Y0-BEE-990

American Men & Women of Science

1992-93 • 18th Edition

The 18th edition of *AMERICAN MEN & WOMEN OF SCIENCE* was prepared by the R.R. Bowker Database Publishing Group.

Stephen L. Torpie, Managing Editor
Judy Redel, Managing Editor, Research
Richard D. Lanam, Senior Editor
Tanya Hurst, Research Manager
Karen Hallard, Beth Tanis, Associate Editors

Peter Simon, Vice President, Database Publishing Group
Dean Hollister, Director, Database Planning
Edgar Adcock, Jr., Editorial Director, Directories

American Men & Women of Science

1992-93 • 18th Edition

A Biographical Directory of Today's Leaders in Physical, Biological and Related Sciences.

Volume 2 • C-F

R. R. BOWKER
New Providence, New Jersey

Published by R.R. Bowker, a division of Reed Publishing, (USA) Inc.

International Standard Book Number

Set:	0-8352-3074-0
Volume I:	0-8352-3075-9
Volume II:	0-8352-3076-7
Volume III:	0-8352-3077-5
Volume IV:	0-8352-3078-3
Volume V:	0-8352-3079-1
Volume VI:	0-8352-3080-5
Volume VII:	0-8352-3081-3
Volume VIII:	0-8352-3082-1

International Standard Serial Number: 0192-8570
Library of Congress Catalog Card Number: 6-7326
Printed and bound in the United States of America.

8 Volume Set

ISBN 0-8352-3074-0

9 780835 230742

Contents

Advisory Committee

Dr. Robert F. Barnes
Executive Vice President
American Society of Agronomy

Dr. John Kistler Crum
Executive Director
American Chemical Society

Dr. Charles Henderson Dickens
Section Head, Survey & Analysis Section
Division of Science Resource Studies
National Science Foundation

Mr. Alan Edward Fechter
Executive Director
Office of Scientific & Engineering Personnel
National Academy of Science

Dr. Oscar Nicolas Garcia
Prof Electrical Engineering
Electrical Engineering & Computer Science Department
George Washington University

Dr. Charles George Groat
Executive Director
American Geological Institute

Dr. Richard E. Hallgren
Executive Director
American Meteorological Society

Dr. Michael J. Jackson
Executive Director
Federation of American Societies for Experimental Biology

Dr. William Howard Jaco
Executive Director
American Mathematical Society

Dr. Shirley Mahaley Malcom
Head, Directorate for Education and Human Resources Programs
American Association for the Advancement of Science

Mr. Daniel Melnick
Sr Advisor Research Methodologies
Sciences Resources Directorate
National Science Foundation

Ms. Beverly Fearn Porter
Division Manager
Education & Employment Statistics Division
American Institute of Physics

Dr. Terrence R. Russell
Manager
Office of Professional Services
American Chemical Society

Dr. Irwin Walter Sandberg
Holder, Cockrell Family Regent Chair
Department of Electrical & Computer Engineering
University of Texas

Dr. William Eldon Splinter
Interim Vice Chancellor for Research,
Dean, Graduate Studies
University of Nebraska

Ms. Betty M. Vetter
Executive Director, Science Manpower Comission
Commission on Professionals in Science & Technology

Dr. Dael Lee Wolfe
Professor Emeritus
Graduate School of Public Affairs
University of Washington

Preface

American Men and Women Of Science remains without peer as a chronicle of North American scientific endeavor and achievement. The present work is the eighteenth edition since it was first compiled as *American Men of Science* by J. Mckeen Cattell in 1906. In its eighty-six year history *American Men & Women of Science* has profiled the careers of over 300,000 scientists and engineers. Since the first edition, the number of American scientists and the fields they pursue have grown immensely. This edition alone lists full biographies for 122,817 engineers and scientists, 7021 of which are listed for the first time. Although the book has grown, our stated purpose is the same as when Dr. Cattell first undertook the task of producing a biographical directory of active American scientists. It was his intention to record educational, personal and career data which would make "a contribution to the organization of science in America" and "make men [and women] of science acquainted with one another and with one another's work." It is our hope that this edition will fulfill these goals.

The biographies of engineers and scientists constitute seven of the eight volumes and provide birthdates, birthplaces, field of specialty, education, honorary degrees, professional and concurrent experience, awards, memberships, research information and adresses for each entrant when applicable. The eighth volume, the discipline index, organizes biographees by field of activity. This index, adapted from the National Science Foundation's Taxonomy of Degree and Employment Specialties, classifies entrants by 171 subject specialties listed in the table of contents of Volume 8. For the first time, the index classifies scientists and engineers by state within each subject specialty, allowing the user to more easily locate a scientist in a given area. Also new to this edition is the inclusion of statistical information and recipients of theNobel Prizes, the Craaford Prize, the Charles Stark Draper Prize, and the National Medals of Science and Technology received since the last edition.

While the scientific fields covered by *American Men and Women Of Science* are comprehensive, no attempt has been made to include all American scientists. Entrants are meant to be limited to those who have made significant contributions in their field. The names of new entrants were submitted for consideration at the editors' request by current entrants and by leaders of academic, government and private research programs and associations. Those included met the following criteria:

1. Distinguished achievement, by reason of experience, training or accomplishment, including contributions to the literature, coupled with continuing activity in scientific work;

   or

2. Research activity of high quality in science as evidenced by publication in reputable scientific journals; or for those whose work cannot be published due to governmental or industrial security, research activity of high quality in science as evidenced by the judgement of the individual's peers;

   or

3. Attainment of a position of substantial responsibility requiring scientific training and experience.

This edition profiles living scientists in the physical and biological fields, as well as public health scientists, engineers, mathematicians, statisticians, and computer scientists. The information is collected by means of direct communication whenever possible. All entrants receive forms for corroboration and updating. New entrants receive questionaires and verification proofs before publication. The information submitted by entrants is included as completely as possible within the boundaries of editorial and space restrictions. If an entrant does not return the form and his or her current location can be verified in secondary sources, the full entry is repeated. References to the previous edition are given for those who do not return forms and cannot be located, but who are presumed to be still active in science or engineering. Entrants known to be deceased are noted as such and a reference to the previous edition is given. Scientists and engineers who are not citizens of the United States or Canada are included if a significant portion of their work was performed in North America.

The information in AMWS is also available on CD-ROM as part of *SciTech Reference Plus*. In adition to the convenience of searching scientists and engineers, *SciTech Reference Plus* also includes *The Directory of American Research & Technology*, *Corporate Technology Directory*, sci-tech and medical books and serials from *Books in Print* and *Bowker International Series*. *American Men and Women Of Science* is available for online searching through the subscription services of DIALOG Information Services, Inc. (3460 Hillview Ave, Palo Alto, CA 94304) and ORBIT Search Service (800 Westpark Dr, McLean, VA 22102). Both CD-Rom and the on-line subscription services allow all elements of an entry, including field of interest, experience, and location, to be accessed by key word. Tapes and mailing lists are also available through the Cahners Direct Mail (John Panza, List Manager, Bowker Files 245 W 17th St, New York, NY, 10011, Tel: 800-537-7930).

A project as large as publishing *American Men and Women Of Science* involves the efforts of a great many people. The editors take this opportunity to thank the eighteenth edition advisory committee for their guidance, encouragement and support. Appreciation is also expressed to the many scientific societies who provided their membership lists for the purpose of locating former entrants whose addresses had changed, and to the tens of thousands of scientists across the country who took time to provide us with biographical information. We also wish to thank Bruce Glaunert, Bonnie Walton, Val Lowman, Debbie Wilson, Mervaine Ricks and all those whose care and devotion to accurate research and editing assured successful production of this edition.

Comments, suggestions and nominations for the nineteenth edition are encouraged and should be directed to The Editors, *American Men and Women Of Science*, R.R. Bowker, 121 Chanlon Road, New Providence, New Jersey, 07974.

Edgar H. Adcock, Jr.
Editorial Director

Major Honors & Awards

Nobel Prizes

Nobel Foundation

The Nobel Prizes were established in 1900 (and first awarded in 1901) to recognize those people who "have conferred the greatest benefit on mankind."

1990 Recipients

Chemistry:

Elias James Corey

Awarded for his work in retrosynthetic analysis, the synthesizing of complex substances patterned after the molecular structures of natural compounds.

Physics:

Jerome Isaac Friedman
Henry Way Kendall
Richard Edward Taylor

Awarded for their breakthroughs in the understanding of matter.

Physiology or Medicine:

Joseph E. Murray
Edward Donnall Thomas

Awarded to Murray for his kidney transplantation achievements and to Thomas for bone marrow transplantation advances.

1991 Recipients

Chemistry:

Richard R. Ernst

Awarded for refinements in nuclear magnetic resonance spectroscopy.

Physics:

Pierre-Gilles de Gennes*

Awarded for his research on liquid crystals.

Physiology or Medicine:

Erwin Neher
Bert Sakmann*

Awarded for their discoveries in basic cell function and particularly for the development of the patch clamp technique.

Crafoord Prize

Royal Swedish Academy of Sciences
(Kungl. Vetenskapsakademien)

The Crafoord Prize was introduced in 1982 to award scientists in disciplines not covered by the Nobel Prize, namely mathematics, astronomy, geosciences and biosciences.

1990 Recipients

Paul Ralph Ehrlich
Edward Osborne Wilson

Awarded for their fundamental contributions to population biology and the conservation of biological diversity.

1991 Recipient

Allan Rex Sandage

Awarded for his fundamental contributions to extragalactic astronomy, including observational cosmology.

Charles Stark Draper Prize

National Academy of Engineering

The Draper Prize was introduced in 1989 to recognize engineering achievement. It is awarded biennially.

1991 Recipients

Hans Joachim Von Ohain
Frank Whittle

Awarded for their invention and development of the jet aircraft engine.

National Medal of Science

National Science Foundation

The National Medals of Science have been awarded by the President of the United States since 1962 to leading scientists in all fields.

1990 Recipients:

Baruj Benacerraf
Elkan Rogers Blout
Herbert Wayne Boyer
George Francis Carrier
Allan MacLeod Cormack
Mildred S. Dresselhaus
Karl August Folkers
Nick Holonyak Jr.
Leonid Hurwicz
Stephen Cole Kleene
Daniel Edward Koshland Jr.
Edward B. Lewis
John McCarthy
Edwin Mattison McMillan**
David G. Nathan
Robert Vivian Pound
Roger Randall Dougan Revelle**
John D. Roberts
Patrick Suppes
Edward Donnall Thomas

1991 Recipients

Mary Ellen Avery
Ronald Breslow
Alberto Pedro Calderon
Gertrude Belle Elion
George Harry Heilmeier
Dudley Robert Herschbach
George Evelyn Hutchinson**
Elvin Abraham Kabat
Robert Kates
Luna Bergere Leopold
Salvador Edward Luria**
Paul A. Marks
George Armitage Miller
Arthur Leonard Schawlow
Glenn Theodore Seaborg
Folke Skoog
H. Guyford Stever
Edward Carroll Stone Jr
Steven Weinberg
Paul Charles Zamecnik

National Medal of Technology

U.S. Department of Commerce,
Technology Administration

The National Medals of Technology, first awarded in 1985, are bestowed by the President of the United States to recognize individuals and companies for their development or commercialization of technology or for their contributions to the establishment of a technologically-trained workforce.

1990 Recipients

John Vincent Atanasoff
Marvin Camras
The du Pont Company
Donald Nelson Frey
Frederick W. Garry
Wilson Greatbatch
Jack St. Clair Kilby
John S. Mayo
Gordon Earle Moore
David B. Pall
Chauncey Starr

1991 Recipients

Stephen D. Bechtel Jr
C. Gordon Bell
Geoffrey Boothroyd
John Cocke
Peter Dewhurst
Carl Djerassi
James Duderstadt
Antonio L. Elias
Robert W. Galvin
David S. Hollingsworth
Grace Murray Hopper
F. Kenneth Iverson
Frederick M. Jones**
Robert Roland Lovell
Joseph A. Numero**
Charles Eli Reed
John Paul Stapp
David Walker Thompson

*These scientists' biographies do not appear in *American Men & Women of Science* because their work has been conducted exclusively outside the US and Canada.

**Deceased [Note that Frederick Jones died in 1961 and Joseph Numero in May 1991. Neither was ever listed in *American Men and Women of Science*.]

Statistics

Statistical distribution of entrants in *American Men & Women of Science* is illustrated on the following five pages. The regional scheme for geographical analysis is diagrammed in the map below. A table enumerating the geographic distribution can be found on page xvi, following the charts. The statistics are compiled by tallying all occurrences of a major index subject. Each scientist may choose to be indexed under as many as four categories; thus, the total number of subject references is greater than the number of entrants in *AMWS*.

All Disciplines

	Number	Percent
Northeast	58,325	34.99
Southeast	39,769	23.86
North Central	19,846	11.91
South Central	12,156	7.29
Mountain	11,029	6.62
Pacific	25,550	15.33
TOTAL	**166,675**	**100.00**

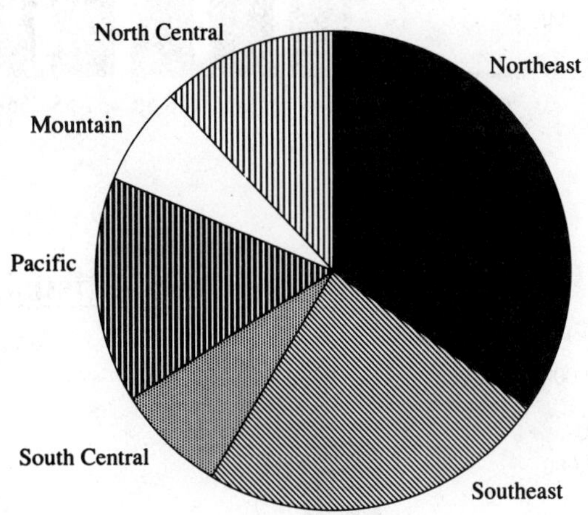

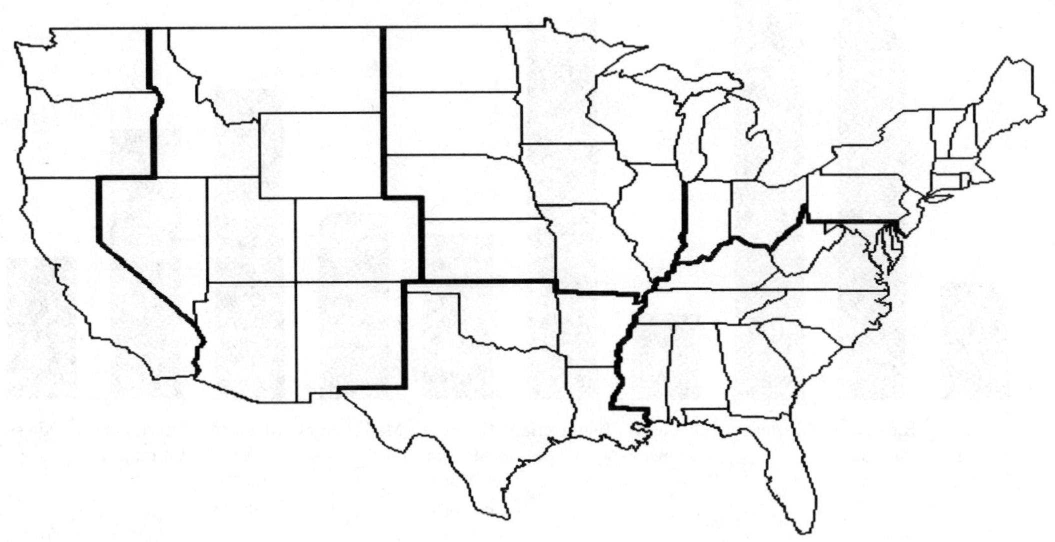

Age Distribution of American Men & Women of Science

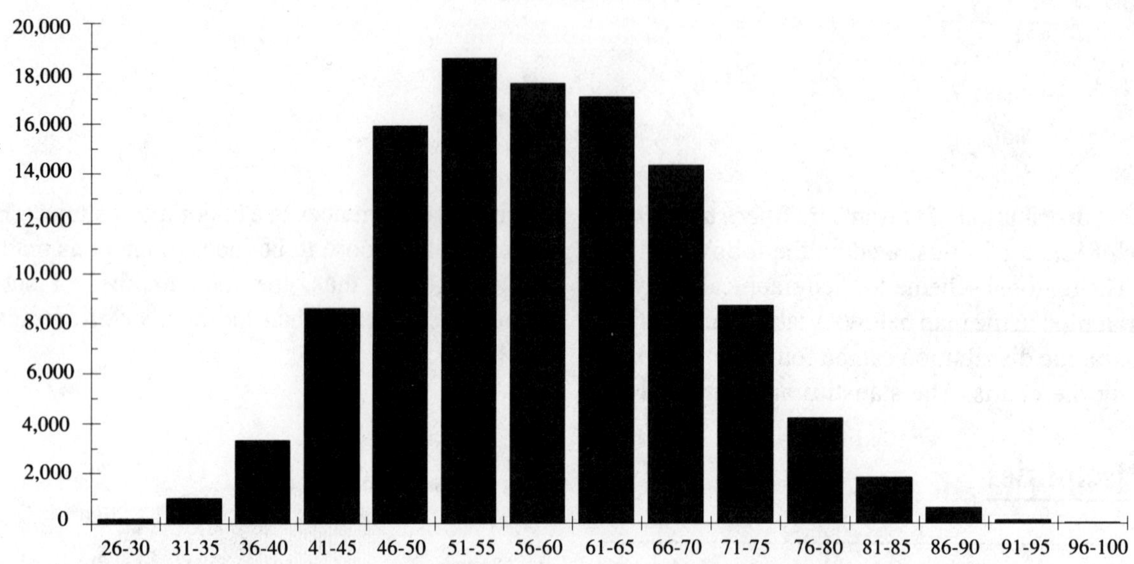

Number of Scientists in Each Discipline of Study

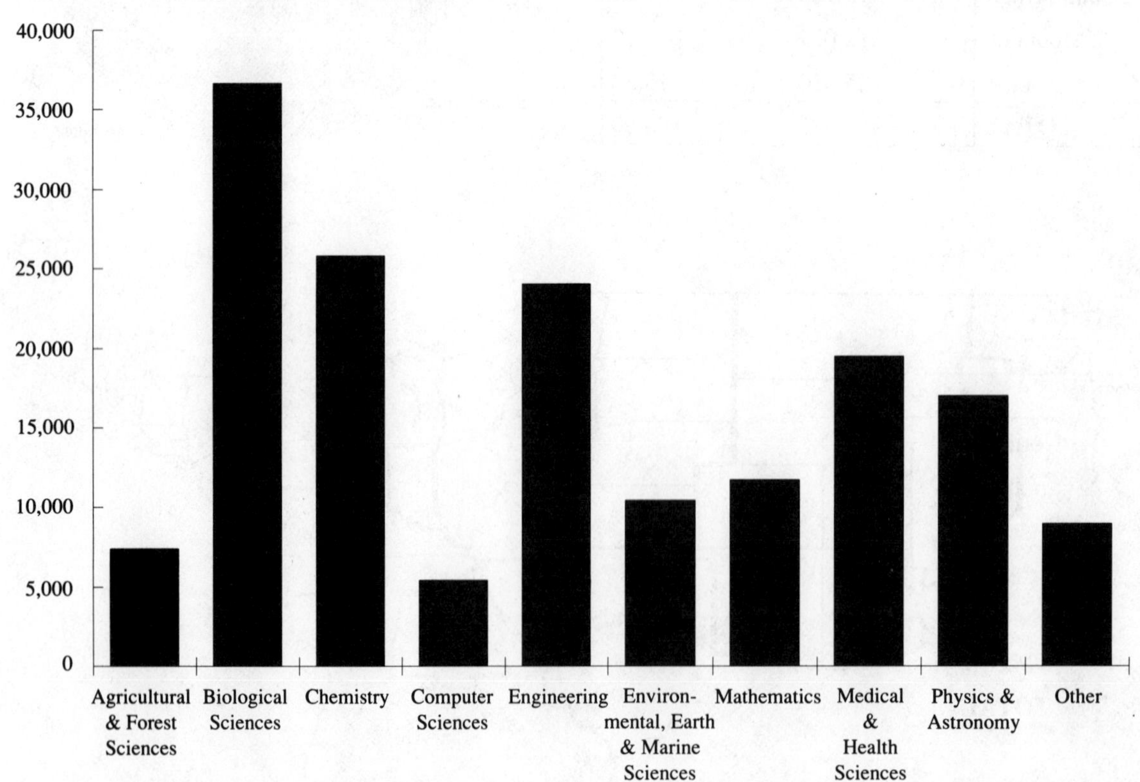

Agricultural & Forest Sciences

	Number	Percent
Northeast	1,574	21.39
Southeast	1,991	27.05
North Central	1,170	15.90
South Central	609	8.27
Mountain	719	9.77
Pacific	1,297	17.62
TOTAL	**7,360**	**100.00**

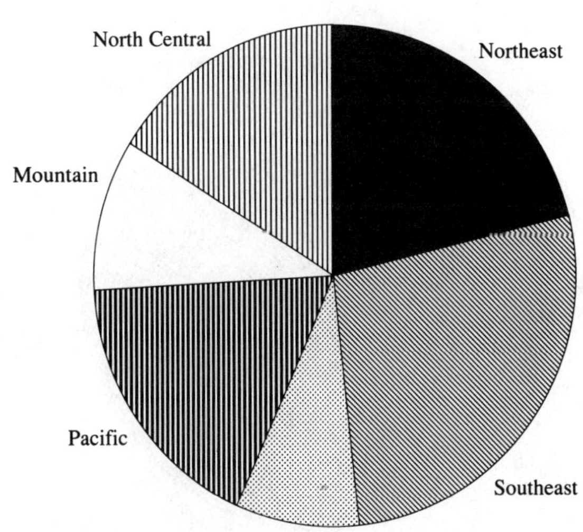

Biological Sciences

	Number	Percent
Northeast	12,162	33.23
Southeast	9,054	24.74
North Central	5,095	13.92
South Central	2,806	7.67
Mountain	2,038	5.57
Pacific	5,449	14.89
TOTAL	**36,604**	**100.00**

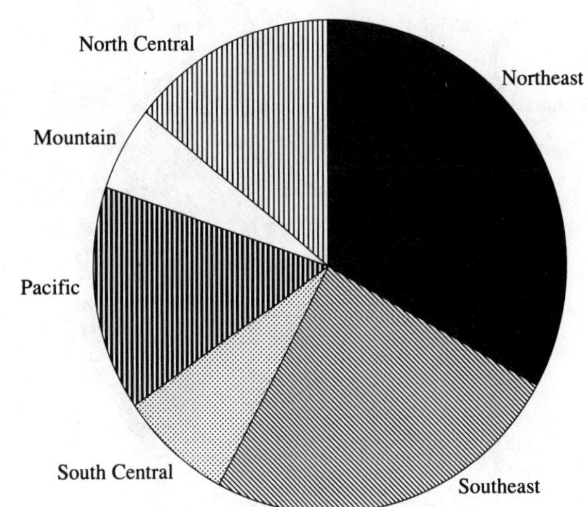

Chemistry

	Number	Percent
Northeast	10,343	40.15
Southeast	6,124	23.77
North Central	3,022	11.73
South Central	1,738	6.75
Mountain	1,300	5.05
Pacific	3,233	12.55
TOTAL	**25,760**	**100.00**

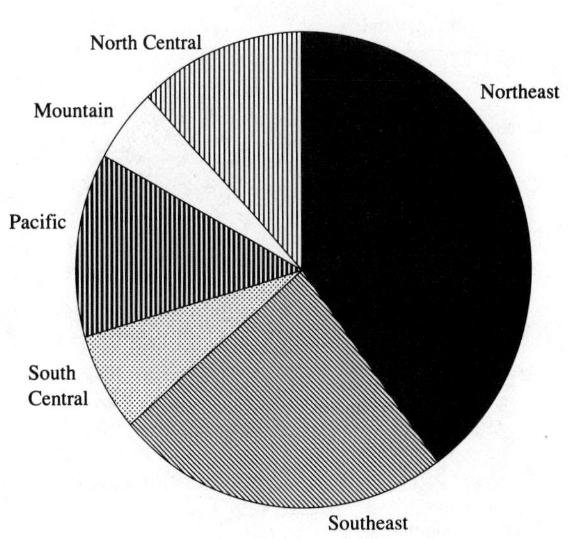

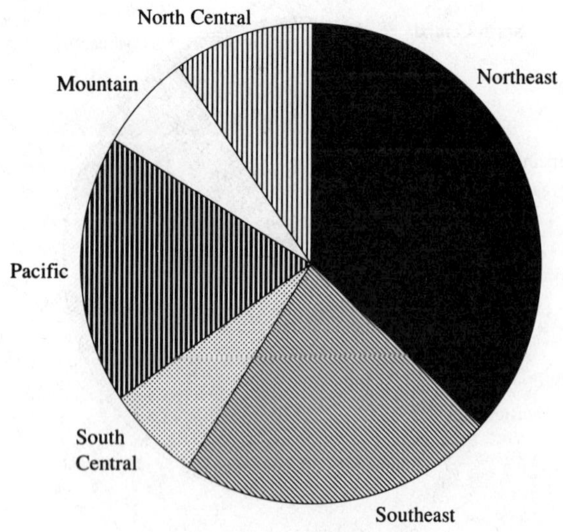

Computer Sciences

	Number	Percent
Northeast	1,987	36.76
Southeast	1,200	22.20
North Central	511	9.45
South Central	360	6.66
Mountain	372	6.88
Pacific	976	18.05
TOTAL	**5,406**	**100.00**

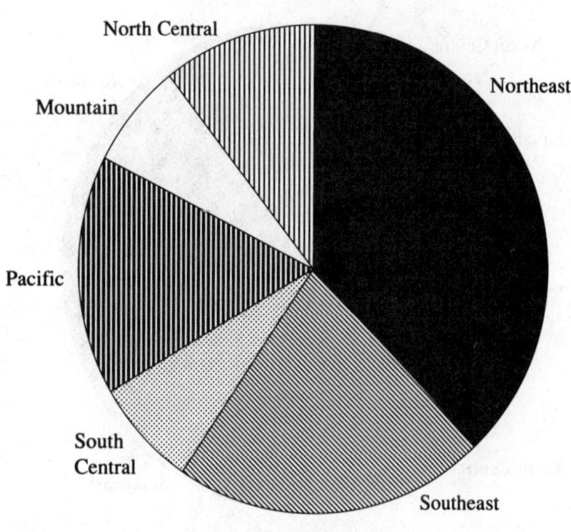

Engineering

	Number	Percent
Northeast	9,122	38.01
Southeast	5,202	21.68
North Central	2,510	10.46
South Central	1,710	7.13
Mountain	1,646	6.86
Pacific	3,807	15.86
TOTAL	**23,997**	**100.00**

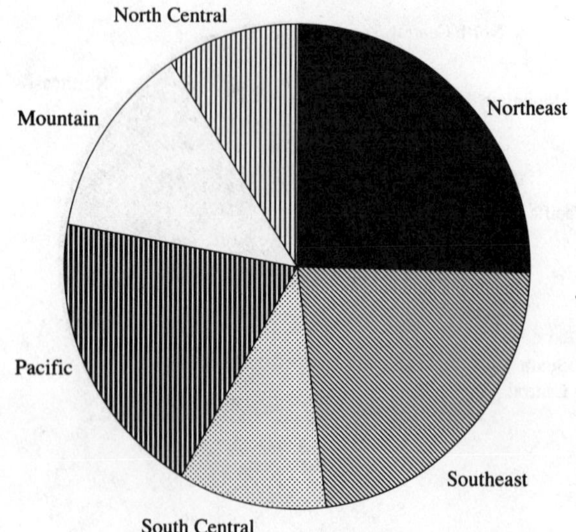

Environmental, Earth & Marine Sciences

	Number	Percent
Northeast	2,657	25.48
Southeast	2,361	22.64
North Central	953	9.14
South Central	1,075	10.31
Mountain	1,359	13.03
Pacific	2,022	19.39
TOTAL	**10,427**	**100.00**

Mathematics

	Number	Percent
Northeast	4,211	35.92
Southeast	2,609	22.26
North Central	1,511	12.89
South Central	884	7.54
Mountain	718	6.13
Pacific	1,789	15.26
TOTAL	**11,722**	**100.00**

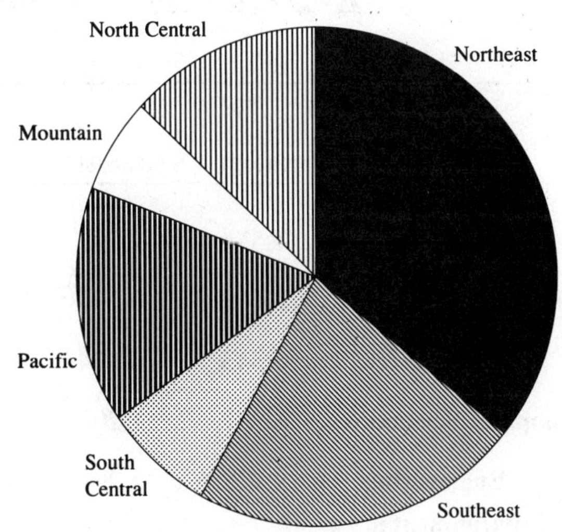

Medical & Health Sciences

	Number	Percent
Northeast	7,115	36.53
Southeast	5,004	25.69
North Central	2,577	13.23
South Central	1,516	7.78
Mountain	755	3.88
Pacific	2,509	12.88
TOTAL	**19,476**	**100.00**

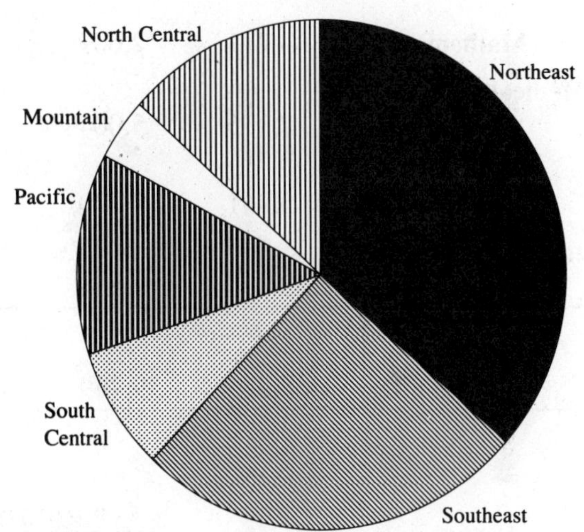

Physics & Astronomy

	Number	Percent
Northeast	5,961	35.12
Southeast	3,670	21.62
North Central	1,579	9.30
South Central	918	5.41
Mountain	1,607	9.47
Pacific	3,238	19.08
TOTAL	**16,973**	**100.00**

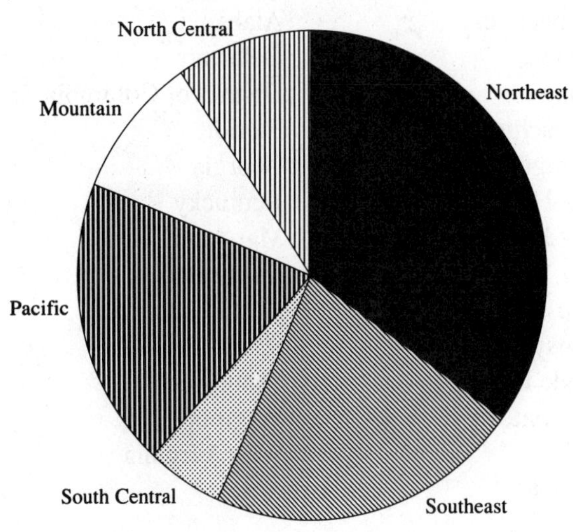

Geographic Distribution of Scientists by Discipline

	Northeast	Southeast	North Central	South Central	Mountain	Pacific	TOTAL
Agricultural & Forest Sciences	1,574	1,991	1,170	609	719	1,297	**7,360**
Biological Sciences	12,162	9,054	5,095	2,806	2,038	5,449	**36,604**
Chemistry	10,343	6,124	3,022	1,738	1,300	3,233	**25,760**
Computer Sciences	1,987	1,200	511	360	372	976	**5,406**
Engineering	9,122	5,202	2,510	1,710	1,646	3,807	**23,997**
Environmental, Earth & Marine Sciences	2,657	2,361	953	1,075	1,359	2,022	**10,427**
Mathematics	4,211	2,609	1,511	884	718	1,789	**11,722**
Medical & Health Sciences	7,115	5,004	2,577	1,516	755	2,509	**19,476**
Physics & Astronomy	5,961	3,670	1,579	918	1,607	3,238	**16,973**
Other Professional Fields	3,193	2,554	918	540	515	1,230	**8,950**
TOTAL	**58,325**	**39,769**	**19,846**	**12,156**	**11,029**	**25,550**	**166,675**

Geographic Definitions

Northeast
Connecticut
Indiana
Maine
Massachusetts
Michigan
New Hampshire
New Jersey
New York
Ohio
Pennsylvania
Rhode Island
Vermont

Southeast
Alabama
Delaware
District of Columbia
Florida
Georgia
Kentucky
Maryland
Mississippi
North Carolina
South Carolina
Tennessee
Virginia
West Virginia

North Central
Illinois
Iowa
Kansas
Minnesota
Missouri
Nebraska
North Dakota
South Dakota
Wisconsin

South Central
Arkansas
Louisiana
Texas
Oklahoma

Mountain
Arizona
Colorado
Idaho
Montana
Nevada
New Mexico
Utah
Wyoming

Pacific
Alaska
California
Hawaii
Oregon
Washington

Sample Entry

American Men & Women of Science (AMWS) is an extremely useful reference tool. The book is most often used in one of two ways: to find more information about a particular scientist or to locate a scientist in a specific field.

To locate information about an individual, the biographical section is most helpful. It encompasses the first seven volumes and lists scientists and engineers alphabetically by last name. The fictitious biographical listing shown below illustrates every type of information an entry may include.

The Discipline Index, volume 8, can be used to easily find a scientist in a specific subject specialty. This index is first classified by area of study, and within each specialty entrants are divided further by state of residence.

CARLETON, PHYLLIS B(ARBARA), b Glenham, SDak, April 1, 30. m 53, 69; c 2. ORGANIC CHEMISTRY. *Educ:* Univ Notre Dame, BSc, 52, MSc, 54, Vanderbilt Univ, PhD(chem), 57. *Hon Degrees:* DSc, Howard Univ, 79. *Prof Exp:* Res chemist, Acme Chem Corp, 54-59, sr res chemist, 59-60; from asst to assoc prof chem 60-63, prof chem, Kansas State Univ, 63-72; prof chem, Yale Univ, 73-89; CONSULT, CARLETON & ASSOCS, 89-. *Concurrent Pos:* Adj prof, Kansas State Univ 58-60; vis lect, Oxford Univ, 77, consult, Union Carbide, 74-80. *Honors & Awards:* Gold Medal, Am Chem Society, 81; *Mem:* AAAS, fel Am Chem Soc, Sigma Chi. *Res:* Organic synthesis, chemistry of natural products, water treatment and analysis. *Mailing Address:* Carleton & Assocs 21 E 34th St Boston MA 02108

Abbreviations

AAAS—American Association for the Advancement of Science
abnorm—abnormal
abstr—abstract
acad—academic, academy
acct—Account, accountant, accounting
acoust—acoustic(s), acoustical
ACTH—adrenocorticotrophic hormone
actg—acting
activ—activities, activity
addn—addition(s), additional
Add—Address
adj—adjunct, adjutant
adjust—adjustment
Adm—Admiral
admin—administration, administrative
adminr—administrator(s)
admis—admission(s)
adv—adviser(s), advisory
advan—advance(d), advancement
advert—advertisement, advertising
AEC—Atomic Energy Commission
aerodyn—aerodynamic
aeronaut—aeronautic(s), aeronautical
aerophys—aerophsical, aerophysics
aesthet—aesthetic
AFB—Air Force Base
affil—affiliate(s), affiliation
agr—agricultural, agriculture
agron—agronomic, agronomical, agronomy
agrost—agrostologic, agrostological, agrostology
agt—agent
AID—Agency for International Development
Ala—Alabama
allergol—allergological, allergology
alt—alternate
Alta—Alberta
Am—America, American
AMA—American Medical Association
anal—analysis, analytic, analytical
analog—analogue
anat—anatomic, anatomical, anatomy
anesthesiol—anesthesiology
angiol—angiology
Ann—Annal(s)
ann—annual
anthrop—anthropological, anthropology
anthropom—anthropometric, anthropometrical, anthropometry
antiq—antiquary, antiquities, antiquity
antiqn—antiquarian
apicult—apicultural, apiculture
APO—Army Post Office
app—appoint, appointed
appl—applied
appln—application
approx—approximate(ly)
Apr—April
apt—apartment(s)
aquacult—aquaculture
arbit—arbitration
arch—archives
archaeol—archaeological, archaeology
archit—architectural, architecture
Arg—Argentina, Argentine
Ariz—Arizona
Ark—Arkansas
artil—artillery
asn—association
assoc(s)—associate(s), associated
asst(s)—assistant(s), assistantship(s)
assyriol—Assyriology
astrodyn—astrodynamics
astron—astronomical, astronomy
astronaut—astonautical, astronautics
astronr—astronomer
astrophys—astrophysical, astrophysics
attend—attendant, attending
atty—attorney
audiol—audiology
Aug—August
auth—author
AV—audiovisual
Ave—Avenue
avicult—avicultural, aviculture

b—born
bact—bacterial, bacteriologic, bacteriological, bacteriology
BC—British Colombia
bd—board
behav—behavior(al)
Belg—Belgian, Belgium
Bibl—biblical
bibliog—bibliographic, bibliographical, bibliography
bibliogr—bibliographer
biochem—biochemical, biochemistry
biog—biographical, biography
biol—biological, biology
biomed—biomedical, biomedicine
biomet—biometric(s), biometrical, biometry
biophys—biophysical, biophysics
bk(s)—book(s)
bldg-building
Blvd—Boulevard
Bor—Borough
bot—botanical, botany
br—branch(es)
Brig—Brigadier
Brit—Britain, British
Bro(s)—Brother(s)
byrol—byrology
bull—Bulletin
bur—bureau
bus—business
BWI—British West Indies

c—children
Calif—California
Can—Canada, Canadian
cand—candidate
Capt—Captain
cardiol-cardiology
cardiovasc—cardiovascular
cartog—cartographic, cartographical, cartography
cartogr—cartographer
Cath—Catholic
CEngr—Corp of Engineers
cent—central
Cent Am—Central American
cert—certificate(s), certification, certified
chap—chapter
chem—chemical(s), chemistry
chemother—chemotherapy
chg—change
chmn—chairman
citricult—citriculture
class—classical
climat—climatological, climatology
clin(s)—clinic(s), clinical
cmndg—commanding
Co—County
co—Companies, Company
co-auth—coauthor
co-dir—co-director
co-ed—co-editor
co-educ—coeducation, coeducational
col(s)—college(s), collegiate, colonel
collab—collaboration, collaborative
collabr—collaborator
Colo—Colorado
com—commerce, commercial
Comdr—Commander

commun—communicable, communication(s)
comn(s)—commission(s), commissioned
comndg—commanding
comnr—commissioner
comp—comparitive
compos—composition
comput—computation, computer(s), computing
comt(s)—committee(s)
conchol—conchology
conf—conference
cong—congress, congressional
Conn—Connecticut
conserv—conservation, conservatory
consol—consolidated, consolidation
const—constitution, constitutional
construct—construction, constructive
consult(s)—consult, consultant(s), consultantship(s), consultation, consulting
contemp—contemporary
contrib—contribute, contributing, contribution(s)
contribr—contributor
conv—convention
coop—cooperating, cooperation, cooperative
coord—coordinate(d), coordinating, coordination
coordr—coordinator
corp—corporate, corporation(s)
corresp—correspondence, correspondent, corresponding
coun—council, counsel, counseling
counr—councilor, counselor
criminol—criminological, criminology
cryog—cryogenic(s)
crystallog—crystallographic, crystallographical, crystallography
crystallogr—crystallographer
Ct—Court
Ctr—Center
cult—cultural, culture
cur—curator
curric—curriculum
cybernet—cybernetic(s)
cytol—cytological, cytology
Czech—Czechoslovakia

DC—District of Columbia
Dec—December
Del—Delaware
deleg—delegate, delegation
delinq—delinquency, delinquent
dem—democrat(s), democratic
demog—demographic, demography
demogr—demographer
demonstr—demontrator
dendrol—dendrologic, dendrological, dendrology
dent—dental, dentistry
dep—deputy
dept—department
dermat—dermatologic, dermatological, dermatology
develop—developed, developing, development, developmental
diag—diagnosis, diagnostic
dialectol-dialectological, dialectology
dict—dictionaries, dictionary
Dig—Digest
dipl—diploma, diplomate
dir(s)—director(s), directories, directory
dis—disease(s), disorders
Diss Abst—Dissertation Abstracts
dist—district
distrib—distributed, distribution, distributive
distribr—distributor(s)
div—division, divisional, divorced
DNA—deoxyribonucleic acid
doc—document(s), documentary, documentation
Dom—Dominion
Dr—Drive
E—east
ecol—ecological, ecology
econ(s)—economic(s), economical, economy
economet—econometric(s)
ECT—electroconvulsive or electroshock therapy
ed—edition(s), editor(s), editorial
ed bd—editorial board
educ—education, educational
educr—educator(s)
EEG—electroencephalogram, electroencephalographic, electroencephalography
Egyptol—Egyptology
EKG—electrocardiogram
elec—elecvtric, electrical, electricity
electrochem-electrochemical, electrochemistry
electroph—electrophysical, electrophysics
elem—elementary
embryol—embryologic, embryological, embryology
emer—emeriti, emeritus
employ—employment
encour—encouragement
encycl—encyclopedia
endocrinol—endocrinologic, endocrinology
eng—engineering
Eng—England, English
engr(s)—engineer(s)
enol—enology
Ens—Ensign
entom—entomological, entomology
environ-environment(s), environmental
enzym—enzymology
epidemiol—epideiologic, epidemiological, epidemiology
equip—equipment
ERDA—Energy Research & Development Administration
ESEA—Elementary & Secondary Education Act
espec—especially
estab—established, establishment(s)
ethnog—ethnographic, ethnographical, ethnography
ethnogr—ethnographer
ethnol—ethnologic, ethnological, ethnology
Europ—European
eval—evaluation
Evangel—evangelical
eve—evening
exam—examination(s), examining
examr—examiner
except—exceptional
exec(s)—executive(s)
exeg—exegeses, exegesis, exegetic, exegetical
exhib(s)—exhibition(s), exhibit(s)
exp—experiment, experimental
exped(s)—expedition(s)
explor—exploration(s), exploratory
expos—exposition
exten—extension

fac—faculty
facil facilities, facility
Feb—February
fed—federal
fedn—federation
fel(s)—fellow(s), fellowship(s)
fermentol—fermentology
fertil—fertility, fertilization
Fla—Florida
floricult—floricultural, floriculture
found—foundation
FPO—Fleet Post Office
Fr—French
Ft—Fort

Ga—Georgia
gastroenterol—gastroenterological, gastroenterology
gen—general
geneal—genealogical, genealogy
geod—geodesy, geodetic
geog—geographic, geographical, geography
geogr—geographer
geol—geologic, geological, geology
geom—geometric, geometrical, geometry
geomorphol—geomorphologic, geomorphology
geophys—geophysical, geophysics
Ger—German, Germanic, Germany
geriat—geriatric
geront—gerontological, gerontology
GES—Gesellschaft
glaciol—glaciology
gov—governing, governor(s)
govt—government, governmental
grad—graduate(d)
Gt Brit—Great Britain
guid—guidance
gym—gymnasium
gynec—gynecologic, gynecological, gynecology

handbk(s)—handbook(s)
helminth—helminthology
hemat—hematologic, hematological, hematology
herpet—herpetologic, herpetological, herpetology
HEW—Department of Health, Education & Welfare
Hisp—Hispanic, Hispania
hist—historic, historical, history
histol—histological, histology
HM—Her Majesty
hochsch—hochschule
homeop—homeopathic, homeopathy
hon(s)—honor(s), honorable, honorary
hort—horticultural, horticulture
hosp(s)—hospital(s), hospitalization
hq—headquarters

ABBREVIATIONS

HumRRO—Human Resources Research Office
husb—husbandry
Hwy—Highway
hydraul—hydraulic(s)
hydrodyn—hydrodynamic(s)
hydrol—hydrologic, hydrological, hydrologics
hyg—hygiene, hygienic(s)
hypn—hypnosis

ichthyol—ichthyological, ichthyology
Ill—Illinois
illum—illuminating, illumination
illus—illustrate, illustrated, illustration
illusr—illustrator
immunol—immunologic, immunological, immunology
Imp—Imperial
improv—improvement
Inc—Incorporated
in-chg—in charge
incl—include(s), including
Ind—Indiana
indust(s)—industrial, industries, industry
Inf—infantry
info—information
inorg—inorganic
ins—insurance
inst(s)—institute(s), institution(s)
instnl—institutional(ized)
instr(s)—instruct, instruction, instructor(s)
instrnl—instructional
int—international
intel—intellligence
introd—introduction
invert—invertebrate
invest(s)—investigation(s)
investr—investigator
irrig—irrigation
Ital—Italian

J—Journal
Jan—January
Jct—Junction
jour—journal, journalism
jr—junior
jurisp—jurisprudence
juv—juvenile

Kans—Kansas
Ky—Kentucky

La—Louisiana
lab(s)—laboratories, laboratory
lang—language(s)
laryngol—larygological, laryngology
lect—lecture(s)
lectr—lecturer(s)
legis—legislation, legislative, legislature
lett—letter(s)
lib—liberal
libr—libraries, library
librn—librarian
lic—license(d)
limnol—limnological, limnology
ling—linguistic(s), linguistical
lit—literary, literature
lithol—lithologic, lithological, lithology

Lt—Lieutenant
Ltd—Limited

m—married
mach—machine(s), machinery
mag—magazine(s)
maj—major
malacol—malacology
mammal—mammalogy
Man—Manitoba
Mar—March
Mariol—Mariology
Mass—Massechusetts
mat—material(s)
mat med—materia medica
math—mathematic(s), mathematical
Md—Maryland
mech—mechanic(s), mechanical
med—medical, medicinal, medicine
Mediter—Mediterranean
Mem—Memorial
mem—member(s), membership(s)
ment—mental(ly)
metab—metabolic, metabolism
metall—metallurgic, metallurgical, metallurgy
metallog—metallographic, metallography
metallogr—metallographer
metaphys—metaphysical, metaphysics
meteorol—meteorological, meteorology
metrol—metrological, metrology
metrop—metropolitan
Mex—Mexican, Mexico
mfg—manufacturing
mfr—manufacturer
mgr—manager
mgt—management
Mich—Michigan
microbiol—microbiological, microbiology
micros—microscopic, microscopical, microscopy
mid—middle
mil—military
mineral—mineralogical, mineralogy
Minn—Minnesota
Miss—Mississippi
mkt—market, marketing
Mo—Missouri
mod—modern
monogr—monograph
Mont—Montana
morphol—morphological, morphology
Mt—Mount
mult—multiple
munic—municipal, municipalities
mus—museum(s)
musicol—musicological, musicology
mycol—mycologic, mycology

N—north
NASA—National Aeronautics & Space Administration
nat—national, naturalized
NATO—North Atlantic Treaty Organization
navig—navigation(al)
NB—New Brunswick
NC—North Carolina
NDak—North Dakota
NDEA—National Defense Education Act
Nebr—Nebraska

nematol—nematological, nematology
nerv—nervous
Neth—Netherlands
neurol—neurological, neurology
neuropath—neuropathological, neuropathology
neuropsychiat—neuropsychiatric, neuropsychiatry
neurosurg—neurosurgical, neurosurgery
Nev—Nevada
New Eng—New England
New York—New York City
Nfld—Newfoundland
NH—New Hampshire
NIH—National Institute of Health
NIMH—National Institute of Mental Health
NJ—New Jersey
NMex—New Mexico
No—Number
nonres—nonresident
norm—normal
Norweg—Norwegian
Nov—November
NS—Nova Scotia
NSF—National Science Foundation
NSW—New South Wales
numis—numismatic(s)
nutrit—nutrition, nutritional
NY—New York State
NZ—New Zealand

observ—observatories, observatory
obstet—obstetric(s), obstetrical
occas—occasional(ly)
occup—occupation, occupational
oceanog—oceanographic, oceanographical, oceanography
oceanogr—oceanographer
Oct—October
odontol—odontology
OEEC—Organization for European Economic Cooperation
off—office, official
Okla—Oklahoma
olericult—olericulture
oncol—oncologic, oncology
Ont—Ontario
oper(s)—operation(s), operational, operative
ophthal—ophthalmologic, ophthalmological, ophthalmology
optom—optometric, optometrical, optometry
ord—ordnance
Ore—Oregon
org—organic
orgn—organization(s), organizational
orient—oriental
ornith—ornithological, ornithology
orthod—orthodontia, orthodontic(s)
orthop—orthopedic(s)
osteop—osteopathic, osteopathy
otol—otological, otology
otolaryngol—otolaryngological, otolaryngology
otorhinol—otorhinologic, otorhinology

Pa—Pennsylvania
Pac—Pacific
paleobot—paleobotanical, paleontology
paleont—paleontology

Pan-Am—Pan-American
parisitol—parasitology
partic—participant, participating
path—pathologic, pathological, pathology
pedag—pedagogic(s), pedagogical, pedagogy
pediat—pediatric(s)
PEI—Prince Edward Islands
penol—penological, penology
periodont—periodontal, periodontic(s)
petrog—petrographic, petrographical, petrography
petrogr—petrographer
petrol—petroleum, petrologic, petrological, petrology
pharm—pharmacy
pharmaceut—pharmaceutic(s), pharmaceutical(s)
pharmacog—pharmacognosy
pharamacol—pharmacologic, pharmacological, pharmacology
phenomenol—phenomenologic(al), phenomenology
philol—philological, philology
philos—philosophic, philosophical, philosophy
photog—photographic, photography
photogeog—photogeographic, photogeography
photogr—photographer(s)
photogram—photogrammetric, photogrammetry
photom—photometric, photometrical, photometry
phycol—phycology
phys—physical
physiog—physiographic, physiographical, physiography
physiol—physiological, phsysiology
Pkwy—Parkway
Pl—Place
polit—political, politics
polytech—polytechnic(s)
pomol—pomological, pomology
pontif—pontifical
pop—population
Port—Portugal, Portuguese
Pos:—Position
postgrad—postgraduate
PQ—Province of Quebec
PR—Puerto Rico
pract—practice
practr—practitioner
prehist—prehistoric, prehistory
prep—preparation, preparative, preparatory
pres—president
Presby—Presbyterian
preserv—preservation
prev—prevention, preventive
prin—principal
prob(s)—problem(s)
proc—proceedings
proctol—proctologic, proctological, proctology
prod—product(s), production, productive
prof—professional, professor, professorial
Prof Exp—Professional Experience
prog(s)—program(s), programmed, programming
proj—project(s), projection(al), projective
prom—promotion
protozool—protozoology
Prov—Province, Provincial
psychiat—psychiatric, psychiatry
psychoanal—psychoanalysis, psychoanalytic, psychoanalytical
psychol—psychological, psychology
psychomet—psychometric(s)
psychopath—psychopathologic, psycho pathology
psychophys—psychophysical, psychophysics
psychophysiol—psychophysiological, psychophysiology
psychosom—psychosomtic(s)
psychother—psychoterapeutic(s), psychotherapy
Pt—Point
pub—public
publ—publication(s), publish(ed), publisher, publishing
pvt—private

Qm—Quartermaster
Qm Gen—Quartermaster General
qual—qualitative, quality
quant—quantitative
quart—quarterly
Que—Quebec

radiol—radiological, radiology
RAF—Royal Air Force
RAFVR—Royal Air Force Volunteer Reserve
RAMC—Royal Army Medical Corps
RAMCR—Royal Army Medical Corps Reserve
RAOC—Royal Army Ornance Corps
RASC—Royal Army Service Corps
RASCR—Royal Army Service Corps Reserve
RCAF—Royal Canadian Air Force
RCAFR—Royal Canadian Air Force Reserve
RCAFVR—Royal Canadian Air Force Volunteer Reserve
RCAMC—Royal Canadian Army Medical Corps
RCAMCR—Royal Canadian Army Medical Corps Reserve
RCASC—Royal Canadian Army Service Corps
RCASCR—Royal Canadian Army Service Corps Reserve
RCEME—Royal Canadian Electrical & Mechanical Engineers
RCN—Royal Canadian Navy
RCNR—Royal Canadian Naval Reserve
RCNVR—Royal Canadian Naval Volunteer Reserve
Rd—Road
RD—Rural Delivery
rec—record(s), recording
redevelop—redevelopment
ref—reference(s)
refrig—refrigeration
regist—register(ed), registration
registr—registrar
regt—regiment(al)
rehab—rehabilitation
rel(s)—relation(s), relative
relig—religion, religious
REME—Royal Electrical & Mechanical Engineers
rep—represent, representative
Repub—Republic
req—requirements
res—research, reserve
rev—review, revised, revision
RFD—Rural Free Delivery
rhet-rhetoric, rhetorical
RI—Rhode Island
Rm—Room
RM—Royal Marines
RN—Royal Navy
RNA—ribonucleic acid
RNR—Royal Naval Reserve
RNVR—Royal Naval Volunteer Reserve
roentgenol—roentgenologic, roentgenological, roentgenology
RR—Railroad, Rural Route
Rte—Route
Russ—Russian
rwy—railway

S—south
SAfrica—South Africa
SAm—South America, South American
sanit—sanitary, sanitation
Sask—Saskatchewan
SC—South Carolina
Scand—Scandinavia(n)
sch(s)—school(s)
scholar—scholarship
sci—science(s), scientific
SDak—South Dakota
SEATO—Southeast Asia Treaty Organization
sec—secondary
sect—section
secy—secretary
seismog—seismograph, seismographic, seismography
seismogr—seismographer
seismol—seismological, seismology
sem—seminar, seminary
Sen—Senator, Senatorial
Sept—September
ser—serial, series
serol—serologic, serological, serology
serv—service(s), serving
silvicult—silvicultural, silviculture
soc(s)—societies, society
soc sci—social science
sociol—sociologic, sociological, sociology
Span—Spanish
spec—special
specif—specification(s)
spectrog—spectrograph, spectrographic, spectrography
spectrogr—spectrographer
spectrophotom—spectrophotometer, spectrophotometric, spectrophotometry
spectros—spectroscopic, spectroscopy
speleol—speleological, speleology
Sq—Square
sr—senior
St—Saint, Street(s)
sta(s)—station(s)
stand—standard(s), standardization
statist—statistical, statistics
Ste—Sainte
steril—sterility

ABBREVIATIONS

stomatol—stomatology
stratig—stratigraphic, stratigraphy
stratigr—stratigrapher
struct—structural, structure(s)
stud—student(ship)
subcomt—subcommittee
subj—subject
subsid—subsidiary
substa—substation
super—superior
suppl—supplement(s), supplemental, supplementary
supt—superintendent
supv—supervising, supervision
supvr—supervisor
supvry—supervisory
surg—surgery, surgical
surv—survey, surveying
survr—surveyor
Swed—Swedish
Switz—Switzerland
symp—symposia, symposium(s)
syphil—syphilology
syst(s)—system(s), systematic(s), systematical

taxon—taxonomic, taxonomy
tech—technical, technique(s)
technol—technologic(al), technology
tel—telegraph(y), telephone
temp—temporary
Tenn—Tennessee
Terr—Terrace
Tex—Texas
textbk(s)—textbook(s)
text ed—text edition
theol—theological, theology
theoret—theoretic(al)
ther—therapy
therapeut—therapeutic(s)
thermodyn—thermodynamic(s)
topog—topographic, topographical, topography
topogr—topographer
toxicol—toxicologic, toxicological, toxicology
trans—transactions
transl—translated, translation(s)
translr—translator(s)
transp—transport, transportation
treas—treasurer, treasury
treat—treatment
trop—tropical
tuberc—tuberculosis
TV—television
Twp—Township

UAR—United Arab Republic
UK—United Kingdom
UN—United Nations
undergrad—undergraduate
unemploy—unemployment
UNESCO—United Nations Educational Scientific & Cultural Organization
UNICEF—United Nations International Childrens Fund
univ(s)—universities, university
UNRRA—United Nations Relief & Rehabilitation Administration
UNRWA—United Nations Relief & Works Agency
urol—urologic, urological, urology
US—United States
USAAF—US Army Air Force
USAAFR—US Army Air Force Reserve
USAF—US Air Force
USAFR—US Air Force Reserve
USAID—US Agency for International Development
USAR—US Army Reserve
USCG—US Coast Guard
USCGR—US Coast Guard Reserve
USDA—US Department of Agriculture
USMC—US Marine Corps
USMCR—US Marine Corps Reserve
USN—US Navy
USNAF—US Naval Air Force
USNAFR—US Naval Air Force Reserve
USNR—US Naval Reserve
USPHS—US Public Health Service
USPHSR—US Public Health Service Reserve
USSR—Union of Soviet Socialist Republics

Va—Virginia
var—various
veg—vegetable(s), vegetation
vent—ventilating, ventilation
vert—vertebrate
Vet—Veteran(s)
vet—veterinarian, veterinary
VI—Virgin Islands
vinicult—viniculture
virol—virological, virology
vis—visiting
voc—vocational
vocab—vocabulary
vol(s)—voluntary, volunteer(s), volume(s)
vpres—vice president
vs—versus
Vt—Vermont

W—west
Wash—Washington
WHO—World Health Organization
WI—West Indies
wid—widow, widowed, widower
Wis—Wisconsin
WVa—West Virginia
Wyo—Wyoming

Yearbk(s)—Yearbook(s)
YMCA—Young Men's Christian Association
YMHA—Young Men's Hebrew Association
Yr(s)—Year(s)
YT—Yukon Territory
YWCA—Young Women's Christian Association
YWHA—Young Women's Hebrew Association

zool—zoological, zoology

American Men & Women of Science

C

CABANA, ALDEE, b Beloeil, Que, July 20, 35; m 58; c 4. PHYSICAL CHEMISTRY, MOLECULAR SPECTROSCOPY. *Educ:* Univ Montreal, BSc, 58, MSc, 59, PhD(chem), 62. *Prof Exp:* Fel chem, Princeton Univ, 61-63; from asst prof to assoc prof, 63-71, dean, Fac Sci, 78-85, PROF CHEM, UNIV SHERBROOKE, 71-, RECTOR, 85- *Concurrent Pos:* Pres, Syndicate Prof Univ Sherbrooke, 75-76; mem, Nat Res Coun Can, 81-86; bd gov, Inst Can Bankers, 87- *Honors & Awards:* Gerhard Herzberg Award, Spectros Soc Can, 76; Acrhambault Prize, Asn Can-French Advan Sci, 83. *Mem:* Fel Chem Inst Can; Spectros Soc Can; Can Asn Physicists; assoc mem Can Soc Chem Eng; Sigma Xi. *Res:* Infrared and Raman spectra of molecular crystals; high resolution infrared spectroscopy. *Mailing Add:* Rector Univ Sherbrooke Sherbrooke PQ J1K 2R1 Can

CABANA, VENERACION GARGANTA, b Lopez, Quezon, Philippines, Jan 9, 42; US citizen. LIPOPROTEIN METABOLISM, ACUTE PHASE RESPONSE. *Educ:* Philippine Union Col, BS, 64; Univ Chicago, cert, 69; Univ Ill, MS, 72, PhD(immunol), 80. *Prof Exp:* Res assoc, Univ Wash, 72-75; from instr to asst prof immunol, Rush Med Col, 80-83; asst prof med, Univ Chicago, 84-87; assoc prof med, Rush Med Col, 87-89; RES ASSOC & ASST PROF, UNIV CHICAGO, 89- *Concurrent Pos:* Guest lectr, Univ Autonoma Montemorelos, Mex, 83-85. *Mem:* Am Asn Immunologists; Am Asn Clin Chem; Am Heart Asn Coun Atherosclerosis. *Res:* Lipoprotein metabolism and the acute phase response, that is alterations of lipids and apolipoproteins in the plasma and response to tissue destructive processes brought about by infections and noninfectious agents. *Mailing Add:* 504 N Richmond St Westmont IL 60559

CABANES, WILLIAM RALPH, JR, b Memphis, Tex, Nov 13, 32; m 62; c 1. ORGANIC CHEMISTRY, POLYMER CHEMISTRY. *Educ:* Univ Tex, Austin, BA, 53, MA, 55, PhD(org chem), 57. *Prof Exp:* Res fel chem, Univ Ariz, 62-64; lectr org chem, Univ Minn, Duluth, 64-65; ASSOC PROF ORG CHEM, UNIV TEX, EL PASO, 65- *Concurrent Pos:* R A Welch grant, 67-70. *Mem:* Fel AAAS; NY Acad Sci. *Res:* Organic polymers; polyampholytes. *Mailing Add:* Dept Chem Univ Tex El Paso TX 79968

CABASSO, ISRAEL, b Jerusalem, Israel, Nov 17, 42. POLYMER CHEMISTRY. *Educ:* Hebrew Univ Jerusalem, BS, 66, MS, 68; Weizmann Inst Sci, PhD(polymer chem), 73. *Prof Exp:* Asst res assoc & teacher chem, Hebrew Univ Jerusalem, 66-68; res asst, Dept Plastic, Weizmann Inst Sci, 68-73; fel, Gulf South Res Inst, 73-74, sr investr polymer chem, 74-75, group leader polymer chem & head & mgr, Polymer Dept, 76-80; PROF & DIR, POLYMER RES INST, STATE UNIV NY SYRACUSE, 80-; CO-DIR MEMBRANE CTR, SYRACUSE UNIV, 87- *Concurrent Pos:* Prof sci, Technicon High Sch, 68-72; sr res assoc, Weizmann Inst Sci, 73-; adj prof polymer dept, Univ Southern Miss, 77-; clin assoc prof biomat dept, La State Univ Med Ctr, New Orleans, 78-81. *Honors & Awards:* Sec Award, Am Chem Soc, 79. *Mem:* Am Chem Soc; Israel Chem Soc; NY Acad Sci. *Res:* Polymer alloys and blends; hollow fiber and synthetic polymeric membrane; transport phenomena, gas separation hemodialysis, reverse osmosis, pervaporation, RO, OF, fuel cells and batteries; polymer materials for artificial organs; self extinguish materials; polymers containing polymer ligard; ion exchangers. *Mailing Add:* Polymer Res Inst ESF State Univ NY Syracuse NY 13210

CABASSO, VICTOR JACK, b Port Said, Egypt, June 21, 15; US citizen; m 48; c 2. VIROLOGY. *Educ:* Lycee Francais, Egypt, BA, 33; Hebrew Univ, Israel, MS, 38; Sorbonne & Univ Algiers, ScD(bact), 41; Am Bd Microbiol, dipl, 61. *Prof Exp:* Spec investr pub health & hyg, Off Foreign Relief & Rehab Opers, US Dept State, Tunis, 43; chief bacteriologist, Greece & Mid East, UNRRA, 44-46; res virologist, Lederle Labs, 46-58; head virus immunol res dept, Lederle Labs, Am Cyanamid Co, 58-67; dir microbiol res dept, Cutter Labs, 67-69, assoc dir res microbiol, 69-74, dir, 74-76, vpres res & develop, 76-80; CONSULT & LECTR, 80- *Concurrent Pos:* Assoc mem, Pasteur Inst, Tunis; mem comt on stand methods for vet microbiol, Nat Acad Sci; mem working comt microbiol, Nat Comt Clin Lab Stand; mem subcomt rabies, Agr Bd, Nat Res Coun. *Mem:* AAAS; fel NY Acad Sci; fel Am Acad Microbiol; Am Pub Health Asn; Am Soc Microbiol; Am Vet Med Asn. *Res:* Virology; viral immunology; vaccine development; plasma fractions. *Mailing Add:* 490 Tharp Drive Moraga CA 94556-2544

CABBINESS, DALE KEITH, b Binger, Okla, May 22, 37; m 58; c 2. ANALYTICAL CHEMISTRY. *Educ:* Southwestern State Col, BS, 60; Univ Ark, MS, 65; Purdue Univ, PhD(chem), 70. *Prof Exp:* Chemist, Dow Chem Co, 60-61; res scientist, 69-75, res group leader, 75-81, ASSOC DIR, CONOCO, INC, 81- *Mem:* Am Chem Soc; Sigma Xi. *Res:* Chemical methods of analysis using kinetic techniques, continuous flow procedures, automation, thermal analysis, environmental analysis; analytical techniques, chromatography, mass spectrometry. *Mailing Add:* Conoco Inc PO Box 1267 Ponca City OK 74603

CABEEN, SAMUEL KIRKLAND, b Easton, Pa, Jan 22, 31. CHEMISTRY, INFORMATION SCIENCE. *Educ:* Lafayette Col, BA, 52; Syracuse Univ, MS, 54. *Prof Exp:* Asst librn, Am Metal Climax Inc, 56-58; librn, Ford Instrument Co Div, Sperry Rand Corp, 58-64; asst dir, 64-68, DIR, ENG SOCS LIBR, 68- *Mem:* Spec Libr Asn; Am Soc Info Sci; Am Libr Asn. *Mailing Add:* Eng Socs Libr 345 E 47th St New York NY 10017-2304

CABIB, ENRICO, b Genoa, Italy, Jan 11, 25; m 55; c 3. BIOCHEMISTRY, ENZYMOLOGY. *Educ:* Univ Buenos Aires, PhD(chem), 51. *Prof Exp:* Investr, Inst Biochem Invest, 49-53; vis investr biochem, Col Physicians & Surgeons, Columbia Univ, 53-54; investr, Inst Biochem Invest, 55-58; investr, Sch Sci, Univ Buenos Aires, 58-67; vis scientist, 67-69, SR RES BIOCHEMIST, NAT INST DIABETES, DIGESTIVE & KIDNEY DIS, 69- *Concurrent Pos:* Instr, Sch Sci, Univ Buenos Aires, 50-53; career investr, Nat Coun Sci & Tech Invests, 60-67; ed, Arch Biochem & Biophys, 70-72 & J Biol Chem, 75-79 & 82-87. *Mem:* Am Soc Microbiol; Soc Complex Carbohydrates; Fedn Am Sci; Am Soc Biol Chemists. *Res:* Metabolism of carbohydrates, particularly studies on the biosynthesis of di- and polysaccharides from sugar nucleotides; biosynthesis of yeast cell wall; regulation of glycogen synthesis in yeast and muscle. *Mailing Add:* Nat Inst Diabetes & Digestive & Kidney Dis Bldg 10 Rm 9N-115 Bethesda MD 20892

CABLE, CHARLES ALLEN, b Akeley, Pa, Jan 15, 32; m 55; c 2. ALGEBRA, NUMBER THEORY. *Educ:* Edinboro State Col, BS, 54; Univ NC, Chapel Hill, MEd, 59; Pa State Univ, PhD(math), 69. *Prof Exp:* Teacher, high sch, NY, 54-55 & Pa, 57-58; from instr to asst prof math, Juniata Col, 59-67; assoc prof, 69-75, chmn dept, 70-90, PROF MATH, ALLEGHENY COL, 75-,. *Concurrent Pos:* Mem bd gov, Math Asn Am, 81-84, assoc ed, Focus, 82-85. *Mem:* Am Math Soc; Math Asn Am. *Res:* Group rings where ring is a finite field and group is Abelian; combinatorics and graph theory. *Mailing Add:* Dept Math Allegheny Col Meadville PA 16335

CABLE, JOE WOOD, b Murray, Ky, Feb 17, 31; m 50; c 3. SOLID STATE PHYSICS. *Educ:* Murray State Univ, AB, 52; Fla State Univ, PhD(chem), 55. *Prof Exp:* PHYSICIST, OAK RIDGE NAT LAB, 55- *Mem:* Am Phys Soc. *Res:* Neutron scattering studies of magnetic materials. *Mailing Add:* Oak Ridge Nat Lab PO Box 2008 Oak Ridge TN 37831

CABLE, PETER GEORGE, b New York, NY, May 5, 36; m 66. SONAR ENGINEERING. *Educ:* Haverford Col, BA, 58; Columbia Univ, MA, 60; Univ Md, PhD(physics), 66. *Prof Exp:* Physicist, Naval Underwater Systs Ctr, 66-68; res asst prof physics, Univ Md, 69-70; physicist, 71-76, SUPVR PHYSICIST, NAVAL UNDERWATER SYSTS CTR, 76- *Concurrent Pos:* Assoc fac, Mitchell Col, 67-69; adj fac, Hartford Grad Ctr, 77-81. *Mem:* Am Phys Soc; Acoust Soc Am; Sigma Xi. *Res:* Application of statistical methods to the characterizations of underwater acoustic transmission phenomena, including multipath fluctuations, scattering from rough boundaries and complex objects; acoustical noise and reverberation. *Mailing Add:* 315 Four Mile River Rd Old Lyme CT 06371

CABLE, RAYMOND MILLARD, b Campton, Ky, Apr 22, 09; m 36; c 4. ZOOLOGY. *Educ:* Berea Col, AB, 29; NY Univ, MS, 30, PhD, 33. *Hon Degrees:* ScD, Berea Col, 55. *Prof Exp:* Assoc prof biol, Berea Col, 33-35; from asst prof to prof, 35-75, EMER PROF PARASITOL, PURDUE UNIV, 75- *Concurrent Pos:* Guggenheim fel, Univ PR, 51-52; Deutsche Gesellschaft fel parasitol. *Honors & Awards:* Rudolph Leuckart Medal. *Mem:* AAAS; Am Soc Zool; Am Micros Soc; Am Soc Parasitol(vpres, 58, pres, 64); NY Acad Sci. *Res:* Parasitology. *Mailing Add:* 820 Carrolton West Lafayette IN 47906

CABOT, JOHN BOIT, II, b Denver, Colo, Aug 16, 46; m; c 1. NEUROANATOMY, NEUROPHYSIOLOGY. *Educ:* Univ Del, BEA, 69; Univ Va, PhD(physiol), 76. *Prof Exp:* Alfred Sloan Fel physiol & neuroanat, Dept Physiol, Univ Va, 76-79; ASSOC PROF NEUROBIOL, DEPT NEUROBIOL & BEHAV, STATE UNIV NY, STONY BROOK, 79- *Concurrent Pos:* Prin investr, Neural Mechanisms Cent Cardiovasc Control, Nat Heart, Lung & Blood Inst, 79-; ad hoc mem, Exp Cardiovasc Study Sect, NIH, 81-82, mem, Exp Cardiovasc Study Sect, Div Res Grants, 83-87. *Mem:* Soc Neurosci; AAAS; Sigma Xi. *Res:* Anatomical, physiological and pharmacological characteristics of central nervous system pathways influencing peripheral cardiovascular function; identification of chemically definable inputs to the thoracolumbar sympathetic preganglionic neurons. *Mailing Add:* Dept Neurobiol & Behav Life Sci Bldg State Univ NY Stony Brook NY 11794

CABOT, MYLES CLAYTON, b Boston, Mass, Oct 23, 48; m. LIPID BIOCHEMISTRY, LIPID METABOLISM IN CELL SIGNAL TRANSDUCTION. *Educ:* Western Carolina Univ, BS, 70, MA, 72; Hebrew Univ, Hadassah Med Sch, Jerusalem, PhD(biochem), 76. *Prof Exp:* Damon Runyon-Walter Winchell Cancer Fel, 76-78; scientist, 78-80, sr staff scientist, Oak Ridge Assoc Univs, 80-84; sr scientist, W Alton Jones Cell Sci Ctr Inc, 84-91; CONSULT. *Concurrent Pos:* Consult, USPHS, 78; adj fac mem, Univ Vt Med Sch, Burlington, Vt, Clarkson Univ, Potsdam, NY & miner res inst, Plattsburgh State Univ, Chazy, NY. *Mem:* FASEB. *Res:* Structure and function of biological membranes; lipid metabolism in cancer cells; membrane modification; tumor promotion; role of lipids in cellular differentiation and agonist-induced cell signal transduction; phospholipase D. *Mailing Add:* 1121 Harvard Santa Monica CA 90403

CABRAL, GUY ANTONY, b Furnas, Port, Dec 15, 38; US citizen; m 69; c 3. VIROLOGY, ELECTRON MICROSCOPY. *Educ:* Univ Mass, BA, 67; Univ Conn, MS, 70, PhD(parasitol), 74. *Prof Exp:* Fel virol, Baylor Col Med, 74-76, asst prof, 76-78; ASST PROF VIROL, DEPT MICROBIOL, MED COL VA, VA COMMONWEALTH UNIV, 78- *Mem:* Am Soc Microbiol; Am Soc Parasitol; AAAS; Electron Micros Soc Am; Sigma Xi. *Res:* Viral oncogenesis; analysis of hepatitis type B subunit vaccines; non-A and non-B hepatitis research. *Mailing Add:* Dept Microbiol Med Col Va Sta Box 678 Richmond VA 23298

CABRERA, BLAS, b Paris, France, Sept 21, 46; US citizen; m 72; c 3. EXPERIMENTAL LOW TEMPERATURE & PARTICLE PHYSICS, ASTROPHYSICS. *Educ:* Univ Va, BS, 68; Stanford Univ, PhD(physics), 75. *Prof Exp:* Res assoc, 75-80, asst prof, 80-86, ASSOC PROF PHYSICS, STANFORD UNIV, 86- *Mem:* Sigma Xi; Am Phys Soc. *Res:* Utilize superconductivity and cyrogenics for precision measurements in condensed matter; particle physics and astrophysics; experiments to measure Cooper pair mass in superconductors and search for new particles in cosmic rays (such as magnetic monupules and weakly interacting massive particles). *Mailing Add:* Dept Physics Stanford Univ Stanford CA 94305

CABRERA, EDELBERTO JOSE, b Pinar del Rio, Cuba, Nov 5, 44; US citizen; m 64; c 2. IMMUNOLOGY, IMMUNOPHARMACOLOGY. *Educ:* Univ Ill, Urbana-Champaign, BS, 67, MS, 68, PhD(zool), 72. *Prof Exp:* Res asst malaria, Univ Ill, Urbana-Champaign, 69-72; res scientist immunol, Univ NMex, 72-77; GROUP LEADER, NORWICH-EATON PHARMACEUT, 77- *Mem:* AAAS; Fedn Am Soc Exp Biol; Am Asn Immunologists; Asn Med Lab Immunologists. *Res:* Immunopharmacology and immunotoxicology of drugs; development of radioimmunoassays and enzyme immunoassays. *Mailing Add:* Norwich-Eaton Pharmaceut Inc PO Box 191 Norwich NY 13815

CABRI, LOUIS J, b Cairo, Egypt, Feb 23, 34; Can citizen; m 59; c 3. MINERALOGY, GEOCHEMISTRY. *Educ:* Univ Witwatersrand, BSc, 54, Hons, 55; McGill Univ, MSc, 61, PhD(geol, geochem), 65. *Prof Exp:* Geologist, Josan, SA, WAfrica, 56, Sierra Leone Mineral Syndicate, 57-58 & New Amianthus Mines, SAfrica, 59; sci officer mineral, Dept Mines & Tech Surv, Can, 64-65, RES SCIENTIST, DEPT ENERGY, MINES & RES CAN, 66- *Concurrent Pos:* Sci ed, Can Mineralogist, 75-82, ed, Plantinum-Group Elements: Mineral, Geol, Recovery, 81. *Honors & Awards:* Waldemar Lindgren Citation, Soc Econ Geol, 66; Queen's Silver Jubilee Medal, 77. *Mem:* Mineral Soc Am; Mineral Asn Can; Geol Asn Can; Can Inst Mining & Metall; Prof Inst Pub Serv Can; Soc Econ Geologists. *Res:* Phase equilibrium research in sulfide and sulfide-type systems and applications to mineralogical problems and to genesis of ore deposits; distribution of trace precious metals in ores and concentrates; platinum-group elements, gold and silver; benefication and economic aspects of ore minerals. *Mailing Add:* 514 Queen Elizabeth Driveway Ottawa ON K1S 3N4 Can

CACCIAMANI, EUGENE RICHARD, JR, b Nyack, NY; m 61; c 3. COMMUNICATIONS THEORY-DIGITAL MODULATION-PSK, INFORMATION THEORY. *Educ:* Union Col, Schenectady NY, BEE, 58; Catholic Univ, MEE, 62, PhD(elec Eng), 72. *Prof Exp:* R&D Comput US Air Force, 58-62; sr mem tech staff, comp Systs RCA Corp, 62-65; mgr modulation tech commun, Comsat Labs, 65-37; vpres eng commun, Am Satellite Corp, 73-81; SR VPRES, COMMUN, M/A-COM TELECOM/ HUGHES NETWORK SYSTS, 81- *Concurrent Pos:* Adj prof, George Wash Univ Continuing Educ Eng Dept, 75-; lectr/consult, Technol Transfer Instit, 77-; ed adv, Telecommun Prod & Technol, 83- *Honors & Awards:* Interface Award, 84. *Mem:* Fel Inst Elec & Electronics Engrs. *Res:* Digital satellite communications and data networks, development and patent holder of major satellite systems including SCPC and IDMA, architect of numerous private corporate digital networks. *Mailing Add:* 11409 Rolling House Rd Rockville MD 20852

CACELLA, ARTHUR FERREIRA, polymer chemistry; deceased, see previous edition for last biography

CACERES, CESAR A, b Puerto Cortes, Honduras, Apr 9, 27. MEDICINE, COMPUTER SCIENCE. *Educ:* Georgetown Univ, BS, 49, MD, 53. *Prof Exp:* Intern med, Boston City Hosp, Mass, 53-54, resident, 54-55; resident, New Eng Med Ctr, 55-56; fel cardiol, George Washington Univ, 56-60, from asst clin prof to assoc prof med, 60-69, prof & chmn dept, 69-71; PRES, CLIN SYSTS ASSOC, 71- *Concurrent Pos:* Chief, Med Systs Develop Lab, USPHS, 60-69; prof elec eng, Univ Md, College Park, 71-76. *Honors & Awards:* Dept Health, Educ & Welfare Super Serv Awards, 63 & 66. *Mem:* Soc Advan Med Systs (past pres); Int Health Eval Asn (past pres); Asn Advan Med Instrumentation (past pres). *Res:* Electrocardiography; cholesterol; cardiology; medical diagnosis; computers in medicine; clinical engineering; medical standards. *Mailing Add:* 1759 Q St NW Washington DC 20009

CACIOPPO, JOHN T, b Marshall, Tex, June 16, 51; m 81; c 1. SOCIAL PSYCHOLOGY, PSYCHOPHYSIOLOGY. *Educ:* Univ Mo, BS, 73; Ohio State Univ, MA, 75, PhD(psychol), 77. *Prof Exp:* Asst prof psychol, Univ Notre Dame, 77-79; from asst prof to assoc prof, 79-85, PROF PSYCHOL, UNIV IOWA, 85- *Concurrent Pos:* Prin investr, NSF, 79-, U I Old Gold Fel, 80, NIH, 80-81 & Hewlett Packard, 87; assoc ed, Psychophysiol, 84-; mem exec comt, Am Psychol Asn, 85-88 & Bd Dir, Soc Psychophysiol Res, 85-88; vis fel, dept psychol, Yale Univ, 86. *Honors & Awards:* Phelps Scholar Award, Univ Mich, 88. *Mem:* Fel Am Psychol Asn; fel Acad Behav Med Res; fel Int Org Psychophysiol; Soc Psychophysiol Res; Soc Exp Social Psychol; AAAS; Sigma Xi; Psychonomic Soc; Asn Appl Psychphysiol & Biofeedback. *Res:* Psychophysiology; communication and persuasion. *Mailing Add:* Dept Psychol 142 Townshend Hall Ohio State Univ 1885 Neil Ave Mall Columbus OH 43210-1222

CADA, GLENN FRANCIS, b Columbus, Nebr, Nov 2, 49; m 72; c 3. FISHERY BIOLOGY, AQUATIC ECOLOGY. *Educ:* Univ Nebr, BS, 71, PhD(zool), 77; Colo State Univ, MS, 73. *Prof Exp:* Res technician limnol, Univ Nebr, 69-71; res asst aquatic ecol, Colo State Univ, 71-73; res asst ichthyol, Univ Nebr, 73-76, teaching asst ichthyol & limnol, 76-77; RES STAFF MEM AQUATIC ECOL, OAK RIDGE NAT LAB, 77- *Mem:* Am Fisheries Soc; Sigma Xi; fel Am Inst Fisheries Res Biologists. *Res:* Ichthyoplankton ecology; evironmental impacts of energy technologies, including steam-electric power plants, hydroelectric power, and coal conversion processes; salmonid ecology. *Mailing Add:* Oak Ridge Nat Lab PO Box 2008 Oak Ridge TN 37831

CADA, RONALD LEE, b Clarkson, Nebr, June 10, 44; m 66; c 2. PUBLIC HEALTH, MICROBIOLOGY. *Educ:* Colo State Univ, BS, 66, MS, 68; Univ Tex, DrPH, 74. *Prof Exp:* Microbiologist lab, Colo Dept Health, 68-70, consult, 70-72; asst prof dis control, Sch Pub Health, Univ Tex, 74-79; dir med specialties, Univ Iowa, 81-82, ext specialist, admin, 79-85; DIR LABS, COLO DEPT HEALTH, 85- *Concurrent Pos:* Lab consult, Community Health Asn, 75-76. *Mem:* Am Pub Health Asn; Conf Pub Health Labs; Asn State & Territorial Pub Health Dirs. *Res:* Methods and evaluation of health laboratory quality assurance practice; development of infection control practice and rural health delivery systems; continuing education for rural laboratory personnel. *Mailing Add:* Colo Dept Health 4210 E 11th Ave Denver CO 80220

CADDELL, JOAN LOUISE, b New York, NY, Jan 16, 27. PEDIATRICS, PEDIATRIC CARDIOLOGY. *Educ:* Col Women Univ Pa, BA & Sch Med, MD, 53; Natl Bd Med Examrs, Cert Med, 54; Am Bd Pediat, cert pediat, 58 & cert pediat cardiol, 62. *Prof Exp:* Rotating intern med, Philadelphia Gen Hosp, 53-54; resident pediat, Children's Hosp Philadelphia, 54-57; fed cardiol instr pediat, Yale Univ Sch Med, New Haven, 59-62; hon lectr pediat, Makerere Univ, Kampala, Uganda,62-63, Rockefeller Found, Ibadou, Nigeria, 63-65; NIH res fel, MALAN Res, Chiang Mai, Thailand, 69-71; res prof pediat & adolescent med, St Louis Univ Sch Med, 80-83; guest researcher, 82-88, ADJ SCIENTIST, NAT INST CHILD HEALTH, NIH, 88- *Concurrent Pos:* Physician to Tarascan Indians, Childrens Hosp Philadelphia Sponsorship, 58; physician, Soc Protect Children from Cruelty, 58-59; prin investr res grants, Am Heart Asn, NIH & St Louis Univ, 72-82. *Mem:* Am Col Cardiol; Soc Exp Biol & Med; Am Inst Nutrit; Am Soc Clin Nutrit; Am Pediat Soc. *Res:* Magnesium metabolism as it applies to mothers, infants and children, with experimental studies in animal models; diagnosis of magnesium deficiency in cardiac, pulmonary and renal aspects and its anatomical and biochemical consequences. *Mailing Add:* Oral Roberts Univ Sch Med 7777 S Lewis Ave Tulsa OK 74171

CADE, JAMES ROBERT, b San Antonio, Tex, Sept 26, 27; m 53; c 6. INTERNAL MEDICINE. *Educ:* Univ Tex, MD, 54. *Prof Exp:* Fel, Cornell Univ, 58-61; assoc prof med, 61-72, chief med sch, 72-80, PROF RENAL MED, UNIV FLA, 72- *Mem:* Am Fedn Clin Res; Am Soc Clin Invest; Am Physiol Soc. *Res:* Renal, electrolyte and exercise physiology. *Mailing Add:* Dept Med Col Med Univ Fla Gainesville FL 32610

CADE, RUTH ANN, b Benton, Miss, Nov 17, 37; m 74; c 1. ENGINEERING, APPLIED MATHEMATICS. *Educ:* Miss State Univ, BS, 63; Univ Ala, MA, 68, PhD(eng math), 69. *Prof Exp:* Chem engr, Southern Res Inst, 63-64; instr eng mech, Univ Ala, 67-68; from asst prof math to prof, 68-88, DIR & PROF ENG TECHNOL, UNIV SOUTHERN MISS, 88- *Mem:* Am Soc Eng Educ. *Res:* Women and minority participation in science and engineering; algorithm analysis; special functions including legendre functions and Mehler- Dirichlet integrals. *Mailing Add:* 361 W Lake Rd Hattiesburg MS 39402

CADE, THOMAS JOSEPH, b San Angelo, Tex, Jan 10, 28; m 52; c 5. ORNITHOLOGY. *Educ:* Univ Alaska, AB, 51; Univ Calif, Los Angeles, MA, 55, PhD(zool), 58. *Prof Exp:* Assoc zool, Univ Calif, Los Angeles, 55-58; NSF fel, Mus Vert Zool, Univ Calif, Berkeley, 58-59; from asst prof to prof zool, Syracuse Univ, 59-67; prof ornith, Col Agr, State Univ NY, Cornell Univ & res dir, Lab Ornith, 67-88, mem staff instr, Res & Exten at Ithaca, 72-88; BIOL PROF, BOISE STATE UNIV, 88- *Concurrent Pos:* NSF grants, 60-65 & sr fel, Transvaal Mus, SAfrica, 65-66; USPHS grant, 63-66. *Mem:* Am Soc Mammal; fel Am Ornith Union. *Res:* Population and physiological ecology; behavior of birds and mammals. *Mailing Add:* Peregrine Fund 5666 W Flying Hawk Lane Boise ID 83703

CADE, WILLIAM HENRY, b San Antonio, Tex, July 5, 46; m 72; c 1. BEHAVIOR GENETICS, SEXUAL SELECTION. *Educ:* Univ Tex, Austin, BA, 68, MA, 73, PhD(zool), 76. *Prof Exp:* From asst prof to assoc prof, 78-85, PROF BIOL, BROCK UNIV, 85-, DEAN, FAC MATH & SCI, 89- *Concurrent Pos:* Can deleg, Int Ethological Conf, 78 & 81. *Mem:* AAAS; Soc Study Evolution; Animal Behav Soc; Entom Soc Can; Entom Soc Am. *Res:* Insect sexual behavior; selective agents influencing behavior include predators and parasites, female mating preferences and factors affecting male competition; behavior genetics of sexual behavior; gypsy moth-plant-fungus interactions. *Mailing Add:* Dept Biol Sci Brock Univ St Catharines ON L2S 3A1 Can

CADENHEAD, DAVID ALLAN, b Tillicoultry, Scotland, Aug 5, 30. PHYSICAL CHEMISTRY. *Educ:* Univ St Andrews, BSc, 53; Bristol Univ, PhD(chem), 57. *Prof Exp:* Can Nat Defence fel, Royal Mil Col, Can, 57-59; fel, Alfred Univ, 59-60; asst prof, 60-64, assoc prof, 64-77, res assoc prof biochem, 70-77, PROF CHEM, STATE UNIV NY BUFFALO, 77-, ASSOC DEAN, FAC NATURAL SCI & MATH, 87- *Concurrent Pos:* State Univ NY Buffalo rep, Univs Space Res Asn, 74-; vis prof, Univ Calif, San Francisco, 76, Tech Univ Munich & Univ Provence, 83, Univ Paris VI, 84. *Honors & Awards:* Alexander von Humboldt Sr US Scientist Award, 83. *Mem:* Am Chem Soc; The Chem Soc. *Res:* Adsorption and porosity; monomolecular and multilayer film studies; surface characterization of phospholipids; molecular interactions in mixed monolayers; cell membrane structure. *Mailing Add:* Dept Chem 112 Acheson Hall State Univ NY Chemistry Rd Buffalo NY 14214

CADET, GARDY, b Port-a-Piment, Haiti, Nov 13, 55; m; c 1. MATERIAL FABRICATION, TRACE ELEMENT ANALYSIS. *Educ:* Brooklyn Col, MS, 82; City Univ New York, PhD(chem), 87. *Prof Exp:* Teaching asst chem, Brooklyn Col, 82-87; prin mem tech staff analytical chem, 87-90, MEM TECH STAFF CHEM, AT&T BELL LABS, 90- *Honors & Awards:* R&D 100 Award, 89. *Mem:* Am Chem Soc; Mat Res Soc; Nat Orgn Black Chemists & Chem Engrs. *Res:* Development and construction of line monitoring systems for process analytical chemistry; on-demand synthesis of precursor gases used in the fabrication of III-V semiconductor materials and devices; nuclear analytical chemistry; thin film fabrication. *Mailing Add:* 225 Austen Rd Orange NJ 07050

CADIGAN, ROBERT ALLEN, b Glen Falls, NY, June 10, 18; m 44; c 4. GEOLOGY, EXPLORATION GEOCHEMISTRY. *Educ:* Univ of Puget Sound, AB, 47; Pa State Univ, MS, 52. *Prof Exp:* Geologist, 48-74, res geologist, 74-82, geologic consult, US Geol Surv, 82-84; RETIRED. *Concurrent Pos:* Adv tech coord comt, Nat Hydrogeochem Surv, US Dept Energy, 75-77; mem, Nat Adv Bd, Am Security Coun; prin investr, Uranium Resource Eval Prog, Delta, Utah Quadrangle, 77-80. *Mem:* Geol Soc Am; Am Soc Econ Paleont & Mineral. *Res:* Textural and geochemical properties of sedimentary rocks; geochemical exploration techniques applied to detecting areas in the United States containing economically important uranium ore deposits; geochemistry of radioactive mineral springs. *Mailing Add:* 9125 W Second Ave Lakewood CO 80226

CADLE, RICHARD DUNBAR, b Cleveland, Ohio, Sept 23, 14; m 40; c 3. PHYSICAL CHEMISTRY. *Educ:* Western Reserve Univ, AB, 36; Univ Wash, PhD(chem), 40. *Prof Exp:* Teaching fel, Univ Wash, 36-38; res chemist, Procter & Gamble Co, 40-47; unit head, Naval Ord Test Sta, Calif, 47-48; sr phys chemist, Stanford Res Inst, 48-54, mgr atmospheric chem physics, 54-63; prog scientist, Nat Ctr Atmospheric Res, 63-66, head, Chem Dept, 66-73, proj scientist, 73-79; CONSULT, 79- *Concurrent Pos:* Asst lectr, Univ Cincinnati, 41-43; consult, Army Chem Corps, 55; vis prof, Ripon Col, 66; mem tech adv comt, Denver Regional Air Pollution Control Agency, 67-72; consult, Gulf South Res Inst, 68-70; mem adv comt air pollution, chem & physics, USPHS, 69-73; panel mem polar meteorol, Nat Acad Sci-Nat Res Coun, 69-73, panel mem atmospheric chem, 70-75; adj prof, Pa State Univ, 72-75. *Mem:* Sigma Xi; Am Chem Soc; Am Geophys Union. *Res:* Properties of dilute solutions of colloidal electrolytes; preparation and identification of aerosols; chemical kinetics; atmospheric chemistry. *Mailing Add:* 4415 Chippewa Dr Boulder CO 80303

CADLE, STEVEN HOWARD, b Cincinnati, Ohio, Feb 19, 46; m 68; c 2. ANALYTICAL CHEMISTRY, ENVIRONMENTAL CHEMISTRY. *Educ:* Univ Colo, BA, 67; State Univ NY Buffalo, PhD(chem), 72. *Prof Exp:* Asst prof chem, Vassar Col, 72-73; assoc sr res chemist, 74-77, sr res scientist, 77-81, SR STAFF RES SCIENTIST & SECT MGR, GEN MOTORS RES LABS, 81- *Honors & Awards:* Young Author Award, Electrochem Soc, 75. *Mem:* Am Chem Soc. *Res:* Analytical methods for trace atmospheric constituents, acid deposition and automotive exhaust emissions. *Mailing Add:* 122230 Hiawatha Dr Utica MI 48087

CADMAN, EDWIN CLARENCE, b Bandon, Ore, May 14, 45; m 68; c 3. BIOCHEMICAL PHARMACOLOGY, CANCER RESEARCH. *Educ:* Stanford Univ, BA, 67; Univ Ore, MD, 71. *Prof Exp:* Resident, Stanford Med Sch, 71-74; fel oncol, Med Sch, Yale Univ, 74-76, from asst prof to assoc prof, 76-83, co-chief, sect med oncol, 81-83; Am Cancer Soc prof clin oncol, 85, PROF MED & DIR CANCER RES INST, UNIV CALIF, SAN FRANCISCO, 83- *Honors & Awards:* Fac Res Award, Am Cancer Soc, 80 & 82. *Mem:* Am Fedn Clin Res; Am Asn Cancer Res; Am Soc Clin Oncol; Am Soc Clin Invest. *Res:* Biochemical modulation and selective killing of cancer cells; transfer of drug resistance. *Mailing Add:* Dept Med Yale Univ 333 Cedar St New Haven CT 06510

CADMAN, THEODORE W, b Osceola Mills, Pa, Feb 28, 40; m 62; c 2. CHEMICAL ENGINEERING. *Educ:* Carnegie-Mellon Univ, BS, 62, MS, 64, PhD(chem eng), 65. *Prof Exp:* From asst prof to assoc prof, Univ Md, 65-73, chmn, dept chem & nuclear eng, 78-84, prof chem eng, College Park, 73-85, PROF CHEM ENG, UNIV MD, BALTIMORE COUNTY, 85- *Concurrent Pos:* Consult eng, ENSCI, Inc, 72- *Mem:* Am Inst Chem Engrs; Am Chem Soc. *Res:* Computer simulation of chemical, biochemical, and petrochemical processes, process dynamics and process control. *Mailing Add:* 9110 St Andrews Pl College Park MD 20740

CADMUS, ROBERT R, b Little Falls, NJ, June 16, 14; m 41; c 2. PREVENTIVE MEDICINE, MEDICAL ADMINISTRATION. *Educ:* Col Wooster, AB, 36; Columbia Univ, MD, 40. *Hon Degrees:* DSc, Col Med & Dent NJ, Newark, 71. *Prof Exp:* Dir prof serv patients, Vanderbilt Clin-Columbia-Presby Hosp, NY, 45-48; asst dir, Univ Hosps, Cleveland, Ohio, 48-50; dir, NC Mem Hosp, 50-62; prof hosp admin, Univ NC, 52-66, chmn dept, 62-66; consult dir, NC Mem Hosp & res dir, Educ & Res Found, 62-66; prof prev med & pres, Col Med & Dent, NJ, Newark, 66-71; dir, Med Ctr Southeastern Wis, 71-75; exec dir & med dir, Found Med Care Eval Southeastern Wis, 75-79; RETIRED. *Concurrent Pos:* Consult, Nat Inst Arthritis & Metab Dis, 60-64 & Div Hosp & Med Facilities, USPHS, 63; mem comt, Clin Res Ctr, Div Res Facilities & Resources, 64-67; consult, Vet Admin Hosp, East Orange, NJ & USPHS Hosp, Staten Island, NY, 66-71; mem nat adv coun, Gen Med Sci, NIH; mem regional adv group, Health Serv & Ment Health Admin-Regional Med Prog; mem comprehensive health planning coun, Health Serv & Ment Health Admin; clin prof, prev med, Med Col Wis; med columnist, Woman's World, 90- *Mem:* Am Hosp Asn; Am Asn Healthcare Consult. *Res:* Administrative medicine. *Mailing Add:* 8851 N Tennyson Dr Milwaukee WI 53217

CADMUS, ROBERT RANDALL, JR, b New York, NY, Oct 30, 46; m 70; c 2. OBSERVATIONAL OPTICAL ASTRONOMY. *Educ:* Swarthmore Col, BA, 68; Univ Wis-Madison, MA, 70, PhD(physics), 77. *Prof Exp:* Res asst physics, Univ Wis-Madison, 68-77; res assoc, Dept Physics & Astron, Univ NC, 77-78; Asst prof, 78-84, ASSOC PROF PHYSICS, GRINNELL COL, 84- *Concurrent Pos:* Vis asst prof physics, Iowa State Univ, 80-81. *Mem:* Am Phys Soc; Sigma Xi; Aston Soc Pac; Am Astron Soc. *Res:* Photometric and spectroscopic studies of semiregular variable stars and development of associated instrumentation. *Mailing Add:* Dept Physics Grinnell Col Grinnell IA 50112-0806

CADOGAN, KEVIN DENIS, b New York, NY, Feb 17, 39; m 61; c 4. PHYSICAL CHEMISTRY. *Educ:* Manhattan Col, BS, 60; Cornell Univ, PhD(phys chem), 66. *Prof Exp:* Res assoc chem, Cornell Univ, 66-67; res assoc chem, Lab Chem Biodynamics, Univ Calif, Berkeley, 67-69; asst prof, 69-74, ASSOC PROF CHEM, CALIF STATE COL, HAYWARD, 74- *Concurrent Pos:* NIH fel, 67-69; Fulbright-Hays lectr, US-UK Educ Comn, 70-71; Fulbright lectr, Inst Educ Technol, Univ Surrey, 70-71. *Mem:* Am Chem Soc. *Res:* Electronic spectroscopy and solid state photochemistry in aqueous and hydrocarbon solutions at cryogenic temperatures; environmental chemistry; induced and natural chemical cycles in the biosphere; scarce chemical research utilization; scientific utopia. *Mailing Add:* Dept Chem Calif State Univ Hayward CA 94542

CADOGAN, W(ILLIAM) P(ATRICK), b Revere, Mass, Nov 2, 19; m 43; c 5. CHEMICAL ENGINEERING. *Educ:* Mass Inst Technol, SB, 41, SM, 46, ScD(chem eng), 48. *Prof Exp:* Chem engr, Standard Oil Co, Ind, 48-51; vpres, Processes Res Inc, 51-56; mgr chem res & develop, Am Mach & Foundry Co, 56-63; vpres res, Emhart Corp, 63-85, dir res, 80-89; RETIRED. *Concurrent Pos:* Adj prof eng, Univ Hartford, 89- *Mem:* AAAS; Sigma Xi; Am Chem Soc; Am Inst Chem Engrs. *Res:* Adsorption of hydrocarbons; glass container forming; packaging machinery; food processing. *Mailing Add:* 35 Fox Den Rd West Simsbury CT 06092

CADORET, REMI JERE, b Scranton, Pa, Mar 28, 28; m 50, 69; c 4. PSYCHIATRY. *Educ:* Harvard Univ, AB, 49; Yale Univ, MD, 53. *Prof Exp:* Intern rotating, Robert Packer Hosp, 53-54; res assoc parapsychol, Parapsychol Lab, Duke Univ, 56-58; from asst prof to assoc prof physiol, Med Sch, Univ Man, 58-65; resident psychiat, Sch Med, Wash Univ, 65-68, from asst prof to assoc prof, 68-73; PROF PSYCHIAT, MED SCH, UNIV IOWA, 73- *Mem:* Psychiat Res Soc. *Res:* Genetics of psychiatric illness; nosology of psychiatric conditions; personality; psychiatric epidemiology. *Mailing Add:* Dept Psychiat 500 Newton Rd Iowa City IA 52242

CADOTTE, JOHN EDWARD, b Minn, Jan 28, 25; m 48; c 5. ORGANIC CHEMISTRY. *Educ:* Univ Minn, MS, 51. *Prof Exp:* Res chemist, Merck & Co, Inc, 51-52, Clinton Foods, Inc, 52-56 & Wood Conversion Co, 56-65; polymer researcher, 65-71, chemist, 66-75, sr chemist, N Star Div, Midwest Res Inst, 75-78; chief chemist, Filmtec Corp, 78-88; RETIRED. *Mem:* Am Chem Soc. *Res:* Adhesives; starch; thermosetting resins; Fourdrinier formed board products and coatings; reverse osmosis membranes. *Mailing Add:* 5637 Scenic Dr Minnetonka MN 55343

CADWALLADER, DONALD ELTON, b Buffalo, NY, June 14, 31; m 60; c 3. PHARMACY, ANALYTICAL CHEMISTRY. *Educ:* Univ Buffalo, BS, 53; Univ Ga, MS, 55; Univ Fla, PhD(pharm), 57. *Prof Exp:* Interim res prof, mil med supply agency contract, Univ Fla, 57-58; res assoc pharm, Sterling-Winthrop Res Inst, 58-60; group leader, White Labs, Inc, 60-61; from asst prof to assoc prof, 61-68, PROF PHARM, UNIV GA, 68-, HEAD DEPT, 77- *Honors & Awards:* Lunsford-Richardson Pharm Award, 57. *Mem:* Fel AAAS; Am Pharmaceut Asn; fel Acad Pharmaceut Sci; fel Am Asn Pharmaceut Sci. *Res:* Behavior of erythrocytes in various solvent systems; solubility and dissolution of drugs; stability of drugs; drug manufacturing procedures and processes; biopharmaceutics. *Mailing Add:* Col Pharm Univ Ga Athens GA 30602

CADY, BLAKE, b Washington, DC, Dec 27, 30; m 60; c 3. SURGERY, ONCOLOGY. *Educ:* Amherst Col, AB, 53; Cornell Univ, MD, 57. *Prof Exp:* Fel surg, Sloan-Kettering Inst & Med Col, Cornell Univ, 65-67; PROF SURG, MED SCH, HARVARD UNIV, 67- *Concurrent Pos:* Surgeon, New England Deaconess Hosp, Boston. *Mem:* AMA; Am Col Surgeons; Soc Surg Oncol (pres, 89); Soc Head & Neck Surgeons; Soc Surg Alimentary Tract; Am Asn Endocrine Surg. *Res:* Clinical interest in human tumors and cancers. *Mailing Add:* 110 Francis St Boston MA 02215

CADY, FOSTER BERNARD, b Middletown, NY, Aug 5, 31; c 4. BIOSTATISTICS, AGRONOMY. *Educ:* Cornell Univ, BS, 53; Univ Ill, MS, 56; NC State Univ, PhD, 60. *Prof Exp:* Soil scientist, Agr Res Serv, USDA, NC State Univ, 57-60; assoc prof statist, Iowa State Univ, 60-68; prof, Univ Ky, 68-71; prof biol statist, Cornell Univ, 71-89; RES SCIENTIST, WINROCK INT, 90- *Mem:* Am Soc Agron; Biomet Soc. *Res:* Statistical methods and experimental design in the agricultural and biological sciences. *Mailing Add:* PO Box 9942 Arlington VA 22209

CADY, GEORGE HAMILTON, b Lawrence, Kans, Jan 10, 06; m 29; c 2. INORGANIC CHEMISTRY, FLUORINE CHEMISTRY. *Educ:* Univ Kans, AB, 27, AM, 28; Univ Calif, PhD(chem), 30. *Prof Exp:* Asst prof chem, Univ S Dak, 30-31; instr inorg chem, Mass Inst Technol, 31-34; res chemist, gen labs, US Rubber Prod, Inc, 34-35 & Columbia Chem Div, Pittsburgh Plate Glass Co, 35-38; from asst prof to prof, 38-72, chmn dept, 61-65, EMER PROF CHEM, UNIV WASH, 72- *Concurrent Pos:* Sect leader, Manhattan Dist Proj, Columbia Univ, 42-43; G N Lewis Mem lectr, 67. *Honors & Awards:* Res Flourine Chem Award, Am Chem Soc, 72. *Mem:* AAAS; Am Chem Soc. *Res:* Preparation and properties of fluorides; hypochlorous acid and hypochlorites; determination of rare gases; composition of gas hydrates. *Mailing Add:* Dept Chem BG-10 Univ Wash Seattle WA 98195

CADY, HOWARD HAMILTON, b Sioux City, Iowa, Jan 15, 31; m 57; c 4. PHYSICAL CHEMISTRY OF HIGH EXPLOSIVES. *Educ:* Univ Wash, BS, 53; Univ Calif, PhD(chem), 57. *Prof Exp:* Asst, Radiation Lab, Univ Calif, 54-57; staff mem, 57-74, sect leader, 74-81, PROJ LEADER, LOS ALAMOS NAT LAB, 81- *Mem:* AAAS; Am Chem Soc; Am Crystallog Asn. *Res:* General physical chemistry; crystallography; phase diagrams; chemistry of explosives. *Mailing Add:* Los Alamos Nat Lab Group M-1 MS C920 Los Alamos NM 87545

CADY, JOHN GILBERT, b Seneca Falls, NY, Jan 30, 14; m 40. SOILS. *Educ:* Syracuse Univ, BS, 36; Univ Wis, MS, 38; Cornell Univ, PhD(soils), 41. *Prof Exp:* Asst soils, Univ Wis, 36-38 & Cornell Univ, 38-42; instr agron, Univ Idaho, 42-43; soil scientist, Mil Geol Unit, US Geol Surv, 44-46; soil scientist, Bur Plant Indust, Soils & Agr Eng, USDA, 46-51 & Soil Surv Lab, Soil Conserv Serv, 52-69; LECTR SOIL SCI, JOHNS HOPKINS UNIV, 69- *Concurrent Pos:* Soil mineralogist, Soil Classification Mission, Econ Coop Admin-Belgian Congo Nat Inst Agron Studies, 51-52. *Mem:* AAAS; fel Geol Soc Am; fel Am Soc Agron. *Res:* Soil genesis; weathering; soil mineralogy; relation of soil characteristics to geomorphology, land use and natural vegetation. *Mailing Add:* Dept Geog & Environ Eng Johns Hopkins Univ Baltimore MD 21218

CADY, K BINGHAM, b Chicago, Ill, Mar 18, 36; c 4. NUCLEAR ENGINEERING. *Educ:* Mass Inst Technol, SB, 56, PhD(nuclear eng), 62. *Prof Exp:* Marine engr, Shipbuilding Div, Bethlehem Steel Co, 56-58, nuclear engr, 58-59; nuclear engr, Jackson & Moreland, Inc, 59-62; asst prof, 62-67, ASSOC PROF NUCLEAR SCI & ENG, CORNELL UNIV, 67- *Concurrent Pos:* Consult, Knolls Atomic Energy Lab, 66-69; nuclear engr, US Atomic Energy Comn, 69-70. *Mem:* Am Nuclear Soc. *Res:* Fast breeder reactor accidents and transients; energy systems analysis; nuclear reactor safety. *Mailing Add:* Hall-Appl Physics Main Campus Cornell Univ 210 Clark Ithaca NY 14853

CADY, LEE (DE), JR, b St Louis, Mo, Nov 22, 27; m 65; c 2. CARDIOLOGY, INDUSTRIAL MEDICINE. *Educ:* Wash Univ, AB, 47, MA, 48, MD, 51; Yale Univ, MPH, 57, DrPH, 59. *Prof Exp:* Intern, Vet Admin Hosp & Baylor Univ, 51-52; epidemic intel officer, USPHS, 52-54; health officer, Int Coop Admin, 54-56; from asst to assoc prof phys med & rehab, NY Univ, 59-64; prof biomath & chmn dept, Univ Tex, Houston, 64-67; staff dir, Comt Emergency Med Care for Los Angeles County, 70-72; ADJ PROF MED, UNIV SOUTHERN CALIF, 67-; CHIEF CARDIOPULMONARY LAB & ACTG MED DIR, LOS ANGELES COUNTY OCCUP HEALTH SERV, 72- *Concurrent Pos:* USPHS spec res fel, 58-61. *Mem:* Fel AAAS; fel Am Col Cardiol; fel Am Pub Health Asn; Sigma Xi. *Res:* Cardiovascular disease; epidemiology; occupational medicine. *Mailing Add:* 1718 Warwick Rd San Marino CA 91108

CADY, PHILIP DALE, b Elmira, NY, June 26, 33; m 55; c 5. CIVIL ENGINEERING. *Educ:* Pa State Univ, BS, 56, MS, 64, PhD(civil eng), 67. *Prof Exp:* Engr, Esso Standrad Oil Co, 56-57 & Lago Oil & Transport Co, 57-62; res asst civil eng, 62-67, from asst prof to assoc prof, 67-75, PROF CIVIL ENG, PA STATE UNIV, 75- *Concurrent Pos:* Secy comt, A2-E1, Hwy Res Bd, Nat Acad Sci-Nat Res Coun, 68- 74; chmn comt 345, Am Concrete Inst, 71-74; consult to pub, indust & govt, 72-; Distinguished Alumni Prof Eng, Pa State Univ, 88- *Honors & Awards:* Thompson Award, Am Soc Testing & Mat, 68. *Mem:* Fel Am Soc Civil Engrs; fel Am Concrete Inst; Am Soc Testing & Mat. *Res:* Durability of portland cement concrete and concrete aggregates. *Mailing Add:* Dept Civil Eng 212 Sackett Bldg Pa State Univ University Park PA 16802

CADY, WALLACE MARTIN, b Middlebury, Vt, Jan 29, 12; m 42; c 3. GEOLOGY. *Educ:* Middlebury Col, BS, 34; Northwestern Univ, MS, 36; Columbia Univ, PhD(geol), 44. *Prof Exp:* Asst geol, Northwestern Univ, 34-36; asst, Columbia Univ, 38-40; substitute tutor, Brooklyn Col, 40-41; from jr geologist to prin geologist, US Geol Surv, 39-63, res geologist 63-85. *Concurrent Pos:* Coun Int Exchange Scholars lectr probs mod tectonics, Voronezh State Univ, USSR, 75. *Mem:* Fel AAAS; Soc Econ Geol; fel Geol Soc Am; Geochem Soc; fel Am Geophys Union. *Res:* Geology of northwestern Washington; regional geology of New England and adjacent Quebec; regional geology of southwestern Alaska; geotectonics; structural geology; metamorphic stratigraphy; archean geology of southwestern Montana. *Mailing Add:* 8585 W Dakota Ave Apt 322 Lakewood CO 80226

CADZOW, JAMES A(RCHIE), b Niagara Falls, NY, Jan 3, 36; m 58; c 4. ELECTRICAL ENGINEERING. *Educ:* Univ Buffalo, BS, 58; State Univ NY, Buffalo, MS, 63; Cornell Univ, PhD(elec eng), 64. *Prof Exp:* Engr, US Army Res & Develop Labs, 58-59; design engr, Bell Aerosysts Corp, 59-61; asst engr, Cornell Aeronaut Labs, 61-62; prof control & commun eng, State Univ NY Buffalo, 64-77; prof, Va Polytech Inst & State Univ, 77-; RES PROF, DEPT ELEC & COMPUT ENG, ARIZ STATE UNIV. *Mem:* Inst Elec & Electronics Engrs. *Res:* Control and signal processing theory; digital filter theory; functional analysis and numerical algorithms; signal and system identification. *Mailing Add:* Dept Elec Eng Vanderbilt Univ Nashville TN 37235

CAENEPEEL, CHRISTOPHER LEON, b South Bend, Ind, July 1, 42; m 71; c 2. CHEMICAL ENGINEERING. *Educ:* Univ Notre Dame, BS, 64, MS, 64, PhD(chem eng), 70. *Prof Exp:* Chemist, O'Brien Paints, South Bend, Ind, 62-64; teaching asst chem eng, Univ Notre Dame, 64-65 & 66-67; engr, Standard Oil Co, Ohio, 65-66; res asst chem eng, Univ Notre Dame, 67-69; prof chem eng, Calif State Polytech Univ, Pomona, 70-80; WITH AMOCO RES CTR, 80- *Concurrent Pos:* Fac fel, NASA-Am Soc Eng Educ, 71-72; consult, Occidental Res Corp, 73- *Honors & Awards:* Intergovt Personnel Act appt, Environ Protection Agency, Durham, NC, 75-76. *Mem:* Am Soc Eng Educ; Am Inst Chem Engrs; Am Chem Soc; Sigma Xi. *Res:* Reactor design; hydro metallurgy; mass transport; mathematical modelling. *Mailing Add:* 3801 W Temple Ave Pomona CA 91768

CAESAR, PHILIP D, b New Haven, Conn, Mar 13, 17; m 40; c 2. PETROLEUM CHEMISTRY. *Educ:* Yale Univ, BS, 39; Univ Ill, MS, 48, PhD(chem), 50. *Prof Exp:* Chemist, res dept, Paulsboro Lab, Socony Mobil Oil Co, Mobil Res & Develop Corp, NJ, 39-45, res assoc, 45-56, supvr chem res, 56-60 & catalysis res, 60-64, mgr prod res sect, cent res div, 64-70, mgr explor process res group, Process Res & Tech Serv Div, 70-77; RETIRED. *Mem:* AAAS; Am Chem Soc. *Res:* Thiophene chemistry; composition and properties of petroleum; catalysis research; petroleum processing; fuel cell applications. *Mailing Add:* Box 546 St John VI 00831

CAFARELLI, ENZO DONALD, b White Plains, NY, Feb 15, 42; div; c 1. EXERCISE PHYSIOLOGY. *Educ:* East Stroudsburg State Col, BS, 69, MEd, 70; Univ Pittsburgh, PhD(exercise physiol), 74. *Prof Exp:* Vis asst fel, John B Pierce Found Lab & fel environ physiol, Yale Univ, 74-77; ASSOC PROF, DEPT PHYS EDUC & FAC SCI, YORK UNIV, TORONTO, 77- *Mem:* Am Col Sports Med; Can Asn Sports Sci; Soc Neurosci. *Res:* Motor and sensory processes in the neuromuscular system. *Mailing Add:* 346 Bethune Col York Univ 4700 Keele St Toronto ON M3J 1P3 Can

CAFASSO, FRED A, b New York, NY, Feb 25, 30; m; c 2. MATERIALS CHEMISTRY. *Educ:* New York Univ, PhD(phys chem), 59. *Prof Exp:* Group leader, 66-69, sect head, 69-74, assoc div dir, 74-83, DEP DIV DIR, ARGONNE NAT LAB, 83-, SR CHEMIST, 76- *Mem:* Am Chem Soc; AAAS; Res Soc Am; NY Acad Sci. *Res:* Physical and materials chemistry; liquid metals chemistry; high temperature spectroscopy; solution thermodynamics; magnetic properties of solids; electrode processes; alkali metal polyether solutions. *Mailing Add:* 1033 Fair Oaks Oak Park IL 60302

CAFFARELLI, LUIS ANGEL, b Buenos Aires, Argentina, Dec 8, 48; m; c 3. MATHEMATICS. *Educ:* Univ Buenos Aires, MS, 68, PHD(math), 72. *Prof Exp:* Fel math, Univ Minn, 73-75, asst prof, 75-77, assoc prof, 77-79, prof, 79-80; prof math, Courant Inst, New York Univ, 80-82, Univ Chicago, 83-86; PROF MATH, INST ADVAN STUDY, 86- *Honors & Awards:* Bochet Prize, Am Math Soc, 84; Plus XII Medal, Vatican Acad Sci, 88. *Mem:* Nat Acad Sci; AAAS. *Res:* Problems of shape and energy optimization in mathematics and mechanics. *Mailing Add:* Inst Adv Study Sch Math Princeton NJ 08540

CAFFEY, HORACE ROUSE, b Grenada, Miss, Mar 24, 29; m 54; c 4. AGRONOMY, PLANT BREEDING. *Educ:* Miss State Univ, BS, 51, MS, 55; La State Univ, PhD(agron), 59. *Prof Exp:* Asst, Miss State Univ, 54-55; agronomist, Miss Rice Grower Asn, 55-57; res assoc, Agron Dept, La Exp Sta, 57-58; assoc agronomist, Delta Br Exp Sta, 58-62, prof & supt, Rice Exp Sta, La State Univ, 62-70; assoc dir, La Agr Exp Sta, 70-79, vchancellor admin & dir, Int Progs, 79-80, vchancellor, Int Prog, Ctr Agr Sci & Rural Develop, 80-81, chancellor, Univ at Alexandria, 81-84, CHANCELLOR, AGR CTR, LA STATE UNIV, BATON ROUGE, 84- *Concurrent Pos:* Int consult. *Mem:* Am Soc Agron. *Res:* Rice breeding; fertilization and cultural practices; agricultural research and extension administration. *Mailing Add:* Agr Ctr La State Univ PO Box 25203 Baton Rouge LA 70894-5203

CAFFEY, JAMES E, b Rockdale, Tex, May 5, 34; m 60; c 3. WATER RESOURCES, CIVIL ENGINEERING. *Educ:* Tex A&M Univ, BS, 55, MS, 56; Colo State Univ, PhD(civil eng), 65. *Prof Exp:* Instr civil eng, Arlington State Col, 59-61; asst civil engr, Colo State Univ, 61, res asst, 61-65; from asst prof to prof civil eng, Univ Tex, Arlington, 65-74; mgr urban resources dept, Planning Div, Turner, Collie & Braden, Inc, Houston, Tex, 74-76; water resources engr, Rady & Assocs, Inc, 76-80, assoc, 80-85; pres, Caffey & Morrison Inc, 85-88; PRES, CAFFEY ENG INC, 88- *Concurrent Pos:* Engr, Freese, Nichols & Endress, Tex, 65-66, consult, 68-69; dir, Am Water Resources Asn, 73-75. *Mem:* Fel Am Soc Civil Engrs; Nat Soc Prof Engrs; Am Geophys Union; fel Am Water Resources Asn; Sigma Xi; Am Water Works Asn. *Res:* Classical and stochastic hydrology, including groundwater, surface water and hydrometeorological problems; water resources planning and management from both supply and quality aspects; hydraulic engineering to complement hydrological interests; flood plain management. *Mailing Add:* 1506 Wagonwheel Trail Arlington TX 76013-3140

CAFFREY, JAMES LOUIS, b Camden, NJ, July 2, 46; m 73. CARDIOVASCULAR PHYSIOLOGY, ENDOCRINOLOGY. *Educ:* Rutgers Univ, BA, 68; Univ Va, PhD(physiol), 73. *Prof Exp:* Post doctoral fel, Colo State Univ, 73-77; ASSOC PROF PHYSIOL, UNIV NTEX, TEX COL OSTEOP MED, 77- *Mem:* AAAS; Am Physiol Soc; Endocrine Soc; Shock

Soc; Soc Exp Biol & Med. *Res:* Stress; adrenal medulla-catecholamines; endogeneous opiates-enkephalines; exercise; role of endogenous opiates in the circulatory system. *Mailing Add:* Dept Physiol Univ NTex Tex Col Osteop Med Ft Worth TX 76107

CAFFREY, ROBERT E, b Scranton, Pa, Mar 9, 20; m 49; c 5. METALLURGY. *Educ:* Pa State Univ, BS, 49, PhD(metall), 55. *Prof Exp:* Res assoc, Pa State Univ, 51-52, instr metall, 52-53; mem tech staff, 54-58, SUPVR, BELL TEL LABS, 58- *Concurrent Pos:* Lectr, Lehigh Univ, 61-63. *Honors & Awards:* Jerome N Behrman Award, 49. *Mem:* Am Soc Metals; Am Inst Mining, Metall & Petrol Engrs; Electrochem Soc; Am Chem Soc; fel Am Inst Chemists; Sigma Xi. *Res:* High temperature application of metals and alloys in electron devices; insulators and metal deposition by chemical vapor deposition. *Mailing Add:* 250 S White Oak Annville PA 17003-1838

CAFLISCH, EDWARD GEORGE, b Union City, Pa, Dec 6, 25; m 49; c 3. ELECTRONICS. *Educ:* Allegheny Col, BS, 48; Ohio State Univ, PhD(org chem), 54. *Prof Exp:* Chemist, Plaskon Div, Libbey-Owens-Ford Glass Co, 48-49; asst & instr org chem, Ohio State Univ, 49-52; sr chemist, Solvents & Intermediates Div, Union Carbide Corp, 54-83; asst prof, WVa State Col, 80-83. *Mem:* Am Chem Soc. *Res:* Synthesis of aldehydes; chlorocarbons; fluorocarbons; polymerization of butadiene, pentadienes, and styrene; development of gas and liquid chromatography equipment; petrochemicals; organic analytical research. *Mailing Add:* 5119 Moody Dr Kingsport TN 37664

CAFLISCH, GEORGE BARRETT, b Columbus, Ohio, Mar 26, 52; m 79; c 4. LIGHT SCATTERING, GEL PERMEATION CHROMATOGRAPHY. *Educ:* WVa Univ, BS, 73; Univ Wis-Madison, PhD(phys chem), 79. *Prof Exp:* Res chemist, 79-84, sr res chemist, 84-89, prin res chemist, 89-90, RES ASSOC, CHEM DIV RES LAB, EASTMAN KODAK, CO, 90- *Mem:* Am Chem Soc. *Res:* Polymer characterization; gel permeation chromatography; quasielastic light scattering; surface tension and wetting; barrier properties of polymeric materials; small angle neutron scattering. *Mailing Add:* Tenn Eastman Co Bldg 150 Kingsport TN 37662

CAFLISCH, ROBERT GALEN, b Charleston, WV, March 19, 56. MICROSCOPIC MODELING, PHASE TRANSITIONS. *Educ:* WVa Univ, BS(physics) & AB(math), 78; Mass Inst Technol, PhD(physics), 84. *Prof Exp:* Physicist, Schlumberger-Doll Res Ctr, 84-86; PHYSICIST, UNIV RI, 86- *Mem:* Am Phys Soc; Am Math Soc. *Res:* Microscopic modeling of a variety of materials systems in order to obtain thermodynamic properties; phase transitions; materials systems including absorbed layers, liquid crystals, spin glasses, ferroelectrics, intercalated graphite and others. *Mailing Add:* 6413 Neff St Houston TX 77074

CAFLISCH, RUSSEL EDWARD, b Charleston, WVa, Apr 29, 54; m 84. MATHEMATICAL FLUID MECHANICS, KINETIC THEORY. *Educ:* Mich State Univ, BS, 75; NY Univ, MS, 87, PhD(math), 88. *Prof Exp:* Asst prof math, Stanford Univ, 79-82; from asst prof to prof, NY Univ, 83-89; PROF MATH, UNIV CALIF, LOS ANGELES, 89- *Concurrent Pos:* Prin investr, Ctr Anal Nonlinear & Heterogeneous Media, 87-89; vis mem, NY Univ, 88-89. *Mem:* Soc Indust & Appl Math; Am Phys Soc; Am Math Soc. *Res:* Analysis and computation of incompressible flows, especially vortex dynamics; singularities in fluid flows; analysis and computation for rarefied gas dynamics; Monte Carlo methods. *Mailing Add:* Dept Math Univ Calif Los Angeles CA 90024-1555

CAFRUNY, EDWARD JOSEPH, b New Castle, Pa, Dec 17, 24; m 48; c 3. PHARMACOLOGY. *Educ:* Ind Univ, AB, 50; Syracuse Univ, PhD, 55; Univ Mich, MD, 60. *Prof Exp:* From instr to assoc prof pharmacol, Med Sch, Univ Mich, 55-65; prof pharmacol, Sch Med, Univ Minn, 65-68; prof & chmn dept pharmacol & exp therapeut, Med Col Ohio, 68-73; pres, Sterling-Winthrop Res Inst, 73-77; prof pharmacol, Med Col Wis, 77-78; prof pharmacol & dean, Grad Sch Biomed Sci, 78-87, DISTINGUISHED PROF, UNIV MED & DENT NJ, 87- *Concurrent Pos:* Consult, Coun on Drugs, Am Med Asn, 63, 64; mem pharmacol & exp therapeut study sect, NIH, 64-68 & anesthesiol training grant comt, 68-71; mem sci adv bd, Pharmaceut Mfg Asn Found, Inc, 68-, pharmacol test comt, Nat Bd Med Examrs, 69-73 & comt on probs of drug safety, Nat Acad Sci-Nat Res Coun, 70-; adj prof pharmacol, Med Col, Cornell Univ. *Mem:* AAAS; Am Soc Pharmacol & Exp Therapeut; NY Acad Sci; Am Soc Nephrol; Am Heart Asn. *Res:* Renal pharmacology; mechanism of action of diuretic agents. *Mailing Add:* Med Sci Bldg C-671 100 Bergen St Newark NJ 07103-2757

CAGAN, ROBERT H, b Brooklyn, NY, Apr 8, 38; m 68; c 2. BIOCHEMISTRY. *Educ:* Northeastern Univ, BS, 60; Harvard Univ, PhD(biochem), 66. *Prof Exp:* USPHS fel, Nobel Med Inst, Karolinska Inst, Sweden, 66-68; from asst prof to assoc prof biochem, Monell Chem Senses Ctr, Sch Med, Univ Pa, 68-79; assoc dir, Res Life Sci, 83-90, DIR FRAGRANCE TECHNOL, COLGATE-PALMOLIVE CO, 90- *Concurrent Pos:* Res chemist, Vet Admin Med Ctr, Philadelphia, 68-83; lectr, Int Symp Food Intake Chem Senses, Fukuoka, Japan, 76, Gordon Res Conf Taste & Olfaction, NH, 78 & 81, Joint Cong, Europ Chemoreception Res Orgn-Int Symp Olfaction & Taste, Noordwijkerhout, Neth, 80, Europ Chemoreception Symp, Jerusalem, Israel, 81, Int Conf Comp Physiol, Crans, Switz, 82; chmn panel/workshop, Winter Conf Brain Res, Colo, 76, 78, 81 & 84 & Idaho, 79; gen chmn, Int Symp Biochem Taste & Olfaction, Philadelphia, 80; mem, Commun Sci Study Sect, NIH, 80; contrib ed, Nutrit Rev, 78-83, rev bd, Amer J Otolaryngology, 78-; adj prof biochem, Sch Med, Univ Pa, 89- *Honors & Awards:* Byron Riegel lectr, Northwestern Univ, 79; Manheimer lectr, Monell Ctr, 88; First Ann Award for Res Excellence, Asn Chemoreception Sci, 89. *Mem:* Fel AAAS; Am Soc Biochem & Molecular Biol; Swed Biochem Soc; Soc Exp Biol Med; Am Soc Neurochem. *Res:* Biochemical basis of physiological function; enzymology of phagocytic cells; biochemical mechanisms of taste and olfaction; membrane receptor mechanisms; neurochemistry of gamma-aminobutyric acid; neural development in tissue culture; research management; life sciences; skin surfactant interactions; sensory evaluation. *Mailing Add:* Colgate-Palmolive Co Res & Develop Div 909 River Rd Piscataway NJ 08855-1343

CAGGIANO, JOSEPH ANTHONY, JR, b Brooklyn, NY, July 30, 38; m 64; c 3. NEOTECTONICS. *Educ:* Allegheny Col, BS, 60; Syracuse Univ, MS, 66; Univ Mass, PhD(geol), 77. *Prof Exp:* Instr earth sci & geomorphol, Div Natural Sci, Elmira Col, 64-67; instr phys & hist geol & geomorphol, Dept Geol, Temple Univ, 67-68; asst, Dept Geol, Univ Mass, 68-72; proj geologist, D'Appolonia Consult Engrs, 73-77; STAFF GEOLOGIST & UNIT MGR, ROCKWELL HANFORD OPER, 77- *Concurrent Pos:* Res geologist field mapping, US Geol Surv, Boston, Mass, 70-73. *Mem:* Geol Soc Am; Asn Eng Geologists; Soc Econ Paleontologists & Mineralogists; Seismol Soc Am; Am Quaternary Asn. *Res:* Geologic analyses of late Cenozoic sediments, landforms, structures, contemporary seismicity and deformation to determine the tectonic evolution of an area and its potential impact on engineered facilities. *Mailing Add:* 330 Snyder Rd Richland WA 99352

CAGLE, FREDRIC WILLIAM, JR, chemistry; deceased, see previous edition for last biography

CAHILL, CHARLES L, b El Reno, Okla, Feb 23, 33; m 54; c 3. PHYSICAL CHEMISTRY, BIOCHEMISTRY. *Educ:* Okla Baptist Univ, AB, 55; Univ Okla, MS, 57, PhD(biochem), 61. *Prof Exp:* From asst prof to prof chem, Oklahoma City Univ, 61-71, chmn dept, 63-71, assoc dean col arts & sci, 67-71; PROVOST & VCHANCELLOR ACAD AFFAIRS & PROF CHEM, UNIV NC, WILMINGTON, 71- *Concurrent Pos:* NIH res grant, 62-70. *Mem:* AAAS; Am Chem Soc; Endocrine Soc; NY Acad Sci. *Res:* Physicochemical characterization of synthetic polymers and proteins. *Mailing Add:* 2216 Parham Dr Wilmington NC 28403

CAHILL, GEORGE FRANCIS, JR, b New York, NY, July 7, 27; m 49; c 6. METABOLISM. *Educ:* Yale Univ, BS, 49; Columbia Univ, MD, 53; Harvard Univ, MA, 66. *Prof Exp:* Intern med, Peter Bent Brigham Hosp, 53-54, from jr resident to sr resident, 54-58; from asst prof to assoc prof med, Harvard Med Sch, 58-70; dir res, Howard Hughes Med Inst, 78-85, vpres, 85-89; prof, 70-90, EMER PROF MED, HARVARD MED SCH, 90-; PROF BIOL SCI, DARTMOUTH COL, 90- *Concurrent Pos:* Res fel biol chem & Nat Res Coun fel med sci, Harvard Univ, 55-57; sr assoc, Peter Bent Brigham Hosp, 59-64, physician, 68-82, sr physician, 82-; dir, Elliott P Joslin Res Lab, 64-78. *Honors & Awards:* Goldberger Award, AMA; Banting Award; Gairdner Award Can; Lilly Prize. *Mem:* AAAS; Am Soc Clin Invest; Am Physiol Asn; Am Clin & Climat Asn; Endocrine Soc; Am Acad Arts & Sci; Asn Am Physicians. *Res:* Metabolism of carbohydrate, lipid and protein. *Mailing Add:* Upton Pond Stoddard NH 03464

CAHILL, JONES F(RANCIS), b Middletown, Ohio, Dec 8, 20; m 44; c 3. AERODYNAMICS. *Educ:* Univ Notre Dame, BS, 42. *Prof Exp:* Aeronaut res scientist aerodyn, Langley Res Ctr, Nat Adv Comt Aeronaut, 42-55; aerodyn engr aerodyn design & develop, Flight Sci Div, Lockheed-Ga Co, 55-69, sr staff scientist aerospace res, 69-74, flight sci mgr, C-141B Prog, 74-78, sr staff specialist, 78-86, mgr, 86-87; CONSULT, 87- *Mem:* Assoc fel Am Inst Aeronaut & Astronaut. *Res:* Experimental research on sophisticated high lift systems; scaling problems in transonic aerodynamics associated with shock-boundary layer interactions and shock induced separation. *Mailing Add:* 226 Whitlock Dr SW Marietta GA 30064

CAHILL, KEVIN M, b New York, NY, May 6, 36; m 61; c 5. MEDICINE, TROPICAL MEDICINE. *Educ:* Fordham Univ, AB, 57; Cornell Univ, MD, 61; Univ London & Royal Col Physicians, dipl trop med & hyg, 63; Am Bd Med Microbiol, dipl, 67; Am Bd Prev Med, dipl, 70. *Hon Degrees:* LLD, Iona Col, 79, LeMoyne Col, 79; DSc, NY Med Col, 80, St John's Univ, 81; LHD, Villanova Univ, 84, Fordham, 85, Univ Cent Am, 87. *Prof Exp:* Head, epidemiol dept & dir trop med, US Naval Med Res Unit, Cairo, Egypt, 63-65; clin assoc prof med, NY Med Col, 65-67; DIR, TROP DIS CTR, LENOX HILL HOSP, NY, 65-; PROF PUB HEALTH & PREV MED, NJ COL MED, 74-; PROF & CHMN, ROYAL COL SURGEONS IN IRELAND, 69- *Concurrent Pos:* Lectr, Univ Cairo & Univ Alexandria, 63-65; consult, UN Health Serv & USPHS, 65-; mem sci adv bd, Am Found Trop Med, 66-75; prof int health & trop med & chmn dept, Royal Col Surgeons, Ireland, 69-; chmn, Health Planning Comn & Health Res Coun, NY State, 75-80, asst to gov health affairs, 75-80; sr mem, New York City Bd Health, 80- *Honors & Awards:* Colles Medal, Royal Col Surgeons, 70. *Mem:* Fel Am Col Chest Physicians; Am Pub Health Asn; Am Soc Trop Med & Hyg; Royal Soc Trop Med & Hyg; AMA; Asn Am Med Col; Am Asn Public Health; Am Col Prev Med. *Res:* Tropical medicine. *Mailing Add:* 850 Fifth Ave New York NY 10021

CAHILL, LAURENCE JAMES, JR, b Frankfort, Maine, Sept 21, 24; m 49; c 3. SPACE PHYSICS. *Educ:* US Mil Acad, BS, 46; Univ Chicago, SB, 50; Univ Iowa, MS, 56, PhD(physics), 59. *Prof Exp:* Asst prof physics, Univ NH, 59-62; chief physics, Hqs NASA, 62-63; from assoc prof to prof physics, Univ NH, 63-68, dir space sci ctr, 67-68; dir Space Sci Ctr, 68-77, PROF PHYSICS, UNIV MINN, MINNEAPOLIS, 77- *Concurrent Pos:* Chmn, US-USSR Comt Co-op Space Res Geomagnetism, 63; mem working group, Int Year Quiet Sun, Int Comt Space Res, 63-64; consult, NASA, 63- *Mem:* Fel Am Geophys Union; Am Phys Soc. *Res:* Investigation of earth's magnetic field by rocket and satellite experiment. *Mailing Add:* Space Sci Ctr Univ Minn Minneapolis MN 55455

CAHILL, PAUL A, b Akron, Ohio, May 21, 59; m 81; c 1. CHEMISTRY GENERAL. *Educ:* Rice Univ, BA, 81; Univ Ill, PhD(chem), 86. *Prof Exp:* SR MEM TECH STAFF, SANDIA NAT LAB, 86- *Concurrent Pos:* Fel, Nat Sci Found, 81. *Mem:* Am Chem Soc; Int Soc Optical Eng. *Res:* Organic nonlinear optical materials. *Mailing Add:* 13824 Brush Pl NE Albuquerque NM 87123-4728

CAHILL, RICHARD WILLIAM, b Utica, NY, Dec 30, 31; m 68; c 3. PHYSICAL CHEMISTRY, POLYMER CHEMISTRY. *Educ:* Niagara Univ, BS, 58; Purdue Univ, PhD(phys chem), 63. *Prof Exp:* RES SCIENTIST CHEM, LOCKHEED MISSILES & SPACE CO, INC, 63- *Mem:* Am Chem Soc; Sigma Xi. *Res:* The fracture toughness of polymeric materials using the

Charpy impact method to obtain the ductile-to-brittle transition temperature and correlation of their chemistry to predict aging trends. *Mailing Add:* Lockheed Missiles & Space Co Dept 6219 Bldg 551 Box 3504 Sunnyvale CA 94088

CAHILL, THOMAS A, b Paterson, NJ, Mar 4, 37; m 65; c 2. NUCLEAR PHYSICS, ATMOSPHERIC CHEMISTRY & PHYSICS. *Educ:* Holy Cross Col, BA, 59; Univ Calif, Los Angeles, MA, 61, PhD(physics), 65. *Prof Exp:* Asst prof in residence physics, Univ Calif, Los Angeles, 65-66; NATO fel, Ctr Nuclear Studies, Saclay, France, 66-67; from asst prof to assoc prof, Univ Calif, Davis, 67-76, dir, Inst Ecol, 72-75, dir, Crocker Nuclear Lab, 80-90, PROF PHYSICS, UNIV CALIF, DAVIS, 76- *Concurrent Pos:* Orgn Am States fel, Chile, 68; Nat Defense Educ Act fel, 59-62; NATO fel, France, 66-67. *Mem:* Am Chem Soc; Air Pollution Control Asn; Am Phys Soc; Am Asn Aerosol Res. *Res:* application of nuclear and radiation techniques to ecological problems; atmospheric chemistry and physics especially visibility. *Mailing Add:* Dept Physics Univ Calif Davis CA 95616

CAHILL, VERN RICHARD, b Tiro, Ohio, May 5, 18; m 46; c 3. MEAT SCIENCE. *Educ:* Ohio State Univ, BSc, 41, MSc, 42, PhD(meat), 55. *Prof Exp:* prof animal sci, Ohio State Univ, 46-88; RETIRED. *Concurrent Pos:* Consult & res, Inst Food Technol, Brazil, 75; consult meat process, USDA, 79-81; consult pork processing, Taiwan, 88. *Honors & Awards:* Fulbright Lectr, 89. *Mem:* Am Soc Animal Sci; Am Meat Sci Asn; Inst Food Technologists. *Res:* Carcass yield of edible portion; meat tenderness and preservation; comminuted meat products. *Mailing Add:* 133 Aldrich Rd Columbus OH 43214

CAHIR, JOHN JOSEPH, b Scituate, Mass, Oct 8, 33; m 62; c 4. METEOROLOGY, CLIMATOLOGY. *Educ:* Pa State Univ, BS, 61, PhD(meteorol), 71. *Prof Exp:* Asst meteorol, Pa State Univ, 61-63; observer, briefer & forecaster, US Weather Bur, 63-64; teacher & researcher, 64-71, PROF METEOROL, PA STATE UNIV, 71-, ASSOC DEAN, COL EARTH AND MINERAL SCI, 80- *Concurrent Pos:* Mem, comt eval res & develop of Nat Environ Satellite Serv, Nat Acad Sci-Nat Res Coun Comt Atmospheric Sci, 76; consult, Dept Earth & Gen Sci, Jackson State Univ, 76-77; ed, Monthly Weather Rev, Am Meteorol Soc, 77-80; prin investr, Nat Earth Satellite Serv res projs, Off Naval Res, US Nat Weather Serv; George Haltiner res chair metereol, US Naval Post Grad Sch, Monterey, Calif, 84; mem info systs ports & harbors comt, Nat Res Coun, 85; chmn, Weather forecasting & anal comt, Am Meteorol Soc, 79-80 & undergrad award comt, 86, chair, 87; Comn Atmos Sci, WMO, 86-; adv, Earth Sci Adv Coun, Univ Space Res Asn, 87-; counr, Am Meteorol Soc, 86-89, serv comt, 88-91. *Honors & Awards:* Appl Meteorol Award, Nat Weather Asn. *Mem:* Fel Am Meteorol Soc; Am Geophys Union; Royal Meteorol Soc; Sigma Xi; Nat Weather Asn (pres-elect, 80-81, pres, 81-82). *Res:* Synoptic meteorology and local forecasting; jet stream dynamics; cloudiness feedback on climate change; use of real time computer graphics systems in forecasting. *Mailing Add:* Pa State Univ 103 Deike Bldg University Park PA 16802

CAHN, ARNO, b Cologne, Germany, Sept 6, 23; nat US; m 50; c 3. ORGANIC CHEMISTRY, SURFACTANTS & DETERGENTS. *Educ:* Queen's Univ, Ont, BSc, 46; Purdue Univ, PhD(chem), 50. *Prof Exp:* Asst, Purdue Univ, 46-48; sr res chemist, Household Prod Div, Lever Bros Co, 50-53, prin res chemist, 53-56, res assoc, 56-58, chief, Org Sect, 58-63, chief, Detergent Liquids Prod Develop Sect, 63-65, res mgr, Chem & Physics Dept, 65-66, develop mgr household prod, 66-73, dir develop, 73-80, vpres tech develop, 80-81; CONSULT, ARNO CAHN CONSULT SERV INC, 81- *Honors & Awards:* Medal Merit, Am Oil Chemists Soc. *Mem:* Am Chem Soc; Am Oil Chemists Soc; fel Am Inst Chemists; Sigma Xi. *Res:* Kinetics of pyridinium salt formation; synthesis of labelled compounds; structure and physical properties of surface-active agents. *Mailing Add:* Arno Cahn Consult Serv Inc 72 E Allison Ave Pearl River NY 10965

CAHN, DAVID STEPHEN, b Los Angeles, Calif, Jan 12, 40; m 71; c 3. GENERAL ENVIRONMENTAL SCIENCES. *Educ:* Univ Calif, Berkeley, BS, 62, MS, 64, DEng(mineral eng), 66. *Prof Exp:* Engr, Homer Res Labs, Bethlehem Steel Corp, 66-68; group leader, Tech Ctr, Am Cement Corp, 68-71; dir, environ control, Amcord, Inc, 71-77, vpres, 77-80; vpres regulatory matters, Calmat Co, 80-90; SR VPRES, CALIF PORTLAND CEMENT CO, 90- *Mem:* Fel AAAS; Am Inst Mining, Metall & Petrol Engrs; Am Inst Chem Engrs; Am Soc Testing & Mat; Air & Waste Mgt Asn. *Res:* Comminution; mixing of particulate solids; operations research; mineral dressing; cementitious composite systems; solid waste utilization; stationary source emissions; monitoring of stationary sources, reclamation of surface mined lands. *Mailing Add:* Calif Portland Cement Co 2025 E Financial Way Glendora CA 91740

CAHN, JOHN W(ERNER), b Ger, Jan 9, 28; nat US; m 50; c 3. PHYSICAL METALLURGY. *Educ:* Univ Mich, BS, 49; Univ Calif, PhD(chem), 52. *Hon Degrees:* ScD, Northwestern Univ, 90. *Prof Exp:* Instr, Inst Study Metals, Univ Chicago, 52-54; mem res lab, Gen Elec Co, 54-64; prof metall, Mass Inst Technol, 64-78; SR FEL, MAT SCI & ENG LAB, NAT INST STANDARDS & TECHNOL, 77- *Concurrent Pos:* Guggenheim fel, 60; chmn, Gordon Conf Phys Metall, 64; vis prof, Israel Inst Technol, 71-72. *Honors & Awards:* Acta Metall Gold Medal, Am Soc Metals, 77, Savveur Award; Von Hippel Award, Mat Res Soc. *Mem:* Nat Acad Sci; Mat Res Soc; Am Inst Mining, Metall & Petrol Engrs; Am Soc Metals; Am Acad Arts & Sci. *Res:* Thermodynamics; metallurgical kinetics; phase transformations; surfaces. *Mailing Add:* 223-A 153 Nat Inst Standards & Technol Gaithersburg MD 20899

CAHN, JULIUS HOFELLER, b Chicago, Ill, Oct 29, 19; m 43, 51; c 2. ASTROPHYSICS. *Educ:* Yale Univ, BA, 42, MS, 47, PhD(physics), 48. *Prof Exp:* Jr engr, Signal Corps Labs, NJ, 42-43; lab asst, Yale Univ, 46-47; asst, gas discharge proj, US Navy, 47-48; asst prof physics, Univ Nebr, 48-50; prin physicist, Battelle Mem Inst, 50-51, asst div consult, 51-59; assoc prof physics & elec eng, 59-70, from assoc prof to prof, physics & astron, 70-85, EMER PROF, UNIV ILL, URBANA-CHAMPAIGN, 85- *Concurrent Pos:* Consult, Boulder Labs, Nat Bur Standards. *Mem:* Am Phys Soc; Am Astron Soc; Royal Astron Soc; Int Astron Union. *Res:* High current electrical discharges in gases; effects of electrostatic interaction on the electronic velocity distribution function; continuous x-ray absorption; antenna impedance; friction; nondestructive testing; solid state physics; plasma physics; stellar evolution; planetary nebulae; galactic absorption distribution; stellar distances. *Mailing Add:* Dept Astron Univ Ill 1002 W Green St Urbana IL 61801

CAHN, PHYLLIS HOFSTEIN, b New York, NY, Sept 22, 28; m 47; c 2. FISH BIOLOGY, ANIMAL BEHAVIOR. *Educ:* NY Univ, BA, 48, MS, 51, PhD(biol), 57. *Prof Exp:* Teaching asst biol, Newark Col Arts & Sci, Rutgers Univ, 48-50; from instr to prof, Stern Col, Yeshiva Univ, 58-69; assoc prof, 68-69, PROF MARINE SCI, C W POST COL, LONG ISLAND UNIV, 69-, CHMN DEPT, 74-; AT NAT SCI FDN DIV, CELL BIOSCI. *Concurrent Pos:* Res fel, Ichthyol Dept, Am Mus Natural Hist, 57-64. *Mem:* AAAS; Am Soc Ichthyologists & Herpetologists; Am Fisheries Soc; Animal Behav Soc. *Res:* Fish sensory systems, particularly behavior and physiology. *Mailing Add:* Long Island Univ C W Post Col Brookville NY 11548

CAHN, ROBERT NATHAN, b New York, NY, Dec 20, 44; m 65; c 2. ELEMENTARY PARTICLES. *Educ:* Harvard Univ, BA, 66; Univ Calif, Berkeley, PhD(physics), 72. *Prof Exp:* Postdoctoral, Stanford Linear Accelerator Ctr, 72-73; asst res prof, Univ Wash, 73-76; asst prof, Univ Mich, 76-78; assoc res prof, Univ Calif, Davis, 78-79; div fel, 79-82, SR SCIENTIST, LAWRENCE BERKELEY LAB, 82- *Concurrent Pos:* Vis fel, Europ Orgn Nuclear Res, 82 & 85; div assoc ed, Phys Rev Lett, 86-89. *Mem:* Fel Am Phys Soc. *Res:* Theory of elementary particles and their interactions. *Mailing Add:* Theory Group Lawrence Berkeley Lab Berkeley CA 94720

CAHNMANN, HANS JULIUS, b Munich, Ger, Jan 27, 06; nat US; m 45; c 3. BIOCHEMISTRY. *Educ:* Univ Munich, Lic Pharm, 30, PhD(chem), 32. *Prof Exp:* Instr chem, Univ Munich, 31-33 & Univ Paris, 33-36; dir res, Labs, Crinex, 36-39; res assoc org chem, French Pub Health Serv, 39-40; res assoc biochem, Univ Aix Marseille, 40-41; dir res & develop, Biochem Prod Corp, NY, 41-42; res assoc, Mt Sinai Hosp, 42-44; sr chemist, Givaudan-Delawanna, 44-47 & William R Warner, 47-49; sr biochemist, 50-75, EMER SCIENTIST, NIH, 76- *Concurrent Pos:* Fel, Japan Soc Promotion Sci, Kyoto Univ, Japan, 71 & Nat Sci Found, 77. *Honors & Awards:* Parke-Davis Award, Am Thyroid Asn, 87. *Mem:* Am Chem Soc; Am Thyroid Asn. *Res:* Biochemistry. *Mailing Add:* 5430 Beech Ave Bethesda MD 20814

CAHOON, GARTH ARTHUR, b Vernal, Utah, Dec 23, 24; m 48; c 4. HORTICULTURE. *Educ:* Utah State Univ, BS, 50; Univ Calif, Los Angeles, PhD(hort sci), 54. *Prof Exp:* Res asst, Dept Floricult & Ornamental Hort, Univ Calif, Los Angeles, 50-54; asst horticulturist, Univ Calif, Riverside, 54-63; assoc prof, 63-67, PROF HORT, OHIO STATE UNIV & OHIO AGR RES & DEVELOP CTR, 67-, ASSOC CHMN, 83- *Concurrent Pos:* US AID consult, India, 68 & 70; mem, Coun Soil Test & Plant Anal; hort consult, Food & Agr Orgn, Somalia, Africa, 76-80 & Belize, Cent Am, 84. *Honors & Awards:* Laurie Award, Am Soc Hort Sci, 56. *Mem:* Fel Am Soc Hort Sci; Int Soc Hort Sci; Am Soc Enol & Viticult. *Res:* Pomology and viticulture; physiology and nutrition. *Mailing Add:* Dept Hort Ohio Agr Res & Develop Ctr Wooster OH 44691

CAHOON, JOHN RAYMOND, b Calgary, Alta, Aug 29, 39; m 65; c 3. MATERIALS SCIENCE, ENGINEERING. *Educ:* Univ Alta, BSc, 61, MSc, 63, PhD(metall eng), 66. *Prof Exp:* Fel metall, Mellon Inst, Carnegie-Mellon Univ, 66-68; from asst prof to assoc prof mech eng, 68-73, PROF MECH ENG, METALL SCI LAB, UNIV MAN, 73-, HEAD DEPT, 75- *Mem:* Fel Am Soc Metals; Can Inst Mining & Metall. *Res:* Solidification of metals; diffusion in metals; composite materials; corrosion. *Mailing Add:* Dept Mech Eng Univ Man Winnipeg MB R3T 2N2 Can

CAHOON, SISTER MARY ODILE, b Houghton, Mich, July 21, 29. CELL PHYSIOLOGY, RADIATION BIOLOGY. *Educ:* DePaul Univ, BS, 54, MS, 58; Univ Toronto, PhD(cell physiol), 61. *Prof Exp:* From instr to assoc prof, 54-67, chmn dept, 61-73, acad dean, 63-69 & 76-81, chmn, Div Natural Sci, 71-76, PROF BIOL, 67-, DEAN & SR VPRES, COL ST SCHOLASTICA, 81- *Concurrent Pos:* Res assoc, McMurdo Sta, Antarctica, 74; Sci Fac Fel, NSF. *Mem:* Am Soc Zoologists. *Res:* Radiation effects on growth of bacterial cells; calcium variations during crustacean intermolt cycle; ultraviolet resistance of bacterial spores; cold adaptation in metabolism of antarctic fauna. *Mailing Add:* Col St Scholastica Duluth MN 55811

CAHOY, ROGER PAUL, b Colome, SDak, Feb 19, 27; m 50; c 4. ORGANIC CHEMISTRY. *Educ:* Dakota Wesleyan Univ, BA, 50; Univ SDak, MA, 52; Univ Nebr, PhD(chem), 56. *Prof Exp:* Develop chemist. silicone prod dept, Gen Elec Co, 56-60; res chemist, Spencer Chem Co, 60-63, sr res chemist, 63-77, supvr, 77-80, sr adv, Gulf Oil Chem Co, 80-82; MGR LAB SERV, PBI GORDON CORP, 82- *Mem:* Am Chem Soc. *Res:* Organometallic compounds; catalysis; biological active organic compounds; industrial chemicals. *Mailing Add:* 8100 W 72 Terr Overland Park KS 66204-1732

CAI, JIN-YI, b Jan 23, 61; Chinese citizen. COMPLEXITY THEORY. *Educ:* Temple Univ, MA, 83; Cornell Univ, PhD(computer sci), 86. *Prof Exp:* Asst prof computer sci, Yale Univ, 86-89; ASST PROF COMPUTER SCI, PRINCETON UNIV, 89- *Concurrent Pos:* NSF res grant, 87-89; NSF presidential young investr, 90; ed, Int J Found Computer Sci, 90- *Mem:* Asn Comput Mach; Am Math Soc; Math Asn Am. *Res:* Theory of computational complexity; theory of polynomial time-space computable functions and the relationship among complexity classes. *Mailing Add:* Dept Computer Sci Princeton Univ Princeton NJ 08544-2087

CAILLE, ALAIN EMERIL, b St Jean, Que, Jan 17, 45; m 66; c 2. PHASE TRANSITIONS, MAGNETISM. *Educ:* Univ Montreal, BSc, 67; McGill Univ, MSc, 69, PhD(physics), 71. *Prof Exp:* Asst, Col France, Paris, 71-72; prof physics, Univ Que, 72-74; PROF PHYSICS, UNIV SHERBROOKE,

74-, VPRES RES, 89- *Concurrent Pos:* Mem coun, Fr Can Asn Advan Sci, 71-72, Natural Sci & Eng Res Can, Fonds poer formation de chercheur et á la recherche du Quebec. *Mem:* Can Asn Physics (vpres, 84-); Fr Can Asn Advan Sci; Natural Sci & Eng Res Coun Can; Int Union Pure & Appl Physics. *Res:* Static and dynamic properties of magneto-structural phase transitions in low dimensional solid state systems; theory and experiments (with collaborators) of ferromagnetic and antiferromagnetic systems. *Mailing Add:* Dept Physics Univ Sherbrooke Sherbrooke PQ J1K 2R1 Can

CAILLE, GILLES, b Montreal, Que, May 5, 35; m 57; c 4. MEDICINAL CHEMISTRY. *Educ:* Univ Montreal, BSc, 59, MSc, 65, PhD(med chem), 69. *Prof Exp:* Monitor instrumental analysis & pharmacog, 63-67, prof instrumental analysis, 67-69, RES PROF PHARMACOL, UNIV MONTREAL, 69- *Concurrent Pos:* Res fel metab psychotrope, Joliette Res Inst, 69-71, co-dir, 71-74; mem comt pharm, St Charles Hosp, Joliette, 69-75; consult, Burroughs Wellcome & Co, 70-76 & Santa Cabrini Hosp, 73-; pres, Biopharm Inc, 71-73. *Mem:* Am Acad Clin Toxicol; Fr Soc Toxicol. *Res:* Metabolism of psychotrope; clinical pharmacology and analytical toxicology. *Mailing Add:* Dept Pharmacol Fac Med Univ Montreal Montreal PQ H3C 3J7 Can

CAILLEAU, RELDA, b San Francisco, Calif, Feb 1, 09. CELL PHYSIOLOGY. *Educ:* Univ Calif, AB, 30, MA, 31; Univ Paris, DSc, 37. *Prof Exp:* Fr Nat Res fel, Pasteur Inst, Paris, 37-40; asst microbiol vitamin assays, Univ Calif, 41-43; assoc nutritionist & home economist, USDA, 43-46; in charge res, Nat Ctr Sci Res, Ministry Educ, France, 47-49; jr asst res biochemist, Univ Calif, Berkeley, 50-53, asst res biochemist tissue cult, 53-55, res assoc oncol & assit res biochemist, cancer res inst, Med Sch, Univ Calif, San Francisco, 55-59, assoc res biochemist, 59-70; res assoc med, Univ Tex, M D Anderson Hosp & Tumor Inst, 70-87; mem staff, John Muir Cancer & Aging Inst, 87-90; CONSULT, 90- *Mem:* Tissue Cult Asn. *Res:* Effects of various growth media and chemotherapeutic agents upon normal and malignant mouse and human cells in vitro and their chromosomes; establishment of human breast carcinoma cell lines from pleural effusions. *Mailing Add:* 1298 Albina Ave Berkeley CA 94706

CAILLIET, GREGOR MICHEL, b Los Angeles, Calif, June 19, 43; m 86; c 1. ICHTHYOLOGY, MARINE ECOLOGY. *Educ:* Univ Calif, Santa Barbara, BA, 66, PhD(biol), 72. *Prof Exp:* Staff res assoc ichthyol, Univ Calif, Santa Barbara, 71-72; asst prof, 72-77, assoc prof, 77-80, PROF ICHTHYOL, MOSS LANDING MARINE LABS, 80- *Mem:* Am Soc Icthyologists & Herpetologists; Am Fisheries Soc; Ecol Soc Am; Am Elasmobrance Soc (pres, 80); Estuarine Res Fed. *Res:* Ecology of marine fishes; age, growth and feeding habits of marine fishes, ecology of marine nekton; midwater ecology; fishery biology. *Mailing Add:* Moss Landing Marine Labs PO Box 450 Moss Landing CA 95039

CAILLOUET, CHARLES W, JR, b Baton Rouge, La, Dec 15, 37; m 77; c 4. FISHERIES SCIENCE. *Educ:* La State Univ, BS, 59, MS, 60; Iowa State Univ, PhD(fishery biol), 64. *Prof Exp:* Asst prof biol, Univ Southwestern La, 64-67; assoc prof, Rosenstiel Sch Marine & Atmospheric Sci, Univ Miami, 67-72; supvry fishery biologist, 72-74, fishery biologist, Res Admin, Gulf Coastal Fisheries Ctr, Galveston Facil, 74-77, chief, Environ Res & Aquacult Div, 77-85, SUPVRY FISHERY BIOLOGIST, NAT MARINE FISHERIES SERV, GALVESTON LAB, 77-, CHIEF, LIFE STUDIES DIV, 85- *Concurrent Pos:* Adj grad fac, Univ Houston, 75-88; vis mem grad fac, Tex A&M Univ, Galveston, 80-, lectr, 82-88. *Mem:* AAAS; Am Fisheries Soc; fel Am Inst Fishery Res Biol; Sigma Xi. *Res:* Directs Kemp's ridley sea turtle head start research project in the Gulf of Mexico. *Mailing Add:* Nat Marine Fisheries Serv Galveston Lab 4700 Ave U Galveston TX 77551-5997

CAIN, CARL, JR, b Chattanooga, Tenn, Jan 6, 31; m 61; c 2. ANALYTICAL CHEMISTRY, CHEMICAL ENGINEERING. *Educ:* Univ Chattanooga, BS, 52; Univ Tenn, MS, 54, PhD(chem), 56. *Prof Exp:* Proj engr, Cramet, Inc, 56-58; res engr, Develop Lab, Combustion Eng, Inc, 58-61; anal res chemist, Rock Hill Lab, Chemetron Corp, 61-62; staff chem engr, 62-70, supvr, Chem Sect, 70-78, SUPVR, CHEM & METAL ENG GROUP, TENN VALLEY AUTHORITY, 78- *Mem:* Am Chem Soc; Am Soc Mech Eng; Am Soc Testing & Mat. *Res:* Power plant analytical chemistry, especially automatic monitors and controllers; corrosion of metals in aqueous environment; industrial water conditioning. *Mailing Add:* 6201 Vance Rd Chattanooga TN 37421

CAIN, CHARLES ALAN, b Tampa, Fla, Mar 3, 43; m 78; c 3. BIOENGINEERING. *Educ:* Univ Fla, BEE, 65; Mass Inst Technol, MSEE, 67; Univ Mich, PhD(elec eng), 72. *Prof Exp:* Mem tech staff, Bell Labs, Inc, 65-68; from asst prof to assoc prof, 72-83, PROF ELEC & COMPUT ENGR & CHMN BIOENG FAC, UNIV ILL, URBANA, 83- *Mem:* Sr mem Inst Elec & Electronics Engrs; Radiation Res Soc; Bioelectromagnetics Soc; NAm Hyperthemia Group. *Res:* Biological effects and medical applications of ultrasound and microwave energy; mathematical modeling of physiological systems. *Mailing Add:* Dept Elec & Comput Eng Univ Mich EECS Bldg 1301 Beal Ave Ann Arbor MI 48109

CAIN, CHARLES COLUMBUS, b Fields, La, Oct 29, 15; m 39; c 2. SOIL SCIENCE. *Educ:* La State Univ, BS, 37; Iowa State Univ, MS, 38, PhD, 56. *Prof Exp:* Jr soil survr, Soil Conserv Serv, USDA, 38-41; asst soil technologist, 41; instr agron & asst agronomist, La State Univ, 41-46; assoc prof, Univ Southwestern La, 46-53, head dept gen agr, 55-65, prof agron, 53-75, dir tech studies, Educ Ctr, New Iberia, 65-67, dir freshman div, Univ, 67-75; CONSULT AGR, 75- *Mem:* Am Soc Agron; Soil Sci Soc Am. *Res:* Soil microscopy; soil morphology and genesis. *Mailing Add:* 720 Girard Park Dr Lafayette LA 70503

CAIN, CHARLES EUGENE, b Ellerbe, NC, June 22, 32; m 58. ORGANIC CHEMISTRY. *Educ:* Univ NC, BS, 54; Duke Univ, MA, 57, PhD(chem), 59. *Prof Exp:* Jr res chemist, Shell Chem Corp, 54; res chemist, E I du Pont de Nemours & Co, 58-60; chmn dept, 60-80, PROF CHEM, MILLSAPS COL, 60- *Mem:* Am Chem Soc; Sigma Xi. *Res:* Reactions of organometallic compounds; reaction mechanisms by relative rates and stereochem effects; electronic versus stereochemical effects in monomers and polymers; analysis of human brain hydrocarbons. *Mailing Add:* Dept Chem Millsaps Col Jackson MS 39210

CAIN, DENNIS FRANCIS, b Republican City, Nebr, June 4, 30; m 61; c 5. BIOCHEMISTRY, SCIENCE ADMINISTRATION. *Educ:* Creighton Univ, BS, 52; Georgetown Univ, MS, 56; Univ Pa, PhD(biochem), 60. *Prof Exp:* NIH fel, 60-61; assoc biochem, Sch Vet Med, Univ Pa, 61-63, from asst prof to assoc prof, 63-68; biochemist, Surg Neurol Br, Nat Inst Neurol Dis & Stroke, 68-74, scientist adminr, Spec Progs, Sci Rev Br, Div Res Grants, NIH, 74-78; prog dir rev, Div Cancer Res Resources & Ctrs, 78-80, chief, Grants Rev Br, Div Extramural Activ, 80-85, SPEC ASST TO ASSOC DIR, CANCER THER EVAL PROG, NAT CANCER INST, NIH, 85- *Concurrent Pos:* Res grant, Nat Inst Neurol Dis & Stroke. *Mem:* AAAS; Am Chem Soc; Sigma Xi. *Res:* Mechanochemistry of muscular contraction; metabolism of phosphorus, nucleic acids and protein in brain tissues. *Mailing Add:* Nat Cancer Inst NIH Exec Plaza Bldg Rm 741 Bethesda MD 20892

CAIN, GEORGE D, b Pittsburgh, Pa, June 1, 40; m 66; c 2. PARASITOLOGY, BIOCHEMISTRY. *Educ:* Sterling Col, BS, 62; Purdue Univ, MS, 64, PhD(biol), 68. *Prof Exp:* Fel, Univ Mass, 68-70; from asst prof to assoc prof, 70-81, chmn dept, 80-84, assoc dean, Col Liberal Arts, 84-87, PROF BIOL, UNIV IOWA, 81- *Concurrent Pos:* Vis res scholar, Univ New Mexico Sch Med, 79-80. *Mem:* AAAS; Am Soc Parasitol; Am Soc Zoologists; Nat Speleol Soc; Am Soc Trop Med Hyg; Sigma Xi. *Res:* Biochemistry of parasite membranes; genetic structure of parasites; lipid metabolism in parasites. *Mailing Add:* Dept Biol Univ Iowa Iowa City IA 52242

CAIN, GEORGE LEE, JR, b Wilmington, NC, Jan 13, 34; m 56; c 1. MATHEMATICS. *Educ:* Mass Inst Technol, BS, 56; Ga Inst Technol, MS, 62, PhD(math), 65. *Prof Exp:* Mathematician, Lockheed-Ga Co, 56-60; from instr to asst prof, 60-68, asst dir, Sch Math, 73-78, assoc prof, 68-80, PROF MATH, GA INST TECHNOL, 80- *Mem:* Soc Indust & Appl Math; Math Asn Am; Am Math Soc. *Res:* Topology; analysis. *Mailing Add:* Sch Math Ga Inst Technol Atlanta GA 30332

CAIN, J(AMES) ALLAN, b Isle of Man, Brit, July 23, 35; div; c 2. GEOLOGY. *Educ:* Univ Durham, BSc, 58; Northwestern Univ, MS, 60, PhD(geol), 62. *Prof Exp:* From instr to asst prof geol, Western Reserve Univ & Case Inst Technol, 61-66; assoc prof, Univ RI, 66-71, chmn dept, 67-83 & 84-87, actg assoc dean, Col Art & Sci, 83-84, actg chmn, Geol Dept, 89-90, actg chmn, Music Dept, 90-91, PROF GEOL, UNIV RI, 71-; STATE GEOLOGIST, 85- *Concurrent Pos:* Res Corp res grant, 62-65; NSF res grant, 64-66; US Navy Underwater Sound Lab grant, 68; hon res assoc, dept geol sci, Harvard Univ, 73; US Geol Surv res grant, 75, 86-87; sabbatical leave, Union Oil, Calif & Hanna Mining Co, 80, US Bur of Mines, 87. *Mem:* Geol Soc Am; Nat Asn Geol Teachers; Am Asn Univ Profs. *Res:* Mineral resources; environmental geology. *Mailing Add:* Dept Geol Univ RI Kingston RI 02881

CAIN, JAMES CLARENCE, b Kosse, Tex, Mar 19, 13; m 38; c 4. INTERNAL MEDICINE, GASTROENTEROLOGY. *Educ:* Univ Tex, BA, 33, MD, 37; Univ Minn, MMS, 47; Am Bd Internal Med, Am Bd Gastroenterol, dipl. *Prof Exp:* Intern, Protestant Episcopal Hosp, Philadelphia, Pa, 37-39; instr path, Med Sch, Univ Tex, Galveston, 39-40; instr path, 40-41 & 46-48, CONSULT MED, MAYO CLIN, 48-, PROF MED, MAYO GRAD SCH MED, UNIV MINN, MINNEAPOLIS, 72- *Concurrent Pos:* From asst prof to clin prof med, Mayo Grad Sch Med, Univ Minn, Minneapolis, 54-72; chmn, Combined Comt of Nat Adv Comt to Selective Serv for Selection of Doctors, Dentists, Vet & Allied Med Personnel & Nat Health Resources Adv Comt, 68-69, adv to dir, Selective Serv Syst, 69-70; mem, Nat Adv Heart Coun, Minn State Bd Med Exam & Nat Adv Comt to Health Manpower Comn; chmn, Fed Use of Health Manpower Panel; civilian consult to Surgeon Gen. *Honors & Awards:* Billings Gold Medal Award, AMA, 63. *Mem:* AAAS; AMA; Am Col Physicians; Am Asn Hist Med; Pan-Am Med Asn; Sigma Xi. *Res:* Lymphatic system; liver and intestinal tract; gastric ulcers. *Mailing Add:* 1915 65th St NE Rochester MN 55904-8910

CAIN, JAMES THOMAS, b Pittsburgh, Pa, May 2, 42; m 71. ELECTRICAL ENGINEERING. *Educ:* Univ Pittsburgh, BS, 64, MS, 66, PhD(elec eng), 70. *Prof Exp:* Asst instr elec eng, Univ Pittsburgh, 66-70; asst prof, 70-77, ASSOC PROF ELEC ENG, UNIV PITTSBURGH, 77- *Concurrent Pos:* Mem tech staff, Bell Tel Labs, 65; consult, var div of Westinghouse Elec Corp, 72-; chmn, Ethics Adv Bd, Inst Electronics & Elec Eng, 84-85 & 87-89, chmn, Strategic Planning Comt. 89-90, pres, Computing Sci Accrediation Bd, 88-89, vpres publ, Inst Electronics & Elec Div VIII Delegate/Dir, 90-; chmn, Mini/Micro Computer Tech Comt, Computer Soc, 80-82, assoc ed, Inst Electronic & Elec Eng Micro Magazine, 80-82, assoc-ed-in-chief, 83-85, co-chmn, Model Prog Computer Sci & Eng proj, 81-82, Educ Activities, 81-85 & 88-90, vchmn, Ethics Adv Bd, 82-83, bd gov, 83-84, Audit Comt, 83-84, Finance Comt, 84-87, rep dir, Computing Sci Accrediation Bd, 84-89, first vpres & acting pres, 86, vpres publ, 86-87, mem, Publ Bd, 86-90; mem Model Curricula Task Force, Joint Computer doc, Asn Comput Mach, 88-90. *Honors & Awards:* Develop Model Curricula Award, Computes Soc, 78, Outstanding Contrib to Computer Sci Award, 83, Meritorious Serv Award, 85, Outstanding Contrib Computer Sci & Eng Educ Award, 89. *Mem:* Inst Elec & Electronic Engrs; Am Soc Eng Educ (treas, 79-81); Asn Comput Mach; Sigma Xi. *Res:* Systems analysis and algorithm development for real time computing systems; interactive computing systems including robotics and biomedical systems. *Mailing Add:* Two Oakhurst Lane Pittsburgh PA 15215-1572

CAIN, JEROME RICHARD, b Piqua, Ohio, Apr 26, 47; m 82. ALGOLOGY. *Educ:* Miami Univ, BA, 69, MS, 72; Univ Conn, PhD(algol), 75. *Prof Exp:* Scientist, Tech Serv Div, Am Can Co, 69-70; teaching asst bot, Miami Univ, 70-72; res asst, Inst Water Resources, Univ Conn, 72-75; from asst prof to assoc prof, 75-84, PROF BOT, ILL STATE UNIV, 84- *Mem:* Phycol Soc Am; Sigma Xi. *Res:* Algal ecology, nutrition and morphology; environmental regulation of sexual and asexual reproduction in microalgae; algal bioassays for nutrient potential in freshwater ecosystems; effects of toxic substances on growth and reproduction of algae. *Mailing Add:* Dept Biol Sci Ill State Univ Normal IL 61761

CAIN, JOHN MANFORD, b Morrison, Ill, June 10, 32; m 55; c 2. RESOURCE MANAGEMENT. *Educ:* Univ Wis, BS, 55, MS, 56, PhD(soil sci), 67. *Prof Exp:* Soil scientist, Soil Conserv Serv, USDA, 58-60; teacher, Mich Jr High Sch, 60-61; soil scientist, Soil Conserv Serv, USDA, 61-65; asst prof soil sci, Wis State Univ, River Falls, 66-67; water resource planner, 67-83, SOIL SPECIALIST, DEPT NATURAL RESOURCES, STATE WIS, 83- *Mem:* Soil & Water Conserv Soc; Am Soc Agron. *Res:* Use of soil survey materials and techniques for nonagricultural interpretations. *Mailing Add:* Off Tech Servs Box 7921 Madison WI 53707

CAIN, JOSEPH CARTER, III, b Georgetown, Ky, Oct 31, 30; m 68; c 5. PLANETARY GEOLOGY. *Educ:* Univ Alaska, BS, 52, PhD, 57. *Prof Exp:* Physicist, Geophys Inst, Univ Alaska, 52-57, asst prof geophys res, 57-59; physicist, Fields & Particles Br, NASA, Goddard Space Flight Ctr, 59-63, head, World Magnetic Surv Sect, Fields & Plasmas Br, 63- 71, physicist, Lab Space Physics, 71-72, Applns Div, Geophys Br, 73-74; physicist, Off Geochem & Geophys, Br Theoret & Appl Geophys, US Geol Surv, Reston, Va, 74-75 & Denver, Colo, 75-83, Off Earthquakes, Volcanoes & Eng, Br Global Seismol & Geomagnetism, Denver, Colo, 83-86; res assoc, Coop Inst Res Environ Sci, Boulder, Colo, 86; vis res scientist, 87-89, ASSOC DEPT GEOL, FLA STATE UNIV, 90- *Concurrent Pos:* Chmn, Div 1, Internal Magnetic Fields, Int Asn Geomagnetism & Aeronomy, 73-75; pres, Sect Geomagnetism & Paleomagnetism, Am Geophys Union, 74-76. *Mem:* fel Am Geophys Union; Int Asn Geomag & Aeronomy; Sigma Xi. *Res:* Magnetic field experiments from spacecraft; geomagnetism; magnetospheric and earth physics; numerical modelling of fields. *Mailing Add:* Dept Geol B-160 Fla State Univ Tallahassee FL 32306-3026

CAIN, LAURENCE SUTHERLAND, b Washington, DC, Feb 4, 46; m 71; c 2. SOLID STATE PHYSICS. *Educ:* Wake Forest Univ, BS, 68; Univ Va, MS, 70, PhD(physics), 73. *Prof Exp:* Res assoc physics, Univ NC, Chapel Hill, 73-76, lectr, 76-78; DOE fac res partic, Oak Ridge Nat Lab, 86-87; asst prof, 78-85, chmn physics 88, ASSOC PROF PHYSICS, DAVIDSON COL, 85- *Mem:* Am Phys Soc; Am Asn Physics Teachers; Sigma Xi; AAAS. *Res:* Elastic, mechanical, and defect properties of solids through the use of ultrasonic, internal friction, diffusion, and conductivity measurements; optical properties of laser materials. *Mailing Add:* Dept Physics Davidson Col Davidson NC 28036

CAIN, STANLEY A, b Deputy, Ind, June 19, 02. APPLIED BIOLOGY. *Prof Exp:* At dept appl biol, Univ Calif, Santa Cruz, 82; RETIRED. *Mem:* Nat Acad Sci. *Mailing Add:* 109 Oak Knoll Santa Cruz CA 95060

CAIN, STEPHEN MALCOLM, b Lynn, Mass, Oct 4, 28; m 51; c 2. PHYSIOLOGY. *Educ:* Tufts Col, BS, 49; Univ Fla, PhD(physiol), 59. *Prof Exp:* Physiologist, Chem Corps Med Lab, 53-56; asst physiol, Johns Hopkins Univ, 56; asst, Univ Fla, 56-59, trainee, 59; physiologist, US Air Force Sch Aerospace Med, 59-71, head sect, 60-71, chief respiratory sect, 68-71; PROF PHYSIOL & ASSOC PROF MED, UNIV ALA, BIRMINGHAM, 71-, PROF ANESTHESIOL, 87-, PROF PEDIAT, 90- *Mem:* Am Physiol Soc; Sigma Xi; Int Soc Oxygen Transport Tissue (treas, 89-94); Am Thoracic Soc; Am Heart Asn. *Res:* Pulmonary physiology; blood gas transport; tissue hypoxia. *Mailing Add:* Dept Physiol & Biophysics Univ Ala Birmingham AL 35294

CAIN, WILLIAM AARON, b Leesville, La, Apr 30, 39; m 59; c 2. IMMUNOLOGY, MICROBIOLOGY. *Educ:* Northwestern State Univ La, BS, 61, MS, 64; La State Univ, PhD(immunol), 66. *Prof Exp:* Res fel, Univ Minn, 66-67, USPHS res fel pediat, 67-69; ASSOC PROF MICROBIOL & IMMUNOL & ADJ ASSOC PROF MED & PEDIAT, HEALTH SCI CTR, UNIV OKLA, 69- *Concurrent Pos:* Immunol consult, Okla Med Res Found, 69-71; allergy immunol consult. *Mem:* Am Soc Microbiol; Am Thoracic Soc. *Res:* Allergy and immediate hypersensitivity; cell mediated immunity and anergy. *Mailing Add:* Dept Microbiol & Immunol Univ Okla Health Sci Ctr Box 26901 Oklahoma City OK 73190

CAIN, WILLIAM F, b Orange, Calif, Apr 7, 37. ISOTOPE GEOCHEMISTRY. *Educ:* Loyola Univ, Los Angeles, BA, 58; Univ Detroit, MS, 65; Univ Santa Clara, STM, 70; Univ Calif, San Diego, PhD(chem), 75. *Prof Exp:* From asst prof to assoc prof, 74-86, chair, Dept Chem, 80- 84, CHAIR, DEPT CHEM & BIOCHEM, LOYOLA MARYMOUNT UNIV, 88-, PROF, 86- *Mem:* Am Chem Soc; AAAS; Sigma Xi. *Res:* Secular atmospheric C-14 fluctuations as recorded in tree ring cellulose; trace elements in environmental materials. *Mailing Add:* Dept Chem & Biochem Loyola Marymount Univ 7101 W 80th St Los Angeles CA 90045-2699

CAINE, DRURY SULLIVAN, III, b Selma, Ala, June 9, 32; m 61; c 2. ORGANIC CHEMISTRY. *Educ:* Vanderbilt Univ, BA, 54, MS, 56; Emory Univ, PhD(chem), 61. *Prof Exp:* Fel org chem, Columbia Univ, 61-62; from asst prof to prof chem, Ga Inst Technol, 74-83; CHMN CHEM DEPT, UNIV ALA, 83- *Mem:* Am Chem Soc; Brit Chem Soc. *Res:* Synthesis of natural products; chemistry of enolate anions and organolithium reagents. *Mailing Add:* Univ Ala PO Box H Tuscaloosa AL 35487

CAINE, T NELSON, b Barrow, Eng, Dec 17, 39; m 66. GEOMORPHOLOGY, HYDROLOGY. *Educ:* Univ Leeds, BA, 61, MA, 62; Australian Nat Univ, PhD(geomorphol), 66. *Prof Exp:* Lectr geog, Univ Canterbury, 66-67; assoc prof geomorphol, 68-80, PROF GEOG, INST ARCTIC & ALPINE RES, UNIV COLO, BOULDER, 80- *Mem:* Glaciol Soc; Am Geophys Union; Am Water Resources Asn; Australian & NZ Asn Advan Sci; Asn Am Geographers. *Res:* Evolution of hillslopes in the alpine environment; snow and glacier hydrology and water resources of the alpine area. *Mailing Add:* Dept Geog Univ Colo Boulder CO 80309

CAIRD, JOHN ALLYN, b Brooklyn, NY, Dec 24, 47. PHYSICS, LASERS. *Educ:* Rutgers Univ, BSEE, 69; Univ Calif, Los Angeles, MSPh, 71; Univ Southern Calif, PhD(physics), 75. *Prof Exp:* Fel & mem tech staff lasers, Hughes Aircraft Co, 69-75, staff physicist, 75-76; fel laser spectros, Chem Div, Argonne Nat Lab, 76-78; sr scientist, Bechtel Nat Inc, 78-81; mem staff appl photochem, Los Alamos Nat Lab, 81-84; PHYSICIST, LAWRENCE LIVERMORE NAT LAB, 84- *Concurrent Pos:* NSF res fel, 76-77. *Mem:* Inst Elec & Electronics Engrs; Am Phys Soc. *Res:* Experimental molecular laser isotope separation; development and spectroscopic analysis of laser devices, atomic structure of lanthanide and actinide elements in crystals and gaseous complexes, electronic transition probabilities and dynamics of energy transfer processes. *Mailing Add:* Lawrence Livermore Nat Lab MS L-493 PO Box 808 Livermore CA 94550

CAIRNCROSS, ALLAN, b Orange, NJ, May 27, 36; m 77; c 4. PHYSICAL ORGANIC CHEMISTRY, PHOTOIMAGING SCIENCE. *Educ:* Cornell Univ, BA, 58; Yale Univ, MS, 59, PhD(org chem), 63. *Prof Exp:* ORG RES CHEMIST, CENT RES DEPT, E I DU PONT DE NEMOURS & CO, INC, 64- *Mem:* Am Chem Soc. *Res:* Stability of tropylium ions; thermal isomerization of tropilidenes; cycloaddition reactions of bicyclobutanes; substituent effects on tropilidene-norcaradiene equilibria, organocopper chemistry and structure, heterocycles, and polymers; photoimaging. *Mailing Add:* 24 Cinnamon Dr RD 3 Hockessin DE 19707-1349

CAIRNS, ELTON JAMES, b Chicago, Ill, Nov 7, 32; m 59, 74; c 3. ELECTROCHEMISTRY, PHYSICAL CHEMISTRY. *Educ:* Mich Col Mining & Technol, BS(chem) & BS(chem eng), 55; Univ Calif, Berkeley, PhD(chem eng), 59. *Prof Exp:* Chemist, US Steel Corp, 53; soils res chemist, Corps Eng, US Army, 54-55; asst thermodyn, Univ Calif, 57-58; phys chemist electrochem, Gen Elec Res Lab, 59-66; phys chemist & group leader, Argonne Nat Lab, 66-70, sect head, 70-73; asst head, electrochem dept, Gen Motors Res Labs, Warren Mich, 73-78; HEAD APPL SCI DIV, LAWRENCE BERKELEY LAB, 78-, ASSOC LAB DIR, 78-; PROF CHEM ENG, UNIV CALIF, BERKELEY, 78- *Concurrent Pos:* Consult, US Dept Interior & US Dept Defense, 69-70, US Dept Defense, 73-78, & NASA, 71, Mat Res Coun, 73; mem subpanel, US Dept Com Panel on Elec Powered Vehicles, 67; mem org comt, Intersoc Energy Conversion Eng Conf, 69, steering comt, 70-, gen chmn, Conf, 76, 90; mem adv bd, Advances Electrochem & Electrochem Eng, 75-; chmn, Energy Conversion, Am Inst Chem Engrs; mem panel, Nat Acad Sci Comt Motor Vehicle Emissions, 72-73; sci deleg, NATO Conf Air Pollution, Eindhoven, 71; div ed, J Electrochem Soc, 69-90; Case Centennial Scholar & lectr, Case Western Reserve Univ, 80; mem, Comt Battery Mat Technol, Nat Acad Sci, 81-82 & Comt Electrochem Aspects Energy Conserv & Prod, 85-86; regional ed, Electrochemica Acta, 85- *Honors & Awards:* Francis Mills Turner Award, Electrochem Soc, 63; Croft lectr, Univ Mo, 79. *Mem:* AAAS; Electrochem Soc (vpres, 86-89, pres 89-90); Am Chem Soc; Am Inst Chem Eng; fel Am Inst Chem; Int Soc Electrochem (vpres, 85-88). *Res:* Electrochemical energy conversion; thermodynamics; transport phenomena; molten salts; liquid metals; surface chemistry. *Mailing Add:* Appl Sci Div Lawrence Berkeley Lab One Cyclotron Rd Berkeley CA 94720

CAIRNS, JOHN, JR, b Conshohocken, Pa, May 8, 23; m 44; c 4. PROTIST COMMUNITY DYNAMICS. *Educ:* Swarthmore Col, AB, 47; Univ Pa, MS, 49, PhD(protozool), 53. *Prof Exp:* From asst cur to cur limnol, Acad Natural Sci, Philadelphia, 48-66, res assoc, 68; prof zool, Univ Kans, 66-68; res prof, 68-70, UNIV DISTINGUISHED PROF BIOL, VA POLYTECH INST & STATE UNIV, 70-, DIR CTR FOR ENVIRON STUDIES, 70- *Concurrent Pos:* Vis prof, Rocky Mountain Biol Lab, 61-63 & 85, Univ Mich Biol Sta, 64-83; lectr, Dept Educ, Temple Univ, 62-63; trustee, Rocky Mountain Biol Lab, 62-72, vpres, 71-72, sr res investr, 84-91; gov bd, Am Inst Biol Sci, 76-79; mem panel fresh-water aquatic life, Nat Acad Sci; chmn, Comt Toxic Chem, Water Pollution Control Fedn, 78-80; pres, Sci Adv Bd, US Environ Protection Agency, 79-83; mem, Environ Studies Bd, Nat Res Coun, 80-83. *Honors & Awards:* Presidential Commendation, 71; Dudley Award, Am Soc Testing & Mat, 78; Super Achievement Award, US Environ Protection Agency, 80; Founder's Award, Soc Environ Toxicol & Chem, 81; Icko Iben Award, Am Water Resources Asn, 84; Morrison Medal, Am Chem Soc, 84; UN Environ Prog Medal, 88; Award of Excellance, Am Fisheries Soc, 89. *Mem:* Nat Acad Sci; fel AAAS; Am Water Resources Asn; Nat Asn Environ Prof; Am Micros Soc (pres, 80); fel Am Acad Arts & Sci. *Res:* Ecotoxicology; restoration ecology; ecosystem hazard evaluation and risk analysis; environmental effects of genetically-engineered microorganisms; fresh water protist colonization dynamics. *Mailing Add:* Univ Ctr Environ & Hazardous Mat Studies Va Polytech Inst & State Univ Blacksburg VA 24061-0415

CAIRNS, JOHN MACKAY, b Scranton, Pa, Dec 3, 12; m 48; c 4. EMBRYOLOGY. *Educ:* Hamilton Col, BA, 35; Univ Rochester, MS, 37; Washington Univ, PhD(embryol), 41. *Prof Exp:* Instr, Washington Univ, 41-42; instr, Univ Tex, 46-52, asst prof biol & embryol, 48-52; asst prof histol & embryol, Sch Med, Univ Okla, 52-56; engr, Bell Aircraft Corp, 56-57; sr cancer res scientist, Roswell Park Mem Inst, 57-67, assoc cancer res scientist, 67-80; RETIRED. *Mem:* Am Soc Zoologists; Am Asn Anatomists; Soc Develop Biol; Int Soc Develop Biol. *Res:* Experimental embryology; regional differentiation in mesoderm; application of control system concepts. *Mailing Add:* Mill St RD 1 Springville NY 14141

CAIRNS, STEPHEN DOUGLAS, b Port Sulphur, La, Nov 26, 49; m 72; c 2. SYSTEMATICS, ZOOGEOGRAPHY. *Educ:* La State Univ, BS, 71; Univ Miami, MS, 73, PhD(biol oceanog), 76. *Prof Exp:* Res assoc, 79-85, fel zool, 81-85, ASSOC CURATOR, SMITHSONIAN INST, 85- *Concurrent Pos:* NSF Grad fel,71-73. *Res:* Descriptive systematics and zoogeogrphy of deep-water western Atlantic, Antarctic, and Pacific Scleractinia and Sylasterina (Coelenterata); worldwide revision and phylogenetic analysis of stylasterine genera. *Mailing Add:* Dept Invert Zool NHB-163 W-329 Smithsonian Inst Washington DC 20560

CAIRNS, THEODORE L, b Edmonton, Alta, July 20, 14; nat US; m 40; c 4. ORGANIC CHEMISTRY. *Educ:* Univ Alta, BSc, 36; Univ Ill, PhD(org chem), 39. *Hon Degrees:* LLD, Univ Alta, 70. *Prof Exp:* Lab asst, Univ Ill, 37-39; instr org chem, Univ Rochester, 39-41; res chemist, E I du Pont de Nemours & Co, 41-45, res supvr, 45-51, lab dir, 51-61, dir basic sci, 62-66, dir res, 66-67, asst dir cent res dept, 67-71, dir cent res & develop dept, 71-79; RETIRED. *Concurrent Pos:* Regents prof, Univ Calif, Los Angeles, 65-66; Fuson lectr, Univ Nev, 68; mem, President Nixon's Sci Policy Task Force, 69, Gov Coun Sci & Technol, 69-72, President's Sci Adv Comt, 70-73 & President's Comt on Nat Medal Sci, 74-75; chmn, Off Chem & Chem Technol, Nat Res Coun, 79-82, mem, Bd Toxicol & Environ Health Hazards, 80-83. *Honors & Awards:* Am Chem Soc Award, 68; Synthetic Org Chem Mfrs Asn US Medal, 68; Perkin Medal, Am Sect, Soc Chem Indust, 73; Cresson Medal, Franklin Inst, 74. *Mem:* Nat Acad Sci. *Res:* Stereochemistry of biphenyls; chemistry of polyamides; reactions of acetylene and carbon monoxide; cyanocarbon chemistry. *Mailing Add:* Two Ridge Run Rd Chadds Ford PA 19317-9332

CAIRNS, THOMAS W, b Hutchinson, Kans, Nov 13, 31; m 59; c 2. MATHEMATICS. *Educ:* Okla State Univ, BS, 53, MS, 55, PhD(math), 60. *Prof Exp:* Head dept, 69-76, chmn div math sci, 72-80, PROF MATH, UNIV TULSA, 69- *Mem:* Math Asn Am; Am Math Soc; Soc Indust & Appl Math. *Res:* Computing; biomathematics; time series analysis. *Mailing Add:* Dept Math Sci Univ Tulsa 600 S Col Ave Tulsa OK 74104

CAIRNS, WILLIAM LOUIS, b St Catharines, Ont, Oct 18, 42. BIOCHEMICAL ENGINEERING, PETROLEUM MICROBIOLOGY. *Educ:* Univ Guelph, BSA, 65; Iowa State Univ, PhD(biochem), 70. *Prof Exp:* Asst prof chem, Univ Ark, Fayetteville, 70-76; res assoc, Max Planck Inst Cell Biol, Heidelberg, 76-78; lectr plant sci, 78-79, ASST PROF BIOCHEM ENG, UNIV WESTERN ONT, 79-, RES ASSOC, 78- *Mem:* Am Chem Soc; AAAS; Am Soc Photobiol; Sigma Xi. *Res:* Microbial production of surface active agents; biochemical basis of circadian rhythms. *Mailing Add:* 402 Victoria St London ON N5Y 4A9 Can

CAIS, RUDOLF EDMUND, b Prague, Czech, Nov 12, 47; US citizen; m 71; c 2. POLYMER CHEMISTRY. *Educ:* Univ Queensland, Australia, BSc Hons, 70, PhD(polymer chem), 75. *Prof Exp:* Consult, 76, MEM TECH STAFF POLYMER CHEM, BELL LABS, 77- *Mem:* Am Chem Soc; Am Phys Soc; AAAS. *Res:* Nuclear magnetic resonance studies of the structure and dynamics of synthetic macromolecules; polymerization and copolymerization mechanisms. *Mailing Add:* 3-43-10 00KA Minami-Ku, Kanagawa 232 Yokohama Japan

CAKMAK, AHMET SEFIK, b Izmir, Turkey, Aug 14, 34; m 71; c 2. EARTHQUAKE ENGINEERING. *Educ:* Robert Col Istanbul, BSE, 57; Princeton Univ, MSE, 58, MA, 60, PhD(eng mech), 62. *Prof Exp:* Asst solid mech, Columbia Univ, 60-62; res assoc, 62-63, from asst prof to assoc prof civil eng, 69-71, chmn dept, 71-80, PROF CIVIL ENG, PRINCETON UNIV, 71- *Concurrent Pos:* Assoc dean, Sch Engr & Appl Sci, Princeton Univ, 81-87. *Mem:* Am Soc Civil Engrs; Am Soc Mech Eng; Soc Rheol. *Res:* Continuum mechanics; statics and dynamics of inelastic media; elastic and viscoelastic systems; earthquake engineering. *Mailing Add:* Sch Eng & Appl Sci Princeton Univ Princeton NJ 08544

CALABI, EUGENIO, b Milano, Italy, May 11, 23; nat US; m 52; c 1. MATHEMATICS. *Educ:* Mass Inst Technol, BS, 46; Univ Ill, AM, 47; Princeton Univ, PhD(math), 50. *Prof Exp:* Asst & instr, Princeton Univ, 47-51; asst prof math, La State Univ, 51-55; from asst prof to prof, Univ Minn, 55-64; prof, 64-69, chmn dept, 73-76, THOMAS A SCOTT PROF MATH, UNIV PA, 69- *Concurrent Pos:* Mem, Inst Advan Study, 58-59; Guggenheim fel, 62-63. *Mem:* Nat Acad Sci; Am Math Soc. *Res:* Differential geometry of complex manifolds. *Mailing Add:* Dept Math Univ Pa 209 S Third-33 St Philadelphia PA 19104-6394

CALABI, LORENZO, b Milan, Italy, Apr 11, 22; nat US; m 49; c 4. APPLIED MATHEMATICS. *Educ:* Swiss Fed Inst Technol, dipl, 46; Univ Milan, PhD(math), 47; Univ Strasbourg, PhD(math), 51. *Prof Exp:* Res fel, Inst Advan Math, Italy, 51-52; asst prof, Boston Col, 52-56; res assoc, Parke Math Labs, Inc, 55-56, assoc dir res, 56-74; tech dir, Solo Corp, 74-78; PRES, PARKE MATH LABS, 78- *Concurrent Pos:* Fulbright travel award, 52; lectr, Holy Cross Col, 53-54; assoc prof, Boston Col, 56-59, lectr, 59-62. *Mem:* Asn Comput Mach; Am Math Soc; Soc Indust & Appl Math; Ital Math Union. *Res:* Computer sciences applications. *Mailing Add:* Nine Moreland Ave Newton MA 02159

CALABI, ORNELLA, US citizen. IMMUNOLOGY, CHEMOTHERAPY. *Educ:* Royal Univ Milan, Italy, BSc, 36; Hebrew Univ, Jerusalem, MSc, 45; Univ Chicago, Msc, 51; Harvard Univ, DSc(microbiol), 57. *Prof Exp:* Res asst trop med, Hebrew Univ, Jerusalem, 40-45; res asst med, Univ Chicago, 48-49, res asst bact & parasitol, 49-51; instr bact, Brandeis Univ, 53-54; instr microbiol & in-chg trop med, Sch Med, Yale Univ, 56-57; res assoc microbiol, Dept Metall, Mass Inst Technol, 57-58; res assoc path, Children's Hosp, Harvard Med Ctr, Boston, 59-60; instr microbiol, Seton Hall Col Med, NJ, 60-61; res assoc path, New York Med Col, 61-63; microbiologist, Walter Reed Army Inst Res, Washington, DC, 64-70, US CIVIL SERV CAREER SCIENTIST, 67- *Concurrent Pos:* Vis scientist, Queen Elizabeth Col, Univ London, 71-72; adj dir, Cantonal Inst Bact & Serol, Lugano, Switz, 73- 74; consult toxicol, Nat Cancer Inst, NIH, Bethesda, Md, 76- *Mem:* NY Acad Sci; AAAS; fel Am Acad Microbiol. *Res:* Pathogenesis of disease; blood and enteric bacterial infections; relapsing fever, shigellosis, melioidosis; humoral and cellular immunity, phagocytosis; chemotherapy, synergism and antagonism of drugs and of oncolytic agents; toxicity tests; mathematical models of drug interaction and their application in biological systems. *Mailing Add:* 5100 Dorsey Ave Kenwood MD 20815

CALABRESE, ANTHONY, b Providence, RI, Feb 25, 37; m 63; c 2. MARINE ECOLOGY. *Educ:* Univ RI, BS, 59; Auburn Univ, MS, 62; Univ Conn, PhD(zool, ecol), 69. *Prof Exp:* FISHERY BIOLOGIST, NAT OCEANIC & ATMOSPHERIC ADMIN, NAT MARINE FISHERIES SERV, 62- *Mem:* Am Fisheries Soc; Nat Shellfisheries Asn (secy-treas, 74-76, vpres, 76-77, pres elect, 77-78, pres, 78-79); Estuarine Res Fedn. *Res:* Development of biological information concerning the effect of pollutants on marine organisms, including shellfish, fin fish and crustaceans, to provide a basis for environmental management. *Mailing Add:* Nat Marine Fisheries Serv Northeast Fisheries Ctr Milford CT 06460

CALABRESE, CARMELO, b Kansas City, Mo, Apr 21, 29; m 54; c 2. ELECTRICAL ENGINEERING. *Educ:* Univ Mo-Columbia, BS, 55, MS, 56, PhD, 65. *Prof Exp:* From instr to prof elec eng, Univ Mo-Columbia, 55-75; dir elec syst res, 75-85, dir electronic commun, 85-90, TECH CONSULT, CONSOL EDISON CO NEW YORK, INC, 90- *Concurrent Pos:* Res & develop fel, Elec Boat/Gen Dynamics, 66-67; Ford Found grant eng, 66; NSF travel grant, 66. *Mem:* Inst Elec & Electronic Engrs. *Res:* Electric power systems; energy conversion. *Mailing Add:* Cent Eng Consol Edison Co New York Inc Four Irving Pl Rm 1445-S New York NY 10003

CALABRESE, DIANE M, b Erie, Pa, Apr 1, 49; m 80. ENTOMOLOGY. *Educ:* Gannon Col, BS, 71; Univ Conn, MS, 74, PhD(entomol), 77. *Prof Exp:* Instr biol, Kans State Univ, 77-78; lectr, Univ Tex, El Paso, 78-80; asst prof, Trinity Col, 80-81; dir, Florence Jones Reineman Wildlife Sanctuary, 81-84, asst prof biol ecol & gen biol, Dickinson Col, 81-87; Noyes res fel, Bunting Inst Radcliffe Res & Study Ctr, 87-88; CONSULT, PAPILLONS DIVERSIFIED ENDEAVORS, 87- *Concurrent Pos:* Ed, newsletter, Women in Entomol, 81-; res assoc, Mo Coord Bd Higher Educ, 90- *Mem:* Soc Syst Zool; Entomol Soc Am; Am Soc Zoologists. *Res:* Problems of an evolutionary-ecological nature in systematics; character congruence among ontogenetic, cytogenetic and adult morphological characters in the Gerridae; land use philosophy and reserve management; neoteny as an evolutionary agent. *Mailing Add:* Papillons Diversified Endeavors 22 Anderson Ave Columbia MO 65203-2673

CALABRESE, JOSEPH C, b Cleveland, Ohio; m 77; c 1. CRYSTALLOGRAPHY, COMPUTING. *Educ:* Univ Wis, PhD(chem), 72. *Prof Exp:* Fel, Univ Milan, 71-72 & Univ Wis, 72-79; CHEMIST, E I DU PONT DE NEMOURS, CO, INC, 80- *Mem:* Am Chem Soc; Am Crystallog Asn. *Res:* Crystallography; crystallographic computing and methodology; organometallic chemistry. *Mailing Add:* Exp Sta PO Box 80228 E I du Pont de Nemours & Co Inc Wilmington DE 19880-0228

CALABRESE, PHILIP G, b Chicago, Ill, Feb 21, 41; m 72; c 4. PROBABILITY LOGIC, REASONING WITH UNCERTAINTY. *Educ:* Ill Inst Technol, BS, 63, MS, 66, PhD(math), 68. *Prof Exp:* Instr math, Ill Inst Technol, 67-68; asst prof, Naval Postgrad Sch, 68-70; asst prof math, Calif State Col, Bakersfield, 70-75; lectr, Humboldt State Univ, 75-79; sr staff engr, Decision Sci Inc, 79-82; analyst-programmer, Orincon Corp, 82-83; sr analyst, Logicon Corp, 83-89; SR RES ASSOC, NAT RES COUN, 90- *Concurrent Pos:* Instr, Rosary Col, 66-67; adj prof math, San Diego State Univ, 80-; instr, Univ Calif, San Diego, 83. *Mem:* Math Asn Am. *Res:* An algebraic synthesis of the foundations of logic and probability; a mathematical theory of ultimate physical quanta; cosmology. *Mailing Add:* 2911 Luna Ave San Diego CA 92117

CALABRESE, RONALD LEWIS, b Peekskill, NY, Sept 24, 47; m 79; c 1. NEUROBIOLOGY. *Educ:* Cornell Univ, BS, 69; Stanford Univ, AM, 70, PhD(neurobiol), 75. *Prof Exp:* ASST PROF NEUROBIOL, DEPT CELLULAR & DEVELOP BIOL, HARVARD UNIV, 78- *Mem:* Soc Neurosci; AAAS. *Res:* Identification and analysis of neural circuits which generate and coordinate rhythmic behaviors at the cellular level. *Mailing Add:* Biol Dept Emory Univ Atlanta GA 30322

CALABRESI, PAUL, b Milan, Italy, Apr 5, 30; US citizen; m 54; c 3. PHARMACOLOGY, CHEMOTHERAPY. *Educ:* Yale Univ, BA, 51, MD, 55; Am Bd Internal Med, dipl, 64. *Prof Exp:* Intern, Harvard Med Serv, Boston City Hosp, Mass, 55-56, asst resident, 58-59; proj assoc, Univ Wis, 56-58; from instr to assoc prof med & pharmacol, Yale Univ, 60-68; PROF MED SCI, BROWN UNIV, 68-, CHMN DEPT MED, 74-; CLIN PROF PHARMACOL, COL PHARM, UNIV RI, KINGSTON, 77- *Concurrent Pos:* Field investr, Nat Cancer Inst, NIH, 56-60, mem, Cancer Chemother Collab Prog Rev Comt, 65-66, bd sci counrs, 83-87, mem, Pharmacol-Toxicol Rev Comt, Nat Inst Gen Med Sci, 67-70, Exp Therapeut Study Sect, 72-76, chmn, 75-76; res fel, Dept Med, Yale Univ, 59-60, head, Div Clin Pharmacol & Chemother & dir, Clin Pharmacol Res Ctr, 65-67; Burroughs Wellcome scholar clin pharmacol, 64-68; Eleanor Roosevelt Int Cancer fel, Am Cancer Soc, 66-67; vis scientist, Univ Lausanne, Switz, 66-67; physician-in-chief & chmn, Dept Med, Roger Williams Gen Hosp, Providence, RI, 68-91, vpres acad affairs, 77-91; mem, Res Coun & Drug Res Bd, Nat Acad Sci, 68-75; mem, Sci Group on Eval & Testing of Drugs for Mutagenicity: Principles & Probs, WHO, 71, consult, Study Group Hycanthone, 71 & Study Group on Eval & Use of Immunosuppressive Agents, 71; counr, Environ Mutagen Soc, 71-74; chief of med, Women & Infants Hosp RI, 74-80; consult, Miriam Hosp, Mem Hosp, Providence Vet Admin Med Ctr, RI Hosp & St Joseph's Hosp, 74-; mem, Sci Prog Comt, Am Col Physicians, 75-78, Clin Pharmacol Comt, 77-82, chmn, 78-82; vis prof numerous univs, 77-; mem, Sci & Pub Affairs Comt, Am Asn Cancer Res, 83-; mem, Med Oncol Training Prog, Am Soc Clin Oncol, 83- *Mem:* Inst Med-Nat Acad Sci; Am Soc Hemat; Am Soc

Pharmacol & Exp Therapeut; Am Soc Clin Oncol (pres, 69-70); Am Fedn Clin Res; Am Asn Clin Res; Am Bd Internal Med (secy-treas, 82-84); Am Cancer Soc; master Am Col Physicians; Am Soc Clin Pharmacol & Therapeut. *Res:* Medical oncology; hematology; virology; immunology; metabolism of nucleic acids; autoimmune disease; bone marrow function and physiology; antimetabolites in viral and cancer chemotherapy. *Mailing Add:* Dept Med Brown Univ Box G Providence RI 02912

CALABRESI, PAUL, b Naples, Italy, Jan 27, 07; nat; m 40. ANATOMY. *Educ:* Cath Univ Am, BA, 31; George Washington Univ, MA, 40; Cambridge Univ, PhD(anat), 55. *Prof Exp:* From instr to prof anat, Sch Med, George Washington Univ, 39-74; vis prof anat, Univ PR Med Ctr, 74-76; PROF MED SCI, BROWN UNIV, 76- *Concurrent Pos:* Vis lectr, Nat Naval Med Ctr, 47-48, Am Univ, 47-48, Washinton Hosp Ctr, 58, Armed Forces Inst Path, 58, Casualty Hosp, 59 & Nat Univ Athens, 69; consult, Naval Med Res Inst, 47-51; examr, Am Bd Orthop Surg, 48-; lectr, Columbian Hosp for Women & George Washington Hosp, 48-49; lectr & supvr, Trinity Col, Cambridge Univ, 53-55; vis prof, Queen's Univ, Belfast, 69; guest lectr, Prince Georges Hosp, Md. *Honors & Awards:* Cross of Eloy Alfaro, Panama. *Mem:* AAAS; Soc Exp Biol & Med; Am Asn Anatomists; Asn Am Med Cols; Anat Soc Gt Brit & Ireland. *Res:* Embryology; pathology. *Mailing Add:* Dept Med Roger Williams Gen Hosp 825 Chalkstone Providence RI 02908

CALAHAN, DONALD A, b Cincinnati, Ohio, Feb 23, 35; m 59; c 4. ELECTRICAL ENGINEERING. *Educ:* Univ Notre Dame, BS, 57; Univ Ill, MS, 58, PhD(elec eng), 60. *Prof Exp:* From instr to assoc prof elec eng, Univ Ill, 59-65; prof, Univ Ky, 65-66; PROF ELEC ENG, UNIV MICH, 66- *Res:* Investigation of theoretical limitations in the design of electrical networks and of the use of computers in network design. *Mailing Add:* Dept Elec & Comput Eng Univ Mich 500 S State St Ann Arbor MI 48104

CALAMARI, TIMOTHY A, JR, b New Orleans, La, Nov 12, 36; m 64; c 3. TEXTILE CHEMISTRY. *Educ:* Loyola Univ, La, BS, 58; La State Univ, MS, 61, PhD(chem), 63. *Prof Exp:* RES CHEMIST, SOUTHERN UTILIZATION RES & DEVELOP LAB, USDA, NEW ORLEANS, 63- *Mem:* Am Asn Textile Chemists & Colorists; Sigma Xi; Am Chem Soc; Am Asn Clin Chemists. *Res:* Additive finishing techniques for the development of dimensionally stable cotton fabric; liquid ammonia based treatments for cotton textiles; new fire retardant finishes for cotton and cotton blends; expert on the proper disposal procedures for toxic and dangerous waste chemicals. *Mailing Add:* 1016 Rosa Ave Metairie LA 70005

CALAME, GERALD PAUL, b Lelocle, Switz, Nov 27, 30; US citizen; m 59; c 1. PHYSICS & ASTROPHYSICS, OPTICS. *Educ:* Col of Wooster, BA, 53; Harvard Univ, AM, 55, PhD(physics), 60. *Prof Exp:* Theoret physicist, Knolls Atomic Power Lab, Gen Elec Co, 59-61; from asst prof to assoc prof nuclear eng, Rensselaer Polytech Inst, 61-66; assoc prof, 66-69, chmn dept, 78-82, PROF PHYSICS, US NAVAL ACAD, 69- *Concurrent Pos:* Consult, Knolls Atomic Power Lab, 61-66. *Mem:* AAAS; Am Asn Physics Teachers. *Res:* Computer modeling of laser sytems. *Mailing Add:* Dept Physics US Naval Acad Annapolis MD 21402

CALAME, KATHRYN LEE, b Leavenworth, Kans, Apr 23, 40; m 62; c 2. GENE REGULATION, ONCOGENE EXPRESSION. *Educ:* Univ Mo, BS, 62; George Washington Univ, MS, 65, PhD(biochem), 75. *Prof Exp:* Fel molecular biol, Dept Biochem, Univ Pittsburgh, 75-78 & Calif Inst Technol, 78-80; asst prof, Sch Med, Univ Calif, Los Angeles, 80-85, assoc prof biol chem, 85-87, prof biol chem, 87-88; PROF MICROBIOL, COLUMBIA UNIV, COL PHYSICIANS & SURGEONS, 88- *Honors & Awards:* Stohlman Award, Leukemia Soc, 89. *Mem:* Sigma Xi; Am Soc Microbiol; AAAS; Am Soc Biol Chemists. *Res:* Regulation of immunoglobin gene expression during B lymphocyte development; mechanism of action of transcriptional enhancer elements; regulation of normal and translocated oncogene expression. *Mailing Add:* Dept Microbiology Columbia Univ Col Physicians & Surgeons 701 W 168th St New York NY 10032

CALANDRA, JOSEPH CARL, b Chicago, Ill, Mar 17, 17; m 44; c 4. TOXICOLOGY, BIOCHEMISTRY. *Educ:* Ill Inst Technol, BS, 38; Northwestern Univ, PhD(chem), 42, MD, 50; Am Bd Clin Chem, dipl, 55; Am Bd Indust Hyg, dipl, 62. *Prof Exp:* Instr chem, Ill Inst Technol, 38; from instr chem to assoc prof path, 38-53, chmn dept path, 49-53, prof, 53-87, EMER PROF PATH, MED SCH, NORTHWESTERN UNIV, 87- *Concurrent Pos:* NIH grant, 58-63; dir toxicol, Indust Bio-Test Labs, Inc, 53-59; pres, Indust Biotest Labs, Inc, 69-76; med dir, NALCO, 67-78. *Mem:* Fel AAAS; fel Am Soc Clin Pharmacol & Chemother; AMA; Am Indust Hyg Asn; Soc Toxicol. *Res:* Occupational medicine; experimental pathology. *Mailing Add:* 4630 Elm Terrace Skokie IL 60076

CALARCO, PATRICIA G, b Lincoln, Nebr; div; c 3. VIROLOGY, OOGENESIS. *Educ:* George Wash Univ, BS, 61; Univ Ill-Urbana, MS, 65, PhD(cell biol), 68. *Prof Exp:* Postdoctoral fel, USPHS, Univ Ill, 68-69, Dept Biol Struct, Univ Wash, Seattle, 69-70; acting asst prof, Zool Dept, Univ Wash, Seattle, 70-71; from asst prof to assoc prof, 71-85, PROF ANAT DEPT, UNIV CALIF, SAN FRANCISCO, 85- *Concurrent Pos:* Mem, NSF adv comt physiol, cell & molecular biol, 79-83; mem, NIH, human embryol & develop study sect, 86-88; prin invest, NSF, NIH, Rockefeller, March of Dimes grants; Chair, Women in Cell Biol, 87; nominating comt, AAAS, 87-91, pres, 91. *Honors & Awards:* Res Career Develop Award, NIH, 79-84; Chancellor's Award Pub Serv, 79. *Mem:* Am Soc Cell Biol; Electron Micros Soc Am (pres, 91); fel AAAS; Am Asn Anatomists; Soc Develop Biol. *Res:* The role of the cell surface; extra cellular matrix and endogenous retroviruses in mammalian development, including fertilization, differentiation and implantation utilizing morphological, immunological, biochemical and molecular techniques. *Mailing Add:* Dept Anat Univ Calif San Francisco CA 94143

CALARESU, FRANCO ROMANO, b Divaccia, Italy, Apr 12, 31; Can citizen; m 58; c 2. PHYSIOLOGY. *Educ:* Univ Milan, MD, 53; Univ Alta, PhD, 64. *Prof Exp:* Intern, Columbus Hosp, Chicago, 54-55, resident med, 55-56; resident radiother, Jefferson Med Col, 56-58; demonstr physiol, Univ Sask, 58-60; demonstr, Univ Alta, 60-64; scientist, Gatty Marine Lab, Scotland, 64-65; asst prof physiol, Univ Alta, 65-67; assoc prof, 67-72, PROF PHYSIOL, UNIV WESTERN ONT, 72- *Concurrent Pos:* Nat Res Coun Can fel, 58-60; Nat Heart Found Can fel, 60-65; Med Res Coun Can scholar, 65-67; Med Res Coun Can vis prof, Inst Cardiovasc Res, Univ Milan, 74-75. *Mem:* AAAS; Am Physiol Soc; Can Physiol Soc; Soc Neurosci. *Res:* Neural autonomic control. *Mailing Add:* Dept Physiol Univ Western Ont London ON N6A 5C1 Can

CALAVAN, EDMOND CLAIR, b Scio, Ore, Jan 13, 13; m 38; c 4. CITRUS PATHOLOGY. *Educ:* Ore State Col, BS, 39, MS, 41; Univ Wis, PhD(plant path), 45. *Prof Exp:* Jr plant pathologist, 45-47, from asst plant pathologist to assoc plant pathologist, 47-59, plant pathologist, Citrus Exp Sta, 59-80, prof, 62-80, EMER PROF PLANT PATH & PLANT PATHOLOGIST, UNIV CALIF, 80- *Concurrent Pos:* Citrograph res award, 66; chmn, Int Orgn Citrus Virologists, 72-75. *Honors & Awards:* Fel, Am Phytopath Soc. *Mem:* Int Soc Citriculture; Int Orgn Mycoplasmologists; fel Am Phytopath Soc; Int Orgn Citrus Virologist; Sigma Xi. *Res:* Diseases of citrus; fungal diseases; viral diseases; regulatory plant pathology; mycoplasma diseases of plants; developed pathogen free citrus clones for California; discovered and cultured the mollicute, Spiroplasma citri, that cause stubborn disease of citrus and many other plant species. *Mailing Add:* Dept Plant Path Univ Calif Riverside CA 92521-0122

CALBERT, HAROLD EDWARD, b Edinburg, Ind, Mar 20, 18; m 42; c 3. BIOCHEMISTRY. *Educ:* Allegheny Col, BA, 39; Univ Wis, MS, 47, PhD(dairy & food technol, biochem), 48. *Prof Exp:* From instr to assoc prof dairy & food industs, 48-60, PROF FOOD SCI & CHMN DEPT, UNIV WIS-MADISON, 61- *Concurrent Pos:* Chmn environ sci training comt, USPHS. *Mem:* AAAS; Am Dairy Sci Asn; Inst Food Technologists; Am Fisheries Soc. *Res:* Dairy and food chemistry; food plant sanitation; product development; aquaculture. *Mailing Add:* 5514 Barton Rd Madison WI 53711

CALBO, LEONARD JOSEPH, b New York, NY, Feb 17, 41; m 64; c 2. INDUSTRIAL ORGANIC CHEMISTRY. *Educ:* Manhattan Col, BS, 62; Seton Hall Univ, MS, 64, PhD(org chem), 66; Univ Conn, MBA, 73. *Prof Exp:* Res chemist, Am Cyanamid Co, 66-76; MKT MGR, CATALYSTS & ION EXCHANGE REAGENTS, KING INDUST INC, NORWALK, CT, 76- *Mem:* Am Chem Soc; Am Mgt Asn. *Res:* Surface coatings; cross-linking agents; thermoset amino and polyester resins; development of new resins and cross-linking agents for electrocoating application; catalysts for curing amino resins. *Mailing Add:* King Indust Inc Science Rd PO Box 588 Norwalk CT 06852-0588

CALCAGNO, PHILIP LOUIS, b New York, NY, Feb 27, 18; m 51. PEDIATRICS. *Educ:* Univ Ga, BS, 40; Georgetown Univ, MD, 43. *Prof Exp:* Rotating intern, Morrisannia City Hosp, New York, 44, resident pediat & internal med, 44-45; chief resident contagious dis, Willard Packer Hosp, 45-46; pediatrician, Floating Hosp, 48; res asst, Children's Hosp, Buffalo, 48-52, assoc pediat, Sch Med, Univ Buffalo, 52-54, from asst prof to assoc prof, 54-62; PROF PEDIAT & CHMN DEPT, GEORGETOWN UNIV, 62- *Mem:* Soc Pediat Res; Am Acad Pediat; NY Acad Sci; Am Pediat Soc; Soc Exp Biol & Med. *Res:* Renal function in infants; physiological studies of the postsurgical state; neurosensory disorders in mental retardation. *Mailing Add:* Georgetown Univ Hosp 3800 Reservoir Rd N Washington DC 20007

CALCATERRA, ROBERT JOHN, b Lincoln, Nebr, Oct 17, 42; m 72; c 3. CHEMICAL ENGINEERING. *Educ:* Univ Nebr, BS, 64, MS, 66; Wash Univ, DSc, 72. *Prof Exp:* Res engr process develop, Monsanto Co, 65-69 & Alcoa, 71-73; res engr, Amoco Chem, Standard Oil Co, Ind, 73-76, sr res engr res planning, 76-80; dir res & develop, Adolph Coors CO, 80-89; EXEC DIR, BOULDER TECHNOL INC, 89- *Concurrent Pos:* Affil prof, Dept Chem Eng, Wash Univ, 73-76; adj prof, Univ Colo, 91- *Mem:* Sigma Xi; Am Inst Chem Engrs; Am Soc Eng Educ; Nat Soc Prof Engrs; Technol Transfer Soc. *Res:* Managing entrepreneurial new business start up companies; engineering economics; transport; reaction engineering; mass transfer; business planning; technology evaluation and forecasting. *Mailing Add:* 1825 Viewpoint Rd Boulder CO 80303

CALCOTE, HARTWELL FORREST, b Meadville, Pa, May 20, 20; m 43; c 4. COMBUSTION, PLASMA CHEMISTRY. *Educ:* Cath Univ Am, BACh, 43; Princeton Univ, PhD(phys chem), 48. *Prof Exp:* Instr electronics, Princeton Univ, 43-45, res asst combustion, 45-48; phys chemist, Exp Inc, 48-51, head thermokinetics dept, 52-54, dir res, 55-56; PRES & DIR RES, AEROCHEM RES LABS, INC, 56- *Concurrent Pos:* Mem papers comt, 8th-22nd Int Symp on Combustion; mem adv panel, phys chem div, Nat Bur Standards, 66-69; mem, Joint Army-Navy-Air Force Tables Thermochem Working Group, 66-71; vpres, Combustion Inst, 85- *Mem:* AAAS; Am Chem Soc; Am Phys Soc; Am Inst Aeronaut & Astronaut; Combustion Inst; Sigma Xi. *Res:* Physical chemistry of combustion processes and electrical discharges; electrical properties of flames; radar interference effects in rocket exhausts; instrumentation; soot formation in flames; chemical synthesis in combustion systems and in low temperature; electrical discharges. *Mailing Add:* AeroChem Res Labs Inc PO Box 12 Princeton NJ 08542

CALDECOTT, RICHARD S, b Vancouver, BC, Apr 15, 24; nat US; m 47; c 3. GENETICS. *Educ:* Univ BC, BSA, 46; State Col Wash, MS, 48, PhD(radiation genetics), 51. *Prof Exp:* Res asst, State Col Wash, 46-49, res assoc, 51; asst prof, Univ Nebr, 51-53; assoc radiobiologist, Brookhaven Nat Lab, 53-54; geneticist, res br, Agr Res Serv, USDA, 54-60 & US Atomic Energy Comn, Washington, DC, 60-63; assoc prof, grad sch, 55-63, prof, 63-65, PROF GENETICS & DEAN COL BIOL SCI, UNIV MINN, ST PAUL, 65-; GENETICIST, CROPS RES DIV, AGR RES SERV, USDA, 63- *Concurrent Pos:* Mem, Nat Acad Sci-Nat Res Coun subcomt on radiation biol; vpres & mem bd dirs, Fresh Water Biol Res Found; mem bd trustees, Argonne Univs Asn, 70-80 & St Paul Sci Mus. *Mem:* AAAS; Genetics Soc Am; Radiation Res Soc; Am Inst Biol Sci; Sigma Xi. *Res:* Radiobiology; biophysics; cytogenetics. *Mailing Add:* Off Pres 202 Morrill Hall Univ Minn Minneapolis MN 55455

CALDER, CLARENCE ANDREW, b Baker, Ore, Oct 30, 37; m 61; c 5. EXPERIMENTAL MECHANICS, WAVE PROPOGATION. *Educ:* Ore State Univ, BSc(ME), 60; Brigham Young Univ, MS, 62; Univ Calif, Berkeley, PhD(mech eng), 69. *Prof Exp:* Engr, Boise Cascade Corp, 60-61; proj engr, Sandia Corp, 62-64; asst prof mech eng, Wash State Univ, 69-74; res engr, Lawrence Livermore Nat Lab, 74-78; ASSOC PROF MECH ENG, ORE STATE UNIV, 78- *Concurrent Pos:* Consult for var co, 70- & Lawrence Livermore Nat Lab, 78-; vis sr lectr, Univ Auckland, NZ, 85-86. *Honors & Awards:* Tatnal Award, Soc Exp Mech, 81. *Mem:* Am Soc Mech Eng; Am Soc Eng Educ; Soc Exp Mech (pres, 87-88). *Res:* Applications of laser induced stress waves and ultrasonics; piezoelectric transducer development; sports mechanics; development of composite smart structures. *Mailing Add:* Dept Mech Eng Ore State Univ Corvallis OR 97331

CALDER, DALE RALPH, b St Stephen, NB, Apr 16, 41; m 65; c 2. INVERTEBRATE ZOOLOGY, SYSTEMATICS. *Educ:* Acadia Univ, BSc, 64; Col William & Mary, MA, 66, PhD(marine Sci), 68. *Prof Exp:* Assoc marine scientist, Va Inst Marine Sci, 69-73; assoc marine scientist, Marine Resources Res Inst, 73-81; CUR-IN-CHG, INVERTEBRATE ZOOL DEPT, ROYAL ONT MUS, 81- *Concurrent Pos:* Nat Res Coun Can fel, 68-69; asst prof marine sci, Univ Va, 69-73, Col william & Mary, 70-73; asst prof marine biol, Col Charleston, 74-81; assoc prof zool, Univ Toronto, 83- *Mem:* Linnean Soc London; Soc Antiq Scotland; Hydrozoan Zoologists Asn. *Res:* Taxonomy, ecology and life history of marine invertebrates, particularly the Hydrozoa and Scyphozoa; ecology of marine benthos. *Mailing Add:* Dept Invert Zool Royal Ont Mus 100 Queens Park Toronto ON M5S 2C6 Can

CALDER, JOHN ARCHER, b Baltimore, Md, Oct 18, 42; m 64. CHEMICAL OCEANOGRAPHY, ORGANIC GEOCHEMISTRY. *Educ:* Southern Methodist Univ, BS, 64; Univ Ill, Urbana, MS, 65; Univ Tex, Austin, PhD(chem), 69. *Prof Exp:* Asst prof oceanog, Fla State Univ, 69-73, assoc prof oceanog, 73-77; MEM STAFF, OFF MARINE POLLUTION ASSESSMENT, NAT OCEANIC & ATMOSPHERIC ADMIN, 77- *Mem:* Geochem Soc; Am Soc Limnol & Oceanog; Am Chem Soc. *Res:* Chemistry of the organic compounds in water and sediment; chemistry of the stable isotopes of carbon and oxygen; biological isotope effects; impacts and management of ocean disposal of wastes. *Mailing Add:* 1820 Reedie Dr Silver Spring MD 20902-3552

CALDER, PETER N, b Springhill, NS, April 9, 38; m 64; c 3. ROCK MECHANICS. *Educ:* NS Tech Col, BEng, 63; Queens Univ, MSc, 67, PhD(mining eng), 70. *Prof Exp:* Drilling & Blasting Supvr, Iron Ore Co, Can, 63-65; PROF ROCK MECH & OPEN PIT MINING, QUEENS UNIV, 70-, HEAD, DEPT MINING ENG, 80- *Concurrent Pos:* Consult mining, 70-; chmn rock mech, Underground Blasting Sub-Comt, 77- *Honors & Awards:* Leonard Gold Medal, Eng Inst Can, 70. *Mem:* Can Inst Mining & Metall Engrs; Soc Mining Engrs. *Res:* Computer applications in open pit design; geotechnical aspects of mine design; rock slope reinforcement; drilling and blasting studies. *Mailing Add:* 117 Seaforth Rd Kingston ON K7M 1E1 Can

CALDER, WILLIAM ALEXANDER, III, b Cambridge, Mass, Sept 2, 34; m 55; c 2. PHYSIOLOGICAL ECOLOGY. *Educ:* Univ Ga, BS, 55; Wash State Univ, MS, 63; Duke Univ, PhD(zool), 66. *Prof Exp:* Instr zool, Duke Univ, 66-67, res assoc, 67; asst prof, Va Polytech Inst, 67-69; assoc prof, 69-74, PROF, DEPT ECOL & EVOLUTIONARY BIOL, UNIV ARIZ, 74- *Concurrent Pos:* Mem instrnl staff, Rocky Mountain Biol Lab, 71, 72, 75, 76, 78, 85 & 87; vis prof, Univ New South Wales, 77 & 83. *Mem:* AAAS; Am Ornith Union; Cooper Ornith Soc; Fedn Am Scientists; Am Soc Zool; Ecol Soc Am; Am Physiol Soc; Am Inst Biol Sci; Soc Conserv Biol. *Res:* Physiology of temperature regulation and respiration; environmental physiology; microclimate, heat and water budgets of birds; allometry; population biology of hummingbirds; ecological scaling. *Mailing Add:* Dept Ecol & Evolutionary Biol Univ Ariz Tucson AZ 85721

CALDERON, ALBERTO PEDRO, b Mendoza, Arg, Sept 14, 20; wid; c 2. MATHEMATICS, CIVIL ENGINEERING. *Educ:* Univ Buenos Aires, dipl civil eng, 47; Univ Chicago, PhD(math), 50. *Prof Exp:* Assoc prof, Ohio State Univ, 50-53; vis mem, Inst Advan Study, Princeton, 53-55; assoc prof math, Mass Inst Technol, 55-59, prof, 72-75; prof, 59-68, Louis Block prof, 68-72, chmn dept, 70-72, univ prof math, 75-85, EMER PROF, UNIV CHICAGO, 85- *Concurrent Pos:* Hon prof, Univ Buenos Aires, 75-; assoc ed, J Functional Anal, J Differential Equations & Advan in Math; invited lectr, Int Gong Mathematicians, Moscow, 66. *Honors & Awards:* Nat Medal of Sci, 91; Latin Am Prize Math, IPCLAR, Santa Fe, Arg, 69; Bocher Prize, 78; Colloquim Lectr, Am Math Soc, 66. *Mem:* Nat Acad Sci; fel Am Acad Arts & Sci; Acad Nacional Ciencias Exactas; Real Acad Ciencias; Acad Sci Inst France; Third World Acad Sci; Acad Sci Latin Am. *Mailing Add:* Dept Math Univ Chicago 5734 S University Ave Chicago IL 60637

CALDERON, CALIXTO PEDRO, b Mendoza, Arg, Dec 29, 39. REAL ANALYSIS, HARMONIC ANALYSIS. *Educ:* Univ Buenos Aires, MA, 65, PhD(math), 69. *Prof Exp:* From asst prof to assoc prof Math, Univ Buenos Aries, 69-72 asst prof, Univ Minn, Minneapolis, 72-74; asst prof, Univ Minn, Minneapolis, 72-74; assoc prof, 76-81, PROF MATH, UNIV ILL, CIRCLE CAMPUS, 81- *Concurrent Pos:* Specialist, Dept Sci Affairs, Pan Am Union, 65-66; vis asst prof, Univ Minn, 69-70; vis assoc prof, Rice Univ, 79. *Mem:* Am Math Soc; Arg Inst Math. *Res:* Real analysis; harmonic analysis; differentiation theory. *Mailing Add:* Math Dept Univ Ill Chicago Circle Chicago IL 60680

CALDERON, NISSIM, b Jerusalem, Israel, Apr 1, 33; US citizen; m 61; c 2. POLYMER CHEMISTRY, ORGANOMETALLIC CHEMISTRY. *Educ:* Hebrew Univ, Jerusalem, MSc, 58; Univ Akron, PhD(polymer chem), 62. *Prof Exp:* Sr res chemist, 62-67, sect head, 67-75, MGR RES & ADMIN, GOODYEAR TIRE & RUBBER CO, 75- *Mem:* Am Chem Soc. *Res:* Developer of the olefin metathesis reaction of homogeneous catalysts; discoverer of the feasibility of cleavage of carbon-to-carbon double bonds by transition metal catalysts. *Mailing Add:* 1766 Brookwood Dr Akron OH 44313-5067

CALDERONE, JULIUS G, b Detroit, Mich, Aug 2, 28. MEDICAL BACTERIOLOGY, MEDICAL MYCOLOGY. *Educ:* San Jose State Univ, BA, 50; Univ Calif, Los Angeles, PhD(microbiol), 64. *Prof Exp:* Med microbiologist, Tulare County Health Dept, Calif, 51-53; clin lab technologist, US Navy, 53-55; med microbiologist, Orange County Health Dept, Calif, 55-58; asst med microbiol, Univ Calif, Los Angeles, 59-63, fel, 64-65; from asst prof to assoc prof, 65-73, PROF MED MICROBIOL, CALIF STATE UNIV, SACRAMENTO, 73- *Mem:* Am Soc Microbiol; Sigma Xi; Wildlife Dis Asn. *Res:* Bacterial and fungal diseases of animals which may be transmitted to humans. *Mailing Add:* 4133 Crondall Dr Sacramento CA 95825

CALDERONE, RICHARD ARTHUR, b Niles, Ohio, Sept 3, 42; m 70; c 4. MICROBIOLOGY. *Educ:* Ohio Univ, Athens, BS, 65; WVa Univ, PhD(microbiol), 70. *Prof Exp:* Fel mycol, WVa Univ, 70-71; asst prof biol, Washington & Jefferson Col, 71-73; fel med mycol, Skin & Cancer Hosp, Temple Univ, 73-74; asst prof, 74-79, assoc prof, 79-88, PROF MICROBIOL DEPT, GEORGETOWN UNIV, 88- *Mem:* Am Soc Microbiol; Med Mycol Soc Am. *Res:* Pathogenesis of endocarditis caused by the yeast Candida albicans, using a rabbit model to study the immune response and how Candida attaches to damaged heart valve endothelium. *Mailing Add:* Microbiol Dept Med Dent Bldg Georgetown Univ 3900 Reservoir Rd NW Washington DC 20007

CALDERWOOD, KEITH WRIGHT, b Henefer, Utah, July 28, 24; m 50; c 5. GEOLOGY. *Educ:* Brigham Young Univ, BS, 50, MS, 51. *Prof Exp:* Geologist, Phillips Petrol Co, 51-56, staff geologist, 57-59, dist geologist, 59-69; consult geologist, 69-76, PRES, CALDERWOOD & MANGUS, INC, 71- *Mem:* Asn Prof Geol Scientists; Am Asn Petrol Geologists. *Res:* Petroleum and mineral exploration and development. *Mailing Add:* Calderwood & Mangus Inc 7900 Honeysuckle Anchorage AK 99502

CALDINI, PAOLO, CLINICAL ANESTHESIA, PULMONARY PHYSIOLOGY. *Educ:* Univ Florence, Italy, MD, 52. *Prof Exp:* ASSOC PROF CLIN ANESTHESIOL, MED SCH, TUFTS UNIV, 81-; CHIEF ANESTHESIOL, NEWTON-WELLESLEY HOSP, 81- *Res:* Cardiovascular physiology. *Mailing Add:* 60 Possum Rd Newton MA 02193

CALDITO, GLORIA C, b Manila, Phillippines, Aug 30, 44. SAMPLE SURVEY METHODS & DATA ANALYSIS, LINEAR MODELS. *Educ:* Univ Santo Tomas, BS, 65; Univ Philippines, Quezon City, MS, 81, Univ Pittsburgh, MA, 86, PhD(biostatist), 90. *Prof Exp:* Instr math & statist, Univ Philippines, 67-72, from asst prof to assoc prof statist, 72-84; teaching asst math & biostatist, Univ Pittsburgh, 84-90; ASST PROF BIOSTATIST, COL MED, NE OHIO UNIVS, 90- *Mem:* Am Statist Asn; Biometric Soc. *Res:* Application of statistical theory and methodology in the biomedical sciences; statistical modelling of disease processes. *Mailing Add:* PO Box 95 Rootstown OH 44272

CALDWELL, AUGUSTUS GEORGE, b Belleville, Can, Apr 7, 23; US citizen; m 48; c 4. SOIL CHEMISTRY. *Educ:* Univ Toronto, BSA, 46, MSA, 48; Iowa State Univ, PhD(soil fertil), 55. *Prof Exp:* Soil surv scientist, Ont Agr Col, 46-52, lectr soil fertil res, 52-54; from asst prof to assoc prof soil chem, Tex A&M Univ, 54-62; PROF AGRON, LA STATE UNIV, BATON ROUGE, 62- *Concurrent Pos:* Tropical soils consult, Africa, 84, 86. *Mem:* Am Soc Agron; Soil Sci Soc Am; Am Chem Soc. *Res:* Soil fertility; soil organic phosphorus; soil tests for phosphorus; plant response to fertilizers; soil characterization; economics of fertilizer use; soil mineralogy; alleviation of acid soil infertility using combination of lime and gypsum to treat surface and subsoil acidity; cultivar tolerance of soil acidity. *Mailing Add:* Dept Agron La State Univ Baton Rouge LA 70803

CALDWELL, CARLYLE GORDON, b Little Rock, Ark, Mar 13, 14; m 40; c 1. PLANT CHEMISTRY. *Educ:* Iowa State Col, BS, 36, PhD, 40. *Prof Exp:* Mem staff, 40-47, res dir, 47-55, vpres res, 55-62, dir, 62-66, exec vpres, 66-68, PRES, NAT STARCH & CHEM CORP, 69-, CHIEF EXEC OFFICER, 75-; DIR, RES CORP, 73- *Concurrent Pos:* Chmn, Nat Starch & Chem Corp, 78- *Honors & Awards:* Claude S Hudson Award, Am Chem Soc, 65; Alsberg-Schoch Award, Am Asn Cereal Chemists, 72. *Mem:* Am Chem Soc; Indust Res Inst; Am Inst Chemists; Asn Res Dirs. *Res:* Fractionation of starch; starch derivatives; polymers. *Mailing Add:* 96 Stone Run Rd Bedminster NJ 07921

CALDWELL, CHRISTOPHER STERLING, b Washington, DC, June 22, 42; m 66; c 1. COMPUTATIONAL FLUID DYNAMICS, NUMERICAL ANALYSIS. *Educ:* Johns Hopkins Univ, BA, 65; Univ Md, MA, 68, PhD(math), 71. *Prof Exp:* Mathematician, Naval Oceanog Off, 70-72; sr mathematician, 72-80, FEL MATHEMATICIAN, BETTIS ATOMIC POWER LAB, 80- *Mem:* Am Math Soc; Soc Indust & Appl Math; Am Nuclear Soc; Sigma Xi. *Res:* Computer simulation of fluid flow. *Mailing Add:* 340 McCombs Rd Venetia PA 15367

CALDWELL, DABNEY WITHERS, b Charlottesville, Va, Mar 26, 27; div; c 4. GEOLOGY. *Educ:* Bowdoin Col, AB, 49; Brown Univ, MA, 53; Harvard Univ, PhD(geol), 59. *Prof Exp:* Geologist, US Geol Surv, 52-54; from instr to asst prof geol, Wellesley Col, 55-67; ASSOC PROF GEOL, BOSTON UNIV, 67-, CHMN GEOL, 81- *Mem:* AAAS; Am Geophys Union; Sigma Xi. *Res:* Glacial geology and fluvial processes in geomorphology; environmental geology; surficial geologic mapping, glacial geology, in the wildlands of northern Maine; groundwater and surface water studies in the northeastern states, in the southern United States and Europe. *Mailing Add:* Dept Geol Boston Univ 725 Commonwealth Ave Boston MA 02215

CALDWELL, DANIEL R, b Oakland, Calif, Feb 29, 36; m 61; c 2. MICROBIOLOGY, BIOCHEMISTRY. *Educ:* Reed Col, BA, 58; Univ Md, MS, 65, PhD(dairy sci), 69. *Prof Exp:* Microbiologist, Agr Res Serv, USDA, Md, 60-69; asst prof microbiol, 69-75, ASSOC PROF MICROBIAL PHYSIOL, UNIV WYO, 75- *Mem:* AAAS; Am Soc Microbiol; Sigma Xi. *Res:* Anaerobic bacterial nutrition and metabolism; rumen microbiology; porphyrin biosynthesis; bacterial carbohydrate fermentation; carbon dioxide fixation; mineral nutrition; anaerobic electron transport; cytochromes in ruminal bacteria. *Mailing Add:* Dept Molecular Biol Univ Wyo Univ Sta Box 3944 Laramie WY 82071

CALDWELL, DAVID KELLER, vertebrate biology; deceased, see previous edition for last biography

CALDWELL, DAVID ORVILLE, b Los Angeles, Calif, Jan 5, 25; m 84; c 2. EXPERIMENTAL ELEMENTARY PARTICLE PHYSICS, PARTICLE ASTROPHYSICS. *Educ:* Calif Inst Technol, BS, 47; Univ Calif, Los Angeles, MA, 49, PhD(physics), 53. *Prof Exp:* From instr to assoc prof physics, Mass Inst Technol, 54-63; vis assoc prof, Princeton Univ, 63-64; lectr, Univ Calif, Berkeley, 64-65, assoc dir, Intercampus Inst Res Particle Accelerators, 76-83, dir, 83-87, PROF PHYSICS, UNIV CALIF, SANTA BARBARA, 65-, ASSOC DIR, INTERCAMPUS INST RES PARTICLE ACCELERATORS, 87- *Concurrent Pos:* Consult, Radiation Lab, Univ Calif, 50-51, 65-70, Am Sci & Eng, 59-60, Inst Defense Analysis, 60-67 & Defense Commun Planning Group, 66-70; NSF fel, Univ Calif, Los Angeles & Edigenossiche Tech Univ, 53-54; physicist, Radiation Lab, Univ Calif, 57, 58 & 64-67; NSF sr fel, 60-61 & Ford Found fel, 61-62; Guggenheim fel, Europ Orgn Nuclear Res, 71-72; mem high energy physics adv panel, Energy Res & Develop Agency & Dept Energy, 74-78. *Honors & Awards:* Alexander Von Humboldt Distinguished Scientist Award, 87. *Mem:* Fel Am Phys Soc. *Res:* Two-photon physics; double beta decay; dark matter searches; hadronic spectroscopy; detector development. *Mailing Add:* Dept Physics Univ Calif Santa Barbara CA 93106

CALDWELL, DOUGLAS RAY, b Lansing, Mich, Feb 16, 36; m 61; c 3. OCEANOGRAPHY. *Educ:* Univ Chicago, AB, 55, BS, 57, MS, 58, PhD(physics), 64. *Prof Exp:* NSF fel, Cambridge Univ, 63-64; res geophysicist, Inst Geophys & Planetary Physics, Univ Calif, San Diego, 64-68; from asst prof to assoc prof, 68-77, prof Dept Oceanog, 77-, ASSOC DEAN COL OCEANOG, 83-, DEAN COL OCEANOG, ORE STATE UNIV, 84- *Res:* Physics of fluids; hydrodynamics and magnetohydrodynamics; stability theory; measurements in the oceans. *Mailing Add:* Ocean Admin Bldg 104 Ore State Univ Corvallis OR 97331-5503

CALDWELL, E GERALD, research administration, for more information see previous edition

CALDWELL, ELWOOD F, b Gladstone, Man, Apr 3, 23; nat US; m 49, 79; c 2. NUTRITION. *Educ:* Univ Man, BSc, 43; Univ Toronto, MA, 49, PhD(nutrit), 53; Univ Chicago, MBA, 56. *Prof Exp:* Chemist, Lake of Woods Milling Co, 43-48; res chemist, Can Breweries Ltd, 48-49; chief chemist, Christie, Brown & Co, 49-51; res assoc nutrit, Univ Toronto, 51-53; group leader processing & packaging res, Quaker Oats Co, 53-62, asst dir foods res, 62-69, dir res & develop, 70-72; prof food sci & nutrit & head dept, Univ Minn, 72-86, exec assoc to dean, Col Agr, 86-88; DIR, SCI SERV, AM ASN CEREAL CHEM, 88- *Concurrent Pos:* Chmn bd dirs, Dairy Qual Control Inst, Inc, St Paul, Minn, 73-88; sci coord, Am Asn Cereal Chem, 86-88; dir, Nat Prog short courses, tech food subjects. *Honors & Awards:* Cert Appreciation Patriotic Civilian Serv, Dept Defense, Army Mat Command, 70. *Mem:* Am Home Econ Asn; Can Inst Food Sci & Technol; fel Inst Food Technol; fel Am Asn Cereal Chem. *Res:* Food chemistry and technology; experimental, clinical and community nutrition; educational and research administration. *Mailing Add:* Am Assoc Cereal Chem 3340 Pilot Knob Rd St Paul MN 55121

CALDWELL, FRED T, JR, b Hot Springs, Ark, May 12, 25; m 47; c 2. SURGERY. *Educ:* Baylor Univ, BS, 46; Washington Univ, MD, 50. *Prof Exp:* Assoc prof surg, Univ Hosp, State Univ NY Upstate Med Ctr, 58-67; PROF SURG, MED CTR, UNIV ARK, LITTLE ROCK, 67- *Concurrent Pos:* Res fel, Washington Univ, 57-58; Cancer Soc clin fel, 57-58; surg consult, Little Rock Vet Admin Hosp, 67- *Mem:* Am Col Surg; Am Physiol Soc; Soc Univ Surg; Am Surg Asn. *Res:* Energy balance following trauma, particularly after thermal burns; pathogenesis of cholesterol gall stones. *Mailing Add:* Univ Ark Med Ctr 4301 W Markham St Little Rock AR 72204

CALDWELL, GLYN GORDON, b St Louis, Mo, Jan 14, 34; m 60; c 3. EPIDEMIOLOGY, PUBLIC HEALTH. *Educ:* St Louis Univ, BS, 60; Univ Mo, MS, 62, MD, 66. *Prof Exp:* Intern med, USPHS Hosp, Brighton, Mass, 66-67; resident internal med, Cleveland Metrop Hosp, 69-71; res biologist virol, Ecol Invest Prog, USPHS, 67-68, actg chief viral dis sect, 68-69, chief oncol & teratology activ, 71-73, asst chief cancer br, Cancer & Birth Defects Div, 73-74, chief field invest sect, 74-77, dep chief cancer br, Bur Epidemiol, 74-77, chief cancer br, Cancer Res Chronic Dis Div, Bur Epidemiol, 77-80, chief cancer Br, Chronic Dis Div, Ctr Environ Health, Ctr Dis Control, 80-85; asst dir, Ariz Dept Health, 85-87, dep dir, 87-90; DIR, TULSA CITY-COUNTY HEALTH DEPT, 90- *Concurrent Pos:* Asst prof microbiol, Univ Kans Med Ctr, 67-73; lectr, Univ Mo Med Ctr, Kans City, 71-73; asst prof prev med epidemiol, Emory Univ, 74-85; asst prof community med, Univ Ariz, 85-86; asst prof, Univ Okla, 90. *Honors & Awards:* Commendation Medal, USPHS. *Mem:* Am Soc Microbiol; NY Acad Sci; Am Soc Prev Oncol; Int Asn Comp Res Leukemia & Related Dis; Sigma Xi; Am Pub Health Asn; AAAS; Nat Asn County Health Officers. *Res:* Cancer epidemiology; tumor virology; hematology; oncology. *Mailing Add:* Tulsa City-County Health Dept 4616 E 15th St Tulsa OK 74112

CALDWELL, J(OSEPH) J(EFFERSON), JR, b Weatherford, Tex, Oct 22, 15; m 40; c 3. MICROWAVE ELECTRONICS. *Educ:* Tex Tech Col, BS, 37; Mass Inst Technol, ScM, 38. *Prof Exp:* Mem staff, Sperry Gyroscope Co, 38-46, Int Tel & Tel Res Lab, 46-50 & Hughes Aircraft Co, 50-59; researcher strategic systs, Boeing Airplane Co, 59-60; researcher electrooptics, TRW Inc, 60-80; SR STAFF ENGR, SPACE & COMPUT GROUP, HUGHES AIRCRAFT CO, INC, 80- *Concurrent Pos:* Res assoc, Stanford Univ, 38, 41 & 43. *Mem:* Sr mem Inst Elec & Electronics Engrs; Am Phys Soc; Sigma Xi. *Res:* Microwave klystrons, magnetrons and high-powered traveling wave tubes; instrument landing, precision microwave bombing systems and microwave communication systems; atomic frequency standards; laser pumps; electric propulsion; laser target designator; electro-optical systems. *Mailing Add:* 26751 Eastvale Rd Rolling Hills CA 90274

CALDWELL, JERRY, immunogenetics, molecular genetics, for more information see previous edition

CALDWELL, JOHN JAMES, b Winnipeg, Man, Aug 4, 44; m 66; c 3. PLANETARY ATMOSPHERES, ASTRONOMY. *Educ:* Univ Man, BSc Hons, 65; Univ Western Ont, MSc, 66; Univ Wis- Madison, PhD(astron), 71. *Prof Exp:* Res assoc, Univ Wis-Madison, 71-72; res assoc astrophys, Princeton Univ, 72-77; adj asst prof atmospheric sci, 77-83, assoc prof astron, State Univ NY, Stony Brook, 83-86; PROF PHYSICS, YORK UNIV, 86- *Concurrent Pos:* Vis prof, Rutgers Univ, 73-77; interdisciplinary scientist space telescope, Marshall Space Flight Ctr, NASA, 78-; vis prof, Univ Hawaii, 81; dir, Space Astrophys Lab, Inst Space & Terrestrial Sci, 87- *Mem:* Am Astron Soc; AAAS. *Res:* Theoretical planetary physics; infrared and radio observation of planets; space astronomy; ultraviolet spectrophotometry of plants; planetary aurorae. *Mailing Add:* Physics Dept York Univ 4700 Keele St North York ON M3J 1P3 Can

CALDWELL, JOHN R, b Middletown, Conn, Oct 11, 18; m 47; c 5. INTERNAL MEDICINE. *Educ:* Lafayette Col, BA, 40; Temple Univ, MD, 43; Am Bd Internal Med, dipl, 55. *Prof Exp:* Med practitioner, Del, 47-50; staff physician, Med Clins, Henry Ford Hosp, Detroit, 52-54, physician-in-charge, Div Hypertension, 54-67, chief, Hypertension Sect, 67-80, staff physician, 80-88, EMER MED STAFF PHYSICIAN, NEPHROL & HYPERTENSION DIV, HENRY FORD HOSP, DETROIT, 88- *Concurrent Pos:* Fel coun arteriosclerosis & coun high blood pressure res, Am Heart Asn; clin assoc prof internal med, Med Sch, Univ Mich. *Mem:* Fel Am Col Physicians; AMA; Am Soc Internal Med. *Res:* Causes, effects and treatment of high arterial pressure; renal artery stenosis; pheochromocytoma; primary aldosteronism; pyelonephritis; social and psychologic factors favoring development of hypertension; compliance to treatment. *Mailing Add:* Nephrol & Hypertension Div Henry Ford Hosp 2799 W Grand Blvd Detroit MI 48202

CALDWELL, JOHN THOMAS, b Mercedes, Tex, June 26, 37; m 61; c 2. EXPERIMENTAL NUCLEAR PHYSICS, NON-DESTRUCTIVE ASSAY TECHNOLOGY. *Educ:* Rice Univ, BA, 59; San Jose State Univ, MA, 61; Univ Calif, Davis, PhD(appl sci), 67. *Prof Exp:* Res scientist nuclear physics, Lawrence Livermore Lab, 61-69; res scientist nuclear physics, Los Alamos Sci Lab, 69-87; CHIEF SCIENTIST & PRES, PAJARITO SCI CORP, 87- *Concurrent Pos:* Consult, Am Soc Testing & Mat Standards comt, non-destructive assay; adv, NDA/NDE working group, Dept Energy. *Honors & Awards:* IR 100 Award, Ind Res Develop Coun, 83. *Mem:* Am Phys Soc. *Res:* Photonuclear physics, photofission, spontaneous fission, nuclear safeguards and nuclear radiation detection technology and detector development; five patents; technical administration. *Mailing Add:* 322 Kimberly Los Alamos NM 87544

CALDWELL, KARIN D, BIOCHEMICAL SEPARATION & CHARACTERIZATION. *Educ:* Univ Uppsala, Sweden, PhD(biochem), 68, FilDr, 76. *Prof Exp:* ASSOC PROF BIOCHEM RES, UNIV UTAH, 85-, DIR, CTR BIOPOLYMERS & INTERFACES, 86- *Concurrent Pos:* Adj assoc prof chem, Univ Utah, 86-, assoc prof, Dept Bioeng. *Res:* Particle and polymer characterization. *Mailing Add:* 2480 Merrill Eng Bldg Univ Utah Salt Lake City UT 84112

CALDWELL, LARRY D, b Manton, Mich, July 13, 32; m 59; c 4. ECOLOGY. *Educ:* Mich State Univ, BSc, 54, MS, 55; Univ Ga, PhD(zool), 60. *Prof Exp:* Asst prof biol, Southeastern La Col, 59-61; from asst prof to assoc prof, 61-69, PROF BIOL, CENT MICH UNIV, 69- *Mem:* AAAS; Am Soc Mammalogists; Am Ornith Union; Wildlife Soc; Sigma Xi. *Res:* Population and physiological ecology; mammalian population phenomena; competition in rodent populations; lipid levels and bird migration. *Mailing Add:* Brooks Hall Cent Mich Univ Mt Pleasant MI 48858

CALDWELL, MARTYN MATHEWS, b Denver, Colo, June 28, 41; div. PLANT ECOLOGY, PLANT PHYSIOLOGY. *Educ:* Colo State Univ, BS, 63; Duke Univ, PhD(bot), 67. *Prof Exp:* From asst prof to assoc prof, 67-75, PROF ECOL, UTAH STATE UNIV, 75- *Concurrent Pos:* NSF fel, Innsbruck, Austria, 68-69; chmn biol prog, Climatic Impact Assessment Prog, Dept Transp, 73-75; mem nat comt photobiol, Nat Acad Sci, 74-78 & Comt Impacts Stratospheric Change, 75-80; vis prof, Univ Wuerzburg, 82; Willard Gardner res award, Utah Acad Sci, Arts & Lett, 83; D Wynne Thorne Res Award, Utah State Univ, 85; prog dir, Pop Biol & Physiol Ecol, Nat Sci Found, Washington, DC, 87-88; mem, Core Proj Planning Comt, Int Geosphere-Biosphere Prog, 90- & Study comt, Sci Comt on Probs of Environ, 90- *Honors & Awards:* Alexander von Humboldt Sr US Scientist Award, WGermany, 81. *Mem:* Ecol Soc Am; Am Soc Plant Physiologists; Brit Ecol Soc; Australian Soc Plant Physiologists; Soc Range Mgt; fel AAAS. *Res:* Physiological ecology of plants under stress in arid and tundra environments; plant photosynthesis, transpiration, water relations, root growth and carbon balance; effects of solar ultraviolet radiation on plants. *Mailing Add:* Ecol Ctr Utah State Univ Logan UT 84322

CALDWELL, MELBA CARSTARPHEN, b Augusta, Ga, May 4, 21; m 60; c 2. VERTEBRATE ZOOLOGY, COMMUNICATIONS SCIENCE. *Educ:* Univ Ga, BS, 41; Univ Calif, Los Angeles, MA, 63. *Prof Exp:* Fishery aide marine zool, US Fish & Wildlife Serv, Ga, 56-59, fishery res biologist, 59-60; staff res assoc, Allan Hancock Found, Univ Southern Calif, 63-67; asst cur & assoc dir res, Marineland of Fla, 67-69; res instr, 70-76, ASSOC RES SCI, INST ADVAN STUDY COMMUN PROCESSES, UNIV FLA, 76-; ASSOC DIR RES, MARINELAND OF FLA, 83- *Concurrent Pos:* Res assoc, Los Angeles County Mus Natural Hist, 61-80; field assoc, Fla State Mus, 75- *Mem:* Fel AAAS. *Res:* Odontocete cetaceans, especially communication, life history and other aspects of general biology; systematics and distribution of marine fishes and odontocete cetaceans. *Mailing Add:* 9507 Ocean Shore Blvd Marineland Florida St Augustine FL 32086

CALDWELL, MICHAEL D, b Spartanburg, SC, Aug 29, 43. METABOLISM. *Educ:* Med Univ SC, MD, 68; Vanderbilt Univ, PhD(physiol), 80. *Prof Exp:* ASSOC PROF SURG, BROWN UNIV, 82-; DIR, SURG METAB LAB & NUTRIT SUPPORT SERV, RI HOSP, 82-, DIR, SURG RES, 86- *Mailing Add:* Dept Surg RI Hosp 593 Eddy St Providence RI 02902

CALDWELL, ROBERT WILLIAM, b Brunswick, Ga, Dec 27, 42; m 65; c 2. PHARMACOLOGY, PHYSIOLOGY. *Educ:* Ga Inst Technol, BS, 65; Emory Univ, PhD(basic health sci), 69. *Prof Exp:* Pharmacologist, Div Med Chem, Walter Reed Army Inst Res, 70-72; from asst prof to assoc prof, 72-79, PROF PHARMACOL, CTR HEALTH SCI, UNIV TENN, MEMPHIS, 79- *Mem:* AAAS; Am Soc Pharmacol & Exp Therapeut; Soc Exp Biol & Med; Soc Neurosci. *Res:* Central nervous system control of hypertension; study of novel cardiac glycosides. *Mailing Add:* 419 Stone Wall St Memphis TN 38112

CALDWELL, ROGER LEE, b Los Angeles, Calif, May 12, 38; m 66; c 2. BIOCHEMISTRY, PLANT PATHOLOGY. *Educ:* Univ Calif, Los Angeles, BS, 61; Univ Ariz, PhD(chem), 66. *Prof Exp:* Nat Acad Sci res assoc, US Food & Drug Admin, Washington, DC, 66-67; from asst prof to assoc prof plant path, 62-80, PROF SOILS, WATER & ENG, UNIV ARIZ, 80-, DIR COUN ENVIRON STUDIES, 74- *Mem:* Am Chem Soc; Am Phytopath Soc; Am Inst Biol Sci; Coun Agr Sci & Technol; Soil Conserv Soc Am. *Res:* Energy alternatives; environmental interactions of agriculture and society; effects of air pollution on plants; pesticide alternatives. *Mailing Add:* 1625 E Kleindale Rd Tucson AZ 85719

CALDWELL, SAMUEL CRAIG, electrical engineering; deceased, see previous edition for last biography

CALDWELL, SLOAN DANIEL, b Lincolnton, NC, May 26, 43; m 69; c 4. AQUATIC ENTOMOLOGY, INVERTEBRATE ZOOLOGY. *Educ:* Western Carolina Univ, BS, 64; Univ Tenn, MS, 66; Univ Ga, PhD(entom), 73. *Prof Exp:* From instr to assoc prof, 69-81, PROF BIOL, GA COL, 81- *Concurrent Pos:* Consult, Ga Power Co, 72-73. *Mem:* NAm Benthological Soc. *Res:* Population, trophic structure and taxonomy of benthic aquatic insects. *Mailing Add:* Dept Biol Ga Col Milledgeville GA 31061

CALDWELL, STEPHEN E, b Columbus, Ohio, Sept 17, 46; m 78. PHYSICS. *Educ:* Ohio State Univ, BSc, 69, MSc, 73, PhD(physics), 74. *Prof Exp:* Mem staff physics, US Army Ballistics Res Lab, 74-76; STAFF MEM PHYSICS, LOS ALAMOS SCI LAB, 76- *Mem:* Am Phys Soc. *Res:* Laser fusion; gas centrifuges. *Mailing Add:* Los Alamos Sci Lab Mail Stop 10410 Los Alamos NM 87545

CALDWELL, WILLIAM GLEN ELLIOT, b Millport, Scotland, July 25, 32; m 61; c 3. GEOLOGY. *Educ:* Glasgow Univ, BSc, 54, PhD, 57. *Prof Exp:* Asst lectr geol, Glasgow Univ, 56-57; spec lectr, Univ Sask, 57-58, from asst prof to prof geol, 58-, head dept geol sci, 72-; VPRES, UNIV WESTERN ONT. *Mem:* Paleont Soc; Brit Geol Asn; fel Geol Soc Am; Brit Paleont Asn; fel Geol Asn Can (pres, 80-81); fel Royal Soc Can. *Res:* Stratigraphy and paleontology of Cretaceous System. *Mailing Add:* Steven Lawson Bldg Univ Western Ont London ON N6A 5B8 Can

CALDWELL, WILLIAM V, b Boyd, Tex, Sept 3, 17; m 53; c 2. MATHEMATICS. *Educ:* Tex Christian Univ, BA, 51; Univ Mich, MA, 56, PhD(math), 60. *Prof Exp:* Analytical engr, transonic lab, Mass Inst Technol, 51-54; asst prof math, Univ Del, 59-61; from asst prof to assoc prof, 61-68, chmn dept, 66-78, PROF MATH, UNIV MICH, FLINT, 68- *Mem:* AAAS; Am Math Soc; Math Asn Am. *Res:* Vector spaces and algebras of light interior functions; theory of quasiconformal functions. *Mailing Add:* 2119 E Second St Flint MI 48503

CALE, ALBERT DUNCAN, JR, b Windsor, NC, Sept 18, 28; m 54; c 4. MEDICINAL CHEMISTRY. *Educ:* Elon Col, BS, 56; Univ SC, MS, 58. *Prof Exp:* Res chemist, 58-63, sr res chemist, 63-75, res assoc med, 75-81, GROUP LEADER & MGR, A H ROBINS, INC, 81- *Mem:* Am Chem Soc. *Res:* Small ring heterocycles. *Mailing Add:* Rte 8 Box 150 Mechanicsville VA 23111

CALE, WILLIAM GRAHAM, JR, b Philadelphia, Pa, Dec 10, 47; m 74. ECOLOGY. *Educ:* Pa State Univ, BS, 69; Univ Ga, PhD(ecol), 75. *Prof Exp:* Consult ecosyst modeling, Colo State Univ, 74-75; from asst prof to prof environ sci, Univ Tex, Dallas, 75-88, head, Environ Sci Prog, 85-88, assoc dean & col master, 87-88; DEAN, COL NATURAL SCI & MATH, UNIV IND, PA, 88- *Mem:* Am Inst Biol Sci; Int Soc Ecol Modelling; Ecol Soc Am. *Res:* Ecosystem modeling and systems analysis; theoretical ecology; uncertainty analysis. *Mailing Add:* 305 Weyand & Hall Ind Univ Indiana PA 15705

CALE, WILLIAM ROBERT, b Paris, Ont, May 4, 13; m 46; c 2. INORGANIC CHEMISTRY, CHEMICAL ENGINEERING. *Educ:* Univ Toronto, BASc, 35, MASc, 36. *Prof Exp:* Asst to works chemist, Elec Reduction Co Can, Ltd, Erco Industs Ltd, 36-38, works chemist, 38-53, tech serv mgr, 53, mgr res tech serv, 54-55, mgr planning & mkt res div, 56-58, mgr eng serv div, 58-62, mgr chlorate develop dept, 63-64, mgr patents & inventions dept, 64-68, mgr patents & info, 68-78, consult, 78-80; RETIRED. *Concurrent Pos:* Mem exec bd, Can Comt Int Water Pollution Res, 75-77; chmn, Environ Comt, Can Mfrs Asn, 78-80; past pres, Can Man or Chem Spec Asn. *Mem:* Fel Chem Inst Can. *Res:* Production processes and applications of phosphorus compounds; chlorates and chlorine dioxide; long-range planning market research and patents; environmental affairs. *Mailing Add:* 53 Greenbrook Dr Toronto ON M6M 2J8 Can

CALEDONIA, GEORGE ERNEST, b Boston, Mass, Nov 9, 41; m 65; c 3. PHYSICAL CHEMISTRY, CHEMICAL PHYSICS. *Educ:* Northeastern Univ, AB, 65, MS, 67. *Prof Exp:* Prin scientist gas phase chem, Avco Everett Res Lab, Avco Corp, 67-73; mgr environ sci, 73-80, vpres res, 80-89, PRES RES, PHYS SCI INC, 89- *Mem:* Am Phys Soc; Am Chem Soc; Am Geophy Union. *Res:* Gas phase kinetics; electron excitation phenomena; absorption and transfer of radiative emission; laser physics; discharge physics; pollutant detection and monitoring; reentry physics; aeronomy. *Mailing Add:* 83 Centre Lane Milton MA 02186

CALEHUFF, GIRARD LESTER, b Williamsport, Pa, Oct 13, 25; m 48; c 5. FLUID MECHANICS, MATHEMATICS. *Educ:* Pa State Univ, BS, 49, MS, 52. *Prof Exp:* Res asst, Pa State Univ, 49-54, asst prof eng res, 54-56; group leader, Westvaco Corp, 56-61, res dir, 61-63, paper mill supt, 63-68, tech dir, 68, res dir, 69-70, asst mgr, 70-75, mill mgr, 75-79, mgr facil planning, 79-90; CONSULT, 90- *Mem:* Am Tech Asn Pulp & Paper Indust; Am Soc Mech Engrs; Am Inst Aeronaut & Astronaut; Can Pulp & Paper Asn. *Res:* Turbulent boundary layers in adverse pressure gradients; turbulence measurements in liquids and gases with application to underwater ordnance; dilute fluid suspensions of papermaking fibers; computer control of paper machines. *Mailing Add:* 46 Thunder Lake Rd Wilton CT 06897

CALENDAR, RICHARD, b Hackensack, NJ, Aug 2, 40; m 69; c 2. MOLECULAR BIOLOGY. *Educ:* Duke Univ, BS, 62; Stanford Univ, PhD(biochem), 67. *Prof Exp:* Fel microbial genetics, Karolinska Inst, Sweden, 67-68; from asst prof to assoc prof, 69-76, PROF MOLECULAR BIOL, UNIV CALIF, BERKELEY, 76- *Concurrent Pos:* Alexander von Humboldt fel, Max Planck Inst Biochem, Munich, Ger, 73; Guggenheim fel, Karolinska Inst, Sweden, 79-80. *Mem:* Am Soc Microbiol. *Res:* Gene control, and DNA replication of bacterial viruses. *Mailing Add:* Dept Molecular & Cell Biol 401 Barker Hall Univ Calif Berkeley CA 94720

CALESNICK, BENJAMIN, b Philadelphia, Pa, Dec 27, 15; m 45; c 1. PHARMACOLOGY. *Educ:* St Joseph's Col, Pa, BS, 38; Temple Univ, AM, 41; Hahnemann Med Col, MD, 44. *Prof Exp:* Lab asst pharmacol, 40-41, assoc, 46-57, asst prof, 57-62, dir human pharmacol, 57-70, PROF PHARMACOL MED, HAHNEMANN MED COL, 62-, DIR DIV HUMAN PHARMACOL, 70- *Concurrent Pos:* Vis instr, Univ Pa, 50; lectr, Women's Med Col Pa, 51-52; chief hypertension clin, St Joseph Hosp, 58- *Mem:* Am Soc Pharmacol & Exp Therapeut; Am Col Clin Pharmacol (pres, 76-78); NY Acad Sci; Soc Exp Biol & Med; Soc Toxicol. *Res:* Human and clinical pharmacology. *Mailing Add:* Div Human Pharmacol Hahnemann Univ Sch Med Philadelphia PA 19102

CALEY, WENDELL J, JR, b Philadelphia, Pa, Jan 16, 28; m 52; c 5. PHYSICS. *Educ:* Houghton Col, BS, 50; Univ Rochester, MS, 59; Temple Univ, PhD(physics), 63. *Prof Exp:* Develop engr, Eastman Kodak Co, 53-59; from instr to asst prof physics & head dept, Gordon Col, 61-66; from assoc prof to prof physics, Eastern Nazarene Col, 66-82, head dept, 67-74; sr staff engr, Memodyne Corp, Needham Heights, Mass, 82-83; sr engr, Codex, Canton, Mass, 84-86; SR SCIENTIST, DELPHAX SYSTS, RANDOLPH, MASS, 86- *Concurrent Pos:* Referce, Am J Physics, 75-81. *Mem:* Am Asn Physics Teachers; Soc Imaging Sci & Technol. *Res:* Hypervelocity projectiles; optics; laser light scattering in liquids; corona and charge generation in ionographic printheads; issued 2 patents. *Mailing Add:* 17 Canton Rd Quincy MA 02171

CALHOON, DONALD ALAN, b Toledo, Ohio, July 20, 50. MICROBIOLOGY. *Educ:* Va Polytech Inst & State Univ, BS, 72, PhD(microbiol), 79; Univ NH, MS, 75. *Prof Exp:* Physical sci aide, Serol Lab, Fed Bur Invest, 72-73; FEL, ORAL BIOL DEPT, SCH DENT, STATE UNIV NY, BUFFALO, 79- *Mem:* Am Soc Microbiol; AAAS; Int Asn Dent Res; Am Asn Dent Res; Sigma Xi. *Res:* Characterization of periodontopathic bacteria including isolation, biochemical characterization, plasmid isolation, cellular fatty acid and soluble protein analyses of these organisms. *Mailing Add:* Silco Inc Box 218 Le Sage WV 25537

CALHOON, ROBERT ELLSWORTH, b Los Angeles, Calif, Dec 29, 38; m 69. QUANTITATIVE GENETICS. *Educ:* San Diego State Univ, AB, 61, MS, 67; Purdue Univ, PhD(genetics), 72. *Prof Exp:* Fel, Purdue Univ, 72-73; asst prof, 73-80, ASSOC PROF BIOL, QUEENS COL, NY, 80- *Mem:* Sigma Xi. *Res:* Genetic selection and correlated response of quantitative characters in Tribolium and Drosophila. *Mailing Add:* Dept Biol Queens Col Flushing NY 11367

CALHOON, STEPHEN WALLACE, JR, b Morrow Co, Ohio, Oct 21, 30; m 52; c 3. ANALYTICAL CHEMISTRY. *Educ:* Houghton Col, BS, 53; Ohio State Univ, MSc, 58, PhD(chem), 63. *Prof Exp:* From instr to assoc prof chem, Houghton Col, 56-64, prof, 64-78, head dept, 71-78; dean, Cent Wesleyan Col, 78-87; VPRES ACAD AFFAIRS, CENT WESTERN COL, 87- *Concurrent Pos:* Asst instr, Ohio State Univ, 60-63. *Mem:* Am Asn Higher Educ. *Res:* Simultaneous analysis of carbon, hydrogen and nitrogen in organic compounds; effects of impurities on properties of ultrapure metals; cardiac-pacemaker electrode decomposition product analysis. *Mailing Add:* Central Wesleyan Col 533 College St PO Box 436 Central SC 29630

CALHOUN, BERTRAM ALLEN, b Petoskey, Mich, May 30, 25; m 48; c 5. MAGNETISM. *Educ:* Univ Man, BSc, 47; Wesleyan Univ, MA, 48; Mass Inst Technol, PhD(physics), 53. *Prof Exp:* Asst grad div appl math, Brown Univ, 48-49; asst, Lab Insulation Res, Mass Inst Technol, 49-53; res engr, Westinghouse Res Labs, 53-56; res staff mem, Res Ctr, 56-66, SR PHYSICIST, DEVELOP LAB, IBM CORP, 66- *Mem:* AAAS; fel Am Phys Soc. *Res:* Magnetic properties of ferrites and garnets. *Mailing Add:* 20255 Glasgow Dr Saratoga CA 95070

CALHOUN, CALVIN L, b Atlanta, Ga, Jan 7, 27; m 48; c 1. CLINICAL NEUROLOGY, ANATOMY. *Educ:* Morehouse Col, BS, 48; Atlanta Univ, MS, 50; Meharry Med Col, MD, 60. *Prof Exp:* Instr biol, Morehouse Col, 50-51; from instr to assoc prof anat & actg chmn dept, 51-72, PROF ANAT & CHMN DEPT, MEHARRY MED COL, 72-, ASSOC PROF MED & DIR DIV NEUROL, 66- *Concurrent Pos:* Resident fel neurol, Univ Minn, 62-65, Nat Inst Neurol Dis & Blindness res fel, 65-66; resource consult, Elem Curric, Minneapolis City Schs, 65-66. *Mem:* AAAS; Am Acad Neurol; Am Asn Anatomists; Nat Med Asn. *Res:* Microscopic anatomy; electron microscopic evaluation of the ultra structure of the evolution of experimental cerebral infarction. *Mailing Add:* Dept Anat Meharry Med Col 1005 18th Ave N Nashville TN 37208

CALHOUN, DAVID H, b Chattanooga, Tenn, Nov 9, 42; m 75; c 1. MICROBIOLOGY, BIOCHEMISTRY. *Educ:* Birmingham Southern Col, BA, 65; Univ Ala, Birmingham, PhD(microbiol), 69. *Prof Exp:* NIH fels, Baylor Col Med, 69-71 & Univ Calif, Irvine, 71-72, instr microbiol, 72-73; from asst prof to assoc prof microbiol, Mt Sinai Sch Med, 83-87; PROF, DEPT CHEM, CITY COL NEW YORK, 87- *Concurrent Pos:* NSF res grant, Mt Sinai Sch Med, 74-76, 87-; NIH res grant, 75-87, Am Cancer Soc, 85-90, NSF, 87- *Honors & Awards:* Irma T Hirschl Career Scientist Award, 80. *Mem:* Am Soc Microbiol; Sigma Xi; Am Soc Biol Chem; NY Acad Sci. *Res:* Control of gene expression. *Mailing Add:* Dept Chem City Col New York Comvent Ave & 138th St New York NY 10031

CALHOUN, FRANK GILBERT, b Charleston, WVa, Mar 13, 39; m 66; c 3. TROPICAL AGRICULTURE & AGRONOMY. *Educ:* Ohio State Univ, BS, 61, MS, 68; Univ Fla, PhD(soil sci), 71. *Prof Exp:* Asst prof soil sci & dir, Soil Characterization Lab, Univ Fla, 71-75, assoc prof & chief party, El Salvad orProj, 75-79; prof Tropical Soils, Tex A&M Univ, 79-86; PROF INT AGRON, OHIO STATE UNIV, 86- *Concurrent Pos:* proj develop officer, Int Progs, Tex A&M Univ, 79-; prin investr soil mgt res, WAfrica, 81-; team leader, Burma Agr Prod Prog, 87- *Mem:* Am Soc Agron; Soil Sci Soc Am; Sigma Xi; Int Soc Soil Sci. *Res:* Properties and management of soils in the semi-arid tropics of Africa, meso-America, and Burma. *Mailing Add:* Dept Agron The Ohio State Univ Columbus OH 43210

CALHOUN, JOHN BUMPASS, b Elkton, Tenn, May 11, 17; m 42; c 2. ECOLOGY. *Educ:* Univ Va, BS, 39; Northwestern Univ, MS, 42, PhD(zool), 43. *Prof Exp:* Instr biol, Emory Univ, 43-44; instr zool & ornith, Ohio State Univ, 44-46; res assoc parasitol, Johns Hopkins Univ, 46-49; NIH spec fel, Jackson Mem Lab, 49-51; psychologist, US Army Med Serv Grad Sch, 51-54; psychologist, Ment Health Intramural Res Prog, NIMH, 54-71, chief sect, Res Behav Systs, Lab Brain Evolution & Behav, Div Biol & Biochem Res, 71-87. *Mem:* Ecol Soc Am; Wildlife Soc; Am Soc Mammal; Am Soc Naturalists; Soc Gen Systs Res. *Res:* Twenty-four hour activity rhythms; vertebrate ecology and social behavior; natural selection; ecology of the Norway rat; zoogeography; population and mental health; dialogue among scientists; theory of emotion and motivation; environmental design; information and conceptual evolution. *Mailing Add:* 5705 Cheshire Dr Bethesda MD 20814

CALHOUN, JOHN C, JR, b Betula, Pa, Mar 21, 17; c 4. PETROLEUM ENGINEERING. *Educ:* Pa State Univ, BS, 37, MS, 41, PhD(petrol & natural gas eng), 46. *Hon Degrees:* DSc, Ripon Col, Wis, 75. *Prof Exp:* Eng trainee, Shell Oil Co, Okla, 37; asst petrol eng, Pa State Univ, 37-45; prof petrol & natural gas eng & head dept, Pa State Univ, 50-55; dean eng & dir eng exp sta & eng & exten serv, Tex A&M Univ, 55-57, vpres eng, Col Syst, 57-60, v chancellor for develop, univ syst, 60-63; sci adv, US Dept Interior, DC, 63-65; acting dir, Water Resources Res, 64; vpres progs, 65-71, dean geosci, 69-71, sea grant prog dir, 69-72; distinguished prof petrol eng, Univ Syst Tex A&M, 65-83, vpres acad affairs, 71-77, exec vchancellor progs, 77-80, dep chancellor eng, 80-83, dir, Crisman Inst Petrol Reservoir Mgt, 84-87, EMER DEP CHANCELLOR ENG, UNIV SYST TEX A&M UNIV, 83- *Concurrent Pos:* Exec dir & pres, Gulf Univs Res Corp, 66-69; chmn spec study group sonic boom in rel to man, US Dept Interior, 67-68, chmn marine affairs action group, 70; mem, comt mineral sci & technol, Nat Acad Sci-Nat Res Coun, 67-69, naval studies bd, 74-79; chmn comt oceanog, Nat Acad Sci, 67-70, chmn ocean sci affairs bd, 70-73; chmn bd, Univ Corp Atmos Res, 68-71; mem, Presidential Task Force Oceanog, 69; chmn, President's Santa Barbara Oil Spill Panel & Panel on Union Oil Lease, 69; chmn marine affairs action group, 70; coun mem, Tex Coastal & Marine Coun, 72-83; mem res coord panel, Gas Res Inst, 77-82; dir, Tex Petrol Res Comt, 78-82; consult, Eng Educ Korea, World Bank, 78-85, petrol educ, Peoples' Repub China, 86-87; chmn, Tech Adv Comt, Gulf Res & Develop Corp, 82-84; sea grant award, 84; mem, Adv Comt Mining & Mineral Resources, US Dept Interior, 87-; consult, Coun Int Educ Exchange, 89-90. *Honors & Awards:* Degolyer Distinguished Serv Medal, Soc Petrol Engrs, 82, Distinguished Lectr, 61-84. *Mem:* Nat Acad Eng; hon mem Am Inst Mining, Metall & Petrol Engrs; hon mem Am Soc Eng Educ (vpres, 68-71, pres, 74); Soc Petrol Engrs (pres, 64); Marine Technol Soc (pres, 76-77); fel AAAS; Sigma Xi; hon mem Am Inst Mining, Metall & Petrol Engrs. *Res:* Petroleum and natural gas production; reservoir engineering; core analysis; behavior of porous materials. *Mailing Add:* Petrol Eng Dept Tex A&M Univ Syst College Station TX 77843-3116

CALHOUN, MILLARD CLAYTON, b Philadelphia, Pa, Aug 25, 35; m 54; c 7. ANIMAL NUTRITION. *Educ:* Univ Del, BS, 58, MS, 60; Univ Conn, PhD(animal nutrit), 67. *Prof Exp:* Res asst animal sci, Univ Del, 58-60; res asst, Univ Conn, 61-66; asst prof animal nutrit, Univ Del, 67-68; asst prof, 68-71, ASSOC PROF ANIMAL SCI, TEX A&M UNIV, 71- *Concurrent Pos:* consult ruminant nutrit & physiology, US AID, Argentina, 72. *Mem:* Am Soc Animal Sci; Am Inst Nutrit; Sigma Xi. *Res:* Endocrine factors in bovine ketosis; fat soluble vitamins and their interrelationships in the bovine; vitamin A and cerebrospinal fluid dynamics; primary acute hypovitaminosis A; feeder lamb nutrition and management; toxic range plants. *Mailing Add:* 1705 Paseo Devaca St San Angelo TX 76901

CALHOUN, NOAH ROBERT, b Clarendon, Ark, Mar 23, 21; m 50; c 2. ORAL SURGERY, DENTAL RESEARCH. *Educ:* Howard Univ, DDS, 48; Tufts Univ, MSD(oral surg), 55. *Prof Exp:* Oral surgeon, Vet Admin Hosp, Tuskegee, Ala, 50-52,53-55,57-65,chief dent serv,55-57, Kessler AFB, Miss,52-53; oral surgeon, Vet Admin Hosp, Washington, DC, 65-74, chief dental serv, 69-82; prof, 68, PROF ORAL SURG, COL DENT, HOWARD UNIV, 82- *Concurrent Pos:* Mem, Adv Comt, Am Bd Oral Surgeons, 67-71; prof lectr, Dent Col, Georgetown Univ; consult, DC Dent Examr Oral Surgeons, 70; mem, Nat Cancer Control Comt, 74 & Cancer Training Control Grant Comt, NIH, 72; consult, Am Bd Oral & Maxillofacial Surg, 81; prof lectr, Georgetown Dent Sch, 82- *Honors & Awards:* Audio Visual Award, Am Soc Oral Surgeons, 73 & 74. *Mem:* Inst Med, ASOS; Am Dent Asn; Am Soc Oral & Maxillofacial Surg; fel Am Col Dent; fel Int Col Dent; Sci Res Soc; Sigma Xi. *Res:* Zinc dental caries, bone healing lefort one osteotomy. *Mailing Add:* 1413 Leegate Rd NW Washington DC 20012

CALHOUN, RALPH VERNON, JR, b Quincy, Fla, Feb 8, 49; m 71; c 3. GUIDANCE & CONTROL, APPLIED PHYSICS. *Educ:* Univ WFla, BS, 71; Fla State Univ, MS, 74, PhD(physics), 75. *Prof Exp:* Instr physics, Drake Univ, 75-77; physicist guid, Air Force Armament Lab, 77-82; physicist & vpres, Tactical Systs, Inc, 82-86; CHIEF SCIENTIST & VPRES, SVERDRUP TECHNOL, TEAS GROUP, 86- *Mem:* Am Phys Soc. *Res:* Guidance and control for tactical weapons; radar; millimeterwave phenomonology; electromagnetic scattering. *Mailing Add:* 265 Shalimar Dr Shalimar FL 32579

CALHOUN, WHEELER, JR, b Columbus, NMex, Nov 19, 16; m 42; c 3. AGRONOMY. *Educ:* Ore State Univ, BS, 46, MS, 53. *Prof Exp:* Res asst seed prod, 48-51, asst prof res farm oper, 55-65, assoc prof new crop res, 65-77, ASSOC PROF AGRON, ORE STATE UNIV, 77- FARM SUPT, 51- *Concurrent Pos:* Mem sci adv comt, Nat Flax Seed Inst, 63-; oilseed crop specialist, Khuzestan Water & Power Authority, Iran, 72-74. *Mem:* Crop Sci Soc Am; Am Soc Agron; Am Soc Oil Chemists; Int Asn Mechanization Field Exp. *Res:* New crop adaptation; agronomic production; variety testing of grass and legumes for seed production potentials. *Mailing Add:* 2655 NW Highland Dr Space No 87 Corvallis OR 97330

CALIGIURI, JOSEPH FRANK, b Columbus, Ohio, Feb 13, 28; m 48; c 4. CONTROL SYSTEMS. *Educ:* Ohio State Univ, BS, 49, MS, 51. *Prof Exp:* Asst proj engr, Sperry Gyroscope Div, Sperry Rand Corp, 51-52, proj engr, 52-56, sr engr, 56-58, eng sect head, 58-60, eng dept head, 60-63, asst chief engr, 63-66, chief engr, 66-69; vpres eng, Guid & Control Systs Div, 69-71, pres, 71-77, company vpres, 74-77, sr vpres, 77-81, GROUP EXEC, LITTON INDUSTS, 77-, EXEC PRES, 81- *Mem:* Am Inst Aeronaut & Astronaut; Inst Elec & Electronics Engrs; Am Inst Navig. *Res:* Inertial navigation and guidance systems for marine, aircraft, missile and space; gyroscopes, accelerometers; gimbal and strapdown platforms, electronics and computers; augmented navigation systems through use of optium filters. *Mailing Add:* Litton Indust Inc 360 N Crescent Dr Beverly Hills CA 90210

CALIGIURI, ROBERT DOMENIC, b San Francisco, Calif, July 29, 51; m 74; c 2. ENVIRONMENTALLY ASSISTED CRACKING. *Educ:* Univ Calif, Davis, BS, 73; Stanford Univ, MS, 74, PhD(mat sci), 77. *Prof Exp:* Engr, Failure Analysis Assocs, 74-77; metallurgist, Lawrence Livermore Nat Lab, 77-78; Mat scientist & prog mgr, Sri Int, 78-87; MGR, MECH ENG, FAILURE ANALYSIS ASSOC, 87- *Concurrent Pos:* prof metall engr, State Calif; vis scientist, Tsukuba Res Ctr, Japan, 88. *Mem:* Am Soc Metals; Am Inst Mining, Metall & Petrol Engrs; Am Defense Preparedness; Sigma Xi. *Res:* The relationship between microstructure, thermomechanical processing variables and the mechanical physical properties of metals; mechanisms of environmentally assisted cracking in metals; mechanisms of high strain rate deformation and fracture in metals and ceramics; finite element modeling of transient structural response. *Mailing Add:* Failure Anal Assoc 149 Commonwealth Dr Menlo Park CA 94025

CALIGUIRI, LAWRENCE ANTHONY, b McKees Rocks, Pa, Aug 10, 33; m 66; c 2. IMMUNOLOGY, VIROLOGY. *Educ:* Bethany Col, WVa, BS, 55; Loyola Univ, Ill, MD, 58; Am Bd Pediat, dipl, 63; Am Bd Allergy Immunol, dipl, 83. *Prof Exp:* Intern, Med Ctr, Univ Pittsburgh, 58-59; from resident to chief resident pediat, Children's Hosp, Pittsburgh, 59-62; guest investr animal virol, Rockefeller Univ, 64-66, res assoc, 66-68, from asst prof to assoc prof, 68-73, adj prof animal virol, 73-83, vis prof virol, 80-81; ADJ PROF MICROBIOL GENETICS & BIOCHEM & CLIN PROF PEDIAT, UNIV PITTSBURGH, 81- *Concurrent Pos:* Consult microbiologist, Albany Med Ctr Hosp, 73-81; prob microbiol & immunol & chmn dept, Albany Med Col, 73-81; res scholar, Am Cancer Soc, 80-81; vis fel, Columbia Univ, 80-81. *Mem:* Soc Gen Microbiol; Am Soc Cell Biologists; Am Soc Microbiol; Am Acad Pediat; Am Asn Immunol; Am Acad Allergy Immunol; Am Col Allergy & Immunol. *Res:* Virus-cell interactions and immune defense mechanisms. *Mailing Add:* Allergic Dis & Asthma Assocs 3801 McKnight East Dr Pittsburgh PA 15237

CALINGAERT, PETER, b New York, NY, Aug 12, 31; m 63; c 2. COMPUTER-SUPPORTED COOPERATIVE WORK, DISTRIBUTED COMPUTER SYSTEMS. *Educ:* Swarthmore Col, BA, 52; Harvard Univ, AM, 54, PhD(appl math), 55. *Prof Exp:* Res assoc bus data processing, Comput Lab, Harvard Univ, 55-56, res fel & instr appl math, 56-57, asst prof, 57-62; systs planner, Int Bus Mach Corp, 62-63, systs eval mgr, 63-66; comput-related instr systs planning mgr, Sci Res Assocs, 66-68, PROF COMPUT SCI, UNIV NC, CHAPEL HILL, 68- *Concurrent Pos:* Nat lectr,

Asn Comput Mach, 69-70 & 73-75; vpres, Microelectronics Ctr NC, 82-83. *Mem:* AAAS; Asn Comput Mach; Fedn Am Sci; Inst Elec & Electronic Engrs; AAUP. *Res:* Harnessing computer technology to facilitate intellectual collaboration by professionals; remedying deficiencies in conventional operating-system support for collaborative applications; communication of technical information; computer science education. *Mailing Add:* Computer Sci Dept Chapel Hill NC 27599-3175

CALINGER, RONALD STEVE, b Aliquippa, Pa, Apr 6, 42; m 74; c 2. HISTORY OF MATHEMATICS, LEONHARD EULER. *Educ:* Ohio Univ, AB, 63; Univ Pittsburgh, MA, 64; Univ Chicago, PhD(hist sci), 71. *Prof Exp:* From instr to assoc prof hist, Rensselaer Polytech Inst, 69-85, dean, Undergrad Col, 82-85; dean arts & sci, 85-87, ASSOC PROF HIST, CATH UNIV AM, 85- *Honors & Awards:* Henry Schumann Prize, 68. *Mem:* Hist Sci Soc; Math Asn Am; Soc Social Studies of Sci. *Res:* History of mathematics on the European continent from the late 17th to the 20th century; scientific biography; sociocultural context of modern science. *Mailing Add:* 12806 Lacy Dr Silver Spring MD 20904

CALIO, ANTHONY JOHN, b Philadelphia, Pa, Oct 27, 29; m 71; c 4. SPACE SCIENCES. *Educ:* Univ Pa, BA, 53. *Hon Degrees:* DSc, Wash Univ, St Louis, 74. *Prof Exp:* Scientist, Bettis Atomic Power Div, Westinghouse Elec Corp, 56-59; mgr nuclear physics sect, Am Mach & Foundry Co, Va, 59-61; exec vpres & mgr opers, Mt Vernon Res Co, 61-63; mem electronic res task group, NASA, Washington, DC, 63-64, chief res eng, Electronics Res Ctr, Boston, 64-65 & Manned Space Sci Prog Off, 65-67, asst dir planetary progs, Off Space Sci & Applns, 67-68, dep dir projs, Sci & Applns Div, Johnson Space Ctr, 68-69, dir sci & applns, 69-75, dep assoc adminr space sci, 75-77, assoc adminr space & terrestrial appln, NASA Hq, Washington, DC, 77-81; dept adminr, Nat Oceanic & Atmospheric Admin, 81-84; undersecy Oceans & Atmospheric, Dept Com, 84-87; sr vpres, Planning Res Corp, McLean, Va, 87-90; VPRES, HUGHES INFO TECH CO, RESTON, VA, 91- *Concurrent Pos:* Sloan fel, Stanford Univ Grad Sch Bus, 74-75; mem bd dirs, Ctr Community Design & Res, Rice Univ, 75-77. *Honors & Awards:* Except Serv Medal, NASA, 69, Except Sci Achievement Medal, 71 & Distinguished Serv Medal, 73 & 81. *Mem:* Fel Am Astronaut Soc; Am Geophys Union; fel Am Inst Aeronaut & Astronaut. *Mailing Add:* Hughes Information Tech Co Suite 100 Reston VA 22090

CALISHER, CHARLES HENRY, b New York, NY, July 14, 36; m 79; c 4. MICROBIOLOGY. *Educ:* Philadelphia Col Pharm & Sci, BSc, 58; Univ Notre Dame, MS, 61; Georgetown Univ, PhD(microbiol), 64. *Prof Exp:* Chief cell develop unit, Microbiol Assoc, Inc, Md, 61-65; chief isolation & serol lab, Ctr Dis Control, 65-69, res microbiologist, 69-74; RES MICROBIOLOGIST, USPHS, 74-, CHIEF, ARBOVIRUS REFERENCE BR, VBDD, 73- *Concurrent Pos:* Mem subcomt interrelationships among catalogued arboviruses, Am Comt Arthropod-borne Viruses, 69-, comt chmn, 82-83, 84-85; mem arboviruses comt, Res Reference Reagents Bd, Nat Inst Allergy & Infectious Dis, 70-; adj assoc prof, Univ NC, 71-73 & Ga State Univ, 71-75; mem comn viral infections, Armed Forces Epidemiol Bd, Dengue Task Force, 70-73; proj officer, PL 480, Arboviruses in Yugoslavia, 71-; consult lab virol, WHO, 71-; mem reagents comt, Pan Am Health Orgn, 72-75 & expert comt dengue in Caribbean, 72-76; fac affil, Dept Microbiol, Colo State Univ, 75- *Mem:* Am Asn Immunologists; AAAS; Am Soc Microbiol; Am Soc Trop Med & Hyg; Sigma Xi; Royal Soc Trop Med Hyg. *Res:* Murine viruses; in vitro growth of tumor cells; cancer viruses; arbovirus ecology and epidemiology; arbovirus taxonomy and classification. *Mailing Add:* US Pub Health Serv PO Box 2087 Ft Collins CO 80522

CALKIN, MELVIN GILBERT, b New Glasgow, NS, May 20, 36; m 61; c 4. PHYSICS. *Educ:* Dalhousie Univ, BSc, 57, MSc, 58; Univ BC, PhD(physics), 61. *Prof Exp:* With Naval Res Estab, Defence Res Bd Can, 61-62; from asst prof to assoc prof, 62-72, chmn, dept physics, 82-85, PROF PHYSICS, DALHOUSIE UNIV, 72- *Concurrent Pos:* Vis prof, Univ Sussex, 71-72 & Univ Bristol, 86. *Res:* Classical and quantum mechanics. *Mailing Add:* Dept Physics Dalhousie Univ Halifax NS B3H 4H6 Can

CALKIN, PARKER E, b Syracuse, NY, Apr 27, 33; m 55; c 2. GLACIAL GEOLOGY. *Educ:* Tufts Univ, BS, 55; Univ BC, MSc, 59; Ohio State Univ, PhD(geol), 63. *Prof Exp:* Res geologist, Tufts Univ, 60-61 & Inst Polar Studies, Ohio State Univ, 61-62; from instr to assoc prof, 63-75, PROF GEOL, STATE UNIV NY BUFFALO, 75- *Concurrent Pos:* Vis res scholar, Scott Polar Res Inst, Cambridge Univ, 70, 86 & Inst Arctic & Alpine Res, Univ Colo, 79. *Honors & Awards:* Antarctica Serv Medal, NSF. *Mem:* Geol Soc Am; Glaciol Soc; Am Quaternary Asn. *Res:* Geomorphology and glacial geology, particularly in northeastern United States and in polar areas. *Mailing Add:* Dept Geol Sci State Univ NY 4240 Ridge Lea Rd Buffalo NY 14226

CALKINS, CARROL OTTO, b Sioux Falls, SDak, June 4, 37; m 59; c 2. ENTOMOLOGY, ECOLOGY. *Educ:* SDak State Univ, BS, 59, PhD(entom), 74; Univ Nebr, MS, 64. *Prof Exp:* Res entomologist, Forage Insect Lab, USDA, Lincoln, Nebr, 60-64 & Northern Grain Insects Res Lab, Brookings, SDak, 64-72, res entomologist, Insect Attractants, Behav & Basic Biol Res Lab, Agr Res Serv, USDA, 72-77; head, entom sect, Seibersdorf Lab, Int Atomic Energy Agency, Vienna, Austria, 77-80; RES LEADER & RES ENTOMOLOGIST INSECT ATTRACTANTS, BEHAV & BASIC BIOL RES LAB, AGR RES SERV, USDA, 80-; ASSOC PROF, UNIV FLA, 80- *Concurrent Pos:* Instr, Univ Nebr, 61-64 & SDak State Univ, 64-72; adj assoc prof, Univ Fla, 80- *Mem:* Ecol Soc Am; Entom Soc Am; Sigma Xi; AAAS; Int Orgn Biol Control Noxious Animals and Plants. *Res:* Ecological basis of insect behavior, population dynamics, distribution and abundance; mating behavior, ethology, quality control of mass-reared insects, Tephritidae and Coleoptera. *Mailing Add:* Biol Res Lab-USDA PO Box 14565 Gainesville FL 32604

CALKINS, CATHERINE E, ANTIBODY REGULATION. *Educ:* Purdue Univ, PhD(cell immunol), 72. *Prof Exp:* PROF IMMUNOL, JEFFERSON MED COL, THOMAS JEFFERSON UNIV, 82- *Mem:* Sigma Xi; Am Asn Immunologists. *Res:* Induction of supressor cells. *Mailing Add:* Jefferson Med Col 1020 Locust St Philadelphia PA 19107

CALKINS, CHARLES RICHARD, b Racine, Wis, May 30, 21; m 44; c 3. CHEMISTRY. *Educ:* Lawrence Col, BA, 42, MS, 47, PhD, 49. *Prof Exp:* Corp dir res & develop sect, Riegel Paper Corp, 51-65, vpres res & develop, 65-72, pres, Riegel Prod Corp, 72-74; exec vpres, Kerr Glass Mfg Corp, 74-77; vpres environ affairs, Am Paper Inst, 77-79; vpres, Kiray Forest Indust, 79-80; vpres, Santa Fe Energy Co, 81-85; RETIRED. *Mem:* AAAS; Am Chem Soc; fel Tech Asn Pulp & Paper Indust; Am Inst Chemists. *Res:* Development of specialty and packaging papers. *Mailing Add:* 6039 Schlentz Hill Rd Pipersville PA 18947

CALKINS, HARMON ELDRED, b Ann Arbor, Mich, July 11, 12; m 38; c 3. BACTERIOLOGY. *Educ:* Transylvania Col, AB, 33; Univ Ky, MS, 37; Univ Pa, PhD(bact), 41. *Prof Exp:* Asst chem & biol, Transylvania Col, 30-33; asst bact, Univ Pa, 38-41 & Johnson Found, 39-41; res bacteriologist, Upjohn Co, 41-48; asst prof bact, Univ Ga, 48-54; assoc prof, SDak State Col, 54-64; assoc prof & assoc bacteriologist, Agr Exp Sta, Univ Idaho, 64-68; assoc prof biol, 68-74, chmn div Natural Sci & Math, 74-80, PROF BIOL, PAUL QUINN COL, 74- *Concurrent Pos:* Technician, Bur Animal Indust, USDA, 34-35. *Res:* Serological reactivity of protein monolayers; virucidal activity of germicides; development of influenza virus vaccine; bovine liver abscess. *Mailing Add:* Div Natural Sci & Math Paul Quinn Col Waco TX 76704

CALKINS, JOHN, b Deming, NMex, Oct 21, 26; m 46; c 3. RADIATION BIOLOGY, PHOTOBIOLOGY. *Educ:* Tex Western Col, BS, 51; Univ Houston, MS, 59; Univ Tex, PhD(biol, physics), 63. *Prof Exp:* From instr to assoc prof, Univ KY, 63-84, PROF RADIATION MED, 84-87; BIOMED PHYSICS BEPT, KING GAISAL SPECIALLIST HOSP & RES CTR, RIYUDH, SAUDI ARABIA, 87- *Mem:* Am Soc Photobiol; Radiation Res Soc; Environ Mutagen Soc. *Res:* Mechanisms and systems for DNA repair; effects of solar ultraviolet radiation; solar ultraviolet photoecology. *Mailing Add:* Dept Radiation Med Univ Ky 800 Rose St Lexington KY 40536

CALKINS, RUSSEL CROSBY, b Cedar Falls, Iowa, Dec 31, 21; m 76; c 3. ANALYTICAL CHEMISTRY. *Educ:* Univ Northern Iowa, BA, 48; Univ Wis, MS, 51, PhD(chem), 53. *Prof Exp:* Instr chem, Univ Northern Iowa, 48-49; from asst to res asst, Univ Wis, 49-53; chemist, Dow Chem Co, 53-59; sr res chemist, 59-69, STAFF RES CHEMIST, KAISER ALUMINUM & CHEM CORP, 69-82; RETIRED. *Mem:* AAAS; Soc Appl Spectros; Am Chem Soc; Sigma Xi. *Res:* Exchange properties of nickel complexes; polymer chemistry; analytical research. *Mailing Add:* 141 St Patricks Dr Danville CA 94526-5153

CALKINS, WILLIAM GRAHAM, b Chicago, Ill, May 29, 26; m 48; c 4. INTERNAL MEDICINE, GASTROENTEROLOGY. *Educ:* Univ Mich, BS, 46, MD, 50. *Prof Exp:* Intern med, US Naval Hosp, St Albans, NY, 50-51; resident, Gen Hosp, Kansas City, Mo, 51-52; resident, Med Ctr, 54-56, from instr to asst prof, Sch Med, 56-64, ASSOC PROF MED, SCH MED, UNIV KANS, 64- *Concurrent Pos:* Staff physician, Vet Admin Hosp, Kansas City, 61-64, chief med serv, 64-70, staff physician, 70- *Mem:* Fel Am Col Physicians; Am Gastroenterol Asn; Am Pancreatic Asn. *Mailing Add:* 10523 Reeder Overland Park KS 66214

CALKINS, WILLIAM HAROLD, b Toronto, Ont, May 28, 18; US citizen; m 43; c 2. INDUSTRIAL CHEMISTRY, POLYMER CHEMISTRY. *Educ:* Univ Calif, BS, 40, PhD(chem), 47. *Prof Exp:* Res chemist, E I Du Pont de Nemours & Co, 47-53, sr res supvr, 53-58, res mgr, 58-68, assoc lab dir, plastics dept, 68-75, mgr feedstocks res, Energy & Mat Dept, 75-77, res mgr & prin consult, cent res dept, 78-85; ADJ PROF DEPT CHEM ENG, UNIV DEL, 85-; CONSULT FUEL SCI, 85- *Concurrent Pos:* Adv comt, Chem Div, Oak Ridge Nat Lab, 76-80, Ctr Res Sulfur in Coal, Ill, 84-85, Consortium Coal Liquefaction, 5 univs, 88-; chair, Del Sect, 74, Fuel Chem Div, Am Chem Soc, 90, Gordon Res conf fuel sci, 88; mem & chmn, Peer Rev Comts, Dept Energy, 83, 86, 90, Ames Lab, Univ Iowa, 88, Argonne Nat Lab, 89. *Honors & Awards:* Glenn Award, Fuel Div, Am Chem Soc; Almquist Lectr, Univ Idaho. *Mem:* Fel Am Chem Soc; fel AAAS; fel Sigma Xi. *Res:* Petrochemicals; coal liquefaction and gasification; heterogeneous catalysis; coal structure and geochemistry; coal pyrolysis; routes to chemicals and intermediates from coal, petroleum, and natural gas. *Mailing Add:* 118 Cambridge Dr Windsor Hills Wilmington DE 19803

CALL, EDWARD PRIOR, b Kent, Ohio, Dec 24, 26; m 54; c 3. REPRODUCTIVE PHYSIOLOGY, AGRICULTURE. *Educ:* Ohio State Univ, BSc, 51; Kans State Univ, PhD(animal breeding), 67. *Prof Exp:* Asst herdsman artificial insemination, Cent Ohio Breeding Asn, 50-51, technician, 51-52; asst prof, 52-58, res asst, 58-63, assoc prof, 63-79, PROF DAIRY SCI & EXTEN SPECIALIST, KANS STATE UNIV, 79- *Mem:* Am Dairy Sci Asn. *Res:* Morphology and histology of bovine genitalia; etiology of ovarian dysfunction and reproductive failure; cyclic changes of peripheral hypophyseal and gonadal hormones in the female bovine; management of dairy cattle; estrous synchronization of cattle. *Mailing Add:* Animal Sci Call Hall Kans State Univ Manhattan KS 66506-1600

CALL, JUSTIN DAVID, b Salt Lake City, Utah, Aug 7, 23; m 51; c 3. CHILD PSYCHIATRY, PEDIATRICS. *Educ:* Univ Utah, BA, 44, MD, 46. *Prof Exp:* Pvt practr, 46-47; intern pediat, Albany Hosp, NY, 47-48; pediat path res, Children's Hosp, Boston, Mass, 48-49; resident pediat, NY Hosp, 49-51; child neuropsychiatrist, Emma Pendleton Bradley Home, 51; resident psychiat, Strong Mem Hosp, 52; from instr to assoc prof, Univ Calif, Los Angeles, 54-68, PROF CHILD & ADOLESCENT PSYCHIAT & PROF, DEPT PEDIAT, UNIV CALIF, IRVINE, 68- *Concurrent Pos:* Mem, Adv Bd, Little Village Nursery Sch, Los Angeles, Exceptional Children's Found; mem, Clin Res Proj Rev Comt, NIMH, 68-72; chmn, Ment Health Adv Bd, Los Angeles County Head 24 Proj, 69-70; training & supervising psychoanalyst, Los Angeles Psychoanalysis Soc & Inst. *Honors & Awards:* Jacques Brien Award. *Mem:* Am Psychiat Asn; Am Acad Child Psychiat; Am Acad Pediat; Am Psychoanal Asn. *Res:* Preventive psychiatry; personality development; language acquisition; infant psychopathology; psychoanalysis. *Mailing Add:* 1958 Galaxy Dr Newport Beach CA 92660

CALL, PATRICK JOSEPH, b Portland, Ore, Sept 28, 49; m 74, 88. PHYSICS. *Educ:* Reed Col, BA, 71; Oxford Univ, DPhil(physics), 74. *Prof Exp:* Mem tech staff mat processing, David Sarnoff Res Ctr, RCA Labs, 75-77; sr scientist thin films develop, Solar Energy Res Inst, 77-82; vpres, Eng Ref Technol, 82-86; gen mgr, US West Optical Pub, 86-88; pres, Earth Info, 88-89, CHIEF OPER OFFICER, MAXWELL MULTI MEDIA, 91- *Concurrent Pos:* Mem tech staff, Domestic Policy Rev Solar Energy, 78. *Mem:* Am Phys Soc. *Res:* Development of thin films and coatings for solar energy applications; analysis of surface and interface properties of thin films. *Mailing Add:* 875 Seventh St Boulder CO 80302

CALL, REGINALD LESSEY, b Rigby, Idaho, Apr 23, 26; m 50; c 3. PHYSICS. *Educ:* Brigham Young Univ, BS, 51; Univ Utah, PhD(physics), 58. *Prof Exp:* Mem tech staff, Bell Labs, 58-68; assoc prof elec eng, Univ Ariz, 68-89; RETIRED. *Concurrent Pos:* Asst dean, Col Eng, Univ Ariz. *Res:* Solid state physics; photovoltaics; solar energy. *Mailing Add:* 1959 Omar Dr Tucson AZ 85704

CALLAGHAN, EUGENE, economic geology, for more information see previous edition

CALLAGHAN, OWEN HUGH, b Johannesburg, SAfrica, Oct 2, 27; m 59; c 3. IMMUNOCHEMISTRY, NEUROSCIENCES. *Educ:* Univ Witwatersrand, MSc, 56; Univ Sheffield, PhD(biochem), 58. *Prof Exp:* Vis scientist, Oxford Univ, 58; res assoc immunol, 59-60, from fac assoc to assoc prof, 61-70, PROF BIOCHEM, CHICAGO MED SCH, 70- *Concurrent Pos:* NIH res career develop award, 63. *Mem:* Brit Biochem Soc; Am Asn Immunologists. *Res:* Enzymology; immunology; histamine release; neurochemistry. *Mailing Add:* Biochem Dept Chicago Med Sch 3333 Green Bay Rd N Chicago IL 60064

CALLAHAM, MAC A, b Ft Payne, Ala, Aug 30, 36; m 60; c 4. WILDLIFE BIOLOGY, FISHERIES. *Educ:* Univ Ga, BS, 58, PhD(wildlife fisheries), 68; George Peabody Col, MA, 61, EdS, 63. *Prof Exp:* Assoc prof biol, Belmont Col, 61-63; assoc prof, 63-72, PROF BIOL, NORTH GA COL, 72-, HEAD DEPT, 72- *Mem:* AAAS; Am Fisheries Soc. *Res:* Fishery management; fish blood proteins; water quality. *Mailing Add:* Dept Biol NGa Col Dahlonega GA 30597

CALLAHAM, ROBERT ZINA, b San Francisco, Calif, May 24, 27; m 49; c 2. RESEARCH PLANNING & EVALUATION, SCI-TECH INFORMATION SYSTEMS. *Educ:* Univ Calif, BS, 49, PhD(bot), 55. *Prof Exp:* Res forester ecol, Bur Entom & Plant Quarantine, Berkeley Forest Insect Lab, USDA, 49-54; geneticist, Calif Forest & Range Exp Sta, 54-58, leader, Inland Empire Res Ctr, Intermountain Forest & Range Exp Sta, 58-60 & Inst Forest Genetics, 60-63, asst dir, Pac Southwest Forest & Range Exp Sta, 63-64, chief, Br Forest Genetics Res, 64-66, asst to dep chief res, 66-70, dir, Div Forest Insects & Dis Res, 70-73, dir, forest environ res staff, 73-76, Dir, Pac Southwest Forest & Range Exp Sta, US Forest Serv, Berkeley, Calif, 76-83; prog coord, Wildland Resources Ctr, Univ Calif, Berkeley, 83-90; RETIRED. *Concurrent Pos:* Mem exec bd, Int Union Forestry Res Orgns, 67-81; ed, Silvae Genetica, 62-66; asst exec dir, Int Union Socs Foresters, 70-74, mem exec bd, 75-79; Cong fel, 72-73; res mgt consult, 83- *Honors & Awards:* Fel, Int Union Socs Foresters, 85; Fel, Soc Am Foresters, 86. *Mem:* Hon Mem, Int Union Forestry Res Orgn,; fel Int Union Socs Foresters. *Res:* Resistance of pine species and species hybrids to bark beetles and other forest insects; pine hybridization and improvement; geographic variation in forest trees; forestry research administration; renewable resources technical information systems; management of research and extension. *Mailing Add:* 26 St Stephens Dr Orinda CA 94563

CALLAHAN, CLARENCE ARTHUR, b Bay City, Tex, Sept 29, 43; div; c 2. ECOLOGY, FISHERIES BIOLOGY. *Educ:* Univ Southwestern La, BS, 66; Auburn Univ, MS, 68; Purdue Univ, PhD(entom), 76. *Prof Exp:* AQUATIC BIOLOGIST LIMNOL, TOXICOL & ECOL ASSESSMENT, US ENVIRON PROTECTION AGENCY, 72- *Mem:* Am Fisheries Soc; Ecol Soc Am; Sigma Xi. *Res:* Ecotoxicology; hazardous waste assessment; ecological assessment. *Mailing Add:* Corvallis Environ Res Lab 200 SW 35th St Corvallis OR 97333

CALLAHAN, DANIEL, b Washington, DC, July 19, 30; m 54; c 6. BIOMEDICAL ETHICS. *Educ:* Yale Univ, BA, 52; Georgetown Univ, MA, 57; Harvard Univ, PhD(philos), 65. *Hon Degrees:* DSci, Col Med & Dent NJ, 81; DH, Univ Colo, 90. *Prof Exp:* Exec ed, Commonwealth Mag, 61-68; staff assoc, The Pop Coun, 69-70; DIR, THE HASTINGS CTR, 69- *Concurrent Pos:* Consult, Pres Comn Pop Growth & the Am Future, 70-71, Comt S-Hemoglobinopathies, Nat Acad Sci, 72, Comn Recombinant DNA Res, NIH, 76 & Med Ethics Comn, Am Col Physicians, 79-; adv bd, Technol in Soc, 81- *Honors & Awards:* Thomas More Medal, Thomas More Asn, 70. *Mem:* Inst Med-Nat Acad Sci; AAAS; Am Philos Asn. *Res:* Biomedical ethics; science and society; social and ethical problems of sicence. *Mailing Add:* The Hastings Ctr 255 Elm Rd Briarcliff Manor NY 10510

CALLAHAN, HUGH JAMES, b Philadelphia, Pa, July 9, 40; m 61; c 5. IMMUNOCHEMISTRY, GLYCOPROTEIN CHEMISTRY. *Educ:* St Joseph Univ, BS, 62; Ill Inst Technol, MS, 67; Thomas Jefferson Univ, PhD(biochem), 70. *Prof Exp:* Res assoc, Evanston Hosp, Ill, 62-67; asst prof, 72-76, ASSOC PROF BIOCHEM, THOMAS JEFFERSON UNIV, 76- *Mem:* Am Asn Immunol; Sigma Xi. *Res:* Immunochemical characterization of lymphocyte membrane receptors; physical-chemical studies of antigen-antibody interactions. *Mailing Add:* Dept Biochem Jefferson Med Col Philadelphia PA 19107

CALLAHAN, JAMES LOUIS, b Cleveland, Ohio, Sept 25, 26; m 49; c 2. INORGANIC CHEMISTRY. *Educ:* Baldwin-Wallace Col, BS, 50; Western Reserve Univ, MS, 54, PhD, 57. *Prof Exp:* From tech res specialist to supvr catalysis res, Standard Oil Co, Ohio, 50-70, mgr chem & catalysis res, 70-75, sr consult, 75-85; RETIRED. *Mem:* Am Chem Soc. *Res:* Catalysis; catalytic conversion of hydrocarbons; solid state chemistry; petrochemicals. *Mailing Add:* 124 Azalea Tr Leesburg FL 34748

CALLAHAN, JAMES THOMAS, b Appalachia, Va, Dec 29, 45; m 68; c 3. RESEARCH ADMINISTRATION, ECOLOGY. *Educ:* Univ SC, BS, 68; Univ Ga, PhD(zool), 72. *Prof Exp:* Asst prog dir, 72-76, ASSOC PROG DIR, ECOSYST STUDIES, NSF, 76- *Concurrent Pos:* Lectr, George Mason Univ, 78, Univ Va, 90-; comnr, Int Conv Adv Comn, 79-81. *Mem:* Ecol Soc Am; NAm Benthological Soc; AAAS. *Res:* Analysis of ecological systems; ecological modeling; nutrient cycles; primary consumer insect populations; science policy; endangered biota; environmental assessment. *Mailing Add:* Div Biotic Systs & Resources 1800 G St NW Washington DC 20550

CALLAHAN, JEFFREY EDWIN, b Boston, Mass, Sept 24, 43; m 67; c 2. PHYSICAL OCEANOGRAPHY. *Educ:* US Naval Acad, BS, 65; Johns Hopkins Univ, MA, 69, PhD(phys oceanog), 71. *Prof Exp:* Weapons officer, 71-74, mil asst to dir sci & eng, US Naval Oceanog Off, 74-76, cmndg officer, USS AFFRAY, 77-78; PRIN ANALYST, ANALYSIS & TECHNOL, INC, 78- *Mem:* Marine Technol Soc; Am Geophys Union; Sigma Xi; Am Geophys Union. *Res:* Large-scale oceanic circulation and its effect on the distributions of properties in the ocean, particularly polar oceans and their influence on abyssal circulation world-wide. *Mailing Add:* PRB Assoc 47 Airport View Dr Hollywood MD 20636

CALLAHAN, JOHN EDWARD, b Buffalo, NY, Feb 26, 41; m 65; c 2. GEOLOGY, ECONOMIC GEOLOGY. *Educ:* State Univ NY, Buffalo, BA, 63, MED, 65; Univ NC, Chapel Hill, MS, 68; Queen's Univ, Ont, PhD(geol), 73. *Prof Exp:* From asst prof to assoc prof, 70-80, PROF GEOL, APPALACHIAN STATE UNIV, 80- *Concurrent Pos:* Geologist, US Geol Surv, 78-83. *Mem:* Can Inst Mining & Metall; Asn Explor Geochemists; Nat Assoc Geol Teachers; Soc Econ Geologists. *Res:* Economic geology and exploration geochemistry. *Mailing Add:* Dept of Geol Appalachian State Univ Boone NC 28608

CALLAHAN, JOHN JOSEPH, b Taylor, Pa, Apr 1, 25; m 53. ANALYTICAL CHEMISTRY, PHYSICAL PROPERTIES. *Educ:* Johns Hopkins Univ, BS, 54. *Prof Exp:* Phys chemist, Physiochm Res Div, Chem Res & Develop Lab, US Dept Army, 54-60, asst, Phys Methods Br, 60-62, chief, Agents Reactions Section, 62-74, actg chief, Phys Chem Br, Chem Res Div, Chem Lab, 74-77, chief, Phys Org Br, Chem Div & res dir, Chem Res & Develop Ctr, Aberdeen Proving Ground, 77-85; SR SCIENTIST, OPTIMETRICS, INC, ANN ARBOR, 85- *Mem:* Am Chem Soc; Res Soc Am. *Res:* Physical chemical characterization and modification of properties of organic compounds; reaction and mechanistic studies of organophosphorus compounds; thermal properties of materials and thermochemistry; measurement of physical properties; instrumental analysis; purification; reaction rates and mechanisms. *Mailing Add:* 2310 Cider Mill Rd Baltimore MD 21234-2506

CALLAHAN, JOHN WILLIAM, b Welland, Ont, July 9, 42; m 69; c 2. BIOCHEMISTRY, NEUROCHEMISTRY. *Educ:* Univ Windsor, BSc, 65, MSc, 66; McGill Univ, PhD(biochem), 70. *Prof Exp:* Med Res Coun Can fel, Univ Calif, Los Angeles, 70-72; asst prof pediat, 73-81, asst prof biochem, 75-81, ASSOC PROF PEDIAT, FAC MED, UNIV TORONTO, 81-, ASSOC PROF BIOCHEM, 81- *Concurrent Pos:* Med Res Coun Can scholar, 73-78; investr neurosci, Res Inst, Hosp Sick Children, 72- *Mem:* NY Acad Sci; Can Biochem Soc; Am Soc Neurochem; Can Soc Clin Invest; Soc Pediat Res. *Res:* Isolation and characterization of lysosomal hydrolases, especially those enzymes involved in the lysosomal storage diseases; the study of the altered gene products in these diseases. *Mailing Add:* Neurosci Div Hosp Sick Children 555 University Ave Toronto ON M5G 1X8 Can

CALLAHAN, JOSEPH THOMAS, b Concord, NH, Mar 31, 22; m 43; c 4. ALLUVIAL GEOLOGY, GROUND WATER CONTAMINATION. *Educ:* Univ NH, BS, 49; Univ Ariz, MS, 51. *Prof Exp:* Geologist, Ground Water Br, US Geol Surv, 51-55, dist geologist, Ga, 55-61, from assoc chief to chief, Ground Water Br, 61-66; geologist-ground water tech adv, US AID, 66-71; chief underground waste disposal invests, Water Resources Div, US Geol Surv, 71, chief groundwater br, 71-72, regional hydrologist, 72-77; consult hydrol, Tetra Tech Int, 77-79; proj mgr, Manila Groundwater Develop, Electrowatt Eng Serv, 79-83; hydrogeologist, Barangay Water Proj, Sheladia Assocs, Inc-USAID-Manila Dept State, 83-86, CONSULT HYDROGEOLOGIST, 86- *Honors & Awards:* Meritorious Serv Award, US Dept Interior, 74; Fed Exec Inst, 74. *Mem:* Geol Soc Am; Nat Water Well Asn; Int Asn Hydrogeology; Am Inst Hydrol. *Res:* Hydrogeologic investigations of tropic zone islands, and in Korea, Japan, Oman, Saudi Arabia and the Philippines. *Mailing Add:* 6320 W Del Monico Lane Glendale AZ 85302-4442

CALLAHAN, KEMPER LEROY, plant pathology; deceased, see previous edition for last biography

CALLAHAN, KENNETH PAUL, b Pasadena, Calif, Dec 15, 43. INORGANIC CHEMISTRY. *Educ:* Univ Santa Clara, BS, 65; Univ Calif, Riverside, PhD(inorg chem), 69. *Prof Exp:* Scholar inorg chem, Dept Chem, Univ Calif, Los Angeles, 70-74; asst prof chem, Brown Univ, 74-79; sr res chemist, Occidental Res Corp, 79-82; SR RES FEL, CATALYTICA, 82- *Mem:* Am Chem Soc; Royal Soc Chem; Sigma Xi; NY Acad Sci. *Res:* Metalloborane synthesis; synthesis, spectroscopy and optical activity of coordination compounds; design and synthesis of new ligands; organic chemistry of carboranes; metal-sulfur chemistry; early transition metal chemistry; homogeneous and heterogeneous catalysis; hydroformulation; immobilized catalysts; acidic catalysts and zeolites; alkane chemistry. *Mailing Add:* 898 E Estates Dr Cupertino CA 95014

CALLAHAN, LESLIE G, JR, b Pocomoke City, Md, July 27, 23; m 44; c 2. MILITARY OPERATIONS RESEARCH, SYSTEMS ENGINEERING. *Educ:* US Mil Acad, BS, 44; Univ Pa, MS, 51, PhD(elec eng), 61. *Prof Exp:* Res & develop coordr electronic systs, Continental Army Command, US Army, Ft Monroe, Va, 52-55, tech opers officer, Univ Mich, 58-61, exec officer, Army Res Off, NC, 61-63, dir aviation electronics, Avionics Lab, Ft

Monmouth, 65-67; commanding officer, Harry Diamond Labs, Washington, DC, 67-69; prof, 69-85, EMER PROF INDUST ENG, GA INST TECHNOL, 85- *Concurrent Pos:* Fac fel, US Gen Accounting Off, 75-76. *Mem:* Opers Res Soc Am. *Res:* Systems engineering applied to the modeling, simulation, analysis and design of man-machine systems in the areas of command and control, air defense, surveillance and avionics. *Mailing Add:* Dept Indust & Systs Eng Ga Inst Technol Atlanta GA 30332

CALLAHAN, LLOYD MILTON, b Hobart, Okla, Mar 28, 34; m 86; c 4. AGRONOMY. *Educ:* Okla State Univ, BS, 59, MS, 61; Rutgers Univ, PhD(agron), 64. *Prof Exp:* From asst prof to assoc prof agron, 64-71, assoc prof ornamental hort, 71-78, PROF ORNAMENTAL HORT, UNIV TENN, KNOXVILLE, 78- *Mem:* Am Soc Agron; Weed Sci Soc Am; Crop Sci Soc Am; Sigma Xi. *Res:* Turfgrass management; cellular morphology and anatomy; plant physiology; herbicide phytotoxicity to turfgrasses; disease; DNA fingerprinting. *Mailing Add:* Dept Ornamental Univ Tenn Hort & Landscape Design Knoxville TN 37901-1071

CALLAHAN, MARY VINCENT, b Bridgeport, Conn, July 2, 22. CHEMISTRY, INSTRUMENTATION. *Educ:* Col Notre Dame, Md, AB, 43; Cath Univ Am, MS, 45, PhD(chem), 66. *Prof Exp:* Teacher, Inst Notre Dame High Sch, 44-45; teacher, Prep Sch, Col Notre Dame, Md, 46-48, instr chem, Col, 48-50; teacher, NY High Sch, 50-54; from instr to prof chem, Col Notre Dame, Md, 54-82, chmn dept chem, 64-82; lectr chem, Catholic Univ, 83-88; lectr chem, Montgomery Col, 88; RETIRED. *Concurrent Pos:* Vis prof & chairperson chem, Summer Sch, Cath Univ Am, 65-; hazardous mat control specialist, Cath Univ, 88- *Mem:* Am Chem Soc; Soc Appl Spectros. *Res:* Free radicals; pyrolysis of hexachloropropylene; instrumental analysis; infrared spectroscopy; analytical chemistry; molecular structure; amino acids; peptides; complexes. *Mailing Add:* 4701 N Charles St Baltimore MD 21210

CALLAHAN, PHILIP SERNA, b Ft Benning, Ga, Aug 29, 23; m 49; c 4. ENTOMOLOGY. *Educ:* Univ Ark, BA & MS, 53; Kans State Col, PhD(entom), 56. *Prof Exp:* From asst prof to assoc prof entom, La State Univ, 56-62; prof entom, Univ Ga, 62-69; entomologist, Southern Grain Insects Res Lab, Agr Res Serv, 62-69, ENTOMOLOGIST, INSECT ATTRACTANTS, BEHAV & BASIC BIOL RES LAB, USDA, UNIV FLA, 69- *Mem:* Optical Soc Am; Am Ornith Union; Entom Soc Am; NY Acad Sci; fel Explorers Club; Sigma Xi. *Res:* Ecology and behavior of Lepidoptera; insect biophysics with special reference to theories of infrared and microwave electromagnetic attraction between insects and host plants; methods of insect biotelemetry; nonlinear coherent radiation in living systems. *Mailing Add:* Insect Attractants USDA Univ Fla Gainesville FL 32604

CALLAHAN, THOMAS WILLIAM, b Akron, Ohio, Apr 24, 25; m 46; c 2. ELECTRICAL ENGINEERING. *Educ:* Purdue Univ, BS, 47. *Prof Exp:* Staff elec engr, Aluminum Co Am, 47-55, works chief elec eng, 56-64, div mgr, 64-73, from asst dir to assoc dir, Alcoa Labs, 73-81, opers dir, 81-82; RETIRED. *Mem:* Indust Res Inst; Sigma Xi. *Res:* Research and development for customer and corporate products and processes in the aluminum and related industries; research and development related to all aspects of corporate processes, equipment and control systems. *Mailing Add:* 106 Hollywood St Lehigh Acres FL 33936

CALLAHAN-COMPTON, JOAN REA, b San Francisco, Calif, Oct 17, 48; m 80; c 2. TECHNICAL AND SCIENCE WRITING. *Educ:* Univ Calif, Berkeley, AB, 72; Univ Ariz, MS, 75, PhD(zool), 76. *Prof Exp:* Res assoc gen biol, Univ Ariz, 77-80; adj res assoc psychol, Univ Ga, 80-85; SR BIOLOGIST, TETRA TECH INC, 90- *Concurrent Pos:* Theodore Roosevelt Mem grant, Am Mus Natural Hist, 76, 90-91; consult, NSF grant, Dept Psychol, Univ Ga, 78-79; consult, Calif State Parks & Recreation Dept, 91-92. *Mem:* Nat Asn Sci Writers; Am Soc Mammalogists. *Res:* Biogeography of small mammals; speciation in Sciuridae; books on applied science. *Mailing Add:* PO Box 3140 Hemet CA 92343

CALLAN, CURTIS, b North Adams, Mass, Oct 11, 42; m 65; c 2. THEORETICAL PHYSICS. *Educ:* Haverford Col, BA, 61; Princeton Univ, PhD(physics), 64. *Prof Exp:* Asst prof physics, Harvard Univ, 67-69; mem, Inst Advan Study, 69-72; prof, 72-86, HIGGINS PROF PHYSICS, PRINCETON UNIV, 86- *Mem:* Nat Acad Sci; Am Phys Soc; Am Acad Arts & Sci. *Res:* Quantum field theory with emphasis on applications to elementary particle physics. *Mailing Add:* Princeton Univ Jadwin Hall Princeton NJ 08544

CALLAN, EDWIN JOSEPH, b Floral Park, NY, Nov 29, 22; m 46; c 2. ATOMIC PHYSICS, TECHNICAL MANAGEMENT. *Educ:* Manhattan Col, BS, 43; Ohio State Univ, MSc, 60. *Prof Exp:* Physicist & chief thermal res sect, Concrete Res Div, Corps Engrs, Miss, 46-53; res & develop adminstr, Directorate Res, Wright Air Develop Ctr, 53-56, chief plans & analyst, USAF Aerospace Res Labs, 56-68, sci adv, Aerospace Res Labs, 68-75; CONSULT, 75- *Concurrent Pos:* Scholar, Dublin Inst Advan Studies, 64-65; mem conf comt, Nat Conf Admin Res, 72-75; chmn, LOM Ltd. *Mem:* Am Phys Soc; Am Chem Soc. *Res:* Atomic transition rates; thermal and elastic properties of concreting materials; radiation shielding; research management; atomic and ionic energy level calculations and methods; materials science engineering; software systems. *Mailing Add:* 4139 Windcross Lane Orlando FL 32809

CALLANAN, JANE ELIZABETH, b Kearny, NJ, Oct 26, 26. SOLIDS, FLUIDS FUEL TECHNOLOGY. *Educ:* Col St Elizabeth, BS, 47; Fordham Univ, MS, 60, PhD(phys chem), 67. *Prof Exp:* Instr-assoc prof chem, Col St Elizabeth, 66-78; res assoc, Univ Mich, 78-80; res chemist, Dept Energy, 80; RES CHEMIST, NAT BUR STANDARDS, US DEPT COM, 81- *Concurrent Pos:* Sr visitor, Inorg Chem Lab, Oxford Univ, Eng, 77-78. *Mem:* Am Chem Soc; Am Inst Chemists; Am Inst Chem Engrs; Sigma Xi. *Res:* Thermophysical properties of complex solids and fluids, particularly those related to the development of fossil fuel technology; development of standards for calorimetry. *Mailing Add:* Ctr Chem Eng 773-10 325 Broadway Boulder CO 80303

CALLANAN, MARGARET JOAN, b Washington, DC, July 31, 26. SCIENCE EDUCATION. *Educ:* Trinity Col, Washington, DC, AB, 48; Cath Univ Am, MS, 50. *Prof Exp:* Chemist, NIH, 50-58; res asst, Nat Res Coun, 58-61; prog officer, NSF, 61-83; CONSULT, 83- *Concurrent Pos:* Fel educ for pub mgt, Stanford Univ, 72-73. *Mem:* Fel AAAS. *Res:* Physico-chemical properties of proteins, nucleic acids and protamines; science education; public understanding of science; finding ways to increase the number of women with science careers. *Mailing Add:* 730 24th St NW Apt 719 Washington DC 20037

CALLANTINE, MERRITT REECE, b Hammond, Ind, Jan 6, 36; m 56; c 2. PHYSIOLOGY, ENDOCRINOLOGY. *Educ:* Purdue Univ, BS, 56, MS, 58, PhD(endocrinol), 61. *Prof Exp:* Teaching asst biol sci, Purdue Univ, 58-60; head sect endocrinol, Parke-Davis Res Labs, Mich, 61-70; dir sci & regulatory affairs, 70-72, sr clin investr, Mead Johnson Res Ctr, 72-76; med-legal liaison, Lilly Res Labs, 76-88; VPRES, PHARMACEUT RES, BIOPHYSICA FOUND, 88- *Concurrent Pos:* Vis assoc prof, Hahnemann Med Col, 73-78. *Mem:* Endocrine Soc; Am Physiol Soc; Soc Exp Biol & Med; Soc Study Reproduction; AAAS; Sigma Xi. *Res:* Endocrinology of reproduction, especially control of gonadotrophin secretion, ovarian function and uterine contractility; mechanism of hormone action; nonsteroidal antifertility agents; hormone antagonists; clinical investigation, antineoplastic agents, antibiotics, estrogen replacement therapy and agents to control uterine contractility. *Mailing Add:* 7855 Bellakaren Pl La Jolla CA 92037

CALLARD, GLORIA VINCZ, b Perth Amboy, NJ, July 20, 38; m 62; c 3. COMPARATIVE ENDOCRINOLOGY. *Educ:* Tufts Univ, BS, 59; Rutgers Univ, MS, 63, PhD(zool), 64. *Prof Exp:* Endocrine biochemist, Personal Prod Co Div, Johnson & Johnson, 64-67; lectr, Col William & Mary, 67-72; asst prof, Tufts Univ, 72-73; res assoc & lectr, Boston Univ, 74-75; res assoc obstet & gynec, Harvard Med Sch, 75-79, asst prof, 79-81; assoc prof, 81-85, PROF BIOL, BOSTON UNIV, 85- *Concurrent Pos:* Co-ed, J Exp Zool; mem, Adv Panel, Reg Biol, NSF, 83-86; mem bd ed, Steroids Div, Gen Comt Edocrinol. *Mem:* Endocrine Soc; fel AAAS; Soc Study Reproduction; Am Soc Zoologists; Soc Neuroscience; NY Acad Sci. *Res:* Reproductive and neuroendocrinology; estrogen synthesis and actions in brain and pituitary; testicular steroidogenesis and regulation of spermatogenesis. *Mailing Add:* Boston Univ Five Cummington St Boston MA 02215

CALLAS, GERALD, b Beaumont, Tex, Oct 14, 32; m 67; c 2. ANATOMY, ELECTRON MICROSCOPY. *Educ:* Lamar State Col, BS, 59; Univ Tex, Galveston, MA, 62, PhD(anat), 66, MD, 67. *Prof Exp:* Intern, John Sealy Hosp, 67-68; asst prof anat, 68-75, ASSOC PROF ANAT, MED BR, UNIV TEX, 75- *Mem:* Am Asn Anatomists. *Res:* Electron microscopy of the lung secondary to altered thyroid function; fine structural changes occurring in the adrenal gland in relation to varying functions of the thyroid gland. *Mailing Add:* Dept Anat Univ Tex Med 301 University Blvd Galveston TX 77550

CALLAWAY, CLIFFORD WAYNE, b Easton, Md, May 28, 41; m 77; c 1. WEIGHT RELATED MEDICAL CONDITIONS, BROAD-BASED MEDICAL POLICY CONSULTING. *Educ:* Univ Delaware, BA, 63; Northwestern Univ, MD, 67. *Prof Exp:* Intern rotating-O, Chicago Wesley Mem Hosp, 67-68, resident internal med, Chicago Wesley Mem/Northwestern, 68-69; med officer, USN, 69-71; resident internal med, Mayo Grad Sch Med, 71-73, advan clin resident endocrinol, Mayo Clin, 75-76, consult endocrinol & metab, 78-86; Mayo Found scholar/res assoc, Joslin Res Lab, 76-78; ASSOC CLIN PROF MED, GEORGE WASHINGTON UNIV, 86- *Concurrent Pos:* Chair, Nutrit Educ Comt, Nutrit Coord Comt, NIH, 79-80, med officer, Nutrit Prog, Nat Inst Arthritis & Metab Dis, NIH, 79-80; actg exec secy, Nutrit Coord Comt, Off Dis Prev & Health Promo, US Dept Health & Human Serv, 80; dir, Nutrit Clin & Nutrit Consult Serv, Mayo Clin, 81-86, Lipid Clin, 82-86 & Ctr Clin Nutrit, Med Ctr, George Washington Univ, 86-88; prev cardiol acad awared, Nat Heart, Lung & Blood Inst, NIH, 86-88; sr consult, Food & Nutrit Bd, Nat Res Coun-Nat Acad Sci, 87- 88; pvt pract, internal med, endocrinol, metab & clin nutrit, 88-; chmn, Olestra Sci Rev Coun, Procter & Gamble, 88-; mem, Dietary Guidelines Adv Comt, USDA/Dept Health & Human Serv, 89-90; med adv bd, NMC Homecare, Nat Med Care, Inc, 89-; health care leader, Resetting Am Table, Am Inst Wine & Food, 90-; spec expert, Working Group Obesity, Nat Inst Diabetes & Digestive & Kidney Dis, NIH, 91-; chmn, Pub Info & Educ Comt, Med Soc DC, 91-; mem, bd dirs, Am Bd Nutrit & bd gov, Am Osler Soc. *Mem:* Hon mem Am Dietetic Asn; Am Bd Nutrit; Am Inst Nutrit; Am Soc Clin Nutrit; Am Osler Soc. *Res:* Human obesity, especially its heterogenecity and the biologically determined responses to semi-starvation; dietary guidelines and national nutrition policy. *Mailing Add:* 2112 F St NW Washington DC 20037

CALLAWAY, ENOCH, III, b La Grange, Ga, July 12, 24; m 48; c 2. PSYCHIATRY. *Educ:* Columbia Univ, AB, 44, MD, 47. *Prof Exp:* Intern med, Grady Mem Hosp, Emory Univ, 47-48; resident psychiat, Worcester State Hosp, Mass, 48-49; from instr to assoc prof, Psychiat Inst, Univ Md, 52-58; prof psychiatry, Univ Calif, San Francisco, 59-; AT BRAIN-BEHAVIOR RES CTR, SONOMA DEVELOP CTR, UNIV CALIF AT ELDRIDGE. *Concurrent Pos:* USPHS res career investr, 54-58; mem psychopharmacol study sect, NIMH, 63-67, mem alcohol & alcohol probs study sect, 68-72. *Mem:* Am Col Neuropsychopharmacol; Soc Biol Psychiat (vpres, 81); Soc Psychophysiol Res (pres, 81). *Res:* Behavioral neurophysiology as related to psychiatric problems; psychopharmacology. *Mailing Add:* One Mt Tiburon Belvedere-Tiburon CA 94920

CALLAWAY, JASPER LAMAR, b Cooper, Ala, Apr 5, 11; m 41; c 3. DERMATOLOGY. *Educ:* Univ Ala, BS, 31; Duke Univ, MD, 32. *Prof Exp:* Fel & instr dermat, Univ Pa, 35-37; prof dermat, Sch Med, Duke Univ, 37-88; RETIRED. *Concurrent Pos:* Consult, Vet Admin Hosp, USPHS & Surgeon Gen, US Air Force; mem, Nat Adv Serol Coun; pres, Am Bd Dermat & Syphilol, 58-59; mem, Spec Adv Group, Vet Admin; mem, Cutaneous Dis Sect, Nat Res Coun. *Mem:* Am Dermat Asn (secy, pres, 58-59); Soc Invest Dermat (pres, 55-56); fel Am Col Physicians; Am Asn Prof Dermatologists (pres, 65); Am Acad Dermat. *Res:* Syphilology; manual clinical mycology. *Mailing Add:* 26 Stoneridge Circle Durham NC 27705

CALLAWAY, JOSEPH, b Hackensack, NJ, July 1, 31; m 49; c 3. THEORETICAL PHYSICS. *Educ:* Col William & Mary, BS, 51; Princeton Univ, MA, 53, PhD(physics), 56. *Prof Exp:* Asst prof physics, Univ Miami, 54-60; from assoc prof to prof physics, Univ Calif, Riverside, 60-67; prof, 67-76, chmn dept, 70-73, BOYD PROF PHYSICS, LA STATE UNIV, BATON ROUGE, 76- *Concurrent Pos:* Consult, res labs, Philco Corp, 61-67; sr vis fel, Imp Col, Univ London, 74-75; mem exec comt, APS Div Condensed Matter Physics, 77-79. *Mem:* AAAS; fel Am Pys Soc; fel Inst Physics; Europ Phys Soc. *Res:* Electron-atom scattering; electronic structure theory; theory of magnetism in solids; atomic clusters. *Mailing Add:* Dept Physics & Astron La State Univ Baton Rouge LA 70803-4002

CALLCOTT, THOMAS ANDERSON, b Columbia, SC, Jan 13, 37; m 61; c 3. SOLID STATE PHYSICS, SURFACE PHYSICS. *Educ:* Duke Univ, BS, 58; Purdue Univ, MS, 62, PhD(physics), 65. *Prof Exp:* Res assoc physics, Purdue Univ, 65; mem tech staff, res div, Bell Tel Labs, 65-68; asst prof, 68-72, assoc prof, 72-78, PROF PHYSICS, UNIV TENN, KNOXVILLE, 78- *Concurrent Pos:* Consult, Oak Ridge Nat Lab, 68-; NSF res grants, 69-77 & 82-; dir, soft x-ray spectros proj, Nat Synch Light Source. *Mem:* Fel Am Phys Soc. *Res:* Soft x-ray spectroscopy of solids; optical and photoemission measurements on solids; physics of solid surfaces. *Mailing Add:* Physics Dept Univ Tenn 401 Nielsen Physics Bldg Knoxville TN 37996

CALLE, CARLOS IGNACIO, b Medellin, Antioquia, July 30, 45; m 71; c 1. NUCLEAR REACTION THEORY, NUMERICAL ANALYSIS. *Educ:* Univ Antioquia, BS, 71; Western Mich Univ, MA, 73; Ohio Univ, PhD(physics), 84. *Prof Exp:* Instr, 81-84, from asst prof to assoc prof, 84-91, PROF PHYSICS, SWEET BRIAR COL, 91- *Mem:* Am Phys Soc; NY Acad Sci. *Res:* Theoretical nuclear physics studies; microscopic theory of nuclear reactions; non-local, energy-dependent optical potential calculations of nucleon elastic scattering from complex nuclei; numerical analysis; nuclear structure calculations. *Mailing Add:* Dept Physics Sweet Briar Col Sweet Briar VA 24595

CALLE, LUZ MARINA, b Titiribi, Antioquia, Sept 30, 47; US citizen; m 71; c 1. ELECTROCHEMICAL CORROSION TESTING, ELECTROCHEMICAL IMPEDANCE MEASUREMENTS. *Educ:* Univ Antioquia, BS, 71; Western Mich Univ, 73; Ohio Univ, PhD(chem), 80. *Prof Exp:* Postdoctoral researcher chem, Ohio Univ, 80-81; asst prof, Stockton State Col, 81; asst prof, 81-87, ASSOC PROF CHEM, RANDOLPH-MACON WOMAN'S COL, 87- *Concurrent Pos:* Fac fel, NASA, 89-90. *Mem:* Am Chem Soc; Nat Asn Corrosion Engrs. *Res:* Electrochemical impedance measurements for evaluating and predicting the performance of metal alloys; electrochemical impedance data analysis; cancer research; biophysical chemistry. *Mailing Add:* Chem Dept Randolph-Macon Woman's Col Lynchburg VA 24503

CALLEN, EARL ROBERT, b Philadelphia, Pa, Aug 28, 25; m 50; c 4. SOLID STATE PHYSICS. *Educ:* Univ Pa, AB, 48, MA, 51; Mass Inst Technol, PhD(physics), 54. *Prof Exp:* Physicist, Nat Security Agency, 55-59; with US Naval Ord Lab, 59-68; prof physics, Am Univ, 68-87; consult, US/Japan Sci & ONR/Tokyo, 87-90; EMER PROF PHYSICS, AM UNIV, 87- *Mem:* Fel Am Phys Soc. *Res:* Magnetostriction; amorphous magnetism; technology transfer and international development. *Mailing Add:* 9110 LeVelle Ct Chevy Chase MD 20815

CALLEN, HERBERT BERNARD, b Philadelphia, Pa, July 1, 19; m 45; c 2. STATISTICAL MECHANICS. *Educ:* Temple Univ, BS, 41, AM, 42; Mass Inst Technol, PhD(physics), 47. *Prof Exp:* Physicist, Kellex Corp, New York, NY, 44-45; electronic res, Princeton Univ, 45; res assoc, Mass Inst Technol, 45-48; from asst prof to prof, 48-85, EMER PROF PHYSICS, UNIV PA, 85- *Concurrent Pos:* Consult, Sperry Rand Corp, 51-84; mem adv panel physics, NSF, 66-69, chmn, 69; chmn fac sen, Univ Pa, 70-71; Guggenheim fel, 72; mem, Comn Statist Mech, Int Union Pure & Appl Physics, 72-78, chmn 75-78. *Honors & Awards:* Elliott Cresson Medal, 84. *Mem:* Nat Acad Sci; fel Am Phys Soc. *Res:* Theoretical physics; theory of solid state; thermodynamics; irreversible statistical mechanics. *Mailing Add:* Dept of Physics Univ of Pa Philadelphia PA 19104

CALLEN, JAMES DONALD, b Wichita, Kans, Jan 31, 41; m 61; c 2. NUCLEAR ENGINEERING. *Educ:* Kans State Univ, BS, 62, MS, 64; Mass Inst Technol, PhD(appl plasma physics), 68. *Prof Exp:* NSF postdoctoral fel, Inst Advan Study, Princeton Univ, 68-69; asst prof aeronaut & astronaut, Mass Inst Technol, 69-72; staff mem, Oak Ridge Nat Lab, 72-75, head, Theory Sect, Fusion Energy Div, 75-79; KERST PROF NUCLEAR ENG PHYS & PHYSICS, UNIV WIS-MADISON, 79- *Concurrent Pos:* Consult, Oak Ridge Nat Lab, 69-72; Jet Proj, Abingdon, Eng, 86-87. *Honors & Awards:* Guggenheim Fel, 86-87. *Mem:* Nat Acad Eng; Am Phys Soc; Am Nuclear Soc. *Res:* Plasma physics, mainly magnetic fusion; nuclear reactor engineering. *Mailing Add:* Dept Nuclear Eng & Eng Phys ERB Univ Wis 1500 Johnson Dr Madison WI 53706

CALLEN, JOSEPH EDWARD, b Moulton, Iowa, Mar 24, 20; m 44; c 2. ORGANIC CHEMISTRY, RESEARCH ADMINISTRATION. *Educ:* Univ Iowa, BS, 42, MS, 43, PhD(org chem), 46. *Prof Exp:* Asst chem, Univ Iowa, 42-43; res chemist, Nat Defense Res Comt, 43-45; res chemist, Proctor & Gamble Co, 46-60, assoc dir res, 60-77; RETIRED. *Concurrent Pos:* Instr, Univ Iowa, 44. *Mem:* Am Chem Soc. *Res:* Synthetic detergents; fats and oils; fluorescent dyes; infrared and ultraviolet spectroscopy; amines; gas chromatography. *Mailing Add:* 3825 Mariners Walk No 621 Cortez FL 34215

CALLENBACH, JOHN ANTON, b Merchantville, NJ, Apr 9, 08; m 33; c 3. ENTOMOLOGY. *Educ:* Univ Wis, BS, 30, MS, 31, PhD(econ entom), 39. *Prof Exp:* Asst econ entom, Univ Wis, 30-33; Va Smelting Co fel, Va Truck Exp Sta, 33-36; instr econ entom, Univ Wis & agent, Bur Entom & Plant Quarantine, USDA, 36-42; asst prof entom, Univ Idaho, 42-44; asst state entomologist, Mont, 44-46; asst prof zool & entom, Mont State Col, 46-48, assoc prof, 48-53; state entomologist, NDak, & prof agr entom & chmn dept, 53-57, assoc dean agr & assoc dir agr exp sta, 57-69, prof entom, 69-74, chmn dept, 69-73, EMER PROF ENTOM, N DAK STATE UNIV, 74- *Mem:* Entom Soc Am. *Res:* Truck crop and cereal crop insects; mosquitoes. *Mailing Add:* 606 Ninth St S Fargo ND 58103-2639

CALLENDER, JONATHAN FERRIS, b Los Angeles, Calif, Nov 7, 44; div; c 4. GEOLOGY, TECTONICS. *Educ:* Calif Inst Technol, BS, 66; Harvard Univ, AM, 68, PhD(geol), 75. *Prof Exp:* Teaching fel geol, Harvard Univ, 67-70, res asst mineral, 70; from asst prof to prof geol, Univ NMex, 72-90; dir, NMex Mus Natural Hist, 84-90; GEN MGR, ADRIAN BROWN CONSULTS, 90- *Concurrent Pos:* Consult, Pub Serv Co, NMex, 73-74 & Sandia Labs, 74-88; vis staff scientist, Los Alamos Sci Lab, 74-78; proj corresp, US Geodynamics Comt, 75-90; pres, NMex Geol Soc, 77; panel mem, Gorleben nuclear disposal site, Fed Repub Ger, 78-79; geol consult, 90- *Mem:* Fel Geol Soc Am; Sigma Xi; Am Geophys Union; Am Asn Petrol Geologists; AAAS. *Res:* Structural geology and petrology; tectonics; nuclear waste disposal in geologic materials; geothermal exploration; geotectonics. *Mailing Add:* 155 S Madison St Suite 240 Denver CO 80209-3014

CALLENDER, WADE LEE, b Kingsville, Ohio, May 27, 26; m 50; c 5. PHYSICAL CHEMISTRY. *Educ:* Col of Wooster, AB, 48; Univ Rochester, PhD(chem), 51. *Prof Exp:* Res chemist, Shell Oil Co, 51-66; sr staff res chemist, Koninklijke-Shell Lab, Amsterdam, 66-68 & Shell Oil Co, 68-73, res assoc, Shell Develop Co, 73-88; RETIRED. *Mem:* Catalysis Soc; Am Chem Soc; Am Soc Testing & Mat. *Res:* Mechanism and kinetics of catalytic reactions; analog and digital computing; catalytic reforming; Ziegler-Natta catalysis. *Mailing Add:* 7718 Westwind Lane Houston TX 77071

CALLENS, EARL EUGENE, JR, b Memphis, Tenn, Mar 8, 40; m 60; c 10. AEROSPACE ENGINEERING, FLUID DYNAMICS. *Educ:* Ga Inst Technol, BSAE, 62, MSAE, 64; Von Karman Inst Fluid Dynamics, Belg, dipl, 67; Univ Tenn, PhD(aerospace eng), 76. *Prof Exp:* Res engr, 68-77, eng supvr res & testing, Sverdrup Corp, 77-78, asst br mgr, ARO Inc, 78-80; asst br mgr, Calspan Field Servs, 81-83; ASSOC PROF MECH ENG, LA TECH UNIV, 88- *Honors & Awards:* Space Shuttle Award, Am Inst Aeronaut & Astronaut, 84; HH Arnold Award, Am Inst Aeronaut & Astronaut, 74; Outstanding Prof, Am Soc Mech Eng, 90. *Mem:* Assoc fel Am Inst Aeronaut & Astronaut; Sigma Xi; Am Soc Mech Engrs; Am Soc Eng Educ; Am Defense Preparedness Asn. *Res:* Aerothermodynamics; hypersonic flow; wake phenomenology; hypervelocity erosion; technical management. *Mailing Add:* 204 Lilinda Dr Ruston LA 71270

CALLERAME, JOSEPH, b New York, NY, Feb 9, 50; m 72; c 2. PHYSICS, ACOUSTICS. *Educ:* Columbia Col, BA, 70; Harvard Univ, MA, 71, PhD(physics), 75. *Prof Exp:* Res assoc, Mass Inst Technol, 75-77; sr res scientist physics, 77-80, mgr, Infrared Detector Lab, 83-89, MGR, MICROWAVE ULTRA SONICS LAB, RAYTHEON CO, 80-, ASST GEN MGR RAYTHEON RES DIV, 89- *Mem:* Inst Elect & Electronic Engrs. *Res:* Surface acoustic wave devices; transducer design; interaction of x-rays with matter; infrared detectors, superconductivity. *Mailing Add:* Raytheon Co Res Div 131 Spring St Lexington MA 02173

CALLERY, PATRICK STEPHEN, b San Jose, Calif, Nov 30, 44; m 65; c 3. MEDICINAL CHEMISTRY. *Educ:* Univ Utah, BS, 68; Univ Calif, PhD(pharmaceut chem), 74. *Prof Exp:* Fel chem, 73-74, asst prof to assoc prof, 74-90, PROF CHEM, UNIV MD, 90- *Concurrent Pos:* BI-4 study sect, NIH, 87-90. *Mem:* Am Chem Soc; Am Soc Mass Spectrometry; Am Asn Col Pharm. *Res:* Biomedical applications of mass spectrometry. *Mailing Add:* Sch Pharm Univ Md 20 N Pine St Baltimore MD 21201

CALLEWAERT, DENIS MARC, b Detroit, Mich, Feb 20, 47; m 71; c 2. CELLULAR IMMUNOLOGY, NATURAL IMMUNITY. *Educ:* Univ Detroit, BS, 69; Wayne State Univ, PhD(biochem), 73. *Prof Exp:* Fel immunol, Med Sch, Wayne State Univ, 73-74; from asst prof to assoc prof, 74-85, PROF CHEM, OAKLAND UNIV, 85-, DIR, INST BIOCHEM & BIOTECHNOL, 89-; PRES, OXFORD BIOMED RES, INC, 83-; PRES, PROTEINS INT, 86- *Concurrent Pos:* Consult, Dept Clin Path, William Beaumont Hosp, 89- *Mem:* Am Asn Immunologists; Am Chem Soc; AAAS; Sigma Xi. *Res:* Molecular mechanism of cell-mediated cytotoxicity reactions mediated by natural killer and activated killer cells; mechanisms involved in regulation of cell-mediated immune reactions; novel methods for drug delivery using monoclonal antibodies and liposomes. *Mailing Add:* Inst Biochem & Biotechnol Oakland Univ Rochester MI 48309

CALLIER, FRANK MARIA, b Antwerp, Belg, Nov 27, 42; m 69; c 2. FEEDBACK CONTROL THEORY, OPTIMAL CONTROL. *Educ:* Univ Ghent, Belg, dipl elec eng, 66; Swiss Fed Inst Technol, cert nuclear eng, 67; Univ Calif, Berkeley, MSc, 70, PhD(elec eng), 72. *Prof Exp:* Asst prof, Dept Elec Eng & Computer Sci, Univ Calif, 72-73; res assoc, Belg Nat Sci Found, Univ Ghent, 73-74; from assoc prof to prof, 74-86, PROF, DEPT MATH, FAC UNIV DE NAMUR, BELG, 86- *Concurrent Pos:* Assoc ed, Systs & Control Lett, 84-90. *Mem:* Fel Inst Elec & Electronic Engrs. *Res:* Applied mathematics in engineering; automatic control theory; system theory; multivariable feedback systems; linear quadratic optimal control; author of more than 30 publications and one book. *Mailing Add:* Fac Univ Nd De La Paix Eight Rempart De La Vierge Namur 5000 Belgium

CALLIHAN, ALFRED DIXON, b Scarbro, WVa, July 20, 08; m 82. PHYSICS. *Educ:* Marshall Col, AB, 28; Duke Univ, AM, 31; NY Univ, PhD(physics), 33. *Hon Degrees:* DSc, Marshall Univ, 64. *Prof Exp:* From asst to instr physics, Marshall Col, 26-29; asst, Duke Univ, 29-30 & NY Univ, 30-34; tutor, City Col New York, 34-37, from instr to asst prof, 37-48; res physicist, Manhattan Dist Proj, Div War Res, Columbia Univ, 41-45; res physicist, Nuclear Div, Union Carbide Corp, 45-73, consult, 73-84; CONSULT, US ARMY COMBAT SYSTS TEST ACTIV, 73- & MARTIN MARIETTA CORP, 84- *Concurrent Pos:* Ed, Nuclear Sci & Eng, 65-83; mem, Nuclear Standards Mgt Bd, Am Nat Standards Inst, 57-87; admin judge, Atomic Safety & Licensing Bd panel, US Nuclear Regulatory Comn, 63- *Honors & Awards:* Exceptional Serv Award, Am Nuclear Soc, 83, Standards Serv Award, 88. *Mem:* Fel Am Phys Soc; fel Am Nuclear Soc; Sigma Xi. *Res:* Reactor physics. *Mailing Add:* 102 Oak Lane Oak Ridge TN 37830

CALLIHAN, CLAYTON D, b Midland, Mich, June 13, 19; m 57; c 2. CHEMICAL ENGINEERING. *Educ:* Mich State Univ, BS, 54, PhD, 57. *Prof Exp:* Lab technician, Dow Chem Co, Mich, 37-44, sr lab technician, 46-50, res engr, 57-60, sr res engr, La, 60-63; assoc prof, 63-71, PROF CHEM ENG, LA STATE UNIV, BATON ROUGE, 71- *Concurrent Pos:* Consult, 63- *Res:* High polymers and their preparation from cellulose; conversion of cellulosic wastes and microbial proteins. *Mailing Add:* 9524 Greenbriar Dr Baton Rouge LA 70815

CALLIHAN, DIXON, b Scarbro, WVa, July 20, 08; m 30, 82. NUCLEAR CRITICALITY SAFETY. *Educ:* Marshall Univ, AB, 28; Duke Univ, MA, 31; NY Univ, PhD(physics), 33. *Hon Degrees:* DSc, Marshall Univ, 64. *Prof Exp:* From instr to asst prof physics, City Col, NY, 34-47; mem staff, Div War Res, Columbia Univ, 42-45 & Union Carbide Corp, 45-73; ADMIN JUDGE, US NUCLEAR REGULATORY COMN, 63- *Concurrent Pos:* Vchmn & mem, Nuclear Standards Bd, Am Nat Standards Inst, 57-88; chmn, Standards Comt, Am Nuclear Soc, 60-91; mem, Reactor Safety Comt, US Army, 63-; ed, Am Nuclear Soc Res J, 64-84. *Honors & Awards:* Except Serv Award, Am Nuclear Soc, 83, Standards Serv Award, 89; Meritorious Serv Award, Am Nat Standards Inst, 87. *Mem:* Fel Am Nuclear Soc; fel Am Phys Soc; Sigma Xi. *Res:* Development of media for separation of the uranium isotopes; application of nuclear data to the nuclear criticality safety of processes with fissionable materials; nuclear reactors. *Mailing Add:* 102 Oak Lane Oak Ridge TN 37830

CALLIHAN, ROBERT HAROLD, b Spokane, Wash, Oct 19, 33; m 54; c 2. AGRONOMY, WEED SCIENCE. *Educ:* Univ Idaho, BS, 57; Ore State Univ, MS, 61, PhD(crop sci), 73. *Prof Exp:* Supt agron, Ore State Univ, 57-60, instr, 60-67; exten specialist potato prod, 67-70, asst exten prof, 69- 70, asst res prof agron, 70-75, ASSOC RES PROF AGRON & WEED SCI, UNIV IDAHO, 75- *Concurrent Pos:* Consult & expert witness, US Dept Justice, DOE, 67-; prof, nat & state comts, regional & local levels; mem, US Army Security Agency, 54-56. *Honors & Awards:* Meritorious Honors Award in Agr, Polish Govt, 77. *Mem:* Weed Sci Soc Am; Am Soc Agron; Crop Sci Soc Am; Potato Asn Am; Europ Asn Potato Res; Int Weed Sci Soc; Soil Sci Soc Am. *Res:* Weed biology and control in range, potatoes, pulses, forages, forests and sugar beets; herbicides, herbigation; crop loss. *Mailing Add:* Plant Sci Dept Univ Idaho Moscow ID 83843

CALLIS, CLAYTON FOWLER, b Sedalia, Mo, Sept 25, 23; m 49; c 2. INORGANIC CHEMISTRY. *Educ:* Cent Col, Mo, AB, 44; Univ Ill, MS, 46, PhD(inorg chem), 48. *Prof Exp:* Instr physics & chem, Cent Col, Mo, 44-45; asst, Univ Ill, 45-47; res chemist, Gen Elec Co, 48-50; res chemist, Monsanto Co, Ala, 51-52 & Ohio, 52-54, res group leader, 54-57 & Mo, 57-59, sr res group leader, 59-60, asst dir res, 60-62, mgr res, 62-69, dir res & devel, Inorg Chem Div, 69-71, Detergents & Fine Chem Div, 71-73 & Detergent & Phosphate Div, 73-75, dir technol planning & eval, 75-77, dir environ opers & technol planning, 77-82 & dir environ opers, Monsanto Fibers & Intermediates Co, Monsanto Co, 82-85; VPRES, CHELAN ASSOCS, 85- *Concurrent Pos:* Civilian with Atomic Energy Comn, 48-50; alt, Indust Res Inst, 75-; mem bd dirs, Am Chem Soc, 77-, chmn bd, 82-83. *Honors & Awards:* St Louis Award, Am Chem Soc, 71. *Mem:* Sigma Xi; Am Inst Chemists; Am Chem Soc; AAAS. *Res:* Chemistry of phosphorus and its compounds; surfactants; detergents; antimicrobials; dentifrices. *Mailing Add:* Two Holiday Lane St Louis MO 63131

CALLIS, JAMES BERTRAM, b Riverside, Calif, Sept 4, 43; m 68; c 1. SPECTROSCOPY, CANCER DIAGNOSIS. *Educ:* Univ Calif, Davis, BS, 65; Univ Wash, PhD(phys chem), 70. *Prof Exp:* Fel, Dept Chem, Univ Wash, 70-72 & Dept Biophys, Univ Pa, 72-73; res asst prof, 75-78, DIR ANAL SERV, DEPT CHEM, UNIV WASH, 78-, RES ASSOC PROF, 80- *Mem:* Am Chem Soc; AAAS; Soc Appl Spectros. *Res:* Analytical chemistry; development of new types of instrumentation, especially for molecular luminescence and mass spectroscopy; biophysical chemistry; development of new instrumentaton for spectroscopic measurements on living entities. *Mailing Add:* Chem Dept BG-10 Univ Wash Seattle WA 98195

CALLIS, JERRY JACKSON, b Parrott, Ga, July 28, 26. VETERINARY MEDICINE. *Educ:* Ala Polytech Inst, DVM, 47; Purdue Univ, MS, 49. *Hon Degrees:* DSc, Purdue Univ, 79, Long Island Univ, Southampton Col, 80. *Prof Exp:* Vet, Brucellosis Test Lab, Purdue Univ, 48-49, State Vet Res Inst, Holland, 49-51 & Animal Dis Sta, Md, 52-53; in charge res opers, 53-56, asst dir, 56-62, dir, 63-86, SR SCI ADV, PLUM ISLAND ANIMAL DIS CTR, 86- *Honors & Awards:* USDA Award, 57; XII Int Vet Cong Prize, Am Vet Med Asn, 74; Am Meat Inst Award, 80. *Mem:* AAAS; Am Vet Med Asn; Tissue Cult Asn; US Animal Health Asn; Am Soc Virologists. *Res:* Tissue culture; virology and immunology; foot-and-mouth disease. *Mailing Add:* Plum Island Animal Dis Ctr Box 848 Greenport NY 11944

CALLIS, PATRIK ROBERT, b Ontario, Ore, Mar 17, 38; m 61; c 2. TWO-PHOTON SPECTROSCOPY, MOLECULAR ORBITAL THEORY. *Educ:* Ore State Univ, BS, 60; Univ Wash, PhD(phys chem), 65. *Prof Exp:* Fel biophys chem, Calif Inst Technol, 66-68; PROF PHYS CHEM, MONT STATE UNIV, 68- *Mem:* Am Chem Soc; Am Soc Photobiol. *Res:* Electronic structure studies of substituted aromatic hydrocarbons, nucleic acids, and amino acids by ultraviolet and laser two-photon spectroscopy in conjunction with computed results from quantum chemical schemes at various levels; emphasis is on establishing correct transition density patterns. *Mailing Add:* Dept Chem Mont State Univ Bozeman MT 59715

CALLISON, GEORGE, b Blue Rapids, Kans, June 1, 40; m 62; c 2. EVOLUTIONARY BIOLOGY, VERTEBRATE MORPHOLOGY. *Educ:* Kans State Univ, BS, 62; Univ Kans, MA, 65, PhD(zool), 69. *Prof Exp:* Asst prof zool & asst cur vert paleont, SDak Sch Mines & Technol, 67; asst prof, 69-72, assoc prof, 73-77, PROF BIOL, CALIF STATE UNIV, LONG BEACH, 78- *Concurrent Pos:* Res assoc vert paleont Los Angeles County Mus County Mus Natural Hist, 69-; consult, Badlands Nat Monument Master Plan, Nat Park Serv, 69 & Fruita Paleont Area, Bur Land Mgt, 77-, Orange County Ralph B Clark Paleont Park, 86, Dinamation Int Corp, 86-; NSF res grants, 77-78, 79-81, & 84-87; Nat Geographic Soc grants, 83-85; Earthwatch grants, 85-87. *Mem:* Sigma Xi; Soc Vert Paleont; Paleont Soc. *Res:* Evolutionary interrelationships of fossil tetrapods from Mesozoic rocks; comparative anatomical and functional morphological concepts. *Mailing Add:* Dept Biol Calif State Univ Long Beach CA 90840

CALLOW, ALLAN DANA, b Somerville, Mass, Apr 9, 16; m; c 5. SURGERY. *Educ:* Tufts Col, BS, 38, MS, 48, PhD, 52; Harvard Univ, MD, 42; Am Bd Surg, dipl, 52. *Hon Degrees:* DSc, Tufts Univ, 87. *Prof Exp:* From instr to prof surg, Med Sch, Tufts Univ, 48-69, vchmn, dept surg & chief gen & vascular surg, 71-77; surgeon, 54-87, CONSULT VASC SURG, NEW ENG MED CTR HOSP, 87- *Concurrent Pos:* Trustee, Tufts Univ, 71-77 & Civic Educ Found; dir, Med Eng Assocs; consult, Surg Gen, Med Dept, US Navy; chmn coun bd of overseers, Tuft Univ. *Honors & Awards:* Clemson Award, 88. *Mem:* Asn Am Med Cols; fel Am Col Surg; Am Heart Asn; Soc Vascular Surg (pres, 86); Int Cardiovasc Soc (secy-gen, 68-77, pres, 77-79, pres, NAm Chap, 74-75); Am Surg Asn; Am Physiol Soc. *Res:* Cardiovascular disease; peripheral vascular surgery, vascular biology, gene therapy prothesis. *Mailing Add:* Wash Univ Sch Med 4960 Audubon Box 8109 St Louis MO 63110

CALLOWAY, DORIS HOWES, b Canton, Ohio, Feb 14, 23; m 81; c 2. EDUCATION ADMINISTRATION. *Educ:* Ohio State Univ, BS, 43; Univ Chicago, PhD(nutrit), 47; Am Bd Nutrit, dipl, 51. *Prof Exp:* Intern dietetics, hosp, Johns Hopkins Univ, 44; res dietician, dept med, Univ Ill, 45; consult nutrit, Med Assocs Chicago, 48-51; nutritionist, QM Food & Container Inst, 51-58, head metab lab, 58-59, chief nutrit br, 59-61; chmn dept food sci & nutrit, Stanford Res Inst, 61-64; provost, prof & cols, 81-87, PROF NUTRIT, UNIV CALIF, BERKELEY, 63- *Concurrent Pos:* Assoc ed, Nutrit Rev, 62-68; mem ed bd, J Nutrit, 67-72, Environ Biol & Med, 69-79, J Am Dietetic Asn, 74-77 & Interdisciplinary Sci Rev, 75-90, Mem bd dirs, Am Bd Nutrit, 68-71; mem panel, White House Conf Food, Nutrit & Health, 69; mem bd trustees, Nat Coun Hunger & Malnutrit in US, 69-71; mem Nat Acad Sci-Nat Res Coun Food & Nutrit Bd, 72-75, 87-90; mem vis comt, Mass Inst Technol, 72-74; mem expert adv panel on nutrit, WHO, 72-; mem adv coun, Nat Inst Arthritis, Metab & Digestive Dis, NIH, 74-77; consult Nutrit Div, Food & Agr Orgn, UN, 74-75; mem adv coun, Nat Inst Aging, NIH, 78-81; mem bd trustees, Centro Internae de Mejoramiento de Maizy Trigo, 83-89; mem bd dir, Winrock Int Inst Agr Develop, 86-90; mem, tech adv comt, Consultative Group Int Agr Res, 89- *Honors & Awards:* Elvehjem Award, Am Inst Nutrit, 86. *Mem:* Inst Med-Nat Acad Sci; fel Am Inst Nutrit (pres, 82-83); Human Biol Coun. *Res:* Human nutrition; protein and energy. *Mailing Add:* Dept Nutrit Sci Univ Calif Berkeley CA 94720

CALLOWAY, E DEAN, b Louisville, Miss, Sept 25, 24; m 52. PHYSICAL CHEMISTRY. *Educ:* Millsaps Col, BS, 48; Univ Ala, MS, 54, PhD(chem), 56. *Prof Exp:* Asst prof chem, Delta State Col, 54-56; assoc prof, Birmingham-Southern Col, 56-58 & Millsaps Col, 58-60; supvr, Chemstrand Co Div, Monsanto Co, Ala, 60-64; asst prof chem, Memphis State Univ, 64-66; assoc prof, 66-69, PROF CHEM, BIRMINGHAM-SOUTHERN COL, 69- *Mem:* Am Chem Soc. *Res:* Ultraviolet absorption and molecular spectroscopy. *Mailing Add:* 27 Ridgewood Dr Columbus MS 39701

CALLOWAY, JEAN MITCHENER, b Indianola, Miss, Dec 18, 23; m 52; c 2. NUMBER THEORY. *Educ:* Millsaps Col, BA, 44; Univ Pa, MA, 49, PhD(math), 52. *Prof Exp:* Asst instr math, Millsaps Col, 44; instr, McCallie Sch, 44-47; asst instr, Univ Pa, 47-52; from asst prof to assoc prof, Carleton Col, 52-60, actg chmn dept, 53-54 & 58-59; chmn dept, 60-73, PROF MATH, KALAMAZOO COL, 60- *Concurrent Pos:* Mem, Inst Advan Study, 59; res assoc, Stanford Univ, 68-69. *Mem:* Am Math Soc; Math Asn Am. *Mailing Add:* Dept Math Kalamazoo Col Kalamazoo MI 49007

CALNE, DONALD BRIAN, b London, Eng, May 4, 36; m 65; c 3. NEUROLOGY, PHARMACOLOGY. *Educ:* Oxford Univ, BSc, 58, MS & BCh, 61, DM, 68. *Prof Exp:* Resident physician med, St Thomas Hosp, London, 63-64; resident physician neurol, Nat Hosps Nervous Dis, London, 65-66 & Univ Col Hosp, London, 67-69; tenured staff physician, Royal Postgrad Med Sch, London, 70-74; clin dir neurol, Nat Inst Neurol & Commun Dis & Stroke, NIH, 74-81, chief exp therapeut br, 76-81; HEAD, DIV NEUROL, UNIV HOSP-UNIV BC SITE, VANCOUVER, 81- *Concurrent Pos:* Mem adv bd, Advan Neurol, 73-, Am Parkinson's Dis Asn, 86-, Dystonia Med Res Found, 82-, Nat Parkinson Found & Parkinson Found Can; consult neurologist, Nat Naval Med Ctr, Bethesda & Walter Reed Army Med Ctr, Washington, DC, 75-80. *Honors & Awards:* Killam Prize; Germania Rossetto Prize. *Mem:* Am Neurol Asn; Am Acad Neurol; Can Neurol Soc. *Res:* Pharmacological study of mechanisms of synaptic function in the central nervous system in health and disease; use of drugs for the manipulation of synaptic transmission in order to benefit neurological disorders. *Mailing Add:* Univ BC 2211 Wesbrook Mall Vancouver BC V6T 1W5 Can

CALNEK, BRUCE WIXSON, b Manchester, NY, Jan 29, 32; m 54; c 2. VETERINARY VIROLOGY. *Educ:* Cornell Univ, DVM, 55, MS, 56. *Prof Exp:* Actg asst prof poultry dis, Cornell Univ, 56-57; assoc prof vet sci, Univ Mass, 57-61; PROF AVIAN DIS, CORNELL UNIV, 61-, CHMN DEPT AVIAN & AQUATIC ANIMAL MED, 77- *Concurrent Pos:* Nat Cancer Inst fel virol, Univ Calif, Berkeley, 67-68; vis scientist oncol, Houghton Poultry Res Sta, Eng, 74-75; cancer res fel, Int Union Against Cancer, 74-75; consult, Virol Study Sect, NIH, 75- *Mem:* Am Vet Med Asn; Am Asn Avian Pathologists; World Vet Poultry Asn. *Res:* Viral oncology, with particular emphasis on the pathogenesis of avian neoplastic diseases of chickens induced by DNA-containing viruses, known as Marek's disease, and RNA-containing viruses causing lymphoid leukosis. *Mailing Add:* NY State Vet Col Ithaca NY 14850

CALO, JOSEPH MANUEL, b Newark, NJ, Nov 9, 44; m 68; c 2. CHEMICAL ENGINEERING, PHYSICAL CHEMISTRY. *Educ:* Newark Col Eng, BS, 66; Princeton Univ, AM, 68, PhD(chem eng), 70. *Hon Degrees:* AM, Adeundem, Brown Univ, 83. *Prof Exp:* Atmospheric physicist, Air Force Cambridge Res Labs, Hanscom AFB, 74; res engr, Exxon Res & Eng Co, Florham Park, 74-76; asst prof, Dept Chem Eng, Princeton Univ, 76-81; assoc prof, 81-89, PROF DIV ENG, BROWN UNIV, 89- *Concurrent Pos:* Consult. *Mem:* Am Inst Chem Engrs; Am Chem Soc; Sigma Xi. *Res:* Chemical reactor/reaction engineering; atmospheric physics and chemistry; carbon reactivity. *Mailing Add:* Div Eng Box D Chem Eng Prog Brown Univ Providence RI 02912

CALPOUZOS, LUCAS, b Detroit, Mich, Oct 20, 27; m 50; c 2. PLANT PATHOLOGY. *Educ:* Cornell Univ, BS, 50; Harvard Univ, AM, 52, PhD(biol), 55. *Prof Exp:* Plant pathologist, R T Vanderbilt Co, 50-51; instr bot, Harvard Univ, 52-53; plant pathologist, Fed Exp Sta, PR, USDA, 55-62; NSF sr fel, Bristol Univ, 62-63; res plant pathologist, Crops Res Div, Agr Res Serv, USDA, 63-67; from assoc prof to prof plant path, Univ Minn, St Paul, 67-71; prof plant sci & head Dept Plant & Soil Sci, Univ Idaho, 71-80; prof plant sci & dean Sch Agr & Home Econ, 80-88, PROF, SCH AGR, CALIF STATE UNIV, CHICO, 88- *Mem:* Am Assoc Adran Sci. *Res:* Crop loss estimates; chemical control of plant diseases; fungus physiology. *Mailing Add:* Sch Agr Calif State Univ Chico CA 95929

CALTAGIRONE, LEOPOLDO ENRIQUE, b Valparaiso, Chile, Mar 1, 27; nat US; m 53; c 3. ENTOMOLOGY. *Educ:* Univ Chile, ErAgron, 51; Univ Calif, PhD(entom), 60. *Prof Exp:* Entomologist, Chilean Ministry Agr, 49-62; from jr entomologist to assoc entomologist, 62-75, PROF & ENTOMOLOGIST, UNIV CALIF, BERKELEY, 75- *Concurrent Pos:* John Simon Guggenheim fel, 57-59; lectr, Sch Agr, Cath Univ Chile, 60-62; consult, USAID. *Mem:* AAAS; Entom Soc Am; Am Inst Biol Sci. *Res:* Biological control of agricultural insect pests; biology of parasitic Hymenoptera; taxonomy of Hymenoptera copidosomatini. *Mailing Add:* 6600 Gatto St El Cerrito CA 94530

CALTRIDER, PAUL GENE, b Mineral Wells, WVa, Jan 14, 35; m 56; c 3. MICROBIAL PHYSIOLOGY. *Educ:* Glenville State Col, BS, 56; WVa Univ, MS, 58; Univ Ill, PhD(plant path, microbiol), 62. *Prof Exp:* Sr microbiologist, Eli Lilly & Co, 62-66, mgr antibiotic qual control & tech serv, 66-76, mgr antibiotic fermentation technol, 76-80, DIR ANTIBIOTIC TECH SERV, ELI LILLY & CO, 80-, DIR FERMENTATION PROD DEVELOP & TECH SERV. *Mem:* Am Soc Microbiol; Soc Indust Microbiol; Am Chem Soc. *Res:* Physiology and biochemistry of microorganisms, especially carbohydrate metabolism, antibiotic fermentations and biosynthesis of antibiotics. *Mailing Add:* Eli Lilly & Co Corp Ctr Indianapolis IN 46285-0002

CALUB, ALFONSO DEGUZMAN, b Aringay, La Union, Philippines, Aug 1, 38; m; c 2. PLANT BREEDING, AGRONOMY. *Educ:* Univ NH, PhD(plant sci), 72. *Prof Exp:* Res asst, Univ Philippines, 60-64, instr, 64-68; res asst, Univ Ky, 68-69 & Univ NH, 69-72; fel, Univ Nebr, 72-74; dir res & sr plant breeder, Alexandria Seed Co, La, 74-81; SR RICE BREEDER, RING AROUND RES, TEX, 81-, STATION MGR, 83- *Concurrent Pos:* Exec comt mem, Rice Tech Working Group, 84-86. *Mem:* Am Soc Agron; Crop Sci Soc Am; Soil Sci Soc Am; Sigma Xi; Am Registry Cert Prof Agron Crops Soils. *Res:* Research, development and production of male sterile lines, maintainer lines, restorer lines, and hybrids of rice; patents on hybrid rice production. *Mailing Add:* Ring Around Prod PO Box 810 East Bernard TX 77435

CALUSDIAN, RICHARD FRANK, b Watertown, Mass, Feb 6, 35; m 62; c 3. THEORETICAL PHYSICS. *Educ:* Harvard Univ, BA, 57; Univ NH, MS, 59; Boston Univ, PhD(physics), 65. *Prof Exp:* Physicist, Mat Res Agency, US Dept Army, 65, fel, Natick Labs, 65-66; assoc prof, 66-69, chmn dept, 66-87, PROF PHYSICS, BRIDGEWATER STATE COL, 69- *Mem:* Am Asn Physics Teachers. *Res:* Electron-phonon interaction; theory of electrical and thermal conductivity. *Mailing Add:* Dept Physics Bridgewater State Col Bridgewater MA 02324

CALVANICO, NICKOLAS JOSEPH, b New York, NY, Aug 18, 36; m 59; c 2. IMMUNOLOGY, PROTEIN CHEMISTRY. *Educ:* City Col New York, BS, 58; DePaul Univ, MS, 60; Univ Vt, PhD(biochem), 64. *Prof Exp:* Trainee immunol, Div Exp Med, Col Med, Univ Vt, 64-65; res instr med, Sch Med, State Univ NY Buffalo, 68-73; asst prof immunol, Mayo Found Sch Med, Univ Minn, 73-77; res health scientist, res serv, Vet Admin 78-85; from asst prof to prof, Dept Med, 79-86, PROF, DEPT DERMAT, MED COL, WIS, 88- *Concurrent Pos:* Arthritis Found fel, Sch Med, State Univ NY Buffalo, 65-68; consult, Erie County Labs, 68-73; indust consult, 73-77; Ed Bd Immunol Invest; dir res, Zeus Sci, NJ, 86-88. *Mem:* NY Acad Sci; Am Asn Immunol. *Res:* Hypersensitivity lung disease; immunological mechanisms causing allergic inflammatory lung disease; purification of environmental allergens of animal and microbiol origin; autoimmune disease; protein purification. *Mailing Add:* Med Col Wis MFRC 8701 Watertown Place Rd Milwaukee WI 53226

CALVELLI, THERESA A, IMMUNOLOGY. *Educ:* Cornell Univ, PhD(leukemic cells), 77. *Prof Exp:* ASST PROF IMMUNOL, ALBERT EINSTEIN COL MED, 83- *Res:* Pediatric Acquired Immune Deficiency Syndrome. *Mailing Add:* Albert Einstein Col Med 1300 Morris Park Ave Rm F-405 Bronx NY 10461

CALVER, JAMES LEWIS, b Pontiac, Mich, June 15, 13; m 45, 70; c 2. MINERALOGY, PHYSICAL GEOLOGY. *Educ:* Univ Mich, AB, 36, MS, 38, PhD(geol), 42. *Prof Exp:* Asst geol, Univ Mich, 35-39; instr, Univ Wichita, 40-42; asst prof, Univ Mo, 42-43; assoc geologist, Tenn Valley Authority, 43-47; geologist, Fla Geol Surv, 47-57; comnr mineral resources & state geologist, Va Div, 57-78; RETIRED. *Honors & Awards:* Past Pres & Hon Mem, Asn of Am State Geologists. *Mem:* Mineral Soc Am; Geol Soc Am; Am Inst Mining, Metall & Petrol Engrs; Am Asn Petrol Geologists; Am Geophys Union. *Res:* Economic geology. *Mailing Add:* 1202 Partridge Lane Charlottesville VA 22901

CALVERLEY, JOHN ROBERT, b Hot Springs, Ark, Jan 14, 32; m 53; c 1. NEUROLOGY. *Educ:* Univ Ore, BS, 53, MD, 55. *Prof Exp:* Resident neurol, State Univ Iowa Hosps, 56; fel med, Mayo Found, 57, resident neurol, 57-59; neurologist, Wilford Hall US Air Force Hosp, San Antonio, Tex, 60-62, chief neurol serv, 62-64; from asst prof to assoc prof, 64-70, chief div, 66-73, PROF NEUROL, MED BR, UNIV TEX, GALVESTON, 70-, CHMN DEPT, 73- *Concurrent Pos:* Asst examr, Am Bd Neurol & Psychiat, 65-77, dir, 77-85; nat consult neurol, Surgeon Gen, US Air Force, 73- *Mem:* Am Neurol Asn; Am Acad Neurol; Asn Univ Prof Neurol. *Res:* Medical education. *Mailing Add:* Dept Neurol Univ Tex Med Br Galveston TX 77550

CALVERT, ALLEN FISHER, b San Diego, Calif, Aug 1, 27; m 55; c 4. BIOCHEMICAL GENETICS. *Educ:* Univ San Francisco, BS, 52; Univ Calif, PhD(biochem), 62; Am Bd Clin Chem, dipl, 72. *Prof Exp:* Fel biochem, Univ Hawaii, 62-64; biochemist, Clin Invest Ctr, US Naval Hosp, Oakland, Calif, 64-65; BIOCHEMIST, MED GENETICS DIV, HOWARD UNIV, 65-, ASSOC PROF PEDIAT & CHILD HEALTH, DEPT PEDIAT, COL MED,74- *Mem:* Am Soc Human Genetics; Am Asn Clin Chem. *Res:* Inborn errors of metabolism; screening for children's diseases of genetic origin. *Mailing Add:* Dept Pediat Col Med Howard Univ Washington DC 20059

CALVERT, DAVID VICTOR, b Chaplin, Ky, Feb 26, 34; m 57; c 2. SOIL & SOIL SCIENCE, AGRONOMY. *Educ:* Univ Ky, BS, 56, MS, 58; Iowa State Univ, PhD(soil fertil), 62. *Prof Exp:* Asst soils chemist, Citrus Exp Sta, 62-67, from assoc prof to prof soil chem, 67-78, DIR, AGR RES CTR, UNIV FLA, 78- *Concurrent Pos:* Off collabr, Soil & Water Conserv Div, Agr Res Serv, USDA, 69-; consult, Jamaican Sch Agr, 70, John F Kennedy Space Ctr, NASA, 74-, Coun Agr Sci & Technol, 75, Univ Beijing, Peoples Repub China, 87 & US Army Corps Engrs, 88; coun, Agr Sci & Technol, 75. *Mem:* AAAS; Am Soc Agron; Sigma Xi; Soc Citricult; Coun Agr Sci & Technol; Soil Sci Soc Am; Am Hort Soc. *Res:* Chemistry of soil phosphorus; soil-plant tracer studies with Nitrogen-fifteen isotope; soil chemistry-fertility, drainage and irrigation with citrus. *Mailing Add:* Agr Res & Educ Ctr Box 248 Univ Fla Ft Pierce FL 34954

CALVERT, JACK GEORGE, b Inglewood, Calif, May 9, 23; m 46; c 2. PHOTOCHEMISTRY, ATMOSPHERIC CHEMISTRY. *Educ:* Univ Calif, Los Angeles, BS, 44, PhD(chem), 49. *Prof Exp:* Nat Res Coun Can fel, 49-50; from asst prof to prof, 50-74, chmn dept, 64-68, Kimberly prof, 74-81, EMER PROF CHEM, OHIO STATE UNIV, 81-; SR SCIENTIST, NAT CTR ATMOSPHERIC RES, 82- *Concurrent Pos:* Chmn, Nat Air Pollution Manpower Develop Adv Comt, 68-69, Nat Adv Comt Conserv Found, 69-72, Comt on Aldehydes, Nat Acad Sci, 79-81 & Environ Chem & Physics Rev Panel, Environ Protection Agency, 80-; Air Pollution Res Grants Adv Comt, Environ Protection Agency, 70-72; mem, Chem & Physics Adv Comt on Air Pollution, 73-76, Comt on Health Effects of Air Pollutants, Nat Acad Sci, 73-74, USA-USSR Joint Comt on Atmospheric Modelling & Aerosols, 75- & vis comt of Brookhaven Nat Lab, Dept Energy & Environ, 76-; Simon Guggenheim Mem Fel, 77-78; chmn, Comt Atmospheric Transport Chem Transformation Acid Precipitation, NAS, 82-83. *Honors & Awards:* Innovative Res in Environ Sci & Technol Award, Am Chem Soc, 81; Frank A Chambers Award, Air Pollution Control Asn, 86. *Mem:* Fel AAAS; Am Chem Soc; fel Am Inst Chemists. *Res:* Atmospheric chemistry; photochemistry; reaction kinetics; formation and decay mechanisms of reactive transients (molecules and free radicals) which are important in the chemistry of the atmosphere; pathways of acid generation in the troposphere. *Mailing Add:* Div Atmospheric Chem Nat Ctr Atmospheric Res Boulder CO 80307

CALVERT, JAMES BOWLES, b Columbus, Ohio, May 28, 35. MICROCOMPUTERS, CIRCUIT DESIGN. *Educ:* Mont State Col, BS, 56, MS, 59; Univ Colo, PhD(physics), 63. *Prof Exp:* Res physicist, 59-63, from lectr to asst prof physics, 62-66, assoc prof astron, 76-80, assoc prof physics, 66-80, ASSOC PROF PHYSICS & ENG, UNIV DENVER, 82- *Mem:* Am Inst Physics; Inst Elec & Electronic Engrs; Am Asn Physics Teachers; Nat Soc Prof Engrs; Am Soc Eng Educ. *Res:* Energy transfer in molecular collisions; fluctuations in light beams; coherence theory; atomic structure. *Mailing Add:* Dept Eng Univ Denver Denver CO 80208-0177

CALVERT, JAY GREGORY, b Ponca City, Okla, Sept 12, 59; m 83; c 1. VIROLOGY. *Educ:* Eastern Ill Univ, BS, 81; Purdue Univ, PhD(genetics), 88. *Prof Exp:* Teaching asst ecol, cell biol, anat, bot & genetics, Dept Biol Sci, Purdue Univ, 81-85, grad instr genetics, 85-88; postdoctoral fel, Dept Microbiol, Univ Guelph, 88-90; POSTDOCTORAL FEL, AVIAN DIS & ONCOL LAB, AGR RES SERV, USDA, 90- *Mem:* Am Soc Virol; AAAS. *Res:* Development of recombinant vaccines to protect commercial poultry flocks from oncogenic viruses; studies on the structure and function of poxvirus genes; replication of RNA viruses and the mode of action of interferon. *Mailing Add:* Avian Dis & Oncol Lab USDA-Agr Res Serv 3606 E Mount Hope Rd East Lansing MI 48823

CALVERT, JON CHANNING, b Sonora, Calif, May 17, 41; m 70; c 2. MEDICAL ADMINISTRATION, GERIATRICS. *Educ:* Stanford Univ, AB, 63; Baylor Univ, MS, MD & PhD(med), 68. *Prof Exp:* Asst prof anat & cell biol, Baylor Univ, 70-73; from asst prof to prof family pract, Med Col Ga, Augusta, 73-82, chmn dept, 76-81; PROF FAMILY MED & CHMN, ORAL ROBERTS UNIV, 82-, ASSOC DEAN CLIN SCI, SCH MED, 84- *Concurrent Pos:* Mem, Gov's Joint Bd Family Pract, 76-81; chmn, Ga Fedn Family Pract Residency Progs, 76-77; mem, Ga Dept Human Resources Adv Coun Phys Health Needs Children & Youth, 76-77; med dir, Tri-County Health Syst, Inc, Augusta, Ga, 81-82; spec adv, White House Conf Aging, 81; chief, Dept Family & Community Med, City of Faith, Tulsa, Okla, 82-; mem, Family Med Deleg, China, 83; med dir, HMO Okla, Tulsa, 85. *Mem:* Am Acad Family Physicians; AAAS; AMA; Am Geriat Soc. *Res:* Ambulatory medical practice; medical and paramedical education. *Mailing Add:* 10522 S 66th E Ave Tulsa OK 74133

CALVERT, OSCAR HUGH, b Dallas, Tex, Oct 28, 18; m 44; c 4. PLANT PATHOLOGY. *Educ:* Okla State Univ, BS, 43; Univ Wis, MS(mycol), 45, PhD(plant path), 48. *Prof Exp:* Asst bot & plant path, Okla State Univ, 42-44; asst, Penicillin Proj, US Army & Dept Bot, Univ Wis, 44-45 & Dept Plant Path, 45-48; asst plant pathologist, Dept Plant Physiol & Path, Agr & Mech Col Tex, 48-52, plant pathologist, USDA, 52-54; consult plant path & nursery retail sales mgr, Lambert Landscape, 54-58; asst prof field crops, 58-67, from asst prof to prof, 67-89, EMER PROF PLANT PATH, UNIV MO-COLUMBIA, 89- *Concurrent Pos:* Vis prof, Res Inst Plant Protection, Budapest, Hungary, 69-70; proj leader soybean dis, Off Rural Develop, SuWon, S Korea, 78. *Mem:* AAAS; Mycol Soc Am; Am Phytopath Soc; Sigma Xi. *Res:* Diseases of field crops; aflatoxin in zea mays (corn). *Mailing Add:* Dept Plant Path Univ Mo 108 Waters Hall Columbia MO 65211

CALVERT, RALPH LOWELL, b Howard, Kans, Jan 10, 10; wid; c 2. MATHEMATICS. *Educ:* Southwestern Col, Kans, AB, 37; Univ Ill, AM, 38, PhD(math, astron), 50. *Prof Exp:* From instr to assoc prof math & astron, Utah State Univ, 40-49; asst prof, Univ Wyo, 49-51; res mathematician, Sandia Corp, 51-73; RETIRED. *Res:* Systems analysis. *Mailing Add:* 7917 Ivanhoe Ave La Jolla CA 92037

CALVERT, RICHARD JOHN, b Columbus, Ohio, Feb 23, 55; m 85; c 1. NUTRITION & CANCER, GASTROINTESTINAL FUNCTION. *Educ:* Duke Univ, AB, 77, Sch Med, MD, 81. *Prof Exp:* Resident, Dept Med, 81-84, Georgetown Univ Hosp; res fel, Dept Surg, 84-86, Univ Pa Hosp; MED RES NUTRITIONIST, US FOOD & DRUG ADMIN, EXP BR, 86- *Concurrent Pos:* Physician, Dept Mil Med, Nat Naval Med Ctr, Bethesda, Md, 90- *Mem:* Am Inst Nutrit; Am Soc Clin Nutrit; AAAS. *Res:* Effects of dietary fibers on intestinal cell proliferation; enzmatics activity; colonic cancer; bile acid metabolism and gastrointestinal transit time; effects of partially absorbed dietary starches on gastrointestinal function. *Mailing Add:* 5805 N Nineth Rd Arlington VA 22205

CALVERT, THOMAS W(ILLIAM), b Dunaskin, Scotland, Apr 12, 36; div; c 2. COMPUTER SCIENCES. *Educ:* Univ London, BSc, 57; Wayne State Univ, MSEE, 64; Carnegie Inst Technol, PhD(elec eng), 67. *Prof Exp:* Elec design engr, Imp Chem Indust, Ltd, Eng, 57-60; electronic engr, Canadair Ltd, Can, 60-61; lectr elec eng, Western Ont Inst Technol, 61-64; instr, Wayne State Univ, 64-65; instr, Carnegie-Mellon Univ, 66-67, asst prof elec eng & bioeng, 67-69, assoc prof elec eng & bioeng & chmn biotech prog, 69-72; dean, Fac Interdisciplinary Studies, 77-85, vpres, Res & Info Systs, 85-90, PROF KINESIOL & COMPUT SCI, SIMON FRASER UNIV, 77-; PRES, SCI COUN BC, 90- *Mem:* Inst Elec & Electronic Engrs; Asn Comput Mach; Can Med & Biol Eng Soc; Comput Sci Asn. *Res:* Computer graphics; in particular simulation and animation of human movement; modelling the nervous system; computer vision; systems theory in physiology. *Mailing Add:* Sci Coun BC 4710 Kingsway Suite 800 Burnaby BC V5H 4M2 Can

CALVET, JAMES P, b Washington, DC, Nov 6, 45. GENETICS. *Educ:* Univ Conn, PhD(cell biol & genetics), 75. *Prof Exp:* ASST PROF BIOCHEM & MOLECULAR BIOL, MED CTR, UNIV KANS, 81- *Mem:* AAAS; Am Soc Cell Biol. *Res:* Regulation of gene expression. *Mailing Add:* Dept Biochem Med Ctr Univ Kans Kansas City KS 66103

CALVIN, CLYDE LACEY, b Winlock, Wash, June 22, 34; m 60; c 2. PLANT ANATOMY. *Educ:* Wash State Univ, BS, 60; Purdue Univ, MS, 62; Univ Calif, Davis, PhD(bot), 66. *Prof Exp:* NIH fel, Univ Calif, Santa Barbara, 65-66; asst prof bot, Calif State Col, Long Beach, 66-67; assoc prof biol, Ore Col Educ, 67-68; ASSOC PROF BIOL, PORTLAND STATE UNIV, 68- *Mem:* Bot Soc Am; Int Soc Plant Morphol. *Res:* Plant anatomy and ultrastructure; tissue relations between dicotyledonous parasites and their hosts. *Mailing Add:* Dept Biol Portland State Univ Box 751 Portland OR 97207

CALVIN, HAROLD I, b Nov 1, 1936; m; c 2. CELL BIOLOGY, GAMETOGENESIS. *Educ:* Columbia Col, BA, 57; Columbia Univ, PhD(biochem), 64. *Prof Exp:* Res scientist genetic develop, Col Physicians & Surgeons, Columbia Univ, 76-86; res scientist, Dept Biochem & Nutrit, Va Polytech Inst & State Univ, 87-88; ASST PROF, DEPT OPTHAL, UNIV MED & DENT NJ, 88-, ADJ PROF, DEPT BIOCHEM & MOLECULAR BIOL, 90- *Mem:* Am Soc Cell Biol; Asn Res Vision & Opthal. *Res:* Testis biochemistry; lens biochemistry. *Mailing Add:* Dept Biochem & Molecular Biol Univ Med & Dent NJ Newark NJ 07103

CALVIN, LYLE D, b Nebr, Apr 12, 23; m 52; c 3. STATISTICS. *Educ:* Parsons Jr Col, AA, 43; Univ Chicago, BS, 48; NC State Univ, BS, 47, PhD, 53. *Prof Exp:* Asst, NC State Univ, 47-50,; biometrician, G D Searle & Co, 50-52; asst statistician, NC State Univ, 52-53; assoc prof, Ore State Univ, 53-57, chmn dept, 62-81, dir, Surv Res Ctr, 73-85, prof statist, 57-88, dean, grad sch, 81-88, EMER PROF & DEAN, ORE STATE UNIV, 89- *Concurrent Pos:* Vis prof, Univ Edinburgh, 67 & Univ Cairo, 71-72; mem, Epidemiol & Biomet Training Comt, Nat Inst Gen Med Sci, 68-72; mem, Am Statist Asn comt, Bur of Census, 71-77; chmn, Comt Pres Statist Soc, 85-91; pres, Biomet Soc, W NAm region, 64-65. *Mem:* Fel AAAS; Biomet Soc (secy, 80-84); fel Am Statist Asn; Int Statist Inst. *Res:* Experimental design and analysis; sampling methods. *Mailing Add:* Dept Statist Ore State Univ Corvallis OR 97331

CALVIN, MELVIN, b St Paul, Minn, Apr 8, 11; m 42; c 3. ORGANIC CHEMISTRY, INORGANIC CHEMISTRY. *Educ:* Mich Col Mines, BS, 31; Univ Minn, PhD(chem), 35. *Hon Degrees:* Numerous from US & foreign univs & cols, 55-79. *Prof Exp:* Rockefeller grant, Univ Manchester, 35-37; instr chem, Univ Calif, Berkeley, 37-40, from asst prof to prof, 41-71, dir, Bio-Org Div, Lawrence Radiation Lab, 46-80, dir, Lab Chem Biodynamics, 60-80, prof molecular biol, 63-80, UNIV PROF CHEM, UNIV CALIF, BERKELEY, 71- *Concurrent Pos:* Off investr, Nat Defense Res Comt, Univ Calif, 41-44, Manhattan Dist Proj, 44-45; mem US deleg, Int Conf Peaceful Uses Atomic Energy, Geneva, Switz, 55; mem joint comt, Int Union Pure & Appl Chem, 56; chem adv comt, US Air Force Off Sci Res, 51-55; vchmn exec coun, Armed Forces-Nat Res Coun Comt Bio-Astronaut, 58-62; mem, President's Sci Adv Comt, 63-65; bd sci counr, NIMH, 68-71; vchmn comt sci & pub policy, Nat Acad Sci, 71, chmn, 72-75; mem energy res adv bd, Dept Energy, 81-85; bd dirs, Dow Chem Co, 64-75 & 77-81. *Honors & Awards:* Nobel Prize in Chem, 61; Sugar Res Found Prize, 50; Flintoff Medal & Prize, The Chem Soc, 55; Hales Award, Am Soc Plant Physiol, 56; Richards Medal, Am Chem Soc, 56, Award Nuclear Appln in Chem, 57, Nichols Medal, 58; Davy Medal, Royal Soc, 64; Gibbs Medal, 77; Gold Medal, Am Inst Chemists, 79; Oesper Prize, Am Chem Soc, 81; Priestly Medal, 78; Feodor Lynen Medal, 83; Sterling B Hendricks Medal, 83; Nat Medal Sci, 89. *Mem:* Nat Acad Sci; AAAS; Am Chem Soc (pres-elect, 70, pres, 71); Fel Am Phys Soc; Am Soc Biol Chem; Royal Soc London. *Res:* Physical chemistry; photosynthesis and chemical biodynamics; plant physiology; chemical evolution; molecular biology; chemical and viral carcinogenesis; solar energy conversion (biomass and artificial photosynthesis). *Mailing Add:* Dept Chem Univ Calif Berkeley CA 94720

CALVIN, WILLIAM HOWARD, b Kansas City, Mo, Apr 30, 39. NEUROPHYSIOLOGY, EVOLUTION. *Educ:* Northwestern Univ, BA,61; Univ Wash, PhD(physiol & biophys), 66. *Prof Exp:* Asst physics, Northwestern Univ, 61; from instr neurol surg & physiol-biophys to asst prof neurol surg, 67-73, assoc prof neurol surg & biol, 74-86, AFFIL ASSOC PROF BIOL, UNIV WASH, 86- *Concurrent Pos:* Nat Acad Sci travel grants to int cong, 66 & 71; NIH grant, 71-84; NIH sr int fel, 78-79; vis prof neurobiol, Hebrew Univ Jerusalem, 78-79. *Mem:* Int Brain Res Orgn; Soc Neurosci; NY Acad Sci; Int Soc Human Ethology; AAAS; Inst Elec & Electronic Engrs; Lang Origins Soc. *Res:* Neuronal processes involved in repetitive discharge, coding of information, integration at single cell level; cortical circuitry for rapid sequencing; law of large number implications for fine sensory discriminations; evolution of the hominid brain, especially its four-fold enlargement in the last 2.5 million years, but also its sensorimotor circuitry and its adaptations for language; tool use. *Mailing Add:* 1543 17th Ave E Univ Wash R1-20 Seattle WA 98112

CALYS, EMANUEL G, b Samos, Greece, Jan 33; US citizen; m 63. ANALYSIS & FUNCTIONAL ANALYSIS. *Educ:* Washburn Univ, BS, 61; Univ Kans, MA, 64; Univ Ky, PhD, 68. *Prof Exp:* PROF MATH, WASHBURN UNIV, 68- *Concurrent Pos:* NSF grant, Northwestern Univ, 63; Nat Defense Educ Act fel, 61-63. *Mem:* Am Math Soc; Math Asn Am. *Mailing Add:* Dept Math Washburn Univ Topeka Topeka KS 66621

CALZONETTI, FRANK J, b Audubon, NJ, Jan 25, 49; m 72; c 2. RESOURCE MANAGEMENT, RESEARCH ADMINISTRATION. *Educ:* Wayne State Univ, BA, 72, MA, 74; Okla Univ, PhD(geog), 77. *Prof Exp:* Res assoc geog, Sci & Pub Policy Prog, Okla Univ, 75-78; assoc prof geog, chmn dept & res assoc prof, Regional Res Inst, 78-87, ASSOC DEAN RES & GRAD STUDIES, COL ARTS & SCI, WVA UNIV, 87- *Concurrent Pos:* Chmn, Energy Specialty Group, Asn Am Geographers, 84-86; proj dir, WVa Elec Export Study, 84-88 & EDA Indust Location Proj. *Mem:* Asn Am Geographers; AAAS. *Res:* Energy and regional development; role of energy prices and availabiltiy in regional economics and industrial location. *Mailing Add:* Dept Geol & Geog WVa Univ Morgantown WV 26506

CAMACHO, ALVRO MANUEL, b Trinidad, WI, Mar 1, 27; US citizen; m 56; c 4. PEDIATRIC ENDOCRINOLOGY. *Educ:* Creighton Univ, 46-49, MD, 53. *Prof Exp:* Intern, Charity Hosp, New Orleans, La, 53-54; resident pediat, Med Br Hosps, Univ Tex, 55-58; fel pediat endocrinol, Children's Hosp, Ohio State Univ, 58-60; fel, Johns Hopkins Hosp, Baltimore, 60-61, instr pediat, 61-63; asst prof, Sch Med, Wayne State Univ, 63-64; from asst prof to assoc prof, 64-72, PROF PEDIAT, COL MED, UNIV TENN, MEMPHIS, 72-, CHIEF SECT PEDIAT ENDOCRINOL, 64-; HD SECT PD ENDOCRIN, 65-81. *Honors & Awards:* Lederle Med Fac Award, 66. *Mem:* Am Fedn Clin Res; Endocrine Soc; Soc Pediat Res; Am Pediat Soc. *Mailing Add:* 848 Adams Ave Memphis TN 38103

CAMARDA, HARRY SALVATORE, b New York, NY, Sept 23, 38. PHYSICS. *Educ:* New York Univ, BS, 63; Columbia Univ, MA, 65, PhD(physics), 70. *Prof Exp:* Res assoc physics, Nevis Labs, Columbia Univ, 70-71; res assoc, Nat Bur Standards, 71-73; sr physicist, Lawrence Livermore Lab, 74-79; PROF PHYSICS, PA STATE UNIV, DELAWARE CO CAMPUS, 79- *Concurrent Pos:* Nat Bur Standards res award, Nat Res Coun, 71; consult, Lawrence Livermore Lab, 79- *Mem:* Am Phys Soc; Sigma Xi; NY Acad Sci. *Res:* Statistical properties of excited nuclei; neutron resonance interactions; low energy nuclear optical model. *Mailing Add:* Pa State Univ Delaware Co Campus Media PA 19063

CAMBEL, ALI B(ULENT), b Merano, Italy, Apr 9, 23; nat US; m 46; c 4. ENERGY TECHNOLOGY & POLICY. *Educ:* Robert Col, Istanbul, BS, 42; Calif Inst Technol, MS, 46; Univ Iowa, PhD(mech eng), 50. *Prof Exp:* From instr to asst prof mech eng, Univ Iowa, 47-53; from assoc prof to prof, Northwestern Univ, 53-61, Walter P Murphy distinguished prof, 61-68, dir gas dynamics lab, 53-66, chmn dept, 57-66; dean col eng, Wayne State Univ, 68-70, exec vpres acad affairs, 70-72; corp vpres, Gen Res Corp, Washington, DC, 72-74; dep asst dir, NSF, 74-75; chmn dept civil, mech & environ eng, George Washington Univ, 75-80, prof, 75-88, dir off energy prog, 75-88, EMER PROF ENG & APPL SCI, GEORGE WASHINGTON UNIV, 88- *Concurrent Pos:* Lectr & chmn symp, Nat Acad Sci, 64; dir res & eng support div, Inst Defense Anal, 66-67, vpres res, 67-68; staff dir, President's Interdept Energy Study; adv & consult to Advan Res Proj Agency, Dept Defense & Aeronaut Systs Div, Off Sci & Technol; indust consult; mem eng sci adv comt, Air Force Off Sci Res. *Honors & Awards:* Pendray Award, Am Inst Aeronaut & Astronaut, 59; McGraw Res Award, Am Soc Eng Educ, 60; NSF Excellence Award, 75. *Mem:* Am Soc Mech Engrs; Am Inst Aeronaut & Astronaut; Am Soc Eng Educ; Sigma Xi. *Res:* Magnetogasdynamics; energetics; social change; nonlinear, nonequilibrium processes in dissipative structures; chaos theory. *Mailing Add:* 6155 Kellogg Dr McLean VA 22101

CAMBEY, LESLIE ALAN, b Sydney Australia, May 15, 34; m 56; c 4. PHYSICS. *Educ:* Univ New SWales, BS, 55, PhD(physics), 60. *Prof Exp:* Fel physics, McMaster Univ, 60-62; sr res scientist, Nuclide Corp, Pa, 62-65, dir res, 65-67; mgr custom prod, CEC/Anal Instruments Div, Bell & Howell Co, Calif, 67-70; vpres, Mat Res Corp, 70-77, sr vpres, 77-81, exec vpres, 81-83, pres & chief operating officer, 83-85; PRES, CAMBEY & WEST, INC, PEARL RIVER, NY, 86- *Mem:* Inst Elec & Electronic Engrs. *Res:* Atomic mass determinations; ion sputtering; ion optics; mass spectroscopy. *Mailing Add:* 100 Gottlieb Dr Pearl River NY 10965

CAMBIER, JOHN C, b Springfield, Ill, June 29, 48. TRANSMEMBRANE SIGNALLING IN LYMPHOCYTES. *Educ:* Univ Iowa, PhD(microbiol & immunol), 75. *Prof Exp:* Asst prof microbiol & immunol, Duke Univ, 78-83; PROF MICROBIOL & IMMUNOL, NAT JEWISH CTR IMMUNOL & RESPIRATORY MED, DENVER, COLO, 83- *Mem:* Am Asn Immunologists; Soc Analytical Cytol. *Mailing Add:* Div Immunol Nat Jewish Ctr Immunol & Respiratory Med 1400 Jackson St Denver CO 80206

CAMBURN, MARVIN E, b Stockbridge, Mich, April, 1, 38; m 59; c 2. BUSINESS ADMINISTRATION & FINANCE. *Educ:* Albion Col, BA, 60; Univ Detroit, MA, 64; Mich State Univ, PhD, 71; Ill Benedictine Col, MBA, 87. *Prof Exp:* Head dept math, Mercyhurst Col, 74-76, chair sci div, 75-77, dir inst develop, 76-78; dean fac & instr, 77-88, PROF BUS ADMIN & FINANCE, ILL BENEDICTINE COL, 88- *Concurrent Pos:* NSF Res Participation Grant IIT, 68; Mich State Univ Natural Sci scholar, 69-71. *Mem:* Math Asn Am; Math Soc Am; Nat Asn Accountants. *Mailing Add:* 6402 New Albany Rd Lisle IL 60532

CAME, PAUL E, b Dover, NH, Feb 25, 37; m 60; c 2. MICROBIOLOGY, VIROLOGY. *Educ:* St Anselm's Col, AB, 58; Univ NH, MS, 60; Hahnemann Med Col, PhD(microbiol), 64. *Prof Exp:* Instr microbiol, Med Units, Univ Tenn & res virologist, St Jude Hosp, 63-65; res assoc biophys cytol, Rockefeller Univ, 65-67; sr virologist, 67-70, mgr dept virol, 70-74, assoc dir microbiol sci & virol, Schering-Plough Corp, 74-75; head dept microbiol, Sterling-Winthrop Res Inst, 75-; PARTNER, HEINDRICK STRUGGES CORP, 91- *Concurrent Pos:* Res, Hem Res, Inc, -91. *Mem:* AAAS; NY Acad Sci; Am Asn Immunologists; Am Asn Cancer Res; Am Soc Microbiol. *Res:* Mechanisms of cellular resistance to viruses; mechanisms of virus oncogenicity; chemotherapy of virus and microbial diseases. *Mailing Add:* 125 S Wacker Floor 28 Chicago IL 60606

CAMENZIND, MARK J, b Palo Alto, Calif, Nov 17, 56. ONLINE INSTRUMENTATION & ANALYSIS FOR SUB-PARTS-PER BILLION IMPURITIES IN PURE CHEMICALS, PROBLEM SOLVING FOR SEMICONDUCTOR PROCESSING. *Educ:* Mass Inst Technol, SB, 78; Univ Calif, Berkeley, PhD(inorg chem), 83. *Prof Exp:* Postdoctoral fel bioinorg chem, Univ BC, 83-86; res chemist pharmaceut, Salutar, Inc, Sunnyvale, Calif, 86-87; RES CHEM ANALYTICAL CHEM, BALAZS ANALYTICAL LAB, 87- *Concurrent Pos:* Comt mem, Comt on Electronics, Filtration & Plastic Piping, Am Soc Testing & Mat, 90- *Mem:* Am Soc Testing & Mat; Am Chem Soc. *Res:* Analytical methods for ultrapure water and chemicals for semiconductor, nuclear power, environmental and chemical industries. *Mailing Add:* Balazs Analytical Lab 1380 Borregas Ave Sunnyvale CA 94089

CAMERINI-OTERO, RAFAEL DANIEL, b Buenos Aires, Arg, Jan 15, 47; US citizen; m 69. MOLECULAR BIOLOGY, HUMAN GENETICS. *Educ:* Mass Inst Technol, BS, 66; NY Univ, MD, 73, PhD(microbiol), 73. *Prof Exp:* Intern pediat, Bellevue Hosp Med Ctr, NY Univ, 73-74; res assoc molecular biol, NIH, 74-77, sr investr, molecular biol lab, Human Biochem Genetics Sect, Nat Inst Arthritis, Metab & Digestive Dis, 77-79, sr investr, Genetics & Biochem Br, 79-82, CHIEF, MOLECULAR GENETICS SECT, GENETICS & BIOCHEM BR, NAT INST DIABETES & DIGESTIVE KIDNEY DIS, NIH, 82-, CHIEF, GENETICS & BIOCHEM BR, 84- *Mem:* Am Chem Soc; Biophys Soc; Am Soc Biol Chem. *Res:* Chromatin and chromosome structure; physical biochemistry of macromolecules; pediatrics; human and medical genetics; biochemistry of homologous recombination; gene targeting. *Mailing Add:* Genetic & Biochem Br Nat Inst Diabetes & Digestive Kidney Dis NIH Bldg 10 Rm 9D15 Bethesda MD 20892

CAMERMAN, ARTHUR, b Vancouver, BC, Apr 12, 39. PHARMACEUTICAL CHEMISTRY, PHARMACOLOGY. *Educ:* Univ BC, BSc Hons, 61, PhD(chem), 64. *Prof Exp:* Fel molecular biol, Royal Inst Great Brit, 64-66; sr fel & res assoc biol struct & pharm, 67-71, res asst prof med neurol & pharmacol, 71-75, res assoc prof, 75-81, RES PROF, UNIV WASH, 81- *Concurrent Pos:* Vis investr, Howard Hughes Med Inst, 71-72; affil, Child Develop & Mental Retardation Ctr, 76-, alcohol & drug abuse Inst, Univ Wash, 77-; Klingstein sr fel neurosci, 83-86. *Honors & Awards:* Epilepsy Res Award, Int League Against Epilepsy, 83. *Mem:* AAAS; Am Crystallog Asn; Am Chem Soc; Am Soc Neurochem; Am Soc Pharmacol & Exp Therapeut. *Res:* Stereochemical basis of drug and biological molecule action; use of crystallography to determine three-dimensional molecular conformations of anti-epileptic and anti-cancer drugs. *Mailing Add:* Div Neurol RG-20 Univ Wash Seattle WA 98195

CAMERMAN, NORMAN, b Vancouver, BC, Apr 12, 39. MOLECULAR BIOLOGY, X-RAY CRYSTALLOGRAPHY. *Educ:* Univ BC, BSc, 61, PhD(chem), 64. *Prof Exp:* Nat Res Coun Can overseas fel, Royal Inst Gt Brit, 64-66; Med Res Coun Can prof asst, 67-69, asst prof, 69-75, assoc prof, 75-79, PROF BIOCHEM, UNIV TORONTO, 79- *Concurrent Pos:* Med Res Coun Can scholar, 69-73. *Mem:* AAAS; Am Crystallog Asn. *Res:* Molecular basis of biological activity; molecular structure determinations of drugs, hormones and other biologically-active substances; correlation of structure and actions. *Mailing Add:* Dept Biochem Univ Toronto One Kings College Circle Toronto ON M5S 1A8 Can

CAMERO, ARTHUR ANTHONY, b Tampa, Fla, June 29, 47; m 88; c 1. PROCESS DEVELOPMENT & ENGINEERING. *Educ:* Univ Fla, BS, 69; Princeton Univ, MA, 72, PhD(chem eng), 75. *Prof Exp:* Engr, Monsanto Co, 70; engr, Exxon Res & Eng Co, 75-76, res engr, 76-78, sr engr, 78-80; sr process engr, Fluor Engrs & Constructors, 80-83; res engr, 83-87, SR RES ENGR, SHELL DEVELOP CO, 87- *Mem:* Am Inst Chem Engrs; Am Chem Soc. *Res:* Coal gasification; coal liquefaction; kinetics; catalysis; petroleum refining; process develop. *Mailing Add:* 7810 Braesdale Lane Houston TX 77071

CAMERON, ALASTAIR GRAHAM WALTER, b Winnipeg, Can, June 21, 25; US citizen; m 55. THEORY. *Educ:* Univ Man, BSc, 47; Univ Sask, PhD(physics), 52. *Hon Degrees:* AM, Harvard Univ, 73; DSc, Univ Sask, 77. *Prof Exp:* Asst prof phys, Iowa State Col, 52-54; from asst to sr res officer, Atomic Energy Can, Ltd, 54-61; sr scientist, Goddard Inst Space Studies, 61-66; prof space phys, Belfer Grad Sch Sci, Yeshiva Univ, 66-73; PROF ASTRON, HARVARD UNIV, 73- *Concurrent Pos:* Vis lectr phys, Yale Univ, 62-68; adj prof phys, NY Univ, 63-65; consult, Nat Aeronaut & Space Admin, 66-; mem vis comt, Bartol Res Found, Franklin Inst, 74-87; chmn, Nat Acad Sci-NRC Space Sci Bd, 76-82. *Honors & Awards:* Bronze Medal, Univ Liege, 59; J Lawrence Smith Medal, Nat Acad Sci, 88; Harry H Hess Medal, American Geophysics Union, 89. *Mem:* Nat Acad Sci; Am Acad Arts & Sci; Royal Soc Can; Am Astron Soc; Am Geophys Union. *Res:* Theoretical research in nuclear physics, astrophysics, and planetary sciences, including nuclear astrophysics, stellar formation and evolution of the solar system and of planets. *Mailing Add:* Harvard Col Observ 60 Garden St Cambridge MA 02138

CAMERON, ALEXANDER MENZIES, b Yonkers, NY, Sept 5, 30; m 52; c 3. VETERINARY PATHOLOGY. *Educ:* Colo State Univ, BS, 61, DVM, 62; Univ Calif, Davis, PhD(comp path), 70. *Prof Exp:* Vet clinician small animals, pvt pract, NMex, 62-63 & Calif, 64-66; field vet dis eradication, USDA, Calif, 63-64; asst prof vet path, Col Vet Med, Univ Ill, 70-76; mem, Staff Path Serv Proj, Nat Ctr Toxicol Res. 76-79, dir, Div Path, 80-82; VET PATHOLOGIST, MCNEIL PHARMACEUT, 82- *Concurrent Pos:* Assoc prof path, Med Sch, Univ Ark, 76-79. *Mem:* Am Col Vet Pathologists; Am Vet Med Asn; Int Acad Path; Soc Toxicol Path. *Res:* Development of mammary neoplasms in animals; toxicologic pathology. *Mailing Add:* Drug Safety Eval McNeil Pharmaceut Spring House PA 19477

CAMERON, BARRY WINSTON, b NS, Feb 7, 40; US citizen; m 68, 79; c 2. PALEOECOLOGY. *Educ:* Rutgers Univ, AB, 62; Columbia Univ, AM, 65, PhD(geol), 68. *Prof Exp:* Teaching asst geol, Columbia Univ, 62 64; from instr to prof geol & chmn dept, Boston Univ, 67-81; dept head, 81-86, PROF GEOL, ACADIA UNIV, WOLFVILLE, NS, 81- *Concurrent Pos:* NSF grants, 70-72, 77-79 & Nat Sci & Eng Res Coun Can grants, 82, 83-85; vis asst prof geol, Colgate Univ, summers 67, 68 & vis assoc prof, Boston State Col, 79. *Mem:* Sigma Xi; Paleont Soc; Geol Soc Am; Soc Econ Paleont & Mineral; Am Asn Petrol Geol; Can Soc Petrol Geol. *Res:* Trace fossils; calcareous algae; paleoecology, sedimentation and stratigraphy; cyclic sedimentation; origin and development of barrier islands; Triassic-Jurassic invertebrate and vertebrate fossils of the Fundy Basin of Nova Scotia; Carboniferous stratigraphy and paleontology of Nova Scotia; holocene peat deposits of Eastern New England. *Mailing Add:* Dept Geol Acadia Univ Wolfville NS B0P 1X0 Can

CAMERON, BRUCE FRANCIS, b Damariscotta, Maine, Sept 10, 34; m 57; c 2. PHYSICAL BIOCHEMISTRY, HEMATOLOGY. *Educ:* Harvard Univ, BS, 56; Univ Pa, MD, 60, PhD(biochem), 62. *Prof Exp:* Vis res assoc chem, Am Univ Beirut, 62-63; res assoc, Univ Ibadan, 63-65; sr staff scientist, New Eng Inst Med Res, 66-69; assoc prof med, Sch Med, Univ Miami, 69-78; sr scientist, Papanicolaou Cancer Res Inst, 71-78; assoc prof pediat med, Sch Med, Univ Cincinnati, 78-86; vpres, prod develop, Comvid, 86-89; DIR TECHNOL, LASERFLEX TECHNOL, 89- *Concurrent Pos:* NIH fel, 60-63; Am Heart Asn advan res fel, 67-69; vis scientist, Am Univ Beirut, 65; investr, Howard Hughes Med Inst, 69-71; adj prof life sci, New Eng Inst Grad Sch, 71-74; estab investr, Am Heart Asn, 72-77, mem, Coun on Basic Sci. *Mem:* AAAS; Am Heart Asn; Am Soc Hemat; Soc Analytic Cytol; NY Acad Sci; Am Soc Biol Chemists. *Res:* Microcomputers in medical training and patient eduction; membrane physiology; studies in sickle cell anemia; biophysical instrumentation with microcomputer systems. *Mailing Add:* Laser Flex Technol 11811 NW Freeway Suite 250 Houston TX 77060

CAMERON, DAVID GEORGE, b Streaky Bay, South Australia, March 13, 47; Can citizen; m 69; c 3. VIBRATIONAL SPECTROSCOPY. *Educ:* Monash Univ, Australia, BSc, 67; LaTrobe Univ, BSc, Hons, 68, PhD, 72. *Prof Exp:* Fel, 72-74, staff, Nat Res Coun, Can, 74-84; ASSOC ED, APPL SPECTROS, 82- *Honors & Awards:* Coblentz Award, 83. *Mem:* Soc Appl Spectroscopy; Coblentz Soc. *Res:* Analytical infrared and Raman spectroscopy; data reduction in analytical spectroscopy. *Mailing Add:* OBP America Res Ctr 4440 Warrensville Center Rd Cleveland OH 44128

CAMERON, DAVID GLEN, b Great Falls, Mont, Dec 6, 34; m 61; c 2. POPULATION GENETICS, EVOLUTIONARY BIOLOGY. *Educ:* Princeton Univ, AB, 57; Stanford Univ, PhD(biol), 66. *Prof Exp:* Res assoc environ med, US Army Med Res Labs, Ft Knox, Ky, 57-59; asst prof genetics, Univ Alta, 64-69; assoc prof zool & head dept zool & entom, 70-74, ASSOC PROF BIOL & GENETICS, MONT STATE UNIV, 74- *Concurrent Pos:* Secy & trustee, Rocky Mountain Biol Labs, Crested Butte, Colo, 68-74. *Mem:* AAAS; Am Inst Biol Sci; Genetics Soc Am; Soc Study Evolution; Am Soc Human Genetics; Sigma Xi. *Res:* Ecological, populational, genetic and evolutionary aspects of genetic polymorphisms in natural populations of small rodents, European rabbits, elk and trout; blood groups serum proteins isoenzymes; human genetics. *Mailing Add:* Dept Biol Mont State Univ Bozeman MT 59717

CAMERON, DON FRANK, b El Paso, Tex, Aug 13, 47; m 69; c 2. MALE INFERTILITY, TESTICULAR FUNCTION. *Educ:* Univ South, BA, 69; Med Univ SC, MS, 72, PhD(anat), 77. *Prof Exp:* Instr anat path, Greenville Tech Educ Ctr, Allied Health Sch, 71-72; instr, 76-80, ASST PROF ANAT, COL MED, UNIV FLA, 80-, DOCTORAL RES FAC, 84- *Concurrent Pos:* Prin res investr, Col Med, Univ Fla, NIH, 82- *Mem:* SEastern Electron Micros Soc; AAAS; Am Asn Anatomists; Sigma Xi; Am Soc Andrology; Am Fed Clin Res; Am Asn Clin Anatomist; fel Soc Electron Micros. *Res:* Male infertility associated with associated with varicocele, pituitary adenoma and diabetes; elucidation of mechanisms of spermatogenesis with special emphasis on sertoli cell structure and function; etiology of disrupted sperm production. *Mailing Add:* Dept Anat Box MDC-6 Univ SFla 12901 Bruce B Downs Blvd Tampa FL 33612

CAMERON, DONALD FORBES, b Edmonton, Alta, Aug 19, 20, m 44; c 4. ANESTHESIOLOGY. *Educ:* Univ Alta, BA, 47, MD, 49; FRCPS(C) cert. *Prof Exp:* Lectr pharmacol, Univ Alta, 50-62, asst prof pharmacol & anesthesia & asst dean fac med, 62-65, assoc dean, 65-74, assoc prof, 65-73, prof anesthesia, 73-88, dean fac med, 74-88; RETIRED. *Concurrent Pos:* Mem, Med Coun Can, 62-, pres, 74-75; mem bd gov, Univ Hosp, 74-; mem gen coun, Can Med Asn, 75- *Honors & Awards:* Order of Brit Empire. *Mem:* Fel Am Col Anesthesiol. *Mailing Add:* 9015 Saskatchewan Dr Edmonton PQ T6G 2B2 Can

CAMERON, DOUGLAS EWAN, b New Brunswick, NJ, Apr 5, 41; c 5. TOPOLOGY. *Educ:* Miami Univ, BA, 63; Univ Akron, MS, 65; Va Polytech Inst & State Univ, PhD(topol), 70. *Prof Exp:* Instr math, Community & Tech Col, Univ Akron, 65-66 & Va Polytech Inst & State Univ, 66-69; from instr to assoc prof, 69-80, PROF MATH, UNIV AKRON, 80- *Concurrent Pos:* US Acad Scis grant, Seklov Inst, Moscow, USSR, 79-81. *Mem:* Am Math Soc; Math Asn Am. *Res:* Properties of maximal and minimal topologies, specifically covering axioms; history of topology in the Soviet Union. *Mailing Add:* Dept Math Scis Univ Akron Akron OH 44325

CAMERON, DOUGLAS GEORGE, b Eng, Mar 11, 17; c 6. MEDICINE. *Educ:* Univ Sask, BSc, 37; McGill Univ, MD & CM, 40; Oxford Univ. *Prof Exp:* Res asst, Nuffield Dept Clin Med, Oxford Univ, 48; sr med res fel, Nat Res Coun Can, 49-52; assoc physician, 51-57, asst dir, 52-57, DIR & PHYSICIAN-IN-CHIEF, UNIV CLIN, MONTREAL GEN HOSP, 57-; PROF MED, FAC MED, McGILL UNIV, 57-, CHMN DEPT, 74- *Concurrent Pos:* Asst prof med, McGill Univ, 49-57, chmn dept, 64-69. *Honors & Awards:* Officer, Order Can, 78. *Mem:* Fel Am Col Physicians; Am Soc Clin Invest; Royal Col Physicians & Surgeons Can (pres, 78); Asn Am Physicians; Am Clin & Climat Asn. *Mailing Add:* 227 Portland Ave Mont Royal Montreal PQ H3R 1V3 Can

CAMERON, DUNCAN MACLEAN, JR, b St Albans, Vt, May 15, 31; m 55; c 3. ZOOLOGY, ECOLOGY. *Educ:* Univ Maine, BS, 54, MEd, 60; Univ Calif, Davis, PhD, 69. *Prof Exp:* Teacher, Pub Schs, Maine & Ohio, 56-61; instr bot, Univ Maine, 61-62; res zoologist, Univ Calif, Davis, 62-64, teaching asst zool, 64-66, assoc, 66-67, lectr, 67-68; fel, Univ BC, 68-69; asst prof biol, 69-73, dir, Div Natural Sci, 80-89, ASSOC PROF BIOL, YORK UNIV, 73- *Concurrent Pos:* Vis prof entom, Univ Maine, 78-79. *Mem:* Ecol Soc Am; Am Soc Mammal; Sigma Xi. *Res:* Population ecology of small mammals; mammalian hibernation; ecological processes, especially competition; ecology of pitcher-plants and their insect associates. *Mailing Add:* Dept Biol York Univ Toronto ON M3S 1P3 Can

CAMERON, EDWARD ALAN, b Brockville, Ont, July 4, 38; m 64; c 2. FOREST ENTOMOLOGY, FOREST PEST MANAGEMENT. *Educ:* Ont Agr Col, BSA, 60; Univ Calif, Berkeley, MS, 67, PhD(entom), 74. *Prof Exp:* Entomologist, Commonwealth Inst Biol Control, 60-65; from asst prof to assoc prof, 70-82, PROF ENTOM, PA STATE UNIV, UNIVERSITY PARK, 82- *Concurrent Pos:* Collabr, USDA, 71-81; gastdozent, Entomologisches Inst, Zurich, Switzerland, 79-80. *Honors & Awards:* Speaker Serv Am Chem Soc, 82- *Mem:* Entom Soc Am; Int Orgn Biol Control; Int Soc Chem Ecol; Sigma Xi. *Res:* Bionomics, ecology and control of forest insects; gypsy moth management; assessment of environmental impact of pesticides; aerial spray technology. *Mailing Add:* Dept Entom 106 Patterson Pa State Univ University Park PA 16802

CAMERON, EDWARD ALEXANDER, b Manly, NC, Nov 10, 07; m 37; c 1. MATHEMATICS. *Educ:* Univ NC, AB, 28, AM, 29, PhD(math), 36. *Prof Exp:* From instr to prof, 29-72, EMER PROF MATH, UNIV NC, CHAPEL HILL, 72- *Concurrent Pos:* Ford Found fel, 51-52; NSF sci fac fel, 65-66; consult, NSF. *Mem:* Am Math Soc; Math Asn Am (treas, 68-72). *Res:* Algebra. *Mailing Add:* 404 Laurel Hill Rd Chapel Hill NC 27514

CAMERON, EUGENE NATHAN, b Atlanta, Ga, Aug 10, 10; c 3. ECONOMIC GEOLOGY. *Educ:* NY Univ, BS, 32; Columbia Univ, AM, 34, PhD(geol), 39. *Prof Exp:* Asst geol, NY Univ, 30-35; asst, Columbia Univ, 36-37, lectr, 37-39, instr, 39; assoc geologist, US Geol Surv, 42-44, geologist, 46, sr geologist, 46-50; from assoc prof to prof econ geol, Univ Wis-Madison, 47-70, chmn dept geol, 55-60, Van Hise distinguished prof econ geol, 70-80, EMER PROF GEOL, UNIV WIS-MADISON, 80- *Concurrent Pos:* Geologist, US Geol Surv, 37-39; asst, State Geol & Natural Hist Surv, Conn, 41-42; consult, NASA, 62-63 & 64-67; deleg, Int Comn Ore Micros, 62-64; mem, Comt Mineral Resources & Environ, 73-75; mem, US Nat Comt Geol, 74-77; vis prof, Carleton Col, 82; consult, Wis Ctr Space Automation & Robotics, 86- *Honors & Awards:* Penrose Medal, Soc Econ Geologists, 85. *Mem:* Fel Geol Soc Am; fel Mineral Soc Am; Soc Econ Geologists (secy, 61-69, vpres, 69, pres, 74). *Res:* Mineralogy; economic geology of pegmatite mineral deposits; internal structure of granitic pegmatites; optical properties of ore minerals; chromite deposits; lunar samples; investigation of origin of chromite deposits in the Bushveld Complex; analysis of mineral resource problems; lunar resources of Helium-3. *Mailing Add:* Dept Geol & Geophys Univ Wis Madison WI 53706

CAMERON, GUY NEIL, b San Francisco, Calif, May 1, 42. ECOLOGY, POPULATION BIOLOGY. *Educ:* Univ Calif, Berkeley, BA, 63; Calif State Col, Long Beach, MA, 65; Univ Calif, Davis, PhD(zool), 69. *Prof Exp:* Asst res zoologist, Univ Calif, Berkeley, 69-71; from asst prof to assoc prof, 71-84, PROF BIOL, UNIV HOUSTON, 84- *Concurrent Pos:* Consult, Gulf Univs Res Consortium, 73, Dames & Moore Environ Engrs, 74 & 75; assoc ed, J Mammal, 80-84; ed, Mammalian Species, 90- *Mem:* Ecol Soc Am; Am Soc Naturalists; Am Soc Mammalogists; Brit Ecol Soc; Soc Study Evolution. *Res:* Investigation of interactions between herbivores and plants; contribution of decomposer organisms to community dynamics; role of behavior and nutrition in population dynamics. *Mailing Add:* Dept Biol Univ Houston Houston TX 77204-5513

CAMERON, H RONALD, b Oakland, Calif, June 30, 29; m 57; c 2. PLANT PATHOLOGY. *Educ:* Univ Calif, BS, 51; Univ Wis, PhD(plant path), 55. *Prof Exp:* Agt, USDA, Univ Wis, 53-55; from asst plant pathologist to assoc plant pathologist, 55-70, PLANT PATHOLOGIST, ORE STATE UNIV, 70- *Concurrent Pos:* NSF fel, Stanford Univ, 62-63; NATO sr fel, East Malling Res Sta, Eng, 72; pres, Univ Fac Senate, Ore State Univ, 85. *Honors & Awards:* Earl Price Award, Res Agr. *Mem:* Am Phytopath Soc (treas, 83-89). *Res:* Virus and bacterial diseases of fruit and nut trees; genetics of pathogenicity. *Mailing Add:* Dept Bot & Plant Path Ore State Univ Corvallis OR 97331

CAMERON, IRVINE R, b Fredericton, NB, June 24, 19; m 45; c 2. POLYMER CHEMISTRY. *Educ:* Univ NB, BS, 41; Univ Toronto, MS, 42. *Prof Exp:* From sci officer to dir eng planning & coord, Chief Tech Serv Br, Can Forces Hq, 70-72; dir gen mat develop planning, 72-76; dir, Gen Res & Develop Serv, Dept Nat Defence, Can, 76-80; RETIRED. *Mem:* Fel Chem Inst Can; assoc fel Can Aeronaut & Space Inst; Am Inst Aeronaut & Astronaut; Prof Inst Pub Serv Can; Can Soc Chem Eng. *Res:* High explosives; propellants for guns and rockets; development of composite rocket propellants of polyurethane type resulting in the Black Brant series of sounding rockets; equipment development; military applications. *Mailing Add:* Four Riopelle Ct Kanata ON K2K 1J1 Can

CAMERON, IVAN LEE, b Los Angeles, Calif, Jan 4, 34; m 55; c 3. ZOOLOGY, ANATOMY. *Educ:* Univ Redlands, BS, 56; Univ Southern Calif, MS, 59; Univ Calif, Los Angeles, PhD(anat), 63. *Prof Exp:* Biologist, Oak Ridge Nat Lab, 63-65; from asst prof anat to assoc prof anat, 65-74, assoc prof cell biol, 68-74, PROF ANAT & PROF CELL BIOL, UNIV TEX HEALTH SCI CTR, SAN ANTONIO, 74- *Mem:* Am Asn Anatomists; Am Soc Cell Biol; Am Asn Cancer Res; Sigma Xi. *Res:* Nuclear-cytoplasmic environmental interactions; chemical changes during the cell-growth-duplication cycle; feedback control of cellular proliferation. *Mailing Add:* 110 Cliffside Dr San Antonio TX 78231

CAMERON, J A, b Pittsburg, Okla, Oct 7, 29; m 53; c 2. MEDICAL MICROBIOLOGY. *Educ:* Maryville Col, BS, 51; Univ Tenn, MS, 55, PhD, 58. *Prof Exp:* Asst, Univ Tenn, 56-58; from instr to assoc prof bact, 58-73, PROF, DEPT MOLECULAR & CELL BIOL, UNIV CONN, 73- *Concurrent Pos:* Res assoc, Calif Inst Technol, 68-69. *Mem:* AAAS; Am Soc Microbiol; Soc Gen Microbiol; Sigma Xi. *Res:* Microbial degradation of polymers. *Mailing Add:* Box U-44 Univ Conn Storrs CT 06269

CAMERON, JAMES N, b Hanover, NH, May 28, 44; m 63; c 2. PHYSIOLOGY, BIOMED INSTRUMENTATION. *Educ:* Univ Wis, Madison, BS, 66; Univ Tex, Austin, PhD(zool), 69. *Prof Exp:* NIH fel, Univ BC, 69-71; from asst prof to assoc prof zoophysiol, Inst Arctic Biol, Univ Alaska, 71-75; from asst prof to assoc prof, 75-84, PROF ZOOLOGY & MARINE STUDIES, MARINE SCI INST, UNIV TEX, 84- *Concurrent Pos:* Secy, Div Comp Physiol & Biochem, Am Soc Zool, 78-79; assoc ed, J Exp Zoology, 78-80 & Marine Biol Lett, 80-85; Nat Sci Found panelist, Regulatory Biol, 80-82; vis scientist, Max-Planck Inst Exp Med, Gottingen, Fed Repub Germany, 82; Int Union Biol Sci, DCPB Prog Comt, 83- *Mem:* Am Soc Zoology; Soc Exp Biol. *Res:* Physiology of respiration, circulation and ion regulation; acid-base regulation; comparative physiology. *Mailing Add:* Dept Marine Sci Univ Tex Austin TX 78712

CAMERON, JAMES WAGNER, b Falls City, Nebr, Apr 23, 13; m 38; c 2. PLANT GENETICS, PLANT BREEDING. *Educ:* Univ Mo, AB, 38; Harvard Univ, AM, 43, PhD(genetics), 47. *Prof Exp:* Asst genetics, Harvard Univ, 41-42; res assoc, Off Sci Res & Develop Contract, 43-45; asst genetics, Harvard Univ, 45-46; from jr geneticist to assoc geneticist, 47-60, lectr hort sic, 60-64, geneticist, 60-80, prof 64-80, EMER PROF, HORT SCI CITRUS RES CTR, UNIV CALIF, RIVERSIDE, 80- *Concurrent Pos:* Sr scholar, East-West Ctr, Univ Hawaii, 62-63; Fulbright lectr, Aegean Univ, Turkey, 70-71; Fulbright travel grant, Palermo Univ, Italy, 76. *Mem:* Am Soc Hort Sci; Sigma Xi; Am Genetic Asn. *Res:* Breeding and genetics of citrus and maize. *Mailing Add:* 4195 Stratus Ct South Salem OR 97302

CAMERON, JOHN, b Glasgow, Scotland, Aug 12, 39. METALLURGICAL ENGINEERING, PYROMETALLURGY. *Educ:* Royal Col Sci & Technol, Scotland, ARCST, 62, DRC, 64; Univ Strathclyde, PhD(metall), 66. *Prof Exp:* Fel, Nat Res Coun Can, NS, 66-68; res assoc, Univ BC, 68-70; asst prof, 70-77, ASSOC PROF, QUEEN'S UNIV, ONT, 77- *Mem:* Am Soc Metals; Am Inst Mining, Metall & Petrol Engrs; Asn Brit Iron & Steel Inst; Asn Brit Inst Metals; Can Inst Mining & Metal. *Res:* Electroslag remelting; vacuum arc remelting; electroslag welding; physical properties of slags, oxides of refractory metals; pyrometallurgy; extractive processes; radioactive waste management; pyrometallurgical processes. *Mailing Add:* 69 Silver Kingston ON K7M 2P6 Can

CAMERON, JOHN ALEXANDER, b Toronto, Ont, May 19, 36; m 58; c 3. NUCLEAR PHYSICS. *Educ:* Univ Toronto, BA, 58; McMaster Univ, PhD(physics), 62. *Prof Exp:* NATO fel nuclear orientation, Clarendon Lab, Oxford Univ, 62-64; from asst prof to assoc prof, 64-74, PROF PHYSICS & ASSOC DEAN SCH GRAD STUDIES, McMASTER UNIV, 74-

Concurrent Pos: Sloan Found fel, 67-68; vis scientist, Cent Res Inst Physics, Hungary, 70; mem fac sci, Univ Paris, 71. *Mem:* Can Asn Physicists. *Res:* Atomic beams; nuclear orientation; angular correlation studies; hyperfine structure in atoms and solids; nuclear moments; nuclear reactions and spectroscopy. *Mailing Add:* Dept Physics McMaster Univ Hamilton ON L8S 4L8 Can

CAMERON, JOHN RODERICK, b Wis, Apr 21, 22; m 47; c 2. MEDICAL PHYSICS. *Educ:* Univ Chicago, BS, 47; Univ Wis, MS, 49, PhD(nuclear physics), 52. *Prof Exp:* Asst prof, Univ Sao Paulo, 52-54; proj assoc, Univ Wis, 54-55; asst prof nuclear physics, Univ Pittsburgh, 55-58; from asst prof to prof, 58-86, dir, Biomed Eng Ctr, 69-76, dir, Med Physics Div, 74-81, Farmington Daniels prof, 79-86, chmn dept, 81-85, EMER PROF MED PHYSICS, UNIV WIS-MADSION, 86- *Concurrent Pos:* Secy-gen, Int Orgn Med Physics, 69-76; consult, Vet Admin Hosp, Madison, Atomic Energy Comn, Washington, Int Atomic Energy Asn, Vienna, Austria, Bur Radiation Health, Fed Drug Admin, 71- *Honors & Awards:* Coolidge Award, Am Asn Physicists Med, 80. *Mem:* Am Phys Soc; Am Asn Physicist in Med (pres, 68). *Res:* Radiation dosimetry using thermoluminescence; physics of diagnostic radiology; physics of bones; physics of the brain; general applications of physics to medicine. *Mailing Add:* 118 N Breese Terr Madison WI 53705

CAMERON, JOHN STANLEY, b Chicago, Ill, Aug 23, 52; m 88; c 1. CARDIOVASCULAR PHYSIOLOGY, COMPARATIVE PHYSIOLOGY OF FISH. *Educ:* Col William & Mary, BS, 74; Univ Mass, Amherst, MS, 77 & PhD(physiol), 80. *Prof Exp:* Postdoc res fel, Albany Med Col, 79-81; res instr & res asst prof pharmacol, Univ Miami Sch Med, 81-84; asst prof, 84-90, ASSOC PROF BIOL SCI, WELLESLEY COL, 90- *Concurrent Pos:* prin investr, NIH, NSF & Am Heart Asn Grants, 80-; postdoc res fel, Nat Heart Lung Blood Inst, 81; vis asst prof pharmacol, Univ Miami Sch Med, 87-88. *Mem:* Am Physiol Soc; Biophys Soc; Am Soc Zoologists; Am Heart Asn; Sigma Xi. *Res:* Cellular electrophysiological alerations underlying cardiac rhythm disturbances in heart disease; ion channels in adaptation of vertebrates to hypoxia and hypothermia. *Mailing Add:* Dept Biol Sci Wellesley Col Wellesley MA 02181-8289

CAMERON, JOSEPH MARION, b Edinboro, Pa, Apr 6, 22; m 46; c 4. MATHEMATICAL STATISTICS. *Educ:* Univ Akron, BS, 42; NC State Col, MS, 47. *Prof Exp:* Math statistician, 47-63, chief statist eng lab, 63-68, chief off measurement serv, Nat Bur Standards, 68-77; CONSULT STATIST, 77- *Mem:* Inst Math Statist; fel Am Statist Asn; Biomet Soc; fel AAAS. *Res:* Application of statistics in physical sciences. *Mailing Add:* 12502 Gould Rd Wheaton MD 20906

CAMERON, LOUIS MCDUFFY, b Richmond, Va, Dec 16, 35; m 59. APPLIED PHYSICS. *Educ:* Univ Richmond, BS, 57; George Washington Univ, MS, 62; Georgetown Univ, PhD(physics), 66. *Prof Exp:* Physicist, NIH, 58-61 & Naval Res Lab, Washington, DC, 61-66; physicist, Ft Belvoir, Va, 66-70, supvry physicist, 70-75, assoc dir, 75-80, dir, night vision & electro optics lab, 80-; DIR, NIGHT VISION & ELECTRO-OPTICS LAB, HARRY DIAMOND LABS. *Concurrent Pos:* Edison Mem training scholar, Naval Res Lab, Washington, DC, 64-65. *Honors & Awards:* Ann Award, Night Vision Lab, 75; Res & Develop Achievement Award, US Army, 75. *Mem:* Am Phys Soc. *Res:* Liquid crystal physics; basic research including thermodynamics, light scattering, x-ray, microscope; development of a set of common modules utilized in infrared thermal imaging systems for all of the Department of Defense; electro-optics. *Mailing Add:* BDM Int 7915 Jones Branch Dr McLean VA 22102

CAMERON, MARGARET DAVIS, b Montgomery, Tex, Apr 25, 20. ORGANIC CHEMISTRY. *Educ:* Tex Women's Univ, BA, 42; Univ Houston, MS, 48; Tulane Univ, PhD(chem), 51. *Prof Exp:* Fulbright scholar & Ramsay mem fel, Univ Leeds, 51-52; USPHS fel, Ohio State Univ, 52-53; res chemist, Monsanto Co, 53-56; from assoc prof to prof chem, 56-72, REGENTS PROF CHEM, LAMAR UNIV, 72- *Mem:* Am Chem Soc; The Chem Soc. *Res:* Chemistry of acetylenic and heterocyclic compounds. *Mailing Add:* 4060 Howard St Beaumont TX 77705

CAMERON, ROBERT ALAN, b Edmonton, Alta, Oct 19, 26; m 52; c 3. EXPLORATION GEOLOGY & GEOPHYSICS. *Educ:* Dalhousie Univ, BSc, 48; Univ Toronto, MASc, 53; McGill Univ, PhD(geol), 56. *Prof Exp:* Lectr geol, Univ NB, 52; geologist, Imperial Oil Ltd, Alta, 52-53 & Malartic Gold Fields Ltd, Que, 56-57; chief geologist, E Malartic Mines Ltd, 57-62; area supvr, McIntyre Porcupine Mines Ltd, 62; asst prof mining, NS Tech Col, 62-68; prin, Sudbury Campus, Cambrian Col, 68-69; ASSOC PROF GEOL, LAURENTIAN UNIV, 69- *Mem:* Fel Geol Asn Can; Can Inst Mining & Metall. *Res:* Geological and geophysical exploration methods. *Mailing Add:* Dept Geol Laurentian Univ Ramsey Lake Rd Sudbury ON P3E 2C6 Can

CAMERON, ROBERT HORTON, mathematics; deceased, see previous edition for last biography

CAMERON, ROSS G, b 1949. CARCINOGENESIS, BIOCHEMISTRY. *Educ:* Univ Western Ont, MD, 72; Univ Toronto, PhD(path & biochem), 80. *Prof Exp:* ASST PROF PATH, DEPT TOXICOL, UNIV TORONTO, 75- *Res:* Experimental pathology. *Mailing Add:* Med Sci Bldg Rm 6212 Univ Toronto Toronto ON M5S 1A8

CAMERON, ROY (EUGENE), b Denver, Colo, July 16, 29; div; c 2. SCIENCE ADMINISTRATION, ENVIRONMENTAL SCIENCE. *Educ:* Wash State Univ, BS, 53 & 54; Univ Ariz, MS, 58, PhD(plant sci), 61. *Prof Exp:* Res chemist, Hughes Aircraft Co, 55-56; sr scientist, Jet Propulsion Lab, Calif Inst Technol, 61-68, mem tech staff, 68-73; pres, Darwin Res Inst, Calif, 73-74; dep dir land reclamation lab, Environ Impact Studies Div, 75-77, dir, energy resources training & develop, Reclamation Lab, Environ Impact Studies Div, Argonne Nat Lab, 77-85; SR STAFF SCIENTIST, LOCKHEED ESCO, 86-, LESC QUAL ASSURANCE OFFICER, 90- *Concurrent Pos:* Consult, Med Sch, Baylor Univ, 66-68; Jet Propulsion Lab team leader, Antarctic Exped, 66-74; vis scientist, Am Soc Agron, 67-69; prin investr, NSF res grant Antarctic microbial ecol, 69-74; Smithsonian Inst study grant Moroccan desert, 69-70; consult, Ecol Ctr, Utah State Univ, 70-72; Dept Com grants, Native Am energy/environ, 78-80; mem, Dept Agr Nat Agr Res & Extension Users Adv Bd, 86-90; bioremediation lab mgr, AWC-Lockheed ESCO, 91-; coord scientist, USDA Soil & Water Qual Projs, Poland, Yugoslavia & Egypt. *Honors & Awards:* Environ Protection Agency Award of Excellence, 88. *Mem:* AAAS; World Future Soc; Soil Sci Soc Am; Am Inst Biol Sci; Soil & Water Conserv Soc; Am Geophys Union; Am Chem Soc; Sigma Xi. *Res:* Desert soil microbiology; blue-green algae; microbial ecology; soil science; polar ecology; environmental monitoring; coal mine reclamation; Native American energy and environmental training; general environmental sciences; numerous publications; global climate change, agroecosystems and groundwater monitoring; discoverer of world's farthest south bacteria, algae, fungi and lichens. *Mailing Add:* Lockheed ESCO 1050 E Flamingo Rd, Suite 126 Las Vegas NV 89119

CAMERON, T STANLEY, b Ft Monmouth, NJ, Aug 6, 42; m 69; c 2. X-RAY CRYSTALLOGRAPHY. *Educ:* Oxford Univ, UK, BA, 65, MA & DPhil(crystallog), 68. *Prof Exp:* Asst prof chem, New Univ Ulster, UK, 71-76; assoc prof, 76-81, PROF CHEM, DALHOUSIE UNIV, 81-, ASSOC DEAN ARTS & SCI, 85- *Res:* The study of the factors influencing the formation of crystals. *Mailing Add:* Dept Chem Dalhousie Univ Halifax NS B3H 4J3 Can

CAMERON, WILLIAM EDWARD, b Long Beach, Calif, Jan 13, 50; m 81; c 1. CONTROL OF RESPIRATION, DEVELOPMENT OF MOTOR SYSTEMS. *Educ:* Northern Ariz Univ, BS, 72, MS, 75; Univ Ariz, PhD, 79. *Prof Exp:* res fel, Div Neurosurg, Univ Tex Houston, 79-80; Res asst prof, Dept Med Univ Wash, 73-84; fel, Dept Physiol & Biophys, Univ Wash, 81-83; ASST PROF, DEPT ANAT & PEDIAT, UNIV PA, PITTSBURGH, 84- *Concurrent Pos:* Ad hoc mem, Neurosurg Study Sect, NIH, 85-, site vis team, 86, resp appl physiol, 86, prin investr, Develop of control of diaphram & upper airways, 87- *Mem:* Soc Neurosci; Am Physiol Soc. *Res:* The properties of the nerve cells that control the contraction of the mammalian diaphram and upper airways; neuroanatomical and neurophysiological techniques are used to describe the course of postnatal development in these neuron populations in order to identify those time when the system might fail; such failures may account for mortality of Sudden Infant Death and Apnea of prematurity. *Mailing Add:* Div Neonatol Magee-Womens Hosp Forbes Ave & Halket St Pittsburgh PA 15235

CAMERON, WINIFRED SAWTELL, b Oak Park, Ill, Dec 3, 18; m 53; c 2. ASTRONOMY, PLANETOLOGY. *Educ:* Northern Ill Univ, BE, 40; Ind Univ, MA, 52. *Prof Exp:* Asst astron, Weather Forecasts, Inc, 42-46 & 49-50; instr, Mt Holyoke Col, 50-51; astron res asst, US Naval Observ, 51-58; aerospace technologist, Lab Space Physics, 59-71, aerospace technologist & acquisition scientist, Nat Space Sci Data Ctr, Goodard Space Flight Ctr, NASA, 71-84. *Honors & Awards:* Asteroid 1575 named Winifred, Minor Planets Ctr, 52; Spec Act Award, NASA, 66, Apollo Achievement Award, 70, Exceptional Contrib to Educ Award, 71, Qual Increase Award, 73, & Spec Achievement Award, 77. *Mem:* Int Astron Union; Am Astron Soc; Am Geophys Union; Int Asn Planetology (vpres, 85-); Sigma Xi. *Res:* Measurements of sunspots; lunar surface and scientific aspects research; scientific results of Mercury program of manned spaceflight; science data acquisition and analyses for Data Center of lunar and planetary surface features and lunar transient phenomena. *Mailing Add:* 200 Rojo Dr Sedona AZ 86336

CAMHI, JEFFREY MARTIN, neurophysiology, animal behavior, for more information see previous edition

CAMIEN, MERRILL NELSON, b Redlands, Calif, Dec 10, 20; m 44; c 2. BIOLOGICAL CHEMISTRY. *Educ:* Univ Calif, Los Angeles, AB, 43, MA, 45, PhD(microbiol), 48; Univ Calif, Irvine, MS, 85. *Prof Exp:* Res assoc biochem, Univ Calif, Los Angeles, 48-50; Fulbright res scholar, Univ Liege, 50-51; jr res chemist, Univ Calif, Los Angeles, 51-53, from asst res chemist to assoc res chemist, 53-62; res chemist, Vet Admin Ctr, Los Angeles, 62-63; res assoc, Mt Sinai Hosp, 64-65, sr scientist, 65-68; res physiologist, Dept Med, Ctr Health Sci, Univ Calif, Los Angeles, 68-70; res biochemist & head biochem & microbiol, Nucleic Acid Res Inst, Int Chem & Nuclear Corp, 69-70, head biol, microbiol & parasitol, 70-71, head neuropharmacol, 71-73; res assoc, Univ Calif, Irvine, 73-76, asst res biochemist, 76-84; RETIRED. *Mem:* Fel AAAS; Am Soc Microbiol; Am Soc Biol Chem; Soc Exp Biol & Med; Am Chem Soc; NY Acad Sci. *Res:* Studies relating to physical biochemistry of bacteriophage DNAs. *Mailing Add:* 1606 Warwick Lane Newport Beach CA 92660

CAMIENER, GERALD WALTER, b Detroit, Mich, Aug 15, 32; m 53; c 3. BIOCHEMISTRY, MICROBIOLOGY. *Educ:* Wayne State Univ, AB, 54; Mass Inst Technol, PhD(biochem), 59. *Prof Exp:* Teaching asst microbiol, Wayne State Univ, 53-54; res biochemist, Upjohn Co, 59-72; pres, ICN Life Sci Group, ICN Pharmaceut, 72-76; PRES, AM RES PROD CO, 76- *Mem:* Am Soc Biol Chemists; Am Asn Clin Chemists; AAAS; Am Chem Soc. *Res:* Biosynthesis of thiamine; isolation and purification of enzymes; tissue culture and antitumor antibiotics; enzyme inhibitors; immunology; immunosuppression; delayed hypersensitivity. *Mailing Add:* 26700 Hurlingham Rd Beachwood OH 44122-2451

CAMILLO, VICTOR PETER, b Bridgeport, Conn, Jan 23, 45; m 66; c 1. PURE MATHEMATICS. *Educ:* Univ Bridgeport, BA, 66; Rutgers Univ, PhD(math), 69. *Prof Exp:* Asst prof math, Univ Calif, Los Angeles, 69-70; asst prof, 70-74, ASSOC PROF MATH, UNIV IOWA, 74- , PROF. *Mem:* Am Math Soc. *Res:* Theory of rings. *Mailing Add:* Dept Math Univ Iowa Iowa City IA 52240

CAMIZ, SERGIO, b Roma, Italy, Dec 22, 46; m 90; c 3. DATA ANALYSIS STATISTICS. *Educ:* Univ Roma, Itlain laurea, 69. *Prof Exp:* Prof algebra-mech, Univ Calabria, 75-77; PROF MATH-ARCHITECTS, UNIV ROMA, 77- *Concurrent Pos:* Consult, Somea, 78-83; prof computer sci, Am Univ Rome, 90- *Mem:* Am Math Soc; Am Statist Asn; Asn Comput Mach; Int Asn Veg Sci; Int Asn Statist Comput. *Res:* Applications of data analysis and statistics to natural sciences, ecology, sociology and archaeology; development of software for scientific applications. *Mailing Add:* Via San Francesco a Ripa 57 Rome I-00153 Italy

CAMMARATA, ARTHUR, physical organic chemistry, for more information see previous edition

CAMMARATA, JOHN, b Brooklyn, NY, Oct 6, 22; m 54; c 2. ELECTRICAL ENGINEERING. *Educ:* Brooklyn Polytech Inst, BEE, 43; Hofstra Univ, MBA, 61. *Prof Exp:* Res asst, Brooklyn Polytech Inst, 42; engr, Gen Elec Co, NY, 43-46; develop engr, Arma Div, Am Bosch Arma, Garden City, 46-50, asst chief engr, 50-61, mgr prod reliability, 61-64, works mgr mfg opers, 64-65, dir, 65-66; gen mgr, Barden Leemath Div, Barden Corp, 66-68; mgr indust eng, Int Tel & Tel Corp, 68-77; CONSULT ENGR, 77- *Mem:* Fel Inst Environ Sci; Inst Elec & Electronics Engrs; Nat Soc Prof Engrs; Am Mgt Asn. *Res:* Servomechanisms; gun fire control systems; analogue computers; missile guidance systems; electromagnetic compatibility. *Mailing Add:* 45 Murray Hill Terr Marlboro NJ 07746-1756

CAMMARATA, PETER S, b Chicago, Ill, Dec 26, 20; m 52; c 3. BIOCHEMISTRY, PHYSICAL ORGANIC CHEMISTRY. *Educ:* Univ Chicago, BS, 43, MS, 47; Univ Wis, PhD(physiol chem), 51. *Prof Exp:* Res assoc cancer res, Univ Chicago, 47-49; asst prof biochem, Yale Univ, 52-54; sr chemist, 54-55, biochem supvr, 56-62, nat prod supvr, 62-64, asst dir biochem, 64-70, dir biochem res div, 70-74, RES ADV, G D SEARLE & CO LABS, 75- *Honors & Awards:* Am Inst Chemists Award, 39. *Mem:* Am Chem Soc. *Res:* Enzymology; sulfated polysaccharides; proteases; prostaglandins; biochemical pharmacology and endocrinology; medicinal chemistry. *Mailing Add:* 8033 N Lorel Ave Skokie IL 60077-2428

CAMMEN, LEON MATTHEW, b Corning, NY, July 28, 49; m 82; c 1. MARINE BENTHIC ECOLOGY, MARINE ECOLOGICAL ENERGETICS. *Educ:* Brown Univ, ScB, 71; NC State Univ, MS, 74, PhD(zool), 78. *Prof Exp:* Res assoc, bot dept, NC State Univ, 77-78; vis fel, Marine Ecol Lab, Bedford Inst Oceanog, NS, 78-79 & Inst Ecol & Genetics, Univ Aarhus, Denmark, 79-80; res assoc, Skidaway Inst Oceanog, 80-81; vis scientist, Inst Biol, Odense Univ, Denmark, 81; RES SCIENTIST, BIGELOW LAB OCEAN SCI, 81- *Mem:* Estuarine Res Fedn; Int Soc Meiobenthologists. *Res:* The cycling of nutrients and energy in marine benthic ecosystems, concentrating on pathways from detritus to invertebrates. *Mailing Add:* Nat Sea Grant Col Prog 1335 East-West Hwy R-OR1 Silver Spring MD 20910

CAMOUGIS, GEORGE, b Concord, Mass, May 10, 30; m 61; c 3. ENVIRONMENTAL RESEARCH, ENVIRONMENTAL TOXICOLOGY. *Educ:* Tufts Univ, BS, 52; Harvard Univ, MA, 57, PhD(biol), 58. *Prof Exp:* Asst comp anat, Tufts Univ, 50-52; lab asst parasitol, Harvard Univ, 55; from asst prof to assoc prof physiol, Clark Univ, 58-64; sr neurophysiologist, Astra Pharmaceut Prod, Inc, 64-66, head neuropharmacol sect, 66-68; PRES & RES DIR, NEW ENG RES, INC, 68- *Concurrent Pos:* Res scientist, NY State Dept Health, 60; panel mem undergrad sci equip prog, NSF, 64 & 65; affil assoc prof, Clark Univ, 64-68, affil prof, 68-79; mem corp, Bermuda Biol Sta Res, Inc, 68-; adj prof, Worcester Polytech Inst, 70-82; adj prof toxicol, Sch Vet Med, Tufts Univ, 81- *Mem:* Soc Environ Toxicol Chem; Biophys Soc; Am Physiol Soc; NY Acad Sci; AAAS. *Res:* Environmental assessments; toxic materials; natural resources; public health; risk assessment and management; professional training. *Mailing Add:* Seven Wheeler Ave Worcester MA 01609

CAMP, ALBERT T(ALCOTT), b Jamestown, NY, Jan 6, 20; m 42, 85; c 7. CHEMICAL ENGINEERING. *Educ:* Yale Univ, BEng, 41; Mass Inst Technol, MS, 56. *Prof Exp:* Res chemist, Hercules Powder Co, 41-50; chem engr & head propellant develop br, US Naval Ord Test Sta, 50-55 & Propellants Div, 55-59; dir, Propellants Div, Lockheed Propulsion Co, 59-64; dir develop, US Navy Propellant Plant, 64-65, dir res & develop, 65-67; assoc tech dir develop & technol, 67-69, dir sci & eng dept, 69-71, head fleet support dept, 71-72, actg tech dir, 72-73; spec asst to tech dir, Naval Ord Sta, 74-75, chief technologist, 75-80, CONTRACTOR CONSULT PROPELLANTS, PROPULSION & EXPLOSIVES TECHNOL, NAVAL SURFACE WEAPONS CTR, 81-; PRES, BRENTLAND GOAT DAIRY CORP, 77- *Concurrent Pos:* Alfred Sloanfel, Mass Inst Technol, 55-56; consult, US Navy & Thiokol Corp, 80-91. *Mem:* Am Chem Soc; Am Defense Prep Asn; Sigma Xi; Am Inst Aeronaut & Astronaut; AAAS. *Res:* Mechanism of solid propellant burning; catalysis of solid propellant burning; paint and varnish chemistry; synthesis of new nitrate esters; polymer chemistry; chemistry of propellants and explosives; ed-in-chef, Weapons J; Ordance Technol; goat milk chemistry; counsellor to physicians on cow milk allergies and use of pasteurized goat dairy products for patients. *Mailing Add:* Rt 1 Box 1278 Welcome MD 20693

CAMP, BENNIE JOE, b Greenville, Tex, Mar 19, 27; m 52; c 2. BIOCHEMISTRY, TOXICOLOGY. *Educ:* ETex State Univ, BS, 49; Tex A&M Univ, MS, 53, PhD(biochem), 56. *Prof Exp:* From asst prof to assoc prof biochem, 56-66, PROF BIOCHEM & BIOPHYS & VET PHYSIOL & PHARMACOL, TEX A&M UNIV, 66- *Concurrent Pos:* Consult in clin chem & environ; cert prof chemist, Am Inst Chem, 72-; cert lab dir, Am Bioanal Bd, 72- *Mem:* Am Chem Soc; hon fel Am Col Vet Toxicol; Am Inst Chemists; Sigma Xi; hon fel Am Acad Vet & Comp Toxicol. *Res:* Chemistry of poisonous plants and toxicology of environmental pollutants. *Mailing Add:* 2704 Burton Dr Bryan TX 77802-2409

CAMP, DAVID CONRAD, b Atlanta, Ga, Aug 20, 34; m 61; c 2. NUCLEAR STRUCTURE, NUCLEAR CHEMISTRY. *Educ:* Emory Univ, BA, 55; Ind Univ, MA, 58, MS, 59, PhD(physics), 63. *Prof Exp:* mgr, Safeguards Technol Prog, 79-89, SR STAFF PHYSICIST, NUCLEAR CHEM DIV, LAWRENCE LIVERMORE NAT LAB, UNIV CALIF, 63-, ASSOC DIV LEADER, 89- *Concurrent Pos:* Mem staff, Inter-Univ Reactor Inst, Delft Technol Univ, 71-72; consult, Mobil Res & Develop Corp, 74 & Environ Protection Agency, 80-83; adj asst prof radiol, Sch Med Univ Calif, 74-87; instr appl sci, Univ Calif, Davis at Livermore, 77-; instr astron, Chabot Col, 87-89. *Honors & Awards:* IR-100 Award, 76, 82. *Mem:* AAAS; Am Phys Soc; Inst Nuclear Mat Mgt; Am Nuclear Soc; Planetary Soc. *Res:* Res and develop program management; nuclear instrumentation develop; multi-element, multi-technique intercomparison studies; design of Si(Li), and HPGe detector systems, x-ray fluorescence and gamma-ray spectrometry systems; x-ray, gamma-ray spectroscopy; nuclear safeguards instrumentation. *Mailing Add:* Lawrence Livermore Lab L-232 Univ Calif Livermore CA 94550

CAMP, DAVID THOMAS, b Toledo, Ohio, Nov 26, 37; m 60; c 3. PROCESS CONTROL. *Educ:* Carnegie Inst Technol, BS, 58, MS, 60, PhD(chem eng), 63. *Prof Exp:* Develop engr photog processing equip, Eastman Kodak Co, 59-60; instr metall eng, Carnegie Inst Technol, 62-63, asst prof, 63-65; from asst prof to assoc prof chem eng, Univ Detroit, 65-74, prof & chmn dept, 74-76; sr process specialist, 76-80, systs assoc, 80-90, SR SYSTEMS ASSOC, DOW CHEM CO, 90- *Mem:* Am Inst Mining, Metall & Petrol Engrs; Am Inst Chem Engrs; Am Soc Eng Educ; Accreditation Bd for Eng & Technol. *Res:* Process engineering and design; process control (principal research area). *Mailing Add:* 5516 Whitehall Midland MI 48640

CAMP, EARL D, b Magazine, Ark, June 12, 18; m 48; c 1. PLANT PATHOLOGY. *Educ:* Tex Tech Univ, BS, 41; Univ NMex, MS, 43; Univ Iowa, PhD(bot), 52. *Prof Exp:* War res staff & investr med mycol, Columbia Univ, 43-44; from instr to prof, 45-85, dept chmn, 59-71, EMER PROF BIOL, TEX TECH UNIV, 85- *Mem:* Fel AAAS. *Res:* Plant morphology and pathology; developmental anatomy of woody monocots. *Mailing Add:* 3104 23rd St Lubbock TX 79410

CAMP, ELDRIDGE KIMBEL, b Tamaqua, Pa, Sept 6, 15; m 47; c 2. ELECTROCHEMISTRY. *Educ:* Pa State Univ, BS, 38. *Prof Exp:* Apprentice exec, Collins & Aikman Corp, 39-40, foreman dyer, 40-42; res chemist metallic corrosion, Westinghouse Res Labs, 46-56; staff engr elec contacts, Prod Develop Lab, Int Bus Machine Corp, 56-57, mgr, Finishes & Corrosion Lab, 57-59, mem tech staff, Mat & Processes Labs, 59-60; dir res & mem tech staff, Chem & Refining Co, Inc, 60-81, sr scientist & consult bd dir, 69-78; Consult, 80-82; RETIRED. *Mem:* Am Electroplaters Soc; Am Chem Soc; Emer mem Electrochem Soc; Am Soc Test & Mat; Nat Gardening Asn. *Res:* Electrode processes applied to electrodeposition; anodizing; electropolishing; galvanic corrosion; surface studies; the study of mechanical and physical properties of precious metal film deposits and processes. *Mailing Add:* 20 Reder Rd Northfield CT 06778

CAMP, FRANK A, III, b Dallas, Tex, Feb 25, 47; m 70; c 2. AQUATIC ECOLOGY. *Educ:* NTex State Univ, BA, 69, MS, 71; Va Polytech Inst & State Univ, PhD(zool), 75. *Prof Exp:* Sr aquatic biologist, WAPORA Inc, 74-75, dir, Ill Div, 75-76; vpres, Jack McCormick & Assocs, 76-78; SR PROJ MGR, TERA CORP, 78- *Concurrent Pos:* Environ eng consult. *Mem:* AAAS; Sigma Xi. *Res:* Benthic diversity studies; continuous flow and static algal bioassays; bioassays at all trophic levels. *Mailing Add:* 3340 Cross Bend Plano TX 75023

CAMP, FREDERICK WILLIAM, b Washington, DC, July 1, 34; m 57; c 3. CHEMICAL ENGINEERING. *Educ:* Columbia Univ, AB, 56, BS, 57; Princeton Univ, MSE, 61, PhD(chem eng), 62. *Prof Exp:* Res engr, Process Develop, Sun Oil Co, 62-75; mgr synfuels technol, Sun Co, 75-83, mgr energy resources technol, 83-88; SR PRIN RES ENGR, SUN REFINING & MKT CO, 88- *Mem:* Am Chem Soc; Soc Petrol Eng; Am Soc Qual Control. *Res:* Athabasca tar sands; development and mathematical modeling of petroleum and petrochemical processes; conversion of coal and oil shale to synthetic fuels; technology administration. *Mailing Add:* 53 Line Rd Malvern PA 19355

CAMP, HANEY BOLON, b Rome, Ga, May 29, 38; m 59; c 3. ENTOMOLOGY, TOXICOLOGY. *Educ:* Auburn Univ, BS, 60, MS, 63; Univ Calif, Riverside, PhD(entom), 68. *Prof Exp:* Regist coordr, Agr Div, 69-71, toxicologist, 71-72, dir regist & toxicol, 72-75, dir biochem, 75-78, spec asst to vpres res & develop, 78-79, VPRES RES & DEVELOP, CIBA-GEIGY CORP, 79- *Mem:* Soc Toxicol; Am Chem Soc; Entom Soc Am; Sigma Xi. *Res:* Insecticide chemistry; toxicology; mode of action of pesticides; environmental impact; metabolism in plants and animals; governmental regulation of pesticides. *Mailing Add:* Ciba-Geisy Corp PO Box 18300 Greensboro NC 27419

CAMP, MARK JEFFREY, b Toledo, Ohio, Dec 19, 47; m 85; c 1. PALEOECOLOGY, MALACOLOGY. *Educ:* Univ Toledo, BSc, 70, MSc, 72; Ohio State Univ, PhD(geol), 74. *Prof Exp:* Asst prof geol, Earlham Col, 74-76; ASSOC PROF GEOL, UNIV TOLEDO, 76- *Mem:* Nat Asn Geol Teachers. *Res:* Comprehensive study of Pleistocene lacustrine deposits in Ohio, Michigan and Indiana, with emphasis on ecologic relationships of fauna and flora, distribution of species, glacial history, stratigraphy and origin of sediments. *Mailing Add:* Dept Geol Univ Toledo Toledo OH 43606

CAMP, PAMELA JEAN, b Little Rock, Ark, Nov 24, 54. CARBOHYDRATE METABOLISM, ZEIN DEGRADATION. *Educ:* Hendrix Col, BA, 76; Univ Ark, MS, 78; NC State Univ, PhD(bot), 82. *Prof Exp:* Teaching fel, Univ Mo, 82-84 & Univ Dayton, 84-85; ASST PROF BIOL & BIOCHEM, DAVIDSON COL, 85- *Concurrent Pos:* Vis lectr, Univ NC, 82. *Mem:* Am Soc Plant Physiologists; Am Inst Biol Sci. *Res:* Zein degradation in corn seeds and corn proteases. *Mailing Add:* Dept Biol Davidson Col PO Box 1719 Davidson NC 28036

CAMP, PAUL R, b Middletown, Conn, Dec 29, 19; m 58; c 3. SOLID STATE PHYSICS. *Educ:* Wesleyan Univ, BA, 41; Harvard Univ, MA, 47; Pa State Col, PhD(physics), 51. *Prof Exp:* Physicist, US Naval Res Lab, 41-44; instr, Wesleyan Univ, 47-48; physicist, Res Lab, Radio Corp Am, 51-53; Ford intern physics, Reed Col, 53-54; from asst prof to assoc prof, Polytech Inst Brooklyn, 54-61; chief, Mat Res Br, Cold Regions Res & Eng Lab, 61-63, physicist at large, 63-65; staff physicist, Comn Col Physics, Univ Mich, 65-67; head, dept physics, 67-73, PROF PHYSICS, UNIV MAINE, ORONO, 73- *Concurrent Pos:* Consult, Rand Corp, 56-61. *Mem:* Int Glaciol Soc; Am Phys Soc; Am Asn Physics Teachers; Sigma Xi. *Res:* Physics of ice. *Mailing Add:* Dept Physics Univ Maine Bennett Hall Orono ME 04473

CAMP, RONALD LEE, b Indianapolis, Ind, Apr 16, 44; m 72. PHYSICAL ORGANIC CHEMISTRY. *Educ:* Univ Mich, BS, 66; Mass Inst Technol, PhD(org chem), 71. *Prof Exp:* Res chemist polymer chem, Union Carbide Corp, 70-73; RES SUPVR ORG CHEM RES & DEVELOP, BASF WYANDOTTE CORP, 73- *Mem:* Am Chem Soc. *Res:* Synthesis of organic pesticides; process research and development for agricultural chemicals; polymer synthesis; surfactant synthesis. *Mailing Add:* PO Box 444 Gross Ile MI 48138-1849

CAMP, RONNIE WAYNE, b Toccoa, Ga, Nov 25, 43; m 65; c 2. GAS ADSORPTION & DESORPTION ISOTHERMS OF POROUS SOLIDS, PARTICLE SIZE & DENSITY OF POROUS SOLIDS. *Educ:* Ga Inst Technol, BEE, 65, MSEE, 68. *Prof Exp:* Sr res engr, Ga Tech Res Inst, 65-70; DIR DEVELOP, MICROMERITICS INSTRUMENT CORP, 70- *Mem:* Inst Elec & Electronics Engrs; Am Soc Testing & Mat. *Res:* Improved instrumental methods for gas adsorption isotherm measurement; particle size measurement; density of solids; pore size distribution determination in solids; electronic systems of instruments and sensors. *Mailing Add:* Micromeritics Instrument Corp One Micromeritics Dr Norcross GA 30093

CAMP, RUSSELL R, b Corning, NY, Mar 26, 41; m 65. PLANT PATHOLOGY, ELECTRON MICROSCOPY. *Educ:* Baldwin-Wallace Col, BS, 64; Miami Univ, MA, 66; Univ Wis, PhD(bot), 70. *Prof Exp:* Asst prof, 70-75, Prof Biol, GORDON COL, 75- *Mem:* AAAS; Am Inst Biol Sci; Mycol Soc Am; Bot Soc Am; Am Phytopath Soc. *Res:* Electron microscopy of host parasite relations of fungal pathogens and their respective plant host. *Mailing Add:* Gordon Col Dept Biol 255 Grapevine Rd Wenham MA 01984

CAMPAGNONI, ANTHONY THOMAS, b Boston, Mass, May 9, 41; m 68. MOLECULAR BIOLOGY. *Educ:* Northeastern Univ, AB, 64; Ind Univ, PhD(chem), 68. *Prof Exp:* Res assoc develop biol, Inst Cancer Res, 68-70; from asst prof to prof biochem, Univ Md, Col Park, 70-85; PROF NEUROSCI, UNIV CALIF, LOS ANGELES, SCH MED, 85- *Concurrent Pos:* Neurol study sect, NIH, 82-85 & 85-86; prog chair, Am Soc Neurochem, 84-85; coun, Am Soc Neurochem, 87-91; chief ed, Develop Neurosci, 87-; dep chief ed, J Neurochem, 90- *Honors & Awards:* Jacob Javits award, Nat Inst Neurol & Commun Dis & Stroke, 85. *Mem:* Am Soc Neurochem; Am Soc Biochem & Molecular Biol; Soc Neurosci; Int Soc Neurochem. *Res:* Genetic and epigenetic factors regulating the expression of nervous system-specific genes. *Mailing Add:* Mental Retardation Res Ctr NPI Rm 47-448 UCLA Ctr Health Sci Los Angeles CA 90024

CAMPAIGNE, ERNEST EDWIN, b Chicago, Ill, Feb 13, 14; m 41; c 3. MEDICINAL CHEMISTRY. *Educ:* Northwestern Univ, BS, 36, MS, 38, PhD(biochem), 40. *Prof Exp:* Lab asst, Northwestern Univ, 36-40, res fel chem, 41-42; instr, Bowdoin Col, 40-41; instr prev med & pub health, Sch Med, Univ Tex, 42-43; from instr to prof, 43-79, EMER PROF CHEM, IND UNIV, BLOOMINGTON, 79- *Concurrent Pos:* Assoc biochemist, M D Anderson Hosp Cancer Res, Univ Tex, 42-43; with Comt Med Res, 44; res partic, Oak Ridge Inst Nuclear Studies, 55; consult, Off Surgeon Gen, Walter Reed Army Inst Res, 59-62; vis lectr, Univ Calif, San Francisco, 62; consult, NIH, 60-64, Am Chem Soc, mem Coun Pub Comt, 60-62 & Policy Comt, 62-64, chmn, Med Div, 68; chmn med comn, Int Union Pure & Appl Chem, 69-73; consult, Drug Develop Comt, Div Cancer Treatment, Dept Health Educ & Welfare, 75-79. *Mem:* AAAS; Am Chem Soc; fel NY Acad Sci; Chem Soc. *Res:* Heterocyclic chemistry; central nervous system agents; antimetabolites; physiologically active and organic sulfur compounds. *Mailing Add:* Dept Chem Ind Univ Bloomington IN 47405

CAMPANA, RICHARD JOHN, b Everett, Mass, Dec 5, 18; m 45; c 2. FOREST PATHOLOGY. *Educ:* Univ Idaho, BSF, 43; Yale Univ, MF, 47, PhD(forest path), 52. *Prof Exp:* Instr forestry, Pa State Univ, 47; asst prof bot, NC State Univ, 47-49; asst plant pathologist, USDA, 49-52; from asst plant pathologist to assoc plant pathologist, Ill Nat Hist Surv, 52-58; head dept bot & plant path, 58-68 & 82-83, prof, 58-84, EMER PROF BOT & FOREST PATH, UNIV MAINE, 85- *Concurrent Pos:* Consult, numerous munic, parks, bot gardens & arboreta, 57-; vis prof, Col Environ Sci & Forestry, State Univ NY, 67; guest botanist, Brookhaven Nat Labs, 67-68; ed, Phytopath News, 70-76; chmn, Orono Conserv Comn, 70-78; pres, Maine Conserv Comn, 72-73; pres, Northeastern Div, Am Phytopath Soc, 75. *Mem:* Fel Am Phytopath Soc; Can Phytopath Soc; Int Soc Arborists (pres, 66-67); Sigma Xi; Am Forestry Asn; Am Soc Conserv Arborists; Int Soc Arboriculture; Int Union Forestry Res Orgn. *Res:* Diseases of forest shade and ornamental trees and woody shrubs. *Mailing Add:* Dept Bot Univ Maine Orono ME 04469-0118

CAMPANELLA, SAMUEL JOSEPH, b Washington, DC, Dec 26, 26; m 58; c 1. ELECTRONICS, COMMUNICATIONS. *Educ:* Cath Univ Am, BS, 50, DSc(elec eng), 65; Univ Md, College Park, MS, 57. *Prof Exp:* Electronics scientist sonar, Naval Res Lab, Washington, DC, 50-53; electronics engr commun, MELPAR Inc, Falls Church, 53-57, mgr electronics res lab, 57-64, mgr electronics res & develop ctr, 64-67; mgr signal processing lab, Cosmat Labs, 67-73, dir commun processing labs, 73, exec dir, 73-83, VPRES & CHIEF SCIENTIST, COMSAT LABS, 83- *Concurrent Pos:* Lectr, George Wash Sch Continuing Educ. *Mem:* Fel Inst Elec & Electronics Engrs; fel AAAS; Am Inst Aeronaut & Astronaut; Sigma Xi. *Res:* Advanced communication satellite technology in time division multiple access; frequency division multiple access; digital speech interpolation; source encoding of speech and video; modulation techniques and on-board communications processing. *Mailing Add:* 18917 Whetstone Circle Gaithersburg MD 20879

CAMPARO, JAMES CHARLES, b Newark, NJ, Mar 15, 56; m 79; c 2. ATOMIC & MOLECULAR PHYSICS. *Educ:* Columbia Univ, BA, 77, MA, 78, MPhil & PhD(chem), 81. *Prof Exp:* MEM TECH STAFF, AEROSPACE CORP, 81- *Mem:* Am Phys Soc. *Res:* Atomic physics; optical pumping in atomic clocks. *Mailing Add:* MS M2-253 Aerospace Corp PO Box 92957 Los Angeles CA 90009

CAMPBELL, ADA MARIE, b Jewell, Iowa, Apr 1, 20. FOOD CHEMISTRY. *Educ:* Iowa State Univ, BS, 42, MA, 45; Cornell Univ, PhD(food sci, nutrit), 56. *Prof Exp:* Instr food & nutrit, NDak State Univ, 45-50; asst home economist exp sta, NMex State Univ, 50-52; actg asst prof food & nutrit, Univ Calif, Los Angeles, 52-53; asst foods, Cornell Univ, 53-56; asst prof food & nutrit, Univ Calif, Los Angeles, 56-61; assoc prof, 61-67, prof food sci, Univ Tenn, 67-80; RETIRED. *Mem:* AAAS; Am Asn Cereal Chem; Inst Food Technol; Am Oil Chem Soc. *Res:* Food lipids. *Mailing Add:* 1431 Cherokee Trail-Apt 94 Knoxville TN 37920

CAMPBELL, ALAN, b Alexandria, Egypt, Aug 28, 44; Can & Brit citizen; m 71; c 2. ECOLOGY, ZOOLOGY. *Educ:* McGill Univ, BSc, 67; Univ Man, MSc, 69; Simon Fraser Univ, PhD(biol), 74. *Prof Exp:* Fel stored prod insects, Agr Can, 73-75; res proj dir tick res, Dept Biol, Acadia Univ, 75-78; res scientist lobster biol, 78-88, RES SCIENTIST MARINE INVERTEBRATE FISHERIES, DEPT FISHERIES & OCEANS, CAN, 88- *Mem:* Can Soc Zoologists; Entom Soc Am; Entom Soc Can; Sigma Xi; Crustaceans Soc. *Res:* population biology of marine invertebrates eg abalone, sea urchins, seoducks. *Mailing Add:* Fisheries & Oceans Can Pac Biol Sta Manaimo BC V9R 5U6 Can

CAMPBELL, ALFRED DUNCAN, biochemisty, for more information see previous edition

CAMPBELL, ALICE DEL CAMPILLO, b Santurce, PR, May 30, 28; m 58; c 2. BIOCHEMISTRY. *Educ:* Columbia Univ, AB, 47; NY Univ, MS, 53; Univ Mich, PhD(biochem), 60. *Prof Exp:* Asst biochem, Pub Health Res Inst, NY, 47-48; dept pharmacol, NY Univ, 48-54; instr, Sch Med, Univ PR, 54-56; res assoc biol, Univ Rochester, 60-68; res assoc, 68-77, SR RES ASSOC BIOL SCI, STANFORD UNIV, 77- *Mem:* Am Chem Soc; Sigma Xi; fel Am Inst Chem. *Res:* Enzymology; intermediary metabolism; lipid metabolism; carboxylation reactions; bacterial and physiological genetics; biotin formation in bacteria. *Mailing Add:* Dept Biol Sci Stanford Univ Stanford CA 94305

CAMPBELL, ALLAN BARRIE, b Winnipeg, Man, Mar 28, 23; m 50; c 1. PLANT BREEDING. *Educ:* Univ Man, BSA, 44, MSc, 48; Univ Minn, PhD(genetics, plant breeding), 54. *Prof Exp:* res scientist wheat breeding, Can Dept Agr, 49-88; RETIRED. *Mem:* Royal Soc Can. *Res:* Breeding improved varieties of common wheat with special emphasis on resistance to stem rust and leaf rust. *Mailing Add:* 492 McNaughton Ave Winnipeg MB R3L 1S4 Can

CAMPBELL, ALLAN MCCULLOCH, b Berkeley, Calif, Apr 27, 29; m 58; c 2. MICROBIAL GENETICS. *Educ:* Univ Calif, BS, 50; Univ Ill, MS, 51, PhD(bact), 53. *Hon Degrees:* DSc, Univ Chicago, 78 & Univ Rochester, 81. *Prof Exp:* Instr bact, Sch Med, Univ Mich, 53-57; res assoc genetics, Carnegie Inst, 57-58; from asst prof to prof biol, Univ Rochester, 58-68; PROF BIOL, STANFORD UNIV, 68- *Concurrent Pos:* Fel, Nat Found Inst Pasteur, 58-59; USPHS res career award, 62-68; lectr, Found Microbiol ,70-71, 87-88; assoc ed, Virology, 63-69, Ann Rev Genetics, 69-84, ed, 84-, Gene, 80-; mem, Genetics Panel, Nat Sci Found, 73-76; Spec ed, Evolution, 85- *Honors & Awards:* Riecher Lectr, Univ Ariz, 87. *Mem:* Nat Acad Sci; fel AAAS; Am Soc Microbiol; Am Soc Nat; Am Soc Virologists; Genetics Soc Am. *Res:* Genetics of bacteriophage; lysogeny; biochemical genetics. *Mailing Add:* Dept Biol Sci Stanford Univ Stanford CA 94305

CAMPBELL, ARTHUR B, b Ann Arbor, Mich, Sept 3, 43; m 71; c 1. RADIATION EFFECTS, SINGLE EVENT UPSETS. *Educ:* Union Col NY, BS, 65; Univ Del, MS, 67, PhD(physics), 71. *Prof Exp:* Fel, Univ Salford, Eng, 71-72 & McMaster Univ, Can, 72-75; res physicist, Bur Mines, US Govt, 75-79; RES PHYSICIST, NAVAL RES LAB, 79- *Mem:* Inst Elec & Electronic Engrs. *Res:* Radiation effects in materials including semiconductor devices and metals; ion implantation and single event upsets in semiconductor memories. *Mailing Add:* Code 6613 Naval Res Lab Washington DC 20375-5000

CAMPBELL, BENEDICT JAMES, b Philadelphia, Pa, Oct 17, 27; m 79; c 2. BIOCHEMISTRY. *Educ:* Franklin & Marshall Col, BS, 51; Northwestern Univ, PhD(biochem), 56. *Prof Exp:* Res biochemist, Glidden Co, 55-58 & Armour Co, 58-60; assoc prof, 60-67, chmn dept, 73-77, PROF BIOCHEM, SCH MED, UNIV MO-COLUMBIA, 67- *Mem:* AAAS; Am Chem Soc; NY Acad Sci; Am Soc Biol Chemists; Biophys Soc. *Res:* Biochemical research; graduate education. *Mailing Add:* Dept Biochem Univ Mo Med Ctr Columbia MO 65201

CAMPBELL, BERNERD EUGENE, b Wooster, Ohio, Nov 1, 38; m 66; c 3. HIGH ENERGY LASERS. *Educ:* Otterbein Col, BS, 61. *Prof Exp:* Researcher physics, US Army Missile Command, Redstone Arsenal, 61-63; RES SCIENTIST LASER TECHNOL, BATTELLE COLUMBUS, BATTELLE MEM INST, 64- *Mem:* Sigma Xi; Am Optical Soc. *Res:* High energy laser device technology and laser applications; co-inventer 3 US patents; laser exafs; directing radiation; suppressing radiation. *Mailing Add:* Battelle Columbus 505 King Ave Columbus OH 43201

CAMPBELL, BONITA JEAN, b Annapolis, Md, Sept 15, 43; div; c 1. INDUSTRIAL ENGINEERING, MANAGEMENT SCIENCE. *Educ:* Colo State Univ, BS, 67; Univ Redlands, MS, 73; Pepperdine Univ, MBA, 73; Univ Calif, Los Angeles, PhD(opers res), 79. *Prof Exp:* Assoc engr, Robinson Eng, 65-68; indust engr, Kaiser Steel Corp, 68-72; acct mgr, Info Systs Div, Gen Elec, 72-73; asst prof quant methods, Pepperdine Univ, 73-75; from asst prof to assoc prof eng, 75-82, dir, Women's Eng Prog, 75-82, prog eng & dept chmn, 82-87, ASSOC VPRES, CALIF STATE UNIV, NORTHRIDGE, 87- *Concurrent Pos:* Consult statistician, Thomas J Dudley & Assoc, 73-75 & L E Chuba Res Assoc, 75-76; adj prof, Pepperdine Univ, 75-77; proj dir, NSF grants, 77-83; mem tech staff & proj engr, Aerospace Corp, 77-87. *Honors & Awards:* Am Soc Eng Educ Award, 79. *Mem:* Am Inst Indust Engrs; Soc Women Engrs; Asn Women Sci; fel Inst Advan Eng. *Res:* Time series analysis, particularly parameter estimation; statistical quality control, particularly sampling designs and plans. *Mailing Add:* Acad Affairs Calif State Univ Northridge CA 91330

CAMPBELL, BONNALIE OETTING, b Springfield, Mo, Aug 21, 33; m 60; c 2. PHYSIOLOGY, ENDOCRINOLOGY. *Educ:* Southwest Mo State Univ, AB, 55; Northwestern Univ, MS, 58, PhD(physiol), 64. *Prof Exp:* Instr biochem, Univ Houston, 66-71, instr physiol, 71-73, ASST PROF PHYSIOL, BAYLOR COL MED, 73- *Concurrent Pos:* NASA grants, 70-; consult endocrinol, Vet Admin Hosp, Houston, 73- *Mem:* AAAS; Am Soc Zoologists; assoc Am Physiol Soc. *Res:* Adrenocorticotropic hormone; circadian rhythms. *Mailing Add:* 2628 Fernwood Houston TX 77005

CAMPBELL, BRUCE (NELSON), JR, b Northampton, Mass, Apr 21, 31; m 56; c 4. BIOCHEMISTRY, ORGANIC CHEMISTRY. *Educ:* Williams Col, AB, 52; Univ Conn, PhD(chem), 58. *Prof Exp:* Asst instr chem, Univ Conn, 55-56; from asst prof to prof chem, MacMurray Col, 57-73, chmn dept chem, 72-73; chmn dept, 73-78, PROF CHEM STATE UNIV NY POTSDAM, 73- *Concurrent Pos:* NSF sci fac fel, Mich State Univ, 65-66; vis prof, Dept Biochem, Univ Ill, Urbana, 73, vis res prof, 80-81; vis prof, Cornell Univ, 87-88. *Mem:* Sigma Xi; AAAS; Am Chem Soc. *Res:* Enzyme isolation and reactions; micelle catalysis; theoretical and synthetic organic chemistry; innovation and computers in chemical education. *Mailing Add:* Chemistry Dept State Univ New York Potsdam NY 13676

CAMPBELL, BRUCE CARLETON, b Trenton, NJ, May 15, 49; m; c 2. HOST-PLANT RESISTANCE, INSECT-PLANT INTERACTIONS. *Educ:* Rutgers Univ, BA, 71, MS, 76; Univ Calif, Davis, PhD(entomol), 80. *Prof Exp:* RES ENTOMOLOGIST, USDA, 80- *Concurrent Pos:* Vis scientist, Dept Bact, Univ Calif, Davis, 86-87. *Mem:* Entomol Soc Am; Phytochem Soc NAm; AAAS. *Res:* Host-plant resistance with respect to the effects of plant natural products on insect development, metabolism and feeding behavior. *Mailing Add:* USDA-WRRC 800 Buchanan St Berkeley CA 94710

CAMPBELL, BRUCE HENRY, b Madison, SDak, Oct 27, 40; m 63; c 4. ANALYTICAL CHEMISTRY. *Educ:* Univ Kans, BS, 62; Univ SDak, MA, 65; Univ Tex, PhD(chem), 68. *Prof Exp:* From asst prof to assoc prof chem, Univ Southern Miss, 67-72; fel, Clarkson Col Technol, 72-74; mgr res anal serv, J T Baker Chem Co, 74-75, sr res chemist, 75-81; sr res chemist, 81-85, group leader, 85-87, PRIN RES SCIENTIST, AM CYANAMID CO, STAMFORD, CONN, 87- *Concurrent Pos:* Ed, Critical Rev Analytical Chem, 75-84. *Mem:* Am Chem Soc; Electrochem Soc; Chem Notation Asn. *Res:* Applied and theoretical electrochemistry; general analytical chemistry. *Mailing Add:* Am Cyanamid Co 1937 W Main Stamford CT 06904

CAMPBELL, C(OLIN) K(YDD), b St Andrews, Scotland, May 3, 27; Can & British citizen; m 54; c 3. ELECTRICAL ENGINEERING, MATERIALS SCIENCE. *Educ:* Univ St Andrews, BSc Hons, 52, PhD(physics), 60; Mass Inst Technol, SM, 53; Univ Dundee, DSc (eng & appl sci), 84. *Prof Exp:* Commun engr, Diplomatic Wireless Serv, Eng, 46-47 & Foreign Off, 47; commun engr, Brit Embassy, Washington, DC, & Brit deleg to UN, NY, 47-48; design engr, Atomic Instrument Co, Mass, 54-57; proj engr, A Kusko Inc, 57; from asst prof to prof elec eng, 60-80, chmn dept, 65-69, prof, 60-89, EMER PROF ELEC & COMPUTER ENG, MCMASTER UNIV, 89- *Concurrent Pos:* Mem, Brit deleg, Meeting of Nobel Physics Prizewinners, Bavaria, 59; vis res prof, Univ BC, 70; exchange scientist, Canada/USSR Acad Sci, Kiev, Vilnius, 76. *Honors & Awards:* Eadie Medal, Royal Soc Can, 83. *Mem:* Fel Royal Soc Can; Sigma Xi; fel Eng Inst Can; fel Royal Soc Arts; fel Inst Elec & Electronics Engrs. *Res:* Thin film physics; synthesis of thin film electronic filter devices; ceramics, power electronics; low temperature physics and superconductivity; masers and lasers; surface acoustic wave devices; author of book on signal processing using surface acoustic wave devices. *Mailing Add:* Dept Elec & Comput Eng McMaster Univ 1280 Main St W Hamilton ON L8S 4L7 Can

CAMPBELL, CARL WALTER, b Decatur, Ill, Jan 10, 29; m 51; c 5. HORTICULTRE, PLANT PHYSIOLOGY. *Educ:* Ill State Norm Univ, BSEd, 51; Kans State Col, MS, 52; Purdue Univ, PhD(plant physiol), 57. *Prof Exp:* Plant physiologist, Agr Mkt Serv, USDA, 57-60; asst horticulturist, 60-66, assoc horticulturist, 66-70, prof, 70-88, EMER PROF HORT, TROP RES & EDUC CTR, UNIV FLA, 88- *Mem:* Int Soc Hort Sci; Int Am Soc Trop Hort; Am Soc Hort Sci. *Res:* Plant growth regulators; post-harvest physiology of fruits; horticultural and physiological aspects of selection, propagation and production of tropical and subtropical fruit crops. *Mailing Add:* Trop Res & Educ Ctr Univ Fla 18905 SW 280th St Homestead FL 33031-3314

CAMPBELL, CARLOS BOYD GODFREY, b Chicago, Ill, July 27, 34; m 79; c 5. BIOEFFECTS, BIOELECTROMAGNETICS. *Educ:* Univ Ill, BS, 55, MS, 57, MD, 63, PhD(anat), 65. *Prof Exp:* Surgical intern, Presby-St Lukes Hosp, Chicago, Ill, 63-64; neuroanatomist, Walter Reed Army Inst Res, Washington, DC, 64-67; from asst prof to assoc prof anat & physiol, Ctr Neural Sci, Ind Univ, Bloomington, 67-74; resident physician, Dept Neurol, Los Angeles County-Univ Southern Calif Med Ctr, 73-74; assoc clin prof anat, Calif Col Med, Univ Calif, Irvine, 75-77, resident physician, dept radiol sci, 75-77; vis assoc biol, Calif Inst Technol, 76-77; prof anat & head dept, Sch Med, Univ PR, 77-79; RES NEUROLOGIST, WALTER REED ARMY INST RES, 79- *Concurrent Pos:* Adj prof psychol, Univ Md, 90-; lectr, Univ Md, dept psychol, 82-90, res prof neurol, Uniformed Serv Univ Health Sci, 83-; adj prof anat, Georgetown Univ, 80-90. *Mem:* Am Asn Anat; Am Soc Zoologists; Am Primatol Soc; Soc Neurosci; Int Primatol Soc; Bioelectromagnetics Soc. *Res:* Nervous system bioeffects of non-ionizing radiation; comparative neuroanatomy; systematic zoology; primate nervous systems; primate evolution. *Mailing Add:* Div Neuropsychiat Walter Reed Army Inst Res Washington DC 20307-5100

CAMPBELL, CATHERINE CHASE, b New York, NY, July 1, 05; m 30; c 1. PALEONTOLOGY, GEOMORPHOLOGY. *Educ:* Oberlin Col, BA & MA, 27; Radcliffe Col, MA, 30, PhD(micropaleont), 32. *Prof Exp:* Instr geol, Mt Holyoke Col, 27-29; tech writer meteorol, Long Range Weather Forecasting Unit, US Army Air Force, 43-45; tech ed rocket proj, Calif Inst Technol, 45-46; underwater ord, US Naval Ord Test Sta, 47-51, supvr pub ed, 51-61; geologist, US Geol Surv, 61-90; RETIRED. *Mem:* Fel Geol Soc Am; Geosci Info Soc; Asn Earth Sci Ed; Soc Tech Commun. *Res:* Environmental geology. *Mailing Add:* 906 Comstock 1333 Jones St San Francisco CA 94109

CAMPBELL, CATHY, statistics, for more information see previous edition

CAMPBELL, CHARLES EDGAR, b Kalamazoo, Mich, Oct 29, 21; m 55; c 4. SOFTWARE SYSTEMS, OPTICS. *Educ:* Univ Mich, BSE, 44, MS, 47. *Prof Exp:* Process engr, Tenn Eastman Corp, Eastman Kodak, 44-46; asst prof physics, Nebr State Col, Chadron, 50-51; res engr, Sci Lab, Ford Motor Co, 51-59; res physicist, Cornell Aeronaut Lab, 59-70; prin engr, 70-80, SR ASSOC PROGRAMMER, LOCKHEED ENG & MGT SERV CO, 80- *Mem:* Fel Optical Soc Am; Soc Photo-Optical Instrumentation Engrs; Am Inst Aeronaut & Astronaut; Am Soc Photogram. *Res:* Validity of Selwyn's equation for image resolution; advantages of using polarization in remote sensing; simulation of the shuttle's dynamics. *Mailing Add:* 1638 Seagate Houston TX 77062

CAMPBELL, CHARLES EDWIN, b Columbus, Ohio, Dec 11, 42; m 65; c 2. MANY-BODY THEORY, QUANTUM FLUIDS. *Educ:* Ohio State Univ, BS, 64; Washington Univ, St Louis, PhD(physics), 69. *Prof Exp:* Res assoc physics, Univ Wash, 69-71 & Stanford Univ, 71-73; from asst prof to assoc prof, 73-81, dir of grad studies, 82-83, head, 83-86, PROF PHYSICS, UNIV MINN, MINNEAPOLIS, 81-; actg dep dir, Theoret Physics Inst, 87-88; CONSULT, 89- *Concurrent Pos:* Humboldt fel, Inst Theoret Physics, Univ Cologne, 81-82; AWV Sabbatical fel, Los Alamos Nat Lab, 89-90. *Mem:* Am Phys Soc. *Res:* Many-body theory; theory of quantum fluids; physical absorption; structure of classical fluids; electronic structure of solids. *Mailing Add:* Sch Physics & Astron Univ Minn 116 Church St Minneapolis MN 55455

CAMPBELL, CHARLES J, b Nanton, Alta, Can, Nov 25, 15; US citizen. FISH BIOLOGY, FISHERIES MANAGEMENT. *Educ:* Wash State Univ, BS, 38. *Prof Exp:* Jr fishery biologist, US Forest Serv, 39-40; from field biologist to chief basin invests, Fishery Div, Ore Wildlife Comn, 41-59, chief fisheries, US Bur Land Mgt, 59-75, asst chief fisheries, Ore Dept Fish & Wildlife, 75-78, environ specialist, 78-79; environ scientist, VTN Oregon, Inc, 79-82. *Concurrent Pos:* Consult, Campbell Craven Environ, 82- *Mem:* Am Fisheries Soc (pres, 73); Am Inst Fisheries Res Biol; Wildlife Soc. *Res:* Fishery biology, management and administration; ecology of fish. *Mailing Add:* 921 SW Cheltenham St Portland OR 97201

CAMPBELL, CHARLOTTE CATHERINE, b Winchester, Va, Dec 4, 14. MEDICAL MYCOLOGY. *Educ:* George Wash Univ, BS, 51; Am Bd Med Microbiol, dipl. *Hon Degrees:* DSc, Lowell Tech Inst, 72 & Shenandoah Univ, 91. *Prof Exp:* Technician, Dept Bact, Walter Reed Army Inst Res, 41-43, bacteriologist, 43-46, med mycologist, 46-49, chief, Mycol Sect, 49-62; from assoc prof to prof med mycol, Sch Pub Health, Harvard Univ, 62-73; prof med mycol & chmn dept med sci, Sch Med, Southern Ill Univ, 73-77; co-prin investr, US/USSR Prog Microbiol, Am Soc Microbiol, 77-80; RETIRED. *Concurrent Pos:* Consult, Vet Admin-US Armed Forces Comn Histoplasmosis & Coccidioidomycosis, 54-62 & Mid Am Res Unit, CZ, 57-62; assoc ed, Sabouraudia; assoc in med, Peter Bent Brigham Hosp, 63-73; mem sci adv bd, Gorgas Mem Inst; mem panel rev skin test antigens, Food & Drug Admin; vis prof, Latin Am Prog, Am Soc Microbiol, 79-84. *Honors & Awards:* Rhoda Benham Medallion for Meritorious Contrib Med Mycol, 78; Isham Award, 79; Lucille K Georg Medallion for Distinguished Contrib Med Mycology. *Mem:* Fel AAAS; Am Pub Health Asn; Am Thoracic Soc; Am Acad Microbiol; Med Mycol Soc of the Americas (pres, 70); hon mem Am Soc Microbiol. *Res:* Antigenic analysis of systemic mycotic agents; epidemiology and ecology of histoplasmosis; serologic diagnosis and chemotherapy of systemic mycoses. *Mailing Add:* 120 Pembroke St Boston MA 02118

CAMPBELL, CLARENCE L, JR, b Indianapolis, Ind, Sept 24, 21; m. VETERINARY MEDICINE. *Educ:* Ohio State Univ, DVM, 45. *Prof Exp:* Vet pvt pract, Ill, 45; field vet, Fla Livestock Sanit Bd, Fla Dept Agr, 45-48, asst state vet, 48-52, state vet & secy, 53-61, STATE VET & DIR DIV ANIMAL INDUST, FLA DEPT AGR & CONSUMER SERV, 61- *Concurrent Pos:* Pres, Nat Assembly State Vet, 56-57. *Honors & Awards:* Meritorious Serv Award, USDA, 62; Cert Serv Award, Am Vet Med Asn, 68. *Mem:* US Animal Health Asn (pres, 65-66); Am Vet Med Asn; Am Asn Equine Practitioners. *Res:* Regulatory veterinary medicine. *Mailing Add:* Fla Dept Agr & Consumer Serv Rm 328 Mayo Bldg Tallahassee FL 32399-0800

CAMPBELL, CLEMENT, JR, b Milton, Pa, Oct 22, 30; m 66; c 1. CHEMISTRY. *Educ:* Bucknell Univ, BS, 51. *Prof Exp:* Res chemist, Pyrotechnics Lab, 51-74 & Explosives Div, 74-77, RES CHEMIST, ENERGETIC MAT DIV, US ARMY ARMAMENT RES, DEVELOP & ENG CTR, PICATINNY ARSENAL, NJ, 77- *Mem:* AAAS; Am Chem Soc; Sigma Xi; NAm Thermal Anal Soc. *Res:* Pre-ignition and ignition reactions of explosive, propellant and pyrotechnic materials; reactions of powdered metals; decomposition of inorganic oxidants; thermoanalysis; gas chromatography/mass spectrometry applied to energetic materials; characterization of environmental pollutants from ammunition plants. *Mailing Add:* 29 Cory Rd Flanders NJ 07836

CAMPBELL, CLYDE DEL, b Wheeling, WVa, Apr 1, 30; m 56; c 1. ORGANIC CHEMISTRY, BIOCHEMISTRY. *Educ:* West Liberty State Col, AB & BS, 53; NC State Col, MS, 55; WVa Univ, PhD(biochem, org chem), 58. *Prof Exp:* Asst biol chem, NC State Col, 53-55 & WVa Univ, 55-58; instr chem, West Liberty State Col, 58-61; sr res chemist, Mobay Chem Co, WVa, 61-68; chmn div sci & math, 68-70, assoc acad dean, 70-73, dean admin, 73-, PRES, WEST LIBERTY STATE COL. *Mem:* AAAS; Am Chem Soc. *Res:* Isolation and identification of chlorophyll and carotenoid pigments; nitrogen metabolism in ruminants; function of vitamin B12 in nitrogen metabolism; carbohydrate analysis of foodstuffs; polyurethanes; isocyanates; plastics and synthetic resins; intermediary metabolism. *Mailing Add:* Liberty Oaks West Liberty State Col West Liberty WV 26074

CAMPBELL, COLIN, b Washington, DC, June 24, 27; m 52; c 3. OBSTETRICS & GYNECOLOGY, MEDICAL ADMINISTRATION. *Educ:* Stanford Univ, AB, 49; Temple Univ, EdM, 67; McGill Univ, MD, CM, 53. *Prof Exp:* Instr obstet & gynec, Temple Univ, 61-64; from asst prof to prof obstet & gynec, Univ Mich, Ann Arbor, 64-78, from asst dean to assoc dean, 72-78; prof & dean, obstet & gynec, Sch Primary Med Care, Univ Ala, Huntsville, 78-83; PROF OBSTET & GYNEC, PRES & DEAN, COL MED, NORTHEASTERN OHIO UNIV COL MED, ROOTSTOWN, 83- *Concurrent Pos:* Consult, Wayne County Gen Hosp, 65-78; examr, Am Bd Obstet & Gynec, 69-76. *Mem:* Fel Am Col Obstet & Gynec; Asn Am Med Cols. *Res:* Erythroblastosis fetalis; medical education. *Mailing Add:* NEOUCOM PO Box 95 Rootstown OH 44272

CAMPBELL, DAN NORVELL, b Mt Enterprise, Tex, Sept 10, 28; m 51; c 2. ANALYTICAL CHEMISTRY, HEALTH SCIENCE. *Educ:* Southern Methodist Univ, BS, 52, MS, 53; Am Acad Indust Hyg, cert, 81. *Prof Exp:* From res chemist to sr res chemist, 55-75, PROCESS SPECIALIST, MONSANTO CO, 70-, INDUST HYG ANALYSIS SPECIALIST, 76- *Mem:* Am Chem Soc; Am Ind Hyg Asn. *Res:* Chromatographic separations in analytical chemistry involving both gas and liquid chromatography. *Mailing Add:* RR 1 Box 218XX Mt Enterprise TX 75681-9799

CAMPBELL, DAVID KELLY, b Long Beach, Calif, July 23, 44; m 67; c 2. NONLINEAR PHENOMENA, QUANTUM & CLASSICAL. *Educ:* Harvard Col, AB, 66; Cambridge Univ, PhD(theoret physics & applied math), 70. *Prof Exp:* Res assoc theoret high energy physics, Univ Ill, 70-72; mem, Inst Advan Study, Princeton, 72-74; staff mem, Theoret Div, 74-85, DIR, CTR NONLINEAR STUDIES, LOS ALAMOS NAT LAB, 85- *Concurrent Pos:* Fel, Ctr Advan Study, Univ Ill, 70-72; Oppenheimer fel, Los Alamos Sci Lab, 74-77; distinguished lectr, Dept Energy-AWU, 90. *Mem:* Fel Am Phys Soc; fel AAAS. *Res:* Nonlinear phenomena in field theory and condensed matter; synthetic metals including conducting polymers and change transfer salts; novel condensed matter materials, including high temperature superconductors. *Mailing Add:* Ctr Nonlinear Studies MS B258 Los Alamos Nat Lab Los Alamos NM 87545

CAMPBELL, DAVID OWEN, b Merriam, Kans, Nov 11, 27; m 54; c 3. RADIOCHEMISTRY, INORGANIC CHEMISTRY. *Educ:* Univ Kansas City, BA, 47; Ill Inst Technol, PhD(chem), 53. *Prof Exp:* CHEMIST, OAK RIDGE NAT LAB, 53- *Honors & Awards:* Special Award, Am Nuclear Soc, 81. *Mem:* Am Chem Soc; Am Nuclear Soc. *Res:* Decontamination of nuclear equipment; molten salt and fluoride volatility fuel processing; transuranium element chemistry and chemistry of protactinium; nuclear reactor fuel reprocessing; radioactive waste treatment, nuclear reactor safety. *Mailing Add:* 102 Windham Rd Oak Ridge TN 37830

CAMPBELL, DAVID PAUL, b Seattle, Wash, May 12, 44; m 69; c 1. GENETICS. *Educ:* Western Wash State Col, BA, 67; Wash State Univ, PhD(genetics), 76. *Prof Exp:* Asst prof genetics, 75-, ASSOC PROF BIOL SCI, CALIF STATE POLYTECH UNIV, POMONA. *Mem:* AAAS; Am Soc Plant Physiologists; Int Plant Tissue Cult Asn. *Res:* Ultrastructural changes in germinating seed tissues of the Solanaceae. *Mailing Add:* 733 Camphor Way Upland CA 91786

CAMPBELL, DONALD BRUCE, b New South Wales, Australia, 42. PLANETARY SCIENCES. *Educ:* Univ Sydney, BS, 63, MS, 65; Cornell Univ, PhD(astron), 71. *Prof Exp:* Res assoc astron, Cornell Univ, 71-73; mem staff, Haystack Observ, Northeast Radio Observ Corp, 73-74; from res assoc to sr res addoc, 74-79,from assoc dir to dir, Arecibo Observ, 79-87, PROF ASTRON, CORNELL UNIV, 88-; CONSULT, 87- *Mem:* Am Astron Soc; Am Geophys Union; Int Astron Union. *Res:* Investigation of solar system bodies with emphasis on their radar scattering properties. *Mailing Add:* 426 Space Sci Bldg Cornell Univ Ithaca NY 14853

CAMPBELL, DONALD EDWARD, characterization of glasses & ceramics; deceased, see previous edition for last biography

CAMPBELL, DONALD L, b Waverly, Iowa, July 16, 40. INORGANIC CHEMISTRY. *Educ:* Iowa State Univ, BS, 62; Univ Ill, Urbana, PhD(inorg chem), 69. *Prof Exp:* Chemist, Liquid Carbonic Div, Gen Dynamics Corp, 62-65; asst prof, 69-78, ASSOC PROF CHEM, UNIV WIS-EAU CLAIRE, 78- *Mem:* Am Chem Soc; AAAS. *Res:* Coordination chemistry of the lanthanide ions; thermodynamic parameters for complex formation and factors effecting complex stability; heavy metals in the environment. *Mailing Add:* Dept Chem Univ Wis Eau Claire WI 54701

CAMPBELL, DONALD R, b Youngstown, Ohio, Oct 12, 30; m 56; c 2. ANALYTICAL CHEMISTRY, RADIOCHEMISTRY. *Educ:* Univ Akron, BS, 53. *Prof Exp:* Jr res chemist, 53-59; sr res chemist, 60-68, res scientist, 69-71, GROUP LEADER, GENCORP INC, 71- *Mem:* Am Chem Soc: Am Inst Chemists. *Res:* Analytical chemistry of high polymers; application of radiochemical techniques to elucidation of the structure and composition of polymers and to the determination of functional groups in polymers; synthesis of isotopically-substituted compounds. *Mailing Add:* 275 N Portage Path No 8E Akron OH 44329

CAMPBELL, DOUGLAS ARTHUR, molecular genetics, for more information see previous edition

CAMPBELL, DOUGLAS MICHAEL, b San Pedro, Calif, May 4, 43; m 66; c 4. MATHEMATICAL ANALYSIS, COMPUTER SCIENCE. *Educ:* Harvard Univ, BS, 67; Univ NC, Chapel Hill, PhD(math), 71. *Prof Exp:* From asst prof to assoc prof, 71-79, PROF MATH, BRIGHAM YOUNG UNIV, 79- *Concurrent Pos:* Russian translr, Am Math Soc, 71-; vis prof, Univ Md, 73, Univ Mich, 78 & Univ Utah, 85; reviewer, Math Rev, 74-, Comput Rev, 83; assoc ed, Math Magazine, 85-; nat lectr, Sigma Xi, 85- *Honors & Awards:* Maeser Award, 81; Fulbright Award, India, 83. *Mem:* Am Math Soc; Sigma Xi; Math Asn Am; Asn Comput Mach. *Res:* Geometric function theory; grammar checkers. *Mailing Add:* Comput Sci Dept Brigham Young Univ Provo UT 84602

CAMPBELL, EARL WILLIAM, b Bowling Green, Ohio. HEMATOLOGY. *Educ:* Harvard Col, BA, 58; Univ Rochester, MD, 62. *Prof Exp:* USPHS fel, Sch Med, Univ Utah, 67-68; instr med, WVa Univ, 68-69, asst prof, Med Ctr, 69-70; pvt pract, Md, 70-73; asst prof, Mich State Univ, 73-74, assoc prof med, 74-78; PROF MED, MED COL OHIO, 78- *Concurrent Pos:* Investr, Acute Leukemia Group B, 68-72, Nat Polycythemia Rubra Vera Group, 73- & Mich Hemophilia Home Care Proj, 74-; consult hemat, Baltimore Cancer Res Inst, Nat Cancer Inst, 71-72; dir, Hemophilia Clin, Lansing Area, 74- *Mem:* AMA; fel Am Col Physicians; Am Soc Hemat. *Res:* Clinical investigation of polycythemia and related myeloproliferative diseases with the Polycythemia Rubra Vera Group; projects on antithrombin III, platelet factor IV. *Mailing Add:* Dept Med Med Col Ohio CS PO Box 10008 Toledo OH 43699

CAMPBELL, EDWARD CHARLES, physics; deceased, see previous edition for last biography

CAMPBELL, EDWARD J, MARITIME DESIGN. *Educ:* Northwestern Univ, BS & MBA. *Hon Degrees:* Two honorary degrees. *Prof Exp:* PRES & CHIEF EXEC OFFICER, NEWPORT NEWS SHIPBUILDING, 79- *Concurrent Pos:* Nat chmn, Shipbuilders Coun Am; mem, Defense Policy Adv Comt on Trade; mem, Com Adv Comt, Nat Oceanic & Atmospheric Admin; bd managers, Am Bur Shipping. *Honors & Awards:* Adm Chester Nimitz Award, US Navy League. *Mem:* Nat Acad Eng; Soc Naval Architects & Marine Engrs. *Res:* Designing, building and servicing nuclear powered ships and conventionally powered ships. *Mailing Add:* Off Pres Newport News Shipbuilding 4101 Washington Ave Newport News VA 23607

CAMPBELL, EDWARD J MORAN, b Yorkshire, Eng, Aug 31, 25; Can citizen; m 54; c 4. GENERAL PHYSIOLOGY. *Educ:* London Univ, BSc, 64, MD, 51, PhD(sci), 54 & FRCP, 85. *Prof Exp:* Asst prof med, Middlesex Hosp London, 55-61; sr lectr med, Royal Postgrad Med Sch, 61-68; PROF MED, MCMASTER UNIV, 68- *Concurrent Pos:* John Hopkins Scholar, 70. *Mem:* Am Physiol Soc; Am Thoracic Soc; Asn Am Physician; fel Royal Soc Can. *Res:* Respiratory muscles; mechanism of dyspnea; controlled oxygen therapy; respiratory disturbances of acid and base; response of breaking to mechanical loads. *Mailing Add:* Dept Med Rm 3N51B McMaster Univ Med Ctr Box 2000 Sta A Hamilton ON L8N 3Z5 Can

CAMPBELL, EDWIN STEWART, b Ada, Ohio, Aug 18, 26; m 49. CHEMICAL PHYSICS. *Educ:* Johns Hopkins Univ, AB, 45; Univ Mich, MSc, 48; Univ Calif, PhD, 51. *Prof Exp:* Instr quant & qual anal, St Martin's Col, 48; asst, Univ Calif, 48-51; lectr & fel, Univ Southern Calif, 51-52; proj assoc, Naval Res Lab, Univ Wis, 52-55; asst prof chem, 55-61, ASSOC PROF CHEM, NY UNIV, 62- *Mem:* AAAS; Am Phys Soc; Fedn Am Scientists; NY Acad Sci. *Res:* Energetics of hydrogen bonding, its implications for structure and properties; structure of liquid and solid water; algorithms for nonadditive and multiple energies. *Mailing Add:* Dept Chem NY Univ New York NY 10003

CAMPBELL, FERRELL RULON, b Afton, Wyo, Nov 14, 37; m 70; c 1. ANATOMY. *Educ:* Utah State Univ, BS, 60, MS, 63; Univ Chicago, PhD(anat), 66. *Prof Exp:* From instr to asst prof anat, Stanford Univ, 66-73; ASSOC PROF ANAT, UNIV LOUISVILLE, 73- *Mem:* Am Asn Anat. *Res:* Hematology. *Mailing Add:* Dept Anat Sch Med Univ Louisville Louisville KY 40292

CAMPBELL, FINLEY ALEXANDER, b Kenora, Ont, Jan 5, 27; m 53; c 3. GEOLOGY. *Educ:* Univ Man, BSc, 50; Queen's Univ, Ont, MA, 56; Princeton Univ, PhD(geol), 58. *Prof Exp:* Geologist, Prospectors Airways Co, Ltd, 50-56; explor geologist, Mining Corp Can, 56-58; asst prof geol, Univ Alta, 58-62, assoc prof, 62-65; head dept, 65-69, vpres, Capital Resources, 69-71, acad vpres, 71-76, prof geol, 76-84, vpres Priorities & Planning, 84-88, EMER PROF GEOL, UNIV CALGARY, 88- *Concurrent Pos:* Vpres, Capital Resources, Univ Calgary, 69-71. *Honors & Awards:* Queen's Jubilee Medal. *Mem:* Royal Soc Can; Geol Asn Can; Mineral Asn Can; Mineral Soc Am; Can Soc Petrol Geologists; Soc Econ Geologists; Can Inst Mining & Metall. *Res:* Economic geology; mineralogy; petrology; geochemistry. *Mailing Add:* 3408 Benton Dr NW Calgary AB T2L 1W8 Can

CAMPBELL, FRANCIS JAMES, b Toledo, Ohio, July 29, 24; m 48; c 7. RADIATION CURING, RADIATION DAMAGE. *Educ:* Univ Toledo, BS, 48. *Prof Exp:* HEAD DIELEC EFFECTS SECT, RADIATION SCI DIV, US NAVAL RES LAB, 58- *Concurrent Pos:* US del, Int Electrotechnol Comn, 64-86; chmn awards comt, Dielec & Elec Insulation Soc, Inst Elec & Electronic Engrs. *Honors & Awards:* Res Publ Award, NRL Dirs, 82. *Mem:* Sigma Xi; Am Chem Soc; fel Inst Elec & Electronic Engrs; Am Soc Testing & Mat; fel Am Inst Chemists. *Res:* Radiation curing adhesives and composites; safety and endurance of electrical insulation under combined environments and stresses-aerospace vehicle wiring, utility power cable, nuclear power plants. *Mailing Add:* Div 4654 Naval Res Lab Washington DC 20375

CAMPBELL, GARY HOMER, IMMUNOSEROLOTY, MONOCLONAL ANTIBODIES. *Educ:* Univ Okla, PhD(microbiol & immunol), 72. *Prof Exp:* RES MICROBIOLOGIST, MALARIA BR, DIV PARASITIC DIS, CTR DIS CONTROL, 82- *Res:* Immunology of malaria parasites. *Mailing Add:* Ctr Dis Control 1600 Clifton Rd Bldg 22 C Atlanta GA 30333

CAMPBELL, GARY THOMAS, b Granite City, Ill, Nov 11, 46; m 68; c 2. NEUROENDOCRINOLOGY. *Educ:* Wash Univ, BS, 68; Northwestern Univ, PhD(biol sci), 72. *Prof Exp:* Res biologist, Monsanto Chem Co, 68; res assoc biol sci, Northwestern Univ, 69, teaching asst, 70-72; instr physiol, Med Col Va, 72-73, asst prof, 73-75; asst prof physiol, Univ Nebr Med Ctr, Omaha, 75-76; mem fac, Dept Physiol, Med Col Va Health Sci Ctr, 76-; assoc prof, Dept Physiol, Univ Nebr Med Ctr; PROF, DEPT ANAT, MED SCH, UNIV SC. *Mem:* AAAS; Sigma Xi; Am Soc Zoologists. *Res:* Description of the cytoarchitecture in brain-pituitary control systems using immunohistochemical methods. *Mailing Add:* Dept Anat Med Sch Univ SC Garners Ferry Rd Columbia SC 29208

CAMPBELL, GAYLON SANFORD, b Blackfoot, Idaho, Aug 20, 40; m 64; c 9. AGRICULTURAL METEOROLOGY, SOIL PHYSICS. *Educ:* Utah State Univ, BS, 65, MS, 66; Wash State Univ, PhD(soils), 68. *Prof Exp:* Captain, US Army Atmospheric Sci Lab, 68-70; from asst prof to assoc prof biophys & soils, 70-80, PROF SOILS, WASH STATE UNIV, 80- *Concurrent Pos:* Consult, Pac Northwest Labs, Battelle Mem Inst, 73-; sr vis fel, Sci Res Coun, Eng, 77-78; vis prof, Univ Nottingham, England, 84-85. *Mem:* Am Meteorol Soc; fel Soil Sci Soc Am; fel Am Soc Agron. *Res:* Evapotranspiration, plant and soil water measurement, computer models of evaporation, transpiration and water uptake by plants; energy budgets of plants and animals. *Mailing Add:* Dept Agron & Soils Wash State Univ Pullman WA 99164-6420

CAMPBELL, GEORGE MELVIN, b Prospect, Pa, May 14, 29; m 59; c 2. PHYSICAL CHEMISTRY, ANALYTICAL CHEMISTRY. *Educ:* Hiram Col, BA, 54; Vanderbilt Univ, MS, 56, PhD(chem), 63. *Prof Exp:* Mem health physics staff, Argonne Nat Lab, 56-58; instr pub health, Univ Minn, 58-60; MEM CHEM STAFF, LOS ALAMOS SCI LAB, UNIV CALIF, 63- *Mem:* Am Chem Soc; fel Am Inst Chemists; Sigma Xi; AAAS. *Res:* Use of super fluorinating agents in actinide chemistry; thermodynamic properties of plutonium compounds; photochemistry of plutonium compounds. *Mailing Add:* 2800 Arizona Los Alamos NM 87544

CAMPBELL, GEORGE S(TUART), b Sauquoit, NY, Nov 29, 26; m 51; c 2. NAVAL ARCHITECTURE. *Educ:* Rensselaer Polytech Inst, BS, 47, BAE, 49; Calif Inst Technol, MS, 51, PhD(aeronaut), 56. *Prof Exp:* Aeronaut res scientist, Nat Adv Comt Aeronaut, 47-53; sr scientist, Hughes Aircraft Co, 54-63; head dept aerospace eng, Univ Conn, 63-71, prof, 63-88, emer prof, 88-; RETIRED. *Mem:* Am Inst Aeronaut & Astronaut; Soc Comput Simulation; Soc Naval Architects & Marine Engrs. *Res:* Aerodynamics; ship dynamics. *Mailing Add:* 73 Hillyndale Rd Storrs CT 06278

CAMPBELL, GEORGE WASHINGTON, JR, b Loma Linda, Calif, Sept 22, 19; m 42; c 4. INDUSTRIAL CHEMISTRY. *Educ:* Univ Calif, Los Angeles, AB, 42; Univ Southern Calif, MS, 47, PhD, 51. *Prof Exp:* Asst, Manhattan Proj, Univ Chicago, 42-43; rubber compounder, Goodyear Tire & Rubber Co, Calif, 43-45; lab assoc chem, Univ Southern Calif, 45-47, asst, 47-51, res assoc, Off Naval Research Proj, 51-52; assoc prof chem, Univ Houston, 53-57, chief investr, Off Ord Res Proj, 52-57; sr res chemist, US Borax Res Corp, 57-63, res supvr, 63-71, mgr, Pilot Plant Res Dept, 71-77, mgr process res, 77-84; RETIRED. *Concurrent Pos:* From instr to prof, George Pepperdine Col, 45-53, head dept, 45-53. *Mem:* Am Chem Soc; fel Am Inst Chemists; Am Soc Testing & Mat; Nat Asn Corrosion Eng. *Res:* Boron chemistry; corrosion; process development. *Mailing Add:* PO Box 2000-47 Mission Viejo CA 92690-2000

CAMPBELL, GERALD ALLAN, b Cincinnati, Ohio, May 30, 46; m 67. ORGANIC POLYMER CHEMISTRY. *Educ:* Univ Cincinnati, BS, 67; Ohio State Univ, PhD(org chem), 71. *Prof Exp:* sr res chemist, 71-77, LAB HEAD POLYMER CHEM, EASTMAN KODAK CO, 77- *Mem:* Am Chem Soc. *Res:* Synthesis of monomers and their subsequent polymerizations to materials of interest in photographic science. *Mailing Add:* 699 Hightower Way Webster NY 14580-2511

CAMPBELL, GILBERT SADLER, b Toronto, Ont, Jan 4, 24; US citizen; m 47, 61; c 6. SURGERY. *Educ:* Univ Va, BA, 43, MD, 46; Univ Minn, MS, 49, PhD(surg), 54. *Prof Exp:* Intern, Univ Minn Hosp, 46, asst physiol, Med Sch, 47-48, instr, 48-49, chief resident surg, Hosp, 53-54, from instr to asst prof, 54-58; prof & chief thoracic surg, Med Ctr Okla & chief surgeon, Vet Admin Hosp, 58-65; prof surg & head dept med ctr, Univ Ark, Little Rock, 65-83; RETIRED. *Concurrent Pos:* Markle scholar, Univ Minn, 54-58. *Honors & Awards:* Horsley Prize, 54. *Mem:* Soc Univ Surg; Am Asn Thoracic Surg; Soc Thoracic Surgeons; Am Physiol Soc; Am Surg Asn. *Res:* Cardiovascular surgery; pulmonary physiology. *Mailing Add:* 66 River Ridge Rd Little Rock AR 72207

CAMPBELL, GRAHAM HAYS, b Houston, Tex, Aug 17, 36; m 60; c 2. COMPUTER SCIENCES. *Educ:* Rice Univ, BA, 57; Yale Univ, MS, 58; Univ Calif, Berkeley, PhD(physics), 65. *Prof Exp:* From asst physicist to assoc physicist appl math, 66-69, COMPUTER SCIENTIST, BROOKHAVEN NAT LAB, 69- *Res:* Computer networks; symbolic algebraic manipulation by computers; operating systems theory. *Mailing Add:* Dept Appl Math Brookhaven Nat Lab Upton NY 11973

CAMPBELL, GRAHAM LE MESURIER, b London, Eng, Nov 25, 41; US citizen; m 68; c 1. DEVELOPMENTAL CYTOLOGY, NEUROIMMUNOLOGY. *Educ:* Cambridge Univ, BA, 64; Univ Pa, PhD(biol), 70. *Prof Exp:* Res asst mammalian cytogenetics, Med Res Coun, Harwell, UK, 64-66; res assoc develop cytol, Wistar Inst, 71-76; DIR VIROL IMMUNOL & BIOL, CANCER INFO DISSEMINATION & ANAL CTR, FRANKLIN INST, 76- *Res:* Functional analysis of purified populations isolated from central nervous systems and neuro systems. *Mailing Add:* 8203 St Martins Lane Philadelphia PA 19118

CAMPBELL, GREGORY AUGUST, b Providence, RI, Sept 18, 41; m 62; c 2. POLYMER EXTRUDER ANALYSIS, MICROCELLULAR FOAM NUCLEATION. *Educ:* Univ Maine, BS, 64, MS, 66, PhD(chem eng), 69. *Prof Exp:* Res engr, Gen Motors Res Labs, 68-70, sr res engr, 70-74, staff res engr, 74-78, sr staff res engr, 78-81; supvr, Polymer Fabrication, Mobil Chem Res & Develop, 81-84; ASSOC PROF CHEM ENG, CLARKSON UNIV, 84- *Concurrent Pos:* Consult, Mobil Chem Corp & Eastman Kodak, 86-, Mobay, 88-90 & Becton Dickinson & Co Res Ctr, 90- *Mem:* Am Inst Chem Engrs; Soc Plastics Engrs; Int Polymer Processing Soc; AAAS. *Res:* Interaction of polymer process parameters and material science on the end use properties of polymer materials; new methods to monitor and control the property development. *Mailing Add:* Chem Eng Dept Clarkson Univ Potsdam NY 13699

CAMPBELL, HALLOCK COWLES, b Cortland, NY, June 4, 10; m 36, 63; c 2. CHEMISTRY. *Educ:* Hamilton Col, BS, 32; Harvard Univ, AM, 34, PhD(chem), 36. *Prof Exp:* Master in chg chem, Browne & Nichols Sch, Mass, 35-38; instr, Queens Col, NY, 38-43; res chemist, Arcos Corp, 43-45, assoc dir res & eng, 45-56, dir res, 56-68, dir res & technol, 68-73; mgr educ, Am Welding Soc, 73-76; CONSULT WELDING EDUC, 76- *Concurrent Pos:* Asst lectr, Eve Tech Sch, Temple Univ, 45-55, co-dir metall, 55-61, dir, 61-73. *Honors & Awards:* Elihu Root Fel Science, 32. *Mem:* AAAS; Am Chem Soc; Am Welding Soc; fel Am Soc Metals; fel Am Inst Chemists. *Res:* Kinetics of thermal explosions; arc welding electrodes; properties of low alloy and high alloy weld metal; thermal explosion of ethylazide gas. *Mailing Add:* 746 Fiddlewood Rd Vero Beach FL 32963

CAMPBELL, HAROLD ALEXANDER, b Zion, Ill, June 27, 09; m 38; c 3. BIOCHEMISTRY. *Educ:* Univ Ill, BS, 35; Univ Wis, PhD(agr chem), 39. *Prof Exp:* Asst, Exp Sta, Univ Wis, 35-39; res chemist & mgr, Cent Labs, Gen Foods Corp, 39-61; vis investr, Walker Lab, Sloan-Kettering Inst Cancer Res, 62-69; RES ASSOC ONCOL, MCARDLE LAB CANCER RES, UNIV WIS-MADISON, 69- *Mem:* Am Chem Soc. *Res:* Isolation of dicumarol hemorrhagic agent in sweet clover disease and its use for the treatment of humans as an anticoagulant; the relationship of L-asparaginase enzyme activity to the antileukemia activity of certain preparations including guinea pig serum; isolation and assay of biological materials of medical significance. *Mailing Add:* 5113 St Cyr Middleton WI 53562

CAMPBELL, IAIN MALCOLM, b Glasgow, Scotland, June 15, 40; Brit citizen; m 73; c 1. BIOCHEMISTRY, ANALYTICAL CHEMISTRY. *Educ:* Glasgow Univ, BS, 62, PhD(chem), 65. *Prof Exp:* Vis asst prof, 67-69, asst prof, 69-74, ASSOC PROF BIOCHEM, UNIV PITTSBURGH, 74- *Concurrent Pos:* Merck Found fac develop award, 70-71. *Mem:* The Chem Soc; Am Chem Soc; Am Soc Mass Spectrometry; Am Oil Chemist's Soc; AAAS. *Res:* Control and biological function of plant and fungal secondary metabolism; natural product biosynthesis; biomass conversion; mode of drug action; radio gas chromatography; gas chromatography/mass spectrometry; liquid chromatography/mass spectrometry. *Mailing Add:* 773 Fruithurst Dr Pittsburgh PA 15228-2535

CAMPBELL, IAN HENRY, geology, for more information see previous edition

CAMPBELL, JACK JAMES RAMSAY, b Vancouver, BC, Mar 29, 18; m 42; c 4. BACTERIOLOGY. *Educ:* Univ BC, BSA, 39; Cornell Univ, PhD(bact), 44. *Prof Exp:* Asst bact, Cent Exp Farm, Ottawa, 39-40; asst, Queen's Univ (Ont), 44-46; from assoc prof to prof dairying, 46-65, prof & head dept, 65-82, EMER PROF MICROBIOL, UNIV BC, 83- *Concurrent Pos:* Vis lectr, Johns Hopkins Univ, 52; res assoc, Univ Ill, 54. *Honors & Awards:* Can Soc Microbiol Award, 66; Centennial Medal, Govt Can, 67; Harrison Prize, Royal Soc Can, 69. *Mem:* AAAS; Am Soc Microbiol; Can Soc Microbiol (pres, 74-75); Royal Soc Can. *Res:* Bacterial physiology. *Mailing Add:* Dept Microbiol Univ BC Vancouver BC V6T 1W5 Can

CAMPBELL, JAMES, b Glasgow, Scotland, July 18, 07; m 54; c 4. PHYSIOLOGY, BIOCHEMISTRY. *Educ:* Univ Toronto, BA, 30, MA, 32, PhD, 38. *Prof Exp:* Res assoc physiol, McGill Univ, 32-33; res assoc, Sch Hyg, 33-40, from asst prof to assoc prof, 40-58, prof physiol, 58-72, spec lectr, 72-77, EMER PROF, UNIV TORONTO, 77- *Mem:* Am Physiol Soc; Am Chem Soc; Am Diabetes Asn; Can Biochem Soc; Brit Biochem Soc. *Res:* Proteolytic enzymes; digestion; nutrition; hormonal control of metabolism; growth hormone; insulin. *Mailing Add:* 54 Summerhill Gardens Toronto ON M4T 1B4 Can

CAMPBELL, JAMES A, b Chipley, Fla, Apr 12, 28; m 55; c 5. BIOLOGY, ECOLOGY. *Educ:* Fla Agr & Mech Univ, BS, 51, MEd, 56; Pa State Univ, DEd(biol sci), 62; Vanderbilt Univ, MDiv, 72. *Prof Exp:* Sci teacher, Roulhac High Sch, 51-56; div chmn biol, Rosenwald Jr Col, 58-64; PROF, TENN STATE UNIV, 64-; DIR, NATURAL SCI & MATH DEPT, AM BAPTIST COL, 65- *Mem:* AAAS; Nat Asn Biol Teachers. *Res:* Fresh water ecology; influence of selected abiotic factors on population density; environmental stress. *Mailing Add:* 2707 Bronte Ave Nashville TN 37216-3705

CAMPBELL, JAMES ALEXANDER, b Guelph, Ont, Oct 10, 13; m 39; c 2. FOOD SCIENCE, NUTRITION. *Educ:* Univ Toronto, BSA, 36; McGill Univ, MSc, 38, PhD(agr chem), 47. *Prof Exp:* Agr asst animal nutrit, Chem Div, Sci Serv, Can Dept Agr, Ottawa, 38-41, chemist, Vitamin & Physiol Res Lab, 41-48; chief, Vitamins & Nutrit Lab, Dept Nat Health & Welfare, 48-62, dir res labs, 63-67, sr sci adv foods, 67-71, actg dir, Nutrit Bur, Food & Drug Directorate, 71-73, dep dir, Caribbean Food & Nutrit Inst, 73-76; NUTRIT CONSULT, 76- *Concurrent Pos:* Past vpres & dir, Prof Inst Pub Serv Can, 52-56; consult, Protein Adv Group, WHO-Food & Agr Orgn-UNICEF, 61-62; vis prof food technol & nutrit, Am Univ Beirut, 62-63; treas, Int Union Nutrit Sci, 75-85; Int Network Food Data Systs, Infood, 83- *Honors & Awards:* Harvey W Wiley Award, Asn Off Analytical Chem, 66; Earle Willard McHenry Award, Can Soc Nutrit Sci, 82. *Mem:* Fel Am Inst Nutrit; Animal Nutrit Res Coun; fel Chem Inst Can; Nutrit Soc Can (past pres). *Res:* Chemical, microbiological and biological assays for vitamins; studies with rats; protein evaluation; physiological availability of drugs and vitamins; evaluation of drugs in oral prolonged action dosage forms; food chemistry; nutrition surveys; food and nutrition policy; diet in celiac disease. *Mailing Add:* 1785 Riverside Dr Suite 2204 Ottawa ON K1G 3T7 Can

CAMPBELL, JAMES ARTHUR, physical chemistry; deceased, see previous edition for last biography

CAMPBELL, JAMES B, b Fraserburgh, Scotland, Sept 16, 39; m 66; c 1. MEDICAL MICROBIOLOGY. *Educ:* Aberdeen Univ, BSc, 62; Univ Alta, PhD(biochem), 65. *Prof Exp:* Asst biochem, Univ Alta, 62-65, fel, 65-66; trainee virol, Wistar Inst, 66-67, res asst, 67-68; from asst prof to assoc prof virol, 68-82, PROF MICROBIOL, UNIV TORONTO, 82- *Concurrent Pos:* Spec lectr, Sch Hyg, Univ Toronto, 67-68; consult virol & biochem & lab dir, Extendicare Diag Serv, Div Extendicare, Can Ltd, Willowdale, Ont, 71-81. *Mem:* Am Soc Microbiol; Can Soc Microbiol; Can Col Microbiol; NY Acad Sci. *Res:* Pathogenesis and biochemistry of animal viruses; serodiagnosis of virus infections; immunization of wildlife against rabies; adenovirus-vectored recombinant vaccines. *Mailing Add:* Dept Microbiol Fac Med Univ Toronto Toronto ON M5S 1A1 Can

CAMPBELL, JAMES EDWARD, b Bay City, Mich, July 16, 17; m 42; c 2. COMPUTER SCIENCE, THEORETICAL CHEMISTRY. *Educ:* Univ Minn, BChE, 40. *Prof Exp:* Lab metallurgist, Saginaw Malleable Iron Div, Gen Motors Corp, 40-41, metallurgist, Res Labs, 41-45; sr metallurgist, Battelle Mem Inst, 45-75; RETIRED. *Concurrent Pos:* Consult metall info, Battelle Mem Inst, 75-77; consult ed, Am Soc Metals, 78-79; secy, Comt E24, Am Soc Testing & Mat. *Honors & Awards:* Award of Merit, Am Soc Testing & Mat, 74. *Mem:* Fel Am Soc Testing & Mat; Am Soc Metals Int; Am Chem Soc. *Res:* Metallurgical research on high strength materials including steels, aluminum alloys, and titanium alloys; fracture toughness of high strength materials; properties of materials at cryogenic temperatures; heat treatments for high strength and toughness; theory for interatomic bonding in molecules and crystal lattices. *Mailing Add:* 3520 S Hamilton Rd Columbus OH 43232-5637

CAMPBELL, JAMES FRANKLIN, b Meridian, Miss, Oct 25, 41; m; c 2. AERONAUTICAL ENGINEERING. *Educ:* Miss State Univ, BS, 63; Va Polytech Inst & State Univ, MS, 68, PhD(aerospace eng), 73. *Prof Exp:* Aerospace engr, 63-72, aeronaut res engr, 72-75, head, appl aerodyn group, 75-81, asst head, Nat Transonic Facil, Aerodyn Br, Langley Res Ctr, 81-84, SR AEROSPACE ENGR, NASA, 84- *Concurrent Pos:* Panel mem, fluid dynamics, Adv Group Aerospace Res & Dev, 87-; sr aerospace eng, High Reynolds Number Aerody Br, 84- *Honors & Awards:* NASA Lifting Body Res Award, 70 & Viking Team Award, 77; Space Shuttle Flag Plague Award, Am Inst Aeronaut & Astronaut, 84; NASA Group Achievement Award, NASA, 82. *Mem:* Assoc fel Am Inst Aeronaut & Astronaut. *Res:* Fluid and flight mechanics; aerodynamics. *Mailing Add:* Mail Stop 294 NASA Langley Res Ctr Hampton VA 23665

CAMPBELL, JAMES FULTON, b Philadelphia, Pa, Nov 24, 32; m 55, 87; c 3. HEALTH SCIENCES, SCIENCE WRITING. *Educ:* Rutgers Univ, BSc, 55; Univ Va, MA, 60, PhD(psychol), 64. *Prof Exp:* Res assoc autonomic psychophysiol, Fels Res Inst, 62-64, USPHS fel psychophysiol, 64-65; asst prof, Antioch Col, 63-65; res assoc psychol, McGill Univ, 65-68, asst prof, 68-73; ASSOC PROF PSYCHOL, CARLETON UNIV, 73- *Concurrent Pos:* Proj officer, Ont Educ Commun Authority, 75-79. *Mem:* Am Asn Study Headache; Int Headache Soc. *Res:* Management of psychological factors in rehabilitation and pain; psychological factors in rehabilitation. *Mailing Add:* Dept Psychol Carleton Univ Colonel By Dr Ottawa ON K1S 5B6 Can

CAMPBELL, JAMES L, b Los Angeles, Calif, Sept 28, 24; m 60; c 3. INVERTEBRATE ZOOLOGY. *Educ:* Univ Calif, Berkeley, AB, 49, MA, 51; Univ Calif, Los Angeles, PhD(zool), 67. *Prof Exp:* Assoc prof, 55-77, PROF BIOL, LOS ANGELES VALLEY COL, 77- *Mem:* AAAS. *Res:* Malacology, study of digestive system of Haliotis cracherodii; echinoderm biology, study of echinoid hoemol system; freshwater primary productivity; marine invertebrates. *Mailing Add:* 5704 Columbus Ave Van Nuys CA 91411

CAMPBELL, JAMES NICOLL, b St Thomas, Ont, June 15, 30; m 54; c 4. MICROBIOLOGY, BIOCHEMISTRY. *Educ:* Univ Western Ont, BA, 51; Univ BC, BA, 55, MSc, 57; Univ Chicago, PhD(microbiol), 60. *Prof Exp:* From asst prof to assoc prof, 60-69, PROF MICROBIOL, UNIV ALTA, 69- *Mem:* Am Soc Microbiol; Can Biochem Soc. *Res:* Interaction of bacteria with plant and animal host species; isolation and characterization of bacterial enzymes. *Mailing Add:* Dept Microbiol Univ Alta Edmonton AB T6G 2M7 Can

CAMPBELL, JAMES STEWART, b Bear River, NS, June 10, 23; m 47; c 5. PATHOLOGY. *Educ:* Dalhousie Univ, BSc, 43, MD, 47; Am Bd Path, dipl, 52; Royal Col Path, fel, 71; FRCP(C). *Prof Exp:* Asst path, Med Sch, Tufts Univ, 49-52, instr, 52-53; from asst prof to prof path, 53-74, prof & head dept, 79-81, CLIN PROF PATH, UNIV OTTAWA, 81- *Concurrent Pos:* From jr asst pathologist to asst pathologist, New Eng Med Ctr, 49-53; registr, Registry for Tissue Reactions to Drugs, Ottawa, 70-74; consult pathologist, Can Tumor Ref Ctr, 58-79 & Nat Defence Med Ctr, Ottawa, 61-87; pathologist, Ottowa Gen Hosp, 53-74, head, Dept Lab Med, 79-81; prof path & chmn dept, Mem Univ Nfld, 74-79; chmn dept lab med, Gen Hosp St John's, 76-79; head path sect, Toxicol Res Div, Health & Welfare Can, 81; vis prof path, Mem Univ Newfoundland, 87. *Mem:* Int Acad Path; Can Med Asn; Can Asn Path; fel Am Col Path; Am Assoc Path. *Res:* Toxicologic and gynecologic pathology; Reproductive failure and endometriosis related to environmental immunotoxicants using PCB-treated rhesus model; gastric carcinoma in Newfoundland and Labrador: high incidence reflects environments and traditional diets; IUD's can precipitate ovarian pregnancy. *Mailing Add:* Path Sect Toxicol Res Div Sir Frederick Banting Res Ctr Health & Welfare Can Tunney's Pasture Ottawa ON K1A 0L2 Can

CAMPBELL, JAMES WAYNE, b Highlandville, Mo, Mar 2, 32; div; c 2. COMPARATIVE BIOCHEMISTRY. *Educ:* Southwest Mo State Col, BS, 53; Univ Ill, MS, 55; Univ Okla, PhD(zool), 58. *Prof Exp:* Asst zool, Univ Ill, 53-55 & Univ Okla, 55-56; Nat Acad Sci-Nat Res Coun fel, Johns Hopkins Univ, 58-59; from instr to assoc prof biol, 59-70, chmn biol dept, 74-78, PROF BIOCHEM CELL BIOL, RICE UNIV, 70- *Concurrent Pos:* NIH spec fel, Univ Wis, 64-65; USPHS career develop award, 66-70; consult prev med, NASA Manned Spacecraft Ctr, Houston, 69-71; consult, BMS, NSF, 72-73, & 74-75, DAR, NSF, 78-79, Presidential Young Investr Awards Comt, 83-85, Exp Prog Stimulate Competitive Res, 80, 86 & DMB, NSF, 85; prog dir, regulatory biol, NSF, 73-74, div dir, physiol, cellular & molecular biol, 79-81, NSF Acad Facil Moderization, Phase I & II, 90; Nat lectr, Sigma Xi, 83-85; mem bd dirs, Am Inst Biol Sci, 89-90. *Mem:* Fel AAAS; Biochem Soc, Eng; Am Physiol Soc; Am Soc Zool; Am Soc Biochemists & Molecular Biologists; Am Soc Cell Biol; Sigma Xi; Am Inst Biol Sci. *Res:* Ammonia detoxication-toxicity. *Mailing Add:* Dept Biochem & Cell Biol Rice Univ PO Box 1892 Houston TX 77251

CAMPBELL, JEPTHA EDWARD, JR, b Atlanta, Ga, Sept 16, 23; m 49; c 3. FOOD SCIENCE. *Educ:* Rollins Col, BS, 47; Univ Wis, MS, 49, PhD(biochem, physiol), 51. *Prof Exp:* Proj assoc biochem, Univ Wis, 51-52; group leader radiobiol, Mound Lab, Monsanto Chem Co, 52-55; chief food chem, USPHS, 55-70; chief microbiol, Biochem Br, Div Microbiol, 70-71, ASST DIR, CINCINNATI FOOD RES LABS, FOOD & DRUG ADMIN, 71- *Honors & Awards:* Serv Award, US Dept Health, Educ & Welfare, 62. *Mem:* Am Chem Soc; Am Inst Nutrit; Am Pub Health Asn; Asn Off Agr Chem. *Res:* Nutrition; radiobiology; food technology; analytical food chemistry. *Mailing Add:* 1618 Dell Terr Cincinnati OH 45230

CAMPBELL, JOHN ALEXANDER, b Detroit, Mich, July 7, 40; m 69. PULMONARY MEDICINE, BIOMEDICAL ENGINEERING. *Educ:* Univ Mich, BSE, 62, MSE, 64, PhD(bioeng), 67; Rush Med Col, MD, 74. *Prof Exp:* Chem engr, process develop dept, Parke, Davis & Co, Mich, 64-65; asst attend biomed engr, Presby-St Luke's Hosp, 67-74; resident internal med, Univ Ill Med Ctr, 74-77; FEL, PULMONARY MED, NORTHWESTERN UNIV MED SCH, 77-, INSTR, 78- *Mem:* Am Inst Chem Engrs; Am Thoracic Soc; fel Am Col Chest Physicians; Sigma Xi. *Res:* Mathematical models of physiological systems; patient monitoring, using computers; artificial organs; exercise physiology; computer applications in clinical medicine and research. *Mailing Add:* 11816 Bevenshire Rd Oklahoma City OK 73162

CAMPBELL, JOHN ARTHUR, b Muskogee, Okla, Nov 2, 30; m 53; c 3. PETROLOGY, SEDIMENTARY ROCKS. *Educ:* Univ Tulsa, BGeol, 55; Univ Colo, MS, 57, PhD(geol), 66. *Prof Exp:* From instr to assoc prof geol, Colo State Univ, 57-74; res geologist, US Geol Surv, 74-81; chmn dept, 86-89, PROF GEOL, FT LEWIS COL, 81-; RES GEOLOGIST, US GEOL SURV, 91- *Concurrent Pos:* Gov, State Oil & Gas Conserv Comn, 91- *Mem:* fel Geol Soc Am; Am Asn Petrol Geol; Soc Econ Paleont & Mineral; Int Asn Sedimentologists. *Res:* Petrology, stratigraphy and depositional environment of paleozoic rocks of the southern Rocky Mountains; uranium in paleozoic rocks of the southwestern United States; paleotectonic influence on sedimentation; synsedimentary tectonics; paleozoic rocks of the Southern Rocky Mountains. *Mailing Add:* 195 Aspen Lane Falls Creek Ranch Durango CO 81301

CAMPBELL, JOHN BRYAN, b Fairmont, Nebr, Mar 30, 33; m 56; c 4. ENTOMOLOGY. *Educ:* Univ Wyo, BS, 61, MS, 63; Kans State Univ, PhD(entom), 66. *Prof Exp:* Entomologist, Agr Res Serv, USDA, Univ Nebr-Lincoln, 66-70; assoc prof, 70-78, PROF ENTOM, UNIV NEBR, NORTH PLATTE STA, 78- *Mem:* Entom Soc Am. *Res:* Biology and ecology of rangeland grasshoppers and livestock insects; biology and control of livestock insects in Central Plains. *Mailing Add:* Univ Nebr West Central Ctr Box 46A RR 4 North Platte NE 69101

CAMPBELL, JOHN C(ARL), b Wilsey, Kans, Apr 20, 20; m 41; c 4. OCCUPATIONAL SAFETY, AGRICULTURAL ENGINEERING. *Educ:* Kans State Univ, BS, 47; Ore State Univ, MS, 49. *Prof Exp:* Exten agr engr, Ore State Univ, 48-54 & Univ Ill, 54-55; assoc prof gen eng, Ore State Univ, 56-65, head dept, 66-72, assoc prof indust eng, 73-77, assoc prof indust & gen eng & dir safety, 78-85; RETIRED. *Mem:* Am Soc Safety Engrs; Am Soc Agr Engrs; Am Soc Eng Educ. *Mailing Add:* 21110 N US Hwy 101 Shelton WA 98584

CAMPBELL, JOHN DUNCAN, b Hamilton, Ont, Can, Apr 22, 23; wid; c 2. PALEOBOTANY. *Educ:* McMaster Univ, BA, 44; Univ BC, MA, 49; McGill Univ, PhD, 52. *Prof Exp:* Res officer, Alta Res Coun, 53-84; PRIN, JAYCON RECONNAISSANCE, 84- *Concurrent Pos:* Assoc acad staff, Forest Sci Dept, Univ Alta, 78-86. *Mem:* Asn Prof Engrs Geol & Geophysics Alta; Can Inst Mining & Metall; Am Soc Testing Mat; Int Asn Plant Taxonomists. *Res:* Coal systematics, distribution and occurrence, paleoecology, wood and modern homologues of Cretaceous and Tertiary continental deposits of Central Canada; coal field sampling; relationship between characteristics of coal as determined in a coal chemical laboratory and suitably corrected for effects of the coal maturation process; the botanical or palaeobotanical make-up of the original coal-forming biotic assemblage. *Mailing Add:* Jaycon Reconnaissance 10908 65A St Edmonton AB T6A 2P7 Can

CAMPBELL, JOHN HOWLAND, b Oklahoma City, Okla, Mar 9, 38; m 62; c 2. ANATOMY. *Educ:* Calif Inst Technol, BA, 60; Harvard Univ, PhD(biol), 64. *Prof Exp:* ASSOC PROF ANAT, SCH MED, UNIV CALIF, LOS ANGELES, 64- *Concurrent Pos:* NSF grants, Pasteur Inst, Paris, France, 64-65 & Commonwealth Sci & Indust Res Orgn, Canberra, Australia, 65-66. *Mem:* AAAS; Am Asn Anat; Soc Math Biol; Am Malacological Soc. *Res:* Genetics; molecular evolution; biological pattern formation; immunology. *Mailing Add:* Dept Anat Sch Med 73-235 Ctr Health Sci Univ Calif 405 Hilgard Ave Los Angeles CA 90024

CAMPBELL, JOHN HYDE, b Ithaca, NY, Dec 2, 47; m 68; c 2. PHYSICAL CHEMISTRY. *Educ:* Rochester Inst Technol, BS, 70; Univ Ill, MS, 72, PhD(phys chem), 75. *Prof Exp:* Chemist, Eastman Kodak Co, 67-70; res chemist energy systs, Lawrence Livermore Lab, Univ Calif, 75-80, proj leader, nuclear waste form develop, 80-82, proj leader, Laser Fusion Target Fabrication, 82-85, PROJ LEADER, LASER MAT DEVELOP, LAWRENCE LIVERMORE LAB, UNIV CALIF, 86- *Mem:* Am Phys Soc; Am Chem Soc; Am Inst Chem Engrs. *Res:* Development of optical materials for high power lasers; investigation of optical damage in high power lasers; development of waste form for disposal of high-level radioactive wastes; operation and analysis of pilot-scale oil shale retorts; investigation of fundamental reaction chemistry of oil shale retorting; development of laser fusion targets. *Mailing Add:* Lawrence Livermore Lab Univ of Calif PO Box 808 L-482 Livermore CA 94550

CAMPBELL, JOHN MORGAN, SR, b Virden, Ill, Mar 24, 22. PETROLEUM ENGINEERING. *Educ:* Iowa State Univ, BS, 43; Univ Okla, MS, 48, PhD, 51. *Prof Exp:* Develop engr & supvr, E I du Pont de Nemours & Co, 43-46; spec instr, Univ Okla, 46-50; mgr Process Equip Div, Black Sivalls & Bryson, Inc, 51-54; prof petrol eng & chmn dept, 54-64, Erle P Halliburton prof, 64, distinguished prof, 64-70; pres, Int Petrol Inst, 70-78; pres, John M Campbell & Co, 78-; RETIRED. *Concurrent Pos:* Pres, John M Campbell & Co, 68-70; consult, over twenty companies & Secy Navy. *Mem:* Nat Acad Eng; Am Inst Chem Eng; Am Chem Soc; Am Inst Mining, Metall & Petrol Engrs. *Res:* PVT behavior of hydrocarbon systems; adsorption; flow of fluids in porous media; engineering economics. *Mailing Add:* John M Campbell & Co 1215 Crossroads Blvd Norman OK 73072

CAMPBELL, JOHN R(OY), b Goodman, Mo, June 14, 33; m 54; c 3. DAIRY SCIENCE, ANIMAL SCIENCE. *Educ:* Univ Mo, BS, 55, MS, 56, PhD(nutrit), 60. *Prof Exp:* From asst prof to prof dairy husb, Univ Mo-Columbia, 70-77; assoc dean agr & prof dairy sci, Univ Ill, Urbana, 77-83, dean Col Agr, 83-88; PRES, OKLA STATE UNIV, 88- *Concurrent Pos:* Southern Ice Cream Mfrs fel, 55-56; Danforth assoc, 70; teaching fel, Nat Asn Col & Teachers Agr, 73. *Honors & Awards:* Award Hon, Am Dairy Sci Asn, 87. *Mem:* AAAS; Am Dairy Sci Asn (pres, 80-81); Am Soc Animal Sci. *Res:* Dairy cattle physiology, especially health, nutrition, production and management; recycling corrugated paper through ruminants in the production of meat and milk. *Mailing Add:* Off Pres Okla State Univ 107 Whitehurst Stillwater OK 74078-0001

CAMPBELL, JOHN RAYMOND, b San Diego, Calif, Feb 6, 44; m 76; c 1. OCEAN ENGINEERING, FLUID MECHANICS. *Educ:* San Diego State Univ, BS, 69; Colo State Univ, PhD(fluid mech), 74. *Prof Exp:* Res assoc, fluid mech & ocean eng, Univ Del, Newark, 74-76; asst prof hydraulic ocean eng, Purdue Univ, 77-78; staff scientist, Los Alamos Sci Lab, 78-81; consult, 80-81, STAFF MEM, FLOW SCI, INC, LOS ALAMOS, NMEX, 81- *Concurrent Pos:* Consult, Midocean Motion Pictures, 79-80, Ft Sill Okla, 80-81, Newark Del Water Comn, 76-77. *Mem:* Inst Elec & Electronics Engrs; Am Meteorol Soc. *Res:* Computer graphics and computer generated movies to visualize physical processes in atmospheric dispersion, superconductors, underground stress field engineering and in oceanographic dispersion modelling. *Mailing Add:* 320 Rim Rd Los Alamos NM 87544

CAMPBELL, JOHN RICHARD, b Pratt, Kans, Jan 16, 32; m 62; c 3. PEDIATRIC SURGERY. *Educ:* Univ Kans, BA, 54, MD, 58; Am Bd Surg, dipl, 64, cert pediat surg, 75, recert, 82. *Prof Exp:* Intern, Univ Pa Hosp, 58-59; surg resident, Med Ctr, Univ Kans, 59-63; asst instr pediat surg, Sch Med, Univ Pa, 65-67; from asst prof to assoc prof, Ore Health Sci Univ, 67-72, actg chmn dept surg, 86-88, secy, Surg Sect, 89-91, PROF SURG & PEDUATM SCH MED, ORE HEALTH SCI UNIV, 72-, CHIEF PEDIAT SURG, 67- *Concurrent Pos:* Surg resident, Children's Hosp Philadelphia, 65-67. *Mem:* Am Pediat Surg Asn; Am Col Surg; Am Acad Pediat; AMA; Pac Asn Pediat Surgeons (pres, 88-89). *Res:* Pediatric surgery and oncology; shock in the newborn infant. *Mailing Add:* Dept Surg Ore Health Sci Univ 3181 SW San Jackson Rd Portland OR 97201

CAMPBELL, JUDITH LYNN, b New Haven, Conn, Mar 24, 43. DNA REPLICATION. *Educ:* Wellesley Col, BA, 65; Harvard Univ, PhD, 74. *Prof Exp:* Res assoc molecular biol, Harvard Med Sch, 74-77; ASSOC PROF CHEM & BIOL, CALIF INST TECHNOL, 77- *Concurrent Pos:* NIH res career develop award, 79-83. *Mem:* Fedn Am Soc Exp Biol. *Res:* Genetic and biochemical studies on the regulation of DNA replication in prokaryotes and eukaryotes; regulation of plasmid replication. *Mailing Add:* Dept Chem & Biol 147-75 Calif Inst Technol Pasadena CA 91125

CAMPBELL, KATHERINE SMITH, b Boston, Mass, Apr 10, 43; m 71; c 1. TIME SERIES. *Educ:* Radcliffe Col, BA, 64; Univ Md, MA, 69; Univ NMex, PhD(math), 79. *Prof Exp:* Asst mem tech staff, Bellcomm, Inc, Wash, DC, 64-66; asst math, Univ Md, 66-68; scientist, EG&G, Los Alamos, NMex, 72-75; STATISTICIAN, LOS ALAMOS NAT LAB, 75- *Concurrent Pos:* Lectr math, Univ NMex, 72. *Mem:* Am Statist Asn; Inst Math Statist. *Res:* Application of statistics in geological exploration; nuclear safeguards and reliability. *Mailing Add:* S-1/MS-600 Los Alamos Nat Lab Los Alamos NM 87545

CAMPBELL, KENNETH B, CARDIAC MECHANICS, HEMODYNAMICS. *Educ:* NIH, PhD(physiol), 78. *Prof Exp:* ASSOC PROF PHYSIOL, WASH STATE UNIV, 76- *Mailing Add:* Dept VCAPP Wash State Univ Pullman WA 99164

CAMPBELL, KENNETH EUGENE, JR, b Jackson, Mich, Nov 4, 43. AVIAN PALEONTOLOGY, HISTORICAL BIOGEOGRAPHY. *Educ:* Univ Mich, BS, 66, MS, 67; Univ Fla, PhD(zool), 73. *Prof Exp:* Asst prof geol & zool, Univ Fla, 75-77; CUR VERT PALEONT, NATURAL HIST MUS, LOS ANGELES, 77- *Mem:* Am Ornith Soc; Wilson Ornith Soc; Cooper Ornith Soc; Soc Vert Paleont; Sigma Xi. *Res:* Avian paleontology and osteology; Cenozoic geology and paleontology of South America. *Mailing Add:* Natural Hist Mus 900 Exposition Blvd Los Angeles CA 90007

CAMPBELL, KENNETH LYLE, b St Paul, Minn, May 3, 48. REPRODUCTIVE ENDOCRINOLOGY, BIOCHEMISTRY. *Educ:* Augsburg Col, BA, 70; Univ Mich, MS, 71, PhD(biochem), 76. *Prof Exp:* Analytical chemist, Fed Water Qual Lab, Duluth, 70; NIH fel ovarian function, Reprod Endocrinol Prog, Dept Path, 76-79, RES INVESTR, REPROD ENDOCRINOL PROG, CTR HUMAN GROWTH & DEVELOP, UNIV MICH, 79- *Concurrent Pos:* Inst for Environ Qual fel, Dept Biochem, Univ Mich, 72-75. *Mem:* Soc Study Reprod. *Res:* Granulosa cell function and structure of the membrana granulosum; biochemical characterization of hormone action and hormone metabolism; toxicological actions of environmental contaminants on reproductive organs. *Mailing Add:* 1128 Michigan Ann Arbor MI 48104-3943

CAMPBELL, KENNETH NIELSEN, medicinal chemistry, research administration; deceased, see previous edition for last biography

CAMPBELL, KENNETH WILFORD, b Ingersoll, Ont, Aug 21, 42; m 65; c 3. RESEARCH MANAGEMENT, BIOTECHNOLOGY. *Educ:* Univ Guelph, BSc, 71, PhD(plant breeding & genetics), 75. *Prof Exp:* res scientist, Agr Can Res Br, 74-81; res mgr, Ciba-Geigy Seeds, 81-84, res mgr, Funk Seeds, 84-85, res mgr, 85-87, RES & DEVELOP MGR, FUNK SEEDS, CIBA-GEIGY, CAN LTD, 87- *Concurrent Pos:* Mem, Barley Subcomt, Can Comt Grain Breeding, 75-84. *Mem:* Can Soc Plant Molecular Biol; Crop Sci Soc Am. *Res:* Breeding superior six-row, and recently two-row barley for the Canadian prairies, with emphasis on malting quality; intercrosses of two-row and six-row types and haploidy; breeding superior barley varieties for Ontario; barley doubled haploid technique; develop new products, techniques for pest and weed control for the Canadian prairies; develop superior corn hybrids for Canada; applications of biotechnology to agricultural industry. *Mailing Add:* Funk Seeds Ciba-Geigy Can Ltd Honeywood Res Sta RR #1 Platteville ON N0J 1S0 Can

CAMPBELL, KEVIN PETER, b Brooklyn, NY, Jan 19, 52; m 74; c 3. CALCIUM TRANSPORT, MUSCLE PHYSIOLOGY. *Educ:* Manhattan Col, BS, 73; Univ Rochester, MS, 76, PhD(biophysics), 79. *Prof Exp:* Fel, Med Res Coun Can, Univ Toronto, 78-81; asst prof, 81-85, assoc prof, 85-88, PROF PHYSIOL, UNIV IOWA, 88- *Concurrent Pos:* Prin investr, Am Heart Asn grant in aid, 82-85. *Mem:* Biophys Soc; NY Acad Sci; AAAS; Sigma Xi. *Res:* Structure and function of the sarcoplasmic reticulum membrane in skeletal and cardiac muscle; molecular mechanisms of calcium movements across excitable membranes. *Mailing Add:* Dept Physiol & Biophys Univ Iowa Iowa City IA 52242

CAMPBELL, KIRBY I, b San Jose, Calif, Jan 12, 33; m 54; c 1. ENVIRONMENTAL HEALTH, TOXICOLOGY. *Educ:* Univ Calif, Davis, BS, 55, DVM, 57; Harvard Univ, MPH, 64. *Prof Exp:* Vet, Butte Vet Hosp, Oroville, Calif, 59-61; proj vet officer, Air Pollution Res Ctr, Univ Calif, Riverside, USPHS, 61-63, actg chief chronic & explor toxicol unit, Health Effects Res Prog, Nat Air Pollution Control Admin, 64-68, dep chief, 68-70, chief vet med, 68-70; chief, Comp & Reprod Toxicity Sect, Toxicol Div, US Environ Protection Agency, Cincinnati, 72-76, chief, Functional Path Br, Lab Studies Div, 76-79, sr toxicologist, Toxicol Div, Health Effects Res Lab, 79-81; RETIRED. *Concurrent Pos:* Mem proj group, Air Pollution Res Adv Comt, Coord Res Coun, 68-72. *Res:* Experimental biology, environmental toxicology, including inhalation; experimental laboratory animal science; mammalian, submammalian, in vivo, in vitro bioassay systems; immune response and defense systems; reproductive function and developmental biology. *Mailing Add:* PO Box 529 Walnut Grove CA 95690

CAMPBELL, LARRY ENOCH, b Brookville, Pa, July 21, 38; m 65; c 3. SOLID STATE PHYSICS. *Educ:* Carnegie Inst Technol, BS, 60, MS, 62, PhD(physics), 66. *Prof Exp:* Appointee physics, Argonne Nat Lab, asst prof, 68-73, ASSOC PROF PHYSICS, HOBART & WILLIAM SMITH COLS, 73- *Concurrent Pos:* Sci asst, Tech Univ Munich, 72-73. *Mem:* Am Phys Soc; Sigma Xi. *Res:* Application of the M-ssbauer effect to solid state and nuclear physics problems. *Mailing Add:* Dept Physics Hobart & William Smith Cols Geneva NY 14456

CAMPBELL, LARRY N, b Oct 6, 46. MATHEMATICS. *Educ:* William Jewell Col, BA, 68; DePauw Univ, MA, 70; Univ Northern Colo, DA, 75. *Prof Exp:* PROF & CHMN DEPT MATH & PHYSICS, SCH OF OZARKS, 75- *Concurrent Pos:* Consult, math dept, Simpson Col, Ind, 88. *Mem:* Math Asn Am. *Mailing Add:* Math/Physics Dept Sch Ozarks Point Lookout MO 65711

CAMPBELL, LAURENCE JOSEPH, b Belmont, WVa, Feb 26, 37; m 71; c 1. LOW TEMPERATURE PHYSICS. *Educ:* Mass Inst Technol, SB & SM, 61; Univ Calif, San Diego, La Jolla, PhD(physics), 65. *Prof Exp:* Res assoc physics, Univ Ill, Urbana, 65-67; STAFF MEM PHYSICS, LOS ALAMOS SCI LAB, 67- *Concurrent Pos:* Vis res scientist, Univ Calif, Berkeley, 79; vis scientist, Acad Sci, USSR, 80-81. *Mem:* Am Phys Soc; AAAS. *Res:* Superfluid hydrodynamics; quantized vortices. *Mailing Add:* LANL B262 Los Alamos NM 87545

CAMPBELL, LINZY LEON, b Panhandle, Tex, Feb 10, 27; m 53. MICROBIOLOGY, BIOCHEMISTRY. *Educ:* Univ Tex, BA, 49, MS, 50, PhD(bact, biochem), 52. *Prof Exp:* Res scientist I marine microbiol, Dept Bact, Univ Tex, 47-50, res scientist II food bact, 50-51; Nat Microbiol Inst res fel plant biochem, Univ Calif, 52-54; from asst prof & asst bacteriologist to assoc prof hort & assoc bacteriologist, State Col Wash, 54-59; assoc prof microbiol, Western Reserve Univ, 59-62; prof microbiol, Univ Ill, Urbana-Champaign, 62-72, dir sch life sci, 71-72, head dept, 63-71; prof microbiol & provost & vpres acad affairs, 72-88, HUGH M MORRIS RES PROF MOLECULAR BIOSCI, UNIV DEL, 88- *Concurrent Pos:* Ed, J Bact, Am Soc Microbiol, 64-65, ed-in-chief, 65-78. *Mem:* AAAS; hon mem Am Soc Microbiol; Am Soc Biol Chem; Am Soc Microbiol (pres, 73-74). *Res:* Microbial metabolism; enzymes; fermentations; food microbiology; thermophilic microorganisms. *Mailing Add:* 400 Morris Univ Del Newark DE 19717-5267

CAMPBELL, LOIS JEANNETTE, b Toledo, Ohio, Nov 16, 23. GEOLOGY. *Educ:* Univ Mich, BS, 44; Ohio State Univ, PhD(glacial geol), 55. *Prof Exp:* Jr geologist statist & eval, Humble Oil & Refining Co, 45-47; from instr to asst prof, 54-75, ASSOC PROF GEOL, UNIV KY, 75- *Mem:* Am Asn Petrol Geologists; Geol Soc Am; Nat Asn Geol Teachers; Sigma Xi. *Res:* Pleistocene geology; late paleozoic invertebrates; geomorphology. *Mailing Add:* Geol Dept Bowman Hall Univ Ky Lexington KY 40506

CAMPBELL, LOUIS LORNE, b Winnipeg, Man, Oct 20, 28; m 54; c 3. INFORMATION THEORY, STATISTICAL COMMUNICATION THEORY. *Educ:* Univ Man, BSc, 50; Iowa State Univ, MS, 51; Univ Toronto, PhD(math), 55. *Prof Exp:* Defence sci serv officer, Defence Res Telecommunications Estab, Can, 54-58; from asst prof to assoc prof math, Univ Windsor, 58-63; assoc prof, 63-67, head, Dept Math & Statist, 80-90, PROF MATH, QUEEN'S UNIV, ONT, 67- *Mem:* Am Math Soc; Soc Indust & Appl Math; Can Math Soc; fel Inst Elec & Electronic Engrs; Statist Soc Can; Can Appl Math Soc; Math Asn Am. *Res:* Information measures; source coding; sampling theorem; satellite communication. *Mailing Add:* Dept Math & Statist Queen's Univ Kingston ON K7L 3N6 Can

CAMPBELL, MALCOLM JOHN, b Wollongong, Australia, Aug 16, 37. SPACE PHYSICS. *Educ:* Univ Sydney, BSc, 58, PhD(physics), 63. *Prof Exp:* Fel space res, Goddard Space Flight Ctr, NASA, Md, 63-65; assoc, Ctr Radiophysics & Space Res, Cornell Univ, 65-72; res assoc, State Univ NY at Syracuse Univ, 72-73; asst prof, 73-80, ASSOC PROF SPACE PHYSICS, WASH STATE UNIV, 80- *Mem:* Am Geophys Union; Am Phys Soc; Sigma Xi. *Res:* Atmospheric physics and chemistry. *Mailing Add:* Air Res Dana Hall 306 Wash State Univ Pullman WA 99164-2910

CAMPBELL, MARY KATHRYN, b Philadelphia, Pa, Jan 20, 39. BIOPHYSICAL CHEMISTRY. *Educ:* Rosemont Col, BA, 60; Ind Univ, PhD(phys chem), 65. *Prof Exp:* Instr radiol sci, Johns Hopkins Univ, 65-68; from asst to assoc prof, 68-81, PROF CHEM, MOUNT HOLYOKE COL, 81- *Concurrent Pos:* Vis scientist, Inst Molecular Biol, Univ Paris, 74-75; vis prof chem, Univ Ariz, 81-82, 88-89. *Mem:* Am Chem Soc; AAAS. *Res:* Specific interactions between proteins and nucleic acids; physical chemistry of nucleic acids and proteins. *Mailing Add:* Dept Chem Mt Holyoke Col South Hadley MA 01075-1461

CAMPBELL, MICHAEL DAVID, b Lancaster, Ohio, Aug 8, 41; m 65; c 5. GROUNDWATER & EXPLORATION GEOLOGY, LANDFILL HYDROGEOLOGY. *Educ:* Ohio State Univ, BA, 66; Rice Univ, MA, 76. *Prof Exp:* Staff geologist & hydrogeologist, Continental Oil Co of Australia, Ltd, 66-69; dist geologist & hydrogeologist, United Nuclear Corp, Wyo, 69-71; co-founder & dir res, NWWA Res FAcil, Columbus, Ohio, 71-76; dir, alt energy, mineral & environ progs, Keplinger & Assocs, Inc, Houston, 76-83; sr partner, Campbell, Foss & Buchanan Inc, Houston, 83-89; corp hyrol consult, Law Eng, Houston, 89-90; REGIONAL DIR GEOSCI, ENSR CONSULT & ENG, HOUSTON, 91- *Concurrent Pos:* Abstr ed, Ground Water, 66-70; tech consult, Water Well J, 71-73; consult, Rice Univ, 73-76; UN tech expert, 76-83; spec lectr, Petroleos Mexicanos, Mexico City, 79; vpres, Norse Windfall Mines Inc, 86-88. *Mem:* Am Inst Prof Geologists; Soc Econ Geologists; Geol Soc Am; Am Inst Mining Engrs; Can Inst Mining & Metall; Asn Geologists Int Develop; Am Inst Hydrogeologists; Asn Ground Water Scientists & Engrs; Asn Eng Geologists. *Res:* Ground water and mineral development; exploration mining; groundwater pollution; geochemistry; geothermal energy; uranium, coal, precious metals, base metals and strategic minerals; resource assessment; geotechnical investigations; mine dewatering; precious metal exploration and economic analysis; expert witness testimony and litigation support. *Mailing Add:* 17419 Sandy Cliffs Dr Houston TX 77090

CAMPBELL, MICHAEL FLOYD, b Sparta, Ill, Nov 5, 42; m 64; c 2. FOOD SCIENCE, AGRICULTURE. *Educ:* Univ Ill, BS, 64, MS, 66, PhD(food sci), 68. *Prof Exp:* Microbiologist, R J Reynolds Indust, 68-70; sr food scientist, Libby McNeill & Libby, 70-71; sr food scientist, A E Staley Mfg Co, 71-73, group leader veg proteins, 73-80, group mgr protein prod develop, 80-82, mgr tech planning & admin, 82-83, tech dir legal, 83-84, dir, patent dept, 84-89, DIR, PATENTS & QUAL ASSURANCE, A E STALEY MFG CO, 89- *Concurrent Pos:* vchmn, Ill State Coun on Nutrit, 77-83; chmn, Sci & Nutrit Comt, Food Protein Coun, 80-82; mem, Soybean Res Adv Inst, USDA, 81-84. *Mem:* Inst Food Technologists. *Res:* Chemicals from carbohydrates; patent and regulatory law; corn sweetners; food starches; food microbiology. *Mailing Add:* A E Staley Mfg Co 2200 E Eldorado Decatur IL 62525

CAMPBELL, MILTON HUGH, b Billings, Mont, Sept 2, 28; m 52; c 4. NUCLEAR WASTE ENGINEERING, ANALYTICAL CHEMISTRY. *Educ:* Mont State Col, BS, 51; Univ Wash, MS, 61. *Prof Exp:* Head analyst chem, Ideal Cement Co, 51-53 & 55; jr chemist, Hanford Atomic Prod Oper, Gen Elec Co, 55-57, anal chemist, 57-59, process chemist, 59-61, sr engr, 61-65; sr engr, Chem Processing Div, Isochem Inc, 65-67, staff engr, 67, mgr separations chem lab, Res & Develop, Atlantic Richfield Hanford Co, 67-73, mgr waste mgt & storage technol, 73-74; staff engr, Exxon Nuclear Co, Inc, 74-83; prin engr, Rockwell Hanford Oper, 83-87; fel engr, Westinghouse Hanford Co, 87-91; CONSULT, 91- *Honors & Awards:* Award of Merit, Am Soc Testing & Mat, 85. *Mem:* Am Chem Soc; fel Am Soc Testing & Mat; Am Nuclear Soc. *Res:* Inorganic ion exchange and liquid-liquid separations; nucleonics; gas chromatography; spectrophotometric analysis; thermochemistry; radioactive material measurements; waste management; safety analysis and environmental analysis; project management; quality assurance. *Mailing Add:* 2119 Beech Richland WA 99352

CAMPBELL, NEIL ALLISON, cell biology, plant physiology, for more information see previous edition

CAMPBELL, NEIL JOHN, b Los Angeles, Calif, Aug 26, 25; Can citizen; m 54; c 2. PHYSICAL OCEANOGRAPHY. *Educ:* McMaster Univ, BSc, 50, MSc, 51; Univ BC, PhD(physics), 55. *Prof Exp:* Phys oceanogr Pac oceanog group, Fisheries Res Bd Can, 55, Atlantic oceanog group, 55-59, oceanogr in chg, 59-63; chief oceanogr, Marine Sci Br, Dept Energy, Mines & Resources, 63-88; dir-gen, Marine Sci & Info Directorate, Dept Fisheries & Oceans, Ont, Can, 76-88; CONSULT, 88- *Concurrent Pos:* Lectr, Dalhousie Univ, 60-63; first vchmn, Intergovt Oceanog Comn, UNESCO, 78-80, 80- *Res:* Arctic and environmental oceanography. *Mailing Add:* 1339 Dowler Ave Ottawa ON K1H 7R8 Can

CAMPBELL, NORMAN E ROSS, b Ft William, Ont, Oct 11, 20; m 47; c 2. MICROBIOLOGY. *Educ:* Ont Agr Col, BSA, 44; Univ Man, MSc, 49, PhD(microbiol), 60. *Prof Exp:* From asst prof to prof microbiol, Univ Man, 69-86, chmn div biol sci, 74-80; RETIRED. *Mem:* Can Soc Microbiol; Arctic Inst NAm; Sigma Xi. *Res:* Soil microbiology; nitrogen fixation in subarctic and arctic soils. *Mailing Add:* 861 Lyon St Winnipeg MB R3T 0G8 Can

CAMPBELL, PAUL GILBERT, b Minneapolis, Minn, Sept 25, 25; m 51. ORGANIC CHEMISTRY. *Educ:* Univ Md, BS, 50; Univ SC, MS, 53; Pa State Univ, PhD(org chem), 57. *Prof Exp:* Chemist, NIH, 50-51; sr chemist, J T Baker Chem Co, 57-59 & Allied Chem Corp, 59-60; CHEMIST, NAT BUR STANDARDS, 60- *Honors & Awards:* Silver Medal, Dept Com, 67. *Mem:* AAAS; Am Chem Soc; Am Soc Test & Mat. *Res:* Organic synthesis; bituminous building materials; organic coatings. *Mailing Add:* 2808 Greenbower Way Ellicott City MD 21043

CAMPBELL, PETER HALLOCK, b Flushing, NY, Mar 16, 40. TAXONOMY & ECOLOGY OF PHYTOPLANKTON. *Educ:* Swarthmore Col, BA, 62; Univ NC, PhD(bot), 73. *Prof Exp:* Teaching asst bot, 69-70, RES ASSOC, UNIV NC, CHAPEL HILL, 72-, AT ENVIRON SCI DEPT. *Res:* Taxonomy and ecology of the phytoplankton of North Carolina lakes, preparation of a guide to these algae. *Mailing Add:* 005 1/2 S Columbia St Chapel Hill NC 27514

CAMPBELL, PRISCILLA ANN, b Mineola, NY, Aug 19, 40; m 72; c 2. IMMUNOLOGY. *Educ:* Colo Col, BA, 62; Univ Colo, MS, 65, PhD(cell biol), 68. *Prof Exp:* Fel immunol, Univ Calif, San Diego, 68-70 & Nat Jewish Hosp, Denver, 70-72; MEM STAFF, DIV BASIC IMMUNOL, DEPT MED, NAT JEWISH HOSP, DENVER, 72- *Concurrent Pos:* Asst prof path, Univ Colo Med Ctr, Denver, 72-78, assoc prof, 78-86, prof, 86-, prof microbiol & immunol, 87-; NIH grant, 74-; mem study sect, NIH, 79-83. *Mem:* AAAS; Am Asn Immunologists; Am Soc Exp Path; Am Soc Microbiol; Soc Leukocyte Biol. *Res:* Cell biology of immune response; resistance to bacterial infection; thymocyte maturation. *Mailing Add:* Dept Med Nat Jewish Ctr Denver CO 80206

CAMPBELL, RALPH EDMUND, b Providence, Utah, Sept 22, 27; m 52; c 5. FOREST SOILS. *Educ:* Utah State Univ, BS, 50, MS, 54. *Prof Exp:* Soil scientist, Exp Sta, SDak State Univ & USDA, 50-54 & NMex Agr Ext Sta, Tucumcari, 54-55, soil scientist, Southern Montana Res Ctr, Agr Res Serv, Mont, 55-64, res soil scientist, 64; res soil scientist, Rocky Mountain Forest & Range Exp Sta, US Forest Serv, 64-85; RETIRED. *Mem:* Soil Sci Soc Am; Am Soc Agron. *Res:* Soil fertility, moisture and management; forest soils problems related to nutrient cycling, sediment movement, water quality and forest climate. *Mailing Add:* 6510 N Longfellow Dr Tucson AZ 85715

CAMPBELL, RAYMOND EARL, b Ranger, Tex, Jan 4, 41; m 65; c 2. HORTICULTURE. *Educ:* Okla State Univ, BS, 63, MS, 66; Kans State Univ, PhD(hort), 72. *Prof Exp:* County exten agent, Coop Exten Serv, Okla State Univ, Coal County, 63-64 & Delaware County, 66-67; instr agr, Friends Univ, 67-68; county exten agent hort, Coop Exten Serv, Kans State Univ, 68-70; exten specialist voc hort, Va Polytech Inst & State Univ, 72-74; exten specialist veg crops, 74-77, ASSOC PROF HORT, OKLA STATE UNIV, 77- *Concurrent Pos:* Dir agr & rural develop progs, Okla State Univ. *Mem:* Am Soc Hort Sci. *Res:* Investigation of environmental influence on vegetable crop production and yield with particular emphasis on temperature and humidity; investigation of applications of artificial soil mixes to greenhouse vegetable growing and vegetable crop alternatives. *Mailing Add:* 245 Agr Hall Okla State Univ Stillwater OK 74078

CAMPBELL, RICHARD DANA, b Oklahoma City, Okla, June 12, 39; m 63; c 2. DEVELOPMENTAL BIOLOGY. *Educ:* Harvard Univ, BA, 61; Rockefeller Inst, PhD(biol), 65. *Prof Exp:* Asst prof organismic biol, 65-70, assoc prof develop & cell biol, 70-74, PROF DEVELOP & CELL BIOL, UNIV CALIF, IRVINE, 74- *Mem:* Am Soc Zool; Am Soc Cell Biol; Soc Develop Biol. *Mailing Add:* Develop & Bio Ctr Univ Calif Irvine CA 92717

CAMPBELL, RICHARD M, b Logan, Ohio, July 29, 30; m 54; c 2. ELECTRICAL ENGINEERING. *Educ:* Ohio State Univ, BEE, 54, MSc, 56, PhD(electron devices), 66. *Prof Exp:* Res asst, Electron Device Lab, Ohio State Univ, 55-56, res assoc, Lab & instr elec eng, Univ, 56-62; staff mem, RCA Labs, 62-63; from instr to asst prof, 63-77, ASSOC PROF ELEC ENG, OHIO STATE UNIV, 77- *Concurrent Pos:* Consult, Electronics Co, 62 & Edmund C Frost & Co, 67. *Mem:* Inst Elec & Electronics Engrs. *Res:* Electron optics problems in television type pickup tubes and microwave tubes. *Mailing Add:* Dept Elec Eng Ohio State Univ 2015 Neil Ave Columbus OH 43210

CAMPBELL, ROBERT A, b Toledo, Ohio, Dec 21, 24; m 49; c 3. PEDIATRICS. *Educ:* Univ Calif, Berkeley, AB, 54; Univ San Francisco, MD, 58; Am Bd Pediat, dipl, 63, cert pediat nephrol, 74. *Prof Exp:* Res asst, Cancer Res Genetics Lab, Univ Calif, Berkeley, 54-58; USPHS fel pediat biochem, 61-63, assoc prof, 63-71, dir, Pediat Renal-Metab Lab, 63-80, PROF PEDIAT, MED SCH, UNIV ORE, 71- *Mem:* AAAS; Am Acad Pediat. *Res:* Metabolic disorders of infancy and childhood; kidney disease. *Mailing Add:* Dept Pediat Ore Health Sci Univ Portland OR 97201

CAMPBELL, ROBERT LOUIS, b Westerville, Ohio, Nov 16, 25; m 49; c 3. NEUROSURGERY. *Educ:* Baldwin-Wallace Col, BS, 45; Ohio State Univ, MD, 49; Am Bd Neurol Surg, dipl, 60. *Prof Exp:* From instr to asst prof, 57-64, PROF NEUROL SURG & DIR MED CTR, IND UNIV, INDIANAPOLIS, 65- *Concurrent Pos:* Consult, Vet Admin Hosp, Indianapolis, 57- & Naval Hosp, Great Lakes, Ill, 63-; mem, Subarachnoid Hemorrhage Study Sect, NIH, 62-; mem consult staff, Wishard Mem Hosp, 65- *Mem:* Cong Neurol Surg; Am Asn Neurol Surg; fel Am Col Surgeons; Soc Neurol Surgeons; AMA; Sigma Xi. *Res:* Spinal cord neurotransmitters; cerebral vasospasm. *Mailing Add:* 8918 W 82nd St Indianapolis IN 46278

CAMPBELL, ROBERT NOE, b Fairmont, Minn, Nov 16, 29; m 54; c 3. PLANT PATHOLOGY. *Educ:* Univ Minn, BS, 52, MS, 54, PhD(plant path), 57. *Prof Exp:* Asst plant path, Univ Minn, 52-54, instr, 54-56; plant pathologist, Forest Prod Lab, 57-59; from asst prof to assoc prof plant path, 59-69, PROF PLANT PATH & PLANT PATHOLOGIST EXP STA, UNIV CALIF, DAVIS, 69- *Honors & Awards:* Campbell Soup Co Award, AAAS, 62. *Mem:* Am Phytopath Soc; AAAS; Am Inst Biol Sci. *Res:* Cause and control of diseases of vegetable crops; fungal transmission of plant viruses. *Mailing Add:* Dept Plant Path Univ Calif Davis CA 95616

CAMPBELL, ROBERT TERRY, b Greenwood, Miss, July 27, 32; m 54; c 2. AGRICULTURAL & FOREST SCIENCES, ANALYTICAL CHEMISTRY. *Educ:* Miss Col, BS, 54; Univ NC, MS, 59. *Prof Exp:* Res chemist, Int Paper, 59-61, mgr, Pulp Group, 62-70, chief, Pulp Res, 71-80, mgr, Technol Planning, 81-84, mgr, Facil Planning, 84-86, MGR, SYSTS TECHNOL, INT PAPER, 87- *Concurrent Pos:* Chmn, Alkaline Pulping Comt, Tech Asn Pulp & Paper Indust, 90-91. *Mem:* Tech Asn Pulp & Paper Indust. *Res:* Development of improved pulping and bleaching processes. *Mailing Add:* 563 Shenandoah Rd W Mobile AL 36608

CAMPBELL, ROBERT WAYNE, b Concordia, Kans, Dec 19, 40; m 60; c 3. POLYMER CHEMISTRY. *Educ:* Kans State Univ, BS, 62; Purdue Univ, PhD(org chem), 66. *Prof Exp:* Res chemist, Phillips Petrol, 66-68, group leader polymer res, 68-73, proj leader explor plastics & fibers, 73-78, sect supvr chem synthesis, 78-80, prod develop mgr, 80-83, sect supvr plastics res, 83, sect supvr adv mat, 84, eng mat br mgr, 84-88, vpres technol, 88-89; VPRES TECHNOL & QUAL, GENCORP POLYMER PROD, 89- *Mem:* Am Chem Soc; Soc Plastics Engrs (secy, 75, treas, 76, pres, 78); Soc Advan Mat & Process Eng; Am Soc Qual Control. *Res:* Synthetic and reaction mechanism aspects of organo-sulfur chemistry; mass spectral fragmentation of sulfonate esters and related materials; polymer synthesis and characterization, including structure-property correlation; sulfur chemical synthesis and applications; specialty chemical commercial development; advanced materials development; specialty polymers and engineered plastics and elastomers development. *Mailing Add:* GenCorp Polymer Prods 350 Springside Dr Akron OH 44333-2475

CAMPBELL, RONALD WAYNE, b Cherryvale, Kans, July 26, 19; m 43; c 3. HORTICULTURE. *Educ:* Kans State Univ, BS, 43, MS, 46; Mich State Univ, PhD(hort), 57. *Prof Exp:* From asst prof to assoc prof, 46-61, head dept hort & forestry, 66-80, PROF HORT, KANS STATE UNIV, 61- *Mem:* AAAS; Am Soc Hort Sci; Am Soc Plant Physiol; Weed Sci Soc Am; Am Pomol Soc; Sigma Xi. *Res:* Physiological studies of influences of agricultural chemicals on plants; winter hardiness and water relations studies; stock and scion relationships. *Mailing Add:* Dept Hort Kans State Univ Manhattan KS 66502

CAMPBELL, RUSSELL HARPER, b Bakersfield, Calif, Apr 20, 28; m 53; c 2. EARTH SCIENCE, ENGINEERING GEOLOGY. *Educ:* Univ Calif, BA, 51. *Prof Exp:* GEOLOGIST, US GEOL SURV, 51- *Concurrent Pos:* Chmn, comt methodologies for predicting mudflow areas, Nat Res Coun, 81-82. *Honors & Awards:* Meritorious Serv Award, US Dept Interior. *Mem:* Fel Geol Soc Am; Am Geophys Union; Soc Econ Paleontologists & Mineralogists. *Res:* Landslide probability and risk distribution; regional structure and stratigraphy of the western Transverse Ranges, California; earthquake hazards; landslides and mudflows; stratigraphy and structure of northwestern Alaska; uranium deposits of southeastern Utah. *Mailing Add:* 12007 Turf Lane Reston VA 22091

CAMPBELL, SAMUEL GORDON, b Oban, Scotland, Dec 10, 33; m 61; c 3. IMMUNOLOGY. *Educ:* Glasgow Univ, BVMS, 56; Toronto Univ, MVSc, 59; Cornell Univ, PhD(vet microbiol), 64. *Prof Exp:* House physician vet med, Vet Sch, Univ Glasgow, 56-57; asst vet microbiol, Ont Vet Col, 57-59; NY State Col Vet Med, Cornell Univ, 61-64; sr lectr, Sch Vet Sci, Melbourne Univ, 64-66; asst prof, NY State Col Vet Med, Cornell Univ, 67-70, dir int educ, 68-74, prof vet microbiol, 70-86, assoc dean, acad affairs, 86-90, DIR INT PROGS, CORNELL UNIV, 90- *Concurrent Pos:* Consult, Vet Educ, Rockefeller Found, Brazil & Peru, 68 & Dept Microbiol, Tuskegee Inst, 73; external examr, Univ Guelph, 73, Melbourne Univ, 74; consult hemoprotozoan dis, USAID, Colombia & Kenya, 75, sheep dis, Indonesia, 87; livestock res, World Bank, Uganda, 90. *Mem:* Royal Col Vet Surg. *Res:* Immunological responses of cattle, sheep, and goats. *Mailing Add:* Col Vet Med Ithaca NY 14853-6401

CAMPBELL, STEPHEN LA VERN, b Bell Plaine, Iowa, Dec 8, 45. MATHEMATICS, APPLIED MATHEMATICS. *Educ:* Dartmouth Col, AB, 67; Northwestern Univ, MS, 68, PhD(math), 72. *Prof Exp:* Instr, Marquette Univ, 69-72; from asst prof to assoc prof, 72-81, PROF MATH, NC STATE UNIV, 81- *Mem:* Soc Indust & Appl Math. *Res:* Singular systems of differential equations and their applications in engineering and biology; bounded linear operators in Banach spaces. *Mailing Add:* Dept Math NC State Univ Raleigh NC 27695-8205

CAMPBELL, STEWART JOHN, b Calgary, Alta, July 22, 47; m 70; c 4. OILSEED CHEMISTRY, OILSEED PROCESSING. *Educ:* Univ Alta, BSc, 69, PhD(chem), 73. *Prof Exp:* Res scientist & plant breeder, Res Br, Agr Can, 73-79; staff mem planning & develop, United Oilseed Prod Ltd, 79-85; VPRES GRAIN DIV, CANBRA FOODS LTD, 85- *Concurrent Pos:* Plant breeder, Can Expert Comt Grain Crops, 73- *Mem:* Chem Inst Can; Agr Inst Can; Am Asn Cereal Chemists; Can Seed Growers Asn; Am Oil Chem Soc. *Res:* Improvement of oilseed and grain crops adapted to western Canada through research in plant breeding, crop production, and oilseed quality. *Mailing Add:* 7535 Hunter Valley Rd NW Calgary AB T2T 3K9 Can

CAMPBELL, SUZANN KAY, b New London, Wis, Apr 19, 43. PHYSICAL THERAPY. *Educ:* Univ Wis-Madison, BS, 65, MS, 68, PhD(neurophysiol), 73. *Prof Exp:* Phys therapist, Cent Wis Colony & Training Sch, 65-68; instr phys ther, Sch Med, Univ Wis-Madison, 68-70, consult, Univ Family Health Serv, 71-72; from asst prof to assoc prof, 72-84, PROF PHYS THER, SCH MED, UNIV NC, CHAPEL HILL, 84-; PROF, DEPT PHYS THER, UNIV ILL, CHICAGO, 88- *Concurrent Pos:* Consult, Div Dis Develop & Learning, Biol Sci Res Ctr, Univ NC, Chapel Hill, 72-; vis lectr phys ther & res fel, Psychophysiol Sect, Dept Pediat & Psychol, Univ Wis-Madison, 81; ed, Phys & Occup Ther in Pediat; Culpeper fel med sci, Univ NC, 85-86; res award, sect pediat, Am Phys Ther Asn, 84. *Honors & Awards:* Golden Pen Award, Am Phys Ther Asn, 78. *Mem:* AAAS; Am Phys Ther Asn; Soc Res Child Develop. *Res:* Infants with central nervous system dysfunction; identification, psychoaffective and sensorimotor development, interaction with mother, and effectiveness of physical therapy in promoting development. *Mailing Add:* Dept Phys Ther Col Assoc Health Prof Univ Ill Chicago 1919 W Taylor St 4th Floor Chicago IL 60680

CAMPBELL, THOMAS COLIN, b Annandale, NJ, Mar 14, 34; m 62; c 5. BIOCHEMISTRY, NUTRITION. *Educ:* Pa State Univ, BS, 56; Cornell Univ, MS, 57, PhD(animal nutrit), 62. *Prof Exp:* Res biologist, Woodard Res Corp, 61-63; res assoc toxicol, Mass Inst Technol, 63-65; from asst prof to prof biochemistry, Va Polytech Inst & State Univ, 65-75; JACOB GOULD SCHURMAN PROF NUTRIT BIOCHEM, DIV NUTRIT SCI, CORNELL UNIV, 75- *Concurrent Pos:* Campus coordr, Philippine Nat Nutrit Prog; NIH res career develop award, 74; sr sci adv, Am Inst Cancer Res; vis scholar, Univ Oxford, Eng. *Honors & Awards:* Roberts Found Nutrit Award. *Mem:* Am Inst Nutrit; Am Soc Pharmacol & Exp Therapeut; Soc Toxicol; Sigma Xi; Am Asn Cancer Res. *Res:* Study of carcinogen metabolism; mechanism of action and human consumption patterns of aflatoxin; nutrition-toxin interactions; environmental health; international nutrition; nutrition and cancer. *Mailing Add:* Div Nutrit Sci Cornell Univ Ithaca NY 14853

CAMPBELL, THOMAS COOPER, b Decatur, Ill, Feb 29, 32; m 57; c 2. PHYSICAL ORGANIC CHEMISTRY, PETROLEUM ENGINEERING. *Educ:* Millikin Univ, BS, 57; Emory Univ, MS, 63; Ga Inst Technol, PhD(chem), 68. *Prof Exp:* Res chemist, Monsanto Co, 67-69, sr res chemist, 69-74; res & develop assoc, PQ Corp, 74-80; MGR, ENHANCED OIL RECOVERY ENG, AMINOIL USA, 80- *Mem:* Soc Petrol Engrs; Am Chem Soc; Sigma Xi. *Res:* Enhanced oil recovery; alkaline waterflooding; sodium silicate applications and technology, surface chemistry; surfactant applications and formulations. *Mailing Add:* 17092 Pleasant Circle Huntington Beach CA 92649-2494

CAMPBELL, THOMAS HODGEN, b Toronto, Ohio, Dec 9, 24. MYCOLOGY. *Educ:* Ohio State Univ, BSc, 46, MSc, 49; Univ Wis, PhD(bot), 52. *Prof Exp:* Instr bot, Univ Tenn, 52-53; biologist, Am Cyanamid Co, NY, 53-54; from asst prof to assoc prof bot, Univ Tenn, 54-61; assoc prof biol, 61-65, head dept, 65-69, prof biol, 65-80, EMER PROF BIOL, COL STEUBENVILLE, 80- *Mem:* AAAS; Mycol Soc Am; Bot Soc Am; Am Phytopath Soc; Mycol Soc France. *Res:* Morphology, cytology and variation of Penicillium chrysogenum; taxonomy and history of mycology. *Mailing Add:* 2022 Cherokee Lane Norman OK 73071

CAMPBELL, TRAVIS AUSTIN, b Washington, DC, Aug 20, 43; m 75. BIOMATHEMATICS. *Educ:* Univ Md, BS, 66, MS, 72, PhD, 80. *Prof Exp:* Agronomist plant breeding statist & weed sci, 67-76, RES AGRONOMIST PLANT BREEDING, USDA, 76- *Mem:* Am Soc Agron; Am Soc Hort Sci. *Res:* New crops breeding; alfalfa breeding. *Mailing Add:* AG Res Serv-Agr Res USDA Rm 339 BG001 Beltsville MD 20705

CAMPBELL, VIRGINIA WILEY, b Denver, Colo, Sept 5, 37; m 57; c 2. MOLECULAR BIOLOGY, IMMUNOLOGY. *Educ:* Univ Colo, BA, 59, MS, 61; Univ Mich, MS & PhD(microbiol), 79. *Prof Exp:* Res assoc, Dept Microbiol, Univ Mich, 79-83, res investr, Med Sch, 83-89; INDUST DEVELOP, UN INDUST DEVELOP ORGN, VIENNA, AUSTRIA, 89- *Mem:* AAAS. *Res:* Gene organization in eucaryotic cells, expression, and regulation in immunoglobulin-producing cells; biosafety; technology transfer to developing countries. *Mailing Add:* Sandgasse 13-2 A-1190 Vienna Austria

CAMPBELL, W(ILLIAM) M(UNRO), b Alvinston, Ont, Jan 21, 15; m 41; c 2. CHEMICAL ENGINEERING. *Educ:* Univ Toronto, BASc, 38; Case Western Reserve Univ, MS, 40; Univ Ill, PhD(chem eng), 50. *Prof Exp:* Asst, Case Western Reserve Univ, 38-40; plant res engr & group leader, Shawinigan Chems, Ltd, 41-46; asst prof chem eng, Queen's Univ, Can, 46-47; instr, Univ Ill, 47-50; group leader, Chem Eng Br, Atomic Energy Can, Ltd, 50-51, head br, 51-56, dir, Chem & Metall Div, 56-65; dir res, Ont Res Found, 65-70; mgr indust appln, Atomic Energy Can, Ltd, 70-71, spec adv to vpres, 71-76; proj dir, W L Wardrop & Assoc Ltd, 76-78; RETIRED. *Mem:* Chem Inst Can. *Res:* Process development based on acetylene; Fischer-Tropsch process; radiochemical processing; solvent extraction; nuclear fuel development; metallurgy. *Mailing Add:* Six Radcliffe Crescent London ON N6H 3X4 Can

CAMPBELL, WALLACE G, JR, b Lockport, NY, July 25, 30; m 57; c 4. PATHOLOGY. *Educ:* Harvard Univ, AB, 53; Cornell Univ, MD, 57. *Prof Exp:* Asst path, Cornell Univ, 58-61, instr, 61-62; from asst prof to assoc prof, 64-71, PROF PATH, EMORY UNIV, 71- *Concurrent Pos:* USPHS trainee path, 59-62. *Mem:* Am Asn Path; Int Acad Path; Am Soc Nephrology; AAAS; Soc Exp Biol Med. *Res:* Experimental hypertension; cardiovascular and renal pathology; coagulation of blood; cellular pathology; arthritis; pneumocystis; vaccinia virus toxicity; endocrine tumors of the pancreas. *Mailing Add:* Dept Path Cornell Univ Med Col 1300 York Ave New York NY 10021

CAMPBELL, WALLACE HALL, b New York, NY, Feb 6, 26; m 56; c 2. GEOMAGNETISM, ATMOSPHERIC PHYSICS. *Educ:* La State Univ, BS, 50; Vanderbilt Univ, MA, 53; Univ Calif, Los Angeles, PhD(physics), 59. *Prof Exp:* Grad res geophysicist, Inst Geophys, Univ Calif, Los Angeles, 55-57, jr res geophysicist, 57-59; asst prof geophys res, Geophys Inst, Univ Alaska, 59-60; group leader ultra low frequency res, Cent Radio Propagation Lab, Nat Bur Stand, 60-65, chief geomagnetism res, Aeronomy Lab, Inst Telecommun Sci & Aeronomy, Environ Sci Serv Admin, 65-67, dir geomagnetism lab, Environ Res Labs, Nat Oceanic & Atmospheric Agency, 67-71, head space magnetism res, 71-73; RESEARCHER EXTERNAL GEOMAGNETIC FIELD & INDUCED CURRENTS, US GEOL SURV, 73- *Concurrent Pos:* Mem various working groups, Int Asn Geomagnetism & Aeronomy, 64- *Honors & Awards:* Boulder Scientist Award, Sigma Xi, 63. *Mem:* Am Geophys Union; AAAS; Soc Terrestrial Magnetism & Elec Japan. *Res:* Upper atmospheric physics; geomagnetic phenomena; natural ultra low frequency field variations; auroral luminosity fluctuations; earth conductivity. *Mailing Add:* 3030 Galena Way Boulder CO 80303

CAMPBELL, WARREN ADAMS, b Berkeley, Calif, June 14, 36; m 58. ASTRONOMY. *Educ:* Willamette Univ, BA, 58; Univ Wis, MS, 60, PhD(astron), 65. *Prof Exp:* Asst prof math, Wash State Univ, 65-70; from asst prof to assoc prof, 70-85, PROF PHYSICS, UNIV WIS, RIVER FALLS, 85- *Mem:* Am Astron Soc. *Res:* Forbidden oxygen and nitrogen lines in the spectra of planetary nebulae. *Mailing Add:* Dept Physics Univ Wis River Falls WI 54022

CAMPBELL, WARREN ELWOOD, b Helena, Mo, Feb 19, 31; m 67; c 2. DATA STORAGE & RETRIEVAL. *Educ:* Univ Tenn, BS, 58, PhD(chem), 62. *Prof Exp:* Res chemist, 62-65, SR RES CHEMIST, TENN EASTMAN CO, EASTMAN KODAK CO, 65- *Mem:* Am Chem Soc; Sigma Xi. *Res:* Real-time operating programs for use with computers. *Mailing Add:* 1935 Lamont St Kingsport TN 37664

CAMPBELL, WILBUR HAROLD, b Santa Ana, Calif, Apr 23, 45; m 81. ENZYMOLOGY, IMMUNOCHEMISTRY. *Educ:* Pomona Col, BA, 67; Univ Wis-Madison, PhD(biochem), 72. *Prof Exp:* Res assoc, Univ Ga, 72-73, Mayo Clin, 73-74 & Mich State Univ, 74-75; asst prof chem, 75-80, assoc prof chem, Col Sci & Forestry, State Univ NY, 80-85; assoc prof biochem, 85-86, PROF BIOCHEM, MICH TECHNOL UNIV, 86- *Concurrent Pos:* Vis prof, Bayreuth Univ, Bayreuth, WGer, 82; vis scientist, Molecular Biol Comput Res Resource; vis scientist, Dana-Farber Cancer Inst; vis scientist, Harvard Sch Pub Health, Harvard Med Sch, 87. *Honors & Awards:* Sigma Xi. *Mem:* Am Chem Soc; Am Soc Plant Physiologists; Japanese Soc Plant Physiologists; Int Soc Plant Molecular Biologists. *Res:* Plant biochemistry; and molecular biology focusing on regulation of nitrogen metabolism using immunochemical approaches; enzyme nitrate reductase; microcomputing for biochemical research; monoclonal antibodies. *Mailing Add:* Dept Biol Sci Mich Technol Univ Houghton MI 49931

CAMPBELL, WILLIAM (ALOYSIUS), developmental anatomy, electron microscopy; deceased, see previous edition for last biography

CAMPBELL, WILLIAM ANDREW, b Paterson, NJ, Jan 29, 06; m 36; c 3. FOREST PATHOLOGY. *Educ:* Pa State Teachers Col, BS, 29; Univ Colo, AM, 31; Pa State Col, PhD(mycol, plant path), 35. *Prof Exp:* Asst forest pathologist, Div Forest Path, Bur Plant Indust, USDA, 36-42, forest pathologist, Guayule Res Proj, Salinas, Calif, 42-46, pathologist, Div Forest Path, Georgia, 46-53, sr pathologist, 53-54, plant pathologist, Southeastern Forest Exp Sta, US Forest Serv, 54-71; prof plant path, Univ Ga, 71-73; CONSULT FORESTRY, 73- *Mem:* Am Phytopath Soc; Mycol Soc Am. *Res:* Diseases of native tree species, particularly root diseases and wood decay; fungi affecting roots of plants. *Mailing Add:* 260 Milledge Heights Athens GA 30606

CAMPBELL, WILLIAM B(UFORD), b Clarksdale, Miss, Nov 23, 35; div. CERAMIC ENGINEERING, MINERALOGY. *Educ:* Ga Inst Technol, BCerE, 58, MSCerE, 60; Harvard Univ, AM, 62; A Ohio State Univ, PhD(mat eng), 67, PGMed, 70; NY Univ Med Sch, PGMed, 80. *Prof Exp:* Asst prof ceramic eng, Ga Inst Technol, 58-59, res asst, Eng Exp Sta, 58-60; res assoc crystal res, Harvard Univ, 59-62; sr scientist, Lexington Labs, Inc, Cambridge, 62, progs dir & vpres, 63-67; from asst prof to assoc prof ceramic eng, Ohio State Univ, 67-74; assoc prof, Dept Eng, Univ Tenn, 74-77; sr partner, Campbell, Churchill, Zimmerman & Assocs, 76-84; pres & chmn, Southeastern Mobility Co, Inc, 77-84; sr mat res engr advisor, AIDE Mgt Resources, Inc, Nuclear Eng Br, Tenn Valley Authority, 86-87, proj engr, Nuclear Eng/Nuclear Power, 87-89; dir, EPRI Ctr Mat Fabrication, 89-90; dir, Innovation & Technol Transfer, Battelle Mem Inst, 90; MANAGING CONSULT, PERFORMANCE CONSULT, INC, 90- *Concurrent Pos:* Res analyst, US Air Force, 65-69; res scientist, Arctic Inst NAm, 60-61; mem, Int Conf Crystal Growth, 66 & task group biomat, Nat Acad Eng; vis scholar, Univ Fla, 70, Miss State Univ, 71 & Va Polytech Inst, 72; mem, Educ Coun, Ga Inst Technol, 72-; off guest, Am Acad Orthop Surgeons, 72-; mem, Rehab Eng Panel, NAS, 75; distinguished lectr, La Tech Univ, 76; Partner, Brae Arden Farms & Brae Arden Ranch, 79-83; consult biomed eng & forensic sci, 84-; trustee, Lakeshore Menal Health Inst, Dept Mental Health, State of Tenn, 86-93; forensic scientist/engr, Med Examr Prog, Commonwealth Ky, 86-, Knox County Med Examr Off, Tenn, 87-; mem, Arts Coalition Citizens with Mental Illness, Knoxville, 88- *Honors & Awards:* Outstanding Young Scientist Award, Sigma Xi, 71; Freeman Award, Am Coun Blind, 76. *Mem:* Fel Am Ceramic Soc; Am Soc Eng Educ; Nat Inst Ceramic Engrs; fel Am Inst Chemists; Nat Soc Prof Engrs; Sigma Xi; Am Soc Testing & Mat; Am Acad Forensic Sci; Asn Advan Med Instrumentation; Am Soc Nondestructive Testing. *Res:* Experimental mineralogy; crystal synthesis; bioceramic materials; pigment degradation mechanisms; thermal effects on reaction kinetics; forensic sciences; three US patents. *Mailing Add:* PO Box 51825 Knoxville TN 37950-1825

CAMPBELL, WILLIAM BRYSON, b Sulphur Springs, Tex, Mar 25, 47; m 75. PHARMACOLOGY. *Educ:* Univ Tex, Austin, BS, 70; Univ Tex Southwestern Med Sch, PhD(pharmacol), 74. *Prof Exp:* Fel pharmacol, Med Col Wis, 74-75, instr, 75-76; from asst prof to assoc prof, 76-90, PROF PHARMACOL, UNIV TEX SOUTHWESTERN MED SCH, 90- *Concurrent Pos:* Coun, high blood pressure res. *Mem:* Am Fedn Clin Res; Am Soc Pharmacol & Exp Therapeut; Am Heart Asn. *Res:* Pharmacology of vasoactive substances and their relationship to the kidney, the adrenal gland and the peripheral vasculature in hypertension. *Mailing Add:* Dept Pharmacol Univ Tex Southwestern Med Sch 5323 Harry Hines Blvd Dallas TX 75235

CAMPBELL, WILLIAM CECIL, b Londonderry, Ireland, June 28, 30; nat US; m 62; c 3. ZOOLOGY. *Educ:* Univ Dublin, BA, 52; Univ Wis, MS, 54, PhD(zool, vet sci), 57. *Prof Exp:* Asst parasitol, Univ Wis, 53-57; res assoc, Merck Inst Therapeut Res, 57-66, dir parasitol, 66-72, dir, Merck, Sharp & Dohme Vet Res & Develop Lab, Australia, 72-73, sr investr, Merck Inst Therapeut Res, 73-76, dir basic parasitol, 77-78, sr dir, 78-84, sr scientist, 84-90; DANA FEL, DREW UNIV, 90- *Concurrent Pos:* Adj prof parasitol, Univ Pa, 77-89, NY Med Col, 85-; mem exec comt, Int Comn Trichinellosis, 72-76, vpres, 76-80, pres, 80-84; mem Steering Comt (filariasis), World Health Orgn, 90-; distinguished vet parasitologist, Am Asn Vet Parasitol. *Honors & Awards:* Discovers Award, Am Pharmaceut Mfrs Asn. *Mem:* Am Soc Parasitol (pres, 87); Am Soc Trop Med & Hyg; Royal Soc Trop Med & Hyg. *Res:* Parasitology; helminthology; chemotherapy. *Mailing Add:* Hall Sci Drew Univ Madison NJ 07940

CAMPBELL, WILLIAM FRANK, b Mt Vernon, Ill, Sept 23, 28; m 54; c 4. RADIATION BOTANY, PLANT PHYSIOLOGY. *Educ:* Univ Ill, BS, 56, MS, 57; Mich State Univ, PhD(radiation bot), 64. *Prof Exp:* Collabr hort, Agr Res Serv, USDA, 64-68, assoc prof plant physiol, 68-77, PROF PLANT SCI, UTAH STATE UNIV, 77- *Mem:* AAAS; fel Am Soc Hort Sci; Am Soc Plant Physiol; Am Soc Agron; Bot Soc Am. *Res:* Effects of ionizing radiation in successive generations on developing and dormant plant embryos; plant growth, physiological responses histo and cyto-chemical and ultrastructural changes as influenced by environmental stresses; ultrastructural plant cytology. *Mailing Add:* Dept Plant Sci Utah State Univ Logan UT 84322-4820

CAMPBELL, WILLIAM H, mathematics, for more information see previous edition

CAMPBELL, WILLIAM HOWARD, b Lakeview, Ore, Aug 13, 42; m 67; c 2. PHARMACOLOGY. *Educ:* Ore State Univ, BS, 65, MS, 68; Purdue Univ, PhD(pharm), 71. *Prof Exp:* Asst prof pharm, Ore State Univ, 71-75; chmn, Dept Pharm Pract, Sch Pharm, Univ Wash, 75-88; DEAN & PROF PHARM, SCH PHARM, AUBURN UNIV, 88- *Concurrent Pos:* Sr investr, Health Serv Res Ctr, Ore Region, Kaiser Found, Portland, 71-; sr consult, Vet Admin Hosp, 75- *Mem:* Am Pub Health Asn; Am Pharmaceut Asn. *Res:* Assessment of drug systems operation within general medical care, specifically adverse drug reactions, drug use review and control, and technological applications. *Mailing Add:* Sch Pharm Auburn Univ Auburn AL 36849-5501

CAMPBELL, WILLIAM J, US citizen. INFORMATION SCIENCE & SYSTEMS. *Educ:* Kutztown State Univ, BA, 75; South Ill Univ, MS, 78. *Prof Exp:* Team chief radioteletype, US Army, 64-68; adv/ed, Vet Affairs Off Newslett, Kutztown Univ, 73-74; regional rep anal chemists, Brandt Assocs, 75-76; proj mgr staff res, Holcomb Res Inst, Butler Univ, 77-78; proj mgr remote sensing training, 78-82, prin investr & syst engr, 83-90, HEAD, DATA MGT SYST FACIL, GODDARD SPACE FLIGHT CTR, NASA, 90- *Concurrent Pos:* Comt mem, Earth Observ Syst Strategic Planning, Goddard Space Flight Ctr, Univs Space Res Asn; mem subcomt, Nat Res Coun-Nat Acad Sci; assoc ed, Am Soc Photogram & Remote Sensing; architect & leader retrieval & visual element, NASA Global Change Technol; consult, Small Bus Innovation Res Submissions. *Mem:* Am Soc Photogram & Remote Sensing. *Res:* Development of intelligent data management value-added services and systems; intelligent user interfaces; geographic information systems; graphical data representation; advanced data structure design; automatic data cataloging; characterization using artificial intelligence and supporting technologies. *Mailing Add:* NASA-Goddard Space Flight Ctr Code 934 Greenbelt MD 20771

CAMPBELL, WILLIAM JACKSON, b Wichita Falls, Tex, Oct 23, 29; m 51; c 3. CLINICAL BIOCHEMISTRY. *Educ:* NTex State Univ, BA, 49; Univ Tex, BS, 52, MS, 53; Ohio State Univ, PhD(phys chem), 60. *Prof Exp:* Chief, Dept Biochem, Walter Reed Army Inst Res, 60-64; prog adminr clin chem, Nat Inst Gen Med Sci, 64-74; exec dir clin chem, Am Asn Clin Chem, 74-81; PRES, STANBIO LAB, INC, 81- *Concurrent Pos:* Lectr, Am Univ, 61-73, prof lectr, 63-65, adj prof, 65-73. *Mem:* Sigma Xi; Am Chem Soc; Am Asn Clin Chemists. *Mailing Add:* Stanbio Lab Inc 2930 E Houston St San Antonio TX 78202-3219

CAMPBELL, WILLIAM JOSEPH, b Washington, DC, July 31, 26; m 48; c 2. PHYSICAL CHEMISTRY. *Educ:* Univ Md, BS, 50, PhD(chem), 56. *Prof Exp:* Asst phys chem, Univ Md, 51; phys chemist, 51-56, supvry phys chemist, 56-62, supvry res chemist, 62-80, RES DIR, US BUR MINES, 80- *Honors & Awards:* Dept Interior Meritorious Serv Award, 62. *Mem:* NY Acad Sci; Am Chem Soc; Soc Appl Spectos; Am Inst Min, Metall & Petrol Eng; fel Am Inst Chem; Sigma Xi. *Res:* Fluorescent x-ray spectroscopy; x-ray diffraction; high temperature physical chemistry; auger spectroscopy; environmental chemistry; mineral particulates; electron optics; chemical-instrumental methods of analysis. *Mailing Add:* 2720 Hambleton Rd Riva MD 21140

CAMPBELL, WILLIAM M, b Philadelphia, Pa, Jan 1, 34; m 58; c 3. CHEMICAL ENGINEERING. *Educ:* Drexel Inst Technol, BS, 56; Univ Houston, MBA, 81. *Prof Exp:* Engr, MW Kellogg Co, 56-71, dir, Proj Eng Dept, 71-74, Process Eng Dept, 74-77, Gen Eng Dept, 77-81; exec vpres, Assoc Kellogg, Can, 81-84; dir offshore technol, 84-87, MGR CLEAN COAL TECHNOL, MW KELLOGG CO, 87- *Mem:* Sigma Xi; Nat Soc Prof Engrs; Am Inst Chem Engrs. *Res:* Development of advanced coal gasification and combustion processes; fluid gas desulfurization and related activities. *Mailing Add:* c/o KLM Mgt One W Loop S Suite 108 Houston TX 77027

CAMPBELL, WILLIAM VERNON, b Chester, SC, May 4, 24; m 47; c 2. ENTOMOLOGY. *Educ:* Miss State Univ, BS, 51, MS, 52; NC State Univ, PhD(entom), 58. *Prof Exp:* Entomologist, Insect Control, Entom Res Br, Agr Res Serv, USDA, Tidewater Field Sta, Va, 52-53 & wheat stem sawfly proj, Cereal & Forage Insects Sect, Entom Res Br, NDak, 53-55; asst entom, 55-58, from asst prof to assoc prof, 59-69, PROF ENTOM, NC STATE UNIV, 70- *Concurrent Pos:* Assoc ed, Peanut Sci, Int Workshop, 80 & 81; consult, Int Crops Res Inst for Semi-Arid Tropics, India, 81. *Mem:* Entom Soc Am; fel Am Peanut Res Educ Soc. *Res:* Biology and control of insects; plant resistance to insects attacking field and forage crops; plant and insect histology; management of arthopods on peanuts in Thailand and Philippines. *Mailing Add:* Dept Entom NC State Univ Box 7613 Raleigh NC 27695

CAMPEANU, RADU IOAN, b Cluj Romania, Mar 21, 49; Can citizen; m 79; c 2. ATOMIC MOLECULAR STRUCTURE. *Educ:* Univ Cluj, BSc, 70, MSc 72; Univ Col, London, PhD(computational, atomic & molecular physics), 77. *Prof Exp:* From asst prof to assoc prof, atomic physic, Dept Physics, Univ Cluj, 72-85; course dir math, ACMS prog, 85-90, ASSOC PROF COMPUTER SCI, YORK UNIV, GLENDON COL, 90- *Concurrent Pos:* Consult, Magna Data Systs, Markham, Ont, 86; staff develop anal, Image Appln Group, IBM Lab, Toronto, 86- 90; adj prof, York Univ, Fac Grad Studies. *Res:* Numerically intensive computing; digital image processing; scientific computing (computational atomic physics). *Mailing Add:* Dept Computer Sci York Univ Glendon Col 2275 Bayview Ave Toronto ON M4N 3M6 Can

CAMPENOT, ROBERT BARRY, b East Orange, NJ, Mar 30, 46. NEUROBIOLOGY. *Educ:* Rutgers Univ, BA, 68; Univ Calif, Los Angeles, MS, 71; Mass Inst Technol-Woods Hole Oceanog Inst, PhD(biol oceanog), 76. *Prof Exp:* res fel neurobiol, Harvard Med Sch, 75-78; asst prof neurobiol, Cornell Univ, 78-87; PROF, DEPT ANAT & CELL BIOL, UNIV ALTA, 87- *Res:* Developmental neurobiology. *Mailing Add:* Dept Anat & Cell Biol Univ Alta Med Sci Bldg Edmonton AB T6G 2H7 Can

CAMPER, NYAL DWIGHT, b Lynchburg, Va, May 12, 39; m; c 2. PLANT PHYSIOLOGY, PLANT BIOCHEMISTRY. *Educ:* NC State Univ, BS, 62, PhD(crop sci), 67. *Prof Exp:* Res asst herbicide physiol, NC State Univ, 62-66; asst prof plant physiol, 66-71, assoc prof, 74-77, PROF PLANT PHYSIOL, CLEMSON UNIV, 77- *Concurrent Pos:* Dir student sci training prog, NSF, 68-70. *Mem:* Sigma Xi; Int Asn Plant Tissue Cult; Am Chem Soc; Am Soc Plant Physiol; Weed Sci Soc. *Res:* Plant tissue and cell culture; physiology and biochemistry of plant growth regulators. *Mailing Add:* Dept Plant Path & Physiol Clemson Univ Clemson SC 29634-0377

CAMPILLO, ANTHONY JOSEPH, b Newark, NJ, June 30, 42; m 69; c 3. AEROSOL OPTICS, NONLINEAR OPTICS. *Educ:* Newark Col Eng, BS, 64; Princeton Univ, MS, 66; Cornell Univ, PhD(elec eng), 73. *Prof Exp:* Res asst, Princeton Univ, 64-66; staff mem, GT&E Labs, Bayside, 66-68; res asst, Cornell Univ, 68-72; physicist, Los Alamos Sci Lab, 72-79; scientist, Brookhaven Nat Lab, 79-81; SECT HEAD, NAVAL RES LAB, 81- *Concurrent Pos:* Adj prof, Dept Chem, Cath Univ Am, 90- *Mem:* Fel Am Phys Soc; Optical Soc Am. *Res:* Nonlinear optics, especially high intensity beam propagation, optical power limiters and quantum dot systems; optical resonant phenomena in microdroplets including cavity quantum electrodynamics; ultrafast spectroscopy; photothermal spectroscopy. *Mailing Add:* Naval Res Lab Code 6546 Washington DC 20375-5000

CAMPION, JAMES J, b Philadelphia, Pa, July 14, 39; m 69. ANALYTICAL CHEMISTRY, PHYSICAL CHEMISTRY. *Educ:* La Salle Col, BA, 61; Univ Pittsburgh, PhD(nonaqueous solutions), 66. *Prof Exp:* Off Saline Water res assoc, Univ Pittsburgh & Mellon Inst, 67-68; asst prof, 68-73, ASSOC PROF ANALYTIC CHEM, STATE UNIV NY COL NEW PALTZ, 73- 68-73. *Mem:* Am Chem Soc; Sigma Xi. *Res:* Solution chemistry of electrolytes and nonelectrolytes in aqueous and nonaqueous solvents. *Mailing Add:* PO Box 286 Tillson NY 12486

CAMPISI, LOUIS SEBASTIAN, b New York, NY, Aug 9, 35; m 63; c 1. INORGANIC CHEMISTRY. *Educ:* City Col New York, BS, 56; Fordham Univ, MS, 60, PhD(inorg chem), 64. *Prof Exp:* Asst prof chem, 62-69, assoc prof, 69-77, PROF CHEM, IONA COL, 78- *Mem:* Am Chem Soc; Sigma Xi. *Res:* Chemistry of uranium and thorium; coordination chemistry. *Mailing Add:* 1103 Sacket Ave Bronx NY 10461

CAMPLING, C(HARLES) H(UGH) R(AMSAY), b Melville, Sask, Nov 30, 22; m 46; c 3. ELECTRICAL ENGINEERING. *Educ:* Queen's Univ, Can, BSc, 44; Mass Inst Technol, SM, 48. *Prof Exp:* Instr math, Queen's Univ, Can, 45-46; asst elec eng, Mass Inst Technol, 46-48, res engr, 48-49; jr res officer, Nat Res Coun Can, 49-50; assoc prof elec eng, Royal Mil Col, 50-55; assoc prof, 55-63, head dept, 67-77, PROF ELEC ENG, QUEEN'S UNIV, ONT, 63- *Honors & Awards:* Ross Medal, Eng Inst Can, 62; Centennial Medal, Inst Elec & Electronics Engrs, 84. *Mem:* Inst Elec & Electronics Engrs; fel Eng Inst Can; Can Soc Elec Eng (pres, 78-80). *Res:* Digital systems. *Mailing Add:* Dept Elec Eng Queen's Univ Kingston ON K7L 3N6 Can

CAMPO, ROBERT D, b New York, NY, Feb 18, 30; m 57; c 5. BIOCHEMISTRY, BIOLOGY. *Educ:* St John's Univ, NY, BS, 52, MS, 57; Rockefeller Inst, PhD(biochem), 63. *Prof Exp:* From instr to asst prof biochem, 63-70, assoc prof, 70-75, PROF ORTHOP SURG, SCH MED, HEALTH SCI CTR, TEMPLE UNIV, 75- *Mem:* Soc Exp Biol & Med; Am Rheumatism Asn; Orthop Res Soc; Sigma Xi. *Res:* Radiation biology; acid mucopolysaccharides; protein polysaccharides of cartilage; bone formation; calcification and degradation enzymes of cartilage; organic matrices of cartilage and bone. *Mailing Add:* 3420 N Broad St Philadelphia PA 19140

CAMPOLATTARO, ALFONSO, b Naples, Italy, Jan 4, 33; US citizen; m 60; c 3. THEORETICAL PHYSICS. *Educ:* Univ Naples, PhD(theoret physics), 59. *Prof Exp:* Res physicist, Univ Naples, 60-65, assoc prof theoret physics, 63-65; NATO fel, Palmer Physics Lab, Princeton Univ, 65-66; vis asst prof, Univ Calif, Irvine, 66-67, actg assoc prof, 67-68; vis assoc prof, 68-70, assoc prof, 70-71, PROF PHYSICS, UNIV MD, BALTIMORE COUNTY, 71- *Concurrent Pos:* Res physicist, Naval Surface Weapon Ctr, White Oak, Silver Spring, Md, 68- *Honors & Awards:* Young Physicist Prize, Ital Physics Soc, 62; US Navy Achievement Awards, 70 & 78. *Mem:* Am Asn Univ Prof; French Soc Phys Chem; AAAS; Am Phys Soc. *Res:* Mathematical aspects of quantum field theory; relativistic astrophysics; combustion theory; detonation theory; solid state physics; particle accelerators. *Mailing Add:* Dept Physics Univ Md Baltimore County Catonsville MD 21228

CAMPONESCHI, EUGENE THOMAS, JR, b Baltimore, Md, June 30, 55; m 80; c 3. MECHANICS OF COMPOSITE MATERIALS. *Educ:* Va Polytech Inst & State Univ, BS, 78, MS, 80; Univ Del, PhD(mech eng), 90. *Prof Exp:* Proj engr, 78-85, SR PROJ ENGR, DAVID TAYLOR RES CTR, 85- *Concurrent Pos:* Consult, Forensic Technol Int, 90- *Honors & Awards:* NAVSEA Medallion, 85. *Mem:* Nat Soc Prof Engrs; Am Soc Testing & Mat; Am Soc Composites; Soc Advan Mat & Process Eng. *Res:* Theoretical mechanics, testing and experimental mechanics, design and failure of fiber-reinforced composite materials; mechanics of materials and structures. *Mailing Add:* 530 Benforest Dr Severna Park MD 21146

CAMPORESI, ENRICO M, ANESTHESIOLOGY, PULMONARY MEDICINE. *Educ:* Univ Milan, Italy, MB, 70. *Prof Exp:* PROF ANESTHESIOL & ASST PROF PHYSIOL, DUKE UNIV, 74- *Mailing Add:* Dept Anesthesiol State Univ NY Health Sci Ctr 750 E Adams St Syracuse NY 13210

CAMRAS, MARVIN, b Chicago, Ill, Jan 1, 16; m 51; c 5. MAGNETISM, ELECTRONICS. *Educ:* Armour Inst Technol, BS, 40; Ill Inst Technol, MS, 42. *Hon Degrees:* LLD, Ill Inst Technol, 68. *Prof Exp:* Physicist, Ill Inst Technol Res Inst, 40-45, sr physicist, 45-59, sr engr, 59-65, sci adv, 65-69, sr sci adv, 69-85, PROF ELEC ENG, IIT INST TECHNOL, 85- *Concurrent Pos:* Engr & draftsman, Delta Star Elec Co, Ill, 39; ed, Trans on Audio, Inst Elec & Electronics Eng, 58-64; chmn, Nat Comt II, Int Electrotech Comn; mem S4 comt, Am Nat Stand Inst; dir, Midwest Acoust Conf, 69-72; mem, Nat Inventors Coun, 72; instr elec eng, Ill Inst Technol, 78. *Honors & Awards:* Distinguished Serv Award, Ill Inst Technol, 48; Scott Medal, 55; Citation, Ind Tech Col, 58; Achievement Award, Inst Elec & Electronics Eng, 58, Consumer Electronics Award, 64 & Broadcasting Papers Award, 64; Info Storage Award, Inst Elec & Electronic Engrs Magnetics Soc, 90; Hon Mem Award, Soc Motion Picture & TV Engrs, 90; John Potts Gold Medal Award, Audio Eng Soc, 69; Washington Award, Western Soc Engrs, 79; Nat Medal Technol, 90. *Mem:* Nat Acad Eng; hon mem Audio Eng Soc; fel AAAS; fel Inst Elec & Electronic Eng; fel Acoust Soc Am; Sigma Xi; fel Soc Motion Picture & TV Engrs. *Res:* Magnetic recording; stereophonic sound; electronics; magnetism; video recording. *Mailing Add:* 560 Lincoln Ave Glencoe IL 60022

CANADY, WILLIAM JAMES, b New York, NY, Dec 8, 24; m 55; c 1. BIOCHEMISTRY. *Educ:* Fordham Univ, BS, 46; George Washington Univ, MS, 50, PhD(biochem), 55. *Prof Exp:* From instr to assoc prof, 58-69, PROF BIOCHEM, MED CTR, WVA UNIV, 69- *Concurrent Pos:* Fel chem, Univ Ottawa, 55-57. *Mem:* Am Chem Soc; Am Soc Biol Chemists. *Res:* Vitamin K; methodology; mechanism of enzyme action; thermodynamics of ionization and solution processes. *Mailing Add:* Med Ctr WVa Univ Morgantown WV 26506

CANALE, RAYMOND PATRICK, b Cortland, NY, Nov 20, 41; m 68; c 2. SANITARY & BIOCHEMICAL ENGINEERING. *Educ:* Syracuse Univ, BS, 64, MS, 66, PhD(sanit eng), 68. *Prof Exp:* From asst prof to assoc prof, 68-77, PROF CIVIL ENG, UNIV MICH, 77- *Mem:* Am Soc Civil Engrs; Water Pollution Control Fedn; Am Chem Soc. *Res:* Biological treatment of domestic and industrial wastes; stream and estuarine pollution; mass transfer in open channel flow; predator-prey relationships in microbial cultures; mathematical ecology. *Mailing Add:* Dept Civil Eng 304 W Eng Bldg Univ Mich Ann Arbor MI 48109-2125

CANALE-PAROLA, ERCOLE, b Frosinone, Italy, Sept 13, 29; US citizen; m 54; c 2. MICROBIOLOGY, BACTERIAL PHYSIOLOGY. *Educ:* Univ Ill, BS, 56, MS, 57, PhD(microbiol), 61. *Prof Exp:* NIH fel, Hopkins Marine Sta, Stanford Univ, 61-63; from asst prof to assoc prof, 63-67, head dept, 84-89, PROF MICROBIOL, UNIV MASS, 73- *Concurrent Pos:* Res grants, NIH, 64-87, NSF, 67-70, US Dept Energy, 81- & US Dept Agr, 85-89; vis scientist, Woods Hole Oceanog Inst, 69, Dept Biol, Amherst Col, 78. *Honors & Awards:* Fel AAAS, 85. *Mem:* Am Soc Microbiol; Am Acad Microbiol; AAAS. *Res:* Bacterial physiology; microbial ecology; spirochetes; sarcinae; bacterial pigments; evolution of microorganisms; cellulose hemicelluloses and pectin degradation by anaerobic bacteria; physiology of bacteria in nutrient-limited environments. *Mailing Add:* Dept Microbiol Univ Mass Amherst MA 01003

CANAL-FREDERICK, GHISLAINE R, b Metz, France, Mar 31, 33; div; c 4. PSYCHOPHYSIOLOGY, HUMAN NEUROPSYCHOLOGY. *Educ:* Univ Md, BS, 71; Howard Univ, MS, 74, PhD(neuropsychol), 78. *Prof Exp:* Psychologist clin & res, Surg Neurol Br, Nat Inst Neurol & Commun Dis & Stroke, 73-76, res psychologist functional neurosurg, clin neurosci, 76-79; STOCKBROKER, LEG, MASON, WOOD & WALKER, 79- *Mem:* Am Psychol Asn; Soc Neurosci; Grad Women Sci; AAAS. *Res:* Human neuropsychology; cerebral organization; cognition; hemispheric asymmetry; age, sex, and hand related factors in cerebral organization. *Mailing Add:* 12025 Wetherfield Lane Potomac MD 20854

CANARY, JOHN J(OSEPH), b Mineola, NY, Jan 9, 25; m 51; c 7. METABOLISM, ENDOCRINOLOGY. *Educ:* St John's Col, NY, BS, 47; Georgetown Univ, MD, 51. *Prof Exp:* NIH trainee metab dis, Georgetown Univ, 54-55; from instr to assoc prof, 56-68, actg dir radioisotope lab, Univ Hosp, 58, dir gen clin res ctr, 73-78, dir, Div & Clins Endocrinol & Metab, 59-84, PROF MED, SCH MED, GEORGETOWN UNIV, 68-, DIR ENDOCRINOL RES, 84- *Concurrent Pos:* Res fel, Georgeton Univ Hosp, NIH, 55-57; clin instr biochem, Sch Med, Georgetown Univ, 57-60, spec lectr, 70-; consult & lectr endocrinol, Nat Naval Med Ctr, 58-82; consult metab, Walter Reed Army Inst Res, 61-81; consult metab & endocrinol, NIH, 76-; mem, Comt Diabetes in Caribbean, WHO/Pan Am Health Orgn; consult traditional med, diabetes & cardiovasc dis, WHO, 79-; co-op invest, Human Nutrit Res Ctr, USDA, Belts, MD, 79- *Mem:* AAAS; Am Fedn Clin Res; Endocrine Soc; Am Soc Bone & Mineral Res; Am Soc Clin Invest. *Res:* Study of the density and chemical composition of human bone in various disease states, particularly osteoporosis and the effects of therapy thereon; body composition and density; adrenocortical hyperfunct. *Mailing Add:* Univ Hosp The Georgetown Univ 3800 Reservoir Rd NW Washington DC 20007

CANAVAN, ROBERT I, b Ridgefield Park, NJ, July 31, 27. COMPUTER SCIENCE. *Educ:* Woodstock Col, Md, AB, 50; NY Univ, PhD(math), 57; Univ Innsbruck, lic theol, 61. *Prof Exp:* Instr math, LeMoyne Col, NY, 51-52; from asst prof to assoc prof, St Peter's Col, NJ, 62-69; assoc prof, 69-74, PROF MATH, MONMOUTH COL, NJ, 74- *Mem:* Am Math Soc; Math Asn Am; Soc Indust & Appl Math; Asn Comput Mach. *Res:* Ordinary and partial differential equations. *Mailing Add:* Dept Computer Sci Monmouth Col West Long Branch NJ 07764

CANAWATI, HANNA N, b Bethlehem, Palestine, Nov 18, 38; m 68; c 3. CLINICAL MICROBIOLOGY, MEDICAL TECHNOLOGY. *Educ:* Damascus Univ, BS, 64; Roosevelt Univ, MS, 71; Chicago Med Sch, PhD(med microbiol), 74. *Prof Exp:* asst prof path, Sch Med, Univ Southern Calif, 75-81, AT DEPT CLIN PATH, UNIV SOUTHERN CALIF; CHIEF MICROBIOL, RANCHO LOS AMIGOS HOSP & JOHN WESLEY COUNTY HOSP, 75- *Concurrent Pos:* Lectr med technol, Masters Prog, Calif State Univ, Dominguez Hills, 76-; microbiol consult, 81- *Mem:* Am Soc Microbiol; Am Acad Microbiol; NY Acad Sci. *Res:* Clinical significance of anaerobes in diabetic patients; bacteriologic study of liver patients spontaneous bacterial pertitonitis; methicillin resistant staph aureus characteristics and drug resistance; in vitro evaluation of cephalosporins against clinically isolated bacteria. *Mailing Add:* 16259 Aurora Crest Dr Whittier CA 90605

CANCRO, MICHAEL P, b Washington, DC, Oct 28, 49. IMMUNOLOGY. *Educ:* Univ Md, PhD(zool), 76. *Prof Exp:* ASSOC PROF PATH, UNIV PA, 79- *Mailing Add:* Dept Path Med Sch Univ Pa 36th & Hamilton Walk Rm 284 G-3 Philadelphia PA 19104

CANCRO, ROBERT, b New York, NY, Feb 23, 32; m 56; c 2. PSYCHIATRY. *Educ:* Fordham Univ, 48-51; State Univ NY, MD, 55, DM Sci, 62;Am Bd Psychiat & Neurol, cert psychiat, 62. *Prof Exp:* Dir alcohol res ward & instr psychiat, State Univ NY Downstate Med Ctr, 62-66; psychiatrist, Menninger Found, 66-69, mem fac, Menninger Sch Psychiat, 67-69; vis assoc, Ctr Advan Study & vis prof, Dept Comput Sci, Univ Ill, Urbana, 69-70; prof psychiat, Univ Conn, Farmington, 70-76; PROF & CHMN DEPT PSYCHIAT, NY UNIV, 76-; DIR, NATHAN S KLINE INST PSYCHIAT RES, 82- *Concurrent Pos:* Consult, Dept Comput Sci, Univ Ill, 67 & Topeka State & Vet Admin Hosps, Topeka, Kans, 67-69. *Honors & Awards:* Fromm-Reichmann Award, 75. *Mem:* AAAS; Am Psychiat Asn; Asn Am Med Cols; AMA; NY Acad Sci. *Res:* Prediction of outcome in schizophrenia; nature of pathology in schizophrenia and in addictions. *Mailing Add:* NYU Med Ctr 550 First Ave New York NY 10016

CANDE, W ZACHEUS, b Amityville, NY, Dec 21, 45; m 77; c 2. CELL BIOLOGY, DEVELOPMENTAL BIOLOGY. *Educ:* Yale Univ, BS, 67; Stanford Univ, PhD(plant physiol), 73. *Prof Exp:* PROF CELL & DEVELOP BIOL, UNIV CALIF, BERKELEY, 76- *Mem:* Am Soc Cell Biol. *Res:* Mechanism of mitosis and meiosis; role of cytoskeleton in plant development. *Mailing Add:* Dept Molecular & Cell Biol 341 Life Sci Addition Univ Calif Berkeley CA 94720

CANDELAS, GRACIELA C, b PR, 22; US citizen. MOLECULAR BIOLOGY, DEVELOPMENTAL BIOLOGY. *Educ:* Univ PR, BS, 44; Duke Univ, MS, 59; Univ Miami, PhD(molecular biol), 66. *Prof Exp:* Instr biol, 51-57, from asst prof to assoc prof, 61-71, PROF BIOL, UNIV PR, RIO PIEDRAS, 71- *Concurrent Pos:* Vis prof biol, Syracuse Univ, 69-71, City Col New York, 74-75 & Rockefeller Univ, 79-80; prof cell & molecular biol, Med Col Ga, 72-74; prin investr, NSF & NIH proj, 83-; mem, Policy Adv Comt, Nat Inst Dent Res, 85-88. *Mem:* Int Soc Develop Biol; Am Soc Cell Biol; Sigma Xi; Int Cell Res Orgn; NY Acad Sci. *Res:* Fibroin synthesis and its regulation using a pair of glands from the spider, Nephila clavipes as a model system; translational level control; modulation by t-RNA and alanine t-RNA genes, and evolution of fibroin synthesis strategies in the spider and silkworm system. *Mailing Add:* Dept Biol Univ PR Rio Piedras PR 00931

CANDER, LEON, b Philadelphia, Pa, Oct 7, 26; m 54; c 2. INTERNAL MEDICINE, CLINICAL PHYSIOLOGY. *Educ:* Temple Univ, MD, 51. *Prof Exp:* Intern, Southern Div, Einstein Med Ctr, 51-52; fel physiol, Dept Physiol & Pharmacol, Grad Sch Med, Univ Pa, 52-54, instr, 54-55, assoc, 55-56; from asst resident to resident med, Beth Israel Hosp, Boston, 56-58; sr instr, Med Sch, Tufts Univ, 58-60; from asst prof to assoc prof, Hahnemann Med Col & Hosp, 60-66, head, Sect Chest Dis, 60-66; prof physiol & internal med & chmn dept, Med Sch, Univ Tex, San Antonio, 66-72; chmn, Dept Med, 72-80, HEAD, SECT CHEST DIS, DAROFF DIV, ALBERT EINSTEIN MED CTR, PHILADELPHIA, 80-; PROF MED, JEFFERSON MED COL, 72-; CLIN PROF MED, HAHNEMANN MED COL, 85- *Concurrent Pos:* Asst lab instr, Sch Auxiliary Med Sci, Univ Pa, 53-55, Am Col Physicians res fel, 54-55; asst, Harvard Med Sch, 57-58; fel, Nat Acad Sci-Nat Res Coun, 55-56; Markle scholar acad med, Hahnemann Med Col & Hosp, 61-66; nat consult, Black Lung Prog, US Dept Labor, 78- *Mem:* Am Fedn Clin Res; Am Physiol Soc; Am Col Physicians. *Res:* Clinical pulmonary physiology; pulmonary disease. *Mailing Add:* 317 Cherry Lane Wynnewood PA 19096

CANDIA, OSCAR A, b Buenos Aires, Arg, Apr 30, 35; m 60; c 3. PHYSIOLOGY, BIOPHYSICS. *Educ:* Univ Buenos Aires, MD, 59. *Prof Exp:* Instr basic physics, Univ Buenos Aires, 60-61, head lab & res assoc biophysics, 62-63; res assoc, Univ Louisville, 64-65, asst prof, 65-68; assoc prof ophthal, 68-77, assoc prof physiol, 78-84, PROF OPHTHAL & PHYSIOL, MT SINAI SCH MED, 84- *Concurrent Pos:* Res fel electrophysiol, Univ Buenos Aires, 60-62; NIH career develop award, 66-71; res assoc, Arg Nat Res Coun, 60-63; mem, Vision Res Prog Comt, Nat Eye Inst, 79-83; assoc ed, Invest Ophthal & Visual Sci. *Honors & Awards:* Alcon Award, 85. *Mem:* Biophys Soc; Asn Res Ophthal; Am Physiol Soc; Asn Res Vision Ophthal. *Res:* Ion transport in biological membranes, models; instrumentation. *Mailing Add:* Dept Ophthal Mt Sinai Sch Med New York NY 10029

CANDIDO, EDWARD PETER MARIO, b Noranda, Que, Mar 28, 46; m 69; c 3. BIOCHEMISTRY, MOLECULAR BIOLOGY. *Educ:* McGill Univ, BSc, 68; Univ BC, PhD(biochem), 72. *Prof Exp:* Fel, Med Res Coun Lab Molecular Biol, Cambridge, Eng, 72-73; from asst prof to assoc prof, 73-85, PROF BIOCHEM, UNIV BC, 85- *Concurrent Pos:* Res grants, Med Res Coun Can, 73-92, NIH, 76-82 & BC Health Care Res Found, 79-91; Med Res Coun Can vis scientist, Pasteur Inst, Paris, 84-85. *Mem:* Can Biochem Soc; Am Soc Biol Chemists. *Res:* Regulation of gene activity in eukaryotic cells; the ubiquitin system; structure and regulation of heat shock genes. *Mailing Add:* Dept Biochem Univ BC Vancouver BC V6T 1W5 Can

CANDY, J(AMES) C(HARLES), b Crickhowell, Wales, Sept 27, 29; US citizen; m 54; c 1. ELECTRONICS ENGINEERING. *Educ:* Univ Wales, BSc, 51, PhD(electronics), 54. *Prof Exp:* Res engr, S Smith & Sons, Eng, 54-56; sr sci officer, Atomic Energy Res Estab, 56-59; res assoc, Dept Physics, Univ Minn, 59-60; MEM TECH STAFF, BELL TEL LABS, 60- *Mem:* Inst Elec & Electronics Engrs. *Res:* High-speed electronic circuits; television transmission. *Mailing Add:* Systs Res Div Bell Tel Labs Inc HO4C 522 Holmdel NJ 07733

CANE, DAVID EARL, b Sept 22, 44. BIO-ORGANIC CHEMISTRY, NATURAL PRODUCTS CHEMISTRY. *Educ:* Harvard Col, AB, 66; Harvard Univ, MA, 67, PhD(chem), 71. *Prof Exp:* NIH fel, 71; res assoc org chem, Swiss Fed Inst Technol, 71-73; from asst prof to assoc prof, 73-80, PROF CHEM, BROWN UNIV, 80- *Concurrent Pos:* Vis assoc prof, Univ Chicago, 80; mem, Bio Organic Natural Prod Study Sect, NIH, 80-84; chmn, Gordon Res Conf Natural Prod, 82; res fel, Alfred P Sloan, 78-82; nat res career develop award, NIH, 78-83; fel, Japan Soc Prom Sci, 83; sr int fel, Fogarty Int Ctr, 89; distinguished vis fel, Christ's Col, Cambridge, 89-90; fel, John Simon Guuggenheim Mem Found, 90; Fulbright scholar, 90. *Honors & Awards:* Ernest Guenther Award, Am Chem Soc, 85; Simonson lectr, Royal Soc Chem, 90-91. *Mem:* Am Chem Soc; Royal Soc Chem; Am Soc Microbiol; Am Soc Pharmacog. *Res:* Biosynthesis of natural products; stereochemistry and synthetic methods. *Mailing Add:* Dept Chem Brown Univ Providence RI 02912

CANE, MARK A, b Brooklyn, NY, Oct 20, 44; m 68; c 2. OCEANOGRAPHY, METEOROLOGY. *Educ:* Harvard Univ, BA, 65, MS, 68; Mass Inst Technol, PhD(meteorol), 76. *Prof Exp:* Math analyst, Comput Appln Inc, 66-70; asst prof math, New Eng Col, 70-72; fel oceanog, Nat Res Coun, 75-76; analyst, Sigma Data Serv Inc, 76-78; earth scientist oceanog, Goddard Space Flight Ctr, NASA, 78-79; from asst to assoc prof oceanog, Mass Inst Technol, 79-84; sr res scientist, 84-87, DOHERTY SR SCIENTIST, LAMONT-DOHERTY GEOL OBSERV, 87- *Concurrent Pos:* Adj asst prof, Dept Geol Sci, Columbia Univ, 77-78, adj assoc prof, 84-88, adj prof, 88- *Mem:* Am Meteorol Soc; Am Geophys Union; Oceanog Soc. *Res:* Dynamics of equatorial ocean circulation; air-sea interaction; el nino; numerical ocean modeling. *Mailing Add:* Lamont-Doherty Geol Observ Palisades NY 10964

CANELLAKIS, EVANGELO S, b Tientsin, China, June 20, 22; nat US; m 48, 75; c 2. BIOCHEMISTRY. *Educ:* Nat Univ Athens, BS, 47; Univ Calif, PhD(biochem), 51. *Prof Exp:* Res asst, Univ Calif, 48-50, asst, 50-51; Nat Found Infantile Paralysis fel, Dept Physiol Chem, Univ Wis, 51-54; Squibb fel pharmacol, Yale Univ, 54-55, from instr to assoc prof, 55-64, PROF PHARMACOL, YALE UNIV, 64- *Concurrent Pos:* Res career develop award, USPHS, 58-63, res career prof, 63-; ed, Biochimica Biophysica Acta, 69. *Mem:* Am Soc Biol Chem; Am Asn Cancer Res; Brit Biochem Soc; Hellenic Chem Soc. *Res:* Amino acid metabolism; pyrimidine metabolism; mechanisms in nucleic acid synthesis; diacridines; relationship of biochemical sites of action of antitumor drugs to their antineoplastic properties; control and regulation of enzymatic activities. *Mailing Add:* Dept Pharmacol Yale Med Sch New Haven CT 06510

CANELLAKIS, ZOE NAKOS, b Lowell, Mass, Sept 7, 27; m 48; c 2. BIOCHEMISTRY. *Educ:* Vassar Col, BA, 47; Univ Calif, MS, 51; Univ Wis, PhD(physiol chem), 54. *Prof Exp:* Asst biochem, Yale Univ, 54-55, instr, 55-59, res assoc, 59-67, asst dean, Grad Sch, 72-77, SR RES SCIENTIST, DEPT PHARMACOL, MED SCH, YALE UNIV, 67- *Res:* Amino acid metabolism; protein and nucleic acid synthesis; polyamines. *Mailing Add:* Dept Pharmacol Yale Univ Med Sch New Haven CT 06510

CANELLOS, GEORGE P, b Boston, Mass, Nov 1, 34; m 58; c 3. INTERNAL MEDICINE, ONCOLOGY. *Educ:* Harvard Univ, AB, 56; Columbia Univ, MD, 60; Am Bd Internal Med, dipl, 68. *Prof Exp:* Intern, Mass Gen Hosp, Boston, 60-61; clin fel, Harvard Med Sch, 61-62; asst resident, Mass Gen Hosp, Boston, 62-63; clin assoc, Nat Cancer Inst, 63-65; sr resident, Mass Gen Hosp, Boston, 65-66; res asst hemat, Royal Postgrad Med Sch London, 66-67; sr investr, Nat Cancer Inst, 67-75, asst chief med br, 73-75, clin dir,

74-75; assoc prof, 75-84, PROF MED, HARVARD MED SCH, 84-; CHIEF DIV MED, SIDNEY FARBER CANCER CTR, 75-; SR ASSOC MED, PETER BENT BRIGHAM HOSP, 75- *Concurrent Pos:* Asst clin prof, Dept Med, Georgetown Univ, 68-74, assoc clin prof, 74-; sr assoc med, 75-84, physician, Peter Bent Brigham Hosp, 84- *Mem:* Am Soc Hemat; Am Fedn Clin Res; Am Asn Cancer Res; Am Soc Clin Oncol; fel Am Col Physicians; Am Soc Clin Invest. *Res:* Cell biology as related to disorders of hemopoiesis; cancer therapy; malignant lymphoma. *Mailing Add:* 116 Sherburn Circle Weston MA 02193

CANFIELD, CARL REX, JR, b Oregon, Ill, May 15, 23; m 52; c 3. ENGINEERING RESEARCH. *Educ:* Ill Inst Technol, BS. *Prof Exp:* Tool designer, Delta Star Elec Co, 42-43; engr, Armour Res Found, 46-51; res engr, Cent Res, Borg-Warner Corp, 51-59, chief engr prod develop, Carne Co, 58-59, dir eng, Marvel-Schebler Prod Div, 59-65, gen mgr, Simms Marvel-Schebler Ltd, London, Eng, 65-68, mgr eng, Warner Elec Brake & Clutch Co, 68-72; vpres res & develop, Schwitzer, Wallace Murray Corp, 72-77, vpres & gen mgr, 77-79, pres, Eng Components Group, Wallace Murray Corp, 79-86; RETIRED. *Concurrent Pos:* Consult, 86- *Mem:* Instrument Soc Am; Soc Automotive Engrs. *Res:* Special instrumentation for road testing of automatic transmissions of cars, trucks and special vehicles; development of wet friction materials for automatic transmissions and fuel injection equipment for cars and trucks; photographic equipment. *Mailing Add:* PO Box 2643 153 E Kelly Jackson WY 83001

CANFIELD, CRAIG JENNINGS, b Pasadena, Calif, May 11, 32; m 54; c 7. CLINICAL PHARMACOLOGY, MALARIOLOGY. *Educ:* Univ Ore, BS, 55, MD, 57. *Prof Exp:* Resident med, Walter Reed Gen Hosp, 60-63; chief dept med clin res, SEATO Med Res Lab, 64-66; fel hemat, Walter Reed Gen Hosp, 68-69, asst dir malaria & chief clin pharm, Walter Reed Army Inst Res, 70-75, dir div med chem, 75-78, dir div exp therapeut, Walter Reed Army Inst Res, 78-84; RES DIR, PHARMACEUT SYSTS INC, 84- *Concurrent Pos:* Prin investr antiparasitic notices of claimed investigational exemption for a new drugs, US Army Med Res & Develop Command, 70-; mem malarial chemother task force, WHO, 75-; prof pharmacol, Uniformed Serv Univ, 77- *Honors & Awards:* Gorgas Medal, 79. *Mem:* Fel Am Col Clin Pharmacol; fel Am Col Physicians; Am Soc Trop Med & Hyg; Am Fedn Clin Res; Am Soc Pharmacol & Exp Therapeut. *Res:* Antiparasitic drug development, especially malaria; hematology; clinical research and pharmacology; diagnosis and treatment of malaria. *Mailing Add:* Pharmaceut Systs Inc 927B N Russell Ave Gaithersburg MD 20879

CANFIELD, EARL RODNEY, b Atlanta, Ga, July 27, 49; m 73; c 1. COMBINATORICS. *Educ:* Brown Univ, ScB, 71; Univ Calif, San Diego, PhD(math), 75. *Prof Exp:* asst prof comput sci, 75-80, ASSOC PROF STATIST & COMPUT SCI, UNIV GA, 80- *Concurrent Pos:* Prin investr, NSF grant, 78-79; vis asst prof comput sci, Univ Calif, San Diego, 78-79. *Res:* Asymptotic and probabilistic methods in combinatorics and analysis of algorithms. *Mailing Add:* Dept Statist & Computer Sci Univ Ga Athens GA 30602

CANFIELD, EARLE LLOYD, b Des Moines, Iowa, Oct 24, 18; m 47; c 3. MATHEMATICS, STATISTICS. *Educ:* Drake Univ, BA, 40; Northwestern Univ, MA, 44; Iowa State Univ, PhD, 50. *Prof Exp:* Teacher high schs, Iowa, 40-46, prin, 43-46; from instr to assoc prof math, 46-58, PROF MATH, DRAKE UNIV, 58-, DEAN GRAD STUDIES, 57- *Concurrent Pos:* Consult, Des Moines Secondary Sch Math Teachers, 51-52; Des Moines consult sch math study group, Yale Univ, 60-61 & Stanford Univ, 61-62; consult, Coun Grad Schs; mem exec comt, Midwestern Asn Grad Schs, 71-75, chmn, 73-74; mem exec comt, Coun Grad Schs in US, 75-76 & 77-80. *Mem:* Math Asn Am. *Res:* Statistical analysis, experimental educational data; mathematical statistics; teaching mathematics, experimental materials. *Mailing Add:* 3820 SW 31st St Des Moines IA 50321

CANFIELD, JAMES HOWARD, b Elmhurst, Ill, Dec 25, 30; m 50; c 2. ORGANIC CHEMISTRY, INFORMATION SCIENCE. *Educ:* Purdue Univ, BSc, 51; Univ Calif, PhD(chem), 54. *Prof Exp:* Res chemist, Elastomers Chem Dept, Jackson Lab, E I du Pont de Nemours & Co, Inc, 56-58; res proj chemist, Whittier Res Lab, Am Potash & Chem Corp, 58-59; res mgr, Magna Prods, Inc, 59-64; propellant specialist, Foreign Tech Div, 65-67, chem specialist, 67-69, tech adv, Info Systs Div, Air Force Syst Command, 70-78, TECH ADV, INFO SERV DIV, FOREIGN TECH DIV, WRIGHT-PATTERSON AFB, 78- *Mem:* Am Chem Soc; The Chem Soc; Am Soc Info Sci; Am Inst Chemists. *Res:* Organic synthesis; general polymer chemistry; organoboron derivatives. *Mailing Add:* 2298 Jacavanda Dr Dayton OH 45431

CANFIELD, RICHARD CHARLES, b Detroit, Mich, Dec 9, 37; m 78; c 4. SOLAR PHYSICS, RADIATIVE TRANSFER. *Educ:* Univ Mich, BS, 59, MS, 61; Univ Colo, PhD(astrogeophysics), 68. *Prof Exp:* Vis scientist, High Altitude Observ, Nat Ctr Atmospheric Res, 68-69; fel, Neth Orgn Sci Res, 69-70; astrophysicist, Sacramento Peak Observ, 70-76; assoc res physicist, 76-80, res physicist, 80-85, ADJ PROF ASTRON, UNIV CALIF, SAN DIEGO, 85-; ASTRONOMER, UNIV HAWAII, 85- *Concurrent Pos:* Mem adv comn, High Altitude Observ Nat Ctr Atmospheric Res, 74-78; mem optical & infrared subcomt, Astron Adv Comt, NSF, 79-; mem comt solar and space physics, Nat Acad Sci, 77-80; mem, Space Sci Bd, Nat Acad Sci, 80-83. *Mem:* Am Astron Soc; Int Astron Union. *Res:* Radiative transfer; application and development of theory of radiative transfer line formation to solar and astrophysical problems, including chromospheric heating, solar flares and solar velocity fields, and quasar emission line regions. *Mailing Add:* Ctr Astrophysics & Space Sci Univ Calif San Diego La Jolla CA 92093-0319

CANFIELD, ROBERT CHARLES, b Forsyth, Mont, Mar, 17, 22; m 56; c 7. CLINICAL RESTORATIVE DENTISTRY, ORAL ANATOMY. *Educ:* Univ Wash, DDS, 51, cert biomed instrumentation, 68. *Prof Exp:* Clin asst, 51-56, clin assoc, 56-67, from asst to assoc prof, 67-74, prof dent, 74-, asst dean regional educ, 79-, AT DEPT NEUROSURG, UNIV WASH. *Concurrent Pos:* Vis prof, Zahnartz Inst, Univ Vienna, Austria, 66; Danforth Assoc, 71; dir student affairs, Univ Wash, 76-79; adj prof neurosurg, Univ Wash, 81; affil, Ctr Res Oral Biol. *Mem:* Int Asn Study Pain; Pierre Fauchard Academie. *Res:* Cerebral responses to tooth pulp stimulation in laboratory animals and man; axonal degeneration patterns in cat brainstem trigeminal nucleus after tooth pulp removal; estimation of sensation magnitude elicited by tooth pulp stimulation in man. *Mailing Add:* 6291 NE North Shore Rd Belfair WA 98528

CANFIELD, ROBERT E, b New York, NY, June 4, 31; m 54; c 3. MEDICINE, ENDOCRINOLOGY. *Educ:* Lehigh Univ, BS, 52; Univ Rochester, MD, 57. *Prof Exp:* From intern to asst resident med, Presby Hosp, 57-59; res assoc, NIH, 59-62; NIH spec res fel, Enzymol Lab, Nat Ctr Sci Res France, 62-63; from asst prof to assoc prof, 63-72, PROF MED, COL PHYSICIANS & SURGEONS, COLUMBIA UNIV, 72-, DIR, CLIN RES CTR, 75- *Mem:* Am Soc Biol Chemists; Am Soc Clin Invest; Endocrine Soc; fel Royal Soc Med; Asn Am Physicians. *Res:* Protein chemistry, especially related to endocrine and metabolism disorders in man; studies of lysozyme, human chorionic gonadotropin, fibrinogen and Pagets disease of bone. *Mailing Add:* Dept Med Col Physicians & Surgeons Columbia Univ 630 W 168th St New York NY 10032-3795

CANFIELD, WILLIAM H, b Oklahoma City, Okla, May 24, 20. SPEECH PATHOLOGY, AUDIOLOGY. *Educ:* Northwestern Univ, BS, 42; Columbia Univ, MA, 50, EdD(speech), 59. *Prof Exp:* Instr speech, Hofstra Col, 50-57; instr speech path, Teachers Col, Columbia Univ, 57-58, res assoc, 58-60, asst prof, 60-63; from assoc prof to prof speech path & audiol, Adelphi Univ, 63-83; RETIRED. *Concurrent Pos:* Consult speech ther, St Barnabas Hosp, Bronx, NY, 59- & St Luke's Hosp, New York, 59- *Mem:* Am Speech & Hearing Asn; Speech Commun Asn; Am Cleft Palate Asn. *Res:* Laryngectomy, cleft palate, Parkinson's disease. *Mailing Add:* 16 E 84th St New York NY 10028

CANHAM, JOHN EDWARD, b Buffalo, NY, Sept 10, 24; m 47; c 7. MEDICINE, NUTRITION. *Educ:* Columbia Univ, MD, 49. *Prof Exp:* Intern, Letterman Gen Hosp, US Army, San Francisco, Calif, 49-50, resident internal med, 51-53, physician, 8th Sta Hosp, Kobe, Japan, 50-51, US Army Hosp, Ft Belvoir, Va, 54-56, chief med serv, US Army Hosp, Wurzburg, Ger, 57-60, chief metab div, US Army Med Res & Nutrit Lab, 61-64, dir res lab, 64-66, comdr, 121st Evacuation Hosp, Ascom, Korea, 66-67, dir res lab, US Army Med Res & Nutrit Lab, 67-73, dir res lab, Letterman Army Inst Res, 73-79; dir clin res, med prod, Cutter Labs, Inc, 79-84; dir Med Affairs, Kabi Vitrum, Inc, 84-90; RETIRED. *Concurrent Pos:* Chief prev med, Wurzburg Med Serv Area, Ger, 57-60; clinician, Interdept Comt Nutrit for Nat Defense, Uruguay Nutrit Surv, 62; affil prof, Colo State Univ, 64-66 & 68-74; US Army liaison rep, Nutrit Study Sect, NIH & Food & Nutrit Bd, Nat Acad Sci-Nat Res Coun, 64-78; US deleg, Far East Conf Nutrit, Manila, Philippines, 67; consult nutrit to Surgeon Gen, US Army, 69-79; mem food formulation res panel, US Dept Defense, 70-78. *Honors & Awards:* Joseph Goldberger Award Clin Nutrit, AMA, 71. *Mem:* Am Inst Nutrit; Am Soc Clin Nutrit; AMA; Asn Mil Surg US; Am Bd Nutrit. *Res:* Nutrition research including applied, clinical and parenteral nutrition, nutrient requirements and nutritional biochemistry; acclimatization, metabolic and nutritional aspects of environment stress. *Mailing Add:* 6835 Wilton Dr Oakland CA 94611

CANHAM, PETER BENNETT, b Toronto, Ont, Apr 26, 41; m 64; c 3. BIOPHYSICS. *Educ:* Univ Toronto, BASc, 62; Univ Waterloo, MSc, 64; Univ Western Ont, PhD(biophys), 67. *Prof Exp:* Proj engr, Can Govt, 62-63; lectr biophys, 67-68, from asst prof to assoc prof, 68-79, PROF BIOPHYS, UNIV WESTERN ONT, 79-, CHMN, DEPT MED BIOPHYSICS, 87- *Mem:* Biophys Soc Can; Int Soc Stereology; Microcirculatory Soc. *Res:* Physics of the blood vessel wall (brain arteries, coronary arteries) and the rheology of mammalian red blood cells. *Mailing Add:* Dept Med Biophysics Univ Western Ont London ON N6A 5C1 Can

CANHAM, RICHARD GORDON, b Arlington, Va, Aug 30, 28; m 52; c 3. PHYSICAL CHEMISTRY. *Educ:* Col William & Mary, BS, 50; Johns Hopkins Univ, MA, 54, PhD, 59. *Prof Exp:* Chemist, Nat Bur Stand, 50-55; from asst prof to assoc prof chem, Col William & Mary, 56-62; assoc prof, Col Charleston, 62-64; assoc prof chem, 64-69, chmn dept phys sci, 71-73, PROF CHEM, OKLA BAPTIST UNIV, 69-, CHMN, DEPT PHYS SCI, 80- *Mem:* Am Chem Soc; Sigma Xi; AAAS. *Res:* Electromotive force of cells; pH in aqueous and nonaqueous media. *Mailing Add:* c/o Okla Baptist Univ Shawnee OK 74801

CANIS, WAYNE F, b Elmira, NY, Aug 30, 39; m 68; c 1. GEOLOGY, ENVIRONMENTAL SCIENCES. *Educ:* Colgate Univ, AB, 61; Univ Mo-Columbia, MA, 63, PhD(geol), 67. *Prof Exp:* Explor geologist, Shell Oil Co, 67-70; assoc prof phys sci, Livingston Univ, 70-80; assoc prof gen sci, 80-85, PROF GEOL, UNIV N ALA, 85- *Mem:* Soc Econ Paleontologists & Minerologists; Am Asn Petrol Geol; Paleont Soc; Geol Soc Am. *Res:* Biostratigraphy; coastal (marine) geology; holothurian sclerites. *Mailing Add:* Dept Physics & Earth Sci Univ N Ala Florence AL 35632-0001

CANIZARES, CLAUDE ROGER, b Tucson, Ariz, June 14, 45; m 68; c 2. X-RAY ASTRONOMY. *Educ:* Harvard Univ, AB, 67, MS, 68, PhD(physics), 72. *Prof Exp:* Mem res staff, Ctr Space Res, Mass Inst Technol, 71-74, from asst prof to assoc prof, 74-84, dep dir, Ctr Space Res, 89-90, PROF PHYSICS, MASS INST TECHNOL, 84-, DIR, CTR SPACE RES, 90- *Concurrent Pos:* Fel Alfred P Sloan Found, 80-84; vis fel, Inst Astron, Univ Cambridge, 81-82; Royal Soc fel, 81-82; Assoc Univ Res Astron vis comt, 81-84; NASA Space Earth Sci Adv Comt, 85-88; prin investr Adv X-ray Astrophy Facil, 85-; mem, Astron Astrophys Survey Comt, Nat Acad Sci, 89-91; NASA fel. *Honors & Awards:* Group Achievement Award, NASA, 80. *Mem:* Int Astron Union; Am Phys Soc; Am Astron Soc; AAAS. *Res:* Optical and X-ray studies of X-ray sources; design and construction of X-ray satellite experiments and spectroscopy. *Mailing Add:* Rm 37-241 Mass Inst Technol Cambridge MA 02139

CANIZARES, ORLANDO, b Havana, Cuba, May 27, 10; nat US; m 38; c 3. DERMATOLOGY. *Educ:* Univ Paris, MD, 35. *Prof Exp:* PROF CLIN DERMAT & SYPHILOL, GRAD MED SCH, NY UNIV, 53-; VIS DERMATOLOGIST & SYPHILOLOGIST, BELLEVUE HOSP, 54- *Concurrent Pos:* Consult, Vet Admin, NY & USPHS Hosp, Staten Island; chief dermat serv, St Vincent's Hosp. *Mem:* AAAS; Am Dermat Asn; fel Am Col Physicians. *Mailing Add:* 210 Grandview Blvd Yonkers NY 10710

CANN, JOHN RUSWEILER, b Bethlehem, Pa, Dec 11, 20; m 46; c 3. BIOPHYSICAL CHEMISTRY, MOLECULAR BIOPHYSICS. *Educ:* Moravian Col, BS, 42; Lehigh Univ, MS, 43; Princeton Univ, MA, 45, PhD(phys chem), 46. *Prof Exp:* Res asst, Manhattan Proj, Princeton Univ, 43-46; res asst, SAM Lab, Carbon & Carbide Chem Corp, 46; res assoc & instr, Cornell Univ, 47; from asst prof to assoc prof, 51-63, PROF BIOPHYS, UNIV COLO HEALTH SCI CTR, DENVER, 63- *Concurrent Pos:* Res fel, Calif Inst Technol, 47-48, sr res fel, 48-50; NIH res grants biophys, Univ Colo Med Ctr, Denver, 51-; USPH spec res fel, Carlsberg Found Biol Inst, Denmark, 61-62; mem planning group biophys mat, Nat Inst Gen Med Students, 65, ad hoc mem biophys & biophys chem B study sect, 67, mem sect A, 76. mem adv panel molecular biol, Div Biol & Med Sci, NSF, 67-70; mem, Molecular & Cellular Biophysics Study Sect, NIH, 81. *Mem:* Fel AAAS; Am Chem Soc; Am Asn Immunol; Am Asn Biol Chemists; Biophys Soc. *Res:* Separation, purification and characterization of proteins; electrophoresis; ultracentrifugation; gel chromatography interaction of proteins with each other and with small molecules; theory of electrophoresis , gel chromatography and ultracentrifugation of reacting macromolecules; CD of proteins and peptides. *Mailing Add:* Dept Biochem Biophys & Genetics Univ Colo Health Sci Ctr Denver CO 80262

CANN, MALCOLM CALVIN, b Yarmouth, NS, Feb 9, 24; m 49. ORGANIC CHEMISTRY, BIOCHEMISTRY. *Educ:* Sir George Williams Col, BSc, 53; McGill Univ, MSc, 55, PhD(biochem), 58. *Prof Exp:* Chemist, Food & Drug Directorate, Can Dept Nat Health & Welfare, 57-66; instr biochemical sci, Eastern Ont Inst Technol, 66-67; MASTER SCH TECHNOL, ALGONQUIN COL, 67- *Mem:* Can Biochem Soc; Can Inst Chem; Sigma Xi. *Res:* Bioassay of corticotropin by means of isolated adrenal tissue, adrenal ascorbic acid and plasma corticosteroids; absorption and toxicity of pesticides in rats of various age groups. *Mailing Add:* 2150 Berwick Ave Ottawa ON K2C 0Y3 Can

CANN, MICHAEL CHARLES, b Schenectady, NY, May 6, 47; m; c 2. ORGANIC CHEMISTRY. *Educ:* Marist Col, BA, 69; State Univ NY Stony Brook, MA, 72, PhD(org chem), 73. *Prof Exp:* NSF fel & assoc instr org chem, Univ Utah, 73-74; lectr org chem, UNiv Colo, Denver, 74-75; asst prof, 75-78, assoc prof, 78-88, PROF ORG CHEM, UNIV SCRANTON, 88- *Mem:* Am Chem Soc. *Res:* The synthesis of aromatic, heterocyclic cations; synthesis of dipeptide salts and sweeteners. *Mailing Add:* Dept Chem Univ Scranton Scranton PA 18510-2192

CANNELL, DAVID SKIPWITH, b Washington, DC, Jan 29, 43; m 67; c 1. THERMAL PHYSICS. *Educ:* Mass Inst Technol, BS, 65, PhD(physics), 70. *Prof Exp:* Asst prof, 70-76, assoc prof, 76-79, PROF PHYSICS, UNIV CALIF, SANTA BARBARA, 79- *Concurrent Pos:* Consult, Dow Chem, 82-85. *Mem:* Fel Am Phys Soc; Guggenheim fel, 78; Sloan Fel, Alfred P Sloan Found, 73-75; Hertz Fel, Hertz Found, 67-70. *Res:* Condensed matter physics, including critical phenomena, laser light scattering and the behavior of systems far from equilibrium. *Mailing Add:* Dept Physics Univ Calif Santa Barbara CA 93106

CANNELL, GLEN H, b Abraham, Utah, Aug 5, 19; m 42, 66; c 6. SOIL PHYSICS. *Educ:* Utah State Univ, BS, 48, MS, 50; Wash State Univ, PhD(agron, soil physics), 55. *Prof Exp:* Soil scientist, Agr Res Serv, USDA, NDak, 54-56; soil physicist, 56-74, prof, 74-90, EMER PROF SOIL PHYSICS, UNIV CALIF, RIVERSIDE, 90- *Concurrent Pos:* Assoc prog dir, Div Pre-Col Educ Sci, NSF, Washington, DC, 69-71. *Mem:* Am Soc Agron; Soil Sci Soc Am; Am Soc Hort Sci; Sigma Xi. *Res:* Soil physics, with emphasis on soil-water-plant relations, soil water movement, soil physical properties and instrumentation for measurement of soil water. *Mailing Add:* 991 Blaine St No 65 Riverside CA 92507

CANNEY, FRANK COGSWELL, b Ipswich, Mass, Oct 8, 20; m 44; c 1. GEOLOGY. *Educ:* Mass Inst Technol, SB, 42, PhD(geol), 52. *Prof Exp:* Res asst, Mass Inst Technol, 49-51; geologist, US Geol Surv, 51-70 & 74-84, chief, Explor Res Br, 70-74; RETIRED. *Concurrent Pos:* Geol consult, 84- *Mem:* Soc Econ Geol; Asn Expl Geochemists (vpres, 70-72); Geol Soc Am; Am Chem Soc. *Res:* Geochemical prospecting for mineral deposits; geochemistry of minor elements in the weathering cycle; remote sensing applied to exploration for mineral deposits. *Mailing Add:* 2954 Routt Circle Lakewood CO 80215

CANNING, T(HOMAS) F, b Boston, Mass, May 26, 27; m 51, 79; c 2. CRYSTALLIZATION, ENVIRONMENTAL. *Educ:* Yale Univ, BE, 50. *Prof Exp:* Prod engr, Merck & Co, Inc, 51-56; develop engr, Nat Res Corp, 56-58; sr res engr, Am Potash & Chem Corp, 58-62, proj engr, 62-65, head crystallization sect, 65-67, head chem eng sect, 67-70, proj mgr, 70-73, mgr plant tech servs, 73-77, mgr eng servs, 78-79, mgr process develop, 80-82, mgr technol licensing, 83-87; mgr environ, Kerr-McGee Chem Corp, 87-90; MGR ENVIRON, NORTH AM CHEM CO, 90- *Mem:* Am Inst Chem Engrs; Am Chem Soc. *Res:* Process improvement and development, design leading to boric acid, salt cake and soda ash processes; process development in liquid-liquid extraction, crystallization, evaporation, heat transfer and solar evaporation. *Mailing Add:* 548 E Dana Ave Ridgecrest CA 93555

CANNIZZARO, LINDA A, b Staten Island, NY, Aug 4, 53. MOLECULAR CYTOGENETICS, GENE MAPPING. *Educ:* St Peter's Col, BS, 75; Fordham Univ, MS, 77, PhD(cytogenetics), 81. *Prof Exp:* Postdoctoral fel path, Dartmouth Med Sch, NH, 81-83; fel human genetics, Cytogenetics & Pediat, Children's Hosp Philadelphia & Univ Pa, 83-84; co-dir clin cytogenetics, M S Hershey Med Ctr, Penn State Univ, 84-86; dir, Gene Mapping Prog, SW Biomed Res Inst, Ariz, 86-89; asst prof molecular cytogenetics, Fels Inst Cancer Res & Molecular Biol, Sch Med, Temple Univ, 89-91; ASST PROF MICROBIOL & IMMUNOL, JEFFERSON CANCER INST, JEFFERSON MED COL, PHILADELPHIA, PA, 91- *Concurrent Pos:* Prin investr, Am Cancer Soc, 88-90. *Mem:* AAAS; Am Soc Human Genetics. *Res:* Determine which genes on specific human chromosomes are responsible for the manifestation and progression of cancer, ie renal cell carcinomas and lung carcinomas; chromosome microdissection; DNA sequences involved in these malignancies are isolated and subjected to further evaluation. *Mailing Add:* Jefferson Cancer Inst Jefferson Alumni Hall 1020 Locust St Philadelphia PA 19107

CANNON, ALBERT, b Charleston, SC, Jan 15, 21; m 43; c 4. CLINICAL PATHOLOGY. *Educ:* Col Charleston, BS, 41; Med Col SC, MD, 49. *Prof Exp:* Intern surg, George Washington Univ Hosp, 49-50; mem surg staff, Charleston Naval Hosp, 50; resident path, US Naval Med Sch, 51-55, instr, 55-57, chief anat path, 57-58; asst prof path, Med Sch, Georgetown Univ, 58-59; asst prof, 59-70, PROF CLIN PATH, MED UNIV SC, 70- *Concurrent Pos:* Lectr lab med, US Naval Hosp, Charleston, SC; med dir, ARC Blood Ctr. *Mem:* AMA; Asn Clin Scientists. *Res:* Anatomic pathology of central nervous system; hematology; histochemistry; clinical chemistry; blood banking. *Mailing Add:* Dept Lab Med Med Univ SC Charleston SC 29403

CANNON, DICKSON Y, b Oklahoma City, Okla, Sept 21, 35; m 56; c 4. EMULSION & SOLUTION RESINS. *Educ:* Davidson Col, NC, BS, 57; Univ Tenn, MS, 59. *Prof Exp:* Tech dir, Conchemco, Inc, Kansas City, Mo, 76-79; res mgr, Reliance Universal, Louisville, Ky, 79-82; VPRES, JONES BLAIR CO, DALLAS, 82- *Mem:* Nat Paint & Coatings Asn; Fedn Coatings Socs. *Res:* Polymers for paint and coatings industry; chemical treatments for formaldehyde removal; paint and coatings formulation. *Mailing Add:* Jones Blair Co 2728 Empire Cent Dallas TX 75235

CANNON, DONALD CHARLES, b Independence, Mo, Nov 14, 34; m 58; c 4. CLINICAL PATHOLOGY, IMMUNOLOGY. *Educ:* Harvard Univ, BA, 56; Univ Chicago, MD, 60, PhD(path), 64. *Prof Exp:* Intern path, Med Ctr, Univ Calif, Los Angeles, 60-61; instr, Univ Chicago, 63-64; instr, Univ NC, 64-65; asst prof, State Univ NY Upstate Med Ctr & asst attend pathologist, State Univ Hosp, 65-68, chief diag reagents sect, Div Biol Stand, NIH, 68-69, asst sect chief clin chem serv, Clin Path Dept, 69-70; asst dir, Bio-Sci Labs, Van Nuys, Calif, 70-71, dir, 71-74; chmn dept, Med Sch, Univ Tex, Houston, 74-80, prof path & lab med, 74-90; PHYSICIAN INTERNAL MED, NUMBER 2 MED PLAZA, 90- *Mem:* AAAS; Acad Clin Lab Physicians & Scientists; AMA; Col Am Pathologists; Am Soc Clin Pathologists. *Res:* Clinical chemistry; blood banking; health care management systems. *Mailing Add:* 1034 Ridgewood Circle Minden LA 71055

CANNON, GLENN ALBERT, b Easton, Md, Apr 11, 40; m 62; c 2. PHYSICAL OCEANOGRAPHY. *Educ:* Drexel Inst, BS, 63; Johns Hopkins Univ, MA, 65, PhD(oceanog), 69. *Prof Exp:* Res asst prof oceanog, Univ Wash, 69-70, res oceangr, Pac Marine Environ Lab, Nat Oceanic & Atmospheric Agency, 70-73; prog dir, NSF, 73-75; RES OCEANOGR, PAC MARINE ENVIRON LAB, NAT OCEANIC & ATMOSPHERIC ADMIN, UNIV WASH, 75-, AFFIL ASSOC PROF OCEANOG, 79- *Concurrent Pos:* Mem adv panel oceanog, NSF, 72-73, 75-76. *Mem:* Am Geophys Union. *Res:* Physical oceanography; estuarine and coastal circulation. *Mailing Add:* NOAA/PMEL/OERD BIN C15700/Bldg 3 7600 Sand Point Way Seattle WA 98115-0070

CANNON, HELEN LEIGHTON, b Wilkinsburg, Pa, Apr 29, 11; div; c 1. GEOCHEMISTRY. *Educ:* Cornell Univ, AB, 32; Univ Pittsburgh, MS, 34. *Prof Exp:* Asst geol, Northwestern Univ, 32-33 & Univ Okla, 34-35; geologist, Oil Geol, Gulf Oil Co, 35-36; geologist minor metal commodity admin, 42-46, geologist geochem prospecting methods, 46-62, geologist explor res, US Geol Surv, 62-88; RETIRED. *Concurrent Pos:* Chmn subcomt geochem environ in health & dis, Nat Res Coun, 69-73, coun mem, AAAS, 69-71, comt coun affairs, 78-81, AAAS, 78-81, chmn Coun & Comt Environ Pub Policy, 73-76. *Honors & Awards:* Meritorious Award, Geol Soc Am, 69; Distinguished Serv Award, Dept Interior, 75. *Mem:* Fel AAAS; fel Geol Soc Am; Asn Explor Geochemists. *Res:* Botanical methods of prospecting; trace element distribution in soils and plants as related to geology, health and disease. *Mailing Add:* US Geol Surv Rte Nine Box 77B Santa Fe NM 87505

CANNON, HOWARD S(UCKLING), b Chicago, Ill, Sept 23, 26; m 47; c 5. METALLURGICAL ENGINEERING, SURFACE ENGINEERING. *Educ:* Carnegie Inst Technol, BS, 50; Rensselaer Polytech Inst, MS, 51, PhD(phys metall), 54. *Prof Exp:* Res engr, Linde Air Prod, 53-55; supvr phys metall, Amco Res, Inc, 55-58; dir phys metall lab, Cent Res & Eng, Continental Can Co, 58-63, dir res inorg mat, Corp Res & Develop, 63-69 & new technol, Metall Div, 70-73, asst to vpres res & eng, 73-76, gen mgr technol & econ, 76-78; dir res, Rasselstein (Steel), W Ger, 78-83; dir, Off Adv Eng Studies, Univ Ill, 84-89; PRES, H C ASSOCS INC, 83- *Mem:* Am Soc Metals; Am Inst Mining, Metall & Petrol Engrs; Soc Mfr Engrs. *Res:* Liquid phase sintering; cooper alloy development; metallic and ceramic coating development; casting research; aluminum and tinplate container technology; steel and aluminum production economics; tinplate and automotive steel research; recycling. *Mailing Add:* 1161 Roxbury Close Rockford IL 61107

CANNON, JERRY WAYNE, b Lambert, Miss, June 24, 42; m 63; c 2. BIOCHEMISTRY. *Educ:* Univ Miss, BSc, 64; Drexel Univ, PhD(chem), 69. *Prof Exp:* Instr biochem, Sch Med, Univ Miss, 68-70; asst res prof, 70-78, assoc prof, 78-84, PROF CHEM, MISS COL, 84- *Mem:* Am Chem Soc. *Res:* Pathway of steroid biosynthesis; synthesis and biological testing of chemical analogs of digitalis. *Mailing Add:* MC Box No 4062 Miss Col Clinton MS 39056

CANNON, JOHN BURNS, b Spartanburg, SC, Mar 5, 48; m 79; c 2. BIOCHEMISTRY, PHARMACEUTICAL CHEMISTRY. *Educ:* Duke Univ, BS, 70; Princeton Univ, PhD(chem), 74. *Prof Exp:* Res chemist, dept chem, Univ Calif, San Diego, 74-76; instr chem, Northern Ill Univ, 77-79; asst prof chem, Cleveland State Univ, 79-84; res chemist, Am Cyanamid Co, Princeton, NJ, 84-; SR PHARMACEUT SCIENTIST, ABBOTT LABS, N CHICAGO, ILL, 87- *Mem:* Am Chem Soc; Sigma Xi; Am Asn Pharmaceut Scientists. *Res:* Interaction of proteins, drugs and porphyrins with liposomes; drug delivery systems; controlled release; targeted drug delivery. *Mailing Add:* Abbott Labs D-495 1400 Sheridan Rd N Chicago IL 60064

CANNON, JOHN FRANCIS, b Monroe, Utah, Dec 14, 40; m 62; c 8. CHEMISTRY EDUCATION. *Educ:* Brigham Young Univ, BS, 65, PhD(phys chem), 69. *Prof Exp:* Fel, Georgetown Univ, 69-70; assoc dir high pressure data ctr, 70-72, assoc dir ctr high pressure res, 72-78, asst prof chem, 78-83, LECTR, BRIGHAM YOUNG UNIV, 83- *Mem:* Am Chem Soc. *Res:* High pressure, high temperature synthesis of inorganic compounds, especially intermetallics and compounds containing lanthanide elements; x-ray crystallography. *Mailing Add:* Dept Chem Brigham Young Univ Provo UT 84602

CANNON, JOHN N(ELSON), b Salt Lake City, Utah, July 27, 27; m 51; c 6. THERMAL SCIENCES, COMBUSTION. *Educ:* Univ Utah, BSME, 52, MS, 55; Stanford Univ, PhD, 65. *Prof Exp:* Aeronaut engr, NAm Aviation, 55-57; from asst prof to assoc prof, 57-68, chmn dept, 62-65, PROF MECH ENG, BRIGHAM YOUNG UNIV, 68- *Concurrent Pos:* Lectr, Univ Southern Calif, 56, Stanford Univ, 61. *Mem:* Am Inst Aeronaut & Astronaut; Am Soc Mech Engrs; Am Soc Eng Educ. *Res:* Fluid mechanics; thermodynamics; heat transfer and combustion. *Mailing Add:* Dept Mech Eng 242CB, Brigham Young Univ Provo UT 84602

CANNON, JOHN ROZIER, b McAlester, Okla, Feb 3, 38; m 57; c 3. MATHEMATICS. *Educ:* Lamar State Col, BA, 58; Rice Univ, MA, 60, PhD(math), 62. *Prof Exp:* Assoc mathematician, Brookhaven Nat Lab, 62-64, NATO fel, 64-65; mem fac math, Purdue Univ, 65-66, prof, 68-69; assoc prof, Univ Minn, Minneapolis, 66-68; prof math, Univ Tex, Austin, 69-81; prof math, Wash State Univ, 81-88; dir, Math Sci Div, Off Naval Res, 86-88; PROF & CHMN MATH, LAMAR UNIV, 88- *Concurrent Pos:* Consult, Mobil Oil Corp, 67-69 & Gen Motors Res Labs, 77-85; vis prof, Tex Tech Univ, 73-74, Colo State Univ, 79, Univ Firenze, 79 & Hahn-Meitner Inst, Berlin, 79. *Mem:* Am Math Soc; Soc Indust & Appl Math; Sigma Xi; Math Asn Am. *Res:* Ordinary and partial differential equations; numerical analysis. *Mailing Add:* Dept Math Lamar Univ Beaumont TX 77710

CANNON, JOSEPH G, b Decatur, Ill, Sept 30, 26; m 60; c 4. MEDICINAL CHEMISTRY. *Educ:* Univ Ill, BS, 51, MS, 53, PhD(pharmaceut chem), 57. *Prof Exp:* From asst prof to assoc prof pharmaceut chem, Univ Wis, 56-62; assoc prof med chem, 62-65, head div med chem & natural prod, 85-90, PROF MED CHEM, UNIV IOWA, 65-, ASST DEAN, GRAD STUDY & RES, COL PHARM, 90- *Mem:* Am Chem Soc; fel Am Asn Pharm Scientists. *Res:* Organic synthesis; structure-activity relationships; nitrogen heterocycles. *Mailing Add:* Col Pharm Univ Iowa Iowa City IA 52242

CANNON, LAWRENCE ORSON, b Logan, Utah, June 11, 35; m 59; c 3. MATHEMATICS. *Educ:* Utah State Univ, BS, 58; Univ Wis, MS, 59; Univ Utah, PhD(math), 65. *Prof Exp:* From instr to asst prof math, Utah State Univ, 61-63; instr, Univ Utah, 63-64; from asst prof to assoc prof, 65-77, head dept, 69-80, PROF MATH, UTAH STATE UNIV, 78- *Concurrent Pos:* Vis prof, Rutgers Univ, 68-69. *Mem:* Am Math Soc; Math Asn Am. *Res:* Upper semicontinuous decompositions of 3-manifolds; wild and tame surfaces; prime number theory. *Mailing Add:* Dept Math Utah State Univ 3900 Logan UT 84322-4125

CANNON, M(OODY) DALE, b Wheeler, Tex, Dec 11, 21; m 47; c 4. AGRICULTURAL ENGINEERING. *Educ:* Okla State Univ, BS, 50; Univ Mo, MS, 53. *Prof Exp:* Exten agr engr, Kans State Col, 53-56; asst prof agr eng & asst agr eng, Univ Agr Exp Sta, Univ Ariz, 56-73, assoc prof agr eng & assoc agr engr, Cotton Res Ctr, 73-; RETIRED. *Concurrent Pos:* Consult. *Mem:* Am Soc Agr Engrs; Coun Agr Sci & Technol. *Res:* Mechanization of cotton production research. *Mailing Add:* 1407 Marilyn Ann Dr Tempe AZ 85281

CANNON, MARVIN SAMUEL, b Toledo, Ohio, Feb 10, 40; m 73. HUMAN ANATOMY, CELL BIOLOGY. *Educ:* Univ Toledo, BS, 60, MS, 65; Ohio State Univ, PhD(human anat), 69. *Prof Exp:* Asst prof biol sci, Capital Univ, 71-73; asst prof anat, Univ Tex Med Br Galveston, 73-76; ASSOC PROF ANAT, TEX A&M UNIV, 76- *Concurrent Pos:* Bremer Found Fund fel, Ohio State Univ, 71-73; Am Heart Asn grant, 80-81; Scott & White Clin & Endowment Fund, 82. *Mem:* Am Asn Anat; Pan Am Asn Anat. *Res:* Microvasculature metabolism; comparative hematology hemopoiesis. *Mailing Add:* Dept Human Anat & Med Neurobiol Col Med College Station TX 77843

CANNON, ORSON SILVER, b Salt Lake City, Utah, Nov 21, 08; m 34; c 5. PHYTOPATHOLOGY. *Educ:* Utah State Col, BS, 35, MS, 37; Cornell Univ, PhD(plant path), 42. *Prof Exp:* Asst plant path, Utah State Agr Col, 33-37; asst exten plant pathologist, Pa State Col, 42-43; head dept, Crop Res Lab, H J Heinz Co, Ohio, 43-48; plant pathologist, Bur Plant Indust, Soils & Agr Eng, USDA, 48-57, head dept bot & plant path, 57-74, EMER PROF BOT & PLANT PATH, UTAH STATE UNIV, 74- *Mem:* AAAS; Am Phytopath Soc; Am Soc Hort Sci; Am Soc Plant Physiol. *Res:* Bacterial wilt of alfalfa; vegetable diseases; mosaic resistance in cucumbers; anthracnose resistance in tomatoes; curly top and wilt resistance in tomatoes. *Mailing Add:* 1407 E 17th N Logan UT 84321

CANNON, PATRICK JOSEPH, b Sioux Falls, SDak, Dec 26, 47; m 78; c 2. POLYMER EXTRUSION-FILM, MACHINE DESIGN. *Educ:* SDak State Univ, BSc, 69; Univ SDak, MBA, 81. *Prof Exp:* Systs engr, Westinghouse Elec Corp, 69-77; eng mgr, 77-89, MGR, BUS DEVELOP, RAVEN INDUSTS, INC, 89- *Concurrent Pos:* Vis prof, Univ SDak, 85 & 87; adj prof, SDak State Univ, 89- *Mem:* Am Inst Aeronaut & Astronaut. *Res:* Development of materials for high altitude, heavy lift, research balloons; development of various plastic film materials for industrial and agricultural applications; invention of controls for manned sport balloons. *Mailing Add:* Raven Industs Inc PO Box 1007 Sioux Falls SD 57117

CANNON, PETER, b Chatham, Eng, Apr 20, 32; m 55; c 4. PHYSICAL CHEMISTRY. *Educ:* Univ London, BSc, 52, PhD(chem), 55. *Prof Exp:* Mem staff, Overseas Chem Dept, Procter & Gamble Co, 55-56; phys chemist, Gen Elec Res Lab, 56-65, mgr opers anal, Gen Elec Info Systs, 65-67; mgr sensors & mat, Gen Elec Mfg & Process Automation Bus Div, 67-69, mgr sensors & microelectronics, 69-72, strategy develop automation & machine tools bus, 72-73; dir new prod develop, 73-75, vpres, Bus Develop, Util & Indust Opers Div, 75-76; STAFF VPRES RES & VPRES, SCI CTR, ROCKWELL INT CORP, 76- *Concurrent Pos:* Adj prof, Polytech Inst Brooklyn, 64-69; lectr, Grad Sch Bus Admin, Univ Va, 65-67. *Mem:* Am Chem Soc; Am Phys Soc; Royal Soc Chem. *Res:* Administrative product development programs and business strategy planning; physics and chemistry of surfaces and the solid state; catalysis; super-pressure reactions, especially diamonds; operations analysis and research, especially mathematical programming. *Mailing Add:* 969 W Maude St Sunnyvale CA 94086-2802

CANNON, PHILIP JAN, b Washington, DC, Nov 15, 40; m 61; c 2. GEOMORPHOLOGY. *Educ:* Univ Okla, BS, 65, MS, 67; Univ Ariz, PhD(geol), 73. *Prof Exp:* Geologist, US Geol Surv, 67-72; res scientist geol, Tex Bur Econ Geol, 72-74; asst prof geol, Univ Alaska, 74-81; chief geologist, Fairbanks, Alaska, 81-83, CHIEF GEOLOGIST, PLANETARY DATA, TECUMSEH, OKLA, 84- *Concurrent Pos:* Mem Sithylemenkat meteorite impact crater discovery group, Alaska, 77. *Honors & Awards:* Ford Bartlett Award, 82. *Mem:* Am Soc Photogram; Sigma Xi; Am Asn Petrol Geologists. *Res:* Geomorphic investigations of Alaska and geologic mapping of coastal areas using side-looking airborne radar imagery; regional geologic explorations using radar and landsat imageries of North America; author of thirty-eight publications in various journals. *Mailing Add:* Planetary Data 302 E Locust Tecumseh OK 74873

CANNON, RALPH S, JR, b York, Pa, Apr 21, 10; m 35; c 1. ECONOMIC GEOLOGY, GEOCHEMISTRY. *Educ:* Princeton Univ, BS, 31, PhD(geol), 35; Northwestern Univ, MS, 33. *Prof Exp:* Mine geologist, NJ Zinc Co, 35-37; res geologist, US Geol Surv, 37-71; RETIRED. *Mem:* Fel Geol Soc Am; fel Soc Econ Geologists. *Res:* Copper resources; isotope geology of lead. *Mailing Add:* 309 Sunrise Ridge Dr Spruce Pine NC 28777-3329

CANNON, RAYMOND JOSEPH, JR, b Hartford, Conn. MATHEMATICS, AIR POLLUTION. *Educ:* Col Holy Cross, AB, 62; Tulane Univ, PhD(math), 67. *Prof Exp:* Asst prof, Vanderbilt Univ, 67-69; asst prof, Univ NC, Chapel Hill, 69-74; assoc prof, Stetson Univ, 74-80; ASSOC PROF MATH, BAYLOR UNIV, 80- *Concurrent Pos:* Off Naval Res fel, Univ Mich, 68-69; vis assoc prof, Baylor Univ, 78-79. *Mem:* Am Math Soc; Math Asn Am. *Res:* Quasiconformal mappings; functions of a complex variable; topological analysis. *Mailing Add:* Dept Math Baylor Univ Waco TX 76798

CANNON, ROBERT H, JR, b Cleveland, Ohio, Oct 6, 23; m 45; c 7. AEROSPACE ROBOTICS, DYNAMICS. *Educ:* Univ Rochester, BS, 44; Mass Inst Technol, ScD(mech eng), 50. *Prof Exp:* Chief scientist, US Air Force, 66-68; prof aeronaut & astronaut & vchmn dept, Stanford Univ, 59-70, dir, Guid & Control Lab, 59-70; US asst secy transp, US Dept Transp, 70-74; prof eng & chmn div eng & appl sci, Calif Inst Technol, 74-80; chmn dept, 79-90, CHARLES LEE POWELL PROF AERONAUT & ASTRONAUT, STANFORD UNIV, 80- *Concurrent Pos:* Chmn, Electronics Res Ctr Adv Group, NASA, 68-70 & Res Adv Subcomt Guid, Control & Navig, 68-70; chmn, Assembly Eng, Nat Res Coun, 74-75 & Energy Eng Bd, 75-81 & mem gov bd, 78-; chmn, sci adv comt, Gen Motors Corp, 79-84, adv bd, Gen Elec Space Sta, 85-, President's Comn Nat Medal Sci, 84-88 & Flight Telerobotic Serv Comn, NASA, 87-; mem, Draper Corp, 75-, Boeing Corp Tech Adv Coun, 84-, Comsat Tech Adv Comn, 85-87, Aerospace & Space Eng bd, 75-79 & 85- *Honors & Awards:* Oldenburger Medal, 88. *Mem:* Nat Acad Eng; fel Am Inst Aeronaut & Astronaut; AAAS. *Res:* Dynamics and automatic control; inertial guidance and automatic flight control of air, water and spacecraft; hydrofoil boats; wave actuated pumps. *Mailing Add:* Dept Aeronautics & Astronautics Stanford Univ Stanford CA 94305

CANNON, ROBERT L, b Kinston, NC, Sept 16, 39; m 69. COMPUTER VISION AND INTELLIGENT SYSTEMS. *Educ:* Univ NC, BS, 61, PhD(comput sci), 73; Univ Wis, Madison, MS, 63. *Prof Exp:* Instr, Philander Smith Col, Ark, 65-66; instr, Wilmington Col, 67-68, asst prof math, 68-69; from asst prof to assoc prof, 73-86, chmn dept, 80-83, PROF COMPUT SCI, UNIV SC, 86- *Concurrent Pos:* Vis prof, Ctr Automation Res, Univ Md, 83-84; vis scientist, IBM fel prog, IBM Sci Ctr, Palo Alto, Calif, 84. *Mem:* Asn Comput Mach; Inst Elec & Electronics Engrs; Int Fuzzy Systs Asn; NAm Fuzzy Systs Asn; Pattern Recognition Soc. *Res:* Development of an integrated system to aid in oil exploration; the system includes a data base of known oil fields which are compared to a prospect using techniques of artificial intelligence and pattern recognition; development of a laboratory for teaching writing skills. *Mailing Add:* Dept Computer Sci Univ SC Columbia SC 29208

CANNON, ROBERT YOUNG, b Boise, Idaho, Sept 11, 17; m 48; c 3. DAIRY SCIENCE. *Educ:* Iowa State Univ, BS, 39; Ohio State Univ, MS, 40; Univ Wis, PhD(dairy indust), 49. *Prof Exp:* PROF FOOD SCI, AUBURN UNIV, 48- *Mem:* Am Dairy Sci Asn; Inst Food Technol; Int Asn Milk, Food & Environ Sanit. *Res:* Dairy foods processing; quality control of foods; food plant sanitation. *Mailing Add:* 525 Forestdale Dr Auburn AL 36830

CANNON, WALTON WAYNE, b Ark, Jan 8, 18; m 48; c 3. ELECTRICAL ENGINEERING. *Educ:* Univ Ill, BSEE, 48, MSEE, 49, PhD(elec eng), 55. *Prof Exp:* Asst microwave magnetrons, Univ Ill, 46-49, res assoc microwave devices, 49-51; from assoc prof to prof elec eng, Univ Ark, 51-61; electronics scientist, Deco Electronics, Va, 61-64; prof elec eng, Va Polytech Inst, 64-69; chmn dept, 69-80, PROF ELEC ENG, WVA UNIV, 69- *Concurrent Pos:* NSF fel, 59-60; consult, Ark Educ TV Asn & Transvideo Corp. *Mem:* Am Soc Eng Educ; Inst Elec & Electronics Engrs; Nat Soc Prof Engrs. *Res:* Radio-frequency mass spectrometry; satellite communication; semiconductor electronics; electronic circuits and systems; communication systems; general electrical engineering. *Mailing Add:* Dept Elec Eng WVa Univ 825 Eng Sci Bldg Morgantown WV 26506

CANNON, WILLIAM CHARLES, b San Diego, Calif, Oct 5, 26; m 60; c 1. PHYSICS, AEROSOL SCIENCE. *Educ:* Wash State Univ, BS, 50; Univ Calif, Los Angeles, MA, 63, PhD(physics), 71. *Prof Exp:* Res engr flight test instrumentation guided missile, NAm Aviation, Downey, 53-55; design engr preliminary anal nuclear reactor res, Atomic Int Div, NAm Aviation Co, Canoga Park, 55-57, sr res eng mat sci, Douglas Aircraft Co, 57-60; res autonetics, Anaheim, 60-62; res asst low temperature physics res, Univ Calif, Los Angeles, 61-71; SR RES SCIENTIST AEROSOL PHYSICS, BIOL DEPT, BATTELLE PAC NORTHWEST LABS, RICHLAND, 72- *Concurrent Pos:* Physicist & consult thermionics res, Jet Propulsion Lab, Pasadena, 69-71. *Mem:* AAAS; Sigma Xi. *Res:* Aerosol science research in instrumentation and measurement methods and in relation of aerosol properties to respiratory deposition and biological effects of aerosol inhalation. *Mailing Add:* 2177 Crestview Ave Richland WA 99352

CANNON, WILLIAM NATHANIEL, b Atlanta, Ga, Oct 15, 27; m 51; c 4. ORGANIC CHEMISTRY. *Educ:* NGa Col, BS, 48; Univ Ga, MS, 50. *Prof Exp:* Chemist, Rohm & Haas Co, Redstone Arsenal, Ala, 50-52; org chemist, 52-57, appln res chemist, 57-66, process chemist, 66-67; sr org chemist, 67-74, RES SCIENTIST, ELI LILLY & CO, 74- *Mem:* Am Chem Soc; Soc Indust Microbiol (secy, 66-67). *Res:* Synthesis of organic compounds for testing and evaluation as agricultural chemicals. *Mailing Add:* PO Box 3216 Breckenridge CO 80424-3216

CANNON, WILLIAM NELSON, JR, b Wilmington, Del, Jan 7, 32; wid; c 3. ENTOMOLOGY. *Educ:* Univ Del, BS, 53, MS, 60; Ohio State Univ, PhD(entom), 63. *Prof Exp:* Sanitarian, Del State Bd Health, 56-58; res assoc entom, Ore State Univ, 63-65; RES ENTOMOLOGIST, FORESTRY SCI LAB, NORTHEASTERN FOREST EXP STA, US FOREST SERV, 65- *Mem:* Entom Soc Am; Soc Pop Ecol. *Res:* Insect-host plant relationships; insect biology; cost-effective integrated pest management strategies. *Mailing Add:* Forestry Sci Lab USDA NE Forest Exp Sta Delaware OH 43015

CANNONITO, FRANK BENJAMIN, b New York, NY, Oct 19, 26; m 53, 77, 85; c 2. ALGEBRA, MATHEMATICAL LOGIC. *Educ:* Columbia Univ, BS, 59, MA, 61; Adelphi Univ, PhD(math), 64. *Prof Exp:* Res mathematician, Res Dept, Grumman Aircraft Eng Corp, 60-62; staff mathematician, Tech Anal Off, Hughes Aircraft Co, 62-64, mgr info sci progs, 64-66; from asst prof to assoc prof math, 66-77, vchmn dept, 69-72, PROF MATH, UNIV CALIF, IRVINE, 77- *Concurrent Pos:* Sr staff mathematician, Hughes Aircraft Co, 66-69; prin investr, Air Force Off Sci Res grant, 66-73; Army Res Off Conf & NSF Conf grants, 69; assoc ed, Info Sci, 68-82. *Mem:* Am Math Soc; Asn Symbolic Logic; London Math Soc. *Res:* Fine degrees of solvability of the word problem and related decision problems in group theory; subgroups of finitely presented groups; algorithms for solvable groups. *Mailing Add:* Dept Math Univ Calif Irvine CA 92717

CANO, FRANCIS ROBERT, b New York, NY, Apr 25, 44; m 69; c 3. MICROBIOLOGY, BIOCHEMISTRY. *Educ:* St John's Univ, NY, BS, 65, MS, 67; Pa State Univ, PhD(microbiol), 70. *Prof Exp:* Res assoc, Inst Microbiol, Rutgers Univ, 71-72; res bacteriologist vaccine res, 72-75, group leader vaccine res, 75-76, mgr biol prod & process improv vaccine develop, 76-79, mgr biol prod & develop, 79-80, MGR BIOL, LEDERLE LABS DIV, AM CYANAMID CO, 80- *Concurrent Pos:* Instr, Rockland Community Col, 74-76; consult, Fisher Diagnostics, 77. *Mem:* Am Soc Microbiol; AAAS. *Mailing Add:* Lederle Labs Pearl River NY 10965

CANO, GILBERT LUCERO, b Mesilla, NMex, Jan 7, 32; m 52; c 4. EXPERIMENTAL ATOMIC PHYSICS. *Educ:* NMex State Univ, BS, 54, MS, 60, PhD(physics), 64. *Prof Exp:* MEM RES STAFF, SANDIA CORP, 64- *Concurrent Pos:* Mem Nat Adv Coun Career Educ, 74-76; sci & energy advisor to gov of NMex, 75-76; mem policy adv comt NMex Mus Natural Hist, 77-82; mem Bd Regents, NMex Inst Mining & Technol, 85- *Mem:* Am Phys Soc. *Res:* Charge spectroscopy of laser-induced blow-off; laser-induced fusion; atomic stopping power of thin metallic films; ion-atom interactions; fast reactor safety. *Mailing Add:* Sandia Labs Div 6454 Albuquerque NM 87115

CANOLTY, NANCY LEMMON, b Washington, Ind, Mar 1, 42; m 68; c 2. NUTRITION. *Educ:* Purdue Univ, BS, 63, MS, 68; Univ Calif, Berkeley, PhD(nutrit), 74. *Prof Exp:* asst prof nutrit, Univ Calif, Davis, 73-80; ASSOC PROF FOODS & NUTRIT, UNIV GA, 80- *Mem:* Inst Food Technologists; Sigma Xi; Am Inst Nutrit; AAAS. *Res:* Nutritional consequences of lithium therapy; protein and energy utilization. *Mailing Add:* Dept Foods & Nutrit Univ Ga Athens GA 30602

CANON, ROY FRANK, b Eagle Pass, Tex, Sept 10, 42; div. MATERIALS SCIENCE. *Educ:* Univ Tex, BES, 64; Univ Tex, Austin, PhD(mech eng), 68. *Prof Exp:* Engr & scientist, Tracor, Inc, Tex, 65-67; asst metallurgist, Argonne Nat Lab, 68-69; engr/scientist, Tracor, Inc, 69-72; tech dir, Turbine Support Div, Chromalloy Am Corp, 72-76, vpres chromalloy compressor technol, 76-81; PRES, R F CANNON INC, 81- *Mem:* Am Soc Metals; Metall Soc. *Res:* Mechanical properties of ceramic nuclear fuel materials; composite materials; development of new repair procedures and protective coatings for gas turbine components; new abradable seals for gas turbine applications. *Mailing Add:* 8917 Willmon Way San Antonio TX 78239

CANONICO, DOMENIC ANDREW, b Chicago, Ill, Jan 18, 30; m 55; c 5. METALLURGY, MATERIAL SCIENCE. *Educ:* Mich Technol Univ, BS, 51; Lehigh Univ, MS, 61, PhD(metall eng), 63. *Prof Exp:* Metallurgist metal joining, Ill Inst Technol Res, 53-58; instr metall, Lehigh Univ, 58-62; engr metal joining, Homer Res Lab, Bethlehem Steel Corp, 62-64; supvr process control, Aerospace Components Div, Atlas Chem, 64-65; metallurgist metal joining, Nuclear Div, Oak Ridge Nat Lab, Union Carbide Corp, 65-74, group leader pressure vessel technol, 74-81; DIR, METALL & MAT LAB, COMBUSTION ENG, INC, 81- *Concurrent Pos:* Consult, Nuclear Regulatory Comn, 73- & Adv Comt on Reactor Safeguards, 75-; mem, Boiler & Pressure Vessel Comt, Am Soc Mech Engrs; mem adv comt, Col Eng, Univ Tenn-Knoxville, Univ Tenn- Chattanooga. *Honors & Awards:* Rene D Wasserman Award, Am Welding Soc, 77; Lincoln Gold Medal, Am Welding Soc, 79; Adams lectr, Am Welding Soc, 83. *Mem:* Am Soc Mech Engrs; Am Welding Soc; fel Am Soc Metals; Sigma Xi; Am Soc Testing & Mat. *Res:* Fracture toughness and weldability of materials for application in energy-producing systems; impact properties; fracture mechanics; creep; fatigue; solidification mechanics; weldment discontinuities; brazing; materials design; ferrous metallurgy. *Mailing Add:* One Windcrest Dr Granby CT 06035

CANONICO, PETER GUY, b Tunis, Tunisia, June 12, 42; US citizen; m 69; c 2. CELL BIOLOGY, TOXICOLOGY. *Educ:* Bucknell Univ, BS, 64; Univ SC, MS, 66; Rutgers Univ, PhD(physiol), 69. *Prof Exp:* res scientist cell physiol, US Army Med Res Inst Infectious Dis, 69-88, chief dept antiviral studies, 80-88; DIR, FREDERICK RES CTR, 88- *Concurrent Pos:* Assoc prof, Hood Col, 76- *Honors & Awards:* Res & Develop Medal, Dept Army, 75. *Mem:* Am Soc Cell Biol; Am Physiol Soc; AAAS; Am Soc Microbiol; Soc Exp Biol & Med. *Res:* Effects of infections and inflammation on cellular functions and physiology and morphology of subcellular organelles; studies of host cellular defense mechanisms to microbial infection; antiviral drug development; toxicology and mechanisms of action of antiviral compounds. *Mailing Add:* Frederick Res Ctr 431 Aviation Way Frederick MD 21701

CANONNE, PERSEPHONE, b Eressos Lesbos, Greece, Jan 17, 32; Can citizen; m 60. ORGANIC & ORGANOMETALLIC CHEMISTRY. *Educ:* Univ Lille, France, BS, 56, PhD(org chem), 59. *Prof Exp:* Asst prof org chem, Univ Lille, France, 59-61; researcher, Dutch Inst Appl Sci Res, Neth, 61-63; from asst prof to assoc prof, 63-72, PROF ORG CHEM, UNIV LAVAL, 72- *Concurrent Pos:* Vis res prof org chem, Univ Ala, 67, Univ Claude Bernard, Lyon, 83 & Univ Montreal, 83. *Mem:* Soc Chem France; Soc Chem Belg; Can Soc Chem; Can Fedn Univ Women. *Res:* Organic chemistry; organic synthesis; development of new reagents and synthetic methods; chemistry of organometallic compounds. *Mailing Add:* Dept de chimie Pavillon Vachon Faculte des sciences et de genie Univ Laval Quebec PQ G1K 7P4 Can

CANRIGHT, JAMES EDWARD, b Delaware, Ohio, Mar 1, 20; m 43; c 4. PALYNOLOGY, PALEOBOTANY. *Educ:* Miami Univ, AB, 42; Harvard Univ, AM, 47, PhD(biol), 49. *Prof Exp:* Teaching fel biol, Harvard Univ, 46-49; from instr to prof bot, Ind Univ, 49-63; chmn dept, 64-72, prof, 64-85, EMER PROF BOT, ARIZ STATE UNIV, 85- *Concurrent Pos:* Am Philos Soc res grant, 53-54; NSF travel grant, Int Bot Cong, Paris, 54, Edinburgh, 64 & Leningrad, 75; res grants, 54-; consult, Mene Grande Oil Co, Venezuela, 58; Guggenheim fel, Indonesia & Malaya, 60-61; vis scientist, US-China coop sci prog, Nat Taiwan Univ, 71; res assoc, Geol Survey of Victoria, Melbourne, Australia, 81. *Mem:* Fel AAAS; Int Orgn Paleobot; Bot Soc Am; Am Inst Biol Sci; Am Asn Stratig Palynologists (pres, 79-80); Int Fed Palynological Socs. *Res:* Tertiary and Cretaceous palynology; paleobotany of the Paleozoic. *Mailing Add:* Dept Bot Ariz State Univ Tempe AZ 85287

CANTE, CHARLES JOHN, b Brooklyn, NY, Oct 31, 41; div; c 3. PHYSICAL CHEMISTRY, SURFACE CHEMISTRY. *Educ:* City Col New York, BS, 63, MA, 65, PhD(phys chem), 67; Iona Col, MBA, 72. *Prof Exp:* Lectr chem, City Col New York, 64-67; phys chemist, Phys Res Lab, Edgewood Arsenal, Dept Army, 67-68, res & develop coordr, 68-69; sr chemist, Tech Ctr, Gen Foods Corp, 69-71, group leader 71-73, mgr new technol res, 75-80, field res mgr, Res Dept, Pet Foods Div, 76-80, mgr Div Tech Res, 80-82; group dir, Cent Res & Eng, 87-90, VPRES RES, ENG & TECHNOL, GEN FOODS USA, 90- *Concurrent Pos:* E I du Pont de Nemours teaching award, 66-67; mgr deserts dir tech res, Gen Foods Corp, 82-84, dir off tech planning, 83-84, dir cent res dept, 86-87. *Mem:* AAAS; Am Chem Soc; NY Acad Sci; Sigma Xi; fel Am Inst Chemists; Inst Food Technologists. *Res:* Colloid chemistry; food chemistry. *Mailing Add:* 442 Washington Ave Pleasantville NY 10570

CANTELO, WILLIAM WESLEY, b Medford, Mass, Sept 11, 26; m 58; c 3. ENTOMOLOGY, INSECT BEHAVIOR. *Educ:* Boston Univ, AB, 48; Univ Mass, MS, 50, PhD(entom), 52. *Prof Exp:* Asst entomologist, Bartlett Tree Res Labs, 52-54, assoc entomologist, 54-55; staff entomologist, US Naval Forces, Marianas, 56-61; entom adv, US Opers Mission, Ministry Agr, Thailand, 61-66; RES ENTOMOLOGIST, USDA, 66- *Mem:* AAAS; Entom Soc Am; Sigma Xi. *Res:* Flies infesting commercial mushroom crops; economic entomology; insect population dynamics and ecology. *Mailing Add:* 11702 Wayneridge St Fulton MD 20759

CANTER, LARRY WAYNE, b Nashville, Tenn, May 25, 39; m 62; c 2. SANITARY ENGINEERING, PUBLIC HEALTH. *Educ:* Vanderbilt Univ, BE, 61; Univ Ill, MS, 62; Univ Tex, PhD(civil eng), 67. *Prof Exp:* Sanit engr, USPHS, 62-65; asst prof civil eng, Tulane Univ, 67-69; from asst prof to assoc prof, 69-76, PROF CIVIL ENG & DIR, UNIV OKLA, 76- *Res:* Water and wastewater treatment. *Mailing Add:* Dept Civil & Environ Eng Univ Okla Norman OK 73019

CANTER, NATHAN H, b Philadelphia, Pa, Nov 17, 42; m 64; c 3. POLYMER PHYSICS. *Educ:* Temple Univ, AB, 63; Princeton Univ, MS, 65, PhD(chem), 66. *Prof Exp:* Res assoc, Corp Res Labs, Exxon Res & Eng Co, 66-80. *Mem:* AAAS; Am Chem Soc; Am Phys Soc; Sigma Xi. *Res:* Statistical and solid state physics; physical chemistry; polymer physics and chemistry. *Mailing Add:* Four Culver Rd Livingston NJ 07039

CANTER, NEIL M, b New York, NY, Jan 9, 56. PRODUCT DEVELOPMENT. *Educ:* Brown Univ, BS, 78; Univ Mich, MS, 79, PhD(chem), 83. *Prof Exp:* Res chemist, Stepan Co, 83-85, supvr esters, prod develop, 85 & indust specialties, 85-88; TECH MGR, MAYCO OIL & CHEM CO, 88- *Mem:* Am Chem Soc; Soc Tribologists & Lubrication Engrs. *Res:* Development of new products and support of business efforts in the industrial lubricant market. *Mailing Add:* Mayco Oil & Chem Co 775 Louis Dr Warminster PA 18974

CANTERINO, PETER JOHN, b Italy, June 4, 20; US citizen; m 59; c 2. PLASTICS PRODUCT DEVELOPMENT, POLYMER PROCESS DEVELOPMENT. *Educ:* Manhattan Col, BS, 43; Univ Cincinnati, MS, 44, DSc(chem), 46. *Prof Exp:* Res assoc, Univ Akron, 46-47; group leader, Nopco Chem Co, 47-51; mgr, Phillips Petrol Co, 51-61; asst dir res, Polymer Chem Div, W R Grace, 61-66; dir res, Plastics Div, Allied Chem, 66-67; mgr plastics res & develop, Mobil Chem Co, 67-77, res assoc, 77-85; CONSULT, MOBIL CHEM CO, POLYMERIX CO & POLYMER PROCESSING INST, STEVENS INST TECHNOL, 85- *Concurrent Pos:* Mem, Educ Comt, Plastics Inst Am, 72-91. *Mem:* Am Chem Soc; Soc Plastics Engrs. *Res:* New or improved polymer processes and products; recycling of consumer and engineering plastic waste; catalysis; synthetic rubber; detergents. *Mailing Add:* 39 Mary Dr Towaco NJ 07082

CANTERNA, RONALD WILLIAM, b Freeport, Pa, May 3, 46; m. GLOBULAR CLUSTERS, PHOTOMETRY. *Educ:* Colgate Univ, BA, 68; Univ Wash, Seattle, PhD(astron), 76; Univ Wyo, BEd, 82. *Prof Exp:* Instr astron, Univ Wash, 76-77; asst prof, La State Univ, 77-79; asst prof, Dept Phys, Colo Col, 84-85; asst prof astron & physics, 79-82, ASSOC PROF, UNIV WYO, 86- *Concurrent Pos:* Spec asst to pres res & develop, Univ Wash, Seattle, 76-77; res fel, Italian Nat Coun Res, 80; asst dir, Wyo Intra Red Observ. *Mem:* Sigma Xi; Asn Astron Educrs; Astron Soc Pac; Am Astron Soc; Am Asn Physics Teachers. *Res:* Properties of globular clusters of the Milky Way galaxy and their role in the chemical evolution and history of galaxy formation; design, development and use of the Washington photometric system for stellar photometry. *Mailing Add:* Dept Physics & Astron Univ Wyo Laramie WY 82071

CANTILLI, EDMUND JOSEPH, b Yonkers, NY, Feb 12, 27; m 48; c 3. TRANSPORTATION ENGINEERING. *Educ:* Columbia Univ, BA, 54, BS, 55; Yale Univ, cert, 57; Polytech Inst Brooklyn, PhD(transp), 72. *Prof Exp:* Sr traffic engr, Port Authority of NY & NJ, 55-61, terminals engr, 61-63, proj planner & engr safety res & studies, 63-65, supvry engr traffic control, 65-67, supvry engr traffic safety, res & studies, 67-69; res assoc transp eng, 69-72, assoc prof, 72-77, PROF TRANSP & SAFETY ENG, POLYTECHNIC UNIV, 77-; PRES TRANSP & ENVIRON SAFETY, EDMUND J CANTILLI, INC, 81-; PARTNER & MGT, PLANNING ENVIRON, TRANSP, URBAN & SAFETY SERV, IMPETUS, 85- *Concurrent Pos:* Pres transp eng, Urbitran Assoc Inc, 73-81; pres & chmn bd transp safety, Inst Safety Transp, 77-; vpres forensic eng, Tech & Med Forensic Consults, Inc, 81-83; pres & chmn bd consult eng, Adeboh Assocs, Inc, 81- *Mem:* Fel Inst Transp Engrs; fel Am Soc Civil Engrs; Am Inst Cert Planners; Am Planning Asn; Am Soc Safety Engrs; Syst Safety soc. *Res:* Transportation safety; traffic safety; transportation system safety; environmental impact assessment, including air, water, noise, social and planning; pedestrian safety. *Mailing Add:* 134 Euston Rd West Hempstead NY 11552

CANTIN, GILLES, b Montreal, Que, Apr 29, 27; US citizen; m 52; c 1. ENGINEERING SCIENCE, MATHEMATICS. *Educ:* Polytech Sch, Montreal, BSc, 50; Stanford Univ, MSc, 60; Univ Calif, Berkeley, PhD(eng sci), 68. *Prof Exp:* Prof math & physics, Tech Sch, Rimouski, 50-52; prof physics, Royal Mil Col, 52-54; prof eng, Univ Col, Ethiopia, 54-59; from asst prof to assoc prof mech eng, 60-71, PROF MECH ENG, NAVAL POSTGRAD SCH, 71- *Concurrent Pos:* Consult, Nat Comt for Ethiopia, Int Geophys Year, 57-58; vis prof, Univ Technol Compiegne, France, 76- *Mem:* Soc Naval Architects & Marine Engrs; Am Soc Mech Engrs; Am Soc Eng Educ. *Res:* Stress analysis; structural mechanics; computer methods; development of finite element methods. *Mailing Add:* 17 via Ladera Monterey CA 93940

CANTIN, MARC, b Que, Aug 7, 33; m 59; c 1. MEDICINE. *Educ:* Laval Univ, MD, 58; Univ Montreal, PhD, 62. *Prof Exp:* Asst, Inst Exp Med & Surg, Univ Montreal, 58-62; instr path, Sch Med, Univ Chicago, 64-65; from asst prof to prof path, Univ Montreal, 65-80; dir, Res Group Hypertension, 84, DIR, LAB PATHOBIOL, CLIN RES INST MONTREAL, 80- *Concurrent Pos:* USPHS fel, 62-63 & grant, 64-66; fel path, Sch Med, Univ Chicago, 62-64; Chicago Heart Asn fel, 63-65 & grant, 64-66; Ill Heart Asn fel, 63-65 & grant, 64-66; Fel Royal Soc of Can, 88. *Mem:* Am Heart Asn; Can Med Asn. *Res:* Juxtaglomerular apparatus; the heart as an endocrine gland; physiopathologic implications. *Mailing Add:* Clin Res Inst Montreal 110 Pine Ave W Montreal PQ H2W 1R7 Can

CANTINO, EDWARD CHARLES, mycology; deceased, see previous edition for last biography

CANTINO, PHILIP DOUGLAS, b Berkeley, Calif, Mar 19, 48; m 80; c 2. ANGIOSPERM SYSTEMATICS. *Educ:* Univ Mich, BS, 70; Harvard Univ, PhD(biol), 80. *Prof Exp:* Botanist, Gray Herbarium, Harvard Univ, 80-81; asst prof, 81-87, ASSOC PROF BOT, OHIO UNI, 87- OHIO UNIV, 81- *Concurrent Pos:* Prin investr, NSF grant, 83-87, 90-92. *Mem:* Sigma Xi; AAAS; Am Soc Plant Taxonomists; Bot Soc Am; Int Asn Plant Taxonomy; Soc Syst Zool. *Res:* Logic and methodology of phylogenetic reconstruction; application of phylogenetic (cladistic) analysis to the systematics of the Labiatae and Verbenaceae. *Mailing Add:* Dept Bot Ohio Univ Athens OH 45701-2979

CANTLEY, LEWIS CLAYTON, b Charleston, WVa, Feb 20, 49. MEMBRANE TRANSPORT, CELL DIFFERENTIATION. *Educ:* WVa Wesleyan Col, BS, 71; Cornell Univ, PhD(phys chem), 75. *Prof Exp:* Fel, Harvard Univ, 75-78, asst prof biochem, fac arts & sci, 78-81, assoc prof, 81-; PROF DEPT PHYSIOL, TUFTS UNIV. *Concurrent Pos:* Instr & scholar, Dreyfus Found, 81; estab investr, Am Heart Asn, 81. *Mem:* Am Heart Asn; Am Soc Biol Chemists; Biophys Soc; Med Found. *Res:* Structures, mechanisms and regulation of proteins which transport ions across eucaryotic cell membranes; investigation of affects of cation fluxes on cell differentiation. *Mailing Add:* Dept Physiol Sch Med Tufts Univ 136 Harrison Ave Boston MA 02111

CANTLIFFE, DANIEL JAMES, b New York, NY, Oct 31, 43; m 65; c 4. PLANT PHYSIOLOGY, VEGETABLE CROPS. *Educ:* Delaware Valley Col, BS, 65; Purdue Univ, MS, 67, PhD(plant physiol), 71. *Prof Exp:* Res asst, Purdue Univ, 65-69; res assoc, Cornell Univ, 69-70; res scientist, Hort Res Inst Ont, 70-74; from asst prof to assoc prof, 74-81, PROF SEED PHYSIOL, UNIV FLA, 81-, CHMN, 84- *Concurrent Pos:* Vis prof, Univ Hawaii, 79-80. *Honors & Awards:* John F Kennedy Award, 65. *Mem:* Fel Am Soc Hort Sci; Am Soc Plant Physiologists; Am Soc Agron; Crop Sci Soc Am; Sigma Xi; Tissue Cult Asn. *Res:* Basic studies and applications with seeds as plant growth units, including the physiology of fruit development, seed formation, seed germination, seed dormancy, seed vigor, and synthetic seed technology. *Mailing Add:* Dept Veg Crops Univ Fla Gainesville FL 32611

CANTLON, JOHN EDWARD, b Sparks, Nev, Oct 6, 21; m 44; c 4. ECOLOGY, RESEARCH ADMINISTRATION. *Educ:* Univ Nev, BS, 47; Rutgers Univ, PhD(bot), 50. *Prof Exp:* Asst prof bot, George Washington Univ, 50-52, assoc prof, 52-53; sr ecologist, Phys Res Lab, Boston Univ, 53-54; from assoc prof to prof bot, Mich State Univ, 54-69, provost, 69-75, vpres res & grad studies, 75-90; CONSULT, 90- *Concurrent Pos:* Mem adv panel environ biol, NSF, 61-64, prog dir environ biol, 65-66, mem adv comt, div environ sci, 66-69, adv comt instnl relations, 70-74; gov bd, Am Inst Biol Sci, 63-66; adv comt health physics, Oak Ridge Nat Lab, 66-69, adv coun, 71-75; exec comt, Div Biol & Agr, Nat Res Coun, 67-71, coun natural resources, 73-81, chmn environ studies bd, 78-81; chmn sci adv bd, Environ Protection Agency, 78-81; vchmn bd, World Resources Inst, 82-; chmn bd, Woods Hole Res Ctr, 89-; mem bd, Cherry Mkt Inst, 90-; mem, Nuclear Waste Technol Rev Bd, 88-; Sci Adv Coun, Neogen Corp, 90- *Mem:* AAAS; Ecol Soc Am (secy, 58-63, vpres, 67-68, pres, 68-69); Am Inst Biol Sci; Bot Soc Am; Am Soc Naturalists. *Res:* Pattern in communities; physiological ecology; Alaskan tundra vegetation; research and academic administration. *Mailing Add:* 1795 Bramble Dr East Lansing MI 48823

CANTONI, GIULIO L, b Milano, Italy, Sept 29, 15; nat US. BIOCHEMISTRY. *Educ:* Univ Milan, MD, 39. *Prof Exp:* Res asst, Dept Biochem & Physiol, Univ Milan, 34-39 & Dept Pharmacol, Oxford, Eng, 40-42; teaching & res asst, Dept Physiol, Med Sch, Univ Mich, 42-43; instr, Dept Pharmacol, Col Med, NY Univ, 43-45; asst prof pharmacol, Long Island Col Med, 45-48; sr fel, Am Cancer Soc, 48-50; assoc prof, Sch Med, Western Reserve, 50-54; CHIEF, LAB GEN & COMP BIOCHEM, NIMH, 54- *Concurrent Pos:* Hon prof, Cath Med Col, Seoul, Korea, 84; vis prof, Col France, Paris, 87 & La Sapienza, Rome, 88. *Mem:* Nat Acad Sci; Am Soc Biol Chemists; Biochem Soc Eng; Am Chem Soc; Harvey Soc; Am Acad Arts & Sci. *Mailing Add:* NIMH Bldg 36 Rm 3D06 Bethesda MD 20205

CANTOR, AUSTIN H, b Liberty, NY, April 1, 42; m; c 3. VITAMIN & MINERAL NUTRITION, POULTRY NUTRITION. *Educ:* Cornell Univ, BS, 64, PhD(nutrit), 74; Teachers Col, Columbia Univ, MA, 65. *Prof Exp:* Res assoc, Cornell Univ, 73-74; asst prof poultry sci, Ohio State Univ, 74-80; ASSOC PROF ANIMAL SCI, UNIV KY, 81- *Concurrent Pos:* Peace Corps Vol, Chile, 65-68. *Mem:* Sigma Xi; Am Inst Nutrit; Poultry Sci Asn. *Res:* Factors affecting vitamin and mineral nutrition in poultry; protein and energy requirements of pullets and laying hens; use of enzymes in diets for improving nutrient utilization. *Mailing Add:* Dept Animal Sci Univ Ky 606 Agr Sci Bldg S Lexington KY 40546-0215

CANTOR, CHARLES ROBERT, b Brooklyn, NY, Aug 26, 42. BIOPHYSICAL CHEMISTRY. *Educ:* Columbia Col, AB, 63; Univ Calif, Berkeley, PhD(chem), 66. *Prof Exp:* From asst prof to prof chem, Columbia Univ, 66-81, prof & chmn human genetics & develop, Col Physicians & Surgeons, 81-89, dep dir educ, 81-85, dep dir biotechnol, Columbia Cancer Ctr, 85-88, Higgins prof, Fac Med, 88-89; dir, Human Genome Ctr, 88-90, SR BIOCHEMIST, CELL & MOLECULAR BIOL DIV, LAWRENCE BERKELEY LAB & PROF MOLECULAR BIOL, UNIV CALIF, BERKELEY, 89-; PRIN SCIENTIST, HUMAN GENOME PROJ, US DEPT ENERGY, 90- *Concurrent Pos:* Alfred P Sloan fel, 69-71; NIH study sect, 71-75; Guggenheim fel, 73-74; Sherman Fairchild Distinguished vis scholar, Calif Inst Technol, 75-76; mem cellular & molecular basis dis rev comt, Nat Inst Gen Med Sci, 77-81; bd trustees, Cold Spring Harbor Labs, 78-85; assoc ed, Ann Rev Biophysics & Biophys Chem, 82- & J Molecular Evolution, 83-; mem, Nomenclature Comn, Int Union Biochem, 84-, US Nat Comt, Int Union Pure & Appl Biophys, 85- & many other sci & adv comts, 85-; Nat Cancer Inst outstanding investr grant, 85. *Honors & Awards:* Eastman Kodak Award, 65; Fresenius Award Chem, 72; Eli Lilly Award, 78; Anal Prize, Clin Chem, 88; Veatch Lectr, Harvard Med Sch, 88; Sol Spiegelman Lectr, Univ Ill, 88; Steinberg/Wylie Lectr, Univ Md, 89; Boyce Thompson Distinguished Lectr, Cornell Univ, 90; Herbert A Sober Award, Am Soc Biochem & Molecular Biol, 90. *Mem:* Nat Acad Sci; Harvey Soc; Am Soc Biol Chem; fel AAAS; Am Acad Arts & Sci. *Res:* Optical properties and conformation of nucleic acids and proteins; structure of the ribosome; mechanism of protein synthesis; affinity labelling; structure of chromatin; photochemical crosslinking; macromolecular assembly; electrophoresis of nucleic acids; gene mapping; chromosome structure and organization. *Mailing Add:* Lawrence Berkeley Lab MS 2-300 One Cyclotron Rd Berkeley CA 94720

CANTOR, DAVID GEOFFREY, b London, Eng, Apr 12, 35; US citizen; m 58; c 2. MATHEMATICS, COMPUTER SCIENCES. *Educ:* Calif Inst Technol, BS, 56; Univ Calif, Los Angeles, PhD(math), 60. *Prof Exp:* Asst prof math, Univ Wash, 62-64; PROF MATH & COMPUT SCI, UNIV CALIF, LOS ANGELES, 64- *Concurrent Pos:* Sloan Found fel, 67; prin investr, NSF, 68- *Mem:* Am Math Soc; Math Asn Am; Soc Indust & Appl Math; Asn Comput Mach; Inst Elec & Electronics Engrs. *Res:* Number theory; combinatorics; algorithms. *Mailing Add:* Dept Math Univ Calif Los Angeles CA 90024

CANTOR, DAVID MILTON, b Grand Rapids, Mich, June 30, 52; m; c 2. ANALYTICAL CHEMISTRY. *Educ:* Mich State Univ, BS, 73; Univ Ill, MS, 75, PhD(chem), 77. *Prof Exp:* chemist anal, Phillips Petrol Co, 77-82; DIR, QA SYSTEMS DEVEL, UPJOHN CO, 87- *Concurrent Pos:* Res scientist, Upjohn Co, 82-87. *Mem:* Am Chem Soc; Soc Appl Spectros. *Res:* Development of diverse computer systems to support quality control functions; analysis and improvement of quality assurance systems and procedures. *Mailing Add:* 14101 Parker Rd Hickory Corners MI 49060-9711

CANTOR, DAVID S, b Harford, Conn, Jan 4, 56; m 85; c 1. DEVELOPMENTAL NEUROPSYCHOLOGY, COMPUTERIZED ELECTROPHYSIOLOGY. *Educ:* Univ Conn, BA, 77; State Univ NY, Stonybrook, MA, 79, PhD(psychol), 82. *Prof Exp:* Instr psychol, State Univ NY, Stony Brook, 79-80; res assoc neurosurg, Sch Med, Univ Md, 81-82, asst res prof human ecol, 82-85, asst res prof psychiat, 86-88, pres, Comprehensive Diag & Treat Serv, 85-90; pres, Mind Inc, 90-91; VPRES, FUNCTIONAL IMAGING ASN, 91- *Concurrent Pos:* Prin investr, NAm Sci Orgn, Sigma Xi, 79-81; proj mgr, Appl Neurosci Inst, 82-85; bd dirs, Md Head Injury Found, 83-86 & Innovative Health Found, 87-; fel, State Univ NY & NY Univ, 80-81; reviewer, Soc Res Child Develop, 82-83. *Mem:* Am Psychol Asn; Soc Res Child Develop; Nat Acad Neuropsychol; AAAS. *Res:* Brain-behavior relationships with specific interests in mapping brain activity to cognitive processes; explore nutritional and environmental pollutants which affect neurochemistry underlying neurophysiology and behavior, primarily across the human age span. *Mailing Add:* 10630 Little Patuxent Pkwy Suite 226 Columbia MD 21045

CANTOR, ENA D, b Montreal, Que, Mar 1, 20; m; c 2. MEDICAL SCIENCES, BIOCHEMISTRY. *Educ:* McGill Univ, BA, 40, BSc, 42, PhD(microbiol, immunol), 68; Harvard Univ, MA, 46. *Prof Exp:* Technician, Royal Victoria Hosp, Can, 42-44; res technician, RI Hosp, Providence, 55-57; res technician, Mass Mem Hosp, 58-59; chief technician, Jewish Gen Hosp, Can, 60-62; technician, McGill Univ, 62-63; mem staff clin virol, Royal Victoria Hosp & McGill Univ, 68-69; chief sanit virol sect, 69-71, chief gen bact sect, 71-79, spec bact anaerobes, Ministry Social Affairs, Montreal, 79-83; chief, proficiency testing prog, Que Pub Health Lab, 83-86, chief, antimicrobial testing, spec testing & res, 86-88, res rev biomed waste disposal viral gastroenteritis, 88-90, IN-CHG, DOC HOSP INFECTION CONTROL, NOSOCOMIAL INFECTIONS, QUE PUB HEALTH LAB, 90- *Mem:* Am Soc Microbiol; Can Asn Clin Microbiol & Infectious Dis; Can Fedn Biol Sci; Can Soc Microbiol; Can Soc Immunol; Am Pub Health Asn; Can Col Microbiologists. *Res:* Enteroviruses; serum proteins; population studies of norms; enterococci, especially chemical and immunological studies; lysogeny of Lancefield group B streptococci; mechanisms of resistance to microbiol agents. *Mailing Add:* Div Bacteriol Laboratoire de Sante Publique du Que Ste-Anne-De-Bellevue PQ H9X 3R5 Can

CANTOR, HARVEY, b New York, NY, Sept 26, 42; m 83; c 2. MEDICINE. *Educ:* Columbia Univ, AB, 63; NY Univ, MD, 67. *Hon Degrees:* MA, Harvard Univ, 79. *Prof Exp:* PROF PATH, MED SCH, HARVARD UNIV, 79- *Concurrent Pos:* Leukemia Soc Scholar, 75; chief, Lab Immunopath, Dana Forber Cancer Inst, Boston, Mass, 79-; Leukemia Soc scholar, 75. *Honors & Awards:* Bordon Prize, 67. *Mem:* Fel Am Soc Immunol; Am Soc Clin Invest; Asn Am Physicians; Am Soc Clin Immunol. *Res:* Definition of the cellular and molecular basis of the immune response. *Mailing Add:* Dept Path Dana Forber Cancer Inst Med Sch Harvard Univ 44 Binney St Boston MA 02115

CANTOR, KENNETH P, b Mt Vernon, NY, 1941. EPIDEMIOLOGY. *Educ:* Oberlin Col, BA, 62; Univ Calif, Berkeley, PhD(biophys), 69; Harvard Sch Pub Health, MPH, 73. *Prof Exp:* Res adminr health effects, US Environ Protection Agency, 73-75; EPIDEMIOLOGIST CANCER EPIDEMIOL, NAT CANCER INST, 75- *Mem:* Soc Epidemiol Res; Am Col Epidemiol; Soc Occup & Environ Health; Int Epidemiol Asn; Int Asn Environ Epidemiol. *Res:* Environmental cancer epidemiology. *Mailing Add:* Environ Epidemiol Br Nat Cancer Inst 443 Executive Plaza N Bethesda MD 20892

CANTOR, MARVIN H, b Brooklyn, NY, Nov 17, 35. CELL PHYSIOLOGY. *Educ:* Boston Univ, AB, 57; Mass Inst Technol, SM, 59; Univ Calif, Los Angeles, PhD(zool), 64. *Prof Exp:* Fel zool, Syst-Ecol Prog, Marine Biol Lab, Woods Hole, 64-65; from asst prof to assoc prof, 65-71, PROF BIOL, CALIF STATE UNIV, NORTHRIDGE, 71- *Mem:* AAAS; Soc Protozool; Am Soc Zool; NY Acad Sci. *Res:* Cell growth and metabolism; cell synchrony; regulation of metabolism. *Mailing Add:* Dept Biol Calif State Univ Northridge CA 91330

CANTOR, STANLEY, b Brooklyn, NY, Sept 23, 29; m 50; c 3. PHYSICAL CHEMISTRY. *Educ:* Tulane Univ, BS, 51, MS, 53, PhD(chem), 55. *Prof Exp:* MEM RES STAFF & GROUP LEADER, ENERGY DIV, OAK RIDGE NAT LAB, 55- *Concurrent Pos:* Exchange fel, Atomic Energy Res Estab, Eng, 63-64. *Mem:* AAAS; Am Chem Soc; Sigma Xi. *Res:* Properties of energy storage materials; thermal analysis; database systems; chemistry of energy systems; information requirements analysis. *Mailing Add:* 898 W Outer Dr Oak Ridge TN 37830

CANTOW, MANFRED JOSEF RICHARD, b Oberhausen, Ger, Mar 21, 26; m 61. HARDWARE SYSTEMS. *Educ:* Univ Mainz, BS, 52, MS, 55, PhD(phys chem), 59. *Prof Exp:* Sr res chemist, Calif Res Corp, 60-66, Chevron Res Co, 66-67; asst dir, Airco Cent Res Labs, 67-71; ADV CHEMIST, IBM CORP, 71- *Concurrent Pos:* Lectr exten, Univ Calif, Berkeley, 62- *Mem:* Am Chem Soc. *Res:* Characterization of polymers; polymer fractionation; physical properties of polymers; polymerization and modification of vinyl polymers. *Mailing Add:* General Products Div IBM 5600 Cottle Rd San Jose CA 95193

CANTRALL, IRVING JAMES, b Springfield, Ill, Oct 6, 09; m 32; c 2. ENTOMOLOGY. *Educ:* Univ Mich, AB, 35, PhD(entom), 40. *Prof Exp:* Asst Orthoptera, mus zool, Univ Mich, 34-37; tech asst, 37-42; jr aquatic biologist, Tenn Valley Authority, 42, asst aquatic biologist, 42-43; asst prof biol, Univ Fla, 46-49; from asst prof to prof, 49-77, EMER PROF ZOOL, UNIV MICH, ANN ARBOR, 78-, EMER CUR INSECTS, MUS ZOOL, 78- *Concurrent Pos:* Mem Univ Mich exped, South & Southwest US, 35, Mex, 41, 53, 59, Guatemala, 56, Cent Am, 61; cur, Edwin S George Reserve, Mus Zool, 49-59, cur insects, 59-77. *Mem:* AAAS; Soc Syst Zool; Ecol Soc Am; Soc Study Evolution; Entom Soc Am. *Res:* Taxonomy of new world Acridoidea. *Mailing Add:* 1531 Las Vegas Dr Ann Arbor MI 48103-5765

CANTRELL, CYRUS D, III, b Bartlesville, Okla, Oct 4, 40; m 72; c 1. NONLINEAR OPTICS, QUANTUM OPTICS. *Educ:* Harvard Univ, AB, 62; Princeton Univ, MA, 64, PhD(physics), 68. *Prof Exp:* From asst prof to assoc prof physics, Swarthmore Col, 67-74; staff mem, Laser Res & Technol Div, Los Alamos Sci Lab, 73-76, assoc group leader, Theoret Chem & Molecular Physics, 76-78, staff mem, Theoret Div Off, 78-79; PROF PHYSICS & ELEC ENG & DIR, CTR APPL OPTICS, UNIV TEX, DALLAS, 80- *Concurrent Pos:* Vis res fel, Princeton Univ, 70-71; adj prof physics, Univ NMex, 76-79; assoc prof, Univ Paris, 80; consult, Los Alamos Nat Lab, 79- *Mem:* Fel Am Phys Soc; fel Optical Soc Am; fel Inst Elec & Electronics Engrs; Sigma Xi. *Res:* Nonlinear interactions of laser light and its applications in physics, chemistry and industrial processing; laser isotope separation; multiphoton excitation of polyatomic molecules; nonlinear optics and pulse propagation; molecular spectroscopy; quantum optics; quantum theory of measurement. *Mailing Add:* Univ Tex at Dallas Richardson TX 75083-0688

CANTRELL, ELROY TAYLOR, b Mobile, Ala, May 10, 43; m 67; c 2. PHARMACOLOGY, CANCER. *Educ:* Ark State Univ, BS, 65; Univ Tenn, MS, 68; Baylor Col Med, PhD(pharmacol, 71. *Prof Exp:* Fel pharmacol, Baylor Col Med, 71-72; res assoc med genetics, M D Anderson Hosp, 72-73; asst prof, 73-76, chmn dept, 73-80, ASSOC PROF PHARMACOL, TEX COL OSTEOP MED, 77- *Concurrent Pos:* Consult, NIH, Environ Protection Agency & Becton Dickinson Res Ctr, 75-79. *Mem:* Am Soc Pharmacol & Therapeut; NY Acad Sci; Am Thoracic Soc. *Res:* The metabolism of chemical carcinogens by human tissues. *Mailing Add:* 4109 Brown Trail Colleyville TX 76034

CANTRELL, GRADY LEON, b Louisville, Ky, Feb 18, 36; m 54, 86; c 1. MATHEMATICS. *Educ:* Univ Louisville, BA, 64; Univ Ky, MS, 66, PhD(math), 68. *Prof Exp:* Asst prof, 68-71, assoc prof, 71- PROF MATH, MURRAY STATE UNIV. *Concurrent Pos:* mem, Ctr Excellence Reservoir Res. *Mem:* Math Asn Am. *Res:* Analytic function theory; continuous functions. *Mailing Add:* Dept Math Murray State Univ Murray KY 42071

CANTRELL, JAMES CECIL, b Palmersville, Tenn, Sept 14, 31; m 54; c 3. TOPOLOGY. *Educ:* Bethel Col, BS, 53; Univ Miss, MS, 55; Univ Tenn, PhD(math), 61. *Prof Exp:* Instr, Dresden High Sch, 53-54; instr math, Univ Tenn, 61-62; from asst prof to assoc prof, 62-70, PROF MATH, UNIV GA, 70-, HEAD DEPT, 74- *Concurrent Pos:* Mem, Inst Advan Study, 66-67; Alfred P Sloan res fel, 66-68. *Mem:* Am Math Soc. *Res:* Topological embeddings of manifolds. *Mailing Add:* Dept Math Grad Univ Ga Athens GA 30602

CANTRELL, JOHN H(ARRIS), b Memphis, Tenn, June 24, 43; m 67; c 1. MICROSTRUCTURE-MATERIAL PROPERTIES RELATIONSHIPS, ELECTRON- ACOUSTIC MICROSCOPY. *Educ:* Univ Tenn, BS, 65, PhD(physics), 76; Univ Cambridge, UK, MA, 88. *Prof Exp:* Nat Res Coun res assoc, 77-79, res physicist, 79-88, SR RES PHYSICIST, LANGLEY RES CTR, NASA, 88- *Concurrent Pos:* Counsult, Oak Ridge Nat Lab, 75-77; adj prof, Dept Physics, Col William & Mary, 83-88; vis scholar, Cavendish Lab, Univ Cambridge, UK, 88-89; overseas fel, Churchill Col, Cambridge, UK, 88-89. *Honors & Awards:* RD-100 Award, 78. *Mem:* Fel Acoust Soc Am; Am Phys Soc; Mat Res Soc; AAAS; Electron Micros Soc Am. *Res:* Acoustoelasticity; ultrasonics; solid state nonlinear acoustics; thermoelasticity; materials characterization; microstructure-material property relationships; scanning electron acoustic microscopy; nondestructive characterization of metal fatigue; mathematical modeling. *Mailing Add:* Langley Res Ctr NASA Mail Stop 231 Hampton VA 23665-5225

CANTRELL, JOHN LEONARD, b Billings, Mont, Feb 7, 39; m 62; c 4. IMMUNOLOGY, IMMUNOGENETICS. *Educ:* Mont State Univ, BS, 67, MS, 69; Univ Calif, Los Angeles, PhD(immunol), 72. *Prof Exp:* Fel, Jackson Lab, Maine, 72-74; asst mem, Okla Med Res Found, 74-76; sr staff fel cancer, Nat Inst Allergy & Infectious Dis, NIH, 76-81; CHIEF IMMUNOL, RIBI IMMUNOCHEM RES, MONT, 81-, VPRES. *Concurrent Pos:* Asst prof, Med Sch, Univ Okla, 74-76; mem clin staff, Univ Hosp & Clin, Univ Okla & Children's Mem Hosp, Oklahoma City, 74-76. *Mem:* Transplantation Soc; Am Asn Cancer Res; Am Asn Immunologists; Sigma Xi; Am Col Radiol. *Res:* Cancer immunology; immunotherapy; immunogenetics of transplantation immunology. *Mailing Add:* Ribi Immunochem Res Inc PO Box 1409 Hamilton MT 59840

CANTRELL, JOSEPH SIRES, b Parker, Kans, July 31, 32; m 58; c 3. PHYSICAL CHEMISTRY, SOLAR PHYSICS. *Educ:* Kans State Teachers Col, AB, 54; Kans State Univ, MS, 57, PhD(phys chem), 61. *Prof Exp:* Res chemist, Procter & Gamble Co, 61-66; mem fac chem, 65-69, assoc prof, 69-85, PROF CHEM, MIAMI UNIV, 85- *Mem:* Electrochem Soc; Am Chem Soc; Am Crystallog Asn; Int Solar Energy Soc; Sigma Xi. *Res:* Chemical kinetics; x-ray crystal structure; mesomorphic structure; thin film and interfacial structure; electron diffraction; chemical applications of solar energy; photoelectrochemical processes; electrochemistry. *Mailing Add:* Dept Chem Miami Univ Oxford OH 45056

CANTRELL, RONALD P, genetics, for more information see previous edition

CANTRELL, THOMAS SAMUEL, b Spartanburg, SC, Aug 29, 38. ORGANIC CHEMISTRY. *Educ:* Univ SC, BS, 58, MS, 59; Ohio State Univ, PhD(chem), 64. *Prof Exp:* NSF fel org chem, Columbia Univ, 64-65; asst prof chem, Rice Univ, 65-70; res chemist, NIH, 70-71; asst prof chem, 71-72, ASSOC PROF CHEM, AM UNIV, 72- *Mem:* Am Chem Soc; Royal Soc Chem. *Res:* Organic photochemistry; nonbenzenoid aromatic compounds. *Mailing Add:* Dept Chem Am Univ Washington DC 20016-8003

CANTRELL, WILLIAM ALLEN, b Everton, Ark, Nov 6, 20; m 45; c 2. PSYCHIATRY. *Educ:* McMurry Col, BS, 40; Univ Tex Med Br Galveston, MD, 43; Am Bd Psychiat & Neurol, cert psychiat, 51. *Prof Exp:* Intern, US Naval Hosp, Corona, Calif, 43-44; resident neuropsychiat, Univ Tex Med Br Hosps, 49-50, asst prof, 50-51; pvt pract, Houston, 51-63; clin prof, 63-68, PROF PSYCHIAT, BAYLOR COL MED, 68- *Concurrent Pos:* Asst prof neuropsychiat, Univ Tex Med Br, 51-54; clin asst prof psychiat, Baylor Col Med, 51-63; from asst chief to sr consult, Psychiat Serv, Methodist Hosp, 52-; clin assoc prof psychiat, Univ Tex Grad Sch Biomed Sci Houston, 59-78; attend staff, Psychiat Serv, Ben Taub Gen Hosp, 63-; asst examr, Am Bd Psychiat & Neurol, 64-; consult psychiat, Social Security Admin, Bur Hearings & Appeals, 64- & Vet Admin Hosp, Houston, 67-; adj prof, Inst Relig, Houston, 68- *Mem:* AAAS; Cent Neuropsychiat Asn (pres, 76-77); fel Am Psychiat Asn; fel Am Col Psychiat. *Res:* Undergraduate medical education, especially methodology and evaluation; doctor-patient relationship as a factor in the quality of health care delivery. *Mailing Add:* Dept Psychiat Baylor Col Med One Baylor Plaza Houston TX 77030

CANTRELL, WILLIAM FLETCHER, b Young Harris, Ga, Oct 29, 16; m 47; c 2. CHEMOTHERAPY. *Educ:* Univ Ga, BS, 38, MS, 39; Univ Chicago, PhD, 49. *Prof Exp:* Instr zool, Univ Ga, 39-40; asst parasitol & pharmacol, Univ Chicago, 41-46; from asst prof to prof pharmacol, Univ Louisville, 49-59; mem staff, Lab Parasite Chemother, NIH, 59-62; from assoc prof to prof, 63-80, EMER PROF PHARMACOL, UNIV TENN, MEMPHIS, 80- *Mem:* Am Soc Pharmacol; Am Soc Parasitol; Am Soc Trop Med & Hyg; Soc Exp Biol & Med. *Res:* Transmission of malaria by mosquitoes; chemotherapy of malaria; mode of action of chemotherapeutic drugs; antigenic variation in trypanosomes; drug resistance. *Mailing Add:* Tenn Ctr Health Sci 874 Union Ave Memphis TN 38103

CANTRILL, JAMES EGBERT, b Frankfort, Ky, Sept 2, 33; m 56; c 6. ORGANIC CHEMISTRY. *Educ:* Univ Notre Dame, BS, 55; Mass Inst Technol, PhD(org chem), 59. *Prof Exp:* Res chemist, Eastman Kodak Co, NY, 59-60; instr chem, St Thomas Moore Col, 60-61, asst prof & chmn dept, 61-63; develop chemist, Plastics Dept, Gen Elec Co, 63-67, mgr res & advan develop, 67-69; mgr res lab, Foster Grant Co, Inc, 69-74, bus develop mgr, 74-77; bus develop mgr petrochem, Am Hoechst Co, NJ, 77-80, TECH RES & DEVELOP, HOECHST CELANESE, SC, 80- *Mem:* Am Chem Soc. *Res:* Polymer synthesis; reaction kinetics and mechanisms. *Mailing Add:* 1400 Hood Rd Greer SC 29611

CANTU, ANTONIO ARNOLDO, b Laredo, Tex, Jan 5, 41. QUANTUM CHEMISTRY, THEORETICAL CHEMISTRY. *Educ:* Univ Tex, Austin, BS, 63, PhD(chem physics), 67. *Prof Exp:* Fel, Univ Alta, 68-71; Orgn Am States fel & vis prof, Inst Physics, Univ Mex, 70; res assoc, Res Coun Alta, 71-72; chemist, Law Enforcement Assistance Admin, US Dept Justice, 72-73; forensic chemist, Bur Alcohol, Tobacco & Firearms, US Dept Treas, 73-83; FBI, 83-86; US SECRET SERVICE, 86- *Honors & Awards:* Forensic Scientist of the Year, 80 (Mid Atlantic Region). *Mem:* Soc Appl Spectros; Am Chem Soc; Sigma Xi. *Res:* Development of techniques to improve conventional methods of ink and paper analysis in the examination of questioned documents; application of statistical pattern recognition techniques to the individualization of physical evidence; improving the detection of fingerprints using optical and chemical techniques; research in protective science. *Mailing Add:* 3013 Seven Oaks Pl Falls Church VA 22042-3155

CANTWELL, EDWARD N(ORTON), JR, b Chicago, Ill, Jan 13, 27; m 48; c 2. FUELS, LUBRICANTS. *Educ:* Northwestern Univ, BSME, 49, MS, PhD(mech eng), 53. *Prof Exp:* Engr, 52-62, div head, 62-79, res mgr, E I Du Pont De Nemours & Co, Inc, 79-84; PRES, CANTWELL CONSULT ASSOCS, INC, 84- *Mem:* Soc Automotive Engrs. *Res:* Combustion of fuels in internal combustion engines; effect of fuels, lubricants and additives on all aspects of engine performance; future engines; measurement and control of vehicle emissions; effect of emissions on the environment; alternate fuels. *Mailing Add:* 2400 Heather Rd W Wilmington DE 19803-2720

CANTWELL, FREDERICK FRANCIS, b Brooklyn, NY, May 22, 41; m 64; c 3. LIQUID CHROMATOGRAPHY, SOLVENT EXTRACTION. *Educ:* Allegheny Col, BA, 63; Univ Iowa, PhD(chem), 72. *Prof Exp:* Sr scientist analytical chem, Endo Labs Inc, E I du Pont de Nemours & Co, Inc, 64-70, asst dir analytical res, 72-74; from asst prof to assoc prof, 75-86, PROF CHEM, UNIV ALTA, 86- *Concurrent Pos:* Nat Res Coun Can res grants, 75- *Mem:* Am Chem Soc; Chem Inst Can. *Res:* Liquid chromatographic research on novel stationary phases including preparation, studies of retention mechanisms and application to drug analysis; solvent extraction applied to flow injection analysis; metal speciation. *Mailing Add:* Dept Chem Univ Alta Edmonton AB T6G 2G2 Can

CANTWELL, JOHN CHRISTOPHER, b St Louis, Mo, Aug 12, 36; m 64; c 3. MATHEMATICS, DIFFERENTIAL TOPOLOGY. *Educ:* St Louis Univ, BS, 57; Univ Notre Dame, PhD(math), 62. *Prof Exp:* Asst, Inst Advan Study, 62-64; asst prof, Univ Iowa, 64-67; from asst prof to assoc prof, 67-73, dept chmn, 86-91, PROF MATH, ST LOUIS UNIV, 73- *Mem:* Asn Mem Inst Advan Study; Am Math Soc; Math Asn Am; Nat Speleol Soc. *Res:* Foliations; morse theory; differential and combinatorial manifolds. *Mailing Add:* 2351 Parkridge St Louis MO 63144

CANVIN, DAVID T, b Winnipeg, Man, Nov 8, 31; m 57; c 4. PLANT PHYSIOLOGY, PLANT BIOCHEMISTRY. *Educ:* Univ Man, BSA, 56, MSc, 57; Purdue Univ, PhD(plant physiol), 60. *Prof Exp:* Res assoc plant sci, Univ Man, 60-63, assoc prof, 63-65; head dept, 80-84, dean, Sch Grad Studies & Res, 84-89, PROF BIOL, QUEEN'S UNIV ONT, 65- *Concurrent Pos:* Secy-treas, Biol Coun Can. *Honors & Awards:* Gold Medal, Can Soc Plant Physiologists. *Mem:* AAAS; Am Soc Plant Physiol; Can Soc Cell Biol; Can Soc Plant Physiol; fel Royal Soc Can. *Res:* Intermediary metabolism in plants; environmental physiology. *Mailing Add:* Dept Biol Queen's Univ Kingston ON K7L 3N6 Can

CANZONIER, WALTER J, b New Brunswick, NJ, Feb 6, 36. BIVALVE SHELLFISH CULTURE, SHELLFISH SANITATION & DEPURATION. *Educ:* St Peter's Col, Jersey City, NJ, BS, 57. *Prof Exp:* Teaching asst biol, Dept Zool, Rutgers Univ, 58-59, res asst, Dept Oyster Culture, 60-67, res assoc, 68-71 & 80-87; researcher, Inst Biol del Mare, CNR, Venezia, 71-77; dir, Coastal Resource Appl Res Lab, Venezia, 77-80; DIR RES & DEVELOP, AQUARIUS ASSOCS, MANASQUAN, NJ, 87- *Concurrent Pos:* Consult, various govt & pvt agencies, Marine Res & Develop, 73-, UNESCO, Paris, Marine Res Facil Design, 78-, Min Fisheries, Eire, Shellfish Sanit, 82-89, consult, SFADCO, Erie, Prog Develop Fisheries, 90-; tech, comt mem, Min Marina Mercantile, Italia, Resource Mgt, 78-80, Min Santà, Italia, Fisheries Prod Control, 78-, Shellfish Sanit, 78-80, tech comt mem, Interstate Shellfish Sanit Conf, 80- *Mem:* Nat Shellfisheries Asn; Soc Invertebrate Path; World Aquacult Soc; Europ Aquacult Soc. *Res:* Shellfish biology and culture; estuarine dynamics; pathology of shellfish; coastal pollution; marine resources evaluation and management; biological and engineering aspects related to bivalve shellfish depuration; author of 46 publications. *Mailing Add:* Aquarius Assocs 44 Cowart Ave Manasquan NJ 08736

CAO, HENGCHU, b Hunan, China, Oct 10, 62. MECHANICAL BEHAVIOR OF ENGINEERING MATERIALS, ENGINEERING FRACTURE MECHANICS. *Educ:* S China Univ Sci & Technol, BS, 82; Univ Calif, Berkeley, MS, 85, PhD(mat sci & eng), 88. *Prof Exp:* Res asst, Ctr Advan Math, Lawrence Berkeley Lab, 84-86; res asst, 86-88, RES ENGR, UNIV CALIF, SANTA BARBARA, 88- *Mem:* Am Ceramic Soc; Ceramic Educ Coun; Mat Res Soc. *Res:* Processing and mechanical behaviors of advanced ceramics, metal-matrix and ceramic-matrix composites; interface fracture mechanisms; evaluation of thin films and coatings; fatigue failure of ferro electric ceramics. *Mailing Add:* Mat Dept Col Eng Univ Calif Santa Barbara CA 93106

CAPALDI, DANTE JAMES, b Windsor, Ont, Sept 4, 57; m 83; c 1. CHEMISTRY, IMMUNOCHEMISTRY. *Educ:* Univ Windsor, BSc, 80, PhD(biochem), 83. *Prof Exp:* Teaching asst & lab instr biochem, Univ Windsor, 80-81, lab coordr & instr introd chem, 81-83; fel, dept internal med, Univ Mich, Ann Arbor, 83-84 & Div Biochem Res, Henry Ford Hosp, Detroit, 84-85; vpres res & develop, Nuclear Diags Inc, 85-87; DIR QUAL ASSURANCE & REGULATORY AFFAIRS, LEECO DIAGS, INC, 87- *Concurrent Pos:* Scholar, Univ Windsor, 82 & 83; adj lectr, dept Natural Sci, Univ Mich, Dearborn, 85-88 & vis scholar, Sch Grad Studies, Ann Arbor, 85-87. *Honors & Awards:* Gold Medal, Soc Chem Indust, 80. *Mem:* Am Asn Clin Chem; Am Chem Soc; Clin Ligand Assay Soc; Can Soc Clin Chemists; Chem Inst Can; Soc Chem Indust; Nat Comt Clin Lab Standards. *Res:* Human erythrocyte membrane surface carbohydrates; development of new immunodiagnostic techniques for thyroid hormones and perioxidase-catalyzed colorimetric assays. *Mailing Add:* 697 Front Rd N Amherstburg ON N9V 2V6 Can

CAPALDI, EUGENE CARMEN, b Philadelphia, Pa, Apr 10, 37; m 63; c 2. ORGANIC CHEMISTRY. *Educ:* Univ Pa, BS, 59; Univ Del, MS, 61, PhD(org chem), 64. *Prof Exp:* Res chemist, Arco Chem Co, 63-66, sr res chemist, 66-76, coorder toxicol, 76-78, chem prod supvr, 78-80, admin mgr, 80-81, mgr toxicol & prod safety, 81-86, mgr Environ, Health & Safety, 86-89, MGR HEALTH RISK MGT, ARCO CHEM CO, DIV ATLANTIC RICHFIELD CORP, 90- *Mem:* Am Chem Soc; Sigma Xi. *Res:* Synthesis of new monomers and chemical intermediates; polyester and polyurethane chemistry; toxicology and product safety. *Mailing Add:* 707 Malin Rd Newtown Square PA 19073

CAPASSO, FEDERICO, b Rome, Italy, June 24, 49. SEMICONDUCTOR PHYSICS. *Educ:* Univ Rome, Italy, Doctorate physics, 73. *Prof Exp:* Staff scientist, Fondazione V Bordoni, 74-76; vis scientist, 76-77, mem tech staff, 77-87, DEPT HEAD, MGT RES, AT&T BELL LABS, 87- *Concurrent Pos:* Distinguished mem tech staff, AT&T Bell Labs, 84; chmn, Solid State Detector Tech Group, Optical Soc Am, 86-87; assoc ed, Photonics Technol Lett, 89-90; mem, External Adv Comt, Beckman Inst, 91-94. *Honors & Awards:* David Sarnoff Award, Inst Elec & Electronic Engrs, 91. *Mem:* Fel Inst Elec & Electronic Engrs; fel Optical Soc Am; fel Am Phys Soc; fel Int Soc Optical Eng. *Res:* Semiconductor physics; transport and optical properties of artificially structured semiconductors and superlattices; pioneered the band structure engineering approach; invented many new electron and optoelectronic devices, including many new solid state photomultipliers and resonant tunneling transistors. *Mailing Add:* AT&T Bell Labs 600 Mountain Ave Rm 7A209 Murray Hill NJ 07974

CAPCO, DAVID G, CELL & DEVELOPMENTAL BIOLOGY. *Educ:* Univ Tex, Austin, PhD(cell biol), 80. *Prof Exp:* ASST PROF CELL BIOL, ARIZ STATE UNIV, 84- *Mailing Add:* Dept Zool Ariz State Univ Tempe AZ 85287-1501

CAPE, ARTHUR TREGONING (TOMMY), b Java Pur, India, Jan 13, 04. METALLURGY. *Educ:* Royal Sch Mines, London, Associateship, 25. *Prof Exp:* Vpres, Coast Metals Inc, 48-88; RETIRED. *Mem:* Fel Am Soc Metals. *Res:* Use of saliva and detecting disease. *Mailing Add:* 1285 Sylvan Rd Monterey CA 93940

CAPE, JOHN ANTHONY, b Helena, Mont, Nov 2, 29; m 62; c 3. SOLID STATE PHYSICS. *Educ:* Carroll Col (Mont), AB, 51; Mont State Univ, MS, 53; Univ Notre Dame, PhD(physics), 58. *Prof Exp:* Prof math & physics, Carroll Col (Mont), 55-56; asst physics, Univ Notre Dame, 57, instr, 57-58; res assoc, Univ Ill, 58-60; res specialist, Atomics Int, 60-64; mem tech staff, Rockwell Int Corp, 64-72, group leader solid state sci, 72-77, prog dir energy conversion, 74-78, proj mgr concentrator solar cells, 78, 87; CHEVRON RES CO, 81- *Concurrent Pos:* Vis res assoc, Stanford Univ, 69. *Mem:* Fel Am Phys Soc. *Res:* Magnetic properties of materials; optical properties of magnetic materials; semiconductors; physical properties of surfaces and interfaces; superconductivity; solar photovoltaic energy conversion. *Mailing Add:* Chevron Res Co PO Box 1627 Richmond CA 94802

CAPE, RONALD ELLIOT, b Montreal, Can, Oct 11, 32; m 56; c 2. BIOCHEMISTRY. *Educ:* Princeton Univ, AB, 53; Harvard Univ, MBA, 55; McGill Univ, PhD(biochem), 67. *Prof Exp:* Pres, Prof Pharmaceut Corp, 57-67, chmn, 67-73; managing partner, Cetus Sci Lab, 71-72, pres, 72-78, CHMN & CHIEF EXEC OFFICER, CETUS CORP, 78- *Concurrent Pos:* Med Res Coun Can Centennial fel, Univ Calif, Berkeley, 67-70. *Mem:* Am Soc Microbiol; Can Biochem Soc; Royal Soc Health; Sigma Xi. *Res:* Mutation and gene manipulation of industrial microorganisms. *Mailing Add:* Cetus Corp 1400 53rd St Emeryville CA 94608-2946

CAPECCHI, MARIO RENATO, b Verona, Italy, Oct 6, 37; US citizen; m 63. CELL BIOLOGY. *Educ:* Antioch Col, BS, 61; Harvard Univ, PhD(biophys), 67. *Prof Exp:* Soc Fels jr fel biophys, Harvard Univ, 66-68, from asst prof to assoc prof biochem, Med Sch, 68-73; PROF HUMAN GENETICS, UNIV UTAH SCH MED, 73- *Concurrent Pos:* Estab investr, Am Heart Asn, 69-72; NIH Career Develop Award, 72-74; Am Cancer Soc Fac Res Award, 74-79. *Honors & Awards:* Am Chem Soc Biochem Award, 69. *Mem:* Nat Acad Sci; Am Biochem Soc; Am Soc Biol Chem. *Res:* Gaining an understanding of how the information encoded in the gene is translated by the cell; expression in Eucaryotic and Procaryotic cells; somatic cell genetics. *Mailing Add:* Dept Biol Univ Utah 342 S Biol Salt Lake City UT 84112

CAPEHART, BARNEY LEE, b Galena, Kans, Aug 20, 40; m 61; c 3. SYSTEM ENGINEERING. *Educ:* Univ Okla, BSEE, 61, MEE, 62, PhD(syst eng), 67. *Prof Exp:* Lectr eng, Northeastern Univ, 64-65; mem tech staff, Aerospace Corp, Calif, 67-68; asst prof indust & syst eng, Grad Eng Educ Syst, Univ Fla, 68-72; assoc prof indust eng, Univ Tenn, Knoxville, 72-73; from assoc prof to prof, 73-87, ASST CHMN, INDUST & SYSTS ENG, UNIV FLA, 87- *Concurrent Pos:* Consult, Martin-Marietta Corp, Fla, US Naval Training Equip Ctr, Orlando, Hicks & Assocs, Orlando, Fla, Casazza, Schultz & Assoc, Washington, DC & Fla Legis; sabbatical, US Dept Energy, Washington, DC, 89-90. *Honors & Awards:* Palladium Medal, Am Asn Eng Socs, 88. *Mem:* Fel AAAS; Inst Elec & Electronics Engrs; fel Inst Indust Engrs; Asn Energy Engrs. *Res:* Digital simulation techniques; energy applications of systems analysis and operations research and energy conservation; energy efficiency. *Mailing Add:* Dept Indust & Systs Eng Univ Fla Gainesville FL 32611

CAPEK, RADAN, b Prague, Czech, Jan 29, 31; Can citizen; m 61; c 2. NEUROPHARMACOLOGY, ELECTROPHYSIOLOGY. *Educ:* Charles Univ, Prague, MD, 56; Czech Acad Sci, PhD(pharmacol), 60. *Prof Exp:* Researcher pharmacol, Inst Pharmacol, Czech Acad Sci, 60-63, head neuropharmacol res unit, 64-69; vis scientist, 69, asst prof, 70-75, ASSOC PROF, DEPT PHARMACOL & THERAPEUT, MCGILL UNIV, MONTREAL, 75- *Concurrent Pos:* Res assoc, dept pharmacol, Col Med, Univ Ill, 62-63; vis scientist, dept pharmacol, Univ Sao Paulo, 63. *Mem:* Pharmacol Soc Can (secy, 80-83, vpres, 83-85, pres, 85-87); Am Soc Pharmacol & Exp Therapeut; Soc Neurosci; Can Col Neuropsychopharmacol. *Res:* Experimental analysis of mechanisms of action of drugs acting on the central nervous system; their action on synaptic transmission, using electrophysiological techniques. *Mailing Add:* Dept Pharmacol & Therapeut McGill Univ 3655 Drummond St Montreal PQ H3G 1Y6 Can

CAPEL, CHARLES EDWARD, b Troy, NY, Dec 26, 22; m 45; c 2. MATHEMATICS. *Educ:* NY State Teachers Col, Albany, BA, 47; Univ Rochester, MA, 50; Tulane Univ, PhD(math), 53. *Prof Exp:* Instr math, Geneseo State Teachers Col, 47-49 & Tulane Univ, 50-51; asst prof, Miami Univ, 53-58; res mathematician, Westinghouse Res Lab, Pa, 58-60; PROF MATH, MIAMI UNIV. 60- *Mem:* Am Math Asn; Math Soc Am; Sigma Xi. *Res:* Inverse limit spaces; functions; fixed points. *Mailing Add:* Math/Stat Dept 213 Bachelor Hall Miami Univ Oxford OH 45056

CAPEN, CHARLES CHABERT, b Tacoma, Wash, Sept 3, 36; m 68. VETERINARY PATHOLOGY. *Educ:* Wash State Univ, DVM, 60; Ohio State Univ, MSc, 61, PhD(vet path), 65; Am Col Vet Pathologists, dipl. *Prof Exp:* Res assoc, 60-62, from instr to prof, 62-72, PROF ENDOCRINOL, COL MED, OHIO STATE UNIV, 72-, CHMN VET PATHOBIOL, 81. *Concurrent Pos:* Consult path, Food & Drug Admin, 73-; mem coun, Am Col Vet Pathologists, 75-81. *Honors & Awards:* Borden Res Award, Am Vet Med Asn, 75. *Mem:* AAAS; Am Vet Med Asn; Endocrine Soc; Int Acad Path; Am Soc Exp Path. *Res:* Comparative and veterinary pathology; endocrine and metabolic diseases; calcium metabolism; ultrastructure of thyroid and parathyroid glands; metabolic bone disease. *Mailing Add:* Dept Vet Pathobiol Ohio State Univ 1925 Coffey Rd Columbus OH 43210

CAPEN, CHARLES FRANKLIN, JR, planetary sciences, education lecturing; deceased, see previous edition for last biography

CAPEN, RONALD L, b Mich, Feb 22, 42; m 85; c 4. PHYSIOLOGY. *Educ:* Univ Calif, Berkeley, PhD(zool), 71. *Prof Exp:* PROF BIOL, DEPT BIOL, COLO COL, 71- *Mem:* Am Physiol Soc. *Res:* Pulmonary blood circulation. *Mailing Add:* Dept Biol Colo Col Cascade Ave Colorado Springs CO 80903

CAPERON, JOHN, b Milford, Utah, Apr 14, 29; m 64; c 3. ECOLOGY, OCEANOGRAPHY. *Educ:* Univ Utah, BS, 52; Scripps Inst Oceanog, Univ Calif, PhD(oceanog), 65. *Prof Exp:* Scientist, Int Bus Mach Corp, 53-59; oceanogr, Supreme Allied Comdr Atlantic Antisubmarine Warfare Res Ctr, Off Naval Res, London, 65-69; assoc prof, 69-74, PROF OCEANOG, UNIV HAWAII AT MANOA, 74- *Mem:* Ecol Asn Am. *Res:* Population dynamics; marine ecology. *Mailing Add:* Whian Whian Via Lismore 2480 N S W Australia

CAPERS, EVELYN LORRAINE, b Los Angeles, Calif, Dec 20, 25; m 58; c 1. CLINICAL MICROBIOLOGY. *Educ:* Univ Calif, Berkeley, BA, 46; Univ Southern Calif, PhD(bact), 71. *Prof Exp:* Med technologist, Childrens Hosp Los Angeles, 47-66; med microbiologist, Martin Luther King Jr Gen Hosp, Los Angeles County Dept Health Serv, 71-74; asst dir microbiol, Bio-Sci Labs, 74-79; clin asst prof path, Sch Med, Univ Southern Calif, 79-84, instr microbiol, 84-86; RETIRED. *Concurrent Pos:* Asst clin prof path, Sch Med, Univ Southern Calif, 73- *Mem:* Am Soc Microbiol; Sigma Xi. *Res:* Diagnostic microbiology and serology. *Mailing Add:* 5503 Verdun Ave Los Angeles CA 40043

CAPETOLA, ROBERT JOSEPH, b Norristown, Pa, Oct 25, 49; m 76; c 3. PHARMACOLOGY. *Educ:* Philadelphia Col Pharm & Sci, BS, 71; Hahnemann Med Col, MS, 74, PhD(pharmacol), 75. *Prof Exp:* Res assoc pharmacol, Hahnemann Med Col, 75-76; sr pharmacologist, Wyeth Lab, Inc, 76-78; scientist, 78-79, sr scientist, 79-81, group leader hypersensitivity dis, 81-83, RES MGR & DIR PHARMACOL, ORTHO PHARMACEUT CORP, 83- *Mem:* Am Soc Pharmacol & Exp Therapeut; Am Rheumatism Asn; Soc Exp Biol & Med; Soc Invest Dermat; Am Chem Soc. *Res:* Therapeutic control of inflammatory and allergic disorders; immunopharmacology; rheumatoid arthritis and asthma; analgesics and dermatopharmacology. *Mailing Add:* Ortho Pharmaceut Corp Rt 202 Raritan NJ 08869

CAPIZZI, ROBERT L, b Philadelphia, Pa, Nov 20, 38; m 65; c 4. ONCOLOGY, CLINICAL PHARMACOLOGY. *Educ:* Temple Univ, BS, 60; Hahnemann Med Col, MD, 64. *Prof Exp:* Fel clin pharmacol, Hahnemann Med Col, 65-66; fel med & pharmacol, 67-69, from asst prof to assoc prof med & pharmacol, Sch Med, Yale Univ, 72-77; prof med & pharmacol & chief, Div Med Oncol, Univ NC, 77-82; DIR, ONCOL RES CTR, CHIEF, SECT HEMAT/ONCOL & CHMN, PIEDMONT ONCOL ASN, BOWMAN GRAY SCH MED, WAKE FOREST UNIV, WINSTON-SALEM, NC, 82- *Concurrent Pos:* Fac develop award clin pharmacol, Pharmaceut Mfrs Asn, 73; investr, Howard Hughes Med Inst, 76. *Mem:* Am Asn Cancer Res; Am Soc Clin Oncol; Am Soc Clin Pharmacol & Therapeut. *Res:* Cancer chemotherapy; clinical pharmacology of anticancer drugs. *Mailing Add:* 3018 Old Clinic Bldg Univ NC Chapel Hill NC 27514

CAPLAN, ARNOLD I, b Chicago, Ill, Jan 5, 42; m 65; c 2. DEVELOPMENTAL BIOLOGY, BIOCHEMISTRY. *Educ:* Ill Inst Technol, BS, 63; Johns Hopkins Univ, PhD(biochem), 66. *Prof Exp:* Fel anat, Johns Hopkins Univ Med Sch, 66-67; fel biochem, Brandeis Univ, 67-68, fel biol, 68-69; ass prof biol, 69-74, assoc prof biol & anat, 74-81, PROF BIOL, CASE WESTERN RESERVE UNIV, 81- *Mem:* AAAS; Soc Develop Biol; Soc Cell Biol. *Res:* Biochemical control of phenotypic expression, especially in chick embryo limb mesodermal cells; biochemical development of muscle, cartilage and bone; stem cell mediated skeletal repair; marrow stroma regeneration. *Mailing Add:* Dept Biol Case Western Reserve Univ Cleveland OH 44106

CAPLAN, AUBREY G, b Pittsburgh, Pa, Oct 22, 23; m 48; c 2. ELECTRICAL DESIGN, TECHNICAL EXPERT WITNESS. *Educ:* Carnegie Inst Technol, BS, 45. *Prof Exp:* Test engr, Douglas Aircraft, Los Angeles, 45-46; power salesman, Duquesne Light Co, Pittsburgh, 46-59; CONSULT ENGR, CAPLAN ENG CO, PITTSBURGH, 59- *Concurrent Pos:* Regional vpres, IES NAm, 75-77. *Mem:* Inst Elec & Electronics Engrs; fel Am Consult Engrs Coun; Construct Specif Inst. *Mailing Add:* 5401 Hobart St Pittsburgh PA 15217

CAPLAN, DANIEL BENNETT, b Boston, Mass, Sept 14, 37; c 3. PEDIATRICS. *Educ:* Brandeis Univ, AB, 58; Tufts Univ, MD, 62. *Prof Exp:* MEM STAFF, CYSTIC FIBROSIS RES CTR, EMORY UNIV. *Mailing Add:* Emory U Cystic Fibrosis Res Ctr 2040 Ridgewood Dr NE Atlanta GA 30322

CAPLAN, DONALD, b Ottawa, Ont, Sept 24, 17; m 41; c 2. SURFACE CHEMISTRY, PHYSICAL METALLURGY. *Educ:* Queen's Univ, Ont, BSc, 40, MSc, 41; Rensselaer Polytech Inst, PhD(metall), 55. *Prof Exp:* Plant chemist & metallurgist, Hull Iron & Steel Foundries Ltd, 41-46; res chemist corrosion chem, Nat Res Coun Can, 46-53; res fel phys chem, Rensselaer Polytech Inst, 53-56; sr res officer corrosion chem, Nat Res Coun Can, 56-82; RETIRED. *Res:* Mechanism and kinetics of high temperature oxidation of metals; corrosion of metals and alloys; materials selection for resistance to corrosion. *Mailing Add:* 426 Simpson Rd Ottawa ON K1H 5A9 Can

CAPLAN, JOHN D(AVID), b Weiser, Idaho, Mar 5, 26; m 52; c 3. CHEMICAL ENGINEERING. *Educ:* Ore State Univ, BS, 49; Wayne State Univ, MS, 55. *Prof Exp:* Head fuels & lubricants dept, 63-67, tech dir basic & appl sci, 67-69, exec dir, Gen Motors Res Labs, 69-87; RETIRED. *Honors & Awards:* Crompton-Lanchester Medal, Brit Inst Mech Engrs, 64. *Mem:* Nat Acad Eng; Am Chem Soc; fel Am Inst Chem Engrs; fel AAAS; fel Soc Automotive Eng. *Res:* Engine fuels and combustion; automotive air pollution. *Mailing Add:* 2512 Loch Creek Way Bloomfield Hills MI 48304-3812

CAPLAN, LOUIS ROBERT, b Baltimore, Md, Dec 31, 36; m 63; c 6. NEUROLOGY, STROKE. *Educ:* Williams Col, BA, 58; Univ Md Sch Med, MD, 62. *Prof Exp:* Intern & jr resident, Boston City Hosp, 62-64; resident neurol, Harvard Neurol Unit, 66-69; fel cardiovasc div, Mass Gen Hosp, 69-70; neurologist, Beth Israel Hosp, 70-78; asst prof neurol, Harvard Med Sch, 71-78; neurologist in chief, Michael Reese Hosp, 78-84; prof neurol, Univ Chicago Sch Med, 81-84; NEUROLOGIST-IN-CHIEF, NEW ENG MED CTR, 84-; PROF NEUROL, TUFTS UNIV, SCH MED, 84- *Concurrent Pos:* Pres, Chicago Neurol Soc, 83-84 & Boston Soc Psychiat & Neurol, 88-89; rep, Coun Med Spec Soc, 82-; practice comt, Am Acad Neurol, 82-; chmn, Stroke Coun, Am Heart Asn, 87- *Mem:* Am Acad Neurol; Am Neurol Asn; Am Heart Asn; Asn Univ Prof Neurol. *Res:* Vetebro-basilae occlusive disease; racial differences in cerebrovascular disease; computer data bases on stroke and diagnosis. *Mailing Add:* Dept Neurol New Eng Med Ctr 750 Washington St Boston MA 02111

CAPLAN, PAUL E, b Far Rockaway, NY, Feb 29, 24; m 73; c 5. CHEMICAL ENGINEERING, INDUSTRIAL HYGIENE. *Educ:* Middlebury Col, AB, 44; Univ Colo, BSChE, 48; Univ Calif, Berkeley, MPH, 49; Univ Mich, Ann Arbor, PE, 58; Am Acad Indust Hyg, dipl, 62; Bd Cert Safety Prof, cert, 71. *Prof Exp:* From asst indust hyg engr to sr indust hyg engr, Bur Occup Health, Calif State Dept Pub Health, 49-71 & adv, Bur Radiol Health, 60-71; res indust hyg eng & dep dir, Div Tech Serv & Div Phys Sci & Eng, Nat Inst Occup Safety & Health, 71-89; RETIRED. *Concurrent Pos:* Mem, Threshold Limits Value Comt, Air Sampling Instruments Comt, Agr Health Comt & Hyg Guides Comt, Am Conf Govt Indust Hyg. *Mem:* Am Conf Govt Indust Hyg; Am Indust Hyg Asn; Nat Soc Prof Engrs. *Res:* Evaluation control and toxicology of hazards involved in use and manufacture of agricultural chemicals, asbestos, lead, silica products, compressed air and other chemicals; development of techniques of air sampling for general environmental hazards evaluation. *Mailing Add:* 1484 Sigma Circle Cincinnati OH 45255

CAPLAN, PAULA JOAN, b Springfield, Mo, July 7, 47; c 2. PSYCHOLOGY OF WOMEN. *Educ:* Harvard Univ, AB, 69; Duke Univ, MA, 71, PhD(psychol), 73. *Prof Exp:* Postdoctoral fel, Neuropsychol Res Unit, Hosp Sick Children, 74-77; clin & res psychologist, Family Court Clin, Clarke Inst Psychiat, 77-80; asst prof, Ont Inst Studies Educ, 80-81, assoc prof, 81-87, head ctr women's studies, 85-87, PROF APPL PSYCHOL, ONT INST STUDIES EDUC, 87-; ASST PROF PSYCHIAT, UNIV TORONTO, 79- *Mem:* Fel Am Orthopsychiat Asn; fel Can Psychol Asn. *Res:* Research methodology; sex differences in intelligence, play, aggression, achievement behavior and spatial abilities; children's learning; hand preference development in infants; child abuse. *Mailing Add:* Dept Appl Psychol Ont Inst Studies Educ Toronto ON M5S 1V6 Can

CAPLAN, PHILIP JUDAH, b Detroit, Mich, May 25, 27; m 52; c 4. SOLID STATE PHYSICS. *Educ:* Yeshiva Univ, BA, 48; Wayne State Univ, MS, 52. *Prof Exp:* Physicist, Electronics Technol & Devices Lab, US Army, 52-88; RETIRED. *Res:* Semiconductor interface defects; electron spin resonance of solids; nuclear magnetic resonance in ferromagnetic materials; dynamic nuclear polarization; optical spectra of laser materials. *Mailing Add:* 105 Bimbler Blvd Ocean NJ 07712

CAPLAN, RICHARD MELVIN, b Des Moines, Iowa, July 16, 29; m 52; c 4. DERMATOLOGY, MEDICAL HUMANITIES. *Educ:* Iowa State Univ, BS, 49; Univ Iowa, MA, 51, MD, 55. *Prof Exp:* From asst prof to assoc prof, 61-69, PROF DERMAT, COL MED, UNIV IOWA, 69-, ASSOC DEAN CONTINUING MED EDUC, 70 - *Mem:* Soc Med Sch Dirs Continuing Med Educ; Am Acad Dermat; Am Dermat Asn. *Mailing Add:* Dept Dermat Univ Iowa Col Med Iowa City IA 52242

CAPLAN, YALE HOWARD, b Baltimore, Md, Dec 27, 41; m 65; c 3. PHARMACEUTICAL CHEMISTRY, PATHOLOGY. *Educ:* Univ Md, BS, 63, PhD(med chem), 68; Am Bd Forensic Toxicol, dipl, 76. *Prof Exp:* Asst pharmaceut chem, Sch Pharm, Univ Md, 64-65; res assoc toxicol & cancer chemother, Sinai Hosp, Baltimore, Md, 68-69, supvr surg res div, 69; asst toxicologist, off chief med examr, State of Md, Baltimore, 69-74; instr toxicol, 72-73, from clin asst prof to clin assoc prof path, 73-85, CLIN PROF PATH, SCH MED, UNIV MD, 85-, DIR GRAD PROG FORENSIC TOXICOL, 74-; CHIEF TOXICOLOGIST, OFF CHIEF MED EXAMR, STATE OF MD, BALTIMORE, 74- *Concurrent Pos:* Am Found Pharmaceut Educ fel, Univ Md, 63-65, NIH fel, 65-68, assoc mem grad fac, 74-77, mem, 77-, adj prof pharmacol & toxicol, Sch Pharm, 86-, adj asst prof biomed chem, 77-79, adj assoc prof, 79-86, adj prof, 86-, mem exec comt, Prog in Toxicol, 84-; asst in surg, dept surg, Sch Med, John Hopkins Univ, Md, 69-70, lectr, div forensic path, Sch Hyg & Pub Health, 73-84, consult, Appl Physics Lab, 73-79; toxicologist, Cent Labs, Assoc Md Pathologists, Ltd, 71-72; prof lectr, dept forensic sci, Col Gen Studies, George Washington Univ, Utah, 75-76; dir toxicol, Md Med Lab, Inc, 77-90; consult, Dept Res Oncol & Cell Biol, Sinai Hosp, 70-74, Allied Chem Co, 75-, Ctr Human Toxicol, Univ Utah, 79-85, Med Diagnostics Inc, 84-85, Ill Dept Law Enforcement, 85, Nat Inst Drug Abuse, 86-90, Armed Forces Inst Path, 87, Nat Transp Safety Bd, 87-90, Toxicol Resource Comt, Col Am Pathologists, 87-, Abbott Labs, 88-90; inspector, Forensic Urine Drug Testing Accreditation Prog, 87-, Nat Lab Cert Prog, Nat Inst Drug Abuse, 88-; sr assoc, Bensinger, DuPont & Assocs, Chicago, 87-90; dir, Forensic Toxicol & Urine Drug Anal Lab, Nat Ctr Forensic Sci, 88-; ed, Forensic Sci Rev, 89-; mem numerous comts, Am Acad Forensic Sci, 76-90, Am Bd Forensic Toxicol, 84-88, Forensic Sci Found, 80-90, Nat Safety Coun, 82-, Soc Forensic Toxicol, 78- *Honors & Awards:* Rolla N Harger Award, Am Acad Forensic Sci, 89. *Mem:* AAAS; fel Soc Forensic Toxicologists (secy, 76-78, vpres, 79-80, pres, 81); fel Am Acad Forensic Sci (treas, 84-86, pres-elect, 86-87, pres, 87-88); Int Asn Forensic Toxicol; fel Am Inst Chemists; Am Chem Soc; Am Asn Clin Chem. *Res:* Analytical forensic and clinical toxicology, particularly development of procedures for analysis and chemical diagnosis in drug related death; experimental toxicology, particularly relationship of toxic concentrations and effects of drugs. *Mailing Add:* 3411 Philips Dr Baltimore MD 21208

CAPLE, GERALD, b International Falls, Minn, Apr 3, 35; m 64. ORGANIC CHEMISTRY. *Educ:* St Olaf Col, BA, 57; Fla State Univ, PhD(org chem), 63. *Prof Exp:* Res assoc chem, Ore State Univ, 63-65, asst prof, 65-66; asst prof, 66-70, ASSOC PROF CHEM, NORTHERN ARIZ UNIV, 70- *Mem:* Am Chem Soc; Sigma Xi. *Res:* Organic synthesis and mechanisms. *Mailing Add:* Northern Ariz Univ Box 5898 Flagstaff AZ 86011

CAPLE, RONALD, b International Falls, Minn, Dec 7, 37; m 59; c 4. ORGANIC CHEMISTRY. *Educ:* St Olaf Col, BA, 60; Univ Mich, MS, 62, PhD(org mechanisms), 64. *Prof Exp:* NSF fel, Univ Colo, 64-65; from asst prof to assoc prof, 65-74, PROF ORG CHEM, UNIV MINN, DULUTH, 74- *Mem:* Am Chem Soc. *Res:* Intermediates in electrophilic additions. *Mailing Add:* Dept Chem Univ Minn Duluth MN 55812

CAPLIN, SAMUEL MILTON, botany, for more information see previous edition

CAPLIS, MICHAEL E, b Ypsilanti, Mich, July 25, 38; m 56; c 7. CLINICAL CHEMISTRY, ANALYTICAL CHEMISTRY. *Educ:* Eastern Mich Univ, BS, 62; Purdue Univ, MS, 64, PhD(biochem), 70. *Prof Exp:* Teacher physics, math & chem, St John High Sch, 58-62; asst biochem, Purdue Univ, 62-67, res asst, 67-69; instr med, 69-80, asst prof biochem, div allied health, Sch Med, Ind Univ, 80-83, ADJ PROF TOXICOL, IND UNIV SCH MED NORTHWEST, 83- *Concurrent Pos:* Anal biochemist, State of Ind, 62-67; toxicologist, Ind Criminal Justice Comn, Region I, 70; dir, Northwest Ind Criminal & Toxicol Lab, 70-83, biochem & toxicol, St Mary Med Ctr, 70-83, Great Lakes Forensic Labs, 83-85. *Mem:* AAAS; Am Asn Clin Chem; Am Chem Soc; Am Acad Clin Toxicol; Am Asn Crim Lab Dirs; Sigma Xi. *Res:* Analytical biochemistry-toxicology; mechanism of action of toxic chemicals and drugs; resistance to toxic action of chemicals and drugs; development of instrumental analytical procedures for study of chemicals, drugs and their metabolites in biological systems; applications of spectrochemical and electrochemical methods. *Mailing Add:* 676 N 50 W Valparaiso IN 46383

CAPLOW, MICHAEL, b New York, NY, May 20, 35; m 59; c 2. BIOCHEMISTRY, ENZYMOLOGY. *Educ:* NY Univ, DDS, 59; Brandeis Univ, PhD(biochem), 63. *Prof Exp:* Asst biochem, Yale Univ, 63-70; assoc prof, 70-73, PROF BIOCHEM, UNIV NC, CHAPEL HILL, 73- *Mem:* AAAS; Am Chem Soc. *Res:* Dynamic properties of the cytoskeleton; chemistry of microtubules. *Mailing Add:* Dept Biochem & Biophysics Univ NC Chapel Hill NC 27599

CAPOBIANCO, MICHAEL F, b Brooklyn, NY, Oct 4, 31; m 65. MATHEMATICS, STATISTICS. *Educ:* Polytech Inst Brooklyn, BChE, 52, MChE, 54, PhD(math), 64; Columbia Univ, MA, 57. *Prof Exp:* Statistician, Am Cyanamid Co, 54-55; from instr to asst prof math, St John's Univ, 55-63; mathematician, Repub Aviation Corp, 63; lectr math, Polytech Inst Brooklyn, 63-64, asst prof, 64-66; assoc prof, 66-71, PROF MATH, ST JOHN'S UNIV, NY, 71- *Concurrent Pos:* Consult, Maimonides Hosp Brooklyn, 66- *Mem:* Am Math Soc; Math Asn Am; Biomet Soc; Soc Indust & Appl Math. *Res:* Statistical decision theory; statistical inference in digraphs; traffic analysis; combinatorics; statistics in the life sciences, especially psychology, economics, sociology, biophysics and cybernetics; tensor analysis and differential geometry. *Mailing Add:* Dept Math St John's Univ 300 Howard Ave Staten Island NY 10301

CAPON, BRIAN, b Wallasey, Eng, Dec 27, 31. BOTANY. *Educ:* La Sierra Col, BA, 58; Univ Chicago, MS, 60, PhD(bot), 61. *Prof Exp:* From asst prof to assoc prof, 61-71, chmn dept, 69-71, PROF BOT, CALIF STATE UNIV, LOS ANGELES, 71- *Res:* Physiology of desert plants. *Mailing Add:* Dept Biol Calif State Univ 5151 State University Dr Los Angeles CA 90032

CAPONE, DOUGLAS GEORGE, b Newark, NJ, April 30, 50; m 79; c 2. MARINE MICROBIOLOGY, MICROBIAL ECOLOGY. *Educ:* Univ Miami, BS, 73, PhD(marine sci), 78. *Prof Exp:* From res asst prof to assoc prof, Marine Sci Res Ctr, State Univ NY, Stony Brook, 70-79; PROF, CHESAPEAKE BIOL LAB, CTR ENVIRON & ESTUARINE STUDIES, UNIV MD, 87- *Mem:* Am Soc Microbiol; Am Soc Limnol & Oceanog; AAAS; Am Chem Soc; Sigma Xi; Oceanog Soc. *Res:* Role and importance of microorganisms in marine ecosystems, particularly in elemental cycles; interactions of the microbiota and environmental pollutants. *Mailing Add:* Chesapeake Biol Lab Univ Md Ctr Environ & Estuarine Studies Solomons MD 20688-0038

CAPONE, JAMES J, b White Plains, NY, Jan 14, 45; m 68; c 2. CLINICAL RESEARCH, REGULATORY AFFAIRS. *Educ:* NY Univ, BS, 67; Fla State Univ, MS, 69, PhD(microbiol), 71. *Prof Exp:* Postdoctoral immunol, Univ Ill, 71; dir res, Kalletad Labs, 71-76; dir qual assurance, Hyland Div, Baxter Health Care, 76-80; dir res, Packard Insts, United Technol, 80-87; dir res audits, G D Searle, 87; PRES CONSULT, DIVERSIFIED BIOMED TECHNOL, 88-; INSTR BIOL SCI, ROOSEVELT UNIV, 89- *Concurrent Pos:* Mem, Nat Comt Clin Lab Standards, 80-84; consult, Advan Qual Systs, 80-90. *Mem:* Regulatory Affairs Professionals Soc; Am Statist Asn; Am Soc Qual Assurance. *Res:* Development of immunochemical metliode and instrumentation for biomedical and clinical research. *Mailing Add:* 1110 Thackeray Dr Palatine IL 60067

CAPONETTI, JAMES DANTE, b Boston, Mass, Mar 15, 32; m 66; c 2. BOTANY, PHARMACY. *Educ:* Mass Col Pharm, BS, 54, MS, 56; Harvard Univ, AM, 59, PhD(biol), 62. *Prof Exp:* From asst prof to assoc prof, 61-84, PROF BOT, UNIV TENN, KNOXVILLE, 84-, ASSOC DEPT HEAD, 86- *Concurrent Pos:* Regist pharmacist, Mass & Tenn. *Mem:* Fel AAAS; Bot Soc Am; Int Soc Plant Morphol; Am Fern Soc (treas, 76-); Tissue Culture Asn; Sigma Xi. *Res:* Morphgenesis and propagation of vascular plants, mostly Pteridophytes, by tissue culture. *Mailing Add:* Dept Bot Univ Tenn Knoxville TN 37996-1100

CAPONIO, JOSEPH FRANCIS, b Canton, Mass, Mar 25, 26; m 57; c 2. BIOCHEMISTRY. *Educ:* St Anselm's Col, AB, 51; Georgetown Univ, PhD(biochem), 59. *Prof Exp:* Sr analyst, Libr Cong, 55-58; chief bibliog div, Off Tech Serv, US Dept Com, 58-61; dir tech info, Defense Document Ctr, Div Sci Area, US Dept Defense, 61-64; sci info officer, NIH, 64-70; assoc dir, Nat Agr Libr, 70-74; dir, Environ Sci Info Ctr, 74-79, dep dir, 79-82, DIR, NAT TECH INFO SERV, DEPT COM, 82- *Concurrent Pos:* Mem off critical tables, Nat Acad Sci-Nat Res Coun, 64- *Honors & Awards:* Silver Medal, Dept Com. *Mem:* Am Chem Soc; Am Soc Info Sci; AAAS. *Res:* Protein and enzyme chemistry; information science. *Mailing Add:* 8417 Ft Hunt Rd Alexandria VA 22308

CAPORALE, LYNN HELENA, b New York, NY, Sept 3, 47. BIOCHEMISTRY, MOLECULAR BIOLOGY. *Educ:* Brooklyn Col, BS, 67; Univ Calif, Berkeley, PhD(molecular biol), 73. *Prof Exp:* Nat Inst Allergy & Infectious Dis fel path, Med Sch, NY Univ, 73-74; assoc researcher leukemia, Sloan Kettering Inst Cancer Res, 75-76; res assoc biochem, Rockefeller Univ, 76-77; asst prof biochem, Sch Med & Dent, Georgetown Univ, 78-84; sr res fel, Merck Sharp & Dohme Res Labs, 84-86, DIR SCI EVAL, MERCK & CO, 87- *Concurrent Pos:* Adj asst prof, Rockefeller Univ, 78-; prin investr, Am Lung Asn res grant, 78-79; NIH res grant, 78-85, spec study sect, 79-84, physiolog chem study sect, 86-90, Nat Res Lab res grant, 82-84; adj fac, Sch Med & Dent, Georgetown Univ, 84-87; Woodrow Wilson hon fel. *Mem:* AAAS; Am Chem Soc; Sigma Xi. *Res:* Peptide hormone antagonists; the complement cascade; evolution of gene families; proteases; peptide synthesis; science policy. *Mailing Add:* Merck & Co Inc RY80M Trl-1101 PO Box 2000 Rahway NJ 07650

CAPORALI, RONALD VAN, b Arnold, Pa, June 22, 36; m 56; c 4. MECHANICAL ENGINEERING, CERAMIC SCIENCE. *Educ:* Pa State Univ, BS, 58, MS, 64, PhD(ceramic sci), 69. *Prof Exp:* Ceramic technologist, Am Glass Res, 58-62; res asst glass struct, Pa State Univ, 62-69; chief glass technologist, Am Glass Res, 69-76. *Concurrent Pos:* Consult glass technol, 76- *Mem:* Am Ceramic Soc; Brit Soc Glass Tech; Soc Soft Drink Technologists; Am Soc Testing & Mat; AAAS; Sigma Xi; Nat Inst Ceramic Engrs. *Res:* Structure and strength of glasses; fracture analysis. *Mailing Add:* RD 1 West Sunbury PA 16061

CAPORASO, FREDRIC, b Jersey City, NJ, May 28, 47; m; c 4. SENSORY EVALUATION, FOOD CHEMISTRY. *Educ:* Rutgers Univ, BS, 69, MS, 72; Pa State Univ, PhD(food sci), 75. *Prof Exp:* Asst prof food sci, Univ Nebr, 75-78; res scientist, med food develop, Am McGaw Div, Am Hosp Supply Corp, 78-80, group leader, food sci, 81-82; CHMN, DEPT FOOD SCI & NUTRIT, CHAPMAN COL, 82-, DIR, FOOD SCI RES CTR, 87- *Concurrent Pos:* Mem exec comt, Sensory Eval Div, Inst Food Technologists, 83-85 & 85-, regional communicator, 83-; chair, Food Sci Adminr Group, 89-91. *Mem:* Inst Food Technologists; Am Meat Sci Asn. *Res:* Sensory evaluation and flavor chemistry of food ingredients and products; medical food product development; muscle food biochemistry; new food product development. *Mailing Add:* Food Sci & Nutrit Dept Chapman Col 333 N Glassell St Orange CA 92666

CAPP, GRAYSON L, b Seattle, Wash, Aug 11, 36; m 59; c 3. BIOCHEMISTRY. *Educ:* Seattle Pac Col, BS, 58; Univ Ore, MS, 62, PhD(biochem), 67. *Prof Exp:* Instr chem, Los Angeles Pac Col, 59-60; res assoc biochem, Ore Primate Ctr, 61-63; NIH fel protein struct, Duke Univ, 66-68; asst prof biochem, 68-70, asst prof chem, 70-73, ASSOC PROF CHEM & PRE-PROF HEALTH SCI ADV, SEATTLE PAC UNIV, 73- *Concurrent Pos:* NIH sr fel, Univ Wash, 69-70. *Mem:* AAAS; Am Chem Soc. *Res:* Protein subunit structure and function with studies of glutamic dehydrogenase and human hemoglobin. *Mailing Add:* Dept Chem Seattle Pac Univ W Nickerson & Third Ave W Seattle WA 98119

CAPP, MICHAEL PAUL, b Yonkers, NY, July 1, 30; m 57; c 4. RADIOLOGY, PEDIATRICS. *Educ:* Roanoke Col, BS, 52; Univ NC, MD, 58; Am Bd Radiol, cert radiol, 62. *Prof Exp:* Intern pediat, Med Ctr, Duke Univ, 58-59, resident radiol, 59-62, assoc, 62-63, from asst prof to assoc prof, 63-70, dir diag radiol, 66-70, asst prof pediat, 68-70; PROF RADIOL, CHMN DEPT & DIR CLIN RADIOL, ARIZ HEALTH SCI CTR, 70-; PRES, UNIV PHYSICIANS INC, 88- *Concurrent Pos:* Lab instr physics, Roanoke Col, 52; teaching asst, Duke Grad Sch Physics, 52-54, radiologist in charge, Pediat Cardiol, Duke Univ Med Ctr, 62-70, dir, Duke Pediat Radiol Prog, 65-70, Med Students Teaching Prog, Diag Radiol, 65-66 & Diag Div, Dept Radiol, 66-70, chief, Pediat & Cardiac Radiol, 63-70; chief of staff, Ariz Health Sci Ctr, 71-73; mem, Univ Ariz Intercol Athletic comt, 81-84, chmn, 84-; guest examr, Am Bd Radiol, 74-, site visit approval residency prog, 76-81, governance comt, 83-85, chmn, exam comt diag radiol, subcomt diag radiol written & oral exam, chmn long range planning comt, 83-84; comt separation diag, ther & nuclear med, Am Col Radiol, 67-70, subcomt resident training, 70-72, comt vis fel, 73-87, comt continuing educ, 74-76, cardiovasc syllabus comt, 79-, comt continuing eval postgrad educ, Comn Diag Radiol, educ delivery systs task force, 81, MR comt govt affairs, 82-85, comt educ & training, 82-83, comt prof self eval & continuing educ & chmn, Sect Cardiovasc II, 82-; pres, coun Cardiovasc Radiol & comt coun affairs, Am Heart Asn, 76-79, res comt, 78- & chmn subcomt C, 85-; res & educ comt, Am Roentgen Ray Soc, 73-76 & chmn, 75-76, chmn publ comt, Am J Roentgenol, 80-87, & chmn search comt pd, 84, adv comt, Lawrence Berkeley Radiation Labs, 76-80; residency review comt, Am Med Asn, 79, rep to Am Bd Radiol, 81-; sci prog comt, Asn Univ Radiologists, 70, mem comt, 70-73, mem award comt, 80, govt affairs comt, NIH, 79-85, sci prog comt, 81-, conjoint comt diag radiol, 80-; vpres, Eastern RAdiol Soc, 73-74, pres, 74-75, chmn adv comt, 79-; diag adv comt, Nat Cancer Inst, 78-80; mem, Cardiovasc Radiol Review Panel, Intersoc Comn Heart Dis Resources, 73; nat comt radiol continuing med educ, Radiol Soc NAm, 78-, sci exhib awards comt, 79-85, bylaws comt, 78-83, pub nfo aadv bd, 83-; assoc ed, W J Med, 80; reviewer, Am J Radiol, Radiol, Radiol Physics, Magnetic Resonance Imaging, Circulation, Invest Radiol, J Western Med, & J Health Care Technol; Toshiba Med Corp grant, 83-88. *Honors & Awards:* Gene Keefe lect, St Josephs Hosp, Mich, 81; 1st Sanford George Bluestein lectr, Yale Univ, 82; Finzi lect, Royal Soc Med, 83; 14th Herbert M Stauffer lect, Temple Univ Med Sch, 84; Wendell G Scott Mem lect, Mallinckrodt Inst Radiol, Wash Univ, 85; Henry J Walton lectr, Univ Md Sch Med, 85; 7th Annual Bernard S Epstein mem lect, Long Island Jewish Med Ctr, 87; Charles E Culpepper lect, Univ Ill, Chicago, 87; Sosman lectr, Harvard Med Sch, 88. *Mem:* Soc Pediat Radiol; AMA; Radiol Soc NAm; Am Roentgen Ray Asn; Am Col Radiol; Asn Univ Radiologists; Soc Chmn Acad Radiol Dept; NAm Soc Cardiac Radiologists. *Res:* Congenital heart disease, particularly left ventricular function; radiology of cardiovascular diseases; radiological physics & imaging; pediatric radiology. *Mailing Add:* Dept Radiol Univ Ariz Health Sci Ctr Tucson AZ 85724

CAPP, WALTER B(ERNARD), b Munhall, Pa, Apr 18, 33; m 65. OPERATIONS RESEARCH. *Educ:* Carnegie Inst Technol, BS, 55, MS, 57. *Prof Exp:* Engr econ eval, Gulf Res & Develop Co, Gulf Oil Corp, 57-60, proj engr opers res, 60-63, sr proj engr, 63-66, sect supvr, Opers Res Sect, 66-71, tech adv, Computation & Commun Serv Dept, 71-75, mgr mgt sci, 76-82, gen mgr info res, 82-85; ORGN CONSULT, CHEVRON CORP, 85- *Mem:* Inst Mgt Sci. *Res:* Management science; economic analysis; statistical analysis; linear programming; computer programming. *Mailing Add:* 152 Ravenhill Rd Orinda CA 94563

CAPPALLO, ROGER JAMES, b Cleveland, Ohio, Sept 30, 49; m 71; c 3. VERY LONG BASELINE INTERFEROMETRY. *Educ:* Mass Inst Technol, SB, 71, PhD(planetary sci), 80. *Prof Exp:* Res assoc, Mass Inst Technol, 80; RES SCIENTIST, HAYSTACK OBSERV, 80- *Mem:* Am Geophys Union. *Res:* Techniques and applications of very long baseline interferometry for geodesy and astronomy; modelling of the moon's rotation. *Mailing Add:* Haystack Observ Rte 40 Westford MA 01886

CAPPAS, C, b Cairo, Egypt, Mar 14, 26; US citizen; m 56; c 1. PHYSICAL CHEMISTRY, MATHEMATICS. *Educ:* Berea Col, BA, 56; Univ Fla, PhD(phys chem), 62. *Prof Exp:* Res assoc phys chem, Princeton Univ, 62, group leader dielectrics, 62-64; assoc prof chem, Oglethorpe Univ, 64-66; assoc prof physics, Chapman Col, 66-67; ASSOC PROF CHEM, UNIV S ALA, 67- *Concurrent Pos:* Consult, anal chem. *Mem:* Am Chem Soc; Sigma Xi. *Res:* Chemical kinetics; determination of physical constants; polymer chemistry; solid state chemistry. *Mailing Add:* Dept Chem Univ S Ala Mobile AL 36688

CAPPEL, C ROBERT, b Connersville, Ind, Nov 17, 42. ORGANIC CHEMISTRY, PHOTOGRAPHIC CHEMISTRY. *Educ:* Ball State Univ, BS, 69; Univ Ill, MS, 72, PhD(chem), 73. *Prof Exp:* Asst chem, Univ Ill, 69-73; SR RES CHEMIST, EASTMAN KODAK CO, 73- *Mem:* Sigma Xi; Soc Photog Scientists & Engrs. *Res:* Investigation and application of organic chemistry to photographic science. *Mailing Add:* 69 Hermitage Rd Rochester NY 14617

CAPPELL, SYLVAIN EDWARD, b Brussels, Belg, Sept 10, 46; US citizen; m 66; c 4. TOPOLOGY. *Educ:* Columbia Univ, 66; Princeton Univ, PhD(math), 69. *Prof Exp:* Princeton nat fel, Princeton Univ, 66-69, from instr to asst prof math, 69-74; assoc prof, 74-78, PROF MATH, COURANT INST, NY UNIV, 78- *Concurrent Pos:* Woodrow Wilson Found fel, 66-67; NSF fel, 66-68; Danforth Found fel, 66-69; vis lectr, Harvard Univ, 70-71; Sloan Found fel, 71-73; vis prof, Weizmann Inst, Israel, 72, Inst Hautes Etudes Sci, 73 & Harvard Univ, 81; Guggenheim Found fel, 89-90. *Mem:* Am Math Soc. *Res:* Manifolds and submanifolds. *Mailing Add:* Courant Inst Math Sci 251 Mercer St New York NY 10012

CAPPELLINI, RAYMOND ADOLPH, b Pittsburgh, Pa, Jan 30, 26; m 50; c 4. PLANT PATHOLOGY. *Educ:* Duquesne Univ, BEd, 50; Pa State Univ, MS, 52; Cornell Univ, PhD, 55. *Prof Exp:* Asst res specialist plant path, 55-61, assoc prof, 61-67, PROF PLANT PATH, RUTGERS UNIV, 67-, CHMN DEPT PLANT PATH, 71- *Mem:* Am Phytopath Soc; Mycol Soc Am; Sigma Xi. *Res:* Market diseases; fungus physiology. *Mailing Add:* Dept Plant Path Rutgers Univ New Brunswick NJ 08903

CAPPS, DAVID BRIDGMAN, b Jacksonville, Ill, Jan 25, 25; m 48; c 3. MEDICINAL CHEMISTRY. *Educ:* Ill Col, AB, 48; Univ Nebr, MS, 50, PhD(org chem), 52. *Hon Degrees:* DSc, Ill Col, 79. *Prof Exp:* Res chemist, Chemstrand Corp, 52-56; from assoc res chemist to res scientist, 56-77, SR SCIENTIST ORG CHEM, RES LABS, PARKE, DAVIS & CO, 77- *Mem:* Am Chem Soc; AAAS. *Res:* Synthesis of experimental drugs for the control of cancer and of infectious diseases, and the study of relationships between molecular structure and biological activities. *Mailing Add:* 1406 Brooklyn Ann Arbor MI 48104

CAPPS, RICHARD H, b Wichita, Kans, July 1, 28; m 55, 75; c 3. THEORETICAL PHYSICS. *Educ:* Univ Kans, AB, 50; Univ Wis, MS, 52, PhD(physics), 55. *Prof Exp:* Fel, Univ Calif, Berkeley, 55-57; actg asst prof physics, Univ Wash, 57-59; fel, Cornell Univ, 58-60; from asst prof to prof, Northwestern Univ, 60-67; PROF PHYSICS, PURDUE UNIV, 67- *Concurrent Pos:* Fulbright & Guggenheim fel, Univ Rome, 62-63; consult, Argonne Nat Lab, Ill, 60-70. *Mem:* Am Phys Soc. *Res:* Theory of the interactions of fundamental particles. *Mailing Add:* Dept Physics Purdue Univ West Lafayette IN 47907

CAPPUCCI, DARIO TED, JR, b Plains, Pa, Aug 19, 41. VETERINARY MEDICINE. *Educ:* Univ Calif, Davis, BS, 63, DVM, 65, MS, 66; Am Bd Vet Pub Health, dipl, 73; Am Registry Cert Animal Scientists, dipl, 75; Univ Calif, San Francisco, PhD(comp path), 76. *Prof Exp:* Head carnivore unit, NIH Animal Ctr, Md, 66-67; vet epidemiologist, Zoonoses Surveillance Unit, Nat Communicable Dis Ctr, Ga, 68; researcher vet med & sci, Independent Invest Studies, 69; vet, Vet Lab Serv, Calif Dept Agr, 69-70; pub health vet, Calif Dept Health, 70-76; pvt pract, 76-78; VET EPIDEMIOLOGIST, FOOD & DRUG ADMIN, MD, 78- *Concurrent Pos:* Mem subcomt pub health, Nat Brucellosis Comt, 68-69; pub health serv med epidemiol preceptor, 80- *Mem:*

Am Vet Med Asn; Am Soc Animal Sci; AAAS; Wildlife Dis Asn; NY Acad Sci; Sigma Xi. *Res:* Reproductive pathophysiology; veterinary public health; comparative medicine; supervision of laboratory animal breeding colony; animal science. *Mailing Add:* 1077 Sanchez St San Francisco CA 94114-3360

CAPRA, J DONALD, b Burlington, Vt, July 20, 37; m 58; c 2. MEDICINE, IMMUNOLOGY. *Educ:* Univ Vt, BS, 59, MD, 63. *Prof Exp:* Intern, St Luke's Hosp, New York, 64, resident, 65; sr surgeon, NIH, 6567; guest investr, Rockefeller Univ, 6769; assoc prof microbiol, Mt Sinai Sch Med, 69-74; PROF MICROBIOL, UNIV TEX HEALTH SCI CTR, DALLAS, 74- *Concurrent Pos:* USPHS fel, 65-67. *Mem:* Am Asn Immunologists; Am Rheumatism Asn. *Res:* Immunogenetics; protein sequences; antibody combining site. *Mailing Add:* Dept Microbiol Univ Tex Southwestern Med Sch 5323 Harry Hines Blvd Dallas TX 75235-9048

CAPRANICA, ROBERT R, b Los Angeles, Calif, May 29, 31; m 58. NEUROBIOLOGY, ELECTRICAL ENGINEERING. *Educ:* Univ Calif, Berkeley, BS, 58; NY Univ, MEE, 60; Mass Inst Technol, ScD(elec eng), 64. *Prof Exp:* Mem tech staff commun res, Bell Tel Labs, NJ, 58-69; assoc prof neurobiol & elec eng, 69-75, PROF NEUROBIOL, BEHAVIOR & ELEC ENG, CORNELL UNIV, 75- *Concurrent Pos:* Ed, J Comp Physiol; NATO sr fel sci, 76-77; Coun Sci & Indust Res, sr fel award, 82-83; partic, Int Cong Physiol Sci, 83. *Mem:* Fel AAAS; Am Soc Zool; fel Acoust Soc Am; Am Physiol Soc; Soc Neurosci; fel Explorers Club; Int Soc Neuroethology (pres elect, 84-); Am Acad Otolaryngol; Am Soc Artificial Internal Organs; Am Soc Ichthyologists & Herpetologists; Am Speech-Lang-Hearing Asn; Int Brain Res Orgn; Int Soc Phonetic Sci; NY Acad Sci. *Res:* Animal sound communication; auditory neurophysiology; neuroethology. *Mailing Add:* Cornell Univ W255 Seeley G Mudd Hall Ithaca NY 14853

CAPRETTA, UMBERTO, b Nereto, Italy, Mar 5, 22; US citizen; m 50; c 2. ANALYTICAL CHEMISTRY. *Educ:* Univ Naples, Dr indust chem, 52. *Prof Exp:* Teacher high sch, Italy, 52-55; lab technician, 55-60, anal chemist, 60-85, SR ENGR, EASTMAN KODAK CO, 85- *Mem:* AAAS; fel Am Inst Chemists; Am Chem Soc; NY Acad Sci; Royal Soc Chem; French Chem Soc; Asn Off Anal Chemists; Am Soc Testing & Mat. *Res:* Instrumental analysis; x-ray emission; polarography; atomic absorption; development of new methods and techniques. *Mailing Add:* 187 River St Rochester NY 14612-4735

CAPRI, ANTON ZIZI, b Czernowitz, Romania, Apr 20, 38; Can citizen; m 60; c 3. FIELD THEORY, MATHEMATICAL PHYSICS. *Educ:* Univ Toronto, BASc, 61; Princeton Univ, MA, 65, PhD(physics), 67. *Prof Exp:* Res physicist, Kimberly-Clark Corp, 61-63; fel, 67-68, vis asst prof theoret physics, 68-69, from asst prof to assoc prof, 69-79, PROF THEORET PHYSICS, UNIV ALTA, 79- *Concurrent Pos:* Sr res fel, Alexander von Humboldt Found, 75-76. *Mem:* Am Phys Soc; Can Asn Physics. *Res:* Higher spin field theories; existence of solutions in quantum field theory; elasticity theory. *Mailing Add:* Theoret Physics Inst Univ Alta Edmonton AB T6G 2J1 Can

CAPRIO, JAMES R, b Niagara Falls, NY, Apr 8, 39; m 58; c 2. ELECTRICAL ENGINEERING, APPLIED MATHEMATICS. *Educ:* State Univ NY, Buffalo, BS, 61, MS, 63; Cornell Univ, PhD(elec eng), 66. *Prof Exp:* Asst electronics engr, Cornell Aeronaut Lab, Cornell Univ, 61-62, assoc electronics engr, 62-66, res electronics engr, 66-67, prin electronics engr, 67-74; ASST MGR, SYSTS ENG DIV, COMPTEK RES, INC, 74- *Concurrent Pos:* Asst prof elec eng, State Univ NY, Buffalo, 67-74; consult prin engr, Cornell Aeronaut Lab, 67- *Mem:* Inst Elec & Electronics Engrs; Soc Indust & Appl Math; Math Asn Am. *Res:* Radar systems and signal processing; statistical communications theory; detection and estimation of signal wave forms and parameters; digital filtering. *Mailing Add:* Comptek Res Inc 110 Broadway Buffalo NY 14203

CAPRIO, JOHN THEODORE, b Norfolk, Va, June 22, 45; m 68. SENSORY PHYSIOLOGY. *Educ:* Old Dominion Univ, BS, 67; Fla State Univ, PhD(physiol), 76. *Prof Exp:* Asst prof, 76-80, ASSOC PROF PHYSIOL, LA STATE UNIV, 80- *Concurrent Pos:* Prin investr, Nat Inst Neurol & Commun Disorders & Stroke grant, 79-81 & 81-84. *Mem:* Soc Neurosci; Sigma Xi; AAAS; Europ Chemoreception Res Orgn. *Res:* Chemical senses; olfaction and taste in aquatic organisms. *Mailing Add:* Dept Zool & Physiol La State Univ Baton Rouge LA 70803

CAPRIO, JOSEPH MICHAEL, b New Brunswick, NJ, Nov 7, 23; m 51; c 3. AGRICULTURAL METEOROLOGY, MICROCLIMATOLOGY. *Educ:* Rutgers Univ, BS, 47, MS, 50; Calif Inst Technol, BS, 48; Utah State Univ, PhD(biometeorol), 70. *Prof Exp:* Agrometeorologist, Am Inst Aerological Res, 50-53; statistician & agronometeorologist, Citrus Exp Sta, Univ Calif, 53-55; from asst prof to assoc prof agr climat, 55-63, PROF AGR CLIMAT, MONT STATE UNIV, 63-; CLIMATOLOGIST, MONT STATE, 77- *Concurrent Pos:* Mem, Tech Comt Western Regional Proj W-48, 57-75, W-148, 76-81, WRCC-47, 82-, W-130, 82- & WRCC- 50, 82-; mem, Tech Comt Great Plains Regional Proj GP-1, 59-61; climat expert, UN World Meteorol Orgn, Iran, 62-63; mem phenology panel, US Int Biol Prog, 67-70; co-invest, Earth Resources Technol Satellite-1, Phenology Satellite Exp, NASA, 72-74; vis scientist, Div Land Use Res, Commonwealth Indust & Sci Orgn, Canberra, Australia, 73-74; mem, UN World Meteorol Orgn Working Group on Methods of Forecasting Agr Crop Develop & Ripening, 75-79; ecologist-meterologist workshop chmn, Panel Biol Indicators Climatic Change, 76; Agristars NASA Prog, 80-84. *Mem:* Am Meteorol Soc; Am Soc Agron; Soil Sci Soc Am. *Res:* Biometeorology; agrometeorology, including statistical climatic analysis; mapping of climatic elements in mountainous areas; study of weather effects on crop production; plant phenology; soil physics; hydrology. *Mailing Add:* Dept Plant & Soil Sci Mont State Univ Bozeman MT 59717

CAPRIOGLIO, GIOVANNI, b Rome, Italy, Aug 9, 32; m 58; c 1. ELECTROCHEMISTRY. *Educ:* Univ Milan, DSc(indust chem), 56. *Prof Exp:* Asst prof electrochem, Univ Modena, 57-59 & Univ Milan, 59-62; consult, Gen Atomic Div, Gen Dynamics Corp, 61-62, staff mem, 62-67; proj mgr, Gulf Gen Atomic, Inc, 67-70; assoc dir lab, Gen Atomic Co, 70-73, mgr mat & chem dept, 73-86; ADJ PROF HAZARDOUS MAT, UNIV CALIF SAN DIEGO, 87- *Mem:* Int Soc Electrochem; Electrochem Soc; Am Nuclear Soc. *Res:* High temperature materials; corrosion; batteries; flue gas desulfurization. *Mailing Add:* 6585 Muirlands Dr La Jolla CA 92037

CAPRIOLI, RICHARD MICHAEL, b New York, NY, Apr 12, 43; m 83. BIOCHEMISTRY, MASS SPECTROMETRY. *Educ:* Columbia Univ, BS, 65, PhD(biochem), 69. *Prof Exp:* Res assoc chem, Purdue Univ, 69-70, from asst prof to assoc prof, 70-75; assoc prof, 75-80, PROF BIOCHEM, MED SCH, UNIV TEX, HOUSTON, 80- *Mem:* Am Soc Biol Chemists; Am Soc Mass Spectrometry. *Res:* Peptide sequencing by mass spectrometry; intermediary metabolism using stable isotopes; mechanisms of enzyme action. *Mailing Add:* Dept Biochem Univ Tex Med Sch PO Box 20708 Houston TX 77225

CAPRIOTTI, EUGENE RAYMOND, b Brackenridge, Pa, June 20, 37; m 60; c 1. ASTRONOMY, PHYSICS. *Educ:* Pa State Univ, BS, 59; Univ Wis, PhD(astron), 62. *Prof Exp:* Res fel astron, Calif Inst Technol, 62-63; asst prof, Univ Calif, Berkeley, 63-64; sr engr, Westinghouse Elec Corp, 64; from asst prof to assoc prof astron, 64-73, PROF ASTRON, OHIO STATE UNIV, 73-, CHMN DEPT, 78-; DIR, PERKINS OBSERV, 78- *Concurrent Pos:* Vis assoc prof, Steward Observ, Univ Ariz, 69-70. *Mem:* Am Astron Soc. *Res:* Active galactic nuclei, quasars, gaseous nebulae and interstellar medium. *Mailing Add:* Dept Astron 5040 Smith Lab Ohio State Univ Columbus OH 43210

CAPRON, ALEXANDER MORGAN, b Hartford, Conn, Aug 16, 44; m; c 2. SCIENCE POLICY. *Educ:* Swarthmore Col, BA, 66; Yale Univ, LLB, 69; Univ Pa, MA, 76. *Prof Exp:* lectr & res assoc, Yale Univ, 70-72, vis lectr law, 76-77; prof law, Univ Pa, 72-82, prof human genetics, 76-82; prof law, ethics & pub policy, Georgetown Univ, 83-84; Topping prof law, med & pub policy, Univ Southern Calif, 85-89; PROF LAW & MED, HENRY W BRUCE UNIV, 89-; CO-DIR, PAC CTR HEALTH POLICY & ETHICS, 90- *Concurrent Pos:* Prin investr, NIH grant, 74-77 & mem, Consensus Develop Conf Caesarean Childbirth, 80; mem, Biomed Res & Develop Adv Panel, Off Technol Assessment, US Cong, 76, chmn, biomed ethics adv comt, 88-; assoc ed, Am J Human Genetics, 77-80; mem, adv comt experimentation, Fed Judicial Ctr, Washington, DC, 78-81; exec dir, Presidents Comn Study Ethical Prob Med & Biomed & Behav Res, 80-83; mem, subcomt on human gene ther, NIH, 84- *Honors & Awards:* Rosenthal Found Award, Am Col Physicians, 83. *Mem:* Inst Med-Nat Acad Sci; fel Inst Soc Ethics & Life Sci; Fel AAAS; Am Soc Law & Med (pres, 88-89). *Res:* Ethical, social and legal implications of developments in the life sciences and the impact of governmental policies and legal rules on medicine and the biomedical and behavorial sciences. *Mailing Add:* Univ Southern Calif Law Ctr Univ Park Los Angeles CA 90089-0071

CAPSTACK, ERNEST, b Fall River, Mass, June 9, 30; m 58; c 2. ORGANIC CHEMISTRY, BIOCHEMISTRY. *Educ:* Mass Inst Technol, BS, 52; Univ RI, MS, 54; Brown Univ, PhD(chem), 59. *Prof Exp:* Res chemist, E I du Pont de Nemours & Co, 58-60; res fel, Steroid Training Prog, Worcester Found Exp Biol, 60-61; res assoc, Clark Univ, 61-62; scientist, Worcester Found Exp Biol, 62-64; vis asst prof, Clark Univ, 62-64; assoc prof chem, 64-66, PROF CHEM, W VA WESLEYAN COL, 66-, CHMN DEPT, 67- *Mem:* Am Chem Soc; Sigma Xi. *Res:* Organic and steroid synthesis; steroid and terpene biosynthesis and metabolism; science and religion; death and dying. *Mailing Add:* Dept Chem WVa Wesleyan Col Buckhannon WV 26201

CAPUTI, ROGER WILLIAM, b Newark, NJ, Jan 9, 35; m 59; c 3. PHYSICAL CHEMISTRY. *Educ:* Calif Inst Technol, BS, 57, PhD(phys chem), 65. *Prof Exp:* Radiol chemist, US Naval Radiol Defense Lab, 57-60; teaching asst, Phys Chem Lab, Calif Inst Technol, 60-64; chemist, US Naval Radiol Defense Lab, 64-68; chemist, Vallecitos Nuclear Ctr, 68-70, mgr prod qual & applns lab, 70-72, sr scientist, Track Etch Develop Lab, 72-76, sr scientist, Core Chem Lab, 76-82, sr scientist, Chem Processes Group, 82-86, SR SCIENTIST, CHEM TECHNOL, VELLECITOS NUCLEAR CTR, GEN ELEC CO, 86- *Mem:* Sigma Xi. *Res:* High temperature studies of nuclear fuel-metal oxide systems; instrumentation; model development of radiation effects on water chemistry in BWR nuclear power plants; development and application of high performance liquid chromatography methods for BWR power plant water analysis; design and development of commercial HPLC systems. *Mailing Add:* Gen Elec Vallecitos Nuclear Ctr Vallecitos Rd Pleasanton CA 94566

CAPUTO, JOSEPH ANTHONY, b Jersey City, NJ, May 10, 40; m 65; c 2. PHYSICAL ORGANIC CHEMISTRY. *Educ:* Seton Hall Univ, BS, 62, MS, 64; Univ Houston, PhD(phys org chem), 67. *Prof Exp:* Fel org chem, Duke Univ, 67-68; from instr to assoc prof org chem, State Univ NY Col Buffalo, 68-77, chmn, dept chem, 74-77; prof chem & dean, Sch Sci, Southwest Tex State Univ, 77-, vpres acad affairs, 80-81; PRES, MILLERSVILLE UNIV, 81- *Mem:* The Chem Soc; Am Chem Soc; Sigma Xi. *Res:* Linear free energy relationships; electronic transmission; reaction mechanisms; diazoalkanes; phosphorus heterocycles; antimalarial compounds. *Mailing Add:* Ten Hemlock Lane Millersville PA 17551

CAPUZZO, JUDITH M, b Manchester, NH, Dec 29, 47; m 72. PHYSIOLOGICAL ECOLOGY, POLLUTION ECOLOGY. *Educ:* Stonehill Col, BS, 69; Univ NH, MS, 71, PhD(zool), 74. *Prof Exp:* Instr biol, Framingham State Col, 74-75; fel, 75-76, investr, 75-76, asst scientist, 76-80, ASSOC SCIENTIST BIOL OCEANOG, WOODS HOLE OCEANOG INST, 80- *Concurrent Pos:* Lectr, Cornell Univ & Univ Pa, 77-; vis prof, Bridgewater State Col, 78-84; participant, Nat Acad Sci Workshop Petrol Marine Environ, 81 & Planning Comt Workshop Land vs Sea Disposal Indust

& Domestic Wastes, 82-83; mem, Nat Acad Sci Ocean Studies Bd, 85-; US rep, Sci Comt Problems Environ, 85-; mem, Mass Marine Fisheries Adv Comn, 85- *Mem:* AAAS; Am Soc Zoologists; Sigma Xi; World Mariculture Soc; Estuarine Res Fedn. *Res:* Comparative physiology of marine larval and postlarval crustaceans including studies of energetics and nutrition; assimilative capacity of the oceans for waste disposal with a consideration of the effects of pollutants on the physiology of marine animals. *Mailing Add:* Woods Hole Oceanog Inst Woods Hole MA 02543

CAPWELL, ROBERT J, b Binghamton, NY, July 6, 40; m; c 2. SPECTROSCOPY. *Educ:* Ohio State Univ, BS, 63; Pa State Univ, MS, 65; Univ Pittsburgh, PhD(phys chem), 70; Rider Col, MBA, 80. *Prof Exp:* Chemist, Gulf Res & Develop Co, 65-68; res scientist phys chem, N L Industs, Inc, 70-72, sect leader phys & analytical chem 72-73, dept head analytical & tech specialist, Cent Res Lab, 73-76, res & develop mgr, Indust Chem Div, 76-81; functional mgr, substrates bus oper, 81-89, DESIGN/DEVELOP & PACKAGING PROG MGR, IBM CORP, 89- *Mem:* Am Chem Soc; Soc Appl Spectros. *Res:* Infrared and Raman spectroscopy; structure of matter; chemical instrumentation; molten salt chemistry; flame retardants; plastics additives; reaction mechanisms; electroless plating; process control; statistical quality control; printed circuit boards; substrates; circuit board packaging; semiconductor packaging; semiconductors. *Mailing Add:* 18 Brittania Dr Danbury CT 06811

CARABATEAS, PHILIP M, b New York, NY, Jan 20, 30; m 60; c 2. ORGANIC CHEMISTRY, MEDICINAL CHEMISTRY. *Educ:* Polytech Inst, Brooklyn, BS, 55; Rensselaer Polytech Inst, PhD(chem), 62. *Prof Exp:* From asst res chemist to res chemist, Sterling Winthrop Res Inst, 55-70, sr res chemist, 70-90, INVESTR ORG CHEM, STERLING RES GROUP, 90- *Mem:* Am Chem Soc. *Res:* Synthesis of nitrogen heterocycles, analgesics, antitussives and antifertility agents. *Mailing Add:* Sterling Res Group Rensselaer NY 12144

CARACENA, FERNANDO, b El Paso, Tex, Mar 13, 36; m 69; c 2. ATMOSPHERIC PHYSICS, PHYSICS. *Educ:* Univ Tex, El Paso, BS, 58; Case Western Reserve Univ, MA, 66, PhD(physics), 68. *Prof Exp:* Atmosphere physicist, White Sands Missile Range, Dept Defense, 58-61, sci programmer, Ballistic Res Labs, Aberdeen Proving Grounds, Md, 59-61; postdoctoral fel, 67-68; asst prof physics, Metrop State Col, 69-76; ATMOSPHERE PHYSICIST, NAT SEVERE STORMS LAB, ENVIRON RES LABS, NAT OCEANIC & ATMOSPHERIC ADMIN, 76- *Concurrent Pos:* Summer Inst Theoret Physics, Univ Colorado, 68; fel, Workshop on Relativistic Wave Equations, Cleveland State Univ, Off Naval Res, 69; fel, Advance Studies Prog, Nat Ctr Atmospheric Res, 76. *Honors & Awards:* NOAA Adminr Award, Work in Windshear, Nat Oceanic & Atmospheric Admin, 84. *Mem:* Am Phys Soc; Sigma Xi; Am Meteorol Soc. *Res:* Objective analysis air safety, especially clear air turbulence, low altitude wind shear, and meteorology of wind-shear-related aircraft accidents; severe local storms, flash floods, and infrared remote sensing; computer analysis of meteorological data, climatology. *Mailing Add:* Nat Oceanic & Atmospheric Admin Environ Res Labs R/E/NSI Boulder CO 80303

CARACO, THOMAS BENJAMIN, b West Palm Beach, Fla, Nov 7, 46; m 79; c 1. ECOLOGY, ANIMAL BEHAVIOR. *Educ:* Univ Rochester, AB, 68; Syracuse Univ, PhD(biol), 77. *Prof Exp:* Fel ecol, Univ Ariz, 77-79; ASST PROF BIOL, UNIV ROCHESTER, 79- *Mem:* Ecol Soc Am; Animal Behav Soc; AAAS. *Mailing Add:* Dept Biol State Univ NY 1400 Washington Ave Albany NY 12222

CARADUS, SELWYN ROSS, b Auckland, NZ, Nov 10, 35; m 59; c 2. MATHEMATICAL ANALYSIS. *Educ:* Univ Auckland, BSc, 57, MSc, 58; Univ Southern Calif, MA, 62; Univ Calif, Los Angeles, PhD(math), 65. *Prof Exp:* Jr lectr math, Univ Auckland, 58-60; from asst prof to assoc prof, 64-76, PROF MATH, QUEEN'S UNIV, ONT, 77- *Mem:* Am Math Soc; Can Math Cong. *Res:* Theory of linear operators in Banach space. *Mailing Add:* Dept Math Queen's Univ Kingston ON K7L 3N6 Can

CARAM, HUGO SIMON, b Buenos Aires, Arg, Mar 14, 45; m 70; c 3. CHEMICAL ENGINEERING. *Educ:* Univ Buenos Aires, BS, 67; Univ Minn, PhD(chem eng), 77. *Prof Exp:* Process engr petrol refining, Shell Compania Argentino de Petroleo Sociedad Anonimol, 70-72; res asst chem eng, Univ Minn, 72-77; from asst prof to assoc prof, 71-85, PROF CHEM ENG, LEHIGH UNIV, 85- *Mem:* AAAS; Am Inst Chem Engrs. *Res:* multiphase flows; reactor analysis; mathematical modeling. *Mailing Add:* Dept Chem Eng Lehigh Univ Bethlehem PA 18015

CARAPELLA, S(AM) C(HARLES), JR, b Tuckahoe, NY, Jan 27, 23. PHYSICAL METALLURGY. *Educ:* Mich State Col, BS, 44; Univ Notre Dame, MS, 48, PhD(phys metall), 50. *Prof Exp:* Metallurgist, Allison Div, Gen Motors Corp, 44; res metallurgist, Asarco, Inc, 50-57, sect head, 57-59, supt high purity metals develop, 59-74 & byprod metals, 74-86; RETIRED. *Concurrent Pos:* Chmn, Tech Comt, Selenium-Tellenium Develop Asn, 72- & tech dir asn, 89. *Mem:* Am Soc Metals. *Res:* Properties of less common metals (selenium and tellurium). *Mailing Add:* 167 Dante Ave Tuckahoe NY 10707

CARAS, GUS J(OHN), b Karitsa, Greece, Jan 20, 29; US citizen; m 58; c 3. CHEMICAL ENGINEERING, INFORMATION SCIENCE. *Educ:* Ga Inst Technol, BS, 58, MS, 67. *Prof Exp:* Process engr, Redstone Div, Thiokol Chem Corp, Ala, 58-61, chem engr, 61-62; chem engr struct & mech lab, Res & Develop Dept, US Army Missile Command, 62-63, info scientist, Redstone Sci Info Ctr, Res & Develop Directorate, 63-67; info scientist pesticides prog, Food & Drug Admin, 67-69, chief tech data unit, Div Community Studies, 69-73, CHIEF STATIST SERV ACTIV, CTR DISEASE CONTROL, HEALTH & HUMAN SERV, 73- *Concurrent Pos:* Instr math, statist & comput sci, Ga State Univ & De Kalb Community Col, 76- *Mem:* Am Soc Info Sci; Am Inst Aeronaut & Astronaut; Am Inst Chem Engrs; Am Statist Asn. *Res:* Research and development in field of solid propellant rockets; analysis and interpretation of scientific and technical literature; design and administration of information storage and retrieval systems; mathematical statistics; computer science. *Mailing Add:* 1659 Clairmont Pl NE Atlanta GA 30329

CARASSA, FRANCESCO, b Napoli, Italy, Mar 7, 22; m 48; c 2. TELECOMMUNICATIONS, RADIO. *Educ:* Politecnico di Torino, Elec Engr, 46. *Prof Exp:* Researcher telecommun, Fabbrica Italiana Magneti Marelli Res Lab, 47-62, dir, 56-62; PROF ELEC COMMUN, POLITECNICO DI MILANO, 62- *Concurrent Pos:* Consult, Gen Tel & Electronics, Italy, 62-78; chmn bd, Siemens Telecommun, 78-80, hon chmn, 80- & chmn, Centro Studi Lab Telecommun, 80-; prin investr, Sirio, 77-84; vis prof, Polytech Inst NY, 84; pres coun, Europ Space Agency, 90- *Mem:* Fel Inst Elec & Electronic Engrs. *Res:* Microwave propagation earth-to-earth and space-to-earth; pulse communication systems; radio-relay systems; satellite communication systems including modulation and demodulation methods; adaptive methods to overcome rain attenuation; communication systems and optical networks. *Mailing Add:* Politecnico Piazza L Da Vinci 32 Milano 20133 Italy

CARASSO, ALFRED SAMUEL, b Alexandria, Egypt, Apr 9, 39; m 64; c 2. APPLIED MATHEMATICS, NUMERICAL ANALYSIS. *Educ:* Univ Adelaide, Australia, BS, 60; Univ Wis-Madison, MS, 64, MA, 65, PhD(math), 68. *Prof Exp:* Asst prof math, Mich State Univ, 68-69; from asst prof to prof math, Univ NMex, 69-81; MATHEMATICIAN, COMPUTING & APPL MATH LAB, NAT INST STANDARDS & TECHNOL, GAITHERSBURG, 82- *Concurrent Pos:* Consult, Los Alamos Nat Lab, 72-81. *Mem:* Soc Indust & Appl Math; Am Math Soc. *Res:* Mathematical and computational analysis of inverse problems and their application in heat conduction, seismology, acoustics, image processing and electromagnetics. *Mailing Add:* Administration A-302 Nat Inst Standards & Technol Gaithersburg MD 20899

CARAWAY, WENDELL THOMAS, b Xenia, Ohio, Nov 4, 20; m 44. BIOCHEMISTRY. *Educ:* Wilmington Col, BS, 42; Miami Univ, MA, 43; Johns Hopkins Univ, PhD(biochem), 50; Am Bd Clin Chem, dipl. *Prof Exp:* Asst chemist, Ohio River Div Labs, US Corps Engr, 43-45, mat engr & head chem lab, 45-46; biochemist, RI Hosp, 50-57; biochemist, Flint Clin Pathologists, 57-85; RETIRED. *Concurrent Pos:* Mem bd dirs, Nat Registry Clin Chem, 67-73. *Honors & Awards:* Ames Award in Clin Chem, 70; Garulet Award, 78. *Mem:* AAAS; Am Chem Soc; Am Asn Clin Chem (pres, 65). *Res:* Kinetics and mechanisms of reactions; serum enzyme assays; ultramicrochemical methods of blood analysis; drug effects on lab tests. *Mailing Add:* 1102 Woodside Dr Flint MI 48503-5341

CARBAJAL, BERNARD GONZALES, III, b New Orleans, La, Feb 15, 33; m 54; c 2. SOLID STATE CHEMISTRY, SEMICONDUCTORS. *Educ:* Univ Minn, PhD(chem), 58. *Prof Exp:* Asst chem, Univ Minn, 54-56; asst prof, Col St Thomas, 57-60; mem tech staff, 60-69, mgr multilevel tech prog, Components Group, 69-73, mgr advan circuit technol, 72-74, mgr design support, 74-76, prog mgr, Automated Array Assembly, 76-80, PROG MGR COMPLEMENTARY METAL-OXIDE SEMICONDUCTOR TECHNOL PROD ENG, TEX INSTRUMENTS, INC, 80- *Mem:* Am Chem Soc. *Res:* Organic semiconductors; thin film polymers; electrical conduction in organic materials; semiconductor processing; metal-oxide-semiconductor devices; thin films; interconnection technology; solar photovoltaic technology; complimentary metal-oxide semiconductor technology. *Mailing Add:* PO Box 655303 MS 3669 Dallas TX 75265-5303

CARBALLO-QUIROS, ALFREDO, plant breeding, quantitative genetics, for more information see previous edition

CARBARY, JAMES F, b June 5, 51. MAGNETOSPHERIC PHYSICS. *Educ:* Univ Ill, BS, 73; Rice Univ, MS, 76, PhD(space physics), 78. *Prof Exp:* Res assoc, Rice Univ, 77-78; staff scientist, Mission Res Corp, 83-84; res assoc, 78-83, STAFF SCIENTIST, GROUP S1G, APPL PHYSICS LAB, JOHNS HOPKINS UNIV, 85- *Mem:* Am Geophys Union; AAAS. *Res:* Research involving magnetospheres of Earth, Jupiter and Saturn; electromagnetic pulse effects; atmospheric radiation. *Mailing Add:* Appl Physics Lab Johns Hopkins Univ Johns Hopkins Rd Bldg 24E-117 Laurel MD 20723-6099

CARBERRY, EDWARD ANDREW, b Milwaukee, Wis, Nov 20, 41; m 67; c 2. INORGANIC CHEMISTRY, ORGANOMETALLIC CHEMISTRY. *Educ:* Marquette Univ, BS, 62; Univ Wis, PhD(inorg chem), 68. *Prof Exp:* PROF CHEM, SOUTHWEST MINN STATE UNIV, 68- *Mem:* Am Chem Soc; Am Sci Glassblowers Soc. *Res:* Preparation, spectroscopic and bonding studies of inorganic and organometallic compounds, especially organosilicon compounds; catenated group IV chemistry emphasizing polysilanes and polygermanes linear and cyclic. *Mailing Add:* Dept Chem Southwest Minn State Univ Marshall MN 56258

CARBERRY, JAMES JOHN, b Brooklyn, NY, Sept 13, 25; m 59; c 2. CHEMICAL ENGINEERING. *Educ:* Univ Notre Dame, BS, 50, MS, 51; Yale Univ, PhD, 57. *Prof Exp:* Chem process engr, Explosives Dept, E I du Pont de Nemours & Co, NJ, 51-53; consult, Olin-Mathieson Chem Corp, Conn, 54-57; sr res engr, Eng Dept, E I du Pont de Nemours & Co, 57-61; from asst prof to assoc prof chem eng, 61-66, PROF CHEM ENG, UNIV NOTRE DAME, 66- *Concurrent Pos:* Consult, Am Oil Co, 62- & Catalytica & Assocs, Inc, 74-; NSF sr fel, Cambridge Univ, 65-66; Hays-Fulbright sr scholar, Univ Rome, 74; Sir Winston Churchill fel, Cambridge Univ, 79. *Honors & Awards:* Yale Eng Asn Award; R H Wilhelm Award, Am Inst Chem Engrs, 76. *Mem:* AAAS; fel NY Acad Sci; fel Am Inst Chem; Am Inst Chem Engrs; Am Chem Soc; Sigma Xi. *Res:* Kinetics; heterogeneous catalysis; heat and mass transfer; nitric acid manufacture; catalytic reactor technology. *Mailing Add:* Dept Chem Eng Univ Notre Dame Notre Dame IN 46556

CARBERRY, JUDITH B, b Oil City, Pa, Mar 10, 36; m 81; c 2. POLLUTION CONTROL. *Educ:* Cornell Univ, BS, 58; Univ Notre Dame, MS, 69, PhD(environ eng), 72. *Prof Exp:* Fel environ eng, Univ Col, London, 72-73; asst prof, Univ Del, 73-78, assoc prof, 78-79; sr fel, The Technion, Israel, 79-80; ASSOC PROF ENVIRON ENG, UNIV DEL, 80- *Concurrent Pos:* Asst dir filtration, NATO Advan Study Inst, 73, partic wastewater, 76, dir & lectr sludge, 79; consult, Planners & Assocs, 73-74, Howard L Robertson, Inc, 75-76, US Naval Ship Res & Develop Ctr, 77-78, Ecolsci, Inc, US Environ Protection Agency, 80-81; guest lectr, Univ Del, 73, Duke Univ, Tulane Univ, Va Polytech, & The Technion, Israel; prin investr grants, US Off Water Res & Technol, US Environ Protection Agency, & NSF, 73-; tech reviewer, US Environ Protection Agency & NSF, 78- *Honors & Awards:* Sigma Xi. *Mem:* Water Pollution Control Fedn; Asn Environ Eng Profs; Soc Women Engrs; Sigma Xi. *Res:* Interfacial phenomena of liquid/particulate reactions and resulting fundamental effects on liquid/solid separation processes; biological particulate transport and adsorption reactions in engineered systems and sediment particle reactions with pollutants in natural systems. *Mailing Add:* Dept Civil Eng Univ Del Newark DE 19716

CARBIENER, WAYNE ALAN, b St Joseph Co, Ind, Feb 21, 36; m 58; c 4. NUCLEAR ENGINEERING. *Educ:* Purdue Univ, BS, 58, MS, 62; Ohio State Univ, PhD(nuclear eng), 75. *Prof Exp:* Res engr nuclear sci, Battelle Columbus Labs, 62-71, sect mgr nuclear eng & sci, 71-75, prog mgr nuclear eng, 76-78; mgr, Off Nuclear Waste Isolation, 78-87, VPRES, PROG MGT, ENERGY SYSTS GROUP, BATTELLE, 87- *Mem:* Am Soc Mech Eng. *Res:* Development and application of technology to isolation of nuclear waste; earth science, nuclear science, geology, engineering and materials science. *Mailing Add:* 4330 Stratton Rd Columbus OH 43220

CARBNO, WILLIAM CLIFFORD, b Tessier, Sask, July 30, 30; m 53; c 1. MATHEMATICS EDUCATION. *Educ:* Univ Sask, BEd, 61, BA, 63, MEd, 64; Univ Toronto, PhD, 76. *Prof Exp:* Asst prof, 67-80, ASSOC PROF MATH EDUC & EDUC PSYCHOL, BRANDON UNIV, 80- *Concurrent Pos:* Chmn sub-comt on math prog, Coun Develop Math Curric, Man, 70-73; mem, Elem Math Curric Comt, 74-; vis prof math educ, Univ Sask, 78-79; dir, Social Sci Fedn Can, 89-91. *Mem:* Can Soc Studies Educ; Can Asn Res Educ. *Res:* Child development; models for memory and cognition. *Mailing Add:* Fac Educ Psych Brandon Univ Brandon MB R7A 6A9 Can

CARBON, JOHN ANTHONY, b Sharon, Pa, Jan 1, 31; m 50; c 2. CENTROMERE DNA, NUCLEIC ACIDS. *Educ:* Univ Ill, BS, 52; Northwestern Univ, PhD(biochem), 55. *Prof Exp:* Res assoc biochem, Med Sch, Northwestern Univ, 55-56; sr res chemist, Org Chem Dept, Abbott Labs, 56-63, res assoc, Biochem Dept, 63-68; prof biochem, 68-90, chmn, Dept Biol Sci, 84-87, AM CANCER SOC RES PROF BIOCHEM, DEPT BIOL SCI, UNIV CALIF, SANTA BARBARA, 90- *Concurrent Pos:* Adj asst prof, Northwestern Univ, 61-68; vis res fel, Dept Biochem, Med Sch, Stanford Univ, 65-66 & 73-74; consult, Abbott Labs, 73-86 & Appl Molecular Genetics, 79-; mem, Study Sect Genetic Basis Dis, NIH, 77-80, Study Sect Genetics, 81 & 86, Howard Hughes Fel Panel, Nat Res Coun, 87; NIH grants, 87- *Honors & Awards:* I M Lewis Lectr, Am Soc Microbiol, 78. *Mem:* Nat Acad Sci; AAAS; Genetics Soc Am; Am Asn Biol Chemists; Am Acad Arts & Sci. *Res:* Nucleic acids; transfer RNA; viral DNA; bacterial plasmids; molecular cloning; yeast molecular genetics; chromosome structure; centromeric DNA. *Mailing Add:* Dept Biol Sci Univ Calif Santa Barbara CA 93106

CARBON, MAX W(ILLIAM), b Monon, Ind, Jan 19, 22; m 44; c 5. NUCLEAR ENGINEERING. *Educ:* Purdue Univ, BS, 43, MS, 47, PhD(heat transfer & thermodyn), 49. *Prof Exp:* Thermodynamacist, Hanford Works, Gen Elec Co, 49-50, pile engr, 50-51, head heat transfer unit, 52-55, head contract eng unit, 55; chief thermodyn sect, Res & Advan Develop Div, Avco Mfg Corp, 55-58; prof mech eng & chmn nuclear eng comt, 59-63, PROF NUCLEAR ENG & ENG PHYSICS & CHMN DEPT, UNIV WIS, MADISON, 63- *Concurrent Pos:* Mem, Adv Comt Reactor Safeguards, US Nuclear Regulatory Comn, 75-87, chmn, 79. *Mem:* AAAS; Am Soc Eng Educ; Am Nuclear Soc. *Res:* Nuclear power; heat transfer. *Mailing Add:* Dept Nuclear Eng & Eng Physics Univ Wis Madison WI 53706

CARBONE, GABRIEL, b New York, NY, Sept 4, 27; m 53; c 1. ENVIRONMENTAL CHEMISTRY. *Educ:* Brooklyn Col, BS, 49. *Prof Exp:* Chemist, Food & Water Chem Lab, New York City Health Dept, 55-67, sr chemist, 67-70, prin chemist & dir food & water chem lab, Bur Labs, 70-88; RETIRED. *Concurrent Pos:* Mem fumigant bd, New York City Health Dept, 70- *Res:* Alternate standards for sharer rapid pasteurization test for dairy products and differentiation of residual and reactivated phosphatase in dairy products; nitrites and nitrates in meat products; trace constituents in potable water. *Mailing Add:* 1848 72nd Brooklyn NY 11204

CARBONE, JOHN VITO, b Sacramento, Calif, Dec 13, 22; m 46; c 3. MEDICINE. *Educ:* Univ Calif, BA, 45, MD, 48. *Prof Exp:* From instr to assoc prof, 51-66, PROF MED, SCH MED, UNIV CALIF, SAN FRANCISCO, 66- *Concurrent Pos:* Giannini fel med, 54-55; consult to Surgeon Gen, US Army, 58 & Letterman Hosp, Travis AFB, San Francisco. *Mem:* Sigma Xi. *Res:* Gastroenterology; metabolic aspects of liver disease. *Mailing Add:* Med Ctr 124 Univ Hosp Univ Calif San Francisco CA 94122

CARBONE, PAUL P, b White Plains, NY, May 2, 31; m 54; c 7. ONCOLOGY. *Educ:* Albany Med Col, MD, 56. *Hon Degrees:* DSc, Albany Med Col. *Prof Exp:* Intern med, USPHS Hosp, Baltimore, Md, 56-57, resident internal med, 58-60; sr investr, Nat Cancer Inst, 60-65, head solid tumor serv, 65-68, chief med br, 68-72, assoc dir med oncol, Div Cancer Treatment, 72-76; Chmn dept human onocol, 77-87, PROF HUMAN ONOCOL & MED & CHIEF CLIN ONOCOL, UNIV WIS, 76-; DIR, WIS CLIN CANCER CTR, 78- *Mem:* Am Soc Hemat; Am Fedn Clin Res; Am Asn Cancer Res; Am Soc Clin Oncol; Am Soc Clin Invest; Asn Am Physicians; Am Col Physicians. *Res:* Research in cancer chemotherapy; clinical trials in cancer. *Mailing Add:* Wis Clin Ctr K4/614 600 Highland Ave Madison WI 53792

CARBONE, RICHARD EDWARD, b Bronx, NY, Sept 26, 44; m 70; c 3. METEOROLOGY, ATMOSPHERIC PHYSICS. *Educ:* NY Univ, BS, 66; Univ Chicago, MS, 69. *Prof Exp:* Analyst opers res, Grumman Aerospace, 68-70; res scientist meteorol, Univ Chicago, 72-74 & Meteorol Res Inc, 74-75, mgr atmospheric physics, 75-76; staff scientist III meteorol, 76-81, mgr Field Observ Facil, 81-89, DIR ATMOSPHERIC TECHNOL DIV, NAT CTR ATMOSPHERIC RES, 89- *Concurrent Pos:* Ed, J Climate & Appl Meteorol, 82-85. *Mem:* Am Meteorol Soc; Sigma Xi. *Res:* Storm kinematics; microphysics; precipitation physics; remote sensing of atmosphere; Doppler radar. *Mailing Add:* 2905 Vassar Dr Boulder CO 80303

CARBONE, ROBERT JAMES, b Hartford, Conn, Aug 17, 30; m 53; c 3. FAMILY PRACTICE MEDICINE, PHYSICS. *Educ:* Univ Conn, BA, 52, MA, 53, PhD, 56; Atonomous Univ Juarez, MD, 78; Am Bd Family Pract, dipl, 83. *Prof Exp:* Asst, Univ Conn, 54-56; physicist, Lincoln Lab, Mass Inst Technol, 56-60 & 62-68 & Bomac Labs, Inc, 61-62; physicist, Philips Res Labs, Technol Univ Eindhoven, 68-69 & Lincoln Lab, Mass Inst Technol, 70-72; alternate group leader, Los Alamos Sci Lab, Univ Calif, 72-76; assoc dir, Hamot Family Med Residency, 84-89; MED DIR, POLYCLIN MED CTR, KLINE FAMILY PRACT, HARRISBURG, PA, 90- *Concurrent Pos:* Family pract resident physician, East Tenn State Univ, Bristol, Tenn, 79-80, Johnson City, 81-82. *Mem:* Am Phys Soc; AMA; fel Am Acad Family Pract; Soc Teachers Family Med. *Res:* Radiofrequency gas discharges; heavy ion-atom interactions; chemical and gaseous lasers; direct current discharges; laser applications in fusion and isotope separation. *Mailing Add:* 2020 Deer Path Rd Harrisburg PA 17110-3471

CARBONELL, JAIME GUILLERMO, b Montevideo, Uruguay, July 29, 53; US citizen; m 75; c 2. COGNITIVE SCIENCE, NATURAL LANGUAGE PROCESSING. *Educ:* Mass Inst Technol, SB, 75; Yale Univ, BS, 76, PhD(comput sci), 79. *Prof Exp:* Res programmer comput sci, Bolt Beranek & Newman, 71-75; res fel, Yale Univ, 76-79; ASST PROF COMPUT SCI, CARNEGIE-MELLON UNIV, 79- *Concurrent Pos:* Consult various orgn, 78- *Mem:* Asn Comput Ling; Asn Comput Mach; Cognitive Sci Soc; NY Acad Sci; Am Asn Artificial Intel. *Res:* Artificial intelligence focusing on computer models of analogical reasoning, planning and counterplanning, natural language processing, man-machine interfaces, modeling human memory and inference processes; knowledge engineering focusing on rule-based expert systems and real time decision-making. *Mailing Add:* Computer Sci Dept Carnegie-Mellon Univ 5000 Forbes Ave Pittsburgh PA 15213

CARBONELL, ROBERT JOSEPH, b El Salvador, Mar 20, 27; m 50; c 5. FOOD SCIENCE. *Educ:* Nat Univ El Salvador, BS, 48; Purdue Univ, MS, 50. *Prof Exp:* Res chemist, Sanit Eng Div, Inst Inter Am Affairs, El Salvador, 45-46, asst head dept chem, Ctr Nat Agron, 46-52, head green coffee processing div, 52-53; assoc prof biochem, Sch Agron, Nat Univ El Salvador, 51-52, assoc prof org anal, Sch Pharmaceut Chem, 51-53; res chemist res & develop food prod, Fleischmann Labs, Standards Brands Inc, 53-54, group leader, 54-57, div head, 57-59, dir, Consumer Prod Res Dept, 59-63, dir, Div Res & Develop, 63-65, mgr, 65-69, dir res, 69-73, corp dir prod & develop planning, 73-74, mgr, Fleischmann Indust Mfg, 74-75, mgr grocery prod mfg, 75, vpres res & develop, 75-81; EXEC VPRES TECHNOL, NABISCO BRANDS INC, 81- *Concurrent Pos:* Mem, Coffee Processing Surv Comn, El Salvador Govt, 52; abstr, Chem Abstr, 53-54. *Mem:* Am Chem Soc; Inst Food Tecnol. *Res:* Composition of fats and oils from tropical species of Salvadorean Flora; nutritional value of coffee pulp; treatment of waste waters from coffee processing plants; effect of crossing upon chemical composition of sunflower oil; coffee bean mucilage. *Mailing Add:* Del Monte Foods Two Alhambra Plaza Coral Gables FL 33134-5254

CARBONI, JOAN M, b Hartford, Conn. CELL BIOLOGY. *Educ:* St Joseph Col, BA, 73; Albert Einstein Col Med, PhD(cell biol), 84. *Prof Exp:* Postdoctoral, Dept Biol, Yale Univ, 84-88; RES SCIENTIST, DEPT CELL/MOLECULAR BIOL, BRISTOL MYERS SQUIBB, 88- *Res:* Taxol, a natural plant product, is a potent inhibitor of cell replication that blocks cells in late G2/M phases of the cell cycle; understanding of the mechanism of action of taxol on microtubules; identification of the taxol-MT binding site. *Mailing Add:* Bristol Myers Squibb Dept 207 Five Research Pkwy Wallingford CT 06492

CARBONI, RUDOLPH A, b Yonkers, NY, Nov 10, 22; m 49; c 4. ORGANIC CHEMISTRY. *Educ:* Columbia Univ, AB, 47, AM, 48; Mass Inst Technol, PhD(org chem), 53. *Prof Exp:* Res chemist, Charles Pfizer & Co, 48-50; res chemist, Cent Res Dept, E I du Pont de Nemours & Co, 53-61, res supvr, Org Chem Dept, 62-67, div head explor intermediates, 67-68, lab mgr, 68-70, res & develop mgr permasep prod, 70-73, tech mgr electronic prod div, 73-82, prin consult, corp staff, 83-84, MGR ACAD RES, ELECTRONICS PROD DEPT, E I DU PONT DE NEMOURS & CO, 85- *Mem:* Am Chem Soc; Am Asn Textile Chem & Colorists. *Res:* Cyanocarbon chemistry; new polynitrogen systems; organofluorine chemistry; antibiotic research; inductive and field effects; dye studies and oxidation catalysts; reverse osmosis; photopolymer systems for printed circuit and microelectronic device fabrication. *Mailing Add:* 312 Center Hill Rd Wilmington DE 19807-1120

CARCHMAN, RICHARD A, BIOSTATISTICS. *Educ:* Downstate Med Sch, State Univ NY, PhD(pharmacol), 72. *Prof Exp:* Prof pharmacol, Med Col Va, 74-90; MGR BIOCHEM RES, PHILIP MORRIS, USA, 90- *Mailing Add:* Philip Morris USA PO Box 26583 Richmond VA 23261

CARD, KENNETH D, b Ont, Jan 23, 37; m 59; c 2. GEOLOGY. *Educ:* Queen's Univ (Ont), BSc, 59; Princeton Univ, MA, 62, PhD(geol), 63. *Prof Exp:* Resident geologist, 63-66, field geologist, 66-74, sr geologist, Ont Div Mines, 74-77; GEOLOGIST, PRECAMBRIAN SUBDIV, GEOL SURV CAN, 77- *Honors & Awards:* Goldrich Medal, Inst Lake Superior Geol, 90. *Mem:* Fel Geol Soc Am; fel Geol Asn Can. *Res:* Regional geology; precambrian stratigraphy, structure and metamorphism; geology and tectonics, Archean Superior Province and Early Proterozoic Southern Province, Canadian Shield. *Mailing Add:* 86 Penfield Dr Kanata ON K2K 1M1 Can

CARD, ROGER JOHN, b Grand Rapids, Mich, Oct 31, 47. HETEROGENEOUS CATALYSIS. *Educ:* Hope Col, BS, 69; Iowa State Univ, MS, 71, PhD(org chem), 75. *Prof Exp:* Asst prof org chem, Univ Kebangsaan Malaysia, 75-77; fel, Bowling Green State Univ, 77-78; res scientist chem, 78-80, SR RES CHEMIST, AMERICAN CYANAMID CO, 80- *Mem:* Am Chem Soc; Am Ceramics Soc. *Res:* Development of new and improved processes for commercial chemical manufacture; preparation of polymer matrix composites; synthesis of ceramic fibers and powders. *Mailing Add:* Am Cyanamid Co 1937 W Main St Stamford CT 06904

CARDARELLI, JOSEPH S, b Bethlehem, Pa, Mar 14, 44; m 68; c 2. MANUFACTURING PROCESS AUTOMATION, COMPUTER INTEGRATED MANUFACTURING. *Educ:* Muhlenberg Col, BS, 66; Univ Wis, MS, 68; Mich State Univ, PhD(chem), 71. *Prof Exp:* Mem res staff, Western Elec Co, 71-76; sr res chemist, Betz Labs, 76-82; mgr qual control, Warner-Lambert Co, 83-86, mgr mfg systs, 86-87, mgr systs technol, 87-89, SR RES ASSOC, WARNER-LAMBERT CO, 89- *Mem:* Am Chem Soc; Am Soc Qual Control. *Res:* Design and implementation of computer integrated manufacturing systems that link process control, materials handling and tracking, quality assurance, manufacturing and packaging and documentation systems to financial, business and planning systems. *Mailing Add:* CP Res & Develop Warner-Lambert Co 175 Tabor Rd Morris Plains NJ 07950

CARDE, RING RICHARD TOMLINSON, b Hartford, Conn, Sept 18, 43; m 74; c 2. INSECT PHEROMONES. *Educ:* Tufts Univ, BS, 66; Cornell Univ, MS, 68, PhD(entom), 71. *Prof Exp:* Fel, NY State Agr Exp Sta, Cornell Univ, 71-75; from asst prof to assoc prof entom, Mich State Univ, 75-82; prof & head entom, 82-87, DISTINGUISHED UNIV PROF, UNIV MASS, 89- *Honors & Awards:* Ciba-Geigy Award, Entom Soc Am, 81. *Mem:* Entom Soc Am; Entom Soc Can; AAAS; Sigma Xi; Soc Study Evolution. *Res:* Insect pheromones and behavior; pheromone identification; use of pheromones in pest management; biosystematics of the Lepidoptera; Kairomones and host finding by parasitoid wasps. *Mailing Add:* Dept Entom Univ Mass Amherst MA 01003

CARDEILHAC, PAUL T, REPRODUCTION, AQUATIC ANIMALS. *Educ:* Univ Pa, PhD(biochem), 64. *Prof Exp:* PROF VET MED, UNIV FLA, 68- *Mailing Add:* Col Vet Med Univ Fla Gainesville FL 32601

CARDELL, ROBERT RIDLEY, JR, b Atlanta, Ga, Nov 11, 31; m 59; c 3. CELL BIOLOGY. *Educ:* Ga Southern Col, BS, 56; Univ Va, MS, 59, PhD(biol), 62. *Prof Exp:* Asst biol, Univ Va, 56-59, instr, 59-60; asst biophys, Edsel B Ford Inst Med Res, 60-62, res assoc, 62-64; res assoc biol & instr, Harvard Univ, 64-67; assoc prof anat, Sch Med, Univ Va, 67-70, prof, 70-79; PROF & CHAIR ANAT & CELL BIOL, COL MED, UNIV CINCINNATI, 79- *Mem:* Am Soc Cell Biol; Am Asn Anat; Endocrine Soc. *Res:* Ultrastructure of cells; morphological action of hormones; cell biology; cellular endocrinology. *Mailing Add:* Dept Anatomy Unic Cincinnati Col Med 231 Bethesda Ave Cincinnati OH 45267

CARDELLI, JAMES ALLEN, DEVELOPMENTAL GENE REGULATION. *Educ:* Univ Wis, PhD(molecular biol), 77. *Prof Exp:* ASST PROF CELL & MOLECULAR BIOL, LA STATE UNIV MED CTR, 85- *Res:* Lysosomal enzyme biosynthesis. *Mailing Add:* Dept Microbiol & Immunol La State Univ Med Ctr 1501 Kings Hwy Shreveport LA 71130

CARDELLO, ARMAND VINCENT, b Stoneham, Mass, Dec 11, 49; m 73; c 3. HUMAN PSYCHOPHYSICS, FOOD ACCEPTANCE. *Educ:* Dartmouth Col, AB, 71; Univ Mass, Amherst, MS, 74, PhD(psychol), 77. *Prof Exp:* Res psychologist, 77-87, SUPVRY RES PSYCHOLOGIST, US ARMY NATICK RES, DEVELOP & ENG CTR, 87- *Concurrent Pos:* Res consult, NIH Res Grant, Harvard Sch Dent Med, 84-86. *Mem:* AAAS; Am Psychol Soc; Eastern Psychol Asn; Inst Food Technologists; Soc Consumer Psychol; Sigma Xi. *Res:* Experimental analysis of the human psychophysical response to oral stimuli, both food and model systems; sensory physiology; functional relationships between sensation and pre- and post-consumatory responses, especially preference, acceptance, choice behavior, and consumption; author of 65 publications. *Mailing Add:* Behav Sci Div US Army Natick Res Develop & Eng Ctr Natick MA 01760-5020

CARDEN, ARNOLD EUGENE, b Birmingham, Ala, Apr 27, 30; m 52; c 6. ENGINEERING. *Educ:* Ala Polytech Inst, BSME, 52; Univ Ala, MS, 56; Univ Conn, PhD(metall), 72. *Prof Exp:* Res engr, Alcoa Res Labs, 56-58; from asst prof to assoc prof eng mech, 58-71, PROF ENG MECH, UNIV ALA, TUSCALOOSA, 71- *Concurrent Pos:* Res partic, Oak Ridge Nat Lab, 59-; NSF fel, Dept Metall, Univ Conn, 69-70; Los Alamos Nat Lab, 68-69, 79-88. *Mem:* Am Soc Mech Engrs; Am Soc Metals; Am Soc Eng Educ; Nat Asn Corrosion Engrs; Sigma Xi. *Res:* Mechanics; mechanics of materials; methods of mechanical testing; material properties; failure analysis. *Mailing Add:* Dept Eng Mech Univ Ala PO Box 870278 Tuscaloosa AL 35487-0278

CARDENAS, CARLOS GUILLERMO, b Laredo, Tex, June 25, 41; m 65; c 2. ORGANIC CHEMISTRY. *Educ:* Univ Tex, BS, 62, PhD(org chem), 65. *Prof Exp:* Asst prof chem, US Naval Postgrad Sch, 65-67; res chemist, Phillips Petrol Co, 67-69, group leader, 69; sr chemist, 69-71, sect mgr, 71-74, mgr res, Glidden-Durkee Div, SCM Corp, 75-85, dir com develop, SCM Org Div, 85-86, DIR LATIN AM & FAR EAST SALES, SCM GLIDCO, 87- *Concurrent Pos:* Off Naval Res grant, 65-67. *Mem:* Am Chem Soc; Royal Soc Chem. *Res:* Olefin synthesis and reaction mechanisms; nuclear magnetic resonance; Diels-Alder reactions; base catalysis; organometallics; terpenes. *Mailing Add:* c-o SCM Glidco Org PO Box 389 Jacksonville FL 32201-0389

CARDENAS, MANUEL, b San Diego, Tex, Sept 18, 42; m 68; c 2. EXPERIMENTAL STATISTICS. *Educ:* Tex A&I Univ, BS, 68, MA, 70; Tex A&M Univ, PhD(statist), 74. *Prof Exp:* Comput programmer, Tex Real Estate Res Ctr, 72-73; asst prof, 74-80, ASSOC PROF STATIST, NMEX STATE UNIV, 80- *Mem:* Am Statist Soc; Biometric Soc. *Res:* Survey sampling and compartmental modeling; the derivation of the distribution of models with time-dependent transition probabilities, particularly as applied to bio-medical sciences. *Mailing Add:* Dept Exp Statist NMex State Univ Box 3130 Las Cruces NM 88003

CARDENAS, RAÚL R, JR, b Galveston, tex, Feb 5, 29; m 61; c 3. EXPERT WITNESS, ENVIRONMENTAL POLLUTION & CHEMISTRY. *Educ:* Univ Tex, Austin, BA, 51; NY Univ, MS, 63, PhD(environ health sci), 70. *Prof Exp:* Res assoc, instr & adj asst prof, Dept Civil Eng, Manhattan Col, 64-66; asst prof civil eng, Dept Civil Eng, NY Univ, 66-72; from asst prof to assoc prof, Dept Civil & Environ Eng, Polytechnic Univ, NY, 72-88; PRES, CARPENTER ENVIRON ASSOCS, INC, NORTHVALE, NJ, 88- *Concurrent Pos:* Prin awardee, US Dept Health, Educ & Welfare, 72; prin investr, Nat Park Serv, 83-84 & 84-85; co-prin investr, NY State Dept Environ Protection, 86; adj prof, Dept Civil & Environ Eng, Polytechnic Univ, 89-, Sch Environ & Occup Health Sci, Hunter Col, 89-; adj assoc prof, Polytechnic Univ, NY, 89-; lectr, Columbia Univ & NY Univ, 89- *Mem:* Water Pollution Control Fedn; Am Soc Microbiol; AAAS; Am Soc Testing & Mat. *Res:* Ultraviolet disinfection of waste waters, nitrification, PAH uptake by clams; lake and pollution studies; expert witness microbiology. *Mailing Add:* 66 Pine Tree Lane Tappan NY 10983

CARDER, DAVID ROSS, b Wayne, Nebr, Aug 14, 40; m 63; c 2. MANAGEMENT SYSTEMS, CIVIL ENGINEERING. *Educ:* Univ Calif, Los Angeles, BS, 63; Stanford Univ, PhD(civil eng), 74. *Prof Exp:* Staff engr consulting, Allied Technol & Capital Inc, 64-65; gen engr construct, Angeles Nat Forest, 65-66, indust engr fire mgt systs res, Riverside Fire Lab, 66-68, indust engr equip develop, San Dimas Equip Develop Ctr, 68-69, indust engr trans systs anal, Transp Systs Planning Proj, 69-71, gen engr mgt systs anal, Prog & Legis, 71-74, PROJ LEADER MULTIRESOURCE MGT RES, ROCKY MOUNTAIN EXP STA, FOREST SERV, USDA, 74- *Concurrent Pos:* Adj fac, Northern Ariz Univ, 75- & Tex Tech Univ, 77- *Mem:* Sigma Xi; Soc Am Foresters. *Res:* Forest management systems engineering; mathematical modeling of the effects of man on forest biophysical processes; communication and display techniques for use in evaluating tradeoffs among management alternatives and in negotiating decisions. *Mailing Add:* Forestry Sci Lab Northern Ariz Univ Flagstaff AZ 86001

CARDER, KENDALL L, b Norfolk, Nebr, Sept 11, 42. PHYSICAL OCEANOGRAPHY. *Educ:* Ore State Univ, MS, 67, PhD(oceanog), 70. *Prof Exp:* PROF OCEANOG, UNIV SFLA, ST PETERSBURG, 69- *Concurrent Pos:* Oceanoptics & polar prog mgr, prog scientist for Nimbus-7 Coastal Zone Color Scanner & Scanning Multichannel Microwave Radiometer Sensors, NASA, 80-82, mem Satellite Ocean Color Sci Working Group, 81-, Imaging Spectrometer Sci Adv Group, Moderate Resolution Imaging Spectrometer Team, High Resolution Imaging Spectrometer Team. *Mem:* Am Geophys Union; Optical Soc Am. *Res:* Ocean optics; distribution and dynamics of marine particulates: their effects on the submarine light field; satellite sensing of ocean variables. *Mailing Add:* Marine Sci Dept Univ SFla 140 Seventh Ave S St Petersburg FL 33701

CARDIASMENOS, APOSTLE GEORGE, b Oakland, Calif. MICROWAVE ENGINEERING, SOLID STATE PHYSICS. *Educ:* Univ Calif, Berkeley, BSEE, 70; Univ Mass, MSEE, 76, PhD(physics & astron), 78. *Prof Exp:* Sr engr microwaves, Five Col Radio Observ, 75-77; sr scientist millimeter waves, Alpha Industs, 77-80, div mgr, millimeter subsysts div, 80-; CONSULT ENGR, MISSILE SYSTS DIV, RAYTHEON CORP. *Concurrent Pos:* Consult microwaves, TRG Div, Alpha Industs, 75-77. *Mem:* Inst Elec & Electronic Engrs. *Res:* Low noise microwave millimeter wave and infrared receivers and devices; masers, lasers and related devices. *Mailing Add:* Consult Eng Raytheonco Missile Sys 50 Apple Hill Dr Tewksbury MA 01876

CARDIFF, ROBERT DARRELL, b San Francisco, Calif, Dec 5, 35; m 62; c 3. PATHOLOGY, CANCER RESEARCH. *Educ:* Univ Calif, Berkeley, BS, 58, PhD(zool), 68; Univ Calif, San Francisco, MD, 62; Am Bd Path, dipl, 69. *Prof Exp:* Rotating intern, Kings County Gen Hosp, Brooklyn, 62-64; instr path & resident anat path, Sch Med, Univ Ore, 64-66; staff pathologist, Dept Neuropsychiat, Walter Reed Army Inst Res, Washington, DC, 68-71; assoc prof, 71-77, PROF PATH, SCH MED, UNIV CALIF, DAVIS, 77-, CHAIR, DEPT PATH, 90- *Honors & Awards:* Triton Res Award, 85; Sadusk Award, 86. *Mem:* AAAS; Am Soc Microbiol; Am Asn Cancer Res; Int Acad Path; Am Asn Path & Bact; Int Asn Breast Cancer Res. *Res:* Mammary tumor systems; breast cancer research; pathology of transgenic animals; diagnostic cytokeratins; lentiviruses. *Mailing Add:* Dept Path Univ Calif Sch Med Davis CA 95616

CARDILLO, FRANCES M, b Rome, NY, Apr 19, 32. BIOLOGY. *Educ:* Col St Rose, BS, 53, MA, 61; St Bonaventure Univ, PhD(biol), 67. *Prof Exp:* Instr bot, Immaculate Conception Jr Col, 66-67; asst prof, Ladycliff Col, 67-69; assoc prof, Quincy Col, 69-71; assoc prof, Ladycliff Col, 71-75; ASSOC PROF BIOL, MANHATTAN COL, 75- *Concurrent Pos:* NSF Acad Year, 69-71. *Mem:* AAAS; Am Inst Biol Sci; Bot Soc Am; Am Fern Soc; Tissue Culture Asn. *Res:* Anatomy of Lycopodium species; biochemical studies of Crossosoma; hybridoma technology. *Mailing Add:* Dept Biol Manhattan Col Mt St Vincent Campus Riverdale NY 10471

CARDILLO, MARK J, b Passaic, NJ, Aug 20, 43. CHEMICAL PHYSICS. *Educ:* Stevens Inst Technol, BS, 64; Cornell Univ, PhD(chem), 70. *Prof Exp:* Res assoc chem, Brown Univ, 69-71; Nat Res Coun Italy fel, Inst Physics, Univ Genoa, 71-72; res assoc, Mass Inst Technol, 72-75; mem staff, 75-81, DEPT HEAD, BELL LABS, 81- *Honors & Awards:* Medard Welch Award, Am Vacuum Soc. *Mem:* Am Chem Soc; fel Am Phys Soc. *Res:* Gas-surface interactions; molecular beams; scattering; reaction kinetics; chemical dynamics on surfaces; high temperature mass spectrometry. *Mailing Add:* Bell Labs 600 Mountain Ave Murray Hill NJ 07974

CARDINAL, JOHN ROBERT, b Flint, Mich, Dec 24, 43; m 70; c 2. MICELLAR SYSTEMS. *Educ:* Univ Mich, BS, 67; Univ Wis, MS 69, PhD(pharm), 73. *Prof Exp:* Res asst, Univ Mich, 66-67; res asst, Univ Wis, 67-71, teaching asst pharmaceut, 70-71; asst prof pharmaceut, Univ Utah, 71-79, assoc prof, 79-82; proj leader, 82-83, MGR, NOVEL DRUG DELIVERY SYSTS, PFIZER INC, 83- *Concurrent Pos:* Vis prof, Upjohn Co, 80; adj assoc prof, Univ Utah, 82- *Mem:* Fel Acad Pharmaceut Sci; Am Chem Soc; AAAS;

Controlled Release Soc; NY Acad Sci. *Res:* Solubilization in micellar systems; mechanisms of cholesterol gallstone formation and of solute permeation through skin; novel drug delivery systems; controlled release drug systems. *Mailing Add:* Cent Res Pfizer Pharmaceut Co Eastern Point Rd Groton CT 06340

CARDINALE, GEORGE JOSEPH, b New York, NY, Mar 30, 36; m 64; c 2. BIOCHEMISTRY. *Educ:* Fordham Univ, BS, 57; Ohio State Univ, PhD(org chem), 65. *Prof Exp:* Fel biochem, Brandeis Univ, 65-67, res assoc, 67-70; asst mem physiol chem, 70-78, asst to dir sci affairs, 78-85, sr scientist, 85-87, ADMIN DIR, ROCHE INST MOLECULAR BIOL, 87- *Concurrent Pos:* Adj assoc prof, Dept Pharmacol & Toxicol, Univ RI, 74- *Mem:* Int Inflammation Club; Sigma Xi; Am Soc Biochem & Molecular Biol; NY Acad Sci; Soc Res Adminr. *Res:* Prolyl hydroxylase and its role in collagen biosynthesis; factors affecting collagen synthesis; role of collagen in atherosclerosis. *Mailing Add:* Roche Inst Molecular Biol Bldg 102 Nutley NJ 07110

CARDINET, GEORGE HUGH, III, b Oakland, Calif, Oct 28, 34; m 57; c 4. COMPARATIVE PATHOLOGY. *Educ:* Univ Calif, BS, 60, DVM, 63, PhD(comp path), 66. *Prof Exp:* From asst prof to assoc prof anat, Neuromuscular Res Lab, Kans State Univ, 66-74; assoc prof, 74-76, PROF ANAT, UNIV CALIF, DAVIS, 76-, PROF PHYS MED & REHAB, 78-, ASSOC DEAN SCH VET MED, 76- ,PROF ANAT. *Concurrent Pos:* NIH fel, 63-66 & 73-74. *Mem:* AAAS; Am Asn Anat; Am Soc Zool; Electron Micros Soc Am; Am Asn Vet Anat. *Res:* Investigations of myopathies in domestic animals including clinical enzymology; enzyme histochemistry and electromicroscopy. *Mailing Add:* Dept Anat 1321 Haring Hall Sch Vet Med Univ Calif Davis CA 95616

CARDIS, ANGELINE BAIRD, b Trenton, NJ, Mar 28, 43; m 62; c 3. ORGANIC CHEMISTRY. *Educ:* Temple Univ, BA, 72, PhD(org chem), 77. *Prof Exp:* Res chemist, Agr Chem Group, FMC Corp, 76-80; sr res chemist polymers res & develop, Mobil Chem Co, 80-82; res assoc, Additives Res Group, 85-89, PROD DIV ADMINR, MOBIL RES & DEVELOP CO, 90. *Concurrent Pos:* Additives lab mgr, Chem Prods Div, Mobil Chem Co, 82-85. *Mem:* Am Chem Soc; AAAS. *Res:* Synthesis of biologically active compounds; structure- activity relationships; alkaloid synthesis and biosynthesis; polyethylene product development; blown and cast film extrusion; synthesis and development of lubricant and fuel additives. *Mailing Add:* 350 W Front St Florence NJ 08518

CARDMAN, LAWRENCE SANTO, b Mt Vernon, NY, Oct 7, 44; m 68; c 3. NUCLEAR PHYSICS, ACCELERATOR PHYSICS. *Educ:* Yale Univ, BA, 66, PhD(physics), 72. *Prof Exp:* Actg instr physics, Electron Accelerator Lab, Yale Univ, 71-72; Nat Acad Sci-Nat Res Coun res fel, Ctr Radiation Res, Nat Bur Standards, 72-73; from asst prof to assoc prof, 73-82, PROF PHYSICS, NUCLEAR PHYSICS LAB, UNIV ILL, 82- *Concurrent Pos:* Vis scientist, Cen Saclay, 80-81, Continuous Electron Beam Accelerator Facil, Newport News, Va, 89-90; prin investr, Nuclear Physics Lab, Univ Ill, 81- *Mem:* Sigma Xi; fel Am Phys Soc. *Res:* Photonuclear physics; elastic and inelastic electron scattering studies of nuclear structure and accelerator physics; polarized electron sources; photocathodes. *Mailing Add:* 909 W Charles St Champaign IL 61821

CARDNER, DAVID V, b Wilmington, Del, June 15, 35; m 61; c 3. CHEMICAL ENGINEERING. *Educ:* Rice Univ, BS, 57, PhD(chem eng), 63. *Prof Exp:* Res engr, Plastics Dept, 63-67, asst div supt process res, 67-68, asst div supt nylon tech, 68-72, asst div supt eng liaison, 72-74, sr eng process comput control, 74-82, TECH ASSOC PROCESS COMPUT CONTROL, E I DU PONT DE NEMOURS & CO, 82- *Concurrent Pos:* Comnr, Sabine River Compact Admin, 81-86 & 89- *Mem:* Am Inst Chem Engrs; Instrument Soc Am; Sigma Xi. *Res:* Computer control of processes; non-linear programming; model studies; process simulations; process development. *Mailing Add:* E I du Pont de Nemours & Co PO Box 1089 Orange TX 77630

CARDON, BARTLEY LOWELL, b Berkeley, Calif, Sept 3, 40. PHYSICS. *Educ:* Univ Calif, Berkeley, AB, 63; Purdue Univ, MS, 66; Univ Ariz, PhD(physics), 77. *Prof Exp:* Res assoc physics, Univ Ariz, 72-77; res fel astrophys, 77-79, res assoc astrophys, Harvard Col Observ, 79-85; STAFF, LINCOLN LAB, MASS INST TECHNOL, 85- *Mem:* Sigma Xi; Optical Soc Am; Am Asn Physics Teachers. *Res:* Beam-foil spectroscopy; spectroscopy of atoms and molecules; experimental determination of atomic and molecular oscillator strengths; study of atomic and molecular processes; infrared phenomenology. *Mailing Add:* MIT/Lincoln Lab PO Box 73 Lexington MA 02173-0073

CARDON, SAMUEL ZELIG, b Chicago, Ill, Nov 24, 18; m 58; c 2. CHEMISTRY. *Educ:* Univ Chicago, BS, 39, MS, 41; Western Reserve Univ, PhD(org chem), 50. *Prof Exp:* Chemist, Harshaw Chem Co, 44-48; org chemist, Lubrizol Corp, 49-51; chemist, Horizons, Inc, 51-53 & Rand Develop Corp, 53-64; secy-treas, 64-80, CHEMIST, GEN TECH SERV, INC, 64, PRES, 80- *Mem:* Am Chem Soc; Am Asn Small Res Co (pres); Sigma Xi. *Res:* Organic and inorganic synthesis; biological regulation and dynamics; chemical carcinogenesis and inhibition; water purification; research management. *Mailing Add:* 13324 Lake Shore Blvd Cleveland OH 44108

CARDONA, MANUEL, b Barcelona, Spain, July 9, 34; US citizen; m 59; c 3. SEMICONDUCTOR PHYSICS, OPTICAL PROPERTIES SOLIDS. *Educ:* Univ Barcelona, Lic Sci, 55; Univ Madrid, Dr Sci, 58; Harvard Univ, PhD(appl sci), 59; Brown Univ, MA, 65. *Hon Degrees:* MA, Brown Univ, 65; Dr Sci, Independent Univ Madrid & Independent Univ Barcelona, 85. *Prof Exp:* Mem tech staff, RCA Lab, Zürich, Switz, & RCA Lab, Princeton, NJ, 61-64; DIR, SOLID STATE PHYSICS, MAX PLANCK INST, STUTTGART, WGER, 71- *Concurrent Pos:* Vis prof, Univ Buenos Aires, 65 & Univ Osaka & Tokyo, 84, Victoria Univ, NZ, 87, Miller vis prof Berkely, UNiv Calif, 88; mem, sci coun, Ger Electron Synchrotron, Hamburg, 75-78; mem, adv bd, Inst Surface Sci, Julich, 77-83; mem, semiconductor comn, Int Union Pure & Appl Physics, 79-84; mem, Comt Condensed Matter, Ger NSF, 80-86; mem bd, Condensed Matter Div, Europ Phys Soc, 90- *Honors & Awards:* N Monturiol Medal, Govt Catalonia, 82; F Isakson Prize, Am Phys Soc, 86; Great Cross of Alfonso X el Salsia, 88; J M Muar Krouland Medal, 89. *Mem:* Nat Acad Sci; fel Am Phys Soc; Ger Physics Soc; Europ Physics Soc. *Res:* Fundamental physical properties of solids, including superconductors and semiconductors; electronic and vibronic structure. *Mailing Add:* Max Planck Inst Solid State Physics Heisenbergstrasse 1 7000 Stuttgart 80 800665 Germany

CARDOSO, SERGIO STEINER, b Belem-Para, Brazil, June 26, 27; m 54; c 4. PHARMACOLOGY, BIOLOGICAL RHYTHMS. *Educ:* Univ Brazil, MD, 52; Univ Sao Paulo, PhD(pharmacol), 64. *Prof Exp:* From asst prof to assoc prof pharmacol, Univ Sao Paulo, 61-68, from asst prof to assoc prof, 67-76, PROF PHARMACOL, SCH MED, UNIV TENN, MEMPHIS, 77- *Mem:* Am Soc Pharmacol & Exp Therapeut. *Res:* Possible role of circadian mitotic rhythms as related to cancer chemotherapy. *Mailing Add:* Dept Pharmacol Univ Tenn Med Sch Memphis TN 38163

CARDUS, DAVID, b Barcelona, Spain, Aug 6, 22; m 51; c 4. CARDIOLOGY, GRAVITATIONAL PHYSIOLOGY. *Educ:* Univ Montpellier, BA, 42, BS; Univ Barcelona, MD, 49. *Prof Exp:* Intern, Hosp Clin, Univ Barcelona, 49-50; resident, Sanitarium of Puig de Olena, Barcelona, 50-53; res assoc physiol, Postgrad Sch Cardiol, Med Sch, Univ Barcelona, 54-55; from instr to asst prof physiol & rehab, 60-65, assoc prof rehab, 65-69, assoc prof physiol, 65-73, PROF PHYSIOL, COL MED, BAYLOR UNIV, 73-, PROF REHAB & HEAD CARDIOPULMONARY LAB, 69-, DIR DIV BIOMATH, 70- *Concurrent Pos:* French Govt res fel, 53-54; Brit Coun res fel, 57; Inst Int Educ fel, Lovelace Found, 57-60; head work tolerance eval unit, Tex Inst Rehab & Res, 60; adj prof math sci, Rice Univ, 70- *Honors & Awards:* August Pi i Sunyer Prize, 68; Gold Medal, 6th Int Cong Phys Med, 72; Elisabeth & Sidney Licht Award, Am Cong Phys Med & Rehab; Medal Narcis Monturiol, Generalitat Catalunya, Spain, 84; Catalunya Enfora Award, Ibero-Am Coop Inst, 87. *Mem:* AAAS; Am Col Cardiol; Am Col Chest Physicians; AMA; NY Acad Sci. *Res:* Experimental exercise and respiratory physiology; mathematical and computer applications to the study of physiological systems; body composition of humans with extensive muscular paralysis; physiology of urinary bladder; benefit cost studies of rehabilitation medicine; physiological effects of microgravity. *Mailing Add:* Tex Inst Rehab & Res 1333 Moursund Ave Houston TX 77030

CARDWELL, ALVIN BOYD, b Lenoir City, Tenn, Oct 16, 02; m 30; c 2. SOLID STATE PHYSICS. *Educ:* Univ Chattanooga, BS, 25; Univ Wis, MS, 27, PhD(physics), 30. *Hon Degrees:* DSc, Univ Chattanooga, 61. *Prof Exp:* Asst physics, Univ Wis, 26-29; from asst prof to assoc prof, Tulane Univ, 30-36; prof, 36-73, head dept physics & physicist in charge, Eng Exp Sta, 37-53, physicist in charge, Agr Exp Sta, 47-53, assoc dean sch arts & sci, 53-55, dir bur gen res, 54-67, head dept physics & physicist in charge, Eng Exp Sta, 57-67, Agr Exp Sta, 57-73, EMER PROF PHYSICS, KANS STATE UNIV, 73- *Concurrent Pos:* Res physicist, Clinton Eng Works, Tenn Eastman Corp, Oak Ridge, Tenn, 43-46. *Mem:* AAAS; fel Am Phys Soc; Am Asn Physics Teachers. *Mailing Add:* Rt 3 Box 310E Kingston TN 37763

CARDWELL, DAVID MICHAEL, b Hickory, NC, June 9, 49; m 82; c 1. NUCLEAR MAGNETIC RESONANCE & MAGNETIC RESONANCE IMAGING METHODS & TECHNIQUES. *Educ:* Univ Ark, BS, 86. *Prof Exp:* RADIO FREQUENCY ENGR, UNIV ARK MED SCI, 86- *Concurrent Pos:* Lectr, Univ Ark, 90- *Mem:* Inst Elec & Electronic Engrs Microwave Tech Soc; Inst Elec & Electronic Engrs Antennas & Propagation Soc; Inst Elec & Electronic Engrs Biomed Eng Soc. *Res:* Design spectroscopy and imaging coils for nuclear magnetic resonance and magnetic resonance imaging research on materials, animals, and humans; design radio frequency hardware for nuclear magnetic resonance and magnetic resonance imaging systems. *Mailing Add:* 2506 S Taylor Little Rock AR 72204

CARDWELL, JOE THOMAS, b Vernon, Tex, Feb 19, 22; m 42; c 2. DAIRY CHEMISTRY. *Educ:* Tex Technol Col, BS, 47, MS, 49; NC State Col, PhD, 56. *Prof Exp:* Vet voc teacher, Knox County Voc Sch, Tex, 47; instr dairy mfg, Tex Technol Col, 47-50; res asst dairy chem, NC State Col, 50-52; from asst prof to assoc prof dairy mfg, 52-64, ASSOC PROF DAIRY SCI, MISS STATE UNIV & DAIRY CHEMIST, AGR EXP STA, 64- *Mem:* Inst Food Technol; AAAS; Am Dairy Sci Asn; Sigma Xi. *Res:* Milk plant sanitation; oxidized flavor development in milk; influence of calcium chloride on yield of cheddar cheese; frozen cultures; chocolate milk; flavor components of foods; products made from delactosed milk; method of detecting NFOM in fluid milk; new products developed from sweet whey. *Mailing Add:* Drawer JC Mississippi State MS 39762

CARDWELL, PAUL H, b Metamora, Mich, Aug 24, 12; m 39; c 4. COLLOID CHEMISTRY. *Educ:* Cent Mich Teacher's Col, AB, 35; Univ Mich, PhD(chem), 41. *Prof Exp:* Chief chemist, Dowell Inc, 41-52, dir lab, 52-54; tech specialist, Dow Chem Co, Mich, 54-65, asst mgr apparatus & instruments, 65-67, mgr apparatus & instruments bus, 67-69; dir res, Deepsea Ventures Inc, 69-75; CONSULT OCEAN NODULE PROCESSING, 75- *Mem:* AAAS; Am Chem Soc; Am Inst Aeronaut & Astronaut; Am Inst Mining, Metall & Petrol Eng; Electrochem Soc. *Res:* Contact angles; colloid chemistry; resins; corrosion inhibitors; chemical treatment of oil wells; reproducible contact angles on reproducible metal surfaces; extractive metallurgy. *Mailing Add:* Zanoni VA 23191

CARDWELL, VERNON BRUCE, b Ft Morgan, Colo, Oct 8, 36; m 54; c 4. AGRONOMY, CROP PHYSIOLOGY. *Educ:* Colo State Univ, BS, 58, MS, 61; Iowa State Univ, PhD(crop prod), 67. *Prof Exp:* Res asst agron, Colo State Univ, 58-60, asst agronomist, 60-64; instr agron, Iowa State Univ, 64-67; from asst prof to assoc prof, 67-78, PROF AGRON, UNIV MINN, ST PAUL, 78- *Honors & Awards:* Agron Educ Award, Am Soc Agron, 84. *Mem:* Am Soc Agron; Crop Sci Soc Am; Coun Agr Sci & Technol; fel Nat Asn Cols & Teachers Agr. *Res:* Seed physiology; seed quality wheat; low temperature germination corn & soybeans; farming systems. *Mailing Add:* 1991 Buford Circle Univ Minn St Paul MN 55108

CARDWELL, WILLIAM THOMAS, JR, b Boulder, Colo, May 27, 17; m 47. ENGINEERING. *Educ:* Calif Inst Technol, BS, 38, MS, 39. *Prof Exp:* Chemist, Standard Oil Co, Calif, 39-41, petrol engr, 41-47, res engr, Chevron Res Co, 47-48, sr res engr, 48-56, res assoc, 56-66, sr res assoc, 66-80; RETIRED. *Mem:* Am Chem Soc; Am Inst Mining, Metall & Petrol Engrs; Acoust Soc Am; Sigma Xi. *Res:* Physical chemistry of drilling fluids; physics and mathematics of oil reservoirs; patents; physics of musical wind instruments. *Mailing Add:* 16731 Ardita Dr Whittier CA 90603

CARELLI, MARIO DOMENICO, b Italy; US citizen. LIPOID METAL FAST REACTORS, THERMAL HYDRAULICS OF LIPOID METALS. *Educ:* Univ Florence, Italy, BSc, 62; Univ Pisa, Italy, PhD(nuclear eng), 66. *Prof Exp:* Postdoctoral fel, Univ Pisa, Italy, 66-67; engr, Atomic Energy Authority, Bologna, Italy, 67-69; sr engr, Westinghouse Advan Reactors Div, 69-73, prin engr, 73-77, fel engr, 77-84, patent chmn, 84-88, mgr, Almr Eng, 88-90, MGR, THERMOFLUID SYSTS & SAFETY, WESTINGHOUSE ADVAN ENERGY SYSTS, 90- *Concurrent Pos:* Adj assoc prof, Energy Resources Prog, Univ Pittsburgh, 78-; prin investr, Strategic Defense Initiative Off, Westinghouse Advan Energy Systs, 87-89. *Mem:* Am Nucelar Soc; Int Asn Hydraulic Res. *Res:* Core design and safety analyses for lipoid metal fast reactors; thermal hydraulics of lipoid metals; advanced nuclear reactors core design for space and terrestrial applications. *Mailing Add:* Westinghouse AES PO Box 158 Madison PA 15663

CAREN, LINDA DAVIS, b Corsicana, Tex, Apr 17, 41; m 63; c 2. MEDICAL MICROBIOLOGY, BIOLOGY. *Educ:* Ohio State Univ, BSc, 62; Stanford Univ, AM, 65, PhD(med microbiol), 67. *Prof Exp:* Tech ed, Tech Info Div, NASA Ames Res Ctr, 72-76; ASST PROF BIOL, SANTA CLARA UNIV, 78- *Concurrent Pos:* Nat Acad Sci fel, 67; consult, Encycl Britannica, 68; instr, DeAnza Col, 77-78; vis lectr, Dept Biol Univ Santa Clara, 74-78, NASA grant, 78-82. *Mem:* Int Soc Study Origins Life; Am Soc Microbiol; AAAS; Sigma Xi; Am Asn Immunologists; Soc Exp & Biol Med; Am Col Toxicol; NY Acad Sci; Reticuloendothelial Soc. *Res:* Role of complement in resistance to disease and in the rejection of skin grafts; use of the enzyme-linked immunosorbent assay to detect anti-influenza antibody; effects of environmental pollutants on immune response and resistance to disease; use of enzyme-linked immunosorbent assay to detect gene dosage effects on antigens in the Duffy, Rhesus and Ss systems; immunotoxicology studies on dimethyl sulfoxide, ethanol, methyl-DOPA. *Mailing Add:* Dept Biol Calif State Univ 18111 Nordhoff St Northridge CA 91330

CAREN, ROBERT POSTON, b Columbus, Ohio, Dec 25, 32; m 63. PHYSICS. *Educ:* Ohio State Univ, BS, 53, MS, 54, PhD(physics), 61. *Prof Exp:* Sr physicist, NAm Aviation Inc, Ohio, 59-60; instr physics, Ohio State Univ, 60-61; res scientist & sr mem res lab, 62-68, mgr infrared progs lab, 68-70, dir eng sci lab, 70-75, dir Palo Alto Res Labs, 75-82, vpres res & develop, 82-87, CORP VPRES SCI & ENG, LOCKHEED MISSILES & SPACE CO, 87- *Concurrent Pos:* Lectr, Univ Santa Clara, chmn tech & opers coun, Aerospace Indust Asn; Deans Adv Bd, Univ Southern Calif, Univ Calif, Davis, Ohio State Univ, Calif Poly San Luis Obisbo. *Honors & Awards:* IR-100 Award, 67. *Mem:* Am Phys Soc; sr mem Inst Elec Eng; assoc fel Am Inst Aeronaut & Astronaut; Am Defense Preparedness Asn; Nat Acad Eng. *Res:* Development of advanced infrared sensor systems and subsystems; radiation heat transfer theory. *Mailing Add:* Lockheed Corp 4500 Park Granada Blvd Calabasas CA 91302

CARES, WILLIAM RONALD, b Dearborn, Mich, 41; c 1. SURFACE CHEMISTRY, CHEMICAL KINETICS. *Educ:* Case Inst Technol, BSChE, 63; Univ Ill, MS, 65, PhD(chem), 69. *Prof Exp:* Res fel, Rice Univ, 69-71; res chemist, US Naval Res Lab, 71-73; res chemist catalysis, Petro-Tex Chem Corp, 73-76; sr res chemist, M W Kellogg Co, 76-87; SR RES ENGR, MOBAY SYNTHETICS CORP, 87- *Mem:* Am Chem Soc; Am Inst Chem Engrs; Royal Soc Chem; Catalysis Soc; Sigma Xi; Am Soc Testing & Mat. *Res:* Kinetics and mechanics of heterogeneous catalytic reactions; flue gas desulfurization chemistry; kinetics of surface reactions; environmental chemistry; chemical engineering. *Mailing Add:* 435 Wycliffe Houston TX 77079-7132

CARESS, EDWARD ALAN, b Columbus, Nebr, Feb 6, 36; m 61; c 3. PHOTOCHEMISTRY, MASS SPECTROMETRY. *Educ:* Dartmouth Col, AB, 58; Univ Rochester, PhD(org chem), 63. *Prof Exp:* Fel, Univ Rochester, 63; res assoc org chem, Mass Inst Technol, 63-65; from asst prof to assoc prof, 65-78, asst dean, Grad Sch Arts & Sci, 71-84, PROF CHEM, GEORGE WASHINGTON UNIV, 78-, ASSOC DEAN, GRAD SCH ARTS & SCI, 85- *Mem:* Am Chem Soc. *Res:* Organic reaction mechanisms; photochemistry of organic compounds; structure determination; mass spectrometric studies. *Mailing Add:* Dept Chem George Washington Univ Washington DC 20052

CARET, ROBERT LAURENT, b Biddeford, Maine, Oct 7, 47; m 69. ORGANIC CHEMISTRY, FLAVOR & FRAGRANCE CHEMISTRY. *Educ:* Suffolk Univ, BA, 69; Univ NH, PhD(org chem), 74. *Prof Exp:* Res assoc, Bio-Res Inst, 67-69; teacher chem, Rumford High Sch, 69-70; teaching asst, Univ NH & Suffolk Univ, 66-72; vis asst prof, 74-75, instr, 75-76, asst prof chem, 76-82, assoc vpres Acad Comput, 85-86, exec asst to pres, 86-87, DEAN NATURAL & MATH SCI, 81-, ASSOC PROF CHEM, 82-, PROVOST & VPRES, TOWSON STATE UNIV, 88- *Concurrent Pos:* Lectr, Suffolk Univ, 72-73; instr, Univ NH, 74; adj prof, Univ Md. *Mem:* Sigma Xi; Am Chem Soc; Am Asn Univ Prof; Am Asn Higher Educ; Am Coun Educ; Coun Col Arts & Sci; Am Asn Univ Adminrs; Am Asn Col Deans; Am Asn State Cols & Univs. *Res:* Stereochemistry, conformational analysis; nuclear magnetic resonance including Carbon-13 nmr and their use in the study of organosulfur compounds; flavor and fragance compounds. *Mailing Add:* Dept Chem Towson State Univ Towson MD 21204

CARETTO, ALBERT A, JR, b Baldwin, NY, May 16, 28; m 60; c 2. NUCLEAR & PHYSICAL CHEMISTRY, SCIENCE EDUCATION. *Educ:* Rensselaer Polytech Inst, BS, 50; Univ Rochester, PhD(chem), 54. *Prof Exp:* Nuclear chemist, Brookhaven Nat Lab, 54-56; nuclear chemist, Univ Calif, Berkeley, 56-57, res chemist, Livermore, 58-59; asst prof chem, 57-58, from asst prof to assoc prof, 59-67, chmn dept, 70-74, PROF CHEM, CARNEGIE-MELLON UNIV, 67- *Concurrent Pos:* Sabbatical Award, Europ Ctr Nuclear Res, 64-65 & Europ Ctr Nuclear Res, Geneva, Switz, 74-75; dir, Pa Gov Sch Sci Carnegie-Mellon, 82- *Mem:* AAAS; Am Chem Soc; Am Phys Soc. *Res:* Nuclear reactions induced with high energy particles; nuclear spectroscopy; radiochemical effects of recoil atoms. *Mailing Add:* Dept Chem Carnegie-Mellon Univ Pittsburgh PA 15213

CAREW, DAVID P, b Monson, Mass, Oct 21, 28; m 51; c 3. PHARMACY. *Educ:* Mass Col Pharm, BS, 52, MS, 54; Univ Conn, PhD(pharm), 58. *Prof Exp:* Asst instr pharmacog, Univ Conn, 54-57; from asst prof to assoc prof, 57-65, actg dean, 84, PROF PHARMACOG, UNIV IOWA, 65-, ASST DEAN, COL PHARM, 75- *Concurrent Pos:* Collab scientist, UN Cannabis Res Prog. *Honors & Awards:* M L Huit Award, 80. *Mem:* Fel AAAS; Am Soc Pharmacog (vpres, 64-65, pres, 65-66); Am Pharmaceut Asn; fel Acad Pharmaceut Sci; Int Plant Tissue Cult Asn; fel Am Asn Pharm Scientists. *Res:* Natural product research; products with therapeutic activity; plant tissue culture and biosynthesis; antibiotics. *Mailing Add:* Col Pharm Univ Iowa Iowa City IA 52242

CAREW, JAMES L, b Lydney, Eng, May 2, 45; c 1. PLEISTOCENE SEA LEVEL, CARBONATE PETROLOGY. *Educ:* Brown Univ, AB, 66; Univ Tex, Austin, MA, 69, PhD, 78. *Prof Exp:* Asst prof geol, Williams Col, Williamstown, Mass, 72-75 & Rensselaer Polytechnic Inst, Troy, NY, 75-77; marine scientist, Williams Col-Mystic Seaport Prog Am Maritime Studies, Mystic Seaport Mus, Conn, 77-80; assoc prof, 81-88, PROF GEOL, COL CHARLESTON, SC, 88- *Concurrent Pos:* Vis asst prof, Conn Col, 78, Univ S Fla, 81; mem SC State Bd Regist Geologists, 86-87, vchmn, 87-88, chmn, 88-89, vchmn, 90-91; convener, Geol Soc Am Penrose Conf, 87; treas, Asn State Bd Geol, 90-91. *Mem:* Geol Soc Am; AAAS; Soc Sed Geol; Paleont Soc; Sigma Xi. *Res:* Stratigraphy and geologic history of Bahamian Islands, particularly San Salvador; use of geologic data to determine sea level high stands over the past several hundred thousand years. *Mailing Add:* Dept Geol Col Charleston Charleston SC 29424

CAREW, JOHN FRANCIS, b Brooklyn, NY, June 26, 37; m 63; c 3. NUCLEAR ENGINEERING, NUCLEAR PHYSICS. *Educ:* St John's Univ, BS, 60, MS, 63; NY Univ, PhD(physics), 68. *Prof Exp:* Physicist, Sperry Gyroscope Co, 64-65; prof physics, St John's Univ, 65-68; physicist, Knolls Atomic Power Lab, Gen Elec, 68-72, mgr monitoring syst, Nuclear Energy Div, 72-76; physicist, 76-79, MGR, LWR SYST, BROOKHAVEN NAT LAB, 79- *Concurrent Pos:* Res assoc, NY Univ, 66-68. *Mem:* Am Nuclear Soc; Am Phys Soc. *Res:* Nuclear energy systems; atomic and nuclear physics. *Mailing Add:* Dept Nuclear Energy Brookhaven Nat Lab Upton LI NY 11973

CAREW, LYNDON BELMONT, JR, b Lynn, Mass, Nov 27, 32; m 60; c 2. ANIMAL NUTRITION, HUMAN NUTRITION. *Educ:* Univ Mass, BS, 55; Cornell Univ, PhD(animal nutrit), 61. *Prof Exp:* Res asst poultry nutrit, Cornell Univ, 55-58, res assoc, 58-59, res asst, 59-61; tech dir, Colombian Nat Poultry Prog & Animal Nutrit Lab, Colombian Agr Prog, Rockefeller Found, Bogota, 61-65; sr res assoc poultry sci, Cornell Univ, 65-66; head poultry res sect, Hess & Clark Div, Richardson Merrill Corp, Ohio, 66-69; assoc prof, 69-75, PROF ANIMAL SCI & HUMAN NUTRIT, UNIV VT, 75- *Concurrent Pos:* Mem tech comt, Animal Nutrit Res Coun, 72-, sci prog mgr, Int Nutrit Proj, Univ Vt, 80-83; USDA regional steering comt, poultry nutrit, 81-88. *Mem:* AAAS; Am Inst Nutrit; Poultry Sci Asn; Soc Exp Biol & Med; Endocrine Soc; Nutrit Educ Soc; Nat Asn Col Teachers Agr; World's Poultry Sci Asn. *Res:* Nutrition-endocrine interactions; nitrate/nitrite toxicity; thyroid function; metabolizable energy; energy metabolism; general poultry nutrition; nutritive properties of fats; Latin American poultry science; dietary fat, amino acids and hormone function; biochemistry. *Mailing Add:* Univ Vt Dept Animal Sci Biores Lab 655 Spear St Burlington VT 05403-6199

CAREW, THOMAS EDWARD, b Columbus, Ga, Dec 18, 43; m 65. ATHEROSCLEROSIS, VASCULAR BIOLOGY. *Educ:* Johns Hopkins Univ, BA, 65; Cath Univ Am, MSE, 67, PhD(bioeng), 71. *Prof Exp:* Res engr, Nat Heart & Lung Inst, 69-72; asst res physiologist, Univ Calif, San Diego, 72-83; asst dir, Deborah Cardiovasc Res Inst, 83-84; asst adj prof med, 85-87, ASSOC ADJ PROF, UNIV CALIF, SAN DIEGO, 87- *Concurrent Pos:* Consult, NSF, 72-; vis prof, Tromso Univ, 77; consult, NIH, 83- *Mem:* Fel Am Heart Asn; Am Physiol Soc; AAAS; NY Acad Sci. *Res:* Studies of lipoprotein metabolism relating to atherosclerosis. *Mailing Add:* Dept Med Univ Calif San Diego La Jolla CA 92093

CAREY, ALFRED W(ILLIAM), JR, b New York, NY, May 5, 24; m 46; c 3. REFRIGERATION. *Educ:* Polytech Inst Brooklyn, BMechE, 49. *Prof Exp:* Res engr, Super Engine Div, White Motors Co, 49-53; prin mech engr, Battelle Mem Inst, 53-55, proj leader mech res, 55-59, asst div chief, 59-61, assoc staff engr prime movers, 61-62; proj mgr combustion res, Cummins Engine Co, Inc, 62-66, mgr, 66-70, mgr advan develop, 70-71, dir advan prod planning, 71-75, dir combustion res, 75-77, sr tech adv res, 77-86, dir advan controls, 80-86; CONSULT, 86- *Mem:* Am Soc Automotive Engrs. *Res:* Heat power; combustion phenomena in compression and spark ignition reciprocating engines; refrigeration, particularly unconventional systems; fuel injection system. *Mailing Add:* 3612 Westenedge Dr Columbus IN 47203

CAREY, ANDREW GALBRAITH, JR, b Baltimore, Md, Apr 11, 32; m 57; c 2. BIOLOGICAL OCEANOGRAPHY. *Educ:* Princeton Univ, AB, 55; Yale Univ, PhD(zool), 62. *Prof Exp:* From asst prof to prof oceanog, 61-87, EMER PROF OCEANOG, ORE STATE UNIV, 87- *Concurrent Pos:* Marshal fel, Denmark, 70; Japan Soc Prom Sci vis prof, Univ Tokyo, 77; vis scientist, Woods Hole Oceanog Inst, 76-77, & Bedford Inst Oceanog, 84-85. *Honors & Awards:* Lindbergh Award, 84. *Mem:* AAAS; Am Soc Limnol & Oceanog; Artic Inst NAm; Ecol Soc Am; Marine Biol Asn UK; Sigma Xi. *Res:* Marine benthic ecology; community, energetics, deep sea; invertebrate zoology; polar ecology. *Mailing Add:* Col Oceanog Ore State Univ Oceanog Admin Bldg 104 Corvallis OR 97331-5503

CAREY, BERNARD JOSEPH, b Pittsburgh, Pa, Feb 28, 41; m 66; c 2. COMPUTER SCIENCE. *Educ:* Univ Pittsburgh, BS, 62; Univ Calif, Santa Barbara, MS, 69, PhD(elec eng comput sci), 71. *Prof Exp:* Design engr, Guid & Control, Litton Industs, 63-68; res asst speech processing, Univ Calif, Santa Barbara, 69-71; asst prof comput sci, Univ Conn, 71-77, assoc prof, 77-; MGR FLEXIBLE AUTOMATION SYSTS, GEN ELEC CORP. *Concurrent Pos:* Consult microprocessing syst, Rogers Corp, 75- *Mem:* Inst Elec & Electronics Engrs. *Res:* Multiprocessor systems design; applications of microprocessors. *Mailing Add:* GE Corp Automation Systs Lab KWD 222 PO Box 8 Schenectady NY 12301

CAREY, CYNTHIA, b Denver, Colo, July 17, 47. PHYSIOLOGICAL ECOLOGY, COMPARATIVE PHYSIOLOGY. *Educ:* Occidental Col, Los Angeles, AB, 69, MA, 70; Univ Mich, PhD(zool), 76. *Prof Exp:* Teaching fel, Univ Mich, 70-74, res asst, 74-76; asst prof, 76-82, ASSOC PROF BIOL, UNIV COLO, 82- *Concurrent Pos:* Vis investr, Univ Peruana Cayetano Heredia, 82-88. *Honors & Awards:* Marcia Brady Tucker Award, Am Ornithologist's Union, 73; A Brazier Howell Award, Cooper Ornith Soc, 75. *Mem:* Am Ornithologist's Union; Am Soc Zoologists; Am Soc Ichthyologists & Herpetologists; Cooper Ornith Soc; fel AAAS; Sigma Xi. *Res:* Physiological adaptation of animals to stressful environments. *Mailing Add:* Dept EPO Biol Univ Colo Boulder CO 80309-0334

CAREY, DAVID CROCKETT, b Montclair, NJ, Oct 2, 39; m 81. HIGH ENERGY PHYSICS, PARTICLE PHYSICS. *Educ:* Mass Inst Technol, BS, 62; Univ Mich, MS, 64, PhD(physics), 67. *Prof Exp:* Technician, Forrestal Res Ctr, Princeton-Penn Accelerator, 59-60; instr physics & math, Upsala Col, 62; res assoc physics, City Col New York, 67-69; PHYSICIST, FERMI NAT ACCELERATOR LAB, 69- *Mem:* AAAS; Am Phys Soc. *Res:* Theoretical and experimental investigations into strong interaction dynamics; charged particle optics; applied mathematics problems of high-energy experimental physics. *Mailing Add:* Fermi Nat Accelerator Lab Batavia IL 60510

CAREY, FRANCIS ARTHUR, b Philadelphia, Pa, May 28, 37; m 63; c 3. ORGANIC CHEMISTRY. *Educ:* Drexel Univ, BS, 59; Pa State Univ, PhD(org chem), 63. *Prof Exp:* NIH fel, Harvard Univ, 63-64; asst prof, 66-71, ASSOC PROF CHEM, UNIV VA, 71- *Mem:* Am Chem Soc. *Res:* Structural and synthetic organic chemistry. *Mailing Add:* Dept Chem Univ Va Charlottesville VA 22903-2454

CAREY, FRANCIS G, b New York, NY, Sept 11, 31; m 59; c 2. BEHAVIOR, PHYSIOLOGY. *Educ:* Harvard Univ, PhD(biol), 60. *Prof Exp:* SR SCIENTIST, WOODS HOLE OCEANOG INST. MASS. 62- *Res:* Behavior, physiology and anatomy of pelagic fish and squids. *Mailing Add:* Woods Hole Oceanog Inst Woods Hole MA 02543

CAREY, GRAHAM FRANCIS, b Cairns, Australia, Nov 14, 44; m 68; c 2. FINITE ELEMENT METHODS, COMPUTATIONAL MECHANICS. *Educ:* Univ Queensland, BS, 65; Univ Wash, MS, 70, PhD, 74. *Prof Exp:* Res fac, Univ Queensland, 66-68; res engr, Boeing Co, 68-70; res prof, Univ Wash, 74-76; PROF, DEPT AEROSPACE ENG, UNIV TEX, AUSTIN, 77- *Concurrent Pos:* Dir Comput Fluid Dynamics Lab, Univ Tex, 86-; adj fel, Minn Supercomput Inst, 86-; ed, Int Jour, Commun Appl Numerical Methods; Eng Found Professorship, Univ Tex. *Mem:* Soc Indust & Appl Math; Am Acad Mech; Soc Eng Sci; Australian Math Soc. *Res:* Finite element methods in computational mathematics and mechanics with specific focus on fluids and transport phenomena, new concepts in supercomputing and finite element algorithms. *Mailing Add:* Dept Aerospace Eng & Eng Mech Univ Tex Austin TX 78712

CAREY, JOHN HUGH, b Windsor, Ont, Jan 28, 47; m 70; c 3. PHOTOCHEMISTRY. *Educ:* Univ Windsor, BSc, 70, MSc, 72; Carleton Univ, PhD(chem), 74. *Prof Exp:* Nat Res Coun Can fel, 74-76, res consult, 76-78, RES SCIENTIST, NAT WATER RES INST, 78- *Mem:* InterAm Photochemry Soc; Sigma Xi. *Res:* Photochemistry of natural waters and of solution-sediment interface; photochemical aspects of pollution and water treatment processes; fate and effects of synthetic organic compounds in natural waters. *Mailing Add:* 578 High Bench Rd Alpine UT 84004-1715

CAREY, JOHN JOSEPH, b Boston, Mass, Dec 10, 11; m 37; c 3. ELECTRICAL ENGINEERING, SAFETY ENGINEERING. *Educ:* Mass Inst Technol, BSEE, 34, MSEE, 53. *Prof Exp:* Elec-mech engr, Panama Canal, 34-41; elec designer, Jackson & Moreland Engrs, 41-42; instr elec eng, Univ NMex, 45; instr, Univ Kans, 45-46; prof, 46-72, EMER PROF ELEC ENG, UNIV MICH, ANN ARBOR, 72- *Concurrent Pos:* Test engr, Gen Elec Co, 39; consult, Commonwealth Assocs, Mich, 52, Gen Elec Co, Ind, 55, Fargo Eng Co, Mich, 55-, Consumers Power Co, 60, Climax Molybdenum Co, Mich, 62- & Mich Consol Gas Co, 63-; eng guid coordr, Engrs Coun Prof Develop, 67-70; Univ Mich rep, Am Power Conf & mem gen & elec prog comts, 67-72; consult elec engr, 72- *Mem:* Nat Soc Prof Engrs; Inst Elec & Electronics Engrs. *Res:* Energy conversion; systems engineering; engineering economics. *Mailing Add:* 3486 Woodland Rd Ann Arbor MI 48104

CAREY, LARRY CAMPBELL, b Coal Grove, Ohio, Nov 5, 33; m 56; c 4. SURGERY. *Educ:* Ohio State Univ, BSc, 55, MD, 59. *Prof Exp:* Intern surg, New York Hosp, Cornell Univ, 59-60; resident, Marquette Integrated Residency Prog, 60-64, chief admin resident, Marquette Univ, 64-65, from asst prof to prof surg, Sch Med, Univ Pittsburgh, 68-74; PROF SURG & CHMN DEPT, COL MED, OHIO STATE UNIV, 75- *Concurrent Pos:* Markle Scholar acad med, 65-70; William S Middleton lectr, Wis State Med Soc, 65; consult, Milwaukee County Gen Hosp, Wis, 65-, Milwaukee Lutheran Hosp, 65-, Columbia Hosp, 65- & St Luke's Hosp, 68-; asst clin prof, Boston Univ, 66-67. *Mem:* AAAS; fel Am Col Surg; AMA; Soc Univ Surg; Soc Surg Alimentary Tract. *Res:* Pancreatic physiology; bioelectric phenomena; shock. *Mailing Add:* 111 S Grant Ave Columbus OH 43215

CAREY, MARTIN CONRAD, b Clonmel, Ireland, June 18, 39; US citizen; div; c 2. GASTROENTEROLOGY, MOLECULAR BIOPHYSICS. *Educ:* Nat Univ Ireland, MB, BCh, BAO, 62, MD, 81, FRCP (I), 85. *Hon Degrees:* DSc, Nat Univ Ireland, 84; AM, Harvard Univ, 89. *Prof Exp:* Clin fel gastroenterol, Boston Univ Hosp, 69-70; res fel & assoc biophys, Sch Med, Boston Univ, 70-73, asst prof med, 73-75; asst prof med, Sch Med, Harvard Univ, 75-79; assoc med gastroenterol, 79-82, physician, Brigham & Womens Hosp, 82-88; assoc prof med, Sch Med, Harvard Univ, 79-88, Lawrence J Henderson, assoc prof health sci & technol, 79-88, PROF MED, SCH MED, HARVARD UNIV, 88-, LAWRENCE J HENDERSON PROF HEALTH SCI & TECHNOL, 88- *Concurrent Pos:* Int Fogarty fel, NIH, 68; fel, Med Res Found, Boston, 72; mem, Nat Inst Arthritis, Metab & Digestive Dis res subcomt, 72-73; fel, J S Guggenheim Found, 74; ad hoc mem, NIH Gen Med A Spec Study Sect, 75-78; assoc ed, J Lipid Res, 78-81; prin investr, Not Inst Arthritis, Diabetes,& Kidney Dis Res grant, 75-; fac mem, Grad Sch Arts & Sci & assoc mem, Dept Physiol & Biophysics, Harvard Univ, 83; co-dir, Core Membrane & Lipid Lab, Harvard Dig Dis Ctr, 84- *Honors & Awards:* McArdle Prize; McArdle, Kennedy, Magennis, Bellingham & Smith Gold Medals; Adolf Windaus Prize, Falk Found, Basel, Switz, 84; Merit Award, NIH, 86; Distinguished Achievement Award, Am Gastroenterol Asn, 90. *Mem:* Am Gastroenterol Asn; Am Soc Study Liver Dis; Am Oil Chem Soc; Am Soc Clin Invest; Biophys Soc; Asn Am Physicians. *Res:* Physical chemistry and pathophysiology of alimentary tract lipids in health and disease; lipidprotein interactions in lipid transport and lipolysis; drug delivery systems, chemistry and physics of micelles, liquid crystals, ordered fluids and emulsions. *Mailing Add:* Harvard Med Sch Brigham & Womens Hosp Boston MA 02115

CAREY, MICHAEL DEAN, b South Bend, Ind, Apr 18, 46; m 67; c 3. ECOLOGY, ANIMAL BEHAVIOR. *Educ:* Wittenberg Univ, BA, 68; Ind Univ, MA, 76, PhD(ecol), 77. *Prof Exp:* Instr zool & ecol, Clark Col, Vancouver, Wash, 76-77; asst prof biol & ecol, Cent Mich Univ, 77-78; ASSOC PROF BIOL, ECOL & ANIMAL BEHAV, UNIV SCRANTON, 78- *Concurrent Pos:* Vis lectr, Univ Minn, ITASCA Sta, 81. *Mem:* Am Ornithologists Union; Cooper Ornith Soc; Ecol Soc Am; Sigma Xi; Wilson Ornith Soc; Soc Field Ornith. *Res:* Breeding sociobiology; evolution of mating systems and territoriality in passerines and insects; empirical field tests of current evolutionary models. *Mailing Add:* Dept Biol Univ Scranton Scranton PA 18510-4625

CAREY, PAUL L, b Arrowsmith, Ill, Nov 4, 23; m 48; c 5. BIOCHEMISTRY. *Educ:* Ill Wesleyan Univ, BS, 48; Kansas State Col, MS, 50; Purdue Univ, PhD(biochem), 58. *Prof Exp:* Jr biochemist, Smith, Kline & French Labs, 50-53; asst, Purdue Univ, 53-54, Walter G Karr fel, 54-57; mgr spec chows lab, Ralston Purina, 58-62; chemist, Penick & Ford, Ltd, 62-65; biochemist, R J Reynolds Co, 65-68, group leader, 68-70; assoc scientist, Ralston Purina Co, 70-78, scientist, 78-82; RETIRED. *Concurrent Pos:* Instr, Winston Salem State Teachers Col, 67. *Mem:* AAAS; Am Chem Soc. *Res:* Nutrition; chemistry of biological materials; process research and development. *Mailing Add:* 10212 Maebern Terr St Louis MO 63127

CAREY, PAUL RICHARD, b Dartford, UK, June 17, 45; Brit & Can citizen. RESONANCE RAMAN IN BIOCHEMISTRY, PROTEIN STRUCTURE & FUNCTION. *Educ:* Univ Sussex, BSc, 66, DPhil(chem), 69. *Prof Exp:* SR RES OFFICER PROTEIN BIOPHYS, INST BIOL SCI, NAT RES COUN, 85-, MGR, CTR PROTEIN STRUCT DESIGN, 87-, HEAD, PROTEIN LAB, INST BIOL SCI, 87- *Concurrent Pos:* Adj prof, Biochem Dept, Univ Ottawa, 87-; mem, Int Adv Comt, Int Conf Raman Spectros, 84-90, Int Conf Lasers in Biol, 87-, Int Network Protein Eng Ctr, 91. *Mem:* Can Chem Soc; Am Chem Soc; Can Biophys Soc. *Res:* Use of resonance Raman spectroscopy in biochemistry; developing resonance Raman, with kinetics and x-ray crystallography, to characterize enzyme-substrate complexes. *Mailing Add:* Inst Biol Sci Nat Res Coun Ottawa ON K1A 0R6 Can

CAREY, THOMAS E, IMMUNOLOGY, CELL BIOLOGY. *Educ:* State Univ NY, Buffalo, PhD(biochem & pharm), 73. *Prof Exp:* ASSOC RES SCIENTIST & DIR CANCER RES LAB, UNIV MICH, 78- *Mailing Add:* Dept Otolaryngol-Head & Neck Surg Cancer Res Lab Univ Mich 6020 KHRI Box 0506 Ann Arbor MI 48109-0506

CAREY, WILLIAM BACON, b Philadelphia, Pa, Dec 6, 26; m 56; c 3. MEDICINE, BEHAVIOR. *Educ:* Yale Univ, BA, 50; Harvard Med Sch, MD, 54; certif, Am Bd Pediat, 60. *Prof Exp:* Rotating intern, Philadelphia Gen Hosp, 54-55, resident pediat, Children's Hosp Philadelphia, 55-57 & 59-60; capt, US Army Med Corp, Ariz, 57-59; instr, dept pediat, Univ Pa, Sch Med, 61-73, assoc, 73-79, from clin asst prof to clin assoc prof, 79-90, CLIN PROF, DEPT PEDIAT, UNIV PA, SCH MED, 90-; DIR, SECT BEHAV PEDIAT, CHILDREN'S HOSP, PHILADELPHIA, 89- *Concurrent Pos:* Res grantee, Am Acad Pediat, 75, 80 & 85; co ed, Develop-Behav Pediat; pvt prac pediat, Media, Pa, 60-89. *Mem:* Am Acad Pediat; Inst Med Nat Acad Sci; Soc Behav Pediat (pres, 90-91); Soc Res Child Develop; Am Pediat Soc. *Res:* Temperament differences in children; co-developer of the Infant Temperament Questionnaire, the Behavioral Style Questionnaire,; the Toddler Temperament Scale, the Middle Child Temperament and the Early Infancy Temperament Questionnaire; relationships of temperament to colic, accidents, sleep disturbances, obesity and behavior. *Mailing Add:* Childrens Hosp Gen Pediat 34th & Civic Ctr Blvd Philadelphia PA 19104

CAREY, WILLIAM DANIEL, b New York, NY, Jan 29, 16; m 44; c 5. SCIENCE POLICY. *Educ:* Columbia Univ, BA, 40, MA, 41; Harvard Univ, MPA, 42. *Hon Degrees:* LHD, Mt Sinai Med Univ. *Prof Exp:* Asst dir, Exec Off US Pres, Bur Budget, 66-69; vpres, Arthur D Little, Inc, 69-74; exec officer & publ, AAAS, 75-87; CONSULT, CARNEGIE CORP, NY, 88- *Concurrent Pos:* Chmn, US aide, US-USSR Bilateral Group Sci Policy, 73-83; trustee, Russell Sage Found & Mitre Corp, 78-88; gov, Ctr Creative Leadership, 79-81; chmn vis comt, US Nat Bur Standards, 80-81; mem, Coun Foreign Relations & Comt Nat Security. *Honors & Awards:* Ralph Coates Roe Medal, Am Soc Mech Engrs, 79; Pub Welfare Medal, Nat Acad Sci, 86.

Mem: hon mem Nat Acad Sci; Inst Med Nat Acad Sci; Nat Acad Pub Admin; Am Soc Pub Admin. *Res:* Carnegie committee on science, technology and government. *Mailing Add:* 3724 Northhampton St NW Washington DC 20015

CARFAGNO, DANIEL GAETANO, b Syracuse, NY, Aug 17, 35; m 58; c 2. PHYSICAL CHEMISTRY, ENVIRONMENTAL MONITORING. *Educ:* Le Moyne Col, NY, BS, 57; Syracuse Univ, PhD(phys chem), 65. *Prof Exp:* Sr res chemist, 67-67, res group leader, 67-69, res specialist, 69-73, GROUP LEADER, ENVIRON LAB, MONSANTO RES CORP, 73- *Mem:* Am Chem Soc. *Res:* Phase equilibria; x-ray diffraction; differential thermal analysis; atmospheric dispersion; physical properties of plutonium and its compounds; isotopic fuels; environmental monitoring for radioactive materials; ground water monitoring. *Mailing Add:* EG&G Mound Appl Technologies Mound Facil PO Box 300 Miamisburg OH 45343

CARFAGNO, SALVATORE P, b Norristown, Pa, Nov 29, 25; m 78; c 1. NUCLEAR PLANT SAFETY, EQUIPMENT AGING. *Educ:* Drexel Univ, BS, 47; Univ Pa, MS, 49; Temple Univ, PhD(physics), 63. *Prof Exp:* head, eng dept, Franklin Res Ctr, Arvin/Calspan Corp, 48-89; TECHNOL CONSULT, 89- *Concurrent Pos:* Adj proj physics, Drexel Evening & Univ Col, 48- *Mem:* Inst Elec & Electronics Engrs; Am Nuclear Soc; Am Phys Soc; Sci Res Soc N Am. *Res:* Nuclear plant aging and life extension; technical evaluation of nuclear plant safety issues; investigation of equipment aging, equipment condition monitoring and equipment qualification methodology. *Mailing Add:* 1616 Riverview Rd Gladwyne PA 19035-1211

CARGILL, B(URTON) F(LOYD), agricultural engineering; deceased, see previous edition for last biography

CARGILL, G SLADE, III, b Atlanta, Ga, Jan 1, 43; m; c 2. APPLIED PHYSICS, MATERIAL SCIENCE. *Educ:* Ga Inst Tech, BA, 65; Harvard Univ, SM, 65, PhD(appl physic), 69. *Prof Exp:* Fac mem, Yale Univ, 70-76; SR MGR STRUCT MAT, INT BUS MACH, TJ WATSON RES CTR, 76- *Mem:* Am Phys Soc; Mat Res Soc (vpres & pres-elect); Am Crystallog Asn. *Res:* Structural studies of amorphous materials, semiconductors, and thin films, especially defect and impurity effects, using x- ray scattering and absorption, electron microscopy, and ion channeling. *Mailing Add:* Res Div TJ Watson Res Ctr PO Box 218 Yorktown Heights NY 10598

CARGILL, ROBERT LEE, JR, b Marshall, Tex, Sept 11, 34; m 65, 80; c 3. CORPORATE MANAGEMENT. *Educ:* Rice Univ, BA, 55; Mass Inst Technol, PhD(org chem), 60. *Prof Exp:* NIH fel org chem, Univ Calif, Berkeley, 60-62; from asst prof to assoc prof chem, Univ SC, 67-73, prof chem, 73-80; adj prof, Rice Univ, 80-84. *Concurrent Pos:* CHIEF EXEC OFFICER, CARGILL INTERESTS, Ltd, 82- *Honors & Awards:* Russell Award for Creative Res, 74. *Mem:* Am Chem Soc; Sigma Xi. *Mailing Add:* PO Box 992 Longview TX 75606-0992

CARGO, DAVID GARRETT, b Pittsburgh, Pa, Dec 28, 24; m 49; c 6. ZOOLOGY. *Educ:* Univ Pittsburgh, BS, 49, MS, 50. *Prof Exp:* Biologist, State of Md, 50-60; res assoc marine biol, Chesapeake Biol Lab, Univ MD, Solomons, 60-91; RETIRED. *Res:* Biology and ecology of marine invertebrates, especially Crustacea and coelenterates. *Mailing Add:* 4965 Mackall Rd St Leonard MD 20685

CARGO, GERALD THOMAS, b Dowagiac, Mich, Mar 2, 30; m 56; c 2. COMPLEX ANALYSIS, INEQUALITIES. *Educ:* Univ Mich, BBA, 52, MS, 53, PhD(math), 59. *Prof Exp:* Asst math, Willow Run Res Ctr, Univ Mich, 55-56; from asst prof to assoc prof, 59-77, PROF MATH, SYRACUSE UNIV, 77- *Concurrent Pos:* Resident res assoc, Nat Acad Sci, 61-62; NSF res grant, 61, 63-66; res analyst, Aerospace Res Labs, 67. *Mem:* Am Math Soc; Math Asn Am; Sigma Xi. *Res:* Boundary behavior of analytic functions; inequalities and convex functions; Hardy classes and Nevanlinna theory; Tauberian theorems; mathematical microbiology; zeros of polynomials. *Mailing Add:* Dept Math 200 Carnegie Bldg Syracuse Univ Syracuse NY 13244-1150

CARHART, RICHARD ALAN, b Evanston, Ill, Aug 30, 39; m 60; c 3. THEORETICAL HIGH ENERGY PHYSICS, MATHEMATICAL PHYSICS. *Educ:* Northwestern Univ, BA, 60; Univ Wis, MA, 62, PhD(physics), 64. *Prof Exp:* Res assoc physics, Univ Wis, 64 & Brookhaven Nat Lab, 64-66; asst prof, 66-70, ASSOC PROF PHYSICS, UNIV ILL, CHICAGO CIRCLE, 70- *Mem:* Am Phys Soc. *Res:* Parity violations in quantum electro-dynamics; nucleon electromagnetic form factors and other topics in theoretical elementary particle physics; nonlinear ordinary and partial differential equations of physics. *Mailing Add:* Dept Physics Box 4348 Univ Ill Chicago Circle Chicago IL 60680

CARICO, JAMES EDWIN, b Galax, Va, Mar 20, 37; m 59; c 2. INVERTEBRATE ZOOLOGY, ARACHNOLOGY. *Educ:* ETenn State Univ, BS, 59; Va Polytech Inst & State Univ, MS, 64, PhD(zool), 70. *Prof Exp:* From asst prof to assoc prof, 64-72, PROF BIOL, LYNCHBURG COL, 72- *Concurrent Pos:* Res grant, Va Acad Sci, 71; res fel arachnology, Harvard Univ, 72. *Mem:* Am Arachnological Soc; Brit Arachnological Soc. *Res:* Taxonomy; ecology; evolution in the spider family Pisauridae; evolution and construction behavior of spider webs. *Mailing Add:* Dept Biol Lynchburg Col Lynchburg VA 24501

CARIM, HATIM MOHAMED, b Hyderabad, India, July 17, 46; m 79; c 3. BIOMEDICAL ENGINEERING, ELECTROCHEMISTRY. *Educ:* Osmania Univ, India, BS, 67, 71; Drexel Univ, MS, 73, PhD(biomed eng), 76. *Prof Exp:* Teaching asst, Drexel Univ, 75-76; SR RES SPECIALIST BIOMED ENG, MED PROD DIV, 3M CO, 77. *Mem:* Electrochem Soc; Soc Biomaterials; Inst Elec & Electronics Engrs; Sigma Xi. *Res:* Biomedical diagnostic techniques; biomaterials; bioelectrodes. *Mailing Add:* 431 W Wentworth Ave St Paul MN 55118-2943

CARITHERS, JEANINE RUTHERFORD, b Boone, Iowa, Sept 26, 33; m 53; c 3. ANATOMY, NEUROENDOCRINOLOGY. *Educ:* Iowa State Univ, BS, 56, MS, 65; Univ Mo, PhD(anat), 68. *Prof Exp:* Vis prof, Univ Louis Pasteur, 76-77; from asst prof to prof vet anat, Iowa State Univ, 68-76, chmn vet anat, 79-89, asst dean, Grad Col, 88-89. *Concurrent Pos:* Asst vpres acad affairs & intern, Iowa State Univ, 87-88. *Mem:* Am Asn Anatomists; Am Asn Vet Anatomists; Soc Neuroscience; Sigma Xi; AAAS. *Res:* Neuroendocrinology, neurosecretion; role of the central nervous system in regulation of adrenal function; development of the hypothalamic neurosecretory system. *Mailing Add:* Dept Vet Anat Iowa State Univ Ames IA 50011

CARL, JAMES DUDLEY, b Centralia, Ill, June 4, 35; m 61. GEOCHEMISTRY. *Educ:* Mo Sch Mines, BS, 57; Univ Ill, MS, 60, PhD(geol), 61. *Prof Exp:* Asst prof geol, Cent Mo State Col, 61-63 & Ill State Univ, 63-68; ASSOC PROF GEOL, STATE UNIV NY COL POTSDAM, 68- *Mem:* Geol Soc Am; Geochemistry_Soc; Mineral Soc Am. *Res:* Petrology; mineralogy; geochemical and structural studies of igneous and metamorphic rocks from coastal Maine and the Northwest Adirondacks, New York; major and trace element variation. *Mailing Add:* Dept Geol State Univ NY Pierrepont Ave Potsdam NY 13676

CARL, PHILIP LOUIS, b Cleveland, Ohio, Dec 6, 39; m 64; c 2. MOLECULAR GENETICS, BIOCHEMISTRY. *Educ:* Harvard Univ, BA, 61; Univ Calif, Berkeley, MS, 63, PhD(biophys), 68. *Prof Exp:* Jane Coffin Childs Mem Fund fel med res, 68-70; asst prof microbiol, Univ Ill, Urbana, 70-77, vis asst prof chem, 77-80; RES ASSOC PROF PHARMACOL, UNIV NC, CHAPEL HILL, 80- *Mem:* AAAS; Am Soc Microbiol; Am Chem Soc. *Res:* Molecular biology of DNA replication carcinogenesis and cancer chemotherapy. *Mailing Add:* Dept Pharmacol FLOB Bldg 231H Univ NC Chapel Hill NC 27514

CARLAN, ALAN J, b New York, NY, Feb 15, 30; m 51; c 3. SOLID STATE PHYSICS, PHYSICAL ELECTRONICS. *Educ:* Brooklyn Col, BA, 51; Worcester Polytech Inst, MS, 57. *Prof Exp:* Physicist, Am Optical Co, Mass, 53-57 & Hoffman Elec, Ill, 57-58; fel, Mellon Inst, 58-61; supvy engr, Syntron Co, Pa, 61-62; pres, Power Components, Inc, 62-66; chief engr, Zener Diodes, Int Rectifier Corp, 66-67; sr tech specialist & group scientist, NAm Rockwell Corp, 67-70; oper mgr, Int Rectifier Corp, 70-73; mgr advan develop, Rockwell Int, 73-76; mem tech staff, Aerospace Corp, 76-86, sr engr specialist, radiation effects, 80-86; proj mgr survivability progs, TRW, 86-89. *Concurrent Pos:* Adv fel, Mellon Inst, 61, vis fac fel, 66; chmn, DNA adv comt on radiation hardness assured mil standard device specif, 84-86, Space Parts Working Group LSI comt, 80-, pres, Carlan Assoc Consults, 87- *Mem:* Inst Elec & Electronics Engrs. *Res:* Investigation and development of improved metal-oxide semiconductors, MNOS integrated circuits and power devices; developed improved manufacturing technologies and processes. *Mailing Add:* 4951 Rock Valley Rd Rancho Palos Verdes CA 90274

CARLAN, AUDREY M, b Brooklyn, NY; m 51; c 3. COMPUTER SCIENCES, EDUCATION ADMINISTRATION. *Educ:* Brooklyn Col, BA, 51; Worcester Polytech Inst, MS, 57. *Prof Exp:* Math physicist, Am Optical Co, 53-57 & TRW, 67-68; chmn dept math & technol, 78-87, PROF MATH & COMPUT SCI, LOS ANGELES SOUTHWEST COL, 68- *Concurrent Pos:* Chair, dist math coun, Los Angeles Community Col Dist, 84-; Gen Mgr, Carlan Assoc Consult, 87- *Res:* Computer Science. *Mailing Add:* Dept Math & Engr Los Angeles Southwest Col 1600 W Imperial Hwy Los Angeles CA 90047

CARLANDER, KENNETH DIXON, b Gary, Ind, May 25, 15; m 39, 75. FISHERIES. *Educ:* Univ Minn, BS, 36, MS, 38, PhD(zool), 43. *Prof Exp:* Lab technician, Work Proj Admin, Univ Minn, 36-38; fishery biologist, Minn State Dept Conserv, 38-46; from asst prof to prof zool, 46-74, distinguished prof, 74-85, EMER PROF, IOWA STATE UNIV, 85- *Concurrent Pos:* Leader, Iowa Coop Fishery Res Unit, 46-65; consult, Ford Found, Egypt, 65-66; vis prof, Satya Wacana Christian Univ, Java Indonesia, 78 & Texas A&M Univ, 82. *Mem:* AAAS; Am Fisheries Soc (pres, 60); Am Soc Limnol & Oceanog; Am Soc Ichthyol & Herpet; Biomet Soc. *Res:* Fishery biology; fish population estimation; limnology; age and growth of fishes. *Mailing Add:* Dept Animal Ecol Iowa State Univ Ames IA 50011

CARLBERG, DAVID MARVIN, b Los Angeles, Calif, Feb 9, 34; m 62; c 2. MOLECULAR BIOLOGY, GENETICS. *Educ:* Univ Calif, Los Angeles, BA, 56, PhD(microbiol), 63. *Prof Exp:* Chemist, Riker Labs, Inc, 56-58; head formula off, Rexall Drug & Chem Co, 58; res engr, McDonnell Douglas Corp, 63-65; mem staff, Hughes Aircraft Co, 65-66; from asst prof to assoc prof, 66-75, PROF MICROBIOL, CALIF STATE UNIV LONG BEACH, 75- *Mem:* Am Soc Microbiol; AAAS. *Res:* Microbial genetics; halobacteria; collection, detection and analysis of microorganisms in air and on surfaces; bio-instrumentation; exobiology. *Mailing Add:* Dept Microbiol Calif State Univ 1250 Bellflower Blvd Long Beach CA 90840

CARLBERG, KAREN ANN, b Seattle, Wash, Nov 15, 49. EXERCISE PHYSIOLOGY, REPRODUCTIVE PHYSIOLOGY. *Educ:* Univ Wash, BS, 72; Univ NMex, MS, 76, PhD(biol), 81. *Prof Exp:* Fel physiol, Univ Fla, Col Med, 81-83; asst prof, 83-88, ASSOC PROF PHYSIOL & ANAT, DEPT BIOL, EASTERN WASH UNIV, 88- *Mem:* Am Physiol Soc; Am Col Sports Med; Sigma Xi; Soc Exp Biol & Med. *Res:* Endocrine aspects of exercise physiology; effects of exercise training on female reproductive function, including menstrual and estrous cycles and fetal development; endocrine, metabolic and carbiovascular responses to physical and mental stress. *Mailing Add:* Dept Biol MS-72 Eastern Washington Univ Cheney WA 99004-2499

CARLE, GLENN CLIFFORD, b San Bernardino, Calif, Aug 17, 36; m 64; c 1. EXOBIOLOGY, PLANETARY ATMOSPHERES. *Educ:* Calif State Polytech Univ, BS, 63. *Prof Exp:* Chemist, 63-80, CHIEF, SOLAR SYST EXPLOR OFF, AMES RES CTR, NASA, 81- *Concurrent Pos:* Co-investr, Viking Biol Gas Exchange Exp, 71-76 & Pioneer Venus Gas Chromatograph,

75-78. *Honors & Awards:* Newcomb Cleveland Award, AAAS, 76. *Mem:* Am Chem Soc. *Res:* Advanced instrument concepts, primarily gas chromatography, for future spacecraft flights which will be used to explore the solar system; major emphasis is on greater sensitivity to volatile molecules. *Mailing Add:* 1503 Meadowlark Lane Sunnyvale CA 94087-4846

CARLE, KENNETH ROBERTS, b Keene, NH, Sept 16, 29; m 57; c 4. PHYSICAL ORGANIC CHEMISTRY. *Educ:* Middlebury Col, AB, 51; Univ NH, MS, 53; Del Univ, PhD(chem), 55. *Prof Exp:* Asst, Gen Chem Lab, Univ NH, 51-52 & Qual Chem Lab, Del Univ, 52-53; res chemist org chem, Am Cyanamid Co, 55-59; assoc prof chem, 59-62, head dept, 62-69, PROF CHEM, HOBART & WILLIAM SMITH COL, 62-, HEAD DEPT, 78- *Concurrent Pos:* Instr, Bridgeport Eng Inst, 57-; Fulbright-Hays lectr, Univ Manila, 66-67; vis res prof, Silliman Univ, Philippines, 74-75. *Mem:* AAAS; Am Chem Soc. *Res:* Organic synthesis; polymers; kinetics; rubber chemistry; structure determination of myeloma proteins; water chemistry. *Mailing Add:* Dept Chem Hobart & William Smith Col Geneva NY 14456

CARLEN, PETER LOUIS, b Edmonton, Alta, July 22, 43; m 70; c 2. NEUROLOGY, NEUROPHYSIOLOGY. *Educ:* Univ Toronto, MD, 67; FRCP(C), 72. *Prof Exp:* Intern, Montreal Gen Hosp, 67-68, resident internal med, 68-69; instr neurophysiol, Dept Zool, Hebrew Univ, Jerusalem, 69-70; resident neurol, Univ Toronto, 70-72; fel neurophysiol neurobiol unit, Hebrew Univ, 72-74; SR PHYSICIAN & HEAD NEUROL PROG, ADDICTION RES FOUND CLIN INST, 74-; STAFF NEUROLOGIST, TORONTO WESTERN HOSP, UNIV TORONTO, 74-, Res assoc, Playfair Neuroscience Unit, Univ Toront,79-, ASSOC PROF, DEPT MED & PHYSIOL, 81- *Mem:* Fel Can Neurol Soc; fel Am Acad Neurol; Soc Neuroscience; AAAS; Can Physiol Soc. *Res:* Neurobiology; acute and chronic effects of psychoactive drugs on neuronal function in man and animal; cellular neurophysiology; neural modelling. *Mailing Add:* 33 Russell St Toronto ON M5S 2S1 Can

CARLEONE, JOSEPH, b Philadelphia, Pa, Jan 30, 46; m 68; c 2. ENGINEERING MECHANICS. *Educ:* Drexel Univ, BS, 68, MS, 70, PhD(appl mech), 72. *Prof Exp:* Eng trainee mech eng, Philadelphia Naval Shipyard, 63-68, mech engr power plants, 68; res assoc composite mat, Drexel Univ, 72-73; CHIEF ENGR MECH, DYNA EAST CORP, 73- *Concurrent Pos:* NDEA fel, Drexel Univ, 68-71, adj prof, 74-75 & 77- *Mem:* Sigma Xi; Am Soc Mech Engrs; Am Defense Preparedness Asn. *Res:* Response of metals to high explosives; shaped charges; ballistics; the mechanics of fiber-reinforced composites; impact of plates and membranes. *Mailing Add:* 19741 Marsala Dr Yorba Linda CA 92686

CARLETON, BLONDEL HENRY, b Portland, Ore, Dec 8, 04; m 35; c 5. PHYSIOLOGY. *Educ:* Univ Ore, BA, 26; Univ Rochester, PhD(physiol), 36. *Prof Exp:* Asst zool, Univ Ore, 26-27; asst physiol, Sch Med, Univ Rochester, 33-36; instr biol & physics, Ga Teachers Col, 36-38; from asst prof to prof zool, 38-74, chmn fac biol, 67-70, EMER PROF ZOOL, UNIV PORTLAND, 74- *Mem:* Sigma Xi; Am Asn Biol Teachers. *Res:* Excitability of amphibian muscle; narcosis and excitability in nerve; oxygen metabolism of muscle and nerve. *Mailing Add:* 6705 N Wilbur Ave Portland OR 97217

CARLETON, HERBERT RUCK, b Rockville Centre, NY, Dec 5, 28; m 51; c 3. OPTICAL PHYSICS. *Educ:* Univ Southern Calif, BA, 58; Cornell Univ, PhD(theoret physics), 64. *Prof Exp:* Test methods engr, Sperry Gyroscope Corp, 49-54; design engr, Gilfillan Bros, Inc, 54-55; sr engr, Canoga Corp, 55-56; prin engr, Bendix Pac Corp, Calif, 56-58; staff mem, Res Ctr, Sperry Rand Corp, 62-67; assoc prof mat sci, 67-73, joint assoc prof mat sci & elec sci, 73-75, chmn dept mat sci & eng, 82-85, JOINT PROF MAT SCI & ELEC ENG, STATE UNIV NY, STONY BROOK, 75- *Concurrent Pos:* Consult, Bendix Pac Corp, 59-60, Defense Systs Dept, Gen Elec Corp, 60-62 & Pro Div, Harris Corp, 82 & 83; vis scientist, IBM San Jose Res Lab, 74, consult, 74-75; vis sci, Scripp's Inst Oceanog, Univ Calif, San Diego, 81. *Mem:* Optical Soc Am; Am Phys Soc; sr mem, Inst Elec & Electronics Eng; Acoust Soc Am. *Res:* Ultrasonic spectroscopy; optical properties of materials; strain-optic properties; non-crystalline solids; elastic properties of crystals; brillouin scattering; hypersonics; ultrasonic amplification in peizoelectric semiconductors; computer simulation of physical properties; dedicated computer system design for simulation. *Mailing Add:* Dept Mat Sci State Univ NY Stony Brook NY 11794

CARLETON, NATHANIEL PHILLIPS, b Burlington, Vt, Mar 16, 29; m 51; c 5. ASTROPHYSICS. *Educ:* Harvard Univ, AB, 51, PhD(physics), 56. *Prof Exp:* Asst prof physics, Harvard Univ, 56-62; PHYSICIST, SMITHSONIAN ASTROPHYS OBSERV, 63- *Mem:* Fel Am Phys Soc; Am Astron Soc. *Res:* Infrared spectroscopy of active galactic nuclei; design of telescopes and instruments. *Mailing Add:* Ctr Astrophys 60 Garden St Cambridge MA 02138

CARLETON, RICHARD ALLYN, b Providence, RI, Mar 15, 31; m 54, 75; c 5. CARDIOLOGY. *Educ:* Dartmouth Col, AB, 52; Harvard Med Sch, MD, 55. *Prof Exp:* Intern med, Boston City Hosp, 55-56, asst resident, 56-57, sr resident, 57-58, sr med resident, Metab Sect, 59-60; from asst prof to prof, Med Sch, Univ Ill, 62-68; assoc dir cardiol, Sect Cardiorespiratory Dis, Rush-Presby-St Luke's Med Ctr, 62-68, dir cardiol, 68-72; prof med, Rush Med Col, 68-72; prof, Univ Calif, San Diego, 72-74; prof med & chmn dept, Dartmouth Med Sch, 74-76; PROF MED, BROWN UNIV MED PROG, 76- *Concurrent Pos:* Res fel, Harvard Med Sch, 58-59; Burton E Hamilton res fel, Cardiol Div, Thorndike Mem Labs, Boston City Hosp, Mass, 58-59; teaching fel, Sch Med, Tufts Univ, 59-60; asst med, Harvard Med Sch, 56-58; from asst attend physician to assoc attend physican, Rush-Presby-St Luke's Med Ctr, 62-67, attend physician, 67-72; chief cardiol, San Diego Vet Admin Hosp, 72-74 & Mem Hosp, Pawtucket, RI, 76-; fel, Coun Cardiol & Epidemiol, Am Heart Asn. *Mem:* Fel Am Col Cardiol; Am Soc Clin Invest; fel Am Col Physicians; Asn Univ Cardiologists. *Res:* Heart disease prevention; cardiac rehabilitation; exercise physiology; coronary disease. *Mailing Add:* Prof Med Brown Univ Mem Hosp Pawtucket RI 02860

CARLEY, CHARLES TEAM, JR, b Greenville, Miss, Dec 27, 32; m 55; c 4. MECHANICAL ENGINEERING. *Educ:* Miss State Univ, BS, 55; Va Polytech Inst, MS, 61; NC State Univ, PhD(mech eng), 65. *Prof Exp:* Instr mech eng, Va Polytech Inst, 58-60; asst prof, Miss State Univ, 60-61; asst, NC State Univ, 61-64; assoc prof, Miss State Univ, 64-68, head dept, 69-90, interim head petrol eng, 90-91, PROF MECH ENG, MISS STATE UNIV, 68- *Mem:* Am Soc Mech Engrs; Am Soc Eng Educ; Nat Soc Prof Engrs. *Res:* Heat transmission; thermodynamics; bioengineering. *Mailing Add:* Dept Mech Eng Miss State Univ Mississippi State MS 39762

CARLEY, DAVID DON, theoretical physics; deceased, see previous edition for last biography

CARLEY, HAROLD EDWIN, b Syracuse, NY, July 8, 42; m 65; c 3. PLANT PATHOLOGY, SOIL SCIENCE. *Educ:* Cornell Univ, BS, 64; Univ Idaho, MS, 66; Univ Minn, St Paul, PhD(plant path), 69. *Prof Exp:* Proj leader agr bactericide-viricide develop, 69-71, group leader fungicides, 71-82, PROD DEVELOP MGR, ROHM AND HAAS CO, 83- *Mem:* Am Phytopath Soc; Sigma Xi; Am Chem Soc. *Res:* Discovery, optomization, and early development of chemical and biological plant disease control agents. *Mailing Add:* 11 Callowhill Rd Chalfont PA 18914-2101

CARLEY, JAMES F(RENCH), b New York, NY, July 16, 23; div; c 3. QUALITY ASSURANCE OIL SHALE ENGINEERING. *Educ:* Cornell Univ, BS, 44, BChE, 47, PhD(chem eng), 51. *Prof Exp:* Instr statist & acct, Cornell Univ, 49-50; res chem engr, Plastics Processing, E I du Pont de Nemours & Co, 50-55; eng ed, Mod Plastics, 56-59; assoc prof chem eng, Univ Ariz, 59-61; tech dir, Prodex Corp, 61-64; develop assoc, Celanese Plastics Co, 64; from assoc prof to prof chem eng, eng design & econ eval, Univ Colo, Boulder, 64-76; staff scientist, oil shale proj, 76-82, composites & polymers group, 82-86, qual assurance, 86-89, STAFF SCIENTIST, POLYMERS SECT, LAWRENCE LIVERMORE NAT LAB, 89- *Concurrent Pos:* Tech ed, Mod Plastics, 82-88. *Mem:* Am Inst Chem Engrs; Fel, Soc Plastics Engrs; Sigma Xi; Am Soc Qual Control. *Res:* Rheology, plastics processing and engineering design of plastics products; quality assurance; fluid flow and heat transfer; fossil fuels technology. *Mailing Add:* Lawrence Livermore Nat Lab PO Box 808 Livermore CA 94550

CARLEY, THOMAS GERALD, b Greenville, Miss, July 3, 35; m; c 2. STRUCTURAL MECHANICS. *Educ:* La State Univ, BS, 58, MS, 61; Univ Ill, PhD(theoret & appl mech), 65. *Prof Exp:* Res engr, Boeing Co, 65-66; asst prof mech eng, Southern Methodist Univ, 66-68; from asst prof to assoc prof, 68-77, PROF ENG SCI & MECH, UNIV TENN, 77- *Concurrent Pos:* Consult, Oak Ridge Nat Lab, 68- *Mem:* Sigma Xi; Am Soc Eng Educ; Am Soc Mech Engrs; Am Acad Mech. *Res:* Structural dynamics; seismic engineering; noise and vibration. *Mailing Add:* Engr Mech Perkins Hall Univ Tenn Knoxville TN 37996

CARLILE, ROBERT ELLIOT, b Admore, Okla, 33; m 56; c 4. PETROLEUM TECHNOLOGY. *Educ:* Univ Tulsa, BS, 58, MS, 60; Tex A&M Univ, PhD(petrol eng), 63. *Prof Exp:* Sr field engr, Kewanee Oil Co, 63-64, assoc prof petrol eng, Univ Mo, Rolla, 64-74, vpres, Petty-Ray Geophy-Geosource, 74-81, admnr, ARAMCO, 81-83; CHMN & PROF PETROL ENG, TEX TECH UNIV, 83- *Concurrent Pos:* Nat Acad Sci Petrol Visitor, Romania, 70 & Yugoslavia, 71; consult, Six Oil Cos, lawyers, 83-; bd dir, Soc Petrol Engrs, 86; Continental fel, 61-63; Exxon fel, 61-63. *Mem:* Soc Petrol Engrs. *Res:* Private analysis of petroleum cores and fluids; computers in petroleum engineering; multi-dimensional and multi-phase simulations of petroleum reservoirs. *Mailing Add:* Petrol Eng Texas Tech PO Box 4099 Lubbock TX 79409

CARLILE, ROBERT NICHOLS, b National City, Calif, Apr 24, 29; m 52; c 3. ELECTRICAL ENGINEERING. *Educ:* Pomona Col, BA, 51; Stanford Univ, MS, 53, EE, 56; Univ Calif, Berkeley, PhD(elec eng), 63. *Prof Exp:* Mem tech staff, Hughes Aircraft Co, 54-57, proj engr, 58-59; assoc prof elec eng, 63-69, PROF ELEC ENG, UNIV ARIZ, 69- *Concurrent Pos:* Vis res prof, Plasma Physics Lab, Princeton Univ, 69-70. *Mem:* Inst Elec & Electronics Engrs; Am Phys Soc. *Res:* Electromagnetic pulse (EMP), phenomenology; fusion plasmas; electric power generation, transmission, and distribution. *Mailing Add:* Dept of Elec Eng Univ of Ariz Tucson AZ 85721

CARLIN, CHARLES HERRICK, b Rockford, Ill, Jan 25, 39; m 65; c 2. ORGANIC CHEMISTRY. *Educ:* Carthage Col, AB, 61; Johns Hopkins Univ, MA, 63, PhD(org chem), 66. *Prof Exp:* From asst prof to assoc prof, 66-75, assoc dean col, 77-79, PROF CHEM, CARLETON COL, 75- *Concurrent Pos:* Sci adv, Food & Drug Admin, 69- *Mem:* AAAS; Sigma Xi. *Res:* Synthetic and mechanistic organic electrochemisty. *Mailing Add:* Dept Chem Carleton Col Northfield MN 55057

CARLIN, HERBERT J, b New York, NY, May 1, 17; m 40; c 2. ELECTROPHYSICS, ELECTRONICS. *Educ:* Columbia Univ, BS, 38, MS, 40; Polytech Inst Brooklyn, DEE, 47. *Prof Exp:* Design engr, Meter Div, Westinghouse Elec Co, NJ, 40-45; instr elec eng, Microwave Res Inst, Polytech Inst Brooklyn, 45-53, res prof, 53-66; dir, Sch Elec Eng, 66-75, J PRESTON LEVIS PROF ENG, CORNELL UNIV, 66- *Concurrent Pos:* Assoc dir, Microwave Res Inst, Polytech Inst Brooklyn, 56-61, head electrophys dept, 61-66; NSF sr res fel, 64-65; mem eval panel, Electromagnetics Dept, Nat Bur Standards; mem vis comt, Lehigh Univ & Univ Pa, 79-; vis prof elec eng, Mass Inst Technol, 72-73 & Univ Genoa, 74; vis res scientist, Ctr Nat d'Etudes des Telecommun, Paris, 79-80; vis prof, Univ Col Dublin, 83; Politecnico ditorino, 87 & 89; Tiaujiu Univ, China, 82; Techuiou, Haifa Israel, 80. *Honors & Awards:* Centennial Medal, Inst Elec & Electronic Engrs, 84. *Mem:* Fel Inst Elec & Electronic Engrs. *Res:* Network theory; microwave components, techniques and measurements. *Mailing Add:* Sch Elec Eng Phillips Hall Cornell Univ Ithaca NY 14853

CARLIN, RICHARD LEWIS, b Boston, Mass, July 28, 35; m 59; c 2. INORGANIC CHEMISTRY. *Educ:* Brown Univ, BS, 57; Univ Ill, MS, 59, PhD, 60. *Prof Exp:* From instr to asst prof chem, Brown Univ, 60-67; from assoc prof to prof chem, Univ Ill, 67-86; CONSULT, 86- *Concurrent Pos:* Vis prof, Kamerlingh Onnes Lab, Leiden. *Mem:* Am Chem Soc; Am Phys Soc. *Res:* Electronic structure of transition metal compounds; synthetic inorganic chemistry; magnetism and spectroscopy; magnetism at low temperatures. *Mailing Add:* Dept Chem Univ Ill Box 4348 Chicago IL 60680

CARLIN, ROBERT BURNELL, b St Paul, Minn, Nov 13, 16; m 52; c 5. ORGANIC CHEMISTRY. *Educ:* Univ Minn, BCh, 37, PhD(org chem), 41. *Prof Exp:* Asst, Univ Minn, 37-40; Lalor Found fel, Univ Ill, 41-42, instr org chem, 42-43; instr, Univ Rochester, 43-46; assoc prof, 46-52, head dept chem, 60-67 & assoc dean, Col Eng & Sci, 67-70, BECKER PROF ORG CHEM, CARNEGIE-MELLON UNIV, 52- *Concurrent Pos:* Consult, Koppers Co, Inc, 52-; Reilly lectr, Univ Notre Dame, 56; chmn, Gordon Res Conf Org Reactions & Processes, 56; mem comt, Off Ord Res, 56-59; eval postdoctoral fel applns, Nat Acad Sci-Nat Res Coun, 55-58, NSF, 55-59 & Air Force Off Sci Res, 66-68. *Mem:* Fel AAAS; Am Chem Soc; Sigma Xi. *Res:* Molecular rearrangements, such as, benzidine and alkyl group migrations; Fischer indole synthesis; polymer synthesis and structure; synthesis of bicyclic N heterocycles. *Mailing Add:* 928 Pa Ave Oakmont PA 15139-1353

CARLIN, RONALD D, b New York, NY, Mar 22, 46; m 76; c 2. CARDIOVASCULAR PHYSIOLOGY, MEDICAL & DENTAL EDUCATION. *Educ:* Fairleigh Dickinson Univ, BS, 67, MS, 69; Columbia Univ, MA, 73, MPhil, 73, PhD(physiol), 75. *Prof Exp:* From asst prof to prof physiol, Col Dent Med, Fairleigh Dickinson Univ, 74-81; res assoc, dept physiol, 79-83, ADJ ASSOC RES SCIENTIST, DEPT PHYSIOL, COL PHYSICIANS & SURGEONS, COLUMBIA UNIV, 83- *Concurrent Pos:* Chmn, Teaching Sect Am Physiol Soc, 86-88, sect rep, Prog Advisatory Comt Soc, 88- *Mem:* Am Physiol Soc; Int Asn Dent Res; Sigma Xi. *Res:* Regional circulations and their regulation during various patho-physiological states; study of current and future graduate and professional school physiology curriculum designs. *Mailing Add:* Col of Dent Med Fairleigh Dickinson Univ 140 Univ Plaza Dr Hackensack NJ 07601

CARLISLE, DAVID BREZ, b Salford, Eng, Mar 12, 26; m 67; c 4. MARINE & FRESH WATER BIOLOGY, GEOCHEMISTRY. *Educ:* Oxford Univ, BA, 47, MS, 51, PhD(zool), 54,. *Hon Degrees:* DSc, Oxford Univ, 63. *Prof Exp:* Zoologist, Marine Biol Asn UK, 51-62; head, Lab Res Div, Anti Locust Res Ctr, 62-69; prof biol & chmn dept, Trent Univ, 69-72; chief, Water Qual Res, Water Sci Group, 72-78, CHIEF SCIENTIST, DEPT ENVIRON, INLAND WATERS DIRECTORATE, GOVT CAN, 78- *Concurrent Pos:* Royal Soc Leverhume Prof Animal Physiology, 66-68; chmn, Isotrace Comt, Univ Toronto, 88- *Mem:* Fel Inst Biol; fel Royal Soc Arts; fel Zool Soc London; fel Am Anthrop Asn; fel Linnean Soc. *Res:* Comparative endocrinology of arthropods; tunicate biology; physiological ecology; environment induced changes in the endocrine systems of arthropods; environmental biology; fluid mechanics; chaos theory; paleogeo chemistry. *Mailing Add:* Environ Can Ottawa ON K1A 0H3 Can

CARLISLE, DONALD, b Vancouver, BC, June 21, 19; m 44; c 4. ECONOMIC GEOLOGY. *Educ:* Univ BC, BASc, 42, MASc, 44; Univ Wis, PhD(geol), 50. *Prof Exp:* Geologist, Consol Mining & Smelting Co, Can, 44-46; lectr geol, 49-50, from instr to prof, 51-89, EMER PROF GEOL & MINERAL RESOURCES, UNIV CALIF, LOS ANGELES, 89- *Concurrent Pos:* Assoc dean grad div, Univ Calif, Los Angeles, 67-74; consult mining co; expert, UN Int Atomic Energy Agency. *Mem:* Fel Geol Soc Am; Soc Econ Geol; Am Inst Mining, Metall & Petrol Engrs; Int Asn Geochemistry & Cosmochemistry. *Res:* Mineral economics and economic theory; mineral deposits and geochemistry; areal studies. *Mailing Add:* Dept Earth & Space Sci Univ Calif Los Angeles CA 90024

CARLISLE, EDITH M, b Vancouver, BC, Can, Jan 5, 22. TRACE ELEMENTS. *Educ:* Univ Wis, PhD(biochem), 50. *Prof Exp:* RESEARCHER & ADJ PROF NUTRIT, SCH PUB HEALTH, UNIV CALIF, LOS ANGELES, 77- *Mem:* AAAS; Am Inst Nutrit. *Mailing Add:* Sch Pub Health Univ Calif Los Angeles CA 90024

CARLISLE, GENE OZELLE, b Bivins, Tex, Feb 11, 39; m 61; c 1. INORGANIC CHEMISTRY. *Educ:* E Tex State Univ, BSc, 61, MSc, 65; N Tex State Univ, PhD(chem), 69. *Prof Exp:* Teacher high sch, Tex, 61-62; instr chem, Texarkana Col, 62-65, asst prof physics, 65-66; instr chem, N Tex State Univ, 66-67; res assoc, Univ NC, 69-70; asst prof chem, 70-74, ASSOC PROF CHEM, W TEX STATE UNIV, 74- *Mem:* Am Chem Soc; Am Phys Soc; Am Inst Physics. *Res:* Synthesis of transition metal complexes and applications of information gained from spectral and magnetic studies of these complexes to structural and bonding problems in coordination chemistry. *Mailing Add:* Dept of Chem W Tex State Univ Canyon TX 79015

CARLISLE, HARRY J, TEMPERATURE REGULATION, FOOD INTAKE. *Educ:* Univ Wash, PhD(physiol & psychol), 64. *Prof Exp:* PROF LAB PHYSIOL PSYCHOL, UNIV CALIF, SANTA BARBARA, 77- *Mailing Add:* Dept Psychol Univ Calif Santa Barbara CA 93106

CARLISLE, KAY SUSAN, b Minn, Nov 9, 45; m 74; c 1. CELL BIOLOGY. *Educ:* Augustana Col, SDak, BA, 67; Colo State Univ, MS, 72. *Prof Exp:* Res asst physiol, Colo State Univ, 67-72; cell biologist sex educ, Fernbank Sci Ctr, 73-77; RES ASSOC, ORE REGIONAL PRIMATE RES CTR, 78- *Mem:* Am Soc Cell Biologists. *Res:* Influence of steroid hormones (estrogen and progesterone) on the female reproductive tract and sexual skin. *Mailing Add:* Ore Primate Res Ctr 505 NW 18th St Beaverton OR 97006

CARLISLE, VICTOR WALTER, b Bunnell, Fla, Oct 3, 22; m 50; c 3. SOIL MORPHOLOGY. *Educ:* Univ Fla, BSA, 47, MS, 53, PhD(soils), 62. *Prof Exp:* Asst soil surveyor, Fla Agr Exp Sta, 47-54; asst soils, Inst Food & Agr Sci, 54-60, from instr to asst prof, 60-67, asst chemist, Agr Exp Sta, 62-67, assoc prof & assoc chemist, 67-74, PROF SOIL SCI, INST FOOD & AGR SCI, UNIV FLA, 74- *Concurrent Pos:* Consult, US AID, Univ Costa Rica, 66, Battelle Mem Inst subcontract, 67 & Jamaica Sch Agr, 70; course coordr, Orgn Trop Studies, Inc, Costa Rica, 68; mem, US AID, Malawi, Africa, 83. *Mem:* Soil Sci Soc Am; Int Soc Soil Sci; Clay Minerals Soc; Am Asn Quaternary Environ. *Res:* Soil genesis, classification and mapping; source of parent materials of soils; weathering; soil mineralogy. *Mailing Add:* 4308 SW 19th Terr Gainesville FL 32608

CARLITZ, LEONARD, b Philadelphia, Pa, Dec 26, 07; m 31; c 2. MATHEMATICS. *Educ:* Univ Pa, AB, 27, AM, 28, PhD(math), 30. *Prof Exp:* Nat Res fel math, Calif Inst Technol, Univ Pa & Cambridge Univ, 30-32; from asst prof to prof, 32-64, JAMES B DUKE PROF MATH, DUKE UNIV, 64- *Concurrent Pos:* Mem, Inst Adv Study, 35-36. *Mem:* Am Math Soc; Math Asn Am; Soc Indust & Appl Math. *Res:* Theory of numbers; arithmetic of polynomials and power series; combinatorial analysis; special functions. *Mailing Add:* Dept Math Duke Univ Durham NC 27706

CARLITZ, ROBERT D, b Durham, NC, June 10, 45. ELEMENTARY PARTICLE PHYSICS. *Educ:* Duke Univ, BS, 65; Calif Inst Technol, PhD(physics), 70. *Prof Exp:* Asst prof physics, Univ Chicago, 72-77; ASSOC PROF PHYSICS, UNIV PITTSBURGH, 77- *Concurrent Pos:* Vis prof, Univ Wash, 76-77, Univ Mich, 79. *Res:* Field theory. *Mailing Add:* Dept Physics 100 Allen Hall Univ Pittsburgh 4200 Fifth Ave Pittsburgh PA 15260

CARLO, WALDEMAR ALBERTO, b Mayaguez, PR, May 20, 52; m 77; c 4. CONTROL RESPIRATORY MUSCLES, PEDIATRICS-NEONATOLOGY & NEONATAL VENTILATION. *Educ:* Univ PR, BS, 75, MD, 77. *Prof Exp:* Intern pediat, Univ Children's Hosp, San Juan, PR, resident, 78-79, chief resident, 79-80; fel neonatology, Rainbow Babies' & Children's Hosp, Cleveland, 80-82; ASST PROF PEDIAT & NEONATOLOGY, CASE WESTERN RESERVE UNIV, 82- *Mem:* Soc Pediat Res; Am Acad Pediat; Am Thoracic Soc; Soc Critical Care. *Res:* Control of breathing and pulmonary physiology in neonatal infants and animals; neonatal ventiliation and high frequency ventilation; lung injury. *Mailing Add:* Div Neonatology 525 New Hillman Bldg University Sta Birmingham AL 35294

CARLOCK, JOHN TIMOTHY, b Paterson, NJ, Nov 7, 51; m 77; c 2. ORGANIC CHEMISTRY. *Educ:* Col St Rose, BA, 73; Brigham Young Univ, PhD(org chem), 77; Med Sch, Univ Okla, MD, 81. *Prof Exp:* Res scientist chem, Continental Oil Co, 77-78; ASSOC PROF MED, HEALTH SCI CTR, UNIV OKLA, 82- *Concurrent Pos:* Fel, Chem Dept, Univ Jubljana. *Honors & Awards:* Coston Award Outstanding Biomed Res, 80. *Mem:* Am Chem Soc; Am Acad Family Practrs; AMA. *Res:* Heterogeneous catalysis synthesis and implementation; heterocyclic photochemistry; organic synthesis; medicine and medicinal chemistry; clinical research. *Mailing Add:* 3819 Dinsmore Castle Dr Columbus OH 43221-4414

CARLON, HUGH ROBERT, b Camden, NJ, Oct 4, 34; m 59, 81; c 4. INFRARED SYSTEMS, CLOUD MICROPHYSICS & IONS. *Educ:* Drexel Univ, BS, 57; George Washington Univ, MS, 71; City Univ Los Angeles, PhD(physics),79. *Prof Exp:* Engr physicist, Edgewood Arsenal, Md, 58-75; physicist chem Res & Develop Lab, 75-77, sr res physicist, Chem Systs Lab, 77-82; sr exchange scientist, Chem Defense Estab, Porton Down, Eng, 82-84; US Army fel, Inst Sci & Technol, Univ Manchester, Eng, 86-87; SR RES PHYSICIST, CHEM RES DEVELOP & ENG CTR, ABERDEEN PROVING GROUND, MD, 84-86 & 87- *Concurrent Pos:* Pres & chmn, Trans-Harford Corp, Bel Air, Md, 69-75 & Harford Serv Corp, 73-78; sr exchange scientist, Chem Defense Estab, Porton Down, Eng, 82-84; army fel, US Army Res & Eng Fel Prog, 85; Inst Sci & Technol, Univ Manchester, Eng, 86-87. *Honors & Awards:* Major Gen Leslie Earl Simon Award, US Army Armament Res & Develop Command, 81; Res & Develop Award, US Army, 82. *Mem:* Optical Soc Am; Am Inst Physics; Sigma Xi; Am Inst Chem Engrs; Int Mensa Soc. *Res:* Engineering physics and atmospheric optics; electromagnetic scattering by aerosols; aerosol physics; molecular structure in atmospheric water vapor; international scientific liaison on military programs; optical countermeasures and smokes; aerosol dispersion by jet engines; chemical hazard and decontamination; chemical treaty verification; operation Desert Storm. *Mailing Add:* 2004 Fallsgrove Way Fallston MD 21047

CARLONE, ROBERT LEO, b Morristown, NJ, Sept 1, 48. DEVELOPMENTAL BIOLOGY, NEUROBIOLOGY. *Educ:* Amherst Col, BA, 70; Univ NH, MSc, 75, PhD(zool), 78. *Prof Exp:* fel, Muscular Dystrophy Asn Can, 78-80; Multiple Sclerosis Soc Can fel, McMaster Univ, 81-; AT DEPT DIOL SCI, BROCK UNIV. *Concurrent Pos:* Chmn, Third Biennial Forum on Regeneration, McMaster Univ, 81. *Mem:* Soc Develop Biol Inc; Tissue Cult Asn; Am Soc Cell Biol. *Res:* Cellular biology of regeneration; neurotrophic interactions involved in amphibian limb regeneration; cell biology and biochemistry of cellular proliferation in regenerating systems. *Mailing Add:* Dept Biol Sci Brock Univ Merrittville Hwy St Catherines ON L2S 3A1 Can

CARLOS, WILLIAM EDWARD, b Brookville, Pa, Jan 18, 49. SOLID-STATE PHYSICS, SEMI-CONDUCTORS. *Educ:* Pa State Univ, BS, 70, MS, 73, PhD(elec eng), 79. *Prof Exp:* RES PHYSICIST, NAVAL RES LABS, 79- *Mem:* Am Phys Soc. *Mailing Add:* Code 6833 Naval Res Lab Washington DC 20375

CARLOTTI, RONALD JOHN, b Martins Ferry, Ohio, Sept 20, 42; m 69; c 4. NUTRITIONAL DESIGN OF HUMAN & PET FOOD & PRODUCT DEVELOPMENT. *Educ:* Ohio State Univ, BS, 64; WVa Univ, MS, 66, PhD(nutrit biochem), 70. *Prof Exp:* Res asst biochem, Col Agr, WVa Univ, 66-70; res assoc enzym, Univ Iowa, 71-72 & 73-74, asst res scientist infant nutrit, 72-73; nutritionist, human nutrit, res dept, Kellogg Co, 74-77; mgr nutrit, res dept, Frito-Lay, Inc, 77-82, prin scientist, 82-85; sr res scientist, Nutrit Dept, Amway Corp, 85-89; DIR FOOD SCI & TECH, COUNTRY

HOME BAKERS, 90- *Mem:* Am Chem Soc; AAAS; Inst Food Technol; Am Asn Cereal Chemists. *Res:* Development of nutritionally designed new bakery products, snack foods, meat replacements and pet foods; development of nutrition labeling, review of nutrition advertising; investigations on sodium, potassium, calcium and other nutrient interactions with essential hypertension in rats; relationship of salt, fat, kind of fat, and level of fat and salt on growth and development of rats and non-human primates; isolation, purification and immunochemical characterization of lactate dehydrogenase from Morris hepatomas. *Mailing Add:* Country Home Bakers 301 28th St SE Grand Rapids MI 49548

CARLQUIST, SHERWIN, b Los Angeles, Calif, July 7, 30. BOTANY. *Educ:* Univ Calif, BA, 52, PhD(bot), 56. *Prof Exp:* Asst prof bot, Grad Sch, Claremont Cols, 56-61, assoc prof, 61-66; prof, 66-77, PROF BOT, CLAREMONT GRAD SCH & POMONA COL, 76-,; PLANT ANATOMIST, RANCHO SANTA ANA BOTANICAL GARDEN, 84- *Honors & Awards:* Gleason Prize, NY Bot Garden, 67; Career award, Bot Soc Am, 77; Int Acad Wood Sci, 87. *Mem:* Bot Soc Am; Am Soc Plant Taxon; Int Asn Wood Anat. *Res:* Comparative anatomy of flowering plants; Compositae, Rapateaceae; problems of insular floras and faunas; wood anatomy; ecological plant anatomy. *Mailing Add:* Rancho Santa Ana Bot Garden Claremont CA 91711-3101

CARLS, RALPH A, b Ringtown, Pa, Aug 9, 38; m 59; c 2. BACTERIOLOGY. *Educ:* Mansfield State Col, BS, 60; Okla State Univ, MS, 64; Univ Wis, PhD(bact), 71. *Prof Exp:* Teacher pub schs, 60-66; prof microbiol, 70-81, PROF BIOL, EDINBORO STATE COL, 81- *Mem:* Am Soc Microbiol. *Res:* Microbial ecology; contamination control and biological indicators. *Mailing Add:* Dept of Biol-Health Serv Edinboro State Col Edinboro PA 16444

CARLSEN, RICHARD CHESTER, b San Francisco, Calif, Feb 5, 40; m 69; c 2. NEUROPHYSIOLOGY. *Educ:* Univ Calif, Berkeley, AB, 63; Univ Ore, PhD(physiol), 73. *Prof Exp:* Res assoc neurophysiol, Duke Univ Med Ctr, 73-76.; asst prof, 76-82, ASSOC PROF HUMAN PHYSIOL, SCH MED, UNIV CALIF, DAVIS, 82- *Concurrent Pos:* Muscular Dystrophy Asn fel physiol & pharmacol, Duke Univ, 74-76; NIH fel, 75. *Mem:* Am Physiol Soc; Soc Neuroscience; Sigma Xi. *Res:* Development and plasticity of synaptic connections in the spinal cord; trophic relationship between nerve and muscle. *Mailing Add:* Dept Human Physiol Univ Calif Sch Med Davis CA 95616

CARLSON, A BRUCE, b Cleveland, Ohio, Jan 31, 37; m 59; c 3. ELECTRICAL & SYSTEMS ENGINEERING. *Educ:* Dartmouth Col, AB, 58; Stanford Univ, MS, 60, PhD(elec eng), 64. *Prof Exp:* ASSOC PROF ELEC ENG, RENSSELAER POLYTECH INST, 63- *Concurrent Pos:* Commun consult, NY State Dept Health, 66-68. *Mem:* Inst Elec & Electronic Engrs; Am Soc Eng Educ. *Res:* Communication systems; engineering education; systems analysis; social context of engineering; electronics. *Mailing Add:* Dept Elec & Systs Eng Rensselaer Polytech Inst 110 Eighth St Troy NY 12180

CARLSON, ALBERT DEWAYNE, JR, b Colfax, Iowa, Dec 8, 30; m 56; c 2. INVERTEBRATE PHYSIOLOGY. *Educ:* Univ Iowa, BA, 52, MS, 59, PhD(zool), 60. *Prof Exp:* From asst prof to assoc prof, 60-80, PROF NEUROBIOL, STATE UNIV NY, STONEYBROOK, 80- *Mem:* Soc Exp Biol; Soc Neuroscience. *Res:* Neural and cellular control of luminescent responses of larval and adult fireflies; neurobiology and behavior of fireflies and aquatic invertebrates. *Mailing Add:* Dept Neurobiol & Behav State Univ NY Stony Brook NY 11794

CARLSON, ALLAN DAVID, b Fargo, NDak, June 19, 39; m 60; c 4. NUCLEAR PHYSICS. *Educ:* Concordia Col, Moorhead, Minn, BA, 61; Univ Wis-Madison, MS, 63, PhD(physics), 66. *Prof Exp:* Staff assoc physics, Gen Atomic Div Gen Dynamics Corp, 66-67; staff mem, Gulf Gen Atomic, 67-70; staff scientist, Gulf Energy & Environ Systs, 70-72; NUCLEAR PHYSICIST, NAT INST STANDARDS & TECHNOL, 72- *Concurrent Pos:* Mem, US Nuclear Cross Sect Adv Comt, 70-72 & Cross Sect Eval Working Group, 72-, chmn standards, 80-; consult, Int Atomic Energy Agency, 83-85; chmn, NEANDC working group B-10 Standard, 89- *Mem:* Am Phys Soc. *Res:* Experimental nuclear physics; neutron cross sections; fluctuations in total neutron cross sections; capture cross sections; resonance parameters; light element, capture and fission cross section standards; neutron reaction mechanisms. *Mailing Add:* RADP B136 Nat Inst Standards & Technol Gaithersburg MD 20899

CARLSON, ARTHUR, JR, b Buffalo, Kans, June 5, 22; m 51; c 3. VETERINARY PHARMACOLOGY. *Educ:* Kans State Univ, DVM, 50; Univ Mo, MS, 67. *Prof Exp:* Pvt pract, Humboldt, Kans, 50-62; res assoc pharmaceut res, Haver-Lockhart Labs, Kansas City, 62-64, assoc dir pharmaceut res, 64-65, dir pharmaceut res, 65-68, vpres pharmaceut res, 68-74; dir qual assurance, Bayvet Corp, Shawnee, Kans, 74-86; pharmaceut consult, 86-87; RETIRED. *Mem:* Am Vet Med Asn. *Res:* Calcium and amino acid metabolism; endotoxin, vaccine adjuvants and prolonged-release oral medications. *Mailing Add:* 9205 W 89th Overland Park KS 66212

CARLSON, ARTHUR STEPHEN, b Brooklyn, NY, Oct 24, 19; m 41; c 2. PATHOLOGY. *Educ:* Brooklyn Col, AB, 41; Cornell Univ, MD, 52; Am Bd Path, Am Bd Nuclear Med, Nat Bd Med Exam, dipl. *Prof Exp:* Asst bacteriologist, Brooklyn Col, 41-42; instr path, Med Col, Cornell Univ, 53-56, lectr, 56-69; assoc pathologist, Community Hop, Glen Cove, 56-74, chief pathologist, 74-, attend-in-charge, Nuclear Med Sect, 81-; RETIRED. *Concurrent Pos:* From intern to resident, NY Hosp, 52-55, provisional asst, 55-57; res fel, NY Heart Asn, 55-56; pres, NY State Bd Med Examr; inspector, Inspection & Acrreditation Prog, Am Asn Blood Banks; clin asst prof path, Med Col, Cornell Univ, 69-77, clin assoc prof, 77- *Mem:* AAAS; Col Am Path; Am Asn Path & Bact; Int Acad Path. *Res:* Pathological and bacteriological biochemistry; morphological pathology. *Mailing Add:* 5002 Club Dr Farming Ridge Reading PA 19606

CARLSON, BILLE CHANDLER, b Jamaica Plain, Mass, June 27, 24; wid; c 2. ELLIPTIC INTEGRALS, HYPERGEOMETRIC FUNCTIONS. *Educ:* Harvard Univ, BA & MA, 47; Oxford Univ, PhD(physics), 50. *Prof Exp:* Staff mem radiation lab, Mass Inst Technol, 43-44; instr physics, Princeton Univ, 50-52, res assoc, 52-54; from asst prof to prof physics, Ames Lab, Iowa State Univ, 54-65, prof physics & math, 65-81, prof math, 81-90; RETIRED. *Concurrent Pos:* Sr res fel math, Calif Inst Technol, 62-63; vis, Poincaré Inst, Univ Paris, 71-72; vis prof, Inst Phys Sci & Technol, Univ Md, 80-81. *Mem:* Fel Am Phys Soc; Am Math Soc; Soc Indust & Appl Math; Math Asn Am. *Res:* Special functions, particularly elliptic integrals and R- functions. *Mailing Add:* 430 Westwood Dr Ames IA 50010

CARLSON, BRUCE ARNE, b St Paul, Minn, Apr 8, 46; m 68; c 2. ORGANIC CHEMISTRY, ORGANOMETALLIC CHEMISTRY. *Educ:* Cornell Univ, AB, 68; Purdue Univ, PhD(org chem), 73. *Prof Exp:* res chemist org chem, Cent Res & Develop Dept, 73-76, res chemist, Chem Dyes & Pigments Dept, 77-78, prod mgr, Methanol Prod Div, 78-82, dist mgr, Chem & Pigments Dept, 82-84, asst plant mgr, Memphis, 84-86, plant mgr, 86-87, bus mgr, Polymer Prod, 87-88, Am dir, Eng Polymers, 88-90, BUS ENG MGR, E I DUPONT DE NEMOURS & CO, INC, 90- *Concurrent Pos:* Dir, Norcom, E I DuPont de Nemours & Co, Inc, 88-90, DePuy- DuPont Orthopaedic, 88-90, Courtney Sanford Fund, Cornell Univ, 80- *Mem:* AAAS; Am Chem Soc; Fedn Am Scientists; Sigma Xi. *Res:* Exploratory chemistry of organoboranes and boron hydrides; heterocyclic chemistry based on HCN and unusual diazo and diazonium ion chemistry; chemistry of chromophores. *Mailing Add:* 2635 Longwood Dr Wilmington DE 19810

CARLSON, BRUCE MARTIN, b Gary, Ind, July 11, 38; m 68; c 2. ANATOMY, DEVELOPMENTAL BIOLOGY. *Educ:* Gustavus Adolphus Col, BA, 59; Cornell Univ, MS, 61; Univ Minn, MD & PhD(anat), 65. *Prof Exp:* From asst prof to assoc prof, 66-75, PROF ANAT, UNIV MICH, ANN ARBOR, 75-, PROF BIOL SCI, 79-, CHMN ANAT & CELL BIOL, 88- *Concurrent Pos:* US Acad Sci exchange fel, Inst Develop Biol, Moscow, USSR, 65-66 & 79 & Inst Physiol, Prague, Czech, 71, 72, 74, 75, 77, 78 & 79; Fulbright fel, Hubrecht Lab, Utrecht, Neth, 73-74; Josiah Macy Jr fel, Univ Helsinki, 81-82; sr fel, Fetzer Inst, Kalamazoo, Mich. *Honors & Awards:* Newcomb-Cleveland Award, AAAS, 74. *Mem:* Int Soc Develop Biol; Am Asn Anat; Am Soc Zool; Am Soc Ichthyologists & Herpetologists; Soc Develop Biol. *Res:* Limb and muscle regeneration; muscle transplantation; aging; limb development and morphogenesis. *Mailing Add:* Dept Anat & Cell Biol Univ Mich Ann Arbor MI 48109

CARLSON, CARL E, b La Crosse, Kans, June 30, 22. PETROLEUM EXPLORATION. *Educ:* Univ Wy, BA, 48, MA, 48. *Prof Exp:* Geol Consult, Mobil Oil, 49-82; INDEPENDENT CONSULT, 82- *Mem:* Fel Geol Soc Am; Am Soc Petrol Geologists. *Mailing Add:* 4734 Oporto Ct San Diego CA 92124

CARLSON, CHARLES MERTON, b Duluth, Minn, Apr 3, 34; wid. THEORETICAL CHEMISTRY. *Educ:* Univ Calif, Riverside, AB, 56; Univ Utah, PhD(chem), 60. *Prof Exp:* Staff scientist, Metall Dept, Brookhaven Nat Lab, NY, 60-62; res specialist, Solid State Physics Group, Boeing Sci Res Labs, 62-70; VIS SCHOLAR, COMPUT SCI GROUP, UNIV WASH, 71- *Mem:* Am Phys Soc; Am Chem Soc; Soc Indust & Appl Math. *Res:* Basic theoretical research in quantum chemistry and statistical mechanics as applied to molecules and solids. *Mailing Add:* 10539 13th Ave NW Seattle WA 98177

CARLSON, CHARLES WENDELL, b Eaton, Colo, May 15, 21; m 43; c 5. ANIMAL NUTRITION. *Educ:* Colo State Univ, BS, 42; Cornell Univ, MSA, 48, PhD(animal nutrit), 49. *Prof Exp:* Asst poultry husb, Cornell Univ, 46-49; from asst prof to prof poultry sci, 49-85, leader poultry res, 67-85, EMER PROF, SDAK STATE UNIV, 85- *Concurrent Pos:* Res assoc, Wash State Univ, 62-63; prin non-ruminant nutritionist, Coop States Res Serv, USDA, Washington, DC, 75-76; Exec vol, Brazil, 81, Dominican Repub, Int Exec Serv Corps, Stamford, Conn, 83 & Port Cur Prince, Haiti, 89; vis prof, Univ Minn, 89-90. *Honors & Awards:* Nat Turkey Fedn Res Award, 61; F O Butler Res Award, 84. *Mem:* Fel AAAS; Am Chem Soc; fel Poultry Sci Asn (pres, 70-71); Am Inst Nutrit. *Res:* Poultry; biochemistry; amino acid and energy metabolism; growth factors; calcification. *Mailing Add:* Dept Animal & Range Sci SDak State Univ Brookings SD 57007-0392

CARLSON, CLARENCE ALBERT, JR, b Moline, Ill, Apr 3, 37; m 58; c 2. FISH BIOLOGY, AQUATIC ECOLOGY. *Educ:* Augustana Col, Ill, AB, 58; Iowa State Univ, MS, 60, PhD(zool), 63. *Prof Exp:* Asst prof biol, Augustana Col, Ill, 62-66; asst prof fishery biol & asst leader NY Coop Fishery Unit, Cornell Univ, 66-72; assoc prof, 72-78, PROF FISHERY BIOL, COLO STATE UNIV, 78- *Concurrent Pos:* Adminr, Colo State Univ Larval Fish Lab, 79-; pres educr sect, Am Fisheries Soc, 83-84. *Mem:* Am Fisheries Soc; Ecol Soc Am; Am Inst Biol Sci. *Res:* Ecology of fishes; river ecology; larval fish biology; environmental effects of energy development. *Mailing Add:* Dept Fishery & Wildlife Biol Colo State Univ Ft Collins CO 80523

CARLSON, CLINTON E, b Tacoma, Wash, July 27, 42; m 68; c 3. SILVICULTURE & CONIFER FORESTS, AIR POLLUTION & FOREST ECOSYSTEMS. *Educ:* Univ Mont, BS, 64, MS, 66, PhD(bot), 78. *Prof Exp:* Plant pathologist, 66-78, RES FORESTER, FOREST SERV, USDA, 79- *Concurrent Pos:* Adj fac, Univ Mont, 72-; pres, Montana Acad Sci, 85-86. *Honors & Awards:* Blue Pencil Award, USDA, 74. *Mem:* Soc Am Foresters. *Res:* Relationships of insects and diseases to their forest environment and the development of silvicultural strategies to deal with those insects and diseases; effect of air pollutants on the forest. *Mailing Add:* NW 600 Old Bentham Lane Florence MT 59833

CARLSON, CURTIS RAYMOND, b Providence, RI, May 22, 45; m 73. VISION, IMAGE PROCESSING. *Educ:* Worcester Polytech, BS, 67; Rutgers Univ, MS, 69, PhD(mech eng), 73. *Prof Exp:* Mem tech staff, 73-81, HEAD IMAGE QUALITY & HUMAN PERCEPTION RES, SARNOFF

RES LABS, RCA, 81- *Mem:* Sigma Xi; Optical Soc Am; Asn Res Vision & Ophthalmology. *Res:* Human visual perception; display image quality; image processing. *Mailing Add:* David Sarnoff Res Ctr Rte 1 Princeton NJ 08540-5300

CARLSON, DALE ARVID, b Aberdeen, Wash, Jan 10, 25; m 48; c 4. SANITARY ENGINEERING, CIVIL ENGINEERING. *Educ:* Univ Wash, BSCE, 50, MSCE, 51; Univ Wis, PhD(civil eng), 60. *Prof Exp:* Engr, City Water Dept, Aberdeen, 51-55; from instr to prof civil eng, Univ Wash, 55-80, chmn dept, 71-76, dean, Col Eng, 76-80 dir solid wastes mgt prog, 69-80, EMER PROF & DEAN, DEPT CIVIL ENG, UNIV WASH, 80-; DEAN, SCH SCI & ENG, SEATTLE UNIV, 90- *Concurrent Pos:* Asst, Wis Alumni Res Found, 60; co-prin investr, US Corps Engrs, states of Ore & Wash, Northwest Pulp & Paper Asn, 60-61 & City of Anacortes, 63-64; prin investr, USPHS, 61-65, Metrop Engrs, Wash, 63-64, Munic of Metrop Seattle, 64-65 & Northwest Pulp & Paper Asn, 64-65; mem, Wash State Bd Health, 64-72, chmn, 71-72; vis prof, Tech Univ Denmark, Copenhagen, 70; vis scientist, Royal Col Agr, Uppsala, Sweden, 76 & 78. *Mem:* Am Soc Civil Engrs; Water Pollution Control Fedn; Am Water Works Asn; Am Asn Prof Sanit Engrs; Int Asn Water Pollution Res; Sigma Xi; Swed Hygiene Asn. *Res:* Biological treatment of domestic and industrial wastes; water quality; solid wastes management; odor removal by soil columns; disinfection for drinking water. *Mailing Add:* Valle Scholar & Scand Exchange Prog 103 Wilson FC-07 Seattle WA 98195

CARLSON, DANA PETER, b Red Wing, Minn, Oct 31, 31; m 56; c 4. ORGANIC CHEMISTRY. *Educ:* Univ Minn, BS, 53; Carnegie Inst Technol, MS, 56, PhD(chem), 57. *Prof Exp:* From res chemist to sr res chemist, 57-71, res assoc, 71-81, RES FEL, E I DU PONT DE NEMOURS & CO, INC, 81- *Mem:* Am Chem Soc. *Res:* Fluorocarbon chemistry; synthesis of monomers and polymers. *Mailing Add:* 13 Brook Lane Chadds Ford PA 19317

CARLSON, DAVID ARTHUR, b Muskegon, Mich, Nov 28, 40; m 63; c 2. ORGANIC CHEMISTRY, ANALYTICAL CHEMISTRY. *Educ:* Kalamazoo Col, BA, 62; Univ Hawaii, MS, 66, PhD(phys org chem), 70. *Prof Exp:* Res chemist, Anaconda Wire & Cable Res Lab, 62-63; RES CHEMIST, USDA, 69- *Mem:* Am Chem Soc; AAAS; Entom Soc Am. *Res:* Biologically active natural products; identification of insect sex attractant hormones and stimulant pheromones in mosquitoes, houseflies and tsetse flies; pesticide analysis; gas chromatography; mass spectrometry. *Mailing Add:* 731 NW 91st St Gainesville FL 32607-1325

CARLSON, DAVID EMIL, b Weymouth, Mass, Mar 5, 42; m 66; c 2. SOLID STATE PHYSICS. *Educ:* Rensselaer Polytech Inst, 63; Rutgers Univ, PhD(physics), 68. *Prof Exp:* Res scientist physics, US Army Nuclear Effects Lab, mem tech staff physics, 70-76; Head photovoltaic device, 77-83, dir res, 83-86, gen mgr, 86-87, VPRES, SOLAREX THIN FILM DIV, RCA LABS, NJ, 87- *Honors & Awards:* Ross Coffin Purdy Award, Am Ceramic Soc, 76; Morris N Liebmann Award, Inst Elec & Electronic Engrs, 84, William R Cherry Award, 88; Walton Clark Medal, Franklin Inst, 86. *Mem:* Am Phys Soc; Inst Elec & Electronics Engrs; Am Vacuum Soc; Sigma Xi. *Res:* Thin film photovoltaic devices. *Mailing Add:* One Buckingham Dr Princeton NJ 08540

CARLSON, DAVID HILDING, b Chicago, Ill, Nov 10, 36; div; c 4. MATHEMATICS. *Educ:* San Diego State Col, AB, 57; Univ Wis, MS, 59, PhD(math), 63. *Prof Exp:* Instr math, Univ Wis, Milwaukee, 62-63; from asst prof to prof math, Ore State Univ, 63-84; AT MATH SCI DEPT, SAN DIEGO STATE UNIV, 84- *Concurrent Pos:* Fulbright-Hays lectr, Univ of Repub Uruguay, 65-66; vis prof, Kent State Univ, 70-71 & Univ Coimbra, Port, 77-78, St Univ Campinas, Braz, 78-79. *Mem:* Am Math Soc; Math Asn Am; Soc Indust Appl Math; Nat Coun Teachers Math. *Res:* Matrix theory, especially inertia theory and location of eigenvalues. *Mailing Add:* Math Sci Dept San Diego State Univ San Diego CA 92182-0314

CARLSON, DAVID L, b Minneapolis, Minn, Sept 3, 36. ELECTRICAL & BIOMEDICAL ENGINEERING. *Educ:* Univ Minn, BSEE, 59; Iowa State Univ, MS, 61, PhD(elec eng), 64. *Prof Exp:* ASSOC PROF ELEC ENG, IOWA STATE UNIV, 64- *Mem:* Inst Elec & Electronics Engrs. *Res:* Electronic instrumentation; respiratory assistance for newborn infants; medical ultrasound. *Mailing Add:* 3430 Southdale Dr Ames IA 50010

CARLSON, DAVID STEN, b New Bedford, Mass, Aug 30, 48; m 70; c 3. ANATOMY, DENTAL RESEARCH. *Educ:* Univ Mass, Amherst, BA, 70, MA, 72, PhD(anthrop), 74. *Prof Exp:* Teaching fel anthrop, Univ Mass, Amherst, 70-74, lectr, 73-74; asst prof anthrop & anat, Wayne State Univ, 74-75; postdoctoral fel, Univ Mich, 75-78, from asst prof to assoc prof anat & cell biol, 78-88, from asst prof to assoc prof anthrop, 78-88, from asst res scientist to assoc res scientist, Ctr Human Growth & Develop, 78-88, assoc prof orthod, 83-88, PROF ANTHROP, 88-, PROF ORTHOD, 88- & RES SCIENTIST, CTR HUMAN GROWTH & DEVELOP, 88- *Concurrent Pos:* Marshall fel, Am Scand Found, 73; assoc, Sch Med, Wayne State Univ, 74-76; scholar, Univ Mich, 75-77. *Mem:* AAAS; Am Asn Phys Anthropologists; Human Biol Coun. *Res:* Analysis of the phylogenetic and ontogenetic factors influencing the form and function of the craniofacial complex. *Mailing Add:* Univ Mich 6217 Dental Ann Arbor MI 48109-1078

CARLSON, DON MARVIN, b Walhalla, NDak, Mar 11, 31; m 51; c 4. GLYCOPROTEINS, GENE EXPRESSION. *Educ:* NDak State Univ, BS, 56; Univ Ill, MS, 58; Mich State Univ, PhD(biochem), 61. *Prof Exp:* Instr, Univ Mich, 63-64; from asst prof to prof biochem, Case Western Reserve Univ, 64-75; head dept biochem, Purdue Univ, 75-81, prof, 81-85; chmn, 85-90, PROF DEPT BIOCHEM & BIOPHYS, UNIV CALIF, DAVIS, 85- *Concurrent Pos:* NIH Physiol Chem Study Sect, 76-80, 83-85; mem, Med Adv Coun, Cystic Fibrosis Found, 86-90. *Mem:* Am Soc Biochem & Molecular Biol (secy, 84-87); Soc Complex Carbohydrates (secy, 74-80, pres, 80-81). *Res:* Chemistry and biosynthesis of glycoproteins; regulation of gene expression. *Mailing Add:* Dept Biochem & Biophys Univ Calif Davis CA 95616

CARLSON, DOUGLAS W, b Jamestown, NDak, Jan 7, 30; m 54; c 2. PHYSICAL CHEMISTRY. *Educ:* Univ NDak, BS, 55; Univ Del, MS, 57, PhD(phys chem), 59. *Prof Exp:* Fel, Princeton Univ, 59-61; res chemist, Minn Mining & Mfg Co, Minn, 61-64; res chemist, E I du Pont de Nemours & Co, Del, 64-67, tech rep, 67-70; dir res, Springfield Res Ctr, Dayco Corp, 70-79; sr vpres, Bandag Inc, 80-82; tech mgr, 82-86, GEN MGR, HYDRIL CO, 86- *Concurrent Pos:* Post doctoral fel, Princeton Univ, 59-61. *Mem:* Am Chem Soc. *Res:* Polymer chemistry; photochemistry; statistics; quality control. *Mailing Add:* 5303 Manor Glen Dr Kingwood TX 77345-1440

CARLSON, DREW E, b Rochester, NY, Feb 9, 48; m 81; c 1. ENDOCRINE & AUTONOMIC PHYSIOLOGY. *Educ:* Cornell Univ, BS, 70; Johns Hopkins Univ, PhD(biomed eng), 76. *Prof Exp:* Fel biomed eng & endocrine physiol, Johns Hopkins Univ, 76-78, asst prof, 78-89; ASST PROF SURG & NEUROBIOL, RI HOSP, BROWN UNIV, 79- *Mem:* AAAS; NY Acad Sci; Endocrine Soc; Soc Neuroscience; Am Physiol Soc; Soc Values Higher Educ. *Res:* Neuroendocrinology and central neural pathways controlling the release of adrenocorticotropin; cardiovascular and metabolic response to trauma; cardiovascular receptors that mediate endocrine and autonomic response to trauma; endocrine and metabolic components in the restitution of blood volume after hemorrhage; modeling of physiological systems. *Mailing Add:* Univ Md MSTF 4-00 Ten S Pine St Baltimore MD 21201

CARLSON, EDWARD H, b Lansing, Mich, Apr 29, 32; m 60; c 3. SOLID STATE PHYSICS. *Educ:* Mich State Univ, BS, 54, MS, 56; Johns Hopkins Univ, PhD(physics), 59. *Prof Exp:* Instr physics, Johns Hopkins Univ, 59-60; NSF fel, State Univ Leiden, 60-61; asst prof, Univ Ala, 61-65; from asst prof to assoc prof, 65-74, PROF PHYSICS, MICH STATE UNIV, 74- *Mem:* Am Phys Soc. *Res:* Rare earth spectra; electron and nuclear magnetic spin resonance; ordered magnetic states; atomic and molecular structure. *Mailing Add:* 3872 Raleigh Dr Okemos MI 48864

CARLSON, ELOF AXEL, b Brooklyn, NY, July 15, 31; m 59; c 5. HISTORY & PHILOSOPHY OF SCIENCE. *Educ:* NY Univ, BA, 53; Ind Univ, PhD, 58. *Prof Exp:* Lectr genetics, Ind Univ, 57-58 & Queen's Univ, 58-60; asst prof zool, Univ Calif, Los Angeles, 60-65, assoc prof, 65-68; prof biol, 68-75, DISTINGUISHED TEACHING PROF BIOL, STATE UNIV NY, STONY BROOK, 75- *Concurrent Pos:* Distinguished vis prof, San Diego State Col, 69, Univ Minn, 74, Univ Utah, 84, Toogaloo Col, 87. *Mem:* Fel AAAS; Genetics Soc Am; Sigma Xi; Am Soc Human Genetics. *Res:* Gene structure and function; chemical and radiation induced mutagenesis; mosaicism in Drosophila and man; history of genetics. *Mailing Add:* Biochem Dept State Univ of NY Stony Brook NY 11794

CARLSON, EMIL HERBERT, b Portland, Ore, Oct 25, 29; m 51; c 4. ORGANIC CHEMISTRY. *Educ:* Willamette Univ, BS, 51; Carnegie Inst Technol, MS, 54, PhD(chem), 56. *Prof Exp:* Asst chem lab, Carnegie Inst Technol, 51-52, asst org qual lab, 52-53, sr asst, 53-54; res chemist, 5S-64, prod specialist org div, Mo, 64-65, mkt analyst, 65-70, res specialist, 70-76, PRIN PROCESS SPECIALIST, AGR DIV, MONSANTO CO, 76- *Concurrent Pos:* Instr org chem, Muscatine Community Col, 79-80. *Mem:* Am Chem Soc; NY Acad Sci; Sigma Xi. *Res:* Agricultural chemicals; functional fluids; intermediate chemicals. *Mailing Add:* Monsanto Co PO Box 473 Muscatine IA 52761-0473

CARLSON, ERIC DUNGAN, b Kansas City, Mo, Jan 19, 29; m 70; c 2. ASTRONOMY. *Educ:* Wash Univ (St Louis), AB, 50; Northwestern Univ, MS, 65, PhD(astron), 68. *Prof Exp:* Mgt trainee printing, R R Donnelley & Sons Co, 56-58; asst to ed-in-chief encycl writing, Consol Bk Publs, 58-61; SR ASTRONR, ADLER PLANETARIUM, 68- *Mem:* Am Astron Soc. *Res:* Peculiar emissionline stars. *Mailing Add:* 2741 Lawndale Evanston IL 60201

CARLSON, ERIC THEODORE, b Westbrook, Conn, Aug 22, 22; m 50; c 1. PSYCHIATRY. *Educ:* Wesleyan Univ, BA, 44; Cornell Univ, MD, 50. *Prof Exp:* Intern int med, Hosp, 50-51, asst res psychiat 51-55, psychiatrist outpatients, 55-56, asst attend psychiatrist, 52-60, assoc attend psychiatrist, 60-70, ATTEND PSYCHIATRIST, HOSP, 70-; CLIN PROF PSYCHIAT, MED COL, CORNELL UNIV, 70- *Concurrent Pos:* Asst psychiat, Med Col, Cornell Univ, 52-53, instr, 53-58, from clin asst prof to clin assoc prof, 58-70. *Mem:* AAAS; Am Psychiat Asn; Am Col Psychiatrists; Int Soc Hist Med; NY Acad Med; Sigma Xi. *Res:* Development of psychiatric thought. *Mailing Add:* Cornell Med Ctr NY Hosp 525 E 68th St New York NY 10021

CARLSON, ERNEST HOWARD, b Seattle, Wash, Dec 23, 33; m 70; c 2. EXPLORATION GEOCHEMISTRY, MINERALOGY. *Educ:* Univ Wash, BSc, 56; Univ Colo, MSc, 60; McGill Univ, PhD(crystal growth), 66. *Prof Exp:* Geologist, US Geol Surv, 58-60; instr geol, Villanova Univ, 65-66; asst prof, 66-75, ASSOC PROF GEOL, KENT STATE UNIV, 75- *Concurrent Pos:* Vis prof, Pahlavi Univ, Shiraz, Iran, 70-71. *Mem:* Am Crystallog Asn; Geol Soc Am; Mineral Soc Am; Sigma Xi; Asn Explor Geochemists. *Res:* Mineralogy; exploration geochemistry; economic geology. *Mailing Add:* Dept Geol Kent State Univ Kent OH 44242

CARLSON, F ROY, JR, b Boston, Mass, Aug 1, 44; m 67. COMPUTER SCIENCE. *Educ:* Univ Rochester, BS, 67; Univ Southern Calif, PhD, 70. *Prof Exp:* Asst prof elec eng & comput sci, 70-76, ASST DEAN, SCH ENG & DIR ENG COMPUT LAB, UNIV SOUTHERN CALIF, 76- *Mem:* Inst Elec & Electronics Engrs; Asn Comput Mach. *Res:* Design of computer based instructional systems, utilizing natural language processing and artificial intelligence techniques for the presentation of instructional material. *Mailing Add:* Sch Eng Comput Sci Sal 200 Univ of Southern Calif Los Angeles CA 90089

CARLSON, FRANCIS DEWEY, b Syracuse, NY, June 29, 21; m 50; c 3. ANIMAL PHYSIOLOGY. *Educ:* Johns Hopkins Univ, AB, 42; Univ Pa, PhD(biophys), 49. *Prof Exp:* Asst, Electroacoustics Lab, Harvard Univ, 42-43, res assoc, 43-46; asst, Univ Pa, 46-49; from instr to assoc prof biophys, 49-60, chmn dept, 56-74, prof biophys, 60- PROF BIOPHYSICS & BIOL

JOHNS HOPKINS UNIV. *Concurrent Pos:* NSF sr fel, 61-62. *Mem:* AAAS; Am Phys Soc; Biophys Soc. *Res:* Mechano-chemistry of muscular contraction; molecular biology; neurophysiology. *Mailing Add:* Dept of Biophys Johns Hopkins Univ Baltimore MD 21218

CARLSON, FREDERICK PAUL, b Aberdeen, Wash, May 26, 38; m 70; c 4. OPTICS, ELECTROMAGNETICS. *Educ:* Univ Wash, BSEE, 60, PhD(elec eng), 67; Univ Md, MS, 64. *Prof Exp:* Instr & control engr elec eng, US Atomic Energy Comn,US Navy, 60-64; res engr, Boeing Aerospace Co, 65-66; from asst prof to prof, Univ Wash, 67-77; vpres, 77-79, actg pres, 77-80, pres, Ore Grad Ctr for Study & Res, 80-87, PROF APPL PHYSICS, ORE GRAD CTR CORP, 77-, PRES, 85- *Concurrent Pos:* Vis scholar, Pac Lutheran Theol Sem, Berkeley, 75-76; vis prof, Stanford Univ, 75-76; consult, Naval Res Adv Comt, 76-79. *Mem:* Optical Soc Am; sr mem Inst Elec & Electronics Engrs; Am Geophys Union. *Res:* Optical data processing; optical computing; biomedical applications of optical data processing and pattern recognition. *Mailing Add:* 11668 Tanglewood Dr Eden Prairie MN 55347

CARLSON, GARY, b Los Angeles, Calif, Mar 6, 28; m 54; c 5. COMPUTER SCIENCES, INTELLIGENT SYSTEMS. *Educ:* Univ Calif, Los Angeles, BS, 56, MA, 58, PhD(indust psychol), 62. *Prof Exp:* Res asst, Western Data Processing Ctr, Univ Calif, Los Angeles, 58-59; sr opers res specialist, Indust Dynamics Dept, Hughes Aircraft Co, Calif, 59-61; with, Info Systs, Inc, 61-63; dir, Comput Res Ctr, Brigham Young Univ, 63-70, dir comput serv, 70-79; pres, Comput Translation Inc, 79-85; PROF INFO MGT, BRINGHAM YOUNG UNIV, 85- *Concurrent Pos:* Mem, Utah State Citizens Comt Rev State Comput Orgn, 68-69. *Mem:* Asn Comput Mach; AAAS. *Res:* Computer administration; computer performance measurement and monitoring. *Mailing Add:* 3289 Mohawk Circle Provo UT 84604

CARLSON, GARY ALDEN, b Hastings, Nebr, Apr 18, 41; m 62, 81; c 3. PHYSICAL CHEMISTRY. *Educ:* Univ Idaho, BS, 63; Univ Calif, Berkeley, PhD(phys chem), 66. *Prof Exp:* SR MEM TECH STAFF PHYS CHEM, SANDIA LABS, 66- *Mem:* Am Chem Soc. *Res:* Coal liquefaction; computer-aided molecular design; pulsed electron beam energy deposition and diagnostics development; medical devices development; liquid metal fast breeder reactor safety research. *Mailing Add:* 10417 A 4th St NE Albuquerque NM 87114

CARLSON, GARY P, b Buffalo, NY, Feb 21, 43; m 68; c 4. TOXICOLOGY, PHARMACOLOGY. *Educ:* St Bonaventure Univ, BS, 65; Univ Chicago, PhD(pharmacol), 69. *Prof Exp:* From asst prof to assoc prof pharmacol, Univ RI, 69-75; assoc prof, 75-80, PROF TOXICOL, PURDUE UNIV, 80-, ASSOC HEAD DEPT, 83- *Mem:* Soc Toxicol; Am Soc Pharmacol & Exp Therapeut; Am Indust Hyg Asn; Soc Risk Analysis. *Res:* Toxicity of pesticides, drugs, chemicals; drug metabolism. *Mailing Add:* Dept Pharmacol & Toxicol Purdue Univ West Lafayette IN 47907

CARLSON, GARY WAYNE, b Idaho Falls, Idaho, Oct 10, 44; m 72; c 4. COMPUTATIONAL PHYSICS, MONTE CARLO TRANSPORT. *Educ:* Univ Utah, BA, 68, PhD(physics), 73. *Prof Exp:* PHYSICIST NUCLEAR PHYSICS, LAWRENCE LIVERMORE LAB, 72- *Res:* Fission cross section measurements; Monte Carlo neutron transport. *Mailing Add:* 817 Hazel St Livermore CA 94550

CARLSON, GEORGE A, b Philadelphia, Pa, 1947; m 74; c 1. IMMUNOGENETICS. *Educ:* Univ PA, BA, 69, Tufts Univ, PhD(physiol), 76. *Prof Exp:* Asst prof, Dept Immunol, Univ Alta, 77-80; staff scientist, Jackson Lab, 80-88; DIR, MCLAUGHLIN RES INST, 88- *Concurrent Pos:* Mem, Immunol & Transplantation Grants Comt, 81-83, Med Res Coun Can, 84-88, Clin Sci Study Sect, NIH. *Mem:* Genetics Soc Am; Am Soc Microbiol; Am Asn Immunol. *Res:* Genetic influences on susceptibility to neurodegenerative disease; hybrid resistence; mitotic recombination in mammalian cells. *Mailing Add:* McLaughlin Res Inst 1625 Third Ave N Great Falls MT 59401

CARLSON, GERALD EUGENE, b Wausa, Nebr, Oct 28, 32; m 54; c 3. ENVIRONMENTAL PHYSIOLOGY, AGRONOMY. *Educ:* Iowa State Univ, BS, 58, MS, 59; Pa State Univ, PhD(forage physiol), 63. *Prof Exp:* Res agronomist, Crop Res Div, Agr Res Serv, USDA, 63-66, res leader, Humid Pasture & Range Invests, 66-72, lab chief, Light & Plant Growth Lab, 72-79, Nat Res Prog leader, Beltsville Agr Res Ctr, 79-83, assoc dir, N Atlantic Area, 83-88; CHMN AGR DEPT, WESTERN ILL UNIV, 88- *Concurrent Pos:* Vis foreign scientist, Japanese Inst Sci & Technol, 70. *Mem:* Fel Am Soc Agron; fel Crop Sci Soc Am. *Res:* Identification of factors limiting light utilization by plants including the effect of temperature and light on photosynthesis, respiration and translocation; research administration. *Mailing Add:* 405 Knoblauch Hall Western Ill Univ Macomb IL 61455

CARLSON, GERALD LOWELL, biological chemistry, for more information see previous edition

CARLSON, GERALD M, b York, Pa, Aug 8, 41. PHARMACOLOGY, PHYSIOLOGY. *Educ:* Univ Pittsburgh, BS, 63; Univ Mich, PhD(pharmacol), 69. *Prof Exp:* Asst prof, Dept Physiol, Univ Pa, 71-76; assoc prof, Dept Pharmacol, Univ Fla, 76-81; ASSOC DIR PROF SERV, SCHERING CORP, 82- *Mem:* Am Physiol Soc. *Res:* Medical/scientific aspects of pharmaceutical marketing (oncology, anti-infectives, gastroenterology, cardiovascular). *Mailing Add:* Schering Corp 2000 Galloping Hill Rd K-5-1 Kenilworth NJ 07033

CARLSON, GERALD MICHAEL, b Clayton, Wash, Sept 26, 46; m 67; c 2. BIOCHEMISTRY. *Educ:* Wash State Univ, BS, 69; Iowa State Univ, PhD(biochem), 75. *Prof Exp:* Res assoc, Inst Enzyme Res, Univ Wis, 75-78; asst prof biochem, Dept Chem, Univ SFla, 78-82; assoc prof biochem, Univ Miss Med Ctr, 83-86; assoc prof, 86-88, PROF BIOCHEM, UNIV TENN, MEMPHIS, 88- *Concurrent Pos:* NIH fel, 76-78. *Mem:* Am Soc Biochem & Molecular Biol; Am Chem Soc. *Res:* Enzymology and protein chemistry. *Mailing Add:* 1432 Harbert Memphis TN 38104

CARLSON, GLENN RICHARD, b New Haven, Conn, Oct 22, 45; m 67; c 2. SYNTHETIC ORGANIC CHEMISTRY. *Educ:* Bates Col, BS, 67; Mich State Univ, PhD(org chem), 71. *Prof Exp:* Res assoc org chem, Univ Chicago, 71-73; res chemist, 73-81, SECT MGR, ROHM & HAAS CO, 81- *Mem:* Am Chem Soc. *Res:* Synthesis of biologically active molecules; development of new synthetic methods in organic chemistry. *Mailing Add:* 305 Britt Rd N Wales PA 19454-2417

CARLSON, GORDON ANDREW, b Jamestown, NY, Jan 11, 17; m 41; c 3. INORGANIC CHEMISTRY. *Educ:* Wittenberg Col, AB, 40; Ohio State Univ, PhD(chem), 44. *Prof Exp:* Res chemist, Columbia Chem Div, Pittsburgh Plate Glass Co, Ohio, 44-50, asst chief chemist, WVa, 50-51 & Columbia-Southern Chem Co, 51-55, asst dir res, 55-62; dept head inorg res, Barberton Chem Tech Ctr, Pittsburgh Plate Glass Co, 62-65, sr res assoc, Chem Div, 65-82; RETIRED. *Mem:* Am Chem Soc; Electrochemistry Soc. *Res:* Chlor-alkali cells; coordination compounds; rare earths; alkalies and chlorine; industrial electrochemistry; metal halides and inorganic chlorinations; chemical metallurgy. *Mailing Add:* 590 Foxwood Blvd Englewood FL 34223-3788

CARLSON, H(ENNING) MAURICE, b Moline, Ill, Feb 13, 16; m 43; c 4. ENVIRONMENTAL SCIENCES. *Educ:* Univ Minn, BS, 39, BME, 43; Univ Louisville, MME, 55; Rutgers Univ, MSc, 75. *Prof Exp:* Teacher, High Sch, Minn, 39-41; asst, Univ Minn, 41-42, instr math, 43-44; res engr, Fuels Div, Battelle Mem Inst, 44-49; from asst prof to assoc prof mech eng, Univ Louisville, 49-57; prof, 57-59, dir eng, 59-62, head, Dept Mech Eng, 57-78, prof, 78-81, EMER PROF MECH ENG, LAFAYETTE COL, 81- *Concurrent Pos:* Mem, Student Develop Comt, Engrs Coun Prof Develop, 60-62, chmn, 63-66, mem ethics comt, 66-72; mem, Northampton County Solid Waste Authority, 80-, bd trutees, Easton YWCA, 81-, bd dirs, shared Housing Resource Ctr, 81-89, Lehigh Valley Community Coun, 82-85, Lehigh Valley adv coun, United Way, 85-91; consult, 81-; coordr, chair prof search comt, Econ Dept, Lafayette Col, 82-83; adv, Lafayette Eve Prog Div, 84-91. *Honors & Awards:* Lincoln Arc Welding Award, 47. *Mem:* Am Soc Mech Engrs; Am Soc Eng Educ; Nat Soc Prof Engrs. *Res:* Engineering education and administration; thermodynamics; fluid mechanics; machine design; consult energy conservation and alternatives; air pollution and control; educational administration. *Mailing Add:* Dept of Mech Eng Lafayette Col Easton PA 18042

CARLSON, HAROLD C(ARL) R(AYMOND), mechanical engineering & metallurgy; deceased, see previous edition for last biography

CARLSON, HAROLD ERNEST, b Staten Island, NY, May 17, 43; m 66; c 2. ENDOCRINOLOGY, METABOLISM. *Educ:* Rensselaer Polytech Inst, BS, 64; Cornell Univ, MD, 68. *Prof Exp:* Intern med, Barnes Hosp, St Louis, Mo, 68-69, resident, 69-70; clin assoc endocrinol, Nat Inst Arthritis & Metab Dis, NIH, Bethesda, Md, 70-72; fel metab, Sch Med, Wash Univ, 72-74; from asst prof to assoc prof med, Sch Med, Univ Calif, Los Angeles, 74-82; from assoc to prof, Sch Med, Univ Mo, Columbia, 82-85; asst chief endocrinol, Wadsworth Vet Admin Hosp, Los Angeles, 74-82; chief, H S Truman Vet Admin Hosp, Columbia, Mo, 82-85; CHIEF ENDOCRINOL, NORTHPORT VET ADMIN HOSP, NY, 85-; PROF MED, STATE UNIV NY SCH MED, STONY BROOK, 85- *Mem:* Endocrine Soc; fel Am Col Physicians; Cent Soc Clin Res; Am Fedn Clin Res; Sigma Xi. *Res:* Regulation of pituitary prolactin secretion and effects of pituitary hormones on calcium metabolism. *Mailing Add:* Med Serv Northport Vet Admin Hosp Northport NY 11768

CARLSON, HARRY WILLIAM, b Philadelphia, Pa, Oct 11, 24; m 45; c 2. AERODYNAMICS, FLUID MECHANICS. *Educ:* Va Polytech Inst, BS, 52. *Prof Exp:* Engr, Langley Res Ctr, NASA, 52-68, sect head aeronaut, Langley Res Ctr, 68-80. *Concurrent Pos:* Mem sonic boom res panel, Interagency Noise Abatement Prog, 68; consult, Kentron Int, 80-88. *Honors & Awards:* Awards, NASA, 66, 68 & 80. *Mem:* Am Inst Aeronaut & Astronaut. *Res:* Design and evaluation of supersonic aircraft aerodynamic configurations, sonic boom generation phenomena and development of prediction and minimization techniques; design and analysis of subsonic wings and flap systems. *Mailing Add:* 122 Dogwood Dr Newport News VA 23606

CARLSON, HARVE J, bacteriology, virology, for more information see previous edition

CARLSON, HERBERT CHRISTIAN, JR, b Brooklyn, NY, May 10, 37; div; c 2. ATMOSPHERIC PHYSICS, SPACE SCIENCE. *Educ:* Cooper Union, BEE, 59; Cornell Univ, MSc, 62, PhD(radio propagation), 65. *Prof Exp:* Res assoc/sr res assoc, Arecibo Observ, Nat Astronomy & Ionosphere Ctr, Cornell Univ, 65-69; res assoc, Dept Space Physics, Rice Univ, 69-70, Dept Geol & Geophys, Yale Univ, 70; head ionosphere dept, Arecibo Observ, Nat Astronomy & Ionosphere Ctr, Cornell Univ, 70-73; sr res scientist atmospheric physics, Inst Phys Sci, Univ Tex, Dallas, 73-77; prog dir aeronomy, NSF, 77-81; chief, ionospheric physics br, 81-88, ACTG CHIEF SCIENTIST, AIR FORCE GEOPHYS LAB, 88- *Concurrent Pos:* Mem, Comt Solar-Terrestrial Res Panel, Nat Acad Sci, 72-75; mem, Sci Adv Comt, Stanford Res Inst Int, 72-77, consult, 72-77; coordr, Incoherant Scattering Working Group, Int Union Radio Union, 73-77; mem, Sci Adv Comt, Nat Astronomy & Ionosphere Ctr, 74-77; mem, Stratosphere Working Group Subcomt, Intergovt Comt Atmosphere Sci, 77-79; mem, Mil Space Syst Tech Plan, 82-84; chair, Air Force Geophys Lab, Civ Policy Bd, 85-86. *Mem:* Am Geophys Union; AAAS; Int Union Radio Sci; Am Inst Aeronaut & Astron. *Res:* Upper atmospheric, radio, and space physics; application of radio and optical techniques, and in situ measurements, to studies of the ionosphere/thermosphere especially thermal balance, chemistry and transport; areas of plasma physics. *Mailing Add:* Air Force Geophys Lab Hanscom AFB MA 01731

CARLSON, IRVING THEODORE, b Colbert, Wash, July 14, 26; m 52; c 2. BREEDING & EVALUATION OF FORAGE CROPS. *Educ:* Wash State Univ, BS, 50, MS, 52; Univ Wis, PhD(agron), 55. *Prof Exp:* Asst wheat breeding, Wash State Univ, 50-52; asst forage grass breeding, Univ Wis, 52-55; asst wheat breeding, Wash State Univ, 55-56; asst prof field crops, NC State Col, 56-60; assoc prof agron, 60-73, PROF AGRON, IOWA STATE UNIV, 73- *Honors & Awards:* Merit Cert, Am Forage & Grassland Coun. *Mem:* Am Soc Agron; Am Forage & Grassland Coun; Crop Sci Soc Am. *Res:* Breeding and evaluation of orchard grass, reed canary grass, smooth bromegrass, tall fescue, switchgrass, big bluestem, indian grass, Canada wild rye, alfalfa and birdsfoot trefoil. *Mailing Add:* Dept Agron Iowa State Univ Ames IA 50011

CARLSON, JAMES ANDREW, b Lewiston, Idaho, Nov 14, 46; m 73. PURE MATHEMATICS. *Educ:* Univ Idaho, BS, 67; Princeton Univ, PhD(math), 71. *Prof Exp:* Asst prof math, Stanford Univ, 71-73 & Brandeis Univ, 73-75; from asst prof to assoc prof, 75-85, PROF MATH, UNIV UTAH, 86- *Concurrent Pos:* Sloan fel, 78-79. *Res:* Algebraic and differential geometry. *Mailing Add:* Dept Math Univ Utah Salt Lake City UT 84112

CARLSON, JAMES C, b Mankato, Minn, Jan 4, 28; div; c 3. RADIOLOGICAL PHYSICS, COMPUTER SYSTEMS DESIGN. *Educ:* Univ Minn, AB, 51, MS, 56. *Prof Exp:* Cancer res scientist, Roswell Park Mem Inst, 54-57; radiol physicist, Hackley Hosp, Muskegon, Mich, 57-86; RETIRED. *Concurrent Pos:* Consult, Vet Admin Hosp, Wood, Wis, 60-76. *Mem:* Radiol Soc NAm; Soc Nuclear Med; Health Physics Soc; Am Asn Physicists Med; Am Asn Med Systs Informatics. *Res:* computer programming for nuclear medicine; hospital computer systems; radiology data management. *Mailing Add:* 3976 Woodmore Dr Greenwood IN 46142

CARLSON, JAMES GORDON, b Port Allegany, Pa, Jan 24, 08; m 36; c 3. CYTOLOGY, RADIOBIOLOGY. *Educ:* Univ Pa, AB, 30, PhD(cytol), 35. *Prof Exp:* Demonstr biol, Bryn Mawr Col, 30-31, instr, 31-35; instr zool, Univ Ala, 35-39, from asst prof to assoc prof, 39-46; sr biologist, NIH, USPHS, Md, 46-47; prof zool, 47-62, ALUMNI DISTINGUISHED SERV PROF, 62-78, EMER PROF, UNIV TENN, KNOXVILLE, 78- *Concurrent Pos:* Rockefeller fel, Carnegie Inst, Cold Spring Harbor Biol Lab & Univ Mo, 40-41; USPHS spec fel, Univ Heidelberg, 64-65; spec consult, USPHS, 43-46, 47-48; consult, Oak Ridge Nat Lab, Tenn, 47-78; mem comt biol & agr fels, Nat Res Coun, 49-52. *Mem:* AAAS (vpres, 55); Radiation Res Soc; Am Soc Cell Biol; Am Inst Biol Sci. *Res:* Effects of fixatives on staining reactions; orthopteran cytology; effects of ultraviolet and ionizing radiations on chromosomes and cell division; effect of chemical agents on chromosomes and cell division. *Mailing Add:* Dept Zool Univ Tenn Knoxville TN 37996-0810

CARLSON, JAMES H, b Cleveland, Ohio, June 10, 35; m 62; c 2. GENETICS. *Educ:* Fenn Col, BS, 58; Ohio State Univ, MS, 60, PhD(genetics), 63. *Prof Exp:* Asst instr zool, Ohio State Univ, 61-63; asst prof biol, Fairleigh Dickinson Univ, 63-66; assoc prof, 66-70, chmn dept, 70-74, PROF BIOL, RIDER COL, 70- *Mem:* AAAS; Genetics Soc Am; Am Genetic Asn. *Res:* Penetrance, expressivity and chromosomal control of a wing venation mutant system in Drosophila melanogaster. *Mailing Add:* Dept of Biol Rider Col 2083 Lawrenceville Rd Lawrenceville NJ 08648

CARLSON, JAMES REYNOLD, b Tulare, Calif, Aug 15, 48; m 75; c 3. CLINICAL MICROBIOLOGY. *Educ:* Calif State Univ, Chico, BA, 74; Univ NC, Chapel Hill, MSPH, 75, PhD(microbiol), 79. *Prof Exp:* Postdoctoral fel microbiol, Sch Med, Univ NC, Chapel Hill, 79-81; asst prof microbiol, Sch Med, Univ Tex, Houston, 81-83; ASSOC PROF MICROBIOL, SCH MED, UNIV CALIF, DAVIS, 83- *Mem:* Am Soc Microbiol; Sigma Xi. *Res:* Development of an HIV vaccine using animal models; development of diagnostic approaches with new molecular biology methods. *Mailing Add:* 306 Anza Ave Davis CA 95616

CARLSON, JAMES ROY, b Windsor, Colo, Mar 9, 39; m 61; c 2. ANIMAL SCIENCE, NUTRITION. *Educ:* Colo State Univ, BS, 61; Univ Wis-Madison, MS, 64, PhD(biochem), 66. *Prof Exp:* Asst prof, 66-71, assoc prof, 71-76, chmn grad prog nutrit, 73-83, PROF ANIMAL SCI, WASH STATE UNIV, 76-, CHMN DEPT ANIMAL SCI, 82- *Mem:* Am Inst Nutrit; Am Soc Animal Sci; Am Thoracic Soc; Soc Exp Biol Med. *Res:* Nutritional biochemistry; chemically-induced lung disease; amino acid and protein metabolism. *Mailing Add:* Dept Animal Sci Wash State Univ Pullman WA 99163

CARLSON, JANET LYNN, b Minneapolis, Minn, Aug 31, 52; m 76; c 1. SYNTHETIC ORGANIC CHEMISTRY. *Educ:* Hamline Univ, St Paul, Minn, BA, 74; Stanford Univ, PhD(org chem), 78. *Prof Exp:* asst prof, 78-87, ASSOC PROF CHEM, MACALESTER COL, ST PAUL, MINN, 87- *Mem:* Am Chem Soc. *Res:* Synthesis of natural products (including pheromones) and herbicides. *Mailing Add:* 1159 Glendon St Maplewood MN 55119

CARLSON, JOHN BERNARD, b Virginia, Minn, Jan 23, 26; m 49. PLANT ANATOMY. *Educ:* St Olaf Col, BA, 50; Iowa State Col, PhD(plant anat), 53. *Prof Exp:* Instr bot, Iowa State Col, 54; from asst prof to assoc prof bot, Univ Minn, Duluth, prof bot, 67-; RETIRED. *Mem:* Bot Soc Am. *Res:* Developmental morphology of soybeans and other plants; floral anatomy of wild rice. *Mailing Add:* 5627 Lester River Rd Duluth MN 55804

CARLSON, JOHN GREGORY, b Minneapolis, Minn, Mar 25, 41; c 2. HUMAN LEARNING & EMOTION, BEHAVIORAL MEDICINE. *Educ:* Univ Minn, BA, 63, PhD(psychol), 67. *Prof Exp:* PROF PSYCHOL, UNIV HAWAII, 67-, CHMN DEPT, 83- *Concurrent Pos:* Vis prof, Univ NMex, 76, Univ Colo, 80, Univ Alaska, 85, Colo State Univ, 86, San Francisco State Univ, 87; pres, Dept Psychol Found, 84-; prog chmn, Biofeedback Soc Am, 85-86; prog chmn, First Int Conf Self Regulation & Health,; NIMH fel, 75-76. *Mem:* Am Psychol Asn; Asn Appl Psychophysiol. *Res:* Behavioral medicine; biofeedback applications; emotion and human learning. *Mailing Add:* Dept Psychol Univ Hawaii Honolulu HI 96822

CARLSON, JOHN W, b Topeka, Kans, Nov 10, 40; m 61; c 2. MATHEMATICS. *Educ:* Kans State Univ, BS, 63, MS, 64; Univ Mo-Rolla, PhD(math), 70. *Prof Exp:* Instr math, Washburn Univ, 64-67; instr, Univ Mo-Rolla, 67-70; asst prof, 70-74, ASSOC PROF, EMPORIA STATE UNIV, 74- *Mem:* Am Math Soc; Math Asn Am; Sigma Xi. *Res:* Topology; quasi-uniform spaces. *Mailing Add:* 1001 Crestview Dr Vermillion SD 57069

CARLSON, JON FREDERICK, b Newport News, Va, July 3, 40; m 66; c 2. ALGEBRA. *Educ:* Old Dom Col, BA, 62; Univ Va, MA, 65, PhD(math), 67. *Prof Exp:* Instr math, Univ Va, 67-68; from asst prof to assoc prof, 68-81, PROF MATH, UNIV GA, 81- *Concurrent Pos:* Sr Fulbright Res Grant, 84. *Mem:* AAAS; Am Math Soc; Math Asn Am. *Res:* Representations of group algebra of finite groups and cohomology of groups. *Mailing Add:* Dept Math Univ Ga Athens GA 30602

CARLSON, KEITH DOUGLAS, b Los Angeles, Calif, Feb 28, 33; m 65. PHYSICAL CHEMISTRY. *Educ:* Univ Redlands, BS, 54; Univ Kans, PhD(chem), 60. *Prof Exp:* Fulbright scholar theoret chem, Oxford Univ, 60-62, hon Am Ramsey fel, 61-62; from asst prof to prof chem, Case Western Reserve Univ, 62-79; SR CHEMIST, ARGONNE NAT LAB, 79- *Concurrent Pos:* Consult, Gen Elec Co, 74-79; mem, High Temperature Sci & Technol Comt, Nat Res Coun, 71-73; chmn, Gordon Res Conf High Temperature Chem, 72. *Mem:* Am Phys Soc; Am Chem Soc; Sigma Xi. *Res:* Solid state physics; organic superconductors; solid state chemistry. *Mailing Add:* Chem Div Argonne Nat Lab Argonne IL 60439

CARLSON, KEITH J, b White Bear Lake, Minn, June 3, 38; m 65; c 2. GEOLOGY. *Educ:* Gustavus Adolphus Col, BS, 60; Iowa State Univ, MS, 62; Univ Chicago, PhD(paleozool), 66. *Prof Exp:* From asst prof to assoc prof, 67-82, PROF GEOL, GUSTAVUS ADOLPHUS COL, 83- *Mem:* Soc Vert Paleont; Paleont Soc; Sigma Xi; Geol Soc Am. *Res:* Vertebrate paleontology. *Mailing Add:* Dept of Geol Gustavus Adolphus Col St Peter MN 56082

CARLSON, KENNETH THEODORE, b Douglas, NDak, June 18, 21; m 49; c 3. CHEMISTRY. *Educ:* Minot State Col, BS, 47; Colo State Col, MA, 53. *Prof Exp:* Instr high schs, NDak, 47-48, 49-51, prin, 51-54; asst prof physics & chem, Mayville State Col, 54-55, from asst prof to assoc prof chem, 55-69, chmn dept sci, 55-85, prof sci, 69-85, emer prof, 85-; RETIRED. *Mem:* Am Chem Soc; Nat Sci Teachers Asn. *Res:* History of chemistry, especially as it pertains to chemical education in the United States. *Mailing Add:* Rte 2 Box 946 San Juan TX 78589

CARLSON, KERMIT HOWARD, mathematics; deceased, see previous edition for last biography

CARLSON, KRISTIN ROWE, b Minneapolis, Minn, July 31, 40. PSYCHOPHARMACOLOGY. *Educ:* Univ Mich, BA, 62; McGill Univ, MA, 63, PhD(psychol), 66. *Prof Exp:* NIMH fel, Univ Waterloo, 66-68, asst prof pharmacol & psychol, 68-75; res asst prof pharmacol, Sch Med, Univ Pittsburgh, 75-77; ASSOC PROF PHARMACOL, SCH MED, UNIV MASS, 77- *Concurrent Pos:* Mem, Pharmacol Sci Rev Comt, Nat Inst Gen Med Sci, 80-84; sabbatical, Univ Helsinki, Finland, 85. *Mem:* Sigma Xi; Am Soc Pharmacol & Exp Therapeut; Soc Neuroscience. *Res:* Effects of psychoactive drugs on behavior and neurochemistry. *Mailing Add:* Dept of Pharmacol Sch of Med Univ of Mass 55 Lake Ave N Worcester MA 01655

CARLSON, LAWRENCE EVAN, b Milwaukee, Wis, Dec 22, 44; c 1. MECHANICAL ENGINEERING, BIOENGINEERING. *Educ:* Univ Wis, BS, 67; Univ Calif, Berkeley, MS, 68, DEng, 71. *Prof Exp:* Asst prof mat eng, Univ Ill, Chicago Circle, 71-74; asst prof eng design, 74-78, ASSOC PROF MECH ENG, UNIV COLO, BOULDER, 78- *Concurrent Pos:* Consult, Prosthetics Res Lab, Northwestern Univ, 71-75; NSF travel grants, Yugoslavia, 72, 75 & res initiation grant, 73-74; NIH res career develop award, 76-81; prin investr, Vet Admin grants, 76-; consult, Ponderosa Assocs, 83- *Honors & Awards:* Ralph R Teetor Award, Soc Automotive Engrs, 76. *Mem:* Am Soc Mech Engrs; Sigma Xi; Rehab Eng Soc NAm. *Res:* Rehabilitation engineering; upper-limb prosthetics; orthotics. *Mailing Add:* Dept Mech Eng Campus Box 427 Univ Colo Boulder CO 80309

CARLSON, LEE ARNOLD, b Rennsaler, Ind, Aug 13, 39. MATHEMATICAL STATISTICS. *Educ:* DePauw Univ, BA, 61; Univ Mich, Ann Arbor, MA, 63. *Prof Exp:* ASSOC PROF, MATH & COMPUTE SCI, VALPARAISO UNIV, 65- *Concurrent Pos:* Vis assoc prof, Cornell Univ, 85-86. *Mem:* Am Math Soc; Inst Elec & Electronics Engrs. *Mailing Add:* Dept Math & Comput Sci Valparaiso Univ Valparaiso IN 46383

CARLSON, LESTER WILLIAM, b Warren, Wis, Sept 12, 33; m 54; c 6. PLANT PATHOLOGY, SILVICULTURE. *Educ:* Carroll Col, Wis, BS, 55; Okla State Univ, MS, 59; Univ Wis, PhD(plant path), 63. *Prof Exp:* Asst prof plant path, SDak State Univ, 63-66; forest pathologist, Forestry Can, 66-74, proj leader, 74-77, prog specialist, tree biol, 77-80, forestry adv, 80-82, dir forest environ res, 82-90, DIR, FOREST PROTECTION RES, FORESTRY CAN, 90- *Mem:* Can Phytopath Soc; Can Inst Forestry; Sigma Xi. *Res:* Plant disease control; nursery cultural practices; tree seedling physiology; forest regeneration and reclamation; forest management and conservation; general environmental sciences; research management. *Mailing Add:* Forestry Can Place Vincent Massey 351 St Joseph Blvd Hull PQ K1A 1G5 Can

CARLSON, LEWIS JOHN, b Valley City, NDak, Oct 4, 24; m 50; c 2. PAPER CHEMISTRY. *Educ:* Jamestown Col, BS, 47; Univ Iowa, MS, 50. *Prof Exp:* Teacher high sch, NDak, 47-48; res chemist, Rayonier, Inc, 50-60; sr chemist, Minn Mining & Mfg Co, 60-62; res chemist, Crown Zellerbach Corp, 62-70, supvr coating appln res, 70-76, mgr coating & process develop, 76-80, mgr paper prod develop, 80-88; vpres technol, James River Corp, 88-89; RETIRED. *Mem:* Tech Asn Pulp & Paper Indust. *Res:* Wood chemistry; carbohydrate; organic synthesis; polymers; paper coatings and treatments. *Mailing Add:* 232 NW 19th Camas WA 98607

CARLSON, MARIAN B, b Princeton, NJ, Oct 19, 52; m 77; c 2. MOLECULAR GENETICS. *Educ:* Harvard-Radcliffe Univ, AB, 73; Stanford Univ, PhD(biochem), 78. *Prof Exp:* Fel genetics, Mass Inst Tech, 78-81; from asst prof to assoc prof, 81-88, PROF GENETICS, COL PHYSICIANS & SURGEONS, 88-- *Res:* Molecular genetics of carbon catabolite repression in yeast. *Mailing Add:* 701 W 168th St New York NY 10032

CARLSON, MARVIN PAUL, b Creston, Iowa, Sept 27, 35; div; c 3. RESOURCE MANAGEMENT, SCIENCE COMMUNICATIONS. *Educ:* Univ Nebr, BS, 57, MS, 63, PhD(geol), 69. *Prof Exp:* Stratigrapher, Univ Nebr, 58, prin geologist, conserv & surv div, 63-84, asst dir, 70-86, PROF, UNIV NEBR, 76- *Concurrent Pos:* Grad Fac fel, Univ Nebr & assoc state geologist, 70-86; Res chmn, Interstate Oil Compact Comn, 71; adv comt mem, Nat Gas Surv, Fed Power Comn, 75. *Mem:* Fel Geol Soc Am; AAAS; Sigma Xi; Am Asn Petrol Geologists. *Res:* Palezoic lithostratigraphy; precambrian techtonics; long range effect of man on natural resource systems. *Mailing Add:* 5542 Shady Creek Ct Apt 10 Lincoln NE 68516

CARLSON, MERLE WINSLOW, b Minneapolis, Minn, June 5, 42. ORGANIC CHEMISTRY, ENVIRONMENTAL CHEMISTRY. *Educ:* Univ Minn, BA, 64; Northwestern Univ, PhD(chem), 69. *Prof Exp:* Res assoc org chem, Univ Wis, 69-70; instr, Northwestern Univ, 70-71; asst prof, State Univ NY, Stony Brook, 71-72; res assoc biochem, Univ Calif-Berkeley, 85-86; asst prof, 72-77, assoc prof org chem, 77-90, PROF ORG CHEM, BUTLER UNIV, 90- *Concurrent Pos:* Chmn phosphate task force, Ind Senate Environ & Ecol Comt, 73-74; co-chmn environ studies prog, Butler Univ, 75-89; pres, Environ Monitoring Serv, Inc, 75-78; bd mem, founder & consult, Wolf Tech Serv, Inc, 76-85. *Mem:* Am Chem Soc; Am Asn Arson Investrs; AAAS; Sigma Xi. *Res:* Synthesis of small-ring organic compounds and cycloaddition reactions; absorption of chlorinated hydrocarbons by plants. *Mailing Add:* Dept of Chem Butler Univ Indianapolis IN 46208

CARLSON, NORMAN ARTHUR, b Geneseo, Ill, Aug 5, 39; m 67; c 1. ORGANIC CHEMISTRY. *Educ:* Augustana Col, Ill, BA, 61; Univ Wis-Madison, MS, 63; Univ Mich, PhD(org chem), 67. *Prof Exp:* CHEMIST, ORG CHEM DEPT, E I DU PONT DE NEMOURS & CO, INC, WILMINGTON, 67- *Mem:* Am Chem Soc. *Mailing Add:* 904 Sharpless Rd Hockessin DE 19707-9378

CARLSON, OSCAR NORMAN, b Mitchell, SDak, Dec 21, 20; m 46; c 3. METALLURGY. *Educ:* Yankton Col, BA, 43; Iowa State Univ, PhD(chem), 50. *Prof Exp:* Prof chem, 50-62, prof mat sci & eng, 62-87, EMER PROF, IOWA STATE UNIV, 87- *Concurrent Pos:* Chmn, Dept Metall & chief div, Ames Lab, Dept Energy, 62-66; vis scientist, Max Planck Inst Metallforschung, Stuttgart, 74-75 & 83; deleg, US-Indo workshop extractive metall, Udaipur, India, 82. *Mem:* Am Chem Soc; Am Soc Metals Int; Am Inst Mining, Metall & Petrol Engrs; Sigma Xi. *Res:* Phase equilibria; preparation of high purity metals; metallurgy of vanadium; mass transport of solutes in metals at high temperatures; effects of interstitial solutes in metals. *Mailing Add:* 223 Met Dev Bldg Iowa State Univ Ames IA 50011

CARLSON, PAUL ROLAND, b St Paul, Minn, Nov 23, 33; m 57; c 2. MARINE GEOLOGY. *Educ:* Gustavus Adolphus Col, BA, 55; Iowa State Univ, MS, 57; Ore State Univ, PhD(oceanog), 67. *Prof Exp:* Res asst eng geol, Soil Eng Lab, Iowa State Univ, 55-57; geologist, US Army Corps Engrs, 57-58; instr geol, Gustavus Adolphus Col, 58-59; teacher gen sci, Cleveland Consol Schs, 59-61; instr geol, Pac Lutheran Univ, 61-63; res assoc geol oceanog, Ore State Univ, 63-67; RES GEOLOGIST, US GEOL SURV, 67- *Mem:* Geol Soc Am; Am Geophys Union; Soc Econ Paleont & Mineral. *Res:* Sedimentology; submarine canyons; processes of sedimentation; estuaries, especially San Francisco Bay; seismic stratigraphy; sedimentologic factors affecting Glacier Bay Fjords; continental margin morphology and sedimentology, Gulf of Alaska and Bering Sea. *Mailing Add:* US Geol Surv 345 Middlefield Rd Menlo Park CA 94025

CARLSON, PHILIP R, b Evanston, Ill, June 2, 31; m 54; c 3. MATHEMATICS. *Educ:* Bethel Col (Minn), BS, 53; Univ Minn, BS, 57, MS, 65, PhD, 71. *Prof Exp:* Teacher high sch, Minn, 57-59; from instr to prof math, Bethel Col, Minn, 60-88; RETIRED. *Concurrent Pos:* Programmed Course in Algebra for High Sch Teachers, 60-66; dir, Job Corps Math Proj Minn Nat Lab, 67-68; vis scholar math, Cambridge Univ (Eng), 73-74, 81-82. *Mem:* Math Asn Am; Nat Coun Teachers Math. *Res:* Mathematics, algebraic theory, statistics. *Mailing Add:* 2018 Thorndale Ave Bethel Col New Brighton MN 55112

CARLSON, RICHARD EUGENE, b Wausa, Nebr, Oct 19, 40; m 60; c 2. AGRONOMY. *Educ:* Univ Nebr, BS, 67; Iowa State Univ, MS, 69, PhD(agr climat), 71. *Prof Exp:* Researcher & teaching asst, Iowa State Univ, 67-68, NDEA fel, 68-71; photointerpreter, Purdue Univ, 71; from asst prof to assoc prof, 71-79, PROF AGR CLIMAT, IOWA STATE UNIV, 79- *Mem:* Am Soc Agron; Crop Sci Soc Am; Soil Sci Soc Am; Am Meteorol Soc. *Res:* Microclimate studies on corn and soybeans with emphasis on moisture stress and photosynthesis; analyses of weather data. *Mailing Add:* Dept of Agron Iowa State Univ Ames IA 50010

CARLSON, RICHARD FREDERICK, b St Paul, Minn, June 19, 36; m 57; c 3. PHYSICS. *Educ:* Univ Redlands, BS, 57; Univ Minn, MS, 62, PhD, 64. *Prof Exp:* Asst res physicist, Univ Calif, Los Angeles, 64, asst prof physics, 64-67; from asst prof to assoc prof, 67-77, coordr dept, 70-72, chmn physics & sci fac, 81-82, PROF PHYSICS, UNIV REDLANDS, 77-, CHMN, 85- *Mem:* Am Phys Soc; Am Sci Affil. *Res:* Experimental nuclear physics; nuclear structure; few nucleon problem. *Mailing Add:* Dept Physics Univ Redlands PO Box 3080 Redlands CA 92373-0999

CARLSON, RICHARD M, b Preston, Idaho, Feb 4, 25. RESEARCH ADMINISTRATION. *Prof Exp:* DIR, RES & TECHNOL LAB, AMES RES CTR, MOFFETT FIELD, CALIF, 76- *Mem:* Nat Acad Eng. *Mailing Add:* Res & Technol Lab Activ Ames Res Ctr Moffett Field CA 94035

CARLSON, RICHARD OSCAR, b Flushing, NY, June 2, 26; m 52; c 3. SEMICONDUCTOR DEVICES, SENSORS. *Educ:* Columbia Univ, AB, 47, AM, 49, PhD(physics), 52. *Prof Exp:* Asst, Columbia Univ, 47-50, physicist, Hudson Labs, 52-54; physicist, Gen Elec Corp Res & Develop Ctr, 54-; RETIRED. *Concurrent Pos:* CONSULT, 87- *Mem:* Fel Am Phys Soc; Electrochemisty Soc; sr mem Inst Elec & Electronics Eng. *Res:* Processing of silicon devices and integrated circuits; experimental transport measurements on pure and doped semiconductor crystals; temperature and humidity sensors; silicon device packaging; electron and gamma irradiation; very large scale integration packaging; solder fatigue. *Mailing Add:* Gen Elec Corp R&D Ctr PO Box 8 Schenectady NY 12301

CARLSON, RICHARD P, b Great Bend, Kans, June 8, 39; m 64; c 2. IMMUNOINFLAMMATORY PHARMACOLOGY. *Educ:* Emporia State Univ, Kans, BA, 62; Hahnemann Med Col, MS, 74; Thomas Jefferson Univ, PhD(physiol), 76. *Prof Exp:* Sci instr biol, physics, chem & geol, Larned High Sch, Kans, 65-66; res biologist, Merck Sharp & Dohme Res Lab, 66-76; sr pharmacologist, Revlon Health Care, 76-79; RES FEL, DEPT EXP THERAPEUT, WYETH-AYERST RES, 79- *Mem:* Sigma Xi; Am Soc Pharmacol & Exp Therapeut; Am Col Rheumatology; Int Asn Inflammation Socs; Inflammation Res Asn. *Res:* Testing of experimental drugs (potential novel antiarthritic and antiallergic) using in vitro and in vivo animal and cellular models which detect antiinflammatory and immunologic activities; effects on cell and humoral immunity and immediate hypersensitivity. *Mailing Add:* Dept Exp Therapeut Wyeth-Ayerst Res CN 8000 Princeton NJ 08543

CARLSON, RICHARD RAYMOND, b Chicago, Ill, Sept 15, 23; m 51; c 3. NUCLEAR PHYSICS. *Educ:* Univ Chicago, PhD(physics), 51. *Prof Exp:* Res assoc physics, Univ Chicago, 51; from asst prof to assoc prof, 51-63, PROF PHYSICS, UNIV IOWA, 63- *Concurrent Pos:* Guggenheim fel, Oxford Univ, 59; mem staff, Los Alamos Sci Lab, Univ Calif, 56; physicist, Oak Ridge Nat Lab, 52. *Mem:* Am Phys Soc. *Mailing Add:* Dept Physics Univ Iowa Iowa City IA 52240

CARLSON, RICHARD WALTER, b Los Angeles, Calif, May 22, 54; m 76. ISOTOPE GEOCHEMISTRY, GEOCHRONOLOGY. *Educ:* Univ Calif, San Diego, BA, 76; Scripps Inst Oceanog, PhD(earth sci), 80. *Prof Exp:* Res asst, Scripps Inst Oceanog, 76-80; res fel, 80-81, MEM STAFF, DEPT TERRESTRIAL MAGNETISM, CARNEGIE INST, 81- *Concurrent Pos:* Mem, Nat Res Coun Panel Explosive Volcanism, 80, NSF Earth Sci Adv Rev Panel, 85- *Mem:* Am Geophys Union; Geol Soc Am; Geochemical Soc. *Res:* Isotope geochemistry; structure of chemical heterogeneity in the earth's mantle; chronology of planetary differentiation; formation of planetary crusts; techniques for high precision isotopic analysis. *Mailing Add:* Dept Terrestrial Magnetism 5241 Broad Branch Rd NW Washington DC 20015

CARLSON, ROBERT BRUCE, b Virginia, Minn, Sept 15, 38; m 60. ENTOMOLOGY. *Educ:* Univ Minn, Duluth, BS, 60; Mich State Univ, MS, 62, PhD(entom), 65. *Prof Exp:* Assoc insect ecologist, NCent Forest Exp Sta, US Forest Serv, 65-67; asst prof entom, 68-71, assoc prof, 71-80, PROF ENTOM, NDAK STATE UNIV, 80- *Concurrent Pos:* Adj prof statist, NDak State Univ, 82- *Mem:* AAAS; Entom Soc Am; Entom Soc Can. *Res:* Ecology; biological control of weeds; applied statistics. *Mailing Add:* 82 22nd Ave N Fargo ND 58102

CARLSON, ROBERT G(USTAV), b Brooklyn, NY, June 6, 28; c 3. METALLURGY, CERAMIC ENGINEERING. *Educ:* Polytech Inst Brooklyn, BME, 51; Rensselaer Polytech Inst, MME, 55; Univ Cincinnati, PhD(metall), 62. *Prof Exp:* Mem staff, CheMet Prog, 51-52, metall engr, Aeronaut & Ord Div, 52-53, asst refractory metals, Res Lab, 53-55, assoc, Lamp Metals & Components Dept, 55-58, mgr advan mat, Advan Engine & Technol Dept, 58-66, mgr composite mat, 66-70, sr composites engr, Mat & Process Technol Lab, 70-77, mgr, Boron Aluminum Fabrication, 77-81, MGR, TURBINE SHROUDS, AIRCRAFT MACH GROUP, MAT & PROCESS TECHNOL LAB, GEN ELEC CO, 81- *Honors & Awards:* Turner Award, 60. *Mem:* Am Inst Mining, Metall & Petrol Engrs; Am Soc Metals; Electrochemical Soc; Sigma Xi. *Res:* Development of composite structural materials; refractory alloys; alkali metal corrosion behavior. *Mailing Add:* 141 Junedale Dr Greenhills OH 45218

CARLSON, ROBERT GIDEON, b Chicago, Ill, Feb 4, 38; m 62; c 2. ORGANIC CHEMISTRY. *Educ:* Univ Ill, BS, 59; Mass Inst Technol, PhD(org chem), 63. *Prof Exp:* From asst prof to assoc prof chem, 63-72, PROF CHEM, UNIV KANS, 72- *Concurrent Pos:* Alfred P Sloan Found res fel, 70-72. *Mem:* Am Chem Soc; Royal Soc Chem. *Res:* Synthetic organic chemistry; structure and synthesis of natural products; highly strained ring systems; photochemical reactions of organic compounds; synthesis of analytical reagents. *Mailing Add:* Dept Chem Univ Kansas Malott Hall Lawrence KS 66045

CARLSON, ROBERT KENNETH, b Chicago, Ill, July 7, 28; m 54; c 2. INORGANIC CHEMISTRY. *Educ:* Northwestern Univ, PhB, 54. *Prof Exp:* Asst chemist, Great Lakes Carbon Corp, 49-50, chemist, 53-57; res chemist, Borg-Warner Corp, 57-59; mat engr, Chance Vought Corp, 59-60, res scientist, 60-62; sr scientist, Res Ctr, Ling-Temco-Vought Corp, 62-63, head mat sci, 63-64, prog mgr, Carbon & Graphite Div, 64-67; vpres & gen mgr, 67-77, PRES & CHIEF EXEC OFFICER, POCO GRAPHITE INC, UNION OIL CO, CALIF, 77- *Concurrent Pos:* Prin investr, US Air Force Indust res grant, 61-62; Chmn Am Carbon Soc, 79- *Mem:* Sigma Xi; Am Chem Soc; fel Am Inst Chemists; Am Carbon Soc. *Res:* Synthesis of new carbon and graphite products; high temperature refractory oxides for rocket and missile application; synthesis of salts of organo phosphoric acids. *Mailing Add:* RR 3 No 210 Decatur TX 76234

CARLSON, ROBERT L, b Gary, Ind, May 22, 24; m 50; c 5. STRUCTURAL MECHANICS. *Educ:* Purdue Univ, BS, 48, MS, 50; Ohio State Univ, PhD(eng), 62. *Prof Exp:* Res engr, Battelle Mem Inst, 50-60, res assoc, 60-62; res engr, US Steel Res Lab, 62-63; res assoc, Dept Aeronaut, Stanford Univ, 63-65, assoc prof, 65-66; prof aerospace eng, 66-88, EMER PROF AEROSPACE ENG, GA INST TECHNOL, 88- *Concurrent Pos:* Lectr, Int Union Theoret & Appl Mech Colloquium, 60, Int Conf on Fatigue & Fatigue Thresholds, 84; consult, EIMAC Div, Varian Assocs, 65-66, US Air Force Flight Dynamics Lab, 68, Pratt & Whitney, 85. *Honors & Awards:* Charles Dudley Medal, Am Soc Testing & Mat, 59. *Mem:* Sigma Xi. *Res:* Phase of structural mechanics which deals with the conditions of external loading and/or temperature environment which lead to failure due to instability and fracture. *Mailing Add:* Sch Aerospace Engr Atlanta GA 30332

CARLSON, ROBERT LEONARD, b Duluth, Minn, Feb 26, 32; m 58; c 2. PHOTOGRAPHIC CHEMISTRY, SURFACE CHEMISTRY. *Educ:* Univ Minn, BA, 58; Univ Ill, PhD(inorg chem), 62. *Prof Exp:* res chemist, 62-70, prod develop mgr, 70-79, dir, qual & develop, 83-85, PROJECT MGR, 3M CO, 86- *Mem:* AAAS; Am Chem Soc. *Res:* Metal ion complexes and molecular complexes in non-aqueous solvents; rheology; photographic chemistry; color photographic imaging systems; surface chemistry. *Mailing Add:* 2243 Berland Pl St Paul MN 55119

CARLSON, ROBERT M, b Cokato, Minn, Oct 30, 40; m 62; c 2. ORGANIC CHEMISTRY. *Educ:* Univ Minn, BChem, 62; Princeton Univ, PhD(chem), 66. *Prof Exp:* Fel, Harvard Univ, 65-66; from asst prof to assoc prof org chem, 66-74, PROF CHEM, UNIV MINN, DULUTH, 74- *Mem:* Sigma Xi; Am Chem Soc. *Res:* Total synthesis of natural products; new synthetic methods; environmental organic chemistry. *Mailing Add:* Dept Chem Univ Minn Duluth MN 55812

CARLSON, ROBERT MARVIN, b Denver, Colo, Mar 18, 32; m 62; c 1. SOIL CHEMISTRY, PLANT NUTRITION. *Educ:* Colo State Univ, BS, 54; Univ Calif, Berkeley, PhD(soil sci), 62. *Prof Exp:* Soil scientist, USDA, 52-54; sr lab technician, Univ Calif, Berkeley, 55-62; prog specialist, Ford Found, Arg, 62-65; asst pomologist, 65-71, ASSOC POMOLOGIST, UNIV CALIF, DAVIS, 71- *Concurrent Pos:* Vis res prof, Nat Univ South, Arg, 62-65. *Mem:* Am Soc Agron; Soil Sci Soc Am; Am Chem Soc. *Res:* Analytical chemistry. *Mailing Add:* 2231 E Third St Duluth MN 55812

CARLSON, ROGER, b St Joseph, Mo, Aug 15, 37; m 60; c 1. MATHEMATICAL STATISTICS. *Educ:* Univ Kansas City, BS, 59, MA, 60; Harvard Univ, PhD(statist), 64. *Prof Exp:* Asst prof, 63-69, ASSOC PROF MATH, UNIV MO-KANSAS CITY, 69- *Mem:* Am Statist Asn. *Res:* Foundations of statistical inference; applications of mathematics to social sciences. *Mailing Add:* Dept Math Univ Mo 5100 Rockhill Rd Kansas City MO 64110

CARLSON, ROGER HAROLD, microbiology, environmental health, for more information see previous edition

CARLSON, ROY DOUGLAS, biophysics, for more information see previous edition

CARLSON, ROY W(ASHINGTON), civil engineering; deceased, see previous edition for last biography

CARLSON, STANLEY DAVID, b St Paul, Minn, Sept 4, 34; m 58, 69; c 5. ENTOMOLOGY, PHYSIOLOGY. *Educ:* Univ Minn, BS, 56; Univ Nebr, MS, 61; Kans State Univ, PhD(entom, physiol), 65. *Prof Exp:* Res entomologist, Stored Prod Insects Res Br, Agr Res Serv, USDA, Kans, 59-64; asst prof entom, Va Polytech Inst, 65-67; NIH spec fel physiol, Karolinska Inst, Sweden, 67-69 & biol, Yale Univ, 69-70; NSF fel, Univ Ill, Urbana, 70-71; from asst prof to assoc prof, 71-76, PROF ENTOM, UNIV WIS-MADISON, 80- *Concurrent Pos:* Entomologist, Colo Dept Agr, 59; USDA res grant, 65-; consult, Panogen Div, Morton Chem Co, 61-62; mem, NEUROSICENCE TRAINING COMT, Univ Wis, 73-; NIH & NSF grants. *Mem:* AAAS; Entom Soc Am; Am Inst Biol Sci; Scand Physiol Soc; Sigma Xi. *Res:* Sensory physiology of insects; physiology of insect vision and visual pigments; neuroanatomy: ultrastructure of compound eye, optic tract, and blood-brain barrier. *Mailing Add:* Dept Entom Univ Wis 1630 Linden Dr Madison WI 53706

CARLSON, STANLEY L, b Glenwood City, Wis, 37. MATHEMATICS. *Educ:* Univ Wis, BS, 61; Univ Northern Iowa, MA, 65; Univ Northern Colo, EdD(math educ), 70. *Prof Exp:* Teacher, Greendale High Sch, 61-65; PROF MATH, UNIV WIS, 65- *Mem:* Am Math Asn. *Mailing Add:* Dept Math Univ Wis Stevens Point WI 54481

CARLSON, THOMAS ARTHUR, b Waterbury, Conn, Apr 1, 28; m 50. MOLECULAR PHYSICS. *Educ:* Trinity Col, Conn, BS, 50; Johns Hopkins Univ, MA, 51, PhD(chem), 54. *Prof Exp:* Sr res staff mem, Oak Ridge Nat Lab, 54-84, corp res fel, 84-88; RETIRED. *Concurrent Pos:* Guggenheim fel, 67; ed, J Electron Spectroscopy, 72-77; consult, 88- *Mem:* fel Am Phy Soc. *Res:* Electron spectroscopy; auger and electron shake-off phenomena; hot atom chemistry; atomic physics; synchrotron radiation. *Mailing Add:* 389 East Dr Oak Ridge TN 37830

CARLSON, TOBY NAHUM, b Brooklyn, NY, Nov 4, 36; m 60; c 2. METEOROLOGY. *Educ:* Mass Inst Technol, BS, 58, MS, 60; Univ London, PhD(meteorol), 65. *Prof Exp:* Res meteorologist, Weather Serv, Inc, on contract to Air Force Cambridge Res Lab, 61-62, Nat Hurricane Res Lab, Nat Oceanic & Atmospheric Admin, Environ Res Lab, 65-74; assoc prof, 74-82, PROF METEOROL, PA STATE UNIV, 82- *Mem:* Am Meteorol Soc. *Res:* Modelling evaportranspiration over vegetation; remote sensing soil moisture. *Mailing Add:* Dept of Meteorol Pa State Univ University Park PA 16802

CARLSON, WAYNE C, b Watseka, Ill, Nov 20, 45; m 71; c 2. AGRONOMY. *Educ:* Univ Ill, BS, 67, MS, 68, PhD(hort), 72. *Prof Exp:* Field res rep, Mobay Corp, 72-75, herbicide prod mgr, 75-78, regional develop mgr, 78-83, mgr prod develop, 83-86, MGR, RES & DEVELOP PLANNING, MOBAY CORP, 86- *Mem:* Weed Sci Soc Am; Coun Agr Sci & Technol. *Res:* Efficacy of products for the control of various pests. *Mailing Add:* Mobay Corp PO Box 4913 8400 Hawthorne Rd Kansas City MO 64120

CARLSON, WAYNE R, b Moline, Ill, Dec 31, 40; m; m 70. CYTOGENETICS. *Educ:* Rockford Col, BA, 62; Ind Univ, Bloomington, MA, 67, PhD(zool), 68. *Prof Exp:* Assoc prof, 68-73, ASSOC PROF BOT, UNIV IOWA, 77- *Concurrent Pos:* NSF res grant, 75-80, 79-83, USDA res grants, 82-86, 86-88 & 89-91. *Res:* Cytogenetics of maize, especially the B chromosome. *Mailing Add:* Dept of Bot Univ of Iowa Iowa City IA 52242

CARLSON, WILLIAM DWIGHT, b Denver, Colo, Nov 5, 28; m 50; c 2. VETERINARY RADIOLOGY, RADIATION BIOLOGY. *Educ:* Colo State Univ, DVM, 52, MS, 56; Univ Colo, PhD(radiol), 58; Am Col Vet Radiol, dipl, 62. *Prof Exp:* Vet, 52-53; asst prof, Col Vet Med, Colo State Univ, 53-55, Am Vet Med Asn fel, 55-57, assoc prof & radiologist, Col Vet Med, 57-62, prof radiol, 62-68, chmn dept radiol & radiation biol, 64-68, dir radiol health animal res lab, Univ-USPHS, 62-68,; pres, Univ Wyo, 68-79; PROF, INTERGOVT PERSONAL ACT, COLO STATE UNIV, 84- *Concurrent Pos:* Civilian adv, US Army Gen Command & Staff Sch, Ft Leavenworth, Kans, 69-72; consult, USPHS, 62-68, Vet Admin Med Facil, 68-70, Am Inst Biol Sci, 69-70, NASA, 69-72 & Surg Gen, US Air Force, 70-76; mem bd comnr, Nat Comn on Accrediting, 69-72; mem nat adv coun health res facil, NIH, 69-73; secy adv comt coal mine safety res, Dept Interior, 71; mem bd visitors, Air Univ, 73-76; vis prof, Colo State Univ, 79-80; chief exec officer, St John's Hosp, Jackson, Wyo, 80-84; res admin, Off Grants & Prog Systs, USDA, Washington, DC, 84- *Mem:* Am Soc Nuclear Med; Am Vet Med Asn. *Res:* Radiation biology, diagnostic & therapeutic radiology, radioactive isotopes; lower animals. *Mailing Add:* 0GPS/CSRS/USDA 323G Aerospace Bldg Washington DC 20250

CARLSON, WILLIAM THEODORE, b Lyons, Nebr, Mar 27, 33; m 60; c 2. INFORMATION SCIENCE. *Educ:* Univ Nebr, BS, 57, MS, 59. *Prof Exp:* Asst & instr agron, Univ WVa, 59-62; asst analyst, Sci Info Exchange, Smithsonian Inst, 62-63, chief plant sci, 63-66, chief agr & appl biol sci, 66-82; pres, Agr & Environ Info Retrieval & Indexing Serv, 83-90; RETIRED. *Concurrent Pos:* Adv, Agr Res Inst-Nat Acad Sci, 66-81. *Mem:* Am Soc Agron; Am Soc Info Sci; hon mem Sigma Xi. *Res:* Forage management; weed control; information storage and retrieval system design; agricultural and plant sciences; environmental biology; water resources; pesticides and agronomy. *Mailing Add:* 335 Coweeta Church Rd Otto NC 28763

CARLSSON, DAVID JAMES, b Hartlepool, Eng, Oct 2, 40; Can citizen; m 65; c 2. FIBER SCIENCE, POLYMER CHEMISTRY. *Educ:* Univ Birmingham, Eng, BSc, 62, PhD(chem), 65. *Prof Exp:* Fel phys chem, 65-67, assoc res officer, 67-77, SR RES OFFICER TEXTILE SCI, NAT RES COUN CAN, 77- *Mem:* Fel Chem Inst Can; Chem Soc. *Res:* Light induced degradation of fiber forming polymers; mechanisms and stabilization; fiber microstructure, characterization and correlation with physical properties. *Mailing Add:* Div Chem Nat Res Coun Can Ottawa ON K1A 0R9 Can

CARLSSON, ERIK, b Vingaker, Sweden, Mar 31, 24; m 51; c 2. RADIOLOGY. *Educ:* Karolinska Inst, Sweden, MD, 52, PhD, 70. *Prof Exp:* Mem staff, Radiol Dept, Karolinska Sjukhuset, Stockholm, Sweden, 54-57, pediat roentgenol, 55-57 & thorax roentgenol, 60-62; from asst prof to assoc prof radiol, Sch Med, Wash Univ, 58-64; assoc prof radiol, head cardiovasc radiol & staff mem, Cardiovasc Res Inst, Med Ctr, 64-68, PROF RADIOL, UNIV CALIF, SAN FRANCISCO, 68- *Concurrent Pos:* NIH training grant & consult, 67-77; dir, Radiol Learning Ctr, Univ Calif, 86, interim chmn, Radiol Dept, 89. *Mem:* Asn Univ Radiol; NY Acad Sci; Sigma Xi; AAAS; Am Heart Asn. *Res:* Diagnostic and cardiovascular radiology. *Mailing Add:* Dept Radiol Univ Calif Sch Med San Francisco CA 94143

CARLSTEAD, EDWARD MEREDITH, b Chillicothe, Mo, Aug 1, 25; m 50; c 3. METEOROLOGY, COMPUTER SCIENCE. *Educ:* Univ Calif, Los Angeles, BS, 46; US Naval Postgrad Sch, MS, 53. *Prof Exp:* Analyst, Joint Weather Bur-Air Force-Navy Anal Ctr, Washington, DC, 49-51; chief analyst, Joint Numerical Weather Prediction Unit, Md, 56-59; res meteorologist, Fleet Numerical Weather Facility, 59-62; officer in chg opers analyst, Pac Command Detachment, Naval Command Systs Support Activity, 62-65; chief sci serv div, Pac Regional Hq, Nat Weather Serv, Environ Sci Serv Admin, 65-71, meteorologist in chg, Nat Weather Serv Forecast Off, Nat Oceanic & Atmospheric Admin, Honolulu, 71-78; chief, Forecast Div, Nat Meteorol Ctr, Nat Oceanic & Atmospheric Admin, Washington, DC, 78-84; METEOROL CONSULT, 84- *Concurrent Pos:* Counr, Am Meteorol Soc, 80-83. *Mem:* Fel Am Meteorol Soc; Sigma Xi; Nat Weather Asn. *Res:* Programming of digital computers to analyze and predict wind and temperature fields in the tropics; improvement of forecasting techniques; physical oceanography; operations analysis. *Mailing Add:* 116 Inverness Lane Ft Washington MD 20744

CARLSTEN, JOHN LENNART, b Minneapolis, Minn, Feb 4, 47; m 78; c 2. NON-LINEAR OPTICS, QUANTUM OPTICS. *Educ:* Univ Minn, BS, 69; Harvard Univ, MA, 71, PhD(physics), 74. *Prof Exp:* Res assoc, Joint Inst Lab Astrophysics, 74-76, sr res assoc, 76-79; asst prof physics, Univ Colo, 76-79; staff mem, Los Alamos Nat Lab, 79-81, assoc group leader, 81-; AT DEPT PHYSICS, MONTANA STATE UNIV, BOZEMAN. *Mem:* Am Phys Soc; Optical Soc Am; Am Asn Phys Teachers. *Res:* Study of spontaneous and stimulated scattering processes, such as stimulated Raman scattering, quantum fluctuations and quantum amplifiers. *Mailing Add:* Dept Physics Montana State Univ Bozeman MT 59717

CARLSTONE, DARRY SCOTT, b Vinita, Okla, May 15, 39; m 62; c 2. PARTICLE PHYSICS. *Educ:* Univ Okla, BS, 61; Purdue Univ, MS, 64, PhD(physics), 68. *Prof Exp:* Asst prof, 67-72, assoc prof, 72-76, PROF PHYSICS, CENT STATE UNIV, OKLA, 76-, CHMN DEPT, 78- *Mem:* Am Phys Soc; Am Asn Physics Teachers; Sigma Xi. *Res:* Theoretical particle physics with special interest in symmetry principles. *Mailing Add:* Dept of Physics Cent State Univ Edmond OK 73034

CARLTON, BRUCE CHARLES, b Burrillville, RI, Aug 3, 35; m 56; c 2. GENETICS. *Educ:* Univ NH, BS, 57; Mich State Univ, MS, 58, PhD(genetics), 60. *Prof Exp:* Asst hort, Mich State Univ, 57-59; USPHS trainee, Stanford Univ, 60-62; from asst prof to assoc prof biol, Yale Univ, 62-71; prof biochem & microbiol, Univ Ga, 71-79, prof molecular & pop genetics, 80-, AT ECOGEN INC. *Concurrent Pos:* USPHS res grant, 63-l NSF res grant, 65-; fel, Silliman Col, 69-71. *Mem:* Am Soc Microbiol. *Res:* Mechanisms of genetic control through studies of mutationally-altered proteins and structure and function of circular DNA elements; correlation between specific plasmias and S-endotoxin production in bacillus thuringiensis. *Mailing Add:* Ecogen Inc 2005 Cabot Blvd W Langhorne PA 19047

CARLTON, DONALD MORRILL, b Houston, Tex, July 20, 37; m 61; c 3. ENVIRONMENTAL SCIENCE, ORGANIC CHEMISTRY. *Educ:* Univ St Thomas, Tex, BA, 58; Univ Tex, Austin, PhD(org chem), 62. *Prof Exp:* Staff mem & group leader polymer sci, Sandia Corp, 62-65; asst dir res, Tracor Inc, 65-69; PRES & BD CHMN, RADIAN CORP SUBSID HARTFORD STEAM BOILER INSPECTION & INS CO, 69- *Concurrent Pos:* Dir, Joy Technologies, Inc, Hartford Steam Boiler Inpection & Ins Co. *Mem:* Nat Coal Coun; Am Chem Soc. *Res:* Energy and environmental science and engineering. *Mailing Add:* Radian Corp PO Box 201088 Austin TX 78720-1088

CARLTON, JAMES THEODORE, b Ft Worth, Tex, Feb 15, 48; m. MARINE BIOLOGY, BIOLOGICAL INVASIONS. *Educ:* Univ Calif, Berkeley, BA, 71; Univ Calif, Davis, PhD(ecol), 79. *Prof Exp:* Sr sci asst invertebrate zool, Calif Acad Sci, 71-73, res assoc, 75-79; vis asst prof biol, Univ Ore-Ore Inst Marine Biol, 78; scholar, NSF Nat Needs, Woods Hole Oceanog Inst, 78-79, guest investr, 79-80, investr, 80-81; marine scientist, Williams Col Mystic Seaport Prog, 82-85; asst prof biol, Ore Inst Marine Biol, Univ Ore, 86-89; DIR, MARITIME STUDIES, WILLIAMS COL, MYSTIC SEAPORT & ASSOC PROF MARINE SCI, WILLIAMSTOWN, MASS, 89- *Concurrent Pos:* Lectr geol & paleobiol, Univ Calif, Davis, 75-76; staff res assoc zool, Bodega Marine Lab-Naval Arctic Res Lab, 76; mem, Working Group Introd & Transfers Marine Organisms, Int Coun Explor Seas, 79-89, chair, 90-; testified Cong Hearings Introduced Species-zebra mussels, 90. *Mem:* AAAS; Am Malacol Union; Ecol Soc Am; Am Soc Zool; Crustacean Soc; Estuarine Res Fedn; Western Soc Malacologists. *Res:* History, biogeography, ecology, mechanisms of dispersal, and environmental impacts and implications of biological invasions of introduced marine, estuarine, and freshwater invertebrates into coastal ecosystems; biogeography of human-dispersed marine and estuarine invertebrates; systematics and biology of marine mollusks. *Mailing Add:* Maritime Studies Prog Williams Col Mystic Seaport Mystic CT 06355

CARLTON, PETER LYNN, b Rochester, Ind, Sept 4, 31; m. PSYCHOLOGY. *Educ:* Grinnell Col, BA, 52; Univ Iowa, MA, 54, PhD(psychol), 55. *Prof Exp:* Teaching asst, Univ Iowa, 52-55; res asst, US Army Med Lab, Fort Knox, 55-57; sr res scientist, Squibb Inst Med Res, 57-62, res assoc psychopharmacol, 62-63; from assoc prof to prof, Dept Psychol, Rutgers Univ, New Brunswick, NJ, 63-68; PROF, DEPT PSYCHIAT, ROBERT WOOD JOHNSON MED SCH, UNIV MED & DENT NJ, 68- *Concurrent Pos:* Fulbright fel, Weizmann Inst, Rehovot, Israel, 71-72; consult, Squibb Inst Med Res, 63-76, Hoffmann-LaRoche, Inc, 75-86 & Lederle Labs, 79-; mem, Neuropsychol Res Rev Comt, NIMH, 66-70, Res Task Force, 72-73, Behav Sci Test Comt, Nat Bd Med Examr, 70-75; assoc ed, Psychopharmacol, 73-80, Neuropsychobiol, 85- *Mem:* Fel Am Psychol Asn; fel AAAS; Behav Pharmacol Soc (pres, 69-70); Am Soc Pharmacol & Exp Therapeut; Soc Neurosci; Asn Acad Psychiat; NY Acad Sci; Res Soc Alcoholism. *Res:* Behavioral pharmacology; compulsive gambling; author of numerous articles. *Mailing Add:* Dept Psychol UMDNJ-Robert Wood Johnson Med Sch 675 Hoes Lane Piscataway NJ 08854-5635

CARLTON, RICHARD WALTER, b Nov 23, 42; US citizen; m 66; c 2. SEDIMENTARY PETROLOGY. *Educ:* Wash State Univ, BS, 65; Ore State Univ, MS, 68, PhD(geol), 72. *Prof Exp:* SR GEOLOGIST, OHIO GEOL SURV, 70- *Mem:* Soc Econ Paleontologists & Mineralogists. *Res:* Mineralogic and petrographic characterization of Middle and Upper Devonian shales in Ohio; prediction of the degree of washability of coal using automated image analysis; cool resource evaluation. *Mailing Add:* 3225 Cimmaron Columbus OH 43221

CARLTON, ROBERT AUSTIN, b Brownsville, Tenn, Apr 30, 27; m 50; c 2. ZOOLOGY. *Educ:* Lambuth Col, BS, 50; George Peabody Col, MA, 51; Ala Polytech Inst, PhD(zool), 58. *Prof Exp:* Instr biol, Northeast Miss Jr Col, 51-54; zool, Ala Polytech Inst, 54-56; assoc prof biol, Delta State Col, 56-64; PROF BIOL, LAMBUTH COL, 64- *Mem:* Nat Wildlife Fed. *Res:* Vertebrate ecology; natural history of vertebrates. *Mailing Add:* Dept Biol Lambuth Col Jackson TN 38301

CARLTON, TERRY SCOTT, b Peoria, Ill, Jan 29, 39; m 60; c 2. THEORETICAL CHEMISTRY. *Educ:* Duke Univ, BS, 60; Univ Calif, Berkeley, PhD(phys chem), 63. *Prof Exp:* From asst prof to assoc prof chem, 63-76, PROF CHEM, OBERLIN COL, 76- *Concurrent Pos:* Vis prof chem, Univ NC, 76. *Mem:* Am Phys Soc; Sigma Xi. *Res:* Intermolecular forces; electron density in atoms. *Mailing Add:* Dept Chem Oberlin Col Oberlin OH 44074

CARLTON, THOMAS A, JR, b Vivian, La, May 3, 27; m 50; c 4. CIVIL ENGINEERING. *Educ:* Tex A&M Univ, BS, 50, MS, 55; Univ Tex, PhD(civil eng), 62. *Prof Exp:* Geotechnical engr, Tex Hwy Dept, 50-53; asst prof civil eng, Lamar State Col, 53-56; asst prof mech eng, Univ Tex, 56-59; assoc prof civil eng, Miss State Univ, 59-61; assoc prof eng mech, Univ Ala, Tuscaloosa, 61-64, prof civil eng, 64-86; county engr, Sumter County, Livingston, Ala, 87-88; RETIRED. *Concurrent Pos:* Consult geotechnical engr, 64- *Mem:* Am Soc Testing & Mat. *Res:* Permeability of clay soils and interaction effects of hazardous wastes on soil properties; development of testing methods and equipment to deal with the specific requirements of permeability research. *Mailing Add:* 10435 Haun Dr Tuscaloosa AL 35405

CARLTON, WILLIAM HERBERT, b Statesboro, Ga, Oct 6, 40; m 62; c 3. MEDICAL PHYSICS. *Educ:* Emory Univ, BS, 62, MS, 64; Rutgers Univ, PhD(radiation biophys), 69; Am Bd Health Physics, dipl, 70; Am Bd Radiol, dipl & cert radiol physics, 73. *Prof Exp:* Lectr radiation sci, Rutgers Univ, 64-69; asst prof, 69-74, ASSOC PROF RADIOL, MED COL GA, 74- *Concurrent Pos:* Consult Colgate-Palmolive Res Lab, 68-69 & Vet Admin Hosp, Augusta, Ga, 70- *Mem:* Am Asn Physicists in Med; Soc Nuclear Med; Health Physics Soc. *Res:* Low energy x-ray absorption; x-ray dose measurements; lacrimal scanning. *Mailing Add:* Dupont Health Protection Bldg 734A Alken SC 29808

CARLTON, WILLIAM WALTER, b Owensboro, Ky, June 17, 29; m 55; c 2. VETERINARY PATHOLOGY, VETERINARY TOXICOLOGY. *Educ:* Univ Ky, BS, 53, MS, 56; Fla Southern Col, BS, 54; Auburn Univ, DVM, 60; Purdue Univ, PhD(vet path), 63; Am Col Vet Path, dipl. *Prof Exp:* Instr path, Purdue Univ, 60-62; asst prof, Mass Inst Technol, 62-65; assoc prof, 65-68, PROF PATH, SCH VET MED, PURDUE UNIV, WEST LAFAYETTE, 68- *Concurrent Pos:* Ed, Fundamental & Appl Toxicol, 80-85. *Mem:* Am Inst Nutrit; Am Asn Avian Path; Am Soc Exp Path; Int Acad Path; Soc Toxicol; Am Soc Vet Pathologists. *Res:* Nutritional and toxicological diseases, especially mycotoxic diseases; auth & co-auth, numerous sci publ. *Mailing Add:* 3716 High Acre Pl West Lafayette IN 47906

CARLUCCI, ANGELO FRANCIS, b Plainfield, NJ, Feb 17, 31. MICROBIAL ECOLOGY, MARINE MICROBIOLOGY. *Educ:* Rutgers Univ, BS, 53, MS, 56, PhD(microbiol), 59. *Prof Exp:* Asst microbiol, Tela RR Co, United Fruit Co, Honduras, 59-61; jr res biologist, Scripps Inst Oceanog, Univ Calif, San Diego, 61-62, postgrad res biologist, 62-64, asst res biologist, Inst Marine Resources, 64-71, assoc res microbiologist, 71-77, RES MICROBIOLOGIST, INST MARINE RESOURCES, UNIV CALIF, SAN DIEGO, 77-, LECTR, 70- *Mem:* AAAS; Am Soc Microbiol; Am Soc Limnol & Oceanog; Sigma Xi. *Res:* General marine microbiology; nitrogen cycle in the sea; vitamins and metabolites in marine ecology; survival of bacteria in seawater; phytoplankton-bacterial interactions; microbiology of sea surface; deep-sea microbiology. *Mailing Add:* Inst Marine Resources A-018 Univ Calif San Diego La Jolla CA 92093

CARLUCCI, FRANK VITO, b New York, NY, Apr 23, 48. LIQUID CHROMATOGRAPHY, AUTOMATED CHEMISTRY. *Educ:* Richmond Col, City Univ New York, BS, 71; Stevens Inst Technol, MS, 76; Rutgers Univ, PhD(food chem), 82. *Prof Exp:* Asst scientist, Schering-Plough Corp, 71-76; sr res scientist, E R Squibb & Sons, Inc, 77-89; MGR, BERLEX LABS, 90- *Concurrent Pos:* Assoc scientist, Hoeschst-Roussel, 77; mgr, Boyle-Midway, 89-90. *Mem:* Am Soc Testing & Mat; Am Inst Chemists. *Res:* Application of high performance liquid chromatography in developing a quantitative assay for specific amines and amino acids found in protein foods and an equation which will index the degree of decomposition. *Mailing Add:* 1371 Thomas Ave N Brunswick NJ 08902

CARLUCCIO, FRANK, b Newark, NJ, Aug 31, 19; m 59; c 1. CHEMICAL ENGINEERING. *Educ:* Newark Tech Sch, AE, 41; Newark Col Eng, BS, 43; Columbia Univ, MS, 47. *Prof Exp:* Jr scientist, Univ Calif, Los Alamos, 44-46; chem engr, FMC Corp, 47-56; chem engr, GAF Corp, 56-60, prog mgr, 60-67, supvr, 67-84; RETIRED. *Mem:* Am Inst Chem Engrs. *Res:* Liquid extraction; high pressure research; process development. *Mailing Add:* 8 Gloria Dr Towaco NJ 07082

CARLUCCIO, LEEDS MARIO, b Leominster, Mass, Sept 12, 36; m 61; c 3. BOTANY, MORPHOLOGY. *Educ:* Mass Col Pharm, BS, 58, MS, 60; Cornell Univ, PhD(paleobot), 66. *Prof Exp:* Instr gen bot, Cornell Univ, 64-66; from asst prof to assoc prof, 66-72, PROF GEN BOT, CENT CONN STATE COL, 72- *Mem:* Bot Soc Am; Sigma Xi; Int Orgn Paleobotany. *Res:* Anatomy and morphology of the progymnosperms of Devonian floras. *Mailing Add:* Dept of Biol Cent Conn State Col New Britain CT 06050

CARLYLE, JACK WEBSTER, b Cordova, Alaska, Feb 23, 33; m 70; c 1. ANALYSIS OF ALGORITHMS. *Educ:* Univ Wash, BA, 54, MS, 57; Univ Calif, Berkeley, MA & PhD(elec eng), 61. *Prof Exp:* Asst prof elec eng, Princeton Univ, 61-63; asst prof eng, 63-68, assoc prof, 68-74, prof & chmn dept, 75-80, PROF COMPUT SCI, UNIV CALIF, LOS ANGELES, 80- *Concurrent Pos:* Vis scientist, Thomas J Watson Res Ctr, IBM, 73. *Mem:* Asn Comput Mach; Inst Elec & Electronics Engrs; Inst Math Statist; Am Math Soc; Soc Indust & Appl Math. *Res:* Computer science theory; stochastic sequential machines; nonparametric methods in signal detection; communication systems. *Mailing Add:* Comput Sci Dept 3731 Boelter Hall Univ Calif Los Angeles CA 90024

CARMACK, MARVIN, b Dana, Ind, Sept 1, 13; m 60. ORGANIC CHEMISTRY. *Educ:* Univ Ill, AB, 37; Univ Mich, MS, 39, PhD(org chem), 40. *Prof Exp:* Asst org chem, Univ Ill, 40-41; Towne instr, Univ Pa, 41-44, from asst prof to prof, 44-53; prof, 53-78, EMER PROF ORG CHEM, IND UNIV, BLOOMINGTON, 78- *Concurrent Pos:* Guggenheim fel, Swiss Fed Inst Technol, 49-50; Fulbright res scholar, Commonwealth Sci & Indust Res Orgn, Melbourne, 60-61. *Honors & Awards:* Marvin Cormack lectr, Ind Univ. *Mem:* AAAS; Am Chem Soc; NY Acad Sci; fel Am Inst Chem. *Res:* Natural products; organic sulfur chemistry; heterocyclic compounds; mammalian pheromones. *Mailing Add:* Dept Chem Ind Univ Bloomington IN 47405

CARMAN, CHARLES JERRY, b Tucumcari, NMex, Nov 14, 38; m 61; c 4. RESEARCH MANAGEMENT, PHYSICAL POLYMER CHEMISTRY. *Educ:* Eastern NMex Univ, BS, 61; Univ Calif, MS, 63. *Prof Exp:* Res chemist, B F Goodrich Co, 63-65, sr res chemist, 65-71, res assoc, 71-77, sr res & develop assoc, 77-79, mgr, Tech Resources Engineered Prod Group, 79-81, mgr Corp Res & Develop Ctr, 81-85, sr mkt mgr, 85-87, DIR MKT GEON VINYL DIV, 87- *Concurrent Pos:* Int health care expert & consult, hemophelia, blood prod, blood safety. *Mem:* Am Chem Soc; Sigma Xi. *Res:* Manage broadbase basic technologies in polymer physics, composites, synthesis; materials research and develop and product development; nuclear magnetic resonance of polymers; relationships of polymer molecular structure and physical properties; nuclear relaxation and polymer molecular motion; electron spin resonance; structure and function of homogeneous and heterogeneous catalysts. *Mailing Add:* 184 Brentwood Dr Hudson OH 44236

CARMAN, GEORGE MARTIN, b Jersey City, NJ, June 2, 50; m 81; c 2. BIOCHEMISTRY, MICROBIOLOGY. *Educ:* William Paterson Col, BA, 72; Seton Hall Univ, MS, 74; Univ Mass, PhD(biochem), 76. *Prof Exp:* Fel biochem, Med Sch Univ Tex, 77-78; PROF FOOD ENZYM, RUTGERS UNIV, 78- *Concurrent Pos:* Robert A Welch Found, Tex, fel, 77-78. *Mem:* Am Soc Microbiol; Sigma Xi; Inst Food Technol; Am Soc Biochem & Molecular Biol. *Res:* Membrane structure and function and phospholipid metabolism. *Mailing Add:* Dept of Food Sci Rutgers Univ New Brunswick NJ 08903

CARMAN, GLENN ELWIN, b Waterloo, Iowa, June 8, 14; m 41; c 2. ENTOMOLOGY. *Educ:* State Univ Iowa, BS, 36; Cornell Univ, PhD(entom), 42. *Prof Exp:* Res entomologist, Rohm & Haas Co, 42-43; from jr entomologist to assoc entomologist, Agr Exp Sta, 43-53, chmn dept, Univ Calif, 63-69, entomologist, Citrus Res Ctr & Agr Exp Sta, 53-81, prof, 60-81, EMER PROF ENTOM, UNIV CALIF, RIVERSIDE, 81- *Concurrent Pos:* Consult, Sunkist Growers & Calif Citrus Qual Coun. *Honors & Awards:* Robert M Howie Award. *Mem:* Fel AAAS; Entom Soc Am. *Res:* Insect toxicology; economic entomology; biology and control of citrus scale insects; evaluation of application equipment; reentry studies; ant and snail control on citrus crops. *Mailing Add:* 5368 Pinehurst Dr Riverside CA 92504

CARMAN, HOWARD SMITH, JR, b Memphis, Tenn, June 8, 59; m 88. CHEMICAL REACTION DYNAMICS, LASER PHOTOPHYSICS. *Educ:* Tex Christian Univ, BS, 81; Rice Univ, MA, 85, PhD(chem), 86. *Prof Exp:* Robert A Welch found, postdoctoral fel, Dept Chem, Rice Univ, 86-87; postdoctoral res assoc, Dept Chem, Univ Tenn, 87-88; RES STAFF, CHEM PHYSICS SECT, OAK RIDGE NAT LAB, 88- *Mem:* AAAS; Am Chem Soc; Am Phys Soc. *Res:* Experimental studies of the dynamics of atom-molecule collisions, electron-molecule collisions, and photoionization and photodissociation processes; experimental studies of the structures and reaction dynamics of molecular clusters. *Mailing Add:* Chem Physics Sect Oak Ridge Nat Lab PO Box 2008 Oak Ridge TN 37831-6125

CARMAN, MAX FLEMING, JR, b Lansing, Mich, Oct 2, 24; m 47; c 2. GEOLOGY. *Educ:* Univ Calif, Los Angeles, AB, 48, PhD(geol), 54. *Prof Exp:* Asst, Univ Calif, Los Angeles, 51-54; prof geol, Petrobras Petrol Co, Brazil, 57-59; asst prof, 54-57, assoc prof geol, 59-64, assoc dean Col Arts & Sci, 61-65, PROF GEOL, UNIV HOUSTON, 64- *Concurrent Pos:* NSF & Fulbright sr res fel, Victoria, NZ, 63. *Mem:* Geol Soc Am; Mineral Soc Am; Geochemisty; Brazilian Geol Soc; Mineral Soc Gt Brit. *Res:* Petrology; petrography; areal and structural geology; optical mineralogy. *Mailing Add:* Dept Geoscis Univ Houston Houston TX 77004-5503

CARMAN, PHILIP DOUGLAS, b Ottawa, Can, Oct 28, 16; m 51; c 3. OPTICS. *Educ:* Univ Toronto, BA, 40; Univ Rochester, MSc, 51. *Prof Exp:* Optical instruments, Res Enterprises Ltd, 40; from jr physicist to sr res officer, Nat Res Coun Can, 41-81; RETIRED. *Concurrent Pos:* Consult optics, 81- *Honors & Awards:* Talbert Abrams Award, Am Soc Photogram, 82. *Mem:* Fel Optical Soc Am; Can Asn Physicists; Am Soc Photogram & Remote Sensing. *Res:* Performance of photographic and photogrammetric systems; optical instrument design and testing. *Mailing Add:* 1332 Snowdon St Ottawa ON K1H 7P4 Can

CARMAN, ROBERT LINCOLN, JR, b Bay Shore, NY, Jan 15, 41; m 64, 75; c 2. LASER PHYSICS, ENERGY & PARTICLE TRANSPORT. *Educ:* Adelphi Col, BA, 62; Harvard Univ, PhD(physics), 68. *Prof Exp:* Res fel SS physics, NSF, 61-62; staff mem physics, Brookhaven Nat Lab, 62-63; staff mem, HE physics, Cambridge Bubble Chamber Group, 62-64; staff mem, SS physics, Lincoln Lab Mass Inst Technol, 64-68; postdoctoral fel nonlinear optics, Gordon McKay Lab, Harvard Univ, 68-70; group leader laser physics, Lawrence Livermore Nat Lab, 70-74; staff mem, sect leader & prog mgr weapons & laser fusion, Los Alamos Nat Lab, 74-84; PROG MGR & SR STAFF SCIENTIST, LASERS, DIRECT ENERGY & PROPULSION PROGS, ROCKETDYNE DIV, ROCKWELL INT, 84- *Concurrent Pos:* Consult, Micronetics, 62-64, Cambridge Res Ctr, NASA, 65-68, Lincoln Lab Mass Inst Technol, 68-70, Maxwell Labs, 80-83, Los Alamos Nat Lab, 84-90 & Lawrence Livermore Nat Lab, 89-90; staff, Jason Summer Studies, Dept Defense, 74; lectr, Nat Acad Sci, USSR, 84; co-founder, pres, chmn bd & vpres mgt, K D C Technol, Inc, Calif, 84-88. *Mem:* Am Phys Soc; Soc Automotive Engrs; Advan Defense Preparedness Asn; Nat Mgt Asn. *Res:* Laser fusion; plasma phenomena; nuclear and conventional weapons design; advanced propulsion systems development; propulsion technology; particle accelerators. *Mailing Add:* 334 W Siesta Ave Thousand Oaks CA 91360

CARMEAN, WILLARD HANDY, b Philadelphia, Pa, Jan 4, 22; m 49; c 3. FOREST SOILS. *Educ:* Pa State Univ, BS, 43; Duke Univ, MF, 47, PhD(forest soils), 53. *Prof Exp:* Res forester soil res & forest surv, Pacific Northwest Forest Exp Sta, 46-51; soil scientist forest soils, Cent States Forest Exp Sta, 53-67, proj leader, NCent Forest Exp Sta, 67-79; assoc prof, 79-84, prof forest soil, EMER PROFF LAKE HEAD UNIV, ONT, 87- *Concurrent Pos:* Forest soils consult, Malaysia, UN Food & Agr Orgn, 74, Rome, 82, Greece, 83, Turkey, 85, Haiti, US/Aid, 82 China, lectures, 87. *Mem:* AAAS; Soc Am Foresters; Soil Sci Soc Am; Ecol Soc Am; Am Soil Conserv Soc; Can Inst Forestry. *Res:* Relations between tree growth and factors of soil and topography. *Mailing Add:* Sch Forestry Lakehead Univ Thunder Bay ON P7B 5E1 Can

CARMEL, RALPH, b Riga, Latvia, Aug 8, 40; US citizen; m 67; c 2. HEMATOLOGY. *Educ:* Yeshiva Univ, BA, 59, BHL, 59; NY Univ, MD, 63. *Prof Exp:* USPHS res fel hemat, Mt Sinai Sch Med, 66-68; res assoc, Aerospace Med Lab, Lackland AFB, 68-70; Wellcome fel, St Mary's Hosp Med Sch, London, 71; from asst prof to assoc prof med, Wayne State Univ, 72-75; chief hemat, Grace Hosp, Detroit, 75; assoc prof, 75-81, PROF MED, 81-, PROF PATHOL, SCH MED, UNIV SOUTHERN CALIF, 86- *Concurrent Pos:* vis investr, Inserm, Hosp Henri Mondor, Creteil, France, 82. *Mem:* Am Soc Hemat; Am Soc Clin Nutrit; Am Inst Nutrit; Am Fedn Clin Res; Am Soc Clin Invest; Int Soc Exp Hemat. *Res:* Megaloblastic anemia and folic acid and vitamin B-12 metabolism, with special interest in the transport of vitamin B-12 and in the proteins binding vitamin B-12. *Mailing Add:* Univ Southern Calif Sch Med 2025 Zonal Ave Los Angeles CA 90033

CARMELI, MOSHE, b June 15, 33; US citizen; m 61; c 3. THEORETICAL PHYSICS. *Educ:* Hebrew Univ, Jerusalem, MSc, 60; Israel Inst Technol, PhD(physics), 64. *Prof Exp:* Lectr physics, Israel Inst Technol, 64; res assoc physics, Lehigh Univ, 64-65; res assoc physics, Univ Md, 65-67, asst prof, 67-68; res physicist, US Air Force, Wright-Patterson AFB, 67-69, sr scientist, 69-72; assoc prof, 72-74, head dept, 73-77, PROF PHYSICS, BEN GURION UNIV, 74-, ALBERT EINSTEIN PROF PHYSICS, 79-, DIR, CTR THEORET PHYSICS, 80- *Concurrent Pos:* Vis prof & mem, Inst Theoret Physics, State Univ NY, Stony Brook, 77-78; vis prof physics, Univ Md, 85-86. *Mem:* Fel Am Phys Soc; AAAS; Am Asn Univ Profs; Sigma Xi; Int Soc Gen Relativity & Gravitation; NY Acad Sci; Israel Physics Soc (vpres, 79-82, pres, 82-85). *Res:* General relativity and gauge theory. *Mailing Add:* Dept Physics Ben Gurion Univ Beer Sheva Israel

CARMEN, ELAINE (HILBERMAN), b New York, NY, Mar 26, 39. PSYCHIATRY. *Educ:* City Col NY, BS, 59; Sch Med, NY Univ, MD, 64; Am Bd Psychiat & Neurol, dipl, 77. *Prof Exp:* Intern, NC Mem Hosp, 64-65, resident, 65-66; staff physician, John Umstead Hosp, 67-69, resident psychiat, 69-72; instr, 72-74, from asst prof to assoc prof, 74-84, asst dir psychiat residence training, 77-84, PROF PSYCHIAT, SCH MED, UNIV NC, CHAPEL HILL, 84-; DIR, EMERGENCY PSYCHIAT SERV, NC MEM HOSP, 84- *Concurrent Pos:* Mem, Panel of Spec Populations, President's Comn Mental Health, 77-78. *Mem:* Fel Am Psychiat Asn; Nat Coalition Women's Ment Health; Asn Acad Psychiat; Nat Women's Health Network. *Res:* Gender & psychotherapy; family violence and violence against women; victims of violence and psychiatric illness. *Mailing Add:* Dept of Psychiat Sch of Med Univ of NC Chapel Hill NC 27514

CARMER, SAMUEL GRANT, b Buffalo, NY, Dec 19, 32; m 60; c 3. BIOMETRICS. *Educ:* Cornell Univ, BS, 54; Univ Ill, Urbana, MS, 58, PhD(agron), 61. *Prof Exp:* Res fel biomath, NC State Col, 61-62; from asst prof to assoc prof, 62-71, PROF BIOMET, UNIV ILL, URBANA-CHAMPAIGN, 71- *Mem:* Fel Am Soc Agron; Am Statist Asn; Biomet Soc; Fel Crop Sci Soc Am. *Res:* Biometrics; teaching graduate level courses and providing individual advice on problems of experimental design; statistical analysis and statistical computing. *Mailing Add:* Dept Agron Turner Hall 1102 S Goodwin Ave Urbana IL 61801

CARMI, SHLOMO, b Cernauti, Romania, July 18, 37; US citizen; m 63; c 3. FLUID MECHANICS, HEAT TRANSFER. *Educ:* Univ Witwatersrand, BSc, 62; Univ Minn, MS, 66, PhD(aeronaut eng), 68. *Prof Exp:* From asst prof to assoc prof mech eng, Col Eng, Wayne State Univ, Detroit, Mich, 68-78, prof, 78-86; PROF & HEAD MECH ENG & MECH DEPT, DREXEL UNIV, PHILADELPHIA, PA, 86- *Concurrent Pos:* Fac res award, Wayne State Univ, 70; sr lectr, Technion, Israel Inst Technol, Haifa, 70-72, vis prof, I Taylor Chair, Mech Eng Dept Technion, 77-78; res specialist, Ford Motor Co & Detroit Edison Co, 73-74, 76-77 & 83; US Army Res Off award, 77-81; grant recipient, US Dept Energy, 77-79 & Dow Chem Co/DOE, 78-79; assoc ed, J Fluids Eng, 81-84; cong fel, sci & tech adv to US Sen, 85-86. *Mem:* Am Phys Soc; Am Soc Eng Educ; Am Soc Mech Engrs. *Res:* Hydrodynamic stability and turbulence by linear and nonlinear theories; eigenvalue bounds; variational and numerical solutions; computer codes for differential equations in fluid mechanics and heat and mass transfer; applied mathematics and modeling; stability studies in fluid, thermal, chemical and biological systems. *Mailing Add:* Dept of Mech Eng & Mechanics Drexel Univ Philadelphia PA 19104

CARMICHAEL, DAVID JAMES, b Casterton, Australia, June 2, 36; m 68; c 1. PROTEIN CHEMISTRY. *Educ:* Univ Melbourne, BAgrSci, 60; Univ Nottingham, PhD(food sci), 66. *Prof Exp:* Technologist, Kraft Foods Ltd, 60-63; demonstr food sci, Univ Nottingham, 63-66; Am Dent Asn fel, Northwestern Univ, 66-67; Helen Hay Whitney Found fel, 67-68; from asst prof to prof dent, Univ Alta, 68-80, dir, res & grad studies, Fac Dent, 76-80; RETIRED. *Concurrent Pos:* Med Res Coun Can grant, 73-76; vis scientist, Med Res Coun Can, 75; vis prof Monash Univ, Clayton, Australia, 75-76. *Mem:* Brit Biochem Soc; Int Asn Dent Res. *Res:* Collagen research; biochemical characterization of normal and lathyritic dentin matrix collagen; immunohistochemical study of collagen fibrillogenesis. *Mailing Add:* Glebe Farm Brace Bridge Heath Lincoln England

CARMICHAEL, DONALD C(HARLES), b Cincinnati, Ohio, May 22, 33; m 56; c 2. SOLAR ENERGY, MATERIALS SCIENCE. *Educ:* Purdue Univ, BS, 55, MS, 57. *Prof Exp:* Metall engr, Allison Div, Gen Motors Corp, 55; prin metall engr, 56-63, asst div chief advan mat develop, 63-65, chief,Mat Appln Div, 66-71, mgr solar photovoltaics, 71-83, SECT MGR, COATINGS & ELECTRONICS MAT, BATTELLE MEM INST, 83- *Concurrent Pos:* Trustee, Am Vacuum Soc, 89. *Mem:* Am Vacuum Soc. *Res:* Development of metallic, ceramic and composite materials; materials compatibility; powder metallurgy; coating technology; vacuum coating; optical and electronic thin films; sputtering; hot-isostatic-pressing process; solar photovoltaic conversion; chemical vapor deposition; thermal spray processes. *Mailing Add:* Battelle Mem Inst 505 King Ave Columbus OH 43201

CARMICHAEL, GREGORY RICHARD, b Marengo, Ill, June, 16, 52; m 75; c 2. CHEMICAL ENGINEERING. *Educ:* Iowa State Univ, BS, 74; Univ Ky, MS, 75, PhD(chem eng), 79. *Prof Exp:* From asst prof to assoc prof, 78-84, PROF CHEM ENG, UNIV IOWA, 85- *Mem:* Am Inst Chem Engrs; Am Chem Soc; Air Pollution Control Asn; Am Meteorol Soc; Sigma Xi; Am Soc Eng Educ; Am Geophys Union. *Res:* Mathematical modeling of the transport, chemical reaction and removal processes affecting the distribution of trace species in the lower troposphere. *Mailing Add:* 2126 Glendale Rd Iowa City IA 52245-3220

CARMICHAEL, HALBERT HART, b St Louis, Mo, Aug 29, 37; m 58; c 2. CHEMICAL KINETICS. *Educ:* Univ Tenn, BS, 59; Univ Calif, PhD(chem), 63. *Prof Exp:* Nat Bur Stand fel, 63-64; asst prof, 64-69, assoc prof, 69-78, PROF CHEM, NC STATE UNIV, 78- *Mem:* Am Chem Soc; Sigma Xi. *Res:* Photochemistry and radiation chemistry of gases. *Mailing Add:* 1001 Marlborough Rd Raleigh NC 27610

CARMICHAEL, HOWARD JOHN, b Manchester, Eng, Jan 17, 50; NZ citizen; m 72. QUANTUM OPTICS, NONLINEAR OPTICS. *Educ:* Univ Auckland, BSc, 71, MSc, 73; Add in Educ: Univ Waikato, PHD, 77. *Prof Exp:* Res assoc, City Col, City Univ NY, 77-78, Univ Tex, Austin, 79-81 & Univ Waikato, 81-82; from asst prof to assoc prof, Univ Ark, 83-89; ASSOC PROF, UNIV ORE, 89- *Concurrent Pos:* Vis lectr, Univ Tex, Austin, 84; vis researcher, Royal Signal & Radar Estab, Malvern, UK, 84; prin investr, NSF, 85-; vis assoc prof, Univ Tex, Austin, 88; vis assoc physics, Calif Inst Technol, 89. *Mem:* Am Phys Soc; fel Optical Soc Am; Am Asn Physics Teachers; Sigma Xi. *Res:* Theoretical study of the quantum mechanical interaction between light and matter, including nonclassical effects such as photon antibunching and squeezing, and cavity quantum electrodynamics; quantum theory of nonlinear optical systems. *Mailing Add:* Dept Physics Univ Ore Eugene OR 97403

CARMICHAEL, IAN STUART, b London, Eng, Mar 29, 30; m 70, 86; c 4. PETROLOGY, GEOCHEMISTRY. *Educ:* Cambridge Univ, BA, 54; Univ London, PhD(geol), 59. *Prof Exp:* Lectr geol, Imp Col, Univ London, 58-63; assoc prof, 65-67, chmn dept geol & geophys, 72-76 & 80-82, assoc dean grad div, 76-78, PROF GEOL, UNIV CALIF, BERKELEY, 67-, ASSOC DEAN GRAD DIV, 85-, ASSOC PROVOST RES, 86- *Concurrent Pos:* Sr foreign scientist, NSF, Univ Chicago, 63; Miller res prof, Miller Inst Sci Res, 67-68. *Honors & Awards:* Bowen Award, Am Geophys Union, 86. *Mem:* Fel Am Geophys Union; fel Mineral Soc Am; fel Mineral Soc Gt Brit & Ireland; Geol Soc Am. *Res:* Origin and cooling history of volcanic rocks. *Mailing Add:* Dept of Geol Univ of Calif Berkeley CA 94720

CARMICHAEL, J W, JR, b Lamesa, Tex, Feb 9, 40. PHYSICAL CHEMISTRY, INORGANIC CHEMISTRY. *Educ:* Eastern NMex Univ, BS, 61; Univ Ill, MS, 63, PhD(phys chem), 65. *Prof Exp:* Asst chem, Univ Ill, 62-65; asst prof, Univ Ark, 65-70; asst prof, 70-73, assoc prof, 73-77, PROF CHEM, XAVIER UNIV LA, 77- *Concurrent Pos:* Pre-Health Professions adv, 74- *Mem:* Am Chem Soc; Nat Asn Minority Med Educ; Royal Soc Chem; Nat Sci Teachers Asn. *Res:* Teaching approaches which make chemistry more accessible to the underprepared. *Mailing Add:* Dept of Chem Xavier Univ Palmetto Pine Sts New Orleans LA 70125

CARMICHAEL, LELAND E, b Huntington Park, Calif, June 15, 30; m 57; c 3. VETERINARY MICROBIOLOGY. *Educ:* Univ Calif, AB, 52, DVM, 56; Cornell Univ, PhD(virol), 59. *Prof Exp:* Asst bact, State Univ NY Vet Col, Cornell Univ, 56-59, res assoc virol, 59-63, asst prof infectious dis & John M Olin Chair, 63-69, PROF VET VIROL & JOHN M OLIN PROF VIROL & VET MICROBIOL, STATE UNIV NY VET COL, CORNELL UNIV, 69- *Concurrent Pos:* NIH grant, 61-81; scientific dir, James A Baker Inst for Animal Health, 75- *Honors & Awards:* Gaines Award, Am Vet Med Asn, 75; Small Animal Res Award, Ralston-Purina, 81. *Mem:* Am Vet Med Asn; NY Acad Sci; US Animal Health Asn; Am Soc Microbiol; Infectious Dis Soc Am; Nat Acad Practice. *Res:* Infectious diseases of domestic animals, principally viral and canine brucellosis; canine parvovirus; immunology; pathogenesis. *Mailing Add:* Vet Virus Res Inst Snyder Hall SUNY Vet Col Cornell Univ Ithaca NY 14850

CARMICHAEL, RALPH HARRY, b Freetown, Ind, Jan 20, 23; m 42; c 1. PHARMACOLOGY. *Educ:* Ind Univ, BS, 49; Butler Univ, MS, 56. *Prof Exp:* Assoc chemist, Lilly Lab Clin Res, Eli Lilly & Co, Indianapolis, Ind, 50-56, supvr clin chem, 56-62, dept head clin chem & hemat, 62-70, dept head drug metab, 70-78, sr biochemist, 78-82; RETIRED. *Mem:* Am Chem Soc; Sigma Xi; NY Acad Sci. *Res:* Drug metabolism. *Mailing Add:* 2732 Parkwood Dr Speedway IN 46224

CARMICHAEL, RICHARD DUDLEY, b High Point, NC, Mar 13, 42; m 67; c 1. MATHEMATICAL ANALYSIS. *Educ:* Wake Forest Col, BS, 64; Duke Univ, AM, 66, PhD(math), 68. *Prof Exp:* Asst prof math, Va Polytech Inst & State Univ, 68-71; assoc prof, 71-80, PROF MATH, WAKE FOREST UNIV, 80- *Concurrent Pos:* Vis lectr, Univ California, Davis, 73; vis assoc prof, Iowa State Univ, 78-79; vis prof, NMex State Univ, 84-85. *Mem:* Am Math Soc; Math Asn Am; Soc Indust & Appl Math. *Res:* Theory of distributions; complex variables. *Mailing Add:* Dept Math PO Box 7388 Wake Forest Univ Winston-Salem NC 27109

CARMICHAEL, ROBERT STEWART, b Toronto, Ont, Jan 11, 42; m 67. GEOPHYSICS, GEOLOGY. *Educ:* Univ Toronto, BASc, 63; Univ Pittsburgh, MS, 64, PhD(earth & planetary sci), 67. *Prof Exp:* Teaching asst geol, Univ Pittsburgh, 63-64 & 66-67; fel geophys, Osaka Univ, Japan, 67-68; geophysicist, Explor & Prod Res Ctr, Shell Develop Co, 69-72; asst prof, Mich State Univ & Univ Mich, Flint, 73-76; assoc prof, 77-82, PROF GEOL, UNIV IOWA, 83- *Concurrent Pos:* Prin investr, NSF & NASA grants; exchange prog, Univ Iceland, 85 & 87; vis prof, Meiji Univ, Tokyo, 88-89. *Honors & Awards:* Group Achievement Award, NASA, 83. *Mem:* Am Geophys Union; Sigma Xi; Soc Explor Geophys; Soc Terrestrial Magnetism & Elec. *Res:* Rock magnetism; computer analysis; properties of earth materials; high-pressure geophysics; exploration geophysics; engineering geophysics. *Mailing Add:* Dept of Geol Univ of Iowa Iowa City IA 52242

CARMICHAEL, RONALD L(AD), b Independence, Mo, Sept 29, 21; m 43; c 2. MINING ENGINEERING. *Educ:* Mo Sch Mines, BS, 44, MS, 47; Colo Sch Mines, ScD(mining eng), 52. *Prof Exp:* Jr mining engr, US Bur Mines, 44-45; mine engr, Rock Island Coal Co, 45-46; instr mining, Mo Sch Mines, 47-48; teaching fel chem mining & metall eng, Ore State Col, 48-49, instr, 49-50; prin mining engr, Battelle Mem Inst, 52-57, proj leader, 57-62, sr metals economist & engr, 62-68; from assoc prof to prof eng mgt, 68-78, prof mining eng, 78-85, PROF EMER, UNIV MO-ROLLA, 85- *Concurrent Pos:* Lectr, Ohio State Univ, 57, 66; consult, Oak Ridge Nat Lab, 80-, AT&T, 82-84. *Mem:* Am Inst Mining, Metall & Petrol Engrs; Sigma Xi. *Res:* Metals and minerals economics; management theory; engineering-economics of the production, transportation refining, processing, fabrication and utilization of metals and minerals. *Mailing Add:* 4022 N River Blvd Independence MO 64050

CARMICHAEL, STEPHEN WEBB, b Detroit, Mich, July 17, 45; div; c 1. HUMAN ANATOMY. *Educ:* Kenyon Col, AB, 67; Tulane Univ, PhD(anat), 71. *Hon Degrees:* DSc, Kenyon Col, 89. *Prof Exp:* Asst prof biol, Delgado Col, 69-71; from instr to assoc prof anat, Sch Med, WVa Univ, 75-82; assoc prof, 82-87, PROF, MAYO MED SCH, 87- *Concurrent Pos:* Fel, Giorgio Cini Found, Milan, 73. *Mem:* Am Asn Anat; Electron Micros Soc Am; Am Soc Cell Biol; Sigma Xi; Soc Neuroscience. *Res:* Morphological aspects of secretion; adrenal medullary cytology. *Mailing Add:* Dept Anat Mayo Clin Med Sci Bldg Three Rochester MN 55905

CARMICHAEL, WAYNE WILLIAM, b Longview, Wash, Aug 22, 47. AQUATIC BIOLOGY & TOXICOLOGY, ENVIRONMENTAL TOXICOLOGY. *Educ:* Ore State Univ, BSc, 69; Univ Alta, MSc, 72, PhD (bot), 74. *Prof Exp:* From asst prof to assoc prof, 76-88, PROF AQUATIC BIOL & TOXICOL, WRIGHT STATE UNIV, DAYTON, OHIO, 88- *Concurrent Pos:* Vis prof, Inst Hydrobiol, Peoples Repub China, 86, dept chem, Univ Hawaii, Honolulu, 86-87; mem, Int Union Pure & Appl Chem Comn, Aquatic Bitoxins Working Group, 82- *Mem:* Int Soc Toxicol; AAAS; N Am Lake Mgt Soc; Soc Int Limnol. *Res:* Aquatic biology/toxicology; environmental toxicology; chemistry; pharmacology; molecular biology; biotechnology of blue-green algae secondary chemicals (biotoxins). *Mailing Add:* Dept Biol Sci Wright State Univ Dayton OH 45435

CARMINES, PAMELA KAY, b Hampton, Va, May 6, 55. RENAL PHYSIOLOGY, MICROCIRCULATION. *Educ:* Longwood Col, BS, 77; Ind Univ, PhD(physiol), 82. *Prof Exp:* res instr, 82-86, RES ASST PROF PHYSIOL & BIOPHYSICS, UNIV ALA, BIRMINGHAM, 86- *Concurrent Pos:* Bd Med Adv, Ala Kidney Found, 86-88. *Honors & Awards:* Award for Excellence in Renal Res, Am Physiol Soc, 81. *Mem:* Am Physiol Soc; Am Soc Nephrology; Int Soc Nephrology; Am Heart Asn. *Res:* regulation of the renal microcirculation; utilizing videometric techniques to evaluate the segmental vascular effects of hormonal vasoactive agents; the effector mechanisms involved in autoregulation of blood flow; the tubuloglomerular feedback response. *Mailing Add:* Dept Physiol Tulane Univ Med Ctr 1430 Tulane Ave New Orleans LA 70112

CARMODY, GEORGE R, b Brooklyn, NY, Mar 29, 38; m 62; c 3. POPULATION GENETICS. *Educ:* Columbia Univ, AB, 60, PhD(zool), 67. *Prof Exp:* Teaching asst zool, Columbia Univ, 61-62; USPHS fel, Univ Chicago, 67-68; asst prof, 69-74, assoc dean sci, 87-89, ASSOC PROF BIOL, CARLETON UNIV, 74- *Concurrent Pos:* Ford Found genetics training grant fel, Univ Chicago, 68-69; vis sr fel, Genetics Dept, Univ Nottingham, Eng, 76-77; guest res animal genetics lab, Nat Inst Environ Health Sci Res Triangle, 82-83; dir, intergrated sci studies, Carleton univ, 82-87; vis prof, Genetics Dept, Univ Hawaii, 89-90. *Mem:* AAAS; Genetics Soc Am; Soc Study Evolution; Genetics Soc Can; Sigma Xi. *Res:* Speciation in Drosophila; protein polymorphisms; maintenance of genetic variability in natural populations; genetic variability in cave-dwelling organisms; DNA sequence variation; forensic applications of DNA typing. *Mailing Add:* Dept Biol Carleton Univ Ottawa ON K1S 5B6 Can

CARMONA, RENE A, b Marseille, France, Aug 6, 47. APPLIED MATHEMATICS. *Educ:* Univ Marseille, PhD(probability & statist), 77. *Prof Exp:* Prof math, Univ St Etienne, France, 78-81; PROF MATH, UNIV CALIF, IRVINE, 81- *Honors & Awards:* Fel, Inst Math Statist. *Mem:* Am Math Soc; Soc Indust & Appl Math. *Mailing Add:* Dept Math Univ Calif Irvine CA 92714

CARMONY, DONALD DUANE, b Indianapolis, Ind, Sept 16, 35; m 61; c 2. HIGH ENERGY PHYSICS. *Educ:* Ind Univ, BS, 56; Univ Calif, Berkeley, PhD(physics), 62. *Prof Exp:* Res physicist, Lawrence Radiation Lab, Univ Calif, Berkeley, 61-62, res physicist & lectr, Univ Calif, Los Angeles, 62-63, res physicist, Univ Calif, San Diego, 63-66; assoc prof, 66-75, PROF PHYSICS, PURDUE UNIV, 75- *Concurrent Pos:* Alexander von Humboldt sr scientist award, 72-73; Sr Fulbright scholar, 79-80. *Mem:* Fel Am Phys Soc. *Res:* Experimental high energy physics; high energy hadron collisions; neutrino interactions; study of pion-pion interactions and strange meson resonances. *Mailing Add:* Dept of Physics Purdue Univ Lafayette IN 47907

CARMOUCHE, L(OUIS) N(ORMAN), b Newton, Kans, Aug 7, 16; m 39; c 3. CHEMICAL ENGINEERING. *Educ:* Univ Kans, BS, 38. *Prof Exp:* Chem engr, Midland Div, Dow Chem Co, 39-40, plant supt, 40-42, tech expert, 44-46, gen mgr, Ludington Div, 46-60, gen mgr, Saginaw Bay Div, 60-63, mgr basic opers, Midland Div, 63-67, gen mgr, Packaging Div, 67-68, mgr, Functional Chem & Serv Dept, 68-69, mgr, Functional Prod & Systs Dept, 69-79, corp prod dir, Exec Dept, 79-81; PRES, TRANSTEL SERV CO, 81- *Concurrent Pos:* Prod supt, Ludington Plant, Dow Magnesium Corp, 42-45. *Mem:* Am Inst Chem Engrs. *Res:* Heat transfer; crystallization; pyrometric processes; drying; management science. *Mailing Add:* 4412 Orchard Midland MI 48640

CARNAHAN, BRICE, b New Philadelphia, Ohio, Oct 13, 33. CHEMICAL ENGINEERING, APPLIED MATHEMATICS. *Educ:* Case Inst Technol, BS, 55, MS, 57; Univ Mich, PhD(chem eng), 65. *Prof Exp:* Lectr chem eng, 60-61, asst dir, Ford Found Proj, 61-63, instr chem eng & biostatist, 63-65, from asst prof to assoc prof, 65-70, asst dir, NSF Proj, 65, PROF CHEM ENG & BIOSTATIST, UNIV MICH, 70- *Concurrent Pos:* Vis prof, Univ Pa, 70; mem, Comt Comput Aids Chem Eng Educ, Nat Acad Eng; chmn, cast div, Am Inst Chem Eng, 80-81. *Honors & Awards:* Comput in Chem Eng Award, Am Inst Chem Eng, 81. *Mem:* AAAS; Am Asn Comput Mach; Am Inst Chem Engrs; Sigma Xi. *Res:* Numerical mathematics; digital computer applications; radiation chemistry. *Mailing Add:* 1605 Kearney Rd Ann Arbor MI 48104

CARNAHAN, CHALON LUCIUS, b Beverly, Mass, Sept 17, 33; m 60; c 2. GROUNDWATER HYDROLOGY. *Educ:* Calif Inst Technol, BS, 55; Univ Calif, Berkeley, MS, 58; Univ Nev, Reno, PhD(hydrol), 75. *Prof Exp:* Radiol chemist, US Naval Radiol Defense Lab, 57-62; sr radiochemist, Hazleton-Nuclear Sci Corp, 62-65, mgr phys chem dept, 65-67; sr assoc scientist, Isotopes, a Teledyne Co, 67-69, scientist & group leader, 69-70, Teledyne Isotopes, 70-71; res assoc, Desert Res Inst, Univ Nev, 71-76; supvry assoc, Environ Sci Assocs, 76-78; STAFF SCIENTIST, LAWRENCE BERKELEY LAB, UNIV CALIF, 78- *Mem:* AAAS; Am Geophys Union; Mat Res Soc. *Res:* Physical and chemical effects of underground nuclear explosions; nuclear reactions; thermodynamics of irreversible processes in flow through porous media; contamination transport in ground water. *Mailing Add:* 14 Camino Sobrante Orinda CA 94563

CARNAHAN, JAMES CLAUDE, b Yonkers, NY, Nov 25, 43; m 71; c 2. POLYMER CHEMISTRY. *Educ:* Columbia Univ, BS, 69; State Univ NY, Albany, PhD(chem), 74. *Prof Exp:* Res fel org chem, Univ Saarlandes, WGer, 74-75; STAFF CHEMIST ORG CHEM, RES & DEVELOP CTR, GEN ELEC CO, 75- *Mem:* Am Chem Soc. *Res:* Liquid and gas chromatography; polymer characterization and molecular weights; physical organic chemistry. *Mailing Add:* 737 St David Lane Schenectady NY 12309-4819

CARNAHAN, JAMES ELLIOT, b Kaukauna, Wis, Jan 26, 20; m 44; c 4. BIOCHEMISTRY, ENVIRONMENTAL BIOLOGY. *Educ:* Univ Wis, BS, 42, MS, 44, PhD(org chem), 46. *Prof Exp:* Asst chem warfare agents, Univ Wis, 42-46; reschemist, E I Du pont de Nemours & Co, 46-56, res Supvr 56-85. *Honors & Awards:* Hoblitzelle Nat Award Res Agr Sci, Tex Res Found, 65. *Mem:* Am Chem Soc; Am Soc Plant Physiol; Am Soc Bio chem; Entom Soc Am; Sigma Xi; Am Soc Neuroscience. *Res:* Organic synthesis; catalytic chemistry; polymer chemistry; biological nitrogen fixation; plant biochemistry; agricultural chemicals, drugs and air pollution effects on plants. *Mailing Add:* 907 Liftwood Dr Wilmington DE 19803

CARNAHAN, JON WINSTON, b Mt Vernon, Ill, Aug 24, 55; m 76; c 2. ANALYTICAL SPECTROMETRY, CHEMICAL ANALYSIS. *Educ:* Southern Ill Univ, Carbondale, BS, 78, MS, 80; Univ Cincinnati, PhD(chem), 83. *Prof Exp:* ASSOC PROF CHEM, NORTHERN ILL UNIV, 83- *Mem:* Am Chem Soc; Soc Appl Spectros; Sigma Xi. *Res:* Analytical atomic spectrometry instrumentation; plasma development and sampling techniques; atomic emission determination of nonmetals in solution; non metal element selective detection for gas, liquid and supercritical fluid chromatography. *Mailing Add:* Dept Chem Northern Ill Univ Dekalb IL 60115

CARNAHAN, ROBERT D, b Pontiac, Mich, June 14, 31; m 53; c 4. METALLURGY, MATERIALS SCIENCE. *Educ:* Mich Tech Univ, BS, 53; Northwestern Univ, PhD(mat sci), 63. *Prof Exp:* Res scientist, Honeywell Res Ctr, Minn, 56-59, res metallurgist, Ord Div, Minneapolis-Honeywell Regulator Co, 59; asst fracture of ceramics, Northwestern Univ, 60-62, resident damping ceramics, 62; mem tech staff, Aerospace Corp, 62-66, staff scientist, 66-68; dir mat sci lab, Universal Oil Prod, 68-73; dir res, Elec & Electronics Lab, Gould Inc, 73-80; sr vpres sci & technol, USG Corp, 80-87; PRES, PARTNERS IN TECHNOL, 88- *Concurrent Pos:* Chmn, Spec Prog Panel Mat Sci, NATO Sci Comt, 82-85; exec vpres, Univ Sci Partners, 89-; pres & chief opers officer, Thixomat, Inc, 90- *Mem:* Am Inst Mining, Metall & Petrol Engrs; Electrochemisty Soc; Inst Elec & Electronic Engrs; Sigma Xi; Am Soc Metals & Mat. *Res:* The relation of structure of metals and nonmetals to mechanical and physical behavior, especially the role of dislocations in the microstrain region and relation to macroplastic flow; dimensional stability and microplastic behavior of solids; gas phase catalysis; solid state sensors; ink jet writing devices; thin film deposition of precision resistors; surface physics and chemistry of solids. *Mailing Add:* Suite 200 Deer Valley Plaza PO Box 2051 Park City UT 84060-2051

CARNAHAN, ROBERT EDWARD, b Kaukauna, Wis, Jan 5, 25; m 50; c 2. ORGANIC CHEMISTRY. *Educ:* Univ Wis, BS, 47; Univ Ill, MS, 48, PhD(org chem), 50. *Prof Exp:* Res chemist, Pfizer, Inc, 50-55, patent agent, 55-60; patent agent, Mead Johnson & Co, 60-69, dir patents, 69-81; DIR PATENTS, BRISTOL-MYERS CO, 81- *Mem:* Am Chem Soc; AAAS. *Res:* Medicinal chemistry. *Mailing Add:* 1327 Timberlake Rd Evansville IN 47710

CARNALL, WILLIAM THOMAS, b Denver, Colo, May 23, 27; m 50; c 3. ACTINIDE CHEMISTRY. *Educ:* Colo State Univ, BS, 50; Univ Wis, PhD(chem), 54. *Prof Exp:* Asst, Univ Wis, 52-54; chemist, 54-75, SR CHEMIST, ARGONNE NAT LAB, 75- *Concurrent Pos:* Sigma Xi fel, Munich, 61-62; vis prof, Tech Univ, Munich, 84. *Mem:* Am Chem Soc; Am Nuclear Soc. *Res:* Chemistry of the actinide elements; theory of lanthanide and actinide element spectra. *Mailing Add:* 5333 S Seventh Ave La Grange IL 60525

CARNES, BRUCE ALFRED, b Duncan, Okla, June 15, 50; m 80; c 3. BIOSTATISTICS, RADIATION BIOLOGY. *Educ:* Univ Utah, BS, 73; Univ Houston, MS, 75; Univ Kans, MA, 80, PhD(statist ecol), 80. *Prof Exp:* Fel biostatist, Argonne Nat Lab, 80-81, res assoc, 81-82, asst biostatistician, 82-87, BIOSTATISTICIAN, ARGONNE NAT LAB, 87- *Mem:* Biometric Soc; Am Statist Asn; Sigma Xi. *Res:* Applied biostatistical issues in ecology and radiation biology with particular interest in the biological effects of low level radiation in whole animal, long term experiments. *Mailing Add:* Argonne Nat Lab 9700 S Cass Ave Bldg 202 Argonne IL 60439

CARNES, DAVID LEE, JR, b Youngstown, Ohio, Mar 16, 46; m 71. COMPARATIVE PHYSIOLOGY. *Educ:* Allegheny Col, BS, 68; Rice Univ, MA, 74, PhD(biol), 75. *Prof Exp:* Res assoc biol, Rice Univ, 74-75; res assoc pharmacol, Univ Mo-Columbia & Harry S Truman Vet Admin Hosp, 75-79; res assoc, Med & Endocrine Div, Children's Hosp Med Ctr & instr oral biol, Harvard Sch Dent Med, 79-84, Dept Pathophysiol, Med Sch, 79-84; ASST PROF, UNIV TEX HEALTH SCI CTR, 84- *Mem:* NY Acad Sci; AAAS; Am Soc Bone & Mineral Res. *Res:* Interrelationship of Parathyroid hormone, calcitonin and vitamin D, as they relate to the control of calcium metabolism. *Mailing Add:* 227 Arcadia Pl San Antonio TX 78209

CARNES, JAMES EDWARD, b Cumberland, Md, Sept 27, 39. CONSUMER ELECTRONICS, SOLID STATE ELECTRONICS. *Educ:* Pa State Univ, BS, 61; Princeton Univ, MA, 67, PhD(electron device physics), 70. *Prof Exp:* mem tech staff,RCA Labs, 69-77, mgr integrated circuit develop, Consumer Electronics Div, 77-78, mgr technol applns, 78-81, dir, New Prod Lab, 81-82, VPRES ENG, CONSUMER ELECTRONICS DIV, RCA, 82- *Concurrent Pos:* Lectr, short course prog, Univ Calif, Los Angeles, 73- *Honors & Awards:* David Sarnoff Outstanding Achievement Award, RCA, 81; Centennial Medal, Inst Elec & Electronics Engrs, 84. *Mem:* Fel Inst Elec & Electronics Engrs. *Res:* Experimental and analytical studies of metal-oxide-silicon integrated circuits, specifically charge-coupled devices including understanding of physics of operation and optimum design for imaging, memory and signal processing applications; application of integrated circuit technology to consumer products; advanced development of consumer electronics video products. *Mailing Add:* David Sarnoff Res Ctr CN 5300 Princeton NJ 08543

CARNES, JOSEPH JOHN, b Rock Island, Ill, Aug 3, 17; m 42; c 5. ORGANIC CHEMISTRY. *Educ:* St Ambrose Col, BS, 39; Univ Iowa, MS, 41, PhD(org chem), 43. *Prof Exp:* Chemist, Nat Defense Res Comt contract, Univ Iowa, 42-43; sr chemist, Am Cyanamid Co, 43-51, group leader, 51-54, sect mgr, 54-59, dir appl res, 59-63, dir contract res, 63-68; dir advan planning, New Eng Inst, 68-69, vpres, 69-75; RETIRED. *Mem:* Fel AAAS; Am Chem Soc; fel Am Inst Chemists. *Res:* Synthetic organic chemistry; process development; ring closure of N-chloroamines; nitrogen mustards; chemical warfare agents; surface active agents; organic phosphorus insecticides; new product development; paper and industrial chemicals; rocket propellant chemistry; energy conversion. *Mailing Add:* 1008 Apollo Beach Blvd No 208 Apollo Beach FL 33572

CARNES, WALTER ROSAMOND, b Winona, Miss, Oct 7, 22; m 46; c 2. ENGINEERING MECHANICS, AEROSPACE ENGINEERING. *Educ:* Ga Inst Tech, BS, 49, MS, 54; Univ Ill, PhD(mech), 64. *Prof Exp:* Instr math, Ga Inst Tech, 50-51, 53-55, from instr to asst prof aeronaut eng, 55-59; assoc prof, 59-63, PROF AEROSPACE ENG & ENG MECH, COL ENG, MISS STATE UNIV, 63-, ASSOC DEAN ENG & DIR INSTR, 70- *Concurrent Pos:* Aircraft struct engr, Lockheed Aircraft pilot, USAF. *Mem:* Nat Soc Prof Engrs; Am Soc Eng Educ. *Res:* Structural dynamics. *Mailing Add:* Col of Eng Miss State Univ Drawer DE Mississippi State MS 39762

CARNES, WILLIAM HENRY, pathology; deceased, see previous edition for last biography

CARNESALE, ALBERT, b Bronx, NY, July 2, 36; m 62; c 2. NUCLEAR ENGINEERING, PUBLIC POLICY. *Educ:* Cooper Union, BME, 57; Drexel Inst Technol, MS, 61; NC State Univ, PhD(nuclear eng), 65 field. *Hon Degrees:* Add in Educ ScD, NJ Inst Technol, 84. *Prof Exp:* Sr engr, Martin Co, Md, 57-62; assoc prof nuclear eng, NC State Univ, 62-69; scientist, US Arms Control & Disarmament Agency, 69-72; prof univ studies, NC State Univ, 72-74; assoc dir prog sci & int affairs, 74-78, PROF PUB POLICY & ACAD DEAN, J F KENNEDY SCH GOVT, HARVARD UNIV, 78- *Concurrent Pos:* Consult, Southern Interstate Nuclear Bd, Ga, 63-69, US Atomic Energy Comn & Res Triangle Inst, NC, 65-69 & US Nuclear Regulatory Comn, 75-77; adv, US Deleg to Strategic Arms Limitation Talks with Soviet Union, 70-72; expert & consult, US Arms Control & Disarmament Agency, 72- & US Dept State, 77-; head US deleg, Int Nuclear Fuel Cycle Eval, 78-80. *Mem:* AAAS; Am Soc Eng Educ; Am Nuclear Soc. *Res:* Nuclear energy policy, nuclear weapons policy and other national and international policy issues having substantial scientific and technological dimensions. *Mailing Add:* J F Kennedy Sch Govt Harvard Univ Cambridge MA 02138

CARNEY, BRUCE WILLIAM, b Guam, Nov 30, 46; US citizen; m 85, 87. ASTRONOMY. *Educ:* Univ Calif, Berkeley, BA, 69; Harvard Univ, AM, 71, PhD(astron), 78. *Prof Exp:* Sci & eng asst, US Army Ballistics Res Lab, 71-74; fel astron, Dept Terrestrial Magnetism, Carnegie Inst Washington, 78-80; asst prof, 80-85, assoc prof astron, 85-89, PROF ASTRON, UNIV NC, CHAPEL HILL, 89- *Concurrent Pos:* Mem, Kitt Peak Nat Observ Users Comt, 80-85 & 88-90, chmn, 83-85; mem, Astron Adv Comt, NSF, 84-87; bd dirs, Astron Soc Pac, 89-92, Optical/Infared Panel, Astron & Astrophysics Survey Comt, 89-90. *Mem:* Am Astron Soc; Astron Soc Pac; Sigma Xi; Int Astron Union. *Res:* Observational aspects of stellar and galactic evolution; particularly Population II systems. *Mailing Add:* Dept Physics & Astron CB No 3255 Phillips Hall Univ NC Chapel Hill NC 27599-3255

CARNEY, D(ENNIS) J(OSEPH), b Charleroi, Pa, Mar 19, 21; m 43; c 5. METALLURGY. *Educ:* Pa State Col, BS, 42; DSc, Mass Inst Technol, 49. *Prof Exp:* Metallurgist, US Steel Corp, Clairton, 42-43, physicist, South Works, 49-50, gen supvr metals res, 50-51, chief develop metallurgist, 51-54, supt elec furnaces, 54-56, open hearth, 56, div supt steel prod, Duquesne Works, 56-59, asst gen supt, Homestead Dist Works, 59-63; gen supt, Duquesne Works, 63-65, vpres long range facil planning, 65-68, vpres appl res, 68-72, vpres res, 72-74; vpres opers, 74-75, exec vpres & dir, 75-76, chief operating officer, 76-77, pres, 76-79, chmn bd, 78-84, CHIEF EXEC OFFICER, WHEELING-PITTSBURGH STEEL CORP, 77-; PRES, INTRA CONTINENTAL CONSULT CORP, 84- *Concurrent Pos:* Metallurgist, Naval Res Lab, 43-45; asst, Carnegie Inst Technol, 46-47 & Mass Inst Technol, 47-49. *Honors & Awards:* Fairless Award, Am Inst

Mining Engrs, 78. *Mem:* Am Inst Mining, Metall & Petrol Engrs; Int & Am Iron & Steel Inst; Soc Metals. *Res:* Gases in steel; hardenability; quenching of steel; sinter; machinability; blast furnace and open hearth studies. *Mailing Add:* 4010 Galt Ocean Dr Apt 710 Ft Lauderdale FL 33308

CARNEY, DARRELL HOWARD, b Boise, Idaho, April 15, 48; m 71; c 2. CELL SURFACES, GROWTH CONTROL. *Educ:* Col Idaho, BS, 70; Univ Conn, PhD(develop biol), 75. *Prof Exp:* Res fel, Univ Calif, Irvine, 75-78; asst prof, 78-82, ASSOC PROF BIOCHEM, UNIV TEX MED BR, 82- *Concurrent Pos:* prin invest grant, 82-, Res Career Develop Award, Nat Cancer Inst, 82- *Honors & Awards:* Res Career Develop Award, Nat Cancer Inst *Mem:* Am Soc Cell Biol; Sigma Xi. *Res:* Molecular events that lead to initiation of cell proliferation; characterization of the thrombin receptor, interaction of thrombin with this receptor, phosphoinositides and transmembrane signals and microtube involvement in initiation. *Mailing Add:* Div Biochem Univ Tex Med Br Galveston TX 77550

CARNEY, EDWARD J, b Rochester, NY, May 15, 29; m 51; c 5. STATISTICS, COMPUTER SCIENCES. *Educ:* Univ Rochester, AB, 51, MS, 58; Iowa State Univ, PhD(statist), 67. *Prof Exp:* Asst factory eng dept, Bausch & Lomb Optical Co, 54-58; staff asst area commun proj, Gen Dynamics Electronics, 58-59; instr indust eng, Iowa State Univ, 59-63, asst prof indust eng & statist, 63-67, assoc prof statist, 67-74, PROF COMPUT SCI & STATIST, UNIV RI, 74- *Mem:* Inst Math Statist; Am Statist Asn; Asn Comput Mach; Sigma Xi. *Res:* Design of experiments; variances of variance component estimates; statistical computations. *Mailing Add:* Dept Computer Sci & Statist Univ RI 264 Tyler Hall Kingston RI 02881

CARNEY, GORDON C, b Glasgow, Scotland, Sept 15, 34; m 60; c 2. INVERTEBRATE PHYSIOLOGY. *Educ:* Univ Durham, BSc, 57, MSc, 60; Univ Minn, PhD(entom), 64. *Prof Exp:* Jr res officer, Atomic Energy Can, Ltd, 63-64, asst res officer, 64-66; asst prof biol, Bowling Green State Univ, 66-69; lectr biol, 69-74, SR LECTR CELL PHYSIOL, BRISTOL POLYTECH, 74- *Res:* Physiological effects of pollutants on aquatic invertebrates; binding of heavy metals by marine algae and its effect on primary production; dietary and environmental uptake of cadmium by Daphnia and other invertebrates; uptake of metals by marine diatoms. *Mailing Add:* Dept Biol Sci Bristol Polytechnic Bristol BS16 1QY England

CARNEY, JOHN MICHAEL, b Chicago, Ill, Jan 27, 47; m 73; c 1. BEHAVIORAL PHARMACOLOGY. *Educ:* St Mary's Col, BS, 68; San Diego State Univ, MS, 70; Univ Mich, PhD(pharmacol), 76. *Prof Exp:* Teaching asst human anat, San Diego State Univ, 68-70 & neurophysiol, Univ Mich, 70-71; fel, Med Col Va, 76-78; ASST PROF PHARMACOL & ADJ ASST PROF PSYCHIAT, UNIV OKLA, 78- *Concurrent Pos:* Nat Inst Drug Abuse fel, 76-78; prin investr, Hoechst-Roussel Pharmaceut Inc, 78-80, consult, Frankfurt, 78-; prin investr, Nat Inst Drug Abuse grant, 78-81; mem, Int Study Group Invest Drugs as Reinforcers. *Mem:* Sigma Xi; Behav Pharmacol Soc. *Res:* Drug abuse; alcoholism; neurobiology of epilepsy; behavioral teratology; mental retardation and drugs. *Mailing Add:* Dept Pharmacol Univ Okla Health Sci Ctr PO Box 26901 Oklahoma City OK 73190

CARNEY, RICHARD WILLIAM JAMES, b Novelty, Mo, June 19, 34; m 57; c 3. ORGANIC & PHARMACEUTICAL CHEMISTRY. *Educ:* McPherson Col, BS, 57; Iowa State Univ, MS, 61, PhD(org chem), 62. *Prof Exp:* Sr chemist, sr res chemist & sr staff chemist, Ciba-Geigy Corp, 62-72, sr staff scientist process res, 72-75, mgr, Chem Develop Res, Pharmaceut Div, 75-82, mgr, Chem Opers Planning & Job Serv, 82-89, dir, Res Statist & Comps Serv, 89-91, DIR, NEW DRUG COORDINATIONS, 91- *Concurrent Pos:* Vis scientist, Basel, 68. *Mem:* Am Chem Soc; Am Inst Chem; NY Acad Sci. *Res:* Diterpenes; chemistry of heterocyclics. *Mailing Add:* 556 Morris Ave Summit NJ 07901

CARNEY, ROBERT GIBSON, dermatology; deceased, see previous edition for last biography

CARNEY, ROBERT SPENCER, b Memphis, Tenn, Aug 9, 45. BIOLOGICAL OCEANOGRAPHY, MARINE BIOLOGY. *Educ:* Duke Univ, BS, 67; Tex A&M Univ, MS, 71; Ore State Univ, PhD(oceanog), 77. *Prof Exp:* Res asst biol oceanog, Ore State Univ, 71-76; fel evolutionary biol, Smithsonian Inst, 76-77, res collabr, 77-81; asst prog dir biol, 78-80, PROG DIR BIOL OCEANOG, NSF, 80-81; MARINE SCI, LA STATE UNIV. *Mem:* Sigma Xi; AAAS. *Res:* Marine benthic ecology; quantitative analysis of ecological data; deep-sea community ecology; functional morphology of deep-sea animals; evolution and systematics of deep-sea holothurians and other echinoderms; radioactive waste disposal. *Mailing Add:* Dept Marine Sci La State Univ-Baton Rouge Baton Rouge LA 70803-7505

CARNEY, ROSE AGNES, b Chicago, Ill. PHYSICS. *Educ:* DePaul Univ, BS, 42, MS, 46; Ill Inst Technol, PhD, 61. *Prof Exp:* Asst metal labs, Univ Chicago, 42-43; instr, Army Spec Training Prog, De Paul, 43-44, instr physics, 44-46; chmn physics & math, St Xavier Col, Ill, 46-48; assoc prof physics, 48-59, prof math & chmn dept, 59-81, chmn, Div Natural Sci, 69-81, PROF MATH, ILL BENEDICTINE COL, 81- *Concurrent Pos:* Consult, Argonne Nat Lab, 62, 63, 64; Chicago regional coordr, Women & Math Prog, Math Asn Am, 85- *Mem:* Am Asn Physics Teachers; Math Asn Am; Sigma Xi. *Res:* Molecular spectroscopy; mathematical physics. *Mailing Add:* Dept of Math Ill Benedictine Col Lisle IL 60532

CARNEY, WILLIAM PATRICK, b Dillon, Mont, July 1, 38; m 65; c 2. HELMINTHOLOGY, ZOONOTIC DISEASES. *Educ:* St Edwards Seminary, Kenmore, Wash, BA, 60; Western Mont Col, Dillon, BS, 62; Univ Mont, Missoula, PhD(zool), 67; Johns Hopkins Univ, MPH, 76. *Prof Exp:* Res assoc parasitol, Minot State Col, NDak, 67-69, Naval Med Res Inst, 69-70; res scientist parasitol, Naval Med Res Unit, Jakarta, 70-74 & Naval Med Res Inst, 74-75, Naval Med Res Unit, Taipei, 76-79; Officer-in-charge infectious dis, Naval Med Res Unit-Jakarta, 79-81; PROG MGR INFECTIOUS DIS, NAVAL MED RES & DEVELOP COMMAND, 81- *Concurrent Pos:* Ed assoc, Chinese J Microbiol, 76-79; vis assoc prof, Nat Taiwan Univ, 77-79, Nat Yang-Ming Med Col, 78-79; consult, Western Pac Region, WHO, 79. *Mem:* Am Soc Parasitologists; Am Micros Soc; Wildlife Dis Asn; Am Malocol Union. *Res:* Epidemiological investigations of tropical infectious diseases of man with emphasis of snail-transmitted diseases in Southeast Asia. *Mailing Add:* 12900 Twinbrook Pkwy Rockville MD 20851

CARNIGLIA, STEPHEN C(HARLES), b San Francisco, Calif, Jan 15, 22; m 47; c 2. MATERIALS SCIENCE, ENGINEERING EDUCATION. *Educ:* Univ Calif, Berkeley, BS, 43, MS, 45, PhD(chem), 54. *Prof Exp:* Instr chem, Marin Col, 46-51; chemist, Radiation Lab, Univ Calif, 51-53 & Calif Res & Develop Co, 53; res dir inorg chem, Westvaco Mineral Prod Div, FMC Corp, 53-56; mgr mat & process res, Atomics Int Div, Rockwell Int Corp, 56-68 & Rocketdyne Div, 68-70; mgr appl chem res & develop, Kaiser Aluminum & Chem Corp, 70-86; LECTR & CONSULT, MAT DISCIPLINES, 86- *Concurrent Pos:* Mem, Nat Acad Sci-Mat Adv Bd Comts Ceramics, 60-67; vis prof, Mat Sci & Eng, Univ Calif, Los Angeles & Davis, 87-90 & Univ Utah, 88; adj prof, Mat Sci & Eng, Univ Calif, Davis, 91- *Honors & Awards:* Ross Coffin Purdy Award, Am Ceramic Soc, 74. *Mem:* Am Catalysis Soc; fel Am Ceramic Soc; Am Ceramic Soc (vpres, 80-81). *Res:* Chemistry, physics, mechanics and preparation of refractory materials; high surface area materials, catalysts and substrates; corrosion engineering. *Mailing Add:* 115 Wilshire Ct Danville CA 94526

CARNOW, BERTRAM WARREN, b Philadelphia, Pa, June 19, 22; m 75; c 6. ENVIRONMENTAL HEALTH, THORACIC DISEASES. *Educ:* NY Univ, BA, 47; Chicago Med Sch, MB & MD, 51. *Prof Exp:* Intern, Cook County Hosp, Chicago, 51-52; resident cardiol, Michael Reese Hosp, Chicago, 52-53, resident clin internal & cardio-pulmonary med, 53-55; physician pvt pract, 55-69; clin assoc, 64-65, asst clin prof, 65-67, from asst prof to assoc prof, 67-70, dir, Div Occup Med, Cook County Hosp, 77-79, dir, Occup Safety & Health, Educ Res Ctr, 77-87, PROF PREV MED & OCCUP MED, DEPT MED, SCH MED, UNIV ILL, 70-, PROF OCCUP & ENVIRON MED, SCH PUB HEALTH, 72- *Concurrent Pos:* Consult & attend physician, Michael Reese Hosp, 55-72 & Univ Ill Hosp, 70-; chest consult, Union Health Serv, 57-76; med dir, Chicago Lung Asn, 69-77; dir, Environ Health Resource Ctr, Ill Inst Environ Qual, 70-78; dir dept Occup & Environ Med, Sch Pub Health, 72-78; pres & sr scientist, Carnow Conibear and Assoc, Occup & Environ Consults, 75; dir, Great Lakes Ctr Occupational Safety & Health, Univ Ill, 76-87. *Mem:* AAAS; Am Thoracic Soc; fel Am Pub Health Asn; fel Am Col Chest Physicians; fel Royal Soc Health; Am Col Toxicol; fel Am Col Epidemiol. *Res:* Effects of environmental hazards on health, including air and water pollution morbidity and mortality; health and energy; occupational diseases including pneumoconiosis, metals, noise, health and toxic waste; medical surveillance hazardous waste workers; toxicology hazardous wastes. *Mailing Add:* Carnow Conibear & Assoc Ltd 333 W Wacker Dr Chicago IL 60606

CARO, LUCIEN G, b Toulon, France, July 5, 28; US citizen; m 54; c 2. MOLECULAR BIOLOGY. *Educ:* Tulane Univ, BS, 57; Yale Univ, PhD(biophys), 59. *Prof Exp:* NSF fel & guest investr cell biol, Rockefeller Inst, 59-61, Helen Hay Whitney fel & res assoc, 61-64; biophysicist, Oak Ridge Nat Lab, 64-70; PROF MOLECULAR BIOL, UNIV GENEVA, 70- *Concurrent Pos:* NSF res grant, 64; vis investr, Inst Molecular Biol, Univ Geneva, Switz, 62-64. *Mem:* AAAS; Swiss Soc Cellular & Molecular Biol; Am Soc Microbiol; Swiss Soc Microbiol. *Res:* Control of DNA replication; plasmids; electron microscopy; autoradiography. *Mailing Add:* Dept Molecular Biol Univ Geneva 30 Quai E Ansermet CH-1211 Geneva 4 Switzerland

CAROFF, LAWRENCE JOHN, b Beaverdale, Pa, Aug 26, 41; c 3. ASTROPHYSICS. *Educ:* Swarthmore Col, BS, 62; Cornell Univ, PhD(appl physics), 67. *Prof Exp:* Res scientist astrophys, 67-81, ACTG DEP DIV CHIEF, SPACE SCI DIV, AMES RES CTR, NASA, 81- *Mem:* Int Astron Union; Am Astron Soc. *Res:* Theoretical research into nature of quasars and active galaxies; cosmology; galactic infrared sources; infrared observations from aircraft and ground observations of galactic H II regions, stars and planets. *Mailing Add:* Code SZB NASA Hq Washington DC 20546

CAROLIN, VALENTINE MOTT, JR, b Sayville, NY, Aug 23, 18; m 50; c 2. INSECT ECOLOGY. *Educ:* Syracuse Univ, BS, 39, MS, 42. *Prof Exp:* Scout, Bur Entom & Plant Quarantine, USDA, NJ, 39; survr, Radio Corp Am Commun, Inc, NY, 39-40; asst, State Univ NY Col Forestry, Syracuse Univ, 40-42; entomologist, Bur Entom & Plant Quarantine, USDA, 46-54, Pac Northwest Forest & Range Exp Sta, US Forest Serv, 54-74, supvry res entomologist, Pac Northwest Forest & Range Exp Sta, 74-76; consult entomologist, 77-86; RETIRED. *Honors & Awards:* Distinguished Tech Commun Award, US Forestry Serv, 79. *Res:* Biological control factors, especially parasites of forest insects; ecology of spruce budworm; western hemlock looper; techniques for field fumigation of European pine shoot moth; associated insects on western conifers; March flies as pollinators. *Mailing Add:* 9030 SE Mill St Portland OR 97216

CAROME, EDWARD F, b Cleveland, Ohio, May 22, 27; m 51, 77; c 8. PHYSICS. *Educ:* John Carroll Univ, BS & MS, 51; Case Inst Technol, PhD(physics), 54. *Prof Exp:* From asst prof to assoc prof, 54-63, PROF PHYSICS, JOHN CARROLL UNIV, 63-; VPRES, EDJEWISE SENSOR PROD, INC, 87- *Concurrent Pos:* Liaison scientist, London Br, Off Naval Res, 68-69; sr res assoc, Hansen Labs Physics, Stanford Univ, 78-79; Underwater Acoust chair, Naval Postgrad Sch, 83-84. *Mem:* Fel Am Phys Soc; fel Acoust Soc Am; fel Inst Elec & Electronic Engrs; Inst Elec & Electronic Engrs Power Eng Soc; Soc Explor Geophysicists. *Res:* Theoretical nuclear structure; theoretical and experimental studies of ultrasonic waveguides; absorption and dispersion of ultrasound in liquids; propagation of acoustic transients; laser induced effects in liquids and solids; acoustically induced effects in optical fibers; optical fiber sensor technology. *Mailing Add:* Edjewise Sensor Products Inc 3450 Green Rd Suite 201 Cleveland OH 44122

CARON, AIMERY PIERRE, b Paris, France, Apr 20, 30; US citizen; m 56; c 1. PHYSICAL CHEMISTRY. *Educ:* Univ Calif, Los Angeles, BS, 55; Univ Southern Calif, MS, 58, PhD(crystallog chem), 62. *Prof Exp:* Lab asst, Univ Southern Calif, 55, US Air Force & Army fel, 62-63; mem res staff phys chem, Space Mat Lab, Northrop Corp, 63-66; asst prof chem, Univ Mass, 66-68; mgr, C & M Caron, Inc, VI, 68-69; admin officer, 69-73, asst to pres, 72-77, ASSOC PROF CHEM, UNIV VI, 73-, DIR COMMUNITY SERVS, 77- *Concurrent Pos:* Dir cont educ & summer sessions, 82- *Mem:* Am Crystallog Asn. *Res:* Determination of molecular structures by x-ray diffraction; inorganic syntheses; infrared spectroscopy; general physical chemistry; scientific computer programming. *Mailing Add:* Univ VI St Thomas VI 00801

CARON, DEWEY MAURICE, b North Adams, Mass, Dec 25, 42; div; c 3. ENTOMOLOGY, APICULTURE. *Educ:* Univ Vt, BS, 64; Univ Tenn, Knoxville, MS, 66; Cornell Univ, PhD(entom), 70. *Prof Exp:* Instr entom, Cornell Univ, 68, admin asst, Dept Entom, 69-70; asst to prof apicult, Univ Md, 70-81, actg chmn entom, 80-81; chmn entolm & appl ecol, 81-84, PROF ENTOM, UNIV DEL, 84- *Concurrent Pos:* Fulbright (Panama), 86-87. *Mem:* Entom Soc Am; Am Inst Biol Sci; Bee Res Asn. *Res:* Pollination ecology; biology and behavior of bees and wasps. *Mailing Add:* Dept Entom & Appl Ecol Univ Del Newark DE 19717

CARON, E(DGAR) LOUIS, b Chicago, Ill, July 31, 22; m 48; c 5. ORGANIC CHEMISTRY, INFORMATION SCIENCE. *Educ:* Tex A&M Univ, BSEE, 44; Bradley Univ, BS, 47; Univ Ill, MS, 48. *Prof Exp:* Res chemist, Upjohn Co, 48-66, sci systs analyst, 66-72, info scientist, 72-87; RETIRED. *Mem:* Am Chem Soc; Inst Elec & Electronics Engrs; Asn Comput Mach; Asn Syst Mgt; Am Inst Chem; Data Processing Mgt Asn. *Res:* Large data base scientific and managerial information storage and retrieval computer systems. *Mailing Add:* 1307 Coolidge Ave Kalamazoo MI 49007-2123

CARON, PAUL R(ONALD), b Fall River, Mass, July 14, 34; m 60; c 1. ELECTRICAL ENGINEERING, COMPUTER ENGINEERING. *Educ:* Bradford Durfee Col Tech, BSEE, 57; Brown Univ, MSc, 60, PhD(eng), 63. *Prof Exp:* Asst elec eng, Brown Univ, 57-63; mem tech staff, Plasma Res Lab, Aerospace Corp, 63-65; assoc prof elec eng, Southeastern Mass Tech Inst, 65-66; staff mem, Electronics Res Ctr, NASA, 66-70; assoc prof, 70-74, PROF ELEC ENG, SOUTHEASTERN MASS UNIV, 74- *Mem:* Inst Elec & Electronics Engrs. *Res:* Electromagnetic theory; antennas; micro processors. *Mailing Add:* Dept of Elec Eng Southeastern Mass Univ North Dartmouth MA 02747

CARON, RICHARD EDWARD, b Pawtucket, RI, April 7, 50; m 74; c 1. INTEGRATED PEST MANAGEMENT. *Educ:* Univ Maine, Orono, BS, 72, NC State Univ, Raleigh, MS, 76, PhD(entom), 81. *Prof Exp:* Res asst, NC State Unv, 76-78; asst prof, 81-86, ASSOC PROF ENTOM, AGR EXT SERV, UNIV TENN, 86- *Mem:* Sigma Xi; Entom Soc Am. *Res:* Corn earworm production on corn relative to survival factors; bollweevil diapause and longevity; effects of soil-applied pesticides; cultivar; planting date on soybean arthropods. *Mailing Add:* 605 Airways Blvd Jackson TN 38301

CARON, RICHARD JOHN, b Montreal, Que, Aug 23, 54; m 76; c 2. CONTINUOUS OPTIMIZATION. *Educ:* Univ Waterloo, BM, 77, MM, 79, PhD(optimization), 83. *Prof Exp:* asst prof, 83-87, ASSOC PROF, UNIV WINDSOR, 87- *Mem:* Math Programming Soc; Can Opers Res Soc. *Res:* Parametric linear and quadratic programming; redundancy in mathematical programming. *Mailing Add:* Dept Math Univ Windsor 401 Sunset Ave Windsor ON N9B 3P4 Can

CARONE, FRANK, b New Kensington, Pa, Nov 28, 27; m 52; c 5. PATHOLOGY. *Educ:* WVa Univ, AB, 48; Yale Univ, MD, 52. *Prof Exp:* Instr, Sch Med, Yale Univ, 59-60; from asst prof to prof, 60-69, MORRISON PROF PATH & DEP CHMN DEPT, SCH MED, NORTHWESTERN UNIV, CHICAGO, 69- *Concurrent Pos:* Life Inst Med Res Fund res fel, 57-59; Markle scholar, 64-69; dir labs, Northwestern Mem Hosp, Chicago, 61- *Mem:* Am Fedn Clin Res; Int Acad Path; Am Soc Exp Path. *Res:* Renal pathophysiology employing micropuncture techniques; light and electron microscopic study of human renal disease. *Mailing Add:* Northwestern Mem Hosp 250 E Superior St Wesley Rm 576 Chicago IL 60611

CAROSELLI, REMUS FRANCIS, b Providence, RI, Oct 4, 16; m 48; c 3. TEXTILE CHEMISTRY. *Educ:* Univ RI, BS, 37. *Prof Exp:* Process control technologist, Ashton Plant, Owens-Corning Fiberglas Corp, 41-46, chief chemist, 46-48, proj mgr textile res, 48-50, asst res mgr, 50-57, lab mgr textile process & prod develop, 57-60, mgr textile prod develop lab, 60-73; PRES, R F CAROSELLI PROD DEVELOP SERV, 74- *Concurrent Pos:* Mem res adv bd, Textile Inst, Princeton Univ. *Mem:* Am Chem Soc; Asn Textile Chemists & Colorists; Textile Res Inst. *Res:* Textiles, plastics, coatings and research and development organization with main emphasis in field of glass fibers. *Mailing Add:* 230 Colonel John Gardner Rd Narragansett RI 02882

CAROSELLI, ROBERT ANTHONY, b Jersey City, NJ. CHEMISTRY. *Educ:* St John's Univ, BS, 66. *Prof Exp:* Group leader, Carter Wallace Pharmaceut, 67-70, regulatory affairs, 70-74; regulatory liaison, Knoll Pharmaceut, 75-78, mgr prod develop, 78-84, dir, 84-87; DIR PROD DEVELOP, CIBA CONSUMER PHARMACEUT, 87- *Mem:* Am Pharmaceut Asn; Inst Food Technologists. *Res:* Development of new or improved dosage forms and manufacturing processes for over the counter pharmaceuticals; collaborate with international facilities to assure smooth technology transfer. *Mailing Add:* Ciba Consumer Pharmaceut Edison NJ 08837

CAROTHERS, OTTO M, algebra; deceased, see previous edition for last biography

CAROTHERS, STEVEN WARREN, b Prescott, Ariz, Dec 19, 43; c 4. ECOLOGY. *Educ:* Northern Ariz, Univ, BS, 66, MS, 69; Univ Ill, Urbana, PhD (ecol), 74. *Prof Exp:* Asst ornithologist, Mus Northern Ariz, 66-67; teaching asst biol, Northern Ariz Univ, 67-68; asst cur zool, Mus Northern Ariz, 69-70; instr ornith, Northern Ariz Univ, 70; teaching asst biol, Univ Ill, 70-71; cur zool, Mus Northern Ariz, 71-74, res ecologist, 74-80; prin, SWC Assocs, 80-84, PRES, SWCA, INC, 84- *Concurrent Pos:* Artist in Residence, Univ Ill, 79; instr, Dept Environ Sci, Univ Va; pvt consult fed burro problems, Dept Navy, consult, commercial & govt contracts; res assoc, Mus Northern Ariz; res scientist, Nat Park Serv; mem, Gov's Comn Ariz Envrion, 84; mem, Southern Bald Eagle Recovery Team. *Mem:* Am Ornithologists Union; Am Soc Mammalogists; Cooper Ornith Soc; Wilson Ornith Soc; Nature Conservancy. *Res:* Work with federal agencies and private institutions at designing proper land-use management plans particularly those affecting non-game wildlife; ecology of Grand Canyon. *Mailing Add:* PO Box 96 Flagstaff AZ 86002

CAROTHERS, ZANE BLAND, b Philadelphia, Pa, Nov 7, 24; m 52; c 2. BOTANY. *Educ:* Temple Univ, BS, 50, MEd, 52; Univ Michigan, PhD(bot), 58. *Prof Exp:* Instr bot, Univ Ky, 57-59; from asst prof to assoc prof, Univ Ill, Urbana Champaign, 59-64, actg chmn dept, 62-63, assoc head dept, 70-72, 83-84, & 88-89, prof bot, 76-83, prof plant biol, 83-91. *Mem:* Brit Bryol Soc; Bot Soc Am; Am Bryol & Lichenol Soc; Int Asn Bryologists; AAAS. *Res:* Ultrastructure of gametogenesis in bryophytes and nonvascular plants; anatomy of vascular plants. *Mailing Add:* 289 Morrill-Plant Biol Univ Ill Urbana-Champaign Urbana IL 61801-3793

CAROVILLANO, ROBERT L, b Newark, NJ, Aug 2, 32; m 52; c 3. SPACE PHYSICS, ASTROPHYSICS. *Educ:* Rutgers Univ, AB, 54; Ind Univ, PhD(physics), 59. *Prof Exp:* From asst prof to assoc prof, 59-67, PROF PHYSICS, BOSTON COL, 67-, CHMN DEPT, 69- *Concurrent Pos:* Vis fac, Mass Inst Technol, 67-68; assoc ed, Cosmic Electrodynamics, 69-72 & Rev Geophys-Space Physics, 75-78. *Mem:* AAAS; Am Phys Soc; Am Geophys Union (secy magnetospheric sect, 70-76; Am Asn Physics Teachers; NY Acad Sci; Sigma Xi. *Res:* Theoretical studies on the solar wind, the magnetosphere, the ionosphere and auroras; plasma physics. *Mailing Add:* Dept of Physics Boston Col Chestnut Hill MA 02167

CAROW, JOHN, b Ladysmith, Wis, Aug 26, 13; m 42; c 2. FORESTRY. *Educ:* Univ Mich, BSF, 37, MF, 38. *Prof Exp:* Field asst forest surv, Appalachian Forest Exp Sta, 37-38, jr forester, 39-42; forest statistician, Am Paper & Pulp Asn, 38-39; shelterbelt asst, US Forest Serv, 39; res forester, Southeastern Forestry Exp Sta, 42-46; from instr to prof, 47-78, EMER PROF FOREST MGT, UNIV MICH, 78- *Mem:* Fel Am Foresters. *Res:* Forest mensuration, inventory techniques, logging cost analysis; forest management. *Mailing Add:* 2027 Hill St Ann Arbor MI 48104

CAROZZI, ALBERT VICTOR, b Geneva, Switz, Apr 26, 25; nat US; m 49; c 2. GEOLOGY. *Educ:* Univ Geneva, MS, 47, DSc, Univ Geneva, 48. *Prof Exp:* Lectr spec geol, 48 & 53, asst prof, 53-57; assoc prof geol, Univ Ill, Urbana-Champaign, 57-59, assoc mem, 69-70, PROF GEOL, CTR ADVAN STUDY, 59- *Concurrent Pos:* Asst vis prof, Univ Ill, 55-56; Am Asn Petrol Geol distinguished lectr, 59; adv, Govt Ivory Coast, Africa, 60-; corresp mem, Int Comt Hist of Geol Sci, 68-; consult adv, Petroleo Brasileiro SAm, Brazil, 69- & Philippine Oil Develop Co, Manila, 70-; Yacimientos Petroliferos Fiscales Arg, 81-; invited prof geol, Fed Univ Ouro Preto, Minas Gerais, Brazil, 83-85; assoc prof geol, Univ Poitiers, France, 85-86; pres, Hist Earth Sci Soc, 84. *Honors & Awards:* Davy Award, Univ Geneva, 49, 54; Plantamour-Prevost Award, 55; Hist Geol Award, Geol Soc Am, 89; Marc-Auguste Pictet Hist Sci Medal, Physics & Nat Hist Soc Geneva, 90. *Mem:* Fel Geol Soc Am; Am Asn Petrol Geol; Soc Econ Paleont & Mineral; Hist Sci Soc; Sigma Xi. *Res:* Sedimentary petrology, such as models of deposition of carbonate rocks and sandstones, experimental studies on porosity in carbonate rocks; history of geology; oil exploration. *Mailing Add:* 245 Nat Hist-Geol Univ of Ill Urbana-Champaign Urbana IL 61801-2999

CARP, GERALD, b New York, NY, Aug 20, 24; m 48; c 2. ELECTRONICS ENGINEERING, PHYSICS. *Educ:* City Col New York, BEE, 48; Polytech Inst Brooklyn, MSEE, 60. *Prof Exp:* Instr elec eng, City Col New York, 48-50; chief weapons effects nuclear weapons, US Army Res & Develop Lab, 50-60; mgr radiation effects opers, Gen Elec Co, 60-66; dir adv develop educ tech, Gen Learning Corp, 66-67; consult info syst, Gen Elec Co, 67-68; group leader sensor-tech, Mitre Corp, 68-74; chief appl technol, Drug Enforcement Admin, 74-76; chief aviation sect res & develop, Fed Aviation Admin, 76-81; PRIN ENGR, MITRE CORP, 81- *Honors & Awards:* Inventors Award, Gen Elec Co, 65. *Mem:* Sigma Xi; Inst Elec & Electronics Engrs. *Res:* Application of nuclear techniques; gamma and neutron interactions; image analysis, pattern recognition to detection of bulk explosives; sensor technology. *Mailing Add:* 8613 Fox Run Potomac MD 20854

CARP, RICHARD IRVIN, b Philadelphia, Pa, May 10, 34; m 60; c 3. MICROBIOLOGY, VIROLOGY. *Educ:* Univ Pa, BA, 55, VMD, 58, PhD(microbiol), 62. *Prof Exp:* Asst virol, Wistar Inst, 58-62, fel, 62-63; virol res dir, Alembic Chem Co, India, 63-64; assoc, Wistar Inst, 64-68; MEM STAFF, INST BASIC RES DEVELOP DISABILITIES, 68-; PROF MICROBIOL, STATE UNIV NY DOWNSTATE MED CTR, 77- *Concurrent Pos:* Vis prof, Div Biochem Virol, Col Med, Baylor Univ, 66-; adj prof, City Univ New York, 69-83. *Mem:* Am Soc Microbiol; Am Soc Virol; AAAS. *Res:* Slow infections of the central nervous system with particular interest in scrapie and the search for the causes of alzheimer disease and amyotrophic lateral sclerosis; model systems of viruses that cause birth defects and mental retardation. *Mailing Add:* Inst Basic Res Develop Disabilities 1050 Forest Hill Rd Staten Island NY 10314

CARPENTER, ADELAIDE TROWBRIDGE CLARK, b Athens, Ga, June 24, 44. GENETICS. *Educ:* NC State Univ, BS, 66; Univ Wash, MS, 69, PhD(genetics), 72. *Prof Exp:* NIH fel cytogenetics, Univ Wis-Madison, 72-74; res assoc, Dept Anat, Duke Univ, 74-75, asst adj prof, 75-76; from asst prof to assoc prof, 76-85, PROF BIOL, UNIV CALIF, SAN DIEGO, 85- *Concurrent Pos:* assoc ed, Genetics, 80- *Mem:* Genetics Soc Am; Am Soc Naturalists; Am Soc Cell Biol; AAAS; Genetics Soc Can. *Res:* Analysis of meiotic and female-sterile mutants in Drosophila melanogaster females and of cell-division mutants by electron microscopy. *Mailing Add:* Dept Biol B-022 Univ Calif San Diego La Jolla CA 92093-0322

CARPENTER, ALDEN B, b Newton, Mass, Feb 24, 36; m 61; c 2. MINERALOGY. *Educ:* Harvard Univ, AB, 57, PhD(geol), 63. *Prof Exp:* From asst prof to assoc prof geol, 63-73, prof geol, Univ Mo-Columbia, 73-81; SR RES ASSOC, CHEVRON OIL FIELD RES CO, LA HABRA, CALIF, 81- *Mem:* Mineral Soc Am; Mineral Asn Can; Geochem Soc; Soc Econ Paleontologists & Mineralogists; Am Asn Petrol Geologists. *Res:* Geochemistry of subsurface waters; diagenesis of sandstones and carbonate rocks. *Mailing Add:* Chevron Oil Field Res Co Box 446 La Habra CA 90633

CARPENTER, ANNA-MARY, b Ambridge, Pa, Jan 14, 16. PATHOLOGY. *Educ:* Geneva Col, BA, 36; Univ Pittsburgh, MS, 37, PhD(microtech), 40; Univ Minn, MD, 58. *Hon Degrees:* DSc, Geneva Col, 68. *Prof Exp:* Res asst, Univ Pittsburgh, 38-40; instr, Moravian Col Women, 41-42; chmn biol curricula, Keystone Col, 42-44; res assoc path dept, Children's Hosp, Pittsburgh, 44-53; lectr mycol, Sch Med, Univ Pittsburgh, 44-53 & Western Reserve Univ, 53-54; from instr to prof anat, Sch Med, Univ Minn, Minneapolis, 54-80; PROF PATH, IND UNIV, 80- *Concurrent Pos:* Mem adv bd, Am Diabetes Asn, 82- *Honors & Awards:* Pioneer Award, Histochem Soc, 88. *Mem:* AAAS; hon mem Histochem Soc (secy, 74-75, treas, 75-79, pres, 80-81); Am Asn Anat; Int Soc Human & Animal Mycol; hon mem Int Soc Stereology (secy-treas, 72-83); Am Diabetes Asn. *Res:* Mycology; histochemistry; quantitation; diabetes. *Mailing Add:* 6424 Hayes Merrillville IN 46410

CARPENTER, BARRY KEITH, b Hastings, Eng, Feb 13, 49; m 74. REACTION MECHANISMS, NEUROCHEMISTRY. *Educ:* Warwick Univ, BSc, 70; Univ Col, Univ London, PhD(chem), 73. *Prof Exp:* NATO fel, Yale Univ, 73-75; from asst prof to assoc prof, 75-85, PROF CHEM, CORNELL UNIV, 85- *Concurrent Pos:* Fel, A P Sloan Found, 80-82, J S Guggenheim Found, 86. *Honors & Awards:* R A Welch lectr, 89; Alexander von Humboldt Award, 90. *Mem:* Royal Soc Chem; Am Chem Soc; AAAS. *Res:* Mechanistic organic and mechanistic organometallic chemistry; neurochemistry. *Mailing Add:* Dept Chem Cornell Univ Ithaca NY 14853-1301

CARPENTER, BENJAMIN H(ARRISON), b Buckhannon, WVa, Apr 5, 21; m 41, 81; c 3. CHEMICAL ENGINEERING, POLLUTION CONTROL. *Educ:* WVa Wesleyan Col, BS, 41; WVa Univ, MS, 61. *Prof Exp:* Res chemist, Union Carbide Corp, 48, mgr qual control systs, 49-53, group leader eng statist, 54-63, eng consult chem div, 64-69; sr engr, Statist Res Div, 69-76, head, Indust Process Studies sect, 77-87, PRES ENG RES APPLN RES TRIANGLE INST, 87- *Mem:* Am Chem Soc; Am Statist Asn; Am Soc Qual Control; Air Pollution Control Asn; Am Iron & Steel Inst; Sigma Xi. *Res:* Quality of life research; new industrial processes; energy efficient processes; industrial pollution control; environmental assessment of industrial processes. *Mailing Add:* 1220 Huntsman Dr Durham NC 27713

CARPENTER, BRUCE H, b Rapid City, SDak, Feb 5, 32; m 52, 75; c 2. PLANT PHYSIOLOGY. *Educ:* Calif State Col, Long Beach, AB, 57, MA, 58; Univ Calif, Los Angeles, PhD(bot), 62. *Prof Exp:* Instr biol, Calif State Univ, Long Beach, 57-59, from asst prof to prof, 62-75, chmn dept, 67-72, assoc vpres Acad affairs-acad personnel, 72-75; provost & acad vpres, Western Ill Univ, 75-82; PRES, EASTERN MONT COL, 82- *Concurrent Pos:* Nat Sci Found res grant, 63-72. *Mem:* AAAS; NY Acad Sci; Am Soc Plant Physiol. *Res:* Plant photoperiodism and Circadian rhythms. *Mailing Add:* Eastern Mont Col 1500 W North 30th St Billings MT 59101

CARPENTER, C(LIFFORD) LEROY, chemical engineering, environmental protection; deceased, see previous edition for last biography

CARPENTER, CAROLYN VIRUS, b Chicago, Ill, Jan 1, 40; m 64; c 3. BIOCHEMISTRY, MICROBIOLOGY. *Educ:* Univ Ill, Urbana, BS, 63; Univ Ill, Chicago Med Ctr, PhD(biochem), 68. *Prof Exp:* Fel biochem, 68-75, res assoc biochem & molecular biol, Northwestern Univ, 75-76; RES SCIENTIST, WEYERHAEUSER CO, 76- *Mem:* Am Chem Soc. *Res:* Mechanism of action of anti-metabolites; specificity profiles of the enzymes involved in cell wall biosynthesis in bacteria; symbiotic nitrogen fixation in non-legumes. *Mailing Add:* 16463 Sixth Ave SW Seattle WA 98166-3501

CARPENTER, CHARLES C J, b Savannah, Ga, Jan 5, 31; m 58; c 3. INFECTIOUS DISEASE. *Educ:* Princeton Univ, AB, 52; Johns Hopkins Univ, MD, 56. *Prof Exp:* From asst prof to prof med, Johns Hopkins Univ, 62-73; chmn, Dept Med, Cases Western Reserve Univ, 73-85; physician-in-chief, Univ Hosps Cleveland, 73-85; PROF MED, BROWN UNIV, 86-, CO-DIR, BROWN INT HEALTH INST, 87- *Concurrent Pos:* Res career develop award, Johns Hopkins Univ, 64-69; mem US deleg, US-Japan Coop Med Sci Prog, 65-, chmn cholera panel, 66-73; mem cholera adv comt, NIH, 66-73; mem expert adv panel bact dis, WHO, 67-; chmn elect, Am Bd Internal Med, 82-83, chmn, 83-84; mem, Adv Coun, Sch Med, Johns Hopkins Univ, 82-, Nat Allergy & Infectious Dis Adv Coun, 85- *Honors & Awards:* Blake Award, Asn Am Physicians, 88. *Mem:* Inst Med Nat Acad Sci; Infectious Dis Soc Am; Asn Profs Med; Asn Am Physicians (secy, 76-80, pres, 87-88); Am Soc Clin Invest. *Res:* Defining the pathophysiology, immunology and optimal means of treating cholera and other dehydrating diarrheal diseases; defining the heterosexual transmission of AIDS in North America. *Mailing Add:* Dept Med Brown Univ Providence RI 02906

CARPENTER, CHARLES CONGDEN, b Denison, Iowa, June 2, 21; m 47; c 3. ZOOLOGY, BEHAVIOR ETHOLOGY. *Educ:* Univ Northern Mich, BA, 43; Univ Mich, MS, 47, PhD(zool), 51. *Prof Exp:* Instr zool, Univ Mich, 51-52 & Wayne State Univ, 52; from instr to prof, 53-88 EMER PROF ZOOL, UNIV OKLA, 88-,. *Concurrent Pos:* NY Zool Soc grant-in-aid, Jackson Hole Res Sta, 51; NSF grant, 56-75; treas, Grassland Res Found, 58-62; mem, Galapagos Int Sci Proj, 64; mem, Sci Adv Comt, Charles Darwin Found for Galapagos Islands; emer cur reptiles & amphibians, Okla Mus Nat Hist, 54. *Honors & Awards:* Regents Award for Superior Accomplishment in Res & Creative Activ, 80; W Frank Blair Eminent Naturalist Award, Southwestern Asn Naturalists, 87. *Mem:* Am Soc Zoologists; Ecol Soc Am; Am Soc Ichthyologists & Herpetologists; Am Soc Mammal; fel Animal Behav Soc (secy, 65-68); fel Herpetologists League (pres, 74 & 75). *Res:* Ecology and behavior of vertebrates; herpetology; dynamics of populations and space relationships of reptiles and amphibians; comparative ecology and behavior. *Mailing Add:* Dept Zool Univ Okla Norman OK 73019

CARPENTER, CHARLES H, b Newark, NJ, July 17, 08; m 36; c 3. CHEMISTRY. *Educ:* Syracuse Univ, BS, 29, MS, 31; Darmstadt Tech Univ, Dr Ing, 33. *Prof Exp:* Asst chem, Carnegie Inst Technol, 34-36; chief chemist, asst dir & tech dir, Herty Found Lab, 36-39; tech dir & gen supt, Southland Paper Mills, 39-47; asst to vpres, NY & Pa Co, 47-50; vpres, White Star Paper Co, 51; CONSULT, 52- *Mem:* Soc Am Foresters; Tech Asn Pulp & Paper Indust; Can Pulp & Paper Asn. *Res:* Cellulose, wood and pulp chemistry. *Mailing Add:* 2345 Wildwood Dr Montgomery AL 36111

CARPENTER, CHARLES PATTEN, b Sellersville, Pa, July 5, 10; m 34; c 2. TOXICOLOGY, BACTERIOLOGY. *Educ:* Franklin & Marshall Col, BS, 31; Univ Pa, AM, 34, PhD(med sci), 37. *Prof Exp:* Instr bact, Hyg Dept, Med Sch, Univ Pa, 36-39, asst prof pub health & prev med, Lab, 39-40; Union Carbide indust fel, Mellon Inst, 40-46, sr fel, Mellon Inst Res, 46-56, from asst admin fel to admin fel, 56-75, adv fel, 75-78, dir, Biotecs Lab, Carnegie Mellon Univ, 79-84; RETIRED. *Mem:* AAAS; Am Chem Soc; Am Soc Toxicol; Am Indust Hyg Asn. *Res:* Toxicity of synthetic organic chemicals; industrial hygiene; acute toxicity of synthetic resins and polymersystems. *Mailing Add:* 339 Woodside Rd Pittsburgh PA 15221

CARPENTER, DAVID O, b Fairmont, Minn, Jan 27, 37; m 61, 85; c 3. NEUROPHYSIOLOGY, BIOPHYSICS. *Educ:* Harvard Univ, BA, 59, MD, 64. *Prof Exp:* Med officer neurophysiol, Lab Neurophysiol, NIMH, 65-72; chmn, Neurol Dept, Armed Fores Radiobiol Res Inst, 73-80; dir, Ctr Labs & Res, 80-85, DEAN SCH PUB HEALTH, STATE UNIV NY, ALBANY & NY STATE DEPT HEALTH, 85- *Concurrent Pos:* Fel neurophysiol, Sch Med, Harvard Univ, 64-65; Moseley fel. *Mem:* Am Physiol Soc; Soc Gen Physiol; Soc Neurosci; Int Brain Res Orgn; Am Pub Health Asn. *Res:* Mechanisms of neuronal cell death; neurotransmitter substances in Aplysia and mammalian nervous systems; neurotoxicology; radiation effects of chlorinated hydrocarbons; supraspinal control mechanisms. *Mailing Add:* Sch Pub Health NY State Dept Health Albany NY 11237

CARPENTER, DELMA RAE, JR, b Salem, Va, Apr 15, 28; m 52; c 3. PHYSICS ENGINEERING. *Educ:* Roanoke Col, BS, 49; Cornell Univ, MS, 51; Univ Va, PhD(physics), 57. *Prof Exp:* Instr physics, Va Mil Inst, 51-54, from asst prof to assoc prof, 55-65, proj dir res labs, 60-68, dep dir, 63-65, dir res labs, 65-85, head dept physics, 69-74, PROF PHYSICS, VA MIL INST, 63- *Concurrent Pos:* Res assoc, US Army Ord Contract, 53, 54, 56 & 60; consult, US Army Res Off, 71-75; trustee, Sci Mus Va, 73-83, chmn bd trustees, 73-78; trustee, Va Mil Inst Found, 77-; vis prof, US Mil Acad, 85-86, Auburn Univ, 86-87. *Honors & Awards:* Pegram Medal, Am Phys Soc; Distinguished Serv Citation, Am Asn Physics Teachers. *Mem:* Fel AAAS; Am Asn Physics Teachers. *Res:* Ordnance development and design; heat transfer in satellite instruments; physics teaching demonstrations. *Mailing Add:* Dept Physics Va Mil Inst Lexington VA 24450

CARPENTER, DEWEY KENNETH, b Omaha, Nebr, June 30, 28; m 55; c 3. PHYSICAL CHEMISTRY, POLYMER CHEMISTRY. *Educ:* Syracuse Univ, BS, 50; Duke Univ, AM, 52, PhD(chem), 55. *Prof Exp:* Res assoc chem, Cornell Univ, 55-56; res assoc & instr, Duke Univ, 56-58; from asst prof to assoc prof chem, Ga Inst Technol, 58-69; assoc prof, 69-74, PROF CHEM, LA STATE UNIV, BATON ROUGE, 74- *Concurrent Pos:* Vis fel, Dartmouth Col, 67-68 & Stanford Univ, 76. *Mem:* Am Chem Soc; Am Sci Affil; Sigma Xi. *Res:* Physical chemistry of high polymers; physical chemistry of macromolecules in solution. *Mailing Add:* Dept Chem La State Univ Baton Rouge LA 70803

CARPENTER, DONALD GILBERT, b Albany, NY, May 23, 27; m 52; c 3. DATA COMMUNICATIONS, SYSTEMS ENGINEERING. *Educ:* Univ Md, BS, 50; Air Force Inst Technol, MS, 60; Iowa State Univ, PhD(nuclear eng), 62; Colo Tech Col, BSEET, 85, BSEE, 86. *Prof Exp:* Assoc prof physics, US Air Force Acad, comdr, 16th & 18th Surveillance Squadrons, US Air Force, chief, Space Surveillance, 53-74, mgr qual assurance & standards, combustion eng, 74-76; prof physics, Chapman Col, 83-84; prof electronics & chmn dept, Colo Tech Col, 84-86; sr engr, Computer Sci Corp, 86-87; prof & dean eng, 87-88; PRIN ENGR & SYS ENG MGR, CONTEL, 88- *Concurrent Pos:* Consult physics, 76-83. *Honors & Awards:* Theodore von Karman Award, Air Force Asn, 72; emer fel, Geront Soc Am, 88. *Mem:* sr mem, Inst Elec & Electronics Eng; fel Geront Soc Am. *Res:* Basic causes of biological aging; quantum mechanical generation of electromagnetic noise; physical bases of psychic phenomena; space physics. *Mailing Add:* 25536 E Hwy 110 Calhan CO 80808-9106

CARPENTER, DWIGHT WILLIAM, b Paducah, Ky, July 25, 36; m 58; c 3. COMPUTER SCIENCES, PHYSICS. *Educ:* Univ Ky, BS, 58; Univ Ill, Urbana-Champaign, MS, 59, PhD(physics), 65. *Prof Exp:* Res assoc physics, Univ Ill, Urbana-Champaign, 64-66; asst prof, Duke Univ, 66-72; prof math & physics & chmn div sci & math, Limestone Col, 72-85; ASSOC PROF COMPUTER SCI, TRANSYLVANIA UNIV, 85- *Mem:* Asn Comput Mach; Am Phys Soc; Am Asn Physics Teachers; Math Asn Am. *Res:* Elementary particle physics. *Mailing Add:* Div of Sci & Math Transylvania Univ Lexington KY 40508

CARPENTER, EDWARD J, b Buffalo, NY, Mar 28, 42; m 67; c 2. BIOLOGICAL OCEANOGRAPHY. *Educ:* State Univ NY Col Fredonia, BS, 64; NC State Univ, MS, 67, PhD(zool), 70. *Prof Exp:* NSF fel biol oceanog, Woods Hole Oceanog Inst, 70-71, from asst scientist to assoc scientist, 71-75; PROF BIOL, STATE UNIV NY STONY BROOK, 75- *Concurrent Pos:* Mem, US-USSR Comt Ocean Pollution, 73-; US-Japanese joint res prog, 79- *Mem:* Am Soc Limnol & Oceanog; Estuarine Res Fedn; Sigma Xi; Phycol Soc Am; AAAS. *Res:* Nitrogen cycling in marine environment; physiology of nitrogen incorporation by algae; nitrogen fixation; primary production of phytoplankton; effects of pollutants on microorganisms. *Mailing Add:* Dept Marine Sci State Univ of NY Stony Brook NY 11794-5000

CARPENTER, EDWIN DAVID, b Great Falls, Mont, Aug 3, 32. ORNAMENTAL HORTICULTURE. *Educ:* Wash State Univ, BS, 57; Mich State Univ, MS, 62, PhD(hort), 64. *Prof Exp:* Sr exp aide, Coastal Wash Res & Exten Unit, Wash State Univ, 57-60; from asst prof to assoc prof, 64-67, chmn hort sect, Dept Plant Sci, 71-87, PROF ORNAMENTAL HORT, UNIV CONN, 77- *Concurrent Pos:* Mem, Northeast Regional Tech Comn, Plant Hort Introd, USDA, 65-80, chmn, 72-79. *Mem:* Am Soc Hort Sci; Am Hort Soc; Int Plant Propagators Soc; Am Soc Bot Gardens & Arboretums; Int Soc Hort Sci. *Res:* Plant anatomy, taxonomy and ecology. *Mailing Add:* Dept Plant Sci Univ Conn 1376 Storrs Rd Storrs CT 06269-4067

CARPENTER, ESTHER, b Meriden, Conn, June 4, 03. ZOOLOGY. *Educ:* Ohio Wesleyan Univ, AB, 25, Univ Wis, MS, 27; Yale Univ, PhD(zool), 32. *Hon Degrees:* DSc, Ohio Wesleyan Univ, 56. *Prof Exp:* Asst zool, Univ Wis, 25-27; lab technician, Yale Univ, 28-29; lab technician biol, Albertus Magnus Col, 30-32; lab technician, Dept Embryol, Carnegie Inst, 32-33; lab technician, 33-34, from instr to prof, 34-63, Myra M Sampson prof, 63-68, EMER PROF ZOOL, SMITH COL, 68- *Concurrent Pos:* Instr, Albertus Magnus Col, 33-34; Howald Scholar, Ohio State Univ, 42-43; res, Strangeways Res Lab, Eng, 53-54, 61; Sophia Smith fel, 73-80. *Mem:* AAAS; Am Soc Cell Biol; Am Soc Zool; Soc Develop Biol; Am Asn Anat; Tissue Cult Soc. *Res:* Experimental embryology; vital staining and transplantation in amphibia; regeneration in Eisenia foetida; tissue culture-differentiation of avian thyroid and femora; differentiation and physiological activities of embryonic thyroid glands in vitro; age changes in rat pituitary. *Mailing Add:* Pennswood Village D-207 Newtown PA 18940

CARPENTER, FRANCES LYNN, b Oklahoma City, Okla, Feb 14, 44. ECOLOGY, EVOLUTION. *Educ:* Univ Calif, Riverside, BA, 66; Univ Calif, Berkeley, PhD(zool), 72. *Prof Exp:* Assoc prof, 72-84, PROF ECOL, UNIV CALIF, IRVINE, 84- *Concurrent Pos:* NSF res grants, 78-80, 81-83 & 84-86. *Mem:* Ecol Soc Am; Soc Study Evolution; Cooper Ornith Soc; fel AAAS. *Res:* Energetics of plant-pollinator coevolved relationships; pollination strategies; behavior and resource partitioning in avian nectar-eaters; territoriality; comparison of generalist and specialist adaptive strategies in plants, birds, insects. *Mailing Add:* Dept of Ecol & Evolutionary Biol Univ of Calif Irvine CA 92717

CARPENTER, FRANK G(ILBERT), b Washington, DC, Mar 26, 20; m 45; c 6. CHEMICAL ENGINEERING. *Educ:* Univ Md, BS, 42; Univ Del, MChE, 46, PhD(chem eng), 49. *Prof Exp:* Res assoc, Nat Bur Stand, 49-63; dir cane sugar ref res proj, USDA, 63-81, res leader sugar processing, Southern Regional Lab, 81-84; RETIRED. *Honors & Awards:* Award, Sugar Indust Technologists, 75. *Mem:* Am Chem Soc; Sugar Indust Technologists. *Res:* Sugar refining research; fluid flow; colorimetry; turbidimetry; adsorption; chromatography; surfaces. *Mailing Add:* 29 Crane St New Orleans LA 70124

CARPENTER, FRANK GRANT, b Toledo, Ohio, Oct 8, 23; m 51; c 2. PHYSIOLOGY. *Educ:* Ohio State Univ, BSc, 48; Columbia Univ, PhD, 51. *Prof Exp:* Asst physiol, Ohio State Univ, 48 & Columbia Univ, 49-51; instr, Univ Rochester, 52-54; asst prof, Med Col, Cornell Univ, 54-57; from asst prof to assoc prof, Dartmouth Med Sch, 57-67; assoc prof, Univ Ala, Birmingham, 67-71, prof pharmacol, 71-88, prof anesthesiol, 73-88; RETIRED. *Concurrent Pos:* Fel, Univ Rochester, 51-54. *Mem:* Am Physiol Soc; Harvey Soc; Am Soc Pharmacol & Exp Therapeut. *Res:* Anesthesia; nerve metabolism; autonomic neuroeffectors. *Mailing Add:* 4131 Stone River Rd Birmingham AL 35213

CARPENTER, FRANK MORTON, b Boston, Mass, Sept 6, 02; m; c 3. ENTOMOLOGY. *Educ:* Harvard Univ, AB, 26, MS, 27, ScD, 29. *Prof Exp:* Nat Res Coun fel zool, 28-31, assoc entom, 31-32, from asst prof to prof, 36-45, chmn dept, 53-59, prof entom & Agassiz prof zool, 36-69, Fisher Prof natural hist, 69-73, asst cur invert paleont, Mus Comp Zool, 32-36, cur fossil insects, 36-73, EMER FISHER PROF NATURAL HIST & EMER AGASSIZ PROF ZOOL, HARVARD UNIV & HON CUR FOSSIL INSECTS, MUS COMP ZOOL, 73- *Concurrent Pos:* Assoc, Carnegie Inst, 31-32. *Honors & Awards:* Paleont Soc Medal, 75. *Mem:* Fel Am Acad Arts & Sci; hon fel Royal Entom Soc,. *Res:* Paleoentomology and insect evolution; North American Neuroptera; Permian insects of Kansas and Oklahoma; Carboniferous insects of North America and Europe. *Mailing Add:* Biol Labs Harvard Univ Cambridge MA 02138

CARPENTER, GAIL ALEXANDRA, b New York, NY, Dec 23, 48; m; c 1. BIOMATHEMATICS. *Educ:* Univ Colo, Boulder, BA, 70; Univ Wis-Madison, MA, 72, PhD(math), 74. *Prof Exp:* Instr appl math, Mass Inst Technol, 74-76; from asst prof to assoc prof, 76-86, PROF MATH, NORTHEASTERN UNIV, 86- *Mem:* Am Math Soc; Int Neural Network Soc; Soc Indust & Appl Math; Soc Neuroscience. *Res:* Neural networks; study of excitable membrane and network phenomena; mathematical biology, using methods of topological dynamics. *Mailing Add:* Ctr For Adaptive Sys Boston Univ 111 Cummington St Boston MA 02215

CARPENTER, GARY GRANT, b San Francisco, Calif, Aug 25, 29. PEDIATRICS. *Educ:* Rutgers Univ, AB, 56; Jefferson Med Col, MD, 60. *Prof Exp:* From instr to asst prof, Sch Med, Temple Univ, 65-68; ASSOC PROF PEDIAT, JEFFERSON MED COL, 68- *Concurrent Pos:* Training fel metab & amino acids, St Christopher's Hosp Children, Pa, 62-63; asst prog dir, clin res ctr, St Christopher's Hosp Children, 65-68. *Mem:* AAAS; Ny Acad Sci; Sigma Xi. *Res:* Amino acid metabolism; cytogenetics. *Mailing Add:* Dept of Pediat Jefferson Med Col Philadelphia PA 19107

CARPENTER, GENE BLAKELY, b Evansville, Ind, Dec 15, 22; m 49; c 2. PHYSICAL CHEMISTRY, CRYSTALLOGRAPHY. *Educ:* Univ Louisville, BA, 44; Harvard Univ, MA, 45, PhD(phys chem), 47. *Prof Exp:* Nat Res fel, Calif Inst Technol, 47-48, res fel, 48-49; from instr to assoc prof, 49-63, prof chem, 68-88, EMER PROF, BROWN UNIV, 88- *Concurrent Pos:* Guggenheim fel, Univ Leeds, 56-57; vis prof, State Univ Groningen, 63-64; Fulbright-Hays lectr, Univ Zagreb, 71-72; vis scientist, Oak Ridge Nat Lab, 80; vis scientist, Univ Gottingen, Fed Repub Ger, 87; consult, Univ Canterbury, Christchurch NZ, 89. *Mem:* Am Chem Soc; Am Crystallog Asn. *Res:* Crystal structure by x-ray diffraction. *Mailing Add:* Dept Chem Brown Univ Providence RI 02912

CARPENTER, GRAHAM FREDERICK, b Apr 16, 44; m; c 2. GROWTH FACTORS, RECEPTORS. *Educ:* Univ RI, BS, 62 & MS, 66; Univ Tenn, PhD(biochem), 74. *Prof Exp:* Res asst prof, dept med, 77-79, from asst prof to assoc prof, 80-86, PROF, DEPT BIOCHEM, VANDERBILT UNIV, 86-, PROF BIOCHEM & MED, SCH MED, 74- *Concurrent Pos:* Investr, Am Heart Asn, 81-86. *Mem:* Am Soc Biochem & Molecular Biol; Am Soc Cell Biol. *Mailing Add:* Vanderbilt Univ Sch Med Dept Biochem Nashville TN 37232

CARPENTER, GRAHAM JOHN CHARLES, b Cardiff, Wales, UK, Jan 3, 39; m 63. METAL PHYSICS, ELECTRON MICROSCOPY. *Educ:* Univ Wales, BSc Hons, 61, PhD(metall), 64. *Prof Exp:* Res demonstr metall, Univ Wales, 63-64; res fel, Cambridge Univ, 64-67; sr res officer mat sci, Chalk River Nuclear Labs, 67-81; SR RES SCIENTIST, CANMET RES LABS, OTTAWA, 81- *Concurrent Pos:* Vis res assoc, Atomic Energy Res Estab, Harwell, 75-76. *Mem:* Micros Soc Can; Electron Micros Soc Am. *Res:* Analytical electron microscopy, physical metallurgy, precipitation phenomena, radiation damage, radiation growth, zirconium alloys, semiconductor devices, ceramics, composite materials, minerals. *Mailing Add:* Metals Technol Lab-CANMET 568 Booth St Ottawa ON K1A 0G1 Can

CARPENTER, HARRY C(LIFFORD), b Richey, Mont, Oct 25, 21; m 52; c 2. CHEMICAL ENGINEERING. *Educ:* Mont State Col, BS, 48, MS, 49. *Prof Exp:* RES CHEM ENGR, PETROL RES CTR, US BUR MINES, 49- *Mem:* AAAS; Am Chem Soc; Am Inst Chem Engrs; Sigma Xi. *Res:* Refining of shale oil by hydrogenation, hydro cracking and other catalytic processes; retorting of oil shale by in situ methods. *Mailing Add:* 1322 Canby Laramie WY 82070

CARPENTER, IRVIN WATSON, JR, b Washington, DC, Nov 29, 23; m 48; c 3. PLANT TAXONOMY. *Educ:* Purdue Univ, BSF, 48, MS, 50, PhD(bot), 52. *Prof Exp:* Instr forestry, Purdue Univ, 52-53; from asst prof to prof bot, Appalachian State Univ, 53-88; RETIRED. *Concurrent Pos:* Cur, Appalachian State Univ Herbarium. *Mem:* Sigma Xi; AAAS. *Res:* Flora of southern Appalachians. *Mailing Add:* 103 Reynolds Rd Boone NC 28607

CARPENTER, JACK WILLIAM, b Worthington, Ohio, July 17, 25; m 50; c 2. THEORETICAL PHYSICS. *Educ:* Mass Inst Technol, BS, 51, MS, 52, PhD(physics), 57. *Prof Exp:* Proj scientist & sr physicist, Allied Res Assoc, Inc, 57-58, chief proj scientist, 58; vpres geophys div & sr physicist, Am Sci & Eng, Inc, Cambridge, 58-69; PRES & DIR, VISIDYNE, INC, MASS, 69- *Res:* Plasma physics; nuclear weapons effects; magnetohydrodynamics; nuclear physics; infrared spectroscopy. *Mailing Add:* Visidyne Inc S Bedford Burlington MA 01803

CARPENTER, JAMES E(DWIN), b Woodsville, NH, Mar 21, 32; m 64; c 3. STRUCTURAL ENGINEERING. *Educ:* Univ Cincinnati, CE, 54; Purdue Univ, MSCE, 60, PhD(struct eng), 65. *Prof Exp:* Engr & inspector, Fay, Spofford & Thorndike, Inc, 57-59; assoc develop engr, Res & Develop Div, Portland Cement Asn, 61-65, develop engr, 65-68, sr res engr, 68-75; mem staff, Concrete Technol Corp, 75-79; PROJ MGR, ANDERSEN-BJORNSTAD-KANE-JACOBS, INC, SEATTLE, WASH, 80- *Mem:* Am Soc Civil Engrs; Am Concrete Inst. *Res:* Structural research in reinforced and prestressed concrete. *Mailing Add:* 16463 Sixth Ave SW Seattle WA 98166

CARPENTER, JAMES E(UGENE), b Syracuse, NY, Jan 19, 21; m 43; c 3. MECHANICAL ENGINEERING, SCIENCE POLICY. *Educ:* Univ Syracuse, BS, 43; Univ Buffalo, MS, 51. *Prof Exp:* Test engr, Res Lab, Curtiss-Wright Corp, 43-44; res engr, Cornell Aeronaut Lab, Inc, 46-50, head struct lab, 50-58, prof mgr, 58-62, br head, 62-64, from asst dept head to dept head, 64-69, sr staff scientist, 69-70; mem comn govt procurement, 70-73; policy analyst, NSF, 73-82; CONSULT, 82- *Res:* Strain gauge and structural research; hypersonic materials. *Mailing Add:* 3412 Mansfield Rd Falls Church VA 22041

CARPENTER, JAMES EDWARD, b Chicago, Ill, Dec 11, 46; m 73; c 4. MATHEMATICS. *Educ:* Ill Benedictine Col, BS, 68; Chicago State Univ, MS, 71. *Hon Degrees:* EdD, Columbia Univ, 78. *Prof Exp:* Teacher math, St Laurence High Sch, 68-74; from asst prof to assoc prof math, St Mary of the Plain Col, 75-81; ASSOC PROF MATH, IONA COL, 81- *Concurrent Pos:* Vis acad, Trinity Col, Dublin, 88-89. *Mem:* Nat Coun Teachers Math; Math Asn Am. *Res:* Investigation of the mathematics and pedagogy of the liberal arts core requirement in mathematics. *Mailing Add:* Dept Math Iona Col New Rochelle NY 10801

CARPENTER, JAMES FRANKLIN, b Brookfield, Mo, Apr 10, 23; m 47; c 1. MATERIALS SCIENCE, PHYSICAL CHEMISTRY. *Educ:* Cent Methodist Col, AB, 48; Northwestern Univ, MS, 50; St Louis Univ, PhD(phys chem), 58. *Prof Exp:* Chemist, Mallinckrodt Chem Works, 50-58; asst mgr res chem, Mallinckrodt Nuclear Corp, 58-60; group leader, United Nuclear Corp, 60-62; SR STAFF ENGR CHEM, McDONNELL DOUGLAS CORP, 62- *Concurrent Pos:* Comt mem nat mat adv bd, Comt Characterization Org Matrix Composites, 78-80. *Mem:* Soc Advan Mat & Proc Eng; Am Inst Aeronaut & Astronaut. *Res:* Physical-organic chemistry; characterization of advanced resin composites and structural adhesives; quality assurance criteria of structural resins based on the physiochemical properties. *Mailing Add:* Sweetwater Ct Lake Ozark MO 65049

CARPENTER, JAMES L(INWOOD), JR, b Fredericksburg, Va, Jan 6, 25; m 47; c 2. INDUSTRIAL ENGINEERING. *Educ:* Col William & Mary, BA, 49, MA, 50. *Prof Exp:* Elec engr, Navy Bur Ships, 50-53 & Newport News Shipbldg & Drydock Co, 53-56; elec engr, Missile Div, Chrysler Corp, 56-57, dept mgr, 57-59, dir plans & prog, Adv Proj Off, Defense Group, 50-61; dir logistics support, Martin Marietta Corp, 61-72, dir systs eng, 72-87; RETIRED. *Concurrent Pos:* Recipient, Greer Award for Contrib to Logistics Mgt, 70. *Mem:* Fel Soc Logistics Engrs (past pres); Syst Safety Soc. *Res:* Engineering analysis of man-machine interface, especially information processing and management. *Mailing Add:* 5214 Alleman Dr Orlando FL 32809

CARPENTER, JAMES WILLIAM, b Shamokin, Pa, Sept 7, 35; m 58; c 4. ORGANIC CHEMISTRY. *Educ:* Lebanon Valley Col, BS, 60; Univ Nebr, MS, 63, PhD(org chem), 65. *Prof Exp:* Instr chem, Univ Nebr, 64-65; CHEMIST, EASTMAN KODAK CO, 65- *Mem:* Soc Photog Sci & Eng. *Res:* Photographic chemistry. *Mailing Add:* 120 Campfire Henrietta NY 14467

CARPENTER, JAMES WOODFORD, b Union, Ky, Jan 6, 22; m 49; c 2. ANIMAL SCIENCE. *Educ:* Univ Ky, BS, 52, MS, 53; Univ Fla, PhD(meat technol), 59. *Prof Exp:* Plant mgr, Univ Fla, 54-59, asst meat scientist, 59-66, assoc meat scientist, 66-71, prof & meat scientist, meat lab, 71-, prof animal sci; RETIRED. *Honors & Awards:* Educ & Cult Medal, Govt S Vietnam, 70. *Mem:* Am Soc Animal Sci; Am Meat Sci Asn. *Res:* Meat animal carcass evaluation, especially quality and palatability factors. *Mailing Add:* PO Box 818 Hawthorne FL 32640

CARPENTER, JOHN HAROLD, b Owatonna, Minn, May 1, 29; m 53; c 2. HIGH TEMPERATURE CHEMISTRY. *Educ:* Macalester Col, BA, 51; Purdue Univ, MS, 53, PhD(chem), 55. *Prof Exp:* Chemist, Univ Calif, 54-55, res chemist, Lawrence Livermore Nat Lab, 55-68; assoc prof chem, 68-71, chmn chem dept, 73-82, PROF CHEM, ST CLOUD STATE UNIV, 71- *Concurrent Pos:* Vis scientist, Los Alamos Nat Lab, 85-86. *Mem:* Am Chem Soc; Am Ceramic Soc; Sigma Xi. *Res:* Structures, vapor pressures and thermodynamics of high melting inorganic compounds. *Mailing Add:* Dept Chem St Cloud State Univ St Cloud MN 56301-4498

CARPENTER, JOHN MARLAND, b Williamsport, Pa, June 20, 35; div; c 4. SOLID STATE PHYSICS, NUCLEAR ENGINEERING. *Educ:* Pa State Univ, BS, 57; Univ Mich, MSE, 58, PhD(nuclear eng), 63. *Prof Exp:* Res assoc nuclear eng, Univ Mich, Ann Arbor, 60-63, fel, Inst Sci & Technol, 63-64, from asst prof to prof, 64-76; vis scientist, 71-72 & 73, SR PHYSICIST & TECH DIR, ARGONNE NAT LAB, 75- *Concurrent Pos:* Consult, Solid State Sci Div, Argonne Nat Lab, 69-71, mem comt intense neutron sources; vis scientist, Lab High Energy Physics, Tsukuba, Japan, 82; adj prof, mem grad faculty, Dept Nuclear Engr, Iowa State Univ, 88-; mem vis comt, Nuclear Eng Dept, Mass Inst Technol, 89- *Mem:* fel Am Phys Soc. *Res:* Neutron inelastic scattering; neutron diffraction; amorphous solids; molecular spectroscopy; nuclear reactor instrumentation and control; neutron thermalization; pulsed moderators; pulsed spallation neutron sources; neutron scattering instrumentation. *Mailing Add:* Argonne Nat Lab Intense Pulsed Neutron Source Argonne IL 60439

CARPENTER, JOHN RICHARD, b Galveston, Tex, May 20, 38; m 58; c 3. ENVIRONMENTAL SCIENCES. *Educ:* Rice Univ, BA, 59; Fla State Univ, MS, 62, PhD geochemical geol 64. *Prof Exp:* Asst geol, Fla State Univ, 63-64; geologist, Fla Geol Surv, 64; US Naval Oceanog Off, 64-66; from asst prof to assoc prof geol, Univ SC, 66-77, actg head dept, 69-70, dir grad studies & asst chmn dept geol, 74-78, PROF GEOL, UNIV SC, 77-, DIR, CTR SCI EDUC, 85- *Concurrent Pos:* Sr col consult, 72-74, sr staff mem, Earth Sci Teacher Prep Proj, 73- *Mem:* Geol Soc Am; Nat Earth Sci Teachers Asn; Nat Asn Geol Teachers. *Res:* Alternative structure modes for earth science education. *Mailing Add:* Ctr Sci Educ Univ SC Columbia SC 29208

CARPENTER, KENNETH HALSEY, b El Dorado Kans, June 22, 39; m 71; c 3. ELECTROMAGNETICS. *Educ:* Kans State Univ, BSEE, 61, MS, 62; Tex Christian Univ, PhD(physics), 66. *Prof Exp:* Columbia Res Corp, 66-68; assoc prof physics, ETenn State Univ, 68-75; Union Carbide Corp, Nuclear Div & Comput Sci Div, Oak Ridge Nat Lab, 75-79; ASSOC PROF ELEC ENG, UNIV MO-ROLLA, 79- *Mem:* Am Phys Soc; Inst Elec & Electronics Engrs; Sigma Xi. *Res:* Plasma diagnostics for magnetic confinement fusion and computing applications to this area. *Mailing Add:* 1513 Westwind Dr Manhattan KS 66502-2434

CARPENTER, KENNETH JOHN, b London, Eng, May 17, 23. NUTRITION. *Educ:* Cambridge Univ, BA, 44, PhD(nutrit), 48,. *Hon Degrees:* DSc, Cambridge Univ, 75. *Prof Exp:* Sci officer nutrit, Rowett Inst, Aberdeen, Scotland, 48-56; reader, Cambridge Univ, 56-77; chmn dept, 81-83, PROF EXP NUTRIT, UNIV CALIF, BERKELEY, 77- *Concurrent Pos:* Kellogg fel, Harvard Univ, 55-56; bursar, Food Res Inst, Mysore, 62. *Mem:* Brit Nutrit Soc; Am Inst Nutrit. *Res:* Availability of vitamins and amino acids in processed foods; history of nutritional science. *Mailing Add:* Dept of Nutrit Sci Univ of Calif Berkeley CA 94720

CARPENTER, KENT HEISLEY, b Williamsport, Pa, Apr 30, 38; m 72; c 2. HYDROMETALLURGY, CHEMICAL PROCESSING. *Educ:* Pa State Univ, BS, 61. *Prof Exp:* Asst res engr, Dept Nuclear Eng, Univ Mich, 61-65; assoc res engr, Conductron Div, McDonnel Douglass, 65-69; owner/mgr, Sports Car Serv, Inc, 69-72; sr proj engr, Thetford Corp, 72-75; sr res assoc, Climax Molybdenum Co, 75-83, sr res engr, Mat Res Ctr, 83-84, MGR COM & TECH SERV, CLIMAX MOLYBDENUM CO, AMAX, INC, GREENWICH, CONN, 84- *Res:* Hydro-and pyro-metallurgical processing of nickel and cobalt ores; reduction of metal oxide powders; crystallization of molybdenum and tungsten salts; elevated temperature and fluid bed processing. *Mailing Add:* 257 Butternut Lane Stamford CT 06903-3834

CARPENTER, LYNN ALLEN, b Cushing, Okla, Apr 25, 43; m; c 3. MICROWAVE ENGINEERING, SAFETY ENGINEERING. *Educ:* Okla State Univ, BS, 64; Univ Ill, Urbana-Champaign, 66, PhD(physics), 71. *Prof Exp:* Res asst elec eng, Univ Ill, 68-71, res assoc, 71-72; asst prof, 72-79, ASSOC PROF ELEC ENG, PA STATE UNIV, 79- *Concurrent Pos:* Res trainee simulation, NASA Manned Spacecraft Ctr, 65; mem staff radar systems, Mass Inst Technol, 68. *Mem:* Am Geophys Soc; Int Sci Radio Union; Inst Elec & Electronics Engrs. *Res:* Development of measurements of the Doppler shift for incoherent-scatter radar systems. *Mailing Add:* 227 Elec Eng E Penn State Univ University Park PA 16802

CARPENTER, MALCOLM BRECKENRIDGE, b Montrose, Colo, July 7, 21; m 49; c 3. ANATOMY, PHYSIOLOGY. *Educ:* Columbia Univ, BA, 43; Long Island Col Med, MD, 47; Am Bd Psychiat & Neurol, dipl, 55. *Prof Exp:* Asst neurol, Columbia Univ, 47 & 48-50, from instr to assoc prof anat, 53-67, prof anat, 62-76; prof anat & neurol, Pa State Univ, 76-78; prof & chmn dept anat, 78-87, PROF ANAT, UNIFORMED SERV UNIV HEALTH SCI, BETHESDA, MD, 87- *Concurrent Pos:* Fel neurol, Columbia Univ, 48-50; Markle scholar med sci, 53-58; surg intern, Bellevue Hosp, NY, 47-48; asst res neurologist, Neurol Inst, NY, 50 & 52-53; consult, Nat Inst Neurol Dis & Blindness, 62-66 & 68-72; mem, Inst Brain Res Orgn; ed, Neurology, 63-72, Am J Anat, 68-74, J Comp Neurol, 71-80, Neurobiology, 71 & Comn Foreign Med Grad, 83; mem, Anat Test Comt, Nat Bd Med Examr, 77-81. *Mem:* Am Asn Anat; Asn Res Nerv & Ment Dis; Am Neurol Asn; Am Acad Neurol; NY Acad Med. *Res:* Neuroanatomic, neurophysiologic and neuropathologic study of motor disturbances, particularly those due to disease of basal ganglia. *Mailing Add:* 380 E Chocolate Ave Hershey PA 17033

CARPENTER, MARTHA STAHR, b Bethlehem, Pa, May 29, 20; m 51; c 1. ASTRONOMY. *Educ:* Wellesley Col, BA, 41; Univ Calif, MA, 43, PhD, 45. *Prof Exp:* Asst astron, Univ Calif, 41-44; instr, Wellesley Col, 45-47; from asst prof to assoc prof, Cornell Univ, 47-54; res grant radio astron, Australian Commonwealth Sci & Indust Res Org, 54-55; res assoc, Ctr Radiophysics & Space Res, Cornell Univ, 55-69; lectr, Univ Va, 69-77, assoc prof astron, 73-; RETIRED. *Mem:* AAAS; Am Astron Soc; Am Asn Variable Star Observers (2nd vpres, 48-49, lst vpres, 49-51, pres, 51-54). *Res:* Galactic structure; radio astronomy. *Mailing Add:* 1101 Hilltop Rd Charlottesville VA 22901

CARPENTER, MARY PITYNSKI, b Detroit, Mich, Feb 20, 26; m 47; c 3. BIOCHEMISTRY. *Educ:* Wayne Univ, BS, 46; Univ Mich, MA, 48, PhD(zool), 52. *Prof Exp:* Res assoc zool, Univ Mich, 51-53; res assoc biochem, Okla Med Res Found, 54-58, biochemist, 58-66; from asst prof to assoc prof biochem, 65-69, PROF BIOCHEM & MOLECULAR BIOL, SCH MED, UNIV OKLA, 69- *Concurrent Pos:* Assoc mem, Okla Med Res Found, 69-76, mem, 76-; mem, Nutrit Study Sect, NIH, 78-82; pres, Okla Acad Sci, 88. *Mem:* AAAS; Am Soc Biol Chem; Am Chem Soc; Am Inst Nutrit; Brit Biochem Soc; Sigma Xi; Am Oil Chemists Soc. *Res:* Role of membrane polyenoic fatty acids and antioxidants on cellular function; function of vitamin E, and its role in testicular differentiation; enzymic oxygenation of unsaturated fatty acids--prostaglandins; eicosanoids. *Mailing Add:* 1218 Cruce Norman OK 73069

CARPENTER, MICHAEL KEVIN, b Port Hueneme, Calif, Sept 21, 54; m. ELECTROCHEMISTRY, PHOTOELECTROCHEMISTRY. *Educ:* Western Wash State Col, BS, 77; Northwestern Univ, MS, 78; Univ Wis-Madison, PhD(inorg chem), 84. *Prof Exp:* Chemist, UOP, Inc, 78-80; STAFF RES SCIENTIST, GEN MOTORS, 84- *Mem:* Am Chem Soc; Electrochemisty Soc. *Res:* Semiconductor electrochemistry; in situ photoelectrochemical characterization; electrochromism; luminescence; battery systems. *Mailing Add:* Gen Motors Res Labs Dept 14 30500 Mound Rd Warren MI 48090-9055

CARPENTER, NANCY JANE, b Detroit, Mich, Nov 15, 46. GENETICS. *Educ:* Albion Col, BA, 68; Univ Mich, MS, 69, PhD(zool), 72. *Prof Exp:* asst prof, 73-79, ASSOC PROF ZOOL, UNIV TULSA, 79-; Asst prof, Univ Tulsa, 73-79, assoc prof zool, 79-; RETIRED. *Concurrent Pos:* Scholar, Univ Mich, 73; assoc clin genetics, Children's Med Ctr, 75-81, assoc dir clin genetics, 81-; clin asst prof pediat, Univ Okla Med Col, Tulsa, 78- *Mem:* Am Soc Human Genetics; Genetics Soc Am; AAAS; Sigma Xi. *Res:* Cytogenetics and clinical genetics. *Mailing Add:* Children's Med Ctr Chapman Inst 5300 E Skelly Dr Tulsa OK 74135

CARPENTER, PAUL GERSHOM, organic chemistry; deceased, see previous edition for last biography

CARPENTER, PHILIP JOHN, b Walnut Creek, Calif, Jan 7, 57. ENVIRONMENTAL GEOPHYSICS, MICROEARTHQUAKE SEISMOLOGY. *Educ:* Univ Minn, Duluth, BS, 79; NMex Inst Mining & Technol, MS, 81, PhD(geosci), 84. *Prof Exp:* Post doctoral seismologist, Los Alamos Nat Lab, Univ Calif, 84-85; ASSOC PROF, GEOL/GEOPHYS, NORTHERN ILL UNIV, 86- *Concurrent Pos:* Geologist, Chevron, USA Inc, San Francisco, 79; res asst, Mobil Field Lab, Mobil Oil Corp, 81; res collabr, Los Alamos Nat Lab, Univ Calif, 86-; researcher, Argonne Nat Lab, Univ Chicago, 89. *Mem:* Am Geophys Union; Geol Soc Am; Asn Eng Geologists; Soc Explor Geophysicists; Asn Groundwater Scientists & Engrs; Seismol Soc Am. *Res:* Application of geophysics to environmental engineering and groundwater problems; microearthquake seismology related to reservoir induced seismicity and geothermal exploration. *Mailing Add:* Dept Geol Northern Ill Univ De Kalb IL 60115-2854

CARPENTER, RAY WARREN, b Berkeley, Calif, Sept 29, 34; m 55; c 3. MATERIALS SCIENCE, SOLID STATE PHYSICS. *Educ:* Univ Calif, Berkeley, BS, 58, MS, 59, PhD(metall), 66. *Prof Exp:* Res asst, Inst Eng Res, Univ Calif, 56-59; sr metallurgist, Aerojet-Gen Nucleonics, 59-65 & Stanford Res Inst, 65-66; sr mem res staff, Oak Ridge Nat Lab, 66-80; dir, Facil High Resolution Electron Micros, 80-85, chair, Grad Prog Sci & Eng Mat, 87-90, PROF, SOLID STATE SCI & ENG, ARIZ STATE UNIV, 80-, DIR, CTR SOLID STATE SCI, 85- *Concurrent Pos:* Dir, Phys Sci Electron Micros Soc Am, 80-82. *Honors & Awards:* Analytical Electron Micros Award, Int Metallog Soc, 76 & 77. *Mem:* Am Inst Mining, Metall & Petrol Engrs; Electron Micros Soc Am (pres, 89). *Res:* Techniques and instrumentation for high resolution analytical electron microscopy and applications to material science and solid state physics; phase transformations and interfaces in metals, ceramics and semiconductors; structural analysis of solids. *Mailing Add:* Ctr Solid State Sci PSB-234 Ariz State Univ Tempe AZ 85287-1704

CARPENTER, RAYMON T, b Topeka, Kans, Jan 14, 29; m 53; c 1. NUCLEAR PHYSICS. *Educ:* Univ Kans, BS, 54, MS, 56; Northwestern Univ, PhD(nuclear physics), 62. *Prof Exp:* Asst instr physics, Univ Kans, 54-56; asst instr physics, Northwestern Univ, 56-58; physicist, Argonne Nat Lab, 58-62; from asst prof to assoc prof physics, 62-82, PROF PHYSICS & ASTRON, 82- *Concurrent Pos:* Guest prof, Univ Heidleberg, 71-; guest sci, Royal Inst Technol, Stockholm, 83-84. *Mem:* AAAS. *Res:* Confinement studies of plasmas; basic laboratory plasma physics; double layers. *Mailing Add:* Dept Physics Univ Iowa Iowa City IA 52242

CARPENTER, RICHARD A, b Kansas City, Mo, Aug 22, 26; m 48; c 3. POLICY ANALYSIS. *Educ:* Univ Mo, BS, 48, MA, 49. *Prof Exp:* Chemist, Shell Oil Co, 49-51; asst mgr, Midwest Res Inst, 51-58; mgr, Callery Chem Co, 58-64; sr specialist sci & tech, Libr Cong, 64-69, chief, Environ Policy Div, Cong Res Serv, 69-72; exec dir comn nat resources, Nat Res Coun, Nat Acad Sci, 72-77; RES ASSOC, ENVIRON & POLICY INST, EAST-WEST CTR, 77- *Concurrent Pos:* Consult, Int Develop Financial Insts. *Mem:* Sigma Xi; Am Chem Soc; Sci Res Soc Am; Ecol Soc Am; Am Sci Affil; Ecol Soc Am. *Res:* Air pollution; environmental chemistry; technology assessment; organizes and conducts multinational interdisciplinary research and methods of environmental assessment in developing countries. *Mailing Add:* 2419 Halekoa Dr Honolulu HI 96821-1039

CARPENTER, RICHARD M, b Cambridge, Mass, Apr 3, 43; m 69; c 3. ELECTRICAL ENGINEERING, APPLIED MATHEMATICS. *Educ:* Tufts Univ, BSEE, 64; Harvard Univ, MS, 66. *Prof Exp:* Aerospace technologist, Electronics Res Ctr, NASA, Mass, 66-70; sr mem tech staff & head comput-aided design, New Prod Line Eng Div, RCA Comput Systs Div, Marlboro, 70-72; chief elec engr, Massa Div, Dynamics Corp Am, 72-75; vpres, 77-90, EXEC VPRES, MASSA PROD CORP, HINGHAM, MASS, 90- *Mem:* Inst Elec & Electronics Engrs. *Res:* Effective design procedures for engineers in a computer-aided circuit design environment; computer-aided circuit design; high speed logic circuit design; development of microprocessor based electo-acoustic control systems. *Mailing Add:* Massa Prod Corp 280 Lincoln St Hingham MA 02043

CARPENTER, ROBERT HALSTEAD, b Pasadena, Calif, June 24, 14; m 41; c 2. GEOLOGY. *Educ:* Stanford Univ, AB, 40, MA, 43, PhD(geol), 48. *Prof Exp:* Geologist, Anaconda Copper Mining Co, Mont, 42-44 & Int Smelting & Ref Co, Utah, 44-46; from asst prof to prof geol, Colo Sch Mines, 47-74; PROF GEOL, UNIV GA, 74- *Concurrent Pos:* Consult geologist, NY & Honduras Rosario Mining Co, Honduras, 49-52, Molybdenum Corp Am, 53-, Thomp Creek Coal & Coke, 53-, Utah Construct Co, 54 & UN Spec Fund Proj, Baldwin Mines, Burma, 63; Fulbright res scholar, italy, 57-58; pres, Int Mineral Eng, 63- *Mem:* Fel Geol Soc Am; Soc Econ Geologists; Am Inst Mining, Metall & Petrol Eng. *Res:* Economic and structural geology; geochemistry of ore deposits. *Mailing Add:* PO Box 632 Elberton GA 30635

CARPENTER, ROBERT JAMES, JR, b Dallas, Tex, Nov 7, 45; m 70; c 1. MATERNAL FETAL MEDICINE, ULTRASONOGRAPHY. *Educ:* Hardin-Simmons Univ, BS, 68; Baylor Col Med, MD, 73, cert, 77. *Prof Exp:* Chief res, 76-77, asst prof obstet & gynec, 77-79, DIR, PRENATAL DIAG CTR, BAYLOR COL MED, 79-; asst prof, 86-87, ASSOC PROF GENETICS, INST MOLECULAR GENETICS, 87- *Concurrent Pos:* Fel obstet ultrasound & fetoscopy, Yale Univ, 77. *Mem:* Am Soc Human Genetics; Nat Perinatal Asn; Am Inst Ultrasound Med; Fetal Med & Surg Soc; Soc Perinatal Obstetricians. *Res:* Premature labor and prenatal diagnosis of fetal malformation with emphasis on in utero therapy. *Mailing Add:* Dept Obstet & Gynec Baylor Col Med 6550 Fannin No 901 Houston TX 77030

CARPENTER, ROBERT LELAND, b St Louis, Mo, June 27, 42; m 65; c 3. AEROSOL PHYSICS, PHARMACOKINETIC MODELING. *Educ:* Univ Mo-Rolla, BS, 65; Univ Tenn, PhD(molecular biol), 75. *Prof Exp:* Res asst plasma physics, Los Alamos Sci Lab, 65-66; engr, McDonnell Douglas Corp, 66-70; sr staff inhalation toxicol, Lovelace Biomed & Environ Res Inst, 75-85; mgr, eng & inhalation, Mantech Inc, 85-89; SR TOXICOLOGIST, ITD, NAVAL MED RES INST, 89- *Concurrent Pos:* Mem comt protection trichothecene micotoxins, Nat Res Coun, 83. *Mem:* Am Asn Aerosol Res; Soc Risk Assessment. *Res:* Inhalation toxicology, source characterization and aerosol generation; aerosol physics. *Mailing Add:* Naval Med Res Inst Wright-Patterson AFB Ohio Dayton OH 45433-6503

CARPENTER, ROBERT RAYMOND, b Connellsville, Pa, Mar 6, 33; m 54; c 2. INTERNAL MEDICINE. *Educ:* Univ Pittsburgh, BS, 54; Univ Rochester, MD, 57. *Prof Exp:* Intern, King County Hosp, 57-58; resident, Sch Med, Univ Wash, 58-60; clin investr & attend physician, Lab Clin Invest, Nat Inst Allergy & Infectious Dis, 62-63, actg chief clin immunol, 63-64; asst prof med, Baylor Col Med, 64-68; from asst prof to assoc prof med & community med, Univ Pittsburgh, 68-72; PROF INTERNAL MED & DIR PRIMARY CARE-COMMUNITY MED, UNIV MICH, ANN ARBOR, 72- *Concurrent Pos:* Markle scholar, 64; consult infectious dis, Georgetown Univ Serv, DC Gen Hosp, 62-63 & Montefiore Hosp, Pittsburgh, 68-71; consult immunol, Vet Admin Hosp, Houston, 64-68, dir immunol & infectious dis, 68; attend physician, Ben Taub Gen Hosp, Houston, 64-68; dir health care, Western Pa Regional Med Prog, Pittsburgh, 68-72. *Mem:* Am Asn Immunol; Am Col Physicians; Am Fedn Clin Res; Am Hosp Asn; Am Col Prev Med. *Res:* Immunology; infectious disease; health care. *Mailing Add:* 201 Oak Dr S Suite 203A Lake Jackson TX 77566

CARPENTER, ROGER EDWIN, b Tucson, Ariz, Oct 13, 35; m 62, 73; c 2. ZOOLOGY. *Educ:* Univ Ariz, BA, 57; Univ Calif, Los Angeles, PhD(zool), 63. *Prof Exp:* Assoc biol, Univ Calif, Riverside, 61-63; from asst prof to assoc prof zool, 63-70, chmn dept, 71-75, PROF ZOOL, SAN DIEGO STATE UNIV, 70- *Mem:* Sigma Xi. *Res:* Environmental physiology of vertebrates. *Mailing Add:* Dept Zool San Diego State Univ San Diego CA 92182

CARPENTER, ROLAND LEROY, b Los Angeles, Calif, Apr 26, 26; m 51; c 1. ASTRONOMY, ASTROPHYSICS. *Educ:* Los Angeles State Col, BA, 51; Univ Calif, Los Angeles, MA, 64, PhD, 66. *Prof Exp:* Electronics technician, Collins Radio Co, 52-55, engr digital commun, 55-57, group supvr, 57-59; res engr, Calif Inst Technol, 59-62, scientist, 62-68; assoc prof physics, Calif State Col, 68-77; MEM STAFF LUNAR & PLANETARY SCI SECT, JET PROPULSION LAB, CALIF INST TECHNOL, 68-; PROF PHYSICS, CALIF STATE UNIV, LOS ANGELES, 77- *Mem:* AAAS; Am Astron Soc; Royal Astron Soc. *Res:* Radar astronomy, especially studies of nearer planets by earth-based radar; galaxies; planetary astronomy. *Mailing Add:* 620 Laurel Lane Monrovia CA 91016

CARPENTER, ROY, US citizen. MARINE CHEMISTRY, GEOCHEMISTRY. *Educ:* Wash & Lee Univ, BS, 61; Univ Calif, San Diego, PhD(chem), 68. *Prof Exp:* Asst prof marine chem & geochemistry, 68-73, ASSOC PROF OCEANOG, UNIV WASH, 73- *Mem:* Am Chem Soc. *Res:* Chemical reactions in the oceans and marine sediments. *Mailing Add:* Dept Oceanog WB-10 Univ of Wash Seattle WA 98195

CARPENTER, SAMMY, b Bolckow, Mo, July 20, 28; m 50; c 2. ORGANIC CHEMISTRY. *Educ:* Northwest Mo State Col, AB & BS, 50; Univ Mo, PhD(chem), 58. *Prof Exp:* Asst chem, Univ Mo, 54-57, asst instr, 58; res chemist, Celanese Corp Am, 58-64; from asst prof to assoc prof, 64-77, PROF CHEM, NORTHWEST MO STATE UNIV, 77-, CHMN DEPT, 68- *Concurrent Pos:* Guest chemist, Nat Bur Stand, 52-54. *Mem:* Am Chem Soc; Sigma Xi; AAAS. *Res:* Substituted styrenes; organometallics and epoxides. *Mailing Add:* Dept Chem Northwest Mo State Col Maryville MO 64468

CARPENTER, STANLEY BARTON, b Searcy, Ark, Aug 29, 37; m 63; c 2. STRESS PHYSIOLOGY, LEAF ANATOMY. *Educ:* Univ Idaho, BS, 59; Univ Wash, MF, 61; Mich State Univ, PhD(tree physiol), 71. *Prof Exp:* Res forester, US Forest Serv, 60-71; from asst prof to assoc prof, Univ, Ky, 71-81; PROF & HEAD DEPT, OKLA STATE UNIV, 81- *Mem:* Soc Am Foresters; Am Forestry Asn. *Res:* Tree physiology research in the areas of surface mine reclamation, stress physiology, and the physiology of hardwood trees; anatomy of leaves. *Mailing Add:* Sch Forestry La State Univ Baton Rouge LA 70803-6202

CARPENTER, STANLEY JOHN, b Mansfield, Ohio, Feb 12, 36; m 63; c 3. CYTOLOGY. *Educ:* Oberlin Col, AB, 58; Univ Iowa, PhD(zool), 64. *Prof Exp:* Trainee path, 64-66, instr anat & cytol, 66-67, asst prof, 67-73, chmn dept, 78-80, ASSOC PROF ANAT, DARTMOUTH MED SCH, 73- *Res:* Cell fine structure; choroid plexus, placenta, embryo. *Mailing Add:* Dept of Anat Dartmouth Med Sch Hanover NH 03755

CARPENTER, STEPHEN RUSSELL, b Kansas City, Mo, July 5, 52; m 79; c 2. ECOSYSTEMS, LIMNOLOGY. *Educ:* Amherst Col, BA, 74; Univ Wis, Madison, MS, 76, PhD(bot), 79. *Prof Exp:* Asst prof biol, Univ Notre Dame, 79-85, assoc prof biol sci, 85-89; BASSETT RES PROF, UNIV WIS, 89- *Mem:* Am Soc Limnol & Oceanog; Brit Ecol Soc; Ecol Soc Am; Sigma Xi; NAm Lake Mgt Soc. *Res:* Lake productivity; Plankton; lake succession; modeling ecological systems; statistical analysis of ecological data; macrophytes. *Mailing Add:* Ctr Limnol Univ Wis Madison WI 53706

CARPENTER, STEVE HAYCOCK, b Cedar City, Utah, May 15, 38; m 60; c 2. SOLID STATE PHYSICS. *Educ:* Univ Utah, BS, 59, PhD(physics), 64. *Prof Exp:* Asst elec & magnetic lab, Univ Utah, 60-62; res physicist, Aerojet-Gen Corp Div, Gen Tire & Rubber Co, 64-65; from lectr to assoc prof & res physicist, 65-74, sr res physicist, 74-76, chmn, Dept Physics, 76-80, PROF, DEPT PHYSICS & METALL, UNIV DENVER, 74- *Concurrent Pos:* Astron head, Physics Div, Denver Res Inst, Univ Denver, 76-77; Invited Fulbright-Hays lectr, Byelo Russian Polytech Inst, Minsk, USSR, 76; panel mem, NSF Reviewer Panel for res invitation grants in mat sci; chmn, NSF Site Visit Team, Mich Tech Univ; reviewer, J Applied Physics; key reader, Metall Transactions; consult, Lawrence Livermore Nat Labs, Rocky Flats, Dunegan-Enduco. *Mem:* Am Phys Soc. *Res:* Defects, including impurities, and their interactions in crystals by means of internal friction measurements; non destructive testing particularly using acoustic emission methods and techniques; explosion welding and powder compaction. *Mailing Add:* Dept of Physics Univ of Denver Denver CO 80210

CARPENTER, STUART GORDON, b Bay City, Mich, Mar 22, 31; m 56; c 5. REACTOR PHYSICS, EXPERIMENTAL DESIGN. *Educ:* Univ Ky, BS, 53; Harvard Univ, AM, 55, PhD(physics), 58. *Prof Exp:* Group leader, Rockwell Int Corp, 57-68; TECH MGR, ARGONNE NAT LAB, 68- *Mem:* Fel Am Nuclear Soc. *Res:* Experimental reactor physics, kinetics and reactivity; design of experiments, analysis and methods development. *Mailing Add:* Argonne Nat Lab Bldg 774 PO Box 2528 Idaho Falls ID 83403-2528

CARPENTER, THOMAS J, b Middlebourne, WVa, Jan 9, 27; m 49; c 3. ORGANIC CHEMISTRY, ANALYTICAL CHEMISTRY. *Educ:* WVa Univ, BS, 49, MS, 51. *Prof Exp:* Plant chemist, Corning Glass Works, Pa, 54-56, supv chem eng, NY, 56-57, mfg eng consult, 57-61, mgr chem & metall eng dept, Tech Staff Div, 61-64; sr mem tech staff, Signetics Corp, 64-65, sect head, Process Improv, 65-66; mgr process technol dept, 66-71, DIR PROCESS TECHNOL, CORNING GLASS WORKS, 71- *Mem:* Am Chem Soc; Am Ceramic Soc; Am Inst Chem Eng; Sigma Xi. *Res:* Chemical process engineering. *Mailing Add:* 2881 County Line Dr Big Flats NY 14814

CARPENTER, WILL DOCKERY, b Moorhead, Miss, July 13, 30; m; c 2. PLANT PHYSIOLOGY. *Educ:* Miss State Univ, BS, 52; Purdue Univ, MS, 56, PhD(plant physiol), 58. *Prof Exp:* Plant biochemist, Inorg Chem Div, Monsanto Chem Co, 58-60, market develop dept, Agr Div, Monsanto Co, 60-75, mgr dept, 68-70, dir dept, 71-75, dir, Prod Develop, 75-76, dir, Environ Oper, 77-80, dir, Environ Mgt, Corp Environ Policy Staff, 80-83, gen mgr, 84-85, VPRES TECHNOL, MONSANTO AGR PROD CO, 86- *Concurrent Pos:* Mem, N Cent Weed Control Conf, pres, 77, hon mem, 82. *Mem:* Fel Weed Sci Soc Am (treas, 75, pres, 80). *Res:* Respiration and carbohydrate metabolism; soil fertility; herbicides; agriculture. *Mailing Add:* Monsanto Agr Co 800 N Lindbergh Blvd St Louis MO 63167

CARPENTER, WILLIAM GRAHAM, b West Liberty, WVa, May 7, 31; m 65; c 3. POLYMER CHEMISTRY. *Educ:* WVa Wesleyan Col, BS, 53; Univ Md, MS, 56, PhD(org chem), 60. *Prof Exp:* Res chemist, Stamford Res Labs, Am Cyanamid Co, 59-63; sr chemist, Plastics Div, Moorehead-Patterson Res Ctr, Am Mach & Foundry Co, 63-64 & Org Div, Cent Res Labs, Interchem Corp, Bloomfield, 64-67; sr chemist, NL Industries Inc, 67-79; tech mgr, Corrosion Eng Div, Pennwalt Corp, 79-85; SR CHEMIST, ASHLAND CHEM CO, 85- *Mem:* Am Chem Soc; Soc Plastics Engrs. *Res:* Fiber process development; polyamides containing phosphorus; addition polymerization; synthesis and evaluation of high temperature polymers; compounding epoxy resins; block copolymers; epoxide polymerization; water soluble polymers, thickeners and retention aids; foundry binder development. *Mailing Add:* 300 Partridge Bend Powell OH 43065

CARPENTER, WILLIAM JOHN, b Pittsburgh, Pa, Sept 15, 27; m 52; c 3. FLORICULTURE. *Educ:* Univ Md, BS, 49; Mich State Univ, PhD(hort), 53. *Prof Exp:* From asst prof to prof hort, Kans State Univ, 53-66; prof hort, Mich State Univ, 68-75; prof & chmn, Dept Ornamental Hort, 75-85, PROF HORT, UNIV FLA, 86- *Concurrent Pos:* Gen Chair, Am Soc Hort Sci, 87. *Mem:* Fel Am Soc Hort Sci. *Res:* Measuring and programming of the greenhouse environment; supplemental lighting of greenhouse crops; floriculture crop physiology; administrative responsibility for Florida's teaching, extension and research programs in ornamental horticulture. *Mailing Add:* Dept Ornamental Hort Univ of Fla Gainesville FL 32611

CARPENTER, ZERLE LEON, b Thomas, Okla, July 21, 35; m 58; c 2. ANIMAL SCIENCE, FOOD SCIENCE. *Educ:* Okla State Univ, BS, 57; Univ Wis, MS, 60, PhD(animal sci), 62. *Prof Exp:* Res assoc, Univ Wis, 58-62; from asst prof to assoc prof, 62-70, PROF ANIMAL SCI, TEX A&M UNIV, 70-, HEAD DEPT, 80-, AT AGR EXTEN SERV. *Mem:* Am Meat Sci Asn; Am Soc Animal Sci; Inst Food Tech; Sigma Xi. *Res:* Determination of histological, biochemical and physical characteristics of beef, pork and lamb muscle as related to quantitative and qualitative components of meat animal species. *Mailing Add:* Tex Agr Exten Serv Tex A&M Univ College Station TX 77843-7101

CARPENTIER, ROBERT GEORGE, b Paris, France, Oct 3, 29. PHYSIOLOGY, CARDIOLOGY. *Educ:* Univ Chile, BS, 49, MD, 57. *Prof Exp:* Instr path & physiol, Col Med, Univ Chile, 57-59, asst prof cardiol, 60-64, from asst prof to prof path & physiol, 65-76; assoc prof, 76-82, PROF PHYSIOL, COL MED, HOWARD UNIV, 82- *Concurrent Pos:* NIH fel cardiac electrophysiol, J E Fogarty Int Ctr Advan Study Health Sci, State Univ NY, Brooklyn, 70-72. *Mem:* Am Physiol Soc; Am Heart Asn; AAAS; NY Acad Sci. *Res:* Cardiovascular physiology, in particular cardiac electrophysiology and pharmacology. *Mailing Add:* Dept Physiol & Biophys 520 W St NW Washington DC 20059

CARPER, WILLIAM ROBERT, b Syracuse, NY, Feb 8, 35; m 66; c 3. NUCLEAR MAGNETIC RESONANCE. *Educ:* State Univ NY Albany, BS, 60; Univ Miss, PhD(chem), 63. *Prof Exp:* Welch Fund fel chem, Texas A&M Univ, 63-65; asst prof, Calif State Col Los Angeles, 65-67; assoc prof, 67-70, PROF CHEM, WICHITA STATE UNIV, 70- *Concurrent Pos:* NIMH spec fel, Univ SFla, 72-73, URRP fel, 80-81, NRC fel, 85, 88-90. *Mem:* Am Chem Soc; The Chem Soc; Asn Biol Chemists; Biophys Soc. *Res:* Structure and function of proteins; chemical dynamics; spectroscopic methods; materials science. *Mailing Add:* Dept of Chem Wichita State Univ Wichita KS 67208

CARPINO, LOUIS A, b Des Moines, Iowa, Dec 13, 27; m 58; c 6. CHEMISTRY. *Educ:* Iowa State Col, BS, 50; Univ Ill, MS, 51, PhD(org chem), 53. *Prof Exp:* Assoc prof org chem, 54-67, PROF CHEM, UNIV MASS, AMHERST, 67- *Concurrent Pos:* Alexander von Humboldt Award, 74. *Mem:* Am Chem Soc; The Chem Soc; Soc Ger Chem; Swiss Chem Soc; Chem Soc Japan. *Res:* Small-ring heterocycles; non-benzenoid aromatic systems; new amino-protecting groups; organo-nitrogen and organo-sulfur chemistry; silica-based organic reagents; peptide synthesis. *Mailing Add:* Dept of Chem Univ of Mass Amherst MA 01003

CARR, ALBERT A, b Covington, Ky, Dec 20, 30; div; c 4. ORGANIC CHEMISTRY, MEDICINAL CHEMISTRY. *Educ:* Xavier Univ, BS, 53, MS, 55; Univ Fla, PhD(org chem), 58. *Prof Exp:* SR ASSOC SCIENTIST, MERRELL-DOW PHARMACEUT, INC, 88- *Mem:* Am Chem Soc. *Res:* Pharmaceuticals; psychotherapeutics; design preparation and characterization of psychotherapeutic, cardiovascular and antiallergy agents. *Mailing Add:* Marion Merrell Dow Inc 2110 E Galbraith Rd Cincinnati OH 45215

CARR, BRUCE R, b Ann Arbor, Mich. OBSTETRICS & GYNECOLOGY, ENDOCRINOLOGY. *Educ:* Univ Mich, BS, 67, MD, 71. *Prof Exp:* Resident obstet & gynec, Univ Tex, 71-75, fel reproductive endocrinol, 78-80, from asst prof to assoc prof, 80-88, PROF OBSTET & GYNEC, SOUTHWESTERN MED CTR DALLAS, UNIV TEX, 88- *Mem:* Am Col Obstet & Gynec; Endocrine Soc; Am Fertil Soc; Soc Gynec Invest; Soc Reproductive Endocrinol. *Res:* Endocrinology of pregnancy; lipoprotein metabolism and cholesterol synthesis in human fetal adrenal and human corpus luteum; regulation of cholesterol synthesis in liver tissue; menopause; GnRH analogs; ovulation induction. *Mailing Add:* Univ Tex Southwestern Med Ctr Dallas 5323 Harry Hines Blvd Dallas TX 75235

CARR, CHARLES JELLEFF, b Baltimore, Md, Mar 27, 10; m 32; c 3. PHARMACOLOGY, TOXICOLOGY. *Educ:* Univ Md, BS, 33, MS, 34, PhD(pharmacol), 37. *Hon Degrees:* DSc, Purdue Univ, 64. *Prof Exp:* From asst prof to prof pharmacol, Univ Md, 37-55; prof, Purdue Univ, 55-57; head pharmacol unit, Psychopharmacol Serv Ctr, NIMH, 57-63; chief sci anal br, Life Sci Div, Army Res Off, chief res & develop hqs, US Dept Army, Va, 63-67; dir life sci res off, Fedn Am Socs Exp Biol, 67-77; exec dir, Food Safety Coun, 77-79, sci counr, 80-82. *Concurrent Pos:* Adj prof, Univ Md, 57-; managing ed, Regulatory Toxicol & Pharmacol, 81-; consult, Nutrition Fedn, 82-84; consult toxicol, 82-; fel Univ Edinburgh, 86. *Mem:* Soc Pharmacol & Exp Therapeut; Am Chem Soc; Am Pharmaceut Asn; Toxicol Forum; Am Col Neuropsychopharmacolog. *Res:* Carbohydrate metabolism; general anesthetic agents; hypotensive alkylnitrites; psychopharmacology; genetic basis for drug metabolic effects; food safety; chemical toxicology. *Mailing Add:* 6546 Belleview Dr Columbia MD 21046

CARR, CHARLES WILLIAM, b Minneapolis, Minn, July 20, 17; m 45; c 2. BIOCHEMISTRY. *Educ:* Univ Minn, BChem, 38, MS, 39, PhD(phys chem), 43. *Prof Exp:* Jr chemist, 39-43, res fel, 43-46, from instr to prof, 46-84, actg head dept, 74-76, assoc head dept, 76-84, EMER PROF BIOCHEM, UNIV MINN, MINNEAPOLIS, 84- *Mem:* Am Chem Soc; Am Soc Biol Chem; Soc Exp Biol & Med; NY Acad Sci. *Res:* Membrane structure and permeability; ion binding with proteins and other biopolymers; ionic effects on enzymes. *Mailing Add:* Dept Biochem Univ Minn Minneapolis MN 55455

CARR, CLIDE ISOM, b Creston, Mont, June 9, 20; m 45; c 3. RESEARCH ADMINISTRATION, TECHNICAL MANAGEMENT. *Educ:* Univ Mont, BA, 42; Univ Calif, PhD(chem), 49. *Prof Exp:* Asst, Univ Calif, 46-49; res chemist, Gen Labs, US Rubber Co, 49-55, Naugatuck Chem, 55-56 & Calif Res Corp, 56-57; res chemist & group leader, Res Ctr, US Rubber Co, 57-60, dept mgr, Fiber Res, 60-61 & Elastomer Res, 61-66, mgr tire eng, 66-69, mgr elastomer res, Uniroyal Res Ctr, Uniroyal Inc, 69-79, dir tire res, 79-81; RETIRED. *Mem:* AAAS; Am Chem Soc; Sigma Xi. *Res:* Management of tire technology and elastomer applications; plastics; fibers; rubber; tires; polymers. *Mailing Add:* 10802 Meade Dr Sun City AZ 85351

CARR, DANIEL OSCAR, b Kansas City, Kans, May 1, 34. BIOCHEMISTRY. *Educ:* Univ Mo-Kansas City, BS, 56; Iowa State Univ, PhD(biochem), 60. *Prof Exp:* From instr to asst prof, 63-69, ASSOC PROF BIOCHEM, UNIV KANS MED CTR, KANSAS CITY, 69- *Concurrent Pos:* USPHS fel, Univ Kans Med Ctr, Kansas City, 60-63. *Mem:* AAAS; Am Chem Soc; Am Soc Biochem & Molecular Biol. *Res:* Mechanisms of enzymic catalysis, roles of vitamins and coenzymes with emphasis upon drug metabolism and clinical biochemistry. *Mailing Add:* Dept Biochem Univ Kans Med Ctr Rainbow Blvd at 39th Ave Kansas City KS 66103

CARR, DAVID HARVEY, b Southport, Eng, Jan 10, 28; Can citizen; m 51; c 1. ANATOMY, CYTOGENETICS. *Educ:* Univ Liverpool, MB, ChB, 50, DSc, Univ Liverpool, 70. *Prof Exp:* Lectr, Univ Western Ont, 58-61, from asst prof to assoc prof, 61-67; assoc prof, 67-70, PROF ANAT, COL HEALTH SCI, MCMASTER UNIV, 70- *Concurrent Pos:* Can Soc Study Fertil res award, 61. *Mem:* Am Asn Anat; Can Fedn Biol Soc; Genetics Soc Am. *Res:* Chromosome studies in clinical syndromes and spontaneous abortions. *Mailing Add:* Dept Anat Col Health Sci McMaster Univ 1200 Main St Hamilton ON L8N 3Z5 Can

CARR, DAVID TURNER, b Richmond, Va, Mar 12, 14; m 79; c 1. INTERNAL MEDICINE, ONCOLOGY. *Educ:* Med Col Va, MD, 37; Univ Minn, MS, 47. *Prof Exp:* Intern, Grady Hosp, Ga, 37-38, asst resident, 38-39; chest serv, Bellevue Hosp, New York, 40-41, chief resident, 41-42; physician, Mt Morris Tuberc Hosp, 42-43; from asst prof to prof med, Mayo Med Sch, 53-79, 64-79, chmn, dept oncol & dir, Mayo Comprensive Cancer Ctr, 74-75; PROF MED, M D ANDERSON HOSP & TUMOR INST, 79- *Concurrent Pos:* Consult, Mayo Clin, 47-79. *Mem:* Am Lung Asn (vpres, 71-72); Am Cancer Soc; fel Am Col Physicians; Am Thoracic Soc (vpres, 63-64); Int Asn Study Lung Cancer (vpres, 74-75, pres, 76, treas, 76-82). *Mailing Add:* M D Anderson Hosp & Tumor Inst Tex Med Ctr Houston TX 77030

CARR, DODD S(TEWART), b Chicago, Ill, Dec 19, 25; m 53; c 2. ELECTRODEPOSITION OF METALS. *Educ:* Loyola Col, BS, 45; Johns Hopkins Univ, MSE, 48, DrEng(chem eng), 50; Rutgers Univ, MBA, 61. *Prof Exp:* Res chemist nickel plating, Int Nickel Co, 49-52; asst dir electrochem res, Bart Mfg Co, 52-56, dir res, Bart Labs & Design, Inc, 56-59; tech asst to vpres, Freeport Nickel Co, 59-60; vpres & tech dir, Bart Mfg Corp, 60-65; vpres & secy, Hill Cross Co, Inc, 65-66; proj mgr chem res, 66-67, mgr, 67-69, MGR CHEM, ELECTROCHEM & PATENTS, INT LEAD ZINC RES ORGN, INC, 69- *Concurrent Pos:* Consult, WHO, 74, 76 & 79. *Honors & Awards:* Silver Medal, Am Electroplaters Soc, 52. *Mem:* Fel Am Inst Chem; NY Acad Sci; sr mem Am Chem Soc; Electrochem Soc; Am Ceramics Soc. *Res:* Organolead chemical applications; zinc oxide stabilization of plastics against ultraviolet light degradation; zinc paint pigments; lead-acid batteries; antifouling coatings for shipbottoms; ceramic foodware safety; electroforming; nickel plating; wood preservatives; patenting and licensing. *Mailing Add:* 1714 Vista St Durham NC 27701-1355

CARR, DONALD DEAN, b Fredonia, Kans, Mar 28, 31; m 55; c 4. GEOLOGY. *Educ:* Kans State Univ, BS, 53, MS, 58; Ind Univ, Bloomington, PhD(geol), 69. *Prof Exp:* Teacher pub sch, 59-60; geologist, Humble Oil & Ref Co, 60-62 & Geosci Div, Tex Instruments Co, 62-63; GEOLOGIST, MINERAL RESOURCES & DATA MGT, IND GEOL SURV, 63- *Concurrent Pos:* Subj ed, Indust Minerals, Encycl Mat Sci & Eng; prof, Ind

Univ. *Honors & Awards:* Hal Williams Hardinge Award, Asn Int Mining Engrs, 88. *Mem:* Geol Soc Am; Am Asn Petrol Geologists; Am Inst Mining, Metall & Petrol Engrs; Am Inst Prof Geol. *Res:* Sedimentology; stratigraphy; geomorphology; economic geology of coal and industrial minerals. *Mailing Add:* Ind Geol Surv 611 N Walnut Grove Bloomington IN 47405

CARR, DUANE TUCKER, b Gunnison, Colo, July 6, 32; m 54; c 2. CHEMISTRY. *Educ:* Western State Col Colo, BA, 54; Purdue Univ, PhD(phys chem), 62. *Prof Exp:* Instr chem, Wabash Col, 60-61; asst prof 61-67, ASSOC PROF CHEM, COE COL, 67- *Concurrent Pos:* Asst prof chem, Agr Col, Haile Selassie Univ, 68-70. *Mem:* AAAS; Am Chem Soc. *Res:* Molecular structure, carbon-13 splittings in fluorine magnetic resonance spectra. *Mailing Add:* Coe Col Cedar Rapids IA 52402-2469

CARR, EDWARD ALBERT, JR, b Cranston, RI, Mar 3, 22; m 52; c 2. CLINICAL PHARMACOLOGY, INTERNAL MEDICINE. *Educ:* Brown Univ, AB, 42; Harvard Univ, MD, 45. *Prof Exp:* Intern, RI Hosp, 45-46; asst resident internal med, Cushing Hosp, 48; instr, Harvard Med Sch, 49-51; resident internal med, Pa Hosp, 51-52; from asst prof to prof internal med & pharmacol, Univ Mich, Ann Arbor, 53-74, dir, Upjohn Ctr Clin Pharmacol, 66-74; prof med & pharmacol & chmn dept pharmacol, Univ Louisville, 74-76; chmn, Dept Pharmacol, 76-88, PROF MED, PHARMACOL & THERAPEUT, STATE UNIV NY, BUFFALO, 76- *Concurrent Pos:* Res fel pharmacol, Harvard Med Sch, 48-49; exchange fel, St Bartholomew's Hosp, London, 52-53; mem, pharmacol Training Comt, Nat Inst Gen Med Sci, 64-66 & Pharmacol-Toxicol Comt, 70-74; mem, revision comt, US Pharmacopeia, 65-75; mem, Pharmacol Comt, Walter Reed Army Inst Res, 67-77, US Joint Comn Prescription Drug Use, 76-80, US Vet Admin Coop Studies Eval Comt, 79-82 & US Army Adv Comt Med Res & Develop, 83-84. *Honors & Awards:* Henry W Elliot Award, Am Clin Pharmacol & Therapeut, 81. *Mem:* Soc Nuclear Med; Am Col Physicians; Am Soc Pharmacol & Exp Therapeut; Am Soc Clin Pharmacol & Therapeut (pres, 74-75); emer mem Am Thyroid Asn; emer mem Endocrine Soc. *Res:* Clinical pharmacology; pharmacology of imaging agents; thyroid; radioisotopes; translation of animal data to human pharmacology. *Mailing Add:* Dept Pharmacol & Therapeut State Univ NY Med Sch Buffalo NY 14214

CARR, EDWARD FRANK, b St Johnsbury, Vt, Aug 18, 20; m 54; c 3. PHYSICS. *Educ:* Mich State Univ, BS, 43, PhD(physics), 54. *Prof Exp:* Asst physics, Mich State Univ, 48-53, instr, 53-54; asst prof, St Lawrence Univ, 54-57; from asst prof to assoc prof, 57-69, PROF PHYSICS, UNIV MAINE, ORONO, 69- *Mem:* Am Phys Soc; Am Asn Physics Teachers. *Res:* Physics of liquid crystals, particularly eletrical properties. *Mailing Add:* Dept of Physics Univ of Maine Orono ME 04473

CARR, EDWARD GARY, b Toronto, Ont, Aug 20, 47; m 87; c 1. CLINICAL CHILD PSYCHOLOGY, DEVELOPMENTAL DISABILITIES. *Educ:* Univ Toronto, BA, 69; Univ Calif, San Diego, PhD(psychol), 73. *Prof Exp:* Adj asst prof psychol, Univ Calif, Los Angeles, 73-76; from asst prof to assoc prof, 76-85, PROF PHYCHOL, STATE UNIV NY, STONY BROOK, 85- *Concurrent Pos:* Fac sponsor, Coun Int Exchange Scholars, 80-82, prof adv bd, May Inst for Autistic Children, 80-; sci comt, Univ Ancona, Italy, 85-, bd dirs, Berkshire Asn Behav Anal, 87-89, prof adv bd, New Eng Ctr for Autism, 88- *Mem:* Am Psychol Asn; Nat Soc Autistic Children. *Res:* Experimental child psychopathology including studies of language disorder; developmental disorders; early social development; analysis and remediation of severe behavior. *Mailing Add:* Dept Psychol State Univ NY Stony Brook NY 11794-2500

CARR, EDWARD MARK, b Warsaw, Poland, Sept 16, 18; nat US; m 44; c 2. ANALYTICAL CHEMISTRY. *Educ:* Univ Minn, BCh, 49, MS, 52. *Prof Exp:* Res chemist process develop, E I du Pont de Nemours & Co, 52-53; res chemist, Toni Div, Gillette Co, 53-56, res supvr anal chem sect, 56-61, res assoc, 61-64; mgr analytical chem group, Noxzema Chem Co, Noxell Corp, 64-69, mgr anal chem, 69-78, sr res scientist, 78-81; RETIRED. *Mem:* Am Chem Soc. *Res:* Analytical chemistry of cosmetic materials and products; physical chemistry of surface active agents; emulsion polymerization. *Mailing Add:* 106 Hanley Dr Grass Valley CA 95949-9438

CARR, FRED K, b Clinton, NC, 1942; m 65; c 2. PHARMACOLOGY, PHARMACEUTICAL CHEMISTRY. *Educ:* NC State Univ, BS, 67; Univ NC, PhD(pharmaceut chem), 75. *Prof Exp:* Nat Res Serv res fel & instr physiol, Albany Med Col, 75-78; asst prof pharmacol, Philadelphia Col Osteop Med, 78-; dir, Off 303, Patent Info, Inc; DIR MKT, PROGRESSIVE INT ELECTRONICS, 90- *Mem:* Am Chem Soc; Am Physiol Soc; Sigma Xi. *Res:* Cardiovascular and biochemical sequelae of circulatory shock. *Mailing Add:* PO Box 2488 Chapel Hill NC 27514

CARR, GEORGE LEROY, b Upperco, Md, Dec 11, 27; m 52; c 3. PHYSICS, SCIENCE EDUCATION. *Educ:* Western Md Col, BS, 48, MEd, 59; Cornell Univ, PhD(sci educ, physics), 69. *Prof Exp:* Teacher high sch, Md, 48-51; chmn dept sci, Milford Mill High Sch, 53-61; instr physics, Cornell Univ, 63-64; chmn dept sci, Pikesville High Sch, Md, 64-65; asst prof educ & physics, Western Md Col, 65-66; from asst prof to assoc prof, 66-70, asst chmn physics dept, 84-87, PROF PHYSICS, UNIV LOWELL, 70-, SERV LAB COORDR, 87- *Concurrent Pos:* Consult, Phys Sci Study Comt, 57-61; adv, Lowell Prof Improv Prog, 69-71; pres, Univ Coun, 81-83. *Mem:* Am Phys Soc; Am Asn Physics Teachers; Am Geophys Union; Sigma Xi. *Res:* Problem-solving in physics; environmental applications of physics; teaching of physics. *Mailing Add:* 2 Gifford Lane Chelmsford MA 01824-2023

CARR, GERALD DWAYNE, b Pasco, Wash, Apr 1, 45; m 77; c 1. SYSTEMATIC BOTANY, EVOLUTION. *Educ:* Eastern Wash State Univ, BA, 68; Univ Wis-Milwaukee, MS, 70; Univ Calif, Davis, PhD(bot), 75. *Prof Exp:* Assoc bot, Univ Calif, Davis, 74-75; asst prof, 75-81, assoc prof, 81-86, PROF BOT, UNIV HAWAII, 86- *Concurrent Pos:* Bot consult. *Mem:* Soc Study Evolution; Am Soc Plant Taxonomists; Bot Soc Am; Int Asn Plant Taxon. *Res:* Biosystematic and evolutionary studies of the Pacific states and Hawaiian tarweeds; biosystematics and cytogenetics of the Hawaiian flora. *Mailing Add:* Dept Bot Univ Hawaii Honolulu HI 96822

CARR, GERALD PAUL, b Denver, Colo, Aug 22, 32; m 79; c 6. HUMAN FACTORS ENGINEERING. *Educ:* Univ Southern Calif, BS, 54; US Naval Postgrad Sch, BS, 61; Princeton Univ, MS, 62. *Hon Degrees:* DSci, Parks Cols, 76. *Prof Exp:* Astronaut, NASA, 66-77; vpres, corp mgr bus develop, Bovay Engrs, 77-80, sr vpres, 80-81; sr consult, Appl Res, Inc, 81-83; mgr, Univ Tex McDonald Observ, 83-85; PRES, CAMUS, INC, 85- *Concurrent Pos:* Adv bd, Eldorado Bank, Tustin, Calif, 77-; mem adv comt, OAST Space Syst Technol, NASA Hq, 86-, Off Space Sci & Appl, Life Sci Strategic Planning Study Comt, 88 & 89; Space Dermat Found, 87; lectr, NASA. *Honors & Awards:* Distinguished Serv Medal, NASA, 74; Gold Space Medal, Fedn Aeronautique Int, 74; De La Vaulx Medal, 74; Haley Astronaut Award, 74; Flight Achievement Award, Am Astronaut Soc, 75. *Mem:* Soc Prof Engrs; fel Am Astron Soc; Nat Space Soc; Soc Exp Text Pilots. *Res:* Space human factors engineering; consult and technical support offered in spacecraft man-machine interface; system design for optimum human productivity in a weightless environment. *Mailing Add:* PO Box 919 Huntsville AR 72740

CARR, HERMAN YAGGI, b Alliance, Ohio, Nov 28, 24; m 59; c 2. CONDENSED STATE PHYSICS. *Educ:* Harvard Univ, BS, 48, MA, 49, PhD(physics), 53. *Prof Exp:* From asst prof to assoc prof, 52-64, PROF PHYSICS, RUTGERS COL, 64- *Concurrent Pos:* Guggenheim fel, 67. *Mem:* Fel Am Phys Soc; Am Asn Physics Teachers. *Res:* Nuclear magnetic resonance; physics of fluids; phase transitions. *Mailing Add:* 928 Tullo Rd Bridgewater NJ 08807

CARR, HOWARD EARL, b Headland, Ala, Sept 16, 15; m 39; c 2. PHYSICS. *Educ:* Auburn Univ, BS, 36; Univ Va, AM, 39, PhD(physics), 41. *Prof Exp:* Asst, Univ Va, 39-40; from asst to assoc prof physics, Univ SC, 41-44; physicist, Ford, Bacon, Davis Corp, Manhattan Dist, Oak Ridge, Tenn, 44; asst prof physics, US Naval Acad, 46-48; from assoc prof to prof, 48-80, head dept, 53-78, EMER PROF PHYSICS, AUBURN UNIV, 80- *Mem:* Fel AAAS; fel Am Phys Soc; Am Asn Physics Teachers; Sigma Xi. *Res:* Isotope separation; thermal diffusion in liquids; mass spectrography; negative ions. *Mailing Add:* 342 Payne St Auburn AL 36830

CARR, IAN, CANCER METASTASIS, ELECTRON MICROSCOPY. *Educ:* Univ Glasgow, PhD(path), 60; Univ Sheffield, MD, 73. *Prof Exp:* PROF PATH, UNIV MAN, 82- *Mailing Add:* Dept Path St Boniface Gen Hosp 409 Tache Ave Winnipeg MB R2H 2A6 Can

CARR, JACK A(LBERT), b Burlington, Ont, Apr 18, 20; m 42; c 2. CHEMICAL ENGINEERING. *Educ:* Univ Toronto, BASc, 41. *Prof Exp:* Mem staff, Dunlop Can Ltd, 40-41, compounder indust rubber goods, 45-49, head foamed latex process control, 49-53; asst mgr, Dunlop Res Ctr, 53-65, asst gen mgr, 65-70, gen mgr, 71-80; RETIRED. *Mem:* Am Chem Soc; Chem Mkt Res Asn; emer mem Indust Res Inst; hon mem Can Res Mgt Asn; fel Chem Inst Can. *Res:* Polymer and polymerization chemistry; polymer physics. *Mailing Add:* 12 Plateau Circle Don Mills ON M3C 1M8 Can

CARR, JAMES DAVID, b Ames, Iowa, Apr 3, 38; m 68; c 2. ANALYTICAL CHEMISTRY. *Educ:* Iowa State Univ, BS, 60; Purdue Univ, PhD(anal chem), 66. *Prof Exp:* Technician gas chromatog, Ivorydale Labs, Procter & Gamble Co, 60; res fel chem, Univ NC, 65-66; from asst prof to assoc prof, 66-88, vchmn dept, 81-83, COORDR FRESHMAN CHEM, UNIV NEBR-LINCOLN, 83-, PROF CHEM, 88- *Concurrent Pos:* Vis assoc prof chem, Purdue Univ, 74-75. *Mem:* Am Chem Soc. *Res:* Kinetics and mechanism of ligand exchange reactions; atom transfer reactions; oxidation kinetics by ferrate VI; water quality analysis and treatment. *Mailing Add:* Dept Chem Univ Nebr Lincoln NE 68588-0304

CARR, JAMES RUSSELL, b Fairfield, Calif, July 30, 57; m 87; c 2. APPLIED ECONOMIC & ENGINEERING GEOLOGY. *Educ:* Univ Nev, Reno, BS, 79; Univ Ariz, MS, 81, PhD(geol eng), 83. *Prof Exp:* Asst prof geol eng, Univ Mo, Rolla, 83-86; asst prof, 86-89, ASSOC PROF GEOL ENG, UNIV NEV, RENO, 89- *Concurrent Pos:* Ed, Int Asn Math Geol Newsletter, 89- *Honors & Awards:* President's Prize, Int Asn Math Geol, 87. *Mem:* Am Soc Photogram & Remote Sensing; Int Asn Math Geol; Asn Eng Geologists; Sigma Xi; Int Soc Rock Mech. *Res:* Applied geostatistics and development of geostatistical software. *Mailing Add:* 30 Bear St Carson City NV 89704

CARR, JAMES W(OODFORD), JR, b Columbus, Miss, Mar 13, 21; m 46; c 2. CHEMICAL ENGINEERING. *Educ:* Miss State Univ, BS, 43; Mass Inst Tech, MS, 47. *Prof Exp:* Head bus & oper serv & safety, Exxon Res & Develop Labs, 47-86; RETIRED. *Concurrent Pos:* Mem Eng Adv Comt, Miss State Univ, 75-, adv coun High Sch Eng Profs, Scottandville Magnet High Sch, Baton Rouge, La, 86-; auth-instr, Short Course in Pilot Plant Technol, La State Univ. *Mem:* Am Inst Chem Engrs; Instrument Soc Am. *Res:* Fluid solids; pilot plant engineering; process control instrumentation and techniques; pilot plant safety. *Mailing Add:* 3272 Svendson Dr Baton Rouge LA 70809

CARR, JAN, b London, Eng, Jan 22, 43; Can citizen; m 66; c 2. ELECTRICAL POWER SYSTEMS. *Educ:* Ryerson Polytech Inst, dipl, 63; Univ Toronto, BASc, 68; Univ Waterloo, MASc, 70, PhD(elec eng), 72. *Prof Exp:* Asst proj eng, Motorola Electronics Inc, 67-68; res engr, Saskatchewan Power Corp, 72-77; consult, 77-80; pres, Amicus Eng Corp, 80-81; ASSOC, ACRES CONSULT SERV LTD. *Concurrent Pos:* Adj prof, Univ Waterloo, 73; consult, 81-; assoc ed, Can J Elec Eng, 82- *Mem:* Inst Elec & Electronics Engrs; Inst Elec Engrs; Can Soc Elec Engrs; Eng Inst Can. *Res:* Instrumentation and simulation of electric power systems particularly with respect to protective relaying; high-voltage direct current transmission and distribution systems. *Mailing Add:* Acres Consult Serv Ltd 5259 Dorchester Rd Box 1001 Niagara Falls ON L2E 6W1 Can

CARR, JEROME BRIAN, b Syracuse, NY, Dec 17, 38; m 61; c 3. TELMATOLOGY, LIMNOLOGY. *Educ:* St Louis Univ, BS, 61; Boston Col, MS, 65; Rensselaer Polytech Inst, PhD(oceanog), 71. *Prof Exp:* Phys oceanogr, US Naval Oceanog Off, Md, 61-62; oceanogr-geophysicist, Sperry

Rand Res Ctr, Mass, 63-66; oceanogr, Gen Dynamics/Electronics, NY, 66, Raytheon Corp, RI, 67-68 & Hazeltine Corp, Braintree, 69-70; vis asst prof oceanog & geol, Purdue Univ, 66-67; environ consult, 70-71; environ scientist, Lowell Technol Inst Res Found, 71-72 & Anal Systs Eng Corp, 72-74; PRES & TECH DIR, CARR RES LAB, INC, 74- *Mem:* AAAS; Am Geophys Union; Soc Mining Metall & Explor; Aquatic Plant Mgt Soc; US Naval Inst; Am Inst Hydrol; Int Peat Soc; NAm Lake Mgt Soc. *Res:* Study of space-time variations of the ocean's thermal structure and relations to underwater acoustics; global tectonics; urban hydrology; developed new lake management techniques using iron and nitrogen control and using biological substitutions, wetland evolution and function; hazardous waste site restoration. *Mailing Add:* 17 Waban St Wellesley MA 02181

CARR, JOHN B, b Olympia, Wash, Aug 29, 37; m 60; c 4. AGRICULTURAL CHEMISTRY MANAGEMENT. *Educ:* St Martin's Col, BSc, 59; Univ Wash, PhD(med chem), 63. *Prof Exp:* NIH fel med chem, Univ Kans, 63-64; org res chemist, Shell Develop Co, 64-77, supvr detergent applications, 77-80, bus researcher, 80-81, mgr org chem, 81-86; MGR ENVIRON SCI, DU PONT AG PROD, 87- *Res:* Lipid metabolism; animal physiology and growth; parasitology; insecticides; herbicides; surfactants. *Mailing Add:* 5 Deer Run Rd Wilmington DE 19807-2403

CARR, JOHN WEBER, III, b Durham, NC, May 16, 23; m 49; c 3. MATHEMATICS. *Educ:* Duke Univ, BS, 43; Mass Inst Technol, MS, 49, PhD(math), 51. *Prof Exp:* Res assoc, Mass Inst Technol, 51-52; res mathematician, Univ Mich, 52-55, asst & assoc prof, 55-59; assoc prof & dir res comput ctr, Univ NC, 59-62, prof, 62-63; assoc prof, 63-66, chmn grad group comput & info sci, 66-73, PROF COMPUT & INFO SCI, MOORE SCH ENG, UNIV PA, 66- *Concurrent Pos:* Lectr math, Univ Mich, 53-55; ed, Comput Reviews, 59-63; vis prof, Jiao Tong Univ, Shanghai, China, 78-79, 81 & NW Telecommun Eng Inst, Xian, China, 78-79; tech consult, UN Develop Prog, Hong Kong Productivity Ctr, Govt Hong Kong, 81; vis prof, Nat Univ Singapore, 85; adj prof, Jiao Tong Univ, Shanghai, NWTI, Xian, China. *Mem:* Am Math Soc; Asn Comput Mach (pres, 57-58); Inst Elec & Electronics Engrs. *Res:* Solution of Schrodinger equation; computing machinery logic and programming theory; numerical analysis; artificial intelligence; problem solving and programming theory. *Mailing Add:* 405 Rancho Arroyo Pkwy Apt 187 Fremont CA 94536

CARR, LAURENCE A, b Ann Arbor, Mich, Mar 21, 42; m 64; c 2. PHARMACOLOGY. *Educ:* Univ Mich, BS, 65; Mich State Univ, MS, 67, PhD(pharmacol), 69. *Prof Exp:* Asst, Mich State Univ, 66; from asst prof to assoc prof, 69-81, PROF PHARMACOL, UNIV LOUISVILLE, 81- *Concurrent Pos:* Fulbright res scholar, 80-81; assoc dean curric, Univ Louisville, Sch Med, 87- *Honors & Awards:* Bristol Lab Award, 65. *Mem:* Am Soc Pharmacol & Exp Therapeut; AAAS; Soc Neuroscience. *Res:* Effects of drugs on the synthesis, release and metabolism of brain catecholamine; interaction of tobacco smoke components with brain neurotransmitters. *Mailing Add:* Dept Pharmacol & Toxicol Univ Louisville Louisville KY 40292

CARR, LAWRENCE JOHN, b De Kalb, Ill, July 4, 39; m 66; c 1. ORGANIC CHEMISTRY, POLYMER CHEMISTRY. *Educ:* St Mary's Col, Minn, BA, 62; Univ Ariz, PhD(org chem), 66. *Prof Exp:* Res chemist, Arco Chem, Atlantic Richfield, 66-71 & A-M Corp, 72-73; group leader chem, 73-79, STAFF SCIENTIST, BORG-WARNER CORP, 79- *Mem:* Am Chem Soc; Sigma Xi. *Res:* Process research on organic and polymeric products; applications research on new chemical products. *Mailing Add:* 1308 Cumberland Circle E Elk Grove Village IL 60007-3804

CARR, MALCOLM WALLACE, anatomy, surgical pathology, for more information see previous edition

CARR, MEG BRADY, b Belvidere, Ill, May 13, 49; m 72. REGRESSION ANALYSIS, NUMERICAL METHODS. *Educ:* Fla State Univ, BA, 71, PhD(appl math), 80; NC State Univ, MA, 72. *Prof Exp:* Grad asst, Nat Ctr Atmospheric Res, 75-80; ASST PROF COMPUT SCI, UNIV OKLA, 80- *Mem:* Am Statist Asn. *Res:* Applied mathematics, numerical methods, statistics, regression analysis and time series analysis; applications of mathematics and statistics to meteorology. *Mailing Add:* Math Dept Univ Okla 601 Elm Norman OK 73019

CARR, MICHAEL H, b Leeds, Eng, May 26, 35; US citizen; m 61; c 1. ASTROGEOLOGY, VULCANOLOGY. *Educ:* Univ London, BSc, 56; Yale Univ, MS, 57, PhD(geol), 60. *Prof Exp:* Res assoc geophys, Univ Western Ont, 60-62; GEOLOGIST, US GEOL SURV, 62- *Concurrent Pos:* Mem, Mariner 9 Mars TV Team, 69-74, Voyager Imaging Team, 76-82 & Galileo Imaging Team, 78-; leader, Viking Mars Orbiter Imaging Team, 70-80 & Mars Observer Interdisciplinary Scientist, 84- *Honors & Awards:* Exceptional Sci Achievement Medal, NASA, 77; Distinguished Serv Award, Dept Interior, 88. *Mem:* AAAS; Geol Soc Am; Am Geophys Union; Am Astron Soc. *Res:* The formation and evolution of solid planetary bodies, mainly Mars & Io; outline of the geologic history, the nature of volcanism and the role of water and ice in surface processes on Mars; volcanism and internal structure of Io. *Mailing Add:* US Geol Surv 345 Middlefield Rd Menlo Park CA 94025

CARR, MICHAEL JOHN, b Portland, Maine, Nov 12, 46; div. GEOLOGY. *Educ:* Dartmouth Col, AB, 69, MS, 71, PhD(geol), 74. *Prof Exp:* Asst prof, 74-85, PROF GEOL, RUTGERS UNIV, 85- *Mem:* Geol Soc Am; Am Geophys Union; Sigma Xi. *Res:* The relation of volcanos and active faults to underthrusting at convergent plate margins, with special emphasis on the Central American convergent plate margin. *Mailing Add:* 155 Emerson Rd Somerset NJ 08873

CARR, NORMAN L(OREN), b Ill, Dec 7, 24; m 48; c 4. CHEMICAL ENGINEERING. *Educ:* Univ Ill, BS, 46; Univ Minn, MS, 50; Ill Inst Tech, PhD(chem eng), 53. *Prof Exp:* Process chem engr, Mallinckrodt Chem Works, 46-48; asst, Univ Minn, 48-49 & Ill Inst Tech, 49-50; res engr, Pure Oil Co, 52-59; sr staff engr, Chem & Minerals Div, Gulf Res & Develop Co, 59-85; CONSULT, 85- *Mem:* Am Chem Soc; Am Inst Chem Engrs. *Res:* Process simulation using analog and digital computers; reactor dynamics and control; fluidization heat transfer; viscosity of natural gases at high pressures; isomerization; kinetics; separations; chemical reaction engineering; coal liquefaction; uranium extraction. *Mailing Add:* 550 Blazier Dr Wexford PA 15090

CARR, PAUL HENRY, b Boston, Mass, May 12, 35; wid; c 5. SOLID STATE PHYSICS. *Educ:* Mass Inst Technol, BS, 57, MS, 61; Brandeis Univ, PhD(solid state physics), 66. *Prof Exp:* Res physicist, Calibration Ctr, Redstone Arsenal, Ala, 61-62; res physicist, 62-68, supvry res physicist, Air Force Cambridge Res Lab, 68-76, BR CHIEF, ELECTROMAGNETICS DIRECTORATE, ROME AIR DEVLOP CTR, 76- *Mem:* Am Phys Soc; Sigma Xi; fel Inst Elec & Electronics Engrs. *Res:* Electron paramagnetic resonance; vacuum standards and measurements; nonlinear phenomena in microwave ultrasonics; microwave phonons; kilomegacycle ultrasonics; microwave frequency elastic surface waves; surface acoustic wave signal processing; monolithic microwave integrated circuits; high temperature superconductivity. *Mailing Add:* 178 Old Billerica Rd Bedford MA 01730

CARR, PETER WILLIAM, b Brooklyn, NY, Aug 16, 44; m 66; c 3. ANALYTICAL CHEMISTRY. *Educ:* Polytech Inst Brooklyn, BS, 65; Pa State Univ, PhD(anal chem), 69. *Prof Exp:* Res assoc path, Med Sch, Stanford Univ, 68-69; from asst prof to assoc prof, Univ Ga, 69-77; assoc prof, 77-81, PROF CHEM, 81-, ASSOC DIR, CTR BIOPROCESS TECHNOL, UNIV MINN, MINNEAPOLIS, 87- *Concurrent Pos:* Consult, Leeds & Northrup Co, 70-78; pres, Symp Anal Chem Pollutants Inc, 74- & Minn Chromatography Forum, 78-; consult, 3M Co, 79- *Honors & Awards:* L S Palmer Award in Chromatography, 84; Benedetti-Pichler Award, 90. *Mem:* Am Chem Soc; Am Asn Clin Chemists; NAm Thermal Anal Soc. *Res:* Bioanalytical chemistry; immobilized enzymes; high performance liquid chromatography; affinity chromatography; thermoanalytical chemistry; clinical chemistry; ion selective electrodes; electroanalytical chemistry. *Mailing Add:* Dept Chem 207 Pleasant St Minneapolis MN 55455

CARR, RALPH W, b St Paul, Minn, Apr 19, 46. COMPUTER SCIENCE. *Educ:* Carlton Col, BA, 68; Univ Wis, PhD(math), 77. *Prof Exp:* Prof math, 77-86, CHMN, COMPUT SCI DEPT, ST CLOUD STATE UNIV, 86- *Mem:* Soc Industr Appl Math; Am Math Soc; Math Asn Am. *Res:* Computer Service Education. *Mailing Add:* 2130 University Ave Madison WI 53705

CARR, RICHARD DEAN, b Columbus, Ohio, June 29, 29; m 53; c 4. DERMATOLOGY. *Educ:* Ohio State Univ, BA, 51, MD, 54. *Prof Exp:* Asst prof dermat, 63-65, dir div, 67-69, ASSOC PROF DERMAT, COL MED, OHIO STATE UNIV, 65- *Mem:* AMA; Am Acad Dermat; Soc Invest Dermat; Am Col Physicians; Am Asn Prof Dermat. *Res:* Clinical dermatology. *Mailing Add:* 1840 Zollinger Rd Columbus OH 43221

CARR, RICHARD J, ULTRASTRUCTURE, VIROLOGY. *Educ:* Univ Ark, PhD(plant path), 83. *Prof Exp:* Fel, Wistar Inst, 83-84; CONSULT, 85- *Mailing Add:* 252 Oak Ridge Ave Summit NJ 07901

CARR, ROBERT CHARLES, b Oakland, Calif, Nov 13, 46. PARTICULATE EMISSIONS CONTROL. *Educ:* Univ Calif, Berkeley, BS, 68, MS, 70. *Prof Exp:* Res asst combustion res, Univ Calif, Berkeley, 68-72; sr engr power plant emissions testing, KVB Eng, 72-74; tech mgr particulate emissions control, 74-82, prog mgr air qual control, 82-84, DIR ENVIRON CONTROL SYSTS, ELEC POWER RES INST, 84- *Mem:* Air Pollution Control Asn. *Res:* Particulate emission control devices for pulverized coal power plants; fabric filtration; electrostatic precipitation; development of particulate measurement instrumentation; nitrogen-oxide emission control; dry sulfuroxide emission control; integrated environmental control systems. *Mailing Add:* PO Box 307 Menlo Park CA 94026

CARR, ROBERT H, b Ames, Iowa, June 3, 35; m 62; c 2. PHYSICS. *Educ:* Cornell Univ, BA, 57; Iowa State Univ, PhD(solid state physics), 63. *Prof Exp:* Res officer physics, Commonwealth Sci & Indust Res Orgn, Australia, 63-64; from asst prof to assoc prof, 64-72, PROF PHYSICS, CALIF STATE UNIV, LOS ANGELES, 72- *Mem:* Am Asn Physics Teachers; Cryogenic Soc Am (past pres). *Res:* Measurement of low temperature characteristics of materials. *Mailing Add:* Dept of Physics Calif State Univ 5151 University Dr Los Angeles CA 90032

CARR, ROBERT J(OSEPH), b Milwaukee, Wis, Mar 27, 31. NUCLEAR CHEMISTRY, ANALYTICAL CHEMISTRY. *Educ:* Univ Calif, Los Angeles, BS, 51; Univ Calif, Berkeley, PhD(chem), 56. *Prof Exp:* From instr to asst prof chem, Wash State Univ, 55-61; chemist, Shell Develop Co, 61-67; prof chem, Col Alameda, 70-90; PROF CHEM, MERRITT COL, OAKLAND, 67-70 & 90- *Mem:* Am Chem Soc. *Res:* Nuclear reactions and spectroscopy; activation analysis; Mossbauer effect spectroscopy; application of radio-isotopes to problems in physical and analytical chemistry. *Mailing Add:* Dept Chem Merritt Col Oakland CA 94619

CARR, ROBERT LEROY, b Pasco, Wash, Sept 2, 40; m 65; c 2. CYTOGENETICS, SYSTEMATIC BOTANY. *Educ:* Eastern Wash Univ, BA, 64; Univ Mich, MA, 66; Ore State Univ, PhD(syst bot), 73. *Prof Exp:* Chmn, dept biol, 76-78, PROF SYST BOT, EASTERN WASH UNIV, 69-, DIR, TURNBULL LAB ECOL STUDIES, 80- *Mem:* Am Asn Plant Taxonomists; Bot Soc Am; Sigma Xi. *Res:* Systematic botany of vascular plants, especially the families Asteraceae and Boraginaceae; cytogenetics of Calycadenia; chemical, morphological and ecological studies in Calycadenia and Hackelia. *Mailing Add:* Dept Biol Eastern Wash Univ Cheney WA 99004

CARR, ROBERT WILSON, JR, b Montpelier, Vt, Sept 7, 34; m 58; c 3. CHEMICAL KINETICS, LASER PHOTOCHEMISTRY. *Educ:* Norwich Univ, BS, 56; Univ Vt, MS, 58; Univ Rochester, PhD(phys chem), 62. *Prof Exp:* From asst prof to assoc prof, 65-76, PROF CHEM ENG, UNIV MINN,

MINNEAPOLIS, 76- *Concurrent Pos:* Res fel, Harvard Univ, 63-65, lectr chem, 64-; asst ed, J Phys Chem, 70-80; NSF fel, Cambridge Univ, 71-72; Hon Ramsay Mem fel, 71-72; Fullbright scholar, 82. *Mem:* Am Chem Soc; Sigma Xi; InterAm Photochemical Soc. *Res:* Gas kinetics; photochemistry; energy transfer; unimolecular reactions; laser induced reactions; chemical reaction engineering. *Mailing Add:* Dept Chem Eng & Mat Sci Univ Minn Minneapolis MN 55455

CARR, ROGER BYINGTON, b Haverhill, Mass, Mar 25, 36; m 60; c 2. ASTRONOMY, PHYSICS. *Educ:* Bucknell Univ, BS, 63; Univ Fla, MS, 65, PhD(physics, astron), 67. *Prof Exp:* From instr to asst prof astron, Carleton Col, 67-72; asst prof physics, Alfred Univ, 72-76, res geophysicist, 76-77; vis asst prof, St John Fisher Col, 78; dir extended educ, State Univ NY Agr & Tech Col, Alfred, 78-84; INT PATENT CLASSIFIER, US PATENT & TRADEMARK OFF, ARLINGTON, VA, 84- *Mem:* Am Astron Soc; Am Meteorol Soc; Sigma Xi. *Res:* Atmospheric extinction; eclipsing binary stars. *Mailing Add:* 121 Jefferson Davis Hwy Bldg CM-2 Rm 904 Arlington VA 22202

CARR, RONALD E, b Newark, NJ, Sept 17, 32; m 57; c 3. OPHTHALMOLOGY, VISUAL PHYSIOLOGY. *Educ:* Princeton Univ, AB, 54; Johns Hopkins Univ, MD, 58; NY Univ, MS, 63. *Prof Exp:* Intern, Cornell-Bellevue Med Serv, 58-59; resident ophthal, Med Ctr, NY Univ, 59-62; clin assoc ophthal, NIH, Md, 63-64, actg assoc ophthalmologist, 64-65; asst prof, 65-69, assoc prof, 69-71, PROF OPHTHAL, MED CTR, NY UNIV, 71- *Concurrent Pos:* Fel ophthal, Med Ctr, NY Univ, 62-63; attend physician, Univ Hosp, NY, 65-; chief ophthal serv, Goldwater Mem Hosp, 67-77; dir retinal clin, Bellevue Hosp, 70- *Mem:* Am Acad Ophthal; Am Col Surg; Asn Res Ophthal; Am Ophthal Soc. *Res:* Clinical applications of visual electrophysiology; hereditary diseases of the retina; drug induced retinal degenerations. *Mailing Add:* Dept of Ophthal NY Univ Med Ctr New York NY 10016

CARR, RONALD IRVING, b Toronto, Ont, May 17, 35; m 68; c 2. IMMUNOLOGY, ALLERGY. *Educ:* Univ Toronto, BA, 58, MD, 62; Rockefeller Univ, PhD(immunol), 69. *Prof Exp:* Sr fel immunol, Nat Jewish Hosp, 69-70; asst prof med, Univ Colo Med Ctr, Denver, 70-77, assoc prof, 78-82; assoc prof, 82-89, ASSOC PROF DEPT MICROBIOL, DALHOUSIE UNIV, 85-, PROF DEPT MED, 89- *Concurrent Pos:* Mem, Pulmonary, Allergy & Clin Immunol Adv Comt, Food & Drug Admin, 73-76 & chmn, 76-77; mem bd of dirs, Lupus Found Am Inc, 77-88; mem staff, dept med, Nat Jewish Hosp Res Ctr, 70-82. *Mem:* Am Asn Immunologists; Am Acad Allergy; Am Col Rheumatol; Assoc Royal Photog Soc; Artistic Fedn Int Art Photog; Soc Clin Immunol; Mucosal Immunol Soc; Can Rheumatism Asn; Can Soc Clin Invest; Can Soc Immunol. *Res:* Mechanisms of food hypersensitivity and possible food related immune disease; factors regulating mucosal immunity; regulatory mechanisms and defects in systemic lupus erythematosus; central nervous system and immune system interactions; neuropeptides and immunoregulation; dietary substances as immunomodulators antibodies to neuropeptides in human disease. *Mailing Add:* Div Rheumatology Halifax Civic Hosp Dalhousie Univ 5938 University Ave Halifax NS B3H 1V9 Can

CARR, RUSSELL L K, b Wakefield, Mich, Apr 26, 26; m 81; c 2. ORGANIC CHEMISTRY. *Educ:* Ohio State Univ, BSc, 51, PhD(chem), 55. *Prof Exp:* From chemist to sr chemist, Hooker Chem Corp, 55-64, res supvr, 64-66, sect mgr res, 66-70, mgr res, 70-83; SUPVR RES, STARKS ASSOC, INC, 84- *Mem:* Am Chem Soc. *Res:* Organic fluorine chemistry; fluorinations; hydrogenations, catalysis; organic phosphorus chemistry; reactions of elemental phosphorus and of esters of phosphorus acids; synthesis of phosphines; polymer chemistry. *Mailing Add:* 69 Layton Ave Eggertsville NY 14226-4229

CARR, SCOTT BLIGH, b Linton, Ky, Oct 11, 34; m 56; c 2. ANIMAL NUTRITION. *Educ:* Western Ky State Col, BS, 56; Univ Ky, MS, 63, PhD(dairy nutrit), 67. *Prof Exp:* Exten specialist forages, 67-71, ASST PROF RUMINANT NUTRIT & SCIENTIST-IN-CHARGE FORAGE TESTING LAB, VA POLYTECH INST & STATE UNIV, 71-, EXTEN SPECIALIST FORAGE EVAL, 77- *Mem:* Am Dairy Sci Asn; Am Soc Animal Sci. *Res:* Forages; dairy and beef nutrition. *Mailing Add:* Dept Dairy Sci Va Polytech Inst Blacksburg VA 24060

CARR, STEPHEN HOWARD, b Dayton, Ohio, Sept 29, 42; m 67; c 2. POLYMER SCIENCE & TECHNOLOGY. *Educ:* Univ Cincinnati, BS, 65; Case Western Reserve Univ, MS, 67, PhD(polymer sci), 70. *Prof Exp:* From asst prof to assoc prof, Northwestern Univ, 70-78, dir, Mat Res Ctr, 84-90, PROF MAT SCI & CHEM ENG, NORTHWESTERN UNIV, 78- *Concurrent Pos:* Assoc surg res, Evanston Hosp, 73- *Honors & Awards:* Ralph R Teeter Award, Soc Automotive Engrs, 80. *Mem:* AAAS; Am Chem Soc; fel Am Phys Soc; Am Soc Metals; Soc Plastic Engrs; Soc Automotive Engrs; Am Inst Chem Engrs. *Res:* Molecular organization in the polymer solid state; electrical polarization and electrical conductivity in polymeric solids; biopolymers; mechanical behavior of polymers. *Mailing Add:* Dept of Mat Sci & Eng Northwestern Univ Evanston IL 60208-9990

CARR, THOMAS DEADERICK, b Ft Worth, Tex, Jan 2, 17; m 61; c 1. PHYSICS, ASTRONOMY. *Educ:* Univ Fla, BS, 37, MS, 39, PhD(physics), 58. *Prof Exp:* Physicist & head blast measurement sect, Ballistic Res Lab, Aberdeen Proving Ground, Md, 40-45; civilian scientist with US Navy Bur Ord, 46; physicist & head antenna & propagation sect & staff mem directorate range develop, Air Force Missile Test Ctr, Patrick Air Force Base, Fla, 50-56; from asst prof to assoc prof physics & astron, Univ Fla, 58-66, assoc chmn, 76-78, chmn astron, 85-88, PROF ASTRON & PHYSICS, UNIV FLA, 79-, DIR, RADIO OBSERV, 85- *Mem:* Am Phys Soc; Am Astron Soc; Am Geophys Union. *Res:* Radio astronomy; geophysics; radio propagation; cosmic radiation; x-ray diffraction; blast measurements; ballistics; guided missile instrumentation. *Mailing Add:* 210 Space Sci Res Bldg Univ Fla Gainesville FL 32611

CARR, TOMMY RUSSELL, b Hardtner, Kans, July 30, 46; m 71; c 2. ANIMAL SCIENCE, FOOD SCIENCE. *Educ:* Kans State Univ, BS, 69, MS, 70; Okla State Univ, PhD(food sci), 75. *Prof Exp:* Res asst animal sci, Kans State Univ, 69-70; instr animal & food sci, Okla State Univ, 70-74; asst prof, 74-80, ASSOC PROF ANIMAL SCI, UNIV ILL, 80- *Mem:* Am Meat Sci Asn; Am Soc Animal Sci; Inst Food Technologists; Sigma Xi. *Res:* Growth and composition of meat animals and in methods used to evaluate compositional differences in both live animals and carcasses. *Mailing Add:* 205B Meat Sci Lab Univ Ill 1503 S Maryland Dr Urbana IL 61801

CARR, VIRGINIA MCMILLAN, b New York, NY, June 24, 44; m 67; c 2. DEVELOPMENTAL NEUROBIOLOGY. *Educ:* Western Reserve Univ, BA, 66; Northwestern Univ, MS, 71, PhD(biol sci), 76. *Prof Exp:* Technician, dept physiol, Western Reserve Univ, 66-67, jr res asst, Case Western Reserve Univ, 67-69; Nat Defense Educ Act fel, 71-73; NIH fel, dept biol, Univ Chicago, 78-80, res assoc, 80-81; asst, dept biol sci, Northwestern Univ, 70-76, res assoc, 76-77, vis scholar, dept biochem, molecular & cell biol, 81-85, asst prof res, 86-90, RES ASSOC DEPT, NEUROBIOL & PHYSIOL, NORTHWESTERN UNIV, 90- *Mem:* AAAS; Soc Develop Biol; Soc Neuroscience; Am Soc Cell Biol. *Res:* Role of cell proliferation, cell death, neuronal-glial and inter-neuronal interactions in developing neural systems; use of immunohistochemical techniques in vitro and in vivo systems. *Mailing Add:* Dept Neurobiol & Physiol Northwestern Univ Evanston IL 60208

CARR, WALTER JAMES, JR, b Knob Noster, Mo, May 6, 18; m 53; c 2. PHYSICS. *Educ:* Univ Mo, Rolla, BS, 40; Stanford Univ, EE, 42, DSc, Carnegie-Mellon Univ, 50. *Hon Degrees:* DSc, Cornegie-Mellon Univ, 50. *Prof Exp:* From physicist to adv physicist, 42-65, mgr theoret physics dept, 65-70, CONSULT MAGNETISM & SUPERCONDUCTIVITY, WESTINGHOUSE RES LABS, 70- *Mem:* Am Phys Soc; Inst Elec & Electronics Engrs. *Res:* Solid state physics; ferromagnetism; superconductivity. *Mailing Add:* Westinghouse Sci & Tech Ctr Beulah Rd Pittsburgh PA 15235

CARR, WILLIAM N, b Thayer, Mo, June 5, 36; m 63; c 1. BIOENGINEERING, BIOMEDICAL ENGINEERING. *Educ:* Carnegie Mellon Univ, BS & MS, 59, PhD(elec eng), 62; Southern Methodist Univ, MBA, 67. *Prof Exp:* Mem tech staff, Tex Instruments, Inc, 62-66; prof electronic sci & dir electronic sci ctr, Southern Methodist Univ, 67-69; pres, Zentron Int, Inc, 69-78; dir eng, Datapoint Corp, 79-86, dir circuit control; CHAIR PROF, DEPT ELEC & COMPUTER ENG & PHYSICS, NJ INST TECHNOL, 86- *Concurrent Pos:* Assoc prof, Col Med, Baylor Univ, 67-78. *Mem:* AAAS; Am Phys Soc; Electrochem Soc; Inst Elec & Electronics Engrs; Soc Nuclear Med. *Mailing Add:* Microelectronics Res Ctr NJ Inst Technol 323 King Blvd Newark NJ 07102

CARRABINE, JOHN ANTHONY, b Cleveland, Ohio, Nov 8, 28; m 51; c 6. X-RAY CRYSTALLOGRAPHY, INORGANIC CHEMISTRY. *Educ:* John Carroll Univ, BS, 51, MS, 53; Case Western Reserve Univ, PhD, 70. *Prof Exp:* Chemist, Thompson Prods, Inc, 51-56; supvr phys res, Brush Beryllium Co, 56-66; asst prof, 66-72, assoc prof, 72-77, PROF CHEM, JOHN CARROLL UNIV, 77-, CHMN, CHEM DEPT, 81- *Concurrent Pos:* Lectr, Eve Col, John Carroll Univ, 56-66. *Mem:* Am Chem Soc. *Res:* Crystal and molecular structures of metal complexes of biochemically important substances; physical metallurgy of beryllium. *Mailing Add:* Dept of Chem John Carroll Univ North Park & Miramar Blvd Cleveland OH 44118-4520

CARRADINE, WILLIAM RADELL, JR, b Austin, Tex, Jan 30, 41; m 63; c 2. CHEMICAL ENGINEERING. *Educ:* Tex Technol Col, BS, 64, MS, 65; Univ Tex, Austin, PhD(chem eng), 70. *Prof Exp:* res engr, Continental Chem, Continental Oil Inc, 69-84; RES ASSOC, VISTA CHEM CO, 84- *Mem:* Am Inst Chem Engrs; Sigma Xi. *Res:* Economic analysis; process simulation; process development. *Mailing Add:* 10601 Bradel Cove Austin TX 78726-1343

CARRADO, KATHLEEN ANNE, b Rome, NY, Apr 24, 60; m 90. MATERIALS CHEMISTRY, CATALYSIS. *Educ:* State Univ NY, Fredonia, BS, 82; Univ Conn, Storrs, PhD(chem), 86. *Prof Exp:* Postdoctoral, 87-89, ASST CHEMIST, ARGONNE NAT LAB, 89- *Mem:* Am Chem Soc; Clay Minerals Soc. *Res:* Solid-state and materials chemistry of heterogeneous catalysts for the petroleum industry; clay mineral and aluminosilicate synthesis, modification, characterization and reactivity; small-angle neutron scattering; x-ray absorption spectroscopy; x-ray powder diffraction exploited. *Mailing Add:* Argonne Nat Lab CHM 200 9700 S Cass Ave Argonne IL 60439-4803

CARRAHER, CHARLES EUGENE, JR, b Des Moines, Iowa, May 8, 41; m 63; c 7. POLYMER CHEMISTRY, PHYSICAL CHEMISTRY. *Educ:* Sterling Col, BA, 63; Univ Mo-Kansas City, PhD(chem), 67. *Prof Exp:* Teaching asst chem, Univ Mo, 63-67; from asst prof to prof chem, Univ SDak, 67-76, dir, Gen Chem Prog, 67-76, chmn, Sci Div, 72-74; prof chem & chmn dept, Wright State Univ, 76-; DEAN SCI, FLA ATLANTIC UNIV. *Concurrent Pos:* Petrol Res Fund grant, 68-; Nat Sci Found grant, 70-73; adv, Sen McGovern, 72-75; Res Fund grant, 73-75; Nat tour speaker, 75-76; Am Chem Soc-Petrol Res Fund grants, 74-82. *Mem:* AAAS; Am Inst Chemists; Sigma Xi (pres-elect, 71-72, pres, 72-73); Am Chem Soc. *Res:* Preparation and characterization of phosphorus containing extended polar polymers and organometallic polymers; practical and theoretical molecular weight distribution determinations; published 30 books and 500 papers. *Mailing Add:* Col Sci Fla Atlantic Univ Boca Raton FL 33431

CARRANO, ANTHONY VITO, b New York, NY, Mar 22, 42; m 64; c 2. GENOME ORGANIZATION, BIOPHYSICS. *Educ:* Rensselaer Polytech Inst, BS, 64; Univ Calif, Berkeley, MB, 70, PhD(biophys), 72. *Prof Exp:* Fel, Div Biol & Med Res, Argonne Nat Lab, Ill, 72-73; biophysicist, Biomed Sci Div, 73-80, sect leader cell biol & mutagenesis, 80-87, SECT LEADER, GENETICS, LAWRENCE LIVERMORE NAT LAB, UNIV CALIF, 87-, DIR HUMAN GENOME CTR, 90- *Concurrent Pos:* Adj prof, Sch Med,

Univ Calif, Davis, 74; assoc ed, Radiation Res, 76-79; vis prof, Univ Calif, Berkeley, 78-79; vis prof, San Jose State Univ, 75-83; mem, Biophys Group, Univ Calif, Berkeley, 76; mem, Genetics Group, Univ Calif, Davis, 77-; mem, Human Genome Coord Comt, Dept Energy, 88-, Joint Adv Comt Human Genome, NIH/Dept Energy, 88- *Honors & Awards:* Environ Mutagen Soc Recognition Award, 86. *Mem:* Am Soc Human Genetics; Genetics Soc Am; Eviron Mutagen Soc (treas, 83-86, pres, 88); AAAS. *Res:* Chromosome organization; mechanisms and significance of mutation and cytogenetic damage; human population cytogenetics; gene mapping. *Mailing Add:* Human Genome Ctr Biomed Sci Div L-452 Lawrence Livermore Nat Lab Livermore CA 94550

CARRANO, RICHARD ALFRED, b Bridgeport, Conn, Oct 1, 40; div; c 2. PHARMACOLOGY, TOXICOLOGY. *Educ:* Univ Conn, BS, 63, MS, 65, PhD(pharmacol & biochem), 67. *Prof Exp:* Res pharmacologist, ICI US, Inc, 66-67, supvr gen pharmacol, 68-74; mgr, pharmacol & toxicol, Adria Labs, Inc, 74-77, dir, preclin res, 77-84; VPRES, PROD DEVELOP, CENTOCOR, INC, 84- *Mem:* Am Chem Soc; Am Soc Pharmacol & Exp Therapeut; Am Pharmaceut Asn; NY Acad Sci. *Res:* Development of new drugs by guiding the studies to demonstrate efficacy and to prove safety in animals for extrapolation to man, administration of the relationships between regulatory and clinical research groups which are necessary to obtain this goal. *Mailing Add:* Dir Prod Dev & Oper Unimed Inc 35 Columbia Rd Somerville NJ 08876-3587

CARRANO, SALVATORE ANDREW, b New Haven, Conn, Feb 2, 15. INORGANIC CHEMISTRY. *Educ:* Yale Univ, BS, 35; Boston Col, MS, 55, PhD(chem), 63. *Prof Exp:* Anal chemist, New Haven Clock Co, 35-38; res chemist, Seymour Mfg Co, 38-53; asst instr chem, Boston Col, 53-55, from asst prof to prof, 56-80, EMER PROF CHEM, FAIRFIELD UNIV, 80-; SR SCIENTIST, RES & DEVELOP DIV, AVCO, 55- *Concurrent Pos:* Teaching fel, Boston Col, 60-63; vis prof, Univ Tenn, 71. *Mem:* Am Chem Soc; Sigma Xi. *Res:* Structure and bonding studies of transition metal complexes utilizing chemical symmetry and group theory. *Mailing Add:* 329 Demarest Dr Orange CT 06477-3404

CARRARA, PAUL EDWARD, b San Francisco, Calif, Sept 16, 47; div. HOLOCENE CLIMATIC CHANGES, GLACIAL GEOLOGY. *Educ:* San Francisco State Col, BA, 69; Univ Colo, MSc, 72. *Prof Exp:* Res assoc, Inst Arctic & Alpine Res, Univ Colo, 72-74; GEOLOGIST, US GEOL SURV, 74- *Mem:* Am Quaternary Asn; Geol Soc Am. *Res:* Holocene and late-Pleistocene glacial geology of the Glacier National Park area, Montana; quaternary geology of the Chewelah, Washington area. *Mailing Add:* US Geol Survey Fed Ctr MS 913 Box 25046 Denver CO 80225

CARRASQUER, GASPAR, b Valencia, Spain, Dec 21, 25; nat US; m 79; c 4. MEDICINE. *Educ:* Univ Valencia, MD, 51. *Prof Exp:* Intern, North Hudson Hosp, NJ, 53; resident internal med, City Hosp New York, 54-55 & Louisville Gen Hosp, Ky, 55-56; from instr to assoc prof med, 59-70, PROF EXP MED, SCH MED, UNIV LOUISVILLE, 71-, ASSOC PHYSIOL, 61- *Concurrent Pos:* Fel, Sch Med, Univ Louisville, 56-59; res fel, Am Heart Asn, 58-60, adv res fel, 60-62; USPHS career develop award, 67-72; estab investr, Am Heart Asn, 62-67. *Mem:* Am Physiol Soc; Am Biophys Soc; Soc Exp Biol & Med; Sigma Xi. *Res:* Ion transport; renal physiology. *Mailing Add:* Health Sci Ctr Univ Louisville Louisville KY 40292

CARRASQUILLO, ARNALDO, b Santa Isabel, PR, July 12, 37; m 63; c 2. MEDICINAL CHEMISTRY, SCIENCE EDUCATION. *Educ:* Univ PR, Rio Piedras, BS, 59, MS, 66; Ohio State Univ, PhD(org chem), 71. *Prof Exp:* Instr chem, Univ PR, Rio Piedras, 59-62, res asst, 62-63; res asst, PR Nuclear Ctr, 63-65 & Ohio State Univ, 70-71; from asst prof to assoc prof, 71-78, PROF CHEM, CATH UNIV PR, 78- *Concurrent Pos:* Biomed res dir, Cath Univ PR, 74-76. *Mem:* Am Chem Soc. *Res:* The synthesis of biologically important molecules, isolation and characterization of biologically active natural products; exploratory organic chemistry; the study and analysis of organic compounds in the biosphere. *Mailing Add:* Catholic Univ PR PO Box 105 Sta 6 Catholic Univ PR Ponce PR 00732

CARRAWAY, CORALIE ANNE CAROTHERS, b Montgomery, Ala; c 2. BIOCHEMISTRY, CELL BIOLOGY. *Educ:* Miss State Univ, BS, 61; Okla State Univ, PhD(biochem), 74. *Prof Exp:* Res asst biochem, Dept of Chem, Univ Ill, 63-66; lab technician, Univ Calif, Berkeley, 67-68; res assoc, Okla State Univ, 74-75; actg asst prof bact, Univ Calif, Los Angeles, 75-76; res assoc, Okla State Univ, 76-78, asst res biochemist, 78-81; asst prof, Dept Oncol, 81-84, ASSOC PROF, DEPT BIOCHEM & MOLECULAR BIOL, SCH MED, UNIV MIAMI, 85- *Concurrent Pos:* Fel Okla Heart Asn, Dept Biochem, Okla State Univ, 77-78. *Mem:* AAAS; Am Soc Cell Biol; Sigma Xi. *Res:* Organization, structure, and interactions of plasma membrane proteins; phosphorylation/dephosphorylation in cellular regulation. *Mailing Add:* Dept Biochem Univ Miami Sch Med R124-MA Miami FL 33101

CARRAWAY, KERMIT LEE, b Utica, Miss, Mar 1, 40; m 62; c 2. BIOCHEMISTRY. *Educ:* Miss State Univ, BS, 62; Univ Ill, PhD(org chem), 66. *Prof Exp:* NIH res fel biochem, Univ Calif, 66-68; from asst prof to assoc prof biochem, Okla State Univ, 68-75, prof biochem, 75-81; PROF & CHMN ANAT, SCH MED, UNIV MIAMI, 81- *Concurrent Pos:* Mem, Molecular Cytol Study Sect, NIH, 75-78. *Mem:* Am Soc Cell Biol; Am Soc Biol Chemists. *Res:* Membrane biochemistry; protein chemistry. *Mailing Add:* Dept Anat Sch Med Univ Miami Miami FL 33136

CARREAU, PIERRE, b Montreal, Que, Nov 14, 39; m 63; c 3. CHEMICAL ENGINEERING, RHEOLOGY. *Educ:* Univ Montreal, BSc, 63, MSc, 65; Univ Wis, PhD(chem eng), 68. *Prof Exp:* Asst chem eng, 64-65, from asst prof to assoc prof chem eng, 68-80, chmn dept, 73-79, PROF, POLYTECH SCH, UNIV MONTREAL, 80- *Mem:* Sr mem Chem Inst Can; Can Soc Chem Engrs; Soc Rheol; Am Inst Chem Engrs. *Res:* Constitutive equations for viscoelastic fluids; transport phenomena applied to polymeric systems. *Mailing Add:* Dept of Chem Eng Univ of Montreal Polytech Sch Montreal PQ H3C 3A7 Can

CARREGAL, ENRIQUE JOSE ALVAREZ, b Padron, Spain, July 2, 32; US citizen; m 60; c 2. MEDICINE, PHYSIOLOGY. *Educ:* Univ Santiago, MD, 54, PhD(med), 60. *Prof Exp:* House doctor, Hosp Clin, Santiago, 54-55; instr physiol, Univ Santiago Med Sch, 55; fel physiol, Span Pharmacol Inst Madrid, 56-59; fel radiol, Univ Madrid, 57-59; fel neurophysiol, Coun Sci Invest, 60; res fel, Inst Med Res, Huntington Mem Hosp, Pasadena, Calif, 60-63; neurophysiologist, Stanford Res Inst, 63-65; assoc prof physiol, Sch Med, 66-69, assoc prof neurosurg, 70-75, anesthesia resident, 75-77, CLIN ASSOC PROF NEUROSURG, SCH MED, UNIV SOUTHERN CALIF, 75- *Concurrent Pos:* Consult, Huntington Mem Hosp, Pasadena, Calif & City of Hope Nat Med Ctr, Duarte. *Mem:* AAAS; Am Physiol Soc; Soc Neuroscience; Am Soc Anesthesiol; Int Soc Study Pain. *Res:* Neurophysiology of pain and sensory mechanism, neural control of respiration; synaptology; anesthesiology. *Mailing Add:* Univ Southern Calif Sch Med Los Angeles CA 90033

CARREIRA, LIONEL ANDRADE, b Worcester, Mass, Nov 1, 44; m 70; c 1. PHYSICAL CHEMISTRY, MOLECULAR SPECTROSCOPY. *Educ:* Worcester Polytech Inst, BS, 66; Mass Inst Technol, PhD(phys chem), 69. *Prof Exp:* Fel chem, Univ Fla, 69-71; res assoc, Univ SC, 71-73; instr, 73-75, ASST PROF CHEM, UNIV GA, 75- *Mem:* Am Chem Soc; Sigma Xi. *Res:* Studies of far infrared and Raman spectra of simple polyatomic molecules having large amplitude oscillations, quasi-linear molecules with anomalously low frequency bending modes and other molecules with unusual vibrational potential functions. *Mailing Add:* Dept of Chem Univ of Ga Athens GA 30601

CARREKER, R(OLAND) P(OLK), JR, b Birmingham, Ala, Aug 6, 25; m 48; c 3. METALLURGICAL ENGINEERING, MATERIALS ECONOMICS. *Educ:* Univ Ill, BS, 45, MS, 47; Rensselaer Polytech Inst, PhD(metall eng), 55. *Prof Exp:* Asst, Res Lab, Gen Elec Corp, 47-52, res assoc, 52-57, metall engr, 57-70, consult mat resources, Mat Resources Oper, 70-73, mgr mat tech & econ anal, 73-83, consult mat resource anal, Corp Purchasing, 83-85; RETIRED. *Mem:* Am Soc Metals; Am Inst Mining, Metall & Petrol Engrs. *Res:* Deformation of metals; continuous casting; process development; economic analysis; mineral ventures evaluation; business systems for strategic planning and forecasting. *Mailing Add:* 13 Buell St Hanover NH 03755

CARREL, JAMES ELLIOTT, b San Pedro, Calif, June 10, 44; m 78; c 2. CHEMICAL ECOLOGY, INSECT BIOCHEMISTRY. *Educ:* Harvard Col, AB, 66; Cornell Univ, PhD(behav biol), 71. *Prof Exp:* From asst prof to assoc prof, 71-86, PROF BIOL SCI, UNIV MO, COLUMBIA, 87- *Concurrent Pos:* Lectr, Dept Zool, Univ Manchester, Eng, 79-80; vis prof, biol sci, Fla State Univ, 81-82, dept chem, State Univ NY, Stony Brook, 90. *Mem:* AAAS; Am Arachnol Soc; Brit Arachnol Soc; Entom Soc Am; Int Soc Chem Ecol; Sigma Xi. *Res:* Chemical defenses of arthropods, spider ecophysiology, chemical ecology, insect biochemistry and molecular biology. *Mailing Add:* Div Biol Sci Univ Mo 209 Tucker Hall Columbia MO 65211

CARRICK, LEE, b Detroit, Mich, Oct 31, 43; m 67. EXPERIMENTAL INFECTIOUS DISEASES. *Educ:* Wayne State Univ, BS, 66; PhD(biol), 73. *Prof Exp:* Technologist, 69-70, res asst, 70-74, instr, 74-76, ASST PROF MICROBIOL, MED SCH, WAYNE STATE UNIV, 76- *Mem:* Soc Exp Biol & Med; Sigma Xi; Am Soc Microbiol; Am Inst Biol Sci; AAAS; Asn Immunol. *Res:* Murine models of opportunistic infections and granulomatous diseases with emphasis on the fractionation and characterization of lympholines and studies of their role in the activation of macrophages for cell-mediated immunity and fibroblasts in chronic inflammation; interactions of macrophages with facultative intracellular bacteria, especially mycobacteria; the molecular basis of whether or not human macrophages permit the intracellular growth of bacilli. *Mailing Add:* Dept Immunol & Microbiol Sch Med Scott Hall Wayne State Univ Detroit MI 48202

CARRICK, WAYNE LEE, b Benton, Ark, Feb 23, 27; m 49; c 5. ORGANIC CHEMISTRY, POLYMER CHEMISTRY. *Educ:* Univ Ark, BS, 52, MS, 53, PhD(phys org chem), 55. *Prof Exp:* Res chemist, 54-56, group leader, 56-65, assoc dir res & develop, Union Carbide Corp, 65-77; VPRES RES & DEVELOP, CHEMPLEX CO, 77- *Mem:* Am Chem Soc; Sigma Xi. *Res:* Reactions mechanisms; kinetics; catalysis; organic compounds of transition metals; high polymers. *Mailing Add:* Box 3601-RFD Long Grove IL 60047

CARRICO, CHRISTINE KATHRYN, b Charlottesville, Va, April 25, 50; m 86. SCIENCE ADMINISTRATION. *Educ:* Hollins Col, BA, 71; Yale Univ, PhD(pharmacol), 76. *Prof Exp:* Fel assoc pharmacol, Yale Univ, 76-77; res assoc, 77-79, health scientist adminr, 79-84, DIR PHARMACOL SCI, NAT GEN MED SCI, NIH, 84- *Mem:* AAAS; Am Soc Pharmacol & Exp Therapeut; NY Acad Sci. *Res:* Molecular pharmacology; clinical pharmacology; drug metabolism and drug disposition. *Mailing Add:* 7238 Wapello Dr Rockville MD 20855-2717

CARRICO, ROBERT JOSEPH, b Mishawaka, Ind, Dec 27, 38; m 76. BIOCHEMISTRY. *Educ:* Purdue Univ, BS, 64; Univ Wis, PhD(biochem), 68. *Prof Exp:* Res asst biochem, Univ Wis, 68-69; fel, Univ Gothenburg, 69-71; res scientist biochem, Miles Labs, Inc, 71-89; RES SCIENTIST BIOCHEM, ENVIRON TEST SYSTS, 90- *Concurrent Pos:* Fel biochem, Albert Einstein Col Med, 71-72. *Res:* Applications of immobilized enzymes in clinical chemistry; activities of modified substrates and cofactors with enzymes; monitoring of therapeutic drugs in blood. *Mailing Add:* 52829 Wildflower Lane Elkhart IN 46514

CARRIER, GEORGE FRANCIS, b Millinocket, Maine, May 4, 18; m 46; c 3. ENGINEERING & APPLIED MATHEMATICS. *Educ:* Cornell Univ, ME, 39, PhD(appl mech), 44. *Prof Exp:* From asst prof to prof eng, Brown Univ, 46-52; Gordon McKay prof mech eng, 52-72, T Jefferson Coolidge prof, 72-88, T JEFFERSON COOLIDGE EMER PROF APPL MATH, HARVARD UNIV, 88- *Concurrent Pos:* Mem, Coun Eng Col, Cornell Univ; assoc ed, J Fluid Mech & Quart Appl Math. *Honors & Awards:* Richards Mem Award, 63 & Timoshenko Medal, 78, Am Soc Mech Engrs; Von

Karman Medal, Am Soc Civil Engrs, 77; Von Karman Prize, Soc Indust & Appl Math, 79; Appl Math & Numerical Analysis Award, Nat Acad Sci, 80; Silver Centennial Medal, Am Soc Mech Engrs, 80; Fluid Dynamics Prize, Am Phys Soc, 84; Dryden Medal, Am Inst Aeronaut, Astronaut, 89; Nat Medal Sci, 90. *Mem:* Nat Acad Eng; Nat Acad Sci; Am Acad Arts & Sci; hon mem Am Soc Mech Engrs; fel Am Acad Arts & Sci; hon fel Brit Inst Math Applns; Am Philos Soc; Soc Indust Appl Math; Sigma Xi; Int Soc Interaction Mech & Math. *Res:* Applied mathematics; hydrodynamics. *Mailing Add:* Pierce Hall 311 Harvard Univ Cambridge MA 02138

CARRIER, GERALD BURTON, b Punxsutawney, Pa, Aug 24, 27; m 49; c 3. MATERIALS SCIENCE, PHYSICS. *Educ:* Bowling Green State Univ, BA & BS, 50; Ohio State Univ, MS, 56. *Prof Exp:* RES PHYSICIST GLASS & CERAMICS, CORNING INC, 52- *Concurrent Pos:* Math instr, Elmira Col. *Res:* Electron microscopy of glasses; glass ceramics; ceramics. *Mailing Add:* Corning Inc Sullivan Park Corning NY 14830

CARRIER, STEVEN THEODORE, b Havre, Mont, Nov 8, 38; m 63; c 2. MEDICAL STATISTICS, BIOSTATISTICS. *Educ:* Calif Polytech State Univ, BS, 67; Tex A&M Univ, PhD(statist), 75. *Prof Exp:* Analyst math, Naval Weapons Ctr, Calif, 67-68; statistician, Am Cyanamid Co, 71-75, group leader statist analyst, Lederle Labs, 75-77; mgr statist sect, Med Oper Dept, Cutter Labs, Berkeley, 77-79; SECT HEAD, BIOSTATIST DEPT, PHARMACEUT PROD DIV RES & DEVELOP, ABBOTT LABS, 79- *Mem:* Drug Info Asn; Am Statist Asn; Biometrics Soc; Soc Clin Trials. *Res:* New statistical methods for the design and analysis of clinical trial data. *Mailing Add:* 35 S Old Creek Rd Vernon Hills IL 60061

CARRIER, W(ILLIAM) DAVID, III, b Allentown, Pa, Dec 21, 43; m 65; c 2. GEOTECHNICAL ENGINEERING. *Educ:* Mass Inst Technol, SB, 65, SM, 66, ScD(civil eng), 68. *Prof Exp:* Lunar soil mech specialist, Johnson Space Ctr, NASA, 68-73; asst chief soils engr, Bechtel, Inc, San Francisco, 73-77; mgr solid waste systs, Woodward-Clyde Consults, 77-78; mgr, Geotechnical Group, 78-80, dir eng, 80-83, PRES, BROMWELL & CARRIER INC, 83- *Concurrent Pos:* Consult geotechnical engr, Lunar & Planetary Inst, Houston, 77-78, 85-; prin investr lunar sample, NASA, 71-72, Fla Inst Phosphate Res, 80-; dir, Lunar Geotech Inst, 90- *Honors & Awards:* Norman Medal, Am Soc Civil Engrs, 72, Middlebrooks Award, 84. *Mem:* Fel Am Soc Civil Engrs; Int Soc Soil Mech & Found Engrs; fel Inst Civil Engrs UK; Am Inst Mining Engrs. *Res:* Geotechnical properties of industrial and mining solid wastes; development of new methods of disposal; physical properties of clays; physical properties of lunar soils. *Mailing Add:* Bromwell & Carrier Inc PO Box 5467 Lakeland FL 33807-5467

CARRIERE, RITA MARGARET, b Toronto, Ont, Can, Apr 25, 30; c 1. HISTOLOGY. *Educ:* McGill Univ, BSc, 50, MSc, 54, PhD, 60. *Prof Exp:* Lectr histol, Univ Montreal, 55-57, asst prof histol, embryol & endocrinol, 57-60; ASST PROF ANAT, STATE UNIV NY DOWNSTATE MED CTR, 60- *Mem:* Sigma Xi; Am Soc Cell Biol. *Res:* Proteolysis in skeletal muscle; secretory processes in rat orbital glands. *Mailing Add:* Anat Dept Box 5 Downstate Med Ctr 450 Clarkson Ave Brooklyn NY 11203

CARRIÈRE, SERGE, b Montreal, Que. MEDICINE. *Educ:* Univ Montreal, BA, 54, MD, 59. *Prof Exp:* INternist, Notre-Dame Hosp, Montreal, 59-62; nephrologist, Harvard Med Sch, Boston, 62-64; from instr to prof med, Univ Montreal, 64-80, prof, Dept Physiol, 80-86, chmn, 80-86, chmn, Dept Med, 86-89, PROF, DEPT MED, UNIV MONTREAL, 86-, DEAN, FAC MED, 89-; MEM, MAISONNEUVE-ROSEMONT HOSP, MONTREAL, 68- *Concurrent Pos:* Vol med, Notre-Dame Hosp, Montreal, 64-66; eligible physician, Maisonneuve-Rosemont Hosp, Montreal, 66-68, mem corp, 70-, mem bd adminr, 70-78, pres, Med Bd, 71-72, chief med, Serv Nephrol, 72-74; consult, Inst Cardiol Monteal, 66; sabbatical, Dept Physiol, Univ Montreal, 75-76. *Honors & Awards:* Queen Elizabeth II Medal, 77; Michel-Sarrazin Award, Club Res Clin, Que, 89; Distinguished Serv Award, Can Soc Clin Invest, 91. *Mem:* Am Soc Clin Invest; Am Fedn Clin Res; Am Soc Nephrol; Int Soc Nephrol; Int Soc Hypertension; Am Soc Physiol. *Mailing Add:* Fac Med Univ Montreal C P 6128 Succ A Montreal PQ H3C 3J7 Can

CARRIGAN, CHARLES ROGER, b Pasadena, Calif, Sept 7, 49; m 76; c 1. FLUID MECHANICS, HEAT TRANSFER. *Educ:* Univ Calif, Los Angeles, BA, 71, MS, 73, PhD(geophys), 77. *Prof Exp:* Res fel, Dept Geod & Geophys, Cambridge Univ, 77-78 & 79, NATO fel, 78-79; res geophysicist, Inst Geophys & Planetary Physics, Univ Calif, Los Angeles, 79-80; mem tech staff, Geophys Res Div, Sandia Nat Labs, 80-89; PHYSICIST, EARTH SCI DEPT, LAWRENCE LIVERMORE NAT LAB, 89- *Mem:* Am Geophys Union; Sigma Xi. *Res:* Natural convection and heat transfer in geophysical and astrophysical systems; laboratory studies of the stability of convecting fluids; fluid mechanics of volcanism and other crystal processes. *Mailing Add:* Earth Sci Dept L-206 Lawrence Livermore Nat Lab Livermore CA 94550

CARRIGAN, P H, JR, hydraulic engineering, hydrology, for more information see previous edition

CARRIGAN, RICHARD ALFRED, b Somerville, Mass, May 11, 06; m 31; c 1. ENVIRONMENTAL CHEMISTRY, SCIENCE ADMINISTRATION. *Educ:* Univ Fla, BS, 32; Cornell Univ, PhD(soil chem), 48. *Prof Exp:* High sch teacher, Fla, 32-33; lab supvr, Magnolia Petrol Co, Tex, 33-38; from asst chemist to assoc chemist exp sta, Univ Fla, 38-45, from assoc biochemist to biochemist, 45-51, prof soils, Col Agr, 48-51; supvr analytical chem, Armour Res Found, Ill, 51-60; prog dir sci facil eval group, Div Inst Prog, NSF, 60-64, staff assoc, 64-70, prog mgr, Div Inst Develop, 70-71, prog mgr, Div Advan Environ Res & Technol, 71-78, prog mgr, Div Prob-focused Res Applns, 78-79, prog dir, 80-82, SR SCI ASSOC, DIV ATMOSPHERIC SCI, NSF, 82- *Concurrent Pos:* Specialist anal chem, Union of Burma Appl Res Inst, 54-55. *Mem:* Fel AAAS; Am Chem Soc; fel Am Inst Chem; Soil Sci Soc Am; Am Geophys Union. *Res:* Chemistry of minor elements in soils; spectrographic analysis; pasture fertility; chemistry of cobalt in soils; spectroscopy of plasmas; federal grant administration. *Mailing Add:* 2475 Virginia Ave NW Apt 304 Washington DC 20037

CARRIGAN, RICHARD ALFRED, JR, b Miami, Fla, Feb 17, 32; m 54; c 2. SCIENCE ADMINSTRATION, SCIENCE COMMUNICATIONS. *Educ:* Univ Ill, BS, 53, MS, 56, PhD(physics), 62. *Prof Exp:* Jr physicist, Firestone Tire & Rubber Co, 53; res physicist, Carnegie Inst Technol, 61-64, asst prof physics, 64-68; dir personnel serv, 72-76, asst dir, 76, asst head, res div, 77-84, PHYSICIST, FERMI NAT ACCELERATOR LAB, 68-, HEAD, OFF RES & TECHNOL APPLN, 84- *Concurrent Pos:* Guest res assoc, Brookhaven Nat Lab, 62-63, guest asst physicist, 63-67; consult, Am Inst Res, 63-64; sr Fulbright fel, Deutsches Elektron-Synchrotron, 67-68. *Mem:* AAAS; Am Phys Soc; Sigma Xi. *Res:* Experimental elementary particle physics; particle scattering; magnetic monopole hypothesis; high energy channeling. *Mailing Add:* Fermi Nat Accelerator Lab Box 500 Batavia IL 60510

CARRIKER, MELBOURNE ROMAINE, b Santa Marta, Colombia, Feb 25, 15; US citizen; m 43; c 4. MALACOLOGY, MARINE INVERTEBRATE BIOLOGY. *Educ:* Rutgers Univ, BS, 39; Univ Wis, PhM, 40, PhD(invert zool), 43. *Hon Degrees:* DSc, Beloit Col, 68. *Prof Exp:* Asst zool, Univ Wis, 39-43; instr, Rutgers Univ, 46-47, asst prof, 47-54; assoc prof, Univ NC, 54-61; fisheries res biologist & chief shellfish mortality prog, Biol Lab, US Bur Com Fisheries, Md, 61-62; dir systs-ecol prog, Marine Biol Lab, 62-72, investr, 72-73; prof, 73-85, EMER PROF, COL MARINE STUDIES, UNIV DEL, 85- *Concurrent Pos:* Mem ornith exped, 34; trustee, Int Oceanog Found, 64-74; mem corp, Marine Biol Lab, 62-78; adv comt biol study of plant site, Conn Yankee Atomic Power Plant, 65-75; adv comt oceanic biol to Off Naval Res, Am Inst Biol Sci, 66-69; adj prof, Univ RI, 65-73 & Boston Univ, 68-73; bioinstrumentation coun, 69-72; mem, President's Fed Water Pollution Control Adv Bd, 69-75; assoc mus comp zool, Harvard Univ, 67-73; res fel, Acad Natural Sci, Philadelphia, 68-80; mem adv bd, Quarterly Rev Biol, 68-83; mem ecol adv comt, Environ Protection Agency Sci Adv Bd, 75- 78; assoc, Del Mus Natural Hist, 75-77. *Mem:* Am Soc Zool; Atlantic Estuarine Res Soc (pres, 61); Nat Shellfisheries Asn (pres, 57-59); Am Malacol Union (pres, 84-85); Partners Americas, Del-Panama Partnership (pres, 86-87). *Res:* Anatomy, histology, behavior, ecology and physiology of gastropods; culture, ecology and microstructure of bivalve larvae; estuarine and marine ecology; mechanisms of penetration of calcareous substrata by invertebrates, especially of boring predatory gastropods; microstructure and chemistry of bivalve shell; tropical shellfish mariculture; estuarine ecology. *Mailing Add:* Col Marine Studies Univ Del Marine Studies Complex Lewes DE 19958-1298

CARRIKER, ROY C, b Spokane, Wash, July 21, 37; m 65. PHYSICS. *Educ:* Wash State Univ, BS, 60; Trinity Col, Conn, MS, 63; Univ Conn, PhD(physics), 68; Harvard Univ, MBA, 76. *Prof Exp:* Sr engr, Pratt & Whitney Aircraft Div, United Aircraft Corp, 60-65, res engr, United Aircraft Res Labs, 65-66, sr res engr, 66-68, supvr optical technol, 68-69, chief electronic instrumentation, 69-70; tech dir, Precision Metals Div, Hamilton Watch Co, 70-72; dir eng & develop, Hamilton Technol, Inc, 72-74; pres, R C Carriker & Assoc, 74-76; GEN MGR, SERMETEL, INC, 76-; VPRES, TELEFLEX, INC, 78-; PRES & CHIEF OPERATING OFFICER, SERMATECH INT INC, 82- *Res:* Solid state physics, especially rare earth-transition metal alloys, transport properties and superconductivity; optics, especially spectroscopy of flames and plasmas, coherent optics and holography; instrumentation techniques. *Mailing Add:* Sermatech Int Inc 155 S Limerick Rd Limerick PA 19468

CARRINGTON, THOMAS JACK, b Amarillo, Tex, June 1, 29; m 56; c 2. GEOLOGY. *Educ:* Univ Ky, BS, 58, MS, 60; Va Polytech Inst, PhD(geol), 65. *Prof Exp:* Petrol consult, Ky, 59; asst prof geol, Birmingham-Southern Col, 61-65, actg chmn dept, 61-62 & 63-65, chmn, 65-67, assoc prof, 65-67; PROF GEOL & HEAD DEPT, AUBURN UNIV, 67- *Concurrent Pos:* Dir, Nat Sci Found undergrad res participation prog & equip grant, Birmingham-Southern Col, 63-64. *Mem:* Geol Soc Am; Nat Asn Geol Teachers; Sigma Xi. *Res:* Stratigraphy, origin and areal distribution of geologic rock formations in the Talladega Group, Alabama; geologic structural development in the Piedmont area of Alabama. *Mailing Add:* 210 Petrie Hall Auburn Univ AL 36849

CARRINGTON, TUCKER, b Cincinnati, Ohio, Oct 19, 27; m 57; c 3. PHYSICAL CHEMISTRY. *Educ:* Univ Va, BS, 48; Calif Inst Technol, PhD(chem), 52. *Prof Exp:* Instr chem, Yale Univ, 52-54; phys chemist, Nat Bur Standards, 56-68, Nat Acad Sci-Nat Res Coun resident res assoc, 56-57; PROF CHEM, YORK UNIV, 68- *Honors & Awards:* Silver Medal, Combustion Inst, 62. *Mem:* Am Phys Soc; Chem Inst Can. *Res:* Energy transfer between resolved quantum states of small molecules in gases. *Mailing Add:* Dept of Chem York Univ Toronto ON M3J 1P3 Can

CARROCK, FREDERICK E, b Utica, NY, Oct 9, 31; m 59; c 1. PHYSICAL CHEMISTRY, ORGANIC CHEMISTRY. *Educ:* Syracuse Univ, BA, 54, MS, 56; State Univ NY, Syracuse, PhD(chem), 59. *Prof Exp:* Asst, State Univ NY, Syracuse, 54-56; res chemist, US Rubber Co, 59-61; supvr lab develop res, Rexall Chem Co, 61-65, asst mgr process develop res, Dart Industs Inc, 65-70, dir prod develop, 70-78; MGR TECH DEVELOP, MOBIL CHEM CO, 78- *Mem:* Am Chem Soc; fel Am Inst Chemists. *Res:* Free radical and polymer chemistry. *Mailing Add:* 151 Albright Lane Paramus NJ 07652-1822

CARROLL, ALAN G, BIOCHEMISTRY, CELL BIOLOGY. *Educ:* Mich State Univ, PhD(zool), 78. *Prof Exp:* ASST PROF EMBRYOL, IND UNIV-PURDUE UNIV, 81- *Mailing Add:* Div Med Oncol Rm 5103A Georgetown Univ Lombardi Cancer Ctr Washington DC 20007

CARROLL, ARTHUR PAUL, b New Orleans, La, Aug 1, 42. PLANT PHYSIOLOGY, PLANT BIOCHEMISTRY. *Educ:* Col Santa Fe, BS, 65; Purdue Univ, PhD(biol), 75. *Prof Exp:* Teacher biol & chmn dept, Archbishop Rummel High Sch, 65-70; instr, Purdue Univ, 75-76; asst prof biol, 76-77, CHMN DEPT SCI, COL SANTA FE, 77-, VPRES ACAD AFFAIRS, 82- *Concurrent Pos:* Consult, Controls for Environ Pollution, 78-; extramural assoc, NIH, 80-81 & NASA-Ames Lab, 81-82. *Mem:* AAAS; Am Soc Plant Physiol. *Res:* Development and reproductive behavior of the green alga, chlorella pyrenoidosa, grown in synchronous cultures. *Mailing Add:* St Michael's High Sch 100 Springs Rd Santa Fe NM 87501

CARROLL, BERNARD JAMES, b Sydney, Australia, Nov 21, 40; m 66; c 2. EXPERIMENTAL PSYCHIATRY, PSYCHOPHARMACOLOGY. *Educ:* Univ Melbourne, BSc, 61, MB, BS, 64, DPM, 69, PhD(endocrinol), 72. *Prof Exp:* Resident med officer, Royal Melbourne Hosp, 65-66, sr house physician, 66-67; clin supvr psychiat, Univ Melbourne, 67-68, med res fel, 68-69; sr res officer, Nat Health & Med Res Coun, 69-70, res fel, 70-71; asst prof psychiat, Univ Pa, 71-73; assoc prof, 73-76, prof psychiat, Univ Mich, Ann Arbor, 76-,; DEPT PSYCHIAT, DUKE UNIV. *Concurrent Pos:* Royal Australian Col Physicians traveling fel endocrinol, 70; Nat Health & Med Res Coun Australia Charles J Martin overseas fel, 71; mem extramural grants comt, Ill Dept Ment Health, 73-, chmn, 76-; consult, Intramural Prog Rev, 75; res scientist, Ment Health Res Inst, Univ MIch, Ann Arbor, 73, assoc dir clin res, 77-, actg chmn, 81- *Mem:* Int Soc Psychoneuroendocrinol; Int Col Neuropsychopharmacology; AAAS; Soc Biol Psychiat. *Res:* Clinical psychobiology of depression and mania; clinical and experimental psychopharmacology; behavioral pharmacology; psychoendocrinology; neuroendocrinology. *Mailing Add:* Dept Pyschiat Duke Univ Med Ctr Box 3322 Durham NC 27710

CARROLL, CLARK EDWARD, b Pittsburgh, Pa, Feb 23, 38. QUANTUM OPTICS, ATOMS IN ELECTROMAGNETIC FIELDS. *Educ:* Calif Inst Technol, BS, 59; Harvard Univ, AM, 65, PhD(physics), 69. *Prof Exp:* Res assoc math, Imperial Col Sci & Technol, 72-73; res assoc appl math, Univ Col Cardiff, 73-74; res assoc physics, Naval Res Lab, 74-76, St Francis Xavier Univ, 76-78, St John Fisher Col, 83-87; res fel chem, Australian Nat Univ, 78-83; vis prof physics, Auburn Univ, 87-88; RES ASSOC PHYSICS, UNIV ROCHESTER, 88- *Mem:* Am Phys Soc. *Res:* Analytic solutions of time-dependent Schrödinger equation for three-state model of atom or molecule driven by two laser beams; related two-state models. *Mailing Add:* Dept Physics & Astron Univ Rochester Rochester NY 14627

CARROLL, DANA, b Palm Springs, Calif, Sept 2, 43; m 66; c 2. MOLECULAR BIOLOGY. *Educ:* Swarthmore Col, BA, 65; Univ Calif, Berkeley, PhD(chem), 70. *Prof Exp:* Fel cell biol, Beatson Inst Cancer Res, 70-72; fel develop biol, Carnegie Inst Washington Dept Embryol, 72-75; asst prof microbiol, 75-81, assoc prof cell viral molecular biol, 81-85, PROF & CO-CHMN BIOCHEM MED SCH, UNIV UTAH, 85- *Mem:* Am Soc Microbiol; Am Soc Biochem & Molecular Biol; AAAS. *Res:* Sequence organization in specific, isolated genes of Xenopus; mechanisms of genetic recombination; xenopus oocyte injection; transposable elements. *Mailing Add:* Dept Biochem Univ of Utah Med Ctr Salt Lake City UT 84132

CARROLL, DENNIS PATRICK, b Minneapolis, Minn, Feb 6, 41; m 67; c 3. ELECTRIC POWER SYSTEMS, COMPUTER SIMULATION. *Educ:* Univ Md, BS, 64; Univ Wis, MS, 65, PhD(elec eng), 69. *Prof Exp:* Electronics eng, US Naval Air Test Ctr, 64-65; asst prof, Univ Wis-Milwaukee, 69-70, Clarkson Univ, 70-73; from assoc prof to prof elec eng, Purdue Univ, 73-85; PROF ELEC ENG, UNIV FLA, 85- *Concurrent Pos:* Exec vpres, Simulation Technologies, Inc, 77-85. *Mem:* Sr mem, Inst Elec & Electronics Engrs. *Res:* Analysis and modeling of electric power systems; computer simulation; power electronics applications. *Mailing Add:* Dept Elec Eng Univ Fla Gainesville FL 32611

CARROLL, DEVINE PATRICK, arachnology, ecology, for more information see previous edition

CARROLL, EDWARD ELMER, b North Bergen, NJ, Feb 13, 30; m 58, 75; c 5. NUCLEAR ENGINEERING. *Educ:* Harvard Univ, AB, 50; Univ Pa, MS, 52, PhD(physics), 60. *Prof Exp:* Fel scientist, Bettis Atomic Power Lab, Westinghouse Elec Corp, 59-66; assoc prof nuclear eng, 66-76, actg chmn dept, 76-80, PROF NUCLER ENG, UNIV FLA, 76- *Concurrent Pos:* Vis staff mem, Los Alamos Sci Lab. *Mem:* Am Phys Soc; Am Nuclear Soc. *Res:* Noise analysis; pulse neutron measurements; nuclear instrumentation. *Mailing Add:* Dept Nuclear Eng Univ Fla Gainesville FL 32611

CARROLL, EDWARD JAMES, JR, b San Diego, Calif, Dec 25, 45; m 68; c 2. DEVELOPMENTAL BIOLOGY, BIOCHEMISTRY. *Educ:* Sacramento State Col, BA, 68; Univ Calif, Davis, PhD(biochem), 72. *Prof Exp:* Fel develop biol, Scripps Inst Oceanog, 72-75; asst prof zool, Univ Md, College Park, 75-76; asst prof, 76-80, PROF BIOL, UNIV CALIF, RIVERSIDE, 86- *Mem:* AAAS; Am Soc Cell Biol; Am Soc Zoologists; Soc Develop Biol. *Res:* Biochemistry of fertilization and activation of embryonic metabolism. *Mailing Add:* Dept of Biol Univ of Calif Riverside CA 92521

CARROLL, F IVY, b Norcross, Ga, Mar 28, 35; m 57; c 2. ORGANIC CHEMISTRY. *Educ:* Auburn Univ, BS, 57; Univ NC, PhD(chem), 61. *Prof Exp:* Sr chemist, 60-67, group leader, 67-71, asst dir, 71-75, DIR ORG & MED CHEM, RES TRIANGLE INST, 75- *Mem:* Am Chem Soc; Chem Soc. *Res:* The design of compounds of potential chemotherapeutic value using molecular modeling-molecular mechanism methods; analgesics, depressants, stimulants, antiradiation agents, antiparasitic agents, anticancer agents and cancer prevention compounds; application of organic physical methods to biological problems, circular dichroism, and hydrogen and carbon 13 nuclear magnetic resonance. *Mailing Add:* Res Triangle Inst PO Box 12194 Research Triangle Park NC 27709-2194

CARROLL, FELIX ALVIN, JR, b High Point, NC, Aug 17, 47; m 72; c 2. ORGANIC CHEMISTRY, PHOTOCHEMISTRY. *Educ:* Univ NC, Chapel Hill, BS, 69; Calif Inst Technol, PhD(org chem), 73. *Prof Exp:* Chemist polymer chem, Burlington Indust Res Ctr, 68-69; from asst prof to assoc prof, 72-86, PROF CHEM, DAVIDSON COL, 86- *Concurrent Pos:* Res grant, Res Corp, 73- & NC Sci, Technol Comt, 75-79; chem exhibit design consult, Discovery Pl Sci Mus, Charlotte, NC, 77; petrol res fund, 82-86; prog chmn, Southeast Regional Meeting of Am Chem Soc, 83; Camille & Henry Dreyfus Found grant, 83-90, NSF grant, 84- 89, Nat Inst Health grant, 88-91. *Mem:* Am Chem Soc; AAAS; Inter-Am Photochemistry Soc. *Res:* Photochemical and photophysical processes; intersystem crossing; heavy atom effects; singlet quenching; exciplexes; kinetics and mechanisms of photochemical reactions; pheromones; history of chemistry; computational chemistry; molecular graphics. *Mailing Add:* Dept Chem Davidson Col PO Box 1749 Davidson NC 28036-1749

CARROLL, FLOYD DALE, b Mt Clare, Nebr, Jan 8, 14; m 44; c 1. ANIMAL HUSBANDRY. *Educ:* Univ Nebr, BS, 37; Univ Md, MS, 39; Univ Calif, PhD(animal nutrit), 48. *Prof Exp:* Prin technician, 48-49, from instr to prof, 49-79, EMER PROF ANIMAL HUSB, UNIV CALIF, DAVIS, 79- *Concurrent Pos:* Fulbright fel, Univ Ceylon, 55-56. *Res:* Vitamin B synthesis in the horse; nutrition in range beef production; beef production under hot climatic conditions; dwarfism in beef cattle; factors affecting beef carcass grades; composition and palatability. *Mailing Add:* 1435 Wake Forest Dr Davis CA 95616

CARROLL, FRANCIS W, b Philadelphia, Pa, Aug 22, 32; wid; c 4. MATHEMATICAL ANALYSIS. *Educ:* St Joseph's Col, Pa, BS, 54; Purdue Univ, MS, 56, PhD(math), 60. *Prof Exp:* Asst prof math, Purdue Univ, 59-60, Mich State Univ, 60 & Univ Wisconsin-Milwaukee, 60-61; from asst prof to assoc prof, 61-75, PROF MATH, OHIO STATE UNIV, 75- *Mem:* Am Math Soc; Math Asn Am. *Res:* Complex analysis; functional equations. *Mailing Add:* Dept Math Ohio State Univ Columbus OH 43210-1174

CARROLL, GEORGE C, b Alton, Ill, Feb 11, 40; m 68; c 2. MYCOLOGY. *Educ:* Swarthmore Col, BA, 62; Univ Tex, PhD(bot), 66. *Prof Exp:* From asst prof to assoc prof, 67-81, PROF BIOL, UNIV ORE, 81- *Concurrent Pos:* Vis prof, Swiss Fed Inst Technol, Inst Gen Bot, 73-74 & Kyoto Univ, Yoshida Col, 88-89. *Mem:* Mycol Soc Am; Brit Mycol Soc. *Res:* Evolution of the fungi; ecology of fungi in terrestrial ecosystems; microbiology of the coniferous forest canopy; ecology and evolution of mutualistic fungal endophytes in leaves of terrestrial plants. *Mailing Add:* Dept Biol Univ Ore Eugene OR 97403

CARROLL, GERALD V, b Meriden, Conn, Apr 9, 21. CRYSTAL OPTICS. *Educ:* Lehigh Univ, BA, 43; Yale Univ, PhD(geol), 53. *Prof Exp:* Instr, Lehigh Univ, 50-51; instr, Trinity Col, Conn, 51-53; asst prof geol, Dartmouth Col, 53-54; assoc prof, Agr & Mech Col, Univ Tex, 54-59; vis assoc prof, Pa State Univ, 59-61; from assoc prof to prof geol, George Washington Univ, 61-83; RETIRED. *Res:* Petrology. *Mailing Add:* 2510 Virginia Ave NW Washington DC 20037

CARROLL, HARVEY FRANKLIN, b New Haven, Conn, Aug 25, 39; m 71; c 1. PHYSICAL CHEMISTRY. *Educ:* Hunter Col, AB, 61; Cornell Univ, PhD(phys chem), 69. *Prof Exp:* Sr chemist, Chem Div, Uniroyal Inc, Conn, 68-69; from asst prof to assoc prof, 70-85, chairperson, 76-79, PROF DEPT PHYS SCI, KINGSBOROUGH COMMUNITY COL, CITY UNIV NY, 85-, DIR, ENG SCI PROG, 87- *Concurrent Pos:* Vis prof, Dept Physical Chem, Hebrew Univ, Jersalem, Israel, 79-80. *Mem:* Am Chem Soc; Sigma Xi. *Res:* High temperature gas phase chemical kinetics; shock tubes; homogeneous isotope exchange reactions in ultra clean systems; unimolecular reactions; chemical and nutrition education. *Mailing Add:* Dept of Phys Sci Kingsborough Community Col Brooklyn NY 11235

CARROLL, J GREGORY, b Philadelphia, Pa, Nov 1, 48; m 73; c 3. EDUCATIONAL PSYCHOLOGY, BEHAVIORAL MEDICINE. *Educ:* Cornell Univ, BA, 70, PhD(educ psychol), 76; Temple Univ, MA, 72. *Prof Exp:* Sr res assoc, Fac Resource Ctr, Univ Cincinnati, 76-77, asst prof biomed commun, Col Med, 76-77; proj dir educ serv & res, Univ Pa Sch Med, 77-80, dir, 80-83, adj asst prof educ, Grad Sch Educ, 77-83; assoc dean & assoc prof health & human develop, Pa State Univ, 83-89; MGR, HEALTH COMMUN & TRAINING, PHARMACEUT DIV, MILES INC, 89- *Concurrent Pos:* Consult, Children's Hosp Philadelphia, 80-82, Mental Health Consortium, 80, Comt Int Med Exchange, 80-82; prog dir, Minority Health Careers Prog, Pa State Univ, 86-89; mem Task Force Med Interview, Soc Gen Internal Med, 81-, Task Force AIDS, 87- *Mem:* AAAS; Am Educ Res Asn; Am Psychol Asn; Asn Study Higher Educ; Asn Am Med Col; Sigma Xi. *Res:* Interpersonal communication skills in the health professions; program evaluation and faculty development in higher education. *Mailing Add:* Pharmaceut Div Miles Inc 400 Morgan Lane West Haven CT 06516

CARROLL, JAMES BARR, b Chicago, Ill, Mar 25, 29; m 52; c 6. SCIENCE ADMINISTRATION, SCIENCE POLICY. *Educ:* Brown Univ, ScB, 52, MS, 57; Univ Conn, PhD(physics), 67. *Prof Exp:* Res engr, United Aircraft Res Labs, 55-68; staff scientist, Royal Prod Co, Litton Industs, 68-72; consult, Off Planning & Mgt, US Environ Protection Agency, 72-74; dir, Off Energy Supply Progs, Fed Energy Admin, 74; CONSULT, 74- *Concurrent Pos:* Lectr, Trinity Col, Conn, 65-66 & Univ Conn, 68. *Mem:* Am Phys Soc. *Res:* Technology assessment; computer modeling techniques; energy research. *Mailing Add:* 25 Welwyn Way Rockville MD 20850

CARROLL, JAMES JOSEPH, b Scranton, Pa, Dec 12, 35; m 60; c 1. BIOCHEMISTRY, CLINICAL CHEMISTRY. *Educ:* Univ Scranton, BS, 57; Pa State Univ, MS, 60, PhD(biochem), 62; Nat Registry Clin Chem, cert; Fairleigh Dickinson Univ, MBA, 77. *Prof Exp:* Scientist, Sandoz Pharmaceut, 62-65; scientist, Warner-Lambert Res Inst, 65-71, sr scientist, 71-73, sr res assoc, Warner-Lambert Co, Inc, 73-76, mgr planning & coord, Warner-Lambert Res Inst, 76-77, prod mgr, Gen Diagnostics Int, Warner-Lambert Int, 77-85; pres, Am Bio Prod Co, 85-87; dir res & develop coagulation & hematol, Baxter Dade, 87-89; DIR RES & DEVELOP HEMOSTASIS, ORTHO DIAG SYSTS, INC, 89- *Concurrent Pos:* Mem adj fac, Dept Mkt, Econ & Finance, Col Bus Admin, Fairleigh Dickinson Univ, Madison, NJ, 77-85. *Mem:* Am Chem Soc; Am Asn Clin Chemists; Asn Clin Scientists; Biomed Mkt Asn. *Res:* Cholesterol metabolism; drug metabolism; creation and development of enzyme assays for use in diagnostics reagents; immuno-enzyme assays for isoenzymes; coagulation; thrombosis; prethrombosis. *Mailing Add:* 48 Cutter Dr Hanover NJ 07936

CARROLL, JOHN MILLAR, b Philadelphia, Pa, Dec 6, 25; m 44; c 7. INDUSTRIAL ENGINEERING, ELECTRONICS. *Educ:* Lehigh Univ, BS, 50; Hofstra Univ, MA, 55; NY Univ, DEngSc, 68. *Prof Exp:* Asst ed, Electronics Mag, 52-54, assoc ed, 54-57, managing ed, 57-64; assoc prof indust eng, Lehigh Univ, 64-68; prof, 68-91, EMER PROF COMPUT SCI, UNIV WESTERN ONT, 91- *Concurrent Pos:* Secy field-test panel, Nat

Stereophonic Radio Comt, 59-60; mem comt info handling, Eng Serv Libr, 64-68; trustee, Eng Index, 68-71; consult, Can Privacy & Comput Task Force, 71 & Comput Security, Royal Can Mounted Police, 75-76. *Mem:* Sr mem Inst Elec & Electronics Engrs. *Mailing Add:* Dept Computer Sci Univ Ont London ON N6A 5B7 Can

CARROLL, JOHN T, III, b Milwaukee, Wis, Apr 5, 60; m 90; c 1. CERAMICS ENGINEERING. *Educ:* NC State Univ, Raleigh, BS, 82, MS, 84, PhD(mech eng), 87. *Prof Exp:* Tech hire, IBM, Res Triangle Park, NC, 80-83; res asst, NC State Univ Precision Eng Ctr, 84-87; tech specialist, Cummins Engine Co, Inc, Columbus, Ind, 87-90; DIR ENG, ENCERATEC, INC, COLUMBUS, IND, 90- *Mem:* Am Soc Precision Eng. *Res:* Develop applications and commercial opportunities for wear resistant, solid, silicon nitride components used in diesel engines, machine tools, and precision mechanical systems. *Mailing Add:* Enceratec Inc 2525 Sandcrest Dr Columbus IN 47203

CARROLL, JOHN TERRANCE, b Hibbing, Minn, Oct 31, 42; m 67; c 1. PARTICLE PHYSICS, COMPUTER SCIENCE. *Educ:* St Mary's Col, Winona, Minn, 64; Univ Wis, Madison, MA, 66, PhD(physics), 71. *Prof Exp:* Res assoc physics, 71-75, MATHEMATICIAN PHYSICS, STANFORD LINEAR ACCELERATOR CTR, 75- *Concurrent Pos:* Sci assoc, Europ Orgn Nuclear Res, 81-82. *Mem:* Am Phys Soc. *Res:* Elementary particle physics; bubble chamber studies of hadronic interactions; development and application of new computer hardware and software techniques for data acquisition and analysis. *Mailing Add:* MS318 Fermi Lab PO Box 500 Batavia IL 60510

CARROLL, KENNETH GIRARD, b Pittsburgh, Pa, Feb 18, 14; m 45; c 1. PHYSICS. *Educ:* Carnegie Inst Technol, BS, 34, MS, 35; Yale Univ, PhD(physics), 39. *Prof Exp:* Instr physics, NC State Col, 39-40; physicist, Nat Adv Comt Aeronaut, Va & Ohio, 40-44, Elastic Stop Nut Corp Am, NJ, 44-46 & res lab, US Steel Corp, 46-54; head physics sect, Int Nickel Co Res Lab, 54-61; sr scientist, Sperry Rand Res Ctr, 61-67 & NASA Electronics Res Ctr, 67-70; vis scientist, Forsyth Dent Ctr, Boston, 70-71; physicist, Argonne Nat Lab, 74-77; CONSULT, 77- *Concurrent Pos:* Nat Acad Sci-Nat Res Coun vis scientist, US Army Natick Labs, 65-66. *Mem:* Am Phys Soc; Electron Micros Soc; Sigma Xi. *Res:* X-ray diffraction and spectroscopy; electron microscopy; physical metallurgy; application of electron beam instruments to biological systems; detection and localization of trace metals in single biological cells. *Mailing Add:* 2424 Pennsylvania Ave NW Apt 701 Washington DC 20037

CARROLL, KENNETH KITCHENER, b Carrolls, NB, Mar 9, 23; m 50; c 3. BIOCHEMISTRY, NUTRITION. *Educ:* Univ NB, BSc, 43, MSc, 46; Univ Toronto, MA, 46; Univ Western Ont, PhD(med), 49. *Prof Exp:* From asst prof to prof med res & actg head dept, 54-68, prof, biochem dept, 68-88, EMER PROF BIOCHEM DEPT, UNIV WESTERN ONT, 88-; CAREER INVESTR, MED RES COUN CAN, 63-, DIR, HUMAN NUTRIT CTR, 90- *Concurrent Pos:* Can Life Ins Off Asn fel, Dept Med Res, Univ Western Ont, 49-52; Merck & Co fel, Dept Chem, Cambridge Univ, 52-53, Agr Res Coun Can fel, 53-54; chmn, comn nutrit & cancer, Int Union Nutrit Sci, 79-89; mem, Comn on Diet & Health Food & Nutrit Bd, US Nat Res Coun, Nat Acad Sci, 86-89; mem, Sci Rev Comt for Revision of Nutrit Recommendations for Can, Health & Welfare, Can, 87-90. *Honors & Awards:* Earle Willard McHenry Award, Can Soc Nutrit Sci, 87. *Mem:* Am Inst Nutrit; Am Oil Chem Soc; Am Soc Biol Chem; Can Biochem Soc; Can Soc Nutrit Sci (secy, 65-67, pres, 78-79); Royal Soc Can; Can Fedn Biol Soc (secy, 67-71). *Res:* Lipid metabolism, cholesterol; atherosclerosis; mammary cancer; polyprenols, dolichol. *Mailing Add:* Dept Biochem Univ Western Ont London ON N6A 5C1 Can

CARROLL, LEE FRANCIS, b Berlin, NH, Oct 14, 37; m 60; c 3. ELECTRICAL ENGINEERING, SCIENCE EDUCATION. *Educ:* Northeastern Univ, BS, 60. *Prof Exp:* Plant elec eng, Fraser Paper, Ltd, 60-64 & Ga Pac Corp, 65-66; maintenance engr, Am Optical Co, 64-65; chief elec eng, Brown Co, Berlin, NH, 66-70 & Wright & Pierce Engrs, 70-73; PROPRIETOR, LEE F CARROLL, PROF ENG, ELEC CONSULT, 73-; PROF & ADJ STAFF, NAT ELEC CODE, NH VOC TECH COL, 87- *Concurrent Pos:* Comnr, Town of Gorham, NH, Water & Sewer Comn, 79-; mem, Bd Trustees, Gould Acad, Bethel, Maine, 80- & NH Dept Transp Appeals Bd, 86- *Honors & Awards:* Lighting Design Award, Illum Eng Soc, 80. *Mem:* Inst Elec & Electronics Engrs; Nat Soc Prof Engrs; Illum Eng Soc; Nat Fire Protection Asn. *Mailing Add:* 43 Evans St PO Box F Gorham NH 03851

CARROLL, MICHAEL M, b Thurles, Ireland, Dec 8, 36; US citizen; m 64; c 2. ROCK MECHANICS, BIOMECHANICS. *Educ:* Nat Univ Ireland, BA, 58, MA, 59; Brown Univ, PhD(appl math), 65. *Hon Degrees:* DSc, Nat Univ Ireland, 79. *Prof Exp:* Res assoc, Div Appl Math, Brown Univ, 64-65; from asst prof to assoc prof, 65-75, PROF APPL MECH, DEPT MECH ENG, UNIV CALIF, BERKELEY, 75-; DEAN, GEORGE R BROWN SCH ENG, RICE UNIV, 88-, PROF MATH SCI & PROF DEPT MECH ENG & MAT SCI, & BURTON J & ANN M MCMURTRY PROFESSORSHIP ENG & MAT SCI, 88- *Concurrent Pos:* Visitor, Sch Theoret Physics, Dublin Inst Advan Studies, 71-72; vis prof, Univ Col Cork, Ireland, 77; consult, Thoratec Labs, Berkeley, Calif, 76- & Terra Tek Inc, Salt Lake City, Utah, 77-; affil mem, Dept Cardiovasc Surg, Pac Med Ctr, 77- *Mem:* Nat Acad Eng; Acoust Soc Am; Soc Rheology; Sigma Xi; fel Am Soc Mech Engrs. *Res:* Continuum mechanics, with emphasis on nonlinear phenomena; mechanics of porous materials; mechanics of oil and gas reservoirs; acoustics; nonlinear optics; biomechanics. *Mailing Add:* George R Brown Sch Eng Rice Univ PO Box 1892 Houston TX 77251

CARROLL, PAUL TREAT, b San Francisco, Calif, Oct 22, 43; m 68; c 2. PHARMACOLOGY, TOXICOLOGY. *Educ:* Univ Calif, Berkeley, AB, 66; San Jose State Univ, MA, 69; Univ Md, PhD(pharmacol), 73. *Prof Exp:* Scientist urol, Alcon Labs, 69-70; fel neurochem, Johns Hopkins Univ, 73-76; asst prof pharmacol, Univ RI, 76-81; ASSOC PROF PHARMACOL, MED SCH, TEX TECH UNIV, 81- *Concurrent Pos:* SmithKline Corp fel, Johns Hopkins Univ, 74-76; NSF neurobiol grant, 78-81. *Mem:* Soc Neuroscience; Am Soc Pharmacol & Exp Therapeut. *Res:* Central cholinergic metabolism. *Mailing Add:* Dept Pharmacol & Exp Therapeut Tex Tech Univ Health Sci Ctr Lubbock TX 79430

CARROLL, RAYMOND JAMES, b Yokohama, Japan, Apr 21, 49, US citizen; m 72. STATISTICS, BIOSTATISTICS. *Educ:* Univ Tex, Austin, BA, 71; Purdue Univ, MS, 72, PhD(statist), 74. *Prof Exp:* From asst prof to prof statist, Univ NC, Chapel Hill, 78-88; PROF & HEAD STATIST, TEX A&M UNIV, 88- *Concurrent Pos:* Co-prin investr, Air Force Off Sci Res, 76-; statist consult, NC State Fisheries Off, 76-78 & Ctr Dis Control, 78-79; vis consult, Nat Heart, Lung & Blood Inst, 80- *Honors & Awards:* Wilcoxon Prize, 85. *Mem:* Fel Inst Math Statist; fel Am Statist Asn; Biomet Soc; Sigma Xi; Int Statist Inst. *Res:* Robustness of statistical inference, transformations, measurement error models. *Mailing Add:* Dept Statist Tex A&M Univ College Station TX 77843-3143

CARROLL, RAYMOND JAMES, b Yokohama, Japan, Apr 21, 49; US citizen; m 72. REGRESSION WEIGHTING & TRANSFORMATION, MEASUREMENT ERROR MODELS. *Educ:* Univ Tex, BA, 71; Purdue Univ, PhD(statist), 74. *Prof Exp:* Prof statist, Univ NC, 74-87; PROF STATIST, TEX A&M UNIV, 87- *Concurrent Pos:* Prin investr, Air Force Off Sci Res Grants, 74-, NIH grants, 89-; vis prof, Univ Wis, 84, Australian Nat Univ, 91; ed, J Am Statist Asn, 87-90. *Honors & Awards:* Wilcoxon Award, Am Soc Qual Control, 85; Comt of Presidents of Statist Socs Award, 88. *Mem:* Am Statist Asn; fel Inst Math Statist; Biometric Soc; fel Int Statist Inst. *Res:* Development and theoretical study of statistical methods for use in medicine, epidemiology, marine sciences, industry and the quality control areas. *Mailing Add:* 9810 Parkwood Dr Bethesda MD 20814

CARROLL, ROBERT BAKER, b Tuscaloosa, Ala, Apr 8, 21; div; c 4. ORGANIC CHEMISTRY, PLANT PATHOLOGY. *Educ:* Univ Ala, BS, 42; La State Univ, BS, 47, MS, 49, PhD(plant path), 51. *Prof Exp:* Asst jr pyrometrist, Ensley-Fairfield Works, US Steel Corp, Ala, 41; asst res chemist, Swann Chem Co, 41-42; asst res chemist, Exp Sta, Hercules Powder Co, Del, 42-43, explosives chemist, Radford Ord Works, Va, 42-43; asst plant physiologist, Boyce Thompson Inst Plant Res, NY, 50-52; microbiologist, Biochem Res Lab, Borden Co, 52-53; independent consult microanal, 53-68; pres, ODV Inc, 68-; AT MED CTR, NEW YORK UNIV. *Mem:* AAAS; Am Ord Asn; Sigma Xi; Weed Sci Soc Am; Am Inst Chemists. *Res:* Food and flavors; pharmaceuticals; pesticides; instrumentation; plant physiology; agronomy; alcoholic beverages; plastics; oils and fats; microanalytical methods using all forms of chromatography; forensic chemistry, especially analysis of drugs and narcotics; field identification of narcotics and dangerous drugs. *Mailing Add:* Path Dept NY Univ 550 First Ave New York NY 10016

CARROLL, ROBERT BUCK, b Wellsburg, WVa, June 27, 40; m 63; c 3. SOYBEAN PATHOLOGY, SOIL-BORNE DISEASES. *Educ:* WVa Univ, BS, 62, MS, 64; Pa State Univ, PhD(plant pathol), 71. *Prof Exp:* Res asst plant pathol, WVa Univ, 62-64; res asst plant pathol, Pa State Univ, 64-70, instr, 70-71; exten specialist, 71-77, ASSOC PROF PLANT PATHOL, UNIV DEL, 77- *Concurrent Pos:* Asst dept chmn, Plant Sci Dept, Univ Del, 77-; assoc ed, Plant Dis J, Am Phytopathol Soc, 81- *Mem:* Am Phytopathological Soc; NY Acad Sci; Sigma Xi. *Res:* Soybean pathology with emphasis on the effect of changing cultural practices (tillage, herbicides, varieties) on the incidence of root and stem diseases. *Mailing Add:* Univ Del 140 Townsend Hall Newark DE 19711

CARROLL, ROBERT GRAHAM, b Darby, Pa, Mar 18, 54; m 84; c 2. CARDIOVASCULAR & RENAL PHYSIOLOGY. *Educ:* Univ Notre Dame, BS, 76; Col Med & Dent NJ, PhD(physiol), 81. *Prof Exp:* Teaching fel physiol, Med Ctr, Univ Miss, 81, instr, 81-84; ASSOC PROF PHYSIOL, SCH MED, E CAROLINA UNIV, 84- *Mem:* Am Physiol Soc; Sigma Xi; Am Heart Asn. *Res:* Cardiovascular and renal physiology; role of the sympathetic nervous system in long term blood pressure regulation; changes in tissue perfusion following hemorrhage; role of dietary proteins in regulating renal function. *Mailing Add:* Sch Med E Carolina Univ Greenville NC 27858-4354

CARROLL, ROBERT J, b Conshohocken, Pa, Aug 24, 28; m 59; c 4. PHYSICAL CHEMISTRY, ELECTRON MICROSCOPY. *Educ:* Drexel Univ, BS, 58. *Prof Exp:* RES CHEMIST, EASTERN REGIONAL RES CTR, USDA, 55- *Mem:* Electron Micros Soc Am; AAAS. *Res:* Application of transmission and scanning electron microscopy to the ultrastructural relationships of skin, leather, milk, meat and meat products; lipid-protein interaction products; immunocytochemical studies of lipid and protein synthesis in lactating mammary tissue. *Mailing Add:* 2740 Clayton St Philadelphia PA 19152

CARROLL, ROBERT LYNN, b Kalamazoo, Mich, May 5, 38; m 87; c 1. VERTEBRATE PALEONTOLOGY. *Educ:* Mich State Univ, BS, 59; Harvard Univ, MA, 61, PhD(biol), 63. *Prof Exp:* Nat Res Coun Can fel, 62-63; NSF fel, 63-64; assoc cur geol, Redpath Mus, McGill Univ, 64-69, assoc prof vert paleont, 69-74, dir, Redpath Mus, 83-90, PROF BIOL, MCGILL UNIV, 74-, CHMN, DEPT BIOL, 90- *Honors & Awards:* Schuchert Award, Paleont Soc, 78. *Mem:* Soc Vert Paleont (vpres, 81-82 & pres, 82-83); Soc Study Evolution; Am Soc Zool; Paleont Soc. *Res:* Anatomy and phylogeny of Carboniferous, Permian and Triassic amphibians and reptiles--labyrinthodonts, microsaurs, captorhinomorphs and eosuchians; origin of lizards and Lissamphibia; macroevolutionary processes evidenced by fossil record. *Mailing Add:* Redpath Mus McGill Univ 859 Sherbrooke St W Montreal PQ H3A 2K6 Can

CARROLL, ROBERT WAYNE, b Chicago, Ill, May 10, 30; m 57, 74, 79; c 2. MATHEMATICS. *Educ:* Univ Wis, BS, 52; Univ Md, PhD(math), 59. *Prof Exp:* Res aeronaut scientist, Nat Adv Comt Aeronaut, Ohio, 52-54; Nat Sci Found fel, 59-60; asst prof math, Rutgers Univ, 60-63, assoc prof, 63-64; assoc prof, 64-67, PROF MATH, UNIV ILL, URBANA-CHAMPAIGN, 67-

Concurrent Pos: Nat Sci Found res grant, 63-70 & 86-87. *Mem:* Math Soc Am. *Res:* Partial differential equations; functional analysis; differential geometry; mathematical physics; Lie groups; author of over 7 books and 130 articles. *Mailing Add:* Dept of Math Univ of Ill Urbana IL 61801

CARROLL, ROBERT WILLIAM, b Geneva, NY, Jan 16, 38; m 80; c 2. PHYSICAL CHEMISTRY. *Educ:* Hobart Col, BS, 59; Fordham Univ, PhD(phys chem), 65. *Prof Exp:* Res chemist, Inmont Corp, 65-68; sr scientist photoconductor res & develop, Optonetics, Inc, 69-70; asst prof chem, Herbert H Lehman Col, 71-78; asst prof chem, Manhattan Col, Mt St Vincent Univ, 79-80; asst prof chem, Fordham Univ, 81-83; CONSULT, 84- *Mem:* Am Chem Soc; Sigma Xi; NY Acad Sci; fel Am Inst Chemists. *Res:* Kinetics of hydrocarbon pyrolysis; field emission and field ionization microscopy; heterogeneous catalysis. *Mailing Add:* RD 3 Box 75 Yorktown Heights NY 10598

CARROLL, SAMUEL EDWIN, b Thamesville, Ont, Dec 20, 28; m; c 2. CARDIOVASCULAR SURGERY, THORACIC SURGERY. *Educ:* Univ Western Ont, BA, 56, MD, 53. *Prof Exp:* Trainee surg, Victoria Hosp, 53-58; registr, Edgware Gen Hosp, London, Eng, 59-60; CARDIOVASC SURGEON, DEPT VET AFFAIRS, WESTMINSTER HOSP, 63-; CHIEF SURGEON, ST JOSEPH'S HOSP, 66-; PROF SURG, UNIV WESTERN ONT, 66- *Concurrent Pos:* Res fel hypothermia, Surg Lab, Children's Hosp Med Ctr, Boston, Mass, 61; Ont Heart Found res fel, 61-, grant in aid, 64-; res asst, St Mark's Hosp, London, Eng, 59-60; examr, Royal Col Physicians & Surgeons Can, 68-74. *Mem:* Fel Am Col Chest Physicians; fel Am Col Surg; NY Acad Sci; Can Cardiovasc Soc; fel Royal Col Physicians & Surgeons Can. *Res:* Hypothermia; small vessel anastomosis; surgical shock. *Mailing Add:* Dept Gen Surg St Joseph's Health Centre London ON N6A 4V2 Can

CARROLL, THOMAS JOSEPH, b Pittsburgh, Pa, April 26, 12. PHYSICS. *Educ:* Univ Pittsburgh, AB, 32; Yale Univ, PhD(physics), 36. *Prof Exp:* Lab asst physics, Yale Univ, 32-36; prof math & physics, Col New Rochelle, 36-41; radio engr, Signal Corps Labs, Ft Monmouth, NJ, 41-43, physicist, Off Chief Signal Officer, Washington, DC, 43-46, Bur Standards, 46-51, Lincoln Lab, Mass Inst Technol, 51-58 & Bendix Radio Co, 58-68; res prof, Dept Elec Eng, George Washington Univ, 68-70; CONSULT, 70- *Concurrent Pos:* Mem, Int Sci Radio Union. *Mem:* Am Phys Soc; Inst Elec & Electronics Engrs; Optical Soc Am; Acoust Soc Am; Am Asn Physics Teachers. *Res:* Microwave radio propagation; twilight scatter propagation; molecular spectra; Faraday effect in molecular spectra. *Mailing Add:* 95 Algonquin Rd Chestnut Hill MA 02167-1001

CARROLL, THOMAS WILLIAM, b Los Angeles, Calif, Aug 22, 32; m 52; c 4. PLANT PATHOLOGY, PLANT VIROLOGY. *Educ:* Calif State Polytech Col, BS, 54; Univ Calif, Davis, MS, 62, PhD(plant path), 65. *Prof Exp:* NIH res plant pathologist, 65-66; from asst prof to assoc prof bot, 66-74, PROF PLANT PATH, MONT STATE UNIV, 74- *Concurrent Pos:* Sigma Xi fac res award, Mont State Univ, 87. *Mem:* Am Phytopath Soc; Am Soc Virol; AAAS; Sigma Xi. *Res:* Plant virology; electron microscopy. *Mailing Add:* Dept Plant Path Mont State Univ Bozeman MT 59717

CARROLL, WALTER WILLIAM, cancer, surgery, for more information see previous edition

CARROLL, WILLIAM J, b Los Angeles, Calif, Nov 23, 23; m; c 4. ENGINEERING. *Educ:* Calif Inst Technol, BS, 48, MS, 49. *Prof Exp:* Indust waste engr, County Engr, County Los Angeles, 49-51; from engr to chief engr, James M Montgomery, Consult Engrs, Inc, 51-65, treas, 56-59, vpres, 59-69, pres & chief exec officer, 69-85, VCHMN BD & CIVIL ENGR, JAMES M MONTGOMERY, CONSULT ENGRS, INC, PASADENA, CALIF, 85- *Concurrent Pos:* Dir, Am Consult Engrs Coun, 73-74, vchmn, Int Eng Comt, 74-78, chmn, Long Range Planning Comt, 80; nat dir, Am Soc Civil Engrs, 76-79; mem, Eng & Environ Comt, World Fedn Eng Orgns, 80-89. *Honors & Awards:* Eng Merit Award, Inst Advan Eng, 76; Cleary Award, Am Acad Environ Engrs, 83. *Mem:* Nat Acad Eng; Am Acad Environ Engrs (vpres, 78, pres, 80); fel Am Soc Civil Engrs (pres-elect, 87-88, pres, 88-89); Am Water Works Asn; Am Geophys Union; fel Inst Advan Eng (treas, 69-70, vpres, 70-71); Int Water Resources Asn; Sigma Xi. *Res:* Water and wastewater system planning and design. *Mailing Add:* James M Montgomery Consult Engrs Inc 250 N Madison Ave PO Box 7009 Pasadena CA 91109-7009

CARRON, NEAL JAY, b Jersey City, NJ, May 27, 41. PHYSICS. *Educ:* Mass Inst Technol, BS, 63; Univ Ill, MS, 65, PhD(physics), 69. *Prof Exp:* Res assoc physics, Rice Univ, Houston, Tex, 69-71; RES STAFF MEM PHYSICS, MISSION RES CORP, SANTA BARBARA, 71- *Mem:* Am Phys Soc. *Res:* Plasma physics; charged particle beams; inertial confinement fusion. *Mailing Add:* Mission Res Corp PO Drawer 719 Santa Barbara CA 93102

CARROZZA, JOHN HENRY, b Bridgeport, Conn, Dec 14, 44; m 77; c 2. MICROBIOLOGY, VETERINARY VACCINES. *Educ:* Fairfield Univ, BS, 66; Univ Conn, PhD(virol), 72. *Prof Exp:* Lab dir vet vaccine, Amerlab Inc, 71-78; dir res & develop vet vaccine, 78-80, GEN MGR, KEEVET LABS, 80- *Mem:* Am Soc Microbiol; AAAS. *Res:* Veterinary vaccines, including those for Marek's Disease, viral arthritis and foul cholera. *Mailing Add:* Keevet Labs PO Box 1706 Anniston AL 36202

CARRUTH, BETTY RUTH, b Comanche, Tex. NUTRITION. *Educ:* Tex Tech Univ, BS, 65, MS, 68; Univ Mo, PhD(human nutrit, sociol), 74. *Prof Exp:* Instr nutrit, Tex Tech Univ, 68-71; res dietitian, Med Ctr & instr med dietetics, Dept Human Nutrit, Foods & Food Systs Mgt, Univ Mo, 74; asst prof nutrit, Univ Minn, St Paul, 74-78; dir nutrit, adolescent health training proj, Dept Pediat, Univ Tex Health Sci Ctr, Dallas, 78-81; HEAD DEPT NUTRIT & FOOD SCI, UNIV TENN, KNOXVILLE, 81- *Mem:* Am Dietetic Asn; Soc Adolescent Med; Sigma Xi. *Res:* determination of significant nutritional and socio-health needs of youth; attitude measurement. *Mailing Add:* 8100 W Cliff Dr Knoxville TN 37909

CARRUTH, JAMES HARVEY, b Baton Rouge, La, Aug 17, 38; m 65; c 2. PURE MATHEMATICS. *Educ:* La State Univ, BS, 61, MS, 63, PhD(math), 66. *Prof Exp:* From asst prof to assoc prof, 66-76, PROF MATH, UNIV TENN, KNOXVILLE, 76- *Concurrent Pos:* Dir, Individualized Retirement Financial Planning Proj, Univ Tenn, 82- *Mem:* Am Math Soc; Math Asn Am. *Res:* Topological semigroups. *Mailing Add:* Dept Math Univ Tenn Knoxville TN 37996

CARRUTH, WILLIS LEE, b Summit, Miss, Feb 21, 09; m 37; c 2. PHYSICAL CHEMISTRY. *Educ:* Asbury Col, BA, 35; Univ Ky, MS, 38. *Prof Exp:* Instr chem, Asbury Col, 35-36 & Univ SDak, 36-38; instr chem & physics, Lewis & Clark Col, 38-39, from asst prof to prof chem, 39-44; prof, Nebr Wesleyan Univ, 44-46; assoc prof math, Col Puget Sound, 46-47, from asst prof to prof chem, 47-58; mem staff, Res & Develop, Appl Physics Corp, 58-74; tech dir, Tinsley Replication Group, Tinsley Labs, Inc, 74-85; RETIRED. *Concurrent Pos:* Registr & admin secy fac, Lewis & Clark Col, 42-44; consult scientist, Varian Assocs, 74-87. *Mem:* Fel AAAS; Am Chem Soc; Optical Soc Am. *Res:* Spectroscopy; instrumental methods of analysis; optics. *Mailing Add:* 2120 Maginn Dr Glendale CA 91202-1130

CARRUTHERS, CHRISTOPHER, b Motherwell, Scotland, Mar 17, 09; nat US; m 40, 59; c 3. CANCER. *Educ:* Syracuse Univ, BS, 33, MS, 35; Univ Iowa, PhD(biochem), 38. *Prof Exp:* Asst chem, Syracuse Univ, 33-35; asst chem, Univ Iowa, 35-38; res assoc, Barnard Free Skin & Cancer Hosp, 42-48; mem div cancer res, Med Sch, Wash Univ, 48-53; assoc cancer res scientist, Rosewell Park Mem Inst, 53-66, prin cancer res scientist, 66-84; RETIRED. *Concurrent Pos:* Res fel, Barnard Free Skin & Cancer Hosp, 38-42; instr, Sch Med, Wash Univ, 41-44; assoc res prof, Med Sch, State Univ, NY Buffalo, 56-79. *Mem:* Fel AAAS; fel NY Acad Sci; Am Soc Biol Chem; Am Asn Cancer Res; Soc Exp Biol & Med; Sigma Xi. *Res:* Polarography in biochemistry and cancer research; biochemistry of epidermis, cell particulates and carcinogenesis; immunology and biochemistry of epidermal proteins and of membranes of various types of rat mammary carcinomas and in rat liver carcinogenesis; metabolism of azocarcinogens by rat liver and distribution of ligandin in azocarcinogen-treated rat liver and azocarcinogen-induced liver tumors; nicotine as a cocarcinogen in mouse skin. *Mailing Add:* 150 Sunset Terrace Orchard Park NY 14127-0094

CARRUTHERS, JOHN ROBERT, b Toronto, Ont, Sept 12, 35; m 57; c 2. MATERIALS SCIENCE. *Educ:* Univ Toronto, BASc, 59, PhD(metall), 66; Lehigh Univ, MS, 61. *Prof Exp:* Mem tech staff semiconductor mat, Bell Tel Labs, 59-63; lectr mat sci, Univ Toronto, 64-65, asst prof, 65-67; mem tech staff crystal chem, Bell Tel Labs, 67-75, head crystal growth & glass res & develop dept, Bell Labs, Murray Hill, NJ, 75-77; dir mat processing, Space Div, NASA, 77-81; proj mgr mat characterization, Solid State Lab, Hewlett-Packard, Palo Alto, Calif, 81-84; DIR, COMPONENTS RES, INTEL CORP, SANTA CLARA, CA, 84- *Concurrent Pos:* Consult, Off Applns, NASA, 74-77; consult, Crystallome (diamond technol), 84-; mem, NASA Space Sta Sci Users TASK FORCE, 84-89; chair, Semiconductor Res Corp, Exec Tech Adv Bd, 87-; mem, Space Studies Bd, Nat Res Coun, 89- *Mem:* AAAS; Am Phys Soc; Sigma Xi; Inst Elec & Electronic Engrs; Mat Res Soc. *Res:* Crystal growth and evaluation, especially influences of fluid convection and phase equilibria on crystal growth, space processing, optical fibers for communications; microelectronics. *Mailing Add:* 22633 Queens Oak Ct Cupertino CA 95014

CARRUTHERS, LUCY MARSTON, b Kansas City, Mo, Jan 20, 37; m 58, 69, 90; c 3. PHYSICAL CHEMISTRY, SCIENTIFIC COMPUTING. *Educ:* Swarthmore Col, BA, 58; Rutgers Univ, MS, 61, PhD(theoret chem), 64. *Prof Exp:* Programmer physics & eng, Mat Sci Ctr, Cornell Univ, 67-69; programmer astronomy, Hale Observ, Pasadena, Calif, 69-70; sr programmer chem, Cornell Univ, 70-73; staff mem weapons physics, TD-Div, Los Alamos Nat Lab, 73-75, staff mem mech & electronics eng, MEE-Div, 75-91; SPECIALIST RES COMPUT, UNIV ARIZ, 87- *Honors & Awards:* Achievement Award, Am Nuclear Soc, 87. *Mem:* Am Nuclear Soc; Asn Comput Mach; Asn Women Sci; Grad Women Sci. *Res:* Computer calculations and analysis, graphics and structural analysis; computing vectoritation, parallelization, and debugging. *Mailing Add:* 2220 E Camino Miraval Tucson AZ 85718

CARRUTHERS, PETER A, b Lafayette, Ind, Oct 7, 35; m 55, 69, 81, 90; c 3. THEORETICAL PHYSICS. *Educ:* Carnegie Inst Technol, BS & MS, 57; Cornell Univ, PhD(theoret physics), 61. *Prof Exp:* From asst prof to prof physics, Cornell Univ, 61-73; leader theoret div, Los Alamos Sci Lab, Univ Calif, 73-80, sr fel, Los Alamos, 80-86; PROF & HEAD, DEPT PHYSICS, UNIV ARIZ, 86- *Concurrent Pos:* NSF fel, 60-61; Sloan res fel, 63-65; vis assoc prof, Calif Inst Technol, 65, vis prof, 69-70 & 77-78; NSF sr fel, Univ Rome, 67-68; mem, Physics Adv Panel, NSF, 75-80, chmn, 78-80, mem, High Energy Physics Adv Panel, 78-; Panel on Pub Affairs, Am Phys Soc; consult, MacArthur Found, SRI Int, 76-81, 80-81 & 84-88, Inst Defense Anal, 85-; Alexander von Humboldt Sr Scientist, 86- *Mem:* Fel Am Phys Soc; fel AAAS; NY Acad Sci. *Res:* Theory of strong interactions of elementary particles; symmetries of elementary particles; quantum theory. *Mailing Add:* Dept Physics Univ Ariz Tucscon AZ 85721

CARRUTHERS, RAYMOND INGALLS, b Pomona, Calif, Jan 4, 51; m 71; c 3. EPIZOOTIOLOGY. *Educ:* Calif Polytech State Univ, BS, 75; Mich State Univ, MS, 79, PhD(entom), 81. *Prof Exp:* Res asst, Mich State Univ, 76-81; asst prof etom, Cornell Univ, 81-85; LEAD SCIENTIST/RES ETOMOLOGIST, USDA, AGR RES SERV, 85- *Honors & Awards:* Wooley Mem Award, 78; Driesbach Mem Award, 81. *Mem:* Entom Soc Am; Entom Soc Can; Soc Invert Path; AAAS. *Res:* Population ecology and systems science; integrated pest management; biological control using insect parasites and pathogens. *Mailing Add:* USDA-Agr Res Serv Fed Plant Soil & Nutrit Lab Cornell Univ Ithaca NY 14853

CARRUTHERS, SAMUEL GEORGE, b Londonderry, Northern Ireland, Sept 18, 45; Brit citizen & Can citizen; m 69; c 4. CARDIOVASCULAR PHARMACOLOGY, CARDIOVASCULAR THERAPEUTICS. *Educ:* Queens Univ, Belfast, MB, BCh, BAO, 69, MD, 75; FRCPS(C), 77; FRCP, 89. *Prof Exp:* Intern & registr, gen internal med, Royal Victoria Hosp, Belfast, 69-73; sr registrar/sr tutor, clin pharmacol, Belfast City Hosp, Royal Victoria Hosp, Queens Univ, Belfast, 73-75; res fel clin pharmacol, Kans Univ Med Ctr, 75-77; from asst prof to prof med & pharmacol, Univ Hosp, Univ Western Ont, 77-88; PROF & HEAD, DEPT MED, DALHOUSE UNIV, HALIFAX, NS, 88- *Concurrent Pos:* Mem, Drug Qual & Therapeut Comt, Ont Ministry Health, 82-86, chmn, 86-88; consult, var pharmaceut co, 78-86. *Honors & Awards:* Piaksky Young Award, Can Soc Clin Pharmacol, 82. *Mem:* Can Soc Clin Invest; Can Hypertension Soc (pres, 90-91); Can Soc Clin Pharmacol(pres,84-86); Brit Pharmacol Soc; Am Soc Clin Pharmacol & Therapeut; Can Soc Internal Med; fel Am Col Physicians. *Res:* Investigation of drug disposition, pharmacological, therapeutic and toxic effects in health and disease; the study of drug interaction; the relationship between drug concentrations and effects; choice of medications for individual patients and society. *Mailing Add:* Dept Med Ste 8-021 Victoria Gen Hosp 1278 Tower Rd Halifax NS B3H 2Y9 Can

CARRYER, HADDON MCCUTCHEN, b Unionville, Mo, Aug 25, 14; m 41; c 3. INTERNAL MEDICINE, ALLERGY. *Educ:* Drake Univ, BA, 35; Northwestern Univ, BM, 38, MS & MD, 39; Univ Minn, PhD(med), 48. *Prof Exp:* Asst med, Mayo Grad Sch, 42-43, from instr to assoc prof, 46-73, PROF MED, MAYO MED SCH, UNIV MINN, 73- *Concurrent Pos:* Consult, Div Med, Mayo Clin, 43- *Mem:* Fel Am Acad Allergy; fel Am Col Physicians; AMA. *Res:* Executive health periodic examinations. *Mailing Add:* 1125 Skyline Dr SW Rochester MN 55902

CARSKI, THEODORE ROBERT, b Baltimore, Md, June 22, 30; m 54; c 4. MEDICINE, MICROBIOLOGY. *Educ:* Johns Hopkins Univ, AB, 52; Univ Md, MD, 56; Am Bd Microbiol, dipl, 65. *Prof Exp:* Med intern, Univ Md, 56-57; sr asst surgeon, Commun Dis Ctr, Ala, 57-60; dir med res, Baltimore Biol Lab Div, B-D Labs, Inc, 60-68; dir microbiol, 68-70, from asst dir to dir, Huntingdon Res Ctr, 70-74, assoc med dir, 74-75, CORP MED DIR, BECTON, DICKINSON & CO, 75- *Mem:* Am Soc Microbiol. *Res:* Fluorescent antibody techniques; mycoplasma; medical immunology; bacteriology and virology; clinical research drugs and diagnostic reagents. *Mailing Add:* Becton Dickinson & Co Box 243 Hunt Valley MD 21030

CARSOLA, ALFRED JAMES, b Los Angeles, Calif, June 6, 19; m 47; c 8. OCEANOGRAPHY, MARINE GEOLOGY. *Educ:* Univ Calif, Los Angeles, AB, 42; Univ Southern Calif, MS, 47; Scripps Inst, Univ Calif, PhD(oceanog), 53. *Prof Exp:* Asst, Univ Southern Calif, 46-47; geophysicist & oceanographer, Electronics Lab, US Navy, 47-60; staff scientist, Lockheed-Calif Co, 60-67, head oceanics div, 67-71; environ res scientist, Southern Calif Coastal Res Proj, 71-72; chief staff scientist, Lockheed Ocean Lab, Lockheed Missiles & Space Co, 72-75; instr, San Diego Mesa & Grossmont Community Col, Univ San Diego & San Diego State Univ, 75-87; RETIRED. *Concurrent Pos:* Instr, San Diego State Col, 56-59, Loyola Univ, 63, Univ San Diego, 59, Univ Calif, Los Angeles, 60-61, 62 & Southern Calif Coastal Res Proj, 71-72. *Mem:* AAAS; Soc Econ Paleont & Mineral; Am Geophys Union; fel Geol Soc Am; Am Soc Limnol & Oceanog. *Res:* Marine sediments; arctic marine seafloor, bathymetry and geomorphology; Seamount sediments; nearshore physical marine processes; micropaleontology; physical oceanography and acoustical oceanography. *Mailing Add:* 3569 Addison St San Diego CA 92106

CARSON, BOBB, b Minneapolis, Minn, July 16, 43. SEDIMENTOLOGY. *Educ:* Carleton Col, BA, 65; Univ Wash, MS, 67, PhD(geol oceanog), 71. *Prof Exp:* from asst prof to assoc prof, 71-82, chmn, 85-91, PROF GEOL, LEHIGH UNIV, 82- *Concurrent Pos:* Consult, Off Marine Geol, US Geol Surv, Woods Hole, Mass, 75-76; sci staff, Deep-sea drilling proj, Leg 57, 77, Leg 136, 91. *Mem:* AAAS; Am Geophys Union; Geol Soc Am; Soc Econ Paleontologists & Mineralogists; Int Asn Sedimentologists. *Res:* Quaternary continental margin sedimentation, northwest pacific ocean; tectonic modification of deep-sea sediments related to subduction; agglomeration of fine-grained sediments. *Mailing Add:* Bldg 31 Dept Geol Sci Lehigh Univ Bethlehem PA 18015

CARSON, BONNIE L BACHERT, b Kansas City, Kans, Aug 11, 40; div; c 1. INFORMATION SCIENCE. *Educ:* Univ NH, BA, 63; Ore State Univ, MS, 66. *Prof Exp:* Teaching asst chem, Ore State Bd Higher Educ, 63-66; lab instr, Univ Waterloo, 68-69; asst abstractor, Chem Abstracts Serv, 69-71, Russian translator, 71-73; asst chemist, 73-75, assoc chemist, 75-80, SR INFO SCIENTIST, MIDWEST RES INST, 80- *Concurrent Pos:* Adminr, Sci-Tech Div, Am Translr Asn, 87-88. *Mem:* Am Chem Soc; NY Acad Sci; Soc Environ Geochem & Health; Am Translr Asn (treas, 89-91); Am Soc Info Sci. *Res:* Assessment and compilation of information on the magnitude of human exposure to, the health effects of, and the environmental fate of toxic chemicals. *Mailing Add:* Midwest Res Inst 425 Volker Blvd Kansas City MO 64110-2299

CARSON, CHESTER CARROL, b Passaic, NJ, Nov 21, 18; m 47; c 2. ANALYTICAL CHEMISTRY. *Educ:* Newark Col Eng, BS, 41; Rensselaer Polytech Inst, MS, 59. *Prof Exp:* Control chemist, Rayon Prod, Celanese Corp Am, Md, 41-42; chemist, Large Steam Turbine-Generator Dept, Gen Elec Co, 46-83; RETIRED. *Honors & Awards:* IR-100 Award, 81. *Mem:* Am Soc Testing & Mat; Am Chem Soc; fel Am Inst Chemists; Sigma Xi. *Res:* Sampling and analysis of atmospheric particulates at gas turbine sites; sampling and analysis of pyrolysates detected in large gas-cooled generators; measurement of hydrogen in steel, of hydrogen, oxygen and nitrogen in steel and metals; problems in gas systems; immediate detection of local overheating in gas-cooled electrical machines; problems in water-cooled systems of generator stators. *Mailing Add:* Currie Ct RD 8 Ballston Spa NY 12020-9112

CARSON, DANIEL DOUGLAS, b Philadelphia, Pa, Aug 7, 53; m; c 4. UTERINE PHYSIOLOGY, EXTRACELLULAR MATRIX INTERACTIONS. *Educ:* Univ Pa, BS, 75; Temple Univ, PhD(microbiol), 80. *Prof Exp:* Postdoctoral biochem, Johns Hopkins Sch Med, 80-83; asst prof, 83-88, ASSOC PROF, BIOCHEM & MOLECULAR BIOL, N D ANDERSON CANCER CTR, UNIV TEX, HOUSTON, 88- *Mem:* Am Soc Cell Biol; Am Chem Soc; Am Soc Biochem & Molecular Biol; Soc Develop Biol. *Res:* Biochemical aspects of implantation and early pregnancy; hormonal control of extracellular matrix and glyco protein expression; cell-cell communication. *Mailing Add:* Dept Biochem & Molecular Biol 1515 Holcombe Blvd MD Anderson Cancer Ctr Houston TX 77030

CARSON, DENNIS A, b New York, NY, May 31, 46; m 70; c 4. RHEUMATOLOGY, BIOCHEMISTRY. *Educ:* Haverford Col, BA, 66; Columbia Univ, MD, 70; Am Bd Internal Med, dipl, 73, dipl rheumatol, 75. *Prof Exp:* Intern & resident, Univ Hosp, San Diego, 70-72; clin assoc chem immunol, Nat Inst Arthritis, Metab & Digestive Dis, NIH, 72-74; from asst mem to mem clin immunol, Scripps Clin & Res Found, 76-89; asst rheumatol, 74-75, PROF MED & DIR, STEIN INST RES AGING, UNIV CALIF, SAN DIEGO, 90- *Concurrent Pos:* Assoc ed, J Arthritis & Rheumatism, 84-88, J Clin Investr, 89-90. *Honors & Awards:* Howley Prize, Arthritis Found, 87. *Mem:* Am Soc Clin Invest; Am Asn Immunologists; Am Asn Physicians. *Res:* Human autoimmune and immunodeficiency diseases, leukemia and lymphoma; arthritis. *Mailing Add:* Dept Med Univ Calif San Diego La Jolla CA 92093-0945

CARSON, EUGENE WATSON, JR, b Cumberland, Va, Mar 27, 39; m 60; c 3. AGRONOMY, PLANT BIOCHEMISTRY. *Educ:* Va Polytech Inst & State Univ, BS, 61, MS, 63; NC State Univ, PhD(soil sci), 66. *Prof Exp:* Va Coun Hwy Invest & Res asst turf & hwy, Va Polytech Inst & State Univ, 60-63; res asst plant physiol, NC State Univ, 63-66; assoc prof, 66-72, PROF AGRON, VA POLYTECH INST & STATE UNIV, 72- *Mem:* Am Soc Agron; Crop Sci Soc Am. *Res:* Plant ecology. *Mailing Add:* 1700 Pratt Dr Blacksburg VA 24060-0506

CARSON, FREDERICK WALLACE, b Quincy, Mass, Mar 18, 40; m 69. ORGANIC CHEMISTRY, BIOCHEMISTRY. *Educ:* Mass Inst Technol, BS, 61; Washington Univ, MA, 63; Univ Chicago, PhD(chem), 65. *Prof Exp:* NIH fel org chem, Princeton Univ, 65-66; asst prof org chem & biochem, Ind Univ, Bloomington, 66-70; asst prof, 70-71, ASSOC PROF ORG CHEM & BIOCHEM, AM UNIV, 72- *Concurrent Pos:* Grants, Washington Heart Asn, 72, NIMH, 73, US Dept Interior, 75, Petrol Res Fund, Biomed Sci & Res Corp. *Mem:* AAAS; Am Chem Soc; Royal Soc Chem; NY Acad Sci. *Res:* Biochemical mechanisms; model enzyme systems; enzyme kinetics; stereochemistry; organosulfur chemistry. *Mailing Add:* Dept of Chem American Univ Washington DC 20016-8003

CARSON, GEORGE STEPHEN, b Lakewood, Ohio, Dec 7, 48; m 69; c 2. MATHEMATICS. *Educ:* Univ Tenn, Knoxville, BS, 70; Univ Calif, Riverside, PhD(math), 75. *Prof Exp:* Lectr math, Calif State Col, San Bernardino, 75-76; mem tech staff, B-1 Div Rockwell, 76-77 & GTE Automatic Elec Labs, Northlake, Ill, 77-78; assoc prin engr, Elec Systs Div, Harris Corp, Melbourne, Fla, 78-81; PRES, GSC ASSOCS INC, REDONDO BEACH, CALIF, 81- *Mem:* Am Math Soc; Math Asn Am; AAAS; Sigma Xi; Inst Elec & Electronics Engrs; Asn Comput Mach. *Res:* Approximation theory; computer graphics; computer systems engineering; digital signal processing. *Mailing Add:* 13254 Jefferson Ave Hawthorne CA 90250

CARSON, GORDON B(LOOM), b High Bridge, NJ, Aug 1, 11; m 37; c 4. INDUSTRIAL & MECHANICAL ENGINEERING. *Educ:* Case Inst Technol, BS, 31; Yale Univ, MS, 32, ME, 38. *Hon Degrees:* DEng, Case Inst Technol, 57; LLD, Rio Grande Col, 73. *Prof Exp:* Instr mech eng, Case Inst Technol, 32-36; engr, Am Ship Bldg Co, 36; from asst prof to assoc prof indust eng, Case Inst Technol, 37-44; mgr eng, Selby, Selby Shoe Co, 44-49, secy of corp, 49-53; dean eng, Ohio State Univ, 53-58, prof indust eng & vpres, 58-71; prof mgt & exec vpres, Albion Col, 71-77; dir finance, Northwood Inst, 77-82; vpres Mich Molecular Inst, 82-88; PRIN, WHITFIELD ROBERT ASSOC, 88- *Concurrent Pos:* Dir res, Cleveland Automatic mach Co, 39-44; chmn tool & die comt, Fifth Regional War Labor Bd, 44-45; chmn, Ohio Selective serv Adv Bd & Tech Personnel, 55-70; vpres, Ohio State Univ Res Found & dir, Cardinal Fund, 67-; vpres, Ga Ctr Automation & Soc,Univ Ga, 69-71; dir, Accuray Corp 59-82. *Mem:* Fel AAAS; fel Am Soc Mech Engrs; Nat Soc Prof Engrs; Am Soc Eng Educ; fel Am Inst Indust Engrs (pres, 56-57). *Res:* Automation, especially as it relates to manufacturing processes; energy use reduction. *Mailing Add:* 5413 Gardenbrook Dr Midland MI 48640

CARSON, HAMPTON LAWRENCE, b Philadelphia, Pa, Nov 5, 14; m 37; c 2. EVOLUTIONARY BIOLOGY. *Educ:* Univ Pa, AB, 36, PhD(zool), 43. *Prof Exp:* Instr zool, Univ Pa, 38-42; from instr to prof, Wash Univ, 43-70; prof, 71-85, EMER PROF GENETICS, UNIV HAWAII, 85- *Concurrent Pos:* Mem Wheelock Exped, Labrador, 34; prof biol, Univ Sao Paulo, 51 & 77; Fulbright res scholar, Univ Melbourne, 61. *Honors & Awards:* Leidy Medal, 85. *Mem:* Nat Acad Sci; Soc Study Evolution (pres, 71); Am Soc Naturalists (pres, 73); AAAS; Am Acad Arts & Sci; Genetics Soc Am (pres, 81). *Res:* Population genetics; genetic systems and their relation to evolution; cytogenetics and evolution of drosophila and other insects. *Mailing Add:* Dept of Genetics Univ of Hawaii Honolulu HI 96822

CARSON, JOHN WILLIAM, b Tallahassee, Fla, Aug 22, 44; m 66; c 1. BULK SOLIDS FLOW. *Educ:* Northeastern Univ, BS, 67; Mass Inst Technol, SM, 68, PhD(mech eng), 71. *Prof Exp:* Res engr, 70-76, vpres, 76-85, PRES, JENIKE & JOHANSON INC, 85- *Mem:* Am Soc Mech Engrs; Am Inst Chem Engrs; Am Soc Civil Engrs. *Res:* Storage and flow of bulk solids; two phase flow; pressures on bin walls; controlled feeding of fine materials. *Mailing Add:* Jenike & Johanson Inc Two Executive Park Dr North Billerica MA 01862

CARSON, JOHNNY LEE, b Asheville, NC, Feb 6, 49; m 73. ENVIRONMENTAL BIOLOGY, CYTOPATHOLOGY. *Educ:* Western Carolina Univ, BS, 71; Univ NC, PhD(bot), 75. *Prof Exp:* RES ASSOC PATH, SCH MED, UNIV NC, CHAPEL HILL, 75-, AT DEPT PEDIAT, INFECTIOUS DIS DIV. *Mem:* Sigma Xi. *Res:* Cytopathological studies of effects of environmental pollutants in mammals; cytology-ecology of algae and fungi. *Mailing Add:* Dept Pediat Infectious Dis Div Univ NC Chapel Hill NC 27514

CARSON, KEITH ALAN, b Roswell, NMex, Sept 4, 52; div. NEUROANATOMY, NEUROCYTOLOGY. *Educ:* NC State Univ, BS, 73; Univ NC, Chapel Hill, PhD(neurobiol), 79. *Prof Exp:* Res fel pharmacol, Duke Univ Med Ctr, 79; res fel neuroanatomy, Harvard Med Sch, 80, instr, 80-81; asst prof, 81-86, ASSOC PROF BIOL SCI, OLD DOMINION UNIV, 86-, DIR, ELECTRON MICROS LAB, 81-, GRAD PROG DIR, BIOMED SCI PHD PROG, 85- *Concurrent Pos:* Asst prof anat & cell biol, Eastern Va Med Sch, 82-86, assoc prof, 86- *Mem:* Soc Neuroscience; Electron Micros Soc Am; Sigma Xi; Am Asn Anatomists. *Res:* Electron microscopy; neuroanatomy; neurocytology; neurohistochemistry. *Mailing Add:* Dept Biol Sci Old Dominion Univ Norfolk VA 23529-0266

CARSON, PAUL LANGFORD, b 42; US citizen; m; c 2. MEDICAL PHYSICS, ULTRASONIC RESEARCH. *Educ:* Colo Col, BS, 65; Univ Ariz, PhD(physics), 72. *Prof Exp:* From instr to assoc prof radiol, Med Ctr, Univ Colo, Denver, 71-81; assoc prof, 81-84, PROF RADIOL, MED SCH, UNIV MICH, 84- *Mem:* fel Am Asn Physicists in Med (pres, 87); fel Am Inst Ultrasound Med (vpres, 78-80); fel Acoust Soc Am. *Res:* Quantitative ultrasound and magnetic resonance imaging; medical imaging system performance and safety. *Mailing Add:* Dept Radiol Univ Mich Hosp Ann Arbor MI 48109-0553

CARSON, PAUL LLEWELLYN, b Ames, Iowa, Mar 27, 19; m 53; c 4. SOIL FERTILITY. *Educ:* Northwest Mo State Teachers Col, BS, 41; Iowa State Col, MS, 47. *Prof Exp:* Teacher pub schs, Mo, 41; agronomist, 48-78, PROF AGRON, S DAK STATE UNIV, 69- *Concurrent Pos:* Adv, Rockefeller Found, Colombia, SA, 67-68; consult, USAID, Pakistan, 76, Brazil, 78 & Botzwana, 81-82. *Mem:* Am Soc Agron; Soil Sci Soc Am. *Res:* Soil management; soil testing to determine fertilizers needed for farmers soils. *Mailing Add:* 808 Christine Ave No 203 Brookings SD 57006

CARSON, RALPH S(T CLAIR), b Durham, NC, Dec 22, 22; m 49; c 2. ELECTRICAL ENGINEERING. *Educ:* Ind Inst Tech, BS, 45; Univ Mich, MS, 52; Univ Ill, PhD(elec eng), 64. *Prof Exp:* Eng asst, Farnsworth Radio & TV Corp, 45-46; instr electronics, Ind Inst Tech, 46-51, chmn, Dept Electronic Eng, 52-60; from res asst to res assoc elec eng, Univ Ill, 61-64; assoc prof, 64-65, PROF ELEC ENG, UNIV MO-ROLLA, 65- *Mem:* Sr mem Inst Elec & Electronics Engrs; Am Soc Eng Educ. *Res:* Electronic devices and circuits. *Mailing Add:* Dept of Elec Eng Univ of Mo Rolla MO 65401

CARSON, ROBERT CLELAND, b Akron, Ohio, Mar 11, 24; m 51; c 4. MATHEMATICS. *Educ:* Purdue Univ, BS, 48, MS, 50; Univ Wis, PhD(math), 53. *Prof Exp:* Asst prof math & statist, Lehigh Univ, 53-57; asst prof math, Western Reserve Univ, 57-58; mathematician, Goodyear Aircraft Corp, Ohio, 58-63; coordr res, 63-71, asst dean grad studies, 68-71, ASSOC PROF MATH, UNIV AKRON, 63- *Mem:* Am Math Soc; Am Statist Asn; Soc Indust & Appl Math. *Res:* Variational methods and stochastic processes. *Mailing Add:* 1573 Maple St Barberton OH 44230

CARSON, ROBERT JAMES, III, b Lexington, Va. GEOLOGY, GEOMORPHOLOGY. *Educ:* Cornell Univ, AB, 63; Tulane Univ, MS, 67; Univ Wash, PhD(geol), 70. *Prof Exp:* Geologist, Texaco Inc, La, 63-67; from asst to assoc prof geol, NC State Univ, 70-74; asst prof, 75-80, ASSOC PROF GEOL & CHMN DEPT, WHITMAN COL, 80- *Concurrent Pos:* Consult, Wash State Dept Ecol & Div Geol & Earth Resources, 69- *Mem:* AAAS; Geol Soc Am; Int Glaciol Soc; Am Quaternary Asn; Sigma Xi. *Res:* Quaternary and environmental geology of the Olympic Peninsula, Washington. *Mailing Add:* Dept Geol Whitman Col Walla Walla WA 99362

CARSON, STEVEN, pharmacology; deceased, see previous edition for last biography

CARSON, STEVEN DOUGLAS, b Bartlesville, Okla, Apr 9, 51; div; c 1. PROTEIN CHEMISTRY, BLOOD COAGULATION. *Educ:* Rice Univ, BA, 73; Univ Tex Med Br, Galveston, PhD(human genetics), 78. *Prof Exp:* Welch undergrad fel, Rice Univ, 70-72; student asst chem, US Environ Protection Agency, 72-73; res technician, Univ Tex Med Br Galveston, 73-78; fel, dept molecular biophysics & biochem, Sch Med, Yale Univ, 78-82, res assoc & lectr, 82; asst prof path, Univ Colo, 82-88; ASSOC PROF PATH & MICROBIOL, UNIV NEBR, 88- *Concurrent Pos:* Fel, NSF, 78-79 & NIH, 79-82. *Mem:* AAAS; Am Soc Human Genetics; NY Acad Sci; Am Soc Biochem & Molecular Biol; Am Heart Asn. *Res:* Protein function and biological regulation such as human biochemical genetics and membrane biochemistry; plasma proteins including clotting factors and lipoproteins, lipid vesicles, tissue factor and monoclonal antibodies; blood coagulation. *Mailing Add:* Dept Path & Microbiol Sch Med Univ Nebr 600 S 42nd St Omaha NE 68198-6495

CARSON, VIRGINIA ROSALIE GOTTSCHALL, b Pittsburgh, Pa; m 60; c 2. PSYCHOPHARMACOLOGY. *Educ:* Calif State Univ, Los Angeles, BA, 60, MA, 65; Univ Calif, Los Angeles, PhD(physiol), 70. *Prof Exp:* Asst clin res chemist, Don Baxter, Inc, 60-64; ment health trainee, Brain Res Inst, Univ Calif, Los Angeles, 65-69, trainee, dept physiol, 69, res trainee, dept pharmacol, 70-71; from asst prof to assoc prof, 71-83, PROF BIOL & CHMN, DIV NATURAL SCI, CHAPMAN COL, 83- *Concurrent Pos:* NIH fel, dept pharmacol, Univ Calif, Los Angeles, 72-74; asst res pharmacologist, dept psychiat & human behav, Univ Calif, Irvine, 72-81, assoc res pharmacologist, 81-; premed adv, Chapman Col, 74-; assoc prof, Southern Calif Col Optom, 79-83. *Mem:* AAAS; Am Psychol Asn; Grad Women Sci; Am Asn Pharmacol & Exp Therapeut; Soc Neuroscience. *Res:* The mechanism of tolerance of the cholinergic system to organophosphates; physiological, psychological and biochemical effects of ethanol. *Mailing Add:* Chapman Col 333 N Glassell St Orange CA 92666

CARSTEA, DUMITRU, b Comuna Paduroiu, Romania, Mar 22, 30; US citizen; m 56; c 4. SOIL CHEMISTRY. *Educ:* Agr Inst, Bucharest, BS, 54; Ore State Univ, MS, 65, PhD(soil chem & clay mineral), 67. *Prof Exp:* Res scientist, Romanian Acad Sci, 54-60; res asst clay mineral & soil chem, Ore State Univ, 61-66; res scientist, proj leader, Can Dept Agr, 66-67; soil scientist, chemist, & hydrologist, US Geol Surv, 67-68, res hydrologist & proj leader, 68-70; environ tech staff & task leader, 74-76, group leader, Mitre Corp, Washington Opers, 76-78; prog mgr & dept head, Environ Sci & Eng, Hittman Corp, 78-80; mgr, environ eng & proj mgr, 80-83, SR CONSULT ENVIRON, ENERGY, NATURAL RESOURCES & PLANNING, SYST DEVELOP CO 83- *Concurrent Pos:* Consult environ & natural resource areas, 76-78; mem Coord Soil Conf, FAO, UN. *Mem:* Am Chem Soc; Int Soc Soil Sci; AAAS; Nat Geophys Soc; Soil Sci Soc Am; NY Acad Sci; Sigma Xi. *Res:* Environmental and energy analysis of coal gasification and coal liquefaction technologies; environmental analysis, regulation, planning and engineering; erosion and sedimentation; water quality and wastewater treatment; soil and hydrological studies; waste management; clay mineralogy; peat development; environmental chemistry and toxicology; land development planning. *Mailing Add:* 13563 Point Pleasant Dr Chantilly VA 22021

CARSTEN, ARLAND L, b Hasting, Minn, Apr 17, 30. RADIOBIOLOGY, HEALTH PHYSICS. *Educ:* Mankato State Col, BSc, 53, MS, 56; Univ Rochester, PhD(biol), 57. *Prof Exp:* From asst to sr assoc radiation biol, Univ Rochester, 55-64; assoc health physicist, 57-62, vis assoc biologist, 62-64, from assoc scientist to scientist, 66-84, SR SCIENTIST, MED RES CTR, BROOKHAVEN NAT LAB, 84-; 73-84, RES PROF PATH, STATE UNIV NY, STONY BROOK, 84 - *Concurrent Pos:* Res fel, Royal Dent Col, Copenhagen, Denmark, 65-66; health physics fel adv, AEC, 60-62; lectr, Am Inst Biol Sci Prog, Med Educ Nat Defense, 61-63; res assoc neurol, Columbia Univ, 64-72; res assoc, Lerner Marine Lab, Bimini, Bahamas; assoc prof, State Univ NY, Stony Brook, 73-84. *Mem:* AAAS; Radiation Res Soc; Health Physics Soc; Am Soc Hemat; Int Soc Exp Hemat; Bioelectromagnetic Soc. *Res:* Acute and late effects of ionizing radiation on mammals, particularly genetic and effects on hematopoietic and nervous tissues; protection against and recovery from radiation injury to man. *Mailing Add:* Bldg 703M Brookhaven Nat Lab Upton NY 11973

CARSTEN, MARY E, b Berlin, Germany, Mar 2, 22; nat US; m 64. SMOOTH MUSCLE, UTERINE PHYSIOLOGY. *Educ:* NY Univ, AB, 46, MS, 48, PhD(biochem), 51. *Prof Exp:* Instr, NY Univ, 52-53; res assoc dept microbiol, Col Physicians & Surg, Columbia Univ, 53-55; asst res physiol chemist, Dept Physiol Chem, 56-61, assoc res biochemist, Depts Biol Chem & Med, 61-63, assoc prof, Depts Physiol, Obstet & Gynec, 63-70, PROF OBSTET & GYNEC, SCH MED, UNIV CALIF, LOS ANGELES, 70- *Concurrent Pos:* Nat Found Infantile Paralysis fel, 54-55; Am Cancer Soc fel, Univ Calif, Los Angeles, 55-57; estab investr, Los Angeles County Heart Asn, 61-64; USPHS res career develop award, 64-69 & 69-74. *Honors & Awards:* Res Award, Los Angeles County Heart Asn, 62, 63, 64. *Mem:* Hon mem Soc Gynecol Invest; Am Soc Biol Chem; Am Chem Soc; NY Acad Sci; Am Physiol Soc; Sigma Xi. *Res:* Ion exchange chromatography; amino acids; protein chemistry; immunochemistry; skeletal, heart and smooth muscle proteins and calcium transport; prostaglandins; myometrial and uterine physiology; calcium control mechanisms in the myometrial cell and the role of phosphoinositol metabolites as second messengers; mechanism of calcium transport in sarcoplasmic reticulum and effects of hormones on calcium transport. *Mailing Add:* 624 N Highland Ave Los Angeles CA 90036

CARSTENS, ALLAN MATLOCK, b Aurora, Ill, Jan 14, 39; m 60; c 3. OPERATIONS RESEARCH. *Educ:* Univ NMex, BS, 61, MS, 63, PhD, 67. *Prof Exp:* From instr to asst prof math, Wash State Univ, 66-70; from asst prof to assoc prof, Mankato State Col, 70-73; sr programmer, 73-76, supv math program, 76-80, mgr sci appl 80-84, MGR DISTRIB DBS DEU, SPERRY CORP, 84- *Concurrent Pos:* Community fac mem, Metro State Univ, Minneapolis, 76-83. *Mem:* Asn Comput Mach; AAAS; Am Math Soc; Math Asn Am; Inst Mgt Sci; Opers Res Soc Am. *Res:* Generalized topological spaces; convergence algebras; products of pretopologies; structures of the lattices of pretopologies, pseudotopologies, limitierungen. *Mailing Add:* 748 NE Ballantyne Lane NE Minneapolis MN 55432

CARSTENS, EARL E, b Rochester, NY, March 26, 50; m 75; c 1. NEUROPHYSIOLOGY, NEUROSCIENCE. *Educ:* Cornell Univ, BS, 72; Univ NC, PhD(neurobiol), 77. *Prof Exp:* Fel, Univ Heidelberg, Ger, 77-80; ASST PROF PHYSIOL, UNIV CALIF, DAVIS, 80- *Mem:* Soc Neuroscience. *Res:* Neurophysiological investigations of neural systems in brain which modulate spinal neurons transmitting information on pain. *Mailing Add:* Dept Animal Physiol Briggs Hall Univ Calif Davis CA 95616

CARSTENS, ERIC BRUCE, b Lethbridge, Alta; m 79. ANIMAL VIROLOGY, INSECT VIROLOGY. *Educ:* Univ Alta, BSc, 68, MSc, 74; Univ Sherbrooke, PhD(virol), 77. *Prof Exp:* Fel virol, Inst Genetics, Univ Cologne, Ger, 77-79; ASSOC PROF VIROL, DEPT MICROBIOL & IMMUNOL, QUEENS UNIV, ONT, 80- *Concurrent Pos:* Alexander von Humboldt res fel. *Mem:* Can Soc Microbiol; Micros Soc Can; Am Soc Microbiol; Am Soc Virol; Can Col Microbiologists. *Res:* Gene structure and function of insect viruses particularly baculoviruses; mode of replication of these viruses so that their genes may be favorably altered leading to better pest control agents. *Mailing Add:* Dept Microbiol & Immunol Queens Univ Kingston ON K7L 3N6 Can

CARSTENS, JOHN C, b Chicago, Ill, Oct 8, 37; m 67. PHYSICS. *Educ:* Monmouth Col, Ill, AB, 59; Univ Mo-Rolla, PhD(physics), 66. *Prof Exp:* Fel, Univ Mo-Rolla, 66-67; asst prof physics, Western Ill Univ, 67-68; PROF PHYSICS & ACTG DIR, CLOUD PHYSICS CTR, UNIV MO-ROLLA, 68- *Concurrent Pos:* Resident res assoc, Argonne Lab, 66-67. *Mem:* Am Meteorol Soc; Am Geophys Union; Am Asn Aerosol Res; Sigma Xi; Am Asn Physics Teachers. *Res:* Mass and heat transport problems in cloud physics; droplet growth. *Mailing Add:* Dept of Physics Norwood Hall Univ of Mo Rolla MO 65401

CARSTENS, ROBERT L(OWELL), transportation engineering, for more information see previous edition

CARSTENSEN, EDWIN L(ORENZ), b Oakdale, Nebr, Dec 8, 19; m 47; c 5. BIOMEDICAL ULTRASOUND, BIOELECTRIC PHENOMENA. *Educ:* Nebr State Teachers Col, BS, 41; Case Western Reserve Univ, MS, 47; Univ Pa, PhD(physics), 55. *Prof Exp:* Asst, Case Western Reserve Univ, 41-42; mem sci staff, Div War Res, Columbia Univ, 42-45, head lab sect, Navy Under-Water Sound Ref Lab, 45-48; res assoc & asst prof, Moore Sch Elec Eng, Univ Pa, 48-55; biophysicist, Ft Detrick, Md, 56-61; from assoc prof to prof elec eng, 61-88, prof biophysics, 81-90, Alfred Gould Yates prof eng, 88-90, SR SCIENTIST, EMER YATES PROF ENG, UNIV ROCHESTER, 90- *Concurrent Pos:* Dir, Biomed Eng, 75-83, Rochester Ctr Biomed Ultrasound, 86- *Honors & Awards:* Holmes Pioneer Award, Am Inst Ultrasound Med Biol, 91. *Mem:* Nat Acad Eng,; fel Inst Elec & Electronics Engrs; Biophys Soc; Biomed Eng Soc; fel Am Inst Ultrasound Med; fel Acoustical Soc Am. *Res:* Ultrasonic and dielectric properties of biological materials; biological effects of ultrasonic and electric fields. *Mailing Add:* Dept Elec Eng Univ Rochester Rochester NY 14627

CARSTENSEN, JENS T(HUROE), b Brooklyn, NY, Jan 9, 26; m 46; c 3. CHEMICAL ENGINEERING. *Educ:* Tech Univ Denmark, MS, 50; Stevens Inst Technol, MS, 64, PhD(phys chem), 67. *Prof Exp:* Develop chemist pharmaceut chem, Lederle Labs, Am Cyanamid Co, 50-53, formulation chemist, 53-55, dept head, Process Improv Lab, 55-57, tech asst to pharmaceut prod mgr, 57-59, dept head encapsulation, 59-60; group leader, Pharmaceut Res Dept, Hoffmann-La Roche, Inc, 60-67; assoc prof pharm, 67-72, PROF PHARM, UNIV WIS-MADISON, 72- *Concurrent Pos:* Assoc prof, Fac Pharm, Univ Paris-Sud, 77-78. *Honors & Awards:* Ebert Award, Acad Pharmaceut Sci, 76, Res Achievement Award, 77; Gent Res Award, Belgium, 78; IPT Award, Acad Pharm Sci, 79. *Mem:* AAAS; NY Acad Sci; Am Pharmaceut Asn; Am Chem Soc. *Res:* Physical pharmacy; kinetics; diffusion controlled processes; tableting and encapsulation operations. *Mailing Add:* Dept Pharm Univ Wis 1150 University Ave Madison WI 53706

CARSWELL, ALLAN IAN, b Toronto, Ont, Oct 4, 33; m 56; c 3. OPTICAL PHYSICS. *Educ:* Univ Toronto, BApplSci, 56, MA, 57, PhD(physics), 60. *Prof Exp:* Nat Res Coun Can fel, Inst Theoret Physics, Amsterdam, 60-61; sr mem sci staff plasma physics, RCA Victor Co, 61-65, dir, Optical & Microwave Physics Lab, 65-68; dir grad prog physics, 71-80, PROF PHYSICS, YORK UNIV, 68-; PRES, OPTECH INC, 74- *Mem:* Fel Royal Soc Can; Can Asn Physicists; Optical Soc Am; Asn Prof Engrs; fel Can Aeronaut & Spaces Inst. *Res:* Laser systems and applications; atmospheric optics; lidar; light scattering. *Mailing Add:* Dept of Physics York Univ Toronto ON M3J 1P3 Can

CART, ELDRED NOLEN, JR, b Irvington, Ky, Jan 6, 34; m 56; c 3. CHEMICAL ENGINEERING. *Educ:* Univ Louisville, BChE, 57; Ohio State Univ, MS, 60. *Prof Exp:* Sect head, Esso Res Labs, Humble Oil & Refining, 60-69; sr staff anal, Corp Planning Dept, Exxon USA, 69-71; sr staff adv, Logistics Dept, Esso 71-76; planning mgr, Govt Res Lab, 76-79, planning mgr, Contract Res Office, 79-81, SR STAFF ADVR, NEW FACIL PROJ, EXXON RES & ENG CO, 81- *Mem:* Am Inst Chem Eng. *Res:* Systems study on future fuels and prime movers for aircraft, marine, railroad, and pipelines; use of coal on residential and commercial sectors. *Mailing Add:* Exxon Res & Eng Co 180 Park Ave Florham Park NJ 07932

CARTER, ANNE COHEN, b New York, NY, Nov 27, 19; m 47; c 2. ENDOCRINOLOGY, BREAST ONCOLOGY. *Educ:* Wellesley Col, BA, 41; Cornell Univ, MD, 44; Am Bd Internal Med, dipl, 58. *Prof Exp:* From instr to asst prof med, Med Col, Cornell Univ, 46-55; from asst prof to prof med, 55-82, VIS PROF, STATE UNIV NY DOWNSTATE MED CTR, 82-; PROF MED, NY MED COL, 82- *Concurrent Pos:* Fel, Russell Sage Inst Path, 51-55; mem, Cancer Clin Training Comt, Nat Cancer Inst, 71-74, mem, Cancer Control Treatment & Rehab Rev Comt, 74-76 & Clin Cancer Prog, Proj Rev Comt, 76-80. *Mem:* Fel AAAS; Soc Exp Biol & Med; Endocrine Soc; Am Soc Clin Oncol; Am Diabetes Asn; Am Fedn Clin Res. *Res:* Endocrinology and metabolism. *Mailing Add:* NY Med Col Dept Med Munger Pavilion Valhalla NY 10595

CARTER, ASHLEY HALE, b Glen Ridge, NJ, June 27, 24; m 72; c 3. ACOUSTICS. *Educ:* Harvard Univ, AB, 45; Brown Univ, ScM, 50, PhD(physics), 63. *Prof Exp:* Res assoc underwater explosions, Woods Hole Oceanog Inst, 46-47; underwater acoustics, res anal group, Brown Univ, 51-53; mem tech staff, 53-65, head eng mech & physics dept, 65-71, head elec protection & interference dept, 71-83, head Ocean Systs Eng Dept, 83-88, RES COORDR, NAVAL SYSTS DIV, AT&T BELL LABS, 88- *Concurrent Pos:* Lectr, Fairleigh Dickinson Univ, 72-75; lectr, Drew Univ, 75-76, adj assoc prof, 76-81, adj prof, 81- *Mem:* Am Phys Soc; Acoust Soc Am. *Res:* Underwater acoustics; propagation of waves in inhomogeneous media; signal processing; electromagnetic interference and protection; mathematical physics. *Mailing Add:* Navy Systs Div Rm 14A431 AT&T Bell Labs Whippany NJ 07981

CARTER, ASHTON BALDWIN, b Philadelphia, Pa, Sept 24, 54; m 83; c 2. DEFENSE & ARMS CONTROL POLICY, TECHNOLOGY POLICY. *Educ:* Yale Univ, BA(medieval hist) & BA(physics), 76; Oxford Univ, PhD(theoret physics), 79. *Prof Exp:* Analyst, Off Technol Assessment, 80-81, Off Secy Defense, 81-82; res fel, Mass Inst Technol, 82-84; from asst prof to assoc prof, 84-88, PROF PUB POLICY, J F KENNEDY SCH GOVT, HARVARD UNIV, 88-, DIR, CTR SCI & INT AFFAIRS, 90- *Concurrent Pos:* Consult, Cong Off Technol Assessment, 82-; mem, Comt Nat Security, Sci & Arms Control, AAAS, 86-, Comt Int Security & Arms Control, Nat Acad Sci, 90-; adj prof, Carnegie-Mellon Univ, 86-; assoc dir, Ctr Sci & Int Affairs, J F Kennedy Sch Govt, Harvard Univ, 88-90; trustee, Mitre Corp, 89-; mem adv comt, Carnegie Comn Sci, Technol & Govt, 89-; mem, Defense Sci Bd, 91-, Panel Nat Security, President's Coun Adv Sci & Technol, 91- *Honors & Awards:* Forum Award, Am Phys Soc, 88. *Mem:* Am Phys Soc; Int Inst Strategic Studies. *Res:* Nuclear weapons and arms control; missile defense, command and control; technology policy. *Mailing Add:* J F Kennedy Sch Govt Harvard Univ 79 J F Kennedy St Cambridge MA 02138

CARTER, BARRIE J, b Dunedin, NZ, Oct 26, 44. CELLULAR BIOLOGY. *Educ:* Univ Otago, PhD, 69. *Prof Exp:* CHIEF, LAB MOLECULAR & CELL BIOL, NAT INST DIABETES, DIGESTIVE & KIDNEY DIS, NIH, 85- *Mem:* Am Soc Microbiol; Am Soc Virol; Am Soc Biochem & Molecular Biol. *Mailing Add:* Lab Molecular & Cellular Biol Nat Inst Diabetes Digestive & Kidney Dis NIH Bldg 8 Rm 301 Bethesda MD 20892

CARTER, CAROL SUE, b San Francisco, Calif, Dec 25, 44; m 70; c 2. ETHOLOGY, REPRODUCTIVE BIOLOGY. *Educ:* Drury Col, BA, 66; Univ Ark, Fayetteville, PhD(zool), 69. *Prof Exp:* Instr biol, Drury Col, BA, 66; NIH trainee, Mich State Univ, 69-70; NIH fel, WVa Univ, 70-71; adj asst prof biol, 70-72; res fel psychopharmacol, Ill Dept Ment Health, 72-74; from asst prof to prof, Dept Psychol, Ecol, Ethol & Evolution, Univ ILL, Champaign 74-85; PROF DEPT ZOOL, UNIV MD, COLLEGE PARK, 85- *Concurrent Pos:* NSF & NIH res grants. *Mem:* Fel AAAS; Soc Neuroscience; Animal Behav Soc; Sigma Xi. *Res:* Mechanisms regulating mammalian reproductive behavior; utilization of endocrine, neuroendocrine and pharmacological techniques to study the physiological basis of behavior; stress & behavior. *Mailing Add:* Dept Zool Rm 42222 Univ Md College Park MD 20742

CARTER, CAROLYN SUE, b Prestonsburg, Ky, May 5, 59. CHEMICAL EDUCATION, SOCIAL CULTURAL ISSUES & SCIENCE EDUCATION. *Educ:* Western Ky Univ, BS, 81; Purdue Univ, MS, 84, PhD(sci educ), 87. *Prof Exp:* ASST PROF SCI EDUC & CHEM, OHIO STATE UNIV, 87- *Concurrent Pos:* Dir, Nat Ctr Sci Teaching & Learning, 90-; consult, Switch Sci Prog, Worthington Sch Dist, 90-; co-prin investr, Dwight D Eisenhower Math & Sci Educ Prog, 90-91. *Mem:* AAAS; Am Chem Soc; Am Educ Res Asn; Nat Asn Res Sci Teaching; Nat Sci Teachers Asn; Sigma Xi. *Res:* Social and cultural factors in science education and how these factors interact to create the contexts in which science teaching, learning and participation take place; beliefs, goals, values and science curriculum. *Mailing Add:* 249 Arps Hall 1945 N High St Columbus OH 43210

CARTER, CHARLEATA A, b Asheville, NC, Dec 6, 60. CANCER BIOLOGY, CELL BIOLOGY. *Educ:* Mars Hill Col, BS, 81; Appalachian State Univ, MA, 82; Clemson Univ, PhD(cell & develop biol), 88. *Prof Exp:* Fel, Univ NC, Chapel Hill, 88-91; IRTA FEL, NAT INST ENVIRON HEALTH SCI, 91- *Mem:* AAAS; Am Asn Cancer Res; Am Soc Zoologists; Metastasis Res Soc; Sigma Xi. *Res:* The involvement of extracellular matrix glycoproteins, cytoskeletal proteins and oncogenes in cellular transformation and metastasis; cancer biology; confocal laser scanning microscopy; electron microscopy; early development in annual fish. *Mailing Add:* Nat Inst Environ Health Sci PO Box 12233 Research Triangle Park NC 27709

CARTER, CHARLES CONRAD, b Seattle, Wash, July 20, 24; m 48; c 4. NEUROLOGY. *Educ:* Reed Col, BA, 46; Univ Ore, MD, 48. *Prof Exp:* Intern, Good Samaritan Hosp, 48-49; resident neurol, Med Sch, Wash Univ, 54-56; from asst prof to prof neurol, Health Sci Ctr, Univ Ore, 62-79, actg head div, 74-75. *Mem:* AMA; Am Acad Neurol. *Res:* Cerebral blood flow. *Mailing Add:* 4238 SW Woodside Circle Lake Oswego OR 97035

CARTER, CHARLES EDWARD, biochemistry; deceased, see previous edition for last biography

CARTER, CHARLES WILLIAMS, JR, b Montpelier, Vt, Nov 25, 45; m 68; c 2. PROTEIN CRYSTALLOGRAPHY. *Educ:* Yale Univ, BA, 67; Univ Calif, San Diego, MS, 68, PhD(biol), 72. *Prof Exp:* Fel chem, Univ Calif, San Diego, 72-73; vis scientist, Med Res Coun Lab Molecular Biol, Cambridge, Eng, 73-74; asst prof, 74-80, ASSOC PROF BIOCHEM & ANAT, UNIV NC, 80-, PROF BIOCHEM & BIOPHYS, 89- *Concurrent Pos:* Prin investr, NIH, 75-, Nat Sci Found, 76-78; Fogarty Sr Int fel, Lure Orsay, France, 86. *Mem:* Biophys Soc; Am Soc Biol Chemists; Am Crystallographic Soc; Am Cancer Soc. *Res:* Structure determination by x-ray crystallography of electron transport proteins and macromolecules responsible for incorporating tryptophan into proteins, including; tryptophanyl-tRNA synthetase and tRNA tryptophan; enzymes of pyrimidine metabolism. *Mailing Add:* Dept Biochem & Biophysics CB No 7260 Univ NC Chapel Hill NC 27599-7260

CARTER, CHARLOTTE, b Birmingham, Ala, Apr 7, 43. GENERAL ANATOMY, SCIENCE EDUCATION. *Educ:* Miles Col, BA, 69; Univ Tenn, MS, 70; Purdue Univ, PhD(biol ed), 83. *Prof Exp:* Instr life sci, Miles Col, 70-80, dean acad affairs, 86-88; teaching asst biol ed, Purdue Univ, 80-81, res assist, 81-83; asst prof biol, 83-86, DIV CHMN MATH & SCI, STILLMAN COL, 88- *Concurrent Pos:* Proposal evaluator, NSF, 80, Curric Mat Develop, 86, Young Scholars Prog, 87 & 88; vis team, Southern Asn Col & Sch, 91. *Mem:* Nat Asn Biol Teachers; Nat Asn Res Sci Teaching; Nat Minority Health Affairs Asn. *Mailing Add:* 4527 18th Ave E No 1308 Tuscaloosa AL 35405

CARTER, CLINT EARL, b Durant, Okla, Apr 28, 41; m 62; c 2. IMMUNOLOGY, PARASITOLOGY. *Educ:* La Sierra Col, BA, 65; Loma Linda Univ, MA, 67; Univ Calif, Los Angeles, PhD(zool), 71. *Prof Exp:* Teaching assoc microbiol, San Bernardino Valley Jr Col, 66-67; teaching asst gen biol & cell & comp physiol, Univ Calif, Los Angeles, 67-68; teaching assoc fel gen zool & cell physiol, Univ Mass, 71-72; from asst prof to assoc prof, 72-86, PROF BIOL, VANDERBILT UNIV, 86- *Mem:* Am Soc Parasitologists; AAAS; Sigma Xi. *Res:* Immunoparasitology and parasite physiology. *Mailing Add:* Dept Biol Vanderbilt Univ Nashville TN 37240

CARTER, DAVID, physics, for more information see previous edition

CARTER, DAVID L, solid state science; deceased, see previous edition for last biography

CARTER, DAVID LAVERE, b Tremonton, Utah, June 10, 33; m 53; c 3. WATER QUALITY. *Educ:* Utah State Univ, BS, 55, MS, 57; Ore State Univ, PhD(soil sci), 61. *Prof Exp:* Soil scientist, Soil & Water Conserv Res Div, 56-60, res soil scientist & line proj leader, 60-65, res soil scientist, 65-71, SUPVRY SOIL SCIENTIST & RES LEADER, AGR RES SERV, USDA, 71- *Honors & Awards:* Soil Conserv Soc Am 50th Ann Award, 85. *Mem:* AAAS; fel Am Soc Agron; fel Soil Sci Soc Am; Int Soc Soil Sci; Soil Conserv Soc Am; Sigma Xi. *Res:* Salt and ion movement through soils; erosion and sediment control on irrigated land; irrigation and drainage water quality; conservation tillage on furrow irrigated land. *Mailing Add:* 523 Eastridge Dr Kimberly ID 83341

CARTER, DAVID MARTIN, b Doniphan, Mo, June 10, 36; m 61; c 3. DERMATOLOGY. *Educ:* Dartmouth Col, AB, 58; Harvard Med Sch, MD, 61; Yale Univ, PhD(biol), 71. *Prof Exp:* Intern med & surg, Med Ctr, Univ Rochester, 61-62, asst resident med, 62-63; surgeon, Venereal Dis Br, Commun Dis Ctr, USPHS, 63-65; resident dermat, Hosp Univ Pa, 65-67; asst prof, 70-73, assoc prof, 73-77, PROF DERMAT, SCH MED, YALE UNIV, 77- *Concurrent Pos:* Med investr, Howard Hughes Med Inst, 70-77. *Mem:* Am Acad Dermat; Soc Invest Dermat; Am Dermat Asn. *Res:* Defenses of cutaneous cells against ultraviolet irradiation. *Mailing Add:* 24 Maple St New Haven CT 06511

CARTER, DAVID SOUTHARD, b Victoria, BC, Mar 25, 26; US citizen; m 49; c 4. MATHEMATICS. *Educ:* Univ BC, BA, 46, MA, 48; Princeton Univ, PhD(math physics), 52. *Prof Exp:* Asst math physics, Princeton Univ, 49-52; mem staff, Los Alamos Sci Lab, Calif, 52-58; NSF res grant math & instr math sci, NY Univ, 57-58; vis assoc prof, Univ Wash, 58 & Univ Calif, Berkeley, 59-61; actg chmn dept, 69-80, PROF MATH, ORE STATE UNIV, 61- *Concurrent Pos:* Consult, Lockheed Missiles & Space Co, 59- *Mem:* Am Math Soc; Sigma Xi. *Res:* Applied analysis; hydrodynamics; statistical mechanics; control and stability theory; numerical analysis and computation. *Mailing Add:* Dept of Math Ore State Univ Corvallis OR 97331

CARTER, DON E, b Norborne, Mo, Jan 6, 26; m 66; c 2. POLYMER TECHNOLOGY, SEPARATION PROCESSES. *Educ:* Univ Mo, BS, 49; NC State Col, MS, 51; Wash Univ, DSc(chem eng), 64. *Prof Exp:* Res engr, 51-59, proj leader, 60-63, sr group leader, 63-69, prin engr, Corp Eng Dept, 69-74, Monsanto fel, 74-82, SR ENG FEL, MONSANTO CO, 82- *Concurrent Pos:* Lectr, Wash Univ, 66-70. *Mem:* Am Inst Chem Engrs; Soc Plastics Engrs; Polymer Processing Inst. *Res:* Polymer process development; design and applications of polymer process machinery; polymer applications; melt crystallization; analytical instrumentation; methods for commercial project evaluation; separations. *Mailing Add:* MCCEng Dept Monsanto Co 800 N Lindbergh Blvd St Louis MO 63167

CARTER, EARL THOMAS, b Baltimore, Md, July 7, 22; m 47; c 3. PHYSIOLOGY. *Educ:* Northwestern Univ, BS, 34, MD, 48, MS, 50; Univ Tex, PhD(physiol), 55. *Prof Exp:* Staff physician, Chicago Munic Tuberc Sanitarium, 50-51; third yr resident med, Chicago Wesley Mem Hosp, 55-56; asst prof physiol & prev med, Ohio State Univ, 56-60; asst prof med, 60-73, PROF PREV MED, MAYO CLIN & MAYO GRAD SCH MED & CHMN DIV, CLIN, 73- *Concurrent Pos:* Staff physician, Univ Hosp & Ohio Tuberc Hosp, 56-60; consult, Mayo Clin, 60- *Mem:* Sigma Xi. *Res:* Environmental medicine and environmental physiology. *Mailing Add:* Preventive Med Div Mayo Clinic Rochester MN 55905

CARTER, EDWARD ALBERT, b New Haven, Conn, Mar 1, 45; m 68; c 1. MACROPHAGE FUNCTION, DRUG & ETHANOL METABOLISM. *Educ:* Sussex, Eng, BSc, 65, PhD(biochem), 71. *Prof Exp:* Assoc med biochem, Harvard Med Sch, 72-84; asst biochem, 72-84, ASSOC BIOCHEMIST, MASS GEN HOSP, 84-; ASSOC PEDIAT BIOCHEM, HARVARD MED SCH, 84-; ASSOC BIOCHEMIST, SHRINER BURNS INST, 84- *Concurrent Pos:* Consult, Danon Electronics, 81. *Mem:* Soc Complex Carbohydrates; Int Soc Biochem Res Alcoholism; Am Burn Asn; Am Asn Study Liver Dis; Am Inst Nutrit; Am Gastroenterol Asn. *Res:* Enterohepatic mucosal immunology; biochemistry; molecular biology. *Mailing Add:* Dept Pediat GI Mass Gen Hosp Boston MA 02114

CARTER, ELMER BUZBY, b Brooklyn, NY, Mar 28, 30; m 57; c 2. COMPUTER SCIENCE. *Educ:* Haverford Col, SB, 53; Fla State Univ, MS, 60, PhD(physics), 62. *Prof Exp:* Res assoc physics, Rice Univ, 62-63, asst prof, 63-65; asst prof, Strasbourg Univ, 65-66; assoc prof physics, Trinity Univ, 66-71, assoc prof, Comput & Info Sci & Urban Studies, 71-84; CONSULT, 84- *Concurrent Pos:* NSF comput sci resident, Syst Develop Corp, Santa Monica, Calif & Sch Archit & Urban Planning, Univ Calif, Los Angeles, 70-71. *Mem:* AAAS; Am Phys Soc; Asn Comput Mach; Sigma Xi. *Res:* Urban studies; data base organization; information systems; systems ethics. *Mailing Add:* 247 E Rosewood San Antonio TX 78212

CARTER, ELOISE, b Waterproof, La, June 9, 45. NUTRITION. *Educ:* Southern Univ, Baton Rouge, BS, 67; Tuskegee Univ, MS, 69; Kans State Univ, PhD(foods & nutrit), 76. *Prof Exp:* Instr nutrit, NC Cent Univ, 69-73, Kans State Univ, 73-76; head dept home econ, 76-85, actg dean, Sch Agr & Home Econ, 85-87, ASSOC DIR, OFF INT PROGS, TUSKEGEE UNIV, 89- *Concurrent Pos:* IPA with USAID/BIFAD & USDA/OICD. *Mem:* Am Dietetic Asn; AAAS; Am Home Econ Asn; Sigma Xi; Asn Women Develop. *Res:* Human nutrition; food eating habits and patterns. *Mailing Add:* Sch Agr & Home Econ Tuskegee Univ Tuskegee AL 36088

CARTER, ELOISE B, b Rome, Ga, Oct 5, 50; m 72; c 2. PLANT ECOLOGY, PLANT EVOLUTION & SYSTEMATICS. *Educ:* Wesleyan Col, AB, 72; Emory Univ, MS, 78, PhD(biol), 83. *Prof Exp:* Instr biol, Agnes Scott Col, 78-83, asst prof, 83-84; asst prof biol, Emory Univ, 84-88; asst prof, 88-91, ASSOC PROF BIOL, OXFORD COL, EMORY UNIV, 91- *Concurrent Pos:* Assoc fac, Biol Dept, Emory Col, Emory Univ, 88-91. *Mem:* Ecol Soc Am; Bot Soc Am; AAAS; Sigma Xi; Asn Women Sci; Asn Biol Lab Educ. *Res:* Evolution and systematics of closely related plant species which are endemic to rock outcrops in the southeastern United States, with particular interest in species of Talinum; developing investigative exercises for undergraduate biology laboratory education. *Mailing Add:* Dept Biol Oxford Col Emory Univ Oxford GA 30267

CARTER, EMILY ANN, b Los Gatos, Calif, Nov 28, 60. SURFACE & INTERFACE SCIENCE. *Educ:* Univ Calif, Berkeley, BS, 82; Calif Inst Technol, PhD(chem), 87. *Prof Exp:* Postdoctoral chem, Univ Colo, Boulder, 87-88; ASST PROF CHEM, UNIV CALIF, LOS ANGELES, 88- *Concurrent Pos:* NSF presidential young investr award, 88; Camille & Henry Dreyfus Found distinguished new fac award, 88. *Honors & Awards:* Innovation Recognition Award, Union Carbide Corp, 89 & 90. *Mem:* Am Chem Soc; Am Phys Soc; Am Vacuum Soc; Sigma Xi; Asn Women Sci. *Res:* Ab initio electronic structure theory; statistical mechanical simulations of semiconductors and metals; thermochemistry and kinetics of adsorbates on surfaces; dynamics of elementary surface chemical reactions; electronic and magnetic properties, bonding, and reactivity of metal clusters and organometallic complexes. *Mailing Add:* Chem & Biochem Dept Univ Calif Los Angeles CA 90024-1569

CARTER, FAIRIE LYN, b Biloxi, Miss, Oct 1, 26. ANALYTICAL CHEMISTRY. *Educ:* Miss State Col for Women, BS, 48; Univ NC, MA, 50. *Prof Exp:* Asst anal chem, Univ NC, 48-50; asst cur limnol, Acad Nat Sci Philadelphia, 50-54; assoc chemist, E Reg Res Lab, USDA, 55-57, res chemist, S Reg Res Lab, 57-65, res chemist, S Forest Exp Sta, USDA, 65-84; CONSULT, 84- *Concurrent Pos:* Vis scientist, CSIRO Termite Group, Canberra, Australia, 86. *Mem:* Am Chem Soc; Sigma Xi. *Res:* Termite biochemistry; termiticides; wood extractives; composition of natural products; amino acids; lipids; chromatography. *Mailing Add:* 576 Comfort Pl Biloxi MS 39530-4434

CARTER, FREDERICK J, b Vernon, NY, Dec 16, 29; m 61; c 5. MATHEMATICS. *Educ:* Le Moyne Col, NY, BS, 56; Univ Detroit, MA, 58. *Prof Exp:* Instr math, Le Moyne Col, NY, 58-63; from instr to asst prof, 63-70, chmn dept, 70-74 & 82-86, ASSOC PROF MATH, ST MARY'S UNIV, TEX, 70-, UNDERGRAD ADV, 74- *Concurrent Pos:* Consult, Alamo Technol, Inc, 83- *Mem:* Am Math Soc; Math Asn Am. *Res:* Integral transforms and distribution theory. *Mailing Add:* Dept Math St Mary's Univ San Antonio TX 78228-8560

CARTER, G CLIFFORD, b Oak Park, Ill, Nov 12, 45; m 66; c 2. ACOUSTICAL SIGNAL PROCESSING. *Educ:* US Coast Guard Acad, BS, 67; Univ Conn, MS, 72, PhD(elec eng), 76. *Prof Exp:* ENGR, US NAVAL UNDERWATER SYSTS CTR, 69- *Concurrent Pos:* Adj fac mem, Univ Conn; adj prof, Kans State Univ. *Mem:* Fel Inst Elec & Electronics Engrs; Am Acoust Soc. *Res:* Signal processing; passive localization and estimation of spectral densities; coherence and time delay. *Mailing Add:* Code 104 Bldg 80 Naval Underwater Systs Ctr New London CT 06320-5594

CARTER, GEORGE EMMITT, JR, b Fayetteville, NC, Jan 18, 46; m 70; c 2. PLANT PHYSIOLOGY. *Educ:* Wake Forest Univ, BS, 68, MA, 70; Clemson Univ, PhD(plant physiol), 73. *Prof Exp:* from asst prof to assoc prof, 73-84, PROF PLANT PHYSIOL, CLEMSON UNIV, 84- *Mem:* Am Soc Plant Physiologists; Sigma Xi; AAAS. *Res:* Mechanism of dormancy in plants; quantitative analysis of plant growth regulators; peach tree physiology; pesticide residue analysis; protein electrophoresis. *Mailing Add:* Dept of Plant Path & Physiol Clemson Univ Clemson SC 29631

CARTER, GEORGE H, b Dobbs Ferry, NY, June 16, 16; m 87; c 3. PSYCHIATRY. *Educ:* Williams Col, BA, 38; Harvard Univ, MD, 43. *Prof Exp:* From instr to asst prof prev med, Sch Med, Boston Univ, 50-53, from instr to asst prof psychiat, 53-59, assoc prof psychiat, 59-86; RETIRED. *Concurrent Pos:* Resident fel psychiat, Harvard Med Sch, 48, teaching fel, 48-49; Commonwealth Fund fel, 66-67; asst, Univ Hosp, 50-55, asst vis physician, 55-59, vis physician, 59-; chmn residency training psychiat, Med Ctr, Boston Univ, 71-74 & curriculum coordr, 74-75, clin dir family ther, 74-75; asst chief inpatient serv psychiat, Bedford Vet Admin Hosp, 75-78, dir family ther, 78-85; neurolinguistic prog master practr, 82. *Mem:* Am Psychiat Asn. *Res:* Psychotherapy; family therapy. *Mailing Add:* Two Mill Pond Dr Charlestown RI 02813

CARTER, GESINA C, b Nootdorp, Netherlands, Dec 15, 39; US citizen; m 62; c 3. SOLID STATE PHYSICS, MAGNETICS. *Educ:* Univ Mich, BS, 60; Carnegie Inst Technol, MS, 62, PhD(physics), 65. *Prof Exp:* Res assoc acoust, Cath Univ, 65-66; physicist, Nat Bur Standards, 66-78; STAFF DIR, SCI & TECH INFO BD, NAT ACAD SCI, 78- *Concurrent Pos:* Vis researcher, Phys Spectrometry Lab, Fac Sci, Univ Grenoble, 75-76; secy, US Nat Comt, Comt Data Sci & Technol, Int Coun Sci Union; past chmn, SIG/Numeric databases, Am Soc Info Sci; chmn Am Soc Testing & Mat, F8.13, Sports Safety & Fencing. *Honors & Awards:* Silver Medal, Dept Com, 77. *Mem:* Am

Phys Soc; Am Inst Physics; Am Soc Info Sci; NY Acad Sci; Am Soc Testing & Mat; Am Soc Metals; Int Asn Hydrogen Energy; AAAS. *Res:* Magnetism in metals; band structure and electronic behavior in metals; transport properties in metals; nuclear magnetic resonance in metals and alloys; thermodynamics and phase diagrams; critical data evaluation; metal-hydrogen systems; numeric databases for sciences; environmental data; scientific data and information policy. *Mailing Add:* Nat Acad of Sci Washington DC 20418

CARTER, GILES FREDERICK, b Lubbock, Tex, Mar 22, 30; m 54; c 3. METALLURGICAL CHEMISTRY, ARCHAEOLOGICAL CHEMISTRY. *Educ:* Tex Tech Univ, BS, 49; Univ Calif, Berkeley, PhD(chem), 53. *Prof Exp:* Asst chem, Univ Calif, 49-50, chemist radiation lab, 50-52; res chemist, E I du Pont de Nemours & Co, 52-63, staff scientist, 63-67; dir state tech serv, 67-70, from assoc prof to prof chem, 70-90, EMER PROF CHEM, EASTERN MICH UNIV, 91- *Mem:* Am Chem Soc; Am Numis Soc; Am Soc Metals; Royal Numis Soc; Soc Archaeol Sci. *Res:* X-ray fluorescence analyses of ancient coins; scientific numismatics. *Mailing Add:* 1303 Grant St Ypsilanti MI 48197

CARTER, HARRY HART, b Ponca, Nebr, Mar 14, 21; m 46; c 7. PHYSICAL OCEANOGRAPHY. *Educ:* US Coast Guard Acad, BS, 43; Scripps Inst Oceanog, MS, 48. *Prof Exp:* Res assoc, Johns Hopkins Univ, 63-68, res scientist, Chesapeake Bay Inst, 68-77; prof, 77-85, EMER PROF MARINE SCI, STATE UNIV NY, STONY BROOK, 85- *Concurrent Pos:* Pres, Hydrocon, Inc, 70-78; prof part-time, Marine Sci, State Univ NY, Stony Brook, 75-77. *Mem:* AAAS; Sigma Xi. *Res:* Estuarine circulation and mixing; coastal oceanography. *Mailing Add:* A-1 Lake Point Dr Pinehurst NC 28374

CARTER, HARVEY PATE, b Friendship, Tenn, Dec 15, 27; m 51; c 1. COMPUTER SCIENCE. *Educ:* David Lipscomb Col, BA, 49; Vanderbilt Univ, MA, 50, PhD(math), 59. *Prof Exp:* Instr math, David Lipscomb Col, 50-51, assoc prof, 53-57; sr mathematician, Oak Ridge Nat Lab, 57-68, asst dir math div, 64-68; assoc prof math, Univ Tenn, Knoxville, 68-70; dir math div, Oak Ridge Nat Lab, 69-73; dir comput sci div, Union Carbide Nuclear Div, Union Carbide Corp, 73-84; DEP DIR, COMPUT & TELECOMMUN, MARTIN MARIETTA ENERGY SYSTS, 84- *Mem:* Asn Comput Mach; Sigma Xi. *Res:* Applications of computer techniques to scientific and engineering problems; management of computer facilities. *Mailing Add:* Martin Marietta Energy Systs PO Box 2003 Oak Ridge TN 37831-7595

CARTER, HERBERT EDMUND, b Morresville, Ind, Sept 25, 10; m 33; c 2. CHEMISTRY, BIOCHEMISTRY. *Educ:* DePauw Univ, AB, 30; Univ Ill, AM, 31, PhD(chem), 34. *Hon Degrees:* ScD, DePauw Univ, 51, Univ Ill, 74, Univ Ind, 74, Univ Ariz, 84; DHL, Thomas Jefferson Univ, 75. *Prof Exp:* From instr to prof chem, Univ Ill, Urbana-Champaign, 32-67, head dept, 54-67, acting dean grad col, 63-65, vchancellor acad affairs, 67-71; coordr interdisciplinary progs, 71-77, head, Dept Biochem, 77-81, RES FEL, OFF ARID LANDS STUDIES, UNIV ARIZ, 81-, SPEC ASST VPRES RES, 81- *Concurrent Pos:* Mem div chem & chem technol, Nat Res Coun, 50-55, mem-at-large, 57-59; mem mgt comt, Gordon Res Conf, 53-56, mem coun, 59-61; chmn biochem study sect, NIH, 54-56, chmn biochem training grant comt, 58-61; mem nat comt, Int Union Pure & Appl Chem, 55-62, chmn, 60-62; mem bd sci counr, Nat Heart Inst, 57-59; mem nat comt, Int Union Biochem, 62-65; mem, President's Comt Nat Medal Sci, 63-66; chmn sect biochem, Nat Acad Sci, 63-66, mem coun, 66-69; mem Nat Sci Bd, 64-76, chmn, 70-74; mem bd trustees, Nutrit Found, 70-84; mem bd trustees, Argonne Univ Assoc, 80-83. *Honors & Awards:* Lilly Award, 43; Nichols Medal, 65; Lipid Chem Award, Am Oil Chem Soc, 66; Kenneth A Spencer Award, 68; Alton E Bailey Award, 70. *Mem:* Nat Acad Sci; AAAS; Am Chem Soc; Am Soc Biol Chemists (pres, 56-57); Am Oil Chem Soc; Am Inst Nutrit; Am Acad Arts Sci. *Res:* Fatty acid metabolism; biochemistry of amino acids; chemistry of streptomycin and other antibiotic substances; biochemistry of complex lipids of plants and animals; structure of bacterial lipopolysaccharides. *Mailing Add:* Off Arid Lands Studies Univ Ariz Tucson AZ 85719

CARTER, HOWARD PAYNE, b Houston, Tex, Sept 9, 21; m 52; c 2. ZOOLOGY. *Educ:* Tuskegee Inst, BS, 41; Columbia Univ, MA, 48; Univ Calif, MA, 57; Univ Wis, PhD(protozool), 64. *Prof Exp:* Instr, 50-54, from asst prof to assoc prof, 54-70, dean, Col Arts & Sci, 68-78, PROF BIOL, TUSKEGEE INST, 70-,. *Concurrent Pos:* Res assoc, Carver Res Found, 58- *Mem:* AAAS; Sigma Xi; Soc Protozool; Am Soc Parasitol; Am Micros Soc. *Res:* Infraciliature of ciliated protozoa; morphology of parasitic and freeliving protozoa; ciliate ultrastructure. *Mailing Add:* PO Box 773 Tuskegee Institute AL 36088

CARTER, HUBERT KENNON, b Athens, Ga, Apr 23, 41; m 62; c 3. EXPERIMENTAL NUCLEAR PHYSICS. *Educ:* Univ Ga, BS, 63; La State Univ, MS, 65; Vanderbilt Univ, PhD(physics), 69. *Prof Exp:* Asst prof physics, Furman Univ, 68-74; MEM FAC RES, UNIV ISOTOPE SEPARATOR, OAK RIDGE ASSOC UNIVS, 72-, SCIENTIST PHYSICS, 73-, DIR, UNIV ISOTOPE SEPARATOR, OAK RIDGE, 85- *Mem:* Am Phys Soc. *Res:* The study of nuclei far from stability with an isotope separator on-line to a heavy-ion accelerator and with fast atom-laser spectroscopy. *Mailing Add:* Oak Ridge Assoc Univ PO Box 117 Oak Ridge TN 37831

CARTER, JACK FRANKLIN, b Lodgepole, Nebr, Oct 1, 19; m 41; c 5. AGRONOMY. *Educ:* Univ Nebr, BS, 41; State Col Wash, MS, 47; Univ Wis, PhD(agron, plant path), 50. *Prof Exp:* Res & teaching fel agron, State Col Wash, 41 & 46-47; res asst agron & plant path, Univ Wis, 47-50; assoc prof & assoc agronomist, 50-59, prof agron & agronomist, 59-87, chmn dept, 60-87, EMER PROF AGRON & AGRONOMIST, NDAK STATE UNIV, 87- *Mem:* Coun Agr Sci & Technol, (pres, 78-79); Am Soc Agron; Crop Sci Soc Am (pres, 72-73). *Res:* Forage crop production and management; pasture research; forage crop diseases. *Mailing Add:* 1345 11th St N Fargo ND 58102

CARTER, JACK LEE, b Kansas City, Kans, Jan 23, 29; m 51; c 3. BOTANY. *Educ:* Emporia State Univ, Kans, BS, 52, MS, 54; Univ Iowa, PhD(bot), 60. *Prof Exp:* Assoc prof biol & head dept, Northwestern Col, Iowa, 55-58 & Simpson Col, 60-62; assoc prof & coordr res & inst grants, Emporia State Univ, Kans, 62-66; assoc dir biol sci curric study, Univ Colo, 66-68, dir, 82-85; PROF BIOL, COLO COL, 68- *Concurrent Pos:* Ed, Am Biol Teacher, 70-74. *Mem:* AAAS; hon mem Nat Asn Biol Teachers; Bot Soc Am; Am Soc Plant Taxonomists; Am Inst Biol Sci. *Res:* Systematic botany and plant geography. *Mailing Add:* PO Box 1251 Silver City NM 88062

CARTER, JAMES CLARENCE, b New York, NY, Aug 1, 27. THEORETICAL PHYSICS. *Educ:* Spring Hill Col, BS, 52; Fordham Univ, MS, 53; Catholic Univ, PhD(physics), 56; Woodstock Col, STL, 59. *Prof Exp:* From instr to asst prof, 60-67, acad vpres, 70-74, ASSOC PROF PHYSICS, LOYOLA UNIV, LA, 67-, PRES, 74- *Concurrent Pos:* NSF exten grant, 63-64; NSF res grant, Loyola Univ, La, 65-71; Nat Acad Sci-Nat Res Coun resident res assoc, Nat Bur Standards, 64-65. *Mem:* Am Phys Soc; Am Asn Physics Teachers; Sigma Xi. *Res:* Nuclear theory and elementary particle theory. *Mailing Add:* President Loyola Univ New Orleans LA 70118

CARTER, JAMES EDWARD, b Great Neck, NY, Nov 3, 29; m 51; c 3. OBSTETRICS & GYNECOLOGY, PATHOLOGY. *Educ:* Univ Vt, BA, 51; NY Med Col, MD, 55. *Prof Exp:* Assoc prof, 70-73, PROF OBSTET & GYNEC & PATH, SCH MED, IND UNIV, INDIANAPOLIS, 73-, ASSOC DEAN STUDENT & CURRICULAR AFFAIRS, 73- *Mem:* AAAS; Am Col Obstet & Gynec; Electron Micros Soc Am; Sigma Xi. *Mailing Add:* OB-GYN Med Sci 164 Sch Med Univ Ind 635 Barnhill Dr Indianapolis IN 46223

CARTER, JAMES EVAN, b Calgary, Alta, Can, Oct 26, 41; m 65; c 4. PHARMACEUTICAL ANALYSIS, PHARMACEUTICAL QUALITY CONTROL. *Educ:* Brigham Young Univ, BS, 66, MS, 69; Univ Utah, PhD(mech chem), 75. *Prof Exp:* Biochemist & prin investr, Utah State Div Health & Utah Community Pesticide Study, 69-73; GROUP LEADER ANAL DEVELOP, AM CRITICAL CARE, DIV AM HOSP SUPPLY CORP, 75- *Mem:* Am Pharmaceut Asn; Am Chem Soc; Asn Off Analytical Chemists. *Res:* Pharmaceutical analysis and control in-vitro; methods development; compatibility studies of drugs in solution; analysis of drugs in feed. *Mailing Add:* PO Box 2321 Flemington NJ 08822-2321

CARTER, JAMES HARRISON, II, b Biloxi, Miss, July 4, 45; m 65; c 2. HEMATOLOGY SYSTEMS, TECHNICAL MANAGEMENT. *Educ:* Ga Inst Technol, BS, 66, PhD(chem), 70. *Prof Exp:* Asst clin chemist, Med Lab Assoc, 70-71; gen mgr, Diag Lab, Biomed Prod Corp, 71-75; lab dir, Vet Diag, Vet Clin Lab, 75-81; sr scientist, 78-84, DIR, RES & DEVELOP, DIAG DIV, COULTER ELECTRONICS, INC, 84- *Concurrent Pos:* Lab consult, 75-; prin investr, Broward County Bd Comn, 81-; fac adv, Walden Univ, 83- *Mem:* Am Asn Clin Chem; Confr Public Health Lab Dir; Am Soc Microbiol; Am Mgt Asn; Am Soc Cell Biol. *Res:* Automated instrument applications to non-human clinical specimens; design and development of hematology reagent and control systems for use with Coulter electronic cell counting; blood cellular components, stabilization and handling; hybridoma cell biology. *Mailing Add:* 740 W 83rd St Hialeah FL 33014

CARTER, JAMES M, b Cass Co, Mich, Dec 7, 21; m 51; c 4. VETERINARY PHYSIOLOGY. *Educ:* Mich State Univ, DVM, 51; Purdue Univ, MS, 63, PhD(physiol), 65. *Prof Exp:* Pvt vet practr, Ind, 51-61; Am Vet Med Asn fel, 61-63, asst, 63-65, asst prof, 65-68, ASSOC PROF VET PHYSIOL, PURDUE UNIV, WEST LAFAYETTE, 68- *Mem:* Am Vet Med Asn; Am Soc Vet Physiol & Pharmacol. *Res:* Renal and respiratory physiology, especially fluid and electrolyte and acid-base balance studies. *Mailing Add:* 1435 Warren Pl Lafayette IN 47905

CARTER, JAMES R, ENDOCRINOLOGY, INTERNAL MEDICINE. *Educ:* Columbia Univ, MD, 59. *Prof Exp:* PROF INTERNAL MED, SCH MED, CASE WESTERN RESERVE UNIV, 78-; DIR, DEPT MED, CLEVELAND METROP GEN HOSP, OHIO, 80- *Mailing Add:* Cleveland Metrop Gen Hosp 3395 Scranton Rd Cleveland OH 44109

CARTER, JAMES RICHARD, b Vicksburg, Miss, Mar 15, 53; m 84; c 2. PLANT SYSTEMATICS, FLORISTICS. *Educ:* Miss State Univ, BS, 75, MS, 78; Vanderbilt Univ, PhD(biol & bot), 84. *Prof Exp:* Asst prof, 84-91, ASSOC PROF BIOL, VALDOSTA STATE COL, 91-, CUR, HERBARIUM, 85- *Concurrent Pos:* Ed, Castanea, 91- *Mem:* Bot Soc Am; Am Soc Plant Taxonomists; Int Asn Plant Taxon. *Res:* Systematics of Cyperaceae, particularly Cyperus and Eleocharis; floristics of southeastern US. *Mailing Add:* Dept Biol Valdosta State Col Valdosta GA 31698

CARTER, JOHN C H, marine biology, fresh water biology; deceased, see previous edition for last biography

CARTER, JOHN HAAS, b Trevorton, Pa, Oct 25, 30; m 54; c 3. OPTOMETRY. *Educ:* Pa State Col Optom, OD, 53; Univ Ind, MS, 59, PhD, 62. *Prof Exp:* Asst div optom, Univ Ind, 57-59, res asst, 59-62; res assoc prof physiol optics, Pa Col Optom, 62-66, res prof, 66-70; dir visual sci div, 71-73, acad dean, 73-77, PROF PHYSIOL OPTICS, NEW ENG COL OPTOM, 70- *Mem:* fel Am Acad Optom. *Res:* Physiological optics; infrared self-recording refractionometer; nature of stimulus to accommodative mechanism of the eye. *Mailing Add:* New Eng Col Optom 424 Beacon St Boston MA 02115

CARTER, JOHN LEMUEL, JR, b Clarksville, Tex, Mar 17, 20; m 47; c 3. NUCLEAR REACTOR COMPUTER CODES, PHYSICS. *Educ:* Baylor Univ, BA, 41; Brown Univ, MSc, 43; Cornell Univ, PhD(physics), 53. *Prof Exp:* Engr sound div, US Naval Res Lab, 43-45; engr, US Naval Underwater Sound Reference Lab, 45-47; asst physics, Cornell Univ, 47-52; engr measurements lab, Gen Elec Co, Mass, 52-55, physicist, Hanford Atomic Prod Oper, 55-58, supvr theoret physics res unit, Hanford Labs, 58-63, tech specialist, 63-64; res assoc, Reactor Physics Dept, Pac Northwest Lab,

Battelle Mem Inst, 65-77; sr engr, Exxon Nuclear Co, 77-85; RETIRED. *Mem:* Am Phys Soc; Am Nuclear Soc; Sigma Xi. *Res:* Nuclear reactor physics; solid state theory; digital computer codes. *Mailing Add:* 1206A NW 133rd St Vancouver WA 98685

CARTER, JOHN LYMAN, b Sisseton, SDak, Mar 14, 34; m 63. INVERTEBRATE PALEONTOLOGY. *Educ:* Univ NDak, BS, 59; Univ Cincinnati, PhD(geol), 66. *Prof Exp:* Res assoc paleont & cur, Univ Ill, Urbana-Champaign, 66-72; assoc cur, 72-81, CUR, CARNEGIE MUS NATURAL HIST, 81- *Mem:* Paleont Soc; Paleont Res Inst; Brit Palaeont Asn; Int Palaeont Asn; Soc Econ Paleont & Mineral; Asn Earth Sci Eds. *Res:* Late Paleozoic Brachiopoda and biostratigraphy. *Mailing Add:* Carnegie Mus Nat Hist 4400 Forbes Ave Pittsburgh PA 15213

CARTER, JOHN NEWTON, b Columbia, Mo, Jan 24, 21; m 48; c 3. AGRONOMY. *Educ:* Univ Mo, BS, 43; Univ Ill, MS, 48, PhD(soil fertil), 50. *Prof Exp:* Spec res asst agron, Univ Ill, 47-50; prin soil scientist, Battelle Mem Inst, 51-55; SOIL SCIENTIST SOIL & WATER CONSERV, AGR RES SERV, USDA, 55- *Mem:* Am Soc Agron; Soil Sci Soc Am; Am Soc Sugar Beet Technologists. *Res:* Sugarbeet nitrogen nutrition; soil fertility; soil nitrogen transformations; soil organic matter; plant nutrition. *Mailing Add:* Snake River Conserv Res Sta USDA-ARS Rt 1 Box 186 Kimberly ID 83341

CARTER, JOHN PAUL, JR, b Charlottesville, Va, July 23, 51. PHARMACEUTICAL CHEMISTRY. *Educ:* Univ Tenn, AB, 73, PhD(chem), 78. *Prof Exp:* Res assoc, Ore State Univ, 78-80; sr scientist, Adria Lab Inc, 81-85; SR SCIENTIST, NOVA PHARMACUET CORP, 85- *Mem:* Am Chem Soc; Royal Inst Chem. *Res:* Cancer chemotheraphy; synthesis of natural products; new synthetic reactions; cardiovascular drugs; anti-emetic therapy; synthesis of natural products; new synthetic reactions. *Mailing Add:* Nova Pharmceut Corp 6200 Freeport Ctr Baltimore MD 21224-2788

CARTER, JOHN PAUL, b Pittsburgh, Pa, Sept 26, 30. CORROSION, ELECTROCHEMISTRY. *Educ:* Univ Pittsburgh, BS, 53. *Prof Exp:* Asst technologist, Res Ctr, US Steel Corp, 55-61; res chemist, US Bur Mines, 61-87; PRES, M&DP ASSOCS INC, 87- *Concurrent Pos:* Chair, T-2E, Nat Asn Corrosion Engrs, 79-85; mem, Nat Acid Precipitation Assessment Prog, 80-87. *Mem:* Am Chem Soc; fel Am Inst Chemists; Am Soc Metals; Am Soc Testing & Mat; Electrochem Soc; Nat Asn Corrosion Engrs (secy, 85-87). *Res:* Corrosion; corrosion abatement; galvanic corrosion between skyward and groundward surfaces of atmospheric panels; author of 80 publications. *Mailing Add:* 3366 Toledo Terr Apt A Hyattsville MD 20782

CARTER, JOHN ROBERT, b Buffalo, NY, Apr 21, 17; m 43; c 2. PATHOLOGY. *Educ:* Hamilton Col, BS, 39; Univ Rochester, MD, 43. *Prof Exp:* Asst path, Univ Iowa, 44, from instr to prof, 44-59; prof path & oncol & chmn dept, Med Ctr, Univ Kans, 59-66; prof path, chmn dept & dir, Inst Path, 66-81, prof path dept orthop, 81-87, EMER PROF PATH, CASE WESTERN RESERVE UNIV, 87- *Concurrent Pos:* Dir path, Univ Hosps, Cleveland, 66-81, dir, Musculoskeletal Path Lab, 81- *Mem:* Am Soc Clin Path; Am Soc Exp Path; Col Am Path; Am Asn Path & Bact (treas, 61-62, secy-treas, 62-65, vpres, 66-67, pres, 67-68); Int Acad Path. *Res:* Orthopedic pathology; musculoskeletal neoplasms. *Mailing Add:* Inst Path Case Western Reserve Univ Cleveland OH 44106

CARTER, JOHN VERNON, b Boise, Idaho, Dec 21, 40; div; c 3. BIOPHYSICAL CHEMISTRY. *Educ:* Whittier Col, BA, 62; Purdue Univ, PhD(org chem), 67. *Prof Exp:* Res scientist org chem, Koppers Co, Inc, Pa, 67-68; res fel, Univ Pittsburgh, 68-70; from asst prof to assoc prof chem, Adams State Col, 70-74; res assoc chem, Univ Minn, Minneapolis, 74-76; asst prof, 76-81, ASSOC PROF HORT SCI, UNIV MINN, ST PAUL, 81- *Mem:* Am Soc Plant Physiol; Sigma Xi; Am Chem Soc. *Res:* Plant stress physiology. *Mailing Add:* Dept of Hort Sci Univ of Minn St Paul MN 55108

CARTER, JOSEPH GAYLORD, b Honolulu, Hawaii, Dec 14, 47; m 69; c 2. PALEOBIOLOGY, PALEONTOLOGY. *Educ:* Univ Kans, BS, 70; Yale Univ, PhD(paleobiol), 76. *Prof Exp:* Asst prof geol, Colgate Univ, 74-76; from asst prof to assoc prof, 76-89, PROF GEOL, UNIV NC, 90- *Concurrent Pos:* Explor geologist, Union Oil Co, 70. *Honors & Awards:* W A Tarr Award. *Mem:* Sigma Xi; Paleont Soc; Paleont Res Inst. *Res:* Ecology and evolution of the Mollusca, especially shell microstructure and evolution of Middle Paleozoic benthic and post-Paleozoic endolithic communities. *Mailing Add:* Mitchell Hall Univ NC Chapel Hill NC 27514

CARTER, JULIA H, EXPERIMENTAL BIOLOGY. *Prof Exp:* PRES & CHIEF EXEC OFFICER, WOOD HUDSON CANCER RES LAB. *Mailing Add:* Wood Hudson Cancer Res Lab Inc 1830 Greenup St Covington KY 41011

CARTER, KENNETH, b Morecambe, Eng, Feb 6, 14; m 40, 70; c 3. MEDICINE, PHARMACEUTICAL CHEMISTRY. *Educ:* Univ London, MRCS & LRCP, 47; Manchester Univ, MPS, 37. *Prof Exp:* Res chemist, Glaxo Labs, Eng, 38-42, med dir, Buenos Aires, 48-50, dir, Qual Control & Prod, 49-50; res liaison officer & secy, Therapeut Res Corp Gt Brit, 46-48; dir develop, Smith Kline & French, Pa, 51-54, med dir & dir develop, Eng, 54-57; sci dir, Ames Co, Ind, 57-58, vpres res & med affairs, 58-62, Miles Labs, Inc, 62-65; vpres & dir med affairs, Syntex Int, 65-71, corp vpres regulatory affairs, Syntex Corp, 71-79; consult, Drug Surveillance Res Unit, Southampton Univ, Eng, 80-81; RETIRED. *Concurrent Pos:* House physician, Postgrad Med Sch, Univ London, 47-48; mem, Royal Col Surgeons; mem, Bd Dirs, Royal Soc Med Found, Inc, 68-81; consult, 81- *Mem:* Fel Pharmaceut Soc Gt Brit; fel Royal Soc Med; Brit Med Asn; Brit Harvein Soc. *Res:* Administration of research, development and medical operations; disciplines necessary for developing new diagnostic and therapeutic agents, such as chemistry, biochemistry, immunology, toxicology, pharmacology, clinical pharmacology and clinical research; world government regulatory affairs. *Mailing Add:* 24612 Olive Tree Lane Los Altos Hills CA 94024

CARTER, KENNETH NOLON, b Columbia, SC, Oct 2, 25; m 54; c 1. ORGANIC CHEMISTRY. *Educ:* Erskine Col, AB, 47; Vanderbilt Univ, MS, 49, PhD(chem), 51. *Prof Exp:* head dept chem, 51-87, Charles A Dana prof, 70-87, EMER PROF,PRESBY COL, 87- *Concurrent Pos:* Mem, local emergency planning comt, Presby Col, 87- *Mem:* Am Chem Soc. *Res:* Organic synthesis; molecular rearrangements. *Mailing Add:* Dept Chem Presby Col Clinton SC 29325

CARTER, LARK POLAND, b Lytton, Iowa, June 26, 30; m 54; c 2. AGRONOMY. *Educ:* Iowa State Univ, BS, 53, MS, 56, PhD(agron), 60. *Prof Exp:* Asst prof agron, Mont State Univ, 60-62, asst dean, Col Agr, 62-65, assoc dean, Col Agr & asst dir, Agr Exp Sta, 65-80; asst dir higher educ, USDA, Washington, DC, 80-81; DEAN, SCH AGR, CALIF POLYTECH STATE UNIV, SAN LUIS OBISPO, 81- *Mem:* Am Soc Agron; Am Asn State Col Agr & Renewable Resources; Sigma Xi. *Res:* Field crop production; forage management; seed certification, grass and legume seed production. *Mailing Add:* Sch Agr Calif Polytech State Univ San Luis Obispo CA 93407

CARTER, LEE, b Rock Island, Ill, Sept 8, 17; m 45; c 3. CHEMICAL ENGINEERING, ENGINEERING ECONOMICS. *Educ:* Purdue Univ, BS, 40; Cornell Univ, MS, 45. *Prof Exp:* Lab asst, Universal Oil Prod Co, 37-38, pilot plant operator, 40-42; mgr tech serv, Va-Carolina Chem Corp, 47; from res engr to mgr, Calgary Off, E B Badger & Sons Co & Stone & Webster, 47-63; mgr eng reports, R W Booker & Assocs, Inc, 63-69, sr vpres & dir proj studies div, 69-72; CONSULT ENGR, 72- *Concurrent Pos:* Mem investment mission, US Dept Com, Colombia, 65. *Mem:* AAAS; fel Am Inst Chem Engrs; Chem Inst Can; fel Brit Inst Petrol; Am Consult Engrs Coun; Fel Brit Inst Chem Engrs; Am Econ Asn. *Res:* Technical-economic investigations of project feasibility; industrial development studies; coal conversion; production of olefins; port development studies; low Btu coal gasification studies aimed at making small users of gas self-reliant in the face of natural gas supply problems; export of Am coal to Holland; technical advice in settlement of claims and in litigation. *Mailing Add:* 622 Belson Ct Kirkwood MO 63122-3073

CARTER, LELAND LAVELLE, b Oberlin, Kans, Nov 27, 37; m 58; c 4. REACTOR PHYSICS, NUCLEAR ENGINEERING. *Educ:* Northwest Nazarene Col, BA, 61; Univ Wash, MS, 64, PhD(nuclear eng), 69. *Prof Exp:* Physicist, Gen Elec Co, 62-63; physicist, Pac Northwest Labs, Battelle Mem Inst, 64-65; nuclear engr, Los Alamos Sci Lab, Univ Calif, 69-73, alt group leader neutron transport, 74-77; prin engr, 77-81, FEL ENGR, WESTINGHOUSE HANFORD CO, 81- *Mem:* Am Nuclear Soc. *Res:* Application of the Monte Carlo method to solve three-dimensional particle transport problems on the digital computer; adjoint simulation, neutron cross-sections, nonlinear radiative transport, criticality, unbiased sampling schemes and shielding for neutron energies 0-50 million electron volts. *Mailing Add:* HO-35 Box 1970 Richland WA 99352

CARTER, LEO F, b Warwick, NY, Dec 20, 39; m 61; c 2. CHEMICAL ENGINEERING. *Educ:* Syracuse Univ, BChE, 61, MChE, 63; Univ Mich, PhD(chem engr), 67. *Prof Exp:* sr specialist, 67-80, SR PROCESS TEHNOL SPEC, MONSANTO CO, SPRINGFIELD, MASS, 80- *Concurrent Pos:* Group leader, explor & fundamental polymer res, Monsanto Co, Springfield, 80-82. *Mem:* Am Chem Soc; Am Inst Chem Engrs; Soc Rheol. *Res:* Multiphase polymer structure-properties relationships; polymerization kinetics and process modeling; rheology of fiber suspensions; laminar flow in porous ducts; mixing in viscous systems. *Mailing Add:* 489 Main St Wilbraham MA 01095-1662

CARTER, LINDA G, b Mansfield, Ohio, Sept 23, 49. SYNTHETIC ORGANIC CHEMISTRY. *Educ:* Case Western Reserve Univ, BA, 71; Purdue Univ, MS, 73; Univ Ill, PhD(org chem), 80. *Prof Exp:* Res chemist, Monsanto Agr Prod Co, 73-76; RES CHEMIST, BIOCHEM DEPT, E I DU PONT DE NEMOURS & CO, 80- *Mem:* Am Chem Soc; Sigma Xi. *Res:* Agrichemical research; metalation chemistry. *Mailing Add:* 822 Starvegut Rd Kenneth Square PA 19348

CARTER, LOREN SHELDON, b Nampa, Idaho, Jan 10, 39; m 68. PHYSICAL CHEMISTRY, INORGANIC CHEMISTRY. *Educ:* Ore State Univ, BS, 61, MS, 66; Wash State Univ, PhD(phys chem), 70. *Prof Exp:* Reactor engr, Phillips Petrol Co, 61-63; from asst prof to assoc prof, 70-79, PROF CHEM, BOISE STATE UNIV, 79- *Mem:* Sigma Xi; Am Chem Soc. *Res:* Surface adsorption on clays; transference number measurements using a centrifuge in organic solvents; hydrocarbons in the environment; chemotaxonomy of cacti. *Mailing Add:* 3401 Tamarack Dr Boise ID 83703

CARTER, MARY EDDIE, b Americus, Ga, March 14, 25. ORGANIC CHEMISTRY. *Educ:* La Grange Col, BA, 46; Univ Fla, MS, 49; Univ Edinburgh, PhD, 56. *Prof Exp:* Instr chem, La Grange Col, 46-47; microscopist, Calloway Mills, 47-48; textile chemist, Southern Res Inst, 49-51; chemist, West Point Mfg Co, 51-53; res chemist, Am Viscose Corp, 56-62, res assoc, FMC Corp, Am Viscose Div, Res & Develop, 62-71; chief textiles & clothing lab, SMNRD, Agr Res Serv, 71-73, dir, Southern Regional Res Ctr, 73-80, ASSOC ADMIN, SCI & EDUC-AGR RES SERV, USDA, 80- *Honors & Awards:* Herty Award, Am Chem Soc, 79. *Mem:* Am Chem Soc; Am Asn Textile Chemists & Colorists; Inter-Soc Color Coun; Fiber Soc; Sigma Xi; fel AAAS. *Res:* Naturally occurring polymers for fiber, food and feed, including safety, nutrition and processing aspects. *Mailing Add:* Rm 302-A Admin Bldg Agr Res Serv USDA Washington DC 20250

CARTER, MARY KATHLEEN, b Franklinton, La, July 11, 22. PHARMACOLOGY. *Educ:* Tulane Univ, BA, 49, MS, 53; Vanderbilt Univ, PhD(pharmacol), 55. *Prof Exp:* Asst pharmacol, Tulane Univ, 49-52; asst, Vanderbilt Univ, 52-53; res assoc, Med Ctr, Univ Kans, 55-57; from instr to prof, 57-84, EMER PROF PHARMACOL, MED SCH, TULANE UNIV, 85- *Concurrent Pos:* USPHS fel, Med Ctr, Univ Kans, 55-57, sr res fel, Med Sch, Tulane Univ, 57-61. *Mem:* Am Soc Pharmacol & Exp Therapeut; Sigma Xi. *Res:* Renal pharmacology; transport of electrolytes and sugars; cholinesterases. *Mailing Add:* Rte 3 Box 724 Kentwood LA 70444

CARTER, MASON CARLTON, b Wash, DC, Jan 14, 33; m 53. PLANT PHYSIOLOGY, FORESTRY. *Educ:* Va Polytech Inst, BS, 55, MS, 57; Duke Univ, PhD(forestry), 59. *Prof Exp:* Asst plant physiol, Va Polytech Inst, 55-56; res forester, Southeastern Forest Exp Sta, USDA, 59-60; asst prof forestry, Auburn Univ, 60-66, assoc prof forestry & bot, 66-70, from alumni assoc prof to alumni prof, 70-73; prof forestry & natural resources & chmn dept, Purdue Univ, West Lafaye tte, 73-85; DEAN, AGR COL, LA STATE UNIV, 85- *Mem:* Am Forestry Asn; Soc Am Foresters; AAAS. *Res:* Mechanisms of herbicidal action; physiology of woody plants. *Mailing Add:* Agr Col La State Univ Rm 142 Ag Admin Bldg Baton Rouge LA 70803

CARTER, MELVIN WINSOR, b Phoenix, Ariz, Jan 22, 28; m 50; c 5. APPLIED STATISTICS, BIOMETRICS. *Educ:* Ariz State Univ, BS, 53; NC State Col, MS, 54, PhD, 56. *Prof Exp:* Asst nutrit, NC State Col, 52-56; asst prof statist, Purdue Univ, 56-58; exp statist, NC State Col, 58-61; assoc prof, 61-66, PROF STATIST, BRIGHAM YOUNG UNIV, 66- *Concurrent Pos:* Vis prof, NC State Univ, 67-68 & Texas A&M Univ, 71; biostatistician, Upjohn Co, 75. *Mem:* Am Statist Asn; Biomet Soc; Sigma Xi. *Res:* Biometrics; nutrition; endocrinology; biochemistry; methodology for the analysis of unbalanced designs. *Mailing Add:* 4621 Ellisbury Dr NE Atlanta GA 30338

CARTER, NEVILLE LOUIS, b Los Angeles, Calif, Aug 21, 34; m 56; c 3. GEOLOGY, GEOPHYSICS. *Educ:* Pomona Col, AB, 56; Univ Calif, Los Angeles, MA, 58, PhD(geol), 63. *Prof Exp:* Res geophysicist, Univ Calif, Los Angeles, 61-63, asst, 63; res geologist, Shell Develop Co, 63-66; assoc prof geol & geophys, Yale Univ, 66-71; prof geophys, earth & space sci, State Univ NY Stony Brook, 71-78; head and prof geophysics, fac assoc, 78-83, PROF GEOPHYSICS, & CTR TECTONOPHYSICS, TEXAS A&M UNIV, 84- *Mem:* Am Geophys Union; Geol Soc Am. *Res:* Tectonophysics; experimental and natural deformation of rocks and minerals. *Mailing Add:* Dept Geophys Tex A&M Univ College Station TX 77843

CARTER, ORWIN LEE, b Geneseo, Ill, Aug 22, 42; m 66; c 2. PHYSICAL CHEMISTRY, IMMUNOLOGY. *Educ:* Univ Iowa, BS, 64; Univ Ill, MS, 65, PhD(x-ray crystallog), 67; Rider Col, MBA, 75. *Prof Exp:* Asst prof phys chem, US Mil Acad, 67-70; from scientist to sr scientist, Rohm & Haas Co, 70-74; dir, Prod Develop, Micromedic Systs, Inc, 74-77; assoc dir, Immunodiagnostics, Becton Dickinson & Co, 77-78, actg dir, Res Ctr, 78-79, vpres, Admin & Planning, 79-80, vpres & gen mgr, Manual Immunoassay, Immunodiagnostics, 80-82; pres, Amersham Corp, 82-86; PRES & CEO, INCSTAR CORP, 86- *Mem:* Am Chem Soc; Am Asn Clin Chemists; Sigma Xi. *Res:* Immunoassay by radioisotope techniques; enzymatic assays in clinical chemistry. *Mailing Add:* 1029 Third Ave S Stillwater MN 55082-5818

CARTER, PAUL BEARNSON, bacteriology; deceased, see previous edition for last biography

CARTER, PAUL RICHARD, b St Louis, Mo, Apr 14, 22; m 44; c 2. MEDICINE, SURGERY. *Educ:* Union Col, Nebr, BA, 44; Loma Linda Univ, MD, 48; Am Bd Surg, dipl; Am Bd Thoracic Surg, dipl. *Prof Exp:* Surg resident, Los Angeles County Gen Hosp, 50-52, 54-56; thoracic surg resident, Olive View Sanitarium, Calif, 56-57; head physician surg, Los Angeles County Gen Hosp, 57-66; assoc prof surg, Univ Calif-Calif Col Med, 66-68; assoc prof in residence, 68-70, prof surg, Col Med, Univ Calif, Irvine, 70-76; CHIEF THORACIC SURG, PETTIS VET ADMIN HOSP, LOMA LINDA, CALIF, 76-; PROF SURG, LOMA LINDA UNIV, 76- *Concurrent Pos:* Fulbright scholar, Oxford Univ, 59; assoc prof surg, Sch Med, Loma Linda Univ, 62-66; chief surg, Rancho Los Amigos Hosp, 66-68; sr attend surgeon, Los Angeles County Gen Hosp & White Mem Hosp, Los Angeles; chief surg, Orange County Med Ctr, 70-76; mem sr staff, Intercommunity Hosp, Covina; mem active staff, Queen of the Valley Hosp, West Covina. *Mem:* Fel Am Col Surg; Am Col Chest Physicians; Am Asn Thoracic Surg; Soc Thoracic Surg; Int Soc Surg. *Res:* Mesenteric vascular occlusion; subphrenic abscess; volvulus of the colon and gall-bladder; pseudotumors of the lung; traumatic thoracic injuries; gallstone obstruction; surgical significance of sternal fracture; rupture of the bronchus; segmental reversal of small intestine; use of the Celestin tube for inoperable carcinoma of the esophagus and cardia; bronchotomy; diaphragmatic hernia; volvulus of stomach. *Mailing Add:* 238 W Badillo St Covina CA 91723-1906

CARTER, PHILIP BRIAN, b Chicago, Ill, July 31, 45; m 69; c 2. MICROBIOLOGY, IMMUNOLOGY. *Educ:* Univ Notre Dame, BS, 67, PhD(microbiol), 71. *Prof Exp:* Assoc fac mem biol, Ind Univ, South Bend, 71; res scientist & proj leader immunol, Ames Res Labs, Miles Labs Inc, 71; res assoc microbial immunol, 71-74, asst mem, 74-78, assoc mem mucosal immunol, Trudeau Inst, 79-81; assoc prof microbiol, Univ Ill, 81-82; assoc vchancellor res, 86-90, dir biotechnol, 86-90, PROF MICROBIOL, SCH VET MED, NC STATE UNIV, 82- *Concurrent Pos:* Fel, Trudeau Inst, 72-74; NIH-USPHS res career develop award, 75-80; guest lectr, Univ Ala, 76-81; consult, Am Asn Accreditation Lab Animal Care, 77-82, bd dirs, 90-92; vis investr, Sch Path, Univ Oxford, 78-79; coun mem, Inst Lab Animal Resources, Nat Res Coun, 78-81; vpres, Res Triangle Br, Am Asn Lab Animal Sci, 88, pres, 89. *Mem:* Asn Gnotobiotics (vpres, 76-77, pres, 77-78); Am Soc Microbiol; Am Asn Lab Animal Sci; Am Thoracic Soc; Reticuloendothelial Soc. *Res:* Immunology of mucosal surfaces; pathogenesis of enteric and respiratory infections; gnotobiology; diseases of laboratory animals. *Mailing Add:* Col Vet Med NC State Univ Box 8401 Raleigh NC 27695-8401

CARTER, R OWEN, JR, b Sligo, La, Aug 18, 15; m 44; c 3. TOXICOLOGY & SAFETY EVALUATION, PHYSICAL CHEMISTRY. *Educ:* Centenary Col, BS, 35; Univ Wis, PhD(phys chem), 39. *Prof Exp:* Sr chemist, Procter & Gamble Co, 39-41; res assoc, Nat Defense Res Comt Proj, Univ Wis, 41-44; sr chemist, Procter & Gamble Co, 44-69, dept head, 69-85; RETIRED. *Concurrent Pos:* Adj asst prof environ health, Univ Cincinnati Col Med, 71-; consult, 85- *Mem:* Am Chem Soc; Am Acad Dermat; Sigma Xi. *Res:* Molecular-kinetic study of proteins; gelatinization of high polymers; fats and oils; soaps and detergents; toxicology and safety evaluation. *Mailing Add:* 6573 Madeira Hills Dr Cincinnati OH 45243-3123

CARTER, RICHARD JOHN, b Brooklyn, NY, Aug 31, 41; m 63; c 2. REGULATORY AFFAIRS. *Educ:* St Francis Col, BS, 62; City Univ NY, MA, 71; Fairleigh Dickinson Univ, MBA, 81. *Prof Exp:* Chemist, Nat Can Corp, 62-67; supvr, FMC Corp, 67-73; mgr, Airwick Indust, 73-76; dir, Boyle Midway, Div Am Home Prod, 76-90; DIR SCI AFFAIRS, RECKITT & COLEMAN HOUSEHOLD PROD, DIV REKITT & COLEMAN US, 90- *Mem:* Am Chem Soc; Soc Cosmetic Chemist; AAAS; Chem Specialites Mfg Asn; Cosmetic Toiletries & Fragrance Asn. *Res:* Development of novel formulae and delivery systems for household pest control. *Mailing Add:* 1655 Valley Rd Wayne NJ 07474

CARTER, RICHARD P(ENCE), b Alderson, WVa, Jan 3, 15; m 39; c 3. CHEMICAL ENGINEERING. *Educ:* Va Polytech Inst, BS, 36; Purdue Univ, PhD(chem eng), 40. *Prof Exp:* Asst, Purdue Univ, 36-40; chem engr, Pilot Plant Develop, Exp Sta, Hercules Inc, Wilmington, Del, 40-42, head pilot plant dept, 42-43, chem engr, 43-45, chief chemist, 45-51, asst plant mgr, 51-59, tech asst to plant mgr, 59-61, tech mgr, NJ, 61-65, sr scientist, Allegany Ballistics Lab, 65-66, div mgr, Polymers Tech Ctr, Hercules Inc, 69-70, oper mgr, 71-80; engr consult plastics processing, 80-; RETIRED. *Mem:* Am Chem Soc; Am Inst Chem Engrs; Am Inst Aeronaut & Astronaut. *Res:* Organic chemical development; synthetic resins development; esterification; distillation; thermal cracking; hydrogenation; general engineering; explosives and solid rocket development; plastics processes. *Mailing Add:* 510 Ruxton Dr Georgian Terr Wilmington DE 19809

CARTER, RICHARD THOMAS, b Portland, Ore, Apr 4, 36; m 57; c 3. PARASITOLOGY, INVERTEBRATE ZOOLOGY. *Educ:* Portland State Univ, BS, 63; Ore State Univ, MA, 67, PhD(parasitol), 73. *Prof Exp:* Instr biol, Portland State Col, 62-63; asst prof, 68-73, ASSOC PROF BIOL, PAC UNIV, 73-, CHMN DEPT, 72- *Mem:* Sigma Xi. *Res:* Investigations into life stage development in Nanophyetus salmincola as related to life history. *Mailing Add:* Dept of Biol Pac Univ Forest Grove OR 97116

CARTER, ROBERT ELDRED, b Minneapolis, Minn, July 14, 23; m 46; c 3. PEDIATRICS, HEMATOLOGY. *Educ:* Univ Minn, BS, 45, MB, 46, MD, 48. *Prof Exp:* From instr to asst prof pediat, Univ Chicago, 56-59; from asst prof to prof, Univ Iowa, 59-67, from asst dean to assoc dean, 61-67; prof pediat, dean & dir, Sch Med, Univ Miss, 67-70; dean med educ prog, 70-74, prof microbiol & pediat, 74-78, EMER PROF, UNIV MINN, DULUTH, 78- *Concurrent Pos:* John & Mary R Markle scholar med sci, 57-62. *Mem:* Soc Pediat Res. *Res:* Biological effects of ionizing radiations; general hematology and bone marrow function. *Mailing Add:* PO Box 2105 Corrales MN 87048

CARTER, ROBERT EMERSON, b Philadelphia, Pa, Feb 3, 20; m 47; c 11. PHYSICS, RADIATION BIOLOGY. *Educ:* Washington Col, BS, 42; Univ Ill, MS, 47. *Prof Exp:* Jr physicist, Off Sci Res & Develop, Purdue Univ, 42-43, instr eng sci & war mgt training, 42-43; jr physicist, Los Alamos Sci Lab, Univ Calif, 43-45; asst nuclear res, Univ Ill, 46-48; mem staff, Los Alamos Sci Lab, 48-63; chmn phys sci dept, Armed Forces Radiobiological Res Inst, 63-75; head, Reactor Sci Div, Inst Resource Mgt, Bethesda, 76-80, proj mgr, US Nuclear Regulatory Comn, 80-87; PRIN PROG SPECIALIST, EG&G IDAHO, INC, 88- *Concurrent Pos:* Consult, Atomic Energy Comn, 47-48, Dept Army, 59-63; instr, Los Alamos Grad Ctr, Univ NMex, 56-63; mem, Subcomt Res Reactors Comt Phys Sci, Nat Acad Sci-Nat Res Coun, 64-69. *Mem:* Am Nuclear Soc; Am Phys Soc; AAAS; Sigma Xi. *Res:* Nuclear physics; nuclear engineering; molecular and atomic physics; mathematical physics; radiation physics; medical physics; radiation biology. *Mailing Add:* 9512 Edgeley Rd Bethesda MD 20814

CARTER, ROBERT EVERETT, b Jamaica, NY, Dec 3, 37; m 62, 75; c 3. PHYSICAL ORGANIC CHEMISTRY. *Educ:* Columbia Col, BA, 58; Calif Inst Technol, PhD(org chem), 62; Gothenburg Univ, Fil Dr, 70. *Prof Exp:* NIH fel gen med sci, Inst Org Chem, Gothenburg Univ, 62-65, lectr org chem, 64-65; consult, A B Hassle Co, 65-67; res assoc, 68-70, docent, Div Org Chem, 70-76, lectr org chem, Lund Inst Technol, 77-87; RES ADV, AB HASSLE, 88- *Concurrent Pos:* Res assoc & proj dir, Comput Org Chem, 80- *Mem:* Am Chem Soc. *Res:* Applications of computers and computer graphics to organic chemistry. *Mailing Add:* AB HASSLE/KF Org Chem Molndal 43183 Sweden

CARTER, ROBERT L(EROY), b Leavenworth, Kans, Aug 22, 18; m 41; c 5. DIRECT ENERGY CONVERSION. *Educ:* Univ Okla, BS, 41; Duke Univ, PhD(physics), 49. *Prof Exp:* Technician, Eastman Kodak Co, 41-42, physicist, Tenn Eastman Corp, 45-46; res asst physics, Duke Univ, 46-49; res engr, Atomics Int Div, NAm Aviation, Inc, 49-53, group leader mat tech, 53-56, nuclear reactor proj engr, 56-59, sr res specialist direct energy conversion, 59-62; assoc prof elec eng, 62-63, prof, 63-88, EMER PROF ELEC & NUCLEAR ENG, UNIV MO-COLUMBIA, 88. *Concurrent Pos:* Vis staff mem, Los Alamos Sci Lab, 68-69 & Oak Ridge Nat Lab, 80; consult, Oak Ridge Nat Lab, 80-88. *Mem:* Am Phys Soc; Am Nuclear Soc; AAAS; Nat Soc Prof Engrs. *Res:* Mass spectrography; microwave spectroscopy; radiation effects; nuclear reactor materials; energy conversion; low level radioactive waste management; applications of superconductivity. *Mailing Add:* Dept of Elec Eng Univ of Mo Columbia MO 65211

CARTER, ROBERT SAGUE, b Poughkeepsie, NY, Nov 15, 25; m 49; c 4. NUCLEAR PHYSICS. *Educ:* Princeton Univ, AB, 48; Harvard Univ, MA, 49, PhD(physics), 52. *Prof Exp:* Assoc physicist, Brookhaven Nat Lab, 52-56; physicist, Westinghouse Res Labs, 56-58; physicist, 58-69, CHIEF REACTOR RADIATION DIV, NAT BUR STANDARDS, 69- *Mem:* Fel Am Nuclear Soc; fel NY Acad Sci; Am Phys Soc. *Res:* Elementary particle physics; neutron physics; reactor engineering. *Mailing Add:* 28432 Mallard Lane Easton MD 21601

CARTER, STEFAN A, b Warsaw, Poland, Mar 25, 28; Can citizen; m 58; c 2. PHYSIOLOGY. *Educ:* Univ Man, MD & BSc, 54, MSc, 56. *Prof Exp:* Lectr, 58-59, from asst prof to assoc prof, 59-77, prof physiol, Fac Med, 77-81, PROF MED, UNIV MAN, 81- *Concurrent Pos:* Clin res fel, NY Hosp-

Cornell Med Ctr, 56-67; Nat Res Coun Can fel physiol, Mayo Found, Univ Minn, 57-58; Man Heart Found grant-in-aid, 59-76; dir cardiovasc sect, Clin Invest Unit, St Boniface Gen Hosp, 58-; dir long term anticoagulant & peripheral vascular dis clin, St Boniface Hosp, 67-; mem coun circulation, Am Heart Asn, 69; St Boniface Gen Hosp Res Found grant-in-aid, 77- *Mem:* Can Physiol Soc; Can Soc Clin Invest; Can Cardiovasc Soc. *Res:* Vasomotor regulation in diabetes; function and elasticity of medium-sized arteries and arterial pressure pulses in health and arterial diseases; arterial occlusive disease in the limbs; biofeedback and vasodilators in Raynaud's diseases. *Mailing Add:* 234 Elm St Winnipeg MB R3M 3P2 Can

CARTER, STEPHEN KEITH, b New York, NY, Oct 30, 37; m 66; c 2. INTERNAL MEDICINE, ONCOLOGY. *Educ:* Columbia Col, AB, 59; New York Med Col, MD, 63. *Prof Exp:* Intern med, Lenox Hill Hosp, 63-64, resident, 64-66, chief resident, 66-67; spec asst to sci dir chemother, Nat Cancer Inst, 67-68, chief cancer ther eval br, 68-73, assoc dir cancer ther eval, 73-74, dep dir div cancer treatment, 74-76; dir, Northern Calif Cancer Prog, Palo Alto, 76-82; SR VPRES PHARMACEUT & MED DEVELOP, PHARMACEUT RES & DEVELOP DIV, BRISTOL-MYERS CO, 82- *Concurrent Pos:* Consult prof med, Stanford Univ, 76-; clin prof med, Univ Calif, San Francisco, 76- *Mailing Add:* Bristol Myers Squibb Co Rte 206 Province Line Rd Lawrenceville NJ 08543-4000

CARTER, THOMAS EDWARD, JR, b Athens, Ga, Jan 23, 53; m 75; c 1. QUANTITATIVE GENETICS, STATISTICS. *Educ:* Univ Ga, BS, 75; NC State Univ, Raleigh, MS, 77, PhD(plant breeding), 80. *Prof Exp:* Res Geneticist Soybean, Agr Res Serv, NC State Univ, USDA, 81-85; DEPT CROP SCI, NC STATE UNIV, 85- *Concurrent Pos:* Prin investr & res grant, Drought Tolerance in Soybeans, 81- *Mem:* Crop Sci Soc Am; Am Soc Agron; Biomet Soc. *Res:* Genetics of drought tolerance in soybean; development of stategies for plant breeding using quantitative genetics; application of plant physiology to plant breeding; efficiency of breeding methods. *Mailing Add:* Dept Crop Sci NC State Univ Box 7620 Raleigh NC 27695

CARTER, TIMOTHY HOWARD, b Los Angeles, Calif, Nov 6, 44; m 76; c 2. CELL BIOLOGY, VIROLOGY. *Educ:* Harvard Col, AB, 66; Princeton Univ, PhD(biochem), 72. *Prof Exp:* Fel, Sch Med, Univ Pa, 72-73, asst prof microbiol, 75-78; fel, Col Physicians & Surgeons, Columbia Univ, 73-75; asst prof biol sci, 78-81, ASSOC PROF BIOL SCI, ST JOHN'S UNIV, 81- *Concurrent Pos:* Prin investr, Nat Cancer Inst, NIH, 78-82, 84- , Nat Oceanog & Atmospheric Asn, 84-87; consult, Bio Glass Serum, Inc, 81-83; mem sci adv bd, Nuclear & Genetic Technol, Inc, 83-85; mem, Ad Hoc Study Sect, Exp Virol, NIH, 85 & rev panels, Minority Biomed Res Support Prog, NIH, 90- *Mem:* Am Soc Biol Chemists; Am Soc Virol; Am Soc Microbiol; Sigma Xi. *Res:* Regulation of gene expression in eukaryotes and their viruses; mechanism of action of tumor promoting substances; effects of aging on biochemical responses to carcinogens and tumor promoters; environmental virology. *Mailing Add:* Dept Biol Sci St John's Univ Jamaica NY 11439

CARTER, WALTER HANSBROUGH, JR, b Winchester, Va, Mar 20, 41; m 63; c 3. BIOSTATISTICS. *Educ:* Univ Richmond, BS, 63; Va Polytech Inst, MS, 66, PhD(statist), 68. *Prof Exp:* Asst prof biomet, 68-72, assoc prof 72-79, PROF BIOSTATIST, MED COL VA, VA COMMONWEALTH UNIV, 79-, CHMN, 84- *Mem:* Sigma Xi; fel Am Statist Asn; Biomet Soc. *Res:* Treatment optimization in combination chemotherapy of cancer; clinical trials; design and analysis of response surface experiments. *Mailing Add:* Dept Biostatist Box 32 Med Col Va Sta Richmond VA 23298

CARTER, WILLIAM ALFRED, b Ada, Okla, Sept 16, 35. ZOOLOGY, ORNITHOLOGY. *Educ:* ECent State Col, BS, 57; Okla State Univ, MS, 60, PhD(zool), 65. *Prof Exp:* Instr biol, Northwestern State Col, Okla, 63-64; from asst prof to assoc prof, 64-71, chmn dept biol, 72-81, PROF BIOL, E CENT OKLA STATE UNIV, 71- *Concurrent Pos:* Res assoc ornith, Okla Mus Natural Hist, Univ Okla, 71-; Okla coordr, US Fish & Wildlife Breeding Bird Surv, 74-; consult, Environ Impact Assessment, Williams Bros Eng Co, 75-78. *Mem:* Am Ornithologists Union; Am Soc Ichthyologists & Herpetologists; Cooper Ornith Soc; Wilson Ornith Soc; Sigma Xi; Soc Study Amphibians & Reptiles. *Res:* Ecology and distribution of birds and herps. *Mailing Add:* Dept Biol E Cent Okla State Univ Ada OK 74820

CARTER, WILLIAM CASWELL, b Waterville, Maine, Jan 16, 17; m 57; c 3. MATHEMATICS. *Educ:* Colby Col, AB, 38; Harvard Univ, PhD(math), 47. *Hon Degrees:* DSc, Univ Newcastle Tyne, 88. *Prof Exp:* Mathematician, Ballistic Res Lab, Aberdeen, 47-52; instr math, Univ Md, 47-51, Johns Hopkins Univ, 51-52; mathematician, Computer Dept, Raytheon Mfg Co, 52-55; dept mgr systs anal, Datamatic, Minn-Honeywell, 55-59; sr engr, T J Watson Res Ctr, Int Bus Mach Corp, 59-61, data systs div, 61-66, staff mem, 66-86; CONSULT, 86- *Concurrent Pos:* Instr, Boston Univ, 52-58; mem, Adv Comt Space & Syst Electronics, NASA, 83-87. *Honors & Awards:* Cert Merit, NASA, 70; Computer Soc Honor Roll, Inst Elec & Electronic Engrs, 74, Computer Soc Award, Contrib Tech Comt, 82. *Mem:* Fel Inst Elec & Electronic Engrs. *Res:* Fault tolerant computer system design and design methods; combinatorial math; computer systems design and analysis; logic design and methods of logic design. *Mailing Add:* PO Box 10 Bailey Island ME 04003

CARTER, WILLIAM DOUGLAS, b Keene, NH, Apr 24, 26; m 50; c 4. GEOLOGY. *Educ:* Dartmouth Col, AB, 49. *Prof Exp:* Geol field asst, Permafrost, US Geol Surv, 48-49, petrol res, 50, geologist uranium explor prog, 51-57, mining geologist & tech adv to Govt Chile, AID, 57-62, commodity geologist, Light Metals & Indust Minerals Br, Resources Res Group, 62-65, geol coordr remote sensing eval & coord staff, 65-67, chmn mineral & land resources working group, Earth Resources Observ Systs Prog, 67-70; asst mgr, Appl Res, Earth Resources Observ Systs Prog, US Geol Surv, 70-74, res scientist, 75-81; CONSULT GEOLOGIST, 82-; PRES, GLOBEX INC, 83- *Concurrent Pos:* Int Union Geol Sci rep to Comt Space Res, Int Coun Sci Unions, 73-; lectr remote sensing, Am Asn Petrol Geologists Continuing Educ Prog, 75-82. *Honors & Awards:* Alan Gordon Award, Am Soc Photogram, 78. *Mem:* Fel Geol Soc Am; Soc Econ Geologists; Am Asn Petrol Geologists; Am Soc Photogram; Soc Explor Geophysicists; hon mem Remote Sensing Soc Latin Am; Geol Soc Nepal. *Res:* Use of satellite data in studies of the tectonics and location of ore deposits in the United States and Andes Mountain region, South America; mineral and energy deposits, chiefly uranium, copper, silica, phosphate and potassium; photogeology; remote sensing; space applications. *Mailing Add:* 2404 Paddock Lane Reston VA 22091-2606

CARTER, WILLIAM EUGENE, b Steubenville, Ohio, Oct 16, 39; m 60; c 3. GEODESY. *Educ:* Univ Pittsburgh, BS, 61; Ohio State Univ, MS, 65; Univ Ariz, PhD(civil eng), 73. *Prof Exp:* Chief astron surv team, 1381st Geod Surv Squad, US Air Force, 61-63, chief, Astron Surv Br, 1st Geod Surv Squad, 66-69, res geodesist lunar laser ranging, Air Force Cambridge Res Lab, 69-72; Lure proj mgr, Inst Astron, Univ Hawaii, 72-76; asst chief, Gravity, Astron & Satellite Br, 77-79, chief, Gravity, Astron & Space Geod Div, 79-82, CHIEF, ADVAN TECHNOL SECT, GEOD RES & DEVELOP LAB, NAT GEOD SURV, 82- *Concurrent Pos:* Mem, Lunar Ranging Exp Team, 75-76; mem, earth rotation working group, Int Astron Union/Int Union Geod & Geophys & prin coordr Very Long Baseline Interferometry, 83-, mem directing bd, Int Earth Rotation Serv, 88- *Mem:* Am Geophys Union; Am Astron Soc; Int Astron Union; Int Asn Geod. *Res:* Application of radio interferometric surveying to the studies of polar motion, earth rotation, crustal deformations, plate motion; research and development of advanced geodetic instrumentation and observational methods. *Mailing Add:* Nat Geodetic Survey N/CG114 Rockville MD 20852

CARTER, WILLIAM G, BIOCHEMISTRY. *Prof Exp:* AT DEPT BIOCHEM & ONCOL, FRED HUTCHINSON CANCER RES CTR, SEATTLE, WASH. *Mailing Add:* Dept Biochem & Oncol Fred Hutchinson Cancer Rec Ctr 1124 Columbia St Seattle WA 98104

CARTER, WILLIAM HAROLD, b Houston, Tex, Nov 17, 38; c 2. OPTICAL PHYSICS. *Educ:* Univ Tex, BS, 62, MS, 63, PhD(elec eng), 66. *Prof Exp:* Res assoc elec eng, Univ Tex, 63-66, instr, 66; contract monitor optics, US Army Mil Intel Corps, 67-69; res assoc physics, Univ Rochester, 69-70; prof elec eng, Univ Nebr-Lincoln, 81-82; RES PHYSICIST OPTICS, OPTICAL SCI DIV, NAVAL RES LAB, 71- *Concurrent Pos:* Asst prof lectr, George Washington Univ, 68-69, assoc prof lectr, 71-75; vis res fel, Univ Reading, Eng, 76-77; assoc ed, J Optical Soc Am, 79-83; instr elec eng, Johns Hopkins Univ, 88- *Mem:* Fel Int Soc Optical Eng; Am Phys Soc; fel Optical Soc Am; Sigma Xi; sr mem Inst Elec & Electronics Engrs. *Res:* Electromagnetic wave propagation; holography; digital image processing; coherence theory; angular spectrum representation; radio astronomy; laser resonater theory; scattering theory; microscopy; electromagnetic beam theory. *Mailing Add:* Code 8304 Naval Res Lab Washington DC 20375

CARTER, WILLIAM WALTON, b Pensacola, Fla, Nov 7, 21; m 45; c 4. PHYSICS. *Educ:* Carnegie Inst Technol, BS, 43; Calif Inst Technol, PhD(physics), 49. *Prof Exp:* Researcher radar, Naval Res Labs, 44-46; staff mem, sect chmn & group leader physics & eng, Los Alamos Sci Labs, 49-59; chief scientist, Missile Command, US Army, 59-67; asst dir defense res & eng, Nuclear Progs, Off Secy Defense, 67-71; tech dir, US Army Electronic Res & Develop Comt, Harry Diamond Labs, 71-84; SR SCIENTIST, PAC-SIERRA RES CORP, 84- *Concurrent Pos:* Proj leader first thermonuclear weapon, 53-56; prin investr, High Power Microwave Prog, US Air Force, 85. *Mem:* Am Phys Soc; AAAS; assoc fel Am Inst Aeronaut & Astronaut. *Res:* Nuclear physics; weapons and their effects; electronics and electronic components; spectroscopy. *Mailing Add:* 1124 Ormond Ct McLean VA 22101

CARTERETTE, EDWARD CALVIN HAYES, b Mt Tabor, NC, July 10, 21; m 55; c 1. PSYCHOACOUSTICS, NEUROPSYCHOLOGY. *Educ:* Univ Chicago, AB, 49; Harvard Univ, AB, 52; Ind Univ, MA, 54, PhD(psychol), 57. *Prof Exp:* Res staff mem acoustics lab, Mass Inst Technol, 51-52; from instr to asst prof psychol, Univ Calif, Los Angeles, 56-63, assoc prof exp psychol, 63-68; vis assoc prof psychol, Univ Calif, Berkeley, 65-66; PROF EXP PSYCHOL & MUSIC, UNIV CALIF, LOS ANGELES, 68-; DIR PACKARD HUMANITIES INST, 86- *Concurrent Pos:* NSF fel, Royal Inst Technol, Stockholm, Sweden & Cambridge Univ, Eng, 60-61; NSF sr fel, Inst Math Studies in Soc Sci, Stanford Univ, 64-65; rev ed, J Auditory Res, 60-69; assoc ed, Perception & Psychophysics, 71- & Music Perception, 81-; mem, Brain Res Inst, Univ Calif, Los Angeles, 74-; distinguished lectr, Am Psychol Asn, 79- *Mem:* Fel Acoust Soc Am; fel AAAS; fel Am Psychol Asn; fel Soc Exp Psychologists (secy-treas, 82-); Psychonomic Soc; Sigma Xi; Scientific Res Soc, 88- *Res:* Psychoacoustics, hearing and speech perception; neuropsychology; mathematical models of cognitive processes. *Mailing Add:* B256 Franz Hall Univ Calif Los Angeles CA 90024

CARTER-SU, CHRISTIN, b Oakland, Calif, May 4, 50. CELL PHYSIOLOGY, HORMONE ACTION. *Educ:* Univ Rochester, PhD(biophys), 78. *Prof Exp:* Fel, Brown Univ, 78-81; asst prof, 81-87, ASSOC PROF PHYSIOL, SCH MED, UNIV MICH, 87- *Mem:* Endocrine Soc; Am Physiol Soc; Am Diabetes Asn; Am Soc Biochem & Molecular Biol. *Res:* Hormonal regulation and mechanism of action of glucose transport systems; growth hormone receptor structure and mechanism of action. *Mailing Add:* Dept Physiol Univ Mich Med Sch Ann Arbor MI 48109-0622

CARTIER, GEORGE THOMAS, b Scranton, Pa, Jan 26, 24; m 46; c 2. APPLIED CHEMISTRY. *Educ:* Haverford Col, AB, 49. *Prof Exp:* Asst org chem, Smith Kline & French, 49-51; tech liaison & mkt develop, Quaker Chem Prods, 51-55; lab dir, Res & Develop Qual Control, A M Collins Div, Int Paper Co, 55-61; consult chemist, Web Processing, 61-65; pres, Keystone Filter Media Co, 65-74. *Concurrent Pos:* Consult, 74- *Mem:* Am Chem Soc; Tech Asn Pulp & Paper Indust; Asn Consult Chemists & Chem Engrs. *Res:* High polymer applications research; decorative and functional paper and paper board coatings research and development; market development; fibrous filter media; high efficiency filter construction. *Mailing Add:* 311 Middle Rd Falmouth ME 04105

CARTIER, JEAN JACQUES, b Beauharnois, Que, Apr 1, 27; m 54. ENTOMOLOGY. *Educ:* Univ Montreal, BA, 48, BS, 52, MS, 53; Kans State Univ, PhD(entom), 56. *Prof Exp:* Res officer, St Jean Res Sta, Can Dept Agr, 53-69, res coordr entom, Res Br, Cent Exp farm, 69-75, asst dir gen, 75-78, dir gen, Eastern region, Can Dept Agr, 78-80, Ont Region, 80-87, Cent Exp Farm, 87-88; RETIRED. *Mem:* Entom Soc Can. *Res:* Crop plants resistance to insects; biology of aphids, particularly aphid biotypes. *Mailing Add:* 12 Glenbrook Way Ottawa ON K2G 0V2 Can

CARTIER, PETER G, b Green Bay, Wis. PHYSICAL CHEMISTRY, POLYMER CHEMISTRY. *Educ:* Lawrence Univ, BA, 68; Cornell Univ, PhD(phys chem), 73. *Prof Exp:* SR CHEMIST, ROHM AND HAAS CO, 73- *Mem:* Am Chem Soc. *Res:* Surface chemistry; the development and characterization of adsorbents and ion exchange resins; reverse osmosis membranes; chromatographic media. *Mailing Add:* Rohm & Haas Spring House PA 19477

CARTLEDGE, FRANK, b Emory, Ga, Aug 26, 38; m 61; c 3. ORGANIC CHEMISTRY. *Educ:* King Col, BA, 60; Iowa State Univ, PhD(org chem), 64. *Prof Exp:* Nat Sci Found res fel organosilicon chem, Univ Sussex, 64-65; fel, Univ Gottingen, 65-66; from asst prof to assoc prof, 66-79, PROF CHEM, LA STATE UNIV, BATON ROUGE, 80- *Mem:* Am Chem Soc; Sigma Xi. *Res:* Synthesis and kinetics of reactions of group four organic compounds; hazardous waste management. *Mailing Add:* Dept of Chem La State Univ Baton Rouge LA 70803-1804

CARTMELL, ROBERT ROOT, b Hagerstown, Ind, Dec 20, 21; m 48; c 3. CATALYTIC CRACKING, REFORMING. *Educ:* Purdue Univ, BSChe, 42, MSChe, 47. *Prof Exp:* Grad asst chem eng, Purdue Univ, 46-47; engr, Oil Field Res, Chevron, 47-48 & Petrol Prod, Petrol Eng Assocs, 48-50; engr petrol refining, Amoco Corp, 52-69, consult petrol refining, 69-86; RETIRED. *Mem:* Am Inst Chem Engrs. *Res:* Eight patents on catalytic cracking. *Mailing Add:* 3701 W 105th Ave Crown Point IN 46307

CARTMILL, MATT, b Los Angeles, Calif; m 71; c 1. PHYSICAL ANTHROPOLOGY. *Educ:* Pomona Col, BA, 64; Univ Chicago, MA, 66, PhD(primate evolution), 70. *Prof Exp:* Assoc anat, Med Ctr, Duke Univ, 69-70, asst prof, 70-74, sociol & anthrop, 70-74, assoc prof anthrop, 74-, prof anat, 81-87, PROF BIOL ANTHROP & ANAT, DUKE UNIV, 87- *Concurrent Pos:* NIH career develop award, 75; co-managing ed, Int J Primatol, 78-79; managing ed, Am J Phys Anthrop, 89-; J S Guggenheim fel, 85-86. *Mem:* Am Anthrop Asn; Am Asn Phys Anthrop; Am Soc Mammal; Primate Soc Gt Brit; Int Primate Soc; Sigma Xi. *Res:* Primate evolution. *Mailing Add:* 3007 Montgomery St Durham NC 27708

CARTON, CHARLES ALLAN, b New York, NY, Feb 28, 20; m 57; c 2. NEUROSURGERY. *Educ:* Yale Univ, BA, 41; Columbia Univ, MD, 44; Am Bd Neurol Surg, dipl, 55. *Prof Exp:* Intern med & surg, Bellevue Hosp, New York, 44-45; asst resident neurol & neuropath, Neurol Inst, NY, 47-48, fel neurosurg, 49, asst resident, 50-51, chief resident, 52; asst prof, Baylor Col Med, 53-56; asst clin prof, Albert Einstein Col Med, 56-58, assoc clin prof, 58-61; assoc clin prof & mem attend staff, 61-74, CLIN PROF NEUROSURG, MED CTR, UNIV CALIF, LOS ANGELES, 74- *Concurrent Pos:* Asst resident, Presby Hosp, New York, 49; chief neurosurg, Vet Admin Hosp, Houston, 53; asst attend neurosurgeon, Jefferson Davis Hosp, 54; asst neurosurgeon, Methodist Hosp & Clin; asst neurosurgeon, M D Anderson Hosp, 55; exec officer, Div Neurosurg, Montefiore Hosp, New York, 56, chief, 58; vis neurosurgeon, Bronx Munic Hosp Ctr, 58; vis neurosurgeon, Morrisania Hosp, 59; adj, Cedars Lebanon Hosp, Los Angeles, 61, asst, 63; asst, Mt Sinai Hosp, 64; attend & co-chief neurosurg, Cedars-Sinai Med Ctr, 71-75. *Mem:* AAAS; Am Asn Neuropathologists; Am Acad Neurol; fel Am Col Surg; AMA. *Res:* Cerebrovascular disease aneurysm and small vessel anastomosis; hydrocephalus. *Mailing Add:* 8631 W 3rd Ste 710 E Los Angeles CA 90048

CARTON, EDWIN BECK, b Baltimore, Md, Apr 27, 27; m 49; c 3. POLYURETHANE FOAMS, POLYOLEFIN DISPERSIONS. *Educ:* Johns Hopkins Univ, BA, 49; Univ Md, MS, 52; Pa State Univ, PhD(org chem), 55. *Prof Exp:* Res chemist, Scott Paper Co, Philadelphia, 55-58, res proj leader, 58-60; mgr, dispersions develop, Cabot Corp, Philadelphia, 60-64, head plastics & corp appl res, Cent Res Ctr, Billerica, Mass, 64-66, mgr, mkt res, Oxides & Pigment Div, Boston, 66-68; mgr, develop, Chomerics, Inc, 68-69; asst dir, res & develop, Bostik Chem Div, Emhart Corp, 69-73, dir, prod develop, 73-76; pres, Sharpe Plastic Prod, Inc, 76-80, Rim Xinde Int, 79-83; vpres, eng & prod develop, Bailey Corp, 83-85. *Concurrent Pos:* Self-employed consult, 85- *Mem:* Fel AAAS; fel Am Inst Chemists; Am Chem Soc. *Res:* Rigid and flexible polyurethane foams, including development of open cell structures; polyolefin dispersions, especially polypropylene; testing crosslinkable olefins with and without fillers; wet strength resins for paper and polymeric paper softeners. *Mailing Add:* 25 Sheffield Rd Newtonville MA 02160

CARTWRIGHT, AUBREY LEE, JR, b Elizabeth City, NC, Jan 13, 52; m; c 3. CELL GROWTH, HORMONE BIOCHEMISTRY. *Educ:* NC State Univ, PhD(nutrit), 82. *Prof Exp:* RES PHYSIOLOGIST, AGR RES SERV, USDA, 82- *Concurrent Pos:* Location coordr, res leader, lead scientist & adj prof, Univ Del. *Mem:* NY Acad Sci; Soc Exp Biol Med; Am Inst Nutrit; Fedn Am Soc Exp Biol; Poultry Sci Asn; Am Soc Animal Sci. *Res:* Develop concepts and principles describing processes that regulate growth of adipose and lean tissue; identify hormonal factors that regulate these processes; identify biochemical and metabolic pathways, and genes that regulate fat and muscle tissue deposition. *Mailing Add:* USDA Agr Res Serv Poultry Res Lab RD 2 Box 600 Georgetown DE 19947

CARTWRIGHT, DAVID CHAPMAN, b Minneapolis, Minn, Dec 2, 37; m 65; c 2. CHEMICAL & ATMOSPHERIC PHYSICS. *Educ:* Hamline Univ, BS, 62; Calif Inst Technol, MS, 63, PhD(chem physics, physics), 68. *Prof Exp:* NATO fel astrophyics & physics, Max Planck Inst Extraterrestrial Physics, 67-68; fel physics, Univ Colo, Boulder, 68-69; mem tech staff, Space Physics Lab, Aerospace Corp, 69-74; staff mem, Theoret Div, Los Alamos Nat Lab, 74-75, dep group leader, Laser Theory Group, 75-76, group leader, 75-80, dep div leader, 80-84, PROG DIR, INERTIAL FUSION & LASER TECHNOL, LOS ALAMOS NAT LAB, 84- *Mem:* Fel Am Phys Soc; Am Geophys Union. *Res:* Electron impact processes in atoms and molecules; spectral properties of simple molecules; auroral and ionospheric processes; gas laser processes; plasma physics of inertial fusion; free-electron lasers. *Mailing Add:* MS-E527 Los Alamos Nat Lab PO Box 1663 Los Alamos NM 87545

CARTWRIGHT, HUGH MANNING, b Hitchin, UK. PHYSICAL CHEMISTRY, SPECTROSCOPY. *Educ:* Univ EAnglia, BSc, 69, PhD(chem), 72. *Prof Exp:* Asst prof Univ Victoria, BC, 73-76, sr lab instr chem, 77-84, LAB OFFICER, OXFORD UNIV, 84- *Concurrent Pos:* Fel, Univ Victoria, BC, 73-74; instr & consult, BC Govt, 78-84. *Res:* chemical education; genetic algorithms; chemometrics. *Mailing Add:* Physical Chem Labs Univ Oxford South Parks Rd Oxford 0X1 3QZ England

CARTWRIGHT, KEROS, b Los Angeles, Calif, July 25, 34; m 55; c 4. HYDROGEOLOGY. *Educ:* Univ Calif, Berkeley, AB, 59; Univ Nev, MS, 61; Univ Ill, PhD(geol), 73. *Prof Exp:* Asst hydrologist, Humboldt River Res Proj, Nev, 59-61; geologist & head, Hydrogeol & Geophysics Sect, 74-84, prin scientist & head, Basic & Environ group, 84-88, HYDROGEOLOGIST, ILL GEOL SURV, 61-, PRIN RES SCIENTIST, 88- *Concurrent Pos:* Vis assoc prof geol, Univ Waterloo, Ont, 76-77; adj prof geol, Northern Ill Univ, 79- & Univ Ill, Urbana- Champaign, 85-; distinguished lectr, Asn Ground-Water Scientists & Engrs, 87. *Honors & Awards:* Birdsall Distinguished Lectr, Geol Soc Am, 87. *Mem:* Am Geophys Union; Geol Soc Am; Am Inst Hydrol; Am Inst Prof Geologists; Int Asn Hydrologists. *Res:* Hydrogeology; author of over 100 publications. *Mailing Add:* 615 E Peabody Dr Champaign IL 61820

CARTWRIGHT, THOMAS CAMPBELL, b York, SC, Mar 8, 24; m 46; c 4. ANIMAL BREEDING. *Educ:* Clemson Col, BS, 48; Tex A&M Univ, MS, 49, PhD, 54. *Prof Exp:* Animal husbandman & geneticist, Agr Exp Sta, 52-58, PROF ANIMAL BREEDING, TEX A&M UNIV, 58- *Mem:* AAAS; Am Genetic Asn; Am Soc Animal Sci. *Res:* Population genetics; genetics of cattle. *Mailing Add:* Dept of Animal Sci Tex A&M Univ College Station TX 77843

CARTY, ARTHUR JOHN, b Hookergate, Eng, Sept 12, 40; m 67; c 3. INORGANIC CHEMISTRY, ORGANOMETALLIC CHEMISTRY. *Educ:* Univ Nottingham, BSc, 62, PhD(chem), 65. *Hon Degrees:* DSc, Univ Rennes, France, 86. *Prof Exp:* Asst prof chem, Mem Univ Nfld, 65-67; from asst prof to assoc prof, 67-75, assoc prof chem, 79-75, PROF CHEM, UNIV WATERLOO, 75-, DEAN RES, 89- *Concurrent Pos:* Royal Soc Nuffield Found fel, 74; actg dir, Guelph-Waterloo Ctr Grad Studies in Chem, 75-76, dir, 76-79; chmn, Chem Grants Selection Comt, Nat Sci & Eng Res Coun, 79-80, group chmn chem, 80-; chmn chem dept, Univ Waterloo, 83-; vpres, Can Soc Chem, 89-90, pres, 90-91. *Honors & Awards:* Alcan Award, Chem Inst Can, 84. *Mem:* Am Chem Soc; Chem Inst Can; fel Royal Soc Can. *Res:* Synthetic and structural inorganic and organometallic chemistry; metal clusters and reactivity patterns for unsaturated molecules in clusters; chemistry of metal acetylides and metal carbonyls; phosphinoacetylenes as ligands and synthons; phosphido bridged bi-and polynuclear complexes; new materials, organometallic polymers with novel conducting and non-linear optical properties. *Mailing Add:* Dept Chem Univ Waterloo Waterloo ON N2L 3G1 Can

CARTY, DANIEL T, b Greenville, Tex, Aug 19, 35; m 58; c 3. ORGANIC CHEMISTRY, POLYMER CHEMISTRY. *Educ:* Univ Calif, Riverside, BA, 61; Univ Hawaii, MS, 63; Stanford Univ, PhD(chem), 68. *Prof Exp:* Chemist, Stanford Res Inst, 63-65; res assoc, Inst Org Chem, Karlsruhe Tech Univ, Ger, 67-68; chemist, Rohm and Haas Co, Bristol, Pa, 68-72; sr res assoc, Fibers Pioneering Res, 72-76; sr res assoc, Plastics Pioneering Res, Pa, 76-77, proj leader textile chem res, 77-79, mkt develop mgr chem specialties, Pacific Region, Plastics Pioneering Res, Calif, 79-83; PROJ LEADER II, NEW PROD DEVELOP, BUS & TECHNOL ACQUISITIONS, CLOROX TECH CTR, PLEASANTON, CALIF. *Concurrent Pos:* Univ Karlsruhe, Ger, post-doctorate, 68. *Mem:* Am Chem Soc. *Res:* Homopolymers and copolymers of acrylates and their alloys with other polymeric systems; chemistry of novel small ring hydrocarbons and valence bond isomers of benzene; chemical modification of nylon and polyester fibers; oxidation chemistry of bleaches; colloid chemistry of aqueous systems; surfactant chemistry. *Mailing Add:* 50 Tyrrel Ct Danville CA 94526

CARUBELLI, RAOUL, b Cordoba, Arg, June 17, 29; US citizen; m 59; c 2. BIOCHEMISTRY. *Educ:* Cordoba Nat Univ, PhD(biochem), 60. *Prof Exp:* Lab instr anal chem pharmaceut drugs & quant anal & biol chem, Cordoba Nat Univ, 53-56; res assoc biochem, Okla Med Res Found, 57-59, biochemist, 60-64, sr investr, 64-65, assoc, 65-67, assoc mem, 67-73; from asst prof to prof biochem, 63-81, PROF BIOCHEM & MOLECULAR BIOL, OKLA MED RES FOUND, SCH, UNIV OKLA, 70-; SR SCIENTIST, DEAN A MCGEE MED EYE INST, 87- *Concurrent Pos:* USPHS res grants, 62-; NIH res career develop award, 68-72; vis investr, Max Planck Inst Virus Res, Ger, 63-64; vis investr, Ger Cancer Res Ctr, Heidelberg, 76-77; res grants, Am Inst Cancer Res, 83-; mem staff, Okla Med Res Found, 73-; adj assoc prof ophthal, Univ Okla Med Sch, 87- *Mem:* AAAS; Sigma Xi; Am Chem Soc; Am Soc Biol Chem; Arg Acad Med Sci. *Res:* Biological and carbohydrate chemistry; chemistry and metabolism of carbohydrates and glycoproteins; dietary modulation of chemical carcinogenesis; effect of free radicals on eye components. *Mailing Add:* Dean A McGee Eye Inst 608 Stanton L Young Blvd Oklahoma City OK 73104-5074

CARUCCIO, FRANK THOMAS, b New York, NY, Sept 7, 35; m 63; c 2. GROUNDWATER GEOLOGY, ENVIRONMENTAL GEOLOGY. *Educ:* City Col NY, BS, 58; Pa State Univ, MS, 63, PhD(geol), 67. *Prof Exp:* Res assoc geol, Pa State Univ, 67-68; asst prof, State Univ NY Col New Paltz, 68-70, assoc prof, 70-71; assoc prof, 71-82, PROF GEOL, UNIV SC, 82- *Concurrent Pos:* State Univ NY Res Found grant, 69; consult geologist, Res

Ctr, Uniroyal, Inc, NY, 69-71; Environ Protection Agency res grants, 73-81, demonstration grant, 75-78, NSF grant, 80-83, Off Surface Mining Reclamation & Enforcement grants, 82-85, 85-88 & Environ Protection Agency grant,85-87; mem, G W Resource & Coal Mining Comt, Nat Res Coun, Nat Acad Sci. *Mem:* Am Geophys Union; Am Water Works Asn; Water Resources Asn; Water Pollution Control Fedn; Soc Environ Geochem & Health; Beijer Inst; Royal Swed Acad Scis; fel Geol Soc Am; AAAS; Nat Water Well Asn. *Res:* Groundwater, its occurrence, movement and quality in relation to the hydrogeologic environment; interactions of pollutants within the hydrogeologic regime and their affects on the ground water potability; surface coal mining; environmental impacts (acid mine drainage). *Mailing Add:* Dept of Geol Univ of SC Columbia SC 29208

CARUOLO, EDWARD VITANGELO, b Providence, RI, Nov 1, 31; m 53; c 4. ANIMAL PHYSIOLOGY, MEDICAL PHYSIOLOGY. *Educ:* Univ RI, BS, 53; Univ Conn, MS, 55; Univ Minn, PhD(physiol), 63. *Prof Exp:* Instr physiol, Univ Minn, 60-63; from asst prof to assoc prof, 63-83, chmn fac physiol, 74-77, PROF EXP PHYSIOL, NC STATE UNIV, 83- *Mem:* Am Dairy Sci Asn; Am Physiol Soc; Nat Mastitis Coun. *Res:* Physiology of milk secretion and ejection in health and disease; general experimental animal physiology; myoepithelial cell occurrence and quantification; aging effect on intestinal motility; milking machine function in disease and health. *Mailing Add:* 1147 Grinnells Animal Health Lab NC State Univ Raleigh NC 27695-7626

CARUSO, FRANK LAWRENCE, b Hackensack, NJ, Nov 18, 49; m 75. PLANT PATHOLOGY, PLANT PHYSIOLOGY. *Educ:* Gettysburg Col, AB, 71; Univ Mass, MS, 75; Univ Ky, PhD(plant path), 78. *Prof Exp:* Res asst plant path, Univ Mass, 72-74, Univ Ky, 74-78; MEM STAFF, DEPT BOT & PLANT PATH, UNIV MAINE, ORONO, 78- *Mem:* Am Phytopathological Soc; Sigma Xi. *Res:* Physiology of diseased plants; bacterial and fungal diseases of fruits. *Mailing Add:* Cranberry Ext Sta PO Box 569 Glen Charlie Road East Wareham MA 02538

CARUSO, FRANK SAN CARLO, b Hartford, Conn, Aug 29, 36; div; c 3. PHARMACOLOGY. *Educ:* Trinity Col, Conn, BS, 58; Univ Rochester, MS, 61, PhD(pharmacol), 63. *Prof Exp:* Sr pharmacologist, Bristol Labs, 63-65, asst dir pharmacol-cardiovasc res, 65-66, coordr biol screening, 67-72, asst dir clin pharmacol, 72-74, dir clin analgesic res, 74-80; dir clin pharmacol, Revlon Health Care Group, 80-85; VPRES RES & DEVELOP, ROBERTS PHARMACEUT CORP, 85- *Mem:* AAAS; NY Acad Sci; Am Soc Clin Pharmacol & Therapeut; Int Asn Study Pain; fel Col Clin Pharmacol. *Res:* Sodium fluoride effects on renal function in dogs; pharmacological and toxicological effects on blood pressure, heart rate and respiration; narcotics and analgesics, narcotics and antagonists; extensive experience in ethical pharmaceutical research and development worldwide, areas of expertise oncology, cardiovascular, endocrine, CNS, and anti-inflammatory. *Mailing Add:* Roberts Pharmaceut Corp Six Indust Way Meridian Ctr III Eatontown NJ 07724

CARUSO, SEBASTIAN CHARLES, b Jamestown, NY, March 7, 26. ENVIRONMENTAL CHEMISTRY. *Educ:* Alfred Univ, BA, 49; Univ Pittsburgh, PhD, 54. *Prof Exp:* Asst, Univ Pittsburgh, 49-50, res asst, 50-54; fel, 54-85, SR FEL & MGR, CARNEGIE-MELLON UNIV, 85- *Mem:* AAAS; Am Chem Soc; Am Soc Testing & Mat; Air Pollution Control Asn. *Res:* Lipid chemistry; micro-isolation and identification of organic substances obtained from natural sources; biochemistry of water pollution; gas chromatography; chemical analysis of surface waters, air, and solid wastes; Environmental Control Processes. *Mailing Add:* 770 Greenfield Ave Pittsburgh PA 15217

CARUTHERS, JOHN QUINCY, b Columbia, Tenn, Feb 28, 13; m 49; c 3. BACTERIOLOGY. *Educ:* Hampton Inst, BS, 33; Iowa State Univ, MS, 41. *Prof Exp:* Instr bact, Atlanta Col Mortuary Sci, 38-62, pres, 53-62; asst prof biol, Spellman Col, 62-81; RETIRED. *Concurrent Pos:* Instr pub schs, Ga, 33-59. *Mem:* Am Soc Microbiol; Nat Sci Teachers Asn. *Res:* Physiological bacteriology; antibiotic resistance in staphylococci. *Mailing Add:* 668 Fielding Ln SW Atlanta GA 30311

CARUTHERS, MARVIN HARRY, b Des Moines, Iowa, Feb 11, 40; c 2. BIOCHEMISTRY, MOLECULAR BIOLOGY. *Educ:* Iowa State Univ, BS, 62; Northwestern Univ, PhD(biochem), 68. *Prof Exp:* Fel biochem, Univ Wis, 68-70, Mass Inst Technol, 70-72; from asst prof to assoc prof, 73-79, PROF BIOCHEM, UNIV COLO, 80- *Concurrent Pos:* Prin investr, various NIH & NSF grants, 73-; career develop award, NIH, 75-80; Guggenheim fel, John Simon Guggenheim Mem Found, 80-81; fac res fel award, Univ Calif. *Mem:* Am Chem Soc; AAAS; Fedn Am Soc Exp Biologists; Sigma Xi. *Res:* Nucleic acid chemistry; biochemistry. *Mailing Add:* Dept Chem & Biochem Univ Colo Campus Box 215 Boulder CO 80309-0215

CARVAJAL, FERNANDO, b San Jose, Costa Rica, June 4, 13; nat US; m 41; c 4. MICROBIOLOGY. *Educ:* Univ Costa Rica, BS, 38; Cornell Univ, MS, 42; La State Univ, PhD(mycol), 43. *Prof Exp:* Res mycologist & dir field studies, Inst del Cafe, San Jose, Costa Rica, 39; res mycologist, Exp Sta, La State Univ, 43-44; sr res mycologist, Schenley Res Labs, 44-49, head Div Microbiol, 49-54; co-head Div Microbiol, Schering Corp, 54-57, head Div Microbiol & vpres, Formet Labs, 57-61; dir res & labs, Arroyo Pharmaceut Corp, 61-72; mgr, fermentation prod, Upjohn Mfg Corp, 72-76, res & develop, 76-78; consult, 79-80; RETIRED. *Concurrent Pos:* Fel, US Off Educ. *Mem:* Sigma Xi; Am Soc Microbiol; Mycol Soc Am; Am Chem Soc. *Res:* Bacteriology; studies on genetics, mutations, physiology and fermentation of fungi and bacteria; plant pathology; antibiotics; discovery of the sexual state of Colletotrichum falcatum; synthesis of steroids; bacteriophages; production of vitamins by microbial fermentation. *Mailing Add:* 4681 Armadillo St Boca Raton FL 33428

CARVAJAL, FERNANDO DAVID, b Buffalo, NY, Aug 28, 44; m 66; c 2. ELECTRONICS ENGINEERING. *Educ:* Univ Colo, Boulder, BSEE, 68. *Prof Exp:* Flight test engr, Mil Flight Test, Cessna Aircraft, 68-69; design engr, Automation Systs Develop, Tex Instruments, 69-71, sr design engr, Circuit Design & Develop, 71-76, eng mgr, Advan Circuits, 76-80 & Linear Circuits, 80-89, SR MEM TECH STAFF, TEX INSTRUMENTS, 89- *Mem:* Inst Elec & Electronics Engrs. *Res:* Led watch circuits; automotive engine control circuits; hall effect sensors; subsurface hall sensor; practical rise and fall time control circuit; band gap temperature correction; granted three US patents. *Mailing Add:* 461 Windmill Lane McKinney TX 75069

CARVALHO, ANGELINA C A, b Azores, Portugal, Sept 26, 38; US Citizen; m 66; c 1. HEMATOLOGY, HEMOSTASIS & THROMBOSIS. *Educ:* Univ Coimbra, Portugal, BS, 62; Lisbon Med Sch, MD, 65, Brown Univ, MA, 85. *Prof Exp:* Resident internal med, Boston Vet Admin Hosp, 69-70; clin & res fel med, Mass Gen Hosp, 71-73; res assoc, Mass Inst Technol, 72-73; head, Spec Coagulation Lab, Mass Gen Hosp, 73-77; assoc hematologist & head, Hemostasis Lab, Mem Hosp, RI, 77-78; CHIEF HEMAT SECT, VET ADMIN MED CTR, PROVIDENCE, RI, 79- *Concurrent Pos:* Asst prof med, Brown Univ, 77-84, assoc prof, 84-; consult, Dept Anesthesia, Mass Gen Hosp, 78-; prin investr, Nat Heart, Lung & Blood Inst, 78-88. *Mem:* Am Heart Asn; Am Soc Hemat; Am Thoracic Soc; Int Soc Thrombosis & Hemostasis; AAAS; NY Acad Sci. *Res:* Thrombosis and hemorrhagic diatheses; the pathogenesis of blood alterations accompanying acute respiratory failure and new diagnostic and therapeutic approaches to detect and treat the thrombotic and hemorrhagic complications of this syndrome. *Mailing Add:* 263 Powell St Stoughton MA 02072

CARVALHO, SERGIO EDUARDO RODRIGUES DE, computer science, for more information see previous edition

CARVELL, KENNETH LLEWELLYN, b N Andover, Mass, May 1, 25. FOREST ECOLOGY. *Educ:* Harvard Univ, BA, 49; Yale Univ, MF, 50; Duke Univ, DF, 53. *Prof Exp:* Assoc forester, Agr Exp Sta & assoc prof silvicult, 53-64, PROF FOREST ECOL & FOREST ECOLOGIST, WVA UNIV, 64- *Res:* Forest regeneration methods and improvement cuttings for hardwood forests; use of chemical herbicides in forest improvement work; ecological effects of herbicides on electric transmission line rights-of-way. *Mailing Add:* Div Forest WVa Univ Morgantown WV 26505

CARVER, CHARLES E(LLSWORTH), JR, b Burlington, Vt, May 27, 22; m 50; c 3. CIVIL ENGINEERING. *Educ:* Univ Vt, BS, 47; Mass Inst Technol, MS, 49, ScD(civil eng), 55. *Prof Exp:* Instr civil eng, Univ Vt, 47; res asst hydrodyn, Mass Inst Technol, 47-49 & 51-55; instr civil eng, Univ Mass, 49-51; sr engr, Glenn L Martin Co, 55-56; res assoc phys oceanog, Woods Hole Oceanog Inst, 56-58; from assoc prof to prof, 58-87, EMER PROF CIVIL ENG, UNIV MASS, 87- *Concurrent Pos:* Lectr, Johns Hopkins Univ, 55-56; consult, Gen Elec Co, 58-60; res & adv develop, Avco Corp, 63. *Mem:* Fel Am Soc Civil Engrs; Sigma Xi. *Res:* Boundary layer control; non-Newtonian flow; oxygen transfer in bubble and spray aeration processes; undersea weapons systems; aircraft weapons systems; wave energy conversion devices. *Mailing Add:* Dept Civil Eng Univ Mass Amherst MA 01003

CARVER, DAVID HAROLD, b Boston, Mass, Apr 18, 30; m 63; c 3. PEDIATRICS, INFECTIOUS DISEASES. *Educ:* Harvard Col, AB, 51; Duke Univ, MD, 55. *Prof Exp:* Asst prof pediat, microbiol & immunol, Albert Einstein Col Med, 63-66; from assoc prof to prof pediat, Sch Med, Johns Hopkins Univ, 73-76, assoc prof microbiol, 67-76; PROF & CHMN DEPT PEDIAT, UNIV TORONTO, 76-; PHYSICIAN-IN-CHIEF, HOSP FOR SICK CHILDREN, TORONTO, 76- *Concurrent Pos:* Res fel pediat, Med Sch, Western Reserve Univ, 56-58; USPHS res fel, Children's Hosp Med Ctr & Harvard Med Sch, 61-63; mem comt infectious dis, Am Acad Pediat, 73-79. *Honors & Awards:* Schaffer Award, Johns Hopkins Univ Hosp, 73; Bain Award, Hosp for Sick Children, 78. *Mem:* Soc Pediat Res; Infectious Dis Soc Am; Am Pediat Soc; Am Acad Pediat; Am Soc Microbiol; Am Soc Virol; Int Soc Infectious Res. *Res:* Virology. *Mailing Add:* Dept Ped CN19 Robert Wood Johnson Med Sch New Brunswick NJ 08903-0019

CARVER, EUGENE ARTHUR, b Randolph, Vt, Dec 28, 44; m 70; c 2. GENETICS, METEORITICS. *Educ:* Univ Md, BS, 67; Univ Chicago, MS, 70, PhD(chem), 74. *Prof Exp:* Res assoc meteoritics, Goddard Space Flight Ctr, NASA, 72-74; ED CHEM, CHEM ABSTR SERV, INC, 74- *Res:* Genetics of dogs with special emphasis on coat colors. *Mailing Add:* 3725 Rome Corners Rd Galena OH 43021

CARVER, GARY PAUL, b Brooklyn, NY, 42; m 75. INTEGRATED CIRCUIT & MICROSTRUCTURE SCIENCE, DISSEMINATION TRANSFER. *Educ:* Clarkson Univ, BS, 63; Cornell Univ, PhD(physics), 70. *Prof Exp:* Res asst, Lab Atomic & Solid State Physics, Cornell Univ, 66-69; res physicist solid state, Naval Surface Warfare Ctr, White Oak Lab, 69-77; physicist, 77-80, mgr, processing res lab, semiconductor electronics div, 80-88, TECH PROG COORDR, MFG TECHNOL CTR, NAT BUR STANDARDS, GAITHERSBURG, MD, 88- *Concurrent Pos:* Vis scientist, Clarendon Lab, Oxford Univ, 76. *Mem:* Am Phys Soc; AAAS; Inst Elec & Electronics Engrs; sr mem Inst Environ Sci. *Res:* Semiconductor materials and processes characterization; correlation of materials and processes with properties of microelectronic semiconductor devices; clean room design and operation; technology transfer and science education; technology dissemination, transfer and technical education. *Mailing Add:* Bldg 225 Rm A331 Nat Bur Standards Gaithersburg MD 20899

CARVER, JAMES CLARK, b Lake Charles, La, Dec 16, 45; m 69; c 2. SURFACE CHEMISTRY, CATALYSIS. *Educ:* Centenary Col La, BS, 67; Univ Tenn, PhD(inorg chem), 72. *Prof Exp:* Res assoc anal chem, Univ Ga, 71-73, elec spectros lab supvr surface characterization, 73-74, lectr anal chem, 74-75; asst prof, Tex A&M Univ, 75-78; staff chemist surface characterization, Exxon Res & Develop Labs, Exxon Corp, 78-; ATTY,

TAYLOR, PORTER, BROOKS & PHILLIPS. *Concurrent Pos:* Prin investr, Petrol Res Fund grant, 76-78, Robert A Welch Found grant, 77-78. *Mem:* Am Chem Soc; Am Vacuum Soc; Am Phys Soc. *Res:* Surface characterization of catalysts using x-ray photoelectron spectroscopy; thermal desorption; secondary ion mass spectroscopy and low energy ion scattering spectrometry. *Mailing Add:* 451 Florida St 8th Floor Baton Rouge LA 70801

CARVER, JOHN GUILL, b Mt Juliet, Tenn, Feb 10, 24; m 56; c 4. PHYSICS ENGINEERING. *Educ:* Ga Inst Technol, BS, 50; Yale Univ, MS, 51, PhD(physics), 55. *Prof Exp:* Res asst physics, Yale Univ, 51-55; sr engr, Aircraft Nuclear Propulsion Dept, Gen Elec Co, 55-56, task leader, 56-60, specialist advan reactor physics develop, Atomic Power Equip Dept, 60-62, mgr fuels & irradiations physics, 62-67; mgr advan res & technol, Space Div, NAm Rockwell Corp, 67-70, supvr advan res, 70-77, supvr electro-optics technol, 77-79, study mgr, Shuttle Systs, Space Div, 79-84, proj mgr, Satellite Systs Div, 84-89, CONSULT, SCI CTR, ROCKWELL INT CORP, 89- *Concurrent Pos:* Chmn, Tech Comt Sensor Systs, Am Inst Aeronaut & Astronaut, 74-75, comt mem, 76-78, mem, Tech Comt Space Sci & Astron, 79-82. *Mem:* Fel AAAS; assoc fel Am Inst Aeronaut & Astronaut; Am Phys Soc; Sigma Xi. *Res:* Particle accelerators; neutron cross sections; nuclear shielding; neutron spectrometry; physics of plutonium fuel cycle; infrared sensor systems; research reactor operation. *Mailing Add:* 3079 Pinewood St Orange CA 92665

CARVER, KEITH ROSS, b Beech Creek, Ky, May 18, 40; m 65; c 1. ELECTRICAL ENGINEERING, ELECTROMAGNETICS. *Educ:* Univ Ky, BS, 62; Ohio State Univ, MS, 63, PhD(elec eng), 67. *Prof Exp:* Res asst elec eng, radio observ, Ohio State Univ, 62-66, res assoc, 66-67; asst prof, Univ Ky, 67-69; assoc prof, NMex State Univ, 66-77, group supvr, E M Res & Develop Group, Phys Sci Lab, 78-81, prof elec eng, 77-; prog mgr, Radar Remote Sensing Systs, Nasa Hq, Washington, DC, 81-; CHMN DEPT ELEC & COMPUT ENG, UNIV MASS, AMHERST. *Concurrent Pos:* Mem, URSI Comn F, 77- *Honors & Awards:* C Holmes MacDonald Award, 76. *Mem:* Am Soc Photogram; Inst Elec & Electronics Engrs. *Res:* Antenna design and measurement microwave remote sensing; radio wave propagation; Remote sensing instrumentation and antenna design and measurement. *Mailing Add:* Dept Elec & Comput Eng Univ Mass Amherst MA 01003

CARVER, MICHAEL BRUCE, b Wales, Mar 3, 41; Can citizen;; c 2. NUMERICAL ANALYSIS. *Educ:* McMaster Univ, Can, BEng, 63; Birmingham Univ, MSc, 65; Ottawa Univ, PhD, 85. *Prof Exp:* Res engr aerodynamics, English Elec Co, 63-65; res engr hydrodynamics, 65-69, simulation analyst numerical math, 69-80, supvry engr computational thermal hydraulics, 80-81, head fuel dynamics sect, 81-83, head appl math dept, 83-86, HEAD THERMAL HYDRAULICS DEPT, ATOMIC ENERGY CAN LTD, 86- *Concurrent Pos:* Lectr, Algonquin Col Tech, 71-78 & Univ Ottawa, 79-; adj prof, McMaster Univ, 87- *Mem:* Asn Prof Engrs Ont; Int Asn Math & Comput Simulation. *Res:* Simulation, numerical analysis, solution of differential equation systems, computational thermal hydraulic analysis of single and two phase flows. *Mailing Add:* Atomic Energy of Can Ltd Chalk River Nuclear Labs Chalk River ON K0J 1J0 Can

CARVER, MICHAEL JOSEPH, b Omaha, Nebr, Apr 4, 23; m 48; c 3. BIOCHEMISTRY. *Educ:* Creighton Univ, BS, 47, MS, 48; Univ Mo, PhD(biochem), 52. *Prof Exp:* Instr chem, Creighton Univ, 48-49; supvr anal dept, Cudahy Labs, 52-56; assoc res prof, 56-66, PROF BIOCHEM, UNIV NEBR MED CTR, OMAHA, 66-, ASST DEAN COL MED, 70- *Mem:* AAAS; Am Chem Soc; Am Inst Chem; Soc Exp Biol & Med; Am Soc Neurochemisty. *Res:* Cytochemistry; electrophoresis; protein chemistry; neurochemistry; psycotropic drugs; mental retardation. *Mailing Add:* Univ Nebr Col Med 602 S 44th Ave Omaha NE 68105

CARVER, O(LIVER) T(HOMAS), b Charleston, WVa, July 14, 24; m 50; c 3. ELECTRICAL ENGINEERING. *Educ:* WVa Univ, BSEE, 49. *Prof Exp:* Engr, 49-55, leader syst design, 55-59, leader, Aerospace Systs Div, 59-61, mgr test systs anal, 61-76, mgr, Ate Systs, RCA Corp, 76- 86. *Concurrent Pos:* Spec lectr, George Washington Univ, 65-66. *Mem:* Am Inst Aeronaut & Astronaut; Inst Elec & Electronics Engrs; Sigma Xi. *Res:* Automatic test systems analysis; advanced test technique and development. *Mailing Add:* 82 Queen Anne Lane Cotuit MA 02635

CARVER, ROBERT E, b Kansas City, Mo, Jan 6, 31; div; c 2. SEDIMENTARY PETROLOGY, HYDROGEOLOGY. *Educ:* Mo Sch Mines, BSMinE, 53; Univ Mo, AM, 59, PhD(geol), 61. *Prof Exp:* Geologist, Bellaire Res Labs, Texaco, Inc, 61-64; from asst prof to assoc prof, 64-80, asst head dept, 71-80, PROF GEOL, UNIV GA, 80- *Concurrent Pos:* Vis prof, Univ Rio Grande do Sul, Brazil, 72; assoc ed, J Sedimentary Petrol, 80-87; consult. *Mem:* Fel Geol Soc Am; Am Asn Petrol Geologists; Soc Econ Paleontologists & Mineralogists. *Res:* Sedimentary petrology; industrial mineralogy; hydrogeology; geology and natural resources of the southeast Atlantic coastal plain. *Mailing Add:* Dept Geol Univ Ga Athens GA 30602

CARY, ARTHUR SIMMONS, b Sacramento, Calif, Nov 30, 25; m 57; c 3. HIGH ENERGY PHYSICS. *Educ:* Fisk Univ, BA, 49, MA, 51; Univ Calif, Riverside, MA & PhD(physics), 69. *Prof Exp:* Instr phys sci, Dillard Univ, 54-56; assoc prof physics, Tenn State Univ, 56-63; from asst prof to assoc prof, Harvey Mudd Col, 69-74; asst prof, 74-80, ASSOC PROF PHYSICS, CALIF POLYTECH STATE UNIV, 80- *Concurrent Pos:* Fulbright lectr, Univ Liberia, WAfrica, 80-81,. *Mem:* Am Phys Soc; Am Asn Physics Teachers; Nat Sci Teachers Asn. *Res:* High energy experimental physics; nuclear emulsion; bubble chamber. *Mailing Add:* Dept of Physics Calif Polytech State Univ San Luis Obispo CA 93407

CARY, HOWARD BRADFORD, b Columbus, Ohio, May 24, 20; m 42; c 2. WELDING ENGINEERING. *Educ:* Ohio State Univ, BIE, 42. *Prof Exp:* Welding engr, Fisher Tank Div, Gen Motors Corp, Mich, 42-44, process engr, 46; welding res engr, Battelle Mem Inst, Marion Power Shovel Co, Marion, Ohio, 46-48, welding engr, 48-51, welding supt, 51-55, asst gen works mgr, 56-58; dir, 59-76, VPRES WELDING SYSTS, HOBART BROS CO, 76-, PRES, HOBART SCH WELDING TECHNOL, 80- *Concurrent Pos:* US deleg, Comn XIV Int Inst Welding; nat pres, Am Welding Soc, 80-81; auth, Modern Welding Technol, copywright 79, Prentice Hall Inc. *Honors & Awards:* A F Davis Silver Medal Award, Am Welding Soc, 63; Samuel Wylie Miller Mem Medal, 72; R O Thomas Mem Award, 88. *Mem:* Am Soc Metals; Am Welding Soc (pres, 80-81); Nat Soc Prof Engrs; Am Soc Mech Engrs. *Res:* Development of automatic metal-arc welding process for steel and aluminum utilizing atomsphere generated outside of the arc; welding education; welding process development. *Mailing Add:* 1492 Surrey Rd Troy OH 45373

CASABELLA, PHILIP A, b Albany, NY, Feb 18, 33; m 60; c 2. PHYSICS. *Educ:* Rensselaer Polytech Inst, BS, 54, MS, 57; Brown Univ, PhD(physics), 59. *Prof Exp:* Res assoc physics, Brown Univ, 59-60; from asst prof to assoc prof, 61-69, chmn dept physics & astron, 68-77, PROF PHYSICS, RENSSELAER POLYTECH INST, 69- *Mem:* AAAS; Sigma Xi; Fel Am Phys Soc. *Res:* Nuclear magnetic resonance and pure quadruple resonance in solids. *Mailing Add:* Dept Physics Rensselaer Polytech Inst Troy NY 12181

CASAD, BURTON M, b Mooreland, Okla, Nov 19, 33; m 54; c 4. CHEMICAL ENGINEERING. *Educ:* Okla State Univ, BS, 55, MS, 57, PhD(chem eng), 60. *Prof Exp:* Process engr, Phillips Petrol Co, 55-56; asst chem eng, Okla State Univ, 56-59; res engr corrosion, 59-63, res group leader, 63-67, sect supvr mat & eng, gas processing, 67-76, SECT DIR RECOVERY SECT, CONTINENTAL OIL CO, 76- *Mem:* Nat Asn Corrosion Engrs; Am Inst Chem Engrs. *Res:* Corrosion of oil and gas well equipment; corrosion; water treating; enhanced oil recovery processes. *Mailing Add:* c/o Conoco 4569 Oslo Pouch Houston TX 77210

CASADABAN, MALCOLM JOHN, b New Orleans, La, Aug 12, 49; m 77; c 2. MOLECULAR GENETICS, GENE EXPRESSION. *Educ:* Mass Inst Technol, SB, 71; Harvard Univ, PhD(microbiol & molecular genetics), 76. *Prof Exp:* Res fel, Stanford Univ, 76-79; asst prof molecular biol, dept biophysics & theoret biol, 80-85, ASSOC PROF, DEPT OF MOLECULAR GENETICS & CELL BIOL, UNIV CHICAGO, 85-; ADV & CO-FOUNDER, THERMOGEN, INC, 89- *Concurrent Pos:* Grad fel, NSF, 71-73; res career develop award, NIH, 81-86; biomed sci study sect, NIH, 86-90. *Mem:* Am Soc Microbiol; AAAS; Genetics Soc Am. *Res:* Gene expression, structures, and function; analysis of genetic control elements by DNA fusion and cloning in prokaryotic and eukaryotic cells; mechanisms of DNA transposition. *Mailing Add:* Dept Molecular Genetics & Cell Biol Univ Chicago 920 E 58th St Chicago IL 60637

CASADY, ALFRED JACKSON, b Milton, Iowa, Feb 16, 16; m 41; c 2. AGRONOMY, PLANT BREEDING. *Educ:* Kans State Univ, BS, 48, MS, 50, PhD(agron), 62. *Prof Exp:* Agronomist plant breeding, Agr Res Serv, USDA, Kans State Univ, 49-78; prof agron, Kans State Univ, 70-78; RETIRED. *Concurrent Pos:* Res agronomist, USDA, 56-78. *Res:* Breeding and selection of cereal crops for improved yield; insect and disease resistance. *Mailing Add:* 105 N Dartmouth Dr Manhattan KS 66502

CASAGRANDE, DANIEL JOSEPH, b Bridgeport, Conn, Jan 25, 45; m 64; c 2. ORGANIC GEOCHEMISTRY. *Educ:* Univ Scranton, BSc, 66; Pa State Univ, PhD(fuel sci), 70. *Prof Exp:* Fel org geochemisty, Univ Calgary, 71; prof org geochemisty, Governors State Univ, 71-79; SR RES SPECIALIST, EXXON PROD RES CO, 79- *Concurrent Pos:* Consult, Village Park Forest South, Ill, 72-, Armour Dial, Inc, 72, Izaak Walton League, 73, Kerr-McGee Corp, 75-78, Shell Oil Co, 76 & Cities Serv Oil Co, 78. *Honors & Awards:* Colin Roscoe Award, Am Chem Soc. *Mem:* Am Chem Soc; Geol Soc Am; The Chem Soc; Geochem Soc; AAAS; Sigma Xi. *Res:* Organic geochemistry of sulfur, metals, amino acids, porphyrins in sediments, petroleum, peat, lignite and coal. *Mailing Add:* 10602 Mt Tipton San Antonio TX 78213

CASAGRANDE, LEO, civil engineering; deceased, see previous edition for last biography

CASAL, HECTOR LUIS, structure indus adducts, spectroscopy low temperatures, for more information see previous edition

CASALI, LIBERTY, b Vivian, WVa, Oct 15, 11. CHEMISTRY. *Educ:* Duke Univ, BS, 33; Univ Colo, PhD(phys chem), 52. *Prof Exp:* Teacher pub schs, WVa, 33-42; asst chemist, Gen Elec Co Labs, Mass, 42-45; asst gen qual phys chem lab, Wheaton Col, Mass, 45-47; instr math, Univ Colo, 48-50, asst chem, 47-51; from instr to asst prof chem, Russell Sage Col, 52-61; assoc prof chem & physics, Winthrop Col, 61-65 & Bridgewater Col, 65-66; from assoc prof to prof, 67-77, EMER PROF CHEM, JAMES MADISON UNIV, 77- *Concurrent Pos:* Abstractor, Chem Abstr. *Mem:* Am Chem Soc; Sigma Xi. *Res:* Heats of reaction of some fluorolefins; molecular spectra; history of science. *Mailing Add:* 722 S Main St Apt B3 Harrisonburg VA 22801

CASALS, JORDI, b Viladrau, Spain, May 15, 11; nat US; m 41; c 1. VIROLOGY. *Educ:* Instituto Nacional, BS, 28; Univ Barcelona, MD, 34. *Prof Exp:* Asst resident physician, Med Sch Hosp, Univ Barcelona, 34-36; asst path, Sch Med, Cornell Univ, 37-38; asst, Dept Path & Bact, Rockefeller Inst, 38-52, virus res prog, Rockefeller Found, 52-69; prof epidemiol, Yale Univ, 69-80, emer prof & sr res scientist, 80-; AT DEPT NEUROL, MT SINAI SCH MED, NEW YORK. *Concurrent Pos:* Consult, Walter Reed Army Med Ctr; mem, Surgeon Gen's Virus Comn, Japan, 47. *Mem:* AAAS; fel Soc Exp Biol & Med; fel NY Acad Sci. *Res:* Arthropod-borne virus infections. *Mailing Add:* Dept Neurol Box 1137 Mt Sinai Sch Med New York NY 10029

CASANI, JOHN R, b Philadelphia, Pa, Sept 17, 32; m; c 4. AERONAUTICS. *Educ:* Univ Pa, BA, 55. *Prof Exp:* Integration engr, Jupiter Radio Inertial Guidance Syst, Jet Propulsion Lab, Calif Inst Technol, Nat Aeronaut & Space Admin, 56-57, accelerometer develop engr, sergeant, 57, payload engr, Pioneers 1 & 4, 58-59, spacecraft syst engr, Rangers 1 & 2, Mariner Mars, 59-65, spacecraft syst engr, Voyager, 65, dep spacecraft syst mgr, Mariner Mars,

66-69, spacecraft syst mgr, 69-70, proj mgr, 70-71, spacecraft syst mgr, Mariner Venus, 71- 74, mgr, Guidance & Control Div, 74-75, mgr, Voyager Proj, 75-77 & Proj Galileo, 77-88, dep asst lab dir, 88-89, ASST LAB DIR, FLIGHT PROJ, JET PROPULSION LAB, CALIF INST TECHNOL, NAT AERONAUT & SPACE ADMIN, CALIF, 89- *Honors & Awards:* Mgt Improvement Award, President US, 74; Astronaut Engr Award, Nat Space Ctr, 81; 1989 Space Flight Award, Am Astronaut Soc, 90. *Mem:* Nat Acad Eng. *Res:* Design of spacecraft systems and their management during specific missions. *Mailing Add:* Jet Propulsion Lab Calif Inst Technol 4800 Oak Grove Dr, MS 180-401 Pasadena CA 91109

CASANOVA, JOSEPH, b Stafford Springs, Conn, May 31, 31; m 56; c 4. ORGANIC CHEMISTRY. *Educ:* Mass Inst Technol, SB, 53; Carnegie Inst Technol, MS, 56, PhD(org chem), 57. *Prof Exp:* Res chemist, E I du Pont de Nemours & Co, Del, 57; res chemist & dir res, Chem Warfare Lab, Army Chem Ctr, 58-59; NIH res fel, Harvard Univ, 59-61; from asst prof to prof chem, Calif State Univ, Los Angeles, 61-85, dean undergrad studies, 78-80, assoc vpres planning & resources, 80-81. *Concurrent Pos:* Vis prof, Ind Univ, 65-66; Fulbright scholar, Dept Org Chem, Univ Lund, Sweden, 70-71. *Mem:* Am Chem Soc; The Chem Soc. *Res:* Sulfonium salts; organic boron compounds; isocyanides; electroorganic chemistry; strained olefins. *Mailing Add:* Dept Chem Calif State Univ 5151 State University Dr Los Angeles CA 90032-4202

CASARETT, ALISON PROVOOST, b New York, NY, Apr 17, 30; c 2. RADIOBIOLOGY. *Educ:* St Lawrence Univ, BS, 51; Univ Rochester, MS, 53, PhD(radiation biol), 57. *Prof Exp:* Res assoc radiation biol, Univ Rochester, 53-58, instr, 58-63; asst prof, 63-69, ASSOC PROF PHYSICAL BIOL, CORNELL UNIV, 69-, ASSOC DEAN GRAD SCH, 73-, VPROVOST, 78- *Mem:* Radiation Res Soc; Sigma Xi. *Res:* Physiological, endocrinological and pathological effects of radiation on mammals. *Mailing Add:* 144 Pine Tree Rd Ithaca NY 14850

CASARETT, GEORGE WILLIAM, b Rochester, NY, Aug 17, 20; m 44; c 2. PATHOLOGY. *Educ:* Univ Rochester, PhD(anat), 52. *Prof Exp:* Jr res assoc, 41-43, asst path, Manhattan Proj, 43-47, chief path unit, Atomic Energy Proj, 47-49, asst chief radiation tolerance sect, 49-52, scientist, 52-62, res assoc radiation ther, 59-68, from instr to assoc prof, 53-59, prof radiation biol, 63-66, chief, Radiation Path Sect, 62-80, PROF RADIATION BIOL & BIOPHYS, SCH MED, UNIV ROCHESTER, 63-, PROF RADIOL, 68-, CHIEF, RADIATION PATH SECT, 62- *Concurrent Pos:* Consult, Surgeon Gen, US Army, 61-70, sci comt effects atomic radiation, UN, 63-, US Armed Forces Radiobiol Res Inst, 64-70 & coun, Am Asn Advan Aging Res; mem, US deleg, UN Conf Peaceful Uses Atomic Energy, Switz, 55, subcomt effects radiation, Nat Acad Sci, 56-64, subcomt nat comt radiation protection, 56-, adv comt to Fed Radiation Coun, 66-70, nat coun radiation protection, 62-, comt on radiation effects, 73- & Int Comn Radiol Protection, 67-70, chmn comt on biol effects of radiation, Nat Acad Sci; mem cancer res training comt, Nat Cancer Inst, 69- *Honors & Awards:* Award & Silver Medal, Am Roentgen Ray Soc, 59; Award, Radiol Soc NAm, 59. *Mem:* AAAS; Am Asn Anat; Am Soc Exp Path; fel Am Geront Soc; fel NY Acad Sci; Sigma Xi. *Res:* Oncology; pathology and hematology of radiations; radioactive substances; radiation biology; gerontology; sterility. *Mailing Add:* Sch Med Univ Rochester 601 Elmwood Ave Rochester NY 14642

CASASENT, DAVID P(AUL), b Washington, DC, Dec 8, 42; m 64; c 5. INFORMATION PROCESSING, RADAR. *Educ:* Univ Ill, Urbana-Champaign, BS, 64, MS, 65, PhD(elec eng), 69. *Prof Exp:* Asst prof, 69-76, PROF ELEC ENG, CARNEGIE-MELLON UNIV, 76- *Concurrent Pos:* Consult, Digital Equip Corp, 70-, Off Naval Res, 75-, Dept Defense, 77-, NSF & Rome Air Develop Ctr, 78-, NASA & Gen Dynamics, 80-; res contract, Off Naval Res, 71-, Air Force Off Sci Res, 75-, NSF, NASA, DARPA. *Honors & Awards:* B Carlton Award, Inst Elec & Electronics Engrs, 76. *Mem:* Fel Inst Elec & Electronics Engrs; fel Optical Soc Am; fel soc Photo-Optical Instrumentation Engrs. *Res:* On-line real-time optical image and signal processing; pattern recognition radar and C3 I signal processing; missile guidance, real-time light modulators; hybrid optical-digital data processing, optical computing; optical neural networks and optical inspection; robotics. *Mailing Add:* Dept Elec Eng Carnegie-Mellon Univ Pittsburgh PA 15213

CASASSA, EDWARD FRANCIS, b Portland, Maine, Nov 10, 24; m 54; c 3. PHYSICAL CHEMISTRY OF MACROMOLECULES. *Educ:* Univ Maine, BS, 45; Mass Inst Technol, PhD(phys chem), 53. *Prof Exp:* Chemist, E I du Pont de Nemours & Co, 45-48; asst, Mass Inst Technol, 49-52; proj assoc chem, Univ Wis, 52-56; fel, 56-59, SR FEL, MELLON INST, 59-, PROF CHEM, CARNEGIE-MELLON UNIV, 67- *Concurrent Pos:* Lectr, Univ Pittsburgh, 56-59; asst ed, J Polymer Sci, 65-69, assoc ed, 69- *Mem:* AAAS; Am Phys Soc; Am Chem Soc. *Res:* Physical chemistry of polymers and proteins; statistical mechanics; light scattering. *Mailing Add:* Mellon Inst 4400 Fifth Ave Pittsburgh PA 15213-2683

CASASSA, ETHEL ZAISER, b Aug 30, 24; m; c 3. PHYSICAL CHEMISTRY. *Educ:* Temple Univ, AB, 44; Columbia Univ, AM, 46, PhD(phys chem), 48. *Prof Exp:* Lab dir, 74-82, from lectr to sr lectr colloids, polymers & surfaces, 75-80, ASSOC PROF CHEM ENG, CARNEGIE-MELLON UNIV, 80- *Mem:* Am Chem Soc; Am Inst Chem Engrs; Sigma Xi; NY Acad Sci. *Res:* Colloid stability; micellization in unusual polar media; solubilization and adsorption phenomena; physical chemistry of polymers; interfacial synthesis of polyesters; mass transport in liquid-liquid systems; surface chemistry of coal water dispersions. *Mailing Add:* Dept Chem Eng Carnegie-Mellon Univ Pittsburgh PA 15213-3890

CASASSA, MICHAEL PAUL, b Pittsburgh, Pa, Sept 1, 56; m 83. PHYSICAL CHEMISTRY. *Educ:* Univ Pittsburgh, BS, 78; Calif Inst Technol, PhD phys chem 84. *Prof Exp:* Res assoc, Calif Inst Technol, 78-83; res fel, Nat Coun-Nat Bur Standards, 83-85 RES CHEMIST, NAT INST STANDARDS & TECHNOL, 85- *Mem:* Am Phys Soc; Am Chem Soc; Am Vaccum Soc. *Res:* Dynamics of molecular excitations; infrared photodissociation spectra of Van der Waals molecules; picosecond vibrational energy flow in weakly bound clusters, in molecules bound to surfaces, and in liquid and solid phase materials. *Mailing Add:* Molecular Physics Div Nat Inst Standards & Technol Bldg 221 Rm B268 Gaithersburg MD 20899

CASAZZA, JOHN ANDREW, b New York, NY, Jan 3, 24; m 49; c 2. TECHNICAL & ECONOMIC SYSTEMS ANALYSIS, INFRASTRUCTURE. *Educ:* Cornell Univ, BEE, 43. *Prof Exp:* Eng, Pub Serv Elec & Gas Co, 46-68, syst planning & develop eng, 68-71, gen mgr, planning & res, 71-74, vpres, planning & res, 74-77; vpres energy consult, Stone & Webster Mgt Consult Inc, 77-79; pres, energy consult, 79-90, CHMN, CASAZZA, SCHULTZ & ASSOCS INC, 91- *Concurrent Pos:* Chmn US Tech Comt, Int Conf of Electronic Researchers on High Tension, 74-81; mem, US/USSR Joint Comn Sci & Tech Coop, 73-80; mem, Energy Policy comt, Inst Elec & Electronic Engrs, 74- & Energy Eng Bd, Nat Res Coun, 90- *Honors & Awards:* Halpern Award for Develop & Use Elec Power Transmission Systs. *Mem:* Fel Inst Elect & Electronic Engrs; Int Conf High-Voltage Elec Power Networks; World Energy Conf. *Res:* Electric power systems including applications of digital controls, in energy storage mechanisms in new forms and types of electricity generator from solar to fusion in fuels and fuel resources and in overall energy system economics. *Mailing Add:* 1901 N Fort Myer Dr Suite 503 Arlington VA 22209

CASBERG, JOHN MARTIN, b LaCrosse, Wis, Nov 26, 34; m 59; c 2. CHEMICAL ENGINEERING. *Educ:* Univ Wis Madison, BS, 56. *Prof Exp:* Develop engr, BF Goodrich Chem, Avon Lake, Ohio, 56-63, sr chem engr, Ky, 63-67; plastics engr, Stauffer Chem, 67-68; SR RES ASSOC, OLIN CORP, 68- *Mem:* Sr mem Soc Plastics Engrs. *Res:* Swimming pool chemicals, including tablets, and devices for feeding these chemicals; granted 12 patents. *Mailing Add:* 370 Sycamore Lane Cheshire CT 06410

CASCARANO, JOSEPH, b Brooklyn, NY, Oct 26, 28; m 54; c 2. CELL PHYSIOLOGY. *Educ:* NY Univ, BA, 50, MS, 53, PhD(biol), 56. *Prof Exp:* US Pub Health Serv fels, Univ Minn, 56-57, NY Univ, 57-58, instr path, Sch Med, 58-59, asst prof, 59-65; assoc prof zool, 65-70, PROF CELL BIOL, UNIV CALIF, LOS ANGELES, 70- *Mem:* AAAS; Am Physiol Soc; Am Soc Cell Biol; Soc Exp Biol & Med; Sigma Xi. *Res:* Cation transport; cell permeability; anaerobic energy metabolism; relation of cell function to energy metabolism; mitochondrial biogenesis; metabolic regulation, heart metabolism; acclimation and adaptation of animals to altitude. *Mailing Add:* Dept of Biol Univ of Calif Los Angeles CA 90024

CASCIANO, DANIEL ANTHONY, b Buffalo, NY, Mar 1, 41; m 64; c 2. CELL BIOLOGY. *Educ:* Canisius Col, BS, 62; Purdue Univ, PhD(cell biol), 71. *Prof Exp:* Res asst tissue cult, Roswell Park Mem Inst, 63-64; res asst cell biol, Purdue Univ, 65-66, asst microbiol, 69; investr biochem, Univ Tenn, 71-73; res biologist, 73-79, DIR, DIV GENETIC TOXICOL, NAT CTR TOXICOL RES, 79- *Concurrent Pos:* Prof Biochem, Univ Arkansas Med Sci. *Mem:* Tissue Culture Asn; AAAS; Environ Mutagen Soc; Genetic Toxicol Asn. *Res:* Development of mammalian somatic cell systems capable of in vitro metabolism of promutagens and procarcinogens. *Mailing Add:* Div Genetic Toxicol Nat Ctr Toxicol Res Jefferson AR 72079

CASCIERI, MARGARET ANNE, b Lancaster, Pa, Feb 9, 51; m 72. GROWTH FACTOR & RECEPTOR RESEARCH, MECHANISMS OF SECRETION. *Educ:* Pa State Univ, BS, 72, MS, 74; Rutgers Univ, PhD(zool/biochem), 82. *Prof Exp:* Staff biochemist, Merck Sharp & Dohme Res Lab, 75-80, res biochemist, 80-82, sr res biochemist, 82-85, res fel, 85-88, ASSOC DIR, MERCK SHARP & DOHME RES LAB, 89- *Mem:* Am Soc Biochem & Molecular Biol; Am Chem Soc; Soc Neuroscience; Endocrine Soc; Am Diabetes Asn; AAAS. *Res:* The structural basis of the interaction of peptides of neurological and endocrinological importance with their receptors; the regulation of the activity of these receptors in disease (ie, diabetes); the mechanisms of regulation of secretion. *Mailing Add:* Merck Sharp & Dohme Res Lab Bldg 80 M-123 Box 2000 Rahway NJ 07065-0900

CASE, CLINTON MEREDITH, b Oregon City, Ore, Nov 20, 40; m 65; c 2. SURFACE PHYSICS, HYDROLOGY. *Educ:* Linfield Col, BA, 63; Univ Nev, Reno, MS, 67, PhD(physics), 70. *Prof Exp:* Res assoc hydrol, Ctr Water Resources Res, Desert Res Inst, Univ Nev, Reno, 70-72, asst res prof, 73-75, assoc res prof, 75-80, res prof hydrol, 80-90; MEM STAFF, DIV ENVIRON PROTECTION, STATE NEV, 91- *Concurrent Pos:* Fulbright scholar, 63. *Mem:* Am Geophys Union; Am Phys Soc; Am Vacuum Soc; Am Asn Physics Teachers; Sigma Xi; NY Acad Sci. *Res:* Statistical mechanics of cooperative phenomena, including surface effects; theoretical groundwater hydrology; experimental solid state physics; theoretical and applied flow in unsaturated porous media. *Mailing Add:* 985 Akard Dr Reno NV 89503

CASE, DAVID ANDREW, b Akron, Ohio, Oct 12, 48; m 80; c 1. PHYSICAL CHEMISTRY. *Educ:* Mich State Univ, BS, 70; Harvard Univ, AM, 72, PhD(chem physics), 77. *Prof Exp:* prof chem, UNIV CALIF, DAVIS, 77-86; MEM, RES INST SCRIPPS CLIN, 86- *Mem:* Am Chem Soc; Biophys Soc; Protein Soc. *Res:* Theoretical chemistry; electronic structures of inorganic compounds and metalloenzymes; dynamics of proteins; relativistic effects in molecular electronic structures. *Mailing Add:* Dept Molecular Biol Res Inst Scripps Clin 10666 N Torrey Pines Rd La Jolla CA 92037-1027

CASE, DENIS STEPHEN, b Sussex, NJ, May 23, 44; c 1. WILDLIFE MANAGEMENT. *Educ:* Hiram Col, BA, 66; Bowling Green State Univ, MS, 70. *Prof Exp:* Proj dir, Cleveland Dept Pub Utilities, 70-73; chief, Div Res, 73-75, WILDLIFE BIOLOGIST, DIV WILDLIFE, OHIO DEPT NATURAL RESOURCES, 75- *Mem:* AAAS; Wildlife Soc. *Res:* Management of native nongame and endangered wildlife species. *Mailing Add:* Div Wildlife Ohio Dept Natural Resources 1840 Belcher Dr Columbus OH 43224

CASE, JAMES B(OYCE), b Lincoln, Ill, Oct 26, 28; wid; c 1. PHOTOGRAMMETRY, GEODESY. *Educ:* Stanford Univ, BS, 50; Ohio State Univ, MS, 57, PhD(geod sci), 59. *Prof Exp:* Res fel glacier mapping, Ohio State Univ & Am Geog Soc, 56-59; prin investr, Ohio State Univ Res Found, 59-60; sr photogrammetrist, Broadview Res Corp, 60-61; prin scientist, Autometric Oper, Raytheon Co, 61-71; supvry geodesist, US Army Topog Command, 71-72; PHYS SCIENTIST, DEFENSE MAPPING AGENCY, 72- *Concurrent Pos:* Ed-in-chief, Photogram Eng & Remote Sensing, 75- *Mem:* AAAS; Brit Cartog Soc; Can Cartog Asn; Can Inst Surv; Sigma Xi; Am Soc Photogram & Remote Sensing; Photogram Soc London. *Res:* Photogrammetry and digital image processing. *Mailing Add:* PO Box 1669 Cedar City UT 84721-1669

CASE, JAMES EDWARD, b Mountain View, Ark, Feb 15, 33; m 55, 72; c 5. GEOLOGY, GEOPHYSICS. *Educ:* Univ Ark, BS, 53, MS, 54; Univ Calif, Berkeley, PhD(geol), 63. *Prof Exp:* Instr geol & math, Lamar State Col, 54-55; geologist, US Geol Surv, 55, geologist & geophysicist, 56-58, geophysicist, 58-65, chief Denver area pub unit & geologist, 66; assoc prof geol & geophys, Tex A&M Univ, 66-69 & Univ Mo, 69-71; GEOPHYSICIST, US GEOL SURV, CALIF, 71- *Mem:* Fel Geol Soc Am; Am Geophys Union; fel, AAAS; Am Asn Petrol Geol; Sigma Xi. *Res:* Gravity; magnetic data; correlation of geophysical data with geologic structure; tectonics of northern South America and Caribbean region; geophysical expression of mafic and ultramafic belts; tectonostratigraphic terranes. *Mailing Add:* US Geol Surv MS 989 345 Middlefield Rd Menlo Park CA 94025

CASE, JAMES FREDERICK, b Bristow, Okla, Oct 27, 26; m 50; c 3. COMPARATATIVE PHYSIOLOGY. *Educ:* Johns Hopkins Univ, PhD(biol), 51. *Prof Exp:* Physiologist avian physiol, USDA, 51-52; insect physiologist, Med Labs, Army Chem Ctr, Md, 55-57, from asst prof to assoc prof zool, Univ Iowa, 57-63; assoc prof, 63-69, PROF NEUROBIOL, UNIV CALIF, SANTA BARBARA, 69- *Mem:* AAAS; Sigma Xi. *Res:* Invertebrate physiology. *Mailing Add:* Dept of Biol Univ of Calif Santa Barbara CA 93106

CASE, JAMES HUGHSON, b Franklinville, NY, May 25, 28; m 51; c 3. AUTOMATA THEORY. *Educ:* Ala Polytech Inst, BS, 50; Tulane Univ, PhD(math), 54. *Prof Exp:* Asst prof math, Univ Utah, 54-59; asst prof, Univ Rochester, 59-61; assoc prof, 61-68, PROF MATH, UNIV UTAH, 68- *Concurrent Pos:* Consult, Gen Dynamics/Electronics, 60-61; vis prof, comput sci, Kans State Univ, 81. *Mem:* Am Math Soc. *Res:* Automata theory; general topology; ethological Models. *Mailing Add:* 2286 Preston St Salt Lake City UT 84106

CASE, JOHN WILLIAM, b Clinton, Iowa, Oct 28, 42; m 68; c 3. RECURSIVE FUNCTION THEORY. *Educ:* Iowa State Univ, Ames, BS, 64; Univ Ill, Urbana, MS, 66, PhD(math), 69. *Prof Exp:* Asst prof comput sci, Univ Kans, Lawrence, 69-73; asst prof, 73-75, ASSOC PROF COMPUT SCI, STATE UNIV NY BUFFALO, 75- *Concurrent Pos:* Vis fel comput sci, Yale Univ, New Haven, Conn, 80-81; vis assoc prof comput sci, NY Univ, New York, 80-81. *Mem:* Asn Comput Mach; AAAS; Am Math Soc; Asn Symbolic Logic; Europ Asn Theoret Comput Sci. *Res:* Recursive function theory in computer science including the theory of inductive inference machines; theoretical applications of self-referential machines or programs; abstract structure of programs; relative succinctness of programs in language hierarchies; biologically motivated automata theory. *Mailing Add:* Comput Sci Dept State Univ of NY - N Campus Buffalo NY 14260

CASE, KENNETH E(UGENE), b Oak Ridge, Tenn, Aug 12, 44; m 66; c 2. QUALITY CONTROL, RELIABILITY ENGINEERING. *Educ:* Okla State Univ, BS, 66, MS, 67, PhD(indust eng), 69. *Prof Exp:* Technician, Aerotron Radio Co, Okla, 59-63; prod engr, Humble Oil Co, Okla, 65; elec engr, Collins Radio Co, Iowa, 66; instr indust eng, Okla State Univ, 67-68; from asst prof to assoc prof indust eng & opers res, Va Polytech Inst & State Univ, 69-74; mgt scientist, GTE Data Serv, Tampa, Fla, 74-75; assoc prof indust eng, 75-78, head indust eng, 80-82, prof, 78-87, REGENTS PROF, SCH INDUST ENG & MGT, OKLA STATE UNIV, 87- *Concurrent Pos:* Mgt consult, Cities Serv Oil Co, Wisc & Okla, 68-69; consult, Boeing Com Airplane Co, Wash, 78, Detroit Diesel Allison Div, Gen Motors, Ind, 79, Mercury Marine, Okla, 80 & Remington Arms, Inc, Ark, Conn, NY, 80-83, Sprague Elec Co, Tex, Calif, 83- 87, AT&T, Okla, NJ, 83-86, Food & Drug Admin, Md, 84, Exxon Chem Co, Tex, La, NJ, 84-, Rockwell, Int, Okla, 84, Ford Motor Co, Okla, Tex, 85-88, Exxon Co, USA, Tex, La, NJ, Calif, 85-, Honeywell, Inc, Colo, 86, Phillips 66 Co, Tex, 87 & Pub Serv Co Okla, 90-; ed, Appl Probability & Statist Dept, Inst Indust Engrs Trans, 77-80; dir conf prog coord, Inst Indust Engrs, 79-82; dir prof regist, Inst Indust Engrs, 81-83, vpres systs eng & tech div, 83-85, pres, 86-87. *Honors & Awards:* Ralph R Teetor Award, Soc Automotive Engrs, 77; George Westinghouse Award, Am Soc Eng Educ, 90; Outstanding Eng Educ, Inst Indust Engrs, 91. *Mem:* Nat Acad Eng; fel Inst Indust Engrs; fel Am Soc Qual Control; Nat Soc Prof Engrs; Am Soc Engr Educ. *Res:* Quality management; process design and improvement; statistical process control in continuous flow systems; economically-based quality control planning; process capability and performance indices. *Mailing Add:* Sch Indust Eng & Mgt Okla State Univ Stillwater OK 74078-0540

CASE, KENNETH MYRON, b New York, NY, Sept 23, 23. PHYSICS. *Educ:* Harvard Univ, SB, 45, MS, 46, PhD, 48. *Prof Exp:* Res assoc, Lawrence Radiation Lab, 49-50; res assoc, Univ Rochester, 50-51; prof, Univ Mich, 51-69; prof, 69-88, EMER PROF PHYSICS, ROCKEFELLER UNIV, 88- *Concurrent Pos:* Adj prof, Inst Nonlinear Studies, Univ Calif, San Diego, 88-, vis prof, 81-82; vis prof, Lawrence Radiation Lab, 56 & 61, Mass Inst Technol, 63-64; scientist, Los Alamos Sci Lab, 44-45; mem Inst Advan Study, Princeton, 48-50 & 56-57; consult, Rand Corp, Ramo-Woolridge Corp, Gen Dynamics, Gen Atomic, Inst Defense Analysis, Stanford Res Inst, Mitre Corp, La Jolla Inst & Phys Dynamics; Guggenheim fel, 63. *Honors & Awards:* Cert Merit, Am Nuclear Soc, 65. *Mem:* Nat Acad Sci; fel Am Phys Soc. *Res:* Neutron diffusion; field theory; relativistic wave equations; orthogonal polynomials. *Mailing Add:* 1429 Calle Altura Univ Calif La Jolla CA 92037

CASE, LLOYD ALLEN, b Ontario, Ore, Jan 1, 43; m 62; c 8. PHYSICS, COMPUTER SCIENCE. *Educ:* Brigham Young Univ, BS, 64, PhD(physics), 68. *Prof Exp:* Teaching asst physics, Brigham Young Univ, 62-68, res asst, 64-68; asst prof, Ind Univ Southeast, 68-71, chmn dept natural sci, 70-72, assoc prof physics, 72-76; res fel, NASA-Ames Lab & Stanford Univ, 71-72; prof & dir comput serv, Ind Univ Southeast, 76-79; mgr, Syst Comput Technol Corp, Malvern, Pa, 79-81; PRES, UNIV NEV SYST & DIR COMPUT SERV, 81- *Concurrent Pos:* NSF fel, Univ Colo, 68; Ind Univ res grant, 71; lectr, Egypt, Belgium. *Mem:* Am Phys Soc; Am Asn Physics Teachers; Sigma Xi; Asn Comput Mach; sr mem Inst Elec & Electronic Engrs. *Res:* Hardware; software; consulting; audits; astrophysics; general relativity. *Mailing Add:* 6901 E Service Rd Hudson CA 95326

CASE, MARVIN THEODORE, b Anna, Ill, Dec 20, 34; m 58; c 2. VETERINARY PATHOLOGY, TOXICOLOGY. *Educ:* Univ Ill, BS, 57, DVM, 59, MS, 64, PhD(vet path), 68; Am Bd Toxicol, dipl, 80. *Prof Exp:* Vet, Ill State Dept Agr, 61-62; from instr to asst prof vet path, Col Vet Med, Univ Ill, 62-69; head path-toxicol sect, Riker Labs, 69-71, mgr path, 71-75, mgr path-toxicol, 75-84; mgr, Clin Develop, Apheresis Prog, 3M Co, 84-87, div scientist, Clin & Reg Affairs, orthod prod, 87-89, DIV VET PATHOLOGIST, 3M PHARMACEUTICALS, 89- *Concurrent Pos:* Dep vet Los Angeles County, Calif & assoc prof comp med, Univ Southern Calif, 69-71. *Mem:* AAAS; Soc Toxicol; Soc Toxicol Pathologists; NY Acad Sci; Int Acad Path. *Res:* Neoplasia; toxicology; animal diseases; teratology. *Mailing Add:* 3M Ctr 270-35-05 St Paul MN 55144

CASE, MARY ELIZABETH, b Crawfordsville, Ind, Dec 10, 25. MICROBIAL GENETICS. *Educ:* Maryville Col, Tenn, BA, 47; Univ Tenn, MS, 50; Yale Univ, PhD(bot), 57. *Hon Degrees:* DSc, Maryville Col, 69. *Prof Exp:* Res assoc genetics, Yale Univ, 57-72; assoc prof zool, 72-79, assoc prof, 80-85, PROF MOLECULAR & POP GENETICS, UNIV GA, 85- *Mem:* Am Soc Microbiol; Am Genetics Soc; Sigma Xi. *Res:* Genetics of microorganisms; neurospora crassa. *Mailing Add:* Genetics Univ Ga Athens GA 30602

CASE, NORMAN MONDELL, b Milton, Ore, Oct 12, 17. HUMAN ANATOMY. *Educ:* Col Med Evangelists, BS, 49; Univ Southern Calif, MS, 54; Loma Linda Univ, PhD(anat), 58. *Prof Exp:* Tech asst anat, Loma Linda Univ, 50-54, from asst instr to assoc prof anat, 54-84; RETIRED. *Mem:* Am Asn Anat; Electron Micros Soc Am; assoc AMA; Royal Micros Soc. *Res:* Electron microscopy of the optic lobe in Octopus. *Mailing Add:* 11635 Vista Lane Yucaipa CA 92399

CASE, ROBERT B, b Columbus, Ohio, July 19, 20; m 61; c 1. CARDIOLOGY, PHYSIOLOGY. *Educ:* Ohio Wesleyan Univ, BA, 43; Mass Inst Technol, BS, 43; Columbia Univ, MD, 48. *Prof Exp:* Res assoc cardiac physiol, Sch Pub Health, Harvard Univ, 52-54; DIR LAB EXP CARDIOL, ST LUKE'S HOSP, 56-; assoc Prof med, 72-79, PROF MED, COL PHYSICIANS & SURGEONS, COLUMBIA UNIV, 79- *Concurrent Pos:* Nat Heart Inst res fel, 52-54; NY Heart Asn sr res fel, 56-61; USPHS res career develop award, 62-67; chief cardiac consult clin, City Health Dept, New York, 62-70; asst clin prof med, Col Physicians & Surgeons, Columbia Univ, 65-72; Attend cardiologist, St Luke's Hosp, 72- *Mem:* Am Physiol Soc; Am Fedn Clin Res. *Res:* Cardiac physiology; coronary artery disease. *Mailing Add:* St Luke's Roosevelt Hosp 421 W 113th St New York NY 10025

CASE, ROBERT OLIVER, b Ft Thomas, Ky, Nov 17, 35. APPLIED MECHANICS, MACHINE DESIGN. *Educ:* US Mil Acad, BS, 58; Univ Ala, Tuscaloosa, MS, 62, PhD(eng mech), 68. *Prof Exp:* Res assoc, Univ Ala, 66-68; asst prof, 68-77, assoc prof, 77-80, PROF MECH ENG, FLA ATLANTIC UNIV, 80-, CHMN DEPT, 77- *Mem:* Soc Exp Stress Analysis. *Res:* Photoelastic investigations of thin-walled structures and non-isotropic structures. *Mailing Add:* Dept Mech Eng Fla Atlantic Univ Boca Raton FL 33431

CASE, RONALD MARK, b Wausau, Wis, Oct 7, 40; m 90; c 2. WILDLIFE ECOLOGY. *Educ:* Ripon Col, AB, 62; Univ Ill, Urbana, MS, 64; Kans State Univ, PhD(biol), 71. *Prof Exp:* Instr ecol, Univ Mo, 71-72; from asst prof to assoc prof, 72-80, PROF WILDLIFE, UNIV NEBR, 80- *Concurrent Pos:* Consult ecol, Midwest Res Inst, 71-72, Mits Kawamoto & Assocs, 72-73. *Mem:* Wildlife Soc; Ecol Soc Am; Am Inst Biol Sci; Am Ornithologists Union. *Res:* Pocket gophers and grasslands; population dynamics and damages to forage; bioenergetics of Bobwhites; study of population status of wild canids, especially swift fox and red fox in Nebraska. *Mailing Add:* Dept Forestry Fisheries & Wildlife Univ Nebr Lincoln NE 68583-0819

CASE, STEVEN THOMAS, b New York, NY, Jan 18, 49; m 71; c 2. CELL BIOLOGY, EUKARYOTIC GENE EXPRESSION. *Educ:* Widner Col, BS, 69; Wilkes Col, MS, 71; Univ Southern Calif, PhD(cellular & molecular biol), 74. *Prof Exp:* Res fel, Karolinska Inst, Stockholm, Sweden, 75-77; res assoc, Yale Univ, 77-78; from asst prof to assoc prof, 79-90, PROF BIOCHEM, UNIV MISS MED CTR, JACKSON, 90- *Concurrent Pos:* Nat Res Serv Award, NIH, 75-78, prin investr, 79-84; prin investr, Am Heart Asn, 82-84, NSF, 85-87 & Off of Naval Res, 87-; ed, Gene, 83- *Honors & Awards:* US Cong Antarctic Serv Medal, NSF, 85. *Mem:* Am Soc Cell Biol; Am Soc Molecular Biol & Biochem; Sigma Xi. *Res:* Learning how physically separate genes are coordinately expressed in differentiated cells; multigene family of Balbiani ring genes that encode secretory proteins in larval salivary glands of Chironomus; assembly of protein biopolymers. *Mailing Add:* Dept Biochem Univ Miss Med Ctr Jackson MS 39216-4505

CASE, TED JOSEPH, b Sioux City, Iowa, July 19, 47. EVOLUTIONARY ECOLOGY. *Educ:* Univ Redlands, BS, 69; Univ Calif, Irvine, PhD(biol), 74. *Prof Exp:* Res assoc & lectr entomol, Univ Calif, Davis, 73-75; asst prof biol, Purdue Univ, 75-78; from asst prof to assoc prof, 78-86, PROF BIOL, UNIV CALIF, SAN DIEGO, 86- *Concurrent Pos:* Vis scholar, Flinders Univ S Australia. *Res:* Evolutionary ecology; island biogeography; lizard ecology. *Mailing Add:* Biol Dept C016 Univ Calif San Diego La Jolla CA 92093

CASE, VERNA MILLER, b Lebanon, Pa, Aug 17, 48; m 76; c 2. ETHOLOGY, ANIMAL BEHAVIOR. *Educ:* Pa State Univ, BS, 70, MS, 72, PhD(zool), 74. *Prof Exp:* ASSOC PROF BIOL, DAVIDSON COL, 74- *Concurrent Pos:* Mem bd dirs, Planned Parenthood Greater Charlotte; pres, Bd Carolina Rapton Ctr; vis prof, MIT-Harvard Med Consortium, 86-87. *Mem:* Animal Behav Soc; AAAS. *Res:* Development and regulation of vertebrate social systems; relationship between social systems, the ecological setting in which they occur and the evolutionary history of the species. *Mailing Add:* Dept Biol Davidson Col Davidson NC 28036

CASE, VERNON WESLEY, b Kitchener, Ont, May 7, 35; m 58; c 3. SOIL FERTILITY. *Educ:* Univ BC, BSA, 58; Cornell Univ, MSc, 61. *Prof Exp:* Res officer res br, Can Dept Agr, 61-68; agr specialist, Res & Develop, Int Minerals & Chem Corp, 68-73; instr soils & plant nutrit, Triton Col, 73-74; mgr Agron Serv Lab, MGR AGRON, INT MINERALS & CHEM CORP, 83- *Mem:* Am Soc Agron; Int Soc Soil Sci; Coun Agr Sci & Technol; Soil Sci Soc Am; Coun Soil & Plant Analysis; Sigma Xi. *Res:* Fertilizer evaluation and plant nutrition research; land reclamation and environmental concerns in agriculture; manager agronomy. *Mailing Add:* 1331 S First St Int Minerals & Chem Corp Terre Haute IN 47808

CASE, WILLIAM BLEICHER, b Elmira, NY, Oct 16, 41; m 80; c 1. PLASMA PHYSICS. *Educ:* Syracuse Univ, BS, 63, MS, 66, PhD(physics), 71. *Prof Exp:* Instr physics, Syracuse Univ, 70-72; vis asst prof, State Univ NY, Binghamton, 74-75; asst prof physics, Hobart & William Smith Cols, 75-80; ASST PROF PHYSICS, GRINNELL COL, 80- *Mem:* Am Phys Soc. *Res:* Electron beam instabilities; microwave generation using electron beams. *Mailing Add:* Dept Physics Grinnell Col Grinnell IA 50112

CASELLA, ALEXANDER JOSEPH, b Taylor, Pa, Aug 10, 39; m 66; c 2. ENERGY PLANNING, ENERGY STUDIES. *Educ:* Villanova Univ, BS, 61; Drexel Inst Technol, MS, 64; Pa State Univ, PhD(physics), 69. *Prof Exp:* Physicist, Frankford Arsenal, Philadelphia, 61-65; asst prof physics, Jacksonville Univ, 69-74; ASSOC PROF ENVIRON STUDIES, SANGAMON STATE UNIV, 74-; PROF & CHAIR ENVIRON, UNIV SAN FRANCISCO, 80- *Concurrent Pos:* Dir Energy Studies, Univ San Francisco, 80-, Consult, vis prof, 83- *Mem:* AAAS; Am Asn Physics Teachers. *Res:* Solar energy; energy technology; policy utility research; energy technology. *Mailing Add:* Dept Environ Studies Sangamon State Univ Springfield IL 62794

CASELLA, CLARENCE J, b New York, NY, Nov 9, 29; m 59; c 5. GEOLOGY. *Educ:* Hunter Col, BA, 56; Columbia Univ, PhD(geol), 62. *Prof Exp:* Lectr geol, Hunter Col, 58-60 & Brooklyn Col, 60-61; asst prof, Villanova Univ, 61-65; asst prof, Dept Earth Sci, 65-68, ASSOC PROF, DIV GEOL, NORTHERN ILL UNIV, 68- *Mem:* AAAS; Geol Soc Am. *Res:* Structure and petrology of metamorphic rocks; lunar geology. *Mailing Add:* Dept of Geol Northern Ill Univ De Kalb IL 60115

CASELLA, JOHN FRANCIS, b Oneida, NY, June 9, 44; m 68; c 2. POLYMER CHEMISTRY. *Educ:* Clarkson Col Technol, BS, 66, MS, 71, PhD(phys chem), 73. *Prof Exp:* res assoc phys chem of proteins, Grad Dept Biochem, Brandeis Univ, 72-74; Res scientist polymer characterization, Ethicon, Inc, 74-80; MEM STAFF, I O LAB CORP, 80- *Concurrent Pos:* Am Cancer Soc fel, 74. *Mem:* Am Chem Soc; AAAS; NY Acad Sci; Sigma Xi. *Res:* Physical chemistry of macromolecules; solution properties of synthetic polymers and how they correlate with processability and final properties of the polymer. *Mailing Add:* 500 Iolab Dr Claremont CA 91711-4881

CASELLA, RUSSELL CARL, b Framingham, Mass, Nov 6, 29; m 52; c 2. PHYSICS. *Educ:* Mass Inst Technol, BS, 51; Univ Ill, MS, 53, PhD(physics), 56. *Prof Exp:* Physicist, Air Force Cambridge Res Ctr, 51-52; asst, Univ Ill, 53-56, from res asst to assoc, 56-58; physicist, Int Bus Mach Res Lab, 58-65; physicist, Nat Bur Standards, 65-69, SOLID STATE & ELEM PARTICLE THEORIST, REACTOR RADIATION THEORY DIV, NAT INST STANDARDS & TECHNOL, 69- *Honors & Awards:* Silver Medal Award, US Dept Com, 73. *Mem:* Am Phys Soc. *Res:* Theoretical solid state physics; theoretical elementary-particle physics. *Mailing Add:* 1485 Dunster Lane Potomac MD 20854

CASEMAN, A(USTIN) BERT, b Phoenix, Ariz, Feb 13, 22; m 46; c 1. CIVIL ENGINEERING, STRUCTURAL ENGINEERING. *Educ:* Utah State Univ, BS, 47, MS, 48;. *Hon Degrees:* Mass Inst Technol, ScD, 61. *Prof Exp:* From instr to asst prof civil eng, Wash State Univ, 48-56; assoc prof, 56-62, Ga Inst Technol, 56-62, prof civil eng, 61-; RETIRED. *Concurrent Pos:* NSF fac fel, 57. *Mem:* Am Soc Civil Engrs; Am Concrete Inst; Sigma Xi. *Res:* Structural theory; design of thin shell structures. *Mailing Add:* 2136 Kodiak Dr NE Atlanta GA 30345

CASERIO, MARJORIE C, b London, Eng, Feb 26, 29; nat US; m 57; c 2. ORGANIC CHEMISTRY. *Educ:* Univ London, BSc, 50; Bryn Mawr Col, MA, 51, PhD(chem), 56. *Prof Exp:* Assoc chemist, Fulma Res Inst, Eng, 52-53; from asst to instr chem, Bryn Mawr Col, 53-56; fel, Calif Inst Tech, 56-64; from asst prof to assoc prof, 65-71, PROF CHEM, UNIV CALIF, IRVINE, 71- *Concurrent Pos:* John S Guggenheim Found fel, 75-76. *Honors & Awards:* Garvan Medal, Am Chem Soc, 75. *Mem:* Am Chem Soc; Royal Soc Chem; Sigma Xi. *Res:* Reaction mechanisms in organic chemistry. *Mailing Add:* 2114 De Mayo Rd Del Mar CA 92014

CASERIO, MARTIN J, b Laurium, Mich, July 18, 16. METALLURGICAL ENGINEERING. *Educ:* Mich Technol Univ, BS, 36, PhD(eng), 63. *Prof Exp:* Group vpres, Elec Components, Gen Motors Corp, 75-81; CONSULT, 81- *Honors & Awards:* Silver Knight Award, Am Mgt Asn, 70; Medal Advan Res, Am Soc Metals, 82; Eli Whittney Mem Award, Soc Mfg Engrs, 84. *Mem:* Fel Am Soc Metals. *Mailing Add:* 2001 Sailfish Point Blvd Apt 307 Stuart FL 34996

CASEY, ADRIA CATALA, b Havana, Cuba, Apr 24, 34; m 67; c 2. CHEMICAL HAZARD ASSESSMENT, MEDICINAL CHEMISTRY. *Educ:* Univ Havana, BS, 56; Univ Miami, MS, 62; Clarkson Col Technol, PhD(org chem), 64. *Prof Exp:* Asst prof chem, Univ Villanueva, Cuba, 56-60; instr, Clarkson Col Technol, 64-65; res chemist, Textile Fibers, E I du Pont de Nemours & Co, Inc Exp Sta, Del, 65-66; mem staff chem, NEng Inst, 66-73; asst prof chem, Univ Bridgeport, Conn, 73-75; sr res chemist, Stauffer Chem Co, 75-77, sr label adminr environ health & safety, 77-81, mgr chem hazard assessment & commun, 81-82, dir, 82-87; PRES, CATALA ASSOC, INC, 87- *Concurrent Pos:* Vpres, Am Conf Chem Labeling; mem bd dirs, Hazardous Mat Adv Coun. *Mem:* Sigma Xi; Am Chem Soc. *Res:* Synthesis; organic synthesis of specialty chemicals; regulatory classification of chemicals. *Mailing Add:* 38 Brickyard Rd Mars PA 16046-9602

CASEY, CAROL A, b Chamberlain, SDak, Dec 10, 53. NITROGEN METABOLISM IN FISH. *Educ:* Augustana Col, Sioux Falls, SDak, BA, 76; Rice Univ, MS, 80, PhD(biol), 81. *Prof Exp:* Res asst, Rice Univ, 76-80; FEL BIOCHEM, UNIV MINN, 80- *Concurrent Pos:* Lab asst biol, Rice Univ, 76-78. *Res:* Obtaining information concerning nitrogen metabolism in fishes specifically on aspects of ammoniagenesis in teleosts as well as osmoregulation in elasmobranchs. *Mailing Add:* 5808 Briggs St Omaha NE 68106

CASEY, CHARLES P, b St Louis, Mo, Jan 11, 42; m 68; c 1. ORGANIC CHEMISTRY. *Educ:* St Louis Univ, BS, 63; Mass Inst Technol, PhD(org chem), 68. *Prof Exp:* Fel org chem, Harvard Univ, 67-68; from asst prof to assoc prof, 68-77, PROF ORG CHEM, UNIV WIS, MADISON, 77- *Honors & Awards:* A C Cope Scholar Award, Am Chem Soc, 88 & Organometallic Chem Award, 91. *Mem:* Am Chem Soc; The Chem Soc. *Res:* Mechanism of organometallic reactions; metal carbene complexes; CO reduction; homogeneous catalysis. *Mailing Add:* Dept Chem Univ Wis Madison WI 53706

CASEY, CLARE ELLEN, b Blenheim, NZ, June 18, 50. LACTATION, TRACE ELEMENTS. *Educ:* Univ Otago, NZ, BHSc, 71, MHSc, 72, PhD(human nutrit), 76. *Prof Exp:* Res assoc, dept pediat, Univ Colo Health Sci, 77-80, 82-86; MACROBERT LECTR HUMAN NUTRIT, UNIV ABERDEEN MED SCH, SCOTLAND, 86- *Concurrent Pos:* Hon res assoc, Rowett Res Inst, Scotland. *Mem:* Am Soc Clin Nutrit; Nutrit Soc; Int Soc Res Human Milk & Lactation; Soc Int Nutrit Res. *Res:* Nutritional epidemiology; nutrition of lactating and pregnant women; production and composition of human milk; nutrition of trace elements and iron; diet in health and disease (CVD, cancer, etc); impact of urbanization on nutrition and health. *Mailing Add:* Dept Med Univ Aberdeen Med Sch Forester Hill Aberdeen AB9-2ZD Scotland

CASEY, DANIEL EDWARD, b West Springfield, Mass, Jan 24, 47. PSYCHOPHARMACOLOGY, MOVEMENT DISORDERS. *Educ:* Univ Va, BA, 69, MD, 72. *Prof Exp:* Resident, Dept Psychiat, Brown Univ, 74-76; from instr to asst prof psychiat, 76-80, dir, Tardive Dyskinesia Clin, 76-82, asst prof, Dept Neurol, 79-83, assoc prof, Dept Psychiat, 80-85, PROF, DEPT PSYCHIAT, ORE HEALTH SCI UNIV, 85- *Concurrent Pos:* Vis res scientist, Sct Hans Hosp, Denmark, 78; assoc investr, Vet Admin Res Career Develop Prog, 76-78, res assoc, 78-81, clin investr, 81-84; collab scientist, Ore Regional Primate Res Ctr, 80- *Mem:* Am Psychiat Asn; Am Col Neuropsychopharmacol; Soc Biol Psychiat; AAAS; Col Int Neuropsychopharmacol. *Res:* Biological psychiatry with emphasis on clinical drug trials and nonhuman primate models of tardive dyskinesia and schizophrenia; neurotransmitter mechanisms of dopamine, acetylcholine, gaba and endorphins. *Mailing Add:* 3250 Southwest Doschdale Dr Portland OR 97201

CASEY, H CRAIG, JR, b Houston, Tex, Dec 4, 34; m 83; c 2. SOLID-STATE ELECTRONICS, OPTICAL PROPERTIES OF SEMICONDUCTORS. *Educ:* Okla State Univ, BS, 57; Stanford Univ, MS, 59, PhD(elec eng), 64. *Prof Exp:* Develop engr, Hewlett-Packard Co, 57-62; mem tech staff, AT&T Bell Labs, 64-79; PROF & CHMN ELEC ENGR, DUKE UNIV, 79- *Concurrent Pos:* Mem, Adv Group Electron Devices, Dept Defense, 75-79; bd dir, Acme Elec Co, 84-91. *Mem:* fel Inst Elec & Electronics Engrs Electron Devices Soc (vpres 86-87, pres 88-89); Am Phys Soc; Am Vacuum Soc; Electrochem Soc. *Res:* Third-fifth compound materials and devices; heterostructure lasers; high-resitivity layers on gallium arsenide; dielectric layers on indiumphosphorus and optoelectronic devices. *Mailing Add:* Dept Elec Eng Duke Univ Durham NC 27706

CASEY, HAROLD W, b Reuter, Mo, Sept 24, 32; m 55; c 2. VETERINARY PATHOLOGY. *Educ:* Univ Mo-Columbia, BS, 54, DVM, 57; Tulane Univ, MPH, 58; Univ Calif, Davis, PhD(comp path), 65; Am Col Vet Pathologists, dipl, 66. *Prof Exp:* US Air Force, 58-80, vet, Nouasseur AFB, 58-59, asst chief vet serv, Morocco, 59-61, res scientist biol div, Hanford Atomic Works, 61-63, chief anat path sect, 65-67, chief cytopath br, 67-70, chief vet path br, Air Force Sch Aerospace Med, 70-71, chief gen vet path br, 71-74, chmn dept vet path, Armed Forces Inst Path, 74-80; dir path, Toxigenics, Inc, 80-82; HEAD, DEPT VET PATH, SCH VET MED, LA STATE UNIV, BATON ROUGE, 82- *Concurrent Pos:* mem, Fac Discussants, Charles Louis Davis, DVM Found, 72-, prog dir, Gross Path Div, 74-80, Adv Bd, 76-; head, WHO Int Ref Ctr Comp Path, 75-80; mem, Comt Histol Classification of Lab Animal Tumors, Inst Lab Animal Resources, Nat Res Coun, 75-79, Carnivores Subcomt, Comt on Animal Models for Res on Aging, NSF, 78-80; coun, Am Col Vet Path, 86-90. *Mem:* Int Acad Path; AAAS; Am Vet Med Asn; Am Col Vet Path (pres-elect, 91). *Res:* Comparative pathology, including ultrastructure and the biological effects of ionizing radiation with special interest in non-human primate pathology. *Mailing Add:* Dept Vet Path La State Univ Baton Rouge LA 70803

CASEY, HORACE CRAIG, JR, b Houston, Tex, Dec 4, 34; m 60, 83; c 2. ELECTRICAL ENGINEERING. *Educ:* Okla State Univ, BS, 57; Stanford Univ, MS, 59, PhD(elec eng), 64. *Prof Exp:* Res & develop engr electronics, Hewlett-Packard Co, 57-62; mem tech staff, Bell Labs, 64-; CHMN DEPT ELEC ENG, DUKE UNIV. *Concurrent Pos:* Mem bd dirs, Acme Elec Corp. *Mem:* Am Phys Soc; fel Inst Elec & Electronic Engrs; Electrochem Soc; Am Vacuum Soc. *Res:* Semiconductor materials and devices; electronic instrumentation. *Mailing Add:* Dept Elec Eng Duke Univ 130 Eng Bldg Durham NC 27706

CASEY, JAMES, b Tipperary, Ireland, Sept 15, 49; m 85; c 3. MECHANICS OF CONTINUOUS MEDIA. *Educ:* Nat Univ Ireland, BE, 71; Univ Calif, Berkeley, PhD(eng sci), 80. *Prof Exp:* Asst prof, 80-85, ASSOC PROF MECH ENG, UNIV HOUSTON, 85- *Mem:* Am Soc Mech Engrs. *Res:* Continuum mechanics; large strain elasticity and plasticity; dynamics; theory and applications of finite plasticity; dynamics of nearly rigid continua. *Mailing Add:* Dept Mech Eng Univ Houston 4800 Calhoun Houston TX 77204

CASEY, JAMES PATRICK, b Syracuse, NY, Aug 5, 15; m 41; c 2. CHEMISTRY. *Educ:* Syracuse Univ, BS, 37; State Univ NY, MS, 47. *Prof Exp:* Head paper serv lab, A E Staley Mfg Co, 37-46; assoc prof pulp & paper mfg, State Univ NY, 46-51; dir tech serv, A E Staley Mfg Co, 51-56, dir res, 56-59; vpres res & develop, Union Starch & Ref Co, 59-67, vpres mkt & develop, Union Div, 67-70, vpres res, Marschall Div, 70-76, CONSULT, MILES LABS, 76- *Mem:* fel Tech Asn Pulp & Paper Indust; emer mem indust Res Inst. *Res:* Paper; starch; textiles; adhesives; consumer products; foods; vegetable oils; chemical technology; enzymes; food cultures; microbiology; fermentation; cereal processing. *Mailing Add:* Many Levels RR 1 Tryon NC 28782

CASEY, JOHN EDWARD, JR, b Cranston, RI, Dec 2, 30; m 56; c 2. CHEMISTRY OF ADVANCED BATTERIES, ORGANIC CHEMISTRY. *Educ:* Providence Col, BS, 52, MS, 57. *Prof Exp:* Control chemist, Allied Chem Corp, 52; asst chem, Providence Col, 55-57; org chemist, Smith Kline & French Labs, 57-64; process chemist, Geigy Chem Corp, RI, 64-66; develop chemist, Rohm and Haas Co, 66-71; mgr process develop, Yardney Elec Corp, 71-76; mgr electrochem progs, GTE LABS, 76-77, mgr prod, Power Sources Ctr, 77-78; dir qual assurance, Altus Corp, 78-80, mgr eng admin, 80-81; owner & dir, Bryant Bur, Mountain View, Calif, 81-82; dir qual assurance, Power Conversion Inc, 82-84; SUPVR POWER SOURCES LAB, LOCKHEED MISSILE & SPACE CO, 84- *Concurrent Pos:* Pvt tutoring, 82- *Mem:* Am Chem Soc. *Res:* Correlation of chemical structure and biological activity; natural products; waste treatment; environmental chemistry; battery technology. *Mailing Add:* 2014 Firethorne Ct Milpitas CA 95035

CASEY, KENNETH L(YMAN), b Ogden, Utah, Apr 16, 35; m 58; c 3. NEUROPHYSIOLOGY, NEUROLOGY. *Educ:* Whitman Col, BA, 57; Univ Wash, MD, 61. *Prof Exp:* Intern, NY Hosp-Cornell Med Ctr, 61-62; res assoc psychol, McGill Univ, 64-66; from asst prof to assoc prof physiol, Univ Mich, 66-69, resident neurol, 71-74, assoc prof physiol & neurophysiol, 74-75, assoc prof neurol, 75-78, PROF NEUROL, UNIV MICH, ANN ARBOR, 78-, PROF PHYSIOL, 75-; CHIEF, NEUROL SERV, VET ADMIN MED CTR, ANN ARBOR, 80- *Concurrent Pos:* NIH Study Sect; Prog award, Bristol-Myers Squibb, 88-93. *Mem:* Am Neurol Asn; Int Asn Study Pain; Am Pain Soc (pres, 84-85); Soc Neurosci; Am Acad Neurol; Sigma Xi; Am Physiol Soc. *Res:* Neurophysiological correlates of behavior; neurophysiology of limbic system; somatosensory neurophysiology and neural mechanism of pain sensation. *Mailing Add:* Chief Neurol Serv 127 Vet Admin Med Ctr 2215 Fuller Rd Ann Arbor MI 48105

CASEY, TIMOTHY M, b Bethlehem, Pa, Jan 13, 49. ENERGETICS. *Educ:* Univ Calif, Los Angeles, 75. *Prof Exp:* PROF ENTOM, COOK COL, RUTGERS UNIV, 77- *Mem:* Am Soc Zoologists; AAAS; Ecol Soc Am; Entom Soc Am. *Mailing Add:* Dept Entom Cook Col Rutgers Univ New Brunswick NJ 08903

CASH, DEWEY BYRON, b Wadley, Ala, Dec 22, 30; m 54; c 2. MATHEMATICAL ANALYSIS. *Educ:* Auburn Univ, BS, 55, MEd, 57, MS, 64. *Prof Exp:* Teacher high schs, Fla & Ga, 55-58; assoc prof, 58-76, PROF MATH, COLUMBUS COL, 76- *Mem:* Math Asn Am. Res Differential equations. *Mailing Add:* Dept Math Columbus Col Columbus GA 31993-2399

CASH, FLOYD LEE, b Wichita Falls, Tex, July 14, 26; m 49; c 2. ELECTRICAL ENGINEERING. *Educ:* Univ Okla, BS, 46; Univ Tex, MS, 51, PhD(elec eng), 55. *Prof Exp:* Instr eng, Midwestern Univ, 47-50; instr elec eng, Univ Tex, 53-54; engr, Elgen Corp, 54-56, res engr, 56-58; dir tech training, Opers Dept, Lane-Wells, 58-59; actg chmn dept, 69-77, PROF ELEC ENG, UNIV TEX, ARLINGTON, 59-, CHMN DEPT, 77- *Concurrent Pos:* Consult, Saturn Electronics Corp, Tex, 61-; consult, Astronaut Div, Ling-Temco-Vought, Inc, 61-63, Aeronaut Div, 62-63 & Electronics Div, 62; consult, Signatrol Corp, Ill, 63- & Hunt Electronics Corp, Tex, 63-; ed, J Soc Prof Well Log Analysts, 62. *Mem:* Sr mem Inst Elec & Electronics Engrs; Soc Petrol Engrs; Am Inst Mining, Metall & Petrol Engrs; Soc Prof Well Log Analysts; Sigma Xi. *Res:* Electromagnetic wave propagation; antennas; radar cross section measurements; oil well log analysis; application of computers to geophysical measurements and interpretations of these measurements. *Mailing Add:* Dept of Elec Eng Univ Tex-Arlington Arlington TX 76019

CASH, ROWLEY VINCENT, b Waterville, NY, June 7, 17; m 51; c 3. HISTORY OF CHEMISTRY. *Educ:* Colgate Univ, AB, 39; Ind Univ, PhD(chem), 52. *Prof Exp:* Asst chem, Ind Univ, 39-42; head dept, Olivet Col, 42-51; from asst prof to prof, 57-81, chmn dept, 69-75, EMER PROF CHEM, CENT CONN STATE UNIV, 81- *Concurrent Pos:* Vis prof, Ariz State Univ, 61 & North Adams State Col, 81. *Mem:* Fel AAAS; Hist Sci Soc; Am Chem Soc; fel Am Inst Chemists; Soc Hist Technol. *Res:* History of chemistry; contributions by Scandinavian chemists. *Mailing Add:* 119 Elizabeth Lane Mt Dora FL 32757

CASH, WILLIAM DAVIS, b Chesnee, SC, Feb 23, 30; m 61; c 3. BIOLOGICAL CHEMISTRY. *Educ:* Univ NC, BS, 51, PhD(pharmaceut chem), 54. *Prof Exp:* Res assoc biochem, Med Col, Cornell Univ, 54, from instr to assoc prof, 56-68; DIR BIOCHEM, CIBA-GEIGY CORP, 68- *Mem:* AAAS; Am Chem Soc; Am Soc Biol Chemists; NY Acad Sci; Sigma Xi. *Res:* Diabetes; prostaglandin biochemistry; platelet biochemistry; neurochemistry. *Mailing Add:* Ciba-Geigy Ltd Pharm Res Dept CH-4002 Basel Switzerland

CASHDOLLAR, KENNETH LEROY, b Pittsburgh, Pa, May 6, 47. DUST EXPLOSIONS, ELECTRO-OPTICS. *Educ:* Dickinson Col, BS, 69; Univ Wis-Madison, MS, 73. *Prof Exp:* Asst res, Yale Univ, 69-71; res assoc, Lowell Technol Inst Res Found, 71-72; proj asst, Univ Wis-Madison, 72-73; RES PHYSICIST, PITTSBURGH RES CTR, BUR MINES, US DEPT INTERIOR, 73- *Concurrent Pos:* Chmn, subcomt E27.05 dusts, Am Soc Testing & Mat, 82- *Mem:* Am Phys Soc; Optical Soc Am; Soc Photo-Optical Instrumentation Engrs; Am Astron Soc; Combustion Inst; Am Soc Testing & Mat. *Res:* Explosibility hazards of dusts and gases; applied electro-optics; multichannel infrared pyrometers and optical dust probes. *Mailing Add:* US Bur Mines Cochrans Mill Rd PO Box 18070 Pittsburgh PA 15236-0070

CASHEL, MICHAEL, b Worthington, Minn, Feb 18, 37; m 62; c 3. BIOCHEMISTRY, GENETICS. *Educ:* Amherst Col, AB, 59; Western Reserve Univ, MD, 63; Univ Wash, PhD(genetics), 68. *Prof Exp:* HEAD, SECT MOLECULAR REGULATION, LAB MOLECULAR GENETICS, NAT INST CHILD HEALTH & HUMAN DEVELOP, 67- *Mem:* Am Soc Biol Chem. *Res:* Regulation of RNA synthesis; transcription specificity of RNA polymerase and nucleotide synthesis; protein synthesis; metabolic dormancy and cell physiology. *Mailing Add:* Molecular Genetics Bldg 6 Rm 335 Nat Inst Child Health Bethesda MD 20892

CASHEN, JOHN F, b West Orange, NJ, Apr 14, 37. SYSTEMS DESIGN. *Educ:* NJ Inst Technol, BS; Univ Calif, Los Angeles, MS & PhD(elec eng). *Prof Exp:* Mem Bell Labs, 60-65 & Hughes Aircraft Co, 65-73; sr res engr, Res & Technol Ctr, Northrop Corp, 73-74, Aircraft Div, 74-75, mgr observables design, 74, vpres advan design & technol, Advan Systs Div, 75-86, VPRES ADVAN PROJS & CHIEF SCIENTIST, B-2 DIV, NORTHROP CORP, PICO RIVERA, CALIF, 86- *Concurrent Pos:* Dep chmn, White House Aeronaut Policy Rev Comt. *Honors & Awards:* Distinguished Serv Medal, Dept Defense, 83. *Mem:* Nat Acad Eng; fel Inst Elec & Electronic Engrs; assoc fel Am Inst Aeronaut & Astronaut. *Res:* Aircraft and military electronics design, development and manufacturing. *Mailing Add:* B2 Div MS T000 Zone AP Northrop Corp 8900 E Washington Blvd Pico Rivera CA 90660

CASHIN, KENNETH D(ELBERT), b Lowell, Mass, May 10, 21; m 44; c 3. CHEMICAL ENGINEERING. *Educ:* Worcester Polytech Inst, BS, 47, MS, 48; Rensselaer Polytech Inst, PhD, 55. *Prof Exp:* From asst prof to prof, 48-63, assoc dean, Eng Sch, 78-81, PROF CHEM ENG, UNIV MASS, 81- *Concurrent Pos:* Consult, Army Chem Corps, 56-57; NSF grants, 57-58, 75-77 & 79-81; adv to establish dept chem eng, Col Petrol & Minerals, Dhahran, Saudi Arabia, 68-70. *Mem:* Am Chem Soc; Am Inst Chem Engrs; Am Soc Eng Educ; Sigma Xi. *Res:* Fluidization of solid particles in gas streams; heat transfer by radiation; spectroscopic studies in fluorine supported flames and other combustion phenomena. *Mailing Add:* 1439 Topeka St Pasadena CA 91104-1551

CASHION, PETER JOSEPH, b Boston, Mass, June 26, 40; m 67; c 4. BIOCHEMISTRY. *Educ:* Boston Col, BSc, 64; Tufts Univ, PhD(biochem), 69. *Prof Exp:* Fel biochem, Univ Wis, 69-70; fel, Mass Inst Technol, 70-72; asst prof, 72-75, ASSOC PROF BIOCHEM, UNIV NB, 75- *Mem:* Am Chem Soc; Can Biochem Soc. *Res:* Nucleic acid chemistry. *Mailing Add:* Dept Biochem Univ New Brunswick College Hill PO Box 4400 Fredericton NB E3B 5A3 Can

CASIDA, JOHN EDWARD, b Phoenix, Ariz, Dec 22, 29; m 56; c 2. TOXICOLOGY. *Educ:* Univ Wis, BS, 51, MS, 52, PhD(entom, biochem), 54. *Prof Exp:* Res asst, Univ Wis, 46-53; med entomologist, Camp Detrick, Md, 53; from asst prof to prof entom, Univ Wis, 54-63; PROF ENTOM & DIR, PESTICIDE CHEM & TOXICOL LAB, UNIV CALIF, BERKELEY, 64- *Concurrent Pos:* Haight travel fel, 58-59; Guggenheim fel, 70-71; vis scholar, Bellagio Study & Conf Ctr, Rockefeller Found, Lake Como, Italy, 78; lectr, Univ New S Wales, Australia, 83, Cornell Univ, 85. *Honors & Awards:* Medal, Seventh Int Cong Plant Protection, Paris, 70; Spencer Award, Am Chem Soc, 78; Jeffery lectr, Univ New S Wales, Australia, 83; Messerger lectr, Cornell Univ, Ithaca, 85. *Mem:* Nat Acad Sci; Am Chem Soc; Entom Soc Am; AAAS. *Res:* Pesticide chemistry; comparative biochemistry. *Mailing Add:* 114 Wellman Hall Univ Calif Berkeley CA 94704

CASIDA, LESTER EARL, JR, b Columbia, Mo, Aug 25, 28; m 53; c 2. MICROBIOLOGY, ECOLOGY. *Educ:* Univ Wis, BS, 50, MS, 51, PhD(bact), 53. *Prof Exp:* Bacteriologist, Abbott Labs, 51 & Pabst Labs, 53-54; res biochemist, Charles Pfizer & Co, Inc, 54-57; asst prof bact, 57-62, assoc prof microbiol, 62-66, PROF MICROBIOL, PA STATE UNIV, 66- *Mem:* Am Soc Microbiol; Am Chem Soc; Brit Soc Gen Microbiol; fel Am Acad Microbiol. *Res:* Microbial physiology and ecology; industrial microbiology. *Mailing Add:* Microbiol Prog S-101 Frear Bldg Pa State Univ University Park PA 16802-6001

CASILLAS, EDMUND RENE, b Westwood, Calif, Nov 24, 38; m 62; c 2. BIOCHEMISTRY. *Educ:* Calif State Univ, BA, 59; Ore State Univ, PhD(biochem), 68. *Prof Exp:* Res assoc biochem, Ore State Univ, 64-68; fel, Ore Regional Primate Res Ctr, 69-71, res scientist reproductive physiol, 71-76; assoc prof, 76-82, PROF CHEM, NMEX STATE UNIV, 82- *Concurrent Pos:* Mem, Pop Res Comt, Nat Inst Child Health & Human Develop, 79-83. *Mem:* Am Soc Biol Chemists; Soc Study Reproduction; Am Chem Soc; Soc Study Fertil; Sigma Xi; Am Soc Andrology. *Res:* Reproductive processes in male animal with emphasis on spermatozoan maturation in the epididymis and regulatory mechanisms involved in spermatozoa physiology and metabolism. *Mailing Add:* Dept of Chem NMex State Univ Box 3C Las Cruces NM 88003

CASJENS, SHERWOOD REID, b Kesley, Iowa, July 10, 45; m 72. BIOCHEMISTRY. *Educ:* Mich State Univ, BS & MS, 67; Stanford Univ, PhD(biochem), 72. *Prof Exp:* ASSOC PROF, DEPT CELLULAR, VIRAL & MOLECULAR BIOL, SCH MED, UNIV UTAH, 74- *Concurrent Pos:* NIH res grant, 75-82, NSF res grant, 79-84. *Res:* Structure and assembly of viruses; genetics and biochemistry of bacteriophage P22 morphogenesis. *Mailing Add:* Dept Cellular Viral & Molecular Biol Univ Utah Med Ctr Salt Lake City UT 84132

CASKEY, ALBERT LEROY, b Wichita, Kans, Nov 26, 31; m 57; c 2. ANALYTICAL CHEMISTRY. *Educ:* Southeast Mo State Col, BS, 53; Iowa State Col, MS, 55, PhD(anal chem), 61. *Prof Exp:* Asst prof chem, Southeast Mo State Col, 58-61, assoc prof, 61-64; ASSOC PROF CHEM, SOUTHERN ILL UNIV, CARBONDALE, 64- *Concurrent Pos:* Chem consult & tech witness legal proc, 60-; NSF grants, 63-77; Off Water Resources Res grants, 65-75. *Mem:* AAAS; Am Chem Soc; Soc Appl Spectros. *Res:* Chelation systems; substituent effects in chelation; water pollution; spectrophotometric methods; ion exchange; inorganic reactions; biological effects of trace substances. *Mailing Add:* Dept of Chem Southern Ill Univ 1506 N Walnut St Carbondale IL 62901-2260

CASKEY, CHARLES THOMAS, b Lancaster, SC, Sept 22, 38; m 60; c 2. HUMAN GENETICS. *Educ:* Duke Univ, MD, 63; Am Bd Internal Med, dipl, 77. *Prof Exp:* Intern & resident, Sch Med, Duke Univ, 63-65; res assoc, Nat Heart & Lung Inst, 65-67, sr investr, 65-70, head med genetics, 70-71; HEAD MED GENETICS, BAYLOR COL MED, 71-, DIR & PROF, INST MOLECULAR GENETICS, 85- *Concurrent Pos:* Investr, Howard Hughes Med Inst, Baylor Col Med, 76-, dir, Robert J Kleberg Jr Ctr Human Genetics, 80-, Med Scientist Training Prog, 82-; fac scholar, Cambridge Univ Med Res Coun, 79-80; dir, ASI Somatic Cell Genetics, NATO, 80-81; mem, Human Genome Steering Comt, Dept Energy, 89-, sci adv bd, Merck Sharp & Dohme Res Labs, 89-; mem sci adv bd, Inst Biosci & Technol, 91- *Mem:* Inst Med-Nat Acad Sci; fel Am Col Physicians; Asn Am Physicians; Am Soc Human Genetics (pres, 90-91); fel AAAS; Am Fedn Clin Res; Am Soc Biochem & Molecular Biol; Am Soc Clin Invest; Fedn Am Socs Exp Biol. *Res:* The mechanism of polypeptide chain determination; somatic cell genetics, inborn error of metabolism and medical genetics. *Mailing Add:* Inst Molecular Genetics Baylor Col Med T809 Houston TX 77030

CASKEY, GEORGE R, JR, b Chicago, Ill, Feb 15, 28; m 52; c 2. MATERIALS SCIENCE. *Educ:* Univ Ill, BS, 50; Mass Inst Technol, SM, 52, PhD(mat sci), 69. *Prof Exp:* Res scientist, E I DuPont de Nemours & Co, 52-64 & 68-75, staff metallurgist, 75-89, FEL METALLURGIST, SAVANNAH RIVER LAB, WESTINGHOUSE SAVANNAH RIVER CO, 89- *Concurrent Pos:* Vis prof mats eng, Va Polytech, 81. *Mem:* Metall Soc; Am Inst Mining, Metall & Petrol Engrs; Sigma Xi. *Res:* Alloy theory and electronic structure; radiation damage in reactor materials; physical metallurgy of uranium; hydrogen solution and transport in metals; hydrogen damage and fracture; helium effects on alloys; nuclear plant aging. *Mailing Add:* 705 Pin Oak Dr Aiken SC 29801

CASKEY, JERRY ALLAN, b Galion, Ohio, Sept 8, 38; m 63; c 3. CHEMICAL ENGINEERING. *Educ:* Ohio Univ BS, 61; Clemson Univ, MS, 63, PhD(chem eng), 65. *Prof Exp:* Chem engr, Dow Chem Co, 65-67, res engr, 67; asst prof chem eng, Va Polytech Inst & State Univ, 67-72; assoc prof, 73-78, PROF CHEM ENG, ROSE-HULMAN INST TECHNOL, 78- *Concurrent Pos:* Vis res prof, Israel Inst Tech, 72. *Mem:* Am Inst Chem Engrs; Am Soc Eng Educ; Int Solar Energy Soc. *Res:* Solar heating technology; water resources engineering. *Mailing Add:* Dept Chem Eng Rose-Hulman Inst 5500 Wabash Ave Terre Haute IN 47803

CASKEY, MICHAEL CONWAY, b Bagley, Minn, Dec 6, 50; m; c 4. SHEEP MANAGEMENT, ADULT EDUCATION. *Educ:* Univ Minn, BS, 72; Mankato State Univ, ME, 76. *Prof Exp:* Instr agr, Univ Minn, 72-74, asst prof, 74-75; INSTR SHEEP MGT, SOUTHWESTERN TECH COL, 75-, DEPT HEAD AGR, 84- *Concurrent Pos:* Consult, Control Data Corp, 82-85; chmn, Nat Sheep Improv Prog, 86-89, Nat Sheep Indust Develop Bd, 86-; rep, US Agr Res Inst, 89- *Mem:* US Animal Health Asn. *Res:* Sheep management; ruminant nutrition; sheep health; genetics and selection; sheep economics; adult education methods. *Mailing Add:* Southwestern Tech Col Pipestone MN 56164

CASLAVSKA, VERA BARBARA, b Chrudim, Czech, Jan 18, 34; US citizen; m 52; c 1. DENTAL RESEARCH, CARIOLOGY. *Educ:* Charles Univ, Prague, MSc, 57, CSc(chem), 65. *Prof Exp:* Res assoc chem, Mining Inst, Czech Acad Sci, 57-65; res assoc mat, Pa State Univ, 66-69; ASST STAFF MEM CHEM, FORSYTH DENT CTR, 69- *Concurrent Pos:* prog chmn, Mineralized Tissue Group, Int Asn Dent Res. *Mem:* Int Asn Dent Res. *Res:* Studies of interactions of topical fluoride with human enamel using in vitro and in vivo models; self-gelling liquid compositions for topical application of medications; dental caries inhibiting compositions; structure and properties of glasses; crystal growth; reduction of ores. *Mailing Add:* 140 The Fenway Boston MA 02115

CASLAVSKY, JAROSLAV LADISLAV, b Tyniste, Czech, Nov 12, 28; m 52; c 1. GROWTH OF SINGLE CRYSTALS AT HIGH TEMPERATURES. *Educ:* Charles Univ, Prague, RNDr, 57; Pa State Univ, PhD(geo chem), 69. *Prof Exp:* Scientist inorg chem, Mining Inst, Czech Acad Sci, 57-62; sr scientist solid state physics, Inst Physics, Prague, 62-65; engr mat sci eng, Wiener Schwachstromwerke, Vienna, 66; res assoc thermodyn, Mat Res Lab, Pa State Univ, 66-69; fel solid state physics, Nat Acad Sci, 69-71; PRIN SCIENTIST SOLID STATE PHYSICS, ARMY MAT TECHNOL LAB, 71- *Concurrent Pos:* Lectr, Mass Inst Technol, 70-71. *Honors & Awards:* Czech Acad Sci Award, 62; Dept Defense Army Res & Develop Award, 79. *Mem:* Am Phys Soc; Am Crystallog Asn; Am Asn Crystal Growth. *Res:* Specialist in the growth and characterization of refractory oxides single crystals; invented the "Optical Differential Thermal Analysis" method and the apparatus; point defects and dislocations in crystals; high temperature equilibria of materials; systems suitable for tunable lasers. *Mailing Add:* 244 East St Lexington MA 02173-2320

CASLER, DAVID ROBERT, b Norwich, NY, May 25, 20; m 42; c 2. PHARMACEUTICS. *Educ:* Union Col, NY, BS, 41. *Prof Exp:* Apprentice, Lathrop's Pharm, 41-42, pharmacist, 42, 46-47; from res asst to res assoc, Sterling-Winthrop Res Inst, 47-61, assoc mem, 61-63, res pharmacist, 63-74, sr res pharmacist, 74-79, group leader, Clin Packaging & Prod Develop Pilot Lab, 64-87, raw mat mgr, 79-87; consult, 87-89; RETIRED. *Concurrent Pos:* Chmn flavor test panel, Sterling-Winthrop Res Inst, 59-67, div narcotics officer, 81-87. *Res:* Flavors; flavor test panel methodology; perfumery; liquid processing and processing equipment; surfactants and emulsions; packaging equipment; clinical packaging. *Mailing Add:* Box 498 RD-2 Rensselaer NY 12144-9802

CASLER, MICHAEL DARWIN, b Green Bay, Wis, Oct 31, 54. GRASS BREEDING, QUANTITATIVE GENETICS. *Educ:* Univ Ill, Urbana, BS, 76; Univ Minn, MS, 79, PhD(plant breeding), 80. *Prof Exp:* ASST PROF AGRON, UNIV WIS-MADISON, 80- *Mem:* Am Soc Agron; Crop Sci Soc Am. *Res:* Breeding cool-season grasses for improved nutritive value and production; utilizing maximum heterozygosity in perennial grasses; evaluating selection systems for improving varietal performance; improving adapted cool-season grass varieties. *Mailing Add:* Dept Agron Univ Wis 1575 Linden Dr Madison WI 53706

CASO, LOUIS VICTOR, b Union, NJ, July 6, 24. HISTOLOGY, IMMUNOLOGY. *Educ:* Manhattan Col, BS, 47; Columbia Univ, AM, 49; Rutgers Univ, BS, 54, PhD(zool), 58. *Prof Exp:* Instr biol sci, Col Pharm, Rutgers Univ, 49-54, asst zool, Univ, 55-58; asst res prof pharmacol, Sch Med, George Washington Univ, 59-60; asst prof anat, Col Med, Ohio State Univ, 61-64; asst prof histol, 64-68, assoc prof, 68-82, prof anat sci, Sch Dent, 82-86, actg chmn dept, 85-86, PROF ANAT, DEPT ANAT, TEMPLE UNIV SCH MED, 86- *Concurrent Pos:* NIH grant, 62-63 & res support grant, 84; Am Cancer Soc grants, 63-64 & 73-74. *Mem:* Am Asn Anat; NY Acad Sci; Tissue Cult Asn; Sigma Xi. *Res:* Effects of viral hemagglutination on human erythrocyte antigens; effects of homologous antiserum on cell growth and morphology; immunology of cancer; action of phytohemagglutinin and neuraminidase on cell growth; antigenicity; cytochemistry; histochemistry and function of thymus gland; immunochemistry of all surface; immunocytochemistry of fibroblast. *Mailing Add:* Sch Dent Temple Univ Philadelphia PA 19140

CASO, MARGUERITE MIRIAM, b Union City, NJ, Mar 2, 19. CHEMISTRY. *Educ:* Col Mt St Vincent, AB, 40; Fordham Univ, MS, 55, PhD(chem), 58. *Prof Exp:* Chemist, Colgate-Palmolive-Peet Co, 41-42; chemist, Nat Starch Prod Co, 42-44; teacher sci, Holy Cross Acad, New York, 46-53; lectr chem, Fordham Univ, 54-55; from instr to assoc prof chem, Col Mt St Vincent, 55-77; vis assoc prof chem, St Joseph's Col, 77-80, assoc prof, 80-; AT COL MOUNT ST VINCENT. *Concurrent Pos:* Dir undergrad res prog, Nat Sci Found, 62- *Mem:* AAAS; Am Chem Soc. *Res:* Analytical and inorganic chemistry; nonaqueous solvents; infrared spectroscopy; complexes. *Mailing Add:* Col of Mount St Vincent Bronx NY 10471-1094

CASOLA, ARMAND RALPH, b Newark, NJ, Feb 21, 18; m 54; c 5. PHARMACEUTICAL CHEMISTRY, MEDICINAL CHEMISTRY. *Educ:* City Col NY, BS, 40; Fordham Univ, MS, 50, PhD(org chem), 56. *Prof Exp:* Res chemist, Am Cyanamid Co, 54-59; sr scientist, Strasenburgh Labs, 59-65; chemist, Div New Drugs, Bur Med, Food & Drug Admin, 65-66, supvry chemist, Div Dent & Surg Adjs, Off New Drugs, 66-70, chief chemist, Bur Drugs, 70-72, supvry chemist, Div Infective Drug Prods, Bur Drugs, 72-90; CONSULT, DRUG INDUST, 91- *Mem:* AAAS; Am Chem Soc; NY Acad Sci; Sigma Xi. *Res:* Synthetic organic chemistry; intermediates; chemotherapeutics; analgesics; hypotensives. *Mailing Add:* 5918 King James Dr Alexandria VA 22310

CASON, JAMES, JR, b Murfreesboro, Tenn, Aug 30, 12; m 35; c 2. ORGANIC CHEMISTRY. *Educ:* Vanderbilt Univ, AB, 34; Univ Calif, MS, 35; Yale Univ, PhD(org chem), 38. *Prof Exp:* Asst & Nat Cancer Inst fel, Harvard Univ, 38-40; instr org chem, DePauw Univ, 40-41; from instr to asst prof, Vanderbilt Univ, 41-45; from asst prof to prof, 45-83, EMER PROF ORG CHEM, UNIV CALIF, BERKELEY, 83- *Mem:* Am Chem Soc; Sigma Xi. *Res:* Structure and synthesis of natural products; branched-chain acids; cyclic reaction intermediates. *Mailing Add:* PO Box 7040 Berkeley CA 94707-7040

CASON, JAMES LEE, b Shongaloo, La, Feb 22, 22; m 46; c 2. DAIRY HUSBANDRY, NUTRITION. *Educ:* La Polytech Inst, BS, 48; Mich State Univ, MS, 50; NC State Col, PhD(dairy husb, nutrit), 56. *Prof Exp:* Actg asst prof dairy husb, La Polytech Inst, 49; instr, Univ Ark, 50-53, asst prof, 53-54; asst prof, Rutgers Univ, 56-59; assoc prof, Univ Md, 59-68, prof, 68-70; PROF AGR & HEAD DEPT, COL PURE & APPL SCI, NORTHEAST LA UNIV, 69- *Mem:* Am Soc Animal Sci; Biomet Soc; Am Dairy Sci Asn; Sigma Xi. *Res:* Dairy cattle nutrition, particularly forage utilization. *Mailing Add:* Dept Agr Northeast La Univ Chem Bldg Monroe LA 71209

CASON, NEAL M, b Chicago, Ill, July 26, 38; m; c 5. HIGH ENERGY PHYSICS. *Educ:* Ripon Col, AB, 59; Univ Wis, MS, 61, PhD(physics), 64. *Prof Exp:* Res assoc high energy physics, Univ Wis, 64-65; from instr to assoc prof, 65-76, dept chmn, 85-89, PROF PHYSICS, UNIV NOTRE DAME, 76- *Concurrent Pos:* Woodrow Wilson fel. *Mem:* Am Phys Soc. *Res:* Charmed particle photo production; meson spectroscopy. *Mailing Add:* Dept Physics Univ Notre Dame Notre Dame IN 46556

CASPARI, ERNST WOLFGANG, developmental genetics, behavioral genetics; deceased, see previous edition for last biography

CASPARI, MAX EDWARD, b Frankfurt-on-Main, Germany, Mar 17, 23; nat US; m 51; c 3. SOLID STATE PHYSICS. *Educ:* Wesleyan Univ, AB, 48; Mass Inst Technol, PhD(physics), 54. *Prof Exp:* Asst insulation res lab, Mass Inst Technol, 48-54; from instr to assoc prof, 54-64, PROF PHYSICS, UNIV PA, 64- *Concurrent Pos:* Chmn dept physics, Univ Pa, 68-73. *Mem:* Am Phys Soc. *Res:* Hyperfine interactions; perturbed angular correlations; magnetism; surfaces. *Mailing Add:* Dept Physics Univ Pa Philadelphia PA 19104

CASPARY, DONALD M, b New York, NY, Sept 1, 43; div; c 3. AUDITORY NEUROSCIENCE, SENSORY NEUROBIOLOGY. *Educ:* Univ Wis, BA, 65; Syracuse Univ, MS, 68; New York Univ, PhD(biol), 71. *Prof Exp:* Fel neurobiol, Nat Inst Neurol & Commun Disorders & Stroke, State Univ NY, Albany, 71-72; asst prof neurobiol, 73-76, prof otolaryngol, Dept Surg, 82-86, PROF PHARMACOL, SCH MED, SOUTHERN ILL UNIV, 76-, PROF OTOLARYNGOL, DEPT SURG, 86- *Concurrent Pos:* Consult, Comt Hearing & Bioacoust, Nat Acad Sci, 73-; NIH prin investr & co-investr, Nat Inst Deafness & Other Commun Disorders, 79-97; Deafness Res Found grants. *Honors & Awards:* Claude Pepper Award, Nat Inst Deafness & other Commun Dis, 90. *Mem:* Acoust Soc Am; Soc Neurosci; AAAS; Sigma Xi; Asn Res Otol. *Res:* Pharmacology; aging. *Mailing Add:* Dept Pharmacol Sch Med Southern Ill Univ PO Box 19230 Springfield IL 62794-9230

CASPER, BARRY MICHAEL, b Knoxville, Tenn, Jan 21, 39; m 61; c 6. THEORETICAL PHYSICS. *Educ:* Swarthmore Col, BA, 60; Cornell Univ, PhD(theoret physics), 66. *Prof Exp:* Consult, Inst Defense Anal, 62-63; instr physics, Cornell Univ, 65-66; from asst prof to assoc prof, 66-77, PROF PHYSICS, CARLETON COL, 77-, DIR, CARLETON TECHNOL POLICY PROJ, 81-; POLICY ADV, OFF SEN PAUL WELLSTONE, US SEN. *Concurrent Pos:* Nat coun mem, Fedn Am Scientist, 71-75 & 80-84, Am Phys Soc, 80-83; chmn, Forum Physics & Soc, Am Phys Soc, 74; res fel, Prog Sci & Int Affairs, Harvard Univ, 75-76; humanities fel, Rockefeller Found, 75-76; NSF fel, Sci Appl to Societal Probs, 76-77; res fel, Prog Sci, Technol & Soc, Mass Inst Technol, 80-81. *Honors & Awards:* Forum on Physics & Soc Prize, Am Phys Soc, 84. *Mem:* Am Phys Soc; AAAS. *Res:* Elementary particle research; arms control research; scientists and public policy; history of science. *Mailing Add:* Off Sen Paul Wellstone US Senate Washington DC 20510

CASPER, JOHN MATTHEW, b Middletown, Pa, Jan 28, 46; m 71; c 1. PHYSICAL CHEMISTRY. *Educ:* Univ Scranton, BS, 67; Univ SC, PhD(chem), 71. *Prof Exp:* Res assoc chem, Univ Md, 71-73; asst prof chem, Polytech Inst NY, 73-80; MEM STAFF, IBM INSTRUMENTS, INC, 80- *Concurrent Pos:* Consult, Church & Dwight Co, Inc, 75- *Mem:* Am Chem Soc; Coblentz Soc; Soc Appl Spectros. *Res:* Application of vibrational spectroscopy to studies of chemical structure and bonding. *Mailing Add:* 56 Lake Dr New Milford CT 06776

CASPERS, HUBERT HENRI, b Oostkamp, Belg, June 5, 29; US citizen; m 59; c 2. SOLID STATE PHYSICS, SPECTROSCOPY. *Educ:* Univ Calif, Los Angeles, BA, 53, MA, 58, PhD(spectros), 62. *Prof Exp:* res physicist & head, Luminescent Mat Sect, US Naval Electronics Lab Ctr, San Diego, 62-89; RETIRED. *Mem:* Am Phys Soc. *Res:* Rare earth solid state spectroscopy; infrared and Raman spectroscopy on solids. *Mailing Add:* 13621 Via Serena Poway CA 92064

CASPERS, MARY LOU, b Wyandotte, Mich. NEUROCHEMISTRY, ENZYMOLOGY. *Educ:* Univ Detroit, BS, 72; Wayne State Univ, PhD(biochem), 77. *Prof Exp:* Asst prof, 77-81, assoc prof biochem, 81-90, PROF BIOCHEM, UNIV DETROIT, 90- *Concurrent Pos:* Vis researcher, Univ Mich, 78; traveling award, NSF, 82; guest researcher, NIMH, 84-85. *Mem:* Am Soc Biol Chemists; Soc Neurosci; Am Chem Soc; Sigma Xi; AAAS. *Res:* Regulation of (sodium and potassium)-adenosinetriphosphatase in neuronal tissues and brain capillary endothelium; ouabain autoradiography of brain sections. *Mailing Add:* Dept Chem Univ Detroit 4001 W McNichols Detroit MI 48221-9987

CASPERSON, LEE WENDEL, b Portland, Ore, Oct 18, 44; m 74; c 4. QUANTUM ELECTRONICS, LASER PHYSICS. *Educ:* Mass Inst Technol, BS, 66; Calif Inst Technol, MS, 67, PhD(elec eng & physics), 71. *Prof Exp:* From asst prof to prof elec eng, Univ Calif, Los Angeles, 71-85; chmn, elec eng, 87-88, PROF ELEC ENG, PORTLAND STATE UNIV, ORE, 83- *Concurrent Pos:* Consult, Northrop Res & Technol Ctr, 73-78 & TRW, 76-83; vis prof, Dept Physics, Univ Auckland, NZ, 81 & Univ Otago, NZ, 90. *Honors & Awards:* Centennial Medal, Inst Elec & Electronics Engrs. *Mem:* Sigma Xi; Optical Soc Am; Inst Elec & Electronic Engrs; Am Soc Eng Educ; fel Optical Soc Am. *Res:* Quantum electronics; laser physics; beam propagation; optical resonators; waveguides; scattering; laser media; laser instabilities. *Mailing Add:* Dept Elec Eng Portland State Univ PO Box 751 Portland OR 97207-0751

CASPI, ELIAHU, b Warsaw, Poland, June 10, 13; US citizen; m 48; c 1. ORGANIC CHEMISTRY, BIOCHEMISTRY. *Educ:* Clark Univ, PhD(org chem), 55. *Prof Exp:* Scientist, Anal Chem, Israel Stand Inst, 43-50; from staff scientist to sr scientist, 51-70, PRIN SCIENTIST, WORCESTER FOUND EXP BIOL, 71- *Concurrent Pos:* US Pub Health Serv res career prog award, 63-72. *Mem:* Fel AAAS; Am Chem Soc; Am Soc Biol Chemists; Royal Soc Chem; Am Heart Asn; Sigma Xi. *Res:* Bioorganic chemistry of natural products; biosynthesis and metabolism of hormones; biosynthesis of sterols in relation to cancer. *Mailing Add:* Worcester Found Exp Biol 222 Maple Ave Shrewsbury MA 01545

CASPI, RACHEL R, b Gdansk, Poland; Israeli citizen. CELLULAR IMMUNOLOGY. *Educ:* Tel-Aviv Univ, Israel, MS, 75; Bar-Ilan Univ, Israel, PhD(immunol), 84. *Prof Exp:* Teaching asst microbiol, Dept Microbiol, Fac Life Sci, Tel-Aviv Univ, 72-75; instr hemat & immunol, Dept Life Sci, Bar-Ilan Univ, 76-84; vis fel, Lab Immunol, 84-86, vis assoc, 87-89, VIS SCIENTIST & ACTG CHIEF, SECT IMMUNOREGULATION, LAB IMMUNOL, NAT EYE INST, NIH, 90- *Honors & Awards:* M Landau Mem Res Award, Israel Acad Sci, 83. *Mem:* Am Asn Immunologists; NY Acad Sci; Asn Res Vision & Ophthal; Israel Immunol Soc. *Res:* Basic mechanisms in immunologically-mediated ocular disease; cellular and molecular immunological approaches to define the sequence of events leading to induction of organ-specific autoimmunity in rodent models of experimental autoimmune uveoretinitis; development of immunotherapeutic strategies. *Mailing Add:* Lab Immunol Nat Eye Inst Bldg 10 Rm 10N222 Bethesda MD 20892

CASS, CAROL E, b Lexington, Ky, Oct 18, 42; m 66. CELL BIOLOGY, BIOCHEMISTRY. *Educ:* Univ Okla, BS, 63, MS, 65; Univ Calif, Berkeley, PhD(cell biol), 71. *Prof Exp:* Fel cancer chemother, 70-73, asst prof, 74-80, ASSOC PROF BIOCHEM, CANCER RES UNIT, UNIV ALTA, 80- *Mem:* Can Biochem Soc; Can Soc Cell Biol; Am Asn Cancer Res; Tissue Cult Asn; NY Acad Sci. *Res:* Nucleoside transport in mammalian cells, biochemical and biological mechanisms of anti-neoplastic agents; mechanisms of resistance to anti-neoplastic agents. *Mailing Add:* Cancer Res Unit McEachern Lab Univ Alta Edmonton AB T6G 2H7 Can

CASS, DAVID D, b Indianapolis, Ind, Sept 1, 38; m 66. EXPERIMENTAL POLLINATION & GAMETE FUSION ANGIOSPERMS. *Educ:* Butler Univ, BS, 61; Univ Okla, PhD(bot), 67. *Prof Exp:* US Pub Health Serv fel, Univ Calif, Berkeley, 66-68, lectr biol, 68-69; from asst prof to assoc prof, 69-80, chmn dept, 79-84, PROF BOT, UNIV ALTA, 80- *Mem:* Can Bot Asn (secy, 78-80); Int Soc Plant Morphologists; Soc Econ Bot; Bot Soc Am; Can Soc Cell Biol; Sigma Xi. *Res:* Experimental pollination; gamete fusion in angiosperms; cell biology of isolated cereal grain male gametes; histochemistry, fluorescence microscopy, scanning and transmission electron microscopy. *Mailing Add:* Dept Bot Univ Alta Edmonton AB T6G 2E9 Can

CASS, GLEN R, b Pasadena, Calif, Apr 18, 47; m 76; c 1. AIR POLLUTION CONTROL, AEROSOL MECHANICS. *Educ:* Univ Southern Calif, BS, 69; Stanford Univ, MS, 70; Calif Inst Technol, PhD(environ eng), 78. *Prof Exp:* Engr, Naval Undersea Ctr, Pasadena, 69; commissioned officer, US Pub Health Serv, 70-73; sr res fel & instr, Calif Inst Technol, 78-79, asst prof environ eng, 79-85, assoc prof, 85-90, PROF ENVIRON ENG & MECH ENG, CALIF INST TECHNOL, 90- *Concurrent Pos:* Consult, S Coast Air Quality Mgt Dist, 78-; mem res screening comt, Calif Air Resources Bd, 80-84, modeling adv comt, 88-; atmosphere chem & phys rev panel, US Environ Protection Agency, 87 & Clean Air Sci Adv Panel Subcomt Visibility, 88-; consult, Health Effects Inst; Nat Res Coun Comt Preserv Hist Documents, 85-86, Comt Haze in Nat Parks, 90-91. *Mem:* Sigma Xi; Am Asn Aerosol Res; Am Chem Soc. *Res:* Air pollution control strategy design; aerosol mechanics; air pollution source characteristics and control technology; visibility; fluid mechanics applied to air quality problems; environmental economics; protection of works of art from damage due to air pollution. *Mailing Add:* Environ Eng Sci Dept Calif Inst Tech Pasadena CA 91125

CASS, THOMAS ROBERT, b San Francisco, Calif, Nov 25, 36; m 59; c 2. ELECTRONICS ENGINEERING. *Educ:* Univ Calif, Berkeley, BS, 58, MS, 61, PhD(metall), 66. *Prof Exp:* NSF fel metall, Univ Paris, Orsay, 65-66; res scientist, 66-68, chief adv mat res & develop, Dept Mat Res, Orlando Div, Martin Aerospace Group, Martin-Marietta Corp, Fla, 68-70; mem res staff, Fairchild Res & Develop Lab, Fairchild Camera & Instrument Corp, 70-71; mem tech staff, Hewlett-Packard Solid State Lab, 71-78, mem tech staff, Hewlett- Packard Intergrated Circuit Lab, 78-86; MEM TECH STAFF, CIRCUIT TECHNOL RES & DEVELOP LAB, 86- *Concurrent Pos:* Lectr mat sci, San Jose State Univ, 73-75; vis prof, Univ Estadual de Campinas, Brazil, 77. *Mem:* Am Phys Soc; Electron Micros Soc Am. *Res:* Characterization of electronic materials and devices; transmission and scanning electron microscopy; x-ray double crystal diffractometry; diffraction contrast theory; ion implantation into semiconductors; sputter deposition of metals and silicon; parametric testing, failure physics and failure analysis of integrated circuits; yield improvement for integrated circuit processes. *Mailing Add:* Hewlett-Packard Labs 3500 Deer Creek Rd Palo Alto CA 94303-0867

CASSADY, GEORGE, b Los Angeles, Calif, Aug 9, 34; m 57; c 3. MEDICINE, PEDIATRICS. *Educ:* Duke Univ, MD, 58. *Prof Exp:* Intern pediat, Duke Univ Med Ctr, 58-59, resident pediat, 60; clin assoc, Med Invest & Genetic Unit, Nat Inst Dent Res, 60-62; sr resident pediat, Children's Hosp Med Ctr, Boston, Mass, 62-63; from asst prof to assoc prof pediat, 64-70, PROF PEDIAT, MED CTR, UNIV ALA, BIRMINGHAM, 70-, DIR NEWBORN DIV, 65-, ASSOC PROF OBSTET & GYNEC, 77- *Concurrent Pos:* Fel pediat cardiol, Duke Hosp, Durham, NC, 59; teaching fel, Harvard Med Sch, 62-63; res fel newborn physiol, Boston Lying-In Hosp & Harvard Med Sch, 63-64; mem bd cert, Am Acad Pediat; chmn maternal & child health comt, State Ala. *Mem:* Am Soc Human Genetics; fel Am Acad Pediat; Soc Pediat Res; Am Fedn Clin Res; NY Acad Sci. *Mailing Add:* Dept Obstet Gynecol Univ Ala Sch Med Univ Station Birmingham AL 35294

CASSADY, JOHN MAC, b Vincennes, Ind, Aug 16, 38; m 59; c 5. ORGANIC CHEMISTRY, MEDICINAL CHEMISTRY. *Educ:* DePauw Univ, BA, 60; Case Western Reserve Univ, MS, 62, PhD(org chem), 64. *Prof Exp:* Res assoc, Case Western Reserve Univ, 64-65; NIH fel, Univ Wis, 65-66; from asst prof to assoc prof, 66-74, assoc dept head, 74-80, PROF MED CHEM & HEAD DEPT, SCH PHARM & PHARMACOL SCI, PURDUE UNIV, 80- *Mem:* AAAS; Am Chem Soc; Royal Soc Chem; fel Acad Pharmaceut Sci; Am Soc Pharmacog; Sigma Xi. *Res:* Isolation and structure elucidation of tumor inhibitors from plants; synthesis of potential tumor inhibitors including pyrimidines, xanthones, acridonea and anthracyclinenes; synthesis of dopaminergic agents. *Mailing Add:* Dept Med Chem Purdue Univ Lafayette IN 47907

CASSAN, STANLEY MORRIS, pulmonary diseases, internal medicine, for more information see previous edition

CASSANOVA, ROBERT ANTHONY, b Columbia, SC, Apr 12, 42; m 64; c 2. SOLAR ENERGY, SPACE TECHNOLOGY & STRATEGIC DEFENSE. *Educ:* NC State Univ, BS, 64; Univ Tenn Space Inst, 67; Ga Inst Technol, PhD(aerospace eng), 75. *Prof Exp:* Res engr, ARO, Inc, 64-67; res engr, Sch Aerospace Eng, 67-77, sr res engr, Eng Exp Sta, 77, ASSOC LAB DIR, ENG EXP STA, GA INST TECHNOL, 79-, DIR, PROG OFF, SOLAR THERMAL ADVAN RES CTR, 81- *Concurrent Pos:* Consult, various energy related private industs, 75- *Mem:* Sigma Xi; AAAS; Solar Thermal Test Facility User's Asn; Int Asn Hydrogen Energy. *Res:* Research and development of advanced energy systems using high temperature solar energy; high temperature materials; advanced strategic defense concepts; space technology. *Mailing Add:* Energy & Mat Sci Lab Ga Tech Res Inst 225 North Ave Atlanta GA 30332

CASSARD, DANIEL W, b E Grand Rapids, Mich, July 30, 23; m 48; c 4. INTERNATIONAL AGRICULTURAL DEVELOPMENT. *Educ:* Univ Calif, Davis, BS, 47, PhD(animal genetics), 52. *Prof Exp:* Teaching & res asst genetics, Univ Calif, Davis & Berkeley, 49-52, instr & asst prof animal sci, Davis, 52-56; prof animal sci, Univ Nevada, Reno, 56-66; animal nutrit adv, IRI Res Inst, Inc, Brazil, 66-69; asst sci dir, Pfizer Int, Inc, 69-77, sci dir, 73-74, vpres agr develop, 74-79, dir, agr res & develop, 79-84; OWNER, CASSARD ASSOCS, TEX, 84- *Concurrent Pos:* Mem, adv comt, Agr Res Serv Western Sheep Breeding Lab, USDA, Idaho, 53-66, bd govs, Am Soc Agr Consult Int, 84-87; dir vocational agr, Dinuba High Sch, Calif, 47-49. *Mem:* Am Soc Animal Sci; Am Inst Biol Sci; Am Soc Agr Consult; Coun Agr Sci & Technol; Am Forage & Grassland Coun. *Res:* Breeding and management of livestock and poultry; development of animal health products and programs. *Mailing Add:* Cassard Assocs Tex 155 McFadden Hut Brownsville TX 78520-8925

CASSEDAY, JOHN HERBERT, b Pasadena, Calif, Aug 11, 34; m 65; c 2. NEUROSCIENCES, PSYCHOLOGY. *Educ:* Univ Calif, Riverside, BA, 60; Ind Univ, MA, 63, PhD(psychol), 70. *Prof Exp:* USPHS trainee, Duke Univ, 70-72, lectr, Dept Psychol, 72-76, asst prof otolaryngol, 72-79, asst prof psychol, 76-80, assoc prof psychol, 80-88, ASSOC MED RES PROF, DEPT SURG, MED CTR, DUKE UNIV, 79-, ASSOC PROF, DEPT NEUROBIOL, MED CTR, 88- *Concurrent Pos:* NSF grant, 72-75 & 77-; NIH grant, 75- *Mem:* AAAS; Acoust Soc Am; Soc Neurosci. *Res:* Hearing; neuroethology. *Mailing Add:* Dept Neurobiol Duke Univ Med Ctr PO Box 3209 Durham NC 27710

CASSEDY, EDWARD S(PENCER), JR, b Washington, DC, Aug 19, 27; m 52; c 4. SCIENCE POLICY, PHYSICS. *Educ:* Union Col, BS, 49; Harvard Univ, SM, 50; Johns Hopkins Univ, DEng, 59. *Prof Exp:* Engr, Potomac Elec Power, 50-51; electronics engr, US Naval Ord Lab, 51-54; res assoc eng, Johns Hopkins Univ, 54-58, res scientist, 58-60; from asst prof to assoc prof electroph, 60-68, PROF ELEC ENG, POLYTECH UNIV, 68- *Concurrent Pos:* Vis staff, Los Alamos Nat Lab, 74-76 & Brookhaven Nat Lab, 79-80. *Mem:* Fedn Am Sci; Am Phys Soc; Inst Elec & Electronics Engrs; Int Asn Energy Economists. *Res:* Antenna and scattering theory; theory of guided waves; nonlinear interaction of waves; electromagnetic theory; plasmas; power systems; energy policies; economics. *Mailing Add:* Polytechnic Univ 333 Jay St New York NY 20895-3136

CASSEL, D KEITH, b Bader, Ill, July 23, 40; m 65; c 2. SOIL PHYSICS. *Educ:* Univ Ill, BS, 63; Univ Calif, Davis, MS, 64, PhD(soil physics), 68. *Prof Exp:* Lab technician, Univ Calif, Davis, 65-68; assoc prof soil physics, NDak State Univ, 68-74; assoc prof, 74-80, PROF SOIL PHYSICS, NC STATE UNIV, 80- *Concurrent Pos:* Tech ed, Soil Sci Soc, Am J, 90- *Mem:* Fel Soil Sci Soc Am; Soil & Water Conserv Soc Am; Int Soc Soil Sci; fel Am Soc Agron. *Res:* Water and solute movement in soils; tillage and modification of hardpan soils; irrigation scheduling; spatial variability of soil properties; soil erosion. *Mailing Add:* Dept of Soil Sci NC State Univ Box 7619 Raleigh NC 27695-7619

CASSEL, DAVID GISKE, b Ainsworth, Nebr, Dec 12, 39; m 66; c 2. PHYSICS, ELEMENTARY PARTICLE PHYSICS. *Educ:* Calif Inst Technol, BS, 60; Princeton Univ, MA, 62, PhD(physics), 65. *Prof Exp:* NSF fel, Europ Orgn Nuclear Res, Geneva, Switz, 65-66; from asst prof to assoc prof, 66-79, PROF PHYSICS, CORNELL UNIV, 79- *Concurrent Pos:* Humboldt Found sr fel, Physics Inst, Bonn Univ, 73-74; sci assoc, DESY, Hamburg, 79-80, 86-87. *Mem:* Am Phys Soc. *Res:* Pi meson form factor; theory of neutron storage rings; neutral K meson decays; symmetries of electromagnetic interactions; neutral K meson photoproduction; inelastic electron scattering; electron-positron annihilation; weak decays of B mesons. *Mailing Add:* Newman Lab Cornell Univ Ithaca NY 14853

CASSEL, DAVID WAYNE, b Toronto, Can, Sept 21, 36; US citizen; m 60; c 3. ALGEBRA. *Educ:* Greenville Col, BS, 59; Syracuse Univ, MA, 62, PhD(math), 67. *Prof Exp:* Instr math, Messiah Col, 62-64; Danforth teacher, Syracuse Univ, 65-67; from asst prof to assoc prof, 67-73, prof math & chmn, Dept Regist, 73-87, DIR COMPUT SERV, MESSIAH COL, 77- *Mem:* Am Math Soc; Math Asn Am. *Res:* Homological algebra; structure of projective modules as it relates to the base ring. *Mailing Add:* Dept Math Messiah Col Grantham PA 17027

CASSEL, J(OSEPH) FRANK(LIN), b Reading, Pa, July 9, 16; m 43; c 4. ORNITHOLOGY, ECOLOGY. *Educ:* Wheaton Col, BS, 38; Cornell Univ, MS, 41; Univ Colo, PhD(zool), 52. *Prof Exp:* From instr to asst prof zool, Colo State Univ, 46-50; from asst prof to prof zool, NDak State Univ, 50-82, chmn dept, 53-63 & 69-77; RETIRED. *Concurrent Pos:* NSF sci fac fel, Mus Comp Zool, Harvard Univ, 63-64. *Mem:* Sigma Xi; Am Ornithologists Union; Am Inst Biol Sci; Soc Study Evolution. *Res:* Population dynamics; speciation; creationism and evolution; birds and mammals of North Dakota. *Mailing Add:* 83 W Boulder St Colorado Springs CO 80903

CASSEL, RUSSELL N, b Harrisburg, Pa, Dec 18, 11; m 62; c 7. HEALTH SCIENCE, COMPUTERS. *Educ:* Millersville State Univ, BS, 37; Penn State Univ, MEd, 39, Univ Southern Calif, EdD, 49; Am Bd Psychol, dipl, 73. *Prof Exp:* Res Psychologist, US Air Force, 41-70; prof educ psychol, Univ Wis, 69-74; DIR, CASSELL PSYCHOL CTR, 74- *Concurrent Pos:* Ed, Educ, Col Student J & Reading Improv. *Mem:* Am Biofeedback Soc; Am Psychol Asn. *Res:* Wrote approximately 400 articles in professional health journals, approximately 30 psychology texts and 12 computer programs in health science. *Mailing Add:* 1362 Santa Cruz C Chula Vista CA 92910

CASSEL, WILLIAM ALWEIN, b Philadelphia, Pa, Mar 25, 24; m 49; c 2. MICROBIOLOGY. *Educ:* Philadelphia Col Pharm & Sci, BS, 46, MS, 47; Univ Pa, PhD(microbiol), 52. *Prof Exp:* Jr bacteriologist, Philadelphia Gen Hosp, 47-48, lab tech asst, 48-50; asst instr microbiol, Univ Pa, 50-51; res assoc, Hahnemann Med Col, 52-53, asst prof, 53-58; assoc prof microbiol, Sch Med, Emory Univ, 58-69, prof microbiol & immunol, 69-87; RETIRED. *Concurrent Pos:* Med fac award, Lederle Labs, 55-58; USPHS res career develop award, 60-65. *Mem:* AAAS; Am Soc Microbiol; Soc Exp Biol & Med; Am Asn Immunol; Tissue Cult Asn; Sigma Xi; NY Acad Sci; Am Asn Univ Prof. *Res:* Variation in viruses; relation of viruses to cancer; microbial cytology; animal virology; oncolytic viruses; genetics; immunity to malignant disease. *Mailing Add:* 2518 Varner Dr NE Atlanta GA 30345

CASSELL, ERIC J, b New York, NY, Aug 2, 28. PUBLIC HEALTH. *Educ:* Queens Col, New York, BS, 50; Columbia Univ, MA, 50; New York Univ Sch Med, MD, 54. *Hon Degrees:* LHD, Med Col Pa, 85. *Prof Exp:* Trainee, US Pub Health Serv, Infectious Dis, Cornell Univ Med Col, 59-61, clin instr, Dept Pub Health, 61-64, clin asst prof, 64-65, consult, 65-66; assoc clin prof community med, Mt Sinai Sch Med, 66, assoc clin prof med, 66, assoc prof Community Med, 67-68, assoc clin prof Community Med, 68-71, LECTR, MT SINAI SCH MED, 71-; CLIN PROF PUB HEALTH, CORNELL UNIV MED COL, 71-, PRACTR INTERNAL MED. *Concurrent Pos:* Intern, Bellevue Hosp, 54-55, asst resident, 55-56, 58-59; asst attending physician, French Hosp, 61-65, attending physician, 65-74; physician to Outpatients, New York Hosp, 61-65, assoc attending physician to In-Patients, 77-84; consult, Dept Preventive Med, Univ Wash, Sch Med, 66-68; prin investr, Robert Wood Johnson Found, grant, 73, Nat Sci Found, grant, 78, Nat Endowment Humanities, 80-85, Nat Libr Med, 80-82, Nat Found Med Educ, 83-86, Louis B Mayer Found, Fundamentals Clin Thinking, 84-86; chmn, Comt Human Rights in Res, Cornell Univ Med Col, 78-80; mem, Comt Ethics, NY State Med Soc, 87-; fel & mem bd dirs, Hastings Ctr. *Mem:* Inst Med-Nat Acad Sci; Am Pub Health Asn; NY Acad Sci; fel Am Col Physicians; AMA. *Res:* Theory of clinical medicine. *Mailing Add:* Mt Sinai Sch Med Fifth Ave & 100th St New York NY 10029

CASSELL, EUGENE ALAN, b Quarryville, Pa, June 11, 34; m 61; c 3. SANITARY ENGINEERING. *Educ:* Pa State Univ, BS, 56; Mass Inst Technol, SM, 58; Univ NC, Chapel Hill, PhD(sanit eng & water resources), 64. *Prof Exp:* Sanit engr, Div Radiol Health, USPHS, 59-61; from asst prof to assoc prof civil eng, Clarkson Col Technol, 64-75; PROF NATURAL RESOURCES, SCH NATURAL RESOURCES, UNIV VT, 75- *Concurrent Pos:* Consult to various consult eng firms, 64-69 & various govt agencies, 65-69. *Mem:* Water Pollution Control Fedn; Am Water Works Asn; Am Soc Civil Engrs. *Res:* Chemical and biological treatment of waste sludges and slurries; flotation; refuse collection optimization. *Mailing Add:* Sch Natural Resources Univ Vt George D Aiken Ctr Burlington VT 05405

CASSELL, GAIL HOUSTON, b Alexander City, Ala, Jan 25, 46; m 67; c 1. MICROBIOLOGY, INFECTIOUS DISEASES. *Educ:* Univ Ala, BS, 69; Univ Ala, Birmingham, MS, 71, PhD(microbiol), 73. *Prof Exp:* Res assoc molecular biol, Univ Ala, 67-68; res asst, Univ Ala, Birmingham, 68-70, instr, 70-73, asst prof, Dept Comp Med, 73-75, from asst prof to assoc prof, 75-82, PROF MICROBIOL, UNIV ALA, BIRMINGHAM, 82-, CHMN, 87- *Concurrent Pos:* NIH grants, 73- & res career develop award, 77-82; mem, Bact/Mycol II Study Sect, Div Res Grants, NIH, 81-91, chmn, 88-90; div counr, Div G, Am Soc Microbiol, 84-89; mem mycoplasmas comt, WHO; chmn, Int Orgn Mycoplasmology, 86-88 & Lab Animal Team, Int Res Prog Comparative Mycoplasmology, 88-; rev panelist, Genetics Panel, Howard Hughes Doctoral Fel Prog Biol Sci, 89-; mem, Ctr Infectious Dis Sci Adv Coun, Ctrs Dis Control, 89-91; mem, Pulmonary Task Force, Nat Bd Med Examiners, 89-90; lectr, Am Soc Microbiol Found, 90-91. *Honors & Awards:* Derreck Edward Award, Int Orgn Mycoplasmology, 80. *Mem:* Am Soc Microbiol; Recticuloendothel Soc; Am Thoracic Soc; Infectious Dis Soc Am; Int Orgn Mycoplasmology. *Res:* Host-parasite relationships in mycoplasmal diseases and the phagocytic cell in host resistance. *Mailing Add:* Dept Microbiol Univ Ala Birmingham Birmingham AL 35294

CASSELMAN, JOHN MALCOLM, b Sydenham, Ont, Apr 25, 40; m 64; c 2. FISHERIES, ZOOLOGY. *Educ:* Ont Agr Col, BSAgr, 64; Univ Guelph, MS, 69; Univ Toronto, PhD(zool), 78. *Prof Exp:* Consult aquatic biol fisheries, 73-74; sr aquatic biologist, James F MacLaren Ltd, 74-76; RES SCIENTIST, FISHERIES BR, RES SECT, ONT MINISTRY NATURAL RESOURCES, 76- *Mem:* Am Fisheries Soc; Am Soc Ichthyologists & Herpetologists; Can Soc Zoologists; Freshwater Biol Asn; Soc Int Limnol. *Res:* Aquatic biology, fisheries, coolwater species, Esocidae; environmental physiology, temperature, photoperiod, nutrition, hormone control, and aquatic acidification; laboratory and natural environment; physiology of growth as related to age and growth determination from calcified tissue; author of 42 scientific publications. *Mailing Add:* Fisheries Res Glenora Fisheries Sta Ont Ministry Natural Resources RR 4 Picton ON K0K 2T0 Can

CASSELMAN, WARREN GOTTLIEB BRUCE, b Vancouver, BC, July 26, 21; m 50; c 2. PHARMACOLOGY, PUBLIC HEALTH ADMINISTRATION. *Educ:* Univ BC, BA, 43, MA, 44; Univ Toronto, MD, 49, PhD(physiol), 52. *Prof Exp:* Asst, Banting & Best Dept Med Res, Univ Toronto, 49-52, res assoc, 52-55, assoc prof, 55-58, assoc prof histol, Dept Anat, 57-58; sr res histochemist, Dept Neuropath, NY State Psychiat Inst, 58-59; assoc mem in chg div cell biol, Inst Muscle Dis, Inc, NY, 59-61; head training sect, Geigy Chem Corp, 61-65; prof pharmacol, Univ Toronto, 65-66; med dir, Pharmaceut Div, Geigy (Can) Ltd, 65-70, head clin pharmacol unit, 70; clin pharmacol adv, Drug Adv Bur, Dept Nat Health & Welfare, Can, 70-72, dir bur, 72-73, dir gen, Drugs Directorate, 73-74, sr adv med, Health Protection Br, 74-80, sr med adv, Int Health Servs, 75-80; prof fac med & dir, Outreach Prog, Health Sci Facil, Univ Western Ont, 80-88; RETIRED. *Concurrent Pos:* Merck fel natural sci, Cytol Lab, Oxford Univ, 53-54; res assoc psychiat, Col Physicians & Surgeons, Columbia Univ, 59-65; Merck lectr, Univs BC, Alta & Sask, 61. *Mem:* Am Physiol Soc; Can Physiol Soc; Pharmacol Soc Can; Can Med Asn. *Res:* pharmacokinetics; histochemistry; primary health care; health planning. *Mailing Add:* 1666 Louise Blvd London ON N6G 2R3 Can

CASSEN, PATRICK MICHAEL, b Chicago, Ill, May 13, 40; m 65; c 1. GAS DYNAMICS, PLANETARY SCIENCE. *Educ:* Univ Mich, BAeroE, 62, MS, 63, PhD(aeronaut & astronaut eng), 67. *Prof Exp:* Res scientist, Space Sci Div, 67-82, actg chief, Space Sci Div, 88-89, CHIEF, THEORET STUDIES BR, AMES RES CTR, NASA, 82-88 & 90- *Concurrent Pos:* Vis scientist, Cornell Univ, 89-90. *Honors & Awards:* Newcomb-Cleveland Award, AAAS, 79; Except Sci Achievement Award, NASA, 80. *Mem:* Am Geophys Union; Am Astron Soc. *Res:* Origin and evolution of the solar system. *Mailing Add:* MS 245-3 Ames Res Ctr NASA Moffett Field CA 94035

CASSEN, THOMAS JOSEPH, b Chicago, Ill. PHYSICAL CHEMISTRY. *Educ:* Polytech Inst Brooklyn, BS, 61, PhD(phys chem), 66. *Prof Exp:* Res assoc chem, Univ Calif, Riverside, 66-68 & Univ Ariz, 68-70; NIH spec fel, 70; coordr, Gen Chem Labs, Univ Ariz, 71; asst prof chem, Univ Ga, 72-80; ASST PROF CHEM, UNIV NC, CHARLOTTE, 80- *Mem:* Am Chem Soc; AAAS. *Res:* Molecular spectroscopy. *Mailing Add:* Dept Chem Univ NC Charlotte Charlotte NC 28223

CASSENS, DANIEL LEE, b Dixon, Ill, Dec 15, 46; m 70; c 2. FORESTRY, WOOD TECHNOLOGY. *Educ:* Univ Ill, Urbana, BS, 68; Univ Calif, Berkeley, MS, 69; Univ Wis-Madison, PhD(forestry), 73. *Prof Exp:* Wood prods technologist, US Forest Prods Lab, 70-73; exten specialist, Coop Exten Serv, La State Univ, 73-77; from asst prof to assoc prof, 77-85, PROF WOOD SCI, PURDUE UNIV, 85- *Concurrent Pos:* Coop ed progs, Nat Hardwood Lumber Asn & Wood & Wood Prod Trade J;; wood prods technologist, TIM Tech Inc, 78- *Mem:* Forest Prods Res Soc; Soc Wood Sci & Technol; Sigma Xi. *Res:* Technology transfer and wood products processing including log and lumber stain; biodeterioration of residential structures; hardwood sawmilling; log and lumber grading; furniture and cabinet manufacturing. *Mailing Add:* Dept Forestry & Natural Resources Purdue Univ West Lafayette IN 47907

CASSENS, PATRICK, b Litchfield, Ill, Oct 21, 38; m 62; c 2. MATHEMATICAL ANALYSIS. *Educ:* St Louis Univ, BS, 60, MS, 62, PhD(math), 66. *Prof Exp:* Teaching asst math, St Louis Univ, 62-64; from instr to asst prof, Univ Mo, St Louis, 64-68; from asst prof to assoc prof math, State Univ NY Col, Oswego, 68-76, assoc dean arts & sci, 73-76; vpres acad affairs, Siena Col, 77-79; prof & vpres acad affairs, Cent State Univ, Okla, 79-83; computer consult, 84-88; vpres acad affairs, Dakota State Univ, 85-86; PROF, MO SOUTHERN STATE COL, 88- *Mem:* AAAS; Am Math Soc; Math Asn Am. *Res:* Summability theory and natural boundaries of functions; 2-metric geometry. *Mailing Add:* PO Box 624 Edmond OK 73083-0624

CASSENS, ROBERT G, b Morrison, Ill, June 10, 37; m 60; c 2. BIOCHEMISTRY. *Educ:* Univ Ill, BS, 59; Univ Wis, MS, 61, PhD(biochem), 63. *Prof Exp:* From asst prof to assoc prof, 64-71, PROF MEAT & ANIMAL SCI, UNIV WIS-MADISON, 71- *Concurrent Pos:* Fulbright grant, Australia, 63. *Mem:* Inst Food Technol; Am Soc Animal Sci. *Res:* Meat science; examination of muscle ultrastructure as effected by rate of postmortem glycolysis; function of zinc in muscle; effect of temperature on rigor mortis and associated changes; histochemistry of fiber types; myoglobin localization; fate of nitrite in cured meat; phosphorylase fluorescent antibody. *Mailing Add:* Muscle Biol Lab Univ Wis Madison WI 53706

CASSERBERG, BO R, b Halsingborg, Sweden, Oct 3, 41; m 63; c 3. PHYSICS. *Educ:* Univ Minn, Minneapolis, BPhys, 64; Princeton Univ, PhD(physics), 68. *Prof Exp:* Asst prof, 68-74, ASSOC PROF PHYSICS, UNIV MINN, DULUTH, 74- *Mem:* Am Asn Physics Teachers. *Res:* Hyperfine structure; magnetic resonance. *Mailing Add:* Dept of Phys Univ of Minn Duluth MN 55812

CASSIDY, CARL EUGENE, b Salineville, Ohio, Dec 4, 24; m 61; c 2. ENDOCRINOLOGY, THYROID DISEASES. *Educ:* Kenyon Col, AB, 46; Western Reserve Univ, MD, 48. *Prof Exp:* Asst, 54-56, clin instr, 56-58, instr, 58-59, sr instr, 59-62, from asst prof to assoc prof, 62-73, CLIN PROF MED, SCH MED, TUFTS UNIV, 73-, PROG DIR, POSTGRAD MED INST, 78- *Concurrent Pos:* Res fel, New Eng Ctr Hosp, 54-56; trainee, 54-56; asst physician, New Eng Ctr Hosp, 56-68, physician, 68-71; physician-in-chief, Baystate Med Ctr, Mass, 72-77; pvt pract, 77- *Mem:* Endocrine Soc; AMA; Am Thyroid Asn; Am Col Physicians. *Res:* Diseases of endocrine glands; antithyroid drugs; radioiodine; hormones of reproductive system; pituitary hormones. *Mailing Add:* PO Box 68 Prudential Ctr Sta Boston MA 02199

CASSIDY, ESTHER CHRISTMAS, b Washington, DC, Aug 5, 33; div; c 3. SCIENCE POLICY PHYSICS, ELECTRICAL ENGINEERING. *Educ:* Manhattanville Col, Purchase, NY, BA, 55. *Prof Exp:* Physicist/proj mgr aerodyn res, Nat Bur Standards, 55-61, physicist/proj mgr high speed elec & optical measurements, 61-66, physicist/proj mgr high voltage elec measurements, 66-73, gen phys scientist cong affairs, 74-75; gen phys scientist, Energy Res & Develop Admin, Dept Energy, 75-78; SUPVRY GEN PHYS SCIENTIST & DIR CONG AFFAIRS, NAT INST STANDARDS & TECHNOL, 78- *Concurrent Pos:* Sci & technol fel, Dept Com, Nat Bur Standards, 73-74. *Honors & Awards:* Silver Medal Outstanding Res Elec Measurements, Dept Com, 70, Invention High-Speed Photographic Shutter Award, Outstanding Performance Award. *Mem:* Sr mem Inst Elec & Electronic Engrs. *Res:* High-speed electro-optical; high voltage electrical measurements; time resolved spectroscopic; high-speed photographic studies of ultra high-temperature gases and electrical insulating materials; more than 20 articles in science & engineering journals; several patents on high-speed photographic shutter. *Mailing Add:* Dir Cong & Legis Affairs Nat Inst Standards & Technol Gaithersburg MD 20899

CASSIDY, HAROLD GOMES, b Havana, Cuba, Oct 17, 06; nat US; m 34. SCIENCE WRITING. *Educ:* Oberlin Col, AB, 30, AM, 32; Yale Univ, PhD, 39. *Hon Degrees:* DSc, St Thomas Inst, 72. *Prof Exp:* Res instr chem, Oberlin Col, 32-33; res chemist, Wm S Merrell Co, Ohio, 33-36; instr chem, Oberlin Col, 36-37; from instr to prof, 38-72, EMER PROF CHEM, YALE UNIV, 72- *Concurrent Pos:* Chmn, Gordon Conf Separation & Purification, 56 & Comt Grants-in-Aid, Nat Exec Bd, Soc Sigma Xi, 69; Nat Sigma Xi lectr, 60, 65 & pres, 77; Ayd lectr, 62; Korzybski Mem lectr, 62; Res Soc spec lectr, 63; Danforth vis lectr, Asn Am Cols Arts Prog, 68 & 71; Sigma Xi centennial lectr, Ohio State Univ, 70; seminar leader libr arts educ, Danforth Workshop, 62-65; sr fel sci, Ctr Advan Studies, Wesleyan Univ, 65-66; consult, Improv Sci Educ in India, 69 & Nat Humanities Fac, 70; prof-at-large, Hanover Col, 72-; Green Honors Chair prof, Tex Christian Univ, 74; assoc ed, Am J Sci. *Mem:* Fel AAAS; Am Chem Soc; Sigma Xi; Fedn Am Soc Exp Biol. *Res:* Oxidation-reduction; chromatography; cybernetics; science education; 8 books and numerous research papers. *Mailing Add:* 605 W Second St Madison IN 47250

CASSIDY, JAMES EDWARD, b Springfield, Mass, Aug 17, 28; m 59; c 2. METABOLISM, ANALYTICAL CHEMISTRY. *Educ:* Univ Mass, BS, 49; Univ Vt, MS, 54; Rensselaer Polytech Inst, PhD(chem), 58. *Prof Exp:* Asst chemist eng exp sta, Univ NH, 49-50; chemist, Army Chem Ctr, 51-52; asst dept chem, Univ Vt, 53-54; asst, Rensselaer Polytech Inst, 54-57; res chemist, Am Cyanamid Co, 58-64; res assoc, Ciba-Geigy Corp, 64-69, group leader, 69-80, sr group leader, 80-87; independent consult, 87; SR RESIDUE CHEM, JELLNEK, SCHWARTZ, CONNOLLY & FRESHMAN, INC, 88- *Mem:* Am Chem Soc; Sigma Xi; NY Acad Sci; Int Soc Study Xenobitics. *Res:* Metabolism and residue of insecticides and herbicides in soil, plants and animals; metabolism of animal health products; separations, instrumentation and tox-kinetic studies. *Mailing Add:* 7846 Somerset Ct Greenbelt MD 20770

CASSIDY, JAMES T, b Oil City, Pa, Sept 10, 30; m 55; c 2. RHEUMATOLOGY. *Educ:* Univ Mich, BS, 53, MD, 55. *Prof Exp:* From instr to assoc prof internal med, 62-73, PROF MED & PEDIAT, MED SCH, UNIV MICH, ANN ARBOR, 73-; PROF & CHMN, DEPT PEDIAT, CREIGTON UNIV, 84- *Concurrent Pos:* Fel, Rackham Arthritis Res Unit, Med Sch, Univ Mich, Ann Arbor, 61-62; Arthritis Found fel, 63-66. *Mem:* Soc Pediat Res; Am Asn Immunol; Am Rheumatism Asn; Am Fedn Clin Res. *Res:* Immunology. *Mailing Add:* Dept Pediat Creighton Univ Sch Med California at 24th Omaha NE 68178

CASSIDY, JOHN J(OSEPH), b Gebo, Wyo, June 21, 30; m 53; c 3. HYDROLOGY, HYDRAULIC ENGINEERING. *Educ:* Mont State Col, BS, 52, MS, 60; Univ Iowa, PhD(fluid mech), 64. *Prof Exp:* Hwy engr, US Bur Pub Roads, 52-53; design engr, Mont State Water Conserv Bd, 55-58; instr fluid mech, Mont State Col, 58-60; mech, Univ Iowa, 60-63; from asst prof to prof civil eng, Univ Mo-Columbia, 63-72, chmn dept, 72-74; dir, State Wash Water Res Ctr, 79-81; asst chief hydraul engr, 74-75, chief hydrol engr, 75-79, chief hydrol eng, 81-85, MGR HYDRAUL & HYDROL, BECHTEL INC, 85- *Concurrent Pos:* Off, US Coast & Geod Surv, 52; eng resident, US Bur Reclamation, 69-70; chmn Am Soc Civil Engrs, Hydraul Div, 78-79, mem, Nat Water Policy Comt, 83-85 & chmn, 84-85; Bechtel fel, 86. *Honors & Awards:* Hunter Rouse Hydraul Eng Lectr, Am Soc Civil Eng, 88. *Mem:* Am Soc Civil Engrs; Am Water Resources Asn; Am Geophys Union; Int Asn Hydraul Res. *Res:* Flood hydrology; reservoir operation studies; water resources; design of hydraulic structures. *Mailing Add:* Bechtel Inc PO Box 3965 San Francisco CA 94119

CASSIDY, MARIE MULLANEY, PHYSIOLOGY, BIOCHEMISTRY. *Educ:* Nat Univ Ireland, PhD(biochem & biophys), 62. *Prof Exp:* PROF PHYSIOL, BIOCHEM & MORPHOL, GEORGE WASHINGTON UNIV MED CTR, 63- *Res:* Structural-functional correlates of biological mechanisms including heart, lung, cardiovascular & gastrointestines; morphology. *Mailing Add:* Dept Physiol George Washington Univ Med Ctr Washington DC 20037

CASSIDY, PATRICK EDWARD, b East Moline, Ill, Nov 8, 37; m 84; c 3. CHEMISTRY. *Educ:* Univ Ill, BS, 59; Univ Iowa, MS, 62, PhD(chem), 63. *Prof Exp:* Fel, Univ Ariz, 63-64; mem tech staff polymer chem, Sandia Corp, 64-66; sr scientist-group leader, Tracor, Inc, 66-69, asst to dir res lab, 69-71; from asst prof to assoc prof, 71-80, PROF CHEM, SOUTHWEST TEX STATE UNIV, 80-, ASSOC VPRES ACAD AFFAIRS, 91- *Concurrent Pos:* Vpres & chief scientist, Tex Res Inst, 75-90. *Mem:* Am Chem Soc; Soc Plastics Engrs; fel Am Inst Chemists. *Res:* Polyphenylene; polyphenylethers; ferrocenes and other organo-metallics; thermal analyses of polymers; epoxy resin modifications; urethanes; high-temperature polymers; adhesives; coupling agents; phenolics; permeation through polymers; heterocyclic polymers; polyimidines; polymer backbone reactions; polyimides; fluorine-containing polymers; polyethers; poly, ether ketones; poly carbonates. *Mailing Add:* Dept Chem Southwest Tex State Univ San Marcos TX 78666

CASSIDY, RICHARD MURRAY, b St John, NB, Oct 28, 44; m 66; c 2. CHROMATOGRAPHY, ELECTROPHORESIS. *Educ:* Univ King Col, BSc, 65; Dalhousie Univ, MSc, 67; McMaster Univ, Phd(anal chem), 70. *Prof Exp:* Fel, Dalhousie Univ, 71-72, res assoc, 72-73; res scientist, Atomic Energy Can Ltd, 73-87; PROF CHEM, UNIV SASK, 88- *Honors & Awards:* Int Ion Chromatography Forum Award, 89. *Mem:* Fel Chem Inst Can. *Res:* Development of new chromatography techniques for the determination of organic and inorganic materials. *Mailing Add:* Chem Dept Univ Sask Saskatoon SK S7N 0W0 Can

CASSIDY, SUZANNE BLETTERMAN, b New York, NY, Jan 12, 44; m 69, 88; c 1. PEDIATRICS, MEDICAL GENETICS. *Educ:* Reed Col, BA, 65; Vanderbilt Univ, MS, 73, MD, 76. *Prof Exp:* Fel med genetics, dept med, Vanderbilt Univ, 76-77; resident pediat, Univ Wash Affil Hosp, 77-79, fel med genetics, Univ Wash, 79-81; from asst prof to assoc prof pediat & med genetics, Univ Conn, 81-88; ASSOC PROF PEDIAT & MED GENETICS, UNIV ARIZ, 88- *Concurrent Pos:* Dir, div human genetics, dept pediat Univ Conn, 85-88, asst prof obstet & gynec, 86-88; mem educ & info comt, Am Soc Human Genetics, 87-90. *Mem:* Am Soc Human Genetics; Am Acad Pediat. *Res:* Delination of human genetic disorders & syndromes; Prader-Willi syndrome. *Mailing Add:* Dept Pediat Sect Genetics/Dysmorphology Univ Ariz Col Med Tucson AZ 85724

CASSIDY, WILLIAM ARTHUR, b New York, NY, Jan 3, 28; m 61; c 3. GEOCHEMISTRY, GEOLOGY. *Educ:* Univ NMex, BS, 52; Pa State Univ, PhD(geochem), 61. *Prof Exp:* Mem staff seismic comput, Superior Oil Co Calif, 52-53; res scientist meteoritics, Lamont Geol Observ, Columbia Univ, 61-67; assoc prof, 68-81, PROF GEOL & PLANETARY SCI, DEPT GEOL & PLANETARY SCI, UNIV PITTSBURGH, 81- *Concurrent Pos:* Prin investr, NSF grants Antarctic Search Meteorites, 76- *Honors & Awards:* Antartic Serv Medal, 77. *Mem:* Am Geophys Union; Meteoritical Soc. *Res:* Origin and evolution of planetary and subplanetary bodies; element abundances and fractionations; meteorites and meteorite craters; lunar & planetary surface phenomena. *Mailing Add:* Dept Geol & Planetary Sci Univ Pittsburgh Pittsburgh PA 15260

CASSIDY, WILLIAM F, b Nome, Alaska, Aug 28, 08. ENGINEERING. *Educ:* Westpoint, BS, 31; State Univ Iowa, MS, 34. *Prof Exp:* Second lieutenant to lieutenant gen chief engr, US Army Corps Engrs, 31-69; RETIRED. *Mem:* Nat Acad Eng. *Mailing Add:* 430 Village Pl Apt 214 Longwood FL 32779

CASSIE, ROBERT MACGREGOR, b Lowville, NY, May 22, 35; m 71; c 2. GEOLOGY STRUCTURAL. *Educ:* St Lawrence Univ, BS, 56; Univ Wis, PhD(geol), 65. *Prof Exp:* Sr res mineralogist, Chem Div, Pittsburgh Plate Glass Co, 62-65, res supvr, Minerals Res, 65-66; asst prof geol, Col Wooster, 66-67; asst prof, 67-73, ASSOC PROF GEOL,COL BROCKPORT, STATE UNIV NY, 73- *Concurrent Pos:* Sr fel, Geophys Lab, Carnegie Inst, 69-70; vis prof, Ind Univ Geol Field Sta, 76- *Mem:* Geol Soc Am; Am Geophys Union; Nat Asn Geol Teachers. *Res:* Petrology and structure of metamorphic and igneous terranes; structure and geomorphology of fold-thrust belts. *Mailing Add:* Dept Earth Sci State Univ NY Brockport NY 14420

CASSIM, JOSEPH YUSUF KHAN, b Khoi, Iran, June 12, 24; m 68; c 1. BIOMACROMOLECULAR PHYSICAL CHEMISTRY. *Educ:* Univ Ill, BA, 51; Univ Chicago, PhD(biophysics), 65. *Prof Exp:* Res & develop physicist, Indust Condenser Corp, 51-58; res fel, Dept Biophysics, Univ Chicago, 65; res biophysicist, Cardiovasc Res Inst, Univ Calif, San Francisco, 65-68; asst prof, Dept Biophysics, 68-71, assoc prof, 71-78, assoc prof, 78-81, PROF, DEPT MICROBIOL, OHIO STATE UNIV, 81- *Mem:* Biophys Soc; Am Soc Photobiol. *Res:* Structure-function studies of biomacromolecules and membranes by means of spectroscopic and spectropolarimetric analysis; purple membrane of Halobacterium halobuim. *Mailing Add:* Dept Microbiol Ohio State Univ 484 W 12th Ave Columbus OH 43210

CASSIN, JOSEPH M, b Lowell, Mass, July 21, 28; m 69. MICROBIAL ECOLOGY. *Educ:* Cath Univ Am, AB, 52; St John's Univ, AM, 58; Howard Univ, MS, 64; Fordham Univ, PhD(microbial ecol), 68. *Prof Exp:* Teacher, sec schs, 52-67; asst prof microbial ecol, Inst Marine Sci, 68-78, asst prof, 68-75, ASSOC PROF BIOL, ADELPHI UNIV, 76- *Concurrent Pos:* Instr, Sci Honors Prog, Sch Eng, Columbia Univ, 59-60; inst dir biol sci curric study, NY Archdiocese Sci Coun, 64-65; res asst microbial ecol, Fordham Univ, 66-78; dir, Adelphi Inst Marine Sci, 76-78 & NSF precol, Inst Marine Biol, 78-79; consult var environ co, Water Qual, Ecosyst Analyst, Aerial Photog. *Mem:* Phycol Soc Am; Ecol Soc Am; Am Inst Biol Sci. *Res:* Phytoplankton physiology and taxonomy. *Mailing Add:* 12 Sherman Rd Glove Cove NY 11542

CASSIN, SIDNEY, b Mass, June 8, 28; m 50; c 4. PHYSIOLOGY. *Educ:* NY Univ, BA, 50; Univ Tex, MA, 54, PhD(physiol), 57. *Prof Exp:* From instr to assoc prof physiol, 57-68, PROF PHYSIOL, COL MED, UNIV FLA, 68- *Concurrent Pos:* NIH spec fel, Nuffield Inst Med Res, Oxford Univ, 62-63, Dept Pediat, Univ London, 79-80. *Mem:* Sigma Xi; Can Physiol Soc; Perinatal Res Soc; Am Physiol Soc; Soc Exp Biol & Med. *Res:* Control of the prenatal pulmonary circulation; control of fetal lung liquid secretions; respiratory physiology; neonatal anoxia; newborn pulmonary circulation; renal physiology. *Mailing Add:* Dept Physiol Col Med Univ Fla Gainesville FL 32601

CASSINELLI, JOSEPH PATRICK, b Cincinnati, Ohio, Aug 23, 40; c 3. ASTROPHYSICS. *Educ:* Xavier Univ, Ohio, BS, 62; Univ Ariz, MS, 65; Univ Wash, PhD(astron), 70. *Prof Exp:* Res asst, Kitt Peak Nat Observ, 63-65; res engr aerospace sci, Boeing Co, Wash, 65-66; res assoc, Joint Inst Lab Astrophys, Colo, 70-72; from asst prof to assoc prof, 72-81, chmn, Astron Dept, 86-89, PROF ASTRON, UNIV WIS-MADISON, 81- *Concurrent Pos:* Vis scientist, Space Res Lab, Astron Inst Utrecht, 75 & Harvard Smithsonian Ctr Astrophys, 81; Fulbright fel, Univ Utrecht, Neth, 85-86. *Mem:* Am Astron Soc; Int Astron Union. *Res:* Theoretical studies of the structure of the expanding stellar atmospheres of hot stars; radiative transfer in stellar atmospheres and studies of stellar coronae and stellar winds; satellite observations of ultraviolet and x-ray emission from hot stars; formation of very massive stars. *Mailing Add:* Washburn Observ Astron Dept Univ Wis 475 N Charter St Madison WI 53706-1582

CASSMAN, MARVIN, b Chicago, Ill, Apr 4, 36; m 72. BIOCHEMISTRY. *Educ:* Univ Chicago, BA, 54, BS, 57, MS, 59; Albert Einstein Med Sch, PhD(biochem), 65. *Prof Exp:* Postdoctoral biochem, Univ Calif, Berkeley, 65-67, asst prof, Santa Barbara, 67-75; adminr biochem, Nat Inst Gen Med Sci, NIH, 75-78, sect chief, 78-84, prog dir biophys & physiol sci, 84-89, DEP DIR, NAT INST GEN MED SCI, NIH, 89- *Concurrent Pos:* Staff mem, Subcomt Sci, Res & Technol, US House Rep, 82-83; sr policy analyst, Off Sci & Technol Policy, White House, 85-86. *Res:* Biochemistry and biophysics. *Mailing Add:* 5608 Beam Ct Bethesda MD 20817

CASSOLA, CHARLES A(LFRED), b Haverhill, Mass, Aug 6, 13; m 46. MATERIALS ENGINEERING. *Educ:* Northeastern Univ, BS, 37. *Prof Exp:* Inspector eng mat, Naval Air Mat Ctr, 40-42; Bur Aeronaut resident rep, Singer Mfg Co, 42-46; mat engr, Naval Air Eng Ctr, 46-56, head plastics br, 57-58, supt high polymer div, 58-70, supt high polymer div, Naval Air Develop Ctr, 70-74; CONSULT, 74- *Mem:* Am Soc Testing & Mat; Soc Aerospace Mat & Process Eng. *Res:* Applied research and development in high polymer materials, primarily those related to fields of structural plastics, elastomers; sealants and textiles as utilized in military aircraft. *Mailing Add:* 1108 Merrick Ave Collingswood NJ 08108

CASSOLA, ROBERT LOUIS, b Brooklyn, NY, June 9, 41; m 65; c 2. NUCLEAR & THEORETICAL PHYSICS, SYSTEMS ANALYSIS. *Educ:* Polytech Inst Brooklyn, BS, 62; Ohio Univ, PhD(nuclear physics), 66. *Prof Exp:* Sr res physicist, Battelle Mem Inst, 66-68; lectr physics, Ohio State Univ, 67-68; fac res fel nuclear physics, Univ Wash, 68-69; staff specialist, 69-77, sr consult systs anal, 77-80, mgr tech resources, 80-84, DIR, TECHNOL PROG, CONTROL DATA CORP, MINNEAPOLIS, 84- *Concurrent Pos:* Fac res fel, Univ Wash, 68-69; lectr physics, Univ Minn, 74- *Honors & Awards:* Dr Charles Kidd Award, Fed Coun Sci & Technol, 68. *Mem:* Sigma Xi; Am Phys Soc. *Res:* Theoretical nuclear physics, especially nuclear reaction theories and nuclear structure; systems analysis, with special emphasis on simulation; analysis of data processing systems which utilize distributed architectures. *Mailing Add:* 4810 W Coventry Rd Minnetonka MN 55345

CASTAGNA, MICHAEL, b Janesville, Wis, Oct 21, 27; m 55; c 4. BIOLOGICAL OCEANOGRAPHY, AQUACULTURE. *Educ:* Fla State Univ, BS, 53, MS, 55. *Prof Exp:* Asst cur marine studios, Marineland, Fla, 55-56; res biologist, US Dept Interior Fish & Wildlife Serv, 56-62; from asst prof to assoc prof, 62-85, PROF, COL WILLIAM & MARY, 85-; SR MARINE SCIENTIST, VA INST MARINE SCI, 62- *Concurrent Pos:* Adj prof, Univ del Norte, Chile; asst prof, Univ Va. *Honors & Awards:* Wallace Award, Nat Shellfisheries Asn; Estuarine Res Fedn Honors. *Mem:* Estuarine Res Fedn (treas, 73-75, secy, 75-77, pres, 77-79); hon mem Nat Shellfisheries Asn (secy-treas, 72-77, vpres, 74-75, pres, 76-77); World Mariculture Soc; Am Malacol Union. *Res:* Larval behavior, mariculture; natural history of mollusks. *Mailing Add:* Rte 1 Box 413 Melfa VA 23410

CASTAGNOLI, NEAL, JR, b Los Angeles, Calif, Sept 6, 36; m 57; c 5. ORGANIC CHEMISTRY, MEDICINAL CHEMISTRY. *Educ:* Univ Calif, Berkeley, BS, 59, MA, 61, PhD(chem), 64. *Prof Exp:* NIH fel, Higher Inst Health, Italy, 64-65; fel, Imp Col, Univ London, 65-67; from asst prof to prof chem & pharmaceut chem, Med Ctr, Univ Calif, San Francisco, 77-, vchmn dept pharmaceut chem & med chem, 77-; PROF CHEM, VET ADMIN TECHNOL CHEM DEPT. *Mem:* Am Chem Soc; Royal Soc Chem; NY Acad Sci. *Res:* Synthesis, metabolism and pharmacological activity of centrally active compounds. *Mailing Add:* VA Technol Chem Dept Blacksburg VA 24061-0212

CASTANEDA, ALDO RICARDO, b Genoa, Italy, July 17, 30; US citizen; m 56; c 3. THORACIC SURGERY, CARDIOVASCULAR SURGERY. *Educ:* San Carlos Univ, Guatemala, MD, 56; Univ Minn, PhD(surg), 63, MS, 64. *Prof Exp:* From instr to prof surg, Univ Minn, Minneapolis, 63-72; prof cardiovasc surg & cardiovasc surgeon-in-chief, PROF SURG, HARVARD MED SCH & CHIEF DEPT CARDIAC SURG, CHILDREN'S HOSP MED CTR, 77- *Concurrent Pos:* Mem adv coun, Cardiovasc Surg Coun, Am Heart Asn, 68. *Mem:* AAAS; Am Asn Thoracic Surg; Am Col Cardiol; Am Col Surg; Am Surg Asn; Sigma Xi. *Res:* Cardiac physiology; extracorporeal circurlation and its biologic effects; combined cardiopulmonary transplantation. *Mailing Add:* Med Ctr Children's Hosp 300 Longwood Ave Boston MA 02115

CASTANER, DAVID, b New York, NY, Aug 4, 34; m 62; c 3. SYSTEMATIC BOTANY. *Educ:* City Col New York, BS, 61; Iowa State Univ, MS, 63, PhD(plant path), 65. *Prof Exp:* Asst plant path, Iowa State Univ, 61-65; from asst prof to assoc prof, 66-73, PROF BOT & CUR HERBARIUM, CENT MO STATE UNIV, 73- *Mem:* Mycol Soc Am; Bot Soc Am; Am Soc Plant Taxon. *Res:* Flora of Johnson County, Missouri; genus Carex. *Mailing Add:* Dept Biol Cent Mo State Univ Warrensburg MO 64093-5053

CASTANERA, ESTHER GOOSSEN, b Winnipeg, Man, July 19, 20; nat US. BIOCHEMISTRY, NUTRITION. *Educ:* Univ Man, BSc, 42; Univ Calif, PhD(animal nutrit), 54. *Prof Exp:* Chemist nitrocellulose & small arms ammunition, Defence Industs, Ltd, Can, 42-44; chemist, Can Breweries Ltd & Graham's Dried Foods, Ltd, 44-45; technician hematol & serol, Can Red Cross Blood Transfusion Serv, 46-48; asst pentosuria, Dept Biol Chem, Sch Med, Creighton Univ, 48-49, Univ Calif, 49-54; asst microenzyme methods, Dept Pharmacol, Sch Med, Univ Wash, 54-55; biochemist cellulose metabolism in rats, Chem Div, US Army Med Nutrit Lab, Colo, 56-57; res biochemist pyridoxine & pregnancy, Univ Calif, 57-58; biochemist res assoc pharmacol, Bio-Med Div, US Radio Defense Lab, 58-60; res biochemist, Univ Calif, Berkeley, 61-65; res chemist, Calif State Dept Pub Health, 65-67; RES BIOCHEMIST, MED CTR, UNIV CALIF, SAN FRANCISCO, 67- *Mem:* AAAS; fel Am Inst Chem; Am Chem Soc; Sigma Xi. *Res:* Radioactive tracers in metabolism; microbiochemical techniques; enzymology. *Mailing Add:* 1417 Grizzly Peak Blvd Berkeley CA 94708

CASTANO, JOHN ROMAN, b New York, NY, June 10, 26; m 51; c 1. GEOLOGY, GEOCHEMISTRY. *Educ:* City Col NY, BS, 48; Northwestern Univ, MS, 50. *Prof Exp:* Asst, Northwestern Univ, 48-50; from geol trainee to sr staff geologist, 50-77, proj leader geochem serv, 75-84, SR STAFF GEOLOGIST, HYDROCARBON CHARGE SYSTS, SHELL DEVELOP CO, 84- *Concurrent Pos:* US Nat Comn Geochem, 77-80. *Mem:* Geol Soc Am; Am Asn Petro Geologists; Soc Econ Paleontologists & Mineralogists; Int Comn Coal Petrol; Soc Org Petrol (pres-elect, 85, pres, 86). *Res:* Petrology and genesis of ancient and recent sand bodies; environment of deposition of sedimentary iron ores; coal petrology; organic geochemistry. *Mailing Add:* 722 Erder Lane Houston TX 77090

CASTATER, ROBERT DEWITT, b Janesville, Wis, Jan 27, 22; m 44; c 2. OPERATIONS ANALYSIS. *Educ:* Milton Col, BS, 47; State Univ Iowa, MS, 52. *Prof Exp:* chief scientist & researcher, Strategic Air Command, Offutt AFB, 57-86; RETIRED. *Concurrent Pos:* Math teacher. *Mem:* Aerospace Industs Asn Am. *Res:* Bomber penetration analysis; war planning; analysis; war games; operational test and evaluation; electronic counter measures. *Mailing Add:* 11421 Hickory Rd Omaha NE 68144

CASTELFRANCO, PAUL ALEXANDER, b Florence, Italy, Oct 16, 21; nat US; m 54; c 2. PLANT PHYSIOLOGY. *Educ:* Univ Calif, AB, 43, MS, 50, PhD(agr chem), 54; Harvard Univ, STB, 57. *Prof Exp:* Asst, Dept Agr Biochem, Univ Calif, 51-54, jr res biochemist, 55-56; USPHS fel biochem, Med Sch, Tufts Col, 57-58; from asst botanist to assoc botanist, 58-65, assoc prof, 65-70, PROF BOT, UNIV CALIF, DAVIS, 70- *Concurrent Pos:* Guggenheim Mem Found fel, 73-74. *Mem:* Am Soc Plant Physiologists; Am Soc Photobiol; Am Soc Biol Chemists. *Res:* Chlorophyll biosynthesis. *Mailing Add:* Dept of Bot Univ of Calif Davis CA 95616

CASTELL, ADRIAN GEORGE, b Formby, Eng, Sept 9, 35; Can citizen; m 61; c 2. SWINE PRODUCTION. *Educ:* Univ Reading, BSc, 59; Univ Hawaii, MS, 62; Univ Alta, PhD(animal nutrit), 67. *Prof Exp:* RES SCIENTIST ANIMAL NUTRIT, AGR CAN RES BR, 67- *Concurrent Pos:* Vis res worker, East of Scotland Col Agr, Edinburgh, 85-86. *Mem:* Brit Soc Animal Prod; Can Soc Animal Sci (secy, 78-81, pres 82-83); Agr Inst Can. *Res:* Swine production, management and nutrition; evaluation of new or potential sources of nutrients for swine. *Mailing Add:* Agr Can Res Sta PO Box 610 Brandon MB R7A 5Z7 Can

CASTELL, DONALD O, b Washington, DC, Sept 19, 35. GASTROENTEROLOGY. *Educ:* George Washington Univ, MD, 60. *Prof Exp:* CHIEF, GASTROENTEROL DIV & PROF MED, BOWMAN GRAY SCH MED, 82- *Mem:* Am Soc Clin Invest; Am Fedn Clin Res; Am Gastroenterol Asn; Am Col Physicians; Am Physiol Soc; Am Asn Study Liver Dis. *Mailing Add:* Bowman Grey Sch Med 300 S Hawthorne Rd Winston-Salem NC 27103

CASTELL, JOHN DANIEL, b Guelph, Ont, Apr 19, 43; m 66, 75, 80; c 6. MARINE SCIENCES, NUTRITIONAL BIOCHEMISTRY. *Educ:* Dalhousie Univ, BSc Hons, 65, MSc, 66; Ore State Univ, PhD(food sci), 70. *Prof Exp:* Fel, Hormel Inst, Univ Minn, 70; res scientist marine nutrit, 71-83, SECT HEAD DIS & NUTRIT, FISHERIES & ENVIRON SCI, DEPT FISHERIES & OCEANS, HALIFAX FISHERIES RES LAB, 83- *Concurrent Pos:* Adj prof biol, Dalhousie Univ, 78-; mem comt, Natural & Renewable Resources, Nat Res Coun, 81-83; mem bd dirs, World Mariculture Soc, 85-86. *Mem:* Nat Shellfish Asn; World Maricult Soc. *Res:* Nutritional requirements of marine species of commercial interest for aquaculture with a main emphasis on lobsters and minor emphasis on oysters and salmonids. *Mailing Add:* 29 Boutilier Lane Dartmouth NS B2X 2H9 Can

CASTELLA, FRANK ROBERT, b Asbury Park, NJ, Nov 9, 36; m 62; c 2. ENGINEERING. *Educ:* City Col New York, BEE, 58; Univ Calif, Los Angeles, MS, 60. *Prof Exp:* Elec eng circuit design, Litton Industs, 60-62; physicist navig systs, Appl Physics Lab, Johns Hopkins Univ, 62-66; elec eng antenna systs, Gen Elec Co, 66-69; elec eng commun systs, Radiation Inc/Harris Corp, 69-72; PHYSICIST RADAR SYSTS, APPL PHYSICS LAB, JOHNS HOPKINS UNIV, 72- *Mem:* Inst Elec & Electronics Engrs. *Res:* Design and analysis of radar systems and communication systems; radar signal processing and tracking systems; detection theory. *Mailing Add:* 16332 Carrs Mill Rd Woodbine MD 21797-8322

CASTELLAN, GILBERT WILLIAM, b Denver, Colo, Nov 21, 24; m 56; c 4. PHYSICAL CHEMISTRY. *Educ:* Regis Col, Colo, BS, 45; Cath Univ, PhD(chem), 49. *Hon Degrees:* ScD, Regis Col, Colo, 67. *Prof Exp:* AEC fel theoret physics, Univ Ill, 49-50; from instr to prof chem, Cath Univ Am, 50-69, asst head dept, 63-65; assoc dean phys sci & eng, grad sch, 69-74, assoc chmn dept, 73-77, 86-87, PROF CHEM, UNIV MD, COLLEGE PARK, 69- *Concurrent Pos:* Consult, Naval Res Lab, 56-63 & Melpar, Inc, 63-67; NSF fel, Max Planck Inst Phys Chem, Gottingen, 62-63. *Mem:* Am Phys Soc; Electrochem Soc; Am Chem Soc. *Res:* Chemical relaxation; electrochemical thermodynamics and kinetics. *Mailing Add:* Dept of Chem Univ of Md College Park MD 20742

CASTELLAN, NORMAN J, b Everett, Wash, Jan 11, 12; wid; c 3. CIVIL ENGINEERING. *Educ:* Univ Colo, BS, 33, MS, 34. *Prof Exp:* Instr civil eng, Univ Colo, 34; sr rodman, US Bur Pub Rd, 34, sr levelman, 35; instrumentman, US Coast & Geod Surv, 34-35; bridge draftsman, Colo State Hwy Dept, 35-36; sr engr, US Bur Reclamation, 36; instr math & civil eng, Colo Sch Mines, 36-39, asst prof, 39-41; asst to dist mgr, War Prod Bd, 41-43, dep regional dir, 47; self employed consult engr, 47-55; assoc prof civil eng, 55-59, head dept civil eng, 60-63, chmn div eng, 62-67, head dept civil eng, 69-76, prof civil eng, 59-80, EMER PROF CIVIL ENG, SACRAMENTO STATE COL, 80- *Honors & Awards:* Western Elec Fund-Am Soc Eng Educ Award, 70-71. *Mem:* Am Soc Civil Engrs; Am Soc Eng Educ. *Res:* Structural analysis. *Mailing Add:* 3113 Calle Verde Ct Sacramento CA 95821

CASTELLANA, FRANK SEBASTIAN, b Flushing, NY, Feb 5, 42. BIOMEDICAL ENGINEERING, MEDICINE. *Educ:* Queens Col, NY, BS, 63; Columbia Univ, BS, 63, MS, 64, EngScD(fluid mech), 69; Albert Einstein Col Med, MD, 76. *Prof Exp:* Res engr, Heat Transfer Res Facil, 68-69, asst prof chem eng, 69-73, ASSOC PROF CHEM ENG, COLUMBIA UNIV, 77-, RES ASSOC, DEPT MED, ST LUKE'S HOSP, 77- *Concurrent Pos:* Med intern, Hosp Ctr, St Luke's Hosp, 76-77. *Mem:* Am Inst Chem Eng; Am Chem Soc. *Mailing Add:* 227 Stuart Rd E Princeton NJ 08540

CASTELLANO, SALVATORE, MARIO, b Trieste, Italy, Sept 28, 25; US citizen; m 53. NUCLEAR MAGNETIC RESONANCE, INDUSTRIAL CHEMISTRY. *Educ:* Liceo Scientifico, Trieste, Italy, dipl, 43, Polytech Milan, Italy, PhD(chem eng), 51. *Hon Degrees:* Libera Docenza, Univ Rome, Italy, 67. *Prof Exp:* Asst prof phys & indust chem, Inst Indust Chem, Polytech Milan, 51-58, dir, Nuclear Magnetic Resonance Lab, 60-62; fel, Mellon Inst, Carnegie-Mellon Univ, 62-75, prof biophys chem, dept Biol Sci, 75-86; PROF MAS SPECTROS, DEPT PHYS CHEM, UNIV PARMA, ITALY, 86- *Concurrent Pos:* Consult, Montecatini Co, 53-58, Assoreni Assocs, 79-81; fel, chem dept, Mass Inst Technol, 58-60; vis prof, Inst Phys Chem, Univ Padova, 74, Univ Pisa & Univ Florence, 83. *Mem:* Int Soc Magnetic Resonance. *Res:* Catalitic studies and industrial developments of oxosyntesis and of other high-pressure syntesis utilizing carbon monoxide and hydrogen gas mixtures; structural studies of chemical and biochemical compounds by means of high resolution nuclear magnetic resonance; mathematical algorithms and computer programs for the analysis of nuclear magnetic resonance spectra. *Mailing Add:* 1425 Chislett St Pittsburgh PA 15202

CASTELLI, VITTORIO, b Rome, Italy, Apr 8, 34; US citizen; m 57; c 1. MECHANICAL ENGINEERING. *Educ:* Columbia Univ, MS, 59, PhD(mech eng), 62. *Prof Exp:* From instr to asst prof mech eng, 59-68, prof, 69-80, ADJ PROF MECH ENG, COLUMBIA UNIV, 80- *Concurrent Pos:* Engr, Franklin Inst, 59-62, sr staff res engr, 62-63, consult, 63-; consult, Conductron Corp, Calif, 64-, AiResearch Corp, Ariz, 64-, Autonetics Div, NAm Aviation, Inc, 64-, Mech Tech Inc, 64-, Rohr Corp, 65-, Xerox Data Systs, 68-, Appl Magnetics Corp, 68-, Comput Peripherals Co, 69-70 & Control Data Corp, 69- *Res:* Fluid mechanics; incompressible and compressible fluid lubrication; applications of computers to engineering problems. *Mailing Add:* Six E Tenth St New York NY 10017

CASTELLI, WALTER ANDREW, b Iquique, Chile, May 5, 29; m 53; c 2. ANATOMY. *Educ:* Univ Chile, DDS, 57; Univ Mich, MS, 64. *Prof Exp:* Instr anat, Concepcion Univ, 57-62, instr anat & res assoc physiopath, 64-66, asst prof anat, 66-68; from asst prof to prof anat, Sch Med, Univ Mich, Ann Arbor, 66-82; asst prof anat, Concepcion Univ, 82-88; PROF ANAT, SCH MED, UNIV MICH, ANN ARBOR, 88- *Concurrent Pos:* W K Kellogg Found fel anat, Univ Mich, 62-64. *Mem:* Int Asn Dent Res; Sigma Xi; Am Asn Clin Anatomists. *Res:* Gross anatomy; vascular studies; head and neck area; experimental tooth transplants in monkeys; periodontal disease studies. *Mailing Add:* Dept of Anat 3725 Med Sci II Univ of Mich Ann Arbor MI 48109

CASTELLI, WILLIAM PETER, b New York, NY, Nov 21, 31; m 61; c 3. EPIDEMIOLOGY, CARDIOLOGY. *Educ:* Yale Col, BS, 53; Cath Univ Louvain, MD, 59. *Prof Exp:* DIR LABS, FRAMINGHAM HEART STUDY, NAT HEART, LUNG & BLOOD INST, NIH, 65- *Concurrent Pos:* Harvard Med Sch fel, 61-65, lectr prev med, 65-; chmn subcomt criteria & methods, Coun Epidemiol, Am Heart Asn, 78- *Res:* To identify those factors associated with subsequent development of cardiovascular disease, such as lipid measures of lipoproteins, hypertension, cigarette smoking, diabetes mellitus overweight and physical activity. *Mailing Add:* 74 White Terr Marlborough MA 01752

CASTELLINO, FRANCIS JOSEPH, b Pittston, Pa, Mar 7, 43; m 65; c 3. BIOCHEMISTRY. *Educ:* Univ Scranton, BS, 64; Univ Iowa, MS, 66, PhD(biochem), 68. *Hon Degrees:* Univ Scranton, LLD, 82. *Prof Exp:* NIH fel biochem, Duke Univ, 68-70; from asst prof to assoc prof, 70-77, PROF BIOCHEM, UNIV NOTRE DAME, 77-, DEAN, COL SCI, 79- *Concurrent Pos:* NIH Res Career Develop grant, 74-77; Camille & Henry Dreyfus Techer/Scholar grant, 74-79. *Mem:* AAAS; NY Acad Sci; Am Soc Biol Chemists; Am Heart Asn. *Res:* Structure function relationships in proteins. *Mailing Add:* Dept Biochem Univ Notre Dame 229 Nieuwland Sci Hall Notre Dame IN 46556

CASTELLION, ALAN WILLIAM, b North Tonawanda, NY, May 1, 34; m 61; c 2. TOXICOLOGY. *Educ:* Univ Buffalo, BS, 56; Univ Utah, PhD(pharmacol), 64. *Prof Exp:* Teaching asst pharmacog, Univ Utah 56-58, pharm, 58-59 & pharmacol, 59-63, res asst, 63-64; sr pharmacologist, Neuropharmacol Sect, Dept Neurol & Cardiol, Res & Develop Div, Smith Kline & French Labs, 66-67, sr scientist, Dept Pharmacol, 67-69; chief, Sect Pharmacol, Norwich Eaton Pharmaceut, 70-77; dir compliance, Inst Prod Develop, 77-82, dir regulatory affairs, 82-84, assoc dir, 84-90, MGR, INST PROD DEVELOP, 90- *Concurrent Pos:* Smith Kline & French res fel, Col Med, Univ Utah, 64-66. *Honors & Awards:* Lunsford Richardson Pharm Award, 64. *Mem:* Sr mem Acad Pharmaceut Sci; Am Soc Pharmacol & Exp Therapeut; Sigma Xi; Regulatory Affairs Prof Soc; Am Pharmaceut Asn; Drug Info Asn; Am Assoc Pharmaceut Sci. *Res:* Pharmaceutical research and development; pharmaceutical regulatory affairs; complaince and quality assurance; toxicology; pharmacy. *Mailing Add:* Bradley Hill Rd Oxford NY 13830

CASTELLO, JOHN DONALD, b Paterson, NJ, May 1, 52; m 82; c 2. FOREST VIROLOGY, FOREST PATHOLOGY. *Educ:* Montclair State Col, BA, 73; Wash State Univ, MS, 75; Univ Wis, PhD(plant path), 78. *Prof Exp:* ASSOC PROF PLANT PATH, MICROBIOL & FOREST PATH, COL ENVIRON SCI & FORESTRY, STATE UNIV NY, 78- *Concurrent Pos:* Vis prof, Institut Fur Pflanzen Krankheiten, Universitat Bonn, 87. *Honors & Awards:* Alexander Von Humboldt Stiftung Award. *Mem:* Am Phytopath Soc. *Res:* Virus and mycoplasma diseases of shade, ornamental and forest tree species in an attempt to relate infection with dieback and decline diseases of major tree species in the northeast. *Mailing Add:* State Univ NY Environ Sci & Forestry Syracuse NY 13210-2788

CASTELLOT, JOHN J, JR, CELL TO CELL INTERACTION. *Educ:* Harvard Univ, PhD(pharmacol), 78. *Prof Exp:* ASST PROF PATH, SCH MED, HARVARD UNIV, 84- *Res:* Vascular endothelial cell, smooth muscle cell interaction; cellular and molecular control of new blood vessel growth during embryo genesis. *Mailing Add:* Anat & Cell Biol Tufts Univ Health Sci Schs 136 Harrison Ave Boston MA 02111

CASTELLUCCI, VINCENT F, b Montreal, Que, July 26, 40. NEUROBIOLOGY. *Educ:* Washington Univ, St Louis, Mo, PhD(neurobiol), 68. *Prof Exp:* ASSOC PROF NEUROSCI, COL PHYSICIANS & SURGEONS, COLUMBIA UNIV, 75- *Mem:* Soc Neurosci; Am Soc Zoologists; Am Physiol Soc; AAAS. *Res:* Cellular & molecular mechanics of learning. *Mailing Add:* Neurobiol & Behav IRCM Clin Res Inst Montreal 110 Pine Ave Montreal PQ H2W 1R7 Can

CASTELNUOVO-TEDESCO, PIETRO, b Florence, Italy, Jan 5, 25; US citizen; m 57; c 2. PSYCHOANALYSIS, PSYCHOSOMATIC MEDICINE. *Educ:* Univ Calif, Los Angeles, BA, 45; Univ Calif, Berkeley, MA, 48; Sch Med, Boston Univ, MD, 52. *Prof Exp:* Physician, Inpatient Unit Psychiat, Boston City Hosp, 58-59; asst prof to prof psychiat, Sch Med, Univ Los Angeles, 59-75; JAMES G BLAKEMORE PROF PSYCHIAT, SCH MED, VANDERBILT UNIV, 75- *Concurrent Pos:* Asst psychiat, Harvard Med Sch, 58-59; chmn, Dept Psychiat, Harbor Gen Hosp, Calif, 59-75; Training & supv analyst, St Louis Psychoanal Inst, 77- *Mem:* fel Am Psychiat Asn; Am Psychoanal Asn; Am Psychosom Soc; fel Am Col Psychiatrists; fel Am Col Psychoanalysts; Int Col Psychosom Med. *Res:* psychotherapy; psychiatric education; author or coauthor of 100 publications. *Mailing Add:* Dept Psychiat Sch Med Vanderbilt Univ 21st Ave S & Garland Nashville TN 37232

CASTEN, RICHARD FRANCIS, b New York, NY, Nov 1, 41; m 64. NUCLEAR PHYSICS. *Educ:* Col of the Holy Cross, BS, 63; Yale Univ, MS, 64, PhD(physics), 67. *Prof Exp:* Fel nuclear physics, Niels Bohr Inst, Univ Copenhagen, 67-69 & Los Alamos Sci Lab, 69-71; from asst physicist to physicist, 71-81, GROUP LEADER, NEUTRON NUCLEAR STRUCTURE GROUP, BROOKHAVEN NAT LAB, 81-, SR SCIENTIST, 82- *Concurrent Pos:* Inst Laue Langevin, Grenoble, France, 76-77; vis scientist, Univ Köln, Germany, 84-85; chmn, NAm Steering Comt Radioactive Beam Accelerator. *Honors & Awards:* Alexander von Humboldt Sr Am Scientist Prize, 83. *Mem:* Sigma Xi; NY Acad Sci; AAAS; Am Phys Soc; Europ Phys Soc; Am Chem Soc. *Res:* Nuclear structure, especially collective modes such as pairing, vibrational, rotational excitations, Coulomb excitation, one and two nucleon transfer reactions, gamma ray deexcitations, nuclei far off stability, Nilsson, Coriolis, interacting Boson approximation models, neutron-proton interaction, symmetries and supersymmetries. *Mailing Add:* Physics Dept Brookhaven Nat Lab Upton NY 11973

CASTEN, RICHARD G, b Philadelphia, Pa, May 14, 43; m 70; c 1. SOFTWARE DEVELOPMENT, APPLIED MATHEMATICS. *Educ:* Temple Univ, AB, 65; Calif Inst Technol, PhD(appl math), 70. *Prof Exp:* Math technician magnetohydrodyn, Gen Elec Co, 64-65; math technician satellite orbits, Aerospace Corp, 66; res fel appl math, Calif Inst Technol, 70; asst prof math, Purdue Univ, West Lafayette, 70-77; MEM TECH STAFF, AEROSPACE CORP, 77- *Mailing Add:* Aerospace Corp 2350 El Segundo Blvd El Segundo CA 90245

CASTENHOLZ, RICHARD WILLIAM, b Chicago, Ill, May 9, 31; m 54, 82; c 2. ECOLOGY, HYDROBIOLOGY. *Educ:* Univ Mich, BS, 52; Wash State Univ, PhD(bot), 57. *Prof Exp:* Asst bot, Wash State Univ, 53-57; from instr to assoc prof, 57-69, PROF BIOL, UNIV ORE, 69- *Concurrent Pos:* John Simon Guggenheim fel, 70-71; Fulbright scholar, Norway, 77-78. *Mem:* Phycol Soc Am; Am Soc Limnol & Oceanog; Am Soc Microbiol; Brit Phycol Soc. *Res:* Physiology and ecology of photosynthetic prokaryotes; biology of thermophilic microorganisms. *Mailing Add:* Dept of Biol Univ of Ore Eugene OR 97403

CASTENSCHIOLD, RENE, b Mount Kisco, NY, Feb 7, 23; m 47; c 3. ELECTRICAL ENGINEERING. *Educ:* Pratt Inst, BEE, 44. *Prof Exp:* Design engr, Gen Elec Co, 44-47; sr prod engr, Am Transformer Co, 47-50; exec eng mgr, Automatic Switch Co, 51-85; PRES, LCR CONSULT ENGRS, PA, 86- *Concurrent Pos:* Lectr, NJ Inst Technol, 67-79; adv, Underwriters Labs, Inc, 73-85; US deleg & chmn US tech adv group, Int Electrotechnical Comn, 81-89. *Honors & Awards:* James H McGraw Award, 86; Achievement Award, Inst Elec & Electronic Engrs, 88; Richard Harold Kaufmann Award, 90. *Mem:* Fel Inst Elec & Electronics Engrs; Nat Soc Prof Engrs; Instrument Soc Am; Nat Elec Mfrs Asn; Can Standards Asn; Nat Fire Protection Asn; Int Asn Elec Inspectors; Nat Acad Forensic Engrs; Am Consult Engrs Coun. *Res:* Research and development of automatic transfer switches and generator controls; author of over 30 papers and articles on product design, ground-fault protection, system coordination and controls for emergency power; nine patents. *Mailing Add:* Lee's Hill Rd New Vernon NJ 07976

CASTENSON, ROGER R, b Galveston, Tex, Aug 6, 43. SCIENCE ADMINISTRATION. *Educ:* Tex A&M Univ, BSAE, 72; Mich State Univ, MBA, 77. *Prof Exp:* Res assoc, Blackland Res Ctr, Temple, Tex, 72-73; mgr pub relations & mem activ, Am Soc Agr Engrs, 73-80; oper mgr, Soc Petrol Engrs, 80-81, bus mgr, 81-86; EXEC VPRES, AM SOC AGR ENGRS, 87- *Mem:* Am Soc Agr Engrs; Soc Petrol Engrs; Sigma Xi. *Res:* Agricultural engineering; management and administration. *Mailing Add:* Am Soc Agr Engr 2950 Niles Rd St Joseph MI 49085

CASTER, KENNETH EDWARD, b New Albany, Pa, Jan 26, 08; m 33. GEOLOGY, PALEONTOLOGY. *Educ:* Cornell Univ, AB, 29, MS, 31, PhD(stratig), 33. *Hon Degrees:* DSc, Univ Cincinnati, 85. *Prof Exp:* Asst entom, Cornell Univ, 28-30, asst geol, 29-30, instr paleont, 30-32, instr geol, 32-35; researcher, Paleont Res Inst, NY, 35-36; cur paleont, Mus, 36-40, fel, Grad Sch, 37-85, from asst prof to prof, 40-78, EMER PROF GEOL, UNIV CINCINNATI, 78- *Concurrent Pos:* Asst head dept sci, NY State Norm Sch, Geneseo, 35-36; Nat Res Coun grant-in-aid, 35-37; trustee, Paleont Res Inst, NY, 39-, vpres, 41-43 & 65, pres, 44, 45, 51-54 & 66; trustee, Cushman Found, 51-56; vis prof awards, US Dept State, 45-47; prof, Sao Paulo, 45-48; mem, Pan Am Cong Mining, Eng & Geol, 47 & US comn geol, 60-64; Guggenheim fel geol studies, SAm, 47-48, SAfrica, 54-55, Australia & NZ, 56; Fulbright vis fel, Univ Tasmania, 55-56, Univ Cologne, 64; German Res Asn grant, 64; US rep, Int Paleont Union, 60-; US off del, Int Geol Cong, 60 & 64. *Honors & Awards:* Gondwana Medal, India, 55; Paleont Soc Medal, 76. *Mem:* AAAS; Am Asn Petrol Geol; fel Paleont Soc (secy, 47-55, vpres, 57, pres, 59); fel Geol Soc Am; Soc Vert Paleont. *Res:* Invertebrate paleontology; paleozoic stratigraphy; paleogeography; Southern Hemisphere historical geology; early echinoderm history; fossil arachnida; problematica. *Mailing Add:* 425 Riddle Rd Cincinnati OH 45220

CASTER, WILLIAM OVIATT, b Topeka, Kans, Dec 7, 19; m 43; c 4. NUTRITION. *Educ:* Univ Wis, BA, 42, MS, 44; Univ Minn, PhD(physiol chem), 48. *Prof Exp:* Asst inorg chem, Univ Wis, 42-43; chemist, Lab Physiol Hyg, Univ Minn, 44-47, asst prof physiol chem, 51-63; biochemist, Nutrit Unit, USPHS, 48-51; from assoc prof to prof, 63-85, EMER PROF NUTRIT, UNIV GA, 86-; ADJ PROF BIOL, GA SOUTHERN UNIV, 89- *Concurrent Pos:* USPHS fel, Nat Heart Inst, 56-61. *Mem:* AAAS; Am Soc Biochem & Molecular Biol; Am Inst Nutrit; Brit Nutrit Soc. *Res:* Human nutrition; water and electrolyte metabolism; radiobiology; chemistry and function of cardiovascular system. *Mailing Add:* Dept Biol Ga Southern Univ LB-8042 Statesboro GA 30460-8042

CASTIGNETTI, DOMENIC, b Boston, Mass, Sept 22, 51; m 75; c 3. SIDEROPHORE METABOLISM, HETEROTROPHIC NITRIFICATION. *Educ:* Merrimack Col, BA, 73; Colo State Univ, MS, 77; Univ Mass, PhD(microbiol), 80. *Prof Exp:* Postdoctoral res biochem, Brandeis Univ, 80-82; asst prof, 82-88, ASSOC PROF BIOL, LOYOLA UNIV, CHICAGO, 88- *Mem:* Am Soc Microbiol; AAAS; Sigma Xi. *Res:* Siderophore metabolism including biodegradation and functioning as disease factors; heterotrophic nitrification including metabolism and bioenergetics. *Mailing Add:* Biol Dept Loyola Univ 6525 N Sheridan Rd Chicago IL 60626

CASTILE, ROBERT G, b Washington, DC, Nov 19, 47. LUNG MECHANICS. *Educ:* Univ Md, MD, 73. *Prof Exp:* ASST PROF PEDIAT, HARVARD MED SCH, 83-; TUFTS NEW ENG MED CTR. *Mem:* Am Thoracic Soc; Am Physiol Soc. *Mailing Add:* 55 Fuller St Dedham MA 02026

CASTILLO, JESSICA MAGUILA, insect pathology, nematology, for more information see previous edition

CASTLE, GEORGE SAMUEL PETER, b Belfast, North Ireland, May 30, 39; Can citizen; m 61; c 3. ELECTRICAL ENGINEERING. *Educ:* Univ Western Ont, BESc, 61, PhD(elec eng), 69; Imp Col, Univ London, DIC, 63; Univ London, MSc, 63. *Prof Exp:* Assoc mem sci staff microwave commun systs, Res & Develop Lab, Northern Elec Co, 63-66; lectr eng sci, 68-69, from asst prof to assoc prof, 69-79, PROF ENG SCI & CHMN, DEPT ELEC ENG, UNIV WESTERN ONT, 79- *Mem:* Inst Elec & Electronics Engrs. *Res:* Application of electrostatic forces to engineering problems. *Mailing Add:* W Ont Univ London ON N6A 5B9 Can

CASTLE, JOHN EDWARDS, b Minneapolis, Minn, June 10, 19; m 42; c 3. APPLIED CHEMISTRY. *Educ:* Carleton Col, BA, 40; Univ Wis, PhD(org chem), 44. *Prof Exp:* Asst chem, Univ Wis, 40-42, tech asst, Off Sci Res & Develop Contract, 42-43 ADJ PROF, COL MARINE STUDIES, UNIV DEL, 79-; chemist cent res dept, E I du Pont de Nemours & Co, Inc, 43-50, res supvr, 50-61, asst lab dir, 61-63, assoc dir mat res, 63-65, lab dir, Electrochem Dept, 65-68, asst res dir, 68-72, mgr environ prods sect, Indust Chem Dept, 72-74, res mgr energy & mat dept, 74-77, environ control mgr cent res dept, 77; consult, 78-79. *Concurrent Pos:* Adj prof, Col Marine Studies, Univ Del, 79- *Mem:* Am Chem Soc; AAAS. *Res:* Contact catalysis; organic synthesis; polymers; biochemistry; chitin, mucopolysaccharides; bioenergy processes. *Mailing Add:* 148 Kendal At Longwood Kennett Square PA 19348-2331

CASTLE, JOHN GRANVILLE, JR, b Buffalo, NY, Sept 9, 24; m 46, 82; c 5. RADIATION DAMAGE, MICROWAVE DEVICES. *Educ:* State Univ NY Buffalo, BA, 47; Yale Univ, PhD(physics), 50. *Prof Exp:* Instr physics, State Univ NY Buffalo, 50-51; assoc res physicist, Cornell Aeronaut Lab, 51-52; res assoc, State Univ NY Buffalo, 53-55; res physicist, Westinghouse Res Lab, 55-58, adv physicist, Westinghouse Res & Develop Ctr, 58-67; prof elec eng & res assoc, Learning Res & Develop Ctr, Univ Pittsburgh, 67-69; dir, Div Natural Sci & Math, Univ Ala, Huntsville, 70-71, prof physics 69-80; MEM TECH STAFF, SANDIA NAT LAB, 80- *Concurrent Pos:* Vis scientist, High Energy Physics Lab, Stanford Univ, 70-71; clin prof therapeut radiol, Med Ctr, Univ Ala, Birmingham, 70-80; consult, US Army Missile Command, 72-80; design, Talking Software, 83- *Mem:* AAAS; Am Phys Soc; Inst Elect & Electronics Engrs. *Res:* Petroleum engineering; solid state physics; medical instrumentation; educational media; microwave and optical properties of solids; infrared properties of gases; biomedical engineering; radiation effects in semiconductor devices. *Mailing Add:* 9801 San Gabriel Rd NE Albuquerque NM 87111

CASTLE, KAREN G, b Elizabeth, NJ, Sept 23, 48; m 71. SPACECRAFT CONTROL, ASTRONOMY. *Educ:* Univ Mich, BS, 70, MS, 72, PhD(astron), 75. *Prof Exp:* Fel physics & instr, Univ Calgary, 75-78; assoc sr comput programmer, Lockheed Electronics Co, 78-81, ENGR, SR LOCKHEED ENG & MGT SERV, LOCKHEED AIRCRAFT CO, 82- & RES SPECIALIST, LOCKHEED MISSILES & SPACE CO. *Concurrent Pos:* Lectr, Downtown Campus, Univ Houston, 81. *Mem:* Am Astron Soc; AAAS; Astron Soc Pac; Sigma Xi; Am Inst Aeronaut & Astronaut; Soc Photo-Optical Instrumentation Engrs. *Res:* Dynamics of mass exchange in binary star systems; orbital mechanics; spacecraft altitude determination and control. *Mailing Add:* Dept 62-11 Bldg 104 Lockheed Missiles-Space Co 1111 Lockheed Way Sunnyvale CA 94089-3504

CASTLE, MANFORD C, b Coeburn, Va, Apr 27, 42; m 69; c 3. PHARMACOLOGY. *Educ:* Berea Col, BA, 65; Univ Kans, PhD(pharmacol & toxicol), 72. *Prof Exp:* Vol chem teacher, Peace Corps, Univ Antioquia, Colombia, 66-68; clin instr pharmacol, Univ Kans, 72-73; staff assoc & NIH fel, Nat Inst Gen Med Sci, 73-75; staff assoc pharmacol, Nat Cancer Inst, 75-76; from asst prof to assoc prof, 78-86, PROF PHARMACOL, EASTERN VA MED SCH, 86- *Concurrent Pos:* Consult, Nat Cancer Inst, 78-80; prin investr, Nat Heart, Lung & Blood Inst, 77-80; dir, Therapeut Drug Monitoring Prog. *Mem:* AAAS; Am Soc Pharmacol & Exp Therapeut; Drug Metab Soc; Int Union Pharmacol; Am Asn Clin Chem. *Res:* Metabolism and disposition of drugs, particularly cardiac glycosides and cancer chemotherapeutic agents; alteration of these processes by other drugs; methods involve use of radioactive tracers and high-pressure liquid chromatography; neurochemistry; neuropharmacology. *Mailing Add:* Dept Pharmacol Eastern Va Med Sch PO Box 1980 Norfolk VA 23501

CASTLE, PETER MYER, b Detroit, Mich, Apr 19, 40; m 61; c 3. CHEMICAL PHYSICS, HIGH TEMPERATURE CHEMISTRY. *Educ:* Univ Mich, BS, 62; Purdue Univ, PhD(phys chem), 70. *Prof Exp:* Vis asst prof phys chem, Purdue Univ, 70; sr engr, 70-80, FEL SCIENTIST PHYS CHEM, WESTINGHOUSE RES LABS, 80- *Concurrent Pos:* Instr night sch, Community Col Allegheny County, Boyce Campus, 74-76; adj prof, Kans State Univ; adv scientist, Winco, 88- *Mem:* Am Soc Mass Spectrometry; Am Chem Soc; Soc Appl Spectros. *Res:* Studies of infrared, visible and near ultraviolet laser photolysis reactions for synthetic applications and understanding of fundamental photophysical processes; high power dye laser development; diode laser spectroscopy. *Mailing Add:* Lawrence Livermore L-471 WINCO SIS Proj PO Box 808 Livermore CA 94550

CASTLE, RAYMOND NIELSON, b Boise, Idaho, June 24, 16; m 37; c 8. CHEMISTRY. *Educ:* Idaho State Univ, BS, 39; Univ Colo, MA, 41, PhD(chem), 44. *Prof Exp:* Asst chem, Univ Colo, 39-42; instr, Univ Idaho, 42-43; instr, Univ Colo, 43-44; res chem, Battelle Mem Inst, 44-46; asst prof chem, Univ NMex, 46-47, pharmaceut chem, 47-50, from assoc prof to prof chem, 50-70, chmn dept, 63-70; prof chem, Brigham Young Univ, 70-81; GRAD RES PROF CHEM, UNIV SFLA, 81- *Concurrent Pos:* Res fel, Univ Va, 52-53; ed, J Heterocyclic Chem; vis prof, Univ SFla, 79. *Honors & Awards:* Chem Award, Int Soc Heterocyclic Soc, 83. *Mem:* Am Chem Soc; fel Royal Soc Chem; Pharmaceut Soc Japan; Int Soc Heterocyclic Chem; Sigma Xi. *Res:* Organic chemistry; optical crystallography; optical crystallographic properties of organic compounds; synthesis of cinnolines; pyridazines; pyridines, polycyclic thiophenes; other related heterocycles of medicinal interest; organic photochemistry; 2D-NMR spectroscopy. *Mailing Add:* Dept Chem Univ SFla Tampa FL 33620-5250

CASTLE, ROBERT O, b Berkeley, Calif, Dec 31, 26; m 54; c 2. GEOLOGY, TECTONICS. *Educ:* Stanford Univ, BS, 48; McGill Univ, MSc, 49; Univ Calif, Los Angeles, PhD(geol), 64. *Prof Exp:* Instr geol, Univ Mass, 49-50; geol field asst, 50-51, geologist, 51-52, GEOLOGIST, US GEOL SURV, 54- *Mem:* Geol Soc Am; Asn Eng Geologists. *Res:* Tectonics engineering geology; igneous petrology; contemporary crustal deformation. *Mailing Add:* 730 Torreya Ct Palo Alto CA 94303

CASTLE, WILLIAM BOSWORTH, clinical medicine, hematology; deceased, see previous edition for last biography

CASTLEBERRY, GEORGE E, b Herring, Okla, Mar 20, 18; m 44. INORGANIC CHEMISTRY. *Educ:* Southwestern State Col, Okla, BS, 39; Univ Okla, MS, 51, PhD(sci educ), 68. *Prof Exp:* Prin & teacher, E Walnut Schs, Okla, 40-41; teacher, Clinton High Sch, 41-42, teacher & asst prin, 45-58; PROF CHEM, SOUTHWESTERN STATE COL, OKLA, 58- *Mem:* Am Chem Soc; Nat Sci Teachers Asn. *Res:* Science education; graduate education of secondary school science teachers of Oklahoma public schools; orthoaminobenzenethiol as a reagent for the gravimetric determination of selenium in compounds. *Mailing Add:* Rt 3 Box 209 Weatherford OK 73096-9303

CASTLEBERRY, RON M, b Meadow, Tex, Oct 14, 41; m 62; c 2. CORN BREEDING, CROP BREEDING. *Educ:* Tex Technol Col, BS, 69; Tex Tech Univ, MS, 70; Univ Nebr, PhD(agron), 73. *Prof Exp:* Res assoc, dept agron, Univ Nebr, 73; corn physiologist, Dekalb-Pfizer Genetics, 73; exec vpres, 84-88, PRES, FFR COOP, 88- *Concurrent Pos:* Assoc ed, Crop Sci, Am Soc Agron, 79-81. *Mem:* Am Soc Agron; Am Soc Plant Physiologists; AAAS; Coun Agr Sci & Technol; Sigma Xi. *Res:* Integrating crop physiology into commercial plant breeding; breeding commercial corn hybrids with improved drought tolerance; coordination and management of commercial breeding programs in forages, soybeans and corn. *Mailing Add:* FFR Coop PO Box 242 Battle Ground IN 47920

CASTLEMAN, ALBERT WELFORD, JR, b Richmond, Va, Jan 7, 36; m 76; c 3. CHEMICAL PHYSICS. *Educ:* Rensselaer Polytech Inst, BChE, 57; Polytech Inst Brooklyn, MS, 63, PhD, 69. *Hon Degrees:* Doktors Honoris Causa, Univ Innsbruck, Austria, 87. *Prof Exp:* Res & develop assoc, High Energy Fuels Div, Olin Mathieson Chem Corp, 57-58; leader, chem res group, Brookhaven Nat Lab, 58-75; prof chem, Univ Colo, 75-82, chmn, chem physics prog, 79-82; prof chem, 82-86, EVAN PUGH PROF CHEM, PA STATE UNIV, 86- *Concurrent Pos:* Consult, adv comt reactor safeguards, US Nuclear Regulatory Comn, 70-80, Chem Mfrs Asn, 75- Nuclear Div, Oak Ridge Nat Lab, 76- & E I DuPont de Nemours, 89-; adj prof atmospheric chem, dept earth & space sci & dept mech, State Univ NY, Stony Brook, 73-75; Coop Inst Res Environ Sci fel, 75-82; Sherman Fairchild distinguished scholar, Calif Inst Technol, 77; guest prof, Physics Inst, Leopold-Franzens Univ, Innsbruck, Austria, 81, 84 & 90; distinguished lectr, Univ Utah, 81, Univ Mass, 87, Inst Sch Mat Sci Technol, Italy, 87; Japan Soc Prom Sci fel, 85; Frontiers in chem lectr, Case Western Reserve Univ, 85; co-ed, Zeitschrift fü r Physik D, 87-; assoc ed, J Phys Chem, 88-91, sr ed, 91-; Fulbright sr scholar, 89-90. *Honors & Awards:* Sr US Scientist von Humboldt Award, 86; Creative Advan Environ Sci & Technol Award, Am Chem Soc, 88. *Mem:* Am Chem Soc; fel Am Phys Soc; Am Geophys Union; Int Asn Meteorol & Atmospheric Physics (secy, 74-75); Sigma Xi; fel AAAS; Am Asn Aerosol Res; NY Acad Sci. *Res:* Molecular properties of small clusters; kinetics of association and unimolecular dissociation reactions; ion-molecule reactions, intramolecular energy transfer, multiphoton ionization; laser photochemistry and photophysics of small clusters; atmospheric chemistry; nucleation phenomena, solvation phenomena. *Mailing Add:* Dept Chem 152 Davey Lab Pa State Univ University Park PA 16802

CASTLEMAN, L(OUIS) S(AMUEL), b St Johnsbury, Vt, Nov 24, 18; c 4. PHYSICAL METALLURGY. *Educ:* Mass Inst Technol, BS, 39, ScD, 50. *Prof Exp:* Metallurgist, Sunbeam Elec Mfg Co, 39-41; asst metall, Mass Inst Technol, 47-50; sr scientist, Westinghouse Atomic Power Div, 50-52, supvry scientist, 53-54; metall specialist, Gen Tel & Electronics Labs, Inc, Sylvania Elec Prod, Inc, 54-64; adj prof, 55-64, PROF PHYS METALL, POLYTECH INST NEW YORK, 64- *Concurrent Pos:* Res assoc, Sch Mines, Columbia Univ, 57-58; chmn fac senate, Polytech Inst New York. *Mem:* Am Inst Mining, Metall & Petrol Engrs; Am Phys Soc; fel AAAS; Sigma Xi. *Res:* Diffusional phenomena; electronic materials; biomaterials. *Mailing Add:* Polytech Inst New York 333 Jay Brooklyn NY 11201

CASTLES, JAMES JOSEPH, JR, b New Brunswick, NJ, Oct 3, 40; m 78; c 3. EDUCATION ADMINISTRATION, MEDICINE. *Educ:* Univ Chicago, BS, 61, MD, 64. *Prof Exp:* Intern-resident med, Columbia-Presby Hosp, NY, 64-66; res assoc biochem, NIH, 66-68; asst prof med, Univ Chicago, 68-73; assoc prof, 73-79, PROF MED & ASSOC DEAN, UNIV CALIF, DAVIS, 79- *Concurrent Pos:* Guggenheim fel, 73. *Mem:* Am Soc Biol Chemists; Am Rheumatism Asn. *Res:* Clinical features of the immune system as it applies to rheumatic disease. *Mailing Add:* Deans Off Sch Med Univ Calif Davis CA 95616

CASTLES, THOMAS R, b St Louis, Mo, Oct 27, 37; m 59; c 3. PHARMACOLOGY. *Educ:* Grinnell Col, BA, 59; Univ Iowa, MS, 62, PhD(pharmacol), 65; Am Bd Toxicol, dipl, 81. *Prof Exp:* From assoc pharmacologist to prin pharmacologist, 65-76, mgr, Pharmaceut Toxicol Lab, Midwest Res Inst, 76-77; MGR TOXICOL LAB, IMP CHEM INDUST, 77- *Mem:* Am Soc Pharmacol & Exp Therapeut; Soc Am Toxicol. *Res:* Thiazide diuretics; aldosterone; carbonic anhydrose inhibitors; pharmacology and toxicology of antimalarial compounds; analgesic tolerance; acute toxicology. *Mailing Add:* Dept Toxicol Stauffer Chem Co 1200 South 47th Richmond CA 94804

CASTLETON, KENNETH BITNER, b Salt Lake City, Utah, July 29, 03; m 31; c 4. SURGERY. *Educ:* Univ Utah, AB, 23, Univ Pa, MD, 27; Univ Minn, PhD(surg), 33; Am Bd Surg, dipl. *Hon Degrees:* LHD, Univ Utah, 75. *Prof Exp:* Instr anat, 33-34, phys diagnosis, 35-36, surg anat, 38-43, assoc clin prof surg, 43-62, prof surg & dean, Col Med, 62-69, vpres med affairs, 69-71, EMER PROF SURG, UNIV UTAH, 71- *Concurrent Pos:* Pvt pract, 33-62. *Mem:* Fel Am Col Surg. *Res:* Clinical surgery; gastroenterology; experimental physiology. *Mailing Add:* 1235 E 200 S Salt Lake City UT 84102

CASTNER, JAMES LEE, b Orange, NJ, June 1, 54; m 81. TROPICAL BIOLOGY, SCIENTIFIC PHOTOGRAPHY. *Educ:* Rutgers Univ, BS, 76; Univ Fla, MS, 83, PhD(entom), 88. *Prof Exp:* SCI PHOTOGR, UNIV FLA, 90- *Concurrent Pos:* Prin investr, Earthwatch, 87-90; res journalist, Feline Press, 88-90. *Mem:* Entom Soc Am; Sigma Xi; Asn Trop Biol. *Res:* Study of the behavior and taxonomy of neotropical leaf-mimicking katydids; camouflage and mimicry; ecotourism; author. *Mailing Add:* Dept Entom Univ Fla Gainesville FL 32611

CASTNER, THEODORE GRANT, JR, b Orange, NJ, June 17, 30; m 67; c 2. SOLID STATE PHYSICS. *Educ:* Cornell Univ, BEng, 53; Univ Ill, MS, 55, PhD(physics), 58. *Prof Exp:* Physicist, Gen Elec Res Labs, 58-63; assoc prof, 63-70, PROF PHYSICS, UNIV ROCHESTER, 70- *Concurrent Pos:* Guggenheim fel, Swiss Fed Inst Technol, 69-70 & Sandia Labs, 76-77; vis scientist, Francis Bitter Nat Magnet Lab; vis prof, Tufts Univ, 83-84. *Mem:* Fel Am Phys Soc; Int Solar Soc. *Res:* Electron spin resonance phenomena in solids; impurities in insulators, semiconductors; cooperative phenomena in magnetic materials; spin-lattice relaxation; dielectric phenomena and metal-insulator transition in semiconductors; magnetoelectric effect; magnetotransport phenomena. *Mailing Add:* Dept Physics & Astron Univ Rochester Rochester NY 14627

CASTO, CLYDE CHRISTY, analytical chemistry, agriculture & food chemistry; deceased, see previous edition for last biography

CASTON, J DOUGLAS, b Ellenboro, NC, June 16, 32; m 58; c 3. BIOCHEMISTRY, EMBRYOLOGY. *Educ:* Lenoir-Rhyne Col, BA, 54; Univ NC, MA, 58; Brown Univ, PhD(biol), 61. *Prof Exp:* From sr instr to assoc prof, 63-75, PROF ANAT, SCH MED, CASE WESTERN RESERVE UNIV, 75- *Concurrent Pos:* Fel develop biol, Carnegie Inst of Wash, 61-63; Am Cancer Soc grants; USPHS grants. *Mem:* Biophys Soc; Am Soc Gen Physiol; Soc Develop Biol. *Res:* Control of metabolic pathways during development; catecholamine synthesis; protein and ribonucleic acid synthesis; folate binding macromolecules. *Mailing Add:* Dept of Anat Sch of Med Case Western Reserve Univ Cleveland OH 44106

CASTON, RALPH HENRY, b Akron, Ohio, Nov 22, 15; m 46; c 3. PHYSICS. *Educ:* Univ Akron, MS, 39; Univ Notre Dame, PhD(physics), 42. *Prof Exp:* Mem staff, Nat Defense Res Comt Proj, Radiation Lab, Mass Inst Technol, 42-45; mem tech staff, Kimberly-Clark Res & Develop Lab, 45-81; RETIRED. *Mem:* Am Phys Soc. *Res:* Physics of rubber; radar; general physics in the paper industry; operations research; computer process control. *Mailing Add:* 638 Stevens St Neenah WI 54956

CASTONGUAY, RICHARD NORMAN, b Worcester, Mass, Dec 22, 39; m 73; c 2. POLYMER APPLICATIONS, MATERIALS & PROCESS ENGINEERING. *Educ:* Col Holy Cross, BS, 62; Cornell Univ, PhD(phys chem), 67. *Prof Exp:* Corp tech dir, MICA Corp, 68-78; mgr, Elec, Electronic & Matrix Develop & Appl Labs, Resins Dept, Ciba Geigy Corp, 78-83; DIR RES & DEVELOP, HICKS & OTIS PRINTS, INC, 84- *Concurrent Pos:* Consult, Gallil Corp, 74-77, Ailtech Div, Cutler Hammer, 77-78 & Am Cyanamid Co, 83-84. *Mem:* Am Phys Soc; Int Soc Hybrid Microelectronics; Soc Advan Mat & Process Eng; Inst Elec & Electronics Engrs. *Res:* Polymer chemistry; polymeric material characteristics including electrical, electronic and structural; resistance materials characteristics. *Mailing Add:* 18 Pond Crest Rd Danbury CT 06811

CASTONGUAY, THOMAS T(ELISPHORE), chemical engineering, physical chemistry; deceased, see previous edition for last biography

CASTOR, CECIL WILLIAM, b Detroit, Mich, Oct 9, 25; m 48; c 3. INTERNAL MEDICINE, RHEUMATOLOGY. *Educ:* Univ Mich, MD, 51; Am Bd Internal Med, dipl, 58. *Prof Exp:* Intern & resident internal med, 51-55; from instr to assoc prof internal med, 55-67, PROF INTERNAL MED, MED SCH, UNIV MICH, ANN ARBOR, 67- *Concurrent Pos:* Arthritis & Rheumatism Found fel, 55-57, sr investr, 57-62; USPHS career res develop award, 63-67; mem study sect gen med, NIH, 70-74; fel comt & prof educ comt, Arthritis Found; distinguished fac lectr, Biomed Res Coun, 91. *Honors & Awards:* Int Geigy Rheumatism Award, 15th Int Cong Rheumatol France, 81; Lee Howley Prize for Arthritis Res, Nat Arthritis Found, 85. *Mem:* Fel Am Col Physicians; Am Soc Clin Invest; Am Fedn Clin Res; Am Rheumatism Asn; Tissue Cult Asn; Sigma Xi; Asn Am Physicians. *Res:* Regulation of connective tissue cells by growth factors (CTAPs), in vitro and in vivo, especially with respect to control of glycosaminohycan and DNA synthesis. *Mailing Add:* Internal Med Arthritis Res Med Sch Univ Mich R4750 Kresge 1 Box 0531 Ann Arbor MI 48109-0531

CASTOR, JOHN I, b Fresno, Calif, Jan 5, 43; m 77; c 2. ASTROPHYSICS. *Educ:* Fresno State Col, BS, 61; Calif Inst Technol, PhD(astron), 67. *Prof Exp:* Res fel physics, Calif Inst Technol, 66-67; res assoc astrophys, Univ Colo, Boulder, 67-69, from asst prof to prof physics & astrophys, 69-81, fel Joint Inst Lab Astrophys, 70-81; PHYSICIST, LAWRENCE LIVERMORE NAT LAB, 81- *Mem:* Fel Royal Astron Soc; Am Astron Soc; Int Astron Union. *Res:* Stellar interiors; pulsating stars; radiative transfer; stellar radiation hydrodynamics. *Mailing Add:* L-23 Lawrence Livermore Nat Lab PO Box 808 Livermore CA 94550

CASTOR, LAROY NORTHROP, b Philadelphia, Pa, Sept 27, 24; m 53; c 4. CELL BIOLOGY, CELL CYCLE. *Educ:* Mass Inst Technol, BS, 48; Univ Pa, PhD(biophys), 54. *Prof Exp:* Am Cancer Soc fel, Univ Toronto, 55; fel, Univ Pa, 56-57; assoc biophysics, Johnson Res Found, 57-63; sr investr cell biol, Biochem Res Found, Del, 63-66; asst mem, 66-78, assoc mem, 78-85, EMER ASSOC MEM, INST CANCER RES, 86- *Res:* Protein synthesis and degradation in relation to control of the cell cycle; cell kinetics; in vitro sensitivity of human tumors to anti-cancer drugs. *Mailing Add:* 3898 Dempsey Lane Huntingdon Valley PA 19006

CASTOR, WILLIAM STUART, JR, b Granville, Ohio, May 23, 26; m 48; c 4. PIGMENTS & FILLERS, PARTICLE-SIZE ASSESSMENT. *Educ:* Northwestern Univ, BS, 47, PhD(chem), 50. *Prof Exp:* Res & develop chemist, Calco Chem Div, Am Cyanamid Co, 50-55, asst Dir Tech Serv, Pigments Div, 55-56, asst plant mgr, Savannah, 57-58, dir res & develop, Pigments Div, 58-61, tech dir, 61-68, dir res & develop, Res Div, 69-72; mgr develop, NJ Zinc Co, Gulf & Western, 74-75, mgr res & develop, 75-79, mgr res & develop & tech serv, Gulf & Western Industs, 80-81, dir res & develop, 82-83; OWNER, PAR TEC LABS, 84- *Concurrent Pos:* Chmn, Educ Comt, Northern Highlands Regional Bd Educ, 67-73; tech consult, 73-74. *Mem:* Am Chem Soc; Am Soc Testing & Mat; Oil & Colour Chemists' Asn; Soc Chem Indust; Tech Asn Pulp & Paper Indust; AAAS; Asn Res Dirs (pres, 77-78). *Res:* Measurement and interpretation of particle size distributions, and methods of controlling and/or modifying them. *Mailing Add:* 111 Schuyler Rd Allendale NJ 07401

CASTORE, GLEN M, DIFFERENTIAL GEOMETRY. *Prof Exp:* Res scientist, David Taylor Naval Res & Develop Ctr, 81-83; res scientist, Nat Bur Standards, 83-84; proj mgr, Control Data, 84-85; sr prin res scientist, Honeywell Systs & Res Ctr, 85-87; PRES, APEIRON INC, 88- *Mem:* Am Math Soc; Soc Indust & Appl Math; Inst Elec & Electronic Engrs. *Res:* Geometric interpretation of quantum theories; robot vision; mathematical models of metal cutting. *Mailing Add:* Apeiron Inc 230 Tenth Ave S Minneapolis MN 55415

CASTRACANE, V DANIEL, b Philadelphia, Pa, Aug 16, 40; m 77; c 2. REPRODUCTIVE PHYSIOLOGY, ENDOCRINOLOGY. *Educ:* Temple Univ, BA, 62; Villanova Univ, MS, 66; Rutgers Univ, PhD(zool), 72. *Prof Exp:* Fel steroid biochem, Worcester Found Exp Biol, 72-73; fel reprod physiol, Sch Med, Case Western Reserve Univ, 73-76; assoc found scientist reproductive endocrinol, Southwest Found Res & Educ, 76-86; PROF & DIR LAB, TEX TECH UNIV, 86- *Mem:* Endocrine Soc; Soc Study Reprod; Am Soc Primatologists. *Res:* Utero-ovarian relationships; pubertal endocrinology; corpus luteum; endocrinology of pregnancy. *Mailing Add:* Tex Tech Univ Health Sci Ctr Dept Obstet & Gynec 1400 Wallace Blvd Amarillo TX 79106

CASTRANOVA, VINCENT, b Trenton, NJ, March 18, 49; m 70; c 1. CELLULAR & RESPIRATORY PHYSIOLOGY, INHALATION TOXICOLOGY. *Educ:* Mt St Marys Col, BS, 70; WVa Univ, PhD(physiol), 74. *Prof Exp:* Fel physiol, WVa Univ, 70-74; fel, Yale Univ, 74-76, res fac, 76-77; from asst prof to assoc prof, 77-85, PROF, WVA UNIV, 85- *Concurrent Pos:* Res physiologist, Nat Inst Occup Safety & Health, 77-83, chief, Biochem Sect, 83-, actg chief, Path Sect, 89- *Mem:* Am Physiol Soc; Sigma Xi; Soc Toxicol. *Res:* Isolation and characterization of lung cells; surfactant release by alveolar type II epithelial cells; excitation-secretion coupling in alveolar macrophages and polymorphonuclear leukocytes; volume regulation by pneumocytes; toxicity of metals and occupational dusts on lung cells; inhalation toxicology. *Mailing Add:* Biochem Sect ALOSH 944 Chestnut Ridge Rd Morgantown WV 26505

CASTRIC, PETER ALLEN, b Pendleton, Ore, Sept 26, 38; m 65; c 1. MICROBIAL PHYSIOLOGY, MOLECULAR BIOLOGY. *Educ:* Ore State Univ, BS, 61; Mont State Univ, PhD(microbiol), 69. *Prof Exp:* Fel microbiol, Dept Bot & Microbiol, Mont State Univ, 64-69, biochem, Dept Biochem & biophysics, Univ Calif, Davis, 69-70 & Dept Biochem, Univ Mich, Ann Arbor, 70-71; from asst prof to assoc prof biol & microbiol, 71-80, PROF MICROBIOL, DEPT BIOL SCI, DUQUESNE UNIV, PITTSBURGH, 80- *Concurrent Pos:* Nat Res Coun sr assoc, Walter Reed Army Inst Res, 85-86. *Mem:* Am Soc Biochem & Molecular Biol; Am Soc Microbiol; AAAS. *Res:* Bacterial cyanide metabolism; molecular biology pseudomonas aeruginosa. *Mailing Add:* 620 N Meadowcraft Ave Pittsburgh PA 15216

CASTRILLON, JOSE P A, b Buenos aires, Arg, Jan 4, 26; US citizen; m 64; c 2. ORGANIC CHEMISTRY, RADIOCHEMISTRY. *Educ:* Univ Buenos Aiers, Dr(chem), 51. *Prof Exp:* Scientist chem, Atanor, SAM, Arg, 50-52, Squibb, SA, 53-55, Arg AEC, 56-64 & PR Nuclear Ctr, 64-73; div head phys sci, PR Nuclear Ctr, 73-76; MEM FAC & HEAD DEPT CHEM, PAN AM UNIV, 76- *Concurrent Pos:* Fel Arg Coun, Univ Colo, 60-61; prof org chem, Univ Asuncion, 71-72. *Mem:* Am Chem Soc; Arg Chem Soc. *Res:* Synthesis of sulfur heterocycles as potential schistosomicides, trypanosomicides and antitumor agents; chemistry of sulfoxides and sulfones; chiral sulfur compounds. *Mailing Add:* Dept Chem Pan Am Univ Edinburg TX 78539-2999

CASTRO, ALBERT JOSEPH, b Santa Clara, Calif, May 13, 16; m 37; c 3. ORGANIC CHEMISTRY. *Educ:* San Jose State Col, AB, 39; Stanford Univ, AM, 42, PhD(org chem), 45. *Prof Exp:* Res chemist, Calif Res Corp, 44-47; asst prof chem, Univ Ariz, 47-49; from asst prof to assoc prof, 49-58, prof chem, 58-85, EMER PROF SAN JOSE STATE UNIV, 85- *Concurrent Pos:* Asst, Univ Santa Clara, 49; consult, Nassau Chems, Inc, 49-53 & Marlyn Co, Inc, 52; vis lectr, Johns Hopkins Univ, 54-55; org chem subcomt, exam comt, Am Chem Soc, 71-86; vis scholar, Stanford Univ, 78; Nat Sci Found reviewer, Chem Div, Chem Synthesis & Anal Sect, 79; adj hoc consult Minority Biomed Support Prog, NIH, 79 & 82, Contracts Rev Br, Develop Therapeut Contract Rev Comt, Nat Cancer Inst, 83-84. *Mem:* Fel Am Inst Chemists; Am Chem Soc; Brit Chem Soc; Sigma Xi. *Res:* Composition of ether solutions of the Grignard reagent; reaction of epoxides with ammonia and metallopyroles; synthesis of arylethylenes; Friedel-Crafts reaction; natural products; chemistry of pyrroles; compounds of possible pharmacological value; reaction of tertiary amines with arylsulfonyl chlorides. *Mailing Add:* Dept Chem San Jose State Univ San Jose CA 95192

CASTRO, ALBERTO, biochemistry, endocrinology; deceased, see previous edition for last biography

CASTRO, ALFRED A, b Buenos Aires, Arg, July 5, 32; US citizen; m 66; c 1. SATELLITE COMMUNICATIONS, SIGNAL PROCESSING. *Educ:* La Plata Univ, Arg, Engr electromech, 56, Engr telecommun, 58; Columbia Univ, MS, 62, EED, 66. *Prof Exp:* Engr, Transradio Int, 56-58 & RCA Int, 58-60; prin engr, Litton Industs, 60-62 & 63-66; teaching asst electronics, Columbia Univ, 62; sect head, Gen Tel-Sylvania, 66-68; mem tech staff, Gen Tel Labs, 69; MGR, MIL COMSAT PROGS, RAYTHEON CO, 69- *Mem:* Inst Elec & Electronics Engrs; Inst Elec & Electronics Engrs Commun Soc. *Res:* Satellite communication systems and equipment; advanced technology, including microwaves, spread spectrum modulation, on board signal processing, adaptive mulling antennas; author of 22 publications. *Mailing Add:* 20 Hunt Rd Sudbury MA 01776

CASTRO, ANTHONY J, b Chicago, Ill, Nov 30, 30; m 81; c 3. ORGANIC POLYMER CHEMISTRY. *Educ:* Univ Chicago, MS, 58, PhD(chem), 62; John Marshall Law Sch, JD, 75. *Prof Exp:* Res supvr polymer chem, Continental Can Co, 62-65; mgr polymer chem, 65-80, ASST DIR RES, ARMAK CO, MCCOOK, 80- *Mem:* AAAS; NY Acad Sci; Am Chem Soc; Soc Plastics Engrs. *Res:* Investigation into areas of unconventional polymer formation; preparation of novel polymeric materials from conventional polymerision techniques; preparation of novel microporous thermoplastics. *Mailing Add:* 2035 Ninth Ave San Francisco CA 94116

CASTRO, CHARLES E, b Santa Clara, Calif, Nov 17, 31; m 53; c 7. BIOCHEMISTRY, INORGANIC CHEMISTRY. *Educ:* San Jose State Col, AB, 53; Univ Calif, Davis, PhD(phys org chem), 57. *Prof Exp:* Res chemist, Shell Develop Co, Calif, 57-60; from asst chemist to assoc chemist, 60-70, CHEMIST & PROF BIOL CHEM, UNIV CALIF, RIVERSIDE, 70- *Concurrent Pos:* Fulbright Scholar, Univ Heidelberg, Ger, 53-54. *Mem:* Am Chem Soc. *Res:* Reactions of organics with transition metal species; anthelmintics; biodehalogenation; iron porphyrin and hemeprotein redox mechanism, nematode semio chemistry. *Mailing Add:* Dept Nematol Univ Calif Riverside CA 92521

CASTRO, GEORGE, b Los Angeles, Calif, Feb 23, 39; m 63; c 4. RESEARCH MANAGEMENT, PHYSICAL CHEMISTRY. *Educ:* Univ Calif, Los Angeles, BS, 60; Univ Calif, Riverside, PhD(chem), 65. *Prof Exp:* Fel chem, Univ Pa, 65-67 & Calif Inst Technol, 67-68; staff mem res, IBM San Jose Res Lab, 68-69, proj leader org photoconductors, 69-73, mgr org solids, 73-75, mgr phys sci, 75-86; MGR SYNCHROTRON STUDIES, IBM ALMADEN RES CTR, 86- *Mem:* Fel Am Phys Soc; Am Chem Soc; AAAS; Hisp Prof Engrs. *Res:* Electronic properties of organic solids. *Mailing Add:* Almaden Res Ctr 650 Harry Rd San Jose CA 95120-6099

CASTRO, GILBERT ANTHONY, b Port Arthur, Tex, Apr 24, 39; m 61; c 2. PHYSIOLOGY, PARASITOLOGY. *Educ:* Lamar State Col, BS, 61; Univ Ark, MS, 63; Univ Tex, PhD(microbiol), 66. *Prof Exp:* NIH fel zool, Univ Mass, 66-68; from asst prof to assoc prof parasitol & lab pract, Med Ctr, Univ Okla, 69-72; assoc prof, 72-77, PROF PHYSIOL, UNIV TEX MED SCH HOUSTON, 77- *Mem:* Am Physiol Soc; Am Soc Parasitol. *Res:* Intestinal physiology; host-parasite relationships; physiology and immunology of host-parasite systems; pathogenesis of gastrointestinal parasites. *Mailing Add:* Dept Physiol Univ Tex Med Sch Houston TX 77225

CASTRO, GONZALO, b Lima, Peru; m 87; c 1. CONSERVATION, BIODIVERSITY. *Educ:* Univ Cayetano Heredia, Lima, BS, 83, MS, 85; Univ Pa, PhD(ecol), 88. *Prof Exp:* Lectr biol, Univ Cayetano Heredia, 83-85; teaching asst, Univ Pa, 85-88; res asst, Acad Natural Sci, Philadelphia, 85-88; postdoctoral fel, Colo State Univ, 88-90; PROG MGR, MANOMET BIRD OBSERV, 90- *Concurrent Pos:* Mem, Pan Am Shorebird Prog, 85- & Eskimo Curlew Adv Group, 91-; reviewer, numerous journals, 85-; vchmn & mem bd dirs, Int Coun Bird Preserv, 90-; consult, Troy Ecol Res Assocs, 90- *Honors & Awards:* Alexander Wilson Prize, Wilson Ornith Soc, 88. *Mem:* AAAS; Am Inst Biol Sci; Soc Conserv Biol; Ecol Soc Am; Cooper Ornith Soc; Am Ornith Union. *Res:* Ecology and conservation of migratory birds in the western hemisphere; social, political, ecological and economic factors and implications of biodiversity conservation in Latin America; global change. *Mailing Add:* WHSRN-Manomet Bird Observ PO Box 936 Manomet MA 02345

CASTRO, PETER, b Mayaguez, PR, July 20, 43; US citizen. MARINE ZOOLOGY, PARASITOLOGY. *Educ:* Univ PR, Mayagüez, BS, 64; Univ Hawaii, MS, 66, PhD(zool), 69. *Prof Exp:* Res assoc, Dept Marine Sci, Univ PR, Mayagüez, 70; instr, Dept Biol Sci, Univ PR, Rio Piedras, 70-71; from asst prof to assoc prof, 72-80, PROF BIOL SCI, CALIF STATE POLYTECH UNIV, POMONA, 80- *Concurrent Pos:* Vis investr, Hopkins Marine Sta, Stanford Univ, 71-82 & Smithsonian Trop Res Inst, 74; lectr, Moss Landing Marine Labs, 73; res dir, Int Prog Spain, Calif State Univ, 81-82; sabbatical leave, Spain, 78 & 86, Italy, 86; Fulbright scholar, Odessa State Univ, USSR, 85. *Mem:* AAAS; Am Soc Zoologists; Sigma Xi. *Res:* Ecological, physiological and behavioral aspects of marine symbioses. *Mailing Add:* Dept Biol Calif Polytechnic Univ Pomona CA 91768

CASTRO, PETER S(ALVATORE), b Chicago, Ill, July 5, 26; m 58; c 1. ELECTRICAL ENGINEERING. *Educ:* Northwestern Univ, BSEE, 51, MSEE, 56, PhD(elec eng), 59. *Prof Exp:* Elec engr, Motorola, Inc, Ill, 51-53; res scientist, Lockheed Res Lab, 59-72; chief engr, Spectrotherm Corp, 72-76; chief engr, Raytek, 76-85; PROJ MGR, VIDEO ENG, SIERRA SCI, 85- *Mem:* AAAS; Inst Elec & Electronics Engrs; NY Acad Sci; Sigma Xi. *Res:* Circuit theory, mainly of distributed parameter networks; analog applications of thin magnetic films; infrared optics and instrumentation-thermography; medical video displays. *Mailing Add:* 901 Madonna Way Los Altos CA 94024-4623

CASTRO, WALTER ERNEST, b Peekskill, NY, Dec 31, 34; m 58; c 3. FLUID & ENGINEERING MECHANICS. *Educ:* State Univ NY Agr & Tech Inst, Delhi, AAS, 54; Ind Inst Technol, BSME, 59; Clemson Univ, MSME, 62; Univ WVa, PhD (eng mech), 66. *Prof Exp:* Instr mech eng, 59-61, asst prof eng mech, 61-62 & 65-67, assoc prof, 67-73, PROF MECH ENG & MECH, CLEMSON UNIV, 73-, ASST DEAN UNDERGRAD STUDIES, COL ENG, 84- *Concurrent Pos:* NASA summer fac fel, Marshall Space Flight Ctr, 68. *Mem:* Am Soc Eng Educ; Am Soc Mech Engrs. *Res:* Experimental mechanics; turbulence; transition; two phase flow; turbulence suppression using high polymer additives; fluid power; hydraulic machinery. *Mailing Add:* Dept Mech Eng Clemson Univ Clemson SC 29631

CASTRONOVO, FRANK PAUL, JR, b Newark, NJ, Jan 2, 40; m 77; c 3. MEDICAL PHYSICS, RADIOPHARMACOLOGY. *Educ:* Rutgers Univ, Newark, BS, 62, New Brunswick, MS, 64; Regist Pharmacist, 65; Johns Hopkins Univ, PhD(radiol sci), 70; Am Bd Radiol, dipl, 78; Am Bd Sci Nuclear Med, dipl, 79. *Prof Exp:* Res sci radiopharmacol, Squibb Inst Med Res, 64-65; asst physicist, 70-80, radiation safety officer, 80-, RADIOPHARMACOLOGIST, MASS GEN HOSP, 80-, asst prof radiol, Harvard Med Sch, 76-86; asst prof, 76-86, ASSOC PROF RADIOL, HARVARD MED SCH, 86- *Concurrent Pos:* Consult med physics & radiopharmaceut sci, Boston hosps, 74-; adj clin prof radiopharmacol, Mass Col Pharm, 76-; vis scientist, Mass Inst Technol, 75- *Honors & Awards:* Nat Res Award, Parenteral Drug Asn, 69. *Mem:* Soc Nuclear Med; Radiopharmaceut Sci Coun; Am Asn Physicists in Med; Am Pharmaceut Asn; Am Soc Hosp Pharmacists. *Res:* Research and development of radiopharmaceuticals; clinical radiopharmacology; quantitative skeletal imaging; ultra-short-lived generator-produced radionuclides; medical physics dosimetry; hospital personnel radiation exposure. *Mailing Add:* Dept of Radiol Mass Gen Hosp Boston MA 02115

CASTRUCCIO, PETER ADALBERT, b New York, NY, Jan 11, 25; m 51. GENERAL EARTH SCIENCES. *Educ:* Univ Genoa, DrEng(elec eng), 46. *Prof Exp:* Asst proj engr, Bendix Radio, Md, 46-50; proj engr, Aircraft Armaments, Inc, 50-55; head preliminary design, Westinghouse Elec Corp, Md, 55-58, dir astronaut, 58-59; tech dir, Aeronca Mfg Co, 59-61; dir adv space progs, Fed Systs Div, IBM Corp, Bethesda, 61-68, dir advan progs, Ctr Sci Studies, Gaithersburg, 68-73; PRES, ECOSYSTS INT, INC, 73- *Concurrent Pos:* Lectr, Space Inst, Univ Md, 59 & 60. *Honors & Awards:* Award, Am Inst Aeronaut & Astronaut, 58. *Mem:* Sr mem Inst Elec & Electronics Engrs; Am Astronaut Soc; Am Inst Aeronaut & Astronaut. *Res:* Space technology, guidance and control, systems architecture, optical and inertial sensors, communications techniques and land survey techniques from ultrahigh flying platforms; environmental modeling. *Mailing Add:* Ecosysts Int Inc PO Box 225 Gambrills MD 21054

CASWELL, ANTHONY H, b Exeter, Eng, Sept 23, 40. CELL BIOLOGY, MUSCLE. *Educ:* Cambridge Univ, Eng, BA, 62; Univ Col Wales, Britain, PhD, 66. *Prof Exp:* Asst prof med, biochem & pharmacol, 70-73, assoc prof, 73-81, PROF PHARMACOL, MED SCH, UNIV MIAMI, 81- *Res:* Mechamism of excitation-contraction coupling in muscle; transverse tubules from skeletal muscle; preparing transverse-tubule vesicles from skeletal muscle with limited contamination from other sources; investigation of transverse-tubules and their interaction with sarcoplasmic reticulum for an understanding of muscle function and contraction. *Mailing Add:* Dept Pharm Sch Med R 189 Univ Miami 1600 NW 10th Ave Miami FL 33101

CASWELL, GREGORY K, b Somerville, NJ, June 13, 47; m 69, 91; c 3. SURFACE MOUNT TECHNOLOGY, MULTICHIP MODULES. *Educ:* Sr technician, Burroughs Corp, 67-73. *Prof Exp:* mem tech staff, Radio Commun Asn-Solid State Technol Ctr, 73-77; chief engr, Tracor Aerospace, 77-86; pres, Caswell Consult, 86-87; DIR, SMT SERV, GALAXY MICROSYSTS, INC, 87- *Mem:* Int Soc Hybrid Microelectronics. *Res:* CMOS and CMOS/SOS radiation hardened process development, input protection networks, advanced surface mount technology, ferrite impregnated printed wiring board technology and multichip modules; over 90 publications in these areas. *Mailing Add:* 1907 Key West Cove Austin TX 78746

CASWELL, HAL, b Los Angeles, Calif, Apr 27, 49; div; c 1. ECOLOGY, MATHEMATICAL BIOLOGY. *Educ:* Mich State Univ, BS, 71, PhD(zool), 74. *Prof Exp:* NSF fel zool, Mich State Univ, 71-74, res assoc, 74-75; from asst prof to assoc prof biol sci, Univ Conn, 75-81; assoc scientist, 81-88, SR SCIENTIST, WOODS HOLE OCEANOG INST, 88- *Concurrent Pos:* Res assoc, Univ Calif, Berkeley, 80-81; vis prof math & ecol, Univ Tenn, 87, distinguished vis prof biol, Univ Miami, 89, vis prof, Univ of the Andes, Merida, Venezuela, 89. *Mem:* Ecol Soc Am; fel AAAS; Population Asn Am; Brit Ecol Soc; Am Soc Naturalists. *Res:* Theoretical population and community ecology; evolutionary demography and life history theory; matrix population models; conservation biology. *Mailing Add:* Biol Dept Woods Hole Oceanog Inst Woods Hole MA 02543

CASWELL, HERBERT HALL, JR, b Marblehead, Mass, May 21, 23; m 48; c 6. ECOLOGY, ORNITHOLOGY. *Educ:* Harvard Univ, SB, 48; Univ Calif, Los Angeles, MA, 50; Cornell Univ, PhD(zool), 56. *Prof Exp:* PROF BIOL, EASTERN MICH UNIV, 55-, HEAD DEPT, 74- *Mem:* Ecol Soc Am; Wilson Ornith Soc; Am Ornith Union; Am Inst Biol Sci; Sigma Xi. *Res:* Terrestrial ecology. *Mailing Add:* 952 Sheridan Ypsilanti MI 48197

CASWELL, JOHN N(ORMAN), b Brooklyn, NY, June 26, 20; m 46; c 2. ELECTRICAL ENGINEERING. *Educ:* Cooper Union, BEE, 41; Brooklyn Polytech Inst, MEE, 49. *Prof Exp:* From asst proj engr to sr proj engr, Sperry Gyroscope Co, Great Neck, 41-52, head eng sect, 52-57, head eng dept, 57-62, mgr Polaris systs opers, 62-67, equip eng mgr Polaris/Poseidon prog Systs Mgt Div, 67-70; prog mgr safeguard, Sperry Corp, 81, 70-73, mgr spec studies, 73-74, poseidon extended operating cycle prog mgr, 74-81, automated mat handling systs prog control mgr,; RETIRED. *Mem:* Am Inst Elec & Electronics Engrs. *Res:* Radar systems requiring precise three coordinate automatic tracking for fire control applications and precise navigation systems. *Mailing Add:* 17730 Villamoura Dr Poway CA 92064

CASWELL, LYMAN RAY, b Omaha, Nebr, Sept 29, 28; m 64; c 3. ORGANIC CHEMISTRY. *Educ:* Ind Univ, BS, 49, MA, 50; Mich State Univ, PhD(org chem), 56. *Prof Exp:* Asst prof chem, Ohio Northern Univ, 55-56; head, Chem Dept, Upper Iowa Univ, 56-61; from asst prof to assoc prof, 61-68, actg chmn dept, 67-70 & 78-79, chmn dept, 79-82, PROF CHEM, TEX WOMAN'S UNIV, 68- *Mem:* AAAS; Am Chem Soc; Hist Sci Soc. *Res:* Aromatic nucleophilic subsitution reactions; cyclic imides and anhydrides; ultraviolet visible and fluorescence spectra; dipole moments. *Mailing Add:* Dept Chem Tex Woman's Univ Denton TX 76204

CASWELL, RANDALL SMITH, b Eugene, Ore, Feb 7, 24; m 45; c 6. PHYSICS. *Educ:* Mass Inst Technol, SB, 47, PhD(physics), 51. *Prof Exp:* Assoc prof physics, Univ Ky, 50-52; res partic solid state physics, Oak Ridge Nat Lab, 52; physicist neutron physics, Nat Bur Standards, 52-69, dep dir, Ctr Radiation Res, 69-78, chief nuclear radiation div, 78-85; CHIEF, IONIZING RADIATION DIV, NAT INST STANDARDS & TECHNOL, 85- *Concurrent Pos:* Adj prof physics, Am Univ, 57-71; mem, Nat Coun Radiation Protection & Measurement, 67-; chmn neutron measurements sect, Adv Comt, Standards Ionizing Radiation Measurement, 69-89; mem, Int Comn Radiation Units & Measurement, 75-, secy, 79-; assoc ed, Radiation Res, 77-80; chmn, sci panel, Comt Interagency Radiation Res & Policy Coord, Off Sci & Technol Policy, 84- *Mem:* Fel Am Phys Soc; Radiation Res Soc. *Res:* Physics and dosimetry; theoretical physics; neutron cross sections. *Mailing Add:* Radiation Phys Bldg C210 Nat Inst Standards & Technol Gaithersburg MA 20899

CASWELL, ROBERT LITTLE, b San Francisco, Calif, Jan 27, 18; m 57; c 1. PESTICIDE CHEMISTRY. *Educ:* Univ Calif, Berkeley, BS, 39. *Prof Exp:* Chemi pesticides, Agr Res Serv, USDA, 41-42 & 46-64, asst chief staff, Off Enforcement Chem, Pesticides Regulation Div, 64-70, Off Pesticides, 70-72, chemist, Criteria & Eval Div, Off Pesticides, Environ Protection Agency, 72-76; CONSULT, 76- *Mem:* AAAS; Am Chem Soc. *Res:* Development of chemical methods of analysis for pesticides. *Mailing Add:* 11626 35th Pl Beltsville MD 20705

CATACOSINOS, PAUL ANTHONY, b New York, NY, Sept 29, 33; m 58; c 3. PETROLEUM GEOLOGY, STRATIGRAPHY. *Educ:* Univ NMex, BA, 57, MS, 62; Mich State Univ, PhD(geol), 72. *Prof Exp:* Explor geologist, Mountain Fuel Supply Co, 62-66; explor geologist, Consumers Power Co, 67-69; from instr to assoc prof geol, Delta Col, 69-80; PROF GEOL, DELTA COL, 80- *Concurrent Pos:* Vpres, Oil & Gas Corp, 82-83. *Honors & Awards:* Bergstein Award, Delta Col, 72 & 88. *Mem:* Am Asn Petrol Geologists; fel Geol Soc Am; Soc Sedimentary Geol; Sigma Xi; Am Astron Soc; fel AAAS. *Res:* Origin and evolution of the Michigan Basin; detailed analysis of the Cambrian and Ordovician stratigraphic sections; Precambrian sediments of Michigan; stratigraphy and petroleum geology of the Michigan Basin. *Mailing Add:* Dept Geol Delta Col University Center MI 48710

CATALANO, ANTHONY WILLIAM, b Brooklyn, NY, Feb 8, 47; m 72. SOLID STATE CHEMISTRY, PHYSICAL CHEMISTRY. *Educ:* Rensselaer Polytech Inst, BS, 68; Brown Univ, PhD(chem), 72. *Prof Exp:* Fel, Div Eng, Brown Univ, 72-74; staff scientist, Inst Energy Conversion, Univ Del, 79-81, mgr mat develop & analysis, 79-81; MEM TECH STAFF, RCA LABS, PRINCETON, NJ, 81- *Mem:* Am Chem Soc; Am Vacuum Soc. *Res:* Solid state chemistry, particularly electrical and optical properties of new materials with application to photovoltaic devices; crystal growth and defect chemistry of zinc phosphorous and development of photovoltaic devices. *Mailing Add:* PO Box 557 Rushland PA 18956-0557

CATALANO, RAYMOND ANTHONY, b Westland, Pa, Feb 21, 38. INVERTEBRATE ZOOLOGY, AQUATIC ENTOMOLOGY. *Educ:* Edinboro State Col, BS, 60; Ind Univ, Pa, MS, 67; Brigham Young Univ, PhD(zool), 78. *Prof Exp:* Educr sci, Avella Area High Sch, Pa, 60-65 & biol, Gateway High Sch, Pa, 65-67; PROF BIOL, CALIF STATE COL, PA, 67- *Concurrent Pos:* Col dir, Marine Sci Consortium, 71-80; res assoc, Ctr Health & Environ Studies, Brigham Young Univ, 75; environ consult, Western Pa

Conserv, 77-; educ consult, Acad Year Biol Inst, NSF, 78-79. *Mem:* NAm Benthological Soc; Am Soc Limnol & Oceanog; Int Asn Theoret & Appl Limnol; Ecol Soc Am. *Mailing Add:* Dept Biol Calif State Col Third St California PA 15419

CATALDI, HORACE A(NTHONY), b Brooklyn, NY, Apr 24, 18; m 51; c 5. ELECTROCHEMISTRY-BATTERIES. *Educ:* City Col New York, BChE, 40; Univ Ill, MS, 47, PhD(chem eng), 49. *Prof Exp:* Physicist, US Navy Dept, 41-43; phys chemist, Manhattan Proj, 43-45; res asst, Univ Ill, 46-49; proj engr, Standard Oil Co, Ind, 50-53; mgr phys chem, proj mgr & mgr pacemaker eng, Gen Elec Co, 53-85; PRES, CATALDI & ASSOCS, INC, 83- *Concurrent Pos:* Consult, mat electrochem, batteries. *Mem:* Electrochem Soc; Am Chem Soc; AAAS. *Res:* Oxidation of metals at high temperatures; stress corrosion; processing high temperature synthetic fibers; clean room materials processing; implantable materials, electrodes, batteries; high-temp solid lubricants; solid state x-ray detectors. *Mailing Add:* 2516 E Menlo Blvd Milwaukee WI 53211

CATALDO, CHARLES EUGENE, b Cardiff, Ala, Feb 12, 27; m 52; c 4. AEROSPACE MATERIALS ENGINEERING. *Educ:* Univ Ala, BS, 50. *Prof Exp:* Metall Tester, US Steel Corp, 50-51; metallogr, US Army Guided Missile Div, 51-53, mat engr, Army Ballistics Missile Agency, 53-60; spec proj engr, NASA-Mat & Processes Labs, 60-75, dep dir, Spec Ex Serv, 75-82; chief, Technol Develop, 82-90, STAFF SCIENTIST, UNITED TECHNOLOGIES CORP, USBI, 90- *Concurrent Pos:* Mem, Civil Serv Bd, Examiners, US Army, 60-64, NASA Mat Adv Comt, 68-78, Am Inst Aeronaut & Astronaut Awards Selection Comt, 81 & Mat Comt, 82. *Mem:* Fel Am Soc Metals; assoc mem Am Inst Aeronaut & Astronaut; Soc Advan Mat & Process Eng; Am Defense Preparedness Asn. *Res:* Conducted pioneering experiments in cryogenic effects on structural materials and hydrogen embrittlement characteristics of alloys at high pressures. *Mailing Add:* 4726 Panorama Dr SE Huntsville AL 35801

CATALDO, DOMINIC ANTHONY, b Altoona, Pa, June 17, 42; m 65; c 1. PLANT PHYSIOLOGY, BIOCHEMISTRY. *Educ:* Ohio State Uni v, BS, 66; Univ Dayton, MS, 68; Yale Univ, MPh & PhD(plant physiol), 72. *Prof Exp:* Instr biol, Univ Dayton, 66-67; res assoc, Dept Agron, Univ Wis, 72-74; res scientist, 74-80, STAFF SCIENTIST PLANT PHYSIOL, BATTELLE-NORTHWEST LABS, 80- *Mem:* Am Soc Plant Physiol; Am Forestry Soc. *Res:* Trace metal metabolism; source-sink relations, nitrogen metabolism, membrane transport mechanisms in plant physiology; kinetics and specificity of root absorption of nutrient and non-nutrient species. *Mailing Add:* 908 S Nelston St Kennewick WA 99336

CATALDO, JOSEPH C, b New York, NY, Apr 1, 37; c 2. WATER RESOURCE ENGINEERING, ENVIRONMENTAL ENGINEERING. *Educ:* City Univ New York, BCE, 60, MSCE, 64, PhD(eng), 69. *Prof Exp:* Naval architech eng, Brooklyn Naval Shipyard, 60-63; lectr, City Univ New York, 63-69, asst prof, 69-71; water resource engr, TAMS, 71-73; environ engr, Dames & Moore, 73-75; asst prof, 75-80, PROF ENG, DEPT CIVIL ENG, COOPER UNION, 75- *Mem:* Am Soc Civil Engrs. *Res:* Hydraulic engineering; sanitary engineering; thermal pollution. *Mailing Add:* Dept Civil Eng Cooper Union Cooper Square New York NY 10003

CATALFOMO, PHILIP, b Providence, RI, Dec 27, 31; m 62; c 2. PHARMACOGNOSY. *Educ:* Providence Col, BS, 53; Univ Conn, BS, 58; Univ Wash, MS, 60, PhD(pharmacog), 63. *Prof Exp:* From asst prof to prof pharmacog, Sch Pharm, Ore State Univ, 63-75, head dept, 66-75; PROF PHARMACOG & DEAN, SCH PHARM, UNIV MONT, 75- *Concurrent Pos:* Am Found Pharmaceut Educ Gustavus A Pfeiffer Mem res fel, 69-70. *Mem:* AAAS; Am Soc Pharmacog; Am Pharmaceut Asn; Acad Pharmaceut Sci; Sigma Xi. *Res:* Investigation of higher plants and fungi for pharmacologically active components; secondary metabolism of marine fungi and mycorrhizal fungi. *Mailing Add:* Univ Sta Univ Wyo Box 3432 Laramie WY 82071

CATALONA, WILLIAM JOHN, b Cleveland, Ohio, Nov 14, 42; m 66; c 1. UROLOGIC ONCOLOGY, TUMOR IMMUNOLOGY. *Educ:* Otterbein Col, BS, 64; Yale Med Sch, MD, 68. *Prof Exp:* Intern surg, Yale New Haven Hosp, 68-69; resident, Univ Calif, San Francisco, 69-70; clin assoc, Surg Br, Nat Cancer Inst, NIH, 70-72; resident urol, Johns Hopkins Hosp, 72-76; assoc prof surg, 76-83, PROF CHIEF UROL SURG, WASH UNIV, 83- *Concurrent Pos:* Asst ed, J Urol, 78- *Honors & Awards:* James Ewing Prize, James Ewing Soc, 72; Grayson Carroll Prize, Am Urol Soc, 74; C E Alken Res Prize, Alken Soc, Bern, Switz, 79; Gold Cystoscope Award, Am Urol Asn, 86. *Mem:* Am Asn Immunologists; Am Asn Cancer Res; Am Urol Asn; Am Col Surg; Soc Int Urol; Am Asn Genitourinary Surgeons; Clin Soc Genitourinary Surgeons. *Res:* Cancer research in the general area of tumor immunology; specific areas of investigation include immunotherapy of bladder cancer; screening and early diagnosis of prostate cancer; genetics of prostate cancer. *Mailing Add:* 4960 Audubon Ave St Louis MO 63110

CATANACH, WALLACE M, JR, b Philadelphia, Pa, Aug 24, 30; m 53; c 4. MECHANICAL ENGINEERING, MECHANICS. *Educ:* Pa State Univ, BS, 52; Bradley Univ, MSME, 58; Lehigh Univ, PhD(mech eng), 67. *Prof Exp:* Res design engr, Caterpillar Tractor Co, 54-59; asst prof mech eng, Lafayette Col, 59-65; res asst mech, Lehigh Univ, 65-67; asst prof, 67-70, ASSOC PROF MECH ENG, LAFAYETTE COL, 70- *Res:* Fracture mechanics; fatigue crack propagation in cylindrical shells; computer aided design using ANSYS and PATRAN software. *Mailing Add:* Dept of Mech Eng Lafayette Col Easton PA 18042

CATANESE, CARMEN ANTHONY, b Niagara Falls, NY, Apr 27, 42; m 67; c 3. ENGINEERING PHYSICS, ELECTRONICS ENGINEERING. *Educ:* Xavier Univ, BS, 64; Yale Univ, MS, 65, PhD(physics), 70. *Prof Exp:* Mem tech staff, RCA Labs, 70-79, group leader, kinescope systs res, 79-81, dir picture tube systs res, 81-88; VPRES, SOLID STATE RES, DAVID SARNOFF RES CTR, SRI INT, 88- *Concurrent Pos:* Consult, NSF, Educ Directorate, 75-76. *Mem:* Am Phys Soc; Soc Info Display; Inst Elec & Electronic Engrs. *Res:* Magnetic materials; solid state physics; display engineering; electron optics. *Mailing Add:* David Sarnoff Res Ctr SRT Int Princeton NJ 08540-6449

CATANIA, PETER J, b Newmarket, Ont, Jan 6, 42; m 62; c 2. RENEWABLE ENERGIES, ENERGY CONSERVATION. *Educ:* Univ Waterloo, BASc, 67; Univ Alta, MSc, 69, PhD(chem eng), 77. *Prof Exp:* Res assoc, Univ Del, 69-71; spec lectr, Univ Regina, 73-74, from asst prof to assoc prof, 74-82, asst dean, 80-86, PROF FAC ENG, UNIV REGINA, SASK, 82- *Concurrent Pos:* Mem, nat proj adv comt & mem bd, western region, Oxfam, 75-76; vis prof, Fed Univ Paraiba, Brazil, 77-79; chmn, Energex '82 & Energex '84, 80-84; dir, Western Can Indust Asn, 84-86; int consult, Guyana govt & Energex '88, 85; chmn, Int Energy Found, 88- *Mem:* Solar Energy Soc Can; Am Inst Chem Engrs; Can Soc Chem Eng; Can Soc Mech Eng; Chem Inst Can. *Res:* International development of energies and energy conservation; international development. *Mailing Add:* Fac Eng Univ Regina Regina SK S4S 0A2 Can

CATANZARO, EDWARD JOHN, b Jamaica, NY, Nov 4, 33; m 62; c 2. GEOCHEMISTRY. *Educ:* Brooklyn Col, BS, 55; Univ Wyo, MA, 57; Columbia Univ, PhD(geochem), 62. *Prof Exp:* Geologist, US Geol Surv, Washington, DC, 62-63; res chemist, Nat Bur Standards, 63-69; asst prof geol, Southampton Col, 69-71; assoc prof, 71-76, PROF CHEM, FAIRLEIGH DICKINSON UNIV, 76- *Concurrent Pos:* Vis sr res assoc, Lamont-Doherty Geol Observ, NY, 70- *Mem:* Geol Soc Am; Geochem Soc; Am Geophys Union. *Res:* Atomic weights of chemical elements; natural isotopic variations in strontium and lead; geochemistry and petrology of Precambrian rocks; mass spectrometric techniques and development; trace metal concentrations and movements in natural waters. *Mailing Add:* 88 Lafayette Ave Park Ridge NJ 07656

CATAPANE, EDWARD JOHN, b Brooklyn, NY, Oct 13, 51. NEUROBIOLOGY, NEUROCHEMISTRY. *Educ:* Fordham Univ, BA, 73, MS, 75, PhD(physiol), 77. *Prof Exp:* From asst prof to assoc prof, 77-85, PROF BIOL, CITY UNIV NEW YORK, 85- *Mem:* Soc Neurosci; AAAS; NY Acad Sci; Am Soc Zoologists. *Res:* Neurobiological studies of the interactions among monoaminergic neurotransmitters and neuroactive peptidergic substances in the nervous system; peripherally innervated organs of invertebrates. *Mailing Add:* Dept Math & Sci City Univ NY Medgar Evers Col 1150 Carroll St Brooklyn NY 11225

CATCHEN, GARY LEE, b Johnstown, Pa, Aug 9, 50; m 74; c 2. HYPERFINE INTERACTIONS, RADIATION DETECTION & MEASUREMENT. *Educ:* Pa State Univ, BS, 71; Columbia Univ, PhD(chem), 79. *Prof Exp:* From res asst to res assoc chem, Columbia Univ, 74-79; res chemist, Conoco, Inc, 79-82; asst prof, 82-90, ASSOC PROF NUCLEAR ENG, PA STATE UNIV, 90- *Mem:* Am Chem Soc; Am Nuclear Soc; Am Ceramic Soc; Am Phys Soc. *Res:* Hyperfine interactions; radiation detection and measurement; health physics; solid-state chemistry. *Mailing Add:* Radiation Sci & Eng Ctr Pa State Univ University Park PA 16802

CATCHINGS, ROBERT MERRITT, III, b Washington, DC, Apr 2, 42; m 70. SOLID STATE PHYSICS. *Educ:* Univ Mich, BS, 64; Wayne State Univ, MS, 66, PhD(physics), 70. *Prof Exp:* ASST PROF PHYSICS, HOWARD UNIV, 70- *Res:* Magnetization, electrical conductivity and nuclear magnetic resonance studies of amorphous solids. *Mailing Add:* 7015 Fitzpatrick Dr Laurel MD 20707

CATCHPOLE, HUBERT RALPH, b London, Eng, May 13, 06; nat US; m 73. ANATOMY. *Educ:* Cambridge Univ, BA, 28, MA, 79; Univ Calif, PhD(physiol), 33. *Prof Exp:* Asst, Univ Calif, 34-35 & Cutter Labs, 35-36; from instr to asst prof physiol, Yale Univ, 36-43, Commonwealth Fund fel, 41-43; assoc path, 46-47, asst prof, 47-51, res assoc prof, 51-61, res prof path, 61-75, PROF HISTOL & EMER PROF PATH, UNIV ILL COL MED, 75- *Concurrent Pos:* USN/USNR, 43-46, NMRI, Bethesda, Md; Vis prof humanities, Rush Univ, 80- *Mem:* Assoc Am Physiol Soc; assoc Endocrine Soc; assoc Brit Soc Endocrinol; Asn Am Anat Soc. *Res:* Physiology of reproduction, aeroembolism and capillary vessels; lactogenic and gonadotrophic hormones; biophysics of ionic distribution in connective tissue and cells; histochemistry of connective tissue. *Mailing Add:* Dept of Histol Univ of Ill Col of Dent Chicago IL 60612

CATE, JAMES RICHARD, JR, b Winters, Tex; m 64; c 2. ENTOMOLOGY. *Educ:* Tex A&M Univ, BS, 67, MS, 68; Univ Calif, Berkeley, PhD(entom), 75. *Prof Exp:* Res asst entom, Tex A&M Univ, 66-68, res assoc, 68-71; res asst, Univ Calif, Berkeley, 71-72, res assoc, 72-74; ASST PROF ENTOM, TEX A&M UNIV, 74- *Mem:* Sigma Xi; Entom Soc Can; Int Orgn Biol Control; Sigma Xi. *Res:* Population ecology and biological control of insect pests of cotton. *Mailing Add:* USDA/CSRS Aerospace Bldg Washington DC 20250-2200

CATE, ROBERT BANCROFT, b Manchester, NH, July 26, 24; m 51. REMOTE SENSING, VISUAL PERCEPTION. *Educ:* Dartmouth Col, BA, 45; NC State Univ, MSc, 60, PhD(soil sci), 70. *Prof Exp:* Admin asst & vconsul, US Foreign Serv, Calcutta, India & Rio de Janeiro, Brazil, 46-52; govt pub rels specialist, San Francisco, 53-56; off mgr, Kaiser Aluminio, Ltd, Belem, Brazil, 56-57; soil chemist, Brit Guiana Soil Surv Proj, Food & Agr Orgn, 61-63; vis assoc prof, NC State Univ & regional dir, Agency Int Develop Int Soil Testing Proj, Brazil 64-73, Guatemala, 74-75 & Colombia, 76; staff scientist, Systs & Serv Div, Lockhead Eng & Mgt Serv Co, Inc, NASA, 77-84, sr res scientist, Technicolor Govt Serv, Inc, 85-86; INDEPENDENT RES, 87. *Concurrent Pos:* NSF coop grad fel, 60. *Res:* Extraction of invariants from remotely sensed data. *Mailing Add:* 303 Coast Blvd 11 La Jolla CA 92037

CATE, RODNEY LEE, b Coleman, Tex, Dec 8, 50; m 72; c 1. PROTEIN CHEMISTRY, TOXICOLOGY. *Educ:* Tarleton State Univ, BS, 73; Ariz State Univ, PhD(chem), 77. *Prof Exp:* Res chemist biochem, Univ Calif, Los Angeles, 77-78; asst prof, 73-83, ASSOC PROF CHEM, MIDWESTERN STATE UNIV, 83- *Concurrent Pos:* Scholar, Univ Calif, Los Angeles, 77-78; consult, N Tex Chem Consults, 78-; sect officer, Am Chem Soc, 81-84. *Mem:* Am Chem Soc; Sigma Xi; Am Asn Advan Sci. *Res:* Enzymology, protein structure and function, protein purification, metabolic regulation, physiology and venom toxicology. *Mailing Add:* Dept Chem Midwestern State Univ 3400 Taft Wichita Falls TX 76308

CATE, THOMAS RANDOLPH, b Nashville, Tenn, Feb 19, 35; m 56; c 5. INFECTIOUS DISEASES. *Educ:* Vanderbilt Univ, BA, 56, MD, 59. *Prof Exp:* Asst prof med, Sch Med, Washington Univ, 66-68; assoc prof med, Duke Univ, 68-75; assoc prof, 75-82, PROF MICROBIOL & MED, BAYLOR COL MED, 82- *Mem:* Infect Dis Soc Am; Am Soc Microbiol; Am Thoracic Soc; Am Fedn Clin Res. *Res:* Pathogenesis of respiratory virus infections with a major focus on mechanisms of resistance. *Mailing Add:* Dept Microbiol Baylor Col Med Houston TX 77030

CATEFORIS, VASILY C, b Athens, Greece, Jan 1, 39; m 63; c 2. GENERAL MATHEMATICS. *Educ:* Univ Wis-Madison, MS, 63, PhD(math), 68. *Prof Exp:* From asst prof to prof, Univ Ky, 68-83; CHMN, STATE UNIV NY, POTSDAM, 87- *Concurrent Pos:* Nat Sci Found fel, 65. *Mem:* Math Asn Am; Math Soc Am. *Mailing Add:* State Univ NY-Potsdam Dept Math Potsdam NY 13676

CATENHUSEN, JOHN, b Tecumseh, Nebr, June 17, 09; m 41; c 2. PLANT ECOLOGY. *Educ:* Univ Wis, PhB, 37, PhM, 39, PhD(bot), 48. *Prof Exp:* Biologist, Arboretum & Wildlife Refuge, Univ Wis, 41-43; asst div mgr, Trop Plantation, Haiti, 43-44; res botanist, Firestone Plantations Co, Liberia, WAfrica, 44-51; assoc prof biol & head dept, Col of Steubenville, 52-55; prof biol & head, Div Sci & Math, Hillsdale Col, 55-81; RETIRED. *Mem:* AAAS; Audubon; Am Ornithologists Union; Wilderness Soc; Nature Conservancy. *Res:* Plant taxonomy; animal ecology; plant breeding; Hevea brasiliensis. *Mailing Add:* 1720 Steamburg Rd Hillsdale MI 49242

CATER, EARLE DAVID, b San Antonio, Tex, Apr 4, 34; m 63, 71; c 1. HIGH TEMPERATURE CHEMISTRY, SOILD STATE CHEMISTRY. *Educ:* Trinity Univ, BS, 54; Univ Kans, PhD(chem), 60. *Prof Exp:* Resident student assoc chem, Argonne Nat Lab, 58-60, res assoc, 60-61; from asst prof to assoc prof, 61-77, PROF CHEM, UNIV IOWA, 78- *Concurrent Pos:* Vis chemist, Oxford Univ, 71; mem ad hoc comt, High Temperature Sci & Technol, Nat Res Coun, 75-77; div ed high temperature mat, J Electrochem Soc, 77-; chmn, Gordon Res Conf High Temperature Chem, 78; vis scientist, Ariz State Univ, 82 & 83. *Mem:* AAAS; Electrochem Soc; Am Chem Soc; Am Asn Univ Prof. *Res:* High temperature physical chemistry; thermodynamics of vaporization processes; mechanisms of reactions at high temperatures; variable valence phenomena in rare earth compounds; electron microscopy of inorganic crystals, particularly of nonstoichiometric compounds; High temperature mass spectrometry. *Mailing Add:* Dept Chem Univ Iowa Iowa City IA 52242-1219

CATER, FRANK SYDNEY, b Chicago, Ill, June 27, 34. MATHEMATICS. *Educ:* Univ Southern Calif, BA, 56, MA, 57, PhD(math), 60. *Prof Exp:* Asst prof math, Univ Ore, 60-65; from asst prof to assoc prof, 65-70, PROF MATH, PORTLAND STATE UNIV, 70- *Concurrent Pos:* Referee, Math Mag, 72-, Real Anal Exchange, Proc Am Math Soc. *Mem:* Am Math Soc; Math Asn Am. *Res:* Analysis and real variables. *Mailing Add:* Dept Math Portland State Univ Portland OR 97207

CATERSON, BRUCE, b Sydney, Australia, Oct 24, 47; m 77; c 2. PROTEOGLYCAN METABOLISM IN CONNECTIVE TISSUES, PATHOGENESIS OF ARTHRITIS & OTHER MUSCULOSKELETAL DISORDERS. *Educ:* Monash Univ, Australia, PhD(biochem), 67, BSc Hons, 70. *Prof Exp:* From instr to asst prof biochem, Univ Ala, Birmingham, 77-82; from assoc prof to prof biochem, WVa Univ, 82-89; PROF SURG, BIOCHEM & BIOPHYS & NORFLEET-RANEY PROF RES ORTHOP, UNIV NC, CHAPEL HILL, 89- *Concurrent Pos:* Benedum distinguishes scholar award biosci & med, WVa Univ, 87-; mem, Pathobiochem Study Sect, NIH, 89-, grant rev comt, Arthritis Found, 89- & Orthop Res & Educ Found, 90- *Mem:* Soc Complex Carbohydrates; Am Soc Biol Chemists; AAAS; Protein Soc; Orthop Res Soc. *Res:* Proteoglycan structure, function and metabolism in connective tissue; proteoglycan metabolism and musculoskeletal disease; osteoarthritis research. *Mailing Add:* Div Orthop Surg Dept Surg Univ NC CB 7055 Burnett-Womack Bldg Chapel Hill NC 27599-7055

CATES, DAVID MARSHALL, b Salisbury, NC, Jan 7, 22; m 48; c 2. POLYMER CHEMISTRY, TEXTILE CHEMISTRY. *Educ:* NC State Col, BS, 49, MS, 51; Princeton Univ, PhD(chem), 55. *Prof Exp:* PROF TEXTILE CHEM, NC STATE UNIV, 55- *Mem:* Am Chem Soc; Fiber Soc; Am Asn Textile Chemists & Colorists. *Res:* High polymers and textile fibers; absorption chemistry. *Mailing Add:* 3 David Clark Lab NC State Univ Raleigh NC 27650

CATES, GEOFFREY WILLIAM, clinical pathology; deceased, see previous edition for last biography

CATES, LINDLEY A, b Chicago, Ill, Nov 20, 32; m 57; c 2. MEDICINAL CHEMISTRY, ORGANIC CHEMISTRY. *Educ:* Univ Minn, BS, 54; Univ Colo, MS, 58, PhD(med chem), 61. *Prof Exp:* Instr pharm, Univ Colo, 58-61; from asst prof to assoc prof, 61-68, PROF MED CHEM, COL PHARM, UNIV HOUSTON, 68-, ASSOC DEAN, 85- *Concurrent Pos:* Prin investr, NIH grants, 62-83; Robert A Welch Found grants, 69-74, 82-88. *Mem:* Am Asn Col Pharm; Am Asn Pharmaceut Scientists. *Res:* Synthesis of organophosphorus compounds as potential chemotherapeutics. *Mailing Add:* Col of Pharm Univ of Houston Houston TX 77004

CATES, MARSHALL L, b Camas, Wash, Dec 17, 42. MATHEMATICAL GROUP THEORIES, COMBINATORIES. *Educ:* Wash State Univ, BA, 65; Univ Ill, MA, 67, PhD(math), 71. *Prof Exp:* PROF MATH, CALIF STATE UNIV, LOS ANGELES, 71- *Concurrent Pos:* Asst anal, Boeing, 65-66; NASA fel, 66-67; Nat Sci Found fel, 69-70. *Mem:* Am Math Soc; Math Asn Am. *Mailing Add:* Dept Math Calif State Univ 5151 State University Dr Los Angeles CA 90032

CATES, VERNON E, b Parsons, Kans, Feb 17, 31; m 66; c 2. ANALYTICAL CHEMISTRY, INORGANIC CHEMISTRY. *Educ:* Kans State Univ, BS, 53, MS, 56, PhD(chem), 62. *Prof Exp:* Teacher high sch, Kans, 56; instr chem, Centenary Col, 56-59; from asst prof to assoc prof, ETex State Univ, 62-67; chmn dept, Dallas Baptist Col, 67-80, prof chem, 67-82; TEACHER, DALLAS INDEPENDENT SCH DIST, 82- *Res:* Gas chromatography; analytical oxidizing agents; analytical methods; teaching innovations. *Mailing Add:* Madison High Sch 3000 Martin Luther King Jr Blvd Dallas TX 75215

CATHCART, JAMES B, b Berkeley, Calif, Nov 22, 17; m 44; c 2. ECONOMIC GEOLOGY & PHSOPHATE DEPOSITS, MINERALOGY OF APATITES. *Educ:* Univ Calif, Berkeley, AB, 39. *Prof Exp:* Teaching fel, Univ Calif, Berkeley, 39-42; FROM GEOLOGIST TO EMER SR RES SCIENTIST, US GEOL SURV, 42- *Honors & Awards:* Meritorious Serv Award, US Dept Interior, 86; Lifetime Achievement Award, Asn Inst Mining Engrs, 88. *Mem:* Fel Geol Soc Am; emer mem Am Asn Petrol Geologists; sr mem Soc Econ Geologists; Soc Econ Paleontologists & Mineralogists. *Res:* Geology of phosphate deposits in the United States, Colombia, Brazil, Saudia Arabia, Antarctica; principally economic geology, stratigraphy and mineralogy. *Mailing Add:* 17225 W 16th Pl Golden CO 80401

CATHCART, JOHN ALMON, b Sparta, Ill, Jan 17, 16; wid; c 2. ORGANIC CHEMISTRY, PATENT SEARCHING. *Educ:* Monmouth Col, Ill, BS, 37; Ohio State Univ, PhD(org chem), 41. *Prof Exp:* Asst chem, Ohio State Univ, 37-41; Anna Fuller Fund fel, Ohio State Univ, 41-42; from instr to asst prof chem & math, Monmouth Col, 42-45; res chemist, Eastman Kodak Co, 45-51, develop engr, 51-60, patent search specialist, 60-64, supvr info sect, Patent Dept, 64-74, patent search specialist, 74-81; RETIRED. *Mem:* Am Chem Soc. *Res:* Polynuclear hydrocarbons; polymers; organic synthesis, patents. *Mailing Add:* 65 Kemphurst Rd Rochester NY 14612

CATHCART, JOHN VARN, b St George, SC, Nov 28, 23; m 66. MECHANICAL PROPERTIES. *Educ:* Clemson Col, BS, 47; Univ Va, PhD(chem), 51. *Prof Exp:* Chemist, phys chem metal surfaces, Oak Ridge Nat Lab, 51-88; RETIRED. *Mem:* AAAS; Am Soc Metals; Electrochem Soc Inc; Am Inst Mining, Metall & Petrol Engrs; fel Am Soc Metals. *Res:* Oxidation of metal surfaces; stress effects during oxidation; diffusion in oxides. *Mailing Add:* 4609 Clinchview Lane Knoxville TN 37931

CATHCART, MARTHA K, b Sturgis, Mich, Dec 3, 51; m 75; c 2. IMMUNOREGULATION. *Educ:* Case Western Reserve Univ, PhD(microbiol), 79. *Prof Exp:* MEM PROF STAFF, RES INST, CLEVELAND CLIN FOUND, 80- *Concurrent Pos:* Adj asst prof biol, Case Western Res Univ, 80- *Honors & Awards:* N Paul Hudson Award, Am Soc Microbiol, 77; William E Lower Award, CCF, 80; Career Woman of Achievement, Young Women Committed Action, 86. *Mem:* Am Asn Immunol; Am Fedn Clin Res; Am Rheumatism Asn; NY Acad Sci; Reticuloendothelial Soc. *Res:* Production of immunosuppressive cytokines by human T lymphocytes; lymphokine that inhibits production of interleukin 2; studies on the mechanisms of inflammatory cell modification of cell and serum lipids including oxidation and transformation to cytotoxic mediators. *Mailing Add:* Cleveland Clinic Found 9500 Euclid Av Cleveland OH 44195

CATHER, JAMES NEWTON, b Carthage, Mo, Mar 17, 31; m 51; c 2. ZOOLOGY. *Educ:* Southern Methodist Univ, BS, 54, MS, 55; Emory Univ, PhD, 58. *Prof Exp:* Instr biol, Emory Univ, 58; from instr to assoc prof zool, Univ Mich, 58-73, from assoc chmn to actg chmn, Div Biol Sci, 76-79, leader, Dept Exp Biol, 75-76, assoc dean, 81-87, PROF ZOOL, UNIV MICH, 73-, DIR, OFF INT PROGS, 87- *Concurrent Pos:* Upjohn fac fel, 65; instr embryol, Marine Biol Lab, 66-67; vis prof, Ore Inst Marine Biol, 69 & 74, Bermuda Biol Sta, 72, Univ Utreclet, 73 & Friday Harbor Lab, 81, Univ Tex, 81 & 87. *Mem:* AAAS; Am Soc Zoologists; Soc Develop Biol; Int Soc Develop Biol. *Res:* Cellular differentiation; molluscan development; invertebrate embryology. *Mailing Add:* Dept Biol Univ Mich Natural Sci Bldg Ann Arbor MI 48109-1048

CATHEY, EVERETT HENRY, geology, for more information see previous edition

CATHEY, JIMMIE JOE, b Whitney, Tex, May 16, 41; m 66; c 2. POWER ELECTRONICS, ELECTRIC MACHINES. *Educ:* Tex A&M Univ, BSEE, 65; Bradley Univ, MSEE, 68, PhD(elec engr), 72. *Prof Exp:* Res engr, Caterpillar Tractor Co, 65-68; sr design engr, Marathon LeTourneau Co, 68-69, proj engr, 72-74, chief elec engr, 74-77, dir appl res, 77-81; instr elec eng, Tex A&M Univ, 69-72; assoc prof, 81-89, PROF ELEC ENG, UNIV KY, 89- *Concurrent Pos:* Vis prof, Univ Ky, 79-80. *Mem:* Inst Elec & Electronic Engrs. *Res:* Novel and conventional power electronic devices utilized by robotics, vehicle traction, and industrial applications. *Mailing Add:* 681 Mount Vernon Dr Lexington KY 40502

CATHEY, LECONTE, b Statesville, NC, Oct 18, 23; m 48; c 5. RADIATION PHYSICS. *Educ:* Davidson Col, BS, 47; Emory Univ, MS, 48; Univ NC, PhD(physics), 52. *Prof Exp:* Sr res physicist, Savannah River Lab, 52-67; PROF PHYSICS, UNIV SC, 68- *Concurrent Pos:* Mem grants rev comt, Dept Health, Educ & Welfare & Environ Protection Agency, 71. *Mem:* AAAS; Am Asn Univ Prof; Sigma Xi; Am Phys Soc; Inst Elec & Electronics Engrs. *Res:* Instrumentation; radiation measurement; Mossbauer spectroscopy. *Mailing Add:* Dept of Physics Univ of S Carolina Columbia SC 29208

CATHEY, WADE THOMAS, JR, b Greer, SC, Nov 26, 37; c 2. IMAGING SYSTEMS, OPTO ELECTRONICS. *Educ:* Univ SC, BS, 59, MS, 61; Yale Univ, PhD(elec eng), 63. *Prof Exp:* Asst elec eng, Univ SC, 59-60 & Yale Univ, 60-62; sr res engr, Autonetics Div, N Am Aviation, Inc, Calif, 62-63, res specialist, 63-64, group scientist, 64-68; from assoc prof to prof, Univ Colo, Denver 68-85, assoc chmn dept, 70-75, chmn dept, 84-85, PROF ELEC & COMPUT ENG,85-, DIR, NSF ENGR RES CTR OPTOELECTRONIC COMPUT SYSTS, UNIV COLO, BOULDER, 87- *Concurrent Pos:* Vis prof, Univ Reading, Eng, 72-73 & Univ Calif, San Diego, 81; consult optical comput, adaptive optics & robot vision, 79- *Mem:* Fel Optical Soc Am; sr mem Inst Elec & Electronic Engrs; Soc Photooptical Instrumentation Engrs. *Res:* Optical computing; optical information processing; holography; imaging theory; laser systems; optical communications systems; coherent optics; adaptive optics; robot vision. *Mailing Add:* 228 Alpine Way Pine Brook Hills Boulder CO 80302

CATHEY, WILLIAM NEWTON, b Pulaski, Tenn, Feb 20, 39; m 63. SOLID STATE PHYSICS. *Educ:* Univ Tenn, BS, 61, MS, 62, PhD(solid state physics), 66. *Prof Exp:* Fel, Nat Res Coun Can, 66-67; asst prof, 67-74, ASSOC PROF PHYSICS, UNIV NEV, RENO, 74-, DEPT CHMN, 79- *Concurrent Pos:* Consult, Reno Metall Ctr, US Bur Mines, Nev. *Mem:* Am Phys Soc; Am Asn Physics Teachers; Sigma Xi. *Res:* Electronic structure of metals using Mossbauer effect; hydrogen storage alloys. *Mailing Add:* Dept of Physics Univ of Nev Reno NV 89507

CATHLES, LAWRENCE MACLAGAN, III, b Brooklyn, NY, Feb 9, 43; m 74; c 2. GEOPHYSICS. *Educ:* Princeton Univ, AB, 65, PhD(geophys), 71. *Prof Exp:* Sr geophysicist, Ledgemont Lab, Kennecott Copper Corp, 71-78; assoc prof geol, Pa State Univ, 78-82; sr res geophys, Chevron Oil Field Res Co, La Habra, 82-86; PROF, CORNELL UNIV, ITHACA, 86- *Mem:* Am Geophys Union; AAAS. *Res:* Earth's viscosity structure inferred from isostatic rebound phenomena; physics and chemistry of copper sulfide leaching from waste dumps; physics and chemistry of hydrothermal systems; basin modeling. *Mailing Add:* Dept Geosci Cornell Univ 2134 Snee Hall Ithaca NY 14853

CATHOU, RENATA EGONE, b Milan, Italy, June 21, 35; US citizen; div. IMMUNOCHEMISTRY, BIOTECHNOLOGY. *Educ:* Mass Inst Technol, BS, 57, PhD(biochem), 63. *Prof Exp:* Res assoc, Mass Inst Technol, 62-65; res assoc, Mass Gen Hosp, 65-70; from asst prof to prof biochem, phys biom & immunol, Sch Med, Tufts Univ, 70-81; PRES, TECH EVAL, 83- *Concurrent Pos:* Lectr, Harvard Med Sch, 69-70; consult, Tufts New England Med Ctr, 71-72; mem, adv panel, molecular biol, NSF, 74-75; vis assoc prof, Chem Dept & Inst Molecular Biol, Univ Calif, Los Angeles, 76-77; mem, bd sci counr, Nat Cancer Inst, NIH, 79-83; co-founder, Clin Assays Inc, Cambridge, Mass, 71, corp secy, 73-75; sr investr, Arthritis Found, 70-75; indust consult, 81-83; sr consult, SRC Assocs, Park Ridge, NJ, 84-; mem adv panels, Small Bus Innov Res Contracts, NIH, 87- *Mem:* Am Asn Clin Chem; AAAS; Am Asn Immunologists; Am Chem Soc; Am Soc Biochem & Molecular Biol; NY Acad Sci; Clin Ligand Assay Soc; Am Inst Chemists. *Res:* Development, scale-up, production and technical evaluations of new therapeutics, biopharmaceuticals, diagnostics, devices and biosensors; physical biochemistry of macro-molecules. *Mailing Add:* Tech Evaluations 430 Marrett Rd Lexington MA 02173

CATIGNANI, GEORGE LOUIS, b Nashville, Tenn, Apr 9, 43; m 72. NUTRITIONAL BIOCHEMISTRY. *Educ:* Vanderbilt Univ, BA, 67, PhD(biochem), 74. *Prof Exp:* Res assoc toxicol, Ctr Environ Toxicol, Dept Biochem, Vanderbilt Univ, 74-75; staff fel nutrit biochem, Lab Nutrit & Endocrinol, Nat Inst Arthritis, Metab & Digestive Dis, 75-78; from asst prof to assoc prof, 78-88, PROF FOOD SCI, NC STATE UNIV, 88- *Mem:* Am Inst Nutrit; AAAS; Inst Food Technol; Sigma Xi; Am Chem Soc; Am Pub Health Asn. *Res:* Mechanism of action of vitamin E; effects of processing on food proteins; in vitro protein digestibility assays; effects of dietary fiber on fat soluble vitamin metabolism. *Mailing Add:* Dept Food Sci 218 Schaub Hall NC State Univ Raleigh NC 27695-7624

CATLETT, DUANE STEWART, b Fremont, Nebr, July 13, 40; m 61; c 2. PROCESS CHEMISTRY, RADIO ANALYTICAL CHEMISTRY. *Educ:* Nebr Wesleyan Univ, BA, 63; Iowa State Univ, PhD(phys chem), 67. *Prof Exp:* Asst prof chem, Minot State Col, 67-68; asst prof, Pac Lutheran Univ, 68-70; staff mem, Los Alamos Nat Lab, 70-74, alt group leader, Mat Technol Group, 74-80, asst to assoc dir, Defense Progs, 80-84, dep group leader, Nuclear Mat Process Technol Group, 84-88, tech dir, Dept Energy, Rocky Flats Area Off, 88-89, MGR, LOS ALAMOS TECHNOL OFF ROCKY FLATS, GOLDEN, COLO, 89- *Concurrent Pos:* Tech ed, Los Alamos Nat Lab's, Defense Sci Jour, 82-84; chmn, Mat Mgt Exec Comt Technol Exchange Group, Dept Energy, 85-88 & PRMP Mgt Oversight Group, 88-89; mem, Plutonium Tech Comt, Dept Energy, 86-89. *Mem:* Am Vacuum Soc; Am Soc Metals. *Res:* Gas-solid kinetics and its theory of the rate-determining processes, such as nucleation and diffusion; thermodynamics of solutions; thin film deposition processes; hot-atom chemistry and radiolysis of condensed media; actinide process chemistry; process analytical chemistry. *Mailing Add:* Los Alamos Nat Lab MS 140 PO Box 1663 Los Alamos NM 87544

CATLIN, AVERY, b New York, NY, Jan 29, 24; m 46; c 4. COMPUTER SCIENCES, SOLID STATE PHYSICS. *Educ:* Univ Va, BEE, 47, MA, 49, PhD(physics), 60. *Prof Exp:* From instr to asst prof elec eng, 49-61, assoc prof mat sci, 61-67, acting chmn dept, 61-62, assoc dean eng & appl sci, 67-74, exec vpres, 74-82, PROF MAT SCI, UNIV VA, 67-, PROF ENG & APPL SCI, 82- *Mem:* AAAS; Am Phys Soc; Inst Elec & Electronics Engrs; Am Inst Mining, Metall & Petrol Engrs; Am Soc Eng Educ. *Res:* Microcomputer applications, computer assisted instruction; properties of thin films; electron microscopy; electronic properties of solids; biomaterials. *Mailing Add:* Dept Computer Sci Univ Va Thornton Hall Charlottesville VA 22903

CATLIN, B WESLEY, b Mt Vernon, NY, June 26, 17; m 54. MICROBIOLOGY, BACTERIAL GENETICS. *Educ:* Univ Calif, Los Angeles, AB, 42, MA, 44, PhD(microbiol), 47. *Prof Exp:* Med lab technician, 39-42; chief lab technologist, Santa Barbara Gen Hosp, 43; teaching asst bact, Univ Calif, Los Angeles, 43-47, res assoc, 47-49; res assoc genetics, Carnegie Inst Wash, 49-50; from asst prof to prof, 50-87, EMER PROF MICROBIOL, MED COL WIS, Milwaukee, 87- *Concurrent Pos:* NIH grantee, 54-87; mem, Neisseriaceae Subcomt, Int Comt Syst Bact, 63-, Am Soc Microbiol, 64-70, secy, 66-78, chmn subcomt Neisseria & Moraxella, 74-78, Taxon Comt, 74-79; mem, Bd Sci Counr, Nat Inst Allergy & Infectious Dis, NIH, Bethesda, Md, 77-80; World Health Org Neisseria Res, Geneva, Switz, 64. *Mem:* AAAS; Am Soc Microbiol; Soc Study Evolution; Genetics Soc Am. *Res:* Neisseriaceae; genetic transformation. *Mailing Add:* 3886 LaJolla Village Dr La Jolla CA 92037-1412

CATLIN, DON HARDT, b New Haven, Conn; m 73; c 2. INTERNAL MEDICINE, CLINICAL PHARMACOLOGY. *Educ:* Yale Univ, BA, 60; Univ Rochester, MD, 65. *Prof Exp:* Resident chief, Med Sch, Univ Calif, Los Angeles, 68-69; maj med, Walter Reed Army Inst Res, 69-72; asst prof, 72-78, ASSOC PROF MED & PHARMACOL, SCH MED, UNIV CALIF, LOS ANGELES, 78- *Concurrent Pos:* Dir, Ctr Study Drug Induced Dis, 76- *Res:* Clinical pharmacology of opiates and endorphins; basic mechanisms of drug dependence; pharmacokinetics and pharmacodynamics. *Mailing Add:* Dept Pharmacol Sch Med 23-278 Ctr Health Sci Univ Calif Los Angeles CA 90024

CATLIN, DONALD E, b Erie, Pa, Apr 29, 36; m 61; c 2. MATHEMATICS. *Educ:* Pa State Univ, BS, 58, MA, 61; Univ Fla, PhD(math), 65. *Prof Exp:* Instr math, Univ Fla, 64-65; asst prof math & statist, 65-72, vchmn dept math, 70-71 & 77-82, ASSOC PROF MATH & STATIST, UNIV MASS, 72- *Concurrent Pos:* Mem, Inst Fundamental Studies. *Mem:* Am Math Soc; Math Asn Am. *Res:* Systems theory; optimal filtering and estimation; marine navigation. *Mailing Add:* Dept Math Univ Mass Amherst MA 01002

CATLIN, FRANCIS I, b Hartford, Conn, Dec 6, 25; m 48; c 3. OTOLARYNGOLOGY. *Educ:* Johns Hopkins Univ, MD, 48, ScD, 59. *Prof Exp:* Intern, Union Mem Hosp, 48-49; intern otolaryngol, Johns Hopkins Hosp, 50 & 52-53, asst resident, 53-54 & 55, otolaryngologist, 56-72; from instr to asst prof otolaryngol, Johns Hopkins Univ, 56-63, asst prof audiol & speech, Sch Hyg & Pub Health, 60-70, assoc prof otolaryngol, Sch Med, 63-72, sci dir, Info Ctr Hearing, Speech & Dis Human Commun, 68-72, assoc prof audiol & speech, Pub Health Admin, 70-72; chief-of-serv, Dept Otolaryngol, St Luke's Episcopal Hosp, Houston, 72-87; PROF OTORHINOLARYNGOL & COMMUN SCI, BAYLOR COL MED, 72-; CHIEF-OF-SERV, DEPT OTOLARYNGOL, TEX CHILDREN'S HOSP, 72- *Concurrent Pos:* Consult, Vet Admin Hosp, Perry Point, 61-72; spec consult, Neurol & Sensory Dis Serv Prog, US Dept Health, Educ & Welfare, 63-64; mem communicative dis res training comt, Nat Inst Neurol Dis & Blindness, 63-65; mem prof adv coun, Nat Easter Seal Soc Crippled Children & Adults, 73-78. *Mem:* AMA; fel Am Acad Opthal & Otolaryngol; fel Am Laryngol, Rhinol & Otol Soc; Am Broncho-Esophagol Asn; Am Speech & Hearing Asn; Am Otological Soc; Sigma Xi. *Res:* Hearing problems in audiology including instrumentation; infectious diseases involving the paranasal sinuses. *Mailing Add:* 13307 Queensbury Lane Houston TX 77079

CATLIN, PETER BOSTWICK, b Ross, Calif, Sept 22, 30; m 52; c 3. PLANT PHYSIOLOGY, POMOLOGY. *Educ:* Univ Calif, BS, 52, MS, 55, PhD(plant physiol), 58. *Prof Exp:* Asst pomologist, 58-64, ASSOC POMOLOGIST, UNIV CALIF, DAVIS, 64-, LECTR POMOL, 69- *Mem:* Am Soc Hort Sci. *Res:* Physiology and biochemistry of plant growth and development; respiratory metabolism; chemical taxonomy. *Mailing Add:* Dept Pomol 1045 Wickson Hall Univ Calif Davis CA 95616

CATLING, PAUL MILES, b London, Eng, Jan 13, 47; m 85; c 1. BOTANY, ECOLOGY. *Educ:* Univ Toronto, BSc, 75 & PhD(bot), 80. *Prof Exp:* Lectr, Univ Toronto, 77-78; res sci, 80-88, CHIEF CUR, AGR CAN, 87- *Concurrent Pos:* Ed comt, Canadian-Field-Naturalist, 85- 88 & J Mex Orchid Soc, 86-88. *Mem:* Canadian Bot Assoc. *Res:* Systematics and ecology of aquatic plants, sedges and other economic plants; systematics and ecology of the orchidacae; phytogeography, diversity, pollination biology. *Mailing Add:* Biosystematics Res Ctr Agr Can, CEF Ottawa Ottawa ON K1A 0C6 Can

CATO, BENJAMIN RALPH, JR, b Belmont, NC, Aug 24, 25; m 48; c 3. MATHEMATICS. *Educ:* Duke Univ, AB, 48, AM, 50. *Prof Exp:* Instr math & physics, Univ Ariz, 50-52; instr math, Univ Md, 52-55; from asst prof to assoc prof, 55-86, EMER PROF MATH, COL WILLIAM & MARY, 86- *Concurrent Pos:* Assoc prof, Summer Inst High Sch Teachers, Nat Sci Found, 59-, assoc dir, 60-68, dir, 68-75. *Mem:* Am Math Soc; Math Asn Am. *Res:* Partial differential equation; integral equations; analysis; linear operators. *Mailing Add:* One Atkins Loop Lake Junaluska NC 28745

CATON, JERALD A, b June 3, 49; m; c 3. THERMODYNAMICS, COMBUSTION ENGINES. *Educ:* Univ Calif, Berkeley, BSME, 72, MSME, 73; Mass Inst Technol, PhD(mech eng), 80. *Prof Exp:* Res engr, Gen Motors Res Lab, Warren, Mich, 73-76; res asst, Mass Inst Technol, Cambridge, 76-79; asst prof, 79-85, ASSOC PROF MECH ENG, TEXAS A&M UNIV, COLLEGE STATION, 85- *Mem:* Am Soc Mech Engrs; Soc Automotive Engrs; Am Soc Eng Educ; Am Chem Soc. *Res:* Thermodynamics, heat transfer, fluid mechanics and internal-combustion engines; alternative fuels for engines; modelling engine and combustion processes; fundamental and applied combustion topics. *Mailing Add:* Dept Mech Eng Tex A&M Univ College Station TX 77843-3123

CATON, RANDALL HUBERT, b Minneapolis, Minn, Aug 19, 42; m 69; c 1. METAL PHYSICS, LOW TEMPERATURE PHYSICS. *Educ:* Univ Minn, BS, 65; Univ Pa, MS, 67; City Univ New York, PhD(physics), 72. *Prof Exp:* Res assoc physics, Univ Ill, 72-74; assoc physicist, Brookhaven Nat Lab, 74-78; ASST PROF PHYSICS, CLARKSON UNIV TECHNOL, 78- *Mem:* Am Phys Soc; Sigma Xi. *Res:* Low temperature physics; electrical properties and heat capacity of metals; magnetic materials; superconductivity; radiation damage in superconductors. *Mailing Add:* 14 Normandy Newport News VA 23606-1516

CATON, ROBERT LUTHER, b Uniontown, Pa, Feb 28, 37; m 69; c 1. METALLURGY & PHYSICAL METALLURGICAL ENGINEERING. *Educ:* Pa State Univ, BS, 59. *Prof Exp:* Res metallurgist, Carpenter Technol, 62-69, supvr stainless res & develop, 69-72, sr metallurgist stainless, 72-75, mgr stainless metall, 75-76, gen mgr alloy develop, 76-83 & new prod develop, 83-86, mgr new prod develop, 86-90; CONSULT, 90- *Mem:* Am Soc Metals. *Mailing Add:* 92 Burning Tree Lane Reading PA 19607

CATON, ROY DUDLEY, JR, b Fresno, Calif, June 7, 30. CHEMICAL EDUCATION. *Educ:* Fresno State Col, BS, 52, MA, 53; Ore State Univ, PhD(chem), 63. *Prof Exp:* Chemist, Chem & Radiol Labs, US Army Chem Ctr, Md, from asst prof to assoc prof, 62-75, PROF CHEM, UNIV NMEX, 75- *Concurrent Pos:* Petrol Res Fund grant, 63-64. *Mem:* Am Chem Soc; Electrochem Soc. *Res:* Chemical education; use of video media in general chemistry instruction. *Mailing Add:* Dept of Chem Univ of NMex Albuquerque NM 87131

CATRAMBONE, JOSEPH ANTHONY, SR, b Chicago, Ill, Sept 21, 24; m 51; c 12. MATHEMATICS, DATA PROCESSING. *Educ:* St Benedict's Col, BS, 47; Univ Maine, MA, 53. *Prof Exp:* Instr math & chem, Damar Acad, 47-50; mathematician adv bd on simulation secretariat, Univ Chicago, 53-54, syst res, 54-56, group leader, Inst Syst Res, 56-58, sr mathematician & asst dir labs appl sci, 58-61; sr scientist sci comput ctr, Serv Bur Corp, 62-63; opers analyst res inst, Ill Inst Technol, 63-68; dir admin data processing, 68-72, asst vpres data processing, 72-78, assoc vpres admin info systs, 78-83, EMER ASSOC VPRES, UNIV ILL, CHICAGO CIRCLE, 83-; VPRES INFO SYSTS, LOYOLA UNIV CHICAGO, 83- *Concurrent Pos:* Instr math, Walton Col Com, 52-, Loyola Univ, 56- & Chicago Jr Col, 57-58; bd dir, Comprehensive Assistance Undergrad Sci Educ, 78-80, vpres, 79 & pres, 80. *Mem:* Asn Comput Math. *Res:* Military weapons systems analyses and evaluations; operations research; computing and information technology. *Mailing Add:* Dir Admin Comp Loyola Univ 820 N Michigan Ave Chicago IL 60611

CATRAVAS, GEORGE NICHOLAS, b Argostoli, Greece, June 22, 16; US citizen. ORGANIC CHEMISTRY, BIOCHEMISTRY. *Educ:* Univ Athens, DCh, 37; Univ Leeds, PhD(org chem), 47; Sorbonne, DSc, 53. *Prof Exp:* Instr org chem, Nat Univ Athens, 37-40; res chemist, Lever Bros & Unilever, Eng, 47-49; in charge res, Nat Ctr Sci Res, France, 50-54; Foreign Oper Admin-Nat Acad Sci fel, Univ Chicago, 54-56, asst prof, 56-63; head biochem res, Technicon Corp, 63-66; proj dir, Molecular Biol Exp, 66-72, chief div neurochem, 72-77, CHMN BIOCHEM DEPT, ARMED FORCES RADIOBIOL RES INST, 77- *Concurrent Pos:* Prof lect, Am Univ, 67- & adj prof, 72- *Mem:* AAAS; Am Soc Biol Chemists; Radiation Res Soc; NY Acad Sci; Royal Soc Chem. *Res:* Intermediary metabolism of lipids; control mechanisms; membranes; action of ionizing radiations on cell constituents; mammalian central nervous system; opiates; microwaves. *Mailing Add:* 8512 Bradmoor Dr Bethesda MD 20817

CATRAVAS, JOHN D, b Athens, Greece, July 24, 50; US citizen; wid; c 1. VASCULAR PHARMACOLOGY, PULMONARY PHARMACOLOGY. *Educ:* Cornell Col, Iowa, BA, 72; Univ Miss, MS, 75, PhD(pharmacol), 78. *Prof Exp:* Postdoctoral fel pulmonary path & pharmacol, Yale Univ, 78-80, res assoc, Dept Anesthesiol, 80-81; from asst prof to assoc prof, 81-90, PROF PHARMACOL & TOXICOL, MED COL GA, 90- *Concurrent Pos:* Prin investr, NIH res grants, 84-; Am Soc Pharmacol & Exp Therapeut travel award, 84; estab investr, Am Heart Asn, 85-90; mem, Toxicol Study Sect, Div Res Grants, NIH, 90-94. *Mem:* Am Thoracic Soc; Soc Toxicol; Am Soc Pharmacol & Exp Therapeut; AAAS; Am Heart Asn; Soc Exp Biol & Med. *Res:* Endothelial cell pathophysiology and pharmacology, focusing on endothelial ectoenzymes, receptors and signal transduction mechanisms. *Mailing Add:* Dept Pharmacol & Toxicol Med Col Ga Augusta GA 30912-2300

CATSIFF, EPHRAIM HERMAN, PHYSICAL CHEMISTRY, POLYMER CHEMISTRY. *Educ:* Pa State Col, BS, 45; Univ Southern Calif, MS, 48; Polytech Inst Brooklyn, PhD(polymer chem), 52. *Prof Exp:* Res asst chem, Princeton Univ, 51-53, res assoc chem, 53-57; chemist, Shell Develop Co, Calif, 57-62; supvr phys properties sect, Res Dept, Thiokol Chem Corp, 62-67, head polymer physics & instrumental res, 67-69, head explor polymer res, 69-74; group leader polymer physics, 74-79, SR SCIENTIST, COMPOSITE PROD RES, CIBA-GEIGY CORP, 79- *Mem:* Am Chem Soc; Soc Rheol; Soc Plastics Engrs; NY Acad Sci; NAm Thermal Anal Soc. *Res:* Creep and stress relaxation of crystalline and amorphous polymers; swelling behavior of filled rubbers; Mullins effect; melt viscometry; epoxy resins characterization; reinforced composites; dynamic mechanical analysis; thermal analysis. *Mailing Add:* 3603 Foxridge Lane Apt D Palos Verdes CA 90274

CATSIMPOOLAS, NICHOLAS, b Athens, Greece, Feb 9, 31; US citizen; m 59; c 3. BIOPHYSICS. *Educ:* Athens Univ, BS, 55 Univ Tenn, MS, 62, PhD(biochem), 64. *Prof Exp:* Biochemist, King Gustaf V Res Inst, Sweden, 59-60; sr res scientist, Res Ctr, Cent Soya Co , Chicago, 65-73; adj assoc prof, Stritch Sch Med, Loyola Univ, 72-73; assoc prof, Mass Inst Technol, 73-80; PROF, SCH MED, BOSTON UNIV, 80- *Mem:* AAAS; Am Soc Biol Chem; Am Chem Soc; Biophys Soc. *Res:* Protein and cell biophysics; electrophoresis. *Mailing Add:* Boston Univ Sch Med 65 Montvale Rd Newton Centre MA 02159

CATT, KEVIN JOHN, b Melbourne, Australia, Sept 24, 32; US citizen; m 69; c 2. ENDOCRINOLOGY, BIOCHEMISTRY. *Educ:* Univ Melbourne, MB, BS, 56, MD, 60; Monash Univ, Australia, PhD(biochem), 67; FRACP, 71. *Prof Exp:* Sr lectr, Dept Med, Monash Univ, 64-68, reader med, 68-69; assoc prof, Endocrine Unit, Dept Med, Med Ctr, Cornell Univ, 69-70; vis scientist, 70-77, CHIEF, SECT HORMONAL REGULATION ENDOCRINOL, NAT INST CHILD HEALTH & HUMAN DEVELOP, NIH, 73-, CHIEF, ENDOCRINOL & REPRODUCTION RES BR, 76- *Concurrent Pos:* Assoc clin prof med, Georgetown Univ, 71-; lectr, Sect Endocrinol, Royal Soc Med, 76; Charles E Culpeper prof, Sch Med, Univ Va, 81. *Honors & Awards:* Eric Sussman Prize Res, Royal Australian Col Physicians, 70. *Mem:* Royal Australian Col Physicians; Coun High Blood Pressure Res; Endocrine Soc; Am Soc Clin Invest; Am Soc Biol Chemists. *Res:* Mechanisms of peptide hormone receptors and regulation of pituitary, gonadal and adrenal functions; regulation of renin-angiotensin system and control of aldosterone secretion. *Mailing Add:* Nat Inst Child Health & Human Develop Endocrinol & Reproduction Res Br Bldg 10 Rm 8C 407 Bethesda MD 20892

CATTANEO, L(OUIS) E(MILE), b Philadelphia, Pa, June 21, 20; m 49; c 5. STRUCTURAL ENGINEERING. *Educ:* Cath Univ Am, BCE, 42. *Prof Exp:* Asst prof civil eng, Cath Univ Am, 46-56; struct res engr, Nat Bur Standards, 56-82; RETIRED. *Mem:* Sigma Xi. *Res:* Structures and materials of construction; live loads on buildings; structural test methods. *Mailing Add:* 5438 MacBeth St Hyattsville MD 20784

CATTANI, EDUARDO, b Rosario, Arg, July 18, 46. ALGEBRAIC & FINANCE GEOMETRY. *Educ:* Univ Buenos Aires, Lic Math, 67; Wash Univ, PhD(math), 72. *Prof Exp:* Courant instr math, NY Univ, 70-73; from asst prof to assoc prof, 73-85, PROF MATH, UNIV MASS, 85- *Concurrent Pos:* Vis prof math, Univ Buenos Aires, 88 & Max Planck Inst, Bonn, 88; corresp mem, CONICET, Arg, 88- *Mem:* Am Math Soc. *Mailing Add:* Dept Math Univ Massachusetts Amherst MA 01003

CATTERALL, JAMES F, b Providence, RI, 1949; m 70; c 2. EUKARYOTIC GENE EXPRESSION, HORMONE-GENE INTERACTION. *Educ:* Colgate Univ, AB, 71; Northwestern Univ, Evanston, MS, 72, PhD(molecular biol), 76. *Prof Exp:* Fel, Dept Cell Biol, Baylor Col Med, 76-78 & res instr, 78-79; asst prof, Rockefeller Univ, 79-84; staff scientist, 79-84, scientist, 84-90, SR SCIENTIST, POP COUN, 90- *Concurrent Pos:* Vis scientist, Lab Molecular Biol, Cambridge, UK, 78; mem, Pop Res Comt, Nat Inst Child Health & Human Develop. *Mem:* Am Soc Cell Biol; Am Soc Microbiol; AAAS; NY Acad Sci; Endocrine Soc. *Res:* Studies of eukaryotic gene structure and regulation of gene expression by steroid hormones; structure-function analysis of glycoprotein horm genes. *Mailing Add:* Pop Coun Rockefeller Univ 1230 York Ave New York NY 10021

CATTERALL, WILLIAM A, b Providence, RI, Oct 12, 40. PHYSIOLOGY. *Educ:* Brown Univ, BA, 68; Johns Hopkins Univ, PhD(physiol chem), 72. *Prof Exp:* Muscular Dystrophy Asn postdoctoral res fel, Lab Biochem Genetics Nat Heart, Lung & Blood Inst, NIH, 72-74, staff fel, 74-76, res chemist, 76-77; assoc prof, 77-81, PROF, DEPT PHARMACOL, SCH MED, UNIV WASH, SEATTLE, 81-, CHMN, 84- *Concurrent Pos:* mem rev panel, Neurobiol, NSF, 80-83; Jacob Javits Neurosci Investr Award, 84; chmn, Gordon Conf Molecular Pharmacol, 85; dir, Grad Prog Neurobiol, Univ Wash, 86-90; assoc, Neurosci Res Prog, 86-; mem bd sci counselors, Nat Heart, Lung & Blood Inst, NIH, 91. *Mem:* Nat Acad Sci; Am Soc Biol Chemists; Am Soc Pharmacol & Exp Therapeut; Am Soc Neurosci; Biophys Soc. *Res:* Author or co-author of over 120 publications. *Mailing Add:* Dept Pharmacol Univ Wash MS SJ-30 Seattle WA 98195

CATTERALL, WILLIAM E(DWARD), b Chicago, Ill, May 29, 20; m 44, 70, 76; c 2. CHEMICAL ENGINEERING. *Educ:* Purdue Univ, BS, 40. *Hon Degrees:* ScD, Mass Inst Technol, 42. *Prof Exp:* Res asst prof chem eng, Univ Pa, 43-46; chem engr, Esso Res & Eng Co, 46-54, sect head, 54-57, eng assoc, 57-63, sr eng assoc, 63-71; RETIRED. *Concurrent Pos:* Consult chem eng, 42-43. *Mem:* Am Chem Soc; Am Inst Chem Engrs. *Res:* Process design and economic evaluation in field of petrochemicals. *Mailing Add:* 5929 East 32nd St Tucson AZ 85711

CATTERSON, ALLEN DUANE, b Denver, Colo, June 26, 29; m 50; c 3. PREVENTIVE MEDICINE, AEROSPACE MEDICINE. *Educ:* Univ Colo, BA, 51, MD, 55; Ohio State Univ, MS, 61. *Prof Exp:* Gen pract, Colo, 58-59; resident aviation med, Ohio State Univ, 59-61; chief resident, Lovelace Found Med Educ & Res, 61-62; actg asst chief, Ctr Med Opers Off, NASA Manned Spacecraft Ctr, 62-63, assoc chief, 63-64, asst to chief, Ctr Med Prog, 64-65, chief flight med br, Ctr Med Off, 65-66, dep dir med res & opers, 66-71; pres, Aerospace Med Consult, 71-76; pres, Airport Med Assocs, 76-83; STAFF PHYSICIAN, KELSEY-SEYBOLD CLIN, 83- *Concurrent Pos:* Mem air traffic controller career comt, US Dept Transp, 69-70. *Honors & Awards:* Melbourne W Boynton Award, Am Astronaut Soc, 71. *Mem:* Fel Aerospace Med Asn; fel Am Col Prev Med. *Res:* Environmental physiology. *Mailing Add:* 4775 Will Clayton Pkway Houston TX 77032

CATTO, PETER JAMES, b Boston, Mass, Mar 23, 43; m 69; c 2. THEORETICAL PLASMA PHYSICS, FUSION. *Educ:* Mass Inst Technol, BS & MS, 67; Yale Univ, PhD(eng & appl sci), 72. *Prof Exp:* Mem sch natural sci, Inst Advan Study, Princeton, NJ, 71-73; asst prof plasma physics, Univ Rochester, 73-78; sr staff scientist, Sci Appln Int Corp, 78-87; SR SCIENTIST, LODESTAR RES CORP, BOULDER, COLO, 87- *Concurrent Pos:* Consult, Fusion Energy Div, Oak Ridge Nat Lab, 74-79 & Magnetic Fusion Energy Prog, Lawrence Livermore Nat Lab, 79-88; vac assoc, Culham Lab, UK, 83 & 87; adv, Off External Affairs, Int Centre Theoret Physics, Trieste, Italy, 88- *Mem:* Fel Am Phys Soc. *Res:* Energetic particle confinement, fast wave minority heating physics, and rf driven fluxes in tokamak plasmas; transport and stability in tokamaks. *Mailing Add:* Lodestar Res Corp 2400 Cent Ave Boulder CO 80301

CATTOLICO, ROSE ANN, b Philadelphia, Pa, July 2, 43; c 1. DEVELOPMENTAL BIOLOGY, MOLECULAR BIOLOGY. *Educ:* Temple Univ, BA, 65, MA, 67; State Univ NY Stony Brook, PhD(develop biol), 73. *Prof Exp:* Res assoc biol, Brookhaven Nat Lab, 67-68; fel biol, McGill Univ, 73-75; from asst prof to assoc prof, 75-86, PROF BOT, UNIV WASH, 86- *Concurrent Pos:* Res grant, Univ Wash, 75 & NSF grant, 76- *Mem:* Soc Develop Biol; Soc Plant Physiologists; Am Soc Cell Biol; Phycological Soc Am; Asn Women Sci. *Res:* Organelle biogenesis; evolution of algal chloroplast DNA; transformation of red algae; rubisco topoisomerase coding site and function in algae. *Mailing Add:* Dept Bot KB-15 Univ Wash Seattle WA 98195

CATTON, IVAN, b Vancouver, BC, June 29, 34; US citizen; m 61; c 3. HEAT TRANSFER, FLUID MECHANICS. *Educ:* Univ Calif, BS, 59, PhD(eng), 66. *Prof Exp:* Engr, Douglas Aircraft Co, 59-62, scientist, 65-67; res engr, 62-65, from asst prof to assoc prof eng, 67-76, PROF ENG, UNIV CALIF, LOS ANGELES, 76- *Concurrent Pos:* Consult, Douglas Aircraft Co, 62-, Rand Corp, 67-, Atomics Int, 70-, mem, Adv Comt Reactor Safety, US Nat Res Coun, 74- & Brookhaven Nat Lab, 82-; ed, J Heat Transfer. *Honors & Awards:* Heat Transfer Mem Award, Am Soc Mech Engrs, 81. *Mem:* Fel Am Soc Mech Engrs; Am Phys Soc; Am Inst Aeronaut & Astronaut; foreign fel Royal Meteorol Soc. *Res:* Thermal stability of fluids; electroconvection in dielectrics; natural convection in confined regions; free surface flows under low gravity conditions; heat transfer in nuclear plants; environmental transport processes; numerical methods; nuclear safety. *Mailing Add:* Univ Calif 48-121 F ENGR IV Los Angeles CA 90024

CATTS, ELMER PAUL, b Elizabeth, NJ, Apr 3, 30; m 52, 79; c 4. MEDICAL ENTOMOLOGY, INSECT ECOLOGY. *Educ:* Univ Del, BSc, 52, MSc, 57; Univ Calif, Berkeley, PhD(parasitol), 63. *Prof Exp:* Asst prof med entom, Univ Del, 62-68, assoc prof med entom & ecol, 68-74, prof entom & ecol, 74-80; PROF & CHMN MED ENTOM & ECOL, WASH STATE UNIV, 80- *Concurrent Pos:* Vis prof, Wash State Univ, 70-71, adj prof wildlife biol, 86-; lectr, Ore Health Sci Univ, 84- *Mem:* Sigma Xi; Entom Soc Am; Wildlife Dis Asn; Am Asn Forensic Sci. *Res:* Myiasis; forensic entomology; behavior and biology of bot flies, tabanids and blow flies. *Mailing Add:* Dept Entom 301 Johnson Hall Wash State Univ Pullman WA 99164

CATURA, RICHARD CLARENCE, b Arkansas, Wis, July 31, 35; m 59; c 3. PHYSICS, X-RAY ASTRONOMY. *Educ:* Univ Minn, BS, 57; Univ Calif, Los Angeles, MS, 59, PhD(physics), 62. *Prof Exp:* Res physicist, Univ Calif, Los Angeles, 62-63; res assoc elem particle physics, Princeton Univ, 63-66; res scientist, 66-73, staff scientist, 73-79, SR STAFF SCIENTIST & LEADER SPACE ASTRON GROUP, LOCKHEED PALO ALTO RES LAB, 79- *Concurrent Pos:* Math consult, Radioisotope Serv, Vet Admin, Los Angeles, Calif, 59-62. *Mem:* Am Phys Soc; Int Astron Union; Am Astron Soc. *Res:* Nuclear physics; transport of high intensity charged particle beams; properties of wide-gap spark chambers; elementary particle physics; solar and cosmic x-ray astronomy; x-ray and extreme ultra violet optics; CCD x-ray image sensors. *Mailing Add:* Lockheed Palo Alto Res Lab D91-30 B256 3251 Hanover St Palo Alto CA 94304

CATZ, BORIS, b Russia, Feb 15, 23; US citizen; m; c 4. MEDICINE. *Educ:* Nat Univ Mex, BS, 41, MD, 47; Univ Southern Calif, MS, 51. *Prof Exp:* Intern, Gen Hosp, Mexico City, 45-46; adj prof, Sch Med, Nat Univ Mex, 47-48; instr, 52-54, asst clin prof, 54-59, ASSOC CLIN PROF MED, UNIV SOUTHERN CALIF, 59- *Concurrent Pos:* Practicing physician, Los Angeles, 51-; chief thyroid clin, Los Angeles County Hosp, 59-69, sr consult, 70- *Mem:* Am Thyroid Asn; fel Am Col Physicians; Soc Exp Biol & Med; Endocrine Soc; fel Am Col Nuclear Med. *Res:* Thyroid disease. *Mailing Add:* 435 N Roxbury Dr Beverly Hills CA 90210

CATZ, CHARLOTTE SCHIFRA, b Paris, France; US citizen. PEDIATRICS, PERINATOLOGY. *Educ:* Univ Buenos Aires, MD, 52. *Prof Exp:* Staff physician, Pediat Serv, Hosp Fernandez, Buenos Aires, Arg, 51-56, supvr, Pediat Tuberc Clin, 55-56; teaching asst pediat, Stanford Univ, 59-61; asst dir, Pediat Out-Patient Clin, Palo Alto Med Ctr, Calif, 63-66; from asst prof to assoc prof pediat, State Univ NY Buffalo, 66-75; pediat med officer, 75-80, CHIEF, PREGNANCY & PERINATOLOGY BR NAT INST CHILD HEALTH & HUMAN DEVELOP, 80- *Concurrent Pos:* Pvt pract pediat, Inst Padua, Buenos Aires, 53-56; actg instr, Sch Med, Stanford Univ, 63-64; assoc attend physician, Children's Hosp, Buffalo, NY, 69-75, clin dir, Birth Defects Ctr, 70-75; Fulbright sr res scholar & guest prof, Univ Rene Descartes, Paris, 73-74; Nat Inst Child Health & Human Develop. *Mem:* Am Acad Pediat; Am Soc Pharmacol & Exp Therapeut; Am Pediat Soc. *Mailing Add:* Pregnancy & Perinatol Br Nat Inst Child Health & Human Develop NIH EPN Bldg 643 Rockville MD 20852

CATZ, JEROME, b New Brunswick, NJ, Dec 11, 32; m 52; c 3. PRODUCT SAFETY, INSTRUMENTATION. *Educ:* Mass Inst Technol, SB & SM, 56, ME, 58, ScD, 60. *Prof Exp:* Asst prof mech eng, Mass Inst Technol, 60-65; assoc prof, 65-70, interim dean, Sch Eng, 72, assoc dean, 72-75, interim dean, 75-77, chmn dept, 78-86, PROF MECH ENG, UNIV MIAMI, 70- *Concurrent Pos:* Consult engr, 54- *Mem:* Am Soc Mech Eng; Nat Soc Prof Engrs; Am Soc Eng Educ; Am Soc Testing & Mat; Am Soc Safety Engrs; Sigma Xi; Instrument Soc Am; Soc Exp Mech. *Res:* Solid ionic arrays as transducers for the measurement of displacement, temperature and heat flux and for use as energy conversion devices. *Mailing Add:* Col Eng Univ Miami PO Box 248294 Coral Gables FL 33124

CAUDILL, REGGIE JACKSON, b Branson, Mo, Oct 20, 49; m 70; c 2. AUTOMATED VEHICLE SYSTEMS. *Educ:* Univ Ala, BSME, 71, MSMH, 73; Univ Minn, PhD(mech eng), 76. *Prof Exp:* Engr dynamic anal, Teledyne-Brown Eng Co, 71-72; res asst, Univ Ala, 72-73, Univ Minn, 73-76; asst prof, Univ Mo-Columbia, 76-77. *Concurrent Pos:* Consult, US Urban Mass Transp Admin & NJ Dept Environ Protection, 78, US Environ Protection Agency, 79 & Transp Systs Ctr, Econ, Inc, 80-81; prin investr, US Dept Transp, 78-80, Nat Sci Found, 80- *Honors & Awards:* Teeter Award, Soc Automotive Engrs. *Mem:* Soc Automotive Engrs; Advan Transit Asn; Am Soc Mech Engrs. *Res:* Design, analysis, and microprocessor implementation of control laws for automated vehicle systems-urban transit, inter-city freight, and industrial material handling; development and integration of computer-aided design, manufacturing, and robotic systems. *Mailing Add:* 20-L Northwood Lake Northport AL 35476

CAUDLE, BEN HALL, b Midlothian, Tex, Apr 27, 23. PETROLEUM ENGINEERING. *Educ:* Univ Tex, BS, 43, PhD(petrol eng), 63. *Prof Exp:* PROF PETROL ENG, UNIV TEX, 63- *Mem:* Nat Acad Eng; hon mem Soc Petrol Eng. *Mailing Add:* Dept Petrol Eng Univ Tex Austin TX 78712

CAUDLE, DANNY DEARL, b Maud, Tex, Oct 8, 37; m 64; c 4. PHYSICAL CHEMISTRY, CORROSION. *Educ:* Centenary Col, BS, 61; Univ Okla, PhD(phys chem), 66. *Prof Exp:* Res asst Chem Res Inst, Univ Okla, 62-66; res scientist, 66-72, corrosion & chem specialist, Prod Dept, 72-74, TECH DEPT, CONOCO INC, 74- *Mem:* Am Chem Soc; Nat Asn Corrosion Eng; Soc Petrol Eng; Sigma Xi; Soc Mining Engrs; Asn Inst Mining Engrs. *Res:* Oil field chemical problems, and environmental engineering. *Mailing Add:* 8835 Sharpview Houston TX 77036

CAUGHEY, DAVID ALAN, b Grand Rapids, Mich, Mar 5, 44; m 69; c 2. AERODYNAMICS, COMPUTATIONAL FLUID DYNAMICS. *Educ:* Univ Mich, BSE, 65; Princeton Univ, MA, 67, PhD(aerospace eng), 69. *Prof Exp:* Scientist aerodynamics, McDonnell Douglas Res Labs, 71-75; from asst prof to assoc prof, 75-84, PROF AEROSPACE ENG, SIBLEY SCH MECH & AEROSPACE ENG, CORNELL UNIV, 84- *Concurrent Pos:* Vis asst prof, Cornell Univ, 74-75; consult, McDonnell Douglas Corp, 75-; vis res scientist, Princeton Univ, 81-82. *Honors & Awards:* Lawrence Sperry Award, Am Inst Aeronaut & Astronaut, 79. *Mem:* Am Inst Aeronaut & Astronaut; Am Phys Soc. *Res:* Numerical techniques for predicting aerodynamic characteristics of flight vehicles, especially in the transonic flow regime. *Mailing Add:* 218 Upson Hall Cornell Univ Ithaca NY 14853

CAUGHEY, JOHN LYON, JR, b Rochester, NY, May 30, 04; m 37; c 1. MEDICAL EDUCATION. *Educ:* Harvard Univ, AB, 25, MD, 30; Columbia Univ, MScD, 35. *Prof Exp:* Intern med, Presby Hosp, New York, 30-32, from asst resident to resident, 32-37; from asst to assoc, Columbia Univ, 35-45; asst dean, 45-48, asst prof med, 45-48, assoc prof clin med, 48-69, prof med & med educ, 69-74, assoc dean, 48-70, dean student affairs, 70-74, EMER PROF MED & MED EDUC, SCH MED, CASE WESTERN RESERVE UNIV, 74-, EMER DEAN STUDENT AFFAIRS, 74-, EMER PROF FAMILY MED, 75- *Concurrent Pos:* Asst physician, Presby Hosp, New York, 37-45; tech aide, Comt Med Res, Off Sci Res & Develop, 43-45. *Honors & Awards:* Abraham Flexner Award, Asn Am Med Col, 74. *Res:* Constitutional medicine; comprehensive health services. *Mailing Add:* 1990 Ford Dr Apt 1208 Cleveland OH 44106

CAUGHEY, THOMAS KIRK, b Scotland, Oct 22, 27; m 52; c 4. APPLIED MECHANICS. *Educ:* Glasgow Univ, BSc, 48; Cornell Univ, MME, 52; Calif Inst Technol, PhD(eng sci), 54. *Prof Exp:* Instr appl mech, 52-54, from asst prof to assoc prof, 55-62, PROF APPL MECH, CALIF INST TECHNOL, 62- *Concurrent Pos:* Consult engr, Jas Howden & Co, Scotland, 49-51, 54-55, NAm Electronics, 55-58, Aeronaut Eng Res Inc, 58-62, Inca Inc, 62-68, Jet Propulsion Lab, Calif Inst Technol, 69- & Tetra-Tech Inc, 70- *Mem:* AAAS; Am Math Soc; Soc Indust & Appl Math; Seismol Soc Am; Sigma Xi. *Res:* Non-linear mechanics; vibrations; acoustics; electronics; applied mathematics; classical physics. *Mailing Add:* 341 S Greenwood Ave Pasadena CA 91107-5018

CAUGHEY, WINSLOW SPAULDING, b Antrim, NH, Nov 9, 26; m 52; c 4. BIOCHEMISTRY. *Educ:* Univ NH, BS, 48, MS, 49; Johns Hopkins Univ, PhD(chem), 53. *Prof Exp:* Sr res asst physiol chem, Med Sch, Johns Hopkins Univ, 53-54, res assoc chem, 54-56; pres, Monadnock Res Inst, 56-59; from asst prof to assoc prof physiol chem, Med Sch, Johns Hopkins Univ, 59-67; prof chem, Univ SFla, 67-69 & Ariz State Univ, 69-73; PROF BIOCHEM & CHMN DEPT, COLO STATE UNIV, 73- *Concurrent Pos:* Chmn subcomt porphyrins, Div Chem & Chem Technol, Nat Acad Sci-Nat Res Coun; Lederle med fac award, 63-66. *Mem:* AAAS; Am Chem Soc; NY Acad Sci; Am Soc Biol Chem; The Chem Soc. *Res:* Inorganic biochemistry; mechanisms of enzymic reactions; hemeproteins and metalloporphyrins; reactions of oxygen in biological systems. *Mailing Add:* Dept Biochem Colo State Univ Ft Collins CO 80523

CAUGHLAN, CHARLES NORRIS, b Pullman, Wash, Jan 20, 15; m 36; c 4. PHYSICAL CHEMISTRY. *Educ:* Univ Wash, BS, 36, PhD(chem), 41. *Prof Exp:* Instr chem, Mont State Col, 41-44; chemist, Eastman Kodak Co, 44-46; from asst prof to assoc prof, 46-51, head dept chem, 67-73, PROF CHEM, MONT STATE UNIV, 51- *Mem:* Am Chem Soc; Am Crystallog Asn. *Res:* Hydrogen bonds in acetoxime; Raman spectroscopy; dielectric properties of polymers and titanium compounds; structures; metal alkoxides; x-ray diffraction; structures of organic phosphates, organic titanates, organic vanadates and natural products. *Mailing Add:* Dept of Chem Mont State Univ Bozeman MT 59715

CAUGHLAN, GEORGEANNE ROBERTSON, b Montesano, Wash, Oct 25, 16; div; c 4. PHYSICS, ASTROPHYSICS. *Educ:* Univ Wash, BS, 37, PhD(physics), 64. *Prof Exp:* From instr to prof, 57-84, EMER PROF PHYSICS, MONT STATE UNIV, 84- *Mem:* Fel Am Phys Soc; Am Astron Soc; Am Asn Physics Teachers; Int Astron Union; Sigma Xi. *Res:* Nuclear astrophysics and the analysis of synthesis of elements in the stars. *Mailing Add:* Dept Physics Mont State Univ Bozeman MT 59717

CAUGHLAN, JOHN ARTHUR, b Pittsfield, Ill, Apr 29, 21; m 42; c 4. ORGANIC CHEMISTRY. *Educ:* Univ Ill, BS, 42; Univ Nebr, MS, 44. *Prof Exp:* Lab foreman, Tenn Eastman Corp, 44-46; res chemist, 46-50, develop chemist, 50-54, head evaluation sect, 54-56, staff asst to dir prod develop, 56-57, asst dir prod develop, 57-60, mgr process chem, 60-62, asst to dir res, 62-67, asst dir res, 67-69, asst dir, Chem Group Res & Develop, 69-72, mgr admin res & develop, Indust Div, 72-76, tech asst, 76-81, TECH ASSOC TO DIR CORP QUAL STAND, MALLINCKRODT INC, 81- *Mem:* Am Chem Soc; Am Pharmaceut Asn. *Res:* Amino ketones; opium alkaloids; organic chemistry; analysis of pharmaceutical chemicals; columbium; tantalum; industrial chemicals. *Mailing Add:* 556 Virginia Ave Webster Groves MO 63119-2795

CAUL, JEAN FRANCES, b Cleveland, Ohio, Aug 19, 15. FOOD SCIENCE. *Educ:* Lake Erie Col, AB, 37; Ohio State Univ, MA, 38, PhD(physiol chem), 42. *Prof Exp:* Res chemist, Borden Co, NY, 42-44; sr proj leader, Arthur D Little, Inc, 44-67; distinguished prof, Kansas State Univ 67-70, prof foods &

nutrit, Col Home Econ, 70-84, head, Sensory Ctr, Food & Nutrit Dept, 84-86; RETIRED. *Concurrent Pos:* Vis instr, Inst Food Sci, Giessen, Ger, 60. *Mem:* Am Chem Soc; fel Inst Food Technol; NY Acad Sci. *Res:* Food technology; flavor measurement; consumer product testing; catfish flavor and preflavoring. *Mailing Add:* 2103 Meadowlark Rd Manhattan KS 66502

CAULDER, JERRY DALE, b Gideon, Mo, Nov 7, 42; m 63; c 1. WEED SCIENCE. *Educ:* Southeast Mo State Univ, BS & BA, 64; Univ Mo, MS, 66, PhD(agron), 69. *Prof Exp:* Res asst weed sci, Univ Mo, 66-70; mkt develop specialist, Monsanto Co, 69-71, mgr, Colombia, SA, 71-73, develop assoc, 73, tech mgr herbicides, 73-74, new prod mgr, 74-84; PRES, MYCOGEN CORP, SAN DIEGO, CALIF, 84- *Res:* Coordination of the discovery, development and manufacture of herbicides and plant growth regulators. *Mailing Add:* 5875 Ciudad Leon Ct San Diego CA 92120-3965

CAULFIELD, DANIEL FRANCIS, b Brooklyn, NY, Aug 4, 35; m 60; c 2. POLYMER CHEMISTRY. *Educ:* Brooklyn Col, BS, 57; Polytech Inst Brooklyn, PhD(chem), 62. *Prof Exp:* Res assoc chem, Polytech Inst Brooklyn, 62; fel Cornell Univ, 62-65; RES CHEMIST, FOREST PROD LAB, US FOREST SERV, 65- *Mem:* Am Chem Soc; Tech Asn Pulp & Paper Indust; fel Int Acad Wood Sci. *Res:* Light scattering; scattering and diffraction of x-rays; structure/properties of paper and cellulose. *Mailing Add:* Forest Prod Lab US Forest Serv Madison WI 53705-2398

CAULFIELD, HENRY JOHN, b Halletsville, Tex, Mar 25, 36; m 66; c 2. OPTICAL COMPUTING, HOLOGRAPHY. *Educ:* Rice Univ, BA, 58; Iowa State Univ, PhD(physics), 62. *Prof Exp:* Staff scientist optics, Tex Instruments, Inc, 62-67; assoc dir night vision, Raytheon Co, 67-68; staff scientist optics, Sperry Rand Res Ctr, 68-72; dir laser technol, Block Eng Inc, 72-77; dir Ctr Optics, Aerodyne Res Inc, 77-85; DIR CTR APPL OPTICS, UNIV ALA, HUNTSVILLE, 85- *Mem:* Fel Optical Soc Am; sr mem Inst Elec & Electronics Engrs; fel Soc Photo-Optical Instrumentation Engrs. *Res:* Optical computing; holography. *Mailing Add:* Ctr Appl Optics Univ Ala Huntsvil Huntsville AL 35899

CAULFIELD, JAMES BENJAMIN, b Minneapolis, Minn, Jan 1, 27; m 50; c 3. CARDIOVASCULAR PATHOLOGY. *Educ:* Miami Univ, BA, 47; Univ Ill, BS, 48, MD, 50. *Prof Exp:* Vis investr, Rockefeller Inst Med Res, 55-56; from instr to asst prof path, Med Ctr, Univ Kans, 56-59; from asst prof to assoc prof path, Harvard Med Sch, 59-75; chmn dept path, Sch Med, Univ SC, 75-85; DEPT PATH, UNIV ALA, BIRMINGHAM, 85- *Concurrent Pos:* USPHS fel, 56-58; asst path, Mass Gen Hosp, 59-69, assoc pathologist, 69-75; pathologist, Shrine Burns Inst, Boston, Mass, 70-75. *Mem:* Int Acad Path; Sigma Xi; Am Asn Pathologists; Am Soc Cell Biol. *Res:* Electron microscopy; spontaneous and induced alterations in fine structure of cells; cardiovascular pathology; collagen matrix of the heart. *Mailing Add:* Dept Path Univ Ala Univ Sta Birmingham AL 35294

CAULFIELD, JOHN P, b Baltimore, Md, Feb 21, 44; div; c 3. PARASITOLOGY, IMMUNOPARASITOLOGY. *Educ:* Loyola Col, BS, 66; Univ Md, MD, 70. *Prof Exp:* Asst prof, 77-81, ASSOC PROF PATH, HARVARD MED SCH, 81-, ASSOC PROF TROP PUB HEALTH, SCH PUB HEALTH, 88- *Concurrent Pos:* Vis prof epidemiol, Shanghai Med Univ, 87. *Mem:* Am Soc Cell Biol; Int Acad Path; Am Acad Path; Am Soc Trop Med Hyg; Am Soc Parasitol. *Res:* Host parasite interactions involving Schistosovia mansoni and human blood cells; mast cell secretion; experimental arthritis. *Mailing Add:* Harvard Med Sch Seeley G Mudd Bldg Rm 517 250 Longwood Ave Boston MA 02115

CAULK, DAVID ALLEN, b Minneapolis, Minn, Sept 24, 50; m 72; c 3. APPLIED MECHANICS. *Educ:* Rensselaer Polytech Inst, BS, 72; Univ Calif, Berkeley, MS, 74, PhD(eng sci), 76. *Prof Exp:* Assoc sr res engr, 76-80, staff res engr, 80-85, SR STAFF RES ENGR ENG MECH, GEN MOTORS RES LABS, 85- *Mem:* Am Soc Mech Engrs. *Res:* Fluid mechanics; metal plasticity; polymer processing; continuum mechanics; metal casting; manufacturing processes. *Mailing Add:* Dept Eng Mech Gen Motors Res Labs Warren MI 48090

CAULTON, M(ARTIN), physics, electronics engineering; deceased, see previous edition for last biography

CAUNA, NIKOLAJS, b Riga, Latvia, Apr 4, 14; US citizen; m 42. ANATOMY, CELL BIOLOGY. *Educ:* Riga Univ, MD, 42; Univ Durham, MSc, 54, DSc, 61. *Prof Exp:* Student demonstr anat, Univ Riga, 35-42, lectr, 42-44; med practitioner, WGer, 44-46; lectr anat, Baltic Univ, Ger, 46-48; from lectr to reader, Durham, Eng, 48-61; prof, 61-84, chmn dept, 75-83, EMER PROF ANAT, SCH MED, UNIV PITTSBURGH, 84- *Concurrent Pos:* Res grants, Royal Soc Eng, 59-61, Am Cancer Soc, 62-63 & USPHS, 62-83. *Mem:* Am Asn Anat; Histochem Soc; Am Soc Cell Biol; Anat Soc Gt Brit & Ireland; fel Royal Micros Soc. *Res:* Development and evolution of tetrapod limbs; development, structure and function of the peripheral receptor organs; control mechanism of the autonomic nervous system; fine structure and functions of the human nasal respiratory mucosa; urticaria. *Mailing Add:* Sch Med Univ Pittsburgh Pittsburgh PA 15261

CAUSA, ALFREDO G, b Montevideo, Uruguay, June 25, 28; US citizen. POLYMER SCIENCE, TEXTILES. *Educ:* Sch Chem & Chem Eng, Montevideo, BSc, 58; Case Inst Technol, MS, 62; Univ Akron, PhD(polymer sci), 68. *Prof Exp:* Chemist textile chem, SAm Subsidiaries, Courtaulds, Ltd, 52-58; res chemist indust fibers, Textile Fibers Div, Can Indust Ltd, 61-64; res chemist polymers, Tarrytown Tech Ctr, Union Carbide Corp, 68-70; PRIN CHEMIST TIRE REINFORCING TECHNOL, GOODYEAR TIRE & RUBBER CO, 70- *Mem:* Am Chem Soc; AAAS. *Res:* Failure modes in tires and other fiber-reinforced composites; fiber fracture; polymer and fiber microstructure; chemistry of fiber finishes and adhesives; chemistry and physics of interfaces. *Mailing Add:* 1255 Ashford Lane Akron OH 44313-6870

CAUSEY, ARDEE, b Baton Rouge, La, July 8, 15; m 38; c 2. CHEMICAL ENGINEERING. *Educ:* Univ La, BS, 36; Univ Mich, MSE, 37. *Prof Exp:* Lab analyst, E I du Pont de Nemours & Co, 37-39, operating supvr camphor plant, 40-46; supvr, Ethyl Corp, 46-56, systs analyst, 56-80; RETIRED. *Mem:* Am Inst Chem Engrs; Am Chem Soc. *Res:* Tetra ethyl lead; oils, fats and wax; organic solvent; fine chemicals; electronic computer. *Mailing Add:* 666 Parlange Dr Baton Rouge LA 70806-1841

CAUSEY, G(EORGE) DONALD, b Baltimore, Md, July 9, 26; m 61; c 4. AUDIOLOGY, SPEECH PATHOLOGY. *Educ:* Univ Md, BA, 50, MA, 51; Purdue Univ, PhD(audiol, speech path), 54. *Prof Exp:* Asst chief audiol clin, DC Health Dept, 54-55; chief acoust res audiol, Vet Benefits Off, 55-64; chief cent audiol & speech path prog, Vet Admin Hosp, 64-82; res prof, Catholic Univ Am, 80-91; RETIRED. *Concurrent Pos:* Res prof, Univ Md, 56-80, dir, Biocommun Lab, 67-80; mem comt hearing, bioacoust & biomech, Nat Acad Sci-Nat Res Coun, 67-82; mem comt hearing aids, Am Nat Standards Inst, 67-82; affil prof eng, Colo State Univ, Ft Collins, 75-78; clin assoc prof surg, Georgetown Univ Med Ctr, 76-82; consult, US Senate, 61-62, Fed Trade Comn, 64-74, Md dept Pub Health, 67-71, US Dept Labor, 77-81; former pres, Biocommon Res Corp, 75-80; nat consult in audiol to US Vet Admin, 84-; pres, Causey Assocs, 87- *Mem:* Fel Am Speech & Hearing Asn; Acoust Soc Am; Sigma Xi; Inst Elec & Electronics Engrs. *Res:* Hearing impairment and measurement techniques; hearing aids; co-patentee of 5 American and 1 British patent. *Mailing Add:* 3504 Dunlop St Chevy Chase MD 20815-5932

CAUSEY, MILES KEITH, b Monroe, La, Dec 23, 40; m 63; c 1. WILDLIFE BIOLOGY. *Educ:* La State Univ, BS, 62, MS, 64, PhD(entom), 68. *Prof Exp:* Asst prof, 68-75, assoc prof, 75-80, PROF ZOOL & ENTOM, AUBURN UNIV, 80- *Mem:* AAAS; Wildlife Soc; Am Soc Mammal. *Res:* Wildlife biology and conservation, especially pesticide-wildlife relationships and environmental degradation. *Mailing Add:* Dept of Zool & Entom Auburn Univ Auburn AL 36849

CAUSEY, WILLIAM MCLAIN, b Cleveland, Miss, Feb 6, 38; c 1. MATHEMATICAL ANALYSIS. *Educ:* Univ Miss, BS, 60, MA, 62; Univ Kans, PhD(math), 66. *Prof Exp:* Asst prof math, Miss State Univ, 66-67; asst prof, Univ Cincinnati, 67-68; assoc prof, 68-73, PROF MATH, UNIV MISS, 73- *Mem:* Am Math Soc; London Math Soc. *Res:* Complex variables; univalent functions. *Mailing Add:* Dept of Math Univ of Miss University MS 38677

CAUTHEN, SALLY EUGENIA, b Montgomery, Ala, Oct 1, 32. BIOCHEMISTRY. *Educ:* Abilene Christian Col, BS, 53; La State Univ, Baton Rouge, MS, 57; Oxford Univ, PhD(biochem), 65. *Prof Exp:* Assoc prof chem, Abilene Christian Col, 56-67; asst prof, Tex Tech Univ, 67-68; PROF CHEM, NORTHEAST LA UNIV, 68- *Mem:* AAAS; Am Chem Soc; Am Soc Microbiol. *Res:* Vitamins and coenzymes, especially biotin, folic acid and vitamin B12; methionine biosynthesis in bacteria; enzymology, especially atropinesterase and methyltransferase; azolesterases related to schizophrenia. *Mailing Add:* Dept of Chem Northeast La Univ Monroe LA 71209

CAUTIS, C VICTOR, b Braila, Romania, June 18, 46; US citizen; m 70; c 1. ELECTROMAGNETIC WAVE PROPAGATION, COMPUTER SCIENCE. *Educ:* Univ Bucharest, Romania, dipl physics, 70; Columbia Univ, New York, MPh, 76, PhD(physics), 77. *Prof Exp:* Res assoc physics, Joint Inst Nuclear Res, Dubna, USSR, 72-73; res asst, Columbia Univ, 74-77; res assoc physics, Stanford Linear Accelerator Ctr, Stanford Univ, 77-81; SR SCIENTIST, TECHNOL COMMUN INT, MOUNTAIN VIEW, CALIF, 81- *Mem:* Am Phys Soc; Inst Elec & Electronics Engrs. *Res:* Electromagnetic wave propagation and radio direction finding. *Mailing Add:* 14360 Debell Rd Los Altos CA 94022

CAVA, MICHAEL PATRICK, b Brooklyn, NY, Feb 13, 26; m 51; c 1. ORGANIC CHEMISTRY. *Educ:* Harvard Univ, BS, 46; Univ Mich, MS, 48, PhD(chem), 51. *Prof Exp:* Fel, Harvard Univ, 51-53; from asst prof to prof chem, Ohio State Univ, 53-65; prof, Wayne State Univ, 65-69, Univ Pa, 69-85; PROF CHEM, UNIV ALA, 85- *Mem:* Am Chem Soc. *Res:* Natural products chemistry; strained ring systems; organosulfur, selenium, and tellurium chemistry. *Mailing Add:* Box 870336 Tuscaloosa AL 35487-0336

CAVAGNA, GIANCARLO ANTONIO, b Milano, Italy, June 3, 38; nat US; m 64; c 2. ORGANIC CHEMISTRY, PAPER CHEMISTRY. *Educ:* Univ Pavia, PhD(org chem), 61. *Prof Exp:* Org res chemist, Inst Carlo Erba Therapeut Res, Italy, 61-63; res chemist, 63-74, sr res chemist, 74-90, RES ASSOC, RES CTR, WESTVACO CORP, 90- *Mem:* Am Chem Soc; Tech Asn Pulp & Paper Indust. *Res:* Organic and colloid chemistry of the papermaking process; chemistry of wood by-products. *Mailing Add:* Westvaco Corp Res Ctr Johns Hopkins Rd Laurel MD 20723

CAVALIER, JOHN F, b Montgomery, WVa, Feb 28, 47; m 70; c 1. MATHEMATICS. *Educ:* WVa Inst Technol, BS, 69; WVa Univ, MS, 72; Va Polytech Inst & State Univ, PhD, 83. *Prof Exp:* PROF MATH, WVA INST TECHNOL, 69- *Mem:* Math Asn Am. *Mailing Add:* PO Box 197 Smithers WV 25186

CAVALIERE, ALPHONSE RALPH, b New Haven, Conn, Jan 23, 37; m 61; c 3. BOTANY. *Educ:* Ariz State Univ, BS, 60, MS, 62; Duke Univ, PhD(mycol), 65. *Prof Exp:* Vis asst prof bot, Duke Univ, 65-66; from asst prof to assoc prof, 66-75, chmn dept, 75-85, PROF BIOL, GETTYSBERG COL, 75- *Concurrent Pos:* Assoc mem, Surtsey Res Soc, Iceland, 65-; mem adv bd, Nat Marine Fisheries Serv, 85- *Mem:* Mycol Soc Am; Am Inst Biol Sci. *Res:* Fungi of Iceland; mycological research on the new volcanic upthrust, Surtsey; marine fungi of Eastern US; marine algae of Bermuda. *Mailing Add:* Dept Biol Gettysburg Col Gettysburg PA 17325

CAVALIERI, ANTHONY JOSEPH, II, b Ft Bragg, NC, Sept 8, 51; m 70, 90; c 1. ENVIRONMENTAL PLANT PHYSIOLOGY. *Educ:* Univ NC, Wilmington, BS, 75; Univ SC, PhD(biol), 79. *Prof Exp:* Fel, Dept Bot, Univ Ill, 80-81; res physiologist, Dept Res Specialists, 81-86, dept dir, 84-90, DIR TECHNOL SUPPORT, PIONEER HI-BRED INT, INC, 90- *Mem:* AAAS; Am Asn Plant Physiologists; Am Soc Agron; Crop Sci Soc Am; Sigma Xi. *Res:* Plant physiology; environmental limitations on crop yield; plant water relations. *Mailing Add:* Plant Breeding Div Pioneer Hi-Bred Int Inc Johnston IA 50131

CAVALIERI, DONALD JOSEPH, b New York, NY, May 5, 42; m 70; c 1. REMOTE SENSING, POLAR OCEANOGRAPHY. *Educ:* City Col New York, BS, 64; Queens Col, New York, MA, 67; NY Univ, PhD(meteorol), 74. *Prof Exp:* Physicist, US Naval Appl Sci Lab, 66-67; from instr to asst prof physics, State Univ NY, 67-70; Nat Res Coun res assoc meteorol, Nat Geophys & Solar-Terrestrial Data Ctr, Nat Oceanog & Atmospheric Admin, 74-76; vis asst prof physics & atmospheric sci, Drexel Univ, 76-77; staff scientist, Systs & Appl Sci Corp, 77-79; phys scientist, Goddard Lab Atmospheric Sci, 79-84, PHYS SCIENTIST, GODDARD LAB HYDROSPHERIC PROCESSES, NASA, 84- *Concurrent Pos:* Assoc ed, J Geophys Res. *Mem:* Am Geophys Union; Am Meteorol Soc; Int Glaciological Soc. *Res:* Passive microwave remote sensing of Arctic and Antarctic sea ice; large scale atmosphere-ocean sea ice interactions. *Mailing Add:* Code 971 Lab Hydrospheric Processes Goddard Space Flight Ctr NASA Greenbelt MD 20771

CAVALIERI, ERCOLE LUIGI, b Milan, Italy, Feb 10, 37. CHEMICAL CARCINOGENESIS. *Educ:* Univ Milan, DSc, 62. *Prof Exp:* Fel, Polytech Zurich, 63-64; lectr & res assoc, Dept Chem, Univ Montreal, 65-67, asst prof, 67-68; res assoc, Melvin Calvin Lab, Lawrence Berkeley Lab, 68-70; asst prof res, Dept Biochem, 71-74, assoc prof, 74-81, PROF RES, DEPT PHARMACEUT SCI, EPPLEY INST RES CANCER, UNIV NEBR MED CTR, 81- *Concurrent Pos:* Proj leader environ carcinogenesis, Sect Polycyclic Hydrocarbons, Pub Health Servs, Nat Cancer Inst, 72-79, prin investr contracts, 79-81; prin investr grants, Nat Cancer Inst, 73-76, 83-86 & 88-91, Nat Inst Environ Health Sci, 79-85; mem sci adv comt, Nat Cancer Inst, 77; prin investr prog proj grant, Nat Cancer Inst, 88-92. *Mem:* Am Chem Soc; AAAS; Am Asn Cancer Res; Fedn Am Scientists; Europ Asn Cancer Res; Soc Free Radical Res; NY Acad Sci. *Res:* Carcinogenesis of polycyclic aromatic hydrocarbons and estrogens with reference to their reaction mechanisms; enzymology of activation; binding to cellular macromolecules; metabolism and tumorigenicity. *Mailing Add:* Eppley Inst Res Cancer Univ Nebr Med Ctr 600 S 42nd St Omaha NE 68198-6805

CAVALIERI, LIEBE FRANK, physical biochemistry, for more information see previous edition

CAVALIERI, RALPH R, b New York, NY, Jan 15, 32; m 57; c 1. ENDOCRINOLOGY, NUCLEAR MEDICINE. *Educ:* NY Univ, BA, 52, MD, 56; Am Bd Internal Med, dipl, 65; Am Bd Nuclear Med, dipl, 73. *Prof Exp:* Intern med, Third Div, Bellevue Hosp, 56-57, resident, 57-59; mem staff nuclear med, US Naval Hosp, Bethesda, Md, 59-61; NATO fel biochem, Nat Inst Med Res, Eng, 61-62; USPHS spec fel nuclear med, Johns Hopkins Univ Hosp, 62-63; CHIEF NUCLEAR MED SERV, VET ADMIN HOSP, 63-; PROF MED & RADIOL, MED CTR, UNIV CALIF, SAN FRANCISCO, 78- *Concurrent Pos:* From asst clin prof to assoc clin prof med & radiol, Med Ctr, Univ Calif, San Francisco, 63-78. *Mem:* Soc Nuclear Med; Asn Am Physicians; Endocrine Soc; Am Fedn Clin Res; Am Soc Clin Invest. *Res:* Thyroid physiology and biochemistry; endocrine control of metabolism; application of radioisotope techniques to medicine. *Mailing Add:* Vet Admin Med Ctr 4150 Clement St San Francisco CA 94121

CAVALLARO, JOSEPH JOHN, b Lawrence, Mass, Mar 18, 32; m 72; c 5. DIAGNOSTIC IMMUNOLOGY. *Educ:* Tufts Col, BS, 52; Univ Mass, Amherst, MS, 54; Univ Mich, Ann Arbor, PhD(virol), 66. *Prof Exp:* Pub health sanitarian, Health Dept, City of Hartford, Conn, 54-55 & 57-61; teaching assoc microbiol, Univ Mass, Amherst, 61-62; res virologist, Med Res Labs, Charles Pfizer & Co, Groton, Conn, 66-67; res assoc virol, dept epidemiol, Virus Lab, Sch Pub Health, Univ Mich, Ann Arbor, 67-71; microbiologist, Diag Immunol Training Sect, Lab Prof Off, 71-86, RES MICROBIOLOGIST, ANAEROBIC BACTERIA BR, HOSP INFECTIONS PROG CTR INFECTIOUS DIS, CTRS DIS CONROL, ATLANTA, GA, 86- *Concurrent Pos:* Consult, PanAm Health Orgn, Columbia, 76 & 77 & Brazil, 77; assoc prof, dept path, Sch Med, Morehouse Col, Ga, 82-; asst prof, dept path & lab med, Sch Med, Emory Univ, Ga, 83- *Mem:* Fel Am Acad Microbiol; Sigma Xi; Am Asn Immunologists; NY Acad Sci; Am Soc Microbiol. *Res:* Laboratory diagnostic immunology/serology test methods; diagnostic anaerobic bacteriology. *Mailing Add:* Ctrs Dis Control 1600 Clifton Rd Bldg 5 Rm 112 Atlanta GA 30333

CAVALLARO, MARY CAROLINE, b Everett, Mass, Feb 2, 32. PHYSICS, SCIENCE EDUCATION. *Educ:* Simmons Col, BS, 54, MS, 56; Ind Univ, EdD, 72. *Prof Exp:* Instr physics & math, Sweet Briar Col, 55-56; instr physics, Simmons Col, 56-58 & Randolph-Macon Woman's Col, 58-59; asst prof, Framingham State Col, 61-63; asst to dean grad studies, 72-78, PROF PHYSICS, SALEM STATE COL, 63-, VCHMN PHYSICS, DEPT CHEM & PHYSICS, 80- *Concurrent Pos:* Boston Univ teaching fel, 59-61, lectr, 61-62. *Mem:* Am Phys Soc; Am Asn Physics Teachers; Nat Sci Teachers Asn. *Res:* Administration in higher education; institutional research graduate education. *Mailing Add:* Dept of Physics Salem State Col Salem MA 01970

CAVALLI-SFORZA, LUIGI LUCA, b Genova, Italy, Jan 25, 22; m 46; c 4. HUMAN POPULATION GENETICS, CULTURAL EVOLUTION. *Educ:* Univ Pavia, Italy, MD, 44; Cambridge Univ, MA, 50. *Hon Degrees:* DSc, Columbia Univ, 80. *Prof Exp:* Asst res genetics, Cambridge Univ, 48-50; lectr genetics & statist, Univ Parma & Univ Pavia, 51-60; prof genetics, Univ Parma, 60-62; prof & dir Genetics Inst, Univ Pavia, 62-70; chmn, 86-90, PROF GENETICS, SCH MED, STANFORD UNIV, CALIF, 70- *Concurrent Pos:* Dir microbiol res, Inst Serum Therapeut, Milan, 50-57; hon fel, Gonville & Caius Col, Cambridge Univ, 82. *Mem:* Foreign assoc Nat Acad Sci; Biomet Soc (pres, 67-68); foreign hon mem Am Acad Arts & Sci; foreign hon mem Japan Soc Human Genetics. *Res:* Bacterial genetics; human population genetics and evolution; cultural evolution; analysis of DNA using restricted enzymes. *Mailing Add:* Dept Genetics Sch Med Stanford Univ Stanford CA 94305-5120

CAVALLITO, CHESTER JOHN, b Perth Amboy, NJ, May 7, 15; m 40; c 3. PHYSIOLOGICAL CHEMISTRY, PHARMACOLOGY. *Educ:* Rutgers Univ, BS, 36; Ohio State Univ, AM, 38, PhD(chem), 40. *Prof Exp:* Asst entom, NJ Exp Sta, 36; res chemist, Goodyear Tire & Rubber Co, 40-41, Winthrop Chem Co, 42-46 & Sterling-Winthrop Res Inst, 46-50; res dir, Irwin, Neisler & Co, 51-63, vpres & dir res, Neisler Labs, Inc, 63-66; prof med chem, Sch Pharm, Univ NC, Chapel Hill, 66-70; exec vpres, Ayerst Labs, 70-78; ADJ PROF MED CHEM, SCH PHARM, UNIV NC, CHAPEL HILL, 78- *Concurrent Pos:* Lectr pharmacol, Univ Ill, 62-66. *Mem:* Fel AAAS; Am Chem Soc; Am Soc Microbiol; fel NY Acad Sci; Acad Pharmaceut Sci; Am Soc Pharmacol Exp Therapeut. *Res:* Medicinals, synthetic and natural; mechanisms of drug action; research administration. *Mailing Add:* 4810 Old Chapel Hill-Hillsborough Rd Hillsborough NC 27278

CAVANAGH, DENIS, b Paisley, Scotland, Dec 27, 23; US citizen; m 51; c 3. OBSTETRICS & GYNECOLOGY. *Educ:* Univ Glasgow, MB, ChB, 52; FRCOG. *Prof Exp:* Mike Hogg Award, Postgrad Sch Med, Univ Tex, 59; from asst prof to prof obstet & gynec, Univ Miami, 59-66; prof & chmn dept, Sch Med, St Louis Univ, 66-77; PROF OBSTET & GYNEC, UNIV S FLA, 77- *Mem:* AAAS; AMA; fel Am Col Surg; fel Am Col Obstet & Gynec; Am Gynec Soc. *Res:* Diagnosis and treatment of gynecological cancer; clinical and laboratory aspects of septic shock; eclamptogenic toxemia. *Mailing Add:* Dept Obstet & Gynec Univ SFla Col Med H Lee Moffitt Cancer Ctr Tampa FL 33620

CAVANAGH, HARRISON DWIGHT, b Atlanta, Ga, July 22, 40; m 64; c 1. VISUAL SCIENCE RESEARCH, CORNEAL SURGERY. *Educ:* Johns Hopkins Univ, AB, 62, MD, 65; Harvard Univ, PhD(biol), 72. *Prof Exp:* Intern pediat, Johns Hopkins Hosp & Univ Sch Med, 65-66, resident & instr ophthal, 69-73; teaching fel biol, Grad Sch Arts & Sci, Harvard Univ, 66-69; fel ophthal, Corneal Unit, Mass Eye & Ear Infirmary, Harvard Med Sch, 75-75, asst prof, 73-76; assoc prof ophthal & corneal surg, Sch Med, Emory Univ, 76-78, F Phinizy Calhoun sr prof & chmn, Ophthal Dept, 78-87; PROF OPHTHAL, SCH MED, GEORGETOWN UNIV, 87- *Concurrent Pos:* Consult, Nat Eye Inst, 75-, mem & chmn Visual Sci Study Sect, 80-84; assoc secy, Am Acad Ophthal, 79-83; exec assoc ed, Ophthal Times, 81-83; exec secy-treas, Asn Res Vision & Ophthal, 82-87; consult, Dept Defense, Ophthal Div, USAF, 82- *Honors & Awards:* Jean Lacerte Lectr, Lavel Univ, Que, 82 & Can Royal Col Surgeons; U K Bochner Lectr, Toronto Univ, 83; Joseph E Koplowitz Lectr, Georgetown Univ, 84. *Mem:* Am Acad Ophthal; fel Am Col Surgeons; Asn Res Vision & Ophthal; Contact Lens Asn Ophthamologists (pres elect, 87-). *Res:* Biochemistry of the Corneal Epithelium; growth regulation. *Mailing Add:* Ctr Sight Sch Med Georgetown Univ 3800 Reservoir Rd NW Washington DC 20007

CAVANAGH, PETER R, b Wolverhampton, England, July 31, 47; m 68, 81; c 4. HUMAN FOOT, HUMAN LOCOMOTION. *Educ:* Loughborough Col, UK, BEd, 69; Royal Free Med Sch, Univ London, PhD(human biomech), 72. *Prof Exp:* Res asst biomech, Royal Free Med Sch, Univ London, 69-72; asst prof biomech, 72-79, assoc prof biomech, 75-81, PROF BIOMECH, PENN STATE UNIV, 81-, PROF & DIR, CTR LOCOMOTION STUDIES, 86. *Concurrent Pos:* Consult, US Olympic Comt, 84-88, NASA, 86-; expert witness falls, footwear & personal injury; vis prof, Manchester Diabetes Ctr, UK, 90-91; chair, Res Comt Foot Coun, 90- *Honors & Awards:* Muybridge Medal, Int Soc Biomech, 87; Wolffe Lectr, Am Col Sports Med, 87. *Mem:* Am Soc Biomech (pres, 86-88); Am Col Sports Med; Int Soc Biomech (coun mem, 88); Am Acad Phys Educ fel; Am Acad Podiat Sports Med; Orthopedic Res Soc; Am Diabetes Assoc; Am Orthop Foot & Ankle Soc. *Res:* Mechanics of lower extermity, principally the foot and its function in human locomotion; diabetic foot, zero-gravity locomotion, shoe research, falls in the elderly and injury preventing research; populations studied; diabetics, the elderly, recreational athletes, and mobility impaired individuals; biomechanics. *Mailing Add:* Ctr Locomotion Studies Rm 10 IM Bldg Pa State Univ University Park PA 16802

CAVANAGH, RICHARD ROY, b Detroit, Mich, Dec 3, 50. MOLECULAR DYNAMICS AT SURFACES,. *Educ:* Wayne State Univ, BA, 72; Harvard Univ, MS, 74, PhD(phys chem), 78. *Prof Exp:* Res fel, 79-80, RES CHEMIST, NAT BUR STANDARDS, 80- *Honors & Awards:* Silver Medal, Dept Commerce, 85, Gold Medal, 90. *Mem:* Am Chem Soc; Am Vacuum Soc; Am Phys Soc. *Res:* Molecular dynamics of surfaces examined by laser diagnostics; energy transfer rates; the pathways and time scales of energy dissipation; laser driver nonthermal reactions. *Mailing Add:* B 248 Chemistry Nat Inst Standard Tech Gaithersburg MD 20899

CAVANAGH, TIMOTHY D, b Berkeley, Calif, Aug 16, 38; m 63; c 2. MATHEMATICS. *Educ:* Calif State Univ, Sacramento, AB, 59, MA, 62; Ohio State Univ, PhD(math educ), 65. *Prof Exp:* From asst prof to assoc prof, 65-73, PROF MATH, UNIV NORTHERN COLO, 73- *Concurrent Pos:* Vis prof, Univ Col, Galway, Ireland, 70-71. *Mem:* Math Asn Am; Nat Coun Teachers Math; Am Educ Res Asn. *Res:* Mathematics education. *Mailing Add:* Dept of Math Univ of Northern Colo Greeley CO 80639

CAVANAH, LLOYD (EARL), b Keytesville, Mo, Sept 18, 19; m 48; c 2. AGRONOMY. *Educ:* Univ Mo, BS, 48, MS, 50. *Prof Exp:* From instr to assoc prof, Univ Mo-Columbia, 48-74, supt dept res farm, 62-77, prof agron, 74-85; RETIRED. *Concurrent Pos:* Exec secy-treas, Mo Seed Improv Asn, Univ Mo, 55-62; dir, Mo Found Seeds; consult, AID Progs Africa, 62-85. *Mem:* Am Soc Agron; Asn Off Seed Certifying Agencies. *Res:* Quality seeds; factors that affect the cleanliness and germination of seeds. *Mailing Add:* 2022 Crestridge Dr Columbia MO 65203

CAVANAUGH, JAMES RICHARD, b Philadelphia, Pa, Sept 17, 34; m 61; c 5. MAGNETIC RESONANCE SPECTROSCOPY. *Educ:* St Joseph's Col, Philadelphia, BS, 56; Columbia Univ, New York, MS, 57, PhD(chem), 60. *Prof Exp:* Res chemist indust res, Mobil Oil Co, Princeton, 60-65; chemist res spectros, 65-72, leader, 72-81, lab chief govt res, 81-88, RES CHEMIST, EASTERN REGIONAL RES CTR, USDA, 88- *Concurrent Pos:* Instr chem, St Joseph's Col, Philadelphia, 67-73. *Mem:* Am Chem Soc; AAAS; Fedn Analytical Chem & Spectros Soc. *Res:* Applications of nuclear magnetic resonance and other spectroscopy techniques to systems of biological and biochemical interest. *Mailing Add:* 27 Meade Rd Ambler PA 19002-5122

CAVANAUGH, ROBERT J, b Scranton, Pa, Nov 11, 42; m 64; c 3. ORGANIC CHEMISTRY. *Educ:* Carnegie-Mellon Univ, BS, 64; Univ Pittsburgh, PhD(org chem), 67. *Prof Exp:* SR RES CHEMIST, E I DU PONT DE NEMOURS & CO, INC, 69- *Mem:* Am Chem Soc. *Res:* Reactions of enamines; catalytic oxidation; polymer synthesis. *Mailing Add:* 18 Chemin Du Pre Du Camp 1228 Plan Les Ouates Plan Les Ouates 1228 Switzerland

CAVARETTA, ALFRED S, b Boston, Mass, 44. SPLINE FUNCTIONS. *Educ:* Dartmouth Col, MA, 66; Univ Wis, PhD(math), 70. *Prof Exp:* Asst prof math, Middlebury Col, 70-71 & Calif Inst Technol, 71-73; PROF, KENT STATE UNIV, 73- *Concurrent Pos:* Vis prof math, Univ Wis, 78. *Mem:* Math Asn Am; Am Math Soc. *Mailing Add:* Dept Math Kent State Univ Kent OH 44242

CAVAZOS, LAURO FRED, b King Ranch, Tex, Jan 4, 27; m 54; c 10. ANATOMY. *Educ:* Tex Tech Col, BA, 49, MA, 51; Iowa State Univ, PhD(physiol), 54. *Prof Exp:* Teaching asst, Tex Tech Col, 49-51; res asst, Iowa State Col, 51-54; from instr to assoc prof anat, Med Col Va, 54-64; prof anat & chmn dept, Tufts Univ, 64-72, assoc dean, Sch Med, 72-73, actg dean, 73-75, dean, 75-80; PROF BIOL SCI & PRES, TEX TECH UNIV & PROF ANAT & PRES, HEALTH SCI CTR, 80. *Honors & Awards:* Lauro F Cavazos Award, 87. *Mem:* Am Asn Anat; Asn Am Med Cols; AAAS; Sigma Xi; Histochem Soc. *Res:* Physiology; histochemistry, electron microscopy and biochemistry of male reproductive system; hormonal factors; fine structure of cells of steroid secretion. *Mailing Add:* 9327 Oak Knolls Lane Houston TX 77078

CAVE, MAC DONALD, b Philadelphia, Pa, May 14, 39; m; c 2. ANATOMY, MOLECULAR BIOLOGY. *Educ:* Susquehanna Univ, BA, 61; Univ Ill, MS, 63, PhD(anat), 65. *Prof Exp:* Am Cancer Soc Swed-Am Exchange fel, Inst Genetics, Univ Lund, 65-66; USPHS fel, Max Planck Inst Biol, 66-67; asst prof anat & cell biol, Sch Med, Univ Pittsburgh, 67-72; assoc prof, 72-79, PROF ANAT, COL MED, UNIV ARK, 79- *Mem:* Am Soc Microbiol; Am Asn Anat; Am Soc Cell Biol. *Res:* The replication and structure of the genetic machinery; synthesis of chromosomal proteins and nucleic acids; organization of genes coding for ribosomal RNA; development of diagnostic and epidemiologic probes for microorganisms. *Mailing Add:* Dept Anat Univ Ark 4301 W Markham Little Rock AR 72205-7199

CAVE, WILLIAM THOMPSON, b Winnipeg, Man, June 8, 17; US citizen; m 41; c 2. PHYSICAL CHEMISTRY. *Educ:* Univ Man, BSc, 39; Oxford Univ, PhD (phys chem), 48. *Prof Exp:* Res chemist, Shawinigan Chems Co, 44-46 & 48-51; res mgr, Cent Res Dept, Monsanto Chem Co, 51-68, tech dir, Monsanto Co, 68-70, dir nuclear opers, Mound Lab, Monsanto Res Corp, 70-82; RETIRED. *Concurrent Pos:* Mem inspection bd, UK & Can, 40-44. *Mem:* Fel AAAS; NY Acad Sci; Am Chem Soc; Soc Appl Spectros. *Res:* Spectroscopy; organic chemistry; instrumentation. *Mailing Add:* 7021 Sulky Lane Rockville MD 20852-4352

CAVELL, RONALD GEORGE, b Sault St Marie, Ont, Oct 15, 38; m 60; c 2. INORGANIC CHEMISTRY. *Educ:* McGill Univ, BSc, 58; Univ BC, MSc, 60, PhD(inorg chem), 62; Cambridge Univ, PhD, 64. *Prof Exp:* From asst prof to assoc prof, 64-74, PROF CHEM, UNIV ALTA, 74- *Honors & Awards:* Alcan Lectr Award, Chem Inst Can, 79. *Mem:* Chem Inst Can; The Chem Soc; Am Chem Soc. *Res:* X-ray and ultraviolet photoelectron spectroscopy; chemistry of transition metal complexes of phosphorus, nitrogen and sulphur ligands; particularly fluorides; chemistry of halogen and perfluoroalkyl derivatives of phosphorus and related compounds. *Mailing Add:* Dept Chem Univ Alta Edmonton AB T6G 2G2 Can

CAVENDER, FINIS LYNN, toxicology, for more information see previous edition

CAVENDER, JAMES C, b Tuxedo, NY, July 27, 36; m 64; c 2. MYCOLOGY. *Educ:* Union Col, BS, 58; Univ Wis, MS, 61, PhD(bot), 63. *Prof Exp:* Res assoc bact, Univ Wis, 63-64; asst prof biol, Wabash Col, 64-69; asst prof, 69-70, ASSOC PROF BOT, OHIO UNIV, 71- *Mem:* AAAS; Bot Soc Am; Mycol Soc Am. *Res:* Ecology, taxonomy and morphology of cellular slime molds. *Mailing Add:* Dept of Bot Ohio Univ Athens OH 45701

CAVENDER, JAMES VERE, JR, b San Antonio, Tex, Oct 10, 22; m 45; c 2. ORGANIC POLYMER CHEMISTRY. *Educ:* Agr & Mech Col, Tex, BSc, 48. *Prof Exp:* Res chemist, Monsanto Chem Co, Mo, 48-51, res chemist, Tex, 51-54, res group leader, 54-74, sr process specialist, Monsanto Co, 74-85; RETIRED. *Mem:* Am Chem Soc. *Res:* Linear polyolefin process development; Ziegler chemistry; organic synthesis; organic medicinals. *Mailing Add:* 1527 18th Ave N Texas City TX 77590-5349

CAVENDER, PATRICIA LEE, b Warren, Ohio, Sept, 26, 50; m 77; c 3. AGRICULTURAL CHEMISTRY. *Educ:* Allegheny Col, BS, 72; Univ Ill, PhD(org chem), 77. *Prof Exp:* Fel, Schering Plough, Inc, 77-79; res chemist, 80-89, SR RES CHEMIST, AGR CHEM DIV, FMC CORP, INC, 89- *Mem:* Am Chem Soc; Sigma Xi. *Res:* Design and synthesis of agrochemicals including insecticides, plant growth regulators and herbicides. *Mailing Add:* FMC Corp Inc Princeton NJ 08543

CAVENEY, STANLEY, b Chester, Eng, Mar 26, 45; m 69; c 1. CELL BIOLOGY, DEVELOPMENTAL BIOLOGY. *Educ:* Univ Witwatersrand, S Africa, BSc, 68; Oxford Univ, DPhil(zool), 71. *Prof Exp:* Asst prof, 73-77, ASSOC PROF ZOOL, UNIV WESTERN ONT, 77- *Mem:* Am Soc Cell Biol; Can Soc Cell Biol; Soc Develop Biol. *Res:* Cell biology of insect development; cellular interactions and communication during insect metamorphosis. *Mailing Add:* Dept of Zool Univ of Western Ont London ON N6A 5B8 Can

CAVENY, LEONARD HUGH, b Atlanta, Ga, Oct 30, 34; m 57; c 5. COMBUSTION, ENERGY CONVERSION. *Educ:* Ga Inst Technol, BME, 56, MS, 60; Univ Ala, PhD(mech eng), 69. *Prof Exp:* Prin engr res & develop, Huntsville Div, Thiokol Corp, 60-67; sr prof staff res, Dept Aerospace & Mech Sci, Princeton Univ, 69-80; prog mgr, energy conversion, US Air Force Off Sci Res, Bolling AFB, Washington, DC, 80-85; DEP DIR, INNOVATIVE & SCI TECHNOL, STRATEGIC DEFENSE INITIATIVE ORGN, PENTAGON, WASHINGTON, DC 85- *Concurrent Pos:* Consult, Huntsville Div, Thiokol Corp, 70-80, Elkton Div, 71-80, US Army Res Off, 69-75, Object Recognition Corp, 76-80, Los Alamos Nat Labs, 78-80 & appl combustion technol, 83-85; chmn, Elec Propulsion Comt, 85-88. *Honors & Awards:* A T Colwell Award, Soc Automotive Engrs, 73; Nat Heat Transfer Award, Am Soc Mech Engrs, 78. *Mem:* Am Inst Aeronaut & Astronaut; Combustion Inst; Am Soc Mech Engrs. *Res:* Experimental and analytical research in combustion, ignition, energy conversion, heat transfer, rocket propulsion and propellants; digital systems, refractory materials and aeroacoustics. *Mailing Add:* 13715 Piscataway Dr Ft Washington MD 20744

CAVERS, PAUL BRETHEN, b Toronto, Ont, Jan 18, 38; m 61; c 4. PLANT ECOLOGY, WEED SCIENCE. *Educ:* Ont Agr Col, BSA, 60; Univ Wales, PhD(weed ecol), 63; Univ London, DIC, 72. *Prof Exp:* Lectr, Univ Western Ont, 63-64, from asst prof to assoc prof, 64-78, actg chmn, 78-79 & 86-87, PROF ECOL, UNIV WESTERN ONT, 78- *Concurrent Pos:* Chmn, subcomt life hist studies, Expert Comt Weeds, Can; grant selection comt pop biol, Nat Res Coun Can, 73-76; vis scientist, Commonwealth Sci & Indust Res, Canberra, Australia, 87-88. *Mem:* Weed Sci Soc Am; Ecol Soc Am; Brit Ecol Soc; Agr Inst Can; Can Bot Asn(pres, 73-74). *Res:* Effects of herbicides and biological control on seed production, viability and dormancy in weeds; seed dispersal and dormancy; seedling establishment; dynamics of seed and plant populations. *Mailing Add:* Dept Plant Sci Univ Western Ont London ON N6A 5B7 Can

CAVERT, HENRY MEAD, b Minneapolis, Minn, Mar 30, 22; m; m 46; c 3. PHYSIOLOGY, MEDICAL SCHOOL & RESEARCH ADMINISTRATION. *Educ:* Univ Minn, MD, 51, PhD(physiol), 52. *Prof Exp:* From asst to assoc prof physiol, 51-68, asst dean med sch, 57-64, PROF PHYSIOL, MED SCH, UNIV MINN, MINNEAPOLIS, 68-, ASSOC DEAN, 64- *Concurrent Pos:* Am Heart Asn res fel, 51-54, estab investr, 54-57; Nat Heart Inst spec res fel biochem & vis prof biochem, Sch Med, Univ Edinburgh, 61-62; mem heart prog proj comt, Nat Heart & Lung Inst, 66-69, consult, 69-; mem, Basic Sci Coun, Am Heart Asn; prin investr, Gen Clin Res Ctr, Univ Minn Med Sch, 78- *Mem:* AAAS; Am Physiol Soc; Asn Am Med Cols; Sigma Xi; Am Med Asn; Am Heart Asn. *Res:* Transport of sugars and amino acids across muscle cell membranes; effects of muscle activity on transmembrane transport; metabolism of myocardium; physician manpower needs and supply; research administration. *Mailing Add:* 3-120 Owre Hall UMHC Box 293 Univ of Minn Med Sch Minneapolis MN 55455

CAVES, CARLTON MORRIS, b Muskogee, Okla, Oct 24, 50; m 84; c 2. THEORETICAL & INFORMATION PHYSICS, QUANTUM OPTICS. *Educ:* Rice Univ, BA, 72; Calif Inst Technol, PhD, 79. *Prof Exp:* Res fel, Calif Inst Technol, 79-81, sr res fel, 82-87; ASSOC PROF ELEC ENG & PHYSICS, UNIV SOUTHERN CALIF, 88- *Honors & Awards:* Einstein Prize, Soc Optical & Quantum Electronics, 90. *Mem:* Am Phys Soc; Optical Soc Am; AAAS; Int Soc Gen Relativity & Gravitation. *Res:* Quantum-mechanical limitations on high-precision measurements; quantum noise theory; theoretical investigations of experimental gravitation; quantum optics; physics of information. *Mailing Add:* Ctr Laser Studies Univ Southern Calif Los Angeles CA 90089-1112

CAVES, THOMAS COURTNEY, b Pryor, Okla, Apr 8, 40; m 64; c 1. THEORETICAL CHEMISTRY, QUANTUM CHEMISTRY. *Educ:* Univ Okla, BS, 62; Columbia Univ, PhD(chem physics), 68. *Prof Exp:* NASA res fel atomic physics, Harvard Col Observ, 68-69, res assoc, 69; asst prof, 69-74, ASSOC PROF CHEM, NC STATE UNIV, 74- *Concurrent Pos:* Vis fel, Gen Elec Res & Develop Ctr, 81-82. *Mem:* Am Phys Soc; Am Chem Soc; Sigma Xi. *Res:* Ab initio calculation of atomic and molecular properties; scattered wave x-alpha calculations on atomic clusters; generalized valance bond calculations on small molecules. *Mailing Add:* 1541 Caswell St Raleigh NC 27608

CAVEY, MICHAEL JOHN, b Elkhorn, Wis, Oct 8, 46. MORPHOLOGY, EMBRYOLOGY. *Educ:* Univ Va, BA, 68; Univ Wash, MS, 71, PhD(zool), 73. *Prof Exp:* Res scientist anat, Sch Med, Univ Southern Calif, 74-76; asst prof, 76-82, ASSOC PROF BIOL SCI, UNIV CALGARY, 82-, CHMN DIV ZOOL, 86-, ADJ ASSOC PROF ANAT, 90- *Concurrent Pos:* NIH fel, 75-76; co-op ed, Cell & Tissue Res, 84-88, ed, 88- *Mem:* Sigma Xi; Am Soc Zoologists; Can Soc Zoologists; Am Asn Anatomists; Am Micros Soc. *Res:* Fine structure and differentiation of contractile tissues and intercellular junctions; morphogenetic movements in marine invertebrates and their subcellular mechanisms; structural and functional assessments of branchial structures in euryhaline decapod crustaceans. *Mailing Add:* Dept Biol Sci 2500 University Dr NW Calgary AB T2N 1N4 Can

CAVIN, WILLIAM PINCKNEY, b Spartanburg, SC, June 2, 25; m 50; c 2. ORGANIC CHEMISTRY. *Educ:* Wofford Col, AB, 45; Duke Univ, AM, 46; Univ NC, PhD(chem), 53. *Prof Exp:* Asst, Duke Univ, 45-46; from instr to prof, 46-62, John M Reeves prof, 62-88, chmn dept, 71-88, EMER JOHN M REEVES PROF CHEM, WOFFORD COL, 88- *Concurrent Pos:* Teaching asst & NSF res fel, Univ NC, 50-53; NSF fac fel-vis prof chem, Brown Univ, 65-66. *Mem:* Am Chem Soc; Sigma Xi. *Res:* Isotope effect of carbon-14 in organic reactions; kinetics and mechanisms of organic reactions. *Mailing Add:* 704 Perrin Dr Spartanburg SC 29302-2404

CAVINESS, BOBBY FORRESTER, b Asheboro, NC, Mar 24, 40; m 61; c 1. COMPUTER ALGEBRA, ANALYSIS OF ALGORITHMS. *Educ:* Univ NC, Chapel Hill, BS, 62; Carnegie-Mellon Univ, MS, 64, PhD(math), 68. *Prof Exp:* Asst prof math, Duke Univ, 67-70; asst prof comput sci, Univ Wis-Madison, 70-75; assoc prof comput sci, Ill Inst Technol, 75-76; from assoc prof to prof math sci, Rensselaer Polytech Inst, 76-81; chairperson comput sci, 81-87, PROF COMPUT SCI & MATH, UNIV DEL, 81- *Concurrent Pos:* NSF grant, 69-70, 71-74 & 76-78; assoc ed, Asn Comput Mach Transactions Math Software, 75-77 & 89-, J Symbolic Computation, 85-; nat lectr, Asn Comput Mach, 81-82; vis res fel, Gen Elec Res & Develop Ctr, 80-81, Hewlett-Packard Equip grants, 82, 84, Syst Develop Found grant, 83; prog dir, NSF, 87-88. *Mem:* Asn Comput Mach; Math Asn Am; Soc Indust & Appl Math; AAAS; Fedn Am Scientists; AAAS. *Res:* Symbolic and algebraic computation; recursive systems for symbolic mathematics; analysis of algorithms; symbolic and algebraic computation, particulary decision procedures for determining closed-form solution for differential equations and research on designing algebraic algorithims for new computer architectures. *Mailing Add:* Dept Comput & Info Sci Univ Del Newark DE 19716

CAVINESS, VERNE STRUDWICK, JR, b Raleigh, NC, July 25, 34; m 62; c 2. NEUROLOGY, NEUROPATHOLOGY. *Educ:* Duke Univ, BA, 56; Oxford Univ, DPhil(exp path), 60; Harvard Univ, MD, 62. *Prof Exp:* Asst prof, 71-76, ASSOC PROF NEUROL, HARVARD MED SCH, 76-; DIR RES NEUROPATH DEVELOP, EUNICE KENNEDY SHRIVER CTR, 76- *Concurrent Pos:* NIH spec res fel, Harvard Med Sch, 69-71; asst neurologist, Mass Gen Hosp, 71-; sr investr, E K Shriver Inst, 71- *Mem:* AAAS; Am Neurol Asn; Am Acad Neurol. *Res:* Developmental neuroanatomy and neuropathology. *Mailing Add:* Dept Neurol Mass Gen Hosp 32 Fruit St Boston MA 02114

CAVITT, STANLEY BRUCE, b Red Oak, Tex, Apr 5, 34; m 65; c 1. PETROLEUM CHEMISTRY. *Educ:* NTex State Col, BA, 56, MS, 57; Univ Tex, PhD(chem), 61. *Prof Exp:* Chemist, Am Oil Co, 61-63; RES ASSOC, TEXACO CHEM CO, INC, 63- *Mem:* Am Chem Soc; The Catalysis Soc. *Res:* Catalyst research; petrochemicals; organic synthesis. *Mailing Add:* Texaco Chem Co Inc PO Box 15730 Austin TX 78761-5730

CAVONIUS, CARL RICHARD, b Santa Barbara, Calif, Dec, 23, 32. RESEARCH ADMINISTRATION. *Educ:* Wesleyan Univ, BA, 53; Brown Univ, MSc, 61, PhD(psychol), 62. *Prof Exp:* USPHS fel, Brown Univ, 62-63; res scientist, Human Sci Res, Inc, 63-65; dir, Eye Res Found, 65-71; von Humboldt fel, Univ Munich, 71-73; J McKeen Cattell Fund fel, Cambridge Univ, 73-74; chief sci officer, Lab Med Phys, Univ Amsterdam, 74-75; PROF NEUROPHYS, UNIV DORTMUND, W GER, 76- *Concurrent Pos:* From asst prof to assoc prof, Sch Med, Univ Md, 64-70; res dir, Inst Pedestrian Res, 73-74; dir, Inst Arbeitsphysiol, 80-81. *Honors & Awards:* Humboldt Award, 71. *Mem:* Am Psychol Asn; fel Optical Soc Am; Human Factors Soc; Exp Psychol Soc; Europ Brain & Behav Soc; fel Am Psychol Asn. *Res:* Sensory coding and processing; human psychophysics, especially visual; human factors and applied physiology. *Mailing Add:* Inst for Occup Physiol Ardeystrasse 67 Dortmund D 4600 Germany

CAWEIN, MADISON JULIUS, hematology, clinical pharmacology; deceased, see previous edition for last biography

CAWLEY, EDWARD PHILIP, b Jackson, Mich, Sept 1, 12; m 39; c 2. DERMATOLOGY. *Educ:* Univ Mich, AB, 36, MD, 40; Am Bd Dermat, dipl, 47. *Prof Exp:* Asst prof dermat, Med Sch, Univ Mich, 48-51; prof, 51-83, EMER PROF DERMAT, SCH MED, UNIV VA, 83-, CHMN DEPT, 51- *Concurrent Pos:* Dir, Am Bd Dermat, 58-67, pres. *Mem:* Am Dermat Asn; Soc Invest Dermat; AMA; NY Acad Sci. *Res:* Medical mycology and dermatopathology. *Mailing Add:* Dept Dermat Univ Va Sch Med Charlottesville VA 22908

CAWLEY, EDWARD T, b Chicago, Ill, Mar 13, 31; m 55; c 5. ECOLOGY. *Educ:* Northern Ill State Teachers Col, BS, 53; Univ Wis, MS, 58, PhD(bot), 60. *Prof Exp:* From asst prof to assoc prof, 60-71, PROF BIOL, LORAS COL, 71-, DIR, ENVIRON RES CTR, 73- *Concurrent Pos:* Mem Iowa State Preserves Adv Bd, 63-75, chmn, 63-68; consult ed, Brown Publ Co. *Mem:* Sigma Xi; Ecol Soc Am. *Res:* Fresh water ecology-diversity indices; ecology of upper Mississippi River. *Mailing Add:* Dept of Biol Loras Col Dubuque IA 52003

CAWLEY, JOHN JOSEPH, b Somerville, Mass, Sept 18, 32; m 56; c 6. PHYSICAL ORGANIC CHEMISTRY. *Educ:* Boston Col, BS, 55; Harvard Univ, MA, 57, PhD(chem), 61. *Prof Exp:* Fel, Univ Wash, 60-61; asst prof, 61-69, ASSOC PROF CHEM, VILLANOVA UNIV, 69- *Concurrent Pos:* Vis prof chem, McMaster Univ, 69. *Mem:* Am Chem Soc; The Chem Soc. *Res:* Chromic acid oxidations; reductions of B-keto systems; epihalohydrin chemistry. *Mailing Add:* Dept of Chem Villanova Univ Villanova PA 19085

CAWLEY, LEO PATRICK, b Oklahoma City, Okla, Aug 11, 22; m 48; c 3. PATHOLOGY, CLINICAL IMMUNOLOGY. *Educ:* Okla State Univ, BS, 48; Univ Okla, MD, 52; Am Bd Path, dipl, 57, cert clin chem, 65, cert blood banking, 73, cert radioisotopic path, 74, Am Bd Nuclear Med, dipl, 76, Am Bd Med Lab Immunol, dipl, 81. *Prof Exp:* Intern path, Wesley Hosp, Wichita, Kans, 52-53, resident, 53-54; resident, Wayne County Gen Hosp, Eloise, Mich, 54-57; clin pathologist & assoc dir labs, 57-69, dir labs, 69-76, dir res & develop, dept lab med, 76-, SCI DIR, WESLEY MED CTR. *Concurrent Pos:* Clin assoc prof path, Sch Med, Univ Kansas, Wichita, 77-80, prof, 80-; sci dir, Wesley Med Res Found. *Mem:* AAAS; Am Soc Clin Pathologists; Am Soc Human Genetics; Am Soc Exp Path; Am Chem Soc. *Res:* Biochemistry; characterization of proteins other than hemoglobin of human erythrocytes, using methods based on column chromatography, electrophoresis, analytic electrophoresis, immunoelectrophoresis and gas chromatography. *Mailing Add:* 7137 E Main Scottsdale AZ 85251-4315

CAWLEY, ROBERT, b Scranton, Pa, Jan 29, 36; m 58; c 4. THEORETICAL PHYSICS, DYNAMICAL SYSTEMS & FRACTALS. *Educ:* Mass Inst Technol, BS, 58; Univ Ill, MS, 60, PhD(physics), 65. *Prof Exp:* Asst prof physics, Clarkson Col Technol, 65-67; res physicist, Nuclear Physics Div, 67-90, PRIN RES SCIENTIST, MATH & INFO SCI BR (CODE R44), NAVAL SURFACE WARFARE CTR, 90- *Mem:* Am Phys Soc; Am Math Soc; Soc Indust & Appl Math; Am Geophys Union. *Mailing Add:* Math & Info Sci Br (Code R44) Naval Surface Warfare Ctr Silver Spring MD 20903-5000

CAWLEY, WILLIAM ARTHUR, b New York, NY, Dec 11, 25; m 51; c 6. ENVIRONMENTAL ENGINEERING. *Educ:* Harvard Univ, AB, 46; Tufts Univ, BS, 50; Mass Inst Technol, MS, 55. *Prof Exp:* Jr sanit engr, Mass Dept Pub Health, 50-52, asst sanit engr, 52-54; sanit engr, Rayonier, Ind, 55-59; tech ed, Scranton Pub Co, 59; staff consult, Res Dept, Mead Corp, 59-63; mgr eng, Sewage & Waste Treatment Dept, Cochrane Div, Crane Co, 63-66; chief pollution control tech br, Fed Water Qual Admin, 66-68, dir, Div Process Res & Develop & Div Water Qual Res, 68-71; dep dir, Off Prog Mgt, Off Res & Develop, US Environ Protection Agency, 71-75, dir, Tech Support Div, 75-78, dep dir, Indust Environ Res Lab, 78-84 & Hazardous Waste Eng Res Lab, 84-87; DIR, GULF COAST HAZARDOUS SUBSTANCE RES CTR, LAMAR UNIV, 87-, RES PROF, 87- *Concurrent Pos:* Ed, J Environ Eng, Div Am Soc Civil Engrs, 73-74. *Honors & Awards:* Bronze Medal, Environ Protection Agency, 83, 84. *Mem:* Am Soc Civil Engrs; Water Pollution Control Fedn; Sigma Xi; Am Acad Environ Engrs; Am Inst Chem Engrs; Air & Waste Mgt Asn. *Res:* Industrial waste treatment; hazardous waste treatment and disposal. *Mailing Add:* Lamar Univ Box 10613 Beaumont TX 77710

CAYEN, MITCHELL NESS, b Montreal, Que, Nov 6, 38; m 67; c 2. DRUG METABOLISM. *Educ:* McGill Univ, BSc, 59, MSc, 61, PhD(agr chem), 65. *Prof Exp:* Res biochemist & head metab sect, Ayerst Labs Res Inc, 65-87, sr res assoc, 80-87, assoc dir, Drug Metab Div, 87-89; SR DIR, DRUG METAB & PHARMACOKINETICS, SCHERING-PLOUGH RES, 89- *Concurrent Pos:* Fel coun arteriosclerosis, Am Heart Asn; ed, Drug Metab Newsletter, 86-; drug metab subsect steering comt, Pharmaceut Mfrs Asn, 87-91. *Mem:* Am Soc Pharmacol & Exp Therapeut; AAAS; NY Acad Sci; Sigma Xi; Int Soc Study Xenobiotics (secy, 86-87, pres, 92-93). *Res:* Lipid metabolism; pathogenesis of atherosclerosis; drug metabolism; pharmacokinetics. *Mailing Add:* 98 Autumn Ridge Rd Bedminster NJ 07921

CAYFORD, AFTON HERBERT, b Hollywood, Calif, Dec 15, 29; m 51; c 4. COMPLEX ANALYSIS, NUMBER THEORY. *Educ:* La Verne Col, BA, 51; Univ Calif, Los Angeles, MA, 58, PhD(math), 61. *Prof Exp:* Mem tech staff, Hughes Aircraft Co, 56-58; instr math, Univ BC, 59-62; res scientist, Jet Propulsion Lab, Calif Inst Technol, 62-63; asst prof, 62-70, ASSOC PROF MATH, UNIV BC, 70- *Mem:* AAAS; Am Math Soc; Math Asn Am; Can Math Soc; Sigma Xi. *Res:* Analytic number theory and properties of certain classes of entire functions of one or several complex variables with special characteristics on domains or finite point sets; growth rate. *Mailing Add:* Dept of Math 121-1984 Mathematics Rd Univ BC Vancouver BC V6T 1Y4 Can

CAYLE, THEODORE, enzymology, industrial microbiology; deceased, see previous edition for last biography

CAYWOOD, STANLEY WILLIAM, JR, b Akron, Ohio, Aug 10, 24; m 59. ORGANIC CHEMISTRY. *Educ:* Harvard Univ, AB, 48, AM, 49; Univ NH, MS, 50; Cornell Univ, PhD(org chem), 53. *Prof Exp:* Res chemist, Elastomers Dept, E I Du Pont de Nemours & Co, Inc, 53-80, res assoc, Polymer Prod Dept, 80-85; RETIRED. *Concurrent Pos:* Vis prof, Clarkson Col Technol, 64-65. *Res:* Polymer synthesis, elastomer testing, evaluation and processing; polymer colloids, synthesis, and characterization. *Mailing Add:* 115 Watford Rd Westgate Farms Wilmington DE 19808

CAYWOOD, THOMAS E, b Lake Park, Iowa, May 9, 19; m 41; c 3. OPERATIONS RESEARCH. *Educ:* Cornell Col, AB, 39; Northwestern Univ, MA, 40; Harvard Univ, PhD(math), 47. *Prof Exp:* Tutor math, Northwestern Univ, 39-40; tutor, Harvard Univ, 41-42, spec res assoc physics, 42-45, asst, 46; sr mathematician & coordr res, Inst Air Weapons Res, Ill, 47-52; supvr opers res, Armour Res Found, 52-53; partner, Caywood-Schiller Assoc, 53-61, Peat, Marwick, Caywood, Schiller & Co, 62-66 & Caywood-Schiller Assoc, 66-70; vpres, Caywood-Schiller Div, A T Kearney Co, 71-78; PROF LECTR MGT SCI & PROD MGT, UNIV CHICAGO, 78- *Concurrent Pos:* Lectr, Sch Bus, Univ Chicago, 53-57 & Sch Bus & Econ, Calif State Univ, Hayward, 80-; consult, Off Asst Secy Defense, 53-60 & opers eval group, US Navy, 60-61; pres, Invests Opers Mex, SA, 59-60; mem alumni bd dir, Cornell Col, 62-65, trustee, 64-, pres bd trustees, 70-72; mem statist comt, Nat Acad Sci-Nat Res Coun, 60-63; mem defense sci bd & chmn ord panel, Off Dir Defense Res & Eng, 60-64; chmn task force, Gun Systs Acquisition, Defense Sci Bd, 75; ed, Opers Res, Opers Res Soc Am, 61-68. *Honors & Awards:* George E Kimball Medal, Opers Res Soc Am, 74. *Mem:* Am Math Soc; Am Inst Indust Eng; Opers Res Soc Am (vpres, 68-69, pres, 69-70); Math Asn Am; Inst Mgt Sci. *Res:* Applied mathematics; operational research; theory of games; probability; evaluation engineering. *Mailing Add:* 704 Argyle Flossmoor IL 60422

CAZEAU, CHARLES J, b Rochester, NY, June 25, 31; m 60; c 2. GEOLOGY. *Educ:* Univ Notre Dame, BS, 54; Fla State Univ, MS, 55; Univ NC, PhD(geol), 62. *Prof Exp:* Explor Geologist, Humble Oil & Refining Co, Tex, 55-56, 57-58; asst prof geol, Clemson Col, 60-63; asst prof, 63-67, ASSOC PROF GEOL, STATE UNIV NY BUFFALO, 67- *Concurrent Pos:* Sigma Xi study grant, 64-65; consult, Union Camp Corp, 74-75. *Mem:* Am Asn Petrol Geologists; Soc Econ Paleontologists & Mineralogists; AAAS; Nat Asn Geol Teachers. *Res:* Detrital mineralogy; recent sediments; Triassic and Pleistocene geology; geoarchaeology of West Mexico; cultural resources. *Mailing Add:* Dept of Geol State Univ of NY Buffalo NY 14214

CAZIER, MONT ADELBERT, b Cardston, Alta, May 27, 11; US citizen; m 55; c 4. SYSTEMATIC ENTOMOLOGY. *Educ:* Univ Calif, BS, 35, PhD(entom), 42. *Prof Exp:* Asst cur entom, Dept Insects & Spiders, Am Mus Natural Hist, 41-43, assoc cur, 46-51, cur, 52-59, chmn dept, 46-59, founder & dir, Southwestern Res Sta, 55-59, resident dir, 50-62; prof zool, Ariz State Univ, 62-81; RETIRED. *Concurrent Pos:* Res entomologist, Univ Calif, 62-64. *Res:* Bionomic entomology; biology; ecology and behavior studies in aculeate Hymenoptera, Coleoptera and Diptera. *Mailing Add:* 3811 S Terrace Rd Tempe AZ 85282

CAZIN, JOHN, JR, b Wheeling, WVa, July 2, 29; m 53; c 3. MICROBIOLOGY, MEDICAL MYCOLOGY. *Educ:* Univ NC, BS, 52, MS, 54, PhD(bact), 57. *Prof Exp:* Instr, 57-58, assoc, 58-59, from asst prof to assoc prof, 59-72, PROF MICROBIOL, UNIV IOWA, 72- *Mem:* Am Soc Microbiol; Mycol Soc Am; Int Soc Human & Animal Mycol; fel Am Acad Microbiol; Med Mycol Soc of the Americas; Sigma Xi. *Res:* Antibiotics and c-albicans; nutrition and sporulation of allescheria; extracellular nucleases from yeasts; virulence and immunology of cryptococus, allescheria and phycomycetes; morphology and physiology of phycomycetes; antigenic peptides of alternaria. *Mailing Add:* Univ Iowa 3-532 Basic Sci Bldg Iowa City IA 52240

CEASAR, GERALD P, b New York, NY, Jan 8, 40; m 67; c 1. SOLID STATE PHYSICS, ELECTRICAL ENGINEERING. *Educ:* Manhattan Col, BS, 62; Columbia Univ, PhD(chem), 67. *Prof Exp:* Air Force Off Sci Res fel chem, Calif Inst Technol, 68-69; NATO & Ramsay Mem fel, Univ Bristol & Oxford Univ, 69; asst prof, Univ Rochester, 69-74; sr scientist, Webster Res Ctr, Xerox Corp, 74-83; res mgr, ARCO Solar, 83-84; RES MGR, BP AM, 84- *Concurrent Pos:* Consult solar energy, Dept of Energy. *Mem:* Am Phys Soc; Am Chem Soc. *Res:* Large area electronics, solar photovoltaic and opto-electronic device technology; amorphous silicon semiconductors; CVD diamond processing and applications; thin film technology, surface and molecular spectroscopies, liquid crystals. *Mailing Add:* BP Am Res Ctr 4440 Warrensville Ctr Rd Cleveland OH 44128

CEBALLOS, RICARDO, b Cadiz, Spain, Jan 12, 30; m 57; c 3. PATHOLOGY, NEUROLOGY. *Educ:* Univ Madrid, MD, 53. *Prof Exp:* Intern, Dept Physiol, Med Sch, Univ Madrid, 47-49, intern, Dept Internal Med, 49-52, rotating intern, 52-53; assoc prof path, Clin Concepcion, Madrid, 55-57; instr, Med Ctr, Univ Ala, 58, asst prof, 58-60; asst pathologist & dir labs, Hotel Dieu Hosp, Kingston, Ont, 60-63; asst dir anat path, 63-64, dir, 64-76, CERT PATHOLOGIST, UNIV HOSP & HILLMAN CLIN, BIRMINGHAM, 76-; PROF PATH & ASSOC PROF NEUROL, MED CTR, UNIV ALA, 69- *Concurrent Pos:* Fel, Menendez & Pelayo Univ, Spain, 53; fel path, Inst Clin & Med Invest, 53-54; Doherty Found fel histochem & path, Med Sch, Univ Ala, 55; Marquesa de Pelayo Found fel, Spain, 56; Med Res Coun Can fel, 62; intern, Dept Path, Gen Hosp, Prov Diputacion of Madrid, Spain, 50-53; secy, Int Cong Internal Med, 56; consult, Vet Admin Hosp, Tuskegee, Ala, 58-60 & 64-67; lectr, Med Sch, Queen's Univ, Ont, 60-63; assoc prof path & instr neurol & med, Med Ctr, Univ Ala, 64-69; consult, WHO. *Mem:* Int Acad Path; AMA; Latin Am Soc Path. *Res:* Pituitary changes in head trauma; neuropathology; hyperparathyroidism; congenital heart disease. *Mailing Add:* 2300 Lane Circle Birmingham AL 35223

CEBRA, JOHN JOSEPH, b Philadelphia, Pa, May 7, 34; m 56; c 4. IMMUNOBIOLOGY, IMMUNOCHEMISTRY. *Educ:* Univ Pa, AB, 55; Rockefeller Inst, PhD(immunochem), 60. *Prof Exp:* Nat Found fel immunol, Weizmann Inst Sci, 60-61; from instr to assoc prof microbiol, Col Med, Univ Fla, 61-67; assoc prof biol, Johns Hopkins Univ, 67-69, prof, 69-79; prof biol, 79-81, chmn dept, 79-83 & 87-90, ANNENBERG PROF NATURAL SCI, UNIV PA, 81- *Concurrent Pos:* NIH career develop award, 64-67; vis prof immunochem, St Mary's Hosp Med Sch, London, 66-67; mem study sect, Nat Inst Allergy & Infectious Dis, 67-71; instr-in-charge physiol course, Marine Biol Lab, Woods Hole, Mass, 72-76; Guggenheim fel & vis scientist, Walter & Eliza Hall Inst Med Res, Melbourne, Australia, 83-84; vis chair prof, Inst Molecular Biol, Acad Sinica, Taipei, Taiwan, Repub of China, 87-88. *Honors & Awards:* Eli Lilly Award Microbiol & Immunol, Am Soc Microbiol, 68. *Mem:* Am Asn Immunologists; Am Soc Microbiol; AAAS. *Res:* Protein chemistry; structure of immunoglobulins; interaction of antigen, antibody and complement components; cellular synthesis of immunoglobulin polypeptide chains. *Mailing Add:* Joseph Leidy Lab Dept Biol Univ Pa Philadelphia PA 19104-6018

CEBULL, STANLEY EDWARD, b Albany, Calif, June 1, 34; m 56; c 3. STRUCTURAL GEOLOGY. *Educ:* Univ Calif, Berkeley, AB, 57, MA, 58; Univ Wash, PhD(geol), 67. *Prof Exp:* Geologist, Tex Petrol Co, Venezuela, 58-62; from asst prof to assoc prof, 67-81, PROF GEOSCI, TEX TECH UNIV, 81- *Mem:* Am Geophys Union; Am Asn Petrol Geologists; fel Geol Soc Am. *Res:* Tectonic geology. *Mailing Add:* Dept of Geosci Texas Tech Univ Lubbock TX 79409

CECCHI, JOSEPH LEONARD, b Chicago, Ill, Mar 29, 47; m 70; c 2. MICROELECTRONIC FABRICATION, THIN-FILM ETCHING & DEPOSITION. *Educ:* Knox Col, AB, 68; Harvard Univ, AM, 69, PhD(physics), 72. *Prof Exp:* Res assoc atomic physics, Argonne Nat Lab, 67; res staff mem plasma physics, 72-79, res physicist, 79-83, PRIN RES PHYSICIST & HEAD, PLASMA PROCESSING, PLASMA PHYSICS LAB, PRINCETON UNIV, 83-, DIR, PROG PLASMA SCI & TECHNOL, 87- *Concurrent Pos:* Vpres & sr scientist, Princeton Sci Consult, Inc, Princeton, NJ, 79-; lectr with rank of prof, Dept Chem Eng, Princeton Univ, 88- *Mem:* Am Phys Soc; Am Vacuum Soc; Electrochem Soc. *Res:* Plasmas for materials processing; plasma etching of silicon semiconductors; atomic processes; materials physics. *Mailing Add:* Dept Chem Eng Princeton Univ Princeton NJ 08544

CECH, CAROL MARTINSON, b Albert Lea, Minn, Dec 26, 47; m 70; c 2. BIOCHEMISTRY, BIOPHYSICAL CHEMISTRY. *Educ:* Grinnell Col, BA, 70; Univ Calif, Berkeley, PhD(chem), 75. *Prof Exp:* Jane Coffin Childs fel biochem, Harvard Univ, 75-77; asst prof chem, Univ Colo, 78-86, sci dir, Somatogenetics Inst, 87-89; DIR, SPEC PROJ, SOMATOGEN, INC, 89- *Mem:* Biophys Soc; Am Soc Biol Chemists. *Res:* RNA polymerase-DNA interactions; gene regulation; blood substitutes. *Mailing Add:* Somatogen Inc 5797 Central Ave Boulder CO 80301

CECH, FRANKLIN CHARLES, forest genetics, for more information see previous edition

CECH, JOSEPH JEROME, JR, b Berwyn, Ill, Dec 5, 43; m 67; c 2. PHYSIOLOGICAL ECOLOGY OF FISHES. *Educ:* Univ Wis-Madison, BS, 66; Univ Tex, Austin, MA, 70, PhD(zool), 73. *Prof Exp:* Assoc fisheries sci, Portland Maine, 73-75; PROF FISHERIES BIOL, UNIV CALIF, DAVIS, 75- *Mem:* Am Inst Biol Sci; Ecol Soc Am; Am Fisheries Soc; Sigma Xi; Estuarine Res Fedn; AAAS; Am Soc Ichthyol Herpet. *Res:* Investigations in the physiological adjustments and adaptations of marine estuarine and freshwater fishes to their environments with emphasis on respiratory, circulatory, ionic and hematological responses to extreme environments or environmental changes. *Mailing Add:* Dept of Wildlife & Fisheries Biol Univ of Calif Davis CA 95616

CECH, ROBERT E(DWARD), b Minneapolis, Minn, Mar 23, 24; m 50; c 3. METALLURGY. *Educ:* Univ Wis, BS, 48, PhD(metall eng), 67; Rensselaer Polytech Inst, MS, 54. *Prof Exp:* Metallurgist, Gen Elec Res Lab, 48-62 & 65-70; instr metall, Univ Wis, 62-65; consult, Gen Motors, 77-85; metallurgist, Watervliet Arsenal, 85-87; RETIRED. *Concurrent Pos:* Consult, 70-77. *Honors & Awards:* Mathewson Award, Am Inst Mining, Metall & Petrol Engrs, 62; Award, Indust Res Mag, 69. *Mem:* Am Inst Mining, Metall & Petrol Engrs; Am Soc Metals; NY Acad Sci. *Res:* Metallurgical reaction kinetics; technology of cobalt-rare earth and ferrite magnets. *Mailing Add:* 196A Hetcheltown Rd Scotia NY 12302

CECH, THOMAS ROBERT, b Chicago, Ill, Dec 8, 47; m 70; c 2. BIOCHEMISTRY, MOLECULAR BIOLOGY. *Educ:* Grinnell Col, BA, 70; Univ Calif, Berkeley, PhD(chem), 75. *Hon Degrees:* DSc, Grinnell Col, 87 & Univ Chicago, 91. *Prof Exp:* Nat Cancer Inst fel molecular biol, dept biol, Mass Inst Technol, 75-77; from asst prof to assoc prof, 78-83, PROF DEPT CHEM, UNIV COLO, BOULDER , 83-; INVESTR, HOWARD HUGHES MED INST, 88- *Concurrent Pos:* Prin investr NIH res grant, 78-; res career develop award, NIH, 80. *Honors & Awards:* Nobel Prize in Chem, 89; Passano Found Young Scientist Award, 84; Harrison Howe Award, 84; Pfizer Award, Enzyme Chem, 85; US Steel Award, Molecular Biol, 87; V D Mattia Award, 87; Newcombe-Cleveland Award, AAAS, 88; Heineken Prize, 88; Gairdner Found Int Award, 88; Louisa Gross Horwitz Prize, 88; Lasker Award, 88; Resenstiel Award, 89; Warren Triennial Prize, 89. *Mem:* Nat Acad Sci; Am Acad Arts & Sci; Am Soc Biochem & Molecular Biol. *Res:* Ribonucleic acid splicing; chromosome telomeres; biologic catalysis by ribonucleic acid. *Mailing Add:* Dept Chem Univ Colo Campus Box 215 Boulder CO 80309-0215

CECIL, DAVID ROLF, b Tulsa, Okla, July 12, 35; m 58; c 1. MATHEMATICS, COMPUTER SCIENCE. *Educ:* Univ Tulsa, BA, 58; Okla State Univ, MS, 60, PhD(math), 62. *Prof Exp:* Sales engr fluid dynamics, Black, Sivalls & Bryson, 57-58; sr res mathematician, Atlantic Ref Co, 62; asst prof math, North Tex State Univ, 62-69; prof, Butler Univ, 69-70; assoc prof, 70-73, chmn dept, 80-85, PROF MATH, TEX A &I UNIV, 73- *Concurrent Pos:* Consult, Region 2, Educ Serv Ctr, Air Force Off Sci Res, Lackland AFB, Tex. *Mem:* Asn Comput Mach; Sigma Xi; Am Statist Soc. *Res:* Program debugging; discrete probability; combinatorics, computer applications. *Mailing Add:* Dept Math Tex A&I Univ Kingsville TX 78363

CECIL, HELENE CARTER, b Tunkhannock, Pa, Jan 25, 33; m 54; c 1. REPRODUCTIVE PHYSIOLOGY. *Educ:* Univ Md, BS, 63, PhD(poultry physiol), 68. *Prof Exp:* RES PHYSIOLOGIST, USDA, 57- *Concurrent Pos:* Bd dirs, Poultry Sci Asn, 83-84. *Mem:* Am Physiol Soc; Poultry Sci Asn (pres, 87-88); AAAS. *Res:* Avian reproductive environmental pollutants on reproduction. *Mailing Add:* Avian Physiol Lab Bldg 262 USDA Beltsville MD 20705

CECIL, SAM REBER, b San Francisco, Calif, Feb 22, 16; m 47; c 2. FOOD SCIENCE. *Educ:* Milligan Col, BS, 37; Univ Ga, MSA, 54. *Prof Exp:* Asst biochem, Sch Med, Vanderbilt Univ, 37-39, asst nutrit, 40-41; from asst food technologist to assoc food technologist, 41-58, food scientist, 58-67, prof, 67-81, EMER PROF FOOD SCI RES, AGR EXP STA, UNIV GA, 81- *Concurrent Pos:* Jr food technologist, Ore State Col, 56-58; mem sci adv coun, Refrigeration Res Found, 74-81. *Mem:* Inst Food Technologists. *Res:* Canning and freezing of fruits and vegetables; storage of canned foods, military and civil defense rations; effects of newer cultural practices on processing and product quality of peanuts. *Mailing Add:* 1119 Maple Dr Griffin GA 30223

CECIL, THOMAS E, b Louisville, Ky, Dec 13, 45; m 71; c 3. GEOMETRY. *Educ:* Col Holy Cross, AB, 68; Brown Univ, PhD(math), 73. *Prof Exp:* Asst prof math, Vassar Col, 73-78; from asst prof to assoc prof, 78-86, PROF MATH, COL HOLY CROSS, 86- *Concurrent Pos:* NSF basic res grants, 75, 76 & 87-91. *Mem:* Am Math Soc; Math Asn Am. *Res:* Taut immersions of manifolds; differential geometry. *Mailing Add:* Dept of Math Col of the Holy Cross Worcester MA 01610

CECILE, MICHAEL PETER, b Chapleau, Ont, Oct 15, 46; m 77; c 1. EARTH SCIENCES, STRATIGRAPHY. *Educ:* Univ Waterloo, BSc, 70; Carleton Univ, MSc, 73, PhD(geol), 76. *Prof Exp:* RES SCIENTIST GEOL, GEOL SURV CAN, 77- *Concurrent Pos:* Fel, Geol Surv Can-Nat Res Coun Can, 76-77. *Mem:* Geol Asn Can; Can Soc Petrol Geologists (pres, 88). *Res:* Regional geology; Paleozoic stratigraphy; economic geology; Protenozoic and Archean geology. *Mailing Add:* Geol Surv Can 3303 33rd St NW Calgary AB T2L 2A7 Can

CEDAR, FRANK JAMES, b Ottawa, Ont, Aug 28, 45. ENVIRONMENTAL LAW. *Educ:* Carleton Univ, BSc, 67, BA, 81; Univ Alta, PhD(synthetic org chem), 73. *Prof Exp:* CHEM EVAL OFFICER, REGISTRATION OF PESTICIDES, PESTICIDES DIR, AGR CAN, 73- *Concurrent Pos:* Exec secy, consult comt indust bio-test pesticide, 81-82; dir, Pesticides Info Div, 86- *Mem:* Chem Inst Can; Am Chem Soc. *Res:* Evaluates data submitted to government by industry for the registration of wood preservatives and industrial biocides. *Mailing Add:* Pesticides Dir Agr Can Carling Ave Ottawa ON K1A 0C5 Can

CEDER, JACK G, b Spokane, Wash, Aug 25, 33; m 55; c 2. PURE MATHEMATICS. *Educ:* Univ Wash, BS, 55, MS, 57, PhD(math), 59. *Prof Exp:* From asst prof to assoc prof, 59-74, PROF MATH, UNIV CALIF, SANTA BARBARA, 74- *Mem:* Am Math Soc. *Res:* Abstract topological spaces; real functions. *Mailing Add:* 1002 E Canon Perdido Santa Barbara CA 93101

CEDERBERG, JAMES W, b Oberlin, Kans, Mar 16, 39; m 67; c 2. MOLECULAR PHYSICS, PHYSICS. *Educ:* Univ Kans, AB, 59; Harvard Univ, AM, 60, PhD(physics), 63. *Prof Exp:* Lectr & res fel physics, Harvard Univ, 63-64; from asst prof to assoc prof, 64-80, PROF PHYSICS, ST OLAF COL, 80- *Concurrent Pos:* NSF sci fac fel, Duke Univ, 69-70; res assoc, Harvard Univ, 76-77; pres & physics cur, Coun on Undergrad Res, 85-88; vis scientist, Am Inst Physics, 87-88. *Mem:* Am Phys Soc; Am Asn Physics Teachers. *Res:* Molecular beams; rotational magnetic moments and magnetic interactions in molecules; molecular hyperfine interactions. *Mailing Add:* Dept Physics St Olaf Col 1520 St Olaf Ave Northfield MN 55057-1098

CEDERGREN, ROBERT J, b Minneapolis, Minn, Nov 27, 39; US & Can citizen. TRNA GENES, BACTERIOLOGY. *Educ:* Univ Minn, BChem, 61; Cornell Univ, PhD(org chem), 66. *Prof Exp:* Postdoctoral biochem, Cornell Univ, 65-66; instr chem, Univ Toronto, 66-67; from asst prof to assoc prof biochem, 67-78, HULL PROF BIOCHEM, UNIV MONTREAL, 79- *Concurrent Pos:* Vis prof biol, Mass Inst Technol, 74-75, vis prof biochem, Univ Mass, Amherst, 83-84. *Mem:* Fel Can Inst Advan Res; Can Soc Biochem; Am Soc Biochem. *Res:* Structure, synthesis, 3-D modeling and evolution of RNA, with special emphasis on the catalytic RNAs. *Mailing Add:* Dept Biochem Univ Montreal Montreal PQ H3C 3J7 Can

CEGLIO, NATALE MAURO, b New York, NY, Sept 6, 44. X-RAY OPTICS, MICROFABRICATION. *Educ:* Columbia Univ, BA, 66, BS, 67; Mass Inst Technol, MS, 69, PhD(physics), 76. *Prof Exp:* Instr physics, Naval Postgrad Sch, 69-73; physicist, 76-80, DEP GROUP LEADER, LAWRENCE LIVERMORE NAT LAB, 81- *Res:* X-ray and thermonuclear burn imaging; time and space resolved x-ray spectroscopy; microfabrication for x-ray optics applications; physics of high temperature, high density plasmas; carbon dioxide gas lasers and gas discharge phenomena. *Mailing Add:* 1177 Avenida Delas Palmas Livermore CA 94550

CEGLOWSKI, WALTER STANLEY, b Newark, NJ, Nov 24, 32; m 64; c 1. MICROBIOLOGY. *Educ:* Univ Vt, BS, 54; Rutgers Univ, MS, 56, PhD(dairy microbiol), 62. *Prof Exp:* Lab asst rickettsial dis, Walter Reed Army Inst Res, DC, 56-58; res asst, Rutgers Univ, 58-62; res biochemist, Immunol Lab, Plum Island Animal Dis Lab, USDA, 62-65; fel dept microbiol, Sch Med, Temple Univ, 65-67, asst prof & dir immunoserol lab, 67-70; assoc prof microbiol, Pa State Univ, 70-79; PROF MICROBIOL & DIR, IMMUNOL LAB, SCH MED, TEMPLE UNIV, 79- *Concurrent Pos:* Vis Scientist, Nat Cancer Inst, 77; vis prof, Univ S Fla Med Sch, 81; mem, Drug Abuse AIDS Res Rev Comt, Nat Inst Drug Abuse, 89- *Mem:* Am Soc Microbiol; Am Asn Immunol; fel NY Acad Sci; Am Asn Clin Chem; Asn Med Lab Immunologists; Int Soc Immunopharmacol. *Res:* Xenobiotics; virus-induced immune suppression; tumor immunology; diagnostic immunology. *Mailing Add:* Dept Microbiol & Immunol Sch Med Temple Univ Philadelphia PA 19140

CEITHAML, JOSEPH JAMES, b Chicago, Ill, May 23, 16; m 42; c 2. BIOCHEMISTRY. *Educ:* Univ Chicago, BS, 37, PhD(biochem), 41. *Prof Exp:* Res assoc biochem, 41-45, from asst prof to assoc prof, 46-58, PROF BIOCHEM, UNIV CHICAGO, 58-, DEAN STUDENTS, DIV BIOL SCI, 51- *Mem:* AAAS; Asn Am Med Cols; Am Soc Biol Chem. *Res:* Isolation of anterior pituitary hormones; metabolism of the malaria parasite; isolation and study of plant enzymes; biochemical genetics. *Mailing Add:* Prof Emer Biochem 2337 W 108th Pl Chicago IL 60643

CELANDER, EVELYN FAUN, b Ottumwa, Iowa, Nov 4, 26; wid; c 3. BIOCHEMISTRY. *Educ:* Drake Univ, BA, 48; Col Osteop Med & Surg, MS, 67. *Hon Degrees:* DSc, Col Osteopth Med & Surg, 78. *Prof Exp:* Reader & res asst, State Univ Iowa & Univ Tex Med Br, 48-55; res technician, Univ Tex Med Br, 55-59, res assoc physiol, 59-61; from instr to asst prof, 61-71, ASSOC PROF BIOCHEM, COL OSTEOP MED & SURG, 71- *Res:* Control and function of fibrinolytic enzyme system in health and disease; biosynthesis of radioactive protein substances; computer technology as applied to information retrieval, data processing and instruction. *Mailing Add:* Dept Biochem Univ Osteop Med & Health Sci 3200 Grand Ave Des Moines IA 50312

CELAURO, FRANCIS L, b Jersey City, NJ, Sept 12, 11; m 40; c 3. MATHEMATICS. *Educ:* NY Univ, AB, 37, AM, 39, PhD(math), 52. *Prof Exp:* Instr math, NY Univ, 37-40; asst prof & chmn, math dept, Loyola Col, 40-43; mathematician, Nat Bur Standards, 43-45; asst prof math, Lehigh Univ, 45-49; prof, E Tenn State Col, 54-57; prof, Cent Mich Univ, 57-62; prof, 62-76, EMER PROF MATH, GEORGE PEABODY COL, VANDERBILT UNIV, 76- *Concurrent Pos:* NSF award, Princeton Univ, 60; consult math, sci & eng. *Mem:* Math Asn Am; Am Asn Univ Professors. *Res:* Applied mathematics; psychology of mathematics learning at college level; factors associated with retention of college mathematics. *Mailing Add:* 1347 Burton Valley Rd Nashville TN 37215

CELENDER, IVY M, b Newark, NJ, Sept 5, 35. NUTRITION. *Educ:* Univ Pittsburgh, PhD(biochem & nutrit), 63. *Prof Exp:* VPRES & DIR NUTRIT, GEN MILLS, 73- *Mailing Add:* 15 Duck Pass Rd St Paul MN 55127

CELESIA, GASTONE G, b Genoa, Italy, Nov 22, 33; m; c 2. NEUROLOGY, NEUROPHYSIOLOGY. *Educ:* Univ Genoa, MD, 59; McGill Univ, MS, 65. *Prof Exp:* Resident neurol, Montreal Neurol Inst, 62-65; from asst prof to prof neurol, Med Ctr, Univ Wis-Madison, 66-76, dir lab EEG & clin neurophysiol, 70-76; prof neurol & vchmn dept, St Louis Univ, 76-79; prof neurol & vchmn dept, Ctr Health Sci, Univ Wis-Madison, 79-83; CHMN DEPT NEUROL, STRITCH SCH MED, LOYOLA UNIV, 83- *Concurrent Pos:* Fel neurophysiol, Med Ctr, Univ Wis-Madison, 60-62 & Montreal Neurol Inst, 64-65; demonstr, McGill Univ, 63-65; clin investr, Vet Admin Hosp, 66-70, consult, 70-76; consult, Cent Wis Colony, 70-76; chief, Neurol Serv, William S Middleton Mem Vet Hosp, 79-; ed-in-chief, Electroencephalography Clin Neurophysiol. *Mem:* Am Acad Neurol; Soc Neurosci; Am Epilepsy Soc; Am EEG Soc; AMA. *Res:* Visual neurosciences; auditory cortex; sensory system. *Mailing Add:* Dept Neurol Stritch Sch Med Loyola Univ 2160 S 1st Ave Maywood IL 60153

CELESK, ROGER A, b Chicago, Ill, Sep 13, 49; m 80; c 1. MICROBIAL ADHERENCE, ELECTRON MICROSCOPY. *Educ:* Univ Ill, BS, 71, Univ Notre Dame, PhD(microbiol), 77. *Prof Exp:* Fel, NIH, 76, Inst Cancer Res, 76-77, & Nat Inst Dent Res, 77-79; ASST PROF BIOL, UNIV DAYTON, 79- *Mem:* Am Soc Microbiol; Am Asn Gnotobiotics; N Am Soc Sigma Xi. *Res:* Microbial ecology and pathogenesis; host-miroflora interactions; bacterial adherence on surfaces; gnotobiotics; electron microscopy; ultrastructural analysis. *Mailing Add:* Miles Pharmaceut 400 Morgan Lane West Haven CT 06516

CELIANO, ALFRED, b Orange, NJ, Aug 8, 28. CHEMICAL KINETICS, PHYSICAL INORGANIC CHEMISTRY. *Educ:* Seton Hall Univ, AB, 49; Catholic Univ, STL, 53; Fordham Univ, MS, 56, PhD(chem), 59. *Prof Exp:* Chmn dept, 59-80, PROF CHEM, SETON HALL UNIV, 59- *Mem:* AAAS; Am Chem Soc. *Res:* The stability and kinetics of formation and substitution of inorganic complexes. *Mailing Add:* Dept of Chem Seton Hall Univ South Orange NJ 07079

CELIK, HASAN ALI, b Bozkir, Turkey. ALGEBRA. *Educ:* Middle East Tech Univ, BS, 64, MS, 65; Univ Calif, Santa Barbara, PhD(math), 71. *Prof Exp:* Asst math, Middle East Tech Univ, 64-66; from asst prof to assoc prof, 71-80, PROF MATH, CALIF STATE POLYTECH UNIV, POMONA, 80- *Mem:* Am Math Soc; Math Asn Am. *Res:* Non-associative algebra including flexible-antiflexible algebras; Jordan-Lie-Alternative Rings. *Mailing Add:* Dept of Math Calif State Polytech Univ Pomona CA 91768

CELINSKI, OLGIERD J(ERZY) Z(DZISLAW), b Warsaw, Poland, Nov 22, 22; nat Can; m 51; c 1. ELECTRICAL ENGINEERING. *Educ:* Polish Univ Col, Dipl Ing, 50; Univ Ottawa, MSc, 61. *Prof Exp:* Develop engr electronics, Airmec Labs, 50-51; digital cts, Comput Devices, Can, 52-54; mail sorting, Post Off Lab, 54-57; asst prof elec eng, Univ Ottawa, 57-75; pres, Northern Energy, 76-81; PRES, INTERACTIVE LEARNING CONTRACTORS INC, 81- *Concurrent Pos:* Univ deleg & consult electronics res, Karachi, Univ Ottawa, 64-66. *Mem:* Brit Inst Elec Engrs. *Res:* Learning processes; computer modeling of human activity; man-machine systems; electrical measurements. *Mailing Add:* 169 Henderson Ave Ottawa ON K1N 7P7 Can

CELIS, ROBERTO T F, b La Paz, Entre Rios, Arg, Apr 22, 30; m 63; c 2. MOLECULAR BIOLOGY, MICROBIOLOGY. *Educ:* Univ Buenos Aires, MD, 57, PhD(biochem), 64. *Prof Exp:* Res assoc microbiol, Univ Buenos Aires, 60-62, lectr, 62-64, asst prof, 64-66; asst prof, 72-75, ASSOC PROF MICROBIOL, MED SCH, NY UNIV, 75- *Concurrent Pos:* Trainee genetics, Med Sch, NY Univ, 68-70, NIH res grant, 73- *Mem:* NY Acad Sci; Am Soc Microbiol. *Res:* Transport of basic amino acids on escherichia coli. *Mailing Add:* Dept Microbiol NY Univ Med Ctr 550 First Ave New York NY 10016

CELLA, RICHARD JOSEPH, JR, b Philadelphia, Pa, Nov 2, 42; m 64; c 1. POLYMER CHEMISTRY, PHYSICAL CHEMISTRY. *Educ:* Univ Del, BS, 64; Cornell Univ, PhD(phys chem), 69. *Prof Exp:* Res chemist, 69-76, tech serv rep, Elastomer Chem Dept, 76-80, sr mkt rep, Polymer Prod Dept, 80-85, AEROSPACE INDUST MGR, E I DU PONT DE NEMOURS & CO, INC, 85- *Res:* X-ray diffraction studies of polymer structure; morphology and solid state characterization of polymers; physical and mechanical properties of elastomers; adhesives; technical sales. *Mailing Add:* One Forest Lane Hockessin DE 19707

CELLARIUS, RICHARD ANDREW, b Oakland, Calif, July 28, 37; m 59; c 2. PHOTOBIOLOGY, ENVIRONMENTAL POLICY. *Educ:* Reed Col, BA, 58; Rockefeller Univ, PhD(biol), 65. *Prof Exp:* USPHS fel, 65-66; asst prof bot, Univ Mich, 66-72; MEM FAC, EVERGREEN STATE COL, 72- *Concurrent Pos:* Res cooperator, US Forest Serv-Pac Northwest Res Sta, 86- *Mem:* AAAS; Sigma Xi; Am Soc Photobiol; Am Soc Plant Physiol; Am Inst Biol Sci. *Res:* Light reactions of photosynthesis, photochemistry, plant and tree physiology, photobiology; environmental policy. *Mailing Add:* Evergreen State Col Olympia WA 98505

CELLER, GEORGE K, b Walbrzych, Poland, Mar 27, 47; US citizen; m 75; c 2. ELECTRONIC MATERIALS, SEMICONDUCTORS. *Educ:* Warsaw Univ, MSc, 69; Purdue Univ, PhD(physics), 76. *Prof Exp:* Res scientist, Physics Dept, Vienna Univ, 69-70; comput scientist, intern Atomic Energy Agency, Vienna, 70; res staff, W Elec Res Ctr, 76-79; mem tech staff, 79-84, RES SUPVR, ADVAN LITHOGRAPHY RES DEPT, AT&T BELL LAB, 84- *Mem:* Fel Am Phys Soc; Mat Res Soc; Electrochem Soc; Inst Elec & Electronic Engrs. *Res:* Interactions of photon and ion beams with semiconductors; structural and electronic properties of materials; nanostructures and x-ray lithography. *Mailing Add:* AT&T Bell Lab 600 Mountain Ave Rm 6F-217 Murray Hill NJ 07974-2070

CELLI, VITTORIO, b Parma, Italy, Aug 13, 36; m 62; c 1. SOLID STATE PHYSICS. *Educ:* Univ Pavia, DSc(physics), 58. *Prof Exp:* Res assoc solid state physics, Univ Ill, 59-61, asst prof, 61-62; asst res physicist, Univ Calif, San Diego, 62-64; lectr & asst theoret physics, Univ Bologna, 64-66; assoc prof physics, 66-69, PROF PHYSICS, UNIV VA, 69- *Concurrent Pos:* Fulbright scholar, 59-62; prof extraordinary, Univ Trieste & guest scientist, Int Ctr Theoret Physics, Trieste, 73-74. *Honors & Awards:* Humboldt Sr US Sci Award, 81, 90. *Mem:* Fel Am Phys Soc; Europ Phys Soc. *Res:* Theoretical physics; theoretical solid state physics; surface science. *Mailing Add:* Dept Physics Univ Va McCormick Rd Charlottesville VA 22901

CELMASTER, WILLIAM NOAH, b Brussels, Belg, June 22, 50; m 78; c 1. PARALLEL COMPUTATION, COMPUTATIONAL PHYSICS. *Educ:* Univ BC, BSc, 71; Harvard Univ, MA, 72, PhD(physics), 77. *Prof Exp:* Postdoctoral physics, Univ Calif, San Diego, 77-79 & Argonne Nat Lab, 79-81; asst prof physics, Northeastern Univ, 81-85 & Univ Ill, 85-86; CHIEF SCIENTIST, PARALLEL COMPUT APPLN, BOLT, BERANEK & NEWMAN, 86- *Concurrent Pos:* Consult, Symbolic, 86. *Mem:* Am Phys Soc; Inst Elec & Electronics Engrs; Soc Indust & Appl Math. *Res:* Theoretical quantum chromodynamics; algorithms and implementations of parallel computation in science; performance analysis and benchmarking of parallel computers. *Mailing Add:* BBN Advan Computers 70 Fawcett St Cambridge MA 02138

CELMER, WALTER DANIEL, b Plymouth, Pa, Sept 13, 25; m 46, 79; c 4. BIO-ORGANIC CHEMISTRY. *Educ:* Bucknell Univ, BS, 47; Univ Ill, PhD(biochem), 50. *Prof Exp:* Res chemist, Pfizer Inc, 50-54, res supvr, 54-61, res mgr, 61-72, res adv, 72-; PRES, MED CHEM ADV SERV. *Concurrent Pos:* Ed, Antimicrobial Agents & Chemother, 72-79. *Mem:* Am Chem Soc; Am Soc Biol Chem; Am Soc Microbiol; Int Platform Asn. *Res:* Discovery from microbial sources of novel chemical entities possessing biological activities, especially antibiotics and to elucidate their structures, stereochemistry, biogenesis, as well as to prepare and study their synthetic modifications. *Mailing Add:* Med Chem Adv Serv Inc 545 Pequot Ave New London CT 06320-4338

CELMINS, AIVARS KARLIS RICHARDS, b Riga, Latvia, Apr 10, 27; m 55, 84; c 3. APPLIED MATHEMATICS, NONLINEAR DATA ANALYSIS. *Educ:* Hannover Tech Univ, BS, 50, MS, 53; Clausthal Tech Univ, PhD(geophys), 57. *Prof Exp:* Mathematician, Seismos GmbH, Ger, 53-57; asst sect head explor geophys, Petrobras-DEPEX, Brazil, 57-58, sect head gravity, 59-61; sr mathematician, Inst Instrumental Math, Bonn, Ger, 61-64; RES MATHEMATICIAN, US ARMY BALLISTIC RES LAB, 64- *Concurrent Pos:* Instr, Univ Del, 65-70. *Mem:* Asn Comput Mach; Europ Asn Explor Geophysicists; Ger Soc Appl Math & Mech; Int Fuzzy Systs Assoc; Am Statist Assoc; NAm Fuzzy Info Processing Soc. *Res:* Numerical mathematics and its applications to technical and physics problems, particularly in the fields associated with fluid dynamics, engineering and exploration geophysics; numerical solution of partial differential equations; nonlinear model fitting to data; fuzzy data analysis. *Mailing Add:* 38 Maple Lane Elkton MD 21921

CELNIKER, SUSAN ELIZABETH, b Culver City, Calif, Mar 13, 54; m 76; c 2. GENETICS, BIOCHEMISTRY. *Educ:* Pitzer Col, BA, 75; Univ NC, PhD(biochem), 83. *Prof Exp:* Postdoctoral fel, 83-86, SR RES FEL GENETICS, CALIF INST TECHNOL, 86- *Mem:* AAAS. *Res:* Developmental biology; genes of the bithorax of Drosophila are regulated and in turn how the products of these genes, the homeoproteins, activate the next set of genes in the developmental heirarchy. *Mailing Add:* Div Biol 156-29 Calif Inst Technol Pasadena CA 91125

CELOTTA, ROBERT JAMES, b New York, NY, Nov 18, 43; m 66; c 2. ATOMIC & MOLECULAR PHYSICS, SURFACE PHYSICS. *Educ:* City Col New York, BS, 64; NY Univ, PhD(physics), 69. *Prof Exp:* Res asst physics, NY Univ, 64-69; res assoc, Joint Inst Lab Astrophys, Univ Colo, 69-71; FEL, PHYSICS LAB, NAT INST STANDARDS & TECHNOL, 71- *Concurrent Pos:* Instr physics, NY Univ, 64-69; ed, Methods of Exp Physics, 81- *Honors & Awards:* E U Condon Award, 80; IR-100 Award, 80 & 85; Silver Medal, Dept Com, 78, Gold Medal, 87; Award for Excellence in Technol Transfer, Fed Lab Consortium, 88. *Mem:* fel Am Phys Soc; AAAS; Am Vacuum Soc. *Res:* Surface physics; atomic and molecular physics; electron polarization phenomena; surface magnetism; electron and tunneling microscopy. *Mailing Add:* Nat Inst Standards & Technol B206 Metrol Bldg Gaithersburg MD 20899

CEMBER, HERMAN, b Brooklyn, NY, Jan 14, 24; m 43; c 2. RADIOBIOLOGY, HEALTH PHYSICS. *Educ:* City Col New York, BS, 49; Univ Pittsburgh, MS, 52, PhD(biophys), 60; Am Acad Environ Engrs, dipl, Am Bd Health Physics, dipl. *Prof Exp:* Res assoc health physics, Grad Sch Pub Health, Univ Pittsburgh, 50-54, from asst prof to assoc prof indust hyg, 54-60; asst prof indust health, Col Med, Univ Cincinnati, 60-64; PROF CIVIL ENG, ENVIRON HEALTH, NORTHWESTERN UNIV, EVANSTON, 64- *Concurrent Pos:* Health physics consult, Carnegie Inst Technol, 52-60; radiol safety officer, Univ Pittsburgh, 52-60; lectr, US Naval Training Prog, Westinghouse Atomic Power Div, 52-55; tech expert occup health, Int Labour Off, Switz, 61-62; vis prof indust health, Col Med, Univ Cincinnati, 64-65; Fulbright vis prof environ health, Hadassah Med Sch, Hebrew Univ Jerusalem, 72-73. *Honors & Awards:* Distinguished Sci Achievement, Health Physics Soc, 90. *Mem:* AAAS; fel Health Physics Soc; Am Acad Environ Engrs; fel Am Pub Health Asn; Am Indust Hyg Asn. *Res:* Biological effects of radiation; experimental lung cancer; environmental radiation and radioactivity. *Mailing Add:* 1800 Kirk St Evanston IL 60202

CEMBROLA, ROBERT JOHN, b New York, NY, April 3, 49; m 77; c 3. POLYMER PHYSICS. *Educ:* Rochester Inst Technol, BS, 72, MS, 73; Univ Mass, MS, 77, PhD(polymer sci), 78. *Prof Exp:* Fel polymer sci, Kyoto Univ, Japan, 78-79; from sr res chemist to group leader polymer physics, Gen Tire & Rubber Co, 79-83, sect head polymer physics & phys testing, 83-85; MGR, ANAL CHEM, DIGITAL EQUIP CORP, 85- *Mem:* Am Chem Soc. *Res:* Solid state physics of polymers; structure-property relationships; rheo-optical study of deformation mechanisms of semi-crystalline polymers; development of analytical methods to support the microelectronics industry. *Mailing Add:* 48 Barbara Jean St Grafton MA 01519

CENCE, ROBERT J, b Cleveland, Ohio, July 16, 30; m 54; c 2. PHYSICS. *Educ:* Univ Calif, Berkeley, AB, 52, PhD(physics), 59. *Prof Exp:* Res assoc physics, Lawrence Radiation Lab, Univ Calif, 59-63, instr, Univ Calif, Berkeley, 62-63; assoc prof, 63-74, PROF PHYSICS & ASTRON, UNIV HAWAII, 74- *Mem:* Am Phys Soc. *Res:* Elementary particle physics. *Mailing Add:* Watanabe Hall Univ Hawaii 2505 Correa Rd Honolulu HI 96822

CENCI, HARRY JOSEPH, b Brooklyn, NY, Oct 2, 30; m 54; c 6. ORGANIC CHEMISTRY, POLYMER CHEMISTRY. *Educ:* Brooklyn Col, BS, 53; Univ Pa, MS, 55, PhD(org chem), 57. *Prof Exp:* Head synthesis lab, 57-74, proj leader indust coatings, 74-77, proj leader, Emulsion Synthesis, 77-79, DEPT MGR, CHEM PROCESS RES, ROHM & HAAS CO, 79- *Mem:* Am Chem Soc. *Res:* Synthesis and application of industrial coatings; polymer synthesis; modifiers for polyvinyl chloride; emulsion and solution polymerization; physical organic chemistry; reactions of cis and trans cyclohexane -1, 3- diols; synthesis and reactions of 1, 2- difluoro-1, 2-dicyanoethylene; photochemistry. *Mailing Add:* 1434 Buttonwood Lane Warminster PA 18974-2502

CENDES, ZOLTAN JOSEPH, b Feffernitz, Austria, May 16, 46; US citizen; m; c 3. NUMERICAL ANALYSIS, ELECTRICAL ENGINEERING. *Educ:* Univ Mich, BSE, 68; McGill Univ, MEng, 70, PhD(elec eng), 73. *Prof Exp:* Prof assoc, Dept Elec Eng, McGill Univ, 72-74; engr, Gen Elec Co, 74-77; pres, Csendes Assoc, Inc, 77-78; engr electromagnetics, Corp Res & Develop, Gen Elec Co, 78-80; assoc prof elec eng, McGill Univ, 80-82; PROF ELEC & COMPUTER ENG, CARNEGIE-MELLON UNIV, 82- *Concurrent Pos:* Co-founder, Ansoft Corp, Pittsburgh, Pa. *Mem:* Inst Elec & Electronic Engrs. *Res:* Numerical methods for partial differential equations; finite element methods; matrix theory; generalized inverses; electromagnetic field theory; electric machine modeling; modeling of microwave circuits; modeling magnetic recording. *Mailing Add:* Dept Elec & Computer Eng Carnegie-Mellon Univ 5000 Forbes Ave Pittsburgh PA 15213

CENEDELLA, RICHARD J, b Pittsburgh, Pa, Jan 12, 39; m 64; c 1. BIOCHEMISTRY, PHARMACOLOGY. *Educ:* Pa State Univ, BS, 61; Jefferson Med Col, PhD(biochem), 66. *Prof Exp:* Res assoc, Sch Med, WVa Univ 65-68, from asst prof to prof pharmacol, 68-76; PROF & CHMN, DEPT BIOCHEM, KIRKSVILLE COL OSTEOP MED, 76- *Res:* Fatty acid metabolism; lipid pharmacology; mechanism of action of hypolipemic drugs; cataracts. *Mailing Add:* Dept of Biochem Kirksville Col Osteop Med Kirksville MO 63501

CENGEL, JOHN ANTHONY, b East Chicago, Ind, July 21, 36; m 62; c 3. PHYSICAL ORGANIC CHEMISTRY. *Educ:* Purdue Univ, BS, 58, PhD(phys chem), 65; Ore State Univ, MS, 59. *Prof Exp:* From proj chemist to sr proj chemist, 65-70, SR RES SCIENTIST, AMOCO CHEM CORP, STAND OIL CO IND, 70- *Concurrent Pos:* NSF grant, 58-59, AEC grant, 63-65. *Mem:* Am Chem Soc. *Res:* Infrared spectroscopy; cellular plastics; viscous polymers; polymer composites; petroleum and gasoline additives; organic acids & anhydrides. *Mailing Add:* Dept Res & Develop PO Box 3011 Amoco Chem Co Naperville IL 60566-7011

CENGELOGLU, YILMAZ, b Turkey, Feb 27, 66; m. NETWORKING & COMMUNICATION, ARTIFICIAL INTELLIGENT. *Educ:* Aegean Univ, BS, 87; Univ Cent Fla, MS, 91. *Prof Exp:* Syst programmer, Bilkent Univ, Turkey, 87-88; syst analyst, GPT-Stromberg Carlson, 89-90; RES ASSOC, INST SIMULATION & TRAINING, 90- *Mem:* Inst Elec & Electronic Engrs; Asn Comput Mach; Nat Soc Prof Engrs. *Res:* Investigate a network of interactive simulators; investigating the feasibility of using open system interconnection network protocol to provide network services for a distributed interactive simulation application. *Mailing Add:* Univ Cent Fla PO Box 678455 Orlando FL 32867

CENKL, BOHUMIL, b Bohómovice, Czech, 34; m 59; c 2. TOPOLOGY ALGEBRA. *Educ:* Charles Univ, Czech, Prom Mat, 58, CSc, 61, RNDr, 67; Czech Acad Sci, Dr Sc, 68. *Prof Exp:* Res assoc, Charles Univ, Czechoslovakia, 61-64 & 66-68; asst prof, Stanford Univ, 65-66; assoc res scientist, Courant Inst, 68-69; PROF, NORTHEASTERN UNIV, 69- *Concurrent Pos:* Vis scholar, Tata Inst, Bombay, 64-65; prin investr, NSF, 68-81. *Mem:* Am Math Soc. *Res:* Algebra topology: tamehomotopy theory, cohom of groups. *Mailing Add:* Dept Math Northeastern Univ 360 Huntington Ave Boston MA 02115

CENKOWSKI, STEFAN, b Wroclaw, Poland, Jan 28, 50; Can citizen; m 81; c 2. DRYING OF FOOD & AGRICULTURAL MATERIALS, RHEOLOGICAL PROPERTIES. *Educ:* Tech Univ, Wroclaw, Poland, BTech, 72, MSc, 74; Agr Univ, Wroclaw, Poland, PhD(drying), 81, ScD, 85. *Prof Exp:* Design engr, Archimedes Manufacture, Wroclaw, Poland, 72-73; res assoc power & mach, Dept Agr Eng, Agr Univ, Wroclaw, Poland, 74-76, res assoc crop preserv, 77-80, asst prof food eng & crop preserv, 81-86; postdoctoral fel food eng & crop preserv, Dept Agr Eng, Univ Sask, 87-88; ASST PROF FOOD ENG, DEPT AGR ENG, UNIV MAN, 88- *Concurrent Pos:* Vis prof, Dept Agr Eng, Agr Univ, Krakow, Poland, 81, Agr Univ, Prague, Czech, 85; postdoctoral fel drying, Silsoe Col, Cranfield Inst Technol, Eng, 83-84; prin investr, Remney Apaines, Brandon, 89, Natural Sci Eng Res Can, 89- *Mem:* Can Soc Agr Eng; Am Soc Agr Eng. *Res:* Effect of food and agricultural materials pretreatment on water activity, sorption hysteresis and their rheological behavior; mathematical modelling and computer simulation of drying processes. *Mailing Add:* Dept Agr Eng Univ Man Winnipeg MB R3T 2N2 Can

CENTER, ELIZABETH M, b Sterling, Ill, Aug 11, 28; m 51; c 1. EMBRYOLOGY. *Educ:* Augustana Col, AB, 50; Stanford Univ, PhD(genetics), 57. *Prof Exp:* Res asst anat, Stanford Univ 51-61, res assoc, 61-69, instr biol sci, 68-71, lectr biol sci, 71-78; from asst prof to assoc prof, 77-84, PROF BIOL, COL NOTRE DAME, CALIF, 84- *Concurrent Pos:* Hon res fel, dept genetics & biomet, Univ Col, London, 84. *Mem:* NY Acad Sci; Am Soc Zoologists; Genetics Soc Am; Am Soc Anat; Sigma Xi; Histochem Soc. *Res:* Developmental genetics. *Mailing Add:* Dept Biol Col Notre Dame Belmont CA 94002

CENTER, ROBERT E, b Breslau, Ger, Nov 12, 35; m 59; c 3. LASERS, CHEMICAL PHYSICS. *Educ:* Univ Sydney, BSc, 56, BE, 58, MEngSc, 59, PhD(gas dynamics), 63. *Prof Exp:* Scientist gas dynamics, Jet Propulsion Lab, Calif Inst Technol, 63-67; prin res scientist, Avco Everett Res Lab, 67-75; dir laser res, 75-79, VPRES RES, MATH SCI NORTHWEST, 79- *Mem:* Am Phys Soc. *Res:* Scattering of fast electrons in gases; vibrational relaxation in anharmonic diatomic molecules under conditions of thermal nonequilibrium; high temperature gas phase kinetics; visible and ultraviolet electric discharge laser development. *Mailing Add:* Spectra Technol Inc 2755 Northup Way Bellevue WA 98004

CENTER, SHARON ANNE, b Eugene, Ore, June 10, 49; m 75. ANIMAL HEPATOLOGY, ANIMAL UROLOGY. *Educ:* Calif State Univ, San Luis Obispo, BS, 71; Univ Calif, Davis, DVM, 75; Am Col Internal Med, dipl, 83. *Prof Exp:* Intern vet med & surg, 75-76, resident vet internal med, 76-77, instr, 80-81, ASST PROF SMALL ANIMAL INTERNAL MED, NY STATE COL VET MED, CORNELL UNIV, 82- *Mem:* Am Col Vet Internal Med; Am Vet Med Asn; Am Animal Hosp Asn; Am Asn Vet Clinicians. *Res:* Improved methods of diagnosing hepatobiliary disease in the domestic species; use of organic anion cholephilic dye plasma clearance and fasting and postprandial values of serum bile acids in animals having hepatobiliary disorders. *Mailing Add:* Small Animal Clin NY State Col Vet Med Cornell Univ Ithaca NY 14853

CENTIFANTO, YSOLINA M, b Panama, Aug 12, 28; US citizen; m 53; c 3. MOLECULAR BIOLOGY. *Educ:* Univ Panama, BS, 51; Western Reserve Univ, MS, 54; Univ Fla, PhD(bact), 64. *Prof Exp:* Asst prof physiol & genetics, Univ Panama, 55-56; res biologist, Kodak Trop Res Lab, Panama, 57; res biologist, Eastman Kodak Res Lab, NY, 58-61; res assoc virol, Ophthal Div, Col Med, Univ Fla 64-65, from instr to assoc prof ophthal & microbiol, 65-77; prof ophthal & microbiol, Col Med, La State Univ, 78-85. *Concurrent Pos:* PROF, TULANE SCH MED, 85-; dir ophthal virol lab, La State Univ, 85- *Mem:* AAAS; Am Chem Soc; Am Soc Microbiol; Asn Res Vision & Ophthal; Am Soc for Virol. *Res:* Viral and bacterial infections of the eye; pathogenesis of disease and its relation to host defense mechanisms, such as interferon and antibody synthesis; studies on recurrent viral infections, herpes simplex virus and ocular disease. *Mailing Add:* La State Univ Eye Ctr 2020 Gravier St B New Orleans LA 70112-2234

CENTNER, ROSEMARY LOUISE, b Newport, Ky, Sept 23, 26. ORGANIC CHEMISTRY. *Educ:* Our Lady of Cincinnati Col, BA, 47; Univ Cincinnati, MS, 49. *Prof Exp:* Libr asst tech libr, 49-52, br librn, 52-56, tech librn, 56-66, mgr tech info serv, 66-72, mgr div info consults, 72-73, mgr, NDA Coord, 73-75, mgr biomed commun, 75-80, mgr tech commun, 80-86, PATENT INFO SPECALIST, PROCTER & GAMBLE CO, 86- *Mem:* Am Chem Soc; Am Med Writers Asn. *Res:* Scientific information; information retrieval; technical translation. *Mailing Add:* 2678 Byrneside Dr Cincinnati OH 45239-6404

CENTOFANTI, LOUIS F, b Youngstown, Ohio, July 25, 43; m 63; c 3. INORGANIC CHEMISTRY. *Educ:* Youngstown State Univ, BS, 65; Univ Mich, MS, 67, PhD(chem), 68. *Prof Exp:* Fel chem, Univ Utah, 68-69; asst prof, Emory Univ, 69-73; sr res chemist, Monsanto Co, 73-; regional adminr, US Dept Energy, 79-81; pres, PPM, Inc, 81-85; SR VPRES, USPCI, 85- *Mem:* Am Chem Soc. *Res:* Synthetic and physical inorganic chemistry. *Mailing Add:* 315 Wilderlake Ct Atlanta GA 30328

CENTOLA, GRACE MARIE, b Utica, NY, July 9, 51; m 78; c 1. ANDROLOGY, INFERTILITY. *Educ:* Utica Col, BS, 73; Georgetown Univ, PhD(anat), 78. *Prof Exp:* Teaching asst anat, Sch Med & Dent, Georgetown Univ, 73-77; instr, Baltimore Col Dent Surg, Univ Md Dent Sch, 77-78, asst prof anat, 78-80; asst prof obstet-gynec, Uniformed Serv Univ Health Sci, Bethesda, Md, 80-84; asst prof obstet-gynec & dir res lab, State Univ NY Upstate Med Ctr, Syracuse, 84-85; asst prof & dir andrology lab, Dept Urol, 85-87, ASST PROF & DIR ANDROLOGY, DEPT OBSTET/GYNEC, UNIV ROCHESTER, 87-91. *Concurrent Pos:* Grad res award, Univ Md Dent Sch, 78-79. *Mem:* Soc Study Reprod; Am Asn Anatomists; Am Fertility Soc; Tissue Cult Soc; Am Col Obstetricians & Gynecologists. *Res:* Andrology; male infertility; sperm motility; cryopreservation of semen; carcinoma cells in vitro. *Mailing Add:* Dept Obstet/Gynec Univ Rochester Med Ctr Box 668 601 Elmwood Ave Rochester NY 14620

CENTONZE, VICTORIA E, b Stamford, Conn, Nov 21, 58. CELL BIOLOGY. *Educ:* Dartmouth Col, PhD(biol), 85. *Prof Exp:* FEL, UNIV WIS, 85- *Mailing Add:* Lab Molecular Biol Univ Wis 1525 Linden Dr Madison WI 53706

CENTORINO, JAMES JOSEPH, marine studies; deceased, see previous edition for last biography

CENTURY, BERNARD, b Chicago, Ill, June 15, 28; m 58; c 4. PHARMACOLOGY. *Educ:* Univ Chicago, PhB, 46, BS & MS, 51, PhD(pharmacol), 53. *Prof Exp:* Biochemist, Biochem Res Lab, Elgin State Hosp, 54-59, res assoc, 59-74, actg dir, 69-74; asst prof biol chem, Univ Ill Col Med, 62-74; mem staff, dept biochem, Michael Reese Med Ctr, 74-77; CLIN CHEMIST, LAB DEPT, ST MARY'S HOSP, WATERBURY, CONN, 77- *Concurrent Pos:* NSF fel pharmacol, Univ Chicago, 53-54. *Mem:* AAAS; Am Chem Soc; Am Inst Nutrit; Am Soc Pharmacol & Exp Therapeut; Am Asn Clin Chem. *Res:* Biochemistry of mental diseases; relationships of dietary lipids to vitamin E deficiency; effect of diet lipids on metabolism of essential fatty acids and tissue compositions; effect of diet lipids on pharmacological responses. *Mailing Add:* St Mary's Hosp Lab Dept 56 Franklin Waterbury CT 06702

CENZER, DOUGLAS, b Detroit, Mich, Nov 15, 47; m 70; c 2. SET THEORY, RECURSION THEORY. *Educ:* Mich State Univ, BS, 68; Univ Mich, PhD(math), 72. *Prof Exp:* From asst prof to assoc prof, 72-87, PROF MATH, UNIV FLA, 87- *Concurrent Pos:* Vis assoc prof, N Tex State Univ, 81-82; mem, MSRI, Berkeley, 89-90. *Mem:* Am Math Soc; Asn Symbolic Logic; Math Asn Am. *Res:* Closure sets and ordinals of inductive definitions; monotone sets & operators; applications to measure theory, topology & computer science; measurable selection; preference orders; Borel isomorphism; Cantor-Bendixson derivative; Turing and Wadge degrees; computational complexity of models, recursive combinatorics. *Mailing Add:* Dept Math 201 Walker Hall Univ Fla Gainesville FL 32611

CEPEDA, JOSEPH CHERUBINI, b Del Rio, Tex, Apr 19, 48. PETROLOGY, VOLCANOLOGY. *Educ:* Univ Tex, Austin, BS, 70, PhD(geol), 77; NMex Inst Mining & Technol, MS, 72. *Prof Exp:* Instr, Appalachian State Univ, 76-77; asst prof, 77-83, ASSOC PROF GEOL, DEPT GEOSCI & FIELD CAMP DIR, W TEX STATE UNIV, 83- *Mem:* Geol Soc Am; Am Geophys Union. *Res:* Field relations and chemistry of igneous and metamorphic rocks; computer applications to geologic problems. *Mailing Add:* Dept of Geosci WTex State Univ Canyon TX 79016

CEPERLEY, DAVID MATTHEW, b Charleston, WVa, Dec 26, 49; m; c 3. STATISTICAL PHYSICS, PHYSICAL CHEMISTRY. *Educ:* Univ Mich, BS, 71; Cornell Univ, PhD(physics), 76. *Prof Exp:* Fel physics, Univ Paris XI, 76-77; fel, Courant Inst, 77-78; staff chemist, Lawrence Berkeley Lab, 78-81, staff scientist, Nat Resource Comput Chem, Lawrence Livermore Nat Lab, 81-87; ASSOC PROF, UNIV ILL, URBANA, CHAMPAIGN, 87- *Mem:* Am Phys Soc; AAAS. *Res:* Computer simulation of many-body systems; particularly quantum and polymeric systems; Monte Carlo Methods. *Mailing Add:* 2114 Lynwood Dr Champaign IL 61821

CEPONIS, MICHAEL JOHN, b Brooklyn, NY, June 25, 16; m 51; c 5. PLANT PATHOLOGY. *Educ:* Cornell Univ, BS, 50, MS, 53. *Prof Exp:* Plant pathologist, Tischler Res Serv, NJ, 53-55; from asst plant pathologist to plant pathologist, 55-63, RES PLANT PATHOLOGIST, AGR RES SERV, USDA, NEW BRUNSWICK, NJ, 63- *Mem:* Am Phytopath Soc; Am Soc Hort Sci. *Res:* Market diseases of fruits and vegetables. *Mailing Add:* 42 Fieldstone Dr Somerville NJ 08876

CEPRINI, MARIO Q, b Jamaica, NY, Sept 30, 25; m 49; c 3. ORGANIC CHEMISTRY, POLYMER & POLYMER ADDITIVES. *Educ:* Queens Col, NY, BS, 47; St John's Univ, NY, MS, 61, PhD(org chem), 67. *Prof Exp:* Chemist, NY State Racing Comn Lab, 47-60; assoc chemist, Hoffmann-LaRoche Inc, NJ, 60-65; sr chemist, Tenneco Chem, Inc, 67-69, group leader metall-org, 70-72, group leader process develop, 72-75, group leader polymer additives, 75-77, sr res scientist, 70-78, lab mgr polymer & additives res, 78-82; mgr environ affairs, Nuodex, Inc, 82-88, dir, 88-91; RETIRED. *Mem:* Soc Plastics Engrs; Am Chem Soc; Sigma Xi. *Res:* Peptide and amino acid chemistry; antibiotics; organic synthesis; polymer additives; process development; polymer synthesis. *Mailing Add:* 513 Ocean Point Ave Cedarhurst NY 11516

CERA, LEE M, b Chicago, Ill, June 24, 50. PHYSICAL PLANT DESIGN, ADMINISTRATION. *Educ:* St Xavier Col, BS, 71; Univ Ill, BA, 73, DVM, 75; Univ Chicago, PhD(path), 91. *Prof Exp:* DIR, MIDWEST PATH STUDY CTR, DAVIS, 83- *Concurrent Pos:* Pathologist, Lincoln Park Zoo, 78-91, mem adv bd, 78-91; dir, Off Animal Care, Univ Chicago, 80-90; assoc prof histol, Univ Chicago, 83-91. *Mem:* Am Vet Med Asn; Sigma Xi; Am Asn Lab Animal Sci. *Res:* Physical plant design. *Mailing Add:* 3421 Golfview Hazel Crest IL 60429

CERAMI, ANTHONY, b Newark, NJ, Oct 3, 40. MEDICAL BIOCHEMISTRY. *Educ:* Rutgers Univ, BS, 62; Rockefeller Univ, PhD(biochem), 67. *Prof Exp:* Asst prof biochem, 69-72, assoc prof & head lab, 72-78, PROF BIOCHEM & HEAD, LAB MED BIOCHEM, ROCKEFELLER UNIV, 78- *Concurrent Pos:* Dean, Rockefeller Univ, 86-91; ed, J Exp Med, 81- *Honors & Awards:* Juv Diabetes Found Award, 78; Luft Award in Diabetes Res, Int Diabetes Found, 88; BCL Award, Asn Clin Biochemists, 89; Connaught-Novo Award, Can Diabetes Asn, 90; Altemeir Award, Surg Infection Soc, 90. *Mem:* Nat Acad Sci; Am Soc Biol Chemists; Am Soc Pharmacol & Exp Therapeut; Am Soc Hemat. *Res:* Diabetes, parasitology. *Mailing Add:* Lab Med Biochem Rockefeller Univ 1230 York Ave New York NY 10021

CERANKOWSKI, LEON DENNIS, b Philadelphia, Pa, July 31, 40; m 63. PHYSICAL CHEMISTRY. *Educ:* Drexel Univ, BS, 63; Princeton Univ, MA, 65, PhD(chem), 69. *Prof Exp:* Chemist, USDA, 61-64; instr phys chem, Princeton Univ, 67-69; SCIENTIST, POLYMER LAB, POLAROID CORP, 69- *Mem:* AAAS; Am Chem Soc; Sigma Xi. *Res:* Physical chemistry of electrolyte and macromolecular solutions. *Mailing Add:* 159 Fiske St Carlisle MA 01741

CERBON-SOLORZANO, JORGE, b Mexico City, Mex, Mar 20, 30; m 54; c 4. MEMBRANES STRUCTURE FUNCTION, SURFACE POTENTIAL MEMBRANE FUNCTION. *Educ:* Nat Sch Biol Sci, BSc, 47, MSc, 57, PhD(microbiol), 63. *Prof Exp:* Asst prof res, Sch Med, Nat Independent Univ Mex, 51-54, prof res, 55-65; prof res, 65-73, head dept, 73-83, BIOL AREA COORDR, CTR ADVAN STUDIES, NAT POLYTECH INST, MEX, 83- *Concurrent Pos:* Invited prof, Sch Med, Univ San Salvador, Cent Am, 59-61; vis prof, Nat Inst Arthritis, Metab & Digestive Dis, NIH, 63-65 & Unilever Res Labs, Eng, 72-73; chmn, Planning Comt Seven, Int Biophysics Cong, 80-81. *Mem:* Mex Soc Biochem (vpres, 75-77, pres, 77-79); Acad Sci Res; Am Soc Microbiol; Pan Am Asn Biochem Socs; Am Soc Biol Chemists; Biophys

Soc. *Res:* Structure-function relationships of membranes; lipid dynamics and their relation to drug membrane interactions; surface potential and membrane phenomena; transport. *Mailing Add:* Dept Biochem C1EA-1PN PO Box 14-740 Mexico City DF 14 07000 Mexico

CERBULIS, JANIS, b Smiltene, Latvia, Dec 5, 13; nat US; m 52; c 3. FOOD CHEMISTRY. *Educ:* Acad Agr, Jelgava, Latvia, PhD(agr), 44; Univ Pa, MS, 57; Rutgers Univ, PhD(biochem), 67. *Prof Exp:* Adminr & instr hort, Baltic Univ, Ger, 47-49; res chemist, Stephen F Whitman & Sons, Pa, 51-55 & Borden Chem Co, 55-56; res chemist, food sci lab, Eastern Regional Res Ctr, Agr Res Serv, USDA, 56-86; EMER RES CHEMIST, FOOD SCI LAB, EASTERN REGIONAL RES CTR, AGR RES SERV, USDA, 87- *Concurrent Pos:* USPHS maintenance grant, 61-63; Am Cancer Soc res grant-in-aid, 63; abstractor, Chem Abstracts, 58-76. *Mem:* Am Chem Soc; Ger Soc Fat Res; Sigma Xi. *Res:* Growing of grass seeds, vegetable seeds and medicinal plants; food value of grass; phenolic resins; earthworm chemical composition; whey composition; lactose chemistry; milk proteins and lipids; milk clotting; milk composition and relationship among milk constituents; lipid composition of beef cattle brain and many wild plant seeds; bialkyl phthalates in foods; extraction of proteins with magnesium and zinc salts; occurrence of diesters of 3-chloro-1,2 propanediol in goats milk, lipids of goats milk; molecular species of chloropropane diol disters and triacylglycerols in milk fat; stereo specific analysis of fatty acid esters of chloropropanediol isolated from goat milk; oxidation of lipids and analysis of oxidation products of lipids. *Mailing Add:* RD Two Box 940 Boyertown PA 19512

CERCONE, NICHOLAS JOSEPH, b Pittsburgh, Pa, Dec 18, 46. COMPUTER SCIENCE. *Educ:* Col Steubenville, BS, 68; Ohio State Univ, MS, 70; Univ Alta, PhD(comput sci), 75. *Prof Exp:* Programmer design automation, Int Bus Mach Corp, 68-69; instr comput sci, Ohio State Univ, 70-71; instr comput, Int Bus Mach Corp, 71-72; asst prof comput sci, Old Dom Univ, 75-76; ASST PROF COMPUT SCI, COMPUT SCI PROG, SIMON FRASER UNIV, 76- *Mem:* Asn Comput Mach; Inst Elec & Electronics Engrs; Royal Photographic Soc. *Res:* Artificial intelligence; natural language processing; representation of knowledge. *Mailing Add:* Comput Sci Dept Simon Fraser Univ Burnaby BC V5A 1S6 Can

CERDA, JAMES J, b Brooklyn, NY, Dec 29, 30; m; c 3. GASTROENTEROLOGY, NUTRITION. *Educ:* Univ Md, MD, 61. *Prof Exp:* PROF & ASSOC CHMN, DEPT MED, UNIV FLA, 76- *Honors & Awards:* Hippocratic Award, 76; Paul Dudley White Award, 89. *Mem:* Am Col Physicians; Am Soc Clin Nutrit; Am Fedn Clin Res; Am Gastroenterol Asn; Am Soc Parenteral & Enteral Nutrit; Am Clin & Climatol Asn. *Res:* Intestinal absorption and malabsorption. *Mailing Add:* Health Ctr Univ Fla Box J214 Gainsville FL 32610

CEREFICE, STEVEN A, b Newark, NJ, Aug 10, 43; m 67. ORGANIC CHEMISTRY, ORGANOMETALLIC CHEMISTRY. *Educ:* Rutgers Univ, BA, 65; Columbia Univ, MA, 66, PhD(org chem), 69. *Prof Exp:* RES CHEMIST ORG CHEM, AMOCO CHEM CORP, 70- *Mem:* AAAS; Am Chem Soc. *Res:* Organic synthesis; photochemistry of hydrocarbons; small ring compounds; bicyclic-polycyclic hydrocarbons; transition metal-catalyzed cycloaddition and electrocyclic reaction of polycyclic hydrocarbons; catalysis and mechanisms studies. *Mailing Add:* 1254 Hearthside Ct Naperville IL 60565-8957

CEREGHINO, JAMES JOSEPH, b Portland, Ore, Oct 27, 37. MEDICINE, NEUROSCIENCES. *Educ:* Portland State Col, BS, 59; Univ Ore Med Sch, MD, 64; Linfield Univ, MS, 71. *Prof Exp:* Intern, Good Samaritan Hosp Med Ctr, 64-65, resident neurol, 65-68; consult, Neurol & Sensory Dis Control Prog, Regional Med Prog Serv & Mental Admin & Dept Health, Educ & Welfare, 68-70; staff neurologist, Epilepsy Br, Nat Inst Neurol & Commun, 70-85; CHIEF, EPILEPSY BR, NIH, DEPT HEALTH & HUMAN SERV, 85- *Concurrent Pos:* Exec secy, serv training subcomt, comt epilepsies, US Pub Health Serv, 68-70; staff comt epilepsies, NIH, 70-85; consult, Comn Control Epilepsy & Consequence, 76-77; mem, Epilepsy Found Am Nat Implementation Task Force, 78-81; captain, US Pub Health Serv, 68-; prof adv bd, Metrop Wash Chap, Epilepsy Found Am, 69-; coun mem, Int League Against Epilepsy, 86-; book rev ed, Epilepsia, 82-86, ed-in-chief, 86- *Honors & Awards:* ILAE Geigy Award, Int League Against Epilepsy, 76; Ambassador Award, Epilepsy Int, 85. *Mem:* Am Epilepsy Soc (pres, 83-84); Am Neurol Asn; Am Acad Neurol; Am Electroencephalographic Soc; Am Med EEG Soc. *Res:* Clinical trials of new anti-epileptic drugs, pharmacology, social and delivery of health care problems in epilepsy. *Mailing Add:* Nat Inst Health Fed Bldg Rm 114 Bethesda MD 20892

CERF, VINTON GRAY, b New Haven, Conn, June 23, 43; m 66; c 2. INFORMATION PROCESSING. *Educ:* Stanford Univ, BS, 65; Univ Calif, Los Angeles, MS, 70, PhD(comput sci), 72. *Prof Exp:* Systs engr, IBM, 65-67; prin programmer, Univ Calif, Los Angeles, 67-72; asst prof elec eng & comput sci, Stanford Univ, 72-76; prin scientist, Defense Adv Res Proj Agency, 76-82; vpres, MCI Digital Info Servs Div, 82-86; VPRES, CORP NAT RES INITIATIVES, 86- *Concurrent Pos:* Consult, 86- *Mem:* Fel Inst Elec & Electronics Engrs; Asn Comput Mach. *Res:* Computers, communication and information processing. *Mailing Add:* 3614 Camelot Dr Annandale VA 22003

CERIMELE, BENITO JOSEPH, b Cincinnati, Ohio, May 11, 36; m 63; c 4. BIOSTATISTICS, BIOMATHEMATICS. *Educ:* Xavier Univ, Ohio, BS, 57, MS, 59; Univ Cincinnati, PhD(math), 63, Ind Univ, MBA, 76. *Prof Exp:* Reactor physicist, Gen Elec Co, Ohio, 57-59; from instr to asst prof math, Xavier Univ, Ohio, 62-66; NIH fel biomath, NC State Univ, 66-68, asst prof, 68-70; mgr, Sci Info Serv, Merrel Dow Pharmaceut, 78-88; sr systs analyst, Lilly Res Labs, 70-78, RES SCIENTIST, LILLY RES CLIN, ELI LILLY & CO, 88- *Concurrent Pos:* Adj assoc prof biostatist, Sch Med, Ind Univ, 76-, lectr, Sch Bus, 85-; adj asst prof biomath, Sch Med, Univ NC, Chapel Hill, 68-70, asst prof biostatist, NC State Univ, 68-70. *Mem:* Biometrics Soc; Am Statist Asn. *Res:* Clinical trials; biomathematical modeling. *Mailing Add:* 659 Williamsburg Ct Greenwood IN 46142

CERINI, COSTANTINO PETER, b Philadelphia, Pa, Nov 19, 31; m 60; c 2. VIROLOGY, IMMUNOLOGY. *Educ:* La Salle Col, BS, 53; Lehigh Univ, MS, 60, PhD(virol), 64. *Prof Exp:* Res virologist, Am Cyanamid Co, 64-70, group leader, 70-81, sr res virologist/proj leader, 81-88, TECH AFFAIRS COORDINATOR, LEDERLE LABS DIV, AM CYANAMID CO, 88- *Mem:* Am Soc Microbiol; NY Acad Sci. *Res:* Development of virus vaccines for human use; indentification and purification of protective antigens from Herpes Simplex Virus; incorporation into vaccines; evaluation in in vivo and in vitro models prior to human trials; vaccine technology assessment. *Mailing Add:* 11 Burnt Dr Pearl River NY 10965

CERKANOWICZ, ANTHONY EDWARD, b Bayonne, NJ, Feb 19, 41; m 66; c 5. COMBUSTION, MECHANICAL ENGINEERING. *Educ:* Stevens Inst Technol, ME, 62, MS, 64, PhD(thermodyn), 70. *Prof Exp:* Res engr combustion, Vitro Labs, 67-69; vpres, Photochem Industs, 69-74; sr staff engr combustion, Exxon Res & Eng Co, 74-87; ASSOC PROF, NJ INST TECHNOL, 87- *Concurrent Pos:* Chmn, Stevens Inst Honor Syst Adv Coun, 75-77. *Mem:* Am Soc Mech Engrs; Combustion Inst; Sigma Xi. *Res:* Pioneering research of techniques for photochemical ignition and enhancement of combustion in unsensitized fuel-air mixtures; analysis and utilization of catalytic combustion techniques in advanced power systems; properties and characteristics of negatively charged liquids; management and control of hazardous waste emissions, particularly complex hydrocarbons, chlorocarbons, and heavy metals. *Mailing Add:* NJ Inst Technol HSMRC-ATC 210 Newark NJ 07102

CERKLEWSKI, FLORIAN LEE, b Danville, Pa, May 28, 49; m 73; c 2. NUTRITION. *Educ:* Pa State Univ, BS, 71; Univ Ill, Urbana, PhD(nutrit), 76. *Prof Exp:* Fel trace metals, Kettering Lab, Univ Cincinnati, 76; asst prof nutrit, Marquette Univ, 76-79; asst prof, 79-84, ASSOC PROF FOODS & NUTRIT, OREGON STATE UNIV, 84- *Mem:* Am Inst Nutrit; Am Home Econ Asn. *Res:* Trace metal nutrition; nutritional biochemistry interactions; nutrition and dental health; fluoride bioavailability. *Mailing Add:* Dept Nutr Food Mtg Oregon State Univ Milam Hall Corvallis OR 97331-5103

CERMAK, J(ACK) E(DWARD), b Hastings, Colo, Sept 8, 22; m 49; c 2. FLUID MECHANICS, WIND ENGINEERING. *Educ:* Colo State Univ, BS, 48, MS, 49; Cornell Univ, PhD, 59. *Prof Exp:* Assoc prof eng mech, Colo State Univ, 47-59, vpres res found, 66-69, pres, 69-72, chmn eng sci major prog, 68-71, prof in charge fluid mech & wind eng prog, 59-85, dir, fluid dynamics & diffusion lab, 63-85, UNIV DISTINGUISHED PROF, COLO STATE UNIV, 86- *Concurrent Pos:* NATO fel, Cambridge Univ, 61-63; lectr, Univ Tex, Cambridge Univ, Israel Inst Technol, Univs Hokkaido, Moscow, Warsaw, Madrid & Marseille, Peking, Nan.ing, Colmbra, 63-; res grants numerous govt agencies and industrial organizations; consult numerous labs & indust co; chmn, Comn on Nat Disasters, Nat Res Coun; pres, Wind Eng Res Coun Inc, 71-85; US ed, Int J Wind Engineering, 74-; pres, Int Asn Wind Eng, 75-79; Sigma Xi nat lectr, 76-78; distinguished lectr, Am Soc Mech Eng, 87- *Honors & Awards:* Freeman Scholar, Am Soc Mech Engrs, 74; Sr Res Award, Am Soc Eng Educ, 87- *Mem:* Nat Acad Eng; hon mem Am Soc Civil Engrs; Am Soc Mech Engrs; Am Soc Eng Educ; Am Acad Mech; Am Meteorol Soc; Am Geophys Union; Am Soc Heating, Refrig & Air Conditioning Engrs; Am Concrete Inst. *Res:* Geophysical fluid mechanics, simulation of atmospheric motions by special wind tunnels, atmospheric diffusion and turbulence; structural aerodynamics; wind engineering; electrokinetics. *Mailing Add:* Col Eng Colo State Univ Ft Collins CO 80523

CERNANSKY, NICHOLAS P, b Pittsburgh, Pa, Mar 10, 46; m 70. COMBUSTION, AIR POLLUTION. *Educ:* Univ Pittsburgh, BS, 67; Univ Mich, MS, 68; Univ Calif, Berkeley, MPH, 73, PhD(mech eng), 74. *Prof Exp:* Develop engr modeling, Westinghouse Power Circuit Breaker Div, 67; air pollution engr vehicle emissions, Nat Air Pollution Control Admin, 68-70; res asst mech eng, Univ Calif, Berkeley, 70-74; from asst prof to assoc prof, 75-85, prof mech eng, 85-87, Raynes prof, 87-88, HESS CHAIR PROF COMBUSTION, DREXEL UNIV, 88- *Concurrent Pos:* Mem technol panel staff, Comt Motor Vehicle Emissions, 74, Diesel Impact Study Comt, Nat Acad Sci, 79-81. *Honors & Awards:* Ralph R Teetor Award, Soc Automotive Engrs, 76,. *Mem:* Air Pollution Control Asn; AAAS; Am Soc Mech Eng; Soc Automotive Engrs; Combustion Inst; Am Inst Aeronaut & Astronaut; Am Soc Eng Educ. *Res:* Analytical and experimental research problems in the areas of fundamental combustion studies, practical combustion systems, propulsion, air pollution, vehicular air pollution, energy and fuel conservation, fuels technology and environmental sciences. *Mailing Add:* Dept Mech Eng Drexel Univ Philadelphia PA 19104

CERNI, TODD ANDREW, b Milwaukee, Wis, Apr 28, 47. CLOUD PHYSICS, RADIATIVE TRANSFER. *Educ:* Marquete Univ, BS, 69, Ind Univ, MS, 71, Univ Ariz, PhD(atmospheric sci), 78. *Prof Exp:* Res assoc, Dept Physics, Univ Wyo, 76-78, res assoc, Dept Atmospheric Sci, 78-79, asst prof atmospheric sci & eng, 79-84; sr physicist, Ophir Corp, 84-88; CHIEF PHYSICIST, K C RES, 89- *Mem:* Am Phys Soc; Am Meteorol Soc. *Res:* Cloud physics; radiative transfer; atmospheric instrumentation. *Mailing Add:* 11231 Main Range Trail Littleton CO 80127

CERNICA, JOHN N, b Romania, May 14, 32; US citizen; m 59; c 5. CIVIL ENGINEERING. *Educ:* Youngstown Univ, BS, 54; Carnegie Inst Technol, MS, 55, PhD(civil eng), 57. *Prof Exp:* Asst prof civil eng & actg head dept, 57-58, assoc prof, 58-61, PROF CIVIL ENG, YOUNGSTOWN STATE UNIV, 61- *Concurrent Pos:* Panelist, NSF; examiner, Ohio State Bd Registr Prof Engrs & Surveyors. *Mem:* Am Soc Eng Educ; Am Soc Civil Engrs; Am Concrete Inst; Nat Soc Prof Engrs. *Res:* Structures and soil mechanics. *Mailing Add:* Dept of Civil Eng Youngstown State Univ 410 Wick Ave Youngstown OH 44555

CERNOSEK, STANLEY FRANK, JR, b Shiner, Tex, Dec 19, 40; m 68; c 2. IMMUNOCHEMISTRY. *Educ:* Pan Am Col, BA, 63; Univ Tex, Austin, PhD(chem), 69. *Prof Exp:* Res assoc peptide chem, Sch Med, Univ Pittsburgh, 68-70; fel biochem, Grad Dept Biochem, Brandeis Univ, 70-73;

asst prof dept biochem, Col Med, Univ Ark, 73-80; res chemist, Nat Ctr Toxicol Res, 78-80; STAFF SCIENTIST, BECKMAN INSTRUMENTS, INC, 80- *Mem:* Am Chem Soc; Royal Soc Chem; Biophys Soc; Sigma Xi. *Res:* Synthetic protein and peptide syntheses; sequential polypeptide syntheses; physical biochemistry; development and applications of radioimmunoassays for chemical carcinogens, hormones and environmental toxicans. *Mailing Add:* Beckman Instruments Inc 200 S Kramer Blvd Brea CA 92621

CERNUSCHI, FELIX, b Montevideo, Uruguay, May 17, 08; m 47; c 2. PHYSICS, ASTROPHYSICS. *Educ:* Univ Buenos Aires, CE, 32; Cambridge Univ, PhD(physics), 38. *Prof Exp:* Res worker, Cordoba Observ, Argentina, 38-39; prof phys sci, Nat Univ Tucuman, Argentina, 39-43; res assoc astrophys, Harvard Col Observ, 44-46; sci adv phys sci, UNESCO, France, 47-48; invited prof physics, Univ PR, 49-50; prof astron, 50-78, prof physics, 55-78, dir dept astron & physics, 55-78, dir fac humanities & sci, 57-78, Univ of the Repub, Uruguay, 57-78; prof, 78-85, EMER PROF PHYSICS, FAC ENG, UNIV BUENOS AIRES, 85- *Concurrent Pos:* Argentine Asn Adv Sci fel, 38-39; Guggenheim fel, 44-46; prof physics & dir fac eng, Univ Buenos Aires, 57-68; mem, Radioastronomy Comn, Univ Buenos Aires, 59-63 & Nat Coun Sci & Tech Invests, Argentina, 59-; invited lectr, Inter-Am Conf Physics, Brazil, 63 & Inter-Am Conf Sci & Technol, DC, 64. *Honors & Awards:* Phys Sci Prize, Buenos Aires City, 65. *Mem:* AAAS; fel Am Phys Soc; emer mem Am Astron Soc; Argentine Nat Acad Sci; Argentine Physics Asn; Columbia Acad Sci; NY Acad Sci; Sigma Xi; hon mem Argentine Math Union. *Res:* Statistical mechanics and its application to liquid state, cooperative phenomena in ordered-disordered transformations in alloys, solid-liquid transitions and astrophysics; interstellar matter; polarization of stellar light; cosmogony; new solutions of the ergodic problem; methods for teaching science and organization of universities; solar energy applications; theory of the birth of giant stars, theory of critical point in the transition liquid-vapor, theory of non-existance of critical point in the solid liquid transition. *Mailing Add:* Av Las Heras 2131 6 Piso Depto A Buenos Aires 1127 Argentina

CERNY, FRANK J, b Dayton, Ohio, June 10, 46; m 68; c 3. EXERCISE PHYSIOLOGY, PULMONARY PHYSIOLOGY. *Educ:* Macalester Col, BA, 68; Univ Wis-Madison, PhD(exercise physiol), 72. *Prof Exp:* Asst prof exercise physiol, Univ Windsor, Ont, 74-78; res asst prof pediat, 76-85, ASST PROF PHYS THERAPY & EXERCISE SCI, STATE UNIV NY, BUFFALO, 86- *Concurrent Pos:* Fel, Univ Freiburg, WGermany, 74; Alexander von Humboldt Found scholar, 75. *Mem:* Am Physiol Soc; Am Col Sports Med. *Res:* Exercise adaptation and therapy in children with cystic fibrosis, asthma, diabetes and congenital heart disease; abdominal muscle function. *Mailing Add:* Dept Phys Therapy & Exercise Sci State Univ Buffalo NY 14214

CERNY, JOSEPH, III, b Montgomery, Ala, Apr 24, 36; m 59, 83; c 2. NUCLEAR CHEMISTRY. *Educ:* Univ Miss, BS, 57; Univ Calif, Berkeley, PhD(nuclear chem), 61; PhD(physics), Univ Tyväskylä, Finland, 90. *Prof Exp:* From asst prof to assoc prof, 61-71, chmn dept, 75-79, PROF CHEM, UNIV CALIF, BERKELEY, 71-, PROVOST RES & DEAN GRAD DIV, 85- *Concurrent Pos:* Consult, US Army Res Off, 63-65 head, nuclear sci div & assoc dir, Lawrence Berkeley Lab, 79-84; Guggenheim fel, Oxford Univ, 69-70; vis fel, Australian Nat Univ, Canberra, 75; Alexander von Humboldt Award, 85. *Honors & Awards:* E O Lawrence Award, US AEC, 74; Nuclear Chem Award, Am Chem Soc, 84. *Mem:* Am Chem Soc; fel Am Phys Soc; Fedn Am Sci; fel AAAS. *Res:* Low energy nuclear science; studies of nuclei far from stability with an on-line mass separator; studies of isobaric analogue states and exotic nuclei; new modes of radioactive decay. *Mailing Add:* 309 Sproul Hall Univ Calif Berkeley CA 94720

CERNY, JOSEPH CHARLES, b Oak Park, Ill, Apr 20, 30; m 62; c 2. UROLOGIC ONCOLOGY, RENAL TRANPLANTATION. *Educ:* Knox Col, BA, 52; Yale Med Col, MD, 56. *Prof Exp:* Intern surg, Med Sch, 56-57, asst resident, 57-58, resident surg, 58-59, resident urol, 59-62, from instr to assoc prof urol, 62-71, CLIN PROF UROL, UNIV MICH, 71-; CHMN DEPT UROL, HENRY FORD HOSP, 71- *Concurrent Pos:* Fel transplant, Peter Bert Brigham Harvard, 58-59. *Mem:* Am Col Surgeons; Am Urol Asn; Am Asn Transplant Surgeons; Am Asn Endorine Surgeons; Nat Kidney Found; Soc Int d'Urologie; Urologic Soc Transplantation & Vascular Surg; Soc Urologic Oncol; Am Fertility Soc. *Res:* Surgical treatment of adrenal gland disorders; comparision of radiation therapy versus surgery in treating prostate cancer; urologic parameters of kidney transplantation; early diagnosis of invasive bladder cancer. *Mailing Add:* Dept Urol Henry Ford Hosp 2799 West Grand Blvd Detroit MI 48202

CERNY, LAURENCE CHARLES, b Cleveland, Ohio, Mar 5, 29; m 55; c 3. BIOPHYSICAL CHEMISTRY. *Educ:* Case Inst Technol, BS, 51, MS, 53; State Univ Ghent, PhD(phys chem), 56. *Prof Exp:* Asst prof chem, John Carroll Univ, 56-60; PROF CHEM, UTICA COL, 60- *Concurrent Pos:* Fel, Med Sch, Univ Minn, 58-59; res assoc hemodynamics, St Vincent Hosp, 59-60; estab investr, Masonic Med Res Lab, 62-67, career develop award, 67-; US-Czech exchange fel, Nat Acad Sci. 67. *Mem:* Fel Am Inst Chem; Biophys Soc; Am Chem Soc; Soc Rheol; Am Heart Asn; Sigma Xi. *Res:* Hemorheology; hemodynamics; polymers; chemical education; plasma expanders. *Mailing Add:* 19 Genesee Ct Utica NY 13502-5149

CERRA, ROBERT FRANK, b Munhall, Pa, Aug 18, 55; m 87. TUMOR CELL MATASTASIS, CELLS SURFACE EXTRACELLULAR MATRIX & LYMPHOCYTE CELL SURFACE. *Educ:* Okla State Univ, PhD(biochem), 81. *Prof Exp:* Asst res scientist, Univ Mich, Ann Arbor, 84-87; ASSOC STAFF INVESTR, HENRY FORD HOSP, DETROIT, 87- *Mem:* AAAS; Am Soc Cell Biol; Sigma Xi. *Res:* Mechanisms of tumor metastasis specifically the cell extracellular matrix in the exterior interaction of lymphocyte with HIV. *Mailing Add:* Div Hemat/Oncol Univ Mich 3700 Upjohn Ctr Ann Arbor MI 48109

CERRETA, KENNETH VINCENT, b Jersey City, NJ, June 12, 42; m 75; c 1. PHARMACOLOGY. *Educ:* Marist Col, BA, 64; Univ Md, MS, 70; Col Med & Dent NJ, PhD(pharmacol), 76. *Prof Exp:* Asst biochemist, Hoffmann-La Roche Inc, 69-70; assoc biochemist, Roche Inst Molecular Biol, 70-73; fel pharmacol, Univ Calif, San Francisco, 76-77; asst prof pharmacol, Sch Med, Univ SDak, 77-; TECH TRAINER, FIELD TRAINING & DEVELOP, SMITH KLINE & FRENCH LABS. *Concurrent Pos:* Lilly Res Inst grant, Univ SDak, 77-78, Smith-Kline Inc grant, 78-79. *Mem:* AAAS; NY Acad Sci; Sigma Xi. *Res:* Neuropharmacology of drugs producing analgesia and sedation, and biochemical factors associated with their mechanism of action. *Mailing Add:* Smith Kline Corp PO Box 7929 One Franklin Plaza 200 N 16th St Philadelphia PA 19102

CERRINA, FRANCESCO, b Moncalvo, Italy, Feb 11, 48; m 84; c 2. ELECTRICAL ENGINEERING. *Educ:* Univ Rome, Italy, PhD(solid state physics), 74. *Prof Exp:* Assoc res, Nat Coun Res, Italy, 74-81; vis scientist, 81-84, ASST PROF ELECTRONIC DEVICES, UNIV WIS-MADISON, 84- *Concurrent Pos:* Res assoc, physics dept, Mont State Univ, Bozeman, 78-79; vis prof, physics dept, Univ Wis, 81-82; affil mem, Ctr X-ray Optics, Lawrence Berkeley Lab, Univ Calif, Berkeley, 84- *Mem:* Am Phys Soc; Am Vacuum Soc; Am Optical Soc; Soc Photo Instrumentation Engrs. *Res:* Applications of x-rays to novel technologies for semiconductor processing (x-ray lithography); theoretical studies of x-ray optical systems for imaging applications in material science. *Mailing Add:* Dept Elec & Comput Eng Univ Wis Madison WI 53706

CERRO, RAMON LUIS, b Santiago del Estero, Arg, Sept 15, 41; m 73; c 3. FLUID MECHANICS, APPLIED MATHEMATICS. *Educ:* Univ Litoral, Arg, BS, 65; Univ Calif Davis, MS, 68, PhD(chem eng), 70. *Prof Exp:* Prof chem eng, Univ Litoral-Arg, 73-80; dir, Inst Design & Develop, 80- 86; PROF CHEM ENG, UNIV TULSA, 87- *Concurrent Pos:* Vis prof chem eng, Univ Minn, 74, 77, 79-80, 86, Univ Calif, Davis, 86; res fel, Nat Res Coun Arg, 76-86. *Mem:* Am Inst Chem Engrs; Sigma Xi; corresp mem Nat Acad Eng Arg. *Res:* Physicochemical hydrodynamics; free surface flows coating flows; secondary oil recovery. *Mailing Add:* Dept Chem Eng Univ Tulsa Tulsa OK 74104

CERRONI, ROSE E, b Weirton, WVa, Mar 29, 30. BIOLOGY, PHYSIOLOGY. *Educ:* Col of Steubenville, BS, 52; Vanderbilt Univ, MA, 55, PhD(biol), 59. *Prof Exp:* Instr nursing sci, Wheeling Hosp Sch Nursing, 52-53; instr microbiol, Vanderbilt Univ, 55-56; NIH fel, Carlsberg Biol Inst, 60-61; res assoc, Temple Univ, 61-62; asst prof biol, West Liberty State Col, 62-67; PROF BIOL, FRANCISCAN UNIV STEUBENVILLE, 66- *Mem:* Am Inst Biol Sci. *Res:* Physiology and ultra structure of the nucleus of acanthamoeba radioautographic studies on tetrahymena pyriformis to determine aspects of the mechanism of heat-induced division sychrony. *Mailing Add:* Dept Biol Univ Steubenville Steubenville OH 43952

CERVENKA, JAROSLAV, b Prague, Czech, Mar 15, 33; US citizen; m 59; c 2. MEDICAL GENETICS, CANCER. *Educ:* Charles Univ, Czech, MD, 58; Czech Acad Sci, CSc(med genetics), 68. *Prof Exp:* Actg chief genetics, Lab Plastic Surg, Prague, 67-68; from asst prof to assoc prof, 68-78, prof med, Med Sch, 78-80, PROF GENETICS, SCH DENT, UNIV MINN, 78-, DIR, DIV CYTOGENETICS & CELL GENETICS, 79- *Concurrent Pos:* Staff consult, Sch Dent, Univ Minn, 66-, mem staff, Grad Sch, 69-, Genetic Clin, Health Sci Ctr, 72- *Mem:* Am Soc Human Genetics; Int Dermatoglyphics Asn; Sigma Xi; Int Found Human Health. *Res:* Cytogenetics of cancer; genetics of congenital abnormalities; clinical cytogenetics. *Mailing Add:* Health & Sci Tower 16-144 Univ of Minn Sch of Dent Minneapolis MN 55455

CERVENY, THELMA JANNETTE, b Prairie Grove, Ark, Sept 25, 49; m 77. CELLULAR PHYSIOLOGY, HEMATOLOGY. *Educ:* Tex Woman's Univ, BS, 71, MS, 72; Mayo Grad Sch Med, PhD(physiol), 84. *Prof Exp:* Instr human physiol, Tex Woman's Univ, 71; instr & course dir biol sci, Air Force Acad, 77-79; prin investr, Armed Forces Radiobiol Res Inst, Defense Nuclear Agency, 83-87; PROG MGR BIOENVIRONMENTAL HAZARDS, AIR FORCE OFF SCI RES, 87- *Concurrent Pos:* Adj Asst Prof physiol, Uniformed Serv Univ Health Sci, 83-, res assoc surg, 88- *Mem:* NY Acad Sci; Am Physiol Soc; Am Astronaut Soc; Aerospace Med Asn; Soc Toxicol. *Res:* Effects of ionizing radiation on endothelial cell function and viability in vitro; effects of radiation on blood brain barrier in vivo. *Mailing Add:* Bolling AFB Bldg 410 Washington DC 20332-6448

CERVONI, PETER, b Jamaica, NY, Mar 4, 31; m 64; c 3. PHARMACOLOGY. *Educ:* St John's Univ, NY, BS, 52; Univ Wash, MS, 55, PhD(pharmacol), 57. *Prof Exp:* Asst pharmacol, Univ Wash, 52-57; pharmacologist, US Army Chem Warfare Labs, 57-59; asst prof pharmacol, Univ Miss, Jackson, 60-61; from instr to asst prof, State Univ NY Brooklyn, 61-66; sr res pharmacologist, Wellcome Res Lab, Burroughs Wellcome & Co, 66-70; sect head cardiovasc/autonomic pharmacol, USV Pharmaceut Corp, 70-71, mgr biol res & develop, 71-73, dir & res scientist, 73-75, dir pharmacol, Pharmaceut Res & Develop, 75-79; HEAD, DEPT CARDIOVASC BIOL RES, MED RES DIV, AM CYANAMID CORP, 79- *Concurrent Pos:* Life Ins Med Res Fund fel, 59-60; adj prof pharmacol, St John's Univ, 75-81, Arnold & Marie Schwartz Col Pharmacy, 82-83, NY Med Col, 82- & Univ Tenn, Med Ctr, 84-; vis prof pharmacol, Ohio State Univ, Med Sch, 75, 78 & 79; counr, Am Soc Pharmacol & Exp Therapeut, 91-94. *Honors & Awards:* Merck Award Dispensing Pharm, 52. *Mem:* Am Heart Asn; Sigma Xi; AAAS; Am Soc Pharmacol & Exp Therapeut; fel NY Acad Sci; fel Coun High Blood Pressure Res. *Res:* Cardiovascular pharmacology. *Mailing Add:* 64 Sheldrake Pl New Rochelle NY 10804

CERWONKA, ROBERT HENRY, b Endicott, NY, Mar 16, 31; m 57; c 2. MARINE ECOLOGY. *Educ:* State Univ NY, Albany AB, 53, MA, 57; Univ Conn, PhD, 68. *Prof Exp:* From instr to assoc prof, 59-69, chmn dept, 79-86, PROF BIOL, STATE UNIV NY COL POTSDAM, 69- *Mem:* Ecol Soc Am; Am Soc Limnol & Oceanog; Am Ornith Union; Sigma Xi. *Res:* Filtering rates of bivalve mollusks. *Mailing Add:* Dept Biol State Univ NY Col Potsdam NY 13676

CESARI, LAMBERTO, b Bologna, Italy, Sept 23, 10; m 39. MATHEMATICAL ANALYSIS. *Educ:* Univ Pisa, PhD(math), 33. *Hon Degrees:* Dr, Univ Perugia, Italy. *Prof Exp:* Asst prof math, Univ Rome, 37-39; assoc prof, Univ Pisa, 39-42; from assoc prof to prof, Univ Bologna, 42-48; vis prof, Inst Adv Study, 48, Univ Calif, 49 & Univ Wis, 50; vis prof, Purdue Univ, 50, prof, 52-60; PROF MATH, UNIV MICH, 60-, R L WILDER PROF, 75- *Honors & Awards:* Russel Lect, Univ Mich, 76. *Mem:* Am Math Soc; Math Asn Am; Soc Indust & Appl Math; Math Union Italy; Sigma Xi. *Res:* Real functions; calculus of variations; surface area theory; asymptotic behavior of differential equations; numerical analysis; ordinary and partial differential equations; optimal control theory; nonlinear analysis. *Mailing Add:* 2021 Washtenaw Ann Arbor MI 48104

CESCAS, MICHEL PIERRE, b Bordeaux, France, Sept 23, 36; m 68; c 2. SOIL CHEMISTRY, SOIL FERTILITY. *Educ:* Laval Univ, BSc, 60; Univ Ill, Urbana, MSc, 65, PhD(agron), 68. *Prof Exp:* Res asst agron, Univ Ill, Urbana, 60-61, 63-68; from asst prof soil chem to prof, 68-77, PROF SOIL SCI, LAVAL UNIV, 77-, HEAD DEPT, 77- *Honors & Awards:* Scarseth's Award, Am Soc Agron. *Mem:* AAAS; Am Soc Agron; Soil Sci Soc Am; Int Soc Soil Sci; Am Chem Soc. *Res:* Electron probe analysis of soils and related materials; phosphorus evolution in soils; rock phosphate transformation with time; analytical chemistry methodology in soil testing and analysis; adsorption phenomena in soils; optimization of fertilization of important crops; use of manures. *Mailing Add:* Fac Agric-Soils Laval Univ Quebec PQ G1K 7P4 Can

CESS, ROBERT D, b Portland, Ore, Mar 3, 33; m 53; c 2. ATMOSPHERIC SCIENCES. *Educ:* Ore State Univ, BS, 55; Purdue Univ, MS, 56; Univ Pittsburgh, PhD(mech eng), 60. *Prof Exp:* Res engr, Westinghouse Res Labs, 56-60, consult, 60-64; assoc prof mech eng, NC State Col, 60-61; assoc prof eng, 61-65, prof, 65-75, prof atmospheric sci, 75-81, LEADING PROF, STATE UNIV NY STONY BROOK, 81- *Concurrent Pos:* NSF res grant, 60-; deleg, US/USSR Bilateral Agreement Environ Protection, 78, 79, 87, 88; assoc ed, J Geophys Res; vis prof, Leningrad State Univ, 80; mem, Carbon Dioxide & Climate Rev Panel, Nat Res Coun, 81-82, 90-91. *Honors & Awards:* Heat Transfer Mem Award, Am Soc Mech Engrs, 77; Spec Achievement Award, NASA, 85 & 88, Except Sci Achievement Medal, 89. *Mem:* AAAS; Am Phys Soc. *Res:* Atmospheric radiation; climate modeling. *Mailing Add:* Inst Terrestrial & Planetary Atmospheres State Univ NY Stony Brook NY 11794

CESSNA, LAWRENCE C, JR, b Cumberland, Md, Feb 1, 39; m 60; c 2. POLYMER PHYSICS, CHEMICAL ENGINEERING. *Educ:* Johns Hopkins Univ, BES, 61; Rensselaer Polytech Inst, PhD(polymer sci), 65. *Prof Exp:* Res assoc polymer sci, Rensselaer Polytech Inst, 65-66; res engr, 66-70, sr res engr, 70-73, res scientist, 73, mgr mat sci div, Hercules Res Ctr, 73-77, asst dir develop projs, 77-80, dir technol mkt, 80-81, dir develop, 81-83, dir qual assurance, 83-85, CORP DIR, PLANNING & ACQUISITIONS, HERCULES INC, 85- *Mem:* Indust Res Inst; Soc Chem Indust; Soc Plastics Engr; Am Chem Soc. *Res:* Ultimate properties of high polymers; physical and mechanical behavior of polymers and polymer based composite materials; chemical engineering fundamentals. *Mailing Add:* 111 Neptune Dr North Star Newark DE 19711-3011

CETRON, MARVIN J, b Brooklyn, NY, July 5, 30; m 55; c 2. RESEARCH MANAGEMENT, INDUSTRIAL ENGINEERING. *Educ:* Pa State Univ, BS, 52; Columbia Univ, MS, 59, cert exec develop, 58; Am Univ, PhD(res & develop mgt), 70. *Prof Exp:* Head performance statist sect, Naval Appl Sci Lab, NY, 52-54, head mgt planning, rev br, 54; eng asst to tech dir opers res, Naval Marine Eng Lab, Md, 56-58, independent res coordr, 58-63, head prog mgt off, 63-64, planning dir, 64-66, head tech forecasting & appraisal, HQ Naval Mat Command, Washington DC, 66-71; FOUNDER & PRES, FORECASTING INT, LTD, 71- *Concurrent Pos:* Ed-in-chief, Technol Assess J, 69-; adj prof, Am Univ & Mass Inst Technol, 70-; mem, Coast Guard Res & Develop Adv Comt. *Honors & Awards:* Armed Forces Mgt Lit Award, 69 & 70; Armed Forces Mgt Award, 70. *Mem:* AAAS; Am Inst Indust Engrs. *Res:* Technological forecasting; operations research in research and development; research and development planning; project selection and resource allocation models; technology assessment. *Mailing Add:* 1001 N Highland St Arlington VA 22210

CEVALLOS, WILLIAM HERNAN, b Quito, Ecuador, Mar 11, 32; US citizen; m 58; c 4. BIOCHEMISTRY, METABOLISM. *Educ:* Mt St Mary's Col, Md, BS, 54; St John's Univ, NY, MS, 56; Georgetown Univ, PhD(biochem), 60. *Prof Exp:* Asst microbiol, St John's Univ, NY, 54-56, lab instr histol, 55-56; asst biochem, Georgetown Univ, 56-59, lab instr chem, Sch Nursing, 57-59; res assoc biochem, Sinai Hosp, Baltimore, Md, 59-64; res assoc, Smith Kline & French Labs, 64-68; res assoc biochem, Div Res, Lankenau Hosp, 68-84; assoc dir clin res, 90 SR CLIN RES MONITOR, WYETH-AYERST RES, 84-, ASST DIR CLIN RES, 87- *Concurrent Pos:* NIH fel, 60-62, grant, 62-64. *Mem:* NY Acad Sci; Am Chem Soc; Am Heart Asn; AAAS; Am Asn Pharmaceut Scientists. *Res:* Lipid metabolism related to atherosclerosis and heart disease; hormonal and chemotherapeutic control of cholesterol metabolism; relationship of circulating and dietary lipids to platelets and thrombosis; pharmaceutical clinical research; clinical pharmacology; pharmacokinetics. *Mailing Add:* Two Warrior Way Hamilton Square NJ 08690

CEVASCO, ALBERT ANTHONY, b New York, NY, Sept 4, 40; m 63; c 2. ORGANIC CHEMISTRY. *Educ:* Manhattan Col, BS, 62; Fordham Univ, PhD(org chem), 68. *Prof Exp:* Asst chem, Fordham Univ, 62-67; res chemist, Explor Res & Develop Dept, Bound Brook, 67-69, res chemist, Decision Making Systs, 69-70, process res chemist, Dyes Tech Dept, 70-75, PRIN RES CHEMIST, AGR RES CTR, AM CYANAMID CO, PRINCETON, NJ, 75- *Mem:* Am Chem Soc; Sigma Xi. *Res:* Heterocyclic syntheses; organic luminescers; dye and brightener intermediates process research and development; process research and development of agricultural organic chemicals, animal health products, pesticides and synthetic pyrethroids. *Mailing Add:* 21 Kingswood Dr Belle Mead NJ 08502

CEZAIRLIYAN, ARED, b Istanbul, Turkey, May 9, 34; US citizen; m 70; c 1. THERMAL PHYSICS, MATERIALS SCIENCE. *Educ:* Robert Col, Istanbul, BSME, 57; Purdue Univ, Lafayette, MSME, 60, PhD, 63. *Prof Exp:* PHYSICIST, NAT INST STANDARDS & TECHNOL, 63- *Concurrent Pos:* Consult, Thermophys Properties Res Ctr, Purdue Univ, 68-; mem, Int Orgn Comt, Europ Thermophys Properties Conf, 74-; chmn, Int Thermophysics Cong, 76-; ed-in-chief, Int J Thermophysics, 80- *Honors & Awards:* Spec Achievement Award, Nat Bur Standards, 77; Gold Medal Award, US Dept Com, 80; Heat Transfer Mem Award, Am Soc Mech Engrs, 81; Gold Medal, High Temperature Soc France, 82; IR-100 Award, 83; Thermal Conductivity Award, Int Thermal Conductivity Conf, 87. *Mem:* AAAS; Am Phys Soc; Am Soc Mech Engrs; Sigma Xi. *Res:* High temperature thermophysics; material properties; calorimetry; high temperature thermometry; high speed pyrometry; optics; transient measurement techniques. *Mailing Add:* Thermophysics Div Nat Inst Standards & Technol Gaithersburg MD 20899

CHA, DAE YANG, b Korea, Sept 25, 36; US citizen; m 65; c 3. CHEMICAL ENGINEERING, SYNTHETIC ORGANIC & NATURAL PRODUCTS CHEMISTRY. *Educ:* Seoul Nat Univ, BS, 59, MSE, 61; Univ Calif, Berkeley, MS, 63; Univ Mich, PhD(chem eng), 68. *Prof Exp:* CHEM ENGR, UPJOHN CO, 66- *Mem:* Am Inst Chem Engrs; Am Chem Soc. *Res:* Heterogeneous catalysis; separation and purification of antibiotics and steroids from fermentation medium; process development on fine chemicals and specialty polymers. *Mailing Add:* 2030 Aberdeen Dr Kalamazoo MI 49008

CHA, SUNGMAN, b Chungpyong, Korea, Mar 1, 28; m 60; c 3. BIOCHEMISTRY, PHARMACOLOGY. *Educ:* Yonsei Univ, Korea, MD, 54; Univ Wis, PhD(pharmacol), 63. *Prof Exp:* Asst pharmacol, Univ Wis, 59-61, trainee, 61-63; from asst prof to assoc prof, 63-76, PROF MED SCI, BROWN UNIV, 76- *Mem:* Am Soc Biol Chem; Am Soc Pharmcaol & Ther; Am Asn Cancer Res. *Res:* Metabolism of nucleotides including analogues; enzymology of cancer cells. *Mailing Add:* Div of Biomed Sci Brown Univ Providence RI 02912

CHABAI, ALBERT JOHN, b Mont, Feb 1, 29; m 58; c 6. PHYSICS. *Educ:* Mont State Col, BS, 51; Lehigh Univ, MS & PhD(physics), 58. *Prof Exp:* Mem physics res staff, 58-65, div supvr, Adv Systs Develop, 65-71, div supvr, Computational Physics & Mech, 71-, DISTINGUISHED MEM TECH STAFF, SANDIA LABS. *Mem:* AAAS. *Res:* Fluid dynamics. *Mailing Add:* 2607 Haines Ave NE Albuquerque NM 87106

CHABNER, BRUCE A, CANCER RESEARCH. *Prof Exp:* DIR, DIV CANCER TREATMENT, NAT CANCER INST, NIH, 82- *Mailing Add:* NIH Nat Cancer Inst Div Cancer Treatment Bldg 31 Rm 3A44 Bethesda MD 20892

CHABOT, BENOIT, b Lachine, Que, Aug 12, 56; m 81; c 2. RNA PROCESSING. *Educ:* Univ Sherbrooke, BSc, 79, MSc, 81; Yale Univ, PhD(molecular biol), 86. *Prof Exp:* Fel mouse genetics, Mt Sinai Hosp Res Inst, Toronto, 86-88; ASST PROF, DEPT MICROBIOL, FAC MED, UNIV SHERBROOKE, 88- *Concurrent Pos:* K M Hunter fel, Nat Cancer Inst Can, 86-88. *Mem:* Int Soc Differentiation Inc; Can Soc Cellular & Molecular Biol. *Res:* Regulation of alternative RNA splicing in mammalian cells. *Mailing Add:* Dept Microbiol Fac Med Univ Sherbrooke Sherbrooke PQ J1H 5N4 Can

CHABOT, BRIAN F, b Orlando, Fla, July 11, 43; m 67; c 2. PHYSIOLOGICAL PLANT ECOLOGY, POPULATION BIOLOGY. *Educ:* Col William & Mary, BS, 65; Duke Univ, PhD(bot & ecol), 71. *Prof Exp:* Asst prof bot, Univ NH, 71-73; from asst prof to assoc prof ecol & systs, Cornell Univ, 73-86, chmn dept, 79-82, assoc dir, Agr Exp Sta & Off Res, 83-90, PROF ECOL & SYSTS, CORNELL UNIV, 86-, DIR, AGR EXP STA & OFF RES, 90- *Concurrent Pos:* NSF presidential young investr award, 85; consult, Exp Prog Stimulate Competitive Res, NSF, 86; prin investr, NEC, 89-91 & 91-95; mem, Exp Stas Comt Orgn & Policy Plant Germplasm Subcomt, 91-93. *Mem:* Fel AAAS; Ecol Soc Am; Am Soc Plant Physiologists; Bot Soc Am. *Res:* Ecology and systematics; physiological plant ecology. *Mailing Add:* Col Agr & Life Sci Cornell Univ 245 Roberts Ithaca NY 14853

CHABRECK, ROBERT HENRY, b Lacombe, La, Mar 18, 33; m 54; c 4. WILDLIFE MANAGEMENT, WETLAND ECOLOGY. *Educ:* La State Univ, Baton Rouge, MS, 57, PhD(bot), 70. *Prof Exp:* Refuge biologist, La Wildlife & Fisheries Comn, 57-59, res biologist, 59-66, res supvr, 66-67; asst leader, La Coop Wildlife Res Unit, 67-72; assoc prof, 72-77, PROF FORESTRY & WILDLIFE MGR, LA STATE UNIV, BATON ROUGE, 77- *Mem:* Wildlife Soc. *Res:* Wetland ecology, especially wetland management, life history studies of American alligator and waterfowl. *Mailing Add:* Sch Forestry Wildlife & Fisheries La State Univ Baton Rouge LA 70803

CHABRIES, DOUGLAS M, b Los Angeles, Calif, Feb 18, 42; m 64; c 6. DIGITAL SIGNAL PROCESSING, ACOUSTIC SCATTERING. *Educ:* Univ Utah, BS, 66; Calif Inst Technol, MS, 67; Brown Univ, PhD(elec eng), 70. *Prof Exp:* Prof computer eng & dept chair elec eng, 83-90, ASST ACAD VPRES, BRIGHAM YOUNG UNIV, 90- *Mem:* Sigma Xi. *Res:* Digital signal processing; acoustic processing; acoustics; underwater scattering; digital imagery; digital image compression. *Mailing Add:* Unit 312 Woodland Hills UT 84653

CHACE, ALDEN BUFFINGTON, JR, b San Francisco, Calif, Dec 13, 39; m 67; c 3. ENVIRONMENTAL OCEAN ACOUSTICS. *Educ:* US Naval Acad, BS, 62; Naval Postgrad Sch, MS, 69; Univ RI, PhD(ocean eng), 75. *Prof Exp:* Asst prof oceanog, Naval Postgrad Sch, 76-79; regional serv, Naval Western Oceanog Ctr, 79-82; opers analyst, Summit Res Corp, 82-83; sr res scientist, Sci Appln Int Corp, 83-84; supvr oceanog eng, Brown & Caldwell, 84-85; PRIN ENGR, BOEING AEROSPACE, 85- *Concurrent Pos:* Submarine officer, US Navy, 62-70, geophysics officer, Fleet Numerical Ocean Ctr, 76; Instr, Leeward Community Col, 81; adj prof oceanog, Naval Postgrad Sch, 84; consult oceanog, 85. *Mem:* Marine Technol Soc; Sigma Xi;

Am Soc Civil Engrs; Am Meteorol Soc. *Res:* Application of ocean environmental acoustics to submarine and anti-submarine warfare; prediction of electromagnetic and sound propagation; gamma ray density testing of marine sediment; ocean wave forecasting; naval submarine operations. *Mailing Add:* 3250 South 164th St Seattle WA 98188-3036

CHACE, FENNER ALBERT, JR, b Fall River, Mass, Oct 5, 08; m 34; c 1. ZOOLOGY. *Educ:* Harvard Univ, AB, 30, AM, 31, PhD(biol), 34. *Prof Exp:* Asst cur mar invert, Mus Comp Zool, Harvard Univ, 34-42, cur Crustacea, 42-46, cur marine invert, US Nat Mus, 46-63; sr zoologist, 63-78, EMER ZOOLOGIST, SMITHSONIAN INST, 78- *Concurrent Pos:* Asst biol, Harvard Univ, 35, Agassiz fel, 35-39, tutor biol, 40-41. *Mem:* AAAS; Crustacean Soc. *Res:* Taxonomy, morphology and distribution of decapod Crustacea. *Mailing Add:* 6 Leland Ct Chevy Chase MD 20815-4906

CHACE, FREDERIC MASON, mining geology; deceased, see previous edition for last biography

CHACE, MILTON A, b Takoma Park, Md, Mar 19, 34; m 58; c 3. DYNAMICS, MECHANICAL ENGINEERING. *Educ:* Cornell Univ, BEP, 57; Univ Mich, MSME, 62, PhD(mech eng), 64. *Prof Exp:* Engr, USAEC, 57-59; mech engr, Atomic Power Develop Assocs, 59-61; staff engr, Systs Develop Div, IBM Corp, 64-67; assoc prof mech eng, 67-75, PROF MECH ENG, UNIV MICH, 75- *Concurrent Pos:* Dir, Comput-Aided Design Lab, 71- & Mech Dynamics Inc, 77- *Mem:* Am Soc Mech Engrs. *Res:* Dynamics of mechanical systems; digital computer simulation of systems; computer aided design. *Mailing Add:* 3265 Maple Rd Ann Arbor MI 48105-9643

CHACKERIAN, CHARLES, JR, b San Francisco, Calif, Feb 6, 35; c 3. PHYSICAL CHEMISTRY, MOLECULAR SPECTROSCOPY. *Educ:* Univ Calif, Berkeley, BS, 58; Univ Wash, PhD(chem), 64. *Prof Exp:* RES SCIENTIST, AMES RES CTR, NASA, 64- *Mem:* Am Phys Soc. *Res:* Shock waves; high temperature chemical kinetics and molecular relaxation; infrared spectroscopy related to planetary and stellar spectroscopy; gas lasers. *Mailing Add:* Ames Res Ctr NASA Moffett Field CA 94035-1000

CHACKO, GEORGE KUTTICKAL, b Trivandrum, India, July 1, 30; US citizen; m 57; c 2. SYSTEMS MANAGEMENT, STATISTICS. *Educ:* Madras Univ, MA, 50; Indian Statist Inst, advan cert, 51; Univ Calcutta, BCom 52; New Sch Soc Res, PhD(econometrics), 59. *Prof Exp:* Consult opers res, Union Carbide Corp, New York, 62-63; mem tech staff, Res Analysis Corp, McLean, Va, 63-65 & MITRE Corp, Washington, DC, 65-67; sr staff scientist, TRW Systs, Washington Opers, 67-70; PROF SYSTS SCI, UNIV SOUTHERN CALIF, 70- *Concurrent Pos:* Fel, Western Mgt Sci Ctr, Univ Calif, Los Angeles, 61; consult space systs anal, Inst Creative Studies, Washington, DC, 68-70; consult comput sci, US Dept of Defense, Indust Col Armed Forces, 70; consult mil opers res, Milcom Systs, Rockville, Md, 71-72; consult info sci, King Res, Rockville, Md, 75-76; consult comput aided mgt controls, On-Line Systems, McLean, Va, 79-82, prin investr adaptive forecasting, 81, prin investr Concomitant Coalitions, 84-85; NSF int sci lectr award, 82; sr Fulbright prof mgt sci, Nat Chengchi Univ, Taipei, Taiwan, 83-85; US Info Agency US sci emissary to Egypt, Burma, India, Singapore, 87; chief sci consult, RJO Enterprises, Inc, Lanham, Md, 88; sr vis res prof, Nat Sci Coun & Nat Chengchi Univ, Taipei, Taiwan, 88-89. *Mem:* Fel AAAS; fel Am Astronaut Soc; Opers Res Soc Am; NY Acad Sci. *Res:* Technology transfer of hi-tech from the US to the Pacific; robotics; artificial intelligence; productivity; program management; US experience for newly industrializing countries; quantitative decision making. *Mailing Add:* 6809 Barr Rd Bethesda MD 20816-1013

CHACKO, GEORGE KUTTY, b Kottarakkara, India, Feb 15, 33; m 62; c 3. BIOCHEMISTRY. *Educ:* Univ Col, Trivandrum, India, BSc, 56; Maharaja's Col, Ernakulam, MSc, 58; Univ Ill, Urbana, PhD(food chem, biochem), 66. *Prof Exp:* Fel biochem, Univ Wash, 66-67; fel, Univ Ariz, 67-68; res asst prof, 68-74, res assoc prof biochem & physiol, 74-77, ASSOC PROF BIOCHEM, MED COL PA, 77- *Concurrent Pos:* Vis prof, Latin Am Biol Sci Ctr Workshop, Caracas, Venezuela, 79, fac med, Xavier Bichat, Univ Paris Seven, Paris, France, 87; invited speaker, Gordon Res Conf, atherosclerosis, 85. *Mem:* Am Soc Biochem & Molecular Biol; AAAS; Sigma Xi. *Res:* Lipid chemistry and biochemistry; lipoprotein-membrane interactions; structural and functional studies on high density lipoprotein receptors. *Mailing Add:* Dept Biochem & Physiol Med Col Pa 3300 Henry Ave Philadelphia PA 19129

CHACKO, ROSY J, b Neendoor, Kerala; India citizen; m 70. MYCOLOGY, FOREST PATHOLOGY. *Educ:* Univ Kerala, BSc, 59; Agra Univ, MSc, 62; Univ Mont, PhD(bot), 72. *Prof Exp:* Lectr, Univ Kerala, 62-67; teaching asst bot, Univ Mont, 68-72; res technician forest path, Col Forestry, Univ Idaho, 73-74, res scientist, 74-77; res technician plant path, Wash State Univ, 77-81; RES TECH PLANT PATH, UNIV ARK, 81- *Concurrent Pos:* Mem mus subcomt, Wash State Univ, 78- *Mem:* Mycol Soc Am. *Res:* Collect and identify fruiting bodies of Ascomycetes and Basidiomycetes that decay timber; make aseptic isolations from sporophores/decay; study the morphology using light and electron microscopic observations. *Mailing Add:* Dept Plant Pathol 217 Plant Sci Bldg Univ Ark Fayetteville AR 72701

CHACKO, SAMUEL K, b Kottarakara, India, Feb 7, 42; US citizen; m 68; c 3. PATHOBIOLOGY, BIOCHEMISTRY. *Educ:* Kerala Vet Col & Res Inst, DVM, 63; Univ Pa, PhD(path), 69. *Prof Exp:* Instr path, Comp Cardiovasc Studies Unit, 65-66, USPHS trainee, 66-69, asscc prof, 74-81, PROF PATH, DEPT PATHOBIOL, SCH VET MED & GRAD SCH ARTS & SCI, UNIV PA, 81- *Concurrent Pos:* Prin investr, Cardiovasc Studies Unit, Univ Pa, 69-75 & Pa Muscle Inst, 72-76; exp pathologist, Molecular Cardiol Sect, Cardiol Br, Nat Heart, Lung & Blood Inst, 75-77; consult, NIH Prog Proj grant, 76; prin investr, Nat Heart, Lung & Blood Inst res grant, 78; chmn, Path Grad Group, Univ Pa, 86; vis prof, Univ Tokyo, 83, Univ Heildelberg, 85. *Mem:* Am Soc Exp Path; Am Soc Cell Biol; Int Soc Heart Res; Sigma Xi; Am Soc Biochem & Molecular Biol. *Res:* Regulation of actomyosin ATPase and contraction in smooth muscle; cytoplasmic filaments in smooth muscle and endothelial cells; contractile protein structure and function in development, growth and aging in cardiac and smooth muscle; urinary bladder hypertrophy. *Mailing Add:* Dept Pathobiol Univ Pa 390 Vet Sch 3800 Spruce Philadelphia PA 19104

CHACON, RAFAEL VAN SEVEREN, b El Salvador, Cent Am, June 10, 31; nat US; wid; c 4. MATHEMATICS. *Educ:* Univ Rochester, BS, 51; Syracuse Univ, PhD(math), 56. *Prof Exp:* Instr, Ohio State Univ, 56-58; asst prof math, Univ Wis, 58-61; assoc prof, Brown Univ, 61-64 & Ohio State Univ, 64-69; prof, Univ Minn, Minneapolis, 69-74; PROF MATH, UNIV BC, 74- *Mem:* Am Math Soc; fel Royal Soc Can. *Res:* Probability theory; ergodic theory; functional analysis. *Mailing Add:* Dept Math Univ BC 2075 Wesbrook Pl Vanoouver BC V6T 1W5 Can

CHADDE, FRANK ERNEST, b Chicago, Ill, June 23, 29; m 53; c 4. ANALYTICAL CHEMISTRY. *Educ:* Univ Ill, BS, 51. *Prof Exp:* Control chemist, Abbott Labs, 51-52, analysis chemist, 54-63, mgr analysis res dept, 63-67, mgr chem control dept, 67-69, mgr Analysis Lab, Abbott Labs, 69-86; RETIRED. *Mem:* Am Soc Qual Control; Am Chem Soc; Acad Pharmaceut Sci. *Res:* Ultra violet spectroscopy; colorimetry; titrimetric analysis; polarography. *Mailing Add:* 2007 Forest Creek Lane Libertyville IL 60048

CHADDOCK, JACK B, b Cameron, WVa, Dec 6, 24; m 55, 73; c 2. MECHANICAL ENGINEERING. *Educ:* Univ SC, BS, 45; Univ WVa, BS, 48; Mass Inst Technol, MS, 49, MechE, 52, ScD(mech eng), 55. *Prof Exp:* From instr to asst prof mech eng, Mass Inst Technol, 50-57; assoc prof, Rensselaer Polytech Inst, 57-59; prof, Purdue Univ, 59-66; prof mech eng, Duke Univ, 66-90, chmn dept, 66-86, dir, Ctr Study Energy Conserv, 75-90, assoc dean, Res & Develop, 87-90, EMER PROF, DUKE UNIV, 90- *Concurrent Pos:* Fulbright lectr, Inst Technol, Helsinki, 55- 56; vis prof, Univ NSW, Australia, 73; vis prof, Bldg Res Estab, Eng, 80; pres & trustee-pres, John B Pierce Found, NY & John B Pierce Lab, New Haven, Conn, 81-; vis sr scientist, Lawrence Berkeley Lab, Univ Calif, Berkeley, 86-87; sr consult, Carrier Corp Res Div, 66; vpres, SUD Assoc, Prof Assocs, 85- *Honors & Awards:* Diamond Key Award, Am Soc Heat, Refrig & Air Cond Engrs, 58, Distinguished Serv Award, 70, E K Campbell Merit Award, 72. *Mem:* Fel Am Soc Heat, Refrig & Air Cond Engrs (pres, 81-82); Int Inst Refrig; fel Am Soc Mech Engrs; Int Solar Energy Soc; Sigma Xi. *Res:* Air conditioning and refrigeration; two-phase flow; building energy systems; solar heating and cooling; heat and mass transfer; editor of two books. *Mailing Add:* Ten Learned Pl Durham NC 27705

CHADDOCK, RICHARD E(ASTMAN), chemical engineering; deceased, see previous edition for last biography

CHADER, GERALD JOSEPH, b Buffalo, NY, Apr 15, 37; c 3. BIOCHEMISTRY. *Educ:* Univ Buffalo, BA, 59; Univ Louisville, PhD(biochem), 66. *Prof Exp:* High sch teacher, NY, 59-60; instr, Med Sch & tutor, Dept Biochem & Molecular Biol, Harvard Univ, 69-71; biochemist, 71-75, CHIEF, SECT RETINAL METAB, NAT EYE INST, 75- *Concurrent Pos:* Fel biochem, Sch Med, Univ Louisville, 66-67; fel biol chem, Harvard Med Sch, 67-69. *Mem:* Am Soc Biol Chemists; Am Soc Neurochem. *Res:* Mechanism of action of vitamin A, involving vitamin-receptor interactions and study of these interactions in relation to retinal function. *Mailing Add:* Lab of Vision Res Nat Eye Inst Bldg 6 Rm 222 Bethesda MD 20892

CHADHA, KAILASH CHANDRA, b Churu, India, July 1, 43; US citizen; m 71; c 3. MAMMALIAN INTERFERONS, HERPES VIRUSES. *Educ:* Univ Rajasthan, India, BSc, 62; Indian Agr Res Inst, New Delhi, MSc, 64; Univ Guelph, PhD(virol), 68. *Prof Exp:* Fel virol, Nat Res Coun, Can, 68-70; cancer res scientist I virol, 70- 72, cancer res scientist II, 73-76, CANCER RES SCIENTIST IV INTERFERONS, ROSWELL CANCER MEM INST, 76- *Concurrent Pos:* Res prof, Niagara Univ, 78-; assoc res prof, dept microbiol, State Univ NY, Buffalo, 79- *Mem:* NY Acad Sci; Am Soc Microbiol; Am Soc Virol; Europ Soc Cancer Res; Int Soc Interferon Res. *Res:* Purification, physico-chemical characterization and mechanism of action of mammalian interferons; gene expression in virally transformed mammalian cells; interferon system in health and disease. *Mailing Add:* Dept Cell & Tumor Biol Roswell Park Cancer Inst Elm & Carlton Sts Buffalo NY 14263

CHADI, JAMES D, b Teheran, Iran, Oct 10, 47; m 82; c 2. SOLID STATE PHYSICS. *Educ:* Cooper Union, BS, 69; Univ Calif, Berkeley, PhD(physics), 74. *Prof Exp:* mem res staff, Xerox Palo Alto Res Ctr, 74-90; MEM RES STAFF, NEC RES INST, 90- *Honors & Awards:* Peter Mark Mem Award, Am Vacuum Soc, 83. *Mem:* Am Phys Soc; Am Vacuum Soc. *Res:* Electronic and structural properties of solids; semiconductor surfaces; defects in semiconductors. *Mailing Add:* NEC Res Inst Four Independence Way Princeton NJ 08540

CHADICK, STAN R, b Mar 27, 41; c 3. MATHEMATICS. *Educ:* Univ Cent Ark, Bs, 63; Univ Ark, MS, 65; Univ Tenn, PhD(math), 69. *Prof Exp:* Teaching asst math, Univ Ark, 63-65; teaching asst math, Univ Tenn, 65-68, lectr, 68-69; from asst prof to prof math, Northwestern State Univ La, 69-83; curric coordr, La Sch Math, Sci & Arts, 83-85; PROF MATH, NORTHWESTERN STATE UNIV, 85-; DIR, LA SCHOLARS' COL, 87- *Concurrent Pos:* Dept head math, Northwestern State Univ La, 78-83; consult, 77-81. *Mem:* Math Asn Am; Nat Coun Teachers Math. *Res:* Author of various scientific articles and books. *Mailing Add:* Dept Math Northwestern State Univ La Natchitoches LA 71457

CHADWICK, ARTHUR VORCE, b Orange, Calif, April 14, 43; m 64; c 2. SEDIMENTOLOGY, PALEOBOTANY. *Educ:* Loma Linda Univ, BA, 65; Univ Miami, PhD(molecular biol), 69. *Prof Exp:* Teaching fel biol, US Nat Defense, 69-70; asst assoc prof biol, Loma Linda Univ, 70-83, assoc prof geol, 80-83; PROF GEOL & BIOL, SOUTHWESTERN ADVENTIST COL, 83- *Concurrent Pos:* Vis prof geol & geophys, Univ Okla, 76-77; res prof geol,

Geosci Res Inst, 82-84. *Mem:* Geol Soc Am; Am Asn Petrol Geologists; Soc Econ Paleontologists & Mineralogists; Am Asn Stratig Palynologists; Sigma Xi. *Res:* Early paleozoic depositional models for southwestern United States; paleozoic spore floras of central United States. *Mailing Add:* 1215 Honeysuckle Dr Keene TX 76059

CHADWICK, DAVID HENRY, industrial organic chemistry; deceased, see previous edition for last biography

CHADWICK, DUANE G(EORGE), b LaGrande, Ore, Jan 24, 25; m 51; c 7. ELECTRICAL ENGINEERING. *Educ:* Utah State Univ, BSEE, 52; Univ Wash, MSEE, 57. *Prof Exp:* Res engr, Boeing Airplane Co, 53-57; from asst prof to assoc prof elec eng, 57-, EMER PROF, UTAH STATE UNIV. *Res:* Electronic control mechanisms; hydrologic instrumentation; solar powered water pumps. *Mailing Add:* 2515 No 1600 E North Logan UT 84321

CHADWICK, GEORGE BRIERLEY, b Vancouver, BC, Can, May 5, 31; m 64; c 1. BUBBLE CHAMBER PHYSICS. *Educ:* Univ British Columbia, BA, 53, MA, 55. *Prof Exp:* Asst physicist, Brookhaven Nat Res, 59-61; res officer, Oxford Nuclear Physics Lab, 61-64; PHYSICIST, STANFORD LINEAR ACCELERATOR CTR, 65- *Mem:* Am Physical Soc. *Res:* Design of positive and negative electron detectors, and use there of for physical studies. *Mailing Add:* SLAC, Bin 94 Stanford CA 94305

CHADWICK, GEORGE F, b Buffalo, NY, July 11, 30; m 52; c 4. PLASTICS CHEMISTRY, CHEMICAL & ENVIRONMENTAL ENGINEERING. *Educ:* Univ Buffalo, BA, 51; Pa State Univ, MA, 56. *Prof Exp:* Res asst petrol, Pa State Univ, 51-54; res scientist, Airco Electronics, 57-64, res supvr, 64-65, develop supvr, 65-67, mgr develop electronics, 67-72, mgr process eng, 73-78; consult eng, Mesch & Assocs, 78-80; sr eng, Agr Chem Div, FMC Corp, 80-82; tech consult, 82-84; sr proc engr, EI Div, WGL Ltd, 84-85; sr proj engr, 85-88, ENG SUPVR, ETHOX CORP, 89- *Mem:* Am Chem Soc; Soc Plastics Engrs; Am Inst Chem Engrs; fel Brit Plastics Inst. *Res:* Plastics; rheology; conductivity of fine dispersed conductive materials in plastics composites; environmental chemistry; materials science engineering. *Mailing Add:* 133 Pin Oak Dr Williamsville NY 14221

CHADWICK, HAROLD KING, b Bay Shore, NY, May 28, 30; m 55; c 5. FISHERIES. *Educ:* Cornell Univ, BS, 52; Univ Mich, MS, 56. *Prof Exp:* FISHERY BIOLOGIST, CALIF DEPT FISH & GAME, 56- *Mem:* Am Fisheries Soc. *Res:* Investigation of Sacramento River striped bass population including harvest and natural mortality rates and strength of year classes; reservoir ecology; fisheries management. *Mailing Add:* Calif Dept of Fish & Game 4001 North Wilson Way Stockton CA 95205

CHADWICK, JUNE STEPHENS, b Fredericton, NB; m 62; c 1. MEDICAL MICROBIOLOGY, PATHOGENESIS. *Educ:* Queen's Univ, BSc, 47, MA, 50; Univ London, PhD, 61. *Prof Exp:* Res scientist, Can Dept Agr, 50-62; from asst prof to assoc prof, 64-78, PROF MICROBIOL, QUEEN'S UNIV, 78- *Mem:* Am Soc Microbiol; Int Soc Develop & Comp Immunol; Soc Invert Path; Am Soc Zoologists. *Res:* Immunity in insects as an aspect of comparative immunology; inductors and mediators of immune effectors in the response to gram negative bacteria; cytotoxic factors. *Mailing Add:* Dept Microbiol & Immunol Queen's Univ Kingston ON K7L 3N6 Can

CHADWICK, NANETTE ELIZABETH, b Providence, RI, July 20, 59. ECOLOGY, BEHAVIOR ETHOLOGY. *Educ:* Univ Calif, Santa Barbara, BA, 80; Univ Calif, Berkeley, PhD(zool), 88. *Prof Exp:* Instr chem, Educ Opportunity Prog, Univ Calif, Santa Barbara, 80-81; instr marine sci, Catalina Island Marine Inst, 81-82; sr instr coral reef ecol, Sch Field Studies, VI, 88-89; POSTDOCTORAL FEL INVERT IMMUNOL, HOPKINS MARINE STA & STANFORD UNIV MED SCH, 90- *Concurrent Pos:* Postdoctoral fel coral reef ecol, Interuniv Inst Eilat & Tel Aviv Univ, Israel, 88-90. *Mem:* Am Soc Zoologists; Ecol Soc Am. *Res:* Behavioral ecology of invertebrates, primarily on cnidarians such as sea anemones, scleractinian corals and corallimorpharians; evolutionary ecology; coral reef ecology; mechanisms of invertebrate behavior. *Mailing Add:* Hopkins Marine Sta Stanford Univ Pacific Grove CA 93950

CHADWICK, RICHARD SIMEON, b Los Angeles, Calif, Nov 23, 41; m 72; c 1. CARDIOVASCULAR DYNAMICS, COCHLEAR MECHANICS. *Educ:* Cornell Univ, BME, 64, MME, 66; Stanford Univ, PhD(aeronaut & astronaut), 71. *Prof Exp:* Mech engr, Norman Eng Co, 64-66; res engr, Marquardt Corp, 66-67; res asst, Dept Aeronaut & Astronaut, Stanford Univ, 69-71; lectr mech eng, Technion-Israel Inst Technol, 71-75, sr lectr, 75-77; adj assoc prof, dept mech & struct, Univ Calif, Los Angeles, 77-80; BIOMED ENGR, NIH, 80- *Concurrent Pos:* Res fel chem eng, Calif Inst Technol, 75-77; vis asst prof fluid mech, dept mech & struct, Univ Calif, Los Angeles, 75-77. *Mem:* Cardiovasc Syst Dynamics Soc. *Res:* Biomechanics; cochlear mechanics; cardiovascular systems dynamics; theoretical models that utilize fluid and solid mechanics concepts to analyze physiological systems. *Mailing Add:* 5420 Alta Vista Rd Bethesda MD 20814

CHADWICK, ROBERT AULL, geology; deceased, see previous edition for last biography

CHADWICK, WALLACE LACY, b Loring, Kans, Dec 4, 1897; m 21; c 2. CIVIL ENGINEERING. *Hon Degrees:* Dr Eng Sci, Univ Redlands, 65. *Prof Exp:* Statistician & plant engr, Calif Alkali Co, 20-21; draftsman, Southern Calif Edison Co, 22-24, div engr, Big Creek-San Joaquin Hydroelec Proj, 24-28, transmission engr, 28-31; from engr to sr engr construct, Colo River Aqueduct, Metrop Water Dist Southern Calif, 31-37, spec assignment, 38; from civil engr to chief civil engr, Southern Calif Edison Co, 37-44, mgr eng dept, 45-50, vpres, 51-62, dir design & construct dams & power sta. *Concurrent Pos:* Chmn, Joint Res Coun Power Plant Air Pollution Control, Los Angeles, 56-62, bd eng consults, James Bay Hydro Proj, Quebec, 71-85, independent panel to rev failure Teton Dam, Idaho, 76, Jubail Rev Bd, Bechtel Group, Saudi Arabia, 77-86, mgt rev bd, S Tex Proj, 82-86; consultdams, Dept Water Resources, Calif, 62-86, Churchill Falls Proj, Labrador, 63-86, Washington Area Rapid Transit Authority, 75-83; mem joint bs environ studies, Nat Acad Sci & Nat Acad Eng, 65-69; pres bd trustees, Univ Redlands, 56-69, chmn exec comt, 69-75; consult engr, 62-90; mem, US Comt, Int Comn Large Dams, 70-86. *Honors & Awards:* Philip T Sprague Award, Instrument Soc Am, 63; Golden Beaver Award, 69; Rickey Medal, Am Soc Civil Eng, 71. *Mem:* Nat Acad Eng; hon mem fel Am Soc Mech Eng; fel Inst Elec & Electronics Engrs; Am Soc Civil Eng (pres, 64-65); Am Concrete Inst. *Mailing Add:* 475 Taylor Dr Claremont CA 91711

CHAE, CHI-BOM, b Seoul, Korea, Sept 25, 40; US citizen; m 69; c 2. GENE REGULATION. *Educ:* Seoul Nat Univ, BS, 67; Univ NC, PhD(biochem). *Prof Exp:* From asst prof to assoc prof, 70-85, PROF BIOCHEM, UNIV NC, 85. *Mem:* Am Asn Biol Chemists. *Res:* Molecular biology; genetic engineering; function and regulation of germ-cell specific histones during spermatogenesis; hormonal regulation of thyroglobulin gene; chromosome structure and cell cycle in yeast. *Mailing Add:* Dept Biochem Univ NC Chapel Hill NC 27599-7260

CHAE, KUN, b Seoul, Korea, Apr 13, 44; m 73; c 2. MEDICINAL CHEMISTRY, ENVIRONMENTAL CHEMISTRY. *Educ:* Seoul Nat Univ, BS, 66; Univ NC, Chapel Hill, MS, 71, PhD(med chem), 73. *Prof Exp:* Vis fel, 73-74, staff fel, 74-77, RES CHEMIST, NAT INST ENVIRON HEALTH SCI, RES TRIANGLE PARK, 78- *Mem:* Am Chem Soc; Sigma Xi. *Res:* Organic synthesis and metabolic studies of lipids; synthesis and analysis of chlorinated aromatics; toxicological and metabolic studies of polychlorinated biphenyls; synthesis and characterization of antigens for immunochemical studies; receptor binding of chlorinated dioxins; receptor binding of estrogens. *Mailing Add:* 5112 Lazywood Lane Durham NC 27712

CHAE, SOO BONG, b Musan, Korea, June 22, 39; m 67; c 2. MATHEMATICS. *Educ:* Seoul Nat Univ, BS, 62; Emory Univ, MA, 63; Univ Rochester, PhD(math), 70. *Prof Exp:* PROF MATH, NEW COL UNIV S FLA, 70- *Mem:* Am Math Soc; Math Asn Am. *Res:* Functional analysis; general topology. *Mailing Add:* Div Nat Sci New Col Univ SFla FL 34243

CHAE, YONG SUK, b Chochiwen, Korea, July 29, 30; m 62; c 2. CIVIL ENGINEERING, ENGINEERING MECHANICS. *Educ:* Dartmouth Col, AB, 56, MS, 57; Univ Mich, PhD(civil eng), 64. *Prof Exp:* Civil engr, Howard, Needles, Tammen & Bergendoff, 60-61; from asst prof to assoc prof, 64-71, PROF CIVIL ENG, RUTGERS UNIV, 71-, ASSOC DEAN, 81- *Concurrent Pos:* Consult, US Army Cold Regions Res & Eng Lab, Mineral Fibre Prod Bur, Continental Oil Co, Singer Co, Dames & Moore, Charles Kupper Int & NL Industries; NATO sr fel sci, Dept State-NSF, 75. *Mem:* Am Soc Civil Engrs. *Res:* Soil mechanics; soil dynamics; foundation engineering; physics of snow, ice and frozen soil; earthquake engineering; environmental effects of construction. *Mailing Add:* Col Eng Rutgers Univ Piscataway NJ 08903

CHAE, YOUNG C, b Seoul, Korea, Mar 31, 32; m 59; c 2. CHEMICAL ENGINEERING, POLYMER SCIENCE. *Educ:* SDak Sch Mines & Technol, BS, 59; Case Western Reserve Univ, PhD(chem eng), 62. *Prof Exp:* Sr res chemist, Monsanto Co, 62-65, res proj leader, 65-66, group leader, 66-70, sr group leader, Org Res Dept, 70-73, com develop mgr, 73-78, mgr com develop, Monsanto New Enterprise Div, 78-80, int mkt dir, Engr Prods Div, 81-88; AREA DIR SALES, JAPAN, 88- *Mem:* Am Chem Soc; Am Inst Chem Engrs; Soc Plastics Indust; Soc Plastics Engrs; Am Mgt Asn. *Res:* Physical characterization of polymers; reinforced plastics; cellular plastics; surface coatings; membrane science; adsorption separation; gas separations; plasticizers; flame retardant polymers and process development. *Mailing Add:* Perma Corp 11444 Lackland Rd St Louis MO 63146

CHAET, ALFRED BERNARD, b Boston, Mass, June 7, 27; m 50; c 3. PHYSIOLOGY. *Educ:* Univ Mass, BS, 49, MS, 50; Univ Pa, PhD(zool), 53. *Prof Exp:* Asst, Univ Pa, 51-53; instr zool, Univ Maine, 53-56; asst prof physiol, Sch Med, Boston Univ, 56-58; from assoc prof to prof biol, Am Univ, 58-66; provost, 67-76, prof biol, 66-, assoc vpres, 76-, ASST VPRES, RES & SPONSOR PROG, UNIV W FLA. *Concurrent Pos:* Res, Marine Biol Lab, Woods Hole, 49, 51-53 & 55-58 & corp mem; res assoc, Boston City Hosp, 56-; NIH spec fel & vis scholar, Scripps Inst, Calif, 64-65; assoc dean sci, Univ WFla, 66-67; chmn, Fla Panhandle Health Systs Agency, Inc. *Mem:* AAAS; Am Soc Zool; Soc Gen Physiol; Am Soc Physiol; Soc Biophys; Sigma Xi. *Res:* Shedding substance in starfish-invertebrate neurohormone; absorption properties of echinoderm tube feet; adhering mechanisms in invertebrates; toxic factor in heat death; thiaminase. *Mailing Add:* Res & Sponsor Prog Univ W Fla Pensacola FL 32514

CHAFETZ, HARRY, b New York, NY, Oct 27, 29; m 55; c 2. ORGANIC CHEMISTRY, MOTOR OIL CHEMISTRY. *Educ:* City Col New York, BS, 50; Pa State Univ, PhD(chem), 54. *Prof Exp:* From chemist to res chemist, 56-61, group leader, 61-70, res assoc, 70-84, SR RES ASSOC, BEACON RES LABS, TEXACO INC, 84- *Mem:* Am Chem Soc. *Res:* Organic synthesis; free radical chemistry; halogenetion and oxidation reactions; lubricant additives; hydrocarbon chemistry. *Mailing Add:* 50A Streit Ave Poughkeepsie NY 12603-2322

CHAFETZ, LESTER, b Providence, RI, Sept 21, 29; m 60; c 2. PHARMACEUTICAL CHEMISTRY. *Educ:* RI Col Pharm & Allied Sci, BS, 51; Univ Wis, MS, 53, PhD(pharmaceut chem), 55. *Prof Exp:* Res assoc anal, Sterling-Winthrop Res Inst, 59-61; chief analytical chemist pharmaceut analysis, Neisler Labs, Inc, 61-64; dir control, drug systs res, Med Ctr, Ark, 64-65; sr scientist appl anal res, Warner-Lambert/Parke-Davis Pharmaceut Res, 65-66, dir apl anal res dept, 66-82, dir phys & anal res, 82-84; pres, Chafetz Assocs, 84-86; PROF PHARM & DIR, THE CTR FOR PHARMACEUT TECHNOL, SCH PHARM, UNIV MO-KANSAS CITY, 87- *Concurrent Pos:* Adj prof, Univ Iowa, 81-86. *Honors & Awards:* Justin L Powers Res Achievement Award Pharmaceut Anal, Am Pharmaceut Asn, 84. *Mem:* AAAS; Am Chem Soc; Acad Pharmaceut Sci; Sigma Xi; Am Assoc Pharm Sci, FAIC, AACP; fel Assoc Inst Chem; Am Asn Chem Pharmacists.

Res: Chemistry of organic medicinals; drug assay methods; drug stability and standards; pharmaceutical dosage forms; quality control of drug products. *Mailing Add:* School of Pharmacy Univ Missouri-Kansas City Kansas City MO 64110-2499

CHAFETZ, MORRIS EDWARD, b Worcester, Mass, Apr 20, 24; m 46; c 3. PSYCHIATRY. *Educ:* Tufts Univ, MD, 48; Am Bd Psychiat & Neurol, dipl, 56. *Prof Exp:* Dir alcohol clin, Mass Gen Hosp, 57-68, dir acute psychiat serv, 61-68, psychiatrist, 64-70, dir clin psychiat serv, 68-70; actg dir, Div Alcohol Abuse & Alcoholism, Nat Inst Alcohol Abuse & Alcoholism, NIMH, 70-71, dir, 71-75; PRES, HEALTH EDUC FOUND, 75- *Concurrent Pos:* Mem ad hoc rev bd res in alcoholism, NIMH, 58-61; mem subcomt alcoholism, Mass Ment Health Planning Proj, 63-; assoc clin prof psychiat, Harvard Med Sch, 68-70. *Honors & Awards:* Maudsley Bequest Lectr, Univ Edinburgh, Scotland, 65; Moses Greeley Parker Lectr, 69. *Mem:* Group Advan Psychiat. *Res:* Treatment, prevention and dynamics of alcoholism and alcohol-related disorders; psychiatric care of urban poor. *Mailing Add:* Health Educ Found 600 New Hampshire Ave NW Suite 452 Washington DC 20037

CHAFFEE, ELEANOR, b Cambridge, Mass, Oct 9, 34. PHYSICAL INORGANIC CHEMISTRY. *Educ:* Mt Holyoke Col, BA, 56; Harvard Univ, MAT, 62; Wellesley Col, MA, 67; Brown Univ, PhD(chem), 71. *Prof Exp:* Res physicist photoconductivity, Stanford Res Inst, 56-57; res physicist phys chem, Arthur D Little, Inc, 59-61; teacher chem, Am High Sch, Lugano, Switz, 62-63 & Lexington High Sch, Mass, 63-65; fel chem, State Univ NY Buffalo, 71-72; RES CHEMIST INORG CHEM, EASTMAN KODAK CO, NY, 72- *Res:* Kinetics and mechanisms of oxidation of substrates, their catalysis by metals, and their complexes; peroxide chemistries. *Mailing Add:* 1786 Lake Rd Webster NY 14580

CHAFFEE, MAURICE A(HLBORN), b Wilkes-Barre, Pa, Jan 10, 37; m 59; c 2. ECONOMIC GEOLOGY, GEOCHEMISTRY. *Educ:* Colo Sch Mines, Geol E, 59; Univ Ariz, MS, 64, PhD(econ geol, mineral), 67. *Prof Exp:* Mine geologist, NJ Zinc Co, Va, 60-62; GEOLOGIST, US GEOL SURV, 67- *Mem:* Soc Econ Geol; Asn Explor Geochem (vpres, 86-88, pres, 88-89). *Res:* Geology and hydrothermal alteration of mineral deposits; trace element chemistry related to mineral deposits; development of new methods and concepts for application of trace element chemistry to mineral exploration. *Mailing Add:* 2968 Pierson Way Lakewood CO 80215

CHAFFEE, ROWAND R J, b El Paso, Tex, Nov 4, 25; m 53; c 6. PHYSIOLOGY, BIOCHEMISTRY. *Educ:* Univ NMex, BS, 46, BA, 51, MS, 52; Harvard Univ, PhD(biol sci), 57. *Prof Exp:* Asst prof zool, Univ Redlands, 58-59; NIH fel cellular physiol, Univ Calif, Berkeley, 59-60; asst prof zool, Univ Calif, Riverside, 60-64; mem staff cellular physiol, Los Alamos Sci Lab, 64-69; PROF ERGONOMICS, UNIV CALIF, SANTA BARBARA, 69- *Mem:* AAAS; Am Soc Mech Engrs; Am Physiol Soc; Soc Exp Biol & Med; Brit Ergonomics Res Soc. *Res:* Biochemistry of hibernating; cold and heat acclimation; cellular physiology; mitosis; enzyme kinetics; primate temperature; altitude acclimation biochemistry. *Mailing Add:* 2527 Bath St Santa Barbara CA 93105

CHAFFEY, CHARLES ELSWOOD, b Montreal, Que, Jan 22, 41; m 71; c 2. CHEMICAL ENGINEERING, RHEOLOGY. *Educ:* McGill Univ, BSc, 61, PhD(chem), 65. *Prof Exp:* PROF CHEM ENG, UNIV TORONTO, 81- *Res:* Rheology of suspensions and reinforced polymer melts. *Mailing Add:* Dept of Chem Eng Univ of Toronto Toronto ON M5S 1A4 Can

CHAFFIN, DON B, b Sandusky, Ohio, Apr 17, 39; m 66; c 1. INDUSTRIAL ENGINEERING, BIOENGINEERING. *Educ:* Gen Motors Inst, BIE, 62; Univ Toledo, MSIE, 64; Univ Mich, PhD(eng), 67. *Prof Exp:* Jr draftsman, Mack Iron Steel Co, Ohio, 55-57; quality control engr, New Departure Div, Gen Motors Corp, Ohio, 60-62, inspection foreman, 62-63; proj engr, Micrometrical Div, Bendix Corp, Mich, 63-64; asst prof phys med, Univ Kans, 67-68; asst prof indust eng, 68-70, assoc prof indust eng & bioeng, 70-77, PROF INDUST & OPERS ENG, UNIV MICH, 77- *Concurrent Pos:* Bioeng grant, Western Elec Co, 67-71, NASA, 70-71, Aerospace Med Res Labs, 70-71 & Nat Inst Occupational Safety & Health, 71-72; consult, Bendix Corp, Mich, 64. *Mem:* Human Factors Soc; Nat Soc Prof Engrs; Am Inst Indust Engrs; Biomed Eng Soc; Brit Ergonomics Res Soc; Sigma Xi. *Res:* Effects and applications of electromyography for bettering human performance; concepts of mechanics to the study of the skeletal-muscle system; expanding the teaching of physiological, neurological, and anatomical concepts as related to the bettering of man-machine systems. *Mailing Add:* Dept Indust Eng Univ Mich Ann Arbor MI 48109

CHAFFIN, ROGER JAMES, electrical engineering, solid state physics; deceased, see previous edition for last biography

CHAFFIN, TOMMY L, b Dallas, Tex, Apr 11, 43; m 65; c 2. ORGANIC CHEMISTRY. *Educ:* Okla State Univ, BS, 65; Univ Ill, PhD(org chem), 69. *Prof Exp:* Sr res chemist, 69-75, tech mgr, 75-80, tech dir, 3M Co, St Paul, 80-; mgr, Sumitomo 3M Ltd, Tokoyo, Japan, 84-87; MANAGING DIR, 3M DENMARK, 87- *Mem:* Am Chem Soc. *Res:* Stereospecific alkylations; anionic block copolymers; Friedel-Crafts chemistry. *Mailing Add:* 3M A/S Fabriksparken 15 Glostrup DK 2600 Denmark

CHAFOULEAS, JAMES G, b Mt Vernon, NY, May 24, 48; m 82; c 2. CELL BIOLOGY, DRUG DISCOVERY. *Educ:* Univ Ariz, BSc & BA, 70, MSc, 73, PhD(genetics), 77. *Prof Exp:* Postdoctoral fel, Dept Cell Biol, Baylor Col Med, 77-80, from instr to asst prof, 80-84; from asst prof to assoc prof, dept physiol, Laval Univ, 84-88, dir, Lab Flow Cytometry, CHUL Res Ctr, 87-89; ASSOC DIR, DEPT BIOCHEM, BIO-MEGA INC, 89- *Concurrent Pos:* Adj prof, dept biochem, Laval Univ, 89-, Univ Montreal, 90- *Mem:* Am Soc Cell Biol; Endocrine Soc; AAAS; Can Fedn Biol Soc. *Res:* Control mechanisms involved in the regulation of cell proliferation; signal transduction and the role of calcium and calmodulin in mediating control of cell cycle progression in normal and transformed cells; study of antiviral drugs. *Mailing Add:* Biochem Dept Bio-Mega Inc 2100 Cunard St Laval PQ H7S 2G5 Can

CHAGANTI, RAJU SREERAMA KAMALASANA, b Samalkot, India, Mar 12, 33; m 66; c 2. GENETICS. *Educ:* Andhra Univ, India, BSc, 54, MSc, 55; Harvard Univ, PhD(biol), 64. *Prof Exp:* Demonstr & res asst bot, Andhra Univ, India, 55-61, lectr, 61-67; mem sci staff, Med Res Coun Radiobiol Unit, Harwell, Eng, 67-71; res assoc & assoc investr, Lab Human Genetics, New York Blood Ctr, 71-76; dir, Lab Cancer Genetics & Cytogenetics, attend geneticist & cytogeneticist, Mem Sloan-Kettering Cancer Ctr, 76-88; from asst prof to assoc prof, 79-88, PROF CELL BIOL & GENETICS, GRAD SCH MED, CORNELL UNIV, 88-; CHIEF, CYTOGENETICS SERV, MEM SLOAN-KETTERING CANCER CTR, 88- *Concurrent Pos:* consult, Lab Human Genetics, New York Blood Ctr, 76-; asst mem, Mem Sloan-Kettering Cancer Ctr, 76-83, assoc mem, 83-87, mem, 87-; assoc prof, Dept Path, Cornell Univ Med Col, 84-88, prof, 88- *Mem:* Am Soc Human Genetics; Genetics Soc Am; Sigma Xi; Harvey Soc. *Res:* Human and mammalian genetics; genetic basis of human cancer. *Mailing Add:* Memorial Sloan-Kettering Cancer Ctr 1275 York Ave New York NY 10021

CHAGNON, ANDRE, b Montreal, Que, Aug 16, 32; m 58; c 2. VIROLOGY, TISSUE CULTURE. *Educ:* Univ Montreal, BA, 53, BSc, 57, PhD(bact), 60. *Prof Exp:* Res asst, 78-80, PROF VIROL, INST ARMAND-FRAPPIER, 80- *Concurrent Pos:* Dir prod & bus opers. *Mem:* Tissue Cult Asn; Fr-Can Asn Advan Sci; Can Soc Microbiol; Soc Cryobiol. *Mailing Add:* Inst Armand-Frappier CP100 Ville de Laval PQ H7N 4Z1 Can

CHAGNON, JEAN YVES, b Quebec, Can, May 27, 34; m 66; c 2. SEISMIC MICROZONATION. *Educ:* Laval Univ, BA, 54; Ecole Polytech, BASc, 58; McGill Univ, MSc, 61, PhD(geol), 65. *Prof Exp:* Engr geol geotech, Quebec Ministry Natural Resources, 64-72, head, Dept Geotech, 72-77; prof eng geol, 77-78, head, Dept Geol, 78-79, PROF ENG GEOL, LAVAL UNIV, 79- *Concurrent Pos:* Mem, Asn Comt Geotech Res, Nat Res Coun Can, 75-81, Subcomt, 79-82 & subcomt, Urgan Eng Terrain Problems, 79-82; consult, Eng Geol Hydro-Quebec, 78-82; assoc ed, Can Geotech J, 77-82. *Mem:* Geol Asn Can; Eng Inst Can; Can Geotech Soc. *Res:* Behavior of sensitive clays and landslides in these clays; seismic microzonation; geotechnical mapping. *Mailing Add:* 778 Francois-Arteau Ste-Foy PQ G1V 3G7 Can

CHAGNON, PAUL ROBERT, b Woonsocket, RI, Nov 11, 29. NUCLEAR PHYSICS. *Educ:* Col of the Holy Cross, BS, 50; Johns Hopkins Univ, PhD(physics), 55. *Prof Exp:* Assoc, Univ Mich, 55, instr, 55-57, asst prof physics, 57-63; from asst prof to assoc prof, 63-69, PROF PHYSICS, UNIV NOTRE DAME, 69- *Mem:* Am Phys Soc. *Res:* Nuclear spectroscopy. *Mailing Add:* Dept of Physics Univ of Notre Dame Notre Dame IN 46556

CHAHINE, MOUSTAFA TOUFIC, b Beirut, Lebanon, Jan 1, 35; US citizen; m 60; c 1. FLUID PHYSICS, ATMOSPHERIC PHYSICS. *Educ:* Univ Wash, BS, 56, MS, 57; Univ Calif, Berkeley, PhD(fluid physics), 60. *Prof Exp:* mem staff, 60-75, mgr Planetary Atmospheres Sect, 75-78, sr res scientist & mgr Earth & Space Sci Div, 78-84, CHIEF SCIENTIST, JET PROPULSION LAB, CALIF INST TECHNOL, 84- *Concurrent Pos:* Consult, USN, 72-76; vis scientist, Mass Inst Technol, 69-70; vis prof, Am Univ, Beirut, 71-72; Regent's lectr, Univ Calif, Los Angeles, 89-90; mem, space & earth sci adv comt, NASA, 82-85, climate res comt, Nat Acad Sci, 85-88, bd dirs, Atmospheric Sci & Climate, 88-; chmn, sci steering group global energy & water cycle exp, World Meteorol Orgn, 88- *Honors & Awards:* Except Sci Achievement Medal, NASA, 69, Outstanding Leadership Medal, 84. *Mem:* Fel Am Inst Physics; Am Meteorol Soc; fel Royal Meteorol Soc; Int Acad Astronaut; Sigma Xi. *Res:* Thermodynamics and statistical fluid physics; strong shock waves and remote sensing of planetary atmospheres; long range numerical weather prediction; environmental sciences, general. *Mailing Add:* Jet Propulsion Lab 180-904 4800 Oak Grove Dr Pasadena CA 91109

CHAI, AN-TI, b Honan, China, 39; m 71; c 2. MOLECULAR SPECTROSCOPY. *Educ:* Nat Taiwan Univ, BS, 61; Kans State Univ, MS, 66, PhD(physics), 68. *Prof Exp:* Asst prof physics, Mich Technol Univ, 68-73; interim asst prof & asst res scientist, Interdisciplinary Ctr for Aeronomy & Atmospheric Sci, Univ Fla, 74-76; GEN PHYSICIST, LEWIS RES CTR, NASA, 76- *Concurrent Pos:* Nat Res Coun adv, Lewis Res Ctr, NASA, 80-; adj assoc prof elec eng & appl physics, Case Western Reserve Univ, 81-83. *Res:* Spectroscopy; atmospheric optics; solar irradiance measurements; standardization and methodology of solar cell measurements; photovoltaic device fabrication and theory; fluid physics under low, or near zero, gravity environment; space power. *Mailing Add:* Lewis Res Ctr NASA 21000 Brook Park Cleveland OH 44135

CHAI, CHEN KANG, b Hopeh, China, Feb 14, 16; m 54; c 2. GENETICS. *Educ:* Army Vet Col, China, DVM, 37; Mich State Col, MS, 49, PhD(animal breeding), 51. *Prof Exp:* Asst, Mich State Col, 49-51; US Dept State fel, 51-52, res fel, 52-55, res assoc, 56, staff scientist, 57-67, SR STAFF SCIENTIST, JACKSON LAB, 67- *Concurrent Pos:* Vis fel, Mass Inst Technol, 52-53; Guggenheim fel, 62-63. *Mem:* Fel AAAS; Am Genetic Asn; Genetics Soc Am; Biomet Soc; Am Asn Phys Anthrop. *Res:* Quantitative genetics; genetic study of endocrine variation; mouse and rabbit genetics; genetic variations in Taiwan aborigines. *Mailing Add:* Jackson Lab Bar Harbor ME 04609

CHAI, HYMAN, b Lithuania, Dec 15, 20; US citizen; m 76; c 2. ALLERGY, IMMUNOLOGY. *Educ:* Univ Witwatersrand, MB, ChB, 43, Clin MD, 51; Royal Col Physicians & Surgeons, DCH, 51, FRCP(E), 74; Am Bd Allergy & Immunol, cert, 75. *Prof Exp:* Intern, Johannesburg Gen Hosp, 44-45; res pediat, Transvaal Mem Hosp, 45-46; fel allergy-immuno, Natl Asthma Ctr, Denver, 62-64, head div med, 64-75, dir med educ, 75-80; assoc prof, 73-80, PROF PEDIAT, SCH MED, UNIV COLO, 80-; SR STAFF PHYSICIAN, DEPT PEDIAT & MED, NAT JEWISH CTR RESPIRATORY IMMUNOL DIS, 80- *Concurrent Pos:* Assoc clin prof pediat, Sch Med, Univ Colo, 73-80, assoc prof, 80-; mem task force, Nat Inst Allergy & Infectious Dis, 77-78; co-investr, Nat Inst Heart, Lung & Blood grants, 75- & Environ Protection Agency grant, 78- *Mem:* Fel Am Acad Allergy; Am Col Allergy; fel Am Thoracic Soc; fel Am Col Chest Physicians. *Res:* Pulmonary physiology of asthma; immunology associated chest disease; allergic process, including immunology. *Mailing Add:* Nat Jewish Ctr For Immunol & Respiratory Med, Dept Pediat 665 S Newport Denver CO 80224

CHAI, WINCHUNG A, b Hunan, China, Aug 21, 39; US citizen; m 69; c 1. MATHEMATICAL ANALYSIS, APPLIED MATHEMATICS. *Educ:* Wittenberg Univ, BA, 60; NY Univ, MS, 64; Polytech Inst Brooklyn, PhD(math), 68. *Prof Exp:* Mathematician, Am Tel & Tel Co, 60-63 & Aerospace Res Ctr, Gen Precision Inc, NJ, 63-68; ASSOC PROF MATH, MONTCLAIR STATE COL, 68- *Concurrent Pos:* Sr consult, Comput Sci & Mgt Info Systs. *Mem:* Am Math Soc; Math Asn Am; Soc Indust & Appl Math; Asn Comput Mach; Inst Elec & Electronics Engrs. *Res:* Applied mathematics and computer science. *Mailing Add:* 6 Keech Briar Lane Pompton Plains NJ 07444

CHAIKEN, IRWIN M, b Providence, RI, May 23, 42. PEPTIDE & PROTEIN CHEMISTRY. *Educ:* Univ Calif, Los Angeles, PhD(biol chem), 68. *Prof Exp:* SR INVESTR, NAT INST DIABETES & DIGESTIVE & KIDNEY DIS, NIH, 73- *Concurrent Pos:* Pres, Int Interest Group Biorecognition Technol, 85- *Mem:* Am Soc Biol Chemsits; NY Acad Sci; AAAS; Sigma Xi. *Mailing Add:* Smith Kline Beecham Sci Lab 709 Swedeland Rd King of Prussia PA 19406-2711

CHAIKEN, JAN MICHAEL, b Philadelphia, Pa, Oct 19, 39; m 39; c 2. MATHEMATICS. *Educ:* Carnegie Inst Technol, BS, 60; Mass Inst Technol, PhD(math), 66. *Prof Exp:* From instr to asst prof math, Cornell Univ, 64-68; researcher, Rand Corp, 68-84; dep area mgr, 84-90, FEL, ABT ASSOC INC, 90-; SR MATHEMATICIAN, LINC, 90- *Concurrent Pos:* Res assoc, Mass Inst Technol, 67-68; adj assoc prof, Univ Calif, Los Angeles, 72-80; assoc ed, Opers Res, 75-77; fac mem, Rand Grand Inst, 81-84; publ serv ed, Interfaces, 86- *Mem:* AAAS; NY Acad Sci; Opers Res Soc Am; Sigma Xi; Inst Mgt Sci. *Res:* Allocation of urban services; models of criminal careers. *Mailing Add:* 66 Birchwood Lane Lincoln MA 01773-4908

CHAIKEN, ROBERT FRANCIS, b Brooklyn, NY, Dec 19, 28; m 53; c 3. CHEMICAL PHYSICS. *Educ:* Univ Ill, BS, 49; Brooklyn Polytech Inst, MS, 58; Univ Calif, Riverside, PhD(chem), 66. *Prof Exp:* Res chemist, Tracerlab Inc, 50-51 & US Testing Co, 51-53; res assoc, George Washington Univ, 53-57; res assoc, Aerojet-Gen Corp, 57-59, tech specialist, 59-61, tech consult, 61-68; sr chem physicist, Stanford Res Inst, 68-70; SUPVRY RES CHEMIST, PITTSBURGH MINING RES CTR, US BUR MINES, 70- *Concurrent Pos:* Lectr, Univ Pittsburgh, 72-76, Pa State Univ, 81-82; adj prof, Univ Pittsburgh, 87-88. *Mem:* AAAS; Am Chem Soc; Combustion Inst (asst treas, 74-78, treas, 85-). *Res:* High temperature and pressure reaction kinetics; theory of combustion and detonation processes; fire and explosion safety; in situ combustion of fossil fuels, coal mine and coal waste fires; solids leaching processes. *Mailing Add:* Pittsburgh Res Ctr US Bur Mines PO Box 18070 Pittsburgh PA 15236

CHAIKIN, LAWRENCE, oral surgery, for more information see previous edition

CHAIKIN, PAUL MICHAEL, b Brooklyn, NY, Nov 14, 45; m 77; c 2. CONDENSED MATTER PHYSICS. *Educ:* Calif Inst Technol, BS, 66; Univ Pa, PhD(physics), 71. *Prof Exp:* From asst prof to prof physics, Univ Calif, Los Angeles, 72-83; prof physics, Univ Pa, 83-88; PROF PHYSICS, PRINCETON UNIV, 88- *Concurrent Pos:* Consult, IBM Res Labs, 75-76; A P Sloan Found fel, 77-81; assoc prof physics, D'orsay Centre, Univ Paris South, 78-79; Res assoc, Exxon Res Labs, 83- *Mem:* Am Phys Soc. *Res:* Superconductivity; magnetism; thermoelectricity; metal-insulator transitions; organic conductors; quasi-one-dimensional compounds, thin metal films, colloids and colloidal crystals. *Mailing Add:* Dept Physics Princeton Univ Princeton NJ 08544

CHAIKIN, PHILIP, b Washington, DC, Dec 23, 48; m 74; c 1. BIOPHARMACEUTICS, PHARMACOKINETICS. *Educ:* Univ Md, BS, 72, PharmD, 77; NJ Med Sch, MD, 87. *Prof Exp:* Scientist, Drug Metab Div, Ortho Pharmaceut Corp, 78-80, sr res scientist, 80-83; med residency, Robert Wood Johnson Sch Med, 87-88; assoc/sr med dir, Pfizer Co, NY, 88-90; DIR CLIN PHARM, BRISTOL-MEYER SQUIBB, PRINCETON, 90- *Concurrent Pos:* Fel pharmacokinetics, Univ Md, 78. *Mem:* Am Soc Hosp Pharmacists; Am Col Clin Pharmacol. *Res:* Preclinical and clinical pharmacokinetics and biopharmaceutics including mechanisms of absorption, distribution, metabolism and excretion of drug substances. *Mailing Add:* Six E Castleton Rd Princeton NJ 08540

CHAIKIN, SAUL WILLIAM, b New York, NY, Dec 25, 21; div; c 2. CHEMISTRY. *Educ:* Brooklyn Col, BA, 43; Univ Chicago, MS, 48, PhD(chem), 48. *Prof Exp:* Asst chem, Toxicity Lab, Univ Chicago, 43-45; res assoc, Univ Calif, Los Angeles, 48-49; asst prof anal chem, Univ WVa, 49-51; sr chemist, Stanford Res Inst, 51-64, mgr surface chem sect, Electronic Mat Dept, 64-68; mgr chem res dept, Memorex Corp, 68-69; dir res, Xidex Corp, 69-76; instr chem, Deanza Col, 77-80; CHEM PATENT SEARCH, SW CHAIKIN & ASSOC, 81- *Concurrent Pos:* Res Corp grantee, 50; consult chem. *Mem:* Am Chem Soc; Sigma Xi; Soc Photog Sci & Eng. *Res:* Development of analytical methods; organic reductions with complex metal hydrides; chelate formation; organoboron chemistry; vapor pressure determination; chromatography; microscopy; surface contamination; permeability; soldering and fluxes. *Mailing Add:* 10480 Ann Arbor Ave Cupertino CA 95014

CHAIRES, JONATHAN BRADFORD, b Riverside, Calif, Apr 1, 50; m 81; c 1. MOLECULAR PHARMACOLOGY. *Educ:* Univ Calif Santa Cruz, BA, 72; Univ Conn, PhD(biophysics), 78. *Prof Exp:* NIH fel, Yale Univ, 79-82; from asst prof to assoc prof, 82-90, PROF BIOCHEM, UNIV MISS MED CTR, 90- *Concurrent Pos:* Jon Humboldt fel, 89-90. *Mem:* Am Chem Soc; Am Soc Biochem & Molecular Biol; Biophys Soc; Sigma Xi. *Res:* Physical biochemistry of nucleic acids and their interactions; molecular pharmacology of anticancer antibiotics. *Mailing Add:* Dept Biochem Univ Miss Med Ctr Jackson MS 39216-4505

CHAISSON, ERIC JOSEPH, b Lowell, Mass, Oct 26, 46; m 76; c 3. SPACE ASTROPHYSICS, RADIO ASTRONOMY. *Educ:* Univ Lowell, BS, 68; Harvard Univ, AM, 69, PhD(astrophys), 72. *Prof Exp:* Res assoc, Harvard-Smithsonian Observ, 72-74; from asst prof to assoc prof astrophys, Harvard Univ, 74-82; prof astrophysics, Haverford Col, 82-86; sr physicist, MIT, 86-87; prof, Wellesley Col, 86-87; SR SCIENTIST & DIR EDUC PROG, SPACE TELESCOPE SCI INST, 87-; ADJ PROF PHYSICS, JOHNS HOPKINS UNIV, 87- *Concurrent Pos:* Nat Acad Sci fel, 72-74; Alfred P Sloan Found res fel, 76-79; chmn pub educ, Harvard-Smithsonian Observ, 78-82; Harlow Shapley vis prof, Am Astron Soc, 79-84; mem sci working group, Extraterrestrial Intel, NASA, 79-90; panel mem, Nat Acad Sci, 88-; educ adv, NSF, 89- *Mem:* AAAS; Am Astron Soc; Am Asn Physics Teachers; Fedn Am Scientists; Int Astron Union; Sigma Xi. *Res:* Gaseous nebulae; interstellar matter; cosmic evolution; extraterrestrial intelligence. *Mailing Add:* Space Telescope Sci Inst Hopkins Homewood Campus Baltimore MD 21218

CHAIT, ARNOLD, b New York, NY, Jan 20, 30; m 65; c 3. MEDICINE, RADIOLOGY. *Educ:* NY Univ, BA, 51; Univ Utrecht, MD, 57; Am Bd Radiol, dipl, 63. *Hon Degrees:* MA, Univ Pa, 75. *Prof Exp:* Intern, Kings County Hosp, NY, 58-59, resident radiol, 59-62; from instr to assoc prof, State Univ NY Downstate Med Ctr, 62-67; from asst prof to prof, 67-76, CLIN PROF RADIOL, UNIV PA, 76- *Concurrent Pos:* Attend physician, Philadelphia Vet Admin Hosp, 69-76; consult, Children's Hosp Philadelphia, 74-76; chmn, Dept Radiol, Grad Hosp, 76-, chmn med staff, 81-83. *Mem:* Soc Uroradiol; Soc Cardiovasc Radiol; Radiol Soc NAm; Am Roentgen Ray Soc; fel Am Col Radiol; Am Heart Asn. *Res:* Cardiovascular radiology. *Mailing Add:* Dept Radiol Grad Hosp 19th & Lombard Sts Philadelphia PA 19104

CHAIT, EDWARD MARTIN, b Brooklyn, NY, May 8, 42; m 66; c 2. ANALYTICAL CHEMISTRY, MASS SPECTROMETRY. *Educ:* Cornell Univ, AB, 64; Purdue Univ, PhD(anal chem), 68. *Prof Exp:* Applns supvr, 67-74; prod mgr instruments, 74-79, res assoc, 79-83, prog mgr, 83-85, bus develop mgr, 85-87, BUS & SALES MGR, MOLECULAR GENETICS, E I DU PONT NEMOURS & CO, INC, 87- *Concurrent Pos:* Ed, J Chem, Biol & Environ Instrumentation; mem adv panel, Ctr Anal Chem, Nat Bur Standards, Chmn Bd Assessment, NIST. *Mem:* Am Soc Mass Spectrometry; Am Chem Soc; Am Soc Testing & Mat; AAAS. *Res:* Organic mass spectrometry; molecular spectroscopy; field ionization phenomena; biomedical gas chromatography/mass spectrometry; environmental analysis; electrophoresis, immunoassay, clinical chemistry; bioanalytical chemistry; biotechnology; business and venture management; technology licensing. *Mailing Add:* Du Pont Co Biotechnology Systs Barley Mill Plaza P24/1356 Wilmington DE 19898

CHAKERIAN, GULBANK DONALD, b Parlier, Calif, Dec 21, 33; m 58; c 2. MATHEMATICS. *Educ:* Univ Calif, Berkeley, AB, 55, PhD(math), 60. *Prof Exp:* Instr math, Calif Inst Technol, 60-63; lectr, 63-64, asst prof, 64-69, PROF MATH, UNIV CALIF, DAVIS, 69- *Mem:* Am Math Soc; Math Asn Am. *Res:* Integral geometry and convex bodies. *Mailing Add:* Dept of Math Univ of Calif Davis CA 95616

CHAKKALAKAL, DENNIS ABRAHAM, b Irinjalakuda, India, Mar 16, 39; m 70. BIOPHYSICS, BIOENGINEERING OF THE MUSCULOSKELETAL SYSTEM. *Educ:* Madras Univ, BSc, 58; Marquette Univ, MS, 62; Washington Univ, PhD(physics), 68. *Prof Exp:* Assoc prof physics, Southern Univ, 69-76; NIH/NRSA fel biophysics & bioeng, Rensselaer Polytech Inst, 76-79, asst prof, 79-80; MEM STAFF, VET ADMIN MED CTR, OMAHA, 80- *Mem:* Orthop Res Soc; Sigma Xi; Bioelec Repair & Growth Soc; AAAS; Soc Biomat; Bioelectromagnetics Soc. *Res:* Many-body theory (physics); biophysical aspects of bone physiology, pathology and fracture healing; electrical stimulation of osteogenesis; bioengineering aspects of the musculoskeletal system relevant to clinical orthopaedics and rehabilitation. *Mailing Add:* Vet Admin Med Ctr 4101 Woolworth Ave Omaha NE 68105

CHAKKO, MATHEW K(ANJHIRATHINKAL), b Kunnamkulam, India, Mar 29, 34; m 65; c 1. ENGINEERING MECHANICS, COMPOSITES MATERIAL SCIENCE. *Educ:* Univ Kerala, BScEng, 55; Syracuse Univ, MME, 61, PhD(mech eng), 65. *Prof Exp:* Inspector factories, Govt Travancore-Cochin, India, 55-56; asst dist engr, Shell Oil Co India, 56-57; trainee, Hindustan Steel, Ltd, 57-58; res asst mech eng, Syracuse Univ, 58-64; res eng, Res & Develop Ctr, B F Goodrich Co, 65-67, sr res eng, 67-70, res assoc, 70-76, sr res & develop assoc, 76-86. *Mem:* Am Chem Soc; Tire Soc. *Res:* Stress analysis; viscous heating and heat transfer in rubber products; composite material science; fatigue of composite and polymeric materials; contact problems in elasticity; contact stress fatigue of metals; metal rolling. *Mailing Add:* 613 McNeil Dr Sagamore Hills OH 44067

CHAKLADER, ASOKE CHANDRA DAS, b Bamrail, India, Sept 1, 30; m 59; c 2. CERAMICS, INORGANIC CHEMISTRY. *Educ:* Univ Calcutta, BSc, 49, Bengal Ceramic Inst, dipl ceramic technol, 51; Univ Leeds, PhD(ceramics), 57. *Prof Exp:* Sr lab asst, Nat Metall Lab, Jamshedpur, India, 51-54; Exhib 1851 sr studentship, Univ Leeds, 57-59; res assoc metall, 59-64, from asst prof to assoc prof, 64-71, PROF METALL, UNIV BC, 71- *Mem:* Inst Ceramics Engrs; Am Soc Metals; Am Ceramic Soc; Can Ceramic Soc; Brit Ceramic Soc; Int Inst Sci & Sintering; Int Acad Ceramists Italy. *Res:* Ceramic composite synthesis and characterization, ceramic composite properties; sintering variables; theory of hot pressing; plasma synthesis. *Mailing Add:* 1788 Acadia Rd Vancouver BC V6T 1R2 Can

CHAKO, NICHOLAS, b Hotove, Albania, Nov 11, 10; nat US; m 52; c 1. MATHEMATICS. *Educ:* Johns Hopkins Univ, PhD(physics), 34; Sorbonne, DSc, 66. *Prof Exp:* Prof math & physics & head dept, State Gym, Albania, 36-37; staff mem, Crufts Lab & tutor physics, Harvard Univ, 38-40; staff mem, Spectros Lab, Mass Inst Technol, 40-41; assoc prof physics, Kans State Univ, 46-47 & Ala Polytech Inst, 47-49; Fulbright exchange prof, State Univ Utrecht, 50-51; guest prof math & physics, Chalmers Univ Technol, Sweden,

51-52; res assoc, Inst Math Sci, NY Univ, 53-56; ASSOC PROF MATH, QUEENS COL, NY, 56- *Concurrent Pos:* Lectr, Ill Inst Technol, 40-41; consult, Balkan Affairs, Off Strategic Serv, DC, 42-45; consult & math physicist, Russell Elec Co, Ill, 42-46; Fulbright grant to Holland, 50-51; lectr, Univ Lund, 52; prof, French Atomic Energy Ctr, Saclay & Univ Paris, 66-67; vis lectr, Laval Univ, 68. *Honors & Awards:* Annual Prize, Royal Soc Eng & Chalmers Alumni Asn Sweden, 52. *Mem:* Am Phys Soc; Inst Elec & Electronics Engrs; Am Math Soc; NY Acad Sci; Acoust Soc Am. *Res:* Absorption of light by organic compounds; geometrical and electron optics; crystal vibrations and acoustics fields; diffraction; special functions; asymptotic integration. *Mailing Add:* 138-10 Franklin Ave Flushing NY 11355

CHAKRABARTI, CHUNI LAL, b Patuakhali, India, Mar 1, 20; m 62; c 1. ANALYTICAL CHEMISTRY, INORGANIC CHEMISTRY. *Educ:* Univ Calcutta, BSc, 41; Univ Birmingham, MSc, 60; Queen's Univ, Belfast, PhD(chem), 62, DSc(chem), 80; FRIC, 63. *Prof Exp:* Supvr chemist, Metal & Steel Factory, Govt India, 41-45; chemist-in-charge, Mines & Indust Dept, Govt Burma, 45-52; chief chemist & mgr, Mineral Resources Develop Corp, 52-59; group leader Noranda Res Ctr, 65; from asst prof to assoc prof anal & inorg chem, 65-76, PROF CHEM, CARLETON UNIV, CAN, 76- *Honors & Awards:* Gerhard Herzberg Award, Spectros Soc Can, 77; Fisher Sci Award, Chem Inst Can, 81; Ionnes Marcus Marci of Kronland Plaque Award, Czech Acad Sci, 84. *Mem:* Fel Chem Inst Can; Am Chem Soc; Royal Soc Chem, Chem Soc, UK. *Res:* Atomic-absorption, atomic-fluorescence and emission spectroscopy; determination of ultratrace elements in air and water; speciation and complexation of trace metals in the natural environment; the effect of heavy-metals-organics interactions on the fixation and release of heavy metals (of geochemical and biochemical interest) in the aquatic environment; electroanalytical techniques for characterization and quantitation of trace metal species. *Mailing Add:* Dept Chem Carleton Univ Colonel By Dr Ottawa ON K1S 5B6 Can

CHAKRABARTI, PARITOSH M, b Apr 1, 40; US citizen; m 68; c 2. ORGANIC CHEMISTRY, PHYSICAL CHEMISTRY. *Educ:* Univ Calcutta, India, BS, 57, MS, 59, DSc(org chem), 64. *Prof Exp:* Fel chem, Ohio State Univ, 64-66 & Univ Hull, Eng, 67-68; asst prof, Mich Technol Univ, 68-69; sr res chemist, FMC Corp, NJ, 69-72; group leader surfactants res, GAF Corp, 72-74, res mgr surfactants, 74-77, tech dir org chem, 77-82; dir res & develop & licensing Chem, 82-86, VPRES, RES & DEVELOP CHEM, PPG INDUST, 87- *Mem:* Am Chem Soc; Indust Res Inst; NY Acad Sci; Am Inst Chemists; Am Asn Immunologists; Coun Chem Res. *Res:* Organic chemicals, inorganic chemicals; surfactants, other fine chemicals and specialty chemicals; analytical chemistry. *Mailing Add:* 1361 Redfern Rd Pittsburgh PA 15241

CHAKRABARTI, SIBA GOPAL, b Rangpur, Bengal, July 1, 23; m 59; c 2. BIOCHEMISTRY. *Educ:* Univ Calcutta, BSc, 45; Rutgers Univ, MS, 60, PhD(biochem), 63. *Prof Exp:* Chemist, Bhartia Elec Steel Co, India, 45-47 & Burn & Co, Ltd, 47-53; assoc chemist, Sam Tour & Co, Inc, NY, 53-54; res assoc biochem, Sloan-Kettering Inst Cancer Res, 54-58; res asst, Rutgers Univ, 58-60; clin chemist, Middlesex Gen Hosp, NJ, 60-62 & St Joseph's Hosp, Hamilton, Ont, 63-65; asst prof dermat & indust health, Med Ctr, Univ Mich, Ann Arbor, 68-73; assoc prof dermat, 73-78, PROF DERMAT, COL MED, HOWARD UNIV, 79- *Concurrent Pos:* NIH res trainee dermat, Univ Mich, 65-68. *Mem:* Am Chem Soc; Soc Invest Dermat; Int Pigment Cell Soc; NY Acad Sci; AAAS. *Res:* Epidermal protein synthesis and characterization of epidermal pre-keratins; control mechanisms in epidermal differentiation; Melanogenesis: chemistry and biology of melanin pigment; metabolism of photoactive compounds; experimental photochemotherapy; chemistry and biology of pigmentation of skin and hair and epidermal differentiation and keratinization. *Mailing Add:* 12806 Stonecrest Dr Silver Spring MD 20904

CHAKRABARTI, SUBRATA K, b Calcutta, India, Feb 3, 41; US citizen; m 67; c 2. OFFSHORE HYDRODYNAMICS, OCEAN ENGINEERING. *Educ:* Jadavpur Univ, India, BS, 63; Univ Colo, Boulder, MS, 65, PhD(eng mech), 68. *Prof Exp:* Asst engr mech eng, Kuljian Corp, 63-64 & Simon Carves, Ltd, 64-65; instr eng design, Univ Colo, 66-67; hydrodynamicist marine res, Chicago Bridge & Iron Co, 68-70, head anal group, 70-79, DIR MARINE RES, CHICAGO BRIDGE & IRON INDUSTS, 79- *Concurrent Pos:* Freeman fel, 79; Tech reviewer, NSF, 80-; assoc ed, Appl Ocean Res, CML, 81-, assoc ed, J Energy Resources Technol, Am Soc Mech Engrs, 83-86; tech seminar, Tex A&M Univ, 82; vis prof, US Naval Acad, 86, 88; ed, J Offshore Mech & Artic Eng, Am Soc Mech Engr, 87-; assoc ed, Marines Structures, 88-; chmn OMAE div, Am Soc Mech Engrs, 87-88. *Honors & Awards:* James Croes Medal, Am Soc Civil Engrs, 74; Ralph James Award, Am Soc Mech Engrs, 85, Achievement Award, 89. *Mem:* Fel Am Soc Civil Engrs; fel AAAS; Sigma Xi; fel Am Soc Mech Engrs. *Res:* Offshore structures; hydrodynamics; ocean wave theories; statistics. *Mailing Add:* 191 E Weller Dr North Plainfield IL 60544-8929

CHAKRABARTI, SUPRIYA, Howrah, W Bengal, India, June 22, 53; m 83; c 1. ULTRAVIOLET INSTRUMENTATION, PLANETARY ATMOSPHERES. *Educ:* Univ Calcutta, India, BE, 75; Univ Calif, MS, 80, PhD, 82. *Prof Exp:* Asst specialist, Space Sci Lab, Univ Calif-Berkeley, 80-82; vis asst prof computer sci, San Francisco State Univ, 82-85; asst res physicist, 82-87, ASSOC RES PHYSICIST, SPACE SCI LAB, UNIV CALIF-BERKELEY, 87- *Concurrent Pos:* Sr fel, Space Sci Lab, Univ Calif-Berkeley, 83-; mem, Spectros-Ctr Eng Develop & Res, NSF-Aeronomy, 85-88, Database, 87-89, Strategic Defense Inst Orgn, Europ technol assessment team, Inst Defense Anal, Washington, DC, 86, Mesosphere & Lower Thermosphere, NASA Space Physics Div, 90, Inner Magnetospheric Imager, Working Group, NASA, 91. *Mem:* Am Geophys Union; Am Inst Physics. *Res:* Experimental studies of the sun-atmosphere-ionosphere relationship; development of novel ultraviolet instruments and experiments for atmospheric studies; spaceborne instrument design; low light level spectroscopic and interferometric studies; atmospheric modelling. *Mailing Add:* Space Sci Lab Univ Calif Berkeley CA 94720

CHAKRABARTY, ANANDA MOHAN, b Sainthia, India, Apr 4, 38; US citizen; m 65; c 2. MICROBIOLOGY, GENETICS. *Educ:* St Xavier's Col, India, BSc, 58; Calcutta Univ, MSc, 60, PhD(biochem), 65. *Hon Degrees:* DSc, Univ Burdwan, 86. *Prof Exp:* Sr sci officer biochem, Calcutta Univ, 64-65; res assoc, Univ Ill, Urbana, 65-71; staff microbiologist, Gen Elec Co, 71-79; prof microbiol, 79-85, DISTINGUISHED UNIV PROF MICROBIOL & IMMUNOL, UNIV ILL MED CTR, 85- *Concurrent Pos:* Adj prof, Dept Biol, State Univ NY, 76-79. *Honors & Awards:* Scientist of the Year Award, Industrial Res Mag, 75, IR-100, 75; Pub Affairs Award, Am Chem Soc, Chicago, 84. *Mem:* Am Soc Microbiol; Am Soc Biol Chemists; Soc Indust Microbiol. *Res:* Evolution and application of hydrocarbon degradative plasmids in Pseudomonas; molecular cloning and genetic engineering with plasmids; genetic basis of hydrocarbon biodegradation; microbial biodegradation of environmental pollutants; pseudomonas infection in cystic fibrosis. *Mailing Add:* Dept of Microbiol Univ of Ill Med Ctr 835 S Wolcott Chicago IL 60612

CHAKRABARTY, MANOJ R, b Bajitpur, Bangladesh, Jan 1, 33; div; c 2. PHYSICAL INORGANIC CHEMISTRY. *Educ:* Univ Calcutta, BS, 51, MS, 54; Univ Toronto, PhD(inorg chem), 62. *Prof Exp:* Instr chem, Bengal Eng Col, India, 55-56; lectr, Indian Sch Mines, 56-57; sessional instr, Univ Alta, 57-59; assoc res scientist mat chem, Ont Res Found, 62-63; from asst prof to assoc prof, 63-69, PROF CHEM, MARSHALL UNIV, 69- *Mem:* Am Chem Soc. *Res:* Coordination compounds; inorganic reaction kinetics; analytical chemistry. *Mailing Add:* Dept of Chem Marshall Univ Huntington WV 25701

CHAKRABORTY, JYOTSNA (JOANA), b Calcutta, India, June 1, 34; m 54; c 1. REPRODUCTIVE PHYSIOLOGY, CELL PHYSIOLOGY. *Educ:* City Col, Univ Calcutta, BS, 54, Sci Col, MS, 56; Inst Nuclear Physics, Calcutta, PhD(biophysics), 62. *Prof Exp:* Res asst cell biol, Inst Nuclear Physics, 60-62; fel, Iowa State Univ, Ames, 62-63; lectr, Inst Nuclear Physics, 65-69; Ford Found fel, Harbor Gen, Med Ctr, Univ Calif, Los Angeles, 69-70; from asst prof to assoc prof, 72-82, PROF CELL & REPRODUCTIVE PHYSIOL, MED COL OHIO TOLEDO, 83- *Concurrent Pos:* Dir ultra structure res, Biophysics Lab, Inst Nuclear Physics, Calcutta, India, 65-69; dir, Electron Micros Lab, Dept Physiol, Med Col Ohio, 70-89; prin investr grant, NIH, 76-80 & 82-85, Stranhan Found, 81-83; vis scholar, Cambridge Univ, Eng, 77. *Mem:* Soc Study Reproduction; AAAS; Am Soc Andrology; Electron Micros Soc India; Electron Micros Soc Am. *Res:* Light and electron microscopic studies of male and female reproductive tracts, mammary glands, spermatozoa, eggs, early embryos and prostate glands; hydridomas and monoclonal antibodies of cell surface antigens. *Mailing Add:* CS #10008 Dept Physiol Med Col Ohio Toledo OH 43699

CHAKRABORTY, RANAJIT, b Calcutta, India, Apr 17, 46; m 74. POPULATION GENETICS, HUMAN GENETICS. *Educ:* Indian Statist Inst, Calcutta, BStatist, 67, MStatist, 68, PhD(biostatist), 71. *Prof Exp:* From lectr to sr lectr statist, Indian Statist Inst, Calcutta, 71-72; vis consult genetics, Pop Genetics Lab, Univ Hawaii, 72-73; res assoc pop genetics, Health Sci Ctr, 73, from asst prof to assoc prof, 73-84, PROF POP GENETICS & HUMAN ECOL, GRAD SCH PUB HEALTH, UNIV TEX, HOUSTON, 84- *Concurrent Pos:* Assoc ed, South Asian Anthrop, 82-, J Indian Anthrop Soc, 82-, Annals Human Biol, 84- & Am J Physics Anthrop, 86- *Mem:* Genetics Soc Am; Am Soc Human Genetics; Am Soc Phys Anthropologists; Am Soc Dermatoglyphics; fel Am Col Epidemiol; Indian Soc Human Genetics; Indian Soc Anthrop; Soc Study Evolution; Indian Statist Inst. *Res:* Statistical methods for genetic determination of quantitative traits; analysis of pedigree data for detection and estimation of familial aggregation of various disorders; mathematical theories of molecular evolution and population dynamics; medico-legal genetics and genetic counseling with DNA markers. *Mailing Add:* Grad Sch Biomed Sci Univ Tex Health Sci Ctr PO Box 20334 Houston TX 77225

CHAKRABURTTY, KALPANA, b India. BIOCHEMISTRY. *Educ:* Univ Calculta, BSc, 61, MSc, 64, PhD(biochem), 68. *Prof Exp:* Asst prof, 73-80, ASSOC PROF BIOCHEM, MED COL WIS, 80- *Mem:* Am Soc Biol Chemists; Sigma Xi. *Res:* Enzymology, chemistry of nucleic acids protein nucleic acid interactions; structure and function of transfer RNA; regulation of ribosomal reactions; yeast genetics and translational control. *Mailing Add:* Dept of Biochem Med Col of Wis Milwaukee WI 53226

CHAKRAVARTHY, SRINIVASARAGHAVAN, b Madras, India, Aug 28, 53; m 82; c 1. STOCHASTIC MODELING, ALGORITHMIC PROBABILITY. *Educ:* Univ Madras, India, BSc, 73, MSc, 75; Univ Del, PhD(opers res), 83. *Prof Exp:* From asst prof to assoc prof, 83-90, PROF MATH & STATIST, DEPT SCI & MATH, GEN MOTORS INST, FLINT, MICH, 90- *Concurrent Pos:* Consult, UPS, Livonia, Mich, 88, Gen Motors, Flint, Mich, 90- *Mem:* Opers Res Soc Am; Math Asn Am; Sigma Xi. *Res:* Stochastic modeling; computational probability; scheduling and production line problems; finite capacity queueing models; reliability theory. *Mailing Add:* Dept Math Eng & Mgt Inst Gen Motors Inst Flint MI 48504-4898

CHAKRAVARTI, KALIDAS, b Gobindapur, India. PHYSICAL CHEMISTRY, SURFACE CHEMISTRY. *Educ:* Univ Calcutta, BS, 57, MS, 59, PhD(chem), 64. *Prof Exp:* Res assoc Univ Calcutta, 64-65; res fel, Univ Minn, Minneapolis, 65-66; res assoc, Lehigh Univ, 67-69; SR SCIENTIST, ALLIED CORP, 69- *Mem:* Am Chem Soc; fel Am Inst Chemists. *Res:* Physical chemistry of macromolecules and biopolymers; colloid and surface chemistry; synthetic fibers and synthetic polymers; fiber surface science; adhesion and bonding of elastomers; fiber finishes. *Mailing Add:* PO Box 450 Hopewell VA 23860-0450

CHAKRAVARTTY, ISWAR C, b Assam, India, March 1, 35. CALCULUS, NUMBER THEORY. *Educ:* Guarwahati Univ, BSC, MSC, 59; Univ Sask, PhD(math), 65. *Prof Exp:* Lectr math, Guarwahati Univ, 65-67; from asst prof to assoc prof, 67-80, PROF MATH, CHAMPLAIN COL, TRENT UNIV, ONT, CAN, 80. *Concurrent Pos:* Lectr math, St Anthony's Col, 59-62; vis prof, Univ WI, Trinidad, 80 & Univ Chulalongkorn, Bangkok, 86. *Mem:* Can Math Cong; Am Math Asn; Am Math Soc; Indian Math Soc; Can Soc Hist & Philos Math. *Res:* Analysis and history of math. *Mailing Add:* Dept Math Trent Univ Champlain Col Peterborough ON K9J 7B8 Can

CHAKRAVARTY, INDRANIL, b New Delhi, India, Jan 7, 54; m 88. INTELLIGENT & HARDWARE SYSTEMS. *Educ:* NY Univ, BEE, 74; Rensselaer Polytech Inst, MEng, 76, PhD(comput & syst eng), 82. *Prof Exp:* Res asst, Image Processing Lab, Rensselaer Polytech Inst, 75-79, instr comput eng, elec, computing systs eng dept, 79-81; prof staff mem, Syst Sci Dept, Schlumberger-Doll Res, 81-85, prog leader, Knowledge-Based Interactive Systs, 85-87, sr res scientist, 88-89, PROG LEADER, MODELING & SIMULATION, SCHLUMBERGER LAB FOR COMPUTER SCI, 89- *Concurrent Pos:* Ed, Comput Reviews, Asn Comput Mach, 85-,. *Mem:* Inst Elec & Electronic Engrs; Asn Comput Mach; Sigma Xi; NY Acad Sci. *Res:* Parallel computer system architectures and algorithms for computer graphics, digital image generation, image processing and robotics. *Mailing Add:* Schlumberger Lab Computer Sci PO Box 200015 Austin TX 78720-0015

CHAKRAVORTY, S(AILENDRA) K(UMER), b Calcutta, India, Jan 5, 22; m 56; c 3. PETROLEUM ENGINEERING. *Educ:* Univ Calcutta, BSc, 42, MSc, 44; Colo Sch Mines, PE, 47; Univ Kans, PhD(petrol geol & eng), 51. *Prof Exp:* Lectr geol, Univ Calcutta, 44-45; instr petrol eng, Univ Kans, 47-48; supvr petrol eng & geol, Dept Mineral Resources, Sask, 51-52, chief oil & gas conserv officer, 52-53; chief petrol engr, New Superior Oils Can Ltd, 54-56, prod mgr, 56-59; sr staff engr, Hudson's Bay Oil & Gas Co, Ltd, 59-76, div petrol engr, 77-80, supvr formation eval, 80-82; supvr formation eval, 83-85, CONSULT ENGR, DOME PETROL LTD, 85- *Mem:* Can Asn Prof Engrs; Am Inst Mining, Metall & Petrol Engrs; Can Inst Mining & Metall; Can Well Logging Soc. *Mailing Add:* 1039 78TH Ave SW Calgary AB T2V 0T9 Can

CHAKRIN, LAWRENCE WILLIAM, b Brooklyn, NY, Oct 21, 38; m 64. PHARMACOLOGY. *Educ:* Long Island Univ, BSc, 62; Univ Minn, PhD(neuropharmacol), 67. *Prof Exp:* USPHS fel, Cambridge, Eng, 67-68; asst dir biol res, Smith Kline & French Labs, 68-77, res pharmacologist, 68-80, assoc dir biol res, 77-80; dir pharmacol, 80, vpres biol, 80-84, exec vpres & chief opers off, 84-85, PRES, STERLING RES GROUP, STERLING-WINTHROP RES INST, 85- *Mem:* Am Soc Pharmacol & Exp Therapeut; Am Soc Neurochem; Am Acad Allergy. *Res:* Respiratory pharmacology; pharmacology of immediate hypersensitivity reactions; central nervous system neuropharmacology; cholinergic mechanisms; cardiovascular pharmacology. *Mailing Add:* Univ City Sci Ctr 3624 Market St Philadelphia PA 19104

CHALABI, A FATTAH, b Mosul, Iraq, Apr 12, 24; m 56. CIVIL ENGINEERING. *Educ:* Univ Baghdad, BSc, 46; Univ Mich, MSc, 52, PhD(civil eng), 56. *Prof Exp:* Supt bldgs, Iraqi Pub Works Dept, 46-49; designer struct eng, Ayres, Lewis, Norris & May, 50-56; asst prof civil eng, Univ Baghdad, 56-59; asst prof mech eng, 59-60, from assoc prof to prof civil eng, 60-78, G I ALDEN PROF ENG, WORCESTER POLYTECH INST, 78- *Concurrent Pos:* Consult, Tippit, Abbott, McCarthy & Straton, Iraq, 56-59; consult engr, 60-81. *Mem:* Am Soc Civil Engrs; Am Concrete Inst; Am Soc Eng Educ. *Res:* Laboratory performance of concrete masonry units made with lightweight aggregate; thermal properties of concrete; construction management and structures information systems; critical varibles influencing the design of large industrial projects. *Mailing Add:* Dept Civil Eng Worcester Polytech Inst 100 Institute Rd Worcester MA 01609

CHALFANT, FOREST EARLE, industrial engineering, for more information see previous edition

CHALFANT, RICHARD BRUCE, b Akron, Ohio, Aug 15, 29; m 53; c 3. ENTOMOLOGY. *Educ:* Univ Akron, BS, 54; Univ Wis, MS, 56, PhD(entom), 59. *Prof Exp:* Asst prof entom, Sch Agr, NC State Univ, 59-66; assoc prof, 66-80, PROF ENTOM, GA COASTAL PLAIN EXP STA, UNIV GA, 80- *Concurrent Pos:* Int agr, Africa. *Mem:* Entom Soc Am. *Res:* Biology, control and management of insects affecting vegetable crops; host-plant resistance; integrated pest management. *Mailing Add:* Ga Coastal Plain Exp Sta Univ of Ga Tifton GA 31793

CHALFIE, MARTIN, CELL DIFFERENTIATION. *Educ:* Harvard Univ, PhD(physiol), 77- *Prof Exp:* ASSOC PROF BIOL SCI, COLUMBIA UNIV, 82- *Res:* Developmental genetics of nerve cells; neuronal degeneration. *Mailing Add:* Dept Biog Sci Columbia Univ Rm 1012 Fairchild New York NY 10027

CHALGREN, STEVE DWAYNE, b Ft Dodge, Iowa, Jan 3, 40; m 62; c 2. MICROBIOLOGY, VIROLOGY. *Educ:* Univ Iowa, BA, 61; Univ Mo, MS, 65, PhD(microbiol, virol), 68. *Prof Exp:* ASSOC PROF BIOL, RADFORD COL, 68- *Concurrent Pos:* Microbiol consult, Labs, Radford Community Hosp, Va, 68- *Mem:* Am Soc Microbiol. *Res:* Diagnostic microbiology. *Mailing Add:* Dept of Biol Radford Univ Univ Station Radford VA 24142

CHALK, DAVID EUGENE, fish & wildlife sciences, for more information see previous edition

CHALK, JOHN, b Sept 13, 22. LINEAR FORMS. *Educ:* Imperial Col Sci & Technol, BSc, 43, PhD, 48; Univ Cambridge, PhD, 51. *Hon Degrees:* DSc, Imperial Col Sci & Technol, 87; ScD. *Prof Exp:* Assoc prof, 60-63, PROF, MATH, UNIV TORONTO, 63- *Concurrent Pos:* Vis fel, Univ Col, London, 65; Sci res fel, Univ Nottingham, 82; vis prof, Imperial Col, London, 86. *Mem:* Fel Royal Soc Can. *Res:* Published numerous articles in various journals. *Mailing Add:* Vanier Col 821 Saint Croix Blvd Montreal PQ H4L 3X9 Can

CHALKLEY, G ROGER, b Sleaford, Eng, June 28, 39; m 62; c 3. BIOCHEMISTRY. *Educ:* Oxford Univ, BA, 61, MA & DPhil(chem), 64. *Prof Exp:* Res fel biol, Calif Inst Technol, 64-67; from asst prof to assoc prof biochem, 67-73, PROF BIOCHEM, ENDOCRINOL & GENETICS, UNIV IOWA, 73- *Res:* Structure and function of chromosomal nucleoproteins; mode of action of steroid hormones; interaction of carcinogens with nuclear material. *Mailing Add:* Dept of Biochem Univ of Iowa Iowa City IA 52242

CHALKLEY, ROGER, b Cincinnati, Ohio, June 21, 31. MATHEMATICS, DIFFERENTIAL EQUATIONS. *Educ:* Univ Cincinnati, ChE, 54; AM, 56, PhD(math), 58. *Prof Exp:* Instr math, Univ Cincinnati, 57-58; mathematician, Oak Ridge Nat Lab, 58-59; asst prof math, Knox Col, Ill, 60-62; from asst prof to assoc prof, 62-79, PROF MATH, UNIV CINCINNATI, 80- *Mem:* Am Math Soc; Math Asn Am. *Res:* Differential equations. *Mailing Add:* Dept Math Univ Cincinnati Cincinnati OH 45221

CHALLICE, CYRIL EUGENE, b London, Eng, Jan 17, 26; m 51; c 2. PHYSICS, BIOPHYSICS. *Educ:* Univ London, BSc, 46, PhD(physics), 49, DSc(biophys), 75; Imp Col, Univ London, ARCS, 46, DIC, 49; Univ Alta, PEng, 74. *Prof Exp:* Biophysicist, Nat Inst Med Res, Eng, 49-52 & Wright-Fleming Inst Microbiol, 52-54; biophysicist & lectr physics, St Mary's Hosp Med Sch, London, 54-57; from asst prof to assoc prof physics, 57-63, head dept, 63-71, vdean, Fac Arts & Sci, 73-76, PROF PHYSICS, UNIV CALGARY, 63- *Concurrent Pos:* NIH fel, 56; NY State Dept Health fel, 58-59; chmn, Nat Comt Biophys, Can, 75-81; Coun,Int Union Pure App Biophysics, 87- *Honors & Awards:* Can Centennial Medal, 67. *Mem:* Fel AAAS; Biophys Soc; Electron Micros Soc Am; fel NY Acad Sci; fel Brit Inst Physics; fel AAAS. *Res:* Structure of biological systems using light and electron microscopy, and electrophysiological methods; structure and function of the heart. *Mailing Add:* Dept Physics Univ Calgary 2500 University Dr Calgary AB T2N 1N4 Can

CHALLIFOUR, JOHN LEE, b Bristol, Eng, June 13, 39; m 67; c 3. MATHEMATICAL PHYSICS. *Educ:* Univ Calif, Berkeley, BA, 60; Cambridge Univ, PhD(theoret physics), 63. *Prof Exp:* Instr math, Princeton Univ, 63-65, lectr, 65-66; asst prof physics, Brandeis Univ, 66-68; assoc prof, 68-78, PROF MATH & PHYSICS, IND UNIV, BLOOMINGTON, 78- *Concurrent Pos:* Vis prof, Univ Göttingen, 70-71, Univ Bielefeld, 75-76, 84 & Univ BC, 80-81; sr Humboldt fel, Univ Göttingen, 87-88. *Mem:* Am Math Soc; fel Am Phys Soc. *Res:* Axiomatic and constructive quantum field theory; theory of distributions; partial differential operators. *Mailing Add:* Dept Phys Ind Univ Bloomington IN 47405

CHALLINOR, DAVID, b New York, NY, July 11, 20; m 52; c 4. FOREST ECOLOGY. *Educ:* Harvard Univ, BA, 43; Yale Univ, MF, 59, PhD(forest ecol), 66. *Prof Exp:* Forestry asst, Conn Agr Exp Sta, 59-60; dep dir, Peabody Mus Natural Hist, Yale Univ, 60-66; spec asst trop biol, Mus Nat Hist, Smithsonian Inst, 66-67, dep dir, 67-69, dir off int activ, 69-71, asst secy sci, 71-87, SCI ADVISOR TO SECY, SMITHSONIAN INST, 88- *Mem:* Soc Am Foresters; Ecol Soc Am; Wildlife Soc; fel AAAS; Sigma Xi. *Res:* Tree-soil interactions in temperate and tropical forest environments. *Mailing Add:* 3117 Hawthorne St NW Washington DC 20008

CHALLONER, DAVID REYNOLDS, b Appleton, Wis, Jan 31, 35; m 58; c 3. INTERNAL MEDICINE, ENDOCRINOLOGY. *Educ:* Lawrence Col, BS, 56; Harvard Univ, MD, 61. *Prof Exp:* Intern, Columbia-Presby Hosp, 61-62, asst resident, 62-63; res assoc, Lab Metab, Nat Heart Inst, 63-65; chief resident, King County Hosp, Univ Wash, 65-66, USPHS spec fel endocrinol, 66-67; from asst prof to prof med & biochem & asst chmn dept med, Sch Med, Ind Univ, Indianapolis, 67-75; prof internal med & dean, Sch Med, St Louis Univ, 75-82; PROF MED & VPRES HEALTH AFFAIRS, UNIV FLA, 82- *Concurrent Pos:* chmn, Pres Comt Nat Medal Sci, 88-90. *Mem:* Inst Med-Nat Acad Sci; Endocrine Soc; Am Physiol Soc; Am Diabetes Asn; Am Fedn Clin Res (pres, 74); Asn Am Physicians; Am Soc Clin Invest; fel AAAS. *Res:* Control mechanisms in intermediary and oxidative metabolism. *Mailing Add:* Univ Fla Box J-14 JHMHC Gainesville FL 32610

CHALMERS, BRUCE, applied physics, metallurgy; deceased, see previous edition for last biography

CHALMERS, JOHN HARVEY, JR, b St Paul, Minn, Mar 5, 40. BIOCHEMICAL GENETICS, MUSICAL ACOUSTICS. *Educ:* Stanford Univ, AB, 62; Univ Calif, San Diego, PhD(biol), 68. *Prof Exp:* NIH-USPHS fel genetics, Univ Wash, 68-71; NIH-USPHS trainee, Univ Calif. Berkeley, 71-73; res fel appl microbiol, Merck Sharp & Dohme Res Labs, 73-75; asst prof biochem, Baylor Col Med, Tex Med Ctr, 76-80, asst prof path, 80-83; sr scientist, Biosyne Corp, 83-87; CONSULT, 87- *Concurrent Pos:* Ed-publ, Xenharmonikon, 73-79, 91-; scholar-in-residence, Ossabaw Island Proj, 79 & Fondazione Rockefeller Cult Ctr, Bellagio, Italy, 80. *Mem:* Sigma Xi; AAAS. *Res:* Biochemical genetics; industrial microbiology; fungal genetics; musical acoustics and experimental music; molecular virology. *Mailing Add:* 9411 Beckford Dr Houston TX 77099-2218

CHALMERS, JOSEPH STEPHEN, b Detroit, Mich, Feb 26, 38; c 3. THEORETICAL PHYSICS. *Educ:* Wayne State Univ, BS, 60, MA, 62, PhD(physics), 67. *Prof Exp:* Assoc prof, 67-80, PROF PHYSICS, UNIV LOUISVILLE, 80- *Concurrent Pos:* Dept Chair, 87-; vis scientist, Arogonne, 82-83. *Mem:* Am Phys Soc. *Res:* Theory of scattering of elementary particles by nuclei; paramagnetic resonance of free radicals. *Mailing Add:* Dept of Physics Univ of Louisville Louisville KY 40292

CHALMERS, ROBERT ANTON, b Wildwood, NJ, Nov 4, 30; m 56; c 2. INTELLIGENT SYSTEMS. *Educ:* Princeton Univ, AB, 52; Northwestern Univ, PhD(physics), 63. *Prof Exp:* STAFF SCIENTIST, LOCKHEED MISSILES & SPACE CO, INC, 63- *Mem:* Am Phys Soc; Inst Elec & Electronic Engrs; Am Asn Artificial Intel. *Res:* Low energy nuclear physics research with electrostatic accelerator; development of computer-oriented laboratory instrumentation; expert systems. *Mailing Add:* Dept 9110 Bldg 203 Lockheed Missiles & Space Co 3251 Hanover St Palo Alto CA 94304

CHALMERS, THOMAS CLARK, b Forest Hills, NY, Dec 8, 17; m 42; c 4. INTERNAL MEDICINE, GASTROENTEROLOGY. *Educ:* Columbia Col, MD, 43; Am Bd Internal Med, dipl, 50. *Prof Exp:* Intern med, Presby Hosp, NY, 43-44; res fel, Malaria Res Unit, Goldwater Mem Hosp, NY Univ, 44-45; resident, 2nd & 4th Med Serv, Harvard Boston City Hosp, 45-47, out-patient physician, 47-48; dir hepatitis study, Comn Liver Dis, Armed Forces

Epidemiol Bd, 51-53; chief med serv, Lemuel Shattuck Hosp, Boston, 55-68; asst chief med dir res & educ, Vet Admin, DC, 68-70; prof med, Sch Med, George Washington Univ, 70-73; pres, Mt Sinai Med Ctr & dean, pres & prof med, Mt Sinai Sch Med, 73-83, EMER PRES & DEAN, MT SINAI MED CTR & SCH MED, 83-; LECTR, HARVARD SCH PUB HEALTH, DEPT EPIDEMIOL & BIOSTATIST, BOSTON UNIV & DEPT MED, SCH MED, TUFTS UNIV, 87- *Concurrent Pos:* Asst, Med Sch, Harvard Univ, 47-49, from instr to asst clin prof, 49-61, lectr, 61-, vis prof, Dept Health Policy & Mgt, Sch Pub Health, 83-84; pvt pract internal med, Cambridge, Mass, 47-53; asst physician, Thorndike Mem Lab, Boston City Hosp, 47-53, assoc vis physician, 2nd & 4th Med Serv, 55-68; jr physician, Mt Auburn Hosp, Cambridge, 47-53, assoc vis physician, 55-68; lectr, Sch Med, Tufts Univ, 55-61 & 87-, prof, 61-68; consult, Faulkner Hosp, Jamaica Plain, Mass, 55-68; assoc staff, New Eng Hosp Ctr, Boston, 55-68; mem, training comt, Nat Heart Inst, 61-65, Spec Rev Panel, Coronary Drug Proj, 66-69, Diet Heart Rev Panel, 68 & policy bd, Urokinase-Streptokinase Pulmonary Embolism Trial, 68-72; mem cancer chemother collab prog rev comt, Nat Cancer Inst, 65-66; mem comt epidemiol & vet follow-up studies, Nat Acad Sci-Nat Res Coun, 65-69, 72-, mem ad hoc comt hepatitis-associated antigen tests, 70-71; mem subcomt on liver, Adv Comt Gen Med, Army Surg Gen Adv Comt, 65-72; mem, sci adv comt, Pharmaceut Mfrs Asn Found, 66-68 & adv comt to fac develop awards in clin pharmacol, 66-70; mem coop studies eval comt, Vet Admin, 70-74; chmn, Nat Coop Crohn's Dis Study Adv Bd, Nat Inst Arthritis, Metab & Digestive Dis, 71-78; mem, Hyper-Immune Gamma Globulin Trials Policy Bd, Nat Heart & Lung Inst, 72-75; chmn, Rev Panel New Drug Regulation, Food & Drug Admin, 75-76; mem, Bd Regents, Nat Libr Med, 78-79; distinguished physician, Boston Va Med Ctr, 86-; bd trustees, Dartmouth-Hitchcock Med Ctr, chmn, 83-88, mem, 88-91; assoc dir, Technol Assessment Group, Harvard Sch Pub Health, 86-; regents vis prof, Dept Statist, Univ Calif, Berkeley, 91. *Mem:* Inst Med-Nat Acad Sci; Am Asn Study Liver Dis (pres, 59); Am Clin & Climat Asn; Am Col Physicians; Am Gastroenterol Asn (pres, 69); Am Acad Arts & Sci; NY Acad Med Res; Clin Trials Soc (pres, 86); Asn Am Physicians; Am Soc Clin Invest. *Res:* Clinical trials; clinical epidemiology. *Mailing Add:* Harvard Sch Pub Health Technol Assessment Group 677 Huntington Ave Boston MA 02115

CHALMERS, WILLIAM, b Edinburgh, Scotland, Nov 16, 05; Can citizen; m 41; c 4. ORGANIC CHEMISTRY. *Educ:* Univ BC, BA, 26, MA, 27; McGill Univ, PhD(sci), 30. *Prof Exp:* Fel chem, Freiberg Univ, Germany, 30-31; pres, Western Chem Indust, Vancouver, 33-68; RETIRED. *Mem:* Am Chem Soc; AAAS. *Res:* Mechanism of polymerization; methacrylic polymers; polymerization of vinyl ethers; fat-soluble vitamins in fish oils; industrial uses of fish by-products; anti-inflammatory activity of the glyceryl ethers of the higher fatty alcohols; alginic acid from seaweed. *Mailing Add:* 4716 Belmont Ave Vancouver BC V6T 1A9 Can

CHALOKWU, CHRISTOPHER ILOBA, b Jos, Nigeria, Dec 30, 52; US citizen; m 86. IGNEOUS-METAMORPHIC PETROLOGY, HIGH TEMPERATURE GEOCHEMISTRY. *Educ:* Northeastern Ill Univ, BS, 78, MS, 80; Miami Univ, Ohio, PhD(geol), 85. *Prof Exp:* Explor geologist, Geneva Pac Corp, 79-80; asst prof, 84-90, ASSOC PROF GEOL, AUBURN UNIV, 90- *Concurrent Pos:* Mem, NSF Rev Panel, 89-; vis Fulbright prof, Univ Ghana, Legon, 90-91, external examr petrol & geochem, 91-; vis res fel, Geol Surv Ghana, 91- *Mem:* Fel Geol Soc London; Geol Soc Am; Mineral Soc Am; Am Geophys Union; Asn Geoscientists Int Develop. *Res:* Magma dynamics and melt migration in layered igneous complexes; petrochemistry and metallization of granitic pegmatites; geothermobarometry; soret crystallization. *Mailing Add:* Dept Geol Auburn Univ Auburn AL 36849

CHALOUD, J(OHN) HOYT, b Omaha, Nebr, Feb 16, 20; m 53; c 2. CHEMICAL ENGINEERING. *Educ:* Iowa State Col, BS(chem eng). *Prof Exp:* Engr process develop, Procter & Gamble, 46-53, sect head tech serv, 53-54, sect head prod develop, 54-58, dept head tech serv, 58-59, assoc dir soap prod develop, 59-69, dir, Explor Develop Div, 69-71, dir, Soap Technol Div, 71-73, dir corp res & develop, prof & regulatory serv, 73-74, dir food prod develop, 74-78, mgr int corp res & develop, 78-81; CONSULT, 82- *Concurrent Pos:* Mem weights & measures adv comt, Nat Bur Standards, 62-65; mem tech comt, Nutrition Found, 74-78; trustee, Food Safety Coun, 76-78. *Mem:* Am Chem Soc; Am Inst Chem Engrs; Am Oil Chem Soc. *Res:* Synthetic detergents; glycerine; fatty acids and alcohols. *Mailing Add:* 204 Poage Farm Rd Wyoming OH 45215

CHALQUEST, RICHARD ROSS, b Denver, Colo, Nov 4, 29; m 54; c 7. VETERINARY MEDICINE, MICROBIOLOGY. *Educ:* Wash State Univ, BS, 51, DVM, 57; Cornell Univ, MS, 59, PhD(path), 60. *Prof Exp:* Asst prof vet microbiol, Wash State Univ, 60-62; res vet, Pfizer Inc, 62-63, mgr agr res & develop, 63-65, dir, 65-72; dir div agr, 72-83, PROF AGR, ARIZ STATE UNIV, 72- *Concurrent Pos:* Mem gov bd, Agr Res Inst, Nat Res Coun, Nat Acad Sci, 67-70. *Mem:* Am Asn Avian Path; Am Vet Med Asn; Poultry Sci Asn; Am Asn Vet Parasitol. *Res:* Agricultural research. *Mailing Add:* Div Agr Ariz State Univ Tempe AZ 85287-3306

CHALUPA, LEO M, b Ger, Mar 28, 45; US citizen; m 66; c 2. NEUROSCIENCES, VISION. *Educ:* Queens Col, BA, 66; City Univ New York, PhD(neuropsychol), 70. *Prof Exp:* Res physiologist psychol, Brain Res Inst, Univ Calif, Los Angeles, 70-75; from asst prof to assoc prof, 75-82, PROF PSYCHOL, UNIV CALIF, DAVIS, 82- *Concurrent Pos:* Fel, Brain Res Inst, Univ Calif, Los Angeles, 70-72; Guggenheim fel, 78; vis scholar, Cambridge Univ, 78-79; vis prof, Inst Neurophysiol, Pisa, Italy; fel-in-residence, Neurosci Inst, New York, 87; mem, NSF Develop Neurosci Grant Rev Panel, 87-88, NIH visual sci study sect, 88- *Honors & Awards:* US-USSR Scientist Exchange Award, Nat Acad Sci, 74. *Mem:* AAAS; Soc Neurosci; Sigma Xi. *Res:* Visual neurophysiology and neuropsychology; plasticity of the visual system. *Mailing Add:* Dept Psychol Univ Calif Davis CA 95616

CHALUPA, WILLIAM VICTOR, b New York, NY, Dec 11, 37; m 60; c 2. ANIMAL NUTRITION. *Educ:* Rutgers Univ, BS, 58, MS, 59, PhD(nutrit), 62. *Prof Exp:* Asst dairy sci, Rutgers Univ, 58-59, asst instr, 59-62, res fel nutrit, 62-63; from asst prof to assoc prof, Clemson Univ, 63-71; mgr rumen metabolic res, Smith Kline Animal Health Prod Div, 71-76; assoc prof, 76-80, PROF NUTRIT, SCH VET MED, UNIV PA, 80- *Concurrent Pos:* Vis scientist, USDA, Md, 69-70; adj assoc prof, Sch Vet Med, Univ Pa, 75- *Honors & Awards:* Am Feed Mfrs Award, 81. *Mem:* AAAS; Am Soc Animal Sci; Am Dairy Sci Asn; Am Inst Nutrit. *Res:* Energy and protein utilization of foodstuffs by ruminant animals; biochemistry of rumen metabolism. *Mailing Add:* Sch Vet Med Univ Pa New Bolton Ctr West Chester PA 19348

CHALUPNIK, JAMES DVORAK, b Bay City, Tex, Nov 10, 30; m 57; c 2. MECHANICS. *Educ:* Tex Tech Col, BSME, 53; Univ Tex, MSEM, 60, PhD(eng mech), 64. *Prof Exp:* Instr mech eng, Univ Tex, 57-58, res asst, 61-64; supvr, Lockheed Missile & Space Co, 58-59, scientist, 59-61; from asst prof to assoc prof mech eng, 64-76, PROF MECH ENG, UNIV WASH, 76- *Concurrent Pos:* Prof eng consult, 64- *Mem:* Am Soc Mech Engrs; fel Acoust Soc Am; Inst Noise Control Eng. *Res:* Acoustics and noise studies; dynamic behavior of materials and structures; mechanical vibrations. *Mailing Add:* 5600 NE 77th Seattle WA 98115

CHALY, NATHALIE, b Louvain, Belgium, March 25, 50; Can citizen. NUCLEAR STRUCTURE AND FUNCTION, CELL ORGANIZATION DURING MITOSIS. *Educ:* Carleton Univ, Ottawa, Can, BSc, 69, MSc, 73; Laval Univ, Quebec City, Can, PhD(biol), 76. *Prof Exp:* Teaching fel virol, McGill Univ, Montreal, Can, 76-78 & Commonwealth Sci & Indust Res Org, Adelaide, Australia, 78-81; res assoc cell biol, Univ Ottawa, Can, 81-85; asst prof, 85-89, ASSOC PROF CELL BIOL, CARLETON UNIV, OTTAWA, 90- *Mem:* Am Soc Cell Biol; Can Soc Cell Biol. *Res:* Investigate the structure and immunocytochemistry of nucleus during the mammalian mitotic cell cycle; role of nucleoskeletal elements in genome organization. *Mailing Add:* Dept Biol Carleton Univ Ottawa ON K1S 5B6 Can

CHAMBERLAIN, A(DRIAN) R(AMOND), b Mich, Nov 11, 29; m 79; c 1. ENGINEERING. *Educ:* Mich State Univ, BS, 51; Wash State Univ, MS, 52; Colo State Univ, PhD, 55. *Hon Degrees:* DEng, Mich State Univ, 71; LHD, Univ Denver, 72. *Prof Exp:* Fulbright grantee, Univ Grenoble, France, 55-56; chief eng sect, Colo State Univ, 57-59, chief eng res & actg dean col eng, 59-61, vpres, 60-66, exec vpres & treas, 66-69, pres, 69-80; pres, Mitchell & Co, 81-85; dir & sr vpres, Simons, LI & Assocs, Inc, 81-88; pres & chmn bd dir, Chemogenetics, Inc, 87-89; EXEC DIR, COLO DEPT HWYS, 87- *Concurrent Pos:* Mem bd trustees, Colo State Univ Res Found, 58-63 & 69-73, pres, 59-60; mem bd dirs, Univ Nat Bank, 62-74, chmn bd, 64-69; mem adv comt environ sci, NSF, 67-69, chmn, 67, mem & chmn comn weather modification, 64-66; mem adv comt on air qual criteria, Nat Air Pollution Control Admin, US Pub Health Serv, 67-70; trustee, Univ Corp Atmos Res, 67-81, chmn bd trustees, 77-79; trustee, Nat Cystic Fibrosis Res Fedn, 71-84; mem bd dir, Nat Ctr Higher Educ Mgt Systs, 74-77, chmn bd, 78-79; pres, State Bd Agr Univs Systs, 78-80; hon prof, El Inst Politecnico, Nacional de Mexico; chmn, Nat Asn Land-Grant Cols & Univs, 80 & mem exec comn, 76-81; mem, NSF Dirs Adv Coun, 78-80; pres, Black Mountain Ranch, Inc, 68-86; mem bd dirs, Mitchell & Co, 68-74 & 79-86, Simons, Li & Assocs, Inc, 80-88 & chmn exec comt, Solaron Corp, 81-88; chmn, Fort Collins-Loveland Airport Authority, 83-85; mem, Bd Dirs, Ft Collins Chambers Com, 85-88; mem exec comt, Western Asn State Hwy & Transp Officials, 87-, Strategic Hwy Res Prog Transp Res Bd, Nat Res Coun, 89-, Transp Res Bd, 91-; vpres, Am Asn State Hwy & Transp Officials, 90- *Mem:* Am Soc Civil Engrs; Sigma Xi. *Res:* Fluid mechanics, hydrology & water resources; financial systems. *Mailing Add:* 4200 Westshore Way Ft Collins CO 80525

CHAMBERLAIN, CHARLES CALVIN, b Evart, Mich, Jan 23, 20; m 46; c 2. ANIMAL NUTRITION. *Educ:* Mich State Univ, BS, 41, MS, 48; Iowa State Univ, PhD, 59. *Prof Exp:* From instr to assoc prof animal nutrit, Univ Tenn, 49-71 & from asst animal husbandman to assoc animal husbandman, 52-71, prof animal nutrit, 71-; RETIRED. *Mem:* Am Soc Animal Sci. *Res:* Cattle and swine. *Mailing Add:* Rte 13 Box 71 Marysville TN 37801

CHAMBERLAIN, CHARLES CRAIG, b Milford, Utah, June 1, 33; m 59; c 4. MEDICAL PHYSICS, RADIOBIOLOGY. *Educ:* Univ Calif, Los Angeles, BA, 59, MA, 61, PhD(med physics), 67. *Prof Exp:* Instr, 67-70, ASST PROF RADIOL, STATE UNIV NY UPSTATE MED CTR, 70-, ASSOC PROF, COL HEALTH RELATED PROFESSIONS, 81- *Mem:* Health Physics Soc; Am Asn Physicists in Med; Am Inst Ultrasound Med. *Res:* Effects of heat and radiation on mammalian cells in culture; physics of diagnostic radiology. *Mailing Add:* Dept Radiol State Univ NY Upstate Med Ctr 750 E Adams St Syracuse NY 13210

CHAMBERLAIN, CRAIG STANLEY, b Sacramento, Calif, Jan 14, 47; m 74. INORGANIC CHEMISTRY, MATERIALS SCIENCE. *Educ:* Calif State Univ, Sacramento, BS, 69, MS, 73; Univ Ill, Urbana, PhD(chem), 78. *Prof Exp:* Forensic chemist, US Army Criminal Invest Lab, 69-71; SR CHEMIST MAT SCI, 3M CO, 78- *Mem:* Am Chem Soc; Sigma Xi; Audobon Soc. *Mailing Add:* 8231 Afton Rd Woodbury MN 55125

CHAMBERLAIN, DAVID LEROY, JR, b Kansas City, Kans, Sept 2, 17; m 45; c 3. ORGANIC CHEMISTRY, CHEMICAL ENGINEERING. *Educ:* Univ Kans, BS, 44, MS, 50; Univ Southern Calif, PhD(org chem), 53. *Prof Exp:* Lab asst, Univ Kans, 43-44; chem engr & tech asst to opers, Rayon Mfg, E I du Pont de Nemours & Co, 44-46; lab asst, Univ Kans, 46-47; res org chemist, Callery Chem Co, 53-56; sr org chemist, Stanford Res Inst, 56-72; res assoc, Nat Forest Prod Asn-Nat Bur Standards, 73-83; RETIRED. *Concurrent Pos:* Consult, 72-73 & 83- *Mem:* Am Chem Soc; fel Am Inst Chem; Sigma Xi. *Res:* Thermal degradation of organic materials; chemistry of organic-inorganic interfaces; chemistry of fire retardant materials; fire test methods for materials. *Mailing Add:* 12209 Pawnee Dr Gaithersburg MD 20760

CHAMBERLAIN, DILWORTH WOOLLEY, b Milford, Utah, June 1, 33; m 58; c 6. ICHTHYOLOGY. *Educ:* Calif State Univ, Los Angeles, BS, 65; Univ Southern Calif, PhD(biol), 80. *Prof Exp:* Res asst, Sch Med, Univ Southern Calif, 59-72, res assoc, 72-74; sci advisor, 74-80, SR SCI ADVISOR ENVIRON PROTECTION, ATLANTIC RICHFIELD CO, 80- *Concurrent Pos:* Chmn fisheries issues task force, Am Petrol Inst, 80- *Mem:* AAAS; Sigma Xi; Soc Petrol Indust Biologists. *Mailing Add:* 1641 Maple Hill Rd Pomona CA 91765

CHAMBERLAIN, DONALD F(RANK), b Wayzata, Minn, July 14, 14; m 40; c 5. CHEMICAL ENGINEERING. *Educ:* Univ Minn, BChE, 36, PhD(chem eng), 40. *Prof Exp:* Asst, Univ Minn, 36-40; res engr, Nat Aniline Div Allied Chem & Dye Corp, 40-46; from asst prof to prof chem eng, Wash Univ, 46-56, vchmn dept, 52-55; adminr, US Govt (classified), 55-76, asst dir, Off Sci Intel, Cent Intel Agency, 63-65, dir, 65-73, inspector gen, 73-76; RETIRED. *Concurrent Pos:* Lectr eng sci & mgt war training, Cornell Univ, 43-45. *Mem:* Am Inst Chem Engrs; Am Chem Soc. *Res:* Foreign science and technology. *Mailing Add:* 9521 Palisades Park Rd Boca Raton FL 33428

CHAMBERLAIN, DONALD WILLIAM, b Green Bay, Wis, Nov 28, 05; m 45; c 3. PLANT PATHOLOGY. *Educ:* St Norbert Col, BA, 29; Univ Wis, MA, 32, PhD(plant path), 43. *Prof Exp:* Instr hort, Univ Wis, 43-45; asst agron, Agr Exp Sta, Univ Ky, 45-46; assoc pathologist, 46-56, PATHOLOGIST, CROPS RES DIV, AGR RES SERV, USDA, 56-; EMER PROF PATH, UNIV ILL, URBANA-CHAMPAIGN, 75- *Concurrent Pos:* From asst prof to prof path, Univ Ill, Urbana-Champaign, 56-75. *Mem:* AAAS; Am Phytopath Soc. *Res:* Bacterial and fungus diseases of soybean; disease resistance; occurrance of races of Pseudomonas glycinea in Illinois; resistance to Phytophthora rot in soybean as expressed in roots or stems. *Mailing Add:* 2022 Boudreau Dr Urbana IL 61801

CHAMBERLAIN, ERLING WILLIAM, b Oslo, Norway, Jan 5, 34; US citizen; m 57. MATHEMATICS. *Educ:* Columbia Univ, AB, 55, MA, 56, PhD(math), 61. *Prof Exp:* From asst prof to assoc prof, 62-70, PROF MATH, UNIV VT, 70- *Concurrent Pos:* NSF res grants, 63-70. *Mem:* Am Math Soc. *Res:* Asymptotic theory of ordinary differential equations in the complex domain, especially with regard to factorization of differential operators. *Mailing Add:* Dept Math Agr Col Univ Vt 85 S Prospect St Burlington VT 05405

CHAMBERLAIN, JACK G, b Detroit, Mich, May 15, 33; m 55; c 4. DEVELOPMENTAL ANATOMY. *Educ:* Occidental Col, BA, 55; San Diego State Col, MA, 57; Univ Calif, Berkeley, PhD(anat), 62. *Prof Exp:* Instr anat, Univ Calif, Berkeley, 62-63; instr, Univ Mich Sch Med, 63-64; asst prof, Univ Calif, San Francisco, 64-72; assoc prof anat, 72-80, PROF ANAT, SCH DENT, UNIV OF PAC, 80-, CHMN DEPT, 72- *Concurrent Pos:* Fac grant, Univ Calif, 62-64; Rackham fac & local cancer res grants, Univ Mich, 63-64, fac grant, 64-65, USPHS grant, 65-70. *Mem:* Soc Develop Biol; Am Asn Anat. *Res:* Normal and abnormal developmental biology, especially pathogenesis of central nervous system abnormalities induced by antivitamins; scanning electron microscopy; effects of fish oils on cardiovascular system in Japanese quail. *Mailing Add:* Dept Anat Univ Pac 2155 Webster St San Francisco CA 94115

CHAMBERLAIN, JAMES LUTHER, b West Chester, Pa, May 16, 25; m 51; c 3. ZOOLOGY. *Educ:* Cornell Univ, BS, 48; Univ Mass, MS, 51; Univ Tenn, PhD(zool), 57. *Prof Exp:* Asst proj leader, WVa Conserv Comn, 48-49; field agt, US Fish & Wildlife Serv, Mass, 51; instr zool, State Teachers Col, NY, 52-53; res assoc, La State Univ, 55-57; assoc prof biol, Randolph-Macon Woman's Col, 57-69; chmn div sci & math, 69-74, PROF BIOL, UTICA COL, 74- *Concurrent Pos:* Res grant, US Forest Serv. *Mem:* Ecol Soc Am; Am Soc Mammal; Am Soc Ichthyologists & Herpetologists; Am Ornithologists Union; Wildlife Soc. *Res:* Marsh ecology; vertebrate ecology. *Mailing Add:* RD 2 Clinton NY 13323

CHAMBERLAIN, JOHN, b Detroit, Mich, Jan 15, 23; m 50; c 3. FLUIDS. *Educ:* Mass Inst Technol, SB, 44. *Prof Exp:* Res engr, Res Lab, Pratt & Whitney Aircraft Div, United Technol Corp, 46-52, asst proj engr, 52-55, proj engr, 55-68, develop engr, 68-87; RETIRED. *Honors & Awards:* George Mead gold medal eng achievement, United Aircraft Corp, 65. *Mem:* Am Inst Aeronaut & Astronaut. *Res:* Combustion and cooling systems for airbreathing engines and rockets. *Mailing Add:* 7300 20th St No 198 Vero Beach FL 32966

CHAMBERLAIN, JOSEPH MILES, b Peoria, Ill, July 26, 23; m 45; c 3. ASTRONOMY. *Educ:* US Merchant Marine Acad, BS, 44; Bradley Univ, BA, 47; Columbia Univ, AM, 50, EdD, 62. *Prof Exp:* Instr nautical sci, US Merchant Marine Acad, 47-50, asst prof astron & meteorol, 50-52; asst astronomer, Am Mus-Hayden Planetarium, NY, 52-53, chmn & astronomer, 53-64; asst dir, Am Mus Natural Hist, 64-68; prof astron, Northwestern Univ, 68-77; DIR, ADLER PLANETARIUM, 68-, PRES, 76- *Concurrent Pos:* Instr, Naval Reserve Officer Sch, NY, 54-55 & Hunter Col, 64-68; consult, Norman Porter & Assocs, NY, 54-; lectr, Nat Artists Corp, NY, 55-57; prof lectr, Univ Chicago, 68-71; vchmn Int Planetarium Dirs Conf, 72-78, chmn, 78-87, vpres, 69-75, pres, 75-87. *Mem:* Int Planetarium Soc; Am Astron Soc; Am Asn Mus (vpres, 71-74, pres, 74-75); Am Polar Soc; Int Astron Union. *Res:* Determination of geodetic coordinates by astronomic methods; planetarium education and administration. *Mailing Add:* Adler Planetarium 1300 S Lake Shore Dr Chicago IL 60605

CHAMBERLAIN, JOSEPH WYAN, b Boonville, Mo, Aug 24, 28; m 49; c 3. AERONOMY, PLANETARY ATMOSPHERES. *Educ:* Univ Mo, AB, 48, AM, 49; Univ Mich, MS, 51, PhD(astron), 52. *Prof Exp:* Proj scientist aurora & airglow, US Air Force Cambridge Res Ctr, 51-53; res assoc Yerkes Observ, Univ Chicago, 53-55, from asst prof to prof, 55-62; assoc dir planetary sci div, Kitt Peak Nat Observ, 62-70, astronr, 70-71; adj prof, 71-76, PROF SPACE PHYSICS & ASTRON, RICE UNIV, 76- *Concurrent Pos:* Mem exec comt, Assembly Math Phys Sci, Nat Res Coun-Nat Acad Sci, 74-78; ed, Rev Geophys & Space Physics, 74-80; sect chmn, Nat Acad Sci, 72-75; assoc ed, Astrophys J, 59-63; dir, Lunar Sci Inst, 71-73; Alfred P Sloan res fel, 61-63. *Honors & Awards:* Helen B Warner Prize, Am Astron. *Mem:* Nat Acad Sci; fel AAAS; Am Astron Soc; Am Phys Soc; fel Am Geophys Union; foreign fel Royal Astron Soc; Am Meteorol Soc. *Res:* Planetary atmospheres; aurora and airglow; aeronomy of the stratosphere; atmospheric pollution; climate. *Mailing Add:* Dept Space Physics & Astron Rice Univ PO Box 1892 Houston TX 77251

CHAMBERLAIN, MALCOLM, organic chemistry, for more information see previous edition

CHAMBERLAIN, NUGENT FRANCIS, b Henderson, Tex, Mar 10, 16; m 43; c 3. SPECTROSCOPY, ELECTRON MICROSCOPY. *Educ:* Agr & Mech Col Tex, ChE, 38. *Prof Exp:* Tech trainee, Humble Oil & Ref Co, 38, from jr chemist to chemist, 39-41, from res chemist to sr res chemist, 41-59, sr res chem engr, 59-61, res specialist, 61-63, res assoc, 63-66, res assoc Esso Res & Eng Co, 66-69, sr res assoc, 69-74, sr res assoc, Baytown Res & Develop Div, 69-82, consult, Exxon Res, 82-84; RETIRED. *Mem:* Am Chem Soc; fel Am Inst Chemists. *Res:* Nuclear magnetic resonance spectroscopy; electron microscopy. *Mailing Add:* PO Box 1298 Wimberley TX 78676-1298

CHAMBERLAIN, OWEN, b San Francisco, Calif, July 10, 20; m 43, 80; c 4. EXPERIMENTAL HIGH ENERGY PHYSICS. *Educ:* Dartmouth Col, BA, 41, Univ Chicago, PhD(physics), 49. *Prof Exp:* From instr to assoc prof, 48-58, PROF PHYSICS, UNIV CALIF, BERKELEY, 58- *Concurrent Pos:* Civilian physicist, Manhattan Dist, Berkeley & Los Alamos, 42-46; Guggenheim fel, 57-58. *Honors & Awards:* Nobel Prize in Physics, 59; Loeb Lectr, Harvard Univ, 59. *Mem:* Nat Acad Sci; fel Am Phys Soc; fel AAAS; fel Am Acad Arts & Sci. *Res:* Fission; Alphaparticle decay; neutron diffraction in liquids; high energy nucleon scattering; antinucleons; author of numerous publications and articles. *Mailing Add:* Dept Physics Univ Calif Berkeley CA 94720

CHAMBERLAIN, PHYLLIS IONE, b Belfast, NY, Oct 22, 38. INORGANIC CHEMISTRY. *Educ:* Houghton Col, BS, 60; State Univ NY Buffalo, PhD(chem kinetics), 68. *Prof Exp:* Teacher pub sch, NY, 60-62; asst prof, 67-75, ASSOC PROF CHEM, ROBERTS WESLEYAN COL, NY, 75- *Concurrent Pos:* Danforth Assoc, 78- *Mem:* Am Chem Soc; Am Sci Affil. *Res:* Kinetics and mechanisms of the reactions of transition metal complexes of organic ligands. *Mailing Add:* Roberts Wesleyan Col 2301 Westside Dr Rochester NY 14624-1997

CHAMBERLAIN, ROBERT GLENN, b Atascadero, Calif, May 8, 39; m 60; c 4. MATHEMATICAL MODELING, SYSTEMS ANALYSIS. *Educ:* Calif Inst Technol, BS, 60, MS, 61. *Prof Exp:* Mech engr spacecraft, US Air Force, 61-64; sr engr, 64-67, MEM TECH STAFF, JET PROPULSION LAB, CALIF INST TECHNOL, 67- *Concurrent Pos:* Mgr SAMICS Develop, Low Cost Solar Array Proj, Jet Propulsion Lab, Calif Inst Technol, 75-85, mgr SDTM software develop, Space Sta Freedom Proj, 85-90. *Mem:* Opers Res Soc Am; AAAS; Inst Mgt Sci; Sigma Xi; Am Inst Aeronaut & Astronaut. *Res:* Space station systems/subsystems design optimization modeling; cost and performance modeling of energy production systems; modeling of the economics of energy system component manufacturing. *Mailing Add:* Jet Propulsion Lab 601-237 4800 Oak Grove Dr Pasadena CA 91109

CHAMBERLAIN, ROY WILLIAM, b Stanton, Calif, July 24, 16; m 44; c 1. MEDICAL ENTOMOLOGY, VIROLOGY. *Educ:* Mont State Col, BS, 42; Johns Hopkins Univ, ScD(parasitol), 49. *Prof Exp:* Chief arbovirus vector lab, 49-67, chief arbovirus infections unit, 67-68, DEP CHIEF VIROL DIV, CTR DIS CONTROL, USPHS, 68- *Mem:* Am Soc Trop Med & Hyg; Sigma Xi; Am Mosquito Control Asn. *Res:* Arthropod transmission of virus diseases of man and animals; behavior of arboviruses in insects and vertebrates. *Mailing Add:* 2311 Echocliff Ct NE Atlanta GA 30345

CHAMBERLAIN, S(AVVAS) G(EORGIOU), b Nicosia, Cyprus, Mar 21, 41; m 65; c 1. ELECTRONICS, SEMICONDUCTOR PHYSICS. *Educ:* Univ Southampton, MSc, 66, PhD(electronics), 68. *Prof Exp:* Proj leader optoelectronics & device physics, Allen Clark Res Ctr, Plessey Co Ltd, Eng, 68-69; asst prof electronics, 69-77, PROF ELEC ENG, UNIV WATERLOO, 77- *Mem:* Inst Elec & Electronics Engrs; assoc mem Brit Inst Elec Engrs; Brit Inst Physics & Phys Soc. *Res:* Solid state semiconductor devices; integrated circuits; device modeling; accurate numerical solutions of p-n junctions and transistors; optoelectronic devices including charge coupled devices. *Mailing Add:* Dept of Elec Eng Univ of Waterloo Waterloo ON N2L 3G1 Can

CHAMBERLAIN, THEODORE KLOCK, b Detroit, Mich, July 18, 30. GEOLOGY, OCEANOGRAPHY. *Educ:* Univ NMex, BS, 52; Scripps Inst Oceanog, Univ Calif, MS, 53, PhD(oceanog), 60. *Prof Exp:* Res marine geologist, Scripps Inst Calif, 56-60; oceanogr, Tokyo Univ Fisheries, Tokyo, 60-62; assoc prof & asst chmn dept oceanog, Univ Hawaii, 62-67; sr oceanogr & mgr sci div, Ocean Sci & Eng Inc, 67-71; dir, Chesapeake Res Consortium, Johns Hopkins Univ, 72-75; head earth resources dept, 75-80, PROF, COLO STATE UNIV, 75- *Concurrent Pos:* Nat Acad Sci-Nat Res Coun fel, 60-62. *Mem:* Am Soc Limnol & Oceanog; AAAS; Royal Siam Soc. *Res:* Littoral processes, physical limnology, marine placers, exploration geology; earth resources development. *Mailing Add:* Dept of Earth Resources Colo State Univ Ft Collins CO 80523

CHAMBERLAIN, WILLIAM MAYNARD, b Montreal, Que, Apr 7, 38; m 61; c 2. AQUATIC BIOLOGY. *Educ:* Univ Toronto, BS, 61, PhD(zool), 68. *Prof Exp:* Res assoc, Limnol Res Ctr, Univ Minn, Minneapolis, 67-69; asst prof, 69-76, ASSOC PROF LIFE SCI, IND STATE UNIV, 76- *Concurrent Pos:* Partic, Gordon Res Conf, 66 & 70 & Int Symp Eutrophication, 67. *Mem:* Am Soc Limnol & Oceanog; Ecol Soc Am. *Res:* Nutrient circulation studies on phosphorus in aquatic ecosystems; assay procedures for available phosphorus; temperature effects on cladoceran populations; nutrient cycles. *Mailing Add:* Dept of Life Sci Ind State Univ Terre Haute IN 47809

CHAMBERLAND, BERTRAND LEO, b Manchester, NH, Mar 17, 34; m 65; c 3. INORGANIC CHEMISTRY. *Educ:* St Anselms Col, AB, 55; Col of the Holy Cross, MS, 56; Univ Pa, PhD(inorg chem), 60. *Prof Exp:* Res chemist, Nat Carbon Co, Ohio, 56-57 & E I du Pont de Nemours & Co, Inc, 60-69; PROF CHEM, UNIV CONN, 69- *Mem:* Am Chem Soc; Royal Soc Chem; Am Ceramic Soc; Am Crystallog Asn; Am Ceramic Soc. *Res:* Inorganic synthesis of molecular and solid state compounds; preparation, characterization and crystal growth of solid state materials; high pressure synthesis and reactions. *Mailing Add:* Dept of Chem U-60 Univ of Conn 215 Glenbrook Rd Storrs CT 06269

CHAMBERLIN, EARL MARTIN, b Cochranville, Pa, Dec 4, 14; m 47; c 7. ORGANIC CHEMISTRY. *Educ:* Philadelphia Col Pharm & Sci, ScB, 36; Boston Univ, AM, 37; Harvard Univ, AM, 44, PhD(org chem), 46. *Prof Exp:* Asst org chem, Merck & Co, 38-40, admin asst to dir res, 40-41, res chemist, 46-47, staff asst to vpres & sci dir, 48-49, res chemist, 49-59, mgr develop res, 59-68, dir, Synthetic Prep Lab, Process Res, Merck & Co, 69-77, sr dir process res, Merch Sharp & Dohme Res Labs, 77-79; RETIRED. *Concurrent Pos:* Lectr, Union Col Cranford, NJ, 78-79. *Mem:* Am Chem Soc; fel Am Inst Chemists. *Res:* Synthetic organic chemistry; alkaloids; fatty acids; quinones; organic therapeutic agents; steroidal hormones; organometallic chemistry. *Mailing Add:* 2028 Hilltop Rd Westfield NJ 07090

CHAMBERLIN, EDWIN PHILLIP, b Louisville, Ky, Jan 7, 41; m 61; c 2. ELECTROMAGNETISM, ATOMIC & MOLECULAR PHYSICS. *Educ:* Tex A&M Univ, BS, 64, PhD(physics), 71. *Prof Exp:* Accelerator engr, Cyclotron Inst, Tex A&M Univ, 71-75; MEM STAFF, LOS ALAMOS NAT LAB, 75- *Mem:* Am Phys Soc. *Res:* Hardware development: calculation, design and modification of ion sources, optics and isotope separators for specific usage. *Mailing Add:* 109 Grand Canyon Los Alamos NM 87544

CHAMBERLIN, HARRIE ROGERS, b Cambridge, Mass, June 13, 20; m; m 46; c 4. PEDIATRICS. *Educ:* Harvard Univ, AB, 42, MD, 45. *Prof Exp:* Instr pediat, Sch Med, Yale Univ, 51-52; from instr to assoc prof, 53-70, prof, 70-84, EMER PROF PEDIAT & DIR FOR DIS OF DEVELOP & LEARNING, CHILD DEVELOP INST, SCH MED, UNIV NC, CHAPEL HILL, 84- *Concurrent Pos:* consult, Div Hosp & Med Facil, USPHS, 65-68. *Mem:* AMA; Am Acad Pediat; assoc Am Acad Neurol; Am Pediat Soc; Am Acad Cerebral Palsy & Develop Med. *Res:* Mental retardation; developmental neurology in the infant and child. *Mailing Add:* Dept Pediat Univ NC 1001 Arrow Head Rd Chapel Hill NC 27514

CHAMBERLIN, HOWARD ALLEN, polymer chemistry, textile chemistry; deceased, see previous edition for last biography

CHAMBERLIN, JOHN MACMULLEN, b Old Hickory, Tenn, Oct 12, 35; m 65; c 1. MATHEMATIC MODELING, NONLINEAR DYNAMICS. *Educ:* Western Ky Univ, BS, 57; Duke Univ, MA, 61, PhD(inorg chem), 64.. *Prof Exp:* Coordr med technol, 65-78, asst to dean, Col Appl Arts & Health, 72-78, ASSOC PROF INORG & PHYS CHEM, WESTERN KY UNIV, 72- *Res:* Mathematical modeling of vapor sorption by coal extracts, DSC oxidation of coal, and CoTSPC monomer-dimer equilibria; investigation of one-dimensional iterated maps. *Mailing Add:* Dept Chem Western Ky Univ Bowling Green KY 42101

CHAMBERLIN, MICHAEL JOHN, b Chicago, Ill, June 7, 37; m 81. NUCLEIC ACIDS, TRANSCRIPTION. *Educ:* Harvard Univ, AB, 59; Stanford Univ, PhD(biochem), 63. *Prof Exp:* Asst prof virol, Univ Calif, 63-67, assoc prof molecular biol, 67-71, assoc prof biochem, 71-73, vchmn dept, 83-88, PROF BIOCHEM, UNIV CALIF, BERKELEY, 73- *Concurrent Pos:* Mem, Physiol Chem Study Sect, NIH, 70-74, Health Molecular Biol Study Sect, 80-84, Am Heart Asn Study Sect, 83-86; ed staff, J Biol Chem, 75-78. *Honors & Awards:* Charles Pfizer Award, Am Chem Soc, 74. *Mem:* Nat Acad Sci; Sigma Xi; Am Soc Biol Chemists; Am Soc Microbiol; fel AAAS. *Res:* Regulation of gene expression, particularly transcription, and factors that determine the specificity of protein-nucleic acid interactions. *Mailing Add:* Dept Biochem Univ Calif Berkeley CA 94720

CHAMBERLIN, RALPH VARY, b Albuquerque, NMex, Apr 22, 56. MAGNETIC PROPERTIES OF MATERIALS, HIGH MAGNETIC FIELDS. *Educ:* Univ Utah, BS, 78; Univ Calif, Los Angeles, MS, 79, PhD(physics), 84. *Prof Exp:* Res asst physics, Univ Utah, 74-78; from res asst to res assoc physics, Univ Calif, Los Angeles, 78-84; res investr physics, Univ Pa, 84-86; ASST PROF PHYSICS, ARIZ STATE UNIV, 86- *Mem:* AAAS; Am Phys Soc. *Res:* Magnetic and electronic properties of low-dimensional systems; slow relaxation in condensed matter. *Mailing Add:* Dept Physics Ariz State Univ Tempe AZ 85287

CHAMBERLIN, RICHARD ELIOT, b Cambridge, Mass, Mar 20, 23; m 53; c 4. MATHEMATICS. *Educ:* Univ Utah, AB, 43; Harvard Univ, AM, 47, PhD(math), 50. *Prof Exp:* Asst prof math, 49-51, from asst prof to assoc prof, 52-70, PROF MATH, UNIV UTAH, 70- *Mem:* Am Math Soc. *Res:* Algebraic topology; critical points of polynomials. *Mailing Add:* 3408 Oakwood St Salt Lake City UT 84109

CHAMBERLIN, WILLIAM BRICKER, III, b Cleveland, Ohio, Aug 8, 43; m 75; c 2. ORGANIC CHEMISTRY. *Educ:* Miami Univ, Oxford, Ohio, BA, 66, MS, 70. *Prof Exp:* Teacher sci, Roosevelt Jr High, Ohio, 68-69; asst instr chem, Miami Univ, Oxford, Ohio, 69-70; chemist, 70-72, PROJ MGR CHEM, LUBRIZOL CORP, 72- *Mem:* Am Chem Soc; Soc Automotive Engrs. *Res:* Development of additives to improve engine lubrication. *Mailing Add:* Lubrizol Corp 29400 Lakeland Blvd Wickliffe OH 44092

CHAMBERS, ALFRED HAYES, b Reading, Pa, Nov 15, 14; m 45. PHYSIOLOGY. *Educ:* Swarthmore Col, AB, 36; Univ Pa, PhD(physiol), 42. *Prof Exp:* Asst instr physiol, Univ Pa, 37-42, instr, 42-45, assoc, 45-47, asst prof, 47-48; from asst prof to assoc prof, 48-70, PROF PHYSIOL & BIOPHYSICS, UNIV VT, 70- *Concurrent Pos:* NIH spec fel, 61, 62. *Mem:* Am Physiol Soc. *Res:* Metabolism; respiration; hearing. *Mailing Add:* Dept Physiol & Biophys Col Med Univ Vt PO Box 121 Jerico VT 05465

CHAMBERS, ANN FRANKLIN, b Evanston, Ill, Apr 4, 48; m 77; c 1. EXPERIMENTAL ONCOLOGY, TUMOR METASTASIS. *Educ:* Duke Univ, BA, 73, PhD(zool), 78. *Prof Exp:* Asst prof, 83-88, ASSOC PROF ONCOL DEPT, UNIV WESTERN ONT, 88-; CAREER SCIENTIST, LONDON REGIONAL CANCER CTR, 83- *Concurrent Pos:* Div head oncol dept, Div Exp Oncol, Univ Western Ont, 89- *Mem:* Am Asn Cancer Res; AAAS; Am Soc Cell Biol; Am Soc Microbiol; Can Soc Cell Biol; Metastasis Res Soc. *Res:* Molecular and cellular mechanisms of tumor progression and metastasis; ras oncogene and metastasis; ras-sensitive and ras-resistant cell lines; genes whose expression is altered by ras; in vivo models for metastasis. *Mailing Add:* London Regional Cancer Ctr 790 Commissioners Rd E London ON N6A 4L6 Can

CHAMBERS, BARBARA MAE FROMM, b Syracuse, NY, Nov 23, 40; m 62; c 4. MATHEMATICS. *Educ:* Univ Ala, BS, 62, MA, 64, PhD(math), 69. *Prof Exp:* Asst math, Univ Ala, 62-69; consult, Washington, DC, 69-70; lectr, Univ Va, Fairfax, 70-71; asst prof math, George Mason Univ, 71-79; ASSOC PROF MATH, NORTHERN VA COMMUNITY COL, 79- *Concurrent Pos:* NASA fel, 63; adv, Northern Va Sci Fair, 72-81. *Mem:* Math Asn Am. *Res:* Analytic function theory and applications. *Mailing Add:* Northern Va Community Col 15200 Neabsco Mills Rd Woodbridge VA 22191

CHAMBERS, CHARLES MACKAY, b Hampton, Va, June 22, 41; m 62; c 4. ACADEMIC ADMINISTRATION, MATHEMATICAL SCIENCES. *Educ:* Univ Ala, BS, 62, MS, 63, PhD(physics), 64; George Washington Univ, JD, 76. *Prof Exp:* Aerospace engr, Marshall Space Flight Ctr, NASA, 62-63; res assoc physics, Univ Ala, 63-64; NSF res fel, Harvard Univ, 64-65; from asst prof to assoc prof math, Univ Ala, 65-69; charter officer & dir, Univ Assocs, Inc, 69-72; assoc dean, George Washington Univ, 72-77; staff assoc & legal adv, Coun Postsec Accreditation, Washington, DC, 77-79, actg pres, 80-81, gen coun, 81-83; PROF MATH, GEORGE WASHINGTON UNIV, 82-; EXEC DIR, AM INST BIOL SCI, 83- *Concurrent Pos:* Dir, NSF and NASA grants Ala math talent search; consult, US Air Force, Salk Inst, US Cong, US Off Educ, NSF, Am Asn Higher Educ, Coun Grad Schs US, US Dept Agr, Amer Coun Educ, Nat League Nursing, Nat Inst Drug Abuse, Law Enforcement Assistance Admin, Fund Improv Postsec Educ & Ctr Mediation Higher Educ; trustee, Southwestern Univ, Washington, DC, 83-87 & BIOSIS, Philadelphia, Pa, 91-; pres, Am Found Biol Sci, 87- *Mem:* AAAS; Am Math Soc; Am Asn Physics Teachers; Sigma Xi; Am Asn Univ Adminrs (pres, 83-84); Am Inst Biol Sci. *Res:* Development and administration of interdisciplinary programs; abstract harmonic analysis; group representation theory; quantitative science in policy studies. *Mailing Add:* 4220 Dandridge Terr Alexandria VA 22309-2807

CHAMBERS, DERRELL LYNN, b Los Angeles, Calif, Feb 3, 34; m 54; c 3. ENTOMOLOGY. *Educ:* Whittier Col, BA, 55; Ohio State Univ, MS, 57; Ore State Univ, PhD, 65. *Prof Exp:* Biol aide entom res div, Whittier Lab, Agr Res Serv, USDA, 55; forestry aide, US Forest Serv, Ohio, 56-57; asst zool & entom, Ohio State Univ, 56-57; entomologist, Entom Res Div, Mexican Fruit Flies Invest Lab, Agr Res Ser, USDA, Mexico City, 57-59, entomologist, Pioneering Res Lab Insect Physiol, Md, 59-61; asst, Sci Res Inst, Ore State Univ, 61-64; sr entomologist, Entom Res Div, Arid Areas Citrus Insects Invest Lab, Agr Res Serv, Calif, 65-67, invests leader, Mex Fruit Flies Invest Lab, 68, Hawaiian Fruit Flies Invest Lab, Honolulu, 69-72, LAB DIR INSECT ATTRACTANTS & BIOL RES LAB, AGR RES SERV, USDA, 72-; PROF ENTOM & NEMATOL, GRAD SCH, ADJ MEM, DEPT ENTOM, UNIV FLA, 88- *Concurrent Pos:* Consult, various govt agencies and progs. *Mem:* Am Inst Biol Sci; Entom Soc Am. *Res:* Insect physiology and behavior. *Mailing Add:* Box 1209 Gainesville FL 32602

CHAMBERS, DONALD A, b New York, NY, Sept 9, 36; div; c 2. BIOCHEMISTRY REGULATORY MECHANISMS. *Educ:* Columbia Univ, PhD(biochem), 72. *Prof Exp:* Asst prof biochem & molecular biol, Univ Calif, San Francisco, 73-74; from asst prof to assoc prof biochem & dermat, Univ Mich, 79; interim head, 85-86, PROF MOLECULAR BIOL & DIR CTR, RES IN PERIODONT DIS ORAL MOLECULAR BIOL, UNIV ILL, 79-, PROF BIOCHEM & DERMAT, ILL, 80-, HEAD DEPT BIOCHEM, COL MED, 86- *Mem:* NY Acad Sci; Am Soc Cell Biol; Am Fed Clin Res; Royal Soc Med; Am Soc Microbiol; Am Soc Biochem & Molecular Biol; Soc Int Dermat; Int Asn Dental Res. *Res:* Biochemistry and molecular biology of disease; bio regulation, cutaneous pathobiology, neuroimmunology, inflammatory and periodontal diseases, host-microbial interactions. *Mailing Add:* Dept Biochem Univ Ill Col of Med 1853 W Polk St Chicago IL 60612

CHAMBERS, DOYLE, b Line, Ark, Mar 24, 18; m 43; c 3. ANIMAL GENETICS, ANIMAL BREEDING. *Educ:* La State Univ, BS, 40, MS, 47; Oklahoma State Univ, PhD(animal breeding), 50. *Prof Exp:* Asst animal sci, La State Univ, 40-42, instr, 45-47; from assoc prof to prof, Okla State Univ, 50-62, assoc dir, Agr Exp Sta, 62-64; dir agr exp sta, La State Univ, Baton Rouge, 64-85, prof animal sci, 77-85; RETIRED. *Concurrent Pos:* Don M Tyler distinguished prof, Okla State Univ, 61. *Mem:* Fel AAAS; Am Soc Animal Sci. *Res:* Inheritance of quantitative traits of economic importance in beef cattle and swine, including growth and carcass traits, maternal traits and efficiency of feed use; dwarfism and cancer eye studies. *Mailing Add:* 2365 Fairway Dr Baton Rouge LA 70809

CHAMBERS, EDWARD LUCAS, b Manhattan, NY, Jan 12, 17; m 54; c 2. CELL PHYSIOLOGY. *Educ:* Princeton Univ, BA, 38; NY Univ, MD, 43. *Prof Exp:* Intern, 2nd Div, Bellevue Hosp, 43-44, asst resident, 4th Div, 44; asst prof anat, Sch Med, Johns Hopkins Univ, 50-52; assoc prof, Sch Med, Univ Ore, 53-54; assoc prof physiol, 54-63, chmn grad prog cellular & molecular biol, 60-71, prof physiol & biochem, 63-73, PROF PHYSIOL & BIOPHYS, SCH MED, UNIV MIAMI, 73- *Concurrent Pos:* Mem corp, Marine Biol Lab, Woods Hole, Mass. *Mem:* Soc Gen Physiol; Am Physiol Soc; Am Soc Cell Biol; Int Soc Develop Biologists. *Res:* Ion exchanges between cell and environment; cell activation and fertilization; cell metabolism; micromanipulation; electrophysiology. *Mailing Add:* Dept Physiol & Biophys Sch Med Univ Miami PO Box 016430 Miami FL 33101

CHAMBERS, FRANK WARMAN, b Washington, DC, Nov 10, 48; m 75; c 3. PLASMA PHYSICS, COMPUTATIONAL PHYSICS. *Educ:* St Joseph's Col, BS, 70; Mass Inst Technol, PhD(physics), 75. *Prof Exp:* Res assoc plasma physics, Mass Inst Technol, 76; PHYSICIST PLASMA PHYSICS, LAWRENCE LIVERMORE LAB, UNIV CALIF, 76- *Mem:* Am Phys Soc. *Res:* Theoretical and experimental research on the interaction of charged particle beams with gases; computer simulation of plasma phenomena; linear stability analysis. *Mailing Add:* L-626 PO Box 808 Livermore CA 94550

CHAMBERS, HOWARD WAYNE, b Buda, Tex, Dec 27, 39. TOXICOLOGY. *Educ:* Tex A&M Univ, BS, 61, MS, 63; Univ Calif, Berkeley, PhD(entom), 66. *Prof Exp:* Res entomologist, Univ Calif, 66-68; from asst prof to assoc prof, 68-79, PROF ENTOM, MISS STATE UNIV, 79- *Mem:* Am Chem Soc; Entom Soc Am. *Res:* Insecticide chemistry; mechanism of action and metabolism of insecticides; insecticide synergists; resistance to insecticides. *Mailing Add:* PO Drawer EM Miss State Univ Mississippi State MS 39762

CHAMBERS, JAMES Q, b Kansas City, Mo, Jan 14, 38; m 64; c 3. ELECTROCHEMISTRY. *Educ:* Princeton Univ, AB, 59; Univ Kans, PhD(chem), 64. *Prof Exp:* Asst prof chem, Univ Colo, 64-69; from asst prof to assoc prof, 69-79, PROF CHEM, UNIV TENN, 79- *Mem:* Am Chem Soc; Electrochem Soc; Mat Res Soc; Soc Electroanal Chem. *Res:* Mechanisms and kinetics of electrode reactions; electrosynthetic methods; electron paramagnetic resonance; ultraviolet and visible spectroscopy of radical ions; biosensors. *Mailing Add:* Dept Chem Univ Tenn Knoxville TN 37916

CHAMBERS, JAMES RICHARD, b Birmingham, Ala, Aug 20, 14; m 39; c 1. ORGANIC CHEMISTRY. *Educ:* Columbia Union Col, BA, 39; Western Reserve Univ, MS, 49; Tex A&M Univ, PhD(org chem), 58. *Prof Exp:* Asst chem, Columbia Union Col, 39-41; prin, Nashville Jr Acad, 41-42; chief chemist, Pennzoil Co, 42-46; head chem dept, Atlantic Union Col, 46-54; instr chem, Southwestern Jr Col, 54-56; asst, Agr & Mech Col Tex, 56-58; head sci dept, Southwestern Jr Col, 58-60; from assoc prof to prof, 60-80, EMER PROF CHEM, WALLA WALLA COL, 80- *Mem:* Am Chem Soc. *Res:* Organic synthesis; organophosphorus chemistry; biochemistry. *Mailing Add:* 22 Tremont College Place WA 99324

CHAMBERS, JAMES ROBERT, b Woodstock, Ont, May 29, 42; m; c 2. QUANTITATIVE GENETICS, POULTRY GENETICS. *Educ:* Univ Guelph, BScAgr, 65, MSc, 67; Univ Wis, PhD(animal genetics), 71. *Prof Exp:* Poultry geneticist, Peel's Poultry Farm, Ltd, 71-74; geneticist, BioBreeding Labs, 74-78; RES SCIENTIST, ANIMAL RES CTR, 78- *Concurrent Pos:* Assoc ed, Poultry Sci Asn, 88-91 & 91-93. *Mem:* Poultry Sci Asn; World Poultry Sci Asn; Am Genetics Asn; Biometrics Soc. *Res:* Genetic selection procedures for joint improvement of growth, feed efficiency and leanness of meat-type chickens; author of one chapter on poultry genetics. *Mailing Add:* Animal Res Ctr Ottawa ON K1A 0C6 Can

CHAMBERS, JAMES VERNON, b Pekin, Ill, Mar 12, 35; c 1. FOOD SCIENCE. *Educ:* Ohio State Univ, BSc, 61, MSc, 66, PhD(food sci), 72. *Prof Exp:* Head microbiologist div qual control, Ross Labs Div, Abbott Labs, 61-68; lab dir, Div Food, Dairies & Drugs, Ohio Dept Agr, 69-71; asst prof food sci, Univ Wis, River Falls, 72-74; exten food scientist, 74-88, PROF FOOD SCI, PURDUE UNIV, 88- *Concurrent Pos:* Consult microbiol food waste treatment & mgt pract, Seiberlings Assoc, 83-, Nal Starch & Chem Corp, 88-, Crystal Creamery & Butter Co, Sacramento, Calif, 88- *Honors & Awards:* Dairy Res Found Award, Inst Food Technologists, Ind. *Mem:* Inst Food Technologists; Am Soc Microbiol: Water Pollution Control Fedn; Am Dairy Sci Asn; Int Asn Milk Food & Environ Sanitarians Inc. *Res:* By-product utilization from cheese manufacturing practices; solvent extraction of lactose from nonfat dry milk; advanced food manufacturing technologies. *Mailing Add:* Smith Hall Rm 101-D Purdue Univ West Lafayette IN 47907

CHAMBERS, JANICE E, b Oakland, CA, Sept 2, 47; m 69; c 2. BIOCHEMICAL TOXICOLOGY, ENVIRONMENTAL TOXICOLOGY. *Educ:* Miss State Univ, PhD(animal physiol), 73; Am Bd Toxicol, dipl, 90. *Prof Exp:* Asst prof biol sci, 80-85, ASSOC PROF BIOL SCI, MISS STATE UNIV, 85- *Concurrent Pos:* Mem, Toxicol Study Sect, NIH; prin investr, res grants. *Mem:* Soc Toxicol; Am Chem Soc; Am Physiol Soc; Soc Neurosci. *Res:* Neurotoxicology of insecticides including neurochemical and behavior studies and insecticide metabolism. *Mailing Add:* Miss State Univ PO Drawer GY Mississippi State MS 39762-5759

CHAMBERS, JOHN MCKINLEY, b Toronto, Ont, Apr 28, 41; m 71. STATISTICS, COMPUTER SCIENCE. *Educ:* Univ Toronto, BSc, 63; Harvard Univ, AM, 65, PhD(statist), 66. *Prof Exp:* Supvr statist, 66-81, head, advan software, 81-83, HEAD, STATIST & DATA ANAL RES, BELL LABS, 83- *Concurrent Pos:* Vis lectr math, Imp Col, Univ London, 66-67; vis lectr statist, Harvard Univ, 69 & Princeton Univ, 71; assoc ed, J Am Statist Asn, 71-74. *Mem:* Fel Am Statist Asn; fel Royal Statist Soc; Asn Comput Mach; Brit Comput Soc; Int Statist Inst. *Res:* Analytical and statistical computing; graphics; software systems; expert software. *Mailing Add:* AT&T Bell Labs Rm 2C282 600 Mountain Ave Murray Hill NJ 07974-2070

CHAMBERS, JOHN WILLIAM, b Richmond, Ky, Nov 7, 29; m 65; c 2. ENDOCRINOLOGY, PHARMACOLOGY. *Educ:* Eastern Ky State Col, BS, 58; Vanderbilt Univ, PhD(pharmacol), 65. *Prof Exp:* From res asst to asst prof pharmacol, Vanderbilt Univ, 64-70; asst prof, Med Col Va, 70-76; assoc prof, 76-78, PROF PHARMACOL & CHMN DEPT, WVA SCH OSTEOPATH MED, 78- *Concurrent Pos:* NIH-USPHS fel, 65-66. *Mem:* AAAS; Am Soc Pharmacol & Exp Therapeut; assoc Am Chem Soc; NY Acad Sci; Sigma Xi. *Res:* Effects of hormones and drugs on amino acid transport and metabolism, on membrane function and on enzyme activity. *Mailing Add:* Dept Pharmacol WVa Sch of Osteopath Med Lewisburg WV 24901

CHAMBERS, KATHLEEN CAMILLE, b Polson, Mont, July 14, 47; m. PSYCHOLOGY. *Educ:* Portland State Univ, BS, 71; Univ Wash, PhD(psychol), 75. *Prof Exp:* From asst prof to assoc prof, Dept Psychol, Portland State Univ, 75-79; asst scientist reproductive biol & behav, Ore Regional Primate Res Ctr, 79-83, assoc scientist, 83-87; ASSOC PROF, DEPT PSYCHOL, UNIV SOUTHERN CALIF, 87- *Concurrent Pos:* Collab scientist, Ore Regional Primate Res Ctr, 76-79 & 87- *Mem:* Animal Behav Soc; Int Primatological Soc; Am Soc Primatologists; Sigma Xi; fel AAAS; Am Psychol Asn; Soc Neurosci. *Res:* Neural and endocrinological determinants of the sex difference in the acquisition and extinction of conditioned food aversions rodent; primate and rodent sexual behavior; neural and endocrinological correlates of sexual behavior in aging primates and rodents. *Mailing Add:* Dept Psychol Seeley Mudd Bldg Univ Southern Calif Los Angeles CA 90089-1061

CHAMBERS, KENTON LEE, b Los Angeles, Calif, Sept 27, 29; m 58; c 2. SYSTEMATICS, EVOLUTION. *Educ:* Whittier Col, AB, 50; Stanford Univ, PhD(biol), 56. *Prof Exp:* Actg instr biol sci, Stanford Univ, 54-55; from instr to asst prof bot, Yale Univ, 56-60; from assoc prof to prof bot, 60-90, cur Herbarium, 60-90, EMER PROF, ORE STATE UNIV, 91- *Concurrent Pos:* Prog dir syst biol, NSF, 67-68. *Honors & Awards:* Merit Award, Bot Soc Am, 90. *Mem:* AAAS; Am Soc Plant Taxonomists (pres, 79); Bot Soc Am; Soc Systematic Zool. *Res:* Taxonomy and biosystematics of angiosperms, especially Compositae; flora of Oregon. *Mailing Add:* Dept Bot Ore State Univ Corvallis OR 97331

CHAMBERS, LARRY WILLIAM, b Hamilton, Ont, Oct 6, 46; m 74; c 1. EPIDEMIOLOGY, HEALTH PROGRAMS EVALUATION. *Educ:* McMaster Univ, BA Hons, 70, MSc, 73; Mem Univ Nfld, PhD(community med), 78. *Prof Exp:* Res asst, Dept Clin Epidemiol & Biostatist, McMaster Univ, 70-71; lectr, Div Community Med, Mem Univ Nfld, 73-76, asst prof, 76-78; asst prof, 78-81, ASSOC PROF HEALTH CARE EVAL & EPIDEMIOL, DEPT CLIN EPIDEMIOL & BIOSTATIST, MCMASTER UNIV, 81- *Concurrent Pos:* Prin investr, Family Pract Nurses Nfld, Nat Health Res & Develop Prog, 73-78, Indexes Health-Prof & Lay Perspectives, 79-81; co-investr, Home Centered Videotaped Coun Rural Parents Hearing Impaired Children, 77-79 & Trial Home Care Chronically Ill Elderly, 79-82; consult, Can Health Surv, 75-, Nfld Cancer Treatment & Res Found, 76-78, Metrop Ment Health Planning Bd Halifax, NS, 77-, Sch Pub Health, Univ NC & Div Family Med, Duke Univ, 78- & Expert Group Qual Care Assessment, Can Physiotherapy Asn & Nat Health & Welfare, 78-; prin investr, Health Care Eval Sem, St John's, Nat Health Res & Develop Prog, 77- & Workshops Rev Health Progs, 79-81; vis prof, Univ Sierra Leone, 80. *Mem:* Can Pub Health Asn; Soc Epidemiol Res; Am Pub Health Asn; Int Epidemiol Asn; Can Asn Teachers Social & Prev Med. *Res:* Development of measures of health status; quantification of quality in health care assessment; cancer epidemiology; research methodology in the health sciences. *Mailing Add:* Dept Biostatist Fac Health Sci McMaster Univ 1200 Main St Hamilton ON L8N 3Z5 Can

CHAMBERS, LEE MASON, b St Marys, WVa, Mar 17, 36; m 60; c 2. ANALYTICAL CHEMISTRY, ELECTROCHEMISTRY. *Educ:* Marshall Univ, BS, 58; Univ Ill, MS, 60, PhD(anal chem), 63. *Prof Exp:* Staff chemist, Ivorydale Tech Ctr, Proctor & Gamble Co, 62-66, group leader res & develop, 66-70, mem staff, Packaging Develop, 76-79, staff analytical chemist & mgr, Paper Testing Lab, 79-86, CHEM TEST DEVELOP, WINSTON HILL TECH CTR, PROCTOR & GAMBLE CO, 87- *Mem:* Am Chem Soc; Sigma Xi. *Res:* Development of chemical and instrumental methods for analysis of consumer products and raw materials including paper products. *Mailing Add:* Proctor & Gamble Co 6100 Center Hill Rd Cincinnati OH 45224

CHAMBERS, RALPH ARNOLD, b Harlan, Ky, Sept 5, 33; m 53; c 2. POLYMER CHEMISTRY. *Educ:* Presby Col, SC, BS, 59; Vanderbilt Univ, PhD(chem), 63. *Prof Exp:* From chemist to sr chemist, Tenn Eastman Co, 62-67, sr res chemist, 67-71, res assoc, 71-75, supt Develop & Control, 75-90, MFG SUPT, TENN EASTMAN CO, 90- *Mem:* Am Chem Soc; Sigma Xi; Tech Asn Pulp & Paper Indust. *Res:* Chemistry of cellulose and its derivatives; polyester chemistry. *Mailing Add:* 2001 Canterbury Rd Kingsport TN 37660

CHAMBERS, RICHARD, b London, Eng, Apr 22, 23; m 56. NEUROLOGY. *Educ:* Oxford Univ, BA, 44, BM & BCh, 47, MA, 48; FRCP(C), 59; FRCP(London), 76. *Prof Exp:* Instr, Univ Toronto, 57-60; from assoc prof to prof, NJ Col Med & Dent, 60-69; PROF NEUROL, JEFFERSON MED COL, 66- *Concurrent Pos:* Fel neurol, Harvard Med Sch, 51-53 & 56; scholar, Royal Col Physicians, Eng, 55-56. *Mem:* Am Acad Neurol; Am Asn Neuropath. *Res:* Cortical physiology; fructose metabolism; virus encephalitis; peripheral neuropathy. *Mailing Add:* Hopkinson Hoiuse 3016 Washington Sq Philadelphia PA 19106

CHAMBERS, RICHARD LEE, b Algona, Iowa, Feb 26, 47; m 78; c 3. GEOPHYSICS, SEISMIC STRATIGRAPHY. *Educ:* Univ Mont, BA, 70, MS, 71; Mich State Univ, PhD(geol), 75. *Prof Exp:* Res scientist chem sedimentology, Great Lakes Environ Res Lab, Dept of Com, 73-80; sr res geologist, Phillips Petrol Co, 80-88; SR RES SCIENTIST, RESERVOIR & RISK ANAL, AMOCO PROD CO RES CTR, TULSA, OKLA, 88- *Mem:* Sigma Xi. *Res:* Probability; Geostatistics, reservoir modelling, 3-D seismics, high resolution p-wave and shear wave seismic data acquisition; seismic interpretation; risk analysis and economic evaluations. *Mailing Add:* 1615 S Nyssa Ave Broken Arrow OK 74012

CHAMBERS, ROBERT J, b Atlanta, Ga, Sept 23, 30; m 74; c 2. ASTRONOMY, INSTRUMENTATION. *Educ:* Univ Wash, BSc, 52; Univ Calif, Berkeley, PhD(astron), 64. *Prof Exp:* From instr to assoc prof, 67-73, PROF ASTRON, POMONA COL, 73-, DIR, BRACKETT OBSERV, 64-, CHMN, PHYSICS DEPT, 88- *Concurrent Pos:* NSF sci fac fel, Univ Calif, Berkeley, 69-70. *Mem:* AAAS; Am Astron Soc; assoc Am Soc Mech Engrs; Optical Soc Am-SPIE. *Res:* Galactic star clusters; astronomical instrumentation. *Mailing Add:* Physics Dept Pomona Col Claremont CA 91711

CHAMBERS, ROBERT ROOD, b Lincoln, Nebr, May 23, 23; m 65; c 3. ORGANIC CHEMISTRY. *Educ:* Univ Nebr, AB, 44; Univ Ill, PhD(org chem), 47; DePaul Univ, JD, 51. *Prof Exp:* Res chemist, Sinclair Res Inc, 47-50, group leader, 50-52, div dir, 52-59, tech mgr, 59-66, vpres, 67-68, pres, 68, vpres res, Sinclair Oil Corp, 68-69, pres, Nuclear Mat & Equip Co, 70-71, Arco Nuclear Co, 71-75 & Arco Med Prod Co, 75-77, pres, Arco Solar Inc, 77-80, pres, Arco Environ & Ardev, 78-80, vpres, Atlantic Richfield Co, 69-82; PRES, SANDHILL SCIENTIFIC INC, 82- *Mem:* Am Chem Soc; Brit Chem Soc. *Res:* Petroleum; petrochemicals. *Mailing Add:* 11439 Laurelcrest Dr Studio City CA 91604-3872

CHAMBERS, ROBERT WARNER, b Oakland, Calif, Oct 27, 24; m 49; c 3. BIOCHEMISTRY, MOLECULAR BIOLOGY. *Educ:* Univ Calif, AB, 49, PhD(biochem), 54. *Prof Exp:* Asst biochem, Univ Calif, 51-54; Res Coun fel, Life Ins Med Res Fund, 54-56; from instr to prof biochem, Sch Med, NY Univ, 56-81; CARNEGIE & ROCKEFELLER PROF BIOCHEM & HEAD DEPT, DALHOUSIE UNIV, HALIFAX, NS, CAN, 81 - *Concurrent Pos:* NSF grant, 58-62; Nat Inst Gen Med Sci grant, 60-78; career scientist, Health Res Coun City New York, 62-72, 78-81; mem subcomt purines & pyrimidines, comt biochem, Nat Acad Sci-Nat Res Coun, 62, chmn, 64; NY Health Res Coun grant, 64-70; mem comt res etiology of cancer, Am Cancer Soc, 66-69; mem postdoctoral fel comt, NSF, 67; exchange scientist, Nat Acad Sci-Nat Res Coun & Polish Acad Sci, 68; grants, Am Cancer Soc, 70-72 & 78-81, Nat Cancer Inst Can, 74-81atives Award, 81-86 & Med Res Coun Can, 86- *Mem:* AAAS; Am Chem Soc; Royal Soc Chem; Am Soc Biol Chem; fel NY Acad Sci; fel Can Inst Chem; Can Biochem Soc; Harvey Soc. *Res:* Photochemistry of nucleic acids; structure-action relationships in transfer RNA; synthesis of oligonucleotides; mutagenesis and carcinogenesis; molecular mechanisms of mutation by carcinogens; precise nature of the carcinogen-induced lesions in DNA that produce mutations and the relationship of these mutations to tumor production. *Mailing Add:* Dept Biochem Dalhousie Univ Sir Charles Tupper Bldg Halifax NS B3H 4H7 Can

CHAMBERS, VAUGHAN CRANDALL, (JR), b Philadelphia, Pa, June 14, 25; m 48; c 4. ORGANIC CHEMISTRY. *Educ:* Swarthmore Col, AB, 47; Mass Inst Technol, PhD(org chem), 50. *Prof Exp:* Chemist, Org Photog Mat, 50-59, sr chemist, 59-60, res assoc, 60-61, res supvr, 61-64, RES MGR PHOTO PROD DEPT, E I DU PONT DE NEMOURS & CO, INC, 64- *Mem:* Am Chem Soc; Soc Photog Sci & Eng; Sigma Xi. *Res:* Physical organic study of small carbocyclic compounds; photography, sensitizing dyes; polymers for use in photographic products. *Mailing Add:* 606 Greenfield Pl Wilmington DE 19809

CHAMBERS, WILBERT FRANKLIN, b Cameron, WVa, Feb 26, 23; m 57; c 3. NEUROANATOMY. *Educ:* WVa Univ, BS, 46, MS, 47; Univ Wis, PhD(anat), 52. *Prof Exp:* Instr anat, Univ Pittsburgh, 49-52; fel, Vanderbilt Univ, 52-54, instr, 54-55; from instr to assoc prof, Univ Vt, 55-67; NIH Fel, Brain Res Inst, UCLA, 61-62; assoc prof, 67-71, PROF ANAT, DARTMOUTH MED SCH, 71- *Mem:* Am Asn Anat; Am Soc Zool. *Res:* Morphology of primate brain; nervous system function in altered endocrine states; the subcommissural organ and water metabolism; effect of amphethamines on the reticular activation response; gonadatrophic centers of the hypothalamus; the thyroid release factors region of the hypothalamus; Rathke's pouch transplants to hypothalamus; growth of small cell carcinoma tissue in brain of the nude mouse. *Mailing Add:* Dept of Anat-Cytol Dartmouth Med Sch Hanover NH 03755

CHAMBERS, WILLIAM EDWARD, b Ravenswood, WVa, Aug 14, 33; m 55; c 3. ANALYTICAL CHEMISTRY. *Educ:* Marshall Univ, BS, 55; Univ Ill, MS, 57, PhD(anal chem), 60. *Prof Exp:* Anal chemist, Parma Res Center, Union Carbide Corp, 59-63, supvr anal div, 63-68, asst dir advan technol projs, Carbon Prod Div, 68-73, dir carbon fiber develop, 73-75, gen mgr res & develop, Parma Tech Ctr, 75-76, VPRES TECHNOL, CARBON PROD DIV, UNION CARBIDE CORP, 76- *Mem:* Soc Appl Spectros; Am Chem Soc; Sigma Xi. *Res:* Flame spectroscopy; nuclear magnetic resonance. *Mailing Add:* 17 Heritage Dr Danbury CT 06810

CHAMBERS, WILLIAM HYLAND, b St Louis, Mo, July 30, 22; m 45; c 4. NUCLEAR SCIENCE. *Educ:* Cornell Univ, BA, 43, MS, 48; Ohio State Univ, PhD(physics), 50. *Prof Exp:* Asst, Cornell Univ, 46-48; asst, Ohio State Univ, 48-49; mem staff & group leader, Los Alamos Nat Lab, 50-77, from asst div leader to assoc div leader, 77-79, dep assoc dir, 79-82; EXEC VPRES, BELL-CHAMBERS ASSOC, INC, 82- *Concurrent Pos:* With AEC Combined Opers Planning Group, Oak Ridge, 67. *Honors & Awards:* Distinguished Assoc Award, US Dept Energy, 82. *Mem:* Fel AAAS; Inst Nuclear Mat Mgt; Am Phys Soc; Am Nuclear Soc. *Res:* Nuclear weapon development; detection of nuclear detonations in space; solar flare x-rays; soft x-ray spectrometry; nuclear safeguards and arms control; nuclear material detection and identification. *Mailing Add:* 336 Andanada Los Alamos NM 87544

CHAMBLEE, DOUGLAS SCALES, b Zebulon, NC, Jan 4, 21; m 49; c 3. AGRONOMY. *Educ:* NC State Univ, BS, 44, MS, 47; Iowa State Univ, PhD(agron), 49. *Prof Exp:* Res instr agron, 43-47, from asst prof to assoc prof, 48-60, PROF AGRON, NC STATE UNIV, 60- *Concurrent Pos:* Mem, NC State Mission, Peru, 61-64; agent div forage crops, USDA. *Mem:* Fel Am Soc Agron. *Res:* Moisture requirements of alfalfa; grass mixtures; fertility and management of permanent pastures; sod-seeding legumes in grass pastures; legume inoculation. *Mailing Add:* 1105 Williams Hall NC State Univ Raleigh NC 27650

CHAMBLISS, CARLSON ROLLIN, b Boston, Mass, July 17, 41. ECLIPSING BINARIES, STELLAR PHOTOMETRY. *Educ:* Harvard Univ, AB, 62; Univ Pa, MS, 64, PhD(astron), 68. *Prof Exp:* Asst prof astron, Georgetown Univ, Washington, DC, 68-70; from asst prof to assoc prof, 70-79, PROF ASTRON, KUTZTOWN UNIV, PA, 79- *Concurrent Pos:* Vis astronomer, Cerro Tolo Inter-Am Observ, 69, Kitt Peak Nat Observ, 75-82 & US Naval Observ, 82-90. *Mem:* Am Astron Soc; Int Astron Union. *Res:* The observation and analysis of the light curves of eclipsing binaries and other types of variable stars. *Mailing Add:* Dept Phys Sci Kutztown Univ Kutztown PA 19530

CHAMBLISS, GLENN HILTON, b Jasper, Tex, Feb 14, 42; m 65; c 1. MICROBIOLOGY, MICROBIAL PHYSIOLOGY. *Educ:* Univ Tex, Austin, BA, 65; Miami Univ, MA, 67; Univ Chicago, PhD(microbiol), 72. *Prof Exp:* Fel, Jane Coffin Childs Mem Fund Med Res, 71-73; fel, Phillipe Found, 73-74; from asst prof to assoc prof, 74-85, PROF BACT, UNIV WIS-MADISON, 85-, PROF CELL & MOLECULAR BIOL, 88- *Concurrent Pos:* Fel microbiol, Inst Microbiol, Univ Paris-Sud, 72-74; dir, Biotechnol Training Prog, Univ Wis-Madison, 89- *Mem:* Am Soc Microbiol; Fedn Europ Biol Socs; Am Chem Soc; Am Acad Microbiol. *Res:* Elucidation of the molecular basis for catabolite repression of gene expression in the gram positive bacterium Bacillus subtilis; elucidation of mechanisms which regulate the synthesis of bacterial enzymes involved in the metabolism of complex carbohydrates such as starch or cellulose. *Mailing Add:* Dept Bact Univ Wis Madison WI 53706

CHAMBLISS, KEITH WAYNE, b Flora, Ill, Dec 16, 26; m 56; c 3. IMMUNOCHEMISTRY. *Educ:* Univ Ill, AB, 48; Univ Ind, AM, 50, PhD(biochem), 52; Am Bd Clin Chem, dipl, 61. *Prof Exp:* Res immunochemist serol, US Army Med Serv Grad Sch, 52-53; res biochemist agr chem, Univ Wyo, 55-56; clin chemist, Toledo Hosp, Ohio, 56-61; sect head immunol, Ames Res Lab Div, Miles Labs Inc, 62-73; dir immunochem res & develop, Am Dade Div, Am Hosp Supply Corp, 73-77, dir appl res, 78-81, sr res assoc oncol res & develop, 81-86; SPECIALIST TOXICOL LAB, SMITH KLINE & BEECHAM, 86- *Mem:* Am Chem Soc; Am Asn Clin Chem. *Res:* Monoclonal antibodies and antibody-antigen reactions applied to diagnostic tests. *Mailing Add:* 7335 Potts Rd Riverview FL 33569

CHAMBLISS, OYETTE LAVAUGHN, b Chapman, Ala, Nov 4, 36; m 61; c 2. VEGETABLE CROPS. *Educ:* Auburn Univ, BS, 58, MS, 62; Purdue Univ, PhD(plant breeding), 66. *Prof Exp:* Res horticulturist, Veg Breeding Lab, Crops Res Div, Agr Res Serv, USDA, 66-70; assoc prof veg crops, 70-78, PROF VEG CROPS, AUBURN UNIV, 78- *Honors & Awards:* Marion Meadows Award, Am Soc Hort Sci, 66, Asgrow Award, 72. *Mem:* Am Soc Hort Sci; Sigma Xi; AAAS; Am Inst Biol Sci. *Res:* Developing insect and disease resistant varieties and investigating nature of resistance in cucurbits, cowpea and tomatoes. *Mailing Add:* Dept Hort Auburn Univ Auburn AL 36849-5408

CHAMBRE, PAUL L, b Kassel, Ger, Aug 7, 18; US citizen; m; c 3. APPLIED MATHEMATICS, ENGINEERING SCIENCE. *Educ:* Univ Calif, Berkeley, BS, 41, PhD(eng), 51; NY Univ, MS, 47. *Prof Exp:* From asst prof to assoc prof math, 53-62, prof math & eng sci, 62-76, PROF MATH NUCLEAR ENG, UNIV CALIF, BERKELEY, 76- *Concurrent Pos:* Consult to US Off Naval Res, 53-; consult, Nat Acad Sci, 81- *Res:* Differential equations; classical analysis; rarified gas dynamics; chemical reactions in flowing systems; neutron transport theory; nuclear waste isolation. *Mailing Add:* Dept Math Univ Calif Berkeley CA 94720

CHAMEIDES, WILLIAM LLOYD, b New York, NY, Nov 21, 49; m 87; c 4. ATMOSPHERIC CHEMISTRY. *Educ:* State Univ NY Binghamton, BA, 70; Yale Univ, MPh, 73, PhD(atmospheric sci), 74. *Prof Exp:* Res investr, Space Physics Res Lab, Univ Mich, 74-75, asst res scientist, 75-76; asst prof physics & atmospheric sci, Univ Fla, 78-80; assoc prof, 80-85, PROF EARTH & ATMOSPHERIC SCI, GA INST TECHNOL, 85-, DIR, SCH EARTH & ATMOSPHERIC SCI, 90- *Concurrent Pos:* Ed, J Geophys Res-Atmospheres, 85-88. *Honors & Awards:* James B MacElwane Award, Am Geophys Union, 83. *Mem:* Am Geophys Union; AAAS. *Res:* Atmospheric chemistry: tropospheric gas-phase and aqueous-phase chemistry, air pollution, global chemical cycles, biospheric-atmospheric interactions and global change. *Mailing Add:* Sch Earth & Atmospheric Sci Ga Inst Technol Atlanta GA 30332

CHAMIS, ALICE YANOSKO, US citizen; m 66; c 3. INFORMATION MANAGEMENT, COMPETITIVE INTELLIGENCE SYSTEMS. *Educ:* McGill Univ, BSc, 59; Case Western Reserve Univ, MS, 62, PhD(info sci), 84. *Prof Exp:* Lit chemist, Alcan, 59-61; libr mgr, B F Goodrich Co, 62-69; asst dir, Cuyahoga County Pub Libr, 70-80; proj mgr, Case Western Reserve Univ, 85-86; asst prof, Col Bus Admin, Kent State Univ, 86-88; PRES, INFO MGT CONSULTS, 88- *Concurrent Pos:* Adj prof, Kent State Univ, 79-86; consult, Info Mgt Consults, 80-85, Database, 83-85; Small Bus Innovation Res Prog Grant, US Dept Health & Human Services, 85. *Mem:* Am Libr Asn; Am Soc Info Sci; Spec Libr Asn. *Res:* Computerized information systems design and implementation; private and commercial online systems; database design and searching; thesaurus and vocabulary control systems; staff training and development; text and image management systems. *Mailing Add:* Info Mgt Consults 24534 Framingham Dr Westlake OH 44145

CHAMIS, CHRISTOS CONSTANTINOS, b Sotira, Greece, May 16, 30; US citizen; m 66; c 3. ENGINEERING MECHANICS. *Educ:* Cleveland State Univ, BCE, 60; Case Western Reserve Univ, MSEM, 62, PhD(mech of solids), 67. *Prof Exp:* Designer indust struct & draftsman consult eng firms, Undergrad Coop Prog, Cleveland, Ohio, 55-60; asst stress & struct anal, Case Western Reserve Univ, 60-62, composite mech, 64-68; res mathematician, B F Goodrich Res Ctr, 62-64; SR AEROSPACE SCIENTIST, LEWIS RES CTR, NASA, 68- *Concurrent Pos:* Tech sem lectr, George Washington Univ, Ga Inst Technol, Univ Akron, Case Western Reserve Univ. *Honors & Awards:* NASA Medal for Except Eng Achievement. *Mem:* Am Soc Civil Engrs; Am Inst Aeronaut & Astronaut; Am Soc Mech Engrs; Soc Advan Mat & Process Eng; Am Soc for Testing & Mat; Soc Exp Mech. *Res:* Advanced structural analysis, composite mechanics, integrated computer programs for structural and stress analysis, structural dynamics and impact, computational structural mechanics. *Mailing Add:* 24534 Framingham Dr Westlake OH 44145

CHAMOT, DENNIS, b New York, NY, June 5, 43; m 74; c 2. TECHNOLOGY POLICY, HUMAN EFFECTS OF NEW TECHNOLOGIES. *Educ:* Polytech Inst Brooklyn, BS & MS, 64; Univ Ill, PhD(org chem), 69; Univ Pa, MBA, 74. *Prof Exp:* Res chemist, E I du Pont de Nemours & Co, Inc, 69-73; asst to exec secy, Coun Unions Prof Employees, Am Fedn Labor & Cong Indust Orgn, 74-77, asst dir, Dept Prof Employees, 77-84, assoc dir, 84-89, EXEC ASST TO PRES, DEPT PROF EMPLOYEES, AM FEDN LABOR & CONG INDUST ORGN, 90- *Concurrent Pos:* Counr, Am Chem Soc, 75-; mem, Labor Res Adv Coun, US Dept Labor; mem, Pub Understanding Sci Adv Comt, NSF, 77-79, chmn, Informal Sci Educ Oversight Comt, 85-86; chmn, Div Prof Rels, Am Chem Soc, 82; travel grant, Swed Inst, 84; mem, Comn Eng & Tech Systs, Nat Res Coun, 85-92, Comt Rev Info Systs Modernization of IRS, 90- & Comt Study Computer Technol & Serv Sector Productivity, 91-; Mary E Switzer mem scholar, Nat Rehab Asn, 89; panel mem, Govt Role Civilian Technol, Nat Acad Sci/Eng, 90- *Honors & Awards:* Plaque, Soc Occup & Environ Health, 82. *Mem:* Sigma Xi; Am Chem Soc; Soc Occup & Environ Health (secy-treas, 78-82). *Res:* Effects of new technologies in the workplace; competitiveness; globalization of technology and technical work; science literacy; science and technology policy; author of numerous publications. *Mailing Add:* Dept Prof Employees Am Fedn Labor & Cong Indust Orgn 815 16th St NW Washington DC 20006

CHAMPAGNE, CLAUDE P, b Sherbrooke, Que, Oct 17, 53; m 81; c 2. FOOD SCIENCE & TECHNOLOGY. *Educ:* Sherbrooke Univ, BSc, 76; Laval Univ, PhD(food sci), 83. *Prof Exp:* Teacher food microbiol, ITA St Hyacinthe, 78-86; RES SCIENTIST, AGR CAN, 87- *Concurrent Pos:* Mem, Exp Comt Food Biotech, Agr Can, 88- *Mem:* Can Inst Food Sci & Technol; Am Soc Microbiol. *Res:* Use of immobilized cell technology for biomass production or food fermentations. *Mailing Add:* Food Res & Develop Ctr 3600 Casavant Blvd W St-Hyacinthe PQ J2S 8E3 Can

CHAMPAGNE, PAUL ERNEST, b Woonsocket, RI, Nov 27, 46; m 68; c 2. COAL PROCESSING TECHNOLOGY, ENERGY PROCESS SYSTEMS DESIGN. *Educ:* Univ RI, BS, 69. *Prof Exp:* Assoc res engr, US Steel Corp Res Ctr, 69-73; res engr tech & econ eval, Ledgemont Lab, Kennecott Copper Corp, 73-74; process res eng, US Steel Corp Res Ctr, 74-75, plant process engr, Coal & Coke Opers, 75-77, plant asst supt coking, 77-78, div engr coal & coking opers, 78-79; mgr coke opers, Semet-Sorway Div, Allied Chem Corp, 79; mgr coal & energy technol, Sci Technol Ctr, Koppers Co, Inc, 80-86; PRES, PROCESS & ENERGY MGT CORP, CONSULT ENG, 86- *Mem:* Am Inst Chem Engrs; Am Inst Mech Engrs; Am Foundrymens Soc. *Res:* Liquid state direct reduction steelmaking; gas/solid fluidized bed direct reduction; fluidized bed design; coal characterization and carbonization; formcoke; e; coal gasification; alternate solid fuels; carbonization research; coal process technology. *Mailing Add:* 1345 Foxwood Dr Monroeville PA 15146

CHAMPE, PAMELA CHAMBERS, b San Francisco, Calif, Aug 29, 45; m 69; c 2. BIOCHEMISTRY, MEDICAL EDUCATION. *Educ:* Stanford Univ, BA, 67; Purdue Univ, MS, 69; Rutgers Univ, PhD(microbiol), 74. *Prof Exp:* Res asst microbiol, Univ Med & Dent NJ, Robert Wood Johnson Med Sch, 69-73, teaching specialist microbiol & biochem, 73-74, from instr to asst prof, 74-84, ASSOC PROF BIOCHEM, UNIV MED & DENT NJ, ROBERT WOOD JOHNSON MED SCH, 84- *Concurrent Pos:* Lectr comprehensive med rev prog, Kuwait, 81-83; mem rev comt, Nat Inst Gen Med Sci, 85-89, reviewers reserve, NIH, 89- *Mem:* AAAS; Sigma Xi; Nat Asn Med Minority Educr; Asn Am Med Col; NY Acad Sci. *Res:* Development of improved methods of medical education, especially pertaining to the presentation of basic science courses, with a focus on biochemistry. *Mailing Add:* Univ Med & Dent NJ Robert Wood Johnson Med Sch 675 Hoes Lane Piscataway NJ 08854

CHAMPE, SEWELL PRESTON, b Montgomery, WVa, Nov 24, 32; m 59, 69; c 2. MOLECULAR BIOLOGY, GENETICS. *Educ:* Mass Inst Technol, SB, 54; Purdue Univ, PhD(biophys), 59. *Prof Exp:* Am Cancer Soc fel, Purdue Univ, 59-61, from asst prof to assoc prof biol, 61-69; PROF MICROBIOL, RUTGERS UNIV, 69- *Concurrent Pos:* NIH res career award, 63- *Res:* Bacteriophage structure and genetics; genetic control of fungal development. *Mailing Add:* 17 Beech Lane Edison NJ 08817

CHAMPINE, GEORGE ALLEN, b Fairmont, Minn, May 16, 34; m 56; c 3. DISTRIBUTED SYSTEMS, COMPUTER GRAPHICS. *Educ:* Univ Minn, BS, 56, MS, 59, PhD(info systs), 75. *Prof Exp:* Dir, Sperry Univac, 58-79; vpres eng, Vydec, Inc, 79-81; SR SCIENTIST, DIGITAL EQUIPMENT CORP, 81- *Concurrent Pos:* Adj prof, Univ Minn, 72-79, Univ Tex, Austin, 84-86, Univ Lowell, 87- *Mem:* Inst Elec & Electronic Engrs Computer Soc; Asn Computer Mach. *Res:* Distributed computer systems. *Mailing Add:* 74 Robert Rd Stow MA 01775

CHAMPION, KENNETH STANLEY WARNER, b Sydney, Australia, Dec 7, 23; m 48; c 4. PHYSICS, ATMOSPHERIC SCIENCES. *Educ:* Univ Sydney, BSc, 44; Univ Birmingham, PhD, 51. *Prof Exp:* Asst lectr physics, Univ Queensland, 46-49; res fel, Australian Nat Univ, 49-52; hon res fel, Univ Birmingham, 51-52; res assoc, Mass Inst Tech, 52-54; asst prof, Tufts Univ, 54-59; sect chief, Space Physics Labs, 59-63, chief, Atmospheric Struct Br, Air Force Cambridge Res Labs, 63-83, CHIEF SCIENTIST, ATMOSPHERIC SCI DIV, AIR FORCE GEOPHYS LABS, 83- *Concurrent Pos:* Res assoc, Comput Ctr, Mass Inst Technol, 56-59; consult, Photochem Lab, Geophys Res Directorate, Air Force Cambridge Ctr, 56-59; vis prof, Univ Adelaide, 64; chmn working group 4, Comt Space Res, Int Coun Sci Unions, 74-79; chmn task group on Comt Int Reference Atmosphere, Comt Space Res, 79-82; chmn sub-comn C4, Comt Space Res, 82-84; chmn comn C, Comt Space Res, 84-86, exec mem, 86-90. *Honors & Awards:* Guenter Loeser lectr, 62. *Mem:* Am Geophys Union; Am Phys Soc; Am Meteorol Soc; fel Brit Phys Soc; Sigma Xi; NY Acad Sci. *Res:* Plasma physics; effects of electromagnetic and magnetic fields; measurement of cross sections for atomic processes; upper atmosphere physics; properties and processes of the atmosphere; model atmospheres. *Mailing Add:* Six Rolfe Rd Lexington MA 02173

CHAMPION, PAUL MORRIS, b Chicago, Ill, Dec 7, 46. PHYSICS, BIOCHEMISTRY. *Educ:* Ohio State Univ, BS, 68; Univ Ill, MS, 70 & PhD(physics), 75. *Prof Exp:* Res assoc solid state physics, Univ Cornell, 75-76 & chem, 76-79; exchange fel, Inst Phys Chem, 79-80; sr res assoc, Univ Cornell, 80-81; asst prof chem, Mass Inst Technol, 81-84; ASSOC PROF PHYSICS, NORTHEASTERN UNIV, 84- *Mem:* Am Phys Soc; Am Chem Soc; Biophys Soc. *Mailing Add:* Phys Dept 106 Dana Northeastern Univ 360 Huntington Ave Boston MA 02115-2299

CHAMPION, WILLIAM (CLARE), b Rockford, Ill, Mar 12, 30; m 63. ORGANIC CHEMISTRY. *Educ:* Univ Ill, BS, 52; Cornell Univ, PhD(org chem), 58. *Prof Exp:* Asst org chem, Cornell Univ, 54-57; fel, Iowa State Univ, 58-59; from asst prof to assoc prof, 59-73, PROF ORG CHEM, COLO COL, 73- *Mem:* Am Chem Soc; Am Crystallog Asn. *Res:* Synthetic organic chemistry; x-ray crystallography. *Mailing Add:* Dept of Chem Colo Col Colorado Springs CO 80903-3294

CHAMPLIN, ARTHUR KINGSLEY, b Portland, Maine, Nov 30, 38; m 66; c 2. REPRODUCTIVE BIOLOGY, GENETICS. *Educ:* Williams Col, BA, 61, MA, 63; Univ Rochester, PhD(biol), 69. *Prof Exp:* NIH trainee biol, Univ Rochester, 65-69; fel biol & reproduction physiol, Jackson Lab, 69-71; asst prof, 71-80, chmn, div natural sci, 80-83, assoc prof biol, 80-87, CHMN DEPT, COLBY COL, 86-, PROF BIOL, 87- *Mem:* Soc Study Reproduction; Am Soc Zoologists; AAAS. *Res:* Genetic and environmental factors affecting mammalian reproduction and early development. *Mailing Add:* Dept Biol Colby Col Waterville ME 04901

CHAMPLIN, KEITH S(CHAFFNER), b Minneapolis, Minn, Aug 20, 30; m 54; c 2. ELECTRICAL ENGINEERING. *Educ:* Univ Minn, BS, 54, MS, 55, PhD(elec eng), 58. *Prof Exp:* Asst, 54-55, from asst prof to assoc prof, 58-66, PROF ELEC ENG, UNIV MINN, MINNEAPOLIS, 66- *Concurrent Pos:* Exchange prof, Sorbonne, 63. *Mem:* Inst Elec & Electronics Engrs; Am Phys Soc; AAAS; Sigma Xi. *Res:* Semiconductors; microwave electronics; solid state devices; fluctuation phenomena. *Mailing Add:* 251 Elec Eng Minneapolis MN 55455

CHAMPLIN, ROBERT L, b Casper, Wyo, Oct 9, 30; m 53; c 3. ENVIRONMENTAL ENGINEERING. *Educ:* Univ Wyo, BS, 59, MS, 61; Harvard Univ, MS, 64, PhD(water chem), 69. *Prof Exp:* From instr to assoc prof, 59-77, PROF CIVIL & ENVIRON ENG, UNIV WYO, 77- *Mem:* Am Chem Soc; Nat Soc Prof Engrs; Am Water Works Asn; Water Pollution Control Fedn; Nat Asn Corrosion Engrs. *Res:* Water quality; pollution; aqueous corrosion. *Mailing Add:* Box 3295 Univ Sta Laramie WY 82071-3295

CHAMPLIN, WILLIAM G, b Rogers, Ark, Sept 10, 23; m 51; c 1. MEDICAL MICROBIOLOGY. *Educ:* Northeastern State Univ, Okla, BS, 48; Univ Ark, MS, 65, PhD(microbiol), 71. *Prof Exp:* SUPV MICROBIOLOGIST, VET ADMIN HOSP, 53- *Concurrent Pos:* Consult, Fayetteville City & Washington Regional Med Ctr, 71-88; guest lectr immunol & vis prof microbiol, Univ Ark, 71-75, 80-81. *Mem:* Am Soc Microbiol; Am Soc Clin Pathologists; Sigma Xi. *Res:* Rapid diagnosis of viral diseases by cell culture and indirect immunofluorescence of clinical material; detection of tumor markers by electroimmunoassay. *Mailing Add:* 3018 Sheryl Ave Fayetteville AR 72703

CHAMPNESS, CLIFFORD HARRY, b Ripley, Surrey, England, Sept 23, 26; Can citizen; m 56; c 2. SEMICONDUCTORS. *Educ:* Imperial Col Sci & Technol, London, BSc, 50; Univ London, MSc, 57; McGill Univ, PhD(physics), 62. *Prof Exp:* Asst exp officer, Armament Res Dept, Ministry of Supply, England, 42-47; res physicist, Res Dept, Metropolitan Vickers Elec Co, 51-57; mem sci staff, Res & Develop Div, Northern Elec Co, Ltd, Montreal, 57-59; head, Solid State Physics Dept, Noranda Res Ctr, Montreal, 62-67; PROF, DEPT ELEC ENG, MCGILL UNIV, 68- *Mem:* Am Phys Soc; Phys Soc London; Can Asn Physicists. *Res:* Optical and electronic properties of semiconductors, Schottky and heterojunctions, photovoltaic cells; study of physical properties of selenium, tellurium and related materials. *Mailing Add:* 135 Prince Rupert Dr Pointe Claire PQ H9R 1M1 Can

CHAMPNEY, WILLIAM SCOTT, b Cleveland, Ohio, Jan 15, 43; m 66; c 3. MOLECULAR GENETICS, RIBOSOMES. *Educ:* Univ Rochester, AB, 65; State Univ NY Buffalo, PhD(biol), 70. *Prof Exp:* Instr microbiol, Col Med, Univ Calif, Irvine, 70-72; asst prof biochem, Univ Ga, 72-79; asst prof genetics, Univ Tex, 79-82; PROF BIOCHEM, E TENN STATE UNIV, 82- *Mem:* Am Soc Biol Chemists; AAAS; Sigma Xi; Am Soc Microbiol; Genetics Soc Am. *Res:* Genetics of bacterial ribosomes; protein-nucleic acid interactions. *Mailing Add:* Biochem Dept E Tenn State Univ Box 19930A Johnson City TN 37614

CHAN, ALBERT SUN CHI, b Kwong Tung, China, Oct 30, 50; m 77; c 1. HOMOGENEOUS CATALYSIS, ORGANOMETALLIC CHEMISTRY. *Educ:* Int Christian Univ, AB, 75; Univ Chicago, MSc, 76, PhD(chem), 79. *Prof Exp:* SR RES CHEMIST, MONSANTO CO, 79- *Mem:* Am Chem Soc; Sigma Xi. *Res:* Organometallic chemistry and its use in homogeneous catalysis; kinetics and mechanisms of homogeneous catalytic reactions; catalytic activation of carbon monoxide and the use of synthesis gas for selective organic reactions. *Mailing Add:* 1233 Danvers Dr St Louis MO 63146

CHAN, ARTHUR WING KAY, b Hong Kong, June 24, 41; m 67; c 1. BIOCHEMISTRY, PHARMACOLOGY. *Educ:* Australian Nat Univ, BSc, 66, PhD(org chem), 69. *Prof Exp:* Res assoc org chem, Wash Univ, 69-71, res assoc pharmacol, Med Sch, 71-73; res scientist III, 73-76, res scientist IV, 76-79, RES SCIENTIST V, RES INST ON ALCOHOLISM, 79- *Concurrent Pos:* Res asst prof pharmacol, State Univ NY Buffalo, 74-81, res assoc prof, 82-; prin investr of res grant, Nat Inst Alcohol Abuse & Alcoholism, 75-77, 84-, NY State Health Res Coun, 76-77, 78-79 & 79-80 & Nat Inst Drug Abuse, 80-81; sponsor postdoctoral fel, Nat Inst Neurol, Commun Dis &

Strokes, 76-78; mem exec comt, NY State Coun Res Scientist. *Mem:* Sigma Xi; Res Soc Alcoholism; Am Soc Pharmacol & Exp Therapeut; Int Soc Biomed Res Alcoholism. *Res:* Alcohol-drug interactions; long-term effects of alcohol/benzodicepine intake; biochemical markers for alcoholism; drug tolerance and dependence. *Mailing Add:* Res Inst on Alcoholism 1021 Main St Buffalo NY 14203

CHAN, BERTRAM KIM CHEONG, b Hong Kong, Nov 28, 37; US citizen; m 69. ELECTROMAGNETIC COMPATIBILITY. *Educ:* Univ New South Wales, Australia, BSc (hons I), 62, MESc, 68; Univ Sydney, Australia, PhD (eng), 66. *Prof Exp:* Res scientist, Australian Atomic Energy Comn Res Estab, 65-67; fel, Univ Waterloo & Atomic Energy Can Ltd, 68-69; res assoc biomed eng, Univ Southern Calif, 69-70; assoc prof math, Loma Linda Univ, 70-71; prof chem & math, Mid East Col, Lebanon, 71-78; prof chem, Atlantic Union Col, 78-80; sr res engr electronic systs eng, Dept Survivability Anal & Parts Develop, Lockheed Missiles & Space Co, Lockheed Corp, 80-84; res specialist, Dept Re-Entry Systs Design, RF & Electrical Design, 85-88, Dept Survivability Analysis & Parts Develop, 88-89, Dept Electronic Systs Integration, 89-90; STAFF ENGR, RES & DEVELOP, ELECTRO MAGNETIC COMPATIBILITY, APPLE COMPUTER INC, 91- *Concurrent Pos:* Vis prof physics, San Jose State Univ, San Jose, 83-87; adj sr lectr elec eng, Cogswell Polytech Col, 87- *Mem:* UK Chartered Engrs. *Res:* Electromagnetic pulse interactions with space vehicles; air mitigation effects on electromagnetic pulses; systems generated electromagnetic pulses; transient analysis of waveforms; computer-aided and experimental analysis of latch-up in integrated circuits; electromagnetic radiation survivability design; development of weapon specifications electromagnetic compatibility EMZ/RFZ, product compliance engineering; calculations of electrical double-layers between charged particles; simultaneous mass transfer and reaction problems; existence and uniqueness solutions analysis. *Mailing Add:* 1534 Orillia Ct Sunnyvale CA 94087-4435

CHAN, BOCK G, b Kwantung, China, June 15, 35; US citizen; m 68; c 1. PHYTOCHEMISTRY. *Educ:* Cornell Univ, PhD(plant physiol), 70. *Prof Exp:* Experimentalist mineral nutrit biochem, dept pomol, Cornell Univ, 60-61 & 63-67, res assoc photosynthesis, dept veg crops, 70-71; fel biochem, 71-73, res plant physiologist, 73-81, supvry plant physiologist, 81-86, res leader, 83-86, RES PLANT PHYSIOLOGIST, WESTERN REGIONAL RES CTR, AGR RES SERV, USDA, 86- *Mem:* Sigma Xi; Phytochem Soc NAm; Entomol Soc Am; Am Chem Soc; AAAS; Am Phytopath Soc; NY Acad Sci. *Res:* Phytochemical basis and the physiology of plant resistance to insects, other pests and pathogens, with special emphasis on the economical crop plants; investigation into enzymology and biosynthetic pathways of the biologically active compounds; biotechnology of antimicrobiol peptides and their application in food protection. *Mailing Add:* 895-Hillside Albany CA 94706

CHAN, CHIU YEUNG, b Hong Kong, Feb 28, 41; US citizen; m 70; c 2. MATHEMATICAL ANALYSIS, APPLIED MATHEMATICS. *Educ:* Univ Hong Kong, BSc, 65; Univ Ottawa, MSc, 67; Univ Toronto, PhD(math), 69. *Prof Exp:* From asst prof to assoc prof, Fla State Univ, 69-81, prof math, 81-83; PROF MATH, UNIV SOUTHWESTERN LA, 82- *Mem:* Soc Indust & Appl Math; Am Acad Mechanics; Am Math Soc. *Res:* Partial differential equations, including Stefan problems; Sturmian theory; mathematical physics, reaction-diffusion problems; quenching phenomena and mathematical modelling; biomathematics and physical mathematics. *Mailing Add:* Dept Math Univ Southwestern LA Lafayette LA 70504-1010

CHAN, CHUN KIN, b Hong Kong, China, Jan 21, 53; m 84; c 2. RELIABILITY ENGINEERING. *Educ:* Chinese Univ Hong Kong, BS, 75; Arizona State Univ, PhD(physics), 80. *Prof Exp:* MEM TECH STAFF, AT&T BELL LABS, 84- *Mem:* Am Phys Soc. *Res:* Very large scale integration reliability. *Mailing Add:* 69 Hazel Ave Livingston NJ 07039

CHAN, DANIEL WAN-YUI, b China, Dec 29, 49; m 76. BIOCHEMISTRY, CLINICAL CHEMISTRY. *Educ:* Univ Ore, BA, 72; State Univ NY, Buffalo, PhD(biochem), 76. *Prof Exp:* Teaching asst biochem, State Univ NY, Buffalo, 72-76, res instr, 76-77; asst prof & assoc dir clin chem, 77-78, co-dir, 79-80, dir clin chem, dept lab med, 81- ASST PROF LAB MED, DEPT PATHOL, AT JOHNS HOPKINS UNIV SCH MED. *Concurrent Pos:* Fel, Erie County Lab, Meyer M Hosp, Buffalo, 76-77. *Mem:* Am Asn Clin Chem; Endocrine Soc; Int Soc Clin Enzym; Nat Acad Clin Biochem; Acad Clin Lab Physicians & Scientists. *Res:* Reproductive and thyroid endocrinology; radioimmunoassay development; therapeutic drug monitoring. *Mailing Add:* Dept Path Johns Hopkins Univ Sch Med 600 N Wolfe St Baltimore MD 21205

CHAN, DAVID S, b Hong Kong, July 23, 40; US citizen; m 75; c 2. PEPTIDE HORMONES, AMINO ACIDS. *Educ:* San Jose State Univ, BA, 64, MS, 70; Univ Southern Miss, PhD(biochem), 73. *Prof Exp:* Teaching asst chem, Univ Southern Miss, 70-73; res fel biol chem, Harvard Univ, 73-75, res assoc, 75-77; prin chemist, 78-83, QUAL ASSURANCE MGR, BECKMAN INSTRUMENTS, INC, 83- *Concurrent Pos:* Vpres, Gemini Sci, Inc, 77- *Mem:* Am Soc Neurochem; Am Soc Qual Control. *Res:* Instrumentation designed for proteins, peptides and nucleotide synthesis, purification and analysis. *Mailing Add:* Beckman Instruments Inc 1050 Page Mill Rd Palo Alto CA 94304

CHAN, EDDIE CHIN SUN, b Singapore, May 13, 31; Can citizen; m. MICROBIOL PHYSIOLOGY. *Educ:* Univ Tex, El Paso, BA, 54, Austin, MA, 57; Univ Md, PhD(microbiol), 60. *Prof Exp:* Nat Res Coun Can fel, 60-62; asst prof microbiol & biochem, Univ NB, 62-65; asst prof, 65-68, ASSOC PROF MICROBIOL & IMMUNOL, McGILL UNIV, 68- *Concurrent Pos:* Nat Res Coun Can & Med Res Coun Can grants. *Mem:* AAAS; Am Soc Microbiol; Can Soc Microbiol. *Res:* Anaerobes of periodontal disease; textbooks in general microbiology. *Mailing Add:* Dept Microbiol & Immunol McGill Univ 3775 University St Montreal PQ H3A 2B4 Can

CHAN, GARY MANNERSTEDT, b Oakland, Calif, Mar 8, 47; m 75; c 3. PEDIATRICS, CALCIUM. *Educ:* Univ Calif, Berkeley, BA, 68; Univ Southern Calif, MD, 72. *Prof Exp:* Resident pediat, Univ Calif, Los Angeles, 75; teaching fel neonatology, Univ Cincinnati, 75-77; asst prof pediat, 77-83, ASSOC PROF HEALTH, UNIV UTAH, 79-, ASSOC PROF PEDIAT & NURSING, 83-, ASSOC PROF PSYCHOL, 84- *Concurrent Pos:* Vis prof nutrit, Nat Dairy Coun, 85-; med dir, Newborn Intensive Care Unit,82- *Mem:* Am Col Nutrit; Am Inst Nutrit; Am Soc Clin Nutrit; Soc Pediat Res; Nat Acad Clin Biochem. *Res:* Calcium and bone mineralization in pediatric subjects; preterm and term infants, children, adolescents and mothers. *Mailing Add:* Dept Pediat Univ Utah 50 N Med Dr Salt Lake City UT 84132

CHAN, HAK-FOON, b Hong Kong, Oct 10, 42; m 68; c 2. AGRICULTURAL CHEMISTRY. *Educ:* Chung Chi Col, Chinese Univ, Hong Kong. dipl sci, 64; Bowling Green State Univ, MA, 68; Univ Mich, PhD(org chem), 71. *Prof Exp:* Fel chem, Univ Rochester, 71-73; sr chemist fungicide & biocide, 73-80, agr prod area mkt mgr, Pac Region, 81-89, AREA DIR, ROHM & HAAS CO, 90- *Concurrent Pos:* Managing dir, Rohm & Haas Co, Hong Kong/People's Repub China. *Mem:* Am Chem Soc; Royal Soc Chem. *Res:* To prepare for evaluation organic compounds which possess fungicidal, bactericidal and biocidal activities. *Mailing Add:* Rohm & Haas Hong Kong Ltd 11/F Dina House 11 Duddell St Hong Kong Hong Kong

CHAN, HARVEY THOMAS, JR, b Astoria, Ore, Mar 5, 40; m 66; c 3. FOOD SCIENCE. *Educ:* Ore State Univ, BSc, 63, PhD(food sci), 69; Univ Hawaii, MSc, 66. *Prof Exp:* Res asst food sci, Univ Hawaii, 63-65 & Ore State Univ, 65-68; FOOD TECHNOLOGIST, AGR RES SERV, USDA, 68- *Concurrent Pos:* Affil grad fac, Dept Food Sci, Univ Hawaii, 69-80, Dept Hort, 80-; consult, Int Exec Serv Corp & Vol Coop Assistance. *Mem:* Inst Food Technologists (secy-treas, 73-74 & 76, chmn-elect, 78-79); Am Chem Soc; Sigma Xi. *Res:* Chemical and biochemical composition of tropical fruits and vegetables; process and product development of tropical fruits and vegetables; changes in nutrients and biochemical constituents during processing. *Mailing Add:* USDA-Agr Res Serv PO Box 4459 Hilo HI 96720

CHAN, JACK-KANG, b Kwangtung, China, Oct 20, 50; US citizen; m 82; c 2. ELECTRICAL ENGINEERING. *Educ:* Univ London, BSc, 74; Chinese Univ Hong Kong, BSc, 75; Polytech Inst NY, PhD(elec eng), 82; Polytech Univ, PhD(math), 90. *Prof Exp:* Microwave engr, Sedco Systs, Raytheon, 79-80; SR MEM TECH STAFF, NORDEN SYSTS, UNITED TECHNOLOGIES, 80- *Concurrent Pos:* Reviewer, Inst Elec & Electronic Engrs Signal Processing Soc, 88; adj lectr, Polytech Univ, 90. *Mem:* Inst Elec & Electronic Engrs; Math Asn Am; Am Math Soc; Inst Elec Eng; Soc Indust & Appl Math. *Res:* Develop new techniques in signal processing for anti-submarine warfare engineering; research in topological measure theory and applied mathematics. *Mailing Add:* Norden Systs 75 Maxess Rd Melville NY 11747

CHAN, JAMES C, b Hong Kong, Nov 20, 37; m 64. MICROBIOLOGY. *Educ:* Int Christian Univ Tokyo, BA, 61; Univ Rochester, PhD(microbiol), 66. *Prof Exp:* Sr res scientist, Squibb Inst Med Res, 66-68; asst prof med, Sch Med, Ind Univ, Indianapolis, 68-71; asst prof biochem virol, Baylor Col Med, 71-72; asst virologist, 72-78, asst prof, 72-79, dir, Virol Prog, 80-81 & 83-85, ASSOC PROF VIROL, CANCER CTR, M D ANDERSON CANCER CTR, UNIV TEX, 79-, CHMN, HYBRIDOMA STUDY GROUP, 80- *Concurrent Pos:* Ind Univ fac res grant & Little Red Door, Inc cancer res grants, 69 & 70; Am Cancer Soc inst grant, 68-70; St Joseph Co, Inc cancer res grant, 71-72; NIH biomed res grant, 78-79; hybridoma res grant, UCF, 80-; Univ Ottawa grant, 83-86; grant, Coun Tobacco Res Inc, 89-91. *Mem:* AAAS; Am Soc Microbiol; Am Asn Cancer Res; Int Leukemia Asn; Am Soc Virol. *Res:* Biochemical changes in cells following infection by tumor viruses, such as RNA tumor viruses; studies of viral involvement in neoplastic diseases in animals and man; monoclonal antibodies to growth factors; tumor cell metastasis. *Mailing Add:* Dept of Virol Univ Tex M D Anderson Cancer Ctr Houston TX 77030

CHAN, JAMES C M, b Hong Kong, Dec 27, 37. PEDIATRIC NEPHROLOGY. *Educ:* McGill Univ, MD, 64. *Prof Exp:* Intern med, McGill Univ, 64-65; resident pediat, Mayo Clin, 65-67, Columbia, 68-70; asst prof pediat, Univ Southern Calif, 70-73; assoc prof child & health develop, Children Nat Med Ctr, George Wash Univ, 73-76; PROF & DIR PEDIAT NEPHROLOGY, MED COL VA, 77-, PROF & VCHMN PEDIAT, 82- *Concurrent Pos:* Fel pediat nephrology, Univ Ore Hosp, 67-68; vis scientist, NIH, 76-77 & 87-88, consult, Nat Heart Lung & Blood Inst, 78-82; chmn med adv bd, Nat Kidney Found, 83-87, med dir, 87- *Mem:* Am Soc Nephrology; Am Soc Pediat Nephrology; Am Acad Pediat; Am Pediat Soc. *Res:* Calcium phosphate growth hormone metabolism in renal tubular acidosis; x-linked hypophosphatemia and chronic renal disease. *Mailing Add:* Dept Pediat Div Health Sci Va Commonwealth Univ Med Col Va Childrens Med Ctr MCV Sta Richmond VA 23298

CHAN, JEAN B, b Toyshan, China, Apr 24, 37; US citizen; m 60; c 2. MATHEMATICS. *Educ:* Univ Chicago, BS, 60, MS, 61; Univ Calif, Los Angeles, PhD(math), 71. *Prof Exp:* Asst prof math, Loyola Univ, Los Angeles, 72-73; from asst prof to assoc prof math, 73-82, chair math dept, 86-89, PROF MATH, SONOMA STATE UNIV, 82-, ACTING CHAIR MATH DEPT, 90- *Mem:* Math Asn Am; Asn Women Math; Am Math Soc. *Res:* Convex sets and geometry. *Mailing Add:* Dept Math Sonoma State Univ 1801 E Cotati Ave Rohnert Park CA 94928

CHAN, JOHN YEUK-HON, b Jan 14, 47; m; c 1. CHEMICAL CARCINOGENESIS, DNA REPAIR & REPLICATION. *Educ:* State Univ NY, Buffalo, PhD(molecular biol), 76. *Prof Exp:* ASST PROF, UNIV TEX M D ANDERSON CANCER CTR, 82- *Mem:* Am Asn Cancer Res. *Res:* Human DNA ligases and polymerases; repair genes. *Mailing Add:* Dept Molecular Path M D Anderson Cancer Ctr Univ Tex 1515 Holcombe Houston TX 77030

CHAN, KAI CHIU, b Canton, China, May 16, 34; US citizen; m 66; c 2. DENTISTRY. *Educ:* Chiba Univ, Japan, BS, 58; Tokyo Med & Dent Univ, DDS, 62; Univ Iowa, MS, 64, DDS, 67. *Prof Exp:* From instr to asst prof, 64-71, assoc prof, 71-76, PROF OPER DENT, COL DENT, UNIV IOWA, 76- *Mem:* Am Asn Dent Schs; Am Dent Asn; Int Asn Dent Res; Acad Oper Dent. *Res:* Burnished amalgam surfaces; retention pins; dental cements. *Mailing Add:* Col of Dent Univ of Iowa Iowa City IA 52242

CHAN, KA-KONG, Hong Kong; m 68; c 4. MEDICINAL CHEMISTRY. *Educ:* Univ NB, PhD(chem), 66. *Prof Exp:* Fel organic chem, Univ NB, 66-67; fel Univ BC, 67-69; res fel biochem, Harvard Univ, 69-70; RES SCIENTIST MED & ORG CHEM, HOFFMAN-LA ROCHE INC, 70- *Mem:* Am Chem Soc. *Res:* Synthesis of natural products; asymmetric synthesis of natural vitamin E; synthesis of antibacterial agents; synthesis of fluorinated retinoids; design and synthesis of enzyme inhibitors. *Mailing Add:* Chem Res Bldg 76 Hoffman-La Roche Inc 340 Nutley NJ 07110

CHAN, KENNETH KIN-HING, b Hong Kong, Nov 30, 40; US citizen; m 72; c 2. PHARMACEUTICS, PHARMACEUTICAL CHEMISTRY. *Educ:* Calif State Univ, San Jose, BS, 64; Univ Calif, Davis, MS, 68; Univ Calif, San Francisco, PhD(pharmaceut chem), 72. *Prof Exp:* Res biochemist, Stanford Univ/Vet Admin Hosp, Palo Alto, 64-66; res assoc pharmaceut chem, 72-73, from asst prof to assoc prof, 73-91, PROF PHARMACEUT, SCH PHARM, UNIV SOUTHERN CALIF, 91- , DIR, PHARMACOANALYTIC LAB, COMPREHENSIVE CTR, 80- *Concurrent Pos:* Prin investr, NIH grants, Nat Cancer Inst, 74-75, 76-79, 81-85, 89-92, co-prin investr, 75-76, 90-93; proj investr, NIH grants, Nat Cancer Inst, Western Cancer Study Group pharmacokinetics, 75-76 & Los Angeles County/Univ Southern Calif Cancer Ctr, 74-77 & 80-; sci adv, US Food & Drug Admin, 83-; mem, Nat Cancer Adv Bd, 91- *Mem:* Am Chem Soc; Am Asn Pharm Sci; Am Asn Cancer Res; Chinese Am Bioscientist Soc Am. *Res:* Drug metabolism; pharmacokinetics; isotope labeling synthesis; cancer chemotherapy; pharmacology of anticancer drugs; drug analysis; nuclear magnetic resonance; mass spectrometry. *Mailing Add:* 605 Indiana Place South Pasadena CA 91030

CHAN, KWING LAM, b Hong Kong, Oct 13, 49; m 73; c 1. COMPUTATIONAL ASTROPHYSICS, SOLAR PHYSICS. *Educ:* Univ Calif, Berkeley, BA, 71; Princeton Univ, PhD(physics), 74. *Prof Exp:* Postdoctoral fel, IBM, Thomas J Watson Res Ctr, 74-76, Calgary Univ, 76-77; lectr physics, Queen's Univ, Ont, 77-80; assoc scientist, 80-86, SR SCIENTIST, APPL RES CORP, 86- *Concurrent Pos:* Vis adj prof, Yale Univ, 90-91. *Mem:* Am Phys Soc; Am Astron Soc; Int Astron Union. *Res:* Numerical study of convection in deep atmospheres; estimation of neutrino mass based on the supernova 1987 A data; magnetic confinement of astrophysical jets; distortion of the cosmic microwave background. *Mailing Add:* Goddard Space Flight Ctr Code 910-2 Greenbelt MD 20771

CHAN, KWOK-CHI DOMINIC, b Hong Kong, July 27, 46; Can citizen; c 2. BEAM INDUCED EFFECTS. *Educ:* Chinese Univ Hong Kong, BSc, 68; Univ Pittsburgh, MSc, 71 & PhD(physics), 75. *Prof Exp:* Sci officer, Atomic Energy Can, 78-86; STAFF MEM, LOS ALAMOS NAT LAB, 86- *Mem:* Am Phys Soc. *Res:* Accelerator designs, mainly on theory and simulations. *Mailing Add:* H825 Los Alamos Nat Lab Los Alamos NM 87545

CHAN, KWOKLONG ROLAND, b Shanghai, China, Oct 18, 33. ATMOSPHERIC CHEMISTRY, ATMOSPHERIC PHYSICS. *Educ:* Univ Hong Kong, BSc, 56; Stanford Univ, MS, 58, PhD(elec eng), 63. *Prof Exp:* Res asst, Radiosci Lab, Stanford Univ, 58-62; asst prof elec eng, Univ Hawaii, 62-63; res scientist, Instrument Div, 63-64, res scientist, Space Sci Div, Ames Res Ctr, 64-80, asst mgr, Upper Atmosphere Res Prog, Washington, DC, 80-81, ASST CHIEF, ATMOSPHERIC EXP BR, AMES RES CTR, NASA, 81- *Mem:* Inst Elec & Electronics Engrs; Am Geophys Union; Sigma Xi; Am Meteorol Soc; Am Inst Aeronaut & Astronaut; AAAS. *Res:* Study of the earth's neutral and ionized atmospheres by experimental technique using ground-based, airborne, balloon-borne or spacecraft instruments; electrical engineering. *Mailing Add:* NASA Ames Res Ctr MS 245-5 Moffett Field CA 94035-1000

CHAN, LAI KOW, b Hong Kong, Nov 5, 40; m 67. STATISTICS. *Educ:* Hong Kong Baptist Col, BSc, 62; Univ Western Ont, MA, 64, PhD(statist), 66. *Prof Exp:* Instr, Univ Toronto, 65-66; from asst prof to prof statist, Univ Western Ont, 66-80; PROF STATIST & HEAD DEPT, UNIV MAN, 80- *Concurrent Pos:* Hon res fel, Univ Col, London, 73-74; vis prof, Univ Umea; vis scholar, Stanford Univ, 85-86. *Mem:* Fel Am Statist Asn; Int Statist Inst; Opers Res Soc Am; fel Inst Math Statist; Statist Soc Can. *Res:* Estimation problems in linear models; statistical quality control; risk assessment of large technological systems. *Mailing Add:* 888 Kilkenny Winnipeg MB R3T 4G3 Can

CHAN, LAURENCE KWONG-FAI, b Hong Kong, Oct 6, 47; US citizen; m 77; c 1. INTERNAL MEDICINE, NEPHROLOGY. *Educ:* Univ Hong Kong, MD, 72; Royal Col Physicians London, MRCP, 77, FRCP; Royal Col Physicians Edinburgh, FRCP, 87; Univ Oxford, PhD(biochem), 82. *Prof Exp:* Res fel biochem, Univ Oxford, 79-81 & Green Col, Oxford, 81-82; clin lectr med, Univ Oxford, 81-83; asst prof, 83-88, ASSOC PROF MED, UNIV COLO SCH MED, 88- *Concurrent Pos:* Vis scholar, Nat Kidney Found, 81; dir dialysis, Health Sci Ctr & dir, Nuclear Magnetic Resonance Lab, Univ Colo, 85- *Mem:* NY Acad Sci; Transplantation Soc; Soc Magnetic Resonance Med. *Res:* Viro nucleomagnetic resonance spectroscopy; renal energy metabolism; clinical renal transplantation. *Mailing Add:* Health Sci Ctr Univ Colo Box C281 4200 E 9th Ave Denver CO 80262

CHAN, LAWRENCE CHIN BONG, b Hong Kong, BCC, Jan 1, 42; US citizen; m 67; c 2. LIPOPROTEIN METABOLISM, ENDOCRINOLOGY. *Educ:* Univ Hong Kong, MB, BS, 66. *Prof Exp:* Intern med & surg, Univ Hong Kong, 66-67; intern rotating, Victoria Hosp, London, Ont, Can, 67-68; resident med, Barnes Hosp, St Louis, Mo, 68-70; res fel biochem, Wash Univ, St Louis, Mo, 70-72; instr med, Sch Med, Vanderbilt Univ, Nashville, Tenn, 72-73; asst prof cell biol & med, 74-78, assoc prof med, 78-83, assoc prof cell biol, 82-86, PROF MED & DIR, LAB MOLECULAR BIOL, DEPT MED, BAYLOR COL MED, 84-, PROF CELL BIOL, 87- *Concurrent Pos:* Consult, Houston Vet Admin Hosp, 74- & res rev comt B, Nat Heart, Lung & Blood Inst, NIH, 82-86; attend physician, Ben Taub Gen Hosp, Houston, Tex, 74-; estab investr, Am Heart Asn, 74-79, deleg, Coun Arteriosclerosis, 81; staff physician, Methodist Hosp, 78-; assoc ed, Metab, 79-86; lectr, Europ Lipoprotein Conf, 83 & 85; metabolism study sect, NIH, 87- *Mem:* Am Soc Clin Invest; Am Soc Cell Biol; Endocrine Soc; Am Diabetes Asn; Am Fedn Clin Res; Am Soc Biochem Molecular Biol. *Res:* The structure, expression and evolution of lipoproteins and lipid-related enzymes; hormonal regulation of gene expression. *Mailing Add:* Baylor Col Med One Baylor Plaza Houston TX 77030

CHAN, LEE-NIEN LILLIAN, b Hong Kong, Sept 28, 41; m 69; c 3. DEVELOPMENTAL BIOLOGY, CELL BIOLOGY. *Educ:* Acadia Univ, BSc, 63; Univ Wis, MS, 66; Yale Univ, PhD(biol), 71. *Prof Exp:* Assoc in res, Yale Univ, 67, res assoc molecular biophys & biochem, 71; res assoc biol, Mass Inst Technol, 71-73; asst prof physiol, Univ Conn Health Ctr, 73-78; ASSOC PROF CELL BIOL, UNIV TEX MED BR GALVESTON, 78- *Concurrent Pos:* NIH res grant, res career develop award. *Mem:* Am Soc Cell Biol; Am Soc Hematol; Soc Develop Biol; Am Soc Biol Chem; AAAS; Int Soc Develop Biol. *Res:* Development of eukaryotes; regulation of specific gene expression; erythropoiesis; differentiation of red cell membrane. *Mailing Add:* Dept of Human Biol Chem & Genetics Univ of Tex Med Br Galveston TX 77550

CHAN, MABEL M, b Hong Kong, Oct, 9, 48; m; c 1. TASTE & NUTRITION, COMMUNITY NUTRITION. *Educ:* Univ Calif, Davis, PhD(nutrit), 75. *Prof Exp:* ASSOC PROF NUTRIT, NY UNIV, 83- *Mem:* Am Inst Nutrit; Am Dietetic Asn; Inst Food Technologists; Sigma Xi; NY Acad Sci; Asn for Chemoreception Sci. *Res:* Computer application in nutrition; Taste and nutrition, community nutrition. *Mailing Add:* Dept Nutrit Food & Hotel Mgt 1201 Educ Bldg NY Univ New York NY 10003

CHAN, MAUREEN GILLEN, b Brooklyn, NY, June 2, 39; m 63. POLYMER CHEMISTRY. *Educ:* Chestnut Hill Col, BS, 61; Stevens Inst Technol, MS, 65. *Prof Exp:* Sr tech aide chem, 62-64, assoc mem tech staff, 64-74, mem tech staff, 74-78, SUPVR, POLYMER STABILIZATION GROUP, BELL LABS, 78- *Mem:* Am Chem Soc; Soc Plastics Engrs. *Res:* Stabilization of high polymers. *Mailing Add:* AT&T Bell Labs Rm 7F-226 Murray Hill NJ 07974

CHAN, MOSES HUNG-WAI, b Xi-an, China, Nov 23, 46; m 72; c 1. LOW TEMPERATURE PHYSICS. *Educ:* Bridgewater Col, BA, 67; Cornell Univ, MS, 70, PhD(exp physics), 74. *Prof Exp:* Asst lectr, Univ Hong Kong, 69-70; res assoc physics, Duke Univ, 73-76; asst prof physics, Univ Toledo, 76-79; from asst prof to prof, 79-90, DISTINGUISHED PROF PHYSICS, PA STATE UNIV, 90- *Concurrent Pos:* Sr res fel, Japan Soc Prom Sci, Univ Tokyo, 82; Guggenheim award, 86-87. *Mem:* Fel Am Phys Soc; Sigma Xi. *Res:* Low temperature properties of quantum fluids and solids, and critical phenomena of fluids; thermodynamic study of phase transition in two and three dimensional systems. *Mailing Add:* Dept Physics Pa State Univ 104 Davey Lab University Park PA 16802

CHAN, PAUL C, b Hong Kong, Mar 15, 36; m 64; c 1. CIVIL ENGINEERING, APPLIED MECHANICS. *Educ:* Chu Hai Col, Hong Kong, BSc, 58; Worcester Polytech Inst, MSc, 62; Tex A&M Univ, PhD(civil eng), 68. *Prof Exp:* Asst prof, 66-77, ASSOC PROF CIVIL ENG, NJ INST TECHNOL, 77- *Mem:* Geol Soc Am; Seismol Soc Am. *Res:* Soil structure interaction; soil behavior under dynamic loadings; stress waves propagation. *Mailing Add:* Dept Civil-Environ Eng NJ Inst Technol 325 High St Newark NJ 07102

CHAN, PETER SINCHUN, b Kwang-tung, China, Dec 1, 38; m 64; c 3. PHARMACOLOGY, BIOCHEMISTRY. *Educ:* Nat Taiwan Univ, BSc, 61; Univ Cincinnati, MSc, 64; Ind Univ, Bloomington, PhD(pharmacol, org chem), 67. *Prof Exp:* Res fel med chem, Univ Mich, 67-68; instr pharmacol, Med Br, Univ Tex, Galveston, 68-69, asst prof, 69-70; res pharmacologist, 70-73, group leader & prin res pharmacologist, 73-75, head dept cardiovasc-renal pharmacol, Lederle Labs Div, 75-76, GROUP LEADER & PRIN RES PHARMACOLOGIST, MED RES DIV, AM CYANAMID CO, 76- *Mem:* Am Soc Pharmacol & Exp Therapeut; Soc Exp Biol & Med; Am Heart Asn; Am Chem Soc; Sigma Xi. *Res:* Cardiovascular and renal pharmacology; pharmacological and biochemical approaches for the search of new drugs and to study mechanisms of drug actions, drug design and regulation of enzyme systems of pharmacologic importance. *Mailing Add:* Cardiovasc CNS Res Sect American Cyanamid Co Lederle Labs Pearl River NY 10965

CHAN, PHILLIP C, b Amoy, China, June 14, 28; US citizen; m 65; c 1. BIOCHEMISTRY. *Educ:* Monmouth Col, Ill, BS, 52; Columbia Univ, MA, 53, PhD(chem), 57. *Prof Exp:* Fel, Sch Med, Johns Hopkins Univ, 57-59; Jane Coffin Childs fel, Max Planck Inst Cell Chem, Ger, 59-60; asst prof, 60-67, assoc prof, 67-76, PROF BIOCHEM, STATE UNIV NY DOWNSTATE MED CTR, 76- *Mem:* Am Chem Soc; Harvey Soc; Am Soc Biol Chem; Sigma Xi. *Res:* Reactive oxygen species in biological systems. *Mailing Add:* Dept Biochem State Univ NY 450 Clarkson Ave Box 8 Brooklyn NY 11203

CHAN, PING-KWONG, b Sun Wei, Kwangtung, China, Oct 22, 49. GENERAL TOXICOLOGY, RISK ASSESSMENT. *Educ:* Northeastern Univ, BSc, 75; Univ Miss Med Sch, PhD(toxicol), 81. *Prof Exp:* Pharmacist, Osco Drug Co, 75-76; TOXICOLOGIST, ROHM & HAAS CO, 80- *Mem:* Sigma Xi; Soc Toxicol; Am Bd Toxicol. *Res:* Toxicology and risk assessment in product development; mechanism of toxicity; metabolism; pharmacokinetics; pesticides toxicology research. *Mailing Add:* Rohm & Haas Co 727 Norristown Rd Spring House PA 19477

CHAN, PO CHUEN, b Ichong, Hupeh, China, May 13, 35; m 61; c 2. CARCINOGENESIS, TOXICOLOGY. *Educ:* Int Christian Univ, Tokyo, BA, 60; Columbia Univ, MA, 63; NY Univ, PhD(biol), 67. *Prof Exp:* Res assoc, Sloan-Kettering Inst Cancer Res, 67-70; assoc mem, Am Health Found, 70-78; cancer res scientist IV, Roswell Park Mem Inst, 78-83; CHEM MGR, NAT INST ENVIRON HEALTH SCI, 83- *Mem:* AAAS; Am Physiol Soc; NY Acad Sci; Tissue Cult Asn; Am Asn Cancer Res. *Res:* Cell production and kinetics; chemical carcinogenesis. *Mailing Add:* Nat Inst Environ Health PO Box 12233 Research Triangle Park NC 27709

CHAN, PUI-KWONG, NUCLEAR PROTEIN, PHOSPHORYLATION. *Educ:* Univ Toronto, PhD(clin biochem), 78. *Prof Exp:* ASST PROF PHARMACOL, BAYLOR COL MED, 78- *Mailing Add:* Dept Pharmacol Baylor Col Med 1 Baylor Plaza Houston TX 77030

CHAN, RAYMOND KAI-CHOW, b Hong Kong, Oct 10, 33; m 60; c 2. PHYSICAL CHEMISTRY. *Educ:* Univ Toronto, BA, 58, PhD(phys chem), 61. *Prof Exp:* Nat Res Coun Can fel, 61-62; asst prof, 62-71, ASSOC PROF CHEM, UNIV WESTERN ONT, 71- *Mem:* Chem Inst Can; NAm Thermal Anal Soc. *Res:* Molecular solid phase transitions under pressure, supercooled liquids and glassy plastic crystals by thermally stimulated depolarization, dielectric properties and differential thermal analysis; air pollution, especially sulfur dioxide limestone/dolomite reaction using thermogravimetry. *Mailing Add:* Dept of Chem Univ of Western Ont London ON N6A 5B9 Can

CHAN, SAI-KIT, b Hong Kong, Jan 4, 41; US citizen; m 69; c 2. PHASE TRANSFORMATIONS, KINETICS OF IRREVERSIBLE PROCESSES. *Educ:* Univ Manchester, Eng, BSc, 63, PhD(physics), 66. *Prof Exp:* Res assoc, Univ Chicago, 66-68; res assoc, Argonne Nat Lab, 68-69; res assoc, Univ Cambridge, Eng, 69-71; sci staff, Max Planck Inst, Gottingen, W Ger, 71-80; PHYSICIST, ARGONNE NAT LAB, 80- *Mem:* Am Phys Soc; Am Ceramic Soc. *Res:* Solid state phase transformations, including critical phenomina, phase separation, nucleation, spinodal decomposition and morphology of precipitates in metals, alloys and ceramic systems. *Mailing Add:* Mat Sci Div Argonne Nat Lab Argonne IL 60439

CHAN, SAMUEL H P, b Nanking, China, Aug 1, 41; m 71. BIOCHEMISTRY, MOLECULAR BIOLOGY. *Educ:* Int Christian Univ, Tokyo, BA, 64; Univ Rochester, PhD(biochem), 69. *Prof Exp:* Asst, Univ Rochester, 64-69; res assoc biochem & fel, Cornell Univ, 69-71; from asst prof to assoc prof, 71-81, PROF BIOCHEM, SYRACUSE UNIV, 81- *Mem:* AAAS; Am Chem Soc; NY Acad Sci; Sigma Xi; Am Soc Biol Chemists. *Res:* Physical, chemical and enzymatic properties of cytochromes and cytochrome oxidase; resolution and reconstitution of inner mitochondrial membrane; mechanism of oxidative phosphorylation; energy metabolism in tumor tissues. *Mailing Add:* Dept Biol Syracuse Univ 130 College Pl Syracuse NY 13244

CHAN, SEK KWAN, b Canton, China, Sept 13, 44; Can citizen; m 40; c 1. EXPLOSIVES HAZARD & SENSITIVITY, PROPELLANT & PYROTECHNICS. *Educ:* Univ NSW, Australia, BE, 67; Sydney Univ, Australia, 68; Univ Toronto, PhD(aerospace), 73. *Prof Exp:* Res physicist, C-I-L Explosives Res Lab, 73-78 & 81-85; sr res scientist, I C I Corp Lab, 78-81; res dir, Space Res Corp, 85-89; RES SCIENTIST, GROUP TECH CENTRE, ICI EXPLOSIVES, 89- *Concurrent Pos:* Vis lectr, Nanjing Univ, China, 83. *Mem:* Am Inst Aeronaut & Astronaut. *Res:* Commercial explosives detonation physics, including sensitivity and hazards; internal ballistics of guns and gun barrel erosion; automobile air bag propellant research and development. *Mailing Add:* ICI Explosives GTC-B McMasterville PQ J3G 1T9 Can

CHAN, SHAM-YUEN, b Hong Kong; US citizen; m 75. MOLECULAR BIOLOGY, CELL BIOLOGY. *Educ:* Hong Kong Baptist Col, BS, 73; Univ Notre Dame, PhD(microbiol), 78. *Prof Exp:* Fel, Lobund Lab, Univ Notre Dame, 78-81, res asst prof, Dept Microbiol, 81-82; res scientist, 82-84, sr res scientist, 84-86, staff scientist, 86-89, SR STAFF SCIENTIST, MILES INC, 89- *Mem:* Am Soc Microbiol; NY Acad Sci; Sigma Xi; AAAS; Tissue Cult Asn. *Res:* molecular expression of factor VIII; serum-free cell culture; chimeric antibodies. *Mailing Add:* Miles Inc PO Box 1986 Berkeley CA 94701

CHAN, SHIH HUNG, b Taiwan, Nov 8, 43; US citizen; m 70; c 2. HEAT TRANSFER, FLUID MECHANICS. *Educ:* Taipei Inst Technol, dipl, 63; Univ Calif, Berkeley, PhD(mech eng), 69. *Prof Exp:* From asst prof to assoc prof mech eng, NY Univ, 69-73; assoc prof, Polytech Inst NY, 73-74; res staff nuclear eng, Argonne Nat Lab, 74-75; from assoc prof to prof mech eng, 75-89, chmn dept, 79-89, WIS DISTINGUISHED PROF & DIR, THERMAL ENG RES LAB, UNIV WIS, MILWAUKEE, 90- *Concurrent Pos:* Prin investr grants, NSF, Off Naval Res, Dept Energy & NIH, 70-; consult, Gen Elec Co, 81, Arrgonne Nat Lab, 75- & Teltech Resources Network, 87- *Mem:* Am Nuclear Soc; Am Soc Mech Eng; Sigma Xi. *Res:* Thermal radiative transfer; radiation properties of gases and nuclear reactor materials; fouling heat transfer; thermal hydraulic analyses of nuclear reactors; two-phase flow; condensation; melting and solidification; liquid metal fuel combustion; multiphase combustion. *Mailing Add:* Dept Mech Eng Univ Wis PO Box 784 Milwaukee WI 53201

CHAN, SHU FUN, b Canton, China, July 27, 39; m 69; c 2. CHEMISTRY. *Educ:* Sun Yat Sen Univ, BSc, 60; Univ Hong Kong, BSc, Hons, 66, PhD(chem), 71; Polytech Univ, NY, MSc, 86. *Prof Exp:* Chemist, Chiap Hwa Mfg Co, Hong Kong, 63-65; teaching asst, Univ Hong Kong, 66-70; lectr inorg chem, Nanyang Univ, 70-78; sr res assoc, 78-80, Brookhaven Nat Lab, NY, 78-80, chem assoc, 80-84; RADIOCHEMIST/NUCLEAR ENGR, LONG ISLAND LIGHTING CO, 85- *Concurrent Pos:* Fel, State Univ NY, Buffalo, 77. *Mem:* Am Chem Soc; Chem Soc London; Am Nuclear Soc. *Res:* Kinetic and mechanistic studies of substitution reactions; electron transfer reactions of coordination compounds; nuclear reactor water chemistry. *Mailing Add:* PO Box 694 Shoreham NY 11786-0694

CHAN, SHU-GAR, b Canton, China, Aug 18, 27; US citizen; m 56. ELECTRICAL ENGINEERING. *Educ:* Univ Wash, Seattle, BS, 52; Columbia Univ, MS, 54; Univ Kans, PhD(elec eng), 64. *Prof Exp:* Lectr elec eng, City Col New York, 54-58; instr, Univ Mich, 58-60 & Univ Kans, 61-62; ASSC PROF, NAVAL POSTGRAD SCH, 64- *Mem:* Inst Elec & Electronics Engrs; Sigma Xi. *Res:* Network theory; computer-aided design; topology. *Mailing Add:* 731 Dry Creek Rd Monterey CA 93940

CHAN, SHUNG KAI, b Hong Kong, Mar 31, 35; m 63. BIOCHEMISTRY. *Educ:* WVa Univ, AB, 56; Univ Wis, PhD(biochem), 62. *Prof Exp:* Asst res chemist, Samuel Robert Noble Found, 58-59; sr res biochemist, Abbott Labs, 62-66; asst prof, 66-70, ASSOC PROF BIOCHEM, MED CTR, UNIV KY, 70- *Mem:* AAAS; Am Chem Soc; Brit Biochem Soc; Am Soc Biol Chem. *Res:* Glycoproteins, structure and function; glycoproteins, biosynthesis and regulation. *Mailing Add:* Dept Biochem Univ Ky Col Med Lexington KY 40536

CHAN, SHU-PARK, b Canton, China, Oct 10, 29; m 56; c 2. ELECTRICAL ENGINEERING. *Educ:* Va Mil Inst, BS, 55; Univ Ill, MS, 57, PhD(elec eng), 63. *Prof Exp:* Instr math & elec eng, Va Mil Inst, 57-59, asst prof math, 61-63; instr, Univ Ill, 60-62, res assoc, co-ord sci lab, 62-63; assoc prof elec eng, 63-68, chmn dept, 69-84, dean eng, 88-89, PROF ELEC ENG, UNIV SANTA CLARA, 68-, NICHOLSON CHAIR PROF, 87- *Concurrent Pos:* NSF res grant, 64-65; conf chmn, Fourth Asilomar Conf Circuits & Systs, 70; ed Int Ser Eng & Scis, Nat Taiwan Univ Press, 73-; spec chair elec eng, Nat Taiwan Univ, 73; overseas consult, Nat Sci Coun, Repub China, 76-; invited lectr, Acad Sinica, Peking, 80; hon prof, Elec Eng Dept, Univ Hong Kong, 80-81, Anhuei Univ, 82- & Schina Univ Sci & Technol, 85-; trustee, W Valley-Mission Community Col Dist Bd, 88- *Mem:* Fel Inst Elec & Electronic Engrs; Am Soc Eng Educ; Am Chem Soc. *Res:* Linear graph theory and its applications to electrical networks; circuit theory; computer-aided analysis and design of linear active networks and systems; author or co-author of 6 books, 50 journal articles and numerous technical reports. *Mailing Add:* Dept of Elec Eng & Comput Sci Univ of Santa Clara Santa Clara CA 95053

CHAN, SIU-KEE, b Canton, China, Sep 11, 36; US citizen; m 66; c 2. FINITE ELEMENT ANALYSIS, PLASTICITY. *Educ:* Cheng Kung Univ, BSME, 59, Univ Va, MMSc, 62, Univ Ill, PhD(mech), 66. *Prof Exp:* Engr, Hayes Int Aircraft Co, 66; FEL ENGR, WESTINGHOUSE RES LABS, 66- *Mailing Add:* 125 Regal Ct Monroeville PA 15146

CHAN, STEPHEN, b Wuchow, China, July 28, 42; c 2. ENDOCRINOLOGY. *Educ:* Univ Hong Kong, BSc, 66, MSc, 68; Univ Hull, PhD(zool), 71. *Prof Exp:* Res assoc, Boston Univ, 71-72; asst prof, Rutgers Univ, 72-77; ASSOC PROF BIOL, STATE UNIV NY BROCKPORT, 77- *Mem:* Soc Study Reproduction; Soc Endocrinol; Soc Exp Biol & Med. *Res:* Mammalian reproduction with emphasis on prenatal biology and aging mechanisms. *Mailing Add:* Dept of Biol Sci State Univ of NY Brockport NY 14420

CHAN, SUNNEY IGNATIUS, b San Francisco, Calif, Oct 5, 36; m 64; c 1. BIOPHYSICAL CHEMISTRY, BIOPHYSICS. *Educ:* Univ Calif, Berkeley, BS, 57, PhD(chem), 60. *Prof Exp:* NSF fel physics, Harvard Univ, 60-61; asst prof chem, Univ Calif, Riverside, 61-63; from asst prof to assoc prof chem physics, 63-68, actg exec officer chem, 77-78, exec officer chem, 78-80, master student houses, 80-83, PROF CHEM PHYSICS, CALIF INST TECHNOL, 68-, PROF BIOPHYS CHEM, 76- *Concurrent Pos:* Consult, Unified Sci Assocs, 63-64, NIH, USPHS, 70-74 & 89-, Procter & Gamble Co, 74-79, Allergan, Irvine, Calif, 83-86 & Cardiovase Devices, 84-86; Merck Sharp & Dohme Res Labs, 74-76 & McGaw Labs, 77-79; Sloan fel, 65-67; Guggenheim Mem fel, 68-69; assoc ed, Ann Rev Magnetic Resonance, 70 & J Am Chem Soc, 83-; Fogarty scholar-in-res, NIH, 87-; dir, Southern Calif Regional Nuclear Magnetic Resonance Facil, 78-88; elected fel, Am Phys Soc, 87; chmn of the fac, 87-89. *Honors & Awards:* Reilly lectr, Univ Notre Dame, 73; Chan mem lectr, Univ Calif, Berkeley, 84. *Mem:* AAAS; Am Phys Soc; Am Chem Soc; Am Soc Biochem & Molecular Biol; Biophys Soc. *Res:* Biological structural determination by various spectroscopic methods; the structure of membrane proteins participating in biological energy conservation and transduction and transmembrane ion translocation; membrane structure and function; bioenegetics. *Mailing Add:* Dept Chem Calif Inst Technol Pasadena CA 91125

CHAN, TAK-HANG, b Hong Kong, June 28, 41; m 69; c 2. ORGANIC CHEMISTRY. *Educ:* Univ Toronto, BSc, 62; Princeton Univ, MA, 63, PhD(chem), 65. *Prof Exp:* Asst prof, 66-71, assoc prof, 72-77, PROF CHEM, McGILL UNIV, 78-, CHMN DEPT, 85- *Concurrent Pos:* Res fel, Harvard Univ, 65-66; Killam fel, 83-85. *Honors & Awards:* Merck, Sharpe & Dohme lectr award, 82. *Mem:* Am Chem Soc; The Chem Soc; Chem Inst Can; AAAS. *Res:* Synthesis of complicated natural products; structural determination of natural products; mechanisms of organic reactions; new synthetic methods with silicon, sulfur and phosphorus compounds. *Mailing Add:* Dept of Chem McGill Univ Montreal PQ H3A 2K6 Can

CHAN, TAT-HUNG, b Hong Kong, Aug 21, 51. AUTOMATA THEORY, COMPUTATIONAL COMPLEXITY. *Educ:* Dartmouth Col, AB, 74; Cornell Univ, MS & PhD(comput sci), 80. *Prof Exp:* Programmer, Asiadata Ltd, Hong Kong, 75-76; asst prof comput sci, Univ Minn, 80-82; asst prof, 82-88, ASSOC PROF, DEPT MATH & COMPUT SCI, STATE UNIV NY, FREDONIA, 88- *Concurrent Pos:* Lectr comput studies, Univ Hong Kong, 85-88. *Mem:* Asn Comput Mach; Soc Indust & Appl Math; Inst Elec & Electronic Engrs, Comput Soc. *Res:* Complexity theory; automata theory; programming logics and program verification; programming languages. *Mailing Add:* Dept Math & Comput Sci State Univ NY Fredonia NY 14063

CHAN, TEH-SHENG, b Taiwan, China. HUMAN GENETICS, MICROBIOLOGY. *Educ:* Nat Taiwan Univ, MD, 63; Yale Univ, PhD(molecular biol), 69. *Prof Exp:* Intern, Nat Taiwan Univ Hosp, 62-63; asst prof physiol, Sch Med, Univ Conn, Farmington, 73-78; ASSOC PROF

MICROBIOL, UNIV TEX MED BR GALVESTON, 78- *Concurrent Pos:* Helen Hay Whitney fel, Rockefeller Univ, 69-71 & Mass Inst Technol, 71-73. *Mem:* Am Soc Microbiol; AAAS. *Res:* Somatic cell genetics; inborn errors of purine metabolism; viral integration in mammalian cells. *Mailing Add:* Dept of Microbiol Univ of Tex Med Br Galveston TX 77550

CHAN, TIMOTHY M, b Hong Kong, Apr 10, 39. METABOLIC REGULATION. *Educ:* Univ Calif, Davis, PhD(biochem), 72. *Prof Exp:* ASSOC PROF PHYSIOL & TOXICOL, UNIV SOUTHERN CALIF, 81- *Mem:* Am Physiol Soc; Am Diabetes Asn. *Mailing Add:* Dept Physiol Inst Toxicol Univ Southern Calif 1985 Zonal Ave Los Angeles CA 90033

CHAN, VINCENT SIKHUNG, b Hong Kong, Sept 1, 49. PLASMA PHYSICS, ELECTRICAL ENGINEERING. *Educ:* Univ Wis-Madison, BSEE, 72, MSEE, 73, PhD(plasma physics), 75. *Prof Exp:* DIV DIR, GEN ATOMICS, 75- *Mem:* Fel Am Phys Soc. *Res:* Radio-frequency heating of fusion plasmas, wave propagation and instabilities in plasmas and transport phenomena in plasmas; high power microwave generation. *Mailing Add:* 13-311 Gen Atomics PO Box 85608 San Diego CA 92186-9784

CHAN, W(AH) Y(IP), b Shanghai, China, Dec 1, 32; nat US; m 61; c 2. REPRODUCTIVE & ENDOCRINE PHARMACOLOGY. *Educ:* Univ Wis, BA, 56; Columbia Univ, PhD(pharmacol), 61. *Prof Exp:* From res assoc biochem to assoc prof pharmacol, 60-76, actg chmn pharmacol, 83-91, PROF PHARMACOL, MED COL, CORNELL UNIV, 76- *Concurrent Pos:* mem basic pharmacol adv comt, Pharmaceut Mfrs Asn Found, 73-80; mem, Pharmacol Study Sect, NIH, 77; consult, Prog Proj Rev, Nat Heart Lung & Blood Inst & NIH, 81. *Honors & Awards:* NIH Res Career Develop Award, 68-73; Irma T Hirschl Career Scientist, 73-77. *Mem:* Soc Exp Biol & Med; Am Soc Pharmacol & Exp Therapeut; Soc Study Reproduction; Harvey Soc; NY Acad Sci; US Pharmacopeia, 85. *Res:* Pharmacology of neurohypophysial hormones and polypeptides; renal pharmacology; reproductive and uterine pharmacology of oxytocin and prostaglandins. *Mailing Add:* Dept of Pharmacol Cornell Univ Med Col 1300 York Ave New York NY 10021

CHAN, WAH CHUN, b Kwangtung, China, Oct 8, 34; Can citizen; m 67; c 4. CONTROL ENGINEERING, OPERATIONS RESEARCH. *Educ:* Nat Taiwan Univ, BSc, 58; Univ NB, MSc, 61; Univ BC, PhD(control), 65. *Prof Exp:* Asst, Univ NB, 59-61; res asst, Univ BC, 64-65; mem sci staff, Res & Develop Labs, Northern Elec Co, 65-67; from asst prof to assoc prof elec eng, 67-77, PROF ELEC ENG, UNIV CALGARY, 77- *Honors & Awards:* 73 Ambrose Fleming Premium, Brit Inst Elec Engrs, 74. *Mem:* Sr mem Inst Elec & Electronics Engrs. *Res:* Optimal control systems; power systems; telecommunication and queueing systems; reliability. *Mailing Add:* Dept Elec Eng Univ Calgary Calgary AB T2N 1N4 Can

CHAN, WAI-YEE, b Canton, China, Apr 28, 50; m 76; c 3. REPRODUCTIVE BIOLOGY, DEVELOPMENTAL MOLECULAR BIOLOGY. *Educ:* Chinese Univ, Hong Kong, BSc 74; Univ Fla, PhD(biochem), 77. *Prof Exp:* Teaching asst biochem, Univ Fla, 74-77; fel, Univ Okla, 77-78, res assoc pediat, 78-79, asst prof pediat & biochem, 79-82, assoc prof pediat biochem & molecular biol, 82-89; PROF PEDIAT BIOCHEM & MOLECULAR BIOL, ANAT & CELL BIOL, GEORGETOWN UNIV, 89- *Concurrent Pos:* Staff affil, GEM Serv, dir, Trace Element Lab & asst sci dir, Biochem Genetics Lab, Okla Children's Hosp, 79-87; consult, Vet Admin Med Ctr, Oklahoma City, Okla, 81-85; consult ed, J Am Col Nutrit, 82-; vis scientist biochem, Univ Wash, 82; affil assoc mem, Okla Med Res Found, 84-87, assoc mem, 87-89. *Mem:* Am Inst Nutrit; NY Acad Sci; Am Soc Biol Chem; Soc Pediat Res; Biochem Soc; Am Soc Human Genetics; Am Inst Immunol. *Res:* Molecular mechanisms of inborn errors of metabolism; molecular reproductive biology. *Mailing Add:* Dept Pediat Georgetown Univ Med Ctr 3800 Reservoir Rd NW Washington DC 20007

CHAN, WING CHENG RAYMOND, b Canton, China, Feb 19, 36; US citizen; m 62; c 3. CHEMICAL ENGINEERING, PROCESS DEVELOPMENT & DESIGN. *Educ:* Univ Calif, BS, 60; Univ Minn, Minneapolis, PhD(chem eng), 64. *Prof Exp:* Res engr, Uniroyal, Inc, 64-65, sr engr, Chem Div, 65-66, sr res engr, 66-69, res scientist, 69-76, ENG ASSOC, UNIROYAL CHEM CO, 76- *Mem:* Am Inst Chem Engrs; Am Chem Soc. *Res:* Computer simulation and optimization of chemical processes; reaction kinetics and reactor engineering; interfacial heat and mass transfer; applied mathematics in chemical engineering; fluid-fluid separation. *Mailing Add:* 250 Sharon Dr Cheshire CT 06410-4239

CHAN, YAT YUNG, b Hong Kong, March 12, 36; Can citizen; m 66; c 1. AERODYNAMICS, FLUID MECHANICS. *Educ:* Nat Taiwan Univ, BSc, 56; Univ Sydney, Australia, MEngSc, 60; Univ Toronto, PhD(aerodynamic eng), 65. *Prof Exp:* Sci officer, Weapons Res Estab, Australia, 60-61; asst res officer, 65-67, assoc res officer, 67-75, SR RES OFFICER, NAT RES COUN, CAN, 75-, CHIEF AERODYNAMICIST, HIGH SPEED AERODYNAMICS LAB, 81- *Concurrent Pos:* Sessional lectr, Carleton Univ, Ottawa, Can, 81- *Mem:* Am Inst Aeronaut & Astronaut; Can Aeronaut & Space Inst. *Res:* Boundary layer theory, computational methods for two and three-dimensional, laminar and turbulent flow; organized structure of jet turbulence with application to jet noise; transonic wind tunnel wall interference; transonic three dimensional flows. *Mailing Add:* Nat Aeronaut Estab Nat Res Coun Can Montreal Rd Ottawa ON R1A 0K6 Can

CHAN, YICK-KWONG, b China, Oct 14, 35. BIOMETRY. *Educ:* Taiwan Prov Col Agr, BS, 55; Univ Minn, MS, 60, PhD(biostatist), 66. *Prof Exp:* Asst prof pub health, Yale Univ, 66-68; assoc prof biomet & head regional statist off, Southeastern Cancer Study Group, Emory Univ, 68-74; asst chief, Coop Stud Prog Coord Ctr, Vet Admin Med Ctr, Perry Point & assoc prof biomet, Univ Md, 74-75; chief, Coop Studies, Prog Coord Ctr, Vet Admin Med Ctr, West Haven & lectr pub health biostatist, 75-87, SPEC ASST TO CHIEF, COOP STUDIES PROG, YALE UNIV, 87- *Mem:* Fel Am Pub Health Asn; Am Statist Asn; Inst Math Statist; Biomet Soc; Soc Clin Trials. *Res:* Application of statistics to clinical trials, epidemiology and bioassays; epidemic simulations. *Mailing Add:* Vet Admin Coop Studies Prog Vet Admin Med Ctr West Haven CT 06516

CHAN, YUN LAI, b Taichung, Taiwan, Dec 3, 41; US Citizen; m 68; c 2. KIDNEY & RENAL PHYSIOLOGY, MEMBRANE BIOPHYSICS. *Educ:* Kaohsiung Med Col, BS, 65; Col Med, Nat Taiwan Univ, MS, 67; Sch Med, Univ Louisville, PhD(pharmacol), 71. *Prof Exp:* Instr renal pharmacol, Sch Med, Univ Louisville, 71-74, physiol, Sch Med, Yale Univ, 74-76; from asst prof to assoc prof, 76-88, PROF PHYSIOL, COL MED, UNIV ILL, 88- *Concurrent Pos:* Res assoc, renal physiol, Max Planck Inst Biophysics, 72-73; prin investr, NIH, 81-88 & Am Heart Asn. *Honors & Awards:* Golden Apple Award. *Mem:* Am Physiol Soc; Am Soc Nephrol; Am Fedn Clin Res; Sigma Xi; NY Acad Sci. *Res:* Neural and hormonal control of renal function; cellular mechanisms of renal tubular transport. *Mailing Add:* Col Med Univ Ill 835 S Wolcott St CM 901 Chicago IL 60680

CHAN, YUPO, US citizen. TRANSORTATION ENGINEERING. *Educ:* Mass Inst Technol, BS, 67, MS, 69, PhD(oper res), 72. *Prof Exp:* Transp consult, Peat Marwick Mitchell & Co, DC, 72-74; asst prof transp & opers res, Pa State Univ, University Park, 74-79; assoc prof transp & policy anal, State Univ NY, Stony Brook, 80-83; assoc prof transp, Wash State Univ, Pullman, Wash, 83-87; PROF & DEP HEAD, DEPT OPER SCI, AIR FORCE INST TECHNOL, 87- *Concurrent Pos:* Hon cong fel, Off Technol Assessment, US Cong, 79-80; mem, Transp Res Bd, Nat Acad Sci. *Mem:* Opers Res Soc Am; Am Soc Civil Engrs. *Res:* Facility location and land use; travel demand forecasting; network routing models; optimization analysis of networks; policy analysis; air transportation; multicriteria decision making. *Mailing Add:* Dept Oper Sci Air Force Inst Technol Wright-Patterson AFB OH 45433

CHANANA, ARJUN DEV, b Lyallpur, Punjab, India, Nov 6, 30; US citizen; m 63; c 2. EXPERIMENTAL PATHOLOGY. *Educ:* Univ Rajasthan, MB & BS, 55; FRCS(E) & FRCS, 60. *Prof Exp:* Internship, residencies & postdoctoral work surg, 55-60; chief resident surg, Bolton Dist Gen Hosp, Eng, 60-63; asst scientist, 63-66, assoc scientist, 66-69, actg head, Div Hemat, 68-69, head scientist, Div Exp Path, 69-77, SR SCIENTIST, MED DEPT, BROOKHAVEN NAT LAB, 79-, CHMN, 86- *Concurrent Pos:* Asst attend physician, Hosp of Med Res Ctr, Brookhaven Nat Lab, 63-65, assoc attend physician, 65-70, head, Div Exp Path, 69-79, attend physician, 70-, chief of medical staff, 74-84; assoc prof path & surg, Health Sci Ctr, State Univ NY Stony Brook, 70-86, res prof path & surg, assoc dean, Sch Med, State Univ NY Stony Brook, 86-; guest prof immunopath, Univ Bern, Switz, 71-72, res consult surg, Nassau County Med Ctr, 70- *Mem:* Am Soc Hemat; Soc Exp Biol & Med; Transplantation Soc; Soc Leakocyte Biol; Am Asn Path; AAAS. *Res:* Pulmonary host-defense mechanisms; respiratory pathophysiology; experimental hematology; transplantation immunology. *Mailing Add:* Med Dept Brookhaven Nat Lab Upton NY 11973

CHANCE, BRITTON, b Wilkes-Barre, Pa, July 24, 13; m 38, 56; c 12. BIOPHYSICS, BIOCHEMISTRY. *Educ:* Univ Pa, BS, 35, MS, 36, PhD(phys chem), 40; Univ Cambridge, PhD(physiol), 42, DSc, 52; Karolinska Inst, Sweden, MD, 62. *Hon Degrees:* DSc, Cambridge Univ, 52, Med Col Ohio, 74, Semmelweise Univ, 76, Hahnemann Med Col, 77, Univ Pa, 85, Univ Helsinki, 90 & Univ Dusseldorf, 91. *Prof Exp:* Actg dir, E R Johnson Found, 40-41, from asst prof to prof, 41-77, univ prof, 77-83, EMER PROF BIOPHYS & PHYS BIOCHEM, SCH MED, UNIV PA, 83-, DIR, INST STRUCT & FUNCTIONAL STUDIES, UNIV CITY SCI CTR, UNIV PA, 82- *Concurrent Pos:* Investr, Off Sci Res & Develop, 41; res assoc, Radiation Lab, Mass Inst Technol, 41-42, group leader, 41-45, assoc div head, 42-45, mem vis comt, 54-56; Guggenheim fel, Nobel Inst, Stockholm & Moleno Inst, 46-48; US Navy consult to attache res, London, 48; comt blood & blood derivatives, Am Red Cross & Nat Res Coun, 48; consult molecular biol, Nat Sci Found, 51-56; vis comt, Bartol Res Found, 55-59; pres sci adv comt, 59-60; coun mem, Nat Inst Alcohol Abuse & Alcoholism, 71-75, Working Group Molecular Control, Nat Cancer Inst, 73-; panel mem, Biotech Res Resources Prog, Div Res Resources, Prin & Resource Invest Meeting, 83; exchange scholar, USSR, 63; foreign fel, Churchill Col, Univ Cambridge, 66; Keilin lectr, 66; exchange scholar, USSR, 63; Redfearn lectr, 70; foreign fel, Churchill Col, Cambridge. *Honors & Awards:* Paul Lewis Award, 50; Morlock Award, 61; Genootschaps Medal, Dutch Biochem Soc, 65; Franklin Medal, 66; Harrison Howe Award, Am Chem Soc, 66, Nichols Award, 70; Heineken Medal, Royal Neth Acad Sci & Lett, 70; Gairdner Award, 72; Post-Cong Festschrift, 73; Semmelweis Medal, 74; Nat Medal Sci, 74; 2nd Julius L Jackson Mem lect, Wayne State Univ, Sigma Xi, 76; Elizabeth Winston Lanier Award, Am Acad Orthop, 86; Gold Medal, Soc Magnetic Resonance Soc, 88. *Mem:* Nat Acad Sci; fel Am Phys Soc; Am Philos Soc (secy, 69, vpres, 84-90); foreign mem Royal Soc London; Am Acad Arts & Sci; Royal Swed Acad Sci; Soc Biol Chemists; fel Inst Radio Engrs; Soc Gen Physiologists; Leopoldina Acad Ger. *Res:* Automatic ship steering; photoelectric control units; radar timing and computing devices; sensitive spectrophotometers; enzyme-substrate compounds and reaction mechanisms of catalases; peroxidases, dehydrogenases; cytochromes; kinetics of multienzyme systems; oscillating enzyme systems; cation accumulation; cell oxygen requirements; non invasive NMR and time resolved optical spectroscopes studies of organ biochemistry. *Mailing Add:* Dept Biochem & Biophys Sch Med Univ Pa Philadelphia PA 19104

CHANCE, C(LAYTON) W(ILLIAM), engineering graphics, for more information see previous edition

CHANCE, CHARLES JACKSON, b Belen, NMex, Apr 18, 14; m 36; c 2. ZOOLOGY, FISHERIES. *Educ:* Berea Col, AB, 35; Univ Tenn, MS, 42. *Prof Exp:* Aquatic biologist, Fisheries Mgt, Tenn Conserv Dept, 40-48; aquatic biologist, Tenn Valley Authority, 48-58, chief aquatic biologist, 58-60, chief Fisheries & Waterfowl Resources Br, 60-79; RETIRED. *Mem:* AAAS; Am Inst Biol Sci; Am Fisheries Soc; Wildlife Soc. *Res:* Fisheries populations and management. *Mailing Add:* 4304 Gaines Rd Knoxville TN 37918

CHANCE, KELLY VAN, b Shamrock, Tex, Jan 19, 47; m 72; c 1. CHEMICAL PHYSICS. *Educ:* Univ Hawaii, BS, 70; Harvard Univ, AM, 72, PhD(chem physics), 77. *Prof Exp:* PHYSICIST, SMITHSONIAN ASTROPHYS OBSERV, HARVARD-SMITHSONIAN CTR ASTROPHYS, 77- *Mem:* Am Chem Soc; Am Geophys Union. *Res:* Molecular spectroscopy; atmospheric chemistry; chemical astrophysics. *Mailing Add:* Ctr For Astrophys 60 Garden St Cambridge MA 02138

CHANCE, KENNETH BERNARD, b New York, NY, Dec 8, 53; m 81; c 4. ENDONDONTICS. *Educ:* Fordham Univ, BS, 75; Case Western Reserve Univ, DDS, 79. *Prof Exp:* Gen pract res gen dent, Jamaica Hosp, 79-80; endodontic resident, Univ Med & Dent NJ, 80-82; attend endodontics, Jamaica Hosp, 80-87; chief endodontics, Kings County Med Ctr, 82-91; asst prof, 82-87, ASSOC PROF ENDODONTICS, UNIV MED & DENT NJ, 87-; ATTEND ENDODONTICS, NCENT BRONX MED CTR, 82-, KINGSBROOK JEWISH MED CTR, 84- *Concurrent Pos:* prin investr, Univ Med & Dent NJ, 84-86, res award, 85, asst dean external affairs, 85-; prin investr, NIH, 89-; consult, Comm Health, State NJ, 90-; fel, Robert Wood Johnson Health Policy Prog, 91-, Pew Nat Dent Leadership Prog, 91-; mem, House Delegates, Nat Dent Asn. *Mem:* Am Asn Dent Schs; Int Asn Dent Res; Nat Dent Asn; Am Asn Endodontists. *Res:* Use of corticosteroids; meticortelone, in the management of pain associated with acute periapical periodontitis. *Mailing Add:* 87 Victory Blvd New Rochelle NY 10804

CHANCE, ROBERT L, b Detroit, Mich, Feb 1, 24; m 52; c 4. ANALYTICAL CHEMISTRY, INORGANIC CHEMISTRY. *Educ:* Wayne State Univ, BS, 48. *Prof Exp:* Sr res chemist, Gen Motors Res Labs, 49-54, group leader inorg analytical chem, 54-61, sr res chemist, 62-78, sr res scientist, 78-81, staff res scientist, 81-85, group leader corrosion res, 83-85; RETIRED. *Res:* General and automotive corrosion research; electrochemical polarization techniques; surface analysis; engine coolants; inhibitor systems; coatings; alloy compositions; metallurgical structure; metal deformation; heat treatment; cavitation phenomena; corrosion surveys; surface modification by ion implantation and laser annealing. *Mailing Add:* 1935 Wiltshire Rd Berkley MI 48072-3315

CHANCE, RONALD E, b Lapeer, Mich, Jan 17, 34; m 61; c 1. BIOCHEMISTRY, METABOLISM. *Educ:* Purdue Univ, BS, 56, MS, 59, PhD(biochem), 62. *Prof Exp:* Asst prof biochem, Purdue, 62-63; sr biochemist, 63-69, res scientist, 69-77, RES ASSOC, ELI LILLY & CO, 78- *Concurrent Pos:* USDA grant, 62-63. *Mem:* AAAS; Am Chem Soc; Am Soc Animal Sci; Sigma Xi. *Res:* Animal nutrition, amino acid requirements and interrelationships in Chinook salmon and weanling pigs; chemistry of protein hormones; isolation, purification, characterization and chemistry of protein hormones. *Mailing Add:* 19303 Flippin Rd Westfield IN 46074

CHANCE, RONALD RICHARD, b Memphis, Tenn, July 24, 47; m 67; c 2. POLYMER CHEMISTRY. *Educ:* Delta State Univ, BS, 70; Dartmouth Col, PhD(chem), 74. *Prof Exp:* Staff physicist, Allied Chem Corp, 74-77, group leader, 77-81, mgr, Allied Corp, 81-85; lab dir, Corp Res, Exxon, 86-90, DIV MGR, PARAMINS TECHNOL, EXXON CHEM, 90- *Mem:* Fel Am Phys Soc; Am Chem Soc. *Res:* Optical and electrical properties of conjugated polymers; polymerization in organized media. *Mailing Add:* Paramins Tech Exxon Chem Linden NJ 07036

CHAND, NARESH, LUNG DISEASES & INFLAMMATION, ALLERGIES & PULMONARY PHARMACOLOGY. *Educ:* Univ Guelph, Can, PhD(pharmacol), 77. *Prof Exp:* ASSOC DIR DEPT PHARMACOL, WALLACE LABS, CARTER-WALLACE INC, 81- *Mailing Add:* Wallace Labs Box One Carter-Wallace Inc Cranbury NJ 08512

CHAND, NARESH, b Dhikana, India, May 10, 51; m 77; c 3. III-V SEMICONDUCTORS, PHYSICS & TECHNOLOGY OF SEMICONDUCTOR DEVICES. *Educ:* Meerut Univ, India, BSc, 70, MSc, 72; Birla Inst Technol & Sci, Pilani, India, MSc, 74; Univ Sheffield, UK, MEng, 80, PhD(elec eng), 83. *Prof Exp:* Jr sci officer, Dept Electronics, Govt India, 74-79; res assoc, Univ Ill, Urbana, 83-85; MEM TECH STAFF, AT&T BELL LABS, 85- *Mem:* Inst Elec & Electronics Engrs; Mat Res Soc; Electrochem Soc. *Res:* Molecular beam epitaxial growth, processing, characterization, and modeling of III-V semiconductors based photonic and high speed electronic devices; gallium arsenide on silicon technology. *Mailing Add:* 461 Park Ave Berkeley Heights NJ 07922

CHANDAN, HARISH CHANDRA, US citizen; m 78; c 2. GLASS SCIENCE. *Educ:* Indian Inst Technol, Kanpur, BTech, 69, MTech, 72; Syracuse Univ, MS, 74; Pa State Univ, PhD(ceramics sci), 80. *Prof Exp:* MEM TECH STAFF, AT&T BELL LABS, 77- *Mem:* Am Ceramic Soc. *Res:* Mechanical behavior and processing of optical fibers; long-term reliability predictions; effect of environment on the strength of optical fibers; new processing techniques to make large preforms to make long lengths of fiber. *Mailing Add:* 1545 Blyth Walk Snellville GA 30278

CHANDAN, RAMESH CHANDRA, b Lahore, Pakistan, July 5, 34; US citizen; m 60; c 3. FOOD TECHNOLOGY, BIOCHEMISTRY. *Educ:* Panjab Univ, India, BSc, 53, Hons, 55, MSc, 56; Univ Nebr, Lincoln, PhD(dairy mfg & chemistry), 63. *Prof Exp:* Lectr chem, Panjab Univ, India, 56-57; lectr dairy chem, Nat Dairy Res Inst, India, 57-59; from asst to assoc dairy sci, Univ Nebr, Lincoln, 59-66; scientist dairy technol, Unilever, Ltd, Eng, 67-69; mgr res & develop, Dairylea Coop, Inc, 70-74; vpres, Whey Prod & Tech Serv, Purity Cheese Co, Anderson Clayton Foods, 74-76; assoc prof, Food Sci & Human Nutrit Dept, Mich State Univ, 76-82; assoc dir, Corp Res & Develop, Land O' Lakes, Minn, 82-84; PRIN SCIENTIST, GEN MILLS, INC, 84- *Honors & Awards:* Nordica Int Award Excel Res, Am Cultured Dairy Prod Inst, 82. *Mem:* Am Dairy Sci Asn; Inst Food Technologists; Am Chem Soc; fel Am Inst Chemists. *Res:* Physico-chemical properties of milk; lipases; lysozymes; lipids; milk protein; microbial chemistry and enzymes; dairy product technology; cultures; cultured products; food product development; whey protein manufacturing technology, quality control, cheese product development; whey fractionation; yogurt technology; health and nutrition. *Mailing Add:* 3257 Rice Creek Rd New Brighton MN 55112

CHANDER, JAGDISH, b Toba Tek Singh, India, Mar 7, 33; m 65; c 3. EXPERIMENTAL NUCLEAR PHYSICS. *Educ:* DAV Col, Jullundur, India, BSc, 52; Panjab Univ, India, BA, 53 & 54; Univ Rajasthan, MSc, 56; Univ Erlangen, Dr rer nat(physics), 61. *Prof Exp:* Demonstr chem, DAV Col, Jullundur, 52-53; demonstr, DSD Col, Gurgaon, 53-54; lectr physics, DAV Col, Jullundur, 56-68; lectr, Birla Sci Col, Pilani, 62 & Panjab Univ, 62-66; asst prof, 66-67, assoc prof, 67-70, PROF PHYSICS & ASTRON, UNIV WIS-STEVENS POINT, 70- *Concurrent Pos:* Res assoc, Siemens Res Lab, Erlangen, Ger, 61; dir, NSF Undergrad Res Partic, 68-71; acad year exten grant col teachers, NSF, 69-70; vis prof, Birla Inst Technol & Sci, Pilani, India, 79. *Mem:* Am Phys Soc; Am Asn Physics Teachers. *Res:* Low energy nuclear physics; semiconductor radiation detectors; teaching of physics. *Mailing Add:* Dept Phys & Astron Univ Wis Stevens Point WI 54481

CHANDER, SATISH, b Muzaffargarh, Pakistan, Dec 25, 37; Canadian citizen; m 61; c 2. VETERINARY PATHOLOGY, DIAGNOSTIC SEROLOGY. *Educ:* Punjab Univ, India, BVSc, 56, MVSc, 60; Vet Sch, Hannover, Ger, DVM, 68; Ont Vet Col, Guelph, PhD(oncol), 71. *Prof Exp:* Res asst, Regional Animal Nutrit Ctr, 59-60; asst res officer, Punjab Agr Univ, 60-65, asst prof path, Vet Sch, 65-66; hematologist, Vet Sch, Hannover, Ger, 68; lectr, Ont Vet Col, 71-72; RES SCIENTIST, AGR CAN, OTTAWA, 72- *Mem:* Can Asn Vet Pathologists; Can Vet Med Asn; Int Asn Comp Res Leukemia & Related Dis. *Res:* Etiology, diagnosis and epizootiology of leukemia in cattle (Bovine Leukosis). *Mailing Add:* 26 Westchester Ct Napean ON K2J 3R1 Can

CHANDLER, A BLEAKLEY, b Augusta, Ga, Sept 11, 26; m; c 2. PATHOLOGY. *Educ:* Med Col Ga, MD, 48. *Prof Exp:* Intern, Baylor Univ Hosp, 48-49; resident path, Univ Hosp, Augusta, Ga, 49-50; from asst prof to assoc prof, 53-62, PROF PATH, MED COL GA, 62-, CHMN DEPT, 75- *Concurrent Pos:* Nat Cancer Inst trainee path, Med Col Ga, 50-51; Commonwealth Fund res fel, Inst Thrombosis Res, Norway, 63-64. *Mem:* Am Asn Hist Med; Int Acad Path; Am Asn Path. *Res:* Thrombosis; experimental and human cardiovascular pathology. *Mailing Add:* Dept Path Med Col Ga Augusta GA 30912-3600

CHANDLER, ALBERT MORRELL, b Pontiac, Mich, July 10, 32; m 59; c 3. BIOCHEMISTRY. *Educ:* Wayne State Univ, BS, 55, PhD(biochem), 62. *Prof Exp:* Pharmacist, Mich, 51-56; res assoc anat, Wayne State Univ, 63-66; from asst prof to assoc prof, 66-73, PROF BIOCHEM, SCH MED, UNIV OKLA, 73- *Concurrent Pos:* NIH fels, Ohio State Univ, 61-63. *Mem:* AAAS; Am Chem Soc; Sigma Xi. *Res:* Plasma protein metabolism; hepatic nucleic acid metabolism; regulation of hexosamine synthesis. *Mailing Add:* Dept Biochem & Molecular Biol Sci Ctr Univ Okla Oklahoma City OK 73190

CHANDLER, ARTHUR CECIL, JR, b Hinton, WVa, Feb 14, 33; m 57; c 4. MEDICINE, OPHTHALMOLOGY. *Educ:* Fla Southern Col, AB, 53; Univ Tenn, MS, 55; Duke Univ, MD, 59; Am Bd Ophthal, dipl, 65. *Prof Exp:* Instr surg, Sch Med, Stanford Univ, 63-65; assoc, 65-66, asst prof, 66-70, ASSOC PROF OPHTHAL, SCH MED, DUKE UNIV, 70-, ASSOC ANAT, 67- *Concurrent Pos:* Consult, Durham Vet Admin Hosp, 65-, chief ophthal, 66-82; consult, Watts Hosp & Sea Level Hosp. *Mem:* AAAS; fel Am Col Surg; AMA; Am Asn Ophthal; Am Optom Asn; Sigma Xi. *Res:* Prevention and treatment of amblyopia ex anopsia. *Mailing Add:* 3901 N Roxboro Rd PO Box 15249 Durham NC 27704

CHANDLER, CARL DAVIS, JR, b Pulaski, Va, Jan 25, 44; m 66; c 2. ANALYTICAL CHEMISTRY. *Educ:* Emory & Henry Col, BS, 65; ETenn State Univ, MA, 67; Va Polytech Inst & State Univ, PhD(chem), 73. *Prof Exp:* ANAL GROUP AREA SUPVR CHEM, RADFORD ARMY AMMUNITION PLANT, HERCULES INC, 67- *Concurrent Pos:* Lectr short course liquid chromatography, Am Chem Soc, 73-; consult, US Army Res Off, Picatinny Arsenal. *Mem:* Am Chem Soc. *Res:* Investigation of column efficiency improvements in gas and liquid chromatography; analysis of propellants and explosives; development of pollution abatement processes and pollution measuring instrumentation specifically for various explosives and propellants. *Mailing Add:* 50 Aldrin St Dublin VA 24084

CHANDLER, CHARLES H(ORACE), b Weihsien, Shantung, China, July 10, 18; div. PHYSICS, ENGINEERING. *Educ:* Col of Wooster, BA, 40; Ohio State Univ, MSc(physics), 46; Worcester Polytech Inst, MS, 84. *Prof Exp:* Mem tech staff, David Sarnoff Res Ctr, 46-54; eng leader, Radio Corp Am, 54-57; chief electronics engr, Gillette Safety Razor Co, 58-60, mgr electronic process develop, 60-62; prin engr, Booz-Allen Appl Res, Inc, 62-64, assoc dir develop & design div, 64-66; sr tech consult, Eng Div, Am Safety Razor Co, 66-71; in charge new prod develop, Micro Switch, Div Honeywell Inc, 71-74; independent consult, Jarrell-Ash, Div Fischer Sci Corp & Ball Bros Res, Inc, 74-77; instrumentation specialist, Centronics Data Comput Corp, 77-78; proj engr, Stanley Tools, Div Stanley Works, Inc, 78-82; proj engr, qual control, M/A-Com, Burlington, Mass, 82-85; RETIRED. *Concurrent Pos:* Cert client instr, Kepner-Tregoe Mgt Course; consult, optical systs mfg opers, 85- *Mem:* Sigma Xi; sr mem Instrument Soc Am; fel AAAS; Inst Defense Prep; Soc Photo-Optical Instrumentation Engrs; Soc Mfg Engrs. *Res:* Propagation in dielectric materials; television circuit development; radar systems and special displays; electronic navigation, bombing, military data-handling systems; applications of electronics and optics in production control systems; manufacturing engineering; bring new technology (including lasers) to bear on inspection and printing of steel tape rules. *Mailing Add:* 1010 Waltham St Lexington MA 02173-8044

CHANDLER, CLAY MORRIS, b McKenzie, Tenn, Nov 2, 27; m 56; c 2. ZOOLOGY. *Educ:* Bethel Col, Tenn, BS, 50; George Peabody Col, MA, 54; Ind Univ, PhD(zool), 65. *Prof Exp:* Asst prof biol, WGa Col, 59-61; prof biol, Bethel Col, Tenn, 65-70; PROF BIOL, MID TENN STATE UNIV, 70- *Concurrent Pos:* Consult environ mgt, planning & eng, Nashville, 77- *Mem:* Sigma Xi; NAm Benthological Soc. *Res:* Ecology and systematics of freshwater triclad Turbellaria; ecology and taxonomy of nematomorpha. *Mailing Add:* Dept Biol Mid Tenn State Univ Murfreesboro TN 37132

CHANDLER, COLSTON, b Boston, Mass, June 7, 39; div; c 3. THEORETICAL PHYSICS, SCATTERING THEORY. *Educ:* Brown Univ, ScB, 61; Univ Calif, Berkeley, PhD(physics), 67. *Prof Exp:* From asst prof to assoc prof, 68-78, PROF PHYSICS, UNIV NMEX, 78- *Concurrent Pos:* Vis prof, Bonn Univ, Germany, 78; fac scientist in residence, Argonne Nat Lab, 84-85; vis scientist, CRIP, Budapest, Hungary, 85- *Mem:* AAAS; Am Phys Soc; Sigma Xi; Am Asn Univ Professors; Int Asn Math Physicists. *Res:* Nonrelativistic quantum mechanical scattering theory. *Mailing Add:* Dept Physics & Astron Univ NMex Albuquerque NM 87131-1156

CHANDLER, DAVID, b Brooklyn, NY, Oct 15, 44; m 66; c 2. CHEMICAL PHYSICS, STATISTICAL MECHANICS. *Educ:* Mass Inst Technol, SB, 66; Harvard Univ, PhD(chem physics), 69. *Prof Exp:* Res chemist, Univ Calif, San Diego, 69-70; asst prof to prof chem, Univ Ill, Urbana, 70-83; prof chem, Univ Pa, 83-85; PROF CHEM, UNIV CALIF, BERKELEY, 86- *Concurrent Pos:* Teaching fel chem, Harvard Univ, 66-69; Alfred P Sload res fel, 72-74; vis assoc prof chem, Columbia Univ, NY, 77-78; vis scientist & consult, IBM Res Ctr, Yorktown Heights, 78 & Oak Ridge Nat Lab, Tenn, 79; Guggenheim fel, 81-82. *Honors & Awards:* Burke lectr, Faraday Div, Royal Soc Chem, 85; Hildebrand Award, Am Chem Soc, 89. *Mem:* Am Inst Physics; fel Am Phys Soc; Am Chem Soc; fel AAAS. *Res:* Theoretical chemistry and statical mechanics; structure, molecular dynamics and quantum processes in liquids; chemical equilibria and kinetics in liquids; aqueous solutions; structure and dynamics of chain molecules. *Mailing Add:* Dept Chem Univ Calif Berkeley CA 94720

CHANDLER, DEAN WESLEY, b Chicago, Ill, June 5, 44. PHYSICAL CHEMISTRY, ANALYTICAL CHEMISTRY. *Educ:* Harvard Univ, BA, 67; Northwestern Univ, PhD(chem), 73. *Prof Exp:* Lectr chem, State Univ NY Albany, 73-74, vis asst prof, 74-75; ASST PROF CHEM, WILLIAMS COL, 75-, ASST DEAN, 78- *Mem:* Am Chem Soc; Am Inst Physics. *Res:* Development of new physical techniques which might prove useful as tools for chemical analysis. *Mailing Add:* 5116 S Woodlawn Chicago IL 60615

CHANDLER, DONALD ERNEST, b San Bernardino, Calif, Nov 22, 25; m 46; c 3. PHYSICS. *Educ:* US Naval Acad, BS, 46; Univ Calif, Los Angeles, MA, 54, PhD(physics), 58. *Prof Exp:* Tech officer, Electronic Systs, Off Naval Res, US Navy, 55-57, physicist, Armed Forces Spec Weapons Proj, 57-59, proj officer, Adv Res Proj Agency, 59-63, sr prog officer, Electronics Lab, 63-66, proj mgr, Advan Res Proj Agency, 66-68; mem tech staff, 68-71, mgr phys sci, Tempo, 71-81, MGR PHYS SCI, GEN ELEC CO, 81- *Mem:* Acoustical Soc Am; Am Geophys Union. *Res:* Propagation of acoustic waves in media exhibiting relaxation; propagation of electromagnetic waves in ionized gases. *Mailing Add:* 921 Arbolado Rd Santa Barbara CA 93103

CHANDLER, DONALD STEWART, b Red Bluff, Calif, Sept 23, 49; m 76; c 3. SYSTEMATIC ENTOMOLOGY, MEDICAL ENTOMOLOGY. *Educ:* Univ Calif, Davis, BS, 71; Univ Ariz, MS, 73; Ohio State Univ, PhD(entom), 76. *Prof Exp:* Entomolgist, Bur Land Mgt, 76-77; sr operator, Mosquito Abatement Dist, Butte County, Calif, 78-81; ENTOMOLOGIST & CUR, UNIV NH, 81- *Honors & Awards:* R E Snodgrass Mem Award, Entom Soc Am, 74. *Mem:* Soc Systematic Zoologists; Sigma Xi; AAAS; Coleopterists Soc. *Res:* Biology, zoogeography and systematics of Anthicidae and Pselaphidae beetle families. *Mailing Add:* Dept Entomology Univ NH Durham NH 03824

CHANDLER, DOUGLAS EDWIN, b Oak Park, Ill, Oct 1, 45; m 69; c 3. CELL BIOLOGY, ELECTRON MICROSCOPY. *Educ:* Univ Rochester, BS, 67; Johns Hopkins Sch Med, MA, 69; Univ Calif, San Francisco, PhD(physiol), 77. *Prof Exp:* From asst prof to assoc prof, 80-90, PROF ZOOL, ARIZ STATE UNIV, 90- *Concurrent Pos:* Prin investr, NSF grants, 81-; vis scholar, Univ Calif, San Diego, 85-86; res career develop awardee, NIH, 85-90; guest ed, J Electron Micros Tech, 89- *Mem:* Am Soc Cell Biol; Soc Develop Biol. *Res:* Fertilization in echinoderms and amphibians; extracellular matrix dynamics; intracellular signals including calcium and protein phosphorylation; freeze-fracture electron microscopy of exocytosis in secretory cells. *Mailing Add:* Dept Zool Ariz State Univ Tempe AZ 85287-1501

CHANDLER, FRANCIS WOODROW, JR, b Milledgeville, Ga, July 25, 43; m 68; c 2. VETERINARY PATHOLOGY. *Educ:* Univ Ga, BS, 66, DVM, 67, PhD(vet path), 73; Am Col Vet Pathologists, dipl. *Prof Exp:* Vet med officer, Venereal Dis Res Lab, Ctr Dis Control, 67-70; resident, Dept Vet Path, Col Vet Med, Univ Ga, 70-73; CHIEF, PATH BR, HOST FACTORS DIV, CTR INFECTIOUS DIS, CENTERS DIS CONTROL, USPHS, 73-; ASSOC CLIN PROF PATH, MED COL GA, 83- *Concurrent Pos:* Consult pathologist, Food & Drug Admin, 77-; collab scientist, Yerkes Primate Res Ctr, Emory Univ, 80- *Mem:* Int Acad Path; NY Acad Sci; Sigma Xi; Am Col Vet Pathologists; Med Mycological Soc Am. *Res:* Animal models of human venereal diseases; pneumocystis carinii pneumonia; mycotic diseases; anemia and neoplasia; ultrastructure of cells; Legionnaires' disease; algal infections (protothecosis); immunohistologic diagnosis of infectious diseases; toxic shock syndrome, acquired immunodeficiency syndrome; in situ nucleic acid hybridization. *Mailing Add:* Dept Path Med Col Ga Sch Med 1120 15th St Augusta GA 30912

CHANDLER, FREDERICK WILLIAM, b Epsom, Surrey, Eng, May 17, 38; Can-Brit citizen; m 69; c 2. PALEOCLIMATES, CLASTIC SEDIMENTOLOGY. *Educ:* Reading Univ, Eng, BSc, 62; London Univ, BSc, 64; Univ Western Ont, PhD(geol), 69. *Prof Exp:* Field geologist proterozoic stratig, Ont Geol Surv, 70-72; RES SCIENTIST SEDIMENTOLOGY, REDBEDS, PALEOCLIMATES & MINERALIZATION IN SEDIMENTS, GEOL SURV CAN, OTTAWA, 72- *Mem:* Geol Asn Can. *Res:* Sedimentology and mineralization; tectonics; red beds of Canada, clastic sedimentology. *Mailing Add:* 223 Roger Rd Ottawa ON K1H 5C5 Can

CHANDLER, HORACE W, b Brooklyn, NY, May 23, 27; m 54; c 2. PHYSICAL CHEMISTRY, CHEMICAL ENGINEERING. *Educ:* Cornell Univ, BChE, 50; NY Univ, MChE, 55; Columbia Univ, DrEngSc, 60. *Prof Exp:* Tech trainee, Gen Chem Div, Allied Chem & Dye Corp, 50-51; chem engr, Gen Aniline & Film Corp, 51-53; res engr, Columbia Mineral Beneficiation Labs, 53-58; tech coordr, Radiation Applns, Inc, 58-60; tech adv to pres, Isomet Corp, 60-68; CHMN DEPT PHYS SCI & MATH, BERGEN COMMUNITY COL, 68- *Concurrent Pos:* Lectr, Stevens Inst Technol, 60. *Mem:* Am Inst Chem Eng; Am Chem Soc; Sigma Xi. *Res:* Radiation, polymer and high temperature chemistry; life support systems. *Mailing Add:* 27 Cedar Ridge Rd New Paltz NY 12561

CHANDLER, J RYAN, b Charleston, SC, July 30, 23; m 41; c 3. SURGERY, OTOLARYNGOLOGY. *Educ:* Duke Univ, MD, 47; Am Bd Otolaryngol, dipl. *Prof Exp:* PROF OTOLARYNGOL, SCH MED, UNIV MIAMI, 52-, CHMN DEPT, 72- *Concurrent Pos:* Consult, Vet Admin Hosp, Miami, Fla, 57 & NIH Commun Dis Training Grant Comt, 60-64. *Mem:* Am Soc Head & Neck Surg; Soc Head & Neck Surg; Am Laryngol Asn; Am Otolaryngol Soc; Am Acad Otolaryngol. *Res:* Otology and head and neck cancer surgery; otolaryngology teaching. *Mailing Add:* Dept Otolaryngol D48 Univ Miami Sch Med 1600 NW Tenth Ave Miami FL 33101

CHANDLER, JAMES HARRY, III, b New Orleans, La, Apr 26, 50; m; c 2. ENVIRONMENTAL CHEMISTRY, QUALITY ASSURANCE. *Educ:* Univ Tenn, Knoxville, BSc, 73, PhD(org chem), 78. *Prof Exp:* Res assoc chem, Int Paper Co, 78-81; chemist, Anal Qual Assurance Off, US Army Environ Hyg Agency, 81-87; PROJ SCIENTIST, & SR INORG DATA REVIEWER, MANTECH ENVIRON TECHNOL INC, 87- *Concurrent Pos:* Qual assurance officer, Region IV, ESAT, 89- *Mem:* Am Chem Soc. *Mailing Add:* 700 Michell Bridge Rd Apt 155 Athens GA 30606

CHANDLER, JAMES MICHAEL, b Wichita, Kans, Sept 30, 43; m 68; c 2. CROP SCIENCE, WEED SCIENCE. *Educ:* WTex State Univ, BS, 65; Okla State Univ, MS, 68, PhD(agron), 71. *Prof Exp:* Res asst, Okla State Univ, 69-71; RES AGRONOMIST WEED RES, SOUTHERN WEED SCI LAB, RES SERV, USDA, 71-; DEPT SOIL & CROP SCI, TEX A&M UNIV. *Mem:* Weed Sci Soc Am; Coun Agr Sci & Technol; Am Soc Agron; Crop Sci Soc Am; Sigma Xi. *Res:* Agronomy and botany related to weed science and the development of principles and practices for economical control of weeds. *Mailing Add:* Dept Soil & Crop Sci Tex A&M Univ College Station TX 77841

CHANDLER, JASPER S(CHELL), b Allen, Nebr, July 21, 11; m 38; c 4. MECHANICAL ENGINEERING. *Educ:* Ga Inst Tech, BS, 34, MS, 36; Pa State Col, PhD(mech eng), 38. *Prof Exp:* Machinist's apprentice, NC & St Louis RR, 29-34; asst, Eng Lab, Ga Inst Technol, 34-36; asst Diesel Lab, Pa State Col, 36-38; res eng, plastics div, Eastman Kodak Co, 38-57, sr res assoc, 58-76, consult, 76-91; RETIRED. *Mem:* Fel Soc Motion Picture & TV Engrs; Soc Photog Sci & Eng. *Res:* Motion picture film handling equipment; cameras; projectors; printers; perforators and special equipment. *Mailing Add:* 185 Dorian Lane Rochester NY 14626

CHANDLER, JERRY LEROY, b Little Falls, Minn, Sept 14, 40; m 69; c 2. BIOMATHEMATICS. *Educ:* Okla State Univ, BS, 63, PhD(biochem), 68. *Prof Exp:* Instr biochem, Okla State Univ, 68-69; chemist & dept leader, Zentrallaor fur Mutigeni Tatsprnfung, 69-72; res assoc biochem, Okla State Univ, 73-74; health standards mgr, Nat Inst Occup Safety & Health, 75-77, sect chief, 77-80, sci advisor, 80-82; CONSULT, 82- *Concurrent Pos:* Fac mem, NIH Found Advan Educ Sci, 81-82. *Mem:* Soc Risk Anal; Environ Mutagen Soc. *Res:* Development of mathematical models of dose-response relationships derived from biochemical mechanisms of action; information systems design; design analysis; risk assessment. *Mailing Add:* 837 Canal Dr McLean VA 22102-1407

CHANDLER, JOHN EDWARD, b Knoxville, Tenn, Feb 26, 41; m 64; c 3. REPRODUCTIVE PHYSIOLOGY. *Educ:* Southern Benedictine Col, BA, 64; Univ Tenn, MS, 69; Va Polytech Inst & State Univ, PhD(animal sci), 77. *Prof Exp:* Asst biologist cancer chemother, Southern Res Inst, 64-66; res assoc reproduction physiol, Comp Animal Res Lab, 67-69; nutrit biochemist amino acid nutrit, Smith Kline Corp, 69-73; asst, Va Polytech Inst, 73-77; asst prof, 77-80, ASSOC PROF REPRODUCTION PHYSIOL, LA STATE UNIV, 80- *Concurrent Pos:* Asst mgr, La Animal Breeders Coop, 77-80, res mgr, 80- *Mem:* Am Soc Animal Sci; Am Dairy Sci Asn; Soc Study Reproduction. *Res:* Biocolloidal aspects of semen with respect to semen preservation, metabolism and interaction with the female environment; heritability of semen quality; experimental design and analysis. *Mailing Add:* Dept Dairy Sci La State Univ Baton Rouge LA 70803

CHANDLER, LOUIS, b Rumania, Jan 15, 22; nat US; m 43. BIOPHYSICS, RADIATION PHYSICS. *Educ:* Univ Chicago, BS, 43; Univ Ill, MS, 47, PhD(biophysics), 54. *Prof Exp:* Instr physics, Univ Chicago, 43-44; physicist, Manhattan Dist, Chicago, Oak Ridge, Calif, 44-46; asst physics, Univ Ill, 46-48; physicist, Anderson Phys Lab, 48-49; chief physicist, Swift & Co, 55-57; from asst prof to assoc prof, Univ Ill, 57-67; prof environ sci, Rutgers Univ, 67-76, dir grad prog radiol health, 68-69, chmn, Dept Radiation Sci, 76-77, prof radiation sci, 67-; RETIRED. *Concurrent Pos:* Pres, Radiation Control, Inc, 59-67; res prof, Univ Tokyo, 63-64; fel adv, AEC, 68-70; first officer dosimetry, Int Atomic Energy Agency, Vienna, 70-72; vis prof, Kyushu Univ Fac Sci, Japan, 78, 83. *Mem:* Am Phys Soc; Radiation Res Soc; Am Asn Physics Teachers; Health Phys Soc; Am Asn Physicists in Med. *Res:* Effect of small dose radiation on cell growth; radiation protection legislation; applications of Mossbauer effect; dosimetry in high level radiation processing and low level environmentally stimulated irradiators. *Mailing Add:* 36 N Ross Hall Blvd Piscataway NJ 08854

CHANDLER, REGINALD FRANK, b Edmonton, Alta, Sept 18, 41; m 62; c 2. NATURAL PRODUCTS CHEMISTRY, ETHNOPHARMACOLOGY & ETHNOBOTANY. *Educ:* Univ Alta, BSc, 62, MSc, 65; Univ Sydney, PhD(pharmaceut chem), 69. *Prof Exp:* Lectr phytochem, Univ Sydney,

65-68; from asst prof to assoc prof, 68-82, PROF PHARM, DALHOUSIE UNIV, 82-, DIR COL PHARM, 89- *Concurrent Pos:* Secy-treas, Asn Fac of Pharm Can, 72-74; expert adv comt on herbs & botanicals, Health Protection Br, Health & Welfare, Can, 85; chmn expert adv comt, Can Drug Scheduling Study Group Can Pharm Asn, 57-87; pres, Asn Fac Pharm Can, 87-88; mem, Can Pharm Asn Nat Drug Adv Comt, 88-, Asn Deans Pharm Can, 89- *Honors & Awards:* B Trevoy Pugsley Lectr, 85. *Mem:* Am Soc Pharmacognosy; Soc Econ Bot; Europ Phytochem Soc. *Res:* Investigation of the medicinal phytochemical and ethnobotanical aspects of maritime flora with particular emphasis on the traditional Micmac and Malecite Indian medicines; examination of herbal and "health foods" for safety and efficiency. *Mailing Add:* Col Pharm Dalhousie Univ Halifax NS B3H 3J5 Can

CHANDLER, RICHARD EDWARD, b Ft Pierce, Fla, Sept 9, 37; m 61; c 1. MATHEMATICS. *Educ:* Fla State Univ, BS, 59, MS, 60, PhD(math), 63. *Prof Exp:* Res assoc, Duke Univ, 63-65; asst prof math, 65-67, assoc prof, 67-72, PROF MATH, NC STATE UNIV, 72- *Mem:* Math Soc; Math Asn Am. *Res:* General topology, especially Hausdorff compactifications. *Mailing Add:* Dept of Math NC State Univ Raleigh NC 27650

CHANDLER, ROBERT FLINT, JR, b Columbus, Ohio, June 22, 07; m 31; c 3. AGRONOMY. *Educ:* Univ Maine, BS, 29; Univ Md, PhD(pomol), 34. *Hon Degrees:* LLD, Univ Maine, 51; DH, Cent Luzon State Univ, 71; ScD, Univ Notre Dame, 71, Univ Philippines, 72, Univ NH, 72 & Univ Md, 75; LittD, Univ Singapore, 71. *Prof Exp:* Horticulturist, State Dept Agr, Maine, 29-31; asst, Exp Sta, Univ Md, 31-34; Nat Res Found fel forestry, Univ Calif, 34-35; from asst prof to prof forest soils, Cornell Univ, 35-47; dean, Col Agr & dir exp sta, Univ NH, 47-50, pres, 50-54; asst dir agr, Rockefeller Found, 54-57, assoc dir agr sci, 57-72, spec field staff mem, Asian Veg Res & Develop Ctr, Taiwan, 72-75; CONSULT INT AGR, 75- *Concurrent Pos:* Vis prof, Agr Mech Col Tex, 40; soil scientist, Rockefeller Found, 46-47; dir, Int Rice Res Inst, Laguna, Philippines & Near East Found; consult rice res & agroforestry. *Honors & Awards:* Gold Medal Award, Govt India, 66; Golden Heart Award, Govt Philippines, 72; Int Agron Award, 72; Presidential End Hunger Award, US Agency Int Develop, 86; World Food Prize, 88. *Mem:* Fel Am Acad Arts & Sci; Am Soc Agron; Crop Sci Soc Am; Soc Int Develop. *Res:* Chemical composition of forest tree leaves and litter; vegetation as soil-forming factor; potassium nutrition of alfalfa; agricultural education and research in the Far East; administration of scientific research. *Mailing Add:* 421 E Minnehaha Ave Clermont FL 34711

CHANDLER, ROBERT WALTER, b Charleston, WVa, Aug 31, 32; m 55; c 3. FLOW CYTOMETRY, CELLULAR IMMUNOLOGY. *Educ:* Morris Harvey Col, Charleston, WVa, BS, 54; Ohio State Univ, MSc, 56; Univ Md, PhD(med microbiol), 67. *Prof Exp:* Res assoc endocrinol, Children's Hosp, Columbus, Ohio, 57-60; instr pediat, Johns Hopkins Univ, Sch Med, Baltimore, Md, 60-62 & 66-67; instr med, Ctr Health Sci, Univ Tenn, 67-69, dir rheumatology res lab, 67-77, from asst prof to assoc prof immunol in med, 69-77, dir gonorrhea res lab, 71-74, dir surg-histocompatibility lab, 77-80, assoc prof surg & immunol, 77-80; ASST TO PATHOLOGIST IN CHARGE HEMAT & IMMUNOL, DEPT PATH, & TECH DIR, CELLULAR IMMUNOL & FLOW CYTOMETRY LABS, BAPTIST MEM HOSP, MEMPHIS, 80- *Concurrent Pos:* Mem patient participation comt, Baptist Mem Hosp, Memphis, 86- *Mem:* Sigma Xi; Am Asn Immunologists; Am Soc Histocompatibility & Immunogenetics; NY Acad Sci; Soc Anal Cytol; AAAS; Am Soc Cell Biol. *Res:* Flow cytometric analyses for DNA aneuploidy in cancer (bladder, leukemia); typing of lymphocytes and other cells in leukemia and lymphoma and acquired immune deficiency syndrome by flow cytometry and monoclonal antibodies. *Mailing Add:* Dept Path Baptist Mem Hosp 1025 Crump Blvd Memphis TN 38104

CHANDLER, ROGER EUGENE, b Elmira, NY, Sept 16, 34; m 56; c 2. ORGANIC CHEMISTRY. *Educ:* Colgate Univ, BA, 56; Mass Inst Technol, PhD(org chem), 61. *Prof Exp:* Res Chemist, Esso Res & Eng Co, 61-64, proj leader detergent additives, 64-66; res coordr, Paramins Div, Enjay Chem Co, 66-67, sect head detergent additives, 67-68; sect head res dept, Esso Petrol Co, Eng, 68-70; head engine testing activity, Enjay Additives Lab, 70-73, sr adv environ conserv & health, 73-78; mgr, Pub Affairs Dept, Exxon Enterprises, 80-82; sr adv, African Affairs, 78-80, asst secy & mgr, Corp Affairs, 82-90, VPRES, TECHNOL & CORP DEVELOP, EXXON CHEM CO, 90- *Concurrent Pos:* Coun, chem res, Indust Res Inst. *Res:* Lubricating oil additives; environmental conservation. *Mailing Add:* Ten Cannon St Norwalk CT 06851

CHANDLER, WILLIAM DAVID, b Brantford, Ont, Jan 29, 40; m 61; c 4. PHYSICAL ORGANIC CHEMISTRY. *Educ:* Queen's Univ, Ont, BSc, 62, PhD(chem), 65. *Prof Exp:* NATO fel, Pa State Univ, 65-66 & Rutgers Univ, 66-67; from asst prof to assoc prof, 67-74, PROF CHEM & HEAD DEPT, UNIV REGINA, 74- *Mem:* fel Chem Inst Can; Am Chem Soc. *Res:* Organic synthesis; diphenyl ether chemistry. *Mailing Add:* Dept Chem Univ Regina Regina SK S4S 0A2 Can

CHANDLER, WILLIAM KNOX, b Oct 13, 33. EXCITATION-CONTRACTION COUPLING, ELECTROPHYSIOLOGY. *Educ:* Univ Louisville, MD, 59. *Prof Exp:* PROF CELLULAR & MOLECULAR PHYSIOL, SCH MED, YALE UNIV, 73- *Mem:* Nat Acad Sci. *Mailing Add:* Dept Cellular & Molecular Biol Sch Med Yale Univ 333 Cedar St New Haven CT 06510

CHANDLER, WILLIS THOMAS, b Argyle, Minn, Feb 15, 23; m 57. METALLURGY. *Educ:* Univ Minn, BMetE, 44, PhD(metall), 55. *Prof Exp:* Asst metall, Manhattan Proj, Univ Chicago, 44 & Mass Inst Technol, 44-45; asst, Univ Minn, 45-48, res assoc, Nuclear Eng Propulsion Aircraft Proj, 48-50; asst prof, Northwestern Univ, 50-53 & Univ Notre Dame, 53-59; sr res engr, 59-62, prin scientist, 62-77, PROJ ENG ADV METALL & FRACTURE MECH, ROCKETDYNE DIV, ROCKWELL INT, 77- *Mem:* AAAS; Am Inst Mining, Metall & Petrol Engrs; Am Soc Metals; Sigma Xi. *Res:* Metallurgy of high temperature and refractory alloys; transformations in metals; hydrogen embrittlement of refractory metals; erosion and corrosion of metals; hydrogen-environment embrittlement. *Mailing Add:* 22714 Margarita Dr Woodland Hills CA 91364

CHANDRA, ABHIJIT, b Calcutta, India, Jan 4, 57; US citizen; m 84; c 2. DESIGN FOR MANUFACTURING, BOUNDARY ELEMENT METHOD. *Educ:* Indian Inst Technol, Kharagpur, BTech(Hons), 78; Univ NB, Fredericton, MS, 80; Cornell Univ, PhD(theoret & appl mech), 83. *Prof Exp:* Sr res engr, Gen Motors Res Labs, 83-85; asst prof, 85-89, ASSOC PROF MFG, UNIV ARIZ, 89- *Concurrent Pos:* Prin investr, NSF, 86-; consult, Goodyear Tire & Rubber Corp, 87-89, Alcoa & Advan Ceramic Res Corp, 90-; NSF presidential young investr, 87. *Honors & Awards:* Achievement Award, J F Lincoln Arc Welding Found, 89. *Mem:* Am Soc Mech Engrs; Sigma Xi. *Res:* Design for manufacturing; forming and machining; analysis, design sensitivity studies and design integration; boundary element method; electronic packaging; thermal design; net shape manufacturing. *Mailing Add:* Dept Aerospace & Mech Eng Univ Ariz Tucson AZ 85721

CHANDRA, DHANESH, b Hyderabad, India, Oct 10, 44. EXTRACTIVE METALLURGY, COMPOSITE MATERIALS. *Educ:* Osmania Univ, India, BEng, 67; Univ Ill, MS, 72; Univ Denver, PhD(metall & mat sci), 76. *Prof Exp:* Mgr planning control, Hyderabad Asbestos Cement Prod, India, 68-70; res asst, Col Eng, 72-76, SR RES METALLURGIST & HEAD STRATEGIC MAT CTR, DENVER RES INST, UNIV DENVER, 76- *Concurrent Pos:* Teaching asst mat characterization, Univ Denver, 75; prin investr grants, US Bur Mine Projs, 76-, Magnasep Corp, 78-79, & Cent Naz Cat Unico Bibl Italiana, Bibl, 76-; consult, Colburn Eng, Denver, 77-, Rocky Mountain Energy Co, Wyoming, 80-; co-investr various projs, Rocky Mountain Energy Co, 77- *Mem:* Am Soc Metals; Am Inst Mining & Metall Engrs; Sigma Xi. *Res:* Extractive metallurgy of metals: nickel, cobalt, chromium, manganese, aluminum and gold; bench scale process development involving kinetic thermodynamic, electron-optical microscopic evaluation; recycling of metals and alloy; unidirectional solidification of metal oxide eutectics for use in electronic materials; fabricating alumium-graphite composites sintering of semi conducting materials and chemical vapor deposition of materials. *Mailing Add:* 1875 Quail Run Rd Reno NV 89523-1827

CHANDRA, G RAM, b India, Feb 10, 33; m. BIOCHEMISTRY. *Educ:* Agra Univ, BS, 51, MS, 53; Univ Alta, PhD(biochem), 62. *Prof Exp:* Res asst biochem, Indian Agr Res Inst, 55-59; scientist, Res Inst Advan Studies, 62-65; scientist, Mich State Univ, 65-66; SCIENTIST, AGR RES SERV, USDA, 66- *Res:* Biological volatiles; senescence; plant nutrition; hormones. *Mailing Add:* USDA B-050 Rm 140 Beltsville MD 20705

CHANDRA, GRISH, b Lucknow, India, Aug 27, 39; US citizen; m 64; c 2. PRECERAMIC POLYMERS, DIELECTRIC FILMS. *Educ:* Univ Lucknow, India, BS, 57, MS, 59; Univ Rajasthan, Jaipur, India, PhD(chem), 63; Univ Sussex, Brighton, UK, DPhil, 67. *Prof Exp:* Lectr inorg chem, KK Col, Lucknow, India, 59-60; res chemist, Dow Corning Corp, UK, 67-76; sr proj chemist, 76-83, assoc res scientist chem, 83-88, SECT MGR CHEM, DOW CORNING CORP, 88- *Mem:* Am Chem Soc; Mat Res Soc. *Res:* Organometallic chemistry of silicon, germanium, titanium, zirconium, platinum and rhodium; ceramic materials for high temperature structural applications; microelectronic packaging. *Mailing Add:* Dow Corning Corp PO Box 994 Midland MI 48686-0994

CHANDRA, JAGDISH, b Hyderabad, India, Oct 11, 35; US citizen; m 61; c 3. MATHEMATICS. *Educ:* Osmania Univ, India, BA, 55, MA, 57; Rensselaer Polytech Inst, PhD(math), 65. *Prof Exp:* Instr math, Rensselaer Polytech Inst, 65-66; res mathematician, US Army Arsenal, Watervliet, 66-70; chief, Appl Math Br, 70-73, assoc dir, 73-74, DIR, MATH DIV, US ARMY RES OFF, DURHAM, 74- *Concurrent Pos:* Adj asst prof, Rensselaer Polytech Inst, 66-70, Union Col, 68-69; adj assoc prof, Duke Univ, 74-80, adj prof, 80- *Mem:* Am Math Soc; Soc Indust & Appl Math; sr mem Inst Elec & Electronic Engrs. *Res:* Mathematical analysis, nonlinear differential and integral equation; operator inequalities; mathematical theories of combustion; control and stability of large-scale systems. *Mailing Add:* US Army Res Off PO Box 12211 Research Triangle Park NC 27709

CHANDRA, KAILASH, b Kanpur, UP, India, Aug 20, 38; m 61; c 3. PHYSICS. *Educ:* Agr Univ, BSc, 56, MSc, 58; Gorakhpur Univ, PhD(physics), 67. *Prof Exp:* Lectr physics, KK Degree Col, India, 58-60; res fel, Coun Sci & Indust Res, 60-62; lectr, Gorakhpur Univ, 62-68; res assoc, Univ Ga, 68-69; assoc prof, 69-73, PROF PHYSICS, SAVANNAH STATE COL, 73-, HEAD DEPT MATH, PHYSICS & COMPUT SCI TECHNOL, 82- *Concurrent Pos:* Vis res fel, Nat Bur Standards, Washington, 78 & NASA Langley Res Ctr, Hampton, Va, 80. *Mem:* Am Asn Physics Teachers; Am Phys Soc; Asn Comput Mach. *Res:* Molecular spectroscopy; laser spectroscopy; competency based education in physics; microcomputers in education. *Mailing Add:* 103 Terrapin Dr Savannah GA 31406

CHANDRA, PRADEEP, b Rajasthan, India, July 4, 44. INTERNAL MEDICINE, HEMATOLOGY. *Educ:* Univ Rajasthan, India, MB, BS, 66; Am Bd Internal Med, dipl, 72, cert hemat, 74. *Prof Exp:* House officer & registr, SMS Med Col Hosp, Univ Rajasthan, 67-68; resident internal med, Wyckoff Heights Hosp, Brooklyn, NY, 68-69; from resident to chief resident, Bronx Lebanon Hosp Ctr, 69-71; clin fel hemat, Long Island Jewish Hosp, New Hyde Park, NY, 71-73; asst prof, 76-80, ASSOC PROF MED, STATE UNIV NY STONY BROOK, 80-; dir educ & assoc attend physician, Dept Oncol, Montefiore Hosp & Med Ctr, Bronx, NY, 80-; MEM STAFF, METHODIST HOSP MED CTR, BROOKLYN, NY. *Concurrent Pos:* Nat Leukemia Asn grants, 75 & 76; assoc scientist, Brookhaven Nat Lab, 74-77, res collabr, 81-, attend physician & prin investr leukemia study, Hosp Med Res Ctr, 74-81, scientist-in-chg, Clin Hemat Lab & Blood Bank, 77-81. *Honors & Awards:* Physician Recognition Award, AMA, 78. *Mem:* Am Soc Hemat; Am Fedn Clin Res; fel Int Soc Hemat; fel Am Col Physicians; Soc Exp Biol & Med. *Res:* Study of lymphocytes in health and disease; special interest in chronic lymphocytic leukemia, measurement of body burden of leukemic cells; growth of lymphocytes and bone marrow cells in cultures; chrono-oncology and oncology. *Mailing Add:* Methodist Hosp Dept Med 506 Sixth St Brooklyn NY 11215

CHANDRA, PURNA, b Khatauli, India, June 23, 29; m 64; c 1. AGRICULTURAL MICROBIOLOGY. *Educ:* Agra Univ, BSc, 49, MSc, 51; Ore State Univ, PhD(bact), 58. *Prof Exp:* Instr bact, Ore State Univ, 58-59; lectr, Univ Bagdad, 59-60; microbiologist, Can Dept Agr Res Sta, Sask, 60-63; asst prof biol, Mt Allison Univ, 63-67; assoc prof, 67-70, PROF BIOL, LAKE SUPERIOR STATE COL, 70- *Concurrent Pos:* Vis scientist, Appl Biochem Div, Dept Sci & Indust Res, New Zealand, 81 & CSIRO Lab Brisbane Australia, 89. *Mem:* Am Soc Microbiol; Soil Sci Soc Am. *Res:* Decomposition of organic matter in soils; nitrogen transformation in soils; biocidal effect on soil microorganisms; soil respiration studies; microbial ecology. *Mailing Add:* Dept of Biol Sci Lake Superior State Univ Sault Ste Marie MI 49783

CHANDRA, SURESH, b Etah, India, July 25, 39; m 64; c 2. FLUID MECHANICS, CHEMICAL ENGINEERING. *Educ:* Allahabad Univ, BSc, 57; Banaras Hindu Univ, BSc, 61; Univ Louisville, MChE, 62; Colo State Univ, PhD(fluid mech), 67. *Prof Exp:* Asst prof mech eng, Univ Miami, 66-71; PROF MECH ENG & CHMN DEPT, A&T STATE UNIV, NC, 71-, DEAN, SCH ENG, 77- *Concurrent Pos:* Partic, Am Soc Eng Educ-Ford Found Residency Prog eng pract, 69. *Mem:* Am Soc Mech Engrs; Am Inst Aeronaut & Astronaut; Am Soc Eng Educ; Sigma Xi. *Res:* Diffusion in turbulent boundary layers; heat transfer in metallic contacts. *Mailing Add:* 6 Westminster Ct Greensboro NC 27410

CHANDRA, SUSHIL, b Varanasi, India, Dec 31, 31; m 55; c 3. AEROSPACE & ATMOSPHERIC PHYSICS. *Educ:* Benares Hindu Univ, BSc, 52, MSc, 54; Pa State Univ, PhD(physics), 61. *Prof Exp:* Lectr physics, U P Col, Varanasi, 54-58; res asst atmospheric physics, Pa State Univ, 58-60, instr physics, 60-61; Nat Acad Sci resident res assoc, Goddard Space Flight Ctr, NASA, Md, 61-64; scientist, Nat Phys Lab, India, 64-66; AEROSPACE TECHNOLOGIST, GODDARD SPACE FLIGHT CTR, NASA, 66- *Mem:* Am Geophys Union. *Res:* Physics of the upper atmosphere; aeronomy; ionosphere. *Mailing Add:* 11901 Blackwood Ct Laurel MD 20708

CHANDRAN, KRISHNAN BALA, b Madurai City, India, May 16, 44; m 72; c 2. BIOMECHANICS, MECHANICAL ENGINEERING. *Educ:* Madras Univ, BS, 63; Wash Univ, MS, 69, DSc(biomed eng), 72. *Prof Exp:* Tool engr, Hindustan Motors Ltd, Calcutta, India, 66-67; asst prof orthop, Biomech Lab, Med Sch, Tulane Univ, 74-78; assoc prof, div mat eng, Col Eng, 78-84, PROF BIOMED ENG, UNIV IOWA, 84- *Honors & Awards:* Borelli Award, Am Soc Biomechanics, 88. *Mem:* Am Acad Mechanics; fel Am Soc Mech Engrs; Am Soc Biomech; Am Heart Asn; Sigma Xi; Biomed Eng Soc. *Res:* Investigations on traumatology of the human head and spine; hemodynamics at arterial curvature sites in relation to the origin of atherosclerosis; left ventricular dynamics with emphasis on the non invasive diagnosis of myocardial infarction; flow dynamics past prosthetic heart valves. *Mailing Add:* Dept Biomed Eng Col Eng Univ Iowa Iowa City IA 52242

CHANDRAN, SATISH RAMAN, b Oct 6, 38; US citizen; m 66; c 2. HUMAN ANATOMY, GENERAL BIOLOGY. *Educ:* Univ Kerala, BS, 55, MS, 58; Univ Ill, Urbana, PhD(entom), 65. *Prof Exp:* Res assoc entom, Univ Ill, Urbana, 65-66; from instr to asst prof biol, Univ Ill, Chicago Circle, 66-72; assoc prof, 72-77, PROF BIOL, KENNEDY-KING COL, 77- *Mem:* Entom Soc Am; Am Inst Biol Sci; Nat Asn Biol Teachers; AAAS. *Res:* Developmental morphology due to thermal stress in mosquitoes; microsomal oxidase activity in insects; toxicity of the photoisomere of cyclodiene insecticides of freshwater animals; lepidopteran morphology. *Mailing Add:* 1648 Western Ave Flossmoor IL 60422

CHANDRAN, V RAVI, b India, May 3, 55. BIOPHARMACEUTICS, PHARMACOKINETICS. *Educ:* Jadaupur Univ, BS, 76, MP, 87; Univ Fl, MS, 83, PhD, 86. *Prof Exp:* Sr res pharmacist, Sterling-Winthrop Res Inst, 86-89; PRES & CHIEF EXEC OFFICER, AM GENERICS, INC, 89- *Concurrent Pos:* Consult, Harvard Med Sch. *Mem:* Am Pharmaceut Asn; Am Asn Pharmaceut Scientists. *Res:* Drug delivery; biopharmaceutics; pharmacokinetics; prodrugs; development of dosage forms; bioavailability. *Mailing Add:* Am Generics Inc PO Box 465 West Sand Lake NY 12196-0465

CHANDRASEKARAN, BALAKRISHNAN, b Lalgudi, India, June 20, 42; m; c 1. INTELLIGENCE SYSTEMS, ARTIFICIAL INTELLIGENCE. *Educ:* Univ Madras, BE, 63; Univ Pa, PhD(elec eng), 67. *Prof Exp:* Eng res specialist, Data Recognition Lab, Philco-Ford Corp, 67-69, eng specialist, Advan Eng & Res, 69, from asst prof to assoc prof, 69-77, PROF COMPUT & INFO SCI, OHIO STATE UNIV, 77- *Concurrent Pos:* Consult, LNK Corp, Md, 69-72 , Lawrence Livermore Nat Labs, 81-83, Smart Syst Technol, McLean, Va, 84-85, Lockheed Corp, Boeing Corp. *Mem:* Fel Inst Elec & Electronic Engrs; Asn Comput Mach; Am Asn Artificial Intel. *Res:* Artificial intelligence, computer graphics, medical diagnosis by computer; intelligent and learning systems; statistical pattern recognition. *Mailing Add:* Dept Comput & Info Sci Ohio State Univ Columbus OH 43210

CHANDRASEKARAN, SANTOSH KUMAR, b Delhi, India. CHEMICAL ENGINEERING, BIOMEDICAL ENGINEERING. *Educ:* Indian Inst Technol, Bombay, BTech, 64; Univ Calif, Berkeley, MS, 65, PhD(chem eng), 71. *Prof Exp:* Res asst chem eng, Univ Calif, Berkeley, 64-65, res & teaching assoc, 67-71; sr engr, E I du Pont de Nemours Inc, 65-67; develop engr, Alza Corp, 71-76, dir systs develop, 74-81, prin scientist, 76-81; dir res & develop, Abcor, Inc, 81-; AT SYNTEX OPHTHALMICS, INC. *Mem:* Am Chem Soc; Am Inst Chem Engrs; Sigma Xi. *Res:* Polymer and pharmaceutical chemistry; design and development of drug delivery systems; ultrafiltration and reverse osmosis. *Mailing Add:* In Site Vision 965 Atlantic Ave Alameda CA 94501

CHANDRASEKHAR, B S, b Bangalore, India, May 24, 28; c 2. SUPERCONDUCTIVITY. *Educ:* Mysore Univ, BSc, 47; Univ Delhi, MSc, 49; Oxford Univ, DPhil(physics), 52. *Prof Exp:* Res assoc physics, Univ Ill, 52-54; res physicist, Westinghouse Res Labs, 54-59, fel physicist, 59-61, sect mgr cryophysics, 61-63; prof physics, 63-67, chmn, physics dept, 65-67 & biol dept & dean sci, 69-70, Perkins prof, 67-87, EMER PROF PHYSICS, CASE WESTERN RESERVE UNIV, 88- *Concurrent Pos:* Vis scientist, Oxford, 54-55; sr vis res fel, Imperial Col, London, 61; consult, Bell Tel Labs, 65, 66 & 68, Argonne Nat Lab, 65-68, Lewis Res Ctr, 66 & NSF Inst Progs, 71; vis prof, Univ Ill, Urbana-Champaign, 77, Tata Inst Fund Res, 80 & Eidgenoessische Tech, Hochsch, 80-81; Fulbright-Hays fel, Imperial Col, London & Univ Cambridge, 78; vis Fulbright prof, Meissner Inst, Munich, Ger, 84-85. *Mem:* Fel Am Phys Soc. *Res:* Liquid helium; superconductivity; electronic properties of solids. *Mailing Add:* Hollerweg 13 D-8038 Groebenzell Germany

CHANDRASEKHAR, PRASANNA, b Bombay, India, Oct 17, 57. THEORETICAL CHEMISTRY, MATERIALS SCIENCE ENGINEERING. *Educ:* Univ Delhi, India, BSc, 78; Concordia Univ, Montreal, Can, MS, 80; State Univ NY, Buffalo, PhD(electroanal chem), 84. *Prof Exp:* Postdoctoral assoc electrochem, Cornell Univ, 84-85; sr res scientist, Phys Sci Ctr, Bloomington, Minn & Defense Systs Div, Horsham, Pa, Honeywell, Inc, 85-87; MGR, ELECTROCHEM PROGS, GUMBS ASSOCS, INC, 87- *Concurrent Pos:* Prin investr for var co. *Mem:* Am Chem Soc; Electrochem Soc; Soc Electroanal Chem; AAAS; Quantum Chem Prog Exchange. *Res:* Dynamic, passive, ultrafast laser shields for use in battlefield; soluble, processibly, environmentally stable, high-conductivity polymeric materials. *Mailing Add:* Gumbs Assocs Inc 11 Harts Lane East Brunswick NJ 08816

CHANDRASEKHAR, S, ARTHRITIS. *Educ:* State Univ NY, PhD(cell biol), 81. *Prof Exp:* Fel, NIH, 81-85; SR BIOLOGIST, ELI LILLY, 85- *Res:* Extra cell matrix; cartilage biology. *Mailing Add:* Connective Tissue Res Lilly Res Labs Eli Lilly & Co Lilly Corp Ctr Indianapolis IN 46285

CHANDRASEKHAR, SUBRAHMANYAN, b Lahore, Brit India, Oct 19, 10; US citizen; m 36. ASTRONOMY, ASTROPHYSICS. *Educ:* Presidency Col, Madras, India, BA, 30; Cambridge Univ, Eng, PhD(theoret physics), 33, ScD, 42. *Hon Degrees:* Numerous from US & foreign univs, 61-91. *Prof Exp:* Fel, Trinity Col, Cambridge Univ, 33-37; res assoc, Univ Chicago, 37-38, from asst prof to prof, 38- 46, distinguished serv prof theoret astrophys, 47-52, Morton D Hull distinguished serv prof, 52-85, EMER MORTON D HULL DISTINGUISHED SERV PROF, UNIV CHICAGO, 86- *Concurrent Pos:* Counr, Am Astron Soc, 51-54 & Am Phys Soc, 64-68; Managing ed, Astrophys J, 52-71. *Honors & Awards:* Nobel Prize in Physics, 83; Bruce Medal, Am Astron Soc Pac, 52; Gold Medal, Royal Astron Soc, 53; Rumford Medal, Am Acad Arts & Sci, 57; Royal Medal, Royal Soc, 62; Nat Medal of Sci, 66; Meghnad Saha Mem Medal, Asiatic Soc, 67; Padma Vibhushan Medal, India, 68; Henry Draper Medal, Nat Acad Sci, 71; Smoluchowski Medal, Polish Phys Soc, 73; Dannie Heineman Prize, Am Phys Soc, 73; Copley Medal, Royal Soc London, 84; R D Birla Mem Award, Indian Physics Asn, 84; Vainu Bappu Mem Award, Indian Nat Acad Sci, 86. *Mem:* Nat Acad Sci; Am Acad Arts & Sci; Am Astron Soc (vpres, 56-58); assoc Royal Astron Soc; Royal Soc; Am Phil Soc; Royal Swedish Acad. *Res:* Internal constitution of stars; white dwarfs; dynamics of stellar systems; theory of stellar atmospheres; radiative transfer; hydrodynamics and hydromagnetics; theory of black holes holes and general relativity; author of ten books. *Mailing Add:* Lab Astrophys & Space Res 933 E 56th St Chicago IL 60637

CHANDRASHEKAR, MUTHU, b Banares, India, Feb 24, 47; m 69. MECHANICAL ENGINEERING, SYSTEMS ENGINEERING. *Educ:* Indian Inst Technol, Kanpur, India, BTech, 69; Univ Waterloo, Ont, MASc, 70, PhD(syst design), 73. *Prof Exp:* Vis prof syst sci, Univ Paraibe, Brazil, 74; vis prof energy systs, Dept Elec Eng & Syst Sci, Mich State Univ, 76; from asst to assoc prof, 73-85, PROF COMPUT DESIGN, UNIV WATERLOO, ONT, 85- *Concurrent Pos:* Consult, Can Centre Inland Waters, Burlington, 74, Can Elec Asn, Montreal, 78 & Nat Res Coun, 78. *Mem:* Int Solar Energy Soc. *Res:* System theory; thermodynamics; networks; energy systems analysis; solar heating simulation; computer aided design of large scale engineering systems; water distribution. *Mailing Add:* Dept Syst Design Eng Univ Waterloo Waterloo ON N2L 3G1 Can

CHANDROSS, EDWIN A, b Brooklyn, NY, Oct 13, 34; m 61; c 2. ORGANIC CHEMISTRY, PHYSICAL CHEMISTRY. *Educ:* Mass Inst Technol, BS, 55; Harvard Univ, MA, 57, PhD(org chem), 60. *Prof Exp:* MEM TECH STAFF CHEM, AT&T BELL LABS, 59-, HEAD, ORG CHEM RES & ENG DEPT, 80- *Concurrent Pos:* Instr, Rutgers Univ, 62-63; prof, Univ Bordeaux I, 88. *Mem:* AAAS; Am Chem Soc; Mat Res Soc. *Res:* Photochemistry; organic materials; specialty polymers; photosensitive materials; adhesion. *Mailing Add:* AT&T Bell Labs 1D246 Murray Hill NJ 07974

CHANDROSS, RONALD JAY, b New York, NY, Mar 21, 35; m 59; c 2. BIOPHYSICS. *Educ:* Polytech Inst Brooklyn, BS, 56; Mass Inst Technol, PhD(phys chem), 61. *Prof Exp:* Sr physicist, Gen Dynamics/Astronaut, 61-63; res chemist, Gen Chem Div, Allied Chem Corp, 63-66; fel biol, Mass Inst Technol, 66-68; res fel, Dept Surg, Mass Gen Hosp, 68-69; res assoc, Lab Reproductive Biol, Univ NC, Chapel Hill, 69-77, res asst prof, 77-84; SR SCIENTIST, UNIV VA, CHARLOTTESVILLE, 84- *Mem:* Am Crystallog Asn; Am Phys Soc; AAAS; Biophys Soc; Sigma Xi. *Res:* X-ray, electron diffraction; electron microscopy; structure of collagen and related substances; membrane structures; steroid structure-activity relationships; protein structure; area detector development. *Mailing Add:* 68 Oak Forest Circle Charlottesville VA 22901

CHANEY, ALLAN HAROLD, b Kerrville, Tex, Dec 11, 23; m 48; c 2. ZOOLOGY. *Educ:* Tulane Univ, BS, 46, MS, 49, PhD(zool), 58. *Prof Exp:* Instr comp anat, Tulane Univ, 50; from asst prof to assoc prof zool, Ark Polytech Col, 53-59; prof & head dept, Del Mar Col, 59-63; from assoc prof to prof, 63-89, EMER PROF BIOL, TEX A&I UNIV, 91- *Mem:* Am Soc Ichthyologists & Herpetologists. *Res:* Herpetology; systematics; marine biology; ornithology; mammalogy; ecology. *Mailing Add:* Dept Biol Tex A&I Univ Kingsville TX 78363

CHANEY, CHARLES LESTER, b Denver, Colo, Dec 21, 30; m 61; c 1. ANALYTICAL CHEMISTRY. *Educ:* Univ Wash, BS, 58. *Prof Exp:* Chemist, US Bur Mines, Nev, 58-60 & Md, 60-62; res asst anal chem, Gen Atomics, 62-67, staff assoc, 67-80, sr scientist, 80-84; OWNER & MGR, SPECTRA CO, 71- *Mem:* Soc Appl Spectros; Am Soc Testing & Mat. *Res:* Design, construct spectroscopy equipment and instrumentation; spectrochemical analysis. *Mailing Add:* 2987 Governor Dr San Diego CA 92122-0204

CHANEY, DAVID WEBB, b Cleveland, Ohio, Dec 19, 15; m 38; c 1. ORGANIC CHEMISTRY. *Educ:* Swarthmore Col, AB, 38; Univ Pa, MS, 40, PhD(org chem), 42. *Prof Exp:* Asst sect leader, Am Viscose Corp, 42-52; sr res group leader, Chemstrand Corp, 52-53, asst dir res, 53-58, exec dir res, 58-60, vpres & exec dir, Chemstrand Res Ctr, Inc, 60-65, tech dir new prod & basic res, 65-67; dean, Sch Textiles, NC State Univ, 67-81, prof textiles, 77-81; RETIRED. *Mem:* AAAS; Am Soc Eng Educ; Am Chem Soc; Am Asn Textile Chem & Colorists. *Res:* Organic synthetic fluorine compounds, especially fluorovinyls; vinyl copolymers; condensation polymers; synthetic fibers. *Mailing Add:* 6000 Sentinel Dr Raleigh NC 27609-3512

CHANEY, GEORGE L, b Coffeyville, Kans, Mar 16, 30; m 52; c 4. MATHEMATICS. *Educ:* Univ Kans, BS, 53, PhD(math educ), 67; Kans State Col Pittsburg, MS, 59. *Prof Exp:* High sch teacher, Kans, 53-56; teacher math, Coffeyville Community Col, 56-61; prof, Kans State Col Pittsburg, 62-64 & 66-68; PROF MATH, OTTAWA UNIV, KANS, 68- *Mem:* Math Asn Am. *Mailing Add:* Dept Math Ottawa Univ Ottawa KS 66067

CHANEY, ROBERT BRUCE, JR, b Helena, Mont, Aug 22, 32; m 63; c 2. AUDIOLOGY. *Educ:* Univ Mont, BA, 58, MA, 60; Stanford Univ, PhD(audiol), 65. *Prof Exp:* Audiologist, Vet Admin, 60-62; res psychologist, US Navy Electronics Lab, Calif, 63-64; from asst prof to assoc prof audiol, 64-77, PROF COMMUN SCI & DISORDERS, UNIV MONT, 77- *Mem:* Acoust Soc Am; Am Speech & Hearing Asn. *Res:* Speech and hearing science; acoustical theory of speech production; neurological basis of speech perception. *Mailing Add:* 14 September Dr Missoula MT 59802

CHANEY, ROBIN W, b Cleveland, Ohio, Dec 13, 38. NUMERICAL ANALYSIS, OPERATIONS RESEARCH. *Educ:* Ohio State Univ, BS, 60, PhD(math), 64. *Prof Exp:* Asst prof math, Western Wash State Col, 64-67 & Univ Calif, Santa Barbara, 67-69; PROF MATH, WESTERN WASH STATE COL, 69- *Concurrent Pos:* NSF fel, 66-67; vis prof, Chalmers Inst Technol, Sweden, 72-73. *Mem:* AAAS; Am Math Soc; Soc Indust & Appl Math; Inst Mgt Sci. *Res:* Convergence analysis of optimization algorithms. *Mailing Add:* Dept Math Comput Sci Western Wash Univ Bellingham WA 98225

CHANEY, STEPHEN GIFFORD, b Ware Co, Pa, Feb 8, 44; m 68; c 1. BIOCHEMISTRY. *Educ:* Duke Univ, BS, 66; Univ Calif, Los Angeles, PhD(biochem), 70. *Prof Exp:* Am Cancer Soc fel microbiol, Sch Med, Washington Univ, 70-72; asst prof, 72-83, ASSOC PROF BIOCHEM, SCH MED, UNIV NC, CHAPEL HILL, 83- *Mem:* Am Chem Soc; Am Asn Chem Res; Am Soc Biochem & Molecular Biol. *Res:* Nucleic acid research; control mechanisms. *Mailing Add:* Dept of Biochem Univ NC FLOB CB No 7260 Chapel Hill NC 27599

CHANEY, WILLIAM R, b McAllen, Tex, Dec 2, 41; m 68; c 2. FORESTRY, PLANT PHYSIOLOGY. *Educ:* Tex A&M Univ, BS, 64; Univ Wis, PhD(tree physiol), 69. *Prof Exp:* Asst prof, 70-73, assoc prof, 73-81, PROF TREE PHYSIOL, PURDUE UNIV, WEST LAFAYETTE, 81- *Concurrent Pos:* Fel, Dept Forestry, Univ Wis, 69-70; vis prof, Univ Col Wales, Aberystwyth, Wales, 77, NC State Univ, 87. *Mem:* Am Soc Plant Physiol; Sigma Xi; Soc Am Foresters; Int Soc Arboriculture; Int Soc Trop Foresters. *Res:* Water relations of woody species; mineland reclamation, mycorrhizae, and tree growth regulators. *Mailing Add:* Dept Forestry & Nat Resources Purdue Univ West Lafayette IN 47907

CHANG, ALBERT YEN, b China, Apr 15, 36; m 64; c 1. BIOCHEMISTRY. *Educ:* Nat Taiwan Univ, BS, 58; Univ Calif, Berkeley, MA, 62; Univ Ill, Urbana, PhD(biochem), 65. *Prof Exp:* PHARMACEUT RES & DEVELOP, STRATEGIC PLANNING & INT LIAISON DIR, UPJOHN CO, 67- *Mem:* AAAS; Am Chem Soc; Am Diabetes Asn; Soc Exp Biol Med; Am Soc Biol Chemists. *Res:* Etiology of diabetes; control of diabetic complications; glycoprotein metabolism; regulation of enzymes in mammalian systems. *Mailing Add:* Pharmaceut Res & Develop Upjohn Co Kalamazoo MI 49001

CHANG, ALFRED TIEH-CHUN, b Shanghai, China, June 22, 42; US citizen; m 70; c 2. REMOTE SENSING, IMAGE PROCESSING. *Educ:* Nat Cheng-Kung Univ, BS, 64; Univ Md, MS, 70, PhD(physics), 71. *Prof Exp:* Programmer, GTE Info Syst, 71-72; res assoc, Nat Res Coun, Nat Acad Sci, 72-73; RES SCIENTIST REMOTE SENSING, GODDARD SPACE FLIGHT CTR, NASA, 74- *Mem:* Am Geophys Union; Am Meteorol Soc; Inst Elec & Electronic Engrs. *Res:* Develop theoretical models to understand the physical processes and interactions of electromagnetic waves with atmosphere and earth surface. *Mailing Add:* Hydrol Sci Br Code 974 NASA Goddard Space Flight Ctr Greenbelt MD 20771

CHANG, AMY Y, CCK RECEPTOR, SIGNAL TRANSDUCTION. *Educ:* Yale Univ, PhD(cell biol), 86. *Prof Exp:* FEL CELL BIOL, YALE UNIV, 86- *Mailing Add:* Dept Human Genetics Yale Univ 333 Cedar St New Haven CT 06510

CHANG, BETTY, b China; US citizen. GERONTOLOGY. *Educ:* Columbia Univ, BS, 61, MAEd, 61; Univ Calif, San Francisco, DNSc, 77. *Prof Exp:* Lectr med-surg nursing, Nursing Sci Prog, Queens Col, 61-68; asst prof family & med-surg nursing, San Francisco State Univ, 68-73; asst prof, 77-81, PROF MED-SURG NURSING, UNIV CALIF, LOS ANGELES, 81- *Mem:* Geront Soc; Am Coun Nurse Researchers. *Res:* Care of the elderly, particularly evaluation of care from the patient's perspective. *Mailing Add:* 1132 Chantilly Rd Los Angeles CA 90077

CHANG, BOMSHIK, b Inchon, Korea, Feb 6, 31; m; c 2. MATHEMATICS. *Educ:* Seoul Nat Univ, BA, 54, MA, 56; Univ BC, PhD, 59. *Prof Exp:* From lectr to asst prof, 58-69, ASSOC PROF MATH, UNIV BC, 69- *Mem:* Am Math Soc; Math Asn Am; Can Math Cong. *Res:* Group theory; Lie algebra. *Mailing Add:* Dept Math Univ BC 2075 Wesbrook Pl Vancouver BC V6T 1Y4 Can

CHANG, BUNWOO BERTRAM, b Kwongtung, China, Oct 9, 47; m 70; c 2. GEOPHYSICS. *Educ:* St John's Univ, Collegeville, Minn, BA, 71; Rice Univ, Houston, MA, 74, PhD(physics), 76. *Prof Exp:* Fel, Div Labs & Res, NY State Dept Health, 76-77, res scientist, 77-80; sr res physicist oil & gas explor, MRD Assoc, Inc, 80-81; vpres geophys res, O'Connor Res Inc, Denver, 82-85; vpres ctr mgr, O'Connor Comput, Inc, Denver & pres, O'Connor Res Inc, Denver, 85-87; PRES, CHANG RES INC, 87- *Concurrent Pos:* co-prin investr, Res Proj, Nat Sci Found, 77-80. *Res:* Development of advanced seismic analysis algorithms and interpretation for oil and gas exploration; modeling inversion calculations and the use of pattern recognition and multivariate statistical analysis techniques. *Mailing Add:* 495 S Federal Blvd Denver CO 80219

CHANG, C HSIUNG, b Taichung, Taiwan, Apr 9, 56. CONTAMINATION CONTROL, FILTRATION PERFORMANCE EVALUATION. *Educ:* Ga Inst Technol, MS, 84. *Prof Exp:* Lab engr, 84-90, CHIEF RES ENGR, HYCON CORP, 90- *Mem:* Nat Fluid Power Asn. *Res:* Develop fluid media for hydraulic systems in fluid power industry. *Mailing Add:* Hycon Corp PO Box 2626 Lehigh Valley PA 18001-2626

CHANG, CATHERINE TEH-LIN, b China. PHOTOCHEMISTRY, PHOTOGRAPHIC CHEMISTRY. *Educ:* Nat Taiwan Univ, BS, 58; Washington Univ, PhD(chem), 64. *Prof Exp:* Res chemist, 64-69, sr res chemist, 69-75, res assoc, 75-83, SR RES ASSOC, IMAGING SYSTMS DEPT, E I DU PONT DE NEMOURS & CO, INC, 83- *Mem:* Am Chem Soc; Soc Photog Scientists & Engrs. *Res:* Stereochemical course of the diazomethane-carbonyl reaction; acid-catalyzed cyclization of farnesol; photopolymerization; photoimaging systems. *Mailing Add:* Exp Sta B352 E I du Pont de Nemours & Co Inc Wilmington DE 19880

CHANG, CHAE HAN JOSEPH, b Seoul, Korea, July 7, 29; US citizen; m 56; c 4. RADIOLOGY. *Educ:* Severance Union Med Col, MD, 53; Am Bd Radiol, dipl, 59; Nagoya Univ, PhD(med sci), 66. *Prof Exp:* Resident radiol, Emory Univ Hosp, 55-58, instr, Sch Med, 58-59; chief, Man Mem Hosp, WVa, 59-63; from assoc prof to prof, Sch Med, WVa Univ, 64-70; PROF RADIOL, SCH MED, UNIV KANS, 70- *Mem:* AMA; fel Am Col Radiol; Radiol Soc NAm; Am Roentgen Ray Soc; Asn Univ Radiol. *Res:* Roentgenological measurement of the right descending pulmonary artery in normal state and pulmonary hypertension; computed tomographic evaluation of the breast. *Mailing Add:* Dept Radiol Univ of Kans Med Ctr Kansas City KS 66103

CHANG, CHARLES HUNG, b Szechwan, China, Apr 4, 24; US citizen; m 56; c 3. SYNTHETIC ORGANIC CHEMISTRY. *Educ:* Nat Cent Polytech Col, China, dipl, 45; Univ Mont, MS, 55; Wayne State Univ, PhD(org chem), 59. *Prof Exp:* Chem engr, Taiwan Fertilizer Co, 47-54; res assoc chem, Wayne State Univ, 58-61; res chemist, GAF Corp, 61-65, res specialist dyestuffs, 65-73; SR RES ASSOC DYESTUFFS, SCOTT ALTHOUSE RES CTR, CROMPTON & KNOWLES CORP, 73- *Mem:* Am Chem Soc; Am Asn Textile Chemists & Colorists. *Res:* Design and synthesis of organic dyestuffs and intermediates. *Mailing Add:* Crompton & Knowles Corp PO Box 341 Reading PA 19603

CHANG, CHARLES YU-CHUN, b Harbin, China, Sept 18, 41; US citizen; m 69; c 1. PHARMACEUTICAL CHEMISTRY. *Educ:* Tamkang Col, Taiwan, BS, 64; Am Univ, MS, 74. *Prof Exp:* Res chemist, Kingdom Pharmaceut Co, Taiwan, 65-67; instr tech & mil Mandarin Chinese, Berlitz Lang Inst, Washington, DC, 67-69; dir qual control, Mifflin McCambridge Co, Riverdale, 69-76; REV CHEMIST, US FOOD & DRUG ADMIN, 76- *Mem:* Nat Asn Pharmaceut Mfrs. *Res:* Large scale purification and disinfection of water; origination of rapid pharmaceutical tests and analyses; activation parameters for hydrogen-deuterium exchanges. *Mailing Add:* 10931 Martingale Ct Rockville MD 20854

CHANG, CHEN CHUNG, b Tientsin, China, Oct 13, 27; nat US; m 51, 77; c 4. MATHEMATICAL LOGIC. *Educ:* Harvard Univ, AB, 49; Univ Calif, PhD(math), 55. *Prof Exp:* Lectr math, Univ Calif, 54-55; instr, Cornell Univ, 55-56; asst prof, Univ Southern Calif, 56-58; from asst prof to assoc prof, 58-64, PROF MATH, UNIV CALIF, LOS ANGELES, 64- *Concurrent Pos:* NSF sr fel, Inst Advan Study, 62-63; Fulbright sr res fel, UK, 66-67; vis fel, All Souls Col, Oxford Univ, 66-67; consult ed, J Symbolic Logic, Asn Symbolic Logic, 68-77; mem US nat comt, Int Union Hist & Philos of Sci, 70-72; ed, Ann Math Logic, 70-76. *Mem:* Asn Symbolic Logic. *Res:* Logic. *Mailing Add:* Dept Math Univ Calif 405 Hilgard Ave Los Angeles CA 90024

CHANG, CHENG ALLEN, sepration science, inorganic stereochemistry, for more information see previous edition

CHANG, CHI, b China, Mar 3, 44; Can citizen; m 73; c 3. SOIL PHYSICS, SOIL PHYSICAL CHEMISTRY. *Educ:* Chung-Hsing Univ, Taiwan, BSc, 67; Univ Man, MSc, 72, PhD(soil physics), 76. *Prof Exp:* Res scientist soil sci, Hydrol Res Div, Environ Can, 76-78; res scientist, Soil Res Inst, 75-76, RES SCIENTIST SOIL SCI, LETHBRIDGE RES STA, AGR CAN, 78- *Mem:* Soil Sci Soc Am; Am Soc Agron; Am Geophys Union. *Res:* Water and nitrate movement in soils under irrigated field conditions; relations of soil physical properties and soil management to salinity of irrigated soils. *Mailing Add:* 4013 Glacier Ave Lethbridge AB T1K 3P2 Can

CHANG, CHI HSIUNG, b Taiwan; US citizen; c 2. BIOCHEMISTRY. *Educ:* Kaohsiung Med Col, BS, 66; Temple Univ, PhD(pharmacol), 75. *Prof Exp:* Res scientist, dept exp therapeut, Roswell Park Mem Inst, 75-79; SR BIOCHEMIST, SOUTHERN RES INST, 79- *Mem:* Am Asn Cancer Res; Am Soc Pharmacol & Exp Therapeut; Sigma Xi. *Res:* Metabolism and metabolic effects of anti-cancer agents; purification and kinetic study of enzyme responsible for DNA synthesis. *Mailing Add:* Southern Res Inst PO Box 55305 Birmingham AL 35255

CHANG, CHI KWONG, b Nanking, China, Dec 25, 47; m 71; c 2. ORGANIC CHEMISTRY, BIOINORGANIC CHEMISTRY. *Educ:* Fu Jen Cath Univ, Taiwan, BS, 69; Univ Calif, San Diego, PhD(chem), 73. *Prof Exp:* Res chemist, Univ Calif, San Diego, 73-75; fel, Univ BC, 75-76; from asst prof to assoc prof, 76-82, PROF CHEM, MICH STATE UNIV, 82- *Concurrent Pos:* Vis scientist, Brookhaven Nat Lab, 79; A P Sloan fel, 80-82; Camille & Henry Dreyfus teacher-scholar, 80-84. *Mem:* Am Chem Soc. *Res:* Synthesis of macrocyclic ligands, porphyrins and metalloporphyrins; models for metalloenzymes; biological oxygen binding and activation; photodynamic cancer therapy. *Mailing Add:* Dept Chem Mich State Univ East Lansing MI 48824

CHANG, CHIA-CHENG, b Tamsui, Taiwan, May 28, 39; US citizen; m 65; c 3. MECHANISM OF CARCINOGENESIS, SOMATIC CELL GENETICS. *Educ:* Chung-Hsing Univ, BS, 62; Univ Mo, PhD(genetics), 71. *Prof Exp:* Postdoctoral investr, Biol Div, Oak Ridge Nat Lab, 71-72; postdoctoral fel, Dept Human Genetics, Med Sch, Univ Mich, 72-74; res assoc, Mich State Univ, 74-75, from asst prof to assoc prof, 75-89, PROF, DEPT PEDIAT & HUMAN DEVELOP, MICH STATE UNIV, 89- *Honors & Awards:* Young Environ Scientist Award, Nat Inst Environ Health Sci, NIH, 78-81. *Mem:* Genetic Soc Am; AAAS; Tissue Culture Asn; Environ Mutagen Soc; Sigma Xi. *Res:* Role of DNA replication and repair in human and mammalian cell mutagenesis; role of gap junctional intercellular communication in tumor promotion; development of sensitive assay systems for environmental mutagens, carcinogens; neoplastic transformation of human breast epithelial cells in vitro. *Mailing Add:* B240 Life Sci Bldg Mich State Univ East Lansing MI 48824

CHANG, CHIEH CHIEN, b Peking, China, July 21, 13; nat US; m 37; c 3. ATOMSPHERIC DYNAMICS, FLUIDS. *Educ:* Nat Northeastern Univ, Peking, BS, 32; Calif Inst Technol, MS, 41, PhD(aeronaut), 50. *Prof Exp:* Instr aeronaut, Tsing Hua Univ, Peking, 34-40, supvr wind tunnel proj, 35-37; res asst, Calif Inst Technol, 40-42; design engr, US Plywood Corp, NY, 42-43; design engr, Glenn Martin Co, Md, 43-46, res engr in charge supersonic aerodyn, 46-47; assoc prof aeronaut, Johns Hopkins Univ, 47-52, contract res dir, 51-52; res prof, Inst Fluid Dynamics & Appl Math, Univ Md, 52-54; prof fluid dynamics, Univ Minn, 54-62; vis prof & dir plasma space sci lab, Cath Univ Am, prof space sci & appl physics & head dept, 63-71, prof aerospace & atmospheric sci, 71-77, prof mech eng, 77-, chmn dept, 78-; RETIRED. *Concurrent Pos:* Lectr, Chinese Air Force Acad, 39-40; mem aeroelasticity comt, Appl Physics Lab, Johns Hopkins Univ, 48-50; consult, Off Sci Res, US Air Force, 51-53, Gen Mills Corp, 55-58, Intercontinental Ballistic Missle Prog, Gen Elec Co, 56-58, Los Alamos Sci Lab, 58-59, Lawrence Radiation Lab, Univ Calif, 59-63 & Goddard Space Flight Ctr, NASA, 62-65; Guggenheim fel, 52-53; sr staff scientist, Phys Res Labs, Aerospace Corp, 61-62; US Atomic Energy Comn rep, Int Conf Plasma Physics & Controlled Nuclear Fusion Res, Salzburg, Austria, Sept, 62; rep, Univ Corp Atmospheric Res. *Honors & Awards:* Gold Medal Award, Repub China, 67; hon res prof, Academia Sinica, People's Repub China, 79. *Mem:* AAAS; Am Phys Soc; fel Am Inst Aeronaut & Astronaut; Am Geophys Union; Am Soc Mech Engrs; Am Meteoral Soc; elected mem Academia Sinica Repub China. *Res:* Inventor of concept of ring vortex cavity reactor for space nuclear propulsion; inventor of superfast thermalization of plasma for controlled thermonuclear fusion reactor; inventor of honeycomb sandwich structures used in flying vehicles; developer of tornado and hurricane modeling; tornado and hurricane dynamics; summer monsoons in Far East, particularly India and China; viscous flow. *Mailing Add:* 8005 Falstaff Rd McLean VA 22101

CHANG, CHIEN-WU, metallurgy & physical metallurgical engineering, for more information see previous edition

CHANG, CHIH-PEI, b Chungking, China, Feb 10, 45; US citizen; m 71; c 3. ATMOSPHERIC DYNAMICS, METEOROLOGY. *Educ:* Nat Taiwan Univ, BS, 66; Univ Wash, PhD(atmospheric sci), 72. *Prof Exp:* From asst prof to assoc prof, 72-82, PROF METEOROL, NAVAL POSTGRAD SCH, 82- *Concurrent Pos:* Prin investr & proj dir, Nat Oceanic & Atmospheric Admin grant, 73- & NSF grant, 75-; adv ed, Papers Meteorol Res, 78-; res consult, Chinese Nat Sci Coun, 77-; assoc ed, J Atmospheric Scis, 83-87; consult, Cent Weather Bur, Taiwan; ed, Monsoon Meteorol, Oxford Univ Press, 85-87. *Honors & Awards:* Clarence Leroy Meisinger Award, Am Meteorol Soc, 83. *Mem:* Fel Am Meteorol Soc; Sigma Xi. *Res:* Tropical and monsoon meteorology, large scale dynamics; numerical weather prediction; operational weather forecasting. *Mailing Add:* Dept Meteorol Code MR/Cp Naval Postgrad Sch Monterey CA 93943-5000

CHANG, CHIN HAO, b Haining, Chekiang, China, July 2, 26; m 61; c 1. ENGINEERING MECHANICS. *Educ:* Univ Taiwan, BS, 53; Va Polytech Inst, MS, 57; Univ Mich, PhD(eng mech), 62. *Prof Exp:* Asst prof eng mech, Univ Ala, Tuscaloosa, 60-64; from assoc prof to prof eng mech, 64-88, EMER PROF ENG MECH, UNIV ALA, 88- *Mem:* Sigma Xi; Am Acad Mech. *Res:* Structural mechanics. *Mailing Add:* 5180 Courton St Alpharetta GA 30202

CHANG, CHIN HSIUNG, b Tainan, Taiwan, May 19, 39; US citizen; m 70; c 2. INORGANIC CHEMISTRY, PHYSICAL CHEMISTRY. *Educ:* Nat Taiwan Univ, BS, 62; Rice Univ, PhD(inorg chem), 67. *Prof Exp:* Sr scientist, Avco Everett Res Lab Inc, 71-73; Staff chemist, Exxon Res & Eng Co, 73-; SR SCIENTIST, RES & TECHNOL, ALLIED-SIGNAL, INC. *Mem:* AAAS; Am Chem Soc; Electrochem Soc; NY Acad Sci; Sigma Xi. *Res:* Solid state chemical researches on materials for electrochemical systems, adsorption and catalysis. *Mailing Add:* Allied-Signal Inc 50 E Algonquin Rd Des Plaines IL 60017-5016

CHANG, CHIN-AN, b Sian, China, June 13, 43; m 66; c 4. PHYSICAL CHEMISTRY, MATERIALS SCIENCE. *Educ:* Chung-Hsing Univ, Taiwan, BS, 63; Colo State Univ, MS, 67; Univ Calif, Berkeley, PhD(phys chem), 70. *Prof Exp:* Vis assoc prof chem, Tsing-Hua Univ, Taiwan, 70-71; res assoc, Tex A&M Univ, 71-72; res fel, Calif Inst Technol, 72-73; chemist, Lawrence Berkeley Lab, Calif, 73-75; RES STAFF MEM, T J WATSON RES CTR, IBM CORP, NY, 75- *Mem:* Am Phys Soc; Mat Res Soc. *Res:* Interface studies of thin films; metal-metal epitaxy on silicon; processing of silicides, metals, polymers and superconductors. *Mailing Add:* T J Watson Res Ctr IBM Corp PO Box 218 Yorktown Heights NY 10598

CHANG, CHIN-CHUAN, b Laiyang, China, Oct 5, 25; m 45; c 3. ENDOCRINOLOGY, REPRODUCTIVE PHYSIOLOGY. *Educ:* Cath Univ Peiping, BS, 48; NY Univ, MA, 61; Univ Wis-Madison, PhD(endocrinol, reprod physiol), 67. *Prof Exp:* Instr biol, Nat Defense Med Col, Taiwan, 55-58; instr zool, Nat Taiwan Univ, 58-63, assoc prof, 63-65; res assoc endocrinol, 69-70, staff scientist, 71-77, SR INVESTR, BIOMED DIV, POP COUN, ROCKEFELLER UNIV, 77- *Concurrent Pos:* Res fel med, Pop Coun, Rockefeller Univ, 67-69. *Mem:* AAAS; Am Fertil Soc; Endocrine Soc; NY Acad Sci; Sigma Xi. *Res:* Relationship between blastocyst and endometrium in process of implantation and requirements of ovarian hormones for formation of deciduomata; endocrine activity of steroids released through dimethylpolysiloxane membrane; more effective methods in contraception. *Mailing Add:* Pop Council Rockefeller Univ York Ave & 66th St New York NY 10021

CHANG, CHING HSONG, molecular endocrinology, reproductive physiology, for more information see previous edition

CHANG, CHING MING, b Nanking, China, Oct 13, 35; m 64; c 2. FLUID PHYSICS, ENGINEERING SCIENCE. *Educ:* Aachen Tech Univ, Dipl Ing, 62, DrIng(fluid physics), 67; State Univ NY, Buffalo, MBA, 85. *Prof Exp:* Res assoc, Shock Tube Lab, Inst Mech, Aachen Tech Univ, 62-64, instr, 64-67, res assoc, 67; vis asst prof eng mech, NC State Univ, 68-70, asst prof, 70-73; sr engr, 73-75, consult, 75, supvr, 75-78, ENG ASSOC, LINDE DIV, UNION CARBIDE CORP, 78- *Concurrent Pos:* Deleg, Int Coun Sci Unions, Madrid, 65, London, 67; asst ed, Plasma Physics, 71-72; adj assoc prof eng sci, aerospace & nuclear eng, State Univ NY Buffalo, 75-79, adj prof eng, 79- *Honors & Awards:* Distinguished Service Award, Nat Soc Prof Engr, Erie-Niagara Chap, 80; Ignace Basinski Award, 84. *Mem:* AAAS; Am Phys Soc; Am Soc Mech Engrs; Nat Soc Prof Engrs. *Res:* Heat transfer; energy conversion; electrohydrodynamics; applied mechanics; electrostatic precipitation; thermal sciences and air pollution; turbomachinery; "expert systems" in mechanical engineering. *Mailing Add:* 171 The Paddock Williamsville NY 14221

CHANG, CHING SHUNG, b China, Dec 19, 47. GEOTECHNICAL ENGINEERING, SOIL MECHANICS. *Educ:* Chen Kung Univ, BS, 69; Univ SC, MS, 71; Univ Calif, Berkeley, PhD(soil mech), 76. *Prof Exp:* Asst prof civil eng, State Univ NY Buffalo, 77-79; from asst prof to assoc prof civil eng, 79-90, PROF CIVIL ENG, UNIV MASS, 90- *Mem:* Am Soc Civil Engrs; Int Soc Soil Mechs & Found Eng. *Res:* Stress/strain behavior of soil; application of computers in geotechnical engineering. *Mailing Add:* Dept Civil Eng Univ Mass Amherst MA 01002

CHANG, CHING-JEN, b Keelung, Taiwan, Feb 2, 41; m 67; c 1. POLYMER SYNTHESIS. *Educ:* Tunghai Univ, Taiwan, BS, 63; Marquette Univ, MS, 67; Univ Calif, Berkeley, PhD(org chem), 71. *Prof Exp:* Res fel chem, Univ Fla, 71-73; sr chemist, 73-81, RES FEL, ROHM & HAAS CO, 81- *Mem:* Am Chem Soc. *Res:* Carbanions; ion-pairs structures; water soluble polymers; hydrophobic association; ionic association; polyurethanes; thickeners; emulsion polymerization; surfactants and dispersants; rheology of polymer solutions and dispersions; acrylic monomers and polymers; binders for nonwovens and textile. *Mailing Add:* Res Labs Rohm & Haas Co Norristown & McKean Rds Spring House PA 19477

CHANG, CHING-JER, b Hsin-chu, Taiwan, Oct 17, 42; c 2. NATURAL PRODUCTS CHEMISTRY, MEDICINAL CHEMISTRY. *Educ:* Nat Taiwan Cheng Kung Univ, BS, 65; Ind Univ, PhD(org chem), 72. *Prof Exp:* Res asst chem, Nat Taiwan Univ, 66-67; teaching asst, NMex Highlands Univ, 68; res & teaching asst, Ind Univ, 68-72; res assoc, 72-73, from asst prof to assoc prof, 73-83, PROF MED CHEM, PURDUE UNIV, WEST LAFAYETTE, 83- *Concurrent Pos:* Mem NIH, Bio-org Nat Prod Study Sect, 86-90; ed adv bd, Jour Nat Prod, 90- *Honors & Awards:* Lion Club Cancer Res Award. *Mem:* Am Chem Soc; The Chem Soc; Am Soc Pharmacog; Phytochem Soc NAm; Am Asn Pharmaceut Sci; Am Asn Cancer Res. *Res:* Isolation, structure elucidation, biosynthesis and partial synthesis of bioactive natural products; interaction of small molecules and drugs with macromolecules; biomedical application of spectroscopy; bioorganic chemistry of molecular recognition and enzyme modeling. *Mailing Add:* Dept of Med Chem & Pharmacog Purdue Univ Sch of Pharm West Lafayette IN 47907

CHANG, CHIN-HAI, b Taiwan, China; US citizen. IMMUNOCHEMISTRY, IMMUNOLOGY. *Educ:* Nat Taiwan Univ, BS, 65; Washington Univ, PhD(develop biochem), 71. *Prof Exp:* Fel physiol chem, Roche Inst Molecular Biol, 71-73; asst prof cancer res, Sch Med, Tufts Univ, 73-76; MGR IMMUNOCHEM RES, SYVA RES INST, PALO ALTO, 76- *Concurrent Pos:* Consult immunochem, Leary Lab, 73-76. *Mem:* AAAS; Int Res Group Carcinoembryonic Proteins; Am Asn Clin Chem; NY Acad Sci; Am Soc Microbiologists. *Res:* Immunodiagnostics; clinical chemistry; biochemistry; tumor immunology; microbiology. *Mailing Add:* Syva Res Inst 3181 Porter Dr Palo Alto CA 94303

CHANG, CHONG-HWAN, b Seoul, Korea, Nov 7, 50; c 2. PROTEIN CRYSTALLOGRAPHY. *Educ:* Seoul Nat Univ, BS, 73, MS, 75; Univ Pittsburgh, PhD(crystallog), 82. *Prof Exp:* Fel protein crystallog, 82-84, asst biophysicist, 84-90, BIOPHYSICIST, BIOL ENVIRON & MED RES DIV,

ARGONNE NAT LAB, 90- *Concurrent Pos:* Adj asst prof, Dept Med Chem & Pharmacog, Col Pharm, Univ Ill, Chicago, 85- *Mem:* Am Crystallog Asn. *Res:* Three dimensional structural analysis of membrane protein, photosynthetic reaction center and immunoglobulin fragments. *Mailing Add:* Biol Environ & Med Res Div Argonne Natl Lab 9700 S Cass Ave Argonne IL 60439-4833

CHANG, CHRISTOPHER TEH-MIN, b Nanking, China, Apr 2, 36; m 69; c 3. GALLIUM ARSENIDE ARSENIC, DIGITAL & LINEAR INTEGRATED CIRCUIT DESIGN. *Educ:* Nat Taiwan Univ, BSEng, 57; Univ Southern Calif, MSEE, 62, PhD(elec eng), 68. *Prof Exp:* Teaching asst physics, Taipei Inst Technol, Taiwan, 59-60; test engr, ALWAC Comput Div, El-tronics, Calif, 61-62; design engr, Appl Res Lab, Calif, 62-63; res asst electromagnetic theory & superconductivity, Univ Southern Calif, 63-68; asst elec engr, High Energy Fac Div, Argonne Nat Lab, 68-73; SR MEM TECH STAFF, DEFENSE SYSTS & ELECTRONICS GROUP, TEX INSTRUMENTS, INC, 73- *Mem:* Am Phys Soc; sr mem Inst Elec & Electronic Engrs. *Res:* Superconductivity, microwave circuits, magnetic bubble device design and material developments; Gallium Arsenide Device and circuit design. *Mailing Add:* Microwave Lab Tex Instruments Inc PO Box 650311 Dallas TX 75265

CHANG, CHU HUAI, b Fukien, China, Oct 1, 17; m 59; c 2. RADIOLOGY. *Educ:* St John's Univ, China, BS, 41, MD, 44. *Prof Exp:* From instr to asst prof radiol, Sch Med, Yale Univ, 54-62; assoc prof & assoc attend radiologist, Med Ctr, 62-67, dir, Radiation Ther Div, 70-85, PROF RADIOL, COL PHYSICIANS & SURGEONS, COLUMBIA UNIV, 67-, ATTEND RADIOLOGIST, COLUMBIA-PRESBY MED CTR, 67- *Concurrent Pos:* Res fel radiol, Sch Med, Univ Calif, 47-49; res fel, Sch Med, Yale Univ, 50-51. *Mem:* AAAS; Sigma Xi; Am Soc Therapeut Radiol; fel Am Col Radiol; Asn Univ Radiol; Am Radium Soc; NY Acad Sci. *Res:* Radiation therapy and radiobiology. *Mailing Add:* 622 W 168th St New York NY 10032

CHANG, CHUAN CHUNG, b Tainan, Formosa, Nov 28, 38; m 63; c 2. ATOMIC & MOLECULAR PHYSICS. *Educ:* Rensselaer Polytech Inst, BS, 62; Cornell Univ, PhD(physics), 67. *Prof Exp:* Res asst physics, Cornell Univ, 63-67; mem tech staff, Bell Labs, 67-83, DISTINGUISHED MEM TECH STAFF, BELL COMMUN RES INC, 83- *Res:* Surface physics; crystallography; electron diffraction; electron spectroscopy; electronics materials; silicon integrated circuit processing; compound semiconductors; superconductivity. *Mailing Add:* 3X1756 Bellcore Red Bank NJ 07701-7020

CHANG, CHUNG-NAN, secretion of protein, for more information see previous edition

CHANG, CHUN-YEN, b Kaoshiung, Taiwan, Oct 12, 37; m 66; c 3. CHEMICAL ENGINEERING. *Educ:* Nat Cheng Kung Univ, BS, 60; Nat Chiao Tung Univ, MS, 62, PhD(electronics), 70. *Prof Exp:* Dir Res, 77-87, dean, 87-90, PROF SEMICONDUCTORS, NAT CHIAO TUNG UNIV, 69-, DEAN ENG, 90-; DIR, NAT NANO-DEVICE LABS, 90- *Mem:* Fel Inst Elec & Electronic Engrs; Am Electromagnetics Acad; Electrochem Soc; Am Inst Physics. *Res:* ULSI physics and technologies; electronic and optoelectronics devices and circuits. *Mailing Add:* Inst Electronics Nat Chiao Tung Univ 1001 University Rd Hsinchu 300 Taiwan

CHANG, CLIFFORD WAH JUN, b Honolulu, Hawaii, July 25, 38. ORGANIC CHEMISTRY. *Educ:* Univ Southern Calif, BS, 60; Univ Hawaii, PhD(chem), 64. *Prof Exp:* Jr chemist, Cyclo Chem Corp, Calif, 59; asst marine chemist, Hawaii Marine Lab, 61; fel chem, Univ Ga, 64-68; from asst prof to assoc prof, 68-79, PROF CHEM, UNIV WEST FLA, 79- *Concurrent Pos:* Vis prof, Univ Okla, 76-77 & Univ Hawaii, 77, 81, 83, 85, 87 & 90. *Mem:* Am Chem Soc; The Chem Soc; Sigma Xi; Am Soc Pharmacog. *Res:* Structural determinations and synthesis of natural products. *Mailing Add:* Dept Chem Univ West Fla Pensacola FL 32514-5751

CHANG, DANIEL P Y, b Shanghai, China, Mar 25, 47; US citizen; m 69; c 2. AIR TOXICS, AIR POLLUTION CONTROL. *Educ:* Calif Inst Technol, BS, 68, MS, 69, PhD(mech eng), 73. *Prof Exp:* Res fel environ health eng, Calif Inst Technol, 73; from asst prof to assoc prof, 73-86, PROF CIVIL ENG, UNIV CALIF, DAVIS, 87- *Concurrent Pos:* Prin investr, Res Initiation Award, NSF, 75-77; co-investr contracts, US Environ Protection Agency, 77-79, proj dir, Air Pollution Area Training Ctr, 78-83; mem res screening comt, Calif Air Resources Bd, 79-83; prin investr, Calif Air Resources Bd, 83-91, US Environ Protection Agency, 86-88; co-investr, Nat Inst Environ Health Sci, 89-91. *Mem:* Air & Waste Mgt Asn; AAAS; Sigma Xi; Am Chem Soc. *Res:* Physico-chemical behavior of aerosols; aerosol effects on health; hazardous wastes incineration and combustion; air toxics emissions and control. *Mailing Add:* Dept Civil Eng Univ Calif Davis CA 95616

CHANG, DAVID BING JUE, b Seattle, Wash, June 27, 35; m 59; c 4. APPLIED PHYSICS. *Educ:* Univ Wash, Seattle, BS, 56; Calif Inst Technol, PhD(physics), 62. *Prof Exp:* Staff physicist, Gen Atomic, 63-66; sr fel, Univ Wash, 66-68; staff physicist, Boeing Co, 68-73; dep asst secy sci & technol, US Dept Commerce, 73-77; mgr res planning, Occidental Res Corp, 77-78, head appl physics, 78-82; chief scientist & asst mgr, Advan Prod Lab, 82-90, DIR TECHNOL, TRAINING & SUPPORT SYSTS GROUP, HUGHES AIRCRAFT CO, 90- *Concurrent Pos:* Res & develop comt, US-Israel, US Govt & US-Japan Sci Comt, 75; fel, Univ Calif, San Diego, 61-63, Univ Wash, Dept Physiol & Biophysics, 66-68; affil investr, Va Mason Res Ctr, 71-82; consult, Firlands Hosp, 72-73; affil assoc prof, Dept Physiol & Biophysics, Univ Wash, 68-82; adj prof, Dept Physics, Univ Calif, Irvine, 81-82; prof physics, Inst Basic Res, Harvard Grounds; mem, Phys Panel, Energy Res Adv Bd, 86, Springer-Verlag Appl & Computational Math. *Mem:* Am Phys Soc; AAAS; Inst Elec & Electronics Engrs. *Res:* Analyses of collective effects in molecules and semiconductors, in fusion and space plasmas, in biophysics and astrophysics; gauge theory of gravitation; development of technology policy; energy technologies; electro optics and lasers; environmental technologies. *Mailing Add:* 14212 Livingston Tustin CA 92680

CHANG, DING, b Kwiechow, China, Sept 2, 40; m 72; c 2. ORGANIC CHEMISTRY, BIOCHEMISTRY. *Educ:* Nat Taiwan Norm Univ, BS, 66; Univ NB, PhD(org chem), 71. *Prof Exp:* Res assoc chem, Univ NB, 71-72; res fel, Inst Biomed Res, Univ Tex, Austin, 72-77; prin res chemist, Bioprods Dept, Beckman Instruments Inc, 77-79; VPRES & MFG DIR, PENINSULA LABS, INC, 79- *Mem:* Am Chem Soc. *Res:* Synthetic chemistry of natural products; alkaloids, peptides and proteins. *Mailing Add:* 611 Taylor Way Belmont CA 94002-4041

CHANG, DONALD CHOY, b Kwangtung, China, Aug 28, 42. BIOPHYSICS, CELL PHYSIOLOGY. *Educ:* Nat Taiwan Univ, BS, 65; Rice Univ, MA, 67, PhD(physics), 70. *Prof Exp:* Res assoc physics, Rice Univ, 70-74; from instr to asst prof, 73-85, ASSOC PROF BIOPHYS, BAYLOR COL MED, 85-; PROF BIOL, HONG KONG UNIV SCI & TECH, 89- *Concurrent Pos:* Welch Found fel biophys, Baylor Col Med, 70-73; adj asst prof physics, Rice Univ, 74-; investr, Marine Biol Lab, Woods Hole, 76-; vis prof biophysics, Beijing Univ, 81; chmn, Int Conf Struct & Function Excitable Cells, 81 & Int Conf Electroporation & Electrofusion, 90. *Mem:* Am Soc Cell Biol; Am Phys Soc; Biophys Soc; AAAS; Soc Neurosci. *Res:* Microscopy; membrane biophysics, mechanisms of nerve excitation; cell fusion, gene transfection by electroporation. *Mailing Add:* Dept Physiol & Molecular Biophys Baylor Col Med Houston TX 77030

CHANG, EDDIE LI, b Nanking, China, Oct 4, 48; US citizen; m 72; c 3. LIGHT-SCATTERING, LIPID MEMBRANES. *Educ:* Antioch Col, BS, 71; Univ Ore, PhD(chem-physics), 77. *Prof Exp:* Res assoc, Dept Chem, Univ Wash, 77-79; Nat Res Coun fel, 79-81, RES PHYSICIST, NAVAL RES LAB, 81- *Concurrent Pos:* Actg sr officer, Off Naval Res, 87-88. *Mem:* Biophys Soc. *Res:* Physical properties of lipid membranes; interaction of lipids with proteins; use of artificial lipid bilayers as membrane models; immunoassays; biosensors; structure-property relationships of archaebacterial lipid membranes. *Mailing Add:* Biomolecular Eng Code 6190 Naval Res Lab Washington DC 20375-5000

CHANG, EDWARD SHI TOU, b Swatow, China, July 6, 40; US citizen; m 80; c 3. ELECTRON-MOLECULE COLLISIONS, SPECTROSCOPY OF RYDBERG STATES. *Educ:* Univ Calif, Riverside, BA, 61, MA, 64, PhD(physics), 67. *Prof Exp:* Res assoc, Goddard Space Flight Ctr, 67-69; res assoc, Univ Chicago, 69-70; from asst prof to assoc prof, 70-84, assist dean, 75-77, PROF PHYSICS, UNIV MASS, AMHERST, 84- *Concurrent Pos:* Sr res fel, Sci Res Coun, Eng, 77-78; summer res scientist, Astron Res Facil, Univ Mass, 80-83; vis prof, Univ Kaiserslavtern, WGer, 83-84 & 87; vis scientist, Inst Theoret Atom & Molecular Physics, Harvard Univ, 90-91. *Mem:* Am Phys Sos; Nat Geog Soc. *Res:* Rydberg states spectroscopy of atoms and molecules and their properties in external fields. *Mailing Add:* Dept Phys Univ Mass Hasbrook Lab Amherst MA 01003

CHANG, ELFREDA TE-HSIN, b Peiping, China, Dec 13, 35; US citizen. THERMODYNAMICS, PHYSICAL CHEMISTRY. *Educ:* Univ Mich, BSE, 57, MS, 59, PhD(phys chem), 63. *Prof Exp:* Mem tech staff chem, 62-82, ENG SPECIALIST, AEROSPACE CORP, 82- *Mem:* Sigma Xi. *Res:* Low and high temperature calorimetry; thermodynamics of gas-condensed-phase equilibria; solubility of gases in propellants; compaitbility of materials with propellants; physical chemistry of propellants. *Mailing Add:* 12610 Havelock Ave Los Angeles CA 90066

CHANG, ELIZABETH B, b Wilmington, Del, Jan 1, 42. COMPUTER SCIENCE. *Educ:* Millersville State Col, BA, 66; Univ Md, MA, 68, PhD(math), 72. *Prof Exp:* From asst prof to assoc prof, 73-83, PROF MATH & COMPUT SCI, CHMN DEPT, HOOD COL, 83- *Concurrent Pos:* NSF fel, Univ Md, 72. *Mem:* Asn Comput Mach; Inst Elec & Electronics Engrs. *Res:* Curriculum development; computer science; computing & mathematical skills for liberal arts students. *Mailing Add:* Dept Math & Comput Sci Hood Col Frederick MD 21701

CHANG, EPPIE SHENG, b Shanghai, China, Aug 20, 46; US citizen; m 68; c 2. BIOCHEMISTRY, NUTRITION. *Educ:* Univ Calif, Berkeley, BA, 67, MA, 71. *Prof Exp:* Res lab asst, Dept Nutrit Sci, Univ Calif, 67-68, res asst, 68-70; asst res scientist, Ames Prod Develop Lab, 71-74, assoc res scientist, 74-77, RES SCIENTIST, AMES BLOOD CHEM LAB, MILES LAB, 77- *Mem:* Am Asn Clin Chem. *Res:* Development of simple clinical, serum tests for diagnostic application. *Mailing Add:* 724 Arastradero Palo Alto CA 94306

CHANG, ERNEST SUN-MEI, b Berkeley, Calif, Dec 7, 50; m 86. ENDOCRINOLOGY, CELL BIOLOGY. *Educ:* Univ Calif, Berkeley, AB, 73; Univ Calif, Los Angeles, PhD(biol), 78. *Prof Exp:* Asst prof, 78-85, ASSOC PROF ANIMAL SCI, UNIV CALIF, DAVIS, 85- *Concurrent Pos:* Am Cancer Soc fel, Univ Chicago, 78; vis prof, Sonoma State Univ, 86- *Mem:* AAAS; World Maricult Soc; Am Soc Zoologists; Crustacean Soc; Nat Shellfisheries Asn. *Res:* Molecular action of insect and crustacean hormones; invertebrate zoology; aquaculture of marine invertebrates. *Mailing Add:* Bodega Marine Lab PO Box 247 Bodega Bay CA 94923

CHANG, FA YAN, b Shantung, China, May 5, 32; m 60; c 3. WEED SCIENCE, PLANT PHYSIOLOGY. *Educ:* Nat Taiwan Univ, BSc, 53; Univ Alta, MSc, 66, PhD(plant sci), 69. *Prof Exp:* Asst agronomist, Taiwan Tobacco Res Inst, Repub China, 54-64; instr weed sci, Univ Alta, 69-70; res assoc, Univ Guelph, 70-74; herbicide eval officer, 74-87, ASSOC DIR, AGR CAN, 87- *Concurrent Pos:* Nat executive, Can Weed Comt, 75-83. *Mem:* Agr Pesticide Soc Can; Weed Sci Soc Am; Can Soc Plant Physiol; Plant Growth Regulator Soc Am. *Res:* Weed control; herbicide physiology and antidotes. *Mailing Add:* Pesticides Directorate 2323 Riverside Second Floor Ottawa ON K1A 0C6 Can

CHANG, FRANK KENG, b Anhwei, China, Feb 12, 22; US citizen; m 62; c 1. SEISMOLOGY, PETROLEUM EXPLORATION. *Educ:* Nat Hunan Univ, BS, 46; St Louis Univ, MS, 58. *Prof Exp:* Petrol engr drilling & production, Chinese Petrol Corp, 46-52; chief seismologist, US Antarctic Res

Proj, Arctic Inst NAm, NSF, 58-60 & res assoc, Geophys & Polar Res Ctr, Univ Wis, 60-61; res geophysicist, Eng Waterways Exp Sta, US Army, 66-; RETIRED. *Concurrent Pos:* Geophys engr seismic data process petrol explor, Tex Instrument, Inc, 66. *Mem:* Seismol Soc Am; Soc Explor Geophysicists; Am Geophys Union; Sigma Xi. *Res:* Earthquake resistant design; site characterization of earthquake ground motions by power spectral densities; strong motion duration, spectral content and predominant period; a quantitative earthquake intensity scale; permanent displacement analysis treating slides an embankment as a rigid block on an inclined plane. *Mailing Add:* 603 Santa Rosa Dr Vicksburg MS 39180

CHANG, FRANK N, PROTEIN SYNTHESIS. *Educ:* Univ Wis-Madison, PhD(molecular biol), 68. *Prof Exp:* PROF BIOL, TEMPLE UNIV, 71- *Mailing Add:* Dept Biol Temple Univ Broad & Montgomery Sts Philadelphia PA 19122

CHANG, FRANKLIN, b Princeton, NJ, Feb 12, 42; m 67, 85; c 3. PHYSIOLOGY, ENTOMOLOGY. *Educ:* Univ Md, BS, 63; Univ Ill, PhD(entom), 69. *Prof Exp:* Asst prof biol, Alma Col, Mich, 69-70; from asst prof to assoc prof, 70-82, PROF ENTOM, UNIV HAWAII, 82-, CHAIR ENTOM DEPT, 91- *Concurrent Pos:* ed bd, Ann Entom Soc Am; examr, ARPE; prin investr. *Mem:* Sigma Xi; AAAS; Entom Soc Am; Am Inst Biol Sci; Am Asn Univ Prof; Am Resistry Prof Entom; Nat Educ Asn. *Res:* Insect biochemistry; endocrinology; reproduction. *Mailing Add:* Dept Entom Univ Hawaii 3050 Maile Way Honolulu HI 96822

CHANG, FRANKLIN SHIH CHUAN, b Nanking, China, Dec 30, 15; US citizen; m 38, 69; c 4. POLYMER CHEMISTRY. *Educ:* Purdue Univ, MS, 49; Univ Md, PhD(chem), 52. *Prof Exp:* Group leader anal & phys chem, Mystik Adhesive Prod, Inc, 52-60; mgr polymer physics, Ingersoll Res Ctr, Borg-Warner Corp, 60-79, staff scientist, 79-80; CONSULT, 80- *Mem:* Am Chem Soc. *Res:* Physical testing of viscoelastic materials; adhesion and adhesives; infrared spectrometry; molecular structure; submicron particle size distribution measurements; polymer physics. *Mailing Add:* 320 S Maple Mt Prospect IL 60018

CHANG, FREDDY WILFRED LENNOX, b Kwang Tung, China, Sept 18, 35; Trinidad & Tobago citizen; m 65; c 3. OPTOMETRY, PHARMACOLOGY. *Educ:* Sir George Williams Univ, BSc, 65; Univ Waterloo, OD, 70; Ind Univ, Bloomington, MS, 73, PhD(physiol optics), 76. *Prof Exp:* Res technician urol, Royal Victoria Hosp, Montreal, 65-67; clinician optom, Drs J D Price & G Grant, Kitchener, Ont, 70; assoc instr optom, Ind Univ, Bloomington, 70-74; asst prof optom, Univ Ala, Birmingham, 74-77, instr pharmacol, 76-77; asst prof optom, 77-80, ASSOC PROF OPTOM & ADJ ASSOC PROF PHARMACOLOGY, MEDICAL SCI PROG, IND UNIV, BLOOMINGTON, 80- *Concurrent Pos:* Am Optom Found fel, 71-74; consult pharmacol, Nat Bd Examr Optom, 74- & Tex State Bd Optom, 78; dir continuing educ, Sch Optom, Univ Ala, Birmingham, 76-77 & Sch Optom, Ind Univ, 77- *Mem:* Am Acad Optom; Asn Res Vision & Ophthal; Am Optom Asn; Sigma Xi. *Res:* Ocular pharmacology, specifically toxicology; physiological disposition of drugs in the eye. *Mailing Add:* Vet Admin Out Patient 425 S Hill St 112C Los Angeles CA 90013

CHANG, FREDERIC CHEWMING, b San Francisco, Calif, Aug 5, 05; m 32; c 3. BIO-ORGANIC CHEMISTRY. *Educ:* Columbia Univ, BA, 27; Harvard Univ, MA, 40, PhD(org chem), 41. *Prof Exp:* From instr to asst prof chem, Lingnan Univ, 30-38, cur dept, 32-38; res assoc, Stanford Univ, 41-42; spec res assoc, Off Sci Res & Develop, Harvard Univ, 42-46; prof chem & chmn dept, Lingnan Univ, 46-51; lectr chem & res assoc path, Univ Tenn, Memphis, 51-59, prof pharmacog, Col Pharm, 59-72, prof biochem, Col Basic Med Sci, 59-76; distinguished res prof biochem, Univ S Ala, 76-82; SR ADJ PROF, HARVEY MUDD COL, 82- *Mem:* Am Chem Soc. *Res:* Naphthoquinones; antimalarials; steroids; medicinal plants; bile acids. *Mailing Add:* Mt San Antonio Gardens 900 E Harrison Ave Apt A22 Pomona CA 91767

CHANG, GEORGE CHUNYI, b Shanghai, China, Aug 23, 35; m 63; c 3. APPLIED MECHANICS, STRUCTURAL ENGINEERING. *Educ:* Taiwan Col Eng, BS, 59; Univ Ill, MS, 62, PhD(struct eng), 66. *Prof Exp:* Struct engr, Severud-Elstad-Krueger & Assocs, 62-63; res asst, Col Eng, Univ Ill, 63-66; res engr, Boeing Airplane Co, Wash, 66-68, res specialist, 68-69; asst prof aerospace eng, US Naval Acad, 69-73, assoc prof, 73-75; br chief energy res, US Dept Energy, 75-79; prof & assoc dean eng, Cleveland State Univ, 79-87; PROF AERONAUT ENG & DEAN GRAD SCH & RES, EMBRY-RIDDLE AERONAUT UNIV, 87- *Concurrent Pos:* Instr, Highline Col, 67-68; lectr, Univ Md, College Park, 69-70; consult, Spacecraft Lab, Commun Satellite Corp, Washington, DC, 69-73, US Dept Energy, 80-82, Pac Northwest Div, Battelle Mem Inst, Richland, 84-86 & US Fed Aviation Admin, 90- *Honors & Awards:* Bunsham Award, Am Vacuum Soc, 86. *Mem:* Am Inst Aeronaut & Astronaut; Am Soc Eng Educ; Am Acad Mech; Am Soc Mech Eng. *Res:* Problems in the areas of applied mechanics, engineering design, computer methods, renewable energy and conservation technology; coatings on turbine blades; integration of flight operations and air traffic management; air traffic control safety. *Mailing Add:* 310 Maple Ave W No 234 Vienna VA 22180

CHANG, GEORGE WASHINGTON, b Madison, Wis, Feb 22, 42; m 69; c 4. INTESTINAL MICROFLORA, INFECTIOUS DISEASE. *Educ:* Princeton Univ, AB, 63, Univ Calif, Berkeley, PhD(biochem), 67. *Prof Exp:* Vis scientist, Lab Molecular Biol, NIH, 67-68; NIH fel biochem, 68-70, asst prof, 70-76, ASSOC PROF FOOD MICROBIOL, NUTRIT SCI DEPT, UNIV CALIF, BERKELEY, 76- *Mem:* Am Soc Microbiol; Inst Food Technologists; Am Inst Nutrit; Am Soc Biochem Molecular Biol. *Res:* Isolation of bacteria producing the major metabolites of the lower intestine; effects of diet and nutritional status on resistance to diarrheal disease; virulence of Salmonella. *Mailing Add:* Dept of Nutrit Sci Univ of Calif Berkeley CA 94720

CHANG, H K, b Shenyang, China, July 9, 40; US citizen; c 2. BIOMEDICAL ENGINEERING. *Educ:* Nat Taiwan Univ, BS, 62; Stanford Univ, MS, 64; Northwestern Univ, PhD(biomed eng), 69. *Prof Exp:* From asst prof to assoc prof eng sci, State Univ NY Buffalo, 69-76; assoc prof, 76-80, prof physiol, biomed eng & med, McGill Univ, 80-84; prof biomed eng & physiol, Univ Southern Calif, 84-90, chmn, biomed eng, 85-90; DEAN, SCH ENG, HONG KONG UNIV SCI & TECH, 90- *Concurrent Pos:* Vis prof, Univ Paris, 81-82; bd dirs, Biomed Eng Soc, 85-88; consult, NIH & Med Res Coun Can; hon prof, Peking Union Med Col & Chinese Acad Med Sci. *Mem:* Am Physiol Soc; Am Thoracic Soc; Biomed Eng Soc (pres, 89-90); Am Inst Chem Engrs; Am Soc Civil Engrs. *Res:* Respiratory physiology with special emphasis on gas transport and blood flow in the lung; critical care of adults and neonates. *Mailing Add:* Sch Eng Hong Kong Univ Sci & Technol 5/F World Shopping Ctr Canton Rd Kowloon Hong Kong

CHANG, HAI-WON, b Seoul, Korea, Apr 27, 29; US citizen; m 56. CHEMISTRY, NEUROCHEMISTRY. *Educ:* Ewha Womans Univ, Korea, BA, 50; Wellesley Col, MA, 56;Columbia Univ, PhD(org chem), 61. *Prof Exp:* Res assoc, Dept Chem, Columbia Univ, 61-63, Brookhaven Nat Lab, 63-64 & Dept Chem, Columbia Univ, 64-69; res assoc, 69-74, asst prof, 75-80, ASSOC PROF NEUROCHEM, COL PHYSICIANS & SURGEONS, COLUMBIA UNIV, 81- *Mem:* Am Soc Biol Chemists; Am Chem Soc; NY Acad Sci; Sigma Xi. *Mailing Add:* Dept Neurol Col Physicians & Surgeons Columbia Univ 630 W 168 St New York NY 10032

CHANG, HAO-JAN, polymer chemistry, organic chemistry, for more information see previous edition

CHANG, HARRY LO, chemical engineering, for more information see previous edition

CHANG, HENRY, b Shanghai, China, Dec 2, 44; US citizen; m 76; c 1. BIOENGINEERING, SCIENCE EDUCATION. *Educ:* Yale Univ, BS, 65; Med Sch, Harvard Univ, MD, 69; diplomate, Am Bd Int Med, 75 & 78. *Prof Exp:* Fel hemat & oncol, Children's Hosp, Boston, 73-76; asst prof, Dept Pediat, New York Hosp, 76-79; asst prof hemat & oncol, Dept Med, Albert Einstein Col Med, 79-83; AT IMMUNOBIOL & TRANSPLANTATION DEPT, NAVAL MED RES INST. *Concurrent Pos:* Vis asst prof, Med Biochem, Rockefeller Univ, 76-78. *Mem:* Am Soc Hematol. *Res:* Blood disorders; red cells; hemoglobin biochemistry; molecular biology; cell growth and differentiation, especially of bone marrow; gene control; pharmacology and drug design. *Mailing Add:* 6506 Landon Lane Bethesda MD 20817

CHANG, HERBERT YU-PANG, b Shanghai, China, Nov 25, 37; m. COMPUTER SCIENCES GENERAL. *Educ:* Univ Ill, BS, 60, MS, 62, PhD(elec eng), 64. *Prof Exp:* MTS-DIR, BELL TEL LABS, 64- *Mem:* Fel Inst Elec & Electronics Engrs. *Res:* Basic and/or applied research in areas of telephone switching systems design, system maintenance techniques, computer software systems and computer aided design systems. *Mailing Add:* AT&T Bell Labs 1200 E Warrenville Rd Naperville IL 60566-7050

CHANG, HOU-MIN, b Chiayi, Taiwan, Aug 29, 38; m 66; c 2. WOOD CHEMISTRY. *Educ:* Nat Taiwan Univ, BS, 62; Univ Wash, MS, 66, PhD(wood chem), 68. *Prof Exp:* Fel, NC State Univ, 68-69, vis asst prof, 69, from asst prof to assoc prof, 70-77, prof wood & paper sci, 77-90, REUBEN B ROBERTSON PROF PULP & PAPER, NC STATE UNIV, 90- *Concurrent Pos:* Sci specialist, Weyerhaeuser Co, 77; vis prof, Univ Tokyo, 81. *Mem:* Am Chem Soc; Am Tech Asn Pulp & Paper Indust; fel Int Acad Wood Sci. *Res:* Species variation in wood lignins; isolation and characterization of cellulase lignin; characterization of residual lignin in kraft pulps of various yield; delignification by oxygen and alkali; lignin biodegradation; decolorization and dechlorination of pulp mill effluents; extended delignification; low chlorine and chlorine-free bleaching. *Mailing Add:* Dept Wood & Paper Sci NC State Univ Raleigh NC 27695-8005

CHANG, HOWARD, b Chiangsu, China, Nov 12, 39. FLUID DYNAMICS, HYDRAULICS. *Educ:* Cheng Kung Univ, Taiwan, BS, 62; Colo State Univ, MS, 65, PhD(fluid mech, hydraul), 67. *Prof Exp:* Asst prof civil eng, Colo State Univ, 67; from asst prof to assoc prof aerospace eng, 67-77, PROF CIVIL ENG, SAN DIEGO STATE UNIV, 77- *Concurrent Pos:* Consult, Rohr Corp, 68- *Honors & Awards:* Outstanding Contrib Aerospace Eng Award, Am Inst Aeronaut & Astronaut, 70. *Mem:* Am Inst Aeronaut & Astronaut; Am Soc Civil Engrs. *Res:* River mechanics; hydraulics; turbulence; hydraulic analogy; potential flows; compressible flows. *Mailing Add:* Dept Civil Eng San Diego State Univ San Diego CA 92182

CHANG, HOWARD HOW CHUNG, b Honolulu, Hawaii, Nov 16, 22; m 52; c 1. THEORETICAL PHYSICS. *Educ:* Calif Inst Technol, BS, 44; Univ Calif, MA, 49; Harvard Univ, PhD(physics), 55. *Prof Exp:* Instr physics, Clarkson Tech Inst, 49-50; instr math, Univ Hawaii, 50-52; mem tech staff, Microwave Lab, Gen Elec Co, 55-57; physicist, Rand Corp, Calif, 57-58 & Hughes Aircraft Co, 58-61; sr math physicist, Stanford Res Inst, 61-76; STAFF SCIENTIST, LOCKHEED MISSILES & SPACE CO, 76- *Concurrent Pos:* Liaison physicist, Off Naval Res, London, 69-70. *Mem:* Am Phys Soc; AAAS. *Res:* Electromagnetic theory; plasma physics; controlled fusion; solar energy; macroscopic applications of superconductivity; energy storage; energy economics. *Mailing Add:* 337 Los Altos Ave Los Altos CA 94022

CHANG, HSIEN-HSIN, b Yuan-lin, China, June 16, 42; m 69, 81; c 2. ORGANIC CHEMISTRY. *Educ:* Nat Taiwan Univ, BS, 65; Univ Miami, MS, 69; Univ Wash, PhD(org chem), 74. *Prof Exp:* Teaching assoc chem, Univ Wash, 73-74; res assoc biochem, molecular & cell biol, Cornell Univ, 74-76; group leader, Quaker Oats Co, 76-77; sr scientist, Pillsbury Co, 77-80; staff chemist, Quaker Oats Co, 80-85; GROUP LEADER, NUTRASWEET CO, 85- *Concurrent Pos:* Dir, Mid-Am Chinese Sci & Technol Asn, 88- *Mem:* Am Chem Soc; Am Oil Chemists Soc; Am Asn Cereal Chemists; Inst Food Technologists. *Res:* Metabolism of coenzymes; synthesis of biologically significant compounds; enzyme modification of food ingredients; microwave interaction with foods; food emulsions; technology for food product development. *Mailing Add:* NutraSweet Co 601 E Kensington Rd Mt Prospect IL 60056

CHANG, HSU, magnetism, electrical engineering; deceased, see previous edition for last biography

CHANG, I-DEE, b Anhwei, China, Mar 21, 22; m 70. AERONAUTICS, MATHEMATICS. *Educ:* Nat Cent Univ, China, BS, 44; Kans State Univ, MS, 55; Calif Inst Technol, PhD(aeronaut, math), 59. *Prof Exp:* Res fel aeronaut, Calif Inst Technol, 59-61; from asst prof to assoc prof, 62-69, PROF AERONAUT & ASTRONAUT, STANFORD UNIV, 70- *Mem:* Am Inst Aeronaut & Astronaut; Am Phys Soc; Biomed Eng Soc. *Res:* Viscous fluid theory; singular perturbation methods; bio-fluid mechanics. *Mailing Add:* 948 Wing Pl Stanford CA 94305

CHANG, IFAY F, b Chung King, China, Apr 4, 42; m 68; c 3. SOLID STATE PHYSICS, ELECTRICAL ENGINEERING. *Educ:* Cheng Kung Univ, Taiwan, BS, 63; Univ RI, MS, 66, PhD(elec eng), 68. *Prof Exp:* Dir res & mgr, Nat Univ Singapore, 85-88; Sr assoc engr, Components Div, IBM Corp, 68-69, staff engr, Vt, 69-70, mem res staff, NY, 70-85, SR MGR, IBM WATSON RES CTR, IBM CORP, 88- *Concurrent Pos:* Vis asst prof, Syracuse Univ, 68-69; vis prof, Univ Tex, Austin, 76-77; consult, IBM Off Prods Div, Austin, 76-77; vpres & bd mem, Chinese Am Acad Prof Asn. *Honors & Awards:* Beatrice Winner Award, Soc Info Display, 89. *Mem:* Am Phys Soc; Inst Elec & Electronics Engrs; fel Soc Info Display; Electrochem Soc; Soc Info Display (pres). *Res:* Lattice dynamics; optical and electrical properties of solids and solid films; impurities and imperfections in solids; mixed crystals; infrared and raman spectroscopy; solid state devices and microelectronics; display and input device and technology; electrochromism; electroluminescence; office systems, multimedia information systems and optical storage technology. *Mailing Add:* IBM Watson Res Ctr PO Box 704 Yorktown Heights NY 10598

CHANG, I-LOK, b Amoy, China, July 9, 43; US citizen; m 70; c 1. MATHEMATICAL ANALYSIS. *Educ:* Calif Inst Technol, BS, 65; Cornell Univ, PhD(math), 71. *Prof Exp:* From instr to asst prof, 70-76, ASSOC PROF MATH, AM UNIV, 76- *Mem:* Am Math Soc; Am Math Asn. *Res:* Functions of a complex variable; numerical analysis. *Mailing Add:* Dept Math Statist Am Univ Washington DC 20016

CHANG, IN-KOOK, b Choonchun, Korea, Aug 24, 43; m 71; c 3. PLANT PHYSIOLOGY. *Educ:* Seoul Nat Univ, BS, 66; Va Polytech Inst & State Univ, MS, 70; Univ Chicago, PhD(biol), 74. *Prof Exp:* Asst prof plant physiol, Va Polytech Inst & State Univ, 75-77; res biologist, Diamond Shamrock Corp, 77-78, sr res biologist, 78-82, res assoc, 82-83; res assoc, SDS Biotech Corp, 83-85; VPRES, INT AGRIPROD CORP, 86- *Mem:* AAAS; Am Soc Plant Physiologists; NY Acad Sci; Weed Sci Soc Am; Am Chem Soc. *Res:* Plant growth and development; effects of growth regulators on crop productivity; plant hormone metabolism; mode of action of herbicides, degradation of herbicides and growth regulators in plants and soils; pesticide and plant growth regulator formulation and evaluation; biorational synthesis of pesticides. *Mailing Add:* 9541 Remington Dr Mentor OH 44060

CHANG, JACK CHE-MAN, b Shanghai, China, Nov 19, 41; m 65; c 2. ANALYTICAL CHEMISTRY. *Educ:* Asbury Col, BA, 61; Univ Ill, Urbana, MS, 63, PhD(chem), 65. *Prof Exp:* Res assoc electrochem luminescence, Mass Inst Technol, 66-67; SR RES CHEMIST, EASTMAN KODAK CO, 67-, DIR, CORP RES LABS, 85- *Mem:* Am Chem Soc; Electrochem Soc; Chinese Am Chem Soc. *Res:* Electrochemistry of organic compounds. *Mailing Add:* 1198 Fox Hollow Webster NY 14580

CHANG, JAE CHAN, b Chong An, Korea, Aug 29, 41; m 65; c 3. HEMATOLOGY, ONCOLOGY. *Educ:* Seoul Nat Univ, MD, 65. *Prof Exp:* Intern med, Ellis Hosp, Schenectady, NY, 65-66; resident internal med, Harrisburg Hosp, Pa, 66-69, fel nuclear med, 69-70; fel hemat-oncol, Med Ctr, Univ Rochester, NY, 70-72; chief hemat-oncol, Vet Admin Hosp, Dayton, Ohio, 72-75; coodr med educ, 76-77, CHIEF HEMAT-ONCOL, GOOD SAMARITAN HOSP, DAYTON, OHIO, 75-, DIR ONCOL UNIT, 77-; CLIN PROF MED, SCH MED, WRIGHT STATE UNIV, DAYTON, OHIO, 80- *Concurrent Pos:* Instr med, Sch Med, Univ Rochester, 70-72; asst clin prof, Col Med, Ohio State Univ, 72-75; assoc clin prof, Sch Med, Wright State Univ, 75-80; co-dir Hematol Lab, Good Samaritan Hosp, Dayton. *Mem:* Fel Am Col Physicians; Am Fedn Clin Res; A Soc Hemat; Am Soc Clin Oncologists; Am Asn Cancer Res; AAAS. *Res:* Clinical research on neoplastic fever using antipyretic agents; psychologic aspects of cancer; biochemical behaviors of neoplasms and chemotherapy of neoplastic diseases. *Mailing Add:* Good Samaritan Hosp 2222 Philadelphia Dr Dayton OH 45406

CHANG, JAMES C, b Shanghai, China, Aug 8, 30; m 70; c 1. PHYSICAL CHEMISTRY, INORGANIC CHEMISTRY. *Educ:* Mt Union Col, BS, 57; Univ Calif, Los Angeles, PhD(chem), 64. *Prof Exp:* Res chemist, E R Squibb & Sons Div, Olin Mathieson Chem Co, 59-62; asst prof chem, State Col Iowa, 64-67; from asst prof to assoc prof, 67-74, actg head dept, 75-77, PROF CHEM, UNIV NORTHERN IOWA, 74- *Concurrent Pos:* Vis scientist, Univ Copenhagen, 69-70, Univ Iowa, 81. *Mem:* Am Chem Soc. *Res:* Inorganic synthesis and substitution reactions of inorganic complex compounds; antimicrobial inorganic complexes. *Mailing Add:* Dept of Chem Univ of Northern Iowa Cedar Falls IA 50613

CHANG, JAW-KANG, b Cholon, South Vietnam, Aug 11, 42; US citizen; m 62. BIOCHEMISTRY. *Educ:* Nat Taiwan Univ, BS, 65; Univ NB, Fredericton, PhD(org chem), 69. *Prof Exp:* Res fel peptide chem, Inst Biomed Res, Univ Tex, Austin, 69-73; sr develop chemist, Dept Bio-Prod, Beckman Instruments Inc, 73-77; res dir, 77-80, VPRES, PENINSULA LABS, INC, 80- *Res:* Synthetic chemistry of natural products; alkaloids, peptides and proteins; design and synthesis of analogs of biological active peptides. *Mailing Add:* Peninsula Labs Inc 611 Taylor Way Belmont CA 94002

CHANG, JEFFREY C F, b Canton, China, Jan 10, 28; US citizen; m 46; c 3. INTELLIGENT SYSTEMS. *Educ:* Western Ill Univ, BS, 65, MS, 66; Univ Ga, PhD(statist & computer sci), 74. *Prof Exp:* From asst prof to assoc prof, 66-86, PROF MATH & COMPUTER SCI, GARDNER-WEBB COL, 86- *Mem:* Asn Comput Mach. *Res:* Pattern recognition for categorical data and artificial intelligence. *Mailing Add:* 143 Woodhill Dr Box 896 Boiling Springs NC 28017

CHANG, JEFFREY PEH-I, b Changteh, China, Oct 10, 17; US citizen; c 2. CELL BIOLOGY. *Educ:* Nat Cent Univ, Chungking, BS, 41; Univ Ill, MS, 46, PhD(zool), 49. *Prof Exp:* From asst prof to assoc prof, Univ Tex Postgrad Sch Biomed Sci, 55-62; from asst biologist to assoc biologist, Univ Tex M D Anderson Hosp & Tumor Inst Houston, 55-64, actg chief sect exp path, 59-64, biologist & prof biol, 64-72; PROF CELL BIOL, UNIV TEX MED BR GALVESTON, 72- *Concurrent Pos:* Spec consult, Nat Cancer Inst, 58-61 & Sch Aerospace Med, Brooks AFB, 62-64; consult, Univ Tex M D Anderson Hosp & Tumor Inst Houston, 72-76; res consult, Nat Sci Coun, Repub China, 74-; academician, Academia Sinica, Repub China. *Mem:* AAAS; Am Soc Cell Biol; Am Soc Exp Path; Am Asn Cancer Res; Histochem Soc. *Res:* Ultrastructural and histochemical studies of cells and tissues; initiation and mechanism of carcinogenesis; tumor production and biology. *Mailing Add:* Div Cell Biol Med Br Univ Tex 3946 Laura Leigh Lane Galveston TX 77546

CHANG, JENNIE C C, b Taiwan, Repub China, Feb 27, 51; US citizen; m; c 1. T CELL IMMUNOLOGY, AUTOIMMUNITY. *Educ:* Nat Taiwan Univ, BS, 72; Univ NTex, MS, 76; Cornell Univ, PhD(immunol), 80. *Prof Exp:* Res assoc immunotoxicol, Cornell Univ, 80-82; postdoctoral fel cellular immunol, Med Ctr, Univ Colo, 82-84, instr, 84-86; asst prof cellular & respiratory immunol, Med Ctr, Univ Ky, 86-90; STAFF SCIENTIST AUTOIMMUNE-DIABETES, IMMUNE RESPONSE CORP, 90- *Mem:* Am Asn Immunologists. *Res:* Past studies include using T cell clone technology, subtypes of marine T cells with regards to lymphokine production, effector function and activation requirement; current studies involve human insulin-dependent diabetes, through the investigation of T cells in pre-diabetic and newly onset diabetic patients. *Mailing Add:* Immune Response Corp 5935 Darwin Ct Carlsbad CA 92008

CHANG, JEN-SHIH, b Tokyo, Japan, Sept 6, 47; Can citizen; m 74; c 2. PLASMA PHYSICS. *Educ:* Musashi Inst Technol, BEng, 69, MEng, 71; York Univ, PhD(space sci), 75. *Prof Exp:* Lectr elec eng, Yomiuri Inst Physics & Technol, 71-72; researcher environ sci, Nat Ctr Sci Res, France, 73-74; proj scientist, York Univ, 75-78, asst prof plasma physics, 78-79; assoc prof, 80-86, PROF ENG PHYS, MCMASTER UNIV, 87- *Concurrent Pos:* Vis prof elec eng, Musashi Inst Tech, 85 & Tokyo Univ, 86, Tokyo Denki Univ, 89. *Mem:* Physics Soc Japan; Inst Elec Engrs Japan; Am Geophys Union; Can Soc Chem Eng; Chem Inst Can. *Res:* Heat, mass, aerosol and charge transport problem in a variable property fluid; electrostatic charging of aerosol particle; plasma diagnostic techniques; glow discharge positive column; lighting to object; two phase flow; corona discharge process. *Mailing Add:* Dept Eng Phys McMaster Univ Hamilton ON L8S 4K1 Can

CHANG, JHY-JIUN, b China, May 29, 44; US citizen; m 68; c 2. SUPERCONDUCTIVITY, SOLID STATE MICROELECTONICS. *Educ:* Nat Taiwan Univ, BS, 66; Case Western Reserve Univ, MS, 69; Rutgers Univ, PhD(physics), 73. *Prof Exp:* Res asst prof physics, Univ Calif, Santa Barbara, 73-76; from asst prof to assoc prof, 77-87, PROF PHYSICS, WAYNE STATE UNIV, 87- *Concurrent Pos:* Prin investr, NSF, 78-86; vis fac, IBM Yorktown, 83, vis assoc prof, Inst Theoret Physics, Santa Barbara, 84 & 90; sr res fel, Calif Inst Technol, 77; consult, STI, Santa Barbara, 88- *Res:* X-ray and electron spectroscopy; phase transition and thermal and dynamic properties of organic conductors; nonequilibrium superconductivity; Josephson effects in superconducting tunnel junctions; grannular superconductors; superconducting devices. *Mailing Add:* Dept Physics Wayne State Univ Detroit MI 48202

CHANG, JOHN H(SI-TEH), b Hopei, China, Mar 9, 37; m 63; c 2. COMPUTER SCIENCE, SYSTEMS PLANNING AND ANALYSIS. *Educ:* Nat Taiwan Univ, BS, 60; Yale Univ, MS, 66, PhD(elec eng), 68. *Prof Exp:* Engr, Int Bus Mach Corp, 63-68, res staff mem, IBM Thomas J Watson Res Ctr, 68-73, sr engr & proj supvr, data processing econ, 73-78, mgr systs eval & data processing econ, 78-84,MGR DATA PROCESSING TECHNOL & ECON & DIR EVAL & PERFORMANCE, IBM CORP, ARMONK, 84- *Concurrent Pos:* Res asst, Yale Univ, 65-68; asst prof, NY Univ, 69-70; lectr, Univ Conn, Stamford, 70-72; adj assoc prof math, Pace Univ, 77-79; adj prof computer sci, Polytech Inst NY, 79-86; adj prof, Info & Commun Syst, GBA, Fordham Univ, 89- *Mem:* Sr mem Inst Elec & Electronic Engrs. *Res:* Investigate problems of data processing technologies and economics for systems evaluation, planning, and demand prediction. *Mailing Add:* IBM Corp Old Orchard Rd Armonk NY 10504

CHANG, JOSEPH YOON, b Malaysia, Oct 22, 52; m 75; c 1. DRUG DISCOVERY, RESEARCH MANAGEMENT. *Educ:* Portsmouth Polytechnic, BS, 74; Univ London, PhD(pharm), 77. *Prof Exp:* Postdoctoral fel, 78-80, res assoc, environ health, Johns Hopkins Univ, 80-81; sr scientist, pharmacol, Wyeth Labs, 81-85, assoc dir, 85-87; DIR, PHARMACOL, WYETH-AYERST RES, 87- *Concurrent Pos:* Dir, Arthritis Found, Md chapter, 80-81. *Mem:* Am Rheumatism Asn; Am Thoracic Soc; NY Acad Sci; Am Soc Pharmacol & Exp Therapeut; Reticuloendothelial Soc. *Res:* Pharmacology and biochemistry of inflammation and allergy; regulation of mediator synthesis and release; antiinflammatory drugs and their effects on the arthritic process; role of macrophages and immune cells on connective tissue biology. *Mailing Add:* Wyeth-Ayerst Res CN 8000 Princeton NJ 08543-8000

CHANG, JOSEPH YUNG, b Nanking, China, Jan 30, 32; nat US; m 63. PHYSICAL CHEMISTRY. *Educ:* Taiwan Col Eng, BS, 53; Univ Notre Dame, MS, 57, PhD(phys chem), 58. *Prof Exp:* Res assoc & fel, Univ Notre Dame, 58-61; proj scientist, Res Div, Philco Corp, 61; assoc res scientist, NY Univ, 61-63; res scientist, 63-76, SR RES SCIENTIST, GRUMMAN AEROSPACE CORP, 76- *Mem:* AAAS; Am Phys Soc; Am Chem Soc; Am Nuclear Soc; Sigma Xi. *Res:* Radiation chemistry and effects on solid state materials; radiation conversion of wastes; solar energy conversion; fusion radiation effects; transient and delayed photolytic and electronic effects of various radiation on optical (including infrared), semiconducting and insulatory materials; radiation effects on GaAs semiconductor devices. *Mailing Add:* 16-D Cold Spring Hills Rd Huntington NY 11743

CHANG, JUANG-CHI (JOSEPH), b Nanking, China, Apr 24, 36. COMMUNICATION SYSTEMS, ELECTRONICS ENGINEERING. *Educ:* Nat Univ Taiwan, BS, 59; Univ BC, MASc, 61; Iowa State Univ, PhD(elec eng), 65. *Prof Exp:* Res assoc elec eng, Res Inst, Univ Ala, Huntsville, 64-65, asst prof, 65-66; specialist, Lockheed-Ga Co, 66-67, staff engr, Lockheed Electronics Co, 67-74; MEM TECH STAFF, THE AEROSPACE CORP, 74- *Mem:* Inst Elec & Electronics Engrs. *Res:* Communication theory; telecommunication systems; satellite communications; microwave devices; multiple beam antennas. *Mailing Add:* The Aerospace Corp 2350 E El Segundo Blvd El Segundo CA 90245

CHANG, JUN HSIN, b Yangmei, Taiwan, Rep China; US citizen; m 70; c 3. CHEMISTRY. *Educ:* Nat Taiwan Norm Univ, BS, 67; Univ Detroit, MS, 71; State Univ NY Buffalo, PhD(org chem), 76. *Prof Exp:* Res assoc chem, Rice Univ, 75-76; RES ASSOC ORG SYNTHESIS, AGR CHEM GROUP, FMC CORP, 76- *Mem:* Am Chem Soc. *Res:* Organic synthesis in natural products, pesticides and pharmaceutical drugs. *Mailing Add:* 39 Cartwright Dr E Princeton Junction NJ 08850-1196

CHANG, JUNG-CHING, b Taipei, Taiwan, Jan 13, 39; m 65; c 1. ANALYTICAL CHEMISTRY, PHYSICAL CHEMISTRY. *Educ:* Tamkang Col, BS, 63; Univ PR, MS, 69; Univ Mo-Kansas City, PhD(phys chem), 75. *Prof Exp:* Chem engr adhesives, Taiwan Sugar Corp, 64-67; res assoc org chem, Univ Ore, 76-77; fel polymer sci, Univ Cincinnati, 77-79; anal chemist, ICN Pharmaceut, Inc, 79-81; RES CHEMIST, ASHLAND CHEM CO, 81- *Mem:* Sigma Xi; NY Acad Sci; Am Chem Soc. *Res:* Thermal analysis; microwave and infrared spectroscopy; chemical kinetics; photochemistry; mass spectrometry; electron spin resonance; photoelectron spectroscopy; x-ray diffraction methods in polymer science; combined gas chromatography and mass spectrometry; liquid chromatography; atomic absorption spectroscopy; mini computers and computer programming; ion chromatography. *Mailing Add:* PO Box 102 Dublin OH 43017-0102

CHANG, KAI, microwave engineering, solid state electronics, for more information see previous edition

CHANG, KAUNG-JAIN, b Fu-Jain Province, China, May 14, 45; m 74; c 2. ROCK MECHANICS, MATERIALS SCIENCE. *Educ:* Cheng KKung Univ, Taiwan, BE, 67; Univ Iowa, ME, 72, DPhil(solid mech), 74. *Prof Exp:* Fel viscoplasticity, Div Mat Eng, Univ Iowa, 75; res assoc creep metals, Dept Theoret & Appl Mech, Cornell Univ, 76-77; SR INVESTR & ASST PROF ROCK MECH, ROCK MECH & EXPLOSIVES RES CTR, UNIV MO-ROLLA, 77- *Mem:* Am Soc Mech Engrs; AAAS. *Res:* Investigation of the fracture phenomena and the constitutive properties of rocks and other materials. *Mailing Add:* 5859 Larboard Lane Agoura Hills CA 91301

CHANG, KENNETH SHUEH-SHEN, b Taipei, Taiwan, Jan 3, 29; m 52; c 5. MICROBIOLOGY, ONCOLOGY. *Educ:* Nat Taiwan Univ, MD, 51; Univ Tokyo, PhD, 60. *Prof Exp:* Asst microbiol, Col Med, Nat Taiwan Univ, 51-55, lectr, 55-59, from assoc prof to prof, 59-67; sr virologist, Flow Labs, 67-69, med officer, Lab Biol, 69-70, HEAD, SECT VIRAL ONCOGENESIS, LAB CELL BIOL, NAT CANCER INST, 70- *Concurrent Pos:* WHO fel, Commonwealth Serum Labs & Dept Microbiol, Univ Melbourne, 56; Nat Acad Sci fel, Dept Med Microbiol & Immunol, Univ Calif, Los Angeles, 62-64; head Bacillus Calmette-Guerin Vaccine Lab, Taiwan Serum & Vaccine Inst, 57-58; head microbiol & serol sect, Dept Clin Path, Nat Taiwan Univ Hosp, 58-67; fel, WHO, 56; postdoctoral fel, Nat Acad Sci, Nat Res Coun, 62-64. *Mem:* Am Asn Cancer Res; Am Soc Microbiol; Tissue Cult Asn; Am Asn Immunologists; Int Asn Comp Res Leukemia & Related Dis. *Res:* Cancer virology, molecular biology and immunology; clinical microbiology. *Mailing Add:* Dir Chang Gung Med Res Ctr Nat Cancer Inst NIH Bldg 37 Rm 1B28 Bethesda MD 20892

CHANG, KUANG-CHOU, b Taipei, Taiwan, Jan 2, 49; m. PHYSICAL ORGANIC CHEMISTRY. *Educ:* Tunghai Univ, Taiwan, BS, 70; Univ Minn, PhD(org chem), 75. *Prof Exp:* Res asst hydrogen bonding, Univ Minn, 71-74; res fel solution kinetics, Brandeis Univ, 74-76; mem staff, Bell Labs, 76-77; scientist, Polaroid Corp, 77-82, SR SCIENTIST, 82- *Res:* Spectroscopic studies of reaction kinetics and structure determination; physics and chemistry of semiconductor photographic science and engineering. *Mailing Add:* Polaroid Corp 750 Main St 4B Cambridge MA 02139

CHANG, KUO WEI, b Shanghai, China, Nov 21, 38; m 63; c 1. BIOMEDICAL ENGINEERING. *Educ:* Nat Taiwan Univ, BS, 60; Univ Cincinnati, MS, 63; Princeton Univ, MA, 67, PhD(aerospace sci), 69. *Prof Exp:* Engr, First Naval Shipyard, Chinese Navy, 60-62; res asst aerospace eng, Univ Cincinnati, 62-64; asst res aerospace sci, Princeton Univ, 64-68; sr scientist physics, 68-71, mgr, Biosensors Dept, 71-75, TECH APPL SCI LABS, GULF & WESTERN CO, 75-; PRES, INDUST & BIOMED SENSORS CORP, 75- *Mem:* Am Phys Soc; Asn Advan Med Instrumentation; Inst Elec & Electronic Engrs. *Res:* Aerospace sciences; plasma physics; magnetohydrodynamics; biophysics; fluid mechanics; kinetic theory of gases; electronics. *Mailing Add:* 32 Buckman Dr Lexington MA 02173

CHANG, KWANG-POO, b Taipei, Taiwan, Nov 12, 42; m 72; c 1. PARASITOLOGY, CELL BIOLOGY. *Educ:* Nat Taiwan Univ, BSc, 65; Univ Guelph, MS, 68, PhD(biol), 72. *Prof Exp:* Fel parasitol, 72-74, from res assoc to asst prof, 74-79, ASSOC PROF, ROCKEFELLER UNIV, 79-; PROF MICROBIOL & IMMUNOL, UNIV HEALTH SCI, CHICAGO MED SCH, 83- *Concurrent Pos:* Consult, WHO, 79, 82 & 88; NIH TMP Study Sect, 87-91. *Honors & Awards:* Seymour Hutner Award, Soc Protozool, 87. *Mem:* Am Soc Microbiol; Am Soc Parasitologists; Am Soc Trop Med Hyg; Am Soc Cell Biol; Soc Protozool. *Res:* Cell and molecular biology of parasitism and symbiosis; applications of concepts and techniques in cell biology, biochemistry, molecular biology and immunology to study protozoan parasites, especially Leishmaniasis; intracellular parasitism in mammalian Leishmaniasis. *Mailing Add:* Univ Health Sci Chicago Med Sch 3333 Green Bay Rd N Chicago IL 60064

CHANG, KWEN-JEN, RECEPTOR MECHANISM, PHARMACOLOGY. *Educ:* State Univ NY, Buffalo, PhD(biochem pharmacol), 72. *Prof Exp:* GROUP LEADER, MOLECULAR BIOL, BURROUGHS WELLCOME CO, 75- *Mailing Add:* Burroughs Wellcome Co 3030 Cornwallis Rd Res Triangle Park NC 27709

CHANG, L(EROY) L(I-GONG), b Honan, China, Jan 20, 36; m 62; c 2. SEMICONDUCTORS, HETEROSTRUCTURES. *Educ:* Nat Taiwan Univ, BS, 57; Univ SC, MS, 61; Stanford Univ, PhD, 63. *Prof Exp:* Mem res staff, Thomas J Watson Res Ctr, Int Bus Mach Corp, 63-68; assoc prof elec eng, Mass Inst Technol, 68-69; mem res staff, 69-75, MER QUANTUM STRUCT, THOMAS J WATSON RES CTR, IBM CORP, 75- *Concurrent Pos:* Adj prof, Brown Univ, 89- *Honors & Awards:* Int Prize for New Mat, Am Phys Soc, 85; David Sarnoff Award, Inst Elec & Electronic Engrs, 90. *Mem:* Fel Am Phys Soc; Am Vacuum Soc; fel Inst Elec & Electronic Engrs; Mat Res Soc; Nat Acad Eng. *Res:* Semiconductor physics, materials and devices. *Mailing Add:* Thomas J Watson Res Ctr IBM Corp PO Box 218 Yorktown Heights NY 10598

CHANG, LAY NAM, b Singapore, June 1, 43; m 67; c 1. QUANTUM FIELD THEORY. *Educ:* Columbia Univ, AB, 64; Univ Calif, Berkeley, PhD(physics), 67. *Prof Exp:* Res assoc, Mass Inst Technol, 67-69 & The Enrico Fermi Inst, Univ Chicago, 69-71; asst prof physics, Univ Pa, 71-78; assoc prof, 78-83, PROF PHYSICS, VA POLYTECH INST & STATE UNIV, 83- *Concurrent Pos:* Vis scientist, Niels Bohr Inst, 74 & Brookhaven Nat Lab, 76, 78, 80, Los Alamos Nat Lab, 81, State Univ NY, Stony Brook, 82, 84, ITP, Santa Barbara, 88. *Res:* Theoretical physics; field theory; particle physics. *Mailing Add:* Physics Dept Va Polytech Inst & State Univ Blacksburg VA 24061

CHANG, LENA, b Yunan, Mainland China, Dec 23, 38; US citizen; m 61; c 2. ACTUARIAL SCIENCE, STATISTICAL MATHEMATICS. *Educ:* Univ Ill, BS, 58, MS, 60, PhD(math), 64. *Prof Exp:* Lectr math, Mich State Univ, 61-62; instr, Univ Ill, Champaign-Urbana, 64-65; asst prof, Univ Ill, Chicago Circle, 65-74; vis assoc prof actuarial sci, Dept Ins & Risk, Temple Univ, 74-76; actuary-statistician, State Rating Bur, Div Ins, Mass, 76-79; INDEPENDENT MATH, ACTUARIAL STATIST, INS CONSULT, CHANG & CUMMINGS, 79- *Concurrent Pos:* Res sci mathematician, Sci Lab, Ford Res Ctr, Ford Motor Co, Mich, 60-61; assoc investr group representation theory, Off Naval Res, 62-64 & NSF res grant, 65-68; expert & consult, US Gen Acct Off, Washington, DC, 74-75; consult actuary, Gordon Assocs, 75-76 & Off State Auditor Gen, Commonwealth of Pa, 76. *Mem:* Am Math Soc; Am Risk & Ins Asn; Am Statist Asn; Sigma Xi. *Res:* Actuarial and mathematical studies of insurance legislation costs; insurance rate filings including auto, worker's compensation products liability and health; loss reserve and investment income; actuarial analysis; operations research; computer science. *Mailing Add:* 4100 Steamboat Bend No 301 Ft Myers FL 33919

CHANG, LEROY L, QUANTUM STRUCTURES. *Educ:* Nat Taiwan Univ, BS, 57; Univ SC, MS, 61; Stanford Univ, PhD(elec eng), 63. *Prof Exp:* Res staff mem, T J Watson Res Ctr, IBM, 63-68; assoc prof elec eng, Mass Inst Technol, 68-69; res staff mem, 69-75, staff to res dir tech planning & control, 84-85, MGR QUANTUM STRUCT, T J WATSON RES CTR, IBM, 75-84 & 85- *Concurrent Pos:* Counr, Mat Res Soc, 82-85; res incentive award, Naval Res Lab, 84. *Honors & Awards:* Int Prize for New Mat, Am Phys Soc, 85. *Mem:* Nat Acad Eng; fel Am Phys Soc; Am Vacuum Soc; fel Inst Elec & Electronics Engrs. *Mailing Add:* T J Watson Res Ctr IBM PO Box 218 Yorktown Heights NY 10598-0218

CHANG, LOUIS WAI-WAH, b Hong Kong, July 1, 44; US citizen; m 68; c 2. EXPERIMENTAL PATHOLOGY. *Educ:* Univ Mass, Amherst, BA, 66; Tufts Univ, MS, 69; Univ Wis-Madison, PhD(path), 72. *Prof Exp:* Instr path, 72-73, dir path lab, Univ Wis-Madison, 72-76, asst prof path, 73-76; assoc prof, 77-80, PROF PATH, UNIV ARK MED SCI, 80-, DIR PATH GRAD PROG, 77-, DIR, EXP PATH, 80- *Concurrent Pos:* NIH & NSF res grants, 73-80; consult, Nat Inst Health, Environ Protection Agency & I J Life Systs, Inc, 79-; Panelist FDA Study, 85; pres, Am Chinese Toxicologist Soc, 87-88; Neuroscience Adv Bd, Off Technol Assessment, US Cong, 88. *Mem:* Am Asn Pathologists; Am Asn Neuropath; Soc Neurosci; Soc Toxicol. *Res:* Environmental toxicology; heavy metal toxicology; experimental pathology on brain, liver, and kidney; histochemistry; electron microscopy; developmental biology and teratology; author of two hundred publications. *Mailing Add:* Dept Path Slot 517 Univ Ark Med Sci 4301 W Markham St Little Rock AR 72205

CHANG, LUCY MING-SHIH, b China, Aug 20, 42; US citizen. BIOCHEMISTRY. *Educ:* Western Reserve Univ, AB, 64, Ind Univ, PhD(biochem), 68. *Prof Exp:* Res assoc biochem, Univ Ky, 68-70, asst prof, 70-72; from asst prof to assoc prof, Univ Conn, 72-77; PROF BIOCHEM, UNIFORMED SERV UNIV HEALTH SCI, 77-, CHMN, 87- *Mem:* AAAS; Am Soc Biol Chem. *Res:* Enzymatic synthesis of DNA in eukaryotic cells. *Mailing Add:* Dept Biochem Uniformed Serv Univ Health Sci 4301 Jones Bridge Rd Bethesda MD 20814

CHANG, LUKE LI-YU, b Honan, China, Sept 18, 35; US citizen; m 59; c 2. MINERALOGY. *Educ:* Nat Taiwan Univ, BS, 57; Univ Chicago, PhD(geophys sci), 63. *Prof Exp:* Sr scientist mineral & ceramic sci, Tem-Pres Res, Inc, 63-67; asst prof geol, Cornell Univ, 67-70; assoc prof geol, Miami Univ, 70-75, prof, 75-81; PROF GEOL & CHMN DEPT, UNIV MD, 81- *Concurrent Pos:* Contrib ed, Phase Diagrams Ceramist, Am Ceramic Soc & Nat Bur Standards, 74- *Mem:* Fel Mineral Soc Am; Am Ceramic Soc; Geochem Soc; Mineral Soc Gt Brit; Can Mineral Soc. *Res:* Mineral synthesis and equilibrium relations in the systems of carbonates, sulfides, and oxides; crystal chemistry of tungstates. *Mailing Add:* Dept of Geol Univ Md College Park MD 20742

CHANG, MEI LING (WU), b Kiangsi, China, Mar 1, 15; US citizen; m 53. NUTRITIONAL BIOCHEMISTRY. *Educ:* Ginling Col, China, BS, 38; Ore State Univ, MS, 49, PhD(nutrit), 51. *Prof Exp:* Res asst histochem of brain, Sch Med, Wash Univ, 51-53; res assoc nutrit, Ore State Univ, 53-54; res assoc nutrit biochem, Univ Ill, 54-62; RES CHEMIST, HUMAN NUTRIT DIV, USDA, 62- *Mem:* Am Inst Nutrit. *Res:* Vitamin metabolism; histochemistry of brain; dietary effect on enzyme system; carbohydrate metabolism. *Mailing Add:* 11329 Frances Dr Beltsville MD 20705

CHANG, MIN CHUEH, physiology; deceased, see previous edition for last biography

CHANG, MING-HOUNG (ALBERT), b Fu Chin, Fu Chen, China, Feb 19, 47; US citizen; m 75; c 2. HARZARDOUS & TOXIC CHEMISTRY, ENVIRONMENTAL BIOCHEMISTRY. *Educ:* Nat Taiwan Univ, BS, 70; Univ Ga, MS, 74, PhD(entom), 80. *Prof Exp:* Res asst entom, Univ Ga, 72-75; marine scientist, Col William & Mary, 75-81; sr syst engr, Comput Sci Corp, 81-85; RES SCIENTIST HAZARDOUS CHEM, BATTELLE COLUMBUS LABS, 85- *Mem:* Entom Soc Am; Marine Technol Soc. *Res:* Protection and decontamination systems for hazardous chemicals; toxic organic residues and petroleum hydrocarbons in marine environments; effect of insecticides on nervous systems in animals. *Mailing Add:* Computer Sci Corps Air Force Flight Test Ctr PO Box 446 Edwards CA 93523-0446

CHANG, MINGTEH, b Fukien, China, Jan 18, 39; m 70; c 3. FOREST HYDROLOGY, WATERSHED MANAGEMENT. *Educ:* Nat Chung-Hsing Univ, Taiwan, BS, 60; Pa State Univ, MS, 68; WVa Univ, PhD(forest hydrol), 73. *Prof Exp:* Watershed technologist, Mountainous Agr Resources Develop Bur, Govt of Taiwan, 61-64; teaching asst surv, Chung-Hsing Univ, 64-67; res assoc hydrol, Water Res Inst, WVa Univ, 73-75; asst prof forest hydrol, 75-80, assoc prof, 80-88, PROF FOREST HYDROL, SCH FORESTRY, STEPHEN F AUSTIN STATE UNIV, 88-, GROUP LEADER FOREST RESOURCES RES, 77- *Mem:* Am Geophys Union; Am Inst Hydrol; Soil & Water Conserv Soc Am; Am Water Resources Asn. *Res:* Quantitative analysis and interpretation of hydrologic and climatologic data; physical and physiologic processes of the soil-plant-atmosphere system in forest environment. *Mailing Add:* Sch Forestry Box 6109 Stephen F Austin State Univ Nacogdoches TX 75961

CHANG, MORRIS, b Chekiang, China, July 10, 31; US citizen; m 53; c 1. ELECTRONICS. *Educ:* Mass Inst Technol, BS, 52, MS, 53; Stanford Univ, PhD(elec eng), 64. *Prof Exp:* Sr engr, Sylvania Elec Prods, Inc, 55-58; mgr germanium develop, 58-61, mgr, Germanium Small Signal Dept, 64-67, vpres, Semiconductor Circuits Div, 67-72, group vpres, Semiconductor Group, 72-75, group vpres, Worldwide Semiconductors Opers, 75-78, group vpres, World Consumer Opers, 78-80, sr vpres corp quality & reliability, Tex Instruments, Inc, 80-; PRES GEN INSTRUMENT CORP, NY. *Mem:* Inst Elec & Electronics Engrs. *Res:* Semiconductor electronics. *Mailing Add:* Ind Tech Res Inst 7th Floor No 106 Ho-Ping E Rd Sec 2 Taipei 10636 Taiwan

CHANG, MOU-HSIUNG, b Taiwan, China, Nov 6, 44; m; c 3. STATISTICAL PHYSICS. *Educ:* Univ Rhode Island, MA, 71, PhD(math), 74. *Prof Exp:* PROF, MATH, UNIV ALA, HUNTSVILLE, 74- *Concurrent Pos:* Vis scientist, Mass Inst Technol, 84-85. *Mem:* Am Math Soc; Soc Indust & Appl Math. *Res:* Stochastic differential equations; interacting diffusions; fluid turbulence. *Mailing Add:* Dept Math Sci Univ Ala Huntsville AL 35899

CHANG, NGEE PONG, b Singapore, Dec 24, 40; m 65; c 2. THEORETICAL HIGH ENERGY PHYSICS. *Educ:* Ohio Wesleyan Univ, BA, 59; Columbia Univ, PhD(physics), 63. *Prof Exp:* Res assoc physics, Columbia Univ, 62-63; res fel, Inst Advan Study, 63-64; res assoc, Rockefeller Univ, 64-65; vis prof, 65-66, PROF PHYSICS, CITY COL NEW YORK, 66- *Concurrent Pos:* Vis prof, Max Planck Inst Physics & Astrophysics, Munich, WGer, 73, Univ Tokyo, Res Inst Fundamental Physics & Kyoto Univ, 74, Ecole Normale Superieure, Nat Ctr Sci Res, Paris, 82 & Nat Lab High Energy, Tsukuba, Japan, 83. *Mem:* Am Phys Soc; fel Am Phys Soc. *Res:* Field theory of weak interactions; symmetries; kinematics at infinite momentum; impact parameter representation; infinite energy scattering; non-abelian gauge theories; quark dynamics; grand unified field theories; asymptotic freedom; renormalization group analysis; dynamical symmetry breaking and its relation with bifurcation; no-scale supergravity; orbifolds; zero cosmological constant; chiral symmetry breaking. *Mailing Add:* Dept Physics City Col New York New York NY 10031

CHANG, PAUL PENG-CHENG, structural engineering, civil engineering, for more information see previous edition

CHANG, PAULINE (WUAI) KIMM, b Shanghai, China, Jan 19, 26; nat US; m 52; c 2. ORGANIC CHEMISTRY. *Educ:* Wellesley Col, BA, 49; Univ Mich, MS, 50, PhD(chem), 55. *Prof Exp:* From res asst to res assoc, Yale Univ, 55-66, sr res assoc pharmacol, 66-83, res scientist, Sch Med, 83-88; RETIRED. *Mem:* Sigma Xi. *Res:* Synthetic organic chemistry; medicinal chemistry; synthesis of labeled compounds. *Mailing Add:* 50 Allendale Dr North Haven CT 06473

CHANG, PEI KUNG (PHILIP), b Shantung Province, China, Jan 9, 36; US citizen; m 66; c 1. FOOD SCIENCE, BIOCHEMISTRY. *Educ:* Nat Taiwan Univ, BS, 60; Colo State Univ, MS, 64; Univ Wis, PhD(food sci), 69; New York Inst Technol, MBA, 82. *Prof Exp:* Res asst nutrient anal green veg, Colo State Univ, 62-64; res asst egg white & yolk protein, Univ Wis, 64-68; res assoc food ingredients, Eastern Res Ctr, Stauffer Chem Co, 68-73; independent food technologist, 68-73; sr assoc, Food Sci Assocs, 84-85; tech consult, food indust, 85-87; tech supvr, Gen Foods, 85-87; MEM TECH STAFF, PEPSICO, 88- *Mem:* Inst Food Technologists; Am Asn Cereal Chemists. *Res:* Egg and milk proteins; developed process cheese emulsifier, whey protein recovery process and egg albumen and whole egg replacers; egg yolk extender for mayonnaise; process for lowering thermogelation temperature of whey protein; process for improving gelation of egg albumen. *Mailing Add:* 232 Coachlight Sq Montrose NY 10548

CHANG, PEI WEN, b China, Apr 26, 23; nat US; m 51; c 3. VETERINARY PATHOLOGY. *Educ:* Mich State Univ, DVM, 51; Univ RI, MS, 60; Yale Univ, PhD, 65. *Prof Exp:* Gen practitioner, Ind, 51-53; area veterinarian, Ind Livestock Sanit Bd, 53-55; asst prof & asst res prof animal path, 55-60, assoc prof, 60-66, PROF ANIMAL PATH, UNIV RI, 66- *Concurrent Pos:* Danforth fel, 61-63. *Mem:* Am Vet Med Asn; Sigma Xi. *Res:* Animal virology; characterization of animal viruses. *Mailing Add:* One Woodward Hall Univ RI Kingston RI 02881

CHANG, PETER HON, bioengineering, for more information see previous edition

CHANG, PING, b Wuhan, China, Oct 22, 60; m 87. OCEANIC WAVES, AIR-SEA INTERACTION. *Educ:* E China Eng Inst, BS, 82; City Col New York, ME, 84; Princeton Univ, MS, 86, PhD(geophys fluid dynamics), 88. *Prof Exp:* Teaching asst mech eng, City Col New York, 83-84; res asst oceanog, Princeton Univ, 84-88; postdoctoral, Joint Inst Study Atmosphere & Ocean, Univ Wash, 88-90; ASST PROF OCEANOG, TEX A&M UNIV, 90- *Mem:* Am Meteorol Soc; Am Geophys Union. *Res:* Geophysical fluid dynamics; equatorial dynamics of oceans and the atmosphere; air-sea interactions; numerical modeling of ocean circulation. *Mailing Add:* Dept Oceanog Texas A&M Univ College Station TX 77843

CHANG, POTTER CHIEN-TIEN, b Canton, China, Mar 21, 34; US citizen; m 61; c 2. STATISTICS. *Educ:* Nat Taiwan Univ, BS, 58; Univ Minn, MS, 66, PhD(biometry), 68. *Prof Exp:* From asst prof to assoc prof, 68-81, prof & head, 81-85, PROF, DIV BIOSTATIST, SCH PUB HEALTH, UNIV CALIF, LOS ANGELES, 81- *Mem:* Am Statist Asn; Biometric Soc. *Res:* Statistical methodology; application of statistics in medical research. *Mailing Add:* Div Biostatist Sch Pub Health Univ Calif 405 Hilgard Ave Los Angeles CA 90024

CHANG, RAYMOND, b Hong Kong, Mar 6, 39; m 68; c 1. PHYSICAL CHEMISTRY. *Educ:* Univ London, BSc, 62; Yale Univ, MS, 63, PhD(phys chem), 66. *Prof Exp:* Res fel, Wash Univ, 66-67; asst prof chem, Hunter Col, 67-68; from asst prof to assoc prof, 68-78, PROF CHEM, WILLIAMS COL, 78- *Mem:* Am Chem Soc; AAAS. *Res:* Electron spin resonance and nuclear magnetic resonance; chemical kinetics of fast reactions; photosynthesis; conformation of proteins. *Mailing Add:* Dept Chem Williams Col Williamstown MA 01267

CHANG, RAYMOND S L, b Tai-Pan, Oct 22, 43. PHARMACOLOGY. *Educ:* Nat Taiwan Univ, BS, 67, MS, 70; Univ Miami, PhD(pharmacol), 77. *Prof Exp:* Postdoctoral fel, Johns Hopkins Univ, 77-79; sr res pharmacologist, 79-83, res fel pharmacol, 83-88, SR RES FEL PHARMACOL, MERCK SHARP & DOHME RES LAB, 88- *Mem:* Am Soc Pharmacol & Exp Therapeut; Soc Neurosci; AAAS. *Res:* Pharmacology; neuroscience. *Mailing Add:* Dept New Lead Pharmacol Merck Sharp & Dohme Res Lab Rm 3025 Bldg 26A West Point PA 19486

CHANG, REN-FANG, b Nanking, China, Jan 14, 38; m 68. THERMODYNAMICS, LASERS. *Educ:* Taiwan Nat Univ, BS, 60; Univ Md, PhD(physics), 68. *Prof Exp:* Res assoc physics, 68-71, asst prof physics & astron, Univ Md, 71-78; RES PHYSICIST, NAT INST STANDARDS & TECHNOL, 78- *Honors & Awards:* Apollo Achievement Award, 69. *Mem:* AAAS; Am Phys Soc; Sigma Xi. *Res:* Measurements of thermophysical properties; theories on properties of binary fluid mixtures near a critical line, critical point phenomena; studies of optical properties of retro reflector. *Mailing Add:* Nat Inst Standards & Technol Bldg 221 Rm A105 Gaithersburg MD 20899

CHANG, RICHARD C(HI-CHENG), b Nanking, China, Jan 19, 18; US citizen; m 41; c 2. MEMBRANE SEPARATIONS, POLYMER CHEMISTRY. *Educ:* Nanking Univ, BS, 40; Syracuse Univ, MChE, 50, PhD(chem eng), 54. *Prof Exp:* Instr chem, Nanking Univ, 40-42; chem engr, Cheng-tu Tannery, China, 42-43; sr chem engr, Chamois Tannery, 43-46; supt Tannery No 2, Taiwan Animal Prod Co, 46-48; res engr, Syracuse Univ, 54-56; sect leader synthetic rubber, Am Synthetic Rubber Corp, 56-64; sr res chem engr, Chemstrand Res Ctr, Inc, Monsanto Co, 64-74; sr res specialist, Monsanto Triangle Park Develop Ctr, 74-81; RETIRED. *Mem:* Am Chem Soc; Am Inst Chem Engr; Sigma Xi. *Res:* Synthetic fibers; leather tanning; synthetic rubber and membrane separation. *Mailing Add:* 516 Emerson Dr Raleigh NC 27609

CHANG, RICHARD KOUNAI, b Hong Kong, June 22, 40; m 61; c 3. SOLID STATE PHYSICS, QUANTUM ELECTRONICS. *Educ:* Mass Inst Technol, BS, 61; Harvard Univ, MS, 62, PhD(solid state physics), 65. *Prof Exp:* Res fel solid state physics, Harvard Univ, 65-66; asst prof, 66-69, assoc prof, 70-76, PROF ENG & APPL SCI, YALE UNIV, 76- *Concurrent Pos:* Consult, Sanders Inc & Sandia Nat Labs. *Mem:* Fel Am Phys Soc; fel Optical Soc Am; Sigma Xi. *Res:* Nonlinear optics; Raman spectroscopy; solid state laser emission; surface science. *Mailing Add:* Appl Physics Yale Univ PO Box 2157 Yale Sta New Haven CT 06520-2157

CHANG, ROBERT SHIHMAN, b China, July 26, 22; nat US; m 51; c 4. VIROLOGY. *Educ:* St John's Univ, China, MD, 46; Harvard Univ, DSc, 52. *Prof Exp:* Assoc prof microbiol, Harvard Univ, 54-68; PROF MED MICROBIOL, SCH MED, UNIV CALIF, DAVIS, 68- , PROF MED MICROBIOL & IMMUNOL. *Concurrent Pos:* Med dir, Davis Free Clins, 74-84; dir, Tissue Typing Lab, Med Ctr, Univ Calif, Davis, 79-84. *Mem:* AAAS; Soc Exp Biol & Med; Am Acad Microbiol; Am Acad Prev Med; Int Soc Antiviral Res; Am Soc Histocompatibility & Immunogenetics. *Res:* Virology; immunology; preventive medicine. *Mailing Add:* Med Micro Univ Calif Davis CA 95616

CHANG, ROBIN, b Hong Kong, June 4, 51; US citizen; m 83; c 2. PROGRAMMABLE LOGIC CONTROLLERS, PARALLEL PROCESSING COMPUTERS. *Educ:* Univ Calif, Berkeley, BS, 72; Princeton Univ, MS & MA, 74, PhD(elec eng & computer sci), 76. *Prof Exp:* Asst instr computer sci, Princeton Univ, 72-73, asst researcher, 72-76; sr engr, Rapistan Corp, 76-78; group leader computer eng, Polaroid Corp, 78-82; PRES, INT PARALLEL MACH, INC, 82- *Concurrent Pos:* Consult, NIH, 74-76 & Lear Siegler Inc, 78-82. *Mem:* Inst Elec & Electronics Engrs; Asn Comput Mach. *Res:* Parallel processing computer architecture, software, supercomputers, algorithms; industrial automation; programmable logic controllers; artificial intelligence. *Mailing Add:* Int Parallel Mach Inc 700 Pleasant St New Bedford MA 02740

CHANG, SHAO-CHIEN, b Mar 1, 30; Can citizen; c 2. MATHEMATICS. *Educ:* Taiwan Norm Univ, BSc, 56; Carleton Univ, Ont, MSc, 63, PhD(math), 68. *Prof Exp:* Teacher, Taipei Chien Kuo High Sch, 55-56; supvr, Chnong Hwa NTS Sch, KL Malaysia, 57-62; sessional lectr math, Carleton Univ, Ont, 63-65; lectr, 65-67, from asst prof to assoc prof, 67-86, PROF MATH, BROCK UNIV, 86- *Concurrent Pos:* Vis prof math, 71-72, Gastprof, Fernuni, 79, 82, 84, 86, 87, 89, 90. *Mem:* Am Math Soc; Math Asn Am; Can Math Cong. *Res:* Mathematical logic, especially syntatical transforms; summability, especially classical and functional analytical method; sequences spaces. *Mailing Add:* Dept of Math Brock Univ St Catharines ON L2S 3A1 Can

CHANG, SHAU-JIN, b Kiangsu, China, Jan 7, 37; m 64; c 2. ELEMENTARY PARTICLE PHYSICS, MATHEMATICAL PHYSICS. *Educ:* Taiwan Univ, BS, 59; Tsing Hua Univ, Taiwan, MS, 61; Harvard Univ, PhD(physics), 67. *Prof Exp:* Mem physics, Inst Advan Study, 67-69; from asst prof to assoc prof, 69-74, PROF PHYSICS, UNIV ILL, URBANA, 74- *Concurrent Pos:* Mem, Inst Advan Study, Princeton, 72-73; Alfred P Sloan fel, 72-74; vis physicist, Fermi Nat Accelerator Lab, 75. *Mem:* fel Am Phys Soc; Chinese Astron Soc. *Res:* Various theoretical topics in quantum field theory; elementary particle physics; nonlinear iterative system. *Mailing Add:* Dept Physics 1110 W Green St Univ Ill Urbana IL 61801

CHANG, SHELDON S L, b Peking, China, Jan 20, 20; m 45, 65; c 3. ELECTRICAL ENGINEERING. *Educ:* Nat Southwest Assoc Univ, China, BS, 42; Tsinghua Univ, MS, 44; Purdue Univ, PhD(elec eng), 47. *Prof Exp:* Design engr, Cent Radio Works, China, 43-44; design engr, Robbins & Myers, Inc, Ohio, 46-47, res & develop engr, 48-52; instr, Purdue Univ, 47-48; from asst prof to prof elec eng, NY Univ, 52-63; PROF ELEC ENG, STATE UNIV NY STONY BROOK, 63- *Concurrent Pos:* Consult, Robbins & Meyers, Inc, 52- & Marine & Air Armament Div, Sperry Gyroscope Co Div, Remington Rand, 56-68; vis Mackay prof, Univ Calif, Berkeley, 69-70. *Mem:* Fel Inst Elec & Electronics Engrs; Am Phys Soc; Am Math Soc. *Res:* Optimal control systems theory; energy conversion; feedback communication systems; electronics. *Mailing Add:* Dept Elec Eng Col Eng State Univ NY Stony Brook NY 11794

CHANG, SHI KUO, b July 17, 44; US citizen; c 2. INFORMATION PROCESSING SYSTEMS, PICTORIAL DATABASE SYSTEMS. *Educ:* Nat Taiwan Univ, BS, 65; Univ Calif, Berkeley, MS, 67, PhD(elec eng & comput sci), 69. *Prof Exp:* Res asst & fel, Electronics Res Lab, dept elec eng & comput sci, Univ Calif, Berkeley, 67-69; res mem, T J Watson Res Ctr, IBM, Yorktown Heights, NY, 69-75; from assoc prof to prof, dept info eng & dir, Info Systs Res Lab, Univ Ill, Chicago Circle, 75-82; PROF & CHMN DEPT ELEC & COMPUT ENG & DIR, INFO SYSTS LAB, ILL INST TECHNOL, 82- *Concurrent Pos:* Asst prof, Sch Elec Eng, Cornell Univ, 70-71; prin investr, NSF, US Army, Off Naval Res & AT&T Found grants & contracts, 71-86; vis prof, dept comput sci, Nat Chiao-Tung Univ, Taiwan & dir, Comput Lab, Inst Math, Acad Sinica, 72-73; consult, Nat Electronics Data Processing Ctr, Repub China, 72-73, IBM, 76 & 84, Comput Ctr, Tan-Run Steel Inc, Kaoshiung, Taiwan, 78-79, Bell Labs, 79-82 & 84, Naval Res Lab, 80-85, Standard Oil Co, 82-83, Honeywell Corp, 84 & R R Donnelley & World Bank China Prog, 85; res fel, Inst Info Sci, Acad Sinica, Repub China; ed-in-chief, Int J Policy Anal & Info Systs, 77-82. *Mem:* Fel Inst Elec & Electronics Engrs; Asn Comput Mach; Chinese Inst Elec Engrs. *Res:* Pictorial information systems; information exchange theory and applications. *Mailing Add:* Knowledge Systs Inst 3420 Main St Skokie IL 60076

CHANG, SHIH-GER, b Taipei, Taiwan, Oct 24, 41; c 1. PHYSICAL CHEMISTRY, ENVIRONMENTAL CHEMISTRY. *Educ:* Cheng Kung Univ, Taiwan, BS, 64, Univ Detroit, MS, 68, Univ Calif, Berkeley, PhD(phys chem), 71. *Prof Exp:* SR CHEMIST, APPLIED SCI DIV, LAWRENCE BERKELEY LAB, 88- *Mem:* Am Chem Soc; AAAS. *Res:* Chemical characterization of air pollutants, and study of chemical reaction of air pollutants; study of surface chemistry of solids; development of fuel gas desulfurization and denitrification process. *Mailing Add:* Div Appl Sci Lawrence Berkeley Lab Bldg 70 Rm 110A Berkeley CA 94720

CHANG, SHIH-YUNG, b Hopei, China, Feb 10, 38; m 64; c 1. QUANTUM CHEMISTRY. *Educ:* Nat Taiwan Univ, BS, 60; Kans State Univ, MS, 65; Univ Wash, PhD(phys chem), 69. *Prof Exp:* Res assoc chem, Univ Calif, Santa Barbara, 69-70; NIH fel, Johns Hopkins Univ, 70-71; mem sci staff, Wolf Res & Develop Corp, 71-73; SR PHYS CHEMIST, ENVIROSPHERE CO, EBASCO SERV INC, 73- *Concurrent Pos:* Adj asst prof, Dept Pharmacol, Mt Sinai Sch Med, City Univ New York, 76-81, Dept Physiol & Biophys, 85-; staff engr, Gibbs & Hill, Inc, 80- *Mem:* Am Phys Soc; Sigma Xi. *Res:* Ab-initio study of molecular wave functions and physical properties; perturbation theory of molecular polarizabilities, force constants and dipole moments; perturbational approach to the determination of molecular electrostatic interaction potential for drug design. *Mailing Add:* 11 Pennsylvania Ave New York NY 10001

CHANG, SHU-PEI, b Sinwui, China, Oct 11, 22; US citizen; m 69; c 1. POLYMER CHEMISTRY, INDUSTRIAL ORGANIC CHEMISTRY. *Educ:* Nat Checkiang Univ, BS, 45; Univ Louisville, PhD(chem), 63. *Prof Exp:* Chem engr & soil scientist, Taiwan Sugar Co, 46-59; res fel, Univ Louisville, 63-64; res chemist, Hort & Spec Crops, Northern Regional Res Ctr, USDA, 64-86; RETIRED. *Mem:* Am Chem Soc; AAAS; Am Oil Chemist's Soc. *Res:* Plasticizers, lubricants and extenders from new seed oils or their derived fatty acids; addition and condensation polymerizations; application of calorimetric, chromatographic and spectroscopic methods to analysis. *Mailing Add:* 309 W Aspen Way Peoria IL 61614

CHANG, SHU-SING, b Shanghai, China, Feb 18, 35; m 60; c 2. PHYSICAL CHEMISTRY, MATERIALS SCIENCE ENGINEERING. *Educ:* Univ Taiwan, BS, 56; Univ Mich, MS, 59, PhD(chem), 62. *Prof Exp:* Chemist, Cent Res Lab, Allied Chem Corp, 61-63; chemist, Inorg Solids Div, Nat Bur Standards, 63-64, CHEMIST, POLYMER DIV, NAT INST STANDARDS & TECHNOL, 64- *Mem:* Am Phys Soc; Am Chem Soc; Am Soc Testing & Mat; Mat Res Soc. *Res:* Thermodynamic properties of globular molecules, plastic crystals, vitreous state and polymers; calorimetry, diffusion, migration, microelectronic packaging, composite processing and automation. *Mailing Add:* Polymers Div Nat Inst Standards & Technol Gaithersburg MD 20899

CHANG, SHUYA, industrial chemistry, for more information see previous edition

CHANG, SIMON H, b June 6, 30; c 3. RECOMBINANT DNA, ENZYMOLOGY. *Educ:* Okla State Univ, PhD(biochem), 65. *Prof Exp:* PROF MOLECULAR BIOL, LA STATE UNIV, 68- *Mailing Add:* Rm 322 Choppin Hall La State Univ Baton Rouge LA 70803

CHANG, STEPHEN SZU SHIANG, b Beijing, China, Aug 15, 18; nat US; m 52. FOOD CHEMISTRY. *Educ:* Nat Chi-nan Univ, BS, 41; Kans State Univ, MS, 49; Univ Ill, PhD(food chem), 52. *Prof Exp:* Res chemist, Universal Pharmaceut Corp, China, 41-44; assoc engr, Nat Resources Comn China, 44-46; supt prod, Chinchow Pulp & Paper Mill, 46-47; res assoc food chem, Univ Ill, 52-55; res chemist, Swift & Co, 55-57; sr res chemist, A E Staley Mfg Co, 57-60; assoc prof food sci, Rutgers Univ, 60-62, chmn dept, 77-86, food sci, 62-88; RETIRED. *Concurrent Pos:* Consult, twenty-nine domestic & foreign co; hon prof, Wuxi Inst Light Indudst, China; chmn bd, Cathay Food Consult Co; indust consult, 88- *Honors & Awards:* Spec Award, Potato Chip Inst Int; Putnam Food Award, Putnam Publ Co; Bailey Award & lipid chem award, Am Oil Chem Soc; Excellence in Res Award, Inst, Inst Food Technologist; Nicholas Appert Medal. *Mem:* Inst Food Technologists; Am Chem Soc; Am Oil Chem Soc (pres, 70); NY Acad Sci. *Res:* Flavor stability of fats and oils; chemical reactions involved in the processing of edible fats and oils; mechanisms of the autoxidation of unsaturated fatty acids; chemistry of food emulsifiers; chemistry of food flavors; isolation and identification of flavor compounds in foods; natural antioxidant, biosynthesis of flavors; fish oils; contributed articles to professional journals; patentee infield. *Mailing Add:* Dept Food Sci Rutgers Univ New Brunswick NJ 08903

CHANG, SUN-YUNG ALICE, b Ci-an, China, Mar 24, 48; m 73; c 2. MATHEMATICAL ANALYSIS. *Educ:* Nat Taiwan Univ, BA, 70; Univ Calif, Berkeley, PhD(math), 74. *Prof Exp:* Asst prof math, State Univ NY Buffalo, 74-75; Hedrick Asst Prof Math, Univ Calif, Los Angeles, 75-77; asst prof math, Univ Md, College Park, 77-80; assoc prof, 80-81, PROF, UNIV CALIF, LOS ANGELES, 81- *Concurrent Pos:* Solan fel, 80-81. *Honors & Awards:* Int Cong Math, 88. *Mem:* Am Math Soc (vpres, 88-90); Women Am Math Soc. *Res:* Investigation of behavior of analytic functions in the complex plane and in several complex variables, approximation of bounded functions by analytic functions; geometric partial differential equations. *Mailing Add:* Dept Math Univ Calif Los Angeles CA 90024

CHANG, TAI MING, b Taiwan, China, Nov 14, 38; m 66; c 2. CHEMICAL ENGINEERING, POLYMER SCIENCE. *Educ:* Nat Taiwan Univ, BS, 61; WVa Univ, MS, 65, PhD(chem eng), 67. *Prof Exp:* Sr res engr, Goodyear Tire & Rubber Co, Akron, 67-70, group leader res & develop, 70-75; eng specialist, 75-81, SR SPECIALIST, MONSANTO CO, 81- *Mem:* Am Chem Soc; Am Inst Chem Engrs. *Res:* Polymerization engineering; solid state polymerization; system simulation and optimization; fluidization; heat and mass transfer; polymer processing; polyvinyl butyral product and process. *Mailing Add:* 50 Canterbury Lane Longmeadow MA 01106-2814

CHANG, TAI YUP, b Korea, Oct 25, 33; m 62; c 2. AIR POLLUTION, QUANTUM CHEMISTRY. *Educ:* Seoul Nat Univ, BS, 58, MS, 60; Univ Wis-Madiosn, PhD(theoret chem), 66. *Prof Exp:* Res fel theoret chem, Univ Wis-Madison, 67 & Harvard Univ, 67-69; res scientist, 69-80, STAFF SCIENTIST, CHEM DEPT, RES STAFF, FORD MOTOR CO, 80- *Mem:* Air Pollution Control Asn; Am Meteorol Soc; Am Chem Soc. *Res:* Atomic and molecular physics; development and application of the perturbation theory; air pollution modeling, urban air quality modeling including atmospheric dispersion and photochemistry; global balance of trace gases. *Mailing Add:* 30091 Mayfair Rd Farmington Hills MI 48331

CHANG, TA-MIN, IMMUNOCHEMISTRY, PROTEIN CHEMISTRY. *Educ:* Univ Calif, Santa Barbara, PhD(biochem), 69. *Prof Exp:* HEAD BIOCHEM LAB, GASTROINTESTINAL UNIT, GENESEE HOSP, 78- *Mailing Add:* Genesee Hosp 224 Alexander St Rochester NY 14607

CHANG, TAO-YUAN, b Taiwan, China, Mar 1, 37; m 63; c 2. OPTOELECTRONIC DEVICES, MOLECULAR BEAM EPITAXY. *Educ:* Nat Taiwan Univ, BS, 59; Stanford Univ, MS, 62; Univ Calif, Berkeley, PhD(elec eng), 66. *Prof Exp:* Actg asst prof elec eng, Univ Calif, Berkeley, 66-67; mem tech staff, 67-84, SUPVR, AT&T BELL LABS, HOLMDEL, NJ, 84- *Mem:* Optical Soc Am; Inst Elec & Electronics Engrs; Am Vacuum Soc. *Mailing Add:* AT&T Bell Labs Rm 4F-429 Crawfords Corner Rd PO Box 3030 Holmdel NJ 07733-1988

CHANG, TA-YUAN, b Kwei-Chow, China, Apr 8, 45; m 72; c 2. LIPID METABOLISM, BIOCHEMICAL REGULATION. *Educ:* Nat Taiwan Univ, BS, 67; Univ NC, Chapel Hill, PhD(biochem), 73. *Prof Exp:* Fel, Wash Univ Med Sch, 73-76; PROF BIOCHEM, DARTMOUTH MED SCH, 76-

Concurrent Pos: Res Career Develop award, NIH. *Mem:* Am Soc Biol Chemists; Am Chem Soc. *Res:* Regulation of cholesterol metabolism in mammalian cells. *Mailing Add:* Dept Biochem Dartmouth Med Sch Hanover NH 03755

CHANG, TE WEN, b Nanchang, China, Oct 12, 20; US citizen; m 52; c 5. INFECTIOUS DISEASES, VIROLOGY. *Educ:* Nat Cent Univ, AB, 41, MD, 45. *Prof Exp:* Resident & asst med, Nat Cent Univ Hosp, 46-49; intern, St Joseph Hosp, Kansas City, 50; res fel virol, Univ Kans, 50-52; fel med, Mass Mem Hosp & Boston Univ, 52-57; instr med, 58-59, asst prof med & microbiol, 60-67, ASSOC PROF MED & COMMUNITY HEALTH, TUFTS UNIV SCH MED, 68- *Concurrent Pos:* Physician, New Eng Med Ctr Hosp, 60; assoc prof sch med, Tufts Univ, 68. *Honors & Awards:* Achievement Award, Am Chinese Med Asn, 87. *Mem:* AAAS; Fedn Clin Invest; Am Soc Microbiol; Infectious Dis Soc Am. *Res:* Treatment and prevention of viral diseases in man; clostridium difficile infections; bacterial toxins. *Mailing Add:* Dept Community Health Tufts Univ Sch Med Boston MA 02111

CHANG, TED T, b Tainan, Taiwan, China, Oct 6, 35; US citizen; m 60; c 3. POLYMER CHEMISTRY, SYNTHETIC INORGANIC & ORGANOMETALLIC CHEMISTRY. *Educ:* Nat Taiwan Univ, BS, 57; Univ Va, MS, 63, PhD(anal chem), 65. *Prof Exp:* Asst lectr anal chem, Nat Chen-Kung Univ, 59-61; postdoctoral, Calif Inst Technol, 65-66; group leader, Wyeth Lab, 71-77; supvr, Celanese Res Co, 77-79; res chemist, 66-71, ASSOC RES FEL ANAL CHEM, AM CYANAMID CO, 79- *Concurrent Pos:* Expert lectr to China, UN, 86. *Mem:* Am Chem Soc; Am Soc Mass Spectrometry; Sigma Xi. *Res:* Utilized mass spectrometry to solve various analytical problems; developed tandem techniques; chromatographic techniques. *Mailing Add:* Am Cyanamid Co 1937 W Main St PO Box 60 Stamford CT 06904-0060

CHANG, THOMAS MING SWI, b Swatow, China, Apr 8, 33; Can citizen; m 58; c 4. BIOTECHNOLOGY, ARTIFICIAL CELLS & ARTIFICIAL ORGANS. *Educ:* McGill Univ, BSc, 57, MD, CM, 61, PhD(physiol), 65; FRCP(C), 72. *Prof Exp:* Intern, Montreal Gen Hosp, 61-62; sessional lectr, McGill, Univ, 64-65, lectr, 65-66, from asst prof to assoc prof, 66-72, dir, Artificial Organ Res Unit, 75-79, PROF PHYSIOL, MCGILL UNIV, 72-, PROF MED, 75-, PROF BIOMED ENG, 90-, DIR, ARTIFICIAL CELLS & ORGANS RES CTR, 79-; CAREER INVESTR, MED RES COUN CAN, 68- *Concurrent Pos:* Med Res Coun Can fel, 62-65, scholar, 65-68; hon prof, Nankai Univ, People's Repub China, 83- *Honors & Awards:* Clemson Award, 80. *Mem:* Am Soc Artificial Internal Organs; Int Soc Artificial Organs; fel Royal Col Physicians Can. *Res:* Inventor of artificial cells; basic and applied research, clinical trial and development of artificial cells for hemoperfusion; artificial kidney, artificial liver, immunosorbent, microencapsultation, blood substitute; immobilized enzymes and multienzymes and cofactor recycling; immobilized cells and microorganisms; artificial blood. *Mailing Add:* Artificial Cells & Organs Res Ctr McGill Univ 3655 Drummond St Montreal PQ H3G 1Y6 Can

CHANG, TIEN SUN (TOM), b Mukden, China, Feb 28, 31; nat US; m 61; c 3. PHYSICS, MECHANICS. *Educ:* Univ Ill, BS, 52, MS, 53 & 54, PhD(theoret & appl mech), 55; Univ Mich, PhD(theoret physics), 63. *Prof Exp:* Asst theoret & appl mech, Univ Ill, 52-54, assoc, 54-55, from instr to asst prof, 55-57; from assoc prof to prof eng mech, Va Polytech Inst, 57-67; NSF prof continuum mech, NC State Univ, 67-76, grad prof nuclear eng & chmn mech prog, 68-76; vis prof & mem staff res, 76-86, DIR, CTR THEORET GEO/COSMO PLASMA PHYSICS, MASS INST TECHNOL, 86- *Concurrent Pos:* Develop engr, Reactor Div, Oak Ridge Nat Lab, 62, res engr, 63, consult, 63-; vis prof, Cambridge Univ, 64-65 & 69, vis lectr, Dept Appl Math & Theoret Physics, 68; lectr & topic organizer, Langley Res Ctr, NASA, 64-65; vis prof, Lehigh Univ, 69, Cornell Univ, 70 & Mass Inst Technol, 71; invited lectr, jointly sponsored by Academia Sinica, Nat Taiwan Univ & Nat Tsing Hua Univ, Taiwan, 70; consult-lectr, adv group aerospace res & develop, NATO, 65-66; ed, Plasma Physics, 67-80; hon res fel, Harvard Univ, 78-79; consult, Nat Magnet Lab, 77-79. *Honors & Awards:* Thompson Award, Am Soc Testing & Mat, 58; Cert Appreciation, Soc Eng Sci, 76, 78. *Mem:* AAAS; Am Geophys Union; fel Am Phys Soc. *Res:* Theoretical and solid state physics; magnetohydrodynamics; plasma physics; fluid mechanics; rheology; radiation gas dynamics; superfluids; hypervelocity impact; nonlinear waves; biomathematics; continuum mechanics; space physics. *Mailing Add:* Mass Inst Technol Br Off PO Box 6 Cambridge MA 02139

CHANG, TIEN-DING, b Chwansha, China, Oct 13, 21; m 51; c 2. GENETICS. *Educ:* Chekiang Univ, BS, 44; Univ Minn, MS, 60; Iowa State Univ, PhD(genetics), 63. *Prof Exp:* Teacher high schs, China, 44-46; agronomist, Taiwan Agr Res Inst, 46-57; cytogeneticist, Dept Med Genetics, Children's Hosp Winnipeg, 64-69; res assoc genetics, Univ Mo-Columbia, 69-74; RES ASSOC CELLULAR & BIOCHEM GENETICS, SLOAN-KETTERING INST CANCER RES, 74- *Concurrent Pos:* Damon Runyon Mem Fund vis investr, Biol Div, Oak Ridge Nat Lab, 63-64; cytogeneticist, Jenkins Found for Res, Salinas, Calif, 73-74. *Mem:* Genetics Soc Am; Genetics Soc Can; Tissue Culture Asn; AAAS. *Res:* Plant and mammalian cytogenetics. *Mailing Add:* 40 Putnam Dr Port Chester NY 10573

CHANG, TIEN-LIN, b Chekiang, China, Nov 22, 43; US citizen; m 71; c 2. ELECTRICAL ENGINEERING, COMPUTER SCIENCE. *Educ:* Nat Taiwan Univ, BS, 65; Rice Univ, MS, 69, PhD(elec eng), 71. *Prof Exp:* Res asst elec eng, Rice Univ, 67-71, fel, 71; MEM TECH STAFF INFO SCI, ROCKWELL INT CORP, 71- *Mem:* Sigma Xi; sr mem Inst Elec & Electronics Engrs. *Res:* Digital signal processing in the areas of filter structure design; applications in image processing and telecommunications; pattern recognition, artificial intelligence, and intelligent machines. *Mailing Add:* 4419 E Emberwood Lane Anaheim CA 92807

CHANG, TIMOTHY SCOTT, b Shaowu, China, May 30, 25; m 55; c 4. POULTRY PATHOLOGY, AVIAN MICROBIOLOGY. *Educ:* Fukien Christian Univ, BA, 46; Duke Univ, MDiv, 51; NC State Univ, BS, 52; Ohio State Univ, MS, 53, PhD(poultry path, avian microbiol), 57. *Prof Exp:* Teacher pub sch, China, 46-48; asst poultry sci, Ohio State Univ, 51-57; dir bact res lab, Whitmoyer Labs, Inc, Rohm and Haas Co, Pa, 57-65; group leader vet microbiol, Vet Res Div, Norwich Pharmacal Co, NY, 65-66, sect chief, 66-69; div mgr diagnostic reagents, Burroughs Wellcome Co, 69-70; dir animal technol dept, S B Penick & Co, CPC Int, Inc, 70-72; assoc prof, 71-77, PROF POULTRY PATH & AVIAN MICROBIOL, MICH STATE UNIV, 77- *Concurrent Pos:* Vis lectr, China, 79, 80, 81, 82, 84, 85, 86, 87, 88, & 89, Malaysia, 87 & 88, Philippines, Indonesia & Thailand, 89; consult, World Bank, 85, Pharmaceut Co, 85-87. *Mem:* Am Poultry Sci Asn; Am Soc Microbiol; World Poultry Sci Asn; Am Asn Avian Pathologists; NY Acad Sci. *Res:* Discovered the function of Bursa of Fabricius in antibody production; immuno-response; poultry diseases; microbial fermentation; antibiotic properties-effects; radiation effects; sanitation-disinfectant; toxicological effects; avian microbiology. *Mailing Add:* Dept Animal Sci Mich State Univ East Lansing MI 48824-1225

CHANG, TSONG-HOW, b Nanking, China, Oct 11, 29; c 3. STATISTICAL MODELING, ENGINEERING ECONOMICS. *Educ:* Nat Taiwan Univ, BS, 53; WVa Univ, MS, 58; Univ Wis, Madison, PhD(mech eng), 72. *Prof Exp:* Sr analyst oper res, Minneapolis Honeywell, 60-61; asst prof indust eng, Miss State Univ, 63-66; asst prof, 69-75, dept chair indust & syst eng, 79-83, ASSOC PROF INDUST ENG, UNIV WIS-MILWAUKEE, 75-, CONSULT STATIST PROCESS CONTROL & DESIGN EXP, 81- *Concurrent Pos:* Adv prof, Shanghai Inst Mech Eng, China, 79- *Mem:* Sigma Xi; Oper Res Soc Am; Am Inst Indust Engrs. *Res:* Statistical modeling and engineering; economic analysis of industrial and manufacturing engineering systems; analysis of survival data for life expectancy predictions and engineering reliability; statistical methods for quality and productivity improvement. *Mailing Add:* Dept Indust & Syst Eng Univ Wis PO Box 784 Milwaukee WI 53201

CHANG, T(AO)-Y(UAN), b Taiwan, China, Mar 1, 37; m 63; c 2. COMPOUND SEMICONDUCTORS, QUANTUM & OPTOELECTRONICS. *Educ:* Nat Taiwan Univ, BS, 59; Stanford Univ, MS, 62; Univ Calif, Berkeley, PhD(elec eng), 66. *Prof Exp:* Engr electronics, Machtronics Inc, Calif, 62; res asst, Univ Calif, Berkeley, 62-66, acting asst prof elec eng, 66-67; MEM TECH STAFF & SUPVR, BELL TEL LABS, 67- *Honors & Awards:* Sr Am Sci Award, Alexander von Humboldt Found, Fed Repub Ger, 79. *Mem:* Sr mem Inst Elec & Electronic Engrs; fel Optical Soc Am; Am Vacuum Soc. *Res:* Instabilities in mirror confined plasmas; infrared mixing in nonlinear crystals; precision spectroscopy of carbon dioxide laser; infrared and far-infrared generation by optical pumping; molecular-beam epitaxy; indium galium arsenide transistors; optoelectronic devices; plasma physics. *Mailing Add:* AT&T Bell Labs 4F429 Holmdel NJ 07733-1988

CHANG, WEN-HSUAN (WAYNE), b Tsingtao, China, Mar 28, 26; m 59; c 2. ORGANIC CHEMISTRY. *Educ:* Fu Jen Univ, China, BSc, 48; Wesleyan Univ, MA, 56; Northwestern Univ, PhD, 59. *Prof Exp:* Assoc engr, Taiwan Agr & Chem Works, Formosa, 49-54; sr res chemist, 58-65, from res assoc to sr res assoc, 65-69, scientist, 69-71, sr scientist, 71-73, dept mgr, 73-84, RES DIR, PIONEER POLYMER RES DEPT, PPG INDUST, 84- *Concurrent Pos:* Lectr modern mgt, 79- *Mem:* Am Chem Soc; NY Acad Sci; Royal Soc Chem. *Res:* Organic chemical synthesis; reaction mechanisms; kinetics; polymer synthesis. *Mailing Add:* PPG Industs PO Box 9 Rosanna Dr Allison Park PA 15101

CHANG, WILLIAM S C, b Kiangsu, China, Apr 4, 31; m 55; c 3. ELECTRICAL ENGINEERING, APPLIED PHYSICS. *Educ:* Univ Mich, BS, 52, MS, 53; Brown Univ, PhD(elec eng), 57. *Prof Exp:* Res assoc & lectr elec eng, Stanford Univ, 57-59; from asst prof to assoc prof, Ohio State Univ, 59-65; prof,Washington Univ 65-79, chmn dept, 65-71, dir, Lab Appl Electronic Sci, 71-79, Samuel Sachs prof,76-79; PROF ELEC ENG, UNIV CALIF, SAN DIEGO, 79- *Mem:* Am Phys Soc; fel Inst Elec & Electronics Engrs; fel Optical Soc Am. *Res:* Quantum electronics; lasers; masers; electromagnetic field theory; infra-red systems; integrated optics. *Mailing Add:* Dept Elec & Comput Eng Univ Calif San Diego La Jolla CA 92093-0407

CHANG, WILLIAM WEI-LIEN, b Taipei, Taiwan, Feb 7, 33; m 65; c 3. ANATOMIC PATHOLOGY, CARCINOGENESIS & ONCOLOGY. *Educ:* Nat Taiwan Univ, MD, 58; Ohio State Univ, MSc(path), 66; McGill Univ, PhD(anat), 70; Am Bd Path, dipl, 78. *Prof Exp:* Asst pharmacol, Nat Taiwan Univ, 60-61; rotating intern med, Buffalo Gen Hosp, 61-62; from asst resident to resident path, Ohio State Univ, 62-66; from asst prof to assoc prof anat, Mt Sinai Sch Med, 70-79; mem fac, Grad Sch, City Univ New York, 72-79; assoc prof 79-85, PROF PATH, MED CTR, WVA UNIV, 85, VCHMN DEPT PATH & DIR AUTOPSY SERV, 89- *Concurrent Pos:* Clin fel, Am Cancer Soc, Dept Path, Ohio State Univ, 64-65; Ont Heart Found fel, Dept Path, Queen's Univ, Ont, 66-67; Nat Cancer Inst res grant, Dept Anat, Mt Sinai Sch Med, 74-77. *Mem:* Am Asn Anat; Am Soc Cell Biol; US & Can Acad Path; AAAS; Am Asn Path. *Res:* Cell population kinetics of colon and salivary glands; chemical carcinogenesis of colon; pathogenesis of human carcinomas, sarcomas, and gastrointestinal diseases, especially colonic carcinogenesis and polypogenesis. *Mailing Add:* Dept Path WVa Univ Health Sci Ctr Morgantown WV 26506

CHANG, WILLIAM Y B, b Amoy, China, June 1, 48; m 77; c 2. PRIMARY PRODUCTIVITY, MODELS ECOSYSTEMS. *Educ:* Univ Pac, MS, 73; Ind Univ, MA, 75, PhD(ecol & math), 79. *Prof Exp:* Res investr ecol & statist, 79-80, asst res scientist, 80-88, ASSOC RES SCIENTIST, GREAT LAKES RES DIV, UNIV MICH, 88-; PROF MGR, DIV INT PROGS, NSF, 88- *Concurrent Pos:* Eigenmann fel biol, Ind Univ, 78, consult statist & comput, 78; consult, Dept Energy & Environ, Brookhaven Nat Lab, 82-83, Dept of the Interior, Water resources Lab Nat Park Serv, 84; res fel, Nat Acad Sci;

vis scholar, Ministry Educ, 86-87, sci consult, Qingdas Environ Res Inst, Ministry Transp, 84, sci advisor, Red Tides Res Ctr, State Oceanic Admin, Peoples Repub of China, 87-92; sci advisor, Yunnan Inst Environ Sci, Yunnan Province, 88-; co-ed, Book Series on environ technol, 82-; mem bd, Directorate Human Dominated Systs, Man & the Biosphere State Dept; distinguished vis scientist, Environ Protection Agency Corvallis Lab & mem vis scientist prog. *Mem:* Am Soc Limnol Oceanog; Int Asn Great Lakes Res; Am Statist Asn. *Res:* Statistical and mathematical models; aquatic ecology and limnology; environmental sciences and aquaculture. *Mailing Add:* Ctr Great Lakes & Aquatic Sci Univ Mich Rm 3113 IST Bldg Ann Arbor MI 48109

CHANG, Y AUSTIN, US citizen; m 56; c 3. MATERIALS CHEMISTRY, THERMODYNAMICS OF DEFECTS. *Educ:* Univ Calif, Berkeley, BS, 54, PhD(metall), 63; Univ Wash, MS, 55. *Prof Exp:* Chem engr, Stauffer Chem Co, Calif, 56-59; res metallurgist, Lawrence Radiation Lab, Univ Calif, Berkeley, 63; sr metall engr, Aerojet-Gen Corp, Calif, 63-67; assoc prof, Univ Wis-Milwaukee, 67-70, chmn dept mat, 71-78, prof mat eng, 70-80, assoc dean res, Grad Sch, 78-80; PROF, DEPT METALL & MINERAL ENG, UNIV WIS-MADISON, 80-, CHMN DEPT, 82- *Concurrent Pos:* Mem bd, Goodwill Residential Community, Inc, Milwaukee, 78-80; mem, Wis Gov Asian Am Adv Coun, 80-82. *Honors & Awards:* Byron Bird Award, 84. *Mem:* Am Inst Mining, Metall & Petrol Engrs; fel Am Soc Metals; Sigma Xi; Nat Asn Corrosion Engrs; Electrochem Soc. *Res:* Therodynamic modelling and phase diagram calculation/predictions of binary and ternary systems; phase stability of magnetic alloys; oxidation of solder alloys and their interaction with solid substrate as applied to electronic packaging; defect structure and therodynamics of IV-VI phases and quaternary metal-sulfur-oxygen systems. *Mailing Add:* Dept Metall & Mineral Eng 1509 University Ave Madison WI 53706

CHANG, YAO TEH, GROWTH OF BONE MARROW CELLS. *Educ:* Ceheeloo Univ, China, MD, 34. *Prof Exp:* PHYSICIAN, NIH, 47- *Mailing Add:* Nat Inst Diabetes & Digestive & Kidney Dis Lab Biochem Pharmacol Bldg Eight Rm 218 Bethesda MD 20892

CHANG, YEW CHUN, b Yinping, China, Oct 10, 41; Can citizen; m 69; c 2. PHYSICAL CHEMISTRY, ORGANIC CHEMISTRY. *Educ:* Univ BC, BSc, 64, MSc, 66; Univ Sussex, DPhil(chem), 68. *Prof Exp:* Sr chemist imaging, Energy Conversion Devices Inc, 70-71, org chem mgr, 71-74; scientist liquid toners, Xerox Res Ctr, Can Ltd, 74-79; sr scientist oil res, Syncrude Can, 79-80; supvr energy res & develop, Arco, 80-82; mat mgr, Amoco, 82-87; PAPER-PIGMENT MGR, COMB ENG, 88- *Concurrent Pos:* Fel, Yale Univ, 68-69 & Univ Calif, Los Angeles, 69-70; hon consult, Indust Adv Coun, Univ Waterloo, 76-79. *Mem:* Assoc mem Am Chem Soc; Soc Photog Scientists & Engrs; Chem Inst Can; Inter-Am Photochem Soc; Am Physics Soc; Tech Asn Pulp & Paper Indust. *Res:* Examined particles; clay in oil emulsions; surfactant dependent on colloids; natural gas components separation and processing; oil sands development; non-fossil fuel energy; thermoelectrics; hydrogen generation and storage; photochemistry and solar energy; pigment-paper chemistry and coating; paper technologies; paper making. *Mailing Add:* James River Corp 349 NW Seventh Ave Camas WA 98607-1999

CHANG, YIA-CHUNG, b Taiwan; US citizen. SUPERLATTICES, SEMICONDUCTORS. *Educ:* Nat Cheng-Kung Univ, Taiwan, BS, 74; Calif Inst Technol, MS, 78 & PhD(physics), 80. *Prof Exp:* Asst prof, 80-86, ASSOC PROF PHYSICS, UNIV ILL, URBANA-CHAMPAIGN, 86-; ASSOC PROF PHYSICS, UNIV ILL, URBANA-CHAMPAIGN, 86- *Concurrent Pos:* Consult, Honeywell Sci Ctr, 83-84, Hughes Res Lab, 85-86, Rockwell Sci Ctr, 87-91 & Bell Commun Res, 88-90. *Mem:* Am Phys Soc. *Res:* Shallow impurities and exciton complexes in semiconductors; optical properties of solids, transport in semiconductors; transport in semiconductors, phonons and polaritons in layered materials; semiconductor surfaces; nonlinear optical properties of solids; phonons and polaritons in layered materials. *Mailing Add:* Dept Physics Univ Ill Urbana-Champaign 1110 W Green Urbana IL 61801

CHANG, YI-HAN, b Peking, China, May 26, 33; US citizen; m 63. BIOCHEMICAL PHARMACOLOGY, IMMUNOLOGY. *Educ:* Stetson Univ, BS, 56; Pa State Univ, PhD(org chem), 61; Univ Conn, PhD(pharmacol), 65. *Prof Exp:* Res chemist, Pfizer, Inc, 60-63, res biochem pharmacologist, 63-68, res supvr, 68-72, group mgr, 72-75; assoc prof, 75-82, exec dir res, Rheumatol Div, 78-83, PROF MED & PHARMACOL, SCH MED, UNIV CALIF, LOS ANGELES, 82- *Mem:* Am Soc Pharmacol & Exp Therapeut; Am Chem Soc. *Res:* Anti-inflammatory and immunosuppressive agents; etiology and pathology of rheumatoid arthritis and hypersensitivity diseases; in vitro, in vivo models of cell-mediated hypersensitivity; drug metabolism; mechanism of action of colchicine. *Mailing Add:* Dept Med-Pharmacol Sch Med Univ Calif 1000 Veterans Ave Los Angeles CA 90024

CHANG, YOON IL, b Pyungbuk, Korea, Apr 12, 42; US citizen; m 66; c 3. NUCLEAR ENGINEERING. *Educ:* Seoul Nat Univ, BS, 64; Tex A&M Univ, ME, 67; Univ Mich, PhD(nuclear sci), 71; Univ Chicago, MBA, 83. *Prof Exp:* Nuclear engr, Nuclear Assurance Corp, 71-74; nuclear engr, 74-76, group leader, 76-77, sect head, 77-78, assoc div dir, 78-84, GEN MGR, IFR PROG, ARGONNE NAT LAB, 84- *Mem:* Am Nuclear Soc. *Res:* Reactor physics; nuclear fuel cycle analysis; reactor core design analysis; advanced reactor concept evaluation. *Mailing Add:* Argonne Nat Lab 9700 S Cass Ave Argonne IL 60439

CHANG, YUNG-FENG, b Taiwan, China, Nov 15, 35; m 68; c 2. BIOCHEMSITRY, MICROBIOLOGY. *Educ:* Nat Taiwan Univ, BS, 58, MS, 60; Univ Pittsburgh, PhD(biochem), 66. *Prof Exp:* Res asst biochem, Nat Taiwan Univ, 61-62; res asst, Univ Pittsburgh, 62-66; res assoc, Sch Med, Univ Md, 66-70, asst prof histol & embryol, Sch Dent, 70-71, from asst prof to assoc prof microbiol, 71-74, PROF BIOCHEM, SCH DENT, UNIV MD, BALTIMORE, 79- *Concurrent Pos:* Prin investr, NIH res grants; vis prof, Walter Reed Army Inst Res, 83. *Mem:* Am Chem Soc; Am Soc Biochem & Molecular Biol; Am Soc Pharmacol & Exp Therapeut; Am Soc Neurochem. *Res:* Lysine metabolism and regulation of enzyme in Pseudomonas putida; Zellweger's syndrome in human genetic disorders; piperolic acid oxidase in peroxisomes and mitochondria in the mammalian brain and liver; lysine metabolism in the mammalian brain; neurochemical and developmental aspects of the metabolic pathways of lysine in the rat; neurotransmitter/neuromodulator activity of lysine metabolites in the brain and its antiepileptic activity. *Mailing Add:* Dept Biochem Univ Md Sch of Dent 666 W Baltimore St Baltimore MD 21201

CHANGNON, STANLEY A, JR, b Donovan, Ill, Apr 14, 28; m 85; c 3. WEATHER MODIFICATION, CLIMATE CHANGE. *Educ:* Univ Ill, BS, 51, MS, 56. *Prof Exp:* Proj dir, Univ Chicago, 53-55; climatologist, Ill State Water Surv, 55-68, head atmospheric sci, 68-79, chief, 80-85, PRIN SCIENTIST, ILL STATE WATER SURV, 85-, DIR, GLOBAL CLIMATE CHANGE PROG, 91- *Concurrent Pos:* Chief ed, J Appl Meteorol, 76-78; mem, Nat Bd Weather Modification, 77-79 & Panel Climate Change, Elec Power Res Inst, 89-; chmn, Adv Comt Atmospheric Sci, NSF, 83-86 & Adv Panel Climate Change, US Environ Protection Agency, 89-; pres, Changnon Climatologist, 85-; res assoc, Univ Rochester, 86-89; counr, Am Meteorol Soc, 90- *Honors & Awards:* Horton Award, Am Geophys Union, 64; Boggess Award, Am Water Resources Asn, 76 & 90; Cleveland Abbe Award, Am Meteorol Soc, 81, Outstanding Appl Sci Award, 91. *Mem:* Fel Am Meteorol Soc; Am Asn State Climatologists (pres, 80-81); Weather Modification Asn (pres, 76-77); fel AAAS. *Res:* Meteorology; effects of weather on agriculture and water; weather modification; climate change. *Mailing Add:* 801 Buckthorn Circle Mahomet IL 61853

CHANGNON, STANLEY ALCIDE, JR, b Donovan, Ill, Apr 14, 28; m 50; c 3. CLIMATOLOGY. *Educ:* Univ Ill, BS, 51, MS, 56. *Prof Exp:* Proj supvr cloud physics, Univ Ill, 52-54; assoc scientist, 55-65, prof climat scientist, 65-70, actg head, 70-71, head, Atmospheric Sci Sect, 71-79, chief, Ill State Water Surv & prof, 75-85, CHIEF SCIENTIST, UNIV ILL, 86-, PROF GEOG. *Concurrent Pos:* NSF res grants, 65-; mem nat comt, Comt Weather Info for Agr, 65; mem, Nat Comt Severe Local Storms & Nat Comt Weather Modification. *Honors & Awards:* Robert Horton Award, Am Geophys Union, 64; Award, Bldg Res Inst, 65; Abby Award, Meteorol Soc, 81; Boggess Award, Am Water Resources Asn, 78, 90. *Mem:* Am Meteorol Soc; AAAS. *Res:* Climatology of Lake Michigan, Illinois and Middle West; severe storms; physical geography; weather modification; agrometeorology; irrigation; urban industrial effects on precipitation; climate change; water resources. *Mailing Add:* Ill State Water Surv 2204 Griffith Dr Champaign IL 61820

CHANIN, LORNE MAXWELL, b Roland, Man, Aug 14, 27; m 50; c 3. PLASMA PHYSICS. *Educ:* Univ Man, BS, 49; Univ NMex, MS, 51; Univ Pittsburgh, PhD(physics), 59. *Prof Exp:* Res engr, Westinghouse Res Labs, 51-59; res scientist & sect head, Honeywell Res Ctr, 59-65; assoc prof, 65-68, PROF ELEC ENG, UNIV MINN, MINNEAPOLIS, 68- *Concurrent Pos:* Consult, US Bur Mines. *Mem:* Am Phys Soc; Inst Elec & Electronics Engrs; Europ Phys Soc. *Res:* Atomic physics; interactions and reactions involving electrons, ions and excited atoms; plasma physics; electrical discharges; gaseous electronics; plasma chemistry. *Mailing Add:* 227 Elec Engr Univ Minn Minneapolis MN 55455

CHANLETT, EMIL T(HEODORE), b New York, NY, Dec 30, 15; m 46; c 2. SANITARY ENGINEERING. *Educ:* City Col New York, BS, 37; Columbia Univ, MS, 39; Am Bd Indust Hyg, Dipl, 62. *Prof Exp:* Asst sanit engr, US Pub Health Serv, 41-43; from asst to assoc prof sanit eng, 46-58, PROF SANIT ENG, UNIV NC, 58- *Concurrent Pos:* W K Kellogg Found field fel, 38-39; spec lectr, Sch Pub Health, Johns Hopkins Univ, 46; consult, Brazilian Govt, Inst Inter-Am Affairs & Spec Pub Health Serv, Amazon River Basin, 48; consult occup & radiol health, State Bd Health, NC; mem, NC Atomic Energy Adv Bd; mem, State Bd Refrig Exam, NC; mem, Sanitarians' Joint Coun, 58-64, chmn, 63, secy, 61 & 64; alternate mem, NC State Bd Sanit Exam, 63-; mem, Am Intersoc Bd Cert Sanitarians, 64. *Mem:* Am Water Works Asn; Water Pollution Control Fedn; Am Indust Hyg Asn; Air Pollution Control Asn; Inter-Am Asn Sanit Engrs (vpres, US sect, 62-64); Sigma Xi. *Res:* Industrial hygiene; radiological health; air pollution control; public health. *Mailing Add:* 622 Greenworld Rd Chapel Hill NC 27514

CHANLEY, JACOB DAVID, b New York, NY, Jan 23, 18; m 42; c 2. BIOCHEMISTRY. *Educ:* City Col New York, BS, 38; Univ Chicago, MS, 40; Harvard Univ, MA, 42, PhD(chem), 44. *Prof Exp:* Asst, Mt Sinai Hosp, 40-41; res worker, War Prod Bd, Columbia Univ, 44-45; res chemist, Mt Sinai Hosp, 45-67; ASSOC PROF BIOCHEM, SCH MED & GRAD SCH, CITY UNIV NEW YORK, 67- *Concurrent Pos:* Fel chem, Harvard Univ, 41-44; Rockefeller traveling fel, Oxford Univ, 49-50; Dazian Found fel, 49-50; sr res assoc, Mt Sinai Hosp, 67-70. *Mem:* AAAS; Am Chem Soc; The Chem Soc; Am Soc Biol Chemists. *Res:* Chemistry; steroid saponins of marine origin; metabolism of adrenergic transmitters; enzyme kinetics. *Mailing Add:* Mt Sinai Sch Med One Gustave Levy Pl New York NY 10029-5291

CHANMUGAM, GANESAR, b Colombo, Ceylon, Oct 24, 39; m 66; c 2. THEORETICAL ASTROPHYSICS. *Educ:* Univ Ceylon, BSc, 61; Cambridge Univ, BA, 63; Brandeis Univ, PhD(physics), 70. *Prof Exp:* Instr physics, Univ Mass, Amherst, 63-64; res fel astrophys, Inst Astrophys, Univ Liege, Belg, 69-71; res assoc, 71-72, from asst prof to assoc prof, 72-83, PROF PHYSICS & ASTRON, LA STATE UNIV, BATON ROUGE, 83- *Concurrent Pos:* Vis fel, Joint Inst Lab Astrophysics, Univ Colo, Boulder, 79-80; vis fel, Space Telescope Sci Inst, 87. *Mem:* Am Phys Soc; Am Astron Soc; Int Astron Union; Royal Astron Soc. *Res:* Theoretical studies of magnetic white dwarfs; neutron stars with application to high energy astrophysics. *Mailing Add:* Dept Physics & Astron La State Univ Baton Rouge LA 70803

CHANNEL, LAWRENCE EDWIN, b Boise, Idaho, Mar 17, 27; m 47; c 5. PHYSICS, ACOUSTICS. *Educ:* Pasadena Col, AB, 50; Univ Calif, Los Angeles, MA, 55. *Prof Exp:* Physicist, US Naval Ord Test Sta, Pasadena, Calif, 53-57; opers analyst, Mass Inst Technol physics eval group, US Dept Navy, Washington, DC, 57-63; res & develop scientist, Lockheed-Calif Co, 63-67, dept mgr, 67-69, div mgr, 69-78, dir, 78-84, vpres, 84-89; CONSULT, CTR NAVAL ANAL, ALEXANDRIA, VA, 89- *Concurrent Pos:* Asst physics, Univ Calif, Los Angeles, 54; instr, Pasadena Col, 64-65 & 69. *Mem:* Acoust Soc Am; Am Inst Aeronaut & Astronaut; Am Helicopter Soc; Oper Res Soc Am. *Res:* Underwater acoustics; military operations research; resource management. *Mailing Add:* 12974 Quaker Hill Rd Nevada City CA 95759

CHANNELL, ROBERT BENNIE, b Gallman, Miss, July 4, 24. SYSTEMATIC BOTANY. *Educ:* Miss State Col, BS, 47, MS, 49; Duke Univ, PhD, 55. *Prof Exp:* Pub sch teacher, 47-49; instr bot, Miss State Col, 49-51; asst, Duke Univ, 51-54; asst cur, Herbarium, 54-55; botanist, Gray Herbarium-Arnold Arboretum, Harvard, 55-57; from asst prof to assoc prof biol, 57-69, chmn dept, 63-73, PROF BIOL, VANDERBILT UNIV, 69- *Mem:* Am Soc Plant Taxon; Int Asn Plant Taxon. *Res:* Conventional and experimental taxonomy of vascular plants, particularly Compositae and Cyperaceae; flora of the Southeastern United States. *Mailing Add:* Dept of Biol Vanderbilt Univ Nashville TN 37240

CHANNIN, DONALD JONES, b Evanston, Ill, Aug 29, 42; c 2. PHYSICS, OPTICS. *Educ:* Case Western Reserve Univ, BS, 64; Cornell Univ, PhD(physics), 70. *Prof Exp:* PROG MGR, SARNOFF RES CTR, RCA LABS, 70- *Concurrent Pos:* Vis prof, Inst Physics & Chem, San Carlos, Univ Sao Paulo, Brazil, 77. *Mem:* Am Phys Soc; Optical Soc Am; Inst Elec & Electronic Engrs. *Res:* Solid state physics; electro-optical phenomenon; liquid crystals; injection lasers; optical communications systems. *Mailing Add:* David Sarnoff Res Ctr CN 5300 Princeton NJ 08543-5300

CHANNON, STEPHEN R, b Washington, DC, Aug 13, 47; m 69; c 1. NONLINEAR DYNAMICS. *Educ:* Ohio State Univ, BS, 70; Rutgers Univ, PhD(physics), 81. *Prof Exp:* Physicist, Fusion Energy Corp, 74-77; consult physicist-programmer, 81-82, DIR THEORY PROG, FUSION ENERGY CORP, 82- *Concurrent Pos:* Dir opers, Consortium Early Testing Advan Feul Fusion, 81-; vis assoc prof, Univ Buenos Aires, 82- *Mem:* Am Physics Soc. *Res:* Nonlinear dynamics; physics of the Mioma concept for controlled fusion (high ion energy plasma confined in a simple magnetic mirror). *Mailing Add:* 546 S Main St Highstown NJ 08520

CHANOCK, ROBERT MERRITT, b Chicago, Ill, July 8, 24; m 48; c 2. PEDIATRICS, VIROLOGY. *Educ:* Univ Chicago, BS, 45, MD, 47. *Hon Degrees:* DSc, Univ Chicago, 77. *Prof Exp:* Nat Res Coun fel, Children's Hosp, Cincinnati, 50-52; asst prof res pediat, Col Med, Univ Cincinnati, 54-56; asst prof epidemiol, Sch Hyg & Pub Health, Johns Hopkins Univ, 56-57; surgeon, USPHS, 57-59, head, respiratory viruses sect, 59-61, CHIEF, LAB INFECTIOUS DIS, NAT INST ALLERGY & INFECTIOUS DIS, NIH, BETHESDA, 68- *Concurrent Pos:* Nat Res Coun fel, 50-51; Nat Found Infantile Paralysis fel, 51-52; sr res fel, USPHS, 56-57; virologist, Children's Hosp DC, 57-; mem Int Nomenclature Comt Myxoviruses, 7th & 8th Int Microbiol Cong; mem, Armed Forces Epidemiol Bd, Comn Acute Respiratory Dis, 60-62, assoc mem, Comn Influenza, 63-74; dir, Int Ref Ctr Lab Mycoplasms, WHO, 62 & mem Expert Adv Panel Virus Dis, 80-85; mem, Int Comt Nomenclature Bacteria, 66; clin prof, Georgetown Univ, 70-71; mem, Nominating Comt, Nat Acad Sci, 79-80; mem, Sci Review Comt, Scripps Clin & Res Found, 86-89. *Honors & Awards:* E Mead Johnson Award Pediat Res, 64; Squibb Award, Infectious Dis Soc Am, 69 & Joseph E Smadel Award, 80; Gorgas Medal, Asn Military Surgeons, 72; Dyer lectr, NIH, 78; Robert Koch Medal, Fed Repub Ger, 81; Virology Prize, ICN Int, 90. *Mem:* Nat Acad Sci; Soc Pediat Res; Am Soc Microbiol; Am Epidemiol Soc; Am Epidemiol; Am Pediat Soc; Am Soc Clin Invest; Soc Exp Biol & Med; Asn Am Physicians; Foreign mem Royal Danish Acad Sci. *Res:* Respiratory and gastrointestinal diseases; virus microbiology. *Mailing Add:* Lab Infectious Dis Nat Inst Allergy & Infectious Dis Bethesda MD 20892

CHANOWITZ, MICHAEL STEPHEN, b Chicago, Ill, May 12, 43; m; c 1. PHYSICS. *Educ:* Cornell Univ, BA, 63, PhD(physics), 71. *Prof Exp:* SR STAFF SCIENTIST, LAWRENCE BERKELEY LAB, 79- *Concurrent Pos:* Div assoc ed, Phys Rev, 88-90. *Mem:* Fel Am Phys Soc. *Res:* Theoretical elementary particle physics. *Mailing Add:* 50A 3115 Lawrence Berkeley Lab Berkeley CA 94720

CHAN-PALAY, VICTORIA, neurosciences, for more information see previous edition

CHANSON, SAMUEL T, Can citizen. COMPUTER COMMUNICATIONS, DISTRIBUTED COMPUTING SYSTEMS. *Educ:* Hong Kong Univ, BSc, 69; Univ Calif, Berkeley, MSc, 71, PhD(elec eng & comput sci), 74. *Prof Exp:* Asst prof elec eng, Purdue Univ, 74-75; asst prof, 75-82, ASSOC PROF COMPUT SCI, UNIV BC, 82-, DIR, DISTRIB SYSTS RES GROUP, 85- *Concurrent Pos:* Consult to indust & govt agencies. *Mem:* Asn Comput Mach; Comput Soc; Inst Elec & Electronic Engrs. *Res:* Protocol testing; local area networks and protocols; distributed operating systems; computer systems performance evaluation. *Mailing Add:* Dept Comput Sci Univ BC Vancouver BC V6T 1W5 Can

CHANT, DONALD A, b Toronto, Ont, Sept 30, 28; m 75; c 4. ENTOMOLOGY. *Educ:* Univ BC, BA, 50, MA, 52; Univ London, PhD(zool), 56. *Hon Degrees:* LLD, Dalhousie Univ, 76, Trent Univ, 83. *Prof Exp:* Res officer, Res Inst, Can Dept Agr, Belleville Ont, 56-60, dir entom & plant path, Res Lab, Vineland, Ont, 60-64; chmn, dept biol control, Univ Calif, Riverside, 64-67; chmn, Dept Zool, Univ Toronto, 67-75, vpres & provost, 75-80, dir, Ctr Toxicol, 80-82; CHMN & PRES, ONT WASTE MGT CORP, 80- *Concurrent Pos:* Mem, Nat Comt Pesticide Use in Agr, Can, 61-64 & subcomt insect control, Nat Res Coun-Nat Acad Sci; chmn, Can Environ Adv Coun, 78-81; chmn sci adv comt, World Wildlife Found, Can, 76-; chmn, Wildlife Toxicol Fund, Can, 85-; sci adv bd, Int Joint Comn on the Great Lakes, 87-89; Can Nat Task Force on Environ & Econ, 87-89. *Honors & Awards:* White Owl Conserv Award, 71; Order Can, 89. *Mem:* Fel Entom Soc Can; Can Soc Zool (pres, 74-75); fel Royal Entom Soc; fel Royal Soc Can. *Res:* Acarology; taxonomy and ecology of predacious phytoseiid mites; principles of predation; biological control. *Mailing Add:* Dept of Zool Univ of Toronto Toronto ON M5S 1A1 Can

CHANTELL, CHARLES J, b Chicago, Ill, May 19, 31; m 57; c 2. VERTEBRATE ANATOMY. *Educ:* Univ Ill, BS, 61; Univ Notre Dame, MS, 63, PhD(biol), 65. *Prof Exp:* ASSOC PROF BIOL, UNIV DAYTON, 65- *Mem:* Soc Vert Paleont; Am Soc Ichthyologists & Herpetologists; Soc Study Amphibians & Reptiles; Sigma Xi. *Res:* Evolution, phylogeny and osteology of the lower vertebrates. *Mailing Add:* Dept of Biol Univ of Dayton Dayton OH 45469

CHANTRY, WILLIAM AMDOR, b Corning, Iowa, Sept 21, 24; m 49; c 3. ADMINISTRATIVE ENGINEERING, FIBER ENGINEERING. *Educ:* Univ Iowa, BS, 49, MS, 51; Cornell Univ, PhD(chem eng), 53. *Prof Exp:* Technologist, Sherwin Williams Co, Ill, 49-50; technologist, Shell Chem Corp, Calif, 53-56, sr technologist, Shell Chem Corp, Tex, 56-59; from res engr to sr res engr, Dacron Res Lab, E I Du Pont de Nemours & Co, Inc, 59-67, res assoc, 67-71, develop assoc, 71-75, develop fel, 75-83, sr res fel, Textile Fibers Dept, Kinston Plant, 83-90; RETIRED. *Mem:* Am Inst Chem Engrs; Sigma Xi; Fiber Soc; Textile Inst. *Res:* Liquid-liquid extraction; heat exchanger design; polymerization kinetics; spinning and drawing on synthetic fibers; fabric finishing; synthetic fibers product development; fiber spinning; polymerization; heat transfer and fluid flow. *Mailing Add:* 1708 St George Pl Kinston NC 28501

CHAO, B(EI) T(SE), b Kiangsu, China, Dec 18, 18; nat US; m 48; c 2. MECHANICAL ENGINEERING, HEAT TRANSFER. *Educ:* Chiao-Tung Univ, BS, 39; Univ Manchester, PhD(mech eng), 47. *Prof Exp:* Engr-in-chg, Small Tools & Gage Div, Cent Mach Works, Nat Resources Comn, China, 41-44; assoc prof, Univ Ill, 53-55, assoc mem, Ctr Advan Study, 63-64, head thermal sci div, dept mech eng, 70-75, prof, 55-87, head dept mech & indust eng, 75-87, EMER PROF MECH ENG, UNIV ILL, URBANA-CHAMPAIGN, 87- *Concurrent Pos:* Spec consult, Scully-Jones & Co, 52-55; mem reviewing staff, Southwest Res Inst, 53-56; spec consult, Chicago Opers Off, AEC, 62; mem reviewing staff, Zentralblatt für Mathematik, Berlin, 70-82; tech ed, J Heat Transfer, Am Soc Mech Engrs, 75-81; mem Westinghouse Awards Comn, Am Soc Eng Educ, 75-81; consult, Argonne Nat Lab, 76-; mem US eng educ deleg to China, Nat Acad Sci, Am Coun Learned Soc & Soc Sci Res Coun, 78; mem, Adv Screening Comn Eng, Fulbright-Hayes Awards Prog, 79-81, chmn, 80-81; mem, Comn on Recommendation for US Army Basic Sci Res, Nat Res Coun, 80-83. *Honors & Awards:* Blackall Award, Am Soc Mech Engrs, 57, Heat Transfer Award, 71; Western Elec Fund Award, Am Soc Eng Educ, 73, Ralph Coats Roe Award, 75, Benjamin Garver Lamme Award, 84; Max Jakob Award, Am Soc Mech Engrs & Am Inst Chem Engrs, 83. *Mem:* Nat Acad Eng; fel Am Soc Eng Educ; Soc Eng Sci; fel Am Soc Mech Engrs; fel AAAS; Academia Sinica. *Res:* Heat and mass transfer; fluid mechanics; multiphase flows. *Mailing Add:* 704 Brighton Dr Urbana IL 61801

CHAO, CHONG-YUN, b Kunming, China, July 5, 30; m 56; c 2. MATHEMATICS. *Educ:* Univ Iowa, BA, 52, MS, 54; Univ Mich, PhD(math), 61. *Prof Exp:* Instr math, Coe Col, 54-56; mathematician, Res Ctr, Int Bus Mach Corp, 61-63; assoc prof math, 63-66, PROF MATH & STATIST, UNIV PITTSBURGH, 66- *Mem:* Am Math Soc; Math Asn Am. *Res:* Algebra, combinations and graph theory. *Mailing Add:* Dept of Math Univ of Pittsburgh 4200 5th Ave Pittsburgh PA 15260

CHAO, EDWARD CHING-TE, b Soochow, China, Nov 30, 19; nat US; m 42; c 3. GEOLOGY. *Educ:* Nat Southwest Assoc Univ, BS, 41; Univ Chicago, PhD(geol), 48. *Prof Exp:* Jr geologist, Geol Surv Szechuan, China, 41-45; fel petrol & geochem, Univ Chicago, 48-49; GEOLOGIST, US GEOL SURV, WASHINGTON, DC, 49- *Concurrent Pos:* Alexander von Humboldt Found Award, sr US scientist, 75. *Honors & Awards:* Wetherill Medal, Franklin Inst, 65; Exceptional Sci Achievement Medal, NASA, 73; Ries Culture Geol Award, Germany, 83. *Mem:* Geol Soc Am; Mineral Soc Am; Geochem Soc. *Res:* Petrology; geochemistry; mineralogy; impact metamorphism; lunar petrology; coal petrology; geology. *Mailing Add:* 11531 Buttonwood Ct Reston VA 22091

CHAO, FU-CHUAN, b Hong Kong, Feb 8, 19; US citizen; m 47; c 3. BIOCHEMISTRY. *Educ:* Lingnan Univ, BA, 41; Univ Calif, Berkeley, PhD(biochem), 51. *Prof Exp:* Res fel biochem, Univ Calif, Berkeley, 51-52, jr res biochemist, 52-53; res instr biochem, Univ Utah, 53-55; res assoc, Stanford Univ, 55-61, biophys chemist, Stanford Res Inst, 61-74; res chemist, Univ San Francisco & Vet Admin Hosp, Palo Alto, 74-78; biochemist, Smithkline Instruments, Inc, 78-84; RETIRED. *Concurrent Pos:* Am Heart Asn fel, 55-57. *Mem:* Am Inst Biol Sci; Am Chem Soc; fel Am Inst Chemists; NY Acad Sci; Sigma Xi. *Res:* Stability of ribonucleoproteins in solution; purification of viruses by chemical methods; prognosis of cancer; metabolism of marijuana; quantitative extraction of polychlorinated biphenyls from milk and blood; immunochemistry. *Mailing Add:* 1524 Channing Ave Palo Alto CA 94303-2801

CHAO, JIA-ARNG, b Hunan, China, July 8, 41; m 67; c 1. PROBABILITY. *Educ:* Nat Taiwan Normal Univ, BS, 64; Nat Tsing Hua Univ, MS, 66; Wash Univ, PhD(math), 72. *Prof Exp:* Teaching & res asst math, Wash Univ, 66-71, instr, 72; asst prof math, Univ Tex, Austin, 72-79; assoc prof, 79-84, PROF MATH, CLEVELAND STATE UNIV, 84-, CHAIRPERSON, 87- *Mem:* Am Math Soc. *Res:* Harmonic analysis, probability. *Mailing Add:* Dept Math Cleveland State Univ Cleveland OH 44115

CHAO, JING, b Zhejiang, China, Nov 7, 24; m 52; c 1. MOLECULAR SPECTROSCOPIC THERMAL DATA & THERMODYNAMIC PROPERTIES OF INORGANIC & ORGANIC SUBSTANCES. *Educ:* Nat Cent Univ, China, BS, 47; Carnegie-Mellon Univ, MS, 60, PhD(phys chem), 61. *Prof Exp:* Asst chem engr, Hsinchu Res Inst, Chinese Petrol Corp, 47-53; assoc chem engr, Union Indust Res Inst, Ministry Econ Affairs, China, 53-57; phys chemist, Thermal Res Lab, Dow Chem Co, 61-69; sr thermodynamicist, 69-73, asst dir molecular thermodyn, 73-78, RES SCIENTIST, THERMODYN RES CTR, TEX A&M UNIV, 78- *Mem:* Am Chem Soc. *Res:* Thermochemistry; chemical thermodynamics; collection, analysis of chemical data and critical evaluation, correlation and estimation of physical and thermodynamic properties for chemical substances. *Mailing Add:* Thermodyn Res Ctr Tex Eng Exp Sta Tex A&M Univ College Station TX 77840

CHAO, JOWETT, b Chekiang, China, Nov 16, 15; nat US; m 39; c 1. PARASITOLOGY. *Educ:* WChina Union Univ, BS, 39; Nat Cent Univ, China, MS, 42; Cornell Univ, PhD(entom), 49. *Prof Exp:* Instr gen zool, WChina Union Univ, 39-40; entomologist, Ministry Agr & Forestry, China, 42-44; teacher chem, Putney Sch, Vt, 47-48; chemist, Vitaminerals, Inc, Calif, 48-54; from asst res zoologist to assoc res zoologist, Univ Calif, Los Angeles, 54-65, res zoologist, 65-77; RETIRED. *Concurrent Pos:* Nat Found Infantile Paralysis fel, 57; lectr, Immaculate Heart Col, Calif, 58-59; Inter-Am Prog Trop Med fel, La State Univ, 61; mem int panel workshop malaria immunol, Walter Reed Army Inst Res, 63; consult, Ore State Univ, 75; Nat Acad Sci exchange scientist, Czechoslovakia and Poland, 78; lectr, Nat Acad Sci, Peoples Repub China, 79. *Mem:* AAAS; Am Soc Trop Med & Hyg; Soc Protozool; Am Soc Parasitol; Am Mosquito Control Asn; Tissue Cult Asn; Soc Invert Path. *Res:* In vitro culture of plasmodium; internal microorganisms; in vitro culture of insect cells and the insect phase of blood parasites; reptilian hemogregarine life cycle and transmission; invertebrate tissue culture. *Mailing Add:* 228 S Palm Dr Beverly Hills CA 90212

CHAO, JULIE, b Tainan, Taiwan, Repub China, Jan 7, 40; US citizen; c 2. HYBRIDOMA, PROTEIN CHEMISTRY. *Educ:* Nat Taiwan Univ, BS, 62; Utah State Univ, MS, 65; Iowa State Univ, PhD(biochem), 70. *Prof Exp:* Pharmacist, Govt Employees' Clin Ctr, 63-65; teaching & res asst biochem, Utah State Univ, 65-67; teaching & res asst biochem, Iowa State Univ, 67-70; teaching fel molecular biol, Univ Conn, 71-74; res assoc molecular biol, 74-75, asst prof, 77-81, ASSOC PROF PHARMACOL, MED UNIV SC, 81-, PRIN INVESTR, 77- *Concurrent Pos:* NIH Res Career Develop award, 79. *Mem:* Am Chem Soc; Am Soc Pharmacol & Exp Therapeut; Am Soc Biol Chem. *Res:* Study the structure, function and regulation of kallikrein gene family enzymes using both protein chemistry and nucleic acid approaches to understand the role of these enzymes in human diseases. *Mailing Add:* Pharmacol Dept Med Univ SC 171 Ashley Ave Charleston SC 29425

CHAO, KWANG CHU, b Chongqing, China, June 7, 25; m 53; c 3. CHEMICAL ENGINEERING. *Educ:* Zhejiang Univ, BS, 48; Univ Wis, MS, 52, PhD(chem eng), 56. *Prof Exp:* Asst eng, Taiwan Alkali Co, 48-51, chem engr, 52-54; res assoc, Univ Wis, 56-57; res engr, Chevron Res Co, 57-63; assoc prof chem eng, Ill Inst Technol, 63-64 & Okla State Univ, 64-68; prof, 68-89, HARRY C PEFFER DISTINGUISHED PROF CHEM ENG, PURDUE UNIV, 89- *Concurrent Pos:* Indust consult, 64-; dir res grants, NSF, Petrol Res Fund, Am Petrol Inst, Gas Processors Asn, Elec Power Res Inst, Gulf Found & Dept Energy. *Mem:* Fel Am Inst Chem Engrs; Am Chem Soc; Sigma Xi. *Res:* Thermodynamics; molecular theory; equilibrium properties of fluids; theory of solutions. *Mailing Add:* Sch of Chem Eng Purdue Univ West Lafayette IN 47907-1283

CHAO, LEE, b Sept 9, 39; m; c 2. GENE CLONING, GENETIC ENGINEERING. *Educ:* Iowa State Univ, PhD(biochem gene functions),70. *Prof Exp:* PROF BIOCHEM, UNIV SC, 74- *Concurrent Pos:* Consult, Amgen, 86- *Mailing Add:* Univ SC 171 Ashley Ave Charleston SC 29425

CHAO, LI-PEN, neurochemistry, biochemistry, for more information see previous edition

CHAO, MOU SHU, b Changsha, China, Nov 20, 24; nat US; m 68. ELECTROCHEMISTRY, ANALYTICAL CHEMISTRY. *Educ:* Nat Cent Univ, China, BS, 47; Univ Ill, MS, 57, PhD(analytical chem), 61. *Prof Exp:* Asst chem, Nat Cent Univ, China, 47-49; technician, Taiwan Fertilizer Co, 50-53, asst engr, 53-55; chemist, Spec Serv Lab, 60-63, res chemist, Electro & Inorg Res Lab, 63-72, res specialist, 72-79, res leader, Inorg Lab, Dow Chem Co, 80-86; CONSULT ASSOC, OMNITECH INT, LTD, 86- *Mem:* Am Chem Soc; Electrochem Soc; Am Inst Chemists; Sigma Xi; Soc Electroanal Chem; NY Acad Sci. *Res:* Chemical instrumentation; electrode kinetics; electrosynthesis, both organic and inorganic. *Mailing Add:* 1206 Evamar Dr Midland MI 48640

CHAO, RAUL EDWARD, b Havana, Cuba, Dec 21, 39; US citizen; m 64; c 2. CHEMICAL ENGINEERING, MANAGEMENT. *Educ:* Univ PR, Mayaguez, BSChE, 61; Johns Hopkins Univ, PhD(chem eng), 65. *Prof Exp:* Instr chem eng, Johns Hopkins Univ, 61-63, res asst, 63-65; proj engr, Exxon Res & Eng Co, NJ, 65-68; chmn, Dept Chem Eng, Univ PR, Mayaguez, 68-77 & Univ Detroit, 77-82; PRES, SYSTEMA GROUP, INC, 82- *Concurrent Pos:* Consult, NASA-Houston, 73-77, Rockwell Int, Columbus, Ohio, 78-, US Dept Energy, 81- & NSF, 86- *Mem:* Am Inst Chem Engrs; Am Chem Soc. *Res:* Polymer processing; heat transfer; reaction engineering; process synthesis emphasizing renewable resources, alternate feedstocks and economic analyses of processes, as well as planning and forecasting; statistical process control; total quality management; design of experiments and montecarlo techniques. *Mailing Add:* 4107 University Dr Coral Gables FL 33146

CHAO, SHERMAN S, b Nanking, China, Aug 16, 47; m 75. CHEMICAL SPECTROSCOPY, TRACE ANALYSIS. *Educ:* Tamkang Univ, BS, 70; Loyola Univ, MS, 74, PhD(anal chem), 77. *Prof Exp:* Res fel, Univ Mo, Columbia, 76-77; sr chemist, Nalco Chem Co, 77-80; supvr, clin anal, 80-85, MGR, ANAL CHEM, 85-, ASST DIR, ANAL CHEM, INST GAS TECHNOL,85- *Concurrent Pos:* Res assoc, Chem Dept, Loyola Univ, 80- *Mem:* Am Chem Soc; Soc Appl Spectros. *Res:* Method development, analytical trouble shooting and analytical basic reseach in fields of energy research, environmental chemistry, trace element analysis, spectroscopy and separation science; analytical laboratory management. *Mailing Add:* 6510 S Wolf Rd Indian Head Park La Grange IL 60525-4369

CHAO, TAI SIANG, b Yangchow, China, Sept 26, 19; US citizen; m 49; c 3. CHEMICAL ENGINEERING, ORGANIC CHEMISTRY. *Educ:* Nat Cent Univ, Chungking, BS, 41; Purdue Univ, MS, 49, PhD(chem eng), 53. *Prof Exp:* Engr, Chungking Battery Plant, Cent Elec Mfg Works, China, 41-47; res assoc chem, Purdue Univ, 53-55; sr res chemist, Archer Daniels Midland Co, 55-61; res chemist, Sinclair Res, Inc, Div Sinclair Oil Corp, 61-69, sr res chemist, 69-74, res assoc, 74-78, SR RES ASSOC, ATLANTIC RICHFIELD CO, 79- *Concurrent Pos:* Wright Air Develop Ctr fel, Purdue Univ, 53-54; Off Naval Res fel, 54-55. *Mem:* Am Chem Soc; fel Am Inst Chemists. *Res:* Engine oils and synthetic lubricants, gasoline, distillates and heavy fuels; additives for lubricants and fuels, especially antioxidants, flow improvers, friction modifiers, organic phosphorus, nitrogen, fluorine and silicon compounds. *Mailing Add:* 2449 Troy Circle Olympia Fields IL 60461-1918

CHAO, TSUN TIEN, b Anhwei, China, July 23, 18; US citizen; m 47; c 3. GEOCHEMISTRY, ANALYTICAL CHEMISTRY. *Educ:* Nat Cent Univ, China, BS, 42; Ore State Univ, MS, 59, PhD(soil chem), 60. *Prof Exp:* Jr soil chemist, Nat Agr Res Bur, China, 44-45, 46-48; assoc agronomist, Taiwan Sugar Exp Sta, 48-52, soil technologist, 52-56; res fel soil chem, Ore State Univ, 56-59, res assoc, 59-61, asst prof, 61-64; from assoc soil chemist to soil chemist, Pineapple Res Inst, 64-66; res chemist, US Geol Surv, 66-88; RETIRED. *Mem:* Soil Sci Soc Am; Am Soc Agron; Geochem Soc; Am Chem Soc; AAAS; Soc Appl Spectros; Asn Explor Geochemists; Clay Minerals Soc. *Res:* Geochemistry of heavy metals in the weathering zone; research in methods of chemical analysis for geochemical exploration. *Mailing Add:* 770 W Ohio Ave Lakewood CO 80226

CHAO, YUH J (BILL), b Taiwan, May 9, 53; US citizen; m 78; c 2. SOLID MECHANICS, FRACTURE MECHANICS. *Educ:* Nat Chang-Kung Univ, BS, 75; Nat Tsing-Hwa Univ, MS, 77; Univ Ill-Urbana, PhD(theoret & appl mech), 81. *Prof Exp:* Mech engr, Stubbs, Overbeck & Assocs, 81-82; asst prof eng mech, Southern Ill Univ, 82-83; asst prof, 84-89, ASSOC PROF MECH ENG, UNIV SC, 89- *Concurrent Pos:* Prin investr, SC Energy Res & Develop Ctr, 85-88 & NSF, 85- *Mem:* Am Soc Mech Engr; Soc Exp Mech; Am Acad Mech; Soc Piping Engrs & Designers; Soc Eng Sci. *Res:* Analysis and design of pressure vessels and piping; fracture mechanics; computer vision applications in mechanical engineering. *Mailing Add:* Dept Mech Eng Univ SC Columbia SC 29208

CHAO, YU-SHENG, b China, Sept 1, 45; m 73; c 2. BIOCHEMISTRY. *Educ:* Nat Taiwan Univ, BS, 69; Univ Miami, MS, 74, PhD(biochem), 77. *Prof Exp:* Fel, Univ Calif, San Francisco, 77-80; SR RES BIOCHEMIST, MERCK SHARP & DOHME RES LAB, 80- *Mem:* Am Soc Biochem & Molecular Biol; Am Heart Asn. *Res:* Effects of pharmacological agents on lipoprotein metabolism. *Mailing Add:* Dept Molecular Pharm & Biochem Merck Sharp & Dohme Res Lab 80 W 250 PO Box 2000 Rahway NJ 07065

CHAPARAS, SOTIROS D, b Lowell, Mass, May 4, 29; m 56; c 2. MICROBIOLOGY, IMMUNOLOGY. *Educ:* Northeastern Univ, BS, 51; Univ Mass, MS, 53; St Louis Univ, PhD(microbiol), 58; Am Bd Med Microbiol, dipl, 74. *Prof Exp:* Instr microbiol, Sch Med, St Louis Univ, 58-59; instr, Sch Med, Univ Southern Calif, 59-60; sr scientist, 60-68, dir myobact & fungal antigens br, 68- 85, CHIEF MYOBACT & CELLULAR IMMUNOL, CTR BIOLOGIES FOOD & DRUG ADMIN, 85-; ASSOC PROF, NC SCH PUB HEALTH, 84- *Concurrent Pos:* Assoc prof, Howard Univ, 62-; fac chmn microbiol & immunol, NIH, 62-87; chmn sci assembly on microbiol & immunol, Am Thoracic Soc, 74-75; panel mem, US-Japan Tuberc Panel, US-Japan Med Coop Prog, 75-80; chmn adv comt, Trudeau Inst, 78-81; lectr, Med Sch, Univ Md, 80-; WHO consult tuberculosis, 81- *Honors & Awards:* Commendation Medal, USPHS, 72, 80. *Res:* Pathogenicity, immunity and hypersensitivity in tuberculosis; standardization of skin test reagents; immunity in cancer; immunology of fungi; genetic engineering. *Mailing Add:* Bur of Biologics Myobact & Fungal Br Food & Drug Admin 8800 Rockville Pike Bethesda MD 20892

CHAPAS, RICHARD BERNARD, b Cleveland, Ohio, Dec 5, 45; m 68; c 3. PHOTOGRAPHIC CHEMISTRY. *Educ:* St Vincent Col, BS, 68; Univ Ill, PhD(chem), 72. *Prof Exp:* sr res chemist, Eastman Kodak Co, 72-78; asst prof chem, Gordon Jr Col, Barnesville, Ga, 78-80; PROD DEVELOP MGR, JOHNSON & JOHNSON, 80- *Concurrent Pos:* Mem adj fac, Rochester Inst Technol, 73-78, Mercer County Community Col, 80- *Honors & Awards:* J & J Entrepreneurial Award. *Mem:* Am Chem Soc; Soc Photog Scientists & Engrs. *Res:* Photographic systems development and analysis; light sensitive materials production; medical devices product development; polymer development and processing; nonwovens development. *Mailing Add:* Two Standfort Ct East Windsor NJ 08520

CHAPATWALA, KIRIT D, b Surat, India; US citizen; m 77; c 2. HEAVY METAL TOXICITY. *Educ:* Gujarat Univ, India, BSc, 70; Miss State Univ, MS, 73, PhD(microbiol & biochem), 78. *Prof Exp:* Tech asst microbiol, Miss State Univ, 71-76, res asst, 72, 73 & 74, res fel, 78; asst prof, 78-80, ASSOC PROF BIOL & MICROBIOL, 80-, CHMN, DIV NATURAL SCI, SELMA UNIV, 87- *Mem:* Am Soc Microbiol; Sigma Xi; Soc Indust Microbiol. *Res:* Effect of cadmium on renal and hepatic gluconeogenic and Adenosine Triphosphate enzymes; study the chemical composition of phage D-11 receptor sites on Mycobacterium phlei; biodegradation of organic compounds. *Mailing Add:* 236 Ramsay Dr Selma AL 36701

CHAPEL, JAMES L, child psychiatry; deceased, see previous edition for last biography

CHAPEL, RAYMOND EUGENE, b Marlow, Okla, Jan 31, 21; m 43; c 3. MECHANICAL ENGINEERING. *Educ:* Okla Agr & Mech Col, BS, 42, MS, 51. *Prof Exp:* From asst prof to assoc prof mech eng, Okla State Univ, 51-68, asst dir eng res, Off Eng Res, 65-67, prof mech & aerospace eng, 68-83, assoc dir eng res, Off Eng Res, 67-83, assoc dean eng, 81-83; RETIRED. *Mem:* Assoc fel Am Inst Aeronaut & Astronaut; Am Soc Eng Educ; Nat Soc Prof Engrs; Sigma Xi. *Res:* Machine analysis and design; aerospace structures. *Mailing Add:* 1024 Graham Stillwater OK 74075

CHAPERON, EDWARD ALFRED, b Burlington, Vt, Sept 24, 30; m 62; c 2. IMMUNOLOGY, MICROBIOLOGY. *Educ:* LeMoyne Col, BS, 57; Marquette Univ, MS, 59; Univ Wis-Madison, PhD(zool), 65. *Prof Exp:* NIH fel immunol, Med Ctr, Univ Colo, Denver, 65-68; asst prof, 68-71, ASSOC PROF MICROBIOL, SCH MED, CREIGHTON UNIV, 71- *Mem:* Am Asn Immunol; Am Soc Microbiol; Reticuloendothelial Soc; Sigma Xi. *Res:* Cellular immunology. *Mailing Add:* 4925 Miami St Omaha NE 68104

CHAPIN, CHARLES EDWARD, b Porterville, Calif, Oct 25, 32; m 58; c 3. STRUCTURAL GEOLOGY, MINERALOGY-PETROLOGY. *Educ:* Colo Sch Mines, Geol Engr, 54, DSc(geochem), 65. *Prof Exp:* Asst prof geol, Univ Tulsa, 64-65; asst prof, NMex Inst Mining & Technol, 65-68, assoc prof & head geosci dept, 68-70; geologist, 70-76, sr geologist, 76-91, DIR, NMEX BUR MINES, 91- *Concurrent Pos:* Adj prof geosci, NMex Inst Mining & Technol, 70-; distinguished lectr, Am Asn Petrol Geol, 85-86. *Honors & Awards:* Van Diest Gold Medal, Colo Sch Mines, 80; Distinguished Res Award, NMex Inst Mining & Technol, 88. *Mem:* Geol Soc Am; Soc Econ Paleontologists & Mineralogists; Am Geophys Union; Sigma Xi; Am Asn Petrol Geol. *Res:* Volcanology; mineral deposits; tectonics. *Mailing Add:* NMex Tech Campus Sta Socorro NM 87801

CHAPIN, DAVID LAMBERT, b Detroit, Mich, May 24, 48; US citizen. NUCLEAR ENGINEERING. *Educ:* Univ Mich, BSE, 70, MSE, 71, PhD(nuclear eng), 74. *Prof Exp:* Researcher fusion reactors, Princeton Plasma Physics Lab, NJ, 74-76; RESEARCHER FUSION REACTORS, WESTINGHOUSE ELEC CORP, 76- *Mem:* Am Nuclear Soc; Am Phys Soc; Sigma Xi. *Res:* Controlled fusion reactor design, especially related to computational nuclear engineering and systems studies. *Mailing Add:* Longue Vue Dr Mt Lebanon PA 15228

CHAPIN, DOUGLAS MCCALL, b Atlanta, Ga, Oct 29, 40; m 62; c 4. NUCLEAR ENGINEERING, ELECTRICAL ENGINEERING. *Educ:* Duke Univ, BS, 62; George Wash Univ, MS, 66; Princeton Univ, PhD(nuclear chem eng), 68. *Prof Exp:* Engr, Naval Reactors Hq, US Atomic Energy Comn, 62-66, MPR Assoc Inc, 68-85; PRIN OFFICER & DIR, MPR ASSOCS INC, 85- *Concurrent Pos:* Prin investr, res prog multidimensional effects during refill & reflood, US Nuclear Regulatory Comn, 74- *Honors & Awards:* Gold Medal, Armed Forces Commun & Electronic Asn & Soc Am Mil Engrs, 62. *Mem:* Am Nuclear Soc; Sigma Xi. *Res:* Power operation; reactor safety, thermal hydraulics, natural circulation, two-phase flow and instrumentation. *Mailing Add:* MPR Assoc Inc 1050 Connecticut Ave NW Washington DC 20036

CHAPIN, DOUGLAS SCOTT, b Muskegon, Mich, July 14, 22; m 44; c 4. PHYSICAL CHEMISTRY. *Educ:* Kans State Col, BS, 44; Ill Inst Technol, MS, 48; Ohio State Univ, PhD(chem), 54. *Prof Exp:* Researcher, Nat Bur Standards, 47; asst, Cryogenic Lab, Ohio State Univ, 48-51; sr cryogenic oper, Herrick L Johnston, Inc, 52; asst, Cryogenic Lab, Ohio State Univ, 53-54; cryogenic engr, Herrick L Johnston, Inc, 54; from asst prof to assoc prof chem, Univ Ariz, 54-66; assoc prog dir grad fels & traineeships, 66-68; head fels & traineeships sect, NSF, 73-77, dir fac oriented progs, 77-78, staff assoc, Sci Personnel Improv Div, 78-82, prog dir grad fel, 82-90, SR SCI ASSOC, NSF, 91-; head fels & traineeships sect, 73-77, dir fac oriented progs, 77-78, staff assoc, Sci Personnel Improv Div, 78-82, PROG DIR GRAD FEL, NSF, 82- *Concurrent Pos:* Staff mem, Lincoln Lab, Mass Inst Technol, 63-64. *Mem:* AAAS; Faraday Soc; Am Chem Soc. *Res:* Low temperature kinetics; ortho-parahydrogen catalysis; separation of ortho-parahydrogen; low temperature thermodynamics of gases absorbed on solids; solid state chemistry of transition metal oxides. *Mailing Add:* Div Sci Personnel Improv NSF Washington DC 20550

CHAPIN, EARL CONE, b Farmington, Ill, Feb 5, 19; m 45; c 3. ORGANIC CHEMISTRY. *Educ:* Univ Ill, BS, 41; Pa State Univ, MS, 42, PhD(org chem), 44. *Prof Exp:* Res chemist, Monsanto Chem Co, 44-50, res group leader, 51-62; chmn dept phys sci, 62-68, dean sch arts & sci, 68-74, PROF CHEM, WESTERN NEW ENGLAND COL, 62- *Mem:* Am Chem Soc. *Res:* Synthetic polymers; organic synthesis. *Mailing Add:* Three Captain Rd Wilbraham MA 01095-1527

CHAPIN, EDWARD WILLIAM, JR, b Baltimore, Md, May 28, 43; c 3. MATHEMATICAL LOGIC, EXPERT MEDICAL SYSTEMS. *Educ:* Trinity Col, Conn, BS, 65; Princeton Univ, MA, 67, PhD(math), 69. *Prof Exp:* asst prof math, Univ Notre Dame, 69-77; CHMN MATH & COMPUT SCI, UNIV MD, EASTERN SHORE, 77- *Mem:* AAAS; Am Math Soc; Math Asn Am; Asn Symbolic Logic; Soc Indust & Appl Math; Asn Comput Mach. *Res:* Algebraic structure of deductive systems; computer-assisted medical diagnosis; stochastic models of set theory; modal logic. *Mailing Add:* Rte 1 Box 81-B Firetower Rd Hebron MD 21830

CHAPIN, F STUART, III, b Portland, Ore, Feb 2, 44; m 66; c 2. PLANT ECOLOGY. *Educ:* Swarthmore Col, BA, 66; Stanford Univ, PhD(biol), 73. *Prof Exp:* Vis instr biol, Univ Javeriana, Colombia, 66-68; instr biol, Stanford Univ, 69-73; from asst prof to prof plant physiol ecol, Univ Alaska, 84-89; PROF ECOL, UNIV CALIF, BERKELEY, 89- *Concurrent Pos:* Guggenheim fel, 79-80. *Mem:* Am Ecol Soc. *Res:* Plant nutritional ecology; adaptations of plants to multiple environmental stresses; plant responses to grazing; nutrient cycling; ecosystem ecology; ecosystem response to global change. *Mailing Add:* Dept Integrative Biol Univ Calif Berkeley CA 94720

CHAPIN, HENRY J(ACOB), b Scranton, Pa, May 26, 08; m 42; c 2. METALLURGY. *Educ:* Haverford Col, BS, 29; Mass Inst Technol, BS, 32. *Prof Exp:* Metallurgist, Am Brake Co, 42-58, sr metallurgist, 58-65; sr metallurgist, Abex Inc, 65-73; RETIRED. *Concurrent Pos:* With Off Sci Res & Develop, 44; consult, 73- *Mem:* Am Soc Metals. *Res:* Ferrous metallurgy; manganese steel; abrasion resistant steels and irons. *Mailing Add:* 61 Oweno Rd Mahwah NJ 07430

CHAPIN, JOAN BEGGS, b St Louis, Mo, Feb 8, 29; div; c 3. SYSTEMATICS. *Educ:* Kans State Univ, BS, 50; La State Univ, MS, 59, PhD(entom), 71. *Prof Exp:* From res asst to assoc prof, 57-82, PROF, DEPT ENTOM, LA STATE UNIV, 82- *Concurrent Pos:* Mem, Standing Comt on Common Names of Insects, Entom Soc Am, 81-87 & chmn, 83-87; local arrangements chmn, ann meeting, Entom Collections Network, 90. *Mem:* Entom Soc Am; Sigma Xi; Coleopterists Soc. *Res:* Systematic studies of Coccinellidae of the Western Hemisphere; taxonomic surveys of Louisiana's insect fauna; preservation, management and improvement of the Louisiana State Univ Insect Collection. *Mailing Add:* Dept Entom La State Univ Baton Rouge LA 70803-1710

CHAPIN, JOHN LADNER, b New York, NY, Sept 1, 16; m 50; c 3. RESPIRATORY PHYSIOLOGY. *Educ:* Denison Univ, AB, 39; Univ Rochester, PhD(physiol), 50. *Prof Exp:* Instr physiol, Sch Med, La State Univ, 50; from instr to asst prof, Sch Med, Univ Colo, 51-65; PROF PHYSIOL, INST CHILD STUDY, UNIV MD, COLLEGE PARK, 65- *Mem:* AAAS; Am Ornithologists Union; Am Physiol Soc; NY Acad Sci; Am Inst Biol Sci. *Res:* Respiratory and environmental physiology; man and his environment; physical aspects of human development. *Mailing Add:* 2400 PArker Ave Wheaton MD 20902

CHAPLER, CHRISTOPHER KEITH, b Des Moines, Iowa, July 19, 40; m 61; c 3. MEDICAL PHYSIOLOGY. *Educ:* Drake Univ, BA, 62, MA, 64; Univ Fla, PhD(physiol), 67. *Prof Exp:* Can Heart Found fel, 67; from asst prof to assoc prof, 68-79, actg head, Dept Physiol, 88-89, PROF PHYSIOL, QUEEN'S UNIV, ONT, 79-, EXEC ASST TO PRIN, 89- *Concurrent Pos:* mem, Spec Study Sect, NIH, 84, Coun Ont Univs, 82-84; consult, cardiovasc panel, Med Res Coun Can, 84-87, Health Effects Inst, 85. *Mem:* Am Physiol Soc; Can Physiol Soc; Can Fedn Biol Sci; Fedn Am Soc Exp Biol. *Res:* Muscle metabolism; cardiovascular physiology; oxygen transport and hypoxia. *Mailing Add:* Dept Physiol Queen's Univ Kingston ON K7L 3N6 Can

CHAPLIN, FRANK S(PRAGUE), b Nova Scotia, May 14, 10; nat US; m 39; c 4. MECHANICAL ENGINEERING. *Educ:* Mass Inst Technol, BS, 32. *Prof Exp:* Stress analyst rail car eng, E G Budd Mfg Co, Pa, 34-39, chief of stress, 39-42 & aviation div, 42-43; prin engr, Artil Div, Off Chief Ord, 43-45; sect chief friction & ballistics res, Franklin Inst Labs, 45-48, assoc dir mech eng, 49-58; consult mech develop, 58-85; RETIRED. *Mem:* Am Soc Mech Engrs; Soc Exp Stress Analysis. *Res:* Friction and lubrication; structures; machine development; planetariums; towers. *Mailing Add:* Rte 2 Box 290A Blounts Creek NC 27814-9610

CHAPLIN, HUGH, JR, b New York, NY, Feb 4, 23; m 45; c 4. MEDICINE. *Educ:* Princeton Univ, AB, 43; Columbia Univ, MD, 47. *Prof Exp:* From intern to resident, Mass Gen Hosp, 47-50; physician in chg, Clin Ctr Blood Bank, NIH, 53-55; from instr to asst prof, 55-62, dir student health serv, 56-57, assoc dean, Med Sch, 57-63, assoc prof & dir, Johnson Insts Rehab, 64-65, PROF MED & PREV MED & KOUNTZ PROF PREV MED, MED SCH, WASH UNIV, 65-; PROF MED & PATH & DIR, BARMAN HOSP BLOOD BANK, 83- *Concurrent Pos:* Fel, Brit Postgrad Med Sch, London, 51-53; Nat Inst Arthritis & Metab Dis trainee, Med Sch, Wash Univ, 55-56; Commonwealth Fund res fel, Wright-Fleming Inst, London, 62-63, Josiah Macy Scholar, 75-76; mem, Am Bd Internal Med, 56. *Mem:* Am Soc Hematol; Am Soc Clin Invest; Am Fedn Clin Res; Asn Am Physicians; Royal Soc Med. *Res:* Hematology; secondary hemolytic anemia. *Mailing Add:* Dept Med & Path Washington Univ Sch Med 660 S Euclid Ave St Louis MO 63110

CHAPLIN, JAMES FERRIS, b Lamar, SC, Nov 10, 20; m 45; c 2. GENETICS. *Educ:* Clemson Col, BS, 48; NC State Col, MS, 52, PhD(plant breeding, genetics), 59. *Prof Exp:* Asst agronomist, Clemson Col, 48-57; res agronomist, Pee Dee Exp Sta, USDA, 57-65, sr scientist, Oxford Tobacco Res Lab, 65-67, leader tobacco breeding & dis invests, Plant Sci Res Div, 67-72, dir, Oxford Tobacco Res Lab, 72-86; prof, 65-86, AGR CONSULT, 86-, EMER PROF CROP SCI, NC STATE UNIV, 86- *Mem:* Fel Am Soc Agron; Am Genetics Asn; fel Crop Sci Soc; Sigma Xi. *Res:* Plant breeding and genetics for disease resistance; chemical constituents and agronomic characteristics; agronomic research. *Mailing Add:* USDA Crops Res Lab Agr Res Serv Oxford NC 27565

CHAPLIN, MICHAEL H, b Olney, Ill, Aug 22, 43; m 65; c 3. PLANT PHYSIOLOGY, NUTRITION. *Educ:* Univ Ky, BS, 65; Rutgers Univ, MS, 66; Mich State Univ, PhD(hort), 68. *Prof Exp:* Prof plant nutrit, Pa State Univ, 85-91; PROF PLANT NUTRIT, IOWA STATE UNIV, 91- *Mem:* Am Soc Hort Sci. *Res:* Nutritional status of economic crops as it affects quality, yield and maturation processes. *Mailing Add:* 106 Hort Dept Iowa State Univ Ames IA 50011

CHAPLIN, NORMAN JOHN, b Hamilton, Ont, Oct 10, 18; US citizen; m 42; c 2. SOLID STATE PHYSICS, ELECTRONICS. *Educ:* Univ Toronto, BASc, 49, MASc, 50. *Prof Exp:* Mem tech staff, Can Radio Mfg Corp, 52-56; mem tech staff, Bell Labs, 56-80; RETIRED. *Mem:* Inst Elec & Electronics Engrs. *Res:* Solidification of alloys; semiconductor devices; reliability physics; integrated circuits; spectrochemical analysis; computers and artificial intelligence. *Mailing Add:* 3155 South Dr Allentown PA 18103

CHAPLIN, ROBERT LEE, JR, b Savannah, Ga, Mar 22, 23; m 56; c 1. PHYSICS. *Educ:* Clemson Univ, BS, 48; NC State Col, MS, 53, PhD(physics), 62. *Prof Exp:* Instr physics, NC State Col, 56-60; res asst, Univ NC, 60-61, fel, 61-62; from asst prof to assoc prof, 62-74, PROF PHYSICS,

CLEMSON UNIV, 74- *Concurrent Pos:* Res partic, Oak Ridge Nat Lab, 63, 74 & 76. *Mem:* Sigma Xi; Am Phys Soc. *Res:* Defect state of metal crystals as produced by electron irradiation damage and removed by thermal annealing. *Mailing Add:* Dept Physics Clemson Univ Clemson SC 29631

CHAPLIN, SUSAN BUDD, physiology, for more information see previous edition

CHAPLINE, GEORGE FREDERICK, JR, b Teaneck, NJ, May 6, 42; m 68; c 1. X-RAY LASERS. *Educ:* Univ Calif, Los Angeles, BA, 61; Calif Inst Technol, PhD(physics), 67. *Prof Exp:* Asst prof physics, Univ Calif, Santa Cruz, 67-69; PHYSICIST, LAWRENCE LIVERMORE LAB, 69- *Honors & Awards:* E O Lawrence Award, US Dept Energy, 83. *Res:* X-ray lasers; strategic defense; relativistic nuclear physics; higher dimensional theories of elementary particles. *Mailing Add:* Dept Physics Lawrence Livermore Lab Livermore CA 94550

CHAPMAN, ALAN J(ESSE), b Los Angeles, Calif, June 22, 25; m 50; c 2. MECHANICAL ENGINEERING. *Educ:* Rice Inst, BS, 45; Univ Colo, MS, 49; Univ Ill, PhD(mech eng), 53. *Prof Exp:* From instr to assoc prof mech eng, 46-58, chmn dept, 55-63 & 64-69, vpres admin, 69-70, dean eng, 75-80, PROF MECH ENG, RICE UNIV, 58- *Concurrent Pos:* Consult, Anderson-Greenwood & Co, 54-, Manned Spacecraft Ctr, NASA, Houston, 63- & Shell Develop Co, 63- *Mem:* Fel Am Soc Mech Engrs; assoc fel Am Inst Aeronaut & Astronaut; Am Soc Eng Educ. *Res:* Heat transfer; fluid dynamics. *Mailing Add:* Dept of Mech Eng Rice Univ PO Box 1892 Houston TX 77251

CHAPMAN, ALAN T, b Columbus, Ohio, Nov 20, 29; m 53; c 3. CERAMIC ENGINEERING. *Educ:* Rutgers Univ, BS, 51; Ohio State Univ, MS, 57, PhD(ceramic eng), 60. *Prof Exp:* Res assoc ceramics, Ohio State Univ, 55-60; ceramic engr, Oak Ridge Nat Lab, Union Carbide Nuclear Co, 60-65; assoc prof ceramic eng, 65-76, HOOD PROF CERAMIC ENG, GA INST TECHNOL, 76- *Concurrent Pos:* Traveling lectr, Oak Ridge Inst Nuclear Studies, 63-64. *Honors & Awards:* B Mifflin Hood Chair of Ceramics. *Mem:* Am Ceramic Soc; Sigma Xi. *Res:* Phase relations in ceramic materials for nuclear applications and single crystal growth of nonmetallic materials. *Mailing Add:* Sch of Ceramic Eng Ga Inst of Technol Atlanta GA 30332

CHAPMAN, ALBERT LEE, b Anderson, Mo, Nov 5, 33; m 56; c 4. ANATOMY. *Educ:* Univ Mo, AB, 56, MS, 59; Univ Nebr, PhD(anat), 62. *Prof Exp:* From instr to assoc prof, 62-74, PROF ANAT, UNIV KANS MED CTR, 74-, DEAN, GRAD STUDIES & RES, 85- *Concurrent Pos:* Spec fel, Viral Lymphoma & Leukemia Br, Nat Cancer Inst, 69- *Mem:* Am Asn Anat; AAAS; NY Acad Sci; Electron Micros Soc Am; Sigma Xi. *Res:* Electron microscopic studies of lymphatic tissues, especially leukemic tissue and virus. *Mailing Add:* Off Grad Studies & Res Univ Kans Med Ctr Kansas City KS 66103

CHAPMAN, ARTHUR BARCLAY, b Windermere, Eng, Oct 25, 08; nat US; m 34; c 3. GENETICS, ANIMAL BREEDING. *Educ:* State Col Wash, BS, 30; Iowa State Col, MS, 31; Univ Wis, PhD(genetics), 35. *Prof Exp:* Asst genetics, Univ Wis, 31-36; Nat Res fel agr, Iowa State Col & Chicago, 36-37; from instr to prof genetics, Univ Wis-Madison, 37-72, prof genetics, meat & animal sci & dairy sci, 72-75, EMER PROF GENETICS & ANIMAL BREEDING, UNIV WIS-MADISON, 75- *Concurrent Pos:* Rockefeller Found res & teaching award, Poland, 60; Fulbright & Guggenheim Mem Found fels, NZ, 66-67 & Midwest Univ Consortium Int Activities-AID, Indonesia, 73; vis prof, Cairo Univ, Egypt, 77. *Honors & Awards:* Animal Breeding & Genetics Award, Am Soc Animal Sci, 68, Morrison Award, 74. *Mem:* Fel AAAS; fel Am Soc Animal Sci (vpres, 63-64, pres, 64-65). *Res:* Animal breeding; genetic effects of irradiation. *Mailing Add:* 1117 Risser Rd Madison WI 53705

CHAPMAN, ARTHUR OWEN, b Heber, Utah, Mar 16, 13; m 41; c 2. HISTOLOGY, PATHOLOGY. *Educ:* Brigham Young Univ, AB, 41; Univ Kans, MA, 49; Univ Nebr, PhD(med sci), 53. *Prof Exp:* Asst instr zool, Univ Kans, 46-47, asst instr anat, 47-49; instr, Col Med, Univ Nebr, 49-53, asst prof, 53-59; assoc prof, 59-67, prof, 67-80, EMER PROF ZOOL, BRIGHAM YOUNG UNIV, 80- *Mem:* Am Asn Anatomists; Sigma Xi. *Res:* Nervous system; irradiation effects on the brain and on embryos; thiamin deficiency and mercury effects on the brain. *Mailing Add:* Brigham Young Univ MLB Mus Provo UT 84602

CHAPMAN, BRIAN RICHARD, b Corpus Christi, Tex, Jan 19, 46; m 67; c 2. ORNITHOLOGY, MAMMALOGY. *Educ:* Tex A&I Univ, BS, 67; Tex Tech Univ, MS, 70, PhD(zool), 73. *Prof Exp:* From asst prof to prof ecol, Corpus Christi State Univ, 82- 90; HERITAGE BIOLOGIST, OKLA BIOL SURV, 90- *Concurrent Pos:* pres, Southwestern Asn Naturalists, 87-88. *Mem:* Wilson Ornith Soc; Am Ornith Union; Cooper Ornith Soc; Am Soc Mammalogists; Sigma Xi. *Res:* Natural history, ecology and distribution of birds and mammals; effects of ectoparasites on the growth and development of juvenile organisms; endangered species research. *Mailing Add:* Okla Natural Heritage Inventory Okla Biol Surv 2001 Priestly Ave Bldg 605 Norman OK 73019-0543

CHAPMAN, CARL JOSEPH, b New York, NY, July 4, 39; m 65. SYSTEMATIC BOTANY. *Educ:* Univ NH, BS, 61, PhD(bot), 65. *Prof Exp:* Fel, Univ NH, 65; instr bot, Univ RI, 65-66; res assoc, Univ Fla, 66-67; from asst prof to assoc prof biol, 67-77, PROF BIOL, CONCORD COL, 77- *Mem:* Bot Soc Am; Sigma Xi. *Res:* Floristic studies; scanning electron microscopy studies of fern spores. *Mailing Add:* Dept of Biol Concord Col Athens WV 24712

CHAPMAN, CARLETON ABRAMSON, b Groveton, NH, Oct 14, 11; m 40; c 1. GEOLOGY. *Educ:* Univ NH, BS, 33; Harvard Univ, AM, 35, PhD(petrog), 37. *Prof Exp:* Asst petrog, Harvard Univ, 34-35, instr, 35-37; instr geol, Univ Ill, Urbana-Champaign, 37-39, assoc, 39-42, from asst prof to prof, 45-48; RETIRED. *Mem:* AAAS; fel Geol Soc Am; fel Mineral Soc Am; Am Geochem Soc; fel Geol Soc London; fel Norweg Geol Soc. *Res:* Petrology; igneous and metamorphic geology; mineralogy; structural geology. *Mailing Add:* 1502 S Race Urbana IL 61801

CHAPMAN, CARLETON BURKE, b Sycamore, Ala, June 11, 15; m 40; c 3. MEDICINE. *Educ:* Davidson Col, AB, 36; Oxford Univ, BA, 38; Harvard Univ, MD, 41, MPH, 44; Am Bd Internal Med, dipl, 48; Am Bd Cardiovasc Dis, 53. *Hon Degrees:* MA, Dartmouth Col, 68; LLD, Davidson Col, 68; DrMeds, Brown Univ, 72. *Prof Exp:* From intern to resident med, Boston City Hosp, Mass, 41-44; from instr to asst prof med, Med Sch, Univ Minn, 47-53; prof, Southwestern Med Sch, Univ Tex, 53-66; dean, Dartmouth Med Sch, 66-73, vpres, 72-73; pres, Commonwealth Fund, New York, 73-80; prof, 80-85, EMER PROF HIST MED, ALBERT EINSTEIN COL MED, BRONX, NY, 85- *Concurrent Pos:* Rockefeller Found fel, Sch Pub Health, Harvard Univ, 44; asst, Harvard Med Sch, 43; consult, Surgeon-Gen, US Army, 44 & US Vet Admin, 47-; fel, Coun Clin Cardiol, Am Heart Asn; vis prof hist med, Johns Hopkins Sch Med, 89- *Mem:* Inst Med-Nat Acad Sci; Am Acad Arts & Sci; Am Soc Clin Invest; fel Am Col Cardiol; Am Heart Asn (pres, 64-65). *Res:* Cardiovascular disease; morals and ethics of bioscience and medicine. *Mailing Add:* Two Allen Lane Hanover NH 03755

CHAPMAN, CLARK RUSSELL, b Palo Alto, Calif, May 13, 45; div; c 1. PLANETARY SCIENCES. *Educ:* Harvard Col, AB, 67; Mass Inst Technol, MS, 68, PhD(planetary sci), 72. *Prof Exp:* Res scientist astrosci, Res Inst, Ill Inst Technol, 71-72; RES SCIENTIST, PLANETARY SCI INST, SCI APPLN, INT CORP, 72- *Concurrent Pos:* Mem, Lunar Sci Inst Coun, Univ Space Res Asn, 75-78; mem sci adv group inner solar syst, NASA, 75-77; mem, Lunar Sci Rev Panel, 75-77, 80-81; mem organizing comt Comn 15, Int Astron Union, 76-, vpres, 79-82, pres, 82-85, mem, Comns 16 & 17, 77-; mem, Comt Planetary Explor, Nat Acad Sci, 77-80, Imaging Team, Galileo Mission, 77-; assoc ed, J Geophys Res, 76-78; columnist, Planetary Report, 80-, mem, Infrared Telescope Opers Working Group, 81-86, solar syst explor mgt coun, 84-88, comet rendezvous asteroid fly by sci working group, 84-85; chmn, Opers Working Group, Planetary Astron Prog, NASA, 85- 88, mem, NASA Planetary Astron Comt, 86-88; mem, Mass Inst Technol Corp Vis Comt, Earth Sci Dept, 86-; chmn organizing comt, Int Conf Near-Earth Asteroids, 91; ed, J Geophys Res-Planets, 91- *Mem:* AAAS; Am Astron Soc (chmn Div Planetary Sci, 82-83); Am Geophys Union; Meteoritical Soc. *Res:* Spectrophotometry of planets, satellites, asteroids, comets and compositional interpretation; analysis of telescopic and spacecraft imagery of planetary atmospheres and surfaces; cratering and impact-erosional processes; planetary accretion; Jupiter's atmospheric circulation; comparative planetology; popularization of planetary science. *Mailing Add:* 6160 N Montebella Tucson AZ 85704

CHAPMAN, DAVID J, b Kingston, WI, Dec 12, 39; m 63; c 1. PLANT BIOCHEMISTRY, PHYCOLOGY. *Educ:* Univ Auckland, BSc, 60; Univ Calif, PhD(marine biol), 65. *Hon Degrees:* DSc, Univ Auckland, 79. *Prof Exp:* Res assoc marine biol, Scripps Inst, Calif, 65-66; res assoc biol, Brookhaven Nat Lab, 66-67; asst prof, Univ Chicago, 68-73; assoc prof, 73-78, PROF BIOL, UNIV CALIF, LOS ANGELES, 78- *Concurrent Pos:* Ger Acad Exchange award, Univ Saarlandes, 70; distnguished vis scientist, Nat Res Coun Can, 79-80. *Mem:* Am Chem Soc; Linnean Soc London; Am Soc Plant Physiol; Bot Soc Am; Phycol Soc Am; Sigma Xi. *Res:* Phycology, algal, chloroplast and natural product biochemistry; chemical taxonomy, phylogeny. *Mailing Add:* Dept of Biol Univ of Calif Los Angeles CA 90024

CHAPMAN, DAVID MACLEAN, b Thunder Bay, Ont, Jan 3, 35. ANATOMY, ZOOLOGY. *Educ:* Univ Man, BSc, 59, MSc, 61; Univ Cambridge, PhD(zool), 64. *Prof Exp:* PROF ANAT, MED COL, DALHOUSIE UNIV, 64- *Mem:* Can Soc Zool; Marine Biol Asn UK; Can Asn Anat. *Res:* Electron microscopy of cnidarians; dermatology. *Mailing Add:* Dept Anat Dalhousie Univ Halifax NS B3H 4H7 Can

CHAPMAN, DEAN R(ODEN), b Ft Sumner, NMex, Mar 8, 22; m 44; c 2. AERONAUTICAL ENGINEERING. *Educ:* Calif Inst Technol, MS, 44, PhD(aeronaut eng), 48. *Prof Exp:* Staff scientist, NASA, 44-46, 48-69, chief, Thermo-&-Gas Dynamics Div, 69-73, dir, astronautics, Ames Res Ctr, 73-80; PROF AERONAUT & ASTRONAUT, & MECH ENG, STANFORD UNIV, 80- *Honors & Awards:* Sperry Award, Am Inst Aeronaut & Astronaut, 52; Rockefeller Pub Serv Award, 59; Exceptional Sci Achievement Award, NASA, 64; Dryden Lectr, 79. *Mem:* Nat Acad Eng; Am Inst Aeronaut & Astronaut. *Res:* Properties and origins of tektites; supersonic aerodynamics; computational and experimental fluid dynamics; base pressure at supersonic velocities. *Mailing Add:* Dept Aeronaut & Astronaut Stanford Univ Stanford CA 94305

CHAPMAN, DEREK D, b Lincoln, Eng, Feb 13, 32; m 58; c 2. ORGANIC CHEMISTRY. *Educ:* Univ Nottingham, BSc, 53, PhD(org chem), 56. *Prof Exp:* Lectr chem, SE Essex Tech Col, Eng, 60-61; sr res chemist, 62-66, RES ASSOC, EASTMAN KODAK CO, 66- *Concurrent Pos:* Fel, Univ Rochester, 56-59, Hull Univ, 59-60 & Univ Rochester, 61-62. *Mem:* Am Chem Soc; Soc Chem Indust; Brit Chem Soc. *Res:* Synthesis of alkaloids; structure determination of antibiotics; synthesis of photographically useful compounds. *Mailing Add:* Seven Andony Lane Rochester NY 14624-4309

CHAPMAN, DOUGLAS GEORGE, b Provost, Alta, Mar 20, 20; nat US; m 43; c 3. BIOMETRICS. *Educ:* Univ Sask, BA, 39; Univ Calif, MA, 40, PhD(math), 49; Univ Toronto, MA, 44. *Prof Exp:* Meteorologist, Meteorol Serv Can, 41-46; asst prof math, Univ BC, 46-48; asst math statist, Univ Calif, 48-49; from asst prof to prof math, Univ Wash, 49-68, dir, Ctr Quantitative Sci, 68-71 & 80-83, dean, Col Fisheries, 71-80; RETIRED. *Concurrent Pos:* Guggenheim fel, Oxford Univ, 54-55; vis prof, NC State Univ, 58-59 & Univ Calif, San Diego, 63-64; adv NPac Fur Seal Comn, 59-; chmn spec study group, Int Whaling Comn, 61-64, sci comn, 65-74; mem, Wash State Census Bur, 65-68; mem comn nat statist, Nat Acad Sci, 71-74 & ocean affairs bd, 72-76; chmn comn sci adv, Marine Mammal Comn, 74-76 & 81-92, chmn comn, 76-81. *Mem:* Fel Inst Math Statist; Biomet Soc; fel Am Statist Asn; Am Fisheries Soc; AAAS. *Res:* Mathematical statistics theory; population estimation; population dynamics. *Mailing Add:* 5843 NE Parkpoint Dr Seattle WA 98115

CHAPMAN, DOUGLAS WILFRED, b Can, Sept 21, 21; nat US; m 53; c 2. ORGANIC CHEMISTRY. *Educ:* Univ Calif, BS, 48; Mass Inst Technol, PhD(chem), 51. *Prof Exp:* Res chemist, Mallinckrodt Inc, 51-65, res assoc, 65-82; RETIRED. *Mem:* Am Chem Soc. *Res:* Organic synthesis; cosmetic pigments; x-ray contrast media. *Mailing Add:* 48 Greendale Dr St Louis MO 63121

CHAPMAN, FLOYD BARTON, horticulture; deceased, see previous edition for last biography

CHAPMAN, GARY ADAIR, b Corvallis, Ore, Aug 30, 37; m 60; c 2. AQUATIC BIOLOGY, WATER POLLUTION. *Educ:* Ore State Univ, BSc, 59, MSc, 65, PhD(fisheries), 69. *Prof Exp:* Agr res technician, USDA, 60-62; RES AQUATIC BIOLOGIST, US ENVIRON PROTECTION AGENCY, 68- *Concurrent Pos:* Team leader, Water Qual Criteria Eval, 80-85 & Sediment Toxicity 83-85; courtesy assoc prof, dept fisheries & wildlife, Oregon State Univ, 81-; actg nat res mgr, Aquatic Biol Effects Acid Precipitation, 82; prin investr, Develop Effluent Bioassay Methods WCoast Marine Organisms, 85- *Honors & Awards:* Superior Serv Medal, US Environ Protection Agency, 72, Except Serv Medal, 76; EPA Commendable Serv Award, 86. *Mem:* Am Fisheries Soc; Sigma Xi; Soc Environ Toxicol & Chem; Am Soc Testing & Mat; Am Inst Fishery Res Biologists. *Res:* Pollution effects on aquatic organisms; salmonid biology and toxicity bioassays; chemistry and toxicity of heavy metals in water; culture and toxicity tests of Pacific marine organisms; dissolved oxygen and bioenergetics of fish. *Mailing Add:* USEPA Hatfield Marine Sci Ctr Newport OR 97365

CHAPMAN, GARY ALLEN, b Bryan, Tex, Mar 25, 38; m 62; c 1. SOLAR PHYSICS. *Educ:* Univ Ariz, BS, 60, PhD(astron), 68. *Prof Exp:* Asst astronr, Univ Hawaii, 68-69; mem tech staff solar physics, Aerospace Corp, 69-77; assoc prof, 77-80, PROF ASTROPHYSICS, CALIF STATE UNIV, NORTHRIDGE, 80-, DIR, SAN FERNANDO OBSERV, 79- *Mem:* AAAS; Am Astron Soc; Am Phys Soc; Sigma Xi; Int Astron Union. *Res:* Fine structure of solar magnetic fields and solar faculae, the effect of faculae on solar oblateness measurements, solar spectroscopy and photometry. *Mailing Add:* Dept of Physics & Astron Calif State Univ 18111 Nordhoff St Northridge CA 91330

CHAPMAN, GARY THEODORE, b Elmwood, Wis, Apr 28, 34; m 56; c 4. FLUID MECHANICS, AERODYNAMICS. *Educ:* Univ Minn, BA, 57; Stanford Univ, MS, 63, PhD(aeronaut, astronaut), 70. *Prof Exp:* Res scientist fluid mech, 57-71, res scientist aerodyn, 71-73, chief, Aerodyn Res Br, 74-78, staff scientist fluid mech, Ames Res Ctr, NASA, 78-89; VIS RES ENG, UNIV CALIF, 89- *Concurrent Pos:* Vis prof aerodyn, Iowa State Univ, 73-74, Univ Fla, 81-82; proj scientist, 79-81. *Mem:* AAAS; Assoc fel Am Inst Aeronaut & Astronaut; Am Soc Eng Educ; Sigma Xi; Soc Indust & Appl Math. *Res:* Basic fluid mechanics; viscous flows including separation; vortices and turbulence; bodies and winged bodies at high angles of attack and flow modeling. *Mailing Add:* Univ Calif 6107 Etcheyerry Hall Berkeley CA 94720

CHAPMAN, GEORGE BUNKER, b Bayonne, NJ, June 10, 25. CYTOLOGY, HISTOLOGY. *Educ:* Princeton Univ, AB, 50, AM, 52, PhD, 53. *Prof Exp:* Asst instr biol, Princeton Univ, 50-52, from asst res to assoc res, 52-56; asst prof zool, Harvard, 56-60; assoc prof anat, Med Col, Cornell Univ, 60-63; chmn dept, 63-90, PROF BIOL, GEORGETOWN UNIV, 63- *Concurrent Pos:* Res biologist, Labs Div, Radio Corp of Am, 53-56. *Mem:* Am Soc Microbiol; Sigma Xi; Am Micros Soc. *Res:* Bacteriophagy; bacterial and general cytology; fine structure of human skin, eye, uterus, and gallbladder; electron microscopy of fish, coelenterates, protozoa and insects. *Mailing Add:* Dept of Biol Georgetown Univ Washington DC 20057

CHAPMAN, GEORGE DAVID, b Windsor, Ont, June 8, 40; m 72; c 2. OPTICAL PHYSICS, ATOMIC PHYSICS. *Educ:* Assumption Univ, BSc, 61; Univ Windsor, MSc, 64, PhD(physics), 65. *Prof Exp:* Overseas fel, Nat Res Coun, Oxford Univ, 65-67; res officer physics, 68-80, SR RES OFFICER PHYSICS, NAT RES COUN, 80- *Concurrent Pos:* Lectr, Carleton Univ, 70-71. *Honors & Awards:* Gold Key, Soc of Chem Industs, 62. *Mem:* Can Asn Physicists; Inst Elec & Electronic Engrs. *Res:* Mass & pressure standards; holography; interferometry; optical data processing; quantum optics; atomic spectroscopy; level crossings; lifetimes. *Mailing Add:* Inst Nat Measurement Standards Nat Res Coun Ottawa ON K1A 0R6 Can

CHAPMAN, HERBERT L, JR, b Kansas City, Mo, July 15, 23; m 46; c 4. ANIMAL NUTRITION. *Educ:* Univ Fla, BSA, 48, MSA, 51; Iowa State Univ, PhD(animal nutrit), 55. *Prof Exp:* Asst animal husbandman, Agr Res Ctr, Agr Exp Sta, Univ Fla, 51-53, from asst animal nutritionist to animal nutritionist, 55-81, dir, 65-81; consult, Maran Grove Corp, 81-90; CONSULT, CHAPMAN & ASSOCS, INC, 87- *Concurrent Pos:* Consult, 63- *Mem:* Fel AAAS; Am Soc Biol Sci; Am Registry Cert Animal Scientists; Am Soc Animal Sci. *Res:* Mineral requirement and interrelations in beef cattle; vitamin interrelations; nutritional requirements of all classes of beef cattle; pasture crop evaluation; chemical residues in beef cattle. *Mailing Add:* PO Box 55 Wauchula FL 33873

CHAPMAN, JACQUELINE SUE, b Vancouver, BC, July 6, 35. HIGH RISK PERINATAL NURSING, DEVELOPMENTAL INTERVENTIONS. *Educ:* Univ BC, BSN, 58; Case Western Reserve Univ, MSN, 67; NY Univ, PhD(nursing), 75. *Prof Exp:* Mem staff, asst head nurse & head nurse, Vancouver Gen Hosp, 58-61; head nurse, Royal Columbian Hosp, BC, 61-61; instr nursing, Univ BC, 61-65; asst prof, Case Western Reserve Univ, 67-69, Cornell Univ, 72-74; assoc prof, 74-79, PROF NURSING, UNIV TORONTO, 79- *Concurrent Pos:* Nat Health Res Scholar, Health & Welfare, Can, 75-81; prin investr, Nat Health Res & Develop Prog, 75-86; cross appointee, Inst Med Sci, Univ Toronto, 80-; nat mem, Sci Rev Comt, Alta Found Nursing Res, 83-86; fel, fac nursing, Univ Calif, San Francisco, 86; vis scholar, fac nursing, Univ Wash, 86. *Mem:* Coun Nurse Researchers. *Res:* Longitudinal follow-up of over 200 prematurely born children up to school age; parental responses to stillbirth, spina bifida and premature births; parental experiences with transfer of their premature infants among level 1, 2, and 3 facilities; effect of an RN antenatal-home- visits-based educational program vs electronic antenatal monitoring on prvention pf pre-term birth; effect of introduction of Als considering based model for caretaking in the NICU on both short and long term developmental and medical outcomes. *Mailing Add:* 50 St George St Toronto ON M5S 1A1 Can

CHAPMAN, JOE ALEXANDER, b Westpoint, Tenn, Oct 23, 19; m 41; c 3. PLANT ECOLOGY. *Educ:* Carson Newman Col, BS, 40; Peabody Col, MA, 47; Univ Tenn, PhD, 57. *Prof Exp:* High sch teacher, 41-42 & 42-44; from asst prof to assoc prof biol, 47-51, prof biol & head dept, Carson Newman Col, 53-; mem staff, Dept Wildlife Mgt, Utah State Univ, Logan, 88-; RETIRED. *Concurrent Pos:* Pres, Appalachian Elec Coop, 79- *Mem:* AAAS; Ecol Soc Am; NY Acad Sci. *Res:* Taxonomy. *Mailing Add:* 727 W Ellis St Jefferson City TN 37760

CHAPMAN, JOHN DONALD, b Estevan, Sask, Can, Feb 18, 41; m 63, 89; c 4. RADIOLOGY. *Educ:* Univ Sask, Saskatoon, BS, 63, MS, 65; Penn State Univ, PhD(biophysics), 67. *Prof Exp:* Asst res officer, Med biophysics Br, Atomic Energy Can Ltd, 68-70, assoc res officer, 71-77; vis prof med physics, Univ Calif, Berkeley, 75-76; assoc clin prof, Univ Alta, 77-80, clin prof, div radiation oncol, 80-86, prof div radiation oncol, dept radiol, 86-90; dir Radiobiol Prog, Cross Cancer Inst, 77-90; DIR TUMOR BIOL & BIOPHYSICS, FOX CHASE CANCER CTR, 90- *Concurrent Pos:* assoc dir, res, Cross Cancer Inst, 83-88; adj prof Radiation Oncol, Univ Pa. *Honors & Awards:* Milford D Schulz Award, Dept Radiation Med, Harvard Med Sch, 83. *Mem:* Radiation Res Soc; Biophysical Soc; Biophysical Soc Can; Brit Inst Radiol; Am Asn Cancer Res. *Res:* The use of radiobiological and cell biophysics techniques for studying resistance of tumor cells to various treatments. *Mailing Add:* Fox Chase Cancer Ctr 7701 Burholme Ave Philadelphia PA 19111

CHAPMAN, JOHN E, b Springfield, Mo, July 5, 31; m 68. PHARMACOLOGY. *Educ:* Southwest Mo State Col, BS(educ) & BS(biol, chem), 54; Univ Kans, MD, 58; Am Bd Clin Pharmacol, dipl. *Hon Degrees:* Dr Hon Causa, Karolinska Inst, Stockholm, Sweden, 87. *Prof Exp:* Instr pharmacol, Med Ctr, Univ Kans, 61-62, asst prof, 62-67, asst acad dean, Med Sch, 63-65, assoc dean, 65-67; assoc dean educ, 67-72, actg dean, 72-73, actg vchancellor med affairs, 73-75, DEAN, VANDERBILT UNIV SCH MED, 75- *Concurrent Pos:* Chmn & bd dirs, Health Educ Media Asn, Coun Deans; mem, Coun Inst Gen Med Sci, NIH; mem, Liason Comt Med Ed; mem, Coun Med Educ, AMA; regent, Am Col Clin Pharmacol. *Mem:* Am Soc Pharmacol & Exp Therapeut; Am Col Clin Pharmacol (vpres & regent); Nat Bd Med Examr. *Res:* Role of humoral agents in central nervous system activity; medical school and medical center administration; medical education. *Mailing Add:* Rm D-3300 Vanderbilt Univ 21st at Garland Ave Nashville TN 37232

CHAPMAN, JOHN FRANKLIN, JR, b Oakdale, Calif, Apr 11, 45; m 68. CLINICAL BIOCHEMISTRY, LABORATORY MEDICINE. *Educ:* San Jose State Col, BA, 68; Calif State Univ, Fresno, MS, 72; Univ NC, Chapel Hill, MPH, 76, DrPH(lab practice), 78. *Prof Exp:* Pub health microbiologist, Fresno County Dept Health, 68-72, sr pub health microbiologist, 72-73, dir, Pub Health Labs, 73-75; fel clin chem, 78-79, ASSOC PROF PATH & ASSOC DIR CLIN CHEM, DEPT PATH, DIV LAB MED, UNIV NC, CHAPEL HILL, 79- *Mem:* Nat Acad Clin Biochem; Asn Clin Lab Physicians & Scientists; Asn Clin Scientists; Am Asn Clin Chem; Sigma Xi. *Res:* High resolution two dimensional electrophoresis in detection of tumor markers and other markers of disease; laboratory determination of amniotic fluid phospholipids in the detection of fetal lung maturity. *Mailing Add:* 2004 Crabtree Lane Chapel Hill NC 27516-9805

CHAPMAN, JOHN JUDSON, b Valdosta, Ga, May 10, 18; m 47; c 4. GEOLOGY. *Educ:* Colo Sch Mines, Geol & Eng, MS, 48; Univ Ill, PhD (geol), 53. *Prof Exp:* Topog engr, US Geol Surv, 41-44 & 46-47; sr geologist, Creole Petrol Corp, 48-51; asst prof geol & head dept, Southern State Col, 53-57, prof geol & chmn div natural sci, 57-68; prof earth sci & head dept, 68-80, prof geol, 80-85, EMER PROF GEOL, WEST CAROLINA UNIV, 85-; CONSULT PETROL GEOLOGIST, MAGNOLIA, ARK, 81- *Mem:* AAAS; Geol Soc Am; Am Asn Petrol Geologists; Nat Asn Geol Teachers; Am Inst Prof Geologists. *Res:* Physical stratigraphy; petroleum geology; cause of life extinctions. *Mailing Add:* 103 W Main St Sylva NC 28779

CHAPMAN, JOHN S, b Sweetwater, Tex, Jan 30, 08; m 32; c 1. MEDICINE. *Educ:* Southern Methodist Univ, BA & BS, 27, MA, 28; Univ Tex, MD, 32. *Prof Exp:* Clin asst prof med, Univ Tenn, 43; clin asst, 43-45, from clin instr to clin assoc prof, 45-52, prof, 52-83, EMER PROF MED, UNIV TEX HEALTH SCI CTR, DALLAS, 83- *Concurrent Pos:* Tuberc consult, Vet Admin Hosp, Dallas, 46; civilian consult, Brooke Army Hosp, 46; consult, Bur Radiol Health & ETex Tuberc Hosp, Tyler; ed-in-chief, Arch Environ Health, AMA, 71-75. *Mem:* AAAS; Am Col Physicians; Am Thoracic Soc. *Res:* Eosinophilic leucocyte; mycobacteria; juvenile tuberculosis. *Mailing Add:* 3606 Lovers Lane Dallas TX 75225

CHAPMAN, JOSEPH ALAN, b Salem, Ore, Apr 28, 42; m 78; c 2. WILDLIFE ECOLOGY. *Educ:* Ore State Univ, BS, 65, MS, 67, PhD(wildlife sci), 70. *Prof Exp:* Wildlife biologist, US Fish & Wildlife Serv, 65-67; res asst, dept fisheries & wildlife, Ore State Univ, 67-69; fac res asst, Natural Resources Inst, Univ Md, 69-70, res asst prof, 70-74, res assoc prof & head, Appalachian Environ Lab, 74-78, prof & head, 78-83; prof Fisheries & Wildlife Sci & dept head, 83-89, PROF & DEAN COL NATURAL RESOURCES, UTAH STATE UNIV, 89- *Concurrent Pos:* Adj prof wildlife sci, Garrett Community Col, 73-; adj assoc prof biol, Frostburg State Col, 73-; guest lectr wildlife ecol, WVa Univ, Univ Md & Univ Ohio, 73- *Mem:* Wildlife Soc; m Soc Mammalogists; Cooper Ornith Soc; Ecol Soc Am; Soc Range Mgt; elected fel Inst Biol, UK; fel Explorers Club. *Res:* Mammalogy. *Mailing Add:* Col Natural Resources Utah State Univ Logan UT 84322-5200

CHAPMAN, JUDITH-ANNE WILLIAMS, b Timmins, Ont, Aug 17, 49; c 1. MEDICAL STATISTICS. *Educ:* Univ Waterloo, BS, 71, PhD(statist), 74. *Prof Exp:* Fel statist, 74-76, STATIST CONSULT CANCER RES, NAT CANCER INST CAN, UNIV WATERLOO, 76- *Mem:* Can Oncol Soc; Am Statist Asn; Can Statist Asn. *Res:* Use of statistics in analysing cancer mortality and incidence data; regional differences in Ontario; different statistics for significance tests; case-control study in bladder cancer. *Mailing Add:* 11 Dayman Ct Kitchener ON N2M 3A1 Can

CHAPMAN, KENNETH REGINALD, b Croydon, Eng, Apr 10, 24; m 46; c 5. ACCELERATOR PHYSICS. *Educ:* Univ London, BSc, 50 & 51, MSc, 60, PhD(nuclear physics), 65. *Prof Exp:* Develop engr, Mullard Radio Valve Co, UK, 40-47; res demonstr physics, Univ Col, Univ Leicester, 51-56; res assoc, Univ Birmingham, 56-66; ACCELERATOR PHYSICIST, FLA STATE UNIV, 66-, PROF, 76- *Mem:* Brit Inst Physics; Brit Inst Elec Eng. *Res:* Accelerator and nuclear physics; ion source research; accelerator development. *Mailing Add:* Dept of Physics Fla State Univ Tallahassee FL 32306-3016

CHAPMAN, KENT M, b Minneapolis, Minn, Oct 28, 28; m 51; c 4. BIOPHYSICS. *Educ:* Univ Minn, BA, 49, MS, 53, PhD(biophys), 62. *Prof Exp:* From asst prof to assoc prof physiol, Univ Alta, 58-65; chmn neurosci sect, Div Biol & Med Sci, 73-76, ASSOC PROF MED SCI, BROWN UNIV, 65- *Mem:* AAAS; Biophys Soc; Am Physiol Soc; Soc Gen Physiologists. *Res:* Neurophysiology; sensory transduction and encoding; electrophysiology; insect mechanoreceptors; vertebrate cardiac mechanoreceptors; biophysical mechanisms of mechanical coupling, mechanoelectric transduction, and sensory encoding in insect cuticular mechanoreceptors; linear transfer functions of compliance, receptor potential, and impulse frequency modulation; thermodynamics and kinetics of electrogenic active ion transport; the role of Onsager reciprocity in active pumping and exchange diffusion in the Na, K-ATPases of red blood cells, epithelia, and neurons. *Mailing Add:* Div Biol & Med Box G-M Brown Univ Providence RI 02912

CHAPMAN, LLOYD WILLIAM, b Pasadena, Calif, Jan 5, 38; m 68; c 2. PHYSIOLOGY. *Educ:* Univ Calif, Los Angeles, BA, 61; Univ Southern Calif, PhD(physiol), 71. *Prof Exp:* From instr to asst prof physiol, Sch Med, Univ Southern Calif, 72-78; ASST PROF PHYSIOL, COL OSTEOP MED OF PAC, 78- *Concurrent Pos:* Consult, Medi Legal Inst, Los Angeles, 72-; vpres, Med Res Found Heart Dis, Los Angeles, 77-; res assoc, Cedars-Sinai Med Ctr, Los Angeles; res assoc, Univ Calif, Riverside. *Mem:* Assoc mem Am Physiol Soc. *Res:* Fluid and electrolyte balance; reflex regulation of circulation. *Mailing Add:* Garfield HS 6167 Topaz St Alta Loma CA 91701

CHAPMAN, LORING FREDERICK, b Los Angeles, Calif, Oct 4, 29; c 3. NEUROSCIENCES, PSYCHIATRY. *Educ:* Univ Nev, BS, 50; Univ Chicago, PhD(psychol & physiol), 55. *Prof Exp:* Res asst, Dept Med, Univ Chicago, 52-55; from asst prof to assoc prof psychol, Dept Med, Col Med, Cornell Univ, 54-61; assoc prof med psychol, Dept Psychiat, Univ Calif, Los Angeles, 61-65; prof, Univ Ore, 65-66 & Georgetown Univ, 66-67; prof behav biol & chmn dept, prof psychiat, human physiol & neurol, 67-79, PROF PSYCHIAT, SCH MED, UNIV CALIF, DAVIS, 79- *Concurrent Pos:* Nat Acad Sci award, 57, 59 & 61; Wilson prize, 58; partic, Skin Conf, 59, 76 & Ciba Found Conf, London, Eng, 60; vis scientist, Univ San Paulo, 59 & Univ Florence, Italy, 79-80; consult, Nat Inst Neurol Dis, 62-, NASA, 73- & Calif Med Facil, Vacaville, 74-; USPHS career develop award, 64; dir res, Fairview Hosp, 65-66; chief lab & behav biol br, NIH, 66-67; mem & res behav biologist, Calif Primate Res Ctr, 67-, mem, Comt, Nat Inst Neurol Dis & Blindness, 67-68 & Res & Training Comt Study Sect, Nat Inst Child Health & Human Develop, 68-73; Commonwealth Fund award & vis scientist, Univ Col, Univ London, 70; vis prof pharmacol & physiol, Fac Med, Botucatu, Brazil, 77; Fogarty Int fel, 79-80. *Mem:* Am Physiol Soc; Soc Neurosci; Am Neurol Asn; Royal Soc Med; Aerospace Med Asn. *Res:* Brain function; biology of behavior; primatology; mental retardation; mental illness and deviant behavior; pain; pleasure; addiction; behavioral and reproductive hazards of psychoactive drugs; coding principles in neural systems; memory; psychopharmacology; neuropeptides; aerospace physiology; positron tomography; growth factors and neurodegenerative diseases. *Mailing Add:* Dept Psychiat Sch Med Univ Calif Davis CA 95616

CHAPMAN, ORVILLE LAMAR, b New London, Conn, June 26, 32; m 55, 81; c 2. ORGANIC CHEMISTRY. *Educ:* Va Polytech Inst, BS, 54; Cornell Univ, PhD(chem), 57. *Prof Exp:* From asst prof to prof chem, Iowa State Univ, 57-74; PROF CHEM, UNIV CALIF, LOS ANGELES, 74- *Concurrent Pos:* Alfred P Sloan Found res fel, 61-; vis res dir, chem & biochem insects, Int Ctr Insect Physiol & Ecology, Nairobi, Kenya, 77-79; mem, Adv Comt Appl Sci & Res Appln Policy, NSF, 77-80; sr US scientist award, Alexander von Humboldt Found, 81. *Honors & Awards:* Founders Prize, Tex Instrument Found, 74; Arthur C Cope Award, Am Chem Soc, 78; Gregory & Freda Halpern Award Photochem, NY Acad Sci, 78; Medal, Havinga Fund Found, 82; NcCoy Award, UCLA, 83. *Mem:* Nat Acad Sci; Am Chem Soc; The Chem Soc; fel AAAS. *Res:* Organic photochemistry; reactive intermediates; natural products; bioorganic chemistry. *Mailing Add:* Dept Chem Univ Calif 405 Hilgard Ave Los Angeles CA 90024

CHAPMAN, PETER JOHN, b Wolverhampton, Eng, Oct 26, 36; m 69; c 2. BIOCHEMISTRY, MICROBIOLOGY. *Educ:* Univ Leeds, BSc, 58, PhD(biochem), 61. *Prof Exp:* Assoc biochem, Univ Ill, Urbana, 61-64; lectr, Univ Hull, 64-66; from asst prof to prof biochem, 66-87, PROF MICROBIOL, UNIV MINN, ST PAUL. *Mem:* AAAS; Am Soc Biol Chemists; Am Soc Microbiol; Soc Gen Microbiol; Soc Indust Microbiol. *Res:* Microbial metabolism of synthetic and naturally-occurring compounds; regulation and organization of genes specifying catabolic enzymes; evolutionary relationships of degradative enzymes. *Mailing Add:* Environ Res Lab US Environ Protection Agency Sabine Island Gulf Breeze FL 32561-3998

CHAPMAN, R KEITH, b Oak Lake, Man, Oct 31, 16; m 42; c 3. INSECT TRANSMISSION OF PLANT PATHOGENS, INSECT RESISTANCE IN HORTICULTURAL CROPS. *Educ:* Ont Agr Col, BSA, 40; Univ Wis, PhD(entom), 49. *Prof Exp:* Lectr entom & zool, Ont Agr Col, 40-45; asst entom, 45-47, from instr to prof, 47-87, EMER PROF ENTOM, UNIV WIS-MADISON, 87- *Concurrent Pos:* Researcher, Cent Am & SAm, Am Cocoa Res Inst; adv, maise borer biol & control, Midwest Univs Consortium for Int Agr/AID Nepal proj. *Honors & Awards:* Award of Merit, Entom Soc Am. *Mem:* Entom Soc Am; Potato Asn Am; Am Phytopath Soc; Coun Agr Sci & Technol. *Res:* Insect transmission of plant disease; vegetable insects and their control; insect resistance in vegetable insects; insecticide resistance in vegetable insects. *Mailing Add:* Dept Entom Univ Wis Col Agr & Life Sci 637 Russell Lab Madison WI 53706

CHAPMAN, RAMONA MARIE, b Columbus, Ky, Sept 17, 45. ONCOLOGY, HEMATOLOGY. *Educ:* Memphis State Univ, BS, 68; Univ Tenn, MD, 70. *Prof Exp:* Intern med, St Pauls Hosp, Dallas, 70-71 & residency, 71-72; residency internal med, Mt Zion Hosp, San Francisco, 72-73; fel hemat & oncol, Scripps Clin & Res Found, 73-76; hon sr registr oncol, St Bartholomews Hosp, London, 76-79; mem staff oncol, US Army, Walter Reed Army Med Ctr, 79-81 & mem hemat, Walter Reed Army Inst Res, 81-82; dir, St John's Regional Oncol Ctr, Mo, 82-84; DIR, CHAPMAN CANCER CTR, 84-; ASST PROF MED, UNIV MO-COLUMBIA, 84- *Concurrent Pos:* Asst prof, Dept Med, Uniformed Serv Univ Health Sci, 79-82; ed bd, Jour Clin Oncol; consult, Jour Am Med Asn. *Honors & Awards:* Physician Recognition Award, Am Med Asn, 86-89. *Mem:* AAAS; Am Women's Med Asn; Am Geriat Soc; AMA; Am Fedn Clin Res; Am Soc Clin Oncol; Am Col Physicians; Am Bds Hemat Oncol; Am Soc Hemat. *Res:* Effects of cancer chemotherapy on sexual function and fertility in humans; possibility of gonadal protection via hormonal gonadal suppression during chemotherapy; current cancer control and cost containment studies. *Mailing Add:* PO Box 2644 Joplin MO 64803

CHAPMAN, RAY LAVAR, b Blackfoot, Idaho. NUCLEAR ENGINEERING, PROPULSION. *Educ:* Brigham Young Univ, BS, 49, MS, 51. *Prof Exp:* Sr dynamics engr stability & control, Gen Dynamics Corp, 51-58; assoc prof nuclear eng, Univ Ariz, 58-61; dept staff engr propulsion, Martin Marietta Corp, 61-73; PRIN PROJ ENGR NUCLEAR ENG, EG&G IDAHO INC, 74- *Concurrent Pos:* Consult, Army Electronic Proving Ground, 58-61. *Res:* Nuclear engineering; technical management; elementary particle physics. *Mailing Add:* EG&G Idaho Inc PO Box 1625 Idaho Falls ID 83415

CHAPMAN, RICHARD ALEXANDER, b Teague, Tex, Sept 24, 32; m 54; c 4. INTEGRATED CIRCUIT TECHNOLOGY. *Educ:* Rice Univ, BA, 54, MA, 55, PhD, 57. *Prof Exp:* Asst physics, Rice Univ, 54-56; physicist, Vallecitos Atomic Lab, Gen Elec Corp, 57-59; physicist, 59-64, mgr mat physics br, 64-71, mgr infrared devices br, 71-76, sr mem tech staff, 76-87, RES FEL, CORP RES & ENG, TEX INSTRUMENTS INC, 76- *Concurrent Pos:* Mem bd gov, Rice Univ, 75-79. *Honors & Awards:* Morton Award, Inst Elec & Electronics Engrs, 87. *Mem:* Fel Am Phys Soc; Inst Elec & Electronics Engrs. *Res:* VLSI CMOS integrated circuit development and characterization; charge coupled devices and infrared imagers; photovoltaic infrared detectors; light emitting diodes; photoluminescence and infrared properties of impurities in semiconductors. *Mailing Add:* 7240 Briarcove Dr Dallas TX 75240

CHAPMAN, RICHARD DAVID, b Atlanta, Ga, June 25, 28; m 55; c 4. CONDENSATION POLYMERIZATION, CRYSTALLINE POLYMERS. *Educ:* Univ Fla, BS, 49; Northwestern Univ, MS, 51, PhD(org chem), 54. *Prof Exp:* Sr chemist, Chemstrand Corp, 53-58, group leader, 58-60, res chemist, Chemstrand Res Ctr, 60-69, res specialist 69-76, sr res specialist, Monsanto Textiles Co, 76-80, sr technol specialist, 80-88, NYLON TECHNOLOGIST, MONSANTO CHEM CO, 89- *Mem:* Am Chem Soc. *Res:* Polymer chemistry; condensation polymers; nylon molding resins. *Mailing Add:* Tech Ctr Monsanto Chem Co PO Box 12830 Pensacola FL 32575

CHAPMAN, ROBERT DALE, b Glendale, Calif, June 4, 55; m 81. ENERGETIC MATERIALS, FLUORINE CHEMISTRY. *Educ:* Univ Calif, Irvine, BA, 77, PhD(chem), 80. *Prof Exp:* Res chemist, Naval Weapons Ctr, China Lake, Calif, 81-82 & Air Force Astronaut Lab, Edwards AFB, Calif, 82-89; sr res chemist, Fluorochem, Inc, Azusa, Calif, 89-91; STAFF SCIENTIST, UNIDYNAMICS/PHOENIX, ARIZ, 91- *Mem:* Am Chem Soc; Sigma Xi. *Res:* Energetic materials chemistry (explosives and propellant ingredients); nitro chemistry; fluorine and difluoramino chemistry; isocyanates; polyphosphazenes; chemistry of noble gases; transition metal chemistry; metalloporphyrin chemistry; kinetics; nuclear magnetic resonance spectroscopy; statistics. *Mailing Add:* Unidynamics/Phoenix PO Box 46100 Phoenix AZ 85063

CHAPMAN, ROBERT DEWITT, b Erie, Pa, July 13, 37; m 64; c 2. ASTROPHYSICS. *Educ:* Pa State Univ, BS, 59; Harvard Univ, PhD(astron), 65. *Prof Exp:* Asst prof astron, Univ Calif, Los Angeles, 64-67; astronomer, NASA, 67-77, assoc chief, Lab Astron & Solar Physics, Goddard Space Flight Ctr, 77-85, user advocate, Office Space Sta, Johnson Space Ctr, 85-86, sr policy analyst, Off Sci & Technol Policy, Exec Off Pres, 86-87, lead scientist, NASA Prog Off, Comput Technol Assoc, 88-90; SR TECH ADV, BENDIX FIELD ENG CORP, 90- *Concurrent Pos:* Vis lectr, Univ Md, 70-75; lectr, Montgomery Col, 81; prog mgr, Utilization Div, Off Space Sta, NASA Hq, 87-88. *Mem:* Am Astron Soc; Int Astron Union. *Res:* Structure of the outermost layers of the sun and stars and related physical problems. *Mailing Add:* 10976 Swansield Rd Columbia MD 21044

CHAPMAN, ROBERT EARL, JR, b Borger, Tex, Jan 11, 41; m 61; c 2. BIOPHYSICS. *Educ:* Yale Univ, MS, 66, PhD, 68. *Prof Exp:* Fel biophys, Univ Calif, San Diego, 68; CHIEF PHOTOG SCIENTIST, UNICOLOR DIV, PHOTOSCI INC, 68-; PRIN ENGR, KAISER OPTICAL SYS. *Concurrent Pos:* Guest lectr, Lansing Community Col, Mich, 81- *Res:* Photographic science; photo chemistry. *Mailing Add:* 9423 Hidden Lake Ct Dexter MI 48130

CHAPMAN, ROBERT MILLS, b Chicago, Ill, Aug 29, 18; m 45, 68; c 1. GEOLOGY. *Educ:* Northwestern Univ, BS, 41. *Prof Exp:* Geologist, Alaskan Br, US Geol Surv, 42-47, in charge Fairbanks off, 47-55, geol res with geochem explor br, Colo, 55-61, in charge off, College, Alaska, 61-69, geologist, Alaskan Geol Br, 70-86; RETIRED. *Concurrent Pos:* Geol res, USGS, 86- *Mem:* Fel Geol Soc Am; Soc Econ Geol; Am Asn Petrol Geol; fel Arctic Inst NAm. *Res:* Alaskan geology; structure and stratigraphy of interior region; geochemical exploration; mineral deposits; history of geologic investigations and the geological survey in Alaska. *Mailing Add:* US Geol Surv MS-904 345 Middlefield Rd Menlo Park CA 94025-3508

CHAPMAN, ROBERT PRINGLE, b Hartland, NB, Sept 12, 26; m 50; c 3. PHYSICS. *Educ:* Mt Allison Univ, BS, 47; McGill Univ, MS, 49; Univ Sask, PhD(physics), 53. *Prof Exp:* Sci officer, Defence Res Estab Atlantic, 53-73, HEAD OCEAN ACOUSTICS SECT, DEFENCE RES ESTAB PAC, 73- *Mem:* Acoustical Soc Am. *Res:* Underwater acoustics; physics of aurora borealis; mass spectroscopy. *Mailing Add:* 1245 Rockcrest Victoria BC V9A 4W4 Can

CHAPMAN, ROGER CHARLES, b New Orleans, La, July 14, 36; m 61; c 3. FORESTRY, STATISTICS. *Educ:* Col William & Mary, BS, 59; Duke Univ, MF, 59; Univ Calif, Berkeley, MA, 65; NC State Univ, PhD(forestry), 77. *Prof Exp:* Res forester, US Forest Serv, 59-66, statistician biomet, 66-69; asst prof biomet, Sch Forestry, Duke Univ, 69-73; from asst prof to assoc prof biomet, Sch Forestry, 80-87, PROF FORESTRY, DEPT NATURAL RESOURCE SCI, WASH STATE UNIV, 87- *Mem:* Am Statist Asn; Biomet Soc. *Res:* Forest growth and yield; estimation theory; forest resources sampling. *Mailing Add:* Dept Natural Resource Sci Wash State Univ Pullman WA 99164-6410

CHAPMAN, ROSS ALEXANDER, b Oak Lake, Man, Dec 10, 13; m 42; c 2. FOOD SCIENCE. *Educ:* Univ Toronto, BSA, 40; McGill Univ, MSc, 41, PhD(chem), 44. *Hon Degrees:* DSc, Univ Guelph, 72. *Prof Exp:* Asst prof, MacDonald Col, McGill Univ, 44-48; head food sect, Food & Drug Div, Dept Nat Health & Welfare, Ottawa, 48-55, head food sect, WHO, Geneva, 55-57, asst dir sci serv, 58-63, asst dir gen food, 63-65, asst dep minister food & drugs, 65-71, spec adv, Off Dep Minister Health, 71-72, dir gen int health serv, 72-73; consult food legis & control, South Pac Comn, New Caledonia, 75-76; consult, Pan Am Health Orgn, Washington, DC, 77-78 & Trinidad, Tobago & Brazil, 80; RETIRED. *Concurrent Pos:* Head Can deleg, UN Comn Narcotic Drugs, 70-73; mem deleg, UN Conf Protocol Psychotropic Substances, 71. *Honors & Awards:* Underwood-Prescott Mem Lectr, Mass Inst Technol, 72. *Res:* Methods of determination of thiamine and riboflavin; ferric thiocyanate methods for fat peroxides and its application to milk powders and fats and oils; methods for determination of antioxidants in fat and oils; behavior of antioxidants; methods for arsenic in fruits and vegetables and tocopherol in butter fat; food and drug legislation. *Mailing Add:* 655 Richmond Rd Unit 48 Ottawa ON K2A 3Y3 Can

CHAPMAN, RUSSELL LEONARD, b Brooklyn, NY, May 30, 46; m 69; c 2. PHYCOLOGY. *Educ:* Dartmouth Col, AB, 68; Univ Calif, Davis, MS, 70, PhD(bot), 73. *Prof Exp:* From asst prof to assoc prof, 73-83, assoc dean, Col Arts & Sci, 79-83, assoc dean, Col Basic Sci, 83-84, PROF BOT, LA STATE UNIV, BATON ROUGE, 83-, CHMN, 88- *Concurrent Pos:* Vis prof, dept molecular cell develop biol, Univ Colo, 84; chmn, phycol sect, Bot Soc Am, 83-85; NSF systematics panel mem, 87-90. *Mem:* Phycol Soc Am (secy, 85-87, vpres, 88, pres, 89); Bot Soc Am; Int Phycol Soc; Brit Phycol Soc; Electron Micros Soc Am. *Res:* Systematics and phylogeny of algae, ribosomal RNA sequencing. *Mailing Add:* Dept Bot La State Univ Baton Rouge LA 70803-1705

CHAPMAN, SALLY, b Philadelphia, Pa, July 28, 46. CHEMISTRY. *Educ:* Smith Col, AB, 68; Yale Univ, PhD(chem), 73. *Prof Exp:* Fel chem, Univ Calif, Irvine, 73-74 & Univ Calif, Berkeley, 75; from asst prof to assoc prof, 75-86, PROF CHEM, BARNARD COL, COLUMBIA UNIV, 86- *Mem:* Am Chem Soc; Am Phys Soc; Sigma Xi. *Res:* Classical trajectory studies of a variety of bimolecular reactive systems; emphasizing energy use and energy disposal and its relationship to the potential energy surface or surfaces. *Mailing Add:* Dept Chem Barnard Col 3009 Broadway New York NY 10027-6598

CHAPMAN, SHARON K, b Hutchinson, Kans, Nov 3, 39. PHARMACOLOGY, MEDICAL SCIENCE. *Educ:* Kans State Col Pittsburg, BS, 61; Univ Fla, PhD(med sci), 70. *Prof Exp:* Asst prof pharmacol, Sch Pharm, Univ Md, 70-72; pharmacologist biochem res, Vet Admin Hosp, Gainesville, Fla, 72-75; ASST PROF PHARMACOL, COL MED, UNIV FLA, 75-; SR CLIN RES SCIENTIST, BURROUGHS WELLCOME CO, 83-, ASSOC HEAD, DEPT INFECT DIS, 88- *Concurrent Pos:* Pharmaceut Mfrs Asn Found res grant, 75-77; grad coordr, Dept Pharmacol, Col Med, Univ Fla, 78-80. *Mem:* Am Soc Pharmacol & Exp Therapeut. *Res:* Biochemical mechanisms of action of drugs and foreign chemicals; chemotherapeutics; polyamine chemistry; cyclic nucleotides; antivirals. *Mailing Add:* Clin Invest Burroughs Wellcome Co 3030 Cornwallis Rd Research Triangle Park NC 27709

CHAPMAN, STEPHEN R, b San Francisco, Calif, Oct 21, 36; m 58; c 2. POPULATION GENETICS, AGRONOMY. *Educ:* Univ Calif, Davis, BS, 59, MS, 63, PhD(genetics), 66. *Prof Exp:* Lab tech agron, Univ Calif, Davis, 60-66; asst prof agron & genetics, Mont Univ, 66-70; from assoc prof to prof, Mont State Univ, 70-77; dir instr, 77-86, from assoc dean to dean & dir, 86-91, PROF AGRON & SOILS, COL AGR SCI, CLEMSON UNIV, 91- *Mem:* Fel Am Soc Agron; Crop Sci Soc. *Res:* Population genetics and ecology of forage grasses. *Mailing Add:* Dept Agron & Soils Clemson Univ Clemson SC 29634

CHAPMAN, THOMAS WOODRING, b Wilkinsburg, Pa, Sept 21, 40; m 66; c 2. CHEMICAL ENGINEERING, EDUCATIONAL ADMINISTRATION. *Educ:* Yale Univ, BE, 62; Univ Calif, Berkeley, PhD(chem eng), 67. *Prof Exp:* Asst prof, 67-77, PROF CHEM ENG, UNIV WIS-MADISON, 78-, ASSOC DEAN ENG, 90- *Concurrent Pos:* Dir, Int Eng Progs, Univ Wis, 89- *Honors & Awards:* Environ Award, Am Inst Chem Engrs, 74. *Mem:* Am Inst Chem Engrs; Am Chem Soc; Am Inst Mining, Metall & Petrol Engrs. *Res:* Mass transfer and separation operations; transport properties; electrochemical and hydrometallurgical processes. *Mailing Add:* Univ Wis Rm 1018 1500 Johnson Dr Madison WI 53706-1687

CHAPMAN, TOBY MARSHALL, b Chicago, Ill, Nov 16, 38; m 61; c 3. BIO-ORGANIC CHEMISTRY, POLYMER CHEMISTRY. *Educ:* Univ Ill, BS, 60; Polytech Inst Brooklyn, PhD(chem), 65. *Prof Exp:* Res fel biol chem, Harvard Med Sch, 65-67; asst prof, 67-74, ASSOC PROF CHEM, UNIV PITTSBURGH, 74- *Mem:* Am Chem Soc; AAAS. *Res:* New methods of peptide and oligonucleotide synthesis; determination of biopolymer structure; vinyl copolymerization; synthesis of phosphorus containing polymers; polyurethane chemistry. *Mailing Add:* 6527 Lilac St Pittsburgh PA 15217-3033

CHAPMAN, VERNE M, b Sacramento, Calif, Oct 4, 38; m 68. GENETICS. *Educ:* Calif State Polytech Col, BS, 60; Ore State Univ, MS, 63, PhD(genetics), 65. *Prof Exp:* Asst prof biol, Millersville State Col, 65-66; fel, Jackson Lab, 66-68; res fel, Yale Univ, 68-72; assoc cancer res scientist, Roswell Park Mem Inst, 72-80; chmn, dept molecular biol, 82-85, ASSOC INST DIR SCI AFFAIRS, STATE UNIV NY BUFFALO, 85-, RES PROF, 86- *Mem:* Soc Develop Biol; Genetics Soc Am; Am Genetics Asn; Am Soc Cell Biol. *Res:* Biochemical genetics and developmental genetics of the mouse. *Mailing Add:* Dept Molecular Biol Roswell Park Mem Inst 666 Elm St Buffalo NY 14263

CHAPMAN, WARREN HOWE, b Chicago, Ill, Oct 30, 25; m 50; c 5. UROLOGY. *Educ:* Mass Inst Technol, BS, 46; Univ Chicago, MD, 52. *Prof Exp:* Intern, St Luke's Hosp, Chicago, 52-53; resident urol, Univ Chicago Clins & instr, Sch Med, 53-57; res assoc, Western Wash State Col, 63-66; clin assoc, dept surg, Univ Wash Affil Hosps, 57-62, clin instr, 62-66, from asst prof to assoc prof urol, 66-73, admin off dept, 66-87, actg chmn, 87-88, PROF UROL, UNIV WASH, 73- *Mem:* AMA; Am Urol Asn; Am Asn Cancer Res; Asn Am Med Cols; Soc Univ Urol. *Res:* Renal and adrenal hypertension; carcinoma of the bladder; microsurgery. *Mailing Add:* Dept Urol RL10 Univ Wash Seattle WA 98195

CHAPMAN, WILLIAM EDWARD, b Miami, Fla, Aug 23, 45; m 75; c 2. PATHOLOGY. *Educ:* Univ Conn, Storrs, BA, 72; Univ Conn, Farmington, MD, 76. *Prof Exp:* Resident path, 76-78, FEL IMMUNOL, DEPT PATH, SCH MED, UNIV CONN, FARMINGTON, 78- *Concurrent Pos:* Fel Mem Sloan Kettering, 82-83; staff pathologist, Beth Israel Med Ctr, 83-86, Vet Hosp, Ashville, 87- *Res:* Babesiosis; immunopathology of the lung. *Mailing Add:* 16N Kensington Rd Asheville NC 28804

CHAPMAN, WILLIAM FRANK, b Hanover, NH, July 26, 44; m 67. GLACIAL GEOLOGY. *Educ:* Univ NH, BS, 66; Univ Mich, MS, 68, PhD(geol), 72. *Prof Exp:* PROF GEOL, SLIPPERY ROCK UNIV, 71- *Mem:* Geol Soc Am; Am Quaternary Asn; AAAS; Nat Asn Geol Teachers; Sigma Xi; Am Meteorol Soc. *Mailing Add:* Dept Geol Slippery Rock Univ Slippery Rock PA 16057

CHAPMAN, WILLIE LASCO, JR, b Chattanooga, Tenn, Dec 17, 28; m 58; c 2. COMPARATIVE PATHOLOGY. *Educ:* Univ Tenn, BS, 50; Auburn Univ, DVM, 57; Colo State Univ, MS, 63; Univ Wis, PhD(vet sci), 68. *Prof Exp:* Pvt pract, 57-62; resident radiol, Colo State Univ, 62-63; asst prof vet med & surg, Sch Vet Med, Univ Ga, 63-64; res fel comp path, Univ Wis, 64-67; asst prof path, 67-71, assoc prof med & surg & head dept, 71-75, assoc prof path, 75-77, PROF PATH, COL VET MED, UNIV GA, 77- *Concurrent Pos:* NIH spec fel, Dept Path & Regional Primate Res Ctr, Univ Wis, 65-67. *Mem:* Am Vet Med Asn; Am Animal Hosp Asn; Am Asn Lab Animal Sci; Int Acad Path. *Res:* Animal models for leishmaniasis; mechanisms of immunity to blood parasites; chemotherapy of leishmaniasis. *Mailing Add:* Dept Path Univ Ga Col Vet Med Athens GA 30602

CHAPNICK, BARRY M, CARDIOVASCULAR PHARMACOLOGY. *Educ:* Univ Chicago, PhD(pharmacol), 71. *Prof Exp:* ASSOC PROF CARDIOVASC PHARMACOL, SCH MED, ST LOUIS UNIV, 80- *Res:* Renal pharmacology. *Mailing Add:* Dept Pharmacol St Louis Univ Sch Med 1402 S Grand St Louis MO 63104

CHAPPEL, CLIFFORD, b Guelph, Ont, Aug 23, 25; m 54, 70; c 4. BIOLOGY. *Educ:* Ont Vet Col, DVM, 50; McGill Univ, MSc, 53, PhD(invest med), 59. *Prof Exp:* Assoc dir biol res, Ayerst Res Labs, 53-65; pres & dir, Bio-Res Labs, Ltd, 65-76; vpres & dir res, Connlab Holdings Ltd, 76-78; pres, FDC Consults, Inc, 78-86; pres, Cantox Inc, 86-90; RETIRED. *Concurrent Pos:* Lectr, Dept Invest Med, McGill Univ. *Mem:* Soc Exp Biol & Med; Am Soc Pharmacol; Pharmacol Soc Can. NY Acad Sci; Am Soc Toxicol; Europ Soc Toxicol; Soc Toxicol Can. *Res:* Pharmacology; toxicology. *Mailing Add:* 1196 Botany Hill Oakville ON L6J 6J5 Can

CHAPPEL, SCOTT CARLTON, b Syracuse, NY, Apr 22, 50; m 76. NEUROENDOCRINOLOGY. *Educ:* Pa State Univ, BS, 72; Univ Md, PhD(physiol), 76. *Prof Exp:* Fel, Ore Regional Primate Res Ctr, 76-78, asst scientist reproduction, 78-79; asst prof, dept obstet-gynec, Univ Pa, 79-; asst prof, Dept Pediat, Univ Mich, C S Mott Childrens Hosp; VPRES & SCI DIR, ARES ADV TECHNOL, 90- *Concurrent Pos:* Res grants, Cammack Trust Fund, 78-79 & NIH, 78-80 & 81-83. *Mem:* Soc Study Reproduction; Endocrine Soc; Am Physiol Soc; NY Acad Sci. *Res:* Reproductive endocrinology and neuroendocrinology. *Mailing Add:* Ares Adv Technol 280 Pond St Randolph MA 02368

CHAPPELEAR, DAVID C(ONRAD), b Dayton, Ohio, Mar 2, 31; m; c 2. CHEMICAL ENGINEERING. *Educ:* Yale Univ, BE, 53; Princeton Univ, MA, 59, PhD(chem eng), 60. *Prof Exp:* Res engr, Plastics Div, Res Dept, Monsanto Co, 53-54 & 56, sr res engr, 60-64, res specialist, 64-65, group supvr, 65, process tech mgr, 65-70, mgr process technol, 70-76, process technol dir, 77, technol dir, 78-80; dir, Corp Res & Develop, Raychem Corp, Menlo Park, 80-82; VPRES, RES & ENG, JOHNSON & JOHNSON, SKILLMAN, NJ, 82- *Concurrent Pos:* Vis lectr, Univ Mass, 61-68, adj assoc prof, 68- *Mem:* Am Inst Chem Engrs; Am Chem Soc; NY Acad Sci; Sigma Xi. *Res:* Polymer product and process development; chemical reactor design; colloidal chemistry; radiation chemistry; flow of dispersed systems; disposable health care products. *Mailing Add:* 51 W Shore Dr Pennington NJ 08534

CHAPPELL, CHARLES FRANKLIN, b St Louis, Mo, Dec 7, 27; m 51; c 3. METEOROLOGY, ATMOSPHERIC PHYSICS. *Educ:* Wash Univ, BS, 49; Colo State Univ, MS, 66, PhD(atmospheric sci), 71. *Prof Exp:* Flight test engr, MacDonnell Aircraft Corp, 50-55; anal & forecasting, Nat Weather Serv, 56-67; res meteorologist, dept atmospheric Sci, Colo State Univ, 67-70; assoc prof meteorol, Utah State Univ, 70-72; res meteorologist, Off Weather Modification, 72-73, sr scientist & dep dir, Atmospheric Physics & Chem Lab, 73-79, actg dir, Off Weather Res & Modification, 79-81, dir, 81-83; dir, Weather Res Prog, Environ Res Labs, Nat Oceanic & Atmospheric Admin, 84-87; head, Appl Sci Group, Res Applications Prog, Nat Ctr Atmospheric Res, 88-89; SR METEOROLOGIST, COOP PROG OPER METEOROL EDUC & TRAINING, UNIV CORP ATMOSPHERIC RES, 90- *Concurrent Pos:* Adj prof, Utah State Univ, 72-74; assoc mem grad fac, Colo State Univ, 73-, vis assoc prof, 74. *Honors & Awards:* Nat Oceanic & Atmospheric Admin Award. *Mem:* Am Geophys Union; fel Am Meteorol Soc; Sigma Xi; Weather Modification Asn; Nat Weather Asn. *Res:* Parameterization of severe convection in mesoscale numerical models, genesis and organization of mesoscale convective systems, and mesoscale modeling of precipitation over mountainous terrain; severe storm prediction; winter storm and snowfall prediction. *Mailing Add:* FW Rm 278 NCAR PO Box 3000 Boulder CO 80307-3000

CHAPPELL, CHARLES RICHARD, b Greenville, SC, June 2, 43; m 68; c 1. MAGNETOSPHERIC PHYSICS. *Educ:* Vanderbilt Univ, BA, 65; Rice Univ, PhD(space sci), 69. *Prof Exp:* Consult, Lockheed Palo Alto Res Lab, 68, assoc res scientist, 68-70, res scientist, 70-73, staff scientist, 73-74; chief, Magnetospheric & Plasma Physics Br, 74-80, chief, Solar Terrestrial Physics Div, Space Sci Lab, 80-87, ASSOC DIR SCI, MARSHALL SPACE FLIGHT CTR, NASA, 87- *Concurrent Pos:* Assoc res physicist, Univ Calif, San Diego, 73-74. *Honors & Awards:* Outstanding Performance & Medal for Exceptional Scientific Achievement, NASA, 80 & 84. *Mem:* Am Geophys Union; AAAS; Int Asn Geomagnetism & Aeronomy. *Res:* Space science; study of low energy particle population of the magnetosphere; the plasmasphere and ionosphere; magnetospheric convection. *Mailing Add:* 2803 Downing Ct Huntsville AL 35801

CHAPPELL, DOROTHY FIELD, b Va, Aug 12, 47. PHYCOLOGY, MARINE BIOLOGY. *Educ:* Longwood Col, Farmville, BS, 69; Univ Va, Charlottesville, MS, 73; Miami Univ Ohio, Oxford, PhD(bot), 77. *Prof Exp:* Teacher biol & phys sci, Prince Edward Found, 69-73; PROF BIOL, WHEATON COL, 77- *Concurrent Pos:* Consult, evaluator & reader, NCent Asn 84-; Fulbright fel, US Educ, NZ, 89-90. *Mem:* AAAS; Phycol Soc Am; Bot Soc Am; Sigma Xi; Electron Micros Soc Am. *Res:* Ultrastructural phycology and systematics; green algae; vascular plant systematics. *Mailing Add:* Dept Biol Wheaton Col Wheaton IL 60187

CHAPPELL, ELIZABETH, b Chicago, Ill, Jan 26, 27; m 73; c 2. PHARMACOLOGY. *Educ:* St Xavier Col, BS, 45; Univ Ill, MS, 47, PhD(pharmacol), 61. *Prof Exp:* Instr sci, St Xavier Col, 47-48; res asst pharmacol, Abbott Labs, 48-53, chem, 56-57, sr pharmacologist, 60-74, regulatory affairs mgr, 74-; RETIRED. *Res:* General pharmacology, including blood coagulation, enzyme induction, antidiabetic drugs and general screening procedures; enzymology; biogenic amines and cyclic amp. *Mailing Add:* 106 Leisureville Blvd Boynton Beach FL 33426-4351

CHAPPELL, GUY LEE MONTY, b Marysville, Ohio, Aug 23, 40; m 62; c 2. RUMINANT NUTRITION, PHYSIOLOGY. *Educ:* Berea Col, BS, 62; Va Polytech Inst, MS, 65, PhD(ruminant nutrit), 66. *Prof Exp:* From asst prof to assoc prof, 66-77, PROF SHEEP EXTEN & RES, UNIV KY, 78- *Mem:* Am Soc Animal Sci; Sigma Xi. *Res:* Factors affecting cellulose digestibility in the ruminant; factors affecting roughage utilization ruminants; high energy rations for early weaned lambs; intensive sheep production; youth progams in animal science; sheep production models. *Mailing Add:* Dept Animal Sci Univ Ky Lexington KY 40506-0512

CHAPPELL, MARK ALLEN, b Agana, Guam, Sept 3, 51. PHYSIOLOGICAL & BEHAVIORAL ECOLOGY. *Educ:* Univ Calif, Santa Cruz, BA, 73; Stanford Univ, PhD(biol), 77. *Prof Exp:* Res biologist, Naval Arctic Res Lab, 78-79; fel biol, Univ Calif, Los Angeles, 79-80, vis prof, 80; asst prof, 80-86, ASSOC PROF BIOL, UNIV CALIF, RIVERSIDE, 86- *Mem:* Ecol Soc Am; Am Soc Zoologists; AAAS; Animal Behav Soc. *Res:* Physiological and behavioral ecology; energetics; temperature regulation; physiological, behavioral and ecological aspects of adaptation to harsh environments. *Mailing Add:* Dept Biol Univ Calif Riverside CA 92521

CHAPPELL, RICHARD LEE, b Buffalo, NY, Mar 9, 38; m 68; c 2. VISION PHARMACOLOGY. *Educ:* Princeton Univ, BSE, 62; Johns Hopkins Univ, PhD(biophys), 70. *Prof Exp:* Test engr nuclear eng, Naval Reactors, US AEC, 62-66; from asst prof to assoc prof, 70-79, PROF BIOL SCI, HUNTER COL, 80- *Concurrent Pos:* Consult, Sci Develop Prog, Inc, 78-; chmn physiol & neurosci sub prog, City Univ NY PLD prog in biol, 86-88; chmn, Dept Biol Sci, Hunter Col, 87-90. *Honors & Awards:* Antartica Geog Feature named in honor, Chappell Peak. *Mem:* AAAS; Asn Res Vision & Ophthal; Inst Elec & Electronic Engrs; Soc Neurosci; Soc Gen Physiologists. *Res:* Electrophysiology; pharmacology; neuroanatomy and information processing with emphasis on the retina and visual systems. *Mailing Add:* Dept Biol Sci Hunter Col 695 Park Ave New York NY 10021

CHAPPELL, SAMUEL ESTELLE, b Abingdon, Va, Apr 18, 31; m 78; c 1. PHYSICS. *Educ:* Va State Col, BS, 52; Pa State Univ, MS, 59, PhD(physics), 62. *Prof Exp:* Physicist radiation physics, 62-72, physicist info specialist, 72-78, PHYSICIST STANDARDS SPECIALIST, NAT BUR STANDARDS, 78- *Concurrent Pos:* Sci fel, Off of President's Spec Trade Rep, 76-77; rep US Dept State, Int Orgn Legal Metrol, 87. *Mem:* Am Phys Soc; AAAS; Sigma Xi; fel Am Soc Testing & Mats. *Res:* Experimental work in ionizing radiation dosimetry and radiation effects on materials; coordinates participation in domestic and international standardization organizations; facilitates international cooperative projects in measurement science and technology. *Mailing Add:* 3812 Garfield St NW Washington DC 20007

CHAPPELL, WILLARD RAY, b Boulder, Colo, Feb 27, 38; m 71; c 2. RISK ANALYSIS, HEALTH EFFECTS. *Educ:* Univ Colo, BA, 62, PhD(physics), 65; Harvard Univ, Am, 63. *Prof Exp:* Fel, Smithsonian Astropys Observ, 65-66; fel, Lawrence Livermore Nat Lab, 66-67; from asst prof to prof physics & astrophys, 67-77, PROF PHYSICS UNIV COLO, DENVER, 77-, DIR, CTR ENVIRON SCI, 79- *Concurrent Pos:* Asst to vpres corp affairs, Sci Applns Inc, 76-77; acad visitor, Imperial Col, Univ London, 83-84; chmn, US Dept Energy Oil Shale Task Force, 78-82; sci adv, Off Gov Colo, 75-77; mem, Dir Life Sci Adv Comt, Los Alamos Nat Lab, 80-83, Nat Res Coun Comt Synfuels Facil Safety, 81-82; prof prev med, Univ Colo Health Sci Ctr, 77-, dir, Molybdenum Proj, 71-78. *Mem:* Am Phys Soc; Soc Environ Geochem & Health (secy/treas, 88-); Nat Asn Environ Prof; Soc Risk Analysis; AAAS. *Res:* Risk analysis of toxic chemicals; health & environmental effects of heavy metals; health & environmental effects of energy production; extrapolation of toxicological data fromanimals to humans. *Mailing Add:* Ctr Environ Sci Campus Box 136 Univ Colo 1200 Lorimer St Denver CO 80202

CHAPPELLE, DANIEL EUGENE, b Washington, DC, Mar 15, 33; m 83; c 3. RESOURCE ECONOMICS. *Educ:* Colo State Univ, BS, 56; Duke Univ, MF, 59; State Univ NY Col Forestry, Syracuse Univ, PhD(forestry econ), 65. *Prof Exp:* Res forester econ, Southeastern Forest Exp Sta, US Forest Serv, 56-61; res asst forestry econ, State Univ NY Col Forestry, Syracuse Univ, 61-64; economist, Pac Northwest Forest & Range Exp Sta, US Forest Serv, 64-66, prin economist, 66-68; assoc prof, 68-71, PROF RESOURCE ECON, MICH STATE UNIV, 71- *Mem:* Asn Environ & Resource Economists; Regional Sci Asn. *Res:* Natural resource economics; regional economics and regional science; economic development. *Mailing Add:* Dept Resource Develop Mich State Univ East Lansing MI 48824-1222

CHAPPELLE, EMMETT W, b Phoenix, Ariz, Oct 24, 25; m 47; c 4. BIOCHEMISTRY, PHOTOBIOLOGY. *Educ:* Univ Calif, BA, 50; Univ Wash, MS, 54. *Prof Exp:* Instr biochem, Meharry Med Col, 50-52; res assoc, Stanford Univ, 55-58; scientist biochem, Res Inst Advan Studies, 58-63; biochemist, Hazleton Labs, 63-66; exobiologist, 66-70, astrochemist, 70-73, Photobiologist, 73-77, AGR REMOTE SENSING SCIENTIST, GODDARD SPACE FLIGHT CTR, NASA, 77- *Concurrent Pos:* Consult, Appl Magnetics Corp, 73-74; NASA fel, Johns Hopkins Univ, 75-77. *Mem:* Am Chem Soc; NY Acad Sci; Am Soc Photobiol; Am Soc Biol Chem. *Res:* Metabolism of iron in mammalian systems; methods for quantitative assay of proteins and amino acids; carbon monoxide utilization by green plants; bioluminescence; methods for microbial determination; interstellar molecules; remote spectral analysis of agricultural crops. *Mailing Add:* 2502 Allendale Rd Baltimore MD 21216

CHAPPELLE, THOMAS W, b Petersburg, Ind, Feb 11, 18; m 52; c 1. LONG RANGE PLANNING & FORECASTING. *Educ:* Univ Cincinnati, ME, 41; George Washington Univ, MEA, 65. *Prof Exp:* Engr, Cincinnati Milling & Grinding Mach, Inc, 41-42 & Atlanta Chem Warfare Procurement Dist, 42-43; asst prof mech eng, chmn com eng dept & coordr co-op plan, Univ Denver, 46-48; personnel dir & asst to contract adminr, Res Labs, Bendix Aviation Corp, 48-51; chief oper capability team & dep chief opers anal, Hqs, Strategic Air Command, 51-55 & 57-58; chief, Off Opers Anal, Hqs, Eighth Air Force, 55-57; br chief, Anal Serv Inc, 58-60, vpres, 60-76; dir prog planning & anal, NASA, 76-77, asst to chief scientist, 77-80, dep dir, Mgt Support Div, 80-87, dir planning integration, Hq, 87-90, MGR, INTERGOVT AFFAIRS, HQ, NASA, 90- *Concurrent Pos:* Dir hist office, HQ, NASA, 82-83. *Mem:* Fel AAAS; Am Astronaut Soc; Nat Space Club. *Res:* Communicating to senior officials of state and local governments NASA's technology advances and activities that could benefit their jurisdictions; managing interagency agreements process and records. *Mailing Add:* Apt A114 6166 Leesburg Pike Falls Church VA 22044

CHAPPELOW, CECIL CLENDIS, JR, b Kansas City, Mo, Apr 12, 28; m 47; c 5. ORGANIC & POLYMER CHEMISTRY. *Educ:* Univ Southern Calif, BA, 51; Univ Mo-Kansas City, MA, 57, PhD, 68. *Prof Exp:* Sr chemist, Chem Div, Midwest Res Inst, 50-66, prin chemist, 66-68, head org & polymeric mat sect, 68-70, head org & polymer chem, 70-73, mgr indust progs, 73-75, head mat sci, 75-85, SR ADV, MIDWEST RES INST, 85- *Concurrent Pos:* Adj prof Clin Dent, Univ Mo, Kansas City, 84- *Mem:* Am Chem Soc; Int Asn Dent Res. *Res:* Synthesis, physicochemical characterization; structure-property correlation; substituted ureas and sulfamides; organometallic monomers and polymers; epoxidation and hydroxylation reactions; hydrogen-fluoride catalyzed condensations; metal chelating agents; biomedical polymer systems; cellulose conversion and utilization; corrosion protective coatings; synthetic polymeric membranes; dental adhesives and composites. *Mailing Add:* 12305 Cherokee Lane Leawood KS 66209

CHAPPELOW, CECIL CLENDIS, III, b Culver City, Calif, July 5, 48; m 78; c 2. CHEMICAL ENGINEERING. *Educ:* Univ Mo-Columbia, BSChE, 70; Univ Calif, Berkeley, PhD(chem eng), 74; Lehigh Univ, MBA, 85. *Prof Exp:* Sr res engr, MPM Div, Pfizer, Inc, 73-77; sr process engr, 77-78, sect mgr res & develop, 78-79, mgr polymer res, 79-84, dept mgr appln & tech serv, 84-86, DIR, POLYMERS TECHNOL, AIR PROD & CHEM, INC, 87- *Mem:* Am Inst Chem Engrs. *Res:* Emulsion and solution polymerization; process and product development and application to adhesives, nonwovens, paper coatings, paints, building products, carpets and textiles; phase equilibria with emphasis on gas-liquid solutions. *Mailing Add:* Air Prod & Chem Inc Allentown PA 18195

CHAPPLE, PAUL JAMES, b St Helier, UK, July 19, 33. CELL BIOLOGY, VIROLOGY. *Educ:* Univ Bristol, BSc, 57, PhD, 60. *Prof Exp:* Res asst, Univ Bristol, Eng, 59-60; res fel, Ministry Agr, Fisheries & Food, Worplesdon, Eng, 60-63; mem sci staff, Common Cold Unit, Med Res Coun, Nat Inst Med Res, Salisbury, 63-66; managing dir & pres, Flow Labs Ltd, Scotland, 66-72, vpres, Int Opers, Flow Labs, Inc, Rockville, 72-74; Dir, W Alton Jones Cell Sci Ctr, Lake Placid, 74-81; PRES, TKI, INC, 81- *Concurrent Pos:* Vis scientist, Microbiol Res Estab, Porton, Eng, 62-63; WHO consult, 64-65; co-investr, United Cerebral Palsy Res & Educ Found Inc, 75-78, Nat Heart, Lung & Blood Inst, 77-78 & Nat Inst Allergy & Infectious Dis, 77-79; prin investr, Nat Heart, Lung & Blood Inst, 77-78 & Nat Inst Aging, 77-80; prog dir, NIH, 78-79; dir, Bank Lake Placid. *Mem:* AAAS; Am Soc Microbiol; Am Soc Cell Biol; Soc Gen Microbiol; Tissue Cult Asn. *Res:* New cell substrate for virus diagnosis and propagation; culture and study of differentiated cells. *Mailing Add:* 242 Twin Creeks Dr Chagrin Falls OH 44022

CHAPPLE, WILLIAM DISMORE, b Boston, Mass, July 24, 36; m 63; c 1. NEUROPHYSIOLOGY, COMPARATIVE PHYSIOLOGY. *Educ:* Harvard Univ, BA, 58; Syracuse Univ, MA, 60; Stanford Univ, PhD(biol), 65. *Prof Exp:* Res biologist, Control Systs Lab, Stanford Res Inst, 63-65; NATO fel, Cambridge Univ & Bristol Univ, 65-66; asst prof, 66-70, assoc prof, 70-80, PROF BIOL, UNIV CONN, 80- *Mem:* Fel AAAS; Soc Neurosci; Brit Soc Exp Biol. *Res:* Comparative neurophysiology; neural basis of patterned movement. *Mailing Add:* Biol Sci Group U-42 Univ Conn 75 N Eaglevill Storrs CT 06268

CHAPUT, RAYMOND LEO, b Manchester, NH, Mar 29, 40; m 62; c 5. RADIATION BIOLOGY, HEALTH PHYSICS. *Educ:* St Anselm Col, BA, 62; Univ Rochester, MS, 65; Univ Tenn, PhD(zool), 67. *Prof Exp:* Prin investr, Armed Forces Radiobiol Res Inst, 67-73; head non ionizing radiation sect, Naval Bur Med & Surg, 73-75; chief nuclear submarine med sect, Naval Undersea Med Inst, 75-78; dir radiation health, Pearl Harbor Naval Shipyard, 78-81; chief, Exp Path Div, Armed Forces Radiobiol Res Inst, 81-83; PROG MGR, NAVY BIOMED RES & DEVELOP, OFF NAVAL TECHNOL, 83- *Concurrent Pos:* Comt mem, Am Nat Standard Inst, 73-75. *Mem:* Health Physics Soc; Sigma Xi. *Res:* Amino acid and metabolism in insect larvae and pupae; effects of ionizing radiation on insect hematocytes; effects of ionizing radiation on animal behavior; neurochemistry and supportive treatment after irradiation using bone marrow and peripheral blood elements. *Mailing Add:* Naval Med Command Washington DC 20372

CHAR, DONALD F B, b Honolulu, Hawaii, Mar 25, 25; m 51; c 5. PUBLIC HEALTH, PEDIATRICS. *Educ:* Temple Univ, MD, 50. *Prof Exp:* Intern, Atlantic City Hosp, 50-51; resident pediat, St Christopher's Hosp Children, 53-56; dir med educ, Kauikeolani Children's Hosp, 56-59 & 62-65; from instr to asst prof pediat, Med Sch, Univ Wash, 59-63; DIR STUDENT HEALTH & PROF PEDIAT, UNIV HAWAII, 65- *Concurrent Pos:* Fel pediat cardiol, St Christopher's Hosp Children, 55-56; res instr, Med Sch, Univ Wash, 62-63; lectr, East-West Training Prog Med Practitioners, Apia, WSamoa, 65; hon consult, Health Ctr, Chinese Univ, Hong Kong, 73; vis prof pediat, Med Sch, Hong Kong Univ, 73; fel, East West Ctr, 80-81; chmn, Hawaii State Bd Health, 83-85; interim dean students, Univ Hawaii, 89-90. *Honors & Awards:* Ruby Rich Burgar Award, 90. *Mem:* Am Acad Pediat; Am Col Health Asn. *Res:* Immune status in adrenalectomized animals; cross cultural health care and foreign students. *Mailing Add:* Student Health Serv Univ Hawaii 1710 East West Rd Honolulu HI 96822

CHAR, WALTER F, b Honolulu, Hawaii, May 27, 20; m 48; c 3. PSYCHIATRY. *Educ:* Temple Univ, MD, 45; Am Bd Psychiat & Neurol, cert psychiat, 52, cert psychoanal, 58, cert child psychiat, 60. *Prof Exp:* Intern, Med Ctr, Temple Univ, 45-46; resident psychiat, Univ Pittsburgh, 48-49; resident, Med Ctr, Temple Univ, 49-52, assoc prof psychiat, Sch Med & dir child psychiat, Med Ctr, 52-62; chmn dept psychiat, 67-69, PROF PSYCHIAT, SCH MED, UNIV HAWAII, 67-, ASSOC CHMN DEPT, 69- *Concurrent Pos:* Pvt pract psychiat & psychoanalyst, 52-; consult psychiat, Child & Family Serv, Honolulu, 62-88, Army Tripler Gen Hosp, 65-88. *Mem:* Am Psychiat Asn; Am Psychoanalyst Asn; Am Acad Child Psychiat; Int Psychoanalyst Asn. *Res:* Evaluation of admission procedure of medical students; transcultural psychiatry; family research; study of normal families from a cross-ethnic point of view. *Mailing Add:* Univ Hawaii Sch Med 1356 Lusitana St Honolulu HI 96813

CHARACHE, PATRICIA, b Newark, NJ, Dec 26, 29; m 51; c 1. MEDICINE. *Educ:* Hunter Col, BA, 52; NY Univ, MD, 57. *Prof Exp:* Intern med, Baltimore City Hosps, 57-58; res assoc immunol, Childrens Hosp & Harvard Univ, 62-64; from instr to asst prof med, 64-73, asst prof microbiol, 70-79, ASSOC PROF MED & LAB MED, JOHNS HOPKINS UNIV, 73- *Concurrent Pos:* USPHS fel res med, Univ Pa, 58-59, Porter fel, 59-60; USPHS fel infect dis, Johns Hopkins Univ, 60-62; res career develop award, Nat Inst Allergy & Infect Dis, 69-73. *Mem:* Am Soc Clin Pharmacol; Reticuloendothelial Soc; Infect Dis Soc Am; Am Soc Microbiol; Fedn Clin Res. *Res:* Immunology and infectious disease, emphasizing bacterial-host interaction; medical microbiology; rapid microbial detection. *Mailing Add:* Dept Lab Med Johns Hopkins Hosp Baltimore MD 21205

CHARACHE, SAMUEL, b New York, NY, Jan 12, 30; m 51; c 1. MEDICINE, HEMATOLOGY. *Educ:* Oberlin Col, BA, 51; NY Univ, MD, 55; Am Bd Internal Med, dipl & cert hemat; Am Bd Clin Path, dipl & cert hemat. *Prof Exp:* Clin assoc, NIH, 56-58; resident med, Hosp Univ Pa, 58-60; from asst prof to assoc prof, 66-78, PROF MED, SCH MED, JOHNS HOPKINS UNIV, 78-, PROF PATH, 81- *Concurrent Pos:* USPHS, 56-58; fel hemat, Johns Hopkins Hosp, 60-62 & 64-66; fel biol, Mass Inst Technol, 62-64. *Mem:* Asn Am Physicians; Am Soc Hemat. *Res:* Hematology. *Mailing Add:* Dept Med Johns Hopkins Hosp Baltimore MD 21205

CHARACKLIS, WILLIAM GREGORY, b Annapolis, Md, Aug 21, 41; m 64; c 2. CHEMICAL & ENVIRONMENTAL ENGINEERING. *Educ:* Johns Hopkins Univ, BES, 64, PhD(environ eng), 70; Univ Toledo, MSChE, 67. *Prof Exp:* Res engr, Olin-Matheson Chem Corp, 64-65; res asst chem eng, Univ Toledo, 65-67; res asst environ eng, Johns Hopkins Univ, 67-70; from asst prof to assoc prof, 70-78, prof environ eng & chmn dept, Rice Univ, 78-; AT DEPT CIVIL ENG, MONT STATE UNIV, BOZEMAN. *Concurrent Pos:* Merck Found fac develop grant, 72; NSF fac sci fel, 76; NSF fac sci fel, Swiss Fed Inst Water Resources & Water Pollution Control, 77-78. *Mem:* Am Inst Chem Engrs; Int Asn Water Pollution Res; Water Pollution Control Fedn. *Res:* Microbial engineering, biofouling, water and wastewater engineering. *Mailing Add:* Dept Civil Eng Mont State Univ Bozeman MT 59717

CHARAP, STANLEY H, b Brooklyn, NY, Apr 21, 32; m 55; c 2. APPLIED MAGNETICS. *Educ:* Brooklyn Col, BS, 53; Rutgers Univ, PhD(physics), 59. *Prof Exp:* Asst physics, Rutgers Univ, 53-55, instr, 57-58; physicist, Res Ctr, Int Bus Mach Corp, 58-64; res scientist, Res Div, Am Stand Corp, 64-65, supvr solid state physics, 65-66, mgr physics & electronics, 66-68; assoc prof, 68-71, assoc dept head, 80-85, actg dept head, 81-82, PROF ELECT & COMPUT ENG, CARNEGIE-MELLON UNIV, 71- *Concurrent Pos:* Sr vis fel, Wales Univ, Cardiff, UK, 76; Bell Lab, Whippany, NJ, summer 73; Ed, Inst Elec & Electronics Engrs Trans on Magnetics, NY, 75-86; gen chmn, Conf on Magnetism & Magnetic Mat, Baltimore, 86, Joint Intermay-Conf, Albuquerque, NMex, 94. *Mem:* fel Inst Elec & Electronic Engrs; Magnetics Soc (secy-treas, 87-88, vpres, 89-90, pres, 91); Sigma Xi. *Res:* Magnetic domains, magnetic recording, magnetic hysteresis models; magnetic fine particles. *Mailing Add:* Dept Elec & Computer Eng Carnegie Mellon Univ Pittsburgh PA 15213

CHARBENEAU, GERALD T, b Mt Clemens, Mich, July 22, 25; m 47; c 3. DENTISTRY. *Educ:* Univ Mich, DDS, 48, MS, 49. *Prof Exp:* Teaching fel dent, Sch Dent, Univ Mich, Ann Arbor, 48-49; from instr to assoc prof, 49-65, PROF DENT, SCH DENT, UNIV MICH, ANN ARBOR, 65-, CHMN DEPT OPER DENT, 69- *Mem:* Am Dent Asn; Int Asn Dent Res; Am Acad Restorative Dent; Acad Oper Dent; Sigma Xi. *Res:* Operative and general restorative dentistry with specific relationship of dental materials to these clinical areas. *Mailing Add:* 1200 Naples Ct Ann Arbor MI 48103

CHARBENEAU, RANDALL JAY, b Ann Arbor, Mich, Oct 5, 50; m 74; c 2. GROUNDWATER POLLUTION, FATE & TRANSPORT MODELING. *Educ:* Univ Mich, BS, 73; Ore State Univ, MS, 75; Stanford Univ, PhD(civil eng), 78. *Prof Exp:* From asst prof to assoc prof, 78-91, PROF CIVIL ENG, UNIV TEX, AUSTIN, 91- *Concurrent Pos:* Dir-at-large, Am Water Resources Asn, 84-87; dir, Ctr Res Water Resources, Univ Tex, 89- *Mem:* Am Geophys Union; Am Soc Civil Engrs; Am Water Resources Asn. *Res:* Groundwater hydraulics and contaminant transport; numerical modeling; radiological assessment; hydraulics. *Mailing Add:* Dept Civil Eng Univ Tex Austin TX 78712

CHARBONNEAU, LARRY FRANCIS, b Faribault, Minn, Aug 14, 39; m 66; c 2. ORGANIC POLYMER CHEMISTRY. *Educ:* Mankato State Col, BS, 64; Univ Ill, Urbana-Champaign, MS, 69, PhD(org chem), 72. *Prof Exp:* Chem technician polymer chem, 3M Co Cent Res, 60-62, chemist photochem, 64-67; assoc sr res chemist org polymer chem, Gen Motors Res Labs, Warren, Mich, 72-77; sr res chemist org polymer chem, Hoechst Celanese Corp, 77-82, develop assoc, 82-87, res assoc/proj leader, Hoechst Celanese Res Div, 87-89, RES SUPVR, ADVAN TECHNOL GROUP, HOECHST CELANESE CORP, 89- *Mem:* Am Chem Soc; NY Acad Sci. *Res:* Synthesis of materials for optoelectronics. *Mailing Add:* 64 Mountainside Rd Mendham NJ 07945-2016

CHARBONNIER, FRANCIS MARCEL, b Monaco, Apr 28, 27; nat US; m 52; c 6. PHYSICS. *Educ:* Polytech Sch Paris, Dipl d'Ing, 49; Univ Wash, PhD(physics), 52. *Prof Exp:* Engr, Militaire de l'Armement, France, 52-55; sr physicist, Linfield Res Inst, 56-57, tech asst to dir, 57-59, asst dir, 59-62; dir res & develop div, Field Emission Corp, 62-64, vpres & res dir, 64-73; ENG MGR, McMINNVILLE DIV, HEWLETT-PACKARD CO, 74- *Mem:* Am Asn Physicists Med; Am Phys Soc; Inst Elec & Electronics Engrs. *Res:* Field emission; electron physics, optics and devices; medical x-ray systems; pulsed radiation sources. *Mailing Add:* Hewlett-Packard Co McMinnville Div 1700 S Baker St McMinnville OR 97128

CHARD, RONALD LESLIE, JR, pediatric oncology & hematology, for more information see previous edition

CHAREST, PIERRE M, b Quebec City, Que, Mar 8, 53. ELECTRON MICROSCOPY, IMMUNOCYTOCHEMISTRY. *Educ:* Laval Univ, BScA, 77, PhD(plant sci), 83. *Prof Exp:* Asst prof, 85-90, ASSOC PROF PLANT ANAT & MOLECULAR BIOL, LAVAL UNIV, 90- *Concurrent Pos:* Postdoctoral fel cell biol, Biozentrum, Basel Univ, Switz, 83-84. *Mem:* Am Soc Cell Biol. *Res:* Molecular study of the plant cytoskeletal proteins and their associated proteins; molecular characterization of the membrane skeletal proteins of the plant cells. *Mailing Add:* Dept Phytology Dav Comtois Laval Univ Ste Foy PQ G1K 7P4 Can

CHARGAFF, ERWIN, b Austria, Aug 11, 05; US citizen; m 29; c 1. BIOCHEMISTRY. *Educ:* Univ Vienna, PhD(chem), 28. *Hon Degrees:* ScD, Columbia Univ, 76; PhD, Univ Basel, 76. *Prof Exp:* Milton Campbell res fel, Yale Univ, 28-30; asst, dept bact & pub health, Univ Berlin, 30-33; res assoc, Pasteur Inst, 33-34; res assoc, 35-38, from asst prof to prof, 38-74, chmn dept, 70-74, EMER PROF BIOCHEM, COL PHYSICIANS & SURGEONS, COLUMBIA UNIV, 74- *Concurrent Pos:* Guggenheim fels, 49 & 57-58; vis prof, Wenner Grens Inst, Univ Stockholm, 49, Univs Rio de Janeiro, Sao Paulo & Recife, 59, Cornell Univ & Univs Naples & Palermo, 66 & Biol Sta, Naples, 69; Harvey lectr, Rockefeller Univ, 56; Plenary Cong lectr, Int Biochem Cong Vienna, 58; vis lectr, Univs Tokyo, Kyoto, Sendai & others, 58; mem comt growth, Nat Res Coun, 52-54; mem adv coun biol, Oak Ridge

Nat Lab, 58-67; Albert Einstein chair, Col France, 65-80. *Honors & Awards:* Pasteur Medal, Paris, 49; Carl Neuberg Medal, 58; Hesup lectr, Columbia Univ, 59; Soc Chem Biol Medal, Paris, 61; Charles Leopold Mayer Prize, Acad Sci, Paris, 63; Dr H P Heineken Prize, Royal Neth Acad Sci, 64; Bertner Found Award, 65; first K A Fonster lectr, Mainz, 68; Miescher Mem lectr, Basel, 69; Gregor Mendel Medal, Halle, Ger, 73; Nat Medal Sci, 74; NY Acad Med Medal, 80. *Mem:* Nat Acad Sci; fel Am Acad Arts & Sci; Am Philos Soc; German Acad Sci; foreign mem Royal Swed Physiol Soc. *Res:* Lipids; lipoproteins; blood coagulation; metabolism of amino acids and inositol; chemistry and biosynthesis of nucleic acids and nucleoproteins; phosphotransferases and other enzymes. *Mailing Add:* 350 Cent Park W Apt 13G New York NY 10025-6547

CHARI, NALLAN C, b Rajahmundry, India, Nov 2, 31; m 54; c 1. CHEMICAL ENGINEERING. *Educ:* Benares Hindu Univ, BS, 54; Univ Mich, MS, 57, ScD(chem eng), 60. *Prof Exp:* Instr chem eng, Benares Hindu Univ, 54-55; asst eng res inst, Univ Mich, 55-60; staff chem engr, Tech Ctr, Owens-Illinois Glass Co, 60-67, chief programmer, 67-70, mgr process comput control, 70-73, mgr systs develop, 73-77, MGR PROCESS ENG & CONTROL, OWENS-ILLINOIS, INC, 77- *Mem:* Am Chem Soc; Am Inst Chem Engrs; Tech Asn Pulp & Paper Indust; Can Pulp & Paper Asn; Nat Soc Prof Engrs. *Res:* Thermodynamics; process dynamics and control; pulp and paper process design and systems engineering study. *Mailing Add:* 1627 Woodhurst Dr Toledo OH 43614

CHARKES, N DAVID, b New York, NY, Aug 13, 31; m 53; c 3. NUCLEAR MEDICINE. *Educ:* Columbia Univ, AB, 52; Washington Univ, MD, 55. *Prof Exp:* USPHS fel arthritis & metab dis, 60-61; assoc radiol, 62-66, clin asst prof, 66, assoc prof radiol & med, 66-71, dir, dept nuclear med, 66-80, PROF RADIOL & ASSOC PROF MED, TEMPLE UNIV HOSP, 71-, RES PROF NUCLEAR MED & PROF MED, 80- *Concurrent Pos:* Dir radioisotope unit, North Div, Albert Einstein Med Ctr, 62-66; Fogarty sr int fel, 76. *Mem:* Soc Nuclear Med; Am Fedn Clin Res; Am Thyroid Asn. *Res:* Diagnostic and therapeutic thyroidology using radioactive iodine; tracer kinetics and compartmental modelling. *Mailing Add:* Temple Univ Hosp Broad & Ontario Sts Philadelphia PA 19140

CHARKEY, LOWELL WILLIAM, AMINO ACID METABOLISM, VITAMINS. *Educ:* Cornell Univ, PhD(amino nutrit), 45. *Prof Exp:* Prof, 36-74, EMER PROF BIOCHEM, COLO STATE UNIV, 74- *Mailing Add:* 828 Gregory Rd Ft Collins CO 80523

CHARKOUDIAN, JOHN CHARLES, b Springfield, Mass, July 29, 41; m 76; c 2. CHEMISTRY, POLYMER SCIENCE. *Educ:* Bates Col, BS, 63; Babson Col, MBA, 67; Boston Univ, MS, 67; Va Polytech Inst & State Univ, PhD(phys inorg chem), 70. *Prof Exp:* Sr scientist, Photog Chem, Polaroid Corp, 70-83; sr res assoc, Tufts Univ Sch Med, 83-85; sr res assoc, Kendall Co, 85-90; GROUP MGR, CORE CHEM, MILLIPORE CORP, 90- *Mem:* AAAS; Am Chem Soc; Am Membrane Soc. *Res:* Diffusion in and permeability of polymers; surface energetics and characterization; surface modification of materials; surface science. *Mailing Add:* 76 Pheasant Hill Lane Carlisle MA 01741

CHARLANG, GISELA WOHLRAB, microbial genetics, microbial ecology; deceased, see previous edition for last biography

CHARLAP, LEONARD STANTON, b Wilmington, Del, Aug 1, 38; c 2. MATHEMATICS. *Educ:* Mass Inst Technol, BS, 59; Columbia Univ, PhD(math), 62. *Prof Exp:* Mem, Inst Advan Study, 62-64 & 90-91; from asst prof to assoc prof math, Univ Pa, 64-69; from assoc prof to prof math, State Univ NY Stony Brook, 69-83; MEM RES STAFF, CTR COMMUN RES, PRINCETON, NJ, 83- *Concurrent Pos:* Mem, Dimacs, Rutgers Univ, 90-91. *Mem:* Am Math Soc. *Res:* Differential geometry; differential topology; homological algebra; flat Riemannian manifolds. *Mailing Add:* Ctr Commun Res Thanet Rd Princeton NJ 08540

CHARLES, EDGAR DAVIDSON, JR, b Florence, SC, Apr 26, 43; m 65; c 1. PUBLIC HEALTH. *Educ:* Pembroke State Univ, BS, 66; Univ Ala, Tuscaloosa, MS, 68, PhD(econ), 71. *Prof Exp:* Assoc prof pub health, 76-82, chmn dept health care orgn, 82-83, ASST PROF HOSP & HEALTH ADMIN, UNIV ALA, BIRMINGHAM, 71-, PROF PUB HEALTH, 82- *Concurrent Pos:* WHO fel, 74. *Mem:* Am Econ Asn; Am Pub Health Asn; Health Econ Res Org. *Res:* Health care organization and health economics; cost benefits and cost-effectiveness analysis; economic evaluation of loss of life, limb and injury; evaluation of the economics and efficacy of new technology. *Mailing Add:* Dept Pub Health SOMSPAH Univ Ala Univ Sta Birmingham AL 35294

CHARLES, GEORGE WILLIAM, b Columbus, Ohio, Dec 24, 15; m 39; c 2. PHYSICS. *Educ:* Ohio State Univ, BA, 37, PhD(physics), 47. *Prof Exp:* Asst physics, Ohio State Univ, 38-42; instr, NC State Col, 42-44; physicist, Naval Ord Lab, 44-46; asst prof physics, Univ Okla, 47-52; sr res physicist, Mound Lab, 52-54; PHYSICIST, OAK RIDGE NAT LAB, 54- *Mem:* Optical Soc Am. *Res:* Atomic spectroscopy; physical optics; heat; spectra of columbium and molybdenum in the extreme ultraviolet; spectra of polonium and of rare earths. *Mailing Add:* 840 W Outer Dr Oak Ridge TN 37830

CHARLES, HARRY KREWSON, JR, b Audubon, NJ, May 29, 44; m 70. ELECTRICAL ENGINEERING, SOLID STATE PHYSICS. *Educ:* Drexel Univ, BS, 67; Johns Hopkins Univ, DPhil(elec eng), 72. *Prof Exp:* Engr aerospace technol elec & commun eng, Goddard Space Flight Ctr, NASA, 68; res assoc solid state physics, 72-73; engr sr staff elec eng & solid state physics, 73-79, sect supvr microelectronics group, 79-81, asst group supvr, 81-84, GROUP SUPVR MICROELECTRONICS, APPL PHYSICS LAB, JOHNS HOPKINS UNIV, 85- *Concurrent Pos:* Nat Defense Educ Act fel; Drexel bd trustees scholar. *Mem:* Sr mem Inst Elec & Electronics Engrs; Am Phys Soc; Int Soc Hybrid Microelectronics; Int Solar Energy Soc; Electron Micros Soc Am. *Res:* Electrical engineering, with emphasis on hybrid microelectronics; thin film resistors, integrated circuits and solar cells. *Mailing Add:* 10500 Patuxent Ridge Way Laurel MD 20707

CHARLES, MICHAEL EDWARD, b Leicester, Eng, Dec 20, 35; Can citizen; m 59; c 2. CHEMICAL ENGINEERING, PIPELINE TRANSPORTATION. *Educ:* Imp Col, Univ London, BASc, 57; Univ Alta, MASc, 59, PhD(chem eng), 63. *Prof Exp:* Res officer, Res Coun Alta, 57-61; res engr, Imp Oil Ltd, Calgary, 63-64; from asst prof to assoc prof chem eng, Univ Toronto, 64-71, asst chmn dept, 70-75, chmn dept, 75-85, PROF CHEM ENG, UNIV TORONTO, 71-, VDEAN, FAC APPL SCI & ENGR, 86-, GOV COUN, 89- *Concurrent Pos:* Dir, Chem Eng Res Consults Ltd, 65-, vpres, 70-75; consult, Imp Oil Ltd, 66 & 74, Milltronics Ltd, 67-71, Wolfe Spiral Pipe Co Ltd, 71-73, Atomic Energy Can Ltd, 72- & Olympia-York Construct, 74; sci adv, Worthington Ltd, Can, 70-74, Panarctic Oil Ltd, 75-77 & 84-85, Ontario Ministry of Labor, 81 & Petro Canada Inc, 85; dir, Ontario Laser & Lightwave Res Ctr, 88-, Ontario Ctr Mat Res, 88- & Mfg Res Corp, Ontario, 88-; mem, Inst Chem Sci & Technol, 89- *Mem:* Can Soc Chem Eng; Can Res Mgt Asn; fel Chem Inst Can; Soc Chem Indust. *Res:* Fluid mechanics of complex systems with industrial significance, especially pipeline transport of solids in slurry and capsule form; two-phase gas-liquid flows; continuous particle size determination. *Mailing Add:* Dept Chem Eng Univ Toronto Toronto ON M5S 1A4 Can

CHARLES, R(ICHARD) J(OSEPH), b Elfros, Sask, Sept 8, 25; US citizen; m 50; c 1. METALLURGY. *Educ:* Univ BC, BS, 48, MS, 49; Mass Inst Technol, ScD(metall), 54. *Prof Exp:* Asst prof metall, Mass Inst Technol, 54-56; res assoc, Gen Elec Res Lab, 56-64 & 65-69, actg mgr, Metals & Ceramics Studies Sect, 64-65, mgr, Properties Br, 69-72, MGR, CERAMICS BR, GEN ELEC CO, CORP RES & DEVELOP, 72- *Concurrent Pos:* Chmn, Gordon Conf Glass, 59-60; Coolidge Fel, Gen Elec Co, 74; adj prof ceramics, Mass Inst Technol, 76-, Robert S Williams lectr, 76. *Honors & Awards:* Raymond Award, Am Inst Mining, Metall & Petrol Engrs, 57; George W Morey Award, Am Ceramic Soc, 72. *Mem:* Am Inst Mining, Metall & Petrol Engrs; Am Ceramic Soc. *Res:* Mechanical properties of brittle materials; electrical and physical chemical properties of oxides, silicates, refractory compounds and metals; permanent magnetism and superconductivity. *Mailing Add:* 2224 Niskayuna Dr Schenectady NY 12309

CHARLES, ROBERT WILSON, b Altoona, Pa, Sept 1, 45; m 83. APPLIED GEOLOGY & HYDROLOGY, ENVIRONMENTAL & EARTH SCIENCES. *Educ:* Bucknell Univ, BS, 67; Mass Inst Technol, PhD(geol), 72. *Prof Exp:* Fel, CIW Geophys Labs, 71-72; fel geochem, Univ BC, 73-74; staff mem geochem, Los Alamos Nat Lab, 74-82, assoc group leader, 82-87, sect leader, 87-89, DEP GROUP LEADER, LOS ALAMOS NAT LAB, 89- *Mem:* AAAS; Mineral Soc Am; Sigma Xi. *Res:* Experimental determination of phase equilibria suitable to describe mineral assemblages found in nature; experimentation involves the routine use and development of high pressure-temperature hydrothermal equipment to duplicate natural conditions. *Mailing Add:* Los Alamos Sci Lab INC-7 M-S J-514 Univ Calif Los Alamos NM 87545

CHARLESWORTH, BRIAN, b Brighton, UK, Apr 29, 45; m 67; c 1. POPULATION GENETICS, EVOLUTIONARY THEORY. *Educ:* Univ Cambridge, UK, BA, 66, PhD(genetics), 69. *Prof Exp:* Lectr genetics, Univ Liverpool, 71-74; lectr biol, Univ Sussex, 74-82, reader, 82-84; PROF ECOL & EVOLUTION, UNIV CHICAGO, 85- *Concurrent Pos:* Vis scientist, Nat Inst Environ Health Sci, 77-88; ed, Heredity, 82-84; dept chmn, Univ Chicago, 86-91; assoc ed, Molecular Biol & Evolution & Paleobiol, 89-, Evolution, 91- *Mem:* Genetics Soc Am; Soc Am Naturalists; Soc Study Evolution. *Res:* Evolutionary theory; population genetics; life-history evolution; evolution of genetic and sexual systems; speciation; Drosophila transposable elements and life-history variation. *Mailing Add:* Dept Ecol & Evolution Univ Chicago 1101 E 57th St Chicago IL 60637

CHARLESWORTH, HENRY A K, b Belfast, Northern Ireland, Jan 29, 31. STRUCTURAL GEOLOGY. *Educ:* Cambridge Univ, MA, 53; Univ Glasgow, PhD(geol), 60. *Prof Exp:* Prof geol, St Andrews Univ, 54-55 & Univ Sask, 55-56; PROF GEOL, UNIV ALTA, 56- *Mem:* Fel Geol Soc Am; fel Geol Soc France (vpres, 87-88). *Mailing Add:* Dept Geol Univ Alta Edmonton AB T6G 2M7 Can

CHARLESWORTH, ROBERT K(ORIDON), b Idaho Falls, Idaho, July 26, 23; m 52; c 4. CHEMICAL ENGINEERING. *Educ:* Univ Wash, BS, 44; Univ Wis, MS, 47; Purdue Univ, PhD(chem eng), 51. *Prof Exp:* Instr chem, Idaho State Col, 47-48; from res & develop engr to sr res engr, Dow Chem Co, 51-68, Res Group Leader, 68-72, process engr specialist, 72-82, res assoc 82-85; RETIRED. *Mem:* Am Chem Soc; Am Inst Chem Engrs. *Res:* High polymers; chemical engineering unit operations; chemical process and design computer simulation. *Mailing Add:* 59 Janin Pl Pleasant Hill CA 94523-1581

CHARLET, LAURENCE DEAN, b Danville, Ill, Oct 6, 46; m 69; c 3. BIOLOGICAL CONTROL, HOST-PLANT RESISTANCE. *Educ:* San Diego State Univ, BS, 69; Univ Calif, Riverside, MS, 73, PhD(entom), 75. *Prof Exp:* Res entomologist, Univ Calif, Riverside, 75-78; RES ENTOMOLOGIST, OILSEEDS RES UNIT, USDA, 78- *Concurrent Pos:* Adj prof entom, NDak State Univ, Fargo; mem gov bd, Int Orgn Biol Control, 84-86, 88-92. *Mem:* Int Orgn Biol Control; Entom Soc Am; Entom Soc Can; Acarological Soc Am. *Res:* Bionomics of pine mites; ecology of domestic mites; pest management and biological control of sunflower insect pests. *Mailing Add:* Agr Res Serv USDA Northern Crop Sci Lab PO Box 5677 State Univ Sta Fargo ND 58105-5677

CHARLEY, PHILIP J(AMES), b Melbourne, Australia, Aug 18, 21; US citizen; m 48; c 3. MECHANICAL & METALLURGICAL ENGINEERING, CHEMISTRY. *Educ:* Univ Wis, BS, 43; Univ Southern Calif, MS, 47, PhD, 60. *Prof Exp:* Test engr, Gen Elec Co, 43-44; lectr, Univ Southern Calif, 47-49; proj engr, Standard Oil Co, 48-55; vpres, 55-70, PRES, TRUESDAIL LABS, INC, 70- *Mem:* AAAS; Am Soc Test & Mat; Am Soc Mech Engrs; Am Soc Metals; Am Chem Soc. *Res:* Analytical mechanics; forensic engineering and chemistry. *Mailing Add:* Truesdail Labs 14201 Franklin Ave Tustin CA 92680-7008

CHARLIER, ROGER HENRI, b Antwerp, Belg, Nov 10, 21; nat US; m 58; c 2. GEOLOGY, GEOGRAPHY. *Educ:* Colonial Univ, Belg, BPol & Admin Sci, 40; Free Univ Brussels, MPolSc, 41, MS, 45; Univ Liege, BS(geol) & BS(geog), 43; Univ Erlangen, PhD(phys geog), 47; Indust Col Armed Forces, dipl, 53; McGill Univ, cert, 53; Univ Paris, LittD(cult geog), 57, ScD(geol, oceanog), 58. *Prof Exp:* Prof geog, Col Baudouin, Belg, 41-42; personal student asst geol, Univ Liege, 43-44; press corresp, 45-51; assoc prof geog & chmn dept, Poly Univ, 51-52; prof phys sci & chmn dept, Finch Col, 52-55; chmn dept geol & geog, Hofstra Univ, 55-58; adj prof geol, Univ Paris, 58-59; vis prof educ, Univ Minn, 59-60; prof geol & geog, Parsons Col, 60-61; dir bur educ travel & study abroad, Northeastern Ill Univ 61-63, chmn area earth sci, 62-63, coordr earth sci progs, 63-66, vchmn dept geog & environ studies, 67-71, prof geog, geol & oceanog, 61-86; prof, 70-88, EMER PROF, FREE UNIV BRUSSELS, BELG, 88- *Concurrent Pos:* Dept dir, UNRRA, 46-47; res analyst, US Govt, 47, 49 & 50; bursar, Carnegie Corp, 53; vis lectr, NY Univ, 53-58, Hunter Col, 57-58 & Univ Aix-Marseille, 58-60; consult, Belg Nat Tourist Off, 56-78, World Tourism Orgn, 74-84 & UNESCO, 84; Fr Govt spec fel, 58-59; resident scholar, Northeastern Ill Univ, 62-65; prof extraordinary oceanog, Univ Brussels, Belg, 70-86; vis prof oceanog, Univ Bordeaux I, France, 71-74 & 83-84; Kellogg fel, Kellogg Found, 82; int adv, COLRUYT Corp, Halle, Belg, 82-83; sci adv, HAECON Corp, Ghent, Belg, 85; vpres, The Green Corp, Chicago, 91- *Honors & Awards:* Knight, Order Acad Palms, France, 72 & Order Leopold, Belg, 73. *Mem:* Asn Am Geogr; Nat Asn Geol Teachers; fel Geol Soc Am; Marine Technol Soc; Am Soc Oceanog. *Res:* Alternative energy from the ocean; marine science and environmental education sponsored by UNESCO and Council of Europe; ocean environmental problems in coastal areas; applications of statistics to the domain of the earth sciences; coastal erosion; ocean energies; ocean economics; ocean engineering. *Mailing Add:* 4055 N Keystone Ave Chicago IL 60641

CHARLTON, DAVID BERRY, b Vancouver, BC, Jan 26, 04; nat US; m 30; c 4. CHEMISTRY, BACTERIOLOGY. *Educ:* Univ BC, BA, 25; Cornell Univ, MS, 29; Iowa State Col, PhD(bact), 33. *Prof Exp:* Asst bacteriologist, Portland, Ore, 26-28; instr bact, Ore State Col, 29-31 & Univ Nebr, 31-32; owner & dir, Charlton Labs, 34-71, consult, Mei-Charlton Inc, 71-74; RETIRED. *Concurrent Pos:* Instr bact, Ore State Col, 34-36. *Mem:* AAAS. *Res:* Food bacteriology; sanitary bacteriology; chlorine compounds as germicides. *Mailing Add:* 14420 SW Farmington Rd No 1 Beaverton OR 97005-2538

CHARLTON, GORDON RANDOLPH, b Newport News, Va, Aug 30, 37; div; c 2. RESEARCH ADMINISTRATION, HIGH ENERGY PHYSICS. *Educ:* Ohio State Univ, BSc, 57; WVa Univ, MSc, 60; Univ Md, PhD(physics), 66. *Prof Exp:* CNRS, Ecole Polytechnique, Paris, 66-69; High Energy Physics Div, Argonne Nat Lab, 69-72; Stanford Linear Accelerator Ctr, 72-73; Physics Dept, Univ Toronto, 73-75; physicist, 75-85, SR PHYSICIST, DIV HIGH ENERGY PHYSIC, DEPT ENERGY, 85- *Mem:* Am Phys Soc; AAAS. *Res:* Experimental high energy physics; management and administration of basic research programs in high energy and elementary particle physics at US national laboratories and universities; major accelerator construction and detector fabrication projects. *Mailing Add:* Div High Energy Physics Dept Energy Washington DC 20845

CHARLTON, HARVEY JOHNSON, b Dillwyn, Va, Aug 18, 34; c 3. MATHEMATICS. *Educ:* Va Polytech Inst, BS, 60, MS, 62, PhD(math), 66. *Prof Exp:* Proj physicist, Atomic Energy Div, Babcock Wilcox Co, Va, 57-59; instr math, Va Polytech Inst, 60-66; ASST PROF MATH, NC STATE UNIV, 66- *Mem:* Am Math Soc; Asn Symbolic Logic; Sigma Xi. *Res:* Modern topology. *Mailing Add:* 236 Singleton St Raleigh NC 27606

CHARLTON, JAMES LESLIE, b St Thomas, Ont, Dec 12, 42; m 64; c 2. PHOTOCHEMISTRY, ORGANIC CHEMISTRY. *Educ:* Univ Western Ont, BSc, 65, PhD(chem), 68. *Prof Exp:* Nat Res Coun Can fel, Calif Inst Technol, 68-70; from asst prof to assoc prof, 70-81, PROF CHEM, UNIV MAN, 81- *Mem:* Am Chem Soc; Chem Inst Can. *Res:* Organic synthesis; asymmetric synthesis. *Mailing Add:* Dept Chem Univ Man Winnipeg MB R3T 2N2 Can

CHARM, STANLEY E, b Boston, Mass, Oct 18, 26; m 52; c 3. BIOCHEMICAL ENGINEERING. *Educ:* Univ Mass, BS, 50; Wash State Col, MS, 52; Mass Inst Technol, BS, 55, ScD(food tgchnol), 57. *Prof Exp:* Res chemist, Am Home Foods Div, Am Home Prod Corp, NY, 52-53; asst food technol, Mass Inst Technol, 53-55, instr, 55-57, asst prof nutrit & food sci, 57-63; assoc prof physiol & sci dir, New Eng Enzyme Ctr, Sch Med, Tufts Univ, 63-; AT PENICILLIN ASSAYS. *Concurrent Pos:* NIH res grant, 61-; consult, Gen Foods Corp, Del, 61-66 & Bur Com Fisheries, US Fish & Wildlife Serv, Mass, 63- *Mem:* Inst Food Technol; Soc Rheol; Soc Cryobiol; Am Inst Chem Engrs; Sigma Xi. *Res:* Food engineering; rheology; biomedical engineering. *Mailing Add:* Penicillin Assays 36 Franklin St Malden MA 02148

CHARMBURY, H(ERBERT) BEECHER, b Hanover, Pa, Sept 21, 14; m 38; c 2. COAL PREPARATION. *Educ:* Gettysburg Col, AB, 36; Univ Pa, MS, 37; Pa State Univ, PhD(fuel technol), 42. *Hon Degrees:* DSc, Gettysburg Col, 71. *Prof Exp:* Asst petrol & natural gas, Pa State Univ, 37-39, asst fuel technol, 39-42, from asst prof to assoc prof, 42-50, assoc prof mineral prep eng, 50-53, asst dean planning & develop, Col Earth & Mineral Sci, 71-73, prof, 53-80; RETIRED. *Concurrent Pos:* Secy mines & mineral industs, Commonwealth of Pa, 63-71; consult coal mining, prep & environ probs, 73- *Honors & Awards:* Environ Conserv Award, Am Inst Mining, Metall & Petrol Engrs, 76, Percy Nicholls Award, 77; Environ Conserv Award, Nat Audubon Soc, 77. *Mem:* Am Chem Soc; Am Gas Asn; Am Inst Mining, Metall & Petrol Engrs. *Res:* Mineral preparation engineering; coal preparation. *Mailing Add:* 420 S Corl St Apt 8 State College PA 16801

CHARNES, ABRAHAM, b Hopewell, Va, Sept 4, 17; m 50; c 3. MATHEMATICS, ECONOMICS. *Educ:* Univ Ill, AB, 38, MS, 39, PhD(math), 47. *Prof Exp:* Off Naval Res fel, Univ Ill, 47-48; from asst prof to assoc prof math, Carnegie Inst Technol, 48-52, assoc prof indust admin, 52-55; prof math & dir res dept transportation & indust mgt, Purdue Univ, 55-57; res prof appl math & econ, Northwestern Univ, 65-68; Walter P Murphy prof, 68-78, JESSE H JONES PROF BIOMATH, GEN BUS & MGT SCI, PROF MATH, GEN BUS & COMPUT SCI & DIR, CTR CYBERNETIC STUDIES, UNIV TEX, AUSTIN, 73- *Concurrent Pos:* Ed, J Inst Mgt Sci, 54- *Mem:* Fel AAAS; fel Economet Soc; Opers Res Soc Am; Asn Comput Mach; Inst Mgt Sci (vpres, 58, pres, 60). *Res:* Topological algebra; functional analysis; differential equations; aerodynamics; hydrodynamic theory of lubrication; statistics; extremal methods; game theory; mathematical theory of management science; biomathematics. *Mailing Add.* Dept Math Univ Tex Austin TX 78712

CHARNEY, DENNIS S, b New York, NY, March 31, 51; c 5. PSYCHOPHARMACOLOGY. *Educ:* Rutgers Univ, AB, 73; Pa State Univ, MD, 77. *Prof Exp:* Intern, Hosp St Raphael, New Haven, Conn, 77-78; fel, 78-81, from asst prof to assoc prof, 81-90, PROF PSYCHIAT, YALE UNIV SCH MED, 90-; CHIEF PSYCHIAT SERV, W HAVEN VET ADMIN MED CTR, 88- *Concurrent Pos:* Falk fel, Am Psychiat Asn Coun Res & Develop, 79-81; psychiat consult, Community Health Care Plan, New Haven, Conn, 81-; biol sci training fel & chief resident, Clin Res Unit, 80-81, assoc unit chief, Clin Neurosci Res Unit, 81-82, dir Ribicoff Res Facil, Conn Mental Health Ctr, New Haven, 83-; dir, Affective Disorders CLin, Dana Psychiat Clin, Yale-New Haven Hosp, 81-83. *Honors & Awards:* Mosby Award for Res Excellence, 77; Clin Res Award, Am Acad Clin Psychiat, 86. *Mem:* AAAS; Am Col Neuropsychopharmacol; Soc Neurosci; Am Psychiat Asn; Soc Biol Psychiat. *Res:* Investigation of the neurobiological etiology of major psychiatric disorders and development of navel, more effective psychiatric medications; author of over 100 technical articles, reviews, chapters and books. *Mailing Add:* Psychiat Serv 116A W Haven Vet Admin Med Ctr W Haven CT 06516

CHARNEY, ELLIOT, b New York, NY, June 1, 22; m 47; c 3. CHEMICAL PHYSICS. *Educ:* City Col New York, BS, 42; Columbia Univ, PhD, 56. *Prof Exp:* Res chemist, Manhattan Proj, 42-45, tech adv, 45-48; consult, US AEC, 48-50; consult writer, Kellex Corp, 50-54; res scientist, Lab Phys Biol, 56-72, CHIEF, SECT SPECTROS & STRUCT, 72-, chief, Lab Chem Phys, NIH, 80-81, 84-85. *Concurrent Pos:* Asst, Columbia Univ, 51-55; vis scientist, Univ Oxford, 62-63; vis fac assoc, Univ Oregon, 65 & Dartmouth Col, 74, 88- *Mem:* Am Phys Soc; NY Acad Sci; AAAS. *Res:* Infrared and ultraviolet spectroscopy; optical rotatory dispersion; structure and interactions of molecules in condensed phases; electro-optic properties; biopolymers. *Mailing Add:* NIH Bldg Two Rm B1-03 Bethesda MD 20892

CHARNEY, EVAN, PEDIATRICS. *Educ:* Albert Einstein Col Med, MD, 60. *Prof Exp:* CHMN, DEPT PEDIAT, SCH MED, UNIV MASS, 87- *Mailing Add:* Med Ctr Univ Mass Worcester MA 01655

CHARNEY, MARTHA R, b Pittsburgh, Pa, Sept 28, 42; m 83; c 1. BIOCHEMISTRY. *Educ:* Carnegie Inst Technol, BS, 64; Univ Pittsburgh, PhD(biochem), 69. *Prof Exp:* Res fel biochem, Sloan-Kettering Inst, 69-71; res fel pharmacol, Columbia Univ, 71-74; sr biochemist, USV Pharmaceut Corp, Revlon, Inc, 74-76, sect head drug disposition, 76-80, ASST DIR, DRUG REGULATORY AFFAIRS, REVLON HEALTH CARE GROUP, REVLON, INC, 80- *Mem:* Am Chem Soc; AAAS; Acad Pharmaceut Sci; Am Soc Pharmacol & Exp Therapeut; Drug Info Asn; Regulatory Affairs Prof Soc. *Res:* Development of analytical methods for the analysis of drugs, drug metabolism, pharmacokinetics and biopharmaceutics. *Mailing Add:* 70 Myrtle Blvd Larchmont NY 10538-2344

CHARNEY, MICHAEL, b New York, NY, Aug 6, 11; m 41, 66; c 5. PHYSICAL ANTHROPOLOGY, FORENSIC ANTHROPOLOGY. *Educ:* Univ Tex, Austin, BA, 34; Univ Colo, Boulder, PhD(anthrop), 69; Am Bd Forensic Anthrop, dipl. *Prof Exp:* Chief lab serv clin path, Sta Hosp, Camp Gordon, Ga, 45-46; chief bacteriologist cancer res, Longevity Res Found, NY, 47-48; dir clin path, Hackensack Bio-Chem Lab, NJ, 46-65; asst prof anthrop, Idaho State Univ, 68-72; from assoc prof to prof, 71-76, EMER PROF ANTHROP & LECTR, COLO STATE UNIV, 77-, DIR, CTR HUMAN IDENTIFICATION, 80- *Concurrent Pos:* Co-dir, Forensic Sci Lab, Colo State Univ, 73, affil prof zool, 74-, dir, Forensic Sci Lab; dep coroner, Larimer County, Colo, 75-; consult identification vet human remains, families of POW-MIA's-Vietnam War. *Mem:* Fel Royal Anthrop Inst Gt Brit; Am Asn Phys Anthropologists; Am Acad Forensic Sci; Soc Study Human Biol; Soc Study Biol; Sigma Xi. *Res:* Problems in facial restoration as an aid in identification of human remains; human identification of skeletonized remains by laser-computer-video of skull and photograph in superimposition; selection pressures in human evolution posed by interplay of biology and culture. *Mailing Add:* Forensic Sci Lab Colo State Univ Ft Collins CO 80523

CHARNEY, WILLIAM, b Russia, Jan 10, 18; nat US; m 47; c 2. MICROBIOLOGY. *Educ:* Johns Hopkins Univ, BA, 40; Rutgers Univ, PhD(microbiol), 53. *Prof Exp:* Bacteriologist, Rare Chems, Inc, 46-50; microbiologist, 53-70, assoc dir microbiol develop, 70-73, DIR MICROBIOL DEVELOP, SCHERING CORP, 73- *Mem:* Am Soc Microbiol; Am Chem Soc; NY Acad Sci; Sigma Xi. *Res:* Microbial transformation of steroids; vitamin B-12; antibiotics. *Mailing Add:* 110 Christopher St Montclair NJ 07042

CHAROLA, ASUNCION ELENA, b Arg, Feb 23, 42. ANALYTICAL CHEMISTRY, PHYSICAL CHEMISTRY. *Educ:* Nat Univ La Plata, lic(anal chem), 67, lic(indust chem), 69, Dr(anal chem), 74. *Prof Exp:* Chief anal chem, Nat Univ La Plata, 72-74; fel, NY Univ, 74-76; asst prof chem, Manhattan Col, 78-81; assoc chemist, Metrop Mus Art, 81-85; ASSOC SCIENTIST, ICCROM, ROME, ITALY, 86- *Concurrent Pos:* Warner-Lambert fel, NY Univ, 76- *Mem:* Agr Chem Asn; Am Chem Soc;

AAAS; US-Int Coun Monuments & Sites; Int Inst Conserv. *Res:* Solid state chemistry; polymorphism; x-ray crystallography; electron microscopy; spectroscopy of solids; electroanalytical chemistry. *Mailing Add:* Muratore 840 San Isidro 1642 Argentina

CHARON, NYLES WILLIAM, b Minneapolis, Minn, Sept 13, 43; m 69. MICROBIOLOGY. *Educ:* Univ Minn, BA, 65, MS, 69, PhD(microbiol), 72. *Prof Exp:* Res asst microbiol, Univ Minn, 65-72; fel biol sci, Stanford Univ, 72-74; asst prof, 74-80, ASSOC PROF MICROBIOL, WVA UNIV, 80- *Mem:* AAAS; Am Soc Microbiol; Sigma Xi; Am Leptospirosis Res Conf. *Res:* Late gene regulation in bacteriophage lambda; biochemical and genetic studies of the spirochete Leptospira, their relative sensitivity to ultraviolet-light irradiation, membrane fluidity, structure, means for motility and isoleucine biosynthesis. *Mailing Add:* Dept Microbiol WVa Univ Med Morgantown WV 26506

CHARP, SOLOMON, b Jersey City, NJ, Jan 20, 20; m 44. ANALYSIS OF DESIGNS, FORENSIC ENGINEERING. *Educ:* Univ Pa, BS, 40, MS, 41. *Prof Exp:* Res asst elec eng, Moore Sch Elec Eng, Univ Pa, 41-42, instr, 42-45, res engr, 44-45, res assoc, 45-48; sr staff engr, Labs Res & Develop, Franklin Inst, 48-59; eng mgr & consult scientist, Gen Elec Co, 59-70; PRES, CHARP ASSOCS, INC, 69- *Concurrent Pos:* Consult engr, 69- *Mem:* Sr mem Inst Elec & Electronic Engrs; assoc fel Am Inst Aeronaut & Astronaut; Sigma Xi. *Mailing Add:* 39 Maple Ave Upper Darby PA 19082-1902

CHARPIE, ROBERT ALAN, b Cleveland, Ohio, Sept 9, 25; m 47; c 4. THEORETICAL PHYSICS. *Educ:* Carnegie Inst Technol, BS, 48, MS, 49, DSc(theoret physics), 50. *Hon Degrees:* PhD, Denison Univ, 65. *Prof Exp:* Physicist, Westinghouse Elec Corp, 47-50; physicist, Oak Ridge Nat Lab, 50-55, asst dir, 55-61, dir, Reactor Div, 58-61; mgr advan develop, Union Carbide Corp, 61-63, gen mgr develop dept, 63-64, dir technol, 64-66, pres electronics div, 66-68; pres, Bell & Howell Co, 68-69; pres, Cabot Corp, 69-86, chmn, 87-88; CHMN, AMPERSAND VENTURES, 88- *Concurrent Pos:* Asst, US Mem Seven-Nation Adv Comt, Int Conf Peaceful Uses Atomic Energy, 55, coordr, US Fusion Res Exhib, 58, secy gen adv comt, AEC, 59-63; ed-in-chief, Proc Int Conf, 55; gen ed, Int Monogr Ser on Nuclear Energy, 55-60; ed, J Nuclear Energy, 55-60; mem, Oak Ridge Bd Ed, 57-61; mem adv comn UN sci activities, State Dept, 61; mem panel, Civilian Technol Pakistan, President's Sci Adv Comn, 61, mem panel oceanog, President's Sci Adv Comt, 65; trustee, Carnegie Inst Technol, 62- *Honors & Awards:* Award, US Chamber Com, 55. *Mem:* Nat Acad Eng; fel Am Nuclear Soc; fel Am Phys Soc; fel NY Acad Sci; Sigma Xi. *Res:* Theoretical, nuclear and reactor physics. *Mailing Add:* Ampersand Ventures 55 William St Suite 240 Wellesley MA 02181

CHARRON, MARTIN, b Sherbrooke, Que, Can, June 28, 60. RADIOLOGY, NUCLEAR MEDICINE. *Educ:* Univ Sherbrooke, Can, MD, 83, ABNM, 88. *Prof Exp:* ASST PROF RADIOL, SCH MED, UNIV PITTSBURGH, 90- *Mem:* Soc Nuclear Med; Radiol Soc NAm; Can Med Asn; AMA. *Res:* AIDS; nuclear cardiology; ped nuclear medicine; computers. *Mailing Add:* Div Nuclear Med Univ Pittsburgh DeSoto at O'Hara St Pittsburgh PA 15213

CHARTIER, VERNON L, b Fort Morgan, Colo, Feb 14, 39; m 67; c 1. HIGH VOLTAGE ENGINEERING, ELECTROMAGNETICS. *Educ:* Univ Colorado, BS, 63. *Prof Exp:* Res Engr, Westinghouse Elect Corp, 63-75; Elect Engr, 75-79, CHIEF ENGR, BONNEVILLE POWER ADMINISTRATION, 79- *Concurrent Pos:* Chmn, Corona & Field Effects Subcomt, Inst Elec & Electronic Engrs, 78-82, Transmission & Dist Commun, 87-88, PES Fel Comt, 90-; Adv, CIGRE Study, 68- *Mem:* US Nat Com Int Engr Con; Am Nat Standards Inst; Acoustical Soc Am; fel Inst Elec & Electronic Engrs. *Res:* High voltage transmission lines; DC fields and ions; acoustical noise; corona; electromagnetic interference. *Mailing Add:* 5190 SW Dover Lane Portland OR 97225

CHARTOCK, MICHAEL ANDREW, b Palo Alto, Calif, May 25, 43; m 71; c 1. ECOLOGY, SCIENCE POLICY. *Educ:* Univ Calif, Berkeley, AB, 65; San Jose State Univ, MA, 71; Univ Southern Calif, PhD(biol), 72. *Prof Exp:* From asst prof to assoc prof zool, Univ Okla, 77-85; STAFF SCIENTIST, LAWRENCE BERKELEY LAB, 85- *Concurrent Pos:* Res fel sci & pub policy, Univ Okla, 71-; consult, BDM Corp, 74-75, US Syn Fuels Corp, 81-82. *Mem:* AAAS; Am Geol Union. *Res:* Research planning, technology assessment and policy oriented research in energy development; energy flow in aquatic ecosystems; role of detritus in coral reefs. *Mailing Add:* Planning & Anal Lawrence Berkeley Lab 50A-4112 Berkeley CA 94720

CHARTON, MARVIN, b Brooklyn, NY, May 1, 31; m 55; c 3. PHYSICAL ORGANIC CHEMISTRY, QUANTITATIVE STRUCTURE. *Educ:* City Col New York, BS, 53; Brooklyn Col, MA, 56; Stevens Inst Technol, PhD(chem), 62. *Prof Exp:* Res chemist, Evans Res & Develop Corp, 55-56; instr, 56-61, from asst prof to assoc prof, 61-67, chmn dept, 69-71, PROF CHEM, PRATT INST, 67 - *Concurrent Pos:* Fel, Intrasci Res Found, 69 -; vis prof, Polymer Res Inst, Polytechnic Univ, Brooklyn, NY; mem, Int Group for Correlation Anal in Org Chem. *Mem:* AAAS; Am Chem Soc; Brit Chem Soc; NY Acad Sci; Int Quant Struct-Activ Relationship Soc. *Res:* Linear free energy relationships in organic chemistry; quantitative treatment of proximity effects; quantitative treatment of bioactivity as a function of molecular structure; quantitative structure activity relationships. *Mailing Add:* Dept Chem Pratt Inst Ryerson St Brooklyn NY 11205

CHARTRAND, GARY, b Sault Ste Marie, Mich, Aug 24, 36; m 68. MATHEMATICS. *Educ:* Mich State Univ, BS, 58, MS, 60, PhD(math), 64. *Prof Exp:* From asst prof to assoc prof, 64-70, PROF MATH, WESTERN MICH UNIV, 70- *Concurrent Pos:* Res grants, US Air Force Off Sci Res, Univ Mich, 65-66 & NIMH, Res Ctr Group Dynamics, 66; NSF grant, 68-69; Off Naval Res fel, 70-71; vis math scholar, Univ Calif, Santa Barbara, 70-71; managing ed, J Graph Theory, 75-78; vis prof, San Jose State Univ, 78. *Mem:* Am Math Soc; Math Asn Am. *Res:* Theory of graphs; connectivity and line-connectivity; graphical partitions; traversability; line, total and permutation graphs; planarity; colorability; graphs and matrices; reconstruction of graphs. *Mailing Add:* Dept Math Western Mich Univ Kalamazoo MI 49008

CHARTRAND, MARK RAY, III, b Miami, Fla, Aug 2, 43; div. ASTRONOMY, SATELLITE COMMUNICATIONS. *Educ:* Case Inst Technol, BS, 65; Case Western Reserve Univ, PhD(astron), 70. *Prof Exp:* Asst to dir astron, Ralph Mueller Planetarium, Cleveland Mus Natural Hist, 65-66; asst astronr & dir educ, Am Mus, 70-74, chmn & assoc astron, Hayden Planetarium, 74-80; exec dir, Nat Space Inst, 80-84. *Concurrent Pos:* Adj asst prof, Fordham Univ, Lincoln Ctr Campus, 72-80; consult & educr, Satellite Commun, 84-; fac mem, New Sch, 80-81; chmn satellite status, 87; consult & announcer, Ariane Rocket Launches, 84-86; vpres, Nat Space Soc, 86- *Honors & Awards:* Armand Spitz Lectr; Margret Noble Lectr. *Mem:* AAAS; sr mem Am Astron Soc; fel Brit Interplanetary Soc; Am Inst Aeronaut & Astronaut; Soc Satellite Profs. *Res:* Galactic structure; photoelectric and photographic photometry. *Mailing Add:* 19751 Frederick Rd Suite 349 Germantown MD 20876

CHARTRAND, ROBERT LEE, b Kans City, MO, Mar 6, 28; m 67; c 4. SCIENCE POLICY, SYSTEM DESIGN & SYSTEM SCIENCE. *Educ:* Univ MO, BA, 49, MA, 49. *Prof Exp:* Naval, intel officer, US Navy, 53-59; mem tech staff info syst, Thompson Ramo Wooldridge R-W Div, 59-61; syst analyst infel syst, Fed Syst Div, IBM Corp, 61-64; mkt adv mgr info syst, Planning Res Corp, 64-66; specialist sci technol info policy, Sci Policy Res Div, 66-77, sr specialist sci technol, 77-88, SR FEL SCI TECHNOL, SCI POLICY RES DIV, LIBR CONG, 88- *Concurrent Pos:* vis prof, Grad Sch Libr Info Sci, Univ Calif, Los Angeles, 68; adj prof & vis scholar, Am Univ, 68; consult, George Washington Univ, 75-77, NSF, 77-79, OTA, 79-89, Nat Acad Sci, 81-83; adv, Smithsonian Inst, 86-; vis prof, Grad Sch Libr Info Sci, Univ Pittsburgh, 89-91; mem, Mapping Sci Com & STI Bd, Nat Res Coun, 90-; adj fel, Ctr Strategic & Int Studies, 91- *Honors & Awards:* Test of Tome Award, Interagency Comt on Automated Data Processing, 79; Award of Merit, Am Soc Info Sci, 85, pioneer, 88. *Mem:* Am Soc Info Sci; fel AAAS. *Res:* Authorship of topical areas concerned with roles & impacts of advance information technologies, computer & telecommunication; author/editor of several books and major congressional reports on information technology, business, agriculture, education, STI and emergency management. *Mailing Add:* 430 Widgeon Pt Naples FL 33942

CHARTRES, BRUCE A, b Adelaide, Australia, July 25, 39. ALGEBRA. *Educ:* Univ Adelaide, BS, 53; Univ Sydney, PhD(theoret physics), 56. *Prof Exp:* Lectr physics, Univ Sydney, 56-61; assoc prof appl math, Brown Univ, 61-65; PROF MATH, UNIV VA, 65- *Mem:* Asn Comput Mach; Am Math Soc; Soc Appl Math. *Mailing Add:* Dept Appl Math Univ Va Thornton Hall Charlottesville VA 22903

CHARVAT, F(EDIA) R(UDOLF), b Pilsen, Czech, Mar 11, 31; US citizen; m 56; c 2. CERAMICS. *Educ:* Univ Leeds, BSc, 53; Mass Inst Technol, DSc(ceramics), 56. *Prof Exp:* Res ceramist, Metals Div, Union Carbide Corp, NY, 56-57, sect leader, 57-59, sect mgr, 59-63, supvr crystal prod, Linde Div, Ind, 63-67, mgr res & develop, Crystal Prod Electronic Div, 67-68, mgr opers, Crystal Prod Div, San Diego, 68-69, asst gen mgr, 69-71, gen mgr, 71-77, gen mgr, 71-77, gen mgr electronics div, 77-84, vpres technol, Electronic Div, 84-86; VPRES & GEN MGR, GEN CERAMICS INC, 86- *Mem:* Am Inst Mining, Metall & Petrol Engrs; Am Ceramic Soc; Am Asn Crystal Growth. *Res:* Solid-state materials; thermal properties of dielectrics; properties of liquid oxides; growth of metallic and non-metallic single crystals. *Mailing Add:* Gen Ceramics Haskell NJ 07420

CHARVAT, IRIS, b Decatur, Ill, Sept 29, 40. MYCOLOGY, CELL BIOLOGY. *Educ:* Univ Calif, Santa Barbara, PhD(cell biol), 73. *Prof Exp:* PROF BOT, UNIV MINN, 73-, DIR GRAD STUDIES, 83- *Mem:* Mycol Soc Am; Bot Soc Am; Sigma Xi; Am Soc Cell Biol. *Mailing Add:* 220 Biosci Ctr Univ Minn St Paul MN 55108

CHARVONIA, DAVID ALAN, b Denver, Colo, July 19, 29; m 53; c 2. DYNAMICS, SYSTEMS CONCEPTS. *Educ:* Univ Colo, BS, 51; Purdue Univ, MS, 53, PhD(propulsion), 59. *Prof Exp:* Div mgr & other positions in electronics & space syst, Aerojet-Gen Corp, 61-68; sr staff mem radar syst, ITT Gilfillan, 68; vpres & tech dir res & develop admin, Telluron, 68-72; staff specialist res & develop prog admin, Off Dir Defense Res & Eng, Dept Defense, 72-75, actg asst dir electronic technol, 75, spec asst to dep dir res & advan technol, 75-77; dir, Europ Regional Off, Defense Advan Res Projs Agency, Stuttgart, W Ger, 77-83; DIR, MCLEAN RES CTR, UNISYS DEFENSE SYSTS, 83- *Concurrent Pos:* Mem, Exec Comt Sci & Technol, ADPA & Defense Res & Develop Comt, Inst Elec & Electronics Engrs. *Mem:* Sigma Xi; Inst Elec & Electronics Engrs; Am Defense Preparedness Asn. *Res:* Exploratory development pertinent to military applications. *Mailing Add:* Unisys Defense Syst 8201 Greensboro Dr Suite 1000 Mclean VA 22102

CHARWAT, ANDREW F(RANCISZEK), b Tallin, Estonia, Feb 10, 25; nat US; m 48; c 1. MECHANICAL ENGINEERING. *Educ:* Stevens Inst Technol, ME, 48; Univ Calif, PhD(mech eng), 52. *Prof Exp:* Lectr, Univ Calif, 49-51, instr, 51-52; aerodynamicist, Propulsion Res Corp, 52-54; preliminary designer, Northrop Aircraft, 54-55; assoc prof, 55-63, PROF ENG & APPL SCI, UNIV CALIF, LOS ANGELES, 63- *Concurrent Pos:* Consult, 58-; Fulbright fel, 61-62; assoc prof, Univ Paris, 61-63; Guggenheim fel, 62-63; dir, Educ Abroad Ctr, Lyon & Grenoble, 86-88. *Res:* Aerodynamics and propulsion; heat transfer; energy systems. *Mailing Add:* Sch Eng & Appl Sci Univ Calif Los Angeles CA 90024

CHARYK, JOSEPH VINCENT, b Canmore, Alta, Sept 9, 20; nat US; m 45; c 4. AERONAUTICS. *Educ:* Univ Alta, BSc, 42; Calif Inst Technol, MS, 43, PhD(aeronaut), 46. *Hon Degrees:* LLD, Univ Alta, 64; Dr Ing, Univ Bologna, 74. *Prof Exp:* Res engr, Jet Propulsion Lab, Calif Inst Technol, 43-46, instr aeronaut, 45; from asst prof to assoc prof aeronaut, Princeton Univ, 46-55; dir, Aerophys & Chem Lab, Missile Systs Div, Lockheed Aircraft Corp, 55-56; dir, Missile Technol Lab, Aeronutronic Systs, Inc, Subsid of Ford Motor Co, 56-58, gen mgr, Space Technol Div, 58-59; chief scientist & Asst Secy Res & Develop, US Dept Air Force, 59, Under Secy of Air Force, 60-63; pres & dir, Commun Satellite Corp, 63-79, chief exec officer, 79-83, chmn bd dirs &

chief exec officer, 83-85; RETIRED. *Concurrent Pos:* Gen ed, Aeronaut Pub Prog, Princeton Univ, 51-56; chmn & mem bd dirs, Charles Stark Draper Lab, Inc; vchmn, Nat Telecommun Security Adv Coun, 82-84, chmn, 84. *Honors & Awards:* Lloyd V Berkner Space Utilization Award, Am Astronaut Soc, 67; Distinguished Aviation Aerospace Serv Award, Nat Aviation Club, 73; Guglielmo Marconi Int Award, Marconi Found, 74; Theodore Von Karman Award, Am Inst Aeronaut & Astronaut, 77, Goddard Astronaut Award, 78; Nat Medal of Technol, 87. *Mem:* Nat Acad Eng; Int Acad Astronaut; fel Am Inst Aeronaut & Astronaut; fel Inst Elec & Electronic Engrs; Nat Space Club; Sigma Xi; Nat Inst Social Sci; Armed Forces Commun & Electronics Asn. *Res:* Jet propulsion, space technology and communications. *Mailing Add:* Commun Satellite Corp 950 L'Enfant Plaza SW Washington DC 20024

CHARYULU, KOMANDURI K N, b Hanamkonda, India, May 24, 24; US citizen; m 44; c 6. ONCOLOGY, RADIATION THERAPY. *Educ:* Andhra Univ, BSc, 45, MD, 51; Royal Col Physicians & Surgeons, dipl med radiation therapy, 60, FFR, 61; Am Bd Radiol, dipl radiation therapy, 70; FRCR. *Prof Exp:* Radium registr & asst surgeon radiother, Radium Inst & Cancer Hosp, India, 55-57; tutor radiol, Osmania Med Col, India, 57-58; hon clin asst, London Hosp, Eng, 59-60, registr, 60-61; locum consult, St Mary's Hosp, Portsmouth, Eng, 62; asst prof radiol, Univ Minn Hosps, Minneapolis, 64-67, assoc prof, 67-70; dir radiation ther, Sch Med, Univ Miami, 70-80, prof radiol, 70-80 & adj prof, 81-85; chief, Miami, Fla, 80-89, CHIEF, DEPT RADIATION THER, VET ADMIN MED CTR, JOHNSON CITY, TENN, 89-; CLIN PROF, ETENN STATE UNIV, 90- *Concurrent Pos:* Spec vis res fel, Mem Hosp & Sloan-Kettering Cancer Inst, 62-63. *Mem:* Int Clin Hyperther Soc; Am Soc Therapeut Radiol; fel Am Col Radiol; Am Radium Soc; fel Am Col Radiol. *Res:* Oxygenation of tissues and study of radiation sensitivity; modification of radio sensitivity by heat and microwaves; endocrine relationships in carcinoma; dose distribution in electron and x-ray therapeutic regimens; prostate cancer; brachy therapy techniques. *Mailing Add:* Radiation Ther Dept Vet Admin Med Ctr 114R Johnson City TN 37684

CHASALOW, FRED I, b Newark, NJ, July 10, 42; m 66. ENDOCRINOLOGY. *Educ:* Brandeis Univ, PhD(biochem), 71. *Prof Exp:* Asst prof pediat, State Univ NY, Stony Brook, 84-89; ASSOC PROF PEDIAT, ALBERT EINSTEIN COL MED, BRONX, NY, 89- *Mem:* Am Soc Biol Chem; Endocrine Soc; Soc Pediat Res. *Res:* Biochemical endocrinology; steroid synthesis, structure & function; growth & growth hormone in children with growth failure; digoxin-like materials. *Mailing Add:* Dept Pediat Schneider Childrens Hosp Long Island Jewish Med Ctr New Hyde Park NY 11042

CHASALOW, IVAN G, b New York, NY, Mar 6, 30; m 54; c 6. OPERATIONS RESEARCH. *Educ:* Mass Inst Technol, BS, 51; Columbia Univ, MA, 52, PhD(chem), 57. *Prof Exp:* Opers analyst, opers eval group, Mass Inst Technol, 56-59; opers analyst, Bell Tell Labs, 59-76, opers analyst, Am Bell Int Inc, 76-78, OPERS ANALYST, AT&T BELL TEL LABS, 78- *Mem:* Sigma Xi; Opers Res Soc Am. *Res:* Military operations research and applied research in underwater sound. *Mailing Add:* 59 Addie Lane Whippany NJ 07981

CHASE, ANDREW J(ACKSON), b Sebec Station, Maine, Feb 16, 16; m 44; c 1. CHEMICAL ENGINEERING. *Educ:* Univ Maine, BS, 49, MS, 51. *Prof Exp:* From instr to prof, 50-82, EMER PROF CHEM ENG, UNIV MAINE, 82- *Honors & Awards:* Forest Prod Div Award, Am Inst Chem Engrs, 79; Ashley S Campbell Award, Col Eng & Sci, 82. *Mem:* Am Inst Chem Engrs; Am Tech Asn Pulp & Paper Indust; Am Soc Eng Educ. *Res:* Surface properties of natural fibers; fluid flow (non-Newtonian); pulp and paper technology. *Mailing Add:* 35 Oak St Orono ME 04473

CHASE, ANN RENEE, b San Bernardino, Calif, Dec 28, 54. ORNAMENTAL PLANT DISEASE. *Educ:* Univ Calif, Riverside, BS, 76, PhD(plant path), 79. *Prof Exp:* ASST PROF PLANT PATH, AGR RES CTR, UNIV FL, 79- *Mem:* Am Phytopath Soc. *Res:* Diseases of ornamental plants, especially foliage plants; description of new diseases and investigation of the role of nutrition, light and temperature in plant disease; importance of cultural controls in disease control. *Mailing Add:* Agr Res Ctr 2807 Binion Rd Apopka FL 32703

CHASE, ARLEEN RUTH, b Boston, Mass, Aug 21, 45; m 83; c 1. IMMUNOCHEMISTRY, PHARMACOLOGY. *Educ:* Boston State Col, BS, 67; Northeastern Univ, MS, 71; Boston Univ, PhD(pharmacol), 82. *Prof Exp:* Immunochemist, Leary Labs, Inc, Boston, Mass, 71-72; supvr, Radioimmunoassay Develop Prog, Collab Res Inc, Waltham, Mass, 72-76; proj mgr & sr scientist Immunochem, Instrumentation Lab Inc, Lexington, Mass, 80-86; dir, Standard Scientifics, Inc, Needham, Mass, 87-89; GROUP LEADER, VITEK SYSTS INC, ROCKLAND, MASS, 90- *Concurrent Pos:* Consult, Collab Res Inc, 76-79; prin investr, Pharmaceut Mfg Assoc Found Inc, 79-80. *Mem:* Am Chem Soc; Am Asn Clin Chem; Clin Ligand Assay Soc. *Res:* Pharmacological investigations of hypertension and aging in animal models; development of state of the art nonisotopic immunoassay for application to existing and future instrumentation. *Mailing Add:* 378 Treble Cove Rd Billerica MA 01862

CHASE, CHARLES ELROY, JR, b Lyndonville, Vt, May 16, 29; m 54; c 2. SURFACE ACOUSTIC WAVE DEVICES, ELECTRONICS. *Educ:* Mass Inst Technol, BS, 50; Camridge Univ, PhD(physics), 54. *Prof Exp:* Staff mem physics, Lincoln Lab, Mass Inst Technol, 54-63 & Francis Bitter Nat Magnet Lab, 63-75; pres, Tachisto Inc, 75-79; sr engr, Raytheon Co, 80-89; RETIRED. *Concurrent Pos:* Fulbright award, Univ Leiden, 62-63; adj assoc prof physics, Boston Univ, 70-71. *Mem:* Fel Am Phys Soc. *Res:* Low temperature physics; ultrasonics in solids and liquids; nonlinear optics; plasmas; quantum electronics; laser development; microelectronics; saw devices. *Mailing Add:* 141 Paul Revere Rd Needham MA 02194-1986

CHASE, CLEMENT GRASHAM, b Phoenix, Ariz, Mar 27, 44; m 66; c 2. MANTLE KINEMATICS, PLATE & CONTINENTAL TECTONICS. *Educ:* Calif Inst Technol, BS, 66; Univ Calif, San Diego, PhD(oceanog), 70. *Prof Exp:* NSF fel, Dept Geod & Geophysics, Univ Cambridge, 70-71; from asst prof to assoc prof, dept geol & geophys, Univ Minn, 71-83; PROF, DEPT GEOSCI, UNIV ARIZ, 84-, CHMN, 91- *Concurrent Pos:* Vis assoc prof, Bd Earth Sci, Univ Calif, Santa Cruz, 78-79. *Mem:* Am Geophys Union; fel Geol Soc Am; Sigma Xi. *Res:* Use of geoid and of isotopic systematics to constrain models of mantle kinematics and convection patterns; flexural isostasy and uplift of mountain ranges; interaction of surficial processes and tectonics; modeling large scale erosion and deposition. *Mailing Add:* Geosci Dept Univ Ariz Tucson AZ 85721

CHASE, CURTIS ALDEN, JR, b Palatka, Fla, Mar 20, 36; m 63. CHEMICAL ENGINEERING. *Educ:* Univ Fla, BS, 59; Ill Inst Technol, PhD(chem eng), 69. *Prof Exp:* Engr, Am Oil Co, Ind, 63-65 & Shell Develop Co, 68-77; engr, Intercomp, 77-78; engr, Todd, Dietrich & Chase, Inc, 79-91; ENGR, TCA RESVOIR ENG SERV INC, 91- *Mem:* Am Inst Mining, Metall & Petrol Engrs; Soc Indust & Appl Math; Soc Petrol Eng; Sigma Xi. *Res:* Applied mathematics as related to solving the partial differential equations describing heat and mass transfer processes. *Mailing Add:* 555 S Camino Del Rio A-3 Durango CO 81301

CHASE, DAVID BRUCE, b Quincy, Mass, June 17, 49; m 71; c 1. OPTICAL SPECTROSCOPY. *Educ:* Williams Col, BA, 70; Princeton Univ, PhD(phys chem), 75. *Prof Exp:* RES CHEMIST, E I DU PONT DE NEMOURS & CO, INC, 75- *Honors & Awards:* Williams Wright Award, 89. *Mem:* Soc Appl Spectros; Am Chem Soc; Optical Soc Am; Coblentz Soc (pres). *Res:* Applications of infrared interferometry to industrial problems and development of laser raman microprobe techniques; Fourier Transform Raman Spectroscopy. *Mailing Add:* CRD E328-131A Exp Sta Wilmington DE 19898

CHASE, DAVID MARION, b Denver, Colo, Jan 20, 30; div; c 1. FLOW-INDUCED NOISE, MATHEMATICAL MODELING. *Educ:* Univ Colo, BS, 51; Princeton Univ, AM, 53, PhD(physics), 55. *Prof Exp:* Staff mem, Los Alamos Sci Lab, 54-57; sr physicist, TRG, Inc, Melville, 57-68; sr scientist, Bolt Beranek & Newman, Inc, 68-79; SR SCIENTIST, CHASE INC, 79- *Concurrent Pos:* Vis asst prof, Iowa State Univ, 56. *Mem:* Am Phys Soc; Am Finance Asn; Acoust Soc Am. *Res:* Analytical modeling in fluid dynamics and structural vibration, especially turbulence and acoustics of flow-induced noise in sonar applications; nuclear models and scattering; acoustics; turbulence; stochastic financial economics. *Mailing Add:* Chase Inc 87 Summer St Suite 510 Boston MA 02110

CHASE, FRED LEROY, b Dedham, Mass, Nov 30, 14; m 45; c 2. CHEMISTRY, RESEARCH ADMINISTRATION. *Educ:* Harvard Univ, AB, 37; Mass Inst Technol, ScD(chem eng), 42. *Prof Exp:* Instr chem eng, Mass Inst Technol, 39-40; res chemist, Dewey & Almy Chem Div, 41-44, lab mgr, 44-58, asst dir res, Container & Chem Spec Div, 58-62, mgr compounding tech ctr, Overseas Chem Div, Eng, 62-64, asst dir res, Container & Chem Spec Div, 64, res dir can & drum sealing compounds, Dewey & Almy Chem Div, 64-70, assoc dir res, 70-73, assoc dir res, 73-79, CONSULT CHEMIST, INDUSTRIAL CHEM GROUP, W R GRACE & CO, 80- *Mem:* Am Chem Soc. *Res:* Rubber; colloid chemistry; canning technology; sealing compounds for containers for food preservation and industrial packaging. *Mailing Add:* 30 Lake Shore Dr Arlington MA 02174-1239

CHASE, GARY ANDREW, b New York, NY, Jan 5, 45; m 68, 80; c 2. GENETICS, STATISTICS. *Educ:* Harvard Univ, AB, 66; Johns Hopkins Univ, PhD(statist), 70. *Prof Exp:* NIH fel, Sch Med, 70-71, asst prof med & biostatist, 71-77, assoc prof health serv admin & biostatist, 78-85, PROF MENT HYG & BIOSTATIST, SCH HYG & PUB HEALTH, JOHNS HOPKINS UNIV, 85- *Concurrent Pos:* Consult, Nat Heart & Lung Inst Collab Lipid Res Prog, 72-85, Vet Admin, 88-90, Nat Inst Ment Health, 86-90. *Mem:* Am Statist Asn; Soc Epidemiol Res; Am Pub Health Asn; fel Am Psychopathological Asn. *Res:* Statistical methods in genetics and public health. *Mailing Add:* Dept Ment Hyg Johns Hopkins Med Inst 624 N Broadway Baltimore MD 21205-2179

CHASE, GENE BARRY, b Southampton, NY, Nov 22, 43; m 74; c 3. HISTORY OF MATHEMATICS. *Educ:* Mass Inst Technol, SB, 65; Cornell Univ, MA, 70, PhD(math), 79. *Prof Exp:* Instr math, Houghton Col, 66-67, Wells Col, 69-70; PROF MATH & COMPUTER SCI, MESSIAH COL, 73- *Concurrent Pos:* Adj instr ling, Univ Wash, Seattle, 79; programmer/analyst, Summer Inst Ling, 82-83 & 87-88. *Mem:* Am Asn Artificial Intel; Asn Comput Mach; Nat Coun Teachers Math; Asn Christians Math Sci (pres 89-91). *Res:* Discovering the historical relationship between Christianity and mathematics. *Mailing Add:* Messiah Col Grantham PA 17027-0801

CHASE, GERALD ROY, b Janesville, Wis, Oct 16, 38; m 61, 73; c 4. BIOSTATISTICS, COMPUTER BASED HEALTH SURVEILLANCE. *Educ:* Beloit Col, BS, 61, Stanford Univ, MS, 63, PhD(statist), 66. *Prof Exp:* From asst prof to assoc prof statist & community health, Univ Mo-Columbia, 66-73; vis scientist, environ biomet, Nat Inst Environ Health Sci, 73-74; assoc prof statist & community health, Univ Mo-Columbia, 74-75; BIOSTATISTICIAN-EPIDEMIOLOGIST, DEPT HEALTH, SAFETY & ENVIRON, JOHNS-MANVILLE CORP, 75- *Mem:* Inst Math Statist; Am Statist Asn; Biomet Soc. *Res:* Applications of statistics in health related fields; occupational health surveillance. *Mailing Add:* Johns-Manville Corp PO Box 5108 Denver CO 80217

CHASE, GRAFTON D, b NJ, May 2, 21; m 53; c 3. PHYSICAL CHEMISTRY. *Educ:* Philadelphia Col Pharm, BSc, 43; Temple Univ, MA, 51, PhD(chem), 55. *Prof Exp:* Instr chem, Philadelphia Col Pharm, 46-48; res scientist, Johnson & Johnson, 48-49; from instr to assoc prof, 49-65, PROF CHEM, PHILADELPHIA COL PHARM & SCI, 65-, DIR RADIOCHEM

LABS, 67-, DIR, DEPT CHEM, 81- *Concurrent Pos:* Assoc canning technologist, Crown Can Co, 45-47; consult, Clinica Quintero Venezuela, 55-56; ed, Remington's Pharmaceut Sci; consult, US Food & Drug Admin, 75- *Mem:* Am Chem Soc; Sigma Xi. *Res:* Fundamentals of radiochemistry; investigation of antigen-antibody interactions and radioimmunoassay. *Mailing Add:* Philadelphia Col Pharm 316 E Lancaster Pk Philadelphia PA 19096

CHASE, HELEN CHRISTINA (MATULIC), b New York, NY, Mar 21, 17; m 42. BIOSTATISTICS. *Educ:* Hunter Col, AB, 38; Columbia Univ, MSc, 51; Univ Calif, DrPH, 61. *Prof Exp:* Jr statistician, NY State Dept Health, 48-50, from biostatistician to prin biostatistician, 50-63; chief mortality statist br, Nat Ctr Health Statist, US Dept Health, Educ & Welfare, 63-65, statistician, Off Health Statist Anal, 65-69; dir res, Asn Schs Allied Health Prof, 69-71; staff assoc biostatist, Inst Med, Nat Acad Sci, 71-72; statistician, Off Res & Statist, Social Security Admin, Va, 73-75, dep chief, Epidemiol Studies Br, Bur Radiol Health, HEW, Md, 75-80; CONSULT, FOOD & DRUG ADMIN, 80-81. *Concurrent Pos:* Lectr biostatist, Grad Sch Nursing, Cath Univ Am, 66-76; consult, US Dept Health, Educ & Welfare, 61-62, White House Conf Food, Nutrit & Health, 69, Nat Acad Sci, 70 & Maternal & Child Health Proj, George Washington Univ, 70-73; mem radiation bioeffects & epidemiol adv comt, Food & Drug Admin, 71-75; mem, Task Force Ionizing Radiation, US Interagency, 78. *Mem:* Fel AAAS; fel Am Pub Health Asn; fel Am Statist Asn; Soc Epidemiol Res; Pop Asn Am. *Res:* Public health; health planning; epidemiology; infant mortality. *Mailing Add:* 6417 15th St Alexandria VA 22307

CHASE, HERMAN BURLEIGH, animal genetics, for more information see previous edition

CHASE, IVAN DMITRI, b Syracuse, NY, Feb 1, 43. ETHOLOGY, SOCIOLOGY. *Educ:* Univ SC, BS, 65; Harvard Univ, MA, 70, PhD(sociol), 72. *Prof Exp:* Vis fel math, Dartmouth Col, 71-73, vis scholar math & soc sci, 73-74, vis scholar sociol, 74-75; hon fel zool, Univ Wis-Madison, 75-78; ASSOC PROF SOCIOL, STATE UNIV NY STONY BROOK, 78- *Concurrent Pos:* Fel Soc Sci Res Coun, 71-73. *Mem:* Animal Behav Soc; Am Sociol Asn. *Res:* Social organization in animals and humans; formation and maintenance of dominance hierarchies; cooperative and non-cooperative behavior; resource distribution. *Mailing Add:* Dept Sociol State Univ NY Stony Brook NY 11794

CHASE, JAY BENTON, b Los Angeles, Calif, July 18, 40; m 63; c 2. RESEARCH ADMINISTRATION. *Educ:* Linfield Col, BA, 62, Iowa State Univ, PhD(physics), 70. *Prof Exp:* Jr physicist, Lawrence Livermore Lab, 62-65; physicist, Ames Lab, 66-70; physicist, 70-78, group leader, 78-81, DEP DIV LEADER, LAWRENCE LIVERMORE NAT LAB, 81- *Mem:* Sigma Xi; Am Defense Preparedness Asn. *Res:* Nuclear weapons; continuum mechanics; equations of state; charged particle and neutron transport; radiative transfer. *Mailing Add:* Livermore Nat Lab PO Box 808 L-35 Livermore CA 94550

CHASE, JOHN DAVID, b Detroit, Mich, Sept 24, 20; m 85; c 1. INTERNAL MEDICINE, EDUCATION ADMINISTRATION. *Educ:* Wabash Col, AB, 42; Western Reserve Univ, 45; Am Bd Internal Med, dipl. *Prof Exp:* Mem staff, Vet Admin, 52-73, chief med dir, Vet Admin Cent Off, Washington, DC, 74-78; assoc dean, 78-83, dean, 81-82, EMER DEAN CLIN PROF, SCH MED, UNIV WASH, 83- *Mem:* Inst Med-Nat Acad Sci; Asn Mil Surgeons; Am Col Chest Physicians; Am Col Physicians. *Mailing Add:* 11 Willow Pl Port Angeles WA 98362

CHASE, JOHN DONALD, b Port Williams, NS, Dec 27, 35; m 63; c 1. CHEMICAL ENGINEERING. *Educ:* Acadia Univ, BSc, 56; McGill Univ, BEng, 58; Univ London, PhD(chem eng), 62. *Prof Exp:* Lectr, W Ham Col Tech, Eng, 59-61; sr res asst combustion, Imp Col, London, 63-64; res engr, Cent Res Div, Am Cyanamid Co, 65-72, sr res engr, 73-74; Proj leader, Gulf Oil, Can, 74-; AT ENERGY MINES RESOURCES, CAN. *Concurrent Pos:* Admiralty fel, 63-64. *Mem:* Sr mem Chem Inst Can; Am Inst Chem Engrs; Combustion Inst; Am Chem Soc. *Res:* Flame kinetics; acetylene decomposition; heterogenous combustion; high temperature inorganic synthesis; plasma physics; crystal growth; plasma chemistry; fuel engineering; oxygenated gasoline components. *Mailing Add:* Energy Mines Resources Can 555 Booth St Ottawa ON K1A 0G1 Can

CHASE, JOHN WILLIAM, b Baltimore, Md, May 30, 44; m 67; c 2. BIOCHEMICAL GENETICS, MOLECULAR BIOLOGY. *Educ:* Drew Univ, BA, 66; Johns Hopkins Univ, PhD(biochem), 71. *Prof Exp:* Res fel biol chem, Harvard Med Sch, 71-74, res assoc, 74-75; from asst prof to prof molecular biol, Albert Einstein Col Med, 75-; PROF DEPT CHEM SCH MED, CASE WESTERN RESERVE UNIV. *Concurrent Pos:* NIH fels, 71-72 & 74-75; Am Cancer Soc fel, 72-73. *Mem:* Am Soc Biol Chemists. *Res:* Enzymology of DNA replication, recombination and repair, biochemical and genetic studies of the functions of single stranded DNA binding proteins and nucleases; analysis of structural and functional domains of proteins. *Mailing Add:* Dept Biochem Sch Med Case Western Reserve Univ 2119 Abington Rd Cleveland OH 44106

CHASE, LARRY EUGENE, b Wadsworth, Ohio, Sept 23, 43; m 64; c 2. ANIMAL NUTRITION, ANIMAL PHYSIOLOGY. *Educ:* Ohio State Univ, BS, 66; NC State Univ, Raleigh, MS, 69; Pa State Univ, Univ Park, PhD(animal nutrit), 75. *Prof Exp:* Dairy supvr, NC Dept Agr, Willard, 68-69; res aide animal nutrit & physiol, Pa State Univ, 69-74; asst prof, 75-81, ASSOC PROF ANIMAL SCI, CORNELL UNIV, 81- *Mem:* Am Dairy Sci Asn; Am Soc Animal Sci. *Res:* Improvement of intake and utilization of foodstuffs by ruminants with emphasis on forage utilization and nitrogen metabolism. *Mailing Add:* 272 Morrison Hall Cornell Univ Ithaca NY 14850

CHASE, LLOYD FREMONT, JR, b San Francisco, Calif, Feb 1, 31; m 52, 79; c 3. PHYSICS, SPACE PHYSICS. *Educ:* Stanford Univ, BS, 53, PhD(physics), 57. *Prof Exp:* Res assoc physics, Stanford Univ, 57-58; res scientist, res labs, Lockheed Missiles & Space Co, 58-64, sr staff scientist & sr mem labs, 64-76, mgr nuclear sci, Lockheed Palo Alto Res Lab, 76-86, MGR APPL PHYSICS LAB, LOCKHEED MISSILES & SPACE CO, 86-, DIR PHYS & ELECTRONIC SCI, RES & DEV DIV. *Concurrent Pos:* Res collabr, Brookhaven Nat Lab, 61-62; vis sr res officer, Univ Oxford, 62-63. *Mem:* Am Phys Soc. *Res:* Low energy nuclear physics; nuclear structure and nuclear reaction mechanisms; space physics. *Mailing Add:* 1850 Sand Hill Rd Palo Alto CA 94304

CHASE, LLOYD LEE, b Milwaukee, Wis, Oct 24, 39; m 68; c 2. PHYSICS. *Educ:* Univ Ill, BS, 61; Cornell Univ, PhD(physics), 66. *Prof Exp:* Mem tech staff, Bell Tel Labs, NJ, 66-69; from asst prof to prof physics, Ind Univ, Bloomington, 69-85; SR SCIENTIST, LAWRENCE LIVERMORE LAB, 85- *Mem:* Am Phys Soc. *Res:* Electron spin resonance; optical pumping in solids; Raman scattering; laser research and development; optical materials research. *Mailing Add:* Lawrence Livermore Lab PO Box 808 Livermore CA 94550

CHASE, MERRILL WALLACE, b Providence, RI, Sept 17, 05; m 31, 61; c 3. IMMUNOLOGY, MICROBIOLOGY. *Educ:* Brown Univ, AB, 27, ScM, 29, PhD(immunol), 31. *Hon Degrees:* Md, Univ Munster, 74; ScD, Brown Univ, 77 & Rockefeller Univ, 87. *Prof Exp:* Instr bact, Brown Univ, 31-32; asst immunol, Rockefeller Inst, 32-43, assoc, 43-53, assoc mem & assoc prof, 53-65, prof, 65-76, EMER PROF IMMUNOL, ROCKEFELLER UNIV, 76- *Concurrent Pos:* Mem, sci & educ coun, Allergy Found Am, 55-82; mem, sci adv bd, St Jude Hosp, Memphis, 62-65 & 66-67; consult, Nat Inst Allergy & Infectious Dis, 70-73 & sci adv comt, 74-76; mem, Sci Adv Comt, Ore Regional Primate Ctr, 71-77. *Honors & Awards:* Distinguished Sci Award, Am Acad Allergy, 69. *Mem:* Nat Acad Sci; Am Asn Immunologists (pres, 56-57); Am Soc Microbiol; Am Asn Lab Animal Sci; fel Am Acad Arts & Sci; hon fel Am Acad Allergy; hon fel Am Col Allergists; fel AAAS; NY Acad Sci. *Res:* Hypersensitivity to simple chemical allergens; cellular transfer of contactant and tuberculin hypersensitivities; immunologic unresponsiveness to chemical allergens; native mycobacterial antigens; Kveim antigen in sarcoidosis. *Mailing Add:* Rockefeller Univ 1230 York Ave New York NY 10021-6399

CHASE, NORMAN E, b Cincinnati, Ohio, June 29, 26; m 54; c 2. MEDICINE. *Educ:* Univ Cincinnati, BS, 50, MD, 53. *Prof Exp:* Instr radiol, Col Physicians & Surgeons, Columbia Univ, 59-61, assoc, 61; assoc attend, 61, from asst prof to assoc prof, 64-67, PROF RADIOL, MED CTR, NY UNIV, 67-, CHMN DEPT, 69- *Concurrent Pos:* Dir radiol, Bellevue Hosp, NY, 65-73, assoc dir, 73-; sr consult, Manhattan Vet Admin Hosp, 69- *Mem:* AMA; Am Soc Neuroradiol (secy-treas, 62, pres elect, 71); Asn Univ Radiol; NY Acad Sci; Am Col Radiol. *Res:* Cerebrovascular disease; radiology; neuroradiology. *Mailing Add:* NY Univ Med Ctr 550 First Ave New York NY 10016

CHASE, RANDOLPH MONTIETH, JR, b Brooklyn, NY, Aug 10, 28; m 55; c 3. MEDICINE, IMMUNOLOGY. *Educ:* NY Univ, AB, 50, MD, 58. *Prof Exp:* Asst med, Sch Med, NY Univ, 59-61, instr, 61-62; asst physician, Rockefeller Inst Hosp, 62-64; asst prof, 64-70, ASSOC PROF MED, SCH MED, NY UNIV, 70-, DIR MICROBE LAB, UNIV HOSP, 64- *Concurrent Pos:* Nat Inst Allergy & Infect Dis fel, Rockefeller Inst, 62-64; clin asst, Bellevue Hosp, 62-64, asst attend physician, 62-, asst vis physician, 65-86, assoc vis physician, 86-; consult, Dept Health Educ & Welfare & FDA, 78-86. *Mem:* AAAS; Transplantation Soc; Sigma Xi. *Res:* Infectious disease and allergy, including the effect of antibiotics on streptococcal cell wall, effect of prophylactic antibiotics in RHD and the appearance of given streptoccal strains in SBE; cross reacting antigens existing between bacteria and mammalicin tissues, especially tissue transplants. *Mailing Add:* Dept Med NY Univ 550 First Ave New York NY 10016

CHASE, RICHARD L, b Perth, Australia, Dec 25, 33; m 65; c 3. GEOLOGY. *Educ:* Univ Western Australia, BSc, 56; Princeton Univ, PhD(geol), 63. *Prof Exp:* Geologist, WAustralian Petrol Ltd, 54-55; asst geologist, Geosurv Australia Ltd, 56-57; sr asst geologist, Ministry Mines, Que, 59; geologist, Ministry Mines & Hydrocarbons, Venezuela, 60-61; Ford Found fel, Woods Hole Oceanog Inst, 63-64, asst scientist, 64-68; from asst prof to assoc prof, 68-78, acting head, 90-91, PROF GEOL SCI, UNIV BC, 78-, PROF OCEANOG, 80- *Mem:* Geol Soc Am; Am Geophys Union; Geol Asn Can. *Res:* Petrology and structural geology; origin of oceanic igneous rocks; history of ocean floors; marine geology, geotectonics and petrology in the northeastern Pacific. *Mailing Add:* Dept Geol Sci Univ BC Vancouver BC V6T 2B4 Can

CHASE, RICHARD LYLE, weed science; deceased, see previous edition for last biography

CHASE, ROBERT A, b Keene, NH, Jan 6, 23; m 46; c 3. RECONSTRUCTIVE SURGERY, HUMAN ANATOMY. *Educ:* Univ NH, BS, 45; Yale Univ, MD, 47; Am Bd Surg, dipl, 55; Am Bd Plastic Surg, dipl, 60. *Prof Exp:* Intern, New Haven Hosp, 47-48; asst resident surg path, bact & cancer clin & asst surg path & bact, Sch Med, Yale Univ, 48-49; asst resident, New Haven Hosp, 49-50, sr asst resident surg, 52-53, chief res surgeon, 53-54; res plastic surgeon, Univ Pittsburgh Hosp, 57-59; from asst prof to assoc prof surg, Sch Med, Yale Univ, 59-63; prof surg & chmn dept, Sch Med, Stanford Univ, 63-73; pres & dir, Nat Bd Med Examr, 73-76; EMILE HOLMAN PROF SURG & PROF ANAT, SCH MED, STANFORD UNIV, 72- *Concurrent Pos:* Fel plastic surg, Univ Pittsburgh Hosp, 57-59; asst, Sch med, Yale Univ, 53-54; attend surgeon, US Vet Admin Hosp, West Haven, Conn, 59-62, consult, 62-63; attend surg, Grace New Haven Community Hosp, 59-63; consult, Christian Med Col & Hosp, India, 62 & Vet Admin Hosp, Palo Alto, Calif; mem med staff, Santa Clara County Hosp; mem, Plastic Surg Res Coun; vis prof, dept anat, Harvard Med Sch, 71;

prof surg, Sch Med, Univ Pa, 74-77. *Mem:* Inst Med Nat Acad Sci; Am Surg Asn; fel Am Col Surg; Am Soc Plastic & Reconstruct Surg; Am Soc Surg Hand (pres, 84); Am Asn Plastic Surg. *Res:* Nerve pedicle regeneration in anterior ocular chamber; objective evaluation of palatopharyngeal function; electronic publishing and teaching material. *Mailing Add:* Div Anat Stanford Univ Sch Med Stanford CA 94305

CHASE, ROBERT L(LOYD), b Brooklyn, NY, Mar 19, 26; m 50, 70; c 3. ELECTRONICS. *Educ:* Columbia Univ, BSEE, 45; Cornell Univ, MEE, 47; Univ Uppsala, PhD(elec eng), 73. *Prof Exp:* From assoc head to head instrumentation div, Brookhaven Nat Labs, 47-77; SR ENGR, LINEAR ACCELERATOR LAB, ORSAY, FRANCE, 77- *Mem:* Fel Inst Elec & Electronics Engrs. *Res:* Nuclear instrumentation. *Mailing Add:* Linear Accelerator Lab Bldg 200 Orsay 91 France

CHASE, ROBERT SILMON, JR, b Abington, Pa, June 9, 30; m 55; c 4. VERTEBRATE ZOOLOGY. *Educ:* Haverford Col, AB, 52; Univ Ark, MS, 55; Bryn Mawr Col, PhD(biol), 67. *Prof Exp:* From instr to assoc prof biol, Lafayette Col, 58-68, asst dean, 61-64, dean studies, 68-69, dean col, 69-70 & 75-78, provost, 70-72, PROF BIOL, LAFAYETTE COL, 74-, HEAD DEPT, 78-, CHARLES A DANA PROF BIOL, 90- *Res:* Amphibian development, vertebrate behavior and ecology. *Mailing Add:* Dept Biol Lafayette Col Easton PA 18042

CHASE, RONALD, b Chicago, Ill, Sept 12, 40; m 68; c 2. NEUROBIOLOGY. *Educ:* Stanford Univ, BSc, 62; Mass Inst Technol, PhD(psychol), 69. *Prof Exp:* Asst prof biol, 71-75, ASSOC PROF BIOL, McGILL UNIV, 75- *Mem:* Soc Neurosci. *Res:* Structure and function in snail brains. *Mailing Add:* Dept Biol 1205 Docteur Penfield Ave Montreal PQ H3A 1B1 Can

CHASE, SHERRET SPAULDING, b Toledo, Ohio, June 30, 18; m 43; c 5. GENETICS, BOTANY. *Educ:* Yale Univ, BS, 39; Cornell Univ, PhD(bot cytol, genetics), 47. *Prof Exp:* Assoc prof bot, Iowa State Col, 47-54; res geneticist & mgr foreign seed opers, DeKalb Agr Asn, Inc, 54-66, dir, Dekalb-Italiana, 65-66; Bullard fel, Bot Mus, Harvard Univ, 66-67, Cabot fel, Forest Res, 67-69, res assoc econ bot, Bot Mus, 69-70; PROF BIOL, STATE UNIV NY OSWEGO, 70- *Concurrent Pos:* Pres, Catskill Ctr Conserv & Develop, Inc; dir, Hanford Mills Mus Corp, 73- *Mem:* AAAS; Bot Soc Am; Am Soc Agron; Genetics Soc Am; Sigma Xi. *Res:* Plant breeding; cytotaxonomy of Najas; parthenogenesis in maize; corn breeding; forest genetics. *Mailing Add:* Box 193 Shokan NY 12481

CHASE, THEODORE, JR, b Boston, Mass, Aug 20, 38; m 65; c 2. ENZYMOLOGY. *Educ:* Harvard Univ, AB, 60; Univ Calif, Berkeley, PhD(biochem), 66. *Prof Exp:* Res assoc biol, Brookhaven Nat Lab, 67-69; asst prof, 69-74, ASSOC PROF BIOCHEM & MICROBIOL, RUTGERS UNIV, 74- *Concurrent Pos:* Food Enzym deleg, People's Rep China, 85- *Mem:* Am Soc Microbiol; Am Chem Soc; Sigma Xi; Am Ornith Union; Brit Ornith Union; Am Soc Biochem & Molecular Biol. *Res:* Mechanism of enzyme action; practical utilization of enzymes and enzyme inhibitors; enzymes in fruit ripening. *Mailing Add:* Dept Biochem & Microbiol Cook Col Rutgers Univ Lipman Hall New Brunswick NJ 08903

CHASE, THOMAS NEWELL, b Westfield, NJ, May 23, 32; m 59; c 2. NEUROLOGY, NEUROPHARMACOLOGY. *Educ:* Mass Inst Technol, BS, 54; Yale Univ, MD, 62. *Prof Exp:* Engr, Singer Mfg Co, Conn, 54-55; res technician, Col Physicians & Surgeons, Columbia Univ, 57-58; intern internal med, Yale-New Haven Med Ctr, 62-63; from asst resident to resident neurol, Mass Gen Hosp, 63-66; guest worker, Lab Clin Sci, 66-68, chief neurol unit, NIMH, 68-74, chief exp therapeut, 70-74; chief, Lab Neuropharmacol, 74-76, dir intramural res, 74-81, CHIEF PHARMACOL SECT, NAT INST NEUROL DIS & STROKE, 76- & CHIEF EXP THERAPEUT BR, 83-; ADJ RES PROF, SCH MED, UNIV MD, 87- *Concurrent Pos:* Fel neuropath, Mass Gen Hosp & Harvard Sch Med, 64-65; Nat Inst Neurol Dis & Stroke spec fel, 66-68; clin assoc prof neurol, Sch Med, Georgetown Univ, 71-; res group Huntington's chorea, World Fedn Neurol; mem sci adv bd, Am Nat Atexia Found, Parkinson's Dis Asn & Nat Parkinson's Found. *Honors & Awards:* Winternitz Prize Path, Yale Univ, 60, Ramsay Prize Clin Med, 61; USPHS Meritorious Serv Medal. *Mem:* Am Soc Neurochem; Soc Neurosci; Asn Res Nerv & Ment Dis; Am Neurol Asn; Am Col Neuro-Psychopharmacol; Int Brain Res Orgn; Am Acad Neurol; Int Soc Neurochem. *Res:* Neuropharmacology; clinical and experimental neurology; neurochemistry; neurohumoral mechanisms; research administration. *Mailing Add:* Nat Inst Neurol Dis & Stroke NIH Bldg 10 Rm 5C103 Bethesda MD 20892

CHASE, THOMAS RICHARD, b Rochester, NY, Mar 15, 54. KINEMATICS, ENGINEERING DATABASES. *Educ:* Rochester Inst Technol, BS, 77, MS, 83; Univ Minn, PhD(mech eng), 84. *Prof Exp:* Asst prof mech eng, Univ RI, 83-85; ASST PROF MECH ENG, UNIV MINN, 85- *Mem:* Am Soc Mech Engrs; Am Soc Eng Educ. *Res:* Computer aided mechanism design and kinematics; database design for computer integrated manufacturing; stress analysis of rolled webs. *Mailing Add:* Mech Eng Dept Univ Minn 111 Church St SE Minneapolis MN 55455

CHASE, VERNON LINDSAY, b Baltimore, Md, Mar 20, 20; m 43; c 4. TEXTILE CHEMISTRY. *Educ:* Western Md Col, BA, 41. *Prof Exp:* Res chemist electrochem, Am Smelting & Refining Co, 41-46; res chemist org chem, Ridbo Labs, 47; dir res & develop textile colors, Color & Chem Div, Interchem Corp, 48-57, prog mgr org coatings, Cent Res Labs, 58-66; res assoc polymers, Tech Ctr, J P Stevens & Co, Inc, 67-83; RETIRED. *Mem:* Am Asn Textile Chemists & Colorists. *Res:* Polymeric systems for elastic fabrics; acrylic binders for pigment printing and non-woven fabrics; foam-backcoating; systems for carpet backing and flame retardancy; dyeing systems for new fiber blends. *Mailing Add:* 413 Sitton Mill Rd Rte 6 Seneca SC 29678

CHASE, WILLIAM HENRY, b Montreal, Que, June 15, 27. IMMUNOPATHOLOGY, NEPHROPATHOLOGY. *Educ:* McGill Univ, BSc, 48, MD & CM, 52. *Prof Exp:* Resident path, Vancouver Gen Hosp, 52-56; from asst prof to assoc prof, 58-74, assoc dir labs, 79-89, PROF MATH, UNIV BC, 74-, ACTG HEAD, 90- *Concurrent Pos:* Res fel anat, Univ Chicago, 56-58. *Mem:* Am Asn Path; Can Soc Immunol. *Res:* Electron microscopy, nephropathology, immunophathology. *Mailing Add:* Dept Path Univ of BC Vancouver BC V6T 1W5 Can

CHASENS, ABRAM I, b Woodbine, NJ, Sept 7, 12; m 42. PERIODONTICS, DENTISTRY. *Educ:* Temple Univ, DDS, 36; NY Univ, cert, 52; Am Bd Periodont, dipl, 54; Am Bd Oral Med, dipl, 66. *Prof Exp:* Asst prof periodont & oral med, Col Dent, NY Univ, 53-57; PROF PERIODONT & ORAL MED & CHMN DEPT, SCH DENT, FAIRLEIGH DICKINSON UNIV, 57 , DIR GRAD PERIODONT, 73- *Concurrent Pos:* Consult, Muhlenberg Hosp, Plainfield, NJ, VA Hosp, Lyons, NJ, St Joseph's Hosp, Paterson, NJ & Holy Name Hosp, Teaneck, NJ, 54-; consult, Dent Educ & Hosp Dent Serv Coun; dir, Am Bd Oral Med, 72-80 & Am Bd Periodont, 80-, chmn, 85-86. *Honors & Awards:* Samuel Charles Miller Mem Award, 71; Hirschfeld Medal, Northeastern Soc Periodont, 72; Hirschfus Mem Award, Am Acad Oral Med, 85; Gold Medal, Am Acad Periodont, 86. *Mem:* Fel Am Col Dent; fel Am Acad Oral Med (pres, 67-68); fel Int Col Dentists; fel Am Acad Periodont; fel Royal Soc Health; fel NY Acad Sci. *Res:* Periodontology; oral medicine; occlusion diseases and disturbances of the temporo-mandibular joint; periodontal surgery. *Mailing Add:* 179 Beverly Rd Hawthorne NJ 07506

CHASIN, LAWRENCE ALLEN, b Willimantic, Conn, July 2, 41; m 61, 75; c 2. BIOCHEMICAL GENETICS. *Educ:* Brown Univ, BS, 62; Mass Inst Technol, PhD(biol), 67. *Prof Exp:* Res assoc microbiol, Lab Enzymol, Ctr Nat Sci Res, 66-68; sr instr cell genetics, Univ Colo Med Ctr, 68-70; from asst prof to assoc prof, 70-81, PROF BIOL SCI, COLUMBIA UNIV, 81- *Concurrent Pos:* Mem genetics study sect, Div Res Grants, NIH, 75-79. *Mem:* AAAS; Genetics Soc Am; Am Soc Cell Biol. *Res:* Application of somatic cell genetic techniques of the regulation of gene expression; biochemical characterization of regulatory variants of cultured mammalian cells. *Mailing Add:* Dept Biol Sci Columbia Univ 912 Fairchild New York NY 10027

CHASIN, MARK, b New York, NY, Feb 20, 42; m 63; c 3. BIOCHEMISTRY, ENZYMOLOGY. *Educ:* Cornell Univ, AB, 63; Mich State Univ, PhD(biochem), 67. *Prof Exp:* Sr res investr biochem pharmacol, Squibb Inst Med Res, 67-74; group leader molecular biol, Ortho Res Found, Otho Pharmaceut Corp, 74-75. sect head biochem res, 75-77, dir, Div Biochem Res, 77-78; dir, INPD, Chasin Enterprises, Inc, 78-80, dir, Clin Develop, 80-81, dir, RIS, 81-86; dir, Technol Develop, Nova Pharmaceut Corp, 86-90; VPRES, RES & DEVELOP, PURDUE FREDERICK, 90- *Mem:* AAAS; Am Chem Soc; NY Acad Sci; Am Soc Biol Chemists; Am Soc Pharmacol & Exp Therapeut; Am Soc Clin Pharmacol & Therapeut; Controlled Release Soc; Drug Info Asn. *Res:* Enzymology and enzyme inhibitors; enzymology concerned with 3', 5'-cyclic adenosine monophosphate; biochemical pharmacology; cardiovascular, hypersensitivity, control nervous system and reproductive research; clinical research; drug delivery systems. *Mailing Add:* Three Wayne Ct Englishtown NJ 07726

CHASIN, WERNER DAVID, b Danzig, Feb 29, 32; US citizen; m 63; c 3. OTOLARYNGOLOGY. *Educ:* Harvard Univ, AB, 54; Tufts Univ, MD, 58. *Prof Exp:* Rotating intern, Mt Sinai Hosp, New York, 58-59; resident otolaryngol, Mass Eye & Ear Infirmary, 59-62, asst otolaryngologist, 62-64; chief otolaryngol, Beth Israel Hosp, 64-68; PROF & CHMN DEPT OTOLARYNGOL, SCH MED, TUFTS UNIV, 68-; OTOLARYNGOLOGIST-IN-CHIEF, TUFTS-NEW ENG MED CTR, 68- *Concurrent Pos:* Vis surgeon, Boston City Hosp, 68-; pres & secy-treas, New Eng Otolaryngol Soc, 69-72. *Mem:* Fel Am Acad Ophthal & Otolaryngol. *Mailing Add:* Dept Otolaryngol Tufts New Eng Med Ctr 750 Washington St Box 850 Boston MA 02111

CHASIS, HERBERT, b New York, NY, Nov 9, 05; m 43; c 2. MEDICINE. *Educ:* Syracuse Univ, AB, 26; NY Univ, MD, 30, ScD(med), 37. *Prof Exp:* From instr to assoc prof, 35-64, PROF MED, COL MED, NY UNIV, 64-, ATTEND PHYSICIAN, UNIV HOSP & CLIN, 55- *Concurrent Pos:* Asst vis physician, Bellevue Hosp, 38-43, assoc attend physician, 44-54, attend physician, 57-; chief, Cardiac Clin, French Hosp, 46-64, consult physician, 64-68; consult, St Lukes Hosp, Newburg, NY & Vet Admin, 51-; consult physician, Phelps Mem Hosp, Tarrytown, NY. *Mem:* Am Soc Clin Invest; Am Physiol Soc; Harvey Soc; Soc Exp Biol & Med; fel Am Col Physicians. *Res:* Cardiovascular and renal physiology; physiological and clinical investigation of renal and hypertensive diseases. *Mailing Add:* Dept Med NY Univ Med Sch New York NY 10016

CHASMAN, CHELLIS, b New York, NY, Feb 11, 32. NUCLEAR PHYSICS. *Educ:* Harvard Univ, BA, 53; Columbia Univ, PhD(physics), 61. *Prof Exp:* Instr physics, Yale Univ, 61-63; SR SCIENTIST PHYSICS, BROOKHAVEN NAT LAB, 63- *Mem:* Fel Am Phys Soc. *Mailing Add:* 29 Conscience Cir Setauket NY 11733

CHASON, JACOB LEON, b Monroe, Mich, May 12, 15; m 42; c 3. PATHOLOGY, NEUROPATHOLOGY. *Educ:* Univ Mich, AB, 37, MD, 40. *Prof Exp:* Chmn dept, 64-78, prof path, neuropath, 51-78, clin prof path, Sch Med, Wayne State Univ, 78-87; RETIRED. *Concurrent Pos:* Neuropathologist, Henry Ford Hosp, 78-88; assoc dean, Wayne State Univ Sch Med, 70-72. *Mem:* Am Soc Clin Path; Col Am Path; Am Acad Neurol; Int Acad Path; Am Asn Neuropathologists. *Res:* Pathology of the nervous system. *Mailing Add:* 4862 Keithdale Lane Bloomfield Hills MI 48302

CHASS, JACOB, b Haifa, Israel, Jan 10, 26; US citizen; m 53; c 2. DISPLACEMENT TRANSDUCERS. *Educ:* Technion, BSc, 51; Univ Pa, MSC, 59. *Prof Exp:* Elec engr, RCA, 53-54; chief eng, Control Div, JRC, 57-64; CHIEF ENGR, PICKERING MEASUREMENT & CONTROL, 65-

Mem: Sr mem Instrument Soc Am. *Res:* Development and design of accurate, temperature compensated transducers for linear and rotary displacement control; development of linear phase displacement transducer; 20 patents concerning TV circuit and electromechanical transducers. *Mailing Add:* 70-25 Yellowstone Blvd Forest Hills NY 11375

CHASSAN, JACOB BERNARD, b New York, NY, Oct 16, 16; m 52; c 3. PSYCHOLOGY, STATISTICS. *Educ:* City Col New York, BS, 39; George Washington Univ, MA, 49, PhD, 58. *Prof Exp:* Statistician, Air Tech Serv Command, Wright Field, Ohio, 42-48; statistician, Med Statist Div, Off Surgeon Gen, US Dept Army, 46-47, chief health reports br, 47-49; chief div tuberc, USPHS, 49-50; chief clin eval & follow-up studies unit, Dept Med & Surg, Vet Admin, 50-53; sci analyst, Mass Inst Tech Opers Eval Group, 53-55; chief statistician, St Elizabeths Hosp, DC, 55-60; math statistician, Off Educ, US Dept Health, Educ & Welfare, 60-61; head statistician, Hoffmann-La Roche, Inc, 61-66; dir statist serv, Sandoz, Inc, NJ, 66-67; assoc dir med res planning, Hoffmann-La Roche Inc, 67-69, psychiat res planner, 69-73, clin res scientist, 73-79; DIR, J B CHASSAN CONSULTS, 80- *Concurrent Pos:* Mem fac, USDA Grad Sch; asst clin prof psychiat, George Washington Univ, 61-73, spec lectr psychiat & behav sci, Sch Med, 73-; lectr biostatist, Seton Hall, 64-65; clin assoc prof statist in psychiat, Med Col, Cornell Univ, 71-73; fac mem, NY Ctr Psychoanal Training, 72-; pvt pract psychother. *Mem:* Fel AAAS; fel Am Statist Asn; Am Acad Psychother; Am Asn Marriage & Family Counrs; Math Asn Am. *Res:* Design of clinical research; applied statistics in epidemiology, psychiatry and psychoanalysis; mathematical statistics. *Mailing Add:* 763 Bloomfield Ave Montclair NJ 07042

CHASSON, ROBERT LEE, experimental physics; deceased, see previous edition for last biography

CHASSY, BRUCE MATTHEW, b Ft Jackson, SC, Oct 22, 42; wid; c 2. BIOCHEMISTRY, MOLECULAR BIOLOGY. *Educ:* San Diego State Col, AB, 62; Cornell Univ, NIH fel & PhD(biochem), 66. *Prof Exp:* Fel biochem, Albert Einstein Med Ctr, 65-67; fel biochem, 68-69, RES CHEMIST, NAT INST DENT RES, 69- *Concurrent Pos:* Prof lectr, Am Univ, 69-72. *Mem:* AAAS; NY Acad Sci; Am Soc Biol Chem; Am Soc Microbiol; Soc Indust Microbiol; Inst Food Technologists. *Res:* Enzyme mechanisms and specificity; plasmids; nucleic acids; nucleotides; sugar transport systems in bacteria. *Mailing Add:* Nat Inst Dent Res NIH Bldg 30 Bethesda MD 20892

CHASTAGNER, GARY A, b Woodland, Calif, Sept 9, 48; m; c 2. PLANT PATHOLOGY. *Educ:* Calif State Univ, Fresno, BA, 71; Univ Calif, Davis, MS, 73, PhD(plant path), 76. *Prof Exp:* Res plant pathologist, Univ Calif, Davis, 77-78; asst plant pathologist, 78-83, assoc plant pathologist, 83-89, PLANT PATHOLOGIST, RES & EXTEN CTR, WASH STATE UNIV, 89- *Mem:* Am Phytopath Soc; Sigma Xi; Int Soc Plant Pathology; Int Soc Hort Sci. *Res:* Epidemiology and control of diseases on turf, christmas trees, ornamental bulbs and hybrid poplars. *Mailing Add:* Res & Exten Ctr Wash State Univ Puyallup WA 98371-4998

CHASTAIN, BENJAMIN BURTON, b Tuscaloosa, Ala, Dec 21, 36; m 79; c 2. INORGANIC CHEMISTRY. *Educ:* Birmingham-Southern Col, BS, 56; Columbia Univ, MA, 57, PhD(inorg chem), 67. *Prof Exp:* Assoc chemist, Southern Res Inst, 57-58; pub sch teacher, Ala, 58-59; from instr to assoc prof chem, 59-70, PROF CHEM, SAMFORD UNIV, 70-, HEAD DEPT, 84- *Concurrent Pos:* Mem secretary's adv comt coal mine safety res, US Dept Interior, 71-74; vis assoc, Calif Inst Technol, 78; chair-elect, Div Hist Chem, Am Chem Soc, 91, chair, 92. *Mem:* Am Chem Soc; Sigma Xi; Am Inst Chemists; Hist Sci Soc; Am Asn Advan Sci. *Res:* Synthesis and electronic structures of coordination complexes on transition metals; history of chemistry; chemical education. *Mailing Add:* Dept of Chem Samford Univ Birmingham AL 35229

CHASTAIN, MARIAN FAULKNER, b Sept 9, 22; US citizen; m 56; c 2. FOOD SCIENCE & TECHNOLOGY. *Educ:* Cedar Crest Col, BS, 44; Fla State Univ, MS, 53, PhD(food, nutrit), 55. *Prof Exp:* Jr chemist, Hoffmann-La Roche, Inc, NJ, 44-52; asst prof foods & nutrit, Purdue Univ, 55-56; assoc prof foods & nutrit, Auburn Univ, 56-59, 62-87; RETIRED. *Mem:* AAAS; Am Home Econ Asn; Am Dietetic Asn; Inst Food Technologists; Sigma Xi. *Res:* Improvement of protein value of foods; effects of cooking methods on palatability and nutrient retention; prevention of oxidative changes in stored foods. *Mailing Add:* 1357 Burke Lane Auburn AL 36830

CHASTEEN, NORMAN DENNIS, b Flint, Mich, Oct 6, 41; m 67; c 1. BIOINORGANIC CHEMISTRY. *Educ:* Univ Mich, AB, 65; Univ Ill, Urbana-Champaign, MS, 66, PhD(chem), 69. *Prof Exp:* NIH fel, 69-70; asst prof chem, Lawrence Univ, 70-74; assoc prof, 74-79, PROF CHEM, UNIV NH, 79- *Concurrent Pos:* Chmn, NIH study sect, 86-88, chmn Gordon Conf, 88 NIH Fogarty fel, 79, Fulbright scholar, 88. *Mem:* AAAS; Am Chem Soc; NY Acad Sci; Am Soc Biochem. *Res:* Proteins of iron storage and transport. *Mailing Add:* Dept of Chem Univ of NH Durham NH 03824

CHASZEYKA, MICHAEL A(NDREW), b Youngstown, Ohio, July 28, 20; m 46. SCIENCE COMMUNICATIONS, MECHANICAL & RESEARCH ENGINEERING. *Educ:* Ohio State Univ, BME, 43; Ill Inst Technol, MS, 59. *Prof Exp:* Apprentice engr, US Steel Corp, 46-47; draftsman, Lombard Corp, 47; designer, Youngstown Steel Tank Co, 47; jr engr, Truscon Steel Div, Repub Steel Corp, 47; sr detailer design, Electromotive Div, Gen Motors Corp, 47-51, jr process engr, 51; res engr, Armour Res Found, 53-61; phys sci coordr phys, Mat & Math Sci, Off Naval Res, Chicago, 61-81; RETIRED. *Concurrent Pos:* Organizer & coordr govt groups eval tech qual of independent res & develop progs; consult, 81- *Mem:* Am Soc Mech Engrs; fel Marine Technol Soc. *Res:* Energy conversion; theoretical and applied fluid mechanics; underwater propulsion; numerical methods. *Mailing Add:* 4147 Grove Ave Western Springs IL 60558

CHATEAUNEUF, JOHN EDWARD, b Swampscott, Mass, Apr 19, 57. REACTIVE INTERMEDIATES, PHOTOCHEMISTRY. *Educ:* Salem State Col, BS, 81; Tufts Univ, PhD(chem), 86. *Prof Exp:* Res assoc, Nat Res Coun Can, 86-88; FAC ASST PROF SPECIALIST, UNIV NOTRE DAME, 88- *Mem:* Sigma Xi; Am Chem Soc; Inter-Am Photochem Soc. *Res:* Physical organic chemistry; time-resolved spectroscopies to study reaction intermediates and mechanistic investigation of reactive species. *Mailing Add:* Radiation Lab Univ Notre Dame Notre Dame IN 46556

CHATELAIN, EDWARD ELLIS, sedimentary petrology, for more information see previous edition

CHATELAIN, JACK ELLIS, b Ogden, Utah, July 17, 22; m 46; c 2. THEORETICAL PHYSICS. *Educ:* Utah State Univ, BS, 47, MS, 48; Lehigh Univ, PhD(physics), 57. *Prof Exp:* Instr physics, Univ Wyo, 50-52; physicist, Dugway Proving Ground, 53; instr physics, Lehigh Univ, 53-57; physicist, Phillips Atomic Energy Div, Phillips Petrol Co, 58; from asst prof to prof physics, Utah State Univ, 57-88; RETIRED. *Concurrent Pos:* Physicist, White Sands Proving Ground, 54-55; sci specialist & consult, Edgerton, Germeshausen & Grier, Inc, Nev, 61-63. *Mem:* Am Phys Soc. *Res:* Theoretical aspects of radiation and its interaction with matter. *Mailing Add:* 1657 N 1600 E Utah State Univ Logan UT 84321

CHATENEVER, ALFRED, b New York, NY, May 1, 16; m 40; c 3. PHYSICAL CHEMISTRY, PETROLEUM ENGINEERING. *Educ:* City Col New York, BS, 36; Columbia Univ, MA, 37; NY Univ, PhD(phys chem), 48. *Prof Exp:* Chief chemist & head res & develop sects, Aquatic Chem Labs, NY, 38-42, tech dir, 46-47; sr chemist, Matam Corp, 42-43; res scientist & group leader, Manhattan Dist Proj, 44-46; pvt chem consult, 48-49; res engr & prof petrol eng, Res Inst, Univ Okla, 49-58; sr res assoc, Sinclair Res, Inc, 58-69; INDEPENDENT CONSULT, 69- *Mem:* AAAS; Am Chem Soc; Nat Asn Corrosion Engrs; Soc Petrol Engrs. *Res:* Kinetics of ionic reactions in solution; gaseous diffusion; physical chemical measurements; water conditioning; corrosion; fluid behavior in porous systems; thermal conductivity; free radicals; research evaluation; groundwater contamination. *Mailing Add:* 4615 S Florence Ave Tulsa OK 74105

CHATER, SHIRLEY S, m 59; c 2. NURSING ADMINISTRATION. *Educ:* Univ Pa, BS, 53; Univ Calif, San Francisco, MS, 60; PhD, 64; Mass Inst Technol Sloan Sch Mgt, cert, 82. *Prof Exp:* Asst vchancellor, Acad Affairs, Univ Calif, San Francisco, Sch Nursing, 74-77, vchancellor, Acad Affairs, 77-82, prof, Social & Behav Sci, 73-82; coun assoc, Am Coun Educ, Washington, DC, 82-84; sr assoc, Presidential Search Consult Serv, Asn Gov Bds Univs & Cols, Washington, DC, 84-86; PRES, TEX WOMAN'S UNIV, DENTON, TEX, 86- *Concurrent Pos:* Commr, Am Coun Educ, 77-82. *Mem:* Inst Med-Nat Acad Sci. *Mailing Add:* Tex Woman's Univ PO Box 23925 Denton TX 76204-1925

CHATFIELD, DALE ALTON, b Pontiac, Mich, Apr 5, 47. ANALYTICAL CHEMISTRY. *Educ:* Oakland Univ, BA & MS, 69; Univ NC, Chapel Hill, PhD(anal chem), 75. *Prof Exp:* Assoc chem, Univ Utah, 74-77, asst res prof mat sci, 77-78; assoc prof, 78-82, ASSOC PROF CHEM, SAN DIEGO STATE UNIV, 82- *Mem:* Am Chem Soc; Am Soc Mass Spectrometry. *Res:* Gas analysis by mass spectrometry; thermal degradation of polymeric materials, surface and thermal ionization processes; chemical coal cleaning research. *Mailing Add:* Dept of Chem San Diego State Univ San Diego CA 92182

CHATFIELD, JOHN A, b San Antonio, TX, 43. ANALYSIS & FUNCTIONAL ANALYSIS. *Educ:* Southwest Tex Univ, BA & MA, 66; Univ Tex, PhD(math), 68. *Prof Exp:* From asst prof to assoc prof, 68-78, PROF MATH, SOUTHWEST TEX UNIV, 78- *Mem:* Am Math Soc; Am Math Asn. *Res:* Analyses & Educational Analysis. *Mailing Add:* Dept Math Southwest Tex State Univ Marcos TX 78666

CHATIGNY, MARK A, b Can, Sept 19, 20; nat US; m 45; c 5. BIOENGINEERING. *Educ:* Univ Calif, BSEE, 49. *Prof Exp:* From jr res engr to assoc res engr, Univ Calif, Oakland, 49-62, res engr, Naval Biosci Lab, 62-81, chmn, Dept Environ Biol, 68, asst dir, 73, eng consult, 81-87, consult, 81-91; RETIRED. *Concurrent Pos:* Consult var agencies, CMCL Bio Res & Prodm Cos & univs; mem, Biohazards work group, Nat Cancer Inst, 63-74, Planetary Quarantine Adv Panel, 68-76, Am Inst Biol Sci & Air Sampling Comt & Am Conf Govt Indust Hyg; lectr, Univ Minn, 71-76; bioeng consult, 81- *Mem:* Inst Environ Sci; Int Soc Contamination Control; Int Aerobiol Asn; Am Biol Safety Asn. *Res:* Aerosols; biohazards; contamination control in biological facilities; facility design for biohazards and contamination control; contamination assessment procedures in microbiology laboratories, hospitals and pharmaceutical plants; studies in development and demonstration of medically acceptable energy conservation techniques in hospitals and laboratories. *Mailing Add:* 507 Santa Ynez San Lorenzo CA 94579

CHÂTILLON, GUY, validity of fringe science, for more information see previous edition

CHATLAND, HAROLD, b Hamilton, Can, Nov 13, 11; nat US; m 37; c 3. MATHEMATICS. *Educ:* McMaster Univ, BA, 34; Univ Chicago, MS, 35, PhD(math), 37. *Prof Exp:* Asst prof math, Univ Mont, 39-46 & Ohio State Univ, 46-49; prof, Univ Mont, 49-59, dean col arts & sci, 54, dean fac, 56-57, acad vpres, 57-59; eng specialist, Sylvania Elec Prod, Inc, 59-62; acad dean, Western Wash State Col, 62-64; sr eng specialist, Electronic Defense Labs, 64-68; mem staff, Sylvania Electronics Systs, 68-71; CONSULT, 71- *Mem:* Soc Indust & Appl Math; Math Asn Am; Sigma Xi. *Res:* Number theory; decision-making techniques. *Mailing Add:* 10566 Blandor Way Los Altos Hills CA 94022

CHATO, JOHN C(LARK), b Budapest, Hungary, Dec 28, 29; nat US; m 54; c 3. MECHANICAL ENGINEERING, BIOENGINEERING. *Educ:* Univ Cincinnati, ME, 54; Univ Ill, MS, 55; Mass Inst Technol, PhD, 60. *Prof Exp:* Asst mech eng, Mass Inst Technol, 56-57, from instr to asst prof, 57-64; assoc prof, Univ Ill, Urbana, 64-69, prof mech eng, 69-71, prof mech & bioeng, 71-75, PROF MECH, ELEC, & BIOENG, UNIV ILL, URBANA, 75- *Concurrent Pos:* Engr & consult indust & govt, 54-; NSF fel, Europe, 61-62; Fogarty sr int fel, Fogarty Found, NIH & Inst Biomed Technol, Zürich, Switz, 78-79; vis exchange scholar, US & Hungarian Academics Sci, 78. *Honors & Awards:* Charles Russ Richards Mem Award, Am Soc Mech Engrs & Pi Tau Sigma, 78. *Mem:* Fel Am Soc Mech Engrs; Am Soc Heating, Refrig & Air-Conditioning Eng; Inst Elec & Electronic Engrs; Am Soc Eng Educ; Int Inst Refrig; Sigma Xi. *Res:* Heat transfer and fluid mechanics; bio-medical engineering; two phase flows; environmental controls and refrigeration; cryogenics. *Mailing Add:* Dept Mech & Indust Eng Univ Ill 1206 W Green St Urbana IL 61801

CHATT, ALLEN BARRETT, b Phoenix, Ariz, July 17, 49; m 71. EPILEPTOLOGY, ELECTROPHYSIOLOGY. *Educ:* State Univ NY, Buffalo, BS, 71; Fla State Univ, MS, 74, PhD(psychol), 78. *Prof Exp:* Res assoc, Marine Biomed Inst, Univ Tex Med Br, 77; fel assoc, 79-81, res asst prof, 81-86, RES SCIENTIST, DEPT NEUROL, SCH MED, YALE UNIV, 86-; RES PSYCHOLOGIST, EPILEPSY CTR, VET ADMIN MED CTR, 79- *Concurrent Pos:* Co-prin investr, NIH Grants, 80-86, prin investr, Vet Admin Merit Rev Grant, 80-; consult, Sachem Hill Prof Ctr, 91- *Mem:* Soc Neurosci; Am Epilepsy Soc; NY Acad Sci; AAAS. *Res:* Discrete neocortical model of epilepsy; monitoring of neuronal responses early in seizure development; anatomical tracing of local cortical circuits mediating spread of abnormalities; identification of circuits critical to propagating abnormalities. *Mailing Add:* Sachem Hill Prof Ctr 157 Goose Lane Guilford CT 06437

CHATT, AMARES, b Calcutta, India; Can citizen. NUCLEAR ANALYTICAL METHODS. *Educ:* Univ Calcutta, India, BSc, 65; Univ Roorkee, India, MSc, 67; Univ Waterloo, Can, MSc, 70; Univ Toronto, Can, PhD(nuclear chem), 74. *Prof Exp:* Lectr chem, St Paul's Col & Vivekananda Col, Calcutta, 67-68; from asst prof to assoc prof, 74-85, PROF CHEM, DALHOUSIE UNIV, HALIFAX, 85- *Concurrent Pos:* Consult & adv, Int Atomic Energy Agency, UN, Vienna, 76-; vis prof, Ctr Anal Res & Develop, Univ Colombo, Sri Lanka, 81; vis scientist, Comn Europ Communities, Joint Res Ctr, Ispra Estab, Italy, 82; chmn, Atomic Methods Tech Comt, Am Nuclear Soc, 85- *Honors & Awards:* Gold Medal, Univ Roorkee, India, 67. *Mem:* Am Nuclear Soc; Can Nuclear Soc; fel Chem Inst Can; Can Soc Chem Eng. *Res:* Development of nuclear analytical methods for toxic elements in environmental, biological, bio-medical, epidemiological, and nutritional materials; radioactive waste management; radioactive waste management. *Mailing Add:* Dept Chem Dalhousie Univ Trace Analysis Res Ctr Halifax NS B3H 4H6 Can

CHATTEN, LESLIE GEORGE, b Calgary, Alta, May 10, 20; m 43; c 3. PHARMACEUTICAL CHEMISTRY. *Educ:* Univ Alta, BSc, 47, MSc, 49; Ohio State Univ, PhD(pharmaceut chem), 61. *Prof Exp:* Head pharmaceut chem sect, Food & Drug Directorate, Dept Nat Health & Welfare, Govt Can, 49-61; from assoc prof to prof pharmaceut chem, Univ Alta, 61-83; PRES, QUAL CONTROL CONSULTS, 85- *Concurrent Pos:* Mem comt assay tablets & capsules, Brit Pharmacopoeia, 53-63 & comt org synthetic substances, 58-63; vis scientist, Med Res Coun Can, 70-71. *Mem:* Sigma Xi; fel Chem Inst Can; fel Royal Soc Chem. *Res:* Qualitative and quantitative pharmaceutical chemistry; application of nonaqueous titrimetry to analysis of drugs and pharmaceuticals; identification of organic medicinal agents; polarography and other electroanalytical techniques; absorption spectrophotometry; fluorimetry; high performance liquid chromatography; redox titrimetry. *Mailing Add:* 1109-161 A St White Rock BC V4A 7N3 Can

CHATTERJEE, BANDANA, b Calcutta, India, Apr 20, 51. MOLECULAR ENDOCRINOLOGY, MOLECULAR BIOLOGY. *Educ:* Univ Calcutta, BS, 70, MS, 73; Univ Nebr, PhD(biochem), 77. *Prof Exp:* Res assoc molecular endocrinol, 77-81, ASST PROF BIOCHEM, OAKLAND UNIV, 81- *Concurrent Pos:* Prin investr, NIH, 82-, Am Heart Asn, Mich, 84- *Mem:* Am Soc Biol Chemists; Sigma Xi. *Res:* Age and hormone dependent regulation of hepatic gene expression; hypolipidemic drug induced fatty acid metabolism in peroxisomes. *Mailing Add:* Dept Chem Oakland Univ Rochester MI 48309

CHATTERJEE, NANDO KUMAR, b W Bengal, India, Nov 1, 38; US citizen; m 69; c 2. BIOCHEMISTRY, MOLECULAR BIOLOGY. *Educ:* Univ Calcutta, India, PhD(bot), 62; Univ Nebr, PhD(genetics biochem), 69. *Prof Exp:* Res assoc genetics, Univ Nebr, 65-67, fel biochem, 69-71; guest scientist biochem, Roche Inst Molecular Biol, 71-74; molecular biologist virol, Plum Island Animal Dis Ctr, Agr Res Serv, USDA, 74-75; sr res scientist virol, 75-77, res scientist IV, 77-80, PRIN RES SCIENTIST V, DIV LABS & RES, NY DEPT HEALTH, 81- *Concurrent Pos:* 10th Int Cong Biochem travel grant; res grants, NIH & NY State Health Res Coun, 77-93; adj assoc prof microbiol & immunol, Albany Med Col & Union Univ, 84-; consult NIH grant, Northwestern Univ Med Sch; immunol/rheumatology, Childrens Mem Hosp, 83-85; fel biol, Univ Va, 63-65; adj assoc prof, Sch Biomed Sci, NY State Health Dept, 89-; coun sci & indust res fel, Govt India. *Mem:* Am Soc Biol Chemists; Am Soc Microbiologists; Am Soc Virologists; Sigma Xi; AAAS. *Res:* Molecular biology of viruses, cells; mechanisms of gene expression in uninfected and virus-infected cells using the parameters of RNA and protein synthesis; virus-induced diabetes mellitus; immunopathogentic mechanisms; mw 64000 autoantigen expression in prediabetic state. *Mailing Add:* Div of Labs & Res IDC Empire State Plaza Albany NY 12201

CHATTERJEE, PRONOY KUMAR, b Varanasi, India, Oct 26, 36; m 62; c 2. POLYMER SCIENCE, PHYSICAL CHEMISTRY. *Educ:* Banaras Hindu Univ, BS, 56, MS, 58; Calcutta Univ, PhD(chem), 63, DSc(chem), 74. *Prof Exp:* Sr res asst polymers, Indian Asn Cultivation Sci, Calcutta, 59-63; Nat Acad Sci-Nat Res Coun res assoc chem, Southern Regional Res Labs, USDA, La, 63-65; res assoc polymers, Princeton Univ, 65-66; sr res chemist, Johnson & Johnson, 66-74, mgr, mat res dept, Personal Prod Co Div, 74-87, MGR & RES FEL, MAT SCI, JOHNSON & JOHNSON ABSORBENT TECHNOL, 88- *Honors & Awards:* P B Hofman Res Scientist Award, Johnson & Johnson, 73; Educ Serv Award, Plastic Inst Am, 74. *Mem:* AAAS; Am Chem Soc; fel Am Inst Chemists; Tech Asn Pulp & Paper Indust; Int Confederation Thermal Anal. *Res:* Rubber vulcanization mechanism; analysis of rubber chemicals; polymer characterization; reaction kinetics; thermal analysis; reaction mechanism of polysulfides; polymerization; inter-fiber bonding mechanism of cellulose; characterization and preparation of chemically modified wood pulp. *Mailing Add:* Chicopee Johnson & Johnson 2351 US Rte 130 Dayton NJ 08810

CHATTERJEE, RAMANANDA, b India, Mar 1, 36; m 65; c 3. SOLID STATE PHYSICS. *Educ:* Calcutta Univ, BS, 54, MS, 56, PhD(physics), 63. *Prof Exp.* Fel, Inst Theoret Physics, Univ Alta, 63-65, from asst prof to assoc prof physics, 65-74, PROF PHYSICS, UNIV CALGARY, 74- *Res:* Theoretical physics with emphasis on solving nonlinear partial differential equations by lie group. *Mailing Add:* Dept Physics Univ Calgary Calgary AB T2N 1N4 Can

CHATTERJEE, SAMPRIT, b Calcutta, India, June 3, 38; US citizen; m. STATISTICS, OPERATIONS RESEARCH. *Educ:* Univ Calcutta, BS & MS, 60; Univ Cambridge, DStat, 62; Harvard Univ, PhD(statist), 66. *Prof Exp:* Instr statist, Boston Univ, 62-63; res asst, Harvard, 63-65; from asst prof statist & opers res to assoc prof statist , 66-74, PROF STATIST, NY UNIV, 74- *Concurrent Pos:* Consult, Mass Ment Health Ctr, 63-65, Coun Drug Abuse, 75-76 & Montefiore Hosp, 76-; res fel, Univ Col, London, 73-74; res scholar, Rand Corp, 68-70, Inst Appl Systs Anal, Vienna, 74 & Environ Protection Res Inst, 81-; vis fel, Stat Dept, Govt New Zealand, 80; vis prof, Auckland Univ, 80 & statist, Stanford Univ, 81; assoc ed, J Am Statist Asn; rev ed, J Bus & Econ Statist; vis prof mgt sci, Sloan Sch, MIT, 87-88; vis prof biostatistics, Harvard Sch Pub Health, 87-88. *Mem:* Am Statist Asn; Biomet Soc; Royal Statist Soc; Opers Res Soc Am; AAAS. *Res:* Environmental problems; ecology; linear models; sample survey; systems analysis; public policy; market research; data analysis. *Mailing Add:* Dept Quantitative Anal NY Univ New York NY 10003

CHATTERJEE, SANKAR, b Calcutta, India, May 28, 43; m 71; c 2. PALEONTOLOGY. *Educ:* Jadavpur Univ, India, BS, 62, MS, 64; Calcutta Univ, PhD(geol), 70. *Prof Exp:* Sr lectr geol, Indian Statist Inst, India, 68-75; asst lectr paleont, Univ Calif, Berkeley, 76; asst lectr geol, George Washington Univ, DC, 76-77; fel paleont, Smithsonian Inst, DC, 77-78; asst lectr geol, George Mason Univ, Va, 78-79; asst lectr & cur paleont, 79-84, assoc prof & dir, Antarctic Res Ctr, 84-86, PROF GEOL, TEX TECH UNIV, 86- *Concurrent Pos:* Prin investr, Geol Marie Byrd Land, Antarctic, NSF, Tex Tech, 79-, stratig & paleont, S Victoria Land, Antarctic, NSF, Tex Tech, 79-86, Explor Triassic Vertebrates at WTex, Nat Geog Soc, 80-81, Explor Lufeng Vertebrates, China, Nat Geog Soc, 85 & Seymour Island, Antarctic Peninsula, NSF. *Honors & Awards:* Antarctic Serv Medal, NSF, 82. *Mem:* Soc Vertebrate Paleont; Am Geol Inst; AAAS; Antarctican Soc. *Res:* Fossil vertebrates of Indian Gondwana rocks; Gondwana reassembly; geology of Antarctica; Triassic fossils from Texas and China, terminal cretaceous extinction, Mesozoic birds, pterosaurs and dinosaurs. *Mailing Add:* Mus Tex Tech Univ Lubbock TX 79409

CHATTERJEE, SATYA NARAYAN, b Calcutta, India, Dec 31, 34; US citizen; m 64; c 3. TRANSPLANT SURGERY, BURNS. *Educ:* Univ Calcutta, MBBS, 57; FRCS(G) & FRCS(E), 63; FRCS, 64. *Hon Degrees:* FICS, Int Col Surgeons, 81. *Prof Exp:* Registrar, Univ Edinburgh, 69-73; fel transplant surg, Univ Southern Calif, 73-75; asst prof, Univ Calif, Los Angeles, 75-77, assoc prof, Davis, 77-84, prof surg, 84-88; CONSULT, 88- *Concurrent Pos:* Consult, Vet Med Ctr, Martinez, 77-88; vis prof, Univ Kuwait, 81, Univ Minas Gerais, Brazil, 83, Univ Guadalajara, Mexico, 84 & 85, Univ Minn, 85 & Univ Heidelberg, WGermany, 85. *Mem:* Transplantation Soc; Europ Dial & Transplantation Soc; Asn Acad Surg; Am Soc Transplant Surgeons; Am Burn Asn. *Res:* Transplantation immunology; infections in surgery, especially in immunocompromised hosts; monoclonal antibodies. *Mailing Add:* 2600 Capitol Ave No 307 Sacramento CA 95816

CHATTERJEE, SUBROTO, b Feb 7, 47; m 74; c 2. GLYCOLIPID METABOLISM, DRUG INDUCED NEPHROTOXITY. *Educ:* Univ Toronto, PhD(biochem), 72; Dalhoune Univ, MS; Lucknow Univ, Msc. *Prof Exp:* ASSOC PROF MEMBRANE BIOCHEM, JOHNS HOPKINS UNIV, 75-, DIR DEPT PEDIAT CORE CELL CULTURE FACIL. *Concurrent Pos:* mem bd dirs, Md Affil, Am Heart Asn; chmn, Howard County Div, Am Heart Asn; mem, NIH & NSF Scientific Rev Comt. *Honors & Awards:* Andrew Mellon Award; Tokten Award, UN; All Star Award, Subroto Chatterjel. *Mem:* Fel Am Soc Biol Chem; AAAS; Am Heart Asn. *Res:* Glycolipids; receptors; lipoproteins; bacterial toxins; enzymes; monoclonal antibodies. *Mailing Add:* Johns Hopkins Univ 601 N Wolfe St Baltimore MD 21205

CHATTERJEE, SUNIL KUMAR, b Calcutta, India, Aug 7, 40; m 72; c 2. BIOCHEMISTRY. *Educ:* Presidency Col, Calcutta, BS, 59; Univ Calcutta, MS, 61, PhD(biochem), 66. *Prof Exp:* Fel protein biosynthesis, Univ Pa, 66-68; res assoc, Inst Cancer Res, 68-70; asst res officer, Univ Calcutta, 70-71; res fel, Max-Planck Inst, Ger, 71-72; SR CANCER RES SCIENTIST, ROSWELL PARK MEM INST, 72- *Mem:* AAAS; Am Asn Cancer Res; NY Acad Sci. *Res:* Biochemistry of cancer cells. *Mailing Add:* Roswell Park Cancer Inst 666 Elm St Buffalo NY 14263

CHATTERJI, DEBAJYOTI, b Puri, India, Aug 4, 44; m 68; c 3. METALLURGICAL ENGINEERING, PHYSICAL CHEMISTRY. *Educ:* Utkal Univ, India, BS, 63; Indian Inst Technol, India, BTech, 66; Purdue Univ, MS, 68, PhD(metall eng), 71. *Prof Exp:* Asst lectr metall, Indian Inst Technol, India, 66-67; asst, Purdue Univ, 67-71; vis scientist, Wright-Patterson AFB, Ohio, 71-73; metall engr, Gen Elec Corp & Develop Ctr, 73-74, tech adminr tech liaison, 74-75, mgr electrochem br, 75-79, mgr chem syst & technol lab, 79-80, mgr, inorg mat & struct lab, 80-83; vpres tech activ, 83-

90, chief exec tech activ, 90, MANAGING DIR TECHNOL, BOC GROUP, 90- *Concurrent Pos:* Nat Res Coun resident res fel, 71-73; mem nat battery adv comt, dept of energy, 76-79; mem indust adv bd, NJ Inventors' Hall Fame, 88- *Honors & Awards:* Geisler Award, Am Soc Metals, 80. *Mem:* Electrochem Soc; Am Soc Metals; Sigma Xi. *Res:* High temperature corrosion and oxidation; batteries; fuel cells; thermochemistry; thermodynamics; mass transfer; electrochemistry; coatings; ceramics; superalloys. *Mailing Add:* BOC Group Tech Ctr 100 Mountain Ave Murray Hill NJ 07974

CHATTERTON, BRIAN DOUGLAS EYRE, b Khartoum, Sudan, July 31, 43; Irish & Can citizen; m 70; c 2. PALEONTOLOGY. *Educ:* Trinity Col, Dublin, BA, 65; Australian Nat Univ, PhD(paleont), 70. *Prof Exp:* Sr demonstr paleont, Australian Nat Univ, 69-70; from asst prof to assoc prof, 70-81, PROF PALEONT, UNIV ALTA, 81- *Concurrent Pos:* Exec dir, Can Geoscience Coun, 84-88; McCalla Professorship, Australian Nat Univ. *Mem:* Geol Asn Can; Paleont Soc; Palaeont Asn; Int Palaeont Asn. *Res:* Ontogenetic, paleoecologic and systematic studies of trilobites from Western and Arctic Canada; systematics, paleoecology and biostratigraphy of conodonts from Western and Arctic Canada; Paleoezoic paleontology; rostroconchs, brachiopods. *Mailing Add:* Dept Geol Univ Alta Edmonton AB T6G 2M7 Can

CHATTERTON, NORMAN JERRY, b Mapleton, Idaho, Feb 11, 39; m 63; c 4. PLANT PHYSIOLOGY. *Educ:* Utah State Univ, BS, 66; Univ Calif, MS, 68, PhD(plant physiol), 70. *Prof Exp:* Plant physiologist crops, 70-81, agr admnr, Agr Res Serv, 81-84, RES LEADER & PLANT PHYSIOLOGIST, USDA, 84- *Honors & Awards:* Fel, Am Soc Agron, Crop Sci Soc Am. *Mem:* Am Soc Agron; Crop Sci Soc Am; Am Soc Plant Physiologists. *Res:* Whole plant photosynthesis; photosynthate partitioning and utlization in forage plants; fructan metabolism in temperate grasses. *Mailing Add:* Agr Res Serv USDA Utah State Univ Logan UT 84322-6300

CHATTERTON, ROBERT TREAT, JR, b Catskill, NY, Aug 9, 35; m 56; c 4. ENDOCRINOLOGY, BIOCHEMISTRY. *Educ:* Cornell Univ, BS, 58, PhD(physiol biochem), 63; Univ Conn, MS, 60. *Prof Exp:* Res fel biol chem, Harvard Univ, 63-65; res assoc, Div Neoplastic Med, Montefiore Hosp & Med Ctr, 65-70; from asst prof to assoc prof obstet & gynec, physiol & biol chem, Univ Ill Med Ctr, 70-79; PROF OBSTET & GYNEC, NORTHWESTERN UNIV MED SCH, 79- *Concurrent Pos:* Prin investr, NIH, 72-75, 75-78, 79-81, 79-82, 82-84, 86-91, USAID, 73-75, 74-75, 78-81, 82-86, US Army Res Off, 87-90, 90-93; mem, sci adv comt, Prog Appl Res Fertil Regulation, USAID, 76-82. *Mem:* Am Chem Soc; Endocrine Soc; Soc Study Reproduction; Soc Gynec Invest; Am Physiol Soc; Sigma Xi. *Res:* Physiology, biochemistry and histology of mammary gland development and ovarian function; steroid biosynthesis; contraceptive development. *Mailing Add:* Prentice Women's Hosp Suite 1130 Northwestern Univ Med Sch Chicago IL 60611

CHATTHA, MOHINDER SINGH, b Gurdaspur, India, Oct 15, 40; m 67; c 2. POLYMER CHEMISTRY. *Educ:* Sikh Nat Col, India, BSc, 60; Banaras Hindu Univ, MSc, 63; Tulane Univ, PhD(chem), 71. *Prof Exp:* Teacher sci, Khalsa High Sch, Wadala, India, 60-61; lectr chem, Sikh Nat Col, 63-65; res fel, Coun Sci & Indust Res, New Delhi, 65-67; fel, Tulane Univ, 71-72; Nat Res Coun fel polymer chem, Wright-Patterson AFB, Ohio, 72-73; SR RES SCIENTIST POLYMER CHEM, RES & DEVELOP, FORD MOTOR CO, 73- *Mem:* Am Chem Soc. *Res:* Organophosphorus chemistry; synthesis and chemistry of new polymers of potential use in coatings, adhesives and elastomers. *Mailing Add:* 7875 Seven Mile Rd Northville MI 48167-9126

CHATTORAJ, SATI CHARAN, b WBengal, India, Aug 1, 34; m 61; c 1. REPRODUCTIVE ENDOCRINOLOGY. *Educ:* Univ Calcutta, BS, 54, MS, 56; Boston Univ, PhD(biochem), 65. *Prof Exp:* Asst prof obstet & gynec, Chicago Med Sch, 65-69; from asst prof to assoc prof, 69-85, PROF BIOCHEM OBSTET & GYNEC, SCH MED, BOSTON UNIV, 85- *Concurrent Pos:* Nat Inst Child Health & Human Develop grant, Sch Med, Boston Univ, 74-77; asst dir gynecic endocrinol, Michael Reese Hosp & Med Ctr, 65-69. *Honors & Awards:* Morris L Parker Award, Chicago Med Sch, 67. *Mem:* AAAS; Am Inst Biol Sci; Endocrine Soc; Soc Study Reproduction; Am Chem Soc; Sigma Xi. *Res:* Reproductive endocrinology; control of steroidogenesis during pregnancy; contraception; breast and endometrial cancer. *Mailing Add:* Five Bryant Lane Dover MA 02030-1422

CHATURVEDI, ARVIND KUMAR, b July 29, 47; m; c 2. BIOCHEMISTRY. *Educ:* Gorakpur Univ, BS, 66; Banaras Hindu Univ, MS, 68; Lucknow Univ, PhD(philos), 72. *Prof Exp:* Res fel, Dept Pharmacol & Therapeut, King George's Med Col, Lucknow Univ, India, 68-74; res assoc, Dept Pharmacol, Sch Med, Vanderbilt Univ, 74-77, res instr, 77-78; asst prof, Dept Pharmaceut Sci & Toxicol, Col Pharm, NDak State Univ, Fargo, 78-82, asst toxicologist, Off State Toxicologist, 78-90, assoc prof, 82-90; SUPVR BIOCHEM RES SECT, TOXICOL & ACCIDENT RES LAB, AEROMED RES DIV, CIVIL AEROMED INST, FED AVIATION ADMIN, 90- *Concurrent Pos:* Numerous grants, 79-89; mem fac develop, Eval & Morale Comt, Col Pharm, 79-80, grad student exam comt, Dept Pharmaceut, 79-85, ad hoc comt curric, 82-84, comt study goals & objectives, 83-84, lab & chem safety, 87-89; assoc mem fac, Grad Col, Health Sci Ctr, Univ Okla & adj prof pharmadynamics & tolicol, 90- *Mem:* Indian Pharmacol Soc; Int Soc Biochem Pharmacol; Am Soc Pharmacol & Exp Therapeut; Soc Environ Toxicol & Chem; Soc Exp Biol & Med; Soc Neurosci; Soc Toxicol; Sigma Xi. *Res:* Correlation between inhibition of pyruvic acid oxidation and anticonvulsant activity; anti-inflammatory and anticonvulsant properties of some natural plant triterpenoids; influence of mecamylamine, atropine and nicotine on the cocaine-induced locomotion of mice. *Mailing Add:* Biochem Res Sect AAM-613 Toxicol & Accident Res Lab Aeromed Res Div Civil Aeromed Inst Fed Aviation Admin PO Box 25082 Oklahoma City OK 73125

CHATURVEDI, MAHESH CHANDRA, b Unnao, India, May 3, 40. METALLURGY, MATERIALS SCIENCE. *Educ:* Banaras Hindu Univ, BSc, 60; Univ Sheffield, MMet, 62, PhD(metall), 66. *Prof Exp:* Res fel, 66-69, asst prof, 69-70, assoc prof, 70-75, PROF METALL, METALL SCI LAB, UNIV MAN, 75- *Mem:* Am Soc Metals; Am Inst Mining, Metall & Petrol Engrs; Electron Micros Soc Am; Can Inst Mining & Metall. *Res:* Precipitation strengthening of metallic materials; electron microscopic studies of phase transformations in metals and alloys; alloy development work; corrosion of metals and alloys. *Mailing Add:* 350 Eng Bldg Univ Man Winnipeg MB R3T 2N2 Can

CHATURVEDI, RAM PRAKASH, b India, Dec 15, 31; m 64; c 2. ATOMIC PHYSICS. *Educ:* Agra Univ, BSc, 53, MSc, 55; Univ BC, PhD(nuclear spectros), 63. *Prof Exp:* Lectr physics, Agra Univ, 55-59; fel & lectr, Panjab Univ, India, 63-64; fel, State Univ NY Buffalo, 64-65; assoc prof, 65-70, PROF PHYSICS, STATE UNIV NY COL CORTLAND, 70-, DISTINGUISHED SERV PROF, 88- *Concurrent Pos:* Mem users group, Oak Ridge Nat Lab, 70; mem, Res Participation Team, Brookhaven Nat Lab; grants, NSF, State Univ Res Found, Dept Energy & NRC. *Mem:* Am Phys Soc; Am Asn Physics Teachers; Sigma Xi. *Res:* Nuclear level schemes; ion-atom interaction. *Mailing Add:* Dept of Physics State Univ of NY Col Cortland NY 13045

CHATURVEDI, RAMA KANT, b Kanker, India, July 7, 33; m 58; c 3. BIO-ORGANIC CHEMISTRY. *Educ:* Agra Univ, India, BSc, 54, MSc, 56, PhD(chem), 60. *Prof Exp:* Lectr chem, H B Technol Inst, India, 60-64 & St Stephen's Col, Univ Delhi, 64-65; res assoc, Ind Univ, 65-66; res assoc biochem, Yale Univ, 66-70; assoc prof, 70-76, chmn, dept sci, 77-82, PROF CHEM, GTR HARTFORD COMMUNITY COL, 76-, DIR SCI, MATH & PSYCH DIV, 82- *Concurrent Pos:* Fulbright Award, US Educ Found in India, 65; asst instr, Yale Univ, 70-74; lectr, Cent Conn State Univ, 73 & 75-77. *Mem:* AAAS; Am Asn Univ Prof. *Res:* Mechanism of acyl transfer reactions with special attention to the formation and partitioning of tetrahedral intermediates involved in these reactions. *Mailing Add:* 39 Beacon St Newington CT 06111

CHATY, JOHN CULVER, b San Francisco, Calif, Sept 12, 25; m 51; c 1. CHEMICAL ENGINEERING. *Educ:* Univ Ala, BS, 57; Univ Va, DSc(chem eng), 62. *Prof Exp:* Prod supvr, Calabama Plant, Olin-Mathieson Chem Corp, 50-53, tech asst, 53-54; engr develop, Esso Res & Eng Co, 57-59; engr res & develop, Chem Div, Union Carbide Corp, 61-75, technol mgr, 75-81, large scale pilot plant mgr, Ethylene Oxide-glycol Div, 81-85; RETIRED. *Concurrent Pos:* Mgr custom synthesis contract bus & pilot plant. *Mem:* Am Inst Chem Engrs. *Res:* Application of fundamentals of chemical engineering in development of chemical process facilities; polymer systems. *Mailing Add:* 1001 Rustling Rd South Charleston WV 25303

CHAU, ALFRED SHUN-YUEN, b Hong Kong, Nov 20, 41; Can citizen; m 66; c 1. ANALYTICAL CHEMISTRY. *Educ:* Univ BC, BSc, 61; Carleton Univ, MSc, 66. *Prof Exp:* Pesticide analyst, Dept Agr, Ottawa, 65-70; chemist, 70-73, head, Spec Serv Sect, 73-80, head qual assurance & methods sect, 80-87, CHIEF, QUAL ASSURANCE, DEPT ENVIRON, CAN, 88- *Concurrent Pos:* Sci reviewer, J Asn Off Anal Chemists, 70- & J Agr & Food Chem, 84; chmn, Data Mgt Work Group, Binat Upper Gt Lakes Connecting Channel Studies, 84-; consult, WHO, People's Repub China, 85; chmn, Qual Assurance Adv Comt, Dioxin, 90- *Honors & Awards:* Caledon Award, 80. *Mem:* Can Inst Chem; Asn Off Anal Chemists; fel Chem Inst Can; fel Int Biog Asn; fel Am Biographic Inst. *Res:* Quality assurance program design and research; development of methodology for the analysis and positive confirmation of organic pollutants, particularly biocides, by chemical derivatization-gas chromatographic techniques; development of standard reference materials for environmental analysis. *Mailing Add:* Water Qual Br PO Box 5050 Burlington ON L7R 4A6 Can

CHAU, CHEUK-KIN, b Hong Kong, Sept 25, 41; m 68. SOLID STATE PHYSICS, CRYOGENICS. *Educ:* MacMurray Col, BA, 62; Univ Ill, Urbana, MS, 63, PhD(physics), 68. *Prof Exp:* Vis asst prof, Ill Inst Technol, 68-69, asst prof physics, 69-75; vis assoc prof, 75-76, assoc prof, 76-80, PROF PHYSICS, CALIF STATE UNIV, CHICO, 80- *Concurrent Pos:* Vis prof, Oregon Grad Ctr, 82-83. *Mem:* Am Phys Soc; Am Asn Physics Teachers; Sigma Xi. *Res:* Apply cryogenic technology to investigate heat transport, electrical transport, optics and radiation damages in condensed materials such as insulators, semiconductors, semi-metal and superconductors. *Mailing Add:* Dept of Physics Calif State Univ Chico CA 95926

CHAU, HIN FAI, b Hong Kong, Jan 6, 64. SOLID-STATE MICROWAVE DEVICES, OPTOELECTRONICS & CIRCUITS. *Educ:* Univ Hong Kong, BS, 85; Univ Mich, Ann Arbor, MSE, 86. *Prof Exp:* RES ASST SOLID-STATE DEVICES, UNIV MICH, ANN ARBOR, 86- *Mem:* Inst Elec & Electronic Engrs; Electrochem Soc. *Res:* The design, fabrication and characterization of heterostructure semiconductor devices, applications of strain-layer systems to these devices, and device modeling. *Mailing Add:* Solid-State Electronics Lab Dept Elec Eng & Computer Sci Univ Mich Ann Arbor MI 48109-2122

CHAU, LING-LIE, b Hunan, China, Jan 18, 39; US citizen; c 1. HIGH ENERGY PHYSICS. *Educ:* Nat Taiwan Univ, BS, 61; Univ Calif, Berkeley, PhD(physics), 66. *Prof Exp:* Fel physics, Lawrence Radiation Lab, Berkeley, 67; mem, Inst Advan Study, 67-69; vis fel, Europe Cttr High Energy Physics, 69; from asst physicist to assoc physicist, Brookhaven Nat Lab, 69-74, physicist, 74-86; PROF, UNIV CALIF, 86- *Concurrent Pos:* Adj prof, Nankai Univ, Tianjin, China. *Mem:* Fel Am Phys Soc; AAAS; NY Acad Sci; Int Asn Math Physics. *Res:* Particle physics. *Mailing Add:* Dept Physics Univ Calif Davis CA 95616

CHAU, MICHAEL MING-KEE, b Hong Kong, Nov 14, 47; m 75; c 2. PHYSICAL ORGANIC CHEMISTRY, POLYMER CHEMISTRY. *Educ:* Calif State Univ Fresno, BS, 71; Univ Ill, Urbana, PhD(chem), 75. *Prof Exp:* res assoc org chem, Tex Tech Univ, 75-77; sr res chemist, 77-82, res assoc, 83-86, sr res assoc, PPG Indust, Inc, 86-88; prin scientist, Fluid Systems Div, Allied Signal, Inc, 88-89, MGR, MEMBRANE RES, 89- *Mem:* Am Chem Soc. *Res:* Polymer design, synthesis and characterization; organic synthesis and mechanistic studies; coatings, inks, adhesives, sealants and membranes. *Mailing Add:* 13718 Butano Way San Diego CA 92129-4440

CHAU, THUY THANH, b Cantho, Vietnam, June 8, 44; m 88; c 1. PHARMACOLOGY, PHARMACY. *Educ:* Univ Saigon, BS, 66; Univ Hawaii, MS, 68; Univ NC, Chapel Hill, PhD(pharmacol), 72. *Prof Exp:* Assoc prof pharmacol, Sch Pharm, Univ Saigon, 72-75; res assoc, Med Col Va, 75-77, asst prof pharmacol, 77-83; res assoc, 84-87, sr res assoc, 87-88, SECT HEAD, DEPT EXP THERAPEUT, WYETH-AYERST RES, 88- *Concurrent Pos:* Chmn dept pharmacol, Sch Med, Univ Minh Duc, 72-75; dir qual control, Tenamyd Labs, Vietnam, 73-75. *Mem:* Sigma Xi; Am Soc Pharmacol & Exp Therapeut; NY Acad Sci. *Res:* Agents that affect the central nervous system and their interactions with neurochemical transmitter systems; opiates and their endogenous ligands; inflammation and analgesia. *Mailing Add:* Dept Exp Therapeut-Immunopharmacol Wyeth-Ayerst Res CN 8000 Princeton NJ 08540

CHAU, VINCENT, b Shanghai, China, Oct 8, 52. PROTEIN CHEMISTRY. *Educ:* Univ Va, PhD(biophysics), 79. *Prof Exp:* asst prof biochem, Sch Med, Univ Fla, 81-88; ASSOC PROF PHARMACOL, WAYNE STATE UNIV, 88- *Mailing Add:* Dept Pharmacol Wayne State Univ Sch Med Rm 6321 540 E Canfield Detroit MI 48201

CHAU, WAI-YIN, b Hong Kong, Sept 17, 39; Can citizen; m 68; c 2. RELATIVISTIC ASTROPHYSICS. *Educ:* Univ Hong Kong, BSc, 62; Columbia Univ, PhD(physics), 67. *Prof Exp:* Asst prof astrophys, Brown Univ, 68-69; from asst prof to assoc prof, 69-82, PROF ASTROPHYS, QUEEN'S UNIV, KINGSTON, ONT, 82- *Mem:* Int Astron Union; Can Astron Soc; Am Astron Soc; Am Phys Soc. *Res:* Relativistic astrophysics; gravitational radiation; low mass close binary systems; neutrino astrophysics; numerical cosmology. *Mailing Add:* Head Dept Physics Hong Kong Univ Pokfulam Hong Kong

CHAU, YIU-KEE, b Canton, China, Dec 6, 27; m 56; c 2. CHEMICAL OCEANOGRAPHY, LIMNOLOGY. *Educ:* Lingnan Univ, BS, 49; Univ Hong Kong, MS, 61; Univ Liverpool, PhD(chem oceanog), 65. *Hon Degrees:* DSc, Univ Liverpool. *Prof Exp:* Sci master, Mid Sch, Macau, 50-54; res officer chem oceanog, Fisheries Res Unit, Univ Hong Kong, 54-59; assoc prof chem, Chinese Univ Hong Kong, 59-68; RES SCIENTIST, NAT WATER RES INST, 68-; PROF GEOL, MCMASTER UNIV, 83- *Concurrent Pos:* UNESCO fel, Commonwealth Sci & Indust Res Orgn, Australia, 57; Brit Coun fel, Univ Liverpool, 63-65. *Mem:* AAAS; assoc Royal Australian Chem Inst; Am Chem Soc; Spectros Soc Can; fel Can Inst Chem; Am Soc Limnol & Oceanog. *Res:* Chemical and biological processes of trace metals in the aquatic environment; interaction of metals and organics in natural water; transformation of elements by biotic and abiotic processes; orgarometallic compounds in the environment. *Mailing Add:* Can Ctr for Inland Waters PO Box 5050 Burlington ON L7R 4A6 Can

CHAUBAL, MADHUKAR GAJANAN, b Nasik City, India, May 15, 30; m 66. PHARMACOGNOSY. *Educ:* Univ Poona, BSc, 51; Univ Bombay, BSc, 54; Univ Toronto, MSc, 60; Univ RI, PhD(pharmaceut sci), 64. *Prof Exp:* Demonstr pharm, Dept Chem Technol, Univ Bombay, 56-58; demonstr pharm, Univ Toronto, 58-60; res asst, Univ RI, 60-64; from asst prof to assoc prof, 64-74, PROF MED CHEM, UNIV OF THE PAC, 74- *Mem:* Acad Pharmaceut Sci; Am Pharmaceut Asn; Am Soc Pharmacog; NY Acad Sci; fel Am Inst Chem. *Res:* Phytochemistry of medicinal plants; biosynthesis of plant constituents; essential oils from plants; constituents of the Saururaceae and pharmacology of the constituents. *Mailing Add:* 669 W Euclid Ave Stockton CA 95204

CHAUBEY, MAHENDRA, b Majagawan, India, Oct 13, 48, Can; m 63; c 3. TRANSPORT THEORY, COLLISION THEORY. *Educ:* Banaras Hindu Univ, BSc, 68, MSc, 72; Univ Rochester, MA, 74; State Univ NY Buffalo, PhD(physics), 78. *Prof Exp:* Fel res asst physics, Oxford Univ, 77-79; res assoc, McGill Univ, 79-80; fel, Univ New Brunswick, Fredericton, 80-81; prof physics, 81-84, PROF MATH, DAWSON COL, 84- *Concurrent Pos:* Lectr, McGill Univ, 80. *Mem:* Inst Physics. *Res:* Quantum statistical theories of transport mechanisms in semiconductors and metal-oxide semiconductor layers; cyclotron resonance in inversion layers; collisional effects on the spectra of high dentity gases in interaction with lasers. *Mailing Add:* Math Dept Dawson Col 3040 Sherbrooke St W Montreal PQ H3Z 1A4 Can

CHAUDHARI, ANSHUMALI, b Allahabad, Uttar Pradesh, India, June 22, 47; US citizen; m 73; c 2. BONE PHYSIOLOGY, RENAL PHYSIOLOGY. *Educ:* Univ Lucknow, India, BS, 66, MS, 68, PhD(chem biol), 76. *Prof Exp:* Postdoctoral fel res, Nat Inst Environ Health Sci, Res Triangle Park, NC, 75-77; res assoc res, Wayne State Univ, Detroit, Mich, 77-80; from adj asst prof to adj assoc prof med res & training, Univ Calif, Los Angeles, 80-85 & Irvine, 85-89; CHIEF RES & TRAINING, BIOCHEM SECT, US ARMY INST DENT RES, WALTER REED ARMY MED CTR, WASH, DC, 89- *Concurrent Pos:* Sr res fel, Indian Coun Med Res, 74-75; vis fel, Fogarty Int Ctr, 75-77; adv res fel, Am Heart Asn, Los Angeles, Calif, 80-83 & sr investr, 83-84; adj assoc prof physiol, Baltimore Sch Dent, Univ Md, 90- *Mem:* Am Soc Pharmaceut & Exp Therapeut; Indian Acad Neurosci; Int Soc Pharm; Soc Biol Chemists India. *Res:* Development of synthetic, biodegradable and biocompatible bone regenerating materials; bone inductive proteins; growth factors on bone cell growth. *Mailing Add:* 2836 Schubert Dr Silver Spring MD 20904

CHAUDHARI, BIPIN BHUDHARLAL, b Taloda, India, Aug 31, 35; m 65. MEDICINAL CHEMISTRY. *Educ:* Univ Bombay, BS, 57 & 59; Univ Iowa, MS, 62, PhD(med chem), 65. *Prof Exp:* Anal chemist, Oriental Chem Industs, Bombay, India, 59-60; res assoc chem, Ill Inst Technol, 65; res chemist, Bauer & Black-Polyken Res Ctr, Kendall Co, 65-69; SR RES CHEMIST, MED CHEM SECT, CHEM RES & DEVELOP LAB, ICI AMERICAS INC, 69- *Mem:* Am Chem Soc. *Mailing Add:* 2308 Patwynn Rd Wynnwood Wilmington DE 19810

CHAUDHARI, PRAVEEN, b Ludhiana, India, Nov 30, 37; m 64; c 2. PHYSICAL METALLURGY, PHYSICS. *Educ:* Indian Inst Technol, Kharagpur, BTech, 61; Mass Inst Technol, SM, 63, ScD(metall), 66. *Prof Exp:* Res asst phys metall, Mass Inst Technol, 61-65, res assoc metall, 65-66; mem res staff, 66-80, DIR PHYS SCI DEPT, IBM THOMAS J WATSON RES CER, 81-, VPRES SCI, 82- *Concurrent Pos:* Mem res staff, Danish Atomic Energy Comn, Denmark, 64; mem vis comt phys sci, Univ Chicago, mats sci & eng dept, Mass Inst Technol & Sch Eng, Univ Ill; mem sci adv bd, Hoechest-Celanese Corp; co-chmn prof comt, nat study mats sci eng, Nat Res Coun; mem solid state sci comt, Nat Acad Sci; mem at large gov bd, Am Inst Physics; exec secy, Pres Reagan's Adv Coun Superconductivity, 88; mem, Nat Comn Superconductivity, 89. *Honors & Awards:* Leadership Award, Metall Soc, 86; George E Pake Award, Am Phys Soc, 87. *Mem:* Nat Acad Eng; fel Am Phys Soc; Am Inst Mining Metall & Petrol Engrs; AAAS. *Res:* Amorphous solids; mechanical properties and defects in crystalline solids; quantum transport; superconductivity; magnetic monopoles and neutrino mass experiments. *Mailing Add:* Thomas J Watson Res Ctr IBM Corp PO Box 218 Rte 134 Yorktown Heights NY 10598

CHAUDHARY, RABINDRA KUMAR, b India; Can citizen. MEDICAL MICROBIOLOGY. *Educ:* Bihar Vet Col, BVSc, 58; Guelph Univ, MSc, 65; Univ Toronto, dipl bacteriol, 66; Univ Ottawa, PhD(microbiol), 71. *Prof Exp:* Res scientist microbiol, Bell & Craig Pharmaceut Co, 66-67; res assoc, Univ Ottawa, 71-75; BIOLOGIST, LAB CTR DIS CONTROL, BUR VIRAL DIS, 75- *Mem:* Am Soc Microbiologists. *Res:* Biophysical studies of hepatitis B antigen. *Mailing Add:* 40 Townsend Dr Napean ON K2J 2V4 Can

CHAUDHRI, SAFEE U, b Punjab, India, May 4, 44; US citizen; m 73; c 4. CHROMATOGRAPHY, SPECTROSCOPY. *Educ:* Punjab Univ, Pakistan, BS, 65, MS, 67; Osaka Univ, Japan, PhD(phys chem), 73. *Prof Exp:* Mgr, Tech Serv, Ittehad Chem, 74-76; anal chemist, Richlyn Labs Inc, 77-79; qual control chemist, Connaught Labs Inc, 79-84; DIR QUAL CONTROL & QUAL ANAL, EVSCO PHARMACEUT, IGI INC, 84- *Concurrent Pos:* Sr res scientist, Lederle-Praxis Biol, 90. *Mem:* Fel Am Inst Chemists; Am Chem Soc. *Res:* Analytical method developments and validation procedures for active pharmaceuticals and insecticides entities. *Mailing Add:* 1040 Holmes Ave Vineland NJ 08360

CHAUDHRY, G RASUL, b Multan, Pakistan, June 6, 48; US citizen; m 82; c 4. BIOTECHNOLOGY, ENVIRONMENTAL MICROBIOLOGY & TOXICOLOGY. *Educ:* Univ Agr, Pakistan, BSc Hons, 69, MSc Hons, 71; Univ Man, Can, PhD(microbiol), 80. *Prof Exp:* Res officer biochem, Agr Res Inst, Pakistan, 71-77; res assoc biochem, Med Sch, Georgetown Univ, Washington, DC, 80-82; vis fel molecular biol, NIH, Bethesda, Md, 82-84, sr staff fel, 84-85; asst prof molecular biol, Univ Fla, Gainesville, 85-89; ASSOC PROF MOLECULAR BIOL, OAKLAND UNIV, ROCHESTER, 89- *Concurrent Pos:* Consult biotechnol, Punjab Univ, Pakistan, 88. *Mem:* Am Soc Microbiol; AAAS; Sigma Xi. *Res:* Molecular genetics of hazardous waste-degrading microorganisms; study of microbiological agents in the environment; molecular biology of genetic diseases and disorders. *Mailing Add:* Dept Biol Sci Oakland Univ Rochester MI 48309

CHAUDHURI, TAPAN K, b Calcutta, India, Nov 25, 44; US citizen; m 80; c 4. MEDICAL SCIENCES. *Educ:* Univ Calcutta, MB, BS, 66. *Prof Exp:* House-staff med, Univ Calcutta, 66-67; fel nuclear med, Yale Univ, 67-68; intern internal med, Southside Hosp, 68-69; resident, WVa Univ, 69-70; clin assoc nuclear med, Univ Iowa, 70-71, asst prof, 71-74; assoc prof, 74-79, PROF NUCLEAR MED, EASTERN VA MED SCH, 79-; CHIEF NUCLEAR MED, VET ADMIN MED CTR, 74- *Concurrent Pos:* Res consult, Univ Okla, 70, vis asst prof, 71; reviewer, J Nuclear Med, Gastroenterol & Chest, 74-85, Ann Int Med, 90. *Mem:* Soc Nuclear Med; Radiol Soc NAm; Sigma Xi; Magnetic Resonance Soc; Am Soc Ultrasound Med; Am Physiol Soc; fel Am Col Physicians; fel Am Col Gastroenterologists; fel Am Col Chest Physicians. *Res:* Nuclear medicine. *Mailing Add:* Vet Admin Med Ctr Hampton VA 23667

CHAUDHURI, TUHIN, b Bengal, India, Jan 3, 42; m 69; c 3. NUCLEAR MEDICINE, HEMATOLOGY. *Educ:* Univ Calcutta, MBBS (MD), 64. *Prof Exp:* Intern med, Med Col & Hosp, Univ Calcutta, 64-65; resident surg, RKM Seva Pratishthan, Calcutta, 65-66; res assoc radiol & nuclear med, Yale Univ, 66-67; from asst prof to assoc prof radiol, Univ Iowa, 69-75, assoc dir nuclear med, 69-75; CHIEF NUCLEAR MED, AUDIE MURPHY VET ADMIN MED CTR, UNIV TEX HEALTH SCI CTR, SAN ANTONIO, 75-, PROF RADIOL, 79- *Concurrent Pos:* Fel nuclear med, Yale Univ, 67-68; James Picker fel, Nat Acad Sci-Nat Res Coun, 67-69; fel, Donner Lab, Univ Calif, Berkeley, 68-69; guest scientist, Lawrence Radiation Lab, Berkeley, 68-69; vis prof nuclear Med, Univ Conn. *Honors & Awards:* Outstanding Young Investr, Soc Nuclear Med, 70. *Mem:* AAAS; Am Fedn Clin Res; Biophys Soc; Soc Nuclear Med; fel Am Col Nuclear Physicans; AMA. *Res:* Gastro-intestinal nuclear medicine; Oesophagal and gastric motility; cardio-vascular nuclear medicine; non-invasive cardiac function studies; diagnosis and therapeutic efficiency studies in cancer. *Mailing Add:* Dept Nuclear Med Audie Murphy Mem Vet Admin Hosp San Antonio TX 78284

CHAUDRY, IRSHAD HUSSAIN, b May 2, 1945; m 74; c 3. ENERGY METABOLISM, CELLULAR TRANSPORT. *Educ:* Monash Univ, PhD, 70. *Prof Exp:* prof surg, Sch Med, Yale Univ, 76-86; PROF SURG & PHYSIOL, MICH STATE UNIV, 86- *Concurrent Pos:* mem study sect, NIH & Vet Admin. *Mem:* Am Physiol Soc; Shock Soc; Surg Infection Soc. *Res:* Shock and trauma; ischemia and repusion; general medical science and immunology. *Mailing Add:* Dept Surg Mich State Univ B424 Clin Ctr E Lansing MI 48824-1315

CHAUFFE, LEROY, b Freetown, La, Dec 26, 36; m 69. ORGANIC CHEMISTRY. *Educ:* Xavier Univ, La, BS, 59; Howard Univ, MS, 64; Univ Calif, Davis, PhD(chem), 66. *Prof Exp:* Chemist, Shell Chem Co, 65-66; fel, Univ Calif, Davis, 66-67; asst prof chem, State Col Long Beach, 67-68; asst prof, 68-74, assoc dean instr, 70-71, actg dean grad studies, 71-72, assoc prof, 74-80, PROF CHEM, CALIF STATE UNIV, HAYWARD, 80- *Mem:* Am Chem Soc; Royal Chem Soc. *Res:* Physical organic chemistry; biochemical kinetics and mechanism. *Mailing Add:* Dept of Chem Calif State Univ Hayward CA 94542

CHAUHAN, VED P S, b Meerut, India, Aug 10, 53; m 80; c 2. SIGNAL TRANSDUCTION BIOCHEMISTRY, MEMBRANE BIOCHEMISTRY. *Educ:* Meerut Univ, India, BS, 73; Postgrad Inst Med Educ & Res, MS, 75, PhD(biochem), 81. *Prof Exp:* Res assoc biochem, Univ Southern Calif, Los Angeles, 81-83; res scientist I biochem, NY State Inst Basic Res, 83-86, res scientist II, 86-87, res scientist III, 87-90, RES SCIENTIST IV & HEAD, SIGNAL TRANSDUCTION BIOCHEM LAB, NY STATE INST BASIC RES, NY, 91- *Mem:* Am Soc Biochem & Molecular Biol; Am Soc Neurochem. *Res:* Role of polyphosphoinositides in signal transduction; membrane-toxicity of amyloid beta-protein-a major protein that accumulates in Alzheimer disease; phosphatidate-mediated calcium traversal across bilayer; studies on lipid-lipid and lipid-protein interactions in biological membrane. *Mailing Add:* Dept Neurochem NY State Inst Basic Res in Develop Disabilities Staten Island NY 10314

CHAUVIN, ROBERT S, b West Beekmantown, NY, Nov 20, 20; m 46; c 2. GEOLOGY, GEOGRAPHY. *Educ:* NY Univ, BS, 43; Columbia Univ, MA, 50, PhD(geol, geog), 55. *Hon Degrees:* LLD, Stetson Univ, 84. *Prof Exp:* Head dept geol & geog, 50-68, dean sci, 68-70, assoc prof geol, 50-55, PROF GEOL, STETSON UNIV, 55-, DEAN COL LIB ARTS, 70-, J OLLIE PROF GEOG & GEOL, 84- *Mem:* Am Geog Soc. *Res:* Arctic flora, fauna and glaciation; geology of Florida. *Mailing Add:* Col of Lib Arts Stetson Univ De Land FL 32721

CHAVE, ALAN DANA, b Whittier, Calif, May 12, 53. PHYSICAL OCEANOGRAPHY, SUB-INERTIAL, ELECTRICAL GEOPHYSICS. *Educ:* Harvey Mudd Col, BS, 75; Mass Inst Tech & Woods Hole Oceanog Inst, PhD(oceanog), 80. *Prof Exp:* Postdoctoral geophys, Scripps Inst Oceanog, 80-82; asst & assoc res geophysicist, 82-86; MEM TECH STAFF, AT&T BELL LABS, 86- *Concurrent Pos:* Adj lectr, Scripps Inst Oceanog, 82-86; guest scientist, Los Alamos Nat Lab, 83-; assoc adj prof, Scripps Inst Oceanog, 86-; mem NJ Sea Grant Adv Bd, 87-90; J Robert Oppenheimer fel, Los Alamos Nat Lab, 85-86; assoc ed, J Geophysics Res, 87-90. *Mem:* Am Geophys Union; Inst Elec & Electronic Engrs; AAAS; Am Meteorol Soc. *Res:* Studies of electromagnetic fields induced by ocean currents and their application to understand low frequency ocean variability; investigations of the electrical structure of the sea floor; development of instrumentation for oceanographic studies. *Mailing Add:* AT&T Bell Labs 3L401 600 Mountain Ave Murray Hill NJ 07974

CHAVE, CHARLES TRUDEAU, mechanical & chemical engineering; deceased, see previous edition for last biography

CHAVE, KEITH ERNEST, b Chicago, Ill, Jan 18, 28; m 51; c 2. GEOLOGY. *Educ:* Univ Chicago, PhB, 48, MS, 51, PhD, 52. *Prof Exp:* Asst, State Geol Surv, Ill, 48; res geologist, Calif Res Corp, 52-59; from asst prof to prof geol, Lehigh Univ, 59-67, assoc dir marine sci ctr, 62-67; chmn dept, 70-73, PROF OCEANOG, UNIV HAWAII, 67- *Concurrent Pos:* Trustee, Bermuda Biol Sta, 62-68; Alexander von Humboldt sr scientist, Univ Kiel, 73-74. *Mem:* Fel AAAS; Soc Econ Paleont & Mineral; Geochem Soc; Am Geophys Union; Am Soc Limnol & Oceanog; Sigma Xi. *Res:* Geochemistry; marine geology. *Mailing Add:* Dept Oceanog Univ Hawaii Honolulu HI 96822

CHAVEL, ISAAC, b Louisville, Ky, Apr 2, 39. GEOMETRY. *Educ:* Brooklyn Col, BA, 61; NY Univ, MS, 64; Yeshiva Univ, PhD(math), 66. *Prof Exp:* Teaching asst math, Brooklyn Col, 61-64; asst prof, Univ Minn, 66-70; asst prof, 70-73, assoc prof, 73-80, PROF MATH, CITY COL NEW YORK, 80- *Mem:* Am Math Soc; Soc Indust & Appl Math. *Res:* Interplay of Riemannian geometry with mathematical analysis, especially as relates to the Laplace operator. *Mailing Add:* 699 W 239th St Riverdale NY 10463

CHAVIN, WALTER, b Dec 6, 25. ENDOCRINOLOGY, RADIOBIOLOGY. *Educ:* City Col New York, BS, 46; NY Univ, MS, 49, PhD(zool), 54. *Prof Exp:* Instr biol, City Col New York, 46-47; asst, NY Aquarium, 48-49; instr zool, Univ Ariz, 49-51; res specialist, Am Mus Natural Hist, 51-53; assoc prof biol, 53-68, prof radiol, 74-80, PROF BIOL SCI, WAYNE STATE UNIV, 68- *Concurrent Pos:* Res assoc, Div Biol & Med Res, Argonne Nat Lab, 55, 56, consult, 57-58; NSF sr fel, 60-61; Sigma Xi fac res award, 68; coordr, US-Japan Seminar Responses of Fish to Environ Changes, Tokyo, 70; consult, Great Lakes Lab, Bur Com Fisheries, US Dept Interior; ed bull, Int Pigment Cell Soc, 79-83. *Mem:* AAAS; Am Soc Cell Biol; Am Physiol Soc; fel NY Acad Sci; Radiation Res Soc; Soc Exp Biol & Med; Sigma Xi; Endocrine Soc; Am Soc Zool (treas, 80-82); Am Asn Cancer Res; Int Pigment Cell Soc. *Res:* Basic pigment studies relating to control of melanocyte enzymatic activity, proliferation and development relative to abnormal melanocyte control (melanoma); control mechanisma (endocrine, cellular, biochemical, drug); environmentally stimulated/inhibited mechanisms. *Mailing Add:* Dept Biol Sci Wayne State Univ 5104 Gullen Mall Detroit MI 48202

CHAWLA, MANGAL DASS, b Lyallpur, India, Oct 10, 32; US citizen; m 62; c 3. ROTATING EQUIPMENT DESIGN & TROUBLE SHOOTING, STRUCTURAL ENGINEERING. *Educ:* Punjab Univ, India, BA, 51, BSc, 55; Univ Alta, Can, MSc, 65; Mich State Univ, East Lansing, PhD(mech eng), 69. *Prof Exp:* Asst engr, Punjab PWD, B&R, India, 55-58; asst prof civil eng, Punjab Eng Col, India, 58-63; proj engr, Avco Lycoming, Charleston, 70-71; assoc prof civil eng, La State Univ, Eunice, 71-74; struct engr, Bechtel Corp, Houston, 74-76, rotating equip specialist, 76-80, pipeline engr, 80-82; rotating equip specialist, Gulf Interstate Eng, Houston, 82-84; AEROSPACE ENGR, FLIGHT DYNAMICS LAB, WRIGHT PATTERSON AFB, 84- *Concurrent Pos:* Struct engr, Whittaker Carswell Co Ltd, Can, 64-65; instr, Mich State Univ, East Lansing, 65-69; civil engr, Brighton Eng Co, Lansing, 66-69; from instr to asst prof, Univ Houston, 74-84. *Mem:* Am Soc Mech Engrs; Asn Unmanned Vehicle Systs. *Res:* Robotics for rapid aircraft turnaround; mechanical subsystem integrity program to increase the reliability, maintainability and supportability of the mechanical systems on a fighter type aircraft. *Mailing Add:* 473 Coy Dr Beavercreek OH 45385

CHAWNER, WILLIAM DONALD, b Calif, May 13, 03; wid; c 2. PETROLEUM GEOLOGY, ENERGY MINERALS. *Educ:* Occidental Col, BS, 25; Calif Inst Technol, MS, 34; La State Univ, PhD(geol), 37. *Prof Exp:* Geologist, Atlantic Refining Co, 27-33, State Geol Surv, La, 34-36, Australasian Petrol Co, 36-41, Carter Oil Co, 41-58 & Humble Oil & Refining Co, 58-68; geol consult, 68-75; RETIRED. *Mem:* Am Asn Petrol Geol; fel Geol Soc Am; Am Inst Mining, Metall & Petrol Eng; Am Inst Prof Geol. *Res:* Petroleum geology; stratigraphy; mineral exploration; uranium deposits. *Mailing Add:* 5409 Lee St W Milton FL 32570

CHAYKIN, STERLING, b New York, NY, Sept 18, 29; m 54; c 4. BIOCHEMISTRY. *Educ:* NY Univ, AB, 50; Univ Wash, PhD(biochem), 54. *Prof Exp:* Runyon fel, Harvard, 56-59; from asst prof to assoc prof biochem, 59-69, chmn dept biochem & biophys, 68-70, assoc dean resident instr, Col Agr & Environ Sci, 70-74, PROF BIOCHEM, UNIV CALIF, DAVIS, 74- *Concurrent Pos:* Fulbright fel, Ger, 66-67; Guggenheim fel, 66-67. *Mem:* AAAS; Am Chem Soc; Am Soc Biol Chemists. *Res:* Enzymology; biochemistry of development, vitamins and mammalian genetics. *Mailing Add:* Dept of Biochem & Biophys Univ of Calif Davis CA 95616

CHAYKOVSKY, MICHAEL, b Mayfield, Pa, Sept 19, 34; m 66; c 2. ORGANIC CHEMISTRY, BIOCHEMISTRY. *Educ:* Pa State Univ, BS, 56; Univ Mich, MS, 59, PhD(org chem), 61. *Prof Exp:* Res fel chem, Harvard Univ, 61-64; asst prof, State Univ NY Buffalo, 64-65; res fel, Labs of Chem, Nat Inst Arthritis & Metab Dis, 65-66; sr res chemist, Hoffmann-La Roche Inc, 66-69; res assoc chem, Harvard Med Sch, 69-80; RES CHEMIST, NAVAL SURFACE WEAPONS CTR, 80- *Mem:* Am Chem Soc. *Res:* Synthetic organic chemistry; steroids, terpenes, sulfur ylides and carbanions; photochemistry; organophosphorous chemistry; heterocycles; carbenes; nitrenes; diborane reductions; cancer chemotherapy. *Mailing Add:* Bldg 343-110 Naval Surface Weapon Ctr Silver Springs MD 20903-5000

CHAZOTTE, BRAD NELSON, b Minneola, NY, Apr 15, 54; m 89. MITOCHONDRIAL BIO-ENERGETICS, BIOPHYSICAL CHEMISTRY. *Educ:* Bucknell Univ, BS, 76; Northern Ill Univ, PhD(chem), 81. *Prof Exp:* res assoc, lab cell biol, 82-86, RES ASST PROF, DEPT CELL BIOL & ANAT, SCH MED, UNIV NC, 86- *Mem:* AAAS; Am Chem Soc; Am Soc Cell Biol; Am Soc Biochem & Molecular Biol; Biophys Soc; Sigma Xi. *Res:* Biochemistry; flouresence recovery after photo bleaching; freeze fracture electron microscopy; membrane fusion; scanning electron microscopy; video microscopy & time-resolved spatial photo. *Mailing Add:* Dept Cell Biol & Anat Univ NC CB No 7090 115 Taylor Hall Chapel Hill NC 27599

CHE, STANLEY CHIA-LIN, b Amoy, China, Sept 11, 46; US Citizen; m 72; c 2. FUEL TECHNOLOGY. *Educ:* Cheng Kung Univ, Taiwan, BS, 68; Univ Utah, PhD(fuels eng), 74. *Prof Exp:* Fel coal gasification, Argonne Nat Lab, 74-75; res engr coal conversion, Occidental Petrol Corp, 75-78, sr res engr coal conversion, Occidental res Corp, 79-82; asst dir & prin engr, Res & Develop Div, Kinetics Technol Inc, 83-85, tech mgr, KTI COSORB Proc, 85-87; assoc dir technol develop, Mfg & Technol Conversions Int, 87-88; mgr environ serv, Jack Bryant & Assoc, 88-89; PRIN DEVELOP ENGR, KTI CORP, 89- *Mem:* Am Chem Soc; Am Inst Chem Engrs. *Res:* Development of novel chemistry to improve the yield and quality of coal tar by flash pyrolysis; upgrading of pyrolysis coal tar; pyrolysis of coals; analytical method development for characterization of coal liquids; catalyst deactivation; characterization and upgrading of shale oil; natural gas liquids pyrolysis; carbon-monoxide recovery and purification; waste lube oil rerefining. *Mailing Add:* 10161 Humbolt St Los Alamitos CA 90720

CHEADLE, VERNON IRVIN, b Salem, SDak, Feb 6, 10; m 39; c 1. BOTANY. *Educ:* Miami Univ, AB, 32; Harvard Univ, AM, 34, PhD(biol bot), 36. *Hon Degrees:* LLD, Miami Univ & Univ RI, 64. *Prof Exp:* Austin teaching fel bot, Harvard Univ, 33-36; from instr to asst prof, RI State Col, 36-42, prof & head dept, 42-52, dir grad div, 43-52; botanist, Exp Sta & prof bot, Univ Calif, Davis, 52-62, chmn dept, 52-60, actg vchancellor, 61-62; chancellor, Univ Calif, Santa Barbara, 62-77; RETIRED. *Concurrent Pos:* Fulbright fel, 59. *Mem:* Fel AAAS; fel Am Acad Arts & Sci; Am Soc Plant Taxon; Bot Soc Am (pres, 61); Int Soc Plant Morph; Int Asn Wood Anat. *Res:* Anatomy; morphology of vascular plants. *Mailing Add:* 891 Cieneguitas Rd Santa Barbara CA 93110

CHEAL, MARYLOU, b St Clair Co, Mich, 1926; m 46; c 3. HUMAN VISUAL ATTENTION. *Educ:* Oakland Univ, BA, 69; Univ Mich, Ann Arbor, PhD(psychobiol), 73. *Prof Exp:* Res investr taste regeneration, Dept Zool, Univ Mich, 73-75, res investr taste develop, Dept Oral Biol, Sch Dent, 75-76; asst psychologist, 76-81, assoc psychologist, Neuropsychol Lab, McLean Hosp, 81-83; fac res, Ariz State Univ, 83-86; RES PSYCHOLOGIST, UNIV DAYTON RES INST, HIGLEY, ARIZ, 86- *Concurrent Pos:* Lectr psychol, Univ Mich, 73-76 & Dept Psychiat, Harvard Med Sch, 77-83; res scientist, Develop Rev Comt, Nat Inst Ment Health, 84-85; adj assoc prof, Ariz State Univ, 86-; vis prof, Air Force Systs Command Univ, Williams Air Force Base, Ariz, 86-88; Charles A King res fel, McLean Hosp, Belmont, Mass, 76-77; res bd adv, Am Biog Inst, 86-; adv coun, Int Biog Ctr, 89- *Mem:* Fel AAAS; Sigma Xi; Soc Neurosci; fel Am Psychol Asn; fel Am Psychol Soc; Int Brain Res Orgn. *Res:* Human visual attention; lifespan growth and development; environmental enrichment; reproduction and behavior of gerbils; attention and habituation. *Mailing Add:* Univ Dayton Res Inst PO Box 2020 Higley AZ 85236-2020

CHEATHAM, JOHN B(ANE), JR, b Houston, Tex, June 29, 24; m 47; c 2. MECHANICAL ENGINEERING. *Educ:* Southern Methodist Univ, BS, 48, MS, 53; Mass Inst Technol, ME, 54; Rice Univ, PhD, 60. *Prof Exp:* Test engr, Gen Elec Co, 48; design engr, Link Belt Co, 49-50; engr, Atlantic Ref Co, 50-53; res assoc, Shell Develop Co, 54-63; assoc prof mech eng, 63-66, chmn dept mech & aerospace eng & mat sci, 69-73, PROF MECH ENG, RICE UNIV, 66- *Concurrent Pos:* Lectr, Univ Houston, 60-63; pres, Cheatham Eng Inc, 77-; pres, Techaid Corp, 77-88; founding tech ed, Am Soc Mech Engrs Trans J Energy Resources Tech, 79-80. *Honors & Awards:* Ralph James Award, Am Soc Mech Engrs, 80; Award for Outstanding Basic Res in Rock Mech, US Comt Rock Mech, NRC, 82. *Mem:* Fel Am Soc Mech Engrs; Am Inst Mining, Metall & Petrol Engrs; Am Soc Eng Educ. *Res:* Rock mechanics; plasticity; machine design; oil well drilling; robotics and automation. *Mailing Add:* Dept Mech Eng & Mat Sci Rice Univ PO Box 1892 Houston TX 77251

CHEATHAM, ROBERT GARY, b Guam Naval Base, Mariana Islands, Oct 30, 26; m 57; c 2. MATERIALS SCIENCE ENGINEERING. *Educ:* The Citadel, BS, 47; Mass Inst Technol, SM, 50, ScD, 57. *Prof Exp:* Res asst plastic mat, Plastic Lab-Colloid & Rubber Lab, Mass Inst Technol, 48-54; dir res, WASCO Chem, 54-57; dir, Plastic Div, D S Kennedy & Co, 57-59; pres mat res, Mat Technol Inc, 59-64; prin scientist polymeric mat, United Aircraft Res Lab, 64-67; unit chief chem mat, Boeing Co, 67-77; chief composite mat, Sikorsky Aircraft Div, United Technol Corp, 78-88; PROF MECH ENG, UNIV BRIDGEPORT, CONN, 88- *Concurrent Pos:* Consult, Stanford Res Int, 68-77; fel, Mass Inst Technol Colloid Lab. *Mem:* Sigma Xi; Am Inst Aeronaut & Astronaut; Am Helicopter Soc. *Res:* Development and application of polymeric materials for high performance vehicles in extreme environments; development of required suitable process techniques; published sixty-five technical papers. *Mailing Add:* Seven Beech Tree Circle Trumbull CT 06611

CHEATHAM, THOMAS J, b Campbellsville, Ky, Oct 8, 44; m 66; c 1. MATHEMATICS. *Educ:* Campbellsville Col, BS, 66; Univ Ky, MS, 68, PhD(math), 71. *Prof Exp:* From asst prof to assoc prof math, Samford Univ, 71-77; vis assoc prof, Univ Ky, 77-78; prof math, Samford Univ, 78-81; PROF MATH & COMPUT SCI, WESTERN KY UNIV, 81- *Concurrent Pos:* Samford Univ Res Fund fels, 74, 75 & 77. *Mem:* Am Math Soc. *Res:* Nonsingular ring and module; numerous publications and journals. *Mailing Add:* Dept Comput Sci Western Ky Univ Bowling Green KY 42101

CHEATHAM, WILLIAM JOSEPH, PATHOLOGY. *Educ:* Vanderbilt Univ, MD, 50. *Prof Exp:* Pathologist, Smith County Community Hosp, 77-90; RETIRED. *Mailing Add:* 200 Ridgeway Rd Marion VA 24354

CHEAVENS, THOMAS HENRY, b Dallas, Tex, May 19, 30; m 55; c 3. ORGANIC CHEMISTRY. *Educ:* Univ Tex, BS, 50, PhD(chem), 55. *Prof Exp:* Res chemist, Org Pigments, Res Div, Am Cyanamid Co, 55-56, develop chemist, Org Chem Div, 56, develop chemist, Intermediates & Rubber Chem, 57, group leader, process develop dept, NJ, 58-61, group leader, Refinery Catalysts Group, Conn, 61-67; supvr indust catalysts res, W R Grace & Co, 67-70, dir indust catalysts res dept, 71-75; sr prog mgr, Div Fossil Energy Res, US Energy Res & Develop Admin, 75-77; res dir, 77-80, PRES, QUEST RES INT, INC, 80-; vpres, Improtee Inc, 80-86; ASST PROF, NMEX HIGHLANDS UNIV, 89- *Concurrent Pos:* Fel Univ Tex, Austin, 87-89. *Mem:* Am Chem Soc; AAAS. *Res:* Total synthesis of stemonaceae alkaloids and pseudo quaianolila sesquiterpenes; cyclo additions and radical cyclization reactions. *Mailing Add:* Dept Phys Sc NMex Highlands Univ Las Vegas NM 87701

CHECHIK, BORIS, b Kislovodsk, USSR, Mar 31, 31; Can citizen; m 58; c 1. TUMOR IMMUNOLOGY, IMMUNOCHEMISTRY. *Educ:* Second Med Sch, Moscow, MD, 56; P A Gertzen Oncol Inst, Moscow, PhD(tumor immunol), 63. *Prof Exp:* Gen practitioner med, Hosp N33, Moscow, 56-59; res investr tumor immunol, P A Gertzen Oncol Inst, 59-71 & Hadassah Med Sch, Jerusalem, 71-73; res assoc, Hosp Sick Children, Toronto, 74-76; res investr tumor immunol, 77-81, SR STAFF MEM, MT SINAI HOSP, TORONTO, 81-; ASSOC PROF, UNIV TORONTO, 81- *Mem:* Can Soc Immunol; Can Fedn Biol Soc. *Res:* Immunochemistry of differentiation antigens of human and animal normal and leukemic hematopoietic cells; immunobiology of adenosine deminase; immunohistochemistry and immunochemistry of oligosaccharides. *Mailing Add:* Res Dept Mount Sinai Hosp 600 Univ Ave Toronto ON M5G 1X5 Can

CHECK, IRENE J, b Austria, Aug 8, 46. IMMUNOLOGY. *Educ:* Case Western Reserve Univ, PhD(microbiol), 72. *Prof Exp:* Asst prof, 80-82, ASSOC PROF PATH, EMORY UNIV, 82- *Mem:* Am Bd Med Lab Immunol; Am Asn Immunologist; Am Soc Microbiol; Am Soc Clin Pathologists. *Mailing Add:* 2765 Townley Circle Atlanta GA 30340

CHECKEL, M DAVID, b Coronation, Alta, Nov 1, 54; m 74; c 2. COMBUSTION IN ENGINES, ALTERNATIVE FUELS. *Educ:* Univ Alta, BSc, 76; Univ Cambridge, PhD(eng), 81. *Prof Exp:* Mech engr, Chevron Standard Ltd, 76-78; asst prof, 81-83, ASSOC PROF MECH ENG, UNIV ALTA, 83- *Concurrent Pos:* Prin, M D Checkel Prof Engr, 83-; fac liaison, Can Soc Mech Engrs, 84-87; vis prof, Eng Dept, Cambridge Univ, 87-88; fel commoner, Churchill Col, Cambridge, 87-88. *Honors & Awards:* Ralph R Teetor Educ Award, Soc Automotive Engrs, 85. *Mem:* Can Soc Mech Engrs; Am Soc Mech Engrs; Combustion Inst Can (secy, 91-); Soc Automotive Engrs. *Res:* Turbulent combustion and turbulent mixing applied to internal combustion engines and dispersion of hazards; enhanced ignitions; alternative fuels; fluid flow visualization; image processing; explosion hazards. *Mailing Add:* Mech Eng Univ Alta Edmonton AB T6G 2G8 Can

CHEEKE, PETER ROBERT, b Duncan, BC, Oct 19, 41; m 70; c 2. ANIMAL SCIENCE, NUTRITION. *Educ:* Univ BC, BSA, 63, MSA, 65; Ore State Univ, PhD(animal nutrit), 69. *Prof Exp:* From asst prof to assoc prof, 69-79, PROF ANIMAL SCI, ORE STATE UNIV, 79- *Concurrent Pos:* Ed, J Appl Rabbit Res. *Mem:* Am Inst Nutrit; Am Soc Animal Sci. *Res:* Nutrition of rabbits; pyrrolizidine alkaloid toxicity; nutritional toxicology interrelationships. *Mailing Add:* Dept Animal Sci Ore State Univ Corvallis OR 97331

CHEEMA, MOHINDAR SINGH, b Sialkot, Panjab, India, Jan 15, 29; m 51; c 4. MATHEMATICS. *Educ:* Univ Panjab, BA, 48, MA, 50; Univ Calif, Los Angeles, MA, 60, PhD(math), 61. *Prof Exp:* Res scholar math, Univ Panjab, 51-54, instr, 54-58; jr res mathematician, Univ Calif, Los Angeles, 58-61; from asst prof to assoc prof, 61-68, PROF MATH, UNIV ARIZ, 68- *Mem:* Am Math Soc; Math Asn Am; Asn Comput Mach; Sigma Xi. *Res:* Number theory; pure mathematics; numerical analysis. *Mailing Add:* Dept of Math Univ of Ariz Tucson AZ 85721

CHEEMA, ZAFARULLAH K, b Gakkhar, WPakistan, Apr 21, 34; US citizen; m 62; c 2. ORGANIC CHEMISTRY, PHARMACEUTICAL CHEMISTRY. *Educ:* Panjab Univ, WPakistan, BPharm, 54; Univ Tubingen, Dr rer nat(chem), 57. *Prof Exp:* Res assoc, Univ Kans, 57-58 & Univ Tübingen, 58-59; chmn dept chem, Knoxville Col, 59-63, assoc prof chem, 61-63; adv chemist, Health Dept, Govt Pakistan, 60-61; lectr chem, Univ Del, 63-64; res chemist, Gen Chem Div, Allied Chem Corp, 64-65, sr res chemist agr div, 66, res assoc, Plastics Div, 66-69, tech supvr res, 69-71; group leader, Am Hoechst Corp, 71-75, mgr res, Keuffel & Esser Co, 75-78; VPRES RES, RICHARDSON GRAPHICS CO, 78- *Concurrent Pos:* Consult, Oak Ridge Nat Lab, 62-63; res fel, Univ Tenn, 63; adj prof & chmn chem dept, Fairleigh Dickinson Univ, 66-67. *Mem:* Am Chem Soc; Soc Photog Scientists & Engrs; Tech Asn Graphic Arts; NY Acad Sci. *Res:* Terpenes; conformational analysis; reaction mechanisms such as epimerization and pinnacol rearrangement; synthesis of pesticides; oxidation; photopolymers, diazo, graphic arts and engineering product development. *Mailing Add:* 34 Cottonwood Rd Wellesley MA 02181-3112

CHEER, CLAIR JAMES, b Lakewood, Ohio, May 16, 37; m 62; c 2. ORGANIC CHEMISTRY. *Educ:* Kenyon Col, BA, 59; Wayne State Univ, PhD(org chem), 64. *Prof Exp:* Res asst chem, Parma Res Lab, Union Carbide Corp, 59; res assoc org chem, Frank J Seiler Res Lab, off aerospace res, US Air Force Acad, 64-67; res assoc chem, Univ Ariz, 67-68; from asst prof to assoc prof, 68-83, asst dean, 73-75, PROF CHEM, UNIV RI, 83- *Concurrent Pos:* Vis scholar, Stanford Univ, 75-76; Fulbright res scholar, Univ Thessaloniki, Greece, 86-87. *Mem:* Am Chem Soc; Royal Soc Chem; Am Inst Physics; Am Crystallog Asn; NY Acad Sci. *Res:* Organic reactions and mechanisms; organic synthetic methods; solid state photochemistry; use of x-ray crystallography in solution of organic chemical problems. *Mailing Add:* Dept of Chem Univ of RI Kingston RI 02881

CHEETHAM, ALAN HERBERT, b El Paso, Tex, Jan 30, 28; m 51; c 4. INVERTEBRATE PALEONTOLOGY. *Educ:* NMex Inst Mining & Technol, BS, 50; La State Univ, MS, 52; Columbia Univ, PhD(paleont), 59. *Prof Exp:* From instr to assoc prof geol, La State Univ, 54-66; assoc cur, 66-69, cur paleobiol, 69-87, SR INVERT PALEONTOLOGIST, SMITHSONIAN INST, 87- *Concurrent Pos:* NSF fel, Brit Mus Natural Hist, 61-62; guest prof, Univ Stockholm, Sweden, 64-65; consult prof, La State Univ, 66; mem & chmn geol screening comt, Coun Int Exchange Scholars, 80-83; assoc ed, paleobiology, 76-; ed, fossil invert, 87. *Mem:* Sigma Xi; fel AAAS; Paleont Soc; Soc Sedimentary Geol; Paleont Asn; Int Paleont Asn. *Res:* Evolution, functional morphology and systematics of fossil and living cheilostome bryozoa; multivariate morphometrics. *Mailing Add:* MRC:NHB 121 Smithsonian Inst Washington DC 20560

CHEETHAM, RONALD D, b Duluth, Minn, Oct 8, 43; m 65; c 3. PHYCOLOGY, AQUATIC MICROBIOLOGY. *Educ:* Univ Minn, Duluth, BA, 65; Univ Minn, St Paul, MS, 67; Purdue Univ, PhD(plant ecol & physiol), 70. *Prof Exp:* Asst prof biol & plant physiol, Minn State Univ, 70-73; asst prof, 73-78, ASSOC PROF BIOL & ECOL, WORCESTER POLYTECH INST, 78- *Concurrent Pos:* Agr ecol advisor, New Eng Res, Inc, 73-; advisor, Worcester Consortium Water Quality, 73- *Mem:* AAAS; Nat Wildlife Soc; Am Ecol Soc; Sigma Xi. *Res:* Aquatic ecology; nitrogen relationships of blue-green algae; public health significance of macroinvertebrates in potable waters. *Mailing Add:* Dept Biol Worcester Poly Inst Worcester MA 01609

CHEEVER, ALLEN, b Brookings, SDak, June 4, 32; m 53; c 4. SCHISTOSOMIASIS. *Educ:* Carlton Col, BA, 54; Harvard Med Sch, MD, 58. *Prof Exp:* Intern, Mt Sinai Hosp, New York 58-59, res path, 59-60; res, path, Nat Instit Health, Bethesda, 62-64; INVESTR, LAB PARASITIC DIS, NAT INSTIT HEALTH, BETHESDA, 64-, ASST CHIEF, 69- *Concurrent Pos:* Atten Pathologist, Lab Path, Nat Can Instit, 73-; Res Fel, Eleanor Roosevelt Cancer Res Found, 65. *Mem:* Am Soc Trop Med & Hyg; Royal Soc Trop Med Hyg. *Res:* Research on the pathology and pathogenesis of schistosomiasis in man and experimental animals. *Mailing Add:* Bldg 4 Rm 126 Nat Instit Health Bethesda MD 20892

CHEEVERS, WILLIAM PHILLIP, b Tallulah, La, Nov 10, 41. MOLECULAR BIOLOGY. *Educ:* Univ Colo, BA, 63; Univ Miss, PhD(microbiol), 68. *Prof Exp:* Sci staff molecular virol, Cancer Res Lab, Univ Western Ont, Can, 70-76; asst prof, 76-80, ASSOC PROF, DEPT VET MICROBIOL & PATH, WASH STATE UNIV, 80- *Res:* Molecular mechanisms of viral persistence in retrovirus diseases. *Mailing Add:* Dept Vet Microbiol Wash State Univ Pullman WA 99163

CHEGINI, NASSER, CELLULAR & MOLECULAR BIOLOGY. *Educ:* Univ Southampton, Eng, PhD(cell & molecular biol), 80. *Prof Exp:* ASST PROF, OBSTET & GYNEC, UNIV LOUISVILLE, 81- *Res:* Corpus luteum. *Mailing Add:* Dept Obstet & Gynec Univ Fla Col Med Box J-294 JHMHC Gainesville FL 32610

CHEH, ALBERT MEI-CHU, b New York, NY, Mar 8, 47; m 70; c 2. TOXICOLOGY. *Educ:* Columbia Univ, BA, 67; Univ Calif, Berkeley, PhD(biochem), 74. *Prof Exp:* Res assoc, Univ Ill, 74; res assoc, Univ Minn, 74-77, NIH fel, 77-78, scientist, 78-80; ASST PROF CHEM, AM UNIV, 80- *Mem:* Am Chem Soc; AAAS; Environ Mutagen Soc; Fedn Am Scientists. *Res:* Detection, analysis and study of the formation of mutagenic electrophiles in environmental samples; study of the mechanisms for the mutagenicity and carcinogenicity of natural products. *Mailing Add:* 5610 Glenwood Rd Bethesda MD 20817-6728

CHEH, HUK YUK, b Shanghai, China, Oct 27, 39; m 69; c 2. ELECTROCHEMISTRY, NUCLEAR HEAT TRANSFER. *Educ:* Univ Ottawa, BASc, 62; Univ Calif, Berkeley, PhD(chem eng), 67. *Prof Exp:* Mem tech staff, AT&T Bell Labs, 67-70; from asst prof to prof, 70-82, RUBEN VIELE PROF ELECTROCHEM, COLUMBIA UNIV, 82- *Concurrent Pos:* Vis res prof, Nat Tsing Hua Univ, 77; prog dir, Nat Sci Found, 78-79. *Honors & Awards:* Electrodeposition Res Award, Electrochem Soc, 88; Sci Achievement Award, Am Electroplaters & Surface Finishers Soc, 89. *Mem:* Am Inst Chem Engrs; Electrochem Soc; NY Acad Sci; Am Electroplaters & Surface Finishers Soc; Sigma Xi. *Res:* Heterogeneous catalysis; biengineering; nuclear heat transfer; electrochemical engineering; electrodeposition; composite materials; battery modeling; mass transfer. *Mailing Add:* Dept Chem Eng Columbia Univ New York NY 10027

CHEIN, ORIN NATHANIEL, b New York, NY, Aug 29, 43; m 65; c 2. MATHEMATICS. *Educ:* NY Univ, BA, 64, MS, 66, PhD(math), 68. *Prof Exp:* From asst prof to assoc prof, 68-79, chmn math dept, 88-90, PROF MATH, TEMPLE UNIV, 79-, DIR TEACHING IMPROV CTR, 90- *Concurrent Pos:* Lady Davis vis scientist, 77-78; Am Philos Soc res grant, 87-88. *Mem:* Am Math Soc; Math Asn Am. *Res:* Automorphisms of free and metabelian groups; Moufang loops; combinatorics; math recreations. *Mailing Add:* Dept Math Temple Univ Philadelphia PA 19122

CHEITLIN, MELVIN DONALD, b Wilmington, Del, Mar 25, 29; m 52; c 3. INTERNAL MEDICINE, CARDIOLOGY. *Educ:* Temple Univ, BA, 50, MD, 54. *Prof Exp:* Chief cardiovasc serv, Madigan Gen Hosp, US Army, 60-64, Tripler Gen Hosp, 64-68 & Letterman Gen Hosp, 68-71, asst clin prof med, Univ Calif, San Francisco, 69-71, chief cardiovasc serv, Walter Reed Army Hosp, 71-74; prof, 74-77, clin prof, 77-79, PROF MED, UNIV CALIF, SAN FRANCISO, 79- *Concurrent Pos:* Consult, Queen's Hosp, Honolulu, 65-68 & Letterman Gen Hosp, 75-; fel, Coun Clin Cardiol, Am Heart Asn, 66-68; cardiovasc consult to surgeon gen, Walter Reed Gen Hosp, 71-74; assoc dir cardiopulmonary unit, San Francisco Gen Hosp, 74- *Mem:* AMA; fel Am Col Physicians; fel Am Col Cardiol; Am Heart Asn; Am Fedn Clin Res. *Mailing Add:* San Francisco Rm 5G1 Gen Hosp Sch Med Univ Calif San Francisco CA 94110

CHELAPATI, CHUNDURI V(ENKATA), b Eluru, Andhra State, India, Mar 11, 33. CIVIL ENGINEERING, ENGINEERING MECHANICS. *Educ:* Andhra Univ, India, BE, 54; Indian Inst Sci, dipl, 56; Univ Ill, MS, 59, PhD(civil eng), 62. *Prof Exp:* Asst prof struct eng, Birla Eng Col, India, 56-57; asst civil eng, Univ Ill, 57-62; asst prof, Calif State Col, Los Angeles, 62-65, NSF instnl grant, 63-65; assoc prof, 65-70, vchmn dept, 71-73, chmn dept, 73-80, PROF CIVIL ENG, CALIF STATE UNIV, LONG BEACH, 70- *Concurrent Pos:* Consult, US Naval Civil Eng Lab, Calif, 63-68; vis scholar, Univ Calif, Los Angeles, 68-69; consult & proj mgr, Holmes & Narvar, Inc, Calif, 68-73; consult, Civil Eng Lab, Naval Construct Battalion Ctr, Port Hueneme, Calif, 75- *Mem:* Am Soc Civil Engrs; Seismol Soc Am; Am Soc Eng Educ; Earthquake Engrs Res Inst; Sigma Xi. *Res:* Dynamics of structures subjected to earthquakes and blasts; arching in soils; soil structure interaction; buckling of structures; nuclear reactor structures; earthquake hazard mitigation; professional review courses; seismic design; computer methods. *Mailing Add:* Dept of Civil Eng Calif State Univ Long Beach CA 90840

CHELEMER, HAROLD, b Green Bay, Wis, Nov 28, 28; m 57; c 3. CHEMICAL ENGINEERING, STATISTICS. *Educ:* Univ Mo, BSc, 49, MSc, 51; Univ Tenn, PhD(chem eng), 55. *Prof Exp:* Sr engr, Reactor Eng & Mat Dept, Westinghouse Atomic Power Div, 57-64, fel engr, PWR Plant Div, 64-70; adv engr, Com Nuclear Fuel Div, Westinghouse Energy & Utility Systs, 70-90; RETIRED. *Mem:* Sigma Xi. *Res:* Pressurized water reactor core thermal and hydraulic design analysis; computer program development; statistical evaluation of engineering and manufacturing data; liquid metal and water thermal-hydraulics; thermodynamics. *Mailing Add:* 5800 Wayne Rd Pittsburgh PA 15230

CHELIKOWSKY, JAMES ROBERT, b Manhattan, Kans, June 1, 48; c 3. SOLID STATE PHYSICS. *Educ:* Kans State Univ, BS, 70; Univ Calif, Berkeley, PhD(physics), 75. *Prof Exp:* Mem tech staff, Bell Tel Lab, Murray Hill, NJ, 76-78; asst prof physics, Univ Ore, 78-80; res assoc, Exxon Res & Eng Co, 80-87; PROF MAT SCI, UNIV MINN, 87- *Mem:* Am Phys Soc; fel Am Phys Soc. *Res:* Electronic structure of solids; band structures; optical properties of solids; phase stability of intermetallic alloys; surface properties of semiconductors and metals; cohesion in metals. *Mailing Add:* Dept Chem Eng & Mat Sci Univ Minn Minneapolis MN 55455

CHELLAPPA, RAMALINGAM, b Tanjore, Madras, India, Apr 8, 53; m 83; c 1. SIGNAL PROCESSING, IMAGE PROCESSING. *Educ:* Univ Madras, BE, 75; Indian Inst Sci, ME, 77; Purdue Univ, MSEE, 78, PhD(elec eng), 81. *Prof Exp:* asst prof, 81-86, ASSOC PROF, ELEC ENG, UNIV SOUTHERN CALIF, LOS ANGELES, 81- *Concurrent Pos:* Consult, Northrop Corp, 85, Hughes Aircraft Co, 86-87. *Honors & Awards:* Jawaharlal Nehru Award, Dept Educ, Govt India, 75; Presidential Young Investr Award, NSF, 85. *Mem:* Inst Elec Electronic Engrs; Am Asn Artificial Intelligence. *Res:* Research in basic computer vision problems; pattern recognition and image processing; statistical analysis. *Mailing Add:* 2210 Curtis Ave B Redondo Beach CA 90278

CHELLEW, NORMAN RAYMOND, b Aurora, Minn, June 17, 17; m 43. INORGANIC CHEMISTRY. *Educ:* Univ Minn, BS, 39. *Prof Exp:* Teacher, Breitung Twp Schs Mich, 40-41; chemist & supvr, E I du Pont de Nemours & Co, 41-46; jr chemist, Argonne Nat Lab, 47-49, assoc chemist, 49-80; RETIRED. *Mem:* Sci Res Soc Am; Am Chem Soc; Am Nuclear Soc. *Res:* Research and development of inorganic chemistry; radiochemistry; development of fission product monitors for nuclear reactor systems. *Mailing Add:* 811 Winthrop Ave Joliet IL 60435

CHELLO, PAUL LARSON, b New Haven, Conn, Nov 11, 42. BIOCHEMICAL PHARMACOLOGY. *Educ:* Johns Hopkins Univ, BA, 64; Univ Vt, PhD(pharmacol), 71. *Prof Exp:* Res assoc molecular therapeut, 74-75, ASSOC, MEM SLOAN KETTERING INST CANCER RES, 75- *Concurrent Pos:* asst prof, Grad Sch Med Sci, Cornell Univ, 75- *Mem:* Am Asn Cancer Res; Am Soc Pharmacol & Exp Therapeut; Am Soc Cell Biol; NY Acad Sci; Sigma Xi. *Res:* Cancer chemotherapy; drug transport; mechanisms of drug resistance; mechanism of action of antifolates; folic acid, vitamin B12 and methionine metabolism. *Mailing Add:* 91 Fair St Guilford CT 06437

CHELTON, DUDLEY B(OYD), b Baltimore, Md, July 17, 28; m 47; c 2. CRYOGENICS. *Educ:* Ohio State Univ, BSME, 48; Mass Inst Technol, SM, 49. *Prof Exp:* Instr mech eng, Mass Inst Technol, 49-50; mech engr, Los Alamos Sci Lab, 50-51; chief, Cryogenics Div, Nat Bur Standards, 68-74, mech engr, 51-77, sr eng consult, 74-77; CONSULT, 77- *Honors & Awards:* Gold Medal Award, US Dept Commerce, 53. *Mem:* Sigma Xi. *Res:* Liquefaction of gases; refrigeration techniques; liquid hydrogen technology; cryogenic processes; cryogenic and liquefied natural gas technology; cryogenic safety; slush hydrogen technology. *Mailing Add:* 500 Mohawk Dr #308 Boulder CO 80303

CHEN, ALBERT TSU-FU, b Fukien, China, Sept 2, 37; m 61; c 2. SOIL MECHANICS, APPLIED MATHEMATICS. *Educ:* Taiwan Univ, BSCE, 58; Univ Mass, MSCE, 62; Rensselaer Polytech Inst, PhD(soil mech), 67. *Prof Exp:* Instr, Rensselaer Polytech Inst, 65-67; RES CIVIL ENGR, US GEOL SURV, 67- *Mem:* Earthquake Eng Res Inst; Am Soc Civil Engrs; Sigma Xi. *Res:* Dynamic soil behavior; consolidation theory; response of soil layers during earthquakes. *Mailing Add:* 333 Kellogg Ave Palo Alto CA 94301

CHEN, ALICE TUNG-HUA, b Taiwan, Rep China, Dec 3, 49; div. COMPUTER NETWORK, COMPUTER ARCHITECTURE. *Educ:* Nat Taiwan Univ, BS, 71; State Univ NY Stony Brook, MS, 73, PhD(numerical anal), 75. *Prof Exp:* Vis fel res math, NIH, 75-77; systs engr, Gen Elec Info Serv Co, 77-81; res specialist, Exxon Prod Res Co, 81-86; COMPUT SCIENTIST, LAWRENCE LIVERMORE NAT LAB, 86- *Mem:* Asn Computing Machinery; Inst Elec & Electronics Engrs. *Res:* Engineering analysis and function specifications for new hardware and software of a computer system; mathematical models for computer system analysis; communication and I/O systems for computer network. *Mailing Add:* Lawrence Livermore Nat Lab Livermore CA 94550-6304

CHEN, AN-BAN, b Chiayi, Taiwan, Oct 10, 42; m 69; c 1. SOLID STATE PHYSICS. *Educ:* Taiwan Norm Univ, BS, 66; Col William & Mary, MS, 69, PhD(physics), 71. *Prof Exp:* Res assoc, Col William & Mary, 71-72; res assoc, Case Western Reserve Univ, 72-74; from asst prof to assoc prof, 74-83, PROF PHYSICS, AUBURN UNIV, 83- *Concurrent Pos:* Vis prof EE, Stanford Univ, 84-85. *Mem:* Am Phys Soc. *Res:* Study of electronic structures and related properties in crystalline solids and in disordered condensed matters. *Mailing Add:* Dept of Physics Auburn Univ Auburn AL 36849

CHEN, ANDREW TAT-LENG, b Kedah, Malaysia, June 17, 38; m 70; c 2. CYTOGENETICS, GENETICS. *Educ:* Chinese Univ Hong Kong, BSc, 61; McGill Univ, MSc, 63; Univ Western Ont, PhD(human cytogenetics), 69; Am Bd Med Genetics, dipl, 84. *Prof Exp:* Assoc psychiat, Emory Univ, 69-70, asst prof, 70-72; asst prof genetics & pediat & dir cytogenetics lab, Med Col Va, 72-74; chief genetics lab, 74-85, chief, Genetics Br, 85-89, ASST DIR GENETICS & EXTRAMURAL AFFAIRS, DIV ENVIRON HEALTH LAB SCI, CTR ENVIRON HEALTH, CTR DIS CONTROL, 89- *Concurrent Pos:* Res geneticist, Ga Ment Health Inst, 69-72. *Mem:* AAAS; Tissue Cult Asn; Sigma Xi; Am Soc Human Genetics; Environ Mutagen Soc. *Res:* Human cytogenetics; tissue culture of mammalian cells. *Mailing Add:* Div Environ Health lab Sci Ctr Environ Health Ctr for Dis Control Atlanta GA 30333

CHEN, ANNE CHI, b Ann Arbor, Mich, May 10, 62; m 90. CELL BIOLOGY, SIGNAL TRANSDUCTION. *Educ:* Mass Inst Technol, BS, 84; Univ Calif, Los Angeles, PhD(molecular biol), 90. *Prof Exp:* Res asst cancer res, Whitehead Inst, Mass Inst Technol, Cambridge, 83-84; consult oncogene workshop, Regional Off Western Pac, WHO, 85; postdoctoral fel cancer res, Dana Farber Cancer Inst & Harvard Med Sch, Boston, Mass, 90-91; POSTDOCTORAL FEL CANCER RES, MED COL, CORNELL UNIV, NY, 91- *Mem:* AAAS; Am Asn Cancer Res. *Res:* Mechanisms of carcinogenesis; regulation of cellular growth control; biology of oncogenes; biology of retinoids and their role in embryogenesis. *Mailing Add:* 1400 Commonwealth Ave West Newton MA 02165

CHEN, ANTHONY BARTHEOLOMEW, RECOMBINANT PROTEIN, DIAGNOSTICS. *Educ:* NJ Col Med & Dent, PhD(microbiol), 74. *Prof Exp:* SCIENTIST, GENENTECH INC, 84- *Mailing Add:* Genentech Inc 460 Point San Bruno Blvd South San Francisco CA 94080

CHEN, ANTHONY HING, b China, July 10, 45; US citizen; m 71. FOOD ENGINEERING. *Educ:* Univ Calif, Berkeley, BS, 69; Ohio State Univ, MS, 71, PhD(food sci), 78. *Prof Exp:* Assoc process develop engr, Res & Develop, Miles Lab, Inc, 71-75, supvr process develop, 75-78; mgr process develop, 78-81, DIR NEW TECHNOL, RES & DEVELOP, ANDERSON CLAYTON CO, 81- *Concurrent Pos:* Lectr food chem & chem eng, Univ Tex, Dallas, 79- *Mem:* Am Inst Chem Engrs; Int Food Technologists; Am Oil Chemist Soc. *Res:* Vegetable oil hydrogenation; process development; process optimization; protein texturization; food rheology; fats and oils processing. *Mailing Add:* 4113 Midnight Dr Plano TX 75093

CHEN, ARTHUR CHIH-MEI, b Boston, Mass, Apr 19, 39; m 63; c 2. SYSTEM DESIGN & SCIENCE, COMPUTER SCIENCES. *Educ:* Mass Inst Technol, BS, 61, MS, 62, PhD(elec eng), 66. *Prof Exp:* Elec engr comput memories, 66-75, WBS proj mgr comput tomography, 75-77, mgr, Energy Progs-Energy Syst Management, 77-80, Energy Systs Management Br, 80, mgr, Elec Systs & Technol Lab, 80-82, MGR, INFO SYSTS LAB, GENERAL ELECTRIC CORP RESEARCH & DEVELOP, 83- *Mem:* Fel Inst Elec & Electronics Engrs; Am Asn Artificial Intel. *Res:* Research and development of distributed computer system for the automatic control of electrical power and drive systems, and for medical electronics. *Mailing Add:* Res & Develop Rm 5C11 Gen Elec Corp One River Rd PO Box 8 Schenectady NY 12301-0008

CHEN, B(IH) H(WA), b Foochow, China, Sept 24, 33; m 66; c 2. CHEMICAL ENGINEERING. *Educ:* Nat Taiwan Univ, BSc, 54; McGill Univ, MEng, 61, PhD(chem eng), 65. *Prof Exp:* Asst res officer, Appl Chem Div, Nat Res Coun Can, 65-66; from asst prof to assoc prof, 66-81, PROF CHEM ENG, TECH UNIV NS, 81- *Concurrent Pos:* vis prof, Yokohama Nat Univ, Japan, 86; hon prof, Dalian Univ Sci & Technol, China. *Mem:* Chem Inst Can. *Res:* Transport phenomena. *Mailing Add:* Dept Chem Eng Tech Univ NS Halifax NS B3J 2X4 Can

CHEN, B(ENJAMIN) T(EH-KUNG), b Shanghai, China. CHEMICAL ENGINEERING, PHOTOGRAPHY. *Educ:* Nat Taiwan Univ, BS, 58; Univ Miss, MS, 63; Univ Rochester, PhD(chem eng), 68. *Prof Exp:* Res asst, Dept Chem Eng, Univ Rochester, 65-67; SR RES CHEMIST, MAT COATING & ENG DIV, RES LABS, EASTMAN KODAK CO, 67- *Mem:* Soc Photog Sci & Eng. *Res:* Making and coating photographic emulsions. *Mailing Add:* 30 Hidden Meadow Penfield NY 14526

CHEN, BANG-YEN, b Ilan, Taiwan, Oct 3, 43; m 68; c 3. GEOMETRY. *Educ:* Tamkang Col, Taiwan, BS, 65; Nat Tsinghua Univ, MS, 67; Univ Notre Dame, PhD(math), 70. *Hon Degrees:* DSc, Sci Univ Tokyo, 82. *Prof Exp:* from res assoc to prof, 70-90, UNIV DISTINGUISHED PROF MATH, MICH STATE UNIV, 90- *Concurrent Pos:* Ed, Tamkang J Math, 70-79, Soochow J Math, 78- & Bull Inst Math Acad Sinica, 79-, Algebras, Groups & Geometry, 88- *Mem:* Am Math Soc. *Res:* Differential geometry, global analysis and algebraic geometry. *Mailing Add:* Dept Math Mich State Univ East Lansing MI 48824-1027

CHEN, BENJAMIN P P, b Shanghai, China, Apr 8, 55; c 2. CELLULAR IMMUNOLOGY. *Educ:* Univ Wis, BSc, 76, MSc, 79, PhD(immunol), 82. *Prof Exp:* Res assoc human oncol, Wis Clin Cancer Ctr, 83-87; fel cell biol, Stanford Univ, 87-90; PROJ MGR IMMUNOL, SYSTEMIX, INC, 90- *Concurrent Pos:* Triton fel, Cetus Triton Res Orgn, 83-87; fel, Cancer Res Inst, Inc, 87-90. *Mem:* Am Soc Microbiologist; Am Asn Immunologist; Am Soc Cell Biol. *Res:* Roles of HCA molecules in antigen presentation; mechanism of human antibody production. *Mailing Add:* Systemix 3400 W Bayshore Rd Palo Alto CA 94303

CHEN, BENJAMIN YUN-HAI, b Taipei, Taiwan, May 3, 51; US citizen; m 79; c 2. NAVAL ARCHITECTURE, OCEAN & COASTAL ENGINEERING. *Educ:* Nat Taiwan Univ, BS, 74; Univ Del, MCE, 78, PhD(ocean & coastal eng), 82. *Prof Exp:* Res asst ocean & coastal eng, Dept Civil Eng, Univ Del, 76-82; eng consult, Coastal & Offshore Eng & Res Inc, Newark, Del, 82-83; res ocean engr, Naval Ocean Res & Develop Activ, Stennis Space Ctr, Miss, 83-84; RES NAVAL ARCHITECT, PROPULSOR TECHNOL, DAVID TAYLOR NAVAL SHIP RES & DEVELOP CTR, BETHESDA, MD, 84- *Concurrent Pos:* Regist prof engr, Md. *Mem:* Am Soc Mech Engrs; Am Soc Civil Engrs; Am Geophys Union; Sigma Xi. *Res:* Wave spectral transformation from deepwater to shallow water for the protection of offshore and coastal structures due to hurricane effects; effects of a real sea state on the offshore structures and sea floor; advanced marine propulsor technology. *Mailing Add:* Code 1544 David Taylor Naval Ship Res & Develop Ctr Bethesda MD 20084-5000

CHEN, C(HIH) W(EN), b Changsha, Hunan, China, Jan 2, 22; m 58; c 2. METALLURGY, MATERIALS SCIENCE. *Educ:* Chiaotung Univ, BS, 44; Ill Inst Technol, MS, 50; Columbia Univ, PhD(phys metall), 54. *Prof Exp:* Res assoc, Columbia Univ, 55-57; fel scientist, Westinghouse Res Lab, 57-66; from assoc prof to prof metall, Iowa State Univ, 66-81; staff, Lawrence Livermore Nat Lab, 81-90; RETIRED. *Concurrent Pos:* Metallurgist, Ames Lab, US Atomic Energy Comn, 66-70, sr metallurgist, Ames Lab, US Dept Energy, 71-81; consult, ITT, Metcal, 80-; Campbell fel, Columbia Univ. *Mem:* Am Phys Soc. *Res:* Crystal defects and radiation effects; magnetism and magnetic materials; thin films and amorphous alloys; electron microscopy. *Mailing Add:* Lawrence Livermore Nat Lab L-350 Livermore CA 94550

CHEN, CARL W(AN-CHENG), b Tainan, Taiwan, Feb 22, 36; US citizen; m 66; c 2. SANITARY ENGINEERING, POWER PLANT IMPACTS. *Educ:* Nat Taiwan Univ, BS, 58; Univ Calif, Berkeley, MS, 63, PhD(eng), 67. *Prof Exp:* Jr engr, Taiwan Pub Works Bur, 60-61, asst engr, 61-62; res asst, Univ Calif, Berkeley, 62-63, res specialist, 63-66, teaching asst, 66-67; assoc engr, Water Resources Engrs, Inc, 67-69, sr engr, 69-73; assoc dir, Tetra Tech, 73-75, dir, 75-78, vpres, 78-83; PRES, SYSTECH ENG, INC, 83- *Concurrent Pos:* Vis scholar, The Swedish Univ Agr Sci, Uppsala, Sweden, 80; mem, Pvt Indust Coun, Costa County, Calif. *Mem:* Am Soc Civil Engrs. *Res:* Data analyses and interpretation of water quality-ecologic responses of water resources systems to wastewater discharges, water resources development, and energy development projects such as power plants and coal mining; hydrodynamic-water quality-ecological simulation of Lake Ontario; simulation model of lake-watershed acidifications; simulation model of trees subjected to the real time stresses of ozone, acid deposition, and drought, storm water modeling; one US patent. *Mailing Add:* Systech Eng Inc 3744 Mt Diablo Blvd Lafayette CA 94549

CHEN, CATHERINE S H, b Chungking, China; US citizen. POLYMER CHEMISTRY. *Educ:* Barnard Col, BA, 50; Columbia Univ, MA, 52; Polytech Inst New York, PhD(polymer chem), 55. *Prof Exp:* Res assoc, Dept Chem, Columbia Univ, 54-56, Celanese Res Co, Celanese Corp, 64-70; sr res chemist, Cent Res Div, Am Cyanamid Co, 56-64; res scientist, Corp Res & Develop Lab, Singer Co, 72-75; res assoc, 75-78, mgr resources chem, Cent Res Div, 78-83, PROG LEADER, ENHANCED OIL RECOVERY, MOBIL RES & DEVELOP CORP, 83-, SR RES ASSOC, ZEOLITE CATALYSIS, 85- *Mem:* Am Chem Soc; Soc Petrol Engrs. *Res:* All aspects of polymer science; tertiary oil recovery chemicals; surfactants and mobility control agents, including polymers; zeolite catalysis. *Mailing Add:* 102 Pinegrove Rd Berkeley Heights NJ 07922-1492

CHEN, CHANG-HWEI, b Taiwan, Oct 12, 41; m 70; c 2. PHYSICAL CHEMISTRY, BIOPHYSICS. *Educ:* Nat Taiwan Univ, BS, 64; Univ Conn, PhD(phys chem), 70. *Prof Exp:* Res assoc chem, Univ Pittsburgh, 70-73; fel, 73-74, res scientist I, Wadsworth Ctr Labs & Res, NY State Dept Health, 74-75, res scientist II, 75-76, res scientist III, 77-80, res sceentist IV, 80-86, RES SCIENTIST V, CHEM, WADSWORTH CTR LABS & RES, NY STATE DEPT HEALTH, 87-; ASSOC PROF, DEPT BIOMED SCI & ADJ PROF, DEPT PHYSICS, STATE UNIV NY, ALBANY, 88- *Mem:* Am Chem Soc; Am Soc Biochem & Molecular Biol; Biophys Soc. *Res:* Physical chemical studies of proteins and lipids; membrane active transport, microcalorimetry; thermodynamics; alcohol biophysics. *Mailing Add:* Wadsworth Ctr Labs & Res NY State Dept Health Albany NY 12201-0509

CHEN, CHAO LING, b Tsaotun, Taiwan, Sept 28, 37; m 66; c 1. VETERINARY PHYSIOLOGY, ENDOCRINOLOGY. *Educ:* Nat Taiwan Univ, DVM, 60; Iowa State Univ, MS, 66; Mich State Univ, PhD(neuroendocrinol), 69. *Prof Exp:* Instr histol, Taipei Med Col, 61-63; asst vet, US Naval Med Res Unit, 63-64; res asst reproductive physiol, Iowa State Univ, 64-66; asst neuroendocrinol, Mich State Univ, 66-69; from asst prof to assoc prof physiol, Col Vet Med, Kans State Univ, 69-76; assoc prof, 76-80, PROF REPRODUCTION, UNIV FLA, 81- *Mem:* Soc Study Reproduction; Am Soc Animal Sci; Endocrine Soc; Int Soc Neuroendocrinol; Am Physiol Soc. *Res:* Hypothalamic regulation of the pituitary gonadotropin secretion by hypothalamic stimulation, lesion and sectioning and by radioimmunoassay; transformation of hypothalamic and pituitary cells. *Mailing Add:* Dept LACS J-136 JHMHC Col Vet Med Univ Fla Gainesville FL 32610-0136

CHEN, CHAO-HSING STANLEY, b Keelung, Taiwan, June 30, 33; US citizen; m 66; c 2. ENGINEERING MECHANICS. *Educ:* Nat Taiwan Univ, BS, 56; Univ RI, MS, 61; Univ Wis, PhD(eng mech), 65. *Prof Exp:* Eng scientist, BF Goodrich Co, 65-66, sr eng scientist, 66-77, eng assoc, 77-79, res & develop assoc, 79-84, SR RES & DEVELOP ASSOC, BF GOODRICH CO, 84- *Mem:* Sigma Xi; Tire Soc; Am Soc Composites. *Res:* Composite materials properties and behaviors and their applications to structures of composite such as laminated plates and shells, tire and inflatables. *Mailing Add:* 9921 Brecksville Rd Brecksville OH 44141

CHEN, CHARLES CHIN-TSE, b Taipei, Taiwan, May 22, 29; 86; c 3. GENERAL PHYSICS. *Educ:* Nat Taiwan Univ, BS, 51; Univ Md, PhD(physics), 62. *Prof Exp:* Instr physics, Nat Taiwan Univ, 51-57; asst, Univ Md, 57-62, res assoc, 63; from asst prof to assoc prof, 63-77, PROF PHYSICS, OHIO UNIV, 77- *Concurrent Pos:* Vis prof, Chubu Univ, Japan, 80. *Mem:* Am Phys Soc. *Res:* Condensed matter physics. *Mailing Add:* Dept Physics Ohio Univ Athens OH 45701

CHEN, CHARLES SHIN-YANG, b Changhua, Taiwan, Dec 25, 34; US citizen; m 63; c 2. MECHANICAL ENGINEERING, ENERGY CONVERSION. *Educ:* Purdue Univ, BS, 57; Univ Wash, MS, 65; Univ Wis, PhD(mech eng), 68. *Prof Exp:* Res engr bldg prod & insulation, Johns-Manville Corp, 57-60; sr res engr aircrafts, Boeing Co, 61-65; assoc prof mech eng, Univ Va, 68-72; prog mgr solar energy, NSF, 72-75; prog mgr solar energy & conserv, US Dept of Energy, 75-77; dir, Ctr Eng Res, Univ Hawaii, 77-80; vpres, Pac Opers, Ultrasysts, Inc, 80-83; dir, Alternative Energy, Am Diversified, 84-88; CONSULT, 88- *Concurrent Pos:* NASA grant hybrid rocket, 68-72; US Dept Interior grant water res, 68-72; consult air pollution, US Environ Protection Agency, 70-75; adj prof mech eng, Univ Miami, 75-76. *Honors & Awards:* Ralph R Teetor Award, Soc Automotive Engrs, 69; Dirs Commendation, NSF, 74. *Mem:* Assoc fel Am Inst Aeronaut & Astronaut; Am Soc Mech Engrs; Soc Automotive Engrs; Combustion Inst; Int Solar Energy Soc. *Res:* Alternative energy fields of solar, wind, ocean thermal, biomass and geothermal energy; combustion phenomena of hydrocarbon fuels; energy conservation in transportation; development of automotive gas turbine and Stirling engines; development of alternative fuels. *Mailing Add:* 3407 Quiet Cove Corona del Mar CA 92625

CHEN, CHEN HO, b Kiangsu, China, Sept 9, 29; US citizen; m 60; c 3. PLANT CYTOLOGY, PLANT PHYSIOLOGY. *Educ:* Nat Taiwan Univ, BS, 54; La State Univ, MS, 60; SDak State Univ, PhD(plant sci), 64. *Prof Exp:* Fel, SDak State Univ, 63-64 & Argonne Nat Lab, US Atomic Energy Comn, 64-65; lectr & head biol, Hong Kong Baptist Col, 65-68; from asst prof to assoc prof, 68-75, PROF BIOL & PLANT SCI, SDAK STATE UNIV, 75- *Mem:* Bot Soc Am; Tissue Cult Asn. *Res:* Development of cell and tissue culture techniques for use in breeding monocotyledonous species. *Mailing Add:* Agr Hall Box 2207 Brookings SD 57007

CHEN, CHENG HSUAN, b Kaohsiung, Taiwan, Sept 1, 46; m 71; c 2. APPLIED PHYSICS. *Educ:* Nat Taiwan Univ, BS, 68; Cornell Univ, PhD, 74. *Prof Exp:* Res assoc, Cornell Univ, 74-76; asst prof, Univ Ill, 76-78; MEM TECHNICAL STAFF, AT&T BELL LABS, 78- *Mem:* AAAS; Am Physical Soc. *Res:* Microstructure of high Tc superconductors, electron energy- loss spectroscopy. *Mailing Add:* 171 Pineway New Providence NJ 07974

CHEN, CHENG-LIN, b Mukden, China, Aug 14, 29; m 48; c 3. ELECTRICAL ENGINEERING. *Educ:* Taiwan Univ, BS, 53; Univ Ill, MS, 56, PhD(elec eng), 59. *Prof Exp:* Res assoc gaseous electronics, Univ Ill, 58-59, res asst prof plasma physics, Coord Sci Lab, 59-63; sr res scientist, Westinghouse Res Labs, 63-88; AT BROOKHAVEN NAT LAB. *Mem:* Am Phys Soc. *Res:* Gaseous electronics; plasma physics; atomic physics; magnetohydrodynamics and laser physics. *Mailing Add:* PO Box 625 Upton NY 11973

CHEN, CHENG-LUNG, b Taichung, Taiwan, Nov 1, 31; m 70; c 2. HYDRAULICS, HYDRAULIC ENGINEERING. *Educ:* Nat Taiwan Univ, BS, 54; Mich State Univ, MS, 60, PhD(hydraul), 62. *Prof Exp:* Res assoc civil eng, Mich State Univ, 62-63; assoc res engr, Utah State Univ, 63-64, from asst prof to assoc prof civil eng, 64-66; assoc prof civil eng, Univ Ill, Urbana-Champaign, 66-69; prof civil eng, Utah State Univ, 69-78; HYDROLOGIST, NAT RES PROG, WATER RESOURCES DIV, US GEOL SURV, 78- *Concurrent Pos:* Grad res asst, Mich State Univ, 58-62; prof on detail from Utah State Univ to Gulf Coast Hydrosci Ctr, US Geol Surv, 76-78. *Honors & Awards:* Henry S Kaminsky Award for Outstanding Res Contrib, Sigma Xi, 83. *Mem:* Am Soc Civil Engrs; Am Geophys Union; Int Asn Hydraul Res; Sigma Xi; Am Water Resources Asn; Int Water Resources Asn. *Res:* Sprinkler and surface irrigation; watershed and flood-plain hydraulics; soil moisture and ground water; fluvial hydraulics and sediment transportation; surface water and open-channel hydraulics; soil and thermal pollution; drainage and porous media flow; fluvial processes and river mechanics; hydrosystems analysis; flash floods and debris flows. *Mailing Add:* Water Resources Div-MS 496 US Geol Surv 345 Middlefield Rd Menlo Park CA 94025

CHEN, CHEN-TUNG ARTHUR, b Changhwa, Taiwan, Apr 22, 49; US citizen; m 76; c 2. DESCRIPTIVE CHEMICAL OCEANOGRAPHY, SEAWATER CARBONATE CHEMISTRY. *Educ:* Nat Taiwan Univ, BS, 70; Univ Miami, MS, 74, PhD(oceanog), 77. *Prof Exp:* From asst prof to assoc prof oceanog, Ore State Univ, 77-84; dir oceanog, Inst Marine Geol, 85-89, PROF OCEANOG, SUN YAT-SEN UNIV, 84-, DIR, CTR MARINE SCI RES & DEAN, COL MARINE SCI, 89- *Concurrent Pos:* Mem UNESCO/ICES/SCOR/IAPSO Joint Panel Experts on Oceanog Tables & Standards, 76-, sci comt oceanog res, 84-, exec mem, Int Ocean Technol Cong, 87-; vis researcher chem, Oak Ridge Nat Lab, 80, vis prof oceanog, Univ Paris VI, 85-; researcher, Sci & Technol Adv Group, 88-; chmn, Nat Global Change Comt, 90-; dir bd, Soc Ocean Sci & Technol, 90-; Joint Global Ocean Flux Studies Western Marginal Seas Info Exchange Group, 90-; ed, Sci Monthly J, Taiwan ROC, 91- *Mem:* Am Geophys Union. *Res:* Solution chemistry; high temperature chemistry; chemical oceanography; marine chemistry; water pollution; limnology; global change studies; seawater carbonate chemistry. *Mailing Add:* Inst Marine Geol Nat Sun Yat-Sen Univ Kaohsiung Taiwan

CHEN, CHIA HWA, b Hupei, China; US citizen. PHYSICAL GENERAL, NUCLEAR STRUCTURE. *Educ:* Nat Taiwan Univ, BS, 53, Nat Tsing Hua Univ, 59, State Univ NY, PhD(physics), 64. *Prof Exp:* From asst prof to assoc prof, 64-76, PROF PHYSICS, CALIF STATE UNIV, 76- *Mem:* Sigma Xi; Am Asn Physics Teachers; Math Asn Am. *Res:* Low energy nuclear physics. *Mailing Add:* Dept Physics-Astron Calif State Univ 1250 Bellflower Blvd Long Beach CA 90840

CHEN, CHIADAO, b Peiping, China, Dec 7, 13; nat US; m; c 1. BIOCHEMISTRY. *Educ:* Shanghai Univ, BS, 36; Dresden Tech Univ, Dipl Ing, 39; Berlin Tech, DSc(biochem), 41. *Prof Exp:* Res chemist, Ciba Pharmaceut, Inc, 42-44; sr res fel, Univ Pittsburgh, 44-48; res assoc & asst prof biochem, Univ Notre Dame, 48-49; assoc prof biochem, Med Sch, Northwestern Univ, Chicago, 49-67, prof, 67-83; RETIRED. *Concurrent Pos:* Biochemist & sect chief, Vet Admin Res Hosp, 54-56. *Mem:* AAAS; Am Chem Soc; NY Acad Sci; Am Soc Biol Chem; Brit Biochem Soc. *Res:* Steroid chemistry; oxygenation mechanism; chemical carcinogenesis; liver function. *Mailing Add:* 192 Spruce Ave Menlo Park CA 94025

CHEN, CHI-HAU, b Fukien, China, Dec 22, 37; m 66; c 2. INFORMATION & COMMUNICATION SCIENCES. *Educ:* Nat Taiwan Univ, BS, 59; Univ Tenn, Knoxville, MS, 62; Purdue Univ, PhD(elec eng), 65. *Prof Exp:* Sr res engr, ADCOM, Inc, 65-66; staff mem, Systs Div, AVCO Corp, 66-68; assoc prof, 68-73, chmn dept, 78-81, PROF ELEC ENG, SOUTHEASTERN MASS UNIV, 73- *Concurrent Pos:* Consult, Systs Div, AVCO Corp, Mass, 68-74; pres, Info Res Lab Inc, 83- *Honors & Awards:* Fel, Inst Elec & Electronics Engrs. *Mem:* Am Asn Artificial Intel; fel Inst Elec & Electronics Engrs; Soc Explor Geophysicists; Am Soc Eng Educ; Int Neural Network Soc. *Res:* Pattern recognition; information theory; communication theory and systems; adaptive and learning systems; digital signal processing; image processing and aritifical intelligence; neural network application to sonar classification. *Mailing Add:* Dept Elec & Comput Eng Southeastern Mass Univ North Dartmouth MA 02747

CHEN, CHIN HSIN, b Che-Kiang, China, Feb 7, 43; m 69. SYNTHETIC ORGANIC CHEMISTRY. *Educ:* Tunghai Univ, Taiwan, BS, 64; Okla State Univ, PhD(org chem), 71. *Prof Exp:* Fel org chem, Ohio State Univ, 71-72 & Harvard Univ, 72-73; sr res chemist, 73-80, RES ASSOC SYNTHETIC ORG CHEM, EASTMAN KODAK CO RES LAB, 80- *Mem:* Am Chem Soc; Sigma Xi. *Res:* Exploratory organic synthesis in the field of organosulfur and organophosphorus heterocyclic chemistry. *Mailing Add:* Eastman Kodak Co Res Lab 1669 Lake Ave Rochester NY 14650

CHEN, CHING-CHIH, b Foochow, China, Sept 3, 37; US citizen; m 61; c 3. INFORMATION SCIENCE. *Educ:* Nat Taiwan Univ, BA, 59; Univ Mich, AMLS, 61; Case Western Reserve Univ, PhD(info sci), 74. *Prof Exp:* Serv librn, Univ Mich, 61-62; sci ref librn, Windsor Pub Libr, Ont, 62; ref librn, McMaster Univ, 62-63, head sci librn, 63-64; sr sci librn, Univ Waterloo, 64-65, head librn, Eng, Math & Sci Libr, 65-68; assoc head sci libr, Mass Inst Technol, 68-71; from asst prof to assoc prof, 71-79, asst dean acad affairs, 77-79, PROF LIBR & INFORM SCI & ASSOC DEAN, SIMMONS COL, 79- *Concurrent Pos:* Consult, Unesco, Who, World Bank & others; vpres, Chen & Chen Consults, Inc. *Honors & Awards:* Gaylord Award, Libr & Info Technol Asn, 90. *Mem:* Am Soc Info Sci; Am Libr Asn; fel AAAS; Am Asn Lib Inform Sci Educ. *Res:* Scientific management, especially the application of modern analytic techniques in library problems; systems analysis; biomedical, scientific and technical library and information services and systems; new technology applications, including microcomputers and videodisc technology; author and editor of twenty-two books and numerous scholarly articles. *Mailing Add:* Grad Sch of Libr & Info Sci Simmons Col 300 The Fenway Boston MA 02115-5898

CHEN, CHING-JEN, b Taipei, Taiwan, July 6, 36; m 65; c 2. MECHANICAL ENGINEERING, APPLIED MATHEMATICS. *Educ:* Taipei Inst Technol, Taiwan, Dipl, 57; Kans State Univ, MS, 62; Case Inst Technol, PhD(mech eng), 67. *Prof Exp:* Design engr, Ta-Tong Grinding Co, Taipei, 59-60; res asst heat transfer, Kansas State Univ, 60-62; res asst fluid mech & heat transfer, Case Inst Technol, 62-67; from asst prof to assoc prof, 67-70, PROF MECH ENG, UNIV IOWA, 77-, CHMN, 82- *Concurrent Pos:* US Sr Scientist Award, Alexander von Humboldt Found, Ger, 74-75; hon prof, Wu-Hun Inst Hydraul & Elec Eng, Wu-Hun, China. *Honors & Awards:* Fel, Am Soc Mech Engrs. *Mem:* Am Soc Mech Engrs; Am Inst Aeronaut & Astronaut; Am Soc Eng Educ; Am Phys Soc; Int Asn Hydraul Res; Am Soc Civil Eng; Japan Soc Flow Visualization; Sigma Xi. *Res:* Fluid mechanics and heat transfer such as melting phenomena, two phase flow, turbulent flow, computational methods, heat convection, condensation and boundary layer phenomenon. *Mailing Add:* EB 2216 Mech Eng Univ of Iowa Iowa City IA 52242

CHEN, CHING-LING CHU, b Taipei, Taiwan, May 8, 48; m 76; c 2. MOLECULAR BIOLOBY, ENDOCRINOLOGY. *Educ:* Nat Taiwan Univ, BS, 71; Columbia Univ, MA, 75, PhD(biochem), 79. *Prof Exp:* Fel, Columbia Univ, 79, fel res, 80-82; SCI POP COUN, ROCKEFELLER UNIV, 82-, ASST PROF, 87- *Mem:* Am Chem Soc; Endocrine Soc; NY Acad Sci; AAAS. *Res:* Isolation and characterization of proopimelanocortin (precursor for adrenocorticortropic hormone-B-Endorphin) genes in rodent and human pituitanes; hormonal regulation of expression of these genes in rat pituitary and possible mechanisms involved in the hormone action; intragonadal regulation of the expression of opiomelonocortin and inhibin genes in different testicular cell types. *Mailing Add:* Pop Coun Rockefeller Univ 1230 York Ave New York NY 10021

CHEN, CHING-NIEN, b China, 45; m; c 1. IMAGE RECONSTRUCTION ALGORITHMS. *Educ:* Nat Taiwan Univ, BS, 67; Nat Tsing-Hua Univ, MS, 69; State Univ NY at Stony Brook, PhD(chem), 80. *Prof Exp:* Lectr, Gen Chem Lab, Nat Tsing-Hua Univ, Taiwan, 69-73; fel nuclear magnetic resonance imaging res, State Univ NY at Stony Brook, 80; vis fel, 80-81, expert nuclear magnetic resonance imaging, 81-88, PHYS SCIENTIST, MAGNETIC RESONANCE IMAGING, NIH, 88- *Concurrent Pos:* Lectr gen chem, Fu-Jen Catholic Univ, 68-69. *Mem:* Soc Magnetic Resonance Med; Am Asn Physicists Med; Inst Elec & Electronic Engrs. *Res:* Nuclear magnetic resonance imaging techniques; three-dimensional image reconstruction from plane-integrals; reconstruction algorithms; computer software in nuclear magnetic resonance imaging; radio frequency probes for nuclear magnetic resonance imaging; radio frequency power deposition in human. *Mailing Add:* Bldg 13 Rm 3W-13 Biomed Eng & Instrumentation Br NIH Bethesda MD 20892

CHEN, CHIN-LIN, b Honan, China, Mar 27, 37; m 67; c 2. ELECTRICAL ENGINEERING. *Educ:* Taiwan Univ, BS, 58; NDak State Univ, MS, 61; Harvard Univ, PhD(appl physics), 65. *Prof Exp:* Res fel electronics, Harvard Univ, 65-66; from asst prof to assoc prof, 66-78, PROF ELEC ENG, PURDUE UNIV, 78- *Mem:* Inst Elec & Electronics Engrs; Optical Soc Am. *Res:* Electromagnetic theory; antennas and diffraction of waves; surface acoustic wave devices, integrated optics. *Mailing Add:* Sch Elec Eng Purdue Univ West Lafayette IN 47907

CHEN, CHIOU SHIUN, b Taipei, Taiwan, Jan 22, 38; m 63; c 1. ELECTRICAL ENGINEERING, AUTOMATIC CONTROL SYSTEMS. *Educ:* Taiwan Univ, BS, 60; Univ Rochester, MS, 64, PhD(elec eng), 67. *Prof Exp:* Assoc engr, Taylor Instrument Co, 64-66; Nat Acad Sci-Nat Res Coun res assoc, Ames Res Ctr, NASA, 67-68; from asst prof to assoc prof, 68-78, PROF ELEC ENG, UNIV AKRON, 78-, HEAD DEPT, 83- *Mem:* Inst Elec & Electronics Engr; Sigma Xi. *Res:* Speech signal analysis and synthesis; microprocessor application in digital control; spectral estimation. *Mailing Add:* Dept Elec Eng Univ Akron Akron OH 44304

CHEN, CHI-PO, b Taiwan, China, May 14, 40; US citizen; m 68; c 2. PHARMACOLOGY, PHARMACY. *Educ:* Kaohsiung Med Col, Taiwan, BSc, 63; Queen's Univ, Ont, MSc, 69; Univ Ky, PhD(pharmacol), 73. *Prof Exp:* Chief pharmacist, Heng Hsin Pharmaceut Co, 64-67; res asst pharmacol, Univ Ky, 69-73; res assoc biochem, Vanderbilt Univ, 73-74; asst prof pharmacol, Univ Conn, Storrs, 74-80; ASSOC PROF PHARMACOL, OHIO COL PODIATRIC MED, CLEVELAND, 80- *Concurrent Pos:* Fac fel, Univ Conn, 75. *Mem:* AAAS; NY Acad Sci; Acad Pharmaceut Sci; Am Pharmaceut Asn; Am Soc Pharmacol & Exp Therapeut. *Res:* Transport through biological membranes with special interests in the area of drug transport in isolated liver cells, perfused liver, the choroid plexus and intestine. *Mailing Add:* Dept Pharmacol Ohio Col Podiatric Med 10515 Carnegie Ave Cleveland OH 44106

CHEN, CHI-TSONG, b Taiwan, Jan 7, 36; m 63; c 3. SYSTEMS THEORY, CONTROL SYSTEMS. *Educ:* Univ Calif, Berkeley, PhD(elec eng), 66. *Prof Exp:* From asst prof to assoc prof, 66-74, PROF ELEC ENG, STATE UNIV NY, STONY BROOK, 74- *Concurrent Pos:* Bell Lab, Grumman Aerospace Corp. *Mem:* Fel Inst Elec & Electronic Engrs. *Res:* Systems and control theory; digital signal processing. *Mailing Add:* Dept of Elec Eng State Univ of NY Stony Brook NY 11794

CHEN, CHONG MAW, b Taoyuan, Taiwan; m 66; c 3. PLANT BIOCHEMISTRY, MOLECULAR BIOLOGY. *Educ:* Taiwan Norm Univ, BS, 58; Univ Kans, MA, 64, PhD(plant biochem), 67. *Prof Exp:* Teaching asst biol, Taiwan Norm Univ, 61-62; fel biochem, McMaster Univ, 67-69; res fel, Roche Inst Molecular Biol, 69-71; from asst prof to prof, 71-88, WIS DISTINGUISHED PROF LIFE SCI, UNIV WIS-PARKSIDE, 88- *Concurrent Pos:* Grants, NSF, NIH, USDA & Off Naval Res. *Mem:* AAAS; Am Chem Soc; Am Soc Biol Chemists & Molecular Biol; Am Soc Plant Physiol; Int Soc Plant Molecular Biol. *Res:* Mechanism of action of plant hormones; plant genetic engineering; gene expression. *Mailing Add:* Dept Biol Sci Univ Wis-Parkside Kenosha WI 53141

CHEN, CHUAN FANG, b Tientsin, China, Nov 15, 32; nat US; m 57; c 3. MECHANICAL & AEROSPACE ENGINEERING. *Educ:* Univ Ill, BSc, 53, MSc, 54; Brown Univ, PhD(eng), 60. *Prof Exp:* Res scientist, Hydronautics, Inc, 60-61, sr res scientist, 61-62; asst to chief engr, 62-63; from asst prof to prof mech eng, Rutgers Univ, 63-80, chairperson, Dept Mech, Indust & Aerospace Eng, 76-80; head, 80-89, PROF, DEPT AEROSPACE & MECH ENG, UNIV ARIZ, 80- *Concurrent Pos:* Consult, Hydronautics, Inc, Vitro Labs & Bard Inc; sr visitor, Dept Appl Math & Theoretical Physics, Cambridge Univ, England, 71-72, 87; vis fel, Res Sch Earth Sci, Australian Nat Univ, 78. *Mem:* Assoc fel Am Inst Aeronaut & Astronaut; fel Am Phys Soc; Am Soc Eng Educ; fel Am Soc Mech Eng; fel AAAS. *Res:* Supersonic interference; supercavitating flow past hydrofoils; heat transfer to gas-solid suspensions; stratified flows; hydrodynamic stability; double diffusive convection. *Mailing Add:* Dept Aerospace & Mech Eng Univ Ariz Tucson AZ 85721

CHEN, CHUAN JU, b Tainan, Taiwan, Mar 21, 47; m 72; c 2. POLYMER SCIENCE ENGINEERING. *Educ:* Chung Yuan Univ, BE, 70; Worcester Polytech Inst, MS, 74; Univ Mich, PhD(polymer sci & eng), 80. *Prof Exp:* SR RES SPECIALIST, MONSANTO CO, 80- *Mem:* Sigma Xi; Am Chem Soc; Soc Plastics Engrs. *Res:* Products development in styrenic and alloys polymers; structure-property-processing relationships of polymers. *Mailing Add:* Monsanto Co 730 Worcester St Indian Orchard MA 01151

CHEN, CHUNG WEI, b Hunan, China, Dec 25, 21; US citizen; m 65; c 2. APPLIED STATISTICS. *Educ:* Nat Chengchi Univ, China, BA, 46, LLB, 48; La State Univ, MBA, 54, PhD(statist), 57. *Prof Exp:* Secy, Human Prov Govt, 46-47; judge, Hunan Changsha Dist Ct, 47-48; statistician, Haloid-Xerox, Inc, 56-61, chief statistician, 61-72, mgr statist serv, 72-76, sr consult corp staff, 76-80, PRIN CONSULT, XEROX CORP, 80- *Concurrent Pos:* Instr, Rochester Inst Technol, 59-67. *Mem:* Am Statist Asn; Am Econ Asn. *Mailing Add:* Xerox Corp El Segundo CA 90245

CHEN, CHUNG-HO, b Kaohsiung, Taiwan, Dec 1, 37; US citizen; m 65; c 4. BIOCHEMISTRY. *Educ:* Chung-Hsing Univ, BS, 62; Okla State Univ, PhD(biochem), 69. *Prof Exp:* Res asst pharmacol, Kaohsiung Med Col, 63-64; asst biochem, Okla State Univ, 64-69; ASST PROF OPHTHALMIC BIOCHEM, SCH MED, JOHNS HOPKINS UNIV, 73- *Concurrent Pos:* Fel biochem, Okla State Univ, 69; fel, Sch Med, Johns Hopkins Univ, 69-73, consult ophthal, 72-73; Nat Eye Inst res grant, 75-; Am Diabetes Asn grant, 75- *Honors & Awards:* Mayers Award, 75. *Mem:* AAAS; Am Chem Soc; Asn Clin Scientists; NY Acad Sci; Asn Res Vision & Ophthal. *Res:* Ocular biochemistry; enzymatic activities in mitochondria; metabolic regulations and disorders; angiogenesis; membrane chemistry; active transport. *Mailing Add:* Woods 363 JHH 601 N Wolfe St Baltimore MD 21205-2103

CHEN, CHUNG-HSUAN, b Amoy, China, Jan 4, 48; US citizen; m 74; c 2. LASER PHYSICS, SUPERCONDUCTIVITY. *Educ:* Nat Taiwan Univ, BS, 69; Univ Chicago, MS, 71, PhD(chem), 74. *Prof Exp:* Res asst chem, Univ Chicago, 70-74 & atomic physics, Univ Ky, 74-76; sr scientist, 76-89, GROUP LEADER PHOTOPHYSICS, OAK RIDGE NAT LAB, 89- *Concurrent Pos:* Adj prof, Physics Dept, Vanderbilt Univ & Univ Tenn, 90- *Honors & Awards:* Res & Develop 100 Award, Indust Res Mag, 84 & 87. *Mem:* Am Phys Soc; Am Chem Soc; Inst Elec & Electronics Engrs. *Res:* Atomic, molecular and laser physics; chemical kinetics and chemical dynamics; solid state chemistry; environmental instrumentation development; superconductivity; DNA mapping and sequency; nuclear magnetic resonance. *Mailing Add:* 1812 Plumb Branch Rd Knoxville TN 37932

CHEN, CINDY CHEI-JEN, b Taipei, Taiwan, Repub of China, Nov 8, 53; US citizen; m; c 1. XEROGRAPHY, TONER. *Educ:* Nat Tsing Hua Univ, Taiwan, Repub of China, BS, 76; Polytech Inst, 82, PhD(phys chem), 84. *Prof Exp:* Asst researcher instrumentation Precision Instrument Ctr, Nat Sci Coun, Repub of China, 76-78; teaching asst & instr org & anal, Polytech Inst, NY, 79-81, grad student phys chem, 79-84; postdoctoral fel kinetics, 84-85, mem res staff electrography, 85-90, TECH SPECIALIST & PROJ MGR XEROGRAPHY, XEROX CORP, 90- *Mem:* Am Chem Soc. *Res:* Complexation of crown ether with metal ions in solvent system; interfacial science and technological issues which related to xerography; electrical and mechanical properties of photoreceptor devices for future photoreceptor design; polymer coating formulation and coating process control. *Mailing Add:* Eight Bethnal Green Rochester NY 14625

CHEN, DAVID J, b China, Dec 6, 44; m 71; c 1. GENETIC TOXICOLOGY, CELL GENETICS. *Educ:* Nat Taiwan Univ, BS, 68; Univ Mo, PhD(genetics), 78. *Prof Exp:* Fel, 78-80, MEM STAFF, LOS ALAMOS NAT LAB, 80- *Concurrent Pos:* Prin investr, Los Alamos Nat Lab, 80- *Mem:* Am Soc Cell Biol; Environ Mutagen Soc; Tissue Culture Asn; Radiation Res Soc. *Res:* Study of mechanisms of DNA repair in mammalian cells; development of human cell mutagenesis assays for detection of chemical and physical mutagens-carcinogens; study of mechanisms of mutation induction by radiations. *Mailing Add:* Genetics Group MS886 Life Sci Div Los Alamos Nat Lab Los Alamos NM 87545

CHEN, DAVIDSON TAH-CHUEN, b Wenling, China, Apr 1, 42; US citizen; m 66; c 2. RESEARCH PHYSICIST, REMOTE SENSING. *Educ:* Nat Taiwan Univ, BS, 63; Univ Calif, Berkeley, MS, 66; NC State Univ, PhD(phys oceanog), 72. *Prof Exp:* Consult engr struct dynamics, John A Blume & Assoc Engrs, 66-67; Nat Res Coun resident res assoc phys oceanog & remote sensing, Wallops Flight Ctr, NASA, 72-74; phys oceanogr, 74-81, res physicist, 81-87, phys sci adminr, Off Naval Res, 87-90, PROG MGR, NAVAL RES LAB, 90- *Concurrent Pos:* Organizer, John W Wright Mem Lect, 80-; pres, Chinese-Am Prof Asn Metrop Washington, DC, 87-88. *Mem:* Am Geophys Union. *Res:* Developing mathematical codes, based upon fluid dynamics and physics of electromagnetic waves, for physical parameters which describe important and meaningful geophysical phenomena; developing microwave remote sensors for inferring the measurements of these physical parameters; developing satellite technology. *Mailing Add:* Code 8301 Naval Res Lab Washington DC 20375-5000

CHEN, DI, b Chekiang, China, Mar 15, 29; m 58; c 2. ELECTRICAL ENGINEERING, MAGNETISM. *Educ:* Nat Taiwan Norm Univ, BS, 53; Univ Minn, Minneapolis, MS, 56; Stanford Univ, PhD(elec eng), 59. *Prof Exp:* Teaching asst elec eng, Nat Taiwan Univ, 53-54; asst prof, Univ Minn, Minneapolis, 59-62; sr res scientist, Honeywell Corp Tech Ctr, 62-65, sr prin res scientist, 66-69, staff scientist & group leader, 69-79, fel, 79-80; technol dir, Advan Mem Lab, Magnetic Peripherals, Inc, 80-82, Optical Peripherals Lab, 83-84; exec vpres & chief tech officer, Optotech, Inc, 84-89; VPRES ENG, LITERAL CORP, 90- *Honors & Awards:* Honeywell Sweatt Eng & Scientist Award, 72. *Mem:* Am Phys Soc; fel Inst Elec & Electronics Engrs. *Res:* Ferrites; ferromagnetic resonance of magnetic films; microwave magnetron research; lasers; magneto-optic effects; microwave modulation of light; optical memory; optical communication and integrated optics. *Mailing Add:* 302 Sunbird Cliffs Lane Colorado Springs CO 80919

CHEN, DILLION TONG-TING, plant anatomy, herbs, for more information see previous edition

CHEN, EDWARD CHUCK-MING, b Houston, Tex, July 30, 37; m 60; c 2. ELECTRON CAPTURE DETECTOR. *Educ:* Rice Univ, BA, 59; Univ Houston, PhD(chem), 66. *Prof Exp:* Res chemist, Ashland Chem-United Carbon, 66-68; chief chemist, Signal Chem, Houston, 68-72; asst prof, Univ Houston, Victoria, 72-75, from asst prof to assoc prof, 75-82, PROF CHEM, UNIV HOUSTON, CLEAR LAKE, 82- *Concurrent Pos:* Consult, Goodyear Chem, Bayport, 72-76 & Chem Exchange, 76- *Mem:* Am Chem Soc; Am Inst Chemists. *Res:* Reactions of thermal electrons; negative ion mass spectrometry; industrial chemical processes. *Mailing Add:* 4046 Durness Houston TX 77025

CHEN, EDWIN HUNG-TEH, b Tainan, Taiwan, Aug 23, 34; US citizen; m 66; c 2. BIOSTATISTICS. *Educ:* Nat Taiwan Univ, BS, 57; Mich State Univ, MS, 64; Univ Calif, Los Angeles, PhD(biostatist), 69. *Prof Exp:* Systs analyst statist, Union Am Comput Corp, 64-65; asst res statistician & lectr biostatist, Univ Calif, Los Angeles, 69-72; assoc prof, 72-78, PROF BIOMET, UNIV ILL, CHICAGO, 78- *Mem:* Am Statist Asn; Inst Math Statist; Biomet Soc. *Res:* Data analysis; robust statistical procedures; statistical computing; computer simulation. *Mailing Add:* Sch Pub Health m/c 925 Univ Ill PO Box 6998 Chicago IL 60680

CHEN, ERH-CHUN, b Kwangtung, China, Nov 11, 34; Can citizen; m 64; c 2. CHEMICAL ENGINEERING, ENVIRONMENTAL ENGINEERING. *Educ:* Nat Taiwan Univ, BEng, 57; McGill Univ, MEng, 64, PhD(chem eng), 68. *Prof Exp:* Shift supvr mfg process, Taiwan Fertilizer Corp, 59-62; res engr hydrometall, Que Iron & Titanium Corp, 62-71; res scientist environ pollution, 71-80, DIR WATER POLLUTION CONTROL ENG, FISHERIES & ENVIRON CAN; SR PROG ENGR, CONSERV & PROTECTION, ENVIRON CAN, 80- *Mem:* Chem Inst Can; Can Soc Chem Eng. *Res:* Oil pollution in cold environment; wastewater treatment and industrial pollution. *Mailing Add:* Conserv & Protection Environ Can Ottawa ON K1A 0H3 Can

CHEN, ER-PING, b Kansu, China, May 19, 44; US citizen; m 73; c 2. FRACTURE MECHANICS. *Educ:* Chung-Hsing Univ, Taiwan, BS, 66; Lehigh Univ, MS, 69, PhD(appl mech), 72. *Prof Exp:* From asst prof to assoc prof mech, Lehigh Univ, 72-78; MEM TECH STAFF, SANDIA LABS, 78- *Mem:* Sigma Xi; Am Soc Mech Engrs. *Res:* Dynamic fracture prediction for geological materials; rock mechanics modelling for waste isolation problems; penetration and perforation. *Mailing Add:* Dept Energy Sandia Nat Labs Div 1514 PO Box 5800 Albuquerque NM 87185

CHEN, FANG SHANG, b Taipeh, Formosa, Feb 25, 28; m 61. ELECTRICAL ENGINEERING. *Educ:* Taiwan Univ, BSEE, 51; Purdue Univ, MSEE, 55; Ohio State Univ, PhD(elec eng), 59. *Prof Exp:* Elec engr, Taiwan Power Co, 51-54; asst, Electron Devices Lab, Ohio State Univ, 55-57, res assoc, 57-59; mem tech staff, Bell Labs, 59-88; RETIRED. *Mem:* mem Inst Elec & Electronics Engrs. *Res:* Microwave electron tubes; solid state devices, especially ferrite devices and masers; optical communication devices; Ga As devices. *Mailing Add:* 198 Oakwood Dr New Providence NJ 07974

CHEN, FRANCIS F, b Canton, China, Nov 18, 29; US citizen; m 56; c 3. ENGINEERING PHYSICS, GENERAL PHYSICS. *Educ:* Harvard Univ, AB, 50, MA, 51, PhD(physics), 54. *Prof Exp:* Res assoc physics, Brookhaven Nat Lab, 53-54; res staff mem, Plasma Physics Lab, Princeton Univ, 54-64, res physicist, 64-69, sr res physicist, 69; PROF ENG & APPL SCI, UNIV CALIF, LOS ANGELES, 69- *Concurrent Pos:* Attend physicist, Nuclear Res Ctr, Fontenay, France, 62-63; chmn fusion adv comt, Elec Power Res Inst, 73-75; consult, TRW Inc, 77-87; chmn, Plasma Phys Div, Am Phys Soc, 83. *Mem:* Fel Am Phys Soc; fel Inst Elec & Electronics Engrs; NY Acad Sci; AAAS; Sigma Xi. *Res:* Basic plasma physics; fusion reactors; laser-plasma interactions; plasma accelerators. *Mailing Add:* 56-125B Engr IV Univ of Calif Los Angeles CA 90024

CHEN, FRANCIS HAP-KWONG, b Shanghai, China, Mar 14, 48; US citizen; m; c 1. NOISE & VIBRATIONS, STRUCTURAL ANALYSIS. *Educ:* Lafayette Col, BS, 71; Univ Ill, Urbana, MS, 73, PhD(theoret & appl mech), 76. *Prof Exp:* Assoc sr res engr, 76-79, staff res engr, 79-85, SR STAFF RES ENGR, GEN MOTORS RES LAB, GEN MOTORS CORP, 85- *Concurrent Pos:* Adj prof acoust & eng mech, Penn State Univ, State Col, Pa, 90. *Mem:* Am Soc Mech Engrs; Soc Automotive Engrs; Sigma Xi; Acoust Soc Am. *Res:* Analytical and experimental structural mechanics, with emphasis on: structural dynamics, system modeling acoustic BEM methods and applications of digital signal processing techniques to noise and vibration problems. *Mailing Add:* Eng Mech Dept Gen Motors Res Lab GM Tech Ctr Warren MI 48090-9055

CHEN, FRANCIS W, immunochemistry, for more information see previous edition

CHEN, FRANKLIN M, b Chang-Hua, Taiwan, Sept 19, 46; m 74; c 2. PHYSICAL CHEMISTRY, BIOCHEMISTRY. *Educ:* Taiwan Univ, BS, 70; Princeton Univ, MS, 73, PhD(phys chem), 76. *Prof Exp:* Res assoc enzyme chem, Rutgers Univ, 76-77; res chemist res & develop, Colgate & Palmolive Co, 77-80; Johnson & Johnson, 80-84; KIMBERLY-CLARK, 84- *Mem:* AAAS; Am Chem Soc; Sigma Xi. *Res:* Colloid and surface chemistry. *Mailing Add:* 1820 W Glendale Ave Appleton WI 54915

CHEN, FRED FEN CHUAN, b Hukou, Taiwan, Nov 20, 34; m 60; c 2. CHEMICAL ENGINEERING. *Educ:* Taiwan Norm Univ, BS, 57; Ga Inst Technol, MS, 61; Va Polytech Inst, PhD(chem eng), 67. *Prof Exp:* Jr process engr, Taiwan Fertilizer Co, 57-59; engr, Celanese Fibers Co Div, Celanese Corp Am, 60-62, develop engr, 62-64, sr develop engr, 64; res chem engr, Eastman Kodak Co, 66-68, sr res chem engr, 68-71, res assoc, 72-76, asst supt, 76-79, sr res assoc, 79-81, asst supt, Tenn Eastman Co Div, 81-86, develop fel, Carolina Eastman Co Div, 86-90; DIR, EASTMAN CHEM CO, 90- *Mem:* Am Chem Soc; Am Inst Chem Engrs; Am Asn Textile Technol; Sigma Xi; Soc Plastic Engrs. *Res:* Polymer engineering; rheology; polymerization kinetics; synthetic fiber technology. *Mailing Add:* Eastman Chem Int Ltd No 09-01 World Trade Ctr Singapore Singapore

CHEN, FU-TAI ALBERT, b Miaoli, Taiwan, Oct 26, 44; US citizen; m 72; c 2. CLINICAL DIAGNOSTIC APPLICATIONS CAPILLARY ELECTROPHORESIS. *Educ:* Nat Taiwan Univ, BSc, 67; Univ Notre Dame, PhD(org chem), 72. *Prof Exp:* Res fel path, Harbor Gen Hosp, Univ Calif Los Angeles, 75-77; head chem dept, Bio-Tech, 77-79; res scientist, Beckman Microbics of Beckman Instruments Inc, 79-82; PRIN RES SCIENTIST, PROTEIN & ORG CHEM, BECKMAN ADVAN TECHNOL, BECKMAN INSTRUMENTS INC, 82- *Concurrent Pos:* Post doctoral biochem, Univ Wis-Madison, 72-75; Am Heart Asn advan res fel, Dept Path, Univ Calif Los Angeles, 75 & 76. *Mem:* Am Chem Soc; Am Asn Clin Chem. *Res:* Capillary electrophoretic analysis of proteins; light scattering based macromolecular characterization; reaction kinetics of immunochemical reaction; general protein analysis by capillary electrophoresis and microbore high-performance liquid chromatography. *Mailing Add:* 105 S Cristal Springs Ct Brea CA 92621

CHEN, GOONG, b Kaosiung, Taiwan, July 7, 50; c 2. ANALYSIS & FUNCTIONAL ANALYSIS. *Educ:* Univ Wis, PhD(appl math), 77. *Prof Exp:* Asst prof math, Southern Univ, 77-78; asst prof, 78-82, assoc prof, math, Pa State Univ, 82-87; PROF MATH & AEROSPACE ENG, TEX A&M UNIV, 87- *Concurrent Pos:* Vis prof, IRIA, Versailles, France, Univ Montreal, 83. *Mem:* Am Math Soc; Soc Industr Appl Math. *Res:* Control & stabilization of partial differential numerical solutions; computer & experimental methods. *Mailing Add:* Dept Math Texas A&M Univ College Station TX 77843

CHEN, H R, b Puten, Fukien, China, June 13, 29; m 60; c 3. DNA, RNA. *Educ:* Taiwan Normal Univ, BS, 52; Univ SC, MS, 60; Yale Univ, PhD(biol), 64. *Prof Exp:* Res assoc biol, Yale Univ, 64-66; asst prof, WVa Univ, 66-69; assoc prof, Univ SC, 70-74; programmer & analyst database, Duke Power Co, 78-79; SR SCIENTIST, DNA, RNA & PROTEIN, NAT BIOMED RES FOUND, GEORGETOWN UNIV MED CTR, 80- *Concurrent Pos:* Vis prof, NC State Univ, 76-77. *Res:* Analyzing sequence data of nucleic acids and proteins in the computerized databases. *Mailing Add:* Nat Biomed Res Found Georgetown Univ Med Ctr 3900 Reservoir Rd Washington DC 20007

CHEN, HAO-CHIA, b Taiwan, China, Feb 3, 35; US citizen; m 65; c 2. ENDOCRINOLOGY, STRUCTURAL CHEMISTRY. *Educ:* Nat Taiwan Univ, BS, 57, MS, 59; Emory Univ, PhD(biochem), 64. *Prof Exp:* Guest investr biochem, Rockefeller Univ, 64-65, res assoc, 65-69, asst prof, 69-72; SR INVESTR CHEM, NAT INST CHILD HEALTH & HUMAN DEVELOP, 72, CHIEF, SECT MOLECULAR STRUCTURE & PROTEIN CHEM, 84- *Concurrent Pos:* Joseph B Whitehead fel, Emory Univ Sch Med, 64-65; hon prof, Beijing Med Univ, People's Rep China, 88- *Mem:* Am Soc Biol Chemists; Endocrine Soc; Am Chem Soc; Protein Soc. *Res:* Chemistry of polypeptides important on reproductive biology; structure and function of gonadotrophins; structure-activity and mechanism of action of peptide antibiotics. *Mailing Add:* Endocrinol & Reproduction Res Br NICHD NIH Bldg Six Rm 2A13 Bethesda MD 20892

CHEN, HARRY WU-SHIONG, b Kaohsiung, Taiwan, June 17, 37; m 69; c 2. CELL BIOLOGY, BIOCHEMISTRY. *Educ:* Nat Chengchi Univ, BA, 60; Univ Kans, PhD(biochem), 69. *Prof Exp:* Fel, 69-70, assoc staff scientist, 70-74, staff scientist, 74-79, sr staff scientist, Jackson Lab, 79-84; PRIN INVESTR, CARDIOVASC SCI, MED PROD DEPT, DU PONT CO. *Mem:* Am Soc Biol & Molecular Biol. *Res:* Regulation of cell growth; function of cholesterol in the surface membranes of mammalian cells. *Mailing Add:* Med Prod Dept E I du Pont de Nemours & Co E 400-3227 Exp Sta Wilmington DE 19898

CHEN, HENRY LOWE, b Shanghai, China, Mar 40; US citizen; c 1. PHYSICS. *Educ:* Harvard Col, BA, 63; Johns Hopkins Univ, MA, 69; Univ Md, PhD(physics), 83. *Prof Exp:* ASSOC PROF PHYSICS, TOWSON STATE UNIV, 65- *Mem:* Am Phys Soc. *Res:* Biology and physics. *Mailing Add:* Dept Physics Towson State Univ Baltimore MD 21204

CHEN, HO SOU, b Taiwan, Nov 24, 32; US citizen; c 2. SOLID STATE PHYSICS. *Educ:* Nat Taiwan Univ, BS, 56; Brown Univ, MS, 63; Harvard Univ, PhD(appl physics), 67. *Prof Exp:* Mem tech staff mat, Bell Tel Labs, 68-70; asst prof chem, Yeshiva Univ, 70-71; physicist, Allied Chem Corp, 71-72; MEM TECH STAFF MAT, BELL LABS, 72- *Honors & Awards:* George W Morey Award, Am Ceramic Soc, 78. *Mem:* Fel Am Phys Soc. *Res:* Structure, thermal and physical properties of metallic glasses and quasicrystals. *Mailing Add:* AT&T Bell Labs 600 Mountain Ave Murray Hill NJ 07974

CHEN, HOFFMAN HOR-FU, b Hualien, Taiwan, Aug 5, 41; US citizen; m 73; c 3. CATALYSIS. *Educ:* Tamkang Univ, BS, 63; Univ Tex, Austin, PhD(org chem), 76. *Prof Exp:* Res assoc org chem, Univ Tex, Austin, 76-77 & Okla State Univ, 77-78; ASSOC PROF ORG CHEM & ENVIRON CHEM, GRAMBLING STATE UNIV, 79- *Concurrent Pos:* Vis scientist catalysis, Brookhaven Nat Lab, 82, 83 & 84, Argonne Nat Lab, 85 & Sch Aerospace Med, 86. *Mem:* Am Chem Soc; AAAS. *Res:* Chemistry of the homogeneous and heterogeneous catalysis such as New Fridel-Crafts reactions, Fenton reaction, Fischer-Tropsch reaction; the reactions related to artificial blood as an oxygen carrier. *Mailing Add:* 707 Tarreyton Dr Ruston LA 71270-2736

CHEN, HOLLIS C(HING), b Chekiang, China, Nov 17, 35; m 61; c 2. ELECTRICAL ENGINEERING, APPLIED MATHEMATICS. *Educ:* Nat Taiwan Univ, BSEE, 57; Ohio Univ, MS, 61; Syracuse Univ, PhD(elec eng), 65. *Prof Exp:* Teaching asst physics, Taiwan Univ, 57-60; res asst, Ohio Univ, 60-61; instr, Syracuse Univ, 61-62, res assoc, 62-65, res fel, 65-67; from asst prof to assoc prof, 67-75, PROF ELEC ENG, OHIO UNIV, 75- *Concurrent Pos:* Mem Comn B, Int Sci Radio Union; NSF grant; Ohio Univ res comts grants; chmn, Dept Elec & Comput Eng, Ohio Univ, 84-86. *Mem:* AAAS; Inst Elec & Electronics Engrs; Am Asn Physics Teachers; Optical Soc Am; Am Geophys Union. *Res:* Electromagnetic radiation in a moving or plasma medium; applied mathematics; electromagnetic theory; plasma physics; applied optics. *Mailing Add:* Dept Elec & Comput Eng Ohio Univ Athens OH 45701-2979

CHEN, HSIEN-JEN JAMES, b Kwangtung, China, June 19, 31; m 71; c 1. NEUROENDOCRINOLOGY. *Educ:* Nat Taiwan Normal Univ, BS, 65, Nat Taiwan Univ Med Col, MS, 69; Mich State Univ, PhD(physiol), 76. *Prof Exp:* Res asst physiol, Mich State Univ, 71-76; fel neuroendocrinol, Mt Sinai Hosp, 76-78, dept anat, Univ Tex Health Sci Ctr, 78-79; asst neuroendocrinologist, Dept Neurobiol, Barrow Neurol Inst, St Joseph Hosp Med Ctr, 79-88; RETIRED. *Concurrent Pos:* Adj asst prof, dept anat, Univ Ariz Health Sci Ctr, 79- *Mem:* Endocrine Soc; Am Physiol Soc; Soc Neurosci; Soc Study Reproduction. *Res:* Neuroendocrine control of pituitary thyroid and pituitary testicular function; effects of pituitary and ovarian hormone on mammary cancers; pineal control of reproduction. *Mailing Add:* 2844 E Cortez St Phoenix AZ 85028

CHEN, HSING-HEN, b Yun Kang, Chekiang, China, May 7, 47; m 71; c 3. PHYSICS, ASTRONOMY. *Educ:* Nat Taiwan Univ, BS, 68; Columbia Univ, MA, 70, PhD(physics), 73. *Prof Exp:* Mem, Sch Natural Sci, Inst Advan Study, 73-75; from vis asst prof to assoc prof, 75-85, PROF PHYSICS & ASTRON, UNIV MD, COLEGE PARK, 85- *Mem:* Am Phys Soc. *Res:* Nonlinear plasma physics; solitons; chaos; laser fusions; plasma astrophysics. *Mailing Add:* Laboratory Plaza Univ Md College Park MD 20742

CHEN, HSUAN, b Shanghai, China, Jan 10, 36; US citizen; m 66; c 2. OPTICS, PARTICLE PHYSICS. *Educ:* Nat Taiwan Univ, BS, 57; Univ Minn, PhD(physics), 70; Univ Mich, MS, 75. *Prof Exp:* Assoc prof, 69-80, PROF PHYSICS, SAGINAW VALLEY STATE COL, 80- *Concurrent Pos:* Res assoc physics, Univ Mich, 72, fel, 74; vis res scientist optics, Univ Mich, 77-78, 79 & 80. *Mem:* Am Phys Soc; Am Optical Soc. *Res:* White light transmission holography and optical processes; high energy neutrino physics. *Mailing Add:* 5200 Nakoma Dr Midland MI 48640

CHEN, HUBERT JAN-PEING, b Kiangsu Prov, China, Oct 29, 42; c 1. STATISTICS. *Educ:* Nat Taiwan Univ, BA, 67; Univ Rochester, MA, 71, PhD(statist), 73. *Prof Exp:* Asst teacher statist, Nat Taiwan Univ, 67-69; teaching asst, Univ Rochester, 69-72; lectr, Ohio State Univ, 72-73; asst prof statist, Memphis State Univ, 73-76; asst prof statist, 76-79, dir consult serv, 78-80, ASSOC PROF, UNIV GA, 79- *Concurrent Pos:* Mem staff, Math Rev, 72- *Mem:* Am Statist Asn; Inst Math Statist. *Res:* Estimation of ranked parameters; ranking and selections; multiple comparisons; regression analysis; scientific computations; simulation and Monte Carlo methods. *Mailing Add:* Dept Statist Univ Ga Athens GA 30602

CHEN, INAN, b Tainan, Taiwan, Oct 9, 33; m 59; c 2. AMORPHOUS SEMICONDUCTORS, ELECTROPHOTOGRAPHY. *Educ:* Nat Taiwan Univ, BS, 56; Tsinghua Univ, China, MS, 58; Univ Mich, PhD(nuclear sci), 64. *Prof Exp:* Res assoc nuclear sci, Univ Mich, 64-65; from scientist to sr scientist, 65-71, PRIN SCIENTIST, RES LABS, XEROX CORP, 71- *Concurrent Pos:* Vis scholar, Tokyo Inst of Technol, 87. *Mem:* Am Phys Soc; Soc Photog Scientists & Engrs. *Res:* Theoretical studies of electronic states in solids; theoretical studies of electronic properties of amorphous and molecular solids, and their applications to electrcophotographic processes. *Mailing Add:* Xerox Corp 114-41D 800 Phillips Rd Webster NY 14580

CHEN, ISAAC I H, b Tainan, Taiwan, Mar 2, 29; US citizen; m 54; c 4. MICROCIRCULATION, CARDIOVASCULAR PHYSIOLOGY. *Educ:* Nat Taiwan Univ, BS, 53; NC State Univ, MS, 65, PhD(physics), 75. *Prof Exp:* Teaching asst physics, Nat Cheng-Kung Univ Taiwan, 55-60, instr, 60-63, assoc prof theoret physics, 66-67; teaching asst physics, NC State Univ, 68-74; asst prof math, Miss Indust Col, 74-77; asst prof physics, Rust Col, Miss, 77-78; res assoc physiol, Univ Tenn, 79-81; fel, La State Univ, Alexandria, 81-82, asst prof physiol, med ctr, 82-88, asst prof physics, 88-89, ASSOC PROF PHYSICS, LA STATE UNIV, ALEXANDRIA, 90- *Concurrent Pos:* Fel, Nat Heart, Lung & Blood Inst, NIH, 80-82; biomed res grant, Biomed Res Grant Found, La, 84; LA Inc Res Grant, Am Heart Asn, 84 & 87. *Res:* Hemodynamical alteration in hypertension; investigate and quantitatively characterize some of the important structural alterations which occur in hypertensive microvascular network; measurement of vessel surface area density, wall-lumen ratio by television microscopy with application of quantitative stereology; use of mathematical analysis of non-Newtonian blood flow; role of vascular alteration in hypertension. *Mailing Add:* Physics Div Sci La State Univ Alexandria Alexandria LA 71302-9633

CHEN, I-WEN, b Tokyo, Japan, Aug 3, 34; US citizen; m 64; c 3. RADIOBIOLOGY, BIOCHEMISTRY. *Educ:* Nat Taiwan Univ, BS, 56; Univ Pa, PhD(biochem), 64. *Prof Exp:* From asst prof to assoc prof, 66-80, PROF RADIOL & BIOL LAB, UNIV CINCINNATI, 80-, PROF PATH LAB, MED CTR, 89- *Concurrent Pos:* NIH fel, 64-66; adj affil mem, Jewish Hosp, Cincinnati, 73- *Mem:* AAAS; Radiation Res Soc. *Res:* Biochemical effect of radiation; radioimmunoassays for various hormones. *Mailing Add:* Dept Radiol & Biol ML-568 Med Ctr Univ Cincinnati Cincinnati OH 45267

CHEN, JAMES CHE WEN, b Taipei, Taiwan, Nov 21, 30; m 59; c 3. DEVELOPMENTAL BIOLOGY, BOTANY. *Educ:* Taiwan Prov Norm Univ, BS, 55; Univ Pa, PhD(bot), 63. *Prof Exp:* Asst instr bot, Taiwan, 56-58; asst instr biol, Univ Pa, 61-62; res assoc, Princeton Univ, 63-65; from asst prof to assoc prof, Washington & Jefferson Col, 65-68; ASSOC PROF BOT, DOUGLASS COL, RUTGERS UNIV, 68- *Mem:* AAAS; Soc Develop Biol; Bot Soc Am. *Res:* Investigations of growth and differentiation of plants and quantitative description of growth processes of plants. *Mailing Add:* Dept of Biol Sci Douglass Col Rutgers Univ New Brunswick NJ 08903

CHEN, JAMES L, b Nanking, China, Nov 3, 14; nat US; m 46; c 2. PHARMACEUTICAL CHEMISTRY. *Educ:* Sino-French Univ, China, BS, 39; Purdue Univ, MS, 40, PhD, 47. *Prof Exp:* Chemist, Merck & Co, 42-45; res chemist, Arlington Chem Co, 47-49; res assoc, E R Squibb & Sons, 49-69, res fel, 69-79, sr res fel, 79-82, res consult, 82-84; RETIRED. *Mem:* Am Pharmaceut Asn; NY Acad Sci; fel Am Inst Chem; AAAS. *Res:* Phytochemistry; biological adhesives; pharmaceutical research; analytical chemistry. *Mailing Add:* 30 Fairview Ave East Brunswick NJ 08816

CHEN, JAMES PAI-FUN, b Fengyuan, Taiwan, May 1, 29; nat US; m 64; c 3. BIOCHEMISTRY, IMMUNOLOGY. *Educ:* Houghton Col, BS, 55; St Lawrence Univ, MS, 57; Pa State Univ, PhD(biochem), 62. *Prof Exp:* From instr to assoc prof, Houghton Col, 60-64; res assoc, Div Exp Med, Col Med, Univ Vt, 64-65; res assoc med, Sch Med, State Univ NY Buffalo, 65-68; res asst prof internal med, Univ Tex Med Br Galveston, 68-70, asst prof human genetics, 70-75; sr res assoc, Biomed Res Div, NASA Johnson Space Ctr, 75-76; res assoc prof, Mem Res Ctr, 76-78, assoc prof, 78-84, PROF MED BIOL, UNIV TENN COL MED, KNOXVILLE, 84- *Concurrent Pos:* Mem Thrombosis Coun, Am Heart Asn. *Mem:* Am Soc Biochem & Molecular Biol; Am Asn Immunologists; Int Soc Hematol; Int Soc Thrombosis Hemostasis. *Res:* Immunoassay of biologically active materials; biomarkers of hypercoagulable state; models of intravascular coagulation and fibrinolysis. *Mailing Add:* Med Ctr Univ Tenn 1924 Alcoa Hwy Knoxville TN 37920-6999

CHEN, JAMES RALPH, b Kingston, Jamaica, Sept 22, 39; US citizen; c 3. PARTICLE PHYSICS, ATOMIC PHYSICS. *Educ:* Brandeis Univ, BA, 62; Harvard Univ, MA, 64, PhD(physics), 69. *Prof Exp:* Res asst physics, Harvard Univ, 67-68; res assoc & fel, Univ Pa, 68-69, asst prof, 69-73; from asst prof to assoc prof, 73-85, PROF PHYSICS, STATE UNIV NY COL GENESE, 85- *Concurrent Pos:* Res consult, Los Alamos Sci Lab, 72, spokesman-scientist, Los Alamos Meson Physics Facil, 74-75; spokesman-scientist, Brookhaven Nat Lab, 72-73; State Univ NY Res Found res fels, 74-76; res consult, Ctr Dis Control, 76; NSF panelist, 77 & 81; Alexander von Humboldt fel & vis scientist, Max Planck Inst Nuclear Physics, Heidelberg, WGer, 79-80; US rep, Int Comt Develop Proton-Induced X-Ray Emission, 80- *Mem:* Am Phys Soc; AAAS; Am Asn Physics Teachers. *Res:* Particle physics experiments; atomic physics x-ray spectroscopic calculations and measurements; elemental analysis using proton-induced x-ray emission and synchrotron radiation. *Mailing Add:* Six Elm St Geneseo NY 14454

CHEN, JIANN-SHIN, b Szechuan, China, May 18, 43; m 73; c 1. BACTERIAL REDOX ENZYMES, STRUCTURE & FUNCTION OF GENES & ENZYMES. *Educ:* Nat Taiwan Univ, BS, 67; Purdue Univ, PhD(biochem), 74. *Prof Exp:* Postdoctoral assoc biochem, Purdue Univ, 75-76; from asst prof to assoc prof, 76-90, PROF MICROBIOL, VA POLYTECH INST & STATE UNIV, 90- *Concurrent Pos:* Consult, Procter & Gamble Co, 81-84, NIH, 84-85, USDA, 85, 87 & 89. *Mem:* Am Chem Soc; Am Soc Biochem & Molecular Biol; Am Soc Microbiol; AAAS; Sigma Xi. *Res:* Molecular biology of nitrogen fixation and solvent production in anaerobic bacteria; mechanisms that control the expression of genes for the two processes in Clostridium. *Mailing Add:* Dept Anaerobic Microbiol Va Polytech Inst & State Univ Blacksburg VA 24061-0305

CHEN, JOHN CHUN-CHIEN, b China, Feb 6, 34; US citizen; m 60; c 3. CHEMICAL ENGINEERING, MECHANICAL ENGINEERING. *Educ:* Cooper Union, BChE, 56; Carnegie-Mellon Univ, MS, 58; Univ Mich, PhD(chem eng), 61. *Prof Exp:* Res sr engr thermal sci, Brookhaven Nat Lab, 60-70; prof mech eng, 70-80, Anderson Prof Chem Eng, 81, CHEM ENG DEPT, LEHIGH UNIV. *Concurrent Pos:* Consult, Brookhaven Nat Lab, 75-, Air Prod & Chem, 75-, Exxon Res & Eng Co, 76-78 & Argonne Nat Lab, 77-, Los Alamos Nat Lab, 87- *Honors & Awards:* Melville Medal, Am Soc Mech Engrs, 80; Donald O Kern Award, 88. *Mem:* Am Inst Chem Engrs; Am Soc Mech Engrs; Am Soc Eng Educ. *Res:* Thermo-fluid sciences; boiling heat transfer and two-phase flows; fluidization and heat transfer; nuclear reactor safety. *Mailing Add:* Dept Chem Eng 111 Research Dr Bldg A Lehigh Univ Bethlehem PA 18015-4791

CHEN, JOSEPH CHENG YIH, b Nanking, China, Nov 12, 33; nat US; m 59; c 1. THEORETICAL PHYSICS, THEORETICAL CHEMISTRY. *Educ:* St Anselm's Col, BA, 57; Univ Notre Dame, PhD(theoret chem), 61. *Prof Exp:* Res assoc theoret chem, Brookhaven Nat Lab, 61-63, assoc chemist, 63-65; vis fel, Joint Inst Lab Astrophys, Univ Colo, 65-66; from asst prof to assoc prof physics, 66-74, PROF PHYSICS, UNIV CALIF, SAN DIEGO, 74- *Concurrent Pos:* Vis lectr, Univ Manchester, 64; Nordic Inst Theoret Atomic Physics vis prof, Univ Oslo, 70. *Mem:* Fel Am Phys Soc; Hist Sci Soc. *Res:* Scattering problem; reaction theories; atomic and molecular systems; many body theory; history of science in China. *Mailing Add:* Dept of Physics Univ of Calif La Jolla CA 92093

CHEN, JOSEPH H, b Tientsin, China, Mar 22, 31; m 56; c 2. PHYSICS. *Educ:* St Procopius Col, BS, 54; Univ Notre Dame, PhD(physics), 58. *Prof Exp:* Instr physics, Univ Notre Dame, 57-58; from asst prof to assoc prof, 58-77, PROF PHYSICS, BOSTON COL, 77- *Mem:* Am Phys Soc. *Res:* Dielectric relaxation phenomena; electron spin resonance and optical properties of solids; Mossbauer effect and high pressure studies in solids; amorphous semiconductors; II-VI compounds. *Mailing Add:* Dept of Physics Boston Col Chestnut Hill MA 02167

CHEN, JOSEPH KE-CHOU, b Hupei, China, May 27, 36; m 67; c 2. MEDICAL MICROBIOLOGY. *Educ:* Nat Taiwan Univ, BS, 59; Univ Pittsburgh, PhD(microbiol), 72. *Prof Exp:* Res asst microbiol, Sch Med, Univ Pittsburgh, 66-72, res assoc biochem, 72-74; cancer res scientist viral oncol, Roswell Park Mem Inst, 75-76; sr staff fel, Lab Gen & Comp Biochem, NIMH, 77-79; staff scientist, Bionetic Lab Prods, Litton Bionetic Inc, 79-83; dir, Biotechnology Develop, Bioqual, Inc, 83-86; SR STAFF MICROBIOLOGIST, VET BIOLOGICS, US DEPT AGR, 87- *Mem:* Am Soc Microbiol. *Res:* Immunodiagnosis techniques development; biotechnology. *Mailing Add:* 7901 Declaration Lane Potomac MD 20854

CHEN, JUEI-TENG, b Taiwan, May 9, 38; US citizen; m 66; c 2. LOW TEMPERATURE PHYSICS. *Educ:* Tunghai Univ, Taiwan, BS, 62; Univ Waterloo, MS, 66, PhD(physics), 69. *Prof Exp:* Res assoc physics, Univ Pa, 68-70; from asst prof to assoc prof, 70-79, PROF PHYSICS, WAYNE STATE UNIV, 79- *Concurrent Pos:* Teaching fel, Nat Res Coun, Can, 68 & 69; res grants, NSF, 74-83; consult, Energy Conversion Devices, Inc, 77- *Mem:* Am Phys Soc; Am Asn Univ Professors; AAAS. *Res:* Superconductivity; electron tunneling; microwave absorption and emission; Josephson effects; proximity effect; nonequilibrium effect; multilayers. *Mailing Add:* 3432 Marc Dr Sterling Heights MI 48310

CHEN, JUH WAH, b Shanghai, China, Nov 10, 28; m 58; c 3. CHEMICAL ENGINEERING, ENVIRONMENTAL ENGINEERING. *Educ:* Taiwan Col Eng, BS, 53; Univ Ill, MS, 57, PhD(chem eng, phys chem), 59. *Prof Exp:* Asst prof chem eng, Bucknell Univ, 59-65; from assoc prof to prof thermal & environ eng, 65-85, chmn dept, 70-85, assoc dean, 85-89, DEAN ENG, COL ENG & TECHNOL, SOUTHERN ILL UNIV, CARBONDALE, 89- *Concurrent Pos:* Res consult engr, Upjohn Co, 60-65. *Mem:* Am Inst Chem Engrs; Am Chem Soc; Soc Eng Educ; Am Soc Mech Eng. *Res:* Kinetics; coal conversion; pollution control. *Mailing Add:* Col Eng & Technol Eng Southern Ill Univ Carbondale IL 62901

CHEN, KAN, b Hong Kong, China, Aug 28, 28; nat US; m 53; c 4. TECHNOLOGY PLANNING & ASSESSMENT. *Educ:* Cornell Univ, BEE, 50; Mass Inst Technol, SM, 51, ScD, 54. *Prof Exp:* Asst elec eng, Mass Inst Technol, 52-54; res mgr, Westinghouse Elec Corp, 54-65; res dir, SRI Int, 66-70; prof environ systs eng & elec eng, Univ Pittsburgh, 70-71; dir, PHD Prog, Urban Technol & Environ Planning, 81-87, PROF ELEC ENG & COMPUT SCI, UNIV MICH, 71- *Concurrent Pos:* From lectr to adj prof, Univ Pittsburgh, 55-66; vis prof, Stanford Univ, 62-63; sr lectr, Carnegie Inst Technol, 64-65; Paul Goebel prof, Univ Mich, 71-73. *Mem:* Fel Inst Elec & Electronics Engrs; fel AAAS. *Res:* Systems technology; operations research; automatic control; computer coordinated systems; systems planning; urban and environmental systems; technology management and assessment; science policy; computer integrated manufacturing systems. *Mailing Add:* Col Eng 4112 EECS Bldg Univ Mich Ann Arbor MI 48109-2122

CHEN, KAO, b Nanjing, China, Mar 21, 19; US citizen; m 48; c 3. SPECIAL POWER SYSTEMS FOR COMPUTERS, GROUNDING & SURGE PROTECTION. *Educ:* Jiao Tong Univ, China, BS, 42; Harvard Univ, MS, 48; Polytech Inst Brooklyn, PhD(elec eng), 53. *Prof Exp:* Engr power plant design & study, Jackson & Moreland, Boston, 48-50; relay specialist relay protection eng, Am Power co, NY, 50-52; proj supvr design generating & distr stat, Ebasco Int Corp, NY, 52-55; sr proj engr indust power dist & illuminating,eng, Westinghouse Elec Corp, 56-68, fel engr in charge, 68-83; dir, syst eng for fourteen plants, N A Philips Lighting Corp, 83-86; pres & consult, 81-87, PRES, CARLSONS CONSULT ENGR, INC, 87- *Concurrent Pos:* Vis prof, Fudan Univ, Shanghai, China, 82; mem, US Nat Comt, Int Comn Illumination; comt chmn, Indust Appln Soc, 83-84, dept chmn, 85-87; designated sem leader, Inst Elec & Electronic Engrs, Energy Management, 90-; mem ed bd, Marcel Dekker, Inc, 90- *Honors & Awards:* Centennial Medal, Inst Elec & Electronic Engrs, 84, IUSD Award of Merit, 85; Medal of Honor, Am Biog Inst, 87. *Mem:* Inst Elec & Electronics Engrs-Indust Appln Soc; Nat Soc Prof Engrs; emer mem Illuminating Eng Soc; fel Inst Elec & Electronics Engrs-Power Eng Soc. *Res:* Perfecting the design and development of industrial power distribution and energy saving illuminating systems for industrial plants. *Mailing Add:* 11816 Caminito Corriente San Diego CA 92128

CHEN, KENNETH YAT-YI, b Chunghwa, Taiwan, Rep of China, Feb 12, 41; US citizen; m 67; c 2. ENVIRONMENTAL ENGINEERING, CIVIL ENGINEERING. *Educ:* Nat Taiwan Univ, BS, 63; Univ RI, MS, 66; Harvard Univ, PhD(environ sci & eng), 70. *Prof Exp:* From asst prof to assoc prof, 70-78, assoc prof & dir, 76-78, PROF & DIR ENVIRON ENG PROG, UNIV SOUTHERN CALIF, 78-, CHMN PROG, 80-, PROF CIVIL ENG, 80- *Concurrent Pos:* Consult, var indust orgn & consult co, 70-; prin investr many grants, 70-; chmn joint task group, Atomic Absorption Method Metals, Water Pollution Control Fedn, 76-78; consult, Dept Eng, Argonne Nat Lab, 77-78 & Dept of Com, Off Energy Related Invention, Nat Bur Standards, 77- *Mem:* Am Soc Civil Engrs; Am Chem Soc; Water Pollution Control Fedn; Am Soc Limnol & Oceanog; Am Water Works Asn. *Res:* Chemical behavior of trace contaminants in environment, chemical and biological processes in environmental engineering. *Mailing Add:* Environ Eng Prog Univ of Southern Calif Los Angeles CA 90007

CHEN, KUEI-LIN, operations research, statistical analysis, for more information see previous edition

CHEN, KUN-MU, b Taipei, Taiwan, Feb 3, 33; nat US; m 62; c 4. ELECTRICAL ENGINEERING. *Educ:* Taiwan Univ, BS, 56; Harvard Univ, MS, 58, PhD(appl physics), 60. *Prof Exp:* Res assoc, Radiation Lab, Univ Mich, 60-64; assoc prof, 64-67, PROF ELEC ENG, MICH STATE UNIV, 67- *Concurrent Pos:* Air Force Cambridge Res Labs res grant, 66-; NSF grant, 66-; Army Res Off grant, 76-78; Navy res grant, 78- *Mem:* Fel AAAS; fel Inst Elec & Electronics Engrs; Sigma Xi. *Res:* Electromagnetic theory; antenna theory; applied plasma physics; biological effects of EM waves. *Mailing Add:* Dept of Elec Eng Mich State Univ E Lansing MI 48824

CHEN, KUO-TSAI, mathematics; deceased, see previous edition for last biography

CHEN, KWAN-YU, b Shanghai, China, Aug 29, 30; m 61; c 3. ASTRONOMY. *Educ:* Ill Inst Technol, BS, 53, MS, 56; Univ Pa, PhD(astron), 63. *Prof Exp:* Asst prof astron, 63-68, from asst prof to assoc prof astron & phys sci, 68-78, PROF ASTRON, UNIV FLA, 78- *Mem:* Am Astron Soc; Int Astron Union; Sigma Xi. *Res:* Photometric study of variable stars. *Mailing Add:* Dept Astron Univ of Fla 211 SSRB Bldg Gainesville FL 32611

CHEN, L T, b Tungkang, Taiwan, Jan 3, 38; US citizen; m 69; c 2. HEMATOPOIETIC TISSUE, FETAL DEVELOPMENT. *Educ:* Nat Taiwan Univ, BSc, 60; Univ Alta, PhD(embryol), 68. *Prof Exp:* Fel anat, Sch Med, Johns Hopkins Univ, 70-72, instr, 72-73, asst prof, 73-80; PROF ANAT, UNIV SFLA, 85- *Concurrent Pos:* Prin investr, Nat Sci Found, 77-79, NIH, 78-81 & Univ S Fla, 83-84. *Mem:* Am Soc Cell Biol; Am Asn Anatomists; Int Soc Exp Hemat. *Res:* Structure and function of hematopoietic tissue (spleen, bone marrow); microcirculation of the spleen; hematopoietics of mammalian embryos in vitro; relationship of maternal fever to fetal development and resorption; vitamin B levels during pregnancy. *Mailing Add:* Dept Anat Univ SFla Col Med Tampa FL 33612

CHEN, LAWRENCE CHIEN-MING, b Ping-Yang, China, Oct 9, 33; Can citizen. PHYCOLOGY. *Educ:* Nat Taiwan Normal Univ, BSc, 60; Univ NB, MSc, 67, PhD(bot), 77. *Prof Exp:* Teaching asst biol, Biol Dept, Nat Taiwan Normal Univ, 60-64; res officer hydraubiol, 67-74, RES OFFICER, MARINE BOT, 74-, SR RES OFFICER, INST MARINE BIOSCI, NAT RES COUN CAN, 88- *Concurrent Pos:* Hon res assoc, Univ NB, 88-; hon prof, Inst Oceanology, Acad Sinica, PRC, 88- *Mem:* Can Bot Asn; Int Phycological Soc; Phycological Soc Am; Phycological Soc Japan; Phycological Soc China; Int Asn Plant Tissue Cult. *Res:* Study of the ontogeny, development and growth of marine phycocolloid plants; tissue, cell or protoplast culture of marine plants. *Mailing Add:* Inst Marine Biosci Nat Res Coun 1411 Oxford St Halifax NS B3H 3Z1 Can

CHEN, LESLIE H(UNG), mechanical engineering, for more information see previous edition

CHEN, LILY CHING-CHUN, b China, Feb 17, 46; m 69. NUCLEAR PHYSICS, COMPUTER SCIENCE. *Educ:* Nat Taiwan Univ, BS, 68; Univ Cincinnati, MS, 69; Univ Wis-Madison, PhD(nuclear physics), 74. *Prof Exp:* Task leader comput sci, Comput Sci Corp, 74-78; SR VPRES & COMPTROLLER, GEN SOFTWARE CORP, 77- *Res:* Orbit and attitude determination for satellites; spacecraft navigation; image processing and data base management. *Mailing Add:* Gen Software Corp 6100 Chevy Chase Dr Laurel MD 20707

CHEN, LINDA LI-YUEH HUANG, b Tokyo, Japan, Mar 22, 37; m 61; c 2. BIOCHEMISTRY. *Educ:* Nat Taiwan Univ, BS, 59; Univ Louisville, PhD(biochem), 64. *Prof Exp:* PROF NUTRIT, UNIV KY, 79-, DIR, MULTIDISCIPLINARY PHD PROG NUTRIT SCI. *Concurrent Pos:* Mem, Cancer Res Manpower Rev Comt, Nat Cancer Inst. *Honors & Awards:* Nat Borden Award Except Res Nutrit, 90. *Mem:* Int Asn Vitamin Nutrit Oncol; Gerontol Soc; Am Inst Nutrit; Am Soc Clin Nutrit; NY Acad Sci. *Res:* Biochemical function of vitamin E and its interaction with other nutrients; tissue antioxidant status; nutrition and aging; nutrient-drug interaction; nutrition and cancer. *Mailing Add:* Dept Nutrit & Food Sci Univ Ky Lexington KY 40506-0054

CHEN, LIU, b Hanchow, China, Jan 3, 46; m 69; c 1. PLASMA PHYSICS, SPACE PHYSICS. *Educ:* Nat Taiwan Univ, BS, 66; Wash State Univ, MS, 69; Univ Calif, Berkeley, PhD(elec eng, comput sci), 72. *Prof Exp:* Mem tech staff plasma physics & space physics, Bell Labs, Murray Hill, NJ, 72-74; res assoc plasma physics, Princeton Univ, 74-75, mem res staff, 75-77, res physicist, 77-80, PROF, PRIN RES PHYSICIST & LECTR, PLASMA PHYSICS LAB & DEPT ASTROPHYS SCI, PRINCETON UNIV, 80- *Mem:* Fel Am Phys Soc. *Res:* Theoretical fusion plasma physics; space plasma physics, especially magnetic pulsations and very low frequency emissions. *Mailing Add:* Plasma Physics Lab Forrestal Campus C-Site Princeton Univ Princeton NJ 08544

CHEN, LO-CHAI, b Kwang-Tung, China, Dec 9, 39; m 65. ICHTHYOLOGY. *Educ:* Nat Taiwan Univ, BS, 61; Univ Alaska, MS, 65; Univ Calif, San Diego, PhD(marine biol), 69. *Prof Exp:* From asst prof to assoc prof, 69-77, PROF ZOOL, SAN DIEGO STATE UNIV, 77- *Mem:* Am Soc Ichthyologists & Herpetologists; Sigma Xi. *Res:* Systematics and zoogeography of fishes, especially Sebastes Scorpaenidae; growth and meristic determination in fishes. *Mailing Add:* Dept of Biol San Diego State Univ San Diego CA 92182

CHEN, MAYNARD MING-LIANG, b Shanghai, China, June 14, 50. QUANTUM CHEMISTRY. *Educ:* Univ Calif, Berkeley, BS, 71; Cornell Univ, PhD(chem), 76. *Prof Exp:* SCIENTIST, POLAROID CORP, 80- *Res:* Quantum chemistry programs to help understand and design molecules used in a photographic context. *Mailing Add:* 20 Fresh Pong Pl Cambridge MA 02138

CHEN, MICHAEL CHIA-CHAO, b Shanghai, China, Jan 13, 47; m. PHYSICAL CHEMISTRY, POLYMER SCIENCE. *Educ:* Univ Minn, BChem, 69; Mass Inst Technol, PhD(phys chem), 73. *Prof Exp:* Res assoc, Mass Inst Technol, 73-75; res assoc, Univ Mass, 75-76; SR STAFF CHEMIST, EXXON CHEM CO, 76- *Concurrent Pos:* Sloan res trainee, Mass Inst Technol, 72. *Res:* Morphology, optical and mechanical properties of polymers; polymer physics; polymer characterization; polyethylene and polypropylene film products, structure-function correlation of nucleic acids, proteins and their model compounds; spectroscopy. *Mailing Add:* Exxon Chem Co PO Box 5200 Baytown TX 77522-5200

CHEN, MICHAEL MING, b Hankow, China, Mar 10, 32; m 61; c 3. MECHANICAL ENGINEERING, BIOENGINEERING. *Educ:* Univ Ill, BS, 55; Mass Inst Technol, SM, 57, PhD(mech eng), 61. *Prof Exp:* Sr staff scientist, Res & Adv Develop Div, Avco Corp, Mass, 60-63; asst prof eng & appl sci, Yale Univ, 63-69; assoc prof, NY Univ, 69-73; PROF MECH & INDUST ENG, UNIV ILL, URBANA-CHAMPAIGN, 73- *Concurrent Pos:* Consult, A D Little Co, 56-58, Bell Tel Labs, 66-69 & Argonne Nat Lab, 76- *Mem:* Am Phys Soc; Am Soc Mech Engrs; AAAS; Sigma Xi. *Res:* Fluid mechanics and heat transfer in energy, manufacturing, and bioengineering. *Mailing Add:* 311 Eliot Dr Urbana IL 61801

CHEN, MICHAEL S K, b Taipei, Formosa, May 18, 41; m 66; c 1. CHEMICAL & BIOCHEMICAL ENGINEERING. *Educ:* Nat Taiwan Univ, BS, 63; Kans State Univ, PhD(chem eng), 69. *Prof Exp:* From lectr to asst prof chem eng, Univ Pa, 69-72; sr process engr, Air Prod & Chem, 72-75, staff engr, 75-77, res mgr, 77-78, staff assoc, 78-79, actg res dir, 81, sect head, 79-82, MGR PROCESS EVAL, AIR PROD & CHEM, 82- *Mem:* Am Inst Chem Engrs. *Res:* Wastewater treatments; flue gas scrubbing; gas separation; mass transfer; air separation; membrane seperation. *Mailing Add:* 6981 Yakl'sville Rd Zionsville PA 18092

CHEN, MICHAEL YU-MEN, b Shanghai, China, Jan 7, 41; m 69. MEDICAL SCIENCES. *Educ:* Shanghai Chinese Med Sch, MD, 64. *Prof Exp:* Radiologist, Shanghai Chinese Med Hosp, 64-79; radiologist, Shanghai Yang Poo Dist Tumor Hosp, 79-83; res fel, 83-87, res instr, 87-88, RES ASST PROF RADIOL, BOWMAN GRAY SCH MED, 88- *Mem:* Radiol Soc NAm; Soc Gastrointestinal Radiologists. *Res:* Independent research in evaluation of diagnostic imaging in general radiology, particularly in the gastrointestinal field. *Mailing Add:* Bowman Gray Sch Med 300 S Hawthorne Rd Winston Salem NC 27103

CHEN, MING CHIH, b China, Aug 15, 20; m 43; c 3. ORGANIC CHEMISTRY. *Educ:* Fukien Christian Univ, BS, 42; Univ Buffalo, PhD, 50. *Prof Exp:* Asst, Univ Buffalo, 48-50; res fel, Purdue Univ, 50-52; res assoc, O-Cel-O Div, Gen Mills Inc, 52-57; group leader, Simoniz Co, 57-62; mgr urethane develop, Sheller Labs, Sheller Mfg Corp, 62-67, dir prod develop, 67-73, dir mfg develop, 73-80, dir anal serv, 80-83, TECH DIR PROCESSING, SHELLER GLOBE CORP. *Mem:* Chem Soc; Soc Plastics Engrs; Soc Plastics Indust; Soc Automotive Engrs; NY Acad Sci. *Res:* Cyanogen in the formation of oxamidine; halogenated hydrocarbon for non-inflammable hydraulic fluid and additives; cellulose and its derivatives; organic coating and polymeric foams. *Mailing Add:* Manufacturing Eng Sheller Globe Corp 1641 Porter St Detroit MI 48216-1935

CHEN, MING M, b Fukien, China, Apr 23, 19; US citizen; m 48; c 2. STRUCTURAL DYNAMICS. *Educ:* Wuhan Univ, BS, 41; Univ Ill, MS, 48, PhD(appl mech), 52; Univ Wash, MS, 52. *Prof Exp:* Res engr, airframe struct, Repub Aviation Corp, 52-53; mem res staff aeroelasticity & struct, Mass Inst Technol, 53-60; assoc prof, 60-66, chmn dept, 68-81, PROF AEROSPACE & MECH ENG, BOSTON UNIV, 66- *Concurrent Pos:* Vis prof, Cheng Kung Univ, Taiwan, 66-67, actg dean acad affairs, 67; actg dir, Eng Sci Res Ctr, Taiwan, 67; gen engr, Transp Syst Ctr, Dept Transp, 75-85. *Honors & Awards:* NASA Tech Brief Award, 75. *Mem:* Am Inst Aeronaut & Astronaut; Am Soc Eng Educ; Soc Exp Stress Anal; Am Asn Univ Prof. *Res:* Flight structures; structural dynamics; high temperature structures; collision dynamics. *Mailing Add:* 227 Harvard St Boston MA 02170

CHEN, MING SHENG, b Taipei, Taiwan, Apr 1, 41; US citizen; m 71; c 2. ANTI-INFECTIVE & ANTI-TUMOR RESEARCH. *Educ:* Nat Chung-Shing Univ, Taiwan, BS, 64; Auburn Univ, Ala, PhD(chem), 71. *Prof Exp:* Fel, Papanicolaou Cancer Res Inst, Fla, 71-72; res assoc org chem, Bluffton Col, Ohio, 72-74; res assoc, Purdue Univ, Ind, 74; sr res assoc, Yale Univ, Conn, 74-82; assoc sr investr, Smithkline Animal Health Prod, 82-83; sr res scientist, Bristol-Myers Pharmaceut Res & Develop, 83-85, res fel, 85-86; sr staff res, 86-87, PRIN SCIENTIST, SYNTEX RES, 88- *Concurrent Pos:* Scholar, Leukemia Soc Am, Inc, 81-86; Niaid Aids Study Sect, 88-89. *Mem:* Am Chem Soc; Am Soc Biol Chemists; Am Soc Pharmacol & Exp Therapeut; Am Soc Virol; Inter-Am Soc Chemotherapy; Sigma Xi. *Res:* Mechanisms of action of antiviral, antibacterial and antitumor compounds; enzymology of the enzymes involved in DNA and RNA synthesis. *Mailing Add:* 10025 Orange Ave Cupertino CA 95014

CHEN, MIN-SHIH, b Kiangsu, China, Nov 13, 42; m 69; c 2. THEORETICAL HIGH ENERGY PHYSICS, SOFTWARE DEVELOPMENT. *Educ:* Nat Taiwan Univ, BSc, 64; Yale Univ, MPhil, 67, PhD(physics), 70. *Prof Exp:* Res assoc physics, Brookhaven Nat Lab, 70-72; res assoc, Stanford Linear Accelerator Ctr, 72-74; lectr, 74-75, scholar physics, 75-76, ADJ ASSOC PROF, UNIV MICH, ANN ARBOR, 76-; Software develop mgr, ADP Network Serv, 79-88; SR SOFTWARE CONSULT, DIGITAL EQUIPMENT CORP, 88- *Concurrent Pos:* Prin, Ann Arbor Chinese Sch, 84-86; proj mgt specialist, ADP Network Serv, 76-79. *Mem:* Proj Mgt Inst; Am Phys Soc. *Res:* Phenomenology in multiparticle production and weak interactions; atomic physics and computational science; software development. *Mailing Add:* 2730 Hampshire Rd Ann Arbor MI 48104

CHEN, MO-SHING, b Chekiang, China, Aug 20, 31; m 59; c 2. ELECTRICAL ENGINEERING. *Educ:* Nat Taiwan Univ, BS, 54; Univ Tex, MS, 58, PhD(elec eng), 62. *Prof Exp:* Elec engr, Taiwan Power Co, 54-56; from asst prof to assoc prof, 62-69, PROF ELEC ENG & DIR ENERGY SYSTS RES CTR, UNIV TEX, ARLINGTON, 69- *Concurrent Pos:* Consult, 32 countries. *Honors & Awards:* Power Eng Educr Award, Edison Elec Inst, 76; Western Elec Award, Am Soc Eng Educ, 77; Halliburton Res Award, 83; Centennial Award, Inst Elec & Electronics Engrs, 84. *Mem:* Fel Inst Elec & Electronics Engrs; Am Soc Eng Educr. *Res:* Electric power generation, transmission, and distribution; system analysis; computer applications; power system load modeling and voltage reduction research. *Mailing Add:* Energy Systs Res Ctr Box 19048 Univ Tex Arlington TX 76019

CHEN, NAI Y, b China, Jan 6, 26; US citizen; m 59; c 2. RESEARCH ADMINISTRATION, TECHNICAL MANAGEMENT. *Educ:* Univ Shanghai, BS, 47; La State Univ, MS, 54; Mass Inst Technol, ScD(chem eng), 59. *Prof Exp:* Dep head, Taiwan Sugar Corp, 47-52; res asst, La State Univ, 52-54; res assoc, Mass Inst Technol, 54-60; chem eng sr scientist, Mobil Res & Develop Corp, 60-79, actg mgr, 79-80, mgr, 80-86, SR SCIENTIST, MOBIL RES & DEVELOP CORP, 79-, RES ADV, 86- *Concurrent Pos:* Dir, Chinese Am Chem Soc, 91. *Mem:* Nat Acad Eng; Am Inst Chem Engrs; Sigma Xi; NAm Catalysis Soc. *Res:* Petroleum refining and petrochemical technologies; heterogeneous catalysis; zeolite catalysis; global energy issues; author of more than 75 publications and three books; awarded more than 120 US patents. *Mailing Add:* Mobil Res & Develop Corp PO Box 1025 Princeton NJ 08540

CHEN, NING HSING, b Kweihsien, Kwangsi, China, Apr 21, 18; US citizen; m 57. CHEMICAL ENGINEERING. *Educ:* Nat Chekiang Univ, China, BS, 41; Polytech Inst New York, BChE, 49, DChE, 61; Univ Mo, MS, 50. *Prof Exp:* Chem engr, First Chem Co, China, 41-48; engr, Allied Chem & Dye Corp, 50-54; process engr, Bechtel Corp, 54-56; proj engr, Indust Process Engrs, 56-58; heat transfer engr, M W Kellogg Co, 58-59; develop & res engr, Aerojet Gen Corp, 59-62; mem tech staff, Aerospace Corp, 62-63; staff engr, Lockheed Missiles & Space Co, 63-64; prin engr, Space Booster Div, Thiokol Chem Corp, Brunswick, 64-65; staff engr, Space Div, Chrysler Corp, Ala, 65-66; from assoc prof to prof, 66-87, EMER PROF CHEM ENG, UNIV LOWELL, 87- *Concurrent Pos:* Vis res prof chem eng, Nat Tsing Hua Univ, Taiwan, 76-77; hon lectr, Univ China, Tianjin, 79 & Univ China, Zhejiang, 81; guest prof, Tientsin Univ, Tientsin, China, 79, ZHejiang Univ, Hong Hov, China, 81, East China Inst Chem Technol, Shanghai, China, 82, South China Inst Technol, Guahgzhov China, 83; vis prof Univ Calif Berkeley, Ca, 84. *Mem:* AAAS; Am Chem Soc; Am Inst Chem Engrs; NY Acad Sci; Am Asn Univ Professors; Sigma Xi. *Res:* Heat transfer; thermodynamics; fluid dynamics; transport and physical properties; distillation; leaching; absorption; extraction; rockets and missiles; optimization. *Mailing Add:* Dept Chem Eng Lowell Univ Lowell MA 01854

CHEN, PAUL EAR, applied mechanics, applied mathematics, for more information see previous edition

CHEN, PETER, b Taipei, Taiwan, Nov 20, 41; m 69; c 2. COMPUTER APPLICATION, STATISTICAL APPLICATION. *Educ:* Provincial Chung-Hsing Univ, Taiwan, BS, 66; Univ Minn, MS, 70. *Prof Exp:* Supvr, Weyerhaeuser Co, 73-74; Programmer, data processing, CBC Comput Serv, 75-76; SPECIALIST WOOD PROD, INT PAPER CO, 76- *Mem:* Forest Prod Soc. *Res:* Develop and refine existing procedures and controls, and provide technical service to production facilities to improve efficiency and reduce costs. *Mailing Add:* 3837 Dutton Dr Plano TX 75023

CHEN, PETER PIN-SHAN, DATABASE, SOFTWARE ENGINEEING. *Educ:* Nat Taiwan Univ, BS, 68; Harvard Univ, SM, AM (computer sci), 71, PhD, 73. *Prof Exp:* Prin engr computer sci, Honeywell, 73-74; vis researcher software, Digital Equip Corp, 74; asst prof mgt sci, Mich Inst Technol, 74-78; Sinclair vis prof, 76-77; assoc prof info sys, Univ Calif Los Angeles, 78-82; PROF COMPUTER SCI, HARVARD UNIV, 90. *Concurrent Pos:* Student assoc, IBM, 70; chair prof computer sci, La State Univ, 73. *Res:* Database management, software engineering, automation of office worker and factory. *Mailing Add:* Dept Comput Sci La State Univ Baton Rouge LA 70803-4020

CHEN, PHILIP STANLEY, JR, b St Johns, Mich, July 3, 32; m 55; c 2. PHARMACOLOGY, PHYSIOLOGY. *Educ:* Clark Univ, BA, 50; Univ Rochester, PhD(pharmacol), 54. *Prof Exp:* Res assoc, Atomic Energy Proj, Univ Rochester, 50-54, jr scientist, 54-56; sr asst scientist, Nat Heart Inst, 56-59; asst prof radiation biol & biophys & pharmacol, Univ Rochester, 59-66; Guggenheim fel, Copenhagen, Denmark, 66-67; grants assoc, NIH, 67-68, spec asst to asst dir prog planning & eval, 68-70, chief spec projs br, Off Prog Analysis, 70-71, chief analysis & eval br, Prog Planning & Eval, 71-72, assoc dir prog planning & eval, Nat Inst Gen Med Sci, 72-74, asst dir, 74-83, ASSOC DIR, INTRAMURAL AFFAIRS, NIH, 83- *Concurrent Pos:* NSF fel, Copenhagen, Denmark, 54-55; chmn, Pub Health Serv Patent Policy Bd, 87- *Mem:* Am Chem Soc; Am Physiol Soc; Radiation Res Soc. *Res:* Radioactive tracer techniques in biology; microanalytical chemistry; bone and mineral metabolism; vitamin D; renal excretion; technology transfer. *Mailing Add:* Bldg One Rm 140 Off Dir NIH Bethesda MD 20892

CHEN, PI-FUAY, b Taipei, Taiwan, Aug 17, 30; m 64; c 2. ELECTRICAL ENGINEERING, AUTOMATIC CONTROL SYSTEMS. *Educ:* Taipei Inst Technol, Taiwan, BS, 56; Va Polytech Inst, MS, 63; Univ Va, DSc(elec eng), 69. *Prof Exp:* Asst engr, Taipei Telecommun Off, 56-60; engr, Int Tel & Tel Corp, 63-65, sr engr, 66; asst prof, George Washington Univ, 68-69; RES ENGR ELECTRONICS, RES INST, US ARMY ENGR TOPOG LABS, 69- *Mem:* Inst Elec & Electronics Engrs; Pattern Recognition Soc; Sigma Xi. *Res:* Various automated and semi-automated techniques for aerial and radar image analysis and cartographic feature extraction and recognition using pattern recognition and image understanding principles. *Mailing Add:* 7200 Marine Dr Alexandria VA 22307

CHEN, PING-FAN, b Kiangyin, China, May 13, 17; m 47, 79; c 3. GEOLOGY. *Educ:* Nat Cent Univ, China, BS, 38; Univ Cincinnati, MS, 57; Va Polytech Inst, PhD(geol), 59. *Prof Exp:* From jr to sr geologist, Nat Geol Surv China, 38-46; from sr geologist to chief petrol geol, Chinese Petrol Corp, 46-55; petrol geologist & head petrol div, 60-66, stratigrapher, WVA Geol & Econ Surv, 67-87; ADJ PROF GEOL, WVA UNIV, 75- *Concurrent Pos:* Hwakang prof, Univ Chinese Cult. *Mem:* Sr fel China Acad. *Res:* Geological exploration work for petroleum, ground water, coal and other mineral deposits in China; detail stratigraphic and structural geology studies in Central Appalachian for oil and gas possibilities in Lower Paleozoic rocks; coal research of West Virginia. *Mailing Add:* 1277 Dogwood Morgantown WV 26505

CHEN, PRISCILLA B(URTIS), HOST PARASITE RELATIONSHIP. *Educ:* Univ Pa, PhD(parasitol), 72. *Prof Exp:* Res assoc, Nat Res Coun, 72-74; immunologist, Litton Bionetics, 75; clin asst prof, 76-88, CLIN ASSOC PROF ORAL BIOL, STATE UNIV NY, BUFFALO, 89- *Mem:* Am Asn Immunologists; Am Soc Microbiol; AAAS. *Res:* Cell mediated immune responses; flow cytometry. *Mailing Add:* Oral Biol State Univ NY 3545 Main St Buffalo NY 14214

CHEN, RAYMOND F, b Canton, China, Oct 21, 33. LUMINESCENT SPECTROSCOPY. *Educ:* Cornell Univ, MD, 59; Univ Utah, PhD(biochem), 63. *Prof Exp:* MED OFFICER BIOPHYSICS, NIH, 63- *Mailing Add:* Nat Heart Lung & Blood Inst NIH Bldg 10 Rm 5D05 Bethesda MD 20892

CHEN, ROBERT CHIA-HUA, b Shanghai, China, Oct 26, 46; m 71; c 2. COMPUTER SCIENCE, ELECTRICAL ENGINEERING. *Educ:* Rensselaer Polytech Inst, BEE, 66; Mass Inst Technol, SM, 68; Carnegie-Mellon Univ, PhD(comput sci), 74. *Prof Exp:* Eng programmer comput sci, Burroughs Corp, 68-69; Asst prof comput & info sci, Univ Pa, 74-; CONSULT ENGR, DIGITAL EQUIP CORP. *Concurrent Pos:* Staff engr, Burroughs Corp, 74- *Mem:* Asn Comput Mach; Inst Elec & Electronics Engrs. *Res:* Concurrency and modularity in computer systems; computer interconnects. *Mailing Add:* Digital Equip Corp BXB1-1/E11 85 Swanson Rd Boxborough MA 01719-1326

CHEN, ROBERT LONG WEN, b Shanghai, China, Aug 23, 25; US citizen; m 58; c 2. PLASMA PHYSICS. *Educ:* Nanking Univ, BS, 47; Univ Syracuse, MA, 57, PhD(physics), 60. *Prof Exp:* Res assoc physics, Syracuse Univ, 60-61, asst prof, 61-62; res assoc, Goddard Inst Space Studies, 62-64; assoc prof, 64-70, PROF PHYSICS, EMORY UNIV, 70- *Mem:* Am Phys Soc. *Res:* Nonequilibrium statistical mechanics; plasma theory; molecular biophysics. *Mailing Add:* Dept Physics Emory Univ Atlanta GA 30322

CHEN, RONG YAW, b Taoyuen, Taiwan, Jan 6, 33; m 65; c 2. FLUID MECHANICS, HEAT TRANSFER. *Educ:* Nat Taiwan Univ, BS, 57; Univ Toledo, MS, 63; NC State Univ, PhD(mech eng), 66. *Prof Exp:* Asst engr mech eng, Shieman Dam Comn, 60-61; asst prof math, Atlantic Christian Col, 65-66; PROF MECH ENG, NJ INST TECHNOL, 66-, ACTG ASSOC CHMN, 91- *Concurrent Pos:* Prin investr, Harry Diamond Lab, Dept Army, DC, 74-75 & Army Res Ctr, Triangle Research Park, NC, 75-78; bd dir, NAm Taiwanese Profs Asn, 82-83, chap pres, 84. *Mem:* Am Soc Mech Eng; Sigma Xi. *Res:* Theoretical analysis on deposition of particles in conduits due to gravity and electro-static charges. *Mailing Add:* 323 King's Blvd Newark NJ 07102

CH'EN, SHANG-YI, b China, Mar 4, 10; nat US; m 31; c 5. SPECTROSCOPY. *Educ:* Yenching Univ, BS, 32, MS, 34; Calif Inst Technol, PhD(physics), 40. *Prof Exp:* Asst, Inst Physics, Nat Acad Peiping, 34-37 & Norman Bridge Physics Lab, Calif Inst Technol, 38-39; lectr physics, Yenching Univ, 39-41, asst prof optics & spectros, 41-43, prof optics & chmn dept physics, 43-46; res prof, Inst Physics, Nat Acad Peiping, 46-49; from assoc prof to prof physics, 49-75, EMER PROF PHYSICS, UNIV ORE, 75- *Concurrent Pos:* Res grants, NSF, 52-75, Off Ord Res, 57-60 & Air Force Off Sci Res, 63-70; vis prof, High Pressure Lab, Nat Ctr Sci Res, Bellevue, France, 61 & Clarendon Lab, Oxford, 68; assoc ed, J Quant Spectros & Radiative Transfer, 72-84. *Mem:* Fel Am Phys Soc; Am Asn Physics Teachers; French Phys Soc. *Res:* Atomic spectroscopy; spectral line shape; pressure and temperature effects. *Mailing Add:* 863 Fairway View Dr Eugene OR 97401

CHEN, SHAO LIN, b China, Aug 15, 18; nat US; m 50; c 2. BIOCHEMISTRY. *Educ:* Nanking Univ, BS, 40; Cornell Univ, PhD(plant physiol, biochem), 49. *Prof Exp:* Fel plant biochem, Carnegie Inst, 49-50; res assoc, Am Smelting & Refining Co, 50-52; biochemist, Red Star Yeast Co, 52-56, sr scientist, 56-60; dir microbiol chem lab, 60-65, dir biochem lab, 66-83, DIR, DRUG ID LAB, UNIVERSAL FOODS CORP, 83- *Concurrent Pos:* Res fel, China Found, 43; Rockefeller Found fel, 46-49; vis assoc prof, Univ Wis-Milwaukee, 65-67. *Mem:* AAAS; Am Chem Soc; fel Am Inst Chemists; fel Royal Soc Health; Sigma Xi. *Res:* Fermentation; enzymology; metabolism; cereal chemistry; food technology. *Mailing Add:* 1550 W Cedar Lane Milwaukee WI 53217

CHEN, SHAW-HORNG, b Hsin-Chu, Taiwan, Aug 18, 48; m 81; c 2. OPTICAL MATERIALS, TRANSPORT PHENOMENA. *Educ:* Nat Taiwan Univ, BS(chem eng), 71, MS(chem), 73; Univ Minn, PhD(chem eng), 81. *Prof Exp:* Asst scientist chem eng res, Inst Nuclear Energy Res, Taiwan, 76-78; asst prof chem eng, 81-85, ASSOC PROF CHEM ENG, UNIV ROCHESTER, 85-; SCIENTIST, LAB LASER ENERGETICS, 87- *Mem:* Am Chem Soc; Am Inst Chem Engr; Mat Res Soc. *Res:* Synthesis, characterization and processing of optical polymers for laser applications; fluid mixing with chemical reaction and molecular transport processes. *Mailing Add:* Dept Chem Eng Univ Rochester Gavett Hall No 206 Rochester NY 14627

CHEN, SHEPLEY S, b Taipei, Taiwan, Mar 28, 38; m 67; c 2. BIOLOGY, PLANT PHYSIOLOGY. *Educ:* Nat Taiwan Univ, BSc, 61; Harvard Univ, PhD(biol), 66. *Prof Exp:* Res assoc biochem, Mich State Univ-AEC Plant Res Lab, 65-69; asst prof, 69-73, ASSOC PROF BIOL SCI, UNIV ILL, CHICAGO, 73- *Concurrent Pos:* Jane Coffin Childs Mem Fund med res fel, 65-66; Nat Coun Sci Develop lectr, Repub of China, 67 & 69. *Mem:* Am Soc Plant Physiologists. *Res:* Physiology and biochemistry of seed germination; plant growth hormones; single-cell protein; algal physiology; algae as human food; microbiology. *Mailing Add:* Dept Biol Sci Univ Ill Chicago Chicago IL 60680

CHEN, SHI-HAN, b Che-Kiang, China, June 29, 36; US citizen; m 68; c 2. BIOCHEMICAL GENETICS, MEDICAL GENETICS. *Educ:* Taiwan Norm Univ, BS, 59; Nat Taiwan Univ, MS, 63; Univ Tex, Austin, PhD(zool), 68. *Prof Exp:* Instr biol, Taiwan Norm Univ, 63-64; teaching & res asst, Univ Tex, Austin, 64-68; res assoc & fel, King County Blood Bank, 69-71; res asst prof, 72-74, res assoc prof, 74-87, RES PROF, DEPT PEDIAT, SCH MED, UNIV WASH, 87- *Mem:* Am Soc Human Genetics; Asn Chinese Geneticists Am; Am Soc Clin Res. *Res:* Genetic variation of enzyme systems in man; biochemical causes of immunodeficiency diseases; molecular biology of Hemophilia B. *Mailing Add:* Dept Pediat Univ Wash Sch Med Seattle WA 98195

CHEN, SHIH-FONG, b Taipei, Taiwan, Sept 28, 49; m; c 2. CANCER CHEMOTHERAPY, CANCER BIOCHEMICAL PHARMACOLOGY. *Educ:* Nat Taiwan Univ, BS, 72, MS, 74; Univ RI, PhD(pharmaceut sci), 80. *Prof Exp:* Asst res fel, Inst Chem Acad Sinica, 74-76; res assoc, Brown Univ, 80-83, res asst prof, 83-84; res biochemist, E I du Pont de Nemours & Co, Inc, 84-88, sr res biochemist, 88-90, res assoc, 90, RES ASSOC, DU PONT MERCK PHARMACEUT CO, 91- *Mem:* Am Chem Soc; Am Cancer Soc. *Res:* Biochemical pharmacology evaluation of novel anticancer agents; mechanism of action studies. *Mailing Add:* Glenolden Lab Du Pont Merck Pharmaceut Co Glenolden PA 19036

CHEN, SHIOU-SHAN, b Tao-yuan, Taiwan, Feb 22, 38; m 69; c 2. CHEMICAL ENGINEERING, PHYSICAL CHEMISTRY. *Educ:* Taipei Inst Technol, BS, 58; Calif Inst Technol, MS, 62, PhD(chem eng & chem), 65. *Prof Exp:* Asst prof chem eng, Mass Inst Technol, 65-67 & Tufts Univ, 67-71; process engr, sr process engr & mgr process eng, Velsicol Chem Corp, 71-74; dir, Res & Develop Labs, 74-81, DIR RES & DEVELOP, BADGER CO, INC, 81- *Concurrent Pos:* Ford Found fel eng, 65-67. *Honors & Awards:* Raytheon Excellence in Technol Award. *Mem:* Am Inst Chem Engrs; Sigma Xi. *Res:* Fluid mechanics; heat and mass transfer; thermodynamics; chemical reactors; chemical and petrochemical processes. *Mailing Add:* Badger Co Inc 56 Woodchuck Rd East Weymouth MA 02189

CHEN, SHUHCHUNG STEVE, b Taipei, Taiwan, Sept 18, 50; m 76. ADHESION SCIENCE. *Educ:* Fu-Jen Cath Univ, Taiwan, BS, 74; Mich State Univ, PhD(org chem), 81. *Prof Exp:* Res assoc polymer chem, Univ Md, College Park, 81-83; proj supvr, 83-88, sr proj supvr, 88-91, RES ASSOC POLYMER & ADHESIVE, NAT STARCH & CHEM CO, 91- *Mem:* Am Chem Soc; Soc Advan Mat & Process Eng; Adhesion Soc. *Res:* Synthesis of new monomers and polymers for use in adhesives. *Mailing Add:* 11 Adams Dr Belle Mead NJ 08502

CHEN, SHU-TEN, cell biology, for more information see previous edition

CHEN, SOW-HSIN, b Taiwan, Mar 5, 35; m 61; c 3. EXPERIMENTAL FLUID PHYSICS, RADIATION PHYSICS. *Educ:* Nat Taiwan Univ, BS, 56; Nat Tsing-Hua Univ Taiwan, MS, 58; Univ Mich, MNS, 62; McMaster Univ, PhD(physics), 64. *Prof Exp:* Asst reactor physics, Argonne Nat Lab, 59-60; from asst prof to assoc prof physics, Univ Waterloo, 64-68; res assoc, Harvard Univ, 67-68; from asst prof to assoc prof nuclear eng, 68-74, PROF NUCLEAR ENG, MASS INST TECHNOL, 74- *Concurrent Pos:* Res assoc & fel, Atomic Energy Res Estab, Harwell, Eng, 64-65; vis scientist, Solid State Sci Div, Argonne Nat Lab, 75; distinguished vis prof, Univ Guelph, Can, 81; vis scientist, Lab Leon Brillouin, Centre d'Etudes Nucleaires, Saclay, France, 81; Alexander Von Humboldt US sr scientist award, WGer, 87-88. *Mem:* Fel Am Phys Soc; fel AAAS; Sigma Xi. *Res:* Study of molecular dynamics in solids and fluids by thermal neutron and laser light scattering. *Mailing Add:* 24-211 Nuclear Eng Dept Mass Inst Technol Cambridge MA 02139

CHEN, SOW-YEH, b Chang-Hwa, Taiwan, Aug 28, 39; m 72; c 2. ORAL PATHOLOGY. *Educ:* Nat Taiwan Univ, BMD, 65; Univ Ill, MS, 70, PhD(path), 72; Am Bd Oral Path, dipl. *Prof Exp:* Spec fel, Nat Inst Dent Res, 71-73; from asst prof to assoc prof, 73-84, EMER PROF PATH, SCH DENT, TEMPLE UNIV, 84- *Concurrent Pos:* Res grants, NIH & STRC. *Mem:* Am Acad Oral Path; Int Asn Dent Res; Am Asn Cancer Educ; Sigma Xi; AAAS. *Res:* Ultrastructural study of tumors in the oral region and pathobiology of oral mucosa. *Mailing Add:* 3223 N Broad St Philadelphia PA 19140

CHEN, STANLEY SHIAO-HSIUNG, b Chekiang, China, Feb 24, 37; US citizen; m 59; c 2. ENGINEERING MECHANICS, BIOMECHANICS. *Educ:* Taipei Inst Technol, dipl mech eng, 57; Ohio Univ, MS, 60; Univ Wis-Madison, PhD(eng mech), 67. *Prof Exp:* Exp engr res & develop, A O Smith Corp, Milwaukee, Wis, 60-64; instr eng, Univ Wis, 66-67; from asst prof to assoc prof, 71-76, PROF ENG, ARIZ STATE UNIV, 76- *Concurrent Pos:* Vis prof eng, Chung Cheng Inst Technol, 73-74; consult, FluiDyne Eng Corp, 73-76; dir, SEM-TEC Labs Inc, 76- *Mem:* Sigma Xi; Am Inst Aeronaut & Astronaut; Am Soc Mech Engrs; Aeronaut & Astronaut Soc Repub China. *Res:* Applied mechanics; biomechanics in human major joint replacement such as knee and hip; gait analysis and application in sport medicine. *Mailing Add:* Dept Aerospace Eng & Eng Sci Ariz State Univ Tempe AZ 85287

CHEN, STEPHEN P K, b Shanghai, China, June 19, 42. CHEMISTRY. *Educ:* Chung Chi Col, Hong Kong, Dipl, 63; Univ Wis, PhD(chem), 68. *Prof Exp:* SR RES CHEMIST, EASTMAN KODAK CO, 68- *Mem:* Am Chem Soc; Soc Rheology. *Res:* Rheological behavior of polymers; molecular mobilities in both rubbery and glassy polymers; polymer physical chemistry. *Mailing Add:* 695 Hightower Way Webster NY 14580

CHEN, STEPHEN SHI-HUA, b Taipei, Taiwan, Repub China, Dec 25, 39; US citizen; m 69; c 2. CLINICAL PATHOLOGY. *Educ:* Nat Taiwan Univ, Taipei, MD, 64; Univ Pittsburgh, PhD(biochem), 72. *Prof Exp:* CLIN ASSOC PROF PATH, STANFORD UNIV, 80- *Concurrent Pos:* Dir Clin Labs, Vet Affairs Med Ctr. *Mem:* Am Asn Pathologists; Col Am Pathologists; Am Soc Clin Pathologists. *Res:* Lipid metabolism. *Mailing Add:* Dept Path Stanford Univ Stanford CA 94305

CHEN, STEVE S, b Fukien Prov, cHINA, Feb 1, 44; US citizen; m; c 3. COMPUTER SCIENCE. *Educ:* Nat Taiwan Univ, BS, 66; Villanova Univ, MS, 72; Univ Ill, PhD(comput sci), 75. *Prof Exp:* Proj engr, Burroughs Corp, 75-78 & Floating Pt Systs, Inc, 78-79; chief designer, Cray Res, Inc, 79-83, vpres develop, 83-86 & sr vpres, 86-87; PRES & CHIEF EXEC OFFICER, SUPERCOMPUTER SYSTS, INC, 87- *Mem:* Nat Acad Eng; fel Am Acad Arts & Sci. *Mailing Add:* Off Pres Supercomputer Systs Inc 1414 W Hamilton Ave Eau Claire WI 54701

CHEN, SUNG JEN, b Chia-yi, Taiwan, Mar 31, 39; US citizen; m 71. CHEMICAL ENGINEERING, MINERAL PROCESSING. *Educ:* Nat Taiwan Univ, BS, 62; Calif Inst Technol, MS, 67; Kans State Univ, PhD(chem eng), 71. *Prof Exp:* Syst analyst, Atlantic Richfield Co, 67-69; vpres res & develop, Kenics Corp, 71-81; ASSOC PROF, UNIV LOWELL, 81- *Mem:* Am Inst Chem Engrs; Am Chem Soc. *Res:* Mixing; polymerization and two-phase flow; solvent extraction; polymer processing. *Mailing Add:* Dept Plastics Eng Univ Lowell One University Ave Lowell MA 01854

CHEN, SU-SHING, b Shanghai, China, Feb 11, 39; US citizen; m 72; c 1. INTELLIGENT SYSTEMS. *Educ:* Nat Taiwan Univ, BS, 61; Univ Tenn, MS, 64; Univ Md, PhD(math), 70. *Prof Exp:* Prof math, Univ Fla, 70-85; prof & chair computer sci, Purdue Univ, Indianapolis, 85-86; chair, 86-89, PROF COMPUTER SCI, UNIV NC, CHARLOTTE, 86-; PROG DIR, KNOWLEDGE MODELS, NSF, 91- *Concurrent Pos:* Vis assoc prof, Ga Inst Technol, 78, Univ Md, 79; vis prof, Univ Bonn, WGer, 80, Univ NC, Chapel Hill, 90-; prof lectr, George Washington Univ, 83-85; prog dir Geom Anal, NSF, 83-84, Intelligent Syst, 84-85; adj prof, NC State Univ, 89- *Mem:* Inst Elec & Electronics Engrs; Int Soc Optical Eng. *Res:* Machine intelligence; neural networks; intelligent control; CIM and computer vision. *Mailing Add:* 9610 Fairmead Dr Charlotte NC 28269

CHEN, T R, b Kaohsiung, Taiwan. TISSUE CULTURE, SOMATIC CELL GENETICS. *Educ:* Nat Taiwan Univ, BS, 58; Yale Univ, PhD(biol), 67. *Prof Exp:* Res assoc, Dept Biol, Yale Univ, 67-71; asst prof tissue cult, Grad Sch Biomed Sci, Univ Tex, Houston, 72-77; dir, Cytogenetic Lab, EKS Ctr Mental Retardation Inc, 77-80; CYTOGENETICIST, AM TYPE CULT COLLECTION, ROCKVILLE, MD, 80- *Mem:* Genetic Soc Am; Tissue Cult Asn. *Res:* Cytogenetic aspect of degrees of diversity in human cancer cells in vitro and in vivo; drug induction of karyotypic changes and stemline evoltuion; mapping of cancer genes. *Mailing Add:* Am Type Cult Collection 12301 Parklawn Dr Rockville MD 20852

CHEN, T(IEN) Y(I), b China, Jan 18, 23; nat US. ENGINEERING. *Educ:* St John's Univ, Shanghai, BS, 45; Polytech Inst Brooklyn, MCE, 51; Univ Ill, PhD(civil eng), 54. *Prof Exp:* Res assoc civil eng, Univ Ill, 54, asst prof, 54-58; assoc prof, Dartmouth Col, 58-62; assoc prof, 62-69, PROF CIVIL ENG, OHIO STATE UNIV, 70- *Mem:* Am Soc Civil Engrs; Am Concrete Inst. *Res:* Structural engineering and mechanics; numerical methods of stress analysis. *Mailing Add:* Dept Civil Eng N470 Hitchcock Hall Ohio State Univ Columbus OH 43210

CHEN, TAR TIMOTHY, b Fuching, Fujien, China, June 23, 45; US citizen; m 69; c 2. CATEGORICAL DATA ANALYSIS, CLINICAL TRIAL. *Educ:* Nat Taiwan Univ, BS, 66; Univ Chicago, MS, 69, PhD, 72. *Prof Exp:* Statistician, Ill Bell Telephone Co, Chicago, 71-73; asst prof statist, Calif State Univ, Hawyard, 73-74; biostatician, Upjohn Co, Kalamazoo, Mich, 75-79; asst prof biometrics, Univ Tex, MD Anderson Cancer Ctr, 79-84; sr biostatician, Alcon Labs, Ft Worth, Tex, 84-89; MATH STATISTICIAN, NAT CANCER INST, BETHESDA, MD, 89- *Concurrent Pos:* Vis assoc prof appl maths, Ching-Hsing Univ, Taichung, Taiwan, 74-75; assoc ed biopharmaceut sect, J Am Statist Asn. *Mem:* Am Statist Asn; Soc Epidemiol Res; Sigma Xi. *Res:* Categorical data; missing and incomplete data; misclassification problem; randomized response; clinical trial methodology. *Mailing Add:* 2297 Glenmore Terr Rockville MD 20850-3059

CHEN, TA-SHEN, b Lung-ching, Taiwan, Feb 5, 32; nat US; m 64; c 2. CONVECTIVE HEAT TRANSFER, FLUID MECHANICS. *Educ:* Nat Taiwan Univ, BS, 54; Kans State Univ, MS, 61; Univ Minn, PhD(mech eng), 66. *Prof Exp:* Mech engr, Taiwan Shipbldg Corp, 55-57 & Engalls-Taiwan Shipbldg & Dry Dock Co, 57-59; mech engr, Twin Cities Mining Res Ctr, US Bur Mines, 63-66, res engr, 66-67; res fel, Univ Minn, 66; from asst prof to assoc prof, 67-73, PROF MECH ENG, UNIV MO-ROLLA, 73- *Concurrent Pos:* NSF grants, 69-70, 75-77, 80-83, 84-86 & 87-88; grad coordr, Dept Mech Eng, Univ Mo-Rolla; assoc ed Am Inst Aeronaut & Astronaut, J Thermophysics & Heat Transfer. *Mem:* Fel Am Soc Mech Engrs; Am Soc Eng Educ; Am Inst Aeronaut & Astronaut; Sigma Xi. *Res:* Convective heat and mass transfer; fluid mechanics; natural and mixed convection; wave and thermal instability of flows. *Mailing Add:* Dept Mech & Aerospace Eng Univ Mo Rolla MO 65401

CHEN, THERESA S, b Taipei, Taiwan, Repub China, Oct 26, 44; US citizen; m 72; c 2. BIOCHEMICAL TOXICOLOGY, AGING BIOLOGY. *Educ:* Nat Taiwan Univ, BS, 67; Univ Louisville, PhD(pharmacol), 71. *Prof Exp:* fel toxicol, Univ Chicago, 71-73; res assoc pharmacol, Univ Pittsburgh, 74-77; sr pharmacologist, div Res & Develop, Travenol Labs, 77-79; asst prof, 79-85, ASSOC PROF PHARMACOL, UNIV LOUISVILLE, 85- *Mem:* NY Acad Sci; Am Soc Pharmacol & Exp Therapeut; Soc Toxicol; Soc Exp Biol & Med. *Res:* Biochemical toxicology; role of glutathione in aging toxicology; general and specific toxicity of environmental pollutants. *Mailing Add:* Dept Pharmacol & Toxicol Univ Louisville Louisville KY 40292

CHEN, THOMAS SHIH-NIEN, b Shanghai, China, Feb 28, 36; US citizen; m 66; c 4. PATHOLOGY, HISTORY & PHILOSOPHY OF SCIENCE. *Educ:* Harvard Univ, AB, 55; NY Med Col, MD, 60. *Prof Exp:* Intern path, Univ Chicago Clin, 60-61; captain, US Air Force, Barksdale AFB, La, 61-63; resident path, St Luke's Med Ctr, New York, 63-67; res fel, Namru No 2, Taipei, Taiwan, 67-69; NIH fel liver dis, Med Sch, 69-71, from asst clin prof to assoc prof, 71-84, PROF, UNIV MED & DENT NJ-NJ MED SCH, 85- *Concurrent Pos:* Staff pathologist, Vet Admin Hosp, East Orange, NJ, 71- *Mem:* Am Asn Study Liver Dis; Inst Asn Study Liver; Am Asn Pathologists. *Res:* Liver disease; medical history. *Mailing Add:* Vet Admin Hosp East Orange NJ 07019

CHEN, THOMAS TIEN, b Liaoyang, Liaoning Prov, China, July 1, 47; US citizen; m 73; c 3. REPRODUCTIVE ENDCRINOLOGY, CELL BIOLOGY. *Educ:* Fu Jen Cath Univ, BS, 71; Univ Fla, MS, 73, PhD(animal sci), 75. *Prof Exp:* Fel, Colo State Univ, 75-78, staff fel reprod endocrinol, NIH, 78-79; asst prof, 79-85, ASSOC PROF ZOOL & FAC MEM, BIOTECHNOL, UNIV TENN, 85-, FAC MEM, CELL BIOL, 81- *Concurrent Pos:* Prin investr, NIH, 82-85 & Am Cancer Soc, 86- *Mem:* AAAS; Soc Study Reprod; Endocrine Soc; Am Soc Cell Biol. *Res:* Regulation of luteinizing-hormone receptor in the ovary, hyperprolactinemia and ovarian function; macrophage growth factors and ovarian function; growth factors and uterine function. *Mailing Add:* Dept Zool Univ Tenn Knoxville TN 37996

CHEN, TIEN CHI, b Hong Kong, Nov 12, 28; US citizen; m 67. COMPUTER SCIENCE. *Educ:* Brown Univ, SB, 50; Duke Univ, MA, 52, PhD(physics), 57. *Prof Exp:* Assoc physicist res ctr, 56-58, staff mathematician, 58-59, mathematician data systs div, 59-61, develop engr, 61-62, sr programmer & mgr problem-oriented programming, 62-67, tech staff mem, Advan Comput Systs, 67-68, STAFF MEM RES CTR, INT BUS MACH CORP, 68-; PROF COMPUT SCI & ELECTRONICS & HEAD, UNITED COL, CHINESE UNIV, HONG KONG, 80- *Concurrent Pos:* Vis scientist, Univ Uppsala, 65-66; mgr technol & systs, Int Bus Mach San Jose fel Res Lab, 73; vis prof electronics, Chinese Univ Hong Kong, 79-80. *Mem:* Am Phys Soc; Asn Comput Mach; fel Inst Elec & Electronics Engrs. *Res:* Eigenvalue problems in chemical physics; digital computer design and applications; numerical analysis; computer algorithms; magnetic bubbles. *Mailing Add:* United Col Chinese Univ Hong Kong Shatin New Territories Hong Kong

CHEN, TIMOTHY SHIEH-SHENG, b Taipei, Taiwan, Feb 8, 49; m 75; c 3. CHEMISTRY, SURFACE CHEMISTRY. *Educ:* Cheng Kung Nat Univ, BS, 71, Univ Notre Dame, MS, 76, PhD(chem), 78. *Prof Exp:* Res assoc, Mich Molecular Inst, 78-79; res assoc, 79-82, RES CHEMIST, HC CHEM RES, SAN JOSE UNIV/AMES RES CTR, NASA, 82- *Mem:* Am Chem Soc. *Res:* Surfactans behaviors in aqueous and nonaqueos solutions; radiation effects on polymeric materials; ultra high molecular weight, polymer synthesis and characteriation; elastomer, expoxy, polyimide and composite res; organic synthesis and characterization; inorganic polymer synthesis and characterization; ceramic fiber process and characterization. *Mailing Add:* 223-4 NASA Ames Res Ctr Moffett Field CA 94035

CHEN, TSAI HWA, b Wenling Chekiang, China, Oct 14, 23; US citizen; m 54; c 4. ELECTRONIC SYSTEMS, COMPUTERS. *Educ:* Pei Yang Univ, China, BSEE, 46; Chiao Tung Univ, MSEE, 61; Univ Calif, Los Angeles, MS, 65, PhD(eng), 68. *Prof Exp:* Asst prof & assoc res fel radio & radar var univ in China & Ord Res Inst, China, 47-59; tech specialist comput circuit design, Electronics Div, Nat Cash Regist Co, 63-69; mem tech staff airborne comput, Guid & Control Systs Div, Litton Indust, 69-70; proj engr comput memory, Res Dept, Data Processing Div, NCR Corp, 70-73; sr proj engr automatic test systs, Support Systs Div, Hughes Aircraft Co, 73-79, sr proj engr, Eng Div, Radar Systs Group, 79- 87, SR SCIENTIST & ENGR, TACTICAL SYSTS DIV, RADAR SYSTS GROUP, HUGHES AIRCRAFT CO, 87- *Mem:* AAAS; Inst Elec & Electronics Engrs. *Res:* Automatic electronic test systems and applications of transmission line theory in large electronic systems; radar systems; rf noise. *Mailing Add:* Hughes Aircraft Co PO Box 92426 Bldg R9 MS 9050 Los Angeles CA 90009

CHEN, TSANG JAN, b Taiwan, China, Nov 13, 34; m 61; c 3. POLYMER CHEMISTRY. *Educ:* Nat Taiwan Univ, BS, 57; NDak State Univ, PhD(polymer chem), 67. *Prof Exp:* SR CHEMIST, RES LABS, EASTMAN KODAK CO, 67- *Mem:* Am Chem Soc. *Res:* Emulsion polymerization; anionic polymerization. *Mailing Add:* 475 Warren Ave Rochester NY 14618-4319

CHEN, TSEH-AN, b Shanghai, China, Oct 26, 28; US citizen; m 54; c 4. BOTANY. *Educ:* Nat Taiwan Univ, BS, 51; Univ Wis, MS, 53; Univ NH, PhD(bot), 62. *Prof Exp:* Res analyst bot, Univ NH, 60-62; from asst prof to assoc prof biol, Fairleigh Dickinson Univ, 62-69; assoc prof entom & econ zool, 69-74, PROF PLANT PATHOL, RUTGERS UNIV, 74-, DIR GRAD PROG NEMATOL, 70- *Concurrent Pos:* Res grants, NIH, 64-66, NJ Dept Health, 65 & NSF, 70-; res assoc, Cornell Univ, 63-66; mem, US-Repub China Prog, NSF, 78-82, USDA Competetive, 79-, NSF, 81-82 & sci proj rev bd, Nat Sci Coun, Repub China; mem, Subcomt Mycoplasmas, 84-; chmn, Dept Plant Path, Rutgers Univ, 91. *Mem:* Fel Am Phytopath Soc; Soc Nematologists; Am Soc Microbiol; Int Orgn Mycoplasmologists. *Res:* Interaction of soil fungi and plant parasitic nematodes on development of root rots; mechanism of plant virus transmission by nematodes; myco-plasma-like organisms that cause plant diseases; corn stunt spiro-plasma; nutrition and pathogenesis of spiroplasma; ultrastructures of nematodes; monoclonal antibodies to mycoplasma like organisms. *Mailing Add:* Dept Plant Path Rutgers Univ New Brunswick NJ 08903

CHEN, TSONG MENG, b Yunlin, Taiwan, July 1, 35; US citizen; m 63; c 3. PLANT PHYSIOLOGY. *Educ:* Nat Taiwan Univ, BS, 58, MS, 62; Univ Calif, Davis, PhD(plant physiol), 66. *Prof Exp:* Res assoc plant physiol, Mich State Univ, 66-68 & Univ Ga, 68-70; assoc prof crop physiol, Nat Taiwan Univ, 70-72; res assoc plant physiol, Univ Ga, 72-73; biologist, Union Carbide Corp, 73-76; assoc plant physiol, Mobil Chem Co, 76-81; sr res assoc, FMC Corp, 81-88; DEPT ELEC ENG, UNIV S FLA, 88-; PARTNER, DAN CONSTRUCT CO, PLAINSBORO, NJ, 88- *Mem:* Am Soc Plant Physiologists; Weed Sci Soc Am; Am Agron Soc. *Res:* Natural and synthetic plant growth regulators which will increase the yield of the major agronomic crops. *Mailing Add:* Three Wheatston Ct Princeton Junction NJ 08550-1936

CHEN, TSONG-MING, b Chia-Yi, Taiwan, China, Nov 25, 34; m 64; c 2. SOLID STATE ELECTRONICS, ELECTRICAL ENGINEERING. *Educ:* Nat Taiwan Univ, BS, 56; Univ Minn, PhD(elec eng), 64. *Prof Exp:* From asst prof to assoc prof eng sci, Fla State Univ, 64-72; assoc prof, 72-74, PROF ELEC ENG, UNIV SFLA, 74- *Concurrent Pos:* Tech consult electronic indust. *Mem:* Inst Elec & Electronics Engrs; Int Soc Hybrid Minoelectronics; Am Vacuum Soc. *Res:* Noise in electronic devices; semiconductor electronics. *Mailing Add:* Dept Elec Eng Univ SFla Tampa FL 33620

CHEN, TU, b I-Lan, Taiwan, Mar 19, 35; US citizen; m 61; c 2. MAGNETISM, MATERIALS SCIENCE. *Educ:* Cheng Kung Univ, Taiwan, BS, 58; Univ Minn, Minneapolis, MS, 64, PhD(metall eng, mat sci), 67. *Prof Exp:* Teacher math & physics, Nan-Yiang Girls High Sch, Taiwan, 61-62; staff engr mat sci res, Int Bus Mach Corp, 67-68; prin scientist mat sci, Corp Res Ctr, Northrop Corp, 68-71; res scientist, Xerox Palo Alto Res Ctr, 71-75, prin scientist mat sci, 75-; CHMN BD, CHIEF SCI, KOMAG, INC. *Mem:* Am Asn Crystal Growth; Inst Elec & Electronics Engrs; Am Chem Soc; Mat Res Soc. *Res:* Solid state physics and magnetism; crystal growth, structure and phase equilibria; inorganic chemistry. *Mailing Add:* Komag Inc 591 Yosemite Dr Milpitas CA 95035

CHEN, TUAN WU, b Taiwan, China, Mar 26, 36; m 63; c 2. THEORETICAL PHYSICS, LIGHT SCATTERING. *Educ:* Nat Taiwan Univ, BS, 58; Tsing Hua Univ, Taiwan, MS, 60; Syracuse Univ, PhD(physics), 66. *Prof Exp:* Instr physics, Univ Guelph, 67-68; from asst prof to assoc prof, 68-79, PROF PHYSICS, NMEX STATE UNIV, 79- *Concurrent Pos:* Fel physics, Univ Toronto, 66-68; vis scientist, Los Alamos Sci Lab, 74-81 & Calif Inst Technol, 82-83; vis prof, Nat Tsing Hua Univ, 74-75. *Mem:* Am Phys Soc. *Res:* Formalism of asymptotic quantum field theory in functional derivatives; effect of masses of gauge fields on chiral symmetry; high energy behaviors of scattering processes; nucleus scattering at medium energy; nonperturbative method for bound state problems; light scattering by particles at high energy. *Mailing Add:* Dept Physics NMex State Univ Las Cruces NM 88003

CHEN, TUNG-SHAN, b April 17, 39; US citizen; m 64; c 2. FOOD CHEMISTRY. *Educ:* Nat Taiwan Univ, Taipei, BS, 60; Univ Calif, Berkeley, MS, 64, PhD(comp biochem), 69. *Prof Exp:* From asst to assoc prof, 69-78, PROF FOOD SCI, CALIF STATE UNIV, NORTHRIDGE, 78-, DIR FOOD SCI, NUTRIT & DIETICS, MARILYN MAGARAM CTR, 90- *Concurrent Pos:* Chmn, Food Sci & Nutrit Div, Calif State Univ, Northridge, 70-74; vis assoc prof, Univ Calif, Los Angeles, 74; food eng specialist, Chinese Univ Develop Proj, Nat Acad Sci, 85; sr res fel, Ctr Cancer & Develop Biol, 85- *Honors & Awards:* Joseph Drown Found Res Award, 83. *Mem:* Am Chem Soc; Am Dietetic Asn; fel Am Inst Chemists; Sigma Xi; Inst Food Technologists. *Res:* Effects of processing and packaging on food qualities; stability of vitamins; improvement of methodology for vitamin analyses. *Mailing Add:* Home Econ Dept Calif State Univ Northridge CA 91330

CHEN, VICTOR JOHN, b Kowloon, Hong Kong, Sept 25, 52; US citizen; m 81; c 1. CHEMICAL MODIFICATION OF PROTEINS, METALLO PROTEINS. *Educ:* Grinnell Col, BA, 74; Iowa State Univ, PhD(biochem), 81. *Prof Exp:* Fel, Univ Minn, St Paul, 80-82 & Med Sch, Univ Tex, Houston, 82-84; SR BIOCHEMIST, LILLY RES LABS, 85- *Mem:* Am Soc Biochem & Molecular Biol. *Res:* Characterization of the iron binding site of the oxidase isopenicillin N Synthase by multiple spectioscopic techniques; synthesis and characterization of chemically modified insulin analogs. *Mailing Add:* Lilly Res Labs Lilly Corp Ctr Indianapolis IN 46285

CHEN, W(AI) K(AI), b Nanking, China, Dec 23, 36; m 62; c 2. ELECTRICAL ENGINEERING. *Educ:* Ohio Univ, BSEE, 60, MS, 61; Univ Ill, Urbana-Champaign, PhD(elec eng), 64. *Prof Exp:* Asst electronics, Ohio Univ, 60-61, from asst prof to prof, Ohio Univ, 64-78, distinguished prof elec eng & grad chmn dept, 78-81; asst graph theory, Co-ord Sci Lab, Univ Ill, 62-63, res assoc topology, 64; PROF & HEAD, DEPT ELEC ENG & COMPUT SCI, UNIV ILL, CHICAGO, 81- *Concurrent Pos:* Vis assoc prof, Purdue Univ, 70-71; vis prof, Univ Hawaii, Manoa, 79; hon prof, Nanjing Inst Technol, Chengdu Inst Radio Eng, Northeast Univ Technol, Zhejiang Univ, East China Inst Technol, Nanjing Inst Post & Telecommun, AnHui Univ & Wuhan Univ, 85, Hangzhou Inst Electronic Eng, China, 90; distinguished guest prof, Dept Elec Eng, Chuo Univ, Tokyo, Japan, 87. *Honors & Awards:* Lester R Ford Award, Math Asn Am, 67; Alexander Von Humboldt Award, Fed Repub Germany, 85; Japan Soc for Promotion Sci Fel Award, Tokyo, 86. *Mem:* Fel AAAS; Fel Inst Elec & Electronics Engrs; Math Asn Am; Asn Comput Mach; Tensor Soc; Sigma Xi; Nat Soc Professional Engrs. *Res:* Communication nets; network topology; graph and network theory; switching circuits; broadband matching; active and passive filters; networks and systems. *Mailing Add:* Dept Elec Eng & Comput Sci M/C 154) Univ Ill Box 4348 Chicago IL 60680

CHEN, WAI-FAH, b Chekiang, China, Dec 23, 36; US citizen; m 66; c 3. STRUCTURAL ENGINEERING. *Educ:* Cheng-Kung Univ, BSCE, 59; Lehigh Univ, MS, 63; Brown Univ, PhD(solid mech), 66. *Prof Exp:* From asst prof to prof civil eng, Lehigh Univ, 66-76; prof, 76-80, HEAD DEPT STRUCTURES, PURDUE UNIV, 80- *Concurrent Pos:* Consult, Exxon Prod Res Co, 79-, Karagozian & Case Struct Engrs, 85-, GA Technol, 87-, Skidmore, Owings & Merrill, 87, World Bank, 88-; chmn, Comt Properties Mat, Am Soc Civil Engrs, 79-82, Comt Connections, 80-82 & Comt Stability, 88-90; vis prof civil eng, Stanford Univ, 83 & Univ Kassel, WGermany, 85; mem, Struct Stability Res Coun, Earthquake Eng Res Coun & Coun Tall Bldg & Urban Habitat; US sr scientist award, Alexander von Humboldt Found, 84. *Honors & Awards:* James F Lincoln Award, Arc Welding Found, 71, 74, 81 & 84; T R Higgins lectr, Am Inst Steel Construct, 85; Raymond C Reese Res

Prize, Am Soc Civil Engrs, 85, Shortridge Hardesty Award, 90. *Mem:* Am Soc Civil Engrs; Int Asn Bridge & Struct Eng; Struct Stability Res Coun; Am Acad Mech. *Res:* Structural connections; beam columns; offshore structures; constitutive modeling of metals, soils and concrete; concrete structures during construction; ice mechanics; soil and concrete plasticity; structural stability; polymer and fiber concrete; structural mechanics solid mechanics. *Mailing Add:* Sch Civil Eng Purdue Univ West Lafayette IN 47907

CHEN, WAYNE H(WA-WEI), b Soochow, China, Dec 13, 22; m 57. ELECTRICAL ENGINEERING, MATHEMATICS. *Educ:* Chiao Tung Univ, BS, 44; Univ Wash, Seattle, MS, 49, PhD(math), 52. *Prof Exp:* Eng staff, Applied Physics Lab & Cyclotron Proj, Univ Wash, Seattle, 49-50, assoc dept math, 50-52; from asst prof to prof elec eng, Univ Fla, 52-89, chmn dept, 65-73, dean, Col Eng & dir eng, Indust Exp Sta, 73-88, EMER PROF ELEC ENG, UNIV FLA, 89- *Concurrent Pos:* Vis prof, Nat Univ Taiwan & Chiao Tung Univ, 64 & Univ Carabobo, Valencia, Venezuela, 72; vis scientist & lectr, Nat Acad Sci to USSR, 67; nat chmn, Elec Eng Dept Heads Asn, 69-70; mem accreditation team, Inst Elec & Electronics Engrs-Eng Coun Prof Develop, 68-72; pres, Col Comput-Aided Design-Comput-Aided Mfg Consortium, Inc, 85-87. *Mem:* Am Soc Eng Educ; fel Inst Elec & Electronics Engrs; Am Asn Univ Prof. *Res:* Switching circuits and theory; network theory; applied mathematics. *Mailing Add:* Col Eng Univ Fla Gainesville FL 32611

CHEN, WEN SHERNG, b Taiwan; US citizen. MEDICINAL CHEMISTRY, BIOCHEMISTRY. *Educ:* Taipei Med Col, BS, 66, Univ NC, PhD(med chem), 74. *Prof Exp:* Fel biochem, Dept Chem, Univ Colo, 74-75; fel, Dept Biochem, Col Med, Univ Iowa, 75-79; SR RES SCIENTIST PROTEIN PROD LAB, KRAFT RES & DEVELOP, DART & KRAFT, INC, 79- *Mem:* Am Chem Soc. *Res:* Physicochemical properties of proteins; isolation and purification of enzymes; enzyme kinetics; chemical modification of enzymes and proteins; design and synthesis of new compounds for biological evaluation, structure-activity relationship. *Mailing Add:* Kraft Inc 801 Waukegan Rd Glenview IL 60025-4391

CHEN, WENPENG, b Ta-Chi, Taiwan. PHYSICS, OPTICS. *Educ:* Nat Tsing-Hwa Univ, Taiwan, BS, 70; Univ Pa, PhD(physics), 77. *Prof Exp:* SCIENTIST PHYSICS, MARTIN MARIETTA LABS, 77- *Mem:* Am Phys Soc; Optical Soc Am. *Res:* Metal surface and adsorbed layer; surface plasmon and surface plariton; visible and infrared spectroscope using attentuated total reflection technique; wave guide and electromagnetic wave propagation; interface of metal and semiconductor; gallium arsenide Impatt diodes; IR switch. *Mailing Add:* Martin Marietta Labs 1450 S Rolling Rd Baltimore MD 21227

CHEN, WILLIAM KWO-WEI, b Shanghai, China, July 18, 28; US citizen; m 51; c 2. POLYMER CHEMISTRY, CHEMICAL ENGINEERING. *Educ:* Mass Inst Technol, BSc, 51; Polytech Inst Brooklyn, PhD(chem), 58. *Prof Exp:* Chem engr, Quaker Chem Prod Co, 51-52; chem engr, Am Mach & Foundry Co, 52-53, sr chem engr, 53-54, group leader plastic prod, 54-56, sect mgr desalination & membranes, 56-61, asst mgr chem lab, 61-63, mgr liquid processing lab, 63-64; mgr res & develop, Celanese Plastic Co, 64-68, VPRES AMCEL CO & GEN EXPORT MGR, CELANESE PLASTIC CO, 68- *Mem:* Am Chem Soc; Soc Plastics Eng. *Res:* Membranes, films and battery separators; desalination and liquid processing; plastic and resins; marketing; international trade. *Mailing Add:* 102 Pinegrove Rd Berkeley Heights NJ 07922

CHEN, WINSTON WIN-HOWER, b 1942; m; c 2. NEUROLOGY. *Educ:* Purdue Univ, PhD(phys chem), 73. *Prof Exp:* assoc prof biochem & neurol, Sch Med, Johns Hopkins Univ, 78-90; CONSULT, 90- *Mem:* Am Soc Biochem & Molecular Biol; Am So Neurochem. *Res:* Glycolipid synthesis; lysosomal disorders; Schwann cell metabolism; cell biol peroxisomal disorders; molecular biology of adrenoleukodystro phytrophy. *Mailing Add:* Sch Med 841 Rm 408 Kennedy Inst Johns Hopkins Univ 707 N Broadway St Baltimore MD 21205

CHEN, YANG-JEN, b Tainan, Taiwan, June 16, 47; c 1. BIOFLUIDDYNAMICS, ENERGY. *Educ:* Chung Yuan Christian Col Sci & Eng, BS, 69; La State Univ, MS, PhD(physics). *Prof Exp:* Res assoc physics, Lehigh Univ, 77-78; sr prod develop engr air pollution control technol, C-E Walther, Inc, 78-79; supvr develop & testing air pollution control technol, Combustion Eng Environ Syst Div, 79-80; SR RES ENGR AIR POLLUTION CONTROL TECHNOL, WESTERN PRECIPITATION DIV, JOY MFG CO, 80- *Mem:* Air Pollution Control Asn; Am Phys Soc; Am Asn Aerosol Res. *Res:* Develop and evaluate the emerging air pollution control technologies including precipitator, fabric filter and scrubber; bench scale research; full scale demonstration research project. *Mailing Add:* 24049 Avenida Crescenta Valencia CA 91355

CHEN, YI-DER, b Taipei, Taiwan, May 29, 40; m 69. THEORETICAL CHEMISTRY, BIOPHYSICS. *Educ:* Nat Taiwan Univ, BS, 63; Nat Tsing Hua Univ, MS, 65; Pa State Univ, PhD(chem), 70. *Prof Exp:* Res chemist biophys Univ Calif, Santa Cruz, 69-72; res assoc theoret biol, Nat Inst Arthritis, Diabetes & Digestive & Kidney Dis, NIH, 72-78, res chemist, 78-; SR RES ASSOC, DEPT MED & GERONT, STANFORD UNIV. *Mem:* Biophys Soc; Am Chem Soc; Fedn Am Soc Exp Biol. *Res:* Statistical mechanics of nonspherical molecules; fluctuations and noise in chemical and biological systems; theoretical studies of membrane transports, muscle contractions, conformational transitions in biopolymers, statistical mechanics of ring polymers and circular DNA. *Mailing Add:* Dept Med & Geront Stanford Univ Stanford CA 94305

CHEN, YIH-WEN, b Taichung, Taiwan, Apr 10, 48; m 77; c 2. ELECTRO-ORGANIC SYNTHESIS, SPECIALTY CHEMICALS. *Educ:* Nat Tsing Hua Univ, Shing Chu, Taiwan, BS, 71, MS, 74; Univ Tex Austin, PhD(anal chem), 82. *Prof Exp:* Anal chemist, Develop Ctr Sci & Technol, 71-72; chemist, Taiwan Aluminum Corp, 74-75; asst scientist, Chun Shan Inst Sci & Technol, 75-78; postdoctoral fel, Solar Energy Res Inst, 82-84; res scientist, Enron Corp, 84-86; res chemist, Warner-Jenkinson Co, Subsid Universal Food Corp, 86-87; SR RES & DEVELOP CHEMIST, HILTON-DAVIS CO, SUBSID PMC, INC, 87- *Res:* Electro-organic synthesis of dyes and polymer dyes for food, drug, cosmetics, markers, inks, cleaners, textiles, plastics, and images; electrosynthesis of conducting polymers for batteries, anti- static agents, sensors, and EMI; thin film electrodeposition; solution deposition; photoelectrochemistry on conductors and semiconductors for solar cells. *Mailing Add:* 8052 Leeshore Dr Maineville OH 45039

CHEN, YING-CHIH, b Chung-King, China, July 7, 48; US citizen; m; c 2. SEMICONDUCTOR LASER, OPTOELECTRONICS. *Educ:* Nat Taiwan Univ, BS, 70; Columbia Univ, MA, 74, PhD(physics), 78. *Prof Exp:* Scientist, McDonnell Douglas, 79-83; mem tech staff, GTE Lab, 83-86; PROF, HUNTER COL, CITY UNIV NEW YORK, 86- *Mem:* Am Phys Soc; Optical Soc Am; Inst Elec & Electronics Engrs. *Res:* Laser physics; nonlinear optics; author of over 50 technical papers. *Mailing Add:* 24 Robin Hill Rd Scarsdale NY 10583

CHEN, YOK, b Soochow, China, July 27, 31; US citizen; m 61; c 3. PHYSICS. *Educ:* Univ Wis, BSc, 52; Purdue Univ, PhD(physics), 65. *Prof Exp:* Electron microscopist, Univ Chicago, 53-55; physicist, Hoffman Semiconductor Prod, Evanston, Ill, 55-58; RES PHYSICIST, OAK RIDGE NAT LAB, 65- *Honors & Awards:* IR 100 Award, 75 & 82. *Mem:* Fel Am Phys Soc; fel Am Ceramic Soc. *Res:* Radiation damage in semiconductors; impurities and defects in insulators; radiation effects and ion implantation in oxides; optical and magnetic resonance spectroscopy. *Mailing Add:* Sol Sta Div Bldg 3025 Oakridge Nat Lab PO Box 2008 Oak Ridge TN 37831

CHEN, YOUNG CHANG, b Shin-Chu, Taiwan, Feb 28, 35; US citizen; m 67; c 2. TUMOR VIROLOGY, MICROBIOLOGY. *Educ:* Nat Chung-Hsing Univ Taiwan, BS, 57; Univ Tokyo, Japan, MS, 63; Univ Utah, AM, 68; Univ Southern Calif, PhD(virol), 73. *Prof Exp:* Lectr chem, Taiwan Prov Shin-Chu Sch Polytech, 59-61; sr lectr microbiol, Univ Southern Calif, 73-76; ASST PROF MICROBIOL & VIROL, N TEX STATE UNIV, 76-; MEM STAFF, DEPT MICROBIOL, UNIV TEX, AUSTIN. *Concurrent Pos:* Fel, Nat Cancer Inst, 73-76; consult, Monogram Industs, Inc, 76-; fac res grants, N Tex State Univ, 76-79. *Mem:* Am Soc Microbiol; AAAS. *Res:* Animal virology and oncogenic virology; somatic cell genetics and viral genetics; antiviral and anti-tumor substances; cell cultures and biology of aging; public health and water pollution microbiology. *Mailing Add:* 2325 E Windsor Dr Denton TX 76201

CHEN, YU, b June 8, 21; US citizen; m 51; c 2. MECHANICS. *Educ:* Chiao Tung Univ, BS, 42; Harvard Univ, MS, 47, ScD, 50. *Prof Exp:* Sr develop engr, Burroughs Corp, 52-56; sr res engr, Ford Motor Co, 56-59; assoc prof mech eng, NY Univ, 59-64; assoc prof, 64-66, chmn, dept mech & nat sci, 81-85; prof, 66-78, DISTINGUISHED PROF MECH, RUTGERS UNIV, 78- *Concurrent Pos:* Consult, Bell Tel Labs, Inc, 60-63 & Goddard Space Flight Ctr, 64-67, Arradcom, 73- *Mem:* Fel Am Soc Mech Engrs; Am Soc Eng Educ; Sigma Xi. *Res:* Solid mechanics; elasticity; structural mechanics; space flight dynamics; mechanics of material processing; optimization of materials processing including all types of forming, extrusion, machining etc by mathematical modeling and computer algorithms to simulate the evolution of the process in full details. *Mailing Add:* 99 Windsor Way Berkeley Heights NJ 07922

CHEN, YU WHY, b Nantungchow, China, Apr 1, 10; m 38; c 1. MATHEMATICS. *Educ:* Univ Gottingen, PhD, 34. *Prof Exp:* Prof math, Peking Univ, 36-45; res assoc, NY Univ, 46-49; res fel, Inst Adv Study, 49-50; assoc prof, Univ Okla, 50-52; from assoc prof to prof, Wayne State Univ, 52-65; prof, 65-82, EMER PROF MATH, UNIV MASS, AMHERST, 82- *Mem:* Am Math Asn; Soc Indust & Appl Math; Am Math Soc. *Res:* Partial differential equations and applications. *Mailing Add:* Dept Math Univ Mass Amherst MA 01003

CHEN, YUDONG, b Suzhou, China, Sept 10, 61. TIME SERIES ANALYSIS. *Educ:* Chongking Univ, China, BSE, 82, MSE, 87; Univ Mich, MSE, 89, PhD(mech eng), 91. *Prof Exp:* Asst engr, Suzhou Power Co, China, 82-84; res asst elec eng, Chongking Univ, China, 84-87; RES ASST MECH ENG, DEPT MECH ENG & APPL MECH, UNIV MICH, ANN ARBOR, 87- *Res:* Computer aided design and manufacturing; computer numerical control; machine vision application in manufacturing; range image analysis; machine condition monitoring and diagnosis; software system; circuit design; electrical power system time series and statistics; quality control. *Mailing Add:* 2364 Bishop No 17 Ann Arbor MI 48105

CHEN, YUH-CHING, b Fukien, China, May 20, 30; m 55; c 3. MANAGEMENT SCIENCE. *Educ:* City Univ New York, PhD(math), 66. *Prof Exp:* Asst lectr math, Taiwan Norm Univ, 54-59; asst lectr, Nanyang Univ, Singapore, 59-60; teacher high sch, Malasia, 60-62; asst prof, Univ Minn, Morris, 64-65 & Wesleyan Univ, 66-71; asst prof, 71-72, chmn dept, 74-78 & 87-90, ASSOC PROF MATH, FORDHAM UNIV, 72- *Concurrent Pos:* Vis prof, Beijing Normal Univ, 78-79 & Huanghe Univ, China, 87. *Mem:* Sigma Xi; NY Acad Sci. *Res:* Homology and homotopy theories and their applications and Eigen value problems of Banach space operators. *Mailing Add:* Dept Math Fordham Univ Bronx NY 10458

CHEN, YUNG MING, b China, Dec 30, 35; US citizen; m 62; c 1. APPLIED MATHEMATICS. *Educ:* Univ Md, BS, 56; Drexel Inst, MS, 58; Univ Calif, Berkeley, MA, 60; NY Univ, PhD(appl math), 63. *Prof Exp:* Res engr, Radio Corp Am, NJ, 56-58; asst, Electronics Res Labs, Univ Calif, Berkeley, 58-59; asst, Dept Math, 59-60; asst appl math, Courant Inst Math Sci, NY Univ, 60-63; asst prof, Purdue Univ, 63-65; assoc prof, Univ Fla, 65-67; assoc prof, 67-72, PROF APPL MATH, STATE UNIV NY STONY BROOK, 72- *Mem:* Sigma Xi; Soc Indust & Appl Math. *Res:* Wave propagation; approximation methods in initial and boundary value problems; numerical analysis and stochastic process numerical methods for inverse problems; numerical methods for inverse problems. *Mailing Add:* Dept Appl Math State Univ NY Stony Brook NY 11794

CHENEA, PAUL F(RANKLIN), b Milton, Ore, May 17, 18; m 41; c 2. ENGINEERING MECHANICS. *Educ:* Univ Calif, BS, 40; Univ Mich, MS, 47, PhD(eng mech), 49. *Hon Degrees:* DSc, Rose Polytech Inst, 68; DEng, Purdue Univ, 68; DEngSc, Tri-State Col, 68; DH Clarkson Col Technol, 71; DEng, Drexel Univ, 71. *Prof Exp:* Proj engr, Contractors Pac Naval Air Bases, 40 & US Army Ord Dept, 41; from instr to assoc prof, Univ Mich, 46-52; asst dean & head div eng sci, Purdue Univ, 52-55, assoc dean eng, 56-58, actg head, Sch Elec Eng, 57-58; Webster vis prof elec eng, Mass Inst Technol, 58-59, head, Sch Mech Eng, 59-61, head div math sci, 60-61, vpres acad affairs, 61-63, actg dean, Sch Sci Educ & Humanities, 62-63; vpres acad affairs, Purdue Univ, 63-67; sci dir, Gen Motors Corp, 67-69, vpres res labs, Tech Ctr, 69-83; RETIRED. *Concurrent Pos:* Consult, Gen Motors Corp, E I du Pont de Nemours & Co, Inc & RCA Corp; dir, Comn Eng Educ. *Mem:* Nat Acad Eng; AAAS; Am Soc Mech Engrs; Nat Soc Prof Engrs; fel Am Acad Arts & Sci. *Res:* Administration; continuum physics; applied mechanics; electromagnetic field theory. *Mailing Add:* PO Box 2548 Prescott AZ 86302

CHENEVERT, ROBERT (BERNARD), b St-Cuthbert, Que, Can, Dec 16, 47; m 75; c 2. ORGANIC CHEMISTRY. *Educ:* Univ Montreal, BS, 68, MS, 70; Univ Sherbrooke, PhD(chem), 75. *Prof Exp:* Res fel chem, Harvard Univ, 75-76; from asst prof to assoc prof, 76-85, PROF CHEM, UNIV LAVAL, 85- *Mem:* Am Chem Soc; Chem Inst Can. *Res:* Use of enzymes and microorganisms as reagents in organic synthesis; host-guest chemistry; chemical ecology. *Mailing Add:* Dept Chem Sci & Genie Univ Laval Quebec PQ G1K 7P4 Can

CHENEY, B VERNON, b Salt Lake City, Utah, June 11, 36; m 64; c 6. THEORETICAL CHEMISTRY, NUCLEAR MAGNETIC RESONANCE. *Educ:* Univ Utah, BA, 61, PhD(phys chem), 66. *Prof Exp:* Teaching & res asst, dept chem, Univ Utah, 61-66; res scientist phys & anal chem, 66-84, SR RES SCIENTIST THEORET CHEM, COMPUT CHEM SUPPORT, UPJOHN CO, KALAMAZOO, MICH, 84- *Concurrent Pos:* Vis scientist, Oxford Univ, 86-87. *Mem:* Am Chem Soc; Sigma Xi. *Res:* Application of theoretical chemistry to problems in pharmacology; drug-receptor interactions and drug structure. *Mailing Add:* Comput Chem Support Upjohn Co Kalamazoo MI 49001

CHENEY, CARL D, b Buhl, Idaho, Feb 19, 34; m 72. BEHAVIOR OF ORGANISMS. *Educ:* Utah State Agr Col, BS, 56; Ariz State Univ, MS, 62, PhD(psychol), 66. *Prof Exp:* Asst prof psychol, Eastern Wash State Col, 66-68; PROF PSYCHOL, UTAH STATE UNIV, 68- *Mem:* Asn Behav Anal; Psychonomic Soc. *Res:* Overt behavior of organisms; environment and behavior interactions. *Mailing Add:* Basic Behav Lab Utah State Univ Logan UT 84322

CHENEY, CHARLES BROOKER, b New Haven, Conn, Mar 2, 12; m 34; c 5. OBSTETRICS & GYNECOLOGY. *Educ:* Yale Univ, BA, 34, MD, 41; Am Bd Obstet & Gynec, dipl, 52. *Prof Exp:* Asst surg obstet & gynec, Sch Med, Yale Univ, 41-43, asst obstet & gynec, 46-47, instr, 47-49, clin instr, 49-52, asst clin prof, 52-69, sr clin assoc, 69-74, assoc clin prof, 74-80, CLIN PROF OBSTET & GYNEC, SCH MED, YALE UNIV, 80-; ASST ASSOC CHIEF OBSTET & GYNEC & ATTEND OBSTETRICIAN & GYNECOLOGIST, YALE-NEW HAVEN HOSP, 57- *Concurrent Pos:* Former pres, Conn Med Exam Bd. *Mem:* AMA; Am Col Obstet & Gynec. *Mailing Add:* 104 Huntington St New Haven CT 06511

CHENEY, CLARISSA M, b Sanford, Maine, Oct 22, 46; m 84. DEVELOPMENTAL GENETICS. *Educ:* Goucher Col, AB, 69; Yale Univ, MPhil, 70; Univ Pa, PhD(biol), 79. *Prof Exp:* Teach sci, private high sch, New Haven, Conn & Boston, Mass, 70-74; teaching asst embryol, Marine Biol Lab, Woods Hole, Mass, 78; NIH Nat Res Serv Award Fel, 79-82, res assoc, 82-84, ASST PROF BIOL, JOHNS HOPKINS UNIV, 84- *Concurrent Pos:* Lectr, Goucher Col, 85-86. *Mem:* Am Soc Cell Biol; Soc Develop Biol; Int Soc Develop Biol. *Res:* The role of the cytoskeleton in development; protein defects in Drosophila imaginal disc mutants. *Mailing Add:* Dept Genetics Wash Univ Sch Med 4566 Scott Ave St Louis MO 63110

CHENEY, DARWIN L, b Burley, Idaho, Sept 1, 40; m 64; c 5. NEUROTRANSMITTER DYNAMICS. *Educ:* Stanford Univ, PhD(pharmacol), 71. *Prof Exp:* Dir neurosci res, Ciba-Geigy, Inc, 83-86, dir neurosci & cardiovasc pharmacol, 85-88, exec dir clin opers & statist, 88-89; PRES & CHIEF EXEC OFFICER, FIDIA RES FOUND, 89- *Mem:* Soc Neurosci; Am Soc Hypertension; Am Soc Pharmacol & Exp Therapeut. *Mailing Add:* FIDIA Res Found 1640 Wisconsin Ave NW Suite 2 Washington DC 20007

CHENEY, ELLIOTT WARD, (JR), b Gettysburg, Pa, June 28, 29; m 52; c 3. MATHEMATICS. *Educ:* Lehigh Univ, BA, 51; Univ Kans, PhD(math), 57. *Prof Exp:* Instr math, Univ Kans, 52-56; design specialist & mathematician, Convair-Astronaut Div, Gen Dynamics Corp, 56-59; mem tech staff, Space Tech Labs, Inc, 59-61; asst prof math, Iowa State Univ, 61-62; from asst prof to assoc prof, Univ Calif, Los Angeles, 62-65; assoc prof, 65-66, PROF MATH, UNIV TEX, AUSTIN, 66- *Concurrent Pos:* Vis assoc prof, Univ Tex, 64; guest prof, Univ Lund, 66-67; vis prof, Mich State Univ, 69-70; assoc ed, J Approximation Theory; assoc ed, J Numerical Anal, Soc Indus & Appl Math & Numerical Function Anal & Optimization; sr res fel, Univ Lancaster, Eng, 84. *Mem:* Am Math Soc; Math Asn Am; Soc Indust & Appl Math. *Res:* Approximation theory; linear inequalities; numerical analysis. *Mailing Add:* Dept Math Univ Tex Austin TX 78712

CHENEY, ERIC SWENSON, b New Haven, Conn, Nov 17, 34; wid; c 4. ECONOMIC GEOLOGY. *Educ:* Yale Univ, BS, 56, PhD(geol), 64. *Prof Exp:* Instr sci, Southern Conn State Col, 63-64; asst prof geol, 64-69, ASSOC PROF GEOL, UNIV WASH, 69- *Concurrent Pos:* Consult, Maine Geol Surv, 63-69, Amoco Minerals Co, 70-71, Texasgulf Inc, 72, Urangesellschaft-USA, 74, Continental Oil Co Inc, 75, Chevron Resources Co, 76-77, Wold Nuclear Co, 78, Seattle City Light, 79-80 & B/T Enterprises, 81; vis prof, Stanford Univ, 74, Univ Pretoria, SAfrica, 84-85, Rande Afrikans Univ, SAfrica, 87-89 & 90; pres, Cambria Corp, 81-91. *Mem:* AAAS; Geol Soc Am; Sigma Xi; Geol Soc SAfrica; Soc Econ Geologists. *Res:* Geology of ore deposits; mineral and energy resources; siting of nuclear power plants; geology of Pacific northwest; Precambrian geology southern Africa. *Mailing Add:* Dept Geol Sci Univ Wash Seattle WA 98195

CHENEY, FREDERICK WYMAN, b Ayer, Mass, Jan 17, 35; m 59; c 2. ANESTHESIOLOGY. *Educ:* Tufts Univ, BS, 56, MD, 60. *Prof Exp:* Instr anesthesiol, 64-66, from asst prof to assoc prof, 66-74, PROF ANESTHESIOL, SCH MED, UNIV WASH, 74- *Concurrent Pos:* NIH res fel, Sch Med, Univ Wash, 66-67. *Mem:* Am Physiol Soc; Am Soc Anesthesiologists; Am Thoracic Soc; Soc Critical Care Med. *Res:* Pulmonary edema; pulmonary vasculature; clinical respiratory care. *Mailing Add:* Dept Anesthesiol Mail Stop RN10 Univ Wash Sch Med Seattle WA 98195

CHENEY, HORACE BELLATTI, b Emerson, Iowa, Dec 15, 13; m 40; c 3. SOIL FERTILITY. *Educ:* Iowa State Univ, BS, 35; Ohio State Univ, PhD(soil fertility), 42. *Prof Exp:* Jr soil surveyor, Soil Conserv Serv, USDA, Mo, Iowa & Wis, 35-37; exten assoc agron, Iowa State Univ, 37-39; res assoc, Ohio State Univ, 39-41; exten & res asst prof agron, econ & sociol, Iowa State Univ, 41-43, exten asst prof agron, 43-44, from assoc prof to prof, 45-52; prof & head dept, 52-77, EMER PROF SOILS, ORE STATE UNIV, 77- *Mem:* Fel AAAS; fel Soil Sci Soc Am (pres, 64); fel Soil Conserv Soc Am; fel Am Soc Agron (pres, 73). *Res:* Soil management. *Mailing Add:* Dept Soil Sci Ore State Univ Corvallis OR 97331

CHENEY, JAMES A, b Los Angeles, Calif, Feb 2, 27; m 51, 67, 88; c 8. AERONAUTICAL & ASTRONAUTICAL ENGINEERING. *Educ:* Univ Calif, Los Angeles, BS, 51, MS, 53; Stanford Univ, PhD(eng mech), 63. *Prof Exp:* Asst engr struct mech, Univ Calif, Los Angeles, 51-53; assoc engr, L T Evans Found Eng Co, 53-55; struct engr, Missile & Syst Div, Lockheed Aircraft Corp, 55-58, head strength group, Lockheed Missile & Space Co, 58-59, grad study engr, 59-62; from asst prof to assoc prof 62-80, dir, Ctr Geotech Modeling, 84-89, PROF CIVIL ENG, UNIV CALIF, DAVIS, 80- *Concurrent Pos:* Staff engr & consult, Lockheed Missile & Space Co, 62-66. *Mem:* Am Soc Civil Engrs. *Res:* Structural mechanics; buckling of structures and structural elements; dynamic response of structure and soil; geotechnical centrifuge model testing; structural properties of bone, constitutive relations in soil mechanics; soil-structure interaction. *Mailing Add:* Dept Civil Eng 2108 Bainer Hall Univ Calif Davis CA 95616

CHENEY, MARGARET, b Lawrence, Kans, Jan 5, 55; m 85. INVERSE PROBLEMS. *Educ:* Oberlin Col, BA, 76; Ind Univ, PhD(math), 82. *Prof Exp:* Postdoctoral appl math, Stanford Univ, 82-84; asst prof math, Duke Univ, 84-88; ASSOC PROF MATH, RENSSELAER POLYTECH INST, 88- *Concurrent Pos:* Grad res assoc, Los Alamos Nat Lab, 76-80. *Mem:* Am Math Soc; Soc Indust & Appl Math; Asn Women Math. *Res:* Inverse problems. *Mailing Add:* Dept Math Sci Rensselaer Polytech Inst Troy NY 12180

CHENEY, MONROE G, b Ft Worth, Tex, Mar 10, 19; m 52; c 3. PHYSICS, GEOPHYSICS. *Educ:* Rice Univ, BA, 41; Univ Tex, MA, 50; Columbia Univ, MA, 52. *Prof Exp:* Radio physicist, US Naval Res Lab, DC, 42-47; asst prof physics, Hardin-Simmons Univ, 47-49; physicist, US Bur Mines, Pa, 52; reservoir engr, Anzac Oil Corp, Tex, 52-55 & Socony Mobil, Venezuela, 55-59; asst prof physics, Arlington State Col, 59-67, asst prof physics, Univ Tex, Arlington, 67-89; RETIRED. *Concurrent Pos:* Mem, Byrd Antarctic Exped, 46-47; consult, Anzac Oil Corp, 55-74. *Mem:* Am Inst Mining, Metall & Petrol Eng; Am Geophys Union. *Res:* Ground constants at radio frequency; gas resaturation of oil; seismic surface-waves; fracturing limestone with crude oil; seismic waves in ice; radiowave patterns; water conservation; underground housing; oil secondary recovery. *Mailing Add:* 1217 W Cedar Arlington TX 76012

CHENEY, PAUL DAVID, b Jamestown, NY, Oct 10, 47; m 66; c 2. PHYSIOLOGY, NEUROPHYSIOLOGY. *Educ:* State Univ NY Col Fredonia, BS, 69; State Univ NY Upstate Med Ctr, PhD, 75. *Prof Exp:* Fel physiol, Univ Wash Sch Med, 74-77; res asst prof, Dept Physiol & Biophysics, Univ Wash, 77-78; ASST PROF PHYSIOL, UNIV KANS MED CTR, 78- *Mem:* Soc Neurosci; Sigma Xi. *Res:* Motor control; muscle receptors; spinal cord; role of cerebral cortex and brainstem in movement. *Mailing Add:* Dept Physiol 39th & Rainbow Blvd Kansas City KS 66103

CHENG, ALEXANDER H-D, b Taipei, Taiwan, May 25, 52; US citizen; m 79; c 2. GROUNDWATER, HYDRAULIC FRACTURING. *Educ:* Nat Taiwan Univ, BS, 74; Univ Mo-Columbia, MS, 78; Cornell Univ, PhD(civil eng), 81. *Prof Exp:* Asst prof civil eng, Sch Civil & Environ Eng, Cornell Univ, 81-82 & Dept Civil Eng & Eng Mech, Columbia Univ, 82-85; ASSOC PROF CIVIL ENG, DEPT CIVIL ENG, UNIV DEL, 85- *Concurrent Pos:* Consult, Dowell Schlumberger Inc, 85-; rec secy, Int Asn Boundary Element Method, 90-91. *Mem:* Am Soc Civil Engrs; Am Geophys Union; Am Soc Petrol Engrs; Am Acad Mech; Int Soc Hydrol Sci. *Res:* Poroelasticity in rock mechanics; hydraulic fracturing; numerical modeling of groundwater and contaminant transport; stochastic groundwater theories; chaotic dynamics; boundary element method. *Mailing Add:* Dept Civil Eng Univ Del Newark DE 19716

CHENG, ANDREW FRANCIS, b Princeton, NJ, Oct 15, 51; m 79; c 2. ASTROPHYSICS. *Educ:* Princeton Univ, AB, 71; Columbia Univ, MPhil, 74, PhD(physics), 77. *Prof Exp:* Fel physics, Bell Tel Labs, 76-78; asst prof physics, Rutgers Univ, 78-83; MEM STAFF, APPL PHYS LAB, JOHNS HOPKINS UNIV, 83- *Concurrent Pos:* Mem, Comt Planetary & Lunar Explor, Nat Acad Sci, 87-90; assoc ed, J Geophys Res, 87-89, Geophys Res Lett, 89-; ed, EOS-Transactions Am Geophys Union, 89- *Honors & Awards:* Group Achievement Awards, NASA, 86 & 90. *Mem:* Am Phys Soc; Am Astron Soc; Am Geophys Union; NY Acad Sci. *Res:* Physics of magnetospheres, pulsars, Jupiter, and Saturn. *Mailing Add:* Appl Phys Lab Johns Hopkins Univ Johns Hopkins Rd Laurel MD 20723-6099

CHENG, CHENG-YIN, b Shinpu, Formosa, Jan 29, 30; nat US; m 61. INSTRUMENTAL METHODS OF ANALYSIS, WATER TREATMENT. *Educ:* Berea Col, BS, 55; Univ Ill, MS, 56, PhD(soil chem), 60. *Prof Exp:* Instr chem, Wilson Col, 60-61; from asst prof to assoc prof, Shippensburg State Col, 61-65; from asst prof to assoc prof, Ithaca Col, 65-68; PROF CHEM, E STROUDSBURG UNIV, 68- *Concurrent Pos:* Vis scientist, Peoples Repub China, 85. *Mem:* Am Chem Soc; Sigma Xi; Am Soc Agron. *Res:* Radioisotope technology; instrumental methods of analysis; ion selective electrodes; air and water quality; field study of the effects of culm ash from fluidized boiler on field corn; characterization of anions present in drinking and natural waters. *Mailing Add:* Dept Chem E Stroudsburg Univ East Stroudsburg PA 18301

CHENG, CHIA-CHUNG, b China, May 5, 25; m 53; c 4. ORGANIC CHEMISTRY. *Educ:* Chekiang Univ, BS, 48; Univ Tex, MA, 51, PhD, 54. *Prof Exp:* Res assoc org res, NMex Highlands Univ, 54-57 & Princeton Univ, 57-59; head, Cancer Chemother Sect, Midwest Res Inst, 59-66, head, Med Chem Sect, 66-78; DIR, MID-AM CANCER CTR & PROF PHARMACOL, UNIV KANS MED CTR, 78- *Concurrent Pos:* Mem med chem A study sect, NIH, 73-77; mem, Adv Comt on Clin Investr Chemother & Hemat, Am Cancer Soc, 82-86; mem, Bd Sci Counr, Div Cancer Etiology, Nat Cancer Inst, 82-86; sr fac res award, Univ Kans Med Ctr, 84. *Honors & Awards:* Sci Award, Midwest Res Inst, 74. *Mem:* AAAS; Am Chem Soc; NY Acad Sci; Chem Soc; Sigma Xi; Res Soc Am. *Res:* Synthesis, identification and reaction mechanism study of organic compounds; synthesis and evaluation of anticancer and antimalarial agents antimetabolites, antibiotics, alkylating agents, vitamin analogs and natural products. *Mailing Add:* 10301 Overbrook Leawood KS 66206-2652

CHENG, CHIANG-SHUEI, b Sinchu, Taiwan, Sept 22, 35; US citizen; m 60; c 2. PLASMA PHYSICS. *Educ:* Nat Taiwan Univ, BS, 58; Nat Tsing Hua Univ, MS, 60; Lehigh Univ, PhD(physics), 68. *Prof Exp:* Instr physics, Lehigh Univ, 65-69; assoc prof, 69-73, PROF PHYSICS, E STROUDSBURG UNIV, 73- *Mem:* Am Phys Soc; Sigma Xi. *Res:* Scattering function and transport theory of a plasma. *Mailing Add:* Box 7060 A Rd Seven Northgate Estate Stroudsburg PA 18360

CHENG, CHING-CHI CHRIS, physical chemistry, inorganic chemistry, for more information see previous edition

CHENG, CHUEN HON, b Hong Kong, Oct 4, 50; m 84; c 1. GEOPHYSICS. *Educ:* Cornell Univ, BSc, 73. *Hon Degrees:* ScD(geophys), Mass Inst Technol, 78. *Prof Exp:* res assoc geophys, 78-82, PRIN RES SCIENTIST, MASS INST TECHNOL, 83- *Mem:* Am Geophys Union; Soc Explor Geophysicists; Seismol Soc Am; Acoust Soc Am. *Res:* Borehole geophysics; elastic wave propagation in porous rocks; planetary interiors. *Mailing Add:* E34-454 Mass Inst Technol Cambridge MA 02139

CHENG, CHUNG-CHIEH, b Chungking, China; US citizen; m 69; c 2. SOLAR PHYSICS. *Educ:* Nat Taiwan Univ, BS, 60; Harvard Univ, MA, 63, PhD(astron & astrophys), 70. *Prof Exp:* Res assoc solar physics, Nat Res Coun, 70-72; solar physicist, Naval Res Lab & Ball Brothers Res, 73-79; astrophysicist, Marshall Space Fight Ctr, NASA, 79-81; ASTROPHYSICIST, NAVAL RES LAB, 81- *Concurrent Pos:* Vis scientist, China Acad Sci, 81, 83, & Arcetric Astrophys Observ, Florence, Italy, 82, 85, 87 & 89. *Mem:* Am Astron Soc; Int Astron Union. *Res:* Interpretation of the ultraviolet, extra ultraviolet and x-ray spectra of the sun; theoretical studies of solar activities; applications of magnetohydrodynamic theory and plasma physics to interpretations of solar phenomena. *Mailing Add:* Code 4177 Naval Res Lab Washington DC 20375-5000

CHENG, DAVID, b Chungking, China, July 21, 41; US citizen; c 2. EXPERIMENTAL PHYSICS. *Educ:* Univ Calif, Berkeley, BS, 62, MS, 63, PhD(exp high energy physics), 65. *Prof Exp:* Physicist exp physics, Lawrence Radiation Lab, Univ Calif, 65-67; assoc physicist, Brookhaven Nat Lab, 67-70; mem tech staff, Bell Labs, 70-74; PROJ LEADER, XEROX RES CTR, 74- *Mem:* Optical Soc Am; Am Phys Soc; Inst Elec & Electronics Engrs; Soc Photog Scientists & Engrs; Soc Photo-Optical Instrumentation Engrs. *Res:* High speed and high density optical recording technique; laser physics and applications; integrated circuit pattern generation techniques; laser and CCD image recording; acousto-optical devices and electro optics. *Mailing Add:* 3709 Ortega Ct Palo Alto CA 94303

CHENG, DAVID H(ONG), b I-Shing, China, Apr 19, 20; nat US; m 49; c 2. ENGINEERING MECHANICS. *Educ:* Franco-Chinese Univ, BS, 42; Univ Minn, MS, 47; Columbia Univ, PhD(mech), 50. *Prof Exp:* Instr civil eng, Rutgers Univ, 49-50; struct eng, Ammann & Whitney, 50-52; struct & develop engr, M W Kellogg Co, 53-55; from asst prof to assoc prof, 55-66, dir grad studies, Sch Eng, 77-78, exec off PhD prog eng, Grad Ctr, 77-78, prof civil eng, City Col New York, 66-85, dean, Sch Eng, 79-85; PRES, TECHTRAN INC, 86- *Concurrent Pos:* Am Soc Eng Educ-NASA fel, 64; NASA fel, 64-65; hon res assoc, Harvard Univ, 67-68; consult, M W Kellogg Co & Inst Defense Anal. *Mem:* Am Soc Mech Engrs; Am Soc Civil Engrs. *Res:* Structural dynamics and vibrations; wind engineering and stability of suspended bridges; stress in pressure vessels and pipings for nuclear power. *Mailing Add:* 200 Old Palisade Rd Ft Lee NJ 07024

CHENG, DAVID H S, b Shanghai, China; US citizen; m 49; c 2. SUBMARINE ANTENNAS, COMPUTER GRAPHICS. *Educ:* St John's Univ, BA, 43; Univ Mo, MA, 49, BS, 56, MS, 58, PhD(elec eng), 64. *Prof Exp:* Scientist, NASA-Am Soc Elec Engrs, Goddard Space Flight Ctr, 65, fel, 67 & 68; elec engr, Naval Underwater Systs Ctr, New London Lab, 70-79; from asst prof to assoc prof, PROF ELEC ENG, UNIV MO-COLUMBIA, 57-; ELECTRONICS ENGR, NAVAL UNDERWATER SYSTS CTR, NEW LONDON LAB, 83- *Concurrent Pos:* Vis scientist, Radiation Lab, Univ Mich, 67-68, consult, 80-83. *Mem:* Inst Elec & Electronics Engrs; Am Phys Soc; Am Geophys Union; Electromagnetics Soc; Sigma Xi; Int Sci Radio Union. *Res:* Research and development in connection with submarine antenna systems. *Mailing Add:* Code 3421 Naval Underwater Systs Ctr New London Lab New London CT 06320

CHENG, DAVID KEUN, b China, Jan 10, 18; US citizen; m 48; c 1. ELECTRICAL ENGINEERING, ELECTROMAGNETICS. *Educ:* Chiao Tung Univ, BSEE, 38; Harvard Univ, SM, 44, ScD, 46. *Hon Degrees:* DEng, Nat Chiao Tung Univ, 85. *Prof Exp:* Engr, Cent Radio Corp, China, 38-43; electronics & proj engr, US Air Force Cambridge Res Lab, 46-48; from asst prof to prof, 48-83, EMER PROF ELEC ENG, SYRACUSE UNIV, 84- *Concurrent Pos:* Dir electromagnetics & antennas res proj, Syracuse Univ Res Inst, 49-83; consult, IBM, 52-53, Gen Elec Co, 57-60, Syracuse Res Corp, 61-65 & TRW, 86-87; John Simon Guggenheim Mem Found fel, 60-61; consult ed elec sci, Addison-Wesley Publ Co, 61-78; consult ed elec eng monographs, Intext Educ Pubs, 69-72; centennial prof, Syracuse Univ, 70; Nat Acad Sci exchange scientist, Hungary, 72, Yugoslavia, 74, Poland & Romania, 78; European distinguished lectr, Inst Elec & Electronics Engrs Antennas & Propagation Soc, 75-76; liaison scientist, US Off Naval Res, London Br, 75-76; mem, US Nat Comn Comts, Int Union Radio Sci; hon prof, Beijing Inst Posts & Telecommun, 82-, Northwest Inst Telecommun Eng, 82-, Shanghai Jiao Tong, Univ, 85-, People's Repub China. *Honors & Awards:* Ann Achievement Award, Chinese Inst Engrs, 62. *Mem:* Fel AAAS; fel Inst Elec & Electronics Engrs; fel Brit Inst Elec Engrs; Sigma Xi; Am Soc Eng Educ; NY Acad Sci. *Res:* Electromagnetic theory; synthesis and optimization of antenna arrays; theory and applications of Walsh functions; communication systems. *Mailing Add:* 4620 N Park Ave No 104E Chevy Chase MD 20815-4549

CHENG, EDWARD TEH-CHANG, b Pingtung, Taiwan, Nov 23, 46; US citizen; m 73; c 2. FUSION ENERGY, NUCLEAR SYSTEMS. *Educ:* Nat Tsin Hua Univ, Taiwan, BS, 69, MS, 71; Univ Wis-Madison, MS, 73, PhD(nuclear eng), 76. *Prof Exp:* Asst scientist fusion eng, Univ Wis-Madison, 76-78; STAFF ENGR, GEN ATOMICS, SAN DIEGO, CALIF, 78- *Mem:* Am Nuclear Soc; Sigma Xi. *Res:* Neutronics; radioactivity; fusion energy-related reactor engineering. *Mailing Add:* Gen Atomics PO Box 85608 San Diego CA 92138

CHENG, FRANCIS SHENG-HSIUNG, US citizen. PHYSICAL CHEMISTRY, POLYMER SCIENCE. *Educ:* Cheng-kung Univ, Taiwan, BS, 59; Baylor Univ, PhD(phys chem), 67; Washington Univ, DSc(polymer sci), 70. *Prof Exp:* Eng consult, Robertson & Assocs, Inc, 71-72; res scientist polymer anal & qual control, United Merchants & Mfg, Inc, 72-75; res specialist injection molding processes, Monsanto, Inc, 75-77; PROCESS ENGR & ASST TO VPRES FOREING TRADE, R T VANDERBILT CO, INC, 77- *Concurrent Pos:* NSF res assoc, Wash Univ, 66-67; res engr, Esso Res & Eng Co, 70-71; chem abstractor, Chem Abstr Serv, 65-75. *Honors & Awards:* Chem Abstr Serv Award, 76. *Mem:* Am Chem Soc; Am Inst Chem Engrs. *Res:* Organic systhesis; physical and thermodynamic characterizations of organic and inorganic sulfur compounds and polymeric substances; implementation of computerized fail-proof process and quality control systems. *Mailing Add:* 12 Proctor Dr West Hartford CT 06117-2037

CHENG, FRANK HSIEH FU, b Shanghai, China, Nov 16, 23; US citizen; m 58; c 3. BIOCHEMISTRY, IMMUNOCHEMISTRY. *Educ:* St John's Univ, China, BS, 46; Univ Tenn, MS, 50; Ind Univ, PhD(biochem), 57. *Prof Exp:* Chemist & assoc supt pharmaceut lab, T W Wu & Co, China, 46-49; asst, Ind Univ, 53-56; biochemist, Toledo Hosp Inst Med Res, Ohio, 58-63; asst prof, 64-69, assoc prof, 69-80, PROF RADIOL, UNIV IOWA, 80- *Concurrent Pos:* Fel, Univ Wis, 56-58. *Mem:* AAAS; Am Chem Soc; Am Acad Allergy; Soc Nuclear Med; Sigma Xi. *Res:* Radiobiology. *Mailing Add:* Dept Radiol Dept Nuclear Med Univ Iowa Iowa City IA 52242

CHENG, FRANKLIN YIH, b Shanghai, China, July 1, 36; m 63; c 2. EARTHQUAKE STRUCTURAL ENGINEERING. *Educ:* Cheng Kung Univ, Taiwan, BSc, 60; Univ Ill, Champaign, MSc, 62; Univ Wis-Madison, PhD(civil eng), 66. *Prof Exp:* Struct engr, C F Murphy Architects & Engrs, Ill, 62-63 & Sargent & Lundy Engrs, 63; from asst prof to assoc prof 66-76, PROF STRUCT ENG, UNIV MO-ROLLA, 76-; STRUCT ENGR, LOS ALAMOS NAT LAB, NMEX, 81- *Concurrent Pos:* Res grant, Nat Sci Found, Univ Mo-Rolla, 67-81; hon fel, Univ Wis; hon prof, Harbin Arch & Civil Eng Inst, China & Xian Inst Metall & Construct Eng, China. *Mem:* Am Soc Civil Engrs; Am Soc Eng Educ; Earthquake Eng Res Inst; Sigma Xi; Struct Stability Res Coun. *Res:* Inelastic behavior of earthquake structures; optimum structural design and optimum control of nondeterministic earthquake structures; nonlinear modeling of low-rise reinforced concrete walls; plastic behavior of tall buildings; dynamic instability of stochastic excitations. *Mailing Add:* Dept Civil Eng Univ Mo Rolla MO 65401

CHENG, GEORGE CHIWO, b China, Sept 7, 28; US citizen; m 70; c 2. COMPUTER SCIENCES, BIOMEDICAL ENGINEERING. *Educ:* South China Univ, BA, 53; Mont State Univ, MS, 60; Univ Ga, PhD, 84. *Prof Exp:* Res assoc, Electronics Res Lab, Mont State Univ, 59-61; asst prof elec eng, Southeastern Mass Tech Inst, 61-62; sr res scientist, 63-68, head prod innovation div, Nat Biomed Res Found, 68-71; consult, Toshiba Res & Develop Ctr, Japan, 71-72; assoc prof, Univ Fla, 73-74; exec staff, Comp Sci Corp, 80-82; CONSULT, AUTOMATIC MED DATA PROCESSING & INT TECHNOL EXCHANGE, 74-; MGR PAC STRATEGY & BUS DEVELOP DIR, ELECTRONIC DATA SYSTS CORP, 82- *Concurrent Pos:* Vis lectr, Southeastern Mass Tech Inst, 62-; managing ed, Pattern Recognition Soc J & Comput in Biol & Med J. *Mem:* AAAS; Am Soc Eng Educ; Pattern Recognition Soc. *Res:* Information processing; pattern recognition; microelectronics and billiongate computer design; nervous system simulation; automatic medical data processing; pictorial data processing by computers; innovation management; international technology exchange; awarded US patents on computerized image systems. *Mailing Add:* 1141 Belvoir La Virginia Beach VA 23464-6724

CHENG, H(SIEN) K(EI), b Macau, South China, June 13, 23; nat US; m 56; c 1. FLUID MECHANICS, APPLIED MATHEMATICS. *Educ:* Chiao Tung Univ, BSc, 47; Cornell Univ, MSc, 50, PhD(aeronaut eng), 52. *Prof Exp:* Aerodynamic engr, Bell Aircraft Corp, 52-56; res aerodynamicist, Cornell Aeronaut Lab Inc, Cornell Univ, 56-57, prin aerodynamicist, 57-63; lectr,

Stanford Univ, 63-64; spec lectr, 64-65, PROF AEROSPACE ENG, UNIV SOUTHERN CALIF, 65- *Concurrent Pos:* Consult to McDonnell Douglas Corp, 64-65, Rand Corp, 65-69, Aerospace Corp, 68-71, Northrop Corp, 71-74, Flow Res Corp, 73- 76 & TRW, 86- *Mem:* Nat Acad Eng; Am Phys Soc; Fel Am Inst Aeronaut & Astronaut. *Res:* Theory of subsonic, transonic, supersonic and hypersonic flows; theoretical gas dynamics; geophysical fluid dynamics; mechanics of animal swimming, flying and soaring; singular-perturbation problems in fluid mechanics. *Mailing Add:* Dept Aerospace Eng Univ Southern Calif Univ Park Los Angeles CA 90007

CHENG, HAZEL PEI-LING, b Hong Kong; Can citizen. GASTROENTEROLOGY, CELL BIOLOGY. *Educ:* McGill Univ, BSc, 67, MSc, 69, PhD(anat), 72. *Prof Exp:* Lectr anat, McGill Univ, 72-73; res staff cell biol, Yale Univ, 73-75; asst prof anat, Univ Ill, Urbana, 75-77; from asst prof to assoc prof, 77-85, PROF ANAT, UNIV TORONTO, 85- *Concurrent Pos:* Adj res assoc cell biol, Rockefeller Univ, 73. *Honors & Awards:* Murray L Barr Jr Scientist Award. *Mem:* Am Asn Anatomists; Am Soc Cell Biol; Can Fed Biol Sci. *Res:* Biology of the intestinal epithelium. *Mailing Add:* Dept Anat Univ Toronto Med Sci Bldg Toronto ON M5S 1A8 Can

CHENG, HERBERT S, b Shanghai, China, Nov 15, 29; US citizen; m 53; c 4. FAILURE ANALYSIS, COMPUTER AIDED DESIGN. *Educ:* Univ Mich, BSME, 52; Ill Inst Technol, MS, 57; Univ Pa, PhD(mech eng), 61. *Prof Exp:* Jr mech engr, Int Harvester Co, 52-53; proj engr, Mach Eng Co, 53-56; instr mech eng, Ill Inst Technol, 56-57, Univ Pa, 57-61; asst prof, Syracuse Univ, 61-62; res engr, Mech Technol, Inc, 62-68; assoc prof, 68-74, dir, Ctr Eng Tribology, 84-89, WALTER P MURPHY PROF MECH ENG, NORTHWESTERN UNIV, 74- *Concurrent Pos:* Consult, Gen Motors Res Lab, Warren, Mich, John Deere Tractor Works, Waterloo, Iowa, Nat Bur Standards, Wash, DC, Eaton Co, Southfield, Mich, Crane Packing Co, Morton Grove, Ill & Gleason Works, Rochester, NY, 85. *Honors & Awards:* Nat Award, Soc Tribolotists & Lubrication Engrs, 87; Mayo D Hersey Award, Am Soc Mech Engrs, 90. *Mem:* Nat Acad Eng; Am Gear Mfg Asn; Soc Tribology & Lubrication Engrs (vpres, 85-90); Gear Res Inst (vpres); fel Am Soc Mech Engrs. *Res:* Failure mechanisms in elastohydrodynamic contacts; surface roughness effects in mixed lubrication; contact fatigue; gas lubrication; tribology of ceramics and composites. *Mailing Add:* Northwestern Univ 219 Catalysis Bldg Evanston IL 60208

CHENG, HSIEN, AUTONOMIC & CARDIOVASCULAR PHARMACOLOGY. *Educ:* Vanderbilt Univ, PhD(pharmacol), 71. *Prof Exp:* SR RESEARCHER PHARMACOL, MERRELL DOW RES INST, 81- *Mailing Add:* Merrell Dow Res Inst 2110 E Galbraith Rd Cincinnati OH 45215

CHENG, HSIEN HUA, economic entomology, insect ecology; deceased, see previous edition for last biography

CHENG, HUNG-YUAN, Taipei, Taiwan, May 12, 50; China citizen; m 75; c 1. ELECTROCHEMISTRY, NEUROCHEMISTRY. *Educ:* Nat Tsing Hua Univ, Taiwan, BS, 72, Ohio State Univ, PhD(anal chem), 78. *Prof Exp:* Teaching asst chem, Ohio State Univ, 74-76, res asst, 76-78; res assoc anal & neuro chem, Univ Kans, 78-80; ASST PROF CHEM, UNIV MD, COLLEGE PARK, 80- *Mem:* Am Chem Soc. *Res:* Development of electroanalytical techniques for neurochemical applications, especially in the field of brain research. *Mailing Add:* Smith Kline Beckman Co L950 709 Swedeland Rd King of Prussia PA 19406-2646

CHENG, HWEI-HSIEN, b Shanghai, China, Aug 13, 32; US citizen; m 62; c 2. SOIL CHEMISTRY, BIOCHEMISTRY. *Educ:* Berea Col, BA, 56; Univ Ill, MS, 58, PhD(agron), 61. *Prof Exp:* Res assoc soil chem, Univ Ill, 61-62; soil biochem, Iowa State Univ, 62-63, asst prof, 64-65; Fulbright res scholar & collab soil chem, soils res ctr, State Agr Univ, Belgium, 63-64; from asst prof to assoc prof soils, 65-77, chmn, prog environ sci & regional planning, 77-79, assoc dean grad sch, 82-86, interim chmn, Dept Agron & Soils, 86-87, PROF SOILS, WASH STATE UNIV, 77- *Concurrent Pos:* Guest scientist, Julich Nuclear Res Ctr, Ger, 72-73, 74 & 79-80, Acad Sinica, Repub China, 78 & Fed Agr Res Ctr, Braunschweig, Ger, 80; assoc ed, J Environ Quality, 83- *Mem:* AAAS; Soc Environ Toxicol Chem; fel Soil Sci Soc Am; Am Chem Soc; fel Am Soc Agron; Coun Agr Sci & Technol; Int Soc Chem Ecol. *Res:* Nitrogen fractionation and distribution in soils; application of nitrogen-15 and carbon-14 in soils research; pesticide movement and transformation in environment; soil organic matter turnover; groundwater quality; acid deposition effect. *Mailing Add:* Dept Soil Sci Univ Minn 1991 Upper Buford Circle St Paul MN 55108

CHENG, KANG, b Kwei-Young, China, Feb 17, 46; m 71; c 1. ENDOCRINOLOGY, DIABETES. *Educ:* Nat Taiwan Univ, BS, 68; State Univ NY Albany, MS, 72; Mt Sinai Sch Med, PhD(biochem), 77. *Prof Exp:* Res fel, Sloan-Kettering Inst Cancer Res, 77-79; res asst prof, Dept Pharmacol, Univ Va, 79-85; res fel, 85-90, SR RES FEL, MERCK SHARP & DOHME RES LAB, 90- *Mem:* Sigma Xi. *Res:* Mechanism of insulin action; regulation of growth hormone secretion. *Mailing Add:* Merck Sharp & Dohme Res Lab PO Box 2000 Rahway NJ 07065

CHENG, KIMBERLY MING-TAK, b Hong Kong, Feb 22, 48; Brit & Can citizen; m 76. AVIAN BEHAVIOR, POULTRY BREEDING. *Educ:* Tenn Technol Univ, BS, 69; Southern Ill Univ, MS, 71; Univ Minn, PhD(poultry breeding), 78. *Prof Exp:* Res specialist waterfowl behav, James Ford Bell Mus Natural Hist, Univ Minn, 78-80; asst prof, 80-85, ASSOC PROF POULTRY GENETICS & BEHAV, DEPT ANIMAL SCI, UNIV BC, 85- *Concurrent Pos:* Dir, Quail Genetic Stock Ctr, Univ BC, 82-; short-term adv, CIDA Indo Proj, 89. *Mem:* Animal Behav Soc; Poultry Sci Asn; Sigma Xi; AAAS; Am Ornithologists Union; Soc Conserv Biol. *Res:* Evolution of waterfowl reproductive behavior; genetic changes affecting behavior associated with domestication; early imprinting and mate preference; development of animal models for research; propagation of endangered waterfowl species. *Mailing Add:* Dept Animal Sci Univ BC Vancouver BC V6T 2A2 Can

CHENG, KUANG LIU, b Chuhsien, Chekiang, China, July 9, 19; nat US; m 49; c 2. MECHANICS, PHYSICS. *Educ:* Ordnance Eng Col, China, BS, 47; Va Polytech Inst, MS, 58, PhD(eng mech), 61. *Prof Exp:* From asst to assoc engr, Chinese Govt Arsenal, 47-56; sr engr, Burroughs Res Ctr, 60-64; res physicist, Cent Res Dept, Lord Corp, 64-66, sr physicist, 66-67, res assoc, 67-72, sr tech assoc, 72-75; consult engr, Am Sterilizer Co, 84-85; RETIRED. *Res:* Solid state magnetic ink; superconductivity and cryotronics; thermoelasticity; viscoelasticity; electroelasticity; mechanical properties of high polymers. *Mailing Add:* 2432 W 36th St Erie PA 16506-3566

CHENG, KUANG LU, b Yangchow, China, Sept 14, 19; nat US. ANALYTICAL CHEMISTRY. *Educ:* Northwestern Col, BS, 41; Univ Ill, MS, 49, PhD(soil chem), 51. *Prof Exp:* Fel, Univ Ill, 51-52; microchemist, Com Solvents Corp, 52-53; instr chem, Univ Conn, 53-55; engr, Westinghouse Elec Corp, 55-57; assoc dir res, Metals Div, Kelsey-Hayes Co, 57-59; mem tech staff, Labs, Radio Corp Am, 59-66; PROF CHEM, UNIV MO-KANSAS CITY, 66- *Concurrent Pos:* Elected Titular mem, Div Anal Chem, Int Union Pure & Appl Chem, 69-79; fel, Bd Trustees, Univ Kansas City, 84. *Honors & Awards:* N T Veatch Award, 79. *Mem:* Fel AAAS; Am Chem Soc; Electrochem Soc; Soc Appl Spectros. *Res:* Photoelectron spectroscopy; ion selective electrodes; surface science; positron annihilation spectroscopy; catalysis. *Mailing Add:* Dept Chem Univ Mo Kansas City MO 64110

CHENG, KUANG-FU, nonparametric inference, applied probability, for more information see previous edition

CHENG, KUO-JOAN, b Taiwan, China, Dec 9, 40; m 68; c 2. AGRICULTURAL MICROBIOLOGY. *Educ:* Nat Taiwan Univ, BS, 63; Univ Sask, MS, 66, PhD(microbiol), 69. *Prof Exp:* Res asst dairy & food sci, Univ Sask, 64-66; Nat Res Coun Can fel microbiol, MacDonald Col, McGill Univ, 69-71; RUMINANT MICROBIOLOGIST, ANIMAL SCI SECT, RES STA, CAN DEPT AGR, 71- *Concurrent Pos:* Vis scientist, Rowett Res Inst, Scotland, 78-79; vis prof, Univ Guelph, 81; adj prof microbiol, Univ Alta, 85-90. *Mem:* Am Soc Microbiol; Can Soc Microbiol; Am Soc Animal Sci; Can Soc Animal Sci; Agr Inst Can. *Res:* Coliform bacteria in Canadian dairy products; degradation of rutin and related flavonoids by anaerobic rumen bacteria; studies on the localization and the role of periplasmic enzymes in bacterial cell; microbiology and biochemistry of digestion in the rumen; ecomicrobiology of digestive tracts of ruminant animals; microbial adhesion in nature and disease. *Mailing Add:* Animal Sci Sect Res Sta Can Dept Agr PO Box 3000 Lethbridge AB T1J 4B1 Can

CHENG, KWOK-TSANG, b Hong Kong, May 5, 50, US citizen; m 75. ATOMIC PHYSICS. *Educ:* Chinese Univ Hong Kong, BSc, 71; Univ Notre Dame, PhD(physics), 77. *Prof Exp:* Researcher physics, Argonne Nat Lab, 77-84; LAWRENCE LIVERMORE NAT LAB, 85- *Mem:* Am Phys Soc. *Res:* Relativistic atomic structure calculations; interactions between atoms and radiation fields. *Mailing Add:* Lawrence Livermore Nat Lab PO Box 808 Livermore CA 94550

CHENG, LANNA, b Singapore, Apr 27, 41; m 69. APPLIED PHYCOLOGY, SYMBIOSES. *Educ:* Univ Singapore, BSc, 63, hons, 64, MSc, 66; Oxford Univ, DPhil(entom), 69. *Prof Exp:* res assoc, 70-72, asst res biologist, 72-77, ASSOC RES BIOLOGIST, SCRIPPS INST OCEANOG, UNIV CALIF, SAN DIEGO, 77- *Concurrent Pos:* Nat Res Coun Can fel biol, Univ Waterloo, 68-69; Am Asn Univ Women fel, 69-70. *Mem:* Orgn Trop Studies; fel Royal Entom Soc London; Western Soc Naturalists (pres, 85); Am Soc Limnol & Oceanog; Phycol Soc Am. *Res:* Taxonomic and biological studies of aquatic insects especially Gerridae, Hemiptera; ecology of marine insects; pleuston; animals of the sea-air interface; marine picopleuston algae; polysaccharide-producing algae as soil conditioners; prochloron-didemnid symbiosis; dunaliella and beta-carotene. *Mailing Add:* A-0202 Scripps Inst Oceanog Univ Calif San Diego La Jolla CA 92093

CHENG, LAWRENCE KAR-HIU, b Hong Kong, July 25, 47; c 2. PHARMACEUTICS. *Educ:* Univ Calif, Santa Barbara, BS, 71; State Univ NY Buffalo, PhD(pharmaceut), 75. *Prof Exp:* Anal chemist pharmaceut, Stuart Pharmaceut, ICI US Inc, 75-80; GROUP MGR DRUG METAB DEPT, A H ROBINS CO, 81- *Mem:* Am Asn Pharmaceut Scientists; AAAS; NY Acad Sci; Am Chem Soc. *Res:* Analytical methods development for drugs and metabolites in the biological tissues and dosage forms; pharmacokinetics of drugs and other pharmacologically active chemicals; bioequivalency studies of new formulations in vivo-in vitro biotransformation of drugs. *Mailing Add:* Drug Metab Dept/Merrell Dow Pharm 2110 Galbraith Rd Cincinnati OH 45215-6300

CHENG, LESTER (LE-CHUNG), b China, Apr 20, 44; US citizen; m 69; c 2. THERMAL SCIENCE, BIOENGINEERING. *Educ:* Nat Taiwan Univ, BS, 65; NDak State Univ, MS, 68; Univ Ill, Urbana, PhD(mech), 71. *Prof Exp:* Res assoc mech, Dept Theoret & Appl Mech, Univ Ill, Urbana, 70-71; sr engr analyst mech eng, Sargent & Lundy Engrs, Chicago, 71-74; sr engr, Gen Atomic Co, San Diego, 74-76; asst prof, 76-79, ASSOC PROF MECH ENG, WICHITA STATE UNIV, 79- *Concurrent Pos:* Consult, Dept Theoret & Appl Mech, Univ Ill, Urbana, 74-76; prin investr NSF grant, 77-80; consult, Vet Admin Ctr, Wichita, Kans, 78-; consult, Kans Energy & Environ Lab, Wichita, 78- *Honors & Awards:* Ralph R Teetor Award, Soc Automotive Eng, 80. *Mem:* Am Acad Mech; Sigma Xi; Am Soc Mech Eng; Am Soc Engr Educ. *Res:* Bio-fluid mechanics; mechanical system analysis; energy. *Mailing Add:* 6915 Croyden Circle Wichita KS 67208

CHENG, PAUL J(IH) T(IEN), b Hsilo, Taiwan, Dec 13, 35; m 66; c 1. CHEMICAL ENGINEERING. *Educ:* Nat Taiwan Univ, BS, 58; Univ Utah, PhD(chem eng), 67. *Prof Exp:* Res asst, Univ Utah, 61, 62-67; res engr, 67-68, sr res eng, 68-81, ENG ASSOC, PHILLIPS PETROL CO, 81- *Mem:* Am Inst Chem Engrs; Am Chem Soc. *Res:* Non-Newtonian behaviors of solid-liquid mixtures; reaction behaviors of solid composite propellants; pyrolysis of polymers; carbon black process innovations; improvements, optimization and exploration of new reaction process technology; process models and computer simulation; carbon fibers; advanced ceramics. *Mailing Add:* 2901 SE Greenwood Ct Bartlesville OK 74006

CHENG, PING, b Canton, China; US citizen; c 2. HEAT TRANSFER, FLUID MECHANICS. *Educ:* Okla State Univ, BS, 58; Mass Inst Technol, MS, 60; Stanford Univ, PhD(aeronaut & astronaut), 65. *Prof Exp:* Res assoc aeronaut & astronaut, Stanford Univ, 64-65; res scientist aeronaut & astronaut, NY Univ, 65-67; res assoc, NASA Ames Res Ctr, 67-68; vis prof mech eng, Nat Taiwan Univ, 68-70; assoc prof, 70-74, PROF MECH ENG, UNIV HAWAII, 74-, CHAIR, 89- *Concurrent Pos:* Vis prof petrol eng, Stanford Univ, 76-77; prin investr, NSF grants, 77-; guest prof, Tech Univ Munich. *Honors & Awards:* Chungshan Award, Chungshan Found, Taipei, Taiwan, 69. *Mem:* Fel Am Soc Mech Engrs; Sigma Xi. *Res:* Convective and boiling heat transfer in porous media and geothermal systems; well test analyses and geothermal reservoir engineering; radiative heat transfer and solar energy; ocean thermal energy conversion. *Mailing Add:* Dept Mech Eng Univ Hawaii Manoa Honolulu HI 96822

CHENG, PI-WAN, b Tainan, Taiwan, China, Apr 1, 43; m 73; c 2. GLYCOPROTEINS, LUNG DISEASES. *Educ:* Nat Taiwan Univ, Taiwan, China, BS, 65, MS, 68; Case Western Reserve Univ, PhD(biochem), 74. *Prof Exp:* Res assoc, Case Western reserve Univ, 74-76, asst prof pediat & biochem, 77-82; ASSOC PROF PEDIAT & BIOCHEM, UNIV NC, CHAPEL HILL, 82- *Concurrent Pos:* Prin investr, Environ Protection Agency & NIH, 85-89; Cystic Fibrosis Found, 88-89. *Mem:* Am Soc Biol Chemists; AAAS; Soc Complex Carbohydrates; Soc Chinese Bioscientists Am. *Res:* Biochemistry of pulmonary secretions, specifically the chemistry and biosynthesis of mucous glycoproteins and the control of respiratory tract mucin secretion. *Mailing Add:* Dept Pediat Univ NC 635 Clin Sci Bldg CB 7220 Chapel Hill NC 27599-7220

CHENG, RALPH T(A-SHUN), b Shanghai, China, Sept 26, 38; m 63; c 2. FLUID DYNAMICS, APPLIED MATHEMATICS. *Educ:* Nat Taiwan Univ, BS, 61; Univ Calif, Berkeley, MS, 64, PhD(mech eng), 67. *Prof Exp:* Asst prof mech eng, State Univ, NY Buffalo, 67-77; MEM STAFF, WATER RESOURCES DIV, US GEOL SURV, 77- *Concurrent Pos:* Proj dir, NSF res grant, 69-; consult, Cornell Aeronaut Lab, 68-70. *Mem:* Am Inst Aeronaut & Astronaut; Am Phys Soc; Am Geophys Union; Sigma Xi. *Res:* Fluid dynamics, especially numerical modeling; thermodynamics; numerical solutions of nonlinear differential equations; geophysical fluid dynamics; transport processes in the atmosphere. *Mailing Add:* Water Resources Div US Geol Surv 345 Middlefield Rd MS64 Menlo Park CA 94025

CHENG, RICHARD M H, b Hong Kong; Can citizen. CONTROL ENGINEERING, INDUSTRIAL AUTOMATION. *Educ:* Hong Kong Univ, BSc, 61; Univ Manchester Inst Sci & Technol, MSc, 67; Birmingham Univ, PhD(control eng), 71. *Prof Exp:* Lectr mech eng, Birmingham Univ, Eng, 69-72; assoc prof, 72-78, PROF MECH ENG, CONCORDIA UNIV, CAN, 78-, DIR, CTR INDUST CONTROL, 83- *Concurrent Pos:* Grad engr, Brit Leyland Motor Corp, Longbridge, Birmingham, Eng, 65-66; dir advan systs, Dynamic Sci Ltd, St Laurent, Can, 79-80; consult, var govt bodies & corps including Can Nat Railways. *Mem:* Fel Brit Inst Mech Engrs; Am Soc Mech Engrs; Inst Elec & Electronics Engrs; sr mem Inst Soc Am. *Res:* Applying fluid mechanics, electronics, mini & microcomputer to industrial control & automation problems; pneumatic brake systems of freight trains; microprocessor-based fuel control of small gas turbines; micro-computer based simulation of process control circuits; computer-aided design using interactive colour graphics; digital control of multi-arm robotic workstation; automated guided vehicle using camera vision. *Mailing Add:* Dept Mech Eng Concordia Univ SB306 1455 De Maisonneuve Blvd W Montreal PQ H3G 1M8 Can

CHENG, SHANG I, b Chiete, Chekiang, China, June 5, 20; US citizen; m 41; c 5. CHEMICAL ENGINEERING. *Educ:* Chekiang Univ, BS, 45; Univ Fla, MS, 59, PhD(chem eng), 61. *Prof Exp:* Chem engr, Hsinchu Res Inst, Taiwan, 45-54; supvr res & develop, Union Indust Res Inst, 54-58; develop scientist, B F Goodrich Chem Co, 61-65; from asst prof to prof, 65-89, EMER PROF CHEM ENG, COOPER UNION, 89- *Concurrent Pos:* Consult, Nat Sci Coun, Repub of China, 74- *Honors & Awards:* Award Syst Design, NASA, 74. *Mem:* Am Inst Chem Engrs; Am Chem Soc. *Res:* Mass transfer and reactor engineering; computer simulation; polymerization kinetics; air pollution abatement; solid waste recycle; energy conservation. *Mailing Add:* 841-G Main St Belleville NJ 07109

CHENG, SHEUE-YANN, b Taiwan, China, June 18, 39; US citizen; m 64; c 2. CHEMISTRY, BIOLOGY. *Educ:* Nat Taiwan Univ, BS, 61; Univ Calif, San Francisco, PhD(pharmaceut, chem), 66. *Prof Exp:* Fel pharmaceut chem, Med Ctr, Univ Calif, San Francisco, 66-67 & Ben May Lab Cancer Res, Univ Chicago, 71-72; fel, 73-78, SR CHEMIST, HIH, 78- *Mem:* Endocrine Soc; AAAS; Am Thyroid Asn; Am Soc Cell Biol. *Res:* Interaction of thyroid hormone with transport proteins, mechanism of the transport of thyroid hormones and the thyroid hormone actions. *Mailing Add:* Bldg 37 Rm 4B09 NIH NCI 9000 Rockville Pike Bethesda MD 20892

CHENG, SHUN, b China, July 10, 19; US citizen; m 47. ENGINEERING MECHANICS. *Educ:* Univ Wis-Madison, PhD(eng mech), 59. *Prof Exp:* Asst prof mech, Univ Dayton, 54-56; from instr to assoc prof eng mech, 56-72, PROF ENG MECH, UNIV WIS-MADISON, 72- *Mem:* AAAS; Am Soc Mech Engrs; Am Inst Aeronaut & Astronaut. *Res:* Solid mechanics; structures. *Mailing Add:* Dept Eng Mech 2348 Eng Bldg Univ Wis 1415 Johnson Dr Madison WI 53706

CHENG, SHU-SING, b Kwangtung, China, Sept 7, 23. ORGANIC CHEMISTRY, MICROBIOLOGY. *Educ:* Nat Col Pharm, Nanking, BS, 47; Univ NC, MS, 59, PhD(pharmaceut chem), 61. *Prof Exp:* Res assoc virol, Sch Med, Univ NC, 60-62; res fel steroid biochem, Worcester Found Exp Biol, 62-63; res asst prof med chem & biochem, Univ NC, 63-65; HEAD MICROBIOL RES, KENDALL RES CTR, KENDALL CO, 66- *Mem:* AAAS; Am Chem Soc; Am Pharmaceut Asn; Am Soc Microbiol. *Res:* Design and synthesis of organic compounds of medicinal interest; chemistry and biochemistry of steroids; tissue culture and microbiology; screening of antimicrobial agents; process research in cold sterilization and monitoring devices. *Mailing Add:* 1820 Laurel Ave Hanover Park IL 60103-3318

CHENG, SIN-I, b China, Dec 28, 21; US citizen; m 49; c 3. AERONAUTICS. *Educ:* Chiao Tung Univ, BS, 46; Univ Mich, MS, 49; Princeton Univ, MA & PhD(aeronaut eng), 52. *Prof Exp:* From instr to assoc prof, 51-60, PROF AERONAUT ENG, PRINCETON UNIV, 60- *Concurrent Pos:* Fel, Chinese Ministry Educ, 48-49; Guggenheim fel, 49-51. *Mem:* Assoc fel Am Inst Aeronaut & Astronaut; Combustion Inst. *Res:* Rocketry; combustion; fluid mechanics; propulsion; aerodynamics; computational methods. *Mailing Add:* Dept Mech & Aerospace Eng Princeton Univ Princeton NJ 08544

CHENG, SZE-CHUH, b Soochow, China, Nov 11, 21; US citizen; c 3. NEUROCHEMISTRY, BIOLOGY. *Educ:* Southwest Assoc Univ, China, BSc, 43; Brown Univ, MSc, 49; Univ Pa, PhD(gen physiol), 54. *Prof Exp:* Res asst gen physiol, Tsinghua Univ, Peking, 43-47; res assoc neurophysiol, Rockefeller Univ, 55-60; sr res scientist neurochem, NY State Psychiat Inst, 60-64; sr res scientist neurochem, NY State Res Inst Neurochem & Drug Addiction, 65-71; prof anesthesiol, Sch Med, Northwestern Univ, Chicago, 71-87; RETIRED. *Concurrent Pos:* Fel, McCollum-Pratt Inst, Johns Hopkins Univ, 53-55; Pub Health Serv spec fel, Agr Res Coun Inst Animal Physiol, Babraham, Eng, 64-65. *Mem:* Am Soc Neurochem; Soc Neurosci; Int Soc Neurochem. *Res:* Metabolism and function of nervous tissue including drug effects. *Mailing Add:* 5497 Blossom Terrace Ct San Jose CA 95124

CHENG, TA-PEI, b Shanghai, China; US citizen. THEORETICAL PHYSICS. *Educ:* Dartmouth Col, AB, 64; Rockefeller Univ, PhD(physics), 69. *Prof Exp:* Mem, Inst Advan Study, 69-71; res assoc, Rockefeller Univ, 71-73; from asst prof to assoc prof, 73-78, chmn, Physics Dept, 78-79, PROF PHYSICS, UNIV MO-ST LOUIS, 78- *Concurrent Pos:* Vis assoc prof physics, Princeton Univ, 77-78; mem, Inst Advan Study, 77-78 & 87-88; vis prof physics, Univ Minn, 79-80; vis scientist, Lawrence Berkeley Lab, Univ Calif, 82-83. *Mem:* Am Phys Soc. *Res:* Theory of elementary particles. *Mailing Add:* Dept Physics Univ Mo St Louis MO 63121

CHENG, THOMAS CLEMENT, b Nanking, China, Nov 5, 30; US citizen; c 3. CELLULAR IMMUNITY, ANTIGENICITY OF ZOOPARASITES. *Educ:* Wayne State Univ, BA, 52; Univ Va, MS, 56, PhD(biol), 58. *Prof Exp:* Instr gen biol, Univ Va, 56-58; from instr to asst prof med histol, Univ Md, 58-59; from asst prof to assoc prof parasitol, cell physiol, invert biol & microbiol, Lafayette Col, 59-64; chief, Immunol & Parasitol Sect, Northeast Marine Health Sci Lab, USPHS, Narragansett, RI, 65; from assoc prof to prof biol, Univ Hawaii, Manoa, 65-69; prof & dir, Ctr Health Sci, Lehigh Univ, 69-80; prof & dir marine biomed res, 80-91, PROF CELL BIOL, MED UNIV SC, 91- *Concurrent Pos:* Mem, Surgeon Gen's Comn Food Protection, USPHS, 65-66, Environ Biol & Chem Study Sect, 69-71, Spec Study Sect, Marine Envrion Health, NIH, 73-80, Spec Comt Use Marine Invert Biomed Res, 73-75, Rev Panel, Div Ocean Sci, NSF, 77-78, Rev Panel Div Cell Physiol, 80-83 & Chem Path Study Sect, Nat Cancer Inst, 87; consult, Div Microbiol, Food & Drug Admin, 68-72, Univ Park Press, 69-80, Acad Press, 69-, Int Copper Res Asn, 70-82, Intext Publ, 73-79, Agr Div, Sandoz Pharmaceut, 79-80 & Xytronyx, San Diego, 83-; instr-in-charge, Res Training Prog, NSF, Japan, 68; assoc ed, Exp Parasitol, 69-89; ed-in-chief, J Invert Path, 69- & Current Topics Comp Pathobiol, 71-73; ed, Comp Pathobiol & asst ed, J Parasitol, 73-83; ann lectr, Southwestern Asn Parasitologists, Dallas, 73; distinguished lectr, Tulane Med Sch, 77; NIH sr Fogarty fel, Cambridge Univ, 80-81; Fulbright sr res scholar & dir res, Nat Ctr Sci Res, France, 86-87; invited chmn, VIT Int Cong Parasitol, Paris, 90. *Honors & Awards:* N R Stoll Distinguished Lectr, NJ Soc Parasitologists, 71 & 88; G C Wheeler Distinguished Lectr, Univ NDak, 73. *Mem:* Fel AAAS; Am Micros Soc (vpres, 73-74, pres, 80-81); Am Soc Parasitologists; Soc Invert Path; Int Orgn Marine Path; Am Soc Zoologists; fel Royal Soc Trop Med & Hyg. *Res:* Comparative immunology, toxicology, parasitology and microbiology; lysosomal and cell membrane biochemistry pertaining to phagocytosis and secretion of bioactive molecules; invertebrate tissue culture; biological control of molluscan vectors of pathogens; molluscan physiology; antigenicity of zooparasites during life cycle stages. *Mailing Add:* Dept Anat & Cell Biol Med Univ SC 171 Ashley Ave Charleston SC 29425

CHENG, TSEN-CHUNG, b Shanghai, China, Dec 24, 44; US citizen; m 74; c 1. ELECTRIC POWER SYSTEMS, DIELECTRIC MATERIALS. *Educ:* Mass Inst Technol, BS, 69, MS, 70. *Hon Degrees:* ScD, Mass Inst Technol, 74. *Prof Exp:* From asst prof to assoc prof, 74-84, LLOYD F HUNT PROF & DIR ELEC POWER ENG, UNIV SOUTHERN CALIF, 84- *Concurrent Pos:* Sr consult, Southern Calif Edison Co, 79-, Pac Gas & Elec Co, 80-, Los Angeles Dept Water & Power, 80-; pres, T C Cheng, ScD, Inc, 81-; prin investr, Res Proj, Dept Water & Power, Elec Power Res Inst, Elec Utilities, 75- *Mem:* Inst Elec & Electronics Engrs; Sigma Xi; Conf Int des Grands Reseaux Electriques. *Res:* High voltage insulation and breakdown, dielectrics and materials properties; author or coauthor of over 80 papers. *Mailing Add:* Dept Elec Eng Univ Southern Calif Los Angeles CA 90089-0271

CHENG, TSUNG O, b Shanghai, China, Mar 30, 25; nat US; m; c 2. CARDIOLOGY. *Educ:* St Johns Univ, China, BS, 47; Pa Med Sch, China, MD, 50; Univ Pa, MMedSc, 56; Am Bd Internal Med, dipl, 61; Am Bd Cardiovasc Dis, dipl, 63. *Prof Exp:* Intern, Hosp St Barnabas, NJ, 50-51; resident internal med, Cook County Hosp, Ill, 52-55; asst cardiol, Mass Gen Hosp, 56-57; asst physician, Cardiac Clin, Johns Hopkins Hosp, 57-59; dir, Cardiopulmonary Lab, Brooklyn Hosp & asst prof med, State Univ NY Downstate Med Ctr, 59-70; dir, Cardiovasc Lab & chief cardiol, Vet Admin Hosp, Brooklyn, 66-70; assoc prof, Sch Med, 70-72, dir cardiac, Catheterization Lab, Med Ctr, 72-77, PROF MED, SCH MED, GEORGE WASHINGTON UNIV, 72- *Concurrent Pos:* Fel Northwestern Univ, 54-55; fel cardiol, Sch Med, George Washington Univ, 55-56; res fel cardiopulmonary physiol, Johns Hopkins Univ & Hosp, 57-59, Am Heart Asn advan res fel, 58-59; consult & chief pediat cardiac clin, Cumberland Hosp, 63-66; asst vis physician, Kings County Hosp, 64-70; Coun Clin Cardiol fel, Am Heart Asn, 64-; consult, Beth Israel Hosp, New York, 70-82; chief cardiol, DC Gen Hosp, 71-72; vis prof, Peking Union Med Col, 86-; hon prof, Shanghai Second Med Univ, China, 86, Qingdao Med Col, China, 89; hon consult, Beijing Hosp, China, 89; hon pres, Dandong First Hosp, China, 89,

Shanghai St Luke's Hosp, China, 90; hon dir, Qingdao Cardiovascular Res Inst, China, 90, Guangdong Cardiovascular Inst, China, 90; guest lectr, Chinese Ministry Public Health, China, 90. *Mem:* Fel Am Col Physicians; fel Am Col Cardiol; fel Am Col Chest Physicians; fel Int Col Angiol; fel Am Col Angiol; fel Soc Cardiac Angiography & Interventions. *Res:* Clinical investigations in cardiopulmonary pathophysiology in health and diseases. *Mailing Add:* Dept Med Cardiol George Washington Med Ctr 2150 Penn Ave NW Washington DC 20037

CHENG, TU-CHEN, BIOTECHNOLOGY RESEARCH, RECOMBINANT DNA. *Educ:* Wayne State, PhD(biochem), 72. *Prof Exp:* RES CHEMIST, DIV BIOTECHNOL, US ARMY, 85- *Mailing Add:* SMCCR-RSB Biotech Div-CRDEC Aberdeen Proving Ground MD 21010

CHENG, WILLIAM J(EN) P(U), b Changsha, China, Sept 26, 15; m 54; c 4. CHEMICAL ENGINEERING. *Educ:* Tsing Hua Univ, China, BS, 39; Washington Univ, St Louis, MS, 51. *Prof Exp:* Asst, Tsing Hua Inst, China, 39-40; asst chem engr, China Veg Oil Corp, 40-41, chem engr & plant supt, 42-44; chem engr, arsenal, Chinese Army, 45-47; sr res chem engr, Petrolite Corp, 53-58, group leader, 58-60, head pilot plant, 60-63, mgr eng res, 63-67, DIR ENG, TRETOLITE DIV, PETROLITE CORP, 67- *Mem:* Am Chem Soc; Am Inst Chem Engrs; NY Acad Sci. *Res:* Petroleum and demulsification; chemicals for petroleum production; synthetic resin and high polymers; industrial surfactants; organic unit process; process development, design and optimization. *Mailing Add:* 705 Louwen Dr St Louis MO 63124

CHENG, WU-CHIEH, b Shanghai, China, Aug 11, 22; US citizen; m 63; c 1. PHYSICAL CHEMISTRY, ATOMIC & MOLECULAR PHYSICS. *Educ:* St John's Univ, China, BS, 44; Kans State Col, MS, 49; Ga Inst Technol, PhD(phys chem), 54. *Prof Exp:* From asst prof to prof chem, Union Univ, 55-66, head dept, 58-66; assoc prof, George Peabody Col, 66-72; teacher rank I, Lyman High Sch, Longwood Fla, 72-75; asst prof, 75-76, assoc prof physics, Paine Col, 76-89; PATENT EXAMR, US DEPT COM, 90- *Concurrent Pos:* Vis instr, Ga Inst Technol, 56; chemist, Northern Regional Res Ctr, USDA, 76; fac res partic, Savannah River Lab, 77, Argonne Nat Lab, 82, Oak Ridge Nat Lab, 84; fac app, Rockwell Hanford Opers, 79; fac res fel, Frank J Seiler Res Lab, US Air Force Acad, 86. *Mem:* Am Chem Soc; Sigma Xi; Am Asn Physics Teachers. *Res:* Molecular structure; chemical education; polymer chemistry; ion-exchange equilibria; starch derivatives; metabolism of unsaturated fatty acids; synthetic resins and plastics; laser plasma; molecular spectroscopy. *Mailing Add:* 278 W Wynngate Dr Maarlinez GA 30907

CHENG, YEAN FU, b China, 1924; US citizen. EXPERIMENTAL MECHANICS, PHOTO MECHANICS. *Educ:* Nat Amoy Univ, BS, 46; Univ Wash, MS, 56; Ill Inst Technol, PhD(mech), 61. *Prof Exp:* Sr basic res scientist, Boeing Sci Res Labs, 61-71; res mech engr, Benet Weapons Lab Armament Res, Develop & Eng Ctr, US Army, 71-89; RETIRED. *Mem:* Am Soc Mech Engrs; Sigma Xi; Am Acad Mech. *Res:* Experimental mechanics; photo mechanics. *Mailing Add:* PO Box 774 Latham NY 12110

CHENG, YIH-SHYUN EDMOND, b Shanghai, China, July, 18, 44; US citizen; m 71; c 1. INTERFERON, GENE REGULATION. *Educ:* Nat Taiwan Univ, BS, 67, MS, 70; Purdue Univ, Ind, PhD(bio sci), 76. *Prof Exp:* Res fel, Cold Spring Harbor Lab, 76-79; res assoc, Harvard Univ, 79-80; RES STAFF, E I DUPONT & CO, 80- *Mem:* Am Soc Microbiol; Am Soc Cell Biol; Int Soc Interferon Res. *Res:* Isolation of an interferon-induced guanylate binding protein and cloning its gene for the understanding of the function of this protein and the regulation of its expression. *Mailing Add:* E 328-349 Dupont Exp Sta Wilmington DE 19898

CHENG, YUNG-CHI, b London, Eng, Dec 29, 44; US citizen; m 69; c 2. CHEMOTHERAPY. *Educ:* Tunghai Univ, Taiwan, BSc, 66; Brown Univ, RI, PhD(biochem & pharmacol), 72. *Prof Exp:* Res assoc, Sect Biochem & Pharmacol, Brown Univ, 72; fel res staff, Pharmacol Dept, Sch Med, Yale Univ, 72-73, res assoc asst prof, 73-74; sr cancer res scientist, Dept Exp Therapeut, Roswell Park Mem Inst, 74-76, asst res scientist, 76-77, res scientist, 77-79, from asst prof to assoc prof, 74-79; PROF PHARMACOL & MED, SCH MED, & CANCER CTR, DEPT BIOCHEM & NUTRI, UNIV NC, CHAPEL HILL, 79- *Concurrent Pos:* Am Leukemia Soc Scholar, 76-81; mem, Sci Adv Bd, G D Searle Pharmaceut & Co, 81; NIH mem, Study Sect Exp Therapeut, Nat Cancer Inst, 80-84; Rhoads Mem award, Am Asn Cancer Res, 81- *Mem:* Sigma Xi; Am Microbiol Soc; Am Soc Biol Chemists. *Res:* Anticancer and antiviral chemotherapy. *Mailing Add:* Dept Pharmacol Sch Med Univ NC Chapel Hill NC 27514

CHENG, YUNG-SUNG, b Fong-Shan, Taiwan, Apr 23, 47; US citizen; m 75; c 2. AEROSOL SCIENCE. *Educ:* Nat Taiwan Univ, BA, 69; Syracuse Univ, MS, 73, PhD(chem eng), 76. *Prof Exp:* Res assoc, Sch Med, Univ Rochester, 76-78; aerosol engr, 78-84, SECT SUPVR, LOVELACE BIOMED & ENVIRON RES INST, 84- *Mem:* Am Indust Hyg Asn; Am Asn Aerosol Res; Am Inst Chem Engrs; Soc Aerosol Res. *Res:* Aerosol science and technology; sampling and characterization of emission sources; design and operation of inhalation exposure systems; deposition and retention of inhaled particles; dynamic behavior of fine particles; instrumentation and measurement. *Mailing Add:* Inhalation Toxicol Res Inst PO Box 5890 Albuquerque NM 87115

CHENGALATH, RAMA, b Calicut, India; Can citizen. HYDROBIOLOGY. *Educ:* Univ Kerala, India, BSc, 62, MSc, 64; Univ Waterloo, Can, MSc, 71. *Prof Exp:* Res biologist, Can Aquatic Identification Ctr, 76; special lectr biol, Trent Univ, Ont, 76; prin investr res, Sci Ctr, Supply & Serv Can, 76-80; invert zoologist, 80-86, CURATOR, ZOOL DIV, CAN MUS NATURE, 86- *Mem:* Can Soc Zoologists; Soc Int Limnologists; Can Soc Environ Biologists; Crustacean Soc; Can Soc Limnologists; Int Asn Great Lakes Res. *Res:* Taxonomy, ecology and distribution of rotifera and cladocera. *Mailing Add:* Zool Div Nat Mus Natural Sci PO Box 3663 Sta D Ottawa ON K1P 6P4 Can

CHENIAE, GEORGE MAURICE, b Mounds, Ill, Aug 27, 28; m 52; c 3. PLANT BIOCHEMISTRY. *Educ:* Univ Ill, BS, 50; NC State Col, MS, 57, PhD(plant physiol), 59. *Prof Exp:* Asst, Oak Ridge Nat Lab, 50-52; Nat Sci fel, 59-60; res scientist, Res Inst Advan Study, 60-75; PROF PLANT PSYSIOL & BIOCHEM, UNIV KY, 75- *Honors & Awards:* Charles F Kettering Award, Am Soc Plant Physiologists, 90. *Mem:* Am Soc Plant Physiologists; Am Soc Biol Chemists. *Res:* Lipid metabolism; respiration; photosynthesis. *Mailing Add:* Dept Agron Univ Ky Agron Sci Ctr N Bldg Rm N205 Lexington KY 40546-0091

CHENICEK, ALBERT GEORGE, b Chicago, Ill, Dec 15, 13; m 43; c 3. POLYMER CHEMISTRY. *Educ:* Univ Chicago, BS, 34, PhD(org chem), 37. *Prof Exp:* Res chemist, Columbis Chem Div, Pittsburgh Plate Glass Co, Ohio, 37-41 & Interchem Corp, 41-53; actg dir develop, Standard Coated Prods, Inc, 53-56; mgr res, Stoner-Mudge Co, 56-61; pres, Unifilm Corp, 61-82; RETIRED. *Mem:* AAAS; Am Chem Soc. *Res:* Chlorination of organic compounds; synthetic resins; coated fabrics; protective coatings; inks. *Mailing Add:* 466 Riverside Dr Princeton NJ 08540

CHENIER, PHILIP JOHN, b Chicago, Ill, Nov 9, 43; m 66; c 4. ORGANIC CHEMISTRY, POLYMER CHEMISTRY. *Educ:* St Mary's Col, BA, 65; Loyola Univ, PhD(org chem), 69. *Prof Exp:* Sr res chemist, Gen Mills Chem, 69-70; from asst prof to assoc prof org chem, 70-82, PROF ORG & INDUST CHEM, UNIV WIS, EAU CLAIRE, 82-, CHMN DEPT, 83- *Mem:* Am Chem Soc. *Res:* Strained polycyclic compounds, their synthesis and reactivity, including especially long-range cyclopropane participation; industrial chemistry. *Mailing Add:* Dept Chem Univ Wis Eau Claire WI 54702-4004

CHENOT, CHARLES FREDERIC, b Canton, Ohio, Sept 16, 38; m 62; c 4. SOLID STATE CHEMISTRY. *Educ:* Col Wooster, BA, 60; Univ Cincinnati, PhD(phys chem), 64. *Prof Exp:* ENG SPECIALIST, CHEM & METALL DIV, GTE SYLVANIA INC, 64- *Mem:* Am Ceramic Soc; Am Chem Soc; Electrochem Soc. *Res:* Physical chemical studies of diffusion in the solid state; research and development of solid state luminescent materials including phase equilibrium relationships, solid state reactions and associated luminescence spectroscopy. *Mailing Add:* RD No 3 Box 100 Towanda PA 18848-9529

CHENOWETH, DARREL LEE, b Indianapolis, Ind, Nov 6, 41; m 63; c 3. CONTROL SYSTEMS. *Educ:* Auburn Univ, MS, 64, PhD(elec eng), 69; Gen Motors Inst, BEE, 64. *Prof Exp:* Instr elec eng, Auburn Univ, 67-68; eng specialist, Vought Aeronaut Div, Ling-Temco-Vought Corp, Tex, 69-70; from asst prof to assoc prof, 70-78, dir, Computer Sci & Eng Doctoral Prog, 88-90, PROF ELEC ENG, UNIV LOUISVILLE, 78- *Mem:* Sr mem Inst Elec & Electronics Engrs. *Res:* Digital control systems; image processing and pattern recognition applied to scene matching navigation systems; microcomputer design; digital signal processing. *Mailing Add:* Dept Elec Eng Univ Louisville Louisville KY 40292

CHENOWETH, DENNIS EDWIN, b Charles City, Iowa, Jan 23, 44; m 75; c 2. BIOCOMPATIBILITY, INFLAMMATORY MEDIATORS. *Educ:* Lewis & Clark Col, BS, 66; Northwestern Univ, PhD(biochem), 71; Univ Calif, Irvine, MD, 75. *Prof Exp:* Fel biochem, Univ Calif, Los Angeles, 70-72, resident physician path, Irvine, 75-77 & San Diego, 77-78; asst mem immunol, Scripps Clin Res Found, 78-81; assoc prof path, Univ Calif, San Diego, 81-86; sr res fel immunol, Travenol Labs, Inc, 86-87; VPRES & SCI DIR, FENWAL DIV, BAXTER HEALTHCARE, INC, 87- *Concurrent Pos:* Estab investr, Am Heart Asn, 78; Clin investr, Vet Admin, 83. *Mem:* Am Asn Pathologists; Am Soc Immunologists; Am Soc Microbiol; Am Fedn Clin Res. *Res:* Identification of biocompatible medical devices; delineation of the effects of inflammatory mediators. *Mailing Add:* Baxter Healthcare Inc 3015 S Daimler St Santa Ana CA 92705

CHENOWETH, JAMES MERL, b Dayton, Ohio, May 22, 24; m 49; c 2. MECHANICAL ENGINEERING. *Educ:* Purdue Univ, BS, 49, MS, 50, PhD(mech eng), 52. *Prof Exp:* Sr res engr two-phase flow, C F Braun & Co, 52-59; sr staff heat transfer, Nat Eng Sci Co, 59-66; sr staff mech eng, Survival Systs, Inc, 66-70; asst tech dir, Heat Transfer Res, Inc, 70-87; CONSULT, 87- *Concurrent Pos:* Chmn, Heat Transfer Div, Am Soc Mech Engrs, 88-89. *Mem:* Fel Am Soc Mech Engrs; Sigma Xi; Survival & Flight Equip Asn. *Res:* Heat transfer and fluid mechanics; two-phase flow; flow-induced tube vibration; heat exchanger fouling. *Mailing Add:* 1135 Winston Ave San Marino CA 91108

CHENOWETH, MAYNARD BURTON, clinical pharmacology; deceased, see previous edition for last biography

CHENOWETH, PHILIP ANDREW, b Chicago, Ill, Aug 21, 19; m 52; c 2. GEOLOGY, STRATIGRAPHY. *Educ:* Columbia Univ, BA, 46, MA, 47, PhD, 49. *Prof Exp:* Instr geol, Amherst Col, 49-51; sr geologist, Sinclair Oil & Gas Co, 51-54; assoc prof geol, Univ Okla, 54-60; staff geologist, Sinclair Oil & Gas Co, 60-65, res assoc, Sinclair Res Ctr, 65-68; CONSULT GEOLOGIST, 54-60, 68- *Honors & Awards:* Prof Award, Univ Okla, 55. *Mem:* Fel Geol Soc Am; Am Asn Petrol Geologists; Am Inst Prof Geologists; Am Inst Mining, Metall & Petrol Eng. *Res:* Petroleum and exploration geology. *Mailing Add:* 7010 S Yale Ave No 216 Tulsa OK 74136

CHENOWETH, R(OBERT) D(EAN), b San Angelo, Tex, Oct 17, 26; m 47; c 4. ELECTRICAL ENGINEERING, POWER SYSTEM ANALYSIS. *Educ:* Agr & Mech Col, Tex, BS, 46, MS, 51; Ga Inst Technol, PhD, 55. *Prof Exp:* Student engr, Westinghouse Elec Corp, 46-47; asst prof & res asst, Agr & Mech Col, Tex, 47-52; asst prof elec eng, Ga Inst Technol, 53-55 & Case Inst Technol, 55-60; prof, Univ Mo, Rolla, 60-67; prof elec eng, Tex A&M Univ, 67-89, asst dean eng, 85-89; RETIRED. *Concurrent Pos:* NSF fel, 52-53. *Mem:* Inst Elec & Electronics Engrs; Am Soc Eng Educ. *Res:* Power in electrical engineering; computers; control. *Mailing Add:* PO Box 9915 College Station TX 77842-9915

CHENOWETH, WANDA L, b Salt Lake City, Utah. NUTRITION. *Educ:* Univ Utah, BS, 56; Univ Iowa, MS, 59; Univ Calif, Berkeley, PhD(nutrit), 72. *Prof Exp:* Res dietitian, Univ Iowa Hosps, 57-59 & Mayo Clin, 59-64; head therapeut dietitian, Univ Iowa Hosps, 64-67; teaching assoc nutrit, Univ Calif, Berkeley, 72; from asst prof to assoc prof, 72-82, PROF NUTRIT, MICH STATE UNIV, 82- *Concurrent Pos:* Consult, Diet, Nutrit & Cancer Prog, Nat Cancer Inst, 76-78. *Mem:* Am Dietetic Asn; Soc Nutrit Educ; Inst Food Technol; Am Inst Nutrit. *Res:* Clinical nutrition; effect of diet and nutrition on gastrointestinal function; mineral metabolism; nutrition in the ederly. *Mailing Add:* Dept Food Sci & Human Nutrit Mich State Univ East Lansing MI 48824

CHENOWETH, WILLIAM LYMAN, b Wichita, Kans, Sept 16, 28; m 55; c 4. URANIUM GEOLOGY. *Educ:* Univ Wichita, AB, 51; Univ NMex, MS, 53. *Prof Exp:* Area geologist, Div Raw Mat, AEC 55-58, chief geol engr, Sect Off, Flagstaff, Ariz, 58-62, proj geologist, Resource Appraisal Br, 62-70, Chief, 70-74, staff geologist, US Energy Res & Develop Admin, 75-77, staff geologist, US Dept Energy, 78-83; CONSULT,GEOLOGIST, 84- *Concurrent Pos:* Res assoc, NMex Bureau Mines, 83-; chmn, Nuc Min Comt, Am Asn Petrol Geologists, 83-; contract geologist, US Geol Surv, 86-89. *Mem:* Fel Geol Soc Am; Am Inst Mining, Metall & Petrol Eng; Am Asn Petrol Geologists. *Res:* Uranium geology; Mesozoic stratigraphy; depositional environment of uranium ore deposits. *Mailing Add:* 707 Brassie Dr Grand Junction CO 81506-3911

CHEN-TSAI, CHARLOTTE HSIAO-YU, b Taipei, Taiwan; US citizen. POLYMER COATINGS ON METAL SUBSTRATES, SOLID STATE DEFORMATION OF THERMOPLASTICS. *Educ:* Nat Tsing Hua Univ, Taiwan, BS, 78; Univ Fla, MS, 79; Univ Mass, Amherst, PhD(chem eng & polymer sci eng), 85. *Prof Exp:* TECH SPECIALIST, POLYMER APPL, ALCOA TECH CTR, 85- *Concurrent Pos:* Mem, Plastic Recycling Comt, Soc Plastics Engrs, 90- *Mem:* Soc Plastics Engrs; Am Chem Soc. *Res:* Apply principles of polymer alloys to formulate protective coatings for metal substrates used in packaging, automobile and engineering applications; research processing, structure, performance relationships in solid state oriented thermoplastics, coextruded and injection molded products. *Mailing Add:* Alcoa Tech Ctr Rte 780 Alcoa Center PA 15069

CHEO, BERNARD RU-SHAO, b Nanking, China, May 29, 30; m 57; c 2. ELECTRICAL ENGINEERING, ENGINEERING PHYSICS. *Educ:* Taiwan Univ, BSc, 53; Univ Notre Dame, MS, 56; Univ Calif, Berkeley, PhD(elec eng), 61. *Prof Exp:* Mem tech staff, Bell Tel Labs, Inc, 60-62; from asst prof to assoc prof elec eng, New York Univ, 62-73; PROF ELEC ENG, POLYTECH UNIV, 73- *Mem:* Inst Elec & Electronics Engrs. *Res:* Electromagnetic theory; plasmas. *Mailing Add:* Dept Elec Eng Polytech Univ 333 Jay St Brooklyn NY 11201

CHEO, LI-HSIANG ARIA S, b Taiwan, China; US citizen; c 2. APPLIED MATHEMATICS, ELECTRICAL ENGINEERING. *Educ:* Nat Cheng Kung Univ, BS, 55; Univ Calif Berkeley, MS, 61; NY Univ, PhD(math), 70. *Prof Exp:* Jr engr elec eng, Taiwan Power Co, 55-56; mem tech staff elec eng, Bell Tel Labs, 60-63; asst prof math, NY Inst, 70-72; from asst prof to assoc prof math, 71-78, assoc prof computer sci, 78-79, PROF COMPUTER SCI, WILLIAM PATTERSON COL NY, 79- *Concurrent Pos:* Res & teaching fel, Univ Calif Berkeley, 57-61, NY Univ, 68-70; chairperson computes sci, William Paterson Co NJ, 85-; tech consult, Los Alamos Nat Lab, 89- *Mem:* Asn Comput Mach; Inst Elec & Electronics Engrs Computer Soc; Sigma Xi. *Res:* Numerical solutions to differential integral equations/algorithmic development and implementation of computational schemes programming languages; information management computer science education. *Mailing Add:* Dept Computer Sci William Paterson Col Wayne NJ 07470

CHEO, PETER K, b Nanking, China, Feb 2, 30; US citizen; m 56; c 5. LASERS & INTEGRATED OPTICS. *Educ:* Aurora Univ, BS, 51; Va Polytech Inst, MS, 53; Ohio State Univ, PhD(physics), 64. *Hon Degrees:* DSc, Aurora Univ, 86. *Prof Exp:* Instr physics, Bethany Col, 54-57; asst prof, Aurora Univ, 57-61; mem prof staff, Bell Tel Labs, 63-70; mgr laser appln res, Aerojet-Gen Corp, 70-71; SR PRIN SCIENTIST, UNITED TECHNOLOGIES RES CTR, 71- *Concurrent Pos:* Adj prof elec eng, Hartford Grad Ctr, 78- *Honors & Awards:* Cert Recognition, NASA, 72, 85; United Technologies Award, 82. *Mem:* Fel Optical Soc Am; fel Inst Elec & Electronics Engrs. *Res:* Laser and electrooptics research; IR waveguide devices; non-linear and coherent phenomena; atomic and molecular spectroscopy; fiber optics; submillimeterwaves; optical engineering. *Mailing Add:* Dept Elec & Syst Eng Univ Conn Storrs CT 06269

CHEPENIK, KENNETH PAUL, b Jacksonville, Fla, Mar 14, 38; m 63; c 2. DEVELOPMENTAL BIOLOGY. *Educ:* Univ Fla, BSAdv, 61, MS, 65, PhD(human anat, biochem, physiol), 68. *Prof Exp:* Instr anat, Bowman Gray Sch Med, Wake Forest Univ, 68-70, asst prof, 70-73; assoc prof, 73-84, PROF ANAT, THOMAS JEFFERSON UNIV, 84- *Mem:* Sigma Xi; NY Acad Sci; Teratology Soc; Am Asn Anatomists; Soc Develop Biol; Am Soc Biochem & Molecular Biol; Am Soc Cell Biol. *Res:* Biochemical mechanisms underlying normal and abnormal mammalian embryogenesis. *Mailing Add:* Dept Anat Jefferson Med Col 1020 Locust St Philadelphia PA 19107

CHEPKO-SADE, BONITA DIANE, b Baltimore, Md, Apr 22, 48; m 71; c 2. SOCIAL ORGANIZATION, LIFE HISTORY ANALYSIS. *Educ:* Duke Univ, BA, 71; Univ PR, MS, 77; Northwestern Univ, PhD(phys anthrop), 82. *Prof Exp:* Instr animal behav, Univ PR, Rio Piedras, 75-76; res asst Rhesus monkey social orgn, Dept Anthrop, Northwestern Univ, 77-83, computer specialist, 84-85, instr primate social orgn, 87; instr genetics human biol, Dept Nat Sci, Loyola Univ, Chicago, 88; INSTR MAMMAL, CRANBERRY LAKE BIOL STA, COL ENVIRON SCI & FORESTRY, STATE UNIV NY, 90- *Concurrent Pos:* Instr, Midwest Talent Search, Northwestern Univ, 83-84, vis scientist, 83-89; Guggenheim found fel primate social orgn, N Country Inst Natural Philos Inc, Mexico, NY, 88-90; adj asst prof behav, Col Environ Sci & Forestry, State Univ NY, Syracuse, 89- *Mem:* Animal Behav Soc; Am Soc Zoologists; Soc Study Evolution; Am Soc Mammalogists. *Res:* Research in the dispersal patterns of mammals and the effects of social organization on population genetics; analysis of the life histories of male Rhesus monkeys, examining the effects of dominance rank on mortality, fecundity, and gene flow due to migration patterns; concentration on the process of group formation by fission. *Mailing Add:* RD 3 Box 53 Emery Rd Mexico NY 13114

CHER, MARK, b Buenos Aires, Arg, June 14, 32; nat US; m 56; c 4. PHYSICAL CHEMISTRY. *Educ:* Calif Inst Technol, BS, 54; Harvard Univ, AM, 55, PhD(chem), 58. *Prof Exp:* Instr chem, Univ Calif, Los Angeles, 57-59, asst prof, 59-60; res specialist, Atomics Int, Calif, 60-63; mem tech staff, NAm Aviation Sci Ctr, 63-69; assoc prof chem & chmn dept, Wis State Univ, River Falls, 69-71; sr chemist, Addressograph-Multigraph Corp, 71-74; mgr qual assurance, 74-80, prog mgr, Plume Model Validation Study-Field Measurements, Environ Monitoring & Serv Ctr, 80-84, MEM TECH STAFF, ROCKWELL INT SCI CTR, ROCKWELL INT CORP, 84- *Mem:* Am Chem Soc. *Res:* Fast reaction kinetics; atoms and free radicals in gas phase; application of ultrasonics to chemical kinetics; shock waves and detonations; photochemistry; radiation chemistry; air pollution monitoring; data quality assurance; program management; surface chemistry-lithographic inks; microwave absorption measurements. *Mailing Add:* Rockwell Int Sci Ctr 1049 Camino Dos Rios Thousand Oaks CA 91320

CHERASKIN, EMANUEL, b Philadelphia, Pa, June 9, 16; m 44; c 1. ORAL MEDICINE. *Educ:* Univ Ala, AB, 39, MA, 41, DMD, 52; Univ Cincinnati, MD, 43; Am Bd Oral Med, dipl, 56. *Hon Degrees:* Dr, Univ Sao Paulo, Brazil, 61. *Prof Exp:* Intern med, Hartford Munic Hosp, 43-44; resident, St Mary's Hosp, Evansville, Ind, 46-47; instr anat, Med Col Ala, 48-50, asst prof physiol, 50-52, assoc prof & chmn, dept oral med, Univ Ala Sch Dent, 52-56, dir postgrad studies, 56-57, prof & chmn, div oral surg & med, 56-62, asst prof, dept med, Med Col Ala, 59-79, prof & chmn, dept oral med, Univ Ala Sch Dent, 62-79, EMER PROF, UNIV ALA, BIRMINGHAM, 79- *Concurrent Pos:* Consult oral med, Vet Admin Hosps, 52-79 & Southeastern Area, Vet Admin, 55-64; mem, res comt, Am Acad Dent Med, 55-59, comt residency oral med, 56-58; staff mem, Univ Ala Hosp, 57-79; adj prof community dent, Col Dent Med, Med Univ SC, 77-80; mem, bd dirs, Nat Comt Prev Alcoholism, 80; mem, sci adv comt, Omni, 80- *Honors & Awards:* Samuel Charles Miller mem lectr, Am Acad Dent Med, 64; Rachel Carson Award, Nat Nutrit Foods Asn, 75; Ann Int Award Prev Dent, Am Soc Prev Dent, 77. *Mem:* Am Acad Oral Med; Am Dent Asn; AMA; hon mem Circle Odontol Paraguay; hon men Dominican Odontol Soc; fel Am Geriat Soc; hon mem Chinese Med Asn; hon mem Ital Med-Dent Asn; fel Int Acad Prev Med (hon pres, 74-75); hon mem Int Acad Metabology. *Res:* Predictive medicine; nutrition; metabolism. *Mailing Add:* Park Tower 904/906 2717 Highland Ave S Birmingham AL 35205-1725

CHERAYIL, GEORGE DEVASSIA, b Kothamangalam, India, Dec 17, 29; US citizen; m 57; c 3. BIOCHEMISTRY, NEUROCHEMISTRY. *Educ:* Univ Madras, BSc, 49, Hons, 52, MA, 55; St Louis Univ, PhD(biochem), 62. *Prof Exp:* Demonstr chem, St Xavier's Col, India, 49-50; lectr, Fatima Mata Nat Col, India, 52-54; lectr, Med Col, Univ Mysore, 54-55; chmn chem dept, Andhra Loyola Col, 55-58; ASSOC PROF PATH, MED COL WIS, 62- *Mem:* AAAS; Am Chem Soc; Am Soc Neurochem; Sigma Xi. *Res:* Metabolism of steroids, especially bile acids; lipid metabolism in central nervous system disorders; biochemistry of phospholipids, sphingolipids and glycolipids; levo-dihydroxyphenylalanine and catecholamines; blood lipids in multiple sclerosis. *Mailing Add:* 4160 Kamala Lane Brookfield WI 53045-7438

CHERBAS, PETER THOMAS, b Bryn Mawr, Pa, Mar 26, 46; m 68; c 1. DEVELOPMENTAL BIOLOGY, INSECT ENDOCRINOLOGY. *Educ:* Harvard Col, BA, 67; Harvard Univ, PhD(biol), 73. *Prof Exp:* Res fel genetics, Cambridge Univ, 73-74; res fel, Harvard Univ, 74-77, asst prof biol, 77-81, assoc prof, 81-85; ASSOC PROF, DEPT BIOL, IND UNIV, 85- *Mem:* Genetics Soc Am; Soc Develop Biol; AAAS; Am Soc Microbiol. *Res:* Molecular analysis of the action of steroid hormones especially the insect molting hormones, the ecdysteroids; insect cell culture and its applications in the study of development. *Mailing Add:* Dept Biol Jordan Hall Rm A317 Ind Univ Bloomington IN 47405

CHEREMISINOFF, PAUL N, b New York, NY, Feb 20, 29; m 52; c 2. CHEMICAL ENGINEERING, ENVIRONMENTAL ENGINEERING. *Educ:* Pratt Inst, BChE, 49; Stevens Inst Technol, MS, 52. *Prof Exp:* Chem engr, Joseph Turner & Co, Ridefield, 49-56; plant mgr, Am Polyglas Co, Carlstadt, 56-60; sr chem engr, Celanese Corp, Newark, 60-63 & Tenneco Corp, Garfield, 63-67; mgr environ serv, Englehard Minerals & Chem Co, 67-73; PROF CIVIL & ENVIRON ENG, NJ INST TECHNOL, 73- *Concurrent Pos:* Eng ed, Water & Sewage Works, 76-80; consult engr, 73-; engr editor, Pollution Eng, 80- *Mem:* Fel NY Acad Sci; Am Chem Soc; Sigma Xi; Am Acad Environ Engrs. *Res:* Author and editor of more than 300 technical books and articles on environmental engineering, sciences, biotechnology and electrotechnology; adsorption phenomena and hazardous wastes treatment. *Mailing Add:* 230 Terrace Ave Hasbrouck Heights NJ 07604

CHERENACK, PAUL FRANCIS, b Hazleton, Pa, June 19, 42. MATHEMATICS. *Educ:* Villanova Univ, BS, 63; Univ Pa, PhD(math), 68. *Prof Exp:* Teaching asst, Univ Pa, 63-68; instr, Villanova Univ, 68; asst prof, Ind Univ, Bloomington, 68-74; MEM FAC MATH, UNIV CAPE TOWN, SAFRICA, 74- *Mem:* Am Math Soc; Math Asn Am. *Res:* Nature of singularities on algebraic varieties via their analytic homotopy groups; algebraic geometry. *Mailing Add:* Dept Math Univ Cape Town Private Bag Rondebosch South Africa

CHERIAN, M GEORGE, b Ernakulam, India, Jan 18, 41; Can citizen. TOXICOLOGY OF METALS. *Educ:* Kerala Univ, India, PhD(biochem), 68. *Prof Exp:* Assoc prof pathol, 80-85, PROF PATH, UNIV WESTERN ONT, 85-, PROF PHARMACOL & TOXICOL, 85- *Concurrent Pos:* Mem, Grants Rev Comt Pharmacol & Toxicol, Med Res Coun Can; mem, US Metallobiochem Study Sect, NIH, 80-84, US Environ Health Sci Rev Comt, 87-; vis scientist, Nat Inst Environ Studies, Tskuba, Japan, 88. *Res:* Role of metallothionein in storage of zinc and cysteine in mammalian development; in detoxification of metal toxicity and its potential use as a tumour marker. *Mailing Add:* Dept Path Univ Western Ont London ON N6A 5C1 Can

CHERIAN, SEBASTIAN K, b Palai, India, Sept 23, 38; m 67; c 2. ZOOLOGY. *Educ:* Univ Kerala, BSc, 59; Duquesne Univ, MS, 63; St Bonaventure Univ, PhD(physiol), 67. *Prof Exp:* USPHS res grant, 61-63; teaching asst physiol, Duquesne Univ, 60-63 & zool, St Bonaventure Univ, 63-66; asst prof biol, St Francis Col, Pa, 66-69; asst prof, 69-74, PROF BIOL, JAMESTOWN COL, 74-, ACTG CHMN DEPT, 70- *Mem:* Am Soc Zoologists. *Res:* Ovarian and uterine responses to exogenous hormone administration in the immature rat; concentration and distribution of uterine glycogen in unilaterally pregnant rats. *Mailing Add:* Dept Biol Jamestown Col Jamestown ND 58401

CHERIN, PAUL, b Brooklyn, NY, Oct 14, 34; div; c 3. MATERIALS SCIENCE, ANALYTICAL CHEMISTRY. *Educ:* Brooklyn Col, BS, 55; Polytech Inst Brooklyn, PhD(phys chem), 63. *Prof Exp:* Staff scientist, Int Bus Mach Corp, 60-61; scientist, Xerox Corp, 62-66, sr scientist, 66-76, tech specialist, 77-89; PRES, TECH VENTURES INC, 86- *Mem:* Am Chem Soc; Sigma Xi; Am Phys Soc. *Res:* Relating structural properties of photoreceptors to electrical properties; problem solving. *Mailing Add:* 30 Whitney Ridge Rd No B9 Fairport NY 14450

CHERITON, DAVID ROSS, b Vancouver, BC, Mar 29, 51. COMPUTER SCIENCE, SOFTWARE ENGINEERING. *Educ:* Univ BC, BSc, 73; Univ Waterloo, MMath, 74, PhD(comput sci), 78. *Prof Exp:* asst prof comput sci, Univ BC, 78-; MEM STAFF, DEPT COMPUT SCI, STANFORD UNIV. *Mem:* Asn Comput Mach; Inst Elec & Electronics Engrs. *Res:* Logical design of mini computer operating systems for portability, reliability and configurability. *Mailing Add:* Dept Comput Sci Stanford Univ Stanford CA 94305

CHERKASKY, MARTIN, b Philadelphia, Pa, Oct 6, 11; m 41; c 2. PUBLIC HEALTH. *Educ:* Temple Univ, MD, 36. *Prof Exp:* Pvt pract, 39-40; exec, Home Care Dept, 47, dir med group, 48-51, chief, Div Social Med, 50, dir, 51-75, pres, 75-81, CONSULT, MONTEFIORE HOSP & MED CTR, 81-; ATRAN PROF COMMUNITY HEALTH, ALBERT EINSTEIN COL MED, 67- *Concurrent Pos:* Consult, NY State Joint Hosp Rev & Planning Coun; consult, Comnr Hosps, New York City Dept Hosps, 61-62; mem exec comt, Health Res Coun New York City, 69-70; mem regional health adv bd, Region II, HEW, 70-71; mem, Gov Steering Comt Social Probs, 70-72. *Mem:* Inst Med Nat Acad Sci; NY Acad Med; Am Pub Health Asn; Asn Am Med Cols. *Mailing Add:* Montefiore Med Ctr 111 E 210th St Bronx NY 10467

CHERKIN, ARTHUR, biochemistry; deceased, see previous edition for last biography

CHERKOFSKY, SAUL CARL, b Lynn, Mass, June 2, 42; m 62; c 3. MEDICINAL CHEMISTRY, PHARMACEUTICAL RESEARCH & DEVELOPMENT. *Educ:* Mass Inst Technol, BS, 63; Harvard Univ, MA, 64, PhD(org chem), 67. *Prof Exp:* Res chemist, Cent Res Dept, 66-80, res supvr, Biochem Dept, 80-81, res mgr, 81-83, mgr, med chem, Biomed Prod Dept, 83-86, dir, med chem, 86-87, DIR CHEM & PHARMACEUT DEVELOP, DUPONT MERCK PHARMACEUTICAL CO, 91- *Mem:* Am Chem Soc; AAAS. *Res:* Anthracene photodimers; carbonium ion rearrangements; bicyclobutane synthesis and polymers; aromatic substitutions; heterocyclic chemistry; medicinal chemistry; pharmaceutical development. *Mailing Add:* 1013 Woodstream Dr Ramblewood Wilmington DE 19810

CHERLIN, GEORGE YALE, b New Haven, Conn, Feb 21, 24; m 45; c 2. MATHEMATICS. *Educ:* Rutgers Univ, MSc, 49, PhD(math), 51. *Prof Exp:* Asst instr, Rutgers Univ, 47-51; asst mathematician, Mutual Benefit Life Ins Co, 51-62; vpres-actuary, Nat Health & Welfare Retirement Asn, Inc, 62-72; second vpres & actuary, Mutual Benefit Life Ins Co, 72-77; pres, APL Bus Consults Inc, 77-81, Prov Mutual Life, 86-87, United Ins Co Am, 87-90; RETIRED. *Concurrent Pos:* Consult actuary, STSC, Inc, 81-82, MONY, 82-83, Actuary Mutural Benefit Life, 83-; mem, Conf Actuaries Pub Pract. *Mem:* Casualty Actuarial Soc; Soc Actuaries; Am Acad Actuaries; Sigma Xi; Actvarial Studies Non-Life Ins; Int Actuarial Asn. *Res:* Actuarial science; computer science; mathematical logic; complex variable; logic; general, mathematics. *Mailing Add:* 4703 Lolis Lane Mt Chalis CA 96067

CHERMACK, EUGENE E A, b New York, NY, Aug 31, 34; m 56; c 4. PHYSICAL METEOROLOGY. *Educ:* Queens Col, NY, BS, 56; Univ Wash, BS, 58; NY Univ, MS, 62, PhD(meteorol), 70. *Prof Exp:* Weather officer, US Air Force, 56-59; asst res scientist, NY Univ, 59-62, instr meteorol, 62-67; ASSOC PROF METEOROL, STATE UNIV NY COL OSWEGO, 67- *Mem:* Am Meteorol Soc; Am Geophys Union; Int Asn Gt Lakes Res; Sigma Xi. *Res:* Development of indirect sounding techniques applied to the atmosphere; infrared radiation and atmospheric optics; surface temperature of lakes and rivers. *Mailing Add:* Rd 3 Box 199 Oswego NY 13126

CHERMS, FRANK LLEWELLYN, JR, b Warwick, RI, June 4, 30; m 52; c 2. POULTRY PHYSIOLOGY. *Educ:* Univ RI, BS, 52; Univ NH, MS, 54; Univ Md, PhD, 58. *Prof Exp:* Instr & asst geneticist, Univ NH, 54-55; from asst prof to prof, Univ Wis, 57-69; reproduction phYsiologist, Nicholas Turkey Breeding Farms, Inc, 69-86; RETIRED. *Mem:* AAAS; Poultry Sci Asn; Soc Study Reproduction; World Poultry Sci Asn. *Res:* Improving the reproductive performance of the turkey through breeding and physiology. *Mailing Add:* 19449 Riverside Dr PO Box Y Sonoma CA 95476

CHERN, MING-FEN MYRA, b Soochow, Kiangsu, Apr 19, 46, US citizen; m 69; c 1. GENETIC EPIDEMIOLOGY, DATABASE MANAGEMENT. *Educ:* Nat Taiwan Univ, BS, 67; Univ Minn, MS, 70, PhD(biomet), 73. *Prof Exp:* Fel biomet & health computer sci, Univ Minn, 71-73, res assoc, 73-74, asst prof, 74-80; asst prof biostatist, Sch Pub Health, Univ Calif, Los Angeles, 80-82; mgr software develop sect, 82-86, MGR PROD ASSURANCE DEPT, LOGICON INC, 87- *Concurrent Pos:* Mem tech staff, Logicon, Inc, 80-82. *Mem:* Am Soc Qual Control. *Res:* Family and epidemiological studies of heterogeneous disease such as epilepsy or diabetes; study design and statistical methodology development related to multivariate survival data analysis; linkage analysis and segregation analysis; computer software metrics. *Mailing Add:* Logicon Inc 222 W Sixth St PO Box 471 San Pedro CA 90733-0471

CHERN, SHIING-SHEN, b Kashing, China, Oct 26, 11; US citizen; m 39; c 2. GEOMETRY. *Educ:* Nankai Univ, China, BS, 30; Tsing Hua Univ, China, MS, 34; Univ Hamburg, DSc, 36. *Hon Degrees:* LLd, Chinese Univ Hong Kong, 69; DSc, Univ Chicago, 69, Univ Hamburg, 71 & State Univ NY, Stony Brook, 85; Dr, ETH Zurich, 82, Nankai Univ, 85. *Prof Exp:* Prof math, Tsinghua Univ, China, 37-43; mem, Inst Advan Study, Princeton, NJ, 43-46; prof math & actg dir, Inst Math, Acad Sinica, Nanking, China, 46-49; prof math, Univ Chicago, 49-60; prof math, Univ Calif Berkeley, 60-79; dir, 81-84, EMER DIR, MATH SCI RES INST, BERKELEY, 84-; EMER PROF, UNIV CALIF, BERKELEY, 79- *Concurrent Pos:* Vis prof, Harvard Univ, 52, Eidgenossische Technische Hoehschule, Zurich, 53, Mass Inst Technol, 57; Guggenheim fel, 54-55 & 66-67; colloquium lectr, Am Math Soc, 60; Miller res prof, Univ Calif, Berkeley, 63-64 & 69-70, fac res lectr, 78; hon prof, Peking Univ & Nankai Univ, China, 78, Chiran Univ, Canton & Inst Syst Sci, Acad Sinica, 80 & Grad Sch, 82, Normal Univ, Beijing, Sci & Technol Univ China, Nanjing Univ, Hongchow Univ, Chekiang Univ, E China Normal Univ, 85; Wolf Found Prize, Israel, 83-84; dir, Nankai Inst Math, Tianjin, China, 84-; Humboldt award, Ger, 82; assoc founding fel, Third World Acad Sci, 83-; foreign mem, Royal Soc London, UK, 85-, Accademia dei Lincei, Rome, Italy, 88- & Académie des Sci, Paris, 89-; hon mem, London Math Soc, UK, 86-; corresp mem, Academia Peloritana, Messina, Sicily, Italy, 86- *Honors & Awards:* Chauvenet Prize, Math Asn Am, 70; Nat Medal Sci, 75; Steele Prize, Am Math Soc, 83; Wolf Prize Math, 84. *Mem:* Nat Acad Sci; fel Am Acad Arts & Sci; Am Math Soc (vpres, 62-64); Math Asn Am; Acad Sinica; hon mem Indian Math Soc; corresp mem Brazilian Acad Sci; hon mem, NY Acad Sci; Am Philos Soc. *Res:* Differential geometry; integral geometry; topology. *Mailing Add:* Dept Math Univ Calif Berkeley CA 94720

CHERN, WEN SHYONG, b Chang-Hwa, Taiwan, Mar 19, 41; US citizen; m 72; c 2. ENERGY, ECONOMETRICS. *Educ:* Nat Chung-Hsing Univ, BS, 64; Univ Fla, Gainsville, MS, 69; Univ Calif, Berkeley, MA, 71, PhD(agr econ), 75. *Prof Exp:* Adj asst prof food & resource econ, Univ Fla, 73-74; res economist mkt, Fla Dept Citrus, 73-74; economist & group leader energy econ, Oak Ridge Nat Lab, 74-81; sr economist energy & resource, Lawrence Livermore Nat Lab, 81-82; from assoc prof to prof consumer econ, Univ Ma, 83-87; PROF AGR ECON, OHIO STATE UNIV, 87- *Concurrent Pos:* Vis assoc prof, Nat Chung-Hsing Univ, Taiwan, 78-79; consult, Elec Power Res Inst, 81-82, Univ Calif, Berkeley, 83, Hogan & Hartson, Washington, DC, 84-85 & Systs Sci Inc, Bethesda, Md, 85-; fel, East-West Ctr, Honolulu, Hawaii, 85,86,87. *Mem:* Am Agr Econ Asn; Am Econ Asn; Int Asn Energy Economists; Am Coun Consumer Interests; Am Statist Asn; Coun Prof Assoc Fed Statist. *Res:* Energy demand estimation and forecasting; econometric modelling of energy-economy relationships in developing countries; analysis of impacts of raising energy prices, energy conservation policies on energy demand and energy use patterns; quantitative reseach in the areas of consumer economics, applied consumption and consumer finance; international comparison of food consumption expenditures; estimation of complete demand systems; Econometric model of US soybean market; health concerns; nutrition; fats and oils consumption. *Mailing Add:* 6920 Perry Dr Worthington OH 43085

CHERNA, JOHN C(HARLES), b Budapest, Hungary, Apr 6, 21; Can citizen; m 56; c 4. INDUSTRIAL ENGINEERING, ENGINEERING GRAPHICS. *Educ:* Swiss Fed Inst Technol, Dipl Ing, 47. *Prof Exp:* Lectr mech technol, Swiss Fed Inst Technol, 45-49; indust engr, Paris, France & Montreal, Can, 49-51; lectr mech eng, 51-54, asst prof, 54-58, ASSOC PROF MECH ENG, MCGILL UNIV, 58- *Concurrent Pos:* Training consult, Bathurst Power & Paper Co Ltd, 64- *Honors & Awards:* Ralph R Teetor Award, Soc Automotive Engrs, 79. *Mem:* Soc Automotive Engrs; Am Soc Metals; Eng Inst Can; Can Soc Mech Eng. *Res:* Metal cutting and working. *Mailing Add:* Dept Mech Eng McGill Univ 817 Sherbrooke St W Montreal PQ H3A 2K6 Can

CHERNAK, JESS, b Brooklyn, NY, Aug 30, 28; m 55; c 2. ELECTRICAL ENGINEERING, COMPUTER SCIENCE. *Educ:* Brooklyn Polytech Inst, BEE, 60; NY Univ, MEE, 61. *Prof Exp:* Mem tech staff comput, 60-63, supvr, 63-68, head dept technol, 68-70, dir technol, 70-75, dir, Digital Transmission Lab, 75-79, EXEC DIR, LOOP TRANSMISSION DIV, BELL LABS, 79- *Concurrent Pos:* Mem tech prog comt, Spring Joint Comput Conf, 67-68, steering comt, 69-70. *Mem:* Inst Elec & Electronics Engrs. *Res:* Systems engineering and development of the communications network including cables, electronics, installation, protection and maintenance systems; characterization of solid state components; thin film and magnetic component design; network design; digital transmission system design; digital terminals; digital regenerators; system analysis. *Mailing Add:* Bell Labs Whippany NJ 07981

CHERNESKY, MAX ALEXANDER, b Inglis, Man, Aug 1, 38; m 65; c 1. MEDICAL VIROLOGY. *Educ:* Univ Guelph, BS, 65; Univ Toronto, MS, 67; Univ BC, PhD(virol), 69. *Prof Exp:* From instr to assoc prof pediat, 70-83, PROF PEDIAT & PATH, MCMASTER UNIV, 84-, PROF BLOOD & CARDIOVASC, MED SCI, 88- *Concurrent Pos:* Dir, Hamilton Reg Virol Lab, St Joseph's Hosp, 78- *Mem:* Am Soc Microbiol; Am Soc Trop Med & Hyg; Can Soc Microbiol; Pan Am Group Rapid Viral Diag. *Res:* Pathogenesis of viral infections; rapid viral diagnosis technology. *Mailing Add:* McMaster Univ Fac Health Ctr 1200 Main St W St Joseph Hosp 50 Charlton Ave E Hamilton ON L8N 3Z5 Can

CHERNIACK, NEIL S, b Brooklyn, NY, May 28, 31; m 55; c 3. INTERNAL MEDICINE. *Educ:* Columbia Univ, AB, 52; State Univ NY, MD, 56. *Hon Degrees:* Dr, Karolinska Inst, Stockholm, Sweden. *Prof Exp:* Intern med, Univ Ill Med Ctr, 56-57, res fel, 57-58, resident, 60-62; res fel pulmonary dis, Columbia Univ, 62-64; from asst prof to assoc prof med, Univ Ill Med Ctr, 64-69; assoc prof, 69-73, prof med & assoc dir pulmonary serv, Univ Pa, 73-77; assoc dean, 83-90, PROF & DIR PULMONARY SERV, CASE WESTERN RESERVE UNIV, 77-, VPRES MED AFFAIRS & DEAN, 90- *Concurrent Pos:* Consult, Chicago State Tuberc Sanitarium, Ill, 64-69; assoc attend physician, Cook County Hosp, 65-69; assoc attend physician & sr res assoc, Michael Reese Hosp, 67-69; sr attend physician, Philadelphia Gen Hosp, Pa, 69-76; vis scholar, Karolinska Inst, Stockholm, 76-77; chief pulmonary serv, Cleveland & Brecksville Vet Admin Hosp, 77-; chief pulmonary serv, Univ Hosp, Cleveland, 77-; assoc ed, J Appl Physiol, 81-; assoc ed, J Lab & Clin Med, 86- *Mem:* Am Thoracic Soc; Am Physiol Soc; Am Soc Clin Invest; Am Asn Physicians Bioeng Soc. *Res:* Control of ventilation and circulation; pulmonary disease; oxygen and carbon dioxide stores of the body; bioengineering; sleep disorders. *Mailing Add:* Sch Med Case Western Reserve Univ Cleveland OH 44106-4109

CHERNIACK, REUBEN MITCHELL, b Can, June 15, 24; m 52; c 3. MEDICINE. *Educ:* Univ Man, MD, 48, MSc, 51, FRCP(C). *Hon Degrees:* DSc, Univ Man, 83. *Prof Exp:* Lectr, Dept Physiol & Med Res, Univ Man, 54-56, from asst prof to prof med, 56-76, chmn fac med, 74-78; PROF MED, UNIV COLO, 76-, VCHMN, FAC MED, 78- *Concurrent Pos:* Fel med, Columbia Univ, 52-54; Life Ins fel & Markle scholar, 54; dir, Cardiorespiratory Unit, Winnipeg Gen Hosp, 54, Inhalation Ther Unit, 61, consult physician in respiratory dis, 58, dir respiratory div, Clin Invest Unit; dir, Joint Respiratory Prog, Sanatorium Bd Man & Univ Man; med dir, D A Stewart Ctr Study & Treat Respiratory Dis; consult, Man Rehab Hosp & Munic Hosps, 58; chmn dept med, Nat Jewish Hosp & Res Ctr, 78-84. *Honors & Awards:* Prowse Prize, 52; Drewery Prize, 53. *Mem:* Fel Am Col Physicians; Am Soc Clin Invest; Am Physiol Soc; Can Col Physicians & Surg; Can Soc Clin Invest (secy, 61-63, pres, 63, past pres, 64). *Res:* Internal medicine; respiratory function and diseases. *Mailing Add:* Pulmonary/Physiol Unit 1400 Jackson St Denver CO 80206

CHERNIAK, EUGENE ANTHONY, b Windsor, Ont, Dec 17, 30; m 55; c 2. PHYSICAL CHEMISTRY. *Educ:* Queen's Univ, Ont, BA, 53, MA, 56; Univ Leeds, PhD(radiation chem), 59. *Prof Exp:* Sci master, Pickering Col, Can, 53-55; Nat Res Coun Can fel, 59-60; lectr chem, Carleton Univ, Can, 60-61, asst prof, 61-65; chmn dept, 65-69, PROF CHEM, BROCK UNIV, 65- *Concurrent Pos:* Vis res prof, FYS-KEM Inst, Uppsala Univ, Sweden. *Mem:* AAAS; Int Am Photochem Soc; fel Chem Inst Can; Am Chem Soc. *Res:* Chemical kinetics; photochemistry (including flash photochemistry); kinetic spectroscopy and spectrophotometry; photoionization in polar fluids; mechanisms of tropospheric reactions involving free radicals; computer modelling of complex chemical mechanisms. *Mailing Add:* Dept Chem Brock Univ St Catharines ON L2S 3A1 Can

CHERNIAK, ROBERT, b New York, NY, June 26, 36; m 61; c 2. BIOCHEMISTRY. *Educ:* City Col New York, BS, 59; Duke Univ, PhD(biochem), 64. *Prof Exp:* Arthritis Found fel biochem, Univ Newcastle, 64 & Albert Einstein Col Med, 65; assoc res biologist, Sterling-Winthrop Res Inst Div, Sterling Drug, Inc, 66-68; from asst prof to assoc prof, 68-81, PROF CHEM, GA STATE UNIV, 81- *Res:* Structure and biosynthesis of polysaccharides of biological origin. *Mailing Add:* Dept Chem Ga State Univ University Plaza Atlanta GA 30303

CHERNIAVSKY, ELLEN ABELSON, b Philadelphia, Pa, 1947; m 68; c 2. OPERATIONS RESEARCH. *Educ:* Stanford Univ, BS, 68; Cornell Univ, MS, 71, PhD(opers res), 73. *Prof Exp:* Assoc scientist, Brookhaven Nat Lab, 73-81; opers res analyst, Anal Sci Corp, 81-82; opers res analyst, 82-88, LEAD ENGR, MITRE CORP, 88- *Concurrent Pos:* Adj prof, State Univ NY, Stony Brook, 73-75; consult, Brookhaven Nat Lab, 81-82. *Mem:* Opers Res Soc Am; AAAS. *Res:* Air traffic control automation support; formulation and evaluation of energy system models, particularly those pertaining to hydrocarbon supplies; multiobjective analysis; cost and performance trends over time for tactical aircraft. *Mailing Add:* 5512 Massachusetts Ave Bethesda MD 20816

CHERNIAVSKY, JOHN CHARLES, b Boston, Mass, Feb 15, 47; m 68; c 2. THEORY, SOFTWARE SYSTEMS. *Educ:* Stanford Univ, BS, 69; Cornell Univ, MS, 71, PhD(comput sci), 72. *Prof Exp:* Asst prof comput sci, State Univ NY, Stony Brook, 72-80, assoc prof, 80-81; prog dir theoret comput sci, NSF, 80-84; prof & dept head, Comput Sci Dept, Georgetown Univ, 84-90; PROG DIR, INST INFRASTRUCTURE, NSF. *Concurrent Pos:* Vis fel, Johns Hopkins Univ, 78-79; consult, Nat Bur Standards, 78-; ed, Book Ser Appl Logic & Computer Sci. *Mem:* Asn Comput Mach; Am Math Soc; Soc Indust & Appl Math; AAAS; Asn Symbolic Logic. *Res:* Theoretical computer science; software engineering; mathematical logic. *Mailing Add:* CISE/OCDA Rm 304 NSF 1800 G St NW Washington DC 20550

CHERNICK, MICHAEL ROSS, b Havre de Grace, Md, Mar 11, 47; m 88; c 1. EXPERIMENTAL DESIGN & ANALYSIS, EXPERT SYSTEMS. *Educ:* State Univ NY, Stony Brook, BS, 69, Univ Md, MA, 73; Stanford Univ, MS, 76, PhD(statist), 78. *Prof Exp:* Mathematician, US Army Mat Systs Anal Activity, 69-74; math statistician, Oak Ridge Nat Lab, 78-80; mem tech staff, Aerospace Corp, 80-88; SR MEM TECH STAFF, NICHOLS RES CORP, 88- *Concurrent Pos:* Vis lectr time series anal, Univ Calif, Santa Barbara, 81; lectr statist, Calif State Univ, Fullerton, 82-85; lectr calculus, Calif State Univ, Dominguez Hills, 84-85; lectr math, Calif State Univ, Long Beach, 90-91. *Honors & Awards:* Jacob Wolfowitz Prize, 83. *Mem:* Am Statist Asn; Bernoulli Soc; Inst Math Statist. *Res:* Stochastic processes; time series analysis; extreme value theory; outlier detection; pattern recognition. *Mailing Add:* 18914 Felbar Ave Torrance CA 90504-5718

CHERNICK, SIDNEY SAMUEL, b Winnipeg, Man, Mar 6, 21; US citizen. BIOCHEMISTRY. *Educ:* Univ Calif, Los Angeles, AB, 43; Univ Calif, MA, 45, PhD(physiol), 48. *Prof Exp:* Physiologist, Med Sch, Univ Calif, 48-51; prof pharmacol & physiol, NDak State Col, 51-52; scientist, 52-63, SCIENTIST DIR, NAT INST ARTHRITIS, METAB & DIGESTIVE DIS, 63- *Concurrent Pos:* Vis scientist, Med Clin, Munich, 63-64. *Honors & Awards:* Purkinje Medal, Czech Med Soc, 69. *Mem:* Soc Exp Biol & Med; Am Soc Biol Chem; Sigma Xi. *Res:* Metabolic defects in endocrine and nutritional diseases; in vitro metabolism; diabetes; mechanisms of hormone action. *Mailing Add:* 6703 Melville Pl Chevy Chase MD 20815

CHERNICK, VICTOR, b Winnipeg, Man, Dec 31, 35; m 57; c 4. PEDIATRICS, RESPIROLOGY. *Educ:* Univ Man, MD, 59; Am Bd Pediat, dipl, 65, FRCP(c), 71. *Prof Exp:* Rotating intern, Winnipeg Gen Hosp, Man, 59-60; Nat Inst Neurol Dis & Blindness perinatal fel, Johns Hopkins Univ, 60, from jr asst resident to asst resident pediat, 60-62, univ fel environ med & pediat, 62-64, from instr to asst prof pediat, 64-66; from asst prof to assoc prof pediat & physiol, 66-71, head dept, 71-79, assoc dean fac med, 83-86, PROF PEDIAT, UNIV MAN, 71- *Concurrent Pos:* Resident, Vet Admin Hosp, Baltimore, 64, attend physician, 64-65, consult, 66; chief respiratory dis & perinatal physiol, Children's Hosp, 67-71; pediatrician in chief, Health Sci Ctr, 71-79; head, sect pediatric respirology, Winnipeg Childrens Hosp, 80- *Honors & Awards:* Queen Elizabeth II Scientist Award, 67; Medal, Can Pediat Soc, 70. *Mem:* Fel Am Thoracic Soc; fel Am Acad Pediat; Soc Pediat Res; Am Pediat Asn; Am Physiol Soc. *Res:* Pulmonary physiology; neonatology; development of control of breathing; development of surfactant. *Mailing Add:* Children's Hosp 685 Bannatyne Ave Winnipeg MB R3E 0W1 Can

CHERNICK, WARREN SANFORD, b Providence, RI, Oct 6, 29. PHARMACOLOGY. *Educ:* RI Col Pharm, BS, 52; Philadelphia Col Pharm, MS, 54, DSc, 56. *Prof Exp:* From instr to asst prof pharmacol, Philadelphia Col Pharm, 54-64; from asst prof to assoc prof, 64-68, PROF PHARMACOL & CHMN DEPT, HAHNEMANN MED COL & HOSP, 68- *Concurrent Pos:* Res assoc, Children's Hosp & Med Sch, Univ Pa, 57-64. *Mem:* Am Pharmaceut Asn; Am Soc Pharmacol & Exp Therapeut. *Res:* Salivary secretion; psychopharmacology; cystic fibrosis. *Mailing Add:* Dept Pharmacol Hahnemann Med Univ 230 N Broad St Philadelphia PA 19102

CHERNIN, ELI, medical parasitology, tropical public health; deceased, see previous edition for last biography

CHERNOCK, WARREN PHILIP, b Fall River, Mass, Jan 12, 26; m 47; c 4. METALLURGY, NUCLEAR ENGINEERING. *Educ:* Columbia Univ, BS, 49; NY Univ, MS, 55. *Prof Exp:* Metall engr, Argonne Nat Lab, 49-51; sr metall engr, Sylvania Elec, 51-56; mgr metall, 56-64, mgr nuclear labs, 64-69, dir nuclear labs, 69-74, VPRES DEVELOP, COMBUSTION ENG, INC, 74-; VPRES NUCLEAR SYSTS. *Concurrent Pos:* Adj prof & chmn dept metall, Hartford Grad Ctr, Rensselaer Polytech Inst, 56-; vis assoc prof nuclear eng, Mass Inst Technol, 61-62. *Mem:* Fel Am Nuclear Soc; fel Am Soc Metals; Am Soc Testing & Mat; Metal Properties Coun; Atomic Indust Forum. *Res:* Irradiation effects; nuclear fuels and materials; anisotropy and preferred orientation; corrosion. *Mailing Add:* Combustion Eng Inc 1000 Prospect Hill Rd Windsor CT 06095

CHERNOFF, AMOZ IMMANUEL, b Malden, Mass, Mar 17, 23; m 53; c 3. HEMATOLOGY, GENETICS. *Educ:* Yale Univ, BS, 44, MD, 47. *Prof Exp:* Intern med, Mass Gen Hosp, 47-48; asst resident, Barnes Hosp, St Louis, Mo, 48-49; res fel hemat, Michael Reese Hosp, Chicago, 49-51; from instr to asst prof, Wash Univ, 51-56; assoc prof, Duke Univ, 56-58; res prof, Univ Tenn, Knoxville, 58-78, dir, Mem Res Ctr, 64-77, prof med, Sch Med, 66-78, assoc vice chancellor, Ctr Health Sci, 77-78; dir, Div Blood Dis & Resources, Nat Heart, Lung & Blood Inst, NIH, Bethesda, MD, sci dir, Am Assoc Blood Banks, 88-90; CONSULT, TRANSFUSION MED, 90- *Concurrent Pos:* Asst dir hemat res lab, Michael Reese Hosp, 50-51; Am Col Physicians fel med, Sch Med, Wash Univ, 51-52, special USPHS fel, 52-53; consult, City Hosp, St Louis, 52-56; asst physician, Barnes Hosp, 52-56; prin investr, NIH grants, 54-77; chief hemat sect, Vet Admin Hosp, Durham, NC, 56-58; attend physician, Univ Tenn Mem Res Ctr & Hosp, 58-78; mem cancer chemother study sect, USPHS, 59-63; USPHS res career award, 62-77; med dir, Cystic Fibrosis Found, Atlanta, 75-77; mem, Health Planning Comt, Nat Asn State Univ & Land Grant Col, 78-79; US coordr, prob area 6, US-USSR Exchange Prog Cardiovasc Dis, 79-88; US deleg, Coun Europe Comt Experts Blood Transfusion & Immunohematology, 79-88. *Mem:* Sigma Xi; fel Am Col Physicians; Am Soc Hemat; Int Soc Hemat; Am Soc Clin Invest; Am Asn Blood Banks. *Res:* Hemolytic anemias; abnormal hemoglobins; biochemical genetics; transfusion medicine; author of over 100 publications related to research and research management. *Mailing Add:* 9417 Copenhaver Dr Potomac MD 20854-3025

CHERNOFF, DONALD ALAN, b Philadelphia, Pa, Feb 18, 52; m 75. MATERIALS ANALYSIS, MICROSCOPY. *Educ:* Univ Chicago, BS, 73, PhD(phys chem), 78. *Prof Exp:* Fel chem, Univ Pa, 78-80; proj leader chem physics, BP Am Corp Res, Cleveland, Ohio, 80-88; proj engr II appl math, Boehringer Mannheim Res & Develop, Indianapolis, 88-90; PRES, ADVAN SURFACE MICROS, INC, 90- *Mem:* Am Chem Soc; Am Phys Soc; Am Vacuum Soc; Sigma Xi. *Res:* Materials characterization using scanning probe microscopes; analytical chemistry; molecular dynamics studied by time-resolved laser spectroscopy; surface science. *Mailing Add:* 6009 Knyghton Rd Indianapolis IN 46220-4955

CHERNOFF, ELLEN ANN GOLDMAN, Philadelphia, Pa, May 14, 52; m 75. DEVELOPMENTAL BIOLOGY, NEURAL REGENERATION. *Educ:* The College, Univ Chicago, BA, 73, PhD,(biol), 78. *Prof Exp:* Fel, dept anat, Sch Med, Univ Pa, 78-79; vis scientist, dept biol, Temple Univ, 79-80; sr res assoc, Dept Molecular Biol & Microbiol, Case Western Reserve Univ, 80-83, sr res assoc, Dept Develop Genetics & Anat, Sch Med, 83-86; ASST PROF, DEPT BIOL, INDIANA-PURDUE UNIV, INDIANAPOLIS, 86-

Concurrent Pos: Prin investr, NSF; vis scientist, Eli Lilly Corp, 88. *Mem:* Am Soc Cell Biol; Soc Develop Biol; Asn Women Sci; Women Cell Biol; AAAS. *Res:* Spinal cord regeneration and factors underlying tissue organization during embryonic development. *Mailing Add:* 6009 Knyghton Rd Indianapolis IN 46220-4955

CHERNOFF, HERMAN, b New York, NY, July 1, 23; m 47; c 2. MATHEMATICAL STATISTICS. *Educ:* City Col New York, BS, 43; Brown Univ, ScM, 45, PhD(appl math), 48. *Hon Degrees:* ScD, Ohio State Univ, 83, Israel Inst Technol, 84. *Prof Exp:* Res assoc, Cowles Comn Res Econ, Chicago, 47-49; asst prof statist & math, Univ Ill, 49-52; from assoc prof to prof statist, Stanford Univ, 52-74; prof appl math, Mass Inst Technol, 74-85; PROF STATIST, HARVARD UNIV, 85- *Concurrent Pos:* Fel, Ctr Advan Study Behav Sci, 59-60. *Honors & Awards:* Wilks Medal, 87. *Mem:* Nat Acad Sci; Am Math Soc; Inst Math Statist (pres, 68-69); Am Statist Asn; Am Acad Arts & Sci; Int Statist Inst. *Res:* Statistical problems in econometrics; sequential design of experiments; rational selection of decision functions; large sample theory; pattern recognition. *Mailing Add:* Dept Statist Harvard Univ Cambridge MA 02138

CHERNOFF, PAUL ROBERT, b Philadelphia, Pa, June 21, 42. MATHEMATICS. *Educ:* Harvard Univ, BA, 63, MA, 65, PhD(math), 68. *Prof Exp:* Lectr, 69-71, asst prof, 71-74, assoc prof, 74-80, PROF MATH, UNIV CALIF, BERKELEY, 80- *Concurrent Pos:* Fel NSF, Univ Calif, Berkeley, 68-69; consult, Inst Defense Analyses, Princeton, 79-; vis prof, Dept Math, Univ Pa, 86. *Mem:* Am Math Soc; Math Asn Am; fel AAAS. *Res:* Functional analysis; operator theory; mathematical physics. *Mailing Add:* Dept of Math Univ of Calif Berkeley CA 94720

CHERNOSKY, EDWIN JASPER, b Rosenberg, Tex, May 21, 14; wid; c 3. ENVIRONMENTAL PHYSICS, SPACE PHYSICS. *Prof Exp:* Chemist, Champion Paper Co, Tex, 37-42; observer, Huancayo Magnetic Observ, Peru, Carnegie Inst, 42-45, res geophysicist, 45-46; mem tech staff physics res dept, Naval Ord Lab, Washington, DC, 46-52; physicist & actg chief geomagnetics unit, 52-55; chief geomagnetic activity sect, Air Force Cambridge Res Labs, 55-67, res physicist, 67-76; mem staff, Visidyne, Inc, 76-78; CONSULT, 79- *Concurrent Pos:* Deleg, Int Asn Geomagnetism & Aeronomy, Toronto, 57, Helsinki, 60, Berkeley, 63, Zurich, 67, Madrid, 69, Moscow, 71, Kyoto, 73, Grenoble, 75, Seattle, 77, Canberra, 79, Edinburgh, 81 & mem comn IV, IX & lunar variations, chmn interdiv comn of hist; mem organizing comt & deleg, Int Symp Equatorial Aeronomy, Huaychulo, Peru, 62, San Jose do Campos, Brazil, 65 & Ahmedabad, India, 69. *Mem:* AAAS; Am Geophys Union; Am Phys Soc; Inst Elec & Electronics Engrs; Sigma Xi. *Res:* Solar geomagnetic relationships, morphology of solar activity and geomagnetic variations, characterization of geomagnetic time variations, recurrence phenomena of geomagnetic variations; 22 year geomagnetic cycle, dichotomy in geomagnetic activity; solar-terrestrial physics. *Mailing Add:* 48 Berkley St Waltham MA 02154

CHERNOW, FRED, b Brooklyn, NY, Sept 13, 32; m 56; c 3. ELECTRICAL ENGINEERING. *Educ:* Brooklyn Col, BA, 55; NY Univ, PhD(ejection polarization), 62. *Prof Exp:* Fel solid state, Lab Insulation Res, Dept Elec Eng, Mass Inst Technol, 61-62, Ford asst prof elec eng, 62-64, asst prof, 64-66; assoc prof, 66-70, prof elec eng, Univ Colo, Boulder, 70-79; PRES, CHERNOW COMMUN, 79- *Mem:* Am Phys Soc; Am Vacuum Soc. *Res:* Transport properties of II-VI compounds; thin film research studies of II-VI compounds and insulators; ion implantation of II-VI compounds. *Mailing Add:* 360 Kiowa Pl Boulder CO 80303

CHERNY, WALTER B, b Montreal, Que, Apr 13, 26; nat US; m 55; c 2. OBSTETRICS & GYNECOLOGY. *Educ:* McGill Univ, BSc, 48, MD, CM, 50. *Prof Exp:* Instr obstet & gynec, Sch Med, Duke Univ, 55, assoc, 55-57, from asst prof to prof, 57-70; dir residency, Postgrad Training in Obstet & Gynec, Good Samaritan Hosp, 70-89; prof obstet & gynec, Univ Ariz, 70-91; RETIRED. *Concurrent Pos:* Chief serv obstet, Lincoln Hosp, 58; consult, Watts Hosp, 58-70. *Mem:* AAAS; AMA; Am Col Obstet & Gynec. *Res:* Obstetrics and gynecology affecting physical and emotional health; reproductive physiology. *Mailing Add:* PO Box 629 Scottsdale AZ 85252-0629

CHERRICK, HENRY M, b Brooklyn, NY, Dec 4, 39; m 61; c 1. ORAL SURGERY, ORAL PATHOLOGY. *Educ:* Univ Fla, Gainesville, AA, 61; Med Col Va, Va Commonwealth Univ, DDS, 65; Ind Univ, MSD, 70; Am Bd Oral Path, dipl. *Prof Exp:* Resident oral surg, Univ Cincinnati Med Ctr, 68; chmn sect oral diag, med & path, Sch Dent, Univ Calif, Los Angeles, 72-76, assoc prof, 73-77, chmn, Div Biol Dent Sci, 74-78, prof sect oral diag, med & path, 76-78, asst dean hosp affairs, 77-78; dean & prof, Sch Dent Med, Southern Ill Univ, Edwardsville, 78-81; dean & prof, Col Dent, Univ Nebr Med Ctr, 81-87; PROF & DEAN, SCH DENT, UNIV CALIF, LOS ANGELES, 87- *Concurrent Pos:* Consult oral path & oral surg var hosps & univs, 70-78; ed consult, J Oral Surg, 73- *Mem:* Am Dent Asn; Am Asn Dent Schs; Am Acad Oral Med; Am Soc Oral Surg; Fel Am Soc Oral & Maxillofacial Surg. *Res:* Chemical carcinogenesis; chemical topical chemotherapy. *Mailing Add:* Sch Dent Univ Calif Los Angeles Ctr Health Sci Los Angeles CA 90024

CHERRINGTON, ALAN DOUGLAS, METABOLIC REGULATION. *Educ:* Univ Toronto, PhD(endocrinol), 73. *Prof Exp:* ASSOC DIR, DIABETES RES & TRAINING CTR, VANDERBILT UNIV, 77-, PROF RENAL PHYSIOL & ENDOCRINOL, 81-, VCHMN, DEPT MOLECULAR PHYSIOL & BIOPHYS, 86- *Res:* Regulation of glycogenolysis; gluconeogenesis. *Mailing Add:* Dept Molecular Physiol & Biophys Sch Med Vanderbilt Univ Light Hall Nashville TN 37232

CHERRINGTON, BLAKE EDWARD, b Belleville, Ont, Mar 16, 37; m 60; c 2. ELECTRICAL ENGINEERING, PHYSICS. *Educ:* Univ Toronto, BASc, 59, MASc, 61; Univ Ill, PhD(elec eng), 65. *Prof Exp:* Res assoc elec eng, Univ Ill, Urbana-Champaign, 65-66, from asst prof to assoc prof, 66-74, asst dean eng, 69-70, prof elec & nuclear eng, 74-79; prof & chmn elec eng, Univ Fla, 79-86; ERICSSON PROF ELEC ENG & DEAN ENG & COMPUTER SCI, UNIV TEX, DALLAS, 86- *Concurrent Pos:* Am Coun Educ fel acad admin, 77-78; dir, Div Elec Commun & Systs Eng, NSF, 84-85. *Honors & Awards:* Super Accomplishment Award, NSF, 85. *Mem:* Inst Elec & ElectronicS Engrs; Am Phys Soc; Am Soc Eng Educ; Soc Mfg Engrs; Nat Soc Prof Engrs. *Res:* Gaseous electronics, plasmas and gas lasers; basic electronic and atomic processes occuring in gas lasers and gas discharges. *Mailing Add:* Erik Jonsson Sch Eng & Computer Sci Univ Tex Dallas PO Box 830688 Richardson TX 75083-0688

CHERRY, DONALD STEPHEN, b Paterson, NJ, Sept 23, 43; m 66; c 2. AQUATIC ECOLOGY. *Educ:* Furman Univ, BSc, 65; Clemson Univ, MSc, 70, PhD(zool), 73. *Prof Exp:* Teacher biol & football coach, J L Mann High Sch, SC, 65-68; instr human ecol, Clemson Univ, 72-73; res assoc & fel, 73-74, asst prof biol, 74-76, ASSOC PROF, CTR ENVIRON STUDIES, VA POLYTECH INST & STATE UNIV, 80- *Concurrent Pos:* Consult, Dept Microbiol, Clemson Univ, 73; investr, Facil Use Agreement, Savannah River Proj, 72-75, co-investr, AEC contract, 73-75; consult, Am Elec Power Serv Corp, Canton, Ohio, 74-75, co-investr, 74-76; co-prin investr, Dept Energy, 79-81. *Mem:* Ecol Soc Am; Am Water Works Asn; Int Water Resources Asn. *Res:* Impact of power production discharges upon aquatic food chains in the drainage systems by site-specific field laboratory and field biomonitoring activities; validation of aquatic field and laboratory data of organisms (bacteria, algae, aquatic insects, micro-invertebrates and fish) for hazard evaluation; correlation of physiological-biochemical to ecological mechanisms of fish to heavy metal toxicity responses from power effluents. *Mailing Add:* Dept Biol Va Polytech Inst & State Univ Blacksburg VA 24061

CHERRY, EDWARD TAYLOR, b Gainesboro, Tenn, Nov 12, 41; m 67; c 4. ENTOMOLOGY. *Educ:* Tenn Polytech Inst, BS, 63; Univ Tenn, MS, 66, PhD(entom), 70. *Prof Exp:* Res assoc appl entom, Miss State Univ, 70-71; asst prof agr biol, Univ Tenn, 71-74; res specialist plant protectants, Ciba-Geigy Corp, 74-79; DIR PROD DEVELOP, RES & DEVELOP AGR CHEM GROUP, FMC CORP, 79- *Mem:* Entom Soc Am; Sigma Xi. *Res:* Chemical insecticides both from a potential nature and extension of existing compounds; subject areas of pest management, biological control. *Mailing Add:* FMC Corp 1735 Market St Philadelphia PA 19103

CHERRY, FLORA FINCH, MATERNAL AND CHILD HEALTH, PUBLIC HEALTH. *Educ:* Tulane Univ, MD, 48. *Prof Exp:* ASSOC PROF MATERNAL & CHILD HEALTH, SCH PUB HEALTH & TROP MED, TULANE UNIV, 70- *Mailing Add:* Dept Appl Health Sci Rm 612 Sch Pub Health & Trop Med Tulane Univ 1501 Canal St New Orleans LA 70112

CHERRY, JAMES DONALD, b Summit, NJ, June 10, 30; m 54; c 3. PEDIATRICS, INFECTIOUS DISEASES. *Educ:* Springfield Col, BS, 53; Univ Vt, MD, 57; Am Bd Pediat, dipl, 62; London Sch Hyg & Trop Med, MSc, 82. *Prof Exp:* Intern pediat, Boston City Hosp, 57-58, asst resident, 58-59; resident, Kings County Hosp, Brooklyn, NY, 59-60; instr, Col Med, Univ Vt, 60-61; NIH fel med, Harvard Med Sch & Thorndike Mem Lab, 61-62; from asst prof to assoc prof, Med Sch, Univ Wis, 63-66; from assoc prof to prof pediat, Sch Med, St Louis Univ, 68-73, assoc prof microbiol, 68-73, vchmn dept pediat, 70-73; PROF PEDIAT, SCH MED, UNIV CALIF, LOS ANGELES, 73- *Concurrent Pos:* Asst attend physician, Mary Fletcher Hosp & DeGoesbriand Hosp, Burlington, Vt, 61-62; asst pediat, Boston City Hosp, 61-62; assoc attend physician, Madison Gen Hosp, 62-67; dir, John A Hartford Res Found, 62-67; Markle scholar, 64; mem med staff, Cardinal Glennon Mem Hosp Children, 66-67; vis worker, Common Cold Res Unit & Clin Res Ctr, Salisbury, Eng, 69. *Mem:* Am Soc Microbiol; Am Fedn Clin Res; Am Acad Pediat; AAAS; Soc Pediat Res. *Res:* Clinical manifestations of infectious diseases; vaccines; interaction of infectious agents in the pathogenesis of disease; epidemiology of infectious diseases. *Mailing Add:* Div Infectious Dis Dept Pediat Univ Calif Sch Med Los Angeles CA 90024

CHERRY, JERRY ARTHUR, b Dayton, Tex, Feb 5, 42; m 65; c 3. POULTRY NUTRITION. *Educ:* Sam Houston State Univ, BS, 64; Univ Mo, PhD(poultry nutrit), 72. *Prof Exp:* Asst prof, 72-77, ASSOC PROF NUTRIT, DEPT POULTRY SCI, VA POLYTECH INST & STATE UNIV, 77- *Mem:* Poultry Sci Asn; Sigma Xi; Nutrit Res Coun; AAAS. *Res:* Lipid metabolism; food intake control mechanisms; energy metabolism. *Mailing Add:* Dept Poultry Sci Livestock Poultry Bldg Univ Ga Athens GA 30602

CHERRY, JESSE THEODORE, b St Louis, Mo, Sept 22, 31; m 55; c 4. GEOPHYSICS, SEISMOLOGY. *Educ:* St Louis Univ, BS, 53, MS, 56, PhD(geophys), 60. *Prof Exp:* Instr eng, St Louis Univ, 57-60; group leader explor geophys, Continental Oil Co, 60-63; group leader geophys comput, Lawrence Radiation Lab, 63-71; sr vpres geophys opers, 71-80, SR VPRES, GEOPHYS & SEISMIC PROGS, SYSTS SCI & SOFTWARE, 80- *Mem:* Seismol Soc Am; Am Geophys Union. *Res:* Rock fracturing from explosive sources; exploration geophysics; computer simulation of non linear material behavior; earthquake hazards. *Mailing Add:* 710 Encinitas Blvd Suite 200 Encinitas CA 92024

CHERRY, JOE H, b Newbern, Tenn, June 3, 34; m 55; c 3. PLANT PHYSIOLOGY, BIOCHEMISTRY. *Educ:* Univ Tenn, BS, 57; Univ Ill, MS, 59, PhD(agron, biochem), 61. *Prof Exp:* Res assoc biochem, Seed Protein Pioneering Res Lab, USDA, La, 61-62; from asst prof to assoc prof hort, Purdue Univ, West Lafayette, 62-67, prof hort, 67-89; PROF & HEAD BOT & MICROBIOL, AUBURN UNIV, 89- *Mem:* AAAS; Am Soc Plant Physiol; Am Soc Biochem & Molecular Biol; Sigma Xi. *Res:* Nucleic acid metabolism during seed germination and plant growth; induction of enzymes during cell differentation; effects of ionizing radiation on plants; mechanism of action of plant hormones; heat stress adaptation in plant cells. *Mailing Add:* Dept Bot & Microbiol Auburn Univ Auburn AL 36849

CHERRY, JOHN PAUL, b Rhinebeck, NY, Jan 31, 41; m 64; c 2. PROTEIN CHEMISTRY, MOLECULAR THEORY. *Educ:* Furman Univ, BS, 63; WVa Univ, MS, 66; Univ Ariz, PhD(genetics & biochem), 71. *Prof Exp:* Res chemist, USDA, 71-72; res assoc biochem & biophys, Tex A&M Univ, 72-73; asst prof food sci, Univ Ga, 73-75; supvry res chemist & lab chief, S Region Res Ctr, USDA, New Orleans, 76-82; assoc dir, 82-84, DIR, E REGION RES CTR, AGR RES SERV USDA, 84- *Concurrent Pos:* Mem, Prog Comt, Am Chem Soc, 85-, Chm Agr & Food Chem Div, 85-86; res grant, Univ Ga, 74-75, S Res Ctr, USDA. *Honors & Awards:* Distinguished Serv Award, Agr & Food Chem Div, Am Chem Soc. *Mem:* Inst Food Technologists; Am Chem Soc; Am Asn Cereal Chemists; Am Peanut Res & Educ Asn; Sigma Xi; Am Oil Chem Soc; Am Asn Anal Chem; Planetary Soc. *Res:* Discovery, isolation, fractionation, purification and modification of proteins and related constituents and their interaction chemistry and deterioration from conventional and nonconventional food sources; characterization of biochemical, functional nutritional and organoleptic properties of proteins for use as food and feed ingredients. *Mailing Add:* ERRC Agr Res Serv USDA 600 E Mermaid Lane Philadelphia PA 19118

CHERRY, LEONARD VICTOR, b Los Angeles, Calif, May 3, 23; m 53; c 1. SOLID STATE PHYSICS. *Educ:* City Col New York, BS, 47; Duke Univ, PhD(chem), 53. *Prof Exp:* Prin chemist, Battelle Mem Inst, 51-53; from asst prof to assoc prof chem, Hampton Inst, 53-57; instr & res assoc, Univ Pittsburgh, 57-61; from asst prof to assoc prof, 61-82, PROF PHYSICS, FRANKLIN & MARSHALL COL, 82- *Mem:* AAAS; Am Phys Soc; Am Asn Physics Teachers; Fedn Am Sci; Bioelec Repair & Growth Soc; Sigma Xi; Soc Biomaterials. *Res:* Kerr effect in aromatic fluorine compounds; physicochemical problems connected with photoengraving and electrophotography; properties of intermetallic compounds; Mossbauer effect; electrical properties of bone. *Mailing Add:* 546 State St Franklin & Marshall Col Lancaster PA 17603

CHERRY, S(HELDON), b Winnipeg, Man, Mar 28, 28; m 62; c 2. EARTHQUAKE ENGINEERING, STRUCTURAL DYNAMICS. *Educ:* Univ Man, BSc, 49; Univ Ill, MS, 51; Bristol Univ, PhD(civil eng), 56. *Prof Exp:* Asst, Univ Ill, 49-52; asst prof civil eng, Univ Man, 55-56; from asst prof to assoc prof, 56-69, asst to head dept, 70-78, PROF CIVIL ENG, UNIV BC, 69-, ASSOC DEAN, FAC GRAD STUDIES, 84- *Concurrent Pos:* Sr res fel, Calif Inst Technol, 63-64; chmn, Can Nat Comt Earthquake Eng, 64-75. *Mem:* Seismol Soc Am; Am Soc Civil Engrs; Eng Inst Can; Earthquake Eng Res Inst. *Res:* Structures; applied mechanics; earthquake engineering, particularly dynamic characteristics of structures and influence of site conditions on ground response. *Mailing Add:* Dept Civil Eng Univ BC Vancouver BC V6T 1W5 Can

CHERRY, WILLIAM BAILEY, b Bowling Green, Ky, Apr 27, 16; m 44; c 2. BACTERIOLOGY. *Educ:* Western Ky State Teachers Col, BS, 37; Univ Ky, MS, 42; Univ Wis, PhD(bact), 49; Am Bd Med Microbiol, dipl, 62. *Prof Exp:* Bacteriologist, Univ Ky, 41-43 & Nat Naval Med Ctr, Md, 43-46; asst prof bact, Univ Tenn, 49-51; res microbiologist, Ctr Dis Control, USPHS, 51-81; RETIRED. *Concurrent Pos:* Assoc prof, Sch Pub Health, Univ NC; contract employee, Environ Protection Agency, 84. *Honors & Awards:* Kimble Award, 67; P R Edwards Award, 68; Difco Award, 74; Medal of Excellence, Ctr Dis Control & Citation, Infectious Dis Soc, 81; Becton-Dickinson Award, Am Soc Microbiol, 82. *Mem:* AAAS; Am Soc Microbiol; fel Am Acad Microbiol; Fedn Am Scientists; Union Concerned Scientists. *Res:* Enteric bacteriology; bacteriology of anthrax; listeriosis; Legionnaires' disease; fluorescent antibody techniques. *Mailing Add:* 1484 Brianwood Rd Decatur GA 30033

CHERRY, WILLIAM HENRY, b New York, NY, Oct 9, 19; m 47; c 3. PHYSICS. *Educ:* Mass Inst Technol, BS, 41; Princeton Univ, MA, 48, PhD, 58. *Prof Exp:* RES PHYSICIST, LABS, RCA CORP, 41- *Honors & Awards:* Levy Medal, Franklin Inst. *Mem:* Am Phys Soc; Inst Elec & Electronics Engrs. *Res:* Electrodynamics in magnetron, betatron and velocity modulated tubes; gas and ultra high frequency discharge; colorimetry in color television; time division multiplex, communications systems; information theory; multiplex, color television systems; cable television education systems; secondary electron emission from surfaces bombarded by positrons; time dependent effects; superconductivity; electromechanical phenomena in accelerated reference frames; special relativity. *Mailing Add:* 24 Dempsey Ave Princeton NJ 08540

CHERTKOFF, MARVIN JOSEPH, b Baltimore, Md, Nov 20, 30; m 63. PHARMACEUTICAL CHEMISTRY, PHYSICAL PHARMACY. *Educ:* Univ Md, BS, 51, MS, 54; Purdue Univ, PhD(pharm), 58. *Prof Exp:* Chemist chem warfare labs, Army Chem Ctr, Md, 56; mgr tech unit, Qual Stand Sect, Merck Sharp & Dohme, 58-59, corp trainee, 59-61, mgr qual control, Pharm Prod, 61-63, supt, Sterile Opers, 63-65, qual motivation coordr, 65-67; planning mgr, Hoffmann-La Roche Inc, 67-68; dir bus develop, Givaudan Corp, 68-69, dir, Aroma Chem Div, 69-70, vpres chem div, 70-73; PRES, BIOZEST LABS, 73- *Mem:* Am Chem Soc; Am Pharmaceut Asn; Am Soc Qual Control. *Res:* Quality control and pharmaceutical production; physical pharmacy. *Mailing Add:* 202 Beechwood Dr Ridgewood NJ 07450-2305

CHERTOCK, GEORGE, b New York, NY, Aug 1, 14; m 37; c 2. PHYSICS. *Educ:* City Col New York, BS, 39; George Washington Univ, MA, 43; Cath Univ Am, PhD(physics), 52. *Prof Exp:* Instr physics, City Col New York, 39; phys sci aide to physicist, Bur Standards, 40-46; physicist, Naval Ship Res & Develop Ctr, 46-82; Consult Vector Res Inc, 84-85; RETIRED. *Concurrent Pos:* Lectr, Univ Md, 53-54 & Cath Univ Am, 64-66; sr vis, Dept Appl Math & Theoret Physics, Cambridge Univ, 66-67. *Mem:* NY Acad Sci; fel Acoust Soc Am; Sigma Xi. *Res:* Underwater acoustics; interaction of elastic structures and hydrodynamic pressures; stellar structure. *Mailing Add:* 12900 Bluhill Rd Silver Spring MD 20906

CHERTOK, ROBERT JOSEPH, b Spartanburg, SC, Sept 5, 35; m 59; c 3. PHYSIOLOGY. *Educ:* Univ SC, BS, 57; Univ Miami, PhD(renal physiol), 65. *Prof Exp:* Sr scientist, Lawrence Livermore Lab, Univ Calif, 64-73; assoc prof, Jackson State Univ, 73-75; assoc prof, Comp Animal Res Lab, Oak Ridge Assoc Univs, 76-83; DEAN MED SCI, AM UNIV CARIBBEAN SCH MED, 83- *Mem:* Am Physiol Soc; Am Soc Nephrology; Soc Exp Biol & Med. *Res:* Renal transport; metabolism of environmental pollutants. *Mailing Add:* Am Univ Caribbean Sch Med Box 400 Plymouth Montserrat British West Indies

CHERTOW, BERNARD, b Brooklyn, NY, Dec 30, 19; m 47; c 5. CHEMICAL ENGINEERING. *Educ:* Ill Inst Technol, BS, 42, MS, 43; Mass Inst Technol, ScD(chem eng), 48. *Prof Exp:* Asst chem eng, Mass Inst Technol, 43-44; staff engr, Off Sci Res & Develop, 44-45; instr chem eng & consult climatic res lab, Mass Inst Technol, 45-47, asst prof chem eng & dir Parlin Field Sta, 47-49; develop engr, Bristol-Myers Co, 49-64, mgr chem develop pilot plants, 64-74, gen mgr, Bristol Labs, 74-80, vpres opers, Indust Div, 80-82, sr vpres, 83-85; pres, Galson Res Corp, 85-86; dir chem engr, Galson & Galson Consults, 86-90; RETIRED. *Concurrent Pos:* Consult, Godfrey Cabot Co, 46-47, Tuscarora Chem Works, Inc, 71-74 & Indust Div, Bristol-Myers, 85-86. *Honors & Awards:* Grand Award, Eng Excellence Am Consult Engrs Coun, 88. *Mem:* Am Chem Soc; Am Inst Chem Engrs; Sigma Xi; AAAS; Int Soc Pharm Engrs. *Res:* Manufacture of antibiotics and anticancer agents; adsorption of gas mixtures; process development for antibiotics and other pharmaceuticals; process development for control of hazardous material in the environment. *Mailing Add:* 139 Sunnyside Park Rd Syracuse NY 13214

CHERVENICK, PAUL A, b Pittsburgh, Pa, Apr 20, 32; m 54; c 3. HEMATOLOGY, ONCOLOGY. *Educ:* Univ Pittsburgh, BS, 57, MD, 61. *Prof Exp:* From intern to resident med, Univ Pittsburgh, 61-64; fel hemat, Univ Utah, 64-67; asst prof med, Rutgers Med Sch, 67-69; assoc prof, 69-73, PROF MED, SCH MED, UNIV PITTSBURGH, 73-, DIR, DIV HEMATOL-ONCOL, 79- *Concurrent Pos:* Consult, Nat Cancer Inst, Am Cancer Soc & NIH; Leukemia Soc Am scholar, 70. *Mem:* Am Soc Clin Invest; Am Soc Clin Res; Am Soc Hemat; fel Am Col Physicians. *Res:* Study of factors controlling the proliferation and maturation of blood leukocytes in health and diseases, such as leukemia; proliferation of cells in an in vitro culture system; internal medicine. *Mailing Add:* 922 Scaife Hall Univ Pittsburgh Sch Med 4200 Fifth Ave Pittsburgh PA 15213

CHERY, DONALD LUKE, JR, b Denver, Colo, Sept 16, 37; div; c 1. HYDROLOGIC SYSTEMS MODELING, GEOHYDROLOGY. *Educ:* Univ Ariz, BS, 60; Utah State Univ, MS, 65, PhD(civil & environmental eng), 76. *Prof Exp:* Res hydrol engr, Agr Res Serv, USDA, 65-80; sr hydrologist, Dames & Moore, 80-82; hydrologist, 82-84, SECT LEADER, US NUCLEAR REGULATORY COMN, HYDROLOGY SECT NMSS-HIGH LEVEL WASTE, 84- *Mem:* Am Soc Civil Engrs; Am Geophys Union; AAAS; Sigma Xi; Am Water Resources Asn. *Res:* Watershed surface water flow; complex watershed models. *Mailing Add:* 6607 Brookville Rd Chevy Chase MD 20815

CHERYAN, MUNIR, b Cochin, India, May 7, 46; US Citizen; m 72; c 2. BIOSEPARATIONS, MEMBRANE TECHNOLOGY. *Educ:* Indian Inst Technol, Kharagpur, India, BTech, 68; Univ Wis-Madison, MS, 70, PhD(food sci), 74. *Prof Exp:* Postdoctoral fel food sci, Univ Wis-Madison, 74-75; res assoc, INTSOY/Food Sci, 75-76, from asst prof to assoc prof food eng, 76-85, PROF FOOD & BIOCHEM ENG, UNIV ILL, URBANA, 85- *Concurrent Pos:* Sr fel, Am Inst Indian Studies, New Delhi, India, 83-84; consult, UN Develop Prog-FAO, 85-; vis fel, Univ Western Sydney, Hawkesbury, Australia, 89; vis lectr, Nat Res Coun, Taiwan, 90; soybean utilization res award, Am Soybean Asn/ICI Am, 91. *Honors & Awards:* Archer-Daniels-Midland Award, Am Oil Chemists Soc, 84; Gardner Award, Asn Food Scientists & Technologists, India, 88. *Mem:* Am Inst Chem Engrs; Inst Food Technologists; Am Dairy Sci Asn; NAm Membrane Soc; Am Chem Soc. *Res:* Food and biochemical engineering; membrane separations technology; bioprocessing (reactor design, bioseparations); dairy technology; processing of oilseeds and cereals. *Mailing Add:* Dept Food Sci Univ Ill Urbana IL 61801

CHESBRO, BRUCE W, IMMUNOLOGY, VIROLOGY. *Educ:* Harvard Univ, MD, 68. *Prof Exp:* CHIEF, LAB PERSISTENT VIRAL DIS, NAT INST ALLERGY & INFECTIOUS DIS, NIH, 79- *Mailing Add:* Nat Inst Allergy & Infectious Lab Persistant Viral Dis NIH Hamilton MT 59840

CHESBRO, WILLIAM RONALD, b Cohoes, NY, Oct 6, 28; m 50; c 2. MICROBIOLOGY. *Educ:* Ill Inst Technol, BS, 51, MS, 55, PhD(bact), 59. *Prof Exp:* Plant bacteriologist, Wanzer Dairy, Ill, 51-55; asst bacteriologist, Am Meat Inst Found, 55-59; from asst prof to assoc prof, 59-68, PROF MICROBIOL, UNIV NH, 68- *Concurrent Pos:* NIH res grants, 60- *Mem:* Am Soc Microbiol; Sigma Xi. *Res:* Microbial physiology and pathogenic microbiology. *Mailing Add:* Dept Microbiol Univ NH Durham NH 03824

CHESEBROUGH, HARRY E, b Ludington, Mich, 1909. MECHANICAL ENGINEERING. *Educ:* Univ Mich, BSE, 32; Chrysler Inst Eng, MSE, 34. *Hon Degrees:* DrEng, Univ Mich. *Prof Exp:* Mem staff, Chrysler Corp, 32-58, vpres, 58-73; RETIRED. *Concurrent Pos:* Trustee, Rackham Eng Found, 65-68. *Mem:* Nat Acad Eng; Soc Automotive Engrs (pres, 60). *Mailing Add:* 471 Dunston Rd Bloomfield Hills MI 48304

CHESEMORE, DAVID LEE, b Janesville, Wis, Nov 3, 39; m 61; c 1. WILDLIFE ECOLOGY. *Educ:* Univ Wis-Stevens Point, BS, 61; Univ Alaska, College, MS, 67; Okla State Univ, PhD(wildlife ecol), 75. *Prof Exp:* Res asst, Wildlife Res Unit, Univ Alaska, College, 61-63, res asst, Dept Wildlife Mgt, 63-64; forester, US Peace Corps, Nepalese Forestry Dept, Birganj, 64-65; res biologist, US Fish & Wildlife Serv, Univ Alaska, College, 67-68; res asst, Wildlife Res Unit, Okla State Univ, 68-72; asst prof, 72-77, assoc prof biol, 77-80, PROF BIOL, CALIF STATE UNIV, FRESNO, 80-

Mem: Wildlife Soc; Am Soc Mammalogists; Sigma Xi. *Res:* Population dynamics and ecology of big game; canid ecology; biometrics and computer applications to field ecology; ecology of rare and endangered species; aspects of scientific photography. *Mailing Add:* Dept Biol Calif State Univ Fresno CA 93740

CHESICK, JOHN POLK, b New Castle, Ind, Aug 8, 33; m 56; c 2. PHYSICAL CHEMISTRY. *Educ:* Purdue Univ, BS, 54; Harvard Univ, PhD(chem), 57. *Prof Exp:* From instr to asst prof chem, Yale Univ, 57-62; assoc prof, 62-71, PROF CHEM, HAVERFORD COL, 71-; MEM STAFF, NMR IMAGING, INC. *Mem:* Am Chem Soc; Sigma Xi. *Res:* Chemical kinetics; molecular structure; photochemistry. *Mailing Add:* Haverford Col Haverford PA 19041

CHESKY, JEFFREY ALAN, b Lynn, Mass, May 11, 46; m 70; c 1. MUSCULAR PHYSIOLOGY, GERONTOLOGY. *Educ:* Cornell Univ, AB, 67; Univ Miami, PhD(physiol, biophys), 74. *Prof Exp:* NIH trainee & res instr physiol & biophys, Sch Med, Univ Miami, 74-77; from asst prof to assoc prof, 77-90, PROF, SANGAMON STATE UNIV, 90- *Concurrent Pos:* Adj asst prof, Fla Int Univ, 76-77; from adj asst prof to adj assoc prof, Sch Med, Southern Ill Univ, 84-90, adj prof, 90-; bd dirs, Am Heart Asn, Ill Affil. *Mem:* Fel Geront Soc Am; AAAS; Am Physiol Soc; Am Aging Asn; Sigma Xi. *Res:* Physiology of aging; age related changes in cardiac and skeletal muscle; contractile protein age changes; exercise and aging. *Mailing Add:* Dept Geront Sangamon State Univ Springfield IL 62708

CHESLER, DAVID ALAN, b New York, NY. MEDICAL PHYSICS. *Educ:* Mass Inst Technol, SB, 55, SM, 55, ScD(elec eng), 60. *Prof Exp:* Engr, Gen Tel & Electronics, 60-70; ASSOC PHYSICIST RADIOL, MASS GEN HOSP, 70- *Concurrent Pos:* NIH res fel, Mass Gen Hosp, 70-72. *Mem:* Sigma Xi; Inst Elec & Electronics Engrs. *Res:* Tomography and computer processing of medical x-ray, radionuclide, and magnetic resonance images. *Mailing Add:* Physics Res Lab Mass Gen Hosp Fruit St Boston MA 02114

CHESLEY, LEON CAREY, b Montrose, Pa, May 22, 08; m 34; c 5. BIOCHEMISTRY. *Educ:* Duke Univ, PhD(physiol), 32. *Prof Exp:* Asst biophysicist, Mem Hosp, New York, 32-35; biochemist, Margaret Hague Maternity Hosp, 35-53; from assoc prof to prof, 53-79, EMER PROF OBSTET & GYNEC, STATE UNIV NY DOWNSTATE MED CTR, 79- *Concurrent Pos:* Instr eve session, City Col New York, 33-35. *Honors & Awards:* Distinguished Scientist Award, Soc Gynec Invest; Chesley Award, Distinguished Res in Hypertension in Pregmancy, Int Soc Study Hypertension in Pregnancy. *Mem:* AAAS; Am Physiol Soc; hon mem Am Gynec Soc; Soc Gynec Invest; fel Royal Col Obstetricians & Gynaecologists. *Res:* Toxemias of pregnancy. *Mailing Add:* 45 Dubonnet Rd Valley Stream NY 11581

CHESNEY, CHARLES FREDERIC, CHRONIC TOXICOLOGY, CARCINOGENICITY. *Educ:* Univ Minn, DVM, 70; Univ Wis-Madison, PhD(med path), 73. *Prof Exp:* MGR DEPT PATH & TOXICOL, RIKER LABS, 84- *Mailing Add:* Path & Toxicol Assoc Five Acorn Dr Sunfish Lake MN 55077-1420

CHESNEY, RUSSELL WALLACE, b Knoxville, Tenn, Aug 25, 41; m 68; c 3. PEDIATRICS, NEPHROLOGY. *Educ:* Harvard Col, AB, 63; Univ Rochester, MD, 68. *Prof Exp:* Intern & resident pediat, Johns Hopkins Hosp, Baltimore, 68-70; clin assoc renal biochem, Nat Inst Child Health & Human Develop, NIH, 70-72; sr asst resident pediat, Johns Hopkins Hosp, 72-73; fel nephrology, Montreal Children's Hosp, McGill Univ, 73-74, fel biochem genetics, DeBelle Lab Biochem Genetics, 74-75; from asst prof to prof pediat, Univ Wis, 75-85; prof pediat, Univ Calif-Davis, 85-88; PROF & CHAIR PEDIAT, UNIV TENN, MEMPHIS, 88- *Concurrent Pos:* Med Res Coun Can fel, 74-75. *Honors & Awards:* E Meade Johnson Award, 85. *Mem:* Soc Pediat Res (pres, 87); Am Soc Nephrology; Am Fedn Clin Res; Am Soc Clin Invest; Am Soc Bone Mineral Metab. *Res:* Renal metabolism and transport; pediatric renal metabolic bone disease; renal tubular disease. *Mailing Add:* LeBonheur Children's Med Ctr Madison Ave Memphis TN 38103

CHESNIN, LEON, b New York, NY, Mar 28, 19; m 40; c 4. SOIL CHEMISTRY, PLANT NUTRITION. *Educ:* Univ Ky, BS, 40; Rutgers Univ, PhD(soils), 48. *Prof Exp:* Soil surveyor, Soil Conserv Serv, USDA, Ind, 41-42; chemurgic res supvr, Joseph E Seagram & Sons, Inc, Ky, 42-44; res asst, NJ Exp Sta, 44-47; asst prof, 47-54, assoc prof & agronomist, 54-85, EMER PROF AGRON, UNIV NEBR, LINCOLN, 85- *Concurrent Pos:* Consult fertilizer co; mem, Nat Micronutrient Comt, Am Coun Fertilizer Appln; consult, Fed Drug Admin, US Congress; pres & consult, Chesin Waste Mgt; consult feed lots, packing plants & galvanizing industs. *Honors & Awards:* Environ Quality Award, US Environ Protection Agency, 79 & 80. *Mem:* Am Soc Agron; Soil Sci Soc Am; Int Soil Sci Soc; hon mem Nat Fertilizer Solutions Asn. *Res:* Micronutrients for crop production; micronutrient and major nutrient interrelations for crop growth; nutrition of crop varieties; influence of soil management on nutrient availability; composting, recycling and disposal of animal, municipal and industrial wastes in soil for crop production as influenced by soil management and waste management practices; waste management and recycling in soils. *Mailing Add:* 3520 S 37th St Lincoln NE 68506

CHESNUT, DONALD BLAIR, b Richmond, Ind, Dec 27, 32; m 54; c 3. PHYSICAL CHEMISTRY. *Educ:* Duke Univ, BS, 54; Calif Inst Technol, PhD(chem), 58. *Prof Exp:* Res assoc & instr physics, Duke Univ, 57-58; res chemist, Cent Res Dept, Exp Sta, E I du Pont de Nemours & Co, Inc, 58-65; assoc prof, 65-71, PROF CHEM, DUKE UNIV, 71- *Mem:* Am Phys Soc; Am Chem Soc. *Res:* Quantum mechanics; magnetic resonance. *Mailing Add:* Dept Chem Duke Univ Durham NC 27706

CHESNUT, DONALD R, JR, b Georgetown, Ky, Dec 11, 48; m 78; c 3. BASIN ANALYSIS, PALEOECOLOGY. *Educ:* Univ Ky, BS, 75, MS, 80, PhD(geol), 88. *Prof Exp:* RES GEOLOGIST, KY GEOL SURV, 79-; DIR & PRES, KY MUS NATURAL HIST, 90- *Mem:* Fel Sigma Xi; fel Geol Soc London; Int Asn Sedimentologists; Int Paleont Asn; Geol Soc Am; Paleont Soc; Palaeont Asn; Paleont Res Inst; Soc Econ Paleontologists & Mineralogists. *Res:* Geologic controls over deposition of Carboniferous sedimentary rocks and the interaction of Carboniferous plants and animals to each other and to the environment; presently specialize in the coal-bearing rocks. *Mailing Add:* 231 McDowell Rd Lexington KY 40502

CHESNUT, DWAYNE A(LLEN), b Stephenville, Tex, Mar 8, 36; m 55; c 3. PETROLEUM RESERVIOR ENGINEERING, PHYSICAL CHEMISTRY. *Educ:* Rice Univ, BS, 59, PhD(phys chem), 63. *Prof Exp:* NASA res asst, Rice Univ, 61-63; chemist, Explor & Prod Res, Shell Develop Co, 63-64, res chemist, 64-65, sect leader phys chem, 65-68, staff reservoir engr, Shell Oil Co, 68-73, sr staff reservoir engr, 73-74; pres, Energy Consult Assocs, Inc, 74-81; pres, Critical Resources, Inc, 81-83; owner, Chestnut & Assoc, 83-87; sr hydrologist, Sci Appln Int Corp, 87-90; PHYSICIST, EARTH SCI DEPT, LAWRENCE LIVERMORE LAB & TECH AREA LEADER, YUCCA MOUNTAIN PROJ, 90- *Concurrent Pos:* Consult, Dept Chem, Rice Univ, 63-64 & Lawrence Radiation Lab, 64-65; lectr, Univ Houston, 68; honorarium teacher, Univ Colo, Denver Ctr, 70-72. *Mem:* NY Acad Sci; fel Am Inst Chemists; Int Solar Energy Soc; Am Inst Chem Engrs; Soc Petrol Engrs. *Res:* Enhanced oil and gas recovery, computer applications and mathematical modeling in petroleum reservoir engineering, risk analysis; nuclear waste repository performance analysis. *Mailing Add:* L-202 Lawrence Livermore Lab PO Box 808 Livermore CA 94550

CHESNUT, ROBERT W, IMMUNOLOGY. *Prof Exp:* DIR IMMUNOL, CYTEL CORP. *Mailing Add:* Cytel Corp 11099 N Torrey Pines Rd La Jolla CA 92037

CHESNUT, THOMAS LLOYD, b Pulaski, Miss, June 14, 42; m 61; c 2. ENTOMOLOGY, FRESH WATER ECOLOGY. *Educ:* Miss State Univ, BS, 65, MS, 66, PhD(entom), 69. *Prof Exp:* Res technologist insect ecol, Boll Weevil Res Lab, Agr Res Serv, 64-68; asst prof biol sci, Fla Technol Univ, 69-72, dir fresh water ecol, 70-72; assoc prof & dir ctr environ study & planning, Ga Col, 72-74, dir off res serv, 74-75, dir off grad studies, 75-76, prof biol & dean grad sch, 77-81; VPRES, RES & GRAD STUDIES, OHIO UNIV, ATHENS, 87- *Mem:* Entom Soc Am. *Res:* Effects of competitive displacement between natural populations of insects; biological control of insect populations using insect parasites; effects of eutrophication on the aquatic habitat. *Mailing Add:* Ohio Univ 122 Univ Res Ctr Athens OH 45701

CHESNUT, WALTER G, b Montclair, NJ, July 20, 28; div; c 5. ENGINEERING PHYSICS. *Educ:* Lehigh Univ, BS, 50, MS, 52; Univ Rochester, PhD(physics), 56. *Prof Exp:* Res assoc high energy physics, Cosmotron Dept, Brookhaven Nat Lab, 56-58; sr assoc appl physics, G C Dewey Corp, NY, 58-62; sr physicist, Radio Physics Lab, 62-63, STAFF SCIENTIST, SRI INT, 63- *Concurrent Pos:* Consult, Stanford Res Inst, 58-61 & Los Alamos Sci Lab, 68-; vis lectr elec eng, Stanford Univ, 73. *Mem:* AAAS; Am Phys Soc; Sigma Xi; Am Geophys Union. *Res:* Utilization of radio and radar waves as a probe of the upper atmosphere or space environments. *Mailing Add:* SRI Int 333 Ravenswood Ave Menlo Park CA 94025

CHESON, BRUCE DAVID, b New York, NY, April 6, 46; m 78; c 1. HEMATOLOGY, ONCOLOGYY. *Educ:* Univ Va, BA, 67; Tufts Med Sch, MD, 71. *Prof Exp:* Intern Internal med, Univ Va Hosp, 71-72, resident internal med, 72-74; fel hematol, New England Med Ctr, 74-76; asst prof hemat & oncol, Univ Utah Med Ctr, 77-84; sr invest hemat & oncol, 84-86, HEAD MED SECT, CANCER THERAPY EVAL PROG, NAT CANCER INST, 86- *Concurrent Pos:* ed, J Clin Oncol, 88-, Oncology, 87-; mem, pub rels comt, Am Soc Clin Oncol, 87-, prog comt, 88, educ comt, 87-, chmn, pub rels comt, 90-; mem, Planning Comt, Nat Blood Resources Educ Cont, 87- *Mem:* Am Soc Clin Oncol; Am Soc Hemat; Am Col Physicians; Am Fedn Clin Res. *Res:* Design, coordination and conduct of National Cancer Inst sponsored clinical trials of cancer therapy, with particular emphasis on adult leukemias, lymphomas and related disorders, as well as bone marrow transplantation. *Mailing Add:* NCI-CTEP Executive Plaza North Rm 741 Bethesda MD 20892

CHESS, KARIN V T, b Hobbs, NMex, Dec 17, 39; m 60. MATHEMATICS. *Educ:* Univ Kans, BS, 62, MA, 64, PhD(math), 68. *Prof Exp:* Instr, Univ Kans, 64-65 & 68-69; from asst prof to assoc prof math, Univ Wis-Eau Claire, 69-82; ASSOC PROF MATH, UNIV SOUTHERN IND, 87- *Mem:* Math Asn Am; Sigma Xi; Nat Coun Teachers Math. *Res:* Generalized nilpotent groups. *Mailing Add:* 1508 Audubon Ct Evansville IN 47715

CHESS, LEONARD, b New York, NY, Apr 9, 43; m 64; c 3. RHEUMATOLOGY. *Educ:* Mass Inst Technol, BS, 64; Downstate Med Ctr, State Univ NY, MD, 68. *Prof Exp:* Intern & resident, Columbia-Presby Med Ctr, 68-70; clin assoc, Nat Cancer Inst, Baltimore, Md, 70-72; sr med resident, Mass Gen Hosp, Boston, 72-73; clin & resident fel, Sch Med, Harvard Univ, 72-74, asst prof med, 74-77; res assoc, Sidney Farber Cancer Inst, 74-77, staff physician, 74-77; assoc prof, 77-82, PROF MED, COL PHYSICIANS & SURGEONS, COLUMBIA UNIV, 82-; DIR, DIV RHEUMATOLOGY, COLUMBIA-PRESBY MED CTR, 77- *Concurrent Pos:* Mem, merit rev bd immunol, Vet Admin, 77, immunobiol study sect, NIH, 78, NY State Acquired Immune Deficiency Task Force, 83 & clin immunol comt, WHO, 84; attend physician, Columbia-Presby Med Ctr, 82-; sect ed, J Immunol, 83. *Mem:* Am Col Physicians; Am Fedn Clin Res; Am Asn Immunologists; Am Soc Clin Invest. *Res:* Basic and clinical immunobiology; function and development of T lymphocytes. *Mailing Add:* Col Physicians & Surgeons Columbia Univ 630 W 168th St New York NY 10032

CHESSER, NANCY JEAN, b Albany, NY, Aug 31, 46; m 75. EXPERIMENTAL SOLID STATE PHYSICS. *Educ:* Cornell Univ, BA, 67; State Univ NY Stony Brook, PhD(physics), 72. *Prof Exp:* Asst physicist & instr physics, Ames Lab & Iowa State Univ, 72-73, assoc physicist & asst prof, 73-75; resident res assoc physics, Feltman Res Lab, 75-77; proj scientist, 77-80, prin scientist, B-K Dynamics Inc, 80-83; res dir, 83-89, VPRES, TECHNOL ANALYSIS, DIRECTED TECHNOLOGIES INC, 83- *Mem:* Am Phys Soc; Sigma Xi. *Res:* Investigation of the dynamical properties of various solids through inelastic and quasielastic neutron scattering. *Mailing Add:* 9418 Overlea Dr Rockville MD 20850-3735

CHESSHIRE, GEOFFREY S, b Duncan, BC, Can, Mar 1, 58. COMPUTER SCIENCE. *Educ:* Univ BC, BA, 80; Calif Inst Technol, PhD(appl math), 86. *Prof Exp:* Appln res scientist, Intel Sci Comput, 86-89; RES STAFF SCIENTIST, T J WATSON RES CTR, IBM, 89- *Concurrent Pos:* Vis scientist appl math, Royal Inst Technol, Stockholm, 88- *Mem:* Soc Indust & Appl Math. *Res:* Parallel algorithms for differential equations. *Mailing Add:* T J Watson Res Ctr IBM 88-504 PO Box 218 Yorktown Heights NY 10598

CHESSICK, RICHARD D, b Chicago, Ill, June 2, 31; m 53; c 3. PSYCHIATRY. *Educ:* Univ Chicago, PhB, 49, SB & MD, 54; Calif Coast Univ, PhD(philos), 77. *Prof Exp:* Asst chief psychiat, USPHS Hosp, Lexington, Ky, 58-60; staff psychiatrist, Michael Reese Hosp, Chicago, 60-61; instr psychiat, 60-61; assoc, 61-62, from asst prof to assoc prof, 62-72, PROF PSYCHIAT, NORTHWESTERN UNIV, EVANSTON, 73- *Concurrent Pos:* Pvt pract, 60-; chief, Vet Admin Res Hosp, Chicago, 61-65, assoc dir res training prog, 64; sr attend psychiatrist, Evanston Hosp & Northwestern Mem Hosp, Chicago; adj prof philos, Loyola Univ, 79-87. *Honors & Awards:* Merck Award, 54; Sigmund Freud Award, 89. *Mem:* Fel Am Psychiat Asn; Am Psychosom Soc; fel Am Acad Psychoanal; fel Am Orthopsychiat Asn; Asn Advan Psychother. *Res:* Psychoanalytic psychotherapy. *Mailing Add:* Suite 628 636 Church St Evanston IL 60201

CHESSIN, HENRY, b Cleveland, Ohio, Dec 8, 19; m 50; c 3. PHYSICS. *Educ:* Western Reserve Univ, BS, 47; Purdue Univ, MS, 50; Polytech Inst Brooklyn, PhD(physics), 59. *Prof Exp:* Instr physics, Polytech Inst Brooklyn, 51-57; res physicist, US Steel Corp, 57-64; PROF PHYSICS, STATE UNIV NY ALBANY, 64- *Mem:* Am Crystallog Asn; Am Inst Mining Metall & Petrol Engrs; Am Phys Soc. *Res:* X-ray crystallography; crystal physics. *Mailing Add:* Dept Physics State Univ NY 1400 Wash Ave Albany NY 12222

CHESSIN, HYMAN, b Cleveland, Ohio, Sept 27, 20; m 42; c 4. PHYSICAL CHEMISTRY. *Educ:* Western Reserve Univ, BS, 47, MS, 49, PhD(phys chem), 51. *Prof Exp:* Analytical chemist, Harshaw Chem Co, 41-43, 46-47; asst phys chem, Western Reserve Univ, 47-50; asst prof phys & analytical chem & res assoc, Kenyon Col, 50-52; asst prof phys chem, Univ Ark, 52-54; dir res, Vander Horst Corp, NY, 54-62; sr res chemist, M&T Chem, Inc, Rahway, 62-68, res assoc, 68- 73, sr res assoc, 73-85, sr scientist, 85-86; RETIRED. *Mem:* Am Chem Soc; Electrochem Soc; Am Electroplaters Soc. *Res:* Electrochemistry; physical chemistry. *Mailing Add:* 110 Rochester Dr Brick NJ 08723

CHESSIN, MEYER, b New York, NY, Feb 5, 21; m 45; c 5. PLANT PHYSIOLOGY, STRESS PHYSIOLOGY. *Educ:* Univ Calif, BS, 41, PhD(plant physiol), 50. *Prof Exp:* Asst bot, Univ Calif, 47-48; from instr to assoc prof, 49-61, PROF BOT, UNIV MONT, 61- *Concurrent Pos:* Res fel, USPHS, Rothamsted Exp Sta, Eng, 56-57; travel awards, Int Photobiol Cong, Copenhagen, 60, Oxford, 64; AEC fel, Univ Minn, 64-65; consult, Nat Libr Med, 70; Nat Acad Sci mem sci exchange, Romania, 71, 75 & Bulgaria, 80, 88; Fulbright Lectr, USSR, 83; researcher, Bulgaria, 87. *Mem:* AAAS; Am Phytopath Soc. *Res:* Virology; preformed and induced inhibitors of virus infection in plants; use of preformed inhibitors in determining mechanisms of virus establishment and self recognition in plants. *Mailing Add:* Dept Bot Univ Mont Missoula MT 59812

CHESSON, EUGENE, JR, b Sao Paulo, Brazil, Dec 1, 28; US citizen; m 54; c 2. CIVIL ENGINEERING, FORENSIC ENGINEERING. *Educ:* Duke Univ, BSCE, 50; Univ Ill, MS, 56, PhD(civil eng), 59. *Prof Exp:* Inspection engr oil refinery, Standard Oil Co, Ind, 53; asst civil eng, Univ Ill, 53-56, res assoc, 56-59, from asst prof to assoc prof, 59-66; chmn dept, 66-75, PROF CIVIL ENG, UNIV DEL, 66-; PRES, CHESSON ENG, INC, 80- *Concurrent Pos:* Mem res coun riveted & bolted struct joints, Eng Found. *Honors & Awards:* W E Wickenden Award, Am Soc Eng Educ, 79-80. *Mem:* Am Soc Civil Engrs; Am Soc Eng Educ; Nat Soc Prof Engrs; Am Inst Steel Construct. *Res:* Static and fatique properties of structural joints; behavior of metal structures; failure analysis. *Mailing Add:* 130 Du Pont Hall Univ Del Newark DE 19716

CHESSON, PETER LEITH, b Medindie, SAustralia, Nov 7, 52; m 74. THEORETICAL ECOLOGY, ECOLOGICAL STATISTICS. *Educ:* Univ Adelaide, BSc, 74, PhD(statist & zool), 78. *Prof Exp:* Fel res biologist, biometry & theoret ecol, Univ Calif, Santa Barbara, 77-81; asst prof, Dept Zool, 81-85, ASSOC PROF, DEPT ZOOL & DEPT STATIST, OHIO STATE UNIV, 85-, PROF, DEPT BOT, 87- *Concurrent Pos:* Lectr, Univ Calif, Santa Barbara, 78. *Mem:* Inst Math Statist; Ecol Soc Am; fel AAAS; Biomet Soc. *Res:* Development of stochastic population and behavior models, of an analytical nature, for competition, predation and host-parasitoid relationships; models for animal movements; theory of bivariate distributions; random probability measures. *Mailing Add:* Res Sch Biol Sci Australian Nat Univ PO Box 475 Canberra ACT 2601 Australia

CHESTER, ALEXANDER JEFFREY, b Davenport, Iowa, Feb 4, 50; m 88; c 1. STATISTICAL CONSULTING, ESTUARINE ECOLOGY. *Educ:* Rutgers Col, AB, 72; Univ Wash, MS, 75. *Prof Exp:* Asst biol oceanog, Univ Wash, 72-75, oceanogr, 75-76; oceanogr, Pac Marine Environ Lab, 76-79, BIOMETRICIAN, BEAUFORT LAB, NAT MARINE FISHERIES SERV, NAT OCEANIC & ATMOSPHERIC ADMIN, 79- *Mem:* Am Soc Limnol & Oceanog; Am Soc Ichthyol Herpet. *Res:* Application of statistical methods to problems in estuarine and fisheries ecology, including community analysis, larval fish growth and distribution, production and fate of organic matter, and microzooplankton ecology. *Mailing Add:* NOAA/Beaufort Lab Nat Marine Fisheries Serv Beaufort NC 28516-9722

CHESTER, ARTHUR NOBLE, b Seattle, Wash, Aug 5, 40; m 69. THEORETICAL PHYSICS. *Educ:* Univ Tex, Austin, BS, 61; Calif Inst Technol, PhD(theoret physics), 65. *Prof Exp:* Physicist, Bell Tel Labs, NJ, 65-69; mem tech staff physics, 69-71, head chem laser sect, 71-73, mgr, Laser Dept, 73-75, asst dir, 75-79, assoc dir, Res Labs, 79-80, mgr high speed integrated circuits & asst mgr, Strategic Systs Div, 80-83, mgr, Tatical Eng Div, 83-85, group vpres & mgr, Space & Strategic Systs Div, Hughes Aircraft Co, 85-88; VPRES & DIR, HUGHES RES LAB, 88- *Concurrent Pos:* Consult, Dept Defense Adv Group, 75-79; co-dir, Int Sch Quantum Electronics, Erice, Italy, 80- *Honors & Awards:* Centennial Medal, Inst Elec & Electronics Engrs, 84. *Mem:* AAAS; Am Phys Soc; fel Inst Elec & Electronics Engrs; fel Optical Soc Am; Lasers & Electro-Optics Soc. *Res:* Electro-optic systems; laser physics; microelectronics. *Mailing Add:* Hughes Res Labs 3011 Malibu Canyon Rd Malibu CA 90265

CHESTER, ARTHUR WARREN, b Brooklyn, NY, Jan 9, 40; m 61; c 3. INORGANIC CHEMISTRY. *Educ:* Brooklyn Col, BS, 61; Mich State Univ, PhD(inorg chem), 66. *Prof Exp:* Res chemist, Mobil Res & Develop Corp, 66-70, sr res chemist, 70-75, assoc, 75-78, res assoc, 78-81, mgr explor process res, 81-84, mgr, 84-88, RES SCIENTIST & MGR, CATALYST CHARACTERIZATION, MOBIL RES & DEVELOP CORP, 88- *Mem:* Am Chem Soc. *Res:* Catalysis by inorganic solids; zeolite chemistry and catalysts; cracking catalysts; catalyst characterization; x-ray absorption spectroscopy; aromatics processing; zeolite synthesis. *Mailing Add:* Mobil Res & Develop Corp Paulsboro NJ 08066

CHESTER, BRENT, b New York, NY, Apr 19, 42; m 67; c 2. CLINICAL MICROBIOLOGY. *Educ:* City Col New York, BS, 62; Long Island Univ, MS, 70; NY Univ, PhD(microbiol), 75. *Prof Exp:* Jr bacteriologist, Bur Labs, New York City Dept Health, 65-68; supvr clin microbiol res, Kings County Res Lab, New York, 68-70; sr supvr microbiol, Mt Sinai Serv, Elmhurst Hosp, New York, 70-74; hosp microbiologist, Vet Admin Hosp, Gainesville, Fla, 74-75, HOSP MICROBIOLOGIST, VET ADMIN HOSP MIAMI, 75-; MEM STAFF, SCH MED UNIV MIAMI. *Mem:* Am Soc Microbiol. *Res:* Isolation and identification procedures for bacteria in clinical specimens with emphasis on Yersinia, Klebsiella and nonfermentative Bacilli. *Mailing Add:* Dept Microbiol Rm 138 Sch Med Univ Miami 1600 NW Tenth Ave Miami FL 33101

CHESTER, CLIVE RONALD, b Brooklyn, NY, Apr 6, 30. MATHEMATICS. *Educ:* NY Univ, AB, 50, MS, 51, PhD(math), 55. *Prof Exp:* Asst math, Inst Math Sci, NY Univ, 51-56; instr, Queens Col, NY, 56-61; from asst prof to assoc prof, Polytech Inst Brooklyn, 61-68; ASSOC PROF MATH, NY INST TECHNOL, 68- *Mem:* Am Math Soc. *Res:* Applied mathematics; wave propagation; partial differential equations. *Mailing Add:* 5928 Flushing Ave Maspeth NY 11378-3241

CHESTER, DANIEL LEON, b Albany, Calif, Feb 26, 43. COMPUTER SCIENCES. *Educ:* Univ Calif, Berkeley, BA, 66, MA, 68, PhD(math), 73. *Prof Exp:* Asst prof math, Univ Tex, Austin, 73-76, asst prof comput sci, 73-80; ASST PROF COMPUT SCI, UNIV DEL, 80- *Concurrent Pos:* Vis scientist, T J Watson Res Ctr, Int Bus Mach, 78-79. *Mem:* Asn Comput Mach; Asn Comput Ling; Inst Elec & Electronics Engrs; Am Asn Artificial Intel; Cognitive Sci Soc. *Res:* Software engineering methodology; natural language question answering systems; artificial intelligence. *Mailing Add:* Dept Comput & Info Sci Univ Del 103 SMI Newark DE 19716

CHESTER, EDWARD HOWARD, b New York, NY, Mar 10, 31; m 59; c 1. PULMONARY MEDICINE, RESPIRATORY PHYSIOLOGY. *Educ:* Ohio Wesleyan Univ, BA, 52; NY Univ, MD, 56. *Prof Exp:* Sr instr med, 62-65, from asst prof to assoc prof med, 65-78, from asst prof to assoc prof biomed eng, 66-78, PROF BIOMED ENG & PROF MED, CASE WESTERN RESERVE UNIV, 79- *Concurrent Pos:* Chief pulmonary serv, US Naval Hosp, Philadelphia, 60-62. *Mem:* Am Thoracic Soc; Am Acad Allergy; Am Col Chest Physicians; Am Fed Clin Res; Inst Elec & Electronics Engrs. *Res:* Pulmonary disease; respiratory physiology; pulmonary mechanics; pulmonary gas transport. *Mailing Add:* 15136 Hill Dr Novelty OH 44072

CHESTER, EDWARD M, b Queens Co, NY, Jan 12, 12; m 38; c 2. INTERNAL MEDICINE. *Educ:* NY Univ, BS, 32; State Univ Iowa, MD, 36; Am Bd Internal Med, dipl, 44. *Prof Exp:* From clin instr to sr clin instr, 43-49, from asst clin prof to prof, 49-80, EMER PROF MED, CASE WESTERN RESERVE UNIV, 80- *Concurrent Pos:* Vis in med, Cleveland Metrop Gen Hosp, 59-63. *Mem:* Am Fedn Clin Res; fel Am Col Physicians. *Res:* Vertebral column acromegaly; infarction of the lung; lung abscess secondary to aseptic pulmonary infarction; diabetes and medical education; retinal changes in systemic disease; hypertensive encephalopathy; nervous system, including optic nerve changes, in Vitamin B12 deficient monkeys. *Mailing Add:* PO Box 304 Berea OH 44017

CHESTER, MARVIN, b New York, NY, Dec 29, 30; m; c 3. SOLID STATE PHYSICS, CRYOGENICS. *Educ:* City Col New York, BS, 52; Calif Inst Technol, PhD(physics), 61. *Prof Exp:* From asst prof to assoc prof, 61-74, PROF PHYSICS, UNIV CALIF, LOS ANGELES, 74- *Honors & Awards:* Alexander von Humboldt Award, 74. *Mem:* Am Phys Soc. *Res:* Configurational emf, a transport property in semiconducting materials; second sound, a thermal wave which exists in certain solids at low temperatures; electric field effects on the infrared absorption of silicon; superfluidity in almost-two-dimensional liquid helium; adsorption phenomena. *Mailing Add:* Dept Physics Univ Calif Los Angeles CA 90024

CHESTNUT, ALPHONSE F, b Stoughton, Mass, Nov 20, 17; m 43; c 2. MARINE ECOLOGY, ZOOLOGY. *Educ:* Col William & Mary, BSc, 41; Rutgers Univ, MSc, 43, PhD, 49. *Prof Exp:* Asst zool, Rutgers Univ, 41-43; res assoc oyster culture, NJ Agr Exp Sta, 43-48; specialist, Inst Marine Sci, 48-49, asst to dir, 49-55, from assoc prof to prof, 49-81, dir, 55-80, EMER PROF, MARINE SCI & ZOOL, INST MARINE SCI, UNIV NC, 81- *Mem:* Am Soc Limnol & Oceanog; hon mem Nat Shellfisheries Asn (vpres, 51-53, pres, 53); Sigma Xi; hon mem Atlantic Estuarine Res Soc (secy-treas, 52-53). *Res:* Food and feeding mechanism of lamellibranchia; estuarine ecology of pelecypods. *Mailing Add:* Inst Marine Sci Univ NC Morehead City NC 28557

CHESTNUT, H(AROLD), b Albany, NY, Nov 25, 17; m 44; c 3. ELECTRICAL ENGINEERING, SYSTEMS SCIENCE. *Educ:* Mass Inst Technol, BSEE, 39, MSEE, 40. *Hon Degrees:* DEE, Case Western Reserve Univ, 66; DEng, Villanova Univ, 72. *Prof Exp:* Mgr, systs eng & analysis, 40-66, Info Sci Lab, 66-67 & Systs Eng & Analysis Br, 67-71, PRES, INT STABILITY FOUND, 83- *Concurrent Pos:* Lectr, exten div, Union Col, 47-49; ed, Systs Eng & Analysis Ser, John Wiley & Sons; mem, NRC Comn Sociotech Systs, 75-78; consult syst eng, Res & Develop Ctr, Gen Elec Co, 71-83. *Honors & Awards:* Centennial Award, Inst Elec & Electronics Engrs, 84, R M Emberson Award, 90; Bellman Control Heritage Award, Am Automatic Control Coun, 85; Rufus Oldenburger Medal, Am Soc Mech Engrs, 90. *Mem:* Nat Acad Eng; Int Fedn Automatic Control (pres, 57-59); fel Instrument Soc Am; Am Automatic Control Coun (pres, 62-63); fel AAAS. *Res:* Control systems and systems engineering; systems research in control and other systems; supplemental ways for improving international stability. *Mailing Add:* 1226 Waverly Pl Schenectady NY 12308

CHESTON, CHARLES EDWARD, b Princeton, NJ, Nov 23, 11; m 38; c 2. FORESTRY. *Educ:* Syracuse Univ, BS, 33; Yale Univ, MF, 40. *Prof Exp:* Asst forester, NJ, 33-42; PROF FORESTRY & CHMN DEPT, UNIV SOUTH, TENN, 42- *Concurrent Pos:* Asst, Univ Mich, 57-58; mem, Tenn Conserv Comn, 63- *Mem:* Soc Am Foresters. *Res:* Forest economics; management of hardwood forest lands; rehabilitation of devastated forest lands on the Cumberland plateau. *Mailing Add:* Dept Forestry Univ South Sewanee TN 37375

CHESTON, WARREN BRUCE, b Rochester, NY, Mar 15, 26; m 50; c 4. PHYSICS. *Educ:* Harvard Univ, BS, 47; Univ Rochester, PhD(physics), 51. *Prof Exp:* Asst prof physics, Wash Univ, 51-52; from assoc prof to prof, Univ Minn, Minneapolis, 53-71, dir space sci ctr, 65-68, dean inst technol, 68-71; chancellor, Univ Ill, Chicago Circle, 71-75; ASSOC DIR, WISTAR INST, 75- *Concurrent Pos:* Fulbright lectr, Univ Utrecht, 58-59; dept sci attache, Am Embassy, London, 63-65. *Mem:* Am Phys Soc. *Res:* Theoretical nuclear and meson physics. *Mailing Add:* Wistar Inst 36th & Spruce Sts Philadelphia PA 19104

CHESWORTH, ROBERT HADDEN, b Artesia, Calif, Sept 27, 29; m 52; c 7. NEUTRON RADIOGRAPHY, FOOD IRRADIATION. *Educ:* Univ Calif, Berkeley, BS, 51. *Prof Exp:* Process engr, Gen Elec Co-Hanford Works, 51-54; res engr, Atomics Int, 54-56; opers mgr, div dir & tech dir, Aerojet-Gen Corp, 56-71; mkt mgr, Gen Atomic Co, 71-77; vpres mkt, Norman Eng Co, 77-79; DIV DIR, GA TECHNOLOGIES INC, 79- *Concurrent Pos:* Mem subcomt energy res & technol, NASA, 69-71; consult radioisotope thermoelec generators, Aerojet-Gen, 71-73. *Honors & Awards:* Bosch & Lomb Sci Award. *Mem:* Am Nuclear Soc(secy, 59). *Res:* Development of the first compact, closed-Brayton cycle nuclear power plant in army gas-cooled reactor systems program; development of low-enriched, highly loaded uranium-zirconium hydride research reactor fuel; development of accelerator-driven sub-thermal neutron radiograph system; development of reactor-based production level neutron radiography system. *Mailing Add:* 225 Via Tavira Encinitas CA 92024

CHETSANGA, CHRISTOPHER J, b Chetsanga Village, Zimbabwe, Aug 22, 35; m 70. MOLECULAR BIOLOGY, BIOCHEMISTRY. *Educ:* Pepperdine Univ, BS, 64; Univ Toronto, MS, 67, PhD(biochem & molecular biol), 69. *Prof Exp:* Tutor biochem, Harvard Univ, 70-72; from asst prof to assoc prof, 72-81, PROF BIOCHEM, UNIV MICH, DEARBORN, 81- *Concurrent Pos:* Res fel biochem, Harvard Univ, 69-72. *Mem:* AAAS; Am Soc Biol Chemists; Biophys Soc. *Res:* Mechanisms of repair of DNA modified by alkylating agents. *Mailing Add:* Biochem Dept Univ Zimbabwe Box MP167 Mt Pleasant Harare Zimbabwe

CHEUNG, HARRY, b Oxnard, Calif, May 5, 31; m 54; c 3. CHEMICAL ENGINEERING. *Educ:* Ga Inst Technol, BS, 53, MS, 55. *Prof Exp:* Engr, Union Carbide Corp, 55-59, proj engr, 59-62, sect engr, 62-65, div engr, 65-67, div mgr, 67-69, eng assoc, 69-71, SR ENG ASSOC, LINDE DIV, UNION CARBIDE CORP, 71- *Mem:* Am Chem Soc. *Res:* Distillation; mass exchange; heat transfer; process development; hydrogen purification and production; air separation and cryogenics. *Mailing Add:* Six Heathwood Rd Buffalo NY 14221-4616

CHEUNG, HERBERT CHIU-CHING, b Canton, China, Dec 19, 33; nat US; m 66; c 2. BIOPHYSICS, PHYSICAL BIOCHEMISTRY. *Educ:* Rutgers Univ, AB, 54 & PhD(phys chem,physics), 61; Cornell Univ, MS, 56. *Prof Exp:* Asst chem, Cornell Univ, 54-56; asst scientist, Fundamental Res Labs, US Steel Corp, 56-57; asst instr chem, Rutgers Univ, 58-60; res chemist Res & Develop Dept, Am Viscose Div, FMC Corp, 60-63 & Gen Chem Div Res Labs, Allied Chem Corp, 63-66; sr fel biophys, Cardiovasc Res Inst, Med Ctr, Univ Calif, San Francisco, 66-69; assoc prof biophys, 69-73, from assoc prof to prof biomath, 73-82, head, Biophys Sect, 74-82, assoc prof, 69-82, PROF BIOCHEM, UNIV ALA MED CTR, 82-, SR SCIENTIST, COMPREHENSIVE CANCER CTR, 76-, DIR GRAD PROG BIOPHYS SCI, 78- *Concurrent Pos:* Lectr,eve div, Pa Mil Col, 61-63; USPHS res career develop award, 71-76; mem Physics Biochem Study Sect, NIH, 80-84; sr scientist, Cystic Fibrosis Res Ctr, Univ Ala Med Ctr, 80-88, adj prof physics, Univ Ala Med Ctr, 90- *Mem:* Am Chem Soc; Biophys Soc; Am Soc Biol Chemists; Protein Soc. *Res:* Molecular basis of contractility; relationship between macromolecular conformation and biological function; fluorescence spectroscopy; calcium-binding proteins. *Mailing Add:* Dept Biochem Univ Ala Birmingham AL 35294

CHEUNG, HOU TAK, b Shanghai, China, Sept 14, 50; m 77. LYMPHOCYTE CELL BIOLOGY. *Educ:* Univ Wis, Oshkosh, BS, 72, MS, 74, PhD(med microbiol), 77. *Prof Exp:* Res assoc, Immunobiol Res Ctr, 77-79; ASST PROF MICROBIOL & IMMUNOL, ILL STATE UNIV, 79- *Mem:* Am Asn Immunologists; Reticuloendothelial Soc; Am Soc Microbiol. *Res:* Cell biology of lymphocytes, particularly the molecular mechanism and regulation of lymphocyte mobility; the immunology of aging, especially interested in the mechanism of the age-related decline of the immune system. *Mailing Add:* Dept Biol Sci Ill State Univ Normal IL 61761

CHEUNG, JEFFREY TAI-KIN, b Shanghai, China, Apr 26, 46; US citizen; m 68. PHYSICAL CHEMISTRY, SEMICONDUCTOR PHYSICS. *Educ:* Univ Calif, Los Angeles, BSc, 69; Harvard Univ, PhD(chem), 74. *Prof Exp:* Res staff chem, Oak Ridge Nat Lab, 75-77; MEM TECH STAFF SEMICONDUCTOR PHYSICS, ROCKWELL INT, 77- *Res:* Gas phase kinetics; semiconductor material research. *Mailing Add:* 1050 Camino Flores Thousand Oaks CA 91360

CHEUNG, JOHN YAN-POON, b Hong Kong, Oct 13, 50; US citizen; m 77; c 2. DIGITAL SIGNAL PROCESSING, BIOMEDICAL ENGINEERING. *Educ:* Ore State Univ, BS, 69; Univ Wash, PhD(elec eng), 75. *Prof Exp:* Sr engr elec eng, Boeing Comput Serv, 77-78; asst prof, 78-83, ASSOC PROF ELEC ENG & COMPUT SCI, UNIV OKLA, 83- *Concurrent Pos:* Prin investr, NSF, 81-82, Eng Found, 81-83. *Mem:* Acad Comput Mach; Inst Elec & Electronics Engrs. *Res:* Development of adaptive recursive filters for system identification in digital signal processing problems; nonconventional computer architecture for executive parallel algorithms; computerized ultrasonic equipments for biomedical signal processing. *Mailing Add:* Univ Okla 202 W Boyd Rm 219 Norman OK 73019

CHEUNG, LIM H, b Hong Kong; US citizen. INFRARED SENSOR, SPACE & AIRBORNE INSTRUMENTATION. *Educ:* Calif Inst Technol, BS, 75; Univ Md, PhD(physics), 80. *Prof Exp:* Postdoctoral fel, Harvard-Smithsonian Ctr Astrophys, 80-81; sr res geophysicist, Gulf Oil Co, 81-84 & Standard Oil Co, 84-86; sr res scientist, 86-89, HEAD, OPTICAL PHYSICS LAB, GRUMMAN AEROSPACE CORP, 89- *Mem:* Am Phys Soc; Am Inst Aeronaut & Astronaut; Soc Photo-Optical Instrumentation Engrs. *Res:* Infrared remote-sensing systems; electro-optical sensor instrumentation; image processing. *Mailing Add:* Grumman Aerospace Corp MS A01-26 Bethpage NY 11714

CHEUNG, MO-TSING MIRANDA, b Hong Kong, Nov 13, 42; Can citizen. ANALYTICAL CHEMISTRY, ECOLOGY. *Educ:* McGill Univ, BSc, 63, PhD(anal chem), 69. *Prof Exp:* Chemist org chem, United Aircraft Co Ltd, Quebec, 69-70; lectr, 72-77, SR LECTR ANALYTICAL CHEM, HONG KONG BAPTIST COL, 77- *Concurrent Pos:* Grant org chem, McGill Univ, 70-71, grant anal chem, 75-76; dep supvr & consult, Chem Testing Lab, Hong Kong Baptist Col, 73-75, dir, 76-, publ & res comt grants, 74, 76, 77-79, 81, secy, 76-79, ed, Acad J, 76-78; chmn, Comt Standardization Methodology Water Pollution, 78-82; Univ Outreach Comt grant, United Bd Christian Higher Educ in Asia, 78 & 80; sr res analyst, Dow Chem Hong Kong, Ltd, 80-81. *Mem:* Asian Ecol Soc (vpres, 77-); Asian Women Inst; Hong Kong Chem Soc. *Res:* Development of new analytical methods for the determination of trace metals by high performance liquid chromatography and investigation of the degree of water pollution in Hong Kong Harbour; air pollution; ion selective electrodes. *Mailing Add:* Dept Chem 224 Waterloo Rd 2nd Floor Kowloon Hong Kong

CHEUNG, PAUL JAMES, b Hong Kong, May 6, 42; US citizen; m 69; c 3. MARINE BIOLOGY, FISH DISEASES. *Educ:* Iona Col, BS, 66; NY Univ, MS, 69, PhD(marine biol), 73. *Prof Exp:* Res staff marine 67-76, FISH PATHOLOGIST, OSBORN LABS MARINE SCI, NEW YORK ZOOL SOC, NY AQUARIUM, 76- *Mem:* Am Micros Soc; NY Acad Sci; Am Soc Zoologists; Am Fishery Soc. *Res:* Mass rearing of marine invertebrates; evaluation of anti-fouling paints; fish diseases in salt and fresh-water fishes; biogenesis and biochemistry of barnacle adhesive and neuroendocrine physiology of barnacle; shark parasitology. *Mailing Add:* Osborn Labs Marine Sci W Eighth St Coney Island Brooklyn NY 11224

CHEUNG, PETER PAK LUN, b China, Feb 2, 39; m 65; c 2. DENTAL MATERIALS. *Educ:* Colo State Univ, BS, 64; Okla State Univ, PhD(chem), 67. *Prof Exp:* Instr chem, Okla State Univ, 67-68; investr, N J Zinc Co, 68-69; res chemist, Penwalt Chem Corp, 69-70, sr chemist, 70-80, proj leader, 80-86; CONSULT, 86- *Mem:* Am Chem Soc; Int Asn Dent Res. *Res:* Surface chemistry of titanium dioxide; properties and compositions of dental materials; composite resin filling materials; alginate and silicone impression materials; x-ray opaque materials; cavity varnishes. *Mailing Add:* 1011 Jones Rd Gulph Mills Conshohecken PA 19428

CHEUNG, SHIU MING, b Canton, China, Dec 31, 42; m 73; c 3. MATHEMATICAL STATISTICS, ELECTRICAL ENGINEERING. *Educ:* Polytech Inst NY, PhD(math), 74. *Prof Exp:* MATHEMATICIAN, FED AVIATION ADMIN TECH CTR, US GOVT, 76- *Mem:* Inst Elec & Electronics Engrs. *Res:* Theories, techniques and procedures which can be used to analyze, design and implement digital filters; construction of algorithms that can be used to filter recorded signals. *Mailing Add:* 143 W Rutgers Ct College Park Egg Harbor City NJ 08215

CHEUNG, WAI YIU, b Canton, China, July 15, 33; US citizen; m 62; c 3. BIOCHEMISTRY. *Educ:* Chung Hsing Univ, Taiwan, BS, 56; Univ Vt, MS, 60; Cornell Univ, PhD(biochem), 64. *Prof Exp:* USPHS trainee, 64-67; asst mem, 67-70, assoc mem, 70-71, MEM, DEPT BIOCHEM, ST JUDE CHILDREN'S RES HOSP, 71- *Concurrent Pos:* Prof biochem, Univ Tenn

Ctr Health Sci; USPHS res career develop award, 71-76; fac scholar, Josiah Macy, Jr Found, 79-80. *Honors & Awards:* Gairdner Int Award, 81; Corcoran Award, 84. *Mem:* AAAS; Am Chem Soc; Am Soc Biol Chem. *Res:* Biological regulatory mechanisms; hormonal action; cyclic nucleotides; calcium and calmodulin. *Mailing Add:* Dept Biochem St Jude Children's Res Hosp Memphis TN 38101

CHEVALIER, HOWARD L, b Beaumont, Tex, Sept 14, 31; m 51; c 2. FLIGHT MECHANICS, AIRCRAFT DESIGN. *Educ:* Tex A&M Univ, BS, 57, MS, 71. *Prof Exp:* Res engr, Aro Inc, 56-63; res scientist, NASA Ames Res Ctr, 63-69; from asst prof to prof, Tex A&M Univ, 69-81, Dresser Indust prof aerospace eng, 81-87; RETIRED. *Concurrent Pos:* Prin investr, NASA, Dept Energy, Fed Aviation Admin & indust, 69-81; consult, Challenge Eng Inc, 81- & indust. *Res:* Aerodynamic; wind tunnel testing; dynamics; stability; control and aircraft design; aircraft design research include canard configurations and remotely controlled aircraft. *Mailing Add:* PO Box 67 New Bagden TX 77870

CHEVALIER, PEGGY, b Green Bay, Wis. PLANT PHYSIOLOGY, PLANT BREEDING. *Educ:* Univ Ill, Urbana, BS, 71; Univ Wis-Madison, MS, 76, PhD(agron, plant breeding & genetics), 78. *Prof Exp:* ASSOC PROF AGRON, WASH STATE UNIV, 78- *Mem:* Crop Sci Soc Am; Am Soc Plant Physiologists; Am Soc Agron; Sigma Xi. *Res:* Sucrose, fructan and starch synthesis in leaves, stems and endosperm of crop plants; use of physiological and biochemical tools in plant breeding programs. *Mailing Add:* Dept Agron & Soils Wash State Univ Pullman WA 99164

CHEVALIER, PETER ANDREW, b Chicago, Ill, Mar 4, 40; m 64; c 2. PHYSIOLOGY. *Educ:* Univ Minn, BA, 62, PhD(physiol), 67. *Prof Exp:* Asst prof physiol, Univ Del, 67-73; asst prof, Mayo Med Sch, 73-77; assoc dir, Lung Dis Div, Nat Heart, Lung & Blood Inst, 79-81, DIR, RES & DEVELOP, MEDTRONIC, INC, 81-; ASSOC PROF PHYSIOL & MED, MAYO MED SCH, 77- *Mem:* AAAS; Am Physiol Soc; Am Heart Asn; Am Col Cardiol. *Res:* Respiratory and circulatory physiology; dynamic regional lung mechanics; Roentgen videodensitometry and computer-based three-dimensional reconstruction techniques; cardiac pacing and electrophysiology; prosthetic heart valves. *Mailing Add:* Medtronic Inc 7000 Central Ave NE Minneapolis MN 55432

CHEVALIER, ROBERT LOUIS, b Chicago, Ill, Oct 25, 46; m 70; c 1. PEDIATRIC NEPHROLOGY. *Educ:* Univ Chicago, BS, 68; Univ Chicago, Pritzker, MD, 72. *Prof Exp:* From asst prof to assoc prof, 77-88, chief pediat nephrology, 78-91, PROF PEDIAT, UNIV VA, 88-, VCHMN PEDIAT, 88- *Concurrent Pos:* Estab investr, Am Heart Asn, 83-88; chmn med adv bd, Nat Kidney Found, Va, 87-89; mem gen med B study sect, NIH, 89-91. *Mem:* Am Acad Pediat; Am Pediat Soc; Am Physiol Soc; Am Soc Nephrology; Soc Pediat Res; Am Soc Pediat Nephrology (pres, 91-). *Res:* Developmental renal physiology, with emphasis on the reninangiotensin system, atrial natriuretic peptide, urinary tract obstruction and compensatory renal hypertrophy. *Mailing Add:* Dept Pediat Univ Va MR4-2034 Charlottesville VA 22908

CHEVALIER, ROGER ALAN, b Rome, Italy, Sept 26, 49; US citizen; m 74; c 2. ASTROPHYSICS. *Educ:* Calif Inst Technol, BS, 70; Princeton Univ, PhD(astron), 73. *Prof Exp:* Asst astronr, Kitt Peak Nat Observ, 73-76, assoc astronr, 76-79; assoc prof astron, 79-85, prof, 85-90, CHAIR, DEPT ASTRON, UNIV VA, 85-, W H VANDERBILT PROF ASTRON, 90- *Concurrent Pos:* Counr, Am Astron Soc, 88-91. *Mem:* Am Astron Soc; Int Astron Union. *Res:* Publish research papers in theoretical astrophysics. *Mailing Add:* 1891 Westview Rd Charlottesville VA 22903

CHEVILLE, NORMAN F, b Rhodes, Iowa, Sept 30, 34; m 58; c 4. VETERINARY PATHOLOGY. *Educ:* Iowa State Univ, DVM, 59; Univ Wis, MS, 63, PhD(path), 64. *Hon Degrees:* Univ Liege, DR, 86. *Prof Exp:* Res vet virol, US Army Biol Lab, 59-61; proj assoc path, Univ Wis, 61-63; PATHOLOGIST, NAT ANIMAL DIS LAB, USDA, 63-, PROF VET PATH, IOWA STATE UNIV, 69- *Concurrent Pos:* Res pathologist, Nat Inst Med Res, London, 68. *Mem:* Am Vet Med Asn; Am Col Vet Path (secy-treas, 74-, pres, 81); Conf Res Workers Animal Dis (vpres, 78, pres, 79). *Res:* Pathology of disease of infectious origin and domesticated animals. *Mailing Add:* Nat Animal Dis Lab Ames IA 50010

CHEVONE, BORIS IVAN, b Lynn, Mass, Aug 25, 43; c 2. PHYTOPATHOLOGY. *Educ:* Univ Mass, Amherst, BA, 65, MS, 68; Univ Minn, St Paul, PhD(entom), 74. *Prof Exp:* res rel plant path, Univ Minn, St Paul, 77-80; asst prof, 80-86, ASSOC PROF PLANT PATH & PHYSIOL, VA POLYTECH INST, 86- *Mem:* Am Phytopath Soc; Sigma Xi. *Res:* Physiology and biochemistry of air pollutant stress on plants. *Mailing Add:* Path & Physiol Dept Va Polytech Inst Blacksburg VA 24060-0304

CHEVRAY, RENE, b Paris, France, Feb 6, 37; m 64; c 2. FLUID MECHANICS. *Educ:* Univ Toulouse, BS, 62; Nat Sch Advan Electrotech & Hydraul, Toulouse, dipl Ing, 62; Univ Iowa, MS, 64, PhD(fluid mech), 67. *Hon Degrees:* ScD, Univ Claude Bernard, Lyon, France, 78. *Prof Exp:* Res engr, Worthington Co, France, 63-64; fel & lectr, Johns Hopkins Univ, 67-69; from asst prof to assoc prof fluid mech, 69-79, prof mech eng, State Univ NY, 79-, actg chmn dept, 80-; MEM STAFF, DEPT MECH ENG, COLUMBIA UNIV. *Concurrent Pos:* State Univ NY Res Found grant in aid, 70-71, fac res fel, 71; fac assoc, NSF res grants, 70-71, 73 & 75, prin investr, 71-73, 75-77, 77-79, 78-80 & 79-81; prin investr, Dept Energy contract, 79-82. *Mem:* Am Phys Soc; Fr Eng Asn; Int Asn Hydraul Res; Sigma Xi; NY Acad Sci. *Res:* Turbulence; two-dimensional and axisymmetric turbulent wakes; electronic instrumentation in turbulence research; preferential transport of heat over momentum in intermittent regions of turbulent shear flows; history of science. *Mailing Add:* Dept Mech Eng Columbia Univ 220 Seely W Mudd New York NY 10027

CHEW, CATHERINE S, b Savannah, Ga, July 6, 42; m; c 1. PHYSIOLOGY. *Educ:* Armstrong State Col, BS, 70; Emory Univ, PhD(physiol), 77. *Prof Exp:* Postdoctoral res fel, NIH, 79-81; instr physiol, Emory Univ, 79-81; from asst prof to assoc prof, 81-89, PROF PHYSIOL, MOREHOUSE SCH MED, 89- *Concurrent Pos:* Porter Found vis prof, Spelman Col, 78-84, Marc vis prof, 84-; from adj asst prof to adj assoc prof physiol, Emory Univ, 81-; adj assoc prof biol, Ga State Univ, 89-; assoc ed, Am J Physiol Gastrointestinal & Liver Physiol, 91-; mem, Women in Physiol Comt, Am Physiol Soc, 86-90, AGA Abstract Selection Comt Sect Esophageal, Gastric & Duodenal Dis, 90 & 91. *Mem:* Am Physiol Soc; Am Soc Cell Biol; Gastroenterol Res Group; NY Acad Sci; Am Women Sci; AAAS. *Res:* Physiology; gastroenterology; biology. *Mailing Add:* Dept Physiol Morehouse Sch Med 720 Westview Dr SW Atlanta GA 30310

CHEW, FRANCES SZE-LING, b Los Angeles, Calif, May 11, 48. BIOLOGY, ENTOMOLOGY. *Educ:* Stanford Univ, AB, 70; Yale Univ, PhD(biol), 74. *Prof Exp:* Fel, Dept Biol Sci, Stanford Univ, 74-75; asst prof, 75-81, ASSOC PROF BIOL, TUFTS UNIV, 81- *Concurrent Pos:* Managing ed, J Lepid Soc, 78-81. *Mem:* Soc Study Evolution; Lepidopterists Soc; AAAS; Ecol Soc Am. *Res:* Plant-herbivore interactions, especially ecology and evolution of plants in the Capparales and their herbivores and fungal symbionts-pathogens. *Mailing Add:* Dept Biol Tufts Univ Medford MA 02155

CHEW, FRANK, b San Francisco, Calif, Aug 17, 16; m 46; c 3. OCEANOGRAPHY. *Educ:* Univ Calif, Los Angeles, BA, 43, MA, 50; Univ Miami, PhD(oceanog), 73. *Prof Exp:* Asst prof oceanog, Univ Miami, 52-59; res assoc, Gulf Coast Lab, Miss, 59-62; res scientist, Lockheed-Calif Co, 62-65 & Bissett-Berman Corp, 65-66; RES OCEANOGR, ATLANTIC OCEANOG & METEOROL LAB, NAT OCEANIC & ATMOSPHERIC ADMIN, 66- *Mem:* Am Meteorol Soc; Am Geophys Union. *Res:* Accelerative process in the Florida Current. *Mailing Add:* 901 Sistina Ave Coral Gables FL 33124

CHEW, GEOFFREY FOUCAR, b Washington, DC, June 5, 24; m 45, 71; c 5. THEORETICAL PARTICLE PHYSICS. *Educ:* George Washington Univ, BS, 44; Univ Chicago, PhD(physics), 48. *Prof Exp:* Jr theoret physicist, Los Alamos Sci Lab, NMex, 44-46; Nat Res fel, Univ Chicago, 46-48; theoret physics res radiation lab, Univ Calif, 48-49, asst prof physics, 49-50; from asst prof to prof, Univ Ill, 50-57; theoret group leader, 69-74 & 79-83, chmn dept, 74-78, Miller Prof, 81-82, PROF PHYSICS, 57-, DEAN PHYS SCI, UNIV CALIF, BERKELEY, 86- *Concurrent Pos:* Fulbright lectr, Les Houches, 53, 60 & 65; fel, Churchill Col, Cambridge, 62-63; lectr, Tata Inst, Nainital, 69; vis prof, Princeton Univ, 70-71 & Univ Paris, 83-84; consult, Los Alamos & Brookhaven Nat Labs; CERN sci assoc, 78-79. *Honors & Awards:* Hughes Prize, Am Phys Soc, 62; E O Lawrence Award, 69. *Mem:* Nat Acad Sci; fel Am Phys Soc; Am Acad Arts & Sci. *Res:* Theoretical particle physics; scattering matrix theory; strong interactions; topological bootstrap theory. *Mailing Add:* Ten Maybeck Twin Dr Berkeley CA 94708

CHEW, HEMMING, b China, Sept 10, 33; Can citizen; m 67; c 1. HYDROLOGY & WATER RESOURCES, GEOMORPHOLOGY & GLACIOLOGY. *Educ:* Univ Western Ont, BSc, 67, PhD(phys chem), 72. *Prof Exp:* Sr demonstr chem, Univ Western Ont, 72-74; PHYS SCIENTIST ENVIRON SCI, NAT HYDROL RES INST, ENVIRON CAN, 74- *Mem:* Int Glaciological Soc; Chem Inst Can; Spectros Soc Can. *Res:* Mechanism of solar radiation absorbed by ice which leads to internal deterioration of ice; weakening of ice strength due to ice porosity is one of the major factors that lead to river ice breakup. *Mailing Add:* 11 Innovation Blvd Saskatoon SK S7N 3H5

CHEW, HERMAN W, b China; US citizen. ELECTROMAGNETISM. *Educ:* Univ Chicago, SM, 58, PhD(physics), 61. *Prof Exp:* Res assoc physics, Univ Pa, 61-62; res scientist, Columbia Univ, 62-63; instr, Princeton Univ, 63-64; asst prof, Case-Western Reserv Univ, 64-67; assoc prof, 67-80, PROF PHYSICS, CLARKSON UNIV, 80- *Mem:* Am Phys Soc; Sigma Xi. *Res:* Weak and electromagnetic interactions; scattering theory. *Mailing Add:* Dept Physics Clarkson Univ Potsdam NY 13676

CHEW, JU-NAM, b China, Oct 8, 23; nat US; m 47; c 3. PETROLEUM RESERVOIR ENGINEERING, ENHANCED OIL RECOVERY. *Educ:* Univ Tex, BSChE, 44, MSChE, 47, PhD(chem eng), 53; Univ Mich, MSE, 49. *Prof Exp:* Chemist, Phelps Dodge Refining Corp, 44; tutor & instr chem, Univ Tex, 44-45; asst chem eng, Eng Res Inst, Univ Mich, 46-49, res assoc, 49; res scientist, Univ Tex, 49-50; sr res engr, Mobil Res & Develop Corp, 53-85; CONSULT ENGR, 85- *Concurrent Pos:* Lectr, Univ Tex, Arlington, 57-71, 85, 87 & 89; foreign expert, SW Petrol Inst, Nanchong, Sichuan, China, 87-88. *Mem:* Am Inst Chem Engrs; Am Chem Soc; Sigma Xi; Am Inst Mining Metall & Petrol Engrs. *Res:* Phase equilibria; fluid mechanics; petroleum reservoir engineering; enhanced oil recovery processes; in-situ coal recovery. *Mailing Add:* 1348 Naples Dr Dallas TX 75232-1538

CHEW, KENNETH KENDALL, b Red Bluff, Calif, Oct 29, 33; m 58; c 2. MARINE BIOLOGY. *Educ:* Chico State Col, BA, 55; Univ Wash, MS, 58, PhD(fisheries), 61. *Prof Exp:* Sr fisheries biologist, Fisheries Res Inst, 61-62, from res asst prof to res assoc prof, 62-67, assoc prof, 67-71, PROF FISHERIES, UNIV WASH, 71- *Mem:* Am Fisheries Soc; Nat Shellfisheries Asn; World Mariculture Soc; AAAS. *Res:* Shellfish biology; marine ecology; growth, condition and survival of Pacific oysters; shellfish toxicity studies in Washington and Southeast Alaska; clam and mussel culture studies; ecological baseline studies. *Mailing Add:* Col Fisheries Univ Wash Seattle WA 98195

CHEW, ROBERT MARSHALL, b Wheeling, WVa, Oct 7, 23; m 46; c 3. ECOLOGY. *Educ:* Washington & Jefferson Col, BS, 44; Univ Ill, MS, 46, PhD(zool, animal ecol), 48. *Prof Exp:* Teaching asst zool, Univ Ill, 44-46; asst prof biol, Lawrence Col, 48-52; instr zool, 52-55, from asst prof to assoc prof biol, 55-66, prof, 66-79, EMER PROF BIOL, UNIV SOUTHERN CALIF,

79- *Concurrent Pos:* Consult, Northrop Space Labs, 62-66 & Lab Nuclear Med & Radiation Biol, Univ Calif, Los Angeles, 65-75; mem exec comt, Desert Biome, US/IBP, 73-76; vis scholar, Univ Ariz, 80- *Mem:* AAAS; Ecol Soc Am; Am Soc Mammalogists; Brit Ecol Soc. *Res:* Structure and function of desert ecosystems; ecology of ants; water metabolism. *Mailing Add:* PO Box 306 Portal AZ 85632

CHEW, VICTOR, b Djakarta, Indonesia, July 9, 23; m 49; c 3. MATHEMATICAL STATISTICS, BIOMETRICS. *Educ:* Univ Western Australia, BSc, 44 & 48; Univ Melbourne, BA, 53. *Prof Exp:* Tech officer, Commonwealth Sci & Indust Res Orgn, Melbourne, Australia, 47-49, res officer, Sydney, 53-55; instr math, Univ Melbourne, 49-53; res asst, Univ Fla, 55-56, asst prof, 56-57; asst statistician, NC State Col, 57-60; math statistician, US Naval Weapons Lab, 60-62; sr engr, RCA Serv Co, 62-70; MATH STATISTICIAN, BIOMET SERV STAFF, AGR RES SERV, USDA, 71- *Concurrent Pos:* Lectr, Am Univ, 60-61, 62-63, Johns Hopkins Univ, 61-62, Brevard Eng Col, 63-67 & Fla State Univ, 67-70. *Mem:* Biomet Soc; Am Statist Asn; Int Asn Statist Phys Sci; fel Royal Statist Soc. *Res:* Experimental designs; mathematical modeling; statistical inference; regression analysis; biometry. *Mailing Add:* Biomet Serv Staff Agr Res Serv USDA Univ Fla Gainesville FL 32611

CHEW, WENG CHO, b Kuantan, Malaysia, June 9, 53; m 77; c 2. ELECTROMAGNETIC SCATTERING, ELECTROMAGNETIC INVERSE SCATTERING. *Educ:* Mass Inst Technol, BS, 76, MS, 78, PhD(elec eng), 80. *Prof Exp:* Postdoctoral fel elec eng, Mass Inst Technol, 80-81; mem prof staff electromagnetics, Schlumberger-Doll Res, 81-82, prog leader, 82-84, dept mgr, 84-85; assoc prof, 85-90, PROF ELEC ENG, UNIV ILL, URBANA-CHAMPAIGN, 90- *Concurrent Pos:* Assoc ed, Trans Geosci & Remote Sensing, Inst Elec & Electronic Engrs, 83-88, Int J Imaging Syst & Technol, 89-, guest ed, 90; guest ed, Radio Sci, 85; NSF presidential young investr, 86. *Mem:* Soc Explor Geophysists; sr mem Inst Elec & Electronics Engrs; sr mem Inst Elec & Electronics Engrs, Geosci & Remote Sensing Soc; sr mem Inst Elec & Electronics Engrs, Antennas & Propagation Soc; sr mem Inst Elec & Electronics Engrs, Microwave Theory & Tech Soc; Am Phys Soc. *Res:* Methods to solve the scattering solutions of complicated geometry, in particular, devising fast and efficient algorithm to solve such problems; study of how the geometry of the scatterers can be extracted if only the scattered data of the object is available; work is used in remote sensing, biological sensing and nondestructive testing. *Mailing Add:* Dept Elec & Comp Eng Univ Ill Urbana IL 61801

CHEW, WILLIAM HUBERT, JR, b Macon, Ga, Sept 21, 33; m 57; c 3. INTERNAL MEDICINE, INFECTIOUS DISEASES. *Educ:* Med Col Ga, MD, 58. *Prof Exp:* Asst med, Sch Med, Tufts Univ, 62-63, instr, 63-64; from instr to assoc prof, 64-72, chief infectious dis sect, 67-70, dir & coordr, Physician Augmentation Prog, 70-75, PROF MED, MED COL GA, 72- *Concurrent Pos:* Res fel infectious dis, Pratt Clin, New Eng Ctr Hosp, 63-64; Markle scholar acad med, 64; dir, NIH Training Grant, 67. *Mem:* Am Fedn Clin Res. *Res:* Candida-host defense interactions, specifically what alterations in host defense occur to permit either local or disseminated candidiasis. *Mailing Add:* Sch Med Med Col Ga 1120 15th St Augusta GA 30912

CHEW, WOODROW W(ILSON), b Burlington, Okla, Jan 29, 13; m 39; c 1. CHEMICAL ENGINEERING. *Educ:* NMex State Univ, BS, 36; Okla State Univ, MS, 38. *Prof Exp:* Asst chem & chem eng, Okla State Univ, 36-38; engr & asst engr, Pub Works Admin, Tex, 38-39; instr chem & physics, Northeastern Okla Jr Col, 39-40; asst prof chem eng, La Polytech Inst, 40-41, from asst prof & head dept to prof & head dept, 41-52; chem engr, Monsanto Chem Co, 45; plant engr & consult, Magnolia Petrol Co, 52; prof & head dept, 52-75, EMER PROF CHEM ENG, LA TECH UNIV, 80- *Concurrent Pos:* Pres, Woodrow W Chew & Assocs, Inc, 75- *Mem:* Am Chem Soc; Am Soc Eng Educ; fel Am Inst Chem Engrs. *Res:* Vapor-liquid equilibrium; mutual solubility in ternary systems; chemical plant economic evaluations and designs. *Mailing Add:* Box 3066 Tech Sta Ruston LA 71272-0047

CHEYDLEUR, BENJAMIN FREDERIC, b Grenoble, France, Oct 22, 12; US citizen; m 43; c 5. COMPUTER SCIENCE, ELECTRONICS. *Educ:* Univ Wis, BA, 38. *Prof Exp:* Statistician, US Civil Serv Comn, Washington, DC, 40-42; chief numerical analyst, Appl Math Div, Naval Ord Lab, Silver Spring, Md, 46-54; dir systs analysis comput design, Remington Rand Univac, St Paul, Minn, 55-58; mgr advan prog comput design, Electronic Data Div, RCA, Camden, NJ, 58-60; staff & prin scientist info sci, Philco-Ford, Willow Grove, Pa & Dearborn, Mich, 60-68; prof eng comput sci, Oakland Univ, 68-; MEM STAFF, ELECTRONIC ENG COMPUT SCI. *Concurrent Pos:* Mem aeroballistics comt, Bur Ord, Dept Defense, 52-54; mem comt data lang systs, Asn Comput Mach, 59-60; mem adv comt info retrieval, Moore Sch, Univ Pa, 62-63; mem, Pa Gov Scranton's Tech Adv Comn, 64; contribr, Merrill Flood Report, Cong Comt Automation, 65; mem int adv bd, Ctr Res & Lang Behavior, Univ Mich, 66-68. *Mem:* Asn Comput Mach; Am Soc Info Sci; Inst Elec & Electronics Engrs; Am Math Soc; Pattern Recognition Soc. *Res:* Computer architecture; information retrieval hardware and switching theory. *Mailing Add:* Electronic Eng Comput Sci 3110 Rosanne Lane Drayton Plains MI 48020

CHEZEM, CURTIS GORDON, b Eugene, Ore, Jan 28, 24; m 85; c 2. PHYSICS, NUCLEAR ENGINEERING. *Educ:* Univ Ore, BA, 51, MA, 52; Ore State Univ, PhD(physics), 60. *Prof Exp:* Flight radio officer, Alaska Div, Pan Am World Airways, 42-43, Pac Div, 43-44; chief radio officer, Hammond Steamship Lines, 44-45; telegrapher, Western Union Tel Co, Ore, 46; announcer & engr, Radio Stas KUGN & KASH, 46-51; staff mem, Los Alamos Sci Lab, 52-67; prof physics & nuclear eng, Los Alamos Grad Ctr, NMex, 62-67; chief systs studies br, Off Safeguards & Mat Mgt, US AEC, 67-69; prof nuclear eng & head dept, Kans State Univ, 69-72; dir, Nuclear Activ, Mid South Servs, New Orleans, 72-77; gen mgr, Waterman Inc, Amarillo, Tex, 77-79; pres, Seven Seas Gifts Inc, Nantucket Island, Mass, 79-86; RETIRED. *Concurrent Pos:* Asst, Univ Ore, 50-51 & Ore State Univ, 55-57; reactor prog supvr, AEC, Bogota, Colombia, 63; vis prof, Tex A&M Univ, 66-67; mem, Kans Nuclear Energy Coun, 70- & Kans State Univ rep, Atomic Indust Forum, 69-72; freelance writer, 87- *Res:* Nuclear materials control and safeguards; nuclear Power; interacting critical and sub-critical nuclear systems; computer programming; data analysis; navigation and communications. *Mailing Add:* 3378 Wisteria St Eugene OR 97404-5930

CHHABRA, RAJENDRA S, b India, Mar 4, 39; m 66; c 2. PHARMACOLOGY, TOXICOLOGY. *Educ:* Vet Col, Mhow, India, BVSc & AH, 62; Univ London, PhD(pharmacol), 70; Am Bd Toxicol, dipl. *Prof Exp:* Res asst pharmacol, Vet Col, Mhow, India, 62-64, asst res officer, 64-66; res pharmacologist, Biorex Labs, London, Eng, 66-67; vis assoc pharmacol, 70-73, sr staff fel, Pharmacol Br, 73-77, pharmacologist, Environ Biol & Chem Br, 78-80, SUPVRY TOXICOLOGIST, DIV TOXICOL RES & TESTING, NAT TOXICOL PROG, NAT INST ENVIRON HEALTH SCI, 80- *Mem:* Soc Toxicol; Am Soc Pharmacol & Exp Therapeut. *Res:* Transport and biotransformation of foreign chemicals; species and strain variations in enzymatic biotransformation of xenobiotics; general toxicology; carcinogenesis. *Mailing Add:* Nat Inst of Environ Health Sci Research Triangle Park NC 27709

CHHEDA, GIRISH B, b Kutch, India, Mar 4, 34; m 62. MEDICINAL CHEMISTRY, BIOCHEMISTRY. *Educ:* Univ Bombay, BSc, Hons 55, BSc, 57; Univ Mich, MS, 59; State Univ NY Buffalo, PhD(med chem), 63. *Prof Exp:* Apprentice drug anal, Glaxo Labs, Bombay, India, 56; pharmaceut chemist, Castophene Mfg Co, 57; fel, State Univ NY Buffalo, 63-64; cancer res scientist, 64-65, sr cancer res scientist, 65-68, assoc cancer res scientist, 68-73, PRIN CANCER RES SCIENTIST, ROSWELL PARK MEM INST, 73- *Concurrent Pos:* Res prof, Niagara Univ, 68- & State Univ NY Buffalo, 68. *Mem:* Am Chem Soc; Fedn Am Soc Exp Biol; Am Asn Cancer Res; Am Soc Mass Spectros. *Res:* Anti-metabolites and enzyme inhibitors; synthesis of oligonucleotides; investigation of human urinary nucleic acid constituents. *Mailing Add:* Roswell Park Mem Inst 666 Elm St Buffalo NY 14203-1135

CHI, BENJAMIN E, b Tientsin, China, June 18, 33; US citizen; m 67. COMPUTER OPERATING SYSTEMS, DATA COMMUNICATIONS. *Educ:* Antioch Col, BS, 55; Rensselaer Polytech Inst, PhD(physics), 62. *Prof Exp:* Fel & lectr physics, Western Reserve Univ, 62-63, instr, 63-65; asst prof physics, State Univ NY, Albany, 65-69, chmn dept, 70-73, assoc dir, Comput Ctr, 82-88, dir, Tech & Network Serv, Comput Ctr, 88-90, ASSOC PROF PHYSICS, STATE UNIV NY, ALBANY, 69- , EXEC DIR, COMPUT SERV, 90- *Mem:* AAAS; Inst Elec & Electronics Engrs; Asn Comput Mach; Sigma Xi. *Res:* Campus-wide information systems; data communications networks and protocols; data network security. *Mailing Add:* Selkirk NY 12158

CHI, CHAO SHU, b Naking, China, Sept 12, 36; US citizen; m; c 3. ELECTRONICS, COMPUTER DESIGN. *Educ:* Nat Cheng-Kung Univ, Taiwan, BS, 61; Worcester Polytech Inst, MS, 65, PhD(elec eng), 74. *Prof Exp:* Engr power syst, Northeast Utilities Serv Co, 63-66; prin mem tech staff advan develop, RCA, Camden, NJ & Marlboro, Mass, 66-71; proj engr comput develop, Digital Equip Corp, Mass, 71-77; prin investr data trans, Sperry Corp, 77-78, mgr magnetic res, Res Ctr, 78-; adv develop group mgr, Magnetic Peripherals Inc, 90; SR LAB HEAD, ADVAN IMAGING RES LAB, EASTMAN KODAK CO, 90- *Concurrent Pos:* Reviewer trans magnetics, Inst Elec & Electronics Engrs, 78; Telecommunication Scholar; WPI Grad Scholar. *Mem:* Inst Elec & Electronics Engrs. *Res:* High density magnetic recording process and identificaon of linear and nonlinear distortions, their origins and solutions to advance the art of computer mass storage technology. *Mailing Add:* Eastman Kodak Co Mail Code 02123 Rochester NY 14650-2123

CHI, CHE, b Peking, China, Feb 6, 49; US citizen; m 77; c 1. NEUROSCIENCE. *Educ:* Nat Taiwan Univ, BS, 70; Okla State Univ, MS, 72; Univ Wis, PhD(neurosci), 76, MS, 81. *Prof Exp:* Res asst, Okla State Univ, 71-72 & Univ Wis, 72-76; res fel, Univ Minn, Minneapolis, 76-77; res fel, Univ Wis, 77-78, asst scientist, 79-81; sr software engr, Nicolet Biomed Instruments, Madison, Wis, 82-86, SOFTWARE MGR, NICOLET AUDIODIAGNOSTICS, 86- *Concurrent Pos:* Vis instr, Univ Ore, 75. *Mem:* Sigma Xi; Asn Res Vision & Optom; Soc Neurosci. *Res:* Scanning and transmission electron microscopy (high-voltage electron microscopy) of the house fly ommatidium and first optic neuropile and blood brain barrier; intrafusal muscles of "wobbler" mice; developing computer programs for 3-D reconstruction of neural networks; biomedical software development. *Mailing Add:* Nicolet Instrument 9225 Verona Rd Madison WI 53711-0287

CHI, CHENG-CHING, b Canton, China, Feb 15, 39; US citizen; m 67; c 1. MECHANICAL SYSTEMS ANALYSIS. *Educ:* Nat Taiwan Univ, BS, 62; Kans State Univ, MS, 65; Univ Calif, Berkeley, PhD(appl mech), 69. *Prof Exp:* Engr, Garrett Corp, 69-71, sr engr, 71-74, sr eng specialist, 74-80, eng supvr, 80-88; ENG MGR, ALLIED SIGNAL AEROSPACE CO, 88- *Concurrent Pos:* Lectr, Calif State Univ, Northridge, 71. *Mem:* Soc Automotive Engrs. *Res:* Nonlinear oscillations; dynamic analysis of high speed trains; computer simulations of jet engine dynamics; vibration signature techniques for machine health monitoring and diagnosis; composite structure analysis. *Mailing Add:* 2525 W 190th St Torrance CA 90509

CHI, DAVID SHYH-WEI, b Taiwan, July 7, 43, US citizen; m 82; c 2. CLINICAL IMMUNOLOGY. *Educ:* Nat Chung-Hsing Univ, Taiwan, BS, 65; Univ Tex, MA, 74, PhD(cell biol), 77; Am Bd Bioanal, dipl, 81. *Prof Exp:* Postdoctoral fel & res asst prof, Sch Med, NY Univ, 77-80; assoc prof, 80-86, DIR, CLIN IMMUNOL LAB, COL MED, ETENN STATE UNIV, 80-, ADJ FAC, DEPT MICROBIOL, 81- *Concurrent Pos:* Clin lab dir, Am Bd Bioanal, 81; chief, Div Biomed Res, ETenn State Univ Col Med, 81-, grad fac, Grad Sch, 82- & prof, Dept Internal Med, 86-; consult, J C L Clin Res Corp, Knoxville, Tenn, 84- *Mem:* Fel Asn Med Lab Immunologists; fel Am Acad Microbiol; Int Soc Develop & Comp Immunol; Am Asn Immunologists; Soc Exp Biol & Med; Clin Immunol Soc; Sigma Xi; Am Soc Microbiol; Am Soc Histocompatibility & Immunogenetics. *Res:* Suppressor and cytotoxic T-cells in the immune responsed; lymphokines; effect of brain peptide, Tyr-MIF-1 on the immune function; tumor associated antigens; immune responses to moraxella catarrhalis and Helicobacter pylori. *Mailing Add:* Dept Internal Med ETenn State Univ Col Med Box 21160A Johnson City TN 37614-1000

CHI, DONALD NAN-HUA, b Medan, Indonesia, June 28, 39; US citizen; m 68; c 2. INFORMATION SCIENCES, MINING ENGINEERING. *Educ:* Willamette Univ, BA, 62; Carnegie Inst Technol, MS, 65; Univ Pittsburgh, PhD(math), 70. *Prof Exp:* res physicist, Univ Pittsburgh, 70-75; supvry res physicist numerical anal & aerodyn, Pittsburgh Res Ctr, US Bur Mines, 75-80, supvry systs analyst, 80-90; DIR, SYSTS & NETWORK MGT BR, FOOD & DRUG ADMIN, 91- *Concurrent Pos:* Mem comt Masters & PhD degrees, Dept Mech Eng, Univ Pittsburgh, 73-, vis lectr, Info Sci, 82-; reviewer, Appl Mech Rev, Am Soc Mech Engrs, 74- *Mem:* Soc Indust & Appl Math. *Res:* Transient phenomena of flame propagation in coal mine networks via computer simulation; numerical techniques in solving gas-dynamic and heat transfer problems and in solving nonlinear algebraic equations and nonlinear least square problems; computer applications ranging from process control, data-reduction to administrative applications; data-base management system and management information system. *Mailing Add:* Food & Drug Admin 5600 Fishers Lane HFC-31 Rm 1274 Rockville MD 20857

CHI, JOHN WEN HUA, b Nanking, China, July 20, 34; US citizen; m 58; c 3. NUCLEAR ENGINEERING. *Educ:* Willamette Univ, BA, 58; Carnegie-Mellon Univ, BS, 58, MS, 60; Univ Pittsburgh, PhD(chem eng), 68. *Prof Exp:* Engr coal processes, Res Dept, Consol Coal Co, 56-58; res engr metall processes, Res Div, Jones & Laughlin Steel Corp, 59-62; sr engr heat transfer res, Astronuclear Lab, 62-68, fel engr thermal design, 68-76, ADV ENGR THERMAL DESIGN, FUSION POWER SYSTS DEPT, WESTINGHOUSE ELEC CORP, 76- *Mem:* Am Nuclear Soc; Am Inst Chem Engrs; Am Inst Aeronaut & Astronaut. *Res:* Cryogenic heat transfer; two-phase flow; advanced heat transfer techniques; heat pipes; advanced energy systems; fusion power. *Mailing Add:* Fusion Power Systs Dept Westinghouse Elec Corp Large PA 15236

CHI, L K, b Shanghai, China, Dec 12, 33; US citizen; c 3. SOFTWARE DEVELOPMENT, MATHEMATICAL APPLICATIONS. *Educ:* Nat Taiwan Univ, BS, 56; Univ Md, MA, 60; Drexel Univ, PhD(appl mech), 68. *Prof Exp:* Res scientist, Space Sci, Inc, Waltham, Mass, 62-65; instr math, Community Col Philadelphia, 65-66; asst prof, Widener Col, 66-68; res assoc elec eng, Drexel Univ, 68-69; sr software specialist, RCA, 69-71; prin syst programer, Sperry Univac, 72-77; from asst prof to assoc prof, 77-87, PROF COMPUT SCI, US NAVAL ACAD, 88- *Mem:* Asn Comput Mach; Am Phys Soc. *Res:* Kinetic theory; boundary layer theory; plasma physics; numerical solutions to various physical problems; cosmology. *Mailing Add:* Comput Sci Dept US Naval Acad Annapolis MD 21402

CHI, LOIS WONG, b Foochow, Fukien, China; m 45; c 3. PARASITOLOGY. *Educ:* Wheaton Col, Ill, BS, 45; Univ Southern Calif, MS, 48, PhD, 53. *Prof Exp:* Res assoc & instr, Loma Linda Univ, 52-56; from instr to assoc prof biol, Immaculate Heart Col, 57-66, chmn dept, 63-66; assoc prof, 66-70, chmn dept, 71-72, PROF BIOL SCI, CALIF STATE COL, DOMINGUEZ HILLS, 70- *Concurrent Pos:* Prin investr, NIH grant, 67-75; mem Nat Adv Coun, NIH, 73 & 74; prog dir, Minority Biomed Support Prog, 77-85. *Mem:* Am Soc Parasitol; Am Soc Trop Med & Hyg; AAAS; NY Acad Sci. *Res:* Host and parasite relationship; immunological control Schistosoma japonicam parasites. *Mailing Add:* Dept Biol Sci Calif State Col Dominguez Hills CA 90747

CHI, MICHAEL, US citizen; m 57; c 2. HIGHWAY SAFETY. *Educ:* Univ Tianjin, BSCE, 46; La State Univ, MS, 49; George Washington Univ, DSc, 69. *Prof Exp:* Designer, Preload Enterprises, 50-53; res engr, Nat Bur Standards, 53-58; assoc prof civil eng, Cath Univ Am, 58-64, prof civil & mech eng, 65-72; phys scientist, Nat Hwy Traffic Safety Admin, 72-73; pres, 74-87, CHIEF EXEC OFFICER, CHI ASSOCS, INC, 87- *Concurrent Pos:* Res engr, Fed Hwy Admin, 67-68; consult, Naval Res Lab, 68-69; exec chmn, Southeastern Conf T&A Mech, 74-75. *Mem:* Fel Am Soc Mech Engrs; Nat Acad Forensic Engrs; Nat Soc Prof Engrs. *Res:* Highway safety and vehicle handling, crash and damage mitigation; injury assessment and prevention; crash simulation and avoidance; dynamics of deformable bodies; approximate methods of engineering analysis. *Mailing Add:* PO Box 769 Arlington VA 22216-0769

CHI, MINN-SHONG, b Taichung, Taiwan, Dec 11, 40; US citizen; m 71; c 2. CHEMISTRY, PSYCHOLOGY. *Educ:* Nat Taiwan Univ, BS, 65; Univ Louisville, MS, 69; Univ Mich, PhD(chem), 72. *Prof Exp:* Fel, Univ Mich, 72-74; res chemist org & polymer chem, Witco Chem Corp, 74-77; staff chemist org, Polymer & Solid Propellant Chem, Allegany Ballistics Lab, Hercules, 77-85. *Mem:* Am Chem Soc. *Res:* Polymer chemistry; polyurethanes, poly amino acids, polyesters; organic chemistry; explosive materials; photochemistry; azo compounds. *Mailing Add:* 204 Meriden Dr Newark DE 19711

CHI, MYUNG SUN, b Chun Buk, Korea, Sept 28, 40; m 70; c 2. NUTRITION, NUTRITIONAL BIOCHEMISTRY. *Educ:* Kon Kuk Univ, BS, 66; Univ Minn, MS, 72, PhD(nutrit), 75. *Prof Exp:* Agr trainee animal sci, Malling Agr Col, Denmark, 66-67; res assoc nutrit, Univ Minn, 74-78; asst prof nutrit, Alcorn State Univ, 78-81; assoc prof, 81-88, PROF NUTRIT, LINCOLN UNIV, 89- *Concurrent Pos:* Res training hypertension, Dept Physiol, Univ-Mo, 82. *Mem:* Poultry Sci Asn; Am Inst Nutrit; Am Heart Asn; Am Asn Clin Chem. *Res:* Lipid metabolism in relation with coronary heart disease; mechanism of nutritional role in hypertension. *Mailing Add:* Dept Human Nutrit Lincoln Univ Schweich Hall Jefferson City MO 65101

CHI, TSUNG-CHIN, solid state physics, ultrasonics, for more information see previous edition

CHIA, E HENRY, Oct 14, 40; US citizen; m; c 2. ALLOYS. *Educ:* WGa Col, BA, 64; Ga Inst Technol, MS, 67, PhD(metall), 75. *Prof Exp:* Metallurgist, Southwire Co, 65-70, chief metallurgist, 70- 78, mgr, Metall Res Lab, 78-87; prin res scientist & dir, Mat Develop & Processing Prog, Ga Inst Technol, 87-89; EXEC VPRES & CHEIF OPER OFFICER, AM FINE WIRE CORP, 89- *Concurrent Pos:* Instr indust metall, Carroll Tech Sch, Ga, 68-71; mem, Non-Ferrous Mgt Comt, Wire Asn Int, 80-83, bd dirs, 83-; adj prof, Sch Mat Eng, Ga Inst Technol, 82-; mem, Non-Ferrous Comt, Hetall Soc, 83- *Honors & Awards:* John W Mordica Award, Wire Asn Int, 80. *Mem:* Fel Am Soc Metals; Metall Soc; Sigma Xi; Am Soc Testing & Mat; Int Metallographic Soc; Sr mem Soc Mfg Engrs. *Res:* Alloys development; microelectronic materials; interconnection technology; precious metals manufacturing; hot and cold deformation; microstructural control, property improvement, thermo- mechanical processing and mechanical behavior in general metallurgical research; recipient of numerous patents. *Mailing Add:* Am Fine Wire 907 Dravenwood Dr Selma AL 36701

CHIA, FU-SHIANG, b Shantung, China, Jan 15, 31; m 63; c 2. ZOOLOGY. *Educ:* Taiwan Norm Univ, BS, 55; Univ Wash, MS, 62, PhD(zool), 64. *Prof Exp:* Lab instr biol, Tunghai Univ, 57-58; asst zool, Univ Wash, 58-64; asst prof life sci, Sacramento State Col, 64-66; sr res officer zool, Univ Newcastle, 66-69; assoc prof, 69-75, chmn dept, 78-83, PROF ZOOL, UNIV ALTA, 75-, DEAN FAC GRAD STUDIES & RES, 83- *Mem:* Am Soc Zoologists; AAAS. *Res:* Developmental biology; marine invertebrate zoology. *Mailing Add:* Dept Zool Univ Alta Edmonton ON T6G 2E9 Can

CHIACCHIERINI, RICHARD PHILIP, b Elmira, NY, Mar 21, 43; m 65; c 2. RISK ASSESSMENT, MODELLING. *Educ:* St Bonaventure Univ, BS, 65; NC State Univ, MES, 67; Va Polytech Inst & State Univ, PhD(statist), 72. *Prof Exp:* Jr statistician, Biometry Sect, Nat Ctr Radiol Health, 67-70; trainee, Bur Radiol Health, Food & Drug Admin, 70-72, chief, Statist Sect, Off Radiol Prog, Environ Protection Agency, 72-73, sr statistician, Epidemiol Br, Bur Radiol Health, Food & Drug Admin, 73-79, chief, Statist Sect, 79-82, chief, Ionizing Radiol Br, 82-85, DIR, DIV BIOMET SCI, CTR DEVICES & RADIOL HEALTH, FOOD & DRUG ADMIN, 85- *Concurrent Pos:* Adj prof, Va Polytech Inst & State Univ, 81-; chief scientist officer, USPHS Comn Corps, 87- *Honors & Awards:* Pub Health Serv Commendation Medal, USPHS, 77, Pub Health Serv Unit Commendation, 77, 87, 88 & 90, Pub Health Serv Citation, 85 & 87, Pub Health Serv Outstanding Serv Medal, 87, Surgeon General's Exemplary Serv Medal, 90. *Mem:* Biomet Soc; Am Statist Asn; Am Col Epidemiol. *Res:* Design, analysis, and interpretation of epidemiologic and animal studies on the safety and effectiveness of medical devices and radiation; extension of categorical data methods and modeling risk and benefit situations. *Mailing Add:* Ctr Devices & Radiol Health FDA HFZ-116 5600 Fishers Lane Rockville MD 20857

CHIAKULAS, JOHN JAMES, b Chicago, Ill, Aug 3, 15; m 49; c 1. ANATOMY. *Educ:* Northwestern Univ, BS, 40, MA, 47; Univ Chicago, PhD(zool), 51. *Prof Exp:* Spec lectr zool, Grinnell Col, 46-47; instr biol, Roosevelt Univ, 51-52; asst prof anat, Chicago Col Optom, 52-53; instr, 53-55, assoc, 56-57, from asst prof to prof, 58-83, actg chmn dept, 80-83, EMER PROF ANAT, CHICAGO MED SCH, 83- *Mem:* Int Soc Chronobiol; Am Soc Zool; Am Asn Anat; Soc Develop Biol; Sigma Xi. *Res:* Wound healing; tissue specificity; organ regeneration; biorythmicity. *Mailing Add:* 3347 N Rutherford Avenue Chicago IL 60634

CHIANELLI, RUSSELL ROBERT, b Newark, NJ, May 22, 44; m 71; c 3. INORGANIC CHEMISTRY. *Educ:* Polytech Inst Brooklyn, BS, 70, PhD(chem), 74. *Prof Exp:* Res chemist, 73-76, SR RES CHEMIST & GROUP HEAD CATALYTIC MAT GROUP, EXXON RES & ENG CO, 76- *Mem:* AAAS; NY Acad Sci; Am Chem Soc; Am Phys Soc; Electrochem Soc. *Res:* Physics and chemistry of solids particularly transition metal chalcogenides, metallic non-metals and related compounds which are studied with x-ray crystallography, optical microscopy and spectroscopy. *Mailing Add:* 151 Station Rd Somerville NJ 08876-3537

CHIANG, ANNE, b Canton, China, Oct 3, 42; US citizen; m 67; c 1. PHYSICAL CHEMISTRY, DISPLAY TECHNOLOGY. *Educ:* Nat Taiwan Univ, BS, 64; Univ Southern Calif, PhD(phys chem), 68. *Prof Exp:* Sr chemist photochem & colloid chem, Memorex Corp, 69-71; MEM RES STAFF & PROJ LEADER EXPLOR DISPLAYS, XEROX PALO ALTO RES CTR, 72- *Mem:* Am Chem Soc; Soc Photographic Scientists & Engrs; Soc Info Display. *Res:* Viscoelasticity of polymers; non-aqueous colloid chemistry; photochemistry of coordination compounds; laser annealed silicon devices, very large scale integration, flat panel displays. *Mailing Add:* 10213 Miner Pl Cupertino CA 95014

CHIANG, BIN-YEA, b China. FOOD SCIENCE & TECHNOLOGY. *Educ:* Chung-Hsing Univ, Taiwan, BS, 66; Kansas State Univ, MS, 72, PhD(food sci), 75. *Prof Exp:* Sr res scientist, Modern Maid Food Prod, Inc, 75-81; PROJ LEADER, NABISCO BRAND INC, 81- *Mem:* Am Asn Cereal Chemists; Inst Food Technol. *Res:* Baking technology; flour quality; cereal grains; cake mixes batter; cookies and crackers. *Mailing Add:* Five Hickory Pl Cedar Knolls NJ 07927-1561

CHIANG, CHAO-WANG, b Ann-Hui, China, May 21, 25; m 60; c 4. MECHANICAL ENGINEERING. *Educ:* Chiao Tung Univ, BS, 48; Univ Wis, PhD(mech eng), 60. *Prof Exp:* Mech engr, Taiwan Hwy Bur, 48-50, 52-56, heavy equip div, 50-52; sr engr, Corning Glass Works, NY, 60-64; from assoc prof to prof mech eng, Univ Denver, 69-74; head dept, 74-82, PROF MECH ENG, SDAK SCH MINES & TECHNOL, 82- *Concurrent Pos:* Mem staff, Nat Comn Space Res, 70-71; vis prof, overseas grad prog, Boston Univ, 83-84. *Mem:* Am Soc Mech Engrs; Am Soc Eng Educ; Sigma Xi. *Res:* Diesel engine combustion; heat transfer and viscous flow problems of glass; conductive, convective and boiling heat transfer; solar energy, geothermal energy and thermoscience. *Mailing Add:* SDak Sch Mines & Technol Rapid City SD 57701

CHIANG, CHIN LONG, b Ningpo, China, Nov 12, 16; US citizen; m 45; c 3. STOCHASTIC PROCESSES. *Educ:* Tsing Hua Univ, China, BA, 40; Univ Calif, Berkeley, MA, 48, PhD(statist), 53. *Prof Exp:* Instr pub health, Univ Calif, Berkeley, 53-55, from asst prof to prof biostatist, 55-87, chmn div measurement sci, 70-75, chmn fac, 75-76, EMER PROF BIOSTATIST, UNIV CALIF, BERKELEY, 87- *Concurrent Pos:* Consult, State Dept Health, Calif, 58 & 64, State Dept Hyg, 61-62, NY State Dept Health, 62, Nat Ctr Health Statist, 62-64, 73-75 & 78, Rand Corp, 73; Nat Ctr Health Serv Res, 73-, WHO, 70-88, Nat Inst Neurol Dis & Stroke & Nat Ctr Health Serv Res & Develop, 73- & Bur Manpower, 78; spec res fel, Nat Heart Inst, 59-60; vis asst prof, Univ Mich, 59; vis lectr, Univ Minn, Minneapolis, 60, 61; spec consult, Nat Vital Statist, 62-64; vis assoc prof, Univ NC, 63, vis prof, 70; Fulbright fel, Gt Brit, 64; vis prof, Yale Univ, 65 & 66, Emory Univ, 67, Univ Pittsburgh, 68, Univ Wash, 69, Univ Tex, 73, Vanderbilt Univ, 75, Beijing Med Univ, 82, Tongji Med Univ, 85 & Peking Univ, 87; assoc ed, Biometrics, 70-75, Math Biosci, 75-87 & World Health Statist Quart, 79-88. *Honors & Awards:* Statist Award, Am Pub Health Asn, 81. *Mem:* AAAS; Am Pub Health Asn; Am Statist Asn; Inst Math Statist; Biomet Soc; Int Statist Inst. *Res:* Stochastic studies of the life table; competing risks; illness and death processes; stochastic modelling. *Mailing Add:* Sch Pub Health Univ Calif Berkeley CA 94720

CHIANG, CHWAN K, b China, Jan 18, 43; US citizen; m 79. SOLID STATE PHYSICS. *Educ:* Mich State Univ, PhD(physics), 74. *Prof Exp:* Res assoc, Univ Pa, 74-78; PHYSICIST, NAT BUR STANDARDS, 78- *Mem:* Am Phys Soc; Am Chem Soc; Electrochem Soc. *Res:* Organic conductors; conducting polymers; high technology super conductors. *Mailing Add:* Nat Inst Standards & Technol Bldg 223 Rm A215 Gaithersburg MD 20899

CHIANG, DONALD C, fluid mechanics; deceased, see previous edition for last biography

CHIANG, FU-PEN, b Checkiang, China, Oct 10, 36; m 63, 90; c 2. EXPERIMENTAL MECHANICS, OPTICAL STRESS ANALYSIS. *Educ:* Taiwan Univ, BS, 57; Univ Fla, MS, 63, PhD(mech), 66. *Prof Exp:* Civil engr, Mil Construct Bur, Taiwan, China, 58-59, Shihmen Dam Construct Comn, 59-61; res asst, Univ Fla, 62-66; res fel mech, Cath Univ Am, 66-67; from asst prof to prof eng, 67-87, LEADING PROF MECH ENG & DIR LAB EXP MECH RES, STATE UNIV NY STONY BROOK, 87- *Concurrent Pos:* NSF res grants, 68-, Off Naval Res grants, Dept Defense, 82- & Army Res Off grant, 88-; vis prof, Swiss Fed Inst Technol, Lausanne, 73-74 & sr vis fel Cavendish Lab, Univ Cambridge, Eng, 80-81; ed, Int J Optics & Lasers Eng, 87-; chmn, Int Conf Photometrics & Speckle Metrol, Soc Photo-optical Instrumentation Engrs, 87, 91. *Mem:* Am Acad Mech; AAAS; NY Acad Sci; fel Optical Soc Am; Soc Photo-optical Instrumentation Engrs; fel Soc Exp Mech; Sigma Xi; Am Soc Mech Engrs; Am Soc Metals Int. *Res:* Theory and applications of photoelasticity; moire methods; holographic inteferometry; laser speckle and white light speckle methods for stress analysis; fracture and fatigue of materials. *Mailing Add:* Dept Mech Eng State Univ NY Stony Brook NY 11794-2300

CHIANG, GEORGE C(HIHMING), b Nanking, China, Sept 12, 31; US citizen; m 59; c 3. STRUCTURAL ENGINEERING, ENGINEERING MECHANICS. *Educ:* Nat Taiwan Univ, BS, 54; Univ Southern Calif, MS, 58 & 63; Stanford Univ, PhD(solid mech), 67. *Prof Exp:* Proj engr, Richard R Bradshaw Inc, 59-62, vpres, 62-64; asst prof eng, 67-69, assoc prof eng & chmn fac civil eng & eng mech, 69-74, PROF ENG, CALIF STATE UNIV, FULLERTON, 74- , PROF CIVIL ENG. *Mem:* Am Inst Aeronaut & Astronaut; Am Concrete Inst. *Res:* Stability of elastic thin shells; stability problems involving viscoelastic materials and dynamics of structures. *Mailing Add:* Sch Eng Calif State Univ Fullerton CA 92634

CHIANG, HAN-SHING, b China, July 2, 29; m 62; c 2. COAL MINING TECHNOLOGY, RESPIRABLE DUST CONTROL. *Educ:* China Mining Inst, BS, 52; Grad Sch Beijing Mining Inst, MS, 55; WVa Univ, MS, 83. *Prof Exp:* Instr mining eng, mining dept, Beijing Mining Inst, 61-79; assoc prof mining eng, Grad Sch China Mining Inst, 79-81; vis prof, 81-83, RES PROF MINING ENG, DEPT MINING ENG, WVA UNIV, 83- *Concurrent Pos:* Mining engr, Bur Technol, China Ministry Coal Indust, 65-66, Bur Prod, 72-78; consult ed, Coal Sci & Technol, China, 83-; consult, China Henan Prov Coal Indust Authority, 84-, China Asn Coal Processing & Utilization, 84- *Res:* Longwall mining; ground control; respirable dust control; mining technology. *Mailing Add:* 353 D Comer Bdg Univ WVa Morgantown WV 26506

CHIANG, HUAI C, b Sunkiang, China, Feb 15, 15; m 46; c 3. INSECT ECOLOGY, INSECT CONTROL. *Educ:* Tsing Hua Univ, China, BS, 38; Univ Minn, MS, 46, PhD(entom), 48. *Hon Degrees:* DSc, Bowling Green State Univ, 79. *Prof Exp:* Asst entom, Tsing Hua Univ, China, 38-40, instr, 40-44; asst, 45-48, res fel, 48-53, from asst prof to prof biol, 54-61, prof entom, 61-83, EMER PROF ENTOM, UNIV MINN, ST PAUL, 84- *Concurrent Pos:* Guggenheim fel, 56-57; mem leader numerous sci del, panels, consult, 75-88. *Honors & Awards:* C V Riley Award, Entom Soc Am, 84; Distinguished Serv Award, Am Inst Biol Sci, 79. *Mem:* Hon mem Entom Soc Am; Ecol Soc Am; Can Entom Soc; fel Royal Entom Soc London; Int Asn Ecol. *Res:* Insect biology and ecology. *Mailing Add:* Dept Entom Univ Minn St Paul MN 55108

CHIANG, JOSEPH FEI, b Hunan, China, Feb 22, 39; m 63; c 2. PHYSICAL CHEMISTRY, APPLIED PHYSICS. *Educ:* Tunghai Univ, Taiwan, BS, 60; Cornell Univ, MS, 65, PhD(phys chem), 67. *Prof Exp:* Fel, Cornell Univ, 67-68; from asst prof to assoc prof, 68-79, PROF CHEM, STATE UNIV NY COL, ONEONTA, 79- *Concurrent Pos:* NIH fel, 75; res fel, Harvard Univ, 78, 79, 81; exchange fel, Nat Acad Sci, Hungary, 80, 86; vis prof, Univ Chicago, 83; fac participant, Argonne Nat Lab, 87-; vis fel, Mass Inst Technol, 88 & 90. *Mem:* Am Chem Soc; Am Phys Soc; AAAS; NY Acad Sci. *Res:* Use of electron diffraction technique and spectroscopic techniques to study molecular structures in gas phase; laser spectroscopic studies of Van der Waals molecules and structures; high temperature TC superconductor. *Mailing Add:* Dept Chem State Univ NY Col Oneonta NY 13820-4015

CHIANG, KARL KIU-KAO, statistics, process engineering, for more information see previous edition

CHIANG, KWEN-SHENG, b Shanghai, China, Feb 12, 39. MOLECULAR BIOLOGY. *Educ:* Nat Taiwan Univ, BS, 59; PhD(biochem Princeton Univ, PhD(biochem sci), 65. *Prof Exp:* Res assoc biochem, Princeton Univ, 65-66; asst prof, 66-72, assoc prof biophys, 72-76, ASSOC PROF BIOCHEM & THEORET BIOL, UNIV CHICAGO, 76- *Concurrent Pos:* USPHS res career develop award, 70-75; ed bd, Plant Sci Letters J, 73-; vis prof, Inst Bot, Acad Sinica, 74-75; mem, Comn Genetics, Univ Chicago. *Mem:* Genetics Soc Am; Am Soc Microbiol; Biophys Soc; Am Soc Cell Biol. *Res:* Molecular mechanisms of meiosis and sexual reproduction; molecular biology of cellular organelles; chloroplast and mitochondria; biochemical mechanisms of non-Mendelian genetics. *Mailing Add:* Cell Biol & Genetics Univ Chicago 920 E 58th St Chicago IL 60637

CHIANG, MORGAN S, b Kiangsu, China, Dec 30, 26; Can citizen; m 65; c 1. GENETICS. *Educ:* Nat Taiwan Univ, BSc, 50; McGill Univ, MSc, 59; Tex A&M Univ, PhD(genetics), 65. *Prof Exp:* Asst genetics, Nat Taiwan Univ, 50-55 & lectr, 55-57; res asst, Jackson Mem Lab, 59-61; RES SCIENTIST, RES STA, CAN DEPT AGR, 65- *Mem:* Agr Inst Can; Genetics Soc Can; Can Soc Hort Sci; Sigma Xi; Am Soc Hort Sci. *Res:* Discovery of mutant Careener in mice; breeding cabbage variety resistant to clubroot disease; development of hybrid cabbage variety Chateauguay; cabbage pollen physiology; released clubroot resistant cabbage varieties Acadie, Richelain and Richesse; discovery of mutant "clustered flowers" in cabbage. *Mailing Add:* Can Dept Agr PO Box 457 Res Sta St Jean PQ J3B 6Z8 Can

CHIANG, PETER K, b Hong Kong, China, Oct 20, 41; US citizen; m 67; c 3. BIOCHEMISTRY. *Educ:* Univ San Francisco, BSc, 65; Univ Alta, MSc, 67, PhD(biochem), 71. *Prof Exp:* Fel, John Hopkins Univ, 71-72; vis fel, NIH, 72-74; sr staff fel, NIMH, 74-80, res scientist, 80-81; res chemist, 81-87, CHIEF DEPT APPL BIOCHEM, WALTER REED ARMY INST RES, 88- *Mem:* Am Soc Biol Chemists; Am Soc Pharmacol & Exp Therapeut; Acad Pharm & Pharmaceut Sci; NY Acad Sci; Am Chem Soc. *Res:* Methylation reactions; methylases; s-adenosylmethionine; s-adenosylhomocysteine; inhibitors of methylation; neurobiology; cholinergic system; differentiation. *Mailing Add:* Div Biochem Walter Reed Army Inst & Res Washington DC 20307-5100

CHIANG, S(HIAO) H(UNG), b Soochow, China, Oct 10, 29; US citizen; m 58; c 3. SOLID & FLUID SEPARATION. *Educ:* Nat Univ Taiwan, BS, 52; Kans State Univ, MS, 55; Carnegie Inst Technol, PhD(chem eng), 58. *Prof Exp:* Proj engr chem eng, Linde Co, Union Carbide Corp, 58-60; from asst prof to prof, 60-88, Ernest E Roth prof, 88-90, WILLIAM KEPLER WHITEFORD PROF CHEM ENG, UNIV PITTSBURGH, 90- *Concurrent Pos:* Consult, Tokten Prog, UN, 84, UN Int Develop Orgn, 86; Hon prof, Nanjing Inst Chem Tech, China; dir, Ctr Energy Res, Univ Pittsburgh, 89- *Mem:* Am Inst Chem Engrs; Am Chem Soc; Soc Mining Engrs; Am Soc Eng Educ; Am Filtraction Soc. *Res:* Mass transfer processes; interface phenomena; coal processing; phase equilibrium; energy conversion. *Mailing Add:* Dept Chem & Petrol Eng Univ Pittsburgh Pittsburgh PA 15261

CHIANG, SOONG TAO, b Shanghai, China, Nov 14, 37; US citizen; m 66; c 2. CLINICAL PHARMACOKINETICS. *Educ:* Univ Calif, Los Angeles, BS, 64, MS, 66, PhD(physics), 70. *Prof Exp:* Teaching asst physics, Univ Calif, Los Angeles, 64-70; res assoc, Univ Nfld, 70-71; instr med physics, Thomas Jefferson Univ, 71-73; sr biomathematician biostatist & pharmacokinetics, Wyeth-Ayerst Res, 73-83, mgr biomath, 83-86, assoc dir, 86-88, DIR, CLIN PHARMACOKINETICS, WYETH-AYERST RES, 88- *Mem:* Am Statist Asn; Am Soc Clin Pharmacol & Therapeut. *Res:* Applied multivariate statistical analyses, pharmacokinetics and numerical analyses. *Mailing Add:* Clin Res & Develop Wyeth-Ayerst Res Radnor PA 19087

CHIANG, TAI-CHANG, b Taipei, Rep China, Aug 28, 49; m 79; c 3. SURFACE PHYSICS, PHOTOELECTRON SPECTROSCOPY. *Educ:* Nat Taiwan Univ, BS, 71; Univ Calif, Berkeley, PhD(physics), 78. *Prof Exp:* Postdoctoral, T J Watson Res Ctr, IBM, 78-80; from asst prof to assoc prof, 80-88, PROF PHYSICS, UNIV ILL, URBANA-CHAMPAIGN, 88- *Honors & Awards:* Presidential Young Investr Award, 84. *Mem:* Am Phys Soc. *Res:* Electronic properties and atomic structure of bulk materials, surfaces and interfaces using synchrotron radiation photoemission, scanning tunneling microscopy and related techniques. *Mailing Add:* Dept Physics Univ Ill 1110 W Green Urbana IL 61801

CHIANG, THOMAS M, b Taiwan, Repub China, Mar 30, 40; US citizen; m 74; c 2. BIOCHEMISTRY. *Educ:* Nat Chung-Hsing Univ, BS, 64; Univ Tenn, Memphis, PhD(biochem), 73. *Prof Exp:* Res asst, Nat Chung-Hsing Univ, Taiwan, 65-66, teaching asst microbiol, 67-69; res fel, 73-76, asst prof, 77-84, ASSOC PROF PLATELET-COLLAGEN INTERACTION, UNIV TENN, MEMPHIS, 85-; RES CHEMIST, VET ADMIN MED CTR, MEMPHIS, 73- *Mem:* AAAS; NY Acad Sci; Am Soc Biol Chemists. *Res:* Platelet-collagen interaction; protein phosphorylation and dephosphorylation. *Mailing Add:* 2151 New Meadow Dr Germantown TN 38138

CHIANG, TZE I, b Fuzhou, Fujian, China, Apr 19, 23; US citizen; m 29; c 3. ECONOMIC FEASIBILITY, ECONOMIC ANALYSIS. *Educ:* Fujian Christian Univ, BA, 46; Okla State Univ, MA, 55; Univ Fla, PhD(agr econ),58. *Prof Exp:* Teacher Chinese, Sin-Ding Jr Girls High Sch, 46-47; mem staff, China Textile Indust Inc, 47-53; teaching asst land use, Okla State Univ, 54-55; res asst agr mkt, Univ Fla, 55-58; asst res economist, 58-62, res economist, 62-65, sr res scientist, 58-74, PRIN RES SCIENTIST, ECON DEVELOP LAB, GTRI, GA INST TECHNOL, 74- *Mem:* Am Agr Econ Soc; Forest Prod Res Soc. *Res:* Economic studies concerning manufacturing processes; input-output relationships; investment requirements; financial returns; author of over sixty formal reports. *Mailing Add:* 3165 Frontenac Ct NE Atlanta GA 30319

CHIANG, YUEN-SHENG, b Tsingtao, China, Feb 2, 36. PHYSICAL CHEMISTRY, TECHNICAL MANAGEMENT. *Educ:* Nat Taiwan Univ, BS, 56; Univ Louisville, MChE, 60; Princeton Univ, PhD(phys chem), 64; Rutgers Univ, MBAC, 79. *Prof Exp:* Res assoc phys chem, Princeton Univ, 63; from scientist to sr scientist, Xerox Corp, NY, 64-69; MEM TECH STAFF, RCA LABS, 69- *Mem:* AAAS; Am Chem Soc; Electron Micros Soc Am; Electrochem Soc; Inst Elec & Electronics Engrs. *Res:* Physics and chemistry of surfaces; crystal growth and dislocation studies; electron paramagnetic resonance; electron microscopy; organic semiconductors; metal physics; ultra high vacuum technology. *Mailing Add:* Jeanne Pierre Inc 75 Frost St Westbury NY 11590

CHIAO, JEN WEI, c 3. CELL BIOLOGY OF CANCER CELL GROWTH, BIOCHEMISTRY OF PROTEINS. *Educ:* Southwestern Univ, BS, 66; Univ Ill, MS, 68, PhD(immunol), 71. *Prof Exp:* Fel immunol, Rockefeller Univ, 71-73; assoc cancer res & immunol, Sloan-Kettering Inst Cancer Res, Cornell Univ Grad Sch, NY, 73-83; PROF IMMUNOL & CANCER RES, DEPT MED & IMMUNOL, NY MED COL, VALHALLA, 83-, PROF, DEPT UROL, 85- *Concurrent Pos:* Ad hoc mem NIH study sect, career develop award, 81-86; dir Immunotype Lab, NY Med Col, 84-; mem bd, Cancer Inst, NY Med Col, 90. *Mem:* Am Asn Immunologists; Am Asn Cancer Res; Int Soc Exp Hematol. *Res:* Growth and differentiation controls of human cancer cells; relationship of cancer cells with immune system; specific projects include leukemia cells, prostate cancer cells and immunology of Lyme disease. *Mailing Add:* Dept Med NY Med Col Valhalla NY 10595

CHIAO, RAYMOND YU, b Hong Kong, Oct 9, 40; US citizen; m 68; c 3. PHYSICS. *Educ:* Princeton Univ, AB, 61; Mass Inst Technol, PhD(physics), 65. *Prof Exp:* Asst prof physics, Mass Inst Technol, 65-67; asst prof, 67-70, assoc prof, 70-77, PROF PHYSICS, UNIV CALIF, BERKELEY, 77- *Concurrent Pos:* Alfred P Sloan fel, 67-72. *Mem:* Am Phys Soc. *Res:* Lasers; non-linear optics; spontaneous and stimulated Brillouin scattering; stimulated Raman scattering; self-trapping of optical beams; superconductivity; astrophysics. *Mailing Add:* Dept Physics Univ Calif Berkeley CA 94720

CHIAO, WEN BIN, b Chang Hua, Taiwan, Sept 21, 48; US citizen; m 75; c 2. ORGANIC POLYMER CHEMISTRY. *Educ:* Nat Tsing Hua Univ, BS, 71; Univ Rochester, PhD(chem), 77. *Prof Exp:* Fel, Univ Md, 76-78; proj supvr, 78-81, RES ASSOC, NAT STARCH & CHEM CORP, 81- *Mem:* Am Chem Soc. *Res:* Synthesis of monomers and polymers; water soluble polymers; adhesives; mechanisms of organic reactions. *Mailing Add:* 434 Farmer Rd Bridgewater NJ 08807

CHIAO, YU-CHIH, b Hsin-Chu, Taiwan, Nov 20, 49; US citizen; m 75; c 2. PHYSICAL POLYMER CHEMISTRY. *Educ:* Nat Tsing-Hua Univ, BS, 72; Univ Rochester, PhD(chem), 76. *Prof Exp:* Muscular Dystrophy Asn fel biochem, Johns Hopkins Univ, 76-78; lectr chem, Rutgers Univ, 80-84; res chemist, Phys Polymer Chem, Allied Corp, 85-89; LAB SUPVR, AQUATECH SYST, 89- *Mem:* Am Chem Soc. *Res:* Transport phenomena of membranes; conformational studies of polyelectrolytes and biopolymers; spectroscopic studies of marcomolecules. *Mailing Add:* Aquatech Syst Seven Powder Horn Dr PO Box 4904 Warren NJ 07959-5191

CHIAPPINELLI, VINCENT A, b Pawtucket, RI, Mar 16, 51. NEUROPHARMACOLOGY, DRUG RECEPTORS. *Educ:* Boston Univ, AB, 73; Univ Conn, PhD(neuropharm), 77. *Prof Exp:* Res fel, Dept Pharmacol, Harvard Med Sch, 77-80; ASST PROF NEUROPHARM, DEPT PHARMACOL, SCH MED, ST LOUIS UNIV, 80- *Mem:* Soc Neurosci; AAAS. *Res:* Nicotinic cholinergic receptors in chick autonomic ganglia, both during development and in mature birds; role of neurotransmission in the biochemical development of nervous system. *Mailing Add:* Dept Pharmacol Sch Med St Louis Univ 1402 S Grand Blvd St Louis MO 63104

CHIARAPPA, LUIGI, b Rome, Italy, Dec 12, 25; US citizen; m 51; c 3. PHYTOPATHOLOGY. *Educ:* Univ Florence, Gen Agr Laurea, 50; Univ Calif, PhD(plant path), 58. *Prof Exp:* Agronomist tech comn, Ital Colonization in Chile, 50-51; entomologist, Di Giorgio Corp, 52-55, plant pathologist, Res Dept, 59-62; CHIEF, PLANT PROTECTION SERV, FOOD & AGR ORGN, UN, 62- *Concurrent Pos:* Vis prof, Univ Calif, Davis, 75-76. *Mem:* Am Phytopath Soc; Int Orgn Citrus Virol; Int Coun Viruses Grapevine; Mediter Phytopath Union. *Res:* Diseases of fruit, nut and vine crops; epidemiology; disease loss appraisal. *Mailing Add:* Via San Lucio 38 Rome Italy

CHIARODO, ANDREW, b New York, NY, June 26, 34; m 69. DEVELOPMENTAL BIOLOGY, CELL BIOLOGY. *Educ:* Fordham Univ, AB, 56, MS, 59; Wash Univ, PhD(zool), 63. *Prof Exp:* NIH fel, Med Col, Cornell Univ, 63; from instr to assoc prof biol, Georgetown Univ, 63-73; grants assoc, Div Res Grants, NIH, 73-74; prog dir, Nat Organ Site Progs, 74-80, CHIEF, ORGAN SYSTS PROG, NAT CANCER INST, 80- *Mem:* Soc Develop Biol. *Mailing Add:* 5120 Klingle St NW Washington DC 20016

CHIASSON, LEO PATRICK, b Cheticamp, NS, May 14, 18; m 48; c 5. GENETICS. *Educ:* St Francis Xavier Univ, BA, 38, BSc, 40; Univ Toronto, PhD(genetics), 44. *Hon Degrees:* LLD, St Francis Xavier Univ, 87. *Prof Exp:* Assoc prof biol, 44-49, prof, St Francis Xavier Univ, 49-87; RETIRED. *Concurrent Pos:* Assoc scientist, Fisheries Res Bd Can, 44-55; sr researcher zool, Columbia Univ, 63-64; vis prof, UBC, 78-79. *Mem:* AAAS; Genetics Soc Am; Genetics Soc Can. *Res:* Tomato species hybrids; relative growth in mice; blood groups and dermatoglyphics in Micmac Indians; distribution of scallops; species hybrids of Abies; mutagenic effects of heat in bacteria. *Mailing Add:* 36 Greenwold Dr Antigonish NS B2G 2H8 Can

CHIASSON, ROBERT BRETON, b Griggsville, Ill, Oct 9, 25; m 44; c 8. COMPARATIVE ANATOMY, COMPARATIVE ENDOCRINOLOGY. *Educ:* Ill Col, AB, 49; Univ Ill, MS, 50; Stanford Univ, PhD(biol sci), 56. *Prof Exp:* Spec supvr, Ill State Mus, 50-51; from instr zool to prof biol sci, 51-75, PROF VET SCI, UNIV ARIZ, 75-, RES PROF ANAT, 83-; RES SCIENTIST, ARIZ EXP STA, 75- *Concurrent Pos:* Fulbright lectr, Univ Sci & Technol, Ghana, 69-70; vis scientist, Poultry Res Ctr, Edinburgh, Scotland, 75-76 & Lab Physiol Domestic Animals, Univ Leuven, Belgium, 85. *Mem:* Am Soc Zoologists; Am Physiol Soc; World Asn Vet Anatomists; NY Acad Sci; Am Asn Vet Anatomists. *Res:* Regulation of pituitary function in the chicken and anatomy of vertebrates; anatomy of the eye of doves and skin of snakes. *Mailing Add:* Dept Vet Sci Univ Ariz Tucson AZ 85721

CHIAZZE, LEONARD, JR, b Falconer, NY, June 19, 34; m 54; c 4. BIOSTATISTICS, EPIDEMIOLOGY. *Educ:* Univ Buffalo, BS, 55, MBA, 57; Univ Pittsburgh, ScD(biostatist), 64. *Prof Exp:* Instr statist, Univ Buffalo, 55-57; asst health serv officer, Nat Cancer Inst, 57-60, sr asst health serv officer, 60-63, health serv officer, 63-64, scientist, 64-66; res assoc ctr pop res, 66-68, dir cerebrovasc dis follow-up & surveillance syst, 68-71, assoc prof community med & int health, 68-77, DIR DIV BIOSTATIST & EPIDEMIOL, SCH MED, GEORGETOWN UNIV, 66-, DIR GRAD PROG BIOSTATIST, 71-, PROF COMMUNITY & FAMILY MED, 77- *Concurrent Pos:* Chief, Biomet Br, Nat Cancer Inst, 75-76; consult epidemiol & biostatist to pvt & pub orgn; chief, Epidemiol & Statist, Lombardi Cancer Res Ctr, 82- *Mem:* Fel Am Pub Health Asn; Am Statist Asn; Int Epidemiol Asn; Soc Epidemiol Res; fel Am Col Epidemiol; Soc Occup & Environ Health (pres, 84-86). *Res:* Chronic disease epidemiology, especially cancer; occupational and environmental epidemilogy; morbidity survey and case register methodologies; population research; clinical trials; occupational health surveillance systems. *Mailing Add:* Div Biostatist & Epidemiol Georgetown Univ Sch Med 3750 Reservoir Rd Washington DC 20007

CHIBA, MIKIO, b Miyagi-Ken, Japan, Aug 4, 29; m 56; c 2. ANALYTICAL CHEMISTRY. *Educ:* Hokkaido Univ, BSc, 53, DSc(anal chem), 62. *Prof Exp:* Chemist, Hokkaido Police Hq, 54-56; res chemist, Sci Police Res Inst, 56-64; RES SCIENTIST, CAN DEPT AGR, 64- *Concurrent Pos:* Nat Res Coun Can fel, 62-64; hon res prof, Brock Univ, 73-; vis prof, Tokyo Univ Fisheries, 80. *Honors & Awards:* Caledon Award, 85. *Mem:* Am Chem Soc; Chem Inst Can; Chem Soc Japan; Pest Sci Soc Japan; Asn Off Anal Chemists. *Res:* Method development for pesticide analysis and to find better ways of applying pesticides. *Mailing Add:* Res Sta Agr Can Vineland Station ON L0R 2E0 Can

CHIBNIK, SHELDON, b New York, NY, Dec 20, 25; m 45; c 2. ORGANIC CHEMISTRY. *Educ:* Cornell Univ, AB, 44; Polytech Inst Brooklyn, MS, 51; Temple Univ, PhD(chem), 55. *Prof Exp:* Lab instr chem, Hunter Col, 46-48; from chemist to sr chemist, Nat Lead Co, 48-55, group leader indust finishes, 55-61; sr res chemist, Mobil Chem Co, Edison, 61-70, sr res chemist, Mobil Res & Develop Corp, Paulsboro, 71-88, assoc chem, 75-80, res assoc, 81-86, SR RES ASSOC, MOBIL RES & DEVELOP CORP, 87- *Mem:* Am Chem Soc. *Res:* Polymer chemistry; protective coatings; monomer synthesis; liquid phase oxidations; heterogeneous catalysis; lubricating oil and fuel additives. *Mailing Add:* Seven Glen View Pl Cherry Hill NJ 08034

CHIBURIS, EDWARD FRANK, b Omaha, Nebr, July 31, 33; m 54; c 5. GEOPHYSICS. *Educ:* Tex A&M Univ, BS, 60, MS, 62; Ore State Univ, PhD(geophys), 65. *Prof Exp:* Res geophysicist, Seismic Data Lab, Teledyne, Inc, 65-68 & dir res, 68-69; assoc prof geophys & asst dir, Marine Sci Inst, Univ Conn, 69-77; assoc prof geophys & asst dir, Weston Observ, Boston Col, 77-80; MGR, FLA LABS, TELEDYNE GEOTECH, 80- *Mem:* Am Geophys Union; Seismol Soc Am; Soc Explor Geophys; Sigma Xi. *Res:* Hypocenter location techniques; seismic network and array analyses; focal mechanism studies; gravity and magnetic methods; crustal studies; computer modeling; geophysical data processing. *Mailing Add:* 1964 First St N Texas City TX 77590

CHIBUZO, GREGORY ANENONU, b Abor, Nigeria, July 25, 43; m 72; c 4. VETERINARY ANATOMY. *Educ:* Tuskegee Inst, BS, 68, DVM, 70, MS, 75; Cornell Univ, PhD, 79. *Prof Exp:* From instr to prof anat, Sch Vet Med, Tuskegee Univ, 70-82; PROF VET ANAT & HEAD DEPT, UNIV MAIDUGORI, NIGERIA, 84-, DEAN, FAC VET MED, 83-; PROF ANAT, TUSKEGEE UNIV, 84- *Concurrent Pos:* Consult bio-med studies, Southern Voc Col, Tuskegee, 74-75. *Mem:* World Asn Vet Anatomists; Am Asn Vet Anatomists; Soc Study Reproduction; Am Vet Med Asn; Am Asn Anatomists. *Res:* Influence of neurohumoral substances on uterine motility; effect of progesterone and/or estrogen on the integrity of intra-ovarian vascular growth and distribution at birth, maturity and menopause; central projection of lingual structures. *Mailing Add:* Dept Anat Tuskegee Univ Sch Vet Med Tuskegee AL 36088

CHICHESTER, CLINTON OSCAR, b New York, NY, Feb 11, 25; m 47, 87; c 6. FOOD TECHNOLOGY. *Educ:* Mass Inst Technol, SB, 49; Univ Calif, MS, 51, PhD, 54. *Prof Exp:* From asst prof to prof food technol, Univ Calif, Davis, 53-70, chmn dept, 67-70; prof food & resource chem, 70-77, PROF FOOD SCI & TECHNOL & ASSOC DIR, INT CTR MARINE RESOURCE DEVELOP, UNIV RI, 77-; MEM STAFF, NUTRIT FOUND, INC. *Concurrent Pos:* Mem, NIH, 57-, Coun Foods & Nutrit, AMA, 62- & Space Sci Bd, Nat Acad Sci, 63-; vpres res, Nutrit Found, New York, 72-74. *Honors & Awards:* Bernardo O'Higgins Award, Govt Chile, 69; Medal, Czech Acad Sci; Babcock Hart Award, Inst Food Technologists, 73. *Mem:* Fel Inst Food Technologists; Am Chem Soc; Optical Soc Am; Am Soc Biol Chemists; Am Inst Chem Eng. *Res:* Pigment biochemistry; food processing. *Mailing Add:* PO Box 271 Wakefield RI 02880-0271

CHICHESTER, LYLE FRANKLIN, b Albany, NY, Nov 5, 31; m 54; c 3. ZOOLOGY. *Educ:* Univ Conn, BS, 54, PhD(zool), 68; Cent Conn State Col, MS, 61. *Prof Exp:* From instr to assoc prof biol, 60-74, chmn dept biol sci, 74-83, PROF BIOL, CENT CONN STATE UNIV, 74- *Mem:* Ecol Soc Am; Am Soc Zoologists; Am Malacol Union. *Res:* Distribution, ecology and systematics of terrestrial slugs, especially introduced European species; application of biochemical methods to systematic problems involving mollusks. *Mailing Add:* Dept Biol Sci Cent Conn State Univ New Britain CT 06052

CHICK, THOMAS WESLEY, b Martin, Tenn, May 1, 40; m 72; c 3. PULMONARY PHYSIOLOGY. *Educ:* Univ Cent Ark, BS, 61; Univ Ark, Little Rock, MD, 65. *Prof Exp:* From intern to resident internal med, Univ Tex Southwestern Med Sch Dallas, 65-68, fel pulmonary med, 68-70, from instr to asst prof internal med, 70-72; asst prof, 72-78, ASSOC PROF INTERNAL MED, SCH MED, UNIV NMEX, 78- *Res:* Clinical pulmonary physiology of obstructive airway disease; effects of bronchodilators, oxygen and exercise. *Mailing Add:* Vet Admin Hosp Albuquerque NM 87108

CHICKOS, JAMES S, b Buffalo, NY, Oct 27, 41; m 66; c 3. ORGANIC CHEMISTRY, PHYSICAL ORGANIC CHEMISTRY. *Educ:* Univ Buffalo, BA, 63; Cornell Univ, PhD(org chem), 66. *Prof Exp:* NIH vis fel, Princeton Univ, 66-67; NIH fel & res assoc, Univ Wis, 67-69; from asst prof to assoc prof, 69-86, PROF ORG CHEM, UNIV MO-ST LOUIS, 87- *Concurrent Pos:* Sabatical res leave, Univ Reading, Eng, 86-87. *Mem:* Am Chem Soc. *Res:* Unimolecular reactions; small ring, non-benzenoid aromatics; chemistry for the non-major; stereochemistry, secondary deuterium isotope effects; measurement of low vapor pressures. *Mailing Add:* Dept Chem Univ Mo 8001 Natural Bridge Rd St Louis MO 63121-4401

CHICKS, CHARLES HAMPTON, b Sandpoint, Idaho, Nov 10, 30; m 56; c 4. MATHEMATICS. *Educ:* Linfield Col, BA, 53; Univ Ore, MA, 56, PhD(math), 60. *Prof Exp:* Adv res engr, Sylvania Electronic Defense Labs, Gen Tel & Electronics Corp, 60-62, engr specialist, 62-69; STAFF ENGR, ESL INC, 69- *Concurrent Pos:* Lectr, Univ Santa Clara, 64-86. *Mem:* Am Math Soc. *Res:* Periodic automorphisms on banach algebras; military operations research; arms control and disarmament. *Mailing Add:* 495 Java Dr Sunnyvale CA 94088-3510

CHICO, RAYMUNDO JOSE, b Hernando, Arg, Sept 17, 30; m 59; c 4. MINING GEOLOGY, ECONOMIC GEOLOGY. *Educ:* Univ Cordoba, dipl geol, 53; Mo Sch Mines, MS, 58; Harvard Univ, MA, 63. *Prof Exp:* Asst, Univ Cordoba, 53; geologist, Direccion Gen de Ingenieros, Arg, 54; geologist, Peruvian Mines, Cerro de Pasco Corp, 54-55, off geologist, NY, 58-59; geol engr, Four Corners Uranium Corp, Colo, 56-57; consult, Nat Lead Co, Arg, 59, Int Basic Econ Corp, NY, 60 & Air Force Cambridge Res Labs, 60; geologist, Ltd War Lab, Aberdeen, Md, 63-65; oceanogr, Nat Oceanog Data Ctr, DC, 65-66; consult econ mining & eng geol indust, US & Latin Am, 66-68; PRES, RAYMUNDO J CHICO, INC, 68- *Concurrent Pos:* Guest crystallog lab, Johns Hopkins Univ, 65; US deleg, NATO Advan Study Inst Uranium, London, 72; pres & chmn bd, Am Gold Minerals Corp; corp dir, Altex Oil Corp; pres & chmn bd, Amada Mineral Corp & Northern Iron Ore Mines Ltd, Can. *Mem:* Am Inst Mining Metall & Petrol Eng; Geol Soc Am; Sigma Xi; Am Mining Cong; Am Asn Petrol Geologists. *Res:* Engineering earth sciences; Inter-American mineral industry; applied geology and mining for business development and economic growth; ore genesis; minerals; field economic geology. *Mailing Add:* 9600 E Grand Ct Englewood CO 80111

CHICOINE, LUC, b Montreal, Que, Apr 19, 29; m 53; c 1. PEDIATRICS. *Educ:* Univ Montreal, BA, 48, MD, 53; FRCP(C), 60. *Prof Exp:* From asst prof to assoc prof, 61-71, chmn dept, 75-82, PROF PEDIAT, UNIV MONTREAL, 71- *Concurrent Pos:* Dir, Poison Control Ctr. *Mem:* Am Acad Pediat; Can Med Asn; Can Pediat Soc. *Res:* Pediatric water and electrolyte problems. *Mailing Add:* Ste Justine Hosp 3175 Cote Ste Catherine Montreal PQ H3T 1C5 Can

CHICOYE, ETZER, b Jacmel, Haiti, Nov 4, 26; US citizen; m 54; c 2. FOOD CHEMISTRY. *Educ:* Univ Haiti, BS, 48; Univ Wis-Madison, MS, 54, PhD(food sci), 68. *Prof Exp:* Grader & cup tester, Nat Coffee Bur, Haiti, 48-52; res asst, Univ Wis-Madison, 52-54, 64-67; analytical chemist, Chicago Pharmacal Co, 54-56; res chemist, Julian Labs, Ill, 56-64; chem res supvr, 67-72, MGR RES, MILLER BREWING CO, 72-, DIR RES, 78- *Mem:* Am Chem Soc; Inst Food Technol; Master Brewers Asn Am; Am Soc Brewing Chemists. *Res:* Steroid chemistry; steroid hormones and vitamin D; cholesterol and degradation products of cholesterol in food; brewing chemistry; flavor chemistry. *Mailing Add:* 3939 W Highland Blvd Milwaukee WI 53201

CHIDAMBARASWAMY, JAYANTHI, b Hamsavaram, India, Nov 14, 27; m 46; c 5. MATHEMATICS. *Educ:* PR Col, Kakinada, India, BA, 50; Andhra Univ, India, MA, 53; Univ Calif, Berkeley, PhD(math), 64. *Prof Exp:* Lectr math, Andhra Univ, India, 53-62; teaching asst, Univ Calif, Berkeley, 62-64, instr, 64-65; asst prof, Univ Kans, 65-66; assoc prof, 66-69, PROF MATH, UNIV TOLEDO, 69- *Mem:* Am Math Soc; London Math Soc; Indian Math Soc. *Res:* Theory of numbers. *Mailing Add:* Dept Math Univ Toledo 2801 W Bancroft Toledo OH 43606

CHIDDIX, MAX EUGENE, b Palestine, Ill, Apr 13, 18; m 44; c 2. ORGANIC CHEMISTRY. *Educ:* Ill State Norm Univ, BEd, 40; Univ Ill, PhD(org chem), 43. *Prof Exp:* Res chemist, Gen Aniline & Film Corp, 43-49, group leader, 50-53, res fel, 53-55, prog mgr acetylene derivatives res, 55-60 & chem & polymer res, 60-68; chief chemist, GAF Corp, 68-69, chief proj chemist, 69-82; RETIRED. *Mem:* AAAS; NY Acad Sci; Am Chem Soc. *Res:* Polymers; plastics; resins; polyamide fibers and film; acetylene and textile chemicals; surfactants; alkylphenols; chelating agents; corrosion inhibitors; reactive dyes; lube oil additives; bactericides; fungicides; herbicides; analytical methods; pollution control. *Mailing Add:* 2324 Ptarmigan Dr Three Walnut Creek CA 94595

CHIDSEY, CHRISTOPHER E, b New York, NY, May 30, 57; m 81; c 2. MATERIALS SCIENCE ENGINEERING. *Educ:* Dartmouth Col, BS, 78; Stanford Univ, PhD(chem), 83. *Prof Exp:* Res assoc, Dept Chem, Univ NC, 83-84; MEM TECH STAFF, AT&T BELL LABS, 84- *Mem:* Am Chem Soc; Electrochem Soc; AAAS. *Res:* Interfacial electron transfer; electronic monolayers and multilayers on solids; electrochemical scanning tunneling microscopy. *Mailing Add:* AT&T Bell Labs Rm 1D-356 600 Mountain Ave Murray Hill NJ 07974-2070

CHIDSEY, JANE LOUISE, b Wilkes-Barre, Pa, Apr 1, 08. PHYSIOLOGY. *Educ:* Wellesley Col, AB, 29; Brown Univ, AM, 31; Cornell Univ, PhD(physiol), 34. *Prof Exp:* Demonstr biol, Brown Univ, 29-31; Coxe fel, Yale Univ, 34-35; instr zool, Smith Col, 35-39; from asst prof to prof biol, 39-73, Fund Adv Educ fac fel, 52-53, actg dean, 61-62, EMER PROF BIOL, WHEATON COL, MASS, 73- *Concurrent Pos:* Mem corp, Mt Desert Island Biol Lab. *Mem:* Sigma Xi. *Res:* Carbohydrate and fat metabolism; general animal and cellular physiology. *Mailing Add:* 50 W Bare Hill Rd Harvard MA 01451

CHIEN, ANDREW ANDAI, b Mt Kisco, NY, Mar 6, 64; m. SOFTWARE SYSTEMS, HARDWARE SYSTEMS. *Educ:* Mass Inst Technol, SB, 84, SM, 87, ScD, 90. *Prof Exp:* Res asst, Lab Computer Sci, Mass Inst Technol, 81-90, res scientist, 90; ASST PROF, DEPT COMPUTER SCI, UNIV ILL, 90- *Mem:* Inst Elec & Electronics Engrs; Asn Comput Mach. *Res:* Architecture of high-performance computer systems; programming systems, compilation, runtime support and hardware architecture. *Mailing Add:* Univ Ill Urbana-Champaign 1304 W Springfield Ave Urbana IL 61801

CHIEN, CHIA-LING, b China, Nov 10, 42; US citizen; m 72; c 2. ARTIFICIALLY STRUCTURED SOLIDS, HIGH TC SUPERCONDUCTORS. *Educ:* Tunghai Univ, Taiwan, BS, 65; Carnegie-Mellon Univ, MS, 68, PhD(physics), 73. *Prof Exp:* Res assoc, 73-74, assoc res scientist, 74-75, vis asst prof, 75-76, from asst prof to assoc prof, 76-83, PROF PHYSICS, JOHNS HOPKINS UNIV, 83- *Mem:* Fel Am Phys Soc; Sigma Xi. *Res:* Magnetic, hyperfine interaction, conductivity, structural and superconductivity in artificially structured solids; superlattices and granular solids; high superconductors. *Mailing Add:* Dept Physics Johns Hopkins Univ Baltimore MD 21218

CHIEN, CHIH-YUNG, b Zizhong, Sichuan, China, Aug 5, 39; m 63; c 3. EXPERIMENTAL HIGH ENERGY PHYSICS. *Educ:* Nat Taiwan Univ, BS, 60; Yale Univ, MS, 63, PhD(physics), 66. *Prof Exp:* Asst res physicist, Univ Calif, Los Angeles, 66-67, asst prof in residence, 67-68, asst prof, 68-69; from asst prof to assoc prof, 69-77, PROF PHYSICS, JOHNS HOPKINS UNIV, 77- *Concurrent Pos:* Ed, Sci & Technol Review; hon prof, Nanjing Univ, 80, Beijing Inst Technol, 83, Xinjiang Univ, 84, Nemongu Univ, Hanzhou Univ, Zhejiang Univ, 85, Yunnan Univ, 86 & Lanzhou Univ, China, 87; prog dir, NSF, 86. *Mem:* Am Phys Soc. *Res:* Experimental research on particle physics; development and application of data processing and nuclear physics detection devices. *Mailing Add:* Dept Physics Johns Hopkins Univ Baltimore MD 21218

CHIEN, HENRY H(UNG-YEH), b Shanghai, China, Sept 28, 35; m 61; c 2. CHEMICAL ENGINEERING. *Educ:* Univ Minn, PhD(chem eng), 63. *Prof Exp:* Sr chem eng, 63-67, eng supt, Cent Eng Dept, 67-75 Monsanto fel, 75-80, SR FEL, ENG TECHNOL DEPT, MONSANTO CO, 80- *Mem:* Am Inst Chem Engrs. *Res:* System modeling, simulation, optimization and control; applied mathematics. *Mailing Add:* Eng Tech Dept 800 N Lindbergh Blvd St Louis MO 63167

CHIEN, JAMES C W, b Shanghai, China, Nov 4, 29; US citizen; m 53; c 3. PHYSICAL CHEMISTRY. *Educ:* St John's Univ, China, BS, 49; Univ Ky, MS, 51; Univ Wis, PhD(phys chem), 54. *Prof Exp:* Sr res chemist, Hercules Powder Co, Del, 54-69; PROF CHEM, UNIV MASS, AMHERST, 69- *Mem:* Am Chem Soc. *Res:* Metalloenzymes and metalloproteins; äelectron paramagnetic resonance crystallography; electrical conducting polymers, polymerization catalysts; oxidation, stabilization and flame retarding polymers; radiation chemistry; oxidation; ultraviolet spectroscopy; electron spin resonance. *Mailing Add:* Dept Polymer Sci & Eng Univ Mass Amherst MA 01003

CHIEN, LUTHER C, b China, Dec 30, 23; US citizen; m 49; c 2. CHEMICAL ENGINEERING, ORGANIC CHEMISTRY. *Educ:* Harvard Univ, BS, 44; Mass Inst Technol, MS, 47. *Prof Exp:* Res engr, E I du Pont de Nemours & Co Inc, 47-55, supvr, 55-57, chief supvr, 57-78, res fel, 79-80, res mgr, fluoro chem process res & develop, 81-85, Dupont fel, 85-88; RETIRED. *Mem:* Am Chem Soc; Am Inst Chem Engrs. *Res:* Fluoro chemical processes. *Mailing Add:* 625 McKinley Dr Pitman NJ 08071

CHIEN, PING-LU, b China, Nov 5, 28; m 57; c 3. ORGANIC CHEMISTRY. *Educ:* Nat Taiwan Univ, BS, 52; Univ Kans, PhD(org chem), 64. *Prof Exp:* Chemist, Union Indust Res Inst, 55-59; res assoc med chem, Univ Kans, 64-65; assoc chemist, 65-68, sr chemist, Midwest Res Inst, 68-83; SR RES SCIENTIST, ORTHO PHARM CORP, 84- *Mem:* Am Chem Soc. *Res:* Synthetic organic and medicinal chemistry; absorption spectroscopy; radiolabeling. *Mailing Add:* 356 Gemini Dr Apt 8 Somerville NJ 08876-4971

CHIEN, SEN HSIUNG, b Taiwan, China, Aug 31, 41; m 70; c 1. SOIL CHEMISTRY. *Educ:* Nat Taiwan Univ, BS, 63; Univ NH, MS, 68; Iowa State Univ, PhD(soil chem), 72. *Prof Exp:* Assoc, Iowa State Univ, 72-73 & Wash Univ, 73-75; RES CHEMIST SOIL CHEM, INT FERTILIZER DEVELOP CTR, 75- *Mem:* Soil Sci Soc Am; Am Soc Agron; Int Soil Sci Soc. *Res:* Dissolution of phosphate rock in relation to the utilization of phosphate rock for direct application to soils. *Mailing Add:* Int Fertilizer Develop Ctr PO Box 2040 Muscle Shoals AL 35662

CHIEN, SHU, b Peiping, China, June 23, 31; m 57; c 2. PHYSIOLOGY. *Educ:* Nat Taiwan Univ, MD, 53; Columbia Univ, PhD(physiol), 57. *Prof Exp:* Intern, Taiwan Univ Hosp, 52-53; asst, 54-56, from instr to assoc prof, 56-69, prof physiol, Col Physicians & Surgeons, Columbia Univ, 69-88; PROF BIOENG & MED, UNIV CALIF, SAN DIEGO, 88- *Concurrent Pos:* Pres, Am Chinese Med Soc, 78-79, Microcirculatory Soc, 80-81; counr, Am Physiol Soc, 85; dir, Inst Biomed Sci, Academia Sinica, Taipei, Taiwan, Repub China, 87-88. *Honors & Awards:* Nanci Medal, 80; Fahraeus Award, 81; Landis Award, 83; Merit Award, Nat Heart Lung & Blood Inst, 88; Melville Medal, 90. *Mem:* Microcirculatory Soc; Harvey Soc; Am Physiol Soc; Soc Exp Biol & Med; NY Acad Sci; Sigma Xi; Academia Sinica; Biomed Eng Soc.

Res: Blood viscosity; red cell membrane; microcirculation; blood flow and volume; hemorrhage; endotoxin shock; body fluids; autonomic nervous system; bioengineering; molecular biology. *Mailing Add:* Ames Bioeng Univ Calif San Diego La Jolla CA 92093-0412

CHIEN, SZE-FOO, b China, Aug, 1929; m 61; c 2. MECHANICAL ENGINEERING. *Educ:* Univ Taiwan, BS, 53; Univ Minn, MS, 56, PhD(mech eng), 61. *Prof Exp:* Instr mech eng, Univ Minn, 56-61; RES ENGR, TEXACO BELLAIRE RES LABS, 61- *Concurrent Pos:* Consult, G H Tennant Co & Furcula Co; invited tech consult, UN Develop Prog, 85. *Mem:* Am Soc Mech Engrs; Sigma Xi. *Res:* Non-Newtonian fluid mechanics; oil-well drilling hydraulics and mechanics; multi-phase flow; secondary and tertiary oil recovery; tar-sands oil technology; product research and development; enchanced gas recovery. *Mailing Add:* 5027 S Braeswood Blvd Houston TX 77096

CHIEN, VICTOR, b Shanghai, China, Nov 4, 48; US citizen; m 83. TELECOMMUNICATION, NETWORK ANALYSIS. *Educ:* Clemson Univ, BS, 81, MS, 82, PhD(math), 85. *Prof Exp:* MEM TECH STAFF, BELL COMMUN RES, 86- *Concurrent Pos:* Vis prof math, Col Charleston, 85-86. *Mem:* Soc Indust & Appl Math. *Res:* Computational mathematics, especially in numerical analysis of partial differential equations and parameter estimation in nonlinear diffusion equations; applied mathematics with mathematical modeling and computer simulation. *Mailing Add:* Bellcore 290 W Mt Pleasant Ave Livingston NJ 07039-0486

CHIEN, YIE W, b Taipei, Taiwan, Oct 20, 38; US citizen; m 64; c 2. RATE-CONTROL DRUG DELIVERY, INDUSTRIAL PHARMACY. *Educ:* Kaohsiung Med Col, Taiwan, BSc, 63; Ohio State Univ, PhD(pharmaceut), 72. *Prof Exp:* Res pharmacist, Res & Develop Div, Seven Seas Pharmaceut Corp, Taipei, Taiwan, 63-66, chief pharmacist, Prod Div, 66-67; res investr, Pharmaceut Res & Develop Group, Prod Develop Dept, Searle Labs, G D Searle & Co, Skokie, Ill, 72-74, res scientist, Biopharmaceut Group, Pharmaceut Develop Dept, 75-76, group leader, Pharmaceut Res Group, New Prod Develop Dept, 76-78; sect head formulation develop, pharmaceut res & develop dept, Endo Labs, The DuPont Co, Garden City, 78-81; head dept, 82-88, PARKE-DAVIS PROF PHARMACEUT, RUTGERS STATE UNIV NJ, 81-, DIR, RES CTR, 82- *Concurrent Pos:* Reviewer, prof & sci journals, 73-; mem, subcomt pharmaceut, Parenteral Drug Asn, 80-83, chmn subcomt, res comt, 81-83; consult, pharmaceut & health care corps, 81- & WHO, 89-; vis prof, State Univ NY, Buffalo, 82-83; mem, working group, NJ Gov Comn Sci & Technol, 83-84. *Mem:* Am Pharmaceut Asn; NY Acad Sci; Parenteral Drug Asn; Controlled Release Soc; Am Asn Pharmaceut Scientists. *Res:* Rate-controlled noninvasive, systemic drug administration; application of controlled drug release technology to the development of novel drug delivery systems; biopharmaceutics and pharmacokinetics; quantitative structure-activity relationships; industrial pharmaceutics technology. *Mailing Add:* Controlled Drug Delivery Res Ctr Rutgers Univ Col Pharm PO Box 789 Piscataway NJ 08855-0789

CHIEN, YI-TZUU, b Shanghai, China, Aug 21, 38; m 65; c 3. ELECTRICAL ENGINEERING, COMPUTER SCIENCE. *Educ:* Nat Taiwan Univ, BSEE, 60; Purdue Univ, MSEE, 64, PhD, 67. *Prof Exp:* Mem tech staff, Bell Tel Labs, Inc, NJ, 66-67; from asst prof to assoc prof elec eng, 67-77, PROF ELEC ENG & COMPUTER SCI, UNIV CONN, 77-; DIV DIR, NSF, 86- *Concurrent Pos:* Group leader, Naval Res Lab, Washington, DC. *Mem:* Fel Inst Elec & Electronics Engrs; Asn Comput Mach. *Res:* Automatic pattern recognition; adaptive and learning systems; artificial intelligence. *Mailing Add:* Nat Sci Found Div Info Robotics & Intel Systs 1800 G St NW Washington DC 20550

CHIERI, P(ERICLE) A(DRIANO), b Mokanshan, China, Sept 6, 05; nat US; m 38. AERONAUTICAL & MECHANICAL ENGINEERING. *Educ:* Univ Genoa, Dr Ing, 27; Univ Naples, ME, 27; Univ Rome, Dr AeroE, 28. *Prof Exp:* Naval architect & marine engr, Res & Exp Div Submarines, Ital Navy Yard, La Spezia, 29-31; naval archit marine eng supt ship hulls & engines, Libera Shipping Corp, Trieste & Genoa, 31-35; aeronaut engr & tech adv to govt, Chinese Comn Aeronaut Affairs, 35-37; dir mat test lab & supt tech voc instruct, Chinese Govt Cent Mil Aircraft Factory, Nanchang, 37-39; aeronaut engr & tech writer, Off Air Attache, Ital Embassy, Washington, DC, 39-41; mem fac aeronaut eng, Tri-State Col, 42; aeronaut engr design & develop, Aeronaut Prod, Inc, 43-44; sr aeronaut engr aerodyn & struct stress anal, Eng & Res Corp, 44-46; assoc prof mech eng, Univ Toledo, 46-47; assoc prof, Newark Col Eng, 47-52; prof mech eng & chmn, Dept Mech & Eng, Univ Southwestern La, 52-72; CONSULT ENGR, 72- *Mem:* AAAS; assoc fel Am Inst Aeronaut & Astronaut; Am Soc Mech Engrs; Soc Exp Stress Analysis; Instrument Soc Am. *Res:* Naval architecture; propulsion; turbomachinery; gas dynamics; aerothermodynamics; aeronautical structures; machine design; vibrations; hydrodynamics; marine engineering; internal combustion engines; turbines. *Mailing Add:* 142 Oak Crest Dr Lafayette LA 70503

CHIERICI, GEORGE J, b Napa, Calif, Nov 8, 26; m 56; c 4. PROSTHODONTICS. *Educ:* Univ Pac, DDS, 50. *Prof Exp:* asst clin prof, 63-68, from asst prof to prof, 68-85, ASST DEAN SCH DENT, UNIV CALIF, SAN FRANCISCO, 85- *Mem:* Int Asn Dent Res; Am Cleft Palate Asn. *Res:* Normal and abnormal growth and development; morphologic and physiologic interrelationships in the orofacial complex. *Mailing Add:* 513 Parnassus Rm S-630 Univ Calif Box 0430 San Francisco CA 94143

CHIGA, MASAHIRO, b Tokyo, Japan, Mar 6, 25; nat US; m 62; c 2. PATHOLOGY. *Educ:* Univ Tokyo, MD, 50. *Prof Exp:* Asst, Inst Infectious Dis, Univ Tokyo, 54; resident, Med Ctr, Univ Kans, 54-58; from asst prof to assoc prof path, Sch Med, Univ Utah, 60-69; assoc prof, 69-72, PROF PATH, SCH MED, UNIV KANS, 72- *Concurrent Pos:* Childs Mem Fund fel, 56-57; Nat Found fel, Metab Lab, Univ Utah, 58-60. *Mem:* Int Acad Path; Am Soc Exp Path. *Res:* Viral and rickettsial infection; experimental oncology; enzymes. *Mailing Add:* Dept Path Univ Kans Med Ctr Kansas City KS 66103

CHIGIER, NORMAN, b Frankfort, SAfrica, Aug 2, 33; US citizen; m 60; c 3. FLUIDS. *Educ:* Univ Witwatersrand, BSc, 52; Univ Cambridge, MA, 60, PhD(eng), 61, ScD, 77. *Prof Exp:* Asst res eng, Univ Cambridge, UK, 50-60; engr, Brit Thompson Houston Co, UK, 53-56; sr investr, Int Flame Res Found, 61-63; sr lectr aeronaut, Technion, Haifa, Israel, 64-66; lectr fuel technol, Univ Sheffield, UK, 66-81; sr res assoc aerodyn, NASA Ames Res Ctr, Calif, 70-71; W J BROWN PROF MECH ENG, CARNEGIE MELLON UNIV, 81- *Concurrent Pos:* Consult, Gen Motors Res Lab, 71-79 & Sandia Lab, Livermore, Calif, 76; founding ed, Progress in Energy & Combustion Sci, 75- & Atomization & Spray Technol, 89-; vis prof, Stanford Univ, Calif, 77, Ecole Centrale, Lyon, France, 77-78, Univ Calif, San Diego, 79 & Technion, Haifa, Israel, 88-89; chmn, Inst Liquid Atomization & Spray Systs, USA, 90-91. *Honors & Awards:* L F Moody Award, Am Soc Mech Engrs, 65; Lubbock-Sambrook Award, Inst Fuel, UK, 68 & 75; Tanasawa Award, Int Conf Liquid Atomization & Spray Systs, 88. *Mem:* Am Soc Mech Engrs; assoc fel Am Inst Aeronaut & Astronaut; Int Conf Liquid Atomization & Spray Systs (pres, 91-94); Combustion Inst; Sigma Xi. *Res:* Physics of disintegration of liquids; atomization; spray analysis; laser diagnostic techniques for velocity and particle size measurement; combustion aerodynamics; swirling flows; particle and fluid mechanics interaction. *Mailing Add:* Mech Eng Dept Carnegie Mellon Univ Pittsburgh PA 15213

CHIGNELL, COLIN FRANCIS, b London, Eng, Apr 7, 38; US citizen; m 66; c 2. ORGANIC CHEMISTRY, PHARMACOLOGY. *Educ:* Univ London, BPharm, 59, PhD(med chem), 62. *Prof Exp:* Vis fel, Nat Inst Arthritis, Metab & Digestive Dis, 62-65; vis assoc, Nat Heart & Lung Inst, 65-70, res pharmacologist, 70-77; CHIEF, LAB ENVIRON BIOPHYS, NAT INST ENVIRON HEALTH SCI, 77- *Concurrent Pos:* Res assoc, Nat Inst Gen Med Sci, 66-69. *Honors & Awards:* J J Abel Prize, Am Soc Pharmacol & Exp Therapeut, 73. *Mem:* Am Chem Soc; Am Soc Pharmacol & Exp Therapeut; Am Soc Biol Chemists; Biophys Soc; Soc Exp Biol & Med. *Res:* Spectroscopic studies of drug interactions with biological systems at a molecular level; structure and function of biological membranes and their interaction with drug molecules; mechanisms of drug photo toxicity. *Mailing Add:* Lab Environ Biophys PO Box 12233 Research Triangle Park NC 27709

CHIGNELL, DEREK ALAN, b London, Eng, July 4, 43; m 70; c 3. BIOCHEMISTRY. *Educ:* Kings Col, Univ London, BS, 64, PhD(biophys chem), 68; Wheaton Col, MA, 77. *Prof Exp:* Fel, Univ Calif, Los Angeles, 68-71; asst prof biochem, Univ Dundee, Scotland, 71-75; assoc prof, 75-85, PROF CHEM, WHEATON COL, ILL, 85- *Concurrent Pos:* Lectr, Med Sch, Loyola Univ, 76- *Mem:* Am Sci Affil; Am Chem Soc. *Res:* Enzymatic digestion of lens crystallins. *Mailing Add:* Dept Chem Wheaton Col Wheaton IL 60187-4295

CHIH, CHUNG-YING, b China, Dec 11, 16; nat US; m 55. PHYSICS. *Educ:* Nat Tsing Hua Univ, China, BSc, 37; Univ Calif, Berkeley, PhD(physics), 54. *Prof Exp:* Instr physics, Fukien Med Col, 37-40; assoc prof, Fukien Teachers Col, 40-44; prof, Nat Chi-nan Univ, 44-45 & Kiangsu Col, 45-48; physicist radiation lab, Univ Calif, 48-54; from asst prof to prof physics, Middlebury Col, 54-68; SCI CONSULT, 68- *Concurrent Pos:* NSF res grant, 57-60. *Mem:* Am Phys Soc. *Res:* Neutron proton scattering; elementary particles. *Mailing Add:* PO Box 2556 Noble Sta Bridgeport CT 06608

CH'IH, JOHN JUWEI, b Tsingtao, China, Oct 29, 33; US citizen; m 62; c 1. BIOLOGICAL CHEMISTRY. *Educ:* Southern Ill Univ, BA, 60; Univ Del, MS, 63; Thomas Jefferson Univ, PhD(biochem), 68. *Prof Exp:* Res technician, Biochem Res Found, Newark, Del, 60-63; clin chemist, St Mary's Hosp, Philadelphia, 63-65; teaching asst biochem, Thomas Jefferson Univ, 65-68, instr, 68-69; from sr instr to assoc prof, 69-81, PROF BIOL CHEM, HAHNEMANN UNIV SCH MED, 81- *Concurrent Pos:* Sr int fel, Fogarty Int Ctr, 78-79. *Mem:* Am Soc Biochem & Molecular Biol; Sigma Xi; AAAS; Am Chem Soc; Am Soc Cell Biol; Soc Exp Biol Med. *Res:* Regulatory mechanisms in nucleic acids and protein biosynthesis of the eukaryotic cells; biogenesis of mammalian cell organelles. *Mailing Add:* Dept Biol Chem Hahnemann Univ Sch Med Philadelphia PA 19102-1192

CHIHARA, CAROL JOYCE, b New York, NY, Oct 31, 41; m 64; c 1. DEVELOPMENTAL GENETICS. *Educ:* Univ Calif, Berkeley, BA, 62, PhD(develop genetics), 72; San Francisco State Univ, MA, 67. *Prof Exp:* NIH fel genetics, Cambridge Univ, 72-73; RESEARCHER DEVELOP BIOL, UNIV CALIF, BERKELEY, 74-; MEM STAFF, DEPT BIOL, UNIV SAN FRANCISCO. *Concurrent Pos:* Lectr cell & molecular biol, San Francisco State Univ, 74; vis scientist, Inst Jacques Monod, Nat Ctr Sci Res, 85-86. *Mem:* AAAS; Genetics Soc Am; Soc Develop Biol. *Res:* Genetic control mechanisms in Drosophila as utilized for the control of lasral cuticle proteins; effect of environmental factors and hormones as well as developmental capacities of the discs; mechanisms of active harmone-induced gene activity. *Mailing Add:* Dept Biol Univ San Francisco 2130 Fulton St San Francisco CA 94117

CHIHARA, THEODORE SEIO, b Seattle, Wash, Mar 14, 29; wid; c 5. MATHEMATICAL ANALYSIS. *Educ:* Seattle Univ, BS, 51; Purdue Univ, MS, 53, PhD(math), 55. *Prof Exp:* Asst math, Purdue Univ, 52-55; from asst prof to prof, Seattle Univ, 55-69, actg head dept, 58-59, head, 59-66; prof assoc, Univ Alta, 69-70; vis prof, Univ Victoria, BC, 70-71; chmn dept, 71-77, PROF MATH, PURDUE UNIV, CALUMET, 71- *Mem:* Am Math Soc; Math Asn Am; Soc Indust Appl Math. *Res:* Theory of orthogonal polynomials; moment problems; special functions. *Mailing Add:* Dept Math Purdue Univ Hammond IN 46322

CHIKALLA, THOMAS D(AVID), b Milwaukee, Wis, Sept 9, 35; m 60; c 3. PHYSICAL METALLURGY, CERAMICS. *Educ:* Univ Wis, BS, 57, PhD(metall), 66; Univ Idaho, MS, 60. *Prof Exp:* Res engr plutonium metall, Gen Elec Co, 57-62; sr res scientist ceramics, 64-67, res assoc 67-72, mgr, Ceramics & Graphite Sect, 72-80, mgr nucelar waste technol, 80-83, mgr, Chem Technol Dept, 83-86, ASSOC DIR, FACIL & OPERS, BATTELLE MEM INST, 86- *Concurrent Pos:* Fel, AEC. *Mem:* Am Inst Mining Metall

& Petrol Engrs; fel Am Ceramic Soc; Sigma Xi; AAAS; Am Soc Metals. *Res:* High temperature phase equilibria in actinide oxide and carbide systems; vaporization behavior and thermodynamics of actinide oxides; crystal structures of nonstoichiometric compounds; materials development for nuclear waste management; nuclear fuels development. *Mailing Add:* 2108 Harris Richland WA 99352

CHIKARAISHI, DONA M, b Chicago, Ill, Jan 6, 47; m 88. NEUROBIOLOGY. *Educ:* Univ Calif, San Diego, PhD(biol), 73. *Prof Exp:* Asst prof biochem, Univ Colo, 80-83; asst prof neurol, 83-85, ASSOC PROF NEUROL, TUFTS SCH MED, 85- *Mem:* Soc Neurosci; Am Soc Microbiol; Am Soc Biol Chemists. *Mailing Add:* Neurosci Prog Tufts Sch Med 136 Harrison St Boston MA 02111

CHIKO, ARTHUR WESLEY, b Yorkton, Sask, Feb 15, 38. PLANT VIROLOGY. *Educ:* Univ BC, BSc, 61; Univ Idaho, MS, 67, PhD(plant path), 70. *Prof Exp:* Res officer forest path, Can Dept Forestry, Forest Entom & Path Lab, Fredericton, NB, 61-63; res scientist plant virol, Res Sta, Winnipeg, Manitoba, Agr Can, 70-79, Saanichton Res & Plant Quarantine Sta, Sidney, BC, 80-; RETIRED. *Mem:* Am Phytopath Soc; Can Phytopath Soc. *Res:* Virus diseases of oranamentals. *Mailing Add:* 2-2654 Lancelot Dr Saanichton BC Z0S 1M0 Can

CHILCOTE, DAVID OWEN, b Sask, Can, July 8, 31; m 53; c 2. CROP PHYSIOLOGY. *Educ:* Ore State Univ, BS, 53, MS, 57; Purdue Univ, PhD(agron), 61. *Prof Exp:* From instr to prof farm crops, 53-87, EMER PROF CROP SCI, ORE STATE UNIV, 87- *Mem:* Am Soc Plant Physiol; Am Soc Agron; fel Crop Sci Soc Am. *Res:* Herbicides and growth regulators; crop physiology and ecology. *Mailing Add:* Dept Crop Sci Ore State Univ Corvallis OR 97331

CHILCOTE, MAX ELI, b Bemidji, Minn, Sept 1, 17; m 43; c 3. CLINICAL CHEMISTRY, LABORATORY MEDICINE. *Educ:* Univ Minn, BS, 38, MS, 41; Univ Mich, PhD(biol chem), 44. *Prof Exp:* Asst biochem, Univ Mich, 40-44; instr, Med Sch, Loyola Univ, Ill, 44-46; Nutrit Found res fel, Pa State Col, 46-48; from asst prof to assoc prof biochem, Sch Med State Univ NY Buffalo, 48-60, asst dir biochem lab, 59-66, dir clin biochem, 66-69, assoc dir, 69-70, DIR ERIE COUNTY LABS, 70- *Concurrent Pos:* Clin assoc prof, 60-70, clin prof biochem, 70- & Dept of Pathol, Sch Med, State Univ NY Buffalo, 74- *Honors & Awards:* Educ Award, Am Asn Clin Chem, 75. *Mem:* Am Asn Clin Chem; Can Soc Clin Chem; Am Chem Soc; Acad Clin Lab Physicians & Scientists (pres, 73-75). *Res:* Clinical chemistry. *Mailing Add:* 55 Sherwin Dr Tonawanda NY 14150-4714

CHILCOTE, WILLIAM W, b Washington, Iowa, Mar 6, 18; m 46; c 2. PLANT ECOLOGY. *Educ:* Iowa State Col, BS, 43, PhD(bot), 50. *Prof Exp:* Instr forestry, Iowa State Col, 46-50; asst prof bot & asst ecologist, Agr Exp Sta Ore State Univ, 50-56, assoc prof bot & assoc ecologist, 56-62, prof bot, 62-81; RETIRED. *Concurrent Pos:* Fulbright grant, Finland, 58-59. *Mem:* Ecol Soc Am; Soc Am Foresters. *Res:* Forest and range ecology; autecology; community dynamics. *Mailing Add:* 3610 NW Van Buren Corvallis OR 97330

CHILD, CHARLES GARDNER, III, surgery; deceased, see previous edition for last biography

CHILD, EDWARD T(AYLOR), b Richmond, Va, July 9, 30; m 55; c 5. CHEMICAL ENGINEERING. *Educ:* Yale Univ, BE, 52; Univ Del, MS, 54, PhD(chem eng), 57; Pepperdine Univ, MBA, 72. *Prof Exp:* Res chem engr, Texaco Res Ctr, NY, 56-68, supvr res, 68-73, asst mgr, Texaco Develop Corp, 73-82, MGR LICENSING, TEXACO INC, 83- *Mem:* Fel Am Inst Chem Engrs; Sci Res Soc Am. *Res:* Process development, engineering and licensing of technology for conversion of coal and heavy petroleum residues to clean gaseous products for the production of chemicals and electric power. *Mailing Add:* Texaco Develop Corp 2000 Westchester Ave White Plains NY 10650

CHILD, FRANK MALCOLM, b Jersey City, NJ, Nov 30, 31; m 60; c 3. CELL BIOLOGY. *Educ:* Amherst Col, AB, 53; Univ Calif, PhD(zool), 57. *Prof Exp:* Instr zool, Univ Chicago, 57-60, asst prof, 60-65; assoc prof biol, 65-73, chmn dept, 74-78, PROF BIOL, TRINITY COL, CONN, 73- *Mem:* Soc Protozool; Am Soc Cell Biol; Am Soc Zoologists; Sigma Xi. *Res:* Protozoan physiology; developmental and cellular biology; cilia and flagella. *Mailing Add:* Dept Biol Trinity Col Hartford CT 06106

CHILD, HARRY RAY, b Bedford, Ind, Oct 30, 28; m 54; c 1. SOLID STATE PHYSICS, MAGNETISM. *Educ:* Univ Tex, BS, 56; Univ Tenn, PhD(physics), 65. *Prof Exp:* PHYSICIST, OAK RIDGE NAT LAB, 56- *Mem:* Am Phys Soc. *Res:* Neutron scattering studies of solids, mostly magnetic properties. *Mailing Add:* Solid State Div 7962 6393 Oak Ridge Nat Lab PO Box 2008 Oak Ridge TN 37831-6031

CHILD, JEFFREY JAMES, b Gateshead, Eng, June 26, 36; m 58; c 2. MICROBIOLOGY. *Educ:* Univ Durham, BSc, 58, PhD, 62. *Prof Exp:* Fel, Prairie Regional Lab, Nat Res Coun Can, 62-63; asst lectr biol, Univ Salford, 63-64, lectr microbiol, 64-67; assoc res officer, 67-76, sr res officer, 76-80, ADMINR, NAT RES COUN CAN, OTTAWA, 80- *Concurrent Pos:* Vis scientist, Commonwealth Sci & Indust Res Orgn, Australia, 74-75. *Honors & Awards:* Medal Award, Can Soc Microbiol, 79. *Mem:* Can Soc Microbiol; Brit Soc Gen Microbiol; Am Soc Microbiol. *Res:* Physiology of fungi; microbiological degradation of natural products; symbiotic nitrogen fixation. *Mailing Add:* Triumph 4004 Westbrook Mall Univ BC Campus Vancouver BC V6T 2A3 Can

CHILD, PROCTOR LOUIS, b Brooklyn, NY, Nov 29, 25; m 52; c 3. PATHOLOGY. *Educ:* Long Island Col Med, MS, 49. *Prof Exp:* Intern, St Agnes Hosp, White Plains, NY, US Army, 49-50, gen med officer, 155th Sta Hosp, Japan, 50, battalion & regimental surgeon, 1st Cavalry Div, Korea, 50-51, gen med officer, Army Hosp, Camp Pickett, Va, 51-52, path resident, Fitzsimons Gen Hosp, Denver, 52-56, chief path serv, 5th Gen Hosp, Stuttgart, 56-58, 130th Sta Hosp, Heidelberg, 58-60, chief path, William Beaumont Gen Hosp, El Paso, 60-64, mem geog path div, Armed Forces Inst Path, 64-66, chief viro-path br & asst chief geog path div, 67; assoc prof path, Sch Med, Temple Univ, 68-80; ASSOC PATHOLOGIST, ALLENTOWN GEN HOSP, PA, 80- *Concurrent Pos:* Mem staff, Roxborough Mem Hosp, Philadelphia, 70- *Mem:* AMA; Am Soc Clin Path; Col Am Path; Int Acad Path. *Res:* Geographic pathology and infectious disease, especially viro-pathology and study of hemorrhagic fevers. *Mailing Add:* 1148 Clearwood Dr Allentown PA 18103

CHILD, RALPH GRASSING, b New York, NY, Oct 7, 19; m 44; c 3. MEDICINAL CHEMISTRY. *Educ:* Hofstra Col, BA, 41; George Washington Univ, MA, 48; Univ Iowa, PhD(org chem), 50. *Prof Exp:* Res chemist, Lederle Labs, Am Cyanamid Co, 50-74, sr res chemist, 74-85, RETIRED. *Mem:* Am Chem Soc; Sigma Xi. *Res:* Medicinal organic chemistry; chemotherapy of virus and neoplastic diseases; chemistry of antibacterial compounds; immune response inhibitors and stimulators; antiinflamatory-analgesics. *Mailing Add:* 22541 Tuckahoe Rd Alva FL 33920

CHILDERS, DONALD GENE, b The Dalles, Ore, Feb 11, 35; m 53; c 2. ELECTRICAL ENGINEERING. *Educ:* Univ Southern Calif, BS, 58, MS, 59, PhD(elec eng), 64. *Prof Exp:* Mem tech staff res & develop, Aeronutronic Div, Ford Motor Co, 58-60 & 61-64 & Hughes Aircraft Co, 60-61; asst prof elec eng, Univ Calif, Davis, 64-65; assoc prof, 65-68, PROF ELEC ENG, UNIV FLA, 68- *Concurrent Pos:* Lectr, Univ Southern Calif, 62-64. *Honors & Awards:* George Westinghouse Award, Am Soc Engr Educ, 75; William J Morlock Award Biomed Engr, Inst Elec & Electronics Engrs, 73. *Mem:* Inst Elec & Electronics Engrs; Acoust Soc Am. *Res:* Computer algorithms, signal processing, spectral analysis, speech analysis, synthesis, recognition; biomedical engineering; visual evoked responses; human-machine interaction. *Mailing Add:* Dept Elec Eng/405 CSE Univ of Fla Gainesville FL 32611

CHILDERS, NORMAN FRANKLIN, b Moscow, Idaho, Oct 29, 10; c 4. HORTICULTURE. *Educ:* Univ Mo, BS, 33, MS, 34; Cornell Univ, PhD(pomol), 37. *Prof Exp:* Asst pomol, Cornell Univ, 34-37; asst prof hort, Ohio State Univ, 37-44, assoc, Ohio Exp Sta, 39-44; asst dir & sr plant physiologist, PR Exp Sta, USDA, 44-47; prof, res specialist & chmn, Dept Hort & Forestry, 48-66, Blake prof hort, Agr Exp Sta, 66-80, emer prof, Rutgers Univ, New Brunswick, 80; MEM STAFF, DEPT FRUIT CROPS, UNIV FLA, GAINESVILLE, 80- *Concurrent Pos:* Adj prof, Univ Fla, Gainesville, 81- *Honors & Awards:* Ware Award, Am Soc Hort Sci. *Mem:* AAAS; Am Soc Plant Physiologists; fel Am Soc Hort Sci. *Res:* Photosynthesis; transpiration; respiration; nutrition of fruits and other horticultural plants; tropical and temperate pomology; tropical vegetables; nightshades effects on arthritis. *Mailing Add:* Fruit Crops Dept Univ Fla Gainesville FL 32611

CHILDERS, RAY FLEETWOOD, b Los Angeles, Calif, Apr 16, 45; m 71; c 2. PHARMACEUTICAL CHEMISTRY. *Educ:* Univ Calif, Los Angeles, BS, 67; Ind Univ, PhD(inorg chem), 72. *Prof Exp:* Res assoc biophys chem, Ind Univ, 72-74; sr analytical chemist, 74-78, RES SCIENTIST, ELI LILLY & CO, 79- *Mem:* Am Chem Soc; AAAS; Controlled Release Soc. *Res:* Automated flow analysis of pharmaceuticals; activity and structure of biological molecules using nuclear magnetic resonance; pharmaco-kinetics; development of modified release, oral pharmaceuticals. *Mailing Add:* Eli Lilly & Co Dept IC 747 Bldg 100/2 Indianapolis IN 46285-0002

CHILDERS, RICHARD LEE, b Birmingham, Ala, Dec 10, 30; m 60; c 2. PHYSICS. *Educ:* Presby Col, SC, BS, 53; Univ Tenn, MS, 56, PhD(particle physics), 62. *Prof Exp:* Res assoc physics, Univ Tenn, 61-63; from asst prof to assoc prof, 63-81, PROF PHYSICS, UNIV SC, 82- *Concurrent Pos:* Consult, Neutron Physics Div, Oak Ridge Nat Lab, 62-64; dir, Honors Prog, 67-69. *Mem:* Am Phys Soc; Am Asn Physics Teachers; Am Inst Physics. *Res:* Electron-position colliding beam experiment; acoustics of musical instruments and rooms. *Mailing Add:* Dept Physics Univ SC Columbia SC 29208

CHILDERS, ROBERT LEE, b Parkersburg, WVa, May 3, 36; div; c 2. PHOTOGRAPHIC CHEMISTRY. *Educ:* WVa Univ, BS, 60, MS, 61; Ohio State Univ, PhD(org chem), 65. *Prof Exp:* Sr res chemist, Eastman Kodak Co, 65-71, res assoc, 71-91, TECH ASSOC, KODAK PATHE, FRANCE, 91- *Mem:* Am Chem Soc. *Res:* Chemistry of photographic emulsions and photographic processing. *Mailing Add:* 343 State St NJ 160 Rochester NY 14650

CHILDERS, ROBERT WAYNE, b Ft Worth, Tex, May 25, 37; m 62; c 3. THEORETICAL PHYSICS. *Educ:* Howard Payne Col, BA, 60; Vanderbilt Univ, PhD(physics), 63. *Prof Exp:* Res fel, Argonne Nat Lab, 63-65; asst prof, 65-72, ASSOC PROF PHYSICS, UNIV TENN, KNOXVILLE, 65- *Concurrent Pos:* Consult, Oak Ridge Nat Lab, 66- *Mem:* Am Phys Soc. *Res:* Elementary particle physics; theory of infinitely rising Regge trajectories and local duality; Veneziano model; quantum field theory; symmetries of elementary particles. *Mailing Add:* Dept Physics & Astron Univ Tenn Knoxville TN 37916

CHILDERS, RODERICK W, b Paris, France, 31; c 1. MEDICINE, CARDIOLOGY. *Educ:* Univ Dublin, BA, 53, MD, 54, MA, 58; Am Bd Cardiovasc Dis, cert, 69. *Prof Exp:* Intern med, St Andrew's Hosp, London, Eng, 54-55; intern surg, May Day Hosp, Croydon, 55-56; chief cardiologist, Royal City Dublin Hosp, Ireland, 59-63; asst prof cardiol, 63-69, assoc prof med, 69-74, HEAD HEART STA, SCH MED, UNIV CHICAGO, 66-, PROF MED, 74- *Concurrent Pos:* Fel, Harvard Univ & West Roxbury Vet Admin Hosp, Mass, 58-59; res assoc nutrit, Sch Pub Health, Harvard Univ, 59-62; med tutor cardiol, Med Sch, Univ Dublin, 59-63; cardiac asst, Nat Children's Hosp, Dublin & vis pediat cardiologist, Rotunda Maternity Hosp, 59-63. *Mailing Add:* Box 161 950 E 59th St Chicago IL 60637

CHILDERS, STEVEN ROGER, b Houston, Tex, Sept 28, 50; m 76. NEUROPHARMACOLOGY, NEUROCHEMISTRY. *Educ:* Univ Tex, Austin, BS, 72; Univ Wis, PhD(physiol chem), 76. *Prof Exp:* Fel pharmacol, Sch Med, Johns Hopkins Univ, 76-79; asst prof, 79-85, ASSOC PROF PHARMACOL, COL MED, UNIV FLA, 85- *Mem:* Soc Neurosci; AAAS. *Res:* Neuropharmacology and neurochemistry; characterization of opioid and other neuropeptide systems in mammalian brain. *Mailing Add:* Dept Pharmacol Col Med Univ Fla Gainesville FL 32610

CHILDERS, WALTER ROBERT, b Kelowna, BC, Mar 29, 16; m 53; c 2. PLANT BREEDING, PLANT GENETICS. *Educ:* McGill Univ, BSc, 38; Univ Wis, MS, 47, PhD(plant breeding & genetics), 51. *Prof Exp:* Asst corn & soybeans, Forage Crop Div, 38-40, CHIEF FORAGE SECT, OTTAWA RES STA, CAN DEPT AGR, 46- *Mem:* Am Soc Agron; Agr Inst Can; Can Soc Genetics; Can Soc Agron; Can Phytopath Soc. *Res:* Orchard and brome grass; timothy; cytology and genetics; alfalfa. *Mailing Add:* 232 Remic Ottawa ON K1Z 5W5 Can

CHILDRESS, CHARLES CURTIS, biochemistry, biomedical engineering, for more information see previous edition

CHILDRESS, DENVER RAY, b Alcoa, Tenn, Feb 5, 37; m 57; c 2. PURE MATHEMATICS. *Educ:* Maryville Col, Tenn, BS, 59; Univ Tenn, Knoxville, MMath, 64, EdD(math educ), 75. *Prof Exp:* Teacher math, Powell High Sch, Knox County, Tenn, 59-60, Maryville Jr High Sch, Maryville, Tenn, 60-62 & Maryville High Sch, 62-65; from asst prof to assoc prof, 67-81, PROF MATH, CARSON-NEWMAN COL, 81- *Mem:* Am Asn Univ Prof; Nat Coun Teachers Math; Math Asn Am. *Res:* Factors which influence student ratings of mathematics teaching and teachers, especially attitudes. *Mailing Add:* Rte 1 Box 46 New Market TN 37820

CHILDRESS, DUDLEY STEPHEN, b Cass Co, Mo, Sept 25, 34; m 59; c 2. BIOMEDICAL ENGINEERING. *Educ:* Univ Mo-Columbia, BS, 57, MS, 58; Northwestern Univ, PhD(elec eng), 67. *Prof Exp:* From instr to asst prof elec eng, Univ Mo-Columbia, 59-63; res asst, Physiol Control Syst Lab, Northwestern Univ, Evanston, 64-66, from asst prof to assoc prof elec eng & orthoped surg, 72-77, PROF ELEC ENG, TECHNOL INST & PROF ORTHOPED SURG RES, MED SCH, NORTHWESTERN UNIV, CHICAGO, 77-, DIR, PROSTHETICS RES LAB, 71-, CO-DIR, REHAB ENG PROG, 72- *Concurrent Pos:* Mem comt prosthetics res & develop, Nat Acad Sci-Nat Res Coun, 69-72, mem sub-comt design, 70-73, chmn upper-extremity prosthetics panel, 71-73; Nat Inst Gen Med Sci res career develop award, 70-75; mem appl physiol & bioeng study sect, NIH, 74-78. *Honors & Awards:* Goldenson Award Res in Med & Technol, United Cerebral Palsy Found. *Mem:* AAAS; Inst Elec & Electronics Eng; Biomed Eng Soc; Rehab Eng Soc NAm; Int Soc Prosthetics & Orthotics; Sigma Xi. *Res:* Rehabilitation engineering, design and development of modern technological systems for disabled people and scientific approach to analysis and description of problems of these people. *Mailing Add:* 112 Dupee Pl Wilmette IL 60091

CHILDRESS, EVELYN TUTT, b Joplin, Mo, Feb 8, 26; m 67. MICROBIOLOGY, IMMUNOLOGY. *Educ:* Lincoln Univ, Mo, BS, 47; Univ Mich, MS, 48, MS, 56; Stanford Univ, PhD(med microbiol), 67. *Prof Exp:* Instr biol, Fla Agr & Mech Univ, 48-49; instr, Lincoln Univ, Mo, 49-52, asst prof, 52-63; mem fac, Fullerton Jr Col, 67-69; asst prof, 69-72, assoc prof biol, 72-77, CALIF STATE UNIV, DOMINGUEZ HILLS, 77- *Mem:* Am Soc Microbiol; Sigma Xi; AAAS; Am Asn Univ Prof. *Res:* Aging in the immune system; origin of naturally occuring antibodies, particularly with their specificity and the question of necessity of antigenic stimulation for their appearance; schistosome immunology. *Mailing Add:* 4623 Don Miguel Dr Los Angeles CA 90008

CHILDRESS, JAMES J, b Kokomo, Ind, Nov 17, 42; m 90. COMPARATIVE PHYSIOLOGY, BIOLOGICAL OCEANOGRAPHY. *Educ:* Wabash Col, BA, 64; Stanford Univ, PhD(biol), 69. *Prof Exp:* From asst prof to assoc prof zool, 69-80, PROF UNIV CALIF, SANTA BARBARA, 81- *Concurrent Pos:* Prin investr, NSF grant, 70-72; NSF grants, 70-; ONR grant, 87-90. *Mem:* Fel AAAS; Am Soc Zoologists; Am Soc Limnol & Oceanog; Sigma Xi; Crustacean Soc; Oceanog Soc; Am Geophys Union. *Res:* Ecological physiology of marine invertebrates and fishes; respiratory physiology; deep-sea biology; effects of hydrostatic pressure on organisms; biology of hydrothermal vent animals; physiology of locomotion in fishes and crustaceans. *Mailing Add:* Dept Biol Univ Calif Santa Barbara CA 93106

CHILDRESS, OTIS STEELE, JR, b Richmond, Va, Dec 27, 36; m 62; c 2. ENGINEERING MANAGEMENT, SYSTEMS ENGINEERING. *Educ:* Va Polytech Inst, BSEE, 62. *Prof Exp:* Aero-decelerator engr, Advan Missions Studies Off, Langley Res Ctr, 67-68, sterilization engr bioeng, Viking Proj Off, 68-70, orbiter sci instruments mgr eng mgt, 70-73, Viking orbiter mgr, 73-76, head, Proj Integration Off Systs Eng, Projs Directorate, 76-78, dep prof mgr & chief engr, Rotor Systs Res Aircraft Proj, 78-80, mgr, Propfan Noise Prog, 80-82, MGR, ROTORCRAFT NOISE REDUCTION RES PROG, LANGLEY RES CTR, NASA, 82- *Concurrent Pos:* Exec mem, Orbiter Imaging Team, Mars Atmospheric Water Detection Team & Mars Infrared Thermal Mapping Team, 70-73; Secy Gov-Indust Working Group on Rotorcraft Noise, 82- *Honors & Awards:* Spec Achievement Award, NASA, 81. *Mem:* Am Helicopter Soc. *Res:* Management and engineering research and development efforts in aeronautics and space. *Mailing Add:* 119 National Lane Williamsburg VA 23185

CHILDRESS, SCOTT JULIUS, b Greenville, SC, Apr 6, 26; m 75. ORGANIC CHEMISTRY. *Educ:* Furman Univ, BS, 47; Univ NC, PhD(chem), 51. *Prof Exp:* Res chemist catalysis, Tenn Eastman Co, 51-52; pharmaceut, Wallace & Tiernan, Inc, 52-58; res chemist pharmaceut, 59-60, group leader, 60-61, mgr med chem sect, 61-68, asst to vpres res & develop, 68-74, asst vpres res & develop, 74-85; RETIRED. *Mem:* Fel NY Acad Sci; Am Chem Soc. *Res:* Design, synthesis and testing of organic compounds of possible therapeutic value. *Mailing Add:* 2202 Hopkinson House Philadelphia PA 19106-4152

CHILDRESS, WILLIAM STEPHEN, b Houston, Tex, Oct 5, 34; m 60, 71; c 4. FLUID MECHANICS, APPLIED MATHEMATICS. *Educ:* Princeton Univ, BSE, 56, MSE, 58; Calif Inst Technol, PhD(aeronaut, math), 61. *Prof Exp:* Assoc res Scientist, Jet Propulsion Lab, 61-64; res assoc magneto-fluid dynamics, Courant Inst Math Sci, 64-66, asst prof, 66-70, assoc prof, 70-76, PROF MATH, NY UNIV, 76- *Concurrent Pos:* Assoc, Inst Henri Poincare, Univ Paris, 67-68; Guggenheim fel, 76-77. *Mem:* Am Math Soc; Soc Indust & Appl Math; Sigma Xi. *Res:* Singular perturbation problems in fluid dynamics and applied mathematics; magnetohydrodynamics; dynamo theory of geomagnetism; viscous flow theory; biomathematics. *Mailing Add:* NY Univ 251 Mercer St New York NY 10003

CHILDS, BARTON, b Chicago, Ill, Feb 29, 16; m 50; c 2. PEDIATRICS, GENETICS. *Educ:* Williams Col, AB, 38; Johns Hopkins Univ, MD, 42. *Prof Exp:* Intern, asst resident & resident pediat, Johns Hopkins Hosp, 42-43, 46-48; res fel, Children's Hosp, Boston, 48-49; mem fac, 49-62, prof sch med, 62-81, EMER PROF PEDIAT, JOHNS HOPKINS UNIV, 81- *Concurrent Pos:* Commonwealth Fund fel, Univ Col, Univ London, 52-53; Markle scholar, 53-58; mem, NIH Consult Comts, 59-; Grover F Powers distinguished scholar, 60-62; mem res adv comt, United Cerebral Palsy Found, 60-63; NIH res career award, 62. *Honors & Awards:* Meade Johnson Award Pediat, NIH, 59; Allen Award, Am Soc Human Genetics, 73; Howland Award, Am Pediat Soc, 89. *Mem:* Inst Med Nat Acad Sci; Am Pediat Soc; Soc Pediat Res; Am Acad Pediat; Am Acad Arts & Sci. *Mailing Add:* Dept Pediat Johns Hopkins Hosp Baltimore MD 21218

CHILDS, DONALD RAY, b Lynn, Mass, May 31, 30; m 64; c 2. MATHEMATICAL PHYSICS. *Educ:* Univ NH, BS, 52, MS, 54; Vanderbilt Univ, PhD(physics), 58. *Prof Exp:* Scientist, Westinghouse Elec Corp, 57-59; sr scientist, Allied Res Assocs, 59-60; prin res scientist, Avco-Everett Res Lab, 60-64; sr scientist, Lab Electronics, 64-66 & Quincy Div, Elec Boat Co, 66-69; PHYSICIST, NAVAL UNDERWATER SYST CTR, 69- *Concurrent Pos:* Spec lectr, Northeastern Univ, 67-69; instr, Roger Williams Col, 87-90. *Res:* Multivariate analysis; signal processing; non-linear mechanics; non-linear control systems; non-linear differential equations. *Mailing Add:* 115 Ethel Dr Portsmouth RI 02871-3921

CHILDS, JAMES FIELDING LEWIS, b Tucson, Ariz, Jan 3, 10; m 36; c 4. PLANT PATHOLOGY. *Educ:* Univ Calif, BS, 37, PhD(plant path), 41. *Prof Exp:* Agent, USDA, 41-43, from asst pathologist to prin pathologist, 43-66, res pathologist, Agr Res Serv, 66-76; RETIRED. *Concurrent Pos:* Consult, Egypt, 55, Morroco, 59, Surinam, 63 & Sudan, 64; ed proc, Int Orgn Citrus Virol, 66. *Mem:* Am Phytopath Soc; Int Orgn Citrus Virol. *Res:* Etiology, virus indexing procedures and programs; control of virus diseases; Rio Grande gummosis, its nature and control; citrus blight (YTD, RLD, SHD), its nature and control. *Mailing Add:* 1206 Nottingham St Orlando FL 32803

CHILDS, LINDSAY NATHAN, b Boston, Mass, Apr 17, 40; m 72; c 3. MATHEMATICS. *Educ:* Wesleyan Univ, BA, 62; Cornell Univ, PhD(math), 66. *Prof Exp:* Asst prof math, Northwestern Univ, 66-68; assoc prof math, 71-80, chmn dept math & statist, 81-84, PROF MATH, STATE UNIV NY, ALBANY, 71- *Mem:* Am Math Soc; Math Asn Am. *Res:* Algebra; number theory. *Mailing Add:* Dept Math State Univ NY Albany NY 12222

CHILDS, MARIAN TOLBERT, b Twin Falls, Idaho, Nov 18, 25; m 52; c 4. LIPID METABOLISM. *Educ:* Univ Calif, Berkeley, BS, 46, PhD(nutrit), 50. *Prof Exp:* Teaching & res asst nutrit, Univ Calif, Berkeley, 46-50; asst prof nutrit, Univ Ill, 50-54; lab tech lipid biochem, Univ Wash, 66-68, from actg asst prof to assoc prof, 68-90, EMER ASSOC PROF NUTRIT, UNIV WASH, 91- *Concurrent Pos:* NIH fel nutrit & atherosclerosis, Dept Med, Univ Wash, 76-78. *Mem:* Sigma Xi; Am Inst Nutrit. *Res:* Lipid metabolism, primarily the interaction of nutrients with lipids; phosphatidy choline turnover studies; hyperlipidemia of pregnancy in rats used turnover and removal studies; omega 3 fatty acids, shellfish. *Mailing Add:* Dept Nutrit Sci/ Med DL10 Univ Wash Seattle WA 98195

CHILDS, MORRIS E, b Yellville, Ark, Mar 30, 23; m 52; c 4. MECHANICAL ENGINEERING. *Educ:* Univ Okla, BS, 44; Univ Ill, MS, 47, PhD(mech eng), 56. *Prof Exp:* Res assoc mech eng, Univ Ill, 47-54; from asst prof to assoc prof, 54-61, chmn dept, 73-80, PROF MECH ENG, UNIV WASH, 61-,. *Mem:* Am Soc Eng Educ; fel Am Soc Mech Engrs; Am Inst Aeronaut & Astronaut. *Res:* Thermodynamics; fluid flow; heat transfer; turbulent boundary layer flow; separated flows. *Mailing Add:* 7857 56th Pl NE Seattle WA 98195

CHILDS, ORLO E, b Loa, Utah, Mar, 28, 14; m 45; c 3. GEOLOGY. *Educ:* Univ Utah, BS, 35, MS, 37; Univ Mich, PhD(geol), 45. *Prof Exp:* Instr geol, Weber State Col, 37-42; geologist, Sinclair-Wyo Oil Co, 45-47; asst prof geol, Colgate Univ, 47-49; asst prof geol, Univ Wyo, 49-50; dir explor projs, Phillips Petrol Co, 50-61; br chief geologist marine geol, US Geol Surv, 61-63; pres, Colo Sch Mines, 63-70; vpres res & Univ prof, Tex Tech Univ, 70-79; dir, 79-85, EMER DIR MINING & MINERALS RESOURCES RES INST, UNIV ARIZ, 85- *Concurrent Pos:* Mem, Oil Shale Comn, Dept Interior, 65 & gen tech adv comt, Off Coal Res, 71-75, rev comt, Cong Pub Land Law, 65-68; mem & chmn, Gulf Coast Univ Res Comt, 71-74. *Mem:* Hon mem Am Asn Petrol Geologists (vpres, 61, pres, 65-66); Geol Soc Am; Am Inst Prof Geologists; Am Asn Petrol Geologists (vpres, 61, pres, 65-66). *Res:* Completed a ten year research project for the American Association of Petroleum Geologists involving 450 geologists and 570 sites in the United States, including Alaska. *Mailing Add:* 7020 N Camino Fray Marcos Tucson AZ 85718

CHILDS, RONALD FRANK, b Liss, Eng, Nov 30, 39; m 65; c 2. PHYSICAL ORGANIC CHEMISTRY. *Educ:* Bath Univ Technol, BSc, 66; Univ Nottingham, PhD(org chem), 66, DSc, 84. *Prof Exp:* Fel, Univ Calif, Los Angeles, 66-68; from asst prof to assoc prof chem, McMaster Univ, 68-78, chmn dept, 82-84, dean fac sci, 84-89, PROF CHEM, MCMASTER UNIV, 78-, VPRES RES, 89- *Mem:* Am Chem Soc; Chem Soc; fel Chem Inst Can. *Res:* Physical organic chemistry, particularly thermal and photochemical rearrangements of carbonium ions. *Mailing Add:* Dept Chem McMaster Univ Hamilton ON L8S 4L8 Can

CHILDS, S(ELMA) BART, b Magnolia, Ark, Jan 3, 38; m 70. COMPUTER SCIENCE, ENGINEERING. *Educ:* Okla State Univ, BS, 59, MS, 60, PhD(eng mech), 66. *Prof Exp:* Instr civil eng, Okla State Univ, 61-64; res engr, Space & Info Systs Div, NAm Aviation, Inc, Okla, 64-65; asst prof mech eng, Univ Houston, 65-68, assoc prof & assoc chmn dept, 68-71; prof appl math & comput sci & chmn dept, Speed Sci Sch, Univ Louisville, 71-74; PROF COMPUT SCI, TEX A&M UNIV, 74- *Mem:* Soc Indust & Appl Math; Asn Comput Mach. *Res:* Numerical and applied mathematics; boundary value problems, codes for analysis; computerized typesetting. *Mailing Add:* Indust Eng Dept Tex A&M Univ College Station TX 77843-3112

CHILDS, WILLIAM HENRY, pomology; deceased, see previous edition for last biography

CHILDS, WILLIAM HENRY, b Kingsport, Tenn, June 2, 41; m 67; c 2. ELECTRICAL ENGINEERING. *Educ:* Mass Inst Technol, BS, 62; Iowa State Univ, MS, 65, PhD(elec eng), 70. *Prof Exp:* Electronic engr, Naval Weapons Ctr, China Lake, Calif, 70-74; mem tech staff, COMSAT Labs, Clarksburg, Md, 74-80; mgr, CAD/CAM Amplica, Newbury Park, Calif, 80-82; EXEC VPRES, EESOF INC, WESTLAKE VILLAGE, CALIF, 82- *Mem:* Inst Elec & Electronics Eng; Sigma Xi. *Res:* Developer and inventor of microwave CAD/CAE software. *Mailing Add:* Eesof Inc 5601 Lindero Canyon Rd Westlake Village CA 91362

CHILDS, WILLIAM JEFFRIES, b Boston, Mass, Nov 9, 26; m 51; c 2. ATOMIC PHYSICS. *Educ:* Harvard Univ, AB, 48; Univ Mich, MS, 49, PhD(physics), 56. *Prof Exp:* PHYSICIST, ARGONNE NAT LAB, 56- *Concurrent Pos:* Vis prof, Univ Bonn, Ger, 72-73. *Mem:* Fel Am Phys Soc; Optical Soc Am. *Res:* Atomic-beam magnetic resonance; hyperfine structure; laser spectroscopy; atomic and molecular structure; laser radiofrequency double resonance. *Mailing Add:* Argonne Nat Lab 9700 S Cass Ave Argonne IL 60439

CHILDS, WILLIAM VES, b Cale, Ark, Sept 14, 35; m 62; c 3. FLUORINE CHEMISTRY, ELECTROCHEMISTRY. *Educ:* Southern State Col, BS, 56; Univ Ark, MS, 60, PhD(phys chem), 63. *Prof Exp:* Sr chemist, Phillips Petrol Co, 62-84; DIV SCIENTIST, 3M CO, 84- *Concurrent Pos:* Vis scientist, Univ Tex, 69-70; consult, NIH, Lung & Heart Inst, 75 & NSF, 90. *Mem:* AAAS: Am Chem Soc (treas, 75); Am Inst Chem Engrs. *Res:* Fluorine chemistry; kinetics; computer simulations; application of small, dedicated computers to data acquisition and processing; synthetic electrochemistry; combustion processes; organic syntheses. *Mailing Add:* 1311 Dallager Ct Stillwater MN 55082-4136

CHILDS, WYLIE J, b Columbia, SC, Feb 25, 22; m 83; c 1. WELDING METALLURGY, CORROSION. *Educ:* Rensselaer Polytech Inst, BMetE, 43, MMetE, 45, PhD(metall), 48. *Prof Exp:* Res staff, Mass Inst Technol, 48-49; from assoc prof to prof & dept chmn metall, Lafayette Col, 49-57; from assoc prof to prof metall, Rensselaer Polytech Inst, 57-70; from vpres to pres, Reel Vortex, Inc, 68-74; sr consult, Prog Assocs, Inc, 74-75; prin engr nuclear power, Gen Elec, 76-80; PROJ MGR, ELEC POWER RES INST, 80- *Concurrent Pos:* Mem, bd dirs, Welding Res Coun, 84-, chmn bd, 90- *Mem:* Am Soc Metals; Am Welding Soc. *Res:* Metallurgical engineering; welding metallurgy; nuclear reactor plant materials; corrosion mitigation; cast metals technology. *Mailing Add:* 2922 Belmont Woods Way Belmont CA 94002

CHILENSKAS, ALBERT ANDREW, b Chicago, Ill, Nov 7, 27; m 63; c 1. CHEMICAL ENGINEERING. *Educ:* Univ Ill, BS, 49. *Prof Exp:* Asst chem engr, 49-51 & 53-60, assoc chem engr, 60-75, MGR ADV BATTERY TECHNOL DEVELOP, ARGONNE NAT LAB, 77- *Honors & Awards:* Indust Res-100 Award, Nat Battery Adv Comt, 81. *Mem:* Sigma Xi; AAAS. *Res:* Nuclear fuel reprocessing; high-temperature lithium-chalcogen batteries; lithium/metal sulfide battery technology. *Mailing Add:* Argonne Nat Lab Bldg 205 Argonne IL 60439

CHILGREEN, DONALD RAY, b Jenkins, Ky, Nov 8, 39; m 64; c 3. SOIL MICROBIOLOGY. *Educ:* Marion Col, AB, 64; Kans State Univ, MS, 67, PhD(microbiol), 74. *Prof Exp:* Asst prof biol, Marion Col, 67-68; instr microbiol, Kans State Univ, 68-69; from asst prof to assoc prof, 70-81, PROF BIOL, IND WESLEYAN UNIV, 81- *Mem:* Am Soc Microbiol; Sigma Xi. *Res:* Diversity of the indigenous thermophilic microorganism in prairie soils; growth curves and respiratory activity of thermophilic bacteria from soil. *Mailing Add:* Dept Biol Ind Wesleyan Univ Marion IN 46953

CHILGREN, JOHN DOUGLAS, b New Ulm, Minn, Sept 14, 43; m 83; c 1. MEMBRANE PHYSIOLOGY & FLOW CYTOMETRY. *Educ:* Gonzaga Univ, BS, 65; Wash State Univ, MS, 68, PhD(zoophysiol), 75. *Prof Exp:* Res assoc psychobiol, US Army Human Eng Lab, 68-69; opers & training adminr, Eighth US Army, UN Command, 69-70; asst prof zool, Ore State Univ, 75-78; asst prof physiol, Nat Col Naturopathic Med, 78-82; consult, Independent Health & Fitness, 83-84; res assoc & asst prof, Linfield Col & Lewis & Clark Col, 85-87; FEL RES NEUROIMMUNOL, VET ADMIN MED CTR, PORTLAND, ORE, 87- *Mem:* AAAS; Am Soc Zoologists; Am Ornithologists Union; Cooper Ornith Soc; Am Physiol Soc; Am Col Sports Med. *Res:* Investigation of the mechanism by which ganglcosides induce loss of CD4 molecule from helper T-lymphocytes to determine if ganglcosides or derivatives have the potential to inhibit HIV infectivity. *Mailing Add:* Dept HPE Portland State Univ PO Box 751 Portland OR 97207-0751

CHILIAN, WILLIAM M, CORONARY RESEARCH, NEURAL CONTROL. *Educ:* Univ Mo, PhD(physiol), 80- *Prof Exp:* RES SCIENTIST, UNIV IOWA, 80. *Res:* Microcirculation. *Mailing Add:* Dept Med & Physiol Tex A&M Col 350 Med Sci Bldg College Station TX 77843-1114

CHILINGARIAN, GEORGE V(AROS), b Tiflis, Georgia, USSR, July 22, 29; m 53; c 3. PETROLEUM ENGINEERING & GEOLOGY. *Educ:* Univ Southern Calif, BE, 49, MS, 50, PhD(geol), 56. *Prof Exp:* Proj engr & chief petrol & chem qual control lab, Wright-Patterson AFB, 54-56; from asst prof to assoc prof, 56-70, actg chmn dept, 64-65, PROF PETROL ENG, UNIV SOUTHERN CALIF, 70- *Concurrent Pos:* Pres, Electroosmotics, Inc, 64-67; sr UN consult, 67-; vpres, Int Resources Consult, Inc, 68-72 & Global Oil Corp, 75- *Mem:* NY Acad Sci; Am Soc Eng Educ; Am Inst Mining Metall & Petrol Engrs; Am Asn Petrol Geol; Am Geophys Union. *Res:* Petroleum products and analysis; geochemical methods of exploration for petroleum; carbonate rocks; porosity, permeability and compaction of sediments; drilling fluids and clays; electrokinetics; 35 books, 180 research articles and 350 scientific reviews. *Mailing Add:* Dept Eng University Park Los Angeles CA 90089-1211

CHILSON, OSCAR P, b Little Rock, Ark, Dec 23, 32; m 85. RECEPTORS. *Educ:* Ark State Teachers Col, BS, 55; Univ Ark, MS, 58; Fla State Univ, PhD(chem), 63. *Prof Exp:* From asst prof to assoc prof, 65-76, PROF BIOL, WASH UNIV, 76-, ASSOC PROF BIOL CHEM, SCH MED, 75- *Concurrent Pos:* Asst chmn, Dept Biol, Wash Univ, 78-80, assoc chmn, 80, actg chmn, 81-83; vis worker, Imp Cancer Res Fund, London, 83-84. *Mem:* Am Soc Biochem & Molecular Biol; AAAS. *Res:* Role of proline metabolism in legume nodules; molecular basis of isozyme structure; regulation of purine metabolism in avian erythrocytes (transport, enzymology); cell surface biochemistry. *Mailing Add:* Dept Biol Wash Univ St Louis MO 63130

CHILTON, A(RTHUR) B(OUNDS), nuclear technology; deceased, see previous edition for last biography

CHILTON, BRUCE L, b Buffalo, NY, June 14, 35. MATHEMATICS. *Educ:* Univ Buffalo, BA, 58, MA, 60; Univ Toronto, PhD(math), 62. *Prof Exp:* Asst prof math, State Univ NY Buffalo, 62-68; dean dept, 68-74, ASSOC PROF MATH, STATE UNIV NY COL FREDONIA, 68- *Mem:* Math Asn Am; Am Math Soc. *Res:* Geometry, especially properties of regular and semiregular figures in Euclidean n-spaces. *Mailing Add:* Dept Math State Univ NY Fredonia NY 14063

CHILTON, JOHN MORGAN, inorganic chemistry; deceased, see previous edition for last biography

CHILTON, MARY-DELL MATCHETT, b Indianapolis, Ind, Feb 2, 39; m 66; c 2. MOLECULAR BIOLOGY, BIOCHEMISTRY. *Educ:* Univ Ill, Urbana, BSc, 60, PhD(chem), 67. *Hon Degrees:* Dr Honoris Causa, Univ Louvain. *Prof Exp:* Fel microbiol, Univ Wash, 67-69, fel biochem, 69-70, asst biologist, 71-73, res asst prof biol, 73-77, res assoc prof, 77-79; assoc prof biol, Wash Univ, 79-83; exec dir agr biotechnol, 83-91, VPRES BIOTECH, CIBA-GEIGY BIOTECHNOL FACIL, 91- *Honors & Awards:* Bronze Medal, Am Inst Chemists, 60. *Mem:* Nat Acad Sci; Am Soc Microbiol. *Res:* Crown gall tumorigenesis; bacterial plasmids; plant genome organization; satellite DNA; DNA and RNA hybridization; bacterial genetics. *Mailing Add:* Ciba-Geigy Biotechnol Facil PO Box 12257 Research Triangle Park NC 27709

CHILTON, NEAL WARWICK, b New York, NY, June 24, 21; m 47; c 5. ORAL MEDICINE. *Educ:* City Col New York, BSc, 39; NY Univ, DDS, 43; Columbia Univ, MSc, 46; Am Bd Endodont & Am Bd Periodont, dipl. *Prof Exp:* Intern, Lincoln Hosp, New York, 43; from instr to asst clin prof pharmacol & therapeut, NY Univ, 44-50; res assoc dent & asst prof dent pub health pract, Columbia Univ, 49-54; from asst prof to assoc prof periodont, Sch Dent, Temple Univ, 52-63, assoc prof prev med, Sch Med, 59-66, clin prof periodont, Sch Dent, 63-76, prof oral med, Sch Med, 66-76; res prof oral med & assoc dir, Clin Res Ctr, Sch Dent Med, Univ Pa, 76-80; res consult, 80-84; prof clin res, Univ Pa, 84-86; SR RES SCIENTIST BIOSTATIST & CLIN DENT RES, COLUMBIA UNIV, 86- *Concurrent Pos:* Lectr, Seton Hall Univ, 47-52 & Temple Univ, 52-53; guest lectr, Evans Dent Inst, Univ Pa, 50-; clin prof, Univ Kansas City, 54-55; res assoc, Fac Med, Columbia Univ, 57-70, sr res assoc biostatist, Sch Pub Health & sr res assoc prev med, Sch Dent & Oral Surg, 70-83, sr res scientist biostatist & prev dent, 83-; asst prof, Grad Sch Med, Univ Pa, 57-70, lectr, Sch Dent Med, 70-; asst chief, Bur Dent Health, State Dept Health, NJ; consult, Coun Dent Therapeut, Am Dent Asn & Surgeon Gen, USPHS; mem comt res manpower, Nat Inst Dept Res & mem dent study sect, Div Res Grants, NIH; ed-in-chief, J Pharmacol & Therapeut Dent; res prof, periodont, NY Univ, 79-90. *Mem:* Am Asn Endodont; fel Am Col Dent; Am Acad Periodont; Int Asn Dent Res; Sigma Xi. *Res:* Diseases of the mouth and gums, etiology, pathology, treatment and prevention; dental public health; design and statistical analysis of agents in clinical trial; clinical therapeutics. *Mailing Add:* 2975 Princeton Pike Lawrenceville NJ 08648

CHILTON, WILLIAM SCOTT, b Philadelphia, Pa, Aug 29, 33; m 65; c 2. ORGANIC CHEMISTRY, NATURAL PRODUCTS CHEMISTRY. *Educ:* Duke Univ, BS, 55; Univ Ill, Urbana-Champaign, PhD(org chem), 63. *Prof Exp:* From asst prof to prof chem, Univ Wash, 63-80; vis prof, Wash Univ, St Louis, 80-83; PROF BOT, DEPT BOT, NC STATE UNIV, 84- *Concurrent Pos:* Sci adv, Food & Drug Admin, 67- *Mem:* AAAS; Am Chem Soc; Am Soc Plant Physiologists; Am Phytopath Soc; Am Soc Pharmacog. *Res:* Structure of natural products; new naturally occurring amino acids; biochemistry of crown gall; fungal phytotoxins. *Mailing Add:* 10513 Winding Wood Trail Raleigh NC 27613

CHIMENTI, DALE EVERETT, b Chicago, Ill, July 26, 46; m 70; c 2. ELASTIC WAVE PROPAGATION, NONDESTRUCTIVE EVALUATION. *Educ:* Cornell Col, Iowa, BA, 68; Cornell Univ, MS, 72, PhD(physics), 74. *Prof Exp:* Res assoc, Argonne Nat Lab, 74-76; vis asst prof physics, Univ Tuebingen, WGer, 76-78; res physicist, Air Force Mat Lab, 78-89; RES PROF, DEPT MAT SCI & ENG, JOHNS HOPKINS UNIV, 90- *Concurrent Pos:* Alexander von Humboldt Found fel 76-78. *Mem:* Am Phys Soc; Acoust Soc Am. *Res:* Elastic wave propagation and ultrasonic reflection in composite materials, plates, and layered structures; applications of ultrasonics to nondestructive evaluation. *Mailing Add:* 102 Maryland Hall Johns Hopkins Univ Baltimore MD 21218-2689

CHIMENTI, FRANK A, b Erie, Pa, May 3, 39; m 64; c 2. APPLIED MATHEMATICS. *Educ:* Gannon Col, BA, 61; John Carroll Univ, MS, 63; Pa State Univ, PhD(math), 70. *Prof Exp:* Res asst appl math, Lord Mfg Co, Pa, 63-65 & Ord Res Lab, Pa State Univ, 65-67; asst prof, 69-74, assoc prof math, State Univ NY Col Fredonia, 74-; mem fac, Dept Computer Sci, Calif State Univ, Domingue Hills; DEPT CHMN COMPUTER SCI, LIBERTY UNIV, 89- *Mem:* Math Asn Am; Am Math Soc; Soc Indust & Appl Math; Asn Comput Mach; Inst Elec & Electronics Engrs. *Res:* General topology; convergence of sequences of sets; multivalued functions; convergence formulas; mathematical modeling and simulation; computer graphics; image processing. *Mailing Add:* Liberty Univ PO Box 20000 Lynchburg VA 24506

CHIMES, DANIEL, b Brooklyn, NY, May 23, 21; m 44; c 3. PRESSURE SENSITIVE ADHESIVES, NONWOVEN FABRICS. *Educ:* Polytechnic Inst Brooklyn, BChE, 42. *Prof Exp:* Qual assurance & prod develop, Permacel Tape Corp, Subsid Johnson & Johnson, 42-55; mgr prod develop, Standard Packaging Corp, 55-58, Rubber & Asbestos Corp, 58-63; VPRES RES & DEVELOP, AM WHITE CROSS LABS, INC, 63- *Mem:* Am Chem Soc. *Res:* Design and production of surgical dressings using various plastic, cloth, nonwoven and composite webs; design and formulation of adhesives; equipment necessary to produce the finished product. *Mailing Add:* 30 Fieldstone Dr Livingston NJ 07039

CHIMOSKEY, JOHN EDWARD, b Traverse City, Mich, Apr 15, 37; div; c 2. PHYSIOLOGY. *Educ:* Univ Mich, MD, 63. *Prof Exp:* Intern internal med, Univ Calif, 64; USPHS fel, Harvard Med Sch, 64-66 & Retina Found, Boston, 66-67; assoc prof physiol, Hahnemann Med Col, 69-70; actg instr dermat, Med Ctr, Stanford Univ, 70-71; asst prof bioeng, Univ Wash, 71-75; assoc prof physiol & surg & dir, Taub Labs Mech Circulatory Support, Baylor Col Med, 75-78; dir grad studies physiol, 78-90, PROF PHYSIOL, MICH STATE UNIV, 78-, INTERIM CHAIR PHYSIOL, 89- *Concurrent Pos:* Guest scientist, US Naval Air Develop Ctr, 69-70; NIH spec fel, Hahnemann Med Col, 70; guest lectr, Hahnemann Med Col & Calif Col Podiatric Med, 70-71; NIH spec fel & Dermat Found fel, Stanford Univ, 70-71; NIH grants, 73-78, 79-83, 84-87 & 87-; US deleg, US-USSR Sci Exchange on Artificial Heart Develop, 76, 77; adj assoc prof bioeng, Rice Univ, 75-78; dir cardiovasc sci training grant; Fulbright distinguished scholar, Brazil, 90-91. *Mem:* AAAS; Am Physiol Soc; Soc Exp Biol & Med. *Res:* Cardiovascular physiology; physiology of atrial natriuretic peptides and central nervous system control of arterial pressure. *Mailing Add:* Dept Physiol Mich State Univ East Lansing MI 48824

CHIN, BYONG HAN, b Shanghai, China, Nov 27, 34; US citizen; m 61; c 3. BIOCHEMISTRY. *Educ:* Yonsei Univ, Korea, BS, 57; Univ Hawaii, MS, 64, PhD(hort), 67. *Prof Exp:* Marine biochemist, Hawaii Marine Lab, Univ Hawaii, 63-64; fel, Mellon Inst, Carnegie-Mellon Univ, 67-79; biochem toxicologist, Diamond Shamrock Corp, 79-81; BIOCHEM TOXICOLOGIST, MITRE CORP, 81- *Mem:* Am Chem Soc; Soc Toxicol; Am Asn Clin Chem. *Res:* The metabolism of pesticides by plants and animals; bioassay of poisonous fishes; methodology development in clinical chemistry; evaluate toxicity studies of pesticides and industrial pollutants. *Mailing Add:* Mitre Corp 7525 Colshire Dr McLean VA 22102

CHIN, CHARLES L(EE) D(ONG), b New York, NY, Feb 4, 23; c 1. ENGINEERING MECHANICS, MATERIAL SCIENCE ENGINEERING. *Educ:* Tri-State Col, BS, 41; Polytech Inst Brooklyn, MAE, 48; Harvard Univ, SM, 49, ScD(eng), 65. *Prof Exp:* Engr, Curtiss-Wright Corp, 41-43; sr engr, Chance Vought Aircraft, 43-44; sr res asst aerodyn, Polytech Inst Brooklyn, 46-48; asst prof aeronaut eng, Univ RI, 50-52; head anal res, Jackson & Church Co, 52-55; asst prof aeronaut eng, Boston Univ, 55-58, prof & chmn dept, 58-68; dir eng & res mach div, Borg-Warner Corp, 68-70, res fel chem & plastics group, 70-71; sr eng specialist & mat sci prin, Monsanto Co, 72-85; MEM MECH ENG FAC, UNIV HARTFORD, 86- *Concurrent Pos:* Staff engr & consult, Res & Advan Design Div, Avco Corp, 62-63; res specialist, NAm Aviation, Inc, Calif, 66; adj fac mech eng, Univ Hartford, 73-85. *Mem:* NY Acad Sci; Am Soc Eng Educ; Soc Plastics Engrs. *Res:* Structural analysis; aerodynamics of sweat cooling; aerodynamic ablation of reentry vehicles; stress concentration around holes; aerodynamics of supersonic wings; atmospheric gust velocity determination; rheology of plastics; mechanics of polymer processing; plastics processing equipment and apparatus. *Mailing Add:* Col Eng Univ Hartford 200 Bloomfield Ave W Hartford CT 06117-0395

CHIN, DAVID, b Boston, Mass. CHEMISTRY. *Educ:* Boston Univ, BA, 64; Purdue Univ, PhD(chem), 71. *Prof Exp:* Teaching asst chem, Purdue Univ, 64-68; sr chemist, 70-80, GROUP LEADER, CONSTRUCTION PROD DIV, W R GRACE & CO, 81- *Mem:* Am Chem Soc; Am Ceramics Soc; Am Soc Testing & Mats. *Res:* Develop chemical admixtures for concrete; hardened concrete analysis; technical service work involving admixtures, cement, hardened concrete, concrete trial mixes. *Mailing Add:* W R Grace & Co 62 Whittemore Ave Cambridge MA 02140-1692

CHIN, DER-TAU, b Chekiang, China, Sept 14, 39. ELECTROCHEMCAL ENGINEERING, CORROSION ENGINEERING. *Educ:* Chung Yuan Col Sci & Eng, BS, 62; Tufts Univ, MS, 65; Univ Pa, PhD(chem eng), 69. *Prof Exp:* Process engr, Taiwan Sugar Corp, 62-63; sci programmer, US Air Force Cambridge Res Labs, 65; sr res engr, Res Labs, Gen Motors Corp, 69-75; assoc prof, 75-80, PROF CHEM ENG, CLARKSON UNIV, 80- *Concurrent Pos:* Vis scientist, Brookhaven Nat Lab, 77 & 80; Materials Dept, Ctr Study Elec Energy, City Univ; consult, Mat Dept, Ctr study Elec Eng, Cidade Univ, RJ Brazil, 79, Hooker Chem & Plastics Corp, 80-, Energy Systs Lab, Cupertino, Calif, 81-, Inst Hydrogren Systs, Univ Toranto, 83-, St Joe Minerals Corp, 84-, Pioneer Bank, 84-; vis prof, Eidgenoessische Tech Hochschule, Zurich, Switz, 81 & Univ Calif, Berkeley, 81; consult, Los Alamos Nat Lab, 81-; vis prof, Nat Univ Singapore, 82-87; lectr, Gorden Res Conf Electrochem, 85; vis prof, Wuhan Univ, People's Repub China, 82- & Nat Tsing Hua Univ, Hsinchu, Taiwan, 89. *Mem:* Electrochem Soc; Am Inst Chem Eng; Am Electroplaters Soc; Inst Colloid Surface Sci; Am Chem Soc; Sigma Xi. *Res:* Electrolytic mass transfer; electrochemical study of flow turbulence; electrochemical machining; high current density electrode process; electrochemical waste treatment; potential and current distribution; engineering analysis of electrochemical systems; corrosion; fuel cells; batteries; electroplating. *Mailing Add:* Dept Chem Eng Clarkson Univ Potsdam NY 13676

CHIN, EDWARD, b Boston, Mass, Sept 4, 26; m 52; c 4. BIOLOGY. *Educ:* Harvard Univ, BS, 48; Univ NH, MS, 53; Univ Wash, PhD, 61. *Prof Exp:* Biol aide clam invest, US Fish & Wildlife Serv, 49-51; fishery res biologist king crab studies, 54-55 & gulf shrimp studies, 55-61; asst sci dir, US Prog Biol Int Indian Ocean Exped, 62-65; assoc prof biol, Tex A&M Univ, 65-68; dir biol oceanog prog, NSF, 68-70; assoc dir, Inst Natural Resources, 70-76, assoc prof zool, 70-79, DIR MARINE SCI, UNIV GA, 77-, PROF ZOOL, 79- *Mem:* Brit Marine Biol Asn. *Res:* Marine invertebrates; marine ecology. *Mailing Add:* Marine Sci Ecol Bldg Univ Ga Athens GA 30602

CHIN, GILBERT YUKYU, physical metallurgy, materials science; deceased, see previous edition for last biography

CHIN, HONG WOO, b Seoul, Korea, May 14, 35; US citizen; m 65; c 3. RADIATION ONCOLOGY, NEURO-RADIATION ONCOLOGY. *Educ:* Seoul Nat Univ, MD, 62, PhD(pharmacol), 74; Am Bd Radiol, dipl, 79. *Prof Exp:* Assoc dir, dept internal med, Red Cross Hosp, Seoul, 70-74; instr radiation med, Col Med, Med Ctr, Univ Ky, 79, asst prof radiation oncol & head sect neuro-radiation oncol, 79-86; radiation oncologist, Albert B Chandler Med Ctr, 79-86; clin prof, Univ Missouri, Kansas City, 87-89; prof radiol, Creighton Univ Sch Med, 88-90; assoc prof, 88, PROF RADIOL, LA STATE UNIV, SHREVEPORT, 90-; CHIEF, RADIATION ONCOL, OVERTON BROOKS VA MED CTR, SHREVEPORT, 90- *Concurrent Pos:* Mem fac internal med, Red Cross Nursing Sch, Seoul 70-74; prin investr brain tumor study, Univ Ky, 79-86, Creighton Univ, 88-90, brain tumor & thoracic oncol studies, Overton Brooks VA Med Ctr, 90-; assoc dir, Radarium Found, 86-88; chief, Radiation Oncol Serv, VA Med Ctr, Shreveport, 88; dir, radiation oncol, Creighton Cancer Ctr, Omaha, 88-90. *Mem:* AMA; Am Col Radiol; Am Soc Therapeut Radiol & Oncol; Radiol Soc NAm; NY Acad Sci; AAAS; Sigma Xi. *Res:* Neuro-oncology; brain brachytherapy. *Mailing Add:* 2900 W 124th Terr Leawood KS 66209

CHIN, HSIAO-LING M, b Shanghai, China, Aug 2, 47; m 74; c 2. ORGANIC CHEMISTRY. *Educ:* Nat Taiwan Univ, BS, 68; Univ Southern Calif, PhD(chem), 74. *Prof Exp:* Res chemist, 73-81, sr res chemist, 81-85, PRIN RES CHEMIST ORG SYNTHESIS, STAUFFER CHEM CO, 85- *Mem:* Am Chem Soc. *Res:* Synthesis of novel organic compounds as agricultural chemicals. *Mailing Add:* ICI 1200 S 47th St Richmond CA 94804-1685

CHIN, JANE ELIZABETH HENG, b Augusta, Ga, Nov 20, 33; m 60; c 2. PHARMACOLOGY. *Educ:* Univ Ga, BS, 54; Univ Mich, MS, 56, PhD(pharmacol), 60. *Prof Exp:* USPHS fel neuropharmacol, Univ Ill, 59-60; USPHS fel neuropharmacol, Sch Med, Stanford Univ, 60-68, USPHS fel biosci, 68-71, res assoc, 71-75, sr res assoc pharmacol, 75-87. *Mem:* Res Soc Alcoholism; Am Soc Pharmacol & Exp Therapeut; Int Soc Biomed Res Alcoholism; Biophys Soc. *Res:* Neuropharmacology; pain and analgesics; psychopharmacology; neuropsychology; biophysics; high pressure physiology; drug effects on membranes; drug tolerance. *Mailing Add:* 727 Christine Dr Palo Alto CA 94303

CHIN, JIN H, b Kwangtung, China, Oct 15, 28; nat US; m 60; c 2. COMPUTATIONAL ENGINEERING ANALYSIS. *Educ:* Stanford Univ, BS, 50; Univ Mich, MSE, 51, PhD(chem eng), 55. *Prof Exp:* Asst, Univ Mich, 54, res assoc, 55-57; heat transfer specialist, Flight Propulsion Div, Gen Elec Co, 57-60; staff engr, 60-80, SR STAFF ENGR, LOCKHEED MISSILES & SPACE CO, 80- *Concurrent Pos:* Consult, Armed Forces Spec Weapons Proj, 57 & Dept Aeronaut Eng, Princeton Univ, 59; asst prof, Mech Eng, San Jose State Col, 67. *Mem:* AAAS; Am Inst Aeronaut & Astronaut. *Res:* Light scattering; particle size determination; applied mathematics; radiative transport; aircraft component cooling; boundary layer theory; computer analyses; chemical kinetics; thermodynamics; cryogenics; propellant behaviors; missiles and space vehicle thermal environments; superorbital entry thermal environments; reentry physics; finite element methods; solid rocket propulsion nozzles; computational fluid dynamics. *Mailing Add:* 727 Christine Dr Palo Alto CA 94303

CHIN, LINCOLN, PHARMACOLOGY, TOXICOLOGY. *Educ:* Univ Utah, PhD(pharmacol & toxicol), 59. *Prof Exp:* PROF PHARMACOL & TOXICOL, UNIV ARIZ, TUCSON, 63- *Mailing Add:* Dept Pharmacol & Toxicol Univ Ariz Tucson AZ 85721

CHIN, MAW-RONG, b Taipei, Taiwan, Jan 6, 51; US citizen; m; c 3. MATERIALS SCIENCE ENGINEERING, GENERAL ENGINEERING. *Educ:* Nat Taiwan Univ, BS, 73; Yale Univ, MS, 77, PhD(appl physics), 81. *Prof Exp:* Res & teaching asst elec eng & solid state related courses, Yale Univ, 76-81; sr engr process develop, Westinghouse Res & Develop Ctr, 81-83; develop engr, Semiconductor Prod Ctr, Hughes Aircraft Co, 83-85, mgr, Technol Sect, 85-88, Technol Dept, 88-91, MGR TECHNOL LAB, TECHNOL DEVELOP, SEMICONDUCTOR PROD CTR, HUGHES AIRCRAFT CO, 91- *Concurrent Pos:* Adv for Univ Calif Prof, Micro Prog State Calif, 87-91; mentor prof, Semiconductor Res Corp, 89-91. *Mem:* Inst Elec & Electronics Engrs Electronic Devices & Solid State Electronics; Elec Chem Soc; Int Soc Optical Eng. *Res:* Development of complementary bipolar and complementary metal-oxide Semiconductor process for high speed, low power high resolution analog circuits; development of electrical erasable programmable read only memory combination with digital and analog functions; implementation of total productive maintenance; granted seeveral patents. *Mailing Add:* 19172 Homestead Lane Huntington Beach CA 92646

CHIN, ROBERT ALLEN, b San Francisco, Calif, Oct 3, 50; m 76. CLASSROOM INSTRUCTION. *Educ:* Univ Northern Colo, BA, 74; Ball State Univ, MAE, 75; Univ Md, PhD(technol educ), 86. *Prof Exp:* Lab asst, Dept Indust Arts, Univ Northern Colo, 73-74; grad asst electronics, Dept Indust Educ & Technol, Col Appl Sci & Technol, Ball State Univ, 74-75; instr drafting, Sioux Falls Independent Sch Dist, 75-79; instr mech drawing, Dept Indust, Technol & Occup Educ, Col Educ, Univ Md, 77-86; ASST PROF ENG GRAPHICS, DEPT CONSTRUCT MGT, SCH INDUST & TECHNOL, E CAROLINA UNIV, 86- *Concurrent Pos:* Reviewer, J Indust Teacher Educ, 87-88 & 90-, Eng Design Grpahics J, 88-; mem, Undergrad Studies Comt, Coun Technol Teacher Educ, 88-; actg chmn, Dept Construct Mgt, ECarolina Univ, 89-90; vchair, Res Comt, Nat Asn Indust Technol, 90- *Mailing Add:* Dept Construct Mgt ECarolina Univ Greenville NC 27858-4353

CHIN, SEE LEANG, b Padang Rengas, Malaya, May 24, 42; Can citizen; m 71; c 2. LASERS. *Educ:* Nat Taiwan Univ, BSc, 64; Univ Waterloo, MSc, 66, PhD(physics), 69. *Prof Exp:* Teacher math, Hua Lian High Sch, Taiping, Malaya, 60; fel physics, Laval Univ, 69-70, res assoc, 71-72, from asst prof to assoc prof physics, 72-82, dir, Optics & Laser Res Lab, 81-87, PROF PHYSICS, LAVAL UNIV, 82- *Concurrent Pos:* Sci consult, K A Mace Ltd, Kitchener, Ont, 73-79; vis scientist, Nuclear Study Ctr, Saclay, France, 75; vis scientist, Ctr Interdisciplinary Studies, Univ Bielefeld, WGer, 79-80, Chalk River Nuclear Lab, Can, 78, Nat Res Coun, Ottawa, 87-88; vis prof, Univ Rochester, 88. *Mem:* Can Asn Physicists; Optical Soc Am. *Res:* Multiphoton interaction of intense lasers with atoms and molecules. *Mailing Add:* Dept Physics Laval Univ Quebec PQ G1K 7P4 Can

CHIN, TOM DOON YUEN, b Guangdong, China, May 29, 22; US citizen; m 50; c 1. MICROBIOLOGY. *Educ:* Univ Mich, MD, 46; Tulane Univ, MPH, 50; Am Bd Prev Med & Am Bd Med Microbiol, dipl. *Prof Exp:* Intern, Binghamton City Hosp, 47; intern, Western Pa Hosp, 47-48; resident, Sea View Hosp, 48-49; dir health unit, State Dept Health, La, 50-51; asst chief epidemiol, Kansas City Field Sta, USPHS, 54-64, chief, 64-66, dir ecol invests prog, Ctr Dis Control, Kansas City, 67-73; chmn dept prev med, 84-89, PROF MED & PREV MED, UNIV KANS MED CTR, KANSAS CITY, 73- *Concurrent Pos:* Assoc ed, Am J Epidemiol, 66-87, ed, 88-; control study sect, NIH, 74-78; vis prof, Univ Minn Sch Pub Health, 80-87; mem, Biometry & Epidemiol Contract Rev Comt, 80-83, & Cancer Ctr Support Rev Comt, 84-88. *Mem:* Fel Am Col Prev Med; fel Am Pub Health Asn; Am Epidemiol Soc; Soc Epidemiol Res; NY Acad Sci; Sigma Xi. *Res:* Infectious diseases; epidemiology. *Mailing Add:* Dept Prev Med Univ Kans Med Ctr 39th/ Rainbow Blvd Kansas City KS 66103

CHIN, WILLIAM W, b New York, NY, Nov 20, 47; m 81; c 2. ENDOCRINOLOGY. *Educ:* Columbia Univ, AB, 68; Harvard Univ, MD, 72. *Prof Exp:* Asst prof, 81-84, ASSOC PROF MED, HARVARD MED SCH, 84-; INVESTR, HOWARD HUGHES MED INST, 84-; CHIEF, DIV GENETICS, BRIGHAM & WOMEN'S HOSP, 87- *Concurrent Pos:* Mem staff, Lab Molecular Endocrinol, Mass Gen Hosp. *Honors & Awards:* Bowditch Lectr, Am Physiol Soc; Van Meter Award, Am Thyroid Asn; Young Investr Award, Am Fedn Clin Res. *Mem:* Endocrine Soc; Am Thyroid Asn; Am Fedn Clin Res; Am Physiol Soc; Am Soc Biochem & Molecular Chem; Am Soc Clin Invest. *Res:* Regulation of expression of anterior pituitary gland and central-nervous-system polypeptide hormone genes. *Mailing Add:* Brigham & Women's Hosp 75 Francis St Boston MA 02115

CHIN, YU-REN, b Shanghai, China, Mar 6, 38; m 66; c 3. CHEMICAL ENGINEERING, CHEMISTRY. *Educ:* Nat Taiwan Univ, BS, 60; Purdue Univ, MS, 65, PhD(chem eng), 69. *Prof Exp:* Sr chem engr, Chem Div, Uniroyal, Inc, 68-73; sr chem engr, SRI Int, 73-79; consult, 80-86, SR CONSULT, PROCESS ECON, 86- *Concurrent Pos:* Lectr process design & eval, Taiwan & Beijing, China. *Mem:* Am Chem Soc; Am Inst Chem Engrs; Sigma Xi; Soc Plastics Engrs. *Res:* Process design and development in chemical industries; chemical kinetics; operations research; technoeconomic evaluation of chemical processes for petrochemicals, inorganochemicals, rubber chemicals, thermoplastics and elastomers. *Mailing Add:* 4185 Georgia Ave Palo Alto CA 94306

CHINARD, FRANCIS PIERRE, b Berkeley, Calif, June 30, 18; m 43; c 3. PHYSIOLOGICAL CHEMISTRY, INTERNAL MEDICINE. *Educ:* Univ Calif, AB, 37; Johns Hopkins Univ, MD, 41. *Prof Exp:* Intern, Presby Hosp, New York, 41-42; Nat Res Coun fel, Rockefeller Inst, 45-46, asst Rockefeller Inst Hosp, 46-49; instr med & physiol chem, Sch Med, Johns Hopkins Univ, 49-51, asst prof physiol chem, 51-56, asst prof med, 52-56, assoc prof physiol chem & med, 59-63; prof exp med, Fac Med, McGill Univ, 63-64; prof med, Sch Med, NY Univ, 64-68; prof med & chem dept, Univ Med & Dent, NJ, 68-74, prof exp med, 75-77, prof res med, 77-90, prof physiol, 78-90, DISTINGUISHED PROF RES MED & PHYSIOL, NJ MED SCH, UNIV MED & DENT NJ, 90- *Concurrent Pos:* Markle scholar, 49-54; asst chief med, Baltimore City Hosps, Md, 53-60, physician in chief, 60-63; chief med, Goldwater Mem Hosp, 66-68; adj prof, NY Univ, 68-70; career scientist, Health Res Coun, NY; prin investr grant pulmonary transport & metabolism in vivo NIH, 70; consult, Louis & Artur Lucian Award Comt for Res Circulatory Dis, McGill Univ, 77-; mem & chmn, Pulmonary Dis Adv Comt, Nat Heart, Lung & Blood Inst, NIH, 71-75, chmn, ad hoc rev comt, Spec Ctr Res Prog Adult Respiratory Distress Syndrome, 78 & Pulmonary Vascular Dis, 80-81; pres, Fac Pract Serv, Inc, NJ Med Sch, 86-88; vis scientist, McGill Univ, Med Res Coun Can, 89-90. *Honors & Awards:* Landis Award, Microcirculatory Soc, 78; Lucien Award, McGill Univ, 89. *Mem:* Am Chem Soc; Am Soc Biol Chem; Am Soc Clin Invest; Soc Exp Biol & Med; fel Am Col Physicians; Am Physiol Soc; fel AAAS; fel NY Acad Sci. *Res:* Membrane permeability; renal and pulmonary physiology and metabolism; correlation of in vivo, isolated perfused organ and isolated constituent cell characteristics and properties. *Mailing Add:* Univ Med & Dent NJ Med Sch 185 S Orange Ave Newark NJ 07103-2757

CHIN-BING, STANLEY ARTHUR, b New Orleans, La, Nov 3, 42. ACOUSTICS, ELECTRO-OPTICS. *Educ:* Tulane Univ, BS, 64; Univ New Orleans, MS, 68, PhD(physics), 73. *Prof Exp:* Systs analyst mech eng, Martin Marietta Corp, 74-76; sr engr electro-optics, Space Div, Chrysler Corp, 76-77; RES PHYSICIST ACOUST, NAVAL OCEAN RES & DEVELOP ACTIV, 78- *Concurrent Pos:* Asst prof physics & eng, Univ New Orleans, 75-80. *Mem:* Am Phys Soc; Optical Soc Am; Am Asn Physics Teachers; Am Inst Physics; NY Acad Sci. *Res:* Theoretical underwater acoustics; electro-optical systems; electric and magnetic fields; radiative lifetimes of excited electronic atomic levels. *Mailing Add:* 3619 Bauvais St Metairie LA 70001

CHINCARINI, GUIDO LUDOVICO, b Venice, Italy, Jan 24, 38; c 2. ASTROPHYSICS. *Educ:* Liceo Scientifico G B Genedetti, Maturity in sci, 56; Univ Padua, Italy, PhD(physics), 61. *Prof Exp:* Astronr, Asiago Observ, Univ Padua, 61-68; res assoc, Wesleyan Univ, 69-71; res scientist & engr, McDonald Observ, Univ Tex, 71-74; vis assoc prof, 75 & 76-77, assoc prof, 77-79, PROF PHYSICS & ASTRON, UNIV OKLA, 79- *Concurrent Pos:* Res assoc Lick Observ, Univ Calif, 64-66; astronr, Hoherlist Observ, Univ Bonn, 68; res assoc, Johnson Space Ctr, NASA, Houston; chair prof astron, Univ Bologna, Italy, 76-78; res assoc, European Southern Observ, 77- *Mem:* Int Astron Union; Am Astron Soc; Sigma Xi. *Res:* Extragalactic astronomy; observational cosmology. *Mailing Add:* Dept Physics & Astron Univ Okla 440 W Brooks Norman OK 73019

CHING, CHAUNCEY T K, b Honolulu, Hawaii, July 25, 40; m 62; c 2. TECHNOLOGY TRANSFER, MATHEMATICAL PROGRAMMING. *Educ:* Univ Calif, Berkeley, BA, 62; Univ Calif, Davis, MS, 65, PhD(agr econ), 67. *Prof Exp:* Asst prof, dept resource econ, Univ Nev, Reno 69-72, prof & head div agr & resource econ, 72-80; prof & chmn dept agr resource econ, 80-84, DIR, HAWAII INST TROP AGR & HUMAN RESOURCES, UNIV HAWAII, 84- *Mem:* Am Agr Econ Asn. *Res:* Application of quantitative methods to economic problems of agriculture and resource use. *Mailing Add:* 3050 Maile Way Gilmore 202 Honolulu HI 96822

CHING, HILDA, b Honolulu, Hawaii, June 30, 34; m 60; c 3. FISH AND INVERTEBRATE PARASITOLOGY. *Educ:* Ore State Univ, BA, 56, MS, 57; Univ Nebr, PhD(zool), 59. *Prof Exp:* Asst parasitol, Agr Exp Sta, Univ Hawaii, 59-60; lectr invert zool, 74-77, lectr biol, Douglas Col, 76-79,; res assoc, Dept Zool, Univ BC, 60-79; lectr biol, 79 & 81, ENDOWED CHAIR, SIMON FRASER UNIV, 90- *Concurrent Pos:* Consult, Fish Parasites, 79-84, Envirocon Ltd, 84-, Hydra Enterprises; consult, Women in Sci & Sci Educ for Girls; fac, Morphol-Parasitol, Nat Ref Centre. *Mem:* Am Soc Parasitol; Soc Can Women Sci. *Res:* Trematodes of fishes and birds; fish parasites; life cycles of parasites. *Mailing Add:* PO Box 2184 Vancouver BC V6B 3V7 Can

CHING, JASON KWOCK SUNG, b Honolulu, Hawaii, Dec 18, 40; m 64; c 3. ENVIRONMENTAL SCIENCE. *Educ:* Univ Hawaii, BS, 62; Pa State Univ, MS, 64; Univ Wash, PhD(meteorol), 74. *Prof Exp:* Res asst meteorol, Pa State Univ, 62-64, Woods Hole Oceanog Inst, 64-66 & Univ Wash, 66-70; res meteorologist, Barbados Oceanog & Meteorol Anal Proj, 70-75, METEROROLOGIST, METEOROL DIV, AIR RES LAB, NAT OCEANIC & ATMOSPHERIC ADMIN, ATMOSPHERIC RES, EXPOSURE ASSESSMENT LAB, ENVIRON PROTECTION AGENCY, 75- *Mem:* Am Meteorol Soc. *Res:* Numerical-theoretical modelling and field studies leading to documentation of the dynamics, thermodynamics and transport characteristics in the planetary boundary layer of the atmosphere. *Mailing Add:* Atmospheric Sci Res Div AREAL Environ Protection Agency Research Triangle Park NC 27711

CHING, MELVIN CHUNG HING, b Honolulu, Hawaii, Feb 11, 35; m 65; c 2. ANATOMY. *Educ:* Univ Nebr, AB, 57, MSc, 60; Univ Calif, Berkeley, PhD(anat), 71. *Prof Exp:* Instr anat, Sch Med & Dent, Univ Rochester, 71-73, asst prof, 73-77; assoc prof anat, Med Col Va, 78-82; sr res fel, NIH, 83-84, expert, 85-89; ASST PROF, OHIO STATE UNIV, 89- *Mem:* Am Asn Anat; Sigma Xi; Endocrinol Soc; Neurosci Soc; AAAS; Soc Study Reproduction; Int Neuroendocrine Soc. *Res:* Neuroendocrinology; endocrinology; neuroendocrine control mechanisms; hypothalamic-pituitary-thyroid gonadal axis. *Mailing Add:* Dept Vet Anat Ohio State Univ 1900 Coffey Rd Columbus OH 43210-1092

CHING, STEPHEN WING-FOOK, b Canton, China, May 6, 36; m 64. COMPUTER ENGINEERING, COMPUTER SCIENCE. *Educ:* Nat Univ Taiwan, BSEE, 58; Univ Pa, MSEE, 61, PhD(elec eng), 66. *Prof Exp:* Electronic engr, Burroughs Corp, Pa, 60-64, sr electronic engr, 65-67; res asst appl math, Johnson Res Found, Univ Pa, 64-65; asst prof elec eng, 67-69, ASSOC PROF ELEC ENG, VILLANOVA UNIV, 69-, MEM STAFF, DEPT MATH. *Mem:* Inst Elec & Electronics Engrs; Am Math Soc; Math Asn Am; Soc Indust & Appl Math; Sigma Xi. *Res:* Scientific computation; electronic engineering. *Mailing Add:* Dept Math Villanova Univ Villanova PA 19085

CHING, TA YEN, b Peking, China, Mar 23, 47; m 75; c 2. POLYMER CHEMISTRY. *Educ:* Fu-Jen Univ, BS, 69; Baylor Univ, MS, 72; Univ Calif, Los Angeles, PhD(chem), 76. *Prof Exp:* Res assoc chem, Yale Univ, 76-77; staff chemist, Corp Res & Develop, Gen Elec Co, 77-; MEM STAFF, CHEVRON RES CO. *Mem:* Am Chem Soc; AAAS. *Res:* Organic photochemistry; chemistry of singlet oxygen; photostabilization of engineering plastics; coatings; water soluble polymers; corrosion inhibition; adhesives and sealants. *Mailing Add:* Chevron Res & Technol Co 100 Chevron Way Richmond CA 94802

CHING, TE MAY, b Soochow, China, Jan 9, 23; US citizen; m 46; c 2. PLANT PHYSIOLOGY. *Educ:* Nat Cent Univ, China, BS, 44; Mich State Col, MS, 50; Mich State Univ, PhD(cytol), 54. *Prof Exp:* Asst wood chem, Nat Cent Univ, China, 44-48; asst plant anat, hist & cytol, Mich State Univ, 50-52, asst plant physiol & cytol, 52-54, instr, 54-56; from asst prof to assoc prof seed physiol, 56-71, PROF SEED PHYSIOL, ORE STATE UNIV, 71- *Concurrent Pos:* Researcher, Wenner-Gren Inst, Stockholm, Sweden, 67-68

& Commonwealth Sci & Indust Res Orgn, Canberra, Australia, 79-80. *Mem:* AAAS; Am Soc Plant Physiol; fel Am Soc Agron; fel Crop Sci Soc Am. *Res:* Seed physiology; cytology; lipid metabolism; structure and function of cellular organelles; developmental biology. *Mailing Add:* 3240 Elmwood Dr Corvallis OR 97330

CHING, WAI-YIM, b Shoashing, China, Oct 18, 45; US citizen; m 75; c 2. SOLID STATE PHYSICS, COMPUTATIONAL PHYSICS. *Educ:* Univ Hong Kong, BSc, 69; La State Univ, MS, 71 & PhD(physics), 74. *Prof Exp:* Res assoc & lectr physics, Univ Wis, Madison, 74-78; from asst prof to prof, 78-88, CURATORS' PROF PHYSICS, UNIV MO, KANSAS CITY, 88-, CHMN PHYSICS DEPT, 90- *Concurrent Pos:* Consult, Argonne Nat Lab, 78-82; vis prof, Univ Sci & Technol China, 83-; grantee, US Dept Energy, 79- *Honors & Awards:* NT Veatch Award, Univ Mo, Kansas City, 85 & Curators' prof, 88- *Mem:* Mem, AAAS; Am Phys Soc; Am Vacuum Soc; Mat Res Soc; Sigma Xi; Am Ceramic Soc. *Res:* Theoretical condensed matter physics; electronic, structural, optical, magnetic and superconducting properties of crystalline and non-crystalline solids. *Mailing Add:* Dept Physics Univ Mo Kansas City MO 64110-2499

CHINITZ, WALLACE, b Brooklyn, NY, Mar 13, 35; m 60; c 2. MECHANICAL ENGINEERING. *Educ:* City Col New York, BME, 57; Polytech Inst Brooklyn, MME, 59, PhD(mech eng), 62. *Prof Exp:* Res engr, Fairchild Engine & Airplane Corp, 57-59; sr sci res engr, Plasma Propulsion Proj, Repub Aviation Corp, 59-60; res asst prof mech eng, Polytech Inst Brooklyn, 60-63; proj engr gas dynamic & combustion res, Gen Appl Sci Labs, Inc, 63-67; PROF MECH ENG, SCH ENG & SCI, COOPER UNION, 67- *Mem:* Am Soc Mech Engrs; Am Soc Eng Educ; assoc fel Am Inst Aeronaut & Astronaut; Combustion Inst. *Res:* High-temperature gas dynamics and combustion; thermodynamics; transport processes; air pollution control; energy. *Mailing Add:* Cooper Union Col 51 Astor Pl New York NY 10003

CHINN, CLARENCE EDWARD, b Cheney, Wash, Dec 1, 25; m 53; c 3. CHEMISTRY. *Educ:* Walla Walla Col, BA, 51; Ore State Col, MS, 53, PhD(soils), 56; Univ Tenn, PhD(inorg chem), 69. *Prof Exp:* From asst prof to assoc prof chem & math, Southern Missionary Col, 56-67; from assoc prof to prof chem, Walla Walla Col, 67-81; RETIRED. *Concurrent Pos:* Dairy processing & bacteriology. *Mem:* Am Chem Soc. *Res:* Soil moisture measurement; effect of herbicides on plant enzymes; solvent extraction of metal chelates. *Mailing Add:* 649 SW Third College Place WA 99324

CHINN, HERMAN ISAAC, b Connellsville, Pa, Apr 8, 13; m 45; c 4. BIOCHEMISTRY. *Educ:* Pa State Col, BS, 34; Northwestern Univ, MS, 35, PhD(biochem), 38. *Prof Exp:* Instr biochem, Med Sch, Northwestern Univ, 38-42; prin chemist, Fla State Bd Health, 46-47; chief dept biochem, Sch Aviation Med, 47-55; sci liaison officer, Off Naval Res, London, 55-57; biochemist, Air Force Off Sci Res, 57-60; dep sci attache, Am Embassy, Ger, 60-63; sci officer, Off Int Sci Affairs, US Dept State, DC, 63-65; sci attache, Am Embassy, Tehran, Iran, 65-67; sci officer, US Dept State, 67-70; sci attache, Am Embassy, Stockholm, Sweden, 70-73; sci attache, Am Embassy, Tel Aviv, Israel, 73-75; consult, US Dept State, Washington, DC, 75-78; sr staff scientist, Fed Am Soc Exp Biol, Bethesda, MD, 76-82; RETIRED. *Mem:* AAAS; Soc Exp Biol & Med; Am Physiol Soc; Am Soc Pharmacol & Exp Therapeut; Am Chem Soc. *Res:* Biochemistry of the eye; aviation physiology; motion sickness. *Mailing Add:* 1257 Harmony Circle Weatherford TX 76087

CHINN, LELAND JEW, b Sacramento, Calif, Oct 19, 24; m 59; c 1. MEDICINAL CHEMISTRY. *Educ:* Univ Calif, BS, 48; Univ Wis, PhD(chem), 51. *Prof Exp:* Asst to prof, Univ Wis, 48-51; res chemist, 52-70, group leader, 70-72, res fel, G D Searle & Co, 72-85 & NutraSweet Co, 85-86; PROF, BIOLA UNIV, 86- *Concurrent Pos:* Vis scientist, Univ Southern Calif, 68. *Mem:* AAAS; Am Chem Soc. *Res:* Natural products; stereochemistry of polycyclic compounds; medicinal chemistry. *Mailing Add:* 10333 Lundene Dr Whittier CA 90601-2032

CHINN, PHYLLIS ZWEIG, b Rochester, NY, Sept 26, 41; m 68; c 2. COMBINATORIES & FINITE MATHEMATICS. *Educ:* Brandeis Univ, BA, 62; Harvard Univ, MAT, 63; Univ Calif, San Diego, MA, 66; Univ Calif, Santa Barbara, PhD(math), 69. *Prof Exp:* Teacher jr high sch, Mass, 63-64; instr math, Mass State Col Salem, 64; asst prof math, Towson State Col, 69-75; from asst prof to assoc prof, 75-83, PROF MATH HUMBOLDT STATE UNIV, 84- *Concurrent Pos:* Coordr, Ann Conf Nat Womens Studies Asn, 82; vis prof, Univ Cent Fla, 83-84; Women & Math Speaker's Bur; dir, Redwood Area Math Proj, 88- *Mem:* Math Asn Am; Nat Coun Teachers Math; Asn Women Sci; Asn Women Math. *Res:* Graph reconstruction problems; frequency partition of graphs; means of improving the teaching of mathematics to prospective teachers; graph coloring problems; coding; discovery learning of mathematics; graphical operations and properties; band width in graphs; women in science and math; graph theory models for wildlife management; primal graphs; domination critical graphs. *Mailing Add:* Dept Math Humboldt State Univ Arcata CA 95521

CHINN, STANLEY H F, b Vancouver, BC, Apr 9, 14; m 44; c 2. SOIL MICROBIOLOGY. *Educ:* Iowa State Col, BS, 40, MSc, 42, PhD, 46. *Prof Exp:* Lectr, Univ Sask, 49-51; BACTERIOLOGIST, RES BR, CAN DEPT AGR, 51- *Concurrent Pos:* Adj prof, Univ Sask. *Mem:* Can Soc Phytopath; Can Soc Microbiol. *Res:* Soil microbiology as related to common rootrot of wheat. *Mailing Add:* 2005 Pembina Saskatoon SK S7Z 1C5 Can

CHINNERY, MICHAEL ALISTAIR, b London, Eng, Sept 27, 33; m 64; c 2. GEOPHYSICS, SEISMOLOGY. *Educ:* Cambridge Univ, BA, 57,MA, 61, DSc, 77; Univ Toronto, MA, 59, PhD(geophys), 62; Brown Univ, MA, 67. *Prof Exp:* Geophysicist, Seismol Serv, Ltd, Eng, 57-58 & Hunting Surv Corp, 59; lectr geophys, Univ Toronto, 61-62; instr, Univ BC, 62-63, asst prof, 63-65; res assoc geol & geophys, Mass Inst Technol, 65-66; from assoc prof to prof, Dept Geol Sci, Brown Univ, 66-73; group leader. Lincoln Lab, Mass Inst Technol, 73-82; DIR, NAT GEOPHYS DATA CTR, NAT OCEANIC & ATMOSPHERIC ADMIN, 82- *Concurrent Pos:* Assoc ed, J Geophys Res, Am Geophys Union, 70-72; chmn adv subcomt geophys & geol, NASA, 78-81; chmn study on geophys data & pub policy, Nat Acad Sci, 80-85; mem coord comt, Data Exchange & Data Ctr, Int Comn Lithosphere, 81-; mem, Panel World Data Ctr, Int Coun Sci Unions, 81- *Mem:* Am Geophys Union; Seismol Soc Am; fel Royal Astron Soc. *Res:* Displacements and stresses in faulting; strength of earth's crust; earthquake mechanism; geotectonics; elasticity theory; earthquake risk; seismic discrimination. *Mailing Add:* Nat Geophys Data Ctr NOAA 325 Broadway Boulder CO 80303-3328

CHINNICI, JOSEPH (FRANK) PETER, b Philadelphia, Pa, Oct 12, 43; div; c 3. GENETICS, EVOLUTIONARY BIOLOGY. *Educ:* La Salle Col, AB, 65; Univ Va, PhD(biol), 70. *Prof Exp:* Asst prof, 70-79, ASSOC PROF BIOL, VA COMMONWEALTH UNIV, 79-; ASSOC PROF HUMAN GENETICS, MED COL VA, 81- *Honors & Awards:* Andrew Fleming Award, Univ Va, 70. *Mem:* Genetics Soc Am; AAAS; Soc Study Evolution; Sigma Xi. *Res:* Genetic control of crossing-over in Drosophila melanogaster; genetic control of aflatoxin and caffeine toxicity resistance in Drosophila melanogaster; effects on crossing-over. *Mailing Add:* Dept Biol Va Commonwealth Univ Box 2012 Richmond VA 23284-2012

CHINOWSKY, WILLIAM, b New York, NY, Feb 24, 29; m 50; c 2. PHYSICS. *Educ:* Columbia Univ, AB, 49, AM, 51, PhD(physics), 55. *Prof Exp:* Res assoc physics, Brookhaven Nat Lab, 54-56, assoc physicist, 56-61; assoc prof, 61-67, PROF PHYSICS, UNIV CALIF, BERKELEY, 67- *Mem:* Am Phys Soc. *Res:* High energy physics. *Mailing Add:* Dept Physics Univ Calif Berkeley CA 94720

CHIO, E(DDIE) HANG, b Macao, Portugual, Mar 25, 48; m 73; c 2. ENTOMOLOGY, TOXICOLOGY. *Educ:* Nat Taiwan Univ, BS, 70; Univ Ill, MS, 76, PhD(entom), 77. *Prof Exp:* SR ENTOMOLOGIST, ELI LILLY & CO, 77-, CHMN, ENTOM RES COMT, 79- *Concurrent Pos:* Bioanalytical chemist; fibrinolysis. *Mem:* Entom Soc Am; Am Registry Prof Entomologists. *Res:* Microbial insecticides research. *Mailing Add:* Lilly Res Labs 1200 Kentucky Ave Indianapolis IN 46221

CHIOGIOJI, MELVIN HIROAKI, b Hiroshima, Japan, Aug 21, 39; US citizen; m 60; c 2. ENGINEERING, OPERATIONS RESEARCH. *Educ:* Purdue Univ, BSEE, 61; Univ Hawaii, MBA, 68; George Washington Univ, DBA, 72. *Prof Exp:* Head weapons component div eng, Qual Eval & Eng Lab, US Navy, 65-69, dir weapons eval & eng div test & eval, Naval Ord Systs Command, Wash, DC, 69-73; dir, Off Indust Anal Energy Conserv, Fed Energy Admin, 73-75, dir, Div Commercialization, 75-80, dep asst secy, 80-84, dir, Off Transp Systs, 84-90, CONSTRUCT MGR, OFF NEW PROD REACTORS, US DEPT ENERGY, 90- *Concurrent Pos:* Prof, George Washington Univ, 72-; mem, Md State Adv Comt Civil Rights, 75-79 & Nat Naval Reserve Policy Bd, 77-80; consult ed, Marcel Dekker, Inc Publs, 78- *Mem:* Inst Elec & Electronics Engrs; Nat Soc Prof Engrs; Acad Mgt; Soc Am Mil Engrs; Asn Sci Technol & Innovation (pres). *Res:* Research and development management theory; energy supply and conservation; industrial energy conservation; buildings energy conservation; management of innovation and technology. *Mailing Add:* 15113 Middlegate Rd Silver Spring MD 20905

CHIOLA, VINCENT, b Bayonne, NJ, May 7, 22; m 52; c 2. INORGANIC CHEMISTRY, LUMINESCENT MATERIALS & CHEMICALS. *Educ:* Wagner Col, BS, 47; Univ Tex, MA, 50. *Prof Exp:* Chemist, Gen Aniline & Film Corp, 50-51; asst, Plastics Lab, Princeton Univ, 51; eng chemist, Chem & Metall Div, Phospher Develop Lab & Phospher Pilot Plant, GTE Sylvania Inc, 51-60, from develop eng to adv develop engr, 60-68, sect head, Chem Develop Lab, 68-69, sect head chem & metall div, Phosphor Develop Lab & Pilot Plant & adminr, res & develop contracts, 69-82; CONSULT, 82- *Mem:* AAAS; Am Chem Soc; Electrochem Soc; Nat Contract Mgt Asn. *Res:* Allyl compounds reaction rates; polyurethane polymerization; protective coatings; chemistry of tungsten and molybdenum; electronic grade chemicals; inorganic luminescent chemicals; phosphors. *Mailing Add:* 329 York Ave Towanda PA 18848

CHIONG, MIGUEL ANGEL, internal medicine, cardiology, for more information see previous edition

CHIOTTI, PREMO, physical chemistry; deceased, see previous edition for last biography

CHIOU, C(HARLES), b Kashing, China, July 15, 24; nat US; m 60; c 2. PHYSICAL METALLURGY, MATERIALS SCIENCE. *Educ:* Pei-Yang Univ, China, BS, 48; Mo Sch Mines, MS, 54; Northwestern Univ, PhD, 59. *Prof Exp:* Asst eng, Taiwan Indust & Mining Corp, 48-52; asst, Northwestern Univ, 55-58; mem res staff, Res Ctr, 58-64, adv metallurgist, 64-65, mgr joining technol dept, 66-68, mgr thin film develop, 63-69, adv engr, 69-77, sr engr & mgr inorg mat, 72-75, sr engr & mgr film process develop, 76-81, PROG COORDR, TECHNICAL STAFF, IBM CORP, 81- *Mem:* Am Soc Metals; Am Inst Mining Metall & Petrol Engrs; Am Vacuum Soc; Sigma Xi; Electrochem Soc. *Res:* Vacuum deposited thin films; superconductivity metals; metallurgy, insulation and interconnection technology of integrated circuits; photoconductor materials and electrophotography; thin film technology; ink jet technology; magnetic bubble device fabrication. *Mailing Add:* 6751 Lookout Bend San Jose CA 95120

CHIOU, CARY T(SAIR), b Maioli, Taiwan, Nov 22, 40; US citizen; m 68; c 2. PHYSICAL CHEMISTRY, SURFACE CHEMISTRY. *Educ:* Cheng Kung Univ, Taiwan, BSE, 65; Kent State Univ, MS, 70, PhD(phys chem), 73. *Prof Exp:* Grad asst phys chem, Kent State Univ, 68-73; fel chem kinetics, Brown Univ, 73-74; fel phys chem, Univ Ky, 74-75; res assoc, Ore State Univ, 75-78, from asst prof to assoc prof environ chem, 78-83; RES HYDROLOGIST & CHEMIST, US GEOL SURV, DENVER, 83- *Concurrent Pos:* Environ lectr, UN Develop Prog, Najing Univ, People's Repub China, 83; environ sci & eng lectr, Nat Sci Coun, Nat Taiwan Univ, 87. *Honors & Awards:* Spec Achievement Award, US Geol Surv, 85. *Mem:*

Am Chem Soc; AAAS. *Res:* Environmental dynamics of organic compounds; evaporation from aqueous and nonaqueous systems; partition equilibria in solvent (lipid)-water mixtures; sorptive mechanisms with soils, sediments, and activated carbons; bioconcentration of organic compounds; interactions of organic solutes with dissolved natural and anthrapogenic organic matter. *Mailing Add:* US Geol Surv Box 25046 MS 408 Denver Fed Ct Denver CO 80225

CHIOU, CHII-SHYOUNG, chemical engineering, for more information see previous edition

CHIOU, GEORGE CHUNG-YIH, b Taoyuan, Taiwan, July 11, 34; US citizen; m 61; c 2. PHARMACOLOGY, BIOCHEMISTRY. *Educ:* Nat Taiwan Univ, BS, 57, MS, 60; Vanderbilt Univ, PhD(pharmacol), 67. *Prof Exp:* Pharmacist, William Pharmaceut Works, Taiwan, 60-61; pharmacist, Chinese Air Force Hosp, 61-62; instr pharmacol, China Med Col, Taiwan, 62-64; from res asst to res assoc, Col Med, Vanderbilt Univ, 64-68; fel pharmacol, Univ Iowa, 68-69; from asst prof to prof pharmacol, Col Med, Univ Fla, 69-78; asst dean, 85, assoc dean, 87-90, PROF & HEAD DEPT PHARMACOL, COL MED, TEX A&M UNIV, 78-, DIR, INST OCULAR PHARMACOL & INST MOLECULAR PATHOGENESIS & THERAPEUT, 84- *Concurrent Pos:* NIH health sci advan award pharmacol, Vanderbilt Univ, 67-68; NIH res grant, Univ Fla, 69-71, Nat Inst Neurol Dis & Stroke res grant, 71-74; Nat Cancer Inst res grant, 75-77; Am Cancer Soc res grant, 77-79; Nat Eye Inst res grants, 76-78 & 81-; Cooper Vision Labs res grant, 80-81; Barnes-Hind Hydrocurve res grant, 85-87, Houston Biotechnol Inc, res grant, 86-88; adj prof biotechnol, Baylor Col Med, 86-, adj prof ophthal, 87-; Am Cyanamid res grant, 89-91. *Honors & Awards:* Distinguished Achievement Award Res, Texas A & M Univ, 84. *Mem:* NY Acad Sci; Am Soc Pharmacol & Exp Therapeut; Sigma Xi; Soc Exp Biol Med; Asn Res Vision & Ophthal; Asn Med Sch Pharmacol. *Res:* Autonomic pharmacology; calcium antagonists; neurochemistry; enzymology; structure-activity relationships of cholinergic and cholinolytic agents; enzyme kinetics of acetylcholinesterase and butyrylcholinesterase; action mechanisms of nicotinic responses; nature of cholinergic receptor; cytolysis of neuroblastomas; treatment of glaucoma; aqueous humor dynamics; ocular pharmacology. *Mailing Add:* Dept Med Pharmacol & Toxicol Tex A&M Univ Col Med College Station TX 77843

CHIOU, MINSHON JEBB, b China, Feb 1, 51; m 80; c 2. FIBER SPINNING TECHNOLOGY, REACTION ENGINEERING. *Educ:* Nat Taiwan Univ, BS, 73; Univ Del, PhD(chem eng), 79. *Prof Exp:* Sr scientist, Jaycor, San Diego, Calif, 79-82; RES ASSOC, E I DU PONT DE NEMOURS & CO, INC, RICHMOND, VA, 82- *Mem:* Am Inst Chem Engrs; Am Chem Soc. *Res:* New product/process development with high performance fiber; fiber spinning technology; reaction engineering; process simulation; transport phenomena; fossil energy conversion processes. *Mailing Add:* E I Du pont de Nemours Co Spruance Res Lab PO Box 27001 Richmond VA 23261

CHIOU, WEN-AN, b Nan-chou, Taiwan, June 24, 48; US citizen; m 73; c 2. ELECTRON MICROSCOPY OF MATERIALS, AMORPHOUS & NANOCRYSTALLINE MATERIALS. *Educ:* Chinese Culture Univ, Taiwan, BS, 70; Univ SFla, Tampa, MS; Tex A&M Univ, College Station, Tex, PhD(oceanog), 81. *Prof Exp:* Staff geologist & supvr x-ray lab x-ray diffraction, Reservoirs, Inc, Houston, Tex, 81-83, sr staff geologist & head anal lab x-ray & sem, 83-86; vpres x-ray & sem, Accumin Anal, 86; electron microscopist & mgr tem lab electron micros, 86-91, ASSOC RES PROF ELECTRON MICROS, NORTHWESTERN UNIV, EVANSTON, ILL, 91- *Concurrent Pos:* Consult & vpres, Accum Anal, Inc, Houston, Tex, 81-; adj lectr, Univ Houston, Clear Lake, Tex, 83-86; consult res scientist, Naval Oceanog & Atmospheric Lab, 88-; consult, Parex Environ, Inc, Addison, Ill, 91-; adj assoc prof, Dept Mech & Indust Eng, Marquette Univ, Milwaukee, Wis, 91- *Mem:* Electron Micros Soc Am; Mat Res Soc; Minerals, Metals & Mat Soc; Clay Minerals Soc; Mineral Soc Am; Am Geophys Union. *Res:* Amorphous and nanocrystalline materials, both metallic alloys and inorganic minerals formed by different techniques using transmission electron microscopy and x-ray diffraction techniques; in situ clay mineralogy and fabric by specially designed environmental cell in a transmission electron microscopy; mineral physics; applied clay science. *Mailing Add:* Dept Mat Sci & Eng Northwestern Univ Evanston IL 60208

CHIOU, WIN LOUNG, b Hsinchu, Taiwan, Aug 29, 38; m 63; c 2. PHARMACOLOGY. *Educ:* Nat Taiwan Univ, BS, 61; Univ Calif, San Francisco, PhD(pharmaceut chem), 69. *Prof Exp:* From res assoc to asst prof pharm, Wash State Univ, 69-71; asst prof pharm, 71-73, assoc prof pharm & occup & environ med, 73-75, dir, Clin Pharmacokinetics Lab, Col Pharm, 75-79, PROF PHARM, MED CTR, UNIV ILL, 76- *Concurrent Pos:* Consult, Med Lett Drugs & Therapeut, 74-75; mem pharmacol study sect, NIH, 81. *Mem:* Am Pharmaceut Asn; Am Asn Pharmaceut Scientists. *Res:* Biopharmaceutics and pharmacokinetics of drugs; drug interactions; solid dispersion formulation of dosage forms; blood level monitoring; liquid chromatography in drug analysis; renal function; formulation of dosage forms. *Mailing Add:* Col Pharm M/C 865 Univ Ill Chicago IL 60680

CHIPAULT, JACQUES ROBERT, biochemistry; deceased, see previous edition for last biography

CHIPLEY, ROBERT MACNEILL, b Cincinnati, Ohio, Nov 20, 39; m 67; c 2. ORNITHOLOGY. *Educ:* Yale Univ, BA, 61; Cornell Univ, PhD(ecol), 74. *Prof Exp:* Sci ed, Dover Publ, 65-68; DIR, NATURAL HERITAGE PROG, NATURE CONSERVANCY, 74- *Mem:* Ecol Soc Am; Am Ornithologists Union; Wilson Ornith Soc. *Res:* Inventory and preservation of natural areas; conservation of endangered species; caribbean ornithology. *Mailing Add:* Nature Conservancy 1815 N Lynn St Arlington VA 22209

CHIPMAN, DAVID MAYER, b New York, NY, Oct 7, 40; m 62; c 3. ENZYMOLOGY, BIO-ORGANIC CHEMISTRY. *Educ:* Columbia Univ, BA, 62, PhD(org chem), 65. *Prof Exp:* Nat Acad Sci-Nat Res Coun res fel biophys, Weizmann Inst, 65-66, NIH res fel, 66-67; asst prof chem, Mass Inst Technol, 67-71; sr lectr biochem, Ben Gurion Univ Negev, 71-74, chmn dept biol, 74-77, assoc prof biochem, 74-89, PROF BIOCHEM, BEN GURION UNIV NEGEV, 90- *Concurrent Pos:* Vis assoc prof chem, Univ Ore, 77-78; vis res scientist, Brandeis Univ, 83-84 & CR&D Dept, E I du Pont de Nemours & Co, 90; vis prof med biochem, Moi Univ, Kenya, 91. *Mem:* AAAS; Am Chem Soc; Israel Biochem Soc; Israel Biophys Soc. *Res:* Enzyme mechanisms particularly relationships between protein structure, ligand binding, catalysis and specifity; specific interest in SR Ca-ATPase and acetohydroxy acid synthase; physical organic chemistry; dynamics and control of metabolic pathways. *Mailing Add:* Ben Gurion Univ Negev PO Box 653 Beersheba Israel

CHIPMAN, GARY RUSSELL, b Berlin, Wis, Feb 27, 43; m 67; c 2. ORGANIC CHEMISTRY. *Educ:* Univ Wis, BS, 65; Univ Mich, MS, 67, PhD(org chem), 70. *Prof Exp:* Res chemist, Amoco Chem Corp, 70-76, res chemist, Standard Oil Co, Ind, 76-81, SR RES CHEMIST, AMOCO CORP, 81- *Mem:* Am Chem Soc; Am Soc Mass Spectrometry. *Res:* Product and process research on random olefin copolymers, polyesters, and block copolymers; organic analysis; mass spectrometry. *Mailing Add:* Amoco Res Ctr PO Box 3011 Naperville IL 60566

CHIPMAN, R(OBERT) A(VERY), b Winnipeg, Man, Apr 28, 12; m 38; c 3. HISTORY & PHILOSOPHY OF SCIENCE. *Educ:* Univ Man, BSc, 32; McGill Univ, MEng, 33; Cambridge Univ, PhD(physics), 39. *Prof Exp:* Asst prof physics, Acadia Univ, 38-40; res asst & asst prof, Queen's Univ, Ont, 40-46; assoc prof elec eng, McGill Univ, 46-57; prof elec eng, 57-78, chmn dept, 59-65, EMER PROF ELEC ENG, UNIV TOLEDO, 78- *Mem:* Fel Inst Elec & Electronics Engrs; Soc Hist Technol. *Res:* Electrical measurements at very high frequencies; history of science and technology; dielectric and magnetic materials. *Mailing Add:* 3547 Rushland Ave Toledo OH 43606-2043

CHIPMAN, ROBERT K, b New York, NY, Nov 16, 31; m 54; c 2. ZOOLOGY. *Educ:* Amherst Col, AB, 53; Tulane Univ, MS, 58, PhD(zool), 63. *Prof Exp:* From instr to asst prof biol, State Univ NY Col Plattsburgh, 61-62; from asst prof to assoc prof zool, Univ Vt, 62-68, asst dean grad sch, 67-68; chmn dept, 68-74, PROF ZOOL, UNIV RI, 68- *Mem:* Am Soc Mammal; Am Soc Zool. *Res:* Morphological variation of fish; vertebrate ecology; rodent population dynamics; vertebrate physiological ecology; physiology of reproduction of rodents. *Mailing Add:* Goshen Rd Bradford VT 05033

CHIPMAN, WILMON B, b Reading, Mass, July 6, 32; m 60; c 3. ORGANIC CHEMISTRY, BIOCHEMISTRY. *Educ:* Harvard Univ, AB, 54; Dartmouth Col, AM, 56; Univ Ill, PhD(org chem), 60. *Prof Exp:* Instr chem, Colby Col, 60-62, asst prof, 62-65; assoc prof, 65-67, chmn dept, 65-81, PROF CHEM, BRIDGEWATER STATE COL, 67- *Concurrent Pos:* Vis lectr, NSF Inst, Mt Hermon Sch, 61-64; NSF vis lectr, 63-65; lectr, NSF in-serv inst, Bridgewater State Col, 66-71, 75-78 & 80-81. *Mem:* Am Chem Soc. *Res:* Heterocyclic chemistry; infrared and ultraviolet spectroscopy; transannular interactions; molecular modeling; nuclear magnetic resonance spectrometry; microcomputers in chemistry; computer-aided instruction; copy editing and scientific writing. *Mailing Add:* 64 Pleasant Dr Bridgewater MA 02324

CHIQUOINE, A DUNCAN, b Upland, Pa, May 3, 26; m 50; c 4. CELL BIOLOGY. *Educ:* Swarthmore Col, AB, 47; Cornell Univ, PhD(zool), 52. *Prof Exp:* Asst, Cornell Univ, 47-52; instr anat, Univ Wash, 52-53; from instr to asst prof biol, Princeton Univ, 53-57; from asst prof to assoc prof anat, Wash Univ, 57-67; PROF BIOL & CHMN DEPT, HAMILTON COL, 72- *Mem:* AAAS; Histochem Soc. *Res:* Histochemistry and cytochemistry; mammalian embryology; electron microscopy of mammalian embryos; germ cells. *Mailing Add:* Dept Biol Hamilton Col Clinton NY 13323

CHIRIGOS, MICHAEL ANTHONY, b Wierton, WVa; Sept 14, 24; c 3. BIOCHEMISTRY. *Educ:* Western Md Col, BS, 52; Univ Del, MS, 54; Rutgers Univ, PhD(biochem), 57. *Prof Exp:* Asst bact, Univ Del, 52-54; fel, Nat Heart & Lung Inst, 57-59, pharmaceut chemist, 59-67; head viral chemother sect, Drug Eval Br, 66-67, HEAD VIRUS & DIS MODIFICATION SECT, NAT CANCER INST, 67-, ASSOC BR CHIEF VIRAL BIOL BR, 70-, CHIEF, IMMUNOPHARMACOL SECT, 81- *Mem:* AAAS; Am Asn Cancer Res; Soc Exp Biol & Med. *Res:* Chemotherapy of cancer and oncogenic viruses; oncogenic virology; transport mechanisms in vitro and in vivo; immune modifiers; biological response modifiers. *Mailing Add:* Four Cold Spring Ct Rockville MD 20854

CHIRIKJIAN, JACK G, b Dec 10, 40; US citizen; m 64; c 2. BIOLOGICAL CHEMISTRY. *Educ:* Trenton State Col, BA, 63; Rutgers Univ, MS, 66, PhD(biochem), 69. *Prof Exp:* Nat Cancer Inst fel biochem, Princeton Univ, 69-71, res assoc, 71-72; from asst prof to assoc prof, 72-81, PROF BIOCHEM, SCHS MED & DENT, GEORGETOWN UNIV, 81- *Concurrent Pos:* Sr consult & dir bd, BRL, Inc, 75-83, pharmacia Fine Chem, 74-; Leukemia Soc Am scholar, 75-80. *Mem:* Am Soc Biol Chemists; Am Soc Microbiol; Am Chem Soc; Sigma Xi; Am Asn Univ Professors. *Res:* Studies dealing with protein-nucleic acid interactions using sequence specific endonucleases as model systems; emphasis is placed on structure and function and uses of the various enzymes. *Mailing Add:* Dept Biochem Georgetown Univ Sch Med & Dent Washington DC 20007

CHIRLIAN, PAUL M(ICHAEL), b New York, NY, Apr 29, 31; m 61; c 2. ELECTRICAL ENGINEERING, ELECTRONICS. *Educ:* NY Univ, BEE, 50, MEE, 52, EngScD, 56. *Hon Degrees:* ME, Stevens Inst Technol, 65. *Prof Exp:* Asst elec eng, NY Univ, 50-51, from instr to asst prof, 51-60; from assoc prof to prof, 60-84, ANSON WOOD BURCHARD PROF ELEC ENG, STEVENS INST TECHNOL, 84- *Concurrent Pos:* Consult, Anesthesia

Assoc, 57-58 & Radio Corp Am, 62-63; consult ed, Matrix Publ, Inc, 78- *Mem:* Fel Am Inst Elec & Electronics Engrs; Am Soc Eng Educ; Sigma Xi. *Res:* Electronic circuits; network theory; physical and medical electronics; digital signal processing; filter synthesis; analog and digital electronics. *Mailing Add:* Dept Elec Eng Stevens Inst Technol Hoboken NJ 07030

CHISARI, FRANCIS VINCENT, b New York, NY, April 5, 42; m 67; c 2. IMMUNOPATHOLOGY. *Educ:* Fordham Univ, AB, 63; Cornell Univ Med Col, MD, 68. *Prof Exp:* Fel anat path, Mayo Grad Sch, Rochester, 69-70; staff assoc exp path, NIH, Md, 70-72; resident internal med, Dartmouth Med Sch, 72-73; res fel exp path, 73-75, asst mem molecular immunol, 75-81, assoc mem clin res, 81-88; MEM & HEAD DIV EXP PATH, SCRIPPS CLIN & RES FOUND, 88- *Concurrent Pos:* Consult, Naval Regional Med Ctr, San Diego, 75-; adj assoc prof, Dept Path, Univ Calif, San Diego, 76-87, Diag Immunol Lab, Orange, Calif, 78-83; Adj prof, Dept Path, Univ Calif, San Diego, 87- *Mem:* Am Asn Immunologists; Am Asn Path; fel Am Col Physicians. *Res:* Virology and immunology. *Mailing Add:* Dept Basic & Clin Res Res Inst Scripps Clin 10666 N Torrey Pines Rd LaJolla CA 92037

CHISCON, J ALFRED, b Kingston, Pa, Feb 18, 33; m 69. GENETICS, EVOLUTION. *Educ:* Bloomsburg State Col, BS, 54; Purdue Univ, MS, 56, PhD(biol), 61. *Prof Exp:* From instr to assoc prof, 61-70, PROF BIOL, PURDUE UNIV, 70- *Concurrent Pos:* Carnegie fel, 68-69. *Mem:* AAAS. *Res:* Social impact of biology; molecular evolution; nucleic acid reassociation studies; DNA relationships between primates; rate of nucleotide sequence change during evolution. *Mailing Add:* Dept Biol Sci Purdue Univ Lafayette IN 47907

CHISCON, MARTHA OAKLEY, b Chicago, Ill, Aug 27, 35; m 69; c 2. PLANT CELL BIOLOGY. *Educ:* Western Ill Univ, BSEd, 56; Purdue Univ, PhD(immunobiol), 71. *Prof Exp:* High sch teacher biol, chem & physics, Ill & Alaska, 56-64; instr biol, 64-70, NIH fel immunol, 71-72, asst prof, 72-77, ASSOC PROF BIOL, PURDUE UNIV, 77- *Mem:* Soc Develop Biol; Genetics Soc; Tissue Cult Asn; Sigma Xi; AAAS. *Res:* Plant cell biology, tissue culture and genetics. *Mailing Add:* Dept Biol Life Sci Bldg Purdue Univ West Lafayette IN 47906

CHISHOLM, ALEXANDER JAMES, b Minnedosa, Man, July 28, 41; m 66; c 3. CLOUD PHYSICS. *Educ:* Univ Alta, BSc, 62; McGill Univ, MSc, 66, PhD(radar meteorol), 70. *Prof Exp:* Meteorol officer, Forecast Div, Can Meteorol Serv, 62-66, meteorologist, 66-70, res scientist, Res & Training Div, 70-74, chief, Cloud Physics Res Div, 74-79, dir, Atmospheric Processes Res Br, Atmospheric Res Environ Serv, Toronto, 79-89, SCI ADV, ATMOSPHERIC RES ENVIRON SERV, OTTAWA, 90- *Mem:* Am Meteorol Soc; Can Meteorol Soc. *Res:* Hailstorm airflow; hail suppression concepts; rainfall enhancement of cumuliform clouds by weather modification. *Mailing Add:* Atmospheric Environ Serv Jules Leger Bldg 14th Floor 25 Eddy St Ottawa PQ K1A 0H3 Can

CHISHOLM, DAVID R, b Chippewa Falls, Wis, Sept 24, 23; c 4. INFECTIOUS DISEASES. *Educ:* Univ Minn, BS, 50; Univ Mich, PhD(bacteriol), 58. *Prof Exp:* Microbiologist, Pfizer & Co, 58-64; MICROBIOLOGIST, BRISTOL LABS, 65- *Res:* Immunolgoy and chemotherapy of infectious diseases. *Mailing Add:* 5100 Highbridge N Fayetteville NY 13066

CHISHOLM, DONALD ALEXANDER, b Waltham, Mass, Oct 27, 36; m 58; c 4. METEOROLOGY, APPLIED STATISTICS. *Educ:* Tufts Univ, BS, 58; NY Univ, MS, 69. *Prof Exp:* Res assoc meteorol, Travelers Res Ctr Inc, 61-69; res scientist solar physics, Geomet Inc, 69-71; res physicist meteorol, Air Force Cambridge Res Labs, 71-75, sci adminr geophys, Air Force Systs Command Hq, 75-77, SUPVR PHYSICIST METEOROL, AIR FORCE GEOPHYS LAB, 78- *Concurrent Pos:* USAF rep, Interdept Comt Atmospheric Sci Subgroup, Nat Climate Prog Plan, 76-77. *Mem:* Am Meteorol Soc; Sigma Xi. *Res:* Mesoscale meteorology including sensor development for observations, objective analysis procedures and mesoscale weather prediction based on physical and dynamical processes. *Mailing Add:* 166 Lincoln St Lexington MA 02173

CHISHOLM, DOUGLAS BLANCHARD, b Bay City, Mich, Mar 1, 43. ENGINEERING MECHANICS, FRACTURE MECHANICS. *Educ:* Cornell Univ, BME, 66, MME, 67; George Washington Univ, DSc, 75. *Prof Exp:* Engr scientist vibration shock, Missile & Space Systs Div, McDonnell-Douglas, 66-68; mem tech staff systs analysis, Comput Sci Corp, 68-70; sr mech engr vibration shock, Div Litton Industs, Amecom, 70-71; res mech engr transp struct safety, Fed Hwy Admin, 71-80; chief, Res Div, Off Pipeline Safety, Res & Spec Progs Admin, 80-89, DIR MECH ENG, INST SAFETY ANALYSIS, US DEPT TRANSP, 89- *Concurrent Pos:* Res fel, George Washington Univ, 75-; consult patent develop, Aluminum Corp Am, 76-, Transp Safety, Inc, 76- & Pole-Lite Ltd, 76- *Honors & Awards:* Bronze Medal, Fed Hwy Admin, 76; Award, Indust Res Mag, 77. *Mem:* Nat Soc Prof Engrs; Am Soc Mech Engrs; Soc Exp Stress Anal; Am Soc Metals; Am Soc Testing & Mat. *Res:* Structural and transportation safety; fatigue and fracture mechanics; materials testing and evaluation. *Mailing Add:* 2865 Sutton Oaks Lane Vienna VA 22180

CHISHOLM, JAMES JOSEPH, b Natick, Mass, May 29, 36; m; c 3. OPTICS, SPECTROSCOPY. *Educ:* Boston Col, BS, 58; Univ Rochester, MS, 65. *Prof Exp:* Res physicist, Air Force Cambridge Res Ctr, 57-58; from physicist to prog mgr spectrophotom, Bausch & Lomb Inc, 58-75, dir spectrophotom res & develop, 75-77; dir eng, Hach Chem Co, 77-78, vpres corp eng & develop, 78-79; eng mgr, Anal Chem Lab & Mat Anal Lab, 79-81, QUAL ASSURANCE MGR, HEWLETT PACKARD CO, 81- *Concurrent Pos:* Pres & gen mgr, Spectra Technol Inc, 79- *Mem:* Optical Soc Am; Soc Photo-Optical Instrumentation Engrs; Am Soc Testing & Mat. *Res:* Monochromators, spectrophotometers, radiometry and photometry; color measurements; atomic absorption spectroscopy; ultraviolet and infrared optics; light sources and detectors; interferometry; diffraction gratings; photometric standards; water analysis; RF treatment of cancer in animals; photolithography & micrometrology; semiconductor materials analysis and vision systems. *Mailing Add:* 2730 Logan Dr Loveland CO 80538

CHISHOLM, MALCOLM HAROLD, b Bombay, India, Oct 15, 45; UK citizen; m 82; c 3. ORGANOMETALLIC CHEMISTRY, INORGANIC CHEMISTRY. *Educ:* Queen Mary Col, BSc, 66, PhD(chem), 69. *Hon Degrees:* DSc, London Univ, 81. *Prof Exp:* Asst prof chem, Princeton Univ, 72-78; from assoc prof to prof, 78-85, DISTINGUISHED PROF CHEM, IND UNIV, 85- *Concurrent Pos:* Chmn div inorg chem, Am Chem Soc, 87-88; Guggenheim fel, 86-87; Claire Hall fel, Cambridge Univ, 86- *Honors & Awards:* Corday Morgan Medal, Royal Soc Chem, 81; Buck Whitney Medal, Am Chem Soc, 87, Award Inorganic Chem, 89. *Mem:* Fel AAAS; Royal Soc Chem; Am Chem Soc; fel Royal Soc. *Res:* Author of over 300 publications on inorganic chemistry and organometallic chemistry. *Mailing Add:* 515 Hawthorne Bloomington IN 47401

CHISHOLM, SALLIE WATSON, b Marquette, Mich, Nov 5, 47. AQUATIC ECOLOGY, PHYTOPLANKTON PHYSIOLOGY. *Educ:* Skidmore Col, BA, 69; State Univ NY Albany, PhD(biol), 74. *Prof Exp:* Res assoc phytoplankton ecol, Inst Marine Resources, Univ Calif, San Diego, 74-76; from asst prof to assoc prof, 76-86, PROF ENVIRON ENG, MASS INST TECHNOL, 86- *Concurrent Pos:* Edgerton prof, 78-80; Henry L Doherty prof ocean utilization, Mass Inst Technol, 80-83. *Mem:* Am Soc Limnol & Oceanog; Ecol Soc Am; Phycol Soc Am; Int Asn Limnol; Sigma Xi; Soc Analytical Cytology. *Res:* Phytoplankton ecology; biological oceanography. *Mailing Add:* 48-425 Ralph M Parsons Lab Mass Inst Technol Cambridge MA 02139

CHISHTI, ATHAR HUSAIN, b Aligarh, India, Dec 25, 57; m 87; c 1. MEMBRANE-CYTOSKELETON, MALARIA PARASITE-HOST INTERACTIONS. *Educ:* AM Univ, India, BSc, 78, MSc, 80; Univ Melbourne, Fla, PhD(biochem), 84. *Prof Exp:* Teaching fel cell biol, Harvard Univ, Cambridge, 86-88, res assoc biol, 84-88; ASST PROF MED, SCH MED, TUFTS UNIV, 88-; ASSOC INVESTR BIOMED RES, ST ELIZABETH'S HOSP, BOSTON, 89- *Concurrent Pos:* Biol tutor, Harvard Lowell House, 86-89. *Mem:* Am Soc Hemat; Am Soc Cell Biol. *Res:* Membrane-cytoskeletal interactions; phosphorylation and oncogene products with kinase activity; malaria parasite-erythrocyte membrane interactions; molecular basis of cell shape charge. *Mailing Add:* St Elizabeth's Hosp ACH-4 736 Cambridge St Boston MA 02135

CHISLER, JOHN ADAM, b Daybrook, WVa, Feb 25, 37; m 59; c 2. BACTERIOLOGY, GENETICS. *Educ:* Ohio State Univ, BSc, 59, MSc, 61, PhD(plant path), 62. *Prof Exp:* Asst prof bot, Marshall Univ, 62-65; PROF BIOL & CHMN DIV SCI & MATH, GLENVILLE STATE COL, 65- *Mem:* AAAS; Am Inst Biol Sci; Sigma Xi. *Mailing Add:* Glenville State Col Glenville WV 26351

CHISM, GRADY WILLIAM, III, b Tampa, Fla, June 18, 46; m 72; c 3. FOOD SCIENCE. *Educ:* Univ Fla, BS, 68; Univ Mass, Amherst, PhD(food sci), 73. *Prof Exp:* Fel food sci, Cook Col, Rutgers Univ, 73-74; asst prof, 74-79, ASSOC PROF FOOD SCI, OHIO RES & DEVELOP CTR, 80-, ASSOC PROF, BIOTECHNOL CTR, 87-; ASSOC PROF, BIOTECHNOL CTR, OHIO STATE CLINIC, 87- *Mem:* Inst Food Technologists; Am Chem Soc; Am Soc Hort Sci; Am Soc Plant Physiologists. *Res:* Biochemical control mechanisms of enzymatic processes important in the development and maintenance of quality in plant tissues used as food; enzymatic regulation of cytokinin levels in tomato fruits. *Mailing Add:* Dept Food Sci & Nutrit Ohio State Univ 2121 Fyffe Rd Columbus OH 43210-1009

CHISMAN, JAMES ALLEN, b Ravenna, Ohio, Mar 4, 35; m 78. INDUSTRIAL & ELECTRICAL ENGINEERING. *Educ:* Univ Akron, BS, 58; Univ Iowa, MS, 60, PhD(mgt eng), 63. *Prof Exp:* Res & develop engr, Firestone Tire & Rubber Co, 58-59; instr indust eng, Univ Iowa, 62-63; from asst prof to assoc prof, 63-76, coordr, systs eng prog, 68-73, head, Eng Technol Dept, 74-80, Indust Eng Dept, 83-84, PROF SYST ENG, CLEMSON UNIV, 77-; PRES, CLEMSON INVEST & DEVELOP CO, 64- *Concurrent Pos:* Consult, Consumer Power Co, 63-72, J P Stevens, 74, US Navy, 75, E I du Pont de Nemours & Co, 77-78 & IBM, 82; adj prof, Overseas Grad Mgt Prog, Boston Univ, 80-81, Apple Comput, 87-88, Am-Luso Comn Educ, Portugal, 87; Fulbright fel, Ireland, 87; Ryobi Motor Prod Corp, 89. *Mem:* Am Inst Indust Engrs; Inst Mgt Sci; Am Soc Eng Educ; Sigma Xi. *Res:* Operations research; production control; simulation modeling. *Mailing Add:* Indust Eng Dept Clemson Univ Clemson SC 29631

CHISOLM, GUY M, PHYSIOLOGICAL TRANSPORT PROCESSES, VASCULAR DISEASES. *Educ:* Univ Va, PhD(chem eng), 72. *Prof Exp:* ASSOC PROF BIOMED ENG, CASE WESTERN RESERVE UNIV, 75-; MEM RES STAFF, CLEVELAND CLIN FOUND, 77- *Mailing Add:* Res Inst FF401 Cleveland Clin Found 9500 Euclid Ave Cleveland OH 44195

CHISOLM, JAMES JULIAN, JR, b Baltimore, Md, July 24, 21; m 48; c 2. MEDICINE. *Educ:* Princeton Univ, AB, 44; Johns Hopkins Univ, MD, 46; Am Bd Pediat, dipl, 52. *Prof Exp:* Intern pediat, Johns Hopkins Hosp, 46-47, asst resident, Johns Hopkins Hosp & asst in pediat, Sch Med, Johns Hopkins Univ, 48; sr asst resident, Babies Hosp, New York, 50-51; resident, Johns Hopkins Hosp, 51-52, asst pediat, Sch Med, Johns Hopkins Univ, 53-55, from instr to asst prof, 55-63, ASSOC PROF PEDIAT, SCH MED, JOHNS HOPKINS UNIV, 63-, PEDIATRICIAN, JOHNS HOPKINS HOSP, 52- *Concurrent Pos:* Fel pediat, Sch Med, Johns Hopkins Univ, 51-53; hosp physician, Baltimore City Hosp, 53-56, asst chief hosp physician, 56-61, assoc chief hosp physician, 61-74, sr staff pediat, 74-; mem panel on lead, Nat Res Coun, 70-71; consult, US Environ Protection Agency, 76- *Mem:* AAAS; Soc Pediat Res; Am Acad Pediat; Am Pediat Soc; Am Pub Health Asn; Am Asn Clin Chemists. *Res:* Renal tubular function; biochemical effects of lead poisoning; porphyrin metabolism; environmental lead exposure. *Mailing Add:* Lead Prog Kennedy Inst 707 N Broadway Baltimore MD 21205

CHISWIK, HAIM H, b Russia, Nov 29, 15; nat US; m 42; c 3. METALLURGY. *Educ:* Harvard Univ, ScB & ScD(phys metall), 41. *Prof Exp:* Res engr metall, Battelle Mem Inst, 42-44 & Air Reduction Co, 44-48; sr metallurgist & assoc dir, Metall Div, Atomic Energy, Argonne Nat Lab, 48-81; CONSULT, 81- *Mem:* Fel Am Soc Metals; Am Inst Mining Metall & Petrol Engrs; Sigma Xi. *Res:* Physical metallurgy in atomic energy fields. *Mailing Add:* 1525 Thornwood Dr Downers Grove IL 60516

CHITALEY, SHYAMALA D, b Nasik, India, Feb 15, 18; m 34; c 2. DEVONIAN FLORA, LATE CRETACEOUS FLORA. *Educ:* Univ Nagpur, India, BSc, 42, MSc, 45; Univ Reading, Eng, PhD(paleobot), 55. *Prof Exp:* Lectr bot, Col Sci, Nagpur, India, 48-61; assoc prof, Inst Sci, Bombay, 61-68, assoc prof & chmn dept bot, Nagpur, 68-74, prof & head dept, Bombay, 74-76; CUR & HEAD DEPT PALEOBOT, CLEVELAND MUS NATURAL HIST, 80- *Concurrent Pos:* Govt India res grants, 55-75, travel grants, Int Bot Conf, Edinburgh, 64 & Seattle, 69; chief ed, Botanique, 70-80; mem orgn comt, IV Int Palynological Conf; liaison, Fossil Soc Cleveland Mus Natural Hist, 80- *Honors & Awards:* Invention Prom Bd Govt India Award, 67. *Mem:* Bot Soc Am; Int Orgn Paleobot; Am Asn Stratigraphical Palynology; Am Asn Paleobotanists; Linnean Soc London. *Res:* Devonian flora of Cleveland shale; Deccan intertrappean flora of India of late cretaceous age; plant paleontology; palynology. *Mailing Add:* Cleveland Mus Natural Hist One Wade Oval University Circle Cleveland OH 44106-1767

CHITGOPEKAR, SHARAD SHANKARRAO, b Raichur, India, Jan 3, 38. STATISTICS, OPERATIONS RESEARCH. *Educ:* Osmania Univ, India, BA, 57; Univ Poona, MA, 59; Fla State Univ, PhD(statist), 68. *Prof Exp:* Lectr statist, Osmania Univ, India, 59-63; res asst, Fla State Univ, 63-68; vis asst prof, Univ Wis, 70-71 & 72-74; asst prof quant methods, Univ Ill, Chicago Circle, 74-78; assoc prof, 78-85, PROF QUANT METHODS, ILL STATE UNIV, 85- *Concurrent Pos:* Fulbright Travel Award, 63. *Mem:* Am Statist Asn; Am Inst Decision Sci; Opers Res Soc Am. *Res:* Dynamic programming; Markovian decision processes. *Mailing Add:* Dept Mgt & Quant Methods Ill State Univ Normal IL 61761

CHITHARANJAN, D, b Madras, India, May 25, 40; c 3. ORGANIC CHEMISTRY. *Educ:* Annamalai Univ, Madras, BSc, 60, MSc, 61; Wayne State Univ, PhD(org chem), 69. *Prof Exp:* Asst prof, 68-70, assoc prof, 73-76, PROF CHEM, UNIV WIS-STEVENS POINT, 77-, DIR MED TECHNOL, 74- *Mem:* Am Chem Soc; Sigma Xi. *Res:* Synthesis of 4-amino-4, 6-dideoxy hexoses and unsaturated derivatives of carbohydrates. *Mailing Add:* Dept Chem Univ Wis 421 Prentice St Stevens Point WI 54481

CHITTENDEN, DAVID H, b Chicago, Ill, July 30, 35; m 61; c 2. CHEMICAL ENGINEERING. *Educ:* Ill Inst Technol, BS, 56; Univ Wis, MS, 57, PhD(chem eng), 61. *Prof Exp:* Teaching asst chem eng, Univ Wis, 59-60, instr, 60-61; process engr, Calif Res Corp, 61-63; asst prof chem eng, Univ NH, 63-70, consult eng exp sta, 63-68; sr tech systs analyst, 70-74, SPECIALIST, OCCIDENTAL SYSTS, 75- *Concurrent Pos:* Lectr, Univ Calif, Santa Barbara, 74-75. *Mem:* Am Inst Chem Engrs; Asn Comput Mach. *Res:* Diffusive mass tranfer in solids; computer application in engineering; chemical process simulation. *Mailing Add:* 26436 Evergreen Rd San Juan Capistrano CA 92675

CHITTENDEN, MARK EUSTACE, JR, b Jersey City, NJ, July 30, 39; m 68; c 2. FISHERIES ECOLOGY. *Educ:* Hobart Col, BA, 60; Rutgers Univ, MS, 65, PhD(aquatic biol), 69. *Prof Exp:* Fisheries biologist, NJ Div Fish & Game, 60-64; res fel, Dept Environ Sci, Rutgers Univ, 64-67, res asst, 67-68, res assoc fisheries, 68-69; asst prof marine fisheries, Col William & Mary & Univ Va, 69-72; assoc marine scientist, Va Inst Marine Sci, 69-72; asst prof, 73-77, ASSOC PROF FISHERIES, DEPT WILDLIFE & FISHERIES SCI, TEX A&M UNIV, 77- *Mem:* Am Fisheries Soc; Am Soc Ichthyologists & Herpetologists; AAAS; Am Inst Fishery Res Biologists. *Res:* Marine, estuarine and anadromous fishes and fisheries; life histories, population dynamics, ecology and factors affecting distributions of fishes and fisheries communities; stock assessments and effects of harvesting fishes and statistics. *Mailing Add:* Dept Marine Sci William & Mary Sch Marine Sci Gloucester Point VA 23062

CHITTENDEN, WILLIAM A, b Columbia, Mo, Dec 7, 27. NUCLEAR ENGINEERING. *Educ:* Univ Ill, BS, 50. *Prof Exp:* Mech engr, Sargent Lundy, 52-54, proj mgr, 54-66, proj dir, 66-73, mgr, Mech Dept, 73-77, dir eng, 77-83, dir servs, 84-87, SR PARTNER, SARGENT LUNDY, 87- *Mem:* Nat Acad Eng; fel Am Nuclear Soc; fel Am Soc Mech Engrs; Nat Soc Prof Engr. *Mailing Add:* Sargent Lundy 55 E Monroe St Chicago IL 60603

CHITTICK, DONALD ERNEST, b Salem, Ore, May 3, 32; m 57; c 1. PHYSICAL CHEMISTRY. *Educ:* Willamette Univ, BS, 54; Ore State Univ, PhD(chem), 60. *Prof Exp:* Instr chem, Univ Puget Sound, 58-59, from asst prof to assoc prof, 59-68; prof chem, George Fox Col, 68-79, chmn, Dept Sci & Math, 74-79; dir res, Pyrenco, Inc, 79-82; PRES, CHITTICK & ASSOC, 82- *Concurrent Pos:* Adj prof, Inst Creation Res, 88- *Mem:* Am Chem Soc; Creation Res Soc. *Res:* Photochemistry; electrochemistry; fuels; programmed instruction; biomass gasification; US and foreign patents in biomass gasification. *Mailing Add:* 34295 NE Wilsonville Rd Newberg OR 97132

CHITTICK, K(ENNETH) A, b NJ, Nov 6, 03; m 28. RADIO & TELEVISION ENGINEERING. *Educ:* Rutgers Univ, BS, 25. *Prof Exp:* Radio engr, Gen Elec Co, 25-30; radio engr, Victor Div, RCA Corp, 30-39, mgr, TV eng, 39-41, mgr, airborne radar & TV, 41-45 & TV eng, 45-60, mgr eng admin, Indianapolis Consumer Prod Div, 60-68; CONSULT, 68- *Concurrent Pos:* Mem, Nat TV Systs Comt. *Honors & Awards:* Dumoht Achievement Award, Radio Club Am. *Mem:* Fel Inst Elec & Electronics Engrs; Electronics Industs Asn. *Res:* Television and other electronic devices for the consumer market and industry. *Mailing Add:* 8020 College Ave Indianapolis IN 76240

CHITTIM, RICHARD LEIGH, mathematics; deceased, see previous edition for last biography

CHITTY, DENNIS HUBERT, b Bristol, Eng, Sept 18, 12; m 36; c 3. POPULATION ECOLOGY. *Educ:* Univ Toronto, BA, 35; Oxford Univ, MA, 47, DPhil(zool), 49. *Prof Exp:* From res officer to sr res officer ecol, Bur Animal Pop, 35-61; prof zool, 61-78, EMER PROF ZOOL, UNIV BC, 78- *Concurrent Pos:* Master teacher, Univ BC, 73. *Honors & Awards:* Fry Medallist, Can Soc Zool, 88. *Mem:* Can Soc Zool; fel Royal Soc Can. *Res:* Regulation of numbers in natural populations; control of rats; history and principles of scientific methodology. *Mailing Add:* 1750 Knox Vancouver BC V6T 1S3 Can

CHITWOOD, DAVID JOSEPH, b Baltimore, Md, April 29, 50. NEMATOLOGY. *Educ:* Univ Md, BS, 72, MS, 75, PhD(plant path), 80. *Prof Exp:* Fel nematologist, 81-82, ZOOLOGIST, NEMATOL LAB, AGR RES SERV, USDA, 82- *Concurrent Pos:* Ed chief, J Nematol, 91- *Mem:* AAAS; Am Phytopath Soc; Soc Nematologists; NY Acad Sci; Am Soc Parasitol. *Res:* Biochemistry, physiology and endocrinology of free-living and plant-parasitic nematodes. *Mailing Add:* Nematol Lab Beltsville Agr Res Ctr East Beltsville MD 20705

CHITWOOD, HOWARD, b Creekmore, Ky, Feb 4, 32; m 53; c 2. DIFFERENTIAL EQUATIONS. *Educ:* Carson-Newman Col, BS, 53; Univ Fla, MS, 55; Univ Tenn, PhD, 71. *Prof Exp:* PROF MATH, CARSON-NEWMAN COL, 57- *Mem:* Math Asn Am. *Res:* Differential equations. *Mailing Add:* Carson-Newman Col Jefferson City TN 37760

CHITWOOD, JAMES LEROY, b St Petersburg, Fla, Mar 17, 43; m 64; c 2. ORGANIC CHEMISTRY. *Educ:* Emory Univ, BS, 65; Univ Calif, Berkeley, PhD(org chem), 68. *Prof Exp:* Res chemist, Eastman Kodak Co, 68-69, sr res chemist, 70-72, res assoc, 72-74, div head phys & analytical chem res, 74, staff asst to exec vpres develop, Res & Develop Admin, 74-75, dir Chem Res Div, 76-78, asst div supt, Org Chem Div, 79, asst div head, Chem Div, Res Labs, 80, dir develop, Chem Div, 81-83, VPRES EASTMAN KODAK CO, 83- *Concurrent Pos:* Vpres & dir, res & develop, Eastman Chem Co. *Mem:* AAAS; Am Chem Soc; Sigma Xi; Am Asn Textile Chemists & Colorists; Am Inst Chemists. *Res:* Organic synthesis; reaction mechanisms; radiation induced reactions; structureproperty relationships. *Mailing Add:* 4412 Beechcliff Dr Kingsport TN 37664

CHITWOOD, PAUL H(ERBERT), electronics engineering, computer science, for more information see previous edition

CHIU, ARTHUR NANG LICK, b Singapore, Mar 9, 29; US citizen; m 52; c 2. STRUCTURAL ENGINEERING, WIND ENGINEERING. *Educ:* Ore State Univ, BA & BS, 52; Mass Inst Technol, SM, 53; Univ Fla, PhD(struct eng), 61. *Prof Exp:* From instr to assoc prof civil eng, Univ Hawaii, 53-59; res specialist struct mech, NAm Aviation, Inc, 62-63; assoc prof civil eng, 59-64, chmn dept, 63-66, assoc dean res, training & fels, Grad Div, 72-76, PROF CIVIL ENG, UNIV HAWAII, MANOA, 64- *Concurrent Pos:* Sci fac fel, Nat Sci Found, 59 & 60; prof struct eng, Colo State Univ-Asian Inst Technol, Bangkok, 66-68; vis res scientist, Naval Civil Eng Lab, 76-77; consult, struct engr. *Mem:* Am Soc Civil Engrs; Am Soc Eng Educ; Nat Soc Prof Engrs; Am Concrete Inst; Sigma Xi; Earthquake Eng Res Inst. *Res:* Dynamic response of structures to wind loads and earthquake forces; computer applications to structural analysis; author of various publications on wind and its effect on structures. *Mailing Add:* Dept Civil Eng 2540 Dole St Honolulu HI 96822

CHIU, CHAO-LIN, b Kaohsiung, China, Nov 9, 34; m 65; c 1. HYDRAULICS, HYDROLOGY. *Educ:* Univ Taiwan, BS, 57; Univ Toronto, MS, 61; Cornell Univ, PhD(hydraulics), 64. *Prof Exp:* Asst, Cornell Univ, 61-64; from asst prof to assoc prof civil eng, 64-71, PROF CIVIL ENG, UNIV PITTSBURGH, 72-, CHMN WATER RESOURCES PROG, 69- *Concurrent Pos:* NSF res initiation grant, 65-66; hydrologist, US Geol Surv, 74-78; vis prof, Univ Karlsruhe, Ger, 80; sr Fulbright fel, 80. *Mem:* Am Soc Civil Engrs; Int Asn Hydraul Res; Am Geophys Union. *Res:* Stochastic Hydraulics; open channel hydraulics; three-dimensional mathematical modeling of open channel flow. *Mailing Add:* Dept Civil Eng Univ Pittsburgh Pittsburgh PA 15261

CHIU, CHARLES BIN, b Foochow, China, May 19, 40; m 64; c 1. ELEMENTARY PARTICLE PHYSICS, HIGH ENERGY PHYSICS. *Educ:* Seattle Pac Col, BSc, 61; Univ Calif, Berkeley, PhD(physics), 66. *Prof Exp:* Fel theoret particle physics, Theoret Group, Lawrence Radiation Lab, Calif, 65-67; vis scientist theory div, Europ Orgn Nuclear Res, Switz, 67-68; sr res fel, Cavendish Lab, Cambridge, 68-69; res fel, Calif Inst Technol, 69-70, Tolman sr res fel, 70-71; asst prof physics, 71-74, ASSOC PROF PHYSICS, UNIV TEX, AUSTIN, 74-; RES SCIENTIST, CTR PARTICLE THEORY, 80- *Concurrent Pos:* Co-organizer, Conf Phenomenology Particle Physics, 71; Alexander von Humboldt Found sr US scientist award, Max Planck Inst Physics & Astrophys, Munich, WGer, 78-79. *Mem:* Am Phys Soc; Sigma Xi; Am Asn Univ Prof; AAAS. *Res:* Participation of measurements on particle scattering cross sections; high energy scattering phenomenon; Regge theory; theory of strong interactions and general theory of elementary particle physics. *Mailing Add:* Dept Physics Univ Tex Austin TX 78712

CHIU, CHING CHING, b Manila, Philippines; US citizen; m 73; c 2. ORGANIC CHEMISTRY, CHEMICAL ENGINEERING. *Educ:* Univ Santo Tomas, BS, 63; Univ Calif, Santa Barbara, MA, 65; Iowa State Univ, PhD(org chem), 69. *Prof Exp:* Res chemist high temperature polymer chem, Marshall Lab, E I du Pont de Nemours & Co, Inc, 69; res assoc barnacle cement chem, Univ Akron, 69-70; prof chem, Claflin Col, 70-74; asst prof, Frederick Community Col, 74-75; fel, 75-77, res scientist chemother, Frederick Cancer Res Ctr, 77-80; sr sci assoc, Antibiotic Ref Standards Eval, US Pharmacopeia Conv, Inc, 80-83; dir, regulatory affairs, Danbury Pharmacol Inc, 84-87, vpres, regulating affairs, 87-89; CORP MGR REGULATORY AFFAIRS, ORGANON TEKNIKA CORP, 89- *Concurrent Pos:* Res coordr, Phelps-Stokes Found Consortium Res Training, Claflin Col, 72-73 & 73-74. *Mem:* Am Chem Soc; Am Asn Pharmaceut Scientists; Regulatory Affairs Prof Soc; Am Pharmaceut Asn. *Res:*

Chemotherapy fermentation; thin-layer chromatography and high performance liquid chromatograph assay of antibiotics and pharmaceuticals; pyrolysis of organic compounds; free radical reaction mechanisms; quantitative and qualitative assay of antibiotics and pharmaceuticals; isolation of antibiotics; structural elucidation of natural products; analysis of pharmaceutical products; quality control and quality assurance of pharmaceutical industries; regulating affairs and compliance of brand name, generic drugs and biological products derived from biotechnology; preparation of IND, NDA, ANDA and DMF. *Mailing Add:* Organon Teknika Corp 1330-A Piccard Dr Rockville MD 20850

CHIU, CHU JENG (RAY), b Tokyo, Japan, Mar 13, 34; Can citizen; m 62; c 2. CARDIOVASCULAR THORACIC SURGERY. *Educ:* Nat Taiwan Univ, MD, 59; FRCS (C), 69; McGill Univ, PhD(exp surg), 70; Am Col Surgeons, FACS, 73. *Prof Exp:* From asst prof to assoc prof, 71-81, PROF SURG, MCGILL UNIV, 81, DIR RES SURG CLIN, 83- *Concurrent Pos:* Scholar, Med Res Coun Can, 71-76; attending surgeon, Montreal Gen Hosp, 71-, Royal Victoria Hosp, Montreal, 73-; consult, Montreal Chinese Hosp, 73-; consult, Medtronic Inc, Minneapolis, 84-, Alleghney Singer Res Inst, Pittsburgh, 87-, Purdue Univ William A Hillenbrand Ctr Biomed Eng, Ind, 90-; assoc ed, J Biomat, Artificial Cells & Artifical Organs, 89-; chmn, Exam Bd, Cardiovascular & Thoracic Surg, Royal Col Physicians & Surg, Can, 86-88; mem, Sci Rev Comt, Med Res Coun Can, 88-; vis prof, USSR Acad Med Sci, Bakulev Inst, 89, Univ Cologne, Ger, 89, Tokyo Women's Med Col, 89. *Mem:* Am Asn Thoracic Surg; Soc Univ Surgeons; Coun Cardiovascular Surg, fel Am Heart Asn; NY Acad Sci; Soc Int Surgeons. *Res:* Cardiovascular research related to myocardial protection and cardiac assist devices; shock and trauma endocrinology; experimental surgery. *Mailing Add:* Montreal Gen Hosp 1650 Cedar Ave Montreal PQ H3G 1A4 Can

CHIU, HONG-YEE, b Shanghai, China, Oct 4, 32; m 66; c 3. ASTROPHYSICS. *Educ:* Okla State Univ, BSc, 56; Cornell Univ, PhD(physics), 59. *Prof Exp:* Nat Acad Sci res assoc physics, Theoret Div, NASA, 60-62, physicist, Goddard Inst Space Studies, 62-78; SPACE SCIENTIST, GODDARD SPACE FLIGHT CTR, 78- *Concurrent Pos:* Mem, Inst Advan Study, 59-61; asst prof physics, Yale Univ, 61-62; from adj asst prof to adj assoc prof, Columbia Univ, 62-68; adj prof physics, City Col New York, 66-78. *Honors & Awards:* Except Medal of Sci, NASA,. *Mem:* Inst Elec & Electronics Engrs; Am Phys Soc; Am Astron Soc; fel Royal Astron Soc. *Res:* Nuclear physics; strange particles; stellar evolution; neutrino astrophysics; x-ray astronomy; cosmology; general relativity. *Mailing Add:* Code 910-2 Goddard Space Flight Ctr Greenbelt Rd Greenbelt MD 20771

CHIU, HUEI-HUANG, b Formosa, China, Dec 29, 30; m 61; c 1. AEROSPACE & MECHANICAL ENGINEERING. *Educ:* Taiwan Univ, BS, 54; Kans State Univ, MS, 56; Princeton Univ, PhD(aeronaut eng), 62. *Prof Exp:* Res assoc aeronaut eng, Princeton Univ, 62-63; from asst prof to assoc prof aeronaut & astronaut, New York Univ, 63-74; PROF, UNIV ILL, 74- *Concurrent Pos:* NASA res grant, 65-66; consult proj engr, Curtiss Wright Co, 65-; res grant, Argonne Nat Lab, 75-88; fel NASA & USAF. *Mem:* Am Inst Aeronaut & Astronaut. *Res:* Energy conversion; propulsion; combustion; fluid dynamics; applied mathematics. *Mailing Add:* Dept Mech Eng PO Box 4348 Chicago IL 60680

CHIU, JEN, b China, June 22, 24; US citizen; m 55; c 4. THERMAL METHODS, CHROMATOGRAPHY. *Educ:* Hunan Univ, BS, 46; Univ Ill, MS, 59, PhD(anal chem), 61. *Prof Exp:* Res chemist, E I Dupont de Nemours Co, Inc, 60-65, sr res chemist, 65-72, res assoc, 72-79, res fel, 79-85; pres, CHIU Enterprises, Inc, 85-88; CONSULT, 88- *Concurrent Pos:* Lectr, Thermal Analysis Inst, Soc Plastics Engrs, & Am Chem Soc. *Honors & Awards:* Mettler Award, NAm Thermal Anal Soc, 82. *Mem:* Am Chem Soc; Int Confedn Thermal Analysis; fel NAm Thermal Analysis Soc (vpres, 75-76, pres, 77); Sigma Xi. *Res:* Polymer characterization; materials properties and characterization. *Mailing Add:* Oaklane Manor 1007 S Hilton Rd Wilmington DE 19803

CHIU, JEN-FU, b Taiwan, China, Sept 30, 40; m 70; c 2. BIOCHEMISTRY. *Educ:* Taipei Med Col, Taiwan, BPharm, 64; Nat Taiwan Univ, MSc, 67; Univ BC, PhD(biochem), 72. *Prof Exp:* Teaching asst chem, Taipei Med Col, 65-67; lectr biochem, Chung Shan Med Col, 67-68; proj investr, Univ Tex M D Anderson Hosp & Tumor Inst Houston, 72-74, asst biochemist, 74-75; asst prof biochem, Sch Med, Vanderbilt Univ, 75-78; assoc prof, 78-, PROF BIOCHEM, COL MED, UNIV VT. *Concurrent Pos:* Rosalie B Hite fel, Univ Tex, 72-73. *Mem:* Am Soc Cell Biol; Am Asn Cancer Res; Can Biochem Soc; Biophys Soc; Am Soc Biol Chem; Sigma Xi. *Res:* Role of nuclear nonhistone proteins in the regulation of genetic activity; biochemistry of chromatin; macromolecular mechanism of carcinogenesis; antibodies to tumor associated antigens. *Mailing Add:* Dept Biochem Univ Vt Given Bldg Rm B405 Burlington VT 05405

CHIU, JOHN SHIH-YAO, b Soochow, China, Sept 8, 28; US citizen; m 59; c 2. APPLIED STATISTICS, ECONOMETRICS. *Educ:* Nat Taiwan Univ, BA, 52; Univ Ky, MS, 55; Univ Ill, PhD(econ & statist), 60. *Prof Exp:* Adv economet, Coun Int Econ Coop, Taiwan Cent Govt, 67-68, adv statist, Bur Statist, 67-68; PROF STATIST, GRAD SCH BUS, UNIV WASH, 69- *Mem:* Am Statist Asn. *Res:* Applied statistics in business and economics. *Mailing Add:* Mgt Sci & Statist Univ Wash Seattle WA 98195

CHIU, KIRTS C, b Taipei, Taiwan, Apr 12, 33; US citizen; m 65; c 1. AGRICULTURAL BIOTECHNOLOGY. *Educ:* Nat Taiwan Univ, BS, 57; MS, 59. *Prof Exp:* Res asst, dept agr biochem, Univ Minn, 61-63; res chemist, Elyria Mem Hosp, Ohio, 64-65; chemist & bacteriologist, Dept Pub Works, Jersey City, 68-69; chemist, Zenith Labs, NJ, 69-70, Jacksonville Eng & Testing Co, Fla, 71-72 & Dept Environ Regulation, State of Fla, 73-75; consult, Ecol Div, Telecommun Industs, Inc, Fla, 75-76; civil engr & consult, Kun-Young Chiu & Assocs, 76-86; SCI & ENG CONSULT, 87- *Mem:* Am Chem Soc; AAAS. *Res:* Utilization of biomass material; mycology methane fermentation of industrial and domestic wastes. *Mailing Add:* PO Box 393 Valdosta GA 31603-0393

CHIU, LUE-YUNG CHOW, b Kiang-su, China, Sept 14, 31; m 60; c 2. THEORETICAL CHEMISTRY. *Educ:* Nat Taiwan Univ, BS, 52; Bryn Mawr Col, MA, 54; Yale Univ, PhD(phys chem), 57. *Prof Exp:* Fel atomic physics, Yale Univ, 57-60; res physicist, Radiation Lab, Columbia, 60-62; res assoc, Lab Molecular Struct & Spectra, Univ Chicago, 62-63 & Inst for Studies of Metals, 63-64; res asst prof chem, Cath Univ Am, 64-65; Nat Res Coun-Nat Acad Sci sr resident res assoc, Lab Theoret Studies, Goddard Space Flight Ctr, NASA, 65-67, sr resident res assoc, Astrochem Sect, 67-68; assoc prof, 68-72, PROF PHYS CHEM, HOWARD UNIV, 72- *Mem:* Am Phys Soc; Am Chem Soc. *Res:* Photochemistry; atomic beam magnetic resonances; theoretical studies on magnetic interactions in molecules; atomic and molecular collisions; interaction of radiation with atoms and molecules; energy and excitation transfer between atoms and molecules; solutions for the dissociative states. *Mailing Add:* Dept Chem Howard Univ Washington DC 20059

CHIU, PETER JIUNN-SHYONG, b Miao-Li, Taiwan, June 9, 42; m 67; c 3. PHARMACOLOGY. *Educ:* Taipei Med Col, BS, 64; Nat Taiwan Univ, MS, 66; Columbia Univ, PhD(pharmacol), 72. *Prof Exp:* Res fel nephrology, Sch Med, Univ Pa, 72-74; sr scientist, Schering Corp, 74-77, prin scientist, 77-82, sr prin scientist, 82-85, sect leader, 86-89; RES FEL, CARDIOVASCULAR/CNS PHARMACOL, SCHERING PLOUGH RES, 89. *Concurrent Pos:* Adj assoc prof physiol, Fairleigh-Dickinson Univ, 87-89. *Mem:* Am Soc Pharmacol & Exp Therapeut; Int Soc Nephrology; Am Soc Nephrology; Am Fedn Clin Res; AAAS; Am Heart Asn; NY Acad Sci; Am Physiol Soc. *Res:* Role of endothelin in control and regulation of cardiovascular-renal function; effect of antihypertensive agents on calcium movements in blood vessels; effect of drugs on platelet aggregation; cardiovascular-renal effects of drugs or hormones mediated by C GMP. *Mailing Add:* Schering-Plough Res 60 Orange St Bloomfield NJ 07003

CHIU, TAI-WOO, b China, May 18, 44. PHYSICAL CHEMISTRY. *Educ:* Chinese Univ Hong Kong, BSc, 66; Univ Miami, PhD(phys chem), 71. *Prof Exp:* Res assoc polymers, Mat Res Ctr, Lehigh Univ, 71-73 & Univ Cincinnati, 73-74; phys chemist rubber res, Polyfibron Div, W R Grace & Co, 75-77; SR RES CHEMIST PAPER RES, SCOTT PAPER CO, 77- *Mem:* Am Chem Soc; Tech Asn Pulp & Paper Indust. *Res:* Mastism and efficiency of non-woven binder. *Mailing Add:* Dept Prod Develop Scott Plaza III Philadelphia PA 19113-1509

CHIU, TAK-MING, b Hong Kong, Aug 26, 49; m 78; c 1. LOCALIZED NMR IN VIVO IN HUMANS, SPECTROSCOPIC IMAGING. *Educ:* McMaster Univ, Hamilton, Ont, BSc, 74; Univ Western Ont, London, PhD(phys chem), 84. *Prof Exp:* Postdoctoral appointee, Chem Div, Radiation & Photochem Sect, Argonne Nat Lab, 84-87; res assoc, Dept Neurol, NMR Res Facil, Henry Ford Hosp, Detroit, 87-88; INSTR BIOPHYS, DEPT NEUROL, HARVARD MED SCH & MASS GEN HOSP, 88-; MAGNETIC RESONANCE PHYSICIST, DEPT NEUROL & BRAIN IMAGING CTR, MCLEAN HOSP, BELMONT, 88- *Mem:* Am Chem Soc; AAAS; Soc Magnetic Resonance Med. *Res:* Biomedical applications of magnetic resonance: in vivo nuclear magnetic resonance detection of biochemical metabolites in human brains; lactate editing and water suppression techniques in localized 1H NMR in cerebro-disorders (Alzheimer disease, stroke and temporal lobe epilepsy) and neuropsychiatry (electroconvulsive therapy). *Mailing Add:* Dept Neurol & Brain Imaging Ctr McLean Hosp Belmont MA 02178

CHIU, THOMAS T, b Shanghai, China, Jan 24, 33; m 60; c 4. CHEMICAL ENGINEERING, PHYSICAL CHEMISTRY. *Educ:* Nat Univ Taiwan, BS, 56; Case Inst Technol, MS, 60, PhD(chem eng), 62. *Prof Exp:* Res chem engr, 62-66, sr res engr, 66-77, RES MGR, DOW CHEM CO, 77- *Mem:* Am Inst Chem Engrs; Am Tech Asn Pulp & Paper Indust; Sigma Xi. *Res:* Physical properties of polyelectrolytes and thermal plastics. *Mailing Add:* 304 Mayfield Lane Midland MI 48640-2958

CHIU, TIEN-HENG, b China, May 2, 53; m; c 1. SEMICONDUCTOR PHYSICS, VACUUM TECHNOLOGY. *Educ:* Univ Ill, Urbana-Champaign, PhD(physics), 82. *Prof Exp:* MEM TECH STAFF, AT&T BELL LABS, 83- *Mem:* Am Physics Soc; Am Vacuum Soc. *Res:* Epitaxial growth of III-V compound semiconductors and its applications in optoelectronic devices and photonic switching. *Mailing Add:* AT&T Bell Labs Rm 4B-517 Holmdel NJ 07733-1988

CHIU, TIN-HO, b Kiang Si Prov, China, June 22, 42; m 69; c 2. SURFACE CHEMISTRY. *Educ:* Chung Chi Col, Chinese Univ Hong Kong, BSc, 64; Lehigh Univ, PhD(chem), 72. *Prof Exp:* High sch teacher chem, Fai Yuen Col, Hong Kong & sci teacher biol, South Western Col, Hong Kong, 64-66; teaching asst chem, Lehigh Univ, 67-69, res asst, Ctr Surface & Coatings Res, 69-72; res assoc, AVCO Everett Res Lab, Subsid AVCO Corp, 72-76, sr staff mem microcalorimetry, 76-79; prof mgr, Biomet Instrumentation Lab Inc, 80-84, prog mgr Res & Develop, 84-86; RES ASSOC, POLYMER LAB, CORP RES, ALLIED-SIGNAL INC, 86- *Concurrent Pos:* Adj assoc prof biomed eng, Div Biol & Med, Brown Univ, 85- *Mem:* Am Chem Soc; Am Soc Artifical Internal Organs. *Res:* The interaction of solid/gas, solid/liquid, gas/liquid and protein/surface interaction including surface, biophysical, colloid, corrosion, polymer and pollution chemistry; biodegradable nerve channels. *Mailing Add:* 754 Ridgewood Rd Millburn NJ 07041-1823

CHIU, TONY MAN-KUEN, b China; US citizen; m 73; c 2. CARBOHYDRATE CHEMISTRY. *Educ:* Okla Baptist Univ, BS, 62; Emporia State Univ, MS, 63; Univ Pac, PhD(org chem), 72. *Prof Exp:* Instr chem, Warren Wilson Col, 63-65; asst prof chem, Viterbo Col, 67-69; Damon Runyon res fel, Sloan-Kettering Cancer Res Inst 71-73; res assoc, Univ SC, 73-74; PROF CHEM, FREDERICK COMMUNITY COL, 74- *Mem:* Am Chem Soc; Am Asn Univ Prof. *Res:* Nucleosides and nucleic acids; synthesis of natural products and derivatives. *Mailing Add:* 11216 Green Watch Way North Potomac MD 20878

CHIU, TSAO YI, b Jean Yang, China, Apr 17, 25; US citizen; m 64; c 2. COASTAL ENGINEERING. *Educ:* Nat Taiwan Univ, BS, 52; Univ Fla, MS, 59, PhD(civil & coastal eng), 72. *Prof Exp:* Res asst coastal eng, coastal eng lab, Univ Fla, 59-65, assoc engr, 73-74, assoc prof, eng sci dept, 75-79, prof, coastal eng dept, 79-81; PROF & DIR COASTAL ENG, BEACHES & SHORES RESOURCE CTR, FLA STATE UNIV, 82- *Honors & Awards:* Jim Purpura Award, Fla Shore & Beach Preserv Asn, 79. *Mem:* Int Asn Hydraul Res. *Res:* Beach erosion; hurricane and storm tides and effects; flushing of coastal waters; coastal hydraulic modeling; inlet hydrodynamics; coastal sediment transport. *Mailing Add:* 203 Morgan Bldg Box 5 2035 E Paul Dirak Dr Tallahassee FL 32310

CHIU, VICTOR, b Tianjin, China; US citizen; c 2. MATHEMATICS OF CURVES & SURFACES. *Educ:* Kent State Univ, BS, 60; Cornell Univ, MS, 66, PhD(theoret physics), 70. *Prof Exp:* Prof physics, math & computer sci, Univ Indianapolis, 70-85; SR DEVELOP ENGR, STRUCT DYNAMICS RES CORP, 85- *Mem:* Soc Appl & Indust Math; Inst Elec & Electronics Engrs Computer Soc. *Res:* Mathematical software related to solid modeling; computer-aided design. *Mailing Add:* 2000 Eastman Dr Milford OH 45150

CHIU, WAN-CHENG, b Meihsien, China, Nov 1, 19; US citizen; m 54; c 3. METEOROLOGY. *Educ:* Nat Cent Univ, China, BS, 41; NY Univ, MS, 47, PhD(meteorol), 51. *Prof Exp:* Technician, Fukein Weather Bur, China, 41-42; teacher math, Pungshan Model Sch, China, 42-43; asst teacher meteorol, Nat Cent Univ, China, 44-45; res assoc, NY Univ, 51-54, from assoc meteorologist to meteorologist, 54-60, res scientist, 60-61; actg chmn dept, 71, 81, 84-85, prof meteorol, 61-86, EMER PROF METEOROL, UNIV HAWAII, 87- *Concurrent Pos:* Vis scientist, Nat Ctr Atmospheric Res, 67-68, sr fel, 75; res academician, Chung Hwa Acad Cult, Repub China. *Mem:* Am Meteorol Soc; Am Geophys Union; Royal Meteorol Soc. *Res:* Atmospheric energy and circulation; the momentum and heat transport in the atmosphere; large-scale sea-air interaction. *Mailing Add:* Dept Meteorol Univ Hawaii Honolulu HI 96822

CHIU, YAM-TSI, b Canton, China, Sept 5, 40; US citizen; m 68. ATMOSPHERIC PHYSICS, PLASMA PHYSICS. *Educ:* Yale Univ, BS, 61, MS, 63, PhD(physics), 65. *Prof Exp:* Res assoc elem particle physics, Yale Univ, 65 & Enrico Fermi Inst Nuclear Studies, Univ Chicago, 65-67; mem tech staff, 67-74, staff scientist, 74-79, sr scientist, 79-84, AEROSPACE CORP GROUP LEADER & SR MEM, LOCKHEED PALO ALTO RES LAB, 84- *Concurrent Pos:* Mem Upper Atmosphere Comn, Int Asn Geomagnetism & Aeronomy, 72-73; prin invest contracts, NASA, Dept of Defense, & Dept of Energy, 78- *Mem:* Am Phys Soc; Am Geophys Union. *Res:* Atmospheric dynamics; ionospheric and solar coronal dynamics-nonlinear waves in gas dynamic and magneto gas dynamic media; space plasma physics. *Mailing Add:* Dept 91-20, Lockheed Palo Alto Res Labs 3251 Hanover Street Bldg 225 Palo Alto CA 94304

CHIU, YIH-PING, b Shanghai, China, Nov 3, 33; US citizen; m 64; c 2. MECHANICAL ENGINEERING. *Educ:* Nat Taiwan Univ, BS, 55; Univ Ill, Urbana, MS, 60; Carnegie Inst Technol, PhD(mech eng), 63. *Prof Exp:* Res scientis, SKF Indust, Inc, 62-80; RES CONSULT, TORRINGTON CO, 80- *Honors & Awards:* Alfred Hunt Award, Am Soc Lubrication Engrs, 75. *Mem:* Am Soc Mech Engrs; Soc Tribologists & Lubrication Engrs; Soc Automotive Engrs. *Res:* Rolling bearing analysis; bearing lubrication and friction analysis; roller-cam follower analysis; roller contact stress and fatigue analysis; plastic deformation and residual stress in rolling sliding contact; bearing load rating study; hook joint analysis; ceramic bearings. *Mailing Add:* Torrington Co 59 Field St Torrington CT 06790

CHIU, YING-NAN, b Canton, China, Nov 25, 33; m 60; c 2. PHYSICAL CHEMISTRY. *Educ:* Berea Col, BA, 55; Yale Univ, MS, 56, PhD(phys chem), 60. *Prof Exp:* Fel chem, Columbia Univ, 60-62; res assoc physics, Univ Chicago, 62-64; from asst prof to assoc prof, 64-70, chmn chem dept, 72-81, PROF QUANTUM CHEM, CATH UNIV AM, 70-, O'BRIEN PROF, 77- *Concurrent Pos:* A P Sloan fel, Harvard & Princeton Univs, 69-71. *Honors & Awards:* Hillebrand Prize, 84. *Mem:* Am Chem Soc; Fel Am Phys Soc. *Res:* Spin-spin and spin-orbit interaction in molecules; quantum theory of molecular structure and valence; theory of optical and magnetic resonance spectra, molecular interactions, chemical reaction dynamics, molecular quantum mechanics; Ligand field theory; electron transfer; energy transfer. *Mailing Add:* Dept Chem Cath Univ Am Washington DC 20017-1567

CHIUEH, CHUANG CHIN, b Taipei, Taiwan, Oct 1, 43; m 70; c 2. PHARMACOLOGY. *Educ:* Taipei Med Col, BS, 66; Nat Taiwan Univ, MS, 68; Mich State Univ, PhD(pharmacol), 74. *Prof Exp:* PHARMACOLOGIST BRAIN IMAGING, NIMH, 85- *Concurrent Pos:* Prin investr, NIMH, 88. *Mem:* Am Soc Pharmacol & Exp Therapeut; Soc Neurosci. *Res:* MPTP-induced private model of Parkinsonism; 6-F-L-DOPA/PET imaging of DA neurons; I-IBZM/SPECT imaging of Dz receptors; neurotoxic mechanism of MPTP. *Mailing Add:* Bldg 10 Rm 4N 317 Bldg 10 Rm 2D-52 Bethesda MD 20892-1000

CHIULLI, ANGELO JOSEPH, b New York, NY; m 60; c 4. MICROBIOLOGY. *Educ:* Hofstra Univ, BA, 54; Syracuse Univ, MS, 59. *Prof Exp:* Researcher, microbiol, Brooklyn Bot Gardens, 59-60; microbiologist, 60-61, LAB DIR, ADVAN BIOFACTURES CORP, 61- *Mem:* Am Soc Microbiol; Soc Indust Microbiol. *Res:* Screening microorganisms for therapeutic enzyme; metabolism of commercially valuable microrganisms. *Mailing Add:* Advan Biofactures Corp 35 Wilbur St Lynbrook NY 11563

CHIUTEN, DELIA FUNGSHE, internal medicine, hematology-oncology, for more information see previous edition

CHIVERS, HUGH JOHN, b Frome, Eng, June 26, 32; m 56; c 2. MANAGEMENT ENGINEERING, ATMOSPHERIC SCIENCES. *Educ:* Univ Manchester, BSc, 56, PhD(physics), 59. *Prof Exp:* Res asst radio astron, Univ Manchester, 60-61; proj leader ionospheric physics, Environ Sci Serv Admin, 61-62, sect chief, 62-64, dir, Space Disturbance Monitoring Sta, 65-67, mem staff, Tropospheric Wave Propagation, Colo, 67-68; MEM FAC, DEPT ELEC ENG & COMP SCI, UNIV CALIF, SAN DIEGO, 68-, RES PHYSICIST. *Concurrent Pos:* Mem Comn G & H, Int Sci Radio Union, 61-; owner, La Jolla Sci, 73- *Mem:* Am Geophys Union; fel Explorers Club. *Res:* Earth-space research using scintillations and absorption caused by charged particles from space; very high frequency radio noise instrumentation. *Mailing Add:* 301 S Granados Ave Solano Beach CA 92075

CHIVIAN, ERIC SETH, b Newark, NJ, June 10, 42; m 68; c 3. SURVEY RESEARCH ON TEENAGERS. *Educ:* Harvard Univ, AB, 64, MD, 68. *Prof Exp:* Resident psychiat, Mass Gen Hosp, 77-80; co-founder & treas, Int Physicians Prev Nuclear War, 80-85; CO-DIR INT VIDEOEDUC PROJ, HARVARD MED SCH, 83-, ASST CLIN PROF PSYCHIAT, 87- *Concurrent Pos:* Staff psychiatrist, Mass Inst Technol, 80-; dir, Int Youth Surv, Am Acad Arts & Sci, 87-90. *Honors & Awards:* Nobel Peace Prize, 85. *Mem:* Physicians Social Responsibility; Int Physicians Prev Nuclear War; Am Psychiat Asn. *Res:* First large scale scientific survey of America and Soviet teenagers' attitudes about the future, US-USSR relations and nuclear war. *Mailing Add:* 56 Hawes St Brookline MA 02146

CHIVIAN, JAY SIMON, b Newark, NJ, Mar 17, 31; m 56; c 2. PHYSICS. *Educ:* Franklin & Marshall Col, BS, 52; Lehigh Univ, MS, 54, PhD(physics), 60. *Prof Exp:* Res asst, Brookhaven Nat Lab, 56; instr physics, Lafayette Col, 59-60; res physicist, Texas Instruments Inc, 60-67; res scientist, Ling-Temco-Vought Res Ctr, 67-71; sr scientist, Vought Advan Technol Ctr Inc, 71-81, SR ENG SPECIALIST, LTV AEROSPACE & DEFENSE CO, 81- *Concurrent Pos:* Adj prof & lectr physics, Univ Tex, Dallas, 84-85. *Mem:* Am Phys Soc; sr mem Inst Elec & Electronics Engrs; Sigma Xi (pres, 84-85); Optics Soc Am. *Res:* Laser applications, holography and nonlinear optics; physical electronics; plasma and neutron physics; thermionic energy conversion; signal detection in antisubmarine warfare; thin film physics; infrared detection; quantum electronics. *Mailing Add:* 919 Warfield Way Richardson TX 75080

CHIVUKULA, RAMAMOHANA RAO, b Vijayavada, India, June 20, 33; US citizen; m 57; c 3. MATHEMATICS. *Educ:* Andra Univ, BA, 53, MA, 55, PhD(math), 60; Univ Ill, PhD, 62. *Prof Exp:* Lectr math, Andhra Univ, 53-59 & Univ Mich, 62-63; from asst prof to assoc prof, 63-80, chmn grad comt, 76-80, PROF MATH, UNIV NEBR, LINCOLN, 80- *Mem:* Am Math Soc; Math Asn Am; Indian Math Soc. *Res:* Functional analysis. *Mailing Add:* Dept Math Univ Nebr Lincoln NE 68588

CHIYKOWSKI, LLOYD NICHOLAS, b Garson, Man, July 26, 29; m 53; c 4. ENTOMOLOGY, DISEASE TRANSMISSION. *Educ:* Univ Man, BSA, 53, MSc, 54; Univ Wis, PhD(entomol), 58. *Prof Exp:* Asst entomol, Univ Wis, 54-58; res officer, Plant Res Inst, Can Dept Agr, 58-67, res scientist, Cent Exp Farm, Cell Biol Res Inst, 67-71, res scientist, 71-80, sr res scientist, Cent Exp Farm, Chem & Biol Res Inst, 80-91; RETIRED. *Mem:* Entom Soc Am; Am Phytopath Soc; Entom Soc Can; Int Orgn Mycoplasm; Can Phytopath Soc. *Res:* Leafhopper transmission of plant viruses; leafhopper transmission of plant viruses and mycoplasmas. *Mailing Add:* 17 Larkspur Dr Nepean ON K2H 6K8 Can

CHIZINSKY, WALTER, b Springfield, Mass, Nov 27, 26; m 53; c 3. REPRODUCTION & HUMAN SEXUALITY, SCIENCE EDUCATION. *Educ:* Univ Mass, BS, 49; Univ Chicago, SM, 50; NY Univ, PhD(biol), 60. *Prof Exp:* Instr & assoc prof biol, Bennett Col, NY, 54-67, chmn, dept sci & math, 60-67; assoc prof & prof biol, Briarcliff Col, 67-74, acad dean, 70-74, chmn, dept sci & math, 71-73, distinguished prof sci, 74-77; vpres & dean fac, Stephens Col, 78-79; ASSOC DEAN INSTR, DIV NAT SCI & MATH, BERGEN COMMUNITY COL, 87-, PROF BIOL, 90- *Concurrent Pos:* Shell Merit fel, Stanford Univ, 69; dean, Div Nat Sci & Math, Bergen Community Col, 81- *Mem:* Am Asn Sex Educr Counr & Therapists; Sex Info & Educ Coun US. *Res:* Human sexuality, especially behavior; anatomy, physiology and behavior of sex and reproduction, especially in vertebrates. *Mailing Add:* Div Nat Sci & Math Bergen Community Col Paramus NJ 07652-1595

CHIZZONITE, RICHARD A, PROTEIN BIOCHEMISTRY, MONOCLONAL ANTIBODIES. *Educ:* Univ Rochester, PhD(pharmacol), 78. *Prof Exp:* RES INVESTR, LA ROCHE RES CTR, 82- *Mailing Add:* Ten Barnesdale Rd Clifton NJ 07013

CHLANDA, FREDERICK P, b Poughkeepsie, NY, July 26, 43; m 64; c 2. POLYMER CHEMISTRY. *Educ:* Clarkson Col Technol, BS, 65, PhD(org chem), 70. *Prof Exp:* Res scientist chem, Columbia Univ, 69-70; sr res chemist, 70-80, RES ASSOC, ALLIED CORP, 80- *Mem:* Am Chem Soc. *Res:* Synthetic membranes, polyelectrolytes, and ion exchange. *Mailing Add:* White Meadow Lake 74 Cayuga Ave Rockaway NJ 07866-1012

CHLAPOWSKI, FRANCIS JOSEPH, b Newport, RI, Feb 28, 44; m 65; c 4. CELLULAR MEMBRANES, CELL CULTURE. *Educ:* Univ Mass, BS, 65; Mich State Univ, PhD(zool), 69. *Prof Exp:* Res fel cell biol, Harvard Univ, 69-70; asst prof anat, 70-73, asst dean admin affairs, 76-78, ASSOC PROF BIOCHEM & MOLECULAR BIOL, MED SCH, UNIV MASS, 73-, DIR, INTERDEPT ELECTRON MICROS FACIL, 74-, ACTING CHAIR BIOCHEM & MOLECULAR BIOL, 88- *Concurrent Pos:* Prin investr, NIH res grant, Univ Mass, 70- *Mem:* Am Soc Cell Biol; Tissue Cult Asn; AAAS; NY Acad Sci. *Res:* Growth differentiation and carcinogenesis of urinary bladder epithelium in vitro. *Mailing Add:* Univ Mass Med Ctr 55 Lake Ave N Worcester MA 01605

CHLEBOWSKI, JAN F, b Toledo, Ohio, Dec 20, 43. ENZYMOLOGY, PROTEIN STRUCTURE. *Educ:* St Marys Col, Minn, BA, 65; Case Western Reserve Univ, PhD(chem), 69. *Prof Exp:* Res fel chem, Univ Col, London, 69-71; res assoc biochem, Yale Univ, 71-79; from asst prof to assoc prof, 79-91, PROF, VA COMMONWEALTH UNIV, 91- *Mem:* Am Chem Soc; Biophys Soc; AAAS; Sigma Xi; Am Soc Microbiol; Am Soc Biochem & Molecular Biol. *Res:* Factors controlling and modulation protein-enzyme structure and function employing biophysical and genetic engineering methods; systems include alkaline phosphatase and alpha-2-macroglobulin. *Mailing Add:* Dept Biochem VA Commonwealth Univ Richmond VA 23298-0614

CHMURA-MEYER, CAROL A, b Chicopee, Mass, Sept 22, 39; m 75. STRATIGRAPHY-SEDIMENTATION. *Educ:* Smith Col, BA, 61; Univ NMex, MS, 63; Stanford Univ, PhD(geol), 70. *Prof Exp:* Lectr geol, Muhlenberg Col, 67-72; sr paleontologist, Mobil Oil Corp, 72-74; res geologist, 74-80, sr geologist palynology geol, 80-84, SR RES ASSOC & SECT SUPVR, CHEVRON OIL FIELD RES CO, 84- *Concurrent Pos:* Res assoc geol, Lehigh Univ, 67-72. *Mem:* Sigma Xi; Am Asn Stratig Palynologists; Am Asn Petrol Geologists. *Res:* Palynology and sedimentology; technical management. *Mailing Add:* Chevron Oil Field Res Co PO Box 446 La Habra CA 90633-0446

CHMURNY, ALAN BRUCE, b Oak Park, Ill, May 31, 44; m 66; c 1. ORGANIC CHEMISTRY. *Educ:* Univ Ill, Urbana, BS, 66; Univ Calif, Los Angeles, PhD(org chem), 71. *Prof Exp:* NIH fel, Mass Inst Technol, 72-73; res scientist, Pfizer Inc, Groton, 73-77, sr scientist org chem, 77-81; dir org chem, Bethesda Res Labs, Rockville, Md, 81-82; MGR BIOTECHNOL RES, W R GRACE, COLUMBIA, MD, 82- *Mem:* Sigma Xi; AAAS; Am Chem Soc. *Res:* Application of enzyme catalyzed reactions to the production of fine organic chemicals on an industrial scale. *Mailing Add:* 6211 White Oak Dr Frederick MD 21701

CHMURNY, GWENDOLYN NEAL, b Gulfport, Miss, Oct 22, 37; m 66; c 1. NUCLEAR MAGNETIC RESONANCE, ORGANIC CHEMISTRY. *Educ:* Memphis State Univ, BA, 59; Univ Ill, Urbana, MS, 61, PhD(phys org chem), 66. *Prof Exp:* Res chemist org chem, E I du Pont de Nemours & Co, 65-66; fel nuclear magnetic resonance, Univ Calif, Los Angeles, 67-68 & spectroscopist, 68-71; fel nuclear magnetic resonance, Ohio State Univ, 71-72 & Mass Inst Technol, 72-73; assoc scientist, 75-81, RES SCIENTIST NUCLEAR MAGNETIC RESONANCE, PFIZER, INC, 81-; SCIENTIST NUCLEAR MAGNETIC RESONANCE, NCI-FCRF, 82- *Mem:* Am Chem Soc; Sigma Xi; NY Acad Sci. *Res:* Carbon-13 and proton nuclear magnetic resonance applied to physical organic, medicinal chemistry and biochemistry; instrumentation and adaption of new techniques. *Mailing Add:* 6211 White Oak Dr Frederick MD 21701

CHO, ALFRED CHIH-FANG, b Shanghai, China, Dec 31, 21; US citizen; m 57; c 2. ACOUSTICS, AEROSPACE SCIENCES. *Educ:* Univ Shanghai, BSc, 43; Univ Tex, MA, 50, PhD(physics), 58. *Prof Exp:* Asst engr, Shanghai Tel Co, 44-48; teaching fel physics, Univ Tex, 49-53, spec instr math, 53-57; sr struct engr, Gen Dynamics/Ft Worth, 57-60, sr physicist, 60-62; sr tech specialist acoust & vibration, res & eng, Space Div, NAm Rockwell, 62-63, supvr, 63-67, mem tech staff, 67-69; sr res specialist antisubmarine warfare & shock & vibration dir, Lockheed-Calif Co, 69-71; eng supvr specialist, Litton Ship Syst, 71-72; mem tech staff, Hughes Aircraft, Space & Commun Group, 72-73; sr eng specialist, Space Div, Rockwell Int, 73-86; prin engr & consult, Jet Prop Lab, Syscon Corp, 86-89; PRIN ENGR & CONSULT, MCDONNELL DOUGLAS SSC, 89- *Concurrent Pos:* Adj prof physics, Tex Christian Univ, 58-62; translr, Am Inst Physics, 64-69. *Mem:* Acoust Soc Am; Brit Acoust Soc. *Res:* Acoustics and vibration of aircraft and launch vehicle; dynamic responses of spacecraft; aeromechanics in aeronautics; pulse statistical analysis in architectural acoustics. *Mailing Add:* 6263 Roundhill Dr Whittier CA 90601

CHO, ALFRED Y, b Peking, China, July 10, 37; m 68; c 4. MICROWAVE & OPTOELECTRONIC DEVICES. *Educ:* Univ Ill, BSEE, 60, MS, 61, PHD(elec eng), 68. *Prof Exp:* Tech staff res, Ion Physics Corp, 61-62, TRW-Space Tech Lab, 62-65; res asst elec eng, Univ Ill, 65-68; tech staff res, 68-84, dept head, 84-87, DIR, AT&T BELL LABS, 87- *Concurrent Pos:* Vis prof, Univ Ill, 77-78, adj prof, 78- *Honors & Awards:* Int Prize New Mat, Am Phys Soc, 82; Morris N Liebmann Award, Inst Elec & Electronics Engrs, 82; GaAs Symposium Award, Ford, 86; Heinrich Walker Medal, 86; Solid State Science & Technol Medal, Electrochem Soc, 87; World Mat Congr Award, ASM Int, 88; Gaede-Langmuir Award, Am Vacuum Soc, 88; NJ Gov Thomas Alva Edison Sci Award, 90; Int Crystal Growth Award, Am Asn Crystal Growth, 90. *Mem:* Nat Acad Sci; Nat Acad Eng; Am Acad Arts & Sci; Am Physics Soc; Am Vacuum Soc; Electrochem Soc; NY Acad Sci; Mat Res Soc; Inst Elec & Electronics Engrs. *Res:* Thin film technology called molecular beam epitaxy--this ultra high vacuum process grows semiconductor, insulator, and metal layers measured in depths as thin as a few atoms, used for experimental quantum physics and fabrication of microwave and optoelectronic devices. *Mailing Add:* 6H 422 AT&T Bell Labs Murray Hill NJ 07974

CHO, ARTHUR KENJI, b Oakland, Calif, Nov 7, 28; m 53; c 2. PHARMACOLOGY, ORGANIC CHEMISTRY. *Educ:* Univ Calif, BS, 52; Ore State Univ, MS, 53; Univ Calif, Los Angeles, PhD(chem), 58. *Prof Exp:* Asst res pharmacologist, Univ Calif, Los Angeles, 58-61; res chemist, Don Baxter Inc, 61-65; res pharmacologist, Nat Heart & Lung Inst, 65-70; assoc prof, 70-74, PROF PHARMACOL, UNIV CALIF, LOS ANGELES, 74- *Mem:* Am Chem Soc; Am Soc Pharmacol & Exp Therapeut. *Res:* Drug metabolism; adrenergic mechanisms. *Mailing Add:* Dept Pharmacol Univ Calif Los Angeles CA 90024

CHO, BYONG KWON, b Seoul, Korea, March 29, 44; m 75; c 3. REACTOR DESIGN, CATALYSIS. *Educ:* Seoul Nat Univ, Korea, BS, 68; Univ Alta, Can, MS, 75; Univ Minn, PhD(chem eng), 80. *Prof Exp:* Res officer, Army Chem Lab, Korea, 68-70; proces engr, Korea Explosives Co, 70-71 & Korea Inst Sch & Technol, 71-73; STAFF RES ENGR, GEN MOTORS RES LABS, 79- *Mem:* Am Inst Chem Engrs; Catalysis Soc; Sigma Xi; Am Chem Soc. *Res:* Theoretical and experimental application of catalysis; catalytic kinetics and mathematical modeling technique to emission control systems such as the automobile catalytic converters; chromatography and catalyst poisoning mixing and reaction. *Mailing Add:* 646 Live Oak St Rochester MI 48063-3788

CHO, BYUNG-RYUL, b Seoul, Korea, Feb 3, 26; m 48; c 3. VETERINARY MEDICINE, MICROBIOLOGY. *Educ:* Seoul Nat Univ, DVM, 50; Univ Minn, MS, 59, PhD(vet microbiol), 61. *Prof Exp:* Instr animal infectious dis, Col Vet Med, Seoul Nat Univ, 57-58, from asst prof to assoc prof poultry dis, 61-64; from res assoc to assoc prof, 64-72, PROF AVIAN DIS, WASH STATE UNIV, 76- *Honors & Awards:* P P Levine Award, 77. *Mem:* AAAS; Am Asn Avian Pathologists; Am Soc Microbiologists; Am Vet Med Asn; World Poultry Vet Asn. *Res:* Viral diseases of poultry, particularly in avian tumor research. *Mailing Add:* 15704 42nd Dr NW Stanwood WA 98292

CHO, CHENG T, b Kaohsiung, Taiwan, Dec 2, 37; US citizen; m 68; c 2. PEDIATRICS, INFECTIOUS DISEASES. *Educ:* Kaohsiung Med Col, Taiwan, MD, 62; Am Bd Pediat, Dipl, 69; Univ Kans, PhD(microbiol), 70. *Prof Exp:* From asst prof to assoc prof,pediat & microbiol, 70-78, actg chmn, 78-79, vchmn, 79-81, CHIEF PEDIAT INFECTIOUS DIS, MED CTR, UNIV KANSAS, 72-, PROF PEDIAT & MICROBIOL, 78- *Concurrent Pos:* Pediat resident, Med Ctr, Univ Kans, 65-67, fel infectious dis, 67-70; vis prof, Tri-Serv Gen Hosp & Nat Defense Med Sch, Taiwan, 80, Vet Gen Hosp, Taiwan, 84, Nat Taiwan Univ Hosp, 87, China Med Col, Taiwan, 90. *Mem:* Am Acad Pediat; Soc Pediat Res; Soc Exp Biol & Med; fel Infectious Dis Soc Am; Am Pediat Soc. *Res:* Pathogenesis; immunity and control of viral infection; clinical pediatric infectious diseases. *Mailing Add:* 10215 Howe Lane Leawood KS 66206

CHO, CHUNG WON, b Seoul, Korea, Feb 7, 31; m 58; c 3. MOLECULAR PHYSICS. *Educ:* Seoul Nat Univ, BSc, 53; Univ Toronto, MA, 55, PhD(physics), 58. *Prof Exp:* Asst, Univ Toronto, 58; from asst prof to assoc prof, 58-67, head dept, 76-82, PROF PHYSICS, MEM UNIV NFLD, 67- *Concurrent Pos:* Vis assoc prof, Pa State Univ, 66-68; vis prof, Inst Space & Astronaut Sci, Tokyo, Japan, 84. *Mem:* Can Asn Physicists. *Res:* Lasers; laser spectroscopy; stimulated laser light scattering; pressure-induced infrared absorption; dye lasers; lidar studies. *Mailing Add:* Dept Physics Mem Univ St John's NF A1B 3X7 Can

CHO, EUN-SOOK, b Seoul, Korea, Dec 12, 40; US citizen; m 67; c 3. NEUROPATHOLOGY. *Educ:* Yonsei Univ, MD, 65. *Prof Exp:* Asst prof path, Cornell Univ, 73-75; asst prof, 75-78, ASSOC PROF PATH, UNIV MED & DENT NJ, NJ MED SCH, 78- *Concurrent Pos:* Asst attend pathologist, Mem Sloan-Kettering Cancer Ctr, 73-75; consult neuropathologist, Mem Sloan-Ketering Center Ctr, 81-89. *Mem:* Am Asn Neuropathologists; Am Acad Neurol. *Res:* Morphological evolution of neurological diseases including skeletal muscle diseases, remote effects of cancer and neurotoxicological conditions; morphological and pathogenetical studies on acquired immune deficiency syndrome and related neurological problems; experimental electrolyte-induced demyelination. *Mailing Add:* Dept Path NJ Med Sch 100 Bergen St Newark NJ 07103

CHO, HAN-RU, b Peking, China, Feb 4, 45; m 71; c 2. DYNAMIC METEOROLOGY. *Educ:* Nat Taiwan Univ, BS, 67; Univ Ill, MS, 70, PhD(atmospheric sci), 72. *Prof Exp:* Res assoc meteorol, Lab Atmospheric Res, Univ Ill, 72-74, asst prof, 74; from asst prof to assoc prof meteorol, Dept Physics, 74-82, PROF PHYSICS, DEPT PHYSICS, UNIV TORONTO, 82- *Concurrent Pos:* Chmn, Toronto Ctr, Can Meteorol & Oceanog Soc, 80-82, mem, Sci Comt, 80-83, Educ Comt, 81-88 & Accreditation Comt, 86-89, vpres, 88-89, pres 89-90; mem, grant select comt, space & astron, Natural Sci & Eng Res Coun Can, 82-85, working group boundary layer probs, World Meteorol Orgn, 82-86, sci subvention adv comt, Atmospheric Environ Serv Can, 82, comt precipitation, Am Geophys Union, 83-87, comt Can mesocale meteorol prog, 83-90; consult, long-range transport air pollutants, Ont Hydro, 83-89; assoc ed, Atmosphere-Ocean, 84- *Honors & Awards:* Pres Prize, Can Meteorol & Oceanog Soc, 86. *Mem:* Am Meteorol Soc; Can Meteorol & Oceanog Soc (vpres, 88-89, pres, 89-90). *Res:* Mesoscale atmospheric dynamics; cloud-mean flow interactions; dynamics of tropical disturbances. *Mailing Add:* Dept Physics Univ Toronto Toronto ON M5S 1A7 Can

CHO, KON HO, b Kangwha, Korea, May 3, 37; m 68. PHYSICAL CHEMISTRY, BIOCHEMISTRY. *Educ:* Seoul Nat Univ, BS, 62; Auburn Univ, MS, 65; Princeton Univ, MA & PhD(chem), 70. *Prof Exp:* Proj engr, Hyosung Moolsan Co, Ltd, 62-63; MEM RES STAFF, WESTERN ELEC CO, INC, 69- *Mem:* Am Chem Soc. *Res:* Mechanisms of enzyme catalysis; conformation and conformational changes of proteins; radiation chemistry of polymers. *Mailing Add:* Bell Labs Crawford Corner Rd Holmdel NJ 07733

CHO, SANG HA, b Korea, Nov 25, 36; m 63; c 2. CHEMICAL ENGINEERING. *Educ:* Univ Seoul, BS, 58; Univ Ala, MSE, 61; Princeton Univ, PhD(chem eng), 65. *Prof Exp:* Fel, Univ Toronto, 65-67; ASST PROF CHEM ENG, QUEEN'S UNIV, ONT, 67- *Concurrent Pos:* Nat Res Coun Can res grants, 67- *Mem:* Chem Inst Can; Can Soc Chem Eng. *Res:* Mass and heat transfer in chemical reactors; turbulence and mixing; high temperature reactions in a shock tube. *Mailing Add:* Dept Chem Eng Queen's Univ Kingston ON K7L 3N6 Can

CHO, SOUNG MOO, b Seoul, Korea, Oct 1, 37; US citizen; m 65; c 6. ADVANCED HEAT TRANSFER EQUIPMENT RESEARCH & DEVELOPMENT. *Educ:* Seoul Nat Univ, Korea, BS, 60; Univ Berkeley, Calif, MS, 64, PhD(mech eng), 67. *Prof Exp:* Researcher, Korea Atomic Energy Res Inst, 60-61; res asst, Univ Calif, Berkeley, 62-64, specialist, 64-67; eng specialist, Garrett Corp, Torrance, Calif, 67-69; staff consult, Energy Technol Eng Ctr, Rockwell Int, Conoga Park, Calif, 69-73; mgr, thermal-hydraulic & systs eng, 73-80, DIR ENG, FOSTER WHEELER ENERGY

CORP, NJ, 81- *Concurrent Pos:* Adj prof, Farleigh Dickinson Univ, NJ, 75-87 & Stevens Inst Technol, NJ, 78-; assoc ed, Appl Mech Rev, Am Soc Mech Engrs, 85-88, J Eng Power & Gas Turbines, Am Soc Mech Engrs, 86-; Fulbright scholar, 62. *Mem:* Am Nuclear Soc; fel Am Soc Mech Engrs. *Res:* Liquid metal fast breeder reactor engineering and thermal-hydraulics; advanced heat transfer equipment design; thermal science. *Mailing Add:* Forest Wheeler Energy Corp Eight Peach Tree Hill Rd Livingston NJ 07039

CHO, TAE MOOK, drug addiction, drug receptors, for more information see previous edition

CHO, YONGOCK, b Seoul, Korea, Mar 20, 35; US citizen; m 61; c 2. HISTOLOGY, HEALTH SCIENCE. *Educ:* Ewha Womans Univ, Korea, BS, 55, MD, 59, MS, 61. *Prof Exp:* Asst histol, Dept Anat, Col Med, Ewha Womans Univ, 59-62, instr, 63-68; res assoc anat & physiol, Ind Univ, Bloomington, 68-71; RES ASSOC & ASST PROF ANAT, UNIV CHICAGO, 71- *Concurrent Pos:* Intern, Dept Surg, Pusan Nat Univ, 61; student dir, Med Col, Ewha Womans Univ, 67-68. *Mem:* Am Asn Anatomists; Sigma Xi; Am Soc Cell Biol. *Res:* Structure of hematopoietic organs under normal and abnormal conditions. *Mailing Add:* Dept Anat A100 Univ Chicago Chicago IL 60637

CHO, YOUNG WON, b Seoul, Korea, Mar 3, 31; US citizen; m 59, 72; c 3. CLINICAL PHARMACOLOGY, INTERNAL MEDICINE. *Educ:* Seoul Nat Univ, MD, 56; Emory Univ, MS, 62; Nihon Univ, Tokyo, 72. *Prof Exp:* Instr physiol, Sch Med, Emory Univ, 63-64; chief cardiovasc res lab, Philadelphia Gen Hosp, 64-68; group dir cardiol, William S Merrell Co, Cincinnati, 68-71; assoc prof med & pharmacol, Sch Med, La State Univ, 71-73; asst prof clin pharmacol, Univ NC, Chapel Hill, 73-76; head cardiopulmonary clin res, Burroughs Wellcome Co, 73-76; assoc dir med res, Riker Lab, St Paul, Minn, 76; med dir, Cooper Lab, Inc, 77-; PROF INTERNAL MED & DIR MED THER, COL MED, UNIV S ALA, MOBILE; RES SCIENTIST, EURASIAN CLIN PHARMACOLOGIST INC, 84- *Concurrent Pos:* NIH fel, Emory Univ, 60-64; instr pharmacol, Sch Med, Univ Pa, 64-70; consult, US Eighth Army Korea, 75- & US Vet Admin Hosp, Castle Point, NY, 68-; ed-in-chief, Int J Clin Pharmacol & Biopharm, 77- *Honors & Awards:* Prin Senatum, Int Soc Clin Pharmacol, 78. *Mem:* Fel Am Col Angiol; Int Soc Clin Pharmacol (vpres, 78-); Int Acad Clin Pharmacol (vpres, 78-); Am Soc Pharmacol & Exp Therapeut; fel Am Col Chest Physicians. *Res:* Development of new compounds in cardiopulmonary medicine; clinical pharmacology of cardiopulmonary drugs. *Mailing Add:* 801 Hyobong Bldg 1306-1 Secho Dong Secho Ku Seoul Republic of Korea

CHO, YOUNG-CHUNG, b Kyung-nam, Korea, Nov 19, 40; US citizen; m 72; c 2. PHYSICS, ACOUSTICS. *Educ:* Seoul Nat Univ, Korea, BS, 63; Mass Inst Technol, PhD(physics), 72. *Prof Exp:* Res scientist nuclear eng, Atomic Energy Res Inst, Korea, 65-66; teaching asst physics, Mass Inst Technol, 66-72; assoc res scientist, NY Univ, 73; res assoc acoustics, Mass Inst Technol, 74-75; consult acoustics, Sonotech, Inc, 76-77; vis scientist, Joint Inst Advan Flight Sci, George Washington Univ, 77-79, aerospace engr acoustics & space commun, NASA Lewis Res Ctr, 79-85, AEROSPACE ENGR OPTICS, NASA AMES RES CTR, 85- *Concurrent Pos:* Consult prof elec eng, Stanford Univ, 84-86; Dryden fel. *Mem:* Am Phys Soc; Acoust Soc Am; Sigma Xi; Am Inst Aeronaut & Astronaut; Optical Soc Am; Inst Elec & Electronics Engrs. *Res:* Duct acoustics; optics; image processing. *Mailing Add:* Mail Stop 260-1 NASA Ames Res Ctr Moffett Field CA 94035

CHOATE, JERRY RONALD, b Bartlesville, Okla, Mar 21, 43; m 63; c 1. MAMMALOGY, EVOLUTIONARY BIOLOGY. *Educ:* Kans State Col, Pittsburg, BA, 65; Univ Kans, PhD(zool), 69. *Prof Exp:* Asst prof biol, Univ Conn, 69-71; dir, Mus High Plains, 71-80, assoc prof, 76-80, PROF ZOOL & DIR MUS, FT HAYS STATE UNIV, 80- *Concurrent Pos:* Guest lectr, Yale Univ, 70; spec consult, Snyecol Corp, 70-75, Harvard Univ, 77, US Fish & Wildlife Serv, 77, Asn Systematics Collections, 78-, Am Mus Natural Hist, 78, Ore State Univ, 79, Sunflower Elec Coop, 80, Kans Fish & Game, 81-, Colo Div Wildlife, 78-, Wyo Game & Fish, 83, US Army Corps Engrs, 83, Univ Fla, 83, Carnegie Mus Nat Hist, 83, Humboldt State Univ, 84 & Univ NMex, 84, Univ Ill Press, 86, Brit Broadcasting Corp, 87, Colo Assoc Univ Press, 89, Inst Museum Serv, 90-, NSF, 91; trustee, Am Soc Mammalogists, 84-, Southwestern Asn Naturalists, 86- *Honors & Awards:* Robert L Packard Excellence Ed Award, Southwestern Asn Naturalists, 88; C Hart Merriam Award, Am Soc Mammologists, 88. *Mem:* Am Soc Mammalogists (rec secy, 74-84); Soc Study Evolution; Soc Syst Zool; Southwestern Asn Naturalists (pres, 79-80). *Res:* Systematics and natural history of mammals; speciation and evolutionary biology of insectivores, bats and rodents; biogeography of mammals on the Great Plains. *Mailing Add:* Mus High Plains Ft Hays State Univ Hays KS 67601-4099

CHOATE, WILLIAM CLAY, b Miami, Fla, Sept 24, 34; div; c 1. ELECTRICAL & CHEMICAL ENGINEERING. *Educ:* Univ Fla, BSChE, 57, BSEE, 61, PhD, 66; Univ Wis, MSChE, 58. *Prof Exp:* Sr mem tech staff res & develop, Esso Res & Eng Co, NJ, 59-61; mem tech staff res, Sandia Corp, NMex, 62 & Bell Tel Labs, NJ, 63; mgr control & data systs, 66-70, mgr anal & simulation, 70-73, MGR SYSTS RES, CENT RES LABS, 73-, SR MEM TECH STAFF, TEX INSTRUMENTS, INC. *Res:* Digital methods for the processing of information; image processing; interactive computer graphics; geophysical signal processing. *Mailing Add:* Tex Instruments Inc PO Box 655474 MS 238 Dallas TX 75265

CHOBANIAN, ARAM V, b Pawtucket, RI, Aug 10, 29; m 55; c 3. CARDIOVASCULAR DISEASES, INTERNAL MEDICINE. *Educ:* Brown Univ, AB, 51; Harvard Med Sch, MD, 55. *Prof Exp:* Intern & chief resident med, Univ Hosp, Boston, 55-59; assoc prof, 60-71, prof med, 71-77, PROF PHARMACOL, SCH MED, BOSTON UNIV, 77-, DIR CARDIOVASCULAR INST, MED CTR, 73-, DIR HYPERTENSION CTR, 75-, DEAN, SCH MED, 88- *Concurrent Pos:* NIH cardiovascular fel, Univ Hosp, Boston, 59-62; Nat Heart & Lung Inst grants, Sch Med, Boston Univ, 68-; lectr, Harvard Med Sch, 71; mem coun arteriosclerosis and high blood pressure res, Am Heart Asn; mem cardiovasc & renal adv comt, Food & Drug Admin, 75-80, chmn, 78-80; mem hypertension & arteriosclerosis adv comt, Nat Heart & Lung Inst, 75-79, chmn, 77-79, chmn, 4th Joint Nat Comt Hypertension, 88. *Honors & Awards:* Modern Med Award, 90; Award Merit, Am Heart Asn, 90; First Bristol Myers-Squibb Lifetime Achievement Award, 90. *Mem:* Am Soc Clin Invest; Am Heart Asn; Am Physiol Soc; Am Fedn Clin Res; Asn Am Physicians. *Res:* Hypertension; arterial metabolism. *Mailing Add:* Cardiovasc Inst Med Ctr Boston Univ 80 E Concord St Boston MA 02118

CHOBOTAR, BILL, b Vita, Man, Sept 2, 34; m 56; c 3. ZOOLOGY, ANIMAL PARASITOLOGY. *Educ:* Walla Walla Col, BA, 63, MA, 65; Utah State Univ, PhD(zool), 69. *Prof Exp:* From asst to assoc prof biol, 68-76, PROF BIOL, ANDREWS UNIV, 76- *Concurrent Pos:* Alexander von Humboldt fel, Univ Bonn, 74, 81, 85 & 88. *Mem:* AAAS; Am Soc Parasitol; Soc Protozool; Electron Micros Soc Am. *Res:* Biology, development and pathology of parasitic infections, especially of parasitic protozoa; life cycles, fine structure and immunity of coccidia. *Mailing Add:* Dept Biol Andrews Univ Berrien Springs MI 49104

CHO-CHUNG, YOON SANG, b Korea, June 11, 32; US citizen; m 65; c 2. BIOCHEMISTRY, MOLECULAR BIOLOGY. *Educ:* Seoul Womens Med Sch, Korea, MD, 56; Univ Wis-Madison, MS, 60, PhD(biochem), 63. *Prof Exp:* Res asst biochem, Dept Biochem, Univ Wis, 57-60, McArdle Lab, 60-63, fel, 63-66; res assoc, Dept Biol Sci, Purdue Univ, 67-70; vis scientist, Lab Biochem, 70-74, res biochemist, 74-80, CHIEF CELLULAR BIOCHEM SECT, LAB PATHOPHYSIOL, NAT CANCER INST, BETHESDA, MD, 81. *Mem:* Am Soc Biol Chemists; Am Asn Cancer Res; AAAS; Soc Exp Biol & Med; NY Acad Sci. *Res:* Mechanisms of actions of hormones in metabolic regulation and development; significance and control mechanism of cyclic nucleotides in growth regulation. *Mailing Add:* Lab Pathophysiol Cellular Biochem Sect Nat Cancer Inst NIH Bldg 10 Rm 5B-38 Bethesda MD 20892

CHOCK, DAVID POILENG, b Medan, Indonesia, Sept 22, 43; US citizen; m 69; c 1. ATMOSPHERIC MODELING, AIR QUALITY STATISTICS. *Educ:* Univ Calif Santa Barbara, BA, 65; Univ Chicago, PhD(chem physics), 68. *Prof Exp:* Res assoc statist physics, State Univ NY Buffalo, 68-69, Free Univ Brussels, Belgium, 69-71 & fluid dynamics, Univ Tex Austin, 71-72; sr res scientist environ sci, Gen Motors Res Labs, 72-82, group leader atmospheric modeling, 77-87, sr staff scientist environ sci, 82-89; PRIN RES SCIENTIST CHEM & ENVIRON SCI, RES STAFF, FORD MOTOR CO, 89- *Concurrent Pos:* Mem, Tech Adv Denver Brown Cloud II Study, Colo Dept Pub Health, 90-; leader, Air Pollution Res Adv Comt Modeling Group, Coord Res Coun, 91- *Mem:* Sigma Xi; Am Geophys Union; Am Meteorol Soc. *Res:* Pollutant transport modeling; atmospheric chemistry and physics; numerical methods for the advection equation; extreme value statistics; air quality trend analysis; radiationless transitions, phonous in a magnetic field; hydrogen-bonded ferroelectrics; hydrodynamic stability. *Mailing Add:* Res Staff MD-3083 Ford Motor Co PO Box 2053 Dearborn MI 48121-2053

CHOCK, ERNEST PHAYNAN, b Medan, Indonesia, Oct 27, 37; US citizen; m 65; c 2. PHYSICAL INORGANIC CHEMISTRY, POLYMER CHEMISTRY. *Educ:* Univ Calif, Santa Barbara, MA, 63, PhD(surface chem), 67. *Prof Exp:* Asst chem, Univ Calif, Santa Barbara, 61-66; consult, Bell & Howell Res Ctr, Calif, 66-67; researcher dept physics, Univ Calif, Santa Barbara, 67-68 & Los Angeles, 67-70, res prof, Univ Calif, Los Angeles, 70-83; CONSULT, 83- *Concurrent Pos:* Consult, Hughes Aircraft Co, 83; sr scientist, TRW, 84-88. *Mem:* Am Chem Soc; Am Phys Soc. *Res:* General material science, inorganic syntheses, preparation of single crystals, intermetallic compounds and alloys, thin films and ultrafine particles; physical properties of solids by x-ray spectroscopy, magnetic susceptibility, conductivity, surface adsorption and catalytic activity; polymerization of organic and inorganic molecules; electrochemical systems for batteries; optical materials holography. *Mailing Add:* 1048 24th St Santa Monica CA 90403

CHOCK, JAN SUN-LUM, b Honolulu, Hawaii, Jan 22, 44; m 68. PHYCOLOGY. *Educ:* Univ Hawaii, BA, 67; Univ NH, PhD(bot), 75. *Prof Exp:* Res asst marine phycol, Univ Hawaii, 67-68; inspector, Animal & Plant Health Inspection Serv, Agr Res Serv, USDA, 68-70; ASST PROF BIOL, KING'S COL, PA, 75- *Mem:* AAAS; Phycol Soc Am. *Res:* Marine algal ecology, specifically in the estuarine intertidal environment; productivity of brown algae and salt marsh grass. *Mailing Add:* Eastwood Sch 31 Yellow Cote Rd Oyster Bay NY 11771

CHOCK, MARGARET IRVINE, b Eureka, Calif, Mar 17, 44; m 65; c 2. GEOGRAPHIC INFORMATION SYSTEMS, COMPUTER GRAPHICS & IMAGE PROCESSING. *Educ:* Univ Calif, Santa Barbara, BA, 65, 67, Los Angeles, MA, 78, PhD(comput sci), 82. *Prof Exp:* Comput programmer, Gen Elec, 67-68; sr programmer, Comput Usage Corp, 68-70; res engr, Univ Calif, Los Angeles, 76-81; PRES CONSULT & SYSTS INTEGRATION, MIB CHOCK, 82- *Mem:* Asn Comput Mach; Inst Elec & Electronics Engrs; Independent Comput Consults Asn; Am Cong Surv & Mapping; Am Soc Photogram & Remote Sensing; Inst Mgt Consults. *Res:* Computer science; geographic information systems; simultaneous processing of multidimentional and multivariable image data; developed software system called Picture Database Management System; computer sciences. *Mailing Add:* M I B Chock 1048 24th St Santa Monica CA 90403-4528

CHOCK, P BOON, b Indonesia, Mar 26, 39. KINETICS, MECHANISMS OF ENZYME REGULATION & ACTION. *Educ:* Univ Chicago, PhD(chem), 67. *Prof Exp:* CHIEF, SECT METAB REGULATION, LAB BIOCHEM, NAT HEART, LUNG & BLOOD INST, NIH, 81- *Mailing Add:* Lab Biochem Nat Heart Lung & Blood Inst NIH Bldg 3 Rm 202 Bethesda MD 20892

CHODOROW, MARVIN, b Buffalo, NY, July 16, 13; m 37; c 2. ACOUSTICS, ELECTRONICS. *Educ:* Univ Buffalo, AB, 34; Mass Inst Technol, PhD(physics), 39. *Hon Degrees:* DL, Univ Glasgow, 72. *Prof Exp:* Asst, physics, Mass Inst Technol, 39-40; res assoc physics, Pa State Col, 40-41; instr, City Col, 41-43; res physicist, Gen Elec Industs, Conn, 43; sr proj engr, Sperry Gyroscope Corp, NY, 43-47; assoc prof physics, Stanford Univ, 47-54, chmn dept appl physics, 62-69, dir, Ginzton Lab, 59-78, prof physics & elec eng, 54-75, Barbara Kimballl Browning prof applied physics, 75-78, EMER PROF APPL PHYSICS & ELEC ENG, STANFORD UNIV, 78- *Concurrent Pos:* Vis lectr, Ecole Normale Superieure, Univ Paris, 55-56; Fulbright fel, Univ Cambridge, 62-63; vis res assoc, Univ Col, Univ London, 69-70; consult, Rand Corp & Lincoln Lab; mem, US Nat Comt, Int Sci Radio Union; mem adv comt on USSR & Eastern Europe, Nat Acad Sci, 69-71, chmn, 71-73; vis scholar, Oxford Univ, 76. *Honors & Awards:* Baker Award, Inst Elec & Electronics Engrs, 62, Lamme Medal, 82. *Mem:* Nat Acad Sci; Nat Acad Eng; fel Am Phys Soc; fel Inst Elec & Electronics Engrs; fel Am Acad Arts & Sci. *Res:* Electronic devices and microwave acoustics. *Mailing Add:* Ginzton Lab Stanford Univ Stanford CA 94305

CHODOS, ALAN A, b Montreal, Que, Aug 19, 43; US & Can citizen; m 70; c 2. MIT BAG MODEL, KALUZA-KLEIN THEORIES. *Educ:* McGill Univ, BSc, 64; Cornell Univ, PhD(physics), 70. *Prof Exp:* Res assoc, Univ Pa, Dept Physics, 70-73; res assoc, Mass Inst Technol Ctr Theoret Physics, 73-76; sr res assoc, 76-80, SR RES PHYSICIST, DEPT PHYSICS, YALE UNIV, 80- *Concurrent Pos:* Dir, Ctr Theoret Physics, Yale Univ, 87- *Mem:* Am Phys Soc; AAAS. *Res:* Theory of elementary particles and gravitation; particle phenomenology, especially MIT bag model; Kaluza-Klein theory; study of possible new phases of Quantum electrodynamics. *Mailing Add:* Dept Physics Yale Univ PO Box 6666 New Haven CT 06511

CHODOS, ARTHUR A, b New York, NY, Oct 27, 22; m 43; c 2. GEOCHEMISTRY, ANALYTICAL CHEMISTRY. *Educ:* City Col NY, BS, 43; Polytech Inst Brooklyn, MS, 69. *Prof Exp:* Org chemist, Fed Telecommun Lab, NJ, 43-48; analytical chemist, Gen Serv Admin, Fed Supply Serv, NY, 48-50; chemist spectros, US Geol Surv, 50-51, spectroscopist, Analytical Labs Br, Colo, 51-52; sr spectroscopist, Calif Inst Technol, 52-70, sr res spectroscopist, 70-77, mem prof staff, div geol sci, 77-87; RETIRED. *Concurrent Pos:* Consult, US Geol Surv, 54-64; co-investr lunar samples; ed, Micro News, 84- *Honors & Awards:* Presidential Award, Microbeam Anal Soc, 80, Spec Serv Award, 91. *Mem:* Microbeam Analysis Soc (treas, Electron Probe Analysis Soc Am, 68-91, pres, 71, treas, 77-78); Soc Appl Spectros; Sigma Xi. *Res:* Application of instrumental analysis to problems in geochemistry and petrology; electron microprobe analysis of lunar materials. *Mailing Add:* 302 Acorn Circle Monrovia CA 91016-1807

CHODOS, ROBERT BRUNO, b Gap, Pa, July 12, 18; m 50; c 4. MEDICINE. *Educ:* Franklin & Marshall Col, BS, 39; Univ Pa, MD, 43; Am Bd Internal Med, dipl, 53; Am Bd Nuclear Med, dipl, 72. *Prof Exp:* Intern, US Naval Hosp, Philadelphia, 43-44; asst med, Sch Med, Boston Univ, 47-49, instr, 52-53; from instr to assoc prof, State Univ NY Upstate Med Ctr, 53-69; PROF MED & RADIOL, ALBANY MED COL, 69-; HEAD NUCLEAR MED DIV, RADIOL DEPT, ALBANY MED CTR, 69- *Concurrent Pos:* Fel med, Evans Mem Hosp, Boston, 47-48, asst resident, 48-49; resident, Cushing Vet Admin Hosp, Framingham, 49-50, physician med serv & radioisotope unit, 50-52; resident, Vet Admin Hosp, Boston, 52-53, asst chief med & dir radioisotope unit, 53-57, assoc chief staff, 57-65, chief radioisotope serv, 57-69; from asst attend physician to assoc attend physician, State Univ NY Upstate Med Ctr, 53-59; attend physician med & radiol, Albany Med Ctr, 69-; chief nuclear med serv, Vet Admin Hosp, 69-71, physician, 71-75; assoc ed, J Nuclear Med, 70-75. *Mem:* AMA; fel Am Col Physicians; Am Fedn Clin Res; Soc Nuclear Med. *Res:* Radioisotopes; human iron absorption; pathologic physiology of anemias; human thyroid metabolism; renal function in hypertension; blood volume in man in health and disease. *Mailing Add:* Dept Radiol Albany Med Col 43 New Scotland Ave Albany NY 12208

CHODOS, STEVEN LESLIE, b Los Angeles, Calif, Apr 4, 43; m 67; c 2. POINTING & TRACKING SYSTEM DEVELOPMENT, INFRARED TECHNOLOGY. *Educ:* Univ Calif, Los Angeles, BS, 65, MS, 66, PhD(physics), 71. *Prof Exp:* Mem tech staff, Logicon, Inc, 70-72; staff engr, TRW Systs, Inc, 72-77; CHIEF SCIENTIST, HUGHES AIRCRAFT CO, 77- *Mem:* Am Phys Soc. *Res:* Image processing, tracking, pointing and control systems. *Mailing Add:* 1926 Westridge Terr Los Angeles CA 90049

CHODOSH, SANFORD, b Carteret, NJ, Jan 14, 28; m 50; c 3. PULMONARY MEDICINE, INTERNAL MEDICINE. *Educ:* Univ Va, BA, 48; Johns Hopkins Univ, MD, 52. *Prof Exp:* From asst prof to assoc prof med, Sch Med, Tufts Univ, 62-74; fel pulmonary med, Bosto City Hosp, 58-59, dir, Sputum Lab, 59-85; actg chief staff, 86-87, chief med serv, 84-91, CHIEF PULMONARY CLIN, VET ADMIN OUTPATIENTS CLIN, 79-, CHIEF STAFF, 87-; ASSOC PROF MED, SCH MED, BOSTON UNIV, 74- *Concurrent Pos:* Chmn human studies comt, Boston City Hosp, 73-84, assoc vis physician, 73-; pres, Pub Responsibility Med & Res Inc, 79- *Mem:* fel Am Col Chest Physicians; Am Thoracic Soc; Sigma Xi; Appl Res Ethics Nat Asn. *Res:* Inflammatory process as it relates to bronchopulmonary disease; clinical pharmacology in pulmonary medicine; methods for detecting patients with early chronic bronchial disease. *Mailing Add:* 35 Oak Hill Rd Wayland MA 01778

CHODROFF, SAUL, b Brooklyn, NY, Apr 29, 14; m 39; c 2. CHEMISTRY. *Educ:* Brooklyn Col, BS, 38, MA, 42; Polytech Inst Brooklyn, PhD(org chem), 48. *Prof Exp:* Asst res chemist, Weiss & Downs, Inc, NY, 39-40; res chemist, Nat Oil Prods Co, NJ, 41-48; assoc res dir, Nopco Chem Co, 48-50; res dir, Norda Essential Oil & Chem Co, 50-65, vpres res & develop, Norda Inc, 65-79; CONSULT, 79- *Honors & Awards:* Am Inst Chem Award. *Mem:* Am Chem Soc; Sigma Xi. *Res:* Chemistry of organic sulfur; catalytic oxidation; organic synthesis in vitamins, hormones and pharmaceuticals; essential oils, aromatics, perfumes and flavors. *Mailing Add:* 243 McDonald Ave Brooklyn NY 11218

CHOE, BYUNG-KIL, b Taegu, Korea, Feb 15, 33; m 60; c 2. MICROBIOLOGY & CELL BIOLOGY. *Educ:* Kyungpook Nat Univ, Korea, MD, 58, MS, 60; Ind Univ, Bloomington, PhD(microbiol), 70. *Prof Exp:* Res assoc med microbiol, Med Sch, Kyungpook Nat Univ, Korea, 59-62; res assoc microbial genetics, Karolinska Inst, 62-68; res assoc microbiol, Ind Univ, Bloomington, 70-71; res asst prof immunol, Med Sch, State Univ NY, Buffalo, 71-73; from asst prof to assoc prof immunol & microbiol, Med Sch, Wayne State Univ, 73-83; assoc prof urol, res dir, Univ Toronto, Ont, 83-86; ASSOC PROF UROL, COL MED, UNIV ILL, 86- *Concurrent Pos:* Boswell fel, NY Res Found, 71-72. *Mem:* Am Soc Microbiol; Tissue Cult Asn; Am Soc Cell Biol; fel Am Acad Microbiol. *Res:* Growth regulation of mammalian cells; tumor immunology; surgery. *Mailing Add:* Dept Surg Ill Col Med Box 6998 804 S Wood Chicago IL 60680

CHOE, HYUNG TAE, b Seoul, Korea, Apr 27, 27; m 64; c 2. PLANT PHYSIOLOGY, HORTICULTURE. *Educ:* Seoul Nat Univ, BS, 54, MSc, 57; Ohio State Univ, PhD(plant physiol, hort), 63. *Prof Exp:* Instr agr, UNESCO Fundamental Educ Ctr, Ministry of Educ, Korea, 56-59; from asst prof to assoc prof, 63-72, PROF BIOL, MANKATO STATE UNIV, 72- *Concurrent Pos:* Res biologist & NSF fel, Thimann Labs, Univ Calif, Santa Cruz, 73-74; vis biochemist, NSF res grant, Univ Calif, Davis, 86-87; prof, Minn State Univ, Akita Campus, Japan, 91-92. *Mem:* AAAS; Am Soc Plant Physiol; Am Soc Hort Sci; Int Soc Hort Sci; Bot Soc Am; Sigma Xi; Japanese Soc Plant Physiol. *Res:* Senescence of isolated chloroplasts and metabolic functions during the senescence in oat leaves; the effect of growth regulators, light qualities on the senescence of chloroplasts and leaves of oat seedlings; ethylene and senescence in relation to enzymes. *Mailing Add:* Dept Biol Sci Mankato State Col Mankato MN 56001

CHOGUILL, HAROLD SAMUEL, b Humboldt, Kans, Jan 12, 07; m; c 1. CHEMISTRY. *Educ:* Col of Emporia, AB, 27; Univ Kans, AM, 31, PhD(chem), 38. *Prof Exp:* Teacher high sch, Kans, 27-30, prin, 28-30; instr math, Garden City Jr Col, 32-37; instr chem, Independence Jr Col, 37-43, ground instr, Civilian Pilot Training, 37-43; prof chem, 46-76, EMER PROF CHEM, FT HAYS STATE UNIV, 76- *Concurrent Pos:* NSF res fel, Univ Col, London, 59-60; vis lectr, Div Chem Educ, Am Chem Soc. *Mem:* Am Chem Soc; Sigma Xi. *Res:* Iodination of phenols and ethers; kinetics of deiodination; halogenated nitroparaffins. *Mailing Add:* 602 S Evergreen Chanute KS 66720-2359

CHOI, BYUNG CHANG, b Changsung, Korea, July 24, 40; m 72; c 2. CATALYSIS, COMBUSTION. *Educ:* Tex A&M Univ, BS, 68; Univ Pa, PhD(chem eng), 79. *Prof Exp:* Syst analyst statist, Int Latex Corp, 68-71; res engr, 76-79, ASSOC, MOBIL RES & DEVELOP, 79- *Mem:* Combustion Inst; Am Chem Soc; Am Inst Chem Engrs; AAAS; NY Acad Sci. *Res:* Modeling various refining processes such as reforming, visbreaking and coking; modeling deactivation phenomenon of reforming catalyst; thermodynamics of colloid stability. *Mailing Add:* Mobil Res & Develop Corp Billingsport Rd Paulsboro NJ 08066

CHOI, BYUNG HO, b Hwang Hae Do, Korea, Oct 16, 28; c 2. PATHOLOGY, NEUROPATHOLOGY. *Educ:* Yonsei Univ, Korea, MD, 53, MSD, 63. *Prof Exp:* Asst in path, Yonsei Univ, Korea, 53-54; resident, Inst Path, Western Reserve Univ, 54-57, demonstr, 56-57; from instr to assoc prof, Yonsei Univ, Korea, 59-65; res assoc path, Albany Med Col, 65-67, asst prof, 67-69; assoc prof path, Sch Med, St Louis Univ, 69-72; from assoc prof to prof path Sch Med & Dent, Univ Rochester, 72-81; PROF & DIR NEUROPATH, CALIF COL MED, UNIV CALIF, IRVINE, 81- *Concurrent Pos:* Univ fel, Western Reserve Univ, 57-59; consult pathologist, Cuyahoga County Hosp, Cleveland, 57-59; attend pathologist, Vet Admin Hosp, Albany, NY, 68-69; assoc pathologist, Cardinal Glennon Mem Hosp, St Louis, 69-72; pathologist, Firmin-Desloge Hosp, 69-72; pathologist, Strong Mem Hosp, Rochester, 72-; Neuropathologist-in-chg, Irvine Med Ctr, Univ Calif, 81- *Mem:* Am Soc Clin Path; Col Am Path; Am Asn Neuropath. *Res:* Kinetics of cell proliferation in arterial intimal cells under different experimental conditions; developmental neurobiology; effects of methylmercury and other environmental agents on developing nervous system. *Mailing Add:* Dept Path Col Med Univ Calif Irvine CA 92717

CHOI, DUK-IN, b Inchon, Korea, Apr 30, 36; m 65; c 1. PLASMA PHYSICS. *Educ:* Seoul Nat Univ, Korea, BS, 59; Univ Colo, PhD(physics), 68. *Prof Exp:* Fel statist mech, Univ Brussels, 68-70; res assoc, Ctr Statist Mech, Univ Tex, Austin, 70-73; RES SCI ASSOC PLASMA PHYSICS, FUSION RES CTR, UNIV TEX, AUSTIN, 73-; MEM STAFF, PHYSICS DEPT, KOREA SDV INST SCI. *Mem:* Am Phys Soc. *Res:* Theory and application of plasma physics in the thermonuclear fusion research. *Mailing Add:* Physics Dept Korea Adv Inst Sci PO Box 150 Choagyanghi Seoul Republic of Korea

CHOI, KEEWHAN, statistics, for more information see previous edition

CHOI, KWAN-YIU CALVIN, b Hong Kong, Aug 30, 54; Brit citizen; m 86. SEMICONDUCTORS. *Educ:* Chinese Univ Hong Kong, BS, 77; Rutgers, State Univ NY, MS, 81; Ariz State Univ, PhD(elec eng), 87. *Prof Exp:* ASST RES SCIENTIST, ARIZ STATE UNIV, 87- *Mem:* Inst Elec & Electronics Engrs; Am Phys Soc. *Res:* Optoelectronic devices; epitaxial growth of semiconductor materials. *Mailing Add:* PO Box 1070 Tempe AZ 85280-1070

CHOI, KYUNG KOOK, b Seoul, Korea, Mar 23, 46; m 76; c 2. DESIGN SENSITIVITY ANALYSIS, SHAPE OPTIMIZATION. *Educ:* Yon-Sei Univ, Korea, BS, 70; Univ Iowa, MS, 77, PhD(appl math), 80. *Prof Exp:* Res scientist, Univ Iowa, 81-83, adj asst prof, 81-83, from asst prof to assoc prof, 84-90, PROF MECH ENG, UNIV IOWA, 90-, ASSOC DIR, CTR COMPUTER AIDED DESIGN, 90- *Concurrent Pos:* Fac scholar award, Univ Iowa, 87-90; consult, US Army Tank Automotive Command, 87, Gen Motors Res Lab, Eng Mechs, 87-89, Super Tech Inst, Portugal, 87 & Ford Motor Co, 89-; mem comt, Design Theory & Methodology Am, Soc Mech Engrs, 88-, Multidisciplinary Design Optimization, Am Inst Aeronaut & Astronaut; prin investr NSF, distrib parameter struct design sensitivity anal

& optimization, 84-87, NASA design sensitivity anal & optimization built up structs, 84-86, NASA-TACOM-NASA Indust Univ Coop Res Ctr Simulation & Design Optimization, 87-, NSF, mgt computer aided eng systs for concurrency, 90-92 & Ford, configuration design sensitivity anal for noise, vibration, harshness & safety responses, 90- *Honors & Awards:* Nat Res Award, NIH, 83. *Mem:* Am Soc Mech Engrs; Sigma Xi; Am Inst Aeronauts & Astronauts. *Res:* Development of mathematical theory numerical methods of design sensitivy analysis; optimization of strutural systems with sizing and shape as the design variables; computer aided design. *Mailing Add:* Col Eng Univ Iowa 2132 Eng Bldg Iowa City IA 52242

CHOI, KYU-YONG, b Korea, Nov 21, 53; m 79; c 2. INDUSTRIAL POLYMERIZATION REACTORS. *Educ:* Seoul Nat Univ, BS, 76, MS, 78; Univ Wis-Madison, PhD (chem eng), 84. *Prof Exp:* Asst prof, 84-88, ASSOC PROF CHEM ENG, UNIV MD, 88- *Concurrent Pos:* NSF presidential young investr award, 86. *Mem:* Am Inst Chem Engrs; Am Chem Soc; Soc Plastics Engrs. *Res:* Modeling, optimization and control of industrial polymerization reactors; kinetics of free radical and melt polycondensation polymerization processes; stochastic on-line estimation of nonlinear chemical reactors. *Mailing Add:* Dept Chem Eng Univ Md College Park MD 20742

CHOI, MAN-DUEN, b Nanking, China, June 13, 45; Can citizen; m 72; c 3. OPERATOR THEORY, MATRIX THEORY. *Educ:* Chinese Univ, Hong Kong, BSc, 67; Univ Toronto, MSc, 70, PhD(math), 73. *Prof Exp:* Lectr math, Univ Calif, Berkeley, 73-76; from asst to assoc prof, 76-82, PROF MATH, UNIV TORONTO, 82- *Honors & Awards:* Coxeter-James Lectr, Can Math Soc, 83; Fel, Royal Soc Can. *Mem:* Am Math Soc; Math Asn Am; Can Math Soc. *Res:* Functional analysis; structure problems of operator algebras, operator theory and sums of squares of polynomials. *Mailing Add:* Dept Math Univ Toronto Toronto ON M5S 1A1 Can

CHOI, NUNG WON, b Pyong Yong, Korea, Nov 1, 31; m 62; c 3. EPIDEMIOLOGY, ONCOLOGY. *Educ:* Seoul Nat Univ, MD, 58; Univ Minn, MPH, 61, PhD(epidemiol), 66. *Prof Exp:* Assoc prof social & prev med, 66-72, PROF EPIDEMIOL & CHIEF SECT EPIDEMIOL & BIOSTATIST, FAC MED, UNIV MAN, 72-; DIR EPIDEMIOL & BIOSTATIST, MAN CANCER TREAT & RES FOUND, 68- *Concurrent Pos:* Med fel epidemiol, Univ Minn, 60-65, fel epidemiol & med statist, Mayo Grad Sch Med, 65-66; USPHS traineeship epidemiol, 62-65; Am Pub Health Asn associateship for prep vital & health statist monogr, 65-66; Can Nat Health Dept positionship epidemiol, 66-72; mem adv comt statist studies, Nat Cancer Inst Can, 69-72; mem nat adv comt epidemiol, Dept Nat Health & Welfare, 74- *Mem:* Fel Am Pub Health Asn; Am Acad Neurol; Can Pub Health Asn; NY Acad Sci. *Res:* Epidemiologies of toxoplasmosis; brain tumor, cardiovascular, cerebrovascular and other neurological diseases; cancer of gastrointestinal tract; childhood and female breast and genital tract congenital malformation; cancer epidemiology. *Mailing Add:* Dept Social & Prev Med Univ Man 753 McDermot Ave Winnipeg MB R3E 0W3 Can

CHOI, SANG-IL, b Korea, Sept 1, 31; m 61; c 2. SOLID STATE PHYSICS, THEORETICAL CHEMISTRY. *Educ:* Seoul Nat Univ, BSc, 53; Brown Univ, PhD(chem), 61. *Prof Exp:* Res assoc theoret chem, Univ Chicago, 61-63; chmn dept physics & astron, 82-88, from asst prof to assoc prof, 63-72, PROF PHYSICS, UNIV NC, CHAPEL HILL, 72- *Concurrent Pos:* Consult, Res Ctr, Am Optical Corp, 67-69; guest lectr, Res Inst Fund Physics, Kyoto Univ, 70. *Mem:* Fel Am Phys Soc; Phys Soc Japan. *Res:* Transport phenomena of gases; semiclassical scattering theory; physics of organic solids; theoretical study of ionic crystals and organic crystals. *Mailing Add:* Dept Physics & Astron CB No 3255 Univ NC Chapel Hill NC 27599-3255

CHOI, SOOK Y, b Seoul, Korea, Dec 13, 38; nat US; m 60; c 3. PHYSIOLOGY, BIOCHEMISTRY. *Educ:* Seton Hall Univ, BS, 64; Rutgers Univ, MS, 70, PhDPhD(physiol), 73. *Prof Exp:* ASSOC PROF BIOL, UPSALA COL, 73-, CHMN DEPT. *Mem:* Am Physiol Soc; Sigma Xi; Am Asn Univ Profs. *Res:* Cellular respiration; oxidative phosphorylation of mitochondria. *Mailing Add:* Dept Biol Upsala Col East Orange NJ 07019

CHOI, STEPHEN U S, b Korea, Feb 15, 42; US citizen; m 69; c 4. FLUID MECHANICS, HEAT TRANSFER. *Educ:* Seoul Nat Univ, BS, 64; Univ Tex, Austin, MS, 74; Univ Calif, Berkeley, PhD(mech eng), 78. *Prof Exp:* Assoc investr, Korea Inst Sci & Technol, 69-73; mem nuclear staff, Bechtel Power Corp, 78-79; staff scientist, Lawrence Berkeley Lab, 80-83; MECH ENGR, ARGONNE NAT LAB, 83- *Concurrent Pos:* Prin investr, Argonne Nat Lab, 87-; vis prof, Dept Eng, Purdue Univ, Calumet, 90-; chmn, Int Energy Agency Advan Fluids Experts Group, 90- *Mem:* Am Soc Mech Engrs; Soc Rheology; Am Soc Heating Refrig & Air-Conditioning Engrs. *Res:* Fluid mechanics including friction reduction; heat and mass transfer; heating, ventilating and air-conditioning; thermalhydraulic experiments and analysis; rheology; building energy conservation technology; advanced-fluids for district heating and cooling systems; awarded one US patent. *Mailing Add:* 2725 Valley Forge Lilse IL 60532

CHOI, SUNG CHIL, b Seoul, Korea, Dec 30, 30; m 67. BIOSTATISTICS. *Educ:* Univ Wash, Seattle, BS, 57, MA, 60; Univ Calif, PhD(biostatist), 66. *Prof Exp:* Mathematician, Boeing Co, 59-62; mem tech staff, Aerospace Corp, 63-64; asst prof math, Calif State Col, 64-66; mem sci staff, Measurement Analysis Corp, 66-67; asst prof, Washington Univ, 67-73, assoc prof biostatist, Med Sch, 73-78, prof, 78; PROF BIOSTATIST & NEUROSURG, MED COL VA, 78- *Concurrent Pos:* Consult, NIH, Am Col Surgeons, Ely-Lilly, 78- *Mem:* Am Statist Asn; Biomet Soc. *Res:* Biostatistical methodology; clinical trials. *Mailing Add:* Dept Biostatist Med Col Va Richmond VA 23298

CHOI, TAI-SOON, b Seoul, Korea, Dec 29, 35; US citizen; m 65; c 3. MEDICINE. *Educ:* Seoul Nat Univ, Korea, MD, 61. *Prof Exp:* Resident pediat, Children's Hosp, Buffalo, 67-69; res asst & instr, Dept Pediat, 71-72, CLIN ASST PROF PEDIAT, STATE UNIV NY, BUFFALO, 72-; RES ASST & INSTR DEPT PEDIAT, PHARMACOL & NEONATOLOGY, CHILDREN'S HOSP, BUFFALO, 71-; DIR NURSERY, BUFFALO MERCY HOSP, 73- *Concurrent Pos:* Fel neonatology, Jewish Gen Hosp, Montreal, 69-70 & Hosp Sick Children, Toronto, 70-71. *Res:* Newborn medicine, especially the prevention of Hyaline Membrane disease; utilizing human plasminogen injections immediately after birth to premature infants. *Mailing Add:* Buffalo Mercy Hosp 565 Abbott Rd Buffalo NY 14220

CHOI, WON KIL, b Kangwon, Korea, Feb 15, 43; m 73; c 3. ELECTRONICS ENGINEERING. *Educ:* Yon-Sei Univ, Seoul, Korea, BSEE, 69; Iowa State Univ, MSEE, 74, PhD(elec eng), 78. *Prof Exp:* Asst prof, Elec Eng Dept, Mich Technol Univ, 78-80; STAFF ENGR, S-CUBED, 80- *Mem:* Inst Elec & Electronics Engrs. *Res:* Electromagnetic pulse hardening for various electronics equipments; analysis of electromagnetic field coupling to enclosures, cables, connectors, gaskets, apertures and slots. *Mailing Add:* 13928 Cayucos Ct San Diego CA 92129

CHOI, YE-CHIN, b Chunchon, Korea, Mar 13, 29; m 50; c 5. VIROLOGY. *Educ:* Univ Tex El Paso, BS, 71; La State Univ, PhD(microbiol), 76. *Prof Exp:* Fel virol, Sch Med, Yale Univ, 76-77; res assoc biochem, State Univ NY Stony Brook, 78-80; STAFF ASSOC GENETICS, COLUMBIA UNIV, 80- *Mem:* Am Soc Microbiol. *Res:* Isolation and characterization of ribosome RNA genes of animals; cloning of intact 18S-28S (45 kb) in cosmid vectors. *Mailing Add:* 1631 Anderson Ave Ft Lee NJ 07024

CHOI, YONG CHUN, b Sunchun, Korea, Dec 25, 35; US citizen; c 2. MOLECULAR BIOLOGY, CANCER RESEARCH. *Educ:* Seoul Nat Univ, MD, 59; Univ Rochester, PhD(biochem), 67. *Prof Exp:* From asst prof to assoc prof, Baylor Col Med, 69-88; PROF & CHMN, DEPT PHARMACOL, COL MED UNIV DONG-A, 88- *Concurrent Pos:* Assoc dean, Col Med & Grad Sch, Dong-A Univ, 91- *Mem:* Am Asn Cancer Res; Am Soc Cell Biol; Am Soc Biochem & Molecular Biol. *Res:* Mechanisms of gene expression in cancer cells; molecular biology of nuclear RNAs of cancer cells; structures of eukaryotic RNAs. *Mailing Add:* Dept Pharmacol Dong-A Univ Col Medicine Pusan 602 103 Republic of Korea

CHOI, YONG SUNG, b Korea, Sept 11, 36; US citizen; m 66. IMMUNOBIOLOGY. *Educ:* Seoul Nat Univ, BMS & MD, 61; Univ Minn, Minneapolis, PhD(biochem), 65. *Prof Exp:* Intern & resident pediat, Univ Minn, Minneapolis, 65-67, med fel, 67-70; res assoc biochem, Salk Inst, La Jolla, Calif, 67-69; asst prof pediat, biochem & path, Univ Minn, Minneapolis, 69-73; mem, Sloan-Kettering Inst & prof biol, Slaon-Kettering div, Cornell Univ Grad Sch Med Sci, 73-; DIR, CELLULAR IMMUNOL LAB, OCHSNER MED FOUND. *Concurrent Pos:* NIH spec res fel, 69-70; fac res award, Am Cancer Soc, 70. *Mem:* Harvey Soc; Am Asn Exp Path; Am Asn Immunol; Sigma Xi; NY Acad Sci. *Res:* Cellular and molecular mechanism of human B-lymphocyte differentiation; molecular cloning of lymphokines; lymphokine therapy of cancer. *Mailing Add:* Cellular Immunol Lab Alton Ochsner Med Found 1516 Jefferson Hwy New Orleans LA 70121

CHOIKE, JAMES RICHARD, b Detroit, Mich, Mar 10, 42; m 66; c 3. MATHEMATICS CURRICULUM DEVELOPMENT, HISTORY OF MATHEMATICS. *Educ:* Univ Detroit, BS, 64; Purdue Univ, MS, 66; Wayne State Univ, PhD(math), 70. *Prof Exp:* Mgt scientist, Burroughs Corp, 62-66; NSF trainee math, Wayne State Univ, 66-70; from asst prof to assoc prof, 70-83, PROF MATH, OKLA STATE UNIV, 83- *Concurrent Pos:* Chmn, Okla-Ark Sect, Math Asn Am, 79-80; fac assoc, Danforth Found, 80-86; pres, Okla Coun Teachers Math, 85-87; chmn, fac coun, Okla State Univ, 87-88; mem bd trustees, Okla Sch Sci & Math, 87-; co-proj dir, Teaching Exp Appl Math Proj, 83-86 & Direct Mid Math & Sci Proj, 87-91; mem math comt, Col Bd Equity Agenda Proj, 90- *Mem:* Math Asn Am; Nat Coun Teachers Math; Am Math Soc; Sigma Xi. *Res:* Study of the behavior of analytic functions in a neighborhood of a singularity; mathematical investigations arising from problems in industry and engineering; development of mathematics curriculum; distance learning via satellite telecasts of AP calculus. *Mailing Add:* Math Dept Okla State Univ Stillwater OK 74078

CHOKSI, JAL R, b Bombay, India, Nov 24, 32; m 63; c 3. MEASURE THEORY & INTEGRATION, ERGODIC THEORY. *Educ:* Univ Cambridge, BA, 54; Univ Manchester, PhD(math), 57. *Prof Exp:* Res asst math, Univ Liverpool, 56-58; res fel, Tata Inst Fundamental Res, India, 58-62, fel, 62-66; vis assoc prof math, Univ Ill, Urbana, 66-67; vis assoc prof, Yale Univ, 67-68; assoc prof, 68-76, PROF MATH, MCGILL UNIV, 76- *Mem:* Am Math Soc; Can Math Soc. *Res:* General and topological measure theory; projective limits; completion regularity; baire category results in ergodic theory; spectral theory of measure preserving transformation; ergodic theory on homogeneous measure algebras. *Mailing Add:* Dept Math & Statist McGill Univ Burnside Hall Montreal PQ H3A 2K6 Can

CHOLAK, JACOB, b Elsass, Russia, Aug 23, 00; nat US; m 32; c 2. ANALYTICAL CHEMISTRY. *Educ:* Univ Cincinnati, ChE, 24; Am Acad Sanit Engrs, dipl; Environ Eng Intersoc, dipl, 57; Am Bd Indust Hyg, dipl, 62. *Prof Exp:* Asst appl physiol, Col Med, Univ Cincinnati, 26-40, res assoc, 40-42, from asst prof to prof environ health eng, 42-70, EMER PROF, ENVIRON HEALTH ENG, UNIV CINCINNATI, 70- *Honors & Awards:* Eminent Chemist Award, Am Chem Soc, 55. *Mem:* Am Chem Soc; Am Indust Hyg Asn. *Res:* Industrial hygiene; atmospheric pollution; field surveys; instrumental and physical analytical methods. *Mailing Add:* 3115 S Whitetree Circle Cincinnati OH 45236-1343

CHOLETTE, A(LBERT), b Quebec, Que, Oct 12, 18; m 43. CHEMICAL ENGINEERING. *Educ:* McGill Univ, BEng, 42; Mass Inst Technol, SM, 43, ScD(chem eng), 45. *Prof Exp:* HEAD DEPT CHEM ENG, LAVAL UNIV, 46-, PROF, 66- *Concurrent Pos:* Mem, Corp Prof Engrs Que. *Mem:* Am Inst Chemists; Chem Inst Can; Eng Inst Can. *Res:* Mixing; reactor design; heat transfer; evaporation. *Mailing Add:* 1324 James Le Moine Ave Quebec PQ G1S 1A3 Can

CHOLICK, FRED ANDREW, b Portland, Ore, Mar 27, 50; m 70. GENETICS, PLANT BREEDING. *Educ:* Ore State Univ, BS, 72; Colo State Univ, MS, 75, PhD(genetics), 77. *Prof Exp:* Grad asst plant breeding, Colo State Univ, 72-77; res assoc plant breeding, Ore State Univ, 77-81; ASSOC PROF PLANT BREEDING, SDAK STATE UNIV, 81- *Mem:* Am Soc Agron; Sigma Xi. *Res:* The development of superior cultivars for increased world food production, with emphasis on the genetic control of yielding ability. *Mailing Add:* Dept Plant Sci SDak State Univ Box 2207A Brookings SD 57007

CHOLLET, RAYMOND, b Flushing, NY, Oct 4, 46; m 90; c 3. PLANT BIOCHEMISTRY & PHYSIOLOGY, PHOTOSYNTHESIS RESEARCH. *Educ:* Colgate Univ, AB, 68; Univ Ill, MS, 69, PhD(bot), 72. *Prof Exp:* Res assoc plant physiol, Univ Ill, 71-72; res scientist plant biochem/physiol, E I du Pont de Nemours & Co, 72-77-; assoc prof 77-82, PROF BIOCHEM, UNIV NEBR, LINCOLN, 82-, HEAD SECT MOLECULAR PLANT BIOL, SCH BIOL SCI, 87- *Mem:* Am Soc Plant Physiologists; AAAS. *Res:* Photosynthetic and photorespiratory carbon metabolism in higher plants, involving studies with isolated enzymes, organelles, protoplasts, cells and intact leaf tissue; regulatory protein phosphorylation in higher plants; light/dark regulation of C4 photosynthesis. *Mailing Add:* Dept Biochem Univ Nebr Lincoln NE 68583-0718

CHOLVIN, NEAL R, b Chippewa Falls, Wis, Sept 8, 28; m 57; c 3. DEVICE PRODUCT DEVELOPMENT. *Educ:* Wayne State Univ, BS, 49; Mich State Univ, DVM, 54, MS, 58; Iowa State Univ, PhD, 61. *Prof Exp:* Instr vet surg & med, Mich State Univ, 55-59; asst prof vet physiol & pharmacol, Iowa State Univ, 59-60; assoc prof vet surg & med, Mich State Univ, 61-63; chmn, dept vet anat, pharmacol & physiol, Iowa State Univ, chemn, Biomed Eng Prog, 63-74 & 80-81 & prof, 63-82, prof vet physiol & pharmacol, 63-82; sect mgr, Dept Exp Surg, Ethicon Res Found, Inc, 82-86 & biomed eng & spec proj, 86-87; CONSULT, MED DEVICE PROD DEVELOP, 87- *Concurrent Pos:* NIH postdoc fel, 60-61, Nat Heart Inst fel, 71-72. *Mem:* AAAS; Am Physiol Soc; Am Vet Med Asn; Biomat Soc; Am Heart Asn; Acad Surg Res; Soc Biomat. *Res:* Cardiopulmonary physiology; experimental surgery; medical device development; biocompatibility. *Mailing Add:* 1910 Edgemont Pl W Seattle WA 98199

CHOMA, JOHN, JR, b Sewickley, Pa, Nov 6, 41. ELECTRICAL ENGINEERING. *Educ:* Univ Pittsburgh, BSEE, 63, MSEE, 65, PhD(elec eng), 69. *Prof Exp:* From instr to asst prof elec eng, Univ Pittsburgh, 65-69; assoc prof, Sacramento State Col, 69-71 & Ill Inst Technol, 71-76; sr res engr, TRW Defense & Space Systs Group, 76-81; assoc prof, 81-90, PROF ELEC ENG & ELECTROPHYS, UNIV SOUTHERN CALIF, 90- *Concurrent Pos:* Consult, Nat Acad Sci, 66-67 & Radio Corp Am, Pa, 67-68; co-investr, NSF grant, 68-69, prin investr, 70-71; consult, Allis-Chalmers Corp, Pa, 69-70, Compact Eng, Jet Propulsion Lab, Aerospace & Acrian, 81-; vis assoc prof, Univ Southern Calif, 81- *Mem:* Fel Inst Elec & Electronics Engrs; Am Soc Eng Educ. *Res:* Stability of large signal electronic systems; active device models for computer-aided analysis and design. *Mailing Add:* Dept Elec Eng Univ Southern Calif Univ Park Phe 616 Los Angeles CA 90089-0271

CHOMAN, BOHDAN RUSSELL, microbiology, for more information see previous edition

CHOMCHALOW, NARONG, b Bangkok, Thailand, Aug 26, 35; m 60; c 3. AGRONOMY, PLANT GENETICS. *Educ:* Kasetsart Univ, Bangkok, BS, 57; Univ Hawaii, MS, 61; Univ Chicago, PhD(bot), 64. *Prof Exp:* Instr agr, Kasetsart Univ, 57-59; res asst bot, Univ Chicago, 61-64; asst prof biol, Northern Ill Univ, 64-66; res officer agr, Appl Sci Res Corp Thailand, Bangkok, 66-71; res dir, Agr Prod Res Inst, 71-76; dep gov, Thai Inst Sci Tech Res, 76-83; regional coordr, FAO/RAPA, Int Bd Plant Genetic Resources, 83-87; sr agronomist, UNDP/FAO Proj, 87-89, consult, 89-90; MANAGING DIR, SIAM DHV CONSULT SERV, LTD, 90- *Concurrent Pos:* Lectr bot, biol, genetics, agronomy & soil sci, Kasetsart Univ, 66-89, lectr bot, Chulalongkorn Univ, 66-89; vchmn, Coord Comt Aquatic Weeds, Agr & Biol Br, Nat Res Coun Thailand, 75-83, mem Biol Control Ctr, 77-83, mem sci ed group, 77-83, mem Comt Agr Res, 79-83; chmn, Coord Comt Plant Genetic Resources, Agr & Biol Br, Nat Res Coun Thailand, 77-83; mem, Int Bd Plant Genetic Resources, FAO, Rome, Italy, 78-83, mem exec comt, 80-83, vchmn, 81-83; tech adv, Bd Sci & Technol for Int Develop, US Nat Acad Sci, 81-83, Int Found Sci, Stockholm, Sweden, 85- *Honors & Awards:* Ratnabhorn Medal, 80. *Mem:* Agr Sci Soc Thailand; Sigma Xi; Orchid Soc Thailand; Ornamental Plant Soc Thailand; Sci Soc Thailand. *Res:* Agronomic investigation and plant improvement of kenaf, jute, peanut, sunflower, banana, basil, Japanese mint and other essential oil crops; genetics and cytogenetics research on banana, mint and Ocimum species; winged bean and fast growing nitrogen fixing trees. *Mailing Add:* SIAM DHV Ltd 31 Phayathai Rd Bangkok 10400 Thailand

CHOMPFF, ALFRED J(OHAN), b Malang, Indonesia, Oct 26, 30; m 52; c 2. PHYSICAL CHEMISTRY, POLYMER SCIENCE. *Educ:* Delft Univ Technol, MS, 62, PhD(phys chem), 65. *Prof Exp:* Chemist, TNO Paint Res Inst, Delft, 54-62; res scientist, Delft Univ Technol, 62-66; staff scientist, Sci Res Staff, Ford Motor Co, 66-75; assoc prof chem eng, Univ Southern Calif, 75-79; mgr, Intermedics Intraocular Inc, 79-80; SR RES ASSOC, AVERY INT, 80- *Concurrent Pos:* Consult, self employed, 75-79. *Mem:* Am Chem Soc; Am Soc Rheology; Brit Soc Rheology; Am Inst Physics; Am Inst Chem Engr. *Res:* Viscoelasticity; rubberelasticity; composites; polymerization engineering; kinetics. *Mailing Add:* 17172 Northfield Ln Huntington Beach CA 92647-5533

CHOMSKY, NOAM, b Philadelphia, Pa, Dec 7, 28. LINGUISTICS. *Educ:* Univ Pa, BA, 49, MA, 51, PhD(ling), 55. *Hon Degrees:* LHD, Univ Chicago, 67, Swarthmore Col, 70 & Loyola Univ; DLitt, Visva-Bharati Univ, 80 & Univ Pa, 84. *Prof Exp:* Asst instr, Univ Pa, 50-51; from asst prof to prof, 55-66, Ferrari P Ward prof mod lang & ling, 66-76, INST PROF & PROF LING, MASS INST TECHNOL, 76- *Concurrent Pos:* NSF postdoctoral fel, 58-59; vis fel, Harvard Univ, 62-65; Jeanette K Watson distinguished vis prof humanities, Syracuse Univ, 82; Samuel Weiner distinguished vis prof, Univ Man, Can, 83; William James fel, Am Psychol Soc, 89. *Honors & Awards:* John Locke Lectr, Oxford, Eng, 69; Betrand Russell Mem Lectr, Trinity Col, Cambridge, Eng, 70; Jawaharlal Nehru Mem Lectr, Delhi, India, 72; Whidden Lectr, McMaster Univ, Ont, 75; Huizinga Mem Lectr, Leiden, Neth, 78; Kant Lectr, Stanford Univ, 78; Woodbridge Lectr, Columbia Univ, 78; Lansdowne Lectr, Univ Victoria, BC, Can, 84; Wood Lectr, Univ South, Sewanee, Tenn, 84; Hilldale Lectr, Univ Wis-Madison, 85; CBC Massey Lectr, Toronto, Ont, 88. *Mem:* Fel Nat Acad Sci; fel AAAS; Linguistic Soc Am; fel Am Acad Arts & Sci; Am Acad Polit & Social Sci; Am Philos Asn. *Mailing Add:* Dept Ling Mass Inst Technol Cambridge MA 02139

CHON, CHOON TAIK, b Seoul, Korea, Oct 4, 46; m 71; c 1. SOLID MECHANICS, MECHANICAL ENGINEERING. *Educ:* Seoul Nat Univ, BSME, 70; Brown Univ, MS, 73, PhD(solid mech), 75. *Prof Exp:* Res assoc dynamic plasticity, Brown Univ, 75-76; assoc sr engr dynamic plasticity & mech composite mat, Gen Motors Res Labs, 76-78, sr res engr, 78-80, PRIN RES ENGR STRUCT DYNAMICS & MECH COMPOSITE MAT & COMPUT-AIDED ENG, FORD MOTOR RES STAFF, 80- *Mem:* Am Soc Mech Engrs; Am Acad Mech; Sigma Xi. *Res:* Structural dynamics for elastic-viscoplastic materials with large deformations; mechanics of composite materials, especially for fiber-reinforced plastics; computer-aided engineering. *Mailing Add:* 4074 Butternut Hill Troy MI 48098

CHONACKY, NORMAN J, b Cleveland, Ohio, Oct 14, 39; m 64; c 2. ATMOSPHERIC PHYSICS, COMPUTERS & INSTRUMENTS. *Educ:* John Carroll Univ, BS, 61; Univ Wis, PhD(physics), 67. *Prof Exp:* Asst prof, Southern Conn State Col, 67-74, assoc prof physics, 74-82; res physicist, Rome Air Develop Ctr, 82-87; vis scholar, Tufts Univ, 87-88; RES ASSOC, BOWDOIN COL, 88- *Concurrent Pos:* Vis res scientist, Max Planck Inst Biophys Chem, Gottingen, WGer, 77; vis instr physics, Univ Nebr-Lincoln, 77; vis assoc prof physics, Bates Col, 89-90. *Mem:* Am Asn Physics Teachers; Am Phys Soc; Optical Soc Am. *Res:* Physics of organic and biologically important materials, including water; atmospheric optics; medical applications of physics. *Mailing Add:* Bowdoin Col Dept Physics Brunswick ME 04011

CHONG, BERNI PATRICIA, b West Palm Beach, Fla, Aug 29, 45; div; c 2. SYNTHETIC ORGANIC CHEMISTRY. *Educ:* Swarthmore Col, AB, 67; Univ Mich, PhD(org chem), 71. *Prof Exp:* Fel org synthesis, Cornell Univ, 71-72; fel org chem, Univ Calif, San Diego, 72-73; res chemist fluid process chem, 74-82, RES SECT MGR, ROHM & HAAS CO, 82- *Mem:* Am Chem Soc. *Res:* Synthesis of ion exchange resins and adsorbents; analytical research; pesticide metabolism and environmental fate. *Mailing Add:* 504 Martin Lane Dresher PA 19025

CHONG, CALVIN, b Jamaica, Dec 12, 43; Can citizen; m 66; c 3. AGRICULTURE, HORTICULTURE. *Educ:* McGill Univ, BSc, 68, MSc, 70, PhD(hort), 72. *Prof Exp:* Res asst, Dept Hort, McGill Univ, 68-72, res assoc hort, 73-74; tech info officer, Info Div, Agr Can, 74, res scientist, Res Br, 74-77; from asst prof to assoc prof, Dept Plant Sci, McGill Univ, 80-84; RES SCIENTIST, HORT RES INST ONT, MINISTRY AGR & FOOD, 84- *Concurrent Pos:* Assoc ed, Can J Plant Sci, 78-81, 88-; adj fac mem, Inst In Vitro Cell Biol, Miner Inst, State Univ NY, 82-; assoc grad fac, dept hort sci, Univ Guelph, 88- *Mem:* Int Plant Propagators' Soc; Int Asn Plant Tissue Cult; Agr Inst Can; Can Soc Hort Sci; Am Soc Hort Sci. *Res:* Physiology of ornamental and horticultural crops; propagation of woody species; tissue culture; nursery management and culture. *Mailing Add:* Hort Res Inst Ont Ministry Agr & Food Vineland Station ON L0R 2E0 Can

CHONG, CLYDE HOK HEEN, b Honolulu, Hawaii, Mar 6, 33; m 64; c 3. ANALYTICAL CHEMISTRY, INORGANIC CHEMISTRY. *Educ:* Wabash Col, BA, 54; Mich State Univ, PhD(chem), 58. *Prof Exp:* Asst chem, Mich State Univ, 54-57; group leader develop, 61-68, res specialist, Mound Lab, 68-88; CONSULT, 88- *Mem:* Am Chem Soc; Sigma Xi. *Res:* Reactions and properties of tetralithium peroxy diphosphate tetrahydrate; nuclear chemistry. *Mailing Add:* 448 Wileray Dr Miamisburg OH 45342

CHONG, JAMES YORK, b Hong Kong, Aug 28, 44; Can citizen; m 69; c 3. BIOLOGY. *Educ:* Carleton Univ, BSc, 69, MSc, 73; Univ Man, PhD(bot), 81. *Prof Exp:* Electron microscopist virus & cereal rust fungi, 73-82, RES SCIENTIST CEREAL RUSTS, RES BR, AGR CAN, 83- *Concurrent Pos:* Adj prof, Univ Man, 84- *Honors & Awards:* Gordon J Green Award, Can Phytopath Soc, 85. *Mem:* Can Phytopath Soc; Am Phytopath Soc; Can Microscopical Soc; Sigma Xi. *Res:* Physiologic specialization and control of oat crown rust; ultrastructural and cytochemical studies of the host/rust fungal interface. *Mailing Add:* Res Sta CDA 195 Dafoe Rd Winnipeg MB R3T 2M9 Can

CHONG, JOSHUA ANTHONY, b Kingston, Jamaica, May 15, 43; US citizen; m 68; c 4. SYNTHETIC ORGANIC CHEMISTRY. *Educ:* Univ Calif, Berkeley, BS, 67; Univ Mich, PhD(org chem), 71. *Prof Exp:* Fel org synthesis, Cornell Univ, 71-74; RES FEL CHEM PROCESS RES, ROHM & HAAS CO, 74- *Mem:* Am Chem Soc. *Res:* Synthesis of physiologically active compounds; polymers and monomers synthesis; organic chemicals process research. *Mailing Add:* Rohm & Haas Co Spring House PA 19477

CHONG, KEN PIN, b Linpin, China, Sept 22, 42; US citizen; m 67; c 2. STRUCTURAL ENGINEERING, SOLID MECHANICS. *Educ:* Cheng Kung Univ, Taiwan, BS, 64; Univ Mass, MS, 66; Princeton Univ, MSE, MA & PhD(struct & mech), 69. *Prof Exp:* Sr proj engr res & develop, Nat Steel Prod Co, Bldg Syst, Nat Steel Corp, 69-74; assoc prof, Univ Wyo, 74-79, prof struct & mech, Dept Civil Eng, 79-89, prof & chair, Struct/Solid Mech, 81-89; DIR STRUCT SYSTS & CONST PROCESSES PROG, ENG, NSF, 89- *Concurrent Pos:* Lectr, Dept Civil Eng, Univ Houston, 72-74; prin investr energy res & develop admin, Dept of Energy, 76-; adv, Tech Adv Comt, Am Inst Timber Construct, 77-; hon prof, Univ Hong Kong, 81; ed, J Thin-walled struct. *Honors & Awards:* Dow Outstanding Young Fac Award, Am Soc Eng Educ, 77. *Mem:* Fel Am Soc Civil Engrs; Am Soc Eng Educ; Am Soc Mech Engrs; Am Acad Mech; Soc Exp Mech; Int Elec & Electronics Engrs. *Res:* Solid mechanics and structural engineering; light-gage steel structures; composite structures; static and dynamic mechanical behavior of geomaterials; author of over 100 technical papers and publications and several books. *Mailing Add:* NSF Rm 1108 1800 G St NW Washington DC 20550

CHONG, SHUANG-LING, b Kiangsu, China, July 4, 43; US citizen; m 67; c 2. PHYSICAL CHEMISTRY, ORGANIC CHEMISTRY. *Educ:* Taiwan Nat Cheng Kung Univ, BS, 64; Rutgers Univ, MS, 66, PhD(phys chem), 69. *Prof Exp:* Fel, Rice Univ, 69-72; chemist, Energy Res & Develop Admin, 75-76; res chemist, Dept Energy, 76-83; RES CHEMIST, WESTERN RES INST, 83- *Mem:* Am Chem Soc; Sigma Xi. *Res:* Research in gas phase photochemistry; ion-molecule reaction; fractionation and characterization of organic materials in oil shale and solid wastes from fossil energy processes. *Mailing Add:* 6300 Georgetown Pike HMR-30 Laramie VA 22101-2296

CHOO, THIN-MEIW, b Tapah, Malaysia, May 15, 47; Can citizen; m 77; c 2. BARLEY GENETICS, BARLEY BREEDING. *Educ:* Nat Taiwan Univ, BSc, 71; McGill Univ, PhD(plant breeding), 76. *Prof Exp:* Res assoc, Dept Crop Sci, Univ Guelph, Ont, 76-78; RES SCIENTIST, RES STA, AGR CAN, 78- *Concurrent Pos:* Vis prof, Univ Ky, 83-84. *Mem:* Crop Sci Soc Am; Am Soc Agron; Genetics Soc Can; Can Soc Agron. *Res:* Breeding and tissue culture of barley; quantitative genetics. *Mailing Add:* Res Sta Agr Can PO Box 1210 Charlottetown PE C1A 7M8 Can

CHOONG, ELVIN T, b Jakarta, Indonesia, Oct 20, 32; US citizen; m 60; c 3. WOOD TECHNOLOGY, FORESTRY. *Educ:* Mont State Univ, BS, 56; Yale Univ, FM, 58; State Univ NY Col Forestry, Syracuse, PhD(wood technol), 62. *Prof Exp:* Instr wood sci & technol, State Univ NY Col Forestry, Syracuse, 61-63; asst prof, Humboldt State Col, 64-65; from asst prof to assoc prof, 65-73, PROF FORESTRY, LA STATE UNIV, 73- *Concurrent Pos:* Vis res fel, Commonwealth Sci & Indust Res Orgn, Australia, 72. *Mem:* Forest Prod Res Soc; Soc Wood Sci & Technol. *Res:* Physical and mechanical properties of wood and wood products; nondestructive testings of wood; wood drying and preservation; wood quality; solar and wood energy. *Mailing Add:* Sch Forestry Wildlife & Fisheries La State Univ Baton Rouge LA 70803-6202

CHOONG, HSIA SHAW-LWAN, b China, July 12, 45; m 71; c 3. ORGANIC CHEMISTRY, POLYMER CHEMISTRY. *Educ:* Nat Taiwan Univ, BS, 66; Mass Inst Technol, PhD(chem), 71. *Prof Exp:* Teaching asst chem, Mass Inst Technol, 67-68, res asst, 68-71; sr chemist, Gen Foods Corp, 72-73; MEM TECH STAFF, HEWLETT PACKARD CO, 73- *Mem:* Am Chem Soc; Electrochem Soc; Sigma Xi. *Res:* Liquid crystals for displays and electronic devices; polymers for optical fiber coating and high resolution lithography; electron beam and x-ray lithography. *Mailing Add:* Hewlett Packard Co 3000 Hanover St Palo Alto CA 94303

CHOPPIN, GREGORY ROBERT, b Eagle Lake, Tex, Nov 9, 27; m 51; c 4. INORGANIC CHEMISTRY, NUCLEAR CHEMISTRY. *Educ:* Loyola Univ La, BS, 49; Univ Tex, PhD(chem), 53. *Hon Degrees:* DSc, Loyola Univ, 69, Chalmers Univ, Goteborg Sweden, 85. *Prof Exp:* Mem staff, Radiation Lab, Univ Calif, 53-56; from asst prof to assoc prof, 56-63, chmn dept, 68-77, PROF CHEM, FLA STATE UNIV, 63- *Concurrent Pos:* Vis scientist, Ctr Study Nuclear Energy, Belgium, 62-63; Fulbright lectr, Uruguay, 65 & Portugal, 69; vis prof, Sci Univ Tokyo, 78; Alexander von Humboldt US sr scientist award, 79; vis lectr, USSR Acad Sci, 84, Japanese Soc Prom Sci & Inst Atomic Energy, 85. *Honors & Awards:* Southern Chem Award, Am Chem Soc, 72, Nuclear Chem Award, 85; Nat Award Chem Educ, Mfg Chem Assoc, 79; Seaborg Award Separation Sci, 89. *Mem:* AAAS; Am Chem Soc. *Res:* Nuclear chemistry; physical chemistry of the actinides and lanthanides; environmental behavior of actinides. *Mailing Add:* Dept Chem Fla State Univ Tallahassee FL 32306-3006

CHOPPIN, PURNELL WHITTINGTON, b Baton Rouge, La, July 4, 29; m 59; c 1. VIROLOGY, INTERNAL MEDICINE. *Educ:* La State Univ, MD, 53; Am Bd Internal Med, dipl. *Hon Degrees:* DSc, Emory Univ, 88, La State Univ, 88; DMed, Univ Cologne, 88; DSc, Tulane Univ Sch Med, 89. *Prof Exp:* Intern ward med, 53-54, asst resident, Barnes Hosp & Sch Med, Wash Univ, 56-57; vis investr, Rockefeller Univ, 57-59, res assoc, 59-60, from asst prof to prof virol & med, 60-80, Leon Hess prof, 80-85, vpres acad progs, 83-85, dean grad studies, 85; vpres & chief sci officer, 85-87, PRES, HOWARD HUGHES MED INST, BETHESDA, MD, 87- *Concurrent Pos:* Nat Found fel, 57-59; asst physician, Rockefeller Univ Hosp, 57-60, res assoc physician, 60-62, from assoc physician to physician, 62-70, sr physician, 70-; mem virol study sect, NIH, 68-72, chmn, 75-78, mem, Nat Allergy & Infectious Dis Adv Coun, 80-83; adj prof, Rockfeller Univ, 85-; mem, Bd Sci Consult, Mem Sloan-Kettering Cancer Ctr, 81-86, chmn, 83-84; mem, Comm Life Sci, Nat Res Coun, 82-; mem, Coun Res & Clin Investr, Am Cancer Soc, 83-85; mem, Coun, Inst Med, 86-; mem, Sci Bd, Ernst Klenk Found, 87-; mem, Exec Comt, Inst Med, 88-; assoc ed, J immunol, 68-72, J Supramolecular structure, 72-75. *Honors & Awards:* Waksman Award Excellence, Nat Acad Sci, 84; Bayne-James Lectr, Johns Hopkins Univ Sch Med, 80; Louis Weinstein Lectr, Tufts Univ Sch Med, 84; Campione Lectr, Northwestern Univ Sch Med, 84; Gudakunst Lectr, Univ Mich, 86; Centennial Lectr, Asn Am Physicians, 86; George Boxer Lectr, Univ Med & Dent, 86. *Mem:* Nat Acad Sci; Am Soc Microbiol; Am Soc Cell Biol; Am Asn Immunol; Am Soc Clin Invest; Asn Am Physicians; Sigma Xi; AAAS; Am Soc Virol; Am Inst Biol Sci; Am Acad Arts & Sci; Am Philos Soc; Am Clin& Climatol Asn; Soc Exper Biol & Med. *Res:* Animal virology; myxoviruses and paramyxoviruses; virus multiplication, structure, and pathogenesis. *Mailing Add:* Howard Hughes Med Inst 6701 Rockledge Dr 900 Bethesda MD 20817-1813

CHOPRA, ANIL K, b Peshawar, India, Feb 18, 41; US citizen; m 76. CIVIL ENGINEERING. *Educ:* Banares Hindu Univ, BSc, 60; Univ Calif, Berkeley, MS, 63, PhD(civil eng), 66. *Prof Exp:* Asst prof civil eng, Univ Minn, Minneapolis, 66-67; lectr & asst res engr, Univ Calif, Berkeley, 67-69, from asst prof to assoc prof, 69-76, vchmn, Div Struct Eng & Struct Mech, 80-83, PROF, DEPT ENG, UNIV CALIF, BERKELEY, 76- *Concurrent Pos:* Engr, Standard Vacuum Oil Co, India, 60; design engr, Kaiser Engrs Overseas Corp, India, 61; vchmn, div struct eng & struct mech, Univ Calif, Berkeley, 80-83; mem, US Comt Large Dams, 78-, Nat Res Coun Comt Earthquake Eng Res, 82; mem, Nat Res Coun Comt Natural Disasters, 80-85, vchmn, 81, chmn, 82 & 83, mem, Earthquake Eng Panel, 74-79; consult, 72- *Honors & Awards:* Huber Res Prize, Am Soc Civil Engrs, 75 & Norman Medal, 79, Raymond C Reese Res Prize, 89. *Mem:* Nat Acad Eng; Am Soc Civil Engrs; Earthquake Eng Res Inst; Seismol Soc Am. *Res:* Author of over 160 publications in structural dynamics and earthquake engineering. *Mailing Add:* Dept Civil Eng Univ Calif Berkeley CA 94720

CHOPRA, BALDEO K, b Multan, W Pakistan, Aug 10, 42; m 64; c 1. PLANT PATHOLOGY, MICROBIOLOGY. *Educ:* Benares Hindu Univ, BSc, 60, MSc, 62; Auburn Univ, PhD(plant path, microbiol), 68. *Prof Exp:* Res asst plant path, Univ Allahabad, 62-65; res asst, Auburn Univ, 65-68, tech res asst, 68-69; assoc prof biol, Prairie View Agr & Mech Col, 69-73; actg head dept, 76-77, ASSOC PROF BIOL, J C SMITH UNIV, 73- *Concurrent Pos:* Proj leader, USDA soybean study grant, Prairie View Agr & Mech Col, 69-72; proj dir, Minority Access Res Career Prog, 78-; prin investr, Minority Biomed Support Proj & Antimicrobiol Susceptibility Analysis, 78-81 & A Comp Study Susceptibility Clin Isolates of Bacteria to Cefoxitin, Cefamandole & Cephalothin, 82- *Mem:* Am Phytopath Soc; Am Soc Microbiol. *Res:* Mycology; taxonomy of Aspergilli and Xylaria; interaction of pesticides and soil microflora; nutritional physiology of Fusarium and Aspergillus flavus. *Mailing Add:* Dept Biol Johnson C Smith Univ 100 Betties Ford Rd Charlotte NC 28216

CHOPRA, DEV RAJ, b Jullundur City, Punjab, Apr 14, 30; m 56; c 1. EXPERIMENTAL SOLID STATE PHYSICS, SURFACE PHYSICS. *Educ:* Punjab Univ, MS, 52; Univ Nebr, MA, 60; NMex State Univ, PhD(physics), 64. *Prof Exp:* Demonstr physics, Punjab Univ, 51-52, lectr, 52-58; asst, Univ Nebr, 58-60; res assoc, NMex State Univ, 60-64; assoc prof, 64-71, res grants, 66-68, 70-75, 77-80, 82-88, 88-91, PROF PHYSICS, ETEX STATE UNIV, 71- *Concurrent Pos:* Prin investr, ETex State Univ res award, 70-80 & Robert A Welch Found, 76-91; consult surface sci, mat characterization & nondestructive eval, 79-81, 88-91; Am Chem Soc grant, 85-88, Tex Instruments Inc, 87-88, 89-90, 90-91; Tex Adv Technol Res grant, 86-89. *Honors & Awards:* Distinguished Scientist Award, Sigma Xi, 87. *Mem:* Am Asn Physics Teachers; Am Phys Soc; Sigma Xi; Am Vacuum Soc. *Res:* Soft x-ray spectroscopy; valence band studies of transition and rare-earth elements and their intermetallics using x-ray photoelectron and appearance potential spectroscopy techniques; nondestructive evaluation; material characterization of solid surfaces and thin films; author of several technical publications. *Mailing Add:* Dept Physics ETex State Univ Commerce TX 75428

CHOPRA, DHARAM PAL, b India, Feb 2, 44; m 68; c 1. CELL BIOLOGY. *Educ:* Univ Delhi, BS, 63; Univ London, MS, 67; Univ Newcastle, Eng, PhD(cell biol, path), 71; Samford Univ, Birmingham, MBA. *Prof Exp:* Res assoc cellular develop biol, Univ Newcastle, Eng, 67-71; asst prof dermat, Skin & Cancer Hosp, Temple Univ, 71-74; sr scientist, Southern Res Inst, 74-81, sect head cell biol, 81-89; PROF, INST CHEM TOXICOL, WAYNE STATE UNIV, 89- *Concurrent Pos:* Nat Cancer Inst & Nat Inst Arthritis & Metab Dis grant, Temple Univ, 71-72; grants, Nat Heart, Lung & Blood Inst & Nat Dent Inst. *Mem:* AAAS; Tissue Cult Asn; Am Asn Cancer Res. *Res:* Regulation of growth and differentiation in normal and neoplastic tissues. *Mailing Add:* Wayne State Univ 2727 Second Ave Rm 4000 Detroit MI 48201

CHOPRA, DHARAM-VIR, b Jullundur, India, Oct 15, 30; US citizen; m 69; c 1. STATISTICS, MATHEMATICS. *Educ:* Panjab Univ, MA, 53; Univ Mich, MS, 61, MA, 63; Univ Nebr, PhD(statist), 68. *Prof Exp:* Lectr math, DAV Col, India, 53-59; instr, Univ Nebr, 63-66; asst prof, SC State Col, 66-67; from asst prof to assoc prof, 67-77, chmn dept, 82-83, 85-88, PROF MATH, WICHITA STATE UNIV, 77- *Concurrent Pos:* Statist consult dent sch, Univ Nebr, 64-65. *Mem:* Am Statist Asn; Inst Math Statist; Indian Math Soc; Int Asn Surv Statisticians; Indian Statist Asn. *Res:* Design of experiments and their analysis; combinatorial mathematics; application of statistics to social sciences; psychology. *Mailing Add:* Dept Math & Statist Wichita State Univ Wichita KS 67208-1595

CHOPRA, INDER JIT, b Gujranwala, India, Dec 15, 39; m 66; c 3. ENDOCRINOLOGY, INTERNAL MEDICINE. *Educ:* All India Inst Med Sci, New Delhi, MB, BS, 61, MD, 65; Am Bd Internal Med, dipl, 72, cert endocrinol, 73. *Prof Exp:* Intern, All India Inst Med Sci, New Delhi, 62, resident med, 63-65; res officer, Indian Coun Med Res, New Delhi, 66; registr med, All India Inst Med Sci, 66-67; resident med, Queen's Med Ctr, Honolulu, Hawaii, 67-68; fel endocrinol, Harbor Gen Hosp, Sch Med, Univ Calif, Los Angeles, Torrance, 68-71; from asst prof to assoc prof, 71-78, PROF MED, SCH MED, UNIV CALIF, LOS ANGELES, 78- *Concurrent Pos:* Staff physician, Harbor Gen Hosp, Torrance, 71-72 & Ctr Health Sci, Univ Calif, Los Angeles, 72-; NIH res career develop award, 72-77 & grant, 72- *Honors & Awards:* Van Meter Armour Award, Am Thyroid Asn, 77, Parke-Davis Award, 88; Ernst-Oppenheimer Award, Endocrine Soc, 80. *Mem:* Sigma Xi; Am Soc Clin Invest; Am Thyroid Asn; Endocrine Soc; fel Am Col Physicians; Asn Am Physicians. *Res:* Thyroid physiology and disease; nature of biologically active thyroid hormones, thyroid hormone metabolism, pathogenesis of Graves' disease, nature of thyroid stimulators, radioimmunoassay, pituitary-thyroid axis; endocrine alterations in systemic diseases. *Mailing Add:* Dept Med Ctr Health Sci Univ Calif Sch Med Los Angeles CA 90024

CHOPRA, JOGINDER GURBUX, b Punjab, India, June 3, 32; US citizen. PEDIATRICS, PUBLIC HEALTH. *Educ:* Univ Bombay, MS, 54; Columbia Univ, MPH & MS(nutrit), 61. *Prof Exp:* Resident physician pediat med, NY Polyclin Med Sch & Hosp, 57-58; fel nutrit & metab dis, Tulane Univ, 58-59;

adv nutrit, Pan Am Health Org, WHO, 61-67, reg adv nutrit res, 68-73; act dir nutrit, 73-74, SPEC ASST MED, BUR FOODS, FOOD & DRUG ADMIN, 75- *Concurrent Pos:* Assoc prof clin pediat, George Washington Univ & Georgetown Univ, 70-; consult, Food & Nutrit Bd, Comt Int Nutrit Prog, Subcomt Nutrit & Fertil, Washington, DC & Comt Res, George Washington Univ Med Ctr, 73-74. *Mem:* Am Soc Clin Nutrit; Am Acad Pediat; Am Pub Health Asn; Brit Med Asn. *Mailing Add:* Food & Drug Admin 4265 Embassy Park Dr NW Washington DC 20016

CHOPRA, NAITER MOHAN, b Amritsar, India, Nov 23, 23; US citizen; m 53; c 4. PESTICIDE CHEMISTRY. *Educ:* Punjab Univ, Lahore, BSc, 44, MSc, 45; Trinity Col, Dublin, PhD(chem), 55. *Prof Exp:* Demonstr chem, Forman Christian Col, Lahore, 46-47; asst lectr, Allahabad Agr Inst, India, 47-49; res fel, Univ Toronto, 55-57; res officer, Can Dept Agr Res Sta, Man, 57-65; from assoc prof to prof chem, NC Agr & Tech State Univ, 65-89, dir, Tobacco Res Proj, 67-89; RETIRED. *Concurrent Pos:* Dir & prin investr, US Dept Agr grants, 72-77, 78-81 & 79- *Mem:* Am Chem Soc; Chem Inst Can; Nat Geog Sci. *Res:* Chemistry of pesticides; chemistry of viricides and antibiotics which could be employed in cure of cereal disease; breakdown of pesticides in tobacco and cigarette smokes. *Mailing Add:* 1803 Red Forest Rd Greensboro NC 27410-3034

CHOQUETTE, PHILIP WHEELER, b Utica, NY, Aug 16, 30; m 59; c 2. GEOLOGY. *Educ:* Allegheny Col, BS, 52; Johns Hopkins Univ, MA, 54, PhD(geol), 57. *Prof Exp:* Geologist, US Geol Surv, 56-58; from res geologist to sr res geologist, Marathon Oil Co, 58-84, res assoc, 84-86; from adj prof to res prof, 87-90, RES PROF, UNIV COLO, BOULDER, 90- *Concurrent Pos:* Assoc ed, J Sedimentary Petrol, 76-84 & Geol Soc Am Bull, 78-80, 90-; counr sedimentology, Soc Econ Paleontologists & Mineralogists, 81-83; co-ed, Carbonate Petrol Reservoirs, 85 & Paleokarst, 88; vis prof, State Univ NY, Stony Brook, 87-88. *Mem:* Fel AAAS; fel Geol Soc Am; Am Asn Petrol Geologists; Soc Econ Paleontologists & Mineralogists; Int Asn Sedimentologists. *Res:* Physical stratigraphy, petrology, geochemistry; diagenesis, porosity of sedimentary carbonate rocks. *Mailing Add:* 5111 S Franklin Littleton CO 80121

CHORIN, ALEXANDRE J, b Warsaw, Poland, June 25, 38. MATHEMATICS. *Educ:* Ecole Polytech, Ing Dipl, 61; NY Univ, MS, 64, PhD, 66. *Prof Exp:* Asst res scientist, AEC Appl Math & Comput Ctr, Courant Inst, NY Univ, 66-68, res scientist, 68-69, from asst prof to assoc prof, 69-72; assoc prof, 72-74, PROF MATH, UNIV CALIF, BERKELEY, 74- *Concurrent Pos:* Vis prof, Weizmann Inst Sci, Rehovoth, Israel, 69, Inst Res & Info, Paris, 75-76, Courant Inst Math Sci, NY Univ, 81; vis Miller res prof, Univ Calif, Berkeley, 71-72, res prof, 82-83; fel, Alfred P Sloan Res Found, 72-74; assoc fac researcher, Lawrence Berkeley Lab, 80-, head, Math Dept, 86-; Guggenheim fel, 87-88; prof math, Tel Aviv Univ, 89; distinguished vis prof, Inst Advan Study, 91- *Honors & Awards:* Award Appl Math & Numerical Anal, Nat Acad Sci, 89. *Mem:* Nat Acad Sci; fel Am Acad Arts & Sci. *Res:* Computational math; applied physics. *Mailing Add:* Math Dept Univ Calif Berkeley CA 94720

CHORPENNING, FRANK WINSLOW, b Marietta, Ohio, Aug 17, 13; m 42; c 4. IMMUNOLOGY, MICROBIOLOGY. *Educ:* Marietta Col, AB, 39; Ohio State Univ, MSc, 50, PhD(microbiol), 63. *Prof Exp:* Admin asst typhus comn, US Army, Philippines & Japan, 45-46, bacteriologist, 4th Army Area Lab, 48-49; serologist & dir blood bank, Med Lab, Ger, 52-55; chief clin path & dir blood bank, Brooke Army Hosp, 55-61; from asst prof to prof, 63-81, EMER PROF MICROBIOL, OHIO STATE UNIV, 81- *Concurrent Pos:* Mem coop study group, WHO, 53-55. *Honors & Awards:* USA Typhus Comn Medal. *Mem:* Sigma Xi; fel Am Acad Microbiol; Am Soc Microbiol; Am Asn Immunol; Asn Gnotobiotics. *Res:* Mechanisms and development of immune responses; immunochemical specificity of bacterial antigens. *Mailing Add:* 275 Hyatts Rd Delaware OH 43015

CHORTYK, ORESTES TIMOTHY, b Romania, June 10, 35; US citizen; m 69; c 3. ORGANIC CHEMISTRY. *Educ:* Drexel Univ, BS, 59; Princeton Univ, MS, 61, PhD(org chem), 63. *Prof Exp:* Chemist, USDA Agr Res Serv, Philadelphia, 63-67, res chemist, 68-71, lab chief, Athens, 71-83, RES LEADER, USDA AGR RES SERV, 83- *Concurrent Pos:* Chmn, Tobacco Chemists Res Conf, 84. *Mem:* Am Chem Soc; Phytochem Soc NAm. *Res:* Chemical constituents of tobacco, cigarettes and cigarette smoke; author and co-author of over 240 publications. *Mailing Add:* Phytochem Res Unit USDA Agr Res Serv Box 5677 Athens GA 30613

CHORVAT, ROBERT JOHN, b Chicago, Ill, Aug 16, 42; m 64; c 3. MEDICINAL CHEMISTRY. *Educ:* Ill Benedictine Col, BS, 64; Ill Inst Technol, PhD(org Chem), 68. *Prof Exp:* Sr res investr, G D Searle & Co, 68-74, res scientist I, 74-79, res scientist II, 79-81, sr res scientist & group leader, 81-87; dir med/org chem, DuPont Crit Care, 87-89, res mgr, Cardiovasc Dis Sect/Med Chem, DuPont Co, 89-91, RES MGR CNS CHEM, DUPONT MERCK PHARMACEUT CO, 91- *Concurrent Pos:* Adj assoc prof med chem, Col Pharm, Univ Ill, Chicago, 87-89. *Mem:* Am Chem Soc; fel Am Inst Chemists. *Res:* Four-membered, single phosphorus atom heterocycles; steroids and nucleo-hetero steroids; pyridone heterocycles; oxygenated sterols; mevalonolactone derivatives; antiarrhythmics, antihypertensive calcium antagonists, alpha and beta blockers, cardiotonics, azaphenothiazine CNS agents, antiglaucoma agents, antithrombotics, 4II receptor antagonists, cognition enhancers, antipsychotics. *Mailing Add:* DuPont Merck Pharmaceut Co Exp Sta PO Box 80353 Wilmington DE 19880-0353

CHOU, CHEN-LIN, b Jiangsu, China, Oct 8, 43; US citizen; m 70; c 2. GEOCHEMISTRY, COAL GEOLOGY. *Educ:* Nat Taiwan Univ, BS, 65; Univ Pittsburgh, PhD(geochem), 71. *Prof Exp:* Scholar geochem, Univ Calif, Los Angeles, 71-72, asst res geochemist, 72-75; sr res assoc geol, Univ Toronto, 75-76, asst prof geol, 76-79; fel geol, McMaster Univ, 79-80; asst geologist, 80-84, assoc geologist, 84-88, GEOLOGIST, ILL STATE GEOL SURV, 89- *Concurrent Pos:* Lectr earth sci, Calif State Univ, Fullerton, 73-74; mem, Assoc Comt on Meteorites, Can Nat Res Coun, 78-80; prin investr coal res projs, Ctr Res Sulfur in Coal, Champaign, Ill, 84-90; lectr, Workshop Applns Neutron Activation Anal Earth Sci, Nat Taiwan Univ, Taipei, China, 86, Trace Element Geochem, Academia Sinica, Beijing, 87. *Honors & Awards:* Nininger Award, 71. *Mem:* Geochem Soc; Meteoritical Soc; Am Geophys Union; Geol Soc Am; Mineral Soc; Int Asn Geochem & Cosmochem. *Res:* Geology and geochemistry of coal; coal quality assessment; black shales, coal cleaning methods for removal of ash, sulfur, sodium and chlorine; methods for trace element determinations; general geochemistry and geostatistics. *Mailing Add:* Ill State Geol Surv 615 E Peabody Dr Champaign IL 61820-6964

CHOU, CHING, b Szechwan, China, Oct 23, 41; US citizen; m 68; c 2. ABSTRACT HARMONIC ANALYSIS. *Educ:* Nat Taiwan Univ, BA, 63; Univ Rochester, PhD(math), 67. *Prof Exp:* From asst prof to assoc prof, 67-84, PROF MATH, STATE UNIV NY, BUFFALO, 84- *Mem:* Am Math Soc. *Res:* Abstract harmonic analysis on locally compact groups, especially inveriant means, weakly almost periodic functions; Fourier algebras and Von N-eumann algebras of locally compact groups. *Mailing Add:* Dept Math State Univ NY 106 Diefendorf Hall Buffalo NY 14214

CHOU, CHING-CHUNG, b Taipei, Taiwan, June 25, 32; m 62; c 3. PHYSIOLOGY, GASTROENTEROLOGY. *Educ:* Nat Taiwan Univ, BM, 58; Northwestern Univ, MS, 64; Univ Okla, PhD(physiol), 66. *Prof Exp:* Intern, Nat Taiwan Univ Hosp, Taipei, 57-58; resident radiol, Chinese Air Force Gen Hosp, 59-60; intern, Washington Hosp Ctr, DC, 60-61; resident internal med, Northwestern Univ, 61-64; instr physiol, Univ Okla, 65-66; from asst prof to assoc prof, 66-73, PROF PHYSIOL & MED, MICH STATE UNIV, 73- *Concurrent Pos:* Pres, Splanchnic Circulation Group, 79-; Fulbright scholar to Brazil. *Mem:* Am Fedn Clin Res; Cent Soc Clin Res; Am Physiol Soc; Am Gastroenterol Asn; fel Am Heart Asn. *Res:* Gastrointestinal physiology; cardiovascular physiology; prostaglandins; hemorrhagic shock; intestinal isohemia; adenosine. *Mailing Add:* Dept Physiol Mich State Univ East Lansing MI 48824-1101

CHOU, CHUNG-CHI, b Taiwan, Dec 24, 36; US citizen; m 61; c 3. PHYSICAL CHEMISTRY, CHEMICAL ENGINEERING. *Educ:* Chen-Kung Univ, BS, 59; Baylor Univ, PhD(phys chem), 68. *Prof Exp:* Process supt, Taiwan Sugar Corp, 59-65; teaching asst, Baylor Univ, 65-68; res assoc res & develop, Amstar Corp, 68-72, mgr, Opers Lab, 72-75, mgr process develop, 75-81, MGR TECH DIV, AM SUGAR DIV, AMSTAR CORP, 81- *Concurrent Pos:* Mem exec comt, US Nat Comt Sugar Anal, 78-; referee, Int Comn Uniform Method Sugar Anal, 78-; mem bd dir, Sugar Processing Res Inst, 81- & NT Sugar Trade Lab,, Inc, 81-83; UN consult, China sugar indust, 83. *Honors & Awards:* George & Eleanore Meade Award, Sugar Indust Technologists, 71 & 78. *Mem:* Sigma Xi; Am Chem Soc; Sugar Indust Technologists; NY Acad Sci. *Res:* Thermal energy utilization improvement in unit operations; application of interfaces physical chemical principles to process design and development; analytical method development for sugar analysis and process control. *Mailing Add:* 103A Pidgeon Hill Rd Huntington Station NY 11746

CHOU, CHUNG-KWANG, b Chung-King, China, May 11, 46; US citizen; m 73; c 2. BIOELECTROMAGNETICS. *Educ:* Nat Taiwan Univ, BSEE, 68; Washington Univ, MS, 71; Univ Wash, PhD(elec eng & physiol), 75. *Prof Exp:* Res trainee, Dept Elec Eng, Washington Univ, 69-70, teaching asst, 70-71; res asst, Dept Rehab Med, Univ Wash, 71-75, res assoc, 76, fel, Dept Physiol & Biophysics, Regional Primate Res Ctr, 76-77, asst prof, 77-81, res assoc prof, dept rehab med & Ctr Bioeng & assoc dir, Bioelectromagnetics Res Lav, dept rehab med, 81-85; ASSOC DIR, DEPT RADIATION RES, CITY OF HOPE, DUARTE, CA, 85- *Concurrent Pos:* Consult, Nat Coun Radiation Protection & Measurement, 78-, Los Alamos Nat Lab, 86-; assoc ed, J Bioelectromagnetics, 89-; mem, comt man & radiation, Inst Elec & Electronic Engrs, 90- *Honors & Awards:* Special Decade Award, Int Microwave Power Inst, 81. *Mem:* Fel Inst Elec & Electronic Engrs; Bioelectromagnetics Soc; Int Microwave Power Inst; Sigma Xi; Radiation Res Soc; NAm Hyperthermia Group. *Res:* Medical applications of electromagnetic energy, especially on cancer hyperthermia; biological effects of ionizing and non-ionizing radiation, particularly on the nervous, immune, cardiovascular and reproductive systems. *Mailing Add:* Div Radiation Oncol City of Hope 1500 E Duarte Rd Duarte CA 91010-0269

CHOU, DAVID YUAN PIN, b Shantung, China, Mar 5, 22; US citizen; m 53; c 3. PHYSICAL CHEMISTRY. *Educ:* Tokyo Inst Technol, BE, 48; Ohio State Univ, PhD(soil chem), 54. *Prof Exp:* Assoc prof, St Augustine's Col, 54-56; head dept, 56-83, Glenn Frye prof, 56-88, EMER PROF CHEM, LENOIR RHYNE COL, 88- *Concurrent Pos:* Am Chem Soc Petrol Res Fund fac award advan sci study, 63; res fel, Univ Kans, 64-65. *Mem:* AAAS; Am Chem Soc; fel Am Inst Chem. *Res:* Nonaqueous solvents; phase equilibria; colloidal properties of silicates. *Mailing Add:* 768 Eighth St N E Hickory NC 28601

CHOU, DORTHY T C T, b Taiwan, Feb 6, 40; m 65; c 2. NEUROPHARMACOLOGY. *Educ:* Nat Taiwan Univ, Taipei, BS, 62, MS, 64; Columbia Univ, New York, PhD(pharmacol), 71. *Prof Exp:* Res fel, Dept Pathobiol, Sch Hyg & Pub Health, Johns Hopkins Univ, Baltimore, 71-72; res assoc, Dept Pharmacol, Col Physicians & Surgeons, Columbia Univ, New York, 72-76; proj specialist, 76-79, RES SPECIALIST, DEPT PHYSIOL & PHARMACOL, GEN FOODS CORP TECH CTR, 79- *Concurrent Pos:* Adj asst prof, Dept Pharmacol, NY Med Col, Valhalla, 76-81. *Mem:* Soc Neurosci; Am Soc Pharmacol & Exp Therapeut; AAAS; NY Acad Sci. *Res:* Neurochemical and electrophysiological techniques to study the mechanism and sites of foods or food components' effects on the central nervous system. *Mailing Add:* Gen Foods Corp Tech Ctr 250 North St T22-1 White Plains NY 10625

CHOU, IIH-NAN, b Taiwan, China, Apr 12, 43; US citizen; m 68; c 2. CELL BIOLOGY, ENVIRONMENTAL TOXICOLOGY. *Educ:* Nat Taiwan Univ, Taipei, Taiwan, BS, 66; Univ Ill, Urbana-Champaign, PhD(biochem), 71. *Prof Exp:* from asst prof to assoc prof, 80-88, PROF MICROBIOL, SCH MED, BOSTON UNIV, 88- *Concurrent Pos:* NIH fel, 72-74, Nat Res Serv Award, 74-75; Cancer Res Scholar Award, Mass Div, Am Cancer Soc, 77-80. *Mem:* Am Soc Cell Biol; NY Acad Sci; AAAS; Soc Toxicol. *Res:* Cell biology; cytoskeleton; mechanisms of growth control; environmental toxicology; mechanisms of toxicity of epigenetic environmental agents, including metals, chlorinated pesticides and herbicides; mechanisms of T cell activation. *Mailing Add:* Dept Microbiol Sch Med Boston Univ 80 E Concord St Boston MA 02118

CHOU, JAMES C S, b Kiangsu, China, Jan 13, 20; US citizen; m 48; c 2. MECHANICAL ENGINEERING. *Educ:* Nat Inst Technol, China, BS, 41; Ga Inst Technol, MS, 49; Okla State Univ, PhD(mech eng), 68. *Prof Exp:* Gen design engr, Hawaiian Commercial & Sugar Co, 51-57; from asst prof to prof mech eng, 60-, EMER PROF, UNIV HAWAII. *Mem:* Am Soc Mech Engrs; Am Soc Eng Educ. *Res:* Thermal properties of sea water. *Mailing Add:* Dept Mech Eng Univ Hawaii Manoa Honolulu HI 96822

CHOU, JANICE Y, CELLULAR DIFFERENTIATION. *Educ:* Univ Utah, PhD(biochem), 69. *Prof Exp:* CHIEF, SECT CELLULAR DIFFERENTIATION HUMAN GENETICS BR, NAT INST CHILD HEALTH & HUMAN DEVELOP, NIH, 69- *Res:* Carcinoembryonic proteins; Gene regulation development. *Mailing Add:* Human Genetics Br Bldg 10 Rm 95242 Nat Inst Child Health & Human Develop 9000 Rockville Pike Bethesda MD 20892

CHOU, JORDAN QUAN BAN, b Shanghai, China, Mar 21, 41; m 68; c 2. POWER PLANT CONTROL, PROCESS SIMULATION & TRAINING. *Educ:* Univ New S Wales, Australia, B Eng, 62 & MEng, 64. *Prof Exp:* Teaching fel eng, Univ New SWales, 63-68; res asst, mech eng, Mass Inst Technol, 66-67; res engr, pump & control valves, Masoneilan Div Worthington Ltd, Can, 67-69; head, hydraulic control sec, Atomic Energy Can Ltd, 69-72; design engr specialist, Ont Hydro, 72-76, supervising design engr, 76-79, head, anal & simulation, 79-89, HEAD, PROCESS CONTROL & INSTRUMENTATION, ONT HYDRO, 90- *Concurrent Pos:* Designer, Lucas Indust Equipt Co, Australia, 62-65, lectr, eng mechs, Mohawk Col Appl Arts & Technol, Can, 67-69, eng mgr, simulator proj, Ontar Hydro, 76-, tech adv, Elec Power Res Inst, 78-84, consult, Westinghouse Ltd, 80-81 & CAE Electron, Can, 81-; proj mgr, Hydro Que, Can, 84- & NY State Elec & Gas Co, 88-, prin investr, AI Res & Develop Proj, Precarn Ltd, Can, 88- *Honors & Awards:* Achievement Award, Instrument Soc Am, 87. *Mem:* Fel Instrument Soc Am; Nat Res Coun Can; Soc Computer Simulation Int; Asn Prof Engrs Can. *Res:* Computer simulation, fluid power control, power plant control, nuclear and fossil power plant operator training, simulator project management, and the application of expert system principles to advanced power plant control and on-line diagnostics. *Mailing Add:* 224 McCraney St W Oakville ON L6H 1H7 Can

CHOU, KUO CHEN, b Guangdong, China, Aug 14, 38; US citizen; m 68; c 1. PROTEIN FOLDING & CONFORMATION, INTERNAL MOTION OF BIOMACROMOLECULES & ITS BIOLOGICAL FUNCTIONS. *Educ:* Nanking Univ, People's Repub China, BS, 60, MS, 62; Chinese Acad Sci, PhD(biochem), 76; Kyoto Univ, Japan, DrSci(biophys), 83. *Prof Exp:* Jr scientist, Shanghai Inst Biochem, Chinese Acad Sci, 76-78, assoc prof, 78-79; vis assoc prof, chem ctr, Lund Univ, Sweden, 79-80 & Max-Planck Inst Biophys Chem, Ger, 80; vis assoc prof, Chem Dept, Cornell Univ, 81-83, sr scientist, Baker Lab, 84-85; vis prof, Biophys Dept, Med Ctr, Univ Rochester, 85-86; sr scientist, Protein Eng Group, Eastman Kodak Co, 86-87; SR SCIENTIST, COMPUTATIONAL CHEM, UPJOHN LABS, 87- *Concurrent Pos:* Mem, res bd adv, Am Biog Inst, 89- *Mem:* Fel Am Inst Chemists; NY Acad Sci; Sigma Xi; Am Chem Soc; Biophys Soc; AAAS. *Res:* More than 80 research and review publications in the areas of protein folding and conformation; low-frequency collective motion of biomacromolecules and its biological functions; diffusion-controlled reactions of enzymes; graph theory of chemical kinetics; DNA-drug intercalation; dynamic principles of protein allosteric transition; trigger effects of antibody; structure and function of bacteriorhodopsin; principles of packing and handedness of proteins; cooperativity of hemoglobin; structural prediction and analysis of growth hormones. *Mailing Add:* Computational Chem Upjohn Labs Kalamazoo MI 49001

CHOU, LARRY I-HUI, b Hunan Province, China, May 5, 36; m; c 1. SOLID MECHANICS, FLUID MECHANICS. *Educ:* Cheng Kung Univ, Taiwan, BS, 58; Colo State Univ, MS, 63; Univ Utah, PhD(civil eng), 79. *Prof Exp:* Analytical engr stress, Clark Brothers Co, New York, 66-68; res engr shell stress, Metal Indust Res Inst, Taiwan, 69-72; engr, Nuclear Power Plant, Bechtel Power Co, 80-82, Tools & Prod, 82-85; Larry & Son Analytics, 87-88; CONSULT, 88- *Mem:* Am Soc Mech Engrs. *Res:* Stress analysis in a conical shell theory and technics; prediction of viscoelastic pavement roughness growth. *Mailing Add:* 3343 Puente Ave Baldwin Park CA 91706

CHOU, LIBBY WANG, b China; US citizen; m 63; c 1. ORGANIC CHEMISTRY. *Educ:* Brooklyn Col, BS, 60; Univ Ill, MA, 62; Wayne State Univ, PhD(org chem), 67. *Prof Exp:* PROF CHEM, ALA A&M UNIV, 67- *Mem:* Am Chem Soc. *Res:* Natural products; isolation and identification of compounds from Chinese medicinal herbs. *Mailing Add:* PO Box 185 Normal AL 35762-0185

CHOU, MEI-IN MELISSA LIU, b Taiwan, Oct 4, 47; m 71; c 2. ORGANOMETALLIC CHEMISTRY & ANALYTICAL CHEMISTRY. *Educ:* Nat Taiwan Normal Univ, BS, 70; Mich State Univ, PhD(chem), 77. *Prof Exp:* Asst, Mich State Univ, 73-74, res asst organometallic chem, 74-77; res asoc hydrocarbon shale & coal, Ill State Geol Surv, 78-80; asst organic chemist, 80-82, ASSOC ORGANIC CHEMIST, DEPT ENERGY & NATURAL RESOURCES, ILL INST, 82- *Mem:* Am Chem Soc; Am Asn Petrol Geologist; Geol Soc Am. *Mailing Add:* 615 E Peabody Champaign IL 61820-6917

CHOU, NELSON SHIH-TOON, b Taipei, Taiwan, Oct 17, 35; m 66; c 2. THERAPEUTIC DRUGS. *Educ:* Nat Taiwan Univ, Taipei, BS, 59; Univ Hawaii, Honolulu, MS, 64; Univ Man, Winnipeg, Can, PhD(animal nutrit), 67; Univ Guelph, Ont, Can, DVM, 73. *Prof Exp:* Res asst, med res, US Naval Med Res Unit-II, 61-62; res asst animal nutrit, Univ Man, 64-67; lectr, Ont Vet Col, Univ Guelph, 67-69, asst prof, 69-71; res vet nutrit & animal drugs, Ralston Purina Co, 73-78; dir res animal drugs, Lab Res Enterprises, 79-80; VET MED OFFICER ANIMAL DRUGS, CTR VET MED, FOOD & DRUG ADMIN, 80- *Concurrent Pos:* Vet, New Kingston Vet Hosp, 79. *Mem:* Am Vet Med Asn; Am Inst Nutrit; Am Soc Microbiol; World Maricult Soc; Can Vet Med Asn. *Res:* Animal nutrition such as vitamins and minerals requirements for growing animals; feed additives for livestock production; biotechnology derived therapeutic drugs; veterinary drugs. *Mailing Add:* 12161 McDonald Chapel Dr Gaithersburg MD 20878-2250

CHOU, PEI CHI (PETER), b China, Dec 1, 24; nat US; m 56; c 4. ENGINEERING, MECHANICS. *Educ:* Nat Cent Univ, China, BS, 46; Harvard Univ, MS, 49; NY Univ, PhD(aeronaut eng), 51. *Prof Exp:* Asst, NY Univ, 49-51; struct eng, Devenco Inc, 51-52; aerodynamicist, Repub Aviation Corp, 52-53; from asst prof to prof aerospace eng, 53-73, dir wave propagation res ctr, 67-73, chmn mech & struct advan study group, 69-73, J HARLAND BILLINGS PROF MECH ENG, DREXEL UNIV, 73- *Concurrent Pos:* Sr stress analyst, Budd Co, 55-57; dynamicist, Prewitt Aircraft Corp, 57-58; staff res scientist, Kellet Aircraft Co, 58-62; consult, Alleghany Ballistics Lab, 61-63, Air Force Mat Lab, 66-70, Army Ballistic Res Labs, 70-71 & pres, Dyna East Corp, 71- *Honors & Awards:* Meritorious Award, Air Armament Div, Am Defense Preparedness Asn, 90. *Mem:* Am Soc Mech Engrs; Am Inst Aeronaut & Astronaut; Am Soc Eng Educ; Am Acad Mech; Am Defense Prep Asn; Am Soc Testing Mat. *Res:* impacts; mechanics of composite materials; explosive-metal interaction; mechanical reliability; manufacturing engineering; terminal ballistics, warhead mechanics. *Mailing Add:* Dept Mech Eng & Mech 32nd & Chestnut Sts Philadelphia PA 19104

CHOU, SHELLEY NIEN-CHUN, b Chekiang, China, Feb 6, 24; nat US; m 56; c 3. NEUROSURGERY. *Educ:* St John's Univ, China, BS, 46; Univ Utah, MD, 49; Univ Minn, MS, 54, PhD, 64. *Prof Exp:* Intern, Providence Hosp, 49-50; univ fel, Univ Minn, 50-53, res fel, 53-55, AEC grant, 54-55; clin asst neurosurg, Univ Utah, 56-58; vis scientist, NIH, 59-60; from instr to assoc prof, 60-68, head dept, 74-89, PROF NEUROSURG, MED SCH, UNIV MINNEAPOLIS, 68- *Concurrent Pos:* Spec Study Sec, NIH, Nat Inst Neurol Dis & Stroke, 74- 76; mem, Am Bd Neurol Surg, 74-79; Am Med Asn-FAA, Study Project, 78-79; mem, Gemnac Neurosurg Panel, 79-80; mem, Fed Drug Admin Adv Panel Nervous Syst Devices, 79-82; mem, Residency Rev Comt Neurol Surg, Accreditation Coun Grad Med Educ, 84-90. *Mem:* Soc Neurol Surg; Am Col Surg; Soc Nuclear Med; Neurosurg Soc Am; Am Acad Neurol Surg. *Res:* Experimental neurosurgery and neurophysiology; isotopic tracer investigation in neurophysiology. *Mailing Add:* Dept Neurosurg Univ Minn Med Sch Minneapolis MN 55455

CHOU, SHYAN-YIH, Taipei, Taiwan, Aug 7, 41; m; c 2. NEPHPROLOGY, HYPERTENSION. *Educ:* Nat Taiwan Univ, MD, 66. *Prof Exp:* PHYSICIAN IN CHG NEPHROLOGY LAB, BROOKDALE HOSP MED CTR, BROOKLYN, 80- *Mailing Add:* Dept Nephrology Lab Brookdale Hosp Med Ctr Linden Blvd at Brookdale Plaza Brooklyn NY 11212

CHOU, T(SU) T(EH), b Shanghai, China, Mar 10, 34; m 73; c 1. THEORETICAL HIGH ENERGY COLLISIONS. *Educ:* Nat Taiwan Univ, BS, 56; Tsing Hua Univ, MS, 58; Univ Iowa, PhD(physics), 65. *Prof Exp:* Res assoc, State Univ NY, Stony Brook, 65-70; asst prof physics, Univ Denver, 70-74; from asst prof to assoc prof, 74-83, PROF PHYSICS, UNIV GA, 83- *Concurrent Pos:* Vis scientist, Stanford Univ, 68; res physicist, Denver Res Inst, 70-74,; vis prof, State Univ NY Stony Brook, 82. *Mem:* Fel Am Phys Soc. *Res:* Phenomenology of high-energy hadron-hadron and lepton-hadron collision; theoretical physics. *Mailing Add:* Dept Physics Univ Ga Athens GA 30602

CHOU, TING-CHAO, b Taiwan, Sept 9, 38; nat US; m 65; c 2. PHARMACOLOGY, CHEMOTHERAPHY. *Educ:* Kaohsiung Med Col, Taiwan, BS, 61; Nat Taiwan Univ, MS, 65; Yale Univ, PhD(pharmacol), 70. *Prof Exp:* Teaching asst pharmacol, Col Med, Nat Taiwan Univ, 64-65; res asst, Yale Univ, 65; fel, Sch Med, Johns Hopkins Univ, 69-72; from asst prof to assoc prof pharmacol, Sloan-Kettering Div, Grad Sch Med Sci, Cornell Univ, 73-88; assoc pharmacol, 72-78, assoc mem, 79-88, MEM, SLOAN-KETTERING INST CANCER RES, 88-; PROF PHARMACOL, SLOAN-KETTERING DIV, GRAD SCH MED SCI, CORNELL UNIV, 88- *Concurrent Pos:* Mem ed adv bd, J Nat Cancer Inst & Cancer Biochem, Biophysics. *Mem:* AAAS; Am Asn Cancer Res; Am Soc Pharmacol Exp Therapeut; Am Soc Prev Oncol; Am Soc Biol Chem; Am Soc Biochem Molecular Biol; NY Acad Sci. *Res:* Pharmacology and biochemistry of cancer chemotherapeutic agents; enzyme kinetics; theoretical biology of dose-effect relationships. *Mailing Add:* Sloan-Kettering Inst Cancer Res Lab Pharmacol 1275 York Ave New York NY 10021

CHOU, TSAI-CHIA PETER, b Chia-Yi, Taiwan, Aug 10, 59; m; c 2. COMPUTER VISION, IMAGE PROCESSING. *Educ:* Nat Chiao-Tung Univ, Hsinchu, Taiwan, BS, 81, MS, 83; Univ Md, College Park, PhD(computer sci), 90. *Prof Exp:* Teaching asst computer, Dept Computer Eng, Nat Chiao-Tung Univ, 81-82, res asst image processing, 81-83; syst analyst info syst, Computer Ctr, Chinese Army, 83-85; res asst computer vision, Ctr Automation Res, Univ Md, 85-90; RES ASSOC COMPUTER SIMULATION, CAMBRIDGE RES ASSOCS, 90- *Mem:* Inst Elec & Electronics Engrs Computer Soc; Asn Comput Mach. *Res:* Computer vision for recovering three-dimensional information from two-dimensional images; estimation of optical flow; shape from texture; recovery of three-dimensional motion; perspective scene generation with imagery overlay for flight simulation. *Mailing Add:* Cambridge Res Assocs Inc 1430 Spring Hill Rd Suite 200 McLean VA 22102

CHOU, TSONG-WEN, b Tokyo, Japan, Feb 10, 33; US citizen; m 58; c 4. FERMENTATION, ENVIRONMENTAL SCIENCE. *Educ:* Nat Taiwan Univ, BS, 55; Utah State Univ, PhD(food sci & technol), 70. *Prof Exp:* Res assoc, Wei-Chaun Foods Co, 59-64, Mass Inst Technol, 69-70; res biochemist, Univ Calif, Davis, 71-72; sr microbiologist, Rachelle Labs, Inc, 72-75; SR MICROBIOLOGIST & BIOCHEMIST, SRI INT, 75- *Mem:* Am Chem Soc; Am Soc Microbiologists; Soc Indust Microbiol. *Res:* Biotechnology; fermentation technology; enzyme technology; biodegradation of chemicals; microbial physiology and genetics; food science; environmental science. *Mailing Add:* 991 Edmonds Way Sunnyvale CA 94087

CHOU, TSU-WEI, b Shanghai, China, June 2, 40; m 68; c 3. MATERIALS SCIENCE, SOLID MECHANICS. *Educ:* Nat Taiwan Univ, BS, 63; Northwestern Univ, MS, 66; Stanford Univ, PhD(mat sci), 69. *Prof Exp:* Res assoc, Mat Res Ctr, Allied Chem Corp, 69; from asst prof to assoc prof, 69-78, PROF MECH ENG, UNIV DEL, 78- *Concurrent Pos:* Frederick Gardner Cottrell fel, 70-71; vis scientist, Argonne Nat Lab, 75-76; sr vis res fel, Brit Sci Res Coun, 76; vis prof, Univ Witwatersrand, SAfrica, 77 & Comt Nat Invest Space, Argentina, 81; liaison scientist, Off Naval Res, London, 83, Ger Aerospace Res Estab, Koln, 83, Tongji Univ, China 90 & Tokyo Sci Univ, Japan, 90; ed, Composites Sci & Technol. *Mem:* Am Soc Mech Engrs; Am Ceramic Soc; Am Soc Mat; Am Soc Testing & Mat; Soc Advan Mat & Processing Eng; Mat Res Soc. *Res:* Fiber composite materials; crystal defect theory; fracture and wear of materials; elasticity; plasticity; microbiomechanics; polymer, metal & ceramic materials. *Mailing Add:* Dept Mech Eng Univ Del Newark DE 19716

CHOU, TZI SHAN, b Taipei, Mar 21, 42; m 69; c 3. PHYSICAL CHEMISTRY, NUCLEAR PHYSICS. *Educ:* Nat Taiwan Univ, BS, 64; Univ SC, MS, 68; Univ Calif, Berkeley, PhD(nuclear chem), 74. *Prof Exp:* Asst gen chem, Nat Taiwan Univ, 65-66; res asst comput, Univ SC, 66-68; res asst low temp, Univ Calif, Lawrence Berkeley Lab, 68-74; res assoc x-ray photoelectron, 74-76, assoc chemist fusion vacuum, 76-78, vacuum scientist, 85, ACTG HEAD INTERSECTING STORAGE ACCELERATOR, BROOKHAVEN NAT LAB, 81- *Mem:* Am Vacuum Soc. *Res:* Surface chemistry of fusion reactor environment; ultra high vacuum technology in accelerator; structure of matter; beam profile instrumentation for storage ring; new technology for accelerator. *Mailing Add:* Bldg 725C Nat Synchrotron Light Source Brookhaven Nat Lab Upton NY 11973

CHOU, WUSHOW, b Shanghai, China, Feb 12, 39; m 65; c 2. COMPUTER SCIENCE, ELECTRICAL ENGINEERING. *Educ:* Chen Kung Univ, Taiwan, BS, 61; Univ NMex, MS, 65; Univ Calif, Berkeley, PhD(elec eng), 68. *Prof Exp:* Asst elec eng, Univ NMex, 63-65; asst elec eng, Univ Calif, Berkeley, 66-67, teaching fel, 67-68, actg asst prof, 68-69; sr tech staff mem, Network Anal Corp, 69-71, vpres telecommun, 71-75; PROF COMPUT SCI & ELEC ENG & DIR COMPUT STUDIES, NC STATE UNIV, 76- *Concurrent Pos:* Vis prof, State Univ NY, Stony Brook, 76; consult, Page Commun Eng, 76-; pres, ACK Comput Appln, Inc, 77-; gen chmn, 7th Data Com Symp, 79-81; tech prog chmn, 6th Data Com Symp, 77-79; ed-in-chief, J Telecommunication Networks, 82- *Mem:* Asn Comput Mach; fel Inst Elec & Electronics Engrs. *Res:* Computer communications; network analysis and optimization; computer performance evaluation. *Mailing Add:* Elec Eng NC State Univ Box 7911 Raleigh NC 27695-7911

CHOU, Y(E) T(SANG), b Kingsu, China, Mar 20, 24; US citizen; m 58; c 2. METALLURGY, MATERIALS SCIENCE. *Educ:* Nat Chungking Univ, BS, 45; Carnegie Inst Technol, MS, 54, PhD(math), 57. *Prof Exp:* Jr res metallurgist, Metals Res Lab, Carnegie Inst Technol, 52-54, proj mathematician, 56-57; scientist, Edgar C Bain Lab for Fundamental Res, US Steel Corp, 57-62, supvr scientist, 62-64, sr scientist, 65-68; assoc prof metall & mat sci, 68-70, PROF METALL & MAT SCI, LEHIGH UNIV, 70- *Concurrent Pos:* US Steel Corp res fel, Cambridge Univ, 60-61; vis prof, Mass Inst Technol, 75-76; vis assoc prof, Brown Univ, 64-65. *Honors & Awards:* Blackall Award, Am Soc Mech Eng. *Mem:* Fel Am Soc Metals; Metall Soc Am; Phys Soc. *Res:* Mechanical properties of metals; flow and fracture of metals; theory of dislocations; superconductivity. *Mailing Add:* Dept Mat Sci & Eng Lehigh Univ Bethlehem PA 18015

CHOU, YUNGNIEN JOHN, b Shanghai, China; US citizen; m 90; c 4. POLYMER CHEMISTRY. *Educ:* Cheng Kung Univ, BS, 70; Lehigh Univ, MS, 76, PhD(polymer sci), 78. *Prof Exp:* Sr chemist res & develop, Sun Chem, 78-80; proj chemist res & develop, Dow Corning, 80-82; proj mgr res & develop, Avery, 82-86; vpres res & develop, PCL Corp, 86-88; VPRES RES & DEVELOP, CMART CO, 88- *Concurrent Pos:* Tech dir res & develop, Polymerica, 87- *Mem:* Am Chem Soc; Tech Asn Pulp & Paper Indust; NY Acad Sci; Sigma Xi. *Res:* Consultant in emulsion polymers, pressure sensitive adhesives and release technology; contract lab study to demonstrate feasibility; market investigation on new product development. *Mailing Add:* Cmart 1646 Francis Way Upland CA 91786

CHOUDARY, JASTI BHASKARARAO, b Jalipudi, India, Jan 15, 33; m 73; c 4. REPRODUCTIVE PHYSIOLOGY, ENDOCRINOLOGY. *Educ:* Madras Vet Col, BVSc, 54; Kans State Univ, MS, 64, PhD, 66. *Prof Exp:* Vet surgeon, India, 54-61; res asst reproductive physiol, Kans State Univ, 63-66; Ford Found fel, Univ Kans Med Ctr, Kansas City, 66-68; sect head reproductive physiol, William S Merrell Co Div, Richardson-Merrell, Inc, 68-71; res scientist endocrinol & metab regulation, G D Searle & Co, 71-; MEM STAFF, FOOD & DRUG ADMIN. *Mem:* Soc Study Reproduction; Endocrine Soc; Am Fertil Soc; Am Asn Anat; NY Acad Sci; Sigma Xi. *Res:* In vitro storage of spermatozoa; pituitary-ovarian relationships; ovarian follicular development and atresia; luteotropic mechanisms; luteolytic mechanisms; control of fertility; drugs; endocrinology of pregnancy; biology of gonadal steroids; biosynthesis of progesterone; radioimmunoassays of steroid hormones; slow-reacting substance anaphylaxis and leukotriene antagonism. *Mailing Add:* Food & Drug Admin HFN-180 5600 Fisheries Lane Parklawn Bldg Rm 10-70 Rockville MD 20857

CHOUDARY, PRABHAKARA VELAGAPUDI, b Morampudi, India, Dec 20, 48; m 78; c 1. RECOMBINANT DNA BIO-TECHNOLOGY, GENE THERAPY. *Educ:* Andhra Univ, India, BSc, 66; Univ Bombay, MSc, 69; Indian Inst Sci, India, PhD(biochem & cell biol), 76. *Prof Exp:* Res assoc biochem, Indian Inst Sci, India, 75-76; fel microbiol, Pa State Univ, 76-77; res assoc human genetics, Yale Univ, 77-80; vis scientist genetic eng, Univ Wis-Madison, 80, 81; sr staff fel molecular genetics, NIMH, 80-83 & NIH, 83-86; prof biotechnol & genetic eng, J Nehru Univ, New Delhi, 86-91; DIR, UCD ANTIBODY ENG LAB, UNIV CALIF, DAVIS, 91- *Concurrent Pos:* Prin investr, NIH, 80-; vis scientist, Indian Inst Sci, India, 83 & Univ Calif-San Diego, 84; consult, Agrigenetics Corp, Colo, 81; head, Lab Molecular Genetics, NIMH, 81-83; coordr, Bioinformatics Ctr, J Nehru Univ, New Delhi, India, 87-91. *Mem:* Am Soc Biochem & Molecular Biol; Am Soc Human Genetics; Sigma Xi; Soc Biochemists India; Asn Microbiologists India; Int Soc Neurochem; Int Soc Plant Molecular Biol; Int Brain Res Orgn. *Res:* Molecular biology of human genetic defects; designing and evaluation of recombinant polyvalent vaccines to tropical parasites; engineering genes of biotechnolgical importance from animals, plants, and microorganisms. *Mailing Add:* Dept Entom Univ Calif Davis CA 95616

CHOUDHARY, MANOJ KUMAR, b Bihar, India, May 22, 52; m 74; c 2. TRANSPORT PHENOMENA, NUMERICAL METHODS. *Educ:* Indian Inst Technol, Kharagpur, BTech Hons, 74; State Univ NY, Buffalo, MS, 76; Mass Inst Technol, ScD(mat eng), 80. *Prof Exp:* Res assoc transp phenomena, Dept Mat Sci & Eng, Mass Inst Technol, 80-82; advan engr, Owens-Corning Fiberglas Corp, 82-85, sr engr, 85-89, proj mgr, 89-90, SR TECH STAFF & PROJ MGR, OWENS-CORNING FIBERGLAS CORP, 90- *Concurrent Pos:* Res staff, State Univ NY, Buffalo, 76; consult, Dept Mat Sci & Eng, Mass Inst Technol, 76-82; mem, Bd Rev Metall Trans, Minerals, Metals & Mat Soc, 85- *Mem:* Am Ceramic Soc; Minerals Metals & Mat Soc; Am Inst Chem Engrs; Sigma Xi. *Res:* Mathematical and physical modeling of flow and heat transfer phenomena in electrically-heated melts; dissolution kinetics of refractory in glass melts; stability of transport phenomena in glass melting furnaces; develop novel approaches for statistical process control; application of novel technique such as microwave, plasma, in glass-making process; mathematical modeling of flow, heat transfer and combustion phenomena; author of numerous publications; awarded one patent. *Mailing Add:* Owens-Corning Fiberglas Corp 2790 Columbus Rd Granville OH 43023-1200

CHOUDHURY, ABDUL LATIF, b Dhaka, Bangladesh, Jan 1, 33; m 60; c 2. THEORETICAL HIGH ENERGY PHYSICS. *Educ:* Univ Dhaka, BS, 53, MS, 54; Free Univ Berlin, PhD(theoret physics), 60. *Prof Exp:* Asst to prof physics, Univ Dhaka, 55, sr lectr, 61-66 & 68, reader, 69; asst to prof theoret physics, Free Univ Berlin, 58-60; res fel, Fritz-Haber Inst Ger, 60; Brit Coun-Colombo Plan res asst, Imp Col, Univ London, 60-61; consult, Nat Bur Standards-Univ Dhaka proj, 63-66; assoc prof, 66-73, PROF PHYSICS & MATH, ELIZABETH CITY STATE UNIV, 73- *Concurrent Pos:* Vis physicist, Int Ctr Theoret Physics, Italy, 68; vis scholar, Univ NC, Chapel Hill, 88-89. *Mem:* Am Phys Soc; Am Asn Univ Prof. *Res:* Quantum field theory; atomic physics; symmetry principles and group theoretical approach to particle physics; magnetic properties of the proposed charmed particles. *Mailing Add:* Dept Phys Sci Box 886 Elizabeth City State Univ Elizabeth City NC 27909

CHOUDHURY, DEO C, b Darbhanga, India, Feb 1, 26; m 63; c 1. NUCLEAR PHYSICS. *Educ:* Univ Calcutta, BS, 44, MS, 46; Univ Calif, Los Angeles, PhD(physics), 59. *Prof Exp:* Coun Sci & Indust scholar, Sci Col, Univ Calcutta, 47-52; res fel, Niels Bohr Inst, Copenhagen, 52-55; asst physics, Univ Rochester, 55-56 & Univ Calif, Los Angeles, 56-59; asst prof, Univ Conn, 59-62; assoc prof, 62-67, PROF PHYSICS, POLYTECH UNIV, 67- *Concurrent Pos:* Sabbatical leave, Niels Bohr Inst, Copenhagen, 78-79; vis asst physicist, Brookhaven Nat Lab, 60; vis physicist, Oak Ridge Nat Lab, 62. *Mem:* Am Phys Soc; NY Acad Sci; Indian Phys Soc; AAAS; Sigma Xi. *Res:* Theoretical investigations of nuclear structure and reactions; theory of B-decay; theory of strong interactions in nuclear physics; theory of high energy nuclear scattering. *Mailing Add:* Dept Physics Polytech Univ 333 Jay St Brooklyn NY 11201

CHOUDHURY, P ROY, b Calcutta, India, Aug 27, 30. MECHANICAL ENGINEERING. *Educ:* Univ Wash, BS, 51, BS, 52, MS, 53; Northwestern Univ, PhD(mech eng), 58. *Prof Exp:* Instr mech eng, Northwestern Univ, 56-58; asst prof, 58-76, ASSOC PROF MECH ENG, UNIV SOUTHERN CALIF, 76- *Res:* Thermodynamics of combustion phenomena and aerothermochemistry; heat transfer. *Mailing Add:* Dept Mech Eng 1453 Univ Southern Calif University Park Los Angeles CA 90089

CHOUINARD, GUY, b Montreal, Que, Dec 20, 44. PSYCHOPHARMACOLOGY. *Educ:* Univ Montreal, BA, 64, MD, 68, MSc, 74; FRCP(C), 72. *Prof Exp:* From lectr to assoc prof, dept psychiat, McGill Univ, 73-87; from asst prof res to assoc prof res, 74-86, PROF RES, DEPT PSYCHIAT, UNIV MONTREAL, 86-; PROF DEPT PSYCHIAT, MCGILL UNIV, 87- *Concurrent Pos:* Prof, Nat Inst Sci Res, Univ Que, 73-80; psychiatrist & pharmacologist, res dept, Louis H LaFontaine Hosp, Univ Montreal, 73-; psychiatrist & pharmacologist, Pharmacol Res Unit, Allan Mem Inst, McGill Univ, 73-, head, Clin Psychopharmacol Res Unit, 80-; asst psychiatrist, Allan Mem Inst, Royal Victoria Hosp, 73-77, assoc psychiatrist, 79-87, sr psychiatrist, 87-; grant, Can Found Advan Clin Pharmacol, 80- 83. *Honors & Awards:* Clin Res Prize, Med Assn French Can, 76. *Mem:* Fel Am Psychiat Asn; Can Col Neuropsychopharmacol; Soc Biol Psychiat; Am Col Neuropsychopharmacol; Int Col Neuropsychopharmacol; Med Asn French Can. *Res:* Made contributions to psychopharmacology in anxiety, depression and schizophrenia; discovered that alprazolam had antipanic effect and later introduced clonazepam; introduced clonazepam in treatment of mania; found that tryptophan is most beneficial when given with lithium; reported cases of schizophrenia which showed that antipsychotic drugs could induce a syndrome called "neuroleptic super-sensitivity psychosis; showed rubidium could be beneficial in schizophrenia. *Mailing Add:* Allan Mem Inst 1025 Pine Ave W Montreal PQ H3A 1A1 Can

CHOUINARD, LEO GEORGE, II, b Waterbury, Conn, Oct 18, 49. COMMUTATIVE SEMIGROUP RINGS. *Educ:* Mass Inst Technol, BS, 70; Princeton Univ, PhD(math), 75. *Prof Exp:* Instr, Univ Kans, 74-76; asst prof math, 76-81, ASSOC PROF MATH, UNIV NEBR, LINCOLN, 81- *Mem:* Am Math Soc; Sigma Xi; Math Asn Am. *Res:* Modules over commutative semigroup rings; applications of finite group actions to combinatorial design problems. *Mailing Add:* Dept Math & Statist Univ Nebr Lincoln NE 68508

CHOUKAS, NICHOLAS C, b Chicago, Ill, Sept 5, 23; m 51; c 4. ORAL & MAXILLOFACIAL SURGERY. *Educ:* Loyola Univ, Ill, DDS, 50, MS, 58; Am Bd Oral Surg, dipl, 61. *Prof Exp:* Fel, 53, from instr to assoc prof oral & maxillofacial surg, 56-70, assoc prof oral biol & chmn dept oral surg, 63-69, PROF ORAL BIOL, GRAD SCH & PROF ORAL & MAXILLOFACIAL SURG, SCH DENT, LOYOLA UNIV CHICAGO, 69- *Concurrent Pos:* Attend oral & maxillofacial surgeon, Hines Vet Admin Hosp, 58-60, consult, 60-; NIH teachers training grant, 61, res grant, 61-64; chief oral surg, Loyola Univ Hosp, 69- *Mem:* Am Asn Oral & Maxillofacial Surg; fel Int Asn Oral Surg; fel Am Col Dentists; fel Int Col Dentists; fel Am Col Stomatologic Surgeons; fel Pan Am Med Asn; fel Int Asn Maxillo-Facial Surg; Am Soc Oral Surgeons; Sigma Xi. *Res:* Growth changes of the temporomandibular joint and mandible of the Macca rhesus monkey following various experimentally induced environments; contributed articles to professional journals. *Mailing Add:* 2160 S First Ave Loyola Univ Med Ctr Maywood IL 60153

CHOULES, GEORGE LEW, biochemistry, for more information see previous edition

CHOUNG, HUN RYANG, polymer chemistry, for more information see previous edition

CHOVAN, JAMES PETER, b Bethlehem, Pa, April 16, 48. ENZYMOLOGY, TOXICOLOGY. *Educ:* Clemson Univ, BS, 70; Lehigh Univ, MS, 77, PhD(biochem), 80. *Prof Exp:* LECTR, STATE UNIV NY, STONY BROOK, 84- *Mem:* Am Chem Soc; NY Acad Sci; AAAS. *Res:* Pharmacokinetics; toxicity; metabolism; mechanism of action and development of analyses for drugs; mistagenesis and/or carcinogenesis of specific DNA adducts; HPLC separations development. *Mailing Add:* CIBA Geigy Corp Bldg MR-1 444 Sawmill River Rd Ardsley NY 10502-2605

CHOVER, JOSHUA, b Detroit, Mich, Mar 26, 28; m 52. MATHEMATICS. *Educ:* Univ Mich, PhD(math), 52. *Prof Exp:* Res mathematician, Bell Tel Labs, 52-56; from instr to assoc prof, 56-65, chmn dept, 77-79, PROF MATH, UNIV WIS-MADISON, 65- *Concurrent Pos:* Mem, Inst Advan Study, 55-56. *Mem:* Am Math Soc. *Res:* Probability and analysis. *Mailing Add:* Dept Math Univ Wis Madison WI 53706

CHOVITZ, BERNARD H, b Norfolk, Va, Nov 10, 24; m 49; c 3. GEODESY, CARTOGRAPHY. *Educ:* Col William & Mary, BS, 44; Harvard Univ, MA, 47. *Prof Exp:* Mathematician, Army Map Serv, 48-60, Geod Intel, Mapping Res & Develop Agency, 60-61; prin scientist, Autometric Oper, Raytheon Co, 61-64; geodesist, Off Res & Develop, US Coast & Geod Surv, 64-65; geodesist, Environ Sci Serv Admin, 65-70; geodesist, 70-74, dir, Geod Res & Develop Lab, 74-81, chief geodesist, Nat Oceanic & Atmospheric Admin, 81-87, mgr, Earth Sci Dept, Sci Applns Res, 87-88; MGR, GEODYNAMICS PROG, ST SYSTS CORP, 89- *Concurrent Pos:* Pres sect II, Int Asn Geod, 75-79; pres, Geodesy Sect, Am Geophys Union, 78-80. *Mem:* Math Asn Am; fel Am Geophys Union; Int Asn Geod. *Res:* Mathematical analysis of map projections; determination of size of earth; application of artificial satellites to the determination of the earth's gravitational field and its size and shape. *Mailing Add:* 8813 Clifford Chevy Chase MD 20815

CHOVNICK, ARTHUR, b New York, NY, Aug 2, 27; m 49; c 2. GENETICS. *Educ:* Ind Univ, AB, 49, MA, 50; Ohio State Univ, PhD(genetics), 53. *Prof Exp:* Asst physiol genetics, Ohio State Univ, 51-53; instr zool, Univ Conn, 53-57, asst prof genetics, 57-59; asst dir, Long Island Biol Asn, 59-60, dir, 60-62; PROF GENETICS, UNIV CONN, 62- *Concurrent Pos:* Assoc ed, Genetics, 72-; mem genetics study sect, Div Res Grants, NIH, 72-76. *Mem:* Fel AAAS; Genetics Soc Am (treas, 81-83). *Res:* Gene structure, function, regulation, mutation, mechanism of recombination. *Mailing Add:* Box U-125 Univ Conn 75 N Eaglevil Box U-131 Beach Hall Univ Conn Storrs CT 06268

CHOW, ALFRED WEN-JEN, b Peiping, China, Jan 11, 24; m 57. MEDICINAL CHEMISTRY, ORGANIC CHEMISTRY. *Educ:* Univ Mich, BS, 46; Univ Minn, PhD(med chem), 50. *Prof Exp:* Res chemist, Bjorksten Res Lab, 51 & Kremers Urban Co, 52-56; sr res chemist, Pabst Res Lab, 56-59; sr med chemist, Smith Kline & French Lab, 59-71, SR INVESTR, SMITH KLINE CORP, 71- *Mem:* AAAS; Am Chem Soc; Royal Soc Chem. *Res:* Preparation of organic compounds of antimicrobial and antiviral interest; synthetic organic chemistry; animal health products; ruminant nutrition. *Mailing Add:* Smith Kline Corp L-301 1500 Spring Garden St Philadelphia PA 19101

CHOW, ARTHUR, b Meaford, Ont, Oct 23, 36; m 63; c 1. ANALYTICAL CHEMISTRY. *Educ:* Univ Toronto, BSc, 61, MA, 62, PhD(anal chem), 66. *Prof Exp:* Lectr chem, Univ Toronto, 62-64, instr, 64-66; Nat Res Coun Can fel, 66-68; from asst prof to assoc prof, 68-79, PROF CHEM, UNIV MAN, 79- *Mem:* Am Chem Soc; fel Chem Inst Can. *Res:* Analytical chemistry of the noble metals; analytical uses of polyurethane foam for extraction and separation; analysis of inorganic pollutants. *Mailing Add:* Dept Chem Univ Man Winnipeg MB R3T 2N2 Can

CHOW, BRIAN G(EE-YIN), b Macau, Aug 10, 41; US citizen; m 69; c 2. PHYSICS, ASTRONOMY. *Educ:* Case Western Reserve Univ, PhD(physics), 69; Univ Mich, MBA, 77, PhD(finance), 80. *Prof Exp:* sr res specialist, Pan Heuristics Res & Develop Assocs, 78-89; SR PHYS SCIENTIST, RAND CORP, 89- *Concurrent Pos:* Res grant, Nat Energy Proj, Am Enterprise Inst Pub Policy, 74-75; consult, Sci Appl Inc, 78, Dept Energy, 78-81, Arms Control & Disarmament Agency, 78-81, US Coun Environ Quality, 80, Navy, 83-84, Defense Nuclear Agency, 83-89, Off Undersecy Defense, 85-, Undersecy Defense Policy, 87-88, President's Sci Adv, 88-89; dir, Defense Res & Eng, 90- *Res:* Elementary particle theory; experimental nuclear spectroscopy; national and international nuclear energy policies; naval defense and campaign; cruise missile deployment; space arms agreements; ballistic missile defense; military space policy; space launch policy; missile proliferation. *Mailing Add:* 926 Harvard Santa Monica CA 90403

CHOW, BRYANT, b Peking, China, Dec 24, 36; nat US; m 59; c 2. MATHEMATICS, STATISTICS. *Educ:* Franklin & Marshall Col, BS, 59; Rutgers Univ, MS, 61; Va Polytech Inst, PhD(statist), 66. *Prof Exp:* Instr statist, Va Polytech Inst, 64-65; asst prof appl math statist, Rutgers Univ, 65-69; assoc prof, Univ Southwestern La, 69-75, prof math & statist, 75-90; VPRES, QUAL PROGS, MOBILE CHAMBER COMMERCE, 90- *Mem:* Am Soc Qual Control; Am Statist Asn; Biomet Soc. *Res:* Non-parametric statistics; biometrics. *Mailing Add:* Mobile Chamber Commerce PO Box 2187 Mobile AL 36652

CHOW, CHAO K, b Hong Kong, Dec 28, 28; US citizen; m 53; c 2. ELECTRICAL ENGINEERING. *Educ:* Utopia Univ & Tsing Hua Univ, China, BEE, 49; Cornell Univ, MEE, 50, PhD(elec eng), 53. *Prof Exp:* Asst prof elec eng, Pa State Univ, 53-55; sr staff scientist, Burroughs Corp, Pa, 55-64; mem res staff & mgr, 64-81, MGR COMPUT ENG, T J WATSON RES CTR, IBM, 81- *Concurrent Pos:* Mem fac grad ctr, Pa State Univ, 61-64; vis prof, Mass Inst Technol, 68-69; assoc ed, J Pattern Recognition, 68- *Mem:* Fel Inst Elec & Electronics Engrs; Soc Pattern Recognition. *Res:* Image processing and pattern recognition; switching theory; medical applications; storage hierarchies; computer sciences; recognition systems; medical ultrasound. *Mailing Add:* Watson Res Ctr IBM Corp PO Box 218 Yorktown Heights NY 10598

CHOW, CHE CHUNG, b Shanghai, China, Feb 6, 35; m 60; c 1. ADHESION SCIENCE,PHYSICAL CHEMISTRY. *Educ:* Univ Hong Kong, BSc, 58; Brown Univ, PhD(chem), 64. *Prof Exp:* Asst chem, Brown Univ, 58-63; res chemist, Cent Res Dept, E I du Pont de Nemours & Co, Inc, Del, 63-69, chemist, East Lab, NJ, 69-70; scientist, xerographic technol dept, 70-86, SR TECH SPECIALIST, PROJ MGR, XEROX CORP, 87- *Mem:* Am Chem Soc. *Res:* Chemical reactions in shock waves; molecular energy transfer; growth kinetics of inorganic fibers; electrical and magnetic properties of solids; magneto-optics; laser-magnetic memories and imaging systems; coagulation mechanism of polymer solutions; mechanism of detonation in solid explosives; imaging and reproduction systems; developement of high temperature polymers; fusing systems for fixing toner images. *Mailing Add:* 286 Valley Green Dr Penfield NY 14526-1740

CHOW, CHING KUANG, b Chiayi, Taiwan, May 7, 40; m 67; c 3. NUTRITIONAL BIOCHEMISTRY. *Educ:* Nat Taiwan Univ, BS, 63; Univ Ill, Urbana, MS, 66, PhD(nutrit sci), 69. *Prof Exp:* From res asst to res assoc nutrit biochem, Univ Ill, Urbana, 64-71; assoc res scientist biochem, NY Univ Med Ctr, 69-70; Nat Vitamin Found fel biochem, 71-72, asst res biochemist, Univ Calif, Davis, 72-76; from asst prof to assoc prof, 76-83, PROF, DEPT NUTRIT & FOOD SCI & GRAD CTR TOXICOL, UNIV KY, 83- *Concurrent Pos:* Fulbright sr res scholar, WGermany, 83-84. *Mem:* Am Inst Nutrit; NY Acad Sci. *Res:* Metabolism and function of vitamin E; cellular antioxidant defense mechanism; nutritional and environmental influence pathological processes. *Mailing Add:* Dept Nutrit & Food Sci Univ Ky Lexington KY 40506

CHOW, CHUEN-YEN, b Nanchang, China, Dec 5, 32; m 60; c 3. AEROSPACE ENGINEERING. *Educ:* Univ Taiwan, BS, 55; Purdue Univ, MS, 58; Mass Inst Technol, SM, 61; Univ Mich, PhD(aeronaut & astronaut eng), 64. *Prof Exp:* Fel, Inst Sci & Tech, Univ Mich, 64-65; from asst prof to assoc prof aerospace eng, Univ Notre Dame, 65-68; assoc prof, 68-76, PROF AEROSPACE ENG, UNIV COLO, 76- *Concurrent Pos:* Distinguished vis prof, US Air Force Acad, 79-80. *Mem:* Sigma Xi; Am Inst Aeronaut & Astronaut. *Res:* Radiative gasdynamics; magnetohydrodynamics; hydromagnetic instability; aerodynamics; computational fluid mechanics. *Mailing Add:* Dept Aerospace Eng Sci Univ Colo Boulder CO 80309-0429

CHOW, DONNA ARLENE, b Toronto, Ont, Mar 9, 41; m 63; c 1. TUMOR IMMUNOLOGY, NATURAL RESISTANCE. *Educ:* Univ Toronto, BSc, 63; Univ Man, PhD(immunol), 75. *Prof Exp:* Sr res scientist, Man Inst Cell Biol, 81-90; asst prof, 81-86, ASSOC PROF IMMUNOL, UNIV MAN, 86- *Mem:* Can Soc Immunol. *Res:* Resistance of natural immune defense mediators against tumor development with a focus on natural antibodies and natural killer cells. *Mailing Add:* Immunol Dept Univ Man 795 McDermot Ave Winnipeg MB R3E 0W3 Can

CHOW, IDA, b China; Brazilian citizen. DEVELOPMENTAL NEUROBIOLOGY. *Educ:* McGill Univ, PhD(physiol), 80. *Prof Exp:* Res asst physiol, Jerry Lewis Neuromuscular Res Ctr, Sch Med, Univ Calif, Los Angeles, 84-89; ASST PROF, DEPT BIOL, AM UNIV, WASH, DC, 89- *Mem:* Soc Neurosci; Soc Develop Biol; Am Soc Cell Biol; AAAS. *Res:* Cell-cell interaction during development; cell culture; synaptogenesis; neuromuscular junction; electrical coupling; electrophysiology; neural development. *Mailing Add:* Dept Biol Am Univ 4400 Massachusetts Ave NW Washington DC 20016-8007

CHOW, KAO LIANG, b Tientsin, China, Apr 21, 18; US citizen; m 64. NEUROSCIENCES, NEUROANATOMY. *Educ:* Yenching Univ China, BA, 43; Harvard Univ, PhD(psychol), 50. *Prof Exp:* Asst, Yerkes Labs Primate Biol, 47-54; asst prof physiol, Univ Chicago, 54-60; assoc prof med, 60-65, prof, 65-84, EMER PROF NEUROL, MED SCH, STANFORD UNIV, 84- *Concurrent Pos:* Mem, Int Brain Res Orgn. *Mem:* AAAS; Am Physiol Soc; Soc Neurosci. *Res:* Neurophysiology of learning and vision; neuroanatomy of vision and central nervous system. *Mailing Add:* Dept Neurol Stanford Univ Sch Med Stanford CA 94305

CHOW, LAURENCE CHUNG-LUNG, b Taipei, Taiwan, Feb 8, 43; US citizen; m 67; c 3. BIOMATERIALS, DENTAL RESEARCH. *Educ:* Chen-Kung Univ, Taiwan, BS, 64; Georgetown Univ, PhD(chem), 70. *Prof Exp:* Res assoc, 69-75, CHIEF RES SCIENTIST, DENT CHEM DIV, AM DENT ASN HEALTH FOUND RES UNIT, NAT BUR STANDARDS, 76- *Concurrent Pos:* prin investr, NIH grants, 82-; mem, Nat Inst Health Study Sect, 83-86; pres, Washington, DC Sect Am Asn of Dental Res, 88; chmn, Gordon Res Conf Calcium Phosphates, 89. *Mem:* Am Chem Soc; Int Asn Dent Res. *Res:* Dental research; dental caries prevention; topical fluoridation of teeth; thermodynamics; solution, biomaterials theories; transport of ions through membranes. *Mailing Add:* 20517 Ann Dyke Way Germantown MD 20874-2825

CHOW, LOUISE TSI, b Hunan, China, Sept 30, 43; m 74. MOLECULAR GENETICS, TUMOR VIROLOGY. *Educ:* Nat Taiwan Univ, BS, 65; Calif Inst Technol, PhD(chem), 73. *Prof Exp:* Res fel biochem, Univ Calif Med Ctr, San Francisco, 73-74; res fel chem, Calif Inst Technol, 74-75; res fel, 75, staff investr, 76, sr staff investr, 77-79, sr staff scientist electron microscopy, Cold Spring Harbor Lab, 79-83; BIOCHEMIST, STRONG HOSP, UNIV ROCHESTER, 83- *Mem:* Am Soc Microbiol. *Res:* Electron microscopic studies of gene arrangements in bacterial, viral and eucaryotic chromosomes; adenovirus RNA transcription and splicing patterns. *Mailing Add:* Univ Rochester Strong Hosp PO Box Biochem 601 Elmwood Ave Rochester NY 14642

CHOW, PAO LIU, b Fukien, China, Nov 28, 36; m 65; c 2. PHYSICAL MATHEMATICS. *Educ:* Cheng Kung Univ, Taiwan, BS, 59; Rensselaer Polytech Inst, MS, 64, PhD(mech), 67. *Prof Exp:* Asst prof math, Rensselaer Polytech Inst, 66-67; asst prof, NY Univ, 67-72; assoc prof, 72-77, PROF MATH, WAYNE STATE UNIV, 77-, MATH CHMN, 90- *Concurrent Pos:* Vis scholar, Univ Calif, Berkeley, 78; vis prof, Inst Nat Recherche Infomatique & Automatique, France, 79, Univ Minn, 85, Bielefeld Univ, Ger, 87. *Honors & Awards:* NSF, US Army Res Off; NASA Grants. *Mem:* Am Math Soc; Soc Indust & Appl Math. *Res:* Stochastic differential equations; stochastic control and optimization; stochastic analysis; applied mechanics. *Mailing Add:* Dept Math Wayne State Univ Detroit MI 48202

CHOW, PAUL C, b Peking, China, Aug 1, 26; m 65; c 3. SOLID STATE PHYSICS, THEORETICAL PHYSICS. *Educ:* Univ Calif, Berkeley, BS, 60; Northwestern Univ, PhD(physics), 65. *Prof Exp:* Res assoc physics, Northwestern Univ, 65; res assoc, Univ Southern Calif, 65-66, asst prof, 66-67; res scientist, Univ Tex, 67-68; assoc prof, 68-80, PROF PHYSICS & ASTRON, CALIF STATE UNIV, NORTHRIDGE, 80- *Concurrent Pos:* Vis asst prof, Univ Tex, 70; consult, Control Data Corp, 81-, World Bank, 87, 88. *Mem:* Am Phys Soc. *Res:* Theoretical solid state physics; computer used as a teaching tool; science education. *Mailing Add:* Dept Physics & Astron Calif State Univ Northridge CA 91330

CHOW, RICHARD H, b Vancouver, BC, Sept 6, 24; nat US; m 48; c 4. PHYSICS. *Educ:* Univ BC, BA, 47, MA, 49; Univ Calif, Los Angeles, PhD(nuclear physics), 55. *Prof Exp:* Asst physics, Univ BC, 46-49; from asst to assoc, Univ Calif, Los Angeles, 50-54; asst res officer reactor physics, Atomic Energy Can, Ltd, 54-58; from asst prof to prof, 58-86, EMER PROF PHYSICS, CALIF STATE UNIV, LONG BEACH, 86- *Concurrent Pos:* Richland fac appointment, NWCol & Univ Asn Sci, 68-69. *Mem:* Am Asn Physics Teachers; Am Phys Soc; Am Nuclear Soc; Sigma Xi. *Res:* Reactor physics; nuclear physics; energy sources. *Mailing Add:* Dept Physics Calif State Univ Long Beach CA 90840

CHOW, RITA K, b San Francisco, Calif, Aug 19, 26. NURSING, GERONTOLOGY. *Educ:* Stanford Univ, BS, 50; Case Western Reserve Univ, MS, 55; Columbia Univ, EdD, 68; George Mason Univ, BIS, 83. *Prof Exp:* Nurse consult, Div Nursing, Bur Health Manpower & Educ, NIH, USPHS, Dept Health, Educ & Welfare, Bethesda, Md, 68, asst to dep dir, Health Manpower Utilization Prog, Nat Ctr Health Serv Res & Develop, Arlington, Va, 69, dep dir, Health Care Resources Br, Rockville, Md, 70, spec asst to dep dir, 71-73, asst chief nurse officer, 73, asst to dep dir, Bur Health Serv Res & Eval, 73, dep dir & chief nurse officer, Off Long-Term Care, 73-77, nurse consult, Instnl Standards Br, Dept Health & Human Serv, 77-78, chief, Qual Assurance Br, Baltimore, 78-82, supvry clin nurse, USPHS Indian Hosp, Rosebud, SDak, 82, health sci analyst, Off Technol Assessment, Rockville, Md, 82-84; asst dir nursing & dir patient educ, GWL Hansen's Dis Ctr, Carville, La, 84-89; NURSE DIR, USPHS, FT WORTH, TEX, 89- *Concurrent Pos:* Instr nursing, Col Nursing, Wayne State Univ, 57-58, basic patient care, US Army Med Training Ctr, Ft Sam Houston, Tex, 57; asst ed, Am J Nursing, 61-62, assoc ed, 62-65. *Mem:* Inst Med-Nat Acad Sci; fel AAAS; fel Geront Soc Am. *Res:* Postoperative cardiovascular nursing care; quality of nursing care; assessment of the long-term care patient. *Mailing Add:* 2406 Cypresswood Trail Apt 921 Arlington TX 76014

CHOW, SHIN-KIEN, b Shan-Tan, China, June 12, 35; m 65; c 3. FLUID MECHANICS, MATHEMATICS. *Educ:* Nat Univ Taiwan, BS, 57; Calif Inst Technol, MS, 61; Univ Iowa, PhD(fluid mech), 67. *Prof Exp:* Sr engr, 65-70, supvr, 70-71, mgr, 71-74, adv eng, Westinghouse Res Labs, 74-83, ADV ENGR, WESTINGHOUSE NUCLEAR CTR, 83- *Mem:* Am Soc Civil Engrs; Am Soc Mech Engrs; Am Inst Aeronaut & Astronaut; Inst Elec & Electronics Engrs. *Res:* Hydrodynamics; thermal-hydraulic systems; electromagnetics. *Mailing Add:* Westinghouse Elec Corp Box 355 Pittsburgh PA 15146

CHOW, TAI-LOW, b Yencheng, China, May 18, 37; m 63; c 2. ASTROPHYSICS, COMPUTER MODELING OF CHAOS. *Educ:* Nat Taiwan Univ, BS, 58; Case Western Reserve Univ, MS, 63; Univ Rochester, PhD(physics, astron), 70. *Prof Exp:* Instr physics, Cheng Kung Univ, 60-61; from asst prof to assoc prof, 69-77, PROF PHYSICS, CALIF STATE UNIV, STANISLAUS, 77-, CHMN DEPT, 73- *Mem:* AAAS; Am Phys Soc; Am Astron Soc; Am Asn Physics Teachers. *Res:* Scattering of high-energy electrons by nucleons; x-ray and gamma-ray astronomy; electron gas in ultraintense magnetic fields and the astrophysical applications; high-velocity neutral hydrogen gases at high galactic latitudes; interstellar medium; pulsar and gravitation radiations; computer modeling of chaos. *Mailing Add:* Dept Physics Calif State Univ Turlock CA 95380

CHOW, TAT-SING PAUL, b China, July 26, 53; US citizen. INTEGRATED CIRCUIT PROCESSING. *Educ:* Augustana Col, BA, 75; Columbia Univ, MS, 77; Rensselaer Polytech Inst, PhD(elec eng), 82. *Prof Exp:* INTEGRATED CIRCUITS PROCESS ENGR, RES & DEVELOP, GEN ELEC CO, 77- *Mem:* Inst Elec & Electronics Engrs; Electrochem Soc. *Res:* Thin film processes for metal oxide silicon integrated-circuits; application of refractory metal and metal silicides to microcircuits. *Mailing Add:* 708 Hampshire St Schenectady NY 12309

CHOW, THIEN LIEN, b Malaysia, Nov 25, 44; Can citizen; m 67; c 2. SOIL PHYSICS, MICROMETEOROLOGY. *Educ:* Univ BC, PhD(soil physics), 73. *Prof Exp:* Res assoc hydrol, Univ BC, 73-74; res scientist soil physics, Can Forestry Serv, 74-78; RES SCIENTIST HYDROL, AGR CAN, 78- *Mem:* Soil Sci Soc Am; Int Soc Soil Sci; Am Soc Soil Conserv. *Res:* Soil-water management in crop production; environmental impact of agricultural practices; soil conservation. *Mailing Add:* Res Sta Agr Can PO Box 20280 Fredericton NB E3B 4Z7 Can

CHOW, TSAIHWA JAMES, b Shanghai, China, Oct 13, 24; US citizen. OCEANOGRAPHY, GEOCHEMISTRY. *Educ:* Nat Chiaotung Univ, BS, 46; Wash State Univ, MS, 49; Univ Wash, PhD(anal chem), 53. *Prof Exp:* Instr, Nat Chiaotung Univ, 46-47; res asst, Univ Wash, 50-52, asst, 52-53, res assoc oceanog, 53-55; res fel geochem, Calif Inst Technol, 55-60; MEM STAFF, SCRIPPS INST OCEANOG, 60-, EMER RES, 88- *Concurrent Pos:* Consult, comt biologic effects atmospheric pollutants, Div Med Sci, Nat Res Coun-Nat Acad Sci, 71, mem, Trace Metal Subcomt of the Safe Drinking Water Comt, Assembly Life Sci, 76-77, panel mem, workshop on Lead in the Human Environ, Environ Studies Bd, 78; partic, Standards & Intercomparison Mat, Analytical Qual Control Serv, Int Atomic Energy Agency, Vienna, 72-78; session chmn, FAO/IAEA Joint Symp on Isotope Ratios as Pollutant Source & Behav Indicators, Vienna, Austria, 74; adv, marine prod proj, Inst del Mar del Peru, Lima, Orgn Am States, 76 & marine pollution comt, 81; session chmn & invited speaker, 26th Cong Int Union Pure & Appl Chem, Tokyo, Japan, 77; vis scientist, Acad Sci People's Repub China, 79 & State Oceanic Admin, People's Repub China, 81, 83 & 85. *Mem:* AAAS; Am Geophys Union. *Res:* Microanalytical chemistry; mass spectrometry; chemical oceanography; trace elements in the sea; strontium-calcium ratio in marine organisms; geochemistry of lead isotopes; geochronology; solution chemistry; lead pollution; plutonium in water. *Mailing Add:* Scripps Inst Oceanog Univ Calif La Jolla CA 92093

CHOW, TSENG YEH, b Shanghai, China, July 22, 21. MATHEMATICS. *Educ:* Nat Chiao-Tung Univ, BS, 42; Mass Inst Technol, SM, 48; Cornell Univ, PhD(math), 53. *Prof Exp:* From asst prof to assoc prof math, Rensselaer Polytech Inst, 52-63; assoc prof, 64-67, PROF MATH, CALIF STATE UNIV, SACRAMENTO, 67- *Mem:* Am Math Soc; Math Asn Am. *Res:* Mathematical analysis; applied mathematics. *Mailing Add:* 4091 Fotos Ct Sacramento CA 95820

CHOW, TSU LING, b China, Apr 4, 15; nat US; m 50. ANIMAL VIROLOGY. *Educ:* Nat Cent Univ, China, BVS, 40; Mich State Univ, PhD(animal path), 50. *Prof Exp:* Asst, Nat Cent Univ, China, 40-42; vet, Ministry Agr & Forestry, 42-45; res assoc, Univ Wis, 50-53; from assoc prof to prof, 69-75, EMER PROF PATH & BACT, SCH VET MED, COLO STATE UNIV, 75- *Mem:* Am Vet Med Asn; Am Soc Microbiol; Conf Res Workers Animal Dis; NY Acad Sci; Int Acad Path. *Res:* Virus diseases of domestic animals such as rinderpest, contagious ecthyma, vesicular stomatitis; infectious bovine rhinotracheitis; blue tongue. *Mailing Add:* 923 Valleyview Rd Ft Collins CO 80524

CHOW, TSU-SEN, b China, Nov 8, 39; US citizen; m 67; c 1. POLYMER PHYSICS, MECHANICAL PROPERTIES. *Educ:* Cheng Kung Univ, Taiwan, BS, 62; Rensselaer Polytech Inst, MS, 66; Carnegie-Mellon Univ, PhD(math, mech), 68. *Prof Exp:* Teaching asst thermodyn, Nat Cheng Kung Univ, 63-64; res assoc polymer/composite, Univ NC, Chapel Hill, 68-72; from scientist to sr scientist, 79-85, PRIN SCIENTIST, POLYMER PHYSICS & COMPOSITE MAT, XEROX WEBSTER RES CTR, XEROX CORP, 85- *Concurrent Pos:* Lectr, CEI-Europe Elsevier, 88- *Mem:* Am Phys Soc; Soc Rheology; Am Chem Soc. *Res:* Mechanical and thermal properties of polymers and composites; statistical and continuous mechanics; adhesion and surface sciences; rheology and stability of disperse systems. *Mailing Add:* Six Red Rose Circle Penfield NY 14526-9779

CHOW, WEN LUNG, b Hankow, China, March 6, 24; US citizen; m 52; c 6. GAS DYNAMICS, FLUID DYNAMICS. *Educ:* Nat Cent Univ, BS, 46; Univ Ill-Urbana, MS, 50, PhD(mech eng), 53. *Prof Exp:* Develop engr plastic films, E I Du Pont Co, 53-55; from asst prof to assoc prof gas dynamics, Univ Ill-Urbana, 55-63, prof mech eng & fluid dynamics, 63-88; PROF MECH ENG, FLA ATLANTIC UNIV, 88- *Concurrent Pos:* Asst engr power generation, Hankow Power & Water Works, 46-49; consult, W L Chow Eng Consults, 63-; sr res assoc, Nat Acad Sci, 82; vis prof, Fla Atlantic Univ, 88-89; prin investr, NASA, 69-79, 86-88 & US Army, 76-88. *Mem:* Fel Am Soc Mech Engrs; assoc fel Am Inst Aeronaut & Astronaut; Sigma Xi. *Res:* Fluid dynamic problems in the area of jet and rocket propulsion; open channel flows, flow through valves and orifices. *Mailing Add:* Dept Mech Eng Fla Atlantic Univ Boca Raton FL 33431-0991

CHOW, WEN MOU, computer science; deceased, see previous edition for last biography

CHOW, WENG WAH, b Singapore, April 22, 48; m 79. LASER PHYSICS. *Educ:* Colo State Univ, BS, 68; Univ Ariz, MS, 74, PhD(physics), 75. *Prof Exp:* Physicist, Max-Planck Inst fur Biophys Chemie, 75-77, physicist, Max-Planck Inst fur Plasmaphysik, 77; physicist, Optical Sci Ctr, Univ Ariz, 78-80; assoc prof, Dept Physics & Astron, Univ NMex, 80-86; SR STAFF ENG, HUGHES AIRCRAFT CO, 86- *Mem:* Optical Soc Am; Am Phys Soc; Inst Elec & Electronics Engrs. *Res:* Laser theory; phased laser arrays; applied optics; laser gyros; short wavelength lasers and high power lasers; semiconductor lasers. *Mailing Add:* Div 2531 Sandia Nat Labs PO Box 5800 Albuquerque NM 87185

CHOW, YUAN LANG, b Formosa, May 28, 29; m 58; c 3. ORGANIC CHEMISTRY. *Educ:* Nat Taiwan Univ, BSc, 51; Duquesne Univ, PhD(chem), 57. *Prof Exp:* Engr, Chinese Petrol Co, 51-54; fel, Ill Inst Technol, 57-58; res fel, Royal Inst Technol, Sweden, 58-59, asst lectr, 59-61; res assoc, Imp Col, Univ London, 61-62; lectr, Univ Singapore, 62-63; asst prof org chem, Univ Alta, 63-65; assoc prof, 65-69, PROF CHEM, SIMON FRASER UNIV, 69- *Mem:* Fel Chem Inst Can; Chem Soc; Am Chem Soc; Sigma Xi. *Res:* Photochemistry in solution, synthetic and mechanistic studies of free radical reaction; carcinogen chemistry. *Mailing Add:* Dept Chem Simon Fraser Univ Burnaby BC V5A 1S6 Can

CHOW, YUAN SHIH, b Hupeh, China, Sept 1, 24; m 63; c 3. MATHEMATICS. *Educ:* Chekiang, China, BS, 49; Univ Ill, MA, 55, PhD(math), 58. *Prof Exp:* Res assoc, Univ Ill, 58-59; staff mathematician, Inst Bus Mach Corp, 59-62; vis assoc prof statist, Columbia Univ, 62-63; from assoc prof to prof, Purdue Univ, 63-68; PROF STATIST, COLUMBIA UNIV, 68- *Mem:* Am Math Soc; Soc Indust & Appl Math; Inst Math Statist. *Res:* Probability; Fourier series. *Mailing Add:* 144 Washington Ave Dobbs Ferry NY 10522

CHOW, YUNG LEONARD, b Foochow, Fukien, China, Dec 29, 36; Can citizen; m 72; c 2. STATIC COMPUTATION, DYNAMIC FIELD COMPUTATION. *Educ:* McGill Univ, BE, 60; Univ Toronto, MASc, 61, PhD(elec eng), 65. *Prof Exp:* Res assoc array design, Nat Radio Astron Observ, 64-66; from asst prof to assoc prof, 66-81, PROF ELEC ENG, UNIV WATERLOO, 81- *Concurrent Pos:* Consult, Westlake Village, Calif, Telemus & Commun Res Ctr, Ottawa, Ont. *Mem:* Inst Elec & Electronics Engrs. *Res:* Microstrip; electromagnetic scattering; antenna array for radio astronomy; transient frequency modulation fields; static electric fields; field simplification by optimization routine; microwave integrated circuit; software as a field theory analysis for circuits. *Mailing Add:* Dept Elec Eng Univ Waterloo Waterloo ON N2L 3G1 Can

CHOW, YUTZE, b Shanghai, China, Sept 29, 27; m 54; c 3. THEORETICAL PHYSICS, MATHEMATICS. *Educ:* Chin-Kung Univ, Taiwan, 52; Univ RI, MS, 57; Brandeis Univ, MA, 64, PhD(physics), 65. *Prof Exp:* Res asst elec eng, Univ RI, 55-57; asst prof, Inst Tech Aeronaut, Brazil, 57-61; asst prof elec eng, Univ Waterloo, 61-62; teaching asst physics, Brandeis Univ, 62-63, res asst, 63-64; from asst prof to assoc prof, 64-67, PROF PHYSICS, UNIV WIS-MILWAUKEE, 67- *Concurrent Pos:* Vis assoc prof, Syracuse Univ, 57. *Mem:* Am Phys Soc. *Res:* Quantum field theory and elementary particle symmetries; Lie groups and Lie algebras. *Mailing Add:* Dept Physics Univ Wis Milwaukee WI 53201

CHOWDHRY, UMA, b Bombay, India, Sept 14, 47; m 70. CERAMICS, HETEROGENEOUS CATALYSTS. *Educ:* Inst Sci, Bombay, BSc, 68; Calif Inst Technol, MS, 70; Mass Inst Technol, PhD(mat sci), 76. *Prof Exp:* Fel, Mass Inst Technol, 76-77; res scientist, E I du Pont de Nemours & Co, Inc, 77-80, res supvr, 81-85, res mgr, Cent Res & Develop Dept, 85-88, GROUP LEADER PHYS SCI, E I DU PONT DE NEMOURS & CO, INC, 80-, LAB DIR, ELECTRONICS, 88-, BUSINESS MGR, 91- *Mem:* Am Ceramic Soc; Mat Res Soc. *Res:* Synthesis and characterization of ceramic powders and heterogeneous catalysts; correlation of catalyst microstructure with catalytic properties; defects in inorganic solids; mass transport in ceramics; microstructural evolution during ceramic fabrication; microstructure property relationships in ceramics; novel precursors and synthetic routes to prepare inorganic powders with controlled morphology and purity; the uses of ceramics as materials for microelectronic packaging and in multilayer capacitors; oxide catalysts for selective oxidation processes and metal catalysts for various hydrogenation reactions. *Mailing Add:* DuPont Elec Barley Mill Plaza P30/2370 Wilmington DE 19880-0030

CHOWDHURI, PRITINDRA, b Calcutta, India, July 12, 27; US citizen; m 62; c 4. ELECTRICAL ENGINEERING. *Educ:* Univ Calcutta, BSc, 45, MSc, 48; Ill Inst Technol, MS, 51; Rensselaer Polytech Inst, DEng(eng sci), 66. *Prof Exp:* Jr engr, Lightning Arresters Sect, Westinghouse Elec Corp, 51-52; elec engr, High Voltage Lab, Maschinenfabrik Oerlikon, Switz, 52-53; res engr, High Voltage Res Comn, Swiss Electrotech Soc, 53-56; develop engr, High Voltage Lab, Gen Elec Co, 56-59, elec engr, Advan Technol Labs, 59-62, engr elec invests, Transp Systs Div, 62-75; staff mem, Los Alamos Nat Lab, 75-86; PROF, ELEC ENG, TENN TECH UNIV, 86- *Concurrent Pos:* Lectr, Grad Ctr, Pa State Univ, Behrend Campus, 69-75. *Honors & Awards:* Invention Disclosure Award, Westinghouse Elec Corp, 51. *Mem:* Sr mem Inst Elec & Electronics Engrs; Int Conf Large High-Voltage Elec Systs; fel AAAS; Inst Elec Engrs (Gt Brit). *Res:* Dielectrics; electrical transients; high voltage engineering; power semiconductor devices; power system engineering; four patents. *Mailing Add:* Ctr Electric Power Tenn Tech Univ PO Box 5032 Cookeville TN 38505

CHOWDHURY, AJIT KUMAR, reproductive physiology, reproductive endocrinology, for more information see previous edition

CHOWDHURY, DIPAK KUMAR, b Jhargram, India, Nov 15, 36; US citizen; m 65; c 2. GEOPHYSICS, SEISMOLOGY. *Educ:* Indian Inst Technol, Kharagpur, BS, 56, MTechnol, 58; Tex A&M Col, PhD(geophys), 61. *Prof Exp:* Lectr geophys, Indian Sch Mines & Appl Geol, Dhanabad, 61-63, asst prof, 63-67; Nat Res Coun fel, Univ BC, 67-69; physicist, Shell Develop Co, Tex, 69-70; asst prof geol, 70-73, ASSOC PROF GEOL, IND UNIV, FT WAYNE, 73-, CHMN, DEPT EARTH & SPACE SCI, 74- *Concurrent Pos:* Geophys consult, Law Eng & Testing Co, Ga, 70-74 & Gen Portland Cement, 75-77; seismol consult, Lincoln Lab, MIT, 74 & 76; geophys consult, Pollution Control Systs of IN, 87, Soil Explor Serv, Ind, 89-90. *Mem:* Am Geophys Union; Am Geol Inst; Sigma Xi; Seismol Soc Am. *Res:* Interpretation of self potential data for tabular shaped ore bodies; elastic wave propagations along layers in two-dimensional models; elastic wave velocities and attenuations in rocks and plastics; variations in azimuths and incident angles due to dipping interfaces; studies on earthquake seismology, upper mantle heterogenity, core-mantle boundary; Benioff Zones, source mechanism and magnitudes of earthquakes; tectonics. *Mailing Add:* Dept Earth & Space Sci Ind Univ-Purdue Univ Ft Wayne IN 46805

CHOWDHURY, IKBALUR RASHID, b Dacca, Pakistan, Feb 12, 39; m 67; c 1. SOIL FERTILITY, AGRONOMY. *Educ:* Univ Dacca, BSc, 60, MS, 63; NDak State Univ, PhD(soil fertil), 70. *Prof Exp:* Sr res asst bact, Pak-SEATO Cholera Res Inst, 64; sr lectr soil microbiol, Univ Dacca, 64-65; res asst soil fertil, NDak State Univ, 65-70, fel soil chem, 70-72; asst prof agr, 72-75, ASSOC PROF AGR, LINCOLN UNIV, 75-, ASSOC PROF-PRIN INVESTR PLANT & SOIL SCI & NATURAL RESOURCES COOP RES, DIR INT PROGS, 83- *Mem:* Am Soc Agron; Soil Sci Soc Am; Int Soc Soil Sci. *Res:* Soil testing; soil chemistry; plant physiology; environmental quality; soybean fertilization with respect to oil and protein. *Mailing Add:* Int Progs Off Lincoln Univ 202 New Memorial Hall Jefferson City MO 65101

CHOWDHURY, MRIDULA, b Calcutta, India, Feb 22, 38; m 61; c 2. REPRODUCTIVE BIOLOGY. *Educ:* Presidency Col, Calcutta, BS, 57; Calcutta Univ, MS, 59, PhD(reproductive physiol), 68. *Prof Exp:* NIH fel reproductive biol, Albert Einstein Med Ctr, 69-71; res assoc, Sch Med, Univ Tex, Houston, 71-74, sr res scientist reprod biol, 74-77, asst prof, Dept Reproductive Med & Biol, 77-90. *Mem:* Am Endocrine Soc; AAAS. *Res:* Feedback control of gonadotropins in male with special emphasis on the control mechanism of synthesis of gonadotropins in the pituitary. *Mailing Add:* Dept Lab Med Col Vet Med Kans State Univ Manhattan KS 66506

CHOWDHURY, PARIMAL, b Chittagong, Bangladesh, Dec 31, 40; US citizen; m 71; c 2. BIOCHEMISTRY, TOXICOLOGY. *Educ:* Dacca Univ, Bangladesh, BSc, 60, MSc, 62; McGill Univ, Can, PhD(immunochem & physiol), 70. *Prof Exp:* Lectr chem, BPC Inst, West Bengal, India, 64-65, lectr med chem, Calcutta Med Col, India, 65-67; fel immunochem, McGill Univ, Can, 67-70; instr med, NJ Med Sch, 70-74, asst prof med, 76-80; asst prof, 80-84, asst prof pharmacol intox, 82-86, ASSOC PROF PHYSIOL & BIOPHYS, UNIV ARK MED SCI, 84-, ASSOC PROF PHARMACOL INTOX, 86- *Concurrent Pos:* Co-investr, NIH Funded Res, 70-80, 82-87 & 88-; prin investr, Inst Pilot Study, GRS grant, NJ, 74-77 & Little Rock, Ark, 80-82; vis sci, MARC, 87 & 89; pres-elect, Little Rock chap, Sigma Xi, 90. *Mem:* Am Physiol Soc; AAAS; Am Asn Univ Professors; Sigma Xi; Society Toxicol; Soc Exp Biol Med; Am Gastroenterol; Am Pancreatic Asn. *Res:* Structure-function relationship of proteins specially immunoglobulins; heavy metal toxicity and mechanism of lung injury; cigarette smoking, pulmonary emphysema, its relationship with serum anti-trypsin activity; blood coagulation, gastrointestinal physiology; determining the effect of cigarette smoke components such as cadmium, nicotine and smoke condensate on the release actions of gastrointestinal peptides. *Mailing Add:* Five New Haven Ct Little Rock AR 72207

CHOWN, EDWARD HOLTON, b Kingston, Ont, Feb 24, 32; m 65; c 3. PETROLOGY, ECONOMIC GEOLOGY. *Educ:* Queen's Univ, Ont, BSc, 55; Univ BC, MASc, 57; Johns Hopkins Univ, PhD(geol), 63. *Prof Exp:* Geologist, Dept Natural Resources, Que, 63-66; from asst prof to assoc prof, Loyola Col Montreal, 66-75; assoc prof, 75-79, PROF SCI, UNIV QUE, CHICOUTIMI, 79- *Mem:* Geol Soc Am; Geol Asn Can; Can Inst Mining & Metall; Mineral Asn Can; Sigma Xi. *Res:* Structure and genesis of archean granitoid intrusions; petrology of proterozoic dyke swarms; origin and metamorphism of archean volcanogenic gold deposits. *Mailing Add:* Dept Earth Sci Univ Que Chicoutimi PQ G7H 2B1 Can

CHOWRASHI, PROKASH K, MUSCLE PROTEIN, MYOSIN FILAMENT ASSEMBLY. *Educ:* Jadavpur Univ, India, PhD(chem), 67. *Prof Exp:* RES ASSOC, DEPT ANAT, SCH MED, UNIV PA, 70- *Res:* V-band protein. *Mailing Add:* Dept Anat Sch Med Univ Pa 37th & Hamilton Walk Philadelphia PA 19104

CHOY, DANIEL S J, b Shanghai, China, May 29, 26; US citizen; m 86; c 2. LASERS. *Educ:* Columbia Col, BA, 44, MD, 49. *Prof Exp:* Intern, Meadowbrook Hosp, Ny, 49-50; asst resident, Goldwater Mem Hosp, 50-51; captain, US Air Force, Wright-Patterson AFB, Ohio, 51-52; dir med educ, Elmhurst Gen Hosp, 58-62; chief, tumor serv, French Hosp, 62-76; ASST CLIN PROF, COLUMBIA PRESBY MED CTR, 80-; DIR LASER LAB, ST LUKE'S ROOSEVELT HOSP, 84- *Concurrent Pos:* Attend, Lenox Hill Hosp, 55-; assoc attend, Dept MEd, St Luke's Roosevelt Hosp, 84- *Mem:* Fel Am Col Physicians; AMA; Am & Int Soc Lasers Surg & Med; Optical Soc Am; AAAS. *Res:* Laser applications in medicine; angiogenesis. *Mailing Add:* 170 E 77th St New York NY 10021

CHOY, PATRICK C, b China, June 16, 44; Can citizen; m 75; c 1. LIPID METABOLISM, CARDIOLOGY. *Educ:* McGill Univ, BSc, 69; Univ NDak, MSc, 72, PhD(biochem), 75. *Prof Exp:* Teaching fel biochem, Univ BC, 75-79; from asst prof to assoc prof, 79-86, PROF BIOCHEM, UNIV MAN, 86- *Concurrent Pos:* Scientist, Med Res Coun Can, 84- *Mem:* Am Soc Biol Chemists; Can Biochem Soc; Can Cardiovasc Soc. *Res:* Lipid metabolism in mammalian hearts; pathogenesis of heart failure. *Mailing Add:* Dept Biochem Fac Med Univ Man Winnipeg MB R3E 0W3 Can

CHOY, WAI NANG, b Canton, China, Nov 23, 46. GENETIC TOXICOLOGY. *Educ:* Rutgers Univ, PhD(molecular biol), 76. *Prof Exp:* Fel pediat, Johns Hopkins Hosp, 76-79; instr biophysics, Sch Hygiene & Pub Health, Johns Hopkins Univ, Calif, 79-82; res geneticist, USDA Western Region Res Ctr, 82-84; RES GENETIC TOXICOLOGIST, HASKELL LAB, E I DUPONT DE NEMOURS, 84- *Mem:* Am Soc Cell Biol; Am Soc Human Genetics; Environ Mutagen Soc; AAAS; Tissue Culture ASn; Genetic Toxicol Asn. *Mailing Add:* 35622 Conovan Lane Fremont CA 94536-2533

CHOYKE, WOLFGANG JUSTUS, b Berlin, Ger, July 24, 26; nat US; m 49; c 2. SOLID STATE PHYSICS. *Educ:* Ohio State Univ, BSc, 48, PhD(physics), 52. *Prof Exp:* Res physicist, Westinghouse Res Labs, 52-60, fel physicist, 60-62, adv physicist, 62-77, consult physicist, 78-88; RES PROF PHYSICS, UNIV PITTSBURGH, 88- *Concurrent Pos:* Adj prof physics, Univ Pittsburgh, 74-88; Humboldt Sr US Scientist Award, 90-92. *Mem:* Fel Am Phys Soc; Optical Soc Am; Mat Res Soc; Am Vacuum Soc. *Res:* Experimental nuclear physics; radiative recombination processes; optical absorption; silicon carbide; reflectivity and ellipsometry of metals and semiconductors; ion implantation; ion beam simulation of neutron damage in solids; ultra-high vacuum surface science; Rutherford backscattering/channeling. *Mailing Add:* 5424 Kipling Rd Pittsburgh PA 15217-1038

CHRAMBACH, ANDREAS C, b Breslau, Germany, Apr 23, 27. GEL ELECTROPHORESIS, SEPARATION SCIENCE. *Educ:* Univ Calif, Berkeley, PhD(biochem), 60. *Prof Exp:* Sr investr, 66-83, SECT CHIEF, NIH, 83- *Mem:* Electrophoresis Soc. *Res:* Quantitative gel electrophoresis and electrofocusing of native macromolecular and subcellular particles. *Mailing Add:* NIH Bldg 10 Rm 6C101 Bethesda MD 20892

CHRAPLIWY, PETER STANLEY, b Pulaski, Wis, Oct 20, 23; m 47; c 4. HERPETOLOGY. *Educ:* Univ Kans, BA, 53, MA, 56; Univ Ill, PhD(zool), 64. *Prof Exp:* From instr to asst prof, 60-64, ASSOC PROF BIOL, UNIV TEX, EL PASO, 64- *Mem:* Am Soc Ichthyologists & Herpetologists. *Res:* Taxonomy of modern amphibians and reptiles; biology of desert flora and fauna. *Mailing Add:* 504 Ridgemont Dr El Paso TX 79912

CHRAPLYVY, ANDREW R, b St Louis, Mo, June 5, 50; m 75; c 3. FIBER OPTICS, NONLINEAR OPTICS. *Educ:* Washington Univ, St Louis, BA, 72; Cornell Univ, MS, 75, PhD(physics), 77. *Prof Exp:* Res scientist, Gen Motors Res Lab, 77-80; MEM TECH STAFF, BELL LABS, 80- *Mem:* Am Phys Soc; Optical Soc Am. *Res:* Nonlinear spectroscopy; fiber optics; high resolution infrared spectroscopy; solid state physics. *Mailing Add:* Crawford Hill Labs L135 Box 400 Holmdel NJ 07733

CHRENKO, RICHARD MICHAEL, b Gillette, NJ, July 16, 30; m 63; c 2. SOLID STATE PHYSICS, SPECTROSCOPY. *Educ:* NY Univ, BA, 52; Harvard Univ, MA, 54. *Prof Exp:* Physicist, Knolls Atomic Power Lab, 56, physicist, Res Lab, 56-65, PHYSICIST, RES & DEVELOP CTR, GEN ELEC CO, 65- *Mem:* Mat Res Soc. *Res:* Molecular and solid state spectroscopy; nuclear physics and reactors; physical properties of diamonds; residual stress; molecular beam epitaxy. *Mailing Add:* Gen Elec Res & Develop Ctr PO Box 8 Schenectady NY 12301

CHREPTA, STEPHEN JOHN, b Watervliet, NY, May 19, 21. ANALYTICAL CHEMISTRY. *Educ:* Fordham Univ, MS, 51. *Prof Exp:* Instr, 46-53, ASST PROF CHEM & DEAN, ST BASIL'S COL, 53- *Mem:* NY Acad Sci; AAAS; Am Chem Soc. *Res:* Spectroscopy; x-ray diffraction. *Mailing Add:* 16134 Veronica E Detroit MI 48021-3642

CHRESTENSON, HUBERT EDWIN, b Grandview, Wash, Oct 21, 27; m 47; c 3. MATHEMATICS. *Educ:* State Col Wash, BA, 49, MA, 51; Univ Ore, PhD(math), 53. *Prof Exp:* Instr math, Purdue Univ, 53-54; asst prof, Whitman Col, 54-57; from asst prof to prof, 57-80, EMER PROF MATH, REED COL, 90- *Mem:* Am Math Soc; Math Asn Am. *Res:* Fourier analysis. *Mailing Add:* Dept Math Reed Col 3203 SE Woodstock Blvd Portland OR 97202

CHRETIEN, MAX, b Basel, Switz, Feb 29, 24; m 58; c 1. PHYSICS. *Educ:* Univ Basel, Switz, PhD(physics), 49. *Prof Exp:* Fels, Univ Birmingham, Eng, 51-53 & Columbia Univ, 53-54, from instr to asst prof, 53-58, ASSOC PROF HIGH ENERGY PHYSICS, BRANDEIS UNIV, 58- *Concurrent Pos:* Swiss Nat Fund fel, 63 & 69. *Mem:* Am Physical Soc. *Res:* Theory of elementary particles. *Mailing Add:* Dept Physics Brandeis Univ Waltham MA 02154

CHRETIEN, MICHEL, b Shawinigan, Que, Mar 26, 36; m 60; c 2. ENDOCRINOLOGY. *Educ:* Univ Montreal, BA, 55, MD, 60; McGill Univ, MSc, 62. *Hon Degrees:* DSc, Univ de Liege, 81. *Prof Exp:* Med Res Coun Can fel, McGill Univ, 60-62; resident med, Peter Bent Brigham Hosp, 62-63, clin fel endocrinol, 63-64; Childs Mem Fund fel biochem, Hormone Res Lab, Univ Calif, Berkeley, 64-66, Med Res Coun Can fel, 66-67; SCI DIR & CHIEF EXEC OFF, CLIN RES INST MONTREAL, 67-; PROF MED, UNIV MONTREAL & HOTEL-DIEU HOSP, 67- *Concurrent Pos:* Res fel endocrinol, Hotel Dieu Hosp, 60-62; asst prof exp med, McGill Univ, 67-; fac scholar, Josiah Macy Jr Found, 79-80. *Honors & Awards:* Basic Res Award, Asn Fr Speaking Physicians Can, 71; Clarke Inst Psychiat Award, 77; Medaille Archambault, Fr-Can Asn Advan Sci, 78. *Mem:* Can Soc Clin Invest; Endocrine Soc; NY Acad Sci; Can Biomed Soc; AAAS. *Res:* Biosynthesis and gene expression of endocrine and neural peptides, the new chemistry of the brain. *Mailing Add:* Clin Res Inst Montreal 110 Pine Ave W Montreal PQ H2W 1R7 Can

CHRIEN, ROBERT EDWARD, b Cleveland, Ohio, Apr 15, 30; m 53; c 4. NUCLEAR PHYSICS. *Educ:* Rensselaer Polytech Inst, BS, 52; Case Inst Technol, MS, 55, PhD(physics), 58. *Prof Exp:* Res assoc physics, 57-59, from asst physicist to physicist, 59-72, group leader, Neutron Physics, 70-81, SR PHYSICIST, BROOKHAVEN NAT LAB, 72-, GROUP LEADER, MEDIUM ENERGY PHYSICS, 81- *Concurrent Pos:* Past secy & past chmn nuclear cross sect adv comt, US AEC; mem, Nuclear Data Comt, Nuclear Energy Agency, 72-84, Tech Adv Panel, Los Alamos Meson Physics Facil, 72-76, Nuclear Physics Div, Comt Nuclear Data & Prog Comt, Am Phys Soc, 73, chmn, 78-79 & Nuclear Data Comt, US Dept Energy, 75-85; chmn, Brookhaven Coun; gen ed, Nuclear Data Sci & Technol, Pergamon Press; mem, Adv Panel Nuclear Data Compilation, Nat Acad Sci. *Mem:* Fel Am Phys Soc; fel AAAS; Sigma Xi; NY Acad Sci; Am Chem Soc. *Res:* Neutron physics; neutron resonance parameters; capture gamma rays from neutron resonances; neutron total cross sections; applications of on-line computers in nuclear physics; nuclear detectors; photonuclear reactions; intermediate energy physics; hypernuclear physics; kaon and proton scattering. *Mailing Add:* 51 S Country Rd Bellport NY 11713

CHRISMAN, CHARLES LARRY, b St Joseph, Mo, Mar 1, 41; m 62; c 2. CYTOGENETICS. *Educ:* Univ Mo-Columbia, BS, 67, MS, 69, PhD(cytogenetics), 71. *Prof Exp:* PROF GENETICS & CYTOGENETICS, PURDUE UNIV, 71- *Mem:* Am Soc Animal Sci; Am Genetic Asn; Sigma Xi. *Res:* Effects of chemicals, drugs and environmental stress on mammalian chromosomes; induced polyploidy in freshwater fish; gene transfer in mice and fish; cytogenetics of neoplasia in domestic animals. *Mailing Add:* Dept Animal Sci Lilly Hall Purdue Univ West Lafayette IN 47907

CHRISMAN, NOEL JUDSON, b Visalia, Calif, July 30, 40; m 89; c 1. MEDICAL ANTHROPOLOGY. *Educ:* Univ Calif, Riverside, BA, 62, Univ Calif, Berkeley, PhD(anthrop), 66, MPH, 67. *Prof Exp:* NIMH fel pub health, Sch Pub Health, Univ Calif, Berkeley, 66-67; asst prof anthrop, Pomona Col, 67-73; from asst prof to assoc prof, 73-84, PROF COMMUNITY HEALTH CARE SYSTS, SCH NURSING, UNIV WASH, 84- *Concurrent Pos:* Nat Endowment Humanities Younger Humanist fel, Univ Calif, Berkeley, 71; vis sci, Fred Hutchinson Cancer Res Ctr, Univ Wash, 86-88. *Mem:* Fel Am Anthrop Asn; fel Soc Appl Anthrop; Soc Med Anthrop; Transcultural Nursing Soc. *Res:* Urban studies, particularly the role of social networks in urban adaptation; class and ethnic variation in health seeking behaviors; working class Americans; network analysis. *Mailing Add:* Sch Nursing Univ Wash Seattle WA 98195

CHRISPEELS, MAARTEN JAN, b Kortenberg, Belg, Feb 10, 38; m 66; c 2. PLANT PHYSIOLOGY. *Educ:* Univ Ill, Urbana, PhD(agron), 64. *Prof Exp:* Res asst agron, Univ Ill, 63-64; res assoc plant biochem, Res Inst Advan Studies, 64-65; plant res lab, AEC, 65-67; res assoc microbiol, Purdue Univ, 67; from asst prof to assoc prof, 67-79, PROF BIOL, UNIV CALIF, SAN DIEGO, 79- *Concurrent Pos:* John S Guggenheim Found fel, 73-74; prog mgr, Competitive Res Grant Off, USDA, 79; consult, plant biotech indust. *Mem:* AAAS; Am Soc Plant Physiol; Am Soc Cell Biol. *Res:* Regulation of plant development, especially seed formation; regulation of gene expression, targeting of proteins to vacuoles; plant biotechnology. *Mailing Add:* C-016 Dept Biol Univ Calif San Diego La Jolla CA 92093-0116

CHRISS, TERRY MICHAEL, b Vallejo, Calif, Dec 10, 45. BOUNDARY LAYER TURBULENCE, SEDIMENT TRANSPORT. *Educ:* Univ Calif, Los Angeles, BS, 68, MS, 71, Ore State Univ, PhD(oceanog), 81. *Prof Exp:* Prof geol, Los Angeles Community Col Dist, 69-74; res assoc, dept oceanog, Dalhousie Univ, 81-85; asst prof, Dept Earth & Environ Sci, Wesleyan Univ, 85-88; mgr univ prog, Keithley Asyst Software, 88-90; OWNER, CUSTOM LAB SOFTWARE SYSTS, 88- *Mem:* Am Geophys Union. *Res:* Field investigation of boundary layer turbulence in the region just above the sediment-water interface (viscous sublayer and lower portion of the turbulent logarithmic layer); studies of turbulence near the sea ice-water interface. *Mailing Add:* Custom Lab Software Systs 22 Main St Centerbrook CT 06409

CHRISSIS, JAMES W, b Pittsburg, Pa, Sept 10, 53. MATH PROGRAMMING, OPTIMIZATION. *Educ:* Univ Pittsburgh, BS, 75; Va Polytech Inst, MS, 77, PhD(opers res), 80. *Prof Exp:* Instr indust eng & opers res, Va Polytech Inst & State Univ, 77-80; asst prof indust & mgt systs eng, Univ SFla, 80-87; asst prof, 87-89, ASSOC PROF OPERS RES, AIR FORCE INST TECHNOL, 89- *Concurrent Pos:* Adj prof, Wright State Univ, 88- *Mem:* Sigma Xi; Opers Res Soc Am; Soc Indust & Appl Math; Inst Indust Engrs. *Res:* Math programming and optimization; large-scale systems; algorithmic processes; systems simulation and analysis. *Mailing Add:* Dept Oper Sci AFIT/ENS Wright-Patterson AFB OH 45433-6583

CHRIST, ADOLPH ERVIN, b Reedley, Calif, May 13, 29; m 51; c 4. CHILD PSYCHIATRY. *Educ:* Univ Calif, Berkeley, AB, 51; Univ Calif, San Francisco, MD, 54; Am Bd Psychiat & Neurol, dipl, 62, cert child psychiat, 67; DMSc, State Univ NY, 87. *Prof Exp:* Intern univ hosp, Univ Calif, San Francisco, 54-55; resident psychiat, Langley Porter Neuropsychol Inst, 58-60, child psychiat, 60-62; instr med sch, Univ Wash, 62-65, asst prof, 65-67, dir psychol inst Scivia children, 62-67; asst prof child psychiat & dir inpatient child psychiat, Albert Einstein Col Med, 68-69; dir training clin psychol, 69-72, EXEC DIR DIV CHILD PSYCHIAT, STATE UNIV NY DOWNSTATE MED CTR, 72- *Mem:* Am Psychiat Asn; Am Orthopsychiat Asn; Am Col Psychiat; World Fedn Ment Health. *Res:* Treatment of childhood psychosis, especially milieu treatment; AIDS prevention. *Mailing Add:* 853 Seventh Ave Apt 6C New York NY 10019

CHRIST, DARYL DEAN, b Buffalo Center, Iowa, Nov 3, 42; m 66; c 1. NEUROPHARMACOLOGY, NEUROPHYSIOLOGY. *Educ:* Univ Iowa, BS, 64; Loyola Univ Chicago, PhD(pharmacol), 69. *Prof Exp:* NIH fel, Loyola Univ Chicago, 69-70; from asst prof to assoc prof pharmacol, Sch Med, Univ Ark, Little Rock, 71-83; ASSOC PROF PHARMACOL, SOUTH BEND CTR MED EDUC, IND UNIV SCH MED, 83- *Concurrent Pos:* Vis assoc prof physiol, Sch Med, Kurume Univ, Japan, 81; adj assoc prof biol, Univ Notre Dame, 83- *Mem:* Am Soc Pharmacol & Exp Therapeut; Soc Neurosci. *Res:* Effects of drugs on ganglionic transmission; pharmacology and physiology of neuromuscular transmission in nematodes. *Mailing Add:* Ctr Med Educ Ind Sch Med Univ Notre Dame Notre Dame IN 46556

CHRIST, JOHN ERNEST, b Ipswich, Mass, Aug, 27, 46. CARDIOVASCULAR PHYSIOLOGY. *Educ:* Baylor Col Med, MD, 73, PhD(physiol), 74. *Prof Exp:* PVT PRACT, PLASTIC SURG, 81- *Mailing Add:* Dept Physiol Baylor Col Med 6560 Fannin Suite 1144 Houston TX 77030

CHRISTADOSS, PREMKUMAR, b Coimbpore, India, May 27, 50; m 77; c 1. IMMUNOGENETICS. *Educ:* Madras Univ, S India, MD, 77. *Prof Exp:* Res scientist, Univ Tex Health Sci Ctr, 82-85; ASST PROF PATH, UNIV VT, 85- *Mem:* Am Asn Immunol; Transplantation Soc; Clin Immunol Soc. *Mailing Add:* Microbiol Bldg Univ Tex Med Br Galveston TX 77550-2782

CHRISTAKOS, SYLVIA, CALCIUM BINDING PROTEINS. *Educ:* State Univ NY, Buffalo, PhD(biochem), 73. *Prof Exp:* Asst prof, 80-85, ASSOC PROF BIOCHEM, UNIV MED & DENT NJ, 85- *Res:* Molecular mechanisms of vitamin D action. *Mailing Add:* Dept Biochem Univ Med & Dent NJ 100 Bergen St Newark NJ 07103

CHRISTE, KARL OTTO, b Ulm, Ger, July 24, 36; nat US; m 62; c 3. INORGANIC CHEMISTRY, PHYSICAL CHEMISTRY. *Educ:* Stuttgart Tech Univ, BS, 57, MS, 60, PhD(inorg chem), 61. *Prof Exp:* Teaching asst anal chem, Stuttgart Tech Univ, 58-60, res assoc inorg polymer chem, 60-61; mem res staff, Stauffer Chem Co, 62-67; mem tech staff, Rocketdyne Div, 67-79, MGR EXPLOR CHEM, ROCKWELL INT, 79- *Honors & Awards:* Creative Work Fluorine Chem, Am Chem Soc, 86. *Mem:* Am Chem Soc. *Res:* Inorganic and organic fluorine chemistry; high energy oxidizers and explosives; structural studies; vibrational spectroscopy; low-temperature matrix isolation; chlorinated hydrocarbons; inorganic polymers; boron hydrides; water treatment chemicals. *Mailing Add:* Rocketdyne Div Rockwell Int BA 26 6633 Canoga Ave Canoga Park CA 91303

CHRISTEN, DAVID KENT, b Paris, Ill, Aug 16, 45; m 70; c 3. SOLID STATE PHYSICS. *Educ:* Univ Ill, BS, 67; Mich State Univ, MS, 70, PhD(physics), 74. *Prof Exp:* Res assoc 74-78, RES STAFF PHYSICS, OAK RIDGE NAT LAB, 78- *Mem:* Am Phys Soc; Sigma Xi. *Res:* Low temperature physics; superconductivity; small-angle neutron scattering. *Mailing Add:* 103 Artesia Lane Oak Ridge TN 37830-7818

CHRISTENBERRY, GEORGE ANDREW, b Macon, Ga, Sept 3, 15; m 37; c 3. MYCOLOGY, ACADEMIC ADMINISTRATION. *Educ:* Furman Univ, BS, 36; Univ NC, AM, 38, PhD(bot), 40. *Prof Exp:* Asst, Univ NC, 36-40; asst prof biol, Meredith Col, 40-41, assoc prof & head dept, 41-43, prof, 43; instr pre-flight prog, Furman Univ, 43-44, instr voc appraiser, Vet Guid, 46, assoc prof biol, 46-48, prof & dean men's col, 48-53; pres, Shorter Col, Ga, 53-58; admin dir, Furman Univ, 58-64; prof biol & chmn dept, Ga Col Milledgeville, 64-65, dean, 65-70; pres, Augusta Col, 70-86; RETIRED. *Concurrent Pos:* Carnegie Found grant, 48; bd dir, Asn State Cols & Univs, 73-75; actg vchancellor, Univ Syst Ga, 79-80; emer pres & emer prof biol, Augusta Col, 86- *Mem:* Am Inst Biol Sci; Mycol Soc Am. *Res:* Taxonomic study of Mucorales in southeastern United States. *Mailing Add:* 500 Norwich Way No 12 Augusta GA 30909

CHRISTENSEN, A(LBERT) KENT, b Washington, DC, Dec 3, 27; m 52; c 5. CELL BIOLOGY, ELECTRON MICROSCOPY. *Educ:* Brigham Young Univ, AB, 53; Harvard Univ, PhD(biol), 58. *Prof Exp:* NIH fel, Col Med, Cornell Univ, 58-59; NIH fel, Harvard Med Sch, 59-60, instr anat, 60-61; from asst prof to assoc prof, Sch Med, Stanford Univ, 61-71; prof anat & chmn dept, Sch Med, Temple Univ, 71-78; chmn dept, 78-82, PROF ANAT & CELL BIOL, UNIV MICH MED SCH, 78- *Mem:* Am Soc Cell Biol; Am Asn Anat; Soc Study Reproduction; Electron Micros Soc Am. *Res:* Cell biology of testis and of steroid-secreting cells; reproductive biology; polysomes. *Mailing Add:* Dept Anat & Cell Biol Med Sci II Bldg Univ Mich Med Sch Ann Arbor MI 48109-0616

CHRISTENSEN, ALLEN CLARE, b Lehi, Utah, Apr 14,35; m 58; c 5. AGRICULTURAL DEVELOPMENT. *Educ:* Brigham Young Univ, BS, 57; Univ Calif, Davis, MS, 60; Utah State Univ, PhD(animal nutrit), 79. *Prof Exp:* From asst prof to assoc prof agr, 64-74, actg dean, Col Agr, 74-75, assoc dean, 75-80, dean, 80-85, actg provost & acad vpres, 85-87, PROF AGR, CALIF STATE POLYTECH UNIV, POMONA, 75-, DEAN, COL AGR, 87- *Concurrent Pos:* Consult, USAID, 83-87, Am Agr Int & Lesotho Agr Col, 89; chmn, Panel Human Capital Develop, Jt Comt Agr Res & Develop, 85-87 & Panel Strengten Grant Prog, 83-87. *Mem:* Am Soc Animal Sci; Asn US Univ Dir Int Agr Prog; Poultry Sci Asn. *Res:* Net energy studies in cattle; amino acid nutrition in poultry; agricultural and human capital development. *Mailing Add:* 4355 Wilson Chino CA 91710

CHRISTENSEN, ANDREW BRENT, b Salt Lake City, Utah, Feb 8, 40; c 4. ATMOSPHERIC PHYSICS, AERONOMY SPACE SCIENCES. *Educ:* Univ Utah, BS, 62; Univ Calif, Berkeley, MA, 64; Univ Denver, PhD(physics), 69. *Prof Exp:* Res assoc shock wave physics, Stanford Res Inst, 64-66; res scientist atmospheric & space sci, Univ Tex, Dallas, 69-79; DIR SPACE SCIENTIST LAB, AEROSPACE CORP, EL SEGUNDO, CALIF, 79- *Concurrent Pos:* Prog dir aeronomy, NSF, 83-85. *Mem:* Am Geophys Union; Int Union Geod & Geophys. *Res:* Atmospheric and space sciences; airglow and auroral physics; atomic and molecular physics; radiation transfer; EUV spectroscopy; low light level photometry. *Mailing Add:* PO Box 92957 Los Angeles CA 90009

CHRISTENSEN, BENT AKSEL, b Copenhagen, Denmark, Mar 22, 28; m 58; c 2. CIVIL ENGINEERING, HYDRAULICS. *Educ:* Copenhagen Univ, Filosofikum, 48; Tech Univ Denmark, MS, 51; Univ Minn, PhD(hydraul), 61. *Prof Exp:* Asst prof hydraul, Tech Univ Denmark, 51-55, assoc prof, 55-59, 61-63; assoc prof, 63-65, PROF CIVIL ENG & HYDRAUL, UNIV FLA, 65- *Concurrent Pos:* Govt supvr, Hydraul Lab, Polytech Sch, Lausanne, 53; Danish Mil Acad, NATO, 55-62; mem bd dirs, Fluid Mech Inst, Univ Fla, 65- *Mem:* Am Soc Civil Engrs; Am Soc Eng Educ; Am Water Resources Asn; Danish Inst Civil Engrs; Int Asn Hydraul Res. *Res:* Hydraulics; pipe and open channel flow; water resources engineering; hydrology; sediment transport; fluid dynamics; ground water flow and soil mechanics. *Mailing Add:* Dept Civil Eng Univ Fla Gainesville FL 32611

CHRISTENSEN, BERT EINAR, b Duluth, Minn, Oct 20, 04; m 32; c 3. SYNTHETIC ORGANIC CHEMISTRY. *Educ:* Wash State Univ, BS, 27; Univ Wash, PhD(chem), 32. *Prof Exp:* Chemist, Atmospheric Nitrogen Corp, Syracuse, 27-28; from instr to prof chem, 31-70, chmn dept, 56-70, EMER PROF CHEM, ORE STATE UNIV, 70- *Mem:* AAAS; Am Chem Soc. *Res:* Organic synthesis of quinazoline, purine, pyrimidine and other related heterocyclic compounds. *Mailing Add:* 337 NW 23rd Corvallis OR 97330

CHRISTENSEN, BURGESS NYLES, b San Francisco, Calif, Oct 4, 40; m 61; c 2. NEUROPHYSIOLOGY. *Educ:* Univ Utah, BA, 63, PhD(biophys, bioeng), 67. *Prof Exp:* Fel physiol, Univ Utah, 67-68; fel, Yale Univ, 68-70; asst prof neurosci, Brown Univ, 70-76; ASST PROF PHYSIOL, UNIV TEX MED BR, 76- *Mem:* AAAS; Soc Neurosci. *Res:* Synaptic transmission in the central nervous system. *Mailing Add:* Dept Physiol & Biophys Univ Tex Med Br Galveston TX 77550

CHRISTENSEN, BURTON GRANT, b Waterloo, Iowa, Apr 8, 30; m 61, 81; c 2. BIO-ORGANIC CHEMISTRY, MEDICINAL CHEMISTRY. *Educ:* Iowa State Univ, BS, 52; Harvard Univ, AM, 54, PhD(chem), 56. *Prof Exp:* Chemist, 56-71, asst dir, 71-73, dir synthetic chem res, 73-76, exec dir synthetic chem res, 76-82, vpres, Basic Chem, 82-87, SR VPRES, CHEM, MERCK, SHARP & DOHME RES LABS, 87- *Concurrent Pos:* Adj prof chem, Stevens Inst Tech, Hoboken, NJ. *Honors & Awards:* Thomas A Edison Patent Award; Cecil L Brown lectr. *Mem:* Am Chem Soc; Royal Soc Chem; Am Inst Chemists. *Res:* Organic chemistry, with emphasis on antibacterial synthesis. *Mailing Add:* Merck & Co Inc R 80 M 113 PO Box 2000 Rahway NJ 07065

CHRISTENSEN, CHARLES RICHARD, b Florence, Ala, Oct 18, 38; m 59; c 3. OPTICAL PHYSICS, SOLID STATE PHYSICS. *Educ:* Vanderbilt Univ, BE, 60; Calif Inst Technol, PhD(chem), 66. *Prof Exp:* Res chemist, 67-68; RES PHYSICIST, PHYS SCI LAB, US ARMY MISSILE COMMAND, 69- *Mem:* Am Phys Soc; Am Chem Soc; Optical Soc Am. *Res:* Coherent optics with applications in signal processing and correlation, coherent optical imaging, optical synthetic aperture methods; magnetic resonance and magnetic materials. *Mailing Add:* 1400 Levert Ave Athens AL 35611

CHRISTENSEN, CHRISTIAN MARTIN, b Ft Collins, Colo, Sept 24, 46; m 68; c 2. ECONOMIC ENTOMOLOGY. *Educ:* Rutgers Univ, BS, 68; Purdue Univ, MS, 70, PhD(entom), 74. *Prof Exp:* Livestock & forage entom exten specialist, 74-80, ASST EXTEN PROF ENTOM, UNIV KY, 80- *Mem:* Entom Soc Am. *Res:* Practical application techniques and control methods for controlling livestock insect pests; development and implementation of a practical integrated pest management system for Kentucky alfalfa producers. *Mailing Add:* 352 Ashmoor Dr Lexington KY 40503

CHRISTENSEN, CLARK G, b Spanish Fork, Utah, June 17, 43; m 67; c 5. ASTRONOMY. *Educ:* Brigham Young Univ, BS, 66; Calif Inst Technol, PhD(astron), 72. *Prof Exp:* ASSOC PROF PHYSICS & ASTRON, BRIGHAM YOUNG UNIV, 72- *Mem:* Am Astron Soc; Astron Soc Pac. *Res:* Luminosity function of galaxies; metal poor stars; variable stars; population syntheses of stellar systems. *Mailing Add:* Dept Phys & Astron Brigham Young Univ Provo UT 84602

CHRISTENSEN, CRAIG MITCHELL, b Chicago, Ill, Jan 25, 32; m 58; c 2. CHEMICAL ENGINEERING, INDUSTRIAL ENGINEERING. *Educ:* Univ Ill, Urbana, BS, 53; Cornell Univ, PhD(chem eng), 61. *Prof Exp:* Res engr, B F Goodrich Res Ctr, B F Goodrich Co, Akron, 60-63, sr res engr, 63-66, mgr qual assurance, 66-73, mgr, Inventory & Distrib Methods, 73-88; LECTR, UNIV AKRON, 88- *Mem:* Am Soc Qual Control; Am Chem Soc; Am Inst Chem Engrs; Am Statist Asn. *Res:* Industrial statistics; liquid-liquid extraction; applied statistical design and analysis of experiments; physical distribution. *Mailing Add:* 5919 Bradford Way Hudson OH 44236-3905

CHRISTENSEN, DOUGLAS ALLEN, b Bakersfield, Calif, Dec 14, 39; m 62; c 3. ELECTRICAL ENGINEERING, BIOENGINEERING. *Educ:* Brigham Young Univ, BS, 62; Stanford Univ, MS, 63; Univ Utah, PhD(elec eng), 67. *Prof Exp:* Sr res engr elec eng, Gen Motors Res Lab, 66-70; asst prof, 70-78, ASSOC PROF ELEC ENG & BIOENG, UNIV UTAH, 78- *Concurrent Pos:* NIH fel, Ctr Bioeng, Univ Washington, 71-73. *Mem:* Inst Elec & Electronics Engrs. *Res:* Bioinstrumentation, especially optical and ultrasound interaction in biological material; microwave hyperthermia. *Mailing Add:* Dept Elec Eng Merrill Eng Bldg Univ Utah Salt Lake City UT 84112

CHRISTENSEN, EDWARD RICHARDS, b Salt Lake City, Utah, Dec 21, 24; m 49; c 1. PETROLEUM CHEMISTRY. *Educ:* Univ Utah, BS, 48, MS, 49; Union Col, MS, 74. *Prof Exp:* Chemist, Beacon Res Labs, Texaco Inc, 49-56, group leader fuels process develop, 56-58, admin asst to dir res, 58-59, asst supvr petrochem res, 60-65, supvr, 65-67, res dir chem & process res, 67-75, asst mgr res & tech dept, Beacon Res Labs, 75-77, asst mgr, Port Arthur Res Labs, Tex, 77-78, mgr sci planning, Res Environ & Safety Dept, 78-84, assoc dir & mgr, Texaco Res Ctr, Beacon, 84-85; RETIRED. *Concurrent Pos:* Bus develop officer, Albany Savings Bank, 85-88. *Mem:* Am Chem Soc; Am Inst Chem Eng; Sigma Xi. *Res:* Synthesis and applications testing of chemicals derived from petroleum and development of pilot scale processes for their manufacture; development of new or improved catalysts and processes for the conversion of crude oil to marketable products; development of improved methods and equipment for carrying out process development. *Mailing Add:* RD/5 Pine Ridge Dr Wappingers Falls NY 12590

CHRISTENSEN, ERIK REGNAR, b Copenhagen, Denmark, Apr 17, 43; m 68; c 3. ENVIRONMENTAL ENGINEERING, WATER CHEMISTRY. *Educ:* Tech Univ Denmark, MS, 67; Univ Calif, Irvine, PhD(Environ Eng), 77. *Prof Exp:* Res asst, Danish Defense Res Bd, 67-69; from asst prof to assoc prof applied nuclear physics & reactor physics, Dept Electro-physics, Tech Univ Denmark, 69-74; teaching asst water qual & teaching assoc radioisotope, Dept Chem, Sch Eng, Univ Calif, Irvine, 74-77; asst prof, 77-82, ASSOC

PROF CIVIL ENG, COL ENG & APPL SCI, UNIV WIS-MILWAUKEE, 82- *Concurrent Pos:* consult, Wis Dept Justice, Milwaukee Metrop Sewage Dist. *Mem:* Am Nuclear Soc; Int Asn Water Pollution Res & Control; Asn Environ Eng Professors; Am Geophys Union; Am Soc Civil Engrs. *Res:* Multiple toxicity of various compounds to algal growth; sources and pathways of pollutants (metals and organics) in sediments using, for example, radiometric geochronologic methods; environmental remote sensing. *Mailing Add:* Dept Civil Eng & Mech Univ Wis Milwaukee WI 53201

CHRISTENSEN, GEORGE CURTIS, b New York, NY, Feb 21; m 47; c 4. VETERINARY ANATOMY. *Educ:* Cornell Univ, DVM, 49, MS, 50, PhD(mammal anat, higher educ), 53. *Hon Degrees:* DSc, Purdue Univ, 78. *Prof Exp:* Instr vet anat, Cornell Univ, 49-53; assoc prof, Iowa State Univ, 53-58; prof vet anat & head dept, Purdue Univ, 58-63; dean col vet med & dir vet med res inst, 63-65, VPRES ACAD AFFAIRS, IOWA STATE UNIV, 65- *Concurrent Pos:* Comnr, NIH & NCent Asn Cols & Sec Schs; mem comt educ & res, Nat Acad Sci, 70-72; mem bd dirs, Iowa State Hyg Lab & Ctr Res Libr, Chicago; mem, Iowa Bd Health; mem exec comt, Nat Asn State Univs & Land-Grant Cols, 73-, chmn coun acad affairs, 73-74; chmn, Comt Educ Telecommun, vpres, Mid-Am State Univs Asn, 75-76. *Mem:* Nat Acad Sci; Am Vet Med Asn; Conf Res Workers Animal Dis; Am Asn Vet Anatomists (pres, 63); World Asn Vet Anatomists (vpres, 63-65); Sigma Xi. *Res:* University administration; comparative cardiovascular anatomy and physiology; comparative vasculature of urogenital and central nervous systems; history of veterinary medical education; higher education. *Mailing Add:* 110 Beardshear Hall Iowa State Univ Ames IA 50010

CHRISTENSEN, GERALD M, b Pocatello, Idaho, Dec 23, 28; m 49; c 2. BIOCHEMISTRY, RADIOBIOLOGY. *Educ:* Univ Utah, BS, 51; Emory Univ, PhD, 58. *Prof Exp:* Fel, Virus Lab, Univ Calif, Berkeley, 58-59; sr biochemist, Boeing Co, Wash, 59-64; from asst prof to assoc prof radiol, Univ Wash, 64-80, assoc prof environ sci, 80; PROG MGR, WASTE TANK SAFETY WESTINGHOUSE HANFORD CO, 80- *Res:* Biological effects of ionizing radiation; enzyme and protein chemistry; amino acid and nucleotide metabolism. *Mailing Add:* Westinghouse Hanford Co PO Box 1970 Richland WA 99352

CHRISTENSEN, H(ARVEY) D(EVON), aerospace & mechanical engineering, for more information see previous edition

CHRISTENSEN, HALVOR NIELS, b Cozad, Nebr, Oct 24, 15; m 39; c 3. BIOCHEMISTRY, BIOPHYSICS. *Educ:* Kearney State Col, BS, 35; Purdue Univ, MS, 37; Harvard Univ, PhD(biol chem), 40. *Hon Degrees:* DSc, Univ Nebr Kearney, 90. *Prof Exp:* Fel chem, Harvard Univ, 40-41; res biochemist, Lederle Labs, 41-42; instr biol chem, Harvard Med Sch, 42-44; dir chem labs, Mary-Imogene Bassett Hosp, Cooperstown, NY, 44-47; asst prof biochem, Harvard Med Sch, 47-49; prof biochem & nutrit & head dept, Sch Med, Tufts Univ, 49-55; chmn dept, 55-70, prof, 55-87, EMER PROF BIOCHEM, UNIV MICH, ANN ARBOR, 87-; ADJ PROF PEDIAT, UNIV CALIF SAN DIEGO, 89- *Concurrent Pos:* Dir lab biochem res, Childrens Hosp, Boston, 47-49; Guggenheim fel, Carlsberg Lab, Copenhagen, 52; consult, NIH, 61-68; Nobel guest prof, Univ Uppsala, 68-69; ed, Biochem Biophys Acta & J Biol Chem; consult ed, Nutrit Reviews. *Honors & Awards:* Russel lectr, Univ Mich, 80. *Mem:* Fel Am Acad Arts & Sci; Am Soc Biol Chem; Am Chem Soc; fel Am Inst Nutrit; Biophys Soc; Soc Gen Physiologists. *Res:* Amino acid transport; intravenous amino acid nutrition; peptide metabolism and antibiotics; cystinosis and transport. *Mailing Add:* Pediat Metab Univ Calif San Diego BSB 4006 La Jolla CA 92093-0609

CHRISTENSEN, HOWARD ANTHONY, b Oakland, Calif, Apr 26, 28; m 52. MEDICAL ENTOMOLOGY, MEDICAL PARASITOLOGY. *Educ:* San Francisco State Col, BA, 63, MA, 66; Univ Calif, Davis, PhD(entom), 68. *Prof Exp:* MED ENTOMOLOGIST, GORGAS MEM LAB, 68- *Mem:* Am Soc Parasitol; Am Soc Trop Med & Hyg; Am Mosquito Control Asn; Sigma Xi. *Res:* Bionomics of arthropod vectors of medical importance and the epidemiology of arthropod borne diseases; leishmaniasis. *Mailing Add:* 813 Acacia Lane Davis CA 95616

CHRISTENSEN, HOWARD DIX, b Logan, Utah, Mar 30, 40; m 66; c 3. NEUROPHARMACOLOGY. *Educ:* Univ Nev, Reno, BS, 61; Univ Calif, Los Angeles, PhD(biophys, nuclear med), 66. *Prof Exp:* Fel pharmacol, Univ Calif, San Francisco, 66-68 & Columbia Univ, 68-69; pharmacologist, Res Triangle Inst, 69-74; ASSOC PROF PHARMACOL, SCHS MED & DENT, UNIV OKLA, 74- *Mem:* Am Soc Pharmacol & Exp Therapeut; NY Acad Sci; Sigma Xi; Soc Neurosci; Soc Exp Biol & Med. *Res:* Central nervous system mechanisms of cannabinoids, anticonvulsants, central nervous system depressants and stimulants and contraceptive steroids; antihypertensives and antiglaucoma drugs; correlations between drug development of analytical methodologies; kinetics and physiological or behavioral response. *Mailing Add:* PO Box 26901 Oklahoma City OK 73190

CHRISTENSEN, JAMES, b Ames, Iowa, Jan 4, 32; m 58; c 3. PHYSIOLOGY, GASTROENTEROLOGY. *Educ:* Univ Nebr, BS, 53, MS & MD, 57. *Prof Exp:* From instr to assoc prof, 65-72, PROF INTERNAL MED, UNIV IOWA, 72- *Concurrent Pos:* Lectr, Univ Alta, 65-66; Markle scholar, 65-70; prin investr, NIH res grant, 67- & USPHS career develop award, 69-74; pres, Am Motility Soc, 82-84; assoc ed, Gastroenterol, 84-86; chmn eval comt gastroenterol, Vet Admin Res, 84-85; dir, Digestive Dis Core Ctr, Univ Iowa, 84-88; consult, Proctor & Gamble Inc, 86- *Mem:* Am Fedn Clin Res; Asn Am Physicians; Am Gastroenterol Asn; Am Soc Clin Invest; Am Clin & Climatol Soc. *Res:* Gastrointestinal motility; autonomic and smooth muscle physiology and anatomy. *Mailing Add:* Dept Internal Med Univ Iowa Hosps Iowa City IA 52242

CHRISTENSEN, JAMES HENRY, b Beaver Dam, Wis, Apr 9, 42; m 63; c 2. PROGRAMMABLE CONTROLLERS, DATA COMMUNICATIONS. *Educ:* Univ Wis, BS, 63, MS, 65, PhD(chem eng), 67. *Prof Exp:* Fel eng design, Dartmouth Col, 67-68; asst prof chem eng, Univ Okla, 68-71, assoc prof, 71-76; pres, Strider Syst, Inc, 76-79; sr syst engr, Tex Instruments, Inc, 79-82; mgr advan res & develop, syst div, 82-86, PRIN ENGR, ADVAN ARCHIT DEVELOP, ALLEN-BRADLEY CO, 86- *Concurrent Pos:* Mem, Comput Aids Chem Eng Educ Comt, Nat Acad Sci, 69-73; consult, Chemshare, Ltd, 69-70; fac fel, Ames Lab, US Atomic Energy Comn, 72; tech expert, prog controllers, Int Electrotech Comn, 81- *Res:* Architectural design of programming support environments for industrial automation. *Mailing Add:* Allen-Bradley Co 747 Alpha Dr Highland Heights OH 44143

CHRISTENSEN, JAMES ROGER, b Des Moines, Iowa, Oct 28, 25; m 51; c 2. VIROLOGY. *Educ:* Iowa State Col, BS, 49; Cornell Univ, PhD(biochem), 53. *Prof Exp:* Fel biophysics, Sch Med, Univ Colo, 53-55; from instr to assoc prof, 55-70, actg chmn, 67-70, PROF MICROBIOL, SCH MED & DENT, UNIV ROCHESTER, 70- *Mem:* Am Soc Microbiologists; Genetics Soc. *Res:* Bacteriophage; DNA replication and recombination; gene function; mutual exclusion. *Mailing Add:* Dept Microbiol Univ Rochester Med Ctr Rochester NY 14642

CHRISTENSEN, JOHN, biochemistry; deceased, see previous edition for last biography

CHRISTENSEN, KENNER ALLEN, b Duluth, Minn, July 18, 43; m 68; c 2. NUCLEAR MAGNETIC RESONANCE, ORGANIC CHEMISTRY. *Educ:* Univ Minn-Duluth, BA, 65; Ohio State Univ, MS, 68, PhD(chem), 71. *Prof Exp:* Fel chem, Univ Utah, 70-73; asst prof, Univ Notre Dame, 73-76; nuclear magnetic resonance specialist, Northwestern Univ, 76-; MEM STAFF, DEPT CHEM, UNIV ARIZ, TUCSON. *Mem:* Am Chem Soc; AAAS. *Res:* Development of instrumentation and techniques in nuclear magnetic resonance; studies of organometallic compounds by nuclear magnetic resonance. *Mailing Add:* Dept Chem Univ Ariz Tucson AZ 85721-0002

CHRISTENSEN, LARRY WAYNE, b Elkhart, Ind, Sept 1, 43; m 65; c 3. ORGANIC CHEMISTRY. *Educ:* Goshen Col, BA, 65; Purdue Univ, PhD(chem), 69. *Prof Exp:* Assoc prof, 69-76, PROF CHEM & CHMN DEPT, HOUGHTON COL, 76- *Concurrent Pos:* Vis fac mem, Univ Ariz, 75; vis res assoc, Univ Fla, 81. *Mem:* Am Chem Soc; Am Sci Affil. *Res:* Mechanistic and synthetic organo-sulfur chemistry; applications of mass spectrometry to organic chemistry; organo-metallic transition metal chemistry; development new synthetic methods. *Mailing Add:* Dept Chem Houghton Col Houghton NY 14744

CHRISTENSEN, MARK NEWELL, b Green Bay, Wis, July 16, 30; m 55. GEOLOGY. *Educ:* Univ Alaska, BS, 52; Univ Calif, PhD, 59. *Prof Exp:* From actg instr to instr geol, Univ Calif, Berkeley, 59-60, from asst prof to assoc prof, 60-74; prof geol & geophys, Univ Calif, Santa Cruz, 60-76, chancellor, 74-76; PROF ENERGY & RESOURCES, UNIV CALIF, BERKELEY, 76- *Concurrent Pos:* Asst dean, Col Letters & Sci, Univ Calif, Berkeley, 65-74. *Mem:* AAAS; Geol Soc Am; Am Geophys Union. *Res:* Structural geology; deformation of rocks; stratigraphy; tectonics. *Mailing Add:* 1514 La Loma Ave Berkeley CA 94708

CHRISTENSEN, MARTHA, b Ames, Iowa, Jan 4, 32. MYCOLOGY, ECOLOGY. *Educ:* Univ Nebr, BSc, 53; Univ Wis, MS, 56, PhD(bot), 60. *Prof Exp:* Proj assoc mycol, Univ Wis, 60-63; from asst prof to assoc prof, 63-75, PROF BOT, UNIV WYO, 75- *Honors & Awards:* Weber-Ernst Award, 53. *Mem:* Mycol Soc Am; Ecol Soc Am; Brit Mycol Soc. *Res:* Ecology and taxonomy of soil microfungi. *Mailing Add:* Dept Bot Univ Wyo Laramie WY 82071

CHRISTENSEN, MARY LUCAS, b St Louis, Mo, Oct 18, 37. CLINICAL VIROLOGY. *Educ:* Col Med, Univ Iowa, BA, 59, MS, 61; Sch Med, Northwestern Univ, PhD(microbiol), 74. *Prof Exp:* Asst bact, Col Med, Univ Iowa, 59-61; res virologist, virus res dept, Wyeth Labs, 61-65 & Abbott Labs, 65-69; chief, Virol Lab, Northwestern Univ Med Ctr, 69-71; Nat Cancer Inst fel biochem, Sch Med, Northwestern Univ, 74-76, fel, 76-78; affil prof staff, Northwestern Mem Hosp, 78-83; asst prof path, 78-87, ASSOC PROF CLIN PATH & PEDIAT, SCH MED, NORTHWESTERN UNIV, 87-, DIR VIROL, CHILDREN'S MEM HOSP, 78- *Concurrent Pos:* Pres, Ill Soc Microbiol; counr, Am Soc Microbiol. *Mem:* Am Soc Microbiol; Pan-Am Group Rapid Viral Diag; Tissue Cult Asn. *Res:* Development of rapid laboratory tests for diagnosis of virus disease, including enzyme immunoassays, immunofluorescent antibody tests; role of coxsackieviruses in the pathogenesis of juvenile dermatomyositis; author virology and microbiology textboks. *Mailing Add:* 2300 Children's Plaza Chicago IL 60614

CHRISTENSEN, N(EPHI) A(LBERT), b Provo, Utah, Jan 19, 03; m 29; c 4. CIVIL ENGINEERING, MATHEMATICS. *Educ:* Brigham Young Univ, BS, 25; Univ Wis, BS, 28; Calif Inst Technol, MS, 34, PhD(civil eng), 39. *Prof Exp:* Teacher high sch, Utah, 25-26; prof exact sci, Ricks Col, Idaho, 28-33; instr, Calif Inst Technol, 33-35; hydraul res lab, Soil Conserv Serv, US Dept Agr, 35-38; dean eng, Colo State Univ, 38-48, dir eng div, exp sta, 39-48; dir, Sch Civil Eng, Cornell Univ, 48-68, emer dir, 68-81; pres, Gas Scribbers, Inc, 80-83; ENG CONSULT, EPITAXY INC, 73- *Concurrent Pos:* Spec lectr, Univ Southern Calif, 36; engr, State Hwy Comn, Wis, 27-28; chief engr, Ballistics Res Lab, Aberdeen Proving Ground, Md, 43-44; chief Rocket Res Div, Ord Res & Develop Ctr, 44-45; consult, Brookhaven Nat Lab & Argonne Nat Lab Atomic Energy Comn; office ord res, US Army, 53-54; mem, NY State Flood Control Comn, 54-60; trustee, Cayuga Heights Village, 56; coordr four eng firms to develop comprehensive sewerage plan for Monroe County, NY, 66-68; col admin adv, Near East Found, Rezaiyeh Agr Col, Iran, 68-72. *Mem:* Am Soc Civil Engrs; Am Soc Eng Educ; Am Geophys Union. *Res:* Rocket research; hydraulics; structures; erosion and sedimentation; irrigation. *Mailing Add:* 7801 Pickard Ave NE Albuquerque NM 87110

CHRISTENSEN, NED JAY, b Clarkston, Utah, June 23, 29; m 51; c 3. AUDIOLOGY. *Educ:* Brigham Young Univ, BA, 54, MA, 55; Pa State Univ, PhD, 59. *Prof Exp:* Instr, Pa State Univ, 55-59; asst prof speech & clin supvr, WVa Univ, 59-62; from asst prof to assoc prof, 62-74, asst dir, 62-70, dir speech path-audiol prog, 72-76, PROF SPEECH, UNIV ORE, 74-, COORDR, COM DISORDERS, 76- *Mem:* Am Speech & Hearing Asn. *Res:* Audiological rehabilitation. *Mailing Add:* Dept Speech Path-Audiol Univ Ore Col Educ Eugene OR 97403

CHRISTENSEN, NIKOLAS IVAN, b Madison, Wis, Apr 11, 37; m 60; c 1. GEOPHYSICS, MINERALOGY. *Educ:* Univ Wis, BS, 59, MS, 61, PhD(geol), 63. *Prof Exp:* Res fel geophys, Harvard Univ, 63-64; assoc prof geol, Univ Southern Calif, 64-67; prof geol, Univ Wash, 67-83; PROF GEOPHYS, PURDUE UNIV, 83- *Mem:* Fel Geol Soc Am; Am Geophys Union. *Res:* Elasticity of rocks and minerals; crystal physics, nature of the earth's interior; oceanography. *Mailing Add:* Dept Earth & Atmospheric Sci Purdue Univ West Lafayette IN 47907

CHRISTENSEN, NORMAN LEROY, JR, b Fresno, Calif, Dec 28, 46; m 68; c 2. PLANT ECOLOGY. *Educ:* Calif State Univ, Fresno, BA, 68, MA, 70; Univ Calif, Santa Barbara, PhD(biol), 73. *Prof Exp:* From asst prof to assoc prof, 73-87, PROF BOT, DUKE UNIV, 87- *Mem:* Ecol Soc Am; AAAS; Sigma Xi; Brit Ecol Soc; Soc Wetland Scientists. *Res:* Effects of disturbance on plant community structure and function; effects of fire on community nutrient relations; forest demography. *Mailing Add:* Dept Bot Duke Univ Durham NC 27706

CHRISTENSEN, ODIN DALE, b Duluth, Minn, Dec 12, 47; div. MINERALOGY, GEOCHEMISTRY. *Educ:* Univ Minn, BA, 70; Stanford Univ, PhD(geol), 75. *Prof Exp:* Asst prof geol, Univ NDak, 75-78; res geochemist, Univ Utah Res Inst, 78-81; res geologist, Newmont Explor Ltd, 81-85, regional geologist, 85-88, dir, US Explor, 88-90, VPRES EXPLOR, NEWMONT GOLD CO, 90- *Mem:* Geol Soc Am; Soc Mining Engrs; Asn Explor Geochemists. *Res:* Geochemistry of geothermal systems; exploration geologic techniques. *Mailing Add:* Newmont Gold Co PO Box 669 Carlin NV 89822

CHRISTENSEN, RALPH C(HRESTEN), b San Mateo, Calif, Dec 11, 39; m 78; c 4. HEALTH PHYSICS, MEDICAL PHYSICS. *Educ:* Stanford Univ, BS, 61, MA, 62; Univ Calif, Berkeley, MBiorad, 65, PhD(biophysics), 71. *Prof Exp:* Res fel biophysics, Pa State Univ, 71-73, asst prof radiobiol, Hershey Med Ctr, 73-76; asst prof radiation sci, 76-80, chmn dept health radiation sci, 83-88, ASSOC PROF RADIATION SCI, UNIV KY, 80-, DIV DIR, 88-, ASSOC PROF RADIATION MED, 89- *Concurrent Pos:* Co-ed, Selected Topics in Reactor Health Physics; chmn, Placement Comt, 78-83, chmn, Manpower & Prof Educ Comt, Health Physics Soc, 83-85, 90-91. *Mem:* Health Physics Soc; Am Assoc Physicists Med; Sigma Xi. *Res:* Radiation dosimetry; applied health physics; applied medical physics; radiation biophysics. *Mailing Add:* 125 Westwood Dr Lexington KY 40503-1137

CHRISTENSEN, RICHARD G, b Imperial, Nebr, Nov 7, 38; m 65; c 3. QUANTITATIVE LC-MS, AUTOMATION OF ANALYSIS. *Educ:* Univ Nebr, BS, 60, MS, 64. *Prof Exp:* CHEMIST, NAT INST STANDARDS & TECHNOL, 59- *Mem:* Sigma Xi. *Res:* Apply liquid chromatography-mass spectroscopy to quantitative measurements of organic analytes; develop novel and automatable sample treatment methods; design instruments to effect the above. *Mailing Add:* Six Walker Ave Gaithersburg MD 20877

CHRISTENSEN, RICHARD MONSON, b Idaho Falls, Idaho, July 3, 32; m 58; c 2. COMPOSITE MATERIALS, VISCOELASTICITY. *Educ:* Univ Utah, BSc, 55; Yale Univ, MEng, 56, DEng, 61. *Prof Exp:* Struct engr, Convair Div, Gen Dynamics, 56-58; mem tech staff, TRW Systs, 61-64; asst prof mech eng, Univ Calif, Berkeley, 64-67; staff res engr, Shell Develop Co, 67-74; prof mech eng, Washington Univ, 74-76; MEM TECH STAFF, LAWRENCE LIVERMORE NAT LAB, 76- *Concurrent Pos:* Lectr, Univ Southern Calif, 62-64, Univ Calif, Berkeley, 69-70, 78 & 80, Univ Houston, 73; chmn, Appl Mech Div, Am Soc Mech Engrs, 80-81; mem, US Nat Comt Theoret & Appl Mech, 80- *Mem:* Nat Acad Eng; Soc Rheology; Am Chem Soc; fel Am Soc Mech Engrs. *Res:* Properties of polymers, wave propagation, failure theories, crack kinetics, composite materials. *Mailing Add:* Lawrence Livermore Nat Lab PO Box 808 Livermore CA 94550

CHRISTENSEN, ROBERT LEE, b Orange, NJ, July 23, 29; m 52; c 3. ATOMIC PHYSICS, COMPUTER SCIENCE. *Educ:* Princeton Univ, AB, 50, MA, 54, PhD(physics), 57. *Prof Exp:* Asst reactor dept, Brookhaven Nat Lab, 53; asst physics dept, Princeton Univ, 57-58; physicist res lab, Int Bus Mach Corp, 58-61, eng mgt, 61-65; vpres, Quantum Sci Corp, 66-67 & A G Becker Inc, 68-82; vpres, Chase Investors Mgt Corp, 82-89; PRIN, RLC MGT SERV, 89- *Mem:* Am Phys Soc; Inst Elec & Electronics Engrs; NY Soc Security Analysts. *Res:* Nuclear and atomic resonance; ultra-high vacuum; photoelectric effect; application of digital computers to image and language processing; management information systems, applied statistics and econometrics; technological forecasting. *Mailing Add:* 100 Devoe Rd Chappaqua NY 10514

CHRISTENSEN, S(ABINUS) H(OEGSBRO) (CHRIS), physics; deceased, see previous edition for last biography

CHRISTENSEN, STANLEY HOWARD, b Boone, Iowa, Mar 6, 35; m 56; c 4. SOLID STATE PHYSICS. *Educ:* Iowa State Univ, BS, 57; Cornell Univ, PhD(physics), 63. *Prof Exp:* Jr physicist, Iowa State Univ, 57; asst prof, 63-67, assoc prof, 67-77, PROF PHYSICS, KENT STATE UNIV, 78-, CHAIR, 83- *Mem:* Am Phys Soc; Am Asn Physics Teachers; AAAS. *Res:* Electron paramagnetic resonance; laser light scattering. *Mailing Add:* Dept Physics Kent State Univ Kent OH 44242

CHRISTENSEN, THOMAS GASH, b Richmond, Va, Sept 16, 44; m 69; c 2. CELL BIOLOGY, ULTRASTRUCTURAL PATHOLOGY. *Educ:* Rutgers Univ, BS, 66; Univ Vt, PhD(bot), 71. *Prof Exp:* Res assoc med biochem, Sch Med, Univ Vt, 71-74; res assoc, Mallory Inst Path, 74-85; res assoc pulmonary path, 74-77, asst res prof, 77-80, asst prof, 80-88, ASSOC PROF PATH, SCH MED, BOSTON UNIV, 88-; SR RES ASSOC, MALLORY INST PATH, 85- *Concurrent Pos:* Vis lectr biol, Univ Vt, 71-74; prin investr, Nat Heart, Lung & Blood Inst grant, 78- *Mem:* Sigma Xi; AAAS; Electron Micros Soc Am; Am Soc Cell Biol; Am Thoracic Soc. *Res:* Airway secretory cell biology in health and disease; ultrastructural aspects of cancer cell differentiation. *Mailing Add:* Mallory Inst Path Boston Univ Sch Med 784 Mass Ave Boston MA 02118

CHRISTENSON, CHARLES O, b Oakland, Calif, Sept 17, 36; m 60; c 4. TOPOLOGY. *Educ:* Univ Kans, BA, 58, MA, 60; NMex State Univ, PhD(math), 64. *Prof Exp:* From asst prof to assoc prof, 64-80, PROF MATH, UNIV IDAHO, 80- *Concurrent Pos:* Vis prof math, Col VI, 81-82. *Mem:* AAAS; Am Math Soc; Math Asn Am; Sigma Xi. *Res:* Knot theory; study of the knotting number; topology of manifolds. *Mailing Add:* Dept Math Univ Idaho Moscow ID 83843

CHRISTENSON, DONALD ROBERT, b Terry, Mont, Mar 31, 37; m 59; c 3. SOIL SCIENCE. *Educ:* Mont State Col, BS, 60, MS, 64; Mich State Univ, PhD(soil sci), 68. *Prof Exp:* From asst prof to assoc prof, 68-79, PROF SOIL SCI, MICH STATE UNIV, 79- *Honors & Awards:* Scarseth Award, Am Soc Agron, 67. *Mem:* Am Soc Agron; Am Soc Sugar Beet Technol; Soil Conserv Soc Am. *Res:* Soil-plant nutrient relationships; response of plants to applied nutrients; soil test correlations; mechanisms of nutrient release from soil minerals. *Mailing Add:* Dept Crop & Soil Sci Mich State Univ East Lansing MI 48824

CHRISTENSON, LISA, b Northampton, Mass, Dec 28, 56; m 90. BIOMATERIALS-BIOCOMPATIBILITY, TRANSPLANTATION. *Educ:* Wellesley Col, AB, 78; Brown Univ, ScM, 86, PhD(med sci), 90. *Prof Exp:* Flow cytometer operator, Beth Israel Hosp, Boston, 80-84; teaching asst, Brown Univ, 84-89; scientist, 89-91, MGR REGULATORY SCI, CELLULAR TRANSPLANTS, INC, 91- *Mem:* Am Soc Artificial Internal Organs; Regulatory Affairs Prof Soc; Sigma Xi; Soc Biomat; Soc Neurosci. *Res:* Encapsulation of cells or tissues within a polymeric membrane, enabling allo- or xeno transplantation without immunosuppression; thymic tissue, parathyroid, dopamine-producing and insulin- producing tissue. *Mailing Add:* Cellular Transplants Inc Four Richmond Sq Providence RI 02906

CHRISTENSON, PAUL JOHN, b Watervliet, NY, Aug 13, 21; m 48; c 9. MEDICINE, PREVENTIVE MEDICINE. *Educ:* Siena Col, NY, BS, 43; Marquette Univ, MD, 46; Columbia Univ, MPH, 55. *Prof Exp:* Pvt pract med, NY, 49-52; dir pub health, Tri County Unit, Va, 52-53 & Mich, 53-54; dir pub health, Utica City Dept Health, NY, 55-57; mem med staff, Eaton Labs, Norwich Pharmacol Col, 57-59, dir clin res, 59-60, med dir, Norwich Prod Div, 60-65; med dir, Quinton Div, Merck & Co, Inc, NJ, 65-69; med dir, S E Massengill Co, 69-71; med dir, Semed Pharmaceut, 69-71; med dir, Ottawa Coun Health Dept, 71-79; health officer & med dir, Dist Health Dept No 2, 79-82; pvt pract med, Mich, 82-83; RETIRED. *Concurrent Pos:* Chief med examr, Ottawa Co, Mich; mem Mich Gov comt med manpower. *Mem:* Am Pub Health Asn; Am Col Prev Med; World Med Asn. *Res:* Clinical pharmaceutical research, especially chemotherapeutic agents; public health. *Mailing Add:* 3061 One-Eight Line Rd West Branch MI 48661

CHRISTENSON, PHILIP A, b Storm Lake, Iowa, July 27, 48; m 71; c 2. CHEMISTRY OF FLAVORS & FRAGRANCES. *Educ:* Loras Col, BS, 70; Boston Col, PhD(org chem), 74. *Prof Exp:* Postdoctoral org chem, Univ Calif, Berkeley, 74-76; res scientist, Fritzche, Dodge & Olcott, Inc, 76-79, group leader, Org Synthesis Lab, 79-86, sect mgr, 86-89, div res & analysis serv, 89-91; DIR, SYNTHESIS LAB, GIVAUDAN CORP, 91- *Mem:* Am Chem Soc. *Res:* Synthesis of new materials which possess desirable organoleptic properties; development of new processes or the improvement of existing processes for the manufacture of flavor and fragrance chemicals. *Mailing Add:* Fritzsche Dodge & Olcott 76 Ninth Ave New York NY 10011

CHRISTENSON, ROGER MORRIS, b Sturgeon Bay, Wis, Sept 28, 19; m 47; c 3. CHEMISTRY. *Educ:* Univ Wis, BS, 41, MS, 42, PhD(food chem), 44. *Prof Exp:* Asst, Univ Wis, 41-44; res chemist, 44-52, res supvr, Synthetic Vehicles, 52-56, mgr new prods res, 58-65, div dir resin res, 65-76, assoc dir res, 76-80, DIR RES, PPG INDUSTS, 80- *Mem:* Am Chem Soc; Fedn Socs Paint Technol. *Res:* Polymer chemistry related to coatings and adhesives; thermosetting and thermoplastic acrylic polymers, alkyd and epoxy resins, melamine and urea resins, polymer dispersions, electrocoating. *Mailing Add:* 10534 NW 36th Pl Gainesville FL 32606-7305

CHRISTIAN, CHARLES DONALD, obstetrics & gynecology; deceased, see previous edition for last biography

CHRISTIAN, CHARLES L, b Wichita, Kans, July 10, 26; m 54; c 3. IMMUNOLOGY. *Educ:* Univ Wichita, BS, 49; Western Reserve Univ, MD, 53. *Prof Exp:* From instr to prof med, Col Physicians & Surgeons, Columbia Univ, 58-70; PROF MED, COL MED, CORNELL UNIV, 70-; PHYSICIAN IN CHIEF, HOSP SPEC SURG, 70- *Concurrent Pos:* Consult, USPHS, 63- *Mem:* AAAS; Soc Exp Biol & Med; Am Asn Immunol; Am Rheumatism Asn (pres 77-78). *Res:* Immunochemistry; experimental pathology. *Mailing Add:* Cornell Univ Med Col 535 E 70th St New York NY 10021

CHRISTIAN, CURTIS GILBERT, b Norwich, NY, Nov 16, 17; m 46; c 2. ORGANIC CHEMISTRY. *Educ:* Univ Calif, BS, 49, MS, 50. *Prof Exp:* Res chemist, Ansco Div, Gen Aniline & Film Corp, 50-52 & Union Oil Co, Calif, 52-55; res assoc, Gasparcolor, Inc, 55-58, prod mgr, 58-60; mgr photo prod pilot plant, Minn Mining & Mfg Co, 60-63, mgr photo prod tech serv, 63-66, mgr tech serv, Photo Film Div, 66-70, mgr prof photog prod lab, Photog Prod

Div, 70-72, tech mgr, Dynacolor Subsidiary, 74-78, sr specialist, Photog Prod Div, 72-82; RETIRED. *Mem:* Am Chem Soc; Soc Photog Sci & Eng; Royal Photog Soc; Soc Motion Picture & TV Engrs. *Res:* Photographic chemistry; organic synthesis. *Mailing Add:* 1636 Lilac Lane St Paul MN 55118-1636

CHRISTIAN, DONALD PAUL, b Akron, Ohio, Mar 16, 49; m 70. POPULATION ECOLOGY, PHYSIOLOGICAL ECOLOGY. *Educ:* Mich State Univ, BS, 71, MS, 73, PhD(zool), 77. *Prof Exp:* ASST PROF BIOL, UNIV MINN, DULUTH, 78- *Mem:* Am Soc Mammal; Ecol Soc Am. *Res:* Mammalian population ecology; physiological ecology, especially water metabolism and bioclimatology. *Mailing Add:* 2029 E Eighth St Duluth MN 55812

CHRISTIAN, FREDERICK ADE, b Lagos, Nigeria, July 8, 37; US citizen; m 71; c 2. HELMINTH PHYSIOLOGY, MEDICAL ENTOMOLOGY. *Educ:* Allen Univ, BS, 62; Wayne State Univ, MS, 64; Ohio State Univ, PhD(parasitol), 69. *Prof Exp:* Teaching assoc biol, Wayne State Univ, 62-64; teaching asst & instr zool, Ohio State Univ, 65-69; from asst prof to assoc prof, 69-74, PROF BIOL & DIR, HEALTH RES CTR, SOUTHERN UNIV, 75-, DIR RES, COL SCI, 76- *Concurrent Pos:* Prin investr, USDA, 72-77, NIH, 72- & NSF, 80-; Coun mem, La Univ Marine Consortium, 78-; extra-mural assoc, NIH, 79; chmn, Univ Res Coun, Southern Univ, 79-, Res Incentive Comt, 81- *Mem:* Am Soc Parasitologists; Am Micros Soc; Nat Minority Health Affairs Assoc; Am Inst Biol Sci. *Res:* Physiology of host-parasite relationships of Fasciola hepatica liver fluke of cattle and man; the effects of environmental pollution pesticides on parasites physiology and occurrences. *Mailing Add:* Southern Br Post Off PO Box 9921 Baton Rouge LA 70813

CHRISTIAN, GARY DALE, b Eugene, Ore, Nov 25, 37; m 61; c 2. ANALYTICAL CHEMISTRY. *Educ:* Univ Ore, BS, 59; Univ Md, MS, 62, PhD(anal chem), 64. *Prof Exp:* Res anal chemist, Walter Reed Army Inst Res, 61-67; from asst prof to assoc prof, Univ Ky, 67-72; PROF CHEM, UNIV WASH, 72- *Concurrent Pos:* Asst prof, Univ Md, 65-66; guest lectr, Walter Reed Army Inst Res, 66; consult electroanal chem, 68-; consult, Miles Labs, Inc, 68-72, Beckman Instruments, Inc, 70-84, 88, Westinghouse Hanford Co, 79-83, Technol Dynamics, 83-85, Instrumentation Lab, 85-88 & Porton Diag Inc, 90-91; Fulbright/Hays scholar, 78-79; vis prof, Univ Libre de Bruxelles, 78-79; invited prof, Univ Geneva, 79; comt examr, Grad Rec Exam Chem Test, 84-90; mem, Sci Bd, Int Chem Olympiad, 92; ed-in-chief, Talanta, 89- *Honors & Awards:* Excellence in Teaching Award, Am Chem Soc, 88. *Mem:* Am Chem Soc; Soc Appl Spectros; fel Am Inst Chemists; Soc Electroanalytical Chem. *Res:* Atomic spectroscopy; electroanalytical chemistry; clinical chemistry; flow injection analysis; chromatography detectors; process analysis. *Mailing Add:* Dept Chem Univ Wash Seattle WA 98195

CHRISTIAN, HOWARD J, b Cambridge, Mass, Sept 17, 23; m 45; c 8. PATHOLOGY. *Educ:* Tufts Univ, BS, 49, MD, 52. *Prof Exp:* Intern path, Boston City Hosp, 52-53, resident, 53-56; from instr to assoc prof path, Sch Med, Tufts Univ, 56-74, clin prof, 74-85; pathologist, Carney Hosp, Boston, 57-85; RETIRED. *Concurrent Pos:* Instr, Sch Med, Boston Univ, 54-56; asst pathologist, St Elizabeth's Hosp, Brighton, Mass, 56-57; jr vis pathologist, Lemuel Shattuck Hosp, Boston, 57-; attend pathologist, Vet Admin Hosp, 61-; hon sr lectr, Aberdeen Univ, 63-64; Commonwealth Fund fel, 63-; assoc prof, Boston Univ Sch Med, 71. *Mem:* AMA; Col Am Pathologists. *Res:* Electron microscopy; cancer and diabetes. *Mailing Add:* 37 Elm St Canton MA 02021

CHRISTIAN, JAMES A, b Kansas City, Mo, June 20, 35; m 55; c 4. BOTANY, GENETICS. *Educ:* Univ Mo-Columbia, BS, 58, MA, 62, PhD(biosysts), 71. *Prof Exp:* Teacher, St Clair Sch Dist, Mo, 58-60 & Mehlville Sch Dist, 60-61; asst bot, Univ Mo, 61-64; asst prof biol, Tarkio Col, 64-66; from asst prof to assoc prof, 66-80, PROF BOT, LA TECH UNIV, 80- *Mem:* Am Soc Plant Taxonomists; Int Asn Plant Taxonomists; Sigma Xi; Bot Soc Am; Torrey Bot Club. *Res:* Biosystematics and genetics of the genus Lupinus; taxonomy and phylogeny of vascular plants. *Mailing Add:* Dept Bot & Bact Box 3158 Tech Sta Ruston LA 71270

CHRISTIAN, JERRY DALE, b Eugene, Ore, Nov 25, 37; m 62; c 3. VAPORIZATION PROCESSES, NUCLEAR FUEL DISSOLUTION. *Educ:* Univ Ore, BS, 59; Univ Wash, PhD(phys chem), 65. *Prof Exp:* Sr res scientist & mgr fund res, Battelle Northwest Lab, 65-71; sr res scientist, Westinghouse Hanford Co, 71-72; sr postdoctoral res assoc, NASA Ames Res Ctr, 72-74; assoc proj engr & group leader, Allied Chem Corp, 74-77; scientist & prog mgr, Sci Applications, Inc, 77-78; STAFF MEM, ADV & CONSULT SCIENTIST, IDAHO CHEM PROCESSING PLANT, WESTINGHOUSE IDAHO NUCLEAR CO, 78- *Concurrent Pos:* Teaching staff, Univ Wash Joint Ctr Grad Study, 65-72; consult, Westinghouse Hanford Co, 72-73, Int Atomic Energy Agency, 87-89; prog mgr, Idaho Chem Processing Plant, Westinghouse Idaho Nuclear Co, 78-, consulting scientist, 87- *Mem:* Am Chem Soc. *Res:* Behavior and control of ruthenium during high-level waste solidification; zirconium nuclear fuel dissolution; high-level airborne, and solid radioactive waste management; radioactive off-gas treatment systems; high temperature oxidation of metals; thermodynamics of metal halide vaporization processes. *Mailing Add:* Westinghouse Idaho Nuclear Co Box 4000 MS 5218 Idaho Falls ID 83403-5218

CHRISTIAN, JOE CLARK, b Marshall, Okla, Sept 12, 34; m 60; c 2. MEDICAL GENETICS. *Educ:* Okla State Univ, BS, 56; Univ Ky, MS, 59, PhD(genetics), 60, MD, 64. *Prof Exp:* From intern to resident internal med, Vanderbilt Univ Hosp, 64-66; from asst prof to assoc prof, 66-74, PROF MED GENETICS, MED SCH, IND UNIV, INDIANAPOLIS, 74-, CHMN, 78- *Mem:* Am Soc Human Genetics; Am Fedn Clin Res; Am Oil Chem Soc. *Res:* Quantitative genetics of cardiovascular diseases; clinical genetics; genetics of aging. *Mailing Add:* Dept Med Genetics Ind Univ Med Sch Indianapolis IN 46223

CHRISTIAN, JOHN B, b Marietta, Ga. MATERIALS SCIENCE ENGINEERING. *Educ:* Univ Louisville, BS, 50. *Prof Exp:* Chemist, E I DuPont de Nemours Co Inc, 52-54, Naval Ordnance Lab, 54-55, Mat Engr, Mat Lab, Wright-Patterson AFB, Ohio, 55-84, vpres, Aerospace Lubricants, Inc, Columbus, Ohio, 84-89; PRES, J B CHRISTIAN, INC, JACKSON, OHIO, 89-; SR VPRES, LUBRICATION TECHNOL INC, JACKSON OHIO, 89- *Concurrent Pos:* Consult, private practice, 84- *Mem:* NY Acad Sci; Nat Lubricating Grease Inst; Am Chem Soc; Am Soc Testing & Mat; Soc Tribology & Lubrication Eng. *Res:* Development of synthetic lubricating greases and grease-like lubricating materials; grease lubricating techniques. *Mailing Add:* 666 Omar Circle Yellow Springs OH 45387

CHRISTIAN, JOHN JERMYN, b Scranton, Pa, Apr 12, 17; m 42, 58; c 2. ENDOCRINOLOGY, PATHOBIOLOGY. *Educ:* Princeton Univ, AB, 39; Johns Hopkins Univ, ScD, 54. *Prof Exp:* Asst res pharmacologist, Wyeth Inst Appl Biochem, 48-51; head animal labs, US Naval Med Res Inst, 51-56, physiologist exp med, 56-59; assoc prof comp path, Univ Pa, 59-62; mem div endocrinol, Albert Einstein Med Ctr, 62-69, prof biol sci, 70-87, EMER PROF, STATE UNIV NY BINGHAMTON, 87- *Concurrent Pos:* Res assoc pathobiol, Johns Hopkins Univ, 54-59; assoc dir, Penrose Res Lab, Philadelphia Zool Soc, 59-62; mem ad hoc comt comp path, Nat Res Coun, Nat Acad Sci, 63-70; mem, Endocrine SS, NIH, 58-62. *Honors & Awards:* Mercer Award, Ecol Soc Am, 57. *Mem:* AAAS; Am Asn Path; Am Ornithologists Union; NY Acad Sci; Wildlife Dis Asn (vpres, 63-64, pres, 65-67); Wildlife Soc. *Res:* Relationship of population density and social factors to endocrine adaptive mechanisms; reproduction; adrenal cortex. *Mailing Add:* Dept Biol Sci State Univ NY Binghamton NY 13901-6000

CHRISTIAN, JOSEPH RALPH, b Chicago, Ill, June 15, 20; m 44; c 2. PEDIATRICS. *Educ:* Loyola Univ, Ill, MD, 44; Am Bd Pediat, dipl, 50. *Prof Exp:* Clin asst pediat, Stritch Sch Med, Loyola Univ, Ill, 48-50, clin instr, 50-51, asst clin prof, 51-52, from asst prof to prof, 53-61, asst chmn dept & dir res, 53-61; prof, Col Med, Univ Ill, 61-71; PROF PEDIAT & CHMN DEPT, RUSH MED COL, 71-; CHMN DEPT PEDIAT, PRESBY-ST LUKE'S HOSP, 61- *Concurrent Pos:* Dir med educ & sr pediatrician, Mercy Hosp, 48-61, dir pediat cardiac clin & cardiac in patient serv, 48-61, dir pediat out patient clin, 49-61, dir pediat residency training prog, 54-61; attend pediatrician, Loyola Serv, La Rabida Sanitarium, 48-61; chief pediat, Lewis Mem Hosp, 51-61; sr attend pediatrician, Cook County Hosp, 59-65. *Mem:* Fel Am Acad Pediat; fel Am Col Chest Physicians; fel Am Col Physicians; Am Fedn Clin Res; fel Am Pub Health Asn. *Res:* Infant nutrition; fluid and electrolyte balance; accidental poisoning in children; pediatric cardiology. *Mailing Add:* Three Oak Brook Club Dr E107 Oak Brook IL 60521

CHRISTIAN, LARRY OMAR, b Kalamazoo, Mich, Aug 20, 36; c 3. PHYSICS, SOLAR ENERGY. *Educ:* Albion Col, BA, 58; Univ Ariz, MS, 64, PhD(physics), 66. *Prof Exp:* Res physicist atmospheric physics, White Sands Missile Range, 65-67; asst prof physics, Willamette Univ, 67-68; RESEARCHER SOLAR ENERGY, ENERGY CONVERSION DEVICES INC, 76- *Res:* Materials research with emphasis toward solar devices. *Mailing Add:* 19 Old Mill Dr No 28 Holland MI 49423

CHRISTIAN, LAUREN L, b LaPorte City, Iowa, June 7, 36; m 59; c 2. SWINE BREEDING. *Educ:* Iowa State Col, BS, 58; Univ Wis, MS, 60, PhD(animal breeding), 63. *Prof Exp:* Res asst animal breeding, Univ Wis, 58-63; asst prof animal sci, Univ Tenn, 63-65; from asst prof to assoc prof, 65-74, PROF ANIMAL SCI, IOWA STATE UNIV, 74- *Concurrent Pos:* Consult genetics, Kleen Leen, Inc, 72-85, US Feed Grains Coun, Japan, 73-81, Taiwan, 75-80, Am Soybean Asn, Japan, 83, Poland, 87 & Monterey Farms, Philippines, 87-88; prin investr, USDA-Coop State Res Serv res grant, 76-80; mem, Coun Agr Sci & Technol. *Mem:* Am Soc Animal Sci. *Res:* Swine genetics; porcine stress syndrome; heterosis; crossbreeding systems; selection for improved feed conversion and leg soundness; prediction of body composition. *Mailing Add:* 101 Kildee Hall 225 Kildee Hall Ames IA 50011

CHRISTIAN, PAUL JACKSON, b Barre, Vt, Sept 9, 20; m 46; c 4. SYSTEMATIC ENTOMOLOGY. *Educ:* Wheaton Col, AB, 47; Univ Kans, PhD(entom), 52. *Prof Exp:* Asst instr biol, Univ Kans, 48-51; from asst prof to assoc prof, Univ Louisville, 52-61; assoc prof, Bethel, Col, Minn, 63-84, chmn dept, 63-77, prof biol, 63-84; RETIRED. *Concurrent Pos:* Consult, Louisville & Jefferson County Dept Pub Health, 60-61. *Mem:* Soc Study Evolution; Soc Syst Zool; Sigma Xi. *Res:* Classification of leaf hoppers. *Mailing Add:* 1559 Asbury St St Paul MN 55108

CHRISTIAN, ROBERT ROLAND, b Meriden, Conn, Nov 10, 24. MATHEMATICS. *Educ:* Yale Univ, BS, 47, MA, 49, PhD(math), 54. *Prof Exp:* Instr math, New Haven Col, 46-49; asst instr, Yale Univ, 48-49; instr, Clark Univ, 49-51; from instr to assoc prof math, Univ BC, 54-89; RETIRED. *Concurrent Pos:* Pvt sch teacher, 47; vis lectr, Univ Ill, 59-60; consult math, Int Develop Asn Educ Proj, Inst Educ, Univ Sierra Leone, 71-72. *Mem:* AAAS; Math Asn Am. *Res:* Mathematics education. *Mailing Add:* 3921 W 13th Ave Vancouver BC V6R 2T1 Can

CHRISTIAN, ROBERT VERNON, JR, b Wichita, Kans, Mar 1, 19; m 44; c 3. CHEMISTRY. *Educ:* Munic Univ Wichita, BS, 40; Iowa State Col, PhD(chem), 46. *Prof Exp:* Asst, Nat Defense Res Comt Proj, Iowa State Col, 42-43; from asst prof to assoc prof, 46-60, PROF CHEM, WICHITA STATE UNIV, 60- *Mem:* AAAS; Am Chem Soc. *Res:* Mass spectroscopy; volatile metal chelates; metal dithiocarbamates; trace metal analysis; chemical instrumentation. *Mailing Add:* 2272 N Fountain Wichita KS 67220

CHRISTIAN, ROSS EDGAR, b DuBois, Pa, Nov 1, 25; m 46; c 2. PHYSIOLOGY, GENETICS. *Educ:* Pa State Univ, BS, 47; Univ Wis, MS, 49, PhD(genetics), 51. *Prof Exp:* Asst genetics, Univ Wis, 47-50, instr & agt, Bur Dairy Indust, USDA, 50-51; asst prof animal husb, Wash State Univ, 51-56; from asst prof to assoc prof, 56-67, PROF ANIMAL SCI, UNIV IDAHO, 67- *Mem:* AAAS; Am Soc Animal Sci; Am Dairy Sci Asn; Am Genetic Asn; Sigma Xi. *Res:* Sterility in farm animals; genetics of fertility. *Mailing Add:* Dept Animal Husb Univ Idaho Moscow ID 83844

CHRISTIAN, SAMUEL TERRY, b Huntington, WVa, Dec 4, 37; m 58; c 3. BIOPHYSICS, ORGANIC CHEMISTRY. *Educ:* Marshall Univ, BA, 60; Univ Tenn, PhD(biochem), 66. *Prof Exp:* Res biochemist, Addiction Res Ctr, NIMH, Ky, 68-69, chief biochem pharmacol sect, 69-72; chief Neurochem Sect, Neurosci Prog & assoc prof psychiat & biochem, 72-76, PROF NEUROBIOL PSYCHIAT, ASSOC PROF PHYSIOL & BIOPHYS, MED CTR, UNIV ALA BIRMINGHAM, 76-, SR SCIENTIST COMPREHENSIVE CANCER CTR, 76-, ASSOC PROF PHARMACOL, 81- *Concurrent Pos:* Fel, Nat Defense Emergency Authorization, 61-66; fel med chem, Univ Tenn, Memphis, 66-67; fel pharmacol, Univ Ky, 67-68; consult, N Miss Res Found, 66-72 & Pierce Chem Co, 75-82; adj asst prof, Depts Community Med & Pharmaceut Chem, Univ Ky, 69-72; scientist, Cystic Fibrosis Ctr, Univ Ala Birmingham, 81- *Mem:* Sigma Xi; NY Acad Sci; Am Soc Biol Chem; Am Soc Neurochem; Soc Neuroscience. *Res:* Neurochemistry and molecular pharmacology; investigation of events produced by psychoactive agents on central nervous system macromolecular or subcellular organelles and their relevance to behavioral or physiological parameters; basic neurochemistry of the brain and its relevance to brain function; central nervous system's metabolism of psychoactive agents. *Mailing Add:* Neuropsychiat Res Prog Univ Ala PO Box 190 Birmingham AL 35294

CHRISTIAN, SHERRIL DUANE, b Estherville, Iowa, Sept 28, 31; m 56; c 3. PHYSICAL CHEMISTRY. *Educ:* Iowa State Univ, BS, 52, PhD, 56. *Prof Exp:* From asst prof to prof, 56-69, asst dean, Col Arts & Sci, 63-66, chmn dept chem, 68-69, GEORGE LYNN CROSS RES PROF CHEM, UNIV OKLA, 69- *Concurrent Pos:* Okla Found res award, 56-57; guest prof chem, Univ Oslo, 66-67 & 74-75 & Univ Trondheim, 79-80; res grants, NSF, Off Saline Water, Dept Interior, PRF Res Corp & Dept Energy; pres, CET Res Group, Norman, Okla. *Mem:* Am Chem Soc; Sigma Xi. *Res:* Physical chemistry of molecular complexes; spectral and thermodynamic properties of hydrogen-bonded and charge-transfer complexes; effect of solvents on complex equilibria; effects of pressure on conformational equilibria; surface and colloid chemistry. *Mailing Add:* Dept Chem Univ Okla 620 Parrington Oval Rm 208 Norman OK 73019

CHRISTIAN, WAYNE GILLESPIE, b King City, Mo, Oct 28, 18; m 43; c 2. GEOPHYSICS. *Educ:* WTex State Univ, BS, 39; Univ Denver, MS, 48, EdD(sci ed), 51. *Prof Exp:* Mus technician, WTex State Teachers Col, 36-37, hist geol lab supvr, 36-39, field & lab supvr paleont & archeol, Mus, 39-40; chief computer, Western Geophys Los Angeles, 40-44; supt schs, Mo, 43-44 & 46-48; teacher pub schs, 44-46; dept dir, Colo State Home Children, 51-52; geophysicist, Sun Oil Co, 52-77; coordr environ affairs, Sunoco Energy Develop Co, 77-78, mgr environ affairs, 78-82, sr environ specialist, 82-84; RETIRED. *Concurrent Pos:* Supvr, Ground Water Surv & Mineral Surv, US Dept Interior & State Tex, 37-39; instr & teaching fel, Univ Denver, 48-56; dir, Chaffield Arboretum Denver Bot Gardens, 87- *Mem:* Soc Vert Paleont. *Mailing Add:* 1360 Monaco St Pkwy Denver CO 80220

CHRISTIAN, WOLFGANG C, b Salzburg, Austria, May 7, 49; US citizen; m 78. LASER SPECTROSCOPY, COMPUTER INTERFACING. *Educ:* NC State Univ, Raleigh, BS, 70, PhD(physics), 75. *Prof Exp:* Asst prof physics & math, Merryhurst Col, 75-78; asst prof physics, Allegheny Col, 78-79; asst prof, Earlham Col, 79-82; ASSOC PROF PHYSICS, DAVIDSON COL, 82- *Concurrent Pos:* Fulbright fel, 76; consult, Oak Ridge Nat Lab, 80- *Mem:* Am Phys Soc; Am Asn Physics Teachers. *Res:* Laser spectroscopy; multiphoton ionization and nonlinear processes in dense alkali vapors using high power dye lasers; use of carbon dioxide lasers for molecular relaxation studies of small gaseous molecules; laboratory control. *Mailing Add:* Physics Dept Davidson Col Davidson NC 28036

CHRISTIANO, JOHN G, b Falerna, Italy, Aug 29, 17; nat US; m 43; c 3. APPLIED MATHEMATICS. *Educ:* Univ Pittsburgh, BS, 39, MS, 42, PhD(math), 50. *Prof Exp:* From instr to assoc prof math, Univ Pittsburgh, 42-59; asst prof & actg head dept, Duquesne Univ, 46-47; prof math, Northern Ill Univ, 59-82; RETIRED. *Concurrent Pos:* Pres, NIU chapter Soc Sigma Xi, 67-68, Ill Sect Am Math Asn, 76-77. *Mem:* Fedn Am Scientists; Math Asn Am; Am Math Soc (pres, Ill Sect, 76-77). *Res:* Mathematics; mechanics. *Mailing Add:* 18186 Lake Bend Dr Jupiter FL 33458

CHRISTIANO, PAUL P, b Pittsburgh, Pa, May 12, 42; m 67; c 1. STRUCTURAL DYNAMICS, GEOTECHNICAL ENGINEERING. *Educ:* Carnegie Inst Technol, BS, 64, MS, 65; Carnegie-Mellon Univ, PhD(civil eng), 68. *Prof Exp:* From asst prof to assoc prof civil eng, Univ Minn, 67-74; assoc prof, 74-81, PROF CIVIL ENG, CARNEGIE-MELLON UNIV, 81- *Concurrent Pos:* Assoc dean eng, Carnegie-Mellon Univ, 82-85, head, Dept Civil Eng, 85-88, dean eng, 88-91, provost, 91- *Mem:* Am Soc Civil Engrs; Am Soc Eng Educ; Earthquake Eng Res Inst. *Res:* Mechanics of solids, including problems in vibrations and stability, with applications to structures and foundations, particularly as regards their response to earthquake excitation. *Mailing Add:* Off Provost Warner Hall Carnegie-Mellon Univ Pittsburgh PA 15213

CHRISTIANS, CHARLES J, b Parkersburg, Iowa, Apr 15, 34; m 57; c 4. ANIMAL BREEDING. *Educ:* Iowa State Univ, BS, 55; NDak State Univ, MS, 58; Okla State Univ, PhD(animal breeding), 61. *Prof Exp:* Asst animal husb, NDak State Univ, 56-58 & Okla State Univ, 58-61; from asst prof to assoc prof, Miss State Univ, 61-64; from asst prof to assoc prof, 64-71, PROF ANIMAL HUSB, UNIV MINN, ST PAUL, 71- *Concurrent Pos:* Nat ed & secy treas, Nat Swine Improv Fedn; lectr & consult, World Wide Col Auctioneering; int consult, Zenton, Nat Japanese Pork Producers , Tokyo & Minn Dept Agr; mem nat ed comt & auth, Pork Indust Handbook, Purdue Univ. *Mem:* Am Soc Animal Sci; Sigma Xi; Am Inst Biol Sci. *Res:* Beef breeding; swine breeding; factors affecting various beef carcass traits. *Mailing Add:* Animal Sci & Agr Exten Dept Agr Univ Minn St Paul MN 55108

CHRISTIANSEN, ALFRED W, b Chicago, Ill, Nov 12, 40. ADHESION. *Educ:* Univ Ill, BS, 63; Case Western Reserve Univ, MS, 65, PhD(macromolecular sci), 70. *Prof Exp:* Res fel, Dept Metall, Univ Liverpool, 70-73; res chemist, Plastics Div, Exxon Chem Co, 73-75; CHEM ENGR, FOREST PROD LAB, USDA, 75- *Mem:* Adhesion Soc; Am Chem Soc; Forest Prod Res Soc. *Res:* Acid catalysis of phenolic resins; conversion of carbohydrates to exterior-durable wood adhesives; characterization and testing of wood adhesives. *Mailing Add:* Forest Prod Lab One Gifford Pinchot Dr Madison WI 53705-2398

CHRISTIANSEN, DAVID ERNEST, b Salt Lake City, Utah, Jan 19, 37; m 62; c 5. CHEMICAL ENGINEERING. *Educ:* Univ Utah, BS, 62; Princeton Univ, MA, 64, PhD(chem eng), 67. *Prof Exp:* Vis res fel, Sloan Found, 67-68; staff mem, Los Alamos Sci Lab, 68-74; sr res engr, Stauffer Chem Co, 75-78; CHEM ENGR, LAWRENCE LIVERMORE LAB, 78- *Mem:* Am Chem Soc; Am Inst Chem Engrs; Sigma Xi. *Res:* Turbulent flow and mixing; photon transport; very high pressure phenomena including explosive initiation and detonation; fluid flow; crystallization; process simulation; fluidized bed processing. *Mailing Add:* 455 Navajo Rd Los Alamos NM 87544

CHRISTIANSEN, DONALD DAVID, b New Jersey, June 23, 27; m 50; c 2. SOLID STATE, ELECTRON DEVICES. *Educ:* Cornell Univ, BEE, 50. *Prof Exp:* Engr, CBS Electronics Div, 50-62; ed, Electronic Design, 62-63; sr ed, Circuit Design Eng, 63-66; sr assoc ed, Electronics, 66, sr ed, 66-67, assoc managing ed, 67-68, ed-in-chief, 68-70, mgr planning & develop, McGraw-Hill Publ, 70-71; ED & PUBL, INST ELEC & ELECTRONICS ENGRS SPECTRUM, 71- *Concurrent Pos:* Chmn, Nat Conf Electronics Med, 71. *Honors & Awards:* Citation, Coun Eng & Sci Soc Exec, 77; Medal, Flanders Acad Arts, Sci & Lit, 80; Centennial Medal, Inst Elec & Electronics Engrs, 84, Gruenwald Award, 90. *Mem:* Fel Inst Elec & Electronics Engrs; NY Acad Sci; fel World Acad Art & Sci; Royal Inst; Franklin Inst; Soc History Technol. *Res:* History of technology; the innovative process; management of engineering and science. *Mailing Add:* Inst Elec & Electronics Engrs Spectrum 345 E 47th St New York NY 10017

CHRISTIANSEN, E A, b Shellbrook, Sask, Sept 20, 28; m 59. GEOLOGY. *Educ:* Univ Sask, BSA, 52, MSc, 56; Univ Ill, PhD(geol), 59. *Prof Exp:* Asst res officer, Sask Res Coun, Univ Sask, 59-63, assoc res officer geol, 63-75, adj prof geol sci, 74, prin res scientist, 75-77; CONSULT GEOLOGIST, 77- *Honors & Awards:* Thomas Roy Award, Eng Geol. *Mem:* Fel Geol Soc Am; fel Geol Asn Can; Am Asn Petrol Geol. *Res:* Glacial and groundwater geology; occurrence of groundwater in drift, Tertiary and upper Cretaceous sediments; engineering geology. *Mailing Add:* E A Christiansen Consult Ltd Box 3087 Saskatoon SK S7K 3S9 Can

CHRISTIANSEN, E(RNEST) B(ERT), chemical engineering; deceased, see previous edition for last biography

CHRISTIANSEN, ERIC H, b Jerome, Idaho, Apr 6, 53; m 79; c 4. GEOCHEMISTRY. *Educ:* Brigham Young Univ, BS, 77; Brown Univ, ScM, 78; Ariz State Univ, PhD(geol), 81. *Prof Exp:* Res assoc, US Geol Surv, 81-82; asst prof econ geol, Univ Iowa, 82-86; ASSOC PROF GEOCHEM, BRIGHAM YOUNG UNIV, 86- *Mem:* Am Geophys Union. *Res:* Origin and evolution of igneous rocks and related ore deposits; fluoine-rich rhyolites; large volume dacitic ignimbrites; geology of terrestrial planets. *Mailing Add:* Dept Geol Brigham Young Univ Provo UT 84602

CHRISTIANSEN, FRANCIS WYMAN, b Richfield, Utah, Feb 19, 12; m 33; c 6. GEOLOGY. *Educ:* Univ Utah, BS, 35, MS, 37; Princeton Univ, PhD(struct ecol geol), 48. *Prof Exp:* Mining geologist, Sierra Consol Mines, Inc, Nev, 37-38; instr geol, Univ Utah, 39-40; indust specialist, War Prod Bd, Washington, DC, 42, asst dep dir mining div, 42-43, chief metals sect, 43-46; from asst prof to prof, 46-76, PROF GEOL & GEOPHYS, UNIV UTAH, 76- *Mem:* AAAS; Geol Soc Am; Am Geophys Union; Am Inst Mining Metall & Petrol Engrs; Am Asn Petrol Geologists. *Res:* Structural geology and ore deposit; polygonal yielding of tabular bodies; magasutures of the earth and continental genesis. *Mailing Add:* 2320 Neffs Lane Holloday Salt Lake City UT 84109

CHRISTIANSEN, J(ERALD) E(MMETT), b Hyrum, Utah, Apr 9, 05; m 29; c 2. IRRIGATION, CIVIL ENGINEERING. *Educ:* Utah Agr Col, BS, 27; Univ Calif, MS, 28, CE, 35. *Hon Degrees:* DSc, Utah State Univ, 76. *Prof Exp:* Jr irrig engr, exp sta, Univ Calif, 28-36, asst irrig engr, 36-42; irrig & drainage engr, Regional Salinity Lab, Bur Plant Indus, USDA, 42-46; dean sch eng & tech, 46-57, prof, 57-70, EMER PROF CIVIL ENG, UTAH STATE UNIV, 70- *Concurrent Pos:* Consult, Resettlement Admin, 36, Rocky Ford Irrig Co, 52-54, Food & Agr Orgn, UN, Uruguay, 57, Italy, Spain, Greece, Turkey, Syria & Iraq, 58, Peru, 68, lectr, Arg, 61; consult, US Agency Int Develop, Spain, 60-61, NC State Agr Mission, Peru, 64, Hydrotechnic Corp, NY & Morocco, 64, Spain, 64 & 65, Utah State Univ-Interam Ctr Integral Develop Waters & Lands, Orgn Am States, Venezuela, 65-70, Arg, 67 & 68, Uruguay, 69, Colombia, 69 & 70 & AID & World Bank, Washington, DC, 75; vis prof, Univ Calif, Davis, 57-58; consult eng sci, Peru, 75 & Chas Main Inc, Boston, 75-76. *Honors & Awards:* Outstanding serv award, Irrig Sprinkler Asn, 68; Royce J Tipton Award, Am Soc Civil Engrs, 76. *Mem:* Am Soc Civil Engrs; fel Am Soc Agr Engrs; Nat Soc Prof Engrs; Am Soc Eng Educ. *Res:* Irrigation and drainage; water measurement; irrigation structures; irrigation by sprinkling; drainage investigations; irrigation water requirements; administration. *Mailing Add:* 544 E 500 N Logan UT 84321

CHRISTIANSEN, JAMES BRACKNEY, b Alden, Minn, Mar 14, 11; m 37; c 2. BIOLOGICAL CHEMISTRY. *Educ:* Carroll Col, Wis, BA, 32; Univ Wis, MA, 34, PhD(biol chem), 39. *Hon Degrees:* DSc, Buena Vista Col, 76. *Prof Exp:* Res assoc, Larrowe Div, Gen Mills Inc, Mich, 39-50; prof & head dept chmn, 54-76, EMER PROF CHEM, BUENA VISTA COL, 76- *Concurrent Pos:* Pvt res, 50- *Mem:* Fel AAAS; Am Chem Soc; World Poultry Sci Asn; Poultry Sci Asn; Am Dairy Sci Asn. *Res:* Chemical measurements of vitamins and other nutrient factors; nutritional requirements of poultry and dogs; livestock feed. *Mailing Add:* Dept Chem Buena Vista Col Storm Lake IA 50588-1798

CHRISTIANSEN, JERALD N, b Arimo, Idaho, July 21, 31; m 54; c 4. SIGNAL PROCESSING SYSTEMS, ENGINEERING MECHANICS. *Educ:* Utah State Univ, BS, 53, MS, 55; Stanford Univ, PhD(eng mech), 58. *Prof Exp:* Instr eng mech, Cornell Univ, 54-56; from instr to asst prof, US Air Force Acad, 58-60; res specialist, 60-61, staff scientist, 61-63, asst resident dir, 63-64, sr staff scientist, 64-65, resident mgr, 65-66, mgr data systs, 66-67, consult scientist, 67-70, sr consult scientist & prog mgr, 70-83, TECH CONSULT, SPACE SYST DIV, LOCKHEED MISSILES & SPACE CO, 84- *Concurrent Pos:* Partner, J E Christiansen & Sons, 52-75; NSF fel, 53-54 & 56-58. *Mem:* Am Soc Civil Engrs; Am Inst Aeronaut & Astronaut. *Res:* Structural dynamics; difference equations; data processing; computer utilization for engineering problems; information systems; signal processing systems. *Mailing Add:* 2068 Cynthia Way Los Altos CA 94022

CHRISTIANSEN, JOHN V, b Sept 28, 27. CIVIL ENGINEERING. *Educ:* Univ Ill, BS, 49; Northwestern Univ, MS, 50. *Prof Exp:* Consult, design eng, 49-52; proj engr, Worthington, 52-61; prin & pres, Skilling, Helle, Christiansen, Robertson, 62-82, consult, 83-86; affil prof, Univ Wash, 83-86; CONSULT STRUCT ENG, 87- *Mem:* Nat Acad Eng; fel Am Soc Civil Engrs; fel Am Concrete Inst; Int Asn Shell & Space Struct. *Res:* Author of 18 articles. *Mailing Add:* 7799 Hansen Rd NE Bainbridge Island WA 98110-1614

CHRISTIANSEN, KENNETH ALLEN, b Chicago, Ill, June 24, 24; m 47; c 4. EVOLUTIONARY BIOLOGY, SPELEOLOGY. *Educ:* Boston Univ, BA, 48; Harvard Univ, PhD(biol), 51. *Prof Exp:* Asst prof biol, Am Univ Beirut, 51-54; instr, Smith Col, 54-55; from asst prof to assoc prof, 55-62, PROF BIOL, GRINNELL COL, 62- *Concurrent Pos:* Corresp, Mus Paris; Iowa Gov's Sci Adv Coun, 73-83, 89-91; vis researcher, Laboratoire Souterraine France, 63, 67-68 & Univ Kyoto, Japan, 74; vis prof biol, Nanjing Univ, 90. *Mem:* Fel AAAS; Soc Study Evolution; Soc Syst Zool; fel Nat Speleol Soc; fel Explorers Club; Sigma Xi. *Res:* Taxonomy and evolution; Collembola. *Mailing Add:* Dept Biol Grinnell Col Grinnell IA 50112

CHRISTIANSEN, MARJORIE MINER, b Canton, Ill, Feb 28, 22; m 51; c 1. NUTRITION, BIOCHEMISTRY. *Educ:* Univ NMex, BS, 49, MA, 55; Utah State Univ, PhD(nutrit, biochem), 67. *Prof Exp:* Chemist, Carnegie-Ill Steel Co, Ind, 42-44; control chemist, Blockson Chem Co, Ill, 44-47; asst dietitian, St Joseph Hosp, Albuquerque, NMex, 48-50; instr sci, Regina Sch Nursing, 50-64, instr nutrit, 52-64, proj dir utilization of basic sci prin in solving nursing care probs, 66-69; prof, 69-84, EMER PROF HOME ECON, JAMES MADISON UNIV, VA, 84- *Concurrent Pos:* USPHS div nursing training grant, 66-68; proj dir dietary sem, Va Regional Med Prog proj grant, 73-76. *Mem:* Am Dietetic Asn. *Res:* Serum alpha-tocopherol, cholesterol and lipids in adults on self-selected diets at different levels of polyunsaturated fat. *Mailing Add:* 94 Laurel St Harrisonburg VA 22801

CHRISTIANSEN, MERYL NAEVE, b Gooselake, Iowa, Sept 5, 25; m 50, 83. PLANT PHYSIOLOGY. *Educ:* Univ Ark, BS, 50, MS, 55; NC State Univ, PhD(crop sci), 60. *Prof Exp:* Asst, Univ Ark, 51-54; agronomist, Crops Res Div, USDA, 55-58; asst, NC State Univ, 58-60; plant physiologist, Crops Res Div, USDA, 60-75, plant physiologist & chief plant stress lab, Sci & Educ Admin, 73-80, PLANT PHYSIOLOGIST & DIR PLANT PHYSIOL INST, AGR RES SERV, USDA, 80- *Mem:* Am Soc Plant Physiol; Crop Sci Soc Am; Phytochem Soc NAm; NY Acad Sci. *Res:* Seed germination physiology; environmental influences on seedling development and metabolism. *Mailing Add:* Plant Physiol Inst USDA Beltsville Agr Res Ctr Beltsville MD 20705

CHRISTIANSEN, PAUL ARTHUR, b Mitchell Co, Iowa, June 7, 32; m 55; c 2. BOTANY, PLANT ECOLOGY. *Educ:* Univ Iowa, BA, 59; Univ Ore, MS, 64; Iowa State Univ, PhD(plant ecol), 67. *Prof Exp:* Teacher, Humboldt Community Schs, Iowa, 59-64; from asst prof to assoc prof, 74-81, PROF BIOL, CORNELL COL, 81- *Concurrent Pos:* Vis prof, Ore State Univ, 69-70, Iowa Lakeside Lab, 84-; vis scientist, Iowa State Univ, 90. *Mem:* AAAS; Sigma Xi. *Res:* Establishment of prairie species; management of natural areas. *Mailing Add:* Dept Biol Cornell Col Mt Vernon IA 52314

CHRISTIANSEN, RICHARD LOUIS, b Denison, Iowa, Apr 1, 35; m 56; c 3. ORTHODONTICS, PHYSIOLOGY. *Educ:* Univ Iowa, DDS, 59; Ind Univ, Indianapolis, MSD, 64; Univ Minn, Minneapolis, PhD(physiol), 70. *Prof Exp:* Intern dent, USPHS Hosp, San Francisco, 59-60; chief dent officer, USPHS Outpatient Clin, St Louis, 60-62; NIH trainee, Sch Dent, Ind Univ, Indianapolis, 62-64; staff orthodontist, Oral Med & Surg Br, Nat Inst Dent Res, 64-66; staff lectr orthod, Sch Dent & NIH trainee grad sch, Univ Minn, Minneapolis, 66-70; prin investr, Oral Med & Surg Br, 70-81, chief, Craniofacial Anomalies Prog Br, 73-81, dir extramural prog, Nat Inst Dent Res, 81-82; DEAN & PROF, SCH DENT, UNIV MICH, 82- *Concurrent Pos:* NIH res fel physiol, Univ Minn, Minneapolis, 66-70; mem & originator numerous state of the art planning comt, Nat Inst Dent Res, 69-; vis prof orthod, Sch Dent, Georgetown Univ, 70-82; vis lectr, Univ Md, 70-82; estab, Int Union Sch Oral Health, 85. *Honors & Awards:* Commendation Medal, US Dept Health Human Serv. *Mem:* Am Dent Asn; Am Asn Orthod; Int Asn Dent Res; AAAS; Int Union Physiol Sci; Sigma Xi. *Res:* Craniofacial malformations; oral physiology, especially intra-oral pressures and motor function, hemodynamics of oral-facial tissues, equilibrium of the dentition and biophysics of orthodontic tooth movement. *Mailing Add:* Sch Dent Univ Mich Ann Arbor MI 48109

CHRISTIANSEN, ROBERT GEORGE, b Sangudo, Alta, Apr 23, 24; m 48. ORGANIC CHEMISTRY. *Educ:* Univ Alta, BSc, 46, MSc, 48; Univ Wis, PhD(chem), 52. *Prof Exp:* Lectr org chem, Univ Alta, 46-48; asst, Univ Wis, 48-50; res assoc, Sterling-Winthrop Res Inst, 51-61, sr res assoc & group leader, 61-78, res fel, 78-87; RETIRED. *Mem:* Am Chem Soc. *Res:* Steroids; medicinal chemistry; modified steroidal hormones. *Mailing Add:* One Huckleberry Rd Castleton Hudson NY 12033

CHRISTIANSEN, ROBERT LORENZ, b Kingsburg, Calif, June 13, 35; m 62; c 3. VOLCANOLOGY, IGNEOUS PETROLOGY. *Educ:* Stanford Univ, BS, 56, MS, 57, PhD(geol), 61. *Prof Exp:* Geologist explor geol, Utah Construct & Mining Co, 57-58; geologist mineral, Stanford Res Inst, 60-61; GEOLOGIST, VOLCANIC GEOL & PETROL, US GEOL SURV, 61- *Concurrent Pos:* Coordr geothermal res prog, US Geol Surv, 76-79 & coordr monitoring & sci studies, eruption Mt St Helen's, 80; chief, Br Igneous & Geothermal Processes, 87-91. *Mem:* AAAS; Mineral Soc Am; Am Geophys Union; Geol Soc Am. *Res:* Igneous petrology; volcanology; geology of cordilleran region of the United States; geothermal energy. *Mailing Add:* US Geol Surv 345 Middlefield Rd MS 910 Menlo Park CA 94025

CHRISTIANSEN, ROBERT M(ILTON), b Chicago, Ill, Nov 5, 24; m 52; c 3. CHEMICAL ENGINEERING. *Educ:* Northwestern Technol Inst, BS, 47; Northwestern Univ, MS, 49; Univ Pa, PhD(chem eng), 55. *Prof Exp:* Jr engr, Universal Oil Prod Co, 48; asst, Am Petrol Inst Res Proj, Northwestern Univ, 48-49; jr engr, Shell Develop Co, 49-52; instr chem eng, Univ Pa, 52-54; proj mgr, Owens-Corning Fiberglas Corp, 55-56, mgr physics res lab, 56-59; chief process engr, Stearns-Roger Eng Corp, 59-70, mgr, Environ Sci Div, 70-86; VPRES, STEARNS-ROGER SERV INC, 85- *Concurrent Pos:* Lectr, Dept Chem Eng, Ohio State Univ, 56-57; independent tech consult, proc & environ eng. *Mem:* Air Pollution Control Asn; Am Chem Soc; Am Inst Chem Engrs. *Res:* Fluid and particle mechanics; chemical and metallurgical plant design; industrial plant design for environmental protection; air pollution control; water pollution control; solids waste handling. *Mailing Add:* 4081 S Holly St Englewood CO 80111

CHRISTIANSEN, WALTER HENRY, b McKees Rocks, Pa, Dec 14, 34; m 60; c 2. GAS DYNAMICS, LASER PHYSICS. *Educ:* Carnegie Inst Technol, BS, 56; Calif Inst Technol, MS, 57, PhD(aeronaut & physics), 61. *Prof Exp:* Sr scientist gas dynamics, Jet Propulsion Lab, 61-62 & 64-67; res assoc prof, 67-70, PROF AERONAUT & ASTRONAUT, UNIV WASH, 70- *Mem:* Assoc fel Am Inst Aeronaut & Astronaut; Am Phys Soc. *Res:* Laser and gas physics; associated with high power gas lasers; gas dynamics; rarefied gas flows. *Mailing Add:* Dept Aeronaut & Astronaut FS-10 Univ Wash Seattle WA 98195

CHRISTIANSEN, WAYNE ARTHUR, b Ft Collins, Colo; m 60; c 2. RADIO ASTRONOMY, THEORETICAL ASTROPHYSICS. *Educ:* Univ Colo, BS, 62; Univ Calif, Santa Barbara, MA, 66, PhD(physics), 68. *Prof Exp:* Scientist physics, E G & G Inc, 62-64; fel astrophys, Joint Inst Lab Astrophys, Univ Colo, 68-70; from asst prof to assoc prof, 70-81, PROF ASTRON, UNIV NC, CHAPEL HILL, 81- *Concurrent Pos:* Vis prof, Leiden Univ, 75, Nat Radio Astron Observ, 76 & 78, Univ Ariz, 78-84; Kenan res prof, 82-83. *Mem:* Am Astron Soc; Royal Astron Soc. *Res:* Origin and evolution of radio galaxies and quasars and the interaction between these objects and their surroundings; cosmic ray and radio source particle acceleration. *Mailing Add:* Dept Physics & Astron Univ NC Chapel Hill NC 27514

CHRISTIANSON, CLINTON CURTIS, b Deer Park, Wis, Sept 25, 28; m 52; c 4. ELECTRIC VEHICLES, BATTERY & FUEL CELL SYSTEMS. *Educ:* Univ Minn, BEE & BS, 51. *Prof Exp:* Engr, Aeronaut & Ord Dept, Gen Elec Co, 54-58 & Aircraft Accessory Turbine Dept, 58-63, team leader fuel cell develop, Direct Energy Conversion Oper, 63-65, mgr prototype design, 65-66, mgr design eng, 66-67, mgr appln eng, 67-68; mgr systs & elec eng, Energy Technol Lab, Gould Labs, 68-76, assoc dir energy res, 76-78; mem staff, 78-81, MGR BATTERY RES & DEVELOP, ARGONNE NAT LAB, 81- *Mem:* Inst Elec & Electronics Engrs; Soc Automotive Engrs. *Res:* Power systems; static power converters and inverters; control systems; batteries; fuel cells; thermoelectrics. *Mailing Add:* Argonne Nat Lab 9700 S Cass Ave Argonne IL 60439

CHRISTIANSON, DEANN, JR, b Grand Forks, Wash, Apr 8, 41. MATHEMATICS STATISTICS. *Educ:* Univ NDak, BA, 63; Univ Ariz, MS, 67; Univ Pacific, EdD, 83. *Prof Exp:* Instr, 67-70, dir, Math Resource Ctr, 73-88, PROF MATH, UNIV PAC, 88- *Mem:* Math Asn Am; Nat Asn Develop Educ; Math Sci Network. *Res:* Factors affecting the learning of under prepared college students in mathematics. *Mailing Add:* Dept Math Univ Pac Stockton CA 95211

CHRISTIANSON, DONALD DUANE, b Fertile, Minn, May 26, 31; c 4. FOOD PHYSICAL PROPERTIES. *Educ:* Concordia Col, BA, 55; NDak State Univ, MS, 57. *Prof Exp:* Prin res chemist biochem, Northern Regional Lab, USDA, 57-89; RETIRED. *Mem:* Am Asn Cereal Chemists. *Res:* The formation of complexes between food carbohydrates (starch, hemicelluloses, sugars) and additive hydrophilic colloids, pectic substances, lipids, and proteins as a basis for developing improved texture and stability in processed foods. *Mailing Add:* 1408W Barker Peoria IL 61606-1708

CHRISTIANSON, GEORGE, b Volga, SDak, May 7, 17; m 47; c 2. FOOD SCIENCE. *Educ:* SDak State Col, BS, 39; Univ Tenn, MS, 40; Univ Minn, MS, 51, PhD(biochem), 53. *Prof Exp:* Asst blood lipids, Univ Minn, 41-42, asst dairy chem, 50-53; res chemist cereal chem, Gen Mills, Inc, 46-48; res scientist meats & meat prod, Rath Packing Co, 53-63; res assoc, James Ford Bell Tech Ctr, Gen Mills, Inc, 63-83; consult, Int Nutrition & Genetics Corp, 84-85; CONSULT, VOLINT EXEC SERV CORP, ZAMBIA, 85- *Mem:* Inst Food Technologists. *Res:* Milk stability; meat preservation studies; cereal chemistry; freezing and storage of meats; meat processing equipment; ready to eat cereals; physical chemistry of sugars; confectionary development; cereal snack development. *Mailing Add:* 18210 N 30 Pl Plymouth MN 55447

CHRISTIANSON, LEE (EDWARD), b Dayton, Ohio, May 5, 40; m 63; c 2. MAMMALOGY, ECOLOGY. *Educ:* Univ NDak, BS, 63; Southern Ill Univ, MA, 65; Univ Ariz, PhD(zool), 67. *Prof Exp:* From asst prof to assoc prof, 67-78, PROF BIOL SCI, UNIV PAC, 78-, DEPT CHMN, 84- *Mem:* Am Soc Mammalogists; AAAS; Am Inst Biol Sci. *Res:* Mammalian systematics and ecology. *Mailing Add:* Dept Biol Sci Univ Pac Stockton CA 95211

CHRISTIANSON, MICHAEL LEE, b Chicago, Ill, Aug 5, 1950. PLANT DEVELOPMENTAL BIOLOGY, PLANT CELL & TISSUE CULTURE. *Educ:* Mich State Univ, BS, 72, MS, 73, PhD(bot-genetics), 76. *Prof Exp:* Res fel crop & soil sci, Mich State Univ, 76-79; SR RES BIOLOGIST, ZOECON CORP, 79- *Concurrent Pos:* Guest biol assoc, Brookhaven Nat Lab, 73; sabbatical, PGEC, USDA, 89. *Mem:* Genetics Soc Am; Am Genetics Asn; Bot Soc Am; Soc Develop Biol. *Res:* Developmental genetics of higher plants, both in whole plants and in vitro; mutation and mutational processes, including epigenetic processes. *Mailing Add:* Zoecon Res Inst Sandoz Crop Protection Corp 975 California Ave Palo Alto CA 94304

CHRISTIE, BERTRAM RODNEY, b Moorefield, Ont, Mar 22, 33; m 60; c 3. CROP BREEDING. *Educ:* Ont Agr Col, BSA, 55, MSA, 56; Iowa State Univ, PhD(crop breeding), 59. *Prof Exp:* From asst prof to prof crop sci, Ont Agr Col, Univ Guelph, 59-89; RES SCIENTIST, AGR CAN, 89- *Mem:* Am Soc Agron; Agr Inst Can; Can Soc Genetics. *Res:* Forage crop breeding. *Mailing Add:* Res Sta-Agr Can PO Box 1210 Charlottetown PE C1A 7M8 Can

CHRISTIE, JOHN MCDOUGALL, b Calcutta, India, Dec 4, 31; m 57; c 3. STRUCTURAL GEOLOGY, ELECTRON MICROSCOPY. *Educ:* Univ Edinburgh, BSc, 53, PhD, 56. *Prof Exp:* Instr geol, Pomona Col, 56-58; from asst prof to assoc prof, 58-68, PROF GEOL, UNIV CALIF, LOS ANGELES, 68- *Concurrent Pos:* Guggenheim fel, 64-65; hon fel, Australian Nat Univ, 64-65. *Mem:* Geol Soc Am; Am Geophys Union. *Res:* Structural geology and petrology; electron microscopy of minerals; experimental deformation of minerals and rocks. *Mailing Add:* Dept Earth & Space Sci Univ Calif 405 Hilgard Ave Los Angeles CA 90024

CHRISTIE, JOSEPH HERMAN, b Magnolia, Ark, Aug 30, 37. ANALYTICAL CHEMISTRY, GEOCHEMISTRY. *Educ:* Rensselaer Polytech Inst, BS, 59; La State Univ, MS, 62; Colo State Univ, PhD(chem), 74. *Prof Exp:* Mem tech staff chem, Rockwell Int, 62-69; asst prof, Colo State Univ, 74-75; supvry chemist, 75-82, RES CHEMIST, US GEOL SURV, 82- *Concurrent Pos:* Fac affil chem, Colo State Univ, 75-82. *Mem:* Am Chem Soc; Electrochem Soc; Soc Appl Spectros; Sigma Xi; Soc Electroanalytical Chem. *Res:* Electrochemical and spectroscopic trace analysis; computer applications in analytical chemistry and instrumentation. *Mailing Add:* Br Geochem US Geol Surv MS 973 Box 25046 Denver Fed Ctr Denver CO 80225

CHRISTIE, MICHAEL ALLEN, b Detroit, Mich, Aug 29, 52; m 83; c 1. CHEMICAL PROCESS DEVELOPMENT. *Educ:* Univ Mich, BS, 74; Mass Inst Technol, PhD(org chem), 78. *Prof Exp:* From assoc sr investr to sr investr, Smith, Kline & Frech Labs, 78-81; group leader, Smith Kline Chem, 81-84, from asst dir to assoc dir, 84-89; DIR, SMITH KLINE BEECHAM CHEM, 89- *Mem:* Am Chem Soc; AAAS. *Res:* Synthetic organic chemistry, especially as applied to practical problems; large-scale synthesis of (poly) hetero-(poly)-cyclic compounds, both aliphatic and aromatic. *Mailing Add:* 900 River Rd Conshohocken PA 19428

CHRISTIE, PETER ALLAN, b Englewood, NJ, Feb 2, 40; m 62. ORGANIC POLYMER CHEMISTRY. *Educ:* Juniata Col, BS, 62; Univ Del, PhD(org chem), 67. *Prof Exp:* Res chemist, 67-74, res scientist, 74-76, SR RES SCIENTIST, RES & DEVELOP CTR, ARMSTRONG WORLD INDUSTS, 76- *Mem:* Sigma Xi; Am Chem Soc. *Res:* Heterocyclic synthesis; organophosphorus chemistry; application of nuclear magnetic resonance spectroscopy to stereochemistry; step-growth, chain growth, and ring-opening polymerization; reactions on polymers; organic-inorganic polymer systems; photochemistry; waterborne polymers/coatings. *Mailing Add:* Res & Develop Ctr Armstrong World Indust Lancaster PA 17604

CHRISTIE, PHILLIP, b Stoke-on-Trent, UK, Oct 10, 61; m 90. COMPUTER HARDWARE MODELING, INTERCONNECTION TECHNOLOGY. *Educ:* Univ Durham, UK, BSc, 82, PhD(molecular electronics), 85. *Prof Exp:* ASST PROF ELEC ENG, UNIV DEL, 87- *Concurrent Pos:* Vis scientist, Univ Del, 86-87. *Mem:* Inst Elec & Electronics Engrs. *Res:* Interconnect-limited system design and simulation applied to neural networks, supercomputers, optical interconnections; use of statistical mechanics and fractal theory in large scale system design, especially for neural networks. *Mailing Add:* Dept Elec Eng Univ Del Newark DE 19716

CHRISTIE, ROBERT WILLIAM, b Mineola, NY, Sept 22, 23; m 48; c 5. PATHOLOGY. *Educ:* Norwich Univ, Northfield, Vt, AB, 47; State Univ NY, MD, 51. *Prof Exp:* Surgeon, Norwich Univ, 53-54; gen practice, Green Mt Clin, Northfield, Vt, 53-54; asst pathologist, Med Ctr, Brookhaven Nat Lab, 56-57; assoc pathologist, Ball Mem Hosp, Muncie, Ind, 58-61; ASST PROF PATH, DARTMOUTH MED SCH, 77- *Concurrent Pos:* Asst scientist exp path, Brookhaven Nat Lab, 56-57; pathologist & lab dir, Androscoggin Valley Hosp, 61-, B D Weeks Mem Hosp, 61- & Upper Conn Valley Hosp, 61- *Mem:* Am Soc Exp Path; AAAS; Am Col Nuclear Med; Am Soc Clin Pathologists; Col Am Pathologists. *Res:* Preventive medicine; problem oriented autopsy; clinical pathology; parasitology. *Mailing Add:* Diag Serv Box 500 Lancaster NH 03584

CHRISTIE, WARNER HOWARD, b Brooklyn, NY, Oct 29, 29; m 87; c 3. MASS SPECTROMETRY. *Educ:* Univ Miami, Fla, BS, 51, MS, 53; Univ Fla, PhD(chem), 58. *Prof Exp:* GROUP LEADER MASS SPECTROMETRY, ANALYTICAL CHEM DIV, OAK RIDGE NAT LAB, 59- *Mem:* Am Soc Mass Spectrometry; Sigma Xi. *Res:* Synthesis and reactions of organic fluorine containing materials; isotope exchange reactions of fluorocarbons; mass spectrometry; application of small computers to mass spectrometry; research development and applications of secondary ion mass spectrometry; resonance; photoionization; mass spectrometry. *Mailing Add:* 952 W Outer Dr Oak Ridge TN 37830

CHRISTLIEB, ALBERT RICHARD, b Boston, Mass, July 24, 35; m 60; c 4. DIABETES MELLITUS, HYPERTENSION. *Educ:* Williams Col, BA, 57; Tufts Med Sch, MD, 61. *Prof Exp:* Assoc dir, Hypertension Unit, Peter Bent Brigham Hosp, 68-69, assoc in med, 71-82; vchmn dept med, New Eng Deaconess Hosp, 75-76, chmn, med admin bd, 82-84; med dir, Joslin Clin Div, Joslin Diabetes Ctr, 77-82; PHYSICIAN, NEW ENG DEACONESS HOSP & JOSLIN CLIN DIV, JOSLIN DIABETES CTR, 69-; ASSOC PROF MED, HARVARD MED SCH, 80- *Concurrent Pos:* Prin investr, NIH grant, 70-76 & mem Concensus Group Diabetes & Hypertension, 85-86; lectr, Asia, Europe, Africa & SAm, 70-; mem bd gov, Joslin Clin, 72-82, trustee, Joslin Diabetes Ctr, 77-82; trustee, New Eng Deaconess Hosp, 82-84; consult med, Brigham & Womens Hosp, 82-; co-ed, Joslin Diabetes Mellitus, 85; mem, Coun Epidemiol & Statist, Am Diabetes Asn; fel, Coun High Blood Pressure Res, Am Heart Asn. *Mem:* Fel Am Col Physicians; fel Am Col Cardiol; Am Fedn Clin Res. *Res:* Mechanisms and epidemiology of hypertension in patients with diabetes mellitus, and the relationships between hypertension and micro- and macrovascular disease in these patients. *Mailing Add:* One Joslin Pl Boston MA 02215

CHRISTMAN, ARTHUR CASTNER, JR, b North Wales, Pa, May 11, 22; m 45; c 6. PHYSICS. *Educ:* Pa State Univ, BS, 44, MS, 50. *Prof Exp:* Instr physics, George Washington Univ, 48-51; physicist opers res off, Johns Hopkins Univ, 51-58; sr physicist, Stanford Res Inst, 58-62, head opers res group, 62-64, mgr opers eval dept, 65-66, mgr opers res dept, 66-69, dir opers res dept, 69-71, dir tactical systs, 71-75; sci adv, Hq US Army Training & Doctrine Command, 75-87; PVT CONSULT, 88- *Concurrent Pos:* Consult, US Navy, 50-51. *Mem:* Am Phys Soc; Sigma Xi; Opers Res Soc Am; fel AAAS. *Res:* Operations research; systems analysis; weapons, information, traffic, postal and health systems; analytic modeling; simulation; field experimentation; reconnaissance; surveillance; target acquisition; interdiction; close support; air defense; countermeasures; x-rays. *Mailing Add:* 900 Calle de los Amigos W-8 Santa Barbara CA 93105

CHRISTMAN, DAVID R, b Columbus, Ohio, Oct 14, 23; m 52; c 3. ORGANIC CHEMISTRY, PHARMACEUTICAL CHEMISTRY. *Educ:* Ohio State Univ, BSc, 47; Carnegie Inst Technol, MSc, 50, DSc(chem), 51. *Prof Exp:* Asst, Carnegie Inst Technol, 47-51; assoc chemist, 51-64, CHEMIST, BROOKHAVEN NAT LAB, 64- *Concurrent Pos:* Lectr, Columbia Univ, 58-64. *Mem:* Am Chem Soc; Soc Nuclear Med. *Res:* Organic radioactivity analysis and syntheses; organic radiation chemistry; organic radiopharmaceuticals with isotopes of short half-life; data processing; use of position emission tomography. *Mailing Add:* Box 2896 Setauket NY 11733-0869

CHRISTMAN, EDWARD ARTHUR, b Lakewood, Ohio, Aug 3, 43; m 79. RADIATION PHYSICS, HEALTH PHYSICS. *Educ:* Ohio Univ, BS, 65; Rutgers Univ, MS, 74, PhD(radiation sci), 77. *Prof Exp:* Mech engr aerospace, Missile Syst Div, Avco Corp, Mass, 65-71; instr radiation sci, Busch Campus, 73-76, supv radiol physicist, 77-89, ASSOC GRAD FAC, RUTGERS UNIV, 80-, ASSOC DIR, 89- *Concurrent Pos:* Fel, Lawrence Berkeley Lab, 76-77. *Mem:* Health Physics Soc; Sigma Xi; Radiation Res Soc; Am Bd Health Physics; AAAS. *Res:* Radiation dosimetry, heavy ions, and radiological health and protection. *Mailing Add:* Dept Radiation Environ Health Safety Bldg 4127 Kilmer Piscataway NJ 08854

CHRISTMAN, JUDITH KERSHAW, b Teaneck, NJ, Apr 8, 41; m 59. BIOCHEMISTRY, MOLECULAR BIOLOGY. *Educ:* NY Univ, AB, 62; Columbia Univ, PhD(biochem), 67. *Prof Exp:* Res fel, Nucleic Acid Dept, NY Blood Ctr, 67-71; asst mem dept enzym, Inst Muscle Dis, 71-74; from asst prof to assoc prof, Mt Sinai Sch Med,74-80, res prof, dept pediat, from assoc prof to prof biochem, 77-87; MEM & DIR, MOLECULAR BIOL DIV, MICH CANCER FOUND, 87- *Concurrent Pos:* Assoc ed, Cancer Res, 82- *Mem:* Am Soc Biochem & Molecular Biol; Am Soc Cell Biol; Sigma Xi; Harvey Soc; Am Soc Cancer Res. *Res:* Role of DNA methylation in regulating gene activity and the mechanisms by which tumor promoters and oncogenic viruses enhance and/or fix expression of the transformed phenotype; expression of cloned hepatitis B virus genes in eukaryotic cells. *Mailing Add:* Dept Molecular Biol Mich Cancer Found Detroit MI 48201

CHRISTMAN, LUTHER P, b Summit Hill, Pa, Feb 26, 15; m 39; c 3. NURSING. *Educ:* Temple Univ, BS, 48, EdM, 52; Mich State Univ, PhD(anthrop & sociol), 65. *Hon Degrees:* DHL, Thomas Jefferson Univ, 80. *Prof Exp:* Dir nursing, Yankton State Hosp, SDak, 53-56; nursing consult, Mich Dept Ment Health, 56-63; assoc prof psychol nursing, Univ Mich, 63-67, res assoc, 64-67; dir nursing, Vanderbilt Univ Hosp, 67-72, dean, Sch Nursing, Vanderbilt Univ, 67-72; John L & Helen Kellogg Dean, 72-87 & vpres nursing affairs, Rush Univ, 72-87; prof sociol, Rush-Presby-St Lukes Med Ctr, 72-89; PRES, CHRISTMAN CORNESKY & ASSOC, 89- *Concurrent Pos:* Vis prof, Col NSW, 78, Nordic Sch Pub Health, Sweden, 79 & Univ Lund, Sweden, 81; lectr, Nat Ziekenhuistitut, Netherlands, 79; consult, Univ Mich, Ann Arbor, 79, & Univ Alberta, 80; mem Rev Comt, Psychol Nursing Educ, Dept Health & Human Servs, NIMH, 80-; consult, Rush Univ, 87-; Dean Emeritus, 87- *Honors & Awards:* Coun Specialists Psychiat Mental Health Nursing Award, 80; Edith Moore Copeland Founders Award Creativity, 81; Old Master, Purdue Univ, 85; Jessie Scott Award, Am Nurses Asn, 87. *Mem:* Inst Med-Nat Acad Sci; Am Nurses Asn; NY Acad Sci; AAAS; fel Soc Appl Anthorp; fel Nat Acad Nursing; Nat Acad Pract. *Res:* Nursing therapy in alcoholism: group structure; experimental model to improve patients care; the selective perceptions resulting from training for a vertical division of labor and the effect on organizational coheson. *Mailing Add:* Rte 1 Box 146 Chapel Hill TN 37034

CHRISTMAN, ROBERT ADAM, b Ann Arbor, Mich, May 16, 24; m 53; c 4. GEOLOGY. *Educ:* Univ Mich, BS, 46, MS, 47; Princeton Univ, PhD(geol), 50. *Prof Exp:* Geologist, US Geol Surv, 50-54; asst prof geol, Cornell Univ, 54-60; assoc prof, 60-87, PROF GEOL, WESTERN WASH UNIV, 87- *Mem:* AAAS; Nat Asn Geol Teachers (secy-treas, 88-); Geol Soc Am; Sigma Xi; Nat Sci Teachers Asn. *Res:* Petrology; mineralogy; earth science for teachers. *Mailing Add:* Dept of Geol Western Wash Univ Bellingham WA 98225

CHRISTMAN, RUSSELL FABRIQUE, b June 20, 36; m 58; c 3. CHEMISTRY. *Educ:* Univ Fla, BS, 58, MS, 60, PhD(chem), 62. *Prof Exp:* Res asst prof sanit chem, Univ Wash, 62-66, asst prof civil eng, 66-68, assoc prof appl sci, 68-74, asst to provost & dir div environ affairs, 70-74; PROF ENVIRON SCI, UNIV NC, CHAPEL HILL, 74-, CHMN DEPT ENVIRON SCI & ENG, 77- *Mem:* AAAS; Water Pollution Control Fedn; Am Water Works Asn; Am Soc Limnol & Oceanog. *Res:* Chemical structures of natural product organic materials in water; methods of organic analysis in water samples; mechanisms of colloidal destabilization with hydrolysis products of aluminum III. *Mailing Add:* Dept Environ Sci & Eng Rosenau Hall 201 H Univ NC Chapel Hill NC 27514

CHRISTMAN, STEVEN PHILIP, b Grass Valley, Calif, May 21, 45; m 68. ECOLOGY, HERPETOLOGY. *Educ:* Univ Fla, BS, 71, PhD(zool), 75. *Prof Exp:* Contract biologist, Fla Game & Fish Comn, 75-76; res biologist, US Fish & Wildlife Serv, 76-85; ENVIRON SPECIALIST, DEPT NATURAL RESOURCES, OFF LAND USE PLANNING & BIOL SERV, 85- *Concurrent Pos:* Sci consult, Fla Comn Rare & Endangered Plants & Animals, 73-; Fla rep, Soc Study Amphibians & Reptiles, Legis Alert Comn, 77-; vis asst cur natural sci, Fla State Mus. *Honors & Awards:* Austin Award, Fla State Mus, 74. *Mem:* AAAS; Am Soc Ichthyologists & Herpetologists; Herpetologist's League; Soc Study Amphibians & Reptiles. *Res:* Biogeography; herpetology; wildlife management; natural history. *Mailing Add:* Dept Natural Resources Mail Sta 45 3900 Commonwealth Blvd Tallahassee FL 32399

CHRISTMAS, ELLSWORTH P, b Warrick Co, Ind, Nov 5, 35; m 58; c 3. CROP PRODUCTION, OILSEED CROPS. *Educ:* Purdue Univ, BS, 58, MS, 61, PhD(agr ed, agron), 64. *Prof Exp:* Teacher sec sch, Ind, 58-60; from asst prof to assoc prof, 64-74, asst dir, Ind Coop Ext Serv, 74-89, PROF AGRON, PURDUE UNIV, 74-, EXTEN AGRONOMIST, 89- *Concurrent Pos:* Agronomist, Int Progs Agr, Purdue-Brazil Proj, Brazil, 69-73. *Mem:* Am Soc Agron; Soil Sci Soc Am; Crop Sci Soc Am. *Res:* Teaching methods in agriculture; soil characterization and conservation; crop production with emphasis on oilseed and specialty crops. *Mailing Add:* Agron Dept Lilly Hall Purdue Univ West Lafayette IN 47907

CHRISTNER, JAMES EDWARD, ATHEROSCLEROSIS, CONNECTIVE TISSUE. *Educ:* Univ Mich, PhD(biochem), 64. *Prof Exp:* Assoc prof biochem, Univ Ala, Birmingham, 78-87; Mem Staff, ENVIRON TEST SYSTS. *Mailing Add:* Environ Test Systs PO Box 4659 Elkhart IN 46514

CHRISTOFFERSEN, DONALD JOHN, b Ogema, Wis, July 27, 34; m 51; c 4. ANALYTICAL CHEMISTRY. *Educ:* Wis State Univ-Stevens Point, BS, 56; Univ Wis, PhD(analytical chem), 66. *Prof Exp:* Chemist, Pure Oil Co, 61-63, group leader gas chromatography, 63-65, sr res chemist, 64-69, SUPVR, SPECTRAL ANALYTICAL CHEM, UNION OIL CO, 69- *Mem:* Am Chem Soc; Am Soc Testing Mat. *Res:* Gas chromatography with petroleum oriented applications. *Mailing Add:* 6182 Saddletree Lane Yorba Linda CA 92686-5832

CHRISTOFFERSEN, RALPH EARL, b Elgin, Ill, Dec 4, 37; m 61; c 2. MATHEMATICS, BIOPHYSICS. *Educ:* Cornell Col, BS, 59; Ind Univ, PhD(phys chem), 64. *Hon Degrees:* LLD, Cornell Col. *Prof Exp:* NIH fel quantum chem, Univ Nottingham, 64-65 & Iowa State Univ, 65-66; from asst prof to prof phys chem, Univ Kans, 66-81, from assoc vchancellor to vchancellor, 78-81; pres, Colo State Univ, 81-83; vpres, Biotechnol & Basic Res, Upjohn Co, 83-87, vpres discovery res, 87-89; SR VPRES RES, SMITHKLINE BEECHAM, 89- *Concurrent Pos:* Alfred P Sloan res fel, 71-73; consult, Lawrence Livermore Lab, 78- & Upjohn Co, Mich; bd mem, Keystone Ctr, Keystone Symp, Nat Biotech Policy Bd & Nat Alliance Aging Res. *Honors & Awards:* Am Inst Chemists Award, 59; Outstanding Scientist of the Year Award, Inst Soc Quantum Biol, 81. *Mem:* AAAS; Am Inst Chemists; Am Chem Soc; Am Phys Soc. *Res:* Quantum chemistry; theory of chemical bonds; ab initio calculations on large molecules; development of algorithms and associated software/hardware to perform large-scale numerical calculations; geometric and electronic structural features of chlorophyll and related molecular systems and their relationship to photosynthesis; biophysics; drug discovery. *Mailing Add:* Smithkline Beecham Res 709 Swedeland Rd PO Box 1539 King of Prussia PA 19406

CHRISTOFFERSON, ERIC, b Newburyport, Mass, May 29, 39; m 61; c 2. GEOLOGICAL OCEANOGRAPHY. *Educ:* Princeton Univ, AB, 61; Univ RI, PhD(oceanog), 73. *Prof Exp:* Instr oceanog, Univ RI, 73-74, res assoc, 74-75; ASST PROF GEOL, RUTGERS UNIV, 75-; MEM STAFF, DEPT GEOL/GEOG, HOWARD UNIV. *Mem:* Sigma Xi; Am Geophys Union. *Res:* Geologic history of the Caribbean Sea Basin. *Mailing Add:* Dept Geol Geog Howard Univ Washington DC 20059

CHRISTOFFERSON, GLEN DAVIS, b Tacoma, Wash, Feb 7, 31; m 51; c 2. PHYSICAL CHEMISTRY. *Educ:* Univ Wash, BS, 53; Univ Calif, Los Angeles, PhD(chem), 58. *Prof Exp:* Asst, Univ Calif, Los Angeles, 53-57; res chemist, Calif Res Corp, 57-63, sr res chemist, 64-68, sr res assoc, Chevron Res Co, 68-91; RETIRED. *Mem:* Soc Appl Spectros; Am Chem Soc; Am Crystallog Asn. *Res:* Organic and inorganic crystal structure determination; x-ray emission and absorption spectroscopy; x-ray low angle scattering; electron diffraction and microscopy. *Mailing Add:* 601 Watchwood Rd Orinda CA 94563

CHRISTOPH, FRANCIS THEODORE, JR, b Alexandria, La, Jan 1, 43; m 68. TOPOLOGY. *Educ:* St Peter's Col, NJ, BS, 64; Rutgers Univ, MS, 66, PhD(math), 69. *Prof Exp:* Asst prof, 69-77, ASSOC PROF MATH, TEMPLE UNIV, 77- *Mem:* Math Asn Am; Am Math Soc. *Res:* Decompositions and extensions of topological semigroups; embedding topological semigroups in topological groups. *Mailing Add:* Five Highpoint Rd Perkasie PA 18944

CHRISTOPH, GREG ROBERT, b Chicago, Ill, Dec 10, 49; m 70; c 1. NEUROPHARMACOLOGY, NEUROBIOLOGY. *Educ:* Univ Chicago, AB, 71; Ind Univ, PhD(exp psychol), 76. *Prof Exp:* Assoc instr psychol, Ind Univ, 74-75; NIMH fel neurosci, Princeton Univ, 76-78; res scientist neuropharmacol, 78-81, PRIN INVESTR NEUROBIOL, EXP STA, E I DU PONT DE NEMOURS & CO, INC, 81- *Res:* Behavioral pharmacology of psychoactive drugs; neurophysiology and neuropharmacology of monoamine containing neurons in brain. *Mailing Add:* Stine Haskell Lab E I du Pont de Nemours & Co Inc Newark DE 19711

CHRISTOPHER, JOHN, b Chicago, Ill, Oct 15, 23; m 47; c 3. MATHEMATICS. *Educ:* Knox Col, AB, 46; Univ Ore, MA, 50, PhD(math), 52. *Prof Exp:* Instr math, Univ Ore, 50-52 & Knox Col, 52-54; asst prof, Univ of the Pac, 54-55; instr, Fresno State Col, 55-56; sr mathematician, Electrodata Corp, Calif, 56-58; asst prof math, Sacramento State Col, 58-60; dir comput ctr, Univ Nebr, 60-63; assoc prof, 63-67, PROF MATH, CALIF STATE UNIV, SACRAMENTO, 67- *Mem:* Am Math Soc. *Res:* Number theory and numerical analysis. *Mailing Add:* Dept Math Sacramento State Col 600 Jay St Sacramento CA 95819

CHRISTOPHER, ROBERT PAUL, b Cleveland, Ohio, Apr 27, 32; m 62; c 3. PHYSICAL MEDICINE & REHABILITATION. *Educ:* Northwestern Univ, BS, 54; St Louis Univ, MD, 59. *Prof Exp:* US Off Voc Rehab fel phys med & rehab, Univ Mich, Ann Arbor, 60-63, from instr to asst prof, 63-67; assoc prof, 67-71, PROF PHYS MED & REHAB, UNIV TENN, MEMPHIS, 71-; MED DIR, LES PASSEES REHAB CTR, 70- *Concurrent Pos:* Chief phys med & rehab, Vet Admin Hosp, Ann Arbor, 63-67; consult, St Jude Children's Res Hosp, 67-, Le Bonheur Children's Hosp, 68-, Vet Admin Hosps, Memphis, 67- & Nashville, 70-, Coun Med Educ, AMA, 69- & Comn Accreditation Rehab Facil, 73-; pres, Am Acad Cerebral Palsy & Develop Med, 86-87. *Mem:* Fel Am Acad Phys Med & Rehab; Am Cong Rehab Med; Am Asn Electromyog & Electrodiag. *Res:* Electrodiagnostic studies in collagen vascular disease; habilitation programs for brain damaged children; primary muscle disease. *Mailing Add:* Div Phys Med & Rehab Univ Tenn 800 Madison Ave Memphis TN 38163

CHRISTOPHERSON, WILLIAM MARTIN, b Salt Lake City, Utah, July 2, 16; m 43; c 1. MEDICINE. *Educ:* Univ Louisville, MD, 42. *Prof Exp:* Ewing fel, Mem Cancer Ctr, NY, 49-50; from asst prof to assoc prof, 50-56, chmn dept, 56-74, PROF PATH, SCH MED, UNIV LOUISVILLE, 56- *Concurrent Pos:* Pathologist, Louisville Gen Hosp, 56-; consult, Med Div, Nat Cancer Inst, Vet Admin Hosp & Ireland Army Hosp, Ky; mem adv comt & spec consult, Cancer Control Prog, USPHS. *Mem:* Int Acad Path (past pres); Am Soc Cytol (past pres); Soc Exp Path; Am Asn Cancer Educ (past pres); Am Cancer Soc. *Res:* Cancer. *Mailing Add:* 2211 Cherokee Pkwy Louisville KY 40204

CHRISTOPHOROU, LOUCAS GEORGIOU, b Limassol, Cyprus, Jan 21, 37; m 63; c 2. ATOMIC PHYSICS, MOLECULAR PHYSICS. *Educ:* Nat Univ Athens, BSc, 60; Manchester, dipl adv physics, 61, PhD(physics), 63, DSc(physics), 69. *Prof Exp:* Res physicist, Health Physics Div, Oak Ridge Nat Lab, 63-64; from asst prof to assoc prof, 64-68, FORD FOUND PROF PHYSICS, UNIV TENN, KNOXVILLE, 69-; HEAD ATOMIC MOLECULAR & HIGH VOLTAGE RES STAFF MEM & CORP FEL, 81- *Concurrent Pos:* Consult, Oak Ridge Nat Lab, 64-66, head atomic & molecular radiation physics group, Health Physics Div, 66-76; corresp mem, Acad Athens, 81- *Honors & Awards:* Humboldt Res Award Sr US Scientist, 90. *Mem:* Fel Am Phys Soc; Radiation Res Soc; fel AAAS. *Res:* Radiation physics; chemical physics; low-energy electron-molecule interactions; electron scattering; electron motion in gases; negative ions; gaseous and liquid dielectrics; photophysical processes; pulsed power; radiation detectors; multiphoton ionization in liquids. *Mailing Add:* Atomic Molecular & High Voltage Physics Group Health & Safety Res Div Oak Ridge Nat Lab Box 2008 Oak Ridge TN 37831

CHRISTOU, GEORGE, b Limassol, Cyprus, July 26, 53; Brit citizen; m 83; c 2. BIOINORGANIC CHEMISTRY. *Educ:* Exeter Univ, UK, BSc, 74, PhD(chem), 77. *Prof Exp:* fel, Manchester Univ, 78-79; NATO fel, Stanford Univ, 80, Harvard Univ, 80-82; lectr, Imperial Col, London, 82-83; ASST PROF CHEM, IND UNIV, 83- *Concurrent Pos:* Res fel, Alfred P Sloan Found, 87-89; teacher-scholar, Camille & Henry Dreyfus Found, 87- *Honors & Awards:* Corday Morgan Medal, Royal Soc Chem, 86. *Mem:* Am Chem Soc; Royal Soc Chem. *Res:* Synthesis of models of the manganese site of the water oxidation enzyme in photosynthesis; synthesis of early transition metal-sulfur clusters and their relevance to industrial hydrodesulfurization of crude oils. *Mailing Add:* Dept Chem Ind Univ Bloomington IN 47405

CHRISTY, ALFRED LAWRENCE, b Pittsburgh, Pa, Mar 15, 45; m 67; c 1. PHOTOSYNTHESIS, PESTICIDES & HERBICIDES. *Educ:* Univ Dayton, BS, 67, MS, 69; Ohio State Univ, PhD(bot), 72. *Prof Exp:* Fel plant physiol, Univ Ga, 72-74; sr res biologist, 74-77, res specialist, 77-78, sr res group leader, 78-81, sci fel, Monsanto Co, 81-86; nat prog leader, Weed Sci, Agr Res Serv, USDA, 86-90; DIR RES, CROP GENETICS INT, 90- *Mem:* Am Soc Plant Physiol; Am Soc Agron; Crop Sci Soc Am; Weed Sci Soc Am; Am Chem Soc; Plant Growth Regulator Soc Am. *Res:* Biocontrol of weeds and pests; natural products as herbicides; photosynthesis and translocation in crop plants in relationship to grain yield; development of screening systems for evaluation of plant growth regulators on crop plants; herbicide physiology; weed control. *Mailing Add:* 1453 Hylton Pl Crofton MD 21114-2114

CHRISTY, JAMES WALTER, b Milwaukee, Wis, Sept 15, 38; m 75; c 4. INFRARED RADIOMETRY, PROTO-STARS. *Educ:* Univ Ariz, BS, 65. *Prof Exp:* Astronr, US Naval Observ, 62-82; PHYSICIST, HUGHES AIRCRAFT, 82- *Concurrent Pos:* Astron researcher, Univ Ariz, 82- *Honors & Awards:* Credited with discovering first satellite of Pluto (Charon), Int Astron Union, 78. *Mem:* Am Astron Soc; Int Astron Union. *Res:* Spectral radiometry of infrared radiation sources; asronomical observaton of Hubble's variable nebula; conceptual basis of cosmology. *Mailing Add:* 1720 W Niona Pl Tucson AZ 85704

CHRISTY, JOHN HARLAN, b Ft Smith, Ark, Aug 13, 37; m 60; c 2. MATHEMATICS. *Educ:* Mass Inst Technol, BS, 59; Vanderbilt Univ, PhD(math), 64. *Prof Exp:* Res asst physics, Los Alamos Sci Lab, 59-60; asst prof math, Southwestern at Memphis, 63-66; assoc prof, Hendrix Col, 66-67; PROF MATH & CHMN DEPT MATH & PHYSICS, TEX WOMAN'S UNIV, 67- *Concurrent Pos:* Vis prof, Univ Ark, 64. *Mem:* Math Asn Am; Sigma Xi. *Res:* Topological dynamics; expansive transformation groups. *Mailing Add:* PO Box 22752 Denton TX 76204

CHRISTY, NICHOLAS PIERSON, b Morristown, NJ, June 18, 23; m 47; c 2. MEDICINE. *Educ:* Yale Univ, AB, 45; Columbia Univ, MD, 51; Am Bd Internal Med, 58. *Prof Exp:* from instr to assoc prof med, Columbia Univ, 56-65, assoc clin prof, 65-72, prof med, Col Physicians & Surgeons, 72-79; chmn dept med, Roosevelt Hosp, 70-79; prof med, State Univ NY, Downstate Med Ctr, 79-88, EMER PROF HEALTH SCI CTR, STATE UNIV NY, BROOKLYN, 89- *Concurrent Pos:* Fel Endocrinol, Columbia Univ, 51-54; Markle scholar, 56; asst physician, Hosp, 54-56, asst attend to physician, 60-62, assoc attend physician, 62-90; ed, J Clin Endocrin & Metab, 63-67; chief staff, Brooklyn Vet Admin Med Ctr, 79-; lectr med, Columbia Univ, 79-88, sr lectr med col, 88- *Honors & Awards:* Borden Award, 51; Harold Swanberg Award, Am Med Writers Asn, 89. *Mem:* AAAS; Am Physiol Soc; Soc Exp Biol & Med; Fedn Clin Res; Endocrine Soc (secy-tres, 78-); Am Col Clin Adminr; Am Soc Clin Investr. *Res:* Clinical disorders of the adrenal cortex; adrenal cortical physiology of animals and man; disorders of the anterior pituitary gland; metabolism of estrogens in hepatic disease; medical sciences; science communications. *Mailing Add:* Columbia Presby Med Ctr Black Bldg Rm 1-111 650 W 168th St New York NY 10032

CHRISTY, ROBERT FREDERICK, b Vancouver, BC, May 14, 16; nat US; m 73; c 3. THEORETICAL PHYSICS, THEORETICAL ASTROPHYSICS. *Educ:* Univ BC, BA, 35, MA, 37; Univ Calif, PhD(theoret physics), 41. *Prof Exp:* Instr physics, Ill Inst Technol, 41-42; res assoc, Univ Chicago, 42-43 & AEC, Los Alamos, NMex, 43-46; from asst prof to prof physics, Calif Inst Technol, 46-86, chmn fac, 69-70, actg pres, 77-78, vpres & provost, 70-86, EMER PROF PHYSICS, CALIF INST TECHNOL, 86- *Concurrent Pos:* Chmn comt nat acad to surv risks of nuclear power. *Mem:* Nat Acad Sci; Int Astron Union; Am Phys Soc; Am Astron Soc. *Res:* Cosmic rays; nuclear physics; astrophysics; variable stars. *Mailing Add:* Calif Inst Technol 1201 E California St Pasadena CA 91125

CHRISTY, ROBERT WENTWORTH, b Chicago, Ill, Nov 2, 22. PHYSICS. *Educ:* Univ Chicago, MS, 49, PhD(physics), 53. *Prof Exp:* Consult, Motorola, Inc, 52-53; from instr to assoc prof, Dartmouth Col, 53-62, chmn dept, 63-67 & 78-80, prof physics, 62-80, Appleton prof, 80-87, EMER PROF PHYSICS, DARTMOUTH COL, 87- *Concurrent Pos:* Consult, TRW Space Technol Labs, 58-68, GTE , 83-87. *Mem:* Fel AAAS; Am Phys Soc. *Res:* Ionic crystals; plastic flow; thermoelectric power; color centers; luminescence; thin films; metal optics. *Mailing Add:* Wilder Physics Lab Dartmouth Col Hanover NH 03755

CHRISTY, WILLIAM O(LIVER), b Barberton, Ohio, Oct 4, 07; m 32; c 1. MECHANICAL ENGINEERING. *Educ:* Houghton Col, AB, 28; Mass Inst Technol, BS, 31. *Prof Exp:* Instr, Houghton Col, 29; draftsman, Goodyear Tire & Rubber Co, Akron, 31-35, designer, Goodrich Tire & Rubber Co, 35-36; develop engr, 36-61, SUPT ENG & CONTROL DEPT, E I DU PONT DE NEMOURS & CO, 61- *Res:* Design and development of rayon processing equipment; selection of metals. *Mailing Add:* 2010 Robin Rd North Augusta SC 29841

CHROMEY, FRED CARL, b Philadelphia, Pa, June 30, 18; m 43; c 4. APPLIED PHYSICS. *Educ:* St Josephs Col, Pa, BS, 40; Cornell Univ, PhD(physics), 44. *Prof Exp:* Asst physics, Ind Univ, 40-42 & Cornell Univ, 42-44; res assoc, Mass Inst Technol contract, res physicist, Los Alamos Sci Lab, NMex, 44-46; RES PHYSICIST, E I DU PONT DE NEMOURS & CO, INC, 46- *Concurrent Pos:* Instr univ exten, Purdue Univ, 41-42, supvr studies, 41. *Mem:* Am Phys Soc. *Res:* Measurement of radio-activity; cyclotron construction and operation; cosmic rays; theory of scattering by colored bodies; viscoelasticity; digital computer programs; statistical experimental designs; explosion hazards. *Mailing Add:* Six N Cliffe Dr Wycliffe Wilmington DE 19809

CHRONES, JAMES, b Weyburn, Sask, Sept 6, 25; Can citizen; m 51; c 3. CHEMICAL ENGINEERING. *Educ:* Univ Sask, BSc, 46, MSc, 48; Mass Inst Technol, SM, 50. *Prof Exp:* Res engr, Atomic Energy of Can Ltd, 50-53; M W Kellogg Co Div, Pullman, Inc, 53-55, process engr, 56-58, process mgr, 61-64; sr process engr, Kellogg Int Corp, 58-60; chem engr & head dept design, Underwood McLellan & Assocs, 65; pres, Cambrian Eng Ltd, 66-70; assoc, Hatch Assocs, Ltd, 71-73; vpres & gen mgr, Pullman Kellogg Can, Ltd, 73-79; pres, Dynawest Projs Ltd, 79-80; consult, 80-81; PRES, CHRONES ENG CONSULTS INC, 81- *Concurrent Pos:* Gen mgr, Deuterium of Can Ltd, 70; prof engr, Ontario & Alberta; consult engr, Ontario. *Mem:* Am Inst Chem Engrs; Can Soc Chem Eng; fel Chem Inst Can. *Res:* Heavy water production processes; hydrocarbon pyrolysis for olefins production; synthesis gas preparation; distillation; vegetable oil refining; design of olefins, petrochemical, sodium sulphate, and vegetable oil refining plants; nuclear fuel reprocessing; bitumen and heavy oil upgrading processes; petrochemical feasibility studies; upgrading residue utilization testing. *Mailing Add:* 111 Lord Seaton Rd Willowdale ON M2P 1K8 Can

CHRONIC, HALKA, geology, for more information see previous edition

CHRONIC, JOHN, b Tulsa, Okla, June 3, 21; m 48, 81; c 4. INVERTEBRATE PALEONTOLOGY, GEOLOGY. *Educ:* Univ Tulsa, BS, 42; Univ Kans, MS, 47; Columbia Univ, PhD(geol), 49. *Prof Exp:* Instr geol, Univ Mich, 49-50; from asst prof to prof, 50-80, EMER PROF GEOL, UNIV COLO, BOULDER, 80- *Concurrent Pos:* Fel, Woods Hole Ocean Inst, 56, exchange lectr, Univ Edinburgh, 58-59; NSF lectr, State Univ NY Col Oneonta, 62; prof & chmn dept, Haile Sellassie Univ, 65-66; exchange prof, Univ PR, 78-79; petrol geologist, Keplinger & Co, Tenneco, Houston, 81-86; org paleontol expert, Inner Mongolia, 87. *Honors & Awards:* Distinguished Serv Award, Rocky Mountain Asn Geol, 72. *Mem:* AAAS; Geol Soc Am; Soc Econ Paleontologists & Mineralogists; Am Asn Petrol Geologists; Paleont Soc; Sigma Xi; Geol Soc London; Geol Soc Greece; Geol Soc Malaysia. *Res:* Colorado paleozoic geology; economic geology. *Mailing Add:* 12402 Copperfield Houston TX 77031-3110

CHRONISTER, ROBERT BLAIR, b Huntingdon, Pa, Aug 24, 42; m 68; c 2. NEUROANATOMY, NEUROBIOLOGY. *Educ:* Juniata Col, Pa, BS, 65; Univ Vt, Burlington, PhD(exp psychol), 72. *Prof Exp:* Fel neurosci, Univ Fla, 71-73; asst prof neurol & psychol, 74-76, asst prof, 76-81, ASSOC PROF ANAT, UNIV S ALA, MOBILE, 81- *Mem:* Soc Neurosci; Am Asn Anatomists. *Res:* Organization of basal forebrain structures. *Mailing Add:* Col Med Univ SAla 307 University Blvd Mobile AL 36688

CHRYSANT, STEVEN GEORGE, b Gargaliani, Greece, Feb 22, 34; US citizen; m 69; c 2. CARDIOVASCULAR DISEASES, HYPERTENSION. *Educ:* Univ Athens, MD, 59, PhD(biochem), 68; Am Bd Internal Med. *Prof Exp:* From instr to asst prof med, Stritch Sch Med, Loyola Univ, 70-72; from asst prof to assoc prof med, Health Sci Ctr, Univ Okla, 72-81, dir div hypertension, 76-81; prof med & chief hypertension, Kans Univ Med Ctr, 81-84; PROF, OKLA MED CTR, 84- *Concurrent Pos:* Res assoc nephrology, Hines Vet Admin Hosp, Ill, 70-71, staff physician, 71-72, asst sect chief renal hypertension & in-chg-hemodialysis unit, 72; dir hypertension screening & treatment prog, Oklahoma City Vet Admin Hosp, 72-; fel, Hypertension Res Coun, Am Heart Asn. *Mem:* Am Heart Asn; Am Soc Nephrology; Int Soc Nephrology; fel Am Col Cardiol; fel Am Col Physicians. *Res:* Systemic and renal hemodynamics effects of hypertension in the rat, dog, and man; role of salt in hypertension; role of prostaglandins in hypertension; pharmacotherapy of hypertension. *Mailing Add:* 5850 W Wilshire Oklahoma City OK 73132-6324

CHRYSOCHOOS, JOHN, b Icaria, Greece, Feb 27, 34; nat US; m 64; c 3. PHYSICAL CHEMISTRY, SPECTROSCOPY. *Educ:* Univ Athens, dipl, 57; Univ BC, MSc, 62, PhD(phys chem), 64. *Prof Exp:* Instr chem, Univ BC, 60-64; res fel chem physics, Harvard Univ, 64-65; res assoc biophys, Michael Reese Hosp, 65-66; res assoc phys chem, Ill Inst Technol, 66-67; from asst prof to assoc prof, 67-76, PROF CHEM, UNIV TOLEDO, 76- *Concurrent Pos:* Sr vis fel, Univ Western Ont, 80; Adj prof, Ctr Photochem Sci, Bowling Green State Univ, Bowling Green, Ohio, 86-; invited prof, Univ Crete, 89. *Mem:* AAAS; Am Chem Soc; Radiation Res Soc; Soc Appl Spectros; Am Soc Photobiol; Interam Photochem Soc; Sigma Xi. *Res:* Molecular luminescence; spectroscopy; lasers; flash spectroscopy; radiation chemistry; luminescence of doped glasses and crystals; spectroscopy of biomolecules; spectroscopy of semiconductors; interfacial electron transfer. *Mailing Add:* Dept Chem Univ Toledo 2801 W Bancroft Toledo OH 43606

CHRYSSAFOPOULOS, HANKA WANDA SOBCZAK, b Porto Alegre, Brazil; m 56. ENVIRONMENTAL SCIENCES, CONSULTING ENGINEERING. *Educ:* Univ Rio Grande de Sul, CE, 51, EME, 52; Univ Ill, Urbana-Champaign, MS, 54, PhD(soil mech), 64. *Prof Exp:* Head Res & Soils Lab, Tech Inst Rio Grande do Sul, 52-55; res asst, dept civil eng, Univ Ill, Urbana-Champaign 55-59 & groundwater resources, Ill State Geol Surv, Champaign, 59-60; geotech res engr, Woodward-Clyde-Sherard & Assocs, Kansas City, Mo, 64-65; asst prof civil eng, Univ Calif, Long Beach, 65-67; pvt res, 67-77; sr engr, Dames & Moore, New York, NY & Boca Raton Fla, 78-84; MEM, BRUCHEM GROUP, WAYNESBORO, VA, 85-; OWNER & PRES, HSCE INC, 86- *Concurrent Pos:* Mem, Regional Coun Eng, Archit & Argron, Brazil, 51-; Fulbright scholar, 54; mem, Am Arbit Asn; consult, UN Develop Prog; consult engr, 86-; mem, US Nat Comt, Pan Am Fedn Eng Socs. *Mem:* Fel Am Soc Civil Engrs; Int Soc Soil Mech & Found Engrs; fel Geol Soc Am; Sigma Xi; sr mem Soc Women Engrs; NY Acad Sci; Am Arbitration Asn; Am Consult Engrs Coun. *Res:* Soils and foundation engineering; geotechnical engineering; basic soil mechanics; engineering geology; information retrieval; environmental sciences; technology transfer. *Mailing Add:* 6642 Patio Lane Boca Raton FL 33433-6632

CHRYSSAFOPOULOS, NICHOLAS, b Istanbul, Turkey, Apr 23, 19; US citizen; m 56. ENVIRONMENTAL IMPACT ASSESSMENTS, SITE SELECTIONS. *Educ:* Robert Col, Istanbul, Turkey, BS, 40; Univ Ill, Champaign, MS, 52, PhD(civil eng), 56. *Prof Exp:* Resident engr, Contractors & Govt Turkey, 40-51; res asst soils, Univ Ill, 52-56, asst prof soils & highways, 56-59; prin & exec vpres, Woodward, Clyde, Sherard & Assoc, 59-68; assoc & partner, Dames & Moore, 68-85; VPRES, HSCE, INC, 86- *Concurrent Pos:* Adj prof, Fla Atlantic Univ, 89- *Mem:* Fel Am Soc Civil Engrs; fel Am Consult Engrs Coun; Am Soc Testing & Mat. *Res:* Surface soils establishing a correlation between agricultural soil types and the engineering properties; pavement design for highways and airports. *Mailing Add:* PO Box 6125 Boca Raton FL 33427

CHRYSSANTHOU, CHRYSSANTHOS, b Thessaloniki, Greece, Oct 15, 25; m 58; c 2. EXPERIMENTAL PATHOLOGY. *Educ:* Aristotelean Univ Thessaloniki, MD, 53. *Hon Degrees:* Dipl, Fac Arias Schreiber Central Hosp, Lima, Peru, 78. *Prof Exp:* Pathologist, Gynec-Obstet Clin, Med Sch, Univ Thessalonika, 53-54; Dept Path, Sch Med, NY Univ, 57; fel, Beth Israel Med Ctr, 58-59, res asst path, 59-60, assoc exp path, 60-63, assoc pathologist, assoc dir path acad affairs, 69-76, PATHOLOGIST, DEPT PATH, BETH ISRAEL MED CTR, 68-, ASSOC DIR DEPT, 64-; PROF PATH, MT SINAI SCH MED, CITY UNIV NEW YORK, 80- *Concurrent Pos:* Lectr, Sch Nursing, Beth Israel Hosp, 58-70; vis prof, Claude Bernard Inst, Univ Montreal, 66, Autonomous Univ Guadalajara, Mex, 86; NIH res grants hypertension, 67-72; assoc prof path, Mt Sinai Sch Med, City Univ NY, 67-80; Off Naval Res contracts, 68-82; NASA grants dysbaric dis, 83-89; dir, Off Res, Beth Israel Med Ctr, 83- *Honors & Awards:* Am Soc Clin Pathologists-Col Am Pathologists Award, 75, 78, 81, 84, 87, 90; AMA Award, 67, 72, 74, 78, 81, 84, 87, 90. *Mem:* Soc Exp Biol & Med; Am Soc Exp Path; Am Asn Path & Bact; Am Asn Cancer Res; NY Acad Sci; Aerospace Med Asn; Undersea &

Hyperbaric Med Soc. *Res:* Hypertension; mechanism and prevention of decompression sickness; dysbaric osteonecrosis; pathophysiology of vasoactive polypeptides; blood-brain barrier; endotoxin induced reactions; coagulation-fibrinolysis; laser photocoagulation; 94 publications in scientific journals and books. *Mailing Add:* Beth Israel Med Ctr First Ave at 16th St New York NY 10003

CHRYSSOSTOMIDIS, CHRYSSOSTOMOS, b Alexandria, Egypt, Oct 27, 42; US citizen. SHIP DESIGN, SEA KEEPING. *Educ:* Univ Durham, Eng, BSc, 65; Mass Inst Technol, SM, 67 & PhD(simulation), 70. *Prof Exp:* From asst prof to assoc prof design, 70-82, PROF DESIGN, OCEAN ENG, MASS INST TECHNOL & DIR, SEA GRANT COL, 82- *Concurrent Pos:* Consult, 70- *Honors & Awards:* Linnard Prize, Soc Naval Archit & Marine Engrs, 75. *Mem:* Soc Naval Architects & Marine Engrs; Royal Soc Naval Architects. *Res:* Computer aided design of large scale structures; dynamic behavior and fabrication of marine structure. *Mailing Add:* Mass Inst Technol 77 Mass Ave Rm E38-300 Cambridge MA 02139

CHRZANOWSKI, ADAM, b Cracow, Poland, Dec 22, 32; Can citizen; m 56; c 2. ENGINEERING. *Educ:* Acad Mining & Metall, Kracow, MEng, 56, DEngSc, 62. *Prof Exp:* Sr asst surv, Acad Mining & Metall, Cracow, 55-61, asst prof, 62-66; fel, 64-66, assoc prof, 66-71, PROF SURV, UNIV NB, 71- *Concurrent Pos:* Mem, Mt Kennedy Exped, Yukon, Alaska, 65; vis prof, Eidgenoessische Tech Hochschule, Zurich, 80. *Mem:* Am Cong Surv & Mapping; Can Inst Surv; Int Soc Mine Surv; Am Geophys Union. *Res:* New methods and instruments in engineering and mining surveying; use of laser in surveying; tectonic movements; mining subsidence. *Mailing Add:* Dept Surv Eng Univ NB Col Hill Box 4400 Fredericton NB E3B 5A3 Can

CHRZANOWSKI, THOMAS HENRY, b Irvington, NJ, April 11, 52; m 75; c 3. MICROBIOL ECOLOGY, MYCOLOGY. *Educ:* Bloomfield Col, BA, 74; Univ SC, MS, 76, PhD(microbiol), 81. *Prof Exp:* asst prof, 81-87, ASSOC PROF MICROBIOL, UNIV TEX, ARLINGTON, 87- *Mem:* Am Soc Microbiol; Am Soc Limnol & Oceanog; AAAS; Sigma Xi. *Res:* Quantifying rates of bacterial heterotrophic activity in aquatic environments. *Mailing Add:* Dept Biol Univ Tex Arlington TX 76019

CHU, BENJAMIN PENG-NIEN, b Shanghai, China, Mar 3, 32; m 59; c 3. PHYSICAL CHEMISTRY. *Educ:* St Norbert Col, BS, 55; Cornell Univ, PhD(phys chem), 59. *Prof Exp:* Res asst nuclear eng, Brookhaven Nat Lab, NY, 57; res assoc chem, Cornell Univ, 58-62; from asst prof to assoc prof, Univ Kans, 62-68; prof chem, 68-88, chmn dept, 78-85, PROF MAT SCI & ENG, STATE UNIV NY STONY BROOK, 82-, LEADING PROF CHEM, 88- *Concurrent Pos:* Sloan res fel, 66-68; Guggenheim fel, 68-69; vis prof & fel, Univ New South Wales & Australian Nat Univ, 74, Wayne State Univ & Hokkaido Univ 75, Univ Koln, 77, Beijing Univ, 79, Inst Theoret Physics, Univ Calif, Santa Barbara & Fudan Univ, 82; assoc ed, Mat Letters, 86-89. *Honors & Awards:* Humboldt Award for Sr US Scientist, 77. *Mem:* Am Chem Soc; fel Am Phys Soc; AAAS; Am Crystal Asn; fel Am Inst Chemists. *Res:* Critical phenomena; molecular configuration and dynamics of macromolecules in solution; structure of non-crystalline media; light scattering and small angle x-ray scattering; ion exchange; DNA gel electrophoresis. *Mailing Add:* Dept Chem State Univ NY Stony Brook NY 11794

CHU, BOA-TEH, b Peiping, China, Sept 26, 24. AERONAUTICAL ENGINEERING. *Educ:* Nat Cent Univ, China, BSc, 45; Johns Hopkins Univ, PhD(aeronaut eng), 54. *Prof Exp:* Asst aeronaut eng, Johns Hopkins Univ, 49-54, res staff asst, 54-56; from asst prof to assoc prof eng, Brown Univ, 56-59; PROF ENG & APPL SCI, YALE UNIV, 64- *Res:* Fluid mechanics; thermodynamics; combustion aerodynamics; magnetohydrodynamics; elasticity; viscoelasticity. *Mailing Add:* Dept Eng & Appl Sci Yale Univ New Haven CT 06520

CHU, CHANG-CHI, b Yentai, Shantung, China, Sept 9, 33; m 68; c 1. CROP PHYSIOLOGY, CROP PRODUCTION. *Educ:* Taiwan Chung-Hsing Univ, BS, 54; Kans State Univ, MS, 73; Cornell Univ, PhD(bot), 76. *Prof Exp:* Asst agronomist & prod mgr, Taiwan Sugar Corp, 55-60, res assoc sugar cane, 60-70; res asst agron, Kans State Univ, 71-73; res asst bot, Cornell Univ, 73-76, res support specialist nitrate pollution, 76-77; crops res & develop specialist & biometrician, Agway, Inc, 77-84; PLANT PHYSIOLOGIST, USDA-ARS, 84- *Mem:* Am Soc Agron; Crop Sci Soc Am; Am Soc Hort Sci; Am Soc Plant Physiologists; Coun Agr Sci & Technol. *Res:* Crop production and physiology. *Mailing Add:* USDA ARS 4151 Hwy 86 Brawley CA 92227

CHU, CHAUNCEY C, b Shanghai, China, Oct 17, 24; nat US; m 44; c 2. MECHANICAL ENGINEERING, DESIGN ENGINEERING. *Educ:* Purdue Univ, BSME, 43; Univ Toronto, MASc, 46; Harvard Univ, MS, 47. *Prof Exp:* Sr res assoc, Fabric Res Labs Inc, 51-67; vpres, Kybe Corp, 67-70; mgr & sr staff engr, 70-72, pres, Puerto Rico Br, 78, SR VPRES MFG, WANG LABS, INC, 72-, PRES, TAIWAN BR, 79- *Concurrent Pos:* Sr consult, Cybetronics, Inc, 60-67. *Mem:* Am Soc Mech Engrs; Am Inst Aeronaut & Astronaut. *Res:* Dynamic and thermal behaviors of polymeric re-entry decelerator materials; development of magnetic tape transports and disc-pack drives; mechanical design of electronic calculators and mini-computers. *Mailing Add:* 43 Deerhaven Rd Lincoln MA 01773-1809

CHU, CHIA-KUN, b Shanghai, China, Aug 14, 27; m 86; c 3. APPLIED MATHEMATICS. *Educ:* Chiao-Tung Univ, China, BS, 48; Cornell Univ, MME, 50; NY Univ, PhD(math), 59. *Prof Exp:* Develop engr, Gen Elec Co, 50-53; asst prof mech eng, Stevens Inst Technol, 53-57; assoc prof eng sci, Pratt Inst, 57-59; assoc prof aero eng, NY Univ, 59-63; vis res assoc, Plasma Res Lab, 63-65, assoc prof, 65-68, Dept Appl Physics & Nuclear Engr, 82-84 & 85-88, PROF ENG SCI, COLUMBIA UNIV, 68-, CHMN APPL MATH COMT, 78- *Concurrent Pos:* Guggenheim fel, 71-72; Sherman Fairchld distinguished scholar, Calif Inst Technol, 84; Weilun lectr, Chinese Univ, Hong Kong, 91. *Mem:* Soc Indust Appl Math; fel Am Phys Soc; fel Japanese Soc Prom Sci. *Res:* Fluid dynamics; plasma physics; large-scale scientific computing. *Mailing Add:* Appl Physics Dept Columbia Univ New York NY 10027

CHU, CHIEH, b Chekiang, China, Jan 17, 22; US citizen; m 48; c 3. CHEMICAL ENGINEERING. *Educ:* Chekiang Univ, BS, 47; Univ Wis, MS, 59, PhD(chem eng), 61. *Prof Exp:* Assoc engr, Chinese Petrol Corp, 47-58; res scientist, Sinclair Res, Inc, 61-65, sr res scientist, Sinclair Oil & Gas Co, 65; asst prof eng & appl sci, Univ Calif, Los Angeles, 65-72; res scientist, Getty Oil Co, 72-81, sr res scientist, 81-84; sr res assoc, 84-85, res consult, 85-90, SR RES CONSULT, TEXACO INC, 90- *Mem:* Am Inst Mining Metall & Petrol Engrs. *Res:* Enhanced oil recovery processes, thermal recovery; reservoir engineering and simulation; chemical kinetics; catalysis; combustion; chemical reactor analysis and design. *Mailing Add:* Texaco Inc PO Box 770070 Houston TX 77215-0070

CHU, CHING-WU, b Hodnam, China, Dec 2, 41; US citizen; m 68; c 2. MATERIALS SCIENCE. *Educ:* Cheng-king Univ, BS, 62; Fordham Univ, MS, 65; Univ Calif, La Jolla, PhD(physics), 68. *Hon Degrees:* DSc, Fordham Univ, Northwestern Univ & Chinese Univ Hong Kong, 88. *Prof Exp:* Mem tech staff physics, Bell Labs, Am Tel & Tel, 68-70; from asst prof to prof, Cleveland State Univ, 70-79; PROF PHYSICS, UNIV HOUSTON, 79-, DIR, TEX CTR SUPERCONDUCTIVITY, 87- *Concurrent Pos:* Resident res assoc, Argonne Nat Lab, 73; consult, Bell Labs, 73, 75 & 78, Marshall Space Flight Ctr, 82; vis staff mem, Los Alamos Sci Lab, 75-80; dir, Space Vacuum Epitaxy Ctr, 86-88 & Magnetic Info Res Lab, 84-; dir, Tex Ctr Superconductivity, 87- *Honors & Awards:* Comstock Award, Nat Acad Sci, 88; Phys & Math Award, NY Acad Sci; Leroy Randle Grumman Medal. *Mem:* Fel Am Phys Soc. *Res:* High pressure low temperature studies; superconductivity, magnetism, dielectrics, novel materials, and magnetic recording. *Mailing Add:* Dept Physics Tex Ctr Superconductivity Houston TX 77204-5506

CHU, CHUNG K, b Seoul, Korea, May 18, 41; US citizen; m 73; c 2. VIRAL CHEMOTHERAPY. *Educ:* Seoul Nat Univ, BS, 64; Idaho State Univ, MS, 70; State Univ NY, Buffalo, PhD(med chem), 74. *Prof Exp:* Fel med chem, Sloan-Kettering Inst Cancer Res, 74-75, res assoc, 76-80; asst prof med chem, Idaho State Univ, 80-82; from asst prof to assoc prof med chem, 82-89, DIR DRUG DISCOVERY GROUP, UNIV GA, 86-, PROF MED CHEM, 90- *Concurrent Pos:* Prin investr, NIH res grants. *Mem:* Am Chem Soc; Int Soc Antiviral Res; Am Asn Cols Pharm; AAAS; Am Asn Cancer Res. *Res:* Design and synthesis of anticancer and antiviral agents; heterocyclic carbohydrates; nucleosides chemistry. *Mailing Add:* Col Pharm Univ Ga Athens GA 30602

CHU, CHUNG-YU CHESTER, b Guangzhou, China, Apr 30, 50; US citizen; m 78; c 2. MASS PRODUCTION & PROCESS SYSTEMS, INDUSTRIAL APPLICATIONS OF ENVIRONMENTAL ENGINEERING. *Educ:* Univ Calif, Berkeley, BS, 79. *Prof Exp:* Plant engr mech eng, 79-84, SR ENGR GEN ENG & ENG MGT, GEN MOTORS CORP, 84- *Concurrent Pos:* Mgr & consult, Chesco, 85-86. *Mem:* Nat Soc Prof Engrs. *Mailing Add:* 3469 Leora St Simi Valley CA 93063

CHU, CHUN-LUNG (GEORGE), b Hong Kong, Apr 29, 50; Can citizen; m 80; c 1. POSTHARVEST PHYSIOLOGY. *Educ:* Nat Chung-Hsing, BSc, 72; Univ Guelph, MSc, 77; Wash State Univ, PhD(hort), 80. *Prof Exp:* Teaching asst postharvest physiol, Wash State Univ, 77-78, res asst, 77-80, teaching asst pomol, 80; RES SCIENTIST APPLE STORAGE, ONT MINISTRY AGR & FOOD, 80- *Concurrent Pos:* Mem, Hort res Inst Ont, Vineland Sta. *Honors & Awards:* Joseph Harvey Gourley Award. *Mem:* Am Soc Hort Sci; Sigma Xi; Agr Inst Can. *Res:* Fruit maturity indices; storage and handling techniques to enhance efficiency of apple production and marketing systems. *Mailing Add:* Hort Res Inst Ont Vineland Station ON L0R 2E0 Can

CHU, DANIEL TIM-WO, b Hong Kong; US citizen. ANTI-INFECTIVES. *Educ:* Univ Alta, BS, 67; Univ NB, PhD(org chem), 71. *Prof Exp:* Indust postdoctoral fel, Bristol-Myers, 71-72; sr chemist, Abbott Labs, 72-84, group leader, 85, group leader & assoc res fel, 85-86, proj leader & asst res fel, 86-87, proj leader & res fel, 87-88, sr proj leader & res fel, 88-90, SR PROJ LEADER & SR RES FEL, ABBOTT LABS, 90- *Mem:* Am Chem Soc; Sigma Xi; Am Soc Microbiol; fel Am Inst Chemists. *Res:* Anti-infective research; author of more than 110 research papers and patents; inventor of antibacterials tosufloxacin and temafloxacin. *Mailing Add:* Abbott Labs D-47N Abbott Park IL 60064

CHU, ELIZABETH WANN, b Shanghai, China, Oct 29, 21; US citizen; m 46; c 1. CYTOLOGY, PATHOLOGY. *Educ:* Univ Hong Kong, BS & BM, 46; Shanghai Med Col, MD, 46; Am Bd Path, dipl, 65. *Prof Exp:* Resident internal med, Beekman Hosp, New York, 48-49; resident internal med, Cambridge City Hosp, Mass, 50-51 & path, 51-52; resident, Boston City Hosp, 52-53; med officer cytol, Washington Cytol Unit, 56-58, MED OFFICER CYTOL, PATH LAB, NAT CANCER INST, 58- *Mem:* Am Soc Cytol. *Res:* Exfoliative cytology, its value in experimental carcinogenesis, metastases and endocrine factors. *Mailing Add:* NIH Clin Ctr Rm 2A-19 Bethesda MD 20892

CHU, ERNEST HSIAO-YING, b Haining, Chekiang, China, June 3, 27; US citizen; m 54; c 3. GENETICS. *Educ:* St John's Univ, China, BS, 47; Univ Calif, Berkeley, MS, 51, PhD(genetics), 54. *Prof Exp:* From res asst to res assoc bot, Yale Univ, 54-59, lectr anat, Sch Med, 58-59; biologist, Oak Ridge Nat Lab, 59-72; PROF HUMAN GENETICS, MED SCH, UNIV MICH, ANN ARBOR, 72-, PROF TOXICOL, SCH PUB HEALTH, 85- *Concurrent Pos:* NSF sr fel, 65; prof zool & biomed sci, Univ Tenn, Knoxville, 67-72; Josiah Macy fac scholar, 78; adv prof, Fudan Univ, China, 80-; Am Cancer Soc fac scholar, 85. *Mem:* Genetics Soc Am; Am Soc Human Genetics; Am Soc Cell Biol; Tissue Cult Asn; Environ Mutagen Soc; Radiation Res Soc. *Res:* Somatic cell genetics; mammalian cytogenetics. *Mailing Add:* Dept Human Genetics Univ Mich Med Sch 4708 Med Sci II Box 0618 Ann Arbor MI 48109-0618

CHU, FLORENCE CHIEN-HWA, b China, May 20, 18; nat US; m 43; c 2. RADIOLOGY. *Educ:* Nat Med Col, Shanghai, MD, 42; Am Bd Radiol, dipl, 50. *Prof Exp:* Instr radiol, 53-55, asst prof clin, 56-61, clin asst prof, 61-69, assoc prof, 69-73, clin prof, 73-77, PROF RADIOL, CORNELL UNIV, 77-; CHIEF, DIV RADIATION ONCOL, NY HOSP-CORNELL MED CTR, 86- *Concurrent Pos:* Fel radiol, Mem Hosp, 49-50, clin asst radiation therapist, 50-53, asst attend radiation therapist, 55-65, assoc attend radiation therapist, 65-69, attend radiation therapist, 69-, actg chmn dept radiation ther, 75, chmn dept, 76-84; res assoc, Sloan-Kettering Inst, 55, assoc mem, 77-84; assoc attend radiologist, New York Hosp, 70-74, attend radiologist, 74- *Mem:* Am Col Radiol; Radiol Soc NAm; Am Radium Soc; Am Soc Ther Radiol & Oncol; Am Soc Clin Oncol; Am Roentgen Soc. *Res:* Ionizing radiation; clinical investigation of management of various types of cancers, particularly breast cancer; clinical investigation of efficacies of radaiation therapy in the treatment of cancer. *Mailing Add:* Stitch Radiation Ther Ctr NY Hosp Cornell Med Ctr 525 E 68th St New York NY 10021

CHU, FUN SUN, b China, May 7, 33; m 58; c 3. BIOCHEMISTRY. *Educ:* Nat Chung-Hsin Univ, BS, 54; WVa Univ, MS, 59; Univ Mo, PhD(biochem), 64. *Prof Exp:* Res assoc, Food Res Inst, Univ Chicago, 63-67; from asst prof to assoc prof, Dept Food Sci & Food Res Inst, 67-77, PROF, DEPT FOOD MICROBIOL & TOXICOL, UNIV WIS-MADISON, 78- *Mem:* Am Soc Microbiol. *Res:* Protein chemistry; biochemistry of microorganisms; mycotoxins; biochemistry of microbial toxins. *Mailing Add:* Food Res Inst Univ Wis 1925 Willow Dr Madison WI 53706

CHU, GORDON P K, ceramics, metallurgy, for more information see previous edition

CHU, HORN DEAN, b Tientsin, China, Sept 9, 33; US citizen; m 62. CHEMICAL ENGINEERING. *Educ:* Waseda Univ, Japan, BS, 59, MS, 61; Univ Pa, MS, 63; Univ Ala, PhD(chem eng), 65. *Prof Exp:* Proj engr, Selas Corp of Am, 65-72; adj prof & asst prof food sci, Cook Col, Rutgers Univ, 72-79; sr engr, MacAndrews & Forbes Co, 79-82; PRES, BERKORP, INC, 82- *Mem:* AAAS; Am Inst Chem Engrs; Am Chem Soc; Inst Food Technologists; fel Am Inst Chemists. *Res:* Combustion instability; heat and fluid processing; thermodynamics and kinetics of heterogeneous reaction; heat and mass transfer; immobilized enzyme reactor; energy conservation in food processing; utlization of biomass; food processing development. *Mailing Add:* 105 The Mews Haddonfield NJ 08033

CHU, HSIEN-KUN, b Shanghai, China, Oct 14, 47; m 76; c 2. ORGANIC CHEMISTRY. *Educ:* Nat Taiwan Univ, BS, 70; Vanderbilt Univ, PhD(chem), 76. *Prof Exp:* Vis instr, Univ Tex-Arlington, 76-77; res assoc chem, Tex Christian Univ, 77-80; proj chemist, Dow Corning Corp, 80-83, res specialist, 83-88; SCIENTIST, LOCTITE CORP, 88- *Concurrent Pos:* Fel, Tex Christian Univ, 77-80. *Mem:* Am Chem Soc; Sigma Xi. *Res:* Mechanisms of organic reactions; photochemistry; organosilicon chemistry; sealants and adhesives. *Mailing Add:* Loctite Corp 705 N Mountain Rd Newington CT 06111-1499

CHU, IRWIN Y E, b Taipei, Taiwan, Sept 28, 37; m 66; c 2. PLANT GENETICS. *Educ:* Chung-Shin Univ, BS, 61; Tokyo Univ, MS, 65, PhD(plant genetics), 68. *Prof Exp:* Fel appl genetics, Nat Inst Genetics, Japan, 68-69; assoc res fel plant genetic, Acad Sinica, Taiwan, 69-72, head lab, Inst Bot, 71-72; asst res prof, Univ Utah, 72-74, res assoc, 74-76; sr plant geneticist, Greenfield Res Lab, Eli Lilly & Co, 76-86; PRES & CHIEF EXEC OFFICER, TWYFORD INT INC, 86- *Concurrent Pos:* Vis assoc prof, Inst Food Crops, Chung-Shing Univ, 71; asst ed newslett, Soc Advan Breeding Res in Asia & Oceania, 70-72. *Mem:* Sigma Xi; Genetic Soc Am; Crop Soc Am; Japanese Soc Genetics; Chinese Soc Agron. *Res:* Genetical studies on plant tissue, cell and protoplast culture; application of tissue culture techniques for plant breeding. *Mailing Add:* Twyford Int Inc 15245 Telegraph Rd Santa Paula CA 93060

CHU, J CHUAN, b Tianjin, China, July 14, 19. ELECTRICAL ENGINEERING. *Educ:* Univ Minn, BS, 42; Univ PA, MSEE, 45. *Prof Exp:* Vpres res & develop, Univac Div, Sperry Rand-Univac, 56-62; vpres & asst gen mgr, Honeywell Info Systs, 62-72; sr vpres, NAm Opers, Wang Corp, 72-80; pres, chmn & chief exec officer, Santec Corp, 80-85; CHMN & EXEC OFFICER, COLUMBIA INT CORP, 85- *Honors & Awards:* Comput Pioneer Award, Inst Elec & Electronics Engrs, 82. *Mem:* Fel Inst Elec & Electronics Engrs. *Mailing Add:* Columbia Int Corp Ten Baldwin Circle Weston MA 02193

CHU, JEN-YIH, b Taipei, Taiwan, July 28, 40; m; c 2. PEDIATRICS. *Educ:* Nat Univ Taiwan, MD, 65; Univ Calif, Berkeley, PhD(nutrit), 71. *Prof Exp:* PROF PEDIAT, SCH MED, ST LOUIS UNIV, 83- *Mem:* Soc Pediat Res; Am Hemat Soc; Am Asn Cancer Res. *Res:* Hematology; iron; vitamin E. *Mailing Add:* Dept Pediat St Louis Univ Sch Med 1465 S Grand Ave St Louis MO 63104

CHU, JOSEPH YUNG-CHANG, b Anhwei, China, Aug 16, 40; US citizen; m 67. ORGANIC CHEMISTRY, PHOTOCHEMISTRY. *Educ:* Nat Taiwan Univ, BS, 61; Fla State Univ, MS, 67; Univ Rochester, PhD(org chem), 72. *Prof Exp:* Res chemist, Sinclair Res, Inc, 67-68; assoc scientist, Xerox Corp, 71-73, scientist, 73-78, sr scientist chem, Webster Res Ctr, 78-80, mgr, China Admin, 80-82, explor technol, 82-84 & tech prog, 84-85, dir, China bus develop & gen mgr, 85-88, dir, China & SPac Opers, 88-90, MGR, TECH OPERS & QUAL, STRATEGY & ARCHIT/D&M, XEROX CORP, 90- *Mem:* Soc Photog Scientists & Engrs; Am Mgt Asn. *Res:* Synthesis; thermal and photochemical reactions of organic compounds; electrical and photoconductive properties of organic materials; reprographic materials and processes. *Mailing Add:* Xerox Corp Bldg 0105-47C 800 Phillips Rd Webster NY 14580

CHU, JU CHIN, b Taitsang, China, Dec 14, 19;. CHEMICAL ENGINEERING. *Educ:* Tsing Hua Univ, BSc, 40; Mass Inst Technol, ScD(chem eng), 46. *Prof Exp:* Asst chem, Tsing Hua Univ, 40-43; instr, Yunan Army Med Acad, 42-43; petrol technologist, Shell Chem Corp, Calif, 46; asst prof chem eng, Wash Univ, 46-49; from assoc prof to prof, Polytech Inst Brooklyn, 49-67; tech adv, Strategic Missile Systs Div, NAm Rockwell Corp, 66-70; TRUSTEE & CHMN BD, TECHNOL RESOURCES, INC, 70- *Concurrent Pos:* Consult, Sun Oil Co, 52, Am Cyanamid Co, 53, Rohm and Haas, 55, Curtiss Wright Corp, 55-57, Union Carbide Nuclear Co, 55-58, Argonne Nat Lab & US Dept Agr, 56-61, Gen Elec Co, 58, Space Technol Labs, 59, Rocketdyne, NAm Aviation, 60-62, space info & systs div, 52-53, Aerospace Corp & Erie Resistor Corp, 62, ARO, Inc, 62-64, Huyck Corp, 62-65, W R Grace, 63, Grumman Aircraft Eng Corp, 64, Gen Motors Corp, 65 & Sandia Corp, 66-67; tech dir, Chem Construct Corp, 56-57; global lectr, Univs & Sci Socs, 62-63; prof, Va Polytech Inst & State Univ, 69-72. *Honors & Awards:* Ann Achievement Award, Chinese Inst Eng, 61; Gold Medal, Chinese Educ Ministry, 63. *Mem:* Fel AAAS; Am Chem Soc; Am Petrol Inst; fel Am Inst Chem Engrs; fel Chinese Acad Sci. *Res:* Advanced weapon systems; nuclear technology and hardening; environment control and industrial waste recycle; process design; unit operation. *Mailing Add:* 21 Yorktown Irvine CA 92720

CHU, KAI-CHING, b Szechwan, China, Nov 19, 44; m 72; c 2. APPLIED MATHEMATICS, SYSTEM THEORY. *Educ:* Nat Taiwan Univ, BS, 66; Harvard Univ, MS, 68, PhD(appl math), 71. *Prof Exp:* Res asst appl math, Harvard Univ, 68-71; mathematician, Systs Control, Inc, 71-72; res staff mem appl math, 73-75, mgr social sci group, dept gen sci, T J Watson Res Ctr, Int Bus Mach Corp, 75-78; vis scientist, Stanford Univ, 79-80; RES, COMPUT SCI DEPT, INT BUS MACH CORP, SAN JOSE, 83- *Concurrent Pos:* Assoc ed, Trans on Automatic Control, 71-; prog mgr, Inst Systs Sci, Singapore, 86-88. *Mem:* Sr mem Inst Elec & Electronics Engrs; Opers Res Soc Am. *Res:* Decision and control theories, optimization and estimation techniques; communication networks; computer applications to urban and industrial problems; image processing and computer graphics. *Mailing Add:* Computer Sci Dept 908 Lundy Lane Los Altos CA 94024

CHU, KEH-CHENG, b Feng-Yang, China, May 19, 33; m 63; c 2. MAGNETIC RESONANCE, RADIATION PHYSICS. *Educ:* Nat Taiwan Univ, BS, 55; Univ Mich, MS, 62, PhD(nuclear sci), 67. *Prof Exp:* Instr nuclear sci, Nat Tsing Hua Univ, 60-62; asst prof physics, Western Ill Univ, 68-72, prof, 74-; ASST RES PHYS, UNIV CALIF, LOS ANGELES. *Mem:* Am Phys Soc. *Res:* Electron spin resonance; electron nuclear double resonance; radiation effects in solids. *Mailing Add:* Knudsen Hall 2-130 Univ Calif Los Angeles CA 90024-1547

CHU, KUANG-HAN, b Chekiang, China, Nov 13, 19; m 62; c 1. CIVIL ENGINEERING. *Educ:* Nat Cent Univ, China, BS, 42; Univ Ill, MS, 47, PhD(civil eng), 50. *Prof Exp:* Struct designer, Ammann & Whitney, 50-51 & D B Steinman, 51-55; actg assoc prof civil eng, Univ Iowa, 55-56; from assoc prof to prof, 56-84, EMER PROF CIVIL ENG, ILL INST TECHNOL, 84- *Concurrent Pos:* Vis prof, Shanghai Inst Rwy Technol, 84-85, China Acad Rwy Sci, 85-87. *Honors & Awards:* Collingwood Prize, Am Soc Civil Engrs, 53. *Mem:* Fel Am Soc Civil Engrs; Am Concrete Inst; Int Asn Bridge & Struct Engrs; Sigma Xi. *Res:* Structural analysis and design; stability and dynamics of structures; computer techniques. *Mailing Add:* 730 Wash St Apt 101 San Francisco CA 94108

CHU, KWO RAY, b Hunan, China, Oct 10, 42; US citizen; m 68; c 1. GYROTRON. *Educ:* Cornell Univ, PhD(physics), 72. *Prof Exp:* Res physicist, Sci Applications, Inc, 73-77; supvr res physicist, Naval Res Lab, 77-; MEM STAFF, DEPT PHYSICS, NAT TSING HUA UNIV, TAIWAN. *Mem:* Fel Am Phys Soc; Inst Elec & Electronics Engrs. *Res:* Plasma physics; relativistic electronics. *Mailing Add:* Dept Physics Nat Tsing Hua Univ Hsinchu Taiwan

CHU, MAMERTO LOARCA, b Philippines, Oct 15, 33; US citizen; m 62; c 5. VIBRATIONS, ACOUSTICS. *Educ:* Iloilo City Univ, Philippines, 56; Univ Houston, MSME, 64, PhD(mech eng), 67. *Prof Exp:* Plant engr, Int Steel, 58-67; res assoc, Univ Houston, 67-68, lectr mech eng, 68-69; asst prof, 69-71, assoc prof, 71-80, PROF MECH ENG, UNIV AKRON, 80- *Concurrent Pos:* Prin investr, NIH, 72-76, Babcock & Wilcox Co, 77-79; adj prof, Akron City Hosp, 77- *Mem:* Am Soc Mech Engrs; Sigma Xi; Acoust Soc Am; Am Soc Eng Educ. *Res:* Acoustics and vibrations as applied to both mechanical and bio-engineering systems, such as audiometric studies of pathological joints, diagnostic applications, modal analysis, random data statistical analysis. *Mailing Add:* Mech Eng Dept Univ Akron Akron OH 44325

CHU, NORI YAW-CHYUAN, b Taipei, Taiwan, Mar 31, 39; US citizen; m 67; c 2. EYE GLASS LENS TECHNOLOGY, PHOTOCHROMIC PLASTICS. *Educ:* Cheng-Kung Univ, Tainan, Taiwan, BS, 61; Univ Chicago, PhD(phys chem), 68. *Prof Exp:* Fel chem, Univ Calif, Riverside, 68-71; res assoc chem, Northeastern Univ, 71-73; sr chemist, 73-82, prin chemist chem & mat sci, 82-89, DIR CHEM TECHNOL, AM OPTICAL CORP, 89- *Concurrent Pos:* Pub Health Serv fel, Univ Calif, Riverside, 69-70. *Mem:* Am Chem Soc; NY Acad Sci; Chinese-Am Chem Soc. *Res:* Basic and applied research in photochromic materials; photophysical processes and energy-transfer mechanisms in molecules and solids; synthesis of photosensitive materials; laser eye protection technology. *Mailing Add:* Am Optical Corp Res Ctr 14 Mechanic St Southbridge MA 01550

CHU, PAUL CHING-WU, b Hunan, China, Dec 2, 41; m 68; c 2. PHYSICS. *Educ:* Cheng-Kung Univ, BS, 62; Fordham Univ, MS, 65; Univ Calif San Diego, PhD(physics), 68. *Hon Degrees:* Dr, Chinese Univ Hong Kong, 88, Northwestern Univ, 88, Fordham Univ, 88, State Univ NY, Farmingdale, 89, Fla Int Univ, 89 & Whittier Col, 91. *Prof Exp:* Teaching asst, Fordham Univ, Bronx, NY, 63-65; res asst, Univ Calif, San Diego, 65-68; from asst prof to prof physics, Cleveland State Univ, Ohio, 70-79; PROF PHYSICS, UNIV

HOUSTON, TEX, 79-, DIR, TEX CTR SUPERCONDUCTIVITY, 87- *Concurrent Pos:* Mem tech staff, Bell Labs, Murray Hill, NJ, 68-70; resident res assoc, Argonne Nat Lab, Ill, 72; vis scientist, Hansens Physics Lab, Stanford Univ, Calif, 73; consult, Bell Labs, Murray Hill, NJ, 73, 75, 78, Marshall Space Flight Ctr, NASA, Huntsville, Ala, 82-87; vis staff mem, Los Alamos Sci Lab, NMex, 75-80; teller, Div Solid State Physics, Am Physics Soc, 76; chmn, Organizing Comt, Int Conf High Pressure & Low Temperature Physics, 77; dir, Magnetic Info Res Lab, Univ Houston, 84-88, M D Anderson chair physics, 87, T L L Temple chair sci, 87-; dir, Space Vacuum Epitaxy Ctr, NASA/UH, 86-88, Solid State Physics Prog, NSF, 86-87; mem, Panel High Temperature Superconductivity, Nat Acad Sci, 87; bd dirs, Indust Technol Res Inst, 88-, Coun Superconductivity Am Competitiveness, 89- & Houston Mus Natural Sci & Burke Planetarium, 89-; mem, Int Mat Prize Comt, Am Phys Soc, 88-89; hon prof, Physics Inst, Chinese Acad Sci, 88, Zhongshao Univ, 88, Nankai Univ, 91 & Chinese Univ Sci & Technol, 91; ad hoc Rev Panel on Long-Range Plan Res & Develop Superconductivity, White House, 89; mem, Res Adv Comt, Inst Technol & Strategic Res, 89. *Honors & Awards:* Phys & Math Sci Award, NY Acad Sci, 87; Leroy Randle Grumman Medal for Outstanding Sci Achievement, Grumman Corp, 87; Nat Medal Sci, 88; Comstock Award, Nat Acad Sci, 88; Int Prize for New Mat, Am Phys Soc, 88; Medal for Sci Merit, World Cult Coun, 89; St Martin de Porres Award, 90. *Mem:* Nat Acad Sci; Am Acad Arts & Sci; Third World Acad Sci; fel Am Phys Soc; Academia Sinica. *Res:* Superconductivity; high pressure physics; low temperature physics; magnetism and dielectrics. *Mailing Add:* Tex Ctr Superconductivity Univ Houston Houston TX 77204-5932

CHU, PE-CHENG, b Shanghai, China, Dec 1, 44; m 76; c 1. AIR ICE OCEAN INTERACTION, GEOPHYSICAL FLUIDS & CLIMATE. *Educ:* Nanking Univ, BS, 68; Inst Atmos Phys, MS, 80; Univ Chicago, PhD(geophys sci). *Prof Exp:* Weather forecaster, 68-73, sect leader agrometerol, Wuxi Observ, 74-78; res asst, Chinese Acad Sci, 78-80; res asst, Univ Chicago, 80-84, res assoc, 85-86; ADJ RES PROF OCEANOG, NAVAL POSTGRAD SCH, 86- *Concurrent Pos:* Co prin investr, off naval res, Naval Postgrad Sch, 86- *Mem:* Am Meterol Soc; Int Glaciological Soc; Am Geophys Union; Am Oceanog Soc. *Res:* Marginal ice-zone; ocean surface mixed layer; flow over topography; turbulence; coupled air-ocean models; el nino and southern oscillation; equatorial and coastal dynamics. *Mailing Add:* Naval Postgrad Sch Code 68-CU Monterey CA 93943

CHU, SHERWOOD CHENG-WU, b Shanghai, China, Aug 30, 37; US citizen; m 59; c 1. MATHEMATICS. *Educ:* Harvard Univ, BA, 59; Univ Md, MA, 61, PhD(math), 63. *Prof Exp:* Res asst, Inst Fluid Dynamics & Appl Math, Univ Md, 60-63; Nat Acad Sci-Nat Res Coun resident res assoc, US Naval Ord Lab, 63-64; mem tech staff, Bellcomm, Inc, 64-66; asst prof math, Univ Del, 66-68; mem tech staff, Bellcomm, Inc, Washington, DC, 68-72; res mathematician, NIH, 72-76; gen engr, US Dept Transp, 76-80, dir Off Univ Res, 80-81, dir Off Regulatory Planning, Mat Transp Bureau, 81-85, dep dir, Off Hazardous Mat Transp, 85-86; lead scientist, Mitre Corp, 86-88; dir, Transp Anal, Monitored Retrievable Storage Rev Comn, 88-89; SR PROF STAFF, US NUCLEAR WASTE TECH REV BD, 89- *Concurrent Pos:* Asst prof lectr, George Washington Univ, 65-66; mem comt study pipeline safety, Nat Res Coun, 87-88; consult, bd radioactive waste mgt, Nat Res Coun, 89. *Mem:* Soc Indust & Appl Math; Soc Risk Anal. *Res:* Applied mathematics. *Mailing Add:* 7012 Marbury Rd Bethesda MD 20817

CHU, SHIH I, b Taipei, Taiwan, Jan 8, 43; m 72; c 3. LASER INDUCED CHEMISTRY, CHAOS & FRACTALS. *Educ:* Nat Taiwan Univ, BS, 65, MS, 68; Harvard Univ, PhD(chem physics), 74. *Hon Degrees:* DSc, Nat Tsing Hua Univ, 71. *Prof Exp:* Res assoc, Joint Inst Lab Astrophysics, 74-76; J W Gibbs Lectr atomic physics, Yale Univ, 76-78; from asst prof to prof, 78-90, WATKINS DISTINGUISHED PROF PHYS CHEM, UNIV KANS, 90- *Concurrent Pos:* Vis scientist, Int Bus Machines Res Lab, San Jose, Calif, 74; consult, Ctr Astrophysics, Harvard Univ, 77 & 80; vis assoc, Calif Inst Tech, 81; prin investr, Dept Energy, Chem Sciences, 80-; vis fel, Joint Inst Lab Astrophysics, 85; vis prof, Univ Calif, Berkeley, 87; John Simon Guggenheim fel, 87-88. *Honors & Awards:* Alfred P Sloan Found Award, 80; Olin Petefish Award in Basic Sci, 88. *Mem:* Fel Am Phys Soc; AAAS. *Res:* Theoretical chemistry and molecular astrophysics; atomic and molecular collisions; intense field multiphoton processes; laser-induced chemical dynamics, complex-coordinate method; interstellar chemistry; application of quantum mechanics to chemical, physical and astronomical problems of current interest; classical & quantum chaos; quantum fractals. *Mailing Add:* Dept Chem Univ Kans Lawrence KS 66045

CHU, SHIRLEY SHAN-CHI, b Peking, China, Feb 16, 29; US citizen; m 54; c 3. SOLID STATE SCIENCE. *Educ:* Taiwan Nat Univ, BS, 51; Duquesne Univ, MS, 54; Univ Pittsburgh, PhD(phys chem), 61. *Prof Exp:* Res assoc x-ray crystallog, Crystallog Lab, Univ Pittsburgh, 61-67; asst prof, Southern Methodist Univ, 68-73, assoc prof, 73-81, prof elec eng, 81-88; PROF, UNIV S FLA, 88- *Concurrent Pos:* Prin investr, Res Contracts & Grants, 76-; associateship prog panelist, Nat Res Coun, 82-; mem coun, Discipline Screening Comt, Int Exchange Scholars, 86-89. *Mem:* Am Crystallog Asn; Mat Res Soc; Int Soc Optical Eng; Inst Elec & Electronics Engrs. *Res:* Epitaxial growth and characterization of compounds semiconductor films; compound semiconductor devices; photovoltaic solar energy conversion; crystal structures of organic and organometallic compounds by x-ray diffraction. *Mailing Add:* Dept Elec Eng Univ S Fla Tampa FL 33620-5350

CHU, SHU-HEH W, NUTRITION, BIOCHEMISTRY. *Educ:* Cornell Univ, PhD(nutrit-biochem), 70. *Prof Exp:* ASST PROF BIOCHEM, SCH MED, HARVARD UNIV, 84- *Mailing Add:* Children's Hosp 300 Longwood Ave Boston MA 02115

CHU, SOU YIE, b Taipei, Taiwan, Feb 17, 42; m 67; c 2. DRUG METABOLISM. *Educ:* Nat Taiwan Univ, BS, 64; Univ Ill, Chicago, PhD(pharmaceut chem), 70. *Prof Exp:* Res asst chem pharmacol, Univ Ill Med Ctr, 70, trainee, 70-74; vis fel, Nat Cancer Inst, 72-73, Nat Inst Arthritis, Metab & Digestive Dis spec fel, Roche Inst, 73-74; PHARMACOLOGIST, DRUG METAB DEPT, ABBOTT LABS, 74- *Mem:* Sigma Xi; Am Pharmaceut Asn; Am Asn Pharmaceut Scientists; Am Chem Soc. *Res:* Analysis of drugs in biological fluids; pharmacokinetics. *Mailing Add:* Drug Metab Dept Abbott Labs D463 AP9 Abbott Park IL 60064

CHU, STEVEN, b St Louis, Mo, Feb 28, 48; m 80; c 2. ATOMIC & MOLECULAR PHYSICS. *Educ:* Univ Rochester, AB, 70, BS, 70; Univ Calif, Berkeley, PhD(physics), 76. *Prof Exp:* Postdoctoral fel, Univ Calif Berkeley, 76-78; mem tech staff, AT&T Bell Labs, 78-87, head, Quantum Electronics Res Dept, 83-87; PROF PHYSICS & APPL PHYSICS, STANFORD UNIV, 87-, THEODORE & FRANCES GEBALLE PROF HUMANITIES & SCI, 90- *Concurrent Pos:* Morris Loeb Lectr, Harvard Univ, 87-88; assoc ed, Optics Letters, 88-; chair laser sci, Topical Group, Am Phys Soc, 89-90; mem, Physics Adv Coun, NSF, 90-; vis prof, Col France, 90; chair, Physics Dept, Stanford Univ, 90- *Honors & Awards:* Broida Prize, Am Phys Soc, 87; Richtinger Mem Prize Lectr, Am Phys Soc/Am Asn Physics Teachers. *Mem:* Fel Am Phys Soc; fel Optical Soc. *Res:* Atomic physics; laser spectroscopy of positronium and muonium, laser cooling and trapping of atoms and molecules and applications of these techniques. *Mailing Add:* Physics Dept Stanford Univ Stanford CA 94305

CHU, SUNG GUN, b Seoul, Korea; m 75; c 2. RHEOLOGY, PHYSICAL-POLYMER CHEMISTRY. *Educ:* HanYang Univ, BS, 73; Univ Tex, Austin, MS, 75, PhD(chem), 78. *Prof Exp:* Fel rheology, Carnegie-Mellon Univ, 78-79; res chemist, Polymer Res Inst, Univ Dayton, 79-81; RES MGR ADHESIVE POLYMER, HERCULES RES CTR, 81- *Mem:* Am Chem Soc; Soc Rheology. *Res:* Characterization of polymer; modification of polymers; adhesive formulation development; viscoelastic properties of polymers as well as lower molecular weight resins. *Mailing Add:* Indust Appln Lab Res Ctr Hercules Inc Wilmington DE 19899

CHU, SUNG NEE GEORGE, b Shanghai, China, Sept 11, 47; US citizen; m 74; c 2. OPTOELECTRONIC MATERIALS, HETEROEPITAXY. *Educ:* Nat Taiwan Univ, BS, 70; Univ Rochester, MS, 74, PhD(mat sci), 78. *Prof Exp:* Res assoc mat sci, Univ Rochester, 78-80; mem tech staff, 80-87, DISTINGUISHED MEM TECH STAFF, MATS SCI, AT&T BELL LABS, 87- *Mem:* Am Inst Physics; Mats Res Soc; Electrochem Soc. *Res:* Materials related problems in heteroepitaxy of III-V & II-VI compound semiconductors; optoelectronic devices prepared by MBE, GSMBE, CBE, MOCVD, hydride & chloride VPE & LPE, metal/III-V contact metallurgy, and degredation mechanisms of optoelectronic devices. *Mailing Add:* AT&T Bell Labs 600 Mountain Ave Murray Hill NJ 07974

CHU, TAK-KIN, b Kwang Tung, China, Dec 14, 38; US citizen; m 74; c 2. PHYSICS. *Educ:* Chung Chi Col, Hong Kong, BS, 61; Dartmouth Col, PhD(physics), 69. *Prof Exp:* Instr & res assoc, Univ Conn, 68-76; asst sr researcher, Physics & Eng, Ctr Info & Numerical Data Analysis & Synthesis, Purdue Univ, 76-80; MEM STAFF, NAVAL SURFACE WEAPONS CTR, 80- *Mem:* Am Phys Soc; Sigma Xi. *Res:* Thermal and electrical properties of materials; relationship of these properties to the fundamental physical processes and to the structural constitution of materials. *Mailing Add:* Naval Surface Weapons Ctr Code R45 White Oak Lab Silver Spring MD 20903

CHU, TA-SHING, b Shanghai, China, July 18, 34; m 62; c 2. ELECTRICAL ENGINEERING, APPLIED MATHEMATICS. *Educ:* Univ Taiwan, BSc, 55; Ohio State Univ, MSc, 57, PhD(elec eng), 60. *Prof Exp:* Res assoc, Antenna Lab, Ohio State Univ, 57-61; res assoc, Div Electromagnetic Res, Courant Inst Math Sci, 61-63; mem tech staff, Radio Res Lab, Bell Labs, 63-83; DISTINGUISHED MEM TECH STAFF, COMMUN SYST RES LAB, AT&T BELL LABS, 83- *Concurrent Pos:* Mem, Comn, B&F, Int Sci Radio Union. *Mem:* Fel Inst Elec & Electronics Engrs. *Res:* Surface wave diffraction; precision gain standard; propagation through precipitation; dual-polarization radio transmission; Crawford Hill 7-meter millimeter antenna; optical communication. *Mailing Add:* AT&T Bell Labs PO Box 400 HON-L 225 Holmdel NJ 07733-0400

CHU, TING LI, b Peking, China, Dec 26, 24; m 54; c 3. PHOTOVOLTAICS, ELECTRONIC MATERIALS. *Educ:* Catholic Univ Peking, BS, 45, MS, 48; Washington Univ, St Louis, PhD(phys chem), 52. *Prof Exp:* From asst prof to assoc prof, Duquesne Univ, 52-56; res scientist, fel scientist & mgr electronic mat, Westinghouse Res Labs, 56-67; prof elec eng, Southern Methodist Univ, 67-88; GRAD RES PROF, UNIV S FLA, 88- *Concurrent Pos:* Consult, Westinghouse Res Labs, 67-69, Tex Instruments, 69-75, Monsanto, NCR Corp, & Union Carbide, 75-78, Poly Solar Incorp, 78-88. *Mem:* Electrochem Soc; Inst Elec & Electronics Engrs; Am Soc Eng Educ. *Res:* Electronic materials and devices, including photovoltaic solar energy conversion, growth and characterization of crystals and films, and fabrication and characterization of junction devices, dielectric-semiconductor devices. *Mailing Add:* Dept Elec Eng Univ S Fla Tampa FL 33620

CHU, TSANN MING, b Kaohsiung, Formosa, Apr 18, 38; m 67; c 2. CANCER RESEARCH. *Educ:* Nat Taiwan Univ, BS, 61; NC State Univ, MS, 65; Pa State Univ, PhD(biochem), 67. *Prof Exp:* Clin chemist, Buffalo Gen Hosp, 69-70; sr cancer res scientist & asst dir clin chem, 70-71, prin cancer res scientist, 71-72, assoc chief cancer res scientist & asst clin chem, 72-76, CHMN DIAG IMMUNOL RES & BIOCHEM, ROSWELL PARK MEM INST, 76- *Concurrent Pos:* Res fel biochem, Med Found Buffalo, 67-69; United Health Found Western NY fel, 68-69; asst prof exp path, State Univ NY Buffalo, 71-74, assoc prof, 74-77, prof, 77-; mem, Comt Cancer Immunodiagnosis, Nat Cancer Inst, 78-79, Tumor Immunol Comt, 79-81, NJ Comn Cancer Res Sci Panel, 83-, Cancer Therapeut Prog Proj Rev Comt, 84-88. *Mem:* Am Asn Cancer Res; Am Chem Soc; Am Soc Biochem & Molecular Biol; Am Asn Immunologists; Am Asn Pathologists. *Res:* Tumor antigen and antibody; Bichemical and immunological markers for cancer; experimental immunotherapy for cancer. *Mailing Add:* Roswell Park Mem Inst 666 Elm St Buffalo NY 14263

CHU, VICTOR FU HUA, b Hankow, China, Jan 22, 18; nat US; m 47, 87; c 3. PHYSICAL CHEMISTRY. *Educ:* Cent China Univ, BS, 38; Yale Univ, PhD(phys chem), 50. *Prof Exp:* Chemist, Chungking Saltpeter Ref, 38-39; chemist & plant supt, Kweichow Saltpeter Ref, 39-42; chemist, Hunan Oil Ref, 42-43; plant supt, Pai-Yeh Oil Ref, 43-44; instr chem, Cent China Univ, 44-47; from res chemist to sr res chemist, E I du Pont de Nemours & Co, Inc, 50-65, res assoc, 65-75, res fel, Imaging Dept, 75-82; RETIRED. *Mem:* Am Chem Soc. *Res:* Color photography; conductance of electrolytes; photographic chemistry and systems. *Mailing Add:* 60 N Beretania St No 2201 Honolulu HI 96817

CHU, VINCENT H(AO) K(WONG), b Shanghai, China, Oct 20, 18; m 50; c 1. INORGANIC CHEMISTRY, PHYSICAL CHEMISTRY. *Educ:* Sun Yat-Sen Univ, BSc, 44; Lehigh Univ, MSc, 57, PhD(inorg chem), 62. *Prof Exp:* Asst engr, Cent Indust Res Inst, China, mfg head, Taiwan Camphor Bur & supt res & mfg, Taipei Chem Works, 44-52; ENGR RAW MAT, HOMER RES LABS, BETHLEHEM STEEL CORP, 57- *Concurrent Pos:* Vis prof, Nat Sun Yat-Sen Univ, Taiwan; consult, China Steel Corp, Taiwan, Indust Technol Res Inst. *Mem:* AAAS; Am Chem Soc; Am Inst Mining, Metall & Petrol Engrs; Sigma Xi; Chinese Inst Engrs (pres, 81-83). *Res:* Organo-metal compound; coordination in aprotic media; reduction kinetics of iron ore; iron ore agglomeration; crystal field theory; powder metallurgy; pyrometallurgy; coal gasification; coking coal immprovement. *Mailing Add:* 1310 Woodland Circle Bethlehem PA 18017

CHU, WEI-KAN, b Kunming, Yunnan, China, Apr 1, 40; US citizen; m 66; c 1. HIGH TEMPERATURE SUPERCONDUCTIVITY, THIN FILM PROCESSING. *Educ:* Cheng-Kung Univ, Taiwan, BS, 62; Baylor Univ, Waco, Tex, MS, 65, PhD(physics), 69. *Prof Exp:* Postdoctoral fel, Baylor Univ, 69-72; res fel, Caltech, 72-75; staff, adv & sr engr, Int Bus Mach Corp, 75-81; res prof physics, Univ NC, 81-88; DIST UNIV PROF PHYSICS, UNIV HOUSTON, 89-; DEP DIR, TEX CTR SUPERCONDUCTIVITY, 89- *Honors & Awards:* Sr US Scientist, Alexander von Humboldt-Stiftung Found, 89. *Res:* Ion implantation in semiconductors and superconductors; application of high temperature superconductivity and thin film processing. *Mailing Add:* Tex Ctr Superconductivity Univ Houston Houston TX 77204-5932

CHU, WESLEY W(EI-CHIN), b Shanghai, China, May 5, 36; US citizen; m 60; c 2. COMPUTER SCIENCE, ELECTRICAL ENGINEERING. *Educ:* Univ Mich, BSE, 60, MSE, 61; Stanford Univ, PhD(elec eng), 66. *Prof Exp:* Engr, Comput Dept, Gen Elec Co, Ariz, 61-62 & Int Bus Mach Corp, Calif, 64-66; res engr, Stanford Electronics Labs, 66; mem tech staff, Bell Tel Labs, NJ, 66-69; assoc prof comput sci, 69-75, chair, 88-91, PROF COMPUT SCI, UNIV CALIF, LOS ANGELES, 75- *Concurrent Pos:* Consult govt agencies & pvt industs; ed text bks, advances comp commun, 74, 76 & 77 & database systs, 79; prog chmn, Fourth Data Commun Symp & chmn, Interprocess Commun Workshop, 75; ed, J Comput Networks, 75- & J Comput & Elec Eng, 78-; assoc ed, Inst Elec & Electronics Engrs Trans Comput, 78-80; US prog chmn, 12th Int Conf Very Large Databases, 86; mem, Asn Comput Mach Spec Interest Group on Data Commun. *Honors & Awards:* Meritorious Serv Award, Inst Elec & Electronics Engrs, 83. *Mem:* Inst Elec & Electronics Engrs. *Res:* Computer communications and computer networking, distributed processing, and distributed data bases; knowledge based systems. *Mailing Add:* Dept Comput Sci Univ Calif Los Angeles CA 90024

CHU, WILLIAM HOW-JEN, polymer physics, for more information see previous edition

CHU, WILLIAM TONGIL, b Seoul, Korea, Apr 16, 34; m 62; c 2. RADIATION PHYSICS. *Educ:* Carnegie Inst Technol, BS, 57, MS, 59, PhD(physics), 63. *Prof Exp:* Res assoc high energy physics, Brookhaven Nat Lab, 63-64; asst prof physics, Ohio State Univ, 64-70; asst prof radiol, Sch Med, Loma Linda Univ, 71-75, assoc prof radiol, 75-78, prof radiol, 78-79; scientist III, Div Accelerator & Fusion Res, 79-87, SR SCIENTIST II, RES MED & RADIATION BIOPHYS, LAWRENCE BERKELEY LAB, UNIV CALIF, BERKELEY, 87- *Concurrent Pos:* Res collabr, Brookhaven Nat Lab, 72-73; res consult, Lawrence Berkeley Lab, 76-79; consult, proton med accelerator facil, Loma Linda Univ, 88- *Honors & Awards:* IR 100 Award, 87. *Mem:* Am Phys Soc; Radiation Res Soc; Am Asn Physicists Med; Sigma Xi; Asn Korean Physicists Am (pres, 86). *Res:* Experimental elementary particle physics; radiation physics and radiation biology; radiation physics for biomedical use of accelerated heavy ions; heavy ion imaging. *Mailing Add:* 64-227 Lawrence Berkeley Lab Berkeley CA 94720

CHU, WING TIN, b Hong Kong, Oct 22, 35; Can citizen; m 68; c 2. AERODYNAMICS, ACOUSTICS. *Educ:* Univ BC, BASc, 61; Univ Toronto, MASc, 63, PhD(aerospace eng), 66. *Prof Exp:* Sr res fel aerospace eng, Inst Aerospace Studies, Univ Toronto, 66-69; asst prof, Univ Southern Calif, 69-73, sr res assoc aerospace eng, 73-75; assoc res officer, 75-80, SR RES OFFICER, NAT RES COUN CAN, 80- *Mem:* Fel Acoust Soc Am; Can Acoust Asn. *Res:* Room acoustics and noise control engineering. *Mailing Add:* Nat Res Coun Can Inst Res Construct Montreal Rd Ottawa ON K1A 0R6 Can

CHU, YUNG YEE, b Hangchow, China, Aug 18, 33; m 67. NUCLEAR CHEMISTRY. *Educ:* Nat Taiwan Univ, BSc, 54; Univ Calif, Berkeley, PhD(chem), 60. *Prof Exp:* Res assoc nuclear chem, 59-61, assoc chemist, 61-65, CHEMIST, BROOKHAVEN NAT LAB, 65- *Mem:* Am Chem Soc; Am Phys Soc. *Res:* Nuclear fission; high energy nuclear reactions; nuclear spectroscopy. *Mailing Add:* Chem Dept Brookhaven Nat Lab Upton NY 11973

CHUA, BALVIN H-L, b May 9, 1946; m 75; c 3. HYPERTENSION, CARDIAC HYPERTROPHY. *Educ:* Univ Wis-Madison, PhD(nutrit), 75. *Prof Exp:* Asst prof, Dept Physiol, Penn State Univ, 78-85; ASSOC PROF BIOCHEM, DEPT PATH, WAYNE STATE UNIV, 86- *Mem:* Am Physiol Soc; Am Soc Biochem. *Res:* Molecular basis of cardiac and vascular hypertrophy and molecular biology in hypertension; control of cardiac collagen metabolism. *Mailing Add:* Dept Path Wayne State Univ Scott Hall Rm 9239 Detroit MI 48201

CHUA, KIAN ENG, b Sumatra, Indonesia, Mar 12, 35; Can citizen; m 68; c 2. BIOCHEMISTRY, MICROBIOLOGY. *Educ:* Miami Univ, BA, 59, MA, 63; Univ Toronto, PhD(zool & limnol), 69. *Prof Exp:* Res assoc radiobiol, Bowman Gray Sch Med, 63-64; res assoc immunol, dept microbiol, Miami Univ, 64-65; SUPVR BIOCHEM, LAB ANALYSIS SYSTS, UNIV TORONTO & ROYAL ONT MUS, 72-88; PROF BIOCHEM, IMMUNOL & MICROBIOL, DEPT BIOL SCI, PHARMACEUT SCI & CHEM, SENECA COL, 73- *Concurrent Pos:* Res fel, reproductive biol & immunol, Toronto Gen Hosp, 90; Lab dir, Bay St Infertil Asn, 89- *Mem:* Am Soc Microbiol; NY Acad Sci. *Res:* Interactions of microorganisms, macroinvertebrates and sediments; biochemical systematics of game fishes; autoimmune diseases and recurrent fetal loss. *Mailing Add:* 100 Bedford Toronto ON M5R 2K2 Can

CHUA, LEON O(NG), b Tarlac, Philippines, June 28, 36; m 61; c 4. ELECTRICAL ENGINEERING, MATHEMATICS. *Educ:* Mass Inst Technol, MSEE, 62; Univ Ill, PhD(elec eng), 64. *Hon Degrees:* Dr, Fed Polytech Sch Lausanne, Switz, 83, Univ Tokushima, Japan, 84. *Prof Exp:* Res engr, Data Syst Div, Int Bus Mach Corp, 62; from asst prof to assoc prof elec eng, Purdue Univ, 64-71; assoc prof, 71-72, PROF ELEC ENG & COMPUT SCI, UNIV CALIF, BERKELEY, 72- *Concurrent Pos:* Mem, Comput-Aided Network Design Comt, Inst Elec & Electronics Engrs, 67-68; vchmn, Circuit Theory Group, Inst Elec & Electronics Engrs, 70-72, ed, Trans Circuits & Systs, 73-75, pres-elect, Soc Circuits & Systs, 75-76 & pres, 76-77; sr vis fel, Cambridge Univ, Eng, 82; sr US scientist fel, Japan Soc Prom Sci, 83. *Honors & Awards:* Browder J Thompson Mem Prize, Inst Elec & Electronics Engrs, 67, W R G Baker Prize, 73 & Centennial Medal, 84; Frederick E Terman Award, Am Soc Eng Educ, 74; Alexander von Humboldt Sr US Scientist Award, Tech Univ Munich, W Germany, 83; Guillemin-Cauer Prize, 85. *Mem:* Fel Inst Elec & Electronics Engrs. *Res:* Nonlinear networks; nonlinear device and system modelling; nonlinear dynamics; computer-aided design; author and co-author of numerous publications and holder of five US patents. *Mailing Add:* Dept Elec Eng & Comput Sci Univ Calif Berkeley CA 94720

CHUAN, RAYMOND LU-PO, b Shanghai, China, Mar 4, 24; US citizen; m 51; c 2. ACOUSTICS, AERONAUTICAL & ASTRONAUTICAL ENGINEERING. *Educ:* Pomona Col, BA, 44; Calif Inst Technol, MS, 45, PhD(aeronaut), 53. *Prof Exp:* Res assoc aeronaut, Eng Ctr, Univ Southern Calif, 53-57, dir & adj prof, 57-64; pres, Celestial Res Corp, 64-68; mgr adv technol, Missile Syst Div, Atlantic Res Corp, 68-72; staff asst technol, Celesco Indust Inc, 72-76; staff scientist, Defense Div, Brunswick Corp, 76-88; STAFF SCIENTIST, FEMTOMETRICS, 88- *Concurrent Pos:* Consult, Adv Group Aero Res & Develop, NATO, 60-64 & Nat Aero & Space Admin, 67-86. *Mem:* Am Geophys Union; AAAS. *Res:* Rarefied gasdynamics; vacuum technology; analytical instrumentation for air quality; aerosol technology. *Mailing Add:* PO Box 1183 Hanalei HI 96714

CHUANG, DE-MAW, b Tainan Hsien, Taiwan, Oct 12, 42; US citizen; m 68; c 2. NEUROPHARMACOLOGY, MOLECULAR BIOLOGY. *Educ:* Nat Taiwan Univ, BS, 66; State Univ NY, Stony Brook, PhD(molecular biol), 71. *Prof Exp:* Assoc, Roche Inst Molecular Biol, 71-73; guest res worker, NIMH, 73-74, staff fel, 74-77, chemist, 77-83, GROUP HEAD, NIMH, 83-, SECT CHIEF, 90- *Mem:* Am Soc Pharmacol & Exp Therapeut; Am Soc Biol Chemists; Soc Neuroscience. *Res:* transynaptic regulation of gene expression; mechanisms of action of antidepressant drugs; regulation of receptor-mediated phospholipid turnover; molecular mechanisms of desensitization of receptors for neurotransmitters; mechanisms of actions of neurotoxins. *Mailing Add:* Biol Psychiat Br NIMH Bldg 10 Rm 3N212 Bethesda MD 20892

CHUANG, HANSON YII-KUAN, b Nanking, China, Sept 24, 35; US citizen; m 66; c 2. BIOCHEMISTRY, PATHOLOGY. *Educ:* Nat Taiwan Univ, BS, 58; Univ NC, PhD(biochem), 68. *Prof Exp:* Res asst chem, Acad Sinica, China, 60-63; res assoc path, 72-73, instr path & biochem, 73-74, asst prof path & biochem, Univ NC, Chapel Hill, 74-75; asst prof path, Brown Univ, 75-77; asst prof path, Univ S Fla, 77-79; RES ASSOC PROF PATH & BIOENG, UNIV UTAH, 79-; CONSULT, GULL LABS VIROL & IMMUNOL, 85- *Concurrent Pos:* Res fel physiol chem, Johns Hopkins Univ, 68-71. *Mem:* Am Chem Soc; Am Asn Pathologists; NY Acad Sci; Int Soc Artificial Organs; Am Asn Blood Banks. *Res:* Enzyme and protein isolation, purification and characterization; metabolism and function of biogenic and cholinergic amines; blood enzymology; platelet function in blood; blood-artificial surface interaction; endothelium function in thrombosis and hemostasis; blood banking and hemotherapy; blood sustitutes; immuno assays; momoclonal antibodies; viral diagnostic assays. *Mailing Add:* Pathology Dept Col Med Univ Utah Salt Lake City UT 84132

CHUANG, HENRY NING, b Nanking, China, July 5, 37; m 65; c 3. MECHANICAL ENGINEERING, ENERGY ENGINEERING. *Educ:* Nat Taiwan Univ, BSME, 58; Univ Md, MSAeroE, 62; Carnegie Inst Tech, PhD(mech eng), 66. *Prof Exp:* From instr to assoc prof, 65-78, Univ Energy Coord, 77-80, PROF MECH ENG, UNIV DAYTON, 78-; DIR, ENERGY ANALYSIS & DIAG CTR PROG, 80- *Mem:* Am Soc Mech Engrs; Am Soc Heating Refrig & Air Conditioning Engrs. *Res:* Energy conservation/conversion and energy engineering. *Mailing Add:* Dept Mech Eng Univ Dayton 300 Col Park Ave Dayton OH 45469-0210

CHUANG, KUEI, b Shanghai, China, June 26, 26; US citizen; m 60. ELECTRICAL ENGINEERING. *Educ:* Nat Taiwan Univ, BSE, 50; Univ Mich, MSE, 52, PhD(elec eng), 58. *Prof Exp:* Spec instr elec eng, Wayne State Univ, 54-55; res assoc, 58-59, from asst prof to assoc prof, 59-69, PROF ELEC ENG, UNIV MICH, ANN ARBOR, 69- *Honors & Awards:* First Prize Indust Div, Inst Elec & Electronics Engrs, 59. *Mem:* Inst Elec & Electronics Engrs. *Res:* Control engineering; stochastic processes; optimal control systems. *Mailing Add:* 3316 EESC Bldg Elec Eng Univ Mich Ann Arbor MI 48109-2122

CHUANG, KUEN-PUO (KEN), b Ilan, Formosa, Jan 20, 33; m 63; c 3. FINITE ELEMENT. *Educ:* Nat Taiwan Univ, BS, 55; Univ Ill, MS, 59, PhD(civil eng), 62. *Prof Exp:* Asst prof civil eng, Univ Okla, 62-66; assoc prof, 66-73, PROF CIVIL ENG, LOYOLA MARYMOUNT UNIV, 73- *Concurrent Pos:* Consult, Agbabian Assocs, El Segundo, Calif, 77 & TRW, Redondo Beach, Calif, 74- *Mem:* Am Soc Civil Engrs; Sigma Xi. *Res:* Structural analysis and finite element; code development and applications. *Mailing Add:* Dept Civil Eng 7101 W 80th St Los Angeles CA 90045

CHUANG, MING CHIA, engineering, computer science, for more information see previous edition

CHUANG, RONALD YAN-LI, b Szuchuan, China, Feb 12, 40; m 67; c 3. BIOCHEMISTRY, PHARMACOLOGY. *Educ:* Nat Taiwan Univ, BS, 61; Univ Calif, Davis, MS, 66, PhD(biochem), 71. *Prof Exp:* Res assoc cancer res, Columbia Univ, 71-72; asst prof pharmacol, Med Ctr, Duke Univ, 72-76; asst res biochemist, Calif Primate Res Ctr, Univ Calif, Davis, 76-78; asst prof biochem, Sch Med, Oral Roberts Univ, 78-81; asst prof, 81-84, ASSOC PROF PHARMACOL, UNIV CALIF, DAVIS, 85- *Concurrent Pos:* NIH fel, Col Physicians & Surgeons, Columbia Univ, 71-72; chemist, Vet Admin Hosp, Durham, NC, 73-76; NIH res grant, Med Ctr, Duke Univ, 74-76 & Univ Calif, Davis, 76-78 & 82- *Mem:* AAAS; Am Soc Biol Chemists; Am Asn Cancer Res; Am Soc Pharmacol & Exp Therapeut. *Res:* Study of the control mechanism of gene expression in leukemic cells; study of the biochemical mechanism of the action of antineoplastic agents; molecular cloning of animal viruses. *Mailing Add:* Dept Pharmacol Univ Calif Sch Med Davis CA 95616

CHUANG, STRONG CHIEU-HSIUNG, b Tainan Co, Taiwan, Jan 1, 39; US citizen; m 67; c 2. PAPER & FOOD PRODUCT & PROCESS DEVELOPMENT. *Educ:* Nat Taiwan Univ, BS, 61; Kans State Univ, MS, 67; Purdue Univ, PhD(mech eng), 70. *Prof Exp:* Res assoc, Purdue Univ, 70-71; develop engr, Procter & Gamble Co, 71-76, sr engr, 76-84, sect head, 84-89; SCI ADV TISSUE TECHNOL, SCOTT PAPER, 89- *Mem:* Tech Asn Pulp & Paper Indust; Am Soc Mech Engrs. *Res:* Fluid mechanics, heat transfer modeling and simulation; equipment and process trouble shooting and problem solving; process equipment development and installation; alternate energy utilization and concept development; fluid mechanics. *Mailing Add:* 205 Walnut Ridge Lane Chadds Ford PA 19317

CHUANG, TSAN IANG, b Hsin-Chu, Taiwan, Apr 21, 33; US citizen; m 58; c 3. SYSTEMATIC BOTANY. *Educ:* Taiwan Normal Univ, BS, 56; Nat Taiwan Univ, MS, 59; Univ Calif, Berkeley, PhD(bot), 66. *Prof Exp:* Asst res fel & cur herbarium, Inst Bot, Academia Sinica, Taiwan, 59-62; asst prof, Univ RI, 66-67; asst prof, 67-71, assoc prof bot, 71-77, CUR HERBARIUM, DEPT BIOL SCI, ILL STATE UNIV, 71-, PROF BOT, 77- *Concurrent Pos:* NSF res grants, 72 & 75. *Mem:* Bot Soc Am; Am Soc Plant Taxonomists; Int Asn Plant Taxon; Int Orgn Plant Biosystematists. *Res:* Systematics and evolution of genera cordylanthus, castilleia, orthocarpus; cytotaxonomy of umbelliferae; pollen morphology and its taxonomic significance of hydrophyllaceae, Campanulaceae and scrophulariaceae. *Mailing Add:* Dept Biol Sci Ill State Univ Normal IL 61761

CHUANG, TZE-JER, b Chiayi, Taiwan, July 19, 43; US citizen; m 74; c 2. FRACTURE MECHANICS. *Educ:* Nat Cheng Kung Univ, BSc, 65; Duke Univ, ScM, 70; Brown Univ, PhD(eng), 75. *Prof Exp:* Civil engr, Taiwan Power Co, 66, 68; sr engr, Westinghouse Elec Corp, 74-80; PHYSICIST, NAT INST STANDARDS & TECHNOL, 80- *Mem:* Am Soc Mech Engrs; Am Soc Metals; Am Ceramic Soc. *Res:* Fracture and deformation of crystalline materials, especially crack tip growth processes at both microstructural and atomic (lattics) levels; mechanical properties of materials at elevated temperatures. *Mailing Add:* Ceramics Div Nat Inst Standards & Technol Gaithersburg MD 20899

CHUBB, CHARLES F(RISBIE), JR, b Pittsburgh, Pa, Jan 31, 20; m 48; c 2. ELECTRICAL ENGINEERING. *Educ:* Princeton Univ, BA, 41; Mass Inst Technol, BS, 43; Polytech Inst Brooklyn, MEE, 50. *Prof Exp:* Mem staff, Radiation Lab, Mass Inst Technol, 43-46; proj engr, Sperry Gyroscope Co, 46-50, sr proj engr, 50-53, res engr, 53-54, head eng sect, 54-57, head eng dept, 58-63; vpres, Dynell Electronics Corp, 63-66, sr vpres, 66-78; vpres technol, Norden Styts, Inc, 78-81, mgr shipboard technol, 81-85; SR VPRES, NOVA TECHNOL, INC, 85- *Mem:* Sr mem Inst Elec & Electronics Engrs. *Res:* Servomechanisms and radar. *Mailing Add:* 89 Cabot Ct Unit L Hauppauge NY 11788

CHUBB, CURTIS EVANS, b Fort Worth, Tex, Feb 19, 45; m 76. REPRODUCTIVE BIOLOGY. *Educ:* Okla State Univ, BS, 68; Johns Hopkins Univ, PhD(reprod biol), 78. *Prof Exp:* NIH fel, male reproductive biol, Johns Hopkins Univ, 72-78 & Univ Tex, Austin, 78-80; asst prof cell biol & anat, 80-87, ASSOC PROF CELL BIOL & NEUROSCI, SOUTHWESTERN MED CTR DALLAS, 87- *Concurrent Pos:* Prin investr, Univ Tex Southwestern Med Ctr, Dallas, 81-; res career develop award, NIH, 86-91. *Honors & Awards:* Young Investr Award, Soc Study Reproduction, 77. *Mem:* Soc Study Reproduction; Am Soc Andrology; AAAS; Am Physiol Soc. *Res:* Control strategies of mammalian testis function; research in male-sterile mice to decipher hormonal requirements for spermatogenesis and to study male infertility. *Mailing Add:* Dept Cell Biol & Neurosci Univ Tex Southwestern Med Ctr Dallas TX 75235-9039

CHUBB, FRANCIS LEARMONTH, b Que, June 26, 13; m 44. ORGANIC CHEMISTRY. *Educ:* McGill Univ, BSc, 35; Univ Southern Calif, MSc, 49, PhD, 52. *Prof Exp:* Chemist, Dom Oilcloth & Linoleum Co, 35-46; lab asst, Univ Southern Calif, 46-50; Can Cancer Soc fel, Univ Alta, 51-54; sr chemist, Merck & Co, Ltd, 54-55; dir org chem res, Frank W Horner Ltd, 55-75; RES ASSOC CHEM, McGILL UNIV, 75- *Mem:* Am Chem Soc; fel Chem Inst Can. *Res:* Pharmaceutical chemistry; five and six-membered heterocyclic compounds; six-membered alicyclic compounds. *Mailing Add:* Dept Chem McGill Univ 801 Sherbrooke St W Montreal PQ H3A 2K6 Can

CHUBB, SCOTT ROBINSON, b New York, NY, Jan 30, 53; m 82; c 1. REMOTE SENSING USING MICROWAVES, RELATIVITY. *Educ:* Princeton Univ, BA, 75; State Univ NY, Stony Brook, MA, 78, PhD(physics), 82. *Prof Exp:* Res assoc, Physics Dept, Northwestern Univ, 82-85, Nat Acad Sci/Nat Res Coun, 85-88; res physicist, Sachs-Freeman Assocs, Inc, 88-89; RES PHYSICIST, OFF NAVAL RES, NAVAL RES LAB, 89- *Mem:* Am Phys Soc; Sigma Xi; Am Geophys Union; Am Inst Aeronaut & Astronaut. *Res:* Theoretical studies of the effects of the ocean on radar imagery derived from space-based platforms; effects of space environment on precision time-keeping and the transfer of time between distant points; theoretical guidance to problems involving cold fusion. *Mailing Add:* Naval Res Lab Code 4234 Washington DC 20375-5000

CHUBB, TALBOT ALBERT, b Pittsburgh, Pa, Nov 5, 23; c 4. GEOPHYSICS, ASTROPHYSICS. *Educ:* Princeton Univ, AB, 44; Univ NC, PhD(physics), 50. *Prof Exp:* Head, upper air physics br, Naval Res Lab, 59-81; PRES, RES SYSTS INC, 81- *Honors & Awards:* E O Hulburt Award, Naval Res Lab, 63. *Mem:* Am Geophys Union; Am Astron Soc; Am Phys Soc; Int Solar Energy Soc. *Res:* Aeronomy; optical geophysics; x-ray astronomy; solar thermal processes for energy recovery. *Mailing Add:* 5023 N 38th St Arlington VA 22207

CHUBB, WALSTON, b Washington, DC, July 23, 23; m 51; c 2. MATERIALS ENGINEERING, MATHEMATICAL MODELS OF MATERIALS BEHAVIOR. *Educ:* Harvard Univ, AB, 44; Univ Mo, BS, 48, MS, 49. *Prof Exp:* Asst engr, Brush Beryllium Co, 49-51; res engr, Battelle Mem Inst, 51-57, asst div consult, 57-62, res assoc, 62-66, fel, Dept Mat Eng, 66-72; prin engr, Nuclear Fuel Div, Westinghouse Elec Corp, 72-86; PVT CONSULT, 86- *Mem:* Am Nuclear Soc; Sigma Xi. *Res:* Refractory nuclear fuels and other high temperature materials; mathematical modeling of the behavior of nuclear materials at high temperatures. *Mailing Add:* 3450 MacArthur Dr Murrysville PA 15668

CHUBER, STEWART, b Queens Village, NY, Dec 22, 30; m 53; c 2. PETROLEUM GEOLOGY. *Educ:* Colo Sch Mines, GeolE, 52; Stanford Univ, MS, 53, PhD(geol), 61. *Prof Exp:* Subsurface geologist, Magnolia Petrol Corp, 53-54; field geologist, Mobil Oil Co Can, Ltd, 54-56, party chief, 56; subsurface geologist, Western Div, Mobil Oil Co, Calif, 57-60 & Franco Western Oil Co, Calif & Tex, 61-65; consult geologist, 65-68; div geologist, Buttes Gas & Oil Co, 68-70; consult geologist, 71; vpres, Cantrell, Wheeler, Lewis & Chuber Geologists & Engrs, 71-72; CONSULT, 73- *Mem:* Geol Soc Am; Sigma Xi; Soc Econ Paleontologists & Mineralogists. *Res:* Stratigraphy; stratigraphic nomenclature; geologic history, especially Permian and Pennsylvanian cyclic sedimentation; numerous oil & gas field studies; clastic sedimentation. *Mailing Add:* Drawer J Schulenburg TX 78956

CHUCKROW, VICKI G, b Brooklyn, NY, July 26, 41. MATHEMATICS. *Educ:* City Col NY, BS, 62; NY Univ, MS, 64, PhD(math), 66. *Prof Exp:* Asst prof, 66-72, ASSOC PROF MATH, CITY COL NY, 72- *Concurrent Pos:* NSF grant, 66-67. *Mem:* Am Math Soc; Math Asn Am. *Res:* Riemann surfaces; Schottky groups, a special subclass of Kleinian groups. *Mailing Add:* Dept Math City Col NY Convent 138th St New York NY 10031

CHUDNOVSKY, GREGORY V, b Kiev, USSR, April 17, 52. TRANSCENDENTAL NUMBERS, SOLVABLE MODELS. *Educ:* Kiev State Univ, Dipl Math, 74; Inst Math, Kiev, PhD(math), 75. *Prof Exp:* Res fel, Kiev State Univ, 74-76; RES PROF, NAT CTR SCI RES, 79-; RES ASSOC, DEPT MATH, COLUMBIA UNIV, 78- *Concurrent Pos:* Vis prof, Inst Higher Sci Studies, France, 77-78; John D & Catherine MacArthur Found fel, 81. *Mem:* Am Math Soc; Am Phys Soc; French Math Soc; Math Asn Am. *Res:* Pure mathematics; number theory, especially theory of transcedental numbers; mathematical physics with emphasis on classical and quantum dynamical systems of physical origin. *Mailing Add:* Dept Math Columbia Univ New York NY 10027

CHUDWIN, DAVID S, b Chicago, Ill, July 11, 50; m 83; c 2. PEDIATRIC ALLERGY & IMMUNOLOGY, MICROBIOLOGY. *Educ:* Univ Mich, Ann Arbor, BS, 72, MD, 76. *Prof Exp:* Resident pediat, Univ Wis Hosps, 76-79; fel immunol, Univ Calif, San Francisco, 79-82; ASST PROF IMMUNOL, RUSH MED COL, 82- *Concurrent Pos:* Asst attend physician, Rush-Presby-St Luke's Med Ctr, 82-, co-dir, Allergy-Clin Immunol Prog, 86-87. *Mem:* Am Acad Allergy & Immunol; Am Asn Immunol; Am Soc Microbiol; AAAS; Am Fedn Clin Res. *Res:* Immunologic host defenses against pneumococcal infections; interactions of antibodies, complement, and acute phase proteins in host defense and autoimmune disease. *Mailing Add:* Dept Immunol Rush-Presby-St Luke's Med Ctr 1735 W Congress Pkwy Chicago IL 60612

CHUEH, CHUN FEI, b Chaochow, China, Sept 17, 32; m 61; c 2. CHEMICAL ENGINEERING, THERMODYNAMICS. *Educ:* Nat Taiwan Univ, BS, 55; Kans State Univ, MS, 57; Ga Inst Tech, PhD(chem eng), 62. *Prof Exp:* Sr chem engr, 62-67, sect head, 67-78, process mgr, Halcon Int Inc, 78-81, tech dir, Halcon Sd Group Inc, 81-86; GEN MGR, FLORASYNTH-SHANGHAI COSFRA LTD, 87- *Mem:* Am Chem Soc; Am Inst Chem Engrs; Nat Asn Corrosion Engrs. *Res:* Theory and application of gas chromatography; thermodynamics of phase equilibria; correlation and prediction of physical properties; distillation. *Mailing Add:* 187-16 Cambridge Rd Jamaica NY 11432

CHUEY, CARL F, b Youngstown, Ohio, Mar 19, 44; div; c 1. PLANT TAXONOMY. *Educ:* Youngstown Univ, BS, 66; Ohio Univ, MS, 69. *Hon Degrees:* EdD, Ohio Christian Col, 71. *Prof Exp:* Instr, 67-74; asst prof, 74-81, ASSOC PROF BIOL, YOUNGSTOWN STATE UNIV, 81- *Concurrent Pos:* Cur herbarium, Youngstown State Univ, 68- *Mem:* Am Fern Soc; Am Soc Plant Taxon; Brit Pteridological Soc. *Res:* Pteridophyte flora of Ohio, western Pennsylvania, and West Virginia; pteridophyte specification by induced mutation. *Mailing Add:* Hof Wapiti Kapradina PO Box 146 New Springfield OH 44443

CHUGH, ASHOK KUMAR, b Amritsar, Punjab, Ind, Dec 9, 42; US citizen; m 70; c 3. GEOTECHNICAL ENGINEERING, TECHNICAL ASSISTANCE & TRAINING. *Educ:* Punjab Univ, BS, 63; Univ Minn, MS, 68; Univ Ky, PhD(civil eng), 73. *Prof Exp:* Asst engr, Punjab Pub Works Dept, 63-67; struct engr, Meyer Mfg, Inc, 68 & Watkins & Assocs, Inc, 68-71; res asst struct anal, Univ Ky, 71-73, res assoc, Soils Lab, 73-75; res engr, US Bur Mines, 76-78; TECH SPECIALIST, US BUR RECLAMATION, 78- *Concurrent Pos:* Struct engr, Am Eng Co, 71-75; lectr, Univ Colo, Denver, 83, Boulder, 88 & 90; consult, Cent Water Comn, New Delhi, India, 88; mem adv bd, Int J Numerical & Analytical Methods Geomech; mem consult bd, World Bank & Asian Develop Bank, 88-; instr, Pub Works Dept, Kualalumpur, Malaysia, 89; researcher, Pub Works Res Inst, Tsukuba, Japan, 90; chmn, Comt Methods Numerical Analysis Dams, US Comt Large Dams of Int Comn Large Dams, 91- *Mem:* Fel Am Soc Civil Engrs. *Res:* Engineering analysis of embankment dams for static and dynamic conditions; stability of natural slopes; development and/or adaptations of computer programs to perform engineering analyses of embankment dams. *Mailing Add:* US Bur Reclamation Mail Stop D-3620 PO Box 25007 Denver CO 80225

CHUGH, YOGINDER PAUL, b Multan, India, Oct 6, 40; m 70; c 3. ROCK MECHANICS, PRODUCTION ENGINEERING. *Educ:* Banaras Hindu Univ, BS, 61; Pa State Univ, MS, 68, PhD(mining eng), 71. *Prof Exp:* Mgr coal mining, Andrew Yule Coal Co, 61-65; res asst, Pa State Univ, 65-70; res assoc, Columbia Univ, 71; res engr soil-rock mech, IIT Res Inst, 72-74; planning engr, Amax Coal Co, 74-76; actg chmn, Dept Mining, 80-81, PROF MINING ENG, SOUTHERN ILL UNIV, 77-, DEPT CHMN, 84-; DIR, ILL MINING & MINERAL RESOURCES RES INST, 89- *Concurrent Pos:* Prin investr, US Bur Mines, US Dept Energy, 73-82; consult, US Dept Commerce & Mining Indust, 78-85, Ill Coal Bd; dir, Ill Mining Res Inst, 88. *Mem:* Am Inst Mining Engrs; Am Soc Eng Educ. *Res:* Surface coal mining; roof control in underground excavations; planning improved mining and reclamation operations; assessing roof conditions prior to mining; effects of moisture on roof control; mine subsidence; coal processing; author or coauthor of over 75 publications; 2 patents. *Mailing Add:* Dept Mining Eng Southern Ill Univ Carbondale IL 62966

CHUI, CHARLES KAM-TAI, b Macao, China, May 7, 40; m 64; c 2. MATHEMATICS. *Educ:* Univ Wis-Madison, BS, 62, MS, 63, PhD(math), 67. *Prof Exp:* Asst prof math, State Univ NY Buffalo, 67-70; assoc prof, 70-74, PROF MATH, TEX A&M UNIV, 74- *Concurrent Pos:* Vis prof, Nat Res Coun, Italy, 80-81. *Mem:* Math Asn Am; Am Math Soc; Soc Indust Appl Math; Int Elec & Electronics Engrs. *Res:* Analysis; approximation theory. *Mailing Add:* Dept Math Tex A&M Univ College Station TX 77843

CHUI, DAVID H K, b Hong Kong, Oct 9, 39; Can citizen; m 67; c 3. HEMATOLOGY, GENETICS. *Educ:* Univ Md, BS, 59; McGill Univ, MD & CM, 63; FRCP(c), 70. *Prof Exp:* Intern, Royal Victoria Hosp, Montreal, 63-64, resident internal med, 64-65; resident internal med, Grad Hosp, Univ Pa, 65-67; fel hemat, Columbia-Presby Med Ctr, NY 67-70; instr, Dept Human Genetics & Develop, Col Physicians & Surgeons, Columbia Univ, 69-70; from asst prof to assoc prof, 70-82, PROF, DEPT PATH, MCMASTER UNIV, 82- *Mem:* Am Soc Clin Invest; Am Fedn Clin Res; Am Soc Hemat; Am Asn Pathologists. *Res:* Developmental biology of erythrypoiesis in health and disease; hemoglobin ontogeny and its regulation; molecular genetics of hematological disorders. *Mailing Add:* Dept Path McMaster Univ Med Ctr Hamilton ON L8N 3Z5 Can

CHUI, SIU-TAT, b Hong Kong, Apr 20, 49; US citizen; m; c 1. THEORETICAL SOLID STATE PHYSICS. *Educ:* McGill Univ, BSc, 69; Princeton Univ, PhD(physics), 72. *Prof Exp:* Instr physics, Princeton Univ, 72-73; mem tech staff, Bell Tel Lab, 73-75; asst prof physics, State Univ NY Albany, 75-79; assoc prof, 79-90, PROF, BARTOL FOUND, FRANKLIN INST, UNIV DEL, 90- *Mem:* Am Phys Soc; Sigma Xi. *Res:* Superconductivity; one-dimensional physics; lattice dynamics; metal-insulator transition; surface physics. *Mailing Add:* Bartol Res Inst Univ Del Newark DE 19716

CHULICK, EUGENE THOMAS, b Jackson, Calif, Jan 8, 44; m 66; c 2. TECHNICAL & GENERAL MANAGEMENT, SALES & MARKETING. *Educ:* Univ of the Pac, BS, 65; Wash Univ, PhD(nuclear chem), 69. *Prof Exp:* Res assoc nuclear chem, Cyclotron Inst, Tex A&M Univ, 69-74, instr, Physics Dept, 74; sr res chemist, Babcock & Wilcox, 74-77, supvr radiochem group, 77-79, mgr, Nuclear & Radiochem Sect, 79-80, MGR, WATER TECH BUS DEVELOP, RES & DEVELOP DIV, BABCOCK & WILCOX CTR, 80-; MGR, SALES & MARKETING, POWERSAFETY INT, 89- *Concurrent Pos:* Tex Res Coun fel, 70-; Tech expert, Int Atomic Energy Agency, 73-74; mem subcomt, Nuclear & Radiochem Sect, Nat Acad Sci, 81-84. *Mem:* AAAS; Am Chem Soc; Am Phys Soc; Am Nuclear Soc. *Res:* Delayed neutrons from fission products; nuclear reactions; internal ionization during beta decay; radiochemistry of pressurized water reactors; water technology and management; one US patent. *Mailing Add:* 1121 Running Cedar Way Lynchburg VA 24503

CHULSKI, THOMAS, b Grand Rapids, Mich, Aug 6, 21; m 47; c 6. ANALYTICAL CHEMISTRY. *Educ:* Mich State Univ, PhD(chem), 53. *Prof Exp:* Chemist, Lindsay Chem Co, 47-50; chemist, Upjohn Co, 53-63; ASSOC PROF CHEM, FERRIS STATE COL, 63- *Mem:* AAAS; Am Chem Soc; NY Acad Sci. *Res:* Analytical chemistry of drugs in biological systems; drug dosage in relation to metabolism and mode of degradation. *Mailing Add:* 258 Mill Big Rapids MI 49307-1722

CHUM, HELENA LI, b Sao Paulo, Brazil, Dec 26, 46. SOLAR ENERGY, BIOMASS CONVERSION. *Educ:* Univ Sao Paulo, Brazil, PhD(phys chem), 72. *Prof Exp:* Asst prof phys chem & coord chem, Univ Sao Paulo, 71-78; PRIN SCIENTIST, SOLAR FUELS RES DIV, SOLAR ENERGY RES INST, GOLDEN, COLO, 79- *Honors & Awards:* Award Excellence Technol Transfer, Fed Lab Consortium; R&D 100 Award, 90. *Mem:* Am Chem Soc; Brazilian Chem Soc; AAAS; Electrochem Soc. *Res:* Electrochemistry in aqueous, non-aqueous and molten salt media; biomass characterization and reactivity; thermally regenerative electrochemical systems; reactivity of organic molecules in coordination compounds; kinetics and mechanism of chemical and electrochemical reactions; conversion of biomass-based compounds into fuels, chemicals and materials by thermo-chemical and electrochemical means. *Mailing Add:* Solar Energy Res Inst 1017 Cole Blvd Golden CO 80401

CHUMLEA, WILLIAM CAMERON, b Ft Worth, Tex, Mar 10, 47; m 73; c 2. HUMAN GROWTH, HUMAN NUTRITION. *Educ:* Wash & Lee Univ, BS, 69; Univ Tex, Austin, MA, 76, PhD(anthrop), 78. *Prof Exp:* FELS ASSOC PROF, DIV HUMAN BIOL, DEPT COMMUNITY HEALTH, SCH MED, WRIGHT STATE UNIV, 85- *Concurrent Pos:* Consult, Nat Ctr Health Statist, NIH, Nat Cancer Inst. *Mem:* Soc Study Human Biol; Human Biol Coun; fel Am Col Nutrit; Am Inst Nutrit; Am Col Sports Med. *Res:* Human growth and development; human nutrition; pediatrics; body composition; longitudinal studies; aging. *Mailing Add:* Div Human Biol Sch Med Wright State Univ 1005 Xenia Ave Yellow Springs OH 45387-1698

CHUN, ALEXANDER HING CHINN, b Wahiawa, Hawaii, Jan 15, 28; m 57; c 4. PHYSICAL PHARMACY. *Educ:* Purdue Univ, BS, 54, MS, 56, PhD, 59. *Prof Exp:* Res pharmacist, 58-62, pharmacologist, 62-71, ASSOC RES FEL BIOPHARMACEUT, ABBOTT LABS, 71- *Mem:* Am Pharmaceut Asn; Am Chem Soc; Sigma Xi; fel Acad Pharmaceut Sci; fel Am Asn Pharmaceut Scientists. *Res:* Drug absorption and distribution; chemical pharmacology; application of physical and colloidal chemistry to pharmaceuticals; biopharmaceutics. *Mailing Add:* 1908 Linden Ave Waukegan IL 60087

CHUN, BYUNGKYU, b Korea, Apr 10, 28; US citizen; m 57; c 2. SURGICAL PATHOLOGY. *Educ:* Seoul Nat Univ, MD, 52. *Prof Exp:* From instr to asst prof, 61-69, ASSOC PROF PATH, MED CTR, GEORGETOWN UNIV, 69- *Concurrent Pos:* Consult, Children's Hosp, DC, 65. *Honors & Awards:* Achievement Award, Angiol Res Found, 66. *Mem:* Col Am Path; AMA; Am Oncol Soc. *Res:* Oncology. *Mailing Add:* 3900 Reservoir Rd NW Washington DC 20007

CHUN, EDWARD HING LOY, b Wahiawa, Hawaii, Nov 2, 30. CHEMISTRY. *Educ:* Univ Hawaii, BA, 52, MA, 54; Harvard Univ, PhD(chem), 58. *Prof Exp:* Res assoc biol, Mass Inst Technol, 58-63; res investr chem, Sch Vet Med, Univ Pa, 63-68, res investr, 68-70, asst prof biochem, 70-73; BIOCHEMIST, ABBOTT LABS, 74- *Mem:* Am Soc Microbiol; Sigma Xi. *Res:* Research in diagnostic immunology; molecular structure of biological macromolecules. *Mailing Add:* 252 W Johnson St Philadelphia PA 19144

CHUN, KEE WON, b Korea; US citizen. THEORETICAL PHYSICS. *Educ:* Univ Pa, AB, 51, PhD(physics), 60; Princeton Univ, AM, 55. *Prof Exp:* Res assoc theoret physics, Columbia Univ, 59-62 & Yale Univ, 62-65; assoc prof physics, 65-69, PROF PHYSICS, UNIV NEW HAVEN, 69-, CHMN DEPT, 71- *Mem:* Am Phys Soc; Am Asn Physics Teachers. *Res:* Theoretical physics. *Mailing Add:* Dept Physics Univ New Haven West Haven CT 06516

CHUN, MYUNG K(I), b Seoul, Korea, Mar 19, 32; US citizen; m 62; c 1. QUANTUM ELECTRONICS. *Educ:* Yonsei Univ, Korea, BSEE, 56, MSEE, 58; Yale Univ, MEng, 62; Rensselaer Polytech Inst, PhD(electrophysics), 69. *Prof Exp:* From engr to chief engr, Christian Broadcasting System, Korea, 56-61; physicist, Res & Develop Ctr, Gen Elec Co, Schenectady, NY, 66-69, develop engr, Aerospace Controls Dept, Binghamton, 69-73; sr physicist, Electronics Lab, Gen Elec Co, Syracuse, NY, 73-77; RES PHYSICIST, NAVAL RES LAB, WASH, DC, 77- *Concurrent Pos:* Instr, Yonsei Univ, Korea, 58-61; adj prof, Sch Advan Technol, State Univ NY Binghamton, 69-70. *Mem:* Am Phys Soc; sr mem Inst Elec & Electronics Engrs; Sigma Xi; Korean Scientist & Engrs Asn Am; Optical Soc Am. *Res:* Laser device areas; optically pumped solid-host lasers; Q-switched lasers; laser resonators; nonlinear optics; uv preionized gas discharge lasers; computer modeling of various laser features: resonators, optical pumping, dynamic gain, and gas discharge; new solid state laser materials; semiconductor lasers. *Mailing Add:* 1407 Claves Ct Vienna VA 22182-1605

CHUN, PAUL W, b Repub of Korea, Dec 14, 28; US citizen; m 64; c 1. PHYSICAL BIOCHEMISTRY. *Educ:* Northwestern State Col, Okla, BS, 55; Okla State Univ, MS, 57; Univ Mo-Columbia, PhD(biochem), 65. *Prof Exp:* Chemist, Worcester Found Exp Biol, 61-63; fel biochem, Univ Mo-Columbia, 65-66; fel, 66-67, asst prof, 67-70, assoc prof, 70-80, PROF BIOCHEM & MOLECULAR BIOL, UNIV FLA, 80- *Mem:* AAAS; Am Chem Soc; Am Soc Biol Chemists; Biophys Soc; Sigma Xi. *Res:* Protein-protein interaction; molecular exclusion. *Mailing Add:* Dept Biochem Univ Fla Box J-245-JHMHC Gainesville FL 32610

CHUN, RAYMOND WAI MUN, b Honolulu, Hawaii, Jan 21, 26; m 60; c 3. PEDIATRICS, NEUROLOGY. *Educ:* St Joseph's Col, BS, 51; Georgetown Univ, MD, 55. *Prof Exp:* Intern med, Philadelphia Gen Hosp, Pa, 55-56; resident pediat, Univ Hosp, Georgetown Univ, 56-58; resident neurol, 58-59 & 60-61, from asst prof to assoc prof, 61-72, PROF PEDIAT NEUROL, UNIV WIS-MADISON, 72- *Concurrent Pos:* Fel pediat neurol, Columbia-Presby Hosp, New York, 59-60; consult, Wis Diag Ctr, 61- *Mem:* Fel Am Acad Pediat; Am Acad Neurol. *Res:* Epilepsy; neurophysiology; clinical research; transillumination of skull of infants. *Mailing Add:* Dept Neurol CSC Univ Wis Sch Clin Ctr 600 Highland Ave Madison WI 53792

CHUN, SUN WOONG, b Seoul, Korea, Mar 30, 34; m 60; c 2. CHEMICAL ENGINEERING, THERMODYNAMICS. *Educ:* Ohio State Univ, BChE & MS, 59, PhD(chem eng), 64. *Prof Exp:* Res asst chem thermodyn, Goodyear Res Proj, Ohio State Univ, 59-63, res fel, Univ, 63-64; res scientist, Union Bag-Camp Paper Corp, 64-67; sr res engr, Gulf Res & Develop Co, 67, sect supvr, 69-75; supvr pyrolysis, Dept Energy, Pittsburgh Energy Res Ctr, 75-76, mgr, Process Sci Div, 76-77, dep dir, 77-78, DIR, PITTSBURGH ENERGY TECHNOL CTR, 78- *Honors & Awards:* Benjamin & Lamme Medal, 81; Diamond Jubilee Award, Pittsburgh Sect, Am Inst Chem Eng, 83. *Res:* Petroleum refining; cool liquefaction and synthetic fuels processing. *Mailing Add:* 3304 Hermar Dr Murrysville PA 15668

CHUNG, ALBERT EDWARD, b Jamaica, West Indies, Dec 18, 36; m 58; c 3. BIOCHEMISTRY. *Educ:* Univ West Indies, BS, 57, MS, 59; Johns Hopkins Univ, PhD(biochem), 62. *Prof Exp:* Res fel, Harvard Univ, 62-64; asst prof biochem, Med Ctr, Univ Colo, 64-67; assoc prof, 67-74, PROF BIOCHEM, UNIV PITTSBURGH, 74-, CHMN BIOL SCI DEPT, 84- *Concurrent Pos:* Fulbright sr res scholar, France, 74-75; mem, Pathobiochem Study Sect, NIH, 85-; vis prof, Inst of Molecular Biol Acad Sinicia, Taiwan, 87-88. *Mem:* AAAS; Am Chem Soc; Am Soc Biol Chemists. *Res:* Regulation of enzymes; basement membrane structure and assembly; cellular differentiation. *Mailing Add:* Dept Biol Sci Fac Arts & Sci Univ Pittsburgh Pittsburgh PA 15260-0001

CHUNG, BENJAMIN T, b China; US citizen; m; c 2. HEAT TRANSFER, FLUID MECHANICS. *Educ:* Kans State Univ, MS, 62, PhD(mech eng), 68; Univ Wisc, MS, 65. *Prof Exp:* Res engr mech eng, Allis-Chalmers Mfg Co, 62-69; from asst prof to assoc prof, 70-80, PROF MECH ENG, UNIV AKRON, 80-, DEPT HEAD, MECH ENG, 84- *Concurrent Pos:* Sr eng analyst consult, Babcock & Wilcox, 77-79. *Honors & Awards:* Inventions & Contrib Bd Space Art Award, NASA. *Mem:* Am Soc Mech Eng; Sigma Xi. *Res:* Nonlinear heat transfer with phase change; turbulent and laminar boundary layer flow with heat transfer; radiative view factor computations. *Mailing Add:* Dept Mech Eng Univ Akron Akron OH 44325

CHUNG, CHI HSIANG, electro-optic materials, thin film technology, for more information see previous edition

CHUNG, CHIEN, nuclear waste management, radiation engineering, for more information see previous edition

CHUNG, CHIN SIK, b Taejon, Korea, May 6, 24; nat US; m 57; c 3. HUMAN GENETICS, BIOSTATISTICS. *Educ:* Ore State Univ, BS, 51; Univ Wis, MS, 53, PhD, 57. *Prof Exp:* Asst genetics, Univ Wis, 52-57, res assoc med genetics, 57-61; vis scientist, NIH, 61-64, res biologist, 64-65; chmn, dept pub health sci, Sch Pub Health, Univ Hawaii, 73-77 & 81-85, chmn, Biomed Sci Grad Prog, 75-78, assoc dean acad affairs, 87-88, PROF PUB HEALTH, UNIV HAWAII, 65-, PROF GENETICS, 69-, CHMN DEPT PUB HEALTH SCI, SCH PUB HEALTH, 90- *Concurrent Pos:* Res grants, NSF & NIH, Hawaii, 73-; consult, NIH, 74; mem, Epidemiol Dis Control, NIH, 82-86. *Mem:* Am Soc Human Genetics; Am Statist Asn; Biomet Soc; Am Pub Health Asn; Soc Study Social Biol; AAAS. *Res:* Genetic epidemiology of human conditions, including muscular dystrophy, blood groups, dental caries, oral clefts, periodontal disease, birth defects and interracial crosses; epidemiologic studies of long-term effects of induced abortion; genetic epidemiology; disease prevention. *Mailing Add:* Sch Pub Health Univ Hawaii Honolulu HI 96822

CHUNG, CHOONG WHA, b Korea, Aug 14, 18; nat US; m 49; c 2. MICROBIOLOGY, MEDICINE. *Educ:* Keio Univ, Japan, MD, 46; Rutgers Univ, PhD(microbiol), 53. *Prof Exp:* Asst microbiol, Med Sch, Seoul Nat Univ, 46-49; res fel microbial biochem, Inst Microbiol, Rutgers Univ, 49-53; instr biochem, Dept Pediat, Sch Med, Johns Hopkins Univ, 53-55; instr, Med Sch, Univ Colo, 55-56; instr chem, Ind Univ, 57-60; chief med chem sect, Med Lab Br, Nat Commun Dis Ctr, Atlanta, Ga, 60-67, asst chief dermal toxicity Br, 67-77, RES CHEMIST, DIV TOXICOL, FOOD & DRUG ADMIN, USPHS, 77- *Mem:* AAAS; Am Chem Soc; Sigma Xi; Am Soc Microbiol; Brit Biochem Soc. *Res:* Delayed and immediate hypersensitivity; chemical allergens; photoallergy; skin biochemistry; immunological aspects of neoplasia. *Mailing Add:* 605 Potomac River Rd McLean VA 22102

CHUNG, DAE HYUN, b Jungup, Korea, Dec 6, 34; US citizen; m 63; c 2. GEOPHYSICS, MATERIALS SCIENCE. *Educ:* Alfred Univ, AB, 59, MS, 61; Pa State Univ, PhD(solid state sci), 66. *Prof Exp:* Fel geophys, Mass Inst Technol, 67-68, res assoc geophys, 68-72; prof geophys, 72-74; dir, Weston Observ, Mass, 72-74; SR SCIENTIST & PROG MGR, LAWRENCE LIVERMORE NAT LAB, 74- *Concurrent Pos:* Consult, Lincoln Lab, Mass Inst Technol, 68-72, US Nuclear Regulatory Comn, 78- & US Dept Energy, 79-; mem, Korea Sci & Technol Develop Bd, 69-71, sci adv to minister sci & technol, Korea; ed, Earthquake Notes, Seismol Soc Am, 72-74; coordr geophys prog, dept appl sci, Univ Calif, Livermore, 75-80; liaison mem, Nat Acad Sci, Nat Acad Eng Comts Seismol, Earthquake Eng, Washington, 89- *Honors & Awards:* Geol Surv of Korea Achievement Award, 70. *Mem:* AAAS; Am Geophys Union; Seismol Soc Am; NY Acad Sci; fel Geol Soc Am. *Res:* Seismic wave propagation; physical state and constitution of the earth's interior; elasticity and equation of state; engineering seismology; earthquake hazard analysis; nuclear waste management technology. *Mailing Add:* 4150 Colgate Way Livermore CA 94550

CHUNG, DAVID YIH, b Shanghai, China, Nov 14, 36; m 67; c 1. PHYSICS, SOLID STATE PHYSICS. *Educ:* Nat Taiwan Univ, BSc, 58; Univ BC, MSc, 62, PhD(physics), 66. *Prof Exp:* Res physicist, Heat Div, Nat Bur Standards, 66-67; from asst prof to assoc prof, 67-77, PROF PHYSICS, HOWARD UNIV, 78- *Concurrent Pos:* Vis prof, Dept Appl Physics & Electronics, Univ Durham, Brit Sci Res Coun vis fel, 73-74; vis fel, Cavendish Lab, Univ Cambridge, 78-79. *Mem:* Am Phys Soc. *Res:* Second sound propagation in solids; superfluid flow in liquid helium; ultrasonic waves in solids; optical property of solids and fiber optics sensors; non-linear acoustics in liquids and solids; magnetic properties of solids at low temperature; properties of magnetic fluids. *Mailing Add:* Dept Physics Howard Univ Washington DC 20059

CHUNG, DEBORAH DUEN LING, b Hong Kong, Sept 12, 52; US citizen; m 76. MATERIALS SCIENCE & ENGINEERING, METALLURGY. *Educ:* Calif Inst Technol, BS & MS, 73; Mass Inst Technol, SM, 75, PhD(mat sci), 77. *Prof Exp:* Res asst mat sci, Mass Inst Technol, 73-77; from asst prof to assoc prof metall & mat sci, Carnegie-Mellon Univ, 82-87; PROF AEROSPACE & MECH ENG, STATE UNIV NY, BUFFALO, 87- *Concurrent Pos:* vis scientist, Francis Bitter Nat Magnet Lab, 74-77; prin investr res proj, Air Force Off Sci Res, 78-83, NSF, 80-84 & 89-91, Strategic Hwy Res Prog, 89-91, Dept of Energy, 86-88 & Def Advan Res Proj Agency, 91-94; consult, res & develop ctr, Westinghouse Elec Corp, 78 & 83, Int Bus Mach Corp, NASA, 85, Union Carbide, 85. *Honors & Awards:* Ladd Award, Carnegie-Mellon Univ, 79; Hardy Gold Medal, Am Inst Mining Metall & Petrol Engrs, 80; Teetor Educ Award, Soc Automotive Engrs, 87. *Mem:* Am Inst Mining Metall & Petrol Engrs; Am Soc Metals Int; Soc Automotive Engrs; Am Carbon Soc; Mat Res Soc; Soc Advan Mat & Process Eng. *Res:* Composite materials; carbon and graphite; metal-matrix composites; polymer-matrix composites; cement and concrete; electronic packaging. *Mailing Add:* 3812 Henley Dr Pittsburgh PA 15235

CHUNG, DO SUP, b Inchon, Korea, Mar 20, 35; m 61; c 2. CHEMICAL ENGINEERING, FOOD SCIENCE. *Educ:* Purdue Univ, BS, 58; Kans State Univ, MS, 60, PhD(chem eng, food sci), 66. *Prof Exp:* From instr to assoc prof, 65-80, PROF AGR ENG, KANS STATE UNIV, 80- *Concurrent Pos:* Agr Res Serv, US Dept Agr res contract, 67-70. *Mem:* Am Soc Agr Engrs; Am Inst Chem Engrs; Inst Food Technol; Am Asn Cereal Chemists. *Res:* Adsorption, desorption and absorption of water by cereal products; heat transfer in grain investigations; physical properties of grains and handling of grain for minimizing damage investigations. *Mailing Add:* Dept Agr Eng Seaton Hall Kans State Univ Manhattan KS 66506

CHUNG, ED BAIK, b Seoul, Korea, Mar 16, 28; US citizen; m 58; c 5. MEDICINE, PATHOLOGY. *Educ:* Severance Union Med Col, MD, 51; Georgetown Univ, MS, 56, PhD(path), 58; Am Bd Path, cert anat path & clin path. *Prof Exp:* Resident path, Med Ctr, Georgetown Univ, 54-58, from instr to asst prof, 58-63; prof oncol, 76-79, PROF PATH, COL MED, HOWARD UNIV, 64- *Concurrent Pos:* Attend pathologist, Howard Univ Hosp, 64 -; consult path, Glenn Dale Hosp, Md, 68-74 & Coroner's Off, Washington, DC, 69-71; spec res fel, Nat Cancer Inst, 71-72. *Mem:* Int Acad Path; Am Soc Clin Path; Col Am Path; NY Acad Sci. *Res:* Orthopedic and renal pathology; oncologic pathology; tumor immunology; soft tissue tumors. *Mailing Add:* Dept Path Howard Univ Col Med 520 W St NW Washington DC 20059

CHUNG, EUNYONG, NEUROPHARMACOLOGY, NEUROCHEMISTRY. *Educ:* Rutgers Univ, PhD(nutrit), 74. *Prof Exp:* ASSOC PROF NEUROBIOL, MT SINAI SCH MED, 76- *Res:* Neurobiology. *Mailing Add:* Dept Neurol Mt Sinai Sch Med Fifth Ave & 100 St New York NY 10029

CHUNG, FRANK H, b Kiangsi, China, July 20, 30; US citizen; m 59; c 3. POLYMER TECHNOLOGY & APPLICATIONS, X-RAY TECHNIQUES IN INDUSTRY. *Educ:* Kent State Univ, PhD(phys chem), 68. *Prof Exp:* Sr scientist phys chem, Sherwin-Williams Co, 68-89; TECH DIR APPL CHEM, MARSON CORP, 89- *Concurrent Pos:* Prog chair, Labcon & Chemtech Forum, 82 & 83; consult China, UN, 85; bd dirs, Mid Am Chinese Sci & Technol Asn, 85-89. *Honors & Awards:* Outstanding Serv Award, Soc Appl Spectros, 80. *Mem:* Am Chem Soc; Soc Appl Spectros; Am Crystallogr Asn; Soc Mfg Engrs. *Res:* Industrial research and development in polymer/plastic technology, coatings technology and x-ray techniques; authored over 30 research papers. *Mailing Add:* 130 Crescent Ave Chelsea MA 02150

CHUNG, HO, b Canton, China, Aug 15, 38; US citizen; m 65; c 2. DRUG DEVELOPMENT, MEDICINE & TOXICOLOGY. *Educ:* Univ Wis-Oshkosh, BS, 62; Univ Md, MS, 73, PhD(pharmacol & toxicol), 74. *Prof Exp:* Nuclear med technician, Straub Clin, Hawaii, 62-64; quality control chemist, Western Foam Co, Ariz, 64-69; res asst, Univ Md, 69-70, teaching asst, 70-74, res fel, 73-76; pharmacologist, 74-75, asst chief drug metab, 75-81, chief biochem pharmacol, 81-85, CHIEF BIOCHEM PHARMACOL & TOXICOL, WALTER REED ARMY INST RES, 85- *Concurrent Pos:* Instr phys sci, Dunbarton Col of Holy Cross, 70-72; consult, Chinese Univ, Hong Kong, 82 & Med & Toxicol Assocs, 84-; Oshkosh scholar, Univ Wis, 59-62; Hudock Assocs Law Off, 80-; adj prof pharmacol & toxicol, Univ Md, Baltimore, 88- *Mem:* Am Soc Pharmacol & Exp Therapeut; Am Acad Clin Toxicol; fel Am Col Clin Pharmacol; Am Col Toxicol; Soc Toxicol; Asn Govt Toxicologists; Sigma Xi. *Res:* Biochemical, pharmacological and toxicological studies of new drugs; new drugs development. *Mailing Add:* Dept Pharmacol & Toxicol Walter Reed Army Inst Res Washington DC 20307-5100

CHUNG, HUI-YING, b China, Nov 23, 27; m 60; c 2. ELECTRICAL ENGINEERING. *Educ:* Nat Taiwan Univ, BS, 51; Iowa State Univ, MS, 60, PhD(elec eng), 64. *Prof Exp:* Engr, Taiwan Power Co, Formosa, 51-57; instr & assoc elec eng, Iowa State Univ, 59-65; from asst prof to assoc prof, 65-90, EMER PROF ELEC ENG, UNIV NEBR, LINCOLN, 90- *Mem:* Am Soc Eng Educ. *Res:* Power system analysis; power electronics. *Mailing Add:* 4930 Crest Haven Dr Lincoln NE 68156

CHUNG, JING YAN, b Miao-Li, Taiwan, Oct 31, 39; US citizen; m 67; c 2. ACOUSTICS, FLUID MECHANICS. *Educ:* Nat Taiwan Univ, BS, 64; Kans State Univ, MS, 68; Purdue Univ, PhD, 74. *Prof Exp:* Staff res engr noise control, Gen Motors Res Labs, Gen Motors Corp, Warren, 74-81; SR SPECIALIST, EXXON PROD RES CO, EXXON CORP, HOUSTON, 81- *Mem:* Acoustical Soc Am. *Res:* Digital signal analysis; new measurement techniques in acoustical engineering. *Mailing Add:* 13310 Pebblebrook Houston TX 77079

CHUNG, JIWHEY, b Korea, Feb 26, 36; m 64; c 2. BIOCHEMISTRY, FOOD SCIENCE. *Educ:* Seoul Nat Univ, BS, 61; Univ Ill, MS, 66; Univ Tenn, PhD(biochem), 69. *Prof Exp:* Res assoc, Albert Einstein Med Ctr, 69-71; RES ASSOC BIOCHEM, UNIV CHICAGO, 71- *Concurrent Pos:*

NIH fel, 71-72 & spec res fel, 72-73; Ill & Chicago Heart Asn grant-in-aid, 75. *Mem:* Sigma Xi. *Res:* Lipid metabolism; mechanism of action of lipoprotein lipase and lecithin cholesterol acyl transferase; synthesis and metabolism of plasma; very low density lipoproteins. *Mailing Add:* 1464 Lori Lyn Lane Northbrook IL 60062

CHUNG, KAI LAI, b Shanghai, China, Sept 19, 17; nat US; c 3. MATHEMATICS. *Educ:* Princeton Univ, MA & PhD(math), 47. *Prof Exp:* Instr math, Princeton Univ, 47-48; asst prof, Cornell Univ, 48-50, vis assoc prof, 51-52; from assoc prof to prof, Syracuse Univ, 53-61; PROF MATH, STANFORD UNIV, 61- *Concurrent Pos:* Vis prof, Columbia Univ, 50-51 & 59, Univ Chicago, 56-57, Univ Strasbourg, 68-69 & Swiss Fed Inst Technol, 70; G A Miller vis prof, Univ Ill, 70-71; Guggenheim fel, 75-76; fel, Churchill Col, Cambridge Univ, 76; ed, Zeitschrift für Wahrscheinlichkeitstheorie und Verwandte Gebiete. *Mem:* Am Math Soc; Inst Math Statist. *Res:* Probability. *Mailing Add:* Dept Math Stanford Univ Bldg 380 Rm 383X Stanford CA 94305

CHUNG, KUK SOO, b Kyungpuk, Korea, Nov 15, 35; US citizen. ASTROPHYSICS, PARTICLE PHYSICS. *Educ:* Yale Univ, BA, 59; Princeton Univ, MA, 61, PhD(physics), 69. *Prof Exp:* Asst prof physics, State Univ NY, 62-68; res assoc, Princeton Univ, 69-70; res scientist, Watson Res Ctr, Int Bus Mach Corp, 70-72; resident res assoc astrophys, Columbia Univ, 72-73; from asst prof to assoc prof, 73-80, PROF PHYSICS, LEHMAN COL, 80- *Concurrent Pos:* Consult, McNulty Assoc, 67-69. *Mem:* Am Phys Soc. *Res:* Theoretical work on gravitational radiations from black holes; dispersion-theoretic approach to pion-pion interactions. *Mailing Add:* Herbert Lehman Col CUNY Bedford Park Blvd Bronx NY 10468

CHUNG, KYUNG WON, b Seoul, Korea, Aug 15, 38; US citizen; m 66; c 2. REPRODUCTIVE ENDOCRINOLOGY. *Educ:* Yonsei Univ, Korea, BS, 64, MS, 66; St Louis Univ, MS, 69; Univ Okla, PhD(anat), 71. *Prof Exp:* Instr biol, Yonsei Univ, 66; fel endocrinol, Hershey Med Ctr, Pa State Univ, 71-72; instr, State Univ NY Downstate Med Ctr, 72-75, asst prof anat, 75-77; from asst prof to assoc prof, 77-86, PROF ANAT SCI, COL MED, UNIV OK, 86-, VCHMN, 89- *Mem:* Sigma Xi; Am Asn Anatomists; Endocrine Soc; Soc Exp Biol & Med. *Res:* Reproductive and molecular endocrinology; testicular morphology and physiology; neuroendocrinology; androgen synthesis and metabolism; male infertility; protein and steroid receptor protein and mRNA assays. *Mailing Add:* Dept Anat Sci Univ Okla Health Sci Ctr Box 26901 Oklahoma City OK 73190

CHUNG, OKKYUNG KIM, b Seoul, Korea, Apr 11, 36; m 61; c 2. CEREAL CHEMISTRY. *Educ:* Ewha Womens Univ, Korea, BS, 59; Kans State Univ, MS, 65, PhD(grain sci), 73. *Prof Exp:* Res asst lipids, Kans State Univ, 64-66, res assoc lipids & surfactants, 73-74; res chem lipids & surfactants, US Grain Mkt Res Lab, Agr Res Serv, USDA, 74-87; SUPVRY RES CHEMIST & RES LEADER, GRAIN QUAL & STRUCT RES UNIT, WINTER WHEAT QUAL LAB, USGMRL, ARS, USDA, 87- *Concurrent Pos:* Assoc ed, Cereal Chem, 78- *Mem:* Am Asn Cereal Chemists; Sigma Xi; Am Chem Soc; Am Oil Chem Soc. *Res:* Functionality of wheat flour lipids and lipid related surfactants in breadmaking; interaction of lipids and surfactants with other flour components during processing into bread; fractionation of wheat flour components; characterization of lipids in cereal grains; wheat gluten; wheat kernel hardness; wheat quality. *Mailing Add:* US Grain Mkt Res Lab 1515 College Ave Manhattan KS 66502

CHUNG, PAUL M(YUNGHA), b Seoul, Korea, Dec 1, 29; nat US; m 52; c 2. COMBUSTION, FLUID MECHANICS. *Educ:* Univ Ky, BS, 52, MS, 54; Univ Minn, PhD(mech eng), 57. *Prof Exp:* From instr to asst prof mech eng, Univ Minn, 55-58; aeronaut res scientist, Ames Res Ctr, NASA, 58-61; head fluid dynamics dept, Aerospace Corp, 61-66; head, Energy Eng Dept, 74-79, PROF FLUID MECH, UNIV ILL, CHICAGO, 66-, DEAN, ENG COL, 79- *Concurrent Pos:* Consult, Sandia Lab, 66-74, Argonne Nat Lab, 75-81 & Aerospace Corp, 80- *Honors & Awards:* Baetijer lectr, Princeton Univ, 76. *Mem:* Fel Am Inst Aeronaut & Astronaut; Am Inst Chem Engrs. *Res:* Combustion; turbulence; fluid mechanics; heat transfer; plasma. *Mailing Add:* Dean Col Eng Univ Ill-Chicago Chicago IL 60680

CHUNG, RAYMOND, chemical engineering, for more information see previous edition

CHUNG, RILEY M, b Peoples Repub China; US citizen; m 72; c 2. GEOTECHNICAL ENGINEERING, EARTHQUAKE ENGINEERING. *Educ:* Nat Taiwan Univ, BSCE, 68; Rensselaer Polytech Inst, 70; Northwestern Univ, PhD(civil eng), 76. *Prof Exp:* Engr, Westenhoff & Norvick, 70-72; proj eng, Harza Eng Co, 72-77; srproj eng, Schnabel Eng Asn, 77-79; chief, geotechnical engr group, Nat Bur Standards, 79-85; DIR, DIV GEOTECHNICAL ENGR & HAZARD MITIGATION, NAT ACAD SCI, 85- *Honors & Awards:* Walter P Murphy fel, Northwestern Univ. *Mem:* Am Soc Civil Engrs; Am Soc Testing Mat; Am Soc Agron; Int Soc Soil Mech & Found Engr; Earthquake Engr Res Inst. *Res:* Soils dynamic properties under earthquake loadings; use of soils as the barrier to isolate radioactive wastes in underground geologic waste isolation system; development of a polymer gage for dynamic soil stress measurement; soil liquefaction study. *Mailing Add:* Nat Res Coun 2101 Constitution Ave NW Washington DC 20418

CHUNG, RONALD ALOYSIUS, b Christiana, Jamaica, WI, Sept 30, 36; US citizen; m 63; c 3. FOOD SAFETY & TOXICOLOGY, HUMAN NUTRITION. *Educ:* Col Holy Cross, BSc, 59; Purdue Univ, MS, 61, PhD(food science), 63. *Prof Exp:* Asst food technol, Purdue Univ, 61-63; prof food sci & nutrit & head home econ, Tuskegee Inst, 68-74, prog coordr food sci & nutrit, 74-85, head home econ, 85-86; PROF & HEAD DEPT HOME ECON, UNIV NORTHERN IOWA, 86- *Concurrent Pos:* Res assoc, Carver Res Found, 63-70; consult, 70- *Honors & Awards:* Res Award, Poultry Sci Asn, 66. *Mem:* Fel AAAS; Am Chem Soc; Inst Food Technol; fel Am Inst Chemists; Am Home Econ Asn. *Res:* Lipid and protein metabolism in poultry and livestock; biochemical activity of food additives in vitro and in vivo; new food product development; dietary related toxemia in pregnancy; radiation preservation of foods; food safety risk assessment. *Mailing Add:* Dept Home Econ Latham 235 Univ Northern Iowa Cedar Falls IA 50614-0332

CHUNG, SHIAU-TA, b Taiwan, Aug 21, 34; m 63; c 3. MICROBIAL GENETICS. *Educ:* Chung-Shin Univ, BS, 57; Univ Tokyo, PhD(microbiol), 67. *Prof Exp:* Res asst microbiol, Univ Manitoba, 67-70; res assoc genetics, Univ Mich, 70-73; SR RES SCIENTIST, MICROBIAL GENETICS, UPJOHN CO, 73- *Mem:* Am Soc Microbiol; AAAS; Soc Indust Microbiol. *Res:* Genetics of streptomyces which produced antibiotics. *Mailing Add:* Bioprocess Res & Develop Upjohn Co Kalamazoo MI 49001

CHUNG, STEPHEN, GENE EXPRESSION, MOLECULAR GENETICS. *Educ:* Univ Calif, Berkeley, PhD(molecular biol), 77. *Prof Exp:* ASST PROF BIOCHEM, MED COL, UNIV ILL, 82- *Mailing Add:* Dept Biochem A-302 Med Col Univ Ill 1853 W Polk St Chicago IL 60612

CHUNG, SUH URK, b Pyongyang, Korea, Nov 11, 36; US citizen; m 63, 84; c 2. HIGH ENERGY PHYSICS. *Educ:* Univ Fla, BES, 61; Univ Calif, Berkeley, PhD(physics), 66. *Prof Exp:* Assoc physicist, 66-74, physicist, 74-90, SR PHYSICIST, BROOKHAVEN NAT LAB, 90- *Concurrent Pos:* Vis scientist, Cern, Geneva, Switz, 69-71, 82-83; vis prof, Physics Dept, Univ Geneva, Swit. *Mem:* Am Phys Soc. *Res:* Experimental high energy physics. *Mailing Add:* Dept Physics Brookhaven Nat Lab Upton NY 11973

CHUNG, TING-HORNG, b Pingtung, Taiwan, Mar 9, 45; m 72; c 2. THERMOPHYSICAL PROPERTY OF FLUIDS. *Educ:* Nat Cheng-Kung Univ, Taiwan, BS, 68, MS, 71; Univ Okla, PhD(chem eng), 80. *Prof Exp:* Chem engr, Kaohsiung Refinery, Chinese Petrol Corp, 71-77; res assoc, Sch Chem Eng, Univ Okla, 80-84; RES ENGR, NAT INST PETROL & ENERGY RES, 84- *Mem:* Soc Petrol Engrs. *Res:* Conducting gas (carbon dioxide, nitrogen and hydrocarbon gases) miscible and immiscible displacement for light and heavy oil recovery research; thermophysical property measurements for petroleum fluids; study of fluid properties based on molecular thermodynamics. *Mailing Add:* PO Box 2128 Bartlesville OK 74005

CHUNG, TZE-CHIANG, b Tainan, Taiwan, Nov 28, 53; m 82; c 1. FUNCTIONAL POLYMERS, RIGID-ROD POLYMERS. *Educ:* Chung Yuan Univ, Taiwan, BS, 76, Univ Pa, PhD(chem), 82. *Prof Exp:* Fel, Univ Calif, Santa Barbara, 82-83, res scientist, 83-84; RES CHEMIST, EXXON RES & ENG CO, 84- *Mem:* Am Chem Soc. *Res:* Synthesis of functional polymers and rigid-rod polymers which possess unusual chemical and physical properties, including permeability, comparatibility, dyeability, adhesion and solution rheology. *Mailing Add:* 548 Brittany Dr State College PA 16803-1423

CHUNG, VICTOR, b Vancouver, BC, Sept 7, 40; m 64; c 1. PHYSICS. *Educ:* Mass Inst Technol, SB, 61, SM, 62; Univ Calif, Berkeley, PhD(physics), 66. *Prof Exp:* Asst res physicist, Univ Calif, San Diego, 66-68; asst prof, 68-78, ASSOC PROF PHYSICS, CITY COL NEW YORK, 78- *Mem:* Am Phys Soc. *Res:* Theoretical high energy physics; quantum electrodynamics; medical physics. *Mailing Add:* Dept Physics City Col NY Convent Ave & 138th St New York NY 10031

CHUNG, YIP-WAH, b Hong Kong; m 77; c 2. SURFACE SCIENCE, SOLID STATE PHYSICS. *Educ:* Univ Hong Kong, BSc, 71, MPhil, 73; Univ Calif, Berkeley, PhD(physics), 77. *Prof Exp:* From asst prof to assoc prof, 77-85, PROF MAT SCI & ENG, NORTHWESTERN UNIV, 85- *Mem:* Am Phys Soc; Metall Soc; Mat Res Soc; Am Vacuum Soc; Soc Tribologists & Lubrication Engrs. *Res:* Metal semiconductor interfaces; energy conversion; chemisorption and catalysis; environmental effects on fatigue, tribology. *Mailing Add:* Dept Mat Sci & Eng Northwestern Univ Evanston IL 60208

CHUNG, YOUNG SUP, b Inchon, Korea, Jan 27, 37; Can citizen; m 65; c 2. MOLECULAR GENETICS, BIOTECHNOLOGY. *Educ:* Purdue Univ, BS, 60; Giessen Univ, W Ger, PhD(microbiol), 64. *Prof Exp:* Asst genetics, Frankfurt Univ, W Ger, 64-65; asst & res assoc, Palo Alto Med Res Found, 65-69; from asst prof to assoc prof, 70-78, PROF GENETICS, DEPT BIOL SCI, UNIV MONTREAL, 79- *Concurrent Pos:* Vis prof, Nagoya Univ, Japan & Bochum Univ, Germany, 80-81; consult, Korean Inst Chem Technol, 85- *Mem:* Am Soc Microbiol; Genetic Soc Am; Can Soc Microbiologists; Genetics Soc Can; Environ Mutagen Soc; Fr-Can Asn Advan Sci. *Res:* Bacterial conjugation; genetic recombination; repair of ozone damage; ozone mutagenesis; cloning the crystal protein gene of B thuringiensis. *Mailing Add:* 3241 Appleton Ave Montreal PQ H3S 1L6 Can

CHUNG, YUN C, b Seoul, Korea, Apr 30, 56; m 82; c 2. FIBER OPTIC COMMUNICATION, QUANTUM ELECTRONICS. *Educ:* Hanyang Univ, Korea, BS, 80; Utah State Univ, MS, 85, PhD(elec eng), 87. *Prof Exp:* Grad fel, Discharge Lasers & Applications Group, Los Alamos Nat Lab, 85-87; MEM TECH STAFF, CRAWFORD HILL LAB, AT&T BELL LABS, 87- *Mem:* Inst Elec & Electronics Engrs; Lasers & Electro-optics Soc. *Res:* High-capacity lightwave systems via wavelength-division- multiplexing, particulary, novel techniques for laser frequency stabilization, precise wavelength control and optical channel selection; characterization of multiple-quantum-well devices; surface-emitting lasers. *Mailing Add:* Crawford Hill Lab AT&T Bell Labs HOH-R113 Holmdel-Keyport Rd Holmdel NJ 07733-0400

CHUONG, CHENG-MING, b Taipei, Taiwan, Sept 15, 52; m 79; c 1. CELL ADHESION MOLECULES, MORPHOGENESIS. *Educ:* Nat Taiwan Univ, MD, 78; Rockefeller Univ, PhD(develop & molecular biol), 83. *Prof Exp:* asst prof develop & molecular biol, Rockefeller Univ, 83-87; ASST PROF PATH, UNIV SOUTHERN CALIF. *Mem:* Soc Neurosci; Am Soc Cell Biol; Sigma Xi. *Res:* Molecular basis for embroynic development; embroynic induction; cell adhesion, movement, and differentiation; developmental biology. *Mailing Add:* Univ Southern Calif 2011 Zonal Ave HMR 204 Los Angeles CA 90033

CHUPKA, WILLIAM ANDREW, b Pittston, Pa, Feb 12, 23; m 55; c 2. LASER SPECTROSCOPY & PHOTOPHYSICS. *Educ:* Univ Scranton, BS, 43; Univ Chicago, MS, 49, PhD(chem), 51. *Prof Exp:* Instr chem, Harvard Univ, 51-54; consult, Argonne Nat Lab, 52-54, assoc physicist, 54-68, sr physicist, 68-75; PROF CHEM, YALE UNIV, 75- *Concurrent Pos:* Guggenheim fel, 61-62. *Mem:* Am Chem Soc; fel Am Phys Soc. *Res:* Laser spectroscopy and photophysics; molecular and atomic structure; high temperature thermodynamics; chemical kinetics; photoionization; vacuum ultraviolet spectroscopy; photoelectron spectroscopy; ion-molecular reactions; mass spectrometry. *Mailing Add:* Sterling Chem Lab Yale Univ PO Box 6666 New Haven CT 06511-8118

CHUPP, EDWARD LOWELL, b Lincoln, Nebr, May 14, 27; m 50; c 3. PHYSICS. *Educ:* Univ Calif, Berkeley, AB, 50, PhD(physics), 54. *Prof Exp:* Staff mem physics, Lawrence Radiation Lab, Univ Calif, 54-59; unit chief geospace physics, Aerospace Div, Boeing Co, 59-62; assoc prof physics, 62-67, PROF PHYSICS, UNIV NH, 67- *Concurrent Pos:* Consult, Lawrence Radiation Lab, Univ Calif, 59-63 & Geophys Corp Am, Mass, 63-65; mem, State of NH Radiation Adv Comt, 66; consult, Solar Physics Subcomt, NASA, 67; mem sci balloon panel, Nat Ctr Atmospheric Res, 70; NATO sr res fel, Max Planck Inst, Munich, 70, Alexander von Humboldt sr award, 72-73, Fulbright-Hayes hon sr fel, Max Planck Inst Extraterrestrial Physics, 72-73. *Mem:* Fel Am Phys Soc; Am Geophys Union; Am Asn Physics Teachers; Am Astron Soc; Sigma Xi. *Res:* Cosmic radiation time variations; gamma ray spectroscopy and gamma ray astronomy; neutron and gamma ray detectors; solar flare physics; solar terrestrial relations; measurements of atomic transition probabilities. *Mailing Add:* 44 Mill Pond Rd Durham NH 03824

CHUPP, TIMOTHY E, b Berkeley, Calif, Nov 30, 54. PARTICLE PHYSICS, NUCLEAR & ATOMIC PHYSICS. *Educ:* Princeton Univ, AB, 79; Univ Wash, PhD(physics, 83. *Prof Exp:* Postdoctoral physics, Princeton Univ, 83-85; from asst prof to assoc prof physics, Harvard Univ, 85-91; ASSOC PROF PHYSICS, UNIV MICH, 91- *Concurrent Pos:* Alfred P Sloan Found fel, 87; NSF presidential young investr award, 87. *Mem:* Am Phys Soc. *Res:* Low energy particle physics particularly by study of symmetries accessible with polarization; tests of time reversal, local Lorentz invariance and linearity in quantum mechanics; structure of the neutron. *Mailing Add:* Dept Physics Univ Mich Ann Arbor MI 48109

CHURCH, ALONZO, b Washington, DC, June 14, 03; wid; c 3. MATHEMATICS. *Educ:* Princeton Univ, AB, 24, PhD(math), 27. *Hon Degrees:* DSc, Case Western Reserve Univ, 69, Princeton Univ, 85, State Univ NY, Buffalo, 90. *Prof Exp:* Nat Res fel math, Harvard Univ, 27-28, Univ Gottingen, 28-29 & Univ Amsterdam, 29; from asst prof to prof, Princeton Univ, 29-67; prof philos & math, Univ Calif, Los Angeles, 67-90; RETIRED. *Concurrent Pos:* Ed, J of Symbolic Logic, 36-79. *Mem:* Nat Acad Sci; AAAS; Am Math Soc; Asn Symbolic Logic; Am Acad Arts & Sci; Am Philos Asn. *Res:* Mathematical logic. *Mailing Add:* Dept Philos Univ Calif Los Angeles CA 90024

CHURCH, BROOKS DAVIS, b Youngstown, Ohio, May 6, 18; c 4. MICROBIOLOGY, ENZYMOLOGY. *Educ:* Univ Mich, BS, 47, MS, 52, PhD(bact), 55. *Prof Exp:* Res bacteriologist, Ft Derrick, Md, 49-50; asst med, Univ Chicago, 50-52; sr res assoc, Warner-Lambert Res Inst, 55-60; asst prof, Univ Wash, 60-62; from asst prof to assoc prof microbiol, Univ Minn, 62-66; sr microbiologist, North Star Res Inst, 66-72; sr microbiologist, 72-76, prof biol sci, Denver Res Inst, Univ Denver, 72-77; PRES BIO-SEARCH ASSOC, INC, 81- *Concurrent Pos:* Mem comt safe drinking water, Nat Acad Sci, 76-77; consult, Marathon Oil, 78-, Allied Mills Starches Australia, 80-, Ralston-Purina, 81-, Marion Labs, 81-, Penwalt Corp, 83, Wacon, 85-88. *Mem:* AAAS; Am Soc Microbiol; Sigma Xi. *Res:* Biochemistry of bacterial spores and cell walls; fungal digestion of food processing wastes; biosynthesis of heteropolysaccharides; bioconversion of corn and wheat starch and milk permeate wastes, via selected fungal species growing in continuous fermentation; feeding domestic livestock. *Mailing Add:* 329 W Caley Dr Littleton CO 80120

CHURCH, CHARLES ALEXANDER, JR, b Rock Hill, SC, Oct 29, 32; m. MATHEMATICS. *Educ:* Va Polytech Inst, BS, 57; Duke Univ, PhD(math), 65. *Prof Exp:* Instr math, Roanoke Col, 59-62; asst prof, WVa Univ, 65-66; asst prof biomet, Med Col Va, 66-67; ASSOC PROF MATH, UNIV NC, GREENSBORO, 67- *Mem:* AAAS; Am Math Soc; Math Asn Am. *Res:* Combinatorial mathematics. *Mailing Add:* Dept Math 383 Bus Econ Bldg Univ NC 1000 Spring Garden St Greensboro NC 27412

CHURCH, CHARLES HENRY, b Phoenix, Ariz, May 15, 29; m 56; c 3. ADVANCED MILITARY TECHNOLOGY. *Educ:* Univ Mo-Rolla, BS, 50; Pa State Univ, MS, 51; Univ Mich, PhD(physics), 59. *Prof Exp:* Assayer anal chem, Bradley Mining Co, Idaho, 50; asst physics, Pa State Col, 50-51; jr physicist, Res Labs Div, Gen Motors Corp, 51-52; res assoc, Eng Res Inst, Univ Mich, 56-58; assoc physicist, Cent Res Lab, Crucible Steel Corp, 58-59; adv physicist, Westinghouse Res Labs, 59-68; programmer, DeFonis Advan Res Proj Agency, 68-75; asst dir Army Res & Technol, Off Dep Chief Staff Res, Develop & Acquisition, 75-87, actg tech dir, 82-83, DIR ADVAN CONCEPTS & TECHNOL ASSESSMENT, OFF ASST SECY ARMY, US ARMY, 87- *Concurrent Pos:* Teacher, Univ Calif Far East Exten, 53-54; lectr, Univ Pittsburgh, 58-59 & Carnegie Inst Technol, 64. *Mem:* Soc Automotive Engrs; Am Phys Soc; fel Optical Soc Am; Inst Elec & Electronics Engrs. *Res:* Lasers; atomic and molecular plasmas; optical instrumentation and design; spectroscopy; weapons systems; solid state physics, energy; military vehicle concepts. *Mailing Add:* 903 Waynewood Blvd Alexandria VA 22308-2609

CHURCH, CLIFFORD CARL, geology, paleontology, for more information see previous edition

CHURCH, DAVID ARTHUR, b Berlin, NH, Apr 3, 39; m 63; c 2. PHYSICS. *Educ:* Dartmouth Col, BA, 61; Univ Wash, MS, 63, PhD(physics), 69. *Prof Exp:* Res assoc physics, Univ Bonn, Germany, 69, Univ Mainz, Germany, 69-71 & Univ Ariz, 71-72; physicist, Lawrence Berkeley Lab, Univ Calif, 72-75; from asst prof to assoc prof, 75-84, PROF PHYSICS, TEX A&M UNIV, 84- *Concurrent Pos:* Vis prof, Univ Tenn, 82 & 84-85; vis scientist, Brookhaven Nat Lab, 85 & Argonne Nat Lab, 89; partic guest, Lawrence Livermore Nat Lab, 90. *Mem:* Fel Am Phys Soc. *Res:* Stored multi-charged ion collisions and precision spectroscopy; fast ion coherence spectroscopy, collisional orientation and alignment; laser spectroscopy and x-ray polarization spectroscopy; photoionization of ions. *Mailing Add:* Dept Physics Tex A&M Univ College Station TX 77843

CHURCH, DAVID CALVIN, b Iola, Kans, Nov 1, 25; m 52. ANIMAL NUTRITION, ANIMAL PHYSIOLOGY. *Educ:* Kans State Univ, BS, 50; Univ Idaho, MS, 52; Okla State Univ, PhD(animal nutrit), 56. *Prof Exp:* From asst prof to assoc prof, 56-70, PROF ANIMAL NUTRIT, ORE STATE UNIV, 70-, EMER PROF ANIMAL NUTRIT. *Mem:* Am Soc Animal Sci; Am Dairy Sci Asn. *Res:* Nutrition of the ruminant animal; rumen physiology; feedstuff and forage evaluation. *Mailing Add:* 1215 NW Kline St Corvallis OR 97330

CHURCH, EUGENE LENT, b Yonkers, NY, July 30, 25; m 48; c 2. METROLOGY, SIGNAL PROCESSING. *Educ:* Princeton Univ, AB, 48; Harvard Univ, PhD(physics), 53. *Prof Exp:* Res assoc, Princeton Univ, 48; res assoc, Brookhaven Nat Lab, 50-52; guest scientist, Argonne Nat Lab, 52-55; guest scientist, Brookhaven Nat Lab, 55-59; Secy of Army fel, Bohr Inst, Copenhagen, 59-61; guest scientist, Brookhaven Nat Lab, 61-71; physicist, Frankford Arsenal, US Army, 71-77; RES PHYSICIST, USA ARMAMENT RES DEVELOP & ENG CTR, 77- *Concurrent Pos:* Mem solid state adv panel, Nat Res Coun, 73-78; teaching fel, Harvard Univ, 49-51; consult, Brookhaven Nat Lab, 81-; consult, Lawrence Livermore Nat Lab, 88-; consult, Oak Ridge Nat Lab, 90- *Mem:* Fel Am Phys Soc; sr mem Inst Elec & Electronics Engr; Optical Soc Am; fel Soc Photo-Optical Inst Engrs; Sigma Xi; fel AAAS. *Res:* Experimental and theoretical optics; surface metrology; signal processing. *Mailing Add:* USA Armament Res Devel & Eng Ctr Dover NJ 07801-5001

CHURCH, GEORGE LYLE, b Boston, Mass, Dec 19, 03; m 34; c 1. BOTANY. *Educ:* Mass Col, BS, 25; Harvard Univ, AM, 27, PhD(bot), 28. *Prof Exp:* Teaching fel bot, Harvard Univ, 26-28; instr, Brown Univ, 28-34; from asst prof to prof bot, 34-59, cur herbarium, 40-72, chmn dept bot, 58-66, Stephen Olney prof, 59-72, EMER PROF BOT, BROWN UNIV, 72- *Mem:* AAAS; Bot Soc Am; Soc Develop Biol; Soc Study Evolution; Am Soc Plant Taxonomists. *Res:* Cytology and taxonomy; cytotaxonomy and cytogenetics of Gramineae. *Mailing Add:* N Bay Manor 171 Pleasant View Ave No 331 Smithfield RI 02917-1732

CHURCH, JOHN ARMISTEAD, b Richmond, Va, Apr 3, 37; m 64; c 3. PHYSICAL CHEMISTRY, ORGANIC CHEMISTRY. *Educ:* Univ Va, BA, 59; Lawrence Univ, MS, 61, PhD(org chem), 64. *Prof Exp:* Sr scientist, Princeton Res Ctr, Am Can Co, 64-70, from res assoc to sr res assoc, 70-83; RES ASSOC, COLGATE-PALMOLIVE CO, 83- *Concurrent Pos:* Consult, 83. *Mem:* Am Chem Soc. *Res:* Cellulose, wood and sugar chemistry; fuels and chemicals from biomass; chemical kinetics; paper deterioration; fabric care; pulp, paper, fiber and textile technology; history of applied optics and astronomy. *Mailing Add:* Colgate-Palmolive Co PO Box 1343 Piscataway NJ 08855

CHURCH, JOHN PHILLIPS, b Columbus, Ohio, July 14, 34; m 59; c 2. REACTOR PHYSICS. *Educ:* Univ Cincinnati, ChE, 57; Univ Fla, MSc, 60, PhD(nuclear eng), 63. *Prof Exp:* Chem engr, Nat Cash Register Co, 58-59; res engr, E I Du Pont de Nemours & Co, Inc, 63-77, staff physicist, 77-82, res staff physicist, 82-87, res assoc, 87-89; ADV ENGR, WESTINGHOUSE ELEC CORP, 87- *Mem:* Am Nuclear Soc; Nat Soc Prof Engrs. *Res:* Reactor kinetics; core design; reactor safety analysis; radiation shielding. *Mailing Add:* 1204 Woodbine Rd SW Aiken SC 29802

CHURCH, JOSEPH AUGUST, b Plainfield, NJ, Sept 4, 46; m. ALLERGY, IMMUNOLOGY. *Educ:* Johns Hopkins Univ, BA, 68; NJ Col Med, MD, 72. *Prof Exp:* Fel allergy & immunol, Med Ctr, Georgetown Univ, 74-76; intern-resident pediat, Childrens Hosp Nat Med Ctr, 72-74; from asst prof pediat to assoc prof clin pediat, 76-86, PROF CLIN PEDIAT, SCH MED, UNIV SOUTHERN CALIF, 86- *Concurrent Pos:* Assoc attend physician & consult, Dept Path, Childrens Hosp, Los Angeles, 76- *Mem:* Am Col Allergists; Am Acad Allergy; Am Acad Pediat; Soc Pediat Res. *Res:* Recognition, management and investigation of immune deficiencies; pediatric aids. *Mailing Add:* Dept Allergy, Childrens Hosp 4650 Sunset Blvd Los Angeles CA 90027

CHURCH, KATHLEEN, CHROMOSOMAL STRUCTURE & FUNCTION. *Educ:* Berkeley Univ, PhD(genetics), 66. *Prof Exp:* PROF GENETICS, ARIZ STATE UNIV. *Mailing Add:* Dept Zool Ariz State Univ Tempe AZ 85287

CHURCH, LARRY B, b St Louis, Mo, Apr 19, 39; m 62; c 2. ANALYTICAL CHEMISTRY, NUCLEAR CHEMISTRY. *Educ:* Univ Rochester, BS, 61; Carnegie Inst Technol, MS, 64, PhD(chem), 66. *Prof Exp:* Fel chem, Univ Calif, Irvine, 66-68; sr chemist, 69-75, res chemist, 75-83, sr res chemist, 83-85,; assoc prof chem, Reed Col, 73-80; ANALYTICAL SCIENTIST, TEKTRONIX, INC, 80- *Concurrent Pos:* Consult, Fabric Flammability Res. *Mem:* Am Chem Soc; Electrochem Soc. *Res:* Inorganic trace analysis; surface and material science phosphors and semi-conductor materials. *Mailing Add:* Tektronix Inc DS 50-289 Beaverton OR 97077

CHURCH, LLOYD EUGENE, anatomy, dentistry; deceased, see previous edition for last biography

CHURCH, MARSHALL ROBBINS, b Richmond, Va, Mar 23, 48. CHEMICAL LIMNOLOGY, ENVIRONMENTAL EFFECTS OF ACIDIC DEPOSITION. *Educ:* Univ Va, BA, 71, PhD(environ sci), 80. *Prof Exp:* Phys sci technician, Denver Wildlife Res Ctr, US Dept Interior, 71-72; teaching asst ecol & aquatic ecol, Dept Environ Sci, Univ Va, 72-75; phys scientist, Waterways Exp Sta, US Army Engrs, 77-78; instr, 79, res assoc aquatic ecol, Univ Va, 80-82; ENVIRON SCIENTIST, CORVALLIS RES LAB, ENVIRON PROTECTION AGENCY, 82- *Concurrent Pos:* Courtesy asst prof, Dept Gen Sci, Ore State Univ, 90- *Honors & Awards:* Bronze Medal Commendable Serv, US Environ Protection Agency. *Mem:* Am Soc Limnol & Oceanog; Ecol Soc Am; Am Geophys Union. *Res:* Effects of nutrients and light on primary productivity in aquatic ecosystems; physiology and biochemistry of algae; effects of acidic deposition on the chemistry and ecology of surface waters. *Mailing Add:* Corvallis Environ Res Lab US Environ Protection Agency 200 SW 35 St Corvallis OR 97333

CHURCH, PHILIP THROOP, b Winchester, Conn, Mar 18, 31; m 54; c 3. TOPOLOGY. *Educ:* Wesleyan Univ, BA, 53; Harvard Univ, MA, 54; Univ Mich, PhD, 59. *Prof Exp:* From asst prof to prof, 58-76, FRANCIS H ROOT PROF MATH, SYRACUSE UNIV, 76- *Concurrent Pos:* NSF grants, 59-62 & 63-69; mathematician, Inst Defense Anal, 62-63; mem, Inst Advan Study, 62 & 65-66; NSF sr fel, 65-66; ed topol, Transactions of & Memoirs of Am Math Soc, 74-77, chmn, Ed Comt, 77; vis fel, Princeton Univ, 76; mem coun, Am Math Soc, 74-77, 80; distinguished vis prof, Univ Alta, 86. *Mem:* Am Math Soc; Math Asn Am; Am Asn Univ Prof. *Res:* Singularities of differentiable maps; nonlinear elliptic partial differential equations. *Mailing Add:* Math Dept Syracuse Univ Syracuse NY 13244-1150

CHURCH, RICHARD LEE, b Escondido, Calif. ENVIRONMENTAL ENGINEERING, OPERATIONS RESEARCH. *Educ:* Lewis & Clark Col, BS, 70; Johns Hopkins Univ, PhD(environ systs), 74. *Prof Exp:* Asst prof, 74-80, assoc prof environ eng, Univ Tenn, 80-; MEM STAFF, DEPT GEOG, UNIV CALIF, SANTA BARBARA. *Concurrent Pos:* Consult, LBJ Sch, Univ Tex, 76-, Brookhaven Nat Lab, 76-77 & Oak Ridge Nat Lab, 78- *Mem:* Opers Res Soc Am; Regional Sci Asn; Asn Am Geographers; Water Pollution Control Fedn. *Res:* Modelling of environmental systems engineering problems; location model development for public and private facilities; water resources systems. *Mailing Add:* Dept Geol Univ Calif Santa Barbara CA 93106

CHURCH, ROBERT BERTRAM, b Calgary, Alta, May 7, 37; m 59; c 2. MEDICAL SCIENCES, SCIENCE ADMINISTRATION. *Educ:* Univ Uppsala, Sweden, dipl, 61; Univ Alta, BSc, 62, MSc, 63; Univ Edinburgh, PhD(animal genetics), 65. *Prof Exp:* Fel molecular biol, Univ Wash, 65-67; asst prof biol, Fac Med, Univ Calgary, 67-69, assoc prof biol & morphol sci, 69-70, prof med biochem & head dept, 70-83, assoc dean med res, 83-89; CONSULT, 89- *Concurrent Pos:* Scholar, Royal Soc Can, 61 & Nat Res Coun, 63; fel, Commonwealth Soc, London, 63 & NIH, 65; affil prof, Animal Reprod Lab, Colo State Univ, 71-83, vis prof, 78; vis prof, Inst Molecular Biol, Soviet Acad Sci, 72 & Sch Vet Sci, Murdoch Univ, 77; exec mem, Natural Sci & Eng Res Coun Can, Ottawa, 78-86; dir, Can Inst Advan Res, Toronto, 83-88 & Vet Infectious Dis Org, Sask, 85-90, Connaught Labs, Willowdale, Ont, 79-87, Bio Star Inc, Sask, Can, 86-, Continental Pharma, 89-, Ciba-Geigy Can Ltd, 90- *Honors & Awards:* Klinck lectr, Agr Inst Can, 86. *Mem:* Soc Study Reproduction; AAAS; Genetics Soc Am; Can Biochem Soc; Can Genetics Soc; Genetics Soc Can (treas, 84-). *Res:* The composition and function of the mammalian genome; the complexity of the DNA base sequences in animal genomes, their arrangement and expression in the preimplantation embryo and in specific tissues; the conditions necessary for successful culture of various species of preimplantation embryos, their manipulation, transfer and storage; transfer of specific gene sequences into animals. *Mailing Add:* Dept Med Biochem Univ Calgary 3330 Hosp Dr NW Calgary AB T2N 4N1 Can

CHURCH, ROBERT FITZ (RANDOLPH), b Philadelphia, Pa, Mar 27, 30; div; c 3. ORGANIC CHEMISTRY. *Educ:* Amherst Col, BA, 51; Univ Mich, MS, 61, PhD(chem), 62. *Prof Exp:* Sales serv rep, Am Cyanamid Co, 51-53, chemist, 55-57; instr, Univ Mich, 60-61; res chemist, Am Cyanamid Co, 61-62; res chemist, 63-79, SR RES CHEMIST, LEDERLE LABS, PEARL RIVER, NY, 79- *Concurrent Pos:* Instr, Bridgeport Eng Inst, 56-57 & 62-68. *Mem:* Am Chem Soc; Sigma Xi. *Res:* Chemistry of steroids and other natural products; pharmaceutical and synthetic organic chemistry; chemistry of small ring compounds; process research and development of pharmaceutical products. *Mailing Add:* Eight Fado Lane Cos Cob CT 06807

CHURCH, RONALD L, b Monterey, Calif, Nov 19, 30. ZOOLOGY, POLLUTION BIOLOGY. *Educ:* San Jose State Col, BA, 53; Univ Calif, Berkeley, MA, 61, PhD(zool), 64. *Prof Exp:* Asst prof biol, San Jose State Col, 63-65; asst prof, Univ Nev, Reno, 65-70, res assoc, Desert Res Inst, 68-70; environ specialist, Div Water Qual, State Water Resources Control Bd, 70-72, ENVIRON SPECIALIST, N COAST REGION, CALIF REGION WATER QUALCONTROL BD, 72- *Concurrent Pos:* NSF grant, 65-66; adj prof, Pac Marine Sta, Univ of the Pac, 74-79. *Mem:* AAAS; Ecol Soc Am; Wildlife Soc; Am Ornith Union; Am Soc Mammal. *Res:* Water quality biology; vertebrate biology; animal ecology. *Mailing Add:* 4409 Brookshire Circle Santa Rosa CA 95405

CHURCH, SHEPARD EARLL, (JR), physical chemistry, for more information see previous edition

CHURCH, STANLEY EUGENE, b Oakland, Calif, Sept 13, 43; m 66; c 2. GEOCHEMISTRY. *Educ:* Univ Kans, BS, 65, MS, 67; Univ Calif, Santa Barbara, PhD(geochem), 70. *Prof Exp:* Res asst geochem, Univ Calif, Santa Barbara, 69-70; res assoc, Carnegie-Mellon Univ, 70-71; Nat Res Coun assoc, Johnson Space Ctr, NASA, 71-73; res assoc geochem, Univ Calif, Santa Barbara, 73-75; res assoc geochem, Hasler Res Ctr, Appl Res Lab, Goleta, Ca, 75-79; MEM STAFF, US GEOL SURV BR GEOCHEM, FED CTR, DENVER COLO, 79- *Concurrent Pos:* Leader, Alaska Mineral Resource Assessment Prog; invited UNESCO lectr, China, 84. *Mem:* Fel Geol Soc Am; Am Geophys Union; Geochem Soc; Soc Econ Geologists; Int Asn Geochemists; Asn Exp Geochemistry. *Res:* The application of isotopic and trace element geochemical methods to the problem of the genesis of calc-alkaline magma and the evolution of the earth's mantle; trace element abundance distributions and the effect of micrometeorite bombardment of the lunar soils to form agglutinites; investigations of elemental dispersion patterns and interpretation of these patterns as clues to the occurrance of economic deposits; use of Pb isotopes in ore deposits to correlate lithostrategicyclic terranes in the Cordillera. *Mailing Add:* US Geol Surv Fed Ctr Mail Stop 973 PO Box 25046 Denver CO 80225

CHURCH, WILLIAM RICHARD, b Tonyrefail, Glamorgan, UK, July 10, 36; m 63; c 4. GEOLOGY. *Educ:* Univ Wales, BSc, 57, PhD(geol), 61. *Prof Exp:* Boese fel, Columbia Univ, 61-62; lectr, 62-63, asst prof, 63-68, ASSOC PROF GEOL, UNIV WESTERN ONT, 68- *Mem:* Fel Geol Soc London; fel Geol Asn Can; Geol Soc Am; Can Inst Mining & Metall. *Res:* Geology of northeast Newfoundland, northwest Ireland, Huronian and Grenvillian of northern Ontario; eclogites; ophiolites of the Appalachian and Pan African systems. *Mailing Add:* Dept Geol Univ Western Ont London ON N6A 5B9 Can

CHURCHER, CHARLES STEPHEN, b Aldershot, Eng, Mar 21, 28; Can citizen; m 59; c 3. VERTEBRATE PALEONTOLOGY, MAMMALOGY. *Educ:* Univ Natal, BSc, 50, Hons, 52, MSc, 54; Univ Toronto, PhD, 57. *Prof Exp:* Lectr zool, 57-59, from asst prof to assoc prof zool, 60-70, assoc prof dent, 69-78, assoc dean, Arts & Sci, 75-78, PROF ZOOL, UNIV TORONTO, 70-, PROF DENT, 78- *Concurrent Pos:* Res assoc, Royal Ont Mus, Toronto, 59-; consult, Geol Surv Can, 66-; actg chmn, Urban & Regional Studies, 80-81; actg dir, Mus Studies, 83-85. *Mem:* Am Soc Mammalogists; Soc Vert Paleont; AAAS; Australian Mammal Soc; fel Geol Asn Can. *Res:* Pleistocene mammals, especially Canadian and African. *Mailing Add:* Dept Zool Univ Toronto Toronto ON M5S 1A1 Can

CHURCHILL, ALGERNON COOLIDGE, b Aug 15, 37; US citizen; m 59; c 3. PHYCOLOGY, PLANT ECOLOGY. *Educ:* Harvard Univ, BA, 59; Univ Ore, MS, 63, PhD(biol), 68. *Prof Exp:* From instr to asst prof, 66-74, ASSOC PROF BIOL, ADELPHI UNIV, 74- *Concurrent Pos:* Sr investr, NY Ocean Sci Lab grant, 70-71. *Mem:* AAAS; Phycol Soc Am; Am Inst Biol Sci; Am Soc Limnol & Oceanog; Int Phycol Soc. *Res:* Physiological ecology of algae; growth and differentiation of marine algae; culturing of algae. *Mailing Add:* Dept Biol Adelphi Univ Garden City NY 11530

CHURCHILL, CONSTANCE LOUISE, b Los Angeles, Calif, May 10, 41; c 1. ORGANIC CHEMISTRY. *Educ:* Baylor Univ, BS, 63, PhD(org chem), 69. *Prof Exp:* From asst prof to prof chem, Dakota State Col, 68-77, chmn, Div Sci, Math & Health Serv, 74-77; chmn div sci, math & technol, Burlington County Col, 78-88, dir acad res & comput, 88-91; DEAN SCI & ALLIED HEALTH, OAKTON COMMUNITY COL, 91- *Mem:* AAAS; Am Chem Soc; Sigma Xi. *Res:* Decomposition of ozonides; water pollution. *Mailing Add:* Dean Sci & Allied Health Oakton Community Col 1600 E Golf Rd Des Plaines IL 60016

CHURCHILL, DEWEY ROSS, JR, b Blackwell, Okla, May 19, 26; m 53; c 4. PHYSICS. *Educ:* Univ Kans, BS, 48, MS, 53. *Prof Exp:* Comput-observer geophys, Seismograph Serv Corp, 50-53; physicist, Res & Develop Lab, Phillips Petrol, 54-55; staff mem physics, Los Alamos Sci Lab, 55-60; sr res engr, 60-61, res scientist theoret physics, 61-89, STAFF ENGR, LOCKHEED MISSLES & SPACE CO, 89- *Mem:* AAAS; Am Phys Soc. *Res:* Radiative processes in molecules and atoms; radiative transfer; nuclear weapons effects; shock hydrodynamics; ion ballistics; aeronomy; survivable RF & optical communications. *Mailing Add:* 1711 Karameos Dr Sunnyvale CA 94087

CHURCHILL, DON W, b Seattle, Wash, Feb 5, 30; m 55; c 3. PSYCHIATRY. *Educ:* Lawrence Col, BS, 51; Univ Wis, MS, 56, MD, 57; Am Bd Psychiat & Neurol, dipl psychiat, 63, dipl child psychiat, 66. *Prof Exp:* Instr path, Univ Wis, 53-56; intern, King County Hosp, Seattle, 57-58; resident psychiat, Cincinnati Gen Hosp, 58-60, fel child psychiat, 60-62; from asst prof to assoc prof, 64-73, PROF PSYCHIAT, MED CTR, IND UNIV-PURDUE UNIV, INDIANAPOLIS, 73-, DIR, RILEY CHILD GUIDE CLIN, 73- *Concurrent Pos:* Consult, Ind Boys' Sch, 64-69 & Ind Sch for Deaf, 69- *Mem:* AAAS; Am Psychiat Asn. *Res:* Pathology of connective tissue and histochemistry of mucomucopolysaccharides; childhood psychosis; language development and its relation to performance and cognitive levels; relationship of success-failure level to affect mood and specific behavioral measures; language and learning abilities. *Mailing Add:* Dept Psychiat Ind Univ Sch Med Indianapolis IN 46202

CHURCHILL, FREDERICK CHARLES, b Pittsburgh, Pa, Nov 7, 40; m 70; c 2. DRUG ANALYSIS, CHROMATOGRAPHY. *Educ:* Lafayette Col, BS, 62; Harvard Univ, MA, 64; Univ Ga, Athens, PhD(anal chem), 80. *Prof Exp:* Health serv officer, Tech Develop Labs, Ctr Disease Control, Savannah, Ga, 64-73; sr health serv officer, Bur Trop Diseases, 73-81, SCIENTIST DIR GRADE, INFECTIOUS DIS, CTR DIS CONTROL, ATLANTA, GA, 81- *Concurrent Pos:* Mem, Comn Officers Asn, USPHS; expert adv panel, WHO. *Mem:* Am Chem Soc; Asn Off Anal Chemists; Collab Int Pesticides Anal Coun. *Res:* Carbon-13 Nuclear Magnetic Resonance studies of fungal metabolites; characterization of immunofluorescent dyes; development of High-Performance Liquid Chromatography and capillary Gas-Liquid Chromatography methods (often employing chemical derivatization) for analysis of pesticides, antiparasitic drugs, and their metabolites in body fluids. *Mailing Add:* Control Technol Br Div Parasitic Dis Ctr Infectious Dis Ctrs Dis Control Atlanta GA 30333

CHURCHILL, GEOFFREY BARKER, b Glen Cove, NY, July 10, 50; m 77; c 3. POLYMER SCIENCE, PLASTICS ENGINEERING. *Educ:* Mass Inst Technol, SB & SM, 74. *Prof Exp:* Res engr plastics, 74-79, SR ENGR, MONSANTO CO, 79- *Mem:* Am Chem Soc. *Res:* Polymer application modeling and feasibility studies; injection molding and simulation of polymer flow; incompatible species in polymer systems, polyblends and polymer alloys; extrusion fundamentals and extruder screw design; adhesion and fiber reinforcement of polymers; polymer formulation and product development for use in automobiles and major appliances. *Mailing Add:* Monsanto Co 800 Worcester St Indian Orchard MA 01151-1089

CHURCHILL, HELEN MAR, biology; deceased, see previous edition for last biography

CHURCHILL, JOHN ALVORD, b Boston, Mass, Mar 25, 20; c 1. NEUROLOGY, PEDIATRIC NEUROLOGY. *Educ:* Trinity Col, BS, 42; Univ Pa, MD, 45. *Prof Exp:* Kirby-McCarthy fel, Johnston Found, Univ Pa, 46-50, asst prof neurol, Sch Med, 49-50; mem staff, Hartford Hosp, 50-53; assoc, Henry Ford Hosp, 53-60; chief dept child neurol, Lafayette Clin & assoc prof neurol, Wayne State Univ, 60-67; head sect neuropediat, Perinatal Br, Nat Inst Neurol Dis & Blindness, 67-72; prof neurol & chief child neurol, Sch Med, Wayne State Univ, 72-87; MEM STAFF, QUILLAN-DISCHNER COL MED, E TENN STATE UNIV, 88- *Concurrent Pos:* Res consult, pediat, Guys Hosp & Spastic Soc, UK, 60; mem comt ment retardation, Dept Ment Health State of Mich, 64-67; res fel bot, Mus Natural Hist, Smithsonian Inst, 70-; res fel, Cranbrook Inst Sci, 72- *Mem:* AAAS; Am Acad Neurol; Asn Res Nerv & Ment Dis; Am Acad Cerebral Palsy. *Res:* Relationships of perinatal events to neurological disabilities in the child; botanical toxicology. *Mailing Add:* 813 Forest Ave Johnson City TN 37601

CHURCHILL, LYNN, b Sacramento, Calif, Mar 27, 47. NEUROCHEMISTRY. *Educ:* Univ Houston, BS, 69; Univ Calif, Irvine, PhD(biol sci), 73. *Prof Exp:* Teaching asst psychobiol, Univ Calif, Irvine, 69-73; res assoc pharmacol, Univ Wis-Madison, 74-77, NIH fel pharmacol, 74-77, proj assoc, 77-79, asst scientist, 79-80; res assoc neurobiol & anat, Med Sch, Univ Tex, Houston, 80-81; res asst prof anat, Med Ctr, Univ Kans, Kansas City, 81-86; RES ASST PROF, WASH STATE UNIV, PULLMAN, 86- *Mem:* Soc Neurosci; Am Soc Neurochem; Sigma Xi. *Res:* Modifications in neurotransmitter receptor proteins and in RNA with drug treatments; adaptation of behavior and neurochemistry after destruction of dopanimergic system. *Mailing Add:* Dept VCAPP Wash State Univ Pullman WA 99164-6520

CHURCHILL, MELVYN ROWEN, b London, Eng, June 2, 40; m 66; c 2. INORGANIC CHEMISTRY, CRYSTALLOGRAPHY. *Educ:* Univ London, BSc, 61, PhD(inorg chem), 64. *Prof Exp:* From instr to assoc prof chem, Harvard Univ, 64-71; prof, Univ Ill, Chicago Circle, 71-75; assoc provost, Fac Natural Sci & Math, 76-78, PROF CHEM, STATE UNIV NY BUFFALO, 75-, ACTG CHMN, 81- *Concurrent Pos:* Alfred P Sloan fel, 68-70; assoc ed, Inorg Chem, 70-82. *Honors & Awards:* Corday-Morgan Medal, Chem Soc, London, 76. *Mem:* Am Chem Soc; Am Crystallog Asn; Royal Soc Chem. *Res:* Crystallographic studies on inorganic compounds particularly organometallic and transition metal complexes; synthetic organometallic chemistry. *Mailing Add:* Dept Chem State Univ NY Buffalo NY 14214

CHURCHILL, PAUL CLAYTON, b Carson City, Mich, Feb 23, 41; m 70. PHYSIOLOGY. *Educ:* Univ Mich, Ann Arbor, BS, 63, PhD(physiol), 69. *Prof Exp:* NSF fel pharmacol, Univ Lausanne, 69-70; asst prof physiol, Sch Med, Univ Mich, Ann Arbor, 70-72; asst prof, 72-80, ASSOC PROF PHYSIOL, SCH MED, WAYNE STATE UNIV, 80- *Mem:* Am Physiol Soc; Int Soc Nephrology; Am Soc Nephrology. *Res:* Physiology of kidney, excretory function, transport; control or renin secretion from kidney. *Mailing Add:* Dept Physiol Sch Med Wayne State Univ 540 E Canfield Detroit MI 48201

CHURCHILL, RALPH JOHN, b Pittsburgh, Pa, July 16, 44; m 66; c 2. WATER CHEMISTRY, ENVIRONMENTAL CHEMISTRY. *Educ:* Univ Ky, BS, 66; Univ Houston, MS, 70; Univ Calif, Berkeley, PhD(civil eng), 73. *Prof Exp:* Engr pollution control, Shell Oil Co, 66-71; consult, Eng-Sci Inc, 73-75; group leader water res, Tretolite Div, Petrolite Corp, 75-81; consult, 81-89; VPRES TECH DEPT, PETROLITE CORP, 89- *Mem:* Am Inst Chem Engr; Am Water Works Asn; Water Pollution Control Fedn. *Res:* Water and wastewater investigation; oil-water separation; water and wastewater treatment technology; mineral scale deposition-inhibition; municipal and industrial wastewater management. *Mailing Add:* 18409 Woodland Meadow Dr Glencoe MO 63038

CHURCHILL, STUART W(INSTON), b Imlay City, Mich, June 13, 20; m 46, 74; c 4. COMBUSTION, HEAT TRANSFER. *Educ:* Univ Mich, BSE(chem eng), 42, BSE(math), 42, MSE, 48, PhD(chem eng), 52. *Hon Degrees:* MA, Univ Pa, 72. *Prof Exp:* Tech asst petrol refining, Shell Oil Co, Inc, 42-46; tech supvr, Electrochem Mfg, Frontier Chem Co, 46-47; res asst, Chem & Metall Eng, Univ Mich, 48-49, res assoc, 49-50, instr chem eng, 50-52, asst prof, 52-55, assoc prof, 55-57, prof, 57-67, chmn, Dept Chem & Metall Eng, 62-67; Carl V S Patterson prof, 67-90, EMER CARL V S PATTERSON PROF CHEM ENG, UNIV PA, 90- *Concurrent Pos:* Vis res, Okayama, Japan, 77. *Honors & Awards:* Prof Progress Award, Am Inst Chem Engrs, 65, William H Walker Award, 69, Warren K Lewis Award, 78 & Founders Award, 80; Max Jakob Mem Award, Am Soc Mech Engrs & Am Inst Chem Engrs, 79. *Mem:* Nat Acad Eng; Am Chem Soc; fel Am Inst Chem Engr (vpres, 65, pres, 66); Combustion Inst. *Res:* Radiantly stabilized combustion, natural convection in enclosures, solar heating, migration of moisture in porous materials. *Mailing Add:* 311A Towne Bldg D3 Univ Pa 220 S 33rd St Philadelphia PA 19104

CHURCHWELL, EDWARD BRUCE, b Sylva, NC, July 9, 40; m 64; c 2. ASTRONOMY, SPECTROSCOPY. *Educ:* Earlham Col, BA, 63; Ind Univ, MS, 67, PhD(astrophys), 70. *Prof Exp:* Scientist radio astron, Max Planck Inst Radio Astronomy, 70-77; asst prof, 77-80, PROF ASTRON, UNIV WIS-MADISON, 77- *Concurrent Pos:* Heinrich Hertz Found fel, 70-72. *Mem:* Am Astron Soc; Astronomische Gesellschaft; Int Astron Union; Union Concerned Scientists. *Res:* Interstellar molecular clouds; star formation; molecular and atomic spectroscopy at MM and microwave frequencies, and the physics of HII regions; observational and theoretical, observations of regions of star formation; observational determinations of mass loss rates from the most luminous stars in the galaxy; studies of interstellar chemistry. *Mailing Add:* Washburn Osberv 475 N Charter St Madison WI 53706

CHURG, JACOB, b Dolhinow, Poland, July 16, 10; US citizen; m 42; c 2. PATHOLOGY, ENVIRONMENTAL HEALTH. *Educ:* Univ Wilno, MD, 33, DSc, 36. *Prof Exp:* Asst path, Sch Med, Univ Wilno, 34-36; asst bact, Mt Sinai Hosp, 38, fel path, 41-43, res assoc, 46-62, assoc attend pathologist, 62-81; prof path & res prof Community Med, Mt Sinai Sch Med, 66-81; CONSULT PATHOLOGIST, MT SINAI HOSP & EMER PROF, MT SINAI MED SCH, 81; EMER PROF PATH, MT SINAI MED SCH, 81- *Concurrent Pos:* Resident, Beth Israel Hosp, Newark, NJ, 39-40; pathologist, Barnert Mem Hosp, Paterson, 46-; chmn, US Mesothelioma Ref Panel, Int Union Against Cancer, 65-81, mem, 81-; chmn, Comt Histol Class Renal Dis, WHO, 75-; Lady Davis vis prof, Hebrew Univ, Jerusalem, 76; Ewert R Angus vis prof, Univ Toronto, 83. *Honors & Awards:* John Redman Oliver lectr, Downstate Med Ctr, Brooklyn, 78; Frederick G Germuth lectr, St Louis, 83; Conrad L Pirahi lectr, Columbia-Presbyterian Med Ctr, New York, 85; John P Peters Award, Am Soc Nephrol, Wash, DC, 87. *Mem:* Am Asn Path; NY Acad Med; Int Acad Path; Int Soc Nephrology; Am Soc Nephrol. *Res:* Vascular diseases; renal structure and diseases; pneumoconioses; syndrome of allergic granulomatosis. *Mailing Add:* 711 Ogden Ave Teaneck NJ 07666

CHURKIN, MICHAEL, JR, b San Francisco, Calif, Jan 6, 32; m 60; c 2. GEOLOGY. *Educ:* Univ Calif, Berkeley, BA, 57, MA, 58; Northwestern Univ, PhD, 61. *Prof Exp:* Fel geol, Columbia Univ, 61-62; geologist, Alaskan Geol Br, US Geol Surv, 62-82; GEOLOGIST, ARCO EXPLOR & TECHNOL CO, 82- *Mem:* AAAS; Geol Soc Am; Am Asn Petrol Geologist. *Res:* Stratigraphy & structure of western Cordillera; accretion tectonics of north Pacific and Arctic; geology of USSR; graptolites; integrated basin analysis. *Mailing Add:* 310 Bend Rd Yuba City CA 95991

CHURNSIDE, JAMES H, b Seattle, Wash, Aug 29, 51; m 72; c 2. OPTICS. *Educ:* Whitworth Col, BS, 74; Ore Grad Ctr, PhD(physics), 78. *Prof Exp:* Fel, Ore Grad Ctr, 78-84; MEM TECH STAFF AEROSPACE CORP, 84-; MEM STAFF, NAT OCEANIC & ATMOSPHERIC ADMIN/ERL WAVE PROPAGATION LAB. *Honors & Awards:* Distinguished Authorship Award, US Dept Com, 89. *Mem:* Fel Optical Soc Am; Acoustical Soc Am. *Res:* Statistical optics; optical propagation through the atmosphere; optical remote sensing of the atmosphere. *Mailing Add:* NOA/ERL Wave Propagation Lab R/E/WPI 325 Broadway Boulder CO 80303

CHUSED, THOMAS MORTON, b St Louis, Mo, Mar 29, 40; m 65; c 2. IMMUNOLOGY. *Educ:* Harvard Col, BA, 62; Harvard Med Sch, MD, 67. *Prof Exp:* Clin assoc immunol, Arthritis Br, Nat Inst Arthritis & Metab Dis, 69-71; sr staff fel, 71-74, sr investr immunol, Lab Microbiol & Immunol, Nat Inst Dent Res, 74-78, SR INVESTR IMMUNOL, LAB IMMUNOL, NAT INST ALLERGY & INFECTIOUS DIS, 78- *Mem:* Am Asn Immunologists; Am Rheumatism Asn. *Res:* Immunoregulation; flow cytometry; autoimmune disease. *Mailing Add:* TW II Nat Inst Health Bethesda MD 20852

CHUSID, JOSEPH GEORGE, b Newark, NJ, Aug 23, 14; m 42; c 2. NEUROLOGY. *Educ:* Univ Pa, AB, 34, MD, 38; Am Bd Psychiat & Neurol, dipl. *Prof Exp:* Asst instr neurol, Univ Pa, 41-45; resident & fel, Neuropsychiat Inst, Univ Ill, 46-47; from asst attend neurologist to attend neurologist, 48-59, assoc dir dept neurol & neurosurg, 59-64, chief neurol serv, 64-70, DIR DEPT NEUROL, ST VINCENT'S HOSP & MED CTR, 70- *Concurrent Pos:* Clin assoc prof med, NJ Col Med & Dent, 61-72; assoc clin prof neurol, Col Physicians & Surgeons, Columbia Univ, 65-, assoc attend neurologist, Columbia Presby Med Ctr, 65-77; prof neurol, NY Med Col, 77- *Mem:* Am Physiol Soc; fel Am Acad Neurol; Asn Res Nerv & Ment Dis; Soc Exp Biol & Med; Am Epilepsy Soc. *Res:* Neurological sciences; cortical connections and functions in the monkey; chronic experimental epilepsy; electroencephalographic studies on humans; cerebral angiography; clinical neurology; effects of major cerebral arterial ligations in monkeys; neuropharmacology; effects of metals on the brain. *Mailing Add:* 3390 Emeric Ave Wantagh NY 11793

CHUSID, MICHAEL JOSEPH, b Coral Gables, Fla, Aug 23, 44; m 72; c 2. IMMUNOLOGY, INFECTIOUS DISEASE. *Educ:* Yale Col, BA, 66; Sch Med, Yale Univ, MD, 70. *Prof Exp:* Intern pediat, Yale New Haven Hosp, 70-71, asst resident, 71-72; clin assoc infectious dis, Lab Clin Invest, Bethesda, 72-75; fel, Children's Hosp Med Ctr, Boston, 75-76; from asst prof to assoc prof, 76-84, PROF PEDIAT, MED COL WIS, 84- *Concurrent Pos:* Dir postgrad med educ, Med Col Wis, 78-84, vchmn, Dept Pediat, 84-, epidemiologist, Children's Hosp, Wis, 84-, chief staff, 91- *Mem:* Am Soc Microbiol; Am Fedn Clin Res; AAAS; fel Infectious Dis Soc Am; Soc Pediat Res; Am Pediat Soc. *Res:* Leukocyte physiology. *Mailing Add:* Dept Pediat MCW MACC Fund Res Bldg 8701 Watertown Plank Rd Milwaukee WI 53226

CHUTE, ROBERT MAURICE, b Naples, Maine, Feb 13, 26; m 46; c 2. ENVIRONMENTAL PHYSIOLOGY. *Educ:* Univ Maine, AB, 50; Johns Hopkins Univ, ScD, 53. *Prof Exp:* From instr to asst prof biol, Middlebury Col, 53-58; asst prof, San Fernando Valley State Col, 58-60; assoc prof & chmn, Lincoln Univ, Pa, 60-61; prof & chmn, 61-62, prof biol & chmn dept, 62-80, Dana prof biol, 78-87, PROF BIOL, BATES COL, 88- *Concurrent Pos:* Dir, Bates-Morse Mt Res Area, 78- *Mem:* AAAS; Sigma Xi. *Res:* Human ecology; estimation of cultural impact on lake and coastal ecosystems. *Mailing Add:* Dept Biol Bates Col Lewistown ME 04240

CHUTJIAN, ARA, b New York, NY, Apr 23, 41; m 77; c 1. ATOMIC & MOLECULAR PHYSICS. *Educ:* Brown Univ, BSc, 62; Univ Calif, Berkeley, PhD(chem physics), 66. *Prof Exp:* Res assoc light diffraction, Holography, Bell Labs, 66-67; res assoc molecular spectros, Univ Southern Calif, 67-68; sr scientist, 69-76, MEM TECH STAFF ELECTRON SCATTERING, JET PROPULSION LAB, CALIF INST TECHNOL, 76- *Concurrent Pos:* Calif Inst Technol Pres Fund grant, 73-74 & 77-78; Planetary Atmospheres Br & Astron/Relativity Br, NASA grant, 78- *Mem:* Am Phys Soc. *Res:* Low-energy electron scattering from positive ions; low-energy electron scattering from neutral atoms and molecules; threshold (zero-energy). *Mailing Add:* Jet Propulsion Lab 183-601 4800 Oak Grove Dr Pasadena CA 91109

CHUTKOW, JERRY GRANT, b Denver, Colo, June 14, 33; m 57; c 4. NEUROLOGY, INTERNAL MEDICINE. *Educ:* Univ Chicago, AB, 52, SB, 55, MD, 58. *Prof Exp:* Intern med, Presbyterian Hosp, City New York, 58-59; resident, Univ Chicago Hosp & Clin, Univ Chicago, 59-62, instr, 62-64, resident neurol, 64-67, asst prof, 67; captain to lt colonel, US Army Med Corps, 67-69; asst prof, Mayo Med Sch, Univ Minn, 70-74, assoc prof, 74-77; chmn neurol, 77-83, PROF NEUROL, STATE UNIV NY, BUFFALO, 77- *Concurrent Pos:* Res asst, Argonne Cancer Res Hosp, 62-64; spec fel, Nat Inst Neurol Dis & Stroke, 65-67; fel med, Schweppe Found, 67-69; chief neurol serv, Martin Army Hosp, 67-69; consult, Mayo Clinic, 69-77; dir, Dept Neurol, Erie Co Med Ctr, 79-82, 90-; chief, Neurol Serv, Vet Admin Med Ctr, 79-84; dir, Neuromuscular Dis Clin & Lab, Dept Neurol, SUNY Buffalo, Erie County Med Ctr & Childrens Hosp, 85-, attend physician, Erie County Med Ctr, 78-, Vet Admin Med Ctr, 77-, Childrens Hosp, 78-, consult, Roswell Park Mem Inst, 78-87, Buffalo Gen Hosp, 78-89. *Honors & Awards:* Sigma Xi. *Mem:* Am Neurol Asn; Am Acad Neurol; Am Col Physicians; Soc Neurosci; Am Col Nutrit; Sigma Xi. *Res:* Metabolism of divalent cations in the central and peripheral nervous systems; expert systems in neurology; computer automation of clinical neurology, neuromuscular diseases. *Mailing Add:* Dept Neurol Sch Med State Univ NY Buffalo 462 Grider St Buffalo NY 14215

CHVAPIL, MILOS, b Kladno, Czech, Sept 29, 28; m 53; c 2. PHYSIOLOGICAL CHEMISTRY, EXPERIMENTAL PATHOLOGY. *Educ:* Charles Univ, Prague, MD, 52, DSc(exp path), 66; Czech Acad Sci, PhD(biochem), 55. *Prof Exp:* Scientist, Inst Indust, Hyg & Occup Dis, Prague, 52-68; fel sci, Max Planck Inst Protein & Leather Res, Ger, 68-69; vis prof med, Sch Med, Univ Miami, 69-70; PROF SURG BIOL, COL MED, UNIV ARIZ, 70- *Concurrent Pos:* Consult scientist, Ministry of Light Indust & Ministry of Food Indust, 58-68; fel surg, Med Sch, Univ Ore, 63; assoc prof exp path, Med Sch, Charles Univ, Prague, 65. *Honors & Awards:* Laureate of State Prize, Czech Repub, 66; Ministry of Health Award & Sci Bd, Czech Acad Sci Award, 68. *Mem:* Am Soc Exp Path. *Res:* Connective tissue physiology; biochemistry control of collagen biosynthesis by chelating agents; oxygen effect on collagen synthesis; mechanism of tissue injury; wound management; medical and industrial uses of collejea; hydrojels in medicine; wound dressing, ligament substutites; cartilage regrowth. *Mailing Add:* Dept Surg Univ Ariz Col Med 1501 N Campbell Tucson AZ 85724

CHWANG, ALLEN TSE-YUNG, b Shanghai, China, Nov 7, 44; US citizen; m 68; c 3. FLUID MECHANICS. *Educ:* Chu Hai Col, BS, 65; Univ Sask, MS, 67; Calif Inst Technol, PhD(mech eng), 71. *Prof Exp:* Res fel eng sci, Calif Inst Technol, 71-73, sr res fel, 73-76, res assoc, 76-78; assoc prof, 78-81, PROF MECH ENG, UNIV IOWA, 81- *Concurrent Pos:* Sr vis fel, Dept Appl Math & Theoret Physics, Univ Cambridge, 74-75; Earle C Anthony fel, 69-70; John Simon Guggenheim fel, 74-75; Distinguished summer fac fel, USN-Am Soc Elec Engrs, 88. *Mem:* Sigma Xi; Am Acad Mech; Am Phys Soc; Am Soc Civil Engrs; Am Soc Mech Engrs; Int Asn Hydraulic Res; NY Acad Sci. *Res:* Low-Reynolds-number hydromechanics; nonlinear water waves; hydrodynamic pressures on dams; interaction hydrodynamics; underwater acoustics; general fluid mechanics and applied mathematics. *Mailing Add:* Inst Hydraulic Res Univ Iowa Iowa City IA 52242

CHYLEK, PETR, b Ostrava, Czech, Nov 6, 37; US citizen. ATMOSPHERIC PHYSICS, OPTICS. *Educ:* Charles Univ, Prague, dipl, 66; Univ Calif, Riverside, PhD(physics), 70. *Prof Exp:* Res assoc physics, Ind Univ, Bloomington, 70-72; fel, Nat Ctr Atmospheric Res, Boulder, 72-73; asst prof, State Univ NY Albany, 73-75; ASSOC PROF GEOSCI, PURDUE UNIV, 75-; DEPT PHYSICS, NMEX STATE UNIV, LAS CRUCES, 88- *Mem:* Am Geophys Union; Am Meteorol Soc; Am Optical Soc. *Res:* Light scattering in the atmosphere, effects of air pollution on climate, radiative transfer and atmospheric optics. *Mailing Add:* Dept Physics NMex State Univ Las Cruces NM 88003

CHYNOWETH, ALAN GERALD, b Harrow, Eng, Nov 18, 27; m 50; c 2. SOLID STATE PHYSICS, MATERIALS SCIENCE. *Educ:* King's Col, Univ London, BSc, 48, PhD(physics), 50. *Prof Exp:* Demonstr physics, King's Col, Univ London, 48-50; res fel, Nat Res Coun Can, 50-52; mem tech staff, 53-60, head crystal electronics res dept, 60-65, asst dir mat res, 65-73, dir mat res, 73-76, exec dir, Electronic Device, Process & Mat Div, Bell Labs, 76-83, VPRES, APPL RES, BELL COMMUN RES, 84- *Concurrent Pos:* Surv dir, Nat Acad Sci Comt on Surv Mat Sci & Eng, 70-73; mem & panel chmn, Nat Acad Sci Comt on Mineral Resources & Environ, 73-75; mem, Nat Mat Adv Bd, 76-79; mem, NATO Spec Prog Panel Mat Sci, 77-82, chmn, 78; assoc ed, Ann Rev Mat Sci, 76-79 & Solid State Commun, 75-83; consult, NATA Adv Study Inst Panel, 82- *Honors & Awards:* Baker Prize, Inst Elec & Electronics Engrs, 67. *Mem:* Fel Am Phys Soc; fel Inst Physics London; fel Inst Elec & Electronics Engrs; Am Inst Mining, Metall & Petrol Engrs; Mat Res Soc; NY Acad Sci. *Res:* Transport properties of semiconductors and insulators; electrical breakdown; ferroelectrics; tunnelling; solid state plasmas; materials research; national materials policies; electronic and photonic devices; telecommunications technology. *Mailing Add:* Bell Commun Res 445 South St Morristown NJ 07962-1910

CHYTIL, FRANK, b Prague, Czech, Aug 28, 24; m 49; c 3. BIOCHEMISTRY, PHYSIOLOGY. *Educ:* Col Chem Tech, Prague, Dip Ing, 49, PhD(biochem), 52, CSc(physiol chem), 56. *Prof Exp:* Res biochemist, Charles Univ, Prague, 49-51; Czech Acad Sci res fel biochem, Inst Human Nutrit, Prague, 52-55; sr scientist physiol, Czech Acad Sci, 56-62, sr scientist microbiol, 63-64; sr res fel biochem, Brandeis Univ, 64, sr res assoc, 65-66; from asst prof to assoc prof, 66-75, PROF BIOCHEM, SCH MED, VANDERBILT UNIV, 75- *Concurrent Pos:* Sect head, Southwest Found Res & Educ, 66-72; General Foods chair nutrit, 84-89. *Honors & Awards:* Osborne & Mendal Award, Am Inst Nutit, 83. *Mem:* Fel Am Inst Nutrit; Am Soc Biol Chem; Endocrine Soc; Am Chem Soc; Sigma Xi. *Res:* Metabolic regulations; mechanism of vitamin A action. *Mailing Add:* Dept Biochem Vanderbilt Univ Sch Med Nashville TN 37232-0146

CHYU, MING-CHIEN, HEAT TRANSFER, THERMO-FLUID SCIENCE. *Educ:* Nat Tsing Hua Univ, BS, 77; Iowa State Univ, MS, 79, PhD(mech eng), 84. *Prof Exp:* Asst prof mech eng, Univ Mo-Columbia/Kansas City, 84-87; asst prof, 87-89, ASSOC PROF MECH ENG, TEXAS TECH UNIV, 89- *Concurrent Pos:* Mem, K-10 comt, Am Soc Mech Eng, 85-, Superconductivity Tech Comt, 90-; prin investr, NSF, 85-88, Gen Dynamics Corp, 90; invited scholar, Beijing Polytechnic Univ, 88; summer res fac, Argonne Nat Lab, 89, 90 & 91; consult, Armstrong World Industs, 88-90, Tex Instruments-Dallas, 89-90, Joshi Prod Technologies Inc, 89-, Medvances Inc, 90-, Whirlpool Corp, 90-, Compaq Computers, 90- *Honors & Awards:* Eng Res Initiation Award, NSF, 85; Teetor Educ Award, Soc Automotive Engrs, 89; Award of Excellence, Halliburton Educ Award, 89. *Mem:* Am Soc Mech Engrs; Am Inst Chem Engrs; Am Inst Aeronaut & Astronaut; Am Soc Eng Educ; Soc Automotive Engrs. *Res:* Applied heat transfer; augmentation; industrial heat exchangers; evaporator; two-phase flow; boiling; heating and refrigeration; thermodynamics; energy; fluid flow; thermal control of superconducting system; electronic system; experimental and numerical methods; analytical modeling. *Mailing Add:* Dept Mech Eng Tex Tech Univ Lubbock TX 79409-1021

CHYUNG, DONG HAK, b Pyungyang, Korea, Aug 5, 37. ELECTRICAL & COMPUTER ENGINEERING. *Educ:* Seoul Nat Univ, BS, 59; Univ Minn, MS, 61, PhD(elec eng), 65. *Prof Exp:* Asst prof elec eng, Univ Minn, 65-66 & Univ SC, 66-68; assoc prof, 68-73, PROF ELEC ENG, UNIV IOWA, 73-, PROF COMPUT ENG. *Mem:* Am Math Soc. *Res:* Control theory; computer-based systems. *Mailing Add:* Dept Elec & Comput Eng Univ Iowa Iowa City IA 52242

CHYUNG, KENNETH, b Seoul, Korea, Mar 16, 36; m 65; c 2. METALLURGY. *Educ:* Mich State Univ, BS, 60, PhD(metall), 65. *Prof Exp:* Res fel metall, Mich State Univ, 66; res supvr, 66-85, SR RES ASSOC, CORNING GLASS WORKS, 85- *Mem:* Am Soc Metals; fel Am Ceramic Soc. *Res:* Mechanical twinning of zinc bicrystals; crystal growth; metal fiber-ceramic composite materials; physical and mechanical properties of glass-ceramics: strengthening, fracture, crystallization and grain growth; creep and stress relaxation; ceramic matrix composites; inorganic fiber and whisker reinforcement; refractory glass ceramic matrix materials. *Mailing Add:* Corning Inc Sullivan Park FR 51 Corning NY 14831

CIALDELLA, CATALDO, b Rochester, NY, Aug 26, 26; m 59; c 4. ORGANIC CHEMISTRY, POLYMER CHEMISTRY. *Educ:* Clarkson Col Technol, BS, 50; Case Inst Technol, MS, 54, PhD(org chem), 56. *Prof Exp:* Chemist, Bausch & Lomb Optical Co, 50-52; res chemist, Esso Res & Eng Co, 56-58, res dir, 58-60, dir res & develop, 60-61; vpres, Hysol Div, Dexter Corp, 61-88; RETIRED. *Mem:* Am Chem Soc. *Res:* Polymers, especially epoxy, urethane silicones. *Mailing Add:* PO Box 3451 Pinehurst NC 28374

CIALELLA, CARMEN MICHAEL, b Trenton, NJ, Apr 18, 25; m 51; c 5. NUCLEAR PHYSICS, BALLISTICS. *Educ:* Pa State Univ, BS, 50. *Prof Exp:* Physicist x-ray physics, Nat Bur Standards, 49-53; res physicist nuclear physics, US Geol Surv, 53-57; res physicist, Ballistic Res Labs, 57-79; RETIRED. *Res:* First laboratory experiments to determine neutron and gamma ray transmission through armored vehicles; developed single crystal spectrometer to measure neutron and gamma ray spectra simultaneously. *Mailing Add:* 622 Southgate Rd Aberdeen MD 21001

CIANCANELLI, EUGENE VINCENT, b Beacon, NY, July 6, 39; m 64; c 2. ECONOMIC GEOLOGY, VOLCANOLOGY. *Educ:* Univ Ariz, BS, 63, MS, 65. *Prof Exp:* Geologist, Phelps Dodge Corp, 66-68; chief geologist, Geothermal Resources Int, 68-73; vpres, Can Geothermal Oil Ltd, 73-75; consult geologist, Eugene V Ciancanelli Consult Geologists, 75-79; chmn, Cascadia Pacific Corp, 86-88; PRES, CASCADIA EXPLOR CORP CONSULT GEOLOGISTS, 79- *Concurrent Pos:* Lectr, geol explor methods, First SINO-US Geothermal Conf, Tianjin, China, 81. *Mem:* Geol Soc Am; Am Asn Petrol Geologists; Geothermal Resources Coun; AAAS; Soc Econ Geologists; Soc Mining Engrs. *Res:* Geothermal energy; volcanogenic ore deposits; hydrothermal systems; petrology of volcanic rocks; structural geology of volcanic regions; petroleum geology; compressed air energy storage (CAES). *Mailing Add:* Cascadia Explor Corp 3358 Apostol Rd Escondido CA 92025

CIANCIO, SEBASTIAN GENE, b Jamestown, NY, June 21, 37; m 63; c 2. PHARMACOLOGY, PERIODONTOLOGY. *Educ:* Univ Buffalo, DDS, 61; Am Bd Periodont, dipl, 70. *Prof Exp:* Fel pharmacol & periodont, 63-65, from asst prof to assoc prof periodont & chmn dept, 65-73, assoc prof pharmacol, Sch Med, 65-73, PROF PERIODONT & CHMN DEPT, SCH DENT, STATE UNIV NY BUFFALO, 73-, CLIN PROF PHARMACOL, SCH MED, 73- *Concurrent Pos:* Res grants, United Health Found of Western NY, 65-66 & 70-71, Nat Inst Dent Res, 67-69 & Merrill Nat Labs, 73-79; consult, Vet Admin Hosp, Buffalo, 70-, Pharmaceut Mfrs Asn, 73-76, US Pharmacopae, Nat Formulary, 75-, Am Cyanamid Corp, 78-84, Dupont Co, 81-82 & Erie County Med Ctr; chmn, Coun Dent Therapeut, Am Dent Asn, 76-78; pres, Pharmacol, Toxicol & Therapeut Group, Int & Am Asn Dent Res, 79; dir, Chautaugua Dent Cong, 79; chmn, US Pharm Dent Comt Rev,

80-; ed, Biol Ther in Dent. *Honors & Awards:* Gies Found Award, 88. *Mem:* Am Dent Asn; fel Int Col Dentists; fel Am Acad Periodont (pres elect, 91); Int Asn Dent Res. *Res:* Papain induced changes in rabbit tissues; local hemostasis; principal fibers of the periodontium; plaque control agents; periodontal observations in twins; acid mucopolysaccharides in gingivitis and periodontitis; antibiotics in periodontal therapy. *Mailing Add:* Dept Periodont State Univ NY Sch Dent Buffalo NY 14214

CIANCOLO, GEORGE J, IMMUNOSUPPRESSION OF CANCER. *Educ:* Univ Miami, PhD(microbiol & immunol), 75. *Prof Exp:* Assoc prof med, Med Ctr, Duke Univ & asst investr, Howard Hughes Med Inst, 78-88; EXEC VPRES RES & DEVELOP, MACRONEX INC, 91- *Res:* Chemotactic dysfunction; immunosuppressive retroviral proteins. *Mailing Add:* Macronex Inc 1000 Park 40 Plaza Suite 290 Durham NC 27713

CIAPPENELLI, DONALD JOHN, b Worcester, Mass, Dec 4, 43. CHEMISTRY. *Educ:* Univ Mass, BSc, 66; Brandeis Univ, MSc, 67, PhD(chem), 71. *Prof Exp:* Dir, Chem Dept, Brandeis Univ, 72-77; DIR CHEM DEPT, HARVARD UNIV, 77-; DIR, CHEM DESIGN CORP, 82- *Concurrent Pos:* Chmn bd, CEO Cambridge Lab Consults, Inc. *Mem:* Am Chem Soc; AAAS. *Mailing Add:* Dept Chem Harvard Univ 12 Oxford St Cambridge MA 02138-2902

CIARAMITARO, DAVID A, b Detroit, Mich, Nov 15, 46. FORMULATIONS, PRODUCT DEVELOPMENT ENGINEERING. *Educ:* Oakland Univ, BA, 68; Univ Ariz, PhD(org chem), 74. *Prof Exp:* Fel pharmacol, Health Sci Ctr, Univ Ariz, 74-78; res chemist, 78-80, CHIEF CHEMIST & HEAD RES & DEVELOP, APACHE POWDER CO, 80- *Concurrent Pos:* Mem tech comt, Inst Makers Explosives, 80-81. *Mem:* Am Chem Soc; Soc Explosives Engrs; Soc Mining Engrs. *Res:* New product development in dynamites; water gels; emulsion explosives and detonating cord; organic synthesis, including synthesis of radiolabeled compounds; investigation of biochemical and pharmacologic phenomena in experimental animals; investigation of degredative reactions in lignin model compounds. *Mailing Add:* 2011 Kitts Ridgecrest CA 93555-2735

CIARLONE, ALFRED EDWARD, b Reading, Pa, May 2, 32; m 59; c 2. PHARMACOLOGY, DENTISTRY. *Educ:* Univ Pittsburgh, DDS, 59, PhD(pharmacol), 74. *Prof Exp:* Gen pract resident dent, Vet Admin Hosp, Pittsburgh, 59-60; self-employed dentist, 60-69; from asst prof to assoc prof oral biol & pharmacol, Sch Dent, 73-82, asst prof pharmacol, Sch Med, 73-79, ASSOC PROF PHARMACOL & TOXICOL, SCH MED, MED COL GA, 79-, PROF ORAL BIOL & PHARMACOL, SCH DENT, 82- *Concurrent Pos:* Nat Inst Dent Res trainee fel, 69-73; consult, Doctors Hosp of Augusta, Ga, 77, J Am Dent Asn, 76- & Nat Bd Test Construct Comt, Dent Examr, 78-82; prin investr, Nat Inst Dent Res, 78-81; mem, Adv Panel Dent, rev US Pharmacopeia, 75-80 & rev cycle, 80-85; prof, Sch Grad Studies, Med Col Ga, 82-; consult, Del Labs, Inc, 80-81 & 83-85; consult oral & maxillofacial surg, Eisenhower Med Ctr, Ft Gordon, Ga, 82-86 & 88-91; mem pharmacol, toxicol & therapeut group, Int Asn Dent Res, 82-83, prog chmn, 85-86, pres, 86-87; secy, pharmacol & therapeut sect, Am Asn Dent Sch, 84-85, chmn, 86-87; prin investr, Nat Inst Dent Res, 86-87. *Mem:* Int Asn Dent Res; AAAS; Am Asn Dent Schs; Am Soc Pharmacol & Exp Therapeuts; Am Asn Oral Biologists. *Res:* Author of over 75 publications in scientific journals involving pharmacology in dentistry, including computer programs and co-author of a textbook; binding of tetracyclines to mineralized tissues; permeability of dentin to drugs. *Mailing Add:* Dept Oral Biol & Pharmacol Sch Dent Med Col Ga Augusta GA 30912-1128

CIBILS, LUIS ANGEL, b Yuty, Paraguay, Mar 22, 27; m 61; c 3. OBSTETRICS & GYNECOLOGY. *Educ:* Col San Jose, BA, 42; Univ Paraguay, MD, 50. *Prof Exp:* Scholar, Span Inst Cult, 51-52; vis asst gynec, Univ Paris, 52-53; res fel obstet & gynec, Univ Repub Uruguay, 57-60; res consult, Western Reserve Univ, 60, from instr to asst prof, 60-66; assoc prof, 66-70, PROF OBSTET & GYNEC, UNIV CHICAGO, 70- *Honors & Awards:* Found Prize, Am Asn Obstet & Gynec, 65. *Mem:* Sigma Xi. *Res:* Physiology of reproduction. *Mailing Add:* Dept Obstet & Gynec Univ Chicago Chicago IL 60637

CIBULA, ADAM BURT, biology, entomology; deceased, see previous edition for last biography

CIBULA, WILLIAM GANLEY, b Cleveland, Ohio, Jan 16, 32; m 56; c 5. REMOTE SENSING, AGARIC TAXONOMY. *Educ:* John Carroll Univ, BS, 56, MS, 65; Univ Mass, PhD(bot), 76. *Prof Exp:* Physicist, Wright Patterson AFB, 56-60 & Picker X-ray Corp, 60-68; BOTANIST, NAT SPACE TECHNOL LABS, NASA, 71- *Concurrent Pos:* Adj prof, Univ South Miss, 78- *Mem:* Mycol Soc Am; AAAS; NAm Mycol Asn; Sigma Xi. *Res:* Development, analysis and specification of computer processed land cover classifications from Landsat and other multi-spectral scanners; Austroriparian agaric and bolete taxonomy; chromatography and chemical analysis of fungal pigments and other metabolites to exemplify their taxonomy and phylogeny. *Mailing Add:* 700 Idlewild Dr Picayune MS 39466

CIBULKA, FRANK, b Prague, Czech, May 27, 23; US citizen; m 51; c 1. ELECTRONICS ENGINEERING. *Educ:* Technol Univ Prague, MSc, 49; Czech Acad Sci, PhD(elec eng), 66. *Prof Exp:* Mgr, Power Electronics Dept, Res Inst Elec Eng, Prague, Czech, 52-68; fel engr, Power Electronics Dept, Res & Develop Ctr, Westinghouse Elec Corp, 69-88; CONSULT POWER ELECTRONICS, 89- *Concurrent Pos:* Deleg, TC 22, Int Electrotech Comn, 63-68. *Res:* High power solid-state conversion; development and testing of high power thyristor valves; analysis of transients in electronic power circuits. *Mailing Add:* 2042 Peach Orchard Dr Falls Church VA 22043

CIBULSKY, ROBERT JOHN, b Johnson City, NY, Feb 4, 46; m 72. ENTOMOLOGY, MICROBIOLOGY. *Educ:* Hamilton Col, BA, 68; Auburn Univ, MS, 71, PhD(entom), 75. *Prof Exp:* MGR FIELD RES & DEVELOP, CHEM & AGR PROD DIV, ABBOTT LABS, 75- *Mem:* Sigma Xi; Entom Soc Am; Am Soc Microbiol; Soc Invert Path; Am Soc Hort Sci. *Res:* Development of microbial products for agricultural uses; organisms include bacterial, fungal, and viral entomopathogens, plant pathogen antagonists, and soil inoculants. *Mailing Add:* 919 Dawes St Libertyville IL 60048-3562

CICCARELLI, ROGER N, b Rochester, NY, Dec 23, 34; m 60; c 2. POLYMER CHEMISTRY. *Educ:* St Bonaventure Univ, BS, 56; Syracuse Univ, MS, 58, PhD(org chem), 61; State Univ NY, PhD(org chem), 61. *Prof Exp:* Chemist, Exxon Res & Eng Co, 61-62; scientist, 62-70, SR SCIENTIST, XEROX CORP, 70- *Mem:* Am Chem Soc; Sigma Xi; Chem Soc. *Res:* Synthetic and polymer chemistry; structure versus electrical properties of polymers; toners and carriers; electrostatics; photoconductors; reaction mechanisms; rocket fuels. *Mailing Add:* 145 Hibiscus Dr Rochester NY 14618

CICCOLELLA, JOSEPH A, b Albany, NY, Sept 3, 09; m 46; c 7. NAVIGATION. *Educ:* Rensselaer Polytech Inst, ME, 31, DME, 34. *Prof Exp:* Asst examr, State Civil Serv Comn, NY, 34-35; engr, Fred Page Construct Co, NY, 35-36; field serv, US Lighthouse Serv, 36-39; comn officer, US Coast Guard, 39-61, chief, Testing & Develop Div, 59-61; dir eng signal prod, Amerace-Esna Corp, 61-72; RETIRED. *Concurrent Pos:* Lectr, US Coast Guard Training Sta; mem night visibility comt, Nat Res Coun; secy, US Indust Comt Lighthouse Conf, US delegate, 55, 60, 65, 70 & 75. *Mem:* Am Soc Naval Engrs; Marine Technol Soc. *Res:* Pharology; search and rescue; maritime safety; aids to navigation. *Mailing Add:* 5519 N Main St Bethania NC 27010

CICCONE, PATRICK EDWIN, b Newark, NJ, Nov 20, 44; m 78; c 3. PHARMACOLOGY, PSYCHIATRY. *Educ:* Harvard Col, BA, 66; Univ Pa, MD, 70; Hosp Univ Pa, residency dipl psychiat, 74; Am Bd Psychiat, 76. *Prof Exp:* Staff psychiatrist, Erich Lindemann Ment Health Ctr, 74-75; asst dir clin res & clin investr psychopharmacol, McNeil Labs, Inc, McNeil Consumer Prod Co, 75-77, assoc dir, 78, dir, 79-83; dir develop, Smith Kline Consumer Prod Co, 84, vpres res & develop, 85-86; dir psychiat res, Phila Va Clin, 86-89; VPRES MED AFFAIRS, WARNER-LAMBERT CO, 89- *Concurrent Pos:* Clin assoc psychiat, Mass Gen Hosp, 74-75; dir & staff psychiatrist, Revere Community Ment Health Ctr, 74-75; clin assoc psychiat, Med Sch, Univ Pa, 76-83, clin asst prof, 87-; vis asst prof, Temple Univ, 78-79. *Mem:* Am Psychiat Asn; AMA; Am Soc Clin Pharmacol & Therapeut; Drug Info Asn; NY Acad Sci. *Res:* Clinical reseach with emphasis on CNS drugs. *Mailing Add:* Four Yorkshire Dr Hackettstown NJ 07840

CICERO, THEODORE JAMES, b Niagara Falls, NY, Aug 14, 42; m 66; c 4. BIOCHEMICAL PHARMACOLOGY. *Educ:* Villanova Univ, BS, 64; Purdue Univ, MS, 66, PhD(physiol psychol), 69. *Prof Exp:* Fel neurochem, 68-70, from asst prof to assoc prof, 70-78, assoc prof neurobiol, 76-82, PROF NEUROPHARMACOL & NEUROBIOL, SCH MED, WASH UNIV, 78- *Concurrent Pos:* Consult, Nat Inst Alcoholism & Alcohol Abuse, 71-75 & NIMH, 72-73; consult, Nat Inst Drug Abuse, 77-81. *Mem:* Soc Neurosci; Soc Biol Psychiat; Am Psychopath Asn; Soc Neurochem; Am Soc Biol Chemists; Am Soc Pharmacol & Exp Therapeut. *Res:* Neurochemical, neurobiological and neuroendocrinological correlates of tolerance to and dependence on narcotics and alcohol; developmental endocrinology. *Mailing Add:* Dept Psychiat Sch Med Wash Univ 4940 Audubon Ave St Louis MO 63110

CICERONE, CAROL MITSUKO, Olaa, Hawaii, July 16, 43; m 67; c 1. NEUROPHYSIOLOGY OF THE VISUAL SYSTEM, VISION. *Educ:* Univ Ill, Urbana, BS, 65, MS, 67, Univ Mich, Ann Arbor, PhD(psychol), 74. *Prof Exp:* Res scientist psychol, Univ Mich, Ann Arbor, 76-78; asst prof, 78-79, assoc prof psychol, Univ Calif, San Diego, 79-86; ASSOC PROF PSYCHOL, UNIV COLO, BOULDER, 87- *Mem:* AAAS; Asn Res Vision & Opthal; Optical Soc Am. *Res:* Psychophysical studies of sensitivity regulation and color vision in humans; neurophysiological studies of the vertebrate visual system involving retino-tectal connections in goldfish, retinal degenerations in rats & sensitivity regulation in the mammalian visual system. *Mailing Add:* Dept Psychol CB 345 Univ Colo Boulder CO 80309-0345

CICERONE, RALPH JOHN, b New Castle, Pa, May 2, 43; m 67; c 1. ATMOSPHERIC CHEMISTRY. *Educ:* Mass Inst Technol, SB, 65; Univ Ill, MS, 67, PhD(elec eng & physics), 70. *Prof Exp:* Physicist, US Dept Com, 67; res asst aeronomy, Univ Ill, 67-70; assoc res scientist aeronomy, Space Physics Res Lab, Univ Mich, Ann Arbor, 70-78; assoc res chemist, Ocean Res Div, 78-80, res chemist, Scripps Inst Oceanog, Univ Calif, San Diego, 80-81; sr scientist & dir, Atmospheric Chem Div, Nat Ctr Atmospheric Res, Boulder, Colo, 80-89; DANIEL G ALDRICH PROF & CHAIR, GEOSCI DEPT, UNIV CALIF, IRVINE, 89- *Concurrent Pos:* lectr & asst prof elec eng, Univ Mich, Ann Arbor, 73-75; assoc ed, J Geophys Res, 77-79; ed, J Geophys Res, 79-83; mem comt atmospheric sci, Nat Acad Sci, 80-82; mem bd, atmospheric sci & climate, Nat Acad Sci, 87-89, comn Geosci, Environ & Resources. *Honors & Awards:* Macelwane Award, Am Geophys Union, 79. *Mem:* Nat Acad Sci; fel AAAS; Am Chem Soc; fel Am Meteorol Soc; fel Am Geophys Union. *Res:* Theoretical and experimental studies of the earth's atmosphere. *Mailing Add:* Geosci Dept Phys Sci Univ Calif Irvine CA 92717

CICHELLI, MARIO T(HOMAS), b Baltimore, Md, Jan 28, 20; m 43; c 4. CHEMICAL ENGINEERING. *Educ:* Loyola Col, Md, BS, 40; Johns Hopkins Univ, PhD(chem eng), 45. *Prof Exp:* Process engr, Kellex Corp, New York, 44-45 & Carbide & Carbon Chem Corp, Manhattan Dist Proj, Oak Ridge, 45-46; res eng fel, Mellon Inst, 46-50; res engr, E I Du Pont De Nemours & Co, Inc, 50-53, res supvr & res mgr, 54-79, mgr, Patent Liaison Div, Fibers Dept, 79-84; CONSULT, CHEM ENG, 88- *Mem:* Am Chem Soc; Am Inst Chem Engrs. *Res:* Heat transmission; diffusional separation processes; high temperature technology. *Mailing Add:* Edenridge Ten Chilton Rd Wilmington DE 19803

CICHOWSKI, ROBERT STANLEY, b Lakewood, Ohio, Feb 28, 42; m 66; c 3. HETEROGENEOUS CATALYSIS, CHILDRENS SCIENCE EDUCATION. *Educ:* Purdue Univ, BSChE, 64; State Univ NY Col Ceramics, Alfred Univ, PhD(ceramic sci), 68. *Prof Exp:* Res chemist, Phillips Petrol Co, Okla, 68-71; lectr, 71, asst prof, 71-75, assoc prof, 76-80, PROF CHEM, CALIF POLYTECH STATE UNIV, SAN LUIS OBISPO, 80- *Concurrent Pos:* Actg dir sci, Pac Sci Ctr, Seattle, Wash, 80. *Mem:* Am Chem Soc; Catalysis Soc; Nat Sci Teachers Asn. *Res:* Chemistry of glazes; surface chemistry of solids and solid state chemistry related to heterogeneous catalysis; photoelectrochemistry. *Mailing Add:* Dept Chem Calif Polytech State Univ San Luis Obispo CA 93407

CIER, H(ARRY) E(VANS), b Monroe, La, Sept 16, 12; m 31; c 3. CHEMICAL ENGINEERING. *Educ:* La State Univ, BS, 33. *Prof Exp:* Chem engr develop div, Esso Standard Oil Co, 34-36; res chem engr res & develop div, Humble Oil & Ref Co, 36-42, 46-48, sr res chem engr, 48-55, res specialist, 55-60, sr res specialist, 60-63, res assoc, 64-65; staff adv to vpres chem, Esso Res & Eng Co, 65-66, eng assoc chem raw mat staff, Esso Res Labs, La, 66-67, eng assoc res planning, Baytown Chem Res Labs, 67-71; chief process engr, Jovan Consult Engrs, Tehran, Iran, 71, consult, 72; sr engr, C F Braun & Co, 75; consult engr, 80; RETIRED. *Mem:* Fel Am Inst Chem; Am Inst Chem Engrs. *Res:* Development of petroleum processes; petrochemical processes; research planning. *Mailing Add:* 2027 Lakeview Dr Baytown TX 77520-1031

CIERESZKO, LEON STANLEY, SR, b Holyoke, Mass, July 30, 17; m 43; c 1. BIOCHEMISTRY. *Educ:* Mass State Col, BS, 39; Yale Univ, PhD(physiol chem), 42. *Prof Exp:* Lab asst physiol chem, Yale Univ, 39- 42; res biochemist, Med Res Div, Sharp & Dohme, Inc, Pa, 42-45; instr biol chem, Sch Med, Univ Utah, 45-46; instr chem, Univ Ill, 46-48; from asst prof to prof, 48-80, chmn dept, 69-70, EMER PROF CHEM, UNIV OKLA, 80- *Concurrent Pos:* Fulbright fel, Zool Sta, Naples, Italy, 55-56; res fel, Yale Univ, 59; off partic, US Prog Biol, Int Indian Ocean Exped, 63; fel & vis investr, Friday Harbor Labs, Seattle, 64; consult prof biochem, Sch Med, Univ Okla, 67-79; vis investr, Dept Marine Sci, Univ PR, 70 & 74-75 & 78, Caribbean Res Inst, VI, 71 & Marine Sci Inst, Univ Tex, 74; consult, Caribbean Res Inst, 70-; vis lectr, Col VI, 71; pvt investr, Sea Grant Res Proj, 71-; vis prof, Marine Lab, Univ Tex, Port Aransas, 78. *Mem:* AAAS; Am Chem Soc; Geochem Soc; Int Soc Chem Ecol; Int Soc Reef Studies. *Res:* Comparative biochemistry; chemistry of natural products from marine animals; chemistry of coelenterates and their zooxanthellae; toxic substances of marine origin; biogeochemistry of coral reefs. *Mailing Add:* Dept Chem Univ Okla 620 Parrington Oval 208 Norman OK 73019-0370

CIFONE, MARIA ANN, b 1945; m 81; c 1. DNA DAMAGE, MAMMALIAN MUTAGENESIS. *Educ:* Univ Pa, PhD(cell biol), 75. *Prof Exp:* CELL BIOLOGIST, HAZELTON LABS AM, 77- *Mem:* Am Soc Cell Biol; Environ Mutagen Soc; Tissue Cult Asn. *Res:* Relationship of anchorage independent growth to genetic damage, tumorigenicity and tumor progression. *Mailing Add:* Dept Genetic Toxicol Hazelton Labs Am 5516 Nicholson Lane Kensington MD 20895

CIFONELLI, JOSEPH ANTHONY, b Utica, NY, Mar 19, 16; m 49; c 2. CARBOHYDRATE CHEMISTRY. *Educ:* Univ Minn, PhD(biochem), 52. *Prof Exp:* Res assoc carbohydrate chem, Univ Minn, 53; res assoc pediat, Bobs Roberts Hosp, 53-64, res assoc biochem, La Rabida Inst, 64-; prof pediat, Univ Chicago, 71-; RETIRED. *Concurrent Pos:* Assoc prof pediat, Univ Chicago, 67-71. *Mem:* Am Chem Soc; Am Soc Biol Chemists. *Res:* Biosynthesis of mucopolysaccharides; enzyme chemistry. *Mailing Add:* 637 Golf Lane Lake Barrington IL 60010

CIFTAN, MIKAEL, b Istanbul, Turkey, Aug 12, 35; US citizen; m 57; c 1. THEORETICAL PHYSICS. *Educ:* Robert Col, Istanbul, BSc, 57; Mass Inst Technol, MSc, 59; Duke Univ, PhD(physics), 67. *Prof Exp:* Res physicist, Raytheon Co, 60-65; res assoc physics, Ind Univ, Bloomington, 68-70; DIR THEORET PHYSICS, US ARMY RES OFF-DURHAM, 70- *Concurrent Pos:* Adj prof physics, Duke Univ, 70- *Honors & Awards:* Issai Lefkowitz Award. *Mem:* Am Phys Soc. *Res:* Symmetry principles in physics and representation theory; elementary particle phenomenology; special functions via group theory; theoretical and experimental laser physics, spectroscopy; biomathematical approach to experimental histocompatibility and genetic mapping; theoretical solid state physics; statistical physics. *Mailing Add:* US Army Res Off PO Box 12211 Research Triangle Park NC 27709

CIHONSKI, JOHN LEO, b Butler, Pa, Oct 24, 48; m 72. INORGANIC CHEMISTRY. *Educ:* Slippery Rock State Col, BA, 71; Tex A&M Univ, PhD(chem), 75. *Prof Exp:* Res assoc chem eng, Tex A&M Univ, 75; sr res chemist catalysis, El Paso Prod Co, 75-89; SR RES CHEM & BUS DEVELOP, CATALYTICA, CALIF, 89- *Mem:* Am Chem Soc; AAAS. *Res:* Solid waste disposal by microbal decomposition; industrial heterogeneous catalysis with continued interests in organometallic photochemistry and molecular spectroscopy. *Mailing Add:* Catalytica 430 Ferguson Dr Bldg 3 Mountain View CA 94043-5272

CIMBALA, MICHELE, HORMONE ACTION, GENETIC DEFICIENCY. *Educ:* WVa Univ, PhD(biochem), 77. *Prof Exp:* ASST PROF BIOCHEM, MED CTR, UNIV MASS, 81- *Mailing Add:* 2916 Birchtree Lane Silver Spring MD 20906

CIMBERG, ROBERT LAWRENCE, b New York, NY, Nov 15, 44. MARINE BENTHIC ECOLOGY, ENVIRONMENTAL IMPACTS. *Educ:* Southern Ill Univ, BS, 67; Calif State Univ, MS, 75; Univ Southern Calif, PhD(biol), 78. *Prof Exp:* Res asst, Calif State Univ, Humboldt, 68-69; res asst, Univ Southern Calif, 69-74, lab instr embryol, Catalina Marine Sci Ctr, 71-74; lectr ecol, Univ Calif, Irvine, 78-80; sr marine biologist, VTN Ore Inc, 80-88; MEM STAFF, LUZ DEVELOP & FINANCE CORP, 88. *Concurrent Pos:* Lectr biol, Calif State Univ, Los Angeles, 79. *Mem:* Sigma Xi; AAAS; Am Soc Zoologists. *Res:* Marine benthic ecology from Alaska to Baja California, including ecology of intertidal and subtidal habitats and impacts of environmental factors and pollutants on physiological and community processes. *Mailing Add:* 286 Cervantes Circle Lake Oswego OR 97035

CIMENT, MELVYN, b New York, NY, Sept 23, 41; m 66; c 2. APPLIED MATHEMATICS. *Educ:* Univ Miami, Fla, BS, 62; NY Univ, MS, 64, PhD(math), 68; Am Univ, JD, 78. *Hon Degrees:* JD, Am Univ, 78. *Prof Exp:* From instr to asst prof math, NY Univ, 67-69; asst prof, Univ Mich, 69-70; vis lectr, Tel Aviv Univ, 70-71; asst prof, Univ Mich, 71-72; mathematician, Naval Surface Weapons Ctr, Silver Spring, Md, 72-77; sr appl mathematician, Nat Bur Standards, 77-83; prog dir, Appl Math, 83-86, Computational Math, 86, dep dir, Div Advan Sci Comput, 86-87, DEP DIR, DIV ADVAN SCI COMPUT, NSF, 87- *Concurrent Pos:* Courant Inst Math Sci fel, NY Univ, 62-66; cong sci fel, US Senate Comt Com Sci & Transp, 80-81; fel, Dept Com Sci & Technol, 80-81. *Honors & Awards:* Group Super Achievement Award, Naval Surface Weapons Ctr, 76. *Mem:* Am Math Soc; Soc Indust & Appl Math. *Res:* Numerical solution of partial differential equations; design and analysis higher order compact implicit methods; high performance computing research, resources and technology. *Mailing Add:* 11712 Kemp Mill Rd Silver Spring MD 20902

CIMINI, LEONARD JOSEPH, JR, b Philadelphia, Pa, Apr 19, 56; m 82; c 2. COMMUNICATIONS, DETECTION & ESTIMATION. *Educ:* Univ Pa, BS, 78, MS, 79, PhD(elec eng), 82. *Prof Exp:* Mem tech staff, Mobile Systs Eng Lab, West Long Branch, NJ, 82-85, MEM TECH STAFF, COMMUN SYSTS RES LAB, AT&T BELL LABS, HOLMDEL, NJ, 85- *Concurrent Pos:* adj prof commun theory, Monmouth Col, 83-88; secy, Commun Theory Comt, Inst Elec & Electronics Engrs Commun Soc, 87-89, vchair, 89-, ed mobile commun, Trans on Commun, 89- *Honors & Awards:* A Atwater Kent Prize, 78. *Mem:* Sr mem Inst Elec & Electronics Engrs. *Res:* Control and switching aspects of future wireless communication systems; author of various publications. *Mailing Add:* 11 Appletree Rd Howell NJ 07731

CINADER, BERNHARD, b Vienna, Austria, Mar 30, 19; c 1. IMMUNOGENETICS, GERONTOLOGY. *Educ:* Univ London, BSc, 45, PhD, 48, DSc, 58. *Prof Exp:* Asst, Lister Inst Prev Med, Eng, 45-46, Beit Mem fel, 49-53, Agr Res Coun grantee, 53-56; fel immunochem, Inst Path, Western Reserve Univ, 48-49; prin sci officer, Dept Exp Path, Agr Res Coun Inst Animal Physiol, Eng, 56-58; head subdiv immunochem div biol, Ont Cancer Inst, 58-69; assoc prof, 58-69, dir dept immunol, 71-81, PROF, DEPT CLIN BIOCHEM, UNIV TORONTO, 70-, PROF DEPT IMMUNOL, 82- *Concurrent Pos:* Lectr & medallist, Fr Soc Biol Chemists, 54; joint chmn sect immunochem, Int Cong Biochem, Belgium, 55, dir, Austria, 58; spec lectr, Pasteur Inst, 60; pub lectr, Univ Col, Univ London, 63; vis prof, Univ Alta, 68 & Univ Manitoba, Inst Res Reproduction, Bombay, 81, Inst Microbiol, Mahiotol Univ, Bangkok, 82; Pfizer fel, Inst Clin Res, Montreal, 72; guest lectr, Acads Sci, Czech & Hungary & Acad Med, Rumania, 74; mem comt on antilymphocytic serum, Med Res Coun, chmn grants panel immunol & transplantation, 70; mem expert adv panel on immunol, WHO, 70-, mem task force standardization immune reagents, 73-, Sci & Tech adv group, spec prog res, develop res human reproduction, Switz, 85-90, task force steering comt, Birth Control Vaccines, Geneva, 85, study group aging, 85; mem spec comt clin immunol, Royal Col Physicians & Surgeons; pres, Int Union Immunol Socs, 69-74; chmn immunol comt, Biol Coun Can; mem, Adv Bd, Human Reprod Task Force Steering Comt, WHO, Med Univ SC, Centre Immunol, Buffalo & Amsterdam chmn, Int Union Immunol Socs-WHO Inst, 71-79; mem, Can Sci Deleg to USSR, 75; mem, Can Nat Comt Immunol, 77; alt rep IUIS, Int Coun Sci Unions, 78-, mem 20th Gen Assembly, 84; mem adv bd, Iroquoian Inst, 90-; mem gov coun, Univ Toronto, 82; mem grant panel epidemiol, immunol, microbiol, path & virol, Nat Cancer Inst; chmn organizing comt, Sixth Int Cong Immunol, 83. *Honors & Awards:* Medal, Soc Biol Chem, Paris, 54; Medal, Inst Pasteur, Paris, 60; Enrique E Ecker Lectr Exp Path, Western Reserve Univ, 64; A Harrington Lectr, State Univ NY, Buffalo, 74; Elizabeth II Jubilee Medal, Gov Gen Can, 77; Thomas W Eadie Medal, Royal Soc Can, 82; Karl Landsteiner Cong Medal, Can Soc Immunol, 86; Found Cinader Lectr, Can Soc Immunol, 87; Jan E Purkyne Medal, Czechoslovak Soc Immunol & Allergol, 88. *Mem:* Am Asn Immunologists; Brit Soc Immunologists; Can Soc Immunol (pres, 66-69 & 79-81); fel Royal Soc Can; Can Fedn Biol Socs; fel NY Acad Sci. *Res:* Characterization of two distinct classes of immunoglobulins, first report on tolerance as a factor in determining response to across-reacting molecule and hypothesis of tolerance-steered specificity; analysis of the effects of antibody on catalysis by biologically active molecules, evidence for inhibition by steric hinderance, competition between antibody molecules for overlapping sites on enzyme; discovery of a murine complement defect as a consequence of experiments based on the hypothesis of tolerance-steered immune responsiveness; analysis of polymorphic gerontological changes in different classes of suppressor cells and in thymus precursors; analysis of age-related changes in isotypes, H1 histones, repair capacity dopamine, insulin receptors and adrenoceptors; development of strategies to alter progression of age related changes by dietary and pharmacological interventions, as methods to analyze interdependence and or independence of compartmentalized aging; a natural law of economic correction deduced from the observation that a variety of different systems show a direct correlation between relative magnitude in early life and subsequent rate of age-related decrease. *Mailing Add:* Dept Immunol Med Sci Bldg Univ Toronto Toronto ON M5S 1A8 Can

CINADR, BERNARD F(RANK), b Brecksville, Ohio, June 5, 33; m 61; c 1. CHEMICAL ENGINEERING. *Educ:* Case Inst Technol, BS, 55, PhD(phys chem), 60. *Prof Exp:* Asst polymer chem, Case Inst Technol, 59-60; res chemist, 60-61, from res engr to sr res engr, 61-69, sect leader, 69-76, sr res & develop assoc, chem div, 76-82, RES & DEVELOP FEL, SPECIALTY POLYMERS & CHEM DIV RES CTR, B F GOODRICH CO, 82- *Mem:* Am Inst Chem Engrs; Am Chem Soc. *Res:* Polymerization chemistry; polymerization process research, including monomer purification, polymerization and isolation and drying of polymers; process design and cost estimation. *Mailing Add:* B F Goodrich Co Res & Develop Ctr 9921 Brecksville Rd Brecksville OH 44141-3289

CINCOTTA, JOSEPH JOHN, b Queens, NY, Sept 15, 31; m 56; c 3. ANALYTICAL CHEMISTRY. *Educ:* Columbia Univ, BS, 53; City Univ New York, MS, 66. *Prof Exp:* Anal chemist, Am Molasses Co, 53-59; anal res chemist, Am Cyanamid Co, 59-68 & M W Kellogg Co, 68-69; sr anal chemist, Conoco Inc Co, 69-84; SR ANAL CHEMIST, VISTA CHEM CO, 84- *Mem:* Am Chem Soc; fel Am Inst Chemists. *Res:* Gas chromatography; coulometry; spectrophotometry; atomic absorption spectrometry and liquid chromatography. *Mailing Add:* 5025 Cloudburst Hill Columbia MD 21044

CINK, CALVIN LEE, b Valley City, NDak, March 1, 47; m 70; c 2. BEHAVIORAL ECOLOGY. *Educ:* NDak State Univ, Fargo, BS, 69; Univ Nebr, Lincoln, MS, 71; Univ Kans, Lawrence, PhD(biol), 77. *Prof Exp:* ASSOC PROF BIOL, BAKER UNIV, 76- *Concurrent Pos:* Vis instr, Biol Dept, Univ Nebr at Omaha, 79, Creighton Univ, Omaha, 80; ed, Winter Bird Pop Studies, 80- *Mem:* Ecol Soc Am; Animal Behav Soc; Am Ornithologists Union; Wildlife Soc; Sigma Xi. *Res:* Vertebrate ecology and behavior; structure of bird communities; population ecology; determinants of territory size; philopatry; dispersal; breeding biology; comparative ethology; animal communication including song dialects. *Mailing Add:* Biol Dept Baker Univ Baldwin City KS 66006

CINLAR, ERHAN, b Divrigi, Turkey, May 28, 41; US citizen. MATHEMATICAL STATISTICS, OPERATIONS RESEARCH. *Educ:* Univ Mich, BSE, 63, MA, 64, PhD(indust eng), 65. *Prof Exp:* From asst prof to assoc prof opers res, Northwestern Univ, 65-71; vis prof, Stanford Univ, 71-72; PROF OPERS RES, NORTHWESTERN UNIV, EVANSTON, 72-, PROF MATH, INDUST ENG & MGMT SCI, ENG SCIS & APPL MATH. *Concurrent Pos:* Assoc ed, J Stochastic Processes & Their Applns, Res & Mgt Sci; vis prof statist, Princeton Univ, 79-80. *Mem:* Am Math Soc; fel Inst Math Statist; Inst Mgt Sci. *Res:* Stochastic processes; theory of regeneration; Markov processes and boundary theory; Markov renewal theory; point processes; random measures. *Mailing Add:* Dept Civil Eng Princeton Univ Princeton NJ 08544

CINO, PAUL MICHAEL, b New York, NY, Dec 8, 46; m 73; c 1. INDUSTRIAL MICROBIOLOGY. *Educ:* Hunter Col, BA, 68; Rutgers Univ, MS, 70, PhD(microbiol), 73. *Prof Exp:* Res fel, Waksman Inst Microbiol, Rutgers Univ, 73-75; res investr, 75-81, sr res investr, 81-87, RES GROUP LEADER, E R SQUIBB & SONS INC, 87- *Concurrent Pos:* Guest lectr, Rutgers Univ. *Mem:* Am Soc Microbiol; Soc Indust Microbiol. *Res:* Microbial bioconversions and enzymatic synthesis of microbial products; new screening techniques. *Mailing Add:* Squibb Inst Med Res Georges Rd New Brunswick NJ 08903

CINOTTI, ALFONSE A, b Jersey City, NJ, Jan 1, 23; m 46; c 5. MEDICINE, OPHTHALMOLOGY. *Educ:* Fordham Univ, BS, 43; State Univ NY Downstate, MD, 46; Am Bd Ophthal, dipl. *Prof Exp:* From asst prof surg & ophthal to assoc prof ophthal, 57-73, PROF OPHTHAL & CHMN DEPT, UNIV MED NJ, 73- *Concurrent Pos:* Assoc examr, Am Bd Ophthal, 53-63; dir resident training ophthal, New York Eye & Ear Infirmary, 55-63, asst dir inst ophthal, 57-63, attend ophthalmologist & dir glaucoma, 63-70, consult glaucoma, 70-; chmn eye health screening prog, State of NJ, 58-; dir ophthal, Jersey City Med Ctr, 63-88 & Univ Hosp, Newark, 71-; med dir, Eye Inst NJ, 74-; mem bd joint comn, Allied Health Personnel Ophthal; chmn med adv comt, Nat Soc Prev Blind, NJ. *Mem:* Fel Am Col Surg; Asn Res Vision & Ophthal; Am Asn Univ Prof Ophthal; Am Asn Ophthal (pres, 78-79); Am Acad Ophthal. *Res:* Prostaglandins in mammalian lenses, correlation of lens functional indices with NMR Data, psychological tests in glaucoma; evaluation of new drugs in glaucoma; eye screening methods. *Mailing Add:* Dept of Ophthal Univ Med & Dent NJ Med Sch 100 Bergen St Newark NJ 07103

CINOTTI, WILLIAM RALPH, b Jersey City, NJ, Sept 14, 26; c 2. PROSTHODONTICS, PERIODONTICS. *Educ:* Georgetown Univ, BS, 46, DDS, 51; NY Univ, cert periodont, 64. *Prof Exp:* Intern oral surg, Martland Med Ctr, 51-52; pvt pract, 52-60; from clin instr to clin asst prof dent, Sch Dent, Seton Hall Univ, 60-67; clin assoc prof, 67-68, assoc prof, 68-71, PROF DENT, NJ DENT SCH, UNIV MED & DENT NJ, 71-, CHMN DEPT RESTORATIVE DENT, 85-, ASSOC DEAN INTERDISCIPLINARY & EXTRAMURAL PROGS, 89- *Concurrent Pos:* Dir dept dent serv, Hudson County, NJ, 70-; pres, Hudson County Ment Health Asn, 75-77, pres emer, 77-; mem bd dirs, Ment Health Asn NJ & Hudson County Men Health Asn; pres, NJ State Bd Dent. *Mem:* Am Prosthodont Soc; fel Am Col Dent; Am Geriat Soc; Royal Soc Health; fel Int Col Dent; Am Acad Craneomandibular Dysfunction; Acad Psychosomatic Med. *Res:* Termporomandibular joint dysfunction; desensitization of teeth; psychologic evaluation of dental patients; evaluation of efficacy of denture adhesive and denture cleanser; evaluation of oral appliances in treatment of reflex spasm of the masticatory muscles; evaluation of cost control procedures for the fabrication of overlay complete dentures; evaluation of the analgesic effect of electrical nerve stimulation with that of conservative dental treatment in patients with temporomandibular joint dysfunction; improving prosthetic appliances for geriatric patients. *Mailing Add:* NJ Dent Sch 100 Bergen St Newark NJ 07103

CINQUINA, CARMELA LOUISE, b Philadelphia, Pa, Mar 11, 36. BACTERIOLOGY. *Educ:* West Chester State Teachers Col, BS, 57; Villanova Univ, MS, 63; Rutgers Univ, PhD(bact), 68. *Prof Exp:* Instr biol & phys educ, York Jr Col, 57-61; from instr to asst prof biol, 61-68, assoc prof, 68-69, prof bact, 69-77, PROF BIOL, WEST CHESTER STATE COL, 77- *Mem:* AAAS; Am Soc Microbiol. *Res:* The effect of caffeine on growth in bacterial cells particularly on morphological changes, variation in lipid composition and changes in cyclic adenosine monophosphate phosphodiesterase activity. *Mailing Add:* Dept of Biol West Chester State Univ West Chester PA 19383

CINTI, DOMINICK LOUIS, b Wilkes-Barre, Pa, Dec 16, 39; m 67; c 2. PHYSIOLOGY, PHARMACOLOGY. *Educ:* Univ Scranton, BS, 61; Jefferson Med Col, MS, 65, PhD(physiol), 68; Drexel Inst Technol, MSBmE, 69. *Prof Exp:* Sr med technician, Jefferson Med Col, 62-63, asst, 64-66; instr physiol, Sch Med, Temple Univ, 66-67; Nat Heart Inst fel, 67-69; res assoc pharmacol, Sch Med, Yale Univ, 69-72; vis scientist, Karolinska Inst, Stockholm, 72-73; from asst prof to assoc prof, 73-81, PROF PHARMACOL, HEALTH CTR, UNIV CONN, FARMINGTON, 81- *Mem:* Am Soc for Pharmacol & Exp Therapeut; Am Soc Biol Chemists. *Res:* Hepatic membrane bound enzymes, such as mixed function oxidase system; microsomal electron transport system; microsomal fatty acid elongation system. *Mailing Add:* Dept Pharmacol Univ Conn Health Ctr Farmington CT 06032

CINTRON, CHARLES, b New York, NY, Dec 25, 36. CELL BIOLOGY, BIOCHEMISTRY. *Educ:* City Col New York, BS, 59; Univ Colo, MS, 61; Univ Conn, PhD(biol), 70. *Prof Exp:* Res assoc ophthal, Med Sch, Yale Univ, 62-65; fel, 70-72, asst scientist, 72-75, assoc scientist, 75-80, SR SCIENTIST CELL BIOL, EYE RES INST, RETINA FOUND, 80- *Concurrent Pos:* Res trainee, Univ Conn, 68-70; Seeing Eye Fund fel, 73-74; NIH res grant, 74-; res assoc, Dept Ophthal, Harvard Med Sch, 75-80, asst prof, 80- *Honors & Awards:* Alcon Res Award, 83. *Mem:* Soc Complex Carbohydrates; Asn Res Vision & Ophthal; Am Soc Cell Biol. *Res:* Electron microscopy; biochemistry; physiology of corneal wound healing; development of mammalian cornea. *Mailing Add:* Eye Res Inst of Retina Found 20 Staniford St Boston MA 02114

CINTRON, GUILLERMO B, b San Juan, PR, Mar 28, 42; m 68; c 3. CARDIOLOGY, INTERNAL MEDICINE. *Educ:* Univ PR, BS, 63; Loyola-Stritch, Chicago, MD, 67. *Prof Exp:* Chief coronary care unit, Vet Admin Hosp, PR, 74-83; from asst prof to assoc prof med, Univ PR Sch Med, 75-83; ASSOC PROF MED, COL MED, UNIV SFLA & CHIEF CARDIOL SECT, VET ADMIN HOSP, TAMPA, 83- *Concurrent Pos:* Prin investr, Vet Admin coop study vasodilators, 75-80 & 85-; consult cardiol, US Naval Hosp, Roosevelt Rds, PR, 75-83; mem, Coun Clin Cardiol, Am Heart Asn. *Mem:* Fel Am Col Cardiol; fel Am Col Physicians; Am Heart Asn; hon mem Dominican Repub Cardiol Soc. *Res:* Evaluation of vasodilator therapy in patients with acute myocardial infarction or heart failure; heart transplantation. *Mailing Add:* Cardiol III-A Vet Admin Hosp 13000 Bruce B Downs Blvd Tampa FL 33612

CIOCHON, RUSSELL LYNN, b Altadena, Calif, Mar 11, 48. OLIGOCENE-MIOCENE PRIMATES. *Educ:* Univ Calif, Berkeley, BA, 71, MA, 74, PhD(anthrop), 86. *Prof Exp:* Lectr sociol & anthrop, Univ NC, Charlotte, 78-81; res assoc paleontol, Univ Calif, Berkeley, 81-82; res assoc paleoanthrop, Inst Human Origins, Berkeley, 83-85; res assoc anat, State Univ NY, Stony Brook, 85-87; ASSOC PROF ANTHROPOL, UNIV IOWA, 87- *Mem:* Am Anthrop Asn; Am Asn Physical Anthropologists; AAAS; Soc Syst Zool; Soc Vert Paleont. *Res:* Paleoanthropology, primarily primate paleontology; documenting the Tertiary evolutionary history of the primates through both field collecting and museum analysis; hominid paleontology. *Mailing Add:* Dept Anthropol Univ Iowa 114 Macbride Hall Iowa City IA 52242

CIOFFI, PAUL PETER, b Cervinara, Italy, June 29, 96; nat US; m 26; c 2. MAGNETISM. *Educ:* Cooper Union, BS, 19, EE, 22; Columbia Univ, AM, 24. *Prof Exp:* Mem tech staff, Bell Tel Labs, 17-61; consult, Arnold Eng Co, Ill, 61-65, magnetics technol, 65-82; RETIRED. *Mem:* AAAS; fel Am Phys Soc; Inst Elec & Electronics Engrs; Magnetics Soc; Sigma Xi. *Res:* Magnetic measurements, materials and circuits; ideal magnetic circuit through superconductivity; electromagnets for intense magnetic fields; deleterious effects and control of interstitial impurities in magnetic materials significantly improved magnetic quality; control of flux distribution in magnetic circuits with superconductors reduces dimensional and energy requirements. *Mailing Add:* 132 Kent Pl Blvd Summit NJ 07901

CIONCO, RONALD MARTIN, b Erie, Pa, Dec 4, 34; m 55; c 3. MICROMETEOROLOGY, METEOROLOGY. *Educ:* Pa State Univ, BS, 57; Cornell Univ, MS, 71. *Prof Exp:* Meteorologist, Army Res & Develop, Meteorol Dept, Electronics Command, 57-58; weather forecaster, Army Res & Develop, Signal Corps, Evans Labs, US Army, 58-60; res meteorologist micro-meteorol, Atmospheric Sci Lab, ERDA, 60-70; RES METEOROLOGIST MICRO- & MESO-METEOROL, ATMOSPHERIC SCI LAB, ERADCOM, 71- *Mem:* Am Meteorol Soc; Sigma Xi. *Res:* Micrometeorological modeling of air flow within and above vegetative canopies including their turbulence and energy budget aspects; mesometeorological modeling of the wind field over complex terrain. *Mailing Add:* 1012 Cedardale Dr Las Cruces NM 88001

CIOSEK, CARL PETER, JR, b Montclair, NJ, Oct 9, 43; m 66; c 1. CELL BIOLOGY, RHEUMATOLOGY. *Educ:* Univ Mass, BS, 66; Univ Vt, PhD(med microbiol), 73. *Prof Exp:* Res assoc microbiol, Hazelton Labs, 66-68; res assoc rheumatology, Col Med, Univ Vt, 73-77; fel, 77-79, RES INVESTR RHEUMATOLOGY, INST MED RES, E R SQUIBB & SONS, 79- *Concurrent Pos:* Nat Arthritis Found fel, 75-77. *Mem:* Am Rheumatism Asn; NY Acad Sci; Tissue Cult Asn; Am Soc Microbiol. *Res:* Connective tissue diseases, especially rheumatoid arthritis; inflammation; autoimmunity; cell defense mechanisms; collagenase, especially proteolytic enzymes; prostaglandins, especially cyclic nucleotides; membrane receptors; cell culture; virology. *Mailing Add:* Inst Med Res Box 4000 Princeton NJ 08540

CIOSLOWSKI, JERZY, b Cracow, Poland, July 14, 63; m 89. QUANTUM CHEMISTRY, ELECTRONIC STRUCTURES OF MOLECULES. *Educ:* Jagiellonian Univ, MSc, 85; Georgetown Univ, PhD(chem), 87. *Prof Exp:* Fel chem, Univ NC, Chapel Hill, 87-88; postdoctoral staff mem, Los Alamos Nat Lab, 88-89; ASST PROF, DEPT CHEM, FLA STATE UNIV, 89- *Mem:* Int Soc Math Chem. *Res:* Development of new approaches to calculating the electronic structure of molecules; density functional theory; the methods of moments; interpretation of electronic wavefunctions. *Mailing Add:* Dept Chem Fla State Univ Tallahassee FL 32306-3006

CIPAU, GABRIEL R, b June 18, 41. CHEMICAL ENGINEERING, COMPUTER SCIENCE. *Educ:* Timisoara Polytech Inst, BS, 64, PhD(chem eng), 68; ECarolina Univ, MS, 73. *Prof Exp:* Asst prof chem eng, Timisoara Polytech Inst, 64-70; dir tech serv, 70-78, VPRES PROD & ENG, BURROUGHS WELLCOME CO, 78- *Concurrent Pos:* Res grant, Timisoara Polytech Inst, 64-68. *Mem:* Am Inst Chem Engrs; Am Chem Soc; Am Mgt Asn; Int Soc Pharmaceut Engrs. *Res:* Applications of computers in process control, laboratory automation and engineering design. *Mailing Add:* PO Box 12861 Research Triangle Park NC 27709-2861

CIPERA, JOHN DOMINIK, b Czech, Aug 7, 23. ORGANIC BIOCHEMISTRY. *Educ:* Tech Univ, Czech, Ing, 48; Univ Toronto, MSA, 51; McGill Univ, PhD(chem), 54. *Prof Exp:* Asst chem, Col Forestry, State Univ NY, 55-56; res assoc, Univ Pittsburgh, 56-58; RES SCIENTIST, ANIMAL RES INST, CAN DEPT AGR, 58- *Honors & Awards:* Eddy Found Award, 52. *Mem:* Am Chem Soc; Biochem Soc; Chem Inst Can. *Res:* Organic chemistry of naturally occurring polymers; peptides; glycosaminoglycans; chemistry and physiology of connective tissues; role of organic matrix in calcification processes. *Mailing Add:* 830 Maplecrest Ave Ottawa ON K2A 2Z8 Can

CIPOLLA, JOHN WILLIAM, JR, b Clifton Heights, Pa, Aug 7, 42; m 67. KINETIC THEORY OF GASES, RADIATIVE TRANSFER. *Educ:* Drexel Univ, BSME, 65; Brown Univ, ScM, 67, PhD(eng), 69. *Prof Exp:* Res scientist, Inst Physics, Univ Milan, 69-70; res scientist, Max-Planck Inst Fluid Mech, 70-71; from asst prof to assoc prof, 71-81, PROF MECH ENG, NORTHEASTERN UNIV, 81- *Concurrent Pos:* Res scientist, Max-Planck Inst Fluid Mech, 73-75; vis asst prof, Brown Univ, 78-79, prof res, 85-86. *Mem:* Sigma Xi; Am Phys Soc; Am Soc Mech Engrs; Am Soc Eng Educ. *Res:* Theoretical research on non-equilibrium problem areas in the thermal sciences, particularly with respect to aerosol and interfacial phenomena, kinetic theory and radiative transfer. *Mailing Add:* Dept Mech Eng 271 SN Northeastern Univ Boston MA 02115

CIPOLLA, SAM J, b Chicago, Ill, July 24, 40; m 66; c 2. ATOMIC PHYSICS, NUCLEAR PHYSICS. *Educ:* Loyola Univ Chicago, BS, 62; Purdue Univ, MS, 65, PhD(nuclear physics), 69. *Prof Exp:* Asst, Purdue Univ, 62-69, res assoc, 69; from asst prof to assoc prof, 69-83, PROF PHYSICS, CREIGHTON UNIV, 83- *Concurrent Pos:* NSF grants, 70-72 & 82-83; res partic, Oak Ridge Nat Lab, 71-80; Cottrell Col sci grant, Res Corp, USA, 72-75 & 78-80; consult, Omaha Pub Power Dist, 74- *Mem:* Am Phys Soc; Am Asn Physics Teachers. *Res:* Radioactivity measurement; operation of nuclear instrumentation; nuclear spectroscopy measurements; radioactive source preparation; vacuum technology; ion-induced inner-shell ionization measurements in atoms; radiation dosimetry in nuclear power plants. *Mailing Add:* Dept Physics Creighton Univ Omaha NE 68178

CIPRIANI, CIPRIANO, b Italy, Aug 25, 23; US citizen; m 52; c 5. STRATIGIC PLANNING. *Educ:* Univ Padua & Bologna, Italy, PhD(indust chem), 49. *Prof Exp:* Res chemist, Snia Viscosa, Italy, 48-52 & Courtaulds Can Ltd, Ont, 52-57; sr res chemist, Celanese Corp Am, 57-62; supvr res & develop, Fibers Div, Allied Chem Corp, 62-68, dep sci dir, Allied Chem SA, Brussels, 68-72; prin chemist, Fiber Div, FMC Corp, 72-76; prin chemist, El Paso Polyolefins Co, Paramus, NJ 77-78, mgr prod develop, 78-79, dir, 80-83; CONSULT, 83- *Mem:* Am Chem Soc. *Res:* Ziegler-Natta catalysts for polymerization of olefins; condensation polymerization: polyamides and polyesters; polymers modification, stabilization, processing and applications; project management; technology evalution and tranfer; market analysis; commercialization of new products. *Mailing Add:* Nine Sunderland Dr Morristown NJ 07960-3622

CIPRIANO, LEONARD FRANCIS, b New York, NY, Feb 26, 38; m 62; c 1. SPACE BIOLOGY, MICROGRAVITY SCIENCE. *Educ:* City Col New York, BS, 59; Univ Calif, Berkeley, PhD(physiol), 70. *Prof Exp:* Res physiologist, US Army Res Inst Environ Med, Natick, Mass, 70-72; lab dir cardiovasc & pulmonary physiol, Lovelace Found Med Educ & Res, Albuquerque, NMex, 72-76; asst prof physiol dept, Baylor Univ Col Med, Houston, Tex, 76-77; prof, Antelope Valley Col, Lancaster, Calif, 77-79; spacelab payload scientist, Gen Elec-Matsco, NASA Ames Res Ctr, Calif, 79-83, spacelab proj mgr, 83-86, proj mgr advan planning, 86-88, mgr advan planning, Lockheed, 88-90; CHIEF CORP SCIENTIST, CTR SPACE & ADVAN TECHNOL, WASH, DC, 91- *Concurrent Pos:* Instr, Calif State Col, Bakersfield, 73-75; consult space life sci, physiol & biol, 78-; consult space processing, 84- *Mem:* Am Physiol Soc; Am Inst Astronaut & Aeronaut; Sigma Xi. *Res:* Cellular and systemic physiology; space sciences; science and technical management; cell biology; pulmonary and cardiovascular physiology; budgeting and forecasting space policy. *Mailing Add:* 1460 Middlefield Rd Palo Alto CA 94301

CIPRIANO, RAMON JOHN, b Warsaw, NY. AIR-SEA INTERACTION, CLOUD PHYSICS. *Educ:* State Univ, NY at Geneseo, BA, 67, MS, 69, at Albany, MS, 75, PhD(atmospheric sci), 79. *Prof Exp:* RES ASSOC, ATMOSPHERIC SCI RES CTR, 79- *Concurrent Pos:* Prin investr, 79- *Mem:* Sigma Xi. *Res:* Role played by the oceans in the production of marine aerosol--this aerosol can be highly enriched in various pollutants, and is largely formed by bursting bubbles from breaking waves. *Mailing Add:* Nine Oak St Stonington CT 06378

CIPRIOS, GEORGE, b New York, NY, June 30, 31; m 64. CHEMICAL ENGINEERING. *Educ:* Columbia Univ, BS, 53, MS, 60. *Prof Exp:* Res assoc chem eng, 56-62, sr engr, 62-69, res assoc, 69-77, sr res assoc, Exxon Res & Eng Co, 77-; MEM STAFF, CORP RES SCI LAB, EXXON RES & ENG CO. *Mem:* Am Inst Chem Engrs; Am Chem Soc. *Res:* Energy conversion, particularly fuel cells; chemical and refinery process research and development. *Mailing Add:* RD 1 Box 159 Camila Dr Pittstown NJ 08867-9715

CIRCEO, LOUIS JOSEPH, JR, b Everett, Mass, Aug 31, 34; m 61; c 2. CONSTRUCTION ENGINEERING, GEOTECHNICAL ENGINEERING. *Educ:* US Mil Acad, BS, 57; Iowa State Univ, MS, 61, PhD(civil eng), 63. *Prof Exp:* Res assoc, Lawrence Radiation Lab, Univ Calif, Livermore, 62-64; civil engr, US Army Corps Engrs, Thailand, 65-66; instr nuclear weapons effects & employ, US Army Engr Sch, Ft Belvoir, Va, 66-68; engr adv, Dept Civil Eng, Vietnamese Nat Mit Acad, Dalat, S Vietnam, 68-69; res mgr, Defense Atomic Support Agency, Washington, DC, 69-72; opers res analyst, Nuclear concepts Br, Supreme Hq Allied Powers Europe, Mons, Belg, 75-79; dir, US Army Construct Eng Res Lab, Champaign, Ill, 79-83, Nuclear Security, Survivability & Safety Directorate Hq, Defense Nuclear Agency, Washington, DC, 83-87; DIR, CONSTRUCT RES CTR, GA INST TECHNOL, ATLANTA, 87- *Concurrent Pos:* Prin res scientist, Col Archit, Ga Inst Technol, 87- *Mem:* Am Soc Civil Engrs; Soc Am Mil Engrs; Sigma Xi. *Res:* Construction technology and engineering with particular expertise in geotechnical engineering; nuclear weapons effects, employment concepts and security, survivability and safety; engineering and waste management applications of plasma are technology; technology transfer. *Mailing Add:* 4245 Navajo Trail Atlanta GA 30319-1532

CIRIACKS, JOHN A(LFRED), b Milwaukee, Wis, Mar 10, 36; m 60; c 5. CHEMICAL ENGINEERING, PHYSICAL CHEMISTRY. *Educ:* Univ Wis, Madison, BS, 58; Inst Paper Chem, MS, 64, PhD(phys chem), 67. *Prof Exp:* Process engr, Neenah Mill, Wis, 58-59, Niagara Falls Mill, NY, 59-61 & Coosa River Mill, Ala, 63; res chem engr, res & eng div, Kimberly-Clark Corp, Wis, 67-75, sr scientist & proj leader, Consumer Prod Res & Develop, 75-83; mgr, tech support, 83-89, VPRES, TECH DIV, ENGINEERED SYSTS INT, 89- *Mem:* Tech Asn Pulp & Paper Indust; Air & Water Management Asn. *Res:* Chemistry and physics of wood pulp fibers; electrokinetic phenomena, mainly charge transport from fibrous networks; reductive burning to regenerate sodium salts, sodium sulfate. *Mailing Add:* 1029 Pembrook Dr Neenah WI 54956

CIRIACKS, KENNETH W, b West Bend, Wis, May 7, 38; m; c 4. GEOLOGY. *Educ:* Univ Wis, BS, 58; Columbia Univ, PhD(geol), 62. *Prof Exp:* Res scientist, Pan Am Petrol Corp Res Ctr, Stand Oil Co Ind, 62-65, sr res scientist, 65-71, staff res scientist, Res Ctr, 71-72; proj geologist, Amoco Prod Co, Houston, 72-73, dist geologist, 73-75, div geologist, Amoco Prod Co, Denver, 76-77, chief geologist, Amoco Int Oil Co, Chicago, 77-79, mgr, Dept Explor, Gupco, Cairo, 79-81, mgr regional explor, Amoco Prod Co Int, Houston, 81-89, vpres explor technol, 89-90, VPRES RES, AMOCO PROD CO, TULSA, 90- *Concurrent Pos:* NSF fel, Columbia Univ, 62. *Mem:* Geol Soc Am; Am Paleont Soc; Asn Econ Paleont & Mineral; Brit Paleont Asn; Am Asn Petrol Geologists. *Res:* Late Paleozoic biostratigraphy; taxonomy, evolution and ecology of fossil and living pelecypods; geological aspects of physical and biological processes in modern marine carbonate environments. *Mailing Add:* Amoco Prod Co PO Box 3385 Tulsa OK 74102

CIRIACY, EDWARD W, b Philadelphia, Pa, Feb 12, 24; c 4. FAMILY MEDICINE. *Educ:* Pa State Col, BS, 48; Temple Univ, MD, 52. *Prof Exp:* Intern, Frankford Hosp, Philadelphia, 52-53, resident surg, Frankford & Temple Univ Hosps, 53-54; pvt pract, 54-71; PROF FAMILY PRACT & HEAD DEPT, UNIV MINN MINNEAPOLIS, 71- *Concurrent Pos:* Mem bd & mem res & develop comt, Am Bd Family Pract, 72-75, mem recert exam panel, 74-76; mem adv bd, Mod Med Publ, 74- *Mem:* Am Acad Family Physicians; AMA; Asn Am Med Cols; Pan-Am Med Asn; Soc Teachers Family Med. *Mailing Add:* Dept Fam Pract & Community Health Univ Minn Mayo Mem Bldg Box 381 Minneapolis MN 55455

CIRICILLO, SAMUEL F, b Newark, NJ, Nov 14, 20; m 48; c 7. AUTOMATIC CONTROL SYSTEMS, TERMINOLOGY. *Educ:* Newark Col Eng, BSME, 42. *Prof Exp:* Design engr vibration controls, Gen Elec Co, Schnectady, NY, 42-44, design sect engr heat-air conditioning, Bloomfield, NJ, 45-52; proj engr, pilot plant controls, Kellex Corp, Jersey City, 44-45; chief engr combustion air conditioning, Quiet-Heet Mfg, Newark, 52-58; supvr res & develop, Res-Controls, Ranco Res Ctr, Ranco Inc, Pompano, Fla, 58-61, dir res, Res & Develop Controls, Columbus, Ohio, 61-67, vpres eng res, Res & Develop Admin, 67-76, dir eng admin, Res & Develop Admin, Ranco Controls Div, 76-88; CONSULT, 88- *Concurrent Pos:* Mem, Measurement & Controls Mgt Bd, Am Nat Standards Inst, 67-85; chmn terminology, Am Soc Heating, Refrig & Air Conditioning Engrs, 82-83. *Mem:* Fel Am Soc Heating Refrig & Air Conditioning Engrs; Am Soc Mech Engrs. *Res:* Development of combustion burners and extended fin heat transfer surfaces; use of fluidic sensors and controls for modulating flow in refrigerant circuits; flow controls for air-cooled heat pumps and control systems. *Mailing Add:* 3277 Somerford Rd Columbus OH 43221

CIRIELLO, JOHN, b Sannicandro, Italy, Oct 18, 50; Can citizen; m 77; c 3. NEUROPHYSIOLOGY, NEUROANATOMY. *Educ:* Univ Western Ont, BA, 74, MSc, 77, PhD(physiol), 79. *Prof Exp:* Fel physiol, Can Heart Found, Dept Physiol, Univ Western Ont, 79-80 & McGill Univ, 80-81; from asst prof to assoc prof, 81-89, PROF PHYSIOL, UNIV WESTERN ONT, 89- *Concurrent Pos:* Instr, Dept Zool, Univ Western Ont, 73-75, univ, 74-80, lectr, Dept Physiol, 78-80, sr res scholar, 81-87; instr, Dept Physiol, McGill Univ, 80-81; career investr, Univ Western Ont, 87- *Honors & Awards:* Astra Award, Can Hypertension Soc, 86. *Mem:* Am Physiol Soc; Can Hypertension Soc; Can Physiol Soc; Soc Neurosci; Int Brain Res Orgn; Am Asn Anatomists; Can Asn Anatomists. *Res:* Peripheral and central neural mechanisms in the pathogenesis of hypertension; central regulation of the circulation; autonomic nervous system responses to accute cerebral ischemia. *Mailing Add:* Dept Physiol & Health Sci Ctr Univ Western Ont London ON N6A 5C1 Can

CIRILLO, VINCENT PAUL, b New York, NY, Oct 16, 25; m 49; c 4. BIOCHEMISTRY. *Educ:* Univ Buffalo, BA, 47; NY Univ, MS, 52; Univ Calif, Los Angeles, PhD(biol), 53. *Prof Exp:* Asst biol, Univ Buffalo, 46-47 & NY Univ, 48-50; asst zool, Univ Calif, Los Angeles, 50-53; asst prof prev med & pub health, Sch Med, Univ Okla, 53-56; sr res microbiologist, Anheuser-Busch, Inc, Mo, 56-59; asst prof microbiol, Col Med & Dent, Seton

Hall Univ, 59-62, assoc prof biochem, 62-64; assoc prof, 64-69, PROF BIOCHEM, STATE UNIV NY STONY BROOK, 69- *Concurrent Pos:* Fulbright award, Israel, 78. *Mem:* Am Soc Biol Chemists; Am Soc Microbiol. *Res:* Mechanisms of sugar transport. *Mailing Add:* Dept Biochem State Univ NY Stony Brook NY 11794

CISIN, IRA HUBERT, statistics; deceased, see previous edition for last biography

CISLER, WALKER L(EE), b Marietta, Ohio, Oct 8, 97; m 39; c 2. MECHANICAL ENGINEERING. *Educ:* Cornell Univ, ME, 22. *Hon Degrees:* LLD, Univ Detroit, 55, Wayne State Univ, 57, Marietta Col, 58 & Univ Akron, 61; DE, Univ Mich, 56; DSc, Univ Toledo, 59, Ind Tech Col, 59 & Mich Technol Univ, 64; EE, Stevens Inst Technol, 59. *Prof Exp:* Asst gen mgr, Pub Serv Elec & Gas Co, 43; chief engr power plants, Detroit Edison Co, 43-48, exec vpres, 48-51, pres & dir, 51-64, chmn bd, 64-77; pres, 77-81, CHMN, OVERSEAS ADV ASSOCS INC, 81- *Honors & Awards:* Washington Award; George Westinghouse Gold Medal. *Mem:* Nat Acad Eng; fel Inst Elec & Electronics Engrs; fel Am Soc Mech Engrs; Soc Am Mil Engrs. *Mailing Add:* Book Bldg Suite 3000 1249 Washington Blvd Detroit MI 48226

CISNE, JOHN LUTHER, b Summit, NJ, Apr 27, 47; m 78; c 2. SYNTHETIC STRATIGRAPHY, SEA LEVEL STUDIES. *Educ:* Yale Univ, BS, 69; Univ Chicago, PhD(geol), 73. *Prof Exp:* Asst prof geol, 73-79, assoc prof geol & biol, 79-86, RES AFFIL, INST STUDY CONTINENTS, CORNELL UNIV, 84-, PROF GEOL & BIOL, DEPT GEOL SCI, DIV BIOL SCI, 86- *Concurrent Pos:* Trustee, Paleontol Res Inst, 76-, actg treas, 80, asst treas, 81- *Mem:* Geol Soc Am; Int Paleontol Asn; fel AAAS; Am Geophys Union; Soc Econ Paleontologists & Mineralogists. *Res:* paleoceanography; stratigraphic measurement of sea level change; combined stratigraphic and morphometric study of evolution. *Mailing Add:* Dept Geol Sci Cornell Univ Ithaca NY 14853

CISZEK, TED F, b Midland, Mich, Jan 26, 42; m 64; c 3. CRYSTAL GROWTH, MATERIALS CHARACTERIZATION. *Educ:* Case Inst Technol, BS, 64; Iowa State Univ, MS, 66. *Prof Exp:* Physicist, Dow Corning Corp, 66-71; consult crystal growth, 71-72; assoc physics, IBM Corp, 72-78; PRIN SCIENTIST, SOLID STATE RES BR, SOLAR ENERGY RES INST, GOLDEN, 78- *Mem:* Am Asn Crystal Growth; Am Phys Soc; Electrochem Soc. *Res:* New crystal growth techniques for photovoltaic materials including silicon, copper indium diselenide and indium phosphide and electrical and structural characterization of these materials. *Mailing Add:* Solid State Res Solar Energy Res Inst-Seri 1617 Cole Blvd Golden CO 80401

CITRON, DAVID S, b Altanta, Ga, Jan 8, 20. MEDICINE. *Educ:* Univ NC, BA, 41; Wash Univ, St Louis, Mo, MD, 44. *Hon Degrees:* DHL, Univ NC. *Prof Exp:* Dir family pract residency, 73-84, dir med educ, Charlotte Mem Hosp, 84-87; dir, Health Educ Ctr, Charlotte, 84-87; CONSULT, 87- *Concurrent Pos:* Vchmn, Nat Bd Med Examnrs, 89-, treas, 91-; exec comt, Fedn State Med Bds US. *Mem:* Inst Med-Nat Acad Sci; Am Col Physicians. *Mailing Add:* 8117 Rising Meadow Rd Charlotte NC 28277

CITRON, IRVIN MEYER, b Atlanta, Ga, May 5, 24; m 65; c 1. ANALYTICAL CHEMISTRY, SCIENCE EDUCATION. *Educ:* Hebrew Univ, Jerusalem, 58; Emory Univ, MS, 61; NY Univ, PhD(sci educ), 69. *Prof Exp:* Chem lab asst, Israel Defense Dept Labs, Weizmann Inst Sci, 54-56; res asst inorg chem, Hebrew Univ, Jerusalem, 56-58; res asst analytical chem, Emory Univ, 58-61; asst prof analytical & inorg chem, Troy State Col, 61-62; asst prof, 62-69, asst chmn dept chem, 67-69, assoc prof, 69-74, chmn dept, 75-78, PROF ANALYTICAL & INORG CHEM, FAIRLEIGH DICKINSON UNIV, 74- *Concurrent Pos:* Deleg, Colloquium Spectroscopicum Int, Ottawa, 67 & Spectros Symp Can, 69 & Pittsburgh Conf, Analytical Chem & Appld Spectros, Cleveland, 79; fac res leave appointment, Argonne Nat Lab, Ill, 81-82. *Mem:* Am Chem Soc; Soc Appl Spectros; Am Asn Univ Profs; Sigma Xi. *Res:* Use of organometallic complexes for analytical purposes; methods of teaching science at high school and college levels; rare earth metals analysis by x-ray fluorescence and atomic absorption spectroscopy; study of uv-visible spectra of rare earth metal complexes. *Mailing Add:* 648 Howard St Teaneck NJ 07666

CITRON, JOEL DAVID, b Brooklyn, NY, Apr 19, 41; m 78; c 1. ORGANIC CHEMISTRY, POLYMER CHEMISTRY. *Educ:* Polytech Inst Brooklyn, BS, 62; Univ Calif, Davis, PhD(org chem), 67. *Prof Exp:* Teaching assoc chem, Univ Calif, Davis, 67-68; res chemist, Elastomer Chem Dept, E I du Pont de Nemours & Co, 69-80, sr res chemist, 80-84, res assoc, Polymer Prod Dept, 84-89; PATIENT ASSOC, RES & DEVELOP, E I DUPONT DE NEMOURS & CO, INC, 89- *Mem:* Am Chem Soc. *Mailing Add:* DuPont Res & Develop Exp Sta 301-315 Wilmington DE 19880-0301

CITRON, STEPHEN J, b New York, NY, July 3, 33; m 57; c 3. APPLIED MATHEMATICS, SPACE MECHANICS. *Educ:* Rensselaer Polytech Inst, BS, 54, MS, 55; Columbia Univ, PhD(mech), 59. *Prof Exp:* Asst prof, Div Eng Sci, 59-62, from assoc prof to prof aeronaut eng sci, 62-76, exec asst to head sch aeronaut, astronaut & eng sci, 62-65, asst to vpres acad affairs, 68-69, asst vpres acad affairs, 69-76, PROF MECH ENG, PURDUE UNIV, 76- *Concurrent Pos:* Ford Found fel, Harvard Univ, 60-61; sr staff consult, Hughes Aircraft Corp, 62; consult, Off Manned Space Flight, NASA, 62-63, Res & Develop Div, Avco Corp, 62-65 & GM, 82- *Mem:* Am Soc Mech Engrs; Am Soc Eng Educ; fel Am Inst Aeronaut & Astronaut; Soc Automotive Engrs. *Res:* Automotive engineering, mathematical techniques in engineering problems; control and application of optimization techniques; orbit mechanics and guidance; industrial management; operations research. *Mailing Add:* Sch Mech Eng Purdue Univ West Lafayette IN 47907

CIUFOLINI, MARCO A, b Rome, Italy, June 14, 56; m 84. SYNTHETIC ORGANIC & NATURAL PRODUCTS CHEMISTRY, ORGANIC CHEMISTRY. *Educ:* Spring Hill Col, BS, 78; Univ Mich, PhD(chem), 81. *Prof Exp:* Postdoctoral chem, Univ Mich, 82 & Yale Univ, 82-84; asst prof, 84-90, ASSOC PROF CHEM, RICE UNIV, 90- *Mem:* Am Chem Soc. *Res:* Organic, bio-organic and organometallic chemistry of potential use in synthesis; total synthesis of natural products and of theoretically interesting molecules. *Mailing Add:* Dept Chem Rice Univ PO Box 1892 Houston TX 77251

CIULA, RICHARD PAUL, b Lorain, Ohio, Dec 8, 33; m 59; c 3. ORGANIC CHEMISTRY. *Educ:* Bowling Green State Univ, BA, 55; Univ Calif, MS, 57; Univ Wash, PhD, 60. *Prof Exp:* From asst prof to assoc prof, 60-68, chmn dept, 66-71, PROF CHEM, CALIF STATE UNIV, FRESNO, 68- *Mem:* Am Chem Soc; Royal Chem Soc. *Res:* Synthesis and properties of small ring compounds, particularly in the cyclobutane series; kinetics and mechanism of the nitrile exchange reaction; synthesis of bicyclic amines. *Mailing Add:* Dept Chem Fresno State Col Fresno CA 93726

CIVAN, MORTIMER M, b New York, NY, Nov 13, 34; m 61; c 3. PHYSIOLOGY. *Educ:* Columbia Univ, AB, 55, MD, 59; Am Bd Internal Med, dipl. *Prof Exp:* Intern, Presby Hosp, 59-60, asst resident, 60-62; staff assoc biophys, NIH, 62-64; instr med, Harvard Med Sch, 65-68, assoc, 68-69, asst prof, 69-72; assoc prof physiol, 72-77, PROF PHYSIOL, SCH MED, UNIV PA, 77-, ASSOC PROF MED, 72- *Concurrent Pos:* USPHS clin & res fel, Mass Gen Hosp & Harvard Univ, 64-65 & USPHS spec fel, Weizmann Inst Sci, 70-71; Am Heart Asn grant-in-aid, 71-73; NSF grant, 73-; NIH grant, 74-; asst, Mass Gen Hosp, 65-72; estab investr, Am Heart Asn, 71-76; fac scholar, Macy Found, 78-79; overseas fel, Churchill Col, Cambridge Univ, 78-79; mem, Steering Comt, Cell & Gen Physiol Sect, Am Physiol Soc, 81-83, chmn, 82-83. *Honors & Awards:* Harold Chaffer Mem Lectr, Fac Med, Univ Otago, Dunedin, NZ, 90. *Mem:* Am Soc Clin Invest; Am Physiol Soc; Biophys Soc; Am Soc Nephrology; Soc Gen Physiol (secy, 81-84); fel AAAS. *Res:* Kinetics of muscle contraction; transport solutes and water across membranes. *Mailing Add:* Dept Physiol Richards Bldg Univ Pa Sch Med Philadelphia PA 19104-6085

CIVELLI, OLIVER, b Fribourg, Switzerland, May 8, 49. MOLECULAR BIOLOGY. *Educ:* Swiss Inst Technol, Zurich, Dipl, 73, Dr Natural Sci, 79. *Prof Exp:* Res asst, Swiss Inst Cancer Res, Luasanne, 73-74, Inst Res Molecular Biol, Univ Paris, 74-79; res assoc, Dept Chem, Univ Ore, 79-82. *Res:* Genetic regulation in eucaryotic cells; post transcriptional regulation; gene expression of hormones; neurotransmitters. *Mailing Add:* Ore Health Sci Univ 3181 SW 52M Jackson Park Rd Portland OR 97201

CIVEN, MORTON, b Boston, Mass, June 20, 29; m 54; c 2. BIOCHEMISTRY. *Educ:* Harvard Univ, MSc, 53, PhD(biochem), 57. *Prof Exp:* Fel, Harvard Med Sch, 58-59; USPHS fel, Nat Inst Med Res, London, 59-61; RES BIOCHEMIST, US VET ADMIN HOSP, LONG BEACH, 62- *Concurrent Pos:* Asst prof biochem, Univ Southern Calif, 62-68, adj asst prof, 68-; adj assoc prof physiol, Univ Calif, Irvine, 72- *Mem:* AAAS; Brit Biochem Soc; Am Soc Biol Chemists. *Res:* Mechanism of enzyme induction, especially effects of peptide hormones on target cell membranes; mechanisms of action of gonadotropins; enzymatic regulation of amino acid metabolism; biochemistry of adrenal cells; effect of toxic chemicals on adrenocortical secretion. *Mailing Add:* Med Res Prog Vet Admin Med Ctr 151 Long Beach CA 90822

CIVIAK, ROBERT L, b Brooklyn, NY, Nov 18, 47. NUCLEAR ENERGY POLICY, ARMS CONTROL. *Educ:* Rensselaer Polytech Inst, BS, 68; Univ Pittsburgh, MS, 70, PhD(physics), 74; Univ Chicago, MLS, 77. *Prof Exp:* Fel res assoc, Mat Res Prog, Brown Univ, 74-76; ref librarian, Nat Oceanic & Atmospheric Admin, 77-78; SPECIALIST ENERGY TECHNOL, SCI POL RES DIV, LIBR CONG, 78-, HEAD ADVAN TECH SECT, CONG RES SERV, 85- *Concurrent Pos:* Arms Control Intern, Lawrence Livermore Nat Lab, 88. *Mem:* AAAS; Am Nuclear Soc; Arms Control Asn; Fedn Am Sci. *Res:* Nuclear energy development and national nuclear energy policy; arms control, strategic defense initiative and nuclear weapons testing. *Mailing Add:* 5526 39th St Washington DC 20015

CIVIN, PAUL, b Rochester, NY, April 29, 19; m 39; c 2. MATHEMATICS. *Educ:* Univ Buffalo, BA, 39; Duke Univ, MA, 41, PhD(math), 42. *Prof Exp:* Instr math, Univ Mich, 42-43 & Univ Buffalo, 43-46; from asst prof to assoc prof, 46-57, PROF MATH, UNIV ORE, 57-, ASSOC PROVOST PLANNING, 77- *Concurrent Pos:* Mem, Inst Advan Study, 53-54; vis res prof, Univ Fla, 60-61; vis prof, Copenhagen Univ, 61-62 & 68-69; consult, Pres, Univ Ore, 73- *Mem:* Am Math Soc. *Res:* Fourier series; topology; two-to-one mappings of manifolds; Banach algebra. *Mailing Add:* 2765 Spring Blvd Eugene OR 97403

CIZEK, LOUIS JOSEPH, b New York, NY, Apr 11, 16; m 41; c 2. PHYSIOLOGY. *Educ:* Fordham Univ, BS, 37; Columbia Univ, MD, 41. *Prof Exp:* Intern med serv, Beekman Hosp, NY, 41-42; from instr to asst prof physiol, Col Physicians & Surgeons, Columbia Univ, 46-56, assoc physiol, 56-; RETIRED. *Concurrent Pos:* Managing ed, Proc, Soc Exp Biol & Med. *Mem:* Fel AAAS; Am Physiol Soc; Harvey Soc; Soc Exp Biol & Med (secy-treas); Am Soc Zool. *Res:* Water and electrolyte balance. *Mailing Add:* 180 Griggs Ave Teaneck NJ 07666

CLAASSEN, E(DWIN) J(ACK), JR, b St Joseph, Mo, June 20, 20; m 42; c 3. CHEMICAL ENGINEERING. *Educ:* Mo Sch Mines, BS, 42; Univ Tex, MS, 44, PhD(chem eng), 48. *Prof Exp:* Instr chem eng, Univ Tex, 43-44; res scientist, Bur Indust Chem, 44-50; mgr res & develop, Sid Richardson Carbon Co, 50-69; plant supt, 70-74, ENG DIR, CHAMPION CHEM, INC, 75- *Mem:* Am Inst Chem Engrs; Am Chem Soc; Nat Asn Corrosion Engrs; Soc Petrol Engrs. *Res:* Electric discharge through gases; production and uses of carbon black; production and uses of oil field chemicals. *Mailing Add:* Champion Chem Inc PO Box 4513 Odessa TX 79760

CLAASSEN, RICHARD STRONG, b Ithaca, NY, May 10, 22; m 45; c 3. MATERIALS SCIENCE ENGINEERING. *Educ:* Cornell Univ, AB, 43; Columbia Univ, MA, 47; Univ Minn, PhD(physics), 50. *Prof Exp:* Asst, Substitute Alloy Material Labs, 44-46 & Univ Minn, 47-50; physicist, Sandia Nat Labs, 51-53, suprv, 53-57, mgr phys sci res dept, 57-60, dir phys res, 60-68, dir electronic component develop, 68-75, dir mat & process sci, 75-82, vpres, Livermore Opers, 82-87; RETIRED. *Concurrent Pos:* Chmn Nat Sci Seminar, 63; mem, Rocky Mountain Sci Coun, 61-76, chmn, 65-66; mem solid state sci comt, Nat Acad Sci-Nat Res Coun, 65-78, chmn, 74, mem nat material adv bd, 73-76; panel chmn surv mat sci, Nat Acad Sci, 71; mem, Mat Res Adv Comt, NSF, 79, chmn, 81; chmn, Tech Eval Panel, Dept Energy, 80. *Mem:* Fel Am Phys Soc. *Res:* Physics of solids; research and development administration. *Mailing Add:* 708 Liquid Amber Pl Danville CA 94506-4528

CLABAUGH, STEPHEN EDMUND, b Carthage, Tex, Apr 2, 18; m 45; c 3. GEOLOGY. *Educ:* Univ Tex, BS, 40, MA, 41; Harvard Univ, PhD(geol), 50. *Prof Exp:* Asst geol, Univ Tex, 40-41; geologist, US Geol Surv, 42-54; from asst prof to assoc prof, 47-55, chmn dept, 62-66, prof geol, 55-77, Fred M Bullard prof geol sci, 77-80, EMER PROF, UNIV TEX, AUSTIN, 80- *Concurrent Pos:* Nat Res Coun fel, Harvard Univ, 46-47; Piper prof geol, 58. *Mem:* Fel Geol Soc Am; Am Geophys Union. *Res:* Geology of Montana corundum deposits; tungsten deposits of Osgood Range, Nevada; igneous and metamorphic rocks of Cornudas Peaks; Texas and New Mexico, and Christmas Mountains, Texas; vermiculite deposits and metamorphic rocks, central Texas; volcanic rocks of western Texas and Mexico. *Mailing Add:* Rte One Box 36 Spicewood TX 78669

CLADIS, JOHN BAROS, b Dawson, NMex, June 21, 22; m 47; c 4. SPACE PLASMA PHYSICS, HIGH-ALTITUDE NUCLEAR-BURST PHENOMENOLOGY. *Educ:* Univ Colo, BS, 44; Univ Calif, Berkeley, PhD(nuclear physics), 52. *Prof Exp:* Asst, Univ Colo, 46-47; physicist, Lawrence Radiation Lab, Univ Calif, 48-52, mem res staff, Los Alamos Sci Lab, 52-55; sr staff scientist, Lockheed Missiles & Space Co, 55-65; CONSULT SCIENTIST & SR MEM RES LAB, THEORET SPACE PHYSICS, LOCKHEED PALO ALTO RES LAB, 65- *Mem:* Am Phys Soc; Am Geophys Union; Int Union Geod & Geophys. *Res:* High energy nuclear scattering experiments; nuclear weapons diagnostic measurements; physics of Van Allen and artificial radiation belts; plasma-magnetic field interactions; magnetospheric physics; high-altitude nuclear-burst phenomenology. *Mailing Add:* Dept 91-20 Lockheed Palo Alto Res Lab Bldg 255 3251 Hanover St Palo Alto CA 94304

CLADIS, PATRICIA ELIZABETH RUTH, b Shanghai, China, July 13, 37; US citizen; m 62; c 2. CONDENSED MATTER PHYSICS. *Educ:* Univ BC, BA, 59; Univ Toronto, MA, 60; Univ Rochester, PhD(physics), 68. *Prof Exp:* Meteorologist, Govt Can, 59-62; programmer-analyst, Katz, Casciato & Shapiro, Ltd, Can, 62; instr physics, Western Conn State Univ, 63-64; res asst, Univ Rochester, 64-68, consult, 68; consult, Univ Toronto, 68-69; res fac sci, Lab Physics Solids, Univ Paris-South, Orsay, 69-72,; MEM TECH STAFF, BELL LABS, 72- *Concurrent Pos:* Kreeger Wolf distinguished prof, Northwestern Univ, Evanston, 76; adj prof, dept physics, Univ Del, Newark, 77-78, Kent State Univ, 89-91; vis prof, Life Sci Sect, Ecole Pratigue des Hauts Etudes, Paris, 83, Solid Physics Group, Ecole Normale Superveure, Paris, 85, Solid Physics, Univ Paris-South, Orsay, France, 86, & Dept Nuclear Physics, Weizmann Inst, Rehovot, Israel, 86-; mem, Comt Status of Women in Physics, Am Phys Soc, 80-82, Nominating Comt, 82 & Maria Goeppert-Mayer Award Comt; co-prin investr, NSF Indust/Univ Coop Res Act, 84-88; mem, Comt Hist Physics, Am Inst Physics, 84-; mem, adv bd, Advan Liquid Crystalline Optical Mat Consort; collab res grant, NATO, 90-92. *Mem:* Fel Am Phys Soc; Am Inst Phys Soc; AAAS; NY Acad Sci. *Res:* Non-linear problems of condensed phases; static and dynamic properties; defect structure. *Mailing Add:* AT&T Bell Labs 600 Mountain Ave Murray Hill NJ 07970

CLAERBOUT, JON F, EARTH SCIENCES. *Educ:* Mass Inst Technol, BS, 60, MS, 63, PhD(geophysics), 67. *Prof Exp:* PROF GEOPHYS, STANFORD UNIV, 67- *Concurrent Pos:* Consult, Chevron Oil Field Res Co, 67-73; vis res geologist, Princeton, 72; vis lectr, Sydney Univ, 73; Founder, Stanford Exploration Proj, 73; adv comt, earth sci, MIT Corp, 79. *Honors & Awards:* Medal Award, Soc Explor Geophysicists. *Mem:* Nat Acad Eng; Soc Explor Geophysicists. *Res:* Author of fundamentals of geophysical data processing and imaging the earth's interior. *Mailing Add:* Stanford Explor Proj Stanford Univ Mitchell Bldg Stanford CA 94305-2215

CLAESSENS, PIERRE, b Brussels, Belg, Sept 5, 39; m 68; c 2. ELECTROCHEMISTRY. *Educ:* Univ Louvain, Lic, 63, Dr(electrochem), 67. *Prof Exp:* Asst prof chem, Univ Montreal, 67-68; res chemist, Noranda Res Ctr, 68-70, group leader, 70-73, head dept, 73-81, chief div scientist, 81-89, PRIN SCIENTIST, NORANDA RES CTR, 89- *Honors & Awards:* Sherrit Hydrometall Award, 90. *Mem:* Am Inst Mining & Petrol Engrs; Can Inst Mining & Metall; Electrochem Soc; Nat Asn Corrosion Engrs. *Res:* Electrodeposition of metals; cathodic process; study of the physical properties of solutions. *Mailing Add:* Noranda Tech Ctr 240 Hymus Blvd Pointe Claire PQ H9R 1G5 Can

CLAFF, CHESTER ELIOT, JR, b Brockton, Mass, Apr 17, 28; m 52; c 2. TECHNICAL TRANSLATION. *Educ:* Mass Inst Technol, BS, 50, PhD(org chem), 53. *Prof Exp:* Chemist polymerization, B B Chem Co, Inc, 55-60; gen mgr, Mark Co, 60-64; vpres, M B Claff & Sons, Inc, 64-71; RETIRED. *Concurrent Pos:* Res assoc, Mass Inst Technol; partic, Rubber Reserve Prog, Reconstruction Corp, 53-55; translr engr, 74-78, tech translr, 78- *Mem:* Am Chem Soc; Am Translrs Asn. *Res:* Preparation and reaction of organosodium compounds; leather technology; acrylic polymerization technology. *Mailing Add:* PO Box 2038 Brockton MA 02405

CLAFLIN, ALICE J, b River Falls, Wis, Feb 12, 32. IMMUNOBIOLOGY. *Educ:* Northern State Col, BS, 53; Univ Wis, PhD(med genetics), 70. *Prof Exp:* Instr med, 70-73, res assoc prof surg, 73-85, RES PROF SURG, SCH MED, UNIV MIAMI, 85- *Mem:* Reticuloendothelial Soc; Am Asn Tissue Banks; Tissue Cult Asn. *Res:* Immune mechanisms of tumor-bearing animals; cellular and humoral immune response with immunosuppressive therapy and transplantation immunology associated with tissue culture and antibodies. *Mailing Add:* Dept Surg Univ Miami Sch Med PO Box 520875 Miami FL 33152

CLAFLIN, ROBERT MALDEN, b Flint, Mich, Nov 11, 21; m 57; c 3. VETERINARY PATHOLOGY. *Educ:* Mich State Univ, DVM, 52; Purdue Univ, MS, 56, PhD(vet path), 58. *Prof Exp:* Instr res animal dis, 52-58, assoc prof vet path, 58-59, prof & head dept vet microbiol, 59-85, assoc dean acad affairs, 85-88, EMER ASSOC DEAN, PATH & PUB HEALTH, SCH VET MED, PURDUE UNIV, 88- *Concurrent Pos:* Mem, Conf Res Workers Animal Dis. *Mem:* Am Vet Med Asn; Int Acad Path. *Res:* Etiology, pathology and epizoology of respiratory diseases of swine, particularly atrophic rhinitis and mucosal diseases of cattle. *Mailing Add:* 601 N Shore Dr Luddington MI 49431

CLAFLIN, TOM O, b Ripon, Wis, Apr 1, 39; m 61; c 1. BIOLOGY. *Educ:* Northern State Col, BS, 61; Univ SDak, MA, 63, PhD(zool), 66. *Prof Exp:* From asst prof to assoc prof, 66-74, PROF BIOL, UNIV WIS-LA CROSSE, 69- *Mem:* AAAS; Am Fisheries Soc. *Res:* Ecology of the benthos of river and lake systems. *Mailing Add:* Dept Biol Univ Wis 1725 State St La Crosse WI 54601

CLAGETT, CARL OWEN, BIOCHEMISTRY. *Educ:* Univ Wis, PhD(biochem), 47. *Prof Exp:* Prof biochem, Pa State Univ, 56-78; RETIRED. *Mailing Add:* 908 Hart Circle State College PA 16801

CLAGETT, DONALD CARL, b Madison, Wis, Dec 31, 39; m 68; c 6. POLYMER CHEMISTRY. *Educ:* Pa State Univ, BS, 61; Yale Univ, MS, 63, PhD(chem), 66. *Prof Exp:* Asst res scientist, NY Univ, 67-68; asst prof chem, Northeastern Univ, 68-73; group leader, Dewey & Almy Div, W R Grace & Co, 73-75, sr group leader, 75-78; prod develop specialist, Plastics Group, 78-81, supvr, 81-83, mgr process develop, 83-86, MGR PROCESS SAFETY, GEN ELEC CO, 86- *Mem:* AAAS; Am Chem Soc; Chem Soc; NY Acad Sci. *Res:* Chemistry of small ring organic compounds; chemistry of nucleic acids; chemical mutagens; chemistry of arthropod venoms; rubber latex formulations; powdered coatings; urethane foam systems; thermoset molding compounds; melt-polymerization of thermoplastics, process development. *Mailing Add:* G E Plastics One Plastics Ave Pittsfield MA 01201

CLAGETT, ROBERT P, BUSINESS ADMINISTRATION. *Educ:* Univ Md, BS, 52; Mass Inst Technol, MS, 67. *Prof Exp:* Gen mgr corp eng, AT&T Technol, 71-73, mat & acct mgt, 73-76, prod line planning & mgt, 76-79, Eng Res Ctr, 79-85; DEAN, COL BUS ADMIN, UNIV RI, 85-, EMER PROF. *Concurrent Pos:* Pres, Nat Eng Consortium, 78; chmn, Nat Commun Forum, Nat Electronics Conf, 78; mem, Bd Overseers, Panel Bd Assessment Nat Bur Standards Prog, Nat Res Coun, 86-, chmn, Comt Rev Defense Logistics Agency's Modernization Prog, 87- & Indust Mobilization Comt, Mfg Studies Bd, 88. *Mem:* Nat Acad Eng; Nat Acad Mgt; Soc Mfg Engrs; emer mem Indust Res Inst. *Res:* Manufacturing; author of various publications; granted 3 patents. *Mailing Add:* Univ RI 1301 Ballentine Hall Kingston RI 02881-0802

CLAGUE, DAVID A, b Philadelphia, Pa, Aug 3, 48; m; c 1. IGNEOUS PETROLOGY. *Educ:* Scripps Inst Oceanography, PhD(earth sci), 74. *Prof Exp:* Nat res coun, US Geol Surv, 75-76; asst prof geol, Middleburry Col, 76-79; RES GEOLOGIST, US GEOL SURV, 79- *Mem:* Fel Geol Soc Am; Am Geophys Union. *Mailing Add:* US Geol Surv Mail Stop 999 345 Middlefield Rd Menlo Park CA 94025

CLAGUE, WILLIAM DONALD, b Mobile, Ala, Nov 29, 20; m 44; c 2. SCIENCE EDUCATION, RESEARCH ADMINISTRATION. *Educ:* Bridgewater Col, AB, 41; Univ Va, MEd, 52, EdD, 60. *Prof Exp:* Teacher high sch, Ala, 41-43; from asst prof to assoc prof chem, Bridgewater Col, 43-60, dean students, 52-66, prof natural sci, 60-66; vpres acad affairs, 75-83, prof educ & dean grad & prof studies, 66-75, exec vpres, 83-87, PROF EDUC MGT, UNIV LA VERNE, 87- *Concurrent Pos:* Consult educ, 72- *Mem:* Nat Soc Study Educ. *Res:* Choline; methods of laboratory instruction in college chemistry; sources of teaching personnel for church related colleges; development leading to accreditation of a new institution of higher education in a frontier; educational management. *Mailing Add:* Univ La Verne 1950 Third St La Verne CA 91750

CLAIBORNE, C CLAIR, b Fredonia, Kans, May 30, 52. POLYMER CHEMISTRY, POLYMER ENGINEERING. *Educ:* Univ Kans, BA, 73; Northwestern Univ, PhD(mat sci & eng), 84. *Prof Exp:* Chemotechnician, Sued Chemie AG, Moosburg, WGer, 73-75; lab mgr, Janes Mfg Inc, Kans, 75-76; res asst, Northwestern Univ, 76-80; res chemist, Phillips Petrol Co, Okla, 80-82; SR MAT ENGR, RES & DEVELOP, ABB POWER T&D CO INC, SHARON, PA, 84- *Concurrent Pos:* Mem, Mercer Local Planning Comt, 90- *Honors & Awards:* Tech Achievement Award, Elec Ins Conf, 87. *Mem:* Am Chem Soc; sr mem Soc Plastic Eng; Am Soc Testing & Mat; Am Radio Relay League. *Res:* Advanced insulating materials; composites; dielectric materials; measuring techniques; combustion products; structure-property relationships of high polymers; environmental sciences. *Mailing Add:* 925 Hall Ave Sharon PA 16146-2434

CLAIBORNE, H(ARRY) C(LYDE), b New Orleans, La, July 13, 21; m 49; c 4. NUCLEAR & CHEMICAL ENGINEERING. *Educ:* La State Univ, BS, 41; Univ Tenn, MS, 49. *Prof Exp:* From jr engr to asst chem engr, Tenn Valley Authority, 42-46; instr chem eng, Univ Tenn, 46-49; develop engr, Atomic Energy Comn, 49-52, tech engr, 52-53, develop engr, res staff, Oak Ridge Nat Lab, 53-; RETIRED. *Mem:* Am Nuclear Soc. *Res:* Nuclear waste management and the design of geologic waste repositories. *Mailing Add:* 795 W Outer Dr Oak Ridge TN 37830

CLAIBORNE, LEWIS T, JR, b Holly Grove, Ark, Sept 17, 35; m 62. ACOUSTICS. *Educ:* Baylor Univ, BS, 57; Brown Univ, PhD(physics), 61. *Prof Exp:* Res assoc physics, Brown Univ, 61-62; res physicist, 62-69, br mgr, Cent Res Labs, 69-74, lab dir, Advan Technol Lab, 75-79, LAB DIR, SYSTS COMPONENTS LAB, CENT RES LABS, TEX INSTRUMENTS, INC, 79- *Mem:* Am Inst Physics; Inst Elec & Electronics Engrs. *Res:* Ultrasonic attenuation; lattice-electron interactions in both normal and superconducting metals; surface acoustic wave devices; charge-coupled devices; microwave devices; infrared detectors. *Mailing Add:* 920 Northlake Dr Richardson TX 75080

CLAISSE, FERNAND, b Quebec, Que, Apr 2, 23; m 49; c 2. PHYSICS, METALLURGY. *Educ:* Laval Univ, BSc, 47, PhD(physics), 57. *Prof Exp:* Physicist, Que Dept Natural Resources Labs, 47-54, chief physicist, 54-58; prof phsyics metals, Laval Univ, 58-84; RETIRED. *Concurrent Pos:* Pres, Claisse Sci Corp, Inc, 76- *Honors & Awards:* Prov Que Sci Award, 58. *Mem:* Am Chem Soc; French-Can Asn Adv Sci; Spectros Soc Can; Soc Appl Spectros; Can Inst Metall. *Res:* Order-disorder reaction; phase transformations; diffusion of atoms in metals; twinning in explosively shocked metals; x-ray fluorescence. *Mailing Add:* 2780 BD de Monaco Quebec PQ G1P 3H2 Can

CLAITOR, L(ILBURN) CARROLL, chemical engineering; deceased, see previous edition for last biography

CLAMAN, HENRY NEUMANN, b New York, NY, Dec 13, 30; m 56; c 3. INTERNAL MEDICINE, IMMUNOLOGY. *Educ:* Harvard Univ, AB, 52; NY Univ, MD, 55. *Prof Exp:* Intern, Barnes Hosp, St Louis, Mo, 55-56, asst resident, 56-57; from asst resident to resident, Mass Gen Hosp, Boston, 57-61; from instr to assoc prof, 62-73, assoc dean fac affairs, 69-71, PROF MED & MICROBIOL & IMMUNOL, UNIV COLO MED CTR, DENVER, 73- *Concurrent Pos:* Fel allergy, Sch Med, Univ Colo, 61-62; consult, Fitzsimons Gen Hosp, 68-; mem immunobiol study sect, NIH, 68-72, mem allergy immunol res comn, 73-77; distinguished prof med, microbiol & immunol, Univ Colo, 90. *Honors & Awards:* Heidelberger Lectr, Colo Univ, 74. *Mem:* Fel Am Acad Allergy Immunol; Am Asn Immunol; Asn Am Phys. *Res:* mast cells and fibrosis; scleroderma. *Mailing Add:* Dept Clin Immunol Univ Colo Med Ctr Denver CO 80262

CLAMANN, H PETER, b Berlin, Ger, Nov 18, 39; US citizen; m 67; c 2. NEUROPHYSIOLOGY, BIOMEDICAL ENGINEERING. *Educ:* St Mary's Univ, Tex, BS, 61; Johns Hopkins Univ, PhD(biomed eng), 67. *Prof Exp:* Res physiologist, Walter Reed Army Inst Res, 68-70; res fel neurophysiol, Harvard Med Sch, 70-72, instr physiol, 72-73; asst prof, 73-78, ASSOC PROF PHYSIOL, MED COL VA, 78- *Concurrent Pos:* NIH res grant, 75-81; sabbatical guest fac, Dept Physiol, Univ Zurich, Switz, 83-84. *Mem:* Biomed Eng Soc; Soc Neurosci; Am Physiol Soc; Inst Elec & Electronics Engrs. *Res:* Motor systems neurophysiology; statistical properties of neuronal spike trains; biomedical instrumentation; electromyography. *Mailing Add:* Dept Physiol Univ Bern Buhlplatz 5 3012 Bern Switzerland

CLAMBEY, GARY KENNETH, b Fergus Falls, Minn, Feb 27, 45; m 69; c 2. PLANT ECOLOGY. *Educ:* NDak State Univ, BS, 67, MS, 69; Iowa State Univ, PhD(bot), 75. *Prof Exp:* Instr nat sci, Fergus Falls Jr Col, Minn, 68-69; specialist prev med, US Army Med Dept, 69-71; asst prof, 74-86, ASSOC PROF BOT, NDAK STATE UNIV, 86- *Mem:* AAAS; Am Inst Biol Sci; Ecol Soc Am; Sigma Xi. *Res:* Analysis of ecosystem structure and function, especially of Midwestern forests and wetlands; environmental impact assessment. *Mailing Add:* Dept Bot NDak State Univ Fargo ND 58105

CLANCY, EDWARD PHILBROOK, b Beloit, Wis, July 3, 13; m 43; c 5. PHYSICS. *Educ:* Beloit Col, BS, 35; Harvard Univ, AM, 37, PhD(physics), 40. *Prof Exp:* Instr physics, Harvard Univ, 37-43; asst prof, Hamilton Col, 43-44; res assoc, Underwater Sound Lab, Harvard Univ, 44-45, lectr, 46; from asst prof to assoc prof, 46-57, PROF PHYSICS, MT HOLYOKE COL, 57- *Mem:* Am Phys Soc; Am Asn Physics Teachers; Sigma Xi. *Res:* Radiation physics; optics. *Mailing Add:* 17 Silverwood Terr South Hadley MA 01075

CLANCY, JOHN, b Dungarvan, Ireland, Oct 27, 22; US citizen; m 52; c 6. PSYCHIATRY. *Educ:* Nat Univ Ireland, MB & ChB, 46; FRCPS(C). *Prof Exp:* Intern med, St Vincents Hosp, Dublin, Ireland, 46; pvt pract, 47-51; resident psychiat, Iowa, 51-54; dir psychiat, Union Hosp, Moosejaw, Sask, 55-59; from asst prof to assoc prof, Univ Iowa, 59-66, prof psychiat, 66-88, emer prof, 88-; RETIRED. *Concurrent Pos:* Mem Gov Comn Alcoholism, Iowa, 60-61 & 66-; consult, Vet Admin Hosp, Iowa City, 66-; dir outpatient & psychiat consult serv, Univ Iowa Hosps, 78. *Mem:* AMA; fel Am Psychiat Asn; Am Psychopath Asn; Am Acad Clin Psychiatrists (pres, 87-88). *Res:* Psychopathology and treatment of alcoholism; anxiety disorder; affective disorder. *Mailing Add:* 2003 N Dodge St Iowa City IA 52245

CLANCY, RICHARD L, b Hardy, Iowa, Dec 26, 33; m 56; c 2. PHYSIOLOGY. *Educ:* Univ Minn, BA, 56, MSc, 61; Univ Kans, PhD(physiol), 65. *Prof Exp:* Asst prof physiol, Ohio State Univ, 67-69; PROF PHYSIOL, SCH MED, UNIV KANS, 69- *Concurrent Pos:* Nat Heart Inst fel, 65-67. *Mem:* Am Heart Asn; Am Physiol Soc; Sigma Xi. *Res:* Acid-base and cardiovascular physiology. *Mailing Add:* Dept Physiol Univ Kans Med Ctr Kansas City KS 66103

CLANDININ, DONALD ROBERT, b Vandura, Sask, Jan 19, 14; m 38; c 3. POULTRY NUTRITION. *Educ:* Univ BC, BSA, 35, MSA, 36; Univ Wis, PhD(biochem, poultry), 48. *Hon Degrees:* DSc, Univ Alta, 85. *Prof Exp:* Poultry geneticist, Govt Alta, 36-38; from lectr to assoc prof poultry nutrit, 38-53, prof poultry nutrit, 53-79, EMER PROF, UNIV ALTA, 79- *Honors & Awards:* Queen's Silver Jubilee Medal; McHenry Award. *Mem:* Fel Agr Inst Can; fel Poultry Sci Asn; World Poultry Sci Asn; Can Soc Nutrit Sci. *Res:* Nutrient requirements of chickens and turkeys; factors affecting protein quality. *Mailing Add:* 11131 55th Ave Edmonton AB T6H 0W9 Can

CLANDININ, MICHAEL THOMAS, b Edmonton, Alta, Feb 27, 49; m; c 1. LIPIDS, FAT METABOLISM. *Educ:* Univ Alta, BSc, 70, PhD(plant biochem), 73. *Prof Exp:* Teaching fel med biochem, Univ Calgary, 74-75; asst prof, dept foods & nutrit & med, Univ Toronto, 77-81, assoc prof nutrit sci, Fac Med, 81-84; res assoc, dept biochem, 75-76, PROF, DEPT FOODS & NUTRIT & MED, UNIV ALTA, 84-, PROF, DIV GASTROENTEROL, 84- *Concurrent Pos:* Scholarship, Nat Res Coun, 70-73; Gulf Oil Can grad fel, Asn Univs & Cols Can, 71-72; grad fac, Univ Toronto, 77; scholar, Alta Heritage Found Med Res, 84; mem, Nat Inst Child Health & Human Develop Comt Methods Milk Analysis, subcomt NIH. *Mem:* Am Inst Nutrit; Nutrit Soc Can; Biol Coun Can; Can Asn Univ Teachers; Am Soc Clin Nutrit; Am Oil Chemists Soc; Can Soc Nutrit Sci; Can Biochem Soc. *Res:* Effect of diet fat on membrane structure and function; physiology of human lactation; effect of diet on diabetes; metabolism of fats in human subjects. *Mailing Add:* Div Gastroenterol Dept Foods Nutrit & Med Univ Alta Newton Res Bldg Rm 533 Edmonton AB T6G 2C2 Can

CLANTON, DONALD CATHER, b Belle Fourche, SDak, Dec 22, 26; m 50; c 2. ANIMAL NUTRITION. *Educ:* Colo State Univ, BS, 49; Mont State Univ, MS, 54; Utah State Univ, PhD(animal nutrit), 57. *Prof Exp:* From asst prof to assoc prof, 58-66, PROF ANIMAL SCI, UNIV NEBR, 66- *Mem:* AAAS; Am Soc Animal Sci. *Res:* Ruminant nutrition, particularly nutrition of reproduction; range nutrition. *Mailing Add:* 914 Grande North Platte NE 69101

CLANTON, DONALD HENRY, mathematics; deceased, see previous edition for last biography

CLANTON, THOMAS L, b Omaha, Nebr, Feb 1, 49. RESPIRATORY MUSCLE FATIGUE & CONTROL. *Educ:* Univ Nebr, PhD(physiol), 80. *Prof Exp:* Fel, 80-82, PROF, PULMONARY DIV, OHIO STATE UNIV, 82- *Mem:* Am Physiol Soc; Am Thoracic Soc; Soc Clin Res. *Mailing Add:* Dept Int Med & Physiol 325 N Means Hall Ohio State Univ 1655 Upham Dr Columbus OH 43210

CLAPP, C(HARLES) EDWARD, b Holden, Mass, Aug 29, 30; m 53; c 4. SOIL BIOCHEMISTRY, SOIL ORGANIC MATTER. *Educ:* Univ Mass, BS, 52; Cornell Univ, MS, 54, PhD(soil chem), 57. *Prof Exp:* Asst soil chemist, Cornell Univ, 52-56; org chemist, Agr Res Serv, USDA, 56-61; from asst prof to assoc prof, 61-76, PROF SOIL SCI, UNIV MINN, ST PAUL, 76-; RES CHEMIST, AGR RES SERV, USDA, 61- *Concurrent Pos:* Hon sr res fel, Univ B'Hom, Eng, 88-89, Hebrew Univ, Israel, 89. *Mem:* Fel Am Soc Agron; fel Soil Sci Soc Am; Int Soil Sci Soc; Sigma Xi; Int Humic Substance Soc. *Res:* Chemistry of soil organic matter; clay-organic complexes; electrophoresis; polysaccharide chemistry; ethylenimine chemistry; viscosity; soil structure; sewage, sludge and wastewater chemistry; soil and crop residue management; nitrogen transformation and modeling; nitrogen-15 and carbon-13 in soil-plant-water biosystem; ground water quality of agricultural soils; humic substances; pesticide complexes. *Mailing Add:* 2847 N Griggs Roseville MN 55113

CLAPP, CHARLES H, b Stamford, Conn, Oct 4, 48. BIO-ORGANIC CHEMISTRY, ENZYMOLOGY. *Educ:* Bowdoin Col, AB, 70; Harvard Univ, PhD(chem), 75. *Prof Exp:* Res fel biochem, Brandeis Univ, 75-77; asst prof chem, Brown Univ, 77-; MEM STAFF, CHEM DEPT, BUCKNELL UNIV. *Mem:* Am Chem Soc; AAAS. *Res:* Mechanisms of enzymatic and related chemical reactions; design of specific enzyme inhibitors. *Mailing Add:* Dept Chem Bucknell Univ Lewisburg PA 17837

CLAPP, JAMES R, b Siler City, NC, Sept 3, 31; m 53; c 2. INTERNAL MEDICINE. *Educ:* Univ NC, MD, 57. *Prof Exp:* Intern & resident med, Parkland Mem Hosp, Dallas, Tex, 57-59; investr kidney & electrolytes, NIH, 61-63; assoc, 63-66, asst prof, 66-70, assoc prof internal med, Sch Med 70-77, PROF NEPHROLOGY & ASSOC PROF PHYSIOL, DUKE UNIV MED CTR, 77- *Concurrent Pos:* USPHS trainee, 59-61 & grant, 63-; fel renol, Southwestern Med Sch, Univ Tex, 59-61; estab investr, Am Heart Asn. *Mem:* Am Physiol Soc; Am Fedn Clin Res. *Res:* Renal physiology and pathophysiology. *Mailing Add:* Med-Box 3014 Duke Univ Med Ctr Durham NC 27710

CLAPP, JOHN GARLAND, JR, b Greensboro, NC, Oct 27, 36; m 59; c 3. AGRONOMY. *Educ:* NC State Univ, BS, 59, MS, 61, PhD(crop sci), 69. *Prof Exp:* Asst agr exten agent, NC State Univ, 61-62; exten agronomist, Clemson Univ, 62-63 & NC State Univ, 63-75; SCI AGRONOMIST, ALLIED CHEM CORP, 75- *Concurrent Pos:* Exten agronomist, Nat Soybean Resource Comt, 70- *Honors & Awards:* Geigy Award in Agron, Am Soc Agron, 72; Meritorious Serv Award, Am Soybean Asn, 74. *Mem:* Am Soc Agron. *Res:* Applied on-farm evaluation of fertilizers, herbicides, growth regulators, nematocides, plant population and tillage methods for soybean production. *Mailing Add:* 310 Clapp Farms Rd Greensboro NC 27405

CLAPP, LEALLYN BURR, organic chemistry; deceased, see previous edition for last biography

CLAPP, NEAL K, b Shelby Co, Ind, Oct 14, 28; m 53; c 3. RADIOBIOLOGY, PATHOLOGY. *Educ:* Purdue Univ, BS, 50; Ohio State Univ, DVM, 60; Colo State Univ, MS, 62, PhD, 64. *Prof Exp:* Instr surg, Vet Clin, Colo State Univ, 60-61; EXP PATHOLOGIST, OAK RIDGE NAT LAB, 64-, EXP PATHOLOGIST, OAK RIDGE ASSOC UNIV, 81- *Mem:* AAAS; Am Vet Med Asn; Radiation Res Soc; Am Asn Cancer Res. *Res:* Radiation pathology; chemical carcinogenesis; studying mechanisms involved in development of colitis and colonic carcinama utilizing a primate animal model. *Mailing Add:* 628 Riverbend Rd Clinton TN 37716

CLAPP, PHILIP CHARLES, b Belleville, Ont, Oct 14, 35; US citizen; m 61; c 3. SOLID STATE PHYSICS. *Educ:* Queen's Univ, BS, 57; Mass Inst Technol, PhD(physics), 63. *Prof Exp:* Lectr magnetism, Mass Inst Technol, 63; physicist, Ledgemont Lab, Kennecott Copper Corp, 63-75, head physics

& metall group, 75-77; vis assoc prof mat sci, Mass Inst Technol, 77-78; head dept, 78-83, PROF METALL, UNIV CONN, 83- *Concurrent Pos:* Sr vis scientist, Oxford Univ, 69-70; vis prof, Nat Comn Atomic Energy, Buenos Aires, Arg, 72, Jiaotong Univ, Shanghai, China, 83 & Nat Inst Appl Sci, Lyon, France, 85-86; adj prof physics, Boston Col, 73-78. *Mem:* Metall Soc; Am Phys Soc. *Res:* Alloy research; Martensitic phase transformations; theories of order-disorder phenomena; grain boundary problems. *Mailing Add:* Univ Conn Box U-136 Storrs CT 06268

CLAPP, RICHARD CROWELL, b Boston, Mass, July 25, 15; m 46; c 2. ORGANIC CHEMISTRY. *Educ:* Bowdoin Col, AB, 37; Harvard Univ, MA, 39, PhD(org chem), 41. *Prof Exp:* Sr chemist, Am Cyanamid Co, 41-54; res chemist, US Army Natick Res & Develop Command, 54-74, head, Org Chem Group, 74-79; RETIRED. *Concurrent Pos:* Chemist, Nat Defense Res Comt, 40-41. *Mem:* Fel AAAS; Am Chem Soc; Sigma Xi. *Res:* Synthesis of carcinogenic hydrocarbons; synthesis of chemotherapeutic agents; antitubercular compounds; sulfur compounds; heterocyclic compounds; natural products; cyanogenetic glycosides; laser dyes; anthraquinone dyes. *Mailing Add:* 194 Bacon St Natick MA 01760

CLAPP, RICHARD GARDNER, b Hanover, NH, Mar 3, 11; m 42; c 2. HIGH-VOLTAGE POWER SUPPLY DESIGN. *Educ:* Univ Pa, BSEE, 31, MSEE, 32. *Prof Exp:* Engr, Philco Corp, res div, 34-53, exec engr, 53-57; mem, opers res dept, Philco Corp, 57-65, TRW Int Resistance Co Div, 65-67, Boeing, 67-75; pres Multi-Pro Inc, Mgt Consults, 75-79; teacher elec, RETS, 79; consult engr, Megavolt Inc, 79-81; RETIRED. *Mem:* Fel Inst Elec & Electronics Engrs. *Mailing Add:* 504 Ellis Rd Havertown PA 19083

CLAPP, ROGER EDGE, theoretical physics, biophysics; deceased, see previous edition for last biography

CLAPP, ROGER WILLIAMS, JR, b Tampa, Fla, Aug 31, 29; m 59; c 4. PHYSICS. *Educ:* Davidson Col, BS, 50; Univ Va, MS, 52, PhD(physics), 54. *Prof Exp:* Res physicist, Army Missile Command, Redstone Arsenal, 56-63; asst prof, 63-66, ASSOC PROF PHYSICS, UNIV S FLA, 66- *Mem:* Am Phys Soc; Am Asn Physics Teachers. *Res:* Surface physics; thin films; history of physics. *Mailing Add:* Dept Physics Univ SFla Tampa FL 33620

CLAPP, THOMAS WRIGHT, b Fulton, Ky, Sept 3, 37; m 69; c 1. BOTANY, SOIL SCIENCE. *Educ:* Murray State Univ, BS, 63; NDak State Univ, MS, 65; Tex A&M Univ, PhD(range sci), 68. *Prof Exp:* Asst prof, 67-80, ASSOC PROF BIOL, ST CLOUD STATE UNIV, 80- *Mem:* Soc Econ Bot; Sigma Xi. *Res:* Allelopathy, seed germination; hormonal regulation of plant growth; plant physiology; plant ecology. *Mailing Add:* Rte 1 Oak Heights Addn Cold Springs MN 56320

CLAPP, WILLIAM LEE, b Memphis, Tenn, Feb 16, 43; m 65; c 3. GENERAL CHEMISTRY. *Educ:* Wake Forest Col, BS, 64; Duke Univ, MA, 66, PhD(chem), 69. *Prof Exp:* Res chemist analytical chem, R J Reynolds Tobacco Co, 68-69; chem officer, Weapons Develop & Eng Lab, Edgewood Arsenal, US Army, 69-71; res chemist analytical chem, 71-72, sect head proj mgt, 72-80, mgr anal technol div, 80-83, mgr prod design technol, 83-85, mgr res & develop planning, 85-87, SR STAFF CHEMIST, R J REYNOLDS TOBACCO CO, 87- *Mem:* Am Chem Soc; Sigma Xi; AAAS. *Res:* Chemical processing of tobacco. *Mailing Add:* Res & Develop Dept R J Reynolds Tobacco Co Winston-Salem NC 27102

CLAPPER, DAVID LEE, b Marshalltown, Iowa, Feb 24, 45; m 68; c 1. SURFACE MODIFICATION TO IMPROVE BIOCOMPATIBILITY, CELL BIOLOGY. *Educ:* Iowa State Univ, Ames, BS, 67, PhD(molecular, cellular & develop biol), 79; Univ Nebr, Omaha, MA, 74. *Prof Exp:* Postdoctoral fel, Hopkins Marine Sta, Stanford Univ, 79-83; res assoc, Dept Pharmacol, Duke Univ, 83-84 & Dept Physiol, Univ Minn, 84-86; sr scientist, 86-90, MGR BIOL RES, BIO-METRIC SYSTS INC, 90- *Concurrent Pos:* Mem, Spec Study Sect Mat Sci, NIH, 90- *Mem:* Am Soc Cell Biol; Soc Biomat; AAAS. *Res:* Modification of biomedical device surfaces to improve biocompatibility; extracellular matrix proteins and peptides to improve cell growth and/or tissue integration; antithrombotic agents to improve blood compatibility. *Mailing Add:* Bio-Metric Systs Inc 9924 W 74th St Eden Prairie MN 55344

CLAPPER, MUIR, b Detroit, Mich, May 26, 13; m 62. INTERNAL MEDICINE, CARDIOLOGY. *Educ:* Wayne State Univ, AB, 33, MD, 36, MS, 40. *Prof Exp:* From instr to assoc prof, 40-53, PROF MED, SCH MED, WAYNE STATE UNIV, 53- *Concurrent Pos:* Consult, Dearborn Vet Hosp, 51- & Macomb Hosp, 57-; emer, Harper Hosp, 62- & Detroit Receiving Hosp. *Mem:* AMA; Am Heart Asn; Am Col Physicians; Asn Univ Cardiol; Am Col Cardiol. *Res:* Cardiology. *Mailing Add:* Wayne State Univ Sch Med 540 E Canfield Ave Detroit MI 48201

CLAPPER, THOMAS WAYNE, b McKean, Pa, Oct 15, 15; m 41; c 3. ORGANIC CHEMISTRY. *Educ:* St Vincent Col, BS, 37; Pa State Univ, MS, 38, PhD(org chem), 42. *Prof Exp:* Res chemist, Pharmaceut Div, Calco Chem Div, Am Cyanamid Co, 40-44, from asst chief chemist to chief chemist, Pharmaceut Dept, 44-48, prod mgr, 48-51, tech dir, Atomic Energy Div, 51-52, gen supt, Chem Processing Plant, 52, asst gen mgr, 52-53; asst to gen mgr, Atomic Energy Div, Phillips Petrol Co, 54; plant mgr, Calera Ref, Chem Construct Corp, 54-56; res mgr, Am Potash & Chem Corp, 56-63, tech dir res, 63-68; dir res, Kerr-McGee Corp, 68-75, mgr technol assessment & planning, 75-81; PRES, CLAPPER ENTERPRISES INC, 81- *Mem:* Am Chem Soc; Electrochem Soc; Am Soc Metals; Am Inst Mining Metall & Petrol Engrs. *Res:* Sulfa drugs; vitamins; chemical processing of uranium; cobalt; high energy fuels; electrochemistry; rare earths; boron compounds; maganese metal and compounds. *Mailing Add:* 12104 Camelot Pl Oklahoma City OK 73120

CLAPSHAW, PATRIC ARNOLD, b Ketchikan, Alaska, Sept 13, 37; c 2. TISSUE CULTURE, PROTEIN ISOLATION. *Educ:* Univ Wash, BA, 62; Univ Calif, San Francisco, MA, 72; Univ Colo, PhD(neurochem), 76. *Prof Exp:* Instr, dept molecular, cellular & develop biol, Univ Colo, 73-76; prin investr, Queens Med Ctr, 81-83; prin investr, Issaquah Health Res Inst, 83-85; DIR, SOLOMON PARK RES INST, 85- *Mem:* Am Soc Neurochem; Int Soc Neurochem; AAAS. *Res:* Neurological disease (Amyotrophic Lateral Sclerosis). *Mailing Add:* 15305 NE Seventh Pl Bellevue WA 98007

CLARDY, JON CHRISTEL, b Washington, DC, May 16, 43; m 66; c 2. STRUCTURAL CHEMISTRY, ORGANIC CHEMISTRY. *Educ:* Yale Univ, BS, 64; Harvard Univ, PhD(chem), 69. *Prof Exp:* From instr to prof, Iowa State Univ, 69-78; prof, 78-90, CHMN CHEM DEPT, CORNELL UNIV, 88-, HORACE WHITE PROF CHEM, 90- *Concurrent Pos:* Camille & Henry Dreyfus Found fel, 72; Alfred P Sloan Found fel, 73; Guggenheim fel, 84-85. *Honors & Awards:* Akron Award, Am Chem Soc, 87. *Mem:* Am Chem Soc; Am Crystallog Asn; fel AAAS. *Res:* Application of x-ray diffraction to problems of biological and chemical interest; natural products chemistry. *Mailing Add:* Dept Chem Cornell Univ Ithaca NY 14853-1301

CLARDY, LEROY, b Ft Worth, Tex, July 16, 10; m 38; c 1. PHYSICAL CHEMISTRY. *Educ:* Tex Christian Univ, BS, 31, MS, 34. *Prof Exp:* Analytical chemist, Armour & Co, Tex, 34-36; chief chemist, Terrell's Labs, 36-37; chemist, Swift & Co, 37-43, physicist, 43-70, mgr control eng div, Eng Res Dept, 70-75; CONSULT, 75- *Res:* Application of instrumentation and automatic control systems to meat packing and allied processes. *Mailing Add:* 835 Edgewater Dr Naperville IL 60540-7434

CLARE, DEBRA A, MILK PROTEIN BIOCHEMISTRY, BOVINE MILK. *Educ:* NC State Univ, PhD(biochem), 81. *Prof Exp:* RES ASSOC, DEPT ANIMAL SCI, NC STATE UNIV, 84- *Res:* Isolation and characterization of bovine milk. *Mailing Add:* 105 Honeysuckle Lane Cary NC 27511

CLARE, STEWART, b Montgomery Co, Mo, Jan 31, 13; m 36. ZOOLOGY, BIOCHEMISTRY. *Educ:* Univ Kans, BA, 35; Iowa State Univ, MS, 37; Univ Chicago, PhD(zool), 49. *Prof Exp:* Tech consult, White-Fringed Beetle Proj, Bur Entom & Plant Quarantine, US Civil Serv Comn, 41-42, instr meteorol, Army Air Force Weather Sch, 42-43; res biologist, Midwest Res Inst, Mo, 45-46; mem spec res proj, Univ Mo-Kansas City, Midwest Res Inst & Kansas City Art Inst, 46-49; instr zool, Univ Alta, 49-50, asst prof zool & lectr sci of color, 50-53; asst prof physiol & pharmacol, Kansas City Col Osteop & Surg, Univ Health Sci, 53; lectr, Univ Adelaide, 54-55; sr res officer entom, Ministry Agr & Gezira Res Sta, Sudan Govt, NAfrica, 55-56; sr entomologist, Klipfontein Org Prod Corp, SAfrica, 57; prof biol & head dept, Union Col, Ky, 58-59, chmn div sci, 59-61; prof biol & head dept, Mo Valley Col, 61-62; Buckbee Found prof biol & lectr, Eve Col, Rockford Col, 62-63; prof biochem & chmn dept & mem res div, Kansas City Col Osteop & Surg, Univ Health Sci, 63-67; prof biol, 67-72, dir biol res, 72-74, EMER PROF BIOL, COL EMPORIA, 74-; RES BIOLOGIST & CONSULT, 74- *Concurrent Pos:* Entomologist, US Bur Entom, 37-40; res & study grants, Alta Res Coun, 51-53, Union Col, Ky, 59-61, Mo Valley Col, 61-62, Rockford Col, 62-63, Adirondack Res Sta, 63-66, NIH, 63-65 & Col Emproia, 67-74; consult, Vols for Tech Assistance, 62- & Info Resource, Nat Referral Ctr for Sci & Technol, Libr of Cong, 70-; lectr & consult, Adirondack Res Sta, NY State Univ, Plattsburgh, 62-66; hon fel, Anglo-Am Acad, Cambridge, 80. *Mem:* NY Acad Sci; Am Entom Soc; Brit Asn Advan Sci; Arctic Inst NAm; Nat Asn Biol Teachers. *Res:* Comparative physiology-biochemistry; circulation of the Arthropoda; trace elements in invertebrates; capillary movement in porous materials; gums, extractives and extraneous materials of plants; biometeorology; chromatology; history of science; science of color. *Mailing Add:* 405 NW Woodland Rd Indian Hills in Riverside Kansas City MO 64150-9445

CLARENBURG, RUDOLF, b Utrecht, Holland, May 3, 31; US citizen; m 59; c 1. PHYSIOLOGICAL CHEMISTRY. *Educ:* Univ Utrecht, Drs, 59, DSc(chem), 65. *Prof Exp:* Res physiologist, Univ Calif, Berkeley, 59-66; assoc prof, 66-74, PROF PHYSIOL, KANS STATE UNIV, 74- *Mem:* AAAS; Am Physiol Soc; Soc Exp Biol Med; Sigma Xi. *Res:* Membrane receptors; transport across biological membranes; physiologic significance of asialoglycoprotein receptors in liver; transport of bilirubin and sulfobromophthalein from blood to bile. *Mailing Add:* 438 Shelle Rd Manhattan KS 66502

CLARIDGE, E(LMOND) L(OWELL), b Delaplaine, Ark, June 5, 17; m 39; c 2. FUEL TECHNOLOGY & PETROLEUM ENGINEERING. *Educ:* Mo Sch Mines, BS, 39, MS, 41, Univ Houston, PhD, 79. *Prof Exp:* Jr res chemist, Wood River Res Lab, Shell Oil Co, 41-43, oper asst, Wood River Ref, 43-47, technologist & asst head exp lab, 47-48, res group leader, Houston Res Lab, 48-51, asst chief res chemist, process res supvr, 51-55, sect leader, Royal Dutch Labs, Amsterdam, 55-57, asst chief res technologist, Houston Res Lab, Shell Oil Co, 57-60, sr technologist mfg res dept head off, NY, 60-64, res assoc, 64-66, sr res assoc, 66-70, sr staff res engr, Shell Develop Co, 70-79; dir grad prog Petrol Eng, 79-87, assoc prof chem eng, 79-90, ADJ PROF CHEM ENG, UNIV HOUSTON, 91- *Concurrent Pos:* Petrol eng consult, Gulf Univ Res Consortium & Tod, Dietrich & Chase, 79-, Lyco Energy Co, 89- *Honors & Awards:* Enhanced Oil Recovery Pioneer, Soc Petrol Engrs, 88. *Mem:* AAAS; Am Chem Soc; Am Inst Chem Engrs; Soc Petrol Engrs; Am Petrol Inst; Can Inst Mining Met & Petrol. *Res:* Crude oil recovery processes. *Mailing Add:* 5439 Paisley Houston TX 77096-4025

CLARIDGE, RICHARD ALLEN, b Chicago, Ill, Feb 22, 32; m 52; c 3. CIVIL ENGINEERING. *Educ:* Fla State Univ, BS & BA, 53; Univ Fla, BCE, 59. *Prof Exp:* Assoc engr, Douglas Aircraft Co, 59-63, engr, 63-65; sr engr, McDonnell Douglas Astro Co, 65-82, tech specialist, 82-86, sr tech specialist, 86-89; SR DESIGN ENGR, GEN DYNAMICS SPACE SYST CO, 89- *Concurrent Pos:* Pres, Atlantic Consults Inc, 74-78. *Mem:* Am Soc Civil Engrs; Am Inst Aeronaut & Astronaut; Nat Soc Prof Engrs; Soc Allied Weight Engrs. *Res:* Field repair techniques in aerospace. *Mailing Add:* 1713 Guldahl Dr Titusville FL 32780

CLARK, A GAVIN, b Warrington, Eng, Nov 18, 38; m. MICROBIOLOGY. *Educ:* Univ Edinburgh, BSc, 63, PhD(microbiol), 66. *Prof Exp:* Res asst bact, Med Sch, Univ Edinburgh, 61-62, res asst microbiol, Sch Agr, 63-66; fel soil sci,Univ Alta, 66-68; asst prof, 68-78, ASSOC PROF MICROBIOL, UNIV TORONTO, 78- *Mem:* Brit Soc Appl Bact; Brit Soc Gen Microbiol; Can Soc Microbiol. *Res:* Foodborne infections; pathogenicity of Vibro parahaemolyticus; campylobacter jejuni. *Mailing Add:* Dept Microbiol Univ Toronto St George Campus Toronto ON M5S 1A3 Can

CLARK, ALAN CURTIS, b Springfield, Mass, May 6, 44; m 70; c 2. ORGANIC CHEMISTRY. *Educ:* Bowdoin Col, AB, 66; Ind Univ, PhD(org chem), 70. *Prof Exp:* Vis res assoc chem, Ohio State Univ, 70-71; fel, Univ Cincinnati, 71-73, lectr, 73-74; res chemist, 74-81, GROUP LEADER, LUBRIZOL CORP, 81- *Mem:* Am Chem Soc. *Res:* Organic synthesis; polymer chemistry; lubricant additive chemistry. *Mailing Add:* Lubrizol Corp Wickliffe Plant 29400 Lakeland Blvd Wickliffe OH 44092

CLARK, ALAN FRED, b Milwaukee, Wis, June 29, 36; m 57; c 4. SUPERCONDUCTIVITY, MAGNETISM. *Educ:* Univ Wis, BS, 58, MS, 59; Univ Mich, PhD(nuclear sci), 64. *Prof Exp:* Nat Acad Sci-Nat Res Coun res assoc low temperature physics, 64-66, staff physicist low temperature physics & mat sci, 66-78, chief thermophys properties of solids, 78-80, CHIEF SUPERCONDUCTOR MAGNETIC MEASUREMENTS, NAT BUR STANDARDS, 81- *Concurrent Pos:* Asst prof mat sci, Colo State Univ, Ft Collins, 65-68; tech ed, Rev Sci Instruments, 74-76; Am ed, Cryogenics, 77-; mem adv panel, Joint NASA, US Air Force & Fed Aviation Admin panel titanium combustion, 75- & magnetic energy storage panel, Dept Energy, 77-; chmn, Int Cryog Mat Conf Bd, 79 -; ed, Int Cryogenics Monogr; sr res scientist, Oxford, 84-85, Liaison scientist, ONR, London, 88-89; mem, Int Cryogenic Mat Conf. *Mem:* Am Soc Testing & Mat; Am Phys Soc; Inst Elec & Electronics Engrs. *Res:* Thermophysical properties of materials at low temperatures; transport properties of high purity metals; combustion of metals; critical properties of superconductors. *Mailing Add:* Nat Inst Standards Metrol Bldg B 258-220 Gaithersburg MD 20899

CLARK, ALFRED, b Beverly, Mass, Aug 5, 09; m 35; c 1. PHYSICAL CHEMISTRY, CHEMICAL ENGINEERING. *Educ:* Purdue Univ, BS, 30; Mich State Univ, MS, 32; Univ Ill, PhD(phys chem), 35. *Prof Exp:* Res chemist viscose, North Am Rayon Corp, 35-36; sr res chem catalysis, Battelle Mem Inst, 36-42; div mgr, Res Catalysis, Publicker Com Alcohol Co, 42-44; br mgr res, Phillips Petrol Co, 44-59, sr res scientist, 59-71; prof chem eng, Univ Okla, 71-79; CONSULT, 79- *Honors & Awards:* Mod Pioneer Creative Indust Award, Nat Asn Mfrs, 66; E V Murphree Award, Am Chem Soc, 67. *Mem:* Sigma Xi. *Res:* Heterogeneous catalysis; polymerization; kinetics of catalytic reactions. *Mailing Add:* Eastwood Apts 2141 E Ave-Apt F Rochester NY 14610-0001

CLARK, ALFRED, JR, b Elizabethton, Tenn, May 5, 36; m 60; c 1. BIOPHYSICAL TRANSPORT PROCESSES. *Educ:* Purdue Univ, BS, 58; Mass Inst Technol, PhD(appl math), 63. *Prof Exp:* NSF fel, 63-64; from asst prof to assoc prof, 64-74, chmn dept, 72-77, PROF MECH ENG, UNIV ROCHESTER, 74- *Concurrent Pos:* Vis fel, Joint Inst Lab Astrophys, Univ Colo, 70-71; visitor, Dept Radiation Biol & Biophys, Univ Rochester, 80-81, 84-85. *Mem:* Am Soc Mech Engrs; Microcirc Soc; Am Phys Soc; Int Soc Oxygen Transport Tissue; Math Asn Am; Soc Indust & Appl Math. *Res:* Present research on biophysical transport processes, particularly the mathematical modelling of oxygen transport to tissue. *Mailing Add:* Dept Mech Eng Univ Rochester Rochester NY 14627

CLARK, ALFRED JAMES, b Cherokee, Iowa, July 14, 33. BIOCHEMISTRY, PUBLIC HEALTH & EPIDEMIOLOGY. *Educ:* Iowa State Univ, BS, 56, MS, 62, PhD(animal nutrit), 63. *Prof Exp:* Fel human nutrit, Sch Pub Health, Univ Calif, Los Angeles, 64-66, asst prof nutrit, 66-73; assoc prof, Sch Home Econ, Purdue Univ, 73-77; ASSOC PROF NUTRIT, SCH HOME ECON, AUBURN UNIV, 77- *Mem:* Am Inst Nutrit; Sigma Xi; AAAS. *Res:* Nutrient needs of adolescents; nutritional status of adolescents; absorption of nutrients from the gastrointestinal tract; nutrient transport in chick uterus and in intestinal epithelium. *Mailing Add:* Dept Nutrit & Foods Auburn Univ Auburn AL 36849

CLARK, ALLAN H, b Cincinnati, Ohio, July 16, 35; c 3. MATHEMATICS. *Educ:* Mass Inst Technol, BS, 57; Princeton Univ, MA, 59, PhD(math), 61. *Prof Exp:* From instr to prof math, Brown Univ, 61-75; dean, Sch Sci, Purdue Univ, West Lafayette, 75-85; PRES, CLARKSON UNIV, POTSDAM, NY, 85- *Concurrent Pos:* NSF grant, 62-; vis mem, Inst Advan Study, Princeton Univ, 65-66; vis prof, Math Inst, Aarhus Univ, 70-71; trustee, Univs Res Asn, Inc, 76- *Mem:* Am Asn Higher Educ; Am Coun Educ; Am Math Soc. *Res:* Algebraic topology. *Mailing Add:* Clarkson Univ Potsdam NY 13676

CLARK, ALLEN KEITH, b Bridgeton, NJ, June 25, 33; m 57; c 2. ORGANIC CHEMISTRY. *Educ:* Catawba Col, AB, 55; Univ NC, PhD(org chem), 60. *Prof Exp:* From asst prof to prof, Old Dominion Univ, 62-66, actg chmn dept, 68-69, asst provost, 72-79, chmn dept, 69-72, dep vpres acad affairs, 80-83. *Concurrent Pos:* Sigma Xi res grant-in-aid, 63-64. *Mem:* Am Chem Soc; Sigma Xi. *Res:* Chemistry of ferrocene; aromatic nitroso compounds. *Mailing Add:* Old Dom Univ Norfolk VA 23529

CLARK, ALLEN LEROY, b Delaware, Iowa, Sept 29, 38; m 55, 81; c 3. ECONOMIC GEOLOGY, GEOCHEMISTRY. *Educ:* Iowa State Univ, BS, 61; Univ Idaho, MS, 63, PhD(geol), 68. *Prof Exp:* Instr geol, Univ Idaho, 65-66; geologist, Bear Creek Mining Co, Kennecott Copper Corp, 66-67; res geologist, US Geol Surv, 67-72, chief, Off Resource Analysis, 72-78, sr staff coordr, 78-80; DIR GEN, INT INST RESOURCE DEVELOP, 80- *Concurrent Pos:* Co-investr, Apollo 12-14, NASA, 70-73. *Mem:* Am Asn Petrol Geol; fel Geol Soc Am; Soc Econ Geologists; Int Asn Genesis of Ore Deposits. *Res:* Economics of international development; economic analysis of exploration and resource availability; wallrock alteration and trace element distributions associated with base metal deposits; platinum group metals distribution in ultramafic rocks; structural analysis of lunar samples; international resource assessment; resource data systems. *Mailing Add:* East West Resource Systs Inst 1777 East West Rd Honolulu HI 96848

CLARK, ALLEN VARDEN, b Attleboro, Mass, Nov 1, 41; m 84; c 3. FOOD SCIENCE, TECHNICAL MANAGEMENT. *Educ:* Mass Inst Technol, BS, 63, MS, 65, PhD(food sci & technol), 69. *Prof Exp:* Prin investr, Corp Res & Develop Lab, 68-80, MGR, CITRUS RES & DEVELOP, COCA-COLA FOODS, 80- *Mem:* AAAS; Am Chem Soc; Inst Food Technologists. *Res:* Isolation and characterization of pigments from protein carbonyl browning systems; tea; food processing; citrus processing; product development; management. *Mailing Add:* 424 Timber Ridge Dr Longwood FL 32779

CLARK, ALTON HAROLD, b Bangor, Maine, Oct 10, 39; m 61; c 2. SOLID STATE PHYSICS. *Educ:* Univ Maine, BA, 61; Univ Wis, Madison, MS, 63; Cornell Univ, PhD(physics), 67. *Prof Exp:* Physicist, Sprague Elec Co, 66-68; from asst prof to prof physics, Univ Maine, Orono, 68-81; group leader, Sohi Engineered Mat Co, 84-88. *Concurrent Pos:* Vis scientist, Xerox Palo Alto Res Ctr, 74-75. *Mem:* Am Phys Soc. *Res:* Electrical and optical properties of crystalline and amorphous semiconductors. *Mailing Add:* Standard Oil Co 4440 Warrensville Ctr Rd Cleveland OH 44128

CLARK, ALVIN JOHN, b Oak Park, Ill, Apr 13, 33; div. GENETICS, MOLECULAR BIOLOGY. *Educ:* Univ Rochester, BS, 55; Harvard Univ, PhD(chem), 59. *Prof Exp:* Asst prof bact, Univ Calif, Berkeley, 62-64, asst prof bact & molecular biol, 64-67, from assoc prof to prof molecular biol, 67-89, PROF GENETICS, UNIV CALIF, BERKELEY, 89- *Concurrent Pos:* Am Cancer Soc fel, 59-61; fel, Yale Univ, 62; John Simon Guggenheim Mem Found fel, 69. *Mem:* Am Soc Microbiol; Genetics Soc Am; Sigma Xi. *Res:* Enzymological and genetic analysis of genetic recombination; bacterial conjugation. *Mailing Add:* Dept Molecular & Cell Biol Univ Calif Barker Hall Berkeley CA 94720

CLARK, ANDREW GALEN, b Urbana, Ill, July 4, 54; m 80. EXPERIMENTAL DROSOPHILA, POPULATION GENETICS. *Educ:* Brown Univ, BS, 76; Stanford Univ, PhD(biol sci), 80. *Prof Exp:* Teaching assoc fel, Ariz State Univ, 80-81; vis researcher, Aarhus Univ, Denmark, 82; ASST PROF GENETICS, PA STATE UNIV, 83- *Concurrent Pos:* Res fel, NATO, 82; Marshall fel, Marshall Found, 82. *Mem:* Genetics Soc Am; Soc Study Evol; Am Soc Human Genetics; Am Soc Naturalists. *Res:* Experimental and theoretical assessment of cytoplasmic variation in natural population of Drosphila; theory of the use of restriction site variation in prenatal diagnosis; sex chromosome evolution. *Mailing Add:* Dept Genetics Pa State Univ University Park PA 16802

CLARK, ARMIN LEE, b Huntington, WVa, June 16, 28; m 51; c 2. GEOLOGY. *Educ:* Marshall Univ, BA, 51; Ohio State Univ, MSc, 58; Univ Tenn, PhD(geol), 73. *Prof Exp:* Teacher sci, Cabell & Mingo Counties, WVa Schs, 58-60 & Jefferson County, Ky Schs, 60-61; from instr to assoc prof, 61-78, PROF GEOL, MURRAY STATE UNIV, 79- *Mem:* Geol Soc Am. *Res:* Petrology of the Eocene sediments in western Kentucky and Tennessee. *Mailing Add:* 1504 Oak Dr Murray KY 42071

CLARK, ARNOLD FRANKLIN, b Madison, Wis, Apr 27, 16; c 3. X-RAY DIFFRACTION, EXPERIMENTAL DESIGNS. *Educ:* Swarthmore Col, AB, 37; Ind Univ, AM, 39, PhD(physics), 41. *Prof Exp:* Asst math & physics, Univ Wis, 37-38; asst physics, Ind Univ, 38-41; res fel & physicist, Univ Calif, 41-46; res assoc, Univ Rochester, 46-47, asst prof physics, 47-49; asst prof physics, Carnegie Inst Technol, 49-54; PHYSICIST, LAWRENCE LIVERMORE NAT LAB, UNIV CALIF, 54- *Mem:* Fel Am Phys Soc; Sigma Xi; Int Solar Energy Soc. *Res:* Nuclear physics; engineering physics; design of electromagnetic accelerators; nuclear emulsions; radiation effects; cloud chambers; solar energy research and development; shallow solar ponds; x-ray diffraction with bent crystal spectrometers; monochromatic x-rays from z pinch plasmas. *Mailing Add:* 3892 Madeira Way Livermore CA 94550

CLARK, ARNOLD M, b Philadelphia, Pa, Jan 28, 16; m 53; c 2. GENETICS. *Educ:* Pa State Col, AB, 37; Univ Pa, MA, 39, PhD(zool), 43. *Prof Exp:* Res biologist, Smyth Labs, Philadelphia, 40-46; asst instr zool, Univ Pa, 44-45; instr, Philadelphia Col Pharm, 45-46; from instr to prof, 46-81, EMER PROF BIOL, UNIV DEL, 81-; SCIENTIST, MARINE BIOL LAB, WOODSHOLE, MASS, 83- *Concurrent Pos:* Radiation biologist, Brookhaven Nat Lab, 53-54. *Mem:* AAAS; Am Soc Naturalists; Radiation Res Soc; Am Genetics Soc; Genetic Asn Am; Sigma Xi. *Res:* Genetics of Habrobracon; toxicological studies on insecticides; resins; pharmaceuticals; gene dosage; radiation damage; action of cell poisons; oxygen poisoning; genetics of aging; studies of genetic mosaics in insects and their use in the analysis of development and behavior; analysis of chromosomal aberrations in man. *Mailing Add:* 53 Wilson Rd Woods Hole MA 02543

CLARK, ARTHUR EDWARD, b Scranton, Pa, July 9, 32; m 58; c 4. SOLID STATE PHYSICS. *Educ:* Univ Scranton, BS, 54; Univ Del, MS, 56; Cath Univ, PhD(physics), 60. *Prof Exp:* RES PHYSICIST, US NAVAL SURFACE WEAPONS CTR, 59- *Mem:* AAAS; Am Phys Soc; Sigma Xi. *Res:* Magnetic, elastic and magnetoelastic properties of solids; ultrasonics and hypersonics. *Mailing Add:* 10421 Floral Dr Adelphi MD 20783

CLARK, BARRY GILLESPIE, b Happy, Tex, Mar 5, 38; m 63; c 4. ASTRONOMY. *Educ:* Calif Inst Technol, BS, 59, PhD(astron), 64. *Prof Exp:* Asst scientist, 64-69, SCIENTIST, NAT RADIO ASTRON OBSERV, 69- *Honors & Awards:* Van Biesbroek Award, 91. *Mem:* Am Astron Soc; Int Astron Union. *Res:* Radio astronomy interferometry and array design software and use. *Mailing Add:* Nat Radio Astron Observ PO Box O Socorro NM 87801-0387

CLARK, BENJAMIN CATES, JR, b Knoxville, Tenn. FLAVOR REACTIONS, ANALYTICAL FLAVOR CHEMISTRY. *Educ:* Duke Univ, BA, 63; Emory Univ, MS & PhD(org chem), 67. *Prof Exp:* Fel, Univ Ga, 67-69; prin staff chemist, 69-75, sr res scientist, 75-86, PRIN INVESTR, CORP RES & DEVELOP DEPT, COCA-COLA CO, 86- *Mem:* Am Chem Soc; Inst Food Technol; Am Soc Testing & Mat; Sigma Xi; Royal Soc Chem. *Res:* Analytical organic chemistry; flavor chemistry particularly, terpene chemistry; the reactions of flavors in dilute aqueous systems and micelles; emulsion chemistry; flavor analysis-carbonated beverages, citrus and other essential oils; beverage technology; photochemistry. *Mailing Add:* Corp Res & Develop Dept PO Drawer 1734 Coca-Cola Co Atlanta GA 30301

CLARK, BENTON C, b Oklahoma City, Okla, Aug 4, 37; m 65. GEOCHEMISTRY, BIOPHYSICS. *Educ:* Univ Okla, BS, 59; Univ Calif, MA, 61; Columbia Univ, PhD(biophys), 69. *Prof Exp:* Res asst radiation instrumentation res & develop, Los Alamos Sci Lab, 59-60; assoc electronic speech recognition, Advan Systs Develop Div, Int Bus Mach Corp, 61; sr staff scientist res & develop, Avco Corp, 68-71; SR RES SCIENTIST, MARTIN-MARIETTA CORP, 71- *Concurrent Pos:* Dep team leader, NASA Viking Inorg Chem Team, 72-78. *Honors & Awards:* NASA Pub Serv Award. *Mem:* Am Inst Aeronaut & Astronaut; Am Astron Soc (Planetary Sci); World Space Found; Planetary Soc. *Res:* Radiobiological effects of very soft x-rays; space radiation research; origin of life via comet pond; geochemical analysis of planetary and comet materials; x-ray fluorescence spectrometry; environmental monitoring; human missions to Mars; space station designs; systems design and systems science. *Mailing Add:* 10890 Park Range Rd Littleton CO 80127

CLARK, BILL PAT, b Bartlesville, Okla, May 15, 39. SEMICONDUCTORS, SOLID STATE PHYSICS. *Educ:* Okla State Univ, BS, 61, MS, 64, PhD(physics), 68. *Prof Exp:* Asst physics, Okla State Univ, 61-68; res fel, Dept Theoret Physics, Univ Warwick, Eng, 68-69; sr mem tech staff, Booz Allen Appl Res, 69-70; sr mem tech staff & analyst, Comput Sci Corp, 70-77; sect mgr data qual assurance, NASA Landsat Proj, Comput Sci Technicolor Assocs, Goddard Space Flight Ctr, 77-79, sr staff scientist, Image Processing Oper, 79-82; sr staff scientist, sci & applications opers, 82-84, SR PRIN ENGR & SCIENTIST, OPERS STAFF, COMPUT SCI CORP, LANDSAT OPERS, 84- *Mem:* AAAS; Am Phys Soc; NY Acad Sci; Inst Elec & Electronics Engrs; Soc Photo-Optical Instrumentation Engrs; Int Remote Sensing Soc; Int Soc Photogram & Remote Sensing; Am Soc Photogram & Remote Sensing. *Res:* Quantum field theory; instabilities and transport properties in solids; mathematical models; systems analysis; analysis of electro optical remote sensors; computer simulations of solar cells, injection lasers, and other devices; digital image processing, future instruments and date processing for remote sensing; mathematical models for biochemical systems. *Mailing Add:* 5811 Barwood Pl Columbia MD 21044

CLARK, BRUCE R, b Pittsburgh, Pa, June 17, 41; m 67; c 2. GEOLOGY. *Educ:* Yale Univ, BS, 63; Stanford Univ, PhD(geol), 68. *Prof Exp:* Res assoc geol, Stanford Univ, 67-68; asst prof, 68-73, assoc prof geol, Univ Mich, Ann Arbor, 73-77; dir rock mech, 77-85, PRES, LEIGHTON & ASSOCS, 86- *Concurrent Pos:* Vis lectr, Monash Univ, Australia, 74-75, Univ Calif, Irvine, 90; mem, US Nat Comt on Rock Mech. *Mem:* Fel Geol Soc Am; Am Geophys Union; Int Soc Rock Mech; Asn Eng Geologist; Seismol Soc Am; Sigma Xi. *Res:* Structural geology; rock mechanics; landslide and earthquake hazard evaluation; earthquake prediction, in situ stress measurements in rock. *Mailing Add:* Leighton and Assocs 2121 Alton Pkwy Irvine CA 92714

CLARK, BURR, JR, b Howell, Mich, Jan 20, 24; m 47; c 3. AGRICULTURAL BIOCHEMISTRY. *Educ:* Mich State Univ, BS, 52; Univ NH, MS, 60; WVa Univ, PhD(agr biochem), 66. *Prof Exp:* Self-employed in agr, 47-49 & 52-56; from asst ed to assoc ed, 63-69, SR ED BIOCHEM, CHEM ABSTRACTS SERV, 69- *Mem:* AAAS; Am Chem Soc. *Res:* Volatile fatty acids metabolism in ruminants and forage quality in relation to volatile fatty acids metabolism in ruminants. *Mailing Add:* 200 Larrimer Ave Worthington OH 43085

CLARK, C(HARLES) C(ANFIELD), b Paterson, NJ, Aug 19, 28; m 59; c 2. METALLURGY. *Educ:* Rensselaer Polytech Inst, BMetE, 50. *Prof Exp:* Res metallurgist, Res Lab, Int Nickel Co, 50-57, metallurgist, 57-61, appln engr aerospace & power industs, 61-69, nickel alloy group mgr prod develop, 70-72; proj mgr, Inco, Inc, 72-75; mgr high temperature alloy develop, Climax Molybdenum Co, 75-80; mgr, Market Develop, Amax Tungsten, 80-87; MGR BUS PLANNING, CLIMAX METALS CO, 87- *Mem:* Am Soc Metals; Am Inst Mining Metall & Petrol Engrs; Am Soc Testing & Mat; Am Soc Mech Engrs. *Res:* Properties and uses of alloy and stainless steels; research and development of nickel base alloys; development of superalloys and chromium-molybdenum steels for energy conversion, petroleum, automotive and aerospace industries; manage tungsten research and development program. *Mailing Add:* Climax Metals Co Box 1700 Greenwich CT 06836-1700

CLARK, C ELMER, b Tooele, Utah, Mar 5, 21; m 51; c 4. PHYSIOLOGY, BIOCHEMISTRY. *Educ:* Utah State Univ, BS, 50; Univ Md, MS, 60, PhD(poultry physiol), 62. *Prof Exp:* Asst prof poultry sci, Utah State Univ, 52-57 & Univ Md, 57-61; assoc prof, Utah State Univ, 62-70, asst dir, Agr Exp Sta, 70-75, Assoc Dir, 75-80, prof animal, dairy & vet sci, 70-88; RETIRED. *Mem:* AAAS; Poultry Sci Asn; World Poultry Sci Asn; Soc Exp Biol & Med. *Res:* Neurohumoral factors in ovulation in chickens; environmental-physiology relationships in the avian species. *Mailing Add:* 228 E 970 N Logan UT 84321

CLARK, C SCOTT, b Rochester, NY, Feb 1, 38; m; c 2. PUBLIC HEALTH, ENVIRONMENTAL HEALTH ENGINEERING. *Educ:* Antioch Col, BS, 61; Johns Hopkins Univ, MS, 63, PhD(eng sci), 65. *Prof Exp:* Sanit engr, Ohio River Valley Water Sanit Comn, 65-67; sr res assoc, 67-70, from asst prof to assoc prof, 71-85, PROF ENVIRON HEALTH, UNIV CINCINNATI, 85- *Concurrent Pos:* Consult, var pub & pvt entities; chmn, Safety & Occup Health Comt, Water Pollution Control Fedn. *Mem:* Am Pub Health Asn; Am Water Works Asn; Water Pollution Control Fedn. *Res:* Epidemiologic-Serologic study of health risks of wastewater exposure; sources of lead in pediatric lead absorption; hazardous waste exposures; soil lead abatement. *Mailing Add:* Dept Environ Health, Univ Ctr Kettering Lab, 3223 Eden Ave Cincinnati OH 45267-0056

CLARK, CARL CYRUS, b Manila, Philippines, Apr 23, 24; US citizen; m 47; c 4. BIOPHYSICS. *Educ:* Worcester Polytech Inst, BS, 44; Columbia Univ, PhD(zool), 50. *Prof Exp:* Res assoc physiol & infrared spectrophotom, Med Col, Cornell Univ, 47-51; asst prof zool, Univ Ill, 51-55; head biophys div, Naval Aviation Med Acceleration Lab, 55-61; mgr, Life Sci Dept, Martin Co, 61-66; assoc chief, Sci & Technol Div, Libr of Cong, 66-68; chief, Task Group on Indust Self-Regulation, Nat Comn Prod Safety, 68-70; staff consult prod safety, Prod Eval Technol Div, Nat Bur Standards, 70-72; head dept life sci, Worcester Polytech Inst, 72-74; exec dir, comn advan pub interest orgn, dir, Health Satellite Proj & Community Health Resources Proj, Monsour Med Found, 74-77; occupant packaging staff, Nat Hwy Traffic Safety Admin, 77-90; PRES, SAFETY SYSTS CO, 91- *Concurrent Pos:* From assoc physiol to asst prof, Sch Med, Univ Pa, 55-61; pres, Safety Systs Co, 68-; lectr, Safety Prog, Univ Southern Calif Eastern Off, 78-80 & Systs Safety & Accident Reconstruct, George Wash Univ, 81- *Mem:* AAAS; Systs Safety Soc; Am Inst Aeronaut & Astronaut; Aerospace Med Asn; Human Factors Soc. *Res:* Infrared spectrophotometry and x-ray diffraction studies of biochemicals; microbiospectrophotometry; human centrifuge dynamic flight simulation; airbag restraint development; flight physiology; auto and home safety; information systems and satellite teleconferences; public interest organizations. *Mailing Add:* Safety Systs Co 23 Seminole Ave Baltimore MD 21228

CLARK, CARL HERITAGE, b Los Angeles, Calif, Nov 18, 25; m 48; c 1. PHARMACOLOGY. *Educ:* State Col Wash, BS & DVM, 47; Ohio State Univ, MS, 49, PhD(physiol), 53. *Prof Exp:* Asst, State Col Wash, 45-47; instr physiol, Ohio State Univ, 47-53; assoc head, Dept Animal Dis Res, 60-67, PROF PHYSIOL & PHARMACOL & HEAD DEPT, SCH VET MED, AUBURN UNIV, 53- *Mem:* Am Vet Med Asn; Am Soc Vet Physiologists & Pharmacologists (pres, 61). *Res:* Pharmacology of antibiotics in animals; pharmacology of body fluids and fluid therapy. *Mailing Add:* 700 Kuderna Acres Auburn AL 36830

CLARK, CHARLES AUSTIN, b Owego, NY, Dec 18, 15; m 37; c 4. ORGANIC CHEMISTRY. *Educ:* Cornell Univ, BS, 37. *Prof Exp:* Bacteriologist, NY State Dept Health Labs, 37-42; org res chemist, GAF Corp, 42-50, mgr, Org Prep Lab Unit, Ansco Div, 50-80; CONSULT, 80- *Mem:* Am Chem Soc; Soc Photog Scientists & Engrs; Am Inst Chemists. *Res:* Optical photographic sensitizing dyes photographic color formers, stabilizers and related intermediates. *Mailing Add:* 14 Westwood Ct Binghamton NY 13905

CLARK, CHARLES CHRISTOPHER, b Erie, Pa, Feb 17, 43. BIOCHEMISTRY. *Educ:* Gannon Col, BA, 65; Northwestern Univ, PhD(biochem), 70. *Prof Exp:* Asst prof, 72-77, ASSOC PROF BIOCHEM, MED SCH, UNIV PA, 77-, ORTHOP SURG, ASSOC PROF RES. *Concurrent Pos:* NIH grant, Univ Wash, 70-72 & res career develop award, 75-80. *Mem:* Soc Complex Carbohydrates. *Res:* Structure and biosynthesis of connective tissue macromolecules with particular emphasis on interstitial collagens, and basement membrane collagen and noncollagen glycoprotein(s). *Mailing Add:* Dept Orthop Surg Univ Pa 424 Med Educ Bldg Philadelphia PA 19104-6081

CLARK, CHARLES EDWARD, b Concord, NC, June 5, 23; m 46; c 2. TOOLING & METHODS, MANUFACTURING PROCESS ENGINEERING. *Educ:* NC State Univ, BS, 51. *Prof Exp:* Engr, United Coop, 51-52; mgr tooling & methods, Perfecting Serv Co, 52-60; develop support engr, Atomic Energy Div, Bendix Corp, 60-62; mgr engr lab, Int Staple & Mach, 62-64; plant engr, Charlotte Pipe & Foundry, 64-75; mgr mfg serv, Crompton & Knowles Corp, 75-77; mach design engr, Norton Co, 77-83 & SIEBE North Inc, 83-88; RETIRED. *Concurrent Pos:* Consult engr, Clark Develop Serv & Siebe North Inc, 88- *Mem:* Nat Soc Prof Engrs; Nat Soc Mfg Engrs. *Res:* Construction and start-up of automated glove making facility. *Mailing Add:* 2253 Sharon Rd Charlotte NC 28207

CLARK, CHARLES EDWARD, b Cuba, NY, Dec 30, 23; m 45; c 2. INDUSTRIAL ENGINEERING CONSULTING, HEALTH SYSTEMS ENGINEERING CONSULTING. *Educ:* Univ Ala, BS, 50. *Prof Exp:* Indust engr, BF Goodrich, Tuscaloosa & Los Angeles, 49-53, Kirkhill Rubber Co, Brea, Calif, 53-55; consult indust engr, R R Wilson Consult, Los Angeles, 55-57, Gulf States Paper Co, Tuscaloosa, 57-59, Ernst & Ernst/Ernst & Whinney-Bham, 59-84; CONSULT INDUST ENGR, HEALTH SYSTS, ENG & GEN MGT, C E CLARK & ASSOCS, 84- *Mem:* Inst Indust Engrs; Nat Soc Prof Engrs; Health Care Info & Mgt Syst Soc; Nat Soc Prof Engrs. *Mailing Add:* Eight Norris Circle Cottondale AL 35453

CLARK, CHARLES KITTREDGE, b Berkeley, Calif, Oct 15, 06; m 35; c 2. NATURAL PRODUCTS CHEMISTRY. *Educ:* Stanford Univ, AB, 28, MA, 29; Univ Fla, PhD(chem), 40. *Prof Exp:* Res chemist, Hercules Powder Co, 30-31; res chemist petrochems, Shell Develop Co, 32; res chemist naval stores, USDA, 33-37; res chemist tall oil, Quaker Chem Prod Corp, 40-42; res chemist terpenes, Naval Stores Div, Glidden Co, 43-45; res chemist wood prod, Weyerhaeuser Timber Co, 46-47; res chemist naval stores, Crosby Chem Inc, 48-52; res chemist wood prod, Crossett Co, 53-62; res chemist chem prod div, Union Camp Corp, 62-73; RETIRED. *Mem:* Am Chem Soc. *Res:* Rosin; terpenes and tall oil products. *Mailing Add:* 231 Andover Dr Savannah GA 31405

CLARK, CHARLES LESTER, b San Jose, Calif, Nov 17, 17; m 40; c 3. MATHEMATICS. *Educ:* Stanford Univ, AB, 39, MA, 40; Univ Va, PhD(math), 44. *Prof Exp:* Asst math, Stanford Univ, 39-40; instr, Univ Va, 42-44; from asst prof to prof, Ore State Col, 44-57; dir comput ctr, 61-71, dir inst res, 64-71, head dept, 57-64 & 71-77, PROF MATH, 57-80, EMER PROF & HEAD DEPT MATH, CALIF STATE UNIV, LOS ANGELES, 80- *Concurrent Pos:* Vis prof, Univ Va, 55-56; consult to educ, indust & legislative groups, 58- *Mem:* Sigma Xi; Am Math Soc; Math Asn Am; Asn Comput Mach. *Res:* Topology; analysis; arc reversing transformations; computer science; privacy and security. *Mailing Add:* PO Box 1006 Twain Harte CA 95383

CLARK, CHARLES MALCOLM, JR, b Greensburg, Ind, Mar 12, 38; m 63; c 2. MEDICINE. *Educ:* Ind Univ, AB, 60, MD, 63. *Prof Exp:* Intern, St Vincent's Hosp, Indianapolis, Ind, 63-64; resident internal med, Ind Univ, 64-65; staff assoc metab res, Nat Inst Arthritis & Metab Dis, 67-69; from asst prof to assoc prof med & pharmacol, Med Sch, Ind Univ, Indianapolis, 69-75; res & educ assoc, 69-71, clin investr endocrinol & metab, 71-74, assoc chief staff educ, 74-78, CHIEF DIABETES SERV, VET ADMIN HOSP, 78-; PROF MED & PHARMACOL, MED SCH, IND UNIV, INDIANAPOLIS, 76-, PROF MED, PHARMACOL & TOXICOL. *Concurrent Pos:* Clin fel diabetes, Joslin Clin, Boston, Mass, 65-66, NIH fel, Joslin Res Lab, 66-67; fel med, Peter Bent Brigham Hosp, Boston, 66-67; Robert Woods Johnson health policy fel, 75-76; dir, Diabetes Res & Training Ctr, 77- *Mem:* AMA; Am Diabetes Asn; Endocrine Soc; fel Am Col Physicians; Am Fedn Clin Res. *Res:* Metabolic control mechanisms; regulation of intermediary metabolism in fetal development. *Mailing Add:* Vet Admin 1481 W Tenth St Indianapolis IN 46202

CLARK, CHARLES RICHARD, b Burbank, Calif, Jan 22, 47; m 72; c 2. TOXICOLOGY. *Educ:* Univ Calif, Davis, BS, 73, PhD(toxicol), 77. *Prof Exp:* Indust & genetic toxicol, risk assessment, petrol hydrocarbon toxicol, Inhalation Toxicol Res Inst, 77-82; MGR TOXICOL & PROD SAFETY, UNOCAL CORP, 82- *Concurrent Pos:* Toxicol comt, Am Petrol Inst; dipl, Am Bd Toxicol, 81. *Mem:* Soc Toxicol; Environ Mutagen Soc; Soc Environ Toxicol Chem. *Res:* Toxicological evaluation of materials used in solar energy; mutagenicity evaluation of coal combustion effluents and diesel exhaust emissions; toxicology of shale and petroleum derived distillates. *Mailing Add:* Unocal Corp 1201 W Fifth St PO Box 7600 Los Angeles CA 90051

CLARK, CHARLES WINTHROP, b Minneapolis, Minn, Sept 30, 52; m 74. RADIATION PHYSICS. *Educ:* Western Wash State Col, BA, 74; Univ Chicago, MS, 76, PhD(physics), 79. *Prof Exp:* Res assoc, Univ Chicago, 79; jr res assoc, Daresbury Lab, England, 79-81; post doctoral res assoc, Nat Res Coun, Nat Bureau Standards, 81-83, PHYSICIST, NAT BUREAU STANDARDS, 84-; ADJ PROF, INST PHYS SCI TECH, UNIV MD, 91- *Concurrent Pos:* Consult, Princeton Plasma Physics Lab, 84-89; mem, Nat Res Coun Comt Line Spectra of Elements, Nat Acad Sci, 87-; vis fel, Australian Nat Univ, Canberra, Australia, 86; topical ed, Atomic Spectros, J Optical Soc Am Bd, 87; mem, Exec Comt, Div Atomic Molecular Optical Physics, Am Physics Soc, 90; chief, Electron & Optical Physics Div, Nat Inst Standards & Technol, 90- *Mem:* Am Phys Soc; Optical Soc Am. *Res:* Theoretical atomic, molecular, and optical physics. *Mailing Add:* A251 Physics Bldg Nat Inst Standards & Technol Gaithersburg MD 20899

CLARK, CHESTER WILLIAM, b San Francisco, Calif, July 18, 06; m 30; c 2. PHYSICS. *Educ:* Univ Calif, BS, 27, MS, 29; State Univ Leiden, PhD(physics), 35. *Prof Exp:* Res chemist, Standard Oil Co, Calif, 29-33; instr chem, Univ Calif, 35-37 & San Francisco City Col, 37-41; res assoc, Johns Hopkins Univ, 46-47; low temp consult & physicist, Naval Res Lab, US Navy, Washington, DC 47, sci asst to dir, Ballistic Res Lab, Aberdeen Proving Ground, 47-51, chief res & develop div & dep comdr, Picatinny Arsenal, 51-54, chief ord res & develop, 55-61, dir army res, 61-63, Commanding Gen, US Army-Japan, 63-65; vpres res, Res Triangle Inst, NC, 65-71; special sci & technol asst to ambassador, Am Embassy, Taiwan, 73-75; corp secy & admin dir, Microelectronics Ctr, NC, 80-81; CONSULT, 81- *Concurrent Pos:* Am Comnr, Joint US-Rep China Comn Rural Reconstruction, Taipei, 73-75. *Mem:* AAAS; Am Phys Soc. *Res:* Low temperature specific heats of inorganic substances; superconductivity of metals and alloys; preparation of interstitial alloys; attainment of temperatures below 1K. *Mailing Add:* 3120 Carol Woods Chapel Hill NC 27514

CLARK, CHRISTOPHER ALAN, b Geneva, NY, Jan 27, 49; m 71; c 2. PLANT PATHOLOGY, AGRICULTURE. *Educ:* Cornell Univ, BS, 70, MS, 73, PhD(plant path), 76. *Prof Exp:* Res assoc, NC State Univ, 76-77; from asst prof to assoc prof, 77-87, PROF PLANT PATH, LA STATE UNIV, 87- *Mem:* Am Phytopath Soc; Am Soc Microbiol. *Res:* Ecology of soil-borne plant pathogens; biological control of plant pathogens; post-harvest pathology; sweet potato diseases; bacterial plant pathogens. *Mailing Add:* Dept Plant Path La State Univ Baton Rouge LA 70803-1720

CLARK, CLARENCE FLOYD, b Briceton, Ohio, May 26, 12; m 34; c 1. ZOOLOGY. *Educ:* Miami Univ, BS, 34; Ohio State Univ, MS, 42. *Prof Exp:* Teacher, High Sch, Ohio, 35-37; fish mgt agent, Ohio Div Wildlife, 37-57, asst supvr fish mgt, 57-60, supvr fish invest, 60-63, asst supvr fish mgt, 63-69; asst supvr fisheries res, 69-71, PROF RES & FISH CULT, SCH NATURAL RESOURCES, OHIO STATE UNIV, 71- *Concurrent Pos:* Consult, Environ Consult, 72-75 & Dept Natural Resources, Ohio Div Wildlife, 74-75; consult, environ fisheries & freshwater mussels probs, 72-80; pres, N Cent Asn Am Fish Soc. *Mem:* Fel Am Inst Fishery Res Biol; Am Fisheries Soc; Int Acad Fishery Sci; Am Soc Ichthyologists & Herpetologists; Sigma Xi. *Res:* Management and propagation of northern pike, minnows, muskelluge, creek chubs; status of freshwater naiads in relation to environmental impacts from man made changes in the environment. *Mailing Add:* 2625 E Southern C 214 Tempe AZ 85282

CLARK, CLAYTON, b Hyde Park, Utah, Mar 9, 12; m 33; c 2. ELECTRICAL ENGINEERING. *Educ:* Utah State Univ, BS, 33; Stanford Univ, EE, 47, PhD(elec eng), 58. *Prof Exp:* Instr elec eng, Utah State Univ, 37-40; res engr, Douglas Aircraft, 47; prof elec eng, 48-77, dir eng exp sta & dir ctr res aeronomy, 64-77, EMER PROF ELEC ENG, UTAH STATE UNIV, 77- *Concurrent Pos:* Mem comn III, Int Sci Radio Union. *Honors & Awards:* Centennial Medal, Inst Elec & Electronics Engrs, 84. *Mem:* Am Soc Eng Educ; Am Geophys Union; fel Inst Elec & Electronics Engrs. *Res:* Radio propagation; microwaves. *Mailing Add:* 798 N 1500 E Logan UT 84321

CLARK, CLIFTON BOB, b Ft Smith, Ark, July 8, 27; m 50; c 3. SOLID STATE PHYSICS. *Educ:* Univ Ark, BA, 49, MA, 51; Univ Md, PhD(physics), 57. *Prof Exp:* Asst prof sci & math, Florence State Teachers Col, Ala, 50-51; asst prof physics, US Naval Acad, 51-55; physicist, US Naval Res Lab, 55-56; assoc prof physics, US Naval Acad, 56-57; from assoc prof to prof, Southern Methodist Univ, 57-65, chmn dept, 62-65; head dept, 65-75, PROF PHYSICS, UNIV NC, GREENSBORO, 65- *Concurrent Pos:* Vis prof physics, Fla State Univ, 75-76. *Mem:* Am Asn Physics Teachers; Am Phys Soc; Sigma Xi. *Res:* Electron-lattice interactions in metals; lattice dynamics. *Mailing Add:* Dept Physics & Astron Univ NC Greensboro NC 27412-5001

CLARK, COLIN WHITCOMB, b Vancouver, Can, June 18, 31; m 55; c 3. RESOURCE ECONOMICS, BEHAVIORAL ECOLOGY. *Educ:* Univ BC, BA, 53; Univ Wash, PhD(math), 58. *Prof Exp:* Instr math, Univ Calif, Berkeley, 58-60; from asst prof to assoc prof, 60-69, PROF MATH, UNIV BC, 69- *Concurrent Pos:* Investr, USAF Off Sci Res, 60-67; vis res scholar, Univ Calif, Berkeley, 65-66; vis res prof, NMex State Univ, 70-71 & Cornell Univ, 87; vis scientist, Commonwealth Sci & Indust Res Orgn, Australia, 75-76; Killam res fel, Univ BC, 75-76 & 81-82; prin investr, US Nat Oceanic & Atmospheric Admin, Nat Marine Fisheries Serv, 77-78; regent's lectr, Univ Calif, Davis, 86. *Honors & Awards:* Jacob Biely Res Prize, Univ BC, 77. *Mem:* Soc Indust & Appl Math; Can Appl Math Soc (pres, 81-84); Asn Environ & Resource Economists; Can Soc Theoret Biol; Res Modeling Asn (pres, 88-90); Soc Math Biol; fel Royal Soc Can. *Res:* Spectral theory of partial differential equations; mathematical models in renewable resource economics; behavioral ecology. *Mailing Add:* 9531 Finn Rd Richmond BC V7A 2L4 Can

CLARK, CONNIE, T-CELL MEDIATED IMMUNITY. *Educ:* Univ Minn, PhD(microbiol), 67. *Prof Exp:* RES MICROBIOLOGIST, VET ADMIN MED CTR, 72- *Mailing Add:* Med Res Div Bldg 31 Rm 308 Vet Admin Med Ctr Minneapolis MN 55417

CLARK, CROSMAN JAY, b Jackson, Mich, Mar 6, 25; m 45; c 3. COMPUTER SCIENCES, ELECTROMAGNETISM. *Educ:* Okla State Univ, BA, 46, MS, 48, PhD(math), 53. *Prof Exp:* Res engr, Curtiss-Wright Corp, 48; instr elec eng, Ohio State Univ, 49; instr & res asst, Okla State Univ, 49-52; mathematician, Stanolind Gas & Oil Co, 51; res mathematician, Continental Oil Co, 53-56; mathematician, Lockheed Missile & Space Corp, 56-61; eng specialist, Sylvania Electronic Defense Lab, Gen Tel & Electronics Corp, 61-64; staff scientist, Apparatus Res Dept, Tex Instruments Inc, Dallas, 64-66; dir advan studies, Northrop Corp, Calif, 66-70; vpres res & develop, Underwater Sci, Inc, Calif, 70-72; prof, 73-88, EMER PROF CYBERNETIC SYST, MGT SCI & INFO RESOURCE MGT, SAN JOSE STATE UNIV, 88-; DIR, INTERDISCIPLINARY SYSTS INST, 78- *Concurrent Pos:* Instr, Okla State Univ, 46-48 & 50-53, Foothill Col, 61-63 & Univ Santa Clara, 64; lectr, Calif State Col, Hayward, 71-72; consult, Nat Endowment Arts, 72-75; pres, Intersci Systs, Inc, 72-; lectr, Univ Calif, Santa Cruz, 78-80; lectr, MBA Info Systs Prog, Col Notre Dame, 89- *Mem:* Environ Design Res Asn; Soc Gen Systs Res; Math Asn Am; Soc Indust & Appl Math; Am Inst Decision Sci. *Res:* Topology; set theory; logic; wave theory; automata; operations research; stochastic processes; decision theory; general systems theory; mathematical-statistical model building; behavioral cybernetics; business-management systems; information resource management; decision theory. *Mailing Add:* 15555 Old Ranch Rd Los Gatos CA 95030

CLARK, D L, RESEARCH ADMINISTRATION. *Prof Exp:* EXEC VPRES & GEN MGR, WESTRONICS. *Mailing Add:* PO Box 961003 2441 Northeast Pkwy Ft Worth TX 76161

CLARK, DALE ALLEN, b Munden, Kans, Sept 14, 22; m 48; c 2. BIOCHEMISTRY. *Educ:* Hastings Col, BA, 44; Univ Colo, MA, 47; Univ Utah, PhD(biochem), 50. *Prof Exp:* Asst prof biochem, Sch Med, Univ Okla, 50-52, assoc prof, 53-54; res biochemist, Vet Admin Hosp, Dallas, 54-59; BIOCHEMIST, CLIN SCI DIV, DET-4 ARMSTRONG LAB, BROOKS AFB, 59- *Concurrent Pos:* Grad res prof biochem, Trinity Univ, San Antonio, 64-68; adj prof, Univ Tex Health Sci Ctr, 68-72; adj res prof, Tex A&M Univ, 69-70; proj officer, Cardiovasc Dis Follow-up Study, West Point. *Mem:* Fel AAAS; Am Chem Soc; Aerospace Med Asn; Soc Exp Biol & Med. *Res:* Atherosclerosis; steroid and sterol metabolism. *Mailing Add:* Clin Sci Div DET-4 Armstrong Lab Brooks AFB TX 78235-5301

CLARK, DAVID A, b St Thomas, Ont, Can, Apr 26, 43; m 66, 86; c 3. CELL IMMUNOLOGY. *Educ:* Univ Western Ont, MD, 67; Univ Toronto, PhD(immunol), 77; Royal Col Phys, 74. *Prof Exp:* From asst prof to assoc prof, 76-86, PROF MED, OBSTET & GYNECOL, MCMASTER UNIV, 86- *Concurrent Pos:* MRC fel & scientist. *Mem:* Am Soc Immunol Reproduction (vpres 87-88); Int Soc Immunol Reproduction; Am Asn Immunologists. *Res:* Immunology of reproduction; immunotoxicology. *Mailing Add:* Molec Virol & Immunol Prog/Reproductive Biol McMaster Univ Rm 3V39 1200 Main St W Hamilton ON L8N 3Z5 Can

CLARK, DAVID BARRETT, b Glen Ellyn, Ill, Nov 1, 13; m 48; c 2. NEUROLOGY, NEUROPATHOLOGY. *Educ:* Univ Chicago, PhD(neuroanat), 40, MD, 46. *Prof Exp:* Asst anat, Univ Chicago, 38-47; intern med, Johns Hopkins Hosp, 47-48, asst resident neurol, 48-49; Fulbright lectr neurol, Nat Hosp, Queen Sq, Eng, 50-51; from asst prof to assoc prof neurol & pediat, Johns Hopkins Hosp, 51-65; chmn dept, 65-79, prof, 65-84, EMER PROF NEUROL, UNIV KY, 84-; CHIEF, NEUROL SERV, VET ADMIN CTR, LEXINGTON, KY, 79- *Concurrent Pos:* Fel neurol, Johns Hopkins Hosp, 49-50; attend neurologist, Baltimore City Hosp, 51-59, Johns Hopkins Hosp, 51-65 & Rosewood State Training Sch, 52-65; adv, Epilepsy Found, 66-; Teale lectr, Royal Col Physicians, 67-; mem field study sect, Nat Inst Neurol Dis & Blindness, 58-60, prog proj study sect, 60-64, neurol res & training sect, 65-69 & residency rev comt, 67-74; dir, Am Bd Psychiat & Neurol, 67-79, emer dir, 79-; guest prof, Royal Alexandra Hosp Children, Sydney, Australia, 73 & 79; vis prof, Royal Soc Med, London, Eng, 85. *Honors & Awards:* Teal Lectr, Royal Col Physicians, London, Eng, 67; Sir

Leonard Parsons Lectr, Birmingham, Eng, 73; Litchfield Lectr in residence, Oxford Univ, Eng, 76; Douglas Reye Mem Lectr, Int Cong Child Neurol, Sydney, Australia, 79; MacKeith Lectr, Brit Soc Child Neurologists, London, Eng, 80. *Mem:* Am Acad Neurol; Am Neurol Asn; Am Asn Neuropath; AMA; Royal Soc Med. *Res:* Pediatric neurology; pathology of cerebral birth injuries; developmental defects of the central nervous system. *Mailing Add:* Vet Admin Med Ctr Neurol 127 Lexington KY 40511

CLARK, DAVID DELANO, b Austin, Tex, Feb 10, 24; m 49; c 3. NUCLEAR PHYSICS, APPLIED NUCLEAR SPECTROSCOPY. *Educ:* Univ Calif, Berkeley, AB, 48, PhD(physics), 53. *Prof Exp:* Asst physics, Univ Calif, Berkeley, 48-51, physicist, 53; res assoc, Brookhaven Nat Lab, 53-55; from asst prof to assoc prof eng physics, 55-64, PROF APPL PHYSICS, CORNELL UNIV, 64-, DIR WARD LAB NUCLEAR ENG, 60- *Concurrent Pos:* Euratom fel, Italy, 62; Guggenheim fel, Niels Bohr Inst, Copenhagen, Denmark, 68-69; vis prof, Tech Univ Munich, Ger, 76; vis scientist, Brookhaven Nat Lab, 82; guest researcher, Nat Inst Standards & Technol, 90. *Mem:* Am Phys Soc; Am Nuclear Soc. *Res:* Nuclear structure physics, especially isomers; nuclear instrumentation applications of nuclear methods in non-nuclear research fields; reactor physics. *Mailing Add:* Ward Lab Cornell Univ Ithaca NY 14853

CLARK, DAVID EDWARD, b Marianna, Fla, July 3, 46; m 70. PROCESSING, CHARACTERIZATION TESTING. *Educ:* Univ Fla, BS, 69, MS, 70, PhD(mat sci & eng), 76. *Prof Exp:* Metallurgist, 3M Co, 70-72; grad res asst, Dept Mat Sci & Eng, Univ Fla, 72-76, postdoctoral fel, 76-77, sr postoctoral fel, 77-78, vis asst prof ceramics, 78-82, assoc prof mat sci & eng, 82-86, PROF, DEPT MAT SCI & ENG, UNIV FLA, 86- *Concurrent Pos:* Scientist & vis prof, Royal Inst Technol, Stockholm, Sweden, 82; ed, Mat Sci World, 83- *Mem:* Fel Am Ceramic Soc; Nat Inst Ceramic Engrs (pres); Mat Res Soc; Am Soc Metals. *Res:* Research in ceramics and ceramic processing, characterization and testing; emphasis on microwave processing. *Mailing Add:* Dept Mat Sci & Eng MAE 136 Univ Fla Gainesville FL 32611-2066

CLARK, DAVID ELLSWORTH, b Paso Robles, Calif, Nov 22, 22; m 47; c 2. ORGANIC CHEMISTRY, ORGANIC SYNTHESIS. *Educ:* Univ Redlands, BA, 47; Stanford Univ, MS, 48, PhD(org chem), 53. *Prof Exp:* Instr chem, Fresno State Col, 50-51 & 53-54, from asst prof to prof, 54-65; assoc acad planning, Chancellor's Off, Calif State Cols, 65-67; acad admin internship, Brown Univ, 67-68; assoc vpres acad affairs, 70-90, EMER PROF, CALIF STATE UNIV, FRESNO, 90- *Concurrent Pos:* NSF sic fac fel, Harvard Univ, 62-63. *Mem:* Am Chem Soc. *Res:* Organic synthesis; chemical therapeutics. *Mailing Add:* 1456 E Browning St Fresno CA 93726

CLARK, DAVID LEE, b Gallipolis, Ohio, Feb 17, 42; m 67; c 1. PHYSICS. *Educ:* Ohio State Univ, BS, 65; Univ Minn, PhD(physics), 77. *Prof Exp:* Res assoc, Williams Lab of Nuclear Physics, Univ Minn, 70-76; fel, Stanford Univ, 76-77, asst prof nuclear physics, Dept Physics, 77-80; ASST PROF PHYSICS, DEPT PHYSICS, UNIV ROCHESTER, 80- *Mem:* Am Phys Soc; Sigma Xi. *Res:* Experimental nuclear physics research with emphasis on weak interactions; properties nuclear states; heavy-ion reactions; application of lasers to nuclear physics. *Mailing Add:* Five S Pittsford Hill Lane Pittsford NY 14534

CLARK, DAVID LEE, b Detroit, Mich, Apr 7, 39; m 67; c 3. VESTIBULAR SYSTEM, EYE MOVEMENTS. *Educ:* Kalamazoo Col, BA, 62; Univ Okla, MS, 63; Mich State Univ, PhD(zool), 67. *Prof Exp:* ASSOC PROF ANAT, COL MED, OHIO STATE UNIV, 68- *Concurrent Pos:* Great Lakes Cols Asn teaching fel, Kenyon Col, 67-68. *Mem:* AAAS; Am Anat Asn. *Res:* Effects of vestibular stimulation therapy applied to developmentally delayed children (cerebral palsied, Down's syndrome, autistic, hyperactive); control of eye movements in children. *Mailing Add:* Dept Anat Ohio State Univ 1645 Neil Ave Columbus OH 43210

CLARK, DAVID LEIGH, b Albuquerque, NMex, June 15, 31; m 51; c 4. PALEONTOLOGY. *Educ:* Brigham Young Univ, BS, 53, MS, 54; Univ Iowa, PhD(geol), 57. *Prof Exp:* Geologist, Standard Oil Co, 54; asst geol, Columbia Univ, 54-55; asst, Univ Iowa, 55-57; asst prof, Southern Methodist Univ, 57-59; from asst prof to assoc prof, Brigham Young Univ, 59-63; assoc prof geol, Univ Wis-Madison, 63-68, prof geol & geophys, 68-74, chmn dept, 71-74, assoc dean natural sci, 86-91, W H TWENHOFEL PROF, UNIV WIS-MADISON, 74- *Concurrent Pos:* Sr Fulbright fel & vis prof, Univ Bonn, 65-66. *Mem:* Geol Soc Am; Am Asn Petrol Geologists; Paleont Soc; Soc Econ Paleontologists & Mineralogists. *Res:* Artic Ocean geology and paleoecology; Paleozoic and Mesozoic conodonts. *Mailing Add:* Dept Geol & Geophys Weeks Hall Univ Wis Madison WI 53706

CLARK, DAVID M, b Denver, Co, July 1, 47. TOPOLOGICAL REPRESENTATIONS OF ALGEBRAS & APPLICATIONS TO LOGIC. *Educ:* Emory Col, BA, 67; Emory Univ, PhD(math) 69. *Prof Exp:* Asst prof math, Ga Inst Technol, 69-70; from asst prof to assoc prof math & comput sci, 70-80, PROF MATH & COMPUT SCI, STATE UNIV NY, COL NEW PALTZ, 78-, CHMN DEPT MATH & COMPUT SCI, 85- *Concurrent Pos:* Woodrow Wilson grad fel, 67; Nat Sci Found grad fel, 67-69; vis res assoc, Humboldt res grant Kassel, Gesamthochschule, WGer, 77-78; Nat Sci & Eng Res Coun Can res fel, Univ Waterloo, Ont, 83. *Mem:* Am Math Soc; Int Neural Network Soc. *Res:* Applicatiom to logic. *Mailing Add:* Dept Math & Comput Sci State Univ NY Col New Paltz New Paltz NY 12561

CLARK, DAVID NEIL, b June 24, 53; m; c 2. HAZARDOUS WASTE CHEMISTRY. *Educ:* Clarkson Col Technol, BS, 75; Mass Inst Technol, PhD(inorg chem), 79. *Prof Exp:* Brigade chem officer & nuclear surety officer, 101st Airborne Div, US Army, 79-81, chief mgt support off & chem surety officer, Rocky Mountain Arsenal, 81-82, dep chief, Environ Div, 82-83; res chemist, ITT Res Inst, 83-87; assoc mgr, 87-88, proj mgr, Battelle Mem Inst, 88-89, CHEM SURETY MGR, BATTELLE, TOOELE OPERS, 90- *Mem:* AAAS; Am Chem Soc; Am Defense Preparedness Asn; Asn Chem Officers; Sigma Xi. *Res:* New methods for the decontamination and destruction of hazardous materials including chemical warfare agents and various industrial chemicals. *Mailing Add:* Battelle 11650 Stark Rd Tooele UT 64074

CLARK, DAVID SEDGEFIELD, b St Stephen, NB, Nov 13, 29; m 52; c 3. FOOD MICROBIOLOGY. *Educ:* McGill Univ, MS, 53, PhD(physiol), 57. *Prof Exp:* Lectr agr bact, MacDonald Col, McGill Univ, 53-57; tech sales, Buckman Labs Can, Ltd, 57-58; from asst res officer to assoc res officer, 58-70, sr res officer, Nat Res Coun Can, 70-76; chief, Div Microbial Res, 76-78, dir, Bur Microbiol Hazards, Health Protection Br, Dept Nat Health & Welfare, 78-85; RETIRED. *Concurrent Pos:* Secy-treas, Int Standing Comn Microbiol Specifications Foods, Int Asn Microbiol Socs, 63-80; hon mem expert panel on food microbiol & hyg, WHO, 70-85. *Honors & Awards:* W J Eva Award, Can Inst Food Sci & Technol; Citation Medal, Int Union Microbiologist Soc. *Mem:* Can Soc Microbiol; Can Inst Food Technologists (pres, 75-76). *Res:* Industrial fermentations; bacterial physiology; meat microbiology; microbiol methodology; effect of the gaseous environment on microorganisms. *Mailing Add:* One Kaymar Dr Ottawa ON K1J 7C8 Can

CLARK, DAVID THURMOND, b Topeka, Kans, Aug 3, 25; m 47; c 2. PARASITOLOGY, IMMUNOLOGY. *Educ:* Univ Nebr, BA, 49, MA, 51; Univ Ill, PhD, 55. *Prof Exp:* Asst zool, Univ Nebr, 49-51 & Univ Ill, 51-55; from instr to prof microbiol & pub health, Mich State Univ, 56-65, asst vpres res & develop, 65-69; staff assoc univ sci & develop, Instnl Rels, NSF, Washington, DC, 69-70; Dean Grad Studies & Res, 70-76, PROF BIOL, PORTLAND STATE UNIV, 76- *Concurrent Pos:* Mem subcomt prenatal & postnatal mortality in swine, Comt Animal Health, Nat Res Coun, 59- *Mem:* Am Soc Parasitologists; Am Micros Soc; NY Acad Sci; Soc Protozool. *Res:* Parasites of wildlife; physiological studies on nematodes; immunology of parasitic infections; science policy. *Mailing Add:* 2434 SW Sherwood Dr Portland OR 97201

CLARK, DENNIS RICHARD, b Palo Alto, Calif, Mar 13, 44. TOXICOLOGY, ANALYTICAL CHEMISTRY. *Educ:* Univ Ore, BA, 66; Stanford Univ, PhD(org chem), 70. *Prof Exp:* Fel chem, Stanford Univ, 70-71, fel pharmacol, 71-73; asst dir, Drug Assay Lab, 73-80, CLIN CHEMIST, STANFORD UNIV HOSP, 80-; VPRES, ANALYTICAL SOLUTIONS, INC, 86- *Mem:* Am Asn Clin Chem. *Res:* Development of methods for drug analysis in biological fluids; study of relationships between drug blood levels and clinical effects of drugs. *Mailing Add:* 568 Weddell Dr Suite 5 Sunnyvale CA 94089

CLARK, DONALD ELDON, b Watertown, SDak, Dec 24, 36; m 59; c 2. ENVIRONMENTAL SCIENCES, GEOCHEMISTRY. *Educ:* Univ SDak, AB, 61; Iowa State Univ, PhD(phys chem), 65. *Prof Exp:* Res asst, Ames Lab, AEC, 61-65; res chemist, Miami Valley Labs, Procter & Gamble Co, 65-66; from asst prof to assoc prof chem, Western Ill Univ, 66-74; sr scientist, Westinghouse Hanford Co, Richland, Wash, 74-80; sr chemist, Brookhaven Nat Lab, Upton, NY, 80-82; proj mgr, Battelle, Columbus, Ohio, 82-88, INT REP, EUROPE BATTELLE PAC, NORTHWEST LABS, 88- *Concurrent Pos:* AEC fac appointee, Battelle Pac Northwest Labs, 69-70 & Douglas United Nuclear, 72; vis staff mem, Los Alamos Sci Lab, Univ Calif, 71. *Mem:* Am Chem Soc; Am Nuclear Soc; AAAS; Am Soc Testing & Mat. *Res:* Geological disposal of high-level nuclear wastes; radiation effects on materials; movement of radionuclides in the geomedia; behavior of high saline solutions under high temperatures and pressure; geotechnical aspects of work management. *Mailing Add:* Westinghouse Hanford Co G6-08 PO Box 1970 Richland VA 99352-0539

CLARK, DONALD LYNDON, b Lyndon, Vt, Feb 17, 20; m 44; c 3. APPLIED PHYSICS, ELECTRICAL ENGINEERING. *Educ:* Univ Vt, BS, 43; Univ Rochester, PhD(physics), 52. *Prof Exp:* Elec engr, Stromberg Carlson Co, 43-46; res assoc physics, Univ Rochester, 52; staff mem & group leader, Defense Electronics, Lincoln Lab, Mass Inst Technol, 52-83; RETIRED. *Mem:* AAAS; Am Phys Soc; Am Asn Physics Teachers; Inst Elec & Electronics Engrs; Sigma Xi. *Res:* Radar and optical measurements of hyper-velocity vehicles outside and within the atmosphere, and their implications for strategic offense and defense systems. *Mailing Add:* PO Box 51 West Newbury VT 05085-0051

CLARK, DONALD RAY, JR, b Garrett, Ind, Jan 20, 40. ECOTOXICOLOGY. *Educ:* Univ Ill, BS, 61; Tex A&M Univ, MS, 64; Univ Kans, PhD(zool & ecol), 68. *Prof Exp:* Asst prof wildlife sci, Tex A&M Univ, 68-72; wildlife biologist res, US Fish & Wildlife Serv, Patuxent Wildlife Res Ctr, 72-80, proj leader, 75-80, supvry wildlife biologist & sect leader, 80-85, WILDLIFE BIOLOGIST RES, US FISH & WILDLIFE SERV, PATUXENT WILDLIFE RES CTR, 85- *Mem:* Sigma Xi; AAAS; Am Soc Mammalogists; Ecol Soc Am; Am Soc Ichthyologists & Herpetologists; Herpetologists' League. *Res:* Effects of environmental contaminants on wildlife population, especially mammals and reptiles. *Mailing Add:* Patuxent Wildlife Res Ctr Laurel MD 20708

CLARK, DOUGLAS NAPIER, b New York, NY, Jan 24, 44; m 68, 89. MATHEMATICAL ANALYSIS. *Educ:* Johns Hopkins Univ, AB, 64, PhD(math), 67. *Prof Exp:* Fel math, Univ Wis-Madison, 67-68; asst prof, Univ Calif, Los Angeles, 68-73; assoc prof, 73-77, PROF MATH, UNIV GA, 77- *Concurrent Pos:* Res assoc, Off Naval Res, Nat Res Coun, 67-68; vis prof, Univ Va, 79-80 & State Univ NY, Stony Brook, 81. *Mem:* Am Math Soc. *Res:* Operator theory and its applications to complex variables, chiefly Toeplitz and Hankel matrices and the study of invariant subspaces; analytic functions in polydiscs; interpolation problems. *Mailing Add:* Dept Math Univ Ga Athens GA 30602

CLARK, DUNCAN WILLIAM, b New York, NY, Aug 31, 10; m 43, 71; c 3. MEDICINE, GENERAL PREVENTIVE MEDICINE. *Educ:* Fordham Col, AB, 32; Long Island Col Med, MD, 36. *Prof Exp:* Intern, Brooklyn Hosp, 36-38; resident med, Kings County Hosp, 38-40; dir student health, Long Island Col Med, 41-49, instr med, 42-47, from asst dean to assoc dean, 43-48, asst prof med, actg chmn prev med & dean, 48-50; chmn dept, 51-78, prof, 51-82, EMER PROF PREV MED & COMMUNITY HEALTH, STATE UNIV NY HEALTH CTR, BROOKLYN, 82- *Concurrent Pos:* Fel, Yale Univ, 40-41; traveling fel, WHO, 52; Commonwealth Found fel, 61; vis prof,

Univ Birmingham, 61; consult, Health Serv Res Study Sect, US Pub Health Serv, 61-65 & 73-77, consult & chmn community health res training, 65-69, consult, Nat Ctr Health Serv Res & Develop, 70-81, mem tech adv group on med care effectiveness, 70-72, res scientist, Fel Rev Comt, 70-72; consult, Nat Acad Sci-Nat Res Coun, 65-68; alt deleg, AMA, 84-, deleg, 89- *Mem:* Am Col Prev Med; Am Col Physicians; Harvey Soc; Am Pub Health Asn; NY Acad Med (vpres, 76-78 & pres, 83-84). *Res:* Medical education; public health and medical care; preventive medicine. *Mailing Add:* Dept Prev Med State Univ NY Health Sci Ctr Brooklyn NY 11203

CLARK, EDWARD ALOYSIUS, b Jersey City, NJ, Jan 28, 34; m 55; c 5. THEORETICAL PHYSICS. *Educ:* Col Holy Cross, BS, 55; Fordham Univ, MS, 60, PhD(physics), 66. *Prof Exp:* From instr to assoc prof, 60-70, chmn dept, 66-70, asst to pres, 70-71, acad vpres, 74-75, PROF PHYSICS, LONG ISLAND UNIV, BROOKLYN CTR, 70-, DEAN, COL LIB ARTS & SCI, 71- *Concurrent Pos:* Consult, State NY Dept Educ, 66 -; pres, Long Island Univ, Brooklyn Ctr, 75-86, sr vpres, 86- *Mem:* Am Phys Soc; Am Asn Physics Teachers. *Res:* Calculation of molecular vibration-rotation spectra. *Mailing Add:* Long Island Univ Brooklyn Ctr Brooklyn NY 11201

CLARK, EDWARD MAURICE, b Edinburgh, Scotland, Feb 16, 20; div; c 1. PLANT BREEDING. *Educ:* Univ Minn, BS, 49, MS, 55, PhD(plant breeding), 56. *Prof Exp:* From asst botanist to assoc botanist, Auburn Univ, 56-62, assoc prof bot, 62-89; RETIRED. *Mem:* AAAS; Genetics Soc Am; Am Genetics Asn; Am Phytopath Soc. *Res:* Diseases of forage legumes and grasses; tall fescue endophyte, acremorium coenophialum. *Mailing Add:* 120 Norwood Ave Auburn AL 36830

CLARK, EDWARD SHANNON, b Schenevus, NY, Apr 26, 30. POLYMER CHEMISTRY. *Educ:* Union Col, NY, BS, 51; Univ Calif, PhD(chem), 56. *Prof Exp:* Res phys chemist, E I du Pont de Nemours & Co, Inc, 55-62, sr res chemist, 62-72; PROF, DEPT CHEM & METALL & POLYMER ENG, UNIV TENN, KNOXVILLE, 72- *Concurrent Pos:* Fulbright scholar, Aarhus Univ, 62-63. *Mem:* Fel Am Chem Soc; Am Crystallog Asn; Fel Am Phys Soc; Fel Soc Plastics Eng; Am Inst Chem Engr; Sigma Xi. *Res:* Structure-property relationships in polymers. *Mailing Add:* Dept Polymer Eng Univ Tenn Knoxville TN 37916

CLARK, ELOISE ELIZABETH, b Grundy, Va, Jan 20, 31. BIOCHEMISTRY, BIOPHYSICS. *Educ:* Mary Washington Col, BA, 51; Univ NC, PhD(zool), 58. *Hon Degrees:* DSc, King Col, 76. *Prof Exp:* Instr biol, Women's Col, Univ NC, Greensboro, 52-53; res asst physiol, Univ NC, Chapel Hill, 53-55; from instr to assoc prof biol, Columbia Univ, 59-69; prog dir, Develop Biol Prog, 69-70, prog dir, Biophysic Prog, 70-73, sect head, Molecular Biol Sect, 71-73; div dir, Biol & Med Sci Div, 73-75, dep asst dir, 75-76, asst dir, Biol, Behav & Soc Sci Directorate, NSF, 76-83; VPRES ACAD AFFAIRS & PROF BIOL SCI, BOWLING GREEN STATE UNIV, 83- *Concurrent Pos:* Coun provosts, Ohio Int Univ, 83-; panel informal sci ed, NSF, 85-, panel minority res ctr excellence, 87-; coun, Nat Adv Res Resources, NIH, 87- *Mem:* AAAS; Am Soc Cell Biol; Soc Gen Physiologists (secy, 65-67); Biophys Soc; Sigma Xi; Marine Biol Lab. *Res:* Physical biochemistry of muscle protein and enzymes; science administration. *Mailing Add:* Acad Affairs Bowling Green State Univ Bowling Green OH 43403

CLARK, EUGENIE, b New York, NY, May 4, 22; div; c 4. ZOOLOGY. *Educ:* Hunter Col, BA, 42; NY Univ, MS, 46, PhD(zool), 50. *Prof Exp:* Asst ichthyol, Scripps Inst Oeanog, 46-47 & NY Zool Soc, 47-48; asst animal behavior, Am Mus Natural Hist, 48-49, res assoc, 50-66; exec dir marine biol, Cape Haze Marine Lab, 55-66; assoc prof biol, City Univ New York, 66-67; assoc prof, 69-73, PROF ZOOL, UNIV MD, COLLEGE PARK, 73- *Concurrent Pos:* AEC fel, 50; Fulbright scholar, Egypt, 51; Saxton fel & Breadloaf Writer's fel, 52; instr, Hunter Col, 54; independent res, behav of sharks, numerous int locations, 54-; mem bd trustees, Nat Parks & Conserv Asn, vpres, 75-83; bd dirs, Nat Aquarium, Washington, DC; emer dir, Mote Marine Lab, Sarasota, Fla; invited lectr, numerous int sci meetings & conf, 80- *Honors & Awards:* Dugan Award, Am Littoral Soc, 69; Cousteau Award, 73; Gold Medal, Soc Women Geogr, 75; David B Stone Medal, 81; John Stomeman Marine Environ Award,; Lowell Thomas Award, Explorers Club, 86- *Mem:* Am Soc Ichthyologists & Herpetologists; Soc Women Geogr; fel AAAS; Am Littoral Soc (vpres); Explorers Club. *Res:* Ichthyology; reproductive behavior of fishes; morphology and taxonomy of plectognath fishes; isolating mechanisms of poeciliid fishes; behavior of sharks; Red Sea fishes. *Mailing Add:* Dept Zool Univ Md College Park MD 20742

CLARK, EVELYN GENEVIEVE, b Brooksville, Ky, Jan 8, 22. MICROBIOLOGY. *Educ:* Georgetown Col, AB, 47; Univ Ky, MS, 53. *Prof Exp:* From instr to assoc prof biol, Georgetown Col, 47-85; RETIRED. *Concurrent Pos:* Consult microbiologist, Cent Baptist Hosp, Ky, 53-64; bacteriologist, Ford Mem Hosp, Ky, 67-71; RETIRED. *Mem:* AAAS; Nat Asn Biol Teachers; Am Soc Microbiol; Am Inst Biol Sci. *Res:* Clinical mycology. *Mailing Add:* 1086 Degaris Mill Rd Georgetown KY 40324

CLARK, EZEKAIL LOUIS, b Gomel, Russia, June 29, 12; US citizen; m 33; c 2. CHEMICAL ENGINEERING, FUEL TECHNOLOGY. *Educ:* Northeastern Univ, BSChE, 37. *Prof Exp:* Chemist, Mass State Dept Pub Health, 37-40; jr engr, Stone & Webster Eng Corp, 40-43; sr engr, Cities Serv Ref Corp, La, 43-45; chem engr, US Bur Mines, Pa & DC, 45-54; dir lab, Israel Mining & Industs, 54-56; pvt consult, 56-64; pres, Pressure Chem Co, 64-74; asst div dir, ERDA, 74-77 & Dept Energy, 77-78; INDEPENDENT CONSULT, 78- *Concurrent Pos:* Assoc prof, Univ Pittsburgh, 51-52 & Israel Inst Technol, 55-56; lectr, Cath Univ Am, 53-54 & Pa State Univ, 65-67. *Mem:* Fel Am Inst Chem Engrs; Am Chem Soc. *Res:* Pilot plant experimentation; high pressure technology; coal science; process development; waste disposal. *Mailing Add:* 4200 N Miller Rd Scottsdale AZ 85251

CLARK, FLORA MAE, b Houston Co, Ala, Nov 19, 33. GENETICS. *Educ:* Ala Col, BS, 60, MAT, 65; Univ Tenn, PhD(zool), 70. *Prof Exp:* Teacher biol, Rehobeth High Sch, Ala, 60-62 & Dependent Educ Group, US Army, France, 62-64; instr, Jacksonville State Univ, 65-67; res assoc genetics, Univ Ga, 70-71; assoc prof, 71-81, PROF BIOL, COLUMBUS COL, 81- *Concurrent Pos:* Consult, St Francis Hosp Lab, 75. *Mem:* AAAS; Am Inst Biol Sci. *Res:* Human chromosome abnormalities. *Mailing Add:* Dept Biol Columbus Col Columbus GA 31993

CLARK, FRANCIS JOHN, b Chicago, Ill, May 30, 33; m 81; c 5. NEUROPHYSIOLOGY. *Educ:* Northwestern Univ, BSEE, 56; Purdue Univ, MSEE, 57, PhD(elec eng), 65. *Prof Exp:* Engr, Cook Res Labs, Ill, 57-59; proj mgr, Advan Res Dept, Sunbeam Corp, 59-61; instr elec eng, Purdue Univ, West Lafayette, 64, asst prof, Sch Elec Eng & Dept Vet Anat, 64-68, assoc prof, 68-72; assoc prof, 72-77, PROF PHYSIOL & BIOPHYS, UNIV NEBR MED CTR, OMAHA, 77- *Concurrent Pos:* Nat Inst Neurol Dis & Stroke spec fel, Nobel Inst Neurophysiol, Karolinska Inst, Sweden, 69-71; fel, Lab Physiology, Magdalen Col, Oxford Univ, Eng, 79-80. *Mem:* Am Phys Soc; Soc Neurosci; fel Muscular Dystrophy Asn. *Res:* Mammalian nervous system. *Mailing Add:* Dept Physiol & Biophys Univ Nebr Med Ctr Omaha NE 68198-4575

CLARK, FRANK EUGENE, b St Louis, Mo, Sept 16, 19; m 48. MATHEMATICS. *Educ:* Dartmouth Univ, BA, 41; Duke Univ, MA, 46, PhD(math), 48. *Prof Exp:* Vis instr math, Duke Univ, 47-48; instr, Tulane Univ, 48-50; from asst prof to assoc prof, 50-61, chmn dept, Univ Col, 59-75, PROF MATH, RUTGERS UNIV, 61- *Concurrent Pos:* Vis prof, Univ Nairobi, Kenya, 75-76. *Mem:* Am Math Soc; Math Asn Am. *Res:* Combinatorial problems; linear programming. *Mailing Add:* 820 Shadowlawn Dr Westfield NJ 07090

CLARK, FRANK S, b San Mateo, Calif, Sept 27, 33; m; c 2. ORGANIC CHEMISTRY, TRIBOLOGY. *Educ:* Stanford Univ, BS, 55; Purdue Univ, PhD(org chem), 60. *Prof Exp:* Res chemist, 60-68, res specialist, 68-83, SR RES SPECIALIST, MONSANTO CO, 83- *Mem:* AAAS; Am Chem Soc; Soc Tribologists & Lubrication Engrs. *Res:* Development of high temperature gas turbine engine oils; organic synthesis; synthesis of synthetic lubricants; and hydraulic fluids, boundary lubrication, corrosion and wetting of synthetic lubricants; process development and or synthetic methods for salicylate esters, salicylic acid, aroma esters, asymmetric catalysts, vanillin, 5-aminosalicylic acid, chlorobenzenes and water purification. *Mailing Add:* Corp Res Dept Monsanto Co 500 Monsanto Ave Sauget IL 62206-1198

CLARK, GARY EDWIN, b Lees Summit, Mo, Jan 5, 39. NUCLEAR STRUCTURE. *Educ:* Park Col, BA, 61; Kans State Univ, MS, 66; Iowa State Univ, PhD(nuclear physics), 72. *Prof Exp:* Instr physics, Kans State Univ, 66-; asst prof, Cent Mo State Col, 66-68; teaching & res, Iowa State Univ, 69-72; staff officer, Nat Acad Sci, 72-85; INTERN, NAT BUR STANDARDS, 72- *Mem:* Am Phys Soc; AAAS; Sigma Xi. *Res:* Photnuclear physics; neutron detectors. *Mailing Add:* 7505 Democracy Blvd Apt-123 Bethesda MD 20817

CLARK, GEORGE, neurophysiology; deceased, see previous edition for last biography

CLARK, GEORGE ALFRED, JR, b Camden, NJ, May 6, 36; m 61; c 1. ORNITHOLOGY. *Educ:* Amherst Col, BA, 57; Yale Univ, PhD(biol), 64. *Prof Exp:* Res assoc zool, Univ Wash, 63-64, actg instr, 64-65; from asst prof to assoc prof zool, 65-82, PROF BIOL, UNIV CONN, 82- *Mem:* Soc Study Evolution; Am Ornith Union; Am Soc Naturalists; Cooper Ornith Soc; Wilson Ornith Soc. *Res:* Integumental structure, behavior and evolution of birds. *Mailing Add:* Biol Box U-43 Univ Conn Storrs CT 06269-3043

CLARK, GEORGE BROMLEY, mining engineering, for more information see previous edition

CLARK, GEORGE C(HARLES), b Battle Creek, Mich, June 26, 30. CHEMICAL ENGINEERING. *Educ:* Univ Mich, BSE, 52, MSE, 53, PhD, 60. *Prof Exp:* Engr reservoir eng res, 57-64, res group leader, 66-67, res assoc, 67-80, SR RES ASSOC, CONTINENTAL OIL CO, 81- *Mem:* AAAS; Am Inst Chem Engrs; Soc Petrol Engrs; Sigma Xi. *Res:* Application of mathematics and computers to reservoir engineering; solution of systems of linear and non-linear partial differential equations; enhanced oil recovery; personal computers & workstations. *Mailing Add:* 2309 E Hartford Ponca City OK 74604

CLARK, GEORGE RICHMOND, II, b Princeton, Maine, Mar 23, 38; m 61; c 3. PALEOBIOLOGY, NATURAL DISASTERS. *Educ:* Cornell Univ, AB, 61; Calif Inst Technol, MS, 66, PhD(geobiol), 69. *Prof Exp:* Asst prof geol, Univ NMex, 69-74, State Univ NY Col Geneseo, 74-76 & Edinboro State Col, Pa, 76-77; asst prof, 77-81, assoc prof, 81-86, PROF GEOL, KANS STATE UNIV, 86- *Mem:* AAAS; Am Soc Zoologists; Am Geophys Union; Geol Soc Am; Paleont Soc; Nat Asn Geol Teachers. *Res:* Growth lines; environmental variations in shell and skeletal morphology; invertebrate calcification and shell microstructure; paleopopulation dynamics; natural disasters; geological education. *Mailing Add:* Dept Geol Thompson Hall Kans State Univ Manhattan KS 66506-3201

CLARK, GEORGE WHIPPLE, b Evanston, Ill, Aug 31, 28; m 54; c 2. PHYSICS, X-RAY ASTRONOMY. *Educ:* Harvard Univ, AB, 49; Mass Inst Technol, PhD(physics), 52. *Prof Exp:* From instr to assoc prof, 52-65, PROF PHYSICS, MASS INST TECHNOL, 65- *Concurrent Pos:* Guggenheim fel & Fulbright res scholar, 63; bd dir, AURA, Inc, 78-; mem, Astron Surv Comt, Nat Acad Sci, 79-81, chmn, Panel High Energy Astrophysics; mem, Space Telescope Inst Coun; prin investr, Small Astron Satellite SAS-3 X-ray Observ, 72-82; mem, various comts, Nat Aeronaut & Space Admin & Nat Acad Sci. *Mem:* Nat Acad Sci; Am Astron Soc; Int Astron Union; Am Acad Arts & Sci; Am Phys Soc. *Res:* Cosmic rays; x-ray astronomy. *Mailing Add:* Dept Physics Mass Inst Technol Cambridge MA 02139

CLARK, GLEN W, b Newdale, Idaho, Apr 17, 31; m 51; c 5. PARASITOLOGY, PROTOZOOLOGY. *Educ:* Ricks Col, BS, 56; Utah State Univ, MS, 58; Univ Calif, Davis, PhD(zool), 62. *Prof Exp:* Instr life sci, Am River Jr Col, 62-64; from asst prof to assoc prof, 64-71, prof zool, 71-77, PROF BIOL, CENT WASH UNIV, 77- *Mem:* Soc Protozool; Wildlife Dis Asn. *Res:* Blood parasites of the class Aves; coccidial parasites of reptiles. *Mailing Add:* Dept Biol Cent Wash Univ Ellensburg WA 98926

CLARK, GORDON MEREDITH, b Akron, Ohio, Nov 21, 34; m 57; c 2. OPERATIONS RESEARCH, INDUSTRIAL ENGINEERING. *Educ:* Ohio State Univ, BIE, 57, PhD(opers res), 69; Univ Southern Calif, MSc in IE, 65. *Prof Exp:* Engr mfg, Lamp Div, Gen Elec Co, 60-61; sr reliability engr reliability anal, Rocketdyne Div, NAm Aviation, 61-65; res assoc indust & syst eng, 65-68, from asst prof to assoc prof, 69-79, PROF INDUST & SYSTS ENG, OHIO STATE UNIV, 79- *Concurrent Pos:* Consult, Army Res Off, 70-76, Comput Sci Corp, 70-73 & Battelle Mem Inst, 78-; ed, Phalanx, Newslett Opers Res Soc Am, 71-73, assoc ed, J Opers Res Soc Am, 75-78. *Mem:* Opers Res Soc Am; Soc Comput Simulation; Mil Opers Res Soc; Inst Mgt Sci; Am Inst Indust Engrs; Sigma Xi. *Res:* Development of improved methods for designing and analyzing simulation of complex systems, improved procedures for predicting and analyzing system reliability and more reliable methods for forecasting system performance. *Mailing Add:* 400 Longfellow Ave Worthington OH 43085

CLARK, GORDON MURRAY, b Montreal, Que, July 9, 25; m 80; c 8. RADIATION BIOLOGY. *Educ:* Sir George Williams Univ, BSc, 48; McGill Univ, MSc, 51; Emory Univ, PhD(radiation biol), 54. *Prof Exp:* Asst, McGill Univ, 50-51 & Univ Miami, 51-52; res, Emory Univ, 54-55 & Univ Mich, 55-57; asst prof zool & radiation biol, 57-64, from asst prof to assoc prof radiobiol, 64-85, prof zool, 77-85, EMER PROF ZOOL & RADIOBIOL, UNIV TORONTO, 85- *Mem:* Soc Protozool; Royal Astron Soc Can; Can Genetic Soc; Sigma Xi; Can Soc Zoologists. *Res:* Radiation effects on the cellular level; chemistry of the cell; dose and dose-rate studies in plant and animal cells; fallout studies; hyperbaric oxygen studies; x-ray and gamma ray irradiation; molecular biology; radiation injury and recovery; molecular level; chronic irradiation studies. *Mailing Add:* 310 E Ave Scarborough ON M1C 2W4 Can

CLARK, GRADY WAYNE, b Candler, NC, Nov 29, 22; m 52. PHYSICS. *Educ:* Clemson Col, BS, 44; Univ Va, PhD(physics), 51. *Prof Exp:* Res engr, Univ Va, 51-52, Linde Air Prod, 52-55 & Va Inst Sci Res, 55-58; res engr, 58-59, LAB HEAD, OAK RIDGE NAT LAB, 59- *Mem:* Am Phys Soc; Am Soc Metals; Am Ceramics Soc; Sigma Xi. *Res:* Crystal physics; crystal growth; eutetic solidification; biomagnetism. *Mailing Add:* 107 E Morningside Dr Oak Ridge TN 37830

CLARK, H(AROLD) B(LACK), b Huntingdon, Pa, May 20, 28; m 58. CHEMICAL ENGINEERING. *Educ:* Pa State Univ, BS, 50; Univ Ill, MS, 51, PhD(chem eng), 57. *Prof Exp:* ENGR, E I DU PONT DE NEMOURS & CO, INC, 57- *Mem:* Am Chem Soc; Am Inst Chem Engrs; Tech Asn Pulp & Paper Indust. *Res:* White pigment. *Mailing Add:* 2026 Floral Dr Wilmington DE 19810-3878

CLARK, HARLAN EUGENE, b Bloomington, Ill, July 29, 41; m 67; c 2. CUSTOMIZED SOFTWARE, X-RAY PHYSICS. *Educ:* Univ Ill, BS, 63; Univ Mich, MS, 64, PhD(chem), 68. *Prof Exp:* Sr res chemist, Res Ctr, Sherwin-Williams Co, 68-73, sr scientist, Sherwin-Williams Chem, Chicago, 73-77; programmer & consult, E S Indust, 77-85; SR SOFTWARE ENGR, SIEMENS ENERGY & AUTOMATION, 85- *Mem:* Am Chem Soc. *Res:* Molecular vibrations; infrared and Raman spectroscopy; titanium dioxide pigments; zinc and barium chemicals; x-ray fluorescence; x-ray diffraction. *Mailing Add:* Analytical X-Ray 6300 Enterprise Lane Madison WI 53719-1173

CLARK, HAROLD ARTHUR, b East Jordan, Mich, Apr 10, 10; m 38; c 2. POLYMER CHEMISTRY. *Educ:* Mich State Univ, BS, 31. *Prof Exp:* Analyst, Monolith Portland Midwest Co, 37-40 & Dow Chem Co, 42-43; analytical supt, 43-48, res chemist, 48-53, res group leader, 53-60, res supvr, 60-64, tech dir, Eng Prod Div, 64-69, asst dir corp develop, 69-70, res scientist, 70-75, consult, Dow Corning Corp, 75-81; INDEPENDENT CONSULT, EXPERT WITNESS, 81- *Mem:* Am Chem Soc; Sigma Xi. *Res:* Silicone resins and polymers; abrasion resistant coatings for plastics and solar collectors. *Mailing Add:* 7081 Cedarhurst Dr SW Ft Myers FL 33919

CLARK, HAROLD EUGENE, b Sunderland, Mass, Feb 21, 06; m 38; c 2. PLANT PHYSIOLOGY. *Educ:* Mass Agr Col, BS, 28; Rutgers Univ, MS, 31, PhD(plant physiol), 33. *Prof Exp:* Nat Res Coun fel, Yale Univ & Univ Conn Exp Sta, 33-35; assoc biochemist, Exp Sta, Pineapple Res Inst, Hawaii, 35-38, head dept physiol & soil, 38-46, head dept chem, 46-47; from assoc prof to prof plant physiol, Rutgers Univ, 47-74, emer prof, 74-; RETIRED. *Mem:* Am Soc Plant Physiol; Am Chem Soc; Am Inst Biol Sci. *Res:* Mineral nutrition. *Mailing Add:* 3617 Applewood Dr Freehold NJ 07728

CLARK, HERBERT MOTTRAM, b Derby, Conn, Sept 3, 18. RADIOCHEMISTRY. *Educ:* Yale Univ, BS, 40, PhD(phys chem), 44. *Prof Exp:* Lab asst, Yale Univ, 40-42, instr chem, 42-46; asst prof phys chem, 46-49, assoc prof phys nuclear chem, 49-51, PROF PHYS & NUCLEAR CHEM, RENSSELAER POLYTECH INST, 51- *Concurrent Pos:* Res assoc, Monsanto Chem Co & Clinton Labs, Oak Ridge, 46-47; consult, Union Carbide & Carbon Corp, 55-58 & US AEC, 65-70; mem subcomt radiochem, Comt Nuclear Sci, Nat Acad Sci-Nat Res Coun, 61-72; consult radiol health, USPHS, 67-69; on leave, Nuclear Power Develop Div, US Dept Energy, Nuclear Alternative Syst Assessment Prog & Int Nuclear Fuel Cycle Eval, 78-79. *Mem:* Fel AAAS; fel NY Acad Sci; Am Chem Soc; Am Phys Soc; Am Soc Eng Educ. *Res:* Radiochemistry applied to the environment; radiological health, nuclear power, nuclear medicine; radiochemical separations; chemistry of technetiam. *Mailing Add:* Dept Chem Rensselaer Polytech Inst Troy NY 12180

CLARK, HOWARD CHARLES, b Auckland, NZ, Sept 4, 29; m 54; c 2. INORGANIC CHEMISTRY. *Educ:* Univ Auckland, BSc, 51, MSc, 52, PhD, 54; Cambridge Univ, PhD, 57, ScD(chem), 72. *Hon Degrees:* Dsc, Cambridge Univ, 57. *Prof Exp:* Jr lectr chem, Univ Auckland, 54-55; res fel, Cambridge Univ, 55-57; from asst prof to prof, Univ BC, 57-65; sr prof, Univ Western Ont, 65-76, head dept, 67-76; prof inorg chem & vpres acad, Univ Guelph, 76-86; PRES & VCHANCELLOR, DALHOUSIE UNIV, 86- *Concurrent Pos:* Mem chem grant selection comt, Nat Res Coun Can, 68-70, chmn, 70; chmn comt chem dept chmn, Ont Univs, 69-; consult, E I du Pont de Nemours & Co, Inc, Del, 70-71. *Honors & Awards:* Noranda Lectr Award, Chem Inst Can. *Mem:* Royal Soc Chem; Am Chem Soc; fel Chem Inst Can (pres, 82-83); fel Royal Soc Can. *Res:* Chemistry of inorganic fluorides and organometallic compounds; coordination chemistry; organometallic and coordination compounds. *Mailing Add:* Pres & VChancelor Dalhousie Univ Main Floor A&A Bldg Halifax NS B3H 4H6 Can

CLARK, HOWARD CHARLES, JR, b Wichita, Kans, June 4, 37; m 57; c 5. GEOPHYSICS, GEOENVIRONMENTAL SCIENCE. *Educ:* Univ Okla, BS, 59; Stanford Univ, MS, 65, PhD(geophys), 67. *Prof Exp:* Teaching asst, Stanford Univ, 59-60; instr physics, Kansas City Jr Col, 61-62; res asst geophys, Stanford Univ, 65; instr geol, Menlo Col, 62-66; asst prof, 66-73, ASSOC PROF GEOL, RICE UNIV, 73- *Mem:* Geol Soc Am; Am Geophys Union; Soc Explor Geophys. *Res:* Marine geophysics; paleomagnetism. *Mailing Add:* Dept Geol Rice Univ Box 1892 Houston TX 77251

CLARK, HOWARD GARMANY, b Birmingham, Ala, Feb 25, 28; m 47; c 3. BIOMEDICAL ENGINEERING. *Educ:* Howard Col, AB, 47; Univ Notre Dame, MS, 49; Univ Md, PhD(org chem), 54. *Prof Exp:* Chemist, Chemstrand Corp, 54-59; group leader, Peninsular Chem Res, Inc, 60; sr chemist, Camille Dreyfus Lab, Res Triangle Inst, 60-67; assoc prof textiles, Clemson Univ, 67-68; assoc prof biomed eng, 68-75, chmn biomed eng dept, 79-84, PROF BIOMED ENG & MECH ENG, DUKE UNIV, 75-, PROF BIOCHEM ENG, 84- *Mem:* AAAS; Am Chem Soc. *Res:* Plant cell culture; coagulation of blood. *Mailing Add:* Dept Biomed Eng Duke Univ Durham NC 27706

CLARK, HOWELL R, b Dexter, Ky, Aug 9, 26; m 58; c 2. INORGANIC CHEMISTRY, ANALYTICAL CHEMISTRY. *Educ:* Murray State Univ, BS, 56; Vanderbilt Univ, MS, 58, PhD(inorg chem), 70. *Prof Exp:* Chemist, Shell Oil Co, 58-63; from asst prof to prof chem, Murray State Univ, 63-86; RETIRED. *Concurrent Pos:* Indust consult, 83-91. *Mem:* Am Chem Soc; Sigma Xi. *Res:* Kinetics of catalyzed organic fluoride hydrolysis; factors affecting acid formation in spoils from surface mining. *Mailing Add:* Rte 8 Box 925 Murray KY 42071

CLARK, HUGH, b Pawling, NY, Apr 15, 14; m 39; c 4. EMBRYOLOGY. *Educ:* Clark Univ, AB, 34; Univ Mich, PhD(zool), 41. *Prof Exp:* Asst zool, Univ Mich, 35-39; prof physiol, Des Moines Still Col Osteop, 39-45; from assoc zool to asst prof, Univ Iowa, 45-47; from asst prof to prof, 47-84, assoc dean, Grad Sch, 63-84, assoc vpres grad educ & res, 82-84, EMER PROF ZOOL, UNIV CONN, 84- *Concurrent Pos:* Dir mus reconstruct, Southern Ill State Norm Univ, 39; consult, Conn Technol Park, 84; Secy, Sci & Technol Inst New Eng, 87- *Mem:* AAAS; Sigma Xi. *Res:* Embryology of hemipenis in North American snakes; homogamy in the earthworm; respiration in reptile embryos; nitrogen metabolism in embryonic development; factors in embryonic differentiation. *Mailing Add:* Conn Tech Develop Co PO Box 575 Storrs CT 06268

CLARK, HUGH KIDDER, b St Louis, Mo, Jan 22, 18; m 42; c 2. REACTOR PHYSICS, NUCLEAR CRITICALITY SAFETY. *Educ:* Oberlin Col, AB, 39; Cornell Univ, PhD(phys chem), 43. *Prof Exp:* Res assoc, Radio Res Lab, Harvard Univ, 43-45; res chemist, E I du Pont de Nemours & Co, Inc, 45-62, res assoc, 62-87; RETIRED. *Honors & Awards:* Nuclear Criticality Safety Div Achievement Award, Am Nuclear Soc, 76. *Mem:* Am Chem Soc; fel Am Nuclear Soc. *Res:* X-ray crystallography; radar direction finding; spinning of synthetic fibers; nuclear reactor physics; criticality safety. *Mailing Add:* 225 Lakeside Dr Aiken SC 29803-5819

CLARK, IRWIN, b Boston, Mass, Apr 28, 18; m 49; c 4. BIOCHEMISTRY. *Educ:* Harvard Univ, AB, 39; Columbia Univ, PhD(biochem), 50. *Prof Exp:* Head, Isotope Dept, Merck Inst, 51-59; from asst prof to prof biochem, Col Physicians & Surgeons, Columbia Univ, 59-71; prof biochem & surg, Sch Med, Univ NC, 71-74; PROF SURG(BIOCHEM), ROBERT WOOD JOHNSON MED SCH, UNIV MED & DENT, NJ, 74- *Concurrent Pos:* Sr Fulbright scholar, Cambridge Univ, 50-51; res career develop award, 62-69; consult, Merck, Sharpe & Dohme, 59-60 & Squibb Inst Med Res, 60-62; vis prof, Rice Univ, 63; ed, Proc, Soc Exp Biol & Med, 67-87. *Mem:* Soc Exp Biol & Med; Am Soc Biochem & Molecular Biol; Am Inst Nutrit; Endocrine Soc; Brit Biochem Soc; Sigma Xi; Am Soc Bone & Mineral Res; Orthop Res Soc; AAAS; Am Chem Soc; Am Col Nutrit. *Res:* Metabolism of bone; hormonal effects on bone; RNA and cancer; growth factors. *Mailing Add:* 443 Stelle Ave Plainfield NJ 07060-2326

CLARK, J(OHN) B(EVERLEY), b Port Dalhousie, Ont, July 13, 24; nat US; m 51; c 3. METALLURGICAL ENGINEERING. *Educ:* Univ Toronto, BASc, 48; Carnegie Inst Tech, MS, 51, PhD(metall eng), 53. *Prof Exp:* Res engr, Dow Chem Co, 52-61; sr res scientist sci lab, Ford Motor Co, 61-66; prof metall eng, Univ Mo-Rolla, 66-86, assoc grad dean, 73-79, asst dean mines & metall, 79-81; group leader metall div, Nat Bur Standards, 86-88; RETIRED. *Concurrent Pos:* Assoc prog dir metall & mat prog, NSF, 72-73, consult, 73-76; category ed, Magnesium, 83-88; ed, Bull Alloy Phase Diagrams, 86-88. *Honors & Awards:* Henry Marion Howe Medal, 60. *Mem:* Fel Am Soc Metals; Am Inst Mining Metall & Petrol Engrs; AAAS. *Res:* Physical metallurgy; phase equilibria and precipitation processes in metal alloy systems. *Mailing Add:* 3529 MacArthur Dr Murrysville PA 15668

CLARK, J DESMOND, b London, Eng, Apr 10, 16. ARCHEOLOGY. *Educ:* Cambridge Univ, BA, MA, PhD(anthrop & archeology), ScD, 79. *Prof Exp:* Dir Nat Mus N Rhodesia, 38-61; prof, 61-86, EMER PROF ANTHROP & ARCHEOL, UNIV CALIF, BERKELEY, 86- *Mem:* Nat Acad Sci; Brit Acad Sci; AAAS; Royal Anthropolotical Inst. *Mailing Add:* Dept Anthrop Univ Calif Berkeley CA 94720

CLARK, J(AMES) EDWIN, b Winnsboro, SC, Mar 3, 33; m 56; c 3. CIVIL ENGINEERING. *Educ:* Univ SC, BSCE, 57, ME, 64; Univ NC, PhD(civil eng), 67. *Prof Exp:* Engr, Union-Bag-Camp Paper Corp, 57-60; sr bridge designer, Smith-Pollite & Assoc, 60-62; instr civil eng, Univ SC, 62-63, res asst, 63-64; instr, NC State Univ, 64-65; asst prof, Miss State Univ, 67-70; assoc prof, 70-81, PROF CIVIL ENG, CLEMSON UNIV, 81- *Concurrent Pos:* Consult & expert witness for traffic collisions. *Mem:* Inst Traffic Engrs; Transp Res Bd. *Res:* Elements of highway safety; emergency medical care and theory of traffic flow; traffic accident analysis for local governments. *Mailing Add:* Dept Civil Eng Clemson Univ Clemson SC 29634

CLARK, JAMES ALAN, b Rahway, NJ, June 15, 50; m 74; c 3. GLACIO-ISOSTASY, ROCK MECHANICS. *Educ:* Grinnell Col, BA, 72; Univ Colo, MA, 74, PhD(geol), 77. *Prof Exp:* Fel, Cornell Univ, 77-79; res scientist, Sandia Nat Lab, 79-82; PROF GEOL, CALVIN COL, 82- *Concurrent Pos:* Prin investr, Nat Sci Found grants, 82- *Honors & Awards:* Kirk Bryan Award, Geol Soc Am, 80; Appl Res Award, Nat Rock Mech Asn, 85. *Mem:* Sigma Xi; Am Sci Asn. *Res:* Determining the effects of surface loads, especially glacial loads, upon earth deformation; global postglacial; tilting of ancient shorelining; neotectonics; earth rheology. *Mailing Add:* Geol Dept Calvin Col Grand Rapids MI 49546

CLARK, JAMES BENNETT, b Shamrock, Tex, Aug 24, 23; m 44; c 2. MICROBIAL PHYSIOLOGY. *Educ:* Univ Tex, Austin, BA, 47, MA, 48, PhD(bact), 50. *Prof Exp:* Res asst bact, Univ Tex, 46-48, res scientist bact genetics, 48-50; asst prof biol, Univ Houston, 50-51; prof microbiol, Univ Okla, 51-81; SR RES SPECIALIST, PHILLIPS PETROL CO, 81- *Mem:* Am Soc Microbiol; Am Acad Microbiol. *Res:* Biocorrosion; microbiol enhancement of oil recovery. *Mailing Add:* 2524 Centre Rd Bartlesville OK 74003

CLARK, JAMES DERRELL, b Atlanta, Ga, Mar 8, 37; m 60; c 3. LABORATORY ANIMAL MEDICINE. *Educ:* Univ Ga, DVM, 61, MS, 64; Tulane Univ, DSc(microbiol), 69. *Prof Exp:* Asst prof lab animal med, Sch Med, Tulane Univ, 67-72; res asst vet med, 61-62, DIR LAB ANIMAL MED, COL VET MED, UNIV GA, 72- *Concurrent Pos:* Vet consult, Audubon Park Zoo, New Orleans, 66-72; consult, Am Asn Accreditation Lab Animal Care, 70- *Mem:* Am Vet Med Asn; Am Asn Lab Animal Sci; Am Asn Zoo Vets; Am Asn Vet Med Cols; Am Col Lab Animal Med; Sigma Xi. *Res:* Study the effects of mycotoxins upon the health of laboratory and domestic animals; development of new and innovative teaching methods. *Mailing Add:* Animal Resources Univ Ga Col Vet Med Athens GA 30602

CLARK, JAMES DONALD, b Plainville, Ind, Jan 1, 18; m 44; c 4. PETROLEUM ENGINEERING & TECHNOLOGY. *Educ:* NMex Sch Mines, BS, 47; Univ Okla, MPE, 49. *Prof Exp:* Asst instr petrol eng, Univ Okla, 47-48; petrol engr, Stanolind Oil & Gas Co, 48-52; chief reservoir engr, Union Oil, Co, Calif, 52-66, regional reservoir engr, 66-80; CONSULT PETROL ENGR, 80- *Concurrent Pos:* Mem, Nat Petrol Council Comt Geothermal-Geopressure Studies, 79-80; reservoir eng test, Dept Energy Geothermal-Geopressure Well Testing, Gulf Coast, 80-84; expert eng witness fed & state cts. *Honors & Awards:* Distinguished Serv Award, Soc Petrol Engrs, 80; Soc Petrol Engrs Legion of Honor Award, 88. *Mem:* Soc Petrol Engrs; Soc Petrol Eval Engrs (pres, 73); Soc Prof Well Log Analysts (vpres, 68-69). *Res:* Secondary recovery; oil field evaluation; natural gas and oil reserves; reservoir limit tests, geopressured-geothermal sands. *Mailing Add:* 5423 Queensloch Houston TX 77096-4027

CLARK, JAMES EDWARD, b Elkins, WVa, Nov 19, 26; m 49, 77; c 7. INTERNAL MEDICINE. *Educ:* WVa Univ, AB, 48; Jefferson Med Col, MD, 52; Am Bd Internal Med, dipl, 59. *Prof Exp:* Intern, Jefferson Med Col Hosp, 52-53, resident med, 53-55, chief resident, 55-56, asst, 56-58, instr, 58-62, assoc, 62-64; asst prof clin med, Jefferson Med Col & dir artificial kidney unit & dialysis unit, Hosp, 64-68; CHIEF MED, CROZER-CHESTER MED CTR, 68-; PROF MED, HAHNEMANN MED COL, 69- *Concurrent Pos:* Vis lectr, Univ Pa; lectr, US Naval Hosp, Philadelphia; courtesy med staff, Pa Hosp, 62-; consult, Jefferson Med Col Med Serv, Philadelphia Gen Hosp & Riddle Mem Hosp; chmn nat adv coun, Nat Kidney Dis Found, 62-64; chmn pharm comt, Riddle Mem Hosp, 62-64; dir health serv, Swarthmore Col; med dir, Franklin Mint, Franklin Ctr, Pa; mem bd dirs, Kidney Found Southeastern Pa, Inc. *Mem:* AAAS; fel Am Col Physicians; AMA; NY Acad Sci; Am Heart Asn. *Res:* Kidney disease and electrolyte metabolism. *Mailing Add:* Chief Med Crozer-Chester Med Ctr Chester PA 19013

CLARK, JAMES HENRY, b Earlington, Ky, June 17, 32; m 57; c 2. REPRODUCTIVE PHYSIOLOGY, ENDOCRINOLOGY. *Educ:* Western Ky State Univ, BS, 59; Purdue Univ, MS, 66, PhD(endocrinol), 68. *Prof Exp:* Instr develop biol, Purdue Univ, 64-68, asst prof endocrinol, 70-73; assoc prof, 73-77, PROF CELL BIOL, BAYLOR COL MED, 77- *Concurrent Pos:* NIH fel biochem endocrinol, Univ Ill, Urbana, 68-70; NIH res grants, 70, 73, 76 & 79; Nat Cancer Inst res grant, 77 & 78; Am Cancer Soc res grant, 72-78; mem rev panel for contraceptive devices, NIH, 73, mem endocrinol study sect, 74-78, mem, Clin Sci Study Sect, 85-; mem sci adv bd, Nat Ctr Toxicol Res; mem adv comt, Am Cancer Soc, 85-; pres, Laurentian Hormone Conf, 85- *Mem:* AAAS; Endocrine Soc; Soc Study Reproduction; Sigma Xi. *Res:* Mechanism of steroid hormone action; the control of reproductive function and the control of hormone indued growth. *Mailing Add:* Dept Cell Biol Baylor Col Med Houston TX 77030

CLARK, JAMES MICHAEL, b Danville, Pa, May 14, 38; m 63; c 1. OXYGEN TOLERANCE IN MAN. *Educ:* Univ Penn, MD, 63, PhD(pharmacol), 70. *Prof Exp:* asst prof pharmacol, 73-78, CLIN ASSOC PROF ENVIRON MED PHARMACOL, SCH MED, UNIV PA, 78- *Honors & Awards:* Stoner-Link Award, Undersea & Hyperbaric Med Soc, 90. *Mem:* Am Physiol Soc; Undersea & Hyperbaric Med Soc; Aerospace Med Asn. *Res:* Pulmonary oxygen toxicity. *Mailing Add:* Inst Environ Med Med Ctr Univ Pa Philadelphia PA 19104-6068

CLARK, JAMES ORIE, II, b South Bend, Ind, Nov 3, 50; m 73. FRUIT PROCESSING TECHNOLOGY, FRUIT JUICE & CONCENTRATE TECHNOLOGY. *Educ:* Ore State Univ, Bs, 83. *Prof Exp:* Prod supvr, Roger's Walla Walla, 81; qual control technician, Frito Lay, Inc, 83; qual control mgr, Tree Top, Inc, 88-89; res & develop mgr, 83-88, DIR OPERS, ALLEN FRUIT, INC, 89- *Mem:* Inst Food Technologists; Am Soc Qual Control; Am Chem Soc. *Res:* Food preservatives; natural fruit pigments; flavor chemistry; sensory techniques; fruit juice, concentrate processing technology; enzyme activities; fruit juice clarification, filtration, packaging, process times and temperatures. *Mailing Add:* Allen Fruit Co Inc 500 E Illinois St Newberg OR 97132

CLARK, JAMES RICHARD, b New Brunswick, NJ, Oct 23, 51; m 78; c 2. HORTICULTURE. *Educ:* Rutgers Univ, BS, 73, MS, 75; Univ Calif, PhD(plant physiol), 79. *Prof Exp:* Res asst hort, Rutgers Univ, 73-75; res asst environ hort, Univ Calif-Davis, 75-78; asst prof hort, Mich State Univ, 78-81; asst prof, 81-84, ASSOC PROF ENVIRON HORT, CTR URBAN HORT, UNIV WASH, 84- *Mem:* Am Soc Hort Sci; Int Soc Arboricult; Int Plant Propagator's Soc. *Res:* Growth and development of woody plants, specifically phase change, assimilate partitioning and the development of crown form; plant-environment interactions in urban areas and crown form of deciduous trees. *Mailing Add:* GF-15 Ctr Urban Hort Univ Wash Seattle WA 98195

CLARK, JAMES SAMUEL, b Bellefonte, Pa, Jan 13, 57; m 80; c 2. PALEOECOLOGY, DEMOGRAPHY-POPULATION THEORY. *Educ:* NC State Univ, BS, 79; Univ Mass, MS, 84; Univ Minn, PhD(ecol), 88. *Prof Exp:* Sr scientist, NY State Mus, 88-90; ASST PROF ECOL, DEPT BOT, UNIV GA, 90- *Concurrent Pos:* Prin investr, NSF, 87-91; mem, Prog Adv Comn, Dahlum Conf Fire, 90-92; guest lectr, Off Interdisciplinary Earth Sci, Global Change Inst, 90. *Honors & Awards:* Cooper Award, Ecol Soc Am, 88 & Mercer Award, 91. *Mem:* Ecol Soc Am; Am Quaternary Asn; Sigma Xi; Brit Ecol Soc; Soc Am Naturalists; Forest Hist Soc. *Res:* Dynamics of terrestrial vegetation and environmental change; determination of long-term fire regimes in North America; effects of climate change on fire and forest ecosystems; factors that cause changes in forest properties at different spatial scales. *Mailing Add:* Dept Bot Univ Ga Athens GA 30602

CLARK, JAMES WILLIAM, b Beaumont, Tex, Nov 28, 24; m 49; c 3. DENTISTRY. *Educ:* Univ Tex, DDS, 47; Univ Toronto, dipl, 52; Am Bd Periodont, dipl, 55. *Prof Exp:* Pvt pract, 47-64; assoc prof periodont, Med Ctr, Univ Ala, 64-69; prof periodont & chmn dept, Col Dent, Univ Tenn, Memphis, 69-77; chmn dept periodont, 80-87, prof periodont & dir clin nutrit, 77-90, EMER PROF PERIODONT, MED COL GA SCH DENT, 90- *Concurrent Pos:* Co-founder, Rowe Smith Mem Found; ed-in-chief, Clin Dent, 76-87, consult ed, 87- *Mem:* Am Acad Periodont; Am Dent Asn; Int Asn Dent Res. *Res:* Clinical investigation of etiology and treatment of periodontal disease. *Mailing Add:* 40 Conifer Lane Augusta GA 30909-4509

CLARK, JIMMY DORRAL, b Hobart, Okla, Feb 21, 39. MYCOLOGY. *Educ:* Wayne State Univ, BS, 65, MS, 66; Univ Calif, Berkeley, PhD(bot), 72. *Prof Exp:* Res assoc, Univ Calif, Berkeley, 72-74; asst prof, 75-80, ASSOC PROF BIOL, UNIV KY, 81- *Mem:* Mycol Soc Am; Bot Soc Am; Am Inst Biol Sci; Sigma Xi. *Res:* Genetic control of senescence and agametic cell fusion in the true slime molds and the associated physiological aspects of aging and incompatible reactions. *Mailing Add:* Morgan Sch Biol Sci Univ Ky Lexington KY 40506

CLARK, JIMMY HOWARD, b Sedalia, Ky, Feb 22, 41; m 74; c 1. RUMINANT NUTRITION, NITROGEN METABOLISM. *Educ:* Murray State Univ, BS, 63; Univ Tenn, PhD(animal sci), 67. *Prof Exp:* Res assoc fel ruminant nutrit, Clemson Univ, 67-68; from asst prof to assoc prof, 68-80, PROF NUTRIT & DAIRY SCI, UNIV ILL, 80- *Concurrent Pos:* Mem comt animal nutrit, Nat Acad Sci, 81-86 & mem comt rev, nutrient req dairy cattle, 83-88; mem bd dirs, Am Dairy Sci Asn, 85-88. *Honors & Awards:* Am Feed Mfrs Award, Am Dairy Sci Asn, 80. *Mem:* Am Dairy Sci Asn; Am Inst Nutrit; Nutrit Soc; Am Soc Animal Sci; Sigma Xi; AAAS. *Res:* Ruminant nutrition and metabolism; limiting nutrients for milk production; amino acid metabolism; rumen fermentation; gluconeogenesis and nutritional hormonal interactions; nonprotein nitrogen utilization. *Mailing Add:* Dept Animal Sci Rm 315 Univ Ill 1207 W Gregory Dr Urbana IL 61801

CLARK, JOHN A(LDEN), b Ann Arbor, Mich, July 9, 23; m 45; c 3. MECHANICAL ENGINEERING. *Educ:* Univ Mich, BSE, 48; Mass Inst Technol, SM, 49, ScD(mech eng), 53. *Prof Exp:* Res engr res dept, United Aircraft Corp, 48; asst dept mech eng, Mass Inst Technol, 49-50, from instr to asst prof, 50-57; prof-in-charge, Heat Transfer Lab, Univ Mich, Ann Arbor, 57-65, chmn dept mech eng, 66-74; bd chmn & sr partner, Solarco, Inc, 76-78; pres, 78-80, CONSULT, CENT SOLAR ENERGY RES CORP, 80-; PROF MECH ENG, 57-,EMER PROF, MECH ENG, UNIV MICH, ANN ARBOR, 88- *Concurrent Pos:* Consult, Corps Engrs, US Army, 52-57, E I du Pont de Nemours & Co, 53-59, Westinghouse Atomic Power Dept, 60-, Kelsey-Hayes Corp, 61- & Babcock & Wilcox Co, 64-; guest prof, Inst Thermodyn Technol, Munich Tech, 65-66 & Inst Thermodyn, Tech Univ Berlin, 72-73; NSF sr fel, 65-66. *Honors & Awards:* Gold Medal Award, 56 & Mem Award, Heat Transfer Div, 78, Am Soc Mech Engrs; Cent Medallion, Am Soc Mech Engrs, 81. *Mem:* Am Soc Mech Engrs; fel Am Soc Mech Engrs. *Res:* Heat and mass transfer; thermodynamics; fluid mechanics; temperature measurement; cryogenic heat transfer; solar energy; economics of energy. *Mailing Add:* Dept Mech Eng Univ Mich Ann Arbor MI 48104

CLARK, JOHN F, b Reading, Pa, Dec 12, 20; m 43, 74; c 2. SPACE SYSTEMS APPLICATIONS & TECHNOLOGY, COMMUNICATIONS. *Educ:* Lehigh Univ, BS, 42, EE, 47; George Wash Univ, MS, 46; Univ Md, PhD(physics), 56. *Prof Exp:* Engr, US Naval Res Lab, 42-47; asst prof elec eng, Lehigh Univ, 47-48; unit head ionospheric physics, US Naval Res Lab, 48-54, br head atmospheric elec, 54-58; dir physics & astron progs, NASA, 58-63, dep assoc adminr space sci & appln & chmn space sci steering comt, 63-65, dir, Goddard Space Flight Ctr, 65-76; dir space appln & technol, RCA Corp, Princeton, 76-87; consult space systs & technol, Jet Propulsion Lab, 87-88; navspace prof aerospace eng, US Naval Acad, 88-90; ACAD PROG CHMN & PROF SPACE TECHNOL, SPACEPORT GRAD CTR, FLA INST TECHNOL, TITUSVILLE, FL, 91- *Concurrent Pos:* Lectr, George Wash Univ, 56-58, res assoc atmospheric physics, Grad Coun, 60-66; mem, Joint Comt Atmospheric Elec, Int Union Geod & Geophys, 60-63; mem, US Nat Comt, Int Sci Radio Union, 62-64; indust & prof adv coun, Pa State Univ, 63-65; mem vis comt physics, Lehigh Univ, 66-74, mem Comt Fed Labs, 71-75; mem, Md Gov Sci Adv Coun, 72-76; mem study panel, Off Telecommun, Nat Assembly Eng,76-77; consult ed space technol, Encycl Sci & Technol, McGraw-Hill, 74-; mem, NJ Govt Sci Adv Comt, 80-86; chmn, Fed Commun Comn Adv Comt Prep Int Telecommun Union, 83; sr tech adv, US deleg, Region 2 Broadcasting Satellite Serv Planning Conf, 83; mem, US deleg, Int Telecommun Union Space World Admin Radio Conf, 85 & Frequency Mgt Adv Comt, Dept Com, 84- *Honors & Awards:* Collier Trophy, Nat Aeronaut Asn, 74. *Mem:* Fel Inst Elec & Electronics Engrs; fel Am Astron Soc; fel Explorers Club; fel Am Inst Aeronaut & Astronaut (vpres, 86-90); Int Acad Astronaut; Satellite Broadcasting & Commun Asn. *Res:* Space applications-earth resources survey, meteorology, communications; upper atmospheric physics; ionospheric physics; atmospheric potential gradient and polar conductivities; algebraic ring theory; weather radar and radar beacon development; satellite station keeping. *Mailing Add:* 555 Fillmore Ave No 406 Cape Canaveral FL 32920

CLARK, JOHN HARLAN, b Helena, Mont, Aug 30, 48; m 67; c 2. FISHERIES. *Educ:* Carroll Col, BA, 71; Colo State Univ, MS, 74, PhD(fisheries), 75. *Prof Exp:* Fisheries biologist III, Alaska Dept Fish & Game, 75-79, chief fisheries scientist, 79-84, regional supvr, 84-91. *Mem:* Am Fisheries Soc; Nat Audubon Soc; Sigma Xi. *Res:* Production capabilities of lakes used for nursery areas by sockeye salmon; sport fisheries population dynamics and management. *Mailing Add:* 1300 College Rd Fairbanks AK 99701-1599

CLARK, JOHN JEFFERSON, b Shrewsbury, Mass, Dec 30, 22; div; c 3. VETERINARY PATHOLOGY. *Educ:* Univ Ga, DVM, 53; Univ Minn, PhD(vet path), 59. *Prof Exp:* Res fel vet path, Univ Minn, 53-57; res assoc, 57-70, RES SECT HEAD VET PATH, UPJOHN CO, 70- *Mem:* Am Vet Med Asn; Am Soc Clin Path. *Res:* Cancer embryology, histology, chemotherapy and immunity; immunology, bacteriology and pathology of leptospirosis and other infectious diseases; parasitology of domesticated animals; hematology; clinical chemistry. *Mailing Add:* Upjohn Co Kalamazoo MI 49001

CLARK, JOHN MAGRUDER, JR, b Ithaca, NY, June 10, 32; m 57; c 2. BIOCHEMISTRY. *Educ:* Cornell Univ, BS, 54; Calif Inst Technol, PhD(biochem), 58. *Prof Exp:* From instr to assoc prof, 58-79, PROF BIOCHEM, UNIV ILL, URBANA, 80- *Mem:* Am Soc Biol Chemists; Am Chem Soc. *Res:* Enzymology related to protein biosynthesis. *Mailing Add:* Roger Adams Lab 1209 W Calif St Urbana IL 61801-3731

CLARK, JOHN PETER, III, b Philadelphia, Pa, May 6, 42; m 68; c 2. CHEMICAL ENGINEERING, PROCESS DEVELOPMENT. *Educ:* Univ Notre Dame, BSChE, 64; Univ Calif, Berkeley, PhD(chem eng), 68. *Prof Exp:* Res engr chem eng, Agr Res Serv, USDA, Berkeley, 68-71, indust engr, Hyattsville, Md, 71-72; from asst prof to assoc prof, Va Polytech & State Univ, 72-78; dir process develop, Int Tel & Tel Continental Baking Co, 78-81; pres, Epstein Process Eng, Inc, 81-90, SR VPRES, A EPSTEIN & SONS INT, INC, 90- *Concurrent Pos:* Consult, Gen Foods Corp, 75-78; adv bd, Univ Ill, Univ Calif, Fla State. *Mem:* Am Inst Chem Engrs; Inst Food Technologists. *Res:* Food engineering; food rheology; food extrusion; cellulose hydrolysis; renewable resources; waste water treatment; fermentation. *Mailing Add:* A Epstein & Sons Int, Inc 600 W Fulton St Chicago IL 60606-1199

CLARK, JOHN R(AY), b Chester, WVa, May 4, 18; m 49; c 2. ELECTRICAL ENGINEERING. *Educ:* Purdue Univ, BS, 39, MSE, 41; Ohio State Univ, PhD(elec eng), 52. *Prof Exp:* Asst elec engr, Radio Lab, Purdue Univ, 39-40; res engr, Continental Elec Co, Ill, 40-41; instr elec eng, Purdue Univ, 41-44, from asst prof to assoc prof, 46-58; prin engr, Electronic Corp, 58-60; head dept, 60-70, PROF ELEC ENG, MICH TECHNOL UNIV, 60- *Mem:* Inst Elec & Electronics Engrs; assoc mem Acoust Soc Am; assoc mem Am Soc Eng Educ. *Res:* Radio; electronic circuits; installation and maintenance of fire control radar; a telemetering device. *Mailing Add:* 914 First St Hancock MI 49930

CLARK, JOHN R(OBERT), metallurgy; deceased, see previous edition for last biography

CLARK, JOHN S, soil chemistry, for more information see previous edition

CLARK, JOHN W(OOD), b Jacksonville, Ill, July 17, 22; m 49, 68; c 3. ENGINEERING MECHANICS. *Educ:* Purdue Univ, BS, 46, MS, 47; Univ Pittsburgh, PhD(math), 54. *Prof Exp:* Res Engr, Alcoa Res Labs, 47-52, from asst chief to chief, Eng Design Div, 57-75, mgr, Eng Properties & Design Div, 75-78, tech adv, 78-80; RETIRED. *Honors & Awards:* Rowland Prize, Am Soc Civil Engrs, 57, Res Prize, 58, Croes Medal, 66. *Mem:* Am Soc Civil Engrs; Struct Stability Res Coun; Sigma Xi. *Res:* Strength and design of aluminum structures. *Mailing Add:* 904 Farragut St Pittsburgh PA 15206

CLARK, JOHN WALTER, b Lockhart, Tex, Apr 7, 35; m 73; c 5. THEORETICAL NUCLEAR PHYSICS, CONDENSED MATTER THEORY. *Educ:* Univ Tex, BS, 55, MA, 57; Wash Univ, PhD(physics), 59. *Prof Exp:* Res assoc physics, Wash Univ, 59; NSF fel, Princeton Univ, 59-61; assoc res scientist, Denver Div, Martin Co, 61; NATO fel, 62-63; from asst prof to assoc prof, 63-72, PROF PHYSICS, WASH UNIV, 72- *Concurrent Pos:* Alfred P Sloan Found fel, 65-67; guest prof, Swed Univ Abo, 71-72. *Honors & Awards:* Feenberg Medal, 87. *Mem:* Fel Am Phys Soc. *Res:* Quantum mechanics of many-body systems; nuclear interactions and nuclear structure; neutron stars; quantum fluids and solids; neural networks; quantum control theory. *Mailing Add:* Dept Physics Wash Univ St Louis MO 63130

CLARK, JOHN WHITCOMB, b Walkerton, Ind, Aug 14, 18; m 61. RADIOLOGY. *Educ:* Harvard Univ, MD, 43. *Prof Exp:* From instr to prof radiol, Univ Ill, 48-70; prof radiol, Rush Med Col, 70-88; dir, Computed Topology & Nuclear Magnetic Resonance, 82-88; RETIRED. *Concurrent Pos:* Resident, Presby-St Luke's Hosp, 46-49, attend radiologist, 49-50, assoc radiologist, 50-55, attend radiologist, Div Radiol & Nuclear Med, 55-70; assoc scientist, Argonne Nat Lab, 52-56, consult, 56-62. *Mem:* AMA; fel Am Col Radiol; Radiation Res Soc; Radiol Soc NAm; Am Roentgen Ray Soc. *Res:* Clinical radiology; radiobiology. *Mailing Add:* 3740 N Lake Shore Dr Chicago IL 60613

CLARK, JON D, b Detroit, Mich, Feb 8, 46; m 86; c 2. COMPUTER AIDED SYSTEMS ENGINEERING. *Educ:* Mich State Univ, MA, 68; Eastern Mich Univ, MBA, 72; Case Western Reserve Univ, PhD(mgt decision systs), 77. *Prof Exp:* Asst prof computer info systs, Univ Tex, Dallas, 75-79; prof BCIS, Univ NTex, 75-79; PROF & CHMN COMPUTER INFO SYSTS, COLO STATE UNIV, 89- *Mem:* Asn Comput Mach; Inst Elec & Electronics Engrs; Data Processing Mgt Asn. *Res:* Application of artificial intelligence to software development, evaluation and maintenance; human factors and computer performance evaluation. *Mailing Add:* Computer Info Systs Dept C105 Clark Bldg Col Bus Ft Collins CO 80523

CLARK, JOSEPH E(DWARD), b Philadelphia, Pa, Oct 2, 35; m 59; c 5. POLYMER SCIENCE, PHYSICAL CHEMISTRY. *Educ:* Villanova Univ, MS, 60; Univ Windsor, PhD(polymer chem), 63. *Prof Exp:* Chemist, Res & Develop Div, Villanova Univ, 57-58; sr res chemist, Res Ctr, W R Grace & Co, 63-65; res assoc weathering of plastics, Mfg Chemists Asn, 65-67, prog mgr, 67-69; tech coordr, textile flammability standards, Nat Bur Standards, 69, chief, div textile flammability standards, Off Flammable Fabrics, 70-73; prof eng & fel pub affairs, Princeton Univ, 73-74; assoc adminr fire safety res, US Fire Admin, 75-79; sr policy analyst & exec secy, intergovt sci, eng & technol adv panel, Off Sci & Technol Policy, Exec Off of the President, 79-83; DEP DIR, NAT TECH INFO SERV, 84- *Mem:* Am Chem Soc; Nat Fire Protection Asn; Am Soc Testing & Mat; Technol Transfer Soc; AAAS. *Res:* Natural and synthetic polymer properties, characterization and degradation; materials problems including durability, weatherability and flammability; analysis of science and technology policy issues; information management. *Mailing Add:* 14607 Westbury Rd Rockville MD 20853-2247

CLARK, JULIAN JOSEPH, b Brooklyn, NY, June 21, 35. PSYCHOSOMATIC MEDICINE. *Educ:* Princeton Univ, AB, 56; Columbia Univ, Col Physicians & Surgeons, MD, 60. *Prof Exp:* Sr psychiatrist, Kings Co Hosp, New York City, 71-; CLIN ASSOC PROF PSYCHIAT, DOWNNSTATE MED CTR, STATE UNIV NY, 78- *Concurrent Pos:* Attend physician liaison psychiat, Dept Psychiat & Med, State Univ Hosp, NY, 71- *Mem:* Am Psychosom Asn; Am Psychosom Soc; Sigma Xi. *Res:* Influence of environmental factors on hyperactive behavior in children; study of inflammatory bowel disease in patients; study of hemodialysis and renal transplant patients. *Mailing Add:* 335 E 17th St Brooklyn NY 11226

CLARK, KENNETH COURTRIGHT, b Austin, Tex, Sept 30, 19; m 47; c 2. ATOMIC PHYSICS, MOLECULAR PHYSICS. *Educ:* Univ Tex, BA, 40; Harvard Univ, AM, 41, PhD(physics), 47. *Prof Exp:* Tutor physics, Harvard Univ, 41-42, spec res assoc, Nat Defense Res Comt Proj, Electro-Acoustic Lab, 41-45, instr physics, 47-48; from asst prof to assoc prof, 48-60, PROF PHYSICS, UNIV WASH, 60- *Concurrent Pos:* Actg head physics div & res assoc prof, Geophys Inst, Univ Alaska, 57-58; consult, US Agency Int Develop, India, 64 & 66 & Battelle Northwest Labs, Battelle Mem Inst, 71-; prog dir aeron, NSF, Washington, DC, 69-70; mem adv comt, Geophys Inst, Univ Alaska, 73-77; vis prof space physics, Latrobe Univ, Melbourne, Australia, 79-80. *Mem:* Fel Am Phys Soc; fel Optical Soc Am; Am Asn Physics Teachers; Am Geophys Union; Sigma Xi. *Res:* Auroral physics; optical excitation processes in upper atmosphere; high resolution spectra; laboratory aeronomy. *Mailing Add:* Dept Physics FM-15 Univ Wash Seattle WA 98195

CLARK, KENNETH EDWARD, b Apr 16, 45; m; c 3. CARDIOVASCULAR PHYSIOLOGY, CARDIOVASCULAR PHARMACOLOGY. *Educ:* Univ Iowa, PhD(pharmacol), 75. *Prof Exp:* PROF OBSTET, GYNEC & PEDIAT, UNIV CINCINNATI, 77- *Mem:* Am Physiol Soc. *Res:* Reproductive physiology. *Mailing Add:* Dept Obstet & Gynec Univ Cincinnati 231 Bethesda Ave Cincinnati OH 45267

CLARK, KENNETH FREDERICK, b Liverpool, Eng, Apr 4, 33; m 62. ECONOMIC GEOLOGY. *Educ:* Univ Durham, BSc, 56, Univ NMex, MS, 62, PhD(geol), 66. *Prof Exp:* Geologist, Anglo Am Corp of SAfrica, Ltd, 56-60; asst prof geol sci, Cornell Univ, 66-71; from assoc prof to prof geol, Univ Iowa, 71-80; PROF GEOL SCI, UNIV TEX, 80- *Concurrent Pos:* Consult, Mexican Govt, 69- *Mem:* Geol Soc Am; Mex Geol Soc; Mex Asn Mining Engrs Metallurgists & Geologists; Am Inst Mining Metall & Petrol Eng; Soc Econ Geologists. *Res:* Exploration and development of mineral resources; geophysical and geochemical exploration. *Mailing Add:* Dept Geol Sci Univ Tex El Paso TX 79968

CLARK, KERRY BRUCE, b Woodbury, NJ, Aug 22, 45; div; c 1. INVERTEBRATE ZOOLOGY, MARINE ECOLOGY. *Educ:* Rutgers Univ, BA, 66; Univ Conn, MS, 68, PhD(invert zool), 71. *Prof Exp:* From asst prof to assoc prof, 71-80, PROF BIOL SCI, FLA INST TECHNOL, 80-, CHMN MARINE BIOL DEPT, 87- *Concurrent Pos:* Vpres, Genetics Data Services, Inc, 83- *Mem:* Am Soc Zoologists; Sigma Xi. *Res:* Taxonomy and ecology of opisthobranch molluscs; physiology of cholroplast and algal symbioses; biogeography and thermal physiology; developmental ecology marine inverts; computer systems for genetics. *Mailing Add:* Dept Biol Sci Fla Inst Technol Melbourne FL 32901-6988

CLARK, L(YLE) G(ERALD), b Gratiot Co, Mich, July 12, 24; m 49; c 4. MECHANICAL ENGINEERING. *Educ:* Univ Mich, BS(eng) & BS(math), 48, MS, 49, PhD(eng), 55. *Prof Exp:* Instr eng mech, Univ Mich, 48-55; dir res eng, Aro Equip Corp, 55-56; chmn mech eng, Univ Del, 56-60; PROF AEROSPACE ENG & ENG MECH, UNIV TEX, AUSTIN, 60- *Mem:* Am Soc Mech Engrs; Am Soc Eng Educ. *Res:* Theoretical mechanics; analytical dynamics; vibrations; nonlinear analysis. *Mailing Add:* Dept Eng Mech Univ Tex Austin TX 78712

CLARK, LARRY P, organic chemistry, medicinal chemistry; deceased, see previous edition for last biography

CLARK, LEIGH BRUCE, b Seattle, Wash, Sept 9, 34; c 1. SPECTROCHEMISTRY. *Educ:* Univ Calif, Berkeley, BS, 57; Univ Wash, PhD(chem), 63. *Prof Exp:* Asst prof, 64-72, ASSOC PROF CHEM, UNIV CALIF, SAN DIEGO, 72- *Res:* Molecular spectroscopy; reflection spectroscopy of molecular crystals. *Mailing Add:* Dept Chem B042 Univ Calif San Diego La Jolla CA 92093

CLARK, LELAND CHARLES, JR, b Rochester, NY, Dec 4, 18; m 39; c 4. BIOCHEMISTRY. *Educ:* Antioch Col, BS, 41; Univ Rochester, PhD(biochem), 44. *Prof Exp:* Chmn biochem dept, Fels Res Inst, 44-58; asst prof biochem, Antioch Col, 44-56, prof, 56-58; from assoc prof to prof surg, Med Ctr, Univ Ala, 58-68; PROF RES PEDIAT, CHILDREN'S HOSP RES FOUND, MED COL, UNIV CINCINNATI, 68- *Concurrent Pos:* Sr res assoc surg & pediat, Univ Cincinnati, 55-58; consult, Wright-Patterson AFB, 56-58 & NIH, 61-; NIH res career award, 62-68; vis prof, Cardiovasc Res Inst San Francisco, 67; fel, coun cerebrovascular dis, Am Heart 67-; ed, Symp Oxygen Transport. *Honors & Awards:* Distinguished Lect Award, Am Col Chest Physicians, 75. *Mem:* AAAS; NY Acad Sci; Sigma Xi; Artificial Organs Soc; Am Heart Asn. *Res:* Vitamin, steroid and oxygen metabolism; polarography; cardiovascular disease; hydrogen and oxygen electrodes in diagnosis; ion exchange resins in biology; glucose electrode; surgical monitoring; intermediary metabolism and synthesis of psychotomimetic drugs; fluorocarbon liquid breathing; artificial blood; enzyme electrode. *Mailing Add:* 218 Greendale Ave Cincinnati OH 45220

CLARK, LINCOLN DUFTON, b Andover, Mass, Jan 18, 23; m 49; c 2. MEDICINE. *Educ:* Harvard Univ, MD, 47. *Prof Exp:* Asst instr psychiat, Med Sch, Harvard Univ, 49-50, asst physician, 51-53; from asst prof to assoc prof psychiat, 55-64, dir, Behav Sci Lab, 64-76, asst prof pharmacol & psychol, 76-80, PROF PSYCHIAT UNIV UTAH, 64-, ADJ PROF PHARMACOL & PSYCHOL, 80- *Concurrent Pos:* NIMH res career award, 63-67 & res scientist award, 67-72; asst physician, Mass Gen Hosp, 51-53; sci assoc, Roscoe B Jackson Mem Lab; chmn adv comt preclin psychopharmacol, NIMH, 61-66, res scientist, 68- *Mem:* Am Col Neuropsychopharmacol; Psychiat Res Soc; Sigma Xi. *Res:* Experimental psychiatry; psychopharmacology; animal behavior; Forensic psychiatry. *Mailing Add:* Dept Psychiat Univ Utah Salt Lake City UT 84112

CLARK, LLEWELLYN EVANS, b Brunswick, Maine, July 30, 32; m 56; c 4. REFUSE TO ENERGY. *Educ:* Univ Maine, BS, 55, MS, 56; Univ Colo, PhD(civil eng), 66. *Prof Exp:* From instr to asst prof mech eng, Univ Maine, 55-63; assoc civil eng, Univ Colo, 63-64; assoc prof mech eng, Univ Maine, 65-67; proj engr, Jones Div, Beloit Corp, 67-68, mgr res, 68-79; proj mgr, 79-81, vpres eng, 82-88, FACIL MGR, VICON RECOVERY ASSOC, 81- *Honors & Awards:* Plant Eng Award, Am Inst Plant Engrs, 86. *Mem:* Am Soc Eng Educ; Soc Exp Stress Analysis; Tech Asn Pulp & Paper Inst; Am Soc Mech Eng. *Res:* Research and development in pulp stock preparation equipment; secondary fiber processing; contaminant removal; deinking systems; thermomechanical pulping systems; dewatering characteristics of pulp and food products; project development refuse to energy systems design; development, construction, start-up and operation of resource recovery plants. *Mailing Add:* Vicon Recovery Assoc 500 Hubbard Ave Pittsfield MA 01201

CLARK, LLOYD ALLEN, b North Battleford, Sask, Mar 17, 32; m 55; c 4. MINERAL EXPLORATION, APPLIED GEOCHEMISTRY. *Educ:* Univ Sask, BE, 54, MSc, 55; McGill Univ, PhD(geol), 59. *Prof Exp:* Fel geochem, Carnegie Inst Geophys Lab, 58-60; from asst prof to assoc prof econ geol, McGill Univ, 60-70; chief geochem res & lab div, Kennecott Explor, Inc, 70-76; explor mgr, Sask Mining Develop Corp, 76-81, chief geologist, Saskatoon, 81-84; CONSULT, CLARK GEOL, 85- *Concurrent Pos:* Nat Res Coun Can res grants, 60-70, fel, 65-66; guest investr, Univ Tokyo & Univ Florence, 65-66; pres, Coronation Mines Ltd, 88- *Mem:* Soc Econ Geol; fel Geol Asn Can; fel Mineral Asn Can; Can Inst Mining & Metall; Soc Geol Appl Mineral Depostis; Asn Explor Geochemists; Mineral Asn of Can. *Res:* Phase equilibrium and related studies in synthetic systems; studies of naturally occurring minerals and ores to yield quantitative information about environment of ore formation; exploration programs for uranium and base metals; lithogeochemistry; overburden geochemistry. *Mailing Add:* 7337-145A St Surrey BC V3S 2Y8 Can

CLARK, LLOYD DOUGLAS, b Ventura, Calif, Oct 30, 40. ELECTRICAL ENGINEERING, MATERIALS SCIENCE. *Educ:* Stanford Univ, BS, 63, MS, 67, PhD(mat sci), 70; Carnegie-Mellon Univ, MS, 65. *Prof Exp:* Res scientist, Xerox Corp, 70-71; res assoc mat sci, Stanford Univ, 71-72; res scientist & mgr, Med Ultrasound, Varian Assocs, 80; new prod res & develop mgr, Xerox Corp, 80-84; proprietor, Clerus Labs, 84-89, PROPRIETOR, PAEDIA CORP, 89- *Concurrent Pos:* Consult, Rodder Instrument, 63-71; proprietor & pvt inventor, Clark Systs, 70-84. *Mem:* Inst Elec & Electronics Engrs. *Res:* Medical ultrasound imaging; nuclear magnetic resonance; x-ray computed tomography imaging; xerography. *Mailing Add:* 499 Carolina St San Francisco CA 94107

CLARK, MALCOLM JOHN ROY, b Bournemouth, Eng, May 22, 44; m 68. ENVIRONMENTAL CHEMISTRY. *Educ:* Univ Victoria, BSc, 66; Univ NB, PhD(chem), 71. *Prof Exp:* BR ENVIRON CHEMIST, BC MINISTRY ENVIRONMENT, 71- *Concurrent Pos:* Vis scientist, Univ Victoria, 75-; pres, Commonwealth Sci Ltd. *Mem:* Am Chem Soc; Am Fisheries Soc; Am Soc Limnol & Oceanog; Am Soc Testing & Mat; Chem Inst Can. *Res:* Investigation of various problems in pollution and environmental chemistry, particularly regarding metals, color and dissolved gases; environmental data storage and retrieval; statistical analyses and image processing. *Mailing Add:* 336 Foul Bay Rd Victoria BC V8S 4G7 Can

CLARK, MALCOLM MALLORY, b Palo Alto, Calif, Sept 21, 31; c 5. QUATERNARY GEOLOGY, TECTONICS. *Educ:* Univ Calif, Berkeley, BS, 57; Stanford Univ, PhD(geol), 67. *Prof Exp:* Engr, Temescal Metall Corp, 57-60; lab mgr, Dumont Mfg Co, 60-61; mfg mgr, Monitor Plastics Co, 61-63; GEOLOGIST, US GEOL SURV, 67- *Mem:* Geol Soc Am; Seismol Soc Am; Int Glaciol Soc; Am Geophys Union. *Res:* Geology of active faults; glacial geology; glaciation of the Sierra Nevada, California. *Mailing Add:* US Geol Surv MS 977 345 Middlefield Rd Menlo Park CA 94025

CLARK, MARTIN RALPH, US citizen. REPRODUCTIVE ENDOCRINOLOGY, BIOLOGICAL CHEMISTRY. *Educ:* Univ Mich, Ann Arbor, BS, 72, PhD(biol chem), 76. *Prof Exp:* Fel reproduction endocrinol, Ford Found, 76-77, NIH individual fel reproduction endocrinol, 77-79, asst prof, 79-84, ASSOC PROF OBSTET/GYNEC & BIOCHEM, UNIV MIAMI, 84- *Mem:* AAAS; Soc Study Reproduction; Endocrine Soc; Am Soc Biol Chemists. *Res:* Biochemistry of hormone action; ovulation. *Mailing Add:* Dept Obstet/Gynec Old Clin Bldg Rm 7576 Chapel Hill NC 27599

CLARK, MARY ELEANOR, b San Francisco, Calif, Apr 28, 27. INTERPRETATIONS OF EVOLUTION, USES OF SCIENCE. *Educ:* Univ Calif, Berkeley, AB, 49, MA, 51, PhD(zool), 60. *Prof Exp:* USPHS fel zool, Bristol Univ, 61-63; Sci Res Coun Gt Brit fel, Bristol & Newcastle Univs, 63-66; vis prof zool, Univ Lund, 67; res asst organismic biol, Univ Calif, Irvine, 67-68; NSF fel environ health eng, Calif Inst Technol, 68-69; from asst prof to assoc prof, San Diego State Univ, 69-73, prof biol, 73-89; CUMBIE CHAIR, CONFLICT RESOLUTION, GEORGE MASON UNIV, 90- *Concurrent Pos:* Mem comn, Am Asn State Col & Univs, Proj 2061, AAAS. *Mem:* Fel AAAS; NY Acad Sci; Am Soc Zoologists; Sigma Xi. *Res:* Culture of normal and malignant fibroblasts; polychaete neurosecretion; monoamine histochemistry; regeneration; biochemistry of amino acids in marine invertebrates; osmoregulation; marine pollution and kelp-bed ecology; biophysics of cell osmolytes; human evolution and human behavior; factors in social change. *Mailing Add:* Ctr Conflict Analysis & Resolution George Mason Univ Fairfax VA 22030-4444

CLARK, MARY ELIZABETH, b Pensacola, Fla, Aug 1, 54; m 78. RADIATION PROTECTION. *Educ:* Fla State Univ, BS, 74, PhD(physics), 83; Univ Mich, MS, 75. *Prof Exp:* Instr math, Fla State Univ, 81-87; state health physicist, 87-89, DIR, FLA OFF RADIATION CONTROL, 89- *Concurrent Pos:* Consult, Taylor County, Fla Sch Dist, 84; Fla rep, Conf Radiation Control Prog Dirs, Inc, 89-; mem, Southeast Compact Comn, 90- *Mem:* Am Phys Soc; AAAS. *Mailing Add:* Fla Off Radiation Control 1317 Winewood Blvd Tallahassee FL 32399

CLARK, MARY JANE, b McKeesport, Pa, Sept 18, 25. BIOCHEMISTRY. *Educ:* Univ Pittsburgh, BS, 47, PhD(biochem), 57. *Prof Exp:* Jr chemist, Koppers Co, Inc, Pa, 52-54; res assoc biochem, Sch Pub Health, Univ Pittsburgh, 54-58; res assoc biochem, Col Physicians & Surgeons, Columbia Univ, 58-64; chmn dept chem, 73-78, asst prof, NY Med Col, Flower & Fifth Ave Hosps, 64-68; assoc prof biochem, Jersey City State Col, 68-, assoc prof chem, 73-, Dean Arts & Sci, 78-; MEM STAFF, DEPT CHEM, JERSEY STATE COL. *Mem:* Am Chem Soc. *Res:* Synthesis of peptides and their enzymic hydrolysis; aromatic biosynthesis in bacteria; general intermediary metabolism of amino acids and derivatives; coenzymes in intermediary metabolism; folic acid and leukemia. *Mailing Add:* Dept Chem Jersey City State Col 2039 Kennedy Blvd Jersey City NJ 07305

CLARK, MELVILLE, JR, b Syracuse, NY, Dec 19, 21. PHYSICS, ELECTRICAL ENGINEERING. *Educ:* Mass Inst Technol, SB, 43; Harvard Univ, AM, 47, PhD(physics), 49. *Prof Exp:* Mem staff microwaves, Radiation Lab, Mass Inst Technol, 42-45; mem staff electronics, Los Alamos Sci Lab, 45-46; mem staff reactors, Brookhaven Nat Lab, 49-53; mem staff neutronics, Radiation Lab, Univ Calif, 53-55; assoc prof nuclear eng, Mass Inst Technol, 55-62; sr eng specialist, Appl Res Lab, Sylvania Elec Prod, Inc, 62-64; sr consult scientist, Res & Advan Develop Div, Avco Corp, 64-67; sr scientist, NASA Electronics Res Ctr, 67-70; sr develop engr, Thermo Electron Eng Corp, 70-73; sr technol strategist, Combustion Eng, 73-83; tech expert witness, Pennie & Edmonds, 84-86; PRES, MELVILLE CLARK ASSOCS, 55- *Concurrent Pos:* Lectr, United Shoe Mach Corp, 55-56; consult, Raytheon Mfg Co, 55-58, Arthur D Little, 57-58 & Aerodyne Res, 83-84; dir, Clark Music Co, 48-60, vpres, 57-60; dir, 416 S Salina St Corp, 57-60; vpres & dir, Meldor Corp, 60-65, pres, 65-67; trustee, Inst Sci Res Music, 91- *Mem:* AAAS; Am Phys Soc; Acoust Soc Am; Inst Elec & Electronics Engrs; Am Inst Physics; Sigma Xi; Am Asn Artificial Intel; Chap Assoc Comput Mach; Fusion Power Assocs. *Res:* Microwave radiation; quantum mechanics; nuclear and plasma physics; reactor engineering; neutral particle transport; musical acoustics; ionospheric propagation; speech research; electric space propulsion; auditory perception; electronics; numerical analysis; mathematical methods of physics; software; heat transfer; recipient of 19 patents. *Mailing Add:* Eight Richard Rd Wayland MA 01778-4099

CLARK, MERVIN LESLIE, b Baltimore, Md, May 18, 21; m 49; c 4. MEDICINE. *Educ:* Va Polytech Inst, BS, 42; Northwestern Univ, MD, 48. *Prof Exp:* From asst prof to assoc prof psychiat, 60-69, from asst prof to assoc prof med, 62-69, actg dir, Div Clin Pharmacol, 70-75, adj assoc prof psychiat, 77-79, PROF MED, SCH MED, UNIV OKLA, 69-, ADJ PROF PSYCHIAT, 79- *Concurrent Pos:* Res fel, Exp Therapeut Unit, Sch Med, Univ Okla, 55-56; chief med serv & dir res unit, Cent State Griffin Mem Hosp, Norman, Okla, 56-; mem sci rev panel, Drug Interactions Eval Prog, Am Pharmaceut Asn, 74. *Mem:* Am Soc Pharmacol & Exp Therapeut; Am Soc Clin Pharmacol & Therapeut; fel Am Col Neuropsychopharmacol. *Res:* Clinical pharmacology and therapeutics; psychopharmacology; controlled clinical trials of new antipsychotic agents; metabolism and kinetics of chlorpromazine. *Mailing Add:* 1019 Mockingbird Lane Norman OK 73071

CLARK, MICHAEL WAYNE, b Portsmouth, Ohio, Dec 31, 52; m 80; c 1. YEAST MOLECULAR CELL BIOLOGY, IMMUNOMICROSCOPY. *Educ:* Northeastern Univ Boston, BSc, 77; Univ Calif, Los Angeles, PhD(biol), 82. *Prof Exp:* Postdoctoral fel molecular biol, Biol Dept, Univ Calif, Los Angeles, 82-84 & Biol Dept, CalTech, Pasadena, 84-89; ASST PROF BIOL, MCGILL UNIV, MONTREAL, 89- *Concurrent Pos:* Co-dir, Yeast Chromosome I Sequencing Ctr, McGill Univ, 90- *Mem:* Can Soc Cell Biol; Am Soc Cell Biol; Am Microbiol Soc. *Res:* Determining how the structural organization of the eukaryotic nucleus impacts upon proper nuclear functioning; nucleus structures such as the nucleolus and the chromosomes are examined during the G2 to M transition. *Mailing Add:* Dept Biol McGill Univ Montreal PQ H3A 1B1 Can

CLARK, NANCY BARNES, b Hamden, Conn, July 1, 39; m 61; c 1. COMPARATIVE ENDOCRINOLOGY, COMPARATIVE PHYSIOLOGY. *Educ:* Mt Holyoke Col, BA, 61; Columbia Univ, MA, 62, PhD(endocrinol), 65. *Prof Exp:* Asst prof zool, 65-70, assoc prof biol, 70-76, PROF BIOL, UNIV CONN, 77- *Mem:* Am Soc Bone & Mineral Res; Am Physiol Soc; Am Soc Zool; Endocrine Soc. *Res:* Parathyroid function, and calcium and phosphate regulation in non-mammalian vertebrates; comparative studies thyroid function. *Mailing Add:* Dept Physiol & Neurobiol U-42 Univ of Conn Storrs CT 06269-3042

CLARK, NATHAN EDWARD, b Milford, Conn, Feb 26, 40; m 80. OCEANOGRAPHY, CLIMATOLOGY. *Educ:* Brown Univ, ScB, 62; Mass Inst Technol, PhD(oceanog), 67. *Prof Exp:* Meteorologist, Southwest Fisheries Ctr, Nat Marine Fisheries Serv, Nat Oceanic & Atmospheric Admin, 67-76; ASST RES OCEANOGR, SCRIPPS INST OCEANOG, 77- *Concurrent Pos:* Res assoc, Scripps Inst Oceanog, 74- *Mem:* Am Meteorol Soc; Am Geophys Union; Sigma Xi. *Res:* Large scale air-sea heat transfer processes and fluctuations in the North Pacific Ocean; effects of large scale changes of ocean and atmosphere on northeastern Pacific fisheries. *Mailing Add:* 4149 Miller St San Diego CA 92103

CLARK, NERI ANTHONY, b New Haven, Conn, Apr 19, 18; m 43; c 1. AGRONOMY. *Educ:* Univ Md, BS, 54, PhD(agron), 59. *Prof Exp:* From asst prof to prof, 58-78, EMER PROF AGRON, UNIV MD, COLLEGE PARK, 79- *Concurrent Pos:* Consult, 78- *Mem:* Am Soc Agron. *Res:* Agronomy; forage management. *Mailing Add:* RD 1 Halpin Rd Middlebury VT 05753

CLARK, NIGEL NORMAN, b Durban, SAfrica, Sept 10, 58. PARTICULATE & MULTIPHASE SYSTEMS, INTERNAL COMBUSTION ENGINES. *Educ:* Univ Natal, Durban, SAfrica, BSc, 80, PhD(eng), 85. *Prof Exp:* Res asst prof, Particle Anal Ctr, 84-86, assoc prof, 86-90, PROF MECH & AEROSPACE ENG, WVA UNIV, 90- *Concurrent Pos:* Adj assoc prof, Col Mineral & Energy Resources, WVa Univ, 86-; NSF presidential young investr, 89. *Honors & Awards:* Ralph R Teetor Award, Soc Automotive Engrs, 88. *Mem:* Am Soc Mech Engrs; Soc Automotive Engrs; Fine Particle Soc. *Res:* Multiphase flows and fluidization, including instrumentation for two phase systems; particle shape analysis and description; internal combustion engine testing and fundamentals. *Mailing Add:* Dept Mech & Aerospace Eng WVa Univ Morgantown WV 26506-6101

CLARK, PATRICIA ANN, physical chemistry, for more information see previous edition

CLARK, PATRICIA ANN ANDRE, b Leadville, Colo, Mar 23, 38; m 60; c 1. APPLIED MATHEMATICS. *Educ:* Mass Inst Technol, SB, 61, SM, 64; Univ Rochester, PhD(astrophys), 69. *Prof Exp:* Instr eng, Univ Rochester & instr math, Rochester Inst Technol, 73-74; asst chief, Int Field Year Great Lakes, US Environ Protection Agency, 74-76; instr eng, Univ Rochester, 76-78; VIS PROF, ROCHESTER INST TECHNOL, 78- *Mem:* Am Astron Soc. *Res:* Biomathematics; mathematical modeling of oxygen from blood to tissue, especially skeletal and cardiac muscle. *Mailing Add:* 319 Chelmsford Rd Rochester NY 14618

CLARK, PAUL ENOCH, physical chemistry; deceased, see previous edition for last biography

CLARK, PETER DAVID, b Peoria, Ill, Oct 27, 52; m 78; c 2. COATINGS TECHNOLOGY, CORROSION SCIENCE. *Educ:* Bradley Univ, Peoria, Ill, BS, 74; Univ Wis, Madison, MS, 75. *Prof Exp:* Res asst, Univ Wis, 74-75; proj leader, PPG Industs-Appliance Finishes, 75-84; sr group leader automotive electrocoatings, 84-89, tech mgr indust coatings, 89-91, TECH MGR AUTOMOTIVE COATINGS, BASF CORP, 91- *Concurrent Pos:* Steering comt, Electrocoat Conf, Gardner Publ, 86-; adv comt, Finishing Conf, Soc Mech Engrs, 88- *Mem:* Am Chem Soc; Fedn Socs Coatings Technol. *Res:* Coatings science; polymer science; electro coating; corrosion science; primer technology; water-borne coatings; powder coatings; high solids coatings; specialty coatings; coatings for plastic substrates. *Mailing Add:* 25888 Rutledge Crossing Farmington Hills MI 48335

CLARK, PETER O(SGOODE), b Ottawa, Ont, June 5, 38; m 61; c 2. ELECTRICAL ENGINEERING, PHYSICS. *Educ:* McGill Univ, BEng, 60; Calif Inst Technol, MSc, 61, PhD(elec eng, physics), 64; Pepperdine Univ, MBA, 82. *Prof Exp:* Mem tech staff, Hughes Res Labs, 64-68, tech sect head, 68-69, assoc dept mgr, 69-70, mgr laser dept, 70-73; chief, Laser Div, Defense Advan Res Proj Agency, 73-75, asst dir technol, 75-77; dir, systems group res, TRW, 77-79, mgr, Power Systs Technol Lab, 79-81, mgr, electronics & technol opers, 81-86, vpres & gen mgr, TRW Microwave Inc, 86-87; PRES, FEI MICROWAVE INC, 87- *Concurrent Pos:* Mem adv comt, Group on Electronic Devices, 69-74; mem, Quantum Electronics Coun, 70-75. *Honors & Awards:* Centennial Medal, Inst Elec & Electronics Engrs, 84; Meritorious Civilian Serv Medal, Secy Defense. *Mem:* Am Defense Preparedness Asn; Inst Elec & Electronics Engrs; Asn Old Crows. *Res:* Electromagnetic theory; electron devices; lasers and laser systems. *Mailing Add:* FEI Microwave Inc 825 Stewart Dr Sunnyvale CA 94086

CLARK, RALPH B, b Farmington, Utah, Sept 12, 33; m 57; c 6. PLANT PHYSIOLOGY, MINERAL NUTRITION. *Educ:* Brigham Young Univ, BS, 57; Utah State Univ, MS, 59; Univ Calif, Los Angeles, PhD(plant sci), 62. *Prof Exp:* Res assoc, Ore State Univ, 63; res chemist, NCent Region, Agr Res Serv, USDA, 63-75; MEM FAC, DEPT AGRON, UNIV NEBR LINCOLN, 75- *Concurrent Pos:* Plant physiologist & fel, Agr Res Serv, USDA. *Mem:* AAAS; Am Soc Plant Physiologists; fel Am Soc Agron; fel Crop Sci Soc Am; fel Soil Sci Soc Am. *Res:* Physiology and biochemistry of mineral nutrition and metabolism in sorghum. *Mailing Add:* 102A KCR Dept Agron Univ Nebr Lincoln NE 68583-0817

CLARK, RALPH LEIGH, b East Jordan, Mich, June 2, 08; m 30; c 4. ELECTRICAL ENGINEERING, PHYSICS. *Educ:* Mich State Col, BS, 30, EE. *Prof Exp:* Radio inspector, Radio Div, US Dept Com, Fed Radio Comn & Fed Commun Comn, Detroit & Washington, DC, 30-35, radio engr, Fed Commun Comn, Washington, DC, 35-41; partner, Ring & Clark Consult Eng Firm, 41-42; contract engr, Bur Aeronaut, US Dept Navy, 42, Lt to Comdr, Bur Aeronaut, USNR, 42-46; contract employee, Res & Develop Bd, 46, dir prog div, 46-49; dep asst dir, CIA Scientific Intel, 49-57; off mgr, Stanford Res Inst, Washington, DC, 57-59; asst dir defense res & eng for commun, 59-62; spec asst to dir, Telecommun Mgt, 62-70; telecommun consult, 70-72; dir Wash off, Inst Elec & Electronics Engrs, 72-76, consult, 76-80; MEM INDEPENDENT STUDIES ARMS CONTROL/DISARMAMENT DEFENSE POLICY, 80- *Concurrent Pos:* Mem, Coun Foreign Rels. *Honors & Awards:* Centennial Medal Extraordinary Achievement, Inst Elec & Electronics Engrs, 84. *Mem:* Fel AAAS; life fel Inst Elec & Electronics Engrs. *Res:* Radio wave propagation measurement and frequency measurement; electronics, including radar navigation systems and countermeasures; research administration; telecommunications policy and management. *Mailing Add:* 4307 N 39th St Arlington VA 22207-4605

CLARK, RALPH M, b Stowe, Vt, Dec 12, 26; m 59; c 1. ZOOLOGY. *Educ:* Univ Vt, AB, 50, MEd, 51; Wash State Univ, MS, 61; Univ Mass, PhD(zool), 63. *Prof Exp:* Teacher, Vt High Sch, 51-56, USAF High Sch, Ger, 56-58 & NY High Sch, 58-59; from asst prof to assoc prof, 63-72; PROF BIOL, STATE UNIV NY COL PLATTSBURGH, 72- *Concurrent Pos:* NY Res Found grants-in-aid, 63-66. *Mem:* AAAS; Am Soc Cell Biol; Sigma Xi. *Res:* Developmental biology. *Mailing Add:* One Washington Pl MR 10 Plattsburgh NY 12901

CLARK, RALPH O, b Broken Bow, Nebr, Aug 17, 12; m 34. CHEMISTRY. *Educ:* Univ Nebr, BS, 34, MS, 35. *Prof Exp:* Control chemist, Kendall Ref Co, 36-37; head anal sect, 37-55, asst dir, Gulf Res & Develop Co, 55-78, tech assoc, 61-78; RETIRED. *Mem:* Am Chem Soc. *Res:* Microanalysis; polarography as applied to analysis; chromatography of petroleum products; absorption analysis; analysis by x-ray fluorescence; emission spectroscopy; process analyzers. *Mailing Add:* White Pine Tree Rd Apt 109 Venice FL 34292-4215

CLARK, RAYMOND LOYD, b Tacoma, Wash, Jan 23, 35; m 55; c 4. PLANT PATHOLOGY. *Educ:* Wash State Univ, BS, 57, PhD(plant path), 61. *Prof Exp:* Res plant pathologist, Crops Res Div, Agr Res Serv, UDSA, 61-67, NCent Regional Plant Introd Sta, 65-90, res leader, Plant Introd Res Unit, 84-90, RES LEADER, WESTERN REGIONAL PLANT INTROD STA, AGR RES SERV, USDA, 90- *Mem:* Am Phytopath Soc; Sigma Xi; Am Soc Hort Sci. *Res:* Diplodia stalk rot of corn; tomato fruit rots; leptosphaerulina leafspot on alfalfa; root knot nematode; disease resistance; evaluation of foreign and wild germplasm. *Mailing Add:* Western Regional Plant Introd Sta USDA Agr Res Serv 59 Johnson Hall Pullman WA 99164-6402

CLARK, REGINALD HAROLD, b London, Eng, June 11, 28. CHEMICAL ENGINEERING. *Educ:* Univ London, BSc, 48, PhD(chem eng), 51. *Prof Exp:* Tech supvr, Murgatroyds Salt & Chem Co, Eng, 50-55; from asst prof to assoc prof, 55-69, chmn & head dept, 62-71, PROF CHEM ENG, QUEEN'S UNIV, ONT, 69- *Mem:* Chem Inst Can; Brit Inst Chem Engrs. *Res:* Fluid and process dynamics; mass transfer; chemical engineering processes; environmental studies; solid waste management. *Mailing Add:* Dept Chem Eng Queens Univ Kingston ON K7L 3N6 Can

CLARK, REX L, b Houston, Tex, Nov 9, 41; m 62; c 3. AGRICULTURAL & BIOLOGICAL ENGINEERING. *Educ:* Univ Ark, BS, 64; NC State Univ, MS, 65; Miss State Univ, PhD(eng), 68. *Prof Exp:* Asst prof, 68-76, assoc prof agr eng, Univ Ga, 76-; MEM STAFF, AGR ENG, UNIV GA. *Honors & Awards:* Outstanding Paper Award, Am Soc Agr Engrs, 70. *Mem:* Am Soc Agr Engrs. *Res:* Physical properties, mechanical, electrical, optical and sonic, of biological materials; mechanical harvesting of fruits and vegetables. *Mailing Add:* Agr Eng Driftmier Eng Ctr Univ Ga Athens GA 30602

CLARK, RICHARD BENNETT, b Charleston, WVa, Nov 1, 20; m 49; c 2. CHEMISTRY. *Educ:* WVa Univ, BS, 42; Yale Univ, PhD(chem), 51. *Prof Exp:* from Chemist to sr chemist, Tenn Eastman Co, 50-65, dept supt, Org Chem Devlop, 65-83; RETIRED. *Mem:* Am Chem Soc; Sigma Xi. *Res:* Processes for manufacture of organic chemicals with particular reference to dyestuffs and chemicals used in photographic processing. *Mailing Add:* 206 Willowbend Lane Kingsport TN 37660

CLARK, RICHARD JAMES, b Lockport, NY, July 15, 35; m 58; c 3. VERTEBRATE ZOOLOGY, ORNITHOLOGY. *Educ:* State Univ NY, Buffalo, BS, 59, MS, 63; Cornell Univ, PhD(vert zool), 70. *Prof Exp:* Researcher vert zool, Cornell Univ, 66-71; instr ethol, 71, from asst prof to assoc prof, 71-81, PROF ETHOL ENVIRON BIOL, FIELD NATURAL HIST & BASIC ORNITHOL, YORK COL, PA, 81- *Concurrent Pos:* Fel, NY Unit, Cooperative Wildlife Res Prog, 70-71; publ grants from var wildlife orgn, 71-; F M Chapman Mem grant, Am Mus Natural Hist, 69 & 76; reviewer instrumentation prog, NSF, 80 & Int Prog, 87; invited partic, XVII Congressus Int Ornithologici, Berlin, WGer, 78; hon adv & assoc ed var publ, Raptor Res Ctr, Hyderabad, India; assoc ed, J Raptor Res, 87; convenor rare owls biol & conserv of Third World Conf Birds of Prey, Eilat, Israel, 87; int corresp, Raptor Res Found, 87- *Mem:* Am Ornithologists Union; Wildlife Soc; Ecol Soc Am; Wilson Ornithol Soc; Raptor Res Found (vpres, 83-, pres, 90-); Nat Audubon Soc; Nat Parks & Conserv Asn; World Working Group Birds Prey (coordr); Nature Conservancy. *Res:* Autoecological studies of raptors, especially nocturnal, avifaunal population studies relative to habitat and land-use changes; research use of Federal lands; endangered avian species in Pennsylvania and adjacent states and the world; ecological parameters affecting bird song. *Mailing Add:* Dept Biol York Col Pa York PA 17403-3426

CLARK, ROBERT, b New York, NY, Apr 19, 33; m 55; c 3. TOXICOLOGY. *Educ:* Columbia Col, AB, 54; Columbia Univ, AM, 55, PhD(psychol), 58. *Prof Exp:* Asst psychol, Columbia Univ, 54-58, lectr, 56-58; res psychologist, Walter Reed Army Inst Res, 58-62; dir behav res, Hazleton Labs, Va, 62-63; res pharmacologist, 63-66, res supvr, 66-71, res assoc, Stine Lab, 71-73, asst dir toxicol, 73-74, DIR TOXICOL, ENDOCRINOL LAB, E I DU PONT DE NEMOURS & CO, INC, 74- *Mem:* AAAS; Am Psychol Asn; Soc Toxicol; Am Soc Pharmacol & Exp Therapeut. *Mailing Add:* Dept Pharmaceut Res & Develop Div E I du Pont de Nemours & Co Inc Barley Mill Plaza Wilmington DE 19898

CLARK, ROBERT A, b Oswego, NY, Jan 14, 42; m 65; c 3. INFECTIOUS DISEASES, CELL BIOLOGY. *Educ:* Syracuse Univ, AB, 63; Columbia Univ, MD, 67. *Prof Exp:* Resident physician, Univ Wash, 67-68 & Columbia Presby Med Ctr, 68-69; NIH fel, 69-72; chief resident physician, Univ Wash, 72-73, asst prof med, 73-77; from assoc prof to prof med, Boston Univ, 77-83; PROF MED & DIR DIV INFECTIOUS DIS, UNIV IOWA, 83-, ASSOC CHMN DEPT, 85-; DIR INFECTIOUS DIS SERV, VET ADMIN MED CTR, IOWA CITY, 83- *Concurrent Pos:* Res Career Develop Award, Nat Cancer Inst, NIH, 75-81; dir div infectious dis, Boston Univ Hosp, 77-83; med investr, Vet Admin, 85-91. *Mem:* Fel Am Col Physicians; Am Soc Clin Invest; Asn Am Physicians; fel Infectious Dis Soc Am; Am Fedn Clin Res; Am Asn Immunologists. *Res:* Acute inflammatory response; function of phagocytes; mechanisms of locomotion, secretion, and generation of toxic oxygen radicals. *Mailing Add:* Dept Med SW-54-15-GH Univ Iowa Iowa City IA 52242

CLARK, ROBERT ALFRED, b Boston, Mass, Oct 28, 08; m 37; c 3. PSYCHIATRY. *Educ:* Harvard Univ, AB, 30, MD, 34. *Prof Exp:* Resident physician neurol, Boston City Hosp, 34-35; intern med, Univ Hosps, Cleveland, 35-37; resident physician psychiat, Boston Psychopathic Hosp, 37-39; sr physician, RI State Hosp, 39-42; sr physician, Western Psychiat Inst, Pittsburgh, 42-44, clin dir, 44-55; clin dir, Friends Hosp, 55-58; med dir, 58-70, chief outpatient serv, 70-71, asst to med dir med educ & res, 71-74, psychiatrist, Northeast Community Ment Health Ctr, 74-78; dir student training, Friends Hosp, 68-78, clin prof, Hahnemann Med Col, 74-86; RETIRED. *Concurrent Pos:* Rockefeller fel, C G Jung Inst, Zurich, 48-49, Bollingen fel, 54; from instr to assoc prof, Sch Med, Univ Pittsburgh, 42-55; asst, Harvard Med Sch, 38-39; asst prof, Jefferson Med Col, 61-; clin assoc prof, Hahnemann Med Col, 70-74. *Mem:* fel Am Psychiat Asn; World Fedn Ment Health. *Res:* Psychiatry and religion. *Mailing Add:* 8301 Forest Ave Elkins Park PA 19117

CLARK, ROBERT ALFRED, b Smith Center, Kans, Aug 18, 24; m 52; c 2. METEOROLOGY, CIVIL ENGINEERING. *Educ:* Kans State Univ, BS, 48; Tex A&M Univ, MS, 59, PhD(meteorol), 64. *Prof Exp:* Hydraul engr, US Bur Reclamation, 45-50, 52-60; from assoc prof to prof meteorol, Tex A&M Univ, 60-73; dir, Off Hydrol, Nat Weather Serv, 73-85. *Mem:* Am Meteorol Soc; Am Soc Civil Eng; Am Geophys Union; Am Water Resources Asn. *Res:* Hydrology; hydrometeorology; physical meteorology. *Mailing Add:* 1801 Carter Creek Pkwy Bryan TX 77802

CLARK, ROBERT ARTHUR, b Melrose, Mass, May 3, 23; m 66. APPLIED MATHEMATICS. *Educ:* Duke Univ, AB, 44; Mass Inst Technol, MS, 46, PhD(math), 49. *Prof Exp:* Instr math, Mass Inst Technol, 46-49, res assoc, 49-50; from instr to prof, Case Inst Technol, 50-67, prof, 67-85, EMER PROF MATH, CASE WESTERN RESERVE UNIV, 85- *Concurrent Pos:* Vis asst prof, Mass Inst Technol, 56-57; vis mem, US Army Res Ctr, Madison, Wis, 61-62. *Mem:* Am Math Soc; Math Asn Am; Soc Indust & Appl Math; AAAS; Sigma Xi. *Res:* Elasticity, shell theory; asymptotic theory of differential equations. *Mailing Add:* Dept Math Case Western Reserve Univ Cleveland OH 44106

CLARK, ROBERT BECK, b Rock Springs, Wyo, July 18, 41; m 59; c 4. ELEMENTARY PARTICLE PHYSICS. *Educ:* Yale Univ, BA, 63, MPhil, 67, PhD(physics), 68. *Prof Exp:* Fac assoc physics, Ctr Particle Theory, Univ Tex, Austin, 68-70, asst prof, dept physics, 70-73; from asst prof to assoc prof, 73-87, PROF PHYSICS, TEX A&M UNIV, 87- *Concurrent Pos:* Mem gov bd, Am Inst Physics, 78-88; pres coun, Inter-Am Conf, 85-87; mem exec bd Coun Sci Soc Pres, 87- *Honors & Awards:* Robert N Little Award, 85. *Mem:* Am Phys Soc; Am Asn Physics Teachers (treas, 78-83, vpres, 84 & pres, 86). *Res:* Investigations of the electromagnetic and weak interactions of elementary particles. *Mailing Add:* Dept Physics Tex A&M Univ College Station TX 77843

CLARK, ROBERT EDWARD HOLMES, b July 8, 47; m 70; c 2. ELECTRON IMPACT EXCITATIONS. *Educ:* Frostburg State Col, BS, 69; Pa State Univ, MS, 78, PhD(astron), 80. *Prof Exp:* FEL, LOS ALAMOS NAT LAB, 80- *Mem:* Sigma Xi. *Res:* Theoretical calculations of electron impact excitation cross sections; rates for positively charged atomic ions. *Mailing Add:* 110 Yosemite Dr Los Alamos NM 87544

CLARK, ROBERT H, b Winnipeg, Man, Dec 25, 21; m 43; c 2. HYDROLOGY. *Educ:* McGill Univ, BEng, 43, MEng, 45. *Prof Exp:* Demonstr civil eng, McGill Univ, 43-44; hydraul design engr, Hydraul Div, Dominion Eng Works, Ltd, Montreal, 46-48; lectr civil eng, Univ Man, 48-52, asst prof, 52; chief hydraul engr, Red River Invest, Winnipeg, Can Dept Resources & Develop, 50-53, asst chief water resources br, 53-57; chief hydraul engr, Can Dept Northern Affairs & Nat Resources, 57-66, chief planning div, Dept Energy, Mines & Resources, 66-68, spec adv, Inland Waters Br, 68-74, sr eng adv, dept environ, 74-80; RETIRED. *Concurrent Pos:* Mem, Prairie Prov Water Bd, 53-69; mem, Souris-Red River Eng Bd, 53-; chmn subcomt hydrol, Nat Res Coun Can, 57-66; mem, Greater Winnipeg Floodway Adv Bd, 62-69; secy, Can Nat Comt, Int Hydrol Decade, 64-68; chmn working group on guide & tech regulations, Comn Hydrometeorol, World Meteorol Orgn, 64-72 & mem adv working group, 68-72; chmn, Can Sect, Int Great Lakes Working Comt, 65-; chmn, Atlantic Tidal Power Eng & Mgt Comt, 66-70; chmn, Can Sect, Int Niagara Bd Control, 69-73; mem for Can, Int Lake Superior Bd of Control & Int Niagara Comt, 69-73; chmn, Mgt Comt Bay Fundy Fed Power Reassessment Studies, 75-78; vpres, Comn Hydrol World Meteorol Orgn, 72-76; pres, Comn Hydrol, 76-84; consult, Water Resources, 81- *Mem:* Am Geophys Union; Int Asn Sci Hydrol. *Res:* Snowmelt floods and river flow under ice conditions; water resources planning and development; tidal power engineering. *Mailing Add:* 8567 Sansum Park Dr Sidney BC V8L 4T6 Can

CLARK, ROBERT NEWHALL, b Ann Arbor, Mich, Apr 17, 25; m 49; c 4. ELECTRICAL ENGINEERING, AERONAUTICS & ASTRONAUTICS. *Educ:* Univ Mich, BSE, 50, MSE, 51; Stanford Univ, PhD, 69. *Prof Exp:* Res engr, Honeywell, Inc, Minn, 51-57; from asst prof to prof, 57-66, consult, Appl Physics Lab, 57-77, NSF fac fel, 66-67, PROF ELEC ENG, AERONAUT & ASTRONAUT, UNIV WASH, 57- *Concurrent Pos:* Consult, Boeing Co, Wash, 58-66 & 71-91 & Perspective, Inc, 64-66; gen chmn, Joint Automatic Control Conf, 66; lectr, Stanford Univ, 68; vis scientist, Fraunhofer Gesellschaft Inst Informationsverarbeitung Technik & Biologie, Karlsruhe, Germany, 76-77; guest prof, Univ Duisburg, Germany, 83-84. *Mem:* Am Inst Aeronaut & Astronaut; fel Inst Elec & Electronics Engrs. *Res:* Automatic control system theory; guidance and control of aerospace vehicles; electromechanical servomechanisms; inertial systems and component development; fault detection in systems. *Mailing Add:* Dept Aeronaut & Astronaut Univ Wash FS-10 Seattle WA 98195

CLARK, ROBERT PAUL, b Jackson, Mich, July 19, 35; m 59; c 4. ELECTROCHEMISTRY, BATTERY TECHNOLOGY. *Educ:* Univ Mich, BS, 57; Univ Ill, MS, 59, PhD(anal chem), 62. *Prof Exp:* Staff mem, Power Sources Div, 62-69, mem tech staff, Explor Battery Div, 69-80, supvr, Storage Batteries Div, 80-87, SUPVR, BATTERY DEVELOP DIV, SANDIA NAT LABS, 87- *Mem:* Am Chem Soc; Electrochem Soc. *Res:* Batteries; energy conversion; thermal batteries; power sources; electrochemistry; storage batteries; utility load leveling; electric vehicles; alternative energy sources. *Mailing Add:* Div 2522 Sandia Nat Labs Albuquerque NM 87185

CLARK, ROBERT VERNON, plant pathology, for more information see previous edition

CLARK, ROGER ALAN, b Feb 24, 46; m; c 2. GEOLOGY APPLIED ECONOMIC & ENGINEERING, ENVIRONMENTAL SCIENCES GENERAL. *Educ:* Bloomsburg Univ, BA, 69; State Univ NY, Binghamton, MA, 74; PhD(geol), 81. *Prof Exp:* Geologist, Amoco Oil, 77-80, Dome Petroleum, 80-81, Slawson Oil, 81-86; environ geologist, Recra Environ, 87-89; ENVIRON GEOLOGIST, NUS CORP, 89- *Mem:* Am Asn Petrol Geologist; Soc Econ Paleontologist & Mineralogist. *Mailing Add:* PO Box 22A Bradfordwoods PA 15015

CLARK, ROGER WILLIAM, b Oxford, Nebr, Nov 23, 42; m 67; c 2. MOLECULAR BIOLOGY, GENETICS. *Educ:* Colo State Univ, BS, 65, MS, 67; Univ Ill, PhD(genetics), 71. *Prof Exp:* from instr to assoc prof molecular genetics, Sch Med & Dent, Univ Rochester, 74-81; RES ASSOC PROF, UNIV DENVER, 81- *Concurrent Pos:* NIH fel gen med sci, Fla State Univ, 71-72; Nat Cancer Inst fel, Univ Tex, Houston, 72-74; prin investr, 78- *Mem:* Sigma Xi; AAAS. *Res:* Chromosome structure; biochemistry and biophysics of nucleic acids; DNA replication and gene amplification in mammalian cells. *Mailing Add:* Dept Biol Univ Denver Univ Park Denver CO 80208

CLARK, RONALD DAVID, b Leeds, Eng, July 18, 38; m 64; c 1. ORGANIC CHEMISTRY, ACADEMIC ADMINISTRATION. *Educ:* Univ Leeds, BSc, 59, PhD(chem), 62. *Prof Exp:* Fel, Univ Nebr, 62-65; sci officer chem, Radiochem Ctr, UK Atomic Energy Authority, 65-67; assoc prof chem, Jamestown Col, 67-75; DEAN, SCH NATURAL & SOCIAL SCI, KEARNEY STATE COL, 75- *Mem:* Am Chem Soc; Royal Soc Chem; Sigma Xi. *Res:* Steroid synthesis; synthesis of small ring compounds and isotopically labelled polypeptides. *Mailing Add:* 1617 8th Ave Kearney NE 68847

CLARK, RONALD DUANE, b Hollywood, Calif, Nov 21, 38; m 67; c 4. ORGANIC CHEMISTRY. *Educ:* Univ Calif, Los Angeles, BS, 60; Univ Calif, Riverside, PhD(org chem), 64. *Prof Exp:* Fel, Mich State Univ, 64-65; sr res chemist, Standard Oil Co, Ohio, 65-69; from asst prof to assoc prof, NMex Highlands Univ, 69-78, chmn, Div Sci & Math, 77-83, dean, Sch Sci & Technol, 88-90, PROF CHEM, NMEX HIGHLANDS UNIV, 78-; PRES, CYCAD PROD, 83- *Concurrent Pos:* Pres, SCE Corp, 87-90. *Mem:* Am Chem Soc. *Res:* High tech ceramics and abrasives; optical polishing chemicals; nonlinear optical materials. *Mailing Add:* Sch Sci & Technol NMex Highlands Univ Las Vegas NM 87701

CLARK, RONALD JENE, b Hutchinson, Kans, July 11, 32; m 56; c 2. INORGANIC CHEMISTRY. *Educ:* Univ Kans, BS, 54, PhD(chem), 58. *Prof Exp:* Asst, Univ Kans, 54-57; res chemist, Linde Co, NY, 58-61; instr & assoc, Iowa State Univ, 61-62; from instr to assoc prof, 62-72, PROF CHEM, FLA STATE UNIV, 72- *Mem:* Am Chem Soc; Chem Soc; AAAS; Sigma Xi. *Res:* Metal coordination compounds-phosphorus triflouride substitution products of metal carbonyls; lower oxidation state compounds of transition and actinide elements; materials science; superconductivity. *Mailing Add:* Dept Chem Fla State Univ Tallahassee FL 32306

CLARK, RONALD KEITH, b Los Angeles, Calif, Aug 28, 41; m 60; c 2. PHYSICAL CHEMISTRY. *Educ:* Univ Calif, Riverside, BA, 63, PhD(phys chem), 66. *Prof Exp:* NIH fel, Cornell Univ, 66-67; chemist, Shell Develop Co, 67-74, sr res chemist, 74-79, staff res chemist, 79-89, SR STAFF RES CHEMIST, SHELL DEVELOP CO, 89- *Honors & Awards:* Serv Award Am Petrol Inst, 90. *Mem:* Sigma Xi; Am Chem Soc; Soc Petrol Engrs. *Res:* Statistical thermodynamics and mechanics of nonelectrolyte solutions; theory of phase transitions, particularly in the critical region; oil-well drilling fluids; phase behavior of polymer solutions. *Mailing Add:* Shell Develop Co PO Box 481 Houston TX 77001

CLARK, RONALD ROGERS, b Percy, NH, Jan 7, 35; m 56; c 4. ELECTRICAL ENGINEERING, RADIO ASTRONOMY. *Educ:* Univ NH, BS, 56; Yale Univ, MEng, 57; Syracuse Univ, PhD(elec eng, automatic control), 63. *Prof Exp:* Assoc prof, 57-67, PROF ELEC & COMPUT ENG, UNIV NH, 67- *Mem:* Inst Elec & Electronics Engrs; Am Geophys Union. *Res:* Communication and control systems; meteor wind radar system; digital signal processing; ionospheric propagation and antenna systems. *Mailing Add:* ECE Univ NH Kingsbury Durham NH 03824

CLARK, RUSSELL NORMAN, b Norwood, Ohio, Mar 23, 21; c 7. ORGANIC CHEMISTRY. *Educ:* Xavier Univ, BS, 42; Univ Detroit, MS, 43; Iowa State Univ, PhD(org chem), 45. *Prof Exp:* Tech supt, E I du Pont de Nemours & Co, Inc, 45-60; vpres res & com develop, Celanese Plastic Co, 60-67; exec vpres, Inmont Corp, 67-70; vpres & gen mgr, Polyester Div, 70-75, VPRES TECH, ECUSTA CORP, 75- *Mem:* Am Chem Soc. *Res:* Polymer chemistry; cellulose chemistry; film fabrication; plastics fabrication; aquaculture; hydroponics. *Mailing Add:* 220 N Shore Dr Hendersonville NC 28739

CLARK, SAM LILLARD, JR, b St Louis, Mo, June 9, 26; m 74; c 6. INTERNAL MEDICINE, ANATOMY. *Educ:* Harvard Univ, MD, 49. *Prof Exp:* Intern med, Mass Gen Hosp, 49-50; from instr to assoc prof anat, Wash Univ, 54-68; chmn dept, 68-77, dir, Ctr Educ Resources, 77-84, resident internal med, 82-84, EMER PROF ANAT, MED SCH, UNIV MASS, 68- *Concurrent Pos:* Nat Res Coun fel med sci biochem & nutrit, Med Sch, Vanderbilt Univ, 50-52; Palmer sr res fel; USPHS sr res fel; USPHS career develop award anat, Wash Univ; mgr ed, Am J Anat, 74-80; pvt pract, internal med, 84- *Mem:* Asn Am Med Cols; Am Asn Anat (vpres, 77-79). *Res:* Cellular differentiation; immunology; electron microscopy of tissues, relating structure to function. *Mailing Add:* 15 Merriam Rd Grafton MA 01519

CLARK, SAMUEL FRIEND, b Danville, Ky, Jan 16, 14; m 46; c 2. ORGANIC CHEMISTRY. *Educ:* Univ WVa, AB, 34, MS, 37; Univ NC, PhD(org chem), 39. *Prof Exp:* Asst org chem, Univ WVa, 34-36, Johns Hopkins Univ, 36-37 & Univ NC, 37-39; res chemist, Union Carbide Chem Co, WVa, 39-42, group leader, 42-46; from assoc prof to prof chem, Univ Miss, 46-63, chmn, 55-63; chmn dept chem & phys sci, 63-67, chmn dept chem, 68-77, PROF CHEM, FLA ATLANTIC UNIV, 63- *Concurrent Pos:* Summer sr scientist, Union Carbide Nuclear Co, Tenn, 51-56; temporary chem liaison & field rep, NSF-AID Prog to Asn Cent Am Univs, Costa Rica, 67-68; US AID consult, Pedag Insts & Simon Bolivar Univ, Venezuela, 69. *Mem:* Am Chem Soc. *Res:* Constitution of natural tannins; synthesis of vinyl monomers and polymers, azo dyes and insecticides; reaction kinetics in radiology; molecular rearrangements. *Mailing Add:* Dept Chem Fla Atlantic Univ Boca Raton FL 33431

CLARK, SAMUEL KELLY, b Ypsilanti, Mich, Nov 3, 24; m 51; c 5. APPLIED MECHANICS, MECHANICAL ENGINEERING. *Educ:* Univ Mich, BSE, 46, MSE, 48, PhD(appl mech), 51. *Prof Exp:* Staff engr, Douglas Aircraft Co, 46-47, Borg Warner Corp, 48-50 & Ford Motor Co Sci Lab, 50-52; asst prof, Case Inst Technol, 52-55; PROF APPL MECH, UNIV MICH, 55- *Mem:* Soc Automotive Engrs; Am Soc Mech Engrs; Soc Exp Stress Analysis; Am Soc Testing & Mat. *Res:* Applied mechanics as applied to problems in the rubber industry, with particular attention to pneumatic tires. *Mailing Add:* Dept Mech Eng & Appl Mech Univ Mich Ann Arbor MI 48109

CLARK, SANDRA HELEN BECKER, b Kansas City, Mo, July 27, 38; div; c 2. GEOLOGY. *Educ:* Univ Idaho, BS, 63, MS, 64, PhD(geol), 68. *Prof Exp:* Geologist, Cominco Am, Inc, 66-67; geologist, Alaska Mineral Resources Br, Calif, 67-72, staff geologist, Off Mineral Resources, US Geol Surv, 72-74, equal employ opportunity officer & spec asst to dir, 78-80, GEOLOGIST, EASTERN MINERAL RESOURCES BUR, US GEOL SURV, RESTON, VA, 80- *Concurrent Pos:* Coord staff mem, Alaska Natural Gas EIS Task Force, Dept of Interior, Washington, DC, 74-75, partic, Mgr Develop Prog, 75-76. *Mem:* Geol Soc Am; Soc Econ Geologists; Am Asn Petrol Geologists; Soc Econ Paleontologists & Mineralogists; Asn Women Geoscientists; Asn Geoscientists Int Develop. *Res:* Geologic mapping; field and laboratory studies of structure and petrology of metamorphic and igneous terranes in northern Idaho, east central and south central Alaska; zinc, lead, barite in east-central US; barite commodity specialist. *Mailing Add:* 11151 Timberhead Lane Reston VA 22091

CLARK, SIDNEY GILBERT, b Wolfsburg, Pa, Sept 2, 30; m 57; c 2. ORGANIC CHEMISTRY. *Educ:* Juniata Col, BS, 53; Pa State Univ, MS, 56, PhD(org chem), 58. *Prof Exp:* res chemist, Monsanto Indust Chem Co, 57-62, res specialist, 62-64, group leader sulfonation-sulfation, 64-66, sr res group leader, 66-69, sect mgr detergents & phosphates res & develop, 69-78, proj mgr com develop, 78-82, mgr com develop, 83-85, BUS DEVELOP MGR, MONSANTO CHEM CO, 86- *Mem:* Am Chem Soc; Sigma Xi; Am Oil Chemists' Soc. *Res:* Alkylation; sulfonation; ethoxylation; surfactant intermediates; anionic and nonionic surfactants; enzymes. *Mailing Add:* 12800 Mason Manor St Louis MO 63141

CLARK, STANLEY JOE, b McPherson, Kans, Sept 22, 31; m 58; c 5. AGRICULTURAL ENGINEERING. *Educ:* Kans State Univ, BS, 54, MS, 59; Purdue Univ, PhD(agr eng), 66. *Prof Exp:* Instr agr eng, Purdue Univ, 60-64; asst prof, Colo State Univ, 64-66; assoc prof, 66-75, PROF AGR ENG, KANS STATE UNIV, 75-, DEPT HEAD, 88- *Concurrent Pos:* Actg br chief, Agr Prod & Food Processing Br, Indust Energy Conserv & Solar Applications, Dept of Energy, Washington, DC, 77-78, consult, Dept of Energy & Off Technol Assessment, 78-79. *Mem:* Am Soc Agr Engrs; Am Soc Eng Educ. *Res:* Crop production mechanization; traction and tillage mechanics; alternative energy sources for agricultural production; energy conservation in agricultural production and food processing; farm equipment product liability and safety; farm equipment design. *Mailing Add:* Dept Agr Eng Kans State Univ Manhattan KS 66506

CLARK, STANLEY PRESTON, b Colby, Kans, June 24, 17; m 42; c 4. CHEMICAL ENGINEERING, GENERAL ENGINEERING. *Educ:* Park Col, AB, 39; Univ Kans, BS, 41; Tex A&M Univ, MS, 58. *Prof Exp:* Chemist, Jones-Dabney Paint Co, 41-42; jr chem engr, Tenn Valley Authority, 42-44; chem engr, Cent Soya Co, 44-48; assoc res engr, Food Protein Res & Develop Ctr, Tex A&M Univ, 48-80; process eng consult, 80-82; RETIRED. *Concurrent Pos:* Consult, UN Children's Fund, 63 & 65. *Mem:* Am Inst Chem Engrs; Am Chem Soc; Am Oil Chem Soc; Sigma Xi. *Res:* Processing of oilseeds and products from oilseeds. *Mailing Add:* Rte Four Box 440 College Station TX 77840

CLARK, STANLEY ROSS, b Berwyn, Alta, Mar 8, 37; m 62; c 2. COMPUTER SCIENCE. *Educ:* Univ BC, BASc, 59; Aberdeen Univ, MSc, 61; Univ Manchester, PhD(comput sci), 67. *Prof Exp:* Chief programmer math, Pac Naval Lab, Dept Nat Defence, 62-64; res assoc comput sci, Inst Comput Studies, Univ Man, 67-68, asst prof, 68-69; ASSOC PROF COMPUT SCI, UNIV VICTORIA, BC, 69- *Concurrent Pos:* Nat Res Coun Can grants, 68-70. *Mem:* Asn Comput Mach; Brit Comput Soc. *Res:* Design and implementation of programming languages; simulation studies of computer systems. *Mailing Add:* 1292 Palmer Rd Victoria BC V8T 2H7 Can

CLARK, STEPHEN DARROUGH, b Seattle, Wash, Apr 10, 45; m 68; c 3. INDUSTRIAL WASTE TREATMENT, WATER MANAGEMENT. *Educ:* Seattle Univ, BS, 68; Mass Inst Technol, PhD(org chem), 72. *Prof Exp:* Fel chem, Syntex Res Inc, Calif, 72-73; res chemist, Arapahoe Chem Inc, 73-75, group leader chem, Arapahoe-Newport Div, Syntex Inc, Ind, 75-78, group leader, Process Develop, Arapahoe Chem Inc, 78-79, mgr, 79; dist rep, Nalco Chem Co, 79-82, area mgr, 81-83, dist mgr, 82-90; TECH DIR, BOILER WATER CHEM, 90- *Concurrent Pos:* Microcomputer consult; owner, Computerland, Mt Vernon, Wa; postdoctoral, Syntex Corp, 72. *Mem:* Am Chem Soc; Sigma Xi; Tech Asn Pulp & Paper Indust; Nat Asn Corrosion Engrs. *Res:* Process development research in pharmaceutical intermediates fine chemicals; pilot plant management; secondary waste treatment operation, fuels and combustion; applications for corrosion control; energy management; development and commercialization of new chemistry for boiler and power generating systems. *Mailing Add:* 1088 Roberts Ct Batavia IL 60510

CLARK, STEPHEN HOWARD, b Boston, Mass, May 3, 40; m 65; c 2. FISH BIOLOGY, MARINE SCIENCE. *Educ:* Univ Maine, BS, 66, MS, 68; Univ Miami, PhD(marine sci), 70. *Prof Exp:* Asst prof biol, Grand Valley State Col, 70-73; res fishery biologist, Nat Marine Fisheries Serv, Gulf Coast Fisheries Ctr, Galveston Lab, Galveston, Tex, 73-74; res fishery biologist, 74-81, SUPVRY RES FISHERY BIOLOGIST, NAT MARINE FISHERIES SERV, NORTHEAST FISHERIES CTR, WOODS HOLE LAB, WOODS HOLE, MASS, 82- *Concurrent Pos:* US rep, Shellfish Comt, Int Coun Explor of Sea, 78-82; mem, working group, Int Coun Explor Sea, Homarus, 78-80, Saithe, 78-82, European hake, 78-81, Arctic Fisheries, 83-, Flatfish, 82-, Pandalus, 83-; chief, Georges Bank-Gulf of Maine Resources Invest, 82-85 & Pop Biol Br, 85- *Mem:* Am Fisheries Soc; Nat Shellfisheries Asn; Am Inst Fishery Res Biologists. *Res:* Fishery science; evaluation of the effects of fishing on marine finfish and crustacean stocks; study of environmental influences on abundance and distribution; fishery biology; survey and sampling methods. *Mailing Add:* Nat Marine Fisheries Serv Northeast Fisheries Ctr Woods Hole MA 02543

CLARK, SYDNEY P, JR, b Philadelphia, Pa, July 26, 29; m 63; c 4. GEOPHYSICS. *Educ:* Harvard Univ, AB, 51, MA, 53, PhD(geol), 55. *Prof Exp:* Res fel geophys, Harvard Univ, 55-57; geophysicist, Geophys Lab, Carnegie Inst, 57-62; WEINBERG PROF GEOPHYS, YALE UNIV, 62-, DIR UNDERGRAD STUDIES, 81- *Concurrent Pos:* Fulbright scholar, Australian Nat Univ, 63. *Mem:* Fel Am Geophys Union. *Res:* Terrestrial and lunar heat flow; high pressure phase equilibria; constitution of earth's interior. *Mailing Add:* 78 N Lake Dr Hamden CT 06517

CLARK, T(HELMA) K, b Pa, Aug 11, 30; m 54; c 1. PSYCHOLOGICAL PHYSICS. *Educ:* Ind Univ, Pa, BS, 52; Univ Pittsburgh, PhD(biol & psychol), 79. *Prof Exp:* RES ASSOC, BIOL SCI DEPT, UNIV PITTSBURGH, 80- *Mem:* Soc Neurosci; Parapsychol Asn. *Res:* Organizational patterns in psychokinetic behavior; function of the locus coeruleus-dorsal noradrenaline system. *Mailing Add:* Biol Sci Dept Univ Pittsburgh Pittsburgh PA 15260

CLARK, THOMAS ALAN, b Leicestershire, Eng, Mar 14, 38; m 60; c 2. ASTRONOMY. *Educ:* Univ Leeds, BSc, 59, PhD(cosmic ray physics), 63. *Prof Exp:* Fel physics, Univ Calgary, 62-64, sessional lectr, 64-65; vis scientist, Defence Res Telecommun Estab, Defence Res Bd Can, 65; lectr physics, Univ Col, Univ London, 66-69, tutor, 68-69; from asst prof to assoc prof, 69-81; CO-DIR, ROTHNEY OBSERV, UNIV CALGARY, 72-, EXEC

ASST DEAN SCI, PHYSICS/ASTRON, 80-, PROF PHYSICS, 81- *Concurrent Pos:* Mem, Adv Comt Stratospheric Pollution, Atmospheric Environ Serv, Can, 74-84, bd dir, Westar, 82-, assoc comt space res, Nat Res Coun, Can, 84-89; Killam res fel, Univ Calgary, 79. *Mem:* Royal Astron Soc Can; fel Brit Inst Physics; Am Astron Soc; Can Astron Soc. *Res:* Infrared astronomy; far infrared solar spectral and eclipse studies by balloon- and aircraft-borne intrumentation; stratospheric emission; spectral measurements in the Far Infra Red as an aid in pollution studies; Infra Red Solar measurements. *Mailing Add:* Physics Dept Univ Calgary 2500 University Dr NW Calgary AB T2N 1N4 Can

CLARK, THOMAS ARVID, b Durango, Colo, Aug 23, 39; m 67. RADIO ASTRONOMY, GEODESY. *Educ:* Univ Colo, BS, 61, PhD(astrophys), 67. *Prof Exp:* Staff scientist astron, Boulder Labs, Environ Sci Serv Admin, Colo, 61-66 & NASA Marshall Space Flight Ctr, 66-68; RADIO ASTRONR, NASA GODDARD SPACEFLIGHT CTR, 68-; ASSOC PROF ASTRON, UNIV MD, 69- *Concurrent Pos:* Exec vpres, 74-81, pres, Radioamateur Satellite Corp, 81-; mem study group 2nd radio astron, Int Consultative Radio Comn, 74-79; mem radio astron subcomt, Comt Radio Frequencies, Nat Res Coun-Nat Acad Sci, 74-79; co-chmn serv working group radio astron, Fed Commun Comn, 75-79. *Mem:* Int Astron Union; Am Astron Soc; Int Sci Radio Union; Am Geophys Union; AAAS. *Res:* Development of very long baseline interferometry for high accuracy astronomical and geophysical measurements; development of astronomical instrumentation; radio frequency spectrum management; education. *Mailing Add:* 6388 Guilford Rd Clarksville MD 21029

CLARK, THOMAS HENRY, b London, Eng, Dec 3, 93; m 27; c 1. STRATIGRAPHY, PALEONTOLOGY. *Educ:* Harvard Univ, AB, 17, AM, 21, PhD(geol, paleont), 23. *Prof Exp:* Instr, Harvard Univ, 15-17 & 20-24; from asst prof to assoc prof paleont, 24-29, Logan prof, 29-62, prof, 62-64, dir univ mus, 25-52, chmn dept geol sci, 52-59, EMER PROF, MCGILL UNIV, 64- *Concurrent Pos:* Geologist, Geol Surv Can, 28-31 & 35 & Dept Mines, Que, 38-64; consult, Dept Natural Resources, Que, 64-69; consult geologist, 62-; adv geol, Redpath Mus, 64- *Honors & Awards:* Logan Medal, 71. *Mem:* Paleont Soc; Geol Soc Am; Royal Soc Can; Geol Asn Can (pres, 58-59); Can Inst Mining & Metal. *Res:* Invertebrate paleontology; Paleozoic stratigraphy. *Mailing Add:* Dept Geol Sci McGill Univ Montreal PQ H3A 2A7 Can

CLARK, TREVOR H, b Haviland, Kans, July 16, 09; m 33, 66; c 4. PHYSICS. *Educ:* Friends Univ, AB, 30, Univ Mich, MS, 33. *Prof Exp:* Serv mgr, Geo E Marshall Co, Kans, 29-32; serviceman, Int Radio, Mich, 33-34; lab asst, Univ Mich, 34; engr, Radio Corp Am Mfg Co, NJ, 34-38; engr, Les Laboratoires LMT, Int Tel & Tel Co, France, 38-40, dept head, Fed Telecommunication Labs, NJ, 40-45, div head, 45-47, mgr tech servs, 47-48, mgr eng servs & special projs, 49-51; asst to pres, Fed Tel & Radio Corp, 48-49; assoc dir, Southwest Res Inst, 51-55; asst to eng mgr, Air Arm Div, Westinghouse Elec Corp, 55-61, mgr underwater launch prog, Aerospace Div, 61-64, mgr deep submergence prog, 64-65, mgr prog opers, 65-67, asst div mgr, 67-69, mgr info serv, 70-71; CONSULT, 71- *Concurrent Pos:* Mem metrop adv bd, Md Coastal Zone Prog, 80-; ed, Lookout, Nat Boating Fedn, 83- *Mem:* Fel Inst Elec & Electronics Engrs; Acoustical Soc Am; assoc fel Am Inst Aeronaut & Astronaut; Am Inst Physics. *Res:* High vacuum; sound; thermionics; microwaves; wave propagation; telephony; electron emission; navigation; vacuum tubes; photocells; multipliers; beam tubes; switching systems; direction finders; antennas; communications; countermeasures; radar; research administration; ecology. *Mailing Add:* 684 Blue Crab Cove Annapolis MD 21401

CLARK, VIRGINIA, b Grand Rapids, Mich, Nov 18, 28; m 53. BIOSTATISTICS. *Educ:* Univ Mich, BA, 50, MA, 51; Univ Calif, Los Angeles, PhD(biostatist), 63. *Prof Exp:* Statistician, Gen Elec Co, 51-54; appl mathematician, Econ Res Proj, Harvard Univ, 54-56, res assoc biostatist, Med Sch, 60-61; statistician, Systs Lab Corp, 56-57; from asst prof to assoc prof biostatist, Sch Pub Health, Univ Calif, Los Angeles, 63-74, assoc prof biomath, 71-74, prof biostatist & prof biomath; MEM STAFF, JULES STEIN EYE INST, HEALTH SCI CTR, UNIV CALIF. *Mem:* Inst Math Statist; Biomet Soc; fel Am Statist Asn. *Mailing Add:* Dept Ophthal Jules Stein Eye Inst Univ Calif Ctr Health Sci 100 Stein Plaza Los Angeles CA 90024-7008

CLARK, VIRGINIA LEE, b Washington, DC, Aug 30, 45. MICROBIAL PHYSIOLOGY, MEDICAL MICROBIOLOGY. *Educ:* Carleton Col, BA, 67; Univ Rochester, PhD(microbiol), 77. *Prof Exp:* Fel biol, Mass Inst Technol, 76-78, instr, 79; asst prof, 79-85, ASSOC PROF MICROBIOL, UNIV ROCHESTER, 85- *Mem:* Am Soc Microbiol; AAAS; NY Acad Sci. *Res:* Genetic engineering and molecular biology of pathogenicity in Neisseria gonorrhoeae; molecular biology of antibiotic resistance in Bacteroides fragilis; regulation of translation in prokaryotes. *Mailing Add:* Dept Microbiol Sch Med Univ Rochester Box 672 Rochester NY 14642

CLARK, WALLACE HENDERSON, JR, b LaGrange, Ga, May 16, 24; c 5. MEDICINE, PATHOLOGY. *Educ:* Tulane Univ, BS, 44, MD, 47. *Prof Exp:* From instr to prof path, Sch Med, Tulane Univ, 49-62; asst prof, Harvard Univ, 62-68, assoc clin prof, 68-69; prof path, Sch Med, Temple Univ, 69-78, chmn dept, 74-78; RES PROF DERMAT & PATH, SCH MED, UNIV PA, 78- *Concurrent Pos:* Markle scholar, 54-60; consult pathologist, Orleans Parish Coroner's Off, 50-52, 54-56 & Armed Forces Inst Path, Washington, DC; assoc path, Mass Gen Hosp, 62-68. *Mem:* Am Asn Path & Bact; Am Asn Cancer Res; Am Soc Exp Path; Am Acad Dermat. *Res:* Dermal pathology; electron microscopy; correlation of ultrastructural changes with known changes in cellular function; tumor progression in human neoplastic systems; immunology and fine structure of primary human cutaneous malignant melanomas; induction of animal model of human malignant melanoma in the guinea pig. *Mailing Add:* Dept Dermat Univ Pa Med Educ Bldg GM 36th & Hamilton Walk Philadelphia PA 19104

CLARK, WALLACE LEE, b St Joseph, Mo, June 12, 44; m 80; c 3. ATMOSPHERIC PHYSICS. *Educ:* Univ Colo, BA, 67. *Prof Exp:* PHYSICIST, AERONOMY LAB, NAT OCEANIC & ATMOSPHERIC ADMIN, DEPT COM, 67- *Honors & Awards:* Silver Medal Award, Dept of Com, 79. *Mem:* Am Geophys Union; Am Meteorol Soc. *Res:* Radar studies of atmospheric dynamics. *Mailing Add:* Aeronomy Lab R/E/AL3 Nat Oceanic & Atmospheric Admin 325 Broadway Boulder CO 80303

CLARK, WALLACE THOMAS, III, b Bakersfield, Calif, Sept 9, 1953; m 81; c 2. INSTRUMENTATION DESIGN & INTEGRATION, TRANSIENT ELECTROMAGNETIC PULSE DATA ACQUISITION & REDUCTION. *Educ:* NMex Inst Mining & Technol, BS, 75; Idaho State Univ, MS, 78; Univ Ark Grad Inst Technol, PhD(instrumental sci), 82. *Prof Exp:* Grad teaching asst physics, Idaho State Univ, 76-78; grad res asst, Univ Ark Grad Inst Technol, 78-82; develop assoc III, Y-12 Weapons Plant, Oak Ridge, Tenn, Union Carbide Nuclear Systs & Martin Marietta Energy Systs, 85; mem tech staff, AT&T Bell Labs, 85-86; ARES site scientist, 86-88, TRESTLE CHIEF SCIENTIST, BDM MSC, 88- *Concurrent Pos:* Pres, NMex Inst Mining & Technol Alumni Assn, 87-90. *Mem:* Instrument Soc Am; Optical Soc Am. *Res:* Systems integration, especially high frequency data transducers, (fiber optic) links to digitizers and controlling computers and controllers; fiber optic and laser based transducers; continuous wave and transient pulse electromagnetic phenomenon. *Mailing Add:* BDM Int 1801 Randolph Rd SE Albuquerque NM 87106

CLARK, WALTER ERNEST, b Stuart, Va, Sept 25, 16; m 39; c 3. APPLIED CHEMISTRY. *Educ:* Va Mil Inst, BS, 37; George Washington Univ, MA, 39; Univ Wis, PhD(chem & chem eng), 49. *Prof Exp:* Instr chem, Va Mil Inst, 39-41; asst prof chem eng, Mo Sch Mines, 49-51; sr res chemist, Oak Ridge Nat Lab, 51-56, group leader, 56-79; RETIRED. *Concurrent Pos:* Lectr, Univ Tenn, 63-64; Fulbright lectr, Tribhuvan Univ, Nepal, 67-68. *Mem:* Am Chem Soc. *Res:* Electrochemistry; polarography; corrosion; nuclear fuel processing; nuclear waste disposal. *Mailing Add:* Rte Box 449 Stuart VA 24171

CLARK, WALTER LEIGHTON, III, b Springfield, Pa, Feb 3, 21; m 45; c 1. FOOD SCIENCE. *Educ:* Pomona Col, BA, 42; Georgetown Univ, MS, 46; Cornell Univ, PhD(biochem), 53. *Prof Exp:* Res assoc, Food Sci & Tech Div, NY State Agr Exp Sta, Cornell Univ, 48-50, asst, 50-53, asst prof biochem, Grad Sch Nutrit, 53-56; sr res biologist, Dept Nutrit & Food Technol, Lederle Labs, Agr Ctr, Am Cyanamid Co, 56-58, group leader food res, 58-61, group leader food res & develop, 61-64; tech mgr new prod develop refrig foods, Res Ctr, Pillsbury Co, 65-67; assoc dir res-explor, Quaker Oats Co, Ill, 67-73; CORP DIR SCI & NUTRIT, HUNT-WESSON FOODS, INC, 73- *Concurrent Pos:* Lectr on US Food Legislation & Regulations by Japan External Trade Relations Orgn, Tokyo, 78; adj prof food sci, Chapman Col, Orange, Calif, 79-; consult & lectr, 79- *Mem:* Am Oil Chemists Soc; Am Chem Soc; Inst Food Technologists (pres-elect, 78-79, pres, 79-80); Am Asn Cereal Chemists. *Res:* Teaching food science; food additives; diet and nutrition; protein sources and technology; food fabrication; heavy metals; food regulations; tomato processing; edible fats and oils; research management. *Mailing Add:* 1505 Clearview Lane Santa Ana CA 92705-1501

CLARK, WAYNE ELDEN, b Lehi, Utah, May 2, 43; m 66; c 4. SYSTEMATICS. *Educ:* Brigham Young Univ, BS, 68, MS, 70; Tex A&M Univ, PhD(entom), 75. *Prof Exp:* Res assoc, Tex A&M Univ, 74-75, fel, 76; fel, Smithsonian Inst, 75-76 & Univ Pa, 76-77; sr scientist, Technassociates, Inc, 77; mgr, Gypsy N Tech Info Proj, Libr Cong, 77-78; from asst prof to assoc prof, 83-89, PROF ENTOM, AUBURN UNIV, 89- *Mem:* Coleopterists Soc; Entom Soc Am; Sigma Xi; Soc Syst Zool. *Res:* Taxonomy, phylogeny and biogeography of weevils (insecta, coleoptera, curculionidae); studies on natural history and host plant relationships. *Mailing Add:* Dept Entom Auburn Univ Auburn AL 36849-5413

CLARK, WESLEY GLEASON, b Wadsworth, Ohio, July 1, 33; m 65; c 3. PHARMACOLOGY. *Educ:* Univ Colo, BA, 55, MS, 58; Univ Utah, PhD(pharmacol), 62. *Prof Exp:* From instr to asst prof, 62-72, ASSOC PROF PHARMACOL, UNIV TEX SOUTHWESTERN MED CTR, DALLAS, 72- *Concurrent Pos:* USPHS grant, Nat Inst Allergy & Infectious Dis, 64-66, Nat Inst Neurol Dis & Stroke, 70-78 & Nat Inst Drug Abuse, 79-82. *Mem:* Am Soc Pharmacol & Exp Therapeut; Soc Neurosci. *Res:* Neuropharmacology; effects of drugs on thermoregulation; bacterial pyrogens; food poisoning; vomiting. *Mailing Add:* Dept Pharmacol Univ Tex Southwestern Med Dallas TX 75325-9041

CLARK, WILBURN O, b Farber, Mo, Feb 9, 38; m 64; c 2. ELECTRICAL & SYSTEMS ENGINEERING, INFORMATION SCIENCE. *Educ:* Univ Kans, BSEE, 60, MSEE, 61, PhD(elec eng), 64. *Prof Exp:* Staff consult comput, Gen Elec Co, Ala, 64-66, consult specialist info serv, Ariz, 66-67; from asst prof to prof eng, Ariz State Univ, 67-80; systs engr, USAF, 80-81; prin engr, Motorola, Inc, 81-88; consult, Southtech, Inc, 89-90; CONSULT, 90- *Concurrent Pos:* Consult, Vendo Co, Mo, 62-64 & Goodyear Aerospace Corp, Ariz, 74-80; in-plant training instr, Motorola Inc, Ariz, 67. *Mem:* Inst Elec & Electronics Engrs; Asn Comput Mach; NY Acad Sci; Sigma Xi. *Res:* Translation and programming language specification and implementation; special purpose digital systems; x-band traveling wave tube experimentation instrumentation; systems engineering; structured software analysis, development, and testing. *Mailing Add:* 603 E Carson Dr Tempe AZ 85282

CLARK, WILLIAM ALAN THOMAS, b Hazlemere, Surrey, Eng, June 8, 48. MATERIALS SCIENCE ENGINEERING, SOLID STATE PHYSICS. *Educ:* Univ Liverpool, Eng, BSc, 72; Oxford Univ, DPhil(metall, sci mat), 86. *Prof Exp:* Sci res coun fel metall, Oxford Univ, 76-78; fel metall, Mich Technol Univ, 78-79; from asst prof to assoc prof metall eng, 79-89, PROF MAT SCI & ENG, OHIO STATE UNIV, 89- *Concurrent Pos:* Prin investr, US Dept Energy, 81-85; dir, Electron Optics Fac, Ohio State Univ, 82-87, mem senate, 88-; prin investr, Nat Sci Found, 83-; consult, Libby-Owens Ford Co, 87-; dept ed, Scripta Metallurgica, 88-; mem, Dept Energy Panel Workshop, 88-; Lawrence Livermore Nat Lab Adv Bicrystals, 88- *Honors &*

Awards: CO Bannister Prize, Livermore Metall Soc, 72. *Mem:* Metall Soc; Electron Micros Soc Am; Mat Res Soc; Inst Metallurgists; Am Soc Metals Int. *Res:* Structure and properties of interfaces in crystalline materials; transmission electron microscopy; electron diffraction; determining the physical and mechanical properties of materials. *Mailing Add:* Dept Mat Sci & Eng Ohio State Univ 116 W 19th Ave Columbus OH 43210

CLARK, WILLIAM B, chemical engineering; deceased, see previous edition for last biography

CLARK, WILLIAM BURTON, IV, b Madison, Fla, Nov 19, 47; m 69; c 2. PERIODONTICS, ORAL MICROBIOLOGY. *Educ:* Univ of the South, BA, 69; Emory Univ, DDS, 73; Harvard Univ, DMSc, 79. *Prof Exp:* From asst prof to assoc prof, 77-87, PROF ORAL BIOL & MICROBIOL, UNIV FLA, 87-; DIR RES, PERIODONT DIS RES CTR, 82- *Concurrent Pos:* Vis prof, Tokyo Dental Col, 82-; Nat Inst Dent Res, NIH, 82, res career develop award; prin investr, Periodont Dis Res Ctr-NIH Ctr grant, 85-; vis scientist, Nat Inst Dent Res, NIH, 85. *Res:* Mechanisms of bacterial attachment to teeth and development of vaccines made from bacterial adhesions to prevent or reduce colonization by periodontal disease-associated bacteria; clinical trials for antiplaque and anti-gingivitis agents; antiobiotic therapy in treatment of peridontal disease. *Mailing Add:* Periodont Dis Res Ctr Box J-424 JHMHC Gainesville FL 32610

CLARK, WILLIAM C, b 1901; m 24; c 3. PHARMACOLOGY. *Educ:* Ind Univ, PhD(pharmacol & physiol), 52. *Prof Exp:* Prof physiol & pharmacol, Univ Ill; RETIRED. *Mailing Add:* 616 Park Ave Covington IN 47932

CLARK, WILLIAM CUMMIN, b Greenwich, Conn, Dec 20, 48; m 80; c 2. POLICY ANALYSIS, SOCIOLOGY OF SCIENCE. *Educ:* Yale Univ, BSc, 71; Univ BC, PhD(ecol), 79. *Prof Exp:* Scientist, Inst Energy Anal Oak Ridge, Assoc Univ, 81-84, proj leader, Int Inst Appl Systs Anal, 84-87; SR RES ASSOC, KENNEDY SCH GOVT, HARVARD UNIV, 87-, ASST DIR, CTR SCI & INT AFFAIRS, 90- *Concurrent Pos:* Mem, US Nat Res Coun Comt Global Change, 87-90, bd Atmospheric Sci & Climate, 84-87, comt Appln Ecol Knowledge Environ Problems, 83-86; mem, Social Sci Res Coun Comt Global Change, 89-, Am Asn Adv Sci Comt Sci, Eng & Pub Policy, 91- *Honors & Awards:* MacArthur Prize, 83. *Mem:* Sigma Xi. *Res:* International science and technology policy; policy analysis for environment and resource management; social dimensions of risk assessment; development strategies for third world; global climate impact assessment; sustainable development. *Mailing Add:* Kennedy Sch Govt Harvard Univ 79 Kennedy St Cambridge MA 02138

CLARK, WILLIAM DEAN, b Guthrie, Okla, Feb 1, 36; m 54; c 2. MATHEMATICAL ANALYSIS. *Educ:* Cent State Col, Okla, BSEd, 62; Univ Tex, Austin, MA, 64, PhD(math), 68. *Prof Exp:* NSF Acad Year Inst partic, Univ Tex, Austin, 62-63; assoc prof math, 66-74, res grant, 68-69, PROF MATH, STEPHEN F AUSTIN STATE UNIV, 74- *Concurrent Pos:* NSF res grant, 70-73. *Honors & Awards:* Georg Polya Award, Math Asn Am, 80. *Mem:* Soc Indust & Appl Math; Math Asn Am. *Res:* Summability of series. *Mailing Add:* N2703 N Pecan Nacogdoches TX 75961-2815

CLARK, WILLIAM DENNIS, b Redding, Calif, Jan 20, 48; m 69, 81. PLANT TAXONOMY, NATURAL PRODUCTS CHEMISTRY. *Educ:* Sacramento State Col, BA, 70; Univ Tex-Austin, PhD(bot), 77. *Prof Exp:* Asst prof, 76-81, ASSOC PROF BOT, ARIZ STATE UNIV, 81- *Concurrent Pos:* Guest prof, Univ Heidelberg, 83-84, Univ Calif, Riverside, 89-90; Alexander von Humboldt fel, 83; Nat Sci Found Fel, 89. *Mem:* Bot Soc Am; Am Soc Plant Taxonomists; Phytochem Soc NAm; Int Asn Plant Taxon; AAAS; Am Asn Pharmacog; Int Soc Chem Ecol. *Res:* Chemical taxonomy of flowering plants, especially compositae; chemical ecology of plants; coevolution, especially plant-insect interactions; plant natural products chemistry, especially flavonoids; chemistry of medicinal plants. *Mailing Add:* Dept Bot Ariz State Univ Tempe AZ 85287-1601

CLARK, WILLIAM EDWIN, b Brunswick, Ga, Dec 7, 34; m 60; c 1. MATHEMATICS. *Educ:* Sam Houston State Col, BA, 60; Tulane Univ, PhD(math), 64. *Prof Exp:* Ford Found res fel math, Calif Inst Technol, 64-65; from asst prof to assoc prof, Univ Fla, 65-70; PROF MATH, UNIV S FLA, 70- *Concurrent Pos:* NSF res grant, 67- *Mem:* Am Math Soc; Math Asn Am. *Res:* Arithmetic coding theory. *Mailing Add:* Dept Math Univ SFla Tampa FL 33620

CLARK, WILLIAM GILBERT, b Los Angeles, Calif, May 26, 30; m 59; c 2. MAGNETIC RESONANCE. *Educ:* Stanford Univ, BS, 52; Cornell Univ, PhD(physics), 61. *Prof Exp:* Asst, Stanford Univ, 52-53; asst, Cornell Univ, 54-60; jr res physicist, Univ Calif, San Diego, 60-62, asst res physicist, 62-64; from asst prof to assoc prof, 64-73, PROF PHYSICS, UNIV CALIF, LOS ANGELES, 73- *Concurrent Pos:* Nat Ctr Sci Res fel, Fac Sci, Orsay, France, 69-70, exchange prof, 70; assoc prof, Sci & Med Univ, Univ Grenoble, France, 75-76. *Mem:* Am Phys Soc. *Res:* Low temperature physics; physical properties of pseudo one-dimensional solids at very low temperatures; experimental solid state physics. *Mailing Add:* Dept Physics Univ Calif 405 Hilgard Ave Los Angeles CA 90024

CLARK, WILLIAM GREER, b Waltham, Mass, Aug 6, 45; m 69; c 3. MARINE STOCK ASSESSMENT. *Educ:* Univ Mich, BA, 67; Univ Wash, PhD(fisheries), 75. *Prof Exp:* Fishery officer, Food & Agr Orgn UN, 75-79; res scientist, Wash Dept Fisheries, 83-88; BIOMETRICIAN, INT PAC HALIBUT COMN, SEATTLE, 88- *Concurrent Pos:* Res assoc, Univ Wash, 79- *Mem:* AAAS; Am Fisheries Soc. *Res:* Dynamics of exploited marine populations; mathematical problems in estimating population sizes and parameters. *Mailing Add:* Int Pac Halibut Comn PO Box 95009 Seattle WA 98145-2001

CLARK, WILLIAM HILTON, b Caldwell, Idaho, Dec 17, 44; m 68; c 3. ENTOMOLOGY, ECOLOGY. *Educ:* Col Idaho, BS, 67; Univ Nev-Reno, MS, 71. *Prof Exp:* Teaching asst bot, Univ Nev-Reno, 67-69, res asst ecol, 68 & 72-73; sr environ qual specialist, water pollution, 74-85, sr water qual analyst, non point source pollution, div environ, 85-89, NON POINT SOURCE MONITORING COORDR, IDAHO DEPT HEALTH & WELFARE, 89- *Concurrent Pos:* Consult, Nev Archaeol Surv, 72; asst investr, Res Adv Bd, Univ Nev, 72; prin investr, Sigma Xi, 73-74 & 76, Found Environ Educ, 74, Am Philos Soc, 74, Ctr Field Res & Earthwatch, 79, 80, 81 & 82 & Cactus & Succulent Soc, 84; res assoc, Col Idaho, 76-80, asst dir, O J Smith Mus Natural Hist & adj prof biol, 80-; AAAS rep, Idaho Acad Sci, 84-; pres, Idaho Acad Sci, 91-92; cert ecologist, Ecological Soc Am, 80-; fisheries scientist, Am Fisheries Soc, 85- *Mem:* AAAS; Brit Ecol Soc; Ecol Soc Am; Entomol Soc Am; NAm Benthological Soc; Sigma Xi(pres). *Res:* Entomology; ants of Idaho and Baja California, ant ecology, aquatic biology; pollution ecology; desert ecology; natural history of Baja California; ecology of Yucca, especially in Southwest United States and Mexico; water quality monitoring of agricultural nonpoint source pollution and effects on aquatic biota; nonpoint source pollution monitoring and evaluation. *Mailing Add:* 6305 Kirkwood Rd Boise ID 83709

CLARK, WILLIAM JESSE, b Salt Lake City, Utah, Sept 29, 23; m 51; c 2. LIMNOLOGY, AQUATIC ECOLOGY. *Educ:* Utah State Univ, BS, 50, MS, 56, PhD(aquatic biol), 58. *Prof Exp:* From asst prof to assoc prof biol, 57-68, assoc prof biol & wildlife sci, 68-75, ASSOC PROF WILDLIFE & FISHERIES SCI, TEX A&M UNIV, 75- *Mem:* Am Soc Limnol & Oceanog; Ecol Soc Am; Int Asn Theoret & Appl Limnol; Sigma Xi. *Res:* Ecology of ponds; regional limnology; limnology and ecology of rivers. *Mailing Add:* Dept Wildlife & Fisheries Sci Tex A&M Univ College Station TX 77843

CLARK, WILLIAM KEMP, b Dallas, Tex, Sept 2, 25; c 6. NEUROSURGERY. *Educ:* Univ Tex, BA, 45, MD, 48. *Prof Exp:* From asst prof to prof neurosurg, Univ Tex Health Sci Ctr, Dallas, 56-88, chmn div, 56-84; RETIRED. *Concurrent Pos:* Dir serv neurosurg, Parkland Mem Hosp, 56-88; attend neurol surg, Vet Admin Hosp, 56-90, consult, 64-90; consult, Children's Med Ctr, 56-90; chmn, Am Bd Neurolog Surg, 76-78; consult mneurosurg to surgeon gen, USN, 77-88 & USAF, 81-88; deleg, World Fedn Neurol Surgeons, 81-85; gov, Am Col Surg, 82-85. *Honors & Awards:* Bucy lectr; Ruge lectr; Keegan Mem lectr; Sally Harrington Goldwater Mem lectr; Charles Elsberg lectr. *Mem:* Am Neurol Asn; Am Acad Neurol Surg; Neurol Soc Am; Am Asn Neurol Surgeons (pres, 81-82); fel Am Col Surgeons; Soc Neurol Surgeons (secy, 79-81, pres, 82-83); World Fedn Neurol Surgeons (pres, 85); corresp mem Soc Brit Neurol Surgeons. *Res:* Injuries to the nervous system; transplant of adrenal mediullary tissue in patients with Parkinsonism. *Mailing Add:* 3909 Euclid Ave Dallas TX 75205

CLARK, WILLIAM MELVIN, JR, b Baldwin, Kans, Apr 17, 22; m 45; c 2. MEDICAL EDUCATION ADMINISTRATION. *Educ:* Baker Univ, AB, 46; Univ Chicago, MD, 49; Am Bd Pediat, dipl. *Prof Exp:* From instr to assoc prof, 54-67, PROF PEDIAT, SCH MED, ORE HEALTH SCI UNIV, 67-, DIR GRAD MED EDUC, 77-, ASSOC MED DIR, UNIV HOSPS & CLINS, 72- *Mem:* Am Acad Pediat; Am Acad Neurol; Sigma Xi. *Res:* Pediatric neurology. *Mailing Add:* Health Sci Ctr Univ Ore Portland OR 97210

CLARK, WILLIAM MERLE, JR, b Amarillo, Tex, Apr 23, 37; m 62; c 2. ELECTRICAL ENGINEERING. *Educ:* Stanford Univ, BS, 58; Univ Calif, Berkeley, MS, 66, PhD(elec eng), 68. *Prof Exp:* Engr, Microwave Components Lab, Sylvania Elec Prod Inc, Calif, 58-59; res asst elec eng, Univ Calif, Berkeley, 65-68; from asst prof to assoc prof elec eng, Univ Tex, Austin, 68-73; mem tech staff, Hughes Res Labs, Malibu, 73-, SR MEM TECH STAFF & HEAD SECT, HUGHES RES LABS. *Mem:* Am Phys Soc; Inst Elec & Electronics Engrs. *Res:* Quantum electronics; non-linear optics. *Mailing Add:* Hughes Aircraft Co 3011 Malibu Canyon Rd Malibu CA 90265

CLARK, WILLIAM R, b Detroit, Mich, Aug 18, 38; m. BIOCHEMISTRY. *Educ:* Univ Calif, Los Angeles, BS, 63; Univ Ill, MS, 65; Univ Wash, PhD(biochem), 68. *Prof Exp:* Trainee cellular immunol, Weizmann Inst Sci, Rehovot, Israel, 68-70; asst prof cell biol, 70-74, assoc prof, 74-78, PROF IMMUNOL, UNIV CALIF, LOS ANGELES, 78- *Concurrent Pos:* Career develop award, NIH, 75-80; head, Parvin Cancer Res Labs, 75-; vis scholar, Stanford Univ, 77-78; assoc dir, 78-88, vchmn biol, 84-88, chair, 88- *Mem:* Am Asn Immunologists; Transplantation Soc; Biophys Soc. *Res:* Mechanisms of cell-mediated cytotoxicity; fetal pancreas transplantation in diabetes; genes involved in T cell function. *Mailing Add:* Molecular Biol Inst Univ Calif Los Angeles CA 90024

CLARK, WILLIAM RICHARD, b New Brunswick, NJ, Oct 25, 49; div; c 2. POPULATION DYNAMICS, COMPUTER SIMULATION. *Educ:* Rutgers Univ, BS, 71; Utah State Univ, MS, 74, PhD(ecol), 79. *Prof Exp:* Instr forestry, Rutgers Univ, 74; instr wildlife, Utah State Univ, 78-79; assoc prof wildlife, Univ Montana, 84; from asst prof to assoc prof, 79-90, PROF ECOL, IOWA STATE UNIV, 90- *Mem:* Wildlife Soc; Am Soc Ecol; Am Soc Mammalogists; Sigma Xi. *Res:* Statistical analysis and simulation of wildlife populations, emphasis on mammals and waterfowl; effects of sport exploitation and economic control on populations. *Mailing Add:* Dept Animal Ecol Iowa State Univ 124 Sci II Ames IA 50011

CLARK, WINSTON CRAIG, b Harlan, Ky, Sept 7, 49; m 75; c 1. NEUROSURGERY, NEUROSURGICAL ONCOLOGY. *Educ:* Ga Instit Technol, BS, 71; Ga State Univ, MHN, 73; Univ Miss, PhD(health care admin), 75; Univ Tenn, MD, 82. *Prof Exp:* Instr, health admin, Univ Tenn, 75-76; from instr to asst prof, health admin, Vanderbilt Univ, 76-78; asst prof, health admin, Meharry Med Col, 76-78; med staff fel neurosurg, Nat Instit Health, 83-84; resident, neurosurg, 82-88, res fel, molecular biol, 87-88, ASST PROF, NEUROSURG, UNIV TENN, 88- *Concurrent Pos:* Assoc Semmes Murphey Clin, 88-; head, sect Neurosurg Oncol, Dept Neurosurg, Univ Tenn, 88-; chief, neurosurg, Vet Admin Med Ctr, Memphis, 88- *Mem:* Cong Neurol

Surg; Am Asn Neurol Surg; Soc Res Neurosurgeons; AMA. *Res:* Biology and treatment of malignancies of the central nervous system, growth factors, immune response, molecular biology; clinical research in LAK cells, growth factors and tumor induced lymphocytes. *Mailing Add:* 920 Madison Suite 201 Memphis TN 38103

CLARKE, ALAN R, b Wolverhampton, Eng, July 14, 38. GLASS TECHNOLOGY, GAS CHROMATOGRAPHY. *Educ:* Salford Univ, Eng, ARIC, 61; Sheffield Univ, Eng, PhD(glass technol), 66. *Prof Exp:* Assoc prof chem, Salem Col, WVa, 66-78; PROF ADMIN, WOODBRIDGE CAMPUS, NORTHERN VA COMMUNITY COL, 78- *Mem:* Soc Indust Archaeol. *Res:* Masonry arch bridges and their role in the industrial revolution. *Mailing Add:* Environ Natural Sci Div Northern Va Community Col 15200 Neabsco Mill Rd Woodbridge VA 22191

CLARKE, ALEXANDER MALLORY, b Richmond, Va, Mar 29, 36; m 59; c 3. BIOPHYSICS, BIOMEDICAL ENGINEERING. *Educ:* Va Mil Inst, BS, 58; Univ Va, MS, 60, PhD(physics), 63. *Prof Exp:* Asst prof, 64-69, ASSOC PROF BIOPHYS, MED COL VA, 69- *Concurrent Pos:* Mem subcomt ocular effects, Am Nat Stand Inst, 69-; mem tech comt, 76 & Int Electrotech Comn, 73- *Mem:* Asn Res Vision & Ophthal; Am Phys Soc; Sigma Xi. *Res:* Biomedical instrumentation; physical chemistry of macromolecules; effects of intense optical sources on the pupil and retina; perception of high frequency sound. *Mailing Add:* 7707 Hollins Rd Richmond VA 23229

CLARKE, ALLAN JAMES, b Adelaide, Australia, Oct 10, 49; m 78; c 3. PHYSICAL OCEANOGRAPHY, GEOPHYSICAL FLUID DYNAMICS. *Educ:* Adelaide Univ, BSc, 71, Hons, 72; Cambridge Univ, PhD(oceanog), 76. *Prof Exp:* Res assoc, Mass Inst Technol, 75-77, phys oceanog, Univ Wash, 78-81; from asst prof to assoc prof, 81-90, PROF PHYS OCEANOG, FLA STATE UNIV, 90- *Concurrent Pos:* Guest investr, Woods Hole Oceanog Inst, 79-80, 82, 84-87; prin investr, NSF, 79-, NASA, 81-82; vis scientist, Commonwealth Sci & Indust Res Orgn, Australia, 83, 84 & 88; consult, shelf dynamics & equatorial dynamics. *Mem:* Am Geophys Union; Am Meterol Soc. *Res:* Wind driven and tidal motions on continental shelves; equatorial ocean dynamics; climatic fluctuations; fluctuations in the large scale wind driven circulation of the antartic circumpolar current; wind driven motions near ice edges. *Mailing Add:* Dept Oceanog B-169 Fla State Univ Tallahassee FL 32306-3048

CLARKE, ALLEN BRUCE, b Sask, Can, Sept 8, 27; m 49; c 3. MATHEMATICS, PROBABILITY THEORY. *Educ:* Univ Sask, BA, 47; Brown Univ, MSc, 49, PhD(math), 51. *Prof Exp:* From instr to prof math, Univ Mich, 51-67; chmn dept math & statist, Western Mich Univ, 67-78, dean arts & sci, 78-88, assoc vpres, 88-90, PROF MATH & STATIST, WESTERN MICH UNIV, 67-, PROVOST, 90- *Concurrent Pos:* Vis scholar, Dartmouth & Fla State Univ, 77-78. *Mem:* Math Asn Am; Inst Math Statist. *Res:* Applied probability and statistics. *Mailing Add:* 2016 Greenbriar Dr Kalamazoo MI 49008

CLARKE, ANN NEISTADT, b Philadelphia, Pa, July 27, 46; m 72. PHYSICAL CHEMISTRY, ENVIRONMENTAL ENGINEERING. *Educ:* Drexel Inst Technol, BS, 68; Johns Hopkins Univ, MA, 70 & 71; Vanderbilt Univ, PhD(phys chem), 75. *Prof Exp:* Res chemist, protein chem, Easter Utilization Res & Develop Div, USDA, 64-68; consult engr, Indust Environ, Sheppard T Powell Assocs; res engr, Ctr Indust Water Qual Mgt, 74-75; asst prof environ eng & chem, Vanderbilt Univ, 75-81; proj mgr, Assoc Water & Air Resources Engrs, Inc, 78-81; dir training & prof develop, Recra Res Inc, 81-82; div dir, Aware Inc, 82-89, DIR REMEDIAL TECHNOL DEVELOP DIV, ECHENFELDER INC, 89- *Mem:* Am Chem Soc; Water Pollution Control Fedn; Am Soc Testing & Mat; Sigma Xi; Nat Environ Trainers Asn; Int Asn Water Pollution Res & Control. *Res:* Remedial technologies development for hazardous waste sites; innovative in-situ and ex-situ techniques; advanced treatability testing design. *Mailing Add:* 1408 Franklin Rd Brentwood TN 37027-6804

CLARKE, ANTHONY JOHN, b Kingston-on-Thames, UK Dec 23, 55; Can Citizen; m 79; c 3. PROTEIN CHEMISTRY, ENZYMOLOGY. *Educ:* Univ Waterloo, BSc, 78, MSc, 80, PHD,(chem, biochem), 83. *Prof Exp:* Res assoc, Carlsberg Res Ctr, 83-84; ASST PROF MICROBIOL, UNIV GUELPH, 86- *Mem:* Am Chem Soc; Am Soc Microbiol; Am Soc Biol Chem; Can Soc Microbiol. *Res:* Enzyme structure & function relationships; carbohydrases. *Mailing Add:* Dept Microbiol Univ Guelph Guelph ON N1G 2W1 Can

CLARKE, BENJAMIN L, ENDOCYTOSIS, GLYCOPROTEIN BIOSYNTHESIS. *Educ:* Univ Tex, Galveston, PhD(biochem), 86. *Prof Exp:* FEL BIOCHEM, M D ANDERSON HOSP & TUMOR CTR, 86- *Mailing Add:* Dept Phys & Biophys Univ Ala Med Sch 893 Basic Health Sci Bldg Birmingham AL 35294

CLARKE, BRUCE LESLIE, b Toronto, Ont, Aug 4, 42; m 68; c 2. THEORETICAL CHEMISTRY. *Educ:* Univ Toronto, BSc, 65; Univ Chicago, PhD(chem), 69. *Prof Exp:* Res asst chem, Univ Calif, Santa Cruz, 69-70; asst prof, 70-76, ASSOC PROF CHEM, UNIV ALTA, 76- *Mem:* Am Phys Soc. *Res:* Diagrammatic stability analysis of oscillatory chemical reaction systems; topological properties of self-organizing chemical networks; non-equilibrium statistical mechanics, fluctuations, phase transitions and critical phenomena. *Mailing Add:* Dept Chem Univ Alberta Edmonton AB T6G 2M7 Can

CLARKE, DAVID E, RECEPTOR PHARMACOLOGY. *Educ:* Univ Bradford, UK, PhD(pharmacol), 68. *Prof Exp:* PROF PHARMACOL, UNIV HOUSTON, UNIV PARK, 73- *Mailing Add:* Dept Pharmacol Syntex Res 3401 Hillview Ave Palo Alto CA 94304

CLARKE, DAVID HARRISON, b Jamestown, NY, Aug 14, 30; m 52; c 3. HUMAN PHYSIOLOGY. *Educ:* Springfield Col, BS, 52, MS, 53; Univ Ore, PhD(phys educ), 59. *Prof Exp:* Asst prof phys educ, Univ Calif, Berkeley, 58-64; from assoc prof to prof phys educ, Univ Md, College Park, 64-80, dir grad studies, Dept Phys Educ, 73-80; chmn, Dept Phys Educ, Ind Univ, Bloomington, 80-84; CHMN, DEPT KINESIOLOGY, UNIV MD, COLLEGE PARK, 84- *Concurrent Pos:* Chmn res sect, Nat Col Phys Educ Asn, 65-66; Pres, Res Consortium, Am Asn Health, Phys Educ & Recreation, 78-80. *Mem:* Asn Health, Phys Educ & Recreation; fel Am Acad Phys Educ (pres, 83-85); fel Am Col Sports Med; Nat Asn Phys Educ Higher Educ (pres, 87-90); Int Coun Health, Phys Educ & Recreation. *Res:* Physiology of exercise, especially in the area of muscular fatigue and strength recovery from exercise. *Mailing Add:* Dept Kinesiology Univ Md College Park MD 20742

CLARKE, DONALD DUDLEY, b Kingston, JA, BWI, Mar 20, 30; US citizen; m 53; c 7. BIOCHEMISTRY. *Educ:* Fordham Univ, BS, 50, MS, 51, PhD(org Chem, enzym), 55. *Prof Exp:* Nat Res Coun Can fel, Banting Inst, 55-57; res scientist neurochem, NY State Psychiat Inst, 57-62; assoc prof, 62-70, chmn, dept chem, 78-84, PROF BIOCHEM, FORDHAM UNIV, 70- *Concurrent Pos:* Res assoc biochem, Col Physicians & Surgeons, Columbia Univ, 59-61; adj assoc prof, Fordham Univ, 61-62; spec fel, NIH, 72-73; mem adv comt, NIMH, 73-77 & NIH, 81-84. *Mem:* AAAS; Am Chem Soc; Am Soc Biol Chemists; Int Soc Neurochem; Am Soc Neurochem; Sigma Xi; fel NY Acad Sci. *Res:* Neurochemistry, glutamic acid metabolism and related compounds with special reference to brain; mold metabolites, structure and biosynthesis. *Mailing Add:* Dept Chem Fordham Univ New York NY 10458

CLARKE, DUANE GROOKETT, b Philadelphia, Pa, Jan 7, 18; m 49. CHEMISTRY. *Educ:* Fla Southern Col, BS, 40; Pa State Univ, MS, 42, PhD(org chem), 44. *Prof Exp:* From res chemist to develop chemist, Rohm & Haas Co, 43-67, head, Pollution Abatement Lab, 67-69, asst to mgr, Environ Control Dept, 70-78, tech assoc environ affairs, 78-83; RETIRED. *Mem:* Am Chem Soc; Sigma Xi. *Res:* Heavy hydrocarbons synthesis and properties; synthesis of insecticides, fungicides and monomers; process development; pollution abatement. *Mailing Add:* PO Box 3215 Meadowbrook PA 19046

CLARKE, EDWARD NIELSEN, b Providence, RI, Apr 25, 25; m 49; c 4. APPLICATION SOLAR PHOTOVOLTAICS. *Educ:* Brown Univ, BS, 45, PhD(physics), 51, Harvard Univ, MS, 47, MES, 48. *Prof Exp:* Physicist solid state physics, res lab, Sylvania Elec Prod, 50-56 & Sperry Semiconductor Div, Sperry Rand Corp, 56-59; founder & vpres opers, Nat Semiconductor Corp, 59-64, vpres corp develop & diversification, 64-65; assoc dean fac, Worcester Polytech Inst, 65-74, dir res, 65-86, assoc dean grad studies, 74-86, DIR, CTR SOLAR ELECRTIFICATION, WORCESTER POLYTECH INST, 85-, PROF ENG SCI, 68- *Concurrent Pos:* Consult, Semiconductor Indust; mem, Nat Coun Univ Res Adminr; tri-col res coordr, Clark Univ, Holy Cross Col & Worcester Polytech Inst, 74-84; mgr, Urban Technol Syst Backup Site, 74-78; dir, Eng Res Coun, 76-79 & 85-86. *Mem:* Sigma Xi; Am Phys Soc; Inst Elec & Electronics Engrs; Am Soc Eng Educ. *Res:* Semiconductors; solar photovoltaics; technology transfer. *Mailing Add:* 85 Richards Ave Paxton MA 01612

CLARKE, FRANCIS, b Montreal, Can, July 30, 45; m 89; c 2. OPTIMIZATION, CALCULUS OF VARIATIONS. *Educ:* McGill Univ, BSc, 69; Univ Wash, PhD(math), 73. *Prof Exp:* Prof math, Univ BC, 73-84; DIR, CTR RES MATH, UNIV MONTREAL, 84- *Concurrent Pos:* Vis prof, Univ Paris, Dauphine, 74-75 & Univ Calif, Berkeley, 79-80; Killam res fel, Killam Found, 78-79; assoc ed, J Math Anal & Appl, Can J Math & Bull Can Math Soc, 85- *Honors & Awards:* Coxeter-James lectr, Can Math Soc, 80. *Mem:* Can Math Soc (vpres, 85-); Am Math Soc; Soc Indust & Appl Math; Math Asn Am; fel Royal Soc Can. *Res:* Optimization; optimal contol; calculus of variations; nonsmooth analysis. *Mailing Add:* Ctr Res Math Univ Montreal CP 6128 Sta A Montreal PQ H3C 3J7 Can

CLARKE, FRANK ELDRIDGE, b Brunswick, Md, Dec 26, 13; m 34, 85; c 2. HYDROLOGY. *Educ:* Western Md Col, AB, 35. *Prof Exp:* Head chem process br, Chem Eng Lab, US Naval Eng Exp Sta, 51-57, head chem engr div, 57-61; res engr, US Geol Surv, 61-62, chief water qual res, 62, chief gen hydrol br, 62-65, asst chief hydrologist, 65-67, assoc chief hydrologist, 67-68, asst dir, 68-71, sr scientist, 72-76; dep undersecy, US Dept Interior, 71-72; CONSULT WELL WATER PROBS, US STATE DEPT, 62- *Concurrent Pos:* Consult, Allwater Resources Related, Egypt, 62-79, Japan, 63-72, Pakistan, 64-66, Univ Queensland, 67-, USSR, 76-77. *Honors & Awards:* Max Hecht Award, 64; Cert Award, Gordon Res Conf, 66. *Mem:* AAAS; Am Inst Chem Eng; Am Chem Soc; hon mem Am Soc Testing & Mat (pres, 74-75). *Res:* Geochemical controls of water quality; environmental sciences; corrosion and encrustation processes, particularly in water wells; systems for environmental impact assessment. *Mailing Add:* PO Box 307 Sykesville MD 21784-0307

CLARKE, FRANK HENDERSON, b Newcastle, NB, Can, Dec 6, 27; m 54; c 2. ORGANIC CHEMISTRY, MEDICINAL CHEMISTRY. *Educ:* Univ NB, BSc, 49, MS, 50; Harvard Univ, PhD(org chem), 54. *Prof Exp:* Fel, Columbia Univ, 53-55; sr res chemist, med chem, Schering Corp, 55-62; res supvr, Geigy Chem Corp, 62-65 assoc dir med chem, 65-67, dir med chem, 67-71, dir med chem, 71-80, DISTINGUISHED RES FEL, CIBA-GEIGY CORP, 80- *Concurrent Pos:* Ed-in-chief, Ann Reports Med Chem, Am Chem Soc, 75-78. *Mem:* AAAS; Am Chem Soc; Am Chem Asn. *Res:* Physical properties of medicinal agents; computer programming; designer of molecular models; x-ray crystallography. *Mailing Add:* Rte Three Box 510 Califon NJ 07830

CLARKE, FREDERIC B, III, b Portsmouth, NH, Aug 31, 42; m 62, 90; c 4. PHYSICAL ORGANIC CHEMISTRY, PHYSICAL CHEMISTRY. *Educ:* Wash Univ, AB, 66; Harvard Univ, AM, 68, PhD(chem), 71. *Prof Exp:* Res chemist fire retardants, Monsanto Co, 71-73, prod supvr mkt, 73-74; asst to dir, 74-77, dep dir, 77-78, dir fire res, ctr Fire Res, Nat Bur Standards, 78-80;

PRES, BENJAMIN CLARK ASSOC INC, 80- *Concurrent Pos:* Cong fels, 76-77; mem, bd dirs, Am Soc Testing & Mat. *Mem:* Nat Fire Protection Asn; Am Chem Soc; Am Soc Testing & Mat. *Res:* Fire retardant chemical mechanisms; organophosphorous chemistry; high-temperature reactions; hazard-risk assessment. *Mailing Add:* Benjamin Clarke Assoc Inc 10605 Concord St Suite 501 Kensington MD 20895

CLARKE, FREDERICK JAMES, b Little Falls, NY, Mar 1, 15; m 38; c 3. CIVIL ENGINEERING. *Educ:* US Mil Acad, BS, 37; Cornell Univ MSCE, 40. *Prof Exp:* Mgr, Hanford Eng Works, US Army, 45-47, exec, Sandia Base, NMex, 47-49, chief, Atomic Res & Develop Sect, 52-53, commandant, US Army Eng Sch, Ft Belvoir, 65-66, chief engrs, US Army, 69-73; exec dir, Nat Comn on Water Qual, 73-76; energy & water consult, Tippetts-Abbett-McCarthy-Stratton, 76-83; RETIRED. *Honors & Awards:* Nat Soc Prof Engrs Award, 78. *Mem:* Nat Acad Engrs; Am Soc Civil Engrs; Nat Soc Prof Engrs; Am Acad Environ Eng; Am Pub Works Asn. *Mailing Add:* 9110 Belvoir Woods Pkwy No 116 Ft Belvoir VA 22060-2717

CLARKE, GARRY K C, b Hamilton, Ont, Oct 6, 41; c 1. GLACIOLOGY. *Educ:* Univ Alta, BSc, 63; Univ Toronto, MA, 64, PhD(physics), 67. *Prof Exp:* Assoc prof, 67-76, PROF GEOPHYS, UNIV BC, 76- *Concurrent Pos:* Partic glaciol expeds, Can & Greenland. *Mem:* Soc Explor Geophys; Am Geophys Union; Int Glaciological Soc (pres, 90-93); fel Royal Soc Can. *Res:* Geophysical applications of statistical communication theory; glacier flow theory. *Mailing Add:* Dept Geophys & Astron Univ BC 2075 Wesbrook Mall Vancouver BC V6T 1Z2 Can

CLARKE, GARY ANTHONY, b Washington, DC, May 31, 46; m 75. MICROBIAL PHYSIOLOGY. *Educ:* Indiana Univ, Pa, BS, 68; St Bonaventure Univ, PhD(biol), 73. *Prof Exp:* asst prof, 73-80, ASSOC PROF BIOL, ROANOKE COL, 80- *Mem:* Am Soc Microbiol; Sigma Xi. *Res:* Chemostatic growth of nitrogen fixing bacteria and the effect of environmental variables on this growth; carbon dioxide fixation. *Mailing Add:* Dept Biol Roanoke Col Salem VA 24153

CLARKE, GEORGE, b Readfield, Maine, Mar 10, 15; m 38; c 4. ANALYTICAL CHEMISTRY. *Educ:* Univ Maine, AB, 36. *Prof Exp:* Chemist, Am Cyanamid Co, 43-51, group leader analytical chem, Reactor Testing Sta, Chem Processing Plant, 51-54, res chemist, 54-62, group leader analytical chem, 62-70, group leader analytical chem, Cent Res Div, 70-85; RETIRED. *Mem:* Am Chem Soc; Am Microchem Soc. *Mailing Add:* 57 Pine Hill Ave Stamford CT 06906

CLARKE, GEORGE A, b New York, NY, Apr 4, 33; m 66; c 2. PHYSICAL CHEMISTRY. *Educ:* City Col New York, BS, 55; Pa State Univ, PhD(phys chem), 60. *Prof Exp:* Res assoc theoret chem, Columbia Univ, 60-62; asst prof chem, State Univ NY Buffalo, 62-68; assoc prof, Drexel Inst Technol, 68-70 & Drexel Univ, 70-71; assoc prof chem, Univ Miami, 71-84 & assoc dean, Col Arts & Sci, 78-84; DEAN, SCH ARTS & SCI, CENT CONN STATE UNIV, 84- *Concurrent Pos:* USPHS grants, 64-66; co-sr investr, AEC Proj Grant, 64-68. *Mem:* Am Chem Soc; Am Phys Soc; Sigma Xi. *Res:* Intermolecular interactions in gaseous and condensed media; studies of molecular complexes; inter-and intramolecular energy transfer processes; electrolyte effects on solvent structure and reactive species; approximation methods in quantum chemistry. *Mailing Add:* Sch Arts & Sci Cent Conn State Univ 1615 Stanley St New Britain CT 06050

CLARKE, JAMES, b New York, NY, Jan 17, 27; m 56; c 3. CHEMICAL ENGINEERING, ORGANIC CHEMISTRY. *Educ:* Cornell Univ, BChE, 53; Univ SC, PhD(chem eng), 69. *Prof Exp:* Process engr, Textile Fibers Dept, E I du Pont de Nemours & Co, Inc, 53-58; res engr, Rock Hill Lab, Chemetron Corp, 59-64; gen mgr biochem, Southeastern Biochem Inc, 63-69; pvt consult, 69-71; DIR, CAROSYN RES ASSOC, 71- *Mem:* AAAS; Am Chem Soc. *Res:* Fine chemicals production with specialization in steroid synthesis and pharmaceuticals; industrial waste treatment; modification of steroids to enhance biological activity. *Mailing Add:* 3905 Kenilworth Rd Columbia SC 29205-1539

CLARKE, JAMES HAROLD, b Rockford, Ill, Jan 10, 45; m 72. ENVIRONMENTAL CHEMISTRY & TOXICOLOGY. *Educ:* Rockford Col, BA, 67; Johns Hopkins Univ, PhD(chem), 72. *Prof Exp:* Res assoc chem & environ eng, Vanderbilt Univ, 72-75; sr vpres & pres, Recra Res, Recra Environ & Health Sci, 80-82; proj mgr, Aware, Inc, 75-80, vpres, 82-84, pres & chief exec officer, 84-89, CHMN, PRES & CHIEF EXEC OFFICER, ECKENFELDER, INC, 89- *Concurrent Pos:* Adj asst prof, dept civil & environ eng, Vanderbilt Univ, 83-; adj asst prof, Div Interdisciplinary Toxicol, Univ Ark Med Sci, 87-88; nat rep, Comn Water Chem, Int Union Pure & Appl Chem, 88-; mem bd trustees, Cumberland Mus, 90- *Mem:* Am Inst Chem Engrs; NY Acad Sci; AAAS; Soc Environ Toxicol & Chem; Soc Risk Analysis; Am Mgt Asn. *Res:* Fate and transport of chemicals in the environment; environmental risk assessment; innovative and emerging technologies for hazardous waste site remediation. *Mailing Add:* 1408 Franklin Rd Brentwood TN 37027

CLARKE, JAMES NEWTON, b Montreal, Que, Mar 9, 48; m 90; c 3. COMPUTER SCIENCE. *Educ:* McGill Univ, BSc, 68; Univ Toronto, MSc, 69; Univ Sydney, PhD(physics, radio astron), 75. *Prof Exp:* Res assoc radio astron, Dept Physics, Queen's Univ, 74-76; res fel, Nuffield Radio Astron Labs, Jodrell Bank, 76-77; res assoc radio astron, Dept Astron, 77-80, asst prof, 80-84, SR TUTOR, DEPT COMPUT SCI, UNIV TORONTO, 84- *Mem:* Inst Elec & Electronics Engrs; Asn Comput Mach. *Res:* Computer architecture, operating systems; extraterrestrial intelligence. *Mailing Add:* Dept Comput Sci Univ Toronto Toronto ON M5S 1A4 Can

CLARKE, JOHN, b Cambridge, Eng, Feb 10, 42; m 79; c 1. PHYSICS. *Educ:* Cambridge Univ, BS, 64, MS, 68, PhD(physics), 68. *Prof Exp:* Scholar, 68-69, from asst prof to assoc prof, 69-73, PROF PHYSICS, UNIV CALIF, BERKELEY, 73- *Concurrent Pos:* Alfred P Sloan fel, 70-72; John Simon Guggenheim fel, 78-79. *Honors & Awards:* Charles Vernon Boys Prize for Physics, 77; Fritz London Mem Award, 87. *Mem:* Am Phys Soc; AAAS; fel Royal Soc London. *Res:* Superconductivity; Josephson tunneling application to measurement of low voltages and magnetic fields and detection of electromagnetic radiation; experimental and theoretical study of low frequency electrical noise in solids; geophysics; magnetotellurics. *Mailing Add:* Dept Physics Univ Calif Berkeley CA 94720

CLARKE, JOHN F, b Hempstead, NY, Sept 5, 39; m; c 6. PHYSICS. *Educ:* Fordham Univ, BS, 61; Mass Inst Technol, MS, 64, PhD(nuclear eng), 66. *Prof Exp:* Res staff mem plasma physics, Oak Ridge Nat Lab, 66-73, group leader confinement physics, 73-74, dir thermonuclear div, 74-77; dep dir, Off Fusion Energy, 77-81, ASSOC DIR FUSION ENERGY, OFF ENERGY RES, DEPT ENERGY, 81- *Concurrent Pos:* Mem fusion power coord comt, Div Controlled Thermonuclear Res, Energy Res & Develop Admin, 74-77; mem, Joint US/USSR Fusion Power Coord Comt, 74-77, co-chmn, 81, Fusion Power Coord Comt, US Dept Energy, 77-81 & co-chmn, Joint US/Japan, 81; chmn Int Thermonuclear Exp Reactor Coun, 88. *Honors & Awards:* Distinguished Assoc Award, Energy Res & Develop Admin, 77. *Mem:* Fel Am Phys Soc; fel AAAS. *Res:* Thermonuclear fusion; plasma heating and confinement; MHD equilibrium; neutral particle and heavy ion transport in plasmas. *Mailing Add:* 2601 14th St Washington DC 20009

CLARKE, JOHN FREDERICK GATES, entomology; deceased, see previous edition for last biography

CLARKE, JOHN FREDERICK GATES, JR, b Pullman, Wash, Nov 11, 33; m 56; c 2. ANALYTICAL CHEMISTRY. *Educ:* Wash State Univ, BS, 55; Purdue Univ, MS, 58, PhD(anal chem), 60. *Prof Exp:* Res chemist, 60-70, sr res chemist & group leader gas chromatography sect, 70-78, RES CHEMIST & GROUP LEADER SEPARATIONS & MASS SPECTROSCOPY SECT, HERCULES RES CTR, HERCULES, INC, 78- *Mem:* Am Chem Soc; Sigma Xi. *Res:* Automation of analytical instrumentation, particularly gas chromatography; glass capillary. *Mailing Add:* 2610 Belaire Dr Wilmington DE 19808-3810

CLARKE, JOHN MILLS, b June 7, 49. PLANT PHYSIOLOGY, WHEAT BREEDING. *Educ:* Univ BC, BSc, 71, MSc, 73; Univ Sask, PhD(crop physiol), 77. *Prof Exp:* RES SCIENTIST, AGR CAN RES BR, 77- *Concurrent Pos:* Assoc ed, Can J Plant Sci, 82-84; adj prof, Dept Crop Sci, Univ Sask, 89- *Mem:* Am Soc Agron; Crop Sci Soc Am. *Res:* Wheat breeding; drought physiology and genetics of related traits in wheat. *Mailing Add:* Res Sta Agr Can Box 1030 Swift Current SK S9H 3X2 Can

CLARKE, JOHN ROSS, b Martinsville, Va, Mar 23, 41; m 65; c 2. SOLID STATE ELECTRONICS. *Educ:* Univ Va, BEE, 64, MS, 67, PhD(elec eng), 70. *Prof Exp:* Elec engr, Warrenton Training Ctr, 64-65; physicist, Eastman Kodak Co Res Labs, 70-82; res & develop mgr, BMI, Textron, Lake Park, Fl, 82-87; MEM STAFF, MICRO ELECTRONIC CTR, 87- *Mem:* Am Vacuum Soc. *Res:* Vacuum deposition preparation and characterization of photoconductive films; low pressure chemical vapor deposition of films for integrated circuit manufacture. *Mailing Add:* 316 King George Loop Cary NC 27511

CLARKE, JOSEPH H(ENRY), gas dynamics; deceased, see previous edition for last biography

CLARKE, KENNETH KINGSLEY, b Miami, Fla, June 7, 24; m 45; c 1. POWER MEASUREMENT & CALIBRATION, PHASE MEASUREMENT & CALIBRATION. *Educ:* Stanford Univ, MSc, 48; Polytech Inst Brooklyn, DEE, 59. *Prof Exp:* Fel, Microwave Res Inst, 49-50; asst prof elec eng, Madras Inst Technol, 50-52; lectr elec eng, Univ Ceylon, 52-54; asst prof elec eng, Clarkson Univ, 54-55; from instr to prof elec eng, Polytech Inst Brooklyn, 55-69, head-grad, 67-69; PRES, CLARKE-HESS COMMUN RES, 69- *Concurrent Pos:* Vis prof, Mid East Tech Univ, 61-62; visitor, Accreditation Bd Eng & Technol, 83-88. *Mem:* Inst Elec & Electronics Engrs; AAAS; Sigma Xi; NY Acad Sci. *Res:* Technical articles, patents and two books on transistor and semiconductor circuits, phase measurements, power measurements and phase standards; design and manufacture of wide-band volt-amphere-wattmeters. *Mailing Add:* 300 Riverside Dr New York NY 10025

CLARKE, LILIAN A, b Humboldt, Iowa, Aug 9, 15; m 42. INORGANIC CHEMISTRY. *Educ:* Grinnell Col, BA, 36; Pa State Univ, MS, 39, PhD(inorg chem), 42. *Prof Exp:* Instr chem, Pa State Univ, 37-42; asst prof chem, Villanova Univ, 63-80; CONSULT, BERKS ASSOC, INC, 68- & BASIC INC, 69- *Mem:* AAAS; Am Chem Soc; Sigma Xi. *Res:* Studies of properties of surface active agents; identification of chemical and biological warfare agents; refefining of crankcase oils; water pollution abatement; solvent extraction of lubricating oils; chemistry of carbon black. *Mailing Add:* 121 Fairmount Ave Laurel Springs NJ 08021

CLARKE, LORI A, b New York, NY, Feb 11, 47; m 74; c 3. SOFTWARE ENGINEERING. *Educ:* Univ Rochester, BA, 69; Univ Colo, Phd(conput sci), 76. *Prof Exp:* Programmer, Sch Med, Univ Rochester, 70-71, Nat Ctr Atmospheric Res, 71-75; from asst prof to assoc prof, 76-86, PROF, COMPUT SCI, UNIV MASS, 81- *Concurrent Pos:* Consult var high technol corps, 75-; prin investr, NSF, Darpa & ONR; nat lectr, Asn Comput Mach, 82-84; secy-treas, Sigsoft, 85-88, vchmn, 88-; distinguished visitor, Tech Comt Software Eng, Inst Elec & Electronics Engrs, 88-; chancellor's lectr, 91. *Mem:* Asn Comput Mach; Inst Elec & Electronics Engrs Comput Soc; Spec Interest Group Prog Lang; Spec Interest Group Software Eng. *Res:* Design of a software development environment to support all phases of the software lifecycle; software testing. *Mailing Add:* Dept Comput & Info Sci Univ Mass Amherst MA 01003

CLARKE, LUCIEN GILL, b Eagle Grove, Iowa, Mar 11, 21; m 48. ENGINEERING, ELECTRONICS. *Educ:* Iowa State Col, BS, 42. *Prof Exp:* Engr, Naval Res Labs, 43, physicist & proj officer, Off Naval Res, 46-47; engr, Raytheon Mfg Co, 47-48; asst gen mgr eng, 48-64, exec dir admin, 64-69, SR RES ENGR, SRI INT, 69- *Mem:* Inst Elec & Electronics Engrs. *Res:* Communications engineering; radio electronics; research management. *Mailing Add:* 12550 Lacresta Dr Los Altos CA 94022

CLARKE, MARGARET BURNETT, b Enid, Okla; m 85. BIOCHEMISTRY, CELL BIOLOGY. *Educ:* Mills Col, BA, 65; Univ Calif, Berkeley, PhD(molecular biol), 71. *Prof Exp:* Res assoc biochem, Univ Calif, San Francisco, 72-76; from asst prof to prof, Dept Molecular Biol, Albert Einstein Col Med, 76-88; MEM, PROG MOLECULAR & CELL BIOL, OKLA MED RES FOUND, 88- *Concurrent Pos:* NIH fel, 72-74; Nat Res Serv Award, 75-76; co-investr, NIH prog proj grant, 76-88; estab investr, Am Heart Asn, 77-82; prin investr, Am Cancer Soc Res grant, 78-80 & 90, NIH res grant, 81- *Mem:* Am Soc Cell Biol; Am Soc Biol Chem; Biophys Soc; NY Acad Sci. *Res:* Molecular basis of motility in eukaryotic cells. *Mailing Add:* Prog Molecular & Cell Biol Okla Med Res Found 825 NE 13 St Oklahoma City OK 73104

CLARKE, MICHAEL J, b St Louis, Mo, Dec 30, 46; m 68; c 3. BIO-INORGANIC CHEMISTRY, ELECTROCHEMISTRY. *Educ:* Catholic Univ Am, AB, 68; Stanford Univ, MS, 70, PhD(inorg chem), 74. *Prof Exp:* Instr chem, City Col San Francisco, 70; asst prof chem, Boston Univ, 74-75; asst prof chem, Wheaton Col, 75-76; from asst prof to assoc prof, 76-85, PROF CHEM, BOSTON COL, 85- *Concurrent Pos:* Ed, Structure & Bonding, 80- *Honors & Awards:* Whiting Found Fel, 84. *Mem:* Am Chem Soc. *Res:* Interactions of metal ions with nucleotides and coenzymes; metal-containing anticancer pharmaceuticals; chemistry of ruthenium and technetium. *Mailing Add:* Dept Chem Boston Col Chestnut Hill MA 02167

CLARKE, RAYMOND DENNIS, b King's Lynn, Eng, May 13, 46; m 69; c 2. ECOLOGY, MARINE BIOLOGY. *Educ:* McGill Univ, BSc, 67; Yale Univ, MFS, 69, MPhil, 71, PhD(ecol), 73. *Prof Exp:* PROF BIOL, SARAH LAWRENCE COL, 72- *Concurrent Pos:* Res grants, Am Philos Soc, 74, Nat Geog Soc, 74 & Sigma Xi, 76; mission chief scientist, Nat Oceanic & Atmospheric Admin, 78-81; vis scientist, West Indies Lab, Fairleigh Dickinson Univ, 79-80; vis prof, Univ Mass Field Sta, 84, & 85, West Indies Lab, Fairleigh Dickinson Univ, 86, & 88. *Mem:* Ecol Soc Am; Sigma Xi; Am Inst Biol Sci; Am Philos Soc; Int Soc Reff Studies; NY Acad Sci. *Res:* Population and community ecology of coral reef fishes with emphasis on habitat structure. *Mailing Add:* Dept Biol Sarah Lawrence Col Bronxville NY 10708

CLARKE, RICHARD PENFIELD, b Baltimore, Md, Jan 30, 19; m 46; c 4. PHYSICAL CHEMISTRY. *Educ:* Princeton Univ, AB, 41, MA, 48, PhD(phys chem), 50; Johns Hopkins Univ, MS, 42. *Prof Exp:* Asst, Manhattan Proj, Princeton Univ, 43-45; chemist, Air Reduction Co, 46; asst, Princeton Univ, 47-50; chemist, res dept, Standard Oil Co Ind, 50-51; proj leader, Exp, Inc, 51-56; mgr res, Okonite Co, 56-59; vpres, Hasche Eng Co, Tenn, 59-64; from vpres to pres, Kalamazoo Spice Extraction Co, 64-70; consult chem engr, Fuel Gas Prod & Econ, 70-73; assoc prof chem, Lake Mich Col, 73-88; RETIRED. *Concurrent Pos:* Vpres, Thermo Tile Corp, 53-59, pres, 59-69. *Mem:* Am Chem Soc; NY Acad Sci; Sigma Xi. *Res:* Dielectric increments of amino acid and polypetide solutions; kinetics; decomposition diborane; heterogeneous catalysis; oxidation hydrocarbons; combustion; chemical engineering. *Mailing Add:* 2124 Aberdeen Dr Kalamazoo MI 49008-1728

CLARKE, ROBERT FRANCIS, b Portsmouth, Va, Oct 8, 19; m 47; c 2. ZOOLOGY, ANIMAL BEHAVIOR. *Educ:* Kans State Teachers Col, BSEd, 55, MS, 57; Okla Univ, PhD(zool), 63. *Prof Exp:* Instr educ, Emporia State Univ, 56-58, from instr to prof biol, 58-85, chmn dept, 72-79; RETIRED. *Mem:* Am Soc Ichthyol & Herpet; Herpet Soc; Kroeber Anthrop Soc. *Res:* Display behavior of lizards, particularly the family Iguanidae; ecology of reptiles; color change in gravid lizards. *Mailing Add:* 2331 Arrowhead Dr Emporia KS 66801

CLARKE, ROBERT LAGRONE, b Tullahoma, Tenn, Mar 10, 17; m 43; c 2. PHARMACEUTICAL CHEMISTRY. *Educ:* Ga Inst Technol, BS, 38; Emory Univ, MS, 39; Univ Wis, PhD(org chem), 47. *Prof Exp:* Instr chem, Young Harris Jr Col, Ga, 39-40; org chemist, Sterling-Winthrop Res Inst, 47-81; RETIRED. *Mem:* Am Chem Soc. *Res:* Ketene acetals; nitrogen heterocycles; steroids; mercurial diuretics; alkaloids. *Mailing Add:* 96 Elsmere Ave Apt 4 Delmar NY 12054-3425

CLARKE, ROBERT LEE, b Vermilion, Alta, Apr 17, 22; m 45; c 4. MEDICAL PHYSICS. *Educ:* Univ Alta, BSc, 43; McGill Univ, PhD(physics), 48. *Prof Exp:* Asst physics, Nat Res Coun, Can, 43-45; from asst res officer to assoc res officer, Atomic Energy Can, Ltd, 48-68; PROF PHYSICS, CARLETON UNIV, 68- *Mem:* Can Asn Physicist. *Res:* Nuclear reactions; neutrons, low to medium energy; radiography; medical physics; radiography, ultrasonic imaging. *Mailing Add:* Dept Physics Carleton Univ Colonel By Dr Ottawa ON K1S 5B6 Can

CLARKE, ROBERT TRAVIS, b Brooklyn, NY, Nov 17, 37; m 58; c 4. PALYNOLOGY, PALEONTOLOGY. *Educ:* Univ Okla, BS, 59, MS, 61, PhD(geol), 63. *Prof Exp:* Res geol palynology, Socony Mobil Oil Co, Inc, 63-71; Offshore Explor, Mobil Oil, New Orleans, 71-73; econ & planning, Mobile Res, 73-76, explor training coordr, 76-85, geol & geochem res, 85-87, asst explor mgr, 87-88, EXPLOR TRAINING COORDR, MOBIL RES, DALLAS, 88- *Mem:* Fel, Geol Soc Am; Am Asn Stratig Palynologists (secy-treas, 70-73, pres, 73-74); Am Asn Petrol Geologists; Dallas Geol Soc. *Res:* Research in the field of palynology applicable to resolving problems in stratigraphic zonation and correlation; paleoecologic and environmental interpretations and age determinations. *Mailing Add:* Mobil Res-DRL PO Box 819047 Dallas TX 75381-9047

CLARKE, ROY, b Bury, Eng, May 9, 47; m 73; c 3. PHYSICS, MATERIALS SCIENCE. *Educ:* Univ London, BS, 69, PhD(physics), 73. *Prof Exp:* Res asst, Cavendish Lab, Univ Cambridge, 73-78; James Franck fel physics, James Franck Inst, Univ Chicago, 78-79; PROF PHYSICS, UNIV MICH, 79- *Mem:* Am Phys Soc. *Res:* Structural phase transitions; x-ray scattering; ferroelectrics; disordered materials, intercalation compounds, photovoltaics, artificial heterostructures; molecular beam epitaxy of metals; non-equilibrium growth; magnetic thin films. *Mailing Add:* Dept Physics Univ Mich Ann Arbor MI 48109

CLARKE, ROY SLAYTON, JR, b Philadelphia, Pa, Jan 23, 25; m 51; c 3. GEOCHEMISTRY. *Educ:* Cornell Univ, AB, 49; George Washington Univ, MS, 57, PhD, 76. *Prof Exp:* Chemist, USDA, 49-51; res assoc chem, George Washington Univ, 51-52; chemist, USDA, 52-53; analytical chemist, US Geol Surv, 53-57; chemist, Div Meteorites, Smithsonian Inst, 57-66, assoc cur, 66-70, CUR, NAT MUS NATURAL HIST, 70- *Mem:* AAAS; Am Chem Soc; Mineral Soc Am; Geochem Soc. *Res:* Geochemistry and metallography of meteorites. *Mailing Add:* Div Meteorites Smithsonian Inst Washington DC 20560

CLARKE, STEVEN DONALD, b Brockton, Mass, Dec 22, 48; m 69; c 2. BIOCHEMISTRY, NUTRITION. *Educ:* Univ Maine, BS, 70; Wash State Univ, MS, 72; Mich State Univ, PhD(human nutrit), 76. *Prof Exp:* Instr, Univ Mich, 75-76; asst prof, Johns Hopkins Univ, 76-78; asst prof nutrit, Ohio State Univ, 78-80; ASST PROF NUTRIT, UNIV MINN, 81- *Concurrent Pos:* NIH trainee fel, Mich State Univ, 74-76; mem, Comt Nutrit Surveillance, State Mich, 75-76; assoc, Health & Weight Prog, Johns Hopkins Univ, 77-78; assoc, Proj Prevention, 77- *Mem:* Am Inst Nutrit; Am Oil Chemists. *Res:* Intracellular regulation of carbohydrate and lipid metabolism as affected by dietary fat and carbohydrate intake and composition. *Mailing Add:* Upjohn Co 7921-25-5 Kalamazoo MI 49001

CLARKE, STEVEN GERARD, b Los Angeles, Calif, Nov 19, 49; m 81; c 1. BIOCHEMISTRY. *Educ:* Pomona Col, BA, 70; Harvard Univ, PhD(biochem & molecular biol), 76. *Prof Exp:* Instr biochem & molecular biol, Harvard Univ, 73-74; Miller fel, Univ Calif, Berkeley, 76-78; from asst prof to assoc prof, 78-87, PROF CHEM & BIOCHEM, UNIV CALIF, LOS ANGELES, 87- *Honors & Awards:* Alfred P Sloan Fel, 82. *Mem:* Am Soc Biol Chemists; Protein Soc. *Res:* Protein methylation reactions; biochemistry of aging; macromolecular degradation and repair processes; spontaneous decomposition of proteins. *Mailing Add:* Dept Chem & Biochem 405 Hilgard Ave Los Angeles CA 90024-1569

CLARKE, THOMAS ARTHUR, b Peoria, Ill, Aug 13, 40. OCEANOGRAPHY, ECOLOGY. *Educ:* Univ Chicago, BS, 62; Univ Calif, San Diego, PhD(oceanog), 68. *Prof Exp:* Asst prof oceanog, 68-74, ASST MARINE BIOLOGIST, HAWAII INST MARINE BIOL, 68-, ASSOC PROF OCEANOG, UNIV HAWAII, 74- *Mem:* AAAS; Ecol Soc Am; Am Soc Limnol & Oceanog; Brit Ecol Soc. *Res:* Behavior and population dynamics of pomacentrid fish; shark ecology; fisheries ecology. *Mailing Add:* Dept Oceanog Univ Hawaii Manoa Honolulu HI 96822

CLARKE, THOMAS LOWE, b Miami, Fla, Sept 27, 48; m 73; c 2. ACOUSTIC CURRENT METERS. *Educ:* Fla Int Univ, BS, 73; Univ Va, MS, 75; Univ Miami, PhD(appl math), 82. *Prof Exp:* Mathematician, Marine Geol & Geophys Lab, 75-81, mathematician, Ocean Acoust Div, Atlantic Oceanog & Meteorol Lab, Nat Oceanic & Atmospheric Admin, 81-86; pres, TLA Res, 86-88, RES PROF, INST SIMULATION & TRAINING, UNIV CENT FLA, 88- *Concurrent Pos:* Consult, Gen Oceanics, Inc & Daubin Systs Corp, 85-; lectr, Fla Int Univ, 85. *Mem:* Inst Elec & Electronics Engrs; Acoust Soc Am; Am Geophys Union; Audio Eng Soc. *Res:* Applications of acoustics and optics in simulation and training; machine perception of images and sounds. *Mailing Add:* Inst Simulation & Training Univ Cent Fla Box 25000 Orlando FL 32816-0054

CLARKE, THOMAS ROY, b Montreal, Que, Dec 23, 40; m 78; c 2. AUDIO VISUAL PROGRAMMING. *Educ:* Univ Toronto, BSc, 63, MA, 64, PhD(astron), 69. *Prof Exp:* Asst cur, 68-73, assoc cur astron, 73-75, HEAD, MCLAUGHLIN PLANETARIUM, ROYAL ONT MUS, 76- *Concurrent Pos:* Instr, Div Exten, Univ Toronto, 70-76; consult, Nat Mus Can, 80 & 85; adv, Can Mus Asn, 83-85. *Mem:* Planetarium Asn Can (vpres, 81-83, pres, 83-85); Can Astron Soc; Am Astron Soc; Int Astron Union; Royal Astron Soc Can. *Res:* Communication of astronomy and related sciences; writing and production of planetarium programs. *Mailing Add:* 100 Queens Park Toronto ON M5S 2C6 Can

CLARKE, W T W, b Toronto, Ont, Nov 6, 20; m 47; c 4. INTERNAL MEDICINE. *Educ:* Univ Toronto, MD, FRCP(C), 50. *Prof Exp:* prof med, Univ Toronto, 66-85, emer prof, 85-; RETIRED. *Concurrent Pos:* Dep physician-in-chief, Toronto Gen Hosp, 69-74; chmn, Drug Qual & Therapeut Comt, Govt Ont. *Mem:* Fel Am Col Physicians; Can Med Asn. *Res:* Diabetes; nephrology. *Mailing Add:* 507 Ontario St Toronto Gen Hosp Toronto ON M4X 1M8 Can

CLARKE, WILBUR BANCROFT, b Colon, Panama, July 22, 29; US citizen; m 59; c 1. ORGANIC CHEMISTRY. *Educ:* Xavier Univ, BS, 50, MS, 53; Univ Ind, PhD(chem), 62. *Prof Exp:* Instr chem, Xavier Univ, 50-53; asst neurol, US Army Chem Ctr, Md, 53-55; asst chem, Ind Univ, 56-58; chemist, Northern Regional Labs, 58-59; PROF CHEM, SOUTHERN UNIV, 60-, CHMN DEPT, 70- *Concurrent Pos:* Res grants, NSF & Sigma Xi, 63-64, NIH, 64-66; res specialist, Miss Test Facil, NASA-Gen Elec Co, 66 & 67; fel, La State Univ, 67-68; consult, NSF, 64. *Mem:* AAAS; Am Chem Soc; Brit Chem Soc. *Res:* Heterocyclics; preparation and elucidation of antiviral and anti-carcinogenic agents; infrared spectroscopy. *Mailing Add:* Southern Univ Box 9745 Baton Rouge LA 70813

CLARKE, WILTON E L, b Kape, SAfrica, Jan 20, 42; m; c 3. COMPUTER SCIENCES. *Educ:* Univ Iowa, MA, 67, PhD(math), 75. *Prof Exp:* Instr, math, Ikiza Seminary, 67-72; instr, math, Heldeiberg Col, 76-79; prof, math, Atlantic Union Col, 79-86; PROF, MATH, LOMA LINDA UNIV, 86- *Concurrent Pos:* Adj prof, math, Riverside, 88- *Mem:* Am Math Soc; Math Asn Am; Asn Comput Mach. *Res:* Research in astronomy. *Mailing Add:* Dept Math & Comput Loma Linda Univ Riverside CA 92515

CLARKSON, ALLEN BOYKIN, JR, b Augusta, Ga, July 1, 43; div; c 1. PARASITOLOGY. *Educ:* Univ of the South, BS, 65; Univ Ga, PhD(zool), 75. *Prof Exp:* Instr biol, Univ Ga, 72-74; fel parasitol, Rockefeller Univ, 74-77; ASST PROF PARASITOL, SCH MED, NY UNIV, 77- *Concurrent Pos:* Adv, WHO, 77-81, spec study sects, NIH. *Mem:* AAAS; Am Soc Trop Med & Hyg; Soc Protozoologists. *Res:* Parasitic protozoa, particularly trypanosomatids; chemotherapy related to carbohydrate pathways for energy production; pneumocystis csnuii; polyamine metabolism; response to iron calculators. *Mailing Add:* Dept Parasitol NY Univ Sch Med 341 E 25th St New York NY 10010-2533

CLARKSON, BAYARD D, b New York, NY, July 15, 26; c 4. HEMATOLOGY, ONCOLOGY. *Educ:* Yale Univ, BA, 48; Columbia Univ, MD, 52. *Prof Exp:* From intern to resident, NY Hosp, 52-58; instr clin med, Med Col, Cornell Univ, 58-62, from asst prof to assoc prof med, 62-74; res fel, 58-59, res assoc, 59-61, assoc, 61-65, assoc mem, 65-71, MEM, SLOAN-KETTERING INST CANCER RES, 71-; PROF MED, MED COL, CORNELL UNIV, 74- *Concurrent Pos:* Mem adv comt ther, Am Cancer Soc, 65-68; mem pharmacol B study sect, NIH, 67-71; mem bd trustees, Cold Spring Harbor Lab, 68; mem ed adv bd, Cancer Res, 70; mem chemother adv comt, Nat Cancer Inst, 71-74; attend physician & chief hemat & lymphoma serv, Mem Hosp, NY, 71-; assoc ed, Cancer Res, 73-76 & 76-81. *Mem:* NY Acad Sci; Am Asn Cancer Res (pres, 80-81); fel Am Col Physicians; Am Soc Clin Oncol (pres, 73-74); Am Soc Clin Invest. *Res:* Cancer chemotherapy; leukemia; cell kinetics and regulation of cell growth as related to control of cancer. *Mailing Add:* Mem Sloan-Kettering Cancer Ctr Cornell Univ Med Col 1275 York Ave New York NY 10021

CLARKSON, JACK E, b Provo, Utah, June 17, 36; m 63; c 3. ANALYTICAL CHEMISTRY, NUCLEAR CHEMISTRY. *Educ:* Brigham Young Univ, BS, 60; Univ Calif, Berkeley, PhD(chem), 65. *Prof Exp:* SR CHEMIST, LAWRENCE LIVERMORE NAT LAB, UNIV CALIF, 65- *Mem:* Am Chem Soc. *Res:* Trace impurity analysis by gas chromatography; chromatographic analysis of explosives; gel permeation chromatography of polymers and explosives; helium ionization detector utilization part-per-million analysis; high pressure liquid chromatography; shale oil analysis; capillary column gas chromatography. *Mailing Add:* 1541 Roselli Dr Livermore CA 94550

CLARKSON, MARK H(ALL), b Lafayette Co, Mo, Sept 27, 17; m 42; c 3. AEROSPACE ENGINEERING. *Educ:* Univ Minn, BAeroEng, 39; Univ Tex, MA, 48, PhD, 53. *Prof Exp:* Shop training, Douglas Aircraft Co, Calif, 39-40, stress analyst, 40-41; aerodynamicist, Consol Vultee Aircraft Co, 41-42, Tenn, 42-43, Tex, 43-45; res engr, Defense Res Lab, Univ Tex, 45-51, res mathematician, 52-54; res & develop proj engr, Chance Vought Aircraft, Tex, 53-59, sr scientist, Res Ctr, 59-60; chmn, Aerospace Eng Dept, 61-72, PROF ENG SCI, UNIV FLA, 61- *Concurrent Pos:* Sr Nat Res Coun fel, NASA Ames Res Ctr, 73-74. *Mem:* Assoc fel Am Inst Aeronaut & Astronaut; Sigma Xi. *Res:* Theoretical and experimental work in fluid mechanics and aerodynamics. *Mailing Add:* 231 Aero Bldg Univ Fla Gainesville FL 32603

CLARKSON, ROBERT BRECK, b Buffalo, NY, Apr 19, 43; m 65; c 3. PHYSICAL CHEMISTRY, SURFACE CHEMISTRY. *Educ:* Hamilton Col, BA, 65; Princeton Univ, MA, 68, PhD(chem), 69. *Prof Exp:* Asst prof phys chem, Univ Wis-Milwaukee, 69-76; head, Electron Paramagnetic Resonance Applications Lab, Varian Assocs, 76-81; dir, Lab Molecular Spectros, 82-84, vis asst prof, Dept Chem, Sch Chem Sci, 84-86, ASSOC PROF MED, UNIV ILL, 86-, ASSOC PROF, CLIN VET MED, VET BIOSCI, 90- *Concurrent Pos:* NSF Indust Res Participation Prog, Exxon Res Corp, 76. *Honors & Awards:* Norton Found Prize; A O Beckman Res Award. *Mem:* Am Chem Soc; Am Phys Soc; Catalysis Soc; Soc Magnetic Resonance Med. *Res:* Applications of electron paramagnetic resonance and nuclear magnetic resonance characterization of gas-solid interactions; electron spin echo and electron-nuclear double resonance studies of disordered systems; characterization of magnetic resonance imaging contrast agents. *Mailing Add:* 367 Noyes Lab Box 61 Univ Ill 505 S Mathews Urbana IL 61801

CLARKSON, ROY BURDETTE, b Cass, WVa, Oct 25, 26; m 52; c 3. BOTANY. *Educ:* Davis & Elkins Col, BS, 51; WVa Univ, MA, 54, PhD, 60. *Prof Exp:* Teacher, pub sch, WVa, 51-56; from instr to assoc prof, 56-69, assoc chmn dept, 69-74, actg chmn, 74-75 & 82-83, PROF BIOL, WVA UNIV, 69-, CUR HERBARIUM, 75- *Concurrent Pos:* Pres, WVa Acad Sci, 80, treas, 83- *Mem:* Am Soc Plant Taxonomists. *Res:* Plant taxonomy and geography; chemosystematics. *Mailing Add:* Dept Biol PO Box 6057 WVa Univ Morgantown WV 26506-6057

CLARKSON, THOMAS BOSTON, b Decatur, Ga, June 13, 31; m 50; c 3. VETERINARY MEDICINE. *Educ:* Univ Ga, DVM, 54; Dipl, Am Col Lab Animal Med, 63. *Prof Exp:* Res assoc pharmacol & exp therapeut sect, S E Massengill Co, 54-57; from asst prof to assoc prof exp med & dir vivarium, 57-64, assoc prof lab animal med & head dept, 64-65, PROF & CHMN, DEPT BOWMAN GRAY SCH MED, WAKE FOREST UNIV, 65-, DIR, ARTERIOSCLEROSIS RES CTR, 71- & COMP MED CLIN RES CTR, 89- *Concurrent Pos:* Mem comt coronary artery lesions & myocardial infarctions, Am Heart Asn, 70-; mem sci adv comt, Univ Wash Regional Primate Res Ctr, 71-; mem adv comt, Cerbrovascular Res Ctr, 73-; mem comt vet med sci, Nat Acad Sci-Nat Res Coun, 75-; mem clin sci panel study nat needs biomed & behav res personnel comt & task force animal models atherosclerosis, Nat Acad Sci, 76-; chmn, Task Force on Res Animal Use, Am Heart Asn; vchmn, coun arteriosclerosis, 79-81, chmn, 81-83; chmn arteriosclerosis, hypertension & lipid metabolism, adv comt, Nat Heart Lung & Blood Inst, 83-85; Am Col Lab Animal med, 84-85. *Honors & Awards:* Griffin Award, Am Asn Lab Animal Sci, 77; Charles River Prize, Am Vet Med Asn, 78; Award of Merit, Am Heart Asn, 87; G Lyman Duff Mem Lectr, Am Heart Asn, 85. *Mem:* Nat Acad Sci; Am Asn Advan Lab Animal Sci; Am Asn Pathologists; Am Heart Asn; Am Soc Exp Path; Sigma Xi; Am Vet Med Asn; fel Soc Behav Med; Acad Behav Med Res; Am Soc Primatol. *Res:* Comparative and experimental atherosclerosis, particularly factors affecting susceptibility and resistance to the disease and the mechanisms by which risk factors affect the pathogenesis. *Mailing Add:* Comp Med Res Ctr Medical Center Blvd Winston-Salem NC 27157-1040

CLARKSON, THOMAS WILLIAM, b UK, Aug 1, 32; m 57; c 3. PHARMACOLOGY. *Educ:* Univ Manchester, BSc, 53, PhD(biochem), 56. *Hon Degrees:* MD, Univ Umea, Sweden, 86. *Prof Exp:* Med Res Coun fel, Univ Manchester, 56-57; instr radiation biol, Univ Rochester, 57-61, asst prof, 61-62; sci officer, Med Res Coun, UK, 62-64; sr fel, Weizmann Inst, 64-65; assoc prof biophys, pharmacol, radiation biol & toxicol, 65-71, head, Div Toxicol, 80-86, PROF TOXICOL & PHARMACOL & BIOPHYS, UNIV ROCHESTER, 71-, DIR ENVIRON HEALTH SCI CTR, 86- *Concurrent Pos:* Mem toxicol adv bd, Food & Drug Admin, 75-77; mem toxicol study sect, NIH, 76-77 & WHO, 75, 80, 87, 89; mem water reuse, Nat Acad Sci, 80-; mem FIFRA sci adv panel, 85-89, nat adv environ health sci coun, 85-89; Sterling Drug vis prof, Albany Med Col, NY, 89. *Mem:* Inst Med-Nat Acad Sci; AAAS; Health Physics Soc; Brit Pharmacol Soc; Soc Toxicol; Am Soc Pharmacol & Exp Therapeut. *Res:* Cellular physiology; reabsorption mechanisms in intestine and kidney; heavy metal toxicology; action of metals on cellular level in intestine, kidney and red blood cells. *Mailing Add:* Sch Med Box EHSC Univ Rochester Rochester NY 14642

CLARY, BOBBY LELAND, b Jesup, Ga, Aug 14, 38; m 57; c 3. AGRICULTURAL ENGINEERING. *Educ:* Univ Ga, BSAE, 60; Okla State Univ, PhD(agr eng), 69. *Prof Exp:* Asst prof agr eng, State Univ NY Agr & Tech Col Alfred, 60-65; prof, Polk Jr Col, Fla, 65-66; res asst, 66-68, from asst prof to assoc prof, 68-78, PROF AGR ENG, OKLA STATE UNIV, 78- *Concurrent Pos:* Actg chief agr & food processes br, Energy Res & Develop Admin, Washington, DC, 75-76. *Honors & Awards:* Young Eng of the Year Award, Okla Soc Prof Engrs, 74; Young Educr Award, 75 & Distinguished Young Agr Engr Award, Southwest Region, 76, Am Soc Agr Engrs. *Mem:* Am Soc Agr Engrs; Am Soc Eng Educ; Inst Food Technol; Nat Soc Prof Engrs. *Res:* Heat and mass transfer in biological materials; agricultural energy. *Mailing Add:* Dept Agr Eng Okla State Univ Stillwater OK 74078

CLARY, WARREN POWELL, b Lewellen, Nebr, Sept 8, 36; m 57; c 3. BOTANY. *Educ:* Univ Nebr, BS, 58; Colo State Univ, MS, 61, PhD(bot), 72. *Prof Exp:* Range conservationist, Rocky Mountain Forest & Range Exp Sta, 60-65, plant ecologist, 65-76; proj leader, Southern Forest Exp Sta, 76-77; PROJ LEADER, INTERMOUNTAIN RES STA, 77- *Concurrent Pos:* Adj prof, NAriz Univ, 67-76, Brigham Young Univ, 79-84; assoc ed, Northwest Sci, 84-86. *Mem:* Soc Range Mgt; Sigma Xi. *Res:* Structure and function of riverine-riparian ecosystems and their response to stress by grazing livestock; plant community response to livestock grazing in salt desert shrub ecosystems. *Mailing Add:* Forestry Serv Lab 316 E Myrtle St Boise ID 83702

CLASE, HOWARD JOHN, b Salisbury, Eng, June 14, 38; m 63; c 2. INORGANIC CHEMISTRY. *Educ:* Cambridge Univ, BA, 60, PhD(inorg chem), 63, MA, 65. *Prof Exp:* Fel, McMaster Univ, 63-65; asst chem, Univ Oulu, 65-66; tutorial fel, Univ Sussex, 66-68; asst prof, 68-74, ASSOC PROF CHEM, MEM UNIV NFLD, 74- *Concurrent Pos:* Vis prof, Univ Reading, 84-85. *Mem:* Royal Soc Chem; Chem Inst Can. *Res:* Application of vibrational spectroscopy to problems in inorganic chemistry. *Mailing Add:* Dept Chem Mem Univ Nfld Elizabeth Ave St John's NF A1C 5S7 Can

CLASEN, RAYMOND ADOLPH, b Chicago, Ill, June 28, 26; m 50; c 1. PATHOLOGY. *Educ:* Univ Ill, BS, 50, MD, 52. *Prof Exp:* Intern, Cook County Hosp, 52-53; resident, Presby Hosp, Chicago, 53-57; Nat Inst Neurol Dis & Blindness fel, 57-59, asst prof 59-73, ASSOC PROF PATH, RUSH-PRESBY-ST LUKE'S MED CTR, CHICAGO, 73- *Mem:* AAAS; Am Asn Neuropath; Am Asn Pathologists & Bacteriologists; Am Soc Exp Path; Int Acad Path; Sigma Xi. *Res:* Experimental neuropathology. *Mailing Add:* 1753 W Congress Pkwy Chicago IL 60612

CLASS, CALVIN MILLER, b Baltimore Co, Md, Jan 27, 24; m 48; c 1. NUCLEAR PHYSICS. *Educ:* Johns Hopkins Univ, AB, 43, PhD(physics), 51. *Prof Exp:* Physicist hydrodyn, Nat Adv Comt Aeronaut, 44-46; asst nuclear physics, Johns Hopkins Univ, 49-52; from instr to prof, 52-85, EMER PROF PHYSICS, RICE UNIV, 85- *Concurrent Pos:* Guggenheim fel, 55-56; master, Mary Gibb Jones Col, Rice Univ, 57-65; sr res assoc, Kellogg Lab, Calif Inst Tech. *Mem:* Fel Am Phys Soc. *Res:* Spectroscopy of light and medium nuclei. *Mailing Add:* Dept Physics Rice Univ Box 1892 Houston TX 77001

CLASS, JAY BERNARD, b Baltimore, Md, Apr 14, 28; m 58; c 2. ADHESIVES, ELASTOMERS. *Educ:* Univ Md, BS, 49; Pa State Univ, PhD(org chem), 52. *Prof Exp:* Res chemist, 52-66, res supvr, 66-78, res scientist & group leader, Adhesives Lab, 78-82, group leader, Mat Sci, 82-89, RES ASSOC, RES CTR, HERCULES INC, 89- *Mem:* Am Chem Soc; Sigma Xi; Am Soc Testing & Mat. *Res:* The effect of low molecular weight hydrocarbon and rosin-based resins on the viscoelastic properties of polymer systems; rheology and viscoelasticity of polymer systems; pressure sensitive and hot melt adhesives. *Mailing Add:* 22 Winterbury Circle Wilmington DE 19808

CLATOR, IRVIN GARRETT, b Huntington, WVa, Nov 2, 41. NUCLEAR PHYSICS, EXPLOSIVES. *Educ:* WVa Univ, BS, 63, MS, 65, PhD(physics), 69. *Prof Exp:* From physicist to res physicist, US Naval Weapons Lab, 65-70; asst prof, 70-74, ASSOC PROF PHYSICS, UNIV NC, WILMINGTON, 74-, CHMN DEPT, 71- *Mem:* AAAS. *Res:* Neutron induced reaction in the 10 to 20 mev energy range with medium A nuclei; explosive material properties. *Mailing Add:* Dept Physics Univ NC 601 S Col Rd Wilmington NC 28403

CLATWORTHY, WILLARD HUBERT, b Auxier, Ky, Oct 16, 15; div; c 2. STATISTICS. *Educ:* Berea Col, BA, 38; Univ Ky, MA, 40; Univ NC, PhD(math statist), 52. *Prof Exp:* Asst, Univ Ky, 38-40; prof, Louisburg Col, 40-42; tool designer, Wright Automatic Packing Mach Co, 42 & Bell Aircraft Corp, 42-43; instr math, Wayne State Univ, 46-49; statistician math & probability, Nat Bur Standards, 52-55; statistician, Bettis Atomic Power Div, Westinghouse Elec Corp, 55-62; prof statist, State Univ NY, Buffalo, 62-81; RETIRED. *Mem:* Fel AAAS; Inst Math Statist; fel Am Statist Asn; Int Asn Statist in Phys Sci; fel Royal Statist Soc. *Res:* Mathematics of statistical design of experiments; combinatorial aspects of design of experiments; regression and least squares; analysis of variance. *Mailing Add:* 378 Cottonwood Dr Williamsville NY 14221

CLAUDE, PHILIPPA, b New York, NY, Jan 21, 36; m 75. CELL BIOLOGY, NEUROBIOLOGY. *Educ:* Cornell Univ, BA, 57; Univ Pa, PhD(zool), 68. *Prof Exp:* Instr, Harvard Med Sch, 69-72, prin res assoc neurobiol, 72-75; asst scientist, Wis Rrgional Primate Res Ctr, 75-84; LECTR BIOL, CORE CURRIC, UNIV WIS-MADISON, 76-, MEM NEUROSCI TRAINING PROG, 77-; ASSOC SCIENTIST, WIS REGIONAL PRIMATE RES CTR, 85- *Concurrent Pos:* NIH fel, Harvard Med Sch, 69-72, NIH spec fel, 72-73. *Mem:* Am Soc Cell Biol; AAAS; Soc Neurosci; Electron Micros Soc Am; Int Soc Develop Neurosci. *Res:* neuronal development and neurospecificity; neuronal tissue culture; membrane specializations; cell surface receptors; internalization and processing of growth factors and hormones; neuronal cell culture; electron microscopic autoradiography; surface replicas; scanning electron microscopy. *Mailing Add:* Regional Primate Res Ctr Univ Wis 1223 Capitol Ct Madison WI 53715-1299

CLAUDSON, T(HOMAS) T(UCKER), b Pratt, Kans, Apr 25, 33; m 55; c 4. MECHANICAL ENGINEERING, METALLURGICAL ENGINEERING. *Educ:* Univ Wash, BS, 55; Ore State Univ, MS, 58, PhD(metall eng), 62. *Prof Exp:* Res engr, Gen Elec Co, 55-61, sr res scientist, 61-65; res assoc, Battelle Mem Inst, 65-66, mgr, mech metall unit, 66-68, mgr, FFTF Fuels Dept, 68-70; mgr, reactor mat, Westinghouse Hanford Co, 70-76, mgr, reactor assembly & res & develop dept, 76-79; group vpres, Criton Technol, 79-85; DIR, ENG DEVELOP PROJ, BATELLE MEM INST, 85- *Concurrent Pos:* Adj asst prof, Univ Wash, Joint Ctr Grad Studies, 62-79. *Mem:* Fel Am Soc Metals; Am Nuclear Soc; Am Soc Mech Engrs. *Res:* Effects of neutron irradiation and nuclear reactor environment on the mechanical behavior and physical properties of structural and fuel cladding materials; heat transfer; physical metallurgy. *Mailing Add:* 1915 Sheriden Pl Richland WA 99352

CLAUER, ALLAN HENRY, b Milwaukee, Wis, Nov 3, 36; m 56; c 3. MECHANICAL METALLURGY, PHYSICAL METALLURGY. *Educ:* Univ Wis-Madison, BSc, 58, MSc, 61; Ohio State Univ, Columbus, PhD(metall eng), 68. *Prof Exp:* Instr metall, Dept Mining & Metals, Univ Wis, 59; RES SCIENTIST SECT MGR, PROCESS METALL SECT & TECH LEADER, ADV MAT, BATTELLE MEM INST, 60- *Concurrent Pos:* Vis scientist, Danish AEC Lab, Riso, Denmark, 72-73. *Mem:* Sigma Xi; fel Am Soc Metals; Am Inst Mining Metall & Petrol Engrs. *Res:* High temperature strength properties of metals and ceramics; the relation between strength and microstructure; laser processing of metals utilizing both shockwave and thermal effects; powder metallurgy; hot isostatic processing. *Mailing Add:* Battelle Mem Inst 505 King Ave Columbus OH 43201

CLAUER, C ROBERT, JR, b Wooster, Ohio, Aug 18, 48; m 75. SPACE PLASMA PHYSICS. *Educ:* Miami Univ, Ohio, BA, 70; Univ Calif, Los Angeles, MS, 74, PhD(geophys & space physics), 80. *Prof Exp:* Res geophysicist, Univ Calif, Los Angeles, 73-79, res assoc, 79-80; res assoc, Stanford Univ, 80-85, sr res assoc, 85-90; ASSOC RES SCIENTIST, UNIV MICH, 91- *Concurrent Pos:* Mem, Magnetic Fields Subgroup Coord Data Anal Workshop, Nat Space Flight Data Ctr, 81-86; mem, data syst users working group, NASA, 82-; mem, NASA ad hoc sci steering comt, Nat Space Sci Data Ctr online data catalogue syst, 84-87. *Mem:* Am Geophys Union; Sigma Xi; Int Union Radio Sci; Int Asn Geomagnetism & Aeronomy. *Res:* Solar-terrestrial relationships; solar wind-magnetosphere-ionosphere coupling; magnetospheric current systems and electrodynamics; incoherent scatter radar measurements of ionospheric electric fields and plasma parameters; data management, display and presentation; substorm and storm phenomenology; computer networks in support of scientific activities. *Mailing Add:* Space Physics Res Lab Univ Mich 2455 Hayward Ann Arbor MI 48109-2143

CLAUS, ALFONS JOZEF, b Belg, May 21, 32; US citizen; m 63; c 3. COMPUTER SCIENCES, OCEAN SCIENCES. *Educ:* Univ Ghent, BS, 56; Univ Mich, PhD(eng mech), 61. *Prof Exp:* Mem tech staff commun satellites, 61-65, SUPVR SPACE PROG OCEAN ACOUST & DATA ANALYSIS, BELL LABS, 65- *Res:* Application of computer science and technology to the study of ocean acoustic phenomena. *Mailing Add:* AT&T Bell Labs 1919 S Eads St Suite 300 Arlington VA 22202

CLAUS, GEORGE WILLIAM, b Council Bluffs, Iowa, Aug 15, 36; m 58; c 3. BACTERIAL PHYSIOLOGY. *Educ:* Iowa State Univ, BS, 59, PhD(physiol bact), 64. *Prof Exp:* Bacteriologist & biochemist, US Army Med Unit, Ft Detrick, Md, 64-66; asst prof microbiol, Pa State Univ, University Park, 66-73; asst prof, 73-76, ASSOC PROF MICROBIOL, VA POLYTECH INST & STATE UNIV, 76- *Concurrent Pos:* Consult, Hoffmann-La Roche, Inc, 74-78, 83-85, Monsanto Co, 88- & Williams & Wilkins Co, 90- *Mem:* AAAS; Am Soc Microbiol; Soc Indust Microbiol; Sigma Xi; US Fedn Cult Collections. *Res:* Physiology and fine-structure of acetic acid bacteria; intracytoplasmic membrane development in nonphotosynthetic gram-negative bacteria; limited oxidations by Gluconobacter; detection of opportunistic pathogens in industrial bioreactors. *Mailing Add:* Dept Biol Va Polytech Inst & State Univ Blacksburg VA 24061-0406

CLAUS, RICHARD OTTO, b Baltimore, Md, May 29, 51. ACOUSTOOPTICS, ULTRASONICS. *Educ:* Johns Hopkins Univ, BES, 73, PhD(elec eng), 77. *Prof Exp:* Instr elec eng, Johns Hopkins Univ, 73-76; asst prof, 77-81, ASSOC PROF ELEC ENG, VA POLYTECH INST & STATE UNIV, 81- *Mem:* Inst Elec & Electronics Engrs; Optical Soc Am; Soc Photo-Optical Instrumentation Engrs; Am Soc Eng Educ. *Res:* Applied optics; nondestructive evaluation; interferometry; ultrasonic transducer characterization; optical fiber instrumentation. *Mailing Add:* Elec Eng Dept Whittemore Hall Va Polytech Inst & State Univ Blacksburg VA 24061-0111

CLAUS, THOMAS HARRISON, b Kansas City, Mo, Jan 17, 43; m 66; c 2. OBESITY, DIABETES. *Educ:* Wheaton Col, Ill, BS, 64; Univ Ill, Chicago, PhD(biochem), 70. *Prof Exp:* Res assoc physiol, Sch Med, Vanderbilt Univ, 69-73, asst prof physiol, 73-81; sr res biochemist, 81-84, GROUP LEADER DIABETES RES, LEDERLE LABS, MED RES DIV, AM CYANAMID CO, 84- *Mem:* Brit Biochem Soc; Am Soc Biochem & Molecular Biol. *Res:* Hormonal regulation of carbohydrate metabolism. *Mailing Add:* Am Cyanamid Co Lederle Labs Bldg 134C Rm ll8 Pearl River NY 10965

CLAUSEN, CHRIS ANTHONY, b New Orleans, La, Dec 7, 40; m 62; c 3. INORGANIC CHEMISTRY. *Educ:* La State Univ, Baton Rouge, BS, 63; La State Univ, New Orleans, PhD(inorg chem), 69. *Prof Exp:* Chemist, Standard Oil Co Calif, 63-66; from asst prof to assoc prof, 69-77, PROF CHEM UNIV CENT FLA, 77- *Concurrent Pos:* AEC res grant, 69-70; tour speaker, Am Chem Soc, 74; res scientist, Dow Chem Co, 77. *Honors & Awards:* Excellence in Sci Res Award, Sigma Xi, 69. *Mem:* Am Chem Soc; Soc Appl Spectros. *Res:* The chemistry of aerosol systems; molecular vibrations in the far infrared; coordination chemistry in marine environments; petroleum development and catalysis studies; forensic chemistry. *Mailing Add:* Dept Chem Univ Cent Fla Orlando FL 32816

CLAUSEN, EDGAR CLEMENS, b St Louis, Mo, Dec 15, 51; m 74; c 1. CHEMICAL ENGINEERING. *Educ:* Univ Mo-Rolla, BS, 74, MS, 75, PhD(chem eng), 78. *Prof Exp:* Process design engr, Monsanto Co, 74; asst prof chem eng, Tenn Tech Univ, 77-81; assoc prof, 81-85, PROF CHEM ENG, UNIV ARK, 85- *Concurrent Pos:* Prof, E I du Pont de Nemours & Co, 78; process design consult, Oak Ridge Nat Lab, 80-81. *Mem:* Am Inst Chem Engrs; Am Soc Eng Educ; Sigma Xi; Am Chem Soc. *Res:* Conversion of biomass to chemicals and energy: methane production by anaerobic digestion, conversion of lignocellulosics to sugars by acid hydrolysis, fermentation of hydrolyzates to chemicals, biological conversion of coal gas to chemicals and energy. *Mailing Add:* Dept Chem Eng Univ Ark Fayetteville AR 72701

CLAUSEN, ERIC NEIL, b Ithaca, NY, July 2, 43. GEOMORPHOLOGY. *Educ:* Columbia Univ, BS, 65; Univ Wyo, PhD(geol), 69. *Prof Exp:* From asst prof to assoc prof geol, 68-78, PROF EARTH SCI, MINOT STATE UNIV, 78- *Mem:* Geol Soc Am; Nat Asn Geol Teachers; Am Quaternary Asn; Sigma Xi; AAAS. *Res:* Geomorphic history of northern great plains; earth science education. *Mailing Add:* Dept Earth Sci Minot State Univ Minot ND 58701

CLAUSEN, WILLIAM E(ARLE), b Rochester, Pa, Mar 11, 38; m 64. ENGINEERING MECHANICS. *Educ:* Lehigh Univ, BS, 60; Ohio State Univ, MSc, 61, PhD(eng mech), 65. *Prof Exp:* Asst prof, 65-69, ASSOC PROF ENG MECH, OHIO STATE UNIV, 69- *Concurrent Pos:* Consult, Battelle Mem Inst, 66- *Mem:* Am Inst Aeronaut & Astronaut; Sigma Xi. *Res:* Elasticity; dynamics; composite materials; shells; numerical techniques. *Mailing Add:* Dept Eng Mech Ohio State Univ 155 W Woodruff Ave Columbus OH 43210

CLAUSER, FRANCIS HETTINGER, b Kansas City, Mo, May 25, 13; m 37; c 2. AERONAUTICS. *Educ:* Calif Inst Technol, BS, 34, MS, 35, PhD(aeronaut), 37. *Prof Exp:* Engr in charge aerodyn & design res, Douglas Aircraft Co, 37-46; prof aeronaut & chmn dept, Johns Hopkins Univ, 46-60, prof mech, 60-64; prof eng & vchancellor, Univ Calif, Santa Cruz, 65-69; chmn, Div Eng & Appl Sci, 69-74, Clark B Millikan prof eng, 69-86, EMER CLARK B MILLIKAN PROF ENG, CALIF INST TECHNOL, 86- *Mem:* Nat Acad Eng; Am Acad Arts & Sci; fel Am Inst Aeronaut & Astronaut. *Res:* Aerodynamics; fluid mechanics; non-linear mechanics. *Mailing Add:* Calif Inst Technol Pasadena CA 91125

CLAUSER, JOHN FRANCIS, b Pasadena, Calif, Dec 1, 42; m 64. QUANTUM MECHANICS, EXPERIMENTAL PHYSICS. *Educ:* Calif Inst Technol, BS, 64; Columbia Univ, MA, 66, PhD(physics), 69. *Prof Exp:* Res physicist quantum physics, Univ Calif & Lawrence Berkeley Lab, 69-75, RES PHYSICIST PLASMA PHYSICS, UNIV CALIF & LAWRENCE LIVERMORE LAB, 75- *Mem:* Am Phys Soc. *Res:* Reconciliation of everyday notions of objectivity, space and time with the observed and predicted behavior of quantum mechanical systems; understanding fundamental processes in magnetic-mirror-confined plasmas; development of computer systems for experimental physics research. *Mailing Add:* 975 Murrieta Blvd No 22 Livermore CA 94550

CLAUSER, MILTON JOHN, b Santa Monica, Calif, June 17, 40; m 61; c 2. NUCLEAR ENGINEERING. *Educ:* Mass Inst Technol, SB, 61; Calif Inst Technol, PhD(physics), 66. *Prof Exp:* NSF res fel physics, Munich Tech Univ, 66-67; mem tech staff, Sandia Nat Labs, 67-79, supvr, Advan Reactor Safety Anal, 80-83, lab rep, Dept Energy, 83-84, supvr, Directed Energy Physics, 84-87, supvr radiation & hydrodynamics theory, 87-89, SUPVR, PULSED POWER PROGS DIV, SANDIA NAT LABS, 89- *Concurrent Pos:* Mgr, Laser Guided Electron-Beam Res Prog. *Mem:* Am Phys Soc. *Res:* Ion beam fusion; target behavior; directed energy; inertial fusion. *Mailing Add:* Org No 1201 Sandia Nat Labs PO Box 5800 Albuquerque NM 87185-5800

CLAUSING, A(RTHUR) M(ARVIN), b Palatine, Ill, Aug 17, 36; m 64; c 2. SOLAR ENERGY, HEAT TRANSFER. *Educ:* Valparaiso Univ, BS, 58; Univ Ill, MS, 60, PhD(mech eng), 63. *Prof Exp:* PROF MECH ENG & ASSOC HEAD DEPT MECH & INDUST ENG, UNIV ILL, URBANA-

CHAMPAIGN, 62- *Concurrent Pos:* Consult, Elec Power Res Inst & Off Energy Rel Inventions. *Mem:* Int Solar Energy Soc; Am Soc Mech Engrs; Am Soc Heating Refrig & Air Conditioning Engrs; Sigma Xi; Am Soc Eng Educ. *Res:* Experimental studies of natural and mixed convective heat transfer using a variable temperature, cryogenic facility; convective energy transport within and from buildings; convective losses from solar receivers; quantifying the performance of passive solar heating systems; variable property influences on convective heat transfer; performance monitoring of solar heating systems. *Mailing Add:* Mech Eng Bldg 1206 W Green St Univ Ill Urbana IL 61801

CLAUSON, W(ARREN) W(ILLIAM), b Chicago, Ill, Mar 9, 26; m 63. CHEMICAL ENGINEERING. *Educ:* Ill Inst Technol, BS, 49, MS, 52, PhD(chem eng), 55. *Prof Exp:* Draftsman, NuWay Refrig Co, 46-47; chem engr, L R Kerns Co, 49-51; engr, Bell & Howell Co, 51-52; asst, Chem Eng Dept, Ill Inst Technol, 54-55; asst prof chem eng, Rose Polytech Inst, 55-58; heat transfer specialist rocket appln res sect, Flight Propulsion Lab Dept, Gen Elec Co, Ohio, 58; supvr heat transfer sect, Appl Mech Div, Aerojet-Gen Corp Div, Gen Tire & Rubber Co, 58-65; mgr, Eng Analysis Dept, Wright Aeronaut Div, Curtiss-Wright Corp, 65-68; proj dir, Gries Reprod Co, 68-69; mgr papermaking develop, Scott Paper Co, 69-71; independent consult, 71-74; proj & process engr, Amoco Chem Corp, 74-88; CONSULT, 88- *Concurrent Pos:* Res fel, Inst Gas Technol, 52-54; consult, Aircraft Gas Turbine Div, Gen Elec Co, 56-58; lectr exten, Univ Calif, Berkeley, 59-61 & Sacramento State Col, 61-65. *Mem:* Inst Chem Engrs. *Res:* Thermodynamics; heat transfer; fluid dynamics; applied mathematics. *Mailing Add:* 134 King George Ct-203 Palatine IL 60067

CLAUSS, ROY H, b Ill, Feb 8, 23; m 45; c 3. CARDIOVASCULAR SURGERY. *Educ:* Northwestern Univ, BS, 43, MD, 46; Am Bd Surg, dipl, 56; Am Bd Thoracic Surg, dipl, 57. *Prof Exp:* Instr surg, Col Physicians & Surgeons, Columbia Univ, 55-57; asst, Harvard Med Sch, 57-58; asst prof, Col Med, Univ Cincinnati, 58-60; assoc prof, Sch Med, NY Univ, 60-69; PROF SURG, NEW YORK MED COL, 69- *Concurrent Pos:* Am Trudeau Soc teaching fel, 54-55; USPHS spec fel, Harvard Med Sch, 57-58; consult, Cabrini Health Care Ctr, 66-; attend surgeon, Lenox Hill Hosp, 76-, St Vincent's Hosp, 77-; courtesy surgeon, Doctor's Hosp, 78- vis surgeon, Metrop & Coler Hosps; consult, NY Vet Med Admin Hosp; mem coun circulation, Am Heart Asn. *Mem:* Am Col Chest Physicians; Am Heart Asn; Am Col Surg; Am Asn Thoracic Surg; Int Cardiovasc Soc. *Res:* Cardiovascular and pulmonary surgery and physiology. *Mailing Add:* 1021 Park Ave New York NY 10028

CLAUSSEN, DENNIS LEE, b Pender, Nebr, Sept 23, 41; m 65; c 2. PHYSIOLOGICAL ECOLOGY. *Educ:* Pomona Col, BA, 63; Univ Calif, Riverside, MA, 66; Univ Mont, PhD(zool), 71. *Prof Exp:* Entomologist, Nutrilite Prod Inc, Calif, 66-68; res asst biol control, Univ Calif, Riverside, 68; from asst prof to assoc prof, 71-83, PROF ZOOL, MIAMI UNIV, 83- *Mem:* AAAS; Sigma Xi; Am Soc Zoologists; Ecol Soc Am; Am Soc Ichthyol Herpet; Herpetologists League. *Res:* The metabolism, thermal relations and water relations of arthropods, amphibians and reptiles. *Mailing Add:* Dept Zool Miami Univ Oxford OH 45056

CLAUSSEN, MARK J, b Whitter, Calif, April 20, 52; m 76; c 2. RADIOASTRONOMY. *Educ:* Mo Southern State Col, BS, 74; NMex State Univ, MS 76; Univ Iowa, PhD(physics), 81. *Prof Exp:* Res fel, Calif Inst Technol, 81-84; at Grad Res Ctr Univ Mass, 84-87; RADIOASTRONOMER, NAVAL RES LAB, 87- *Mem:* Am Astron Soc. *Res:* Radio interferometry of molecular lines in star-forming regions and evolved stars; continuum interferometry of galactoc H II regions and planetary nebulae. *Mailing Add:* Naval Res Lab Code 4134C Washington DC 20375

CLAUS-WALKER, JACQUELINE LUCY, b Paris, France, Dec 13, 15; US citizen; m 65. ENDOCRINOLOGY. *Educ:* Univ Paris, BA, 35; Sorbonne, MS, 46; Union Col, BS, 51; Univ Houston, MS, 55; Baylor Univ, PhD(physiol), 66. *Prof Exp:* Lab technician, Robert Packer Hosp, 47-48; chief pharmacist, James Walker Mem Hosp, 51-53; res asst, Univ Tex M D Anderson Hosp & Tumor Inst, 55-58; res asst, 61-66, from instr to asst prof physiol chem, Dept Biochem & Dept Rehab, 66-71, assoc prof physiol, 71-74, asst prof biochem, 71-84, dir Neuroendocrine Lab, 66-84, PROF, DEPT REHAB, BAYLOR COL MED, 84- *Mem:* AAAS; NY Acad Sci; Am Soc Bone & Mineral Res; Endocrine Soc; Am Physiol Soc. *Res:* Pharmacology of spinal cord injury; biochemistry and histology of skin in spinal cord injury; bone and collagen metabolism in man with section of the cervical spinal cord. *Mailing Add:* 5127 Jackwood Houston TX 77096

CLAUSZ, JOHN CLAY, b Hackensack, NJ, Oct 5, 40; m 63; c 2. MYCOLOGY, MICROBIOLOGY. *Educ:* Ohio Wesleyan Univ, BA, 62; Univ NC, Chapel Hill, MA, 66, PhD(bot), 70. *Prof Exp:* Asst prof biol, St Andrews Presby Col, 69-76, State Univ Geneseo, 76-79; asst prof biol, State Univ NY Geneseo, 76-79; from asst prof to assoc prof, 79-88, PROF BIOL, CARROLL COL, WAUKESHA, WIS, 88- *Mem:* Mycol Soc Am; Am Soc Microbiol; Sigma Xi. *Res:* Physiology and ecology of fungi, especially the aquatic fungi; lipids in water molds; heavy metal uptake by water mold Achlya. *Mailing Add:* Dept Biol Carroll Col Waukesha WI 53186

CLAVAN, WALTER, b Philadelphia, Pa, Apr 6, 21; m 45; c 2. ANALYTICAL CHEMISTRY, SPECTROSCOPY. *Educ:* Univ Pa, BS, 42, MS, 47, PhD(analytical chem), 49. *Prof Exp:* Analytical chemist, E J Lavino & Co, 42-44; group leader, analytical dept, King of Prussia Technol Ctr, Penwalt Corp, 49-70, dir analytical serv, 70-74, mgr, analytical chem dept, 74-83, corp analytical scientist, 83-88; RETIRED. *Concurrent Pos:* Lectr, eve div, La Salle Col, 60-70. *Mem:* AAAS; Am Chem Soc; Geochem Soc; Sigma Xi. *Mailing Add:* 1219 W Wynnewood Rd Unit 611 Wynnewood PA 19096

CLAVEAU, ROSARIO, b Chicoutimi, Que, Dec 13, 24; m 51; c 2. HEMATOLOGY, INTERNAL MEDICINE. *Educ:* Chicoutimi Sem, BA, 45; Laval Univ, MD, 50; FRCP(C). *Prof Exp:* MEM DEPT HEMAT, HOPITAL DE CHICOUTIMI, LAVAL UNIV, 60-, HON PROF, FAC MED, 64- *Mem:* Am Soc Hemat; fel Am Col Physicians; Can Med Asn; fel Royal Col Physicians (Can). *Mailing Add:* Hemat Lab Hopital de Chicoutimi Chicoutimi PQ G7H 5H6 Can

CLAVENNA, LEROY RUSSELL, b Joliet, Ill, May 12, 43; m 68; c 3. CHEMICAL ENGINEERING. *Educ:* Univ Ill, BS, 66; Univ Minn, PhD(chem eng), 71. *Prof Exp:* Res engr, solid state chem, Exxon Res & Develop Labs, 71-74, res engr, coal gasification, 74-78, staff engr, 78-80, proj leader gasification & indirect liquefaction, 80-86, RES ASSOC, ADVAN FUEL PROCESSES, EXXON RES & DEVELOP LABS, 86- *Mem:* Am Inst Chem Engrs. *Res:* Surface physics; solid state inorganic chemistry; synthetic fuels; catalysis. *Mailing Add:* Exxon Res & Develop Labs PO Box 2226 Baton Rouge LA 70821-2226

CLAWSON, ALBERT J, b Curtis, Nebr, Feb 15, 24; m 48; c 6. ANIMAL NUTRITION. *Educ:* Univ Nebr, BS, 49; Kans State Univ, MS, 51; Cornell Univ, PhD(nutrit), 55. *Prof Exp:* Asst, Kans State Univ, 49-51; animal husbandman, Exp Sta, North Platte, Nebr, 51-52; asst, Cornell Univ, 52-55; assoc prof, 55-69, PROF ANIMAL SCI, NC STATE UNIV, 69- *Concurrent Pos:* Res fel, Centro Internacional Agricultura Tropical, Cali, Colombia, 72-73. *Mem:* Am Inst Nutrit; Am Soc Animal Sci. *Res:* Nutrient requirements for reproduction in swine; amino acid requirements of pig as determined by manipulation of dietary ingredients; indirect methods of determining live animal composition and factors influencing carcass composition. *Mailing Add:* Rte 1 Box 7 New Hill NC 27562

CLAWSON, DAVID KAY, b Salt Lake City, Utah, Aug 8, 27; m 52; c 2. ORTHOPEDIC SURGERY. *Educ:* Harvard Univ, MD, 52; Am Bd Orthop Surg, dipl, 61. *Prof Exp:* Intern surg, Stanford Univ Hosp, 52-53, resident, 53-54, resident orthop, 54-55; resident, San Francisco City & County Hosp, 54-57; mem bd ed adv, Orthopedic Rev, 72-75; assoc ed, J Bone & Joint Surg, 73-75. resident orthop, Stanford Univ Hosp, 56-57; asst prof surg, Univ Calif, Los Angeles, 58; head div surg & from asst prof to assoc prof surg, Univ Wash, 58-65, prof orthop & chmn dept, 65-75; dean, Col Med, Univ Ky, 75-83, assoc vpres clin affairs, 82-83; EXEC VCHANCELLOR, MED CTR, UNIV KANS, 83- *Concurrent Pos:* Nat Found Infantile Paralysis fels orthop, 55-57 & advan orthop, 57-58; hon sr registr, Royal Nat Orthop Hosp & clin res asst, Univ London, 57-58. *Mem:* Am Acad Orthop Surg; Royal Soc Med; Asn Bone & Joint Surgeons (secy, 72-75, pres, 77); fel Am Geriat Soc; AMA. *Res:* Infections of bone and joints; bone implants; health care delivery; Orthopedic manpower. *Mailing Add:* Med Ctr Univ Kans 39th & Rainbow Blvd Kansas City KS 66103

CLAWSON, ROBERT CHARLES, b South Haven, Mich, May 10, 29; m 58; c 5. HISTOLOGY, EMBRYOLOGY. *Educ:* Spring Hill Col, BS, 50; St Louis Univ, MS, 53; Loyola Univ Chicago, PhD(anat), 62. *Prof Exp:* From instr to asst prof anat, Stritch Sch Med, 64-68; from asst prof to assoc prof, 68-77, PROF ANAT, SCH MED, LA STATE UNIV, SHREVEPORT, 77- *Concurrent Pos:* NIH fel, Loyola Univ Chicago, 62-64. *Mem:* Am Soc Zool; Am Asn Anat. *Res:* Structure of defensive glands of arthropods; gross anatomy. *Mailing Add:* 351 Pennsylvania Ave Shreveport LA 71105

CLAXTON, LARRY DAVIS, b Chattanooga, Tenn, June 17, 46; m 71; c 2. GENETIC TOXICOLOGY, BIOTECHNOLOGY HEALTH. *Educ:* Middle Tenn State Univ, Murfreesboro, BS, 67; Memphis State Univ, MS, 71; NC State Univ, Raleigh, PhD(genetics), 80. *Prof Exp:* Res asst toxicol res, Oak Ridge Nat Lab, 71-72; biologist, genetic toxicol res, Nat Inst Environ Health Sci, Res Triangle Park, NC, 72-77; GENETIC TOXICOLOGIST RES, US ENVIRON PROTECTION AGENCY, 77-; ADJ ASSOC PROF, SCH PUB HEALTH, UNIV NC, CHAPEL HILL, 88- *Concurrent Pos:* Adj asst prof, Sch Pub Health, Univ NC, Chapel Hill, 82-88; chmn task group, Am Soc Testing & Mat, 82-88; counr, Environ Mutagen Soc, 85-87, mem, Critical Issues Comt, 83-84; chmn tech comt, WHO. 87-; adj asst prof, Sch Textiles, NC State Univ, Raleigh, 87- *Honors & Awards:* Bronze Medal, US Environ Prot Agency, 80, Sci & Tech Achievement Award, 83 & 86 & 89; Spec Act & Achievement Award, Alaskan Oil Spill, Environ Protection Agency, 90, Biotechnol Health Res Award, 90. *Mem:* Environ Mutagen Soc; Genotoxicity & Environ Mutagen Soc (pres, 83-85); AAAS; Soc Risk Anal; Air & Waste Mgt Asn. *Res:* Developing genetic toxicology test procedures for assessing complex environmental pollutants; applying computerized structure-activity analyses to environmental chemicals; developing methods to access the health effects of biotechnology. *Mailing Add:* Health Effects Res Lab MD 68 US Environ Protection Agency Research Triangle Park NC 27711

CLAY, CLARENCE SAMUEL, b Kansas City, Mo, Nov 2, 23; m 45; c 4. GEOPHYSICS. *Educ:* Kans State Univ, BA, 47, MS, 48; Univ Wis, PhD(physics), 51. *Prof Exp:* Asst prof physics, Univ Wyo, 50-51; res physicist, Carter Oil Co, 51-55; sr res scientist, Hudson Lab, Columbia Univ, 55-67; PROF GEOL & GEOPHYS, UNIV WIS-MADISON, 68- *Mem:* Fel Acoust Soc Am; Am Geophys Union; Soc Explor Geophys. *Res:* Wave propagation in inhomogeneous media; scattering at rough interfaces; ocean acoustics; electrical geophysics; acoustic measurements of fish populations. *Mailing Add:* Weeks Hall Dept Geol & Geophys Univ Wis 1215 W Dayton Madison WI 53706

CLAY, FORREST PIERCE, JR, b Sutherland, Va, Nov 15, 27. PHYSICS, ELECTRONICS ENGINEERING. *Educ:* Randolph-Macon Col, BS, 48; Univ Va, MS, 50, PhD(physics), 52. *Prof Exp:* With Atlantic Res Corp, 52; asst prof physics, Georgetown Univ, 52-54 & Rutgers Univ, 54-61; from asst prof to assoc prof, 57-61, chmn, Dept Physics, 79-80, PROF PHYSICS, OLD DOM UNIV, 61- *Concurrent Pos:* Vis prof, Randolph-Macon Col, 57-58. *Mem:* Am Phys Soc; Am Asn Physics Teachers; Sigma Xi; Inst Elec & Electronics Engrs. *Res:* Positronium decay; quadrupole mass spectrometry electronics. *Mailing Add:* Dept Physics Old Dom Univ Norfolk VA 23529

CLAY, GEORGE A, b Cambridge, Mass, June 24, 38; m 65; c 4. NEUROPHARMACOLOGY. *Educ:* Dartmouth Col, AB, 61; Boston Univ, MA, 64, PhD, 68. *Prof Exp:* Nat Heart & Lung Inst fel, 67-68, Nat Inst Gen Med Sci Res Assocs Pharmacol Training Prog fel, 68-70; asst prof pharmacol, Bowman Gray Sch Med, 70-72; from res investr to sr res investr, G D Searle & Co, 72-74, group leader, 74-80, sect head, Cent Nerv Syst Pharmacol, 80-, ASST DIR GASTROENTEROL, G D SEARLE & CO. *Mem:* Am Soc Pharmacol & Exp Therapeut; AAAS. *Res:* Biochemical mechanisms of action of drugs affecting the central nervous system. *Mailing Add:* Res & Develop Div G D Searle & Co 4901 Searle Pkwy Skokie IL 60077

CLAY, JAMES RAY, b Burley, Idaho, Nov 5, 38; m 59; c 3. MATHEMATICS. *Educ:* Univ Utah, BS, 60; Univ Wash, MS, 62, PhD(math), 66. *Prof Exp:* Assoc engr, Boeing Co, 60-63; phys scientist, US Govt, 64-66; from asst prof to assoc prof, 66-74, assoc head dept, 69-72, PROF MATH, UNIV ARIZ, 74- *Concurrent Pos:* Guest prof, Univ T06bingen, 72-73, Univ London, 73, Tech Univ, Munich, 79-80, Univ Edinburgh, 80, Univ Stellenbosch, 89 & Bundeswehr Universitaet, Hamburg, 90-91. *Honors & Awards:* Spec Award, Humboldt Found, 72. *Mem:* Am Math Soc. *Res:* Abstract algebra; computer science; algebraic structures arising from endomorphism and mappings of groups. *Mailing Add:* Dept Math Univ Ariz Tucson AZ 85721

CLAY, JOHN PAUL, physical chemistry; deceased, see previous edition for last biography

CLAY, MARY ELLEN, b Freeport, Ohio, July 28, 40. ENTOMOLOGY, BIOLOGY. *Educ:* Muskingum Col, BS, 63; Ohio State Univ, MS, 66, PhD(entom), 69. *Prof Exp:* Res assoc mosquito biol, 69-73, lectr introd biol prog, 73-74, asst prof, 74-79, ASSOC PROF ENTOM, OHIO STATE UNIV, 79- *Mem:* Am Inst Biol Sci; Am Mosquito Control Asn; AAAS; Arctic Inst NAm; Entom Soc Am. *Res:* Insect biology; structural and functional aspects of the mosquito crop; insect diapause; mosquito neuroendocrine system; internal and external factors controlling diapause processes; biology of phychodids. *Mailing Add:* Dept Entomo, Ohio State Univ 1735 Neil Ave Columbus OH 43210

CLAY, MICHAEL M, b Cleveland, Ohio, Aug 10, 20; m 55; c 2. PHARMACOLOGY. *Educ:* Ohio State Univ, BA, 41, PhD(pharmacol), 53; Univ Toledo, BS, 50, MS, 51. *Prof Exp:* Res assoc endocrinol, Ohio State Univ, 52-53; from asst to assoc prof pharmacol, Col Pharm, Columbia Univ, 53-64; guest prof, Med Clin, Univ Münster, 64-68; prof pharmacol, Univ Houston, 68-84; RETIRED. *Mem:* Am Pharmaceut Asn; Am Geront Soc; NY Acad Sci. *Res:* Connective tissue physiology; cardiovascular disease. *Mailing Add:* 1926 Norfolk St No 14 Houston TX 77089

CLAYBAUGH, GLENN ALAN, b Lincoln, Nebr, Dec 10, 27; m 50; c 2. BACTERIOLOGY. *Educ:* Univ Nebr, BSc, 49; Mich State Univ, MSc, 50; Iowa State Univ, PhD(dairy bact), 53. *Prof Exp:* Sr bacteriologist, 53-60, sect leader, 60-63, prod mgr, 63-65, assoc, Mkt Div, 65-69, mkt dir, 69-72, DIR PROF SERV, MEAD JOHNSON NUTRIT DIV, MEAD JOHNSON & CO, 72- *Mem:* AAAS; Am Soc Microbiol; Am Dairy Sci Asn; fel Royal Soc Health; fel Am Pub Health Asn. *Res:* Infant feeding; dairy and food bacteriology; antibiotics and non-sporulating anaerobic bacteria in dairy products; quality control of milk products, infant formulas and other specialized food products; hospital consulting. *Mailing Add:* 1612 Russell Evansville IN 47720-5525

CLAYBERG, CARL DUDLEY, b Tacoma, Wash, Mar 1, 31; m 77; c 2. VEGETABLE CROPS. *Educ:* Univ Wash, BS, 54; Univ Calif, PhD(genetics), 58. *Prof Exp:* Asst genetics, Univ Calif, 54-56; asst geneticist, Conn Agr Exp Sta, 57-61, assoc geneticist, 61-74; assoc prof hort & forestry, 74-77, PROF HORT, KANS STATE UNIV, 77- *Mem:* Am Genetic Asn; Soc Study Evolution; Bot Soc Am; Am Soc Hort Sci. *Res:* Plant genetics, especially in Phaseolus and Cucumis. *Mailing Add:* Dept Hort Kans State Univ Manhattan KS 66506

CLAYBROOK, JAMES RUSSELL, b Cleburne, Tex, Aug 24, 36; m 63. MOLECULAR PHYSIOLOGY. *Educ:* Univ Tex, BS, 57, PhD(chem), 63. *Prof Exp:* Asst prof biochem, Med Sch, Univ Ore, 63-66; asst scientist, Ore Regional Primate Res Ctr, 63-66; NIH res fel microbiol, Univ Ill, Urbana-Champaign, 66-68; researcher, Int Lab Genetics & Biophys, Naples, Italy, 68-69; assoc prof physiol, Med Col Ohio, Toledo, 69-74; OFF GRANTS & CONTRACTS, MED COL GA, 74- *Mem:* AAAS; Soc Gen Physiol; NY Acad Sci. *Res:* Role of nucleic acids in development and cell differentiation. *Mailing Add:* 0ff Grants & Contracts Res Admin Med Col Ga Augusta GA 30912

CLAYCOMB, CECIL KEITH, b Twin Falls, Idaho, Oct 19, 20; m 43; c 2. BIOCHEMISTRY. *Educ:* Univ Ore, PhD(biochem), 51. *Prof Exp:* AEC res asst, Med Sch, 49-51, asst prof, 51-61, head dept, 51-81, asst to pres Health Sci Ctr Minority Student Affairs, 75-78, 80-85, prof, Sch Dent, 61-85, EMER PROF BIOCHEM, ORE HEALTH SCI UNIV, 85. *Concurrent Pos:* Vis res scientist, Inst Dent Res, United Dent Hosp, NSW, Australia, biol sci coordr, 72-76, chmn, Biol Sci Comt, 77-79. *Mem:* AAAS; Am Chem Soc; NY Acad Sci; Int Asn Dent Res. *Res:* Oral collagen metabolism using proline labeled with tritium and/or radiocarbon. *Mailing Add:* 3326 SW 13th Ave Portland OR 97201-2922

CLAYCOMB, WILLIAM CREIGHTON, b Cincinnati, Ohio, Dec 20, 42; c 1. BIOLOGICAL CHEMISTRY. *Educ:* Ind Univ, AB, 66, PhD, 69. *Prof Exp:* Asst prof cell biophys, Baylor Col Med, 72-76; assoc prof, 76-82, PROF BIOCHEM, SCH MED, LA STATE UNIV, 82- *Concurrent Pos:* Res fel biol chem, Harvard Med Sch, 69-72 & NIH fel differentiation, 70-72; estab investr, Am Heart Asn, 75-80. *Mem:* AAAS; Soc Develop Biol; Am Soc Cell Biol; Am Soc Zool; Am Soc Biol Chemists. *Res:* Regulation of cell differentiation and cell proliferation; developmental biology; DNA replication. *Mailing Add:* Biochem Dept La State Univ Med Col 1901 Perdido St New Orleans LA 70112

CLAYMAN, BRUCE PHILIP, b New York, NY, Sept 2, 42; m 62; c 1. SOLID STATE PHYSICS. *Educ:* Rensselaer Polytech Inst, BS, 64; Cornell Univ, PhD(physics), 69. *Prof Exp:* From asst prof to assoc prof, 68-80, assoc dean grad studies, 76-79, actg dean sci, 80, PROF PHYSICS, SIMON FRASER UNIV, 80-, DEAN GRAD STUDIES, 85-,. *Concurrent Pos:* Consult, Xerox Palo Alto Res Ctr, 79-80; vis prof physics, Emory Univ, 81; chmn, triumf bd mgt, 88-90. *Mem:* Am Phys Soc; Can Asn Physicists. *Res:* Far-infrared spectroscopic study of pure and doped materials. *Mailing Add:* Dean Grad Studies Simon Fraser Univ Burnaby BC V5A 1S6 Can

CLAYPOOL, GEORGE EDWIN, b Shenandoah, Iowa, Nov 20, 39; m 62; c 3. GEOCHEMISTRY, GEOLOGY. *Educ:* Colo State Univ, BS, 63; Univ Calif, Los Angeles, PhD(geochem), 74. *Prof Exp:* Lab technician chem, United Testing Labs, Monterey Park, 61-62; chemist geol, Denver Res Ctr, Marathon Oil Co, 63-69; res asst geochem, Inst Geophys & Planetary Physics, Univ Calif, Los Angeles, 69-74; res chemist geochem, US Geol Surv, Denver, 74-88; MOBIL OIL, 87- *Concurrent Pos:* Teaching asst chem, Univ Calif, Los Angeles, 70-71. *Mem:* Geochem Soc. *Res:* Research in geochemistry of sedimentary organic substances and petroleum; investigation of chemical reactions in earth's crust linked to the decomposition of organic matter; geochemistry of light stable isotopes. *Mailing Add:* 520 Keller Springs Rd No 514 Dallas TX 75248

CLAYPOOL, LAWRENCE LEONARD, pomology; deceased, see previous edition for last biography

CLAYTON, ANTHONY BROXHOLME, b Solihull, Eng, Jan 14, 40; m 63; c 4. ORGANIC CHEMISTRY. *Educ:* Univ Aston, ARIC, 62; Univ Birmingham, PhD(org chem), 65. *Prof Exp:* Res assoc, Cornell Univ, 65-67; from res chemist to sr res chemist, 67-89, RES SCIENTIST, HERCULES INC, 89- *Mem:* Am Chem Soc; Royal Soc Chem; Sigma Xi. *Res:* Organic Fluorine chemistry; agricultural chemistry; polymer chemistry; general organic chemistry. *Mailing Add:* Hercules Inc Res Ctr Wilmington DE 19894

CLAYTON, CLIVE ROBERT, b Croydon, Eng, Oct 26, 49. CORROSION SCIENCE, ELECTRON SPECTROSCOPY. *Educ:* Univ Surrey, BSc, 73, PhD(surface chem), 76. *Prof Exp:* Res assoc, 76-78, from asst prof to assoc prof, 78-89, PROF MAT SCI, STATE UNIV NY, STONY BROOK, 89- *Mem:* Electrochem Soc; Inst Metallurgists. *Res:* Corrosion of metals and alloys; electron spectroscopy; surface analysis; modification of surface properties of metals and alloys by ion implantation. *Mailing Add:* Dept Mat Sci & Eng State Univ NY Stony Brook NY 11794

CLAYTON, DALE LEONARD, b Harrisville, Mich, Apr 16, 39; m 61; c 2. ZOOLOGY. *Educ:* Andrews Univ, BA, 62; Loma Linda Univ, MA, 64; Mich State Univ, PhD(zool), 68. *Prof Exp:* Teaching asst biol sci, Mich State Univ, 64-67, dir human biol labs, 67-69, from instr to asst prof physiol, 67-69, asst prof, 70-77, prof biol, Walla Walla Col, 77-81; PROF BIOL & CHMN, DEPT BIOL, SOUTHWESTERN ADVENTIST COL, KEENE, TEX, 81- *Concurrent Pos:* Vis prof, Philippine Union Col, 78-80. *Mem:* AAAS; Int Soc Chronobiol; Sigma Xi; Animal Behav Soc. *Res:* Physiological basis of animal behavior, especially the interaction of internal circadian rhythms, homeostatic mechanisms, development rates and external variables of photoperiod, twilight, temperature and other environmental perturbations. *Mailing Add:* Dept Biol Southwestern Adventist Col Keene TX 76059

CLAYTON, DAVID WALTON, b Leicester, Eng; m 55; c 3. ORGANIC CHEMISTRY, WOOD CHEMISTRY. *Educ:* Univ London, BSc, 45, MSc, 51; Cambridge Univ, BA, 50, PhD(org chem), 53. *Prof Exp:* Res asst chem, Brit Leather Mfrs Res Asn, London, 45-48; Nat Res Coun Can fel, 53-55; sr scientific officer, Radiochem Ctr, United Kingdom Atomic Energy Authority, Amersham, Eng, 55-58; res chemist, Pulp & Paper Res Inst Can, 58-68, dir process res div, 68-76, asst dir res, 77-79, dir res, 79-83, dir strategic planning, 84-90, ASSOC SCI ED, J PULP & PAPER SCI, PULP & PAPER RES INST CAN, 90- *Concurrent Pos:* Mem, Task Force Biotechnol, Ministry State Sci & Technol, Ottawa, 80-81; chmn, IEA Pulp & Paper Exec Comt, 84-90. *Mem:* Royal Soc Chem; Chem Inst Can; Am Chem Soc; Can Pulp & Paper Asn. *Res:* Chemistry of alkaline pulping; chemistry of pulp bleaching; carbohydrate and wood chemistry. *Mailing Add:* PO Box 358 24 Westwood Dr Hudson PQ J0P 1H0 Can

CLAYTON, DONALD DELBERT, b Shenandoah, Iowa, Mar 18, 35; m; c 4. ASTRONOMY, NUCLEAR PHYSICS. *Educ:* Southern Methodist Univ, BS, 56; Calif Inst Technol, MS, 59, PhD(physics), 62. *Prof Exp:* Res fel physics, Calif Inst Technol, 61-63; asst prof space sci, 63-65, from assoc prof to prof physics & space sci, 65-75, A H BUCHANAN PROF ASTROPHYS, RICE UNIV, 75- *Concurrent Pos:* Vis fel, Inst Theoret Astron, Cambridge, UK, 67-72; mem panel astrophys & relativity, astron & physics surv comt, Nat Res Coun, 70-; Alfred P Sloan Found fel, 66-68; fel, Am Phys Soc, 68-; Fulbright, fel, 79-80; fel Meteoritical Soc, 82- *Honors & Awards:* George Darwin lectr, Royal Astron Soc, 81. *Mem:* Fel Am Phys Soc; Meteoritical Soc; Am Astron Soc. *Res:* Nucleosynthesis; space science; stellar evolution; geochemistry; cosmology; origin of solar system. *Mailing Add:* Dept Physics & Astron Clemson Univ Clemson SC 29634

CLAYTON, EUGENE DUANE, b Ravena, Nebr, Mar 20, 23; m 47; c 3. NUCLEAR ENGINEERING, PHYSICS. *Educ:* Whitman Col, BA, 47; Univ Ore, MS, 49, PhD(physics), 52. *Prof Exp:* Res physicist appl res reactors, Gen Elec Co, 51-56, supvr reactor lattice physics, 56-57, supvr critical mass physics, 57-65; mgr criticality res & analysis, 65-81, MGR CRITICAL MASS LAB, BATTELLE-NORTHWEST, 81- *Concurrent Pos:* Res assoc prof, Dept Nuclear Eng, Univ Wash, 64-78. *Honors & Awards:* Nuclear Criticality Safety Achievement Award, Am Nuclear Soc. *Mem:* Fel Am Nuclear Soc; Am Phys Soc. *Res:* Criticality measurements and studies; nuclear criticality safety in fuel recycle operations; criticality safety evaluations; techniques for criticality prevention and control; fissionability and criticality of actinide elements; anomalies of nuclear criticality. *Mailing Add:* 2041 Davison Richland WA 99352

CLAYTON, FRANCES ELIZABETH, b Texarkana, Tex, Nov 6, 22. GENETICS. *Educ:* Tex State Col Women, BA, 44; Univ Tex, MA, 47, PhD(zool, genetics), 51. *Prof Exp:* Teacher, high sch, Ark, 44-45; tutor & fel, Univ Tex, 45-50; instr zool, Univ Ark, 50-51 & Univ Tex, 51-52, Hite fel, 52-53, res scientist, Genetics Found, 53-54; from asst prof to prof, 54-87, EMER PROF ZOOL, UNIV ARK, FAYETTEVILLE, 87- *Concurrent Pos:* Vis colleague, Univ Hawaii, 63-64. *Mem:* Soc Study Evol; fel AAAS; Am Naturalist; Am Genetic Asn; Sigma Xi. *Res:* Cytogenetics in Drosophila species. *Mailing Add:* Dept Zool SE 632 Univ Ark Fayetteville AR 72701

CLAYTON, FRED RALPH, JR, b Knoxville, Tenn, Dec 22, 40; m 62; c 1. ELECTROANALYTICAL CHEMISTRY, CHEMICAL EDUCATION. *Educ:* Univ Tenn, BS, 66, PhD(chem), 72. *Prof Exp:* Chem technician nuclear res, Oak Ridge Nat Lab, Union Carbide Nuclear Co, 59-62; process engr nylon prod, Firestone Synthetic Fibers Co, Firestone Tire & Rubber Co, 64-66; from asst prof to assoc prof, 71-86, PROF CHEM, FRANCIS MARION COL, 86- *Mem:* Am Chem Soc; Sigma Xi; AAAS; Am Inst Chemists. *Res:* Electroanalytical chemistry in non-aqueous solvents; environmental analysis in waste water systems; surfactant and detergent analysis and synthesis. *Mailing Add:* Dept Chem & Physics Francis Marion Col PO Box F7500 Florence SC 29501

CLAYTON, GLEN TALMADGE, b Elmo, Ark, Jan 30, 29; m 50; c 2. PHYSICS. *Educ:* Univ Ark, BS, 53, MS, 54; Univ Mo, PhD, 60. *Prof Exp:* Instr physics, Univ Ark, 53-54; asst prof, William Jewell Col, 54-56; instr, Univ Mo, 56-58; res assoc, Argonne Nat Lab, 58-60; from asst prof to prof, Univ Ark, Fayetteville, 60-72; PROF PHYSICS & DEAN SCH SCI & MATH, STEPHEN F AUSTIN STATE UNIV, 72-, DEAN GRAD SCH. *Mem:* Am Asn Physics Teachers; Am Phys Soc. *Res:* X-ray; neutron diffraction; electronics. *Mailing Add:* RR 6 No 340 Nacogdoches TX 75961

CLAYTON, J(OE) T(ODD), b Etowah, Tenn, Oct 2, 24; m 46; c 3. FOOD ENGINEERING & TECHNOLOGY. *Educ:* Univ Tenn, BSAE, 49; Univ Ill, MS, 51; Cornell Univ, PhD, 62. *Prof Exp:* Asst agr eng, Univ Ill, 50-51, from instr to asst prof, 51-57; asst prof, Univ Conn, 54-55; assoc prof, 57-61, head dept food & agr eng, 66-85, PROF AGR ENG, UNIV MASS, AMHERST, 61-, PROF FOOD ENG, 85- *Concurrent Pos:* NSF fel, Cornell Univ, 60-62; vis prof, Univ Reading, 70-71; NSF-NATO sr fel sci, Univ Reading & Nat Col Food Technol, Eng, 71; Japan Soc Prom Sci fel, Univ Tokyo, 81; vis scientist, US Army Natick Res Develop & Eng Ctr, 84-85. *Honors & Awards:* Elected Fel, Am Soc Agr Engrs, 85. *Mem:* AAAS; Am Soc Agr Engrs; Int Soc Biometeorol; Inst Food Technol; NY Acad Sci. *Res:* Design of food production and preservation structures and processes; food and biological engineering; engineering properties of food and other biological materials. *Mailing Add:* Dept Food Eng Univ Mass Amherst MA 01003

CLAYTON, JAMES WALLACE, b New Westminster, BC, Nov 4, 33; m 57; c 4. FISH & WILDLIFE SCIENCES. *Educ:* Univ BC, BA, 55; Univ Sask, PhD(phys & org chem), 62. *Prof Exp:* Chemist wood pulp bleaching, Res Div, MacMillan & Bloedel, Ltd, BC, 56-58 & cereal protein chem, Grain Res Lab, Can Grain Comn, Can, 62-66; RES SCIENTIST, FRESHWATER INST, CAN DEPT FISHERIES & OCEANS, 67- *Concurrent Pos:* Vis scientist, Dept Biochem, Univ Otago, Dunedin, NZ, 75-76, & Dept Biol, Queen's Univ, Kingston, Ont, Can, 89-90; adj prof, Univ Manitoba. *Mem:* Genetics Soc Am; Genetics Soc Can; Soc Systs Zool; Am Genetic Asn; Soc Study Evolution; Soc Marine Mammal; Sigma Xi. *Res:* Molecular genetics of fish and marine mammals; population genetics; systematics and evolution of freshwater fishes and marine mammals. *Mailing Add:* Freshwater Inst 501 Univ Crescent Winnipeg MB R3T 2N6 Can

CLAYTON, JOE EDWARD, b Tillar, Ark, Sept 17, 32; m 52; c 3. AGRICULTURAL ENGINEERING. *Educ:* Univ Ark, BS(agr) & BS(agr eng), 59; Clemson Univ, MS, 60. *Prof Exp:* Agr engr, Allis-Chalmers Mfg Co, 59; AGR RES ENGR, USDA, 60- *Mem:* Am Soc Agr Engrs. *Res:* Design and field testing of harvesting and farm processing machinery. *Mailing Add:* 1002 SE Third St Belle Glade FL 33430

CLAYTON, JOHN CHARLES (HASTINGS), b Pittston, Pa, June 15, 24; m 68. PHYSICAL INORGANIC CHEMISTRY. *Educ:* St Joseph's Col, Philadelphia, BS, 49; Univ Pa, MS, 50, PhD(chem), 53. *Prof Exp:* Asst chem, Univ Pa, 52, fel, 53-54; sr res chemist, 54-81, FEL SCIENTIST, BETTIS ATOMIC POWER LAB, WESTINGHOUSE ELEC CORP, 81- *Mem:* AAAS; Am Chem Soc; Am Ceramic Soc; Am Nuclear Soc; Sigma Xi; Am Soc Metals; NY Acad Sci. *Res:* Inorganic chemistry of solids; physical and inorganic chemistry of nuclear materials; thermodynamics and material properties; corrosion and hydriding of Zircaloy. *Mailing Add:* Bettis Atomic Power Lab Westinghouse Elec Corp PO Box 79 West Mifflin PA 15122

CLAYTON, JOHN MARK, b Kevil, Ky, Aug 6, 45; m 70; c 3. MEDICINAL CHEMISTRY, MEDICAL RESEARCH. *Educ:* Tenn Technol Univ, BS, 68; Univ Tenn, PhD(pharmaceut sci), 71. *Prof Exp:* Res assoc drug design, Dept Chem, Pomona Col, 71-72; res biologist chem carcinogenesis, Nat Ctr Toxicol Res, Food & Drug Admin, 72-73; clin res assoc, Consumer Opers-Hq, Schering-Plough Co, 74-75, dir clin & regulatory serv, 75-78, vpres qual control, clin & regulatory serv, 78-85, sr vpres, consumer opers, 85-89, SR VPRES SCI & REGULATORY AFFAIRS, SCHERING-PLOUGH HEALTH CARE PROD, 89- *Concurrent Pos:* Asst prof pharmacol, Col Med, Univ Ark, 73-77; asst prof molecular biol, Col Pharm, Univ Tenn, 74; mem res bd adv, Memphis State Univ; NIH fel; assoc prof pharmaceut, Col Pharm, Univ Tenn, 87- *Mem:* Am Chem Soc; Am Acad Dermat; NY Acad Sci; AAAS; Soc Invest Dermat. *Res:* Quantitative structure-activity relationship approach to drug design; contact dermatitis; evaluations of sunscreens. *Mailing Add:* Schering-Plough Health Care Prod 110 Allen Rd Liberty Corner NJ 07938

CLAYTON, JOHN WESLEY, JR, b Philadelphia, Pa, Sept 1, 24. TOXICOLOGY, RESPIRATORY PHYSIOLOGY. *Educ:* Wheaton Col, AB, 48; Univ Pa, AM, 50, PhD(parasitol), 54. *Prof Exp:* Toxicologist, Haskell Lab, E I du Pont de Nemours & Co, Inc, 54-60, asst dir labs, 60-69; dir environ sci lab, Hazleton Labs, TRW Inc, 69-71; dir toxicol ctr, Univ Wis, 71-73; chief toxicol br & dir health effects div, US Environ Protection Agency, 73-74; prof pharmacol, toxicol, microbiol & immunol & dir, Toxicol Prog, 74-90, EMER PROF PHARMACOL, TOXICOL, MICROBIOL & IMMUNOL, UNIV ARIZ, 90- *Concurrent Pos:* Mem tech adv bd, State Air Pollution Control Bd, Va, 70-71 & policy bd, Nat Ctr Toxicol Res, 73-74; consult, Wis Dangerous Substances Bd, 72-73, Reese & Schluechter, Ill, 72-73, Kennecott Copper Corp, 74- & Food & Drug Admin, 74-; mem sci adv bd, Food & Drug Admin; mem adv bd, Inhalation Toxicol Res Inst, Electronic Resources Develop Agency; mem ed bd, Am Indust Hyg Asn, 67-; mem ed bd, Soc Toxicol, 69-, chmn tech comt, 70-71 *Mem:* Am Indust Hyg Asn; Soc Toxicol. *Res:* Toxicology of fluorocarbons, including cardiac effects; action of fluoro-olefins on renal function; pyrolysis products of fluoropolymers. *Mailing Add:* 7762 N Harelson Pl Tucson AZ 85704

CLAYTON, NEAL, b Ripley, Miss, Aug 23, 13; m 40; c 1. SEISMOLOGY, GEOPHYSICS. *Educ:* Miss Col, BA, 34; La State Univ, MS, 37. *Prof Exp:* Eng trainee, Schlumberger Well Surv, Tex, 37; seismic helper, Humble Oil Co, 37, seismic computer, 37-39 & Seismic Explor, Inc, 40-41; assoc physicist, US Navy, 42-43; seismic observer, Magnolia Petrol Co, 43-44; seismic party chief, NAm Geophys Co, 44-46 & Repub Explor Co, 46-51; asst mgr domestic opers, Century Geophys Corp, 51-54; pres & supvr, Liberty Explor Co, 54-56; mem staff & dist geophysicist, Sohio Petrol Co, La, 56-60; supvr, Index Explor Co, Tex, 60-61; Gulf Coast geophysicist, Cosden Petrol Co, 61-64; consult geophysicist, 64-87; RETIRED. *Mem:* Soc Explor Geophys; Am Asn Petrol Geologists. *Res:* Application of seismology to exploration for oil and gas; geophysical prospecting. *Mailing Add:* PO Box 3637 Corpus Christi TX 78404

CLAYTON, PAULA JEAN, b St Louis, Mo, Dec 1, 34; m 58; c 3. PSYCHIATRY. *Educ:* Univ Mich, Ann Arbor, BS, 56; Washington Univ, MD, 60. *Prof Exp:* Intern, St Luke's Hosp, St Louis, 60-61; asst resident & chief resident psychiat, Barnes & Renard Hosps, St Louis, 61-65; from instr to assoc prof psychiat, 65-74, PROF PSYCHIAT, SCH MED, WASH UNIV, 74- *Concurrent Pos:* Consult psychiatrist, Malcolm Bliss Ment Health Ctr, St Louis, 72- & dir training & res, 75; dir, Barnes & Renard Hosp Psychiat Inpatient Serv, 75-81; PROF & HEAD, DEPT PSYCHIAT, UNIV MINN, 81- *Mem:* Fel Am Psychiat Asn; Psychiat Res Soc; Asn Res Nerv Ment Dis; Am Psychopath Asn; Soc Biol Psychiat. *Res:* Studies dealing with nosology, course and treatment of patients with psychiatric diagnosis; also the symptomatology and course of normal bereavement. *Mailing Add:* Dept Psychiat Univ Minn Box 77 Mayo 420 Delaware St SE Minneapolis MN 55455

CLAYTON, RAYMOND BRAZENOR, b Manchester, Eng, Sept 16, 25; m 62; c 2. BIOCHEMISTRY, ENDOCRINOLOGY. *Educ:* Univ Manchester, BSc, 49, MSc, 50, PhD(chem), 52. *Prof Exp:* Res fel org chem, Univ Manchester, 52-53; res fel biochem, Univ Chicago, 53-54, res fel chem, Harvard Univ, 54-55; Imp Chem Indust fel, Oxford Univ, 55-56; res dir, Manchester Cancer Res Trust Fund, 56-58; res fel chem, Harvard Univ, 59-63; assoc prof, 63-68, PROF BIOCHEM, DEPT PSYCHIAT & BEHAV SCI, STANFORD UNIV. *Concurrent Pos:* Hon lectr, Univ Manchester, 56-58; estab investr, Am Heart Asn, 60-65. *Mem:* AAAS; Am Soc Biol Chemists; Endocrine Soc; Am Soc Zoologists; Royal Soc Chem. *Res:* Steroid biosynthesis; comparative aspects of steroid and terpenoid metabolism; genetics of steroid hormone metabolism and hormonal effects; action of steroid hormones in the central nervous system. *Mailing Add:* Biochem Psych Td 114 Stanford Univ Stanford CA 94305

CLAYTON, ROBERT ALLEN, biochemistry; deceased, see previous edition for last biography

CLAYTON, ROBERT NORMAN, b Hamilton, Ont, Mar 20, 30; m 71; c 1. GEOCHEMISTRY. *Educ:* Queen's Univ, Ont, BSc, 51, MS, 52; Calif Inst Technol, PhD(chem), 55. *Hon Degrees:* DSc, McMaster Univ, Can, 90. *Prof Exp:* Res fel geochem, Calif Inst Technol, 55-56; asst prof geochem, Pa State Univ, 56-58; from asst prof to assoc prof chem, Univ Chicago, 58-66, master phys sci col div, assoc dean col & assoc dean phys sci, 69-72, chmn, Dept Geophys Sci, 76-79, prof chem & geophys, 66-80, ENRICO FERMI DISTINGUISHED SERV PROF, UNIV CHICAGO, 80- *Concurrent Pos:* Guggenheim fel, 64-65; Sloan fel, 64-66. *Honors & Awards:* NASA Excep Sci Achievement Medal, 77; George P Merrill Award, Nat Acad Sci, 80; Goldschmidt Medal, Geochem Soc, 81; Leonard Medal, Meteoritical Soc, 82; Cresson Medal, Franklin Inst, 85; Bowie Medal, Am Geophys Union, 87. *Mem:* AAAS; fel Am Geophys Union; fel Meteoritical Soc; fel Am Acad Arts & Sci; fel Royal Soc Can; fel Royal Soc London. *Res:* Natural variations of stable isotope abundances. *Mailing Add:* Enrico Fermi Inst Univ Chicago Chicago IL 60637

CLAYTON, RODERICK KEENER, b Tallin, Estonia, Mar 29, 22; US citizen; m 44; c 2. PHOTOSYNTHESIS, VISION. *Educ:* Calif Inst Technol, BS, 47, PhD(physics, biol), 51. *Prof Exp:* Merck fel, Stanford Univ, 51-52; assoc prof physics, US Naval Post Grad Sch, 52-57; NSF sr fel, 57-58; sr biophysicist, biol div, Oak Ridge Nat Lab, 58-62; vis prof microbiol, Dartmouth Col Med Sch, 62-63; sr investr, C F Kettering Res Lab, 63-66; prof plant biol & appl & eng physics, 66-81, L H Baily prof physics, 81-84, EMER PROF BIOPHYSICS, CORNELL UNIV, 84- *Concurrent Pos:* Lalor fel, Woods Hole Marine Biol Lab, 55; consult, Firestone Res & Develop Lab, 56-57. *Mem:* Nat Acad Sci; Am Acad Arts & Sci; Soc Gen Physiol; Am Soc Plant Physiol; Am Soc Photobiol; Biophys Soc. *Res:* Physical aspects of photosynthesis; biochemistry of photosynthetic bacteria. *Mailing Add:* 4176 Inglewood Blvd Apt 9 Los Angeles CA 90006-5250

CLAYTON, WILLIAM HOWARD, b Dallas, Tex, Aug 16, 27; c 2. PHYSICAL OCEANOGRAPHY. *Educ:* Bucknell Univ, BSc, 49; Tex A&M Univ, PhD(phys oceanog), 56. *Prof Exp:* Provost & dean, Moody Col Marine Sci & Maritime Resources, 74-77, pres, 77-87, EMER PRES, TEX A&M UNIV, GALVESTON, 87- *Mem:* Am Meteorol Soc; Am Geophys Union; Sigma Xi. *Res:* Micrometeorology; numerical analysis; water level variations; air-sea interchange; oceanographic and meteorological instrumentation; machine computational methods. *Mailing Add:* Tex A&M Univ 5222 Denver Dr Galveston TX 77551

CLAYTON-HOPKINS, JUDITH ANN, b Santa Monica, Calif, Sept 17, 39; c 1. ENDOCRINOLOGY. *Educ:* Univ Calif, Los Angeles, BA, 60, MA, 63, PhD(zool), 66. *Prof Exp:* Staff physiologist, Worcester Found Exp Biol, 68-70; res assoc, Dept Biol Sci & Dept Nutrit Sci, Univ Conn, Storrs, 70-71, asst prof physiol, 71-72; dir, radioisotope/endocrine lab, Dept Path & Lab Med, St Francis Hosp & Med Ctr, Hartford, 72-80; mem staff, Abbott Labs, Chicago, 81-83; ASST PROF LAB MED, HEALTH CTR, UNIV CONN, FARMINGTON, 73-; TECHNOL ASSESSMENT, TRAVENOL LABS, DEERFIELD, IL, 83- *Concurrent Pos:* Res assoc NIH grant, Harvard Med Sch & Beth Israel Hosp, Boston, 67-68; consult, Automation Med Lab Sci Rev Comt, NIH, 73-77; consult radioimmunoassay, Bur Med Devices & Diag Prod, Food & Drug Admin, 76-79; ed, Selected Methods Clin Chem, Am Asn Clin Chemists, 74-; chief oncofet antigen & endocrine physiol, Ctr Dis Contrl, Atlanta, Ga, 80-81; Immunol & Ligand Assay, Nat Comt Clin Lab Standards. *Mem:* Am Soc Zool; Endocrine Soc; NY Acad Sci; Am Physiol Soc; Am Asn Clin Chem. *Res:* Radioimmunochemistry; biochemical and molecular mechanisms of hormone action. *Mailing Add:* 100 Nassau Circle Round Lake IL 60073

CLAYTOR, THOMAS NELSON, b Tulsa, Okla, Aug 17, 49; m 69; c 2. ULTRASONIC MICROSCOPY, ACCOUSTIC NOISE ANALYSIS. *Educ:* Okla State Univ, BS, 71; Purdue Univ, MS, 73, PhD(physics), 76. *Prof Exp:* Scientist, Argonne Nat Lab, 77-86; PHYSICIST, LOS ALAMOS NAT LAB, 86- *Mem:* Am Phys Soc; Inst Elec & Electronics Engrs; Sigma Xi; Am Soc Nondestructive Testing. *Res:* Ultrasonic microscopy for materials evaluation; measurements and modeling of the ultrasonic properties of metals and composite materials; development of signal processing techniques for acoustic and ultrasonic noise analysis. *Mailing Add:* 712 Meadow Lane White Rock NM 87544

CLAZIE, RONALD N(ORRIS), b Oakland, Calif, Apr 17, 38; m 61; c 2. CHEMICAL ENGINEERING, INDUSTRIAL & MANUFACTURING ENGINEERING. *Educ:* Univ Calif, BS, 60, PhD, 67; Calif Inst Technol, MS, 61. *Prof Exp:* Process engr, Dow Chem Co, 61-63; asst, Univ Calif, 63-67; mgr chem process develop, Raychem Corp, Menlo Park, Calif, 67-70, supvr chem prod, 70-71, mgr chem process eng, 71-73, mgr eng, Mat Div, 73-74, mgr eng, Wire & Cable Div, 74-76, mgr eng, Thermofit Div, 76-78, tech servs mgr, 78-79, tech mgr, AME Div, 79-80, mfg mgr, 80-81, tech dir, 81-85, prod mgr, Microelectronics Div, 85-88; CHIEF OPERATING OFFICER, CYCLOTOMICS, BERKELEY, CALIF, 89-; VPRES OPER, THERMALUX, RICHMOND, CALIF, 90- *Mem:* Am Inst Chem Engrs; Sigma Xi. *Res:* Product design and application of heat shrinkable plastic tubing and molded parts; microelectronics interconnection in surface mount technology; aesogel process development. *Mailing Add:* 415 Santa Rita Menlo Park CA 94025

CLEALL, JOHN FREDERICK, b Birmingham, Eng, Feb 6, 34; m 56; c 2. ORTHODONTICS. *Educ:* Otago Univ, Dunedin, NZ, BDS, 56, MDS, 60, DDS, 64; FRCP, 67. *Prof Exp:* Prof & chmn orthod, Dept Orthod, Fac Dent & Dept Dent Sci, Fac Grad Studies & Res, Univ Man, 64-74; PROF & HEAD, DEPT ORTHOD, UNIV ILL, CHICAGO 74- *Concurrent Pos:* Dir res exec coordr, Eastman Dent Dispensary, 60-64; res assoc, Res Dept, Winnipeg Children's Hosp, 64-74. *Honors & Awards:* Milo Hellman Res Award, 64. *Mem:* Int Asn Dent Res; Am Asn Orthodontists. *Res:* Bone growth of the craniofacial complex, oral physiology, clinical research, including growth of the craniofacial region in humans and physiologic aspects of muscle function; computer-aided diagnostic systems and orthodontic appliance systems. *Mailing Add:* Dept Orthod Col Dent Univ Ill 801 S Paulina St Chicago IL 60612

CLEARE, HENRY MURRAY, b Dalton, Ga, Aug 5, 28; m 50; c 3. MEDICAL PHYSICS. *Educ:* Ga Inst Technol, BS, 51. *Prof Exp:* Physicist, 51-58, Eastman Kodak Co, res physicist, 58-62, res assoc, 62-79, head, Radiography Lab, 64-78, sr res assoc, Res Labs, 79-86, tech adv, Black & White Div, 78-86; RETIRED. *Concurrent Pos:* Clin prof, Dept Radiol, Univ Mo. *Res:* Image-forming properties of medical and industrial radiographic systems; properties of quantum limited radiographic systems; photographic radiation dosimetry; photographic effects of radiations in space. *Mailing Add:* 24 Beacon Hill Fairport NY 14450-3328

CLEARFIELD, ABRAHAM, b Philadelphia, Pa, Nov 9, 27; m 49; c 2. INORGANIC CHEMISTRY. *Educ:* Temple Univ, BA, 48, MA, 50; Rutgers Univ, PhD(phys chem, crystallog), 54. *Prof Exp:* Assoc chemist, Titanium Alloy Mfg Div, Nat Lead Co, 54-56, sr chemist, 56-58, asst chief chem res, 58-63; from asst prof to prof chem, Ohio Univ, 63-74; assoc prog dir thermodyn, NSF, 74-75; PROF CHEM, TEX A&M UNIV, 76-, DIR, MAT SCI PROG, TEX A&M UNIV, 86- *Concurrent Pos:* Lectr, Niagara Univ, 57-60; consult, Bio-Rad Labs, Tizon Chem Co & Magnesium Elektron, Manchester, Eng, 75-; assoc dean res, Col Sci, 85- *Mem:* Am Chem Soc; Am Crystallog Asn; Mat Res Soc. *Res:* Chemistry of transition metals, especially titanium and zirconium; solid state chemistry; x-ray diffraction and crystal structure, inorganic ion exchangers; heterogeneous catalysis. *Mailing Add:* Dept Chem Tex A&M Univ College Station TX 77843

CLEARY, JAMES WILLIAM, b Evanston, Ill, Apr 13, 26; m 60; c 4. ORGANIC CHEMISTRY, POLYMER CHEMISTRY. *Educ:* Loyola Univ, Ill, BS, 50; State Col Wash, MS, 53, PhD, 56. *Prof Exp:* Sr exp aide agr chem, State Col Wash, 51-53, asst chem, 53-55; res chemist, Phillips Petrol Co, 56-68, sr res chemist, 68-85; PRIN RES CHEMIST, DARTCO MFG CO, 85- *Mem:* AAAS; Am Chem Soc. *Res:* Peppermint oil; a-ketobutyrolactones; synthesis and modification of synthetic rubber; sulfur compounds; polyolefins; polymerization catalysts; condensation polymers; polyamides; polyesters; polyvinyl pyridines; polyphenylene sulfides; polymer blends; polyarylates. *Mailing Add:* 2213 Henry St PO Box 5867 Neenah WI 54957

CLEARY, MARGOT PHOEBE, b Lawrence, Mass, Mar 8, 48. NUTRITIONAL BIOCHEMISTRY. *Educ:* Regis Col, BA, 70; Columbia Univ, MS, 71, MPhil, 73, PhD(nutrit), 76. *Prof Exp:* Instr, Teachers Col, Columbia Univ, 76, NIH fel, Inst Human Nutrit, 76-78; asst prof, dept nutrit & food sci, Drexel Univ, 78-82; asst prof, 83-86, ASSOC PROF, HORMEL INST, UNIV MINN, 86- *Concurrent Pos:* Res assoc, Vassar Col, 77-78. *Mem:* Soc Exp Biol Med; Am Diabetic Assoc. *Res:* Adipose tissue development; obesity; nutrition and women; nutrition, growth and development. *Mailing Add:* Hormel Inst Univ Minn 801 16th Ave NE Austin MN 55912

CLEARY, MICHAEL, b Co Mayo, Ireland, Aug 16, 50. APPLIED MECHANICS. *Educ:* Nat Univ Ireland, BE, 72; Brown Univ, MS, 74, PhD(mech solids & struct), 75. *Prof Exp:* Asst prof, 76-79, ASSOC PROF MECH ENG, MASS INST TECHNOL, 79- *Concurrent Pos:* Indust fel, Marathon Oil Co, 78. *Mem:* Am Soc Mech Engrs; Am Soc Civil Engrs; Soc Eng Sci; Soc Petrol Engrs; Int Soc Rock Mech. *Res:* Constitutive relations for porous fluid-infiltrated media; fracture in geophysical and geotechnical applications; material microstructural modelling; recovery of oil and gas; mining; structural analysis; energy and material resources. *Mailing Add:* Dept Mech Eng Mass Inst Technol 77 Massachusetts Ave Cambridge MA 02139

CLEARY, MICHAEL E, b San Francisco, Calif, Sept 3, 47; m 69; c 4. TECHNICAL MANAGEMENT. *Educ:* Univ Calif, Los Angeles, BS, 69; Ore State Univ, PhD(org chem), 75. *Prof Exp:* Fel, Univ Utah, Cheves Walling, 75-76; res assoc, Spreckels Sugar Co, Amstar Corp, 76-79; sr res chemist, Am Crystal Sugar Co, 79-82, mgr chem res, 82-88; MGR, WESTERN DIV TECH SERV, HOLLY SUGAR CORP, 88- *Concurrent Pos:* Lectr & instr, Beet Sugar Inst, Beet Sugar Develop Found, 82- *Mem:* Am Chem Soc; Royal Soc Chem; Asn Off Anal Chemists. *Res:* Chemical phases of beet sugar manufacture, from soil chemistry to crystal structure of sucrose. *Mailing Add:* Holly Sugar Corp 2000 Powell St Suite 800 Emeryville CA 94608

CLEARY, PAUL PATRICK, b Watertown, NY, July 9, 41; m 63; c 2. MOLECULAR BIOLOGY, MEDICAL MICROBIOLOGY. *Educ:* Univ Cincinnati, BS, 65; Univ Rochester, MS, 69, PhD(microbiol genetics), 71. *Prof Exp:* Assoc prof microbiol & pediat, 72-84, PROF MICROBIOL, UNIV MINN, MINNEAPOLIS, 84- *Concurrent Pos:* Trainee, Stanford Univ, 68-71; trainee biol sci, Univ Calif, Santa Barbara, 71-72; biotechnology consult, Educ Develop, Indonesia, 85, 87, 88, 90. *Mem:* Am Soc Microbiol; Sigma Xi; Infectious Dis Soc Am. *Res:* Regulation of the biotin gene cluster in Escherichia coli; regulation of the arabinose operon in Escherichia coli; genetic determinants for resistance to phagocytosis in group A streptococci; complement specific peptidase from Streptococci; molecular pathogenesis. *Mailing Add:* Dept Microbiol Box 196 Univ Minn Minneapolis MN 55400

CLEARY, STEPHEN FRANCIS, b New York, NY, Sept 28, 36; m 59; c 5. BIOPHYSICS, RADIOBIOLOGY. *Educ:* NY Univ, BSChE, 58, PhD(biophys), 64; Univ Rochester, MS, 60. *Prof Exp:* Res engr sterio-specific polymers, Texus-US Chem Co, 58-59; teaching asst radiation biol, Inst Environ Med, NY Univ, 60-62, res assoc biophys, 62-64; from asst prof to prof biophys, 64-82, PROF PHYSIOL & BIOPHYS, MED COL VA, VA COMMONWEALTH UNIV, 82- *Concurrent Pos:* Consult, Nat Insts Occup Safety & Health Sci, NIH, 84- *Mem:* Biophys Soc; NY Acad Sci; Bioelectromagnetics Soc. *Res:* Biological effects of non-ionizing radiation; lasers, light, microwave and radiofrequency radiation; effects of ionizing radiation on the mammalian eye; structural bonding forces in viruses. *Mailing Add:* Dept Physiol & Biophys Med Col Va Box 551 Richmond VA 23298

CLEARY, TIMOTHY JOSEPH, b Philadelphia, Pa, Aug 8, 42; m 65; c 2. MEDICAL MICROBIOLOGY. *Educ:* Mt St Mary's Col, BS, 64; Univ Cincinnati, MS, 68, PhD(microbiol), 69; Am Bd Med Microbiol, dipl, 74. *Prof Exp:* Asst prof biol, Duquesne Univ, 69-71; fel, Ctr Dis Control, 71-73; asst prof path, 73-78, asst prof microbiol, 75-79, ASSOC PROF PATH, UNIV MIAMI, 78-, ASSOC PROF MICROBIOL, 79- *Mem:* Am Soc Microbiol; Sigma Xi. *Res:* Antimicrobial drug interactions and assays; immuno-assay procedures for the diagnosis of infectious diseases. *Mailing Add:* Dept Microbiol Sch Med Jackson Mem Hosp Univ Miami 1600 NW Tenth Ave Miami FL 33101

CLEARY, WILLIAM JAMES, b St Louis, Mo, Dec 10, 43; m 67; c 1. MARINE GEOLOGY, SEDIMENTARY PETROLOGY. *Educ:* Southern Ill Univ, BA, 65; Duke Univ, MA, 67; Univ SC, PhD(geol), 72. *Prof Exp:* Geologist, Pan Am Petrol Corp, 67-68; asst prof, 72-76, ASSOC PROF GEOL, UNIV NC, WILMINGTON, 76- RES ASSOC MARINE SCI, 74- *Mem:* Geol Soc Am; Sigma Xi; Soc Econ Paleontologists & Mineralogists; Am Asn Geol Teachers. *Res:* Continental margin sedimentation off Southeastern United States; barrier island sedimentation; turbidite sedimentation on the Hatteras Abyssal Plain; deep sea sedimentation on the Balearic Abyssal Plain. *Mailing Add:* Dept Geol Univ NC 601 S Col Rd Wilmington NC 28403

CLEASBY, JOHN LEROY, b Madison, Wis, Mar 1, 28; m 50; c 3. SANITARY ENGINEERING, CIVIL ENGINEERING. *Educ:* Univ Wis, BS, 50, MS, 51; Iowa State Univ, PhD(sanit eng), 60. *Prof Exp:* Inspection engr refinery construct, Standard Oil Co, Ind, 51-52; proj engr, Consoer, Townsent & Assoc, Ill, 52-54; from instr to prof, 54-83, ANSON MARSTON DISTINGUISHED PROF ENG, IOWA STATE UNIV, 83- *Honors & Awards:* Norman Medal, Am Soc Civil Engrs, 80. *Mem:* Nat Acad Eng; Am Water Works Asn; Water Pollution Control Fedn; Am Soc Civil Engrs; Nat

Soc Prof Engrs; Am Soc Eng Educ; Asn Environ Eng Prof; hon mem Am Water Works Asn. *Res:* Water treatment research for economy of plant design and operation, particularly sand and diatomite filtration, softening & flocculation. *Mailing Add:* Dept Civil Eng Iowa State Univ Ames IA 50011

CLEATOR, IAIN MORRISON, b Edinburgh, Scotland, Oct 18, 39; m 61; c 3. SURGERY, MEDICINE. *Educ:* Univ Edinburgh, MBChB, 62; FRCS(E), 66; FRCS, 67; FRCS(C), 72; FACS, 74. *Prof Exp:* Resident surg, Edinburgh Teaching Hosps, 62-72; fel, 70-71, asst prof surg, 72-80, ASSOC PROF SURG, UNIV BC, 80-; DIR, GASTROINTESTINAL CLIN, ST PAUL'S HOSP, VANCOUVER, 74- *Concurrent Pos:* Clin instr surg, Univ Edinburgh, 68-72. *Mem:* Can Asn Gastroenterol. *Res:* Gastrointestinal hormones in relation to physiology, obesity, diabetes; peptic ulcer disease. *Mailing Add:* Dept Gen Surg Univ BC 2194 Health Sci Mall Vancouver BC V6T 1W5 Can

CLEAVER, CHARLES E, b Paris, Ky, Mar 14, 38; m 59; c 3. MATHEMATICS. *Educ:* Eastern Ky Univ, BS, 60; Univ Ky, MS, 63, PhD(math), 68. *Prof Exp:* Instr math, Murray State Univ, 62-64, asst prof, 64-65; from asst prof to assoc prof math, Kent State Univ, 68-83, asst chmn math, 75-76, asst dean, Col Arts & Sci, 76-83; PROF & HEAD, MATH/COMPUTER SCI DEPT, THE CITADEL, 83- *Mem:* Am Math Soc; Math Asn Am. *Res:* Operator theory in Banach spaces; functional analysis. *Mailing Add:* The Citadel Charleston SC 29409

CLEAVER, FRANK L, mathematics; deceased, see previous edition for last biography

CLEAVER, JAMES EDWARD, b Portsmouth, Hants, Eng, May 17, 38; m 64. CANCER. *Educ:* St Catharine's Col, Cambridge Univ, BA, 61, PhD(radiobiol), 64. *Prof Exp:* Res fel neurosurg, Mass Gen Hosp, Boston & surg, Harvard Med Sch, 64-66; asst res biophysicist, Lab Radiobiol, 66-68, from asst prof to assoc prof radiobiol, 68-74, PROF RADIOL, SCH MED, UNIV CALIF, SAN FRANCISCO, 74- *Honors & Awards:* Res Award, Radiation Res Soc, 73. *Mem:* Radiation Res Soc; Photobiol Soc; Am Asn Cancer Res; Nat Coun Radiation Protection & Measurements. *Res:* Effects of ultraviolet light on mammalian cells and mechanisms of recovery from radiation damage; dermatology; mutagenesis; xeroderma pigmentosum; radiobiology of tritium decays. *Mailing Add:* Lab Radiol Univ Calif Sch Med San Francisco CA 94143-0750

CLEAVES, EMERY TAYLOR, b Easton, Pa, May 11, 36; m 60; c 4. PHYSICAL GEOGRAPHY, ENVIRONMENTAL GEOLOGY. *Educ:* Harvard Col, BA, 60; John Hopkins Univ, MA, 64, PhD(geog), 73. *Prof Exp:* Assoc geologist, 63-65, geologist IV, 65-73, asst dir, 73-80, DEP DIR & PRIN GEOLOGIST, MD GEOL SURV, 81- *Mem:* Fel Geol Soc Am. *Res:* Chemical weathering of crystalline rocks; landscape evolution; landform mapping and its application to environmental geology. *Mailing Add:* 3609 Woodholm Dr Jarrettsville MD 21084

CLEBNIK, SHERMAN MICHAEL, b Lynn, Mass, Nov 15, 43. GLACIAL GEOLOGY. *Educ:* Ind Univ, MA, 67, Univ Mass, Amherst, BA, 65, PhD(geol), 75. *Prof Exp:* PROF EARTH SCI, EASTERN CONN STATE UNIV, 73-, ASSOC PROF ENVIRON EARTH SCI. *Mem:* Geol Soc Am; AAAS; Am Asn Univ Prof. *Res:* Glacial geology of eastern Connecticut. *Mailing Add:* Earth Sci Dept Eastern Conn State Univ 83 Windham St Willimantic CT 06226

CLEBSCH, ALFRED, JR, b Clarksville, Tenn, Jan 20, 21; m 48; c 2. HYDROGEOLOGY, HYDROLOGY. *Educ:* Univ Chicago, SB, 47, SM, 48. *Prof Exp:* Geologist mil geol, US Geol Surv, 49-54, hydrologist ground water, 54-73, regional hydrologist mgt, 73-86; RETIRED. *Mem:* Geol Soc Am; Am Geophys Union; AAAS. *Res:* Geothermal energy; ground disposal of radioactive wastes. *Mailing Add:* Fifty Hoyt Lakewood Denver CO 80226

CLEBSCH, EDWARD ERNST COOPER, b Clarksville, Tenn, June 6, 29; m 56; c 3. PLANT ECOLOGY. *Educ:* Univ Tenn, AB, 55, MS, 57; Duke Univ, PhD(bot), 60. *Prof Exp:* Res assoc, 60-63, from asst prof to assoc prof bot, 63-75, PROF BOT & ECOL, UNIV TENN, KNOXVILLE, 75- *Mem:* Fel AAAS; Ecol Soc Am; Bot Soc Am; Am Inst Biol Sci. *Res:* Mineral cycling in Southern Appalachian ecosystems; physiological ecology of Arctic-Alpine plants; flora and vegetation of Tennessee; mountain environments; ecology of the Aleutian Islands; phytoliths. *Mailing Add:* Dept Bot Univ Tenn Knoxville TN 37996-1100

CLEEK, GEORGE KIME, chemistry, for more information see previous edition

CLEEK, GIVEN WOOD, b Warm Springs, Va, Nov 6, 16; m 41; c 3. CHEMISTRY, GLASS TECHNOLOGY. *Educ:* George Washington Univ, BS, 54. *Prof Exp:* Lab apprentice, Nat Adv Comt Aeronaut, 35-36; lab apprentice, Nat Bur Standards, 36-41, glassworker, 46-49, technologist, 49-57, phys chemist, 58-67, res chemist, 67-73; RETIRED. *Concurrent Pos:* Consult, 74-80. *Honors & Awards:* Silver Medal, US Dept Commerce, 73. *Mem:* Am Chem Soc; Am Ceramic Soc; Optical Soc Am; Am Soc Test & Mat. *Res:* Development of special optical glasses having higher refractive indices and special dispersions; development of infrared transmitting glasses; determination of physical properties of glass as a function of chemical composition. *Mailing Add:* 7304 Good Samaritan Ct No 332 El Paso TX 79912

CLEELAND, CHARLES SAMUEL, b Jacksonville, Ill, Sept 23, 38; m 81; c 1. NEUROPSYCHOLOGY, PSYCHOLOGY. *Educ:* Wesleyan Univ, BS, 60; Wash Univ, PhD(psychol), 66. *Prof Exp:* Instr psychiat, Med Sch, Univ Mo, 64-65; from instr to assoc prof, 66-84, PROF NEUROL, MED SCH, UNIV WIS-MADISON, 84- *Mem:* AAAS; Soc Psychophysiol Res; Soc Neurosci. *Res:* Cancer pain, its impact and treatment; behavioral treatment in illness. *Mailing Add:* Dept Neurol 46-530 CSC Univ Wis 600 Highland Madison WI 53792

CLEGG, FREDERICK WINGFIELD, b Atlanta, Ga, Oct 9, 44; div; c 1. ELECTRICAL ENGINEERING, TELECOMMUNICATIONS SYSTEMS. *Educ:* Oakland Univ, BS, 65; Stanford Univ, MS, 67, PhD(elec eng, comput sci), 70. *Prof Exp:* Develop engr eng software, Ford Motor Co, 67; res asst elec eng, Stanford Univ, 67-69, instr, 69-70; asst prof, Univ Santa Clara, 70-75; sect mgr, Software Res & Develop, Hewlett-Packard Co, 75-89; dir software, 89-90, DIR ENG SERV & CAE, RAYNET CORP, MENLO PARK, CALIF, 90- *Concurrent Pos:* Consult various pvt & govt agencies, 67- *Mem:* Sigma Xi; Inst Elec & Electronics Engrs. *Res:* Software engineering; computer architecture; microprocessors; digital systems reliability. *Mailing Add:* 8093 Hyannisport Dr Cupertino CA 95014-4011

CLEGG, JAMES S, b Aspinwall, Pa, July 27, 33; div; c 3. PHYSIOLOGY, BIOCHEMISTRY. *Educ:* Pa State Univ, BS, 58; Johns Hopkins Univ, PhD(biol), 61. *Prof Exp:* Res assoc biol, Johns Hopkins Univ, 61-62; asst prof zool, 62-64, from assoc prof biol to prof biol, Univ Miami, 64-86; PROF ZOOL & DIR BODEGA MARINE LAB, UNIV CALIF, 86- *Concurrent Pos:* Wilson fel, 58-59; Fulbright sr res award, Univ London, 78; CNRS Thiais France 1983; vpres, Nat Asn Marine Labs. *Honors & Awards:* Fel AAAS. *Mem:* AAAS; Am Soc Zoologists; Am Soc Cell Biol; Sigma Xi; Biophys Soc; Soc Cryobiol; Bioelectromagnetics Soc. *Res:* Comparative biochem & biophys; mechanisms of cryptobiosis; properties and role of water in cellular metabolism; cytoplasmic orgn. *Mailing Add:* Bodega Marine Lab Univ Calif PO Box 247 Bodega Bay CA 94923

CLEGG, JOHN C(ARDWELL), b Heber, Utah, Sept 19, 27; m 53; c 8. ELECTRICAL ENGINEERING. *Educ:* Univ Utah, BS, 49, MS, 54, PhD(elec eng, physics), 57. *Prof Exp:* Test engr, Gen Elec Co, 49-50, develop engr, 50-53; mem tech staff, Space Tech Labs, 57-61; from asst prof to assoc prof elec eng, 61-70, PROF ELEC ENG, BRIGHAM YOUNG UNIV, 70- *Concurrent Pos:* Consult, TRW Systs, 62-; mem, Nat Comt Illum Eng Soc. *Mem:* Inst Elec & Electronics Engrs; Am Soc Eng Educ; Sigma Xi; Illum Eng Soc. *Res:* Nonlinear control systems; high velocity projectiles; inertial instruments; ultra-high pressure; clinical instruments; power electronics; reliability and efficiency of lighting systems. *Mailing Add:* 1785 N 1500 East Provo UT 84604

CLEGG, MICHAEL TRAN, b Pasadena, Calif, Aug 1, 41; c 4. POPULATION GENETICS, EVOLUTION. *Educ:* Univ Calif, Davis, BS, 69, PhD(genetics), 72. *Prof Exp:* Res asst genetics, Univ Calif, Davis, 72-76; from instr to asst prof biol, Div Biol Med Sci, Brown Univ, 76-84; from assoc prof bot to prof bot & genetics, Univ Ga, 76-84; DEPT BOT & PLANT SCI, UNIV CALIF, RIVERSIDE, 84- *Concurrent Pos:* Prin investr, NSF grants, 74-90. *Honors & Awards:* Guggenheim Fel, 81-82. *Mem:* Nat Acad Sci; Genetics Soc Am; Am Soc Naturalists; Am Genetic Asn; Sigma Xi; Soc Study Evolution (secy, 79-82). *Res:* Population genetics; plant molecular evolution. *Mailing Add:* Dept Bot & Plant Sci Univ Calif Riverside CA 92521

CLEGG, ROBERT EDWARD, b Providence, RI, July 29, 14; m 41; c 3. BIOCHEMISTRY. *Educ:* Univ RI, BS, 36; NC State Col, MS, 39; Iowa State Univ, PhD(phys chem, nutrit), 48. *Prof Exp:* Assoc prof, 48-54, PROF BIOCHEM, BIOCHEM AGR EXP STA, KANS STATE UNIV, 54- *Mem:* AAAS; Am Chem Soc; Am Soc Biol Chemists. *Res:* Hormone influence on lipoprotein level; application of trace techniques to biochemical problems; enzyme kinetics; in vivo protein formation. *Mailing Add:* Dept Biochem Kans State Univ Manhattan KS 66506

CLEGG, THOMAS BOYKIN, b Emory University, Ga, Jan 6, 40; m 68; c 1. EXPERIMENTAL NUCLEAR PHYSICS. *Educ:* Emory Univ, BA, 61; Rice Univ, MA, 63, PhD(physics), 65. *Prof Exp:* Res assoc physics, Rice Univ, 65 & Univ Wis-Madison, 65-68; from asst prof to assoc prof, 68-76, PROF PHYSICS, UNIV NC, CHAPEL HILL, 76- *Concurrent Pos:* Staff mem, Triangle Univ Nuclear Lab, Durham, 68-; Fulbright grant, 75-76; vis physicist, Ctr Nuclear Res, Saclay, France, 75-76, Ctr Nuclear Res, Bruyeres-le-Chatel, France, 81, 83-84; vis scientist, Inst Fundamental Electronics, Univ Paris, 76; chmn, dept physics, Univ NC, Chapel Hill, 89-; treas, 83-85, chmn-elect, 90-91, bd trustees, Southeastern Univs Res Asn, 79- *Mem:* Fel Am Phys Soc. *Res:* Elastic scattering of polarized protons and deuterons from nuclei; development of polarized ion beams for nuclear physics experiments. *Mailing Add:* Dept Physics Phillips Hall Univ NC Chapel Hill NC 27599-3255

CLEGHORN, ROBERT ALLEN, b Cambridge, Mass, Oct 6, 04; m 32, 77; c 3. PSYCHIATRY. *Educ:* Univ Toronto, MD, 28; Aberdeen Univ, DSc(physiol), 32; FRC Psych, 71. *Hon Degrees:* DSc, McMaster Univ, Hamilton, Ontario. *Prof Exp:* Jr rotating intern, Toronto Gen Hosp, 28-29; demonstr physiol, Aberdeen Univ, 29-32; demonstr med & asst attend physician, Toronto Gen Hosp, 33-46; from asst to prof psychiat, 46-64, prof & chmn dept, 64-70, EMER PROF PSYCHIAT, McGILL UNIV, 71-; MEM STAFF, PSYCHIAT DEPT, SUNNYBROOK HOSP, 78- *Concurrent Pos:* Dir therapeut res lab, Allan Mem Inst, 46-64, dir, 64-70, hon consult, 70-; res assoc, Harvard Univ Med Sch, 53-54; psychiatrist in chief, Royal Victoria Hosp, 64-70. *Mem:* Am Psychiat Asn; Can Physiol Soc; Can Med Asn; Can Psychiat Asn; fel Royal Col Psychiatrists. *Res:* Physiology and clinical aspects of the adrenal cortex; clinical endocrinology; autonomic nervous system in adrenal insufficiency; shock and blood substitutes; physiological correlates and psychoanalytic studies in psychosomatic states and psychopharmacology; study of schizo-affective psychoses and lithium therapy. *Mailing Add:* Dept Psychiat Sunnybrook Med Ctr Univ Toronto 2075 Bayview Ave Toronto ON M4N 3M3 Can

CLELAND, CHARLES FREDERICK, b Indianapolis, Ind, July 1, 39; m 78; c 1. SMALL BUSINESS INNOVATION RESEARCH PROGRAM, PLANT PHYSIOLOGY. *Educ:* Wabash Col, BA, 61; Stanford Univ, PhD(plant physiol), 67. *Prof Exp:* NSF fels, 66-68; Milton Fund res grant, Harvard Univ, 68; asst prof biol, Harvard Univ, 68-73, lectr, 73-74; vis asst prof bot, Univ NC, Chapel Hill, 74-75; plant physiologist, Radiation Biol Lab, Smithsonian Inst, 75-87, USDA, SMALL BUS INNOVATION RES PROG, COORDR, 87- *Concurrent Pos:* NSF res grants, 69 & 72; prof lectr, dept biol,

George Washington Univ, 79-82; NATO grant, 78. *Mem:* Bot Soc Am; Am Soc Plant Physiologists; AAAS; Am Inst Biol Sci; Japanese Soc Plant Physiologists. *Res:* Plant physiology of flowering and biology of Lemnaceae; general agricultural sciences; forests and related resources; plant production and protection; animal production and protection; air, water and soils; food science and nutrition; rural and community development; aquaculture; industrial applications. *Mailing Add:* 9506 Culver Kensington MD 20895

CLELAND, FRANKLIN ANDREW, b St Francis, Kans, Oct 7, 28; m 56; c 2. CHEMICAL ENGINEERING. *Educ:* Tex A&M Univ, BS, 50; Princeton Univ, PhD(chem eng), 54. *Prof Exp:* Engr, Shell Develop Co, 54-62, supvr chem eng res, 62-63, supvr process eng, 63-66, dept head process develop, 66-67, mgr process develop, Synthetic Rubber Div, Shell Chem Co, 67-69, mgr tech dept, Polymer Div, 70-72, mgr chem eng dept, Shell Develop Co, 72-74, mgr physics & systs dept, 74-76, mgr process eng, chem, Shell Oil Co, 76-77, dir corp res & develop eng, 77-84, MGR SYNFUELS RES, DEVELOP & OPER, SHELL DEVELOP CO, 84- *Concurrent Pos:* Chmn, grad prog adv comt, Chem Eng Dept, Tex A&M Univ, 81-84; mem, adv coun, Chem Eng Dept, Princeton Univ, 83-84, chmn, 85-; mem, eval panel, Ctr Chem Eng, Nat Bur Standards, 83-84; mem, Col Eng Adv Bd, Univ Calif, Berkeley, 85-, vis comt, Chem Eng Dept, Univ Wis, 87- *Honors & Awards:* Fel, Am Inst Chem Engrs. *Mem:* Am Inst Chem Engrs; Am Chem Soc. *Res:* Coordination, planning and guidance of engineering research programs aimed at developing improved engineering procedures for use during process development, process design and mechanical design of oil and petrochemical processes. *Mailing Add:* 13 Woodlake Sq 333 PO Box 1380 Houston TX 77063

CLELAND, GEORGE HORACE, b Pasadena, Calif, July 19, 21; m 53; c 1. ORGANIC CHEMISTRY. *Educ:* Occidental Col, BA, 42; Calif Inst Technol, PhD(chem), 51. *Prof Exp:* Res assoc, Nat Defense Res Comn, 43-45; res chemist org chem, Naval Ord Test Sta, 52-54; assoc prof, 54-70, PROF CHEM, OCCIDENTAL COL, 70- *Mem:* AAAS; Am Chem Soc; Sigma Xi. *Res:* Reactions of amino acids; reaction mechanisms. *Mailing Add:* Dept Chem Occidental Col Los Angeles CA 90041

CLELAND, JOHN GREGORY, b Middlesboro, Ky, Feb 10, 46; m 69; c 2. SYNTHETIC FUELS, ENERGY SYSTEMS. *Educ:* Univ Tenn, BS, 70; Univ Ala, Huntsville, MS, 75; NC State Univ, PhD(mech eng), 81. *Prof Exp:* Naval architect, Charleston Naval Shipyard, 69; mech eng asst, US Army Missile Command, 70-72; res teaching asst, Univ Ala, Huntsville, 72-74; sr engr, Sperry Rand Corp, 73-75; sr dept & prog mgr, 75-83, MGR, INDUST APPLNS, RES TRIANGLE INST, 83- *Concurrent Pos:* Consult, Univ Ala, Huntsville, 72-74. *Mem:* Am Inst Chem Eng; Am Soc Mech Engrs; Soc Mfg Engrs; Technol Transfer Soc. *Res:* Fossil energy systems experimentation and modeling; design of measurement devices; studies of industrial environmental problems and control technologies; management or projects and personnel; robotics; aerospace technology transfer. *Mailing Add:* 7008 Branton Dr Apex NC 27502

CLELAND, JOHN W, experimental solid state physics; deceased, see previous edition for last biography

CLELAND, LAURENCE LYNN, b Defiance, Ohio, Oct 22, 39; m 61; c 1. ENGINEERING SCIENCES, PHYSICS. *Educ:* Purdue Univ, BS, 61, MS, 64, PhD(elec eng), 68. *Prof Exp:* Proj engr, Univ Calif, 63-64; asst prof, Purdue Univ, 68-69; group leader, 69-70, div leader, 70-71, dep dept head, 78-79, PROF MGR, NUCLEAR SAFETY, LAWRENCE LIVERMORE LAB, UNIV CALIF, 79- *Concurrent Pos:* Consult, Midwest Appl Sci Corp, 66-69, CTS Microelectronics Inc, 69; NSF grant, 64-68. *Mem:* Sr mem Inst Elec & Electronics Engrs; AAAS; Am Nuclear Soc. *Res:* System and management sciences applied to energy issues; nuclear safeguards, waste management, safety analyses and methodology development. *Mailing Add:* 3575 Deer Crest Dr Danville CA 94506

CLELAND, ROBERT E, b Baltimore, Md, Apr 30, 32; m 57; c 2. PLANT PHYSIOLOGY. *Educ:* Oberlin Col, AB, 53; Calif Inst Technol, PhD(biochem), 57. *Prof Exp:* USPHS fels plant physiol, Lund, 57-58 & King's Col, London, 58-59; asst prof bot, Univ Calif, Berkeley, 59-64; assoc prof, 64-68, PROF BOT, UNIV WASH, 68- *Concurrent Pos:* Guggenheim fel, Univ Leeds, 67-68. *Mem:* Am Soc Plant Physiol (pres, 74-75. *Res:* Mechanism of auxin action; cell extension; gravitropism. *Mailing Add:* Dept Bot KB-15 Univ Wash Seattle WA 98195

CLELAND, ROBERT LINDBERGH, b St Francis, Kans, June 10, 27; m 56; c 4. PHYSICAL CHEMISTRY. *Educ:* Agr & Mech Col Tex, BS, 48; Mass Inst Technol, SM, 51, PhD(phys chem), 56. *Prof Exp:* Res asst chem, Mass Inst Technol, 48-50, res employee, Div Indust Coop, 50-52, res asst, 52-56; res assoc chem, Cornell Univ, 56-58; from asst prof to assoc prof, 60-71, PROF CHEM, DARTMOUTH COL, 71- *Concurrent Pos:* Fulbright res scholar, State Univ Leiden, 58-59; USPHS spec res fel, Retina Found, Mass, 59-60; res fel, Univ Uppsala, 68; vis prof, Univ Strasbourg, 68-69; assoc ed, Macromolecules, 74-76; vis prof, Univ Uppsala, 77-78. *Res:* Physical chemistry of polyelectrolytes. *Mailing Add:* Dept Chem Dartmouth Col Hanover NH 03755

CLELAND, WILFRED EARL, b St Francis, Kans, Aug 10, 37; m 66; c 2. ELEMENTARY PARTICLE PHYSICS. *Educ:* Agr & Mech Col Tex, BS, 59; Yale Univ, MS, 60, PhD(physics), 64. *Prof Exp:* Instr physics, Yale Univ, 63-64; vis scientist, Europ Orgn Nuclear Res, Geneva, 64-67; from asst prof to assoc prof physics, Univ Mass, Amherst, 67-70; assoc prof, 70-78, PROF PHYSICS, UNIV PITTSBURGH, 78- *Concurrent Pos:* NATO fel, 64-65; NSF fel, 65-66. *Mem:* Am Phys Soc. *Res:* Studies of interactions of elementary particles using electronic techniques at high energy accelerators, including high transverse momentum phenomena; lepton identification, and reactions using relativistic heavy ions. *Mailing Add:* Dept Physics Univ Pittsburg 4200 Fifth Ave Pittsburgh PA 15213

CLELAND, WILLIAM WALLACE, b Baltimore, Md, Jan 6, 30; m 67; c 2. BIOCHEMISTRY. *Educ:* Oberlin Col, AB, 50; Univ Wis, MS, 53, PhD(biochem), 55. *Prof Exp:* NSF fel, Univ Chicago, 57-59; from asst prof to assoc prof, 59-66, PROF BIOCHEM, UNIV WIS-MADISON, 66- *Honors & Awards:* Merck Award, 90. *Mem:* Nat Acad Sci; Am Chem Soc; Am Soc Biol Chemists; Am Acad Arts & Sci. *Res:* Use of enzyme kinetics to deduce enzymatic mechanisms. *Mailing Add:* Enzyme Inst Univ Wis 1710 University Ave Madison WI 53705

CLELLAND, RICHARD COOK, b Camden, NY, Aug 23, 21; m 63; c 2. APPLIED STATISTICS. *Educ:* Hamilton Col, BA, 44; Columbia Univ, AM, 49; Univ Pa, PhD, 56. *Prof Exp:* Instr math, Syracuse Univ, 46-47; instr, Hamilton Col, 50-53; from asst prof to assoc prof statist, Univ Pa, 56-66, chmn dept statist & opers res, 66-71, actg dean, Wharton Sch, Univ Pa, 71-72, assoc dean, 75-81, actg assoc provost, 81, DEP PROVOST, UNIV PA, 81-, PROF STATIST, 66- *Mem:* Fel Am Statist Asn; Inst Math Statist; Opers Res Soc Am; Inst Mgt Sci. *Res:* Experimental design; statistical methodology; operations research. *Mailing Add:* 106 Col Hall Univ Pa Philadelphia PA 19104

CLEM, JOHN R, b Waukegan, Ill, Apr 24, 38; m 60; c 2. THEORETICAL SOLID STATE PHYSICS, SUPERCONDUCTIVITY. *Educ:* Univ Ill, Urbana, BS, 60, MS, 62, PhD(physics), 65. *Prof Exp:* Res assoc physics, Univ Md, 65-66; vis res fel, Tech Univ Munich, 66-67; from asst prof to assoc prof, 67-75, chmn dept, 82-85, PROF PHYSICS, IOWA STATE UNIV, 75-, DISTINGUISHED PROF LIB ARTS & SCI, 89- *Concurrent Pos:* Consult, Argonne Nat Lab, 71-76; vis staff mem, Los Alamos Sci Lab, 71-83; Fulbright-Hayes sr res scholar, Inst Solid State Res, Juelich, Ger, 74-75; consult, Brookhaven Nat Lab, 80-81, Watson Res Ctr, Int Bus Mach Corp, 82-85, Allied Signal Aerospace Co, 90; vis scientist, Watson Res Ctr, Int Bus Mach Corp, 85-86; mem, policy comt, J Low Temperature Physics, 89-95. *Honors & Awards:* Award for Sustained Outstanding Res in Solid State Physics, Dept Energy, 88. *Mem:* AAAS; fel Am Phys Soc; Sigma Xi; Am Asn Univ Professors. *Res:* Theoretical research in solid state physics, low-temperature solid state physics; electrodynamic and thermal properties of superconductors subjected to electrical currents and magnetic fields. *Mailing Add:* 2307 Timberland Rd Ames IA 50010-8251

CLEM, LESTER WILLIAM, b Frederick, Md, June 23, 34; m 57; c 5. IMMUNOLOGY, IMMUNOCHEMISTRY. *Educ:* Western Md Col, BS, 56; Univ Del, MS, 60; Univ Miami, PhD(microbiol), 63. *Prof Exp:* From instr to assoc prof microbiol, 64-71, prof immunol & med microbiol, Col Med, Univ Fla, 71-79; PROF MICROBIOL & CHMN DEPT, UNIV MISS MED CTR, 79- *Concurrent Pos:* Res assoc immunol, Variety Children's Res Found, Miami, 63-66; WHO consult immunol, India, 70; instr physiol course, Marine Biol Lab, Woods Hole, 72 & 74. *Mem:* AAAS; Am Asn Immunol; Am Asn Zool. *Res:* Phylogenetic development of immunological competency and immunoglobulin structure and function. *Mailing Add:* Dept Microbiol Med Ctr Univ Mass 2500 N State St Jackson MS 39216

CLEMANS, GEORGE BURTIS, b Huntington, NY, May 11, 38; m 61. ORGANIC CHEMISTRY, BIOCHEMISTRY. *Educ:* Va Polytech Inst, BS, 60; Duke Univ, MA, 63, PhD(chem), 64. *Prof Exp:* Res assoc chem, Ind Univ, 64-65; res assoc chem, Univ Ark, 65-66, asst prof, 66-74, assoc prof, 74-76, PROF CHEM, BOWLING GREEN STATE UNIV, 76- *Mem:* Sigma Xi. *Res:* Stereospecific reactions of dicyclopentadiene derivatives; synthesis of diterpenes. *Mailing Add:* Dept Chem Bowling Green State Univ Bowling Green OH 43403

CLEMANS, KERMIT GROVER, b Adrian, NDak, Apr 14, 21; m 44; c 3. MATHEMATICAL STATISTICS. *Educ:* Jamestown Col, BS, 43; Univ Minn, MA, 48; Univ Ore, PhD(math statist), 53. *Prof Exp:* Instr math, Willamette Univ, 48-50 & Univ Ore, 50-53; math & statist consult, US Naval Test Sta, 53-59; dean div sci & technol, 60-67, prof, 59-67, EMER PROF MATH, SOUTHERN ILL UNIV, EDWARDSVILLE, 67- *Honors & Awards:* Am Stat Asn. *Mem:* Math Asn Am; Am Soc Qual Control; Sigma Xi. *Res:* Extreme value statistics. *Mailing Add:* 33 Chaparral Edwardsville IL 62025

CLEMANS, STEPHEN D, b Gloversville, NY, Apr 1, 39; m 61; c 2. STRUCTURAL CHEMISTRY. *Educ:* Rensselaer Polytech Inst, BS, 61, MS, 64, PhD(org chem), 67. *Prof Exp:* Asst res chemist, Sterling Winthrop Res Inst, Rensselaer, 63-69, res chemist, 69-76, sr res chemist & group leader, 76-77, sect head Analytical Chem, 77-90, RES FEL, STERLING WINTHROP RES INST, RENSSELAER, 90- *Concurrent Pos:* Fel, Harvard Univ, 67. *Mem:* Am Chem Soc; Sigma Xi. *Res:* Pharmaceutical research; structure determination by spectroscopic methods of compounds of pharmaceutical interest. *Mailing Add:* RR 1 Box 1396 Poestenkill NY 12140

CLEMENCE, SAMUEL PATTON, b Knoxville, Tenn, May 23, 39; m 67; c 2. CIVIL ENGINEERING. *Educ:* Ga Inst Technol, BS, 62, MS, 64, PhD(civil eng), 73. *Prof Exp:* Res asst civil eng, Ga Inst Technol, 63-64; Lieutenant construct, US Navy Civil Eng Corps, 64-69; res asst civil & geotech eng, Ga Inst Technol, 69-73; asst prof, Civil Eng Dept, Univ Mo, Rolla, 73-77; ASSOC PROF CIVIL & GEOTECH ENG, CIVIL ENG DEPT, SYRACUSE UNIV, 77-, CHMN DEPT, 82- *Concurrent Pos:* Expert witness, Construct contractors, 77-; consult, O'Brien & Gere, Dames & Moore, & Mobil Oil, 77-; prin investr res contracts, Niagara Mohawk Power Corp, 80- *Mem:* Am Soc Civil Engrs; Int Soc Soil Mech & Found Eng; Asn Eng Geologists; Sigma Xi; Int Soc Rock Mech. *Res:* Uplift capacity of helical anchors; bearing capacity of drilled piers; hydraulic fracturing of soils; hazard waste contaminents in clay. *Mailing Add:* Dept Civil Engr Hinds Hall Syracuse Univ Syracuse NY 13244-1190

CLEMENCY, CHARLES V, b New York, NY, Feb 12, 29; m 54; c 3. GEOCHEMISTRY, CLAY MINERALOGY. *Educ:* Polytech Inst Brooklyn, BS, 50; NY Univ, MS, 58; Univ Ill, PhD(geol), 61. *Prof Exp:* Chemist, Sylvania Elec Prod Inc, 53-58; asst prof, 61-67, ASSOC PROF GEOL,

STATE UNIV NY BUFFALO, 67- *Concurrent Pos:* Ford Found for study grant, Univ Sao Paulo, Brazil, 69; sr exchange scientist, Nat Acad Sci, Czech, 74. *Mem:* Fel Mineral Soc Am; AAAS; Geochem Soc; Clay Minerals Soc; Sigma Xi. *Res:* Analytical geochemistry; rock weathering; low temperature water-rock interactions. *Mailing Add:* 656 Creekside Dr Tonawanda NY 14150

CLEMENS, BRUCE MONTGOMERY, b Houston, Tex, Nov 19, 53; m 81; c 2. AMORPHOUS METALS, MAGNETIC THIN FILMS. *Educ:* Colo Sch Mines, BSc, 78; Calif Inst Technol, MS, 79, PhD(appl physics), 82. *Prof Exp:* Sr res scientist, Dept Physics, Gen Motors Res, 82-87, staff res scientist, 87-88; ASST PROF, DEPT MAT SCI & ENG, STANFORD UNIV, 89- *Concurrent Pos:* Instr physics, Oakland Univ, Rochester, Mich, 83-84. *Mem:* Am Phys Soc; Mat Res Soc; Sigma Xi. *Res:* Structure and properties of amorphous metals and compositionally modulated metal films; solid state reaction to form amorphous alloys; sputter deposition of metal films; magnetic thin films. *Mailing Add:* Dept Mat Sci & Eng Stanford Univ Stanford CA 94305-2205

CLEMENS, CARL FREDERICK, b Elkland, Pa, Nov 24, 24; m 53; c 8. CHEMISTRY. *Educ:* Ohio Univ, BS, 55; Univ Rochester, MBA, 66. *Prof Exp:* Asst biochem, Univ Rochester, 52-53; jr chemist, Haloid Co, 55-56, chemist, Haloid-Xerox, Inc, 56-58, proj chemist, 58-59; sr proj chemist, Xerox Corp, 59-64, scientist, 64-66, mgr instrumental anal, 66-67, mgr mat appln & develop, 67-69, tech prog mgr, Advan Develop Dept, 69-72, technol prog mgr, Xerographic Technol Dept, 72-77, mgr mats processing area, 77-79, mgr environ health & safety area, 79-85; RETIRED. *Concurrent Pos:* Consult. *Mem:* Am Chem Soc; Am Mgt Asn; fel Am Inst Chemists; Am Soc Safety Engrs; Soc Risk Assessment. *Res:* Xerography; application of polymer science to xerographic materials; polymer characterization and analytical chemistry; technical management; risk assessment. *Mailing Add:* 1119 Lake Rd Ontario NY 14519

CLEMENS, CHARLES HERBERT, b Dayton, Ohio, Aug 15, 39; m 83; c 3. MATHEMATICS. *Educ:* Holy Cross Col, AB, 61; Univ Calif, Berkeley, PhD(math), 66. *Prof Exp:* From asst prof to assoc prof math, Columbia Univ, 70-75; assoc prof, 75-76, PROF MATH, UNIV UTAH, 76- *Concurrent Pos:* Sloan fel, 74-75; ed, Pac J Math, 86- *Mem:* Am Math Soc. *Res:* Complex geometry. *Mailing Add:* Dept Math Univ Utah Salt Lake City UT 84112

CLEMENS, DANIEL THEODORE, b Los Angeles, Calif, May 5, 59. RESPIRATORY PHYSIOLOGY, ENERGETICS. *Educ:* Univ Calif, Santa Cruz, BA, 81; Univ Calif, Los Angeles, PhD(biol), 88. *Prof Exp:* Res assoc, Dept Organismal Biol & Anat, Univ Chicago, 89-91; ASST PROF BIOL, WILLIAMS COL, 91- *Concurrent Pos:* Lectr, Biol Sci Col Div, Univ Chicago, 90-91. *Mem:* Am Soc Zoologists; Sigma Xi. *Res:* Ecological physiology focuses on respiration and gas exchange, energetics and temperature regulation in terrestrial vertebrates; blood physiology and ventilatory responses of small birds to high altitude, nocturnal hypothermia in finches and the role of perfusion in the regulation of cutaneous gas exchange in amphibians. *Mailing Add:* Dept Biol Williams Col Williamstown MA 01267

CLEMENS, DAVID HENRY, b Newton, Mass, Nov 8, 31; m 53; c 3. ORGANIC CHEMISTRY. *Educ:* Middlebury Col, AB, 53; Univ Wis, PhD(chem), 57. *Prof Exp:* Head ion exchange & pollution control synthesis, 57-75, proj leader coatings res, 75-76, MGR POLYMERS & RESINS SYNTHESIS, ROHM & HAAS CO, 76- *Mem:* Am Chem Soc; Royal Soc Chem. *Res:* Organic synthesis; vinyl polymers; plasticizers; ion exchange; adsorbents; membrane processes; organic coatings; chemicals for textiles, leather, paper, non-wovens and chemical specialties. *Mailing Add:* 2000 Richard Rd Willowgrove PA 19090-2699

CLEMENS, DONALD FAULL, b Dover, Ohio, Aug 14, 29; m 50; c 5. ALUMINUM POLISHING BATHS, SUGAR ACID CHELANTS. *Educ:* Fla Southern Col, BS, 61; Univ Fla, MS, 63, PhD(chem), 65. *Prof Exp:* Lectr chem, Univ Fla, 65; assoc prof, 65-69, PROF CHEM, E CAROLINA UNIV, 69- *Concurrent Pos:* Vpres, Whitehurst Assocs Consults, 81- *Mem:* Am Chem Soc; Am Inst Chemists; Sigma Xi. *Res:* Use of peat and slurries of peat in fuel oil and alchohol as alternate fuels; development of chemical polishing baths for aluminum; production of sugar acid chelants. *Mailing Add:* 1701 Sulgrave Rd Greenville NC 27834

CLEMENS, EDGAR THOMAS, b July 4, 38; m 74; c 5. PHYSIOLOGY, GASTROENTEROLOGY. *Educ:* Univ Ill, BS, 66; N Mex State, MS, 68; Univ Nebr, PhD(biochem of nutrit), 71; Cornell Univ, PhD(gastroenterol), 75. *Prof Exp:* Sr lectr Vet Med, Cornell Univ, 75-78; instr vet med, Cornell Univ, 78-80; asst prof vet med, New Eng Med Ctr, 80-81; assoc prof vet sci, 81-87, PROF ANIMAL SCI, UNIV NEBR, 87- *Mem:* Am Soc Animal Sci; Am Inst Nutrit; Comp Gastroenterol Soc; Am Soc Vet Physiol & Pharmacol; Am Asn Zool Park & Aquariums. *Res:* Clinical nutrition; gastroenterology; stress physiology. *Mailing Add:* Dept Vet Sci 140 Vet Basic Sci Bldg Univ Nebr E Campus Lincoln NE 68583

CLEMENS, HOWARD PAUL, b Arthur, Ont, May 31, 23; nat US; m; c 4. ZOOLOGY. *Educ:* Univ Western Ont, BS, 46, MS, 47; Ohio State Univ, PhD(zool), 49. *Prof Exp:* Asst zool, Univ Western Ont, 47; from instr to assoc prof, 49-72, PROF ZOOL, UNIV OKLA, 72-, MEM GRAD FAC, 80- *Mem:* AAAS; Am Soc Limnol & Oceanog; Am Fisheries Soc; Am Soc Zoologists; Am Soc Ichthyologists & Herpetologists. *Res:* Limnology; fishery biology; aquatic invertebrates; fish endocrines. *Mailing Add:* 1621 Cruce Norman OK 73069

CLEMENS, JAMES ALLEN, b Windsor, Pa, Feb 4, 41; m 64; c 2. NEUROENDOCRINOLOGY, NEUROPHYSIOLOGY. *Educ:* Pa State Univ, BS, 63, MS, 65; Mich State Univ, PhD(physiol), 68. *Prof Exp:* RES ADV, ELI LILLY & CO, 69- *Concurrent Pos:* Nat Inst Neurol Dis & Blindness fel, Univ Calif, Los Angeles, 68-69; Biochem Endocrinol Study Sect (NIH). *Mem:* AAAS; Endocrine Soc; Soc Neurosci; Int Soc Psychoneuroendocrinol. *Res:* Neural mechanisms that control anterior pituitary hormone secretion; stroke and neurological disease research. *Mailing Add:* Div CNS & Endocrine Res Lilly Res Labs Indianapolis IN 46285

CLEMENS, JON K(AUFMANN), b Sellersville, Pa, May 10, 38; m 59; c 3. VIDEO SYSTEMS, ELECTRICAL ENGINEERING. *Educ:* Goshen Col, BA, 60; Mass Inst Technol, BS & MS, 63, PhD(elec eng), 65. *Prof Exp:* Mem tech staff, 65-75, head Signal Systs Res Group, 75-80, dir, Video-Disc Systs Res, 80-83, vpres, Consumer Electronics Res, David Sarnoff Res Ctr, RCA Corp, 83-87; pres, Chronar Corp, 87-89; SR VPRES SCI GROUP, SRI INT, 90- *Honors & Awards:* Edward Rhine Prize, 80; Vladimirk Zwovkin Award, Inst Elec & Electronics Engrs, 83. *Mem:* Inst Elec & Electronic Engrs; Sigma Xi. *Res:* Video disc; image processing, digital recording of video, audio and data; communication transmission and video systems. *Mailing Add:* SRI Int 333 Ravenwood Ave Menlo Park CA 94025

CLEMENS, LAWRENCE MARTIN, b Chicago, Ill, Nov 14, 37; m 63; c 1. ORGANIC CHEMISTRY. *Educ:* Ill Inst Technol, BS, 59; Carnegie Inst Technol, PhD(org chem), 64. *Prof Exp:* NIH fel chem, Univ Notre Dame, 63-64; res chemist, Archer Daniels Midland Co, 64-69; res specialist, 69-80, res mgr, 80-83, LAB MGR, MINN MINING & MFG CO, 83- *Mem:* Am Chem Soc; Adhesion Soc; Adhesive & Sealant Coun. *Res:* Ultraviolet curing coatings; adhesive chemistry and polymer chemistry; product development and applied chemistry. *Mailing Add:* Minn Mining & Mfg Co 3M Ctr Bldg 236-1B-21 St Paul MN 55144

CLEMENS, ROBERT JAY, b Lansdale, Pa, Jan 17, 57; m 77; c 2. DIKETENE CHEMISTRY, HETEROCYCLIC SYNTHESIS. *Educ:* Ursinus Col, BS, 77; Princeton Univ, MA, 80, PhD(chem), 83. *Prof Exp:* Tech serv engr, Firestone Plastics Co, 77-78; SR RES CHEMIST, RES LABS, EASTMAN CHEM DIV, EASTMAN KODAK CO, 82- *Honors & Awards:* Elias Singer Award, 88. *Mem:* Sigma Xi; Am Chem Soc. *Res:* Synthesis, modification and utilization of small-ring heterocycles; chemistry of diketene and derivatives. *Mailing Add:* PO Box 431 Kingsport TN 37662-0431

CLEMENS, STANLEY RAY, b Souderton, Pa, May 11, 41; m 62; c 4. MATHEMATICS. *Educ:* Bluffton Col, AB, 63; Ind Univ, MA, 65; Univ NC, PhD(math), 68. *Prof Exp:* Instr, Univ NC, 67-68; asst prof, 68-76, ASSOC PROF MATH, ILL STATE UNIV, 76- *Mem:* Am Math Soc; Math Asn Am. *Res:* Topology. *Mailing Add:* Dept Math Bluffton Col Bluffton OH 45817

CLEMENS, WILLIAM ALVIN, b Berkeley, Calif, May 15, 32; m 55; c 4. PALEONTOLOGY. *Educ:* Univ Calif, Berkeley, BA, 54, PhD(paleont), 60. *Prof Exp:* NSF fel, 60-61; from asst prof & asst cur to assoc prof zool & assoc cur fossil higher vert, Univ Kans, 61-67; assoc prof, 67-71, PROF PALEONT, UNIV CALIF, BERKELEY, 71- *Concurrent Pos:* NSF fel, 68-69; John Simon Guggenheim Found fel, 74; Alexander von Humboldt fel, 78-79; vis prof, Miller Inst, 82- *Mem:* Soc Syst Zool; Soc Vert Paleont; Geol Soc Am; Palaeont Asn; Zool Soc London; Sigma Xi. *Res:* Evolution of Mesozoic and Cenozoic mammals. *Mailing Add:* Dept Paleont Univ Calif Berkeley CA 94720

CLEMENS, WILLIAM BRYSON, bacteriology, for more information see previous edition

CLEMENST, JAMES LEE, hydrometallurgy, extractive mettallurgy, for more information see previous edition

CLEMENT, CHRISTINE MARY (COUTTS), b Chatham, Ont; m 75; c 1. ASTRONOMY. *Educ:* Univ Toronto, BSc, 63, MA, 64, PhD(astron), 67. *Prof Exp:* Fel astron, Asiago Astrophys Observ, Italy, 67-68; res assoc, David Dunlap Observ, 69-74, LECTR ASTRON, UNIV TORONTO, 74- *Mem:* Am Astron Soc; Can Astron Soc; Am Asn Variable Star Observers; Royal Astron Soc Can; Can Asn Women Sci. *Res:* Study of variable stars in globular clusters. *Mailing Add:* Dept Astron McLennan Phys Labs Univ Toronto Toronto ON M5S 1A7 Can

CLEMENT, DUNCAN, b Pittsfield, Mass, Oct 22, 17. SCIENCE ADMINISTRATION. *Educ:* Mt St Mary's Col, Md, BS, 40; Harvard Univ, MA, 42, PhD(bot), 48. *Prof Exp:* Econ botanist, Atkins Garden, Harvard Univ, Cuba, 48-53, in chg, 49-53, dir, 53-60, dir, US, 61-63; consult, Off Int Sci Activities, NSF, 62-63, prog dir, 63-64, head sci liaison staff, Costa Rica, 64-68, prof assoc, Off Int Progs, 68-74; sci attache, US Embassy, Madrid, Spain, 74-79; resident dir, Nat Res Coun, NAS, Quito, Ecuador, 83-84; RETIRED. *Mem:* Sigma Xi. *Res:* International science cooperation. *Mailing Add:* 620 SW 70th Ave Pembroke Pines FL 33023

CLEMENT, GERALD EDWIN, b Austin, Minn, Nov 5, 35; m 57; c 2. CLINICAL CHEMISTRY. *Educ:* Univ Minn, BA, 57; Purdue Univ, PhD(org chem), 61; Am Bd Clin Chemists, cert, 75. *Prof Exp:* NIH fel, Northwestern Univ, 61-63; asst prof org chem, Harpur Col, 63-68; assoc prof chem, Kenyon Col, 68-74; CLIN CHEMIST, ALLENTOWN SACRED HEART HOSP, 74- *Concurrent Pos:* Fel, Hahnemann Hosp, 72-74. *Mem:* Am Asn Clin Chemists; Am Chem Soc; Am Soc Biol Chemists. *Res:* Organic reaction mechanisms, especially acid, base and enzyme catalysis. *Mailing Add:* 2027 Greenleaf St Allentown PA 18104

CLEMENT, JOHN REID, JR, b East Spencer, NC, Apr 14, 21; m 45; c 2. PHYSICS. *Educ:* Catawba Col, AB, 43. *Prof Exp:* Proj physicist, low temperature thermomet & calorimet, US Naval Res Lab, 46-53, head, Cryogenic Properties, Matter & Devices Sect, 53-59, asst head, Cryogenics Br, 59-62, head, 62-69, assoc supt, Solid State Div, 69-74, head, High Magnetic Field Facil, 74-80, sr res physicist, 80-83; RETIRED. *Mem:* Fel Am Phys Soc; Sigma Xi. *Res:* High voltage electricity; low temperature thermometry and calorimetry; cryogenic devices; high magnetic fields. *Mailing Add:* 128 Canterberry Dr Salisbury NC 28144

CLEMENT, JOSEPH D(ALE), b Kalamazoo, Mich, Jan 7, 28; m 56; c 2. PHYSICS, NUCLEAR ENGINEERING. *Educ:* Western Mich Univ, BS, 49; Univ Wis, MS, 53, PhD(physics), 57. *Prof Exp:* Jr scientist, inst atomic res, Iowa State Univ, 49-52; sr scientist, Bettis Atomic Power Lab, Westinghouse Elec Corp, 57-59, group leader, Astronuclear Lab, 60-62; supvry engr reactor physics, Nuclear Div, Martin Co, 59-60; dept mgr nuclear eng, Nuclear Mat & Equip Corp, 62-65; assoc prof, 65-68, PROF NUCLEAR ENG, GA INST TECHNOL, 68- *Mem:* Am Phys Soc; Am Nuclear Soc. *Res:* Nuclear physics; interactions of neutrons with complex nuclei; nuclear engineering; reactor physics; shielding; radioisotope utilization; nuclear power economics. *Mailing Add:* Dept Nuclear Eng Ga Inst Technol Atlanta GA 30332

CLEMENT, LORAN T, b Kansas City, Mo, Aug 3, 46. PEDIATRICS, REGULATION OF IMMUNE RESPONSE. *Educ:* Univ Nebr, MD, 73. *Prof Exp:* Assoc prof pediat, Univ Ala, Birmingham, 83-85; ASSOC PROF PEDIAT, UNIV CALIF, LOS ANGELES, 85- *Mem:* Am Asn Immunologists. *Mailing Add:* Dept Pediat Sch Med Univ Calif Los Angeles CA 90024

CLEMENT, MAURICE JAMES YOUNG, b Vancouver, BC, Sept 11, 38; m 75; c 1. THEORETICAL ASTROPHYSICS. *Educ:* Univ BC, BSc, 60, MSc, 61; Univ Chicago, PhD(astrophys), 65. *Prof Exp:* From asst prof to assoc prof, 67-81, PROF ASTRON, UNIV TORONTO, 81 - *Concurrent Pos:* Fel, Princeton Univ, 65-66; Nat Res Coun Can fel, 66-67. *Mem:* Am Astron Soc; Can Astron Soc; Int Astron Union. *Res:* Equilibrium and stability of rotating stars; differential rotation and meridian circulation in stars. *Mailing Add:* Dept Astron Univ Toronto Toronto ON M5S 1A7 Can

CLEMENT, ROBERT ALTON, b Brockton, Mass, Aug 12, 29; m 55; c 4. INDUSTRIAL ORGANIC CHEMISTRY. *Educ:* Mass Inst Technol, BS, 50; Univ Calif, Los Angeles, PhD(chem), 54. *Prof Exp:* Proj assoc chem, Univ Wis, 54-55; from instr to asst prof, Univ Chicago, 55-62; res chemist, E I DuPont de Nemours & Co, Inc, 62-86; SUPVR & CONSULT, 86- *Mem:* Am Chem Soc. *Res:* Synthetic methods in organic chemistry; polymer chemistry. *Mailing Add:* 04470 B-306 E I DuPont de Nemours & Co Inc Wilmington DE 19898

CLEMENT, STEPHEN LEROY, b Ventura, Calif, Aug 25, 44; m 82; c 1. BIOLOGICAL WEED CONTROL, ENTOMOLOGY. *Educ:* Univ Calif, Davis, BS, 67, MS, 72, PhD(entom), 76. *Prof Exp:* Fel & res entomologist pheromones, Univ Calif, Davis, 77; asst prof entom, Ohio Agr Res & Develop Ctr, Ohio State Univ, 77-81; res entomologist, biol control weeds, Agr Res Serv, USDA, Albany, Calif, 81; RES ENTOMOLOGIST, PLANT GERMPLASM RES, AGR RES SERV, USDA, WASH STATE UNIV, 86- *Concurrent Pos:* Environ Protection Agency grant pest mgt, Ohio Agr Res & Develop Ctr, 77-81; mem agr mission US embassy, Rome, Italy, 82-86; affil prof entom, Univ Idaho, 86- *Mem:* Entom Soc Am; Sigma Xi; Int Orgn Biol Control. *Res:* Applied ecology with emphasis on insects, weeds and agronomic crops; biological control; host plant resistance; pest management. *Mailing Add:* Plant Germplasm Res USDA ARS Wash State Univ 59 Johnson Hall Pullman WA 99164-6402

CLEMENT, WILLIAM GLENN, b Denver, Colo, Apr 11, 31; m 52; c 2. GEOPHYSICS, ELECTRICAL ENGINEERING. *Educ:* Stanford Univ, BS, 56, PhD(geophys), 63. *Prof Exp:* Geophysicst, Pan Am Petrol Corp, Tex, 63-65, sr geophysicist, Res Lab, Okla, 65-68, staff res engr, 68-74, sr res engr, Amoco Prod Res Ctr, 74-76; MGR GEOPHYS RES, CITIES SERV OIL CO, 76- *Mem:* Soc Explor Geophys; Europ Asn Explor Geophys. *Res:* Application of communication theory to the analysis of seismic and potential field data. *Mailing Add:* Aramco No 195466 Box 5923 Dhahran 31311 Saudi Arabia

CLEMENT, WILLIAM H, b Johnstown, Pa, Dec 17, 31; m 65; c 2. ORGANIC CHEMISTRY, ENVIRONMENTAL CHEMISTRY. *Educ:* Gettysburg Col, AB, 54; Univ Del, MS, 57, PhD(chem), 60. *Prof Exp:* Chemist, E I du Pont de Nemours & Co, 55 & Pittsburgh Plate Glass Co, 56-57; res chemist, Gulf Oil Corp, 59-62; asst prof chem, Waynesburg Col, 62-63; NSF fel, Univ Buffalo, 63; USDA fel, Univ Cincinnati, 63-65; asst prof, Ithaca Col, 65-70, prof, 70-71; res assoc environ sci, Rutgers Univ, 71-81; ENVIRON CONSULT, 81- *Concurrent Pos:* US Army Chem Corps, 57. *Mem:* Am Chem Soc; Sigma Xi. *Res:* Oxidation reactions and mechanisms; organometallics; catalysis; organic synthesis; environmental chemistry; water quality. *Mailing Add:* 1180 Worthington Heights Blvd Worthington OH 43235-2133

CLEMENT, WILLIAM MADISON, JR, b Rome, Ga, Dec 15, 28; m 50. GENETICS. *Educ:* Univ Ga, BSA, 50; Univ Calif, PhD, 58. *Prof Exp:* Lab technician agron, Univ Calif, 52-57; res geneticist, Forage & Range Br, Agr Res Serv, USDA, 58-65; assoc prof cytol & cytogenetics, Vanderbilt Univ, 65-86, dir undergrad studies, Dept Gen Biol, 86; CONSULT, FORENSIC SCI LAB, UNIV SALA, MOBILE, 89- *Concurrent Pos:* Asst prof, Univ Minn, 61-65. *Mem:* Crop Sci Soc Am; Am Genetics Soc; Genetics Soc Can; Am Soc Naturalists. *Res:* Cytogenetics of medicago sativa and related species; chromosome structure. *Mailing Add:* PO Box 28 Gulf Shores AL 36547

CLEMENTE, CARMINE DOMENIC, b Penns Grove, NJ, Apr 29, 28; m 68. NEUROANATOMY. *Educ:* Univ Pa, AB, 48, MS, 50, PhD, 52. *Prof Exp:* Dir, Brain Res Inst, 76-87, from instr to assoc prof, 52-63, chmn dept, 63-73, PROF ANAT, SCH MED, UNIV CALIF, LOS ANGELES, 63-; PROF SURG ANAT, CHARLES R DREW POSTGRAD MED SCH, LOS ANGELES, CALIF, 63- *Concurrent Pos:* Gianinni Found fel, 53-54; hon res assoc, Univ Col, Univ London, 53-54; consult, Vet Admin Hosp, Sepulveda, Calif & Martin Luther King Mem Hosp; ed, Gray's Anat & Exp Neurol; vis scientist & spec consult, NIH, 58; mem, Biol Stain Comn; mem subcomt neuropath, Nat Acad Sci; mem med adv bd, Bank of Am-Gianinni Found; mem admin bd, Asn Am Med Cols, 73-81; mem bd dirs, King-Drew Med Ctr & Charles R Drew Postgrad Med Sch, 85-; distinguished vis prof, Mich State Univ Med Sch, 86; liaison comt med educ, Am Med Asn & Am Asn Med Clin, 80-86; mem, Nat Bd Med Examrs, 80-88; pres, Asn Anat Chairmen, 71-72. *Honors & Awards:* Res Award, Pavlovian Soc NAm, 68; Japan Soc Prom Sci Res Award, 78; Guggenheim Award, 88; Ann Award in Sci, Nat Paraplegia Found, 73; Martin E Rehfuss Prize & Medal, 87. *Mem:* Inst Med Nat Acad Sci; Am Physiol Soc; Am Acad Cerebral Palsy; Pavlovian Soc Nam (vpres, 70-72, pres, 72-73); Am Asn Anat (vpres, 70-72, pres, 76-77); Am Acad neurol. *Res:* Regeneration of nerve fibers; effects of x-irradiation on brain; neurocytology; basic neurology; sleep and wakefulness. *Mailing Add:* Dept Anat & Cell Biol Univ Calif Sch Med Los Angeles CA 90024

CLEMENTS, BURIE WEBSTER, b Pierce, Fla, Dec 16, 27; m 52; c 3. MEDICAL ENTOMOLOGY. *Educ:* Univ Fla, BS, 54, MSA, 56. *Prof Exp:* Entomologist, stored prod insects, Ga Lab, USDA, 56-58, Savannah, Ga Lab, 58-60; entomologist, Vero Beach Lab, Entom Res Ctr, Div Health, 60-64, entomologist & adminr, John A Mulrennan Sr Res Lab, 64-90; RETIRED. *Mem:* Entom Soc Am; Am Mosquito Control Asn; Sigma Xi; Registry Prof Entomologist. *Res:* Development of effective and economical methods for controlling insects of public health importance. *Mailing Add:* 1421 Airport Road Panama City FL 32405

CLEMENTS, GEORGE FRANCIS, b Colfax, Wash, Apr 17, 31; m 52; c 4. MATHEMATICS. *Educ:* Univ Wis, BSME, 53; Syracuse Univ, MA, 57, PhD(math), 62. *Prof Exp:* Asst math, Syracuse Univ, 55-62; asst prof appl math, 62-68, assoc prof math, 68-76, PROF MATH, UNIV COLO, BOULDER, 76-, FAC GRAD SCH, 80- *Mem:* Am Math Soc. *Res:* Combinatorial theory. *Mailing Add:* Dept Math Box 426 Univ Colo Boulder CO 80309

CLEMENTS, GREGORY LELAND, b Lincoln, Nebr, Apr 5, 49; m 74; c 4. ASTRONOMY. *Educ:* Univ Iowa, BA, 71, MS, 76, PhD(physics), 78. *Prof Exp:* Asst prof physics & astron, Dickinson Col, 78-82; ASSOC PROF PHYSICS & MATH, MIDLAND COL, 83- *Concurrent Pos:* Computer programmer, 82-83. *Mem:* Am Asn Physics Teachers. *Res:* Spectrophotometry of planets, satellites, asteroids and eclipsing binary stars. *Mailing Add:* Midland Lutheran Col Fremont NE 68025

CLEMENTS, JOHN ALLEN, b Auburn, NY, Mar 16, 23; m 49; c 2. LUNG FUNCTION, INTERFACIAL PHENOMENA. *Educ:* Cornell Univ, MD, 47. *Hon Degrees:* Dr, Berue, 90. *Prof Exp:* Res asst physiol, Med Sch, Cornell Univ, 47-49; physiologist, Med Labs, Army Chem Ctr, Md, 51-62, asst chief clin invest br, 52-61; res assoc physiol & assoc prof pediat, 61-64, PROF PEDIAT & AM HEART ASN CAREER INVESTR, SCH MED, UNIV CALIF, SAN FRANCISCO, 64- *Concurrent Pos:* Res assoc physiol, Sch Med, Univ Pa, 52-58; lectr, Sch Med, Johns Hopkins Univ, 55-61; consult, Baltimore City Hosp, 57-61, Roswell Park Mem Inst, 58-61 & Surgeon Gen, USPHS, 64-68; sci counr, Nat Heart & Lung Inst, 72; assoc ed, Am Rev Respiratory Dis, 73-80; assoc ed, Am Jour Physiol, 88-; assoc ed, Annual Rev Physiol, 88- *Honors & Awards:* Modern Med Distinguished Achievement Award, 73; Howard Taylor Ricketts Medal, Univ Chicago, 75; Edward Livingston Trudeau Medal, Am Lung Asn, 82; Gairdner Found Int Award, 83; J Burns Amberson lectr, Am Thoracic Soc, 91. *Mem:* Nat Acad Sci; Am Physiol Soc; hon fel Am Col Chest Physicians; Am Thoracic Soc; fel Royal Col Physicians London; fel AAAS; NY Acad Sci. *Res:* Biophysics; respiration; membrane and cardiovascular physiology; published numerous articles in various journals. *Mailing Add:* Cardiovasc Res Inst Med Ctr Univ Calif San Francisco CA 94143

CLEMENTS, JOHN B(ELTON), b Dillwyn, Va, May 25, 29; m 56; c 2. AIR POLLUTION. *Educ:* Hampden-Sydney Col, BS, 50; Univ Va, PhD(org chem), 55. *Prof Exp:* Res chemist, Summit Res Labs, Celanese Corp Am, 55-60; from res chemist to sr res chemist, Chemstrand Res Ctr, Monsanto Co, 60-68; chief, Bioanal Sect, 68-70, chemist, Off Criteria & Standards, 70-72, chief, Methods Standardization Br, 72-75, chief, Qual Assurance Br, 75-79, dir, Environ Monitoring Div, 79-85, DIR, QUAL ASSURANCE DIV, ENVIRON PROTECTION AGENCY, 85- *Mem:* Am Chem Soc; Am Soc Testing & Mat. *Res:* Measurement of air pollutants; quality assurance for air pollution monitoring data. *Mailing Add:* Environ Protection Agency MO-77 Triangle Park NC 27711

CLEMENTS, LINDA L, b Phoenix, Ariz, Oct 6, 45; m 71; c 2. COMPOSITE MATERIALS, POLYMERIC MATERIALS. *Educ:* Stanford Univ, BS, 67, PhD(mat sci eng), 75; Univ Pa, MSE, 71. *Prof Exp:* Engr, Lawrence Livermore Lab, 74-78, prog mgr, 77-78; proj dir, Advan Res & Applications Corp, Sunnyvale, Calif, 78-81; assoc prof, 81-85, PROF MAT ENG, SAN JOSE STATE UNIV, 85- *Concurrent Pos:* Prin investr, NASA Ames Res Ctr, 78-81, proj dir & prin investr, 81-83; lectr, Mat Eng Dept, San Jose State Univ, 79-81; consult composite mat & polymers, US Dept Justice & var co, 81-; nat adj fac mem, Am Soc Metals Int, 84-; proj dir & prin investr, Aerojet Strategic Propulsion Co, Sacramento, 85-86 & Northrop Corp, Hawthorne, Calif, 86-90; chair, Int Career Develop Comt, Soc Advan Mat & Process Eng Int, 86- *Mem:* Soc Advan Mat & Process Eng Int; Am Soc Metals Int; Am Chem Soc; Am Soc Testing & Mat; Soc Plastics Engrs; Soc Women Engrs. *Res:* High performance materials, emphasizing composite materials, polymers, and mechanical behavior; hazardous materials; materials engineering, with emphasis on composite materials, polymers, and test methodology. *Mailing Add:* Mat Eng Dept San Jose State Univ San Jose CA 95192

CLEMENTS, REGINALD MONTGOMERY, b Vancouver, BC, Apr 13, 40; m 64; c 1. PLASMA PHYSICS. *Educ:* Univ BC, BASc, 63, MASc, 64; Univ Sask, PhD(plasma physics), 67. *Prof Exp:* Fel plasma physics, Univ Alta, 67-68; from asst prof to assoc prof, 68-79, PROF PLASMA PHYSICS, UNIV VICTORIA, 79- *Mem:* Combustion Inst; Soc Automotive Engrs. *Res:* Gas discharge physics; plasma diagnostic methods, specifically electrostatic and rf probes; microwave-plasma interactions; electrical phenomena in combustion; ignition systems, underwater acoustics. *Mailing Add:* Dept Physics Univ Victoria PO Box 1700 Victoria BC V8W 2Y2 Can

CLEMENTS, THOMAS, b Chicago, Ill, June 1, 98. MININGS, ENGINEERING GEOLOGY. *Educ:* Tex Sch Mines, EM, 22; Calif Inst Tech, MA, 23, PhD(geol), 32. *Prof Exp:* Prof geol, Univ Southern Calif, 26-64; GEOLOGIST CONSULT, 63- *Mem:* Fel Geol Soc Am; Am Petrol Geologist. *Mailing Add:* 2171 Vista Del Mar St Los Angeles CA 90068

CLEMENTS, WAYNE IRWIN, b Hillsdale, NJ, Feb 11, 31. FORENSIC ENGINEERING. *Educ:* Newark Col Eng, BSEE, 57, MSEE, 61; Univ Pa, MS, 75. *Prof Exp:* Design engr, Gen Elec, 57-58 & ITT Labs, 58-59; ASSOC PROF ELEC & COMPUTER ENG, NJ INST TECHNOL, 59- *Concurrent Pos:* Consult engr, 64- *Mem:* Inst Elec & Electronics Engrs; Instrumentation Soc Am; Am Asn Univ Professors. *Res:* Design engineering education; forensic engineering. *Mailing Add:* PO Box Five Edison NJ 08818-0005

CLEMENTS, WILLIAM EARL, b Temple, Tex, Mar 8, 42; m 64; c 2. PHYSICS, ATMOSPHERIC PHYSICS. *Educ:* Tex A&I Univ, BS, 65, MA, 66; NMex Inst Mining & Technol, PhD(physics), 74. *Prof Exp:* Asst prof physics, Calif Polytech State Univ, San Luis Obispo, 66-70; PHYSICIST ATMOSPHERIC PHYSICS, LOS ALAMOS NAT LAB, UNIV CALIF, 74- *Mem:* Am Meteorol Soc. *Res:* Complex terrain meteorology. *Mailing Add:* 300 Rover Blvd White Rock NM 87544

CLEMENTSON, GERHARDT C, b Black Earth, Wis, May 3, 17; m 43; c 3. COMPUTER SCIENCES, OPERATIONS RESEARCH. *Educ:* US Mil Acad, BS, 42; Calif Inst Technol, MS, 45; Mass Inst Technol, MSAE, 48, ScD, 50. *Prof Exp:* Asst prof elec eng, US Air Force Inst Technol, 54-55, prof aeronaut & head dept, US Air Force Acad, 55-61; asst to exec vpres-tech, Space & Info Syst Div, NAm Aviation, Inc, 64-68; dir opers res, Auto-Tronix Universal Corp, 68-70; PROF COMPUT & MGT SCI & DIR ACAD COMPUT CTR, METROP STATE COL, 70- *Mem:* Asn Comput Mach; Sigma Xi. *Res:* Guidance and control; education; management sciences. *Mailing Add:* 6423 S Sycamore St Littleton CO 80120

CLEMENTZ, DAVID MICHAEL, b Cleveland, Ohio, Sept 4, 45; m 67; c 2. CLAY MINERALOGY, PETROLEUM ENGINEERING. *Educ:* Univ Ariz, BS, 67; Purdue Univ, MS, 69; Mich State Univ, PhD(clay mineral), 73; Pepperdine Univ, MBA, 80. *Prof Exp:* Res chemist surface chem, 73-76, sr res chemist org geochem, 77-79, petrol prod engr, 80-81, sr res assoc, 82-84, asst pres, 84, mgr res prop, 85-86, MGR, RES ENGR, ENHANCED OIL RECOVERY, CHEVRON OIL FIELD RES CO, 87- *Concurrent Pos:* Mem coun, Clay Minerals Soc, 80-82; mem bd of trustees, LaHabra Children Mus, 86-; pres, LaHabra Area Chamber Com, 87-88. *Mem:* Clay Minerals Soc; Soc Petrol Engrs; Am Chem Soc. *Res:* Enhanced oil recovery; surface chemistry of minerals especially clays; clay organic interactions; petroleum source rock characterization; organic geochemistry; petroleum and reservoir engineering. *Mailing Add:* 238 Canova Close SW Calgary AB T2W 6A6 Can

CLEMETSON, CHARLES ALAN BLAKE, b Canterbury, Eng, Oct 31, 23; m 47; c 4. OBSTETRICS & GYNECOLOGY. *Educ:* Oxford Univ, BM & BCh, 48, MA, 50; FACOG, FRCOG, FRCSC. *Prof Exp:* Res asst obstet, Univ Col Hosp, London, 50-52, lectr obstet & gynec, 56-58; house surgeon, London Hosps, 52-54; registr, Ashton-under-Lyne Hosp, 54-56; asst prof, Univ Sask Hosp, 58-61; asst prof, Med Ctr, Univ Calif, San Francisco, 61-67; lectr maternal health, Univ Calif, Berkeley, 63-67; assoc prof obstet & gynec, State Univ NY Downstate Med Ctr, 67-72, prof, 72-81; dir obstet & gynec, Methodist Hosp Brooklyn, 67-81; PROF OBSTET & GYNEC, TULANE UNIV, 81- *Mem:* Fel Am Col Obstet & Gynec. *Res:* Obstetric and menstrual physiology. *Mailing Add:* 314 Country Club Dr Pineville LA 71360

CLEMMENS, RAYMOND LEOPOLD, b Baltimore, Md, Apr 2, 22; m 52; c 3. MEDICINE, PEDIATRICS. *Educ:* Loyola Col, Md, BS, 47; Univ Md, MD, 51; Am Bd Pediat, dipl, 56. *Prof Exp:* From instr to prof, 54-86, EMER PROF PEDIAT, UNIV MD, BALTIMORE, 86- *Concurrent Pos:* Pediat consult, USPHS & Md State Health Dept, 62-; mem adv comt, Crippled Children's Prog, Md State Health Dept, 62-; mem adv coun ment hyg, State Bd Health & Ment Hyg, 62-67; mem task force, Nat Inst Neurol Dis & Stroke, 65- *Mem:* Fel Am Acad Pediat; Am Pediat Soc; Am Acad Ment Deficiency. *Res:* Handicapped children's diagnostic and evaluation clinic. *Mailing Add:* 522 Goucher Blvd Towson MD 21204

CLEMMONS, DAVID ROBERTS, b Nashville, Tenn, May 19, 47; m 71; c 2. GROWTH FACTORS, CELLULAR GROWTH REGULATION. *Educ:* Davidson Col, BS, 69; Univ NC, MD, 74. *Prof Exp:* Intern med, Mass Gen Hosp, 74-75, jr resident, 75-76; sr resident, John Hopkins Hosp, 76-77; fel endocrinol, Mass Gen Hosp, 77-79; fel endocrinol, 79-80, asst prof, 80-84, ASSOC PROF MED, UNIV NC, 84- *Concurrent Pos:* Prin investr, Nat Inst Aging, 80-, Nat Heart, Lung & Blood Inst, NIH, 81-; consult, Monsanto Inc, 82-, Synergen Inc, 86-; vis scientist, Nat Inst Diabetes, Digestive & Kidney Dis, 85-86. *Mem:* Am Soc Clin Invest; Am Fedn Clin Res; Am Soc Cell Biol; Am Col Physicians; Endocrine Soc; Asn Clin Res Dir. *Res:* The isolation characterization and structural determination of polypeptide growth factors and the proteins that modulate their biologic actions; the hormonal variables that regulate cellular growth and differentiation. *Mailing Add:* 908 Emory Dr Chapel Hill NC 27514

CLEMMONS, JACKSON JOSHUA WALTER, b Beloit, Wis, Mar 24, 23; m 52; c 4. BIOCHEMISTRY, PATHOLOGY. *Educ:* Univ Wis, BS, 48, MS, 49, PhD, 56; Western Reserve Univ, MD, 59; Am Bd Path, dipl, 64. *Prof Exp:* Res assoc path, Univ Wis, 51-56; univ fel, Western Reserve Univ, 57-60, Helen Hay Whitney fel, 60-62; asst prof, 62-64, assoc prof, 64-77, PROF PATH, SCH MED, UNIV VT, 77- *Mem:* Am Soc Exp Path; NY Acad Sci; Int Acad Path; Am Asn Clin Chem; Sigma Xi. *Mailing Add:* Dept Path Univ Vt Med Sch Burlington VT 05401

CLEMMONS, JOHN B, b Rome, Ga, Apr 11, 18; m 47; c 2. MATHEMATICS, PHYSICS. *Educ:* Morehouse Col, AB, 37; Univ Atlanta, MS, 39. *Prof Exp:* Admin prin high sch, Ky, 40-43; asst prin high sch, Md, 43-47; assoc prof, 47-74, PROF MATH, SAVANNAH STATE COL, 74-, HEAD DEPT MATH & PHYSICS, 47- *Concurrent Pos:* Lectr, Univ Southern Calif, 50; chmn acad adv comt to Bd Regents, State of Ga. *Mem:* Am Math Soc; Nat Inst Sci. *Res:* Special properties of convex sets. *Mailing Add:* 647 Hamilton Ct Savannah GA 31401

CLEMMONS, ROGER M, b Springfield, Mo, Aug 10, 49; m. PHYSIOLOGY, ANIMAL. *Educ:* Wash State Univ, DVM, 73, PhD (neurophysiol), 79. *Prof Exp:* Pvt pract, small animal, 73-75; post doctoral fel, Wash State Univ, 75; asst prof, 79-85, ASSOC PROF, COL VET MED, UNIV FLA, 85- *Honors & Awards:* Res Award, Scottish Terrier Club Am, 85. *Mem:* Am Vet Med Asn; Am Vet Neurol Asn; Am Physiol Soc; Am Heart Asn; Int Soc Vet Perinatology. *Res:* Comparative hemostasis; platelet physiology. *Mailing Add:* Col Vet Med Univ Fla Box J-126 JHMHC Gainesville FL 32610

CLEMONS, ERIC K, TELECOMMUNICATIONS TECHNOLOGY. *Educ:* Mass Inst Technol, SB, 70; Cornell Univ, MS, 74, PhD(opers res), 76. *Prof Exp:* Asst prof, 76-81, ASSOC PROF DECISION SCI, WHARTON SCH, UNIV PA, 81- *Concurrent Pos:* Proj dir, Reginald H Jones, Ctr Sponsored Res Proj Info Syst, Telecommun & Bus Strategy, 87-; consult, pvt & pub sectors. *Res:* Strategic information systems; telecommunications technology and its commercial application; information systems in international securities markets; information technology on the distribution channel; several publications. *Mailing Add:* Dept Decision Sci Wharton Sch Univ Pa Philadelphia PA 19104-6366

CLEMONS, RUSSELL EDWARD, b Warner, NH, Oct 1, 30; m 68. ECONOMIC GEOLOGY. *Educ:* Univ NMex, BS, 60, MS, 62; Univ Tex, Austin, PhD(geol), 66. *Prof Exp:* From instr to asst prof geol, Univ Tex, Arlington, 65-69; assoc prof, 69-74, PROF GEOL, NMEX STATE UNIV, 74- *Concurrent Pos:* Consult, NMex Bur Mines & Mineral Resources, 71- *Mem:* Fel Geol Soc Am; Am Asn Petrol Geologists. *Res:* Areal field geology and geologic mapping in Mex and NMex; igneous petrography and associated mineral deposits of plutons in northeastern Mex and southern NMex; carbonate petrography of paleozoic and cretaceous rocks in Southwestern US. *Mailing Add:* Dept Earth Sci NMex State Univ Las Cruces NM 88003

CLEMSON, HARRY C, b Pawtucket, RI, Aug 12, 34; m 57; c 3. BIOCHEMISTRY, CLINICAL CHEMISTRY. *Educ:* Univ RI, BS, 58, PhD(pharmaceut chem), 66; Univ NH, MS, 61. *Prof Exp:* Instr chem, Rochester Inst Technol, 60-63; USPHS fel pharmacol, Yale Univ, 66-68; asst prof med chem, Northeastern Univ, 68-70; HEAD BIOCHEM, LYNN HOSP, 70- *Mem:* AAAS; Am Chem Soc; Am Asn Clin Chemists; Am Inst Chemists; Sigma Xi. *Res:* Diagnostic enzymology; drug analysis. *Mailing Add:* 173 Perkins Row Topsfield MA 01983

CLENCH, MARY HEIMERDINGER, b Louisville, Ky, Jan 18, 32; wid. AVIAN GASTROINTESTINAL MOTILITY, ORNITHOLOGY. *Educ:* Wheaton Col, Mass, BA, 53; Yale Univ, MS, 55 & 59, PhD(biol), 64. *Prof Exp:* Instr biol & conserv, Am Mus Natural Hist, 55-56; instr biol, Wittenberg Col, 56-57; asst zool, Yale Univ, 57-58, instr, 58-59, asst, 60; asst ornith, Peabody Mus, Yale Univ, 59-63; adj cur birds, Fla State Mus, Univ Fla, 80-83; res tech, Vet Admin Med Ctr, Gainesville, Fla, 83-86; res assoc, Univ Va, 86-87; from res asst to asst cur, 63-67, assoc cur ornith, 68-80, RES ASSOC, CARNEGIE MUS NATURAL HIST, 80-; ASST PROF, RES, UNIV TEX MED BR, 87- *Concurrent Pos:* Frank M Chapman Mem Fund grant, Am Mus Natural Hist, 64-65; asst prof, Univ Fla, 67-68; adj assoc prof biol, Univ Pittsburgh, 74-80; dir, Nat Audubon Soc, 78-84; ed, Living Bird, Cornell Univ, 81-82. *Mem:* Fel Am Ornith Union; Wilson Ornith Soc (pres, 87-89); Cooper Ornith Soc. *Res:* Avian physiology (gastrointestinal motility); avian anatomy; pterylography of passerines; passerine systematics and taxonomy; migration patterns; other banding-related studies. *Mailing Add:* Div Gastroenterol Dept Int Med G-64 Univ Tex Med Br Galveston TX 77550

CLENDENIN, JAMES EDWIN, b Paris, Tenn, June 10, 39; m 62; c 1. ACCELERATORS. *Educ:* Univ Va, BEE, 62; Columbia Univ, PhD(physics), 75. *Prof Exp:* Res assoc physics, Yale Univ, 75-79; staff physicist, 79-88, SOURCE SYSTS LEADER, STANFORD LINEAR ACCELERATOR CTR, STANFORD UNIV, 88- *Mem:* Am Phys Soc; Inst Elec & Electronics Engrs. *Res:* Polarized electron sources; high energy physics; accelerator physics. *Mailing Add:* Slac-Bin 66 PO Box 4349 Stanford CA 94309

CLENDENIN, MARTHA ANNE, b Salem, Ohio, Jan 26, 44. NEUROSCIENCES. *Educ:* Med Col Va, Richmond, BS, 65, MS, 70, PhD(anat), 72. *Prof Exp:* NIH fel, Sweden, 72-73; from asst prof to assoc prof anat, Eastern VA Med Sch, 73-78; PROF & CHAIR DEPT PHYS THER, UNIV FLA, GAINESVILLE, 83-, PROF DEPT ANAT, 83- *Honors & Awards:* Dorothy Briggs Mem Award for Res, Am Phys Ther Asn, 72. *Mem:* Soc Neurosci; Am Asn Anatomists; Sigma Xi; Am Phys Ther Asn; AAAS. *Res:* Investigations of normal mechanisms of movement and posture utilizing intra and extracellular recording techniques in animals and electromyography in human subjects. *Mailing Add:* Dept Phys Ther Univ Fla Box J-154 Gainesville FL 32610

CLENDENING, JOHN ALBERT, b Martinsburg, WVa, Mar 6, 32; m 54; c 3. PALYNOLOGY, ORGANIC PETROLOGY. *Educ:* WVa Univ, BS, 58, MS, 60, PhD, 70. *Prof Exp:* Asst coal geologist, WVa Geol & Econ Surv, 60-64, palynologist, 64-66, coal geologist & palynologist, 66-68; geologist, Pan Am Petrol Corp, 68-71; sr geologist, 71-74, staff geologist, 74-75, staff paleontologist, 75-78, geol assoc, 78-80, SR GEOL ASSOC, AMOCO PROD CO, 80- *Concurrent Pos:* Counr, Am Asn Stratig Palynologists, 75-76. *Honors & Awards:* Distinguished Serv Award, Am Asn Stratig Palynologists, 87. *Mem:* Fel Geol Soc Am; Am Asn Stratig Palynologists (vpres, 76-77, secy-treas, 78-82, pres elect, 82-83, pres, 83-84); Soc Org Petrol (pres, 84); Am Asn Petrol Geologists; Asn Petrol Geochem Explorationists. *Res:* Applied non-seismic hydrocarbon exploration techniques. *Mailing Add:* Amoco Prod Co Box 3092 Houston TX 77253

CLENDENNING, LESTER M, b Toronto, Ohio, Feb 4, 33. MATHEMATICAL PHYSICS, RELATIVITY. *Educ:* Milwaukee Sch Engr, BSEE, 54; Univ Pittsburgh, MS, 67, PhD(physics), 71. *Prof Exp:* Elec Engr, RCA, 54-57; from asst to assoc prof, 58-77, PROF PHYSICS, HUMBOLDT STATE UNIV, 77- *Concurrent Pos:* Vis scholar, Relativity Ctr, Univ Tex, 77-78; fac exchange, Calif State Univ, Long Beach, 87-88. *Mem:* Am Math Soc; Planetary Soc; Int Soc Gen Relativity. *Mailing Add:* Dept Physics Humboldt State Univ Arcata CA 95521

CLENDENNING, WILLIAM EDMUND, b Waynesburg, Pa, June 23, 31; m 58; c 4. DERMATOLOGY. *Educ:* Allegheny Col, BS, 52; Jefferson Med Col, MD, 56. *Prof Exp:* Instr dermat, Case Western Reserve Univ, 60-61; sr investr, Dermat Br, Nat Cancer Inst, 61-63; sr instr dermat, Case Western Reserve Univ, 63-66, asst prof, 66-67; clin assoc prof, 67-72, PROF CLIN DERMAT, DARTMOUTH MED SCH, 72- *Concurrent Pos:* Nat Cancer Inst fel, Sch Med, Case Western Reserve Univ, 63-66 & 67-69; mem gen med A study sect, NIH, 65-69; assoc ed, J Invest Dermat, 67-72; mem, Path & Chemother. *Mem:* Am Acad Dermat; Soc Invest Dermat; Am Dermat Asn; Am Soc Dermatopath; Am Fedn Clin Res. *Res:* Mycosis fungoides; contact dermatitis. *Mailing Add:* Dartmouth Med Sch Two Maynard Rd Hanover NH 03755

CLENDENON, NANCY RUTH, b Ahoskie, NC, July 27, 33. NEUROCHEMISTRY. *Educ:* Old Dom Univ, BS, 62; Univ NC, Chapel Hill, PhD(biochem), 68. *Prof Exp:* From instr to assoc prof, 69-88, EMER ASSOC PROF NEUROCHEM, COL MED, OHIO STATE UNIV, 89- *Concurrent Pos:* Res fel neurochem, Ohio State Univ Hosp, 68-69; Ohio State Univ Res Found-Charles R Kistler Mem Found, 69-82. *Mem:* AAAS; Int Soc Neurochem; Am Chem Soc; Soc Neurosci; Am Soc Neurochem; Sigma Xi. *Res:* Cerebral hydrolytic enzyme localization in subcellular fractions of normal and pathologic tissues; combined modalities in the treatment of experimentally induced brain tumors in rats; response of hydrolytic, mitochondrial and plasma membrane enzymes in dog spinal cord and cerebrospinal fluid to experimental injury; fluid and electrolyte alterations in experimental spinal cord trauma. *Mailing Add:* 34 Raymond Dr Arapahoe NC 28510-9747

CLENDINNING, ROBERT ANDREW, b Schenectady, NY, Dec 12, 31; m 76; c 4. ORGANIC CHEMISTRY, POLYMER CHEMISTRY. *Educ:* Union Col, BS, 53; Rensselaer Polytech Inst, PhD(org chem), 59. *Prof Exp:* Asst, Rensselaer Polytech Inst, 53-58; chemist, 58-68, proj scientist, 68-83; res scientist, 86-87, SR RES SCIENTIST, AMOCO PERFORMANCE PROD, INC, 87- *Mem:* Am Chem Soc. *Res:* Synthetic organic chemistry; monomer synthesis and polymerization; mechanism of polymerization reactions; condensation polymerization reactions; polyolefin and olefin copolymers research and development; polymer evaluation. *Mailing Add:* 6310 Stallion Dr Cumming GA 30130

CLERMONT, YVES WILFRED, b Montreal, Can, Aug 14, 26; m 50; c 3. MICROSCOPIC ANATOMY. *Educ:* Univ Montreal, BSc, 49; McGill Univ, PhD(anat), 53. *Prof Exp:* From lectr to assoc prof, 53-63, chmn dept, 75-85, PROF ANAT, MCGILL UNIV, 63- *Concurrent Pos:* Anna Fuller fel, 54-55; Lalor Found award, 62; consult, WHO, Geneva, 72-78; mem adv comt, Ford Found, 76-79; Anat Test Comt, US Nat Bd Med Examr, 79-83, secy, Lucian Award Comt, 80- *Honors & Awards:* S L Siegler Award, Am Fertility Soc, 66; Hopkins-Maryland Award Reprod Biol, 86; Van Campenhout Award, Can Fertil & Andrology Soc, 86; J C B Grant Award, Can Asn Anatomists, 86; Distinguished Andrologist Award, Am Soc Andrology, 88. *Mem:* Am Asn Anat (2nd vpres, 70-73); Soc Study Reproduction; Can Asn Anat; Microsc Soc Can; fel Royal Soc Can; Am Asn Andrology. *Res:* Histology and histophysiology of mammalian male reproductive system; cytological studies of spermatogenesis with light and electron microscopes; ultrastructure and function of the Golgi apparatus of various cell types. *Mailing Add:* Dept Anat 3640 Univ St Montreal PQ H3A 2B2 Can

CLESCERI, LENORE STANKE, b Chicago, Ill, Aug 9, 35; m 57; c 5. APPLIED & ENVIRONMENTAL BIOLOGY. *Educ:* Loyola Univ, BS, 57; Marquette Univ, MS, 61; Univ Wis, PhD(biochem), 63. *Prof Exp:* USPHS res fel, 63-65; asst prof, 66-74, ASSOC PROF MICROBIAL BIOCHEM, RENSSELAER POLYTECH INST, 74- *Concurrent Pos:* Mem, sci adv bd, Environ Protection Agency, 76-84; consult, Life Systs Inc, 81-, Ayerst Lab, 82, Procter & Gamble, 83-84 & NY State Energy Res Authority, 89-91; mem, Nat Acad Sci Water, Sci & Technol bd, 89- *Mem:* Soc Environ Toxicol & Chem; Am Chem Soc; Soc Int Limnol; Am Soc Microbiol; Water Pollution Control Fedn. *Res:* Kinetics of microbial growth; microbial biotransformations; aquatic toxicity testing. *Mailing Add:* Dept Biol Rensselaer Polytech Inst Troy NY 12180-3590

CLESCERI, NICHOLAS LOUIS, b Chicago, Ill, Sept 13, 36; m 57; c 5. ENVIRONMENTAL ENGINEERING. *Educ:* Marquette Univ, BCE, 58; Univ Wis, MSCE, 61, PhD(sanit eng), 64. *Prof Exp:* Foreman construct, A L Clesceri & Co, Ill, 54-58 & Hahn Plumbing Co, Wis, 58; civil engr, sewer eng div, City of Milwaukee, 58-59; NIH res fel sanit eng, Swiss Fed Inst Technol, 63-65; from asst prof to assoc prof environ eng, 65-75, PROF ENVIRON ENG, RENSSELAER POLYTECH INST, 75- *Concurrent Pos:* NSF res grants, 66-67; res & teaching asst, Univ Wis, 61 & 62; mem, Environ Adv Bd, Army Corp Engrs; chmn, Joint Task Group Phosphorus Anal, 16th Ed Standard Methods for Water Pollution Control Fedn. *Mem:* Water Pollution Control Fedn; Am Chem Soc. *Res:* Chemical limnology; problems of eutrophication by nutrient discharges; inorganic nutrition of algae; chemical methods of nutrient removal from wastewaters; acid precipitation research. *Mailing Add:* Dept Environ Eng-102 RI Rensselaer Polytech Inst Troy NY 12181

CLEVELAND, BRUCE TAYLOR, b Boston, Mass, Aug 6, 37; m 64. NUCLEAR CHEMISTRY. *Educ:* Johns Hopkins Univ, PhD(physics), 70. *Prof Exp:* Lectr physics, State Univ NY Buffalo, 70-72; res assoc, Columbia Univ, 72-75; sr res assoc, Brookhaven Nat Lab, 76-78, assoc scientist, 79-80, scientist, 81-85; STAFF MEM, LOS ALAMOS NAT LAB, 85- *Mem:* Am Phys Soc. *Res:* Experimental nuclear physics; studies of double beta decay and solar neutrinos. *Mailing Add:* Los Alamos Nat Lab P-3 MS D449 Los Alamos NM 87545

CLEVELAND, DON W, b Mo, Aug 26, 50. MICROTUBULES, NEUROFILAMENTS. *Educ:* Princeton Univ, PhD(biochem), 77. *Prof Exp:* Assoc prof, 84-88, PROF BIOL CHEM, JOHNS HOPKINS UNIV, 88- *Concurrent Pos:* Ed, Molecular & Cellular Biol; assoc ed, J Cell Biol. *Mem:* Am Soc Cell Biol; Am Soc Biochem & Molecular Biol. *Mailing Add:* Dept Biol Chem Johns Hopkins Univ 725 N Wolfe St Baltimore MD 21205

CLEVELAND, DONALD EDWARD, b Seattle, Wash, May 25, 28; m 55; c 3. CIVIL ENGINEERING. *Educ:* Mass Inst Technol, SB, 49; Yale Univ, cert, 50, MEng, 59; Tex A&M Univ, PhD(civil eng), 62. *Prof Exp:* Res asst, Bur Hwy Traffic, Yale Univ, 50-51, res assoc, 56-59; consult engr, Ramp Bldgs Corp, 54-56; assoc prof civil eng, Tex A&M Univ, 59-64 & Univ Va, 64-65, assoc prof, 65-68, PROF CIVIL ENG, UNIV MICH, 68- *Concurrent Pos:* Mem tech comts, Transp Res Bd, Nat Acad Sci-Nat Res Coun; consult, Detroit & Philadelphia; sr Fulbright res fel, Tech Univ, Aachen, WGer, 71-72; sr Fulbright lectr, Helsinki Tech Univ, Finland, 78-79. *Honors & Awards:* President's Award, Inst Transp Eng, 61. *Mem:* Am Soc Civil Engrs; Opers Res Soc Am; Inst Transp Eng; Sigma Xi. *Res:* Analytic techniques in transportation. *Mailing Add:* PO Box 1938 Ann Arbor MI 48106-1938

CLEVELAND, GREGOR GEORGE, b San Jose, Calif, Sept 21, 48; m 69; c 1. BIOPHYSICS, MAGNETIC RESONANCE. *Educ:* La Tech Univ, BS, 70; Rice Univ, MA, 73, PhD(physics), 75. *Prof Exp:* Instr physics, Univ Houston, Downtown, 75; asst prof, 75-80, ASSOC PROF PHYSICS, UNIV NC, GREENSBORO, 80- *Mem:* Am Phys Soc; Biophys Soc. *Res:* Investigation of the physicochemical state of water and ions in various biological or model systems utilizing pulsed nuclear magnetic resonance. *Mailing Add:* Dept Physics 101 Petty Sci Bldg Univ NC 1000 Spring St Greensboro NC 27412

CLEVELAND, JAMES PERRY, b Charlotte, NC, Feb 20, 42; m 65. ORGANIC CHEMISTRY. *Educ:* Ga Inst Technol, BS, 63, PhD(org chem), 67. *Prof Exp:* NIH fel org chem, Ore State Univ, 69-70; mem staff, Res Labs, 70-77, RES ASSOC, TENN EASTMAN CO, 77- *Mem:* Am Chem Soc; Chem Soc. *Res:* Mechanisms of reactions of organo-phosphorus and organo-sulfur compounds. *Mailing Add:* 4427 Fieldstone Dr Kingsport TN 37664-5031

CLEVELAND, JOHN H, b Bloomington, Ind, Nov 12, 32; m 53; c 3. ECONOMIC GEOLOGY. *Educ:* Univ Ind, BA, 54, PhD(geol), 63; Univ Wis, MS, 61. *Prof Exp:* From asst prof to assoc prof, 62-70, PROF GEOL, IND STATE UNIV, 70- *Mem:* Nat Asn Geol Teachers; Mineral Soc Am; Geol Soc Am; Sigma Xi. *Res:* Geology and geochemistry of magnesite deposits; lead-zinc deposits of the upper Mississippi Valley; economic geology of Indiana; computer utilization. *Mailing Add:* Dept Geol Ind State Univ Terre Haute IN 47809

CLEVELAND, LAURENCE F(ULLER), electrical engineering; deceased, see previous edition for last biography

CLEVELAND, MERRILL L, b Orleans, Ind, Feb 29, 28; m 50; c 1. ENTOMOLOGY. *Educ:* Ind State Teachers Col, BS, 50; Purdue Univ, MS, 54, PhD, 64. *Prof Exp:* Pub sch teacher, 50-52; entomologist, Fruit Insects Sect, Entom Res Div, USDA, 55-68, asst chief fruit insects res br, 68-72, Nat Res Prog Leader, Nat Prog Staff, Fruit & Veg Insects, 72-82, agr res adv, off adminr, Agr Res Serv, 82-86; RETIRED. *Mem:* AAAS; Am Registry Prof Entomologists; Entom Soc Am. *Mailing Add:* 1220 Burton St Silver Spring MD 20910

CLEVELAND, RICHARD WARREN, b Santa Ana, Calif, Dec 4, 24; m 52; c 2. PLANT BREEDING. *Educ:* Univ Calif, BS, 49, PhD(genetics), 53. *Prof Exp:* Asst agron, Univ Calif, 49-53; from asst prof to prof, 53-88, EMER PROF PLANT BREEDING, PA STATE UNIV, STATE COLLEGE, 88- *Mailing Add:* 361 Laurel Lane State College PA 16803

CLEVELAND, WILLIAM GROVER, JR, b Pittsburg, Calif, Aug 27, 51; m 79; c 2. FLUID MECHANICS, INSTRUMENTATION. *Educ:* Gettysburg Col, BA, 73; Univ Md, MS, 79. *Prof Exp:* Physicist res, Nat Bur Standards, 72-79, MECH ENGR RES, NAT INST STANDARDS & TECHNOL, 79- *Mem:* Am Phys Soc. *Res:* Experimental and theoretical fluid mechanics and turbulence research on flow measurement processes; develop automated data acquisition and analysis systems. *Mailing Add:* Nat Inst Standards & Technol Bldg 230 Rm 105 Gaithersburg MD 20899

CLEVELAND, WILLIAM SWAIN, b Sussex, NJ, Jan 24, 43; m 82; c 4. STATISTICS. *Educ:* Princeton Univ, AB, 65; Yale Univ, MS, 67, PhD(statist), 69. *Prof Exp:* Asst prof statist, Univ NC, Chapel Hill, 69-72; MEM TECH STAFF, BELL LABS, 72- *Mem:* Am Statist Asn; Inst Math Statist; Asn Comput Mach; Inst Elec & Electronics Engrs. *Res:* Data analysis; psychophysics; graphical methods in statistics; time series analysis; software reliability; non parametric regression. *Mailing Add:* AT&T Bell Labs 600 Mountain Ave Murray Hill NJ 07924

CLEVENGER, RICHARD LEE, b Columbus, Ind, May 16, 31; m 79; c 1. ORGANIC CHEMISTRY, BIOCHEMISTRY. *Educ:* Univ Ind, BS, 59; Univ Louisville, PhD(org chem), 63. *Prof Exp:* Instr chem, Univ Louisville, 62-63; from asst prof to assoc prof chem, Easts Tex State Univ, 63-81, dir forensic chem, 73-81; consult, Dalchem, 80-82, OWNER & CHIEF EXEC OFFICER, RMEC LABS, 82- *Concurrent Pos:* Head chem, Independence Community Col. *Mem:* Am Chem Soc. *Res:* Thiadiazoles; hydrazones; geometrical isomerism; cyclopentane derivatives; azasteroids; neuro-biochemistry; B vitamins; drug metabolites; forensic chemistry. *Mailing Add:* Independence Community Col Independence KS 67301

CLEVENGER, SARAH, b Indianapolis, Ind, Dec 19, 26. PLANT BIOCHEMISTRY. *Educ:* Miami Univ, AB, 47; Ind Univ, PhD(plant physiol), 57. *Prof Exp:* Asst prof biol, Berea Col, 57-59 & 61-63, Wittenburg Col, 59-60, Eastern Ill Univ, 60-61; from asst prof to prof, 78-85, EMER PROF LIFE SCI, IND STATE UNIV, TERRE HAUTE, 85- *Concurrent Pos:* Contributed, Book Forum, 74-79. *Mem:* Am Soc Plant Taxonomists; Int Asn Plant Taxon; Phytochem Soc NAm (secy, 67-68); Am Inst Biol Sci. *Res:* Flower pigments; gene control of flower pigments and simulation of population biology. *Mailing Add:* 717 S Henderson St Bloomington IN 47401

CLEVENGER, WILLIAM A, ENGINEERING. *Prof Exp:* RETIRED. *Mem:* Nat Acad Eng. *Mailing Add:* 156 Meadow Lark Lane Sequim WA 98382

CLEVER, HENRY LAWRENCE, b Mansfield, Ohio, June 14, 23; m 56; c 1. PHYSICAL CHEMISTRY. *Educ:* Ohio State Univ, BSc, 45, MS, 49, PhD(chem), 51. *Prof Exp:* Jr chemist, Shell Develop Co, 45-47; asst, Res Found, Ohio State Univ, 47-50; instr & res assoc, Duke Univ, 51-54; from instr to assoc prof, 54-65, PROF CHEM, EMORY UNIV, 65-, DIR SOLUBILITY DATA, 81- *Concurrent Pos:* Partic, Oak Ridge Nat Lab, 57; res assoc, Univ Mich, 63-64; res assoc, Polymer Res Inst, Univ Mass, Amherst, 72-73; vis prof, Univ Melbourne, 88. *Mem:* AAAS; Am Chem Soc. *Res:* Thermodynamics; solubility; surface tension; heat capacity; thermal properties; light scattering; compilation and evaluation of solubility and other data. *Mailing Add:* Dept Chem Emory Univ Atlanta GA 30322-2210

CLEVER, LINDA HAWES, b Seattle, Wash; m; c 1. OCCUPATIONAL MEDICINE. *Educ:* Stanford Univ, AB, 62, MD, 65; Am Bd Internal Med, dipl, 71; Am Bd Prev Med, cert occup med, 85. *Prof Exp:* Intern med, Palo Alto Stanford Hosp, 65-66, resident, 66-67 & 69-70; NIH res fel infectious dis, Stanford Univ Sch Med, 67-68; postdocral fel community med, Univ Calif Med Ctr, San Francisco, 68-69; med dir, Diag & Treatment Ctr, St Marys Hosp & Med Ctr, 70-76; CHMN, DEPT OCCUP HEALTH, PAC PRESBY MED CTR, 77- *Concurrent Pos:* Inpatient preceptor, St Marys Med Ctr, 71-76; clin prof, dept med & assoc physician, Univ Calif, San Francisco, Med Ctr; mem, comt med ctr occup health serv & occup med pract comt, Am Occup Med Asn; mem, Gov Adv Comt, Am Col Physicians, 77-, Health Care Delivery Subcomt, 82-84, Publ Policy Comt, 90-, chmn, Credentials Comt, 91-; mem, Bd Health Prom Dis Prev, Inst Med, Nat Acad Sci, 81-87; bd dirs, Western Occup Med Asn, 79-80; adv comt, Univ Calif Occup Health Ctrs, 79-, chmn, 82-, adv bd, Peoples Health Resources, 74-76; mem, consult comt accreditation continuing med educ, Calif Med Asn, 72-74, comt patient care serv, Calif Hosp Asn, 72-77; mem, community health comt, San Francisco Comprehensive Health Planning Coun, 71-73, actg chmn, 72-73, bd dir; Haight Ashbury Health Comt, 71-76. *Honors & Awards:* Lee Ann McCaffrey Mem Lectr, Mayo Clin, 90. *Mem:* Inst Med-Nat Acad Sci; fel Am Col Physicians; Am Col Occup Med; Am Indust Hyg Asn; Am Pub Health Asn. *Res:* Student research grant; stimulation of interferon in humans using inactivated virus; nurse practitioner care of patients with chronic, stable disease. *Mailing Add:* Dept Occup Health Pac Presby Med Ctr PO Box 7999 San Francisco CA 94115

CLEWE, THOMAS HAILEY, b San Francisco, Calif, Sept 9, 25; div; c 3. CONTRACEPTION, MEDICAL RESEARCH. *Educ:* Stanford Univ, BS, 49, MD, 55. *Prof Exp:* Res assoc reprod, Stanford Univ, 55-56; instr anat, Sch Med, Yale Univ, 57-58; asst res prof obstet & gynec & anat, Sch Med, Univ Kans, 58-61; asst res prof obstet & gynec, Sch Med, Vanderbilt Univ, 61-66; res assoc, Div Reprod Physiol, Delta Regional Primate Res Ctr, 66-73; asst clin res dir, Squibb Inst Med Res, 73-75, assoc clin res dir, 75-76; assoc med dir, Syntex Res, 76-84; RETIRED. *Concurrent Pos:* Pop Coun med res fel, Sch Med, Yale Univ, 56-57; assoc prof anat, Tulane Univ, 66-73, clin assoc prof obstet & gynec, Med Sch, 68-73. *Honors & Awards:* Rubin Award, Am Fertil Soc, 59. *Mem:* Am Asn Anatomists; Soc Gynec Invest. *Res:* Structure and function of mammalian oviduct; physiology of early stages of mammalian reproduction; fertility and sterility; comparative reproduction, especially of primates; fertility control devices; physiology of skin; design of clinical investigations. *Mailing Add:* 1450 Oak Creek Dr Apt 308 Palo Alto CA 94304-2065

CLEWELL, DAYTON HARRIS, physics, engineering; deceased, see previous edition for last biography

CLEWELL, DON BERT, b Dallas, Tex, Sept 5, 41; m 68; c 2. BIOCHEMISTRY, MICROBIOLOGY. *Educ:* Johns Hopkins Univ, AB, 63; Ind Univ, Indianapolis, PhD(biochem), 67. *Prof Exp:* Biologist, Univ Calif, San Diego, 69-70; from asst prof to assoc prof, 70-77, res scientist, Dent Res Inst, 80-89, PROF BIOL & MAT SCI, SCH DENT & MICROBIOL, SCH MED, UNIV MICH, ANN ARBOR, 77- *Concurrent Pos:* Nat Cancer Inst fel molecular genetics, Univ Calif, San Diego, 67-69; USPHS res career develop award, 75-80. *Mem:* AAAS; Am Chem Soc; Am Asn Biochem & Molecular Biol; Am Asn Oral Biol; Int Asn Dent Res; Am Soc Microbiol. *Res:* Molecular biology of bacterial plasmids and transposons; molecular genetics in enterococci and streptococci that encode antibiotic resistance and virulence determinants; nucleic acid chemistry. *Mailing Add:* Biol & Mat Sci Univ Mich Sch Dent Ann Arbor MI 48109

CLIATH, MARK MARSHALL, b Palo Alto, Calif, July 6, 35; m 64; c 1. PESTICIDE RESIDUE CHEMISTRY. *Educ:* Univ Calif, Riverside, AB, 63, PhD(soil sci), 78. *Prof Exp:* CHEMIST SOIL-PESTICIDE INTERACTIONS, SOIL & WATER QUAL PROTECTION RES, USDA, RIVERSIDE, 66- *Concurrent Pos:* Consult qual assurance, Southern Calif Edison, 85- *Mem:* Am Chem Soc; Soil Sci Soc Am; Sigma Xi. *Res:* Soil-pesticide interactions; volatilization of pesticides from agricultural soil and water surfaces; vapor behavior of pesticides in soil, water and air; quality assurance-toxic chemical analysis. *Mailing Add:* USDA Soil-Water Dept Soil & Environ Sci Univ Calif Riverside CA 92521

CLIBURN, JOSEPH WILLIAM, b Hazlehurst, Miss, Jan 20, 26; m 50; c 2. ZOOLOGY, BOTANY. *Educ:* Millsaps Col, BS, 47; Univ Southern Miss, MA, 53; Univ Ala, PhD(zool), 60. *Prof Exp:* Instr, pub schs, Miss, 47-53; instr biol, Copiah-Lincoln Jr Col, 53-55; instr, pub schs, Miss, 55-58; assoc prof, 60-64, PROF BIOL, UNIV SOUTHERN MISS, 64- *Concurrent Pos:* Prof zool, Gulf Coast Res Lab, 70- *Mem:* Am Soc Ichthyologists & Herpetologists; Soc Study Amphibians & Reptiles. *Res:* Taxonomy and zoogeography of southeastern amphibians, reptiles and fishes. *Mailing Add:* Dept Biol Univ Southern Miss Box 5018 Hattiesburg MS 39406

CLICK, ROBERT EDWARD, b Wenatchee, Wash, Mar 22, 37; m 56; c 2. IMMUNOBIOLOGY. *Educ:* Wash State Univ, BS, 60; Univ Calif, Berkeley, PhD(biochem), 64. *Prof Exp:* NIH fels, Columbia Univ, 64-65 & Sloan-Kettering Cancer Inst, 65-66; NIH fel, Univ Wis, 66-68, asst prof immunol, 68-72; scientist, Wis Alumni Res Inst, Madison, 72-73; assoc, Sloan-Kettering Cancer Inst, 73-74; ASST PROF IMMUNOL, DEPT MICROBIOL, UNIV MINN, MINNEAPOLIS, 74- *Mem:* AAAS; Soc Develop Biol; Soc Plant Physiol; Am Asn Immunol. *Res:* Genetic control of immune responses, primarily in vitro. *Mailing Add:* Altick Assoc Rte 3 River Falls UT 54022

CLIETT, CHARLES BUREN, b Montpelier, Miss, July 10, 24; m 46; c 2. AEROSPACE ENGINEERING. *Educ:* Ga Inst Technol, BS, 45, MS, 50. *Prof Exp:* Instr aeronaut eng, Miss State Univ, 47-49; asst, Ga Inst Technol, 49-50; from asst prof to assoc prof, 50-60, PROF AEROSPACE ENG, MISS STATE UNIV, 60-, HEAD DEPT, 60- *Mem:* Am Soc Eng Educ; Am Inst Aeronaut & Astronaut; Nat Soc Prof Engrs; Aerospace Dept Chmn Asn. *Res:* Engineering mechanics; stress analysis and structural dynamics. *Mailing Add:* Dept Aerospace Eng Drawer A Mississippi State MS 39762

CLIETT, OTIS JAY, b Athens, Ga, Nov 8, 44. MATHEMATICS. *Educ:* Univ Ga, BS, 66, MA, 68, PhD(math educ), 70. *Prof Exp:* From asst prof to assoc prof, 70-82, PROF MATH, GA SOUTHWESTERN COL, 82- *Mem:* Nat Coun Teachers Math; Math Asn Am. *Mailing Add:* Dept Math Ga Southwestern Col Americus GA 31709

CLIFF, EUGENE M, b Jamaica, NY, Jan 24, 44. AERONAUTICAL & ASTRONAUTICAL ENGINEERING, OPTIONAL TRAJECTORY SHAPING. *Educ:* Clarkson Col, BS, 65; Univ Ariz, MS, 68, PhD(aerospace eng), 70. *Prof Exp:* Asst prof aerospace eng, Univ Ariz, 69-71; from asst prof to assoc prof aerospace & ocean eng, 75-80, PROF AEROSPACE & OCEAN ENG, VA POLYTECH STATE UNIV, 80-, REYNOLDS METALS PROF, 90- *Concurrent Pos:* Vis prof mech, US Mil Acad, 80-81. *Mem:* Am Inst Aeronaut & Astronaut; Inst Elec & Electronics Engrs; Math Asn Am; Soc Indust & Appl Math. *Res:* Distributed perimeter control. *Mailing Add:* Va Polytech Inst & State Univ 215 Randolph Hall Blacksburg VA 24061

CLIFF, FRANK SAMUEL, b Carson City, Nev, Apr 3, 28; m 55; c 3. VERTEBRATE ZOOLOGY. *Educ:* Stanford Univ, AB, 51, PhD, 54. *Prof Exp:* Herpetologist, Sefton-Stanford exped, Gulf of Calif, 52-53; asst comp anat & gen biol, Natural Hist Mus, Stanford Univ, 52, 53-56; from instr to asst prof, Colgate Univ, 56-59; from asst prof to assoc prof, 59-69, PROF COMP ANAT & GEN BIOL, CALIF STATE UNIV, CHICO, 69- *Mem:* Am Soc Ichthyologists & Herpetologists. *Res:* Reptiles of islands adjacent to Baja California and Mexico; reptiles of western North America and Mexico; insular evolution; osteology of reptiles. *Mailing Add:* Dept Biol Calif State Univ Chico CA 95929

CLIFFORD, ALAN FRANK, b Natick, Mass, June 8, 19; m 49; c 2. INORGANIC CHEMISTRY, FLUORINE CHEMISTRY. *Educ:* Harvard Univ, AB, 41; Univ Del, MS, 47, PhD(inorg chem), 49. *Prof Exp:* Anal chemist & lab supvr, Kankakee Ord Works, Ill, 41-43; asst, Manhattan Dist Proj, Radiation Lab, Univ Chicago, 43; anal res chemist, Clinton Labs, Tenn, 43-44; res chemist, Hanford Eng Works, Wash, 44-45; develop chemist, Exp Sta, E I du Pont de Nemours & Co, Del, 45-47; instr chem, Univ Del, 47-49; asst prof inorg chem, Ill Inst Technol, 49-51; from asst prof to assoc prof, Purdue Univ, 53-66; PROF INORG CHEM & HEAD DEPT CHEM, VA POLYTECH INST & STATE UNIV, 66- *Concurrent Pos:* Guggenheim fel, Cambridge Univ, 51-53; mem subcomt solubility data, Comt Equilibrium Data, Int Union Pure & Appl Chem, 73- *Mem:* AAAS; Am Chem Soc; fel NY Acad Sci; Chem Soc; Sigma Xi. *Res:* Rare earth, inorganic fluoride chemistry; hydrogen fluoride system; acid theory; inorganic polymers; oxidations in liquid ammonia; multiple bonding in inorganic compounds; hypofluorites; Mossbauer spectrometry of rare earths and biological materials. *Mailing Add:* Dept Chem Va Polytech Inst & State Univ Blacksburg VA 24061-0212

CLIFFORD, DONALD H, b Burlington, Vt, June 7, 25; m 54; c 4. CANINE BEHAVIOR, EUTHANASIA. *Educ:* Univ Montreal, DVM, 50; Univ Minn, MPH, 55, PhD(vet med), 59; Am Col Lab Animal Med, dipl, 60; Am Col Vet Surgeons, dipl. *Prof Exp:* Intern vet med, Angell Mem Animal Hosp, Boston, Mass, 50-51, resident, 51-52; from instr to assoc prof vet surg, Col Vet Med, Univ Minn, 52-64; assoc prof, div exp biol, Baylor Col Med, Houston, 64-65; assoc prof, 65-71, DIR, DIV LAB ANIMAL MED, MED COL OHIO, TOLEDO, OHIO, 65-, PROF, 77- *Concurrent Pos:* Clin practr, Angell Mem Animal Hosp, Boston, Mass & Rowley Mem Animal Hosp, Springfield, Mass, 50-52; head small animal surg, dept vet surg & radiol, Univ Minn, 52-65; dir, Animal Res Fac, VA Hosp, Houston, Tex, 65-71, Animal Med Lab, Med Col Ohio, Toledo, 71-; consult, Univ Toledo, 76-, NAm Sci Asn, Northwood, Ohio, 77-, Bowling Green State Univ, Ohio, 78-, Cleveland Clinc, 79-, St Vincent Hosp & Med Ctr, Toledo, Ohio, 80-, Am Asn Accreditation Lab Animal Care. *Mem:* Fel Am Col Vet Surgeons; Am Vet Med Asn; Am Asn Lab Animal Sci; Am Asn Lab Animal Practitioners. *Res:* Veterinary surgery, especially comparative restraint and anesthesiology in laboratory, zoological and domestic animals; pathology and surgery of the canine mouth and esophagus; pathogenesis and treatment of achalasia of the esophagus in dogs and cats; effect of acupuncture on the cardiovascular system of dogs; euthanasia of laboratory animals and pit bulls. *Mailing Add:* Div Lab Animal Med Med Col Ohio CS 10008 Toledo OH 43699

CLIFFORD, GEORGE O, b Akron, Ohio, Apr 30, 24; m 48; c 3. INTERNAL MEDICINE, HEMATOLOGY. *Educ:* Tufts Univ, MD, 49; Am Bd Internal Med, dipl, 57. *Prof Exp:* Intern Henry Ford Hosp, Detroit, 49-50; resident med, Detroit Receiving Hosp, 54-55; from instr to assoc prof, Col Med, Wayne State Univ, 55-63; assoc prof, Med Col, Cornell Univ, 63-72; PROF MED & CHMN DEPT, SCH MED, CREIGHTON UNIV, 72- *Concurrent Pos:* Res fel, Detroit Receiving Hosp, 50-52; res fel, Med Sch, Univ NC, 54; Markle scholar, 59; dir blood bank med lab serv, Mem Hosp, Sloan-Kettering Inst, 63-72, dir hemat, 63-70, assoc chmn dept med, 70-72. *Mem:* AMA; Am Fedn Clin Res; Am Soc Hemat; NY Acad Sci; fel Am Col Physicians. *Res:* Clinical and research hematology; hemaglobiopathies; megaloblastic anemias; leukemias. *Mailing Add:* 601 N 30th St Omaha NE 68131

CLIFFORD, HOWARD JAMES, b Binghamton, NY, May 29, 39; m 62; c 3. PHYSICAL CHEMISTRY, MOLECULAR SPECTROSCOPY. *Educ:* Univ NMex, BS, 63; Wash State Univ, PhD(chem), 71. *Prof Exp:* Lab technician, Vet Admin Hosp, Albuquerque, 63-64; instr, Regina Sch Nursing, Univ Albuquerque, 64-66; teaching asst chem, Univ NMex, 66-67; res asst, Wash State Univ, 67-70; assoc prof, 70-76, ADMIN VPRES, UNIV PUGET SOUND, 76-, MEM STAFF, PHYSICS DEPT. *Concurrent Pos:* Lab technician, Med Sch, Univ NMex, 65. *Mem:* AAAS; Sigma Xi; Am Chem Soc. *Res:* Synthesis, purification and spectroscopic investigation of six-coordinated metal complexes containing ions from the second and third transition series group VIII elements; examination of effects due to differences in molecular symmetries using absorption and emission spectroscopy. *Mailing Add:* Dept Physics Univ Puget Sound Tacoma WA 98416

CLIFFORD, HUGH FLEMING, b Warren, Pa, Dec 9, 31; m 61; c 2. LIMNOLOGY, INVERTEBRATE ZOOLOGY. *Educ:* Mich State Univ, BS, 58, MS, 59; Ind Univ, PhD(limnol), 65. *Prof Exp:* Fishery biologist, Mo Conserv Comn, 60-62; from asst prof to assoc prof, 65-78, PROF ZOOL, UNIV ALTA, 78- *Mem:* NAm Benthological Soc; Can Soc Zoologists. *Res:* Stream ecology; ecology of mayflies. *Mailing Add:* Dept ZooL Univ Alta Edmonton AB T6G 2E9 Can

CLIFFORD, PAUL CLEMENT, statistics, quality control, for more information see previous edition

CLIFFORD, STEVEN FRANCIS, b Boston, Mass, Jan 4, 43. WAVE PROPAGATION, GEOPHYSICS. *Educ:* Northeastern Univ, BSEE, 65; Dartmouth, PhD(eng sci), 69. *Prof Exp:* Physicist environ res, 69-70, supvr physicist, wave propagation lab, 70-82, chief, Propation Studies Group, 82-86, DIR, WAVE PROPAGATION LAB, NAT OCEANIC & ATMOSPHERIC ADMIN, BOULDER, 86- *Concurrent Pos:* Assoc ed, J Optical Soc Am, 80-83; topical adv atmospheric optics, Optical Soc Am, 82-85. *Mem:* Fel Optical Soc Am; Int Radio Sci Union; fel Acoust Soc; Am Meteorol Soc; Am Geophys Union; sr mem Inst Elec & Electronics Engrs. *Res:* Engaged in theoretical research in the field of optical acoustical & microwave propagation through turbulent geophysical flows. *Mailing Add:* Environ Res Lab R/E/WP Nat Oceanic & Atmospheric Admin 325 Broadway Boulder CO 80303

CLIFFTON, MICHAEL DUANE, b Hastings, Nebr, Feb 28, 52. SYNTHETIC ORGANIC CHEMISTRY. *Educ:* Univ Nebr, Omaha, 74, Lincoln, PhD(chem), 80. *Prof Exp:* Fel, Ga Inst Technol, 80-81; res chemist, Phillips Petrol Co, 81-84, Union Carbide Corp, 84-86 & Dartco Mfg Inc, 86-88; RES CHEMIST, TENN EASTMAN, 88- *Mem:* Am Chem Soc. *Res:* Engineering plastics; monomer synthesis; polyester film. *Mailing Add:* 1080 Hidden Valley Rd Kingsport TN 37663

CLIFT, WILLIAM ORRIN, b Flint, Mich, Mar 27, 14; m 51; c 2. GEOLOGY, STRATIGRAPHY-SEDIMENTATION. *Educ:* Univ Mich, BS, 38; Columbia Univ, PhD, 56. *Prof Exp:* Stratigrapher, Sinclair Refining Co, Venezuela, 46-49, stratigr-paleontologist, Sinclair Petrol Co, 49, chief geologist, Ethiopia, 49-52, gen supt, 53-56, mgr, Sinclair Somal Corp, Somalia, 56-59, pres, Sinclair & BP Explor Co, NY, 59-62, vpres, Sinclair Int Oil Co, 62-69; vpres, Podesta, Meyers, Rominger & Clift, Inc, 69-74; dir explor sci, Forest Oil Corp, Denver, 74-86; RETIRED. *Mem:* Fel Geol Soc Am; Asn Petrol Geologists; Paleont Soc. *Res:* Eocene stratigraphy and paleontology. *Mailing Add:* 25 Wedge Way Littleton CO 80123

CLIFTON, BRIAN JOHN, b Preston, Lancashire, Eng, Oct 14, 37; US citizen; m 64; c 2. MATERIALS SCIENCE ENGINEERING, ASTRONOMY. *Educ:* Univ Col N Wales, Banger, BSc, 61, PhD(elec eng), 64. *Prof Exp:* Vis asst prof elec eng, Cambridge, Mass, 64-66, STAFF MEM, LINCOLN LAB, MASS INST TECHNOL, 66- *Concurrent Pos:* Fac mem, Grad Sch Eng, Northeastern Univ, Boston, Mass, 69-73; vis fel, Dept Physics, Queen Mary Col, London, UK, 81-83 & Sch Physics, Univ Bath, UK, 84-88; sr US scientist, Alexander von Humboldt, 86; res fel, Max Planck Soc, 86. *Mem:* Sr mem Inst Elec & Electronics Engrs; AAAS. *Res:* Gallium arsenide monolithic integrated circuits; high-speed electronics; low-noise receiver technology for radio astronomy; plasma diagnostics; metrology. *Mailing Add:* 45 Millpond Rd Sudbury MA 01776-2560

CLIFTON, DAVID GEYER, b Pomeroy, Ohio, Mar 20, 24; m 56. PHYSICAL CHEMISTRY. *Educ:* Miami Univ, Ohio, BA, 48, MA, 50; Ohio State Univ, PhD, 55. *Prof Exp:* Res chemist, Film Dept, E I du Pont de Nemours & Co, Inc, 55-56; staff mem, Los Alamos Sci Lab, Univ Calif, 57-64; sr res chemist, Gen Motors Defense Res Labs, 64-68; STAFF MEM, LOS ALAMOS SCI LABS, UNIV CALIF, 68- *Mem:* Am Chem Soc; Am Inst Physics; AAAS; Am Nuclear Soc. *Res:* Thermodynamics; rocket propellant systems; advanced breeder reactor fuels; aerophysics; plutonium chemistry. *Mailing Add:* 352 Cheryl Ave Los Alamos NM 87544-3610

CLIFTON, DAVID S, JR, b Raleigh, NC, Nov 15, 43; m 83; c 4. TECHNOLOGY. *Educ:* Georgia Inst Technology, BIE, 66; Georgia State Univ, MBA, 70, PhD(economics), 80. *Prof Exp:* Dir, Economic Develop Lab, 79-90, DIR, ECON DEVELOP & TECHNOL TRANSFER, GA TECH RES INST, 90- *Concurrent Pos:* Consult, United Nations Industrial Develop Orgn, 82; vis assoc prof, Ga Inst Technol, Sch Indust & Systs Eng, 81. *Mem:* Sigma Xi; Am Econ Asn; Technol Transfer Soc. *Res:* Technology policies and initiatives to stimulate technology- based economic development; cost benefit analysis and methodology and project feasibility analysis, that is, the analysis of new ventures. *Mailing Add:* Ga Tech Res Inst Off Dir Ga Inst Technol Atlanta GA 30332

CLIFTON, DONALD F(REDERIC), b Pella, Iowa, Apr 28, 17; m 52; c 2. METALLURGY. *Educ:* Mich Col Mining & Technol, BS, 40; Univ Utah, PhD(metall), 57. *Prof Exp:* Metallurgist, Globe Steel Tubes Co, 40-45 & Inst Study Metals, Chicago, 46-50; asst prof metall, Univ Ky, 54-57; from asst prof to prof, 57-82, EMER PROF METALL, 82- *Mem:* Am Phys Soc; Am Crystallog Asn; Am Soc Metals. *Mailing Add:* 2166 Randall Flat Rd Moscow ID 83843

CLIFTON, HUGH EDWARD, b Prospect, Ohio, July 29, 34; m 57; c 3. GEOLOGY. *Educ:* Ohio State Univ, BSc, 56; Johns Hopkins Univ, PhD(geol), 63. *Prof Exp:* GEOLOGIST, US GEOL SURV, 63- *Mem:* AAAS; fel Geol Soc Am; Soc Econ Paleont & Mineral; Int Soc Sedimentol. *Res:* Sedimentary petrography; sedimentology. *Mailing Add:* US Geol Surv, Mail Stop 99 345 Middlefield Rd Menlo Park CA 94025

CLIFTON, JAMES ALBERT, b Fayetteville, NC, Sept 18, 23; m 49; c 2. INTERNAL MEDICINE. *Educ:* Vanderbilt Univ, BA, 44, MD, 47; Am Bd Internal Med, dipl, 55; Am Bd Gastroenterol, dipl, 62. *Prof Exp:* Intern, Univ Hosps, Univ Iowa, 47-48, resident dept med, 48-51; mem staff, Vet Admin Hosp, Tenn, 52-53; assoc internal med, 53-54, from asst prof to assoc prof, 54-63, prof med, 63-72, vis prof dept physiol, 64, vchmn dept med, 67-70 & head dept, 70-75, actg head dept, 75-76, ROY J CARVER PROF MED, UNIV IOWA, 73- *Concurrent Pos:* Fel med, Mass Mem Hosp, 55-56; NIH spec res fel, 55-56; attend physician, Vet Admin Hosp, Iowa, 53-; consult, Surgeon Gen, USPHS, 64-; mem, Nat Adv Arthritis & Metab Dis, 70-73; mem exec comt, Am Bd Internal Med, 78-81, chmn, 80-81; sci adv comt, Ludwig Inst Cancer Res, 85-; vis prof gastroenterol, St Mark's Hosp, Univ London, 84-85. *Honors & Awards:* Stengel Award, Am Col Physicians. *Mem:* Inst Med-Nat Acad Sci; Asn Am Physicians; Am Col Physicians (pres, 77-78); Am Gastroenterol Asn (pres, 70-71); AAAS. *Res:* Patho-physiology of the gastrointestinal system; mechanisms of intestinal absorption and colon cancer. *Mailing Add:* Dept Med John Collonton Pavillion Fourth Fl Iowa City IA 52242

CLIFTON, KELLY HARDENBROOK, b Spokane, Wash, July 22, 27; m 49; c 3. CANCER BIOLOGY, ENDOCRINOLOGY. *Educ:* Univ Mont, BA, 50; Univ Wis, MS, 51, PhD(zool), 55. *Prof Exp:* Am Cancer Soc res fel, Children's Cancer Res Found, 55-56, res assoc exp path, 56-59; res fel, Dept Path, Harvard Univ, 57-59; from asst prof to assoc prof, Dept Radiol, Univ Wis-Madison, 59-67, prof radiol & path, 67-75, asst dean pre-med affairs, 72-77, actg chmn path, 77-78; chief res & dir, Radiation Effects Res Found, Hiroshima, Japan, 80-82; PROF HUMAN ONCOL & RADIOL, MED SCH, UNIV WIS-MADISON, 75- *Concurrent Pos:* Nat lectr, Sigma Xi, 90-92. *Mem:* Am Asn Cancer Res; Soc Exp Biol & Med; Am Soc Exp Path; Radiation Res Soc. *Res:* Radiation and endocrine oncogenesis, particularly of breast and thyroid; radiobiology; physiologic feedback mechanisms in general. *Mailing Add:* Dept Human Oncol Med Sch, Univ Wisc K4/330 Clinical Sci Ctr, 600 Highland Ave Madison WI 53792

CLIFTON, RODNEY JAMES, b Orchard, Nebr, July 10, 37; m 58; c 4. SOLID MECHANICS. *Educ:* Univ Nebr, BS, 59; Carnegie Inst Technol, MS, 61, PhD(civil eng), 64. *Prof Exp:* Fel, 64-65, from asst prof to assoc prof, 65-71, PROF ENG, BROWN UNIV, 71-; PROF, RUSH C HAWKINS UNIV, 88- *Concurrent Pos:* NSF sci fac fel, Univ Southampton, Eng, 71-72; consult, Terra Tek Inc, Salt Lake City, 73-; chmn exec comt, Div Eng, Brown Univ, 74-79; consult, Sandia Nat Labs, Albuquerque, 79-; vis prof mat sci & eng, Stanford Univ, 79-80. *Honors & Awards:* Melville Medal, Am Soc Mech Engrs. *Mem:* Am Soc Civil Engrs; fel Am Acad Mech; AAAS; Soc Eng Sci. *Res:* Plate impact theory and experiments; dynamic plasticity; dislocation and dynamic fracture; mechanics of hydraulic fracturing; rock mechanics; numerical methods. *Mailing Add:* Div Eng Brown Univ Providence RI 02912

CLIKEMAN, FRANKLYN MILES, b Havre, Mont, Mar 6, 33; m 58; c 2. NUCLEAR ENGINEERING, NUCLEAR PHYSICS. *Educ:* Mont State Univ, BS, 55; Iowa State Univ, PhD(nuclear physics), 62. *Prof Exp:* Fel physics, Iowa State Univ, 62-63; from asst prof to assoc prof nuclear eng, Mass Inst Technol, 63-70; PROF NUCLEAR ENG, PURDUE UNIV, 70- *Mem:* Am Nuclear Soc; Am Physics Soc; Am Soc Eng Educ. *Res:* Experimental reactor physics; applied radiation; radiation detection; fast reactor physics. *Mailing Add:* Sch Nuclear Eng Purdue Univ West Lafayette IN 47907

CLIMENHAGA, JOHN LEROY, b Delisle, Sask, Nov 7, 16; m 43; c 2. ASTROPHYSICS. *Educ:* Univ Sask, BA, 45, MA, 49; Univ Mich, MA, 56, PhD, 60. *Prof Exp:* Instr physics, Regina Col, 46-48; from asst prof to prof, 49-82, head dept physics, 56-69, dean Fac Arts & Sci, 69-72, EMER PROF PHYSICS, UNIV VICTORIA, BC, 82- *Concurrent Pos:* Mem, Nat Comt Can, Int Astron Union, 67-71; mem scholar comt, Nat Res Coun, 69-72, mem radio & elec eng div adv bd, 71-74. *Honors & Awards:* Minor planet named in honor, Minor Planet Climenhaga, 86. *Mem:* Am Astron Soc; Can Asn Physicists; Royal Astron Soc Can; Int Astron Union. *Res:* Abundance ratio C-12/C-13 in carbon stars; spectral studies of late type supergiants; cometary spectra. *Mailing Add:* Dept Physics & Astron Univ Victoria Victoria BC V8W 3P6 Can

CLINARD, FRANK WELCH, JR, b Winston-Salem, NC, Aug 4, 33; m 68. MATERIALS SCIENCE. *Educ:* NC State Univ, BME, 55, MS, 57; Stanford Univ, PhD(mat sci), 65. *Prof Exp:* Staff mem metallurgist, Sandia Corp, 57-61; mat scientist & prin investr, Los Alamos Nat Lab, Univ Calif, 64-89, sect leader, radiation effects, 78-89, LAB ASSOC, LOS ALAMOS NAT LAB, LOS ALAMOS, 89- *Concurrent Pos:* Vis prof, NMex Tech, 90- *Mem:*

AAAS; Am Soc Metals; fel Ceramic Soc; Am Nuclear Soc; Mat Res Soc. *Res:* Structural defects in crystalline solids; phase transformations; physical properties of rare earths; behavior of crystalline materials at cryogenic temperatures; radiation damage; fusion reactor materials; high-temperature ceramic superconductors. *Mailing Add:* 2940 Arizona Ave Los Alamos NM 87544

CLINE, CARL F, b Detroit, Mich, Nov 25, 28; m 54; c 4. METALLURGY, CERAMICS. *Educ:* Ga Inst Technol, BS, 53; Niagara Univ, MS, 58. *Prof Exp:* Develop engr, Nat Carbon Co, 53-55; prin ceramist, Battelle Mem Inst, 55-56; sr engr, Res Div, Carborundum Co, 56-58; group leader mat res, Chem Div, Lawrence Radiation Lab, Univ Calif, Livermore, 58-70; dep mgr mat dynamics dept, Physics Int Co, 70-72; mgr strength physics dept, Mat Res Ctr, Allied Chem Co, 72-74; SECT LEADER, METALS & CERAMICS DIV, LAWRENCE LIVERMORE NAT LAB, 74-, C GROUP LEADER, PHYSICS DEPT, 86- *Concurrent Pos:* Panel mem, Solid State Sci Comn, 76-; chmn, Tungsten Comt, Nat Acad Sci, 77-78, dynamic compaction metal & ceramic powders, 81; dep div leader, Mat Sci Div, 81-, mem staff, physics dept, Lawrence Livermore Nat Lab, 84- *Honors & Awards:* IR-100 Award, Indust Res Mag, 76. *Mem:* Am Phys Soc; fel Am Ceramic Soc; Am Soc Metals; Sigma Xi. *Res:* Crystal growth of oxides; advanced materials for ceramic armor and cutting tools; ferroelectrics for high strain applications; management of materials research and development; crystal growth; physical ceramics; hard compounds; properties and processing of amorphous metals; high rate forming; rapid solification technology. *Mailing Add:* Physics Dept Lawrence Livermore Nat Lab Livermore CA 94550

CLINE, DOUGLAS, b York, Eng, Aug 28, 34; m 75; c 2. NUCLEAR PHYSICS. *Educ:* Univ Manchester, BSc, 57, PhD(physics), 63. *Prof Exp:* Res fel physics, Univ Manchester, 60-63; res assoc, 63-65, from asst prof to assoc prof, 65-76, PROF PHYSICS, UNIV ROCHESTER, 76- *Concurrent Pos:* Dir, Nuclear Struct Res Lab, mem prog adv comt, Brookhaven Nat Lab, 74-78, Mich State Univ, 81-85, Argonne Nat Lab, 86-89; chmn exec comt, Holifield Fac, Oak Ridge Nat Lab; heavy-ion linear accelerator, Lawrence Berkeley Lab; mem nuclear sci adv comt, Dept Energy/NSF, 83-86, chmn, Gammasphere Steering Comt, 89-91. *Mem:* Fel Am Phys Soc. *Res:* Heavy ion nuclear physics, including Coulomb excitation, electromagnetic moments and transition strengths; transfer reactions, collective model interpretation of nuclear properties; nuclear structure. *Mailing Add:* Nuclear Struct Res Lab Univ Rochester Rochester NY 14627

CLINE, EDWARD TERRY, b Ischua, NY, Aug 20, 14; m 39, 67; c 3. ORGANIC CHEMISTRY, POLYMER CHEMISTRY. *Educ:* Antioch Col, BS, 36; Ohio State Univ, PhD(chem), 39. *Prof Exp:* Asst, Ohio State Univ, 36-39; res org chemist, E I du Pont de Nemours & Co, Inc, 39-66, res assoc, 66-79; RETIRED. *Mem:* Am Chem Soc; Sigma Xi. *Res:* Textile treatments; new textile fibers; polymers; specialty film; chemical development. *Mailing Add:* 209 Lewis Ave PO Box 402 Penney Farms FL 32079

CLINE, GEORGE BRUCE, b McConnellsburg, Pa, Sept 30, 36; m 61. PHYSIOLOGY, ISOZYMIC REGULATION. *Educ:* Juniata Col, BS, 58; State Univ NY, PhD(physiol), 67. *Prof Exp:* Consult molecular anat sect, Oak Ridge Nat Lab, 64-66, res assoc, 66-67; from asst prof to assoc prof, Univ Ala, Birmingham, 67-75, prof biol, 75-79, chmn dept, 75-80, asst prof physiol & biophys, Sch Med, 67-79. *Concurrent Pos:* Consult, Electro-Nucleonics, Inc, NJ, 67-75 & Argonne Nat Lab, 77-79; vis prof, Univ Brussels, 73-74; adj prof, Col Virgin Islands, 82-83; mem adv bd, Wolfson Bioanalyt Ctr, Univ Surrey, Guildford, Gt Brit, 78- *Mem:* World Aquaculture Soc; Soc Invert Path. *Res:* Stock identification and life history studies of scyllarid lobsters and stone crabs from the Gulf of Mexico and the Caribbean Sea; scyllarid lobster, fresh water prawn and stone crab larvae rearing and development in intensive systems; examination of isozymic marker in individual larvae to determine when adult genes are switched on during development. *Mailing Add:* Dept Biol Univ Col Univ Ala Univ Sta Birmingham AL 35294

CLINE, HARVEY ELLIS, b Cambridge, Mass, Aug 15, 40; m 62; c 1. METALLURGY, SOFTWARE SYSTEMS. *Educ:* Mass Inst Technol, BS, 62, MS, 64, PhD(metall), 65. *Prof Exp:* RES METALLURGIST, GEN ELEC RES & DEVELOP CTR, 65- *Concurrent Pos:* Coolidge fel, Gen Elec Corp. *Mem:* Am Inst Mining Metall & Petrol Engrs; Am Inst Physics; Inst Elec & Electronics Engrs. *Res:* Solidification and properties of eutectics; magnetic properties of type II superconductors; semiconductor processing; developed the thermomogration process; algorithms for constructing three dimensional images from computed temography and magnetic resonance; three-dimensional software system. *Mailing Add:* 845 Harris Dr Niskayuna NY 12309

CLINE, J(ACK) F(RIBLEY), b Cadillac, Mich, June 19, 17; m 41; c 2. ELECTRICAL ENGINEERING. *Educ:* Univ Mich, BS, 38, MS, 41, PhD, 50. *Prof Exp:* From instr to asst prof elec eng, Univ Mich, 42-57; sr res eng, Stanford Res Inst, 57-76, staff engr, 76-84; RETIRED. *Concurrent Pos:* Design specialist, Antenna Lab, Douglas Aircraft Co, 53-54. *Mem:* Inst Elec & Electronics Eng. *Res:* Radio transmitters, receivers, antennas; electronic and microwave circuits and measurements; wave propagation; filters; multicouplers; radar; radio navigation; multilateration tracking; signal population analysis; systems evaluation. *Mailing Add:* 3527 Arbutus Dr Palo Alto CA 94303-4415

CLINE, JACK HENRY, b Columbus, Ohio, Feb 27, 27; m 48; c 8. ANIMAL NUTRITION. *Educ:* Ohio State Univ, BS, 50, MS, 52, PhD(animal sci), 56. *Prof Exp:* Asst animal nutrit, Exp Sta, 51-56, from instr to assoc prof, 56-69, PROF ANIMAL SCI, OHIO STATE UNIV, 69- *Concurrent Pos:* With Ohio Agr Res & Develop Ctr, Wooster; consult feed comp. *Mem:* Am Soc Animal Sci; Sigma Xi. *Res:* Ruminant and non-ruminant nutrition; feeding and metabolism studies with sheep and swine; mineral metabolism. *Mailing Add:* 23937 Buena Vista Rd Rockbridge OH 43149

CLINE, JAMES E, b Detroit, Mich, Mar 10, 31; m 53; c 4. EXPERIMENTAL NUCLEAR PHYSICS. *Educ:* Univ Mich, BSE, 53, MS, 54, PhD(physics), 58. *Prof Exp:* Res assoc synchrotron proj, Univ Mich, 54-57; physicist, Atomic Energy Div, Phillips Petrol Co, 57-64, group leader, Decay Schemes Group, 64-66; group leader exp physics, Nat Reactor Test Sta, Aerojet Nuclear Co, 66-70, sect chief, 70-73; lab dir, Nuclear Environ Serv, Div Sci Applns Inc, 73-81, div mgr, 81-83, corp chief scientist, Sci Applns Int Corp, 84-90; PRES, CLINE & ASSOCS, 90- *Concurrent Pos:* Lectr nuclear physics, Idaho State Univ, 58-59 & Univ Idaho, 58-73; asst prof & collab physics, Utah State Univ, 65-73; mem, Int Comt Radiation Metrol. *Mem:* Am Phys Soc; Am Nuclear Soc; Inst Elec & Electronics Eng; Am Soc Testing & Mat; AAAS. *Res:* Gamma ray spectroscopy and automatic data analysis, measurement and study of radioiodine molecular forms in nuclear power plants; transuranics and other non-gamma ray emitting nuclides in Radwaste systems; instrumentation technniques and services for radioactive effluent monitoring; methodology for analyses of bulk radioactive wastes and of coal elemental constituents using non-destructive techniques. *Mailing Add:* Cline & Assocs Inc PO Box 1513 Rockville MD 20849

CLINE, JAMES EDWARD, b Glens Falls, NY, Nov 13, 13; m 37; c 3. OPERATIONS RESEARCH. *Educ:* Cornell Univ, AB, 34; Harvard Univ, AM, 35, PhD(phys chem), 37. *Prof Exp:* Asst photochem, Harvard Univ, 37-41; asst chemist, Tenn Valley Authority, Wilson Dam, Ala, 41-42, assoc chemist, 42-45; chief res chemist, Beacon Co, Boston, 45-48; phys chemist & proj leader, Mass Inst Technol, 48-52; engr specialist, Sylvania Elec, 52-58; eng group leader, Raytheon Mfg Co, 58-61; staff scientist, Kearfott Semiconductor Corp, 61-62; eng specialist, Sylvania Electronic Syst Div, Gen Tel & Electronics Corp, 62-64; eng specialist aerospace technol, Electronic Res Ctr, NASA, 64-70; OPERS RES, US DEPT TRANSP, CAMBRIDGE, 70- *Res:* Instrumentation for vapor trace detection; computer interfacing; microelectronics. *Mailing Add:* 1731 Beacon St No 1011 Brookline MA 02146-4350

CLINE, KENNETH CHARLES, b Ann Arbor, Mich, May 12, 48; m 71; c 1. PLANT MEMBRANE BIOGENESIS. *Educ:* Univ Mich, Ann Arbor, BS & BSE, 71; Univ Colo, Boulder, PhD(biochem), 79. *Prof Exp:* Postdoctoral fel, Dept Bot, Univ Wis-Madison, 79-84; ASST PROF DEPT BIOL, UNIV FLA, 84- *Mem:* Am Soc Plant Physiol; Int Soc Plant Molecular Biol. *Res:* Structure function, and biogenesis of chloroplast membranes; mechamisms of synthesis and incorporation of chloroplast envelope membrane lipids and proteins. *Mailing Add:* Dept Fruit Crops Univ Fla Gainesville FL 32611

CLINE, MARLIN GEORGE, b Bertha, Minn, Dec 31, 09; m 36; c 3. SOIL MORPHOLOGY, SOIL GENESIS. *Educ:* NDak State Col, BS, 35; Cornell Univ, PhD(soils), 42. *Hon Degrees:* DSc, NDak State Univ & Trinity Col, Dublin, 65. *Prof Exp:* Jr soil surveyor, 35-38, assoc soil scientist, 41-42, soil scientist, 44-45, agt correlation soils, USDA, 46-74; from asst prof to prof, 42-74, head dept agron, 63-70, EMER PROF SOIL SCI, CORNELL UNIV, 74- *Concurrent Pos:* Soil scientist, Econ Coop Admin, Brit Africa, 49; Cornell contract, Philippines, 54-56; mem US mission soil & water, USSR, 58. *Honors & Awards:* Cert of Merit, Soil Conserv Serv, USDA. *Mem:* Fel Soil Sci Soc Am; hon mem Am Soc Agron. *Res:* Morphology, genesis and cartography of soils; soil management. *Mailing Add:* 107 Brandywine Rd Ithaca NY 14850

CLINE, MICHAEL CASTLE, b Richmond, Va, Apr 3, 45; m 69; c 1. FLUID DYNAMICS. *Educ:* Va Polytech Inst & State Univ, BS, 67; Purdue Univ, MS, 68, PhD(mech eng), 71. *Prof Exp:* Nat Acad Sci res assoc, Langley Res Ctr, NASA, 71-73; STAFF MEM FLUID DYNAMICS, LOS ALAMOS NAT LAB, 73- *Mem:* Am Inst Aeronaut & Astronaut. *Res:* Computation of high speed, compressible fluid flows. *Mailing Add:* 151 Piedra Loop Los Alamos NM 87544

CLINE, MORRIS GEORGE, b Los Angeles, Calif, Aug 10, 31; m 59; c 6. PLANT PHYSIOLOGY, ECOLOGY. *Educ:* Univ Calif, Berkeley, BS, 53; Brigham Young Univ, MS, 61; Univ Mich, PhD(bot), 64. *Prof Exp:* Asst plant physiologist, Colo State Univ, 64-66; res fel environ plant physiol, Calif Inst Technol, 66-68; asst prof, 68-74, ASSOC PROF BOT, OHIO STATE UNIV, 74- *Mem:* Am Soc Plant Physiologists. *Res:* Effects of temperature, low and high intensity visible and ultraviolet radiation on plants; gravity interaction with mechanisms of apical dominance; effects of soil environment on root growth of mountain shrubs; hormone action and nucleic acid metabolism. *Mailing Add:* Dept Plant Biol Ohio State Univ 1735 Neil Ave Columbus OH 43210

CLINE, RANDALL EUGENE, b Marietta, Ohio, Oct 4, 31; m 56; c 3. APPLIED MATHEMATICS. *Educ:* Marietta Col, BA, 53; Purdue Univ, MS, 55, PhD(math), 63. *Prof Exp:* Res asst & instr math, Purdue Univ, 55-59; res assoc, Inst Sci & Technol, Mich, 59-63, assoc res mathematician, 63-65, res mathematician, 65-68; from assoc prof to prof math & comput sci, 68-77, PROF MATH, UNIV TENN, KNOXVILLE, 77- *Concurrent Pos:* Mem staff, Math Res Ctr, US Army-Univ Wis, 64-65. *Mem:* AAAS; Soc Indust & Appl Math; Sigma Xi. *Res:* Matrix theory; algorithms. *Mailing Add:* 203 Ayres Hall Univ Tenn Knoxville TN 37996

CLINE, SYLVIA GOOD, b Atlantic City, NJ, Dec 27, 28; m 50; c 3. CELL PHYSIOLOGY, CYTOLOGY. *Educ:* Bryn Mawr Col, PhD(biol), 65. *Prof Exp:* Res asst physiol chem, Univ Pa, 50-51; res asst biol, Bryn Mawr Col, 60-64, fel, 65-66; res assoc biochem, Queens Col, NY, 66-67, lectr chem & biochem, 67-68; from asst prof to assoc prof, 68-78, asst dean acad affairs, 73-76, PROF BIOL, QUEENSBOROUGH COMMUNITY COL, 78- *Concurrent Pos:* Guest lectr, Biochem Cytol Lab, Rockefeller Univ, 66-67, Bryn Mawr Col, 84- *Mem:* AAAS; NY Acad Sci; Sigma Xi. *Res:* Glucose metabolism; RNA catabolism; growth of protozoa. *Mailing Add:* Dept Biol Queensborough Community Col Bayside NY 11364

CLINE, THOMAS L, b Peiping, China, May 14, 32; US citizen; m 54; c 3. PHYSICS. *Educ:* Hiram Col, BA, 54; Mass Inst Technol, PhD(physics), 61. *Prof Exp:* Res assoc, Lab Nuclear Sci, Mass Inst Technol, 60-61; PHYSICIST, SPACE SCI LAB, GODDARD SPACE FLIGHT CTR, NASA, 61-, ACTG HEAD, NUCLEAR ASTROPHYS BR, 89- *Concurrent Pos:* actg chief, High Energy Astrophys Prog, NASA, DC, 70-71, 84-85, asst head, Cosmic Radiation Br, 70-82. *Honors & Awards:* Except Sci Achievement Award, NASA, 81; John Lindsay Mem Award, Goddard Space Flight Ctr, NASA, 80. *Mem:* Am Phys Soc; Am Astron Soc; Int Astron Union. *Res:* Cosmic rays; solar particle production; astrophysics; gamma ray astronomy. *Mailing Add:* Goddard Space Flight Ctr NASA Code 661 Greenbelt MD 20771

CLINE, THOMAS WARREN, b Oakland, Calif, May 6, 46; m 86. DEVELOPMENTAL GENETICS. *Educ:* Univ Calif, Berkeley, AB, 68; Harvard Univ, PhD(biochem), 73. *Prof Exp:* Fel develop genetics, Helen Hay Whitney Found, Univ Calif, Irvine, 73-76; from asst to prof biol, Princeton Univ, 76-90; PROF GENETICS, UNIV CALIF BERKELEY, 90- *Mem:* Genetics Soc Am; Soc Develop Biol; AAAS. *Res:* Developmental regulation of gene expression and pattern formation in Drosophila melanogaster with emphasis on oogenesis, sex determination, and X-chromosome dosage compensation. *Mailing Add:* Molecular & Cell Biol Dept Div Genetics Univ Calif Berkeley CA 94720

CLINE, TILFORD R, b Virginia, Ill, Mar 15, 39; m 71; c 2. ANIMAL NUTRITION. *Educ:* Univ Ill, PhD(nutrit), 65. *Prof Exp:* PROF ANIMAL SCI, PURDUE UNIV, 77- *Mem:* Am Soc Animal Sci; Am Inst Nutrit; AAAS. *Res:* Swine nutrition and metabolism. *Mailing Add:* Dept Animal Sci Purdue Univ West Lafayette IN 47907

CLINE, WARREN KENT, b Bluefield, Va, July 28, 21; m 46; c 2. ORGANIC CHEMISTRY. *Educ:* Va Polytech Inst, BS, 42; Ohio State Univ, PhD(chem), 50. *Prof Exp:* Chemist, Gen Chem Defense Corp, 42-43; res chemist, Magnolia Petrol Co, 43-46; asst, Ohio State Univ, 46-50; res org chemist, Olin Mathieson Chem Corp, Olin Corp, 50-59, res assoc, 59-60, supvr res group, Film Opers, 60-69, mgr polymer res group, Film Div, 69-76, sr res assoc, Tobacco Indust Res Group, 76-; RETIRED. *Mem:* Am Chem Soc; Sigma Xi. *Res:* Organic synthesis; polymer chemistry. *Mailing Add:* 67 Deerwood La Brevard NC 28712

CLINE, WILLIAM H, JR, b Elkins, WVa, Dec 28, 40; m 59; c 3. AUTONOMIC NERVOUS SYSTEM, CARDIOVASCULAR. *Educ:* Davis & Elkins Col, BS, 62; WVa Univ, MS, 64, PhD(pharmacol), 65. *Prof Exp:* Fel pharmacol, Sch Med, WVa Univ, 65-66; asst prof, Sch Med, Univ Mo, 66-71, assoc prof, 71-73; dir, Dept Med Sci, Div Pharmacol, 73-79, PROF PHARMACOL, SCH MED, SOUTHERN ILL UNIV, 73- *Concurrent Pos:* Adj prof, Sagamon State Univ, 73-77 & 81-; prof, Dept Physiol, Col Sci, Southern Ill Univ, Carbondale, 74-78. *Mem:* Sigma Xi; NY Acad Sci; AAAS. *Res:* Pharmacology of the involvement of antiotensin II in blood pressure control including interactions with the autonomic nervous system and adrenal medulla in hypertension. *Mailing Add:* Dept Pharmacol Sch Med Southern Ill Univ PO Box 19230 Springfield IL 62784-9230

CLINESCHMIDT, BRADLEY VAN, b Redding, Calif, Dec 11, 41; c 2. NEUROPHARMACOLOGY, RECEPTOROLOGY. *Educ:* Ore State Univ, BS, 64; Univ Wash, PhD(pharmacol), 68. *Prof Exp:* Res assoc pharmacol, Exp Therapeut Br, Nat Heart & Lung Inst, NIH, 69-71, sr staff fel, 71-72; res fel, 72-75, dir neuropsychopharmacol, 75-80, sr dir pharmacol, 80-82, exec dir, 82-84, GROUP DIR, MERCK SHARP & DOHME RES LAB, 84- *Mem:* Am Soc Pharmacol & Exp Therapeut; Soc for Neuroscience. *Res:* Neuropharmacological, behavioral and neurochemical actions of drugs affecting the central nervous system; pharmacological analysis of receptors; autonomic pharmacology. *Mailing Add:* Merck Sharp & Dohme Res Lab West Point PA 19486

CLINGMAN, WILLIAM HERBERT, JR, b Grand Rapids, Mich, May 5, 29; m 51; c 2. PHYSICAL CHEMISTRY. *Educ:* Univ Mich, BS, 51; Princeton Univ, MA & PhD(phys chem), 54. *Prof Exp:* Res chemist, Am Oil Co, 54-57, group leader, 57-59; sect head, Tex Instruments Inc, 59-61, dir, Energy Res Lab, 61-62, mgr, Res Exploitation Dept, 62-64 & Corp Res & Develop Mkt, 64-67; PRES, W H CLINGMAN & CO, 67- *Concurrent Pos:* Vpres technol, Precision Measurement Inc, 85- *Mem:* Am Chem Soc; Inst Elec & Electronics Eng. *Res:* Radiation chemistry; catalysis; organic reaction mechanism; energy conversion; thermodynamics; combustion. *Mailing Add:* 4416 Mc Farlin Dallas TX 75205

CLINNICK, MANSFIELD, b Somerville, NJ, Jan 21, 22; m 42; c 6. MATHEMATICS. *Educ:* Calif State Polytech Col, BS, 47; Univ Calif, Berkeley, AB, 53, MA, 55. *Prof Exp:* Prin programmer, Lawrence Radiation Lab, Univ Calif, 55-58; proj mgr, Broadview Res Corp, 58-60; from asst prof to assoc prof math, Calif State Polytech Col, 60-67; comput scientist, Lawrence Berkeley Lab, Univ Calif, Berkeley, 67-82; mathematician & programmer, Lawrence Livermore Lab, 82-86; RETIRED. *Mem:* Sigma Xi; Math Asn Am; Asn Comput Mach. *Res:* Digital computer programming; algebra. *Mailing Add:* 189 Pier Ave Shell Beach CA 93449-2035

CLINTON, BRUCE ALLAN, b Jackson Heights, NY, Apr 27, 37; m 67; c 1. CELLULAR IMMUNOLOGY, MONOCLONAL ANTIBODY. *Educ:* Hofstra Univ, NY, BA, 61, MA, 62; Rutgers Univ, NJ, PhD(immunol), 69. *Prof Exp:* Instr immunol, Hofstra Univ, 62-63; res fel immunopath, Scripps Clinic Res Found, 69-72; WHO fel cellular immunol, Univ Autonoma De Mexico, Mexico City, 72; res scientist immunol, G D Searle & Co, 72-73, sr res scientist, 73-76, group leader, 76-77; dir immunohematol, 77-81, dir immunology res & develop, Div Am Hosp Supply Corp, AM Dade, 81-83; PRES & CHIEF EXEC OFFICER, TRINITY LABS, RALEIGH, NC, 83- *Concurrent Pos:* Vpres res & develop, Meridian Diagnostics, Cincinnati. *Mem:* Sigma Xi; Am Asn Parasitol; Am Asn Immunol. *Res:* Lymphokines and delayed hypersensitivity publications; immunochemistry of parasites; immune response to parasitic infection and auto immune disease; monoclonal antibody production; allergic response; blood-bank immunology. *Mailing Add:* Trinity Labs Inc 533 Pylon Dr Raleigh NC 27606

CLINTON, GAIL M, b Klamath Falls, Ore, Jan 23, 46. CELL BIOLOGY, VIROLOGY. *Educ:* Univ Calif, San Diego, BA, 69, MS, 71, PhD(virol), 75. *Prof Exp:* Fel, Harvard Med Sch, 75-77, res assoc, 77-79; res assoc, Children's Hosp Med Ctr, Boston, 79-81; ASST PROF, DEPT BIOCHEM, LA STATE UNIV MED CTR, 81- *Mem:* AAAS; Am Soc Microbiol; Am Soc Biol Chemists; Sigma Xi. *Res:* Role of protein modifications in growth regulation of animal cells and viruses. *Mailing Add:* Biochem Dept Ore Health Sci Univ 3181 SW Sam Jackson Park Portland OR 97201

CLINTON, RAYMOND OTTO, b Burbank, Calif, Apr 12, 18; m 38; c 2. ORGANIC CHEMISTRY. *Educ:* Calif Inst Technol, BS, 40; Univ Calif, Los Angeles, MA, 41, PhD(org chem), 43. *Prof Exp:* Jr chemist, C F TenEyck & Co, Calif, 35-38; asst, Calif Inst Technol, 36-40; asst inorg chem, Univ Calif, Los Angeles, 41; chemist, Union Oil Co, Calif, 41-42; instr org chem, Marymount Col, 42; sr res chemist, Gasparcolor, Inc, Calif, 46-47 & Sterling-Winthrop Res Inst, 43-46, group leader, 47-56, head org sect, 56-60, dir res admin, 60-67; asst to chmn, Sterling Drug, Inc, 67-74, dir corp serv, 74-79; CONSULT, 79- *Concurrent Pos:* Chemist, Black Diamond Oil Co, Calif, 39-40; res chemist, Nat Defense Res Comt proj, Univ Calif, Los Angeles, 41-43; adj prof Rensselaer Polytech Inst, 52-62. *Mem:* Am Chem Soc; NY Acad Sci; Royal Soc Chem. *Res:* Chemistry of plant pigments; antitubercular agents; local anesthetics; antimalarials; sulfur-containing amines; anti-virus agents; color photography dyes; flavonones and related plant pigments; synthesis of acids related to vitamin A; steroids; structural activity relationships. *Mailing Add:* 110 W Mora Dr Green Valley AZ 85614

CLINTON, WILLIAM L, b St Louis, Mo, Sept 17, 30; m 52; c 9. THEORETICAL PHYSICS. *Educ:* St Louis Univ, PhD(chem), 59. *Prof Exp:* Teacher chem, St Louis Univ, 55-58; res assoc, Brookhaven Nat Lab, 58-60; asst prof chem, 60-63, from asst prof to assoc prof physics, 64-70, PROF PHYSICS, GEORGETOWN UNIV, 70- *Concurrent Pos:* Consult, Nat Bur Standards, Brookhaven Nat Lab, Oak Ridge Nat Lab & Lawrence Berkeley Lab; prog dir, Condensed Matter Theory, Nat Sci Found. *Mem:* Am Phys Soc; AAAS. *Res:* Application of quantum mechanics to molecular and solid state physics and theoretical chemistry; electronic structure. *Mailing Add:* Dept Physics Georgetown Univ Washington DC 20057

CLIPPINGER, FRANK WARREN, JR, b Appleton, Wis, Oct 27, 25; m 50; c 2. ORTHOPEDIC SURGERY. *Educ:* Drury Col, Mo, AB, 48; Wash Univ, St Louis, MD, 52. *Prof Exp:* From instr to assoc prof, 57-70, PROF ORTHOP SURG, SCH MED, DUKE UNIV, 70- *Concurrent Pos:* Chmn, Prosthetics Res & Develop Comt, Nat Acad Sci, 75-76; med dir, Duke Hosp W-Duke Univ Med Sch, 75-, dir rehab, Duke Univ Med Ctr, 75-; chmn, Orthop Surg Adv Coun, Am Col Surgeons, 73-76, mem bd gov, 75- *Mem:* Am Orthop Asn; Am Acad Orthop Surgeons; Am Col Surgeons; Am Soc Surg Hand; Sigma Xi. *Res:* Research and development of artificial limbs; socket design and sensory feedback mechanism. *Mailing Add:* Dept Surg Duke Univ Med Ctr Durham NC 27710

CLISE, RONALD LEO, b Westernport, Md, Aug 19, 23; c 3. GENETICS, ZOOLOGY. *Educ:* Marietta Col, BA, 49; Mich State Univ, MS, 52; Western Reserve Univ, PhD(genetics), 60. *Prof Exp:* From instr to assoc prof, 55-72, chmn dept, 60-62, PROF BIOL, CLEVELAND STATE UNIV, 72-, MEM STAFF, DEPT QUANTUM ANALYSIS. *Mem:* AAAS. *Res:* Population studies with Drosophila. *Mailing Add:* Dept Quantum Analysis Cleveland State Univ Euclid Ave at E 24th St Cleveland OH 44115

CLITHEROE, H JOHN, b Hornchurch, Eng, Jan 2, 35; m 68; c 1. ENDOCRINOLOGY, GYNECOLOGY. *Educ:* Univ Sheffield, BSc, 59, MIBiol & PhD(med), 62; Royal Soc Health, FRSH, 74. *Prof Exp:* Johnson & Johnson fel, Rutgers Univ, 62-63; NIH fel pharmacol, 63-65, asst prof, 66-70, CLIN ASST PROF OBSTET & GYNEC, NJ COL MED, 70-; PROF BIOL SCI, COL STATEN ISLAND, CITY UNIV NY, 75- *Concurrent Pos:* Consult cytol, St Elizabeth's Hosp, Elizabeth, NJ, 67-75 & St Vincent's Med Ctr, New York, 67-80; assoc prof biol sci, Staten Island Community Col, NY Univ, 70-75. *Mem:* Brit Soc Endocrinol; Brit Inst Biol; Pan-Am Cancer Cytol Soc; NY Acad Sci; World Population Soc. *Res:* Uterine physiology; cytology of reproductive organs; sex selection of offspring; sexual therapy. *Mailing Add:* Dept Biol Sci Col Staten Island City Univ NY Staten Island NY 10301

CLIVER, DEAN OTIS, b Oak Park, Ill, Mar 2, 35; m 60; c 4. VIROLOGY. *Educ:* Purdue Univ, BS, 56, MS, 57; Ohio State Univ, PhD(agr), 60. *Prof Exp:* Fel, Ohio State Univ, 60; resident res assoc virus serol, US Army Chem Corps Biol Labs, Ft Detrick, Md, 61-62; res assoc virol, food res inst & dept microbiol, Univ Chicago, 62-66; from asst prof to assoc prof, 66-76, PROF VIROL, FOOD RES INST & DEPT BACT, UNIV WIS-MADISON, 76- *Concurrent Pos:* Resident res assoc, Nat Acad Sci-Nat Res Coun, 61-62; consult, WHO, 69-, prin investr & head, Collab Ctr Food Virol, Madison Inst Food Technologists, 75- *Mem:* Am Soc Microbiol; Sigma Xi. *Res:* Food research; studies on virus contamination of foods and water; animal virology. *Mailing Add:* Food Res Inst 1925 Willow Dr Univ Wis Madison WI 53706

CLOGSTON, ALBERT MCCAVOUR, b Boston, Mass, July 13, 17; m 41; c 2. SOLID STATE PHYSICS. *Educ:* Mass Inst Technol, SB, 38, PhD(physics), 41. *Prof Exp:* Teaching fel physics, Mass Inst Technol, 38-41, mem staff, radiation lab, 41-46; res physicist, Bell Tel Labs, 46-63, asst dir mat res lab, 63-65, dir, phys res lab, 65-71; vpres res, Sandia Labs, 71-73; exec dir res, physics & acad affairs div, Bell Tel Labs, 73-82; MEM, CTR MAT SCI, LOS ALAMOS NAT LABS, 82- *Concurrent Pos:* Mem comt, Physics & Soc, Am Inst Physics, 67-72, comt on prof concerns, 76-79, adv comt, Corp Assocs, 76-83, chmn, 82; mem, Comt Problems Physics & Soc, Am Phys Soc, 69-70, Applns of Physics, 75-76, chmn, 75, mem, Physics Manpower Panel, 76-79, counr-at-large, 84-87, exec comt, 85 & 86; mem, Solid State Sci Res Eval Group, Air Force Off Sci Res, 69-72, Stanford Synchrotron Radiation Lab Sci Policy Bd, 76-79, external adv comt, Los Alamos Nat Lab Ctr Mat Sci, 81-83, comn, Physics for Develop, Int Union for Pure & Appl Physics, US nat comt, 81-87, external adv comt, Mat Adv Comt, Univ Ill, 82-85, chmn, 83, mem, Air Force Sci Adv Bd, 82-86, chmn, Sci Panel, 83-86; mem, Panel on Condensed Matter, Physics Survey Comt, 73-74, adv bd, Off Phys Sci, 75-78, Comt on Sci & Pub Policy, 76-80, Comt on Educ & Employ of Minority-Group Mems in Sci, 76-78, Comn Human Resources, 77-81, Assembly Math & Phys Sci, 78-81, Comt Int Human Resource Issues, 81, Nat Acad Sci Coun & Exec Comt, 84-87, Nat Res Coun Gov Bd, 84-87, Comt on Sci, Eng & Pub Policy, 87- *Mem:* Nat Acad Sci; fel Am Phys Soc. *Res:* Magnetism; theory of metals; superconductivity; nuclear magnetic resonance; alloys and intermetallic compounds; non-linear quantum physics. *Mailing Add:* Los Alamos Nat Labs PO Box 1663 Los Alamos NM 87545

CLOKE, PAUL LEROY, b Orono, Maine, Feb 6, 29; m 55; c 2. GEOCHEMISTRY. *Educ:* Harvard Univ, AB, 51; Mass Inst Technol, PhD(geol), 54. *Prof Exp:* Res geologist, mining & explor geol, Anaconda Co, 54-57; res fel geochem, Harvard Univ, 57-59; from asst prof to assoc prof, 59-69, PROF GEOCHEM, UNIV MICH, ANN ARBOR, 69- *Concurrent Pos:* Chemist, Dept Sci & Indust Res, Gracefield, NZ, 66-67; ed, The Geochem News, 65-73. *Mem:* AAAS; Soc Econ Geologists; Geochem Soc; Am Inst Mining Metall & Petrol Engrs; Mineral Soc Can; Sigma Xi. *Res:* Application of chemistry to geologic problems, such as hydrothermal solutions, ore deposits, mineral-solution equilibria at low to high temperature and pressure, and recent sediments. *Mailing Add:* 2951 E De Silva Dr Las Vegas NV 89121

CLONEY, RICHARD ALAN, b Port Angeles, Wash, Feb 12, 30; m 52; c 3. DEVELOPMENTAL BIOLOGY, HISTOLOGY. *Educ:* Humboldt State Col, AB, 52, MA, 54; Univ Wash, PhD(zool), 59. *Prof Exp:* NIH fel anat, Sch Med, 59-61, from asst prof to assoc prof zool, 61-72, PROF ZOOL, UNIV WASH, 72- *Concurrent Pos:* Consult, Develop Biol Sect, Educ Develop Ctr, Mass; NSF grant; instr, Fri Harbor Labs, 63, 68-73, 75, 78 & 80; assoc ed, Cell & Tissue Res, 76- *Mem:* Am Soc Zoologists; AAAS; Am Soc Cell Biol. *Res:* Electron microscopic, cinemicrographic and experimental analyses of metamorphosis in ascidians; microfilaments and morphogenesis; fine structure of ascidian larvae; chaetogenesis in polychaetes; structure of chromatophore organs and iridophores in cephalopods; cinemicrography of marine invertebrate larvae. *Mailing Add:* Dept Zool Univ Wash Seattle WA 98195

CLONEY, ROBERT DENNIS, b Boston, Mass, May 6, 27. PHYSICAL CHEMISTRY. *Educ:* Spring Hill Col, BS, 52; Cath Univ Am, PhD(chem), 57; Woodstock Col, STB, 61. *Prof Exp:* Instr, pvt sch, NY, 52-53; res assoc chem, Woodstock Col, 57-62; from instr to asst prof, 62-71, ASSOC PROF CHEM, FORDHAM UNIV, 71- *Mem:* Am Chem Soc; Sigma Xi. *Res:* Quantum chemistry; molecular structure. *Mailing Add:* Dept Chem Fordham Univ Bronx NY 10458

CLONINGER, CLAUDE ROBERT, b Beaumont, Tex, Apr 4, 44; m 69; c 2. PSYCHIATRY, POPULATION GENETICS. *Educ:* Univ Tex, Austin, BA, 62; Washington Univ, St Louis, MD, 70. *Hon Degrees:* MD, Univ Umeå, Sweden, 83. *Prof Exp:* Resident psychiat, Wash Univ, 70-73, from asst prof to assoc prof psychiat, 73-80, assoc prof genetics, 79-80, PROF PSYCHIAT & GENETICS, SCH MED, UNIV WASH, 81-, CHMN, DEPT PSYCHIAT, 89- *Concurrent Pos:* NIMH res scientist award, 75-; vis assoc prof, Pop Genetics Lab, Univ Hawaii, 78-79; consult, Am Psychiat Asn Task Force on Nomenclature, 78-80; vis prof psychiat & genetics, Umea Univ, Sweden, 80; psychiatrist, Jewish Hosp St Louis, 76-; mem res rev comt psychopath & clin biol, NIMH, 80-; assoc ed, Am J Human Genetics, 80-83, J Clin Psychiat, 81-87 & Genetic Epidemiol, 84-; assoc ed, Genetic Epidemiol, 84-; consult, World Health Orgn, 81; mem sci adv coun, Tex Dept, Ment Health & Ment Retardation, 83-84; sci counr to dir of NIAAA, 88-; bd dirs, Res Soc Alcoholism, 87- *Honors & Awards:* Milton Parker Mem Award, Ohio State Univ, 83 & 85. *Mem:* Inst Med-Nat Acad Sci; Am Psychopath Asn; Am Soc Human Genetics; Behav Genetics Asn; Soc Biol Psychiat Social Biolog; Am Psychopath Asn (treas, 84-90, vpres, 90); Am Psychiat Asn. *Res:* Genetic epidemiology of psychiatric disorders and classification of psychiatric disorders. *Mailing Add:* Sch Med Wash Univ 4940 Audubon St St Louis MO 63110

CLOPPER, HERSCHEL, b Winthrop, Mass, May 11, 41; m 63; c 2. CHEMICAL ENGINEERING, QUALITY ENGINEERING. *Educ:* Mass Inst Technol, BS, 62, MS, 63; Rice Univ, PhD(chem eng), 68. *Prof Exp:* Res engr, Plastics Dept, Res & Develop Div, E I du Pont de Nemours & Co, 67-68 & Fluorocarbons Div, 68-70; asst prof chem eng, Cath Univ Am, 70-71; develop engr, Polaroid Corp, 71-77, prod develop engr, 77-81, prin engr, Battery Div, 81-85, qual mgr, precision molding optics, 85-89, qual mgr, electronics mfg, 89-90, TECH MGR, MFG STRATEGY, POLAROID CORP, 90- *Mem:* Am Inst Chem Engrs; Am Chem Soc; Sigma Xi. *Res:* Rheology of and reactor design for non-Newtonian fluids; materials development; web coating technology. *Mailing Add:* Four Ford Lane Framingham MA 01701

CLORE, WALTER JOSEPH, b Tecumseh, Okla, July 1, 11; m 34; c 3. HORTICULTURE. *Educ:* Okla Agr & Mech Col, BS, 33; State Col Wash, PhD, 47. *Prof Exp:* Asst hort, 37-46, assoc horticulturist, 46-51, horticulturist, 51-76, EMER HORTICULTURIST, IRRIG AGR RES & EXTEN CTR, WASH STATE UNIV, 76- *Concurrent Pos:* Consult, viticulture. *Mem:* Am Soc Hort Sci; hon mem Am Soc Enol. *Res:* Production of grapes for juice and wine and asparagus production under irrigation. *Mailing Add:* 1317 Paterson Rd Prosser WA 99350

CLOSE, CHARLES M(OLLISON), b Ilion, NY, Mar 15, 27; m 57; c 3. ELECTRICAL ENGINEERING. *Educ:* Lehigh Univ, BS, 50; Stevens Inst Technol, MS, 53; Rensselaer Polytech Inst, PhD(network theory), 62. *Prof Exp:* Engr, Westinghouse Elec Corp, 50-52; asst, Stevens Inst Technol, 52-54; from instr to assoc prof, 54-67, PROF ELEC ENG, RENSSELAER POLYTECH INST, 67-, ASSOC CURRIC CHMN, ELEC COMPUT & SYSTS ENG & CURRIC CHMN ELEC ENG, 76- *Mem:* Inst Elec & Electronics Engrs; Am Soc Eng Educ. *Res:* Network theory. *Mailing Add:* 18 Berkshire Dr Clifton Park NY 12065

CLOSE, DAVID MATZEN, b Plainfield, NJ, Mar 9, 42; m; c 2. RADIATION BIOLOGY, COMPUTER INTERFACING. *Educ:* Franklin & Marshall Col, AB, 64; WVa Univ, MS, 67; Clark Univ, PhD(physics), 72. *Prof Exp:* Fel physics, Univ Conn, 73-74; fel radiation biol, Univ Rochester, 74-78; PROF PHYSICS, ETENN STATE UNIV, 78-, ADJ PROF, COL MED, 79- *Mem:* Am Phys Soc; fel Radiation Res Soc; Sigma Xi. *Res:* Electron spin resonance and electron-nuclear double resonance studies of radiation damage to nucleic acid constituents; theoretical calculations of free radical structures. *Mailing Add:* Physics Dept ETenn State Univ Box 22060A Johnson City TN 37614

CLOSE, DONALD ALAN, b Tucson, Ariz, Nov 19, 46. NUCLEAR PHYSICS, NUCLEAR SPECTROSCOPY. *Educ:* Hastings Col, BA, 68; Univ Kans, MA, 70, PhD(physics), 72. *Prof Exp:* NSF presidential internship, Los Alamos Sci Lab, Univ Calif, 72-73, staff mem, 73, staff physicist nuclear physics, 73-87, STAFF PHYSICIST NUCLEAR PHYSICS, LOS ALAMOS NAT LAB, UNIV CALIF, 90- *Concurrent Pos:* Consult, Int Atomic Energy Agency, Vienna, Austria, 87-90. *Mem:* Am Phys Soc; Sigma Xi. *Res:* Muonic atoms, gamma-ray spectroscopy and nuclear structure; proton induced x-ray fluorescence; development of assay instrumentation for fissionable material; application of Monte Carlo and discrete ordinates transport calculations to problems in nuclear science and geophysics; determination of uranium enrichment and plutonium isotopics by gamma-ray spectroscopy; qualitative and quantitative measurements of nuclear materials. *Mailing Add:* Los Alamos Nat Lab Group Q2-MS-J562 Los Alamos NM 87545

CLOSE, PERRY, b Chicago, Ill, May 20, 21; m 61; c 2. GENETICS, PHYSIOLOGY. *Educ:* Univ Calif, Berkeley, AB, 47, MA, 48; Univ Tex, PhD(zool), 55. *Prof Exp:* Asst prof biol, Univ Southwestern La, 48-49; asst human genetics, Univ Tex, 54-55; aviation physiologist, med dept, US Naval Air Sta, Va, 55-57 & US Naval Sch Aviation Med, 57-62; mem res staff, Northrop Space Labs, Calif, 62-64; res biologist, Vet Admin Hosp, Long Beach, Calif, 64-65; head life sci, Chrysler Space Div, 65-68; PROF BIOL, CITY COL SAN FRANCISCO, 68- *Mem:* Am Soc Human Genetics; assoc fel Aerospace Med Asn; Soc Study Social Biol. *Res:* Genetics of longevity; hereditary deafness; low pressure and impact patho-physiology and the effects of radiation in combination with aerospace stresses; evaluation of radiotherapy procedures; genetics of mental retardation. *Mailing Add:* 272 Dennis Dr Daly City CA 94015

CLOSE, R(ICHARD) N(ORCROSS), b Philadelphia, Pa, Apr 20, 23; m 52; c 3. ELECTRONICS. *Educ:* Univ Rochester, BS, 43. *Prof Exp:* Proj engr, Radiation Lab, Mass Inst Technol, 43-45; prog dir reconaissance systs, Airborne Instruments Lab Div, Cutler Hammer, Inc, 45-60, dir, Indust Electronics Div, 60-68; progs mgr, Raytheon Co, 68-74, pres, Raytheon Europe Electronics Co, 74-75; PRES, R N CLOSE ASSOCS INC, 75- *Concurrent Pos:* Sci consult, USAF, 45; consult, Air Navig Develop, 50. *Mem:* Inst Elec & Electronics Engrs. *Res:* Military electronics; reconnaissance systems; radar data processing. *Mailing Add:* 85 Highland Circle Wayland MA 01778

CLOSE, RICHARD THOMAS, b New York, NY, Dec 24, 34; m 58; c 7. COMPUTER SCIENCE, ELECTROMAGNETIC THEORY. *Educ:* Iona Col, BS, 56; Cath Univ Am, PhD(physics), 67. *Prof Exp:* Res physicist, Naval Res Lab, Washington, DC, 59-68; assoc prof physics, St Bonaventure Univ, 68-71; dir comput ctr, State Univ NY Agr & Tech Col, Alfred Univ, 71-76; assoc prof, 76-80, PROF COMPUT SCI, US COAST GUARD ACAD, 80- *Concurrent Pos:* Lectr, Univ Md, 60-68; adj fac, Grad Sch, Univ New Haven, 76-80 & Hartford Grad Ctr, 80-88. *Mem:* Data Processing Mgt Asn; Asn Comput Mach; Inst Elec & Electronics Engrs Comput Soc. *Res:* Antenna research with particular emphasis on frequency independent scanning arrays; intense electron beam studies including both theoretical and experimental studies; solution of elliptic partial differential equations using various numerical methods; large-scale programming systems; computer algorithms; computer programming languages. *Mailing Add:* Comput Sci Dept US Coast Guard Acad New London CT 06320-4195

CLOSMANN, PHILIP JOSEPH, b New Orleans, La, July 28, 25; m 56; c 6. PHYSICS, CHEMICAL ENGINEERING. *Educ:* Tulane Univ, BE, 44; Mass Inst Technol, SM, 48; Calif Inst Technol, MS, 50; Rice Inst, PhD(physics), 53. *Prof Exp:* physicist, Shell Develop Co, 53-86; CONSULT, 86- *Mem:* Am Phys Soc; Soc Petrol Engrs; Petrol Soc Can Inst Mining & Metall. *Res:* Thermal recovery of hydrocarbons; oil shale recovery; fluid flow; heat flow; low temperature physics; fluidized solids. *Mailing Add:* 27 Williamsburg Houston TX 77024

CLOSS, GERHARD LUDWIG, b Wuppertal, Ger, May 1, 28. MAGNETIC RESONANCE, ELECTRON TRANSFER. *Educ:* Univ Tubingen, Dipl Chem, 53, PhD(chem), 55. *Prof Exp:* Fel, Harvard Univ, 55-57; from asst prof to prof, 57-74, A A MICHELSON DISTINGUISHED SERV PROF CHEM, UNIV CHICAGO, 74- *Concurrent Pos:* A P Sloan Found fel, 62-66. *Honors & Awards:* Arthur C Cope Award, Am Chem Soc, 71, James Flack Norris Award, 74. *Mem:* Nat Acad Sci; AAAS; Am Chem Soc; Am Acad Arts & Sci. *Res:* Chemistry of reactive intermediates; photosynthesis; porphyrins; magnetic resonance. *Mailing Add:* Dept Chem Univ Chicago Chicago IL 60637

CLOSSON, WILLIAM DEANE, b Remus, Mich, Feb 3, 34; m 69. ORGANIC CHEMISTRY. *Educ:* Wayne Univ, BS, 56; Univ Wis, PhD(org chem), 60. *Prof Exp:* Asst org chem, Univ Wis, 56-60; NSF res fel, Harvard Univ, 60-61; from instr to asst prof, Columbia Univ, 61-66; assoc prof, 66-71, PROF CHEM, STATE UNIV NY ALBANY, 71- *Concurrent Pos:* Res grants, Petrol Res Fund, 62-65, USPHS, 63-66 & 67- & NSF, 65-67; Alfred P Sloan res fel, 68-70; Nat Acad Sci-Nat Res Coun travel grant, IVPAC Cong, Jerusalem, 75; Air Force Off res grant, 76-77. *Mem:* AAAS; Am Chem Soc; Royal Soc Chem. *Res:* Solvolytic reactions of organic compounds; electronic absorption spectra of ketones, esters and alkyl azides; reactions of organic anion radicals; mecuration reactions and silver pi complexes of substituted alkenes. *Mailing Add:* Dept Chem State Univ NY Albany NY 12222

CLOTFELTER, BERYL EDWARD, b Prague, Okla, Mar 23, 26; m 51; c 3. PHYSICS. *Educ:* Okla Baptist Univ, BS, 48; Univ Okla, MS, 49, PhD(physics), 53. *Prof Exp:* Instr math & physics, Okla Baptist Univ, 49-50; res physicist, Phillips Petrol Co, 53-55; assoc prof physics, Univ Idaho, 55-56; from asst prof to assoc prof physics, Okla Baptist Univ, 56-63; assoc prof, 63-68, PROF PHYSICS, GRINNELL COL, 68- *Concurrent Pos:* NSF sci fac fel, 68-69. *Mem:* Am Phys Soc; Am Asn Physics Teachers; AAAS. *Res:* Astrophysics; cosmology; history of science. *Mailing Add:* Dept Physics Grinnell Col Grinnell IA 50112

CLOTHIER, GALEN EDWARD, b Stafford, Kans, Nov 7, 33; m 55; c 3. CELL BIOLOGY, DEVELOPMENTAL BIOLOGY. *Educ:* Fresno State Col, AB, 55; Ore State Univ, MS, 57, PhD(biol), 60. *Prof Exp:* Asst prof zool, Los Angeles State Col, 60-62; from asst prof to assoc prof, 62-68, PROF BIOL, CALIF STATE UNIV, SONOMA, 68- *Mem:* Am Soc Zoologists; AAAS; Sigma Xi. *Res:* Physiology of mitosis; developmental biology of sea urchins. *Mailing Add:* Dept Biol Sonoma State Univ Rohnert Park CA 94928

CLOTHIER, ROBERT FREDERIC, b Pocatello, Idaho, Sept 15, 25. MECHANICAL ENGINEERING, PHYSICS. *Educ:* Univ Southern Calif, BE, 42, MSc, 43; Univ Montpellier, Dr Univ & DSc(physics), 65. *Prof Exp:* Asst chem, Univ Southern Calif, 41-42, asst physics, 42-43, lectr math, 45-46, instr mech eng, 49; instr physics, Ariz State Univ, 43-44; staff tech supvr, Electromagnetic Div, Manhatten Eng Dist, 44-45; instr math, Univ Okla, 46-47; assoc prof mech eng, Auburn Univ, 50-55; sr engr & adminstr, Atomic Power Div, Westinghouse Elec Corp, 55-57; prof eng & in chg nuclear eng, 57-77, chmn dept mech eng, 70-77, PROF MECH ENG, SAN JOSE STATE UNIV, 77- *Concurrent Pos:* Prof indust nuclear eng, Atomic Power Div, Westinghouse Elec Corp, 54, consult, Atomic Power Div, 54-55; consult, US Atomic Energy Comn, 59-; vis prof, London Polytech, 64-65 & 70. *Mem:* Am Soc Mech Engrs; Am Soc Eng Educ. *Res:* Dielectric properties of glass and cellulose materials; electrochemical isolation of isotopes; nuclear engineering; thermodynamics. *Mailing Add:* Dept Mech Eng San Jose State Univ San Jose CA 95192

CLOUD, GARY LEE, b Muskegon, Mich, Sept 8, 37; m 72; c 2. BIOMECHANICS, NONDESTRUCTIVE EVALUATION. *Educ:* Mich Technol Univ, BS, 59, MS, 61; Mich State Univ, PhD(appl mech), 66. *Prof Exp:* From asst prof to assoc prof, 66-79, PROF MECH, DEPT METALL, MECH & MAT SCI, MICH STATE UNIV, 79- *Concurrent Pos:* Lectr mech eng, Univ Zambia, Africa, 69, Juneau Icefield Res Prog, Seattle, Wash, 70-72; sr res assoc, Air Force Mat Lab, Nat Res Coun, 75-76; NSF fel, Dept Physics, Imp Col, Eng, 76-77; consult, numerous co, 78-; ed, Exp Mech, 85-88. *Mem:* Soc Exp Mech; Am Optical Soc; Soc Photo-Optical Instrumentation Engrs; Sigma Xi. *Res:* Developing and applying optical and electronic techniques in measurement of motion and strain in geomechanics, glaciers, caverns; biomechanics, eye, tissue, kinematics; composite materials; fracture; fastening. *Mailing Add:* Dept Metall Mech & Mat Sci Mich State Univ East Lansing MI 48824-1226

CLOUD, JAMES DOUGLAS, electrical engineering, for more information see previous edition

CLOUD, JOSEPH GEORGE, b Phoenixville, Pa, Nov 27, 44; m 70; c 3. OOGENESIS, ENDOCRINOLOGY. *Educ:* WVa Univ, BS, 66; Univ Wis, Madison, MS, 68, PhD(physiol), 74. *Prof Exp:* Asst fel reproductive biol, John Hopkins Univ, 74-76, asst prof, 76-77; ASST PROF ZOOL, UNIV IDAHO, 77- *Mem:* Soc Study Reproduction; Am Soc Zoologists; Sigma Xi. *Res:* Hormonal control of the reinitiation of meiosis in oocytes at the time of ovulation. *Mailing Add:* Dept Biol Sci Col Lett/Sci Univ Idaho Moscow ID 83843

CLOUD, PRESTON, geology, invertebrate paleontology; deceased, see previous edition for last biography

CLOUD, WILLIAM K, seismology, mechanical engineering; deceased, see previous edition for last biography

CLOUD, WILLIAM MAX, b Wilmot, Kans, Mar 27, 23; m 47; c 3. ATOMIC SPECTROSCOPY. *Educ:* Southwestern Col, BA, 47; Univ Wis, MS, 49, PhD(physics), 55. *Prof Exp:* Instr physics, Southwestern Col, 49-50, asst prof physics & counr men, 50-53; from asst prof to assoc prof, Kans State Teachers Col, 55-62; assoc prof, Eastern Ill Univ, 62-66, prof physics & chmn div pre-eng studies, 66-89; RETIRED. *Mem:* Am Asn Physics Teachers; Sigma Xi; Am Phys Soc. *Res:* High resolution atomic spectroscopy using an atomic beam. *Mailing Add:* Dept Physics Eastern Ill Univ Charleston IL 61920

CLOUGH, DAVID EDWARDS, b New Brunswick, NJ, Apr 12, 46; wid; c 2. CHEMICAL ENGINEERING. *Educ:* Case Western Reserve Univ, BS, 68; Univ Colo, MS, 69, PhD(chem eng), 75. *Prof Exp:* Chem engr, E I du Pont de Nemours & Co, 69-72; instr, 75, asst prof, 75-, ASSOC PROF CHEM ENG, UNIV COLO. *Concurrent Pos:* Consult, E I du Pont de Nemours & Co, 72- & Latin Am Scholar Prog, Am Univ, 75- *Mem:* Am Inst Chem Engrs; Instrument Soc Am; Am Soc Eng Educ. *Res:* Optimization and control of chemical processes; application of real time computers; mathematical modeling and applied mathematics in chemical engineering. *Mailing Add:* Dept Chem Eng Univ Colo Campus Box 422 Boulder CO 80309

CLOUGH, FRANCIS BOWMAN, b Boise, Idaho, Feb 4, 24; m 62. ENVIRONMENTAL CHEMISTRY. *Educ:* Univ Wyo, BS, 44; Princeton Univ, MA, 48, PhD(chem), 51. *Prof Exp:* Chemist org res, Distillation Prod, Inc, 44-46; asst, Princeton Univ, 46-50; from instr to asst prof chem, Va Polytech Inst & State Univ, 50-55; from asst prof to assoc prof chem, Stevens Inst Technol, 55-80; CONSULT & PROPRIETOR, CHEM CRITERIA, 81- *Concurrent Pos:* Vis scientist, Nat Ctr Atmospheric Res, 79. *Mem:* AAAS; Am Chem Soc; fel Am Inst Chemists; Air & Waste Mgt Asn. *Res:* Optical rotatory power; solvent interactions; energy transfer in inorganic reactions. *Mailing Add:* 8451 Allison Ct Arvada CO 80005

CLOUGH, G WAYNE, ENGINEERING. *Prof Exp:* DEAN ENG, VA POLYTECH INST. *Mem:* Nat Acad Eng. *Mailing Add:* Dean Eng N Hall Va Polytech Inst Blacksburg VA 24061

CLOUGH, GARRETT CONDE, ecology, conservation, for more information see previous edition

CLOUGH, JOHN WENDELL, b Oak Bluffs, Mass, Jan 3, 42; m 68; c 1. GEOPHYSICS. *Educ:* Northeastern Univ, BS, 65; Univ Wis, Madison, MS, 70, PhD(geophys), 74. *Prof Exp:* Proj assoc geophys, Geophys & Polar Res Ctr, Univ Wis, 74-75; sci dir & asst prof geophys, Ross Ice Shelf Proj, Univ Nebr, Lincoln, 75-79; asst prof geophys, NC State Univ, Raleigh, 79-81; GEOPHYSICIST, MARATHON OIL CO, CASPER, WY, 81- *Mem:* Soc Explor Geophys. *Res:* Radar echo sounding of polar ice thickness; geophysical survey of Ross Ice Shelf, Antarctica; magneto tellurics. *Mailing Add:* 5311 Pine Arbor Dr Houston TX 77066-2550

CLOUGH, RAY WILLIAM, b Seattle, Wash, July 23, 20; m 42; c 3. STRUCTURAL & EARTHQUAKE ENGINEERING. *Educ:* Univ Wash, BS, 42; Calif Inst Technol, MS, 43; Mass Inst Technol, MS, 47, ScD, 49. *Hon Degrees:* DTech, Chalmers Univ Technol, Gothenburg, Sweden, 79 & Norweg Inst Technol, Trondheim, Norway, 82. *Prof Exp:* From asst prof to prof civil eng, 49-83, nishkian prof struct eng, 83-87, EMER PROF, UNIV CALIF, BERKELEY, 87- *Concurrent Pos:* Fulbright fel, res ship vibrations, Norway, 56-57 & 72-73; overseas fel, Churchill Col, Cambridge Univ, 63-64; consult struct dynamics and earthquake engineering. *Honors & Awards:* Res Award, Am Soc Civil Engrs, 61, Howard Award, 70, Moisseiff Award, 80; Newmark Medal, 79; Martin P Korn Award, Prestressed Concrete Inst, 90. *Mem:* Nat Acad Sci; Nat Acad Eng; Seismol Soc Am; Earthquake Eng Res Inst; Am Soc Civil Engrs. *Res:* Application of electronic digital computers to problems of structural analysis; response of structures to earthquake loads. *Mailing Add:* Rm 537 Davis Hall Univ Calif Berkeley CA 94720

CLOUGH, RICHARD H(UDSON), b Springer, NMex, Aug 25, 22; m 45; c 2. CIVIL ENGINEERING, CONSTRUCTION MANAGEMENT. *Educ:* Univ NMex, BS, 43; Univ Colo, MS, 49; Mass Inst Technol, ScD(civil eng), 51. *Prof Exp:* Instr & asst prof, Univ NMex, 46-49, assoc prof, 51-52; res assoc, Mass Inst Technol, 49-51; vpres & owner, Lembke, Clough & King, Inc, 52-57; from assoc prof civil eng to prof & chmn dept, 57-60, dean, col eng, 60-68, PROF CIVIL ENG, UNIV NMEX, 68- *Honors & Awards:* SIR Award, NMex Bldg Br, Assn Gen Contractors Am. *Mem:* Am Soc Civil Engrs; Nat Soc Prof Engrs; Am Soc Eng Educ; Am Soc Testing & Mat. *Res:* Metrication of construction industry; bidding strategies; construction project time control; construction forensics. *Mailing Add:* 1025 Pueblo Solano NW Albuquerque NM 87107

CLOUGH, ROBERT RAGAN, b Sibley, Iowa, Feb 25, 42. COMPUTER SCIENCES GENERAL, SOFTWARE SYSTEMS. *Educ:* Univ Md, AB, 64; Northwestern Univ, MS, 66, PhD(math), 67. *Prof Exp:* Asst prof math, Univ Notre Dame, 67-77; SOFTWARE SPECIALIST, BURROUGHS CORP, 77- *Res:* Algebraic topology; fiber spaces; homotopy theory. *Mailing Add:* 753 Juniper Dr Palatine IL 60067

CLOUGH, ROGER LEE, b Salt Lake City, Utah, Aug 24, 49; m 70. ORGANIC CHEMISTRY. *Educ:* Univ Utah, BA, 71; Calif Inst Technol, PhD(chem), 75. *Prof Exp:* NATO fel, Tech Univ Munich, Ger, 75-76; NSF fel, Univ Calif, Los Angeles, 76-77; STAFF SCIENTIST CHEM, SANDIA LABS, 77- *Mem:* Am Chem Soc. *Res:* Organic chemistry; photochemistry and radiation chemistry; polymer degradation and stabilization; free radical chemistry; molecular conformational studies; nuclear magnetic resonance; transition metal organometallics. *Mailing Add:* Sandia Labs Org 1811 Albuquerque NM 87185

CLOUGH, SHEPARD ANTHONY, b Hanover, NH, July 19, 31; m 57; c 6. ATMOSPHERIC RADIATIVE TRANSFER. *Educ:* Cornell Univ, BEng Phys, 54; Columbia Univ, MA, 58. *Prof Exp:* Res scientist, Air Force Geophysics Lab, 57-87; SR SCIENTIST, ATMOSPHERIC & ENVIRON RES, INC, 87- *Concurrent Pos:* Co-prin investr, Tropospheric Emission Sounder, Earth Orbital Shuttle, NASA, 88- & prin investr, Atmospheric Radiation Measurement Prog, Dept Energy, 90-95; mem, Lippincott Award Comt, Optical Soc Am, 90-93. *Mem:* Optical Soc Am; Soc Indust & Appl Math. *Res:* Radiative transfer in the atmosphere including the effects of molecular absorption and cloud, aerosol and molecular scattering; retrieval of atmospheric state parameters and radiative problems related to climate charge. *Mailing Add:* Atmospheric & Environ Res Inc 840 Memorial Dr Cambridge MA 02139

CLOUGH, STUART BENJAMIN, b Tisbury, Mass, Mar 16, 37; m 61; c 3. PHYSICAL CHEMISTRY, POLYMER CHEMISTRY. *Educ:* Univ Mass, Amherst, BS, 59, PhD(chem), 66; Univ Del, MChE, 61. *Prof Exp:* Chemist, Dewey & Almy Chem Div, W R Grace Co, 61-62; res fel chem, Univ Mass, Amherst, 62-65; chemist, US Army Natick Labs, 65-68 & US Army Mat & Mech Res Ctr, 68-70; from asst prof to assoc prof, 70-78, chmn dept, 81-84, PROF CHEM, UNIV LOWELL, 78- *Mem:* Am Chem Soc; Am Phys Soc; fel Am Col Physicians. *Res:* Structure and physical properties of bulk polymers; modeling of polymers. *Mailing Add:* Dept Chem Univ Lowell Lowell MA 01854-2881

CLOUGH, STUART CHANDLER, b Richmond, Va, July 29, 43; m 68; c 2. ORGANIC CHEMISTRY. *Educ:* Univ Richmond, BS, 65; Univ Fla, PhD(chem), 69. *Prof Exp:* Res assoc chem, State Univ NY Buffalo, 69-71; res assoc, Philip Morris, Inc, 71-73; ASSOC PROF CHEM, UNIV RICHMOND, 73- *Mem:* Am Chem Soc; Sigma Xi. *Res:* Thermal and photochemical rearrangements and reactions of small organic molecules with particular interest in the formation and characterization of reactive intermediates and natural products. *Mailing Add:* 125 Fairwood Dr Richmond VA 23235

CLOUGH, WENDY GLASGOW, b Salem, NJ, Sept 7, 42. ANIMAL VIROLOGY. *Educ:* Harvard Univ, BA, 64; Dartmouth Med Sch, BMS, 66; Harvard Med Sch, MD, 68. *Prof Exp:* Fel molecular biol, Dept Microbiol, Sch Med, Tufts Univ, 68-71; NIH spec fel molecular biol, Dept Biol, John

Hopkins Univ, 71-73; fel biochem, Dept Biochem & Molecular Biol, Harvard Univ, 73-74; res assoc molecular biol, Basic Sci Div, Sidney Faber Cancer Inst, Harvard Med Sch, 74-75, instr, 75-77; asst prof, 77-81, ASSOC PROF MOLECULAR BIOL, DEPT BIOL SCI, MOLECULAR BIOL DIV, UNIV SOUTHERN CALIF, 81- *Concurrent Pos:* Fel, Med Found Boston, 76-77; res career develop award, Nat Cancer Inst, 80. *Mem:* AAAS; Am Soc Microbiol; Sigma Xi; Asn Women Sci. *Res:* Molecular biology and biochemistry of herpes viruses, specifically Epstein-Barr virus; eucaryotic and viral DNA replication; mechanism of action of anti-viral drugs; viral DNA methylation; molecular genetics of bacterial viruses. *Mailing Add:* ACBR 126 Molecular Biol Univ Southern Calif Los Angeles CA 90007

CLOUSER, WILLIAM SANDS, b Atlantic City, NJ, Feb 8, 21; m 51; c 1. ENGINEERING MECHANICS, MATERIALS SCIENCE. *Educ:* Univ NMex, BSME, 55; Univ Wis, MS, 57, PhD(eng mech), 58. *Prof Exp:* Draftsman, Los Alamos Sci Lab, 49-53; asst prof eng mech, Univ Wis, 58-63; staff mem eng mech, 63-90, ASSOC, LOS ALAMOS NAT LAB, 90- *Concurrent Pos:* Vis asst prof & Ford Found res grant, Univ Mich, 60; adv prof, Univ NMex, 71- *Res:* Properties of wood and wood products; graphite and metals behavior; research and development for weapons systems and reactor components; stress analysis. *Mailing Add:* 126 Rover Blvd Los Alamos NM 87544

CLOUTHIER, DENNIS JAMES, b Le Pas, Man, Aug 27, 50; m 84. LASER SPECTROSCOPY, PHOTOPHYSICS. *Educ:* Univ Sask, BSc, 75, PhD(chem), 80. *Prof Exp:* Res assoc, Herzberg Inst Astrophys, Nat Res Coun Can, 79-82; staff combustion scientist, Atomic Energy Can, 83-84; ASST PROF CHEM, UNIV KY, 84- *Mem:* Chem Inst Can; Inter-Am Photochem Soc; Am Chem Soc. *Res:* Laser spectroscopy and the dynamics of the electronic excited states of polyatomic molecules. *Mailing Add:* Dept Chem Univ Ky Lexington KY 40506-0002

CLOUTIER, CONRAD FRANCOIS, b Chateau-Richer, Que, July 11, 48; c 1. BIOLOGICAL CONTROL. *Educ:* Laval Univ, BSc, 70, MSc, 72; Simon Fraser Univ, PhD(entom), 78. *Prof Exp:* Lectr, 76-78, asst prof, 78-81, ASSOC PROF ENTOM, DEPT BIOL, LAVAL UNIV, 81- *Mem:* Entom Soc Can; Entom Soc Am; Can Soc Zoologists. *Res:* Insect parasitoids; aphid biology and ecology; biological control. *Mailing Add:* Dept Biol Fac Sci & Eng Laval Univ Quebec PQ G1K 7P4 Can

CLOUTIER, GILLES GEORGES, b Quebec City, Que, June 27, 28; m 54; c 5. PHYSICS. *Educ:* Laval Univ, BA, 49, BASc, 53; McGill Univ, MSc, 56, PhD(physics), 59. *Prof Exp:* Tech officer, Defence Res Bd Can, 53-54; sr mem sci staff, plasma physics, res labs, RCA Victor Co, Ltd, Can, 59-64; assoc prof physics, Univ Montreal, 64-68; sci dir, Hydro-Quebec Inst Res, 68-71, dir res, 71-74, asst dir, 74-78; pres, Alta Res Coun, 78-83; exec vpres, technol & int affairs, Hydro-Quebec, 83-85; RECTOR, UNIV MONTREAL, 85- *Concurrent Pos:* Mem comn, Int Sci Radio Union, Can, 64-68; assoc comt space res, Nat Res Coun Can, 64-69, mem Coun, 73-77. *Mem:* Sr mem Inst Elec & Electronics Engrs; Am Phys Soc; Can Asn Physicists (pres, 72-73); Royal Soc Can; Asn Univs & Cols Can. *Res:* Plasma physics; microwave optics; electron impact phenomena; electric propulsion; electromagnetic waves and plasmas; arc physics. *Mailing Add:* Recteur Univ Montreal PO Box 6128 Stat A Montreal PQ H3C 3J7 Can

CLOUTIER, JAMES ROBERT, b Opelousas, La, Mar 19, 47; m 76; c 3. GUIDANCE, NAVIGATION & CONTROL TACTICAL MISSILES. *Educ:* Univ Southwestern La, BS, 69; Rice Univ, MA, 74, PhD(math sci), 75. *Prof Exp:* Mathematician Poseidon-Trident fire control, US Naval Surface Warfare Ctr, US Navy, 69-77, mathematician oceanog, US Naval Oceanog Off, 77-83; chief, Inertial Technol, 83-86, Advan Guidance Concepts, 86-87, tech adv, 88-90, SR SCIENTIST, GUIDANCE & CONTROL, ARMAMENT DIRECTORATE WRIGHT LAB, AIR FORCE ARMAMENT LAB, 90- *Concurrent Pos:* Adj asst prof, Univ Fla, 84-; mem, Am Inst Aeronauts & Astronauts Guidance Navig & Control Tech Comt, 87-90; mem, Inst Elec & Electronics Engrs Aerospace Controls Tech comt, 90- *Honors & Awards:* Goldsborough Award, US Naval Oceanog Off, 81. *Mem:* Assoc fel Am Inst Aeronaut & Astronauts; sr mem Inst Elec & Electronics Engrs. *Res:* Basic research and exploratory development in the guidance, navigation and control of tactical missiles; target acceleration modelling; target state estimation; advanced guidance laws; modern bank to turn autopilots; system integration. *Mailing Add:* 520 Golf Course Dr Niceville FL 32578

CLOUTIER, LEONCE, b Quebec City, Que, Feb 5, 28; m 51; c 3. CHEMICAL ENGINEERING. *Educ:* Laval Univ, BScA, 50, DSc(chem eng), 63. *Prof Exp:* Plant engr, Can Packers Ltd, 50-52; design engr nuclear reactor, C D Howe Co Ltd, 52-55; lectr, Laval Univ, 55-59, from assoc prof to prof chem eng, 59-88, dept head, 69-75, secy, fac sci & eng, 80-85; RETIRED. *Concurrent Pos:* Tech consult, Aluminum Co Can Ltd; vis prof, Univ Calif, Berkeley, 75-76. *Honors & Awards:* CIC Prize, 49. *Mem:* Int Asn Hydrogen Energy. *Res:* Efficiency and mechanism of mixing for liquid continuous systems; process instrumentation and control; chemical kinetics; applications of computers to chemical engineering. *Mailing Add:* 880 Lienard Sainte-Foy PQ G1V 2W5 Can

CLOUTIER, PAUL ANDREW, b Opelousas, La, Feb 7, 43; m 64; c 2. SPACE SCIENCE, PLASMA PHYSICS. *Educ:* Univ Southwestern La, BS, 64; Rice Univ, PhD(space sci), 67. *Prof Exp:* Res assoc space sci, 67, from asst prof to assoc prof, 67-76, PROF SPACE PHYSICS & ASTRON, RICE UNIV, 76- *Res:* Experimental research in ionospheric currents and fields, especially rocketborne magnetometers; theoretical research in macroscopic plasma phenomena, especially interaction of the solar wind with planetary atmospheres. *Mailing Add:* Dept Space Physics & Astron Rice Univ Houston TX 77001

CLOUTIER, ROGER JOSEPH, b North Attleboro, Mass, July 25, 30; m 54; c 5. HEALTH PHYSICS. *Educ:* Univ Mass, BS, 56; Univ Rochester, MS, 57. *Prof Exp:* Assoc engr, Westinghouse Elec Corp, 57-59; scientist, Med Div, Oak Ridge Assoc Univ, 59-74, chmn, Spec Training Div, 74-77, dir prof training, 77-89, assoc chmn, Mert Div, 86-89, ASSOC CHMN, EESD, OAK RIDGE ASSOC UNIV, 89- *Mem:* Health Physics Soc (pres, 82-83); Sigma Xi; Soc Risk Analysis. *Res:* Radiation safety and dosimetry; medical use of radioisotopes; environment evaluation. *Mailing Add:* EESD Oak Ridge Assoc Univs Oak Ridge TN 37831-0117

CLOUTMAN, LAWRENCE DEAN, b Pratt, Kans, Nov 5, 44; m 69; c 1. ASTROPHYSICS, FLUID DYNAMICS. *Educ:* Univ Kans, BS, 68; Ind Univ, MA, 71, PhD(astrophys), 72. *Prof Exp:* STAFF MEM NUMERICAL FLUID DYNAMICS, LOS ALAMOS SCI LAB, 72- *Mem:* Am Astron Soc; Sigma Xi. *Res:* Methodology development in numerical fluid dynamics; theoretical performance studies on gas centrifuges and internal combustion engines; theoretical astrophysics. *Mailing Add:* 3968 Fairland Dr Pheasanton CA 94588

CLOVER, RICHMOND BENNETT, b Johnson City, NY, Jan 30, 43; m 66; c 2. MAGNETISM. *Educ:* Cornell Univ, BS, 65; Yale Univ, MS, 67, PhD(appl physics), 69. *Prof Exp:* Mem tech staff, RCA Labs, 69-72; mem tech staff, Hewlett-Packard Lab, 72-73, dept mgr, 72-77; vpres eng & mkt, Intel Magnetics, Inc, 77-; MGR-TECHNOL EXCHANGE, INTEL CORP. *Mem:* Inst Elec & Electronics Engrs. *Res:* Investigation of magnetic bubble domain materials, devices and memory systems. *Mailing Add:* Intel Corp 2625 Walsh Rd SC4-270 Santa Clara CA 95051

CLOVER, WILLIAM JOHN, JR, b St Louis, Mo, Nov 2, 44; m 66; c 2. NONDIFFERENTIABLE & NONCONVEX OPTIMIZATION, DEPENDENT STOCHASTIC PROCESSES & SIMULATION. *Educ:* Ill Inst Technol, BS, 66, MS, 68. *Prof Exp:* Asst prof math, Concordia Univ, 69-73; SR SCIENTIST MATH, HORRIGAN ANALYSIS, 73- *Mem:* Am Math Soc; Math Asn Am; Soc Indust & Appl Math; AAAS; Sigma Xi. *Res:* Models of dependent probabilistic processes that are mixtures of dicrete and continuous subprocesses; algorithms that prescribe optimal inputs. *Mailing Add:* 412 S Seventh Ave Maywood IL 60153

CLOVIS, JAMES S, b Waynesburg, Pa, Aug 14, 37; m 70. PHYSICAL ORGANIC CHEMISTRY. *Educ:* Waynesburg Col, BS, 59; Calif Inst Technol, PhD(chem), 63. *Prof Exp:* Fel, Univ Munich, 62-63; mem staff, Rohm & Haas Co, 63-70, lab head, process chem, 70-74, proj leader, Pollution Control Res, 74-76, res & develop mgr, Indust Chem & Plastics, Europ Region, 76-79, mgr res & develop, Indust Chem, NAm Region, 79-81, dir res, indust chem, 82-83, dir res, New Technol, 83-86, DIR RES AG CHEM & SEPARATION TECHNOL, ROHM & HAAS CO, 86- *Mem:* Am Chem Soc. *Res:* Benzidine rearrangement; 1, 3-dipolar addition; phosphorus chemistry; plastics research; process research; pollution control research; ion exchange research; oil additive research. *Mailing Add:* 207 Pen Valley Terr Morrisville PA 19067

CLOVIS, JESSE FRANKLIN, b Clarksburg, WVa, Jan 31, 21; m 48; c 3. SYSTEMATIC BOTANY. *Educ:* WVa Univ, BSF, 47, MS, 52; Cornell Univ, PhD, 55. *Prof Exp:* Instr bot, Univ Conn, 55-57; from asst prof to assoc prof biol, WVa Univ, 57-72, pre-med adv, 63-70, prof biol, 72-83, emer prof, 83; RETIRED. *Mem:* Bot Soc Am; Am Soc Plant Taxonomists. *Res:* Aquatic plants; speciation. *Mailing Add:* 234 Willowdale Rd Morgantown WV 26505

CLOWER, DAN FREDRIC, b Crystal Springs, Miss, Mar 9, 28; m 51; c 4. COTTON PEST MANAGEMENT, INSECTICIDE RESISTANCE. *Educ:* La State Univ, BS, 49; Cornell Univ, PhD(econ entom), 55. *Prof Exp:* Asst entom, Cornell Univ, 50-55; asst entomologist, 55-58, from assoc prof to prof, 63-85, EMER PROF ENTOM, LA STATE UNIV, BATON ROUGE, 85- *Concurrent Pos:* Consult on cotton pest mgt & insecticide resistance, 85- *Mem:* Entom Soc Am; Sigma Xi. *Res:* Prevention and delay on the development of resistance in various cotton production areas. *Mailing Add:* 232 Ranchette Village Loop Searcy AR 72143

CLOWERS, CHURBY CONRAD, JR, b Little Rock, Ark, Mar 23, 34; m 59; c 2. ANALYTICAL CHEMISTRY. *Educ:* Univ Mo, Kansas City, BS, 57; Univ Mo, Columbia, PhD(analytical chem), 66. *Prof Exp:* Res chemist, Textile Fibers Dept, Nylon Tech Div, E I du Pont de Nemours & Co, Inc, 66-68; analytical res chemist, Anal Chem Dept, 68-83, SR ANALYTICAL CHEMIST II, MERRELL DOW RES CTR, MERRELL DOW PHARMACEUTICALS, INC, 83- *Mem:* Am Chem Soc. *Res:* Pharmaceutical, biochemical and metabolic analytical research; chemical instrumentation; physicochemical studies of drug degradation; chromatographic separations; spectra-structure relationships. *Mailing Add:* Three Revel Ct Cincinnati OH 45217

CLOWES, GEORGE HENRY ALEXANDER, JR, surgical metabolism; deceased, see previous edition for last biography

CLOWES, RONALD MARTIN, b Calgary, Alta, Mar 18, 42; m 68, 85. SEISMOLOGY, TECTONICS. *Educ:* Univ Alta, BSc, 64, MSc, 66, PhD(geophys), 69. *Prof Exp:* Nat Res Coun Can postdoctoral & hon res fel geophys, Australian Nat Univ, 69-70; from asst prof to assoc prof, 70-83, PROF GEOPHYS, UNIV BC, 83- *Concurrent Pos:* Consult, Horton Maritime Explor Ltd, 75-76; vis assoc prof, Inst Geophys, Copenhagen Univ, 79-80 & Lab Geophys, Aarhus Univ, Denmark, 79-80; Can Coun Villam res fel, 87-88. *Honors & Awards:* Soc Explor Geophysicists Award, 68. *Mem:* Am Geophys Union; Soc Explor Geophysicists; Geol Soc Am; Can Soc Explor Geophysicists; Can Geophys Union; Geol Asn Can. *Res:* Structure and properties of the earth's crust and upper mantle from analysis of reflected and refracted seismic waves generated by controlled sources; relationship of seismic results with tectonics and geology. *Mailing Add:* Dept Geophys & Astron Univ BC Vancouver BC V6T 1Z2 Can

CLOWES, ROYSTON COURTENAY, microbial genetics, biotechnology; deceased, see previous edition for last biography

CLOYD, GROVER DAVID, b Mosheim, Tenn, Oct 25, 18; m 85; c 3. VETERINARY MEDICINE. *Educ:* Auburn Univ, DVM, 42. *Prof Exp:* Self employed, Vet Med Pract, 47-55; dir vet serv, Ky Chem Indust, 55-57; asst dir field res pharmaceut prod, Richardson-Merrell Inc, 57-59, dir field res, 59-62, asst dir res, 62-65, sci develop aide gen mgr, 65-68; vet med dir, A H Robins Co, Inc, 68-71, dir vet med, 71-72, dir vet med & consumer prod res & develop, 72-83, regulatory affairs, 83-85; RETIRED. *Mem:* Am Asn Indust Vets (secy, 70-74, pres, 75-76); Am Vet Med Asn. *Res:* Pharmaceutical products research and development, veterinary and human. *Mailing Add:* 104 Fairwood Blvd Fairhope AL 36532

CLOYD, JAMES C, III, b Louisville, Ky, July 1, 48. PHARMACY, CLINICAL PHARMACOLOGY. *Educ:* Purdue Univ, BS, 71; Univ Ky, DPharm, 76. *Prof Exp:* Asst prof, 76-80, ASSOC PROF & CHMN DEPT PHARM PRACT, COL PHARM, UNIV MINN, 81- *Concurrent Pos:* Resident, Hosp Pharm, Univ Hosp, Univ Ky Med Ctr, 73-76; consult, licensure examr, Nat Asn Bds Pharm, 77-; fel parmacokinetics, Dept Pharmaceut, Sch Pharm, Univ Wash, 85-86. *Mem:* Am Asn Cols Pharm; Am Soc Hosp Pharmacists; Am Pharmaceut Asn; Am Epilepsy Soc; Acad Pharmaceut Sci. *Res:* Clinical pharmacology of anticonvulsant drugs; physical-chemical compatibility of drugs in intravenous solutions. *Mailing Add:* Health Sci Unit F Col Pharm Univ Minn 308 Harvard St SE Minneapolis MN 55455

CLUFF, CARWIN BRENT, b Central, Ariz, Feb 20, 35; m 68; c 5. HYDROLOGY. *Educ:* Univ Ariz, BS, 59, MS, 61; Colo State Univ, PhD, 77. *Prof Exp:* Asst engr civil eng, Calif Dept Water Resources, 61-62; res assoc hydrol, Water Resources Res Ctr, Univ Ariz, 62-63, from asst hydrologist to assoc hydrologist, 63-75; expert in hydrol, Food & Agr Orgn, UN, 75-76; ASSOC HYDROLOGIST, WATER RESOURCES RES CTR, UNIV ARIZ, 76- *Mem:* Am Soc Civil Engrs; Am Water Resources Asn. *Res:* Evaporation and seepage control; water harvesting; reuse of municipal waste water; solar energy. *Mailing Add:* Dept Hydrol Univ Ariz Tucson AZ 85721

CLUFF, EDWARD FULLER, b Dedham, Mass, Feb 14, 28; m 56; c 2. ORGANIC CHEMISTRY. *Educ:* Mass Inst Technol, BS, 49, PhD(org chem), 52. *Prof Exp:* Res chemist, 52-69, develop supvr, 69-70, res div head, 70-74, develop supt, 74-77, gen process supt, 77-80, res mgr, E I du Pont de Nemours & Co, Inc, 80-85; CONSULT POLYMER CHEM, 85- *Res:* Polymer chemistry; elastomers, including polyurethanes and fluoroelastomers; hydrocarbon elastomers and olefin polymerization via transition metal catalysts. *Mailing Add:* 706 Hertford Rd Wilmington DE 19803

CLUFF, LEIGHTON EGGERTSEN, b Salt Lake City, Utah, June 10, 23; m 44; c 2. ALLERGY, INFECTIOUS DISEASES. *Educ:* George Wash Univ, MD, 49. *Hon Degrees:* DSc, C Hahneman Med Univ, Long Island Univ, St Louis Univ & George Wash Univ. *Prof Exp:* House officer med, Johns Hopkins Hosp, Baltimore, 49-50; asst res physician, Hosp, Duke Univ, 50-51; asst res physician, Johns Hopkins Hosp, 51-52; asst physician & vis investr, Rockefeller Inst, 52-54; res physician, Johns Hopkins Hosp, 54-55, from instr to prof, 54-66; prof med & chmn dept, Col Med, Univ Fla, Gainesville, 66-76; vpres, Robert Wood Johnson Found, 76-80, exec vpres, 80-86, pres, 86-90; VET ADMIN DISTINGUISHED PHYSICIAN & PROF MED, UNIV FLA, 90- *Concurrent Pos:* Markle scholar, 55-62; consult, Food & Drug Admin; consult ed, Dermatol Dig; chmn training grant comn, Nat Inst Allergy & Infectious Dis; mem, Nat Res Coun-Nat Acad Sci Drug Res Bd; mem, Nat Comn Pharm & Pharm Educ; chmn comt biomed res & med training in the Vet Admin, Nat Acad Sci-Nat Res Coun; expert adv panel on bact dis, WHO. *Honors & Awards:* Ordronaux Award, 49; Res Leader Award, NIH, 62-66. *Mem:* Inst Med-Nat Acad Sci; Am Fedn Clin Res; Am Soc Clin Invest; Soc Exp Biol & Med; Am Acad Allergy; Asn Am Physicians; Am Clin Climat Asn. *Res:* Elucidation of the role of psychologic factors in convalescence from acute infection; studies of the epidemiology of staphylococcal infections; pathogenesis of fever due to bacterial pyrogens; mechanism of host injury in infection; studies of the epidemiology of adverse drug reactions; health care policy. *Mailing Add:* 8851 SW 45th Blvd Gainesville FL 32608

CLUFF, LLOYD STERLING, b Provo, Utah, Sept 29, 33; c 2. EARTHQUAKE TECHNOLOGY. *Educ:* Univ Utah, BS, 60. *Prof Exp:* Geologist, Lottridge Thomas & Assoc, 60; staff geologist, Woodward-Clyde Consult, 60-65, assoc & chief eng geologist, 65-71, vpres, prin & dir, 71-85; MGR, GEOSCI DEPT, PAC GAS & ELEC CO, 85- *Concurrent Pos:* Consult, Venezuelan Pres Earthquake Comn, 67-; mem consult bd, San Francisco Bay Conserv & Develop Comn, 68-73; mem consult panel to President & Secy Interior for Santa Barbara Oil Leak, 69; mem state Calif joint comt, Seismic Safety & Gov Earthquake Coun, 72-75; mem US Geol Surv earthquake adv panel, Nat Acad Sci comt seismol & SRI Int oversight comt earthquake prediction, 75-; mem consult bd, Int Atomic Energy Agency, Vienna. *Honors & Awards:* Hogentogler Award, Am Soc Testing & Mat, 65. *Mem:* Nat Acad Eng; Asn Eng Geologists (pres, 68-69); Earthquake Eng Res Inst; Int Asn Eng Geologists (vpres, 70-74); Seismol Soc Am; Struct Engrs Asn. *Res:* Active faults, earthquake and geologic hazards; engineering geology. *Mailing Add:* Geosci Dept Rm 768 Pac Gas & Elec 215 Market St San Francisco CA 94106

CLUFF, ROBERT MURRI, b Buenos Aires, Argentina, Jan 17, 53; US citizen; c 2. GEOLOGY. *Educ:* Univ Calif, Riverside, BS, 74; Univ Wis-Madison, MS, 76. *Prof Exp:* PRES, DISCOVERY GROUP, INC, 87- *Honors & Awards:* A I Levorsen Award, Am Asn Petrol Geologists, 80. *Mem:* Am Asn Petrol Geologists; Soc Econ Paleontologists & Mineralogists; Soc Petrol Engrs. *Res:* Sedimentology and diagenesis of carbonate rocks; sedimentology, paleoecology and depositional environments of shales; rock-geophysical log response; petrophysics. *Mailing Add:* Discovery Group Inc 535 16th St Suite 900 Denver CO 80202

CLUM, JAMES AVERY, b Sidney, NY, July 7, 37; m 62; c 1. PHYSICAL METALLURGY. *Educ:* Ohio State Univ, BMetE, 60; Carnegie Inst Technol, MSc, 63, PhD(metall eng), 68. *Prof Exp:* Res metallurgist, Battelle Mem Inst, 64-67; fel metall eng, Ohio State Univ, 67-68; res fel metall, Cambridge Univ, 68-69; asst prof phys metall, Univ Wis, Madison, 69-82, assoc dir Univ-Indust Res Prog, 70-82; PROF MECH ENG, STATE UNIV NY, BINGHAMTON, 85-, DIR INST RES ELECTRONIC PACKAGING, 88- *Concurrent Pos:* Ford Motor Co, 73-74, Vanderbilt Univ, 77-78; Ford Found fel, Am Soc Engr Educ, 73-74. *Mem:* AAAS; Am Soc Metals; Am Inst Mining Metall & Petrol Engrs; Am Soc Eng Educ; Am Ceramic Soc; Sigma Xi. *Res:* Surface physics; diffusion; nucleation; solid-solid phase transformations; relation of material properties and microstructure. *Mailing Add:* Mech & Indust Eng Watson Sch Eng State Univ NY Binghamton NY 13902-6000

CLUMP, CURTIS WILLIAM, b Reading, Pa, May 27, 23; m 44; c 2. CHEMICAL ENGINEERING. *Educ:* Bucknell Univ, BS, 47, MS, 49; Carnegie Inst Technol, PhD(chem eng), 54. *Prof Exp:* Res engr, Air Reduction, 47-48; instr, Univ Rochester, 48-50; res engr, Gen Foods, 53-55; from asst prof to prof, 55-88, chmn dept, 69-70, EMER PROF CHEM ENG, LEHIGH UNIV, 88- *Concurrent Pos:* Consult. *Honors & Awards:* Hillman Award, Lehigh Univ. *Mem:* Am Inst Chem Engrs; Am Soc Eng Educ. *Res:* Fluid mechanics; dry blending. *Mailing Add:* Dept Chem Eng Lehigh Univ Bethlehem PA 18015

CLUNIE, THOMAS JOHN, b Racine, Wis, Mar 8, 40; m 70. CHEMICAL ENGINEERING, APPLIED MATHEMATICS. *Educ:* Northwestern Univ, BS, 63; Univ Notre Dame, PhD(chem eng), 68. *Prof Exp:* Engr new investments develop, 67-69, sr proj engr, 69-72, group leader, 72-77, SECT HEAD, EXXON RES & ENG CO, TEX, 77- *Concurrent Pos:* Tech adv, Esso Res & Eng Co/Imp Oil Enterprises Ltd, 68-69. *Mem:* AAAS; Am Chem Soc; Am Inst Chem Engrs; Sigma Xi. *Res:* Unsteady state heat transfer; direct reduction of iron ore; liquified natural gas technology. *Mailing Add:* 2127 Hillside Oak Lane Houston TX 77062

CLUTTER, MARY ELIZABETH, b Charleroi, Pa; div. BOTANY. *Educ:* Allegheny Col, BS, 53; Univ Pittsburgh, MS, 57, PhD(bot), 60. *Hon Degrees:* DSc, Allegheny Col, 86. *Prof Exp:* Res assoc, Yale Univ, 61-73, lectr biol, 65-78, sr res assoc, 73-78; prog dir, NSF, 76-81, sect head, 81-84, sr sci adv, 85-87, div dir, 84-85 & 87-88, ASST DIR, NSF, 88- *Concurrent Pos:* Mem bd dirs, AAAS, 86-90 & Army Sci Bd, 88- *Honors & Awards:* Pres Rank Award, 85; Career Recognition Award, Women Cell Biol, 86. *Mem:* AAAS; Am Soc Cell Biol; Am Soc Plant Physiologists; Int Soc Plant Molecular Biol; Soc Develop Biol; Sigma Xi; Asn Women Sci. *Res:* Function of polytene chromosomes in plant embryo development. *Mailing Add:* NSF Washington DC 20550

CLUTTERHAM, DAVID ROBERT, b Chicago, Ill, Feb 10, 22; m 45, 82; c 6. MATHEMATICS. *Educ:* Cornell Col, BA, 45; Univ Ariz, MS, 48; Univ Ill, PhD(math), 53. *Prof Exp:* Asst math, Univ Ariz, 46-48 & Univ Ill, 48-49, asst digital comput, 50-53; design specialist, Convair Div, Gen Dynamics Corp, 53-59, sect chief comput design, Martin-Orlando, 59-63, mgr, Info Sci Dept, 63-64; mem tech staff, Bunker-Ramo Corp, Calif, 64-65; sr scientist, Radiation Inc, Fla, 65-68; prof math sci & head dept, 68-85, prof computer sci, 86-89, EMER PROF, FLA INST TECHNOL, 89- *Mem:* Am Math Soc; Asn Comput Mach; Inst Elec & Electronics Engrs; Sigma Xi. *Res:* Digital computers and computing techniques. *Mailing Add:* 200 Miami Ave Indialantic FL 32903

CLUXTON, DAVID H, b Martinsville, Ohio, 1943; m. EXPERIMENTAL PHYSICS, SCIENCE EDUCATION. *Educ:* Wilmington Col, AB, 65; Mich State Univ, MS, 67; Kent State Univ, PhD(physics), 72. *Prof Exp:* Asst prof, 72-77, ASSOC PROF PHYSICS, RUSSELL SAGE COL, 77- *Concurrent Pos:* Consult, Gen Elec Res & Develop Ctr, Schenectady, 75-81; NY Sect rep, Am Asn Physics Teachers, 84-92. *Mem:* Am Phys Soc; AAAS; Nat Asn Res Sci Teaching; Am Asn Physics Teachers. *Res:* Applications of physics to biological systems; intensity fluctuation spectroscopy. *Mailing Add:* Four Aspen Lane Clifton Park NY 12065-4807

CLYDE, CALVIN G(EARY), b Springville, Utah, Sept 5, 24; m 48; c 8. GROUND WATER HYDROLOGY, FLUID MECHANICS. *Educ:* Univ Utah, BS, 51; Univ Calif, Berkeley, MS, 52, CE, 53, PhD(civil eng), 61. *Prof Exp:* From instr to assoc prof civil eng, Univ Utah, 53-63; from asst dir to assoc dir, Utah Water Res Lab, 65-77, PROF CIVIL ENG, UTAH STATE UNIV, 63- *Concurrent Pos:* Sr develop engr, Hercules Inc, 61-63, consult, 62-63. *Mem:* Am Soc Civil Engrs; Am Soc Eng Educ; Am Inst Aeronaut & Astronaut; Nat Water Well Asn. *Res:* Turbulence, instrumentation; hydraulics; sediment transport; ground water hydrology; water resources planning; systems analysis; simulation of water resources systems; hydrology; hydroelectric power development. *Mailing Add:* 839 N 1400 E Logan UT 84321

CLYDE, WALLACE ALEXANDER, JR, b Birmingham, Ala, Nov 7, 29; m 53; c 3. PEDIATRICS, INFECTIOUS DISEASES. *Educ:* Vanderbilt Univ, BA, 51, MD, 54. *Prof Exp:* Intern pediat, Univ Hosp, Vanderbilt Univ, 54-55; asst resident, NC Baptist Hosp, 55-56; resident, Univ Hosp, Vanderbilt Univ, 56-57; trainee infectious dis, 61-62, from instr to assoc prof pediat, 62-72, assoc prof bact, 68-72, PROF PEDIAT & BACT, UNIV NC, CHAPEL HILL, 72- *Concurrent Pos:* Nat Inst Allergy & Infectious Dis fel prev med, Case Western Reserve Univ, 59-61, career develop award, 63-73; assoc mem comn acute respiratory dis, Armed Forces Epidemiol Bd, 63-72; mem bact-mycol study sect, Div Res Grants, NIH, 66-70; vis assoc prof path,Yale Univ, 71-72; prin investr, Pediat Pulmonary Specialized Ctr Res, 76-91; spec adv, Pan Am Health Orgn, 81-; chmn-elec, Int Orgn Mycoplasmology, 82-84, chmn, 84-86, 90; vis scientist, Statens Seruminstitut, Copenhagen, Denmark, 88. *Honors & Awards:* Bronze Medal, City of Bordeaux, France, 83. *Mem:* Am Soc Microbiol; Soc Pediat Res; Am Soc Clin Invest; Infectious Dis Soc Am; Am Pediat Soc; Int Orgn Mycoplasmology (secy-gen, 80-82). *Res:* Infectious diseases of children, especially nonbacterial respiratory infections; relationship of Mycoplasmataceae to human disease; respiratory disease pathogenesis. *Mailing Add:* Dept Pediat Sch Med Univ NC CB#7220 535 Burnett Womack Bldg Chapel Hill NC 27599-7220

CLYDESDALE, FERGUS MACDONALD, b Toronto, Ont, Feb 19, 37; m 79; c 2. CHEMISTRY, FOOD SCIENCE. *Educ:* Univ Toronto, BA, 60, MA, 62; Univ Mass, PhD(food sci), 66. *Prof Exp:* Chemist, Can Industs Ltd, 60; physiol chemist, Can Defence Res Med Lab, 62; from fel to asst prof food sci, 66-72, assoc prof, 72-76, prof food sci & nutrit, 76-89, PROF & HEAD DEPT, DEPT FOOD SCI, UNIV MASS, AMHERST, 89- *Honors & Awards:* William V Cruess Award, Inst Food Technologists, 76, Babcock Hart Award, 84, Tanner Lectureship Award, Chicago, 85, Tressler Award, 86, Nicholas Appert Award, 89. *Mem:* AAAS; Inst Food Technologists; Am Chem Soc; Inter-Soc Color Coun; Am Inst Nutrit. *Res:* Basic chemical changes in processed foods and their effect on quality and nutrition; basic color measurement problems involved with foods; chemistry of minerals and its relationship to bioavailability. *Mailing Add:* Dept Food Sci & Nutrit Univ Mass Amherst MA 01003

CLYMER, ARTHUR BENJAMIN, b Cleveland, Ohio, Aug 7, 20; m 53; c 2. CHRONTOLOGY. *Educ:* Oberlin Col, BA, 41; Ohio State Univ, MS, 46; Mass Inst Technol, MS, 48. *Prof Exp:* Math engr, Owens-Ill Glass Co, 49-53; res coordr, Bituminous Coal Res, Inc, 53-55; sr tech specialist, NAm Aviation, Inc, 55-61; environ engr, Ohio Environ Protection Agency, 72-76; simulation engr, Electronic Asn, Inc, 77-81; dir training, Autodynamics, Inc, 81-83; RETIRED. *Concurrent Pos:* Pvt consult, Clymer Technol, 61-72, 76-77 & 83-88. *Honors & Awards:* Sr Sci Simulation Award, Soc Computer Simulation, 70. *Mem:* Am Soc Mech Eng; Soc Computer Simulation; Inst Elec & Electronics Engrs. *Res:* Determination of the formula for the ages at which events can occur in the development and later life of organisms; observation of many events having multimodal events ages; applications to evolution. *Mailing Add:* 32 Willow Dr Apt 1B Ocean NJ 07712

CMEJLA, HOWARD EDWARD, b Milwaukee, Wis, Dec 25, 26; m 48; c 6. PARASITOLOGY, ENTOMOLOGY. *Educ:* Univ Wis, BS, 50, MS, 51, PhD(entom, zool), 54. *Prof Exp:* Collabr, Div Apicult & Biol Control, USDA, Wis, 51-54; parasitologist, Parasitol Dept, Abbott Labs, 54-58, Food & Drug Admin liaison, 58-63; dir regulatory liaison, 63-70, vpres & dir prod planning & develop, 69-79, DIR, RES & DEVELOP ADMIN, AYERST LABS, 79- *Concurrent Pos:* Food & Drug Admin liaison. *Mem:* Am Soc Parasitologists; Entom Soc Am; Am Soc Trop Med & Hyg; Sigma Xi. *Res:* Drugs-product planning and development, and research and development administration; amebiasis; apiculture; bee diseases and colony management. *Mailing Add:* 452 Dorset Dr Cocoa Beach FL 32931-3834

COACHMAN, LAWRENCE KEYES, b Rochester, NY, Apr 25, 26; m 81; c 4. PHYSICAL OCEANOGRAPHY. *Educ:* Dartmouth Col, AB, 48; Yale Univ, MF, 51; Univ Wash, PhD(oceanog), 62. *Prof Exp:* Hydrographer & oceanogr, Dartmouth Col Blue Dolphin Labrador Expeds, 50-55; sr scientist gases in glacier ice, Univ Oslo, 55-57; asst, 57-62, from asst prof to assoc prof, 62-72, PROF OCEANOG, UNIV WASH, 72- *Concurrent Pos:* Sr scientist, Arctic Inst NAm Expeds, Greenland, 58 & 63, Bering & Chukchi Seas, 64, 66 & 69-78, Cent Arctic Ocean, 70-72; mem Saronikos Systs Proj, Athens, 73-75; chmn, US-USSR Oceanog Exchange Deleg, 64; Arctic Inst NAm/McGill Univ vis prof, 67; vis lectr, AARI, Leningrad, 76; PROBES ecosyst study, 76-82; ISHTAR ecosyst study, 84-; vis prof, Universite de Liege, 87-88. *Mem:* Am Geophys Union; fel Arctic Inst NAm. *Res:* Physical oceanography of Arctic Ocean and peripheral seas; oceanography of coastal, fjord and ice-covered waters. *Mailing Add:* Sch Oceanog WB-10 Univ Wash Seattle WA 98195

COAD, BRIAN WILLIAM, b England, Oct 19, 46; m 73; c 1. ICHTHYOLOGY, ZOOGEOGRAPHY. *Educ:* Univ Manchester, Eng, BSc, 70; Univ Waterloo, Ont, MSc, 72; Univ Ottawa, PhD(biol), 76. *Prof Exp:* Assoc prof ichthyol & zool, Pahlavi Univ, Shiraz, Iran, 76-79; res assoc, 79-81, assoc cur, 81-86, CUR FISHES, NAT MUS NATURAL SCI, OTTAWA, CAN, 86- *Mem:* Am Soc Ichthyologists & Herpetologists; Freshwater Biol Asn; Japanese Soc Ichthyol; Indian Soc Ichthyol; French Soc Ichthyol. *Res:* Systematics and zoogeography of the fresh water fishes of Southwest Asia and of North America. *Mailing Add:* 1021 Castle Hill Crescent Ottawa ON K1P 6P4 Can

COADY, LARRY B, b Ottumwa, Iowa, Mar 4, 33; m 56; c 10. ELECTRICAL ENGINEERING. *Educ:* Iowa State Univ, BS, 59, MS, 63, PhD(elec eng), 65. *Prof Exp:* Jr res engr, Collins Radio Co, 59-62; from instr to asst prof, 62-75, ASSOC PROF ELEC ENG, IOWA STATE UNIV, 75- *Mem:* Inst Elec & Electronics Engrs. *Res:* Circuit and systems theory. *Mailing Add:* Dept Elec Eng Iowa State Univ 201 Coover Ames IA 50011

COAKER, A(NTHONY) WILLIAM, b Johannesburg, SAfrica, Oct 8, 27; m 54; c 5. CHEMICAL ENGINEERING. *Educ:* Univ Witwatersrand, BSc, 49, PhD(chem eng), 53. *Prof Exp:* Demonstr chem, Univ Witwatersrand, 49-50; asst works chemist chem eng, Masonite Africa, Ltd, 51-52; asst res officer chem, SACSIR, 52-53; chemist plastics technol, Monsanto Co, 53-56, group leader vinyl plastics, 57-64, group leader plasticizer applns, 64-66, sr group leader, Org Chem Div, 66-71, mkt supvr plasticizers, Monsanto Indust Chem Co, 71-75; vpres, Triple R Indust, 76-77; DIR, MGF SERV, TENNECO CHEMICALS INC, 78- *Mem:* Soc Plastics Engrs; Am Chem Soc; Am Inst Chem Engrs. *Res:* Polymerization; thermoplastics processing; plastisol technology; compounding of vinyl plastics; industrial wastes treatment; rheology. *Mailing Add:* 6726 Chadbourne Dr North Olmstead OH 44070

COAKLEY, CHARLES SEYMOUR, b Washington, DC, July 4, 14; m; c 5. ANESTHESIOLOGY. *Educ:* George Washington Univ, MD, 37; Am Bd Anesthesiol, dipl, 48. *Prof Exp:* From instr to assoc prof, 39-49, PROF ANESTHESIOL & CHMN DEPT, SCH MED, GEORGE WASHINGTON UNIV, 49- *Concurrent Pos:* Consult, Walter Reed Army Med Ctr, Vet Admin Hosp, Wash, DC & NIH; med consult, CARE-Medico. *Mem:* AMA; Am Col Anesthesiol; Asn Am Med Cols; Am Soc Anesthesiol; Asn Univ Anesthetists. *Res:* Use of monitors for anesthetized and critically ill patients; applications of computer of continuous automated monitoring. *Mailing Add:* 5175 Watson St Washington DC 20016

COAKLEY, JAMES ALEXANDER, JR, b Long Beach, Calif, Dec 14, 46; m 71; c 2. REMOTE SENSING. *Educ:* Univ Calif, Los Angeles, BS, 68; Univ Calif, Berkeley, MA, 70, PhD(physics), 72. *Prof Exp:* Scientist climate, Nat Ctr Atmospheric Res, 78-88, PROF, SR RES DEPT, ATMOSPHERIC SCI, ORE STATE UNIV, 88- *Concurrent Pos:* Nat Ctr Atmospheric Res fel, 72-73; res assoc, Nat Acad Sci-Nat Res Coun-Nat Environ Satellite Serv, 73-74; affil prof & vis asst prof, Univ Corp Atmospheric Res, Dept Earth Sci, Iowa State Univ, 76-78; lectr, dept asto-geophysics, Univ Colo, 82-85; assoc ed, J Geophys Res, 81-86; vis scholar, Calif Space Inst, Univ Calif, San Diego & Jet Propulsion Lab, 85-86; assoc ed J of Climate, 88-; ed adv bd, Tellus B, 88; AMS comt on Atmospheric Radiation, 86; chmn comt on Atmospheric Radiation, 88. *Mem:* AAAS; Am Geophys Union; Am Meteorol Soc. *Res:* Climate models; radiative transfer in planetary atmospheres; remote sensing of atmospheres from space. *Mailing Add:* Dept Atmospheric Sci Ore State Univ Corvallis OR 97331-2209

COAKLEY, JOHN PHILLIP, b Nassau, Bahamas, Jan 7, 40; m 68; c 2. SEDIMENTOLOGY, COASTAL ENGINEERING. *Educ:* St Francis Xavier Univ, BSc, 64; Univ Ottawa, MSc, 67; Univ Waterloo, PhD, 85. *Prof Exp:* Sci officer quant geol limnogeol, Geol Surv Can, 66-67; sci officer, 67-72, RES SCIENTIST NEARSHORE SEDIMENTOLOGY, LAKES RES BR, CAN, DEPT ENVIRON, 72- *Concurrent Pos:* Dir environ control, Bahamas Develop Corp, 75-76. *Mem:* Int Asn Great Lakes Res; Sigma Xi; Geol Asn Can. *Res:* Processes and response of coastal systems in the Great Lakes; paleoenvironments and long-term post-glacial evolution of Great Lakes shorelines; tracing contaminated sediment transport. *Mailing Add:* 23 Lower Horning Rd Hamilton ON L8S 3E9 Can

COAKLEY, MARY PETER, b South Amboy, NJ. PHYSICAL CHEMISTRY, INORGANIC CHEMISTRY. *Educ:* Georgian Court Col, AB, 47; Univ Notre Dame, MS, 53, PhD, 55. *Prof Exp:* Parochial high sch teacher, 42-45; TEACHER CHEM & CHMN DEPT, GEORGIAN COURT COL, 46- *Concurrent Pos:* Mem atomic energy proj, Univ Notre Dame, 54-55; res grants, Atomic Energy Comn, 57-60 & 61-63, Petrol Res Fund, 66-68 & NSF, 68-70; res, Univ Calif, Berkeley, 68. *Mem:* Am Chem Soc; NY Acad Sci. *Res:* Nuclear magnetic resonance studies of tin complexes; spectroscopy; ultraviolet and infrared absorption; spectra of metal chelates. *Mailing Add:* Dept Chem Georgian Court Col Lakewood NJ 08701

COALE, ANSLEY J, b Baltimore, Md, Nov 14, 17. DEMOGRAPHY. *Educ:* Princeton Univ, BA, 39, MA, 41, PhD, 47. *Hon Degrees:* DHC, Univ Louvain, 79, Univ Leige, 82; LLD, Univ Penn, 83. *Prof Exp:* From asst prof to prof, 47-86, William Church Osborne prof, 64-86, EMER PROF ECON & PUB AFFAIRS, PRINCETON UNIV, 86-, RES ASSOC, OFF POP RES, 86- *Concurrent Pos:* Fel Brit Acad. *Honors & Awards:* Mindel Sheps Prize Math Demog, 74; Irene Taeuber Award, Pop Asn Am, 83. *Mem:* Nat Acad Sci; fel Am Acad Arts & Sci; fel Am Statist Asn; fel AAAS; Am Philos Soc. *Res:* Mathematical demography. *Mailing Add:* Princeton Univ 21 Prospect Ave Princeton NJ 08544

COALSON, JACQUELINE JONES, b Oklahoma City, Okla, Mar 12, 38; div; c 3. PATHOLOGY. *Educ:* Okla Baptist Univ, BS, 60; Univ Okla, MS, 63, PhD(path), 65. *Prof Exp:* Asst assoc full prof path, Med Ctr, Univ Okla, 68-78; PROF PATH, UNIV TEX MED CTR, SAN ANTONIO, 79- *Concurrent Pos:* Mem, Pulmonary Dis Adv Comt, Div Lung Res, NIH, 76-81, chmn & mem, Res Manpower Rev Comt, Nat Heart, Lung & Blood Inst, 81-86. *Mem:* Am Thoracic Soc; Am Asn Path; Int Acad Pathol; Am Soc Cell Biol. *Res:* Morphologic, morphometric, and immunocytochemical studies of diseased lungs of both human and experimental animals. *Mailing Add:* 96 Shavano Dr San Antonio TX 78231

COALSON, ROBERT ELLIS, b Hobart, Okla, Dec 7, 28; m 62; c 3. ANATOMY. *Educ:* Univ Okla, BS, 49, MS, 51, PhD(med sci), 55. *Prof Exp:* Instr anat, Sch Nursing, Univ Okla, 52-54; instr, Sch Med, Vanderbilt Univ, 57-60; from asst prof to assoc prof, 60-70, PROF ANAT, SCH MED, UNIV OKLA, 70-, ASST PROF PATH, 68- *Concurrent Pos:* Consult, Meharry Med Col, 58-60. *Mem:* Am Soc Zool; Tissue Cult Asn; Sigma Xi. *Res:* Comparative embryology, anatomy and histology; tissue transplantation. *Mailing Add:* Anat Dept Univ Okla Med Ctr PO Box 26901 Oklahoma City OK 73190

COAN, EUGENE VICTOR, b Los Angeles, Calif, Mar 26, 43. ENVIRONMENTAL SCIENCES, MALACOLOGY. *Educ:* Univ Calif, Santa Barbara, AB, 64; Stanford Univ, PhD(biol sci), 69. *Prof Exp:* Dir polit activ, Zero Pop Growth, 69-70; consult, 70-75, major issues specialist, 75-76, ASST CONSERV DIR, SIERRA CLUB, 77- *Concurrent Pos:* Ed, Western Soc Malacol; res assoc, Calif Acad Sci, Santa Barbara Mus Natural Hist & La County Mus Natural Hist; treas, Calif Malacozoological Soc; co-ed, Malacologia. *Mem:* Am Malacol Union; Nat Audubon Soc; Sierra Club. *Res:* History of natural sciences; taxonomy and distribution of northwest American bivalves; taxonomic studies on groups of bivalve mollusks in the eastern pacific; taxa and bibliographics of early west coast malacologists. *Mailing Add:* 891 San Jude Ave Palo Alto CA 94306

COAN, STEPHEN B, b New York, NY, Apr 8, 21; m 42; c 2. ORGANIC CHEMISTRY. *Educ:* Univ Mich, BS, 41; Polytech Inst Brooklyn, MS, 50, PhD(chem), 54. *Prof Exp:* Analytical chemist, Gen Chem Co, 41-42, TNT control chemist, 42-43, res chemist, 46; develop chemist, 46-51, res chemist, 51-54, patents-res liaison, 51-63, from assoc dir to dir, Patent Dept, 63-78, STAFF VPRES LICENSING, SCHERING CORP, 78- *Mem:* Am Chem Soc; Licensing Exec Asn. *Res:* Organic and medicinal research; organic synthesis; pharmaceuticals. *Mailing Add:* 72 Sykes Ave Livingston NJ 07039

COAR, RICHARD J, b Hanover, NH, May 2, 21. ENGINEERING. *Educ:* Tufts Univ, BS, 42. *Prof Exp:* Engr, 42-56, chief engr, Fla res & develop ctr, 56-70, asst gen mgr, 70-71, vpres, 71-76, exec vpres, 76-83, pres, Pratt & Whitney, 83-84; exec vpres, United Technol, 84-86; RETIRED. *Concurrent Pos:* Consult, technol export, Dept Defense; mem, Aeronaut & Space Eng Bd, NASA. *Honors & Awards:* George Westinghouse Gold Medal, Am Soc Mech Engrs. *Mem:* Nat Acad Eng; Am Soc Metals. *Mailing Add:* 2802 SE Dunes Dr Stuart FL 34996

COARTNEY, JAMES S, b Coles Colo, Ill, Sept 3, 38; m 64; c 2. PLANT PHYSIOLOGY, WEED SCIENCE. *Educ:* Eastern Ill Univ, BSEd, 60; Purdue Univ, MS, 63, PhD(plant physiol), 67. *Prof Exp:* Res asst plant physiol, Purdue Univ, 60-66; from asst prof to assoc prof plant physiol, 66-79, ASSOC PROF HORT, VA POLYTECH INST & STATE UNIV, 79- *Mem:* Weed Sci Soc Am; Am Soc Hort Sci; Int Plant Propagators Soc. *Res:* Plant growth regulation and mode of action of herbicides. *Mailing Add:* 206 Locust St Floyd VA 24091

COASH, JOHN RUSSELL, b Denver, Colo, Sept 24, 22; m 48; c 2. GEOLOGY. *Educ:* Colo Col, BA, 47; Univ Colo, MA, 49; Yale Univ, PhD, 54. *Prof Exp:* From asst prof to prof, Bowling Green State Univ, 49-64, dir res, 65-66; assoc prog dir, NSF, 66-68; dean, Sch Arts & Sci, 68-87, PROF GEOL, CALIF STATE COL, BAKERSFIELD, 87- *Concurrent Pos:* NSF vis lectr, 62, 63, 64 & 65; chmn dept, Bowling Green State Univ, 54-64, asst to provost & dir hon prog, 63-65; panelist, NSF, 68- & NIH, 72-73. *Honors & Awards:* RW Webb Award, Nat Asn Geol Teachers, 84. *Mem:* Fel AAAS; fel Geol Soc Am; Am Asn Petrol Geol; Nat Asn Geol Teachers (pres, 81); Nat Sci Teachers Asn. *Res:* Field geology of northern Nevada, central Rocky Mountains and Sierra Nevada; stratigraphic and structural geology; science administration. *Mailing Add:* Calif State Col 9001 Stockdale Hwy Bakersfield CA 93309

COATES, ANTHONY GEORGE, b Staines, Eng, May 20, 36; m 61; c 1. PALEONTOLOGY, STRATIGRAPHY. *Educ:* Univ London, BSc, 59, PhD(geol), 63. *Prof Exp:* Geologist, Jamaican Geol Surv, 62-64; lectr geol, Univ West Indies, 64-67; assoc prof, 67-74, asst vpres Acad Affairs, 86-89, PROF GEOL, GEORGE WASHINGTON UNIV, 74-, ASSOC VPRES ACAD AFFAIRS & RES, 89- *Concurrent Pos:* res assoc, Smithsonian Inst, 68-; sr lectr, Kings Col, London, 72-73; vis prof, Museum Natural Hist, Paris, 75-76. *Mem:* Paleont Soc Am; fel Geol Soc London; Geol Soc France; Geol Soc Am. *Res:* Paleontology, stratigraphy and sedimentation of ordovician of Normandy; cretaceous Caribbean corals and biostratigraphy; Jamaican cretaceous stratigraphy; Biology, Geology of clonal organisms, Isthmus of Panama as a biogeographic barrier. *Mailing Add:* Acad Affairs George Washington Univ Washington DC 20052

COATES, ARTHUR DONWELL, b Steubenville, Ohio, June 14, 28; m 57; c 3. PHYSICAL CHEMISTRY, FUEL SCIENCE. *Educ:* Col Steubenville, BS, 50; Univ Del, MS, 61. *Prof Exp:* Practice engr chem & metall, Wheeling Steel Corp, 50-51; chemist, Ballistic Res Labs, 51-55, chief ignition sect, Combustion & Incendiary Effects Br, 55-57, chief, Spec Prob Sect, Nuclear Physics Br, 57-59, chief, Radiation Damage Sect, Nuclear Physics Br, 60-70, chief, Methodology Sect, Combustion & Incendiary Effects Br, 70-75, tech asst to chief detonation & deflagration dynamics lab, 75-76, tech asst to chief, Terminal Ballistics Div, 76-78, spec asst to dir, 79-86; PRES, SILVER FARMS, INC, 82-; CONSULT, BALLISTIC TECHNOL, COMBUSTION CHEM, INCENDIARY MAT APPLNS & TECH WRITING, 86- *Mem:* AAAS; Am Chem Soc; fel Am Inst Chem; NY Acad Sci; Sigma Xi; Am Defense Preparedness Asn. *Res:* Fuels; combustion; powdered metals; pyrophoric materials; propellants; explosives; mass spectrometry; thin film physics; ignition; chemistry of exothermic reactions; thermal analysis. *Mailing Add:* 311 N Osborn Lane Aberdeen MD 21001

COATES, CLARENCE L(EROY), JR, b Hastings, Nebr, Nov 5, 23; m 43, 69; c 3. COMPUTER ENGINEERING. *Educ:* Univ Kans, BS, 44, MS, 48; Univ Ill, PhD(elec eng), 53. *Prof Exp:* Instr elec eng, Univ Kans, 46-48; from instr to assoc prof, Univ Ill, 48-56; res engr, Info Studies Sect, Gen Elec Res Lab, 56-63; prof elec eng, Univ Tex, Austin, 63-67, prof elec eng & comput sci, 67-71, chmn dept elec eng, 64-66, dir electronics res ctr, 67-71; prof elec eng & dir, Coord Sci Lab, Univ Ill, Urbana-Champaign, 71-73; head, Sch Elec Eng, Purdue Univ, 73-83, prof, 73-88; RETIRED. *Concurrent Pos:* Adj prof, Rensselaer Polytech Inst, 58-63; mem, Comput Elec Eng Comt, Comn Educ, Nat Acad Eng, 68-73, chmn, 71-73. *Mem:* Fel AAAS; fel Inst Elec & Electronics Engrs. *Res:* Switching theory; adaptive systems; computer organization; recognition systems. *Mailing Add:* 26 Stoney Ridge Asheville NC 28804

COATES, DONALD ALLEN, b Sonoma, Calif, Apr 10, 38; c 3. STRATIGRAPHY, ENVIRONMENTAL GEOLOGY. *Educ:* Univ Colo, BA, 61, MS, 64; Univ Calif, Los Angeles, PhD(geol), 69. *Prof Exp:* Geologist, Texaco Inc, 63-64 & CRA Explor, 64-65; res assoc, Inst Polar Studies, Ohio State Univ, 69-71; asst prof geol, Cleveland State Univ, 71-75; GEOLOGIST, US GEOL SURV, 75- *Concurrent Pos:* Adj asst prof, Calif State Univ, Los Angeles, 69, Ohio State Univ, 69-71; consult, 69-75. *Mem:* Geol Soc Am; AAAS; Sigma Xi. *Res:* Stratigraphy of Gondwana glacial deposits in South America and Antarctica; natural burning of coal beds and structure, petrology, paleomagnetism, and relation to landscape development of the baked rocks (clinker) produced; quaternary geology and recent tectonics of the Bengal delta, Bangladesh. *Mailing Add:* US Geol Surv Box 25046 Fed Ctr MS-972 Denver CO 80225

COATES, DONALD ROBERT, b Grand Island, Nebr, July 23, 22; m 44; c 3. GEOLOGY, GEOMORPHOLOGY. *Educ:* Col Wooster, BA, 44; Columbia Univ, MA, 48, PhD(geol), 56. *Prof Exp:* Asst geol, Columbia Univ, 46-48; asst prof & head dept, Earlham Col, 48-51; geologist, Ground Water Br, US Geol Sury, 51-54; chmn dept 54-63, from instr to assoc prof, 54-63, PROF GEOL, STATE UNIV NY BINGHAMTON, 63- *Concurrent Pos:* Party chief, Ind Geol Surv, 49; lectr, Ind Univ, 50; res geologist, Off Naval Res, 54; consult, Chernin & Gold, NY, 55-68; geologist, Gen Hydrol Br, US Geol Surv, 58-60; vis prof, Cornell Univ, 58, 60, 61, 84 & 85; State Univ NY Res Found fels, 61 & 66, grants-in-aid, 62-65; vis prof, Univ Ill, 63; assoc prog dir, NSF, 63-64, consult, 64-69; vis geoscientist for Am Geol Inst, 63-65; consult, US Army Corps Engrs, 65-68, NY State Atty Gen, 65-, US Dept Com, 72-75 & Consol Edison of New York, 75-76; proj dir, NY State Atomic & Space Develop Authority, 75; consult, US Nat Park Serv, 71-77, NY Dept Transp, 76- & Niagara Mohawk Power Corp, 76-; Consult, Town of Islip, 74-77, Town of Vestal, 81-, Town of Norwich, Broome County, 81- & NY Dept Environ Consv, 84-; Empire State Elec Energy Res Corp, 78-, NY State Elec & Gas Corp, 83-85, Town of Vernon, Town of Deerfield, Town of Trenton, 85-, Pollution Enterprises, Inc, 82-87, Town of Nichols, 87- & Town of Rodman, 87-; Susquehanna-Cayuga Groundwater Protection Asn, 85-, Pure Water for Life, Inc, 87-, Fahs Construct, Inc, 87-, Hagopian Eng, Inc, 88 & Picciano Construct, Inc, 88, Aetna Insurance, 87- & Mohonk Preserve, 89- *Mem:* Asn Eng Geologists; Geol Soc Am; Asn Prof Geol Scientists; AAAS. *Res:* Geomorphology; environmental geology and environmental lawsuits; glacial geology of eastern United States; analysis of man's changes of rivers and coasts; evaluation of water and earth surface resources; landslides; geomorphology and engineering; urban, glacial & coastal geomorphology; analysis of contamination of wells; pollutants, including industrial materials, gasoline spills and salinization; glaciation of the Catskill Mountains; landfill investigations; groundwater studies. *Mailing Add:* Dept Geol Sci State Univ NY Binghamton NY 13901

COATES, GARY JOSEPH, b Annapolis, Md, July 20, 47; div; c 1. PASSIVE SOLAR ENERGY DESIGN. *Educ:* NC State Univ, BS, 69, Master Archit. *Prof Exp:* Asst prof environ psychol, design methods & planning strategies, dept design & environ analysis, NY State Col Human Ecol, Cornell Univ, 71-77; assoc prof, 77-84, PROF PASSIVE SOLAR DESIGN & SUSTAINABLE COMMUNITY DESIGN, DEPT ARCHIT & DESIGN, KANS STATE UNIV, 84- *Concurrent Pos:* Dir, Univ Man Appropriate Technol Prog, Manhattan, Kans, 78-83, consult, 83- *Mem:* Soc Bldg Sci Educrs; Am Solar Energy Soc. *Res:* Technical, social, economic, educational and ethical requirements for the creation of a renewable resource-based, sustainable society; community energy planning; ecological design and organic architecture. *Mailing Add:* Dept Archit Kans State Univ Manhattan KS 66506

COATES, GEOFFREY EDWARD, b London, Eng, May 14, 17; m 51; c 2. ORGANOMETALLIC CHEMISTRY, INORGANIC CHEMISTRY. *Educ:* Oxford Univ, BA, 38, BSc, 39, MA, 42; Bristol Univ, DSc(chem), 54. *Prof Exp:* Res chemist, Magnesium Metal Corp, Eng, 40-45; lectr chem, Bristol Univ, 45-53; prof chem & head dept, Univ Durham, 53-68; head dept chem, Univ Wyo, 68-77, prof, 68-79; RETIRED. *Concurrent Pos:* Consult, Rio Tinto Zinc Corp, 47-71; Imp Chem Indust, 55-68; Clarke, Chapman & Co, 55-69 & Ethyl Corp, 57-69. *Mem:* Am Chem Soc; Royal Soc Chem; Royal Inst Chem; Sigma Xi. *Res:* Beryllium chemistry; organometallic compounds. *Mailing Add:* 1801 Rainbow Ave Laramie WY 82070

COATES, JESSE, b Baton Rouge, La, Mar 12, 08; m 39; c 2. CHEMICAL ENGINEERING. *Educ:* La State Univ, BS, 28; Univ Mich, MS & PhD(chem eng), 36. *Prof Exp:* Chemist & treating engr, Nat Lumber & Creosoting Co, 28; chemist, Int Paper Co, 28-29, Meeker Sugar Refining, 30-31 & Punta Alegre Sugar Co, 31; chem engr, Tex Pac Coal & Oil Co, 32-33 & United Gas Pub Service, 33-36; from asst prof to prof, 36-69, chmn dept, 55-67, 69-70, EMER ALUMNI PROF CHEM ENG, LA STATE UNIV, BATON ROUGE, 69- *Concurrent Pos:* Consult, Nat Gas Odorizing Co, 45-56; mem Nat Coun Eng Exam. *Honors & Awards:* Coates Mem Award, Am Chem Soc, 58. *Mem:* Am Chem Soc; fel Am Inst Chem Engrs; fel Am Inst Chemists; Am Soc Eng Educ. *Res:* Thermodynamics of solutions; thermal conductivity of liquids; mass transfer; distillation; evaporation; process development. *Mailing Add:* St James Pl Apt 344 333 Lee Dr Baton Rouge LA 70808

COATES, JOSEPH FRANCIS, b Brooklyn, NY, Jan 3, 29; m 52; c 5. TECHNOLOGY ASSESSMENT, FUTURES RESEARCH. *Educ:* Polytech Inst Brooklyn, BS, 51; Pa State Univ, MS, 53. *Hon Degrees:* DSc, Clarmont Grad Sch, 85. *Prof Exp:* Res chemist, Atlantic Refining Co, Philadelphia, 53-60; chief chemist, Onyx Chem Co, Jersey City, 60-61; staff scientist, Inst Defense Analysis, Arlington, 61-70; proj mgr, NSF, Wash, DC, 70-74; asst to dir, Off Technol Assessment, US Cong, 74-78, sr assoc, 78-79; PRES, J F COATES INC, WASH, DC, 79- *Concurrent Pos:* Adj prof, Am Univ, 71-73 & George Washington Univ, 72- *Mem:* AAAS; World Future Soc; Am Chem Soc. *Res:* Impacts of science and technology on society and on the future. *Mailing Add:* 3738 Kanawha St NW Washington DC 20015

COATES, R(OBERT) J(AY), b Lansing, Mich, May 8, 22; m 46; c 1. PHYSICS, ELECTRICAL ENGINEERING. *Educ:* Mich State Univ, BSEE, 43; Univ Md, MSEE, 48; Johns Hopkins Univ, PhD(physics), 57. *Prof Exp:* Radio engr, US Naval Res Lab, 43-46, electronic scientist, 46-56, head, solar physics sect, 56-59; consult, 59-60, assoc chief, tracking systs div, 60-62, chief, space data acquisition div, 62-63, chief, Advan Develop Div, 63-71, chief, Advan Data Systs Div, 71-74, sr scientist applns, 74-79, PROJ MGR CRUSTAL DYNAMICS, GODDARD SPACE FLIGHT CTR, 79- *Concurrent Pos:* Jr instr, Johns Hopkins Univ, 49-52; deleg, Int Sci Radio Union, 57, 60 & 63. *Honors & Awards:* Exceptional Performance Award, Goddard Space Flight Ctr, 71; NASA Group Achievement Award Viking Lander Biol Instrument, 85; NASA Group Achievement Award Crustal Dynamics Proj, 86. *Mem:* Am Phys Soc; fel Inst Elec & Electronics Engrs; Am Geophys Union; AAAS; Sigma Xi. *Res:* Microwave; radio astronomy; telemetry; satellite tracking; geophysics; satellite communications. *Mailing Add:* 529 Whitingham Dr Silver Spring MD 20771

COATES, RALPH L, b Moroni, Utah, July 24, 34; m 51; c 3. CHEMICAL ENGINEERING. *Educ:* Univ Utah, BSChE, 59, PhD(chem eng), 62. *Prof Exp:* Res assoc chem eng, Univ Utah, 60-62; sr engr, Hercules Powder Co, 62-63; tech specialist propellant combustion, Lockheed Propulsion Co, 63-67; assoc prof, Brigham Young Univ, 67-77, prof chem eng, 77-; COATES INDUST PIPING CO. *Mem:* Am Inst Aeronaut & Astronaut; Am Chem Soc; Am Inst Chem Engrs. *Res:* Combustion; fluid mechanics; heat transfer; thermodynamics. *Mailing Add:* Coates Indust Piping Co 2358 W 12420 S Riverton UT 84065

COATES, ROBERT MERCER, b Evanston, Ill, May 21, 38; m 64. ORGANIC CHEMISTRY. *Educ:* Yale Univ, BS, 60; Univ Calif, Berkeley, PhD(chem), 64. *Prof Exp:* NIH fel, Stanford Univ, 63-65; asst prof, 65-77, PROF CHEM, UNIV ILL, URBANA-CHAMPAIGN, 77- *Concurrent Pos:*

A P Sloan Found fel, 71-73; Guggenheim fel, 80. *Mem:* Am Chem Soc; Royal Soc Chem; Am Soc Pharmacog. *Res:* Synthesis and biosynthesis of natural products; synthetic methods; biogenetic-like rearrangement of terpenes; synthesis and reactions of polycyclic compounds. *Mailing Add:* Chem Dept Univ Ill 1209 W California Urbana IL 61801-3731

COATS, ALFRED CORNELL, b Portland, Ore, Mar 12, 36; m 63; c 2. NEUROPHYSIOLOGY. *Educ:* Stanford Univ, BA, 59; Baylor Univ, MD, 62, MS, 63. *Prof Exp:* From instr to assoc prof, 62-74, PROF PHYSIOL & OTOLARYNGOL, BAYLOR COL MED, 74- *Concurrent Pos:* Dir, Electronystagmography Lab & mem consult staff, Methodist Hosp, 63-, St Luke's Hosp, 73-81 & Hermann Hosp, 75-77; mem consult staff, Ben Tauber Hosp, 65-, Vet Admin Hosp, 69- *Mem:* Am Neuro-Otologic Asn; Soc Neurosci. *Res:* Physiology of peripheral auditory system; clinical vestibulometry; study of balance-and-equilibrium system in humans. *Mailing Add:* Dept Otolaryngol Baylor Col Med Houston TX 77030

COATS, JOEL ROBERT, b Kenton, Ohio, Apr 24, 48; m 71, 88; c 5. TOXICOLOGY. *Educ:* Ariz State Univ, BS, 70; Univ Ill, MS, 72, PhD(entom), 74. *Prof Exp:* Res assoc environ toxicol, Univ Ill, Urbana, 74-76; vis prof, Univ Guelph, Ont, 76-78; from asst prof to assoc prof, 78-86, PROF, TOXICOL, IOWA STATE UNIV, 86- *Concurrent Pos:* Consult & prin investr grants; chmn-elect & prog chmn, Agrochem Div, Am Chem Soc, 91. *Mem:* AAAS; Am Chem Soc; Entom Soc Am; Soc Toxicol; Soc Environ Toxicol & Chem; Pestic Soc Japan. *Res:* Toxicology and chemistry of pesticides, especially insecticides; mode of action, selectivity, metabolism, degradation, uptake mechanisms, environmental fate and effects; natural insecticides; editor of 3 books on pesticides. *Mailing Add:* Dept Entom Iowa State Univ Ames IA 50011

COATS, KEITH HAL, b Ann Arbor, Mich, Nov 14, 34; m 56; c 3. PETROLEUM ENGINEERING, APPLIED MATHEMATICS. *Educ:* Univ Mich, BS, 56, MS, 57 & 58, PhD(chem eng), 59. *Prof Exp:* Lectr chem eng, Univ Mich, 59, asst prof, 59-61; sr res engr, Jersey Prod Res Co, 61-64; res assoc, Esso Prod Res Co, 64-66; assoc prof petrol eng, Univ Tex, Austin, 66-70; BD CHMN, INTERCOMP RESOURCE DEVELOP & ENG, 70-, TECH DIR SCI SOFTWARE. *Concurrent Pos:* Consult, Northern Natural Gas Co, Nebr, 66- & Chevron Res Corp, 67- *Mem:* Am Inst Mining Metall & Petrol Engrs; Am Inst Chem Engrs. *Res:* Computer simulation of oil and gas reservoir performance through numerical solution of partial differential equations. *Mailing Add:* Sci Software Intercomp 10333 Richmond Ave Suite 1000 Houston TX 77042-4122

COATS, RICHARD LEE, b Madill, Okla, Feb 14, 36; m 59; c 4. NUCLEAR PHYSICS, REACTOR PHYSICS. *Educ:* Univ Okla, BS, 59, MS, 63, PhD(nuclear eng), 66. *Prof Exp:* Mem tech staff, Sandia Labs, 66-69, div supvr, Nuclear Eng & supvr, Reactor Containment Safety Studies Div, 69-81; PRES, OMEGA BUILDERS, ALBUQUERQUE, NM, 81- *Mem:* Am Nuclear Soc. *Res:* Experimental and theoretical nuclear reactor physics; coupled reactor dynamics; stochastic reactor kinetics; Monte Carlo reactor physics calculations; advanced reactor safety research. *Mailing Add:* 7178 Cherokee Rd Dexter NM 88230

COATS, ROBERT ROY, b Toronto, Ont, Nov 22, 10; US citizen; m 37; c 3. GEOLOGY. *Educ:* Univ Wash, BS, 31, MS, 32; Univ Calif, PhD(geol), 38. *Prof Exp:* Geologist, Storey County Mines, 37; asst prof geol, Univ Alaska, 37-39; from jr geologist to geologist, US Geol Surv, 39-64, res geologist, 64-81; RETIRED. *Mem:* Fel Mineral Soc Am; Soc Econ Geologists; fel Geol Soc Am; Sigma Xi. *Res:* Tin deposits of Alaska; alteration by hydrothermal solutions; Aleutian volcanoes; geology of the northeastern Great Basin; tectonics and Mesazoic and Tertiary plutonism and vulcanism in northeastern Nevada and their relation to ore deposition. *Mailing Add:* 3130 Trout Gulch Rd Aptos CA 95003

COBB, CAROLUS M, b Lynn, Mass, Jan 22, 22; m 66; c 1. PHYSICAL CHEMISTRY. *Educ:* Mass Inst Technol, SB, 44, PhD(phys chem), 51. *Prof Exp:* Chemist, Tenn Eastman Corp, 44-46 & Ionics, Inc, 51-55; prin scientist, Allied Res Assocs, Inc, 55-60; CHIEF CHEMIST, AM SCI & ENG, INC, CAMBRIDGE, MASS, 60- *Concurrent Pos:* Affil in bioeng, Forsyth Dent Ctr, Harvard Sch Pub Health, 64- *Mem:* Am Chem Soc; Am Phys Soc. *Res:* Titanium and solution chemistry; ion exchange chromatography; atmospheric and physical optics; fluorescent materials; nuclear phenomena; high temperature and electronic materials; thermodynamics; biochemistry. *Mailing Add:* Am Sci & Eng Inc 40 Erie St Cambridge MA 02139

COBB, CHARLES MADISON, b Kansas City, Mo, Sept 20, 40; m 64; c 1. DENTISTRY, PERIODONTICS. *Educ:* Univ Mo-Kansas City, DDS, cert periodont & MS, 64; Georgetown Univ, PhD(anat), 71. *Prof Exp:* Prof periodont, Sch Dent, La State Univ, 71-72; asst prof periodont, 73-74, investr periodont, Inst Dent Res, 73-76, SCH DENT, UNIV ALA, BIRMINGHAM, 74-; clin assoc prof dent, 78-86, PROF PERIODONT, SCH DENT, UNIV MO, KANSAS CITY, 86- *Honors & Awards:* Balant Orban Prize, Am Acad Periodont, 66. *Mem:* Int Asn Dent Res; Am Acad Periodont; Am Dent Asn; Sigma Xi; Pierre Fouchard Acad. *Res:* Ultrastructure and histopathology of periodontal disease. *Mailing Add:* 424 W 67th Terr Kansas City MO 64113

COBB, DONALD D, b Atlantic, Iowa, May 4, 43. PHYSICS. *Educ:* Northern Ill Univ, BS, 65; Univ Iowa, MS, 68, PhD(physics), 70. *Prof Exp:* Sect leader, EG&G, Los Alamos, 70-75; group leader atmospheric sci, 76-89, DIV LEADER SPACE SCI, LOS ALAMOS NAT LAB, 89- *Mem:* Am Phys Soc; Am Geophys Union. *Res:* Aeronomy; nuclear test detection. *Mailing Add:* 3202 Woodland Los Alamos NM 87544

COBB, EMERSON GILLMORE, b Slaughters, Ky, Nov 28, 07; m 29; c 2. ORGANIC CHEMISTRY. *Educ:* Union Col, Ky, AB, 28; Univ Ky, MS, 31; Univ NC, PhD(org chem), 41. *Hon Degrees:* LHD, Union Col, Ky, 61. *Prof Exp:* High sch instr, Ky, 28-29 & 32-40; asst prof chem, La Polytech Inst, 40-42; prof & head dept, Dakota Wesleyan Univ, 42-48; chmn dept, 48-74, prof, 48-78, EMER PROF CHEM, UNIV PAC, 78- *Concurrent Pos:* Fulbright vis lectr, Univ Peshawar, 61-62; vis lectr, Univ Baja Calif & Univ Marine Sci, 74-75; consult, chem educ. *Mem:* Am Chem Soc. *Res:* Natural plant products; protective coatings; constitution of tannins. *Mailing Add:* PO Box 228 Burson CA 95225

COBB, FIELDS WHITE, JR, b Key West, Fla, Feb 16, 32; m 58; c 3. PLANT PATHOLOGY, FOREST PATHOLOGY. *Educ:* NC State Univ, BS, 55; Yale Univ, MF, 56; Pa State Univ, PhD(plant path), 63. *Prof Exp:* Res forester, Southeastern Forest Exp Sta, US Forest Serv, 55-57, plant pathologist, Southern Forest Exp Sta, 57; statist clerk agr econ, Agr Mkt Serv, USDA, 57-58; instr plant path, Pa State Univ, 63; from asst prof to assoc prof, 63-82, PROF PLANT PATH, UNIV CALIF, BERKELEY, 82- *Mem:* Soc Am Foresters; Am Phytopath Soc. *Res:* Diseases of forest trees, particularly rusts and those of roots and the vascular system, their causes, epidemiology, development and control; interactions with insects, air pollutants and activities of humans. *Mailing Add:* Dept Plant Path Univ Calif Berkeley CA 94720

COBB, FREDERICK ROSS, CARDIOVASCULAR DISEASE, HEART DISEASE. *Educ:* Univ Miss, MD, 64. *Prof Exp:* PROF MED, DUKE UNIV, 83- *Mem:* Am Soc Clin Invest. *Res:* Cardiovascular hemodynamics. *Mailing Add:* Dept Med Durham VAMC Durham NC 27705

COBB, GLENN WAYNE, b Jonesboro, La, Dec 1, 36; m 58; c 2. PLANT MORPHOGENESIS, PLANT PATHOLOGY. *Educ:* La Polytech Inst, BS, 58; Purdue Univ, MS, 61, PhD(plant morphol), 62. *Prof Exp:* Asst prof biol, Stephen F Austin State Univ, 62-65; from assoc prof to prof biol, 65-77, PROF BOT, MCNEESE STATE UNIV, 77- *Concurrent Pos:* Fac res grants, 63-64 & 65. *Res:* Plant disease research. *Mailing Add:* Biol & Environ Sci McNeese State Univ Lake Charles LA 70609

COBB, GROVER CLEVELAND, JR, b Atlanta, Ga, Feb 6, 35; m 54; c 2. NUCLEAR PHYSICS. *Educ:* Univ Ga, BS, 56, MS, 57; Univ Va, PhD(physics), 60. *Prof Exp:* Asst prof, 60-69, ASSOC PROF PHYSICS, NC STATE UNIV, 69- *Mem:* Am Phys Soc. *Res:* Neutron scattering; optical spectroscopy; gaseous discharge experiments; nuclear cross sections; plasma oscillations. *Mailing Add:* Dept Physics Box 8202 NC State Univ Raleigh NC 27695-8202

COBB, JAMES C, b Urbana, Ill, Jan 28, 48; m; c 1. COAL. *Educ:* Univ Ill, PhD(geol), 80. *Prof Exp:* Res assoc, Ill Geol Surv, 75-80; head coal & mineral sect, 80-90, ASST STATE GEOLOGIST, KY GEOL SURV, 90- *Concurrent Pos:* Chmn, Coal Div, Geol Soc Am. *Mem:* Fel Geol Soc Am; Soc Mineralogists & Paleontologists; Soc Mining Engrs; Sigma Xi. *Res:* Coal resources; origin of coal. *Mailing Add:* 228 Mining & Mineral Resources Bldg Univ Ky Lexington KY 40506-0107

COBB, JAMES TEMPLE, JR, b Cincinnati, Ohio, Mar 9, 38; m 64; c 2. COAL PROCESSING, ENERGY PROJECT MANAGEMENT. *Educ:* Mass Inst Technol, SB, 60; Purdue Univ, MS, 63, PhD(chem eng), 66. *Prof Exp:* Engr, Esso Res Lab, 67-70; asst prof, 70-75, ASSOC PROF CHEM ENG, UNIV PITTSBURGH, 75- *Concurrent Pos:* Consult, ACV Power Corp, Goodyear Tire & Rubber Co, Gulf Res & Develop Co, US Dept Energy, Cannon Boiler Works & Gas Desulfurization Corp. *Mem:* Am Inst Chem Engrs; Am Soc Eng Educ; Nat Soc Prof Engrs; Asn Energy Engrs; Sigma Xi. *Res:* High temperature fuel cells; warm fog dispersal; chemical process development; enzyme engineering; coal conversion processes; zeolite catalysis; solid waste incineration; beneficial utilization of ash. *Mailing Add:* 141 Deerfield Dr Pittsburgh PA 15235

COBB, JEWEL PLUMMER, b Chicago, Ill, Jan 17, 24; div; c 1. CELL PHYSIOLOGY. *Educ:* Talladega Col, AB, 44; NY Univ, MS, 47, PhD(cell biol), 50. *Hon Degrees:* Numerous from various univs & cols. *Prof Exp:* Instr, Anat Dept & dir, Tissue Culture Lab, Univ Ill, 52-54; instr, res surg, 55-56, asst prof, New York Univ, 56-60; prof, Biol Dept, Sarah Lawrence Col, 60-69; dean & prof zool, Conn Col, 69-76; dean & prof biol sci, Douglass Col, 76-81; pres, 81-90, EMER PRES & PROF BIOL SCI, CALIF STATE UNIV, FULLERTON, 90-, TRUSTEE PROF, 90- *Concurrent Pos:* Comm Acad Affairs, Am Coun Educ, Washington, DC; Comt, Opportunities for Women & Minorities in Sci, AAAS, 73-77; Adv coun, Allergy & Infectious Dis, NIH, 79-80; consult, Off Technol Assessment, US Congress, 76-78; mem of Corp, Marine Biol Inst, Woods Hole, 72-; mem adv bd, Signs, J Women Cult & Soc, 78-86; mem US Dept State Adv Comt on Oceans & Int Environ & Sci Affairs, Washington, DC, 80-; bd dirs, Allied Chem Corp, Morristown, NJ, 80, Nat Conf Christians & Jews, Orange Coun, 82-, Orange Coun Phil Soc, 82-86; mem, Community Develop Comt, Am Cancer Soc, Orange County, 83-84; mem bd dirs, First Interstate Bancorp, 85-, Am Assembly, Bernard Col, 86-, Newport Harbor Art Mus, 89-; mem, Adv Comt Competitive Technol, State Calif Dept Com, 89- *Mem:* Inst Med-Nat Acad Sci; fel NY Acad Sci; fel AAAS; Am Asn Univ Women; Asn Women Sci; Sigma Xi; Tissue Cult Asn; Am Conf Acad Deans; fel NY Acad Sci; Tissue Cult Asn. *Res:* Published 36 articles in the area of factors which influence growth, morphology and genetic expressions of normal and cancer pigment cells; genetics studies in polyoma virus; author or co-author of over 40 publications. *Mailing Add:* Off Pres Admin 815 Calif State Univ 5151 State University Dr Los Angeles CA 90032-8500

COBB, JOHN CANDLER, b Boston, Mass, July 8, 19; m 46; c 4. PREVENTIVE MEDICINE. *Educ:* Harvard Univ, BA, 41, MD, 48; Johns Hopkins Univ, MPH, 54. *Prof Exp:* Asst malaria control, Friends Serv Comt, 41-42; instr maternal & child health, Sch Hyg & Pub Health, Johns Hopkins Univ, 51-54, instr pediat, Sch Med, 51-56 & psychiat, 52-56, asst prof maternal & child health, Sch Hyg, 54-56; area consult, USPHS, Div Indian Health, NMex, 56-60; Johns Hopkins Univ & Ford Found dir med social res proj, Lahore, Pakistan, 60-64; chmn dept, 66-72, prof, 65-84, EMER PROF PREV MED, SCH MED, UNIV COLO, DENVER, 84- *Concurrent Pos:* Mem comt peace educ & family planning, Am Friends Serv Comt, 64-74;

mem comt Indian health, Asn Am Indian Affairs, 65-73; WHO short term consult maternal & child health & family planning, Indonesia, 69-70, sci group adv res methods fertility regulations, Genva, 70 & family health educ, Western Pac Region, 71-72; mem gov sci adv coun, Colo; mem environ coun; mem Nat Med Comm, Planned Parenthood/World Population, 71-73, pres, Am Asn Planned Parenthood Physicians, 72-73, bd mem, Planned Parenthood Fedn Am, 72-73; mem air pollution control comn, Colo, 76-; mem gov's task force Nuclear Energy Plant, 75; mem bd dirs, Rocky Mountain Ctr Environ & Colo Coalition Full Employ, 78-80; mem, Colo Gov Task Force Health Effects of Air Pollution, 78-79; vis prof, preventative med, Sch Pub Health, Guangxi Med Col, Nanning, Guangxi, Rep China; mem bd dirs, Appropriate Rural Technol Asn, 87- *Mem:* AAAS; Int Solar Energy Soc; Am Pub Health Asn. *Res:* Environmental health. *Mailing Add:* PO Box 1403 Corrales NM 87048

COBB, JOHN IVERSON, b Marianna, Fla, Feb 9, 38. TOPOLOGY. *Educ:* Fla State Univ, BA, 60; Univ Wis, MA, 61, PhD(math), 66. *Prof Exp:* Instr math, Racine Ctr, Univ Wis, 66; asst prof, Rutgers Univ, 66-69; asst prof, 69-74, ASSOC PROF MATH, UNIV IDAHO, 74- *Mem:* Am Math Soc; Math Asn Am. *Res:* Point-set topology; piece-wise linear topology. *Mailing Add:* Dept Math Univ Idaho Moscow ID 83843

COBB, LOREN, b Boston, Mass, May 8, 48; m 75; c 1. MATHEMATICAL STATISTICS. *Educ:* Cornell Univ, BA, 70, MA, 71, PhD(sociol), 73. *Prof Exp:* Asst prof sociol, Univ NH, 72-77; fel psychiat, Med Sch, Univ SFla, 77-79; from asst prof to assoc prof biomet, Med Univ SC, 79-88; ASSOC PROF, COMMUNITY MED, UNIV NMEX, 88- *Concurrent Pos:* Res assoc, Cornell Univ, 73-77. *Mem:* Inst Math Statist; Am Statist Asn; Math Assoc Am. *Res:* Estimation and hypothesis testing for stochastic nonlinear dynamical systems; statistical catastrophe theory and its applications in the biological and social sciences. *Mailing Add:* Community Med Sch Med Univ NMex Albuquerque NM 87131

COBB, R M KARAPETOFF, paper chemistry, for more information see previous edition

COBB, RAYMOND LYNN, b Ochelata, Okla, Dec 10, 29; m 66; c 1. ORGANIC CHEMISTRY. *Educ:* Ottawa Univ, BS, 51; Univ Kans, PhD(org chem), 55. *Prof Exp:* Res chemist & res assoc, Phillips Petrol Co, 55-86; sr develop scientist, Union Carbide Co, 87-90; RETIRED. *Mem:* Am Chem Soc. *Res:* Organic nitrogen compounds; catalytic organic processes; thermal reactions; organosulfur compounds; reaction mechanisms; alkylation processes; halogenations; organo silicone surfectants. *Mailing Add:* 334 Live Oak Dr Buda TX 78610-9288

COBB, THOMAS BERRY, b Atlanta, Ga, Nov 4, 39; m 64. PHYSICS, CHEMISTRY. *Educ:* Southern Missionary Col, BA, 60; Univ SC, MS, 63; NC State Univ, PhD(physics), 68. *Prof Exp:* Instr physics, Western Md Col, 63-65; res assoc chem, Univ NC, 68-69; asst prof, 69-74, ASSOC PROF PHYSICS & ASST VPROVOST RES, BOWLING GREEN STATE UNIV, 74- *Mem:* AAAS; Am Asn Physics Teachers; Sigma Xi. *Res:* Nuclear magnetic resonance. *Mailing Add:* 856 Ferndale Bowling Green OH 43402

COBB, WILLIAM MONTAGUE, anatomy, physical anthropology; deceased, see previous edition for last biography

COBB, WILLIAM THOMPSON, b Spokane, Wash, Nov 10, 42; div; c 4. PLANT PATHOLOGY, TOXIC & HAZARDOUS WASTE. *Educ:* Eastern Wash Univ, BA, 64; Ore State Univ, PhD(plant path), 73. *Prof Exp:* Mgr agron, Sun Royal Co, 70-74; sr scientist plant path, Lilly Res Labs, 74-78, res scientist, 78-87; CONSULT TOXIC & HAZARDOUS WASTE & GROUNDWATER CONTAMINATION, 88- *Concurrent Pos:* Plant pathol instr, Columbia Basin Col, 71, 73, 75, & 77. *Mem:* Sigma Xi; Am Phytopath Soc; Weed Sci Soc Am; Am Soc Agron; Ctr Appln Sci Technol. *Res:* Field screening of experimental pesticides; crop fertility and disease interactions; determination of groundwater contamination potential of pesticides. *Mailing Add:* 815 S Kellogg Kennewick WA 99336

COBBAN, WILLIAM AUBREY, b Anaconda, Mont, Dec 31, 16; m 42; c 3. GEOLOGY. *Educ:* Univ Mont, BA, 40; Johns Hopkins Univ, PhD(geol), 49. *Prof Exp:* Geologist, Carter Oil Co, Tulsa, 39-45; PALEONTOLOGIST & STRATIGRAPHER, US GEOL SURV, 46- *Honors & Awards:* Paleont Medal, Paleont Soc Am, 85 R C Moore Paleont Medal, 90; Distinguished Serv Award, Dept of Interior, 86. *Mem:* Soc Econ Paleontologists & Mineralogists; Paleont Soc; Am Asn Petrol Geol; Geol Soc Am; AAAS. *Res:* Upper Cretaceous stratigraphy and paleontology of the Rocky Mountain area. *Mailing Add:* US Geol Surv Fed Ctr Bldg 25 Mail Stop 919 Denver CO 80225

COBBE, THOMAS JAMES, b Cincinnati, Ohio, July 18, 18; m 44; c 2. FOREST ECOLOGY. *Educ:* Univ Cincinnati, BA, 40, MA, 41; Univ Mich, PhD(bot), 53. *Prof Exp:* Asst prof biol, Am Univ, 48-49; instr, Oberlin Col, 49-52; from instr to asst prof, Capital Univ, 52-57; from asst prof to prof, 57-84, EMER PROF BOT, MIAMI UNIV, 84- *Mem:* AAAS; Ecol Soc Am; Am Soc Photogram. *Res:* Secondary forest successions; Dutch elm disease. *Mailing Add:* 108 Tenney Ct Oxford OH 45056

COBBLE, JAMES WIKLE, b Kansas City, Mo, Mar 15, 26; m 49; c 2. PHYSICAL CHEMISTRY, INORGANIC CHEMISTRY. *Educ:* Northern Ariz Univ, AB, 46; Univ Southern Calif, MS, 49; Univ Tenn, PhD(phys chem), 52. *Prof Exp:* Chemist, Oak Ridge Nat Lab, 49-52; postdoc res assoc, Lawrence Berkeley Lab, Univ Calif, 52-54; from asst prof to prof chem, Purdue Univ, 55-73; PROF CHEM & DEAN, GRAD DIV & RES, SAN DIEGO STATE UNIV, 73- *Concurrent Pos:* Instr, Univ Calif, 53; consult, Elec Power Res Inst; Guggenheim fel, 66; vpres, San Diego State Univ Found, 75-; bd vis, US Air Force Air Univ, 84-, chair, 88-90; bd trustees, Calif Western Law Sch, 87- *Honors & Awards:* E O Lawrence Award, AEC, 70; Robert A Welch Found lectr, 71. *Mem:* Am Chem Soc; fel Am Phys Soc; Sigma Xi. *Res:* Radiochemistry; physical-inorganic chemistry; correlation of thermodynamic properties with structures; high temperature solutions; nuclear chemistry; mechanisms of nuclear reactions. *Mailing Add:* Dept Chem San Diego State Univ San Diego CA 92182-0328

COBBLE, MILAN HOUSTON, b St Paul, Minn, Mar 13, 22; m 49; c 4. MECHANICAL ENGINEERING. *Educ:* Univ Mich, BSME, 48, PhD(mech eng), 58; Wayne State Univ, MSME, 52. *Prof Exp:* Res assoc fluids lab, Univ Mich, 55-56; from asst prof to assoc prof, Univ Del, 55-62; prof mech eng, NMex State Univ, 62-84; vis prof, Univ Mich, 85-88; RETIRED. *Concurrent Pos:* Fel, NSF fac, 59-62, Shell, 54-55, Eng Res Inst, 55-56; consult, Lockheed, 84-85; academic specialist, US Info Agency, 90. *Mem:* Fel AAAS; Am Soc Mech Engrs; Soc Rheology; Am Soc Eng Educ; Am Acad Mech; Sigma Xi. *Res:* Nonlinear mathematics; solar energy; heat transfer. *Mailing Add:* 2619 Fairway Las Cruces NM 88001

COBBOLD, R S C, b Worcester, Eng, Dec 10, 31; Can citizen; m 63; c 3. BIOMEDICAL ENGINEERING. *Educ:* Univ London, BSc, 56; Univ Sask, MSc, 61, PhD(elec eng), 65. *Prof Exp:* Asst exp officer electronics, Ministry of Supply, Eng, 49-53; sci officer, Defence Res Bd, Ottawa, Can, 56-59; from lectr to assoc prof elec eng, Univ Sask, 60-66; from assoc prof to prof elec eng, Univ Toronto, 66-75, dir Inst Biomed Eng, 74-83. *Mem:* Inst Elec & Electronics Engrs; Int Fedn Med & Biol Eng; fel Royal Soc Can. *Res:* Medical ultrasonics; biomedical transducers; semiconductor transducers. *Mailing Add:* Inst Biomed Eng Univ Toronto Toronto ON M5S 1A1 Can

COBERLY, CAMDEN ARTHUR, b Elizabeth, WVa, Dec 21, 22; m 46; c 4. CHEMICAL ENGINEERING, ACADEMIC ADMINISTRATION. *Educ:* WVa Univ, BS, 44; Carnegie Inst Technol, MS, 47; Univ Wis-Madison, PhD(chem eng), 49. *Prof Exp:* Chief engr & supvr, Mallinckrodt Chem Works, 49-64; chmn dept, 68-71, assoc dean eng, 71-86, PROF CHEM ENG, UNIV WIS-MADISON, 64- *Mem:* Fel Am Inst Chem Engrs; fel AAAS; Am Soc Eng Educ; Am Chem Soc; Nat Asn Corrosion Engrs; Sigma Xi. *Res:* Chemical engineering transfer operations; chemical plant design. *Mailing Add:* 4114 N Sunset Ct Madison WI 53705

COBERN, MARTIN E, b New York, NY, Dec 25, 46; m 69; c 2. MEASUREMENT-WHILE-DRILLING, OILWELL SURVEYING & LOGGING. *Educ:* Cooper Union, BS, 67; Yale Univ, MPh, 69, PhD(physics), 74. *Prof Exp:* Foreign visitor, Ctr Nuclear Studies, Saclay, 74-75; res assoc physics, Univ Pa, 75-77; sr scientist, Westinghouse Bettis Atomic Power Lab, 77-79; res scientist, NL Industs-Drilling Systs Technol, 79-80, sr res scientist, 80-81; proj mgr, 81-82, mgr physics & analysis, 82-84, dir res, 85-87, VPRES RES, TELECO OILFIELD SERV INC, 87- *Concurrent Pos:* Mem, Well Logging Indust Adv Comt, Univ Houston, 81-87; mem, Technol Comt, Soc Prof Well Log Analysts, 85-, distinguished speaker, 85-87 & 88-89, dir-at-large, 88-89. *Mem:* Am Phys Soc; Soc Petrol Engrs; Am Nuclear Soc; Soc Prof Well Log Analysts. *Res:* New sensors and methods for borehole logging; measurement-while-drilling; directional surveying. *Mailing Add:* 727 Rustic Lane Cheshire CT 06410

COBLE, ANNA JANE, b Raleigh, NC, July 12, 36; m 85. PHYSICAL PROPERTIES OF MEMBRANES. *Educ:* Howard Univ, BS, 58, MS, 61; Univ Ill, Urbana, PhD(biophysics), 73. *Prof Exp:* Instr physics, NC Agr & Tech State Univ, 60-64; assoc, Ctr Biol Nat Syst, Washington Univ, 69-71; lectr, 71-74, ASST PROF PHYSICS & BIOPHYSICS, HOWARD UNIV, 74- *Concurrent Pos:* MARC fac fel, NIH, 79-80. *Mem:* AAAS; Am Asn Physics Teachers; Sigma Xi; Minority Women Sci. *Res:* Biological effects and properties of ultrasound; physical studies of membranes and membrane transport. *Mailing Add:* 1000 Fairview Ave Takoma Park MD 20912

COBLE, HAROLD DEAN, b Burlington, NC, Feb 3, 43; m 65; c 3. WEED SCIENCE, AGRONOMY. *Educ:* NC State Univ, BS, 65, MS, 67; Univ Ill, Urbana, PhD(weed sci), 70. *Prof Exp:* From asst prof to assoc prof, 70-80, PROF WEED SCI, NC STATE UNIV, 80- *Concurrent Pos:* Consult, Union Carbide Corp, 74-77, Monsanto Chem Co, 73- & Velsicol Chem Corp, 76- *Mem:* Am Soc Agron; Weed Sci Soc Am; Plant Growth Regulator Group; Am Peanut Res & Educ Asn; Sigma Xi. *Res:* Weed-crop interactions as related to integrated pest management; herbicide effects on crops and weeds; crop yield enhancement with plant growth regulators. *Mailing Add:* Dept Crop Sci NC State Univ Raleigh NC 27695-7620

COBLE, R(OBERT) L(OUIS), b Uniontown, Pa, Jan 22, 28; m 52; c 5. MATERIALS PROCESSING, CERAMICS. *Educ:* Bethany Col, WVa, BS, 50; Mass Inst Technol, ScD(ceramics), 55. *Prof Exp:* Res asst ceramics, Mass Inst Technol, 50-55; ceramist, Gen Elec Res Lab, 55-60; from asst prof to assoc prof, 60-69, prof ceramics, 69-87, EMER PROF CERAMICS, MASS INST TECHNOL, 87- *Honors & Awards:* Award, Nat Inst Chem Engrs, 60 & Ross Coffin Purdy Award, Am Ceramic Soc, 70; Sosman lectr, Am Ceramic Soc, 79; Humboldt-Stiftung Award, Max-Plank Inst, 84; Frenkel Prize, Int Inst Sci Sintering, 85. *Mem:* Nat Acad Eng; fel Am Ceramic Soc; Nat Inst Ceramic Engrs; AAAS. *Res:* High temperature creep; thermodynamics of solid solutions and defect equilibria in ceramics; diffusion in non-metals; modeling of diffusion-controlled phenomena; creep rupture; powder metallurgy and ceramics processing. *Mailing Add:* Dept Mat Sci & Eng Mass Inst Technol Cambridge MA 02139

COBLER, JOHN GEORGE, b Conneaut Lake, Pa, Sept 15, 18; m 41; c 2. ORGANIC POLYMER CHEMISTRY. *Educ:* Col Wooster, BA, 40. *Prof Exp:* Lab asst, Col Wooster, 38-39; control chemist, Ohio Exp Sta, 39-40; asst, Purdue Univ, 40-41; res chemist, Distillation Prod, Inc, NY, 42-45; lab dir, Bordon Co, NJ, 45-47; head spec prod develop, Bordens Soy Processing Co, 47-49; group leader, Anal Dept, Dow Chem Co, 49-59, tech expert, 59-69, tech consult, 69-70, assoc anal scientist, 70-75, ASSOC SCIENTIST, HEALTH & ENVIRON SCI, DOW CHEM CO, 75- *Concurrent Pos:* Asst, Manhattan Proj, Univ Rochester, 44-45. *Mem:* Am Chem Soc; Sigma Xi; Am Soc Testing & Mat; NY Acad Sci. *Res:* Structure and composition of polymers; stability and degradation of polymers. *Mailing Add:* 2008 Rapanos Dr Midland MI 48640

COBURN, CORBETT BENJAMIN, JR, b Lake Providence, La, Dec 7, 40; m 61; c 1. ENVIRONMENTAL PHYSIOLOGY, PHYSIOLOGICAL ECOLOGY. *Educ:* La Polytech Inst, BS, 62, MS, 64; Univ Southern Miss, PhD(zool), 70. *Prof Exp:* Instr biol, Calif Baptist Col, 64-65 & East Central Jr Col, 66-68; prof, Wesleyan Col, 70-71; asst prof, West Liberty State Col, 71-72; asst prof, 72-80, PROF BIOL, TENN TECHNOL UNIV, 80-; DIR, TECH AQUA BIOL FIELD STA, 84- *Mem:* Am Fisheries Soc; AAAS. *Res:* Vertebrate hematology; metabolic responses of aquatic animals to pollutants; RNA/DNA ratio as affected by season and water quality; environmental assessment of highway impact on organisms in flood plain. *Mailing Add:* Dept Biol Tenn Technol Univ Cookeville TN 38501

COBURN, EVERETT ROBERT, b Manchester, NH, Aug 10, 15; m 39; c 2. ORGANIC CHEMISTRY. *Educ:* Harvard Univ, SB, 38, AM, 40, PhD(org chem), 41. *Prof Exp:* Lilly fel polarog studies of quinones, Harvard Univ, 41-42; res chemist, Nat Defense Res Comt, 42-43; prof chem, Bennington Col, 43-80; RETIRED. *Concurrent Pos:* Lectr, Middlebury Col, 44; consult, Sprague Elec Co, 52- *Mem:* Am Chem Soc. *Res:* Diels-Alder reactions on quinones; polarographic work on quinones and related compounds; incendiary mixtures and design of apparatus for use. *Mailing Add:* 43 Spindle Point Meredith NH 03253-9731

COBURN, FRANK EMERSON, b Toronto, Ont, Apr 25, 12; m 40; c 4. PSYCHIATRY. *Educ:* Univ Toronto, BA, 36, MD, 39; RCPS(C), cert psychiat. *Prof Exp:* Assoc prof psychiat, Univ Iowa, 50-55; assoc prof, 55-58, prof, 58-80, EMER PROF PSYCHIAT, UNIV SASK, 80- *Concurrent Pos:* Baker lectr, Univ Mich, 54; mem, Am Bd Psychiat & Neurol, 47- *Mem:* Fel Am Psychiat Asn; Am Psychosom Soc; Can Psychiat Asn; Can Med Asn. *Res:* Teaching; therapy; community psychiatry. *Mailing Add:* Student Health Centre Univ Sask Saskatoon SK S7N 0W0 Can

COBURN, HORACE HUNTER, b Cambridge, Mass, May 10, 22; m 47; c 3. PHYSICS. *Educ:* Ohio State Univ, BS, 43; Univ Ill, MS, 47; Univ Pa, PhD(physics), 56. *Prof Exp:* Assoc prof physics, Moravian Col, 50-51; from asst prof to assoc prof, 54-69, PROF PHYSICS, NMEX STATE UNIV, 69- *Concurrent Pos:* Physicist, Manhattan Proj, Tenn, 44-46; consult, Los Alamos Sci Lab, 62- & USAID, India, 66 & 69; mem fac, Inst Optics, Univ Rochester, 64-65; consult, USAID, India, 66 & 69. *Mem:* Nat Sci Teachers Asn; Am Asn Physics Teachers; Optical Soc Am; Soc Photo-optical & Instrumentation Engrs. *Res:* Optics. *Mailing Add:* NMex State Univ Box 30001 Dept 3D Las Cruces NM 88003-0001

COBURN, JACK WESLEY, b Fresno, Calif, Aug 6, 32; m 58; c 3. INTERNAL MEDICINE, NEPHROLOGY. *Educ:* Univ Redlands, BS, 53; Univ Calif, Los Angeles, MD, 57; Am Bd Internal Med, dipl, 65, cert nephrology, 74. *Prof Exp:* Intern med, Med Ctr, Univ Calif, Los Angeles, 57-58; asst res physician, Univ Wash Hosp Syst, Seattle, 58-60; assoc res physician, Med Ctr, Univ Calif, Los Angeles, 60-61; sect chief gen med, Wadsworth Hosp, Los Angeles, 68-69; from asst prof to assoc prof, 65-73, prof med, 73-85, ADJ PROF MED, SCH MED, UNIV CALIF, LOS ANGELES, 85-; STAFF PHYSICIAN, VET AFFAIRS WADSWORTH HOSP CTR, 85- *Concurrent Pos:* Nat Inst Arthritis & Metab res fel, Vet Admin Hosp & Univ Calif, Los Angeles, 61-63, clin investr award, Vet Affairs Ctr, 65-67; pvt pract, Nephrol & Internal Med, 65-; chief metab res ward, Vet Affairs Med Ctr, Los Angeles, 67-70; chief, nephrology sect, Wadsworth Hosp Ctr, 70-81; mem coun, Western Asn Physicians, 80-83, Am Soc Bone & Mineral Res, 81-84; Med Investr Award, Vet Affairs Med Ctr, 83-85. *Honors & Awards:* Lederle Award Human Nutrit, Am Inst Nutrit, 81; Frederic C Bartter Award, Am Soc Bone & Mineral Res. *Mem:* Am Fedn Clin Res; AMA; Am Soc Nephrology; fel Am Col Physicians; Am Physiol Soc; Am Asn Physicians; Am Soc Chem Invest. *Res:* pathophysiology of uremia; aluminum toxicity and metabolism; vitamin D and calcium metabolism; metabolic bone diseases. *Mailing Add:* 9400 Brighton Way Beverly Hills CA 90210

COBURN, JOEL THOMAS, b Kenosha, Wis, Sept 3, 55; m 88; c 1. CHROMATOGRAPHY, CHEMOMETRICS. *Educ:* Univ Wis-Milwaukee, BS, 79, MS, 80; Purdue Univ, PhD(analytical chem), 85. *Prof Exp:* Res chemist, BASF Corp, 85-89; SR RES CHEMIST, DOW CHEM CO, 89- *Mem:* Am Chem Soc; Sigma Xi. *Res:* Chromatographic and spectroscopic determination of trace levels of odorous compounds in industrial chemicals; industrial applications of chemometrics; fluorescence spectroscopy; supercritical fluid chromatography. *Mailing Add:* Dow Chem Co Bldg 1897 Midland MI 48667

COBURN, JOHN WYLLIE, b Vancouver, BC, Nov 9, 33; m 66; c 2. PLASMA CHEMISTRY. *Educ:* Univ BC, BASc, 56, MASc, 58; Univ Minn, Minneapolis, PhD(elec eng), 67. *Prof Exp:* Fel physics, Simon Fraser Univ, 67-68; RES STAFF MEM, MAT SCI DEPT, RES LAB, INT BUS MACH CO, 68- *Mem:* Am Vacuum Soc. *Res:* Particle diagnostics in glow discharges; thin film formation by sputtering and by plasma polymerization; physics and chemistry of dry etching processes. *Mailing Add:* IBM Almaden Res Ctr K33/801 650 Harry Rd San Jose CA 95120-6099

COBURN, LEWIS ALAN, b Austin, Tex, Aug 16, 40; m 66; c 1. ANALYSIS. *Educ:* Univ Mich, BS, 61, MS, 62 & PhD(math), 64. *Prof Exp:* Asst prof math, NY Univ, 64-65 & Purdue Univ, 65-66; from asst prof to prof math, Yeshiva Univ, 66-79; PROF & CHAIR, STATE UNIV NY BUFFALO, 79- *Concurrent Pos:* Prin investr, NSF grants, 67- *Mem:* Am Math Soc. *Res:* Author of over 30 research journal articles on operator theory and analysis. *Mailing Add:* Dept Math State Univ NY Buffalo NY 14214

COBURN, MICHAEL DOYLE, b Houston, Tex, Aug 6, 39; div; c 2. ORGANIC CHEMISTRY. *Educ:* Univ Tex, Austin, BS, 62, PhD(chem), 64. *Prof Exp:* STAFF MEM, LOS ALAMOS SCI LAB, UNIV CALIF, 64- *Mem:* Am Chem Soc; Int Soc Heterocyclic Chem. *Res:* Synthesis of energetic organic compounds, predominantly in the heterocyclic field. *Mailing Add:* PO Box 1633 MS 920 Los Alamos NM 87544

COBURN, RICHARD KARL, b Salt Lake City, Utah, Feb 24, 20; m 42; c 16. APPLIED MATHEMATICS. *Educ:* Utah State Univ, BS, 42 & 43; Univ Wash, MS, 56; Univ Ill, MA, 62. *Prof Exp:* Instr math & physics, Pa State Univ, 46-48; prof chem & physics, Ricks Col, 48-58; chmn dept, 58-80, prof math, Brigham Young Univ, Hawaii Campus, 58-87; RETIRED. *Concurrent Pos:* Math consult, Lockheed Airplane Co, 60-62; pres, Hawaii Coun Teachers Math; assoc dir, Hawaii Sci Fair; lectr, Nat Coun Teachers Math. *Mem:* Am Chem Soc; Math Asn Am. *Res:* Analytical methods of solving inequalities of order two and higher. *Mailing Add:* Dept Math Brigham Young Univ Hawaii Campus PO Box 1734 Laie Oahu HI 96762

COBURN, ROBERT A, b Akron, Ohio, Dec 31, 38; m 66; c 2. MEDICINAL CHEMISTRY, PHYSICAL ORGANIC CHEMISTRY. *Educ:* Univ Akron, BS, 60; Harvard Univ, AM, 62, PhD(org chem), 66. *Prof Exp:* Res chemist, US Army Natick Labs, Mass, 65-66; from asst prof to assoc prof, 68-84, PROF MED CHEM, SCH PHARM, STATE UNIV NY, BUFFALO, 84- *Mem:* Am Chem Soc; Royal Soc Chem. *Res:* Synthesis and molecular structure studies of heterocyclic compounds of biological and/or pharmacological significance. *Mailing Add:* 35 Chestnut Hill Lane S Williamsville NY 14221

COBURN, RONALD F, b Grand Rapids, Mich, Dec 10, 31; m 62; c 2. PHYSIOLOGY. *Educ:* Northwestern Univ, BS, 54, MD, 57. *Prof Exp:* Intern, Presby-St Luke's Hosp, Chicago, 57-58; resident internal med, Vet Admin Res Hosp, Chicago, 58-60; from instr to assoc med, 63-66 & asst prof, 67-68, from asst prof to assoc prof physiol, 66-75, PROF PHYSIOL & MED, SCH MED, UNIV PA, 75- *Concurrent Pos:* Fel physiol, Sch Med, Univ Pa, 60-63; mem, Vet Admin Respiratory Syst Res Eval Comt, 69-71. *Mem:* Am Physiol Soc; Am Soc Clin Invest; Am Fedn Clin Res. *Res:* Pulmonary and carbon monoxide physiology; heme catabolism; pulmonary gas exchange; tissue oxygenation; airway physiology. *Mailing Add:* Dept Physiol Univ Pa Sch Med Philadelphia PA 19104

COBURN, STEPHEN PUTNAM, b Orange, NJ, Nov 10, 36. BIOCHEMISTRY. *Educ:* Rutgers Univ, BS, 58; Purdue Univ, MS, 61, PhD(biochem), 64. *Prof Exp:* DIR, DEPT BIOCHEM, FT WAYNE STATE DEVELOP CTR, 63- *Concurrent Pos:* Instr, Ind Univ-Purdue Univ, Ft Wayne; dipl, Am Bd Clin Chem. *Mem:* Am Chem Soc; Am Asn Ment Retardation; Am Asn Clin Chem; Brit Biochem Soc; Am Inst Nutrit. *Res:* Biochemistry of mental retardation and other metabolic diseases; vitamin B6. *Mailing Add:* Ft Wayne State Develop Ctr Dept Biochem 4900 St Joe Rd Ft Wayne IN 46835-3299

COBURN, THEODORE JAMES, b Newton, Mass, June 11, 26; m 49; c 3. SOLID STATE PHYSICS. *Educ:* Ohio State Univ, BSc, 47, PhD(physics), 57. *Prof Exp:* Contract administr, Armaments Br, Off Naval Res, Washington, DC, 53-55; proj engr, Apparatus & Optical Div, 57-60, sr res physicist, 60-80, SR STAFF RES ASSOC, RES LABS, EASTMAN KODAK CO, NY, 80- *Mem:* Inst Elec & Electronics Engrs; Am Phys Soc; Electrochem Soc. *Res:* Surface state physics as applied to electrostatistics; infrared spectroscopy; military applications of infrared; solid state silicon photosensor and integrated circuit very large scale integration fabrication and testing. *Mailing Add:* 156 Pinecrest Dr Rochester NY 14617

COBURN, WILLIAM CARL, JR, b Duluth, Minn, Nov 2, 26; m 51; c 2. PHYSICAL ORGANIC CHEMISTRY. *Educ:* Univ Colo, BA, 48; Fla State Univ, MA, 51, PhD(phys chem), 54. *Prof Exp:* Phys chemist, 54-59, sr phys chemist, 59-67, HEAD MOLECULAR SPECTROS SECT, SOUTHERN RES INST, 67- *Mem:* AAAS; Coblentz Soc; Am Chem Soc; Sigma Xi. *Res:* Kinetics and mechanisms of organic reactions; molecular complexing; hydrogen bonding; theoretical and applied infrared spectroscopy; ultraviolet spectroscopy; nuclear magnetic resonance spectroscopy; mass spectroscopy. *Mailing Add:* Southern Res Inst 2000 Ninth Ave S Birmingham AL 35205

COCANOWER, R(OBERT) D(UNLAVY), b Hoehne, Colo, Dec 6, 20; m 47; c 2. PETROLEUM ENGINEERING. *Educ:* Univ Okla, BS, 46. *Prof Exp:* Chemist, Mercury Refining Co, 38-41; dist mgr, West Co, 46-52, supvr res, 53-65, mgr eng & develop, 65-77; DIR MKT, CRC WIRELINE, 77- *Mem:* Am Inst Mining Metall & Petrol Engrs; Am Petrol Inst; Am Soc Petrol Engrs; Independent Petrol Assoc Am. *Res:* Radioactivity; explosives; detection of radiation in oil well bores; use of radioactive isotopes in tracing fluids in oil wells; design of oil well perforating guns; design of electronic instruments for oil wells. *Mailing Add:* 4001 Hartwood Ft Worth TX 76109

COCCA, M(ICHAEL) A(NTHONY), b Green Island, NY, Mar 27, 25; m 48; c 4. METALLURGY. *Educ:* Rensselaer Polytech Inst, BMet, 51, MMetE, 59. *Prof Exp:* Metallurgist, Res Lab, Gen Elec Co, 51-54, supvr heat treating & surfaces, 54-57, admin asst, 57-59, metallurgist, 59-70, mgr struct mat appln, 70-76, prog mgr, Int Mat Progs, Gas Turbine Dept, 76-81, mgr struct mat eng, Gen Elec Co, 84-86; RETIRED. *Mem:* Am Soc Metals; Am Vacuum Soc (secy-treas, 61-); Am Inst Mining Metall & Petrol Engrs; Am Welding Soc. *Res:* Physical metallurgy of refractory and reactive metals, including high purity atmosphere technology in refining and processing; thin film technology; reactor structural materials; pressure vessel steels; nickel base alloys; process control, forging, casting, fabrication of alloy steels and superalloys. *Mailing Add:* Ten West St Green Island NY 12183

COCCODRILLI, GUS D, JR, b Peckville, Pa, July 28, 45; m 67; c 2. NUTRITION, FOOD SCIENCE & TECHNOLOGY. *Educ:* Pa State Univ, BS, 67, PhD(food sci), 71; Va Polytech Inst, MS, 69. *Prof Exp:* Sr chemist, Gen Foods Corp, 71-74, prof leader nutrit, 74-76, group leader nutrit, 76-80, lab mgr dent res, 80-81, sr lab mgr, nutrit sci, 81-86, dir, 87-89, GROUP DIR, GEN FOODS USA, KRAFT GEN FOODS, 90- *Concurrent Pos:* Adj prof, NY Med Col, 84- *Mem:* Inst Food Technologists; Nutrit Today Soc; Am Inst Nutrit. *Res:* Mineral nutrition research; trace element metabolism; vitamin nutrition; dental health research; cereal nutrition. *Mailing Add:* Gen Foods USA Kraft Gen Foods 555 S Broadway Tarrytown NY 10591

COCEANI, FLAVIO, b Trieste, Italy, Jan 3, 37; m 69; c 2. NEUROPHYSIOLOGY. *Educ:* Univ Bologna, MD, 61, Docent(human physiol), 68. *Prof Exp:* Asst prof, Univ Bologna, 65-66; vis scientist neurochem, Montreal Neurol Inst, 66-68; from asst prof to assoc prof, 68-83, PROF PHYSIOL & PEDIAT, UNIV TORONTO, 83-; PROG DIR, RES INST, HOSP SICK CHILDREN, 77- *Concurrent Pos:* Nat Res Coun fel & res fel, Dept Physiol, Univ Bologna, 61-62; res fel neurophysiol, Montreal Neurol Inst, 62-64; asst scientist, Res Inst, Hosp Sick Children, 68-77. *Honors & Awards:* Lectureship Award, Can Cardiovasc Soc, 81. *Mem:* AAAS; Am Soc Neurochem; Can Soc Clin Invest; NY Acad Sci; Can Physiol Soc. *Res:* Role of prostaglandins in brain function; role of prostaglandins in fetal and neonatal cardiovascular homeostasis. *Mailing Add:* Res Inst Hosp Sick Children Toronto ON M5G 1X8 Can

COCH, NICHOLAS KYROS, b New York, NY, Mar 30, 38; m 71. SEDIMENTOLOGY, ENVIRONMENTAL GEOLOGY. *Educ:* City Col New York, BS, 59; Univ Rochester, MS, 61; Yale Univ, PhD(geol), 65. *Prof Exp:* Asst prof, Southampton Col, Long Island Univ, 65-67; from asst prof to assoc prof, 67-75, PROF GEOL, QUEENS COL, CITY UNIV NY, 76-; PHD FAC EARTH & ENVIRON SCI, GRAD CTR, CITY UNIV NY, 83- *Concurrent Pos:* Consult geologist, Va Div Mining Resources, 65-71; prin investr, NASA grant, 75-77; Nat Oceanog & Atmospheric Admin, 81-83; mem sci adv comt, NY State Disaster Prep Comt, 88- *Mem:* Nat Asn Geol Teachers; fel Geol Soc Am; Am Asn Petrol Geologists; Int Asn Sedimentol; Am Inst Prof Geologists; Soc Econ Paleont & Mining. *Res:* Coastal and estuarine sedimentology; Atlantic Coastal Plain stratigraphy; environmental geology; geology of Hudson & Block Island Estuaries and adjacent waters; determination of sedimentary structures and dispersal patterns in sediments utilizing radiographic techniques; coastal geology; geologic effects of hurricanes. *Mailing Add:* Dept Geol Queens Col Flushing NY 11367

COCHIS, THOMAS, b Boston, Mass, June 24, 36; m 60; c 1. BOTANY, HORTICULTURE. *Educ:* McNeese State Col, BS, 60; La State Univ, MS, 62, PhD(hort, bot), 64. *Prof Exp:* Asst hort, La State Univ, 60-62, instr, 62-64; asst prof bot, Millsaps Col, 64-66; asst prof hort, Univ Fla, 66-68; PROF BOT, JACKSONVILLE STATE UNIV, 68- *Mem:* AAAS; Am Soc Hort Sci; Am Inst Biol Sci; Int Soc Hort Sci. *Res:* Horticultural and botanical research and teaching. *Mailing Add:* Dept Biol Ayers Hall Jacksonville State Univ Jacksonville AL 36265

COCHKANOFF, O(REST), b Homeglen, Alta, Mar 14, 26; m 53; c 2. MECHANICAL ENGINEERING. *Educ:* Univ BC, BASc, 49; Univ Toronto, MASc, 52; Iowa State Univ, PhD(theoret & appl mech), 63, Nat Defense Col Can, dipl, 79. *Prof Exp:* Instr mach design, Univ Toronto, 49-53; from asst prof to assoc prof mech eng, Tech Univ NS, 53-63, prof, 63-76 & 79-87, head dept, 63-76, dean eng, 71-78, EMER PROF MECH ENG, TECH UNIV NS, 91-; PRES, ATLANTIC ENG CONSULT LTD, 88- *Concurrent Pos:* Consult indust, 52- & Defense Res Bd Can, 54- *Mem:* Fel Can Aeronaut & Space Inst; Am Soc Automotive Engrs; fel Can Soc Mech Engrs; fel Eng Inst Can. *Res:* Applied mechanics; experimental methods; fundamental studies and applications of dynamics, stress analysis and fluid mechanics; ocean engineering and aerospace operations. *Mailing Add:* Atlantic Eng Consult 33 Ridgevalley Rd Halifax NS B3P 2E4 Can

COCHRAN, ALLAN CHESTER, b Long Beach, Calif, Jan 23, 42; m 62, 85; c 4. TOPOLOGY. *Educ:* E Cent State Col, BS, 62; Univ Okla, MA, 64, PhD(math), 66. *Prof Exp:* From asst prof to assoc prof, 66-77, PROF MATH, UNIV ARK, FAYETTEVILLE, 77-, VCHMN DEPT, 79- *Mem:* Am Math Soc; Math Asn Am. *Res:* Convergence space theory; theory of topological algebras. *Mailing Add:* SE 301 Dept Math Univ Ark Fayetteville AR 72701

COCHRAN, ANDREW AARON, b W Plains, Mo, Sept 28, 19; m 42; c 3. BIOPHYSICS, QUANTUM PHYSICS. *Educ:* Univ Mo-Rolla, BS, 41, MS, 63. *Prof Exp:* Chem engr, Phillips Petrol Co, Kans, 41-46; org res chemist, Mallinckrodt Chem Works, Mo, 46-52; chemist, Metall Res, Rolla Metall Res Ctr, US Bur Mines, 52-60, supvry phys chemist, 60-62, supvry res chemist, 62-80; RETIRED. *Mem:* Am Inst Mining Metall & Petrol Eng; Sigma Xi. *Res:* The fundamental theory of biophysics; development of new processes for recovering titanium, tin, and manganese from their respective ores; development of processes for treating industrial wastes to recover metals and reduce pollution. *Mailing Add:* Col Hills RFD 4 Box 83 Rolla MO 65401-9366

COCHRAN, BILLY JUAN, b Dec 10, 33; m 55; c 2. AGRICULTURAL ENGINEERING. *Educ:* Miss State Univ, BS, 58; Tex A&M Univ, MS, 62; Okla State Univ, PhD(agr eng), 73. *Prof Exp:* Instr, Tex A&M Univ, 58-64; from asst prof to assoc prof agr eng, La State Univ, 78-89; prof, Irri-Thai Farm Mach Proj, Bangkok, Thailand, 90; SPECIALIST SOLID WASTE, LA STATE UNIV, 90- *Concurrent Pos:* Proj leader, Agr Res Serv, USDA, 64-67, Am Sugar Cane League, 72-73, La Dept Conserv, 76-77, Battelle Columbus Labs, 77- & Dept Energy, 78- *Mem:* Am Soc Agr Engrs; Am Soc Sugarcane Technologists; Int Soc Sugarcane Technologists. *Res:* Development of system and machinery for biomass production as an alternate energy resource; development of systems and machinery for sugarcane production. *Mailing Add:* La Coop Ext Serv Knapp Hall Rm 268E La State Univ Baton Rouge LA 70803-1900

COCHRAN, CHARLES NORMAN, b Pittsburgh, Pa, Mar 24, 25; m 46; c 3. MATERIALS SCIENCE ENGINEERING. *Educ:* Westminster Col, Pa, BS, 45; Ohio State Univ, MS, 47. *Prof Exp:* Scientist, Alcoa Labs, Phys Chem Div, 47-58, sect head, 58-63, mgr, Phys Chem Div, 63-80, mgr technol forecasting, 80-84, tech consult, 84-88; PVT TECH CONSULT, 88- *Honors & Awards:* Dr Rene Wasserman Award, Am Welding Soc, 72; A V Davis Award, Alcoa, 85. *Mem:* Am Chem Soc; Metall Soc; Sigma Xi. *Res:* Oxidation of metals; gas in light metals; gas sorption; high temperature chemistry of aluminum; aluminum smelting; joining of aluminum; energy involvement of aluminum industry; technology forecasting; fuel cells; competition among materials. *Mailing Add:* 22 Oakglen Dr Oakmont PA 15139

COCHRAN, DAVID L(EO), b New York, NY, Dec 17, 29; m 56; c 3. MECHANICAL ENGINEERING. *Educ:* Univ Nev, BSME, 51; Stanford Univ, MSME, 52, PhD(mech eng), 57. *Prof Exp:* Asst mech eng, Stanford Univ, 51-57; sr mech engr, Aerojet-Gen Nucleonics, Gen Tire & Rubber Co, 57-61, asst mgr space power dept, 61-63; SR VPRES & DIR ENG, MB ASSOCS, SAN RAMON, 63- *Mem:* Am Soc Mech Engrs; Am Inst Aeronaut & Astronaut; Am Ord Asn. *Res:* Heat transfer; aerodynamics; mechanics; thermodynamics; research, development and manufacture of passive electronic warfare systems, miniature and rocket signaling and marking systems. *Mailing Add:* 472 Constitution Dr Danville CA 94526

COCHRAN, DAVID LEE, BONE RESORPTION, CONTROL OF CELL GROWTH. *Educ:* Med Col Va, DDS, 81, PhD(biochem), 82; Harvard Univ, MSc, 85. *Prof Exp:* ASST PROF PERIODONT, MED COL VA, 85- *Mailing Add:* 906 Dawnwood Rd Midlothian VA 23113-4461

COCHRAN, DONALD GORDON, b New Hampton, Iowa, July 5, 27; m 52; c 3. INSECT PHYSIOLOGY. *Educ:* Iowa State Univ, BS, 50; Va Polytech Inst, MS, 52; Rutgers Univ, PhD(entom), 55. *Prof Exp:* Entomologist, Chem Ctr, US Army, 55-57; assoc entomologist, Agr Exp Sta, 57-59, assoc prof, 59-64, PROF ENTOM, VA POLYTECH INST & STATE UNIV, 64-, HEAD DEPT, 85- *Mem:* AAAS; Entom Soc Am. *Res:* Physiological and genetical aspects of insect resistance to insecticides; cockroach genetics and cytogenetics; biochemistry of insect excretion. *Mailing Add:* Dept Entom Va Polytech Inst & State Univ Blacksburg VA 24061-0319

COCHRAN, DONALD ROY FRANCIS, b San Francisco, Calif, Oct 6, 26; m 60; c 3. NUCLEAR SCIENCE. *Educ:* Univ Calif, Berkeley, BS, 48; Johns Hopkins Univ, PhD(nuclear chem), 54. *Prof Exp:* Instr chem, Johns Hopkins Univ, 52-54; mem staff physics, 54-68, group leader, 68-73, ASST DIV LEADER, MESON PHYSICS DIV, LOS ALAMOS SCI LAB, 73- *Concurrent Pos:* Mem adv panel on accelerator radiation safety, US AEC, 70-73 & US DOE, 76- *Mem:* Am Phys Soc; Am Inst Physics; Am Nuclear Soc; Am Chem Soc. *Res:* Medium energy physics; nuclear chemistry and physics; accelerators. *Mailing Add:* MS H 830 Los Alamos Nat Lab MPDO PO Box 1663 Los Alamos NM 87545

COCHRAN, GEORGE THOMAS, b Washington, DC, Dec 28, 38; m; c 2. INORGANIC CHEMISTRY, ANALYTICAL CHEMISTRY. *Educ:* Univ Richmond, BS, 60; Univ Tenn, MS, 63; Clemson Univ, PhD(chem), 67. *Prof Exp:* Teacher high sch, Va, 60-61; from asst prof to assoc prof chem, Rollins Col, 67-78; qual assurance staff specialist, Burroughs Wellcome Pharmaceut Co, 78-81, head, Dept Admin & Auditing, 81-82, mgr, Compliance Dept, 83-85, Compliance & Auditing Dept, 85-90, MGR, COMPLIANCE & VALIDATION CERT, BURROUGHS WELLCOME PHARMACEUT CO, 90- *Mem:* Am Chem Soc; Am Soc Qual Control. *Res:* Coordination chemistry of univalent metal ions; high pressure liquid chromatography; organic polarography. *Mailing Add:* Burroughs Wellcome Co Box 1887 Greenville NC 27834

COCHRAN, GEORGE VAN BRUNT, b New York, NY, Jan 20, 32; m 70; c 3. RESEARCH ADMINISTRATION. *Educ:* Dartmouth Col, AB, 53; Columbia Univ, MD, 56, ScD(med, physiol), 67. *Prof Exp:* dir, Orthop Res Lab, St Lukes Hosp, 70-81; dir, Biomech Res Unit, Helen Hayes Hosp, 72-81, DIR, ORTHOP ENG & RES CTR, 81-; PROF CLIN ORTHOP SURG, COL PHYSICIANS & SURGEONS, COLUMBIA UNIV, 81-; ADJ PROF BIOMED ENG, RENSSELAER POLYTECH INST, 81- *Concurrent Pos:* Assoc prof clin orthop surg, Col Physicians & Surgeons, Columbia Univ, 75-81; adj prof biomed eng, Rensselaer Polytech Inst, 75-81. *Honors & Awards:* Nicholas Andry Award, Asn Bone & Joint Surgeons, 75; Kappa Delta Award, Am Acad Orthop Surg, 68. *Mem:* Orthop Res Soc; Am Acad Orthop Surgeons; fel NY Acad Sci; fel Explorers Club (pres, 81-85); Sigma Xi. *Res:* Biomechanics and electrophysiology of bone and soft tissues; gait analysis in musculoskeletal disabilities. *Mailing Add:* Orthop Eng & Res Ctr Helen Hayes Hosp West Haverstraw NY 10993

COCHRAN, HENRY DOUGLAS, JR, b Plainfield, NJ, Sept 13, 43; m 73; c 2. CHEMICAL ENGINEERING, COAL CONVERSION. *Educ:* Princeton Univ, BSE, 65; Mass Inst Technol, SM, 67, PhD(chem eng), 73. *Prof Exp:* Instr chem eng, Mass Inst Technol, 67-68; develop engr, Oak Ridge Nat Lab, 68-69; res asst, Dept Chem Eng, Mass Inst Technol, 69-72; develop engr, 72-74, group leader, 74-77, mgr coal conversion, 77-83, SR RES STAFF, OAK RIDGE NAT LAB, 83- *Mem:* Am Inst Chem Engrs; Sigma Xi; Am Chem Soc; Am Phys Soc. *Res:* catalysis; fluid dynamics; physical properties; vapor-liquid equilibria; pyrolysis; statistical mechanics; physics, fluids; chemistry, thermodynamics. *Mailing Add:* 105 Graceland Rd Oak Ridge TN 37830

COCHRAN, JAMES ALAN, b San Francisco, Calif, May 12, 36; m 58; c 2. APPLIED MATHEMATICS. *Educ:* Stanford Univ, BS, 56, MS, 57, PhD(math), 62. *Prof Exp:* Res mathematician, Stanford Res Inst, 55-58; asst math, Stanford Univ, 58-61; mem tech staff, Bell Tel Labs, 62-65, supvr, Electromagnetic Res Dept, 65-68, supvr, Appl Math & Statist Dept, 69-72; prof math, Va Polytech Inst & State Univ, 72-78; chmn dept, Wash State Univ, 78-84, chmn grad prog statist, 81-84, prof math, Dept Pure & Appl Math, 78-89, CAMPUS DEAN, WASH STATE UNIV, TRI-CITIES, 89- *Concurrent Pos:* Vis prof math, Stanford Univ, 68-69, Wash State Univ, 77 & Univ New S Wales, Australia, 85; foreign scholar math, Nanjing Inst Technol, China, 84; Fulbright fel, 84; Gordon fel comput & math, Deakin Univ, Australia, 85, 87. *Mem:* Am Math Soc; Soc Indust & Appl Math; Math Asn Am; Sigma Xi. *Res:* Applied research in electromagnetic theory and microwave propagation; basic research in special functions, asymptotics, differential and integral equations, and operator theory. *Mailing Add:* Wash State Univ Tri-Cities 100 Sprout Rd Richland WA 99352

COCHRAN, JEFFERY KEITH, b Ramey AFB, PR, Oct 11, 52. DISCRETE-EVENT SIMULATION, HIERARCHIAL DECISION PROCESSES. *Educ:* Purdue Univ, BSE, 73, MSNE, 76, MSIE, 82, PhD(indust eng), 84. *Prof Exp:* Assoc engr, Aerojet Nuclear Co, 73; engr, Stone & Webster Eng

Corp, 74; res asst, NASA, 75-76; staff mem, Los Alamos Nat Lab, 76-80; grad instr, Purdue Univ, 81-84; sr res scientist, Battelle Northwest Labs, 84; asst prof, 84-88, ASSOC PROF, ARIZ STATE UNIV, 88- *Concurrent Pos:* Prin investr, Western Elec Prod, 85, Honeywell, Inc, 85-86, Ariz Dept Transp & Motorola, Inc, 86-87, NSF, 87-92, Inst Mfg & Automations Res, 88-90 & USN, 91-92. *Mem:* Soc Computer Simulation; Opers Res Soc Am; sr mem Soc Mfg Engrs; sr mem Inst Indust Engrs; Asn Comput Mach; Am Soc Eng Educ. *Res:* Intelligent simulation environments; artificial intelligence and knowledge engineering; computer integrated manufacturing; mathematical modeling especially stochastic systems; decision analysis and decision support systems; applied operations research. *Mailing Add:* Indust & Mgt Systs Eng Ariz State Univ Tempe AZ 85287

COCHRAN, JOHN CHARLES, b Akron, Ohio, Feb 10, 35; m 58; c 3. SYNTHETIC INORGANIC & ORGANOMETALLIC CHEMISTRY. *Educ:* Col Wooster, BA, 57; Univ NC, MA, 60; Univ NH, PhD(org chem), 67. *Prof Exp:* Instr chem, Randolph-Macon Women's Col, 60-62; from asst prof to assoc prof, 66-83, chmn dept, 81-87, PROF CHEM, COLGATE UNIV, 83- *Concurrent Pos:* Vis prof, State Univ NY, Albany, 81, Syracuse Univ, 87-88. *Mem:* Am Chem Soc. *Res:* Preparation and reactions of organotin compounds; Reimer-Tiemann reaction; promotion of organic reactions by ultrasound. *Mailing Add:* Dept Chem Colgate Univ Hamilton NY 13346

COCHRAN, JOHN EUELL, JR, b Dawson, Ala, May 22, 44; m 65; c 2. AEROSPACE ENGINEERING, AIRCRAFT ACCIDENT ANALYSIS. *Educ:* Auburn Univ, BAE, 66, MS, 67; Univ Tex, Austin, PhD(aerospace eng), 70; Jones Law Inst, JD, 76. *Prof Exp:* Res engr, Dept Aerospace Eng & Eng Mech, Univ Tex, 69; from instr to assoc prof, 69-80, PROF AEROSPACE ENG, AUBURN UNIV, 80- *Concurrent Pos:* Am Soc Elec Engrs-NASA fel, Marshall Space Flight Ctr, 70 & 71; consult, Northrop Serv Inc, 72-73, Battelle Mem Labs, 74 & 76, aviation litigation, 77-, Accident Prev, Invest & Anal Co, 77-86 & SRS Technologies, 84-89; vis assoc prof eng sci & systs, Univ Va, 75; pres, Eaglemark, Inc, 84-; atty law, 77-; assoc ed, J Guid Control & Dynamics, 89- *Mem:* Assoc fel Am Inst Aeronaut & Astronaut; Nat Soc Prof Engrs; Sigma Xi; Am Astronaut Soc; Am Bar Asn. *Res:* Guidance, control and dynamics of spacecraft and aircraft; stability of nonlinear systems; dynamics of rocket launchers; products liability; astrodynamics; dynamics of towed flight vehicles. *Mailing Add:* Dept Aerospace Eng Auburn Univ Auburn AL 36849-5338

COCHRAN, JOHN FRANCIS, b Saskatoon, Sask, Jan 29, 30; m 57; c 2. SOLID STATE PHYSICS. *Educ:* Univ BC, BASc, 50, MASc, 51; Univ Ill, PhD(physics), 55. *Prof Exp:* Res assoc physics, Univ Ill, 55-56; Nat Res Coun Can fel, Clarendon Lab, Eng, 56-57; asst prof, Mass Inst Technol, 57-65; dean sci, 81-85, PROF PHYSICS, SIMON FRASER UNIV, 65- *Concurrent Pos:* Sloan fel, 58-60. *Mem:* Am Phys Soc; Can Asn Physicists. *Res:* Electronic properties of very pure metals; ferromagnetic metals. *Mailing Add:* Dept Physics Simon Fraser Univ Burnaby BC V5A 1S6 Can

COCHRAN, JOHN RODNEY, b St Joseph, Mo, Feb 7, 20; m 41; c 2. SPEECH PATHOLOGY. *Educ:* Utah State Univ, BS, 49, MS, 50; Univ Utah, PhD(speech path), 59. *Prof Exp:* Asst speech path, Univ Utah, 54-59; assoc prof speech path & psychol, Southwest Tex State Col, 59-61; assoc prof speech & speech path, chmn dept speech path & audiol & dir speech & hearing clin, Eastern NMex Univ, 61-70; prof & dir speech & hearing clin, Univ Alaska, 70-71; assoc prof & dir speech & hearing clin, Lehman Col, 71-73; prof speech & path, Kearney State Col, 73-88; RETIRED. *Mem:* Am Speech & Hearing Asn. *Res:* Communication abilities of air force personnel. *Mailing Add:* 4215 Pony Express Rd Kearney NE 68847

COCHRAN, KENNETH WILLIAM, b Chicago, Ill, Nov 2, 23; m 45; c 2. TOXICOLOGY, PHARMACOLOGY. *Educ:* Univ Chicago, SB, 47, PhD(pharmacol), 50; Am Bd Indust Hyg, cert toxicol. *Prof Exp:* Res asst toxicity lab, Univ Chicago, 48-50, res assoc, US Air Force Radiation Lab & Dept Pharmacol, 50-52; from instr to assoc prof pharmacol, Sch Med, Univ Mich, 52-89, res assoc epidemiol, Sch Pub Health, 52-55, from asst prof to prof epidemiol, 55-89, secy fac, 70-73, EMER PROF EPIDEMIOL, SCH PUB HEALTH, UNIV MICH, ANN ARBOR, 89-, EMER ASSOC PROF PHARMACOL, SCH MED, 89- *Concurrent Pos:* Mem, pharmacol & endocrinol fel rev panel, NIH, 60-64; chair, Toxicol Comt, N Am Mycological Asn, 80-87; exec secy, N Am Mycological Asn, 88- *Mem:* Fel AAAS; Am Soc Microbiol; Am Acad Indust Hyg; Am Soc Pharmacol; Mycol Soc Am; Soc Exp Biol Med; Sigma Xi. *Res:* Virus chemotherapy; toxicology. *Mailing Add:* 3556 Oakwood Ann Arbor MI 48104

COCHRAN, LEWIS WELLINGTON, b Perryville, Ky, Oct 12, 15; m 40; c 2. PHYSICS. *Educ:* Morehead State Col, BS, 36; Univ Ky, MS, 39 & 40, PhD(physics), 52. *Hon Degrees:* DSc, Univ Ky, 82, Ky State Univ, 83. *Prof Exp:* Instr math & physics, Morehead State Col, 39-41 & Cumberland Univ, 41; from asst prof to prof physics, Univ Ky, 46-81, actg head dept, 56-58, assoc dean, 63-66, provost, 65-67, actg dean, Grad Sch, 66-67, dean, Grad Sch & vpres res, 67-70, vpres acad affairs, 70-81; RETIRED. *Mem:* Am Phys Soc; Am Asn Physics Teachers; Sigma Xi. *Res:* Proton induced nuclear reactions; interaction of radiation with matter; experimental nuclear physics; nuclear structure physics; gaseous electronics. *Mailing Add:* 1581 Beacon Hill Rd Lexington KY 40504

COCHRAN, PATRICK HOLMES, b Guthrie Center, Iowa, Oct 14, 37; m 61; c 3. FOREST SOILS. *Educ:* Iowa State Univ, BSc, 59; Ore State Univ, MSc, 63, PhD(soils), 66. *Prof Exp:* Asst soils, Ore State Univ, 64-65; asst prof silvicult, State Univ NY Col Forestry, Syracuse Univ, 65-67; SOIL SCIENTIST, PAC NORTHWEST FOREST & RANGE EXP STA, US FOREST SERV, 67- *Mem:* Am Soc Agron; Soc Am Foresters; Sigma Xi. *Res:* Relationships of physical, thermal and chemical soil properties to the distribution of wild land vegetation and tree growth. *Mailing Add:* 480 SE Airpark Dr Bend OR 97702

COCHRAN, PAUL TERRY, b Sullivan, Ind, Jan 27, 38; m 72; c 2. CARDIOLOGY, MEDICAL EDUCATION. *Educ:* DePauw Univ, BA, 60; Western Reserve Univ, MD, 64; Am Bd Internal Med, dipl; Am Bd Cardiovasc Dis, dipl. *Prof Exp:* Staff cardiologist, Malcolm Grow US Air Force Hosp, 69-71; clin cardiologist, Gallatin Med Group, Downey, Calif, 71-72; asst prof med, 72-77, CLIN ASSOC PROF MED, SCH MED, UNIV NMEX, 77-; CLIN CARDIOLOGIST, SOUTHWEST CARDIOL ASN, 81- *Concurrent Pos:* Dir cardiac diag lab, Bernalillo County Med Ctr, 72-77; staff cardiologist, Vet Admin Hosp, Albuquerque, 72-77; fel coun clin cardiol, Am Heart Asn. *Mem:* Fel Am Col Cardiol; fel Am Col Physicians; Am Heart Asn. *Res:* Clinical investigations of the hemodynamics of heart disease; research in teaching methods in cardiovascular medicine. *Mailing Add:* 1101 Medical Arts Ave NE Bldg 5 Albuquerque NM 87102

COCHRAN, ROBERT GLENN, b Indianapolis, Ind, July 12, 19; m 44; c 1. PHYSICS. *Educ:* Ind Univ, AB, 48, MS, 50; Pa State Univ, PhD(nuclear physics), 57. *Prof Exp:* Res asst cyclotron group, Ind Univ, 47-50; nuclear physicist & group leader chg swimming pool reactor fac, Physics Div, Oak Ridge Nat Lab, 50-54; assoc prof nuclear eng & dir res reactor facil, Pa State Univ, 54-59; head dept & dir nuclear sci ctr, 59-84, PROF NUCLEAR ENG, TEX A&M UNIV, 59- *Concurrent Pos:* Consult, Nat Regulatory Comn, US Air Force, and others. *Mem:* Am Nuclear Soc; Am Phys Soc; Am Soc Eng Educ; Nat Soc Prof Engrs. *Res:* Nuclear physics; decay schemes and isomeric states of short half-life nuclids; nuclear reactor physics and nuclear engineering; power reactors and their environmental effects. *Mailing Add:* Dept Nuclear Eng Tex A&M Univ Col Station TX 77843-3133

COCHRAN, STEPHEN G, b Indianapolis, Ind, Aug 19, 47; m 69; c 3. PHYSICS, COMPUTER SCIENCE. *Educ:* Loyola Univ, BS, 69; Cornell Univ, MS, 73, PhD(physics), 74. *Prof Exp:* PHYSICIST, DEPT ENERGY, LAWRENCE LIVERMORE LAB, UNIV CALIF, 74- *Mem:* Am Phys Soc. *Res:* Continuum dynamics; materials properties; numerical simulation; creation of models of material behavior exhibited in experiments. *Mailing Add:* 1536 Naples Ct Livermore CA 94550

COCHRAN, THOMAS B, b Washington DC, Nov 18, 40; m 71; c 2. NUCLEAR WEAPONS RESEARCH, ARMS CONTROL. *Educ:* Vanderbilt Univ, BE, 62, MS, 65, PhD(physics), 67. *Prof Exp:* Asst prof physics, Naval Postgrad Sch, 67-71; group supvr, Litton Sci Support Lab, Mellonics Div, Litton Industs, 69-71; sr res assoc, Resources Future, 71-73; SR STAFF SCIENTIST, NATURAL RESOURCES DEFENSE COUN, INC, 73- *Honors & Awards:* Szilard Award, Am Phys Soc, 87; Pub Serv Award, Fedn Am Scientists, 87. *Mem:* Fel Am Phys Soc; AAAS; Sigma Xi; Health Physics Soc; Am Nuclear Soc. *Res:* Nuclear weapons research and production, arms control; nuclear weapons proliferation, safeguards, seismic verification national energy research and development policy (principally nuclear energy issues, the breeder reactor and plutonium recycle; radiation exposure standards. *Mailing Add:* NRDC 1350 New York Ave NW Suite 300 Washington DC 20005

COCHRAN, THOMAS HOWARD, Aliquippa, Pa, Apr 6, 40; m 62; c 4. COMBUSTION, PROPULSION. *Educ:* Renesselaer Polytech Inst, BME, 62, Case Inst Technol, MSME, 67. *Prof Exp:* Proj engr exp res, 62-73, Sect head res & tech, 74-79, br chief syst anal, 79-80, br chief elec propulsion & space exp, 81-82, dep div chief, space propulsion, 83-84, DEP DIR, SPACE STA SYST, LEWIS RES CTR, NASA, 84- *Concurrent Pos:* Space Sta Task Force, NASA Hq, Washington, DC, 82-83. *Res:* Technology efforts in electric propulsion; definition of fundamental experiments to be conducted in space; analytical and experimental research on gravitational effects on fundamental combustion and fluid physics phenomena as applied to spacecraft fire safety and propellant handling systems. *Mailing Add:* 31115 Nantucket Row Bay Village OH 44140

COCHRAN, VERLAN LEYERL, b Declo, Idaho, Feb 19, 38; m 69; c 2. SOIL MANAGEMENT, SOIL FERTILITY. *Educ:* Calif State Polytech Univ, BS, 66; Wash State Univ, MS, 71. *Prof Exp:* Soil scientist, Land Mgt & Water Conserv Res Unit, 66-85, SOIL SCIENTIST, SUBARTIC AGR RES UNIT, ARS-USDA, 85- *Concurrent Pos:* Asst soil scientist, Wash State Univ, 73-85; affil fac, Univ Idaho, 81-85 & Univ Alaska, Fairbanks, 85- *Mem:* Am Soc Agron; Soil Sci Soc Am; Sigma Xi; Am Soc Microbiol. *Res:* Soil management and fertility problems associated with reduced tillage systems in small grain production with emphasis on water use efficiency, nutrient cycling, and production of greenhouse gasses. *Mailing Add:* 2370 Debbie Way Fairbanks AK 99709

COCHRAN, WILLIAM RONALD, b Kalamazoo, Mich, May 24, 40. PHYSICS, OPTICS. *Educ:* Univ Calif, Los Angeles, BA, 62, MS, 64, PhD(physics), 69. *Prof Exp:* From asst prof to assoc prof, 69-81, PROF PHYSICS, YOUNGSTOWN STATE UNIV, 81- *Mem:* Am Phys Soc; Optical Soc Am. *Res:* Spectroscopy of ions in crystals; heterodyne spectroscopy of solids; optical physics; photo conductivity. *Mailing Add:* 115 Col Lane Poland OH 44514

COCHRANE, CHAPPELLE CECIL, b Conway, SC, Oct 28, 13; m 40; c 3. ORGANIC CHEMISTRY. *Educ:* Howard Univ, BS, 38, MS, 40, Ohio State Univ, PhD(chem), 51. *Prof Exp:* Assoc, Howard Univ, 40-41; instr chem, Morgan State Col, 41-42 & Cent State Col, 46-47; res chemist, Glidden Co, 51-58, Armour & Co, 58-60, US Army Chem Ctr, 60-62 & Nalco Chem Co, 62-74; from asst prof to assoc prof chem, Chicago State Univ, 74-83; RETIRED. *Concurrent Pos:* Assoc, Ohio State Univ, 51. *Honors & Awards:* Lloyd Hall Award, Howard Univ, 38. *Mem:* Am Chem Soc. *Res:* Polynuclear aromatic hydrocarbons; steroids; fatty acids. *Mailing Add:* Two Forest Blvd Park Forest IL 60466

COCHRANE, DAVID EARLE, b Coldspring, NY, July 16, 44; c 2. PHYSIOLOGY, NEUROBIOLOGY. *Educ:* Cornell Univ, BS, 66; Univ Vt, MS, 68, PhD(physiol & biophys), 71. *Prof Exp:* Res assoc electrophys, Univ Vt, 71-72; NIH fel cellular aspects of secretion, Dept Pharm, Yale Univ, 72-76; from asst prof to assoc prof, 76-90, PROF BIOL, TUFTS UNIV, 90- *Res:* Mast cell physiology, peptides, and cellular aspects of inflammation. *Mailing Add:* Dept Biol Tufts Univ Medford MA 02155

COCHRANE, HECTOR, b Stowmarket, Eng, Mar 16, 40; m 65; c 3. PHYSICAL CHEMISTRY. *Educ:* Univ Nottingham, BSc, 61, PhD(chem), 64; Univ Ill, MBA, 86. *Prof Exp:* Fel fuel sci, Pa State Univ, 64-66; res chemist, 66-73, group leader, 73-80, TECH DIR, CAB-O-SIL RES & DEVELOP, CABOT CORP, 80- *Mem:* Am Chem Soc; Sigma Xi. *Res:* Formation of fine particles in flames; study of the surface chemistry and aggregate morphology of particles and how these properties affect their theological properties in liquids and reinforcement properties in rubber. *Mailing Add:* 1807 Bridgestone Ave Champaign IL 61820

COCHRANE, ROBERT LOWE, b Morgantown, WVa, Feb 10, 31. REPRODUCTIVE PHYSIOLOGY, ENDOCRINOLOGY. *Educ:* WVa Univ, BA, 53; Univ Wis, MS, 54, PhD(genetics), 61. *Prof Exp:* Res asst genetics, Univ Wis, 53-55; laborer, Fur Exp Sta, Petersburg, Alaska, 55; animal husb agent, USDA, 55-61; biologist, US Food & Drug Admin, 61-62; sr res fel primate reproduction, Med Sch, Univ Birmingham, Eng, 62-65; proj assoc, Sch Med, Univ Pittsburgh, 65-66; sr endocrinologist, Lilly Res Labs, 66-80; res assoc, G D Searle & Co, 80-81; laborer & adv, Short's Fur Farm, Granton, Wis, 81-83; res assoc, Marshfield Med Found, Inc, Wis, 83-84; biologist, Northwood Fur Farms, Inc, Cary, Ill, 84; consult, Fur Farming, Food & Agr Org, India, 85; ADJ PROF, DIV ANIMAL & VET SCI, WVA UNIV, 87- *Concurrent Pos:* Res asst zool, Univ Wis, 57-60. *Mem:* AAAS; Am Soc Animal Sci; Am Inst Biol Sci; Soc Study Reproduction; Endocrine Soc; Soc Exp Biol Med. *Res:* Contraception; ovoimplantation; control of corpus luteum and ovary function; immunological tolerance, hormonal control of growth and effects of estrogens on reproduction in mustelids. *Mailing Add:* 404 Junior Ave Morgantown WV 26505

COCHRANE, ROBERT W, b Toronto, Ont, Mar 2, 40; m 64; c 3. MAGNETISM, METALLIC ALLOYS. *Educ:* Univ Toronto, BSc, 63, MA, 65, PhD(physics), 69. *Prof Exp:* Postdoctoral, Yale Univ, 69-71; res assoc physics, McGill Univ, 71-74, reader, 74-78; res assoc, 79-85, assoc prof, 85-88, PROF PHYSICS, DEPT PHYSICS, UNIV MONTREAL, 88- *Concurrent Pos:* Co-chmn, 6th Int Conf Rapidly Quenched Metals, 84-87. *Mem:* Can Asn Physicists; Mat Res Soc. *Res:* Structural, electrical and magnetic properties of metals and alloys; amorphous metals; metallic multilayers-particularly combinations of amorphous and crystalline alloys. *Mailing Add:* Dept Physics Univ Montreal Montreal PQ H3C 3J7 Can

COCHRANE, WILLIAM, b Toronto, Ont, Mar 18, 26; m 51; c 3. PEDIATRICS. *Educ:* Univ Toronto, MD, 49; FRCP(C), 56. *Prof Exp:* Clinician, Hosp Sick Children, Toronto, 56-58; assoc prof pediat, Fac Med, Dalhousie Univ, 58-63, prof & head dept, 63-67; dean fac med, Univ Calgary, 67-73, pres & vchancellor, 74-78; CHMN & CHIEF EXEC OFFICER, CONNAUGHT LABS LTD, 78- *Concurrent Pos:* Dep minister health serv, Govt Alta, 73-74. *Mem:* Soc Pediat Res; fel Am Col Physicians; Can Pediat Soc; Can Soc Clin Invest. *Res:* Metabolic diseases of children; biochemical relationship of protein and amino acid metabolism to mental disease. *Mailing Add:* 15-3203 Rideau Pl SW Calgary AB T2S 2T1 Can

COCHRUM, ARTHUR L, b Kingsville, Tex, Jan 5, 25. GEOLOGY. *Educ:* Univ Tex, MA, 51. *Prof Exp:* PETROL CONSULT, 77- *Mem:* Fel Geol Soc Am; Am Prof Geologists. *Res:* Petroleum. *Mailing Add:* 322 Knipp Rd Houston TX 77024

COCIVERA, MICHAEL, b Pittsburgh, Pa, Jan 21, 37; m 69; c 2. INORGANIC CHEMISTRY, MATERIALS SCIENCE ENGINEERING. *Educ:* Carnegie-Mellon Univ, BSc, 59; Univ Calif, Los Angeles, PhD(phys chem), 63. *Prof Exp:* Mem tech staff chem, Bell Tel Labs, 63-69; assoc prof, 69-74, PROF CHEM, UNIV GUELPH, 74- *Mem:* Am Chem Soc; Chem Inst Can; Mat Res Soc; Electrochem Soc. *Res:* Preparation of thin-film semiconductors by electrochemical deposition; study of the photophysical properties of thin-film semiconductors; study of properties of the interface (solid-solid and solid-liquid). *Mailing Add:* Dept Chem Univ Guelph Guelph ON N1G 2W1 Can

COCK, LORNE M, b Tatamagouche, NS, June 1, 32; m 57; c 2. ANIMAL NUTRITION. *Educ:* McGill Univ, BSc, 54; Univ Wis, MS, 60; Univ Maine, PhD(animal nutrit), 66. *Prof Exp:* From instr to asst prof animal sci, Univ Maine, 65-69; assoc prof, 69-74, PROF ANIMAL SCI, NS AGR COL, 74-, HEAD DEPT, 69- *Mem:* Am Soc Animal Sci; Am Dairy Sci Asn; Agr Inst Can. *Res:* Energy metabolism of ruminants; influence of dietary nitrogen on ruminant heat increment; fasting metabolism of sheep. *Mailing Add:* Dept Animal Sci NS Agr Col PO Box 550 Truro NS B2N 5E3 Can

COCKE, JOHN, b Charlotte, NC, 1925. TECHNICAL CONTRIBUTIONS IN COMPUTING. *Educ:* Duke Univ, BS, 46, PhD(math), 56. *Prof Exp:* Engr, Air Eng Co, 46-49; engr, Gen Elec Co, 49-50; RES STAFF MEM, IBM RES DIV, 56- *Concurrent Pos:* Vis prof, Mass Inst Tech, Elec Eng Dept, 62; vis prof, Courant Inst Math Sci, 68-69; mem, IBM Corp Tech Comt, 77-78. *Honors & Awards:* Nat Medal of Technol, 91; Eckert-Maunchy Award, Am Comput Soc & Inst Elec & Electronics Engrs, 85; Am Turing Award, Asn Comput Mach, 87. *Mem:* Nat Acad Eng; fel Am Acad Arts & Sci. *Res:* Inventor of Reduced Instruction Set Computer (RISC); author of numerous articles published in various journals. *Mailing Add:* IBM Corp TJ Watson Res Ctr PO Box 704 Yorktown Heights NY 10598

COCKERELL, GARY LEE, b Fort Dodge, Iowa, Dec 20, 45; m 69; c 2. ONCOLOGY. *Educ:* Univ Calif, Davis, BS, 68, DVM, 70; Ohio State Univ, PhD(vet path), 76. *Prof Exp:* Vet lab admin officer res, US Army Med Res Inst Infectious Dis, 70-73; veterinarian, McClellan Vet Clinic, 71-73; grad res assoc path, Ohio State Univ, 73-76; asst prof, 76-81, ASSOC PROF, DEPT PATH, NY STATE COL VET MED, CORNELL UNIV, 81- *Concurrent Pos:* Res fel, NIH, 75-76; prin investr, Nat Cancer Inst contract, 76-81; mem bd dirs, Cornell Veterinarian, 81- *Mem:* Am Col Vet Path; Int Asn Comparative Leukemia Res; Am Vet Med Asn; Int Acad Path; Vet Cancer Soc. *Res:* Comparative pathology and the immunobiology of naturally occurring neoplasms in domestic animals; chemical and viral induced neoplasms in domestic and laboratory species of animals. *Mailing Add:* Dept Path Col Vet Med Colo State Univ Ft Collins CO 80523

COCKERHAM, COLUMBUS CLARK, b Mountain Park, NC, Dec 12, 21; m 44; c 3. POPULATION & QUANTITATIVE GENETICS. *Educ:* NC State Col, BS, 43, MS, 49; Iowa State Univ, PhD(animal breeding & genetics), 52. *Prof Exp:* Asst prof biostatist, Sch Pub Health, Univ NC, 52-53; from assoc prof to prof exp statist, 53-72, WILLIAM NEAL REYNOLDS PROF STATIST & GENETICS, NC STATE UNIV, 72-, DISTINGUISHED UNIV PROF, 88-, EMER PROF, 91- *Concurrent Pos:* Prin investr, NIH Grant, 60-63, proj dir, 63-, mem genetics study sect, 65-69; consult, adv comt protocols for safety eval, Food & Drug Admin, 67-69; ed, Theoret Pop Biol, 75-81, assoc ed, 82-; assoc ed, Am J Human Genetics, 78-80. *Honors & Awards:* Outstanding Res Award, NC State Univ; O Max Gardner Award; Superior Service Award, US Dept Agr. *Mem:* Nat Acad Sci; fel AAAS; Biomet Soc; Am Soc Animal Sci; Genetics Soc Am; fel Am Soc Agron; fel Crop Sci Soc Am; foreign hon mem, Genetics Soc Japan. *Res:* Population and quantitative genetic theory; estimation of genetic parameters in populations; development of mating and experimental designs for estimation; selection theory; applications to plant and animal breeding. *Mailing Add:* Dept Statist Box 8203 NC State Univ Raleigh NC 27695-8203

COCKERHAM, LORRIS G(AY), b Denham Springs, La, Sept 27, 35; m 57; c 4. RADIATION TOXICOLOGY, NEUROTOXICOLOGY. *Educ:* La Col, Ba, 57; Colo State Univ, Ms, 73, PhD(physiol toxocol), 79. *Prof Exp:* Instr biol sci, US Air Force Acad, 73-75, asst prof & div chief, 75-77; DIV CHIEF PHYSIOL, ARMED FORCES RADIOBIOL RES INST, DEFENSE NUCLEAR AGENCY, 80-; ASST PROF NEUROPHYSIOL, UNIFORMED SERV UNIV HEALTH SCI, 81- *Concurrent Pos:* Consult, Orgn Concerned Toxicity Agent Orange & Assoc Chlorinated Dioxins, 79-; prin investr, Armed Forces Radiobiol Res Inst, 80-, comdr, Air force Element, 81- *Mem:* Sigma Xi; Am Physiol Soc; NY Acad Sci; Soc Neurosci; Aerospace Med Asn; Soc Toxicol. *Res:* Effect of selected chemicals on radiation induced early transient incapacitation and performance decrement; chemicals of special interest are neurotoxins and their antidotes. *Mailing Add:* PO Box 24318 Little Rock AR 72221-4318

COCKERLINE, ALAN WESLEY, b Toronto, Ont, Oct 2, 26; m 51; c 3. CYTOLOGY. *Educ:* Univ Mich, BS, 52, MS, 53; Mich State Univ, PhD(bot), 61. *Prof Exp:* From asst prof to assoc prof, 61-89, EMER PROF, DIR MED TECHNOL, TEX WOMAN'S UNIV, 75-, TECHNOL PROG, 75- *Concurrent Pos:* Proj dir undergrad res partic, NSF, 64-72. *Mem:* AAAS; Am Chem Soc; Am Asn Plant Physiologists; Mycol Soc Am; Tissue Cult Asn; Sigma Xi. *Res:* Mycotoxins. *Mailing Add:* 2509 Woodhaven Denton TX 76201-2245

COCKETT, ABRAHAM TIMOTHY K, b Maui, Hawaii, Sept 4, 28; c 4. UROLOGY, PHYSIOLOGY. *Educ:* Brigham Young Univ, BS, 50; Univ Utah, MD, 54. *Prof Exp:* Res fel, Dept Med, Univ Southern Calif, 57-58; assoc prof urol, Univ Calif, Los Angeles, 62-69; PROF UROL SURG, SCH MED, UNIV ROCHESTER, 69- *Concurrent Pos:* Chief urol, Harbor Gen Hosp, Torrance, Calif, 62-69; urologist-in-chief, Strong Mem Hosp, Rochester, NY, 69- *Mem:* AAAS; Soc Univ Surgeons; Am Urol Asn; Soc Univ Urologists; Undersea Med Soc. *Res:* Kidney physiology and transplantation; underwater physiology related to problems in decompression sickness and treatment of these alterations; support of man in outer space; studies related to renal, urinary and testicular physiology. *Mailing Add:* Dept Urol Univ Rochester 601 Elmwood Ave Rochester NY 14642

COCKING, W DEAN, b San Diego, Calif, Oct 27, 40;; m. PLANT ECOLOGY. *Educ:* Pomona Col, BA, 62; Cornell Univ, MS, 67; Rutgers Univ, PhD(bot & ecol), 73. *Prof Exp:* Instr, Mohawk Valley Community Col, Utica, 66-69; asst prof biol, 73-79, ASSOC PROF BIOL, JAMES MADISON UNIV, 80- *Concurrent Pos:* Dir, James Madison Univ fire ecol res proj, Shenandoah Nat Park, 75-80, forest ecol res proj, 80-86 & terrestrial mercury accumulation proj, Waynesboro, Va, 83-; mem, educ comt, Am Inst Biol Sci, 81-84. *Mem:* Ecol Soc Am; Brit Ecol Soc; Sigma Xi. *Res:* Ecosystem and physiological ecology; stability of systems in response to stress; ecosystem development; practical application of ecological concepts; plant community pattern in Shenandoah National Park; bioaccumulation of mercury in terrestrial flood plain ecosystems. *Mailing Add:* Dept Biol James Madison Univ Harrisonburg VA 22807

COCKRELL, ROBERT ALEXANDER, b Yonkers, NY, Aug 11, 09; m 33; c 3. DENDROLOGY, WOOD SCIENCE & TECHNOLOGY. *Educ:* Syracuse Univ, BS, 30, MS, 31; Univ Mich, PhD(wood technol), 34. *Prof Exp:* Asst, NY State Col Forestry, Syracuse Univ, 30-32; asst, Sch Forestry & Conserv, Univ Mich, 32-34, jr forester, 34-35; assoc prof forestry, Clemson Col, 35-36; from asst prof to prof, Univ Calif, Berkeley, 36-77, assoc forester, Exp Sta, 45-62, assoc dean, Grad Div Univ, 56-67, secy, Acad Senate, 68-78, 87-88, wood technologist, Exp Sta, 62-77, EMER PROF FORESTRY, UNIV CALIF, BERKELEY, 77- *Concurrent Pos:* Technologist, Forest Prod Lab, US Forest Serv, 42-45; forestry res specialist, US Army, Japan, 50; Orgn Europ Econ Coop sr vis fel, 61. *Mem:* AAAS; Soc Am Foresters; Forest Prod Res Soc; Soc Wood Sci & Technol; Int Asn Wood Anatomists. *Res:* Anatomy of tropical woods; mechanical properties of wood; gluing of wood; wood shrinkage; cell wall structure. *Mailing Add:* Forestry Dept 145 Mulford Hall Univ Calif Berkeley CA 94720

COCKRELL, RONALD SPENCER, b Kansas City, Mo, June 26, 38; m 60; c 2. BIOCHEMISTRY. *Educ:* Univ Mo, BS, 63, BMedS, 64; Univ Pa, PhD(molecular biol), 68. *Prof Exp:* Asst prof, 69-74, ASSOC PROF BIOCHEM, SCH MED, ST LOUIS UNIV, 74- *Concurrent Pos:* USPHS fel, Cornell Univ, 68-69. *Mem:* Am Soc Biochem & Molecular Biol. *Res:* Membrane structure and function; hypertension. *Mailing Add:* Dept Biochem St Louis Univ Sch Med St Louis MO 63104

COCKRUM, ELMER LENDELL, b Sesser, Ill, May 29, 20; m 43; c 3. VERTEBRATE ZOOLOGY. *Educ:* Univ Kans, PhD(zool), 51. *Prof Exp:* Asst cur mammal, Mus Natural Hist, Univ Kans, 46-48, res assoc, 51-52, fel embryol & gen zool, 48-49; from asst prof to assoc prof, 52-60, dir, Desert Biol

Sta, 65-68, mammalogist, Agr Exp Sta 74-80, head, Dept Ecol & Evolutionary Biol, 76-84, PROF ZOOL, UNIV ARIZ, 60-, CUR MAMMALS, 57- *Mem:* AAAS; Am Soc Zool; Am Soc Mammal. *Res:* Mammals of Kansas and Arizona; microtine rodents; life history studies of bats. *Mailing Add:* Dept Zool Univ Ariz Tucson AZ 85721

COCKRUM, RICHARD HENRY, b Culver, Calif, Sept 23, 50; m 78; c 1. ELECTRONIC MATERIALS ENGINEERING. *Educ:* Calif State Polytech Univ, Pomana, BS, 73, ME, 75. *Prof Exp:* Mem tech staff, Jet Propulsion Lab, Calif Inst Technol, 73-80; lectr, 80-82, assoc prof, 82-87, PROF ELEC ENG, CALIF STATE POLYTECH UNIV, POMONA, 87-, DEPT CHAIR, 86- *Concurrent Pos:* Lectr, Calif State Polytech Univ, Pomona, 75-80; mem tech staff, Jet Propulsion Lab, Calif Inst Technol, 80-85, consult, 85-; consult, MD Software, Inc, 81-91. *Mem:* Fel Inst Elec & Electronics Engrs; Am Soc Eng Educ. *Res:* Semiconductor materials, silicon and gallium arsenide, for use in transister manufacturing and solar cell technology; computer software development for blood gas chemistry interpretation; author of numerous publications. *Mailing Add:* Dept Elec & Computer Eng Calif State Polytech Univ Pomona CA 91768-4065

COCKS, FRANKLIN H, b Staten Island, NY, Oct 1, 41; m 66; c 2. MATERIALS SCIENCE, PHYSICAL METALLURGY. *Educ:* Mass Inst Technol, BS, 63, MS, 64, ScD(phys metall), 65. *Prof Exp:* Fulbright fel, Imp Col, Univ London, 65-66; staff scientist, Tyco Labs, Inc, 66-67, sr scientist, 67-70, asst head, Mat Sci Dept, 70-72; assoc prof mech eng & mat sci, 72-75, PROF MECH ENG & MAT SCI, DUKE UNIV, 75- *Concurrent Pos:* Consult, Los Alamos Nat Lab, 79-; vis scholar, Div Appl Physics, Harvard Univ, 81; vis scientist, Div Chem Physics, Commonwealth Sci Indust & Res Orgn, Sydney, Australia, 85. *Mem:* Nat Asn Corrosion Engrs; Am Inst Mining Metall & Petrol Engrs; Brit Inst Metall. *Res:* Corrosion; electronic materials; failure analysis; x-ray diffraction; photovoltaic materials; amorphous materials; positron annihilation; non-destructive testing. *Mailing Add:* Dept Mech Eng & Mat Sci Duke Univ Durham NC 27706

COCKS, GARY THOMAS, b Long Beach, Calif, Mar 27, 43; m 83; c 1. SCIENCE COMMUNICATIONS, BIOCHEMISTRY. *Educ:* Cornell Univ, AB, 64; Univ Calif, Berkeley, PhD(biochem), 71. *Prof Exp:* Res fel, dept microbiol & molecular genetics, Med Sch, Harvard Univ, 71-73; assoc ed, 73-83, SR ASSOC ED, PROCEEDINGS, NAT ACAD SCI, 83- *Mem:* Am Chem Soc; AAAS; Coun Biol Editors. *Res:* Edit papers in biochemistry and related fields. *Mailing Add:* Nat Acad Sci 2101 Constitution Ave Washington DC 20418

COCKS, GEORGE GOSSON, b Sioux City, Iowa, Mar 22, 19; m 42; c 4. CHEMICAL MICROSCOPY. *Educ:* Iowa State Col, BS, 41; Cornell Univ, PhD(chem micros), 49. *Prof Exp:* Chemist, Allison Div, Gen Motors Corp, 41; asst chem micros, Cornell Univ, 46-49; asst chief physics solids div, Battelle Mem Inst, 49-64; assoc prof, 64-81, EMER PROF CHEM MICROS, CORNELL UNIV, 81- *Concurrent Pos:* Vis staff mem, Sci Lab, Los Alamos Nat Lab, 80-81, staff mem, 81-90. *Mem:* AAAS; Am Chem Soc; Optical Soc Am; Electron Micros Soc Am (exec secy, 59-75); Am Soc Metals; Sigma Xi. *Res:* Light and electron microscopy; chemical microscopy; optical crystallography; structure of polymer gels; growth of crystals in gels; crystallization of ice; cement. *Mailing Add:* 1719 Hyland St Bayside CA 95524

COCKSHUTT, E(RIC) P(HILIP), b Brantford, Ont, May 30, 29; m 54; c 5. ENERGY, RESOURCE INDUSTRIES. *Educ:* Univ Toronto, BASc, 50; Mass Inst Technol, SM, 51, ME, 52, ScD(mech eng), 54. *Prof Exp:* Res officer, Nat Res Coun Can, 53-67, prin res officer & sect head, Engine Lab, 67-76, dir, div energy, 75-86, exec dir, eng progs, 86-90; EXEC DIR, CAN MEM COMT, WORLD ENERGY COUN, 91- *Concurrent Pos:* Sessional lectr, Carleton Univ, 59-73. *Honors & Awards:* Casey Baldwin Award, Can Aeronaut & Space Inst, 61 & 65; Turnbull lectr, Can Aeronaut & Space Inst, 76. *Mem:* Fel Can Aeronaut & Space Inst; fel Can Soc Mech Eng; Solar Energy Soc Can Inc; fel Eng Inst Can. *Res:* Energy supply, conversion and end-use of all types. *Mailing Add:* 120 Dorothea Dr Ottawa ON K1V 7C7 Can

COCOLAS, GEORGE HARRY, b Flushing, NY, July 9, 29; m 53; c 4. MEDICINAL CHEMISTRY. *Educ:* Univ Conn, BS, 52; Univ NC, PhD(pharm), 56. *Prof Exp:* Org res chemist, Res Labs, Nat Drug Co, 56-58; from asst prof to assoc prof, 58-73, head div med chem, Sch Pharm, 75-82, PROF MED CHEM, UNIV NC, CHAPEL HILL, 73-, ASSOC DEAN, 82- *Concurrent Pos:* Ed, Am J Pharm Educ, 80- *Mem:* Am Chem Soc; Am Pharmaceut Asn; Am Asn Col Pharm. *Res:* Stereochemistry and biological activity; cholinergic mechanisms. *Mailing Add:* Sch Pharm Univ NC Chapel Hill NC 27599-7360

CODD, EDGAR FRANK, b Portland, Eng, Aug 19, 23; div; c 4. FOUNDATIONS OF COMPUTER SCIENCE. *Educ:* Oxford Univ, BA & MA, 48; Univ Mich, MS, 63, PhD(commun sci), 65. *Prof Exp:* Prog mathematician, IBM, NY, 49-53, head, Multiprogramming Systs, 59-61; mgr, Can Guided Missle Prog, 53-57; consult, 61-85; CHIEF SCIENTIST, CODD & DATE, INC, 85- *Concurrent Pos:* Researcher, Univ Mich, 64-65. *Honors & Awards:* Turing Award, Asn Comput Mach, 81. *Mem:* Nat Acad Eng; fel Brit Comput Soc; Asn Comput Mach; Sigma Xi. *Res:* multiprogramming, relational approach to data base management. *Mailing Add:* Codd & Date Inc 2099 Gateway Pl Suite No 220 San Jose CA 95110-1017

CODD, JOHN EDWARD, b Spokane, Wash, Oct 13, 36; m 64; c 5. TRANSPLANTATION IMMUNOLOGY, THORACIC SURGERY. *Educ:* Gonzaga Univ, BA, 58; St Louis Univ, MD, 63. *Prof Exp:* Asst prof, 71-74, assoc prof, 75-80, prof surg, med sch, St Louis Univ, 80-84,; CHIEF, CARDIOTHORACIC SURG, DE PAUL HEALTH CTR, 84- *Concurrent Pos:* Chief, Unit II Surg, John Cochran Vet Admin Hosp, 71-84. *Mem:* Am Col Surgeons; Soc Transplant Surgeons; Asn Acad Surg. *Res:* Organ preservation for transplantation. *Mailing Add:* Dept Surg St Louis Univ Sch Med 1325 S Grand St Louis MO 63104

CODDING, EDWARD GEORGE, b Ionia, Mich, Jan 17, 42; m 68; c 1. ANALYTICAL CHEMISTRY. *Educ:* Cent Mich Univ, BSc, 65; Mich State Univ, PhD(chem), 71. *Prof Exp:* Asst prof chem, Kent State Univ, 74-77; asst prof chem, Univ Calgary, 77-84. *Mem:* Am Chem Soc; Soc Appl Spectros. *Res:* Application of solid state image sensors as spectrochemical detectors; digital data handling techniques; incorporation of digital and analog instrumentation techniques for the measurement of chemical information. *Mailing Add:* 6231 72nd St NW Calgary AB T3B 3V9 Can

CODDING, PENELOPE WIXSON, b Emporia, Kans, Sept 18, 46. X-RAY DIFFRACTION. *Educ:* Mich State Univ, BSc, 68, PhD(x-ray crystallog), 71. *Prof Exp:* Fel, Dept Biochem, Univ Alta, 71-73, Dept Chem, 73-74; asst prof, Dept Chem, Kent State Univ, 74-76; from res assoc to assoc prof, 76-89, PROF, DEPT CHEM, UNIV, CALGARY, 89- *Mem:* Am Crystallog Asn; Am Chem Soc; Am Inst Physics; fel Can Inst Chem. *Res:* Structural aspects of the activity of drugs, neurotoxins and peptides using single crystal X-ray diffraction; correlation of conformational and pharmacological data to determine the molecular basis for binding to receptors. *Mailing Add:* Dept Chem Univ Calgary Calgary AB T2N 1N4 Can

CODDINGTON, EARL ALEXANDER, b Washington, DC, Dec 16, 20; m 45; c 3. MATHEMATICS. *Educ:* Johns Hopkins Univ, PhD(math), 48. *Prof Exp:* Physicist, Naval Ord Lab, Washington, DC, 42; mathematician, Navy Dept, 42-46; instr, Johns Hopkins Univ, 48-49; instr math, Mass Inst Technol, 49-52; from asst prof to assoc prof, 52-59, chmn dept, 68-71, PROF MATH, UNIV CALIF, LOS ANGELES, 59- *Concurrent Pos:* Fulbright lectr, Univ Copenhagen, Denmark, 55-56, vis prof, 63-64; vis assoc prof, Princeton Univ, 57-58. *Mem:* Math Asn Am; Am Math Soc. *Res:* Differential equations; analysis. *Mailing Add:* Dept Math Univ Calif Los Angeles CA 90024

CODE, ARTHUR DODD, b Brooklyn, NY, Aug 13, 23; m 43; c 4. SPACE ASTRONOMY, ASTROPHYSICS. *Educ:* Univ Chicago, MS, 47, PhD(astron & astrophysics), 50. *Prof Exp:* Asst, Yerkes Observ, Univ Chicago, 46-49; instr, Univ Va, 50; from instr to asst prof astron, Univ Wis, 51-56; assoc prof & mem staff, Mt Wilson & Palomar Observs, Calif Inst Technol, 56-58; prof, 58-69, dir observ, 58-70, JOEL STEBBINS PROF ASTRON, UNIV WIS-MADISON, 69-, DIR, SPACE ASTRON LAB & HILLDALE PROF ASTRON, 86- *Concurrent Pos:* Vis astronr, Radcliffe Observ, Pretoria, SAfrica, 53, Kitt Peak Nat Observ, 67 & 69 & Cerro-Tolulo Inter-Am Observ, 78; mem, Astron Panel Space Sci Bd, Nat Acad Sci, 58, Astron Mission Bd, NASA & many other nat adv panels & comts; prin investr, Wis Exp Package, first Orbiting Astron Observ, 68 & Wis Ultraviolet Photo-Polarimeter, Astro Mission, 90; chmn bd dirs, Asn Univ Res Astron, 76-80; actg dir, Space Telescope Sci Inst, 81-82. *Honors & Awards:* Pub Ser Award, NASA, 70. *Mem:* Nat Acad Sci; Am Astron Soc (vpres, 76-78, pres, 82-84); Int Acad Astronaut; fel Am Acad Arts & Sci. *Res:* Photoelectric photometry of stars and nebulae; stellar spectroscopy; development of instruments; satellite astronomy; author of approximately 120 scientific publications. *Mailing Add:* Washburn Observ Univ Wis 475 N Charter St Madison WI 53706

CODE, CHARLES FREDERICK, b Can, Feb 1, 10; nat US; m 35; c 3. PHYSIOLOGY. *Educ:* Univ Man, MD & BSc, 34; Univ Minn, PhD, 39. *Prof Exp:* Lectr physiol, Univ London, 35-36; asst exp surg, Mayo Grad Sch Med, Univ Minn, 37, from instr to prof physiol, 38-75, dir med educ & res, 66-72; assoc dir & sr res scientist, Ctr Ulcer Res & Educ, Wadsworth Vet Admin Hosp & Univ Calif, Los Angeles, 75-80; PROF MED & SURG, UNIV CALIF, SAN DIEGO, 80- *Concurrent Pos:* Mem staff & consult, Mayo Clin, 40-75; staff physician, Vet Admin Med Ctr, San Diego, 80- *Honors & Awards:* Theobald Smith Award, AAAS, 38; Friedenwald Medal, Am Gastroenterol Asn, 74. *Mem:* Am Physiol Soc; Brit Physiol Soc; Am Soc Pharmacol & Exp Therapeut; Am Soc Clin Invest; Am Gastroenterol Asn. *Res:* Metabolism of histamine, relationship to gastric secretion; hypersensitive state; physiology of gastrointestinal tract; motor action of the alimentary canal, secretion and absorption from the stomach, small and large bowel. *Mailing Add:* Gastroenterol Sect Vet Admin Med Ctr 3350 La Jolla Village Dr San Diego CA 92161

CODEN, MICHAEL H, b New York, NY, Mar 6, 47. COMPUTER LOCAL AREA NETWORKING, FIBER OPTICS. *Educ:* Mass Inst Technol, BS, 67; Columbia Univ, MS, 75; Courant Inst Math Sci, MS, 79. *Prof Exp:* Eng prod mgr, Hewlett Packard Co, Inc, 67-69; eng & mkt mgr, Digital Equip Corp, 69-72; asst vpres, Maher Terminals Incorp, 72-75; div mgr, Exxon Corp, 75-79; PRES, CHMN & CHIEF EXEC OFFICER, CODENOLL TECHNOL CORP, 79- *Concurrent Pos:* Preceptor, Columbia Univ, 74-75; lectr, Farleigh-Dickinson Univ, 79. *Honors & Awards:* Award Merit, Am Chem Soc. *Mem:* AAAS; NY Acad Sci; Inst Elec & Electronics Engrs; Optical Soc Am; Soc Photo-Optical Instrumentation. *Res:* Computer hardware and local area network architecture: data business and multiprocessor; development of semiconductor laser sources and systems; fiber optic components and systems for very high speed data communications, including development of compound semiconductor components, microelectronics and electromagnetic wave theory. *Mailing Add:* Codenoll Technol Corp 1086 N Broadway Yonkers NY 10701

CODERRE, JEFFREY ALBERT, b Southbridge, Mass, Aug 20, 53; m 81; c 1. CANCER RESEARCH, DRUG DESIGN. *Educ:* Southampton Col, Long Island Univ, BS, 75; Yale Univ, PhD(chem), 81. *Prof Exp:* Fulbright scholar geochem, Univ Iceland, 75-76; res fel med, Univ Calif, San Francisco, 81-84; ASST SCIENTIST CANCER RES, BROOKHAVEN NAT LAB, 84- *Mem:* Am Soc Biol Chemists. *Res:* Synthesis and testing of melanoma-specific radiopharmaceuticals; development of tumor-specific, boron-containing compounds for neutron capture therapy; use of radiosensitizers to augment radiation therapy. *Mailing Add:* Med Dept Brookhaven Nat Lab Upton NY 11973

CODINGTON, JOHN F, b Macon, Ga, Feb 9, 20; m 52; c 3. BIOCHEMISTRY. *Educ:* Emory Univ, AB, 41, MA, 42; Univ Va, PhD(org chem), 45. *Prof Exp:* Res asst, Joint Off Sci Res & Develop & Comt Med Res Proj, Univ Va, 43-45; chemist, NIH, Md, 45-49; res biochem, Columbia Univ, 51-55; asst, Sloan-Kettering Inst Cancer Res, 55-59, assoc, 59-67; asst biochemist, 67-74, assoc biochemist, Dept Med, 74-88, consult biochem, Mass Gen Hosp, 85-88; prin res assoc, 67-84, assoc prof biol chem, 84-87, BIOL CHEM & MOLECULAR PHARMACOL, MED SCH, HARVARD UNIV, 87-; SR SCIENTIST, BOSTON BIOMED RES INST, 86- *Concurrent Pos:* Asst prof, Cornell Med Col, 62-67. *Mem:* AAAS; Am Soc Cell Biol; Am Soc Biol Chemists; Am Chem Soc; Soc Complex Carbohydrates. *Res:* Isolation, structures, immunological properties and functions of glycoproteins of tumor cell surfaces; tumor antigens; diagnosis and immunotherapy in human cancer. *Mailing Add:* 1725 Commonwealth Ave West Newton MA 02165

CODISPOTI, LOUIS ANTHONY, b Brooklyn, NY, June 6, 40; m 68. CHEMICAL OCEANOGRAPHY. *Educ:* Fordham Univ, BS, 62; Univ Wash, MS, 65, PhD(oceanog), 73. *Prof Exp:* Oceanogr, US Naval Hydrographic Off, 66-69; actg asst prof chem oceanog, Univ Wash, 73-74, fel, 74-75, prin oceanogr, 75-80; res scientist, Bigelow Lab, 80-87; SR SCIENTIST, MONTERAY BAY AQUARIUM RES INST, 87- *Mem:* Arctic Inst NAm; Am Soc Limnol & Oceanog; Sigma Xi. *Res:* Upwelling and marine identification; upwelling; chemical oceanography of the Arctic Ocean; marine denitrification; global nutrient budgets, cycles and feedback mechanisms; marine carbon dioxide system. *Mailing Add:* 160 Central Ave Pac Grove CA 93950

CODRINGTON, ROBERT SMITH, b Victoria, BC, Dec 11, 25; nat US; m 48; c 1. PHYSICS, BIOENGINEERING & BIOMEDICAL ENGINEERING. *Educ:* Univ BC, BA, 46, MA, 48; Univ Notre Dame, PhD(physics), 51. *Prof Exp:* Asst physics, Univ BC, 45-47; res assoc, Univ Notre Dame, 48-51; asst res specialist, Rutgers Univ, 51-54; physicist, Schlumberger, Ltd, 54-62; mgr eng, Varian Assoc, 62-79, mgr, Res & Develop Instruments Div, 79-84, mgr, res & develop Varian Imaging Systs, 84-88, SR SCIENTIST & CONSULT, VARIAN & SIEMENS, 88- *Honors & Awards:* Nat Telemetry Prize, 62; Petrol Inst Award, 76. *Mem:* Am Phys Soc; Inst Elec & Electronics Engrs; Inst Soc Am; Soc Magnetic Resonance in Med; Int Soc Magnetic Res. *Res:* Magnetic instrumentation; geophysics; radar and telemetry synchronization; high resolution nuclear magnetic resonance; optical instruments; magnetic resonance imaging. *Mailing Add:* Spectros Imaging Systs Varian & Siemens 1120 Auburn St Fremont CA 94538

CODY, D THANE, b St John, NB, June 23, 32; m 63; c 2. OTOLARYNGOLOGY, PHYSIOLOGY. *Educ:* Dalhousie Univ, MD & CM, 57; Univ Minn, PhD(otolaryngol), 66. *Prof Exp:* Asst to staff otorhinolaryngol, Mayo Clin, 62-63, consult, 63-68, from instr to assoc prof otolaryngol & rhinol, Mayo Grad Sch Med, 63-74, prof otolaryngol, Mayo Med Sch, Univ Minn, 74-87, chmn, dept otorhinolaryngol, Mayo Clin, 68-87, chmn, bd gov, Mayo Clin, Jacksonville, Fla, 86-87, EMER PROF OTOLARYNGOL, MAYO MED SCH, UNIV MINN, 88- *Concurrent Pos:* Edward John Noble travel award, Mayo Found, 61; Am Acad Ophthal & Otolaryngol Res Award, 61; pres, Am Bd Otolaryngol, 88. *Honors & Awards:* Dr John Black Award, 55. *Mem:* Fel Am Acad Ophthal & Otolaryngol; fel Am Col Surg; Am Laryngol, Rhinol & Otolaryngol Soc; Asn Res Otolaryngol; Am Otol Soc. *Res:* Labyrinthine otosclerosis. *Mailing Add:* Mayo Jacksonville Clin 4500 San Pablo Rd Jacksonville FL 32224

CODY, GEORGE DEWEY, b New York, NY, May 16, 30. SOLID STATE PHYSICS. *Educ:* Harvard Univ, AB, 52, MA, 54, PhD(physics), 57. *Prof Exp:* Asst, Harvard Univ, 52-57; staff physicist, RCA Labs, 57-69, dir, Solid State Lab, 69-78; sr res assoc, 78-80, SCIENTIFIC ADV, CORP RES LABS, EXXON RES & ENG CO, 80 - *Concurrent Pos:* Harvard Univ Parker fel, Clarendon Lab, Oxford Univ, 57-58; regents prof, Univ Calif, San Diego, 69. *Honors & Awards:* Ballantine Medal, Franklin Inst, 79. *Mem:* Fel Am Phys Soc. *Res:* Low temperature physics; superconductivity; high temperature thermal conductivity; thin films; magnetic properties of metals; optical properties of amorphous semiconductors. *Mailing Add:* Exxon Res & Eng Corp Clinton Twp Rte 22 E Annandale NJ 08801

CODY, JOHN T, b Susquehanna, Pa, Oct 17, 49. CHEMISTRY. *Educ:* Iowa Wesleyan Col, BA, 71; Univ Iowa, MS, 74, PhD(chem, sci educ), 79. *Prof Exp:* Instr biochem, Sch Health Care Sci, USAF, 80-85; vis lectr, dept community med, Univ Okla, 80-84, clin assoc prof, dept family med, 84-85; chief, Qual Control Div, 85-87, Analytical Sci Div, 87-89, DEP DIR, DRUG TESTING LAB, USAF, 89- *Mem:* Am Chem Soc; Am Asn Mass Spectrometry; Am Soc Forensic Sci; Soc Forensic Toxicologists. *Res:* Drug stability; drug analysis procedure; interferences to analytical procedures. *Mailing Add:* AFDTL Brooks AFB TX 78235-5000

CODY, MARTIN LEONARD, b Peterborough, Eng, Aug 7, 41; c 4. COMMUNITY ECOLOGY, ECOLOGICAL BIOGEOGRAPHY. *Educ:* Edinburgh Univ, MA, 63; Univ Pa, PhD(zool), 66. *Prof Exp:* Asst prof zool, 66-72, PROF BIOL, UNIV CALIF, LOS ANGELES, 72- *Concurrent Pos:* Vis prof, Univ Capetown, Univ W Australia, W Univ Can. *Honors & Awards:* John Simon Guggenheim Fel, 78-79. *Mem:* Soc Study Evolution; Ecol Soc Am; Am Soc Naturalists; Cooper Ornith Soc; Bot Soc Am. *Res:* Population biology; community ecology; patterns of competition and coexistence of plant and animal communities; convergent and divergent evolution; ecological biogeography; Mediterranean-climate ecosystems in California, Chile, Europe and South Africa; island biogeography; theoretical ecology. *Mailing Add:* Dept Biol Univ Calif Hilgard Ave Los Angeles CA 90024

CODY, REGINA JACQUELINE, b Steubenville, Ohio, May 13, 43. SPACE SCIENCE. *Educ:* West Liberty State Col, BS, 65; Univ Pittsburgh, PhD(phys chem), 72. *Prof Exp:* Nat Acad Sci-Naval Res Labs res assoc IR laser res, Naval Res Lab, Washington, DC, 72-74; ASTROPHYSICIST PHOTOCHEM, LASER SPECTROS, GODDARD SPACE FLIGHT CTR, NASA, 74- *Mem:* Am Chem Soc; Am Phys Soc; Inter-Am Photochem Soc; Sigma Xi. *Res:* Photochemistry of small molecules; laser spectroscopy, especially laser induced fluorescence spectroscopy; chemistry of comets, interstellar clouds and planetary atmospheres. *Mailing Add:* 8406 Snowden Oaks Pl Montpelier Laurel MD 20708

CODY, REYNOLDS M, b Asheville, NC, Apr 17, 29; m 58; c 2. MICROBIOLOGY. *Educ:* Univ Tenn, BA, 56; Miss State Univ, MS, 61, PhD(microbiol), 64. *Prof Exp:* Instr microbiol, Miss State Univ, 58-61, res asst, 61-64; asst prof bact, 64-65, ASSOC PROF MICROBIOL & BOT, SCH VET MED, AUBURN UNIV, 65- *Mem:* Am Soc Microbiol; Soc Indust Microbiol. *Res:* Biochemistry and physiology of pathogenic microorganisms including viruses; elucidation of metabolic pathways in Serratia indica. *Mailing Add:* Dept Bot & Microbiol Auburn Univ Auburn AL 36849

CODY, TERENCE EDWARD, b Orrville, Ohio, June 3, 38; div; c 2. MOLECULAR BIOLOGY, TOXICOLOGY. *Educ:* Mt Union Col, BS, 60; Case Western Reserve Univ, MS, 68; Ohio State Univ, PhD(bot), 72. *Prof Exp:* Asst ed biochem, Chem Abstracts Serv, 64-68; fel environ health, 72-73, ASST PROF ENVIRON HEALTH, UNIV CINCINNATI, 73- *Concurrent Pos:* Environ Consult. *Mem:* AAAS; Tissue Cult Asn; Sigma Xi. *Res:* The applicability of bioassays and bioindicators for evaluating environmental impacts and health effects of organic chemicals and carcinogens. *Mailing Add:* Inst Environ Health 3223 Eden Ave Univ Cincinnati Med Ctr Cincinnati OH 45267-0056

CODY, VIVIAN, b San Diego, Calif, Jan 28, 43. CRYSTALLOGRAPHY. *Educ:* Univ Mich, BS, 65; Univ Cincinnati, PhD(chem), 69. *Prof Exp:* Res asst chem, Univ Mich, 63-65; teaching asst, Univ Cincinnati, 65-67, NSF fel, 67-69; fel, Univ Mo, St Louis, 69-70; fel, endocrinol trainee, Med Found Buffalo, 70-72, res scientist crystallog, 72-78, assoc res scientist, 79-86, SR RES SCIENTIST, MED FOUND BUFFALO, 86-; ASSOC PROF MED, STATE UNIV NY BUFFALO, 79-, ASSOC RES PROF MED CHEM, 87- *Concurrent Pos:* Fac Res Award, Am Cancer Soc, 85-90; US assoc ed, J Molecular Graphics, Endocrine Res; treas, Am Chem Soc, western NY sect, 88-90. *Mem:* Endocrine Soc; Am Thyroid Asn; Am Crystallog Asn (secy, 88-94); Am Chem Soc; Europ Thyroid Asn; Royal Soc Chem. *Res:* Structure-function analysis of thyroid, flavonoid, and cardioactive agents as well as antifolate antineoplastic agents, using the techniques of x-ray crystallography of protein complexes; computer graphics; computation chemical methods for drug design. *Mailing Add:* Med Found Buffalo 73 High St Buffalo NY 14203

CODY, WAYNE LIVINGSTON, b Albany, NY, May 1, 59. THREE-DIMENSIONAL STRUCTURE ANALYSIS, LIGAND RECEPTOR INTERACTIONS. *Educ:* Allegheny Col, BS, 81; Univ Ariz, PhD(org chem), 85. *Prof Exp:* Res chemist, Eastman Chem Div, Eastman Kodak Co, Kingsport, Tenn, 85-88; res scientist, Immunobiol Res Inst, Johnson & Johnson, Annandale, NJ, 88-90; RES ASSOC, WARNER-LAMBERT/PARKE-DAVIS PHARMACEUT RES, ANN ARBOR, MICH, 90- *Mem:* Am Chem Soc; Am Peptide Soc; Sigma Xi; AAAS; Protein Soc. *Res:* Synthesis, purification of peptides for pharmacological evaluation and drug discovery; design of peptides and modified peptides utilizing structure activity relationships, molecular modeling and solution structures from nuclear magnetic resonance analysis; synthesis of unusual amino acids. *Mailing Add:* Parke-Davis Pharmaceut Res 2800 Plymouth Rd Ann Arbor MI 48106-1047

CODY, WILLIAM JAMES, b Hamilton, Ont, Dec 2, 22; m 50; c 5. BOTANY. *Educ:* McMaster Univ, BA, 46. *Prof Exp:* Botanist, 46-87, cur, Vascular Plant Herbarium, 59-87, RES ASSOC, CAN DEPT AGR, 88- *Mem:* Am Soc Plant Taxon; Int Asn Plant Taxon; Can Bot Asn. *Res:* Floristics of Mackenzie district, Northwest Territories and Yukon Territory; Canadian ferns. *Mailing Add:* 1189 Tara Dr Ottawa ON K2C 2H4 Can

COE, BERESFORD, b Philadelphia, Pa, May 4, 19; m 45; c 3. ORGANIC CHEMISTRY. *Educ:* Earlham Col, AB, 41. *Prof Exp:* Res chemist, Barrett Div, Allied Chem & Dye Corp, 41-46; res chemist, Rohm & Haas Co, 46-81; RETIRED. *Mem:* Am Chem Soc. *Res:* Adhesives technical service. *Mailing Add:* 17029 106th Ave Sun City AZ 85373-1923

COE, CHARLES GARDNER, b Philadelphia, Pa, Nov 13, 50; m 72; c 1. INORGANIC CHEMISTRY, ORGANIC CHEMISTRY. *Educ:* Thiel Col, BA, 72; Carnegie-Mellon Univ, PhD(inorg chem), 76. *Prof Exp:* Teaching asst, Carnegie-Mellon Univ, 72-76; res chemist, 76-81, GROUP LEADER, MAT RES, AIR PROD & CHEM INC, 81- *Mem:* Am Chem Soc; Sigma Xi. *Res:* Synthetic inorganic chemistry; interaction and activation of small molecules by metalions; heterogeneous and homogeneous catalysis. *Mailing Add:* 1381 Walnut Ln Macungie PA 18062

COE, EDWARD HAROLD, JR, b San Antonio, Tex, Dec 7, 26; m 49; c 2. GENETICS. *Educ:* Univ Minn, BS, 49, MS, 51; Univ Ill, PhD(bot), 54. *Prof Exp:* Res fel genetics, Calif Inst Technol, 54-55; res assoc field crops, 55-58, assoc prof, 59-63, PROF AGRON, UNIV MO, 64-; GENETICS, USDA, 55-, RES LEADER, PLANT GENETICS, 77- *Concurrent Pos:* Ed, Maize Genetics Coop Newslett, 75- *Mem:* AAAS; Genetics Soc Am; Am Genetic Asn; Crop Sci Soc Am; Am Soc Naturalists. *Res:* Genetics of maize; genome mapping; extramendelian inheritance; fertilization and development. *Mailing Add:* 206 Heather Lane Columbia MO 65203

COE, ELMON LEE, b Phoenix, Ariz, Mar 6, 31; m 61. BIOCHEMISTRY. *Educ:* Harvard Univ, AB, 52; Univ Calif, Los Angeles, PhD(physiol chem), 61. *Prof Exp:* Res assoc biochem, Univ Ind, 60-61; MEM STAFF, DEPT BIOCHEM, MED SCH, UNIV BEIRUT. *Concurrent Pos:* NIH grant, 64-67. *Res:* Interactions between metabolic pathways and metabolic control mechanisms; carbohydrate metabolism; biochemistry of tumors. *Mailing Add:* 225 Loraine Woods Dr RR 1 Rural Box 17 Macon GA 31210

COE, FREDRIC LAWRENCE, b Chicago, Ill, Dec 25, 36; m 65; c 2. INTERNAL MEDICINE, NEPHROLOGY. *Educ:* Univ Chicago, BA & BS, 57, MD, 61. *Prof Exp:* Dir & chmn renal div, Michael Reese Hosp, 72-81; from asst prof to prof med, 69-80, PROF PHYSIOL, UNIV CHICAGO, 80-; CHIEF, RENAL SECT, UNIV CHICAGO HOSPS, 81- *Concurrent Pos:* USPHS fel, Univ Tex Southwestern Med Sch, Dallas, 67-69; mem renal prog, Michael Reese Hosp, 69-72, chmn prog, 72- *Mem:* Am Physiol Soc; Am Soc Nephrology; Am Fedn Clin Res; Cent Soc Clin Res; Am Soc Clin Invest; Asn Am Physicians. *Res:* Renal physiology; causes of renal calculi; medical education. *Mailing Add:* Dept Med & Physiol Sch Med Univ Chicago 5841 S Maryland Box 28 Chicago IL 60637

COE, GERALD EDWIN, b Granville, Ill, Apr 21, 22; m 41; c 3. GENETICS, CYTOLOGY. *Educ:* Tex Col Arts & Indust, BS, 42; Univ Tex, PhD(bot), 52. *Prof Exp:* Asst, Univ Tex, 52; geneticist, Sci & Educ Admin, USDA, 52-86; RETIRED. *Concurrent Pos:* Sugar Beet Develop Found Cooperator, USDA, 86- *Mem:* Am Soc Sugarbeet Technologists. *Res:* Cytology and genetics in the genus Beta; DNA isolation & transfer. *Mailing Add:* 4300 Maple Pl Beltsville MD 20705

COE, GORDON RANDOLPH, b Cincinnati, Ohio, Aug 10, 33; m 60; c 1. POLYMER CHEMISTRY, PHYSICAL ORGANIC CHEMISTRY. *Educ:* Centre Col, Ky, AB, 55; Univ Tenn, MS, 60; Univ Cincinnati, PhD(inorg chem), 64. *Prof Exp:* Assoc prof chem, Univ Southern Miss, 64-67; chemist, E I du Pont, 67-70 & Joseph E Seagrams, 70-72; PROG MGR POLYMER CHEM, GEN ELEC CO, 72- *Mem:* Soc Plastics Engrs. *Res:* Polymer characterization, stabilizer and additive analysis. *Mailing Add:* 3132 Trinity Rd Louisville KY 40206

COE, JOHN EMMONS, b Evanston, Ill, Sept 1, 31; m 54; c 3. PENTRAXINOLOGY, HEPATOLOGY. *Educ:* Oberlin Col, BA, 53; Hahnemann Med Col, MD, 57. *Prof Exp:* Intern, Univ Ill Res & Educ Hosp, 57-58; resident internal med, Med Ctr, Univ Colo, 58-60; surgeon, Rocky Mountain Lab, Nat Inst Allergy & Infectious Dis, NIH, 60-63; fel path, Scripps Clin & Res Found, 63-65; MED OFFICER, ROCKY MOUNTAIN LAB, NAT INST ALLERGY & INFECTIOUS DIS, NIH, 65- *Concurrent Pos:* Affil prof microbiol & zool, Univ Mont, 74- *Mem:* Am Asn Immunologists. *Res:* Selective induction of antibody formation in immunoglobulin classes of rodents, especially Syrian hamsters; evolution, function of pentraxins (female protein) in various hamster species; estrogen induction of hepatotoxicity and hepatocarcinogenisis. *Mailing Add:* Rocky Mountain Lab Nat Inst Allergy & Infectious Dis NIH Hamilton MT 59840

COE, KENNETH LOREN, b Omaha, Nebr, Apr 16, 27. CHEMISTRY, INFORMATION SCIENCE. *Educ:* Tarkio Col, BA, 49. *Prof Exp:* From asst ed to assoc ed, Chem Abstr, 51-58, head ed dept, 59-61, managing ed, abstr issues, 62-69, ed opers, 69-90; RETIRED. *Mem:* AAAS; Am Chem Soc; Coun Biol Ed; Asn Earth Sci Ed. *Res:* Chemical documentation. *Mailing Add:* 1631 Roxbury Rd Apt D1 Columbus OH 43212

COE, MICHAEL DOUGLAS, b New York, NY, May 14, 29; m 55; c 5. ARCHAEOLOGY. *Educ:* Harvard Univ, AB, 50, PhD, 59. *Prof Exp:* Asst prof anthrop, Univ Tenn, 58-60; from instr to prof anthrop, 60-90, C J MACCURDY PROF ANTHROP, YALE UNIV, 90- *Mem:* Nat Acad Sci; Royal Anthrop Inst; Mex Anthrop Soc; Soc Am Archeol. *Res:* Archaeology and ethnohistory of Mesoamerica; historical archaeology of Northeastern US; history of writing systems. *Mailing Add:* Dept Anthrop Yale Univ PO Box 2114 Yale Sta New Haven CT 06520

COE, RICHARD HANSON, b Stamford, Conn, Jan 29, 20; m 55; c 6. CHEMISTRY. *Educ:* Wesleyan Univ, BA, 41, MA, 42; Stanford Univ, PhD(chem), 47. *Prof Exp:* Chemist, Conn State Water Comn, 41-42; actg instr chem, Stanford Univ, 43, asst, Nat Defense Res Comt, 44; asst res chemist, Calif Res Corp, 45-46; instr chem, Wesleyan Univ, 46-48; sr technologist, Shell Oil Co, 49-64, sr res chemist, 64-67, supvr, 67-69, staff res chemist, 69-71, sr staff res chemist, 71-77, mgr OSH Regulations, Safety & Indust Hyg, 77-82; RETIRED. *Honors & Awards:* Indust Wastes Medal, Fed Sewage & Indust Wastes Asn, 52. *Mem:* AAAS; Am Chem Soc; Am Inst Chem. *Mailing Add:* Star Route, Box 98 Eastsound WA 98245

COE, ROBERT STEPHEN, b Toronto, Ont, Feb 20, 39; m 69; c 2. GEOPHYSICS, GEOCHEMISTRY. *Educ:* Harvard Univ, BA, 61; Univ Calif, Berkeley, MS, 64, PhD(geophys), 66. *Prof Exp:* From asst prof to assoc prof, 68-78, PROF EARTH SCI, UNIV CALIF, SANTA CRUZ, 78- *Concurrent Pos:* Fel, Australian Nat Univ, 66-68. *Mem:* AAAS; Am Geophys Union; Geol Soc Am; Soc Terrestrial Magnetism & Elec Japan. *Res:* Paleomagnetism and tectonics; effects of shear stress on polymorphic transitions in minerals. *Mailing Add:* Dept Earth Sci Univ Calif Santa Cruz CA 95064

COELHO, ANTHONY MENDES, JR, b Danbury, Conn, May 26, 47; m 74. BEHAVIORAL MEDICINE, STRESS. *Educ:* Western Conn State Univ, BS, 70; Univ Tex, Austin, MA, 73, PhD(anthrop), 75. *Prof Exp:* Asst prof anthrop, Tex Tech Univ, 74-75; asst scientist, 75-76, assoc scientist, Dept Cardiopulmonary Dis, 76-86, HEAD BEHAV MED LAB, SOUTHWEST FOUND BIOMED RES, 75-, SCIENTIST, DEPT PHYSIOL & MED, 86- *Concurrent Pos:* Adj assoc prof, Dept Pediat, Univ Tex Health Sci Ctr, San Antonio, 76-, Dept Dent Diag Sci, 84-90 & adj prof, Dept Surg/Neurosurg, 89-; lectr, Dept Behav & Cult Sci, Univ Tex San Antonio, 77-82; bd dirs, Am Soc Primatologists, 82-84; mem, Sci & Prof Liaison Comt, Soc Behav Med, 82-88; prin investr, Med Res Training grant, 84-87, Exercise, Stress & Atherosclerosis grant, 85-90 & Exercise, Stress & Blood Pressure Regulation grant, 89-94; mem, Minority Inst Res Training Prog Comt, Nat Heart, Lung & Blood Inst, 87-90, Res Manpower Rev Comt, 88-93; ed bk rev, Am J Primatology, 89-92; consult, Am Asn Accreditation Lab Animal Care, 90-93. *Mem:* Am Asn Phys Anthrop; Am Soc Primatologists (exec secy, 82-84); Bioelectromagnetics Soc; Human Biol Coun; Nat Coun Univ Res Adminrs; Sigma Xi; Soc Behav Med; Soc Res Adminrs. *Res:* Behavioral medicine; behavioral development, social behavior and stress; exercise, stress and blood pressure regulation; physical growth, development, nutrition and evolution; primate ecology, evolution, time and energy budgets; behavioral aspects of psychosocial stress and disease. *Mailing Add:* Behav Med Lab Southwest Found Biomed Res PO Box 28147 San Antonio TX 78228-0147

COEN, GERALD MARVIN, b Camrose, Alta, Mar 26, 39; m 64; c 2. SOIL SCIENCE, PEDOLOGY. *Educ:* Univ Alta, BSc, 63, MSc, 65; Cornell Univ, PhD(agron), 70. *Prof Exp:* Res scientist, 69-74, biol scientist, 74-86, ACTG UNIT HEAD, AGR CAN, 86- *Concurrent Pos:* Assoc asst prof, Univ Alta, 71-; assoc ed, Can J Soil Sci, 75-77; exec, Can Soc Soil Sci, 80-86, 89-92. *Mem:* Agr Inst Can; Can Soc Soil Sci (pres, 90-91); Am Soc Agron; Int Soc Soil Sci; Soil Conserv Soc Am. *Res:* Soil survey and resource inventories including research on soil genesis, classification and interpretation necessary to carry out the inventory programs. *Mailing Add:* Agr Can Soil Surv 4445 Calgary Trail S Edmonton AB T6H 5R7 Can

COESTER, FRITZ, b Berlin, Ger, Oct 16, 21; nat US; m 52; c 6. PHYSICS. *Educ:* Univ Zurich, PhD(theoret physics), 44. *Prof Exp:* Res, Sulzer Bros, Inc, Switz, 44-46; asst, Univ Geneva, 46-47; from asst prof to prof physics, Univ Iowa, 47-63; SR PHYSICIST, ARGONNE NAT LAB, 63- *Concurrent Pos:* Mem, Inst Advan Study, 53-54. *Mem:* Fel Am Phys Soc; Switz Phys Soc; Europ Phys Soc. *Res:* Quantum theory of fields; theoretical nuclear physics; theoretical physics. *Mailing Add:* Argonne Nat Lab 9700 S Cass Bldg 203 Argonne IL 60439

COETZEE, JOHANNES FRANCOIS, b Bloemfontein, Union SAfrica, Nov 25, 24; m 54; c 1. ANALYTICAL CHEMISTRY. *Educ:* Univ Orange Free State, SAfrica, BSc, 44, MSc, 49; Univ Minn, PhD(chem), 55. *Prof Exp:* Instr chem, Univ Orange Free State, SAfrica, 45-48, lectr, 49-51 & Univ Witwatersrand, 56-57; from asst prof to assoc prof, 57-66, PROF CHEM, UNIV PITTSBURGH, 66- *Concurrent Pos:* Titular mem Int Union Pure & Appl Chem. *Res:* Non-aqueous solutions; electroanalytical chemistry. *Mailing Add:* Dept Chem Univ Pittsburgh Pittsburgh PA 15260

COFER, DANIEL BAXTER, DEVELOPMENT OF WIRE INDUSTRY. *Educ:* Ga Inst Technol, BME, 53. *Prof Exp:* EXEC VPRES, SOUTHWIRE CO, 87- *Honors & Awards:* John W Mordica Med Award, 71. *Mem:* Am Soc Testing Mat; fel Am Soc Metals; Insulated Power Cable Eng Asn. *Res:* Wire development. *Mailing Add:* Southwire Co One Southwire Dr Carrollton GA 30119

COFER, HARLAND E, JR, b Atlanta, Ga, Dec 28, 22; m 45; c 6. ECONOMIC GEOLOGY, MINERALOGY. *Educ:* Emory Univ, AB, 47, MS, 48; Univ Ill, Urbana-Champaign, PhD(geol), 57. *Prof Exp:* Asst prof geol, Emory Univ, 48-58; sr geologist, Indust Chem Div, Am Cyanamid Co, Ga, 58-66; prof geol, 66-88, EMER PROF GEOL & PHYSICS, GA SOUTHWESTERN COL, 88- *Concurrent Pos:* Geologist, Groundwater Div, US Geol Surv, Ga, 53-54; consult, Consol Quarries Inc, Ga, 52-53 & Indust Chem Div, Am Cyanamid Co, 52- *Mem:* Geol Soc Am; Clay Minerals Soc; Asn Ground Water Scientists & Engrs. *Res:* Clay mineral research, especially utilization of kaolin group minerals as sources for alumina; site evaluations for solid waste and liquid waste disposal; x-ray diffraction and electron microscopy of ceramic products. *Mailing Add:* 409 Judy Lane Americus GA 31709

COFFEE, ROBERT DODD, b East Orange, NJ, Dec 18, 20; m 71; c 2. CHEMICAL SAFETY, INSTRUMENTATION. *Educ:* Newark Col Eng, BS, 42; Univ Wis, Madison, MS, 47, PhD(chem eng), 49. *Prof Exp:* Develop engr, Plastics Div, Monsanto Chem Co, Mass, 42-46; develop engr, Eastman Kodak Co, 49-53, sr develop engr, 53-63, supvr tech safety, 63-81, tech asst, 81-83; RETIRED. *Concurrent Pos:* Chmn, Am Soc Testing & Mat, 67, 69, 77 & 79-83; safety consult, US Dept Transp, 71-73; lectr, Nat Safety Coun, 75-78. *Mem:* Emer mem Am Chem Soc. *Res:* Plastics manufacture and processing; equipment design; chemical safety; hazard evaluation and protection. *Mailing Add:* 142 Camberly Pl Penfield NY 14526-2712

COFFEEN, W(ILLIAM) W(EBER), b Champaign, Ill, Aug 13, 14; m 45; c 2. CERAMICS. *Educ:* Univ Ill, BS, 35, MS, 37, prof degree, 49; Rutgers Univ, PhD, 69. *Prof Exp:* Assoc, Univ Ill, 35-37; ceramic engr, Canton Stamping & Enameling Co, Ohio, 37-39; asst prof ceramics eng, Ga Inst Technol, 40-41; res assoc, Porcelain Enamel Inst, Nat Bur Standards, 41-42, glass technologist, 42-45; res supvr, Metal & Thermit Corp, 45-62, dir res M&T Chem, Inc, 63-67, tech adv to pres, 67-68; from assoc prof to prof, 69-84, EMER PROF CERAMIC ENG, CLEMSON UNIV, 84- *Honors & Awards:* Greaves-Walker Award, 80. *Mem:* Fel Am Ceramic Soc (vpres, 68-69); Nat Inst Ceramic Engrs (pres, 63-64). *Res:* Vitreous enamels; optical glass; opacification in vitreous media; electronic ceramics, especially capacitors; tin, antimony and zirconium compounds in ceramics; arc welding equipment; thermite reactions; mineral beneficiation; ceramic prosthetics. *Mailing Add:* 227 Kings Way Clemson SC 29631

COFFEY, CHARLES WILLIAM, II, b Somerset, Ky, July 28, 49; m 71. MEDICAL PHYSICS, HEALTH PHYSICS. *Educ:* Univ Ky, BS, 71, MS, 72; Purdue Univ, PhD(bionucleonics), 75; Am Bd Radiol, cert therapeut radiol physics, 77. *Prof Exp:* Asst med physicist & asst prof clin physics, 75-78, CHIEF CLIN PHYSICS & ASST PROF MED PHYSICS, DEPT RADIATION THER, MED CTR, UNIV KY, 78- *Mem:* Am Asn Physicists Med; Health Physics Soc; Sigma Xi. *Res:* Applied and clinical physics in radiation medicine; high energy electron beams, used in radiotherapy; the use of linear accelerators in radiotherapy; computer techniques in radiotherapy. *Mailing Add:* Dept Radiation Med Rm C-21 Univ Ky Med Ctr Lexington KY 40536

COFFEY, DEWITT, JR, b Gilmer, Tex, Apr 12, 35; m 68. CHEMICAL PHYSICS. *Educ:* Abilene Christian Col, BS(chem), 58; Univ Tex, BS(chem eng), 58, PhD(chem), 67. *Prof Exp:* Res asst drilling & well completion, Mobil Oil Field Res Lab, Tex, 58-61; McBean fel, Stanford Res Inst, 66-68; assoc prof, 68-77, PROF CHEM, SAN DIEGO STATE UNIV, 77- *Concurrent Pos:* Res Corp Frederick Gardner Cottrell grants-in-aid, 69-; consult, Stanford Res Inst, Calif, 69-; San Diego State Univ Found fac res grant, 69-; vis prof, Kyushu Univ, Fukuoka City, Japan, 75. *Mem:* AAAS; Am Chem Soc; Am Phys Soc. *Res:* Infrared laser; microwave spectroscopy; stark-shifted spectroscopy; computational chemistry. *Mailing Add:* Dept Chem San Diego State Univ San Diego CA 92182

COFFEY, DONALD STRALEY, b Bristol, Va, Oct 10, 32; m 53; c 2. PHARMACOLOGY, BIOCHEMISTRY. *Educ:* E Tenn State Univ, BS, 57; Johns Hopkins Univ, PhD(biochem), 64. *Hon Degrees:* DSc, King Col, 77. *Prof Exp:* Chemist, NAm Rayon Corp, Tenn, 55-57; chem engr, Westinghouse Corp, Md, 57-59; actg dir James B Brady Urol Res Lab, Johns Hopkins Hosp, 59-60, instr physiol chem, 65-66, from asst prof to assoc prof pharmacol, 66-74, assoc prof, Oncol Ctr, Sch Med, 73-74, dir, James Buchanan Brady Lab Reproductive Biol, Johns Hopkins Hosp, 64-74, actg chmn, Dept Pharmacol & Exp Therapeut, Sch Med, 73-74, DIR, DEPT UROL RES LABS, JOHNS HOPKINS HOSP, 74-, PROF PHARMACOL & EXP THERAPEUT, 74-, PROF UROL, 75-, DEP DIR ONCOL CTR, SCH MED, JOHNS HOPKINS UNIV & PROF, 86- *Concurrent Pos:* USPHS res career award, 66-72; asst ed, J Molecular Pharmacol, 67-71 & Cancer Res, 81. *Honors & Awards:* D R Edwards Award, 81. *Mem:* Am Soc Pharmacol & Exp Therapeut; Am Soc Biol Chemists; Cell Biol. *Res:* Control of cell replication and DNA synthesis; biochemistry of mammalian nuclei; growth and development of the prostate gland; control of reproduction; cancer chemotherapy. *Mailing Add:* Dept Urol Johns Hopkins Univ Sch Med Marburg 121 600 N Wolfe St Baltimore MD 21205

COFFEY, JAMES CECIL, JR, b Salisbury, NC, July 2, 38; m 57; c 2. ENDOCRINOLOGY, TERATOLOGY. *Educ:* Catawba Col, BA, 63; Univ NC, MSPH, 66, PhD(parasitol, biochem), 70. *Prof Exp:* Res assoc, Dent Res Ctr, 71-72, asst prof pediat & oral biol, Sch Med & Dent Res Ctr, 72-78, ASSOC PROF ORAL BIOL & PEDIAT, SCH DENT & MED, UNIV NC, CHAPEL HILL, 78- *Concurrent Pos:* Fel pediat endocrinol, Sch Med, Univ NC, 70-71. *Mem:* Endocrine Soc. *Res:* Androgen metabolism; physiology of submaxillary glands; cleft palate; muscular dystrophy. *Mailing Add:* Dent Res Ctr Univ NC Chapel Hill NC 27514

COFFEY, JOHN JOSEPH, b Cambridge, Mass, Apr 24, 40. BIOCHEMISTRY, BIOCHEMICAL PHARMACOLOGY. *Educ:* Harvard Univ, AB, 61; Johns Hopkins Univ, PhD(biochem), 67. *Prof Exp:* NSF fel, Virus Lab, Univ Calif, Berkeley, 67-68; BIOCHEMIST, ARTHUR D LITTLE, INC, 68- *Mem:* Am Chem Soc; AAAS. *Res:* Kinetics of regulatory processes; enzyme induction; allosteric properties of proteins; drug distribution and metabolism; protein binding of drugs; pharmacokinetics; cancer chemotherapy. *Mailing Add:* 2200 Corley Dr Villa Sierra Apt 4D Las Cruces NM 88001

COFFEY, JOHN WILLIAM, b Sedalia, Mo, Jan 20, 37; m 63; c 3. BIOCHEMISTRY. *Educ:* Rockhurst Col, BS, 59; Tulane Univ, PhD(biochem), 63. *Prof Exp:* Res fel biochem, Touro Res Inst, 63-65 & Rockefeller Univ, 65-67; asst prof, Tulane Med Sch, 67-69, sr scientist, 69-75; asst group leader, 76-78, group leader, 79-83, RES LEADER, HOFFMANN-LA ROCHE, INC, 83- *Mem:* AAAS; NY Acad Sci; Am Soc Biol Chemists; Sigma Xi; Am Col Rheumatol. *Res:* Arachidome acid metabolism; lysosomal functions; vitamin B12 metabolism; collagen metabolism; inflammation. *Mailing Add:* 152 Washington Ave West Caldwell NJ 07006

COFFEY, MARVIN DALE, b Midvale, Idaho, Apr 25, 30; m 52; c 5. ENTOMOLOGY, ZOOLOGY. *Educ:* Brigham Young Univ, AB, 52, MA, 53; Wash State Univ, PhD(entom), 57. *Prof Exp:* From instr to assoc prof zool & entom, 57-67, chmn dept biol, 65-70, PROF ZOOL & ENTOM, SOUTHERN ORE STATE COL, 67- *Concurrent Pos:* Asst prof, Fresno State Col, 64-65; vis prof, Tex A&M Univ, 69-70 & Univ Ky, 76-77. *Mem:* AAAS; Entom Soc Am; Am Inst Biol Sci. *Res:* Taxonomy, ecology and medical importance of parasitic arthropods and of the Diptera. *Mailing Add:* Dept Biol Southern Ore State Col Ashland OR 97520

COFFEY, RONALD GIBSON, b Monte Vista, Colo, Dec 29, 36; m 59; c 3. BIOCHEMISTRY, PHARMACOLOGY. *Educ:* Colo State Univ, BS, 58; Ore State Univ, PhD(biochem), 63. *Prof Exp:* Res asst biochem, Ore State Univ, 58-63; chief div chem, Fifth US Army Med Lab, St Louis, Mo, 63-65; res assoc biochem, Univ Ore, 65-68; head div biochem, Children's Asthma Res Inst & Hosp, Denver, Colo, 68-73; assoc immunopharmacol, Sloan-Kettering Inst Cancer Res, 73-82; asst prof biol, Cornell Univ Sch Med, Sloan-Kettering Div, 76-82; ASSOC PROF PHARMACOL, COL MED, UNIV S FLA, 82- *Concurrent Pos:* NIH grant, Children's Asthma Res Inst & Hosp, Denver, Colo, 70-73; ACS grant, Univ SFla, 88-91. *Mem:* Am Soc Biochem Molecular Biol; Am Soc Pharmacol Exp Therapeut. *Res:* Regulation of cell membrane enzymes, especially adenylate cyclase, guanylate cyclase, protein kinase C and ATPase by hormonal and pharmacologic agents and its application to immunology, allergy and cancer. *Mailing Add:* Dept Pharmacol Col Med Univ SFla 12901 N Bruce B Downs Blvd Tampa FL 33612

COFFEY, TIMOTHY, b Washington, DC, June 27, 41; m 63; c 3. TECHNICAL MANAGEMENT. *Educ:* Mass Inst Technol, BS, 62; Univ Mich, MS, 63, PhD(physics), 67. *Prof Exp:* Physicist, Air Force Cambridge Res Lab, 64-65; res asst physics, Univ Mich, 65-66; physicist, EG&G, Inc, 66-71; br head, 71-75, div head plasma physics, 75-82, asoc dir res 80-82, DIR RES, NAVAL RES LAB, 82-- *Mem:* Fel Am Phys Soc; AAAS; Sigma Xi. *Res:* Theoretical non-linear mechanics; theoretical plasma physics; relativistic electron beam theory; ionospheric and space plasma theory. *Mailing Add:* Naval Res Lab Code 1001 Washington DC 20375-5000

COFFIN, DAVID L, b Neshameny, Pa, Feb 24, 13; m 40; c 1. ENVIRONMENTAL HEALTH, EXPERIMENTAL PATHOLOGY. *Educ:* Univ Pa, VMD, 38. *Prof Exp:* Asst prof clin path, Univ Pa, 40-46; pathologist, Angell Mem Hosp, 46-56; prof path, Univ Pa, Philadelphia, 57-61; res adv air pollution, Robert Taft Eng Ctr, 61-71; sr res adv air pollution, US Environ Protection Agency, 71-91; VIS RES SCI, UNIV NC, CHAPEL HILL, 91- *Concurrent Pos:* Res assoc path, Harvard Med, Calif, 46-56; dir res path, Caspary Inst, New York City, 56-61; vis prof, Los Alamos Lab, Univ Calif, 78-80; consult mem, Comt Refining Syn Fuels, Nat Acad Sci, 80-81. *Mem:* Am Asn Path & Bact; Soc Toxicol; Int Acad Path; Am Col Vet Path; AAAS. *Res:* Environmental health; interaction of toxicants with infectious agents; influence of environmental factors on pulmonary defense to infectious disease and cancer; role of organics and mineral fibers in elicitation of cancer. *Mailing Add:* Ctr Environ Med & Lung Biol Univ NC Chapel Hill NC 27514

COFFIN, FRANCES DUNKLE, b Philipsburg, Pa; m 51; c 3. SOLID STATE PHYSICS. *Educ:* Wilson Col, AB, 43; Cornell Univ, PhD(phys chem), 50. *Prof Exp:* Chemist, Armstrong Cork Co, 43-45; physicist, Nat Adv Comn Aeronaut, 50-58 & Nat Aeronaut & Space Admin, 58-69. *Mem:* Am Chem Soc. *Res:* Electron diffraction studies on the crystalline nature of the surface of germanium solar cells; influence of orientation on the creep of single crystal wires of zinc. *Mailing Add:* 27038 Bruce Rd Bay Village OH 44140

COFFIN, HAROLD GARTH, b PEI, Can, Nov 7, 38; m 63; c 3. AGRICULTURAL MARKETING, GRAIN POLICY INTERNATIONAL TRADE. *Educ:* McGill Univ, BSc, 62, Univ Conn, MSc, 67, PhD(agr econ), 70. *Prof Exp:* Res economist, PEI Mkt Develop Ctr, 70-71; dir econ res & secy bd, Livestock Feed Bd Can, Montreal, 71-79; ASSOC PROF MKT, CHMN & AGR ECONOMIST, MACDONALD COL, MCGILL UNIV, 79- *Concurrent Pos:* Counr, Can Agr Econ & Farm Mgt Soc, 83-86. *Mem:* Can Agr Econ & Farm Mgt Soc (pres-elect, 87-88); Am Agr Econ Asn; Agr Inst Can. *Res:* Marketing and trade problems in agriculture, including regional competition price discovery, grain and livestock marketing and trade policy. *Mailing Add:* Dept Agr Econ MacDonald Col McGill Univ Box 189 Lakeshore Rd Ste Anne de Bellevue PQ H9X 1C0 Can

COFFIN, HAROLD GLEN, b Nanning, China, Apr 9, 26; US citizen; m 47; c 2. PALEONTOLOGY. *Educ:* Walla Walla Col, BA, 47, MA, 52; Univ Southern Calif, PhD, 55. *Prof Exp:* Head dept biol, Can Union Col, 47-52; res fel, Univ Southern Calif, 52-54; head div sci & math, Can Union Col, 54-56; assoc prof, Walla Walla Col, 56-58, head dept, 58-64; prof zool, Geosci Res Inst, Andrews Univ, 64-65, prof paleont, 65-80; SR RES SCIENTIST, GEOSCI RES INST, LOMA LINDA UNIV, 80- *Mem:* AAAS; Sigma Xi; Geol Soc Am. *Res:* Marine invertebrates; science and religion, especially as related to geology and biology. *Mailing Add:* 11146 Rosarita Dr Loma Linda CA 92354-3208

COFFIN, JOHN MILLER, b Boston, Mass, Apr 20, 44; m 68; c 2. VIROLOGY, MOLECULAR BIOLOGY. *Educ:* Wesleyan Univ, BA, 67; Univ Wis, PhD(molecular biol), 72. *Prof Exp:* Trainee oncol, Univ Wis, 67-72; fel, Inst Molecular Biol, Univ Zurich, 72-75; asst prof, 75-78, assoc prof, 78-, PROF MOLECULAR BIOL, SCH MED, TUFTS UNIV. *Concurrent Pos:* Jane Coffin Childs Mem Fund Med Res fel, 72-74; prin investr, Nat Cancer Inst, 75- & Am Cancer Soc, 75-; fac award, Am Cancer Soc, 78-83, mem virol study sect, 80-; Outstanding Investr Award, Nat Cancer Inst, 87-; bd trustee, Leukemia Soc Am, 87- *Mem:* Am Soc Microbiol. *Res:* Viral oncology; genome structure and genetics of RNA tumor viruses. *Mailing Add:* Dept Molecular & Microbiol Tufts Univ Sch Med 136 Harrison Ave Boston MA 02111

COFFIN, LAURENCE HAINES, b Buenos Aires, Arg, June 4, 33; US citizen; m 58; c 3. THORACIC SURGERY, CARDIAC SURGERY. *Educ:* Mass Inst Technol, BS, 55; Case Western Reserve Univ, MD, 59; Am Bd Surg & Bd Thoracic Surg, dipl, 68. *Prof Exp:* From instr to sr instr thoracic surg, Case Western Reserve Univ, 67-69; assoc prof surg, 69-75, PROF SURG, COL MED, UNIV VT, 75-, CHIEF SECT THORACIC & CARDIAC SURG, 69-, ATTEND & CHIEF THORACIC SERV, MED CTR HOSP VT, 70-, DIR THORACIC & CARDIOVASC SURG, 77- *Concurrent Pos:* Surg intern, Univ Hosps Cleveland, 59-60, resident, 60-61 & 63-67, asst thoracic surgeon, 67-69; chief thoracic surg, Vet Admin Hosp Cleveland, 67-69; mem coun cardiovasc surg, Am Heart Asn, 70- *Mem:* Soc Thoracic Surgeons; Am Acad Surg; fel Am Col Surg; Am Asn Thoracic Surgeons. *Res:* Pathophysiology of burn shock; cardiovascular physiology. *Mailing Add:* One S Prospect St Med Ctr Hosp Vt Burlington VT 05401

COFFIN, LOUIS F(USSELL), JR, b Schenectady, NY, Aug 30, 17; m 43; c 8. MECHANICAL METALLURGY. *Educ:* Swarthmore Col, BS, 39; Mass Inst Technol, ScD, 49. *Prof Exp:* From instr to asst prof mech eng, Mass Inst Technol, 41-49; res assoc, Gen Physics Unit, Knolls Atomic Power Lab, Gen Elec Co, 49-55, mech engr, Res & Develop, Gen Elec Corp, 55-86; CONSULT & DISTINGUISHED RES PROF, MECH ENG DEPT, RENSSELAER POLYTECH INST, 86- *Concurrent Pos:* Adj assoc prof, dept metall, Rensselaer Polytech Inst, 56-59; mem comt, Mat Adv Bd, Nat Acad Sci, 60-62, 65 & 83-84; adj prof, Union Col, 67-80; Gen Elec Coolidge fel, 74; vis fel, Clare Hall, Cambridge Univ, Eng, 76; mem adv comt, dept metall, Univ Pa, 77-80. *Honors & Awards:* Hunt Award, 58; Dudley Award, Am Soc Testing & Mat, 75; Nadai Award, Am Soc Mech Engrs, 79; Albert Sauveur Award, Am Soc Metals, 80; Francis Clamer Medal, Franklin Inst, 84. *Mem:* Nat Acad Eng; fel Am Soc Mech Engrs; Am Inst Mining Metall & Petrol Engrs; fel Am Soc Metals; fel Am Soc Testing & Mat. *Res:* Behavior of materials under stress; lubrication and friction; plasticity; metal working; fatigue. *Mailing Add:* 1178 Lowell Rd Schenectady NY 12308

COFFIN, PERLEY ANDREWS, b Newburyport, Mass, Oct 8, 08; m 39; c 1. RUBBER CHEMISTRY, POLYMER CHEMISTRY. *Educ:* Northeastern Univ, BChE, 31; Mass Inst Technol, MS, 33. *Prof Exp:* Lab asst, Simplex Wire & Cable Co, Mass, 30-31; control chemist, Vultex Chem Co, 33-37; develop chemist, Gen Latex & Chem Corp, 37-42; lab mgr, Gen Tire & Rubber Co, Tex, 43-45; sect head qual control, Gen Latex & Chem Corp, 46-61, lab mgr, 62-78; RETIRED. *Mem:* Am Chem Soc; Am Soc Testing & Mat. *Res:* Polymerization; rubber; synthetic resin and plastics; latex. *Mailing Add:* Four Hammond St Gloucester MA 01930

COFFINO, PHILIP, b New York, NY, Sept 7, 42. CELL BIOLOGY, GENETICS. *Educ:* Univ Calif, Berkeley, BA, 63; Einstein Col Med, PhD, 71, MD, 72. *Prof Exp:* Resident, Cancer Res Inst, 73-74, ASST PROF MICROBIOL, UNIV CALIF, SAN FRANCISCO, 74-, ASST PROF RESIDENCE, DEPT MED, 75-, ASSOC PROF MED & MICROBIOL. *Concurrent Pos:* Am Cancer Soc fac res award; NIH res career develop award. *Honors & Awards:* Trygve Tuve Mem Award, NIH, 77. *Mem:* Am Soc Clin Invest; Am Soc Biol Chemists. *Res:* Genetics of animal cells, mutagenesis; hormone and cyclic nucleotide mediated regulation; growth regulation. *Mailing Add:* Dept Microbiol & Med Univ Calif Third & Parnassus Ave San Francisco CA 94143

COFFMAN, CHARLES BENJAMIN, b Baltimore, Md, Dec 19, 41; m 71; c 2. AGRONOMY. *Educ:* Univ Md, BS, 66, MS, 69, PhD(soil mineral), 72. *Prof Exp:* Asst geol, Univ Md, 66-71, instr 72; RES AGRONOMIST, AGR ENVIRON QUAL INST, AGR RES SERV, USDA, 72- *Concurrent Pos:* Instr geol, Frederick Community Col, 73-77; Ed, Northeastern Weed Soc; chmn, Personnel Comt, Weed Sci Soc Am. *Mem:* Am Soc Agron; Weed Sci Soc Am; Coun Agr sci & Technol. *Res:* Evaluation of potential new herbicides and controlled release formulations of existing herbicides; weed control in no-till agriculture; weed control in sustainable agriculture. *Mailing Add:* AEQI Agr Res Serv USDA Bldg 001 ARC-West Beltsville MD 20705

COFFMAN, CHARLES VERNON, b Hagerstown, Md, Oct 23, 35; m 63; c 2. MATHEMATICS. *Educ:* Johns Hopkins Univ, BES, 57, PhD(math), 62. *Prof Exp:* Assoc engr, Appl Physics Lab, Johns Hopkins Univ, 57; vis mem res staff math, Res Inst Advan Study, Martin Co, Md, 60; from asst prof to assoc prof, 62-71, PROF MATH, CARNEGIE-MELLON UNIV, 71- *Mem:* Am Math Soc. *Res:* Differential equations and functional analysis. *Mailing Add:* Dept Math Carnegie-Mellon Univ 5000 Forbes Ave Pittsburgh PA 15213

COFFMAN, EDWARD G, JR, b Los Angeles, Calif, Aug 16, 34. COMPUTER SCIENCE, OPERATIONS RESEARCH. *Educ:* Univ Calif, Los Angeles, BA, 56, MS, 61, PhD(eng), 66. *Prof Exp:* Mem tech staff, Syst Develop Corp, Santa Monica, 58-65; asst prof, Elec Eng & Comput Sci Dept, Princeton Univ, 66-69; vis prof comput sci, Univ Newcastle Tyne & Univ Durham, Eng, 69-70; prof computer sci, Pa State Univ, 70-76, actg head dept, 73; prof elec eng & computer sci, Columbia Univ, 76-77 & Univ Calif, Santa Barbara, 77-79; MEM TECHNICAL STAFF, AT&T BELL LABS, MURRAY HILL, 79- *Concurrent Pos:* Numerous res & travel grants, NSF & NASA, 67-; vis asst prof, Elec Eng Dept, Polytech Inst Brooklyn, 67-69; chmn, Computer Conf, Inst Elec & Electronics Engrs, 69, Conf System Sci, 84 & Nat Conf, Opers Res Soc Am, 76; assoc ed, J Asn Comp Mach, 69-75 & 79-85, ed-in-chief, 75-79; ed, Soc Indust & Appl Math J Comput, 82-86; mem, numerous comts & conferences, Inst Elec & Electronics Engrs, Opers Res Soc Am, Asn Comput Mach & Int Fedn Info Processing; vis lectr elec eng & computer sci, Princeton Univ, 80. *Honors & Awards:* Outstanding Contrib Award, Asn Comput Mach, 87. *Mem:* Fel Inst Elec & Electronics Engrs; Asn Comput Mach; Int Fedn Info Processing. *Res:* Data structures and algorithms; mathematical models of computer organization and operating systems; computer performance evaluation; quenching theory; combinatorial problems in sequencing and allocation; stochastic optimization; author of numerous publications. *Mailing Add:* Bell Labs 600 Mountain Ave Murray Hill NJ 07974

COFFMAN, HAROLD H, b Overbrook, Kans, Feb 16, 15; m 46; c 4. PETROLEUM WAXES, SYNTHETIC WAXES. *Educ:* Kans State Univ, BS, 40. *Prof Exp:* Res chemist, Bareco Oil Co, 40-42, 46-52, dir res, 52-55; dir, Barnsdall Res Group, Petrolite Corp, 55-66, asst dir, Appln Res Lab, 66-70, mgr qual control, Bareco Div, 70-83; RETIRED. *Mem:* Am Chem Soc; Am Soc Testing & Mat. *Res:* Microcrystalline petroleum waxes in regard to their properties; end use applications and processes of manufacturing. *Mailing Add:* PO Box 766 Barnsdale OK 74002

COFFMAN, JAMES BRUCE, b Cheyenne, Wyo, July 15, 25; m 50; c 4. GEOLOGY. *Educ:* Univ Nebr, BS, 50. *Prof Exp:* Gen mgr geol res, Exxon Prod Res Co, 67-70, explor mgr, Esso Explor, Inc, 70-72, mgr, Prod Dept, Esso Europe, 72-74, exec vpres, Esso Explor & Prod UK, Inc, 73-74, vpres explor res, Exxon Prod Res Co, 74-81; pres & chief oper officer, Ammoil Inc, 81-84; pres & chief exec officer, Aminoil Inc, 84-85; PRES & CHIEF EXEC OFFICER, J B COFFMAN & ASSOC, INC, 85- *Mem:* Am Asn Petrol Geologists; Soc Explor Geophysicists; Am Petrol Inst; Am Geol Inst; Geol Soc Am; Nat Ocean Indust Asn. *Mailing Add:* 511 Barker's Cove Houston TX 77079

COFFMAN, JAY D, b Quincy, Mass, Nov 17, 28; m 55; c 4. INTERNAL MEDICINE. *Educ:* Harvard Col, BA, 50; Boston Univ, MD, 54. *Prof Exp:* From asst to assoc prof, 60-70, PROF MED, MED CTR, BOSTON UNIV, 70- *Mem:* Am Physiol Soc; Am Fedn Clin Res; Am Heart Asn; Am Soc Clin Invests. *Res:* Peripheral vascular physiology and disease. *Mailing Add:* Peripheral Vascular Sect Univ Hosp 88 E Newton St Boston MA 02118

COFFMAN, JOHN W, b El Dorado, Kans, Dec 19, 31; m 54; c 3. PHYSICS. *Educ:* Univ Kans, BS, 54, MS, 56. *Prof Exp:* Res asst, Univ Kans, 54-56; physicist atmospheric acoustics, Missile Geophys Div, White Sands Missile Range, 58-60, supvry physicist, Missile Meteorol Div, 60-63, res atmospheric physicist, Environ Sci Dept, 63-67; MEM STAFF, GODDARD SPACE FLIGHT CTR, NASA, 67- *Mem:* Am Phys Soc. *Res:* Polarizable dielectrics; atmospheric infrared absorption and emission spectra; low frequency sound propagation in the atmosphere. *Mailing Add:* Box 153 Bealsville MD 20839

COFFMAN, MOODY LEE, b Abilene, Tex, July 25, 25; m 47; c 4. THEORETICAL PHYSICS, ACOUSTICS. *Educ:* Abilene Christian Univ, BA, 47; Univ Okla, MA & MS, 49; Tex A&M Univ, PhD(physics), 54. *Prof Exp:* Instr physics, E Tex State Univ, 49-51; instr, Tex A & M Univ, 51-53, instr math, 53-54; sr nuclear engr, Convair, Tex, 54-55; from asst prof to assoc prof physics & math, Abilene Christian Univ, 55-60, head dept, 56-60; sr physicist, Missile & Space Systs Dept, Hamilton Standard Div, United Aircraft Corp, Conn, 60-61; prof physics & head dept, Oklahoma City Univ, 61-69; PROF PHYSICS, CENT STATE UNIV, OKLA, 69- *Concurrent Pos:* Adj prof, Tex Christian Univ, 54-55; adj assoc prof physics, Hartford Grad Ctr, Rensselaer Polytech Inst, 60-61; consult, Convair, 55-57; consult physics, 69-; vpres res, Acoustic Controls, Inc, Tex, 69-; owner & consult, Acoustic By Coffman, 86-; vpres res & develop, Terra Shield Resources, Inc, Okla, 87- *Mem:* Am Phys Soc; Am Math Soc; Am Asn Physics Teachers; Am Geophys Union; Sigma Xi; Math Asn Am; Am Inst Physics. *Res:* Electromagnetic theory and quantum mechanics related to molecular and atomic structure; mechanics of charged particles; geomagnetism; architectural acoustics; six US and two Canadian patents; author of numerous publications. *Mailing Add:* 1832 NW 17th St Oklahoma City OK 73106

COFFMAN, ROBERT EDGAR, b Grosse Pointe Farms, Mich, Jan 5, 31; m 59; c 3. CHEMICAL PHYSICS. *Educ:* Univ Ill, BS, 53; Univ Calif, Berkeley, MS, 55; Univ Minn, PhD(chem physics), 64. *Prof Exp:* Chemist, Hanford Atomic Prod Oper, Gen Elec Co, Wash, 55-56; chemist, Chemet Prog, NY, 56-57, phys chemist, Advan Semiconductor Lab, Semiconductor Prod Dept, 57-60; NSF fel physics, Nottingham Univ, 64-65; asst prof chem, Augsburg Col, 65-67; from asst prof to assoc prof, 67-79, PROF CHEM, UNIV IOWA, 79- *Concurrent Pos:* NATO sr fel, Cambridge Univ, 73; vis prof, Quantum Theory Proj, Univ Fla, Gainesville, Fla, 85. *Mem:* AAAS; Am Phys Soc; Am Chem Soc; Sigma Xi. *Res:* Quantum chemistry; electron paramagnetic resonance in inorganic, metal-ligand and biological molecules. *Mailing Add:* Dept Chem Univ Iowa Iowa City IA 52240

COFFMAN, WILLIAM PAGE, b Vandergrift, Pa, Jan 7, 42; m 65; c 3. LOTIC ECOLOGY. *Educ:* Thiel Col, BS, 63; Univ Pittsburgh, PhD(biol), 67. *Prof Exp:* Hydrobiologist, Karlsruhe, Ger, 68-69; asst prof, 69-74, ASSOC PROF ECOL, UNIV PITTSBURGH, 74- *Res:* Ecology of stream benthos; ecology and taxonomy of aquatic insects, particularly the Dipteran family Chironomidae. *Mailing Add:* Dept Biol Sci A234 Langley Hall Univ Pittsburgh Pittsburgh PA 15260

COFRANCESCO, ANTHONY J, b New Haven, Conn, Feb 24, 10; m 41; c 3. INDUSTRIAL ORGANIC CHEMISTRY. *Educ:* Wesleyan Univ, BA, 33, MA, 34; Yale Univ, PhD(org chem), 39. *Prof Exp:* Res chemist, Calco Chem Co, NJ, 39-44; chief chemist, Arnold & Hofmann, RI, 44-45 & Carwin Chem Co, Conn, 45-50; group leader, 50-54, sect mgr, TPM Dyes, 54-58 & Intermediates, 58-62, mgr, Chem Specialty Sect, GAF Corp, Rensselaer, 62-75; CONSULT INDUST ORG CHEM, 75- *Concurrent Pos:* Res assoc, Biol Dept, State Univ Albany, 83-85. *Mem:* Sigma Xi. *Res:* Anthraquinone intermediates and dyes; industrial organic chemicals; deoxynucleosides and deoxynucleotides. *Mailing Add:* Knight Road Delanson NY 12053

COGAN, ADRIAN ILIE, b Bucharest, Romania, Feb 11, 46; m 70; c 2. ELECTRONICS, SOLID STATE PHYSICS. *Educ:* Polytech Inst, Bucharest, BS, 68, MS, 69, PhD(solid state physics), 75. *Prof Exp:* Res electronics, Inst Semiconductor Res, 69-73, group leader, 73-75, lab chief, 75-77; tech staff mem semiconductor devices, GTE Labs, Waltham, 78-84; assoc prof elec eng, Northeastern Univ, Boston, 82-84; mgr device design, Siliconix Inc, 84-87; ENG DIR, MICROWAVE TECHNOL INC, 87- *Concurrent Pos:* Mem, Nat Romanian Comn for Microwave Applns, 75-76; lectr, Univ Berkeley Ext, 87- *Mem:* Inst Elec & Electronics Engrs; Sigma Xi; Electrochem Soc. *Res:* Microwave semiconductor devices; metal-semiconductor devices and contacts; electrical field analysis in planar microwave structures; semiconductor devices processing. *Mailing Add:* Microwave Technol Inc 4268 Solar Way Freemont CA 94538

COGAN, DAVID GLENDENNING, b Fall River, Mass, Feb 14, 08; m 34; c 2. OPHTHALMOLOGY. *Educ:* Dartmouth Col, AB, 29; Harvard Univ, MD, 32. *Hon Degrees:* DSc, Duke Univ. *Prof Exp:* Asst ophthal, Harvard Med Sch, 34-40, from asst prof to prof ophthalmic res, 40-63, Henry Willard Williams prof ophthal, 63-70, prof, 70-74, actg dir lab, 40-42, dir, Howe Lab Ophthal, 43-74; chief, 76-85, SR MED OFFICER, NEURO-OPHTHAL SECT, NAT EYE INST, NIH, 85-; EMER HENRY WILLARD WILLIAMS PROF OPHTHAL, HARVARD MED SCH, 74- *Concurrent Pos:* Moseley traveling fel, Harvard Med Sch, 37-38; asst ophthal, Mass Eye & Ear Infirmary, 34-40, clin asst, 35-39, from asst surgeon to surgeon, 39-74, dir, Ophthal Labs, 47-74, chief ophthal, 60-66; consult, Los Alamos Med Ctr; mem comt ophthalmic consults, Nat Res Coun; Coun Inst Neurol & Blindness, USPHS; mem coun, Nat Eye Inst, 70-73; ed-in-chief, AMA, Arch of Ophthal, 60-67; med officer USPHS, 74-76. *Honors & Awards:* Hektoen Silver Medal, AMA, 61; Howe Medal Am, Ophthal Soc, 65; Gonin Medal, Int Coun Ophthal; Hon Award, Asn Res. *Mem:* AAAS; AMA; Am Soc Clin Invests; Am Neurol Soc; Can Ophthal Soc; Am Ophtal Soc; Japanese Ophtal Soc; Ger Ophtal Soc Vision & Ophthal. *Res:* Clinical physiology of the eye; neuro-ophthalmology. *Mailing Add:* Bldg 10 Rm 6C 401 NIH Bethesda MD 20892

COGAN, EDWARD J, b Milwaukee, Wis, Jan 18, 25; m 47; c 1. MATHEMATICAL LOGIC. *Educ:* Univ Wis, BA, 46, MA, 48; Pa State Univ, PhD(math philos), 55. *Prof Exp:* Instr math, Pa State Univ, 48-50, 51-55 & Dartmouth Col, 55-57; MEM FAC, SARAH LAWRENCE COL, 57- *Concurrent Pos:* Dir, NSF Insts, 59-64, co-dir, Upward Bound Prog, 66-69; consult, Metrop Sch Study Coun, 59-60; co-dir, Spec Prog, Sarah Lawrence Col, 74-78. *Mem:* Math Asn Am; Asn Symbolic Logic. *Res:* Foundations of mathematics; theory of sets; combinatory logic; automatic programming languages for computers. *Mailing Add:* Dept Math Sarah Lawrence Col Bronxville NY 10708

COGAN, HAROLD LOUIS, b Framingham, Mass, May 30, 31; m 55; c 2. PHYSICAL CHEMISTRY. *Educ:* Univ Mass, BS, 54; Yale Univ, MS, 56, PhD(phys chem), 58. *Prof Exp:* Asst, Yale Univ, 54-56, NSF asst phys chem of electrolytes, 56-57; PRES, HAROLD L COGAN, INC, 57- *Mem:* Am Chem Soc; Sigma Xi. *Res:* Thermodynamic studies of the effect of pressure upon ionic equilibria; pressure and temperature dependence of the dielectric constant of water; use of coaxial cavity resonators for dielectric constant measurements. *Mailing Add:* 2600 Hampshire Blvd Grand Rapids MI 49506

COGAN, JERRY ALBERT, JR, b Flushing, NY, Jan 22, 35; m 58; c 3. CHEMICAL ENGINEERING. *Educ:* Amherst Col, BA, 56; Mass Inst Technol, SM, 58. *Prof Exp:* Instr chem eng, Mass Inst Technol, 57-58; chem engr, E I du Pont de Nemours & Co, Inc, 58-61; chem engr, Deering Milliken Res Corp, 61, prod supvr, Deering Milliken, Inc, 62, new prod mgr, 62, exec vpres, 63, PRES, MILLIKEN RES CORP, 64- *Mem:* Inst Chem Engrs; Am Asn Textile Chem & Colorists. *Res:* Research and development management. *Mailing Add:* 1303 Pinecrest Rd Spartanburg SC 29302

COGAN, MARTIN, NEPHROLOGY, RENAL PHYSIOLOGY. *Educ:* Harvard Univ, MD, 74. *Prof Exp:* ASST PROF MED, UNIV CALIF, SAN FRANCISCO, 77- *Mailing Add:* Vet Admin Med Ctr 4150 Clement San Francisco CA 94121

COGBILL, BELL A, b Marianna, Ark, Mar 18, 09; m 43; c 4. ELECTRICAL ENGINEERING. *Educ:* Univ Tenn, BSEE, 31, MS, 36. *Prof Exp:* Instr elec eng, Univ Tenn, 34-36; engr, Gen Elec Co, 36-64; from assoc prof to prof elec eng, Northeastern Univ, 64-74; RETIRED. *Mem:* AAAS; fel Inst Elec & Electronics Engrs; Am Soc Eng Educ; Nat Soc Prof Engrs. *Res:* Electromagnetic characteristics of large power transformers; power system engineering; electric power circuits; power transformers design. *Mailing Add:* 51 Jackson Rd Wellesley MA 02181

COGBURN, ROBERT FRANCIS, b 1944. MARKOV CHAINS, STOCHASTIC DIFFERENTIAL EQUATIONS. *Educ:* Univ Calif, BA, 56, PhD(math), 60. *Prof Exp:* Asst prof math statist, Univ Berkeley, 60-65; consult, Robert Cogburn, 65-70; assoc prof math, 70-73, PROF MATH, UNIV NMEX, 73- *Concurrent Pos:* Vis asst prof, Princeton Univ, 61-62; lectr, Univ Berkeley, 65-70. *Mem:* Am Math Soc; Inst Math Statist. *Res:* Probability and stochastic processes; general theory of Markov chains in random environment; random evolutions; averaging methods in stochastic differential equations; application to channeling theory. *Mailing Add:* Math Dept Univ NMex Albuquerque NM 87131

COGBURN, ROBERT RAY, b Weatherford, Tex, Mar 3, 35; m 58; c 4. ENTOMOLOGY, STORED PRODUCTS. *Educ:* Univ Tex A&M Univ, BS, 58, MS, 61. *Prof Exp:* Entomologist, Agr Res Serv, USDA, Tex, 59-63, res entomologist, Stored Prod Insects Br, Calif, 63-66, Ga, 66-69, RES ENTOMOLOGIST, AGR RES SERV, USDA, STORED RICE INSECTS LAB, BEAUMONT, TEX, 69- *Mem:* Entom Soc Am. *Res:* Research with insects affecting stored rice. *Mailing Add:* Stored-Rice Insects Lab USDA Rte 7 Box 999 Beaumont TX 77713

COGDELL, THOMAS JAMES, b Quanah, Tex, Aug 19, 34; m 61; c 4. ORGANIC CHEMISTRY. *Educ:* Midwestern Univ, BA, 55; Univ Tex, MA, 62; Harvard Univ, PhD(chem), 65. *Prof Exp:* Chemist, Dow Chem Co, 55-58; mem tech staff, Bell Tel Labs, 65-66; asst prof, 66-74, ASSOC PROF CHEM, UNIV TEX, ARLINGTON, 74- *Mem:* AAAS; Am Chem Soc; Sigma Xi. *Res:* Organic reaction mechanisms; carbonium ion rearrangements; stable free radicals; benzyne intermediates. *Mailing Add:* 2922 Lakeshore Dr Arlington TX 76013

COGEN, WILLIAM MAURICE, b Chicago, Ill, Mar 30, 09; m 41; c 2. GEOLOGY. *Educ:* Calif Inst Technol, BS, 31, MS, 33, PhD(geol), 37. *Prof Exp:* Petrol geologist, Superior Oil Co, Tex, 36-37 & Shell Oil Co, 37-62; consult, 63-67; supvr tech writers, Lockheed Electronics Co, 67-74; GEOL CONSULT, 74- *Mem:* Fel Geol Soc Am; Am Asn Petrol Geologists. *Res:* Heavy minerals of Gulf Coast sediments; mechanics of landslides; geology of Texas Gulf coast; petroleum geology. *Mailing Add:* 11 Hawks Hill Ct Oakland CA 94618

COGGESHALL, NORMAN DAVID, b Ridgefarm, Ill, May 15, 16; m 40; c 4. PHYSICS. *Educ:* Univ Ill, BA, 37, MS, 39, PhD(physics), 42. *Prof Exp:* Asst physics, Univ Ill, 37-41, instr, 42-43; dir, Phys Sci Div, 43-66, vpres, Proc Sci Dept, 66-70, vpres, Explor & Prod, 70-76, vpres, Govt-Technol Coordr, Gulf Res & Develop Co, 76-80; CONSULT, 81- *Concurrent Pos:* Mem adv bd, Nat Bur Stand; mem adv comm, Am Petr Inst. *Honors & Awards:* Recipient of Resolution of Appreciation, Div Refining, Am Petrol Inst; Award, Am Chem Soc, 69; fel, Am Phys Soc. *Mem:* Am Chem Soc; Am Mass Spec Soc; Am Phys Soc. *Res:* Mass, infrared and ultraviolet spectroscopy; molecular physics; general research management; separation processes; process instrumentation. *Mailing Add:* 701 Driftwood Dr Lynn Haven FL 32444

COGGESHALL, RICHARD E, b Chicago, Ill, May 29, 32; m 59; c 4. ANATOMY. *Educ:* Univ Chicago, BA, 51; Harvard Med Sch, MD, 56. *Prof Exp:* Instr anat, Harvard Med Sch, 64-65, assoc, 65-67, from asst prof to assoc prof, 67-71; PROF ANAT, PHYSIOL & BIOPHYS, UNIV TEX MED BR, GALVESTON, 71- *Concurrent Pos:* USPHS career develop award, 66-; NIH fels, Univ Tex Med Br, Galveston, 71-74 & 75-79, NIH grant, 71- *Mem:* Am Asn Anatomists; Am Soc Cell Biol; Am Soc Neurosci; Sigma Xi; AAAS. *Res:* Neurobiology and the structure of the nervous system. *Mailing Add:* Marine Biomed Inst 200 Univ Blvd Galveston TX 77550

COGGIN, JOSEPH HIRAM, b Birmingham, Ala, Feb 4, 38; m 57; c 4. MICROBIOLOGY, VIROLOGY. *Educ:* Vanderbilt Univ, BA, 59; Univ Tenn, MS, 61; Univ Chicago, PhD(midrobiol), 65. *Prof Exp:* Sr bacteriologist, Tenn Dept Pub Health, 59-60; sr res virologist virus & cell biol div, Merck Inst; therapeut res, 65-67; asst prof & virologist, Univ Tenn, Knoxville, 67-68, assoc prof microbiol, 68-73, prof microbiol, 73-77; PROF & CHMN, COL MED, UNIV SALA, 77-, CHMN, DEPT MICROBIOL, & IMMUNOL & PATH. *Concurrent Pos:* Sect chief & consult, Molecular Anat Prog, Tumor transplanation Study, Oak Ridge Nat Labs, 68-; mem immunol sci study sect, Nat Cancer Inst, 75. *Mem:* AAAS; Tissue Cult Asn; Soc Exp Biol & Med; Am Soc Microbiol; Am Asn Immunol. *Res:* Drug resistance in microorganisms; tumor immunology; virology of cancer. *Mailing Add:* Dept Microbiol & Immunol Lab Molecular Biol Col Med Univ SAla Mobile AL 36688

COGGINS, CHARLES WILLIAM, JR, b NC, Nov 17, 30; m 51; c 3. PLANT PHYSIOLOGY. *Educ:* NC State Col, BS, 52, MS, 54; Univ Calif, PhD(plant physiol), 58. *Prof Exp:* Asst plant physiologist, 57-64, assoc plant physiologist, 64-70, chmn dept bot & plant sci, 75-82, PLANT PHYSIOLOGIST, DEPT BOT & PLANT SCI, CITRUS RES CTR, UNIV CALIF, RIVERSIDE, 70- *Honors & Awards:* Am Soc Hort Sci Award, 66 & 84. *Mem:* AAAS; Am Soc Hort Sci; Am Soc Plant Physiologists; Am Inst Biol Sci; Int Soc Citricult; Plant Growth Regulator Soc Am. *Res:* Evaluation of vegetative, reproductive and fruit quality responses of citrus, avocado and other subtropical fruit to plant regulators. *Mailing Add:* Dept Bot & Plant Sci Univ Calif Riverside CA 92521-0124

COGGINS, JAMES RAY, b Denton, NC, Dec 2, 47; m 70; c 2. PARASITOLOGY, ELECTRON MICROSCOPY. *Educ:* E Carolina Univ, BA, 70, MA, 72; Wake Forest Univ, PhD(biol), 75. *Prof Exp:* Res assoc parasitol, Univ Notre Dame, 75-77; ASST PROF ZOOL, UNIV WIS-MILWAUKEE, 77- *Concurrent Pos:* NIH trainee fel, 75-77. *Mem:* Am Soc Parasitologists; Electron Micros Soc Am; Am Micros Soc; AAAS. *Res:* Growth and development of parasitic helminths. *Mailing Add:* Dept Biol Sci Univ Wis Milwaukee WI 53201

COGGINS, LEROY, b Thomasville, NC, July 29, 32; m 56; c 5. VETERINARY VIROLOGY. *Educ:* NC State Col, BS, 55; Okla State Univ, DVM, 57; Cornell Univ, PhD(vet virol), 62. *Prof Exp:* Vet res officer virus res, Ft Detrick, Md, 57-59; asst, NY State Col Vet Med, Cornell Univ, 59-62; res assoc, Cornell Univ, 62-63; vet res officer, EAfrican Vet Res Orgn, USDA, 63-68; prof virol, NY State Col Vet Med, Cornell Univ, 68-80; PROF & HEAD DEPT MICROBIOL, PATH & PARASITOL, SCH VET MED, NC STATE UNIV, 80- *Mem:* Am Vet Med Asn; US Livestock Sanit Asn; Conf Res Workers Animal Dis; Am Asn Equine Practrs; Sigma Xi. *Res:* Virus research; viruses of variola, hog cholera, bovine virus diarrhea and African swine fever and the host response to these agents; equine infectious anemia; equine influenza. *Mailing Add:* 309 Kelso Ct Cary NC 27511

COGGON, PHILIP, b Kirkby, Eng, Mar 22, 42; US citizen; m 65; c 1. FOOD CHEMISTRY. *Educ:* Univ Nottingham, BS, 63, PhD(chem), 66. *Prof Exp:* Res fel chem, Univ Sussex, 66-68; res assoc crystallog, Duke Univ, 68-70; group leader tea res, Thomas J Lipton, Inc, 70-76, mgr bev res & process develop, 76-78; MGR OPER SERV, ROYAL ESTATE TEA CO, 79- *Mem:* Am Chem Soc. *Res:* Organic chemistry of natural products from food sources; process and product development of beverage products. *Mailing Add:* 800 Sylvan Ave Royal Estate Tea Co Englewood Cliffs NJ 07632

COGHLAN, ANNE EVELINE, b Boston, Mass, Mar 29, 27. MICROBIOLOGY. *Educ:* Simmons Col, BS, 48; Boston Univ, MEd, 53; Univ Vt, MS, 57; Univ RI, PhD(biol sci), 65. *Prof Exp:* Instr bact, Colby Jr Col, 49-59; NSF fel, 59-61; chmn dept, 72-78, PROF BIOL, SIMMONS COL, 62-, DEAN SCI, 78- *Mem:* AAAS; Am Soc Microbiol; Am Soc Cell Biol. *Res:* General microbiology; basic bacteriology; immunology; host-parasite relationships. *Mailing Add:* 65 Belcher Circle Milton MA 02186

COGHLAN, DAVID B(UELL), b Cleveland, Ohio, Apr 14, 20; m 42, 69; c 5. CHEMICAL ENGINEERING. *Educ:* Yale Univ, BChE, 41. *Prof Exp:* Chemist, Gen Chem Co, 41-42, TNT mfg foreman, 42-43, res engr, 43-46; res engr, E I du Pont de Nemours & Co, 46-49; head res eng div, 49-57, asst dir res, 57-62, mgr contract res, 62-65, MGR SPEC PROJ, FOOTE MINERAL CO, 65- *Mem:* Soc Indust & Appl Math; Am Inst Chem Engrs; Am Chem Soc; Am Inst Mining Metall & Petrol Engrs. *Res:* Fluosulfonic acid; lithium chemical processes; solar evaporation pond design. *Mailing Add:* 582 Grubbs Mill Rd West Chester PA 19380

COGLEY, ALLEN C, b Grinnell, Iowa, Apr 16, 40; m 63; c 2. AERONAUTICAL ENGINEERING, ATMOSPHERIC SCIENCES. *Educ:* Iowa State Univ, BSAE, 62; Univ Va, MSAE, 64; Stanford Univ, PhD(aeronaut sci), 68. *Prof Exp:* Res scientist, Langley Res Ctr, NASA, 62-63; from asst prof to assoc prof fluid mech, Univ Ill, Chicago Circle, 68-79, prof, 79-83; prof & head, Mech Eng, Univ Ala, Huntsville, 83-86; PROF HEAD, MECH ENG, KANS STATE, 87- *Concurrent Pos:* NASA and NSF res grants. *Mem:* Am Inst Aeronaut & Astronaut; Am Meteorol Soc; Am Soc Eng Educ; Am Soc Mech Engrs. *Res:* Wave propagation; radiative gas dynamics; radiative transfer; fluid mechanics. *Mailing Add:* Dept Mech Eng Durland Hall Kans State Univ Manhattan KS 66506

COGLIANO, JOSEPH ALBERT, b Brooklyn, NY, Mar 4, 30. ORGANIC CHEMISTRY. *Educ:* Polytech Inst Brooklyn, BS, 51; Princeton Univ, MA, 56, PhD, 58; George Washington Univ, MS, 57. *Prof Exp:* Chemist, Nat Bur Stand, 51-53; res assoc, George Washington Univ, 53-54; asst, Princeton Univ, 54-57; RES ASSOC, W R GRACE & CO, 57- *Concurrent Pos:* Staff mem, Nat Acad Sci, 55. *Mem:* Am Chem Soc. *Res:* Organic research and synthesis; process development; organophosphorous chemistry; physico-chemical measurements; permeability; foams-preparation and properties; technical trouble shooting; technical program management; bound enzymes. *Mailing Add:* 7379 Rte 32 Columbia MD 21044-3310

COGSWELL, GEORGE WALLACE, b New York, NY, Feb 8, 23; m 49; c 4. ENVIRONMENTAL ANALYSIS, SYNTHETIC FINE INORGANIC CHEMICALS. *Educ:* City Col New York, BS, 53; Fordham Univ, MS, 55, PhD(org chem), 60. *Prof Exp:* Chemist indust detergents, Colgate-Palmolive Co, NJ, 50-53; lab asst, Fordham Univ, 54-57; res assoc, Dept Pharmacol, Med Col, Cornell Univ, 57-58; sr develop chemist-proj coord, A E Staley Mfg Co, 58-64; sect mgr, A-U Proj, Armour & Co, 64-65; mkt develop mgr, Hooker Chem Co, Inc, 65-68; PRES, WOODBURN ANALYTICAL LAB, 68-; PRES, FIELDING CHEM CO, INC, 78- *Concurrent Pos:* Instr chem, Anderson Col, SC, 74-; consult forensic chem. *Mem:* Am Chem Soc; Am Pharmaceut Asn; Tech Asn Pulp & Paper Indust. *Res:* Synthetic and natural polymers; specialty and fine chemicals; coatings; paper; detergents, ozonolysis; structure activity studies; instrumental and wet analyses. *Mailing Add:* Wallace Corp PO Box 282 Anderson SC 29622-0282

COGSWELL, HOWARD LYMAN, b Susquehanna Co, Pa, Jan 19, 15; m 38; c 1. ORNITHOLOGY, ECOLOGY. *Educ:* Whittier Col, BA, 48; Univ Calif, Berkeley, MA, 51, PhD(zool), 62. *Prof Exp:* Asst prof biol sci, Mills Col, 52-64; from assoc prof to prof, 64-80, EMER PROF BIOL SCI, CALIF STATE UNIV, HAYWARD, 80- *Concurrent Pos:* NSF sci fac fel, 63-64. *Mem:* Am Ornith Union; hon mem Cooper Ornith Soc; Soc Wetland Scientists; Wilson Ornith Soc. *Res:* Habits, phenology and populations of birds of California; habitat distribution and selection in birds; biology of shorebirds in relation to tides; behavior of birds in relation to wetland habitats. *Mailing Add:* 1548 East Ave Hayward CA 94541

COGSWELL, HOWARD WINWOOD, JR, b Sherman, Tex, Apr 22, 23; m 47; c 1. ANALYTICAL CHEMISTRY. *Educ:* Austin Col, BS, 47. *Prof Exp:* Jr chemist, Cities Serv Refining Corp, 48-49, analytical chemist, Cities Serv Res & Develop Co, 49-53, sr analytical chemist, 53-56, sect leader, 56-60, sr res chemist, 58-60, res assoc, 60-61; head analytical res group, Petro-Tex Chem Corp, 61-77; SUPT, ANALYTICAL SERV & QUAL CONTROL, DENKA CHEM CORP, 77- *Mem:* Am Chem Soc. *Res:* Catalytic petroleum processing; catalyst reactivation and development; research, design, and construction in gas chromatography; analytical instrumentation and methods development. *Mailing Add:* 526 Shawnee St Houston TX 77034

COHAN, CHRISTOPHER SCOTT, b New York, NY, Sept 25, 52; m 74; c 2. NEURONAL REGENERATION, GROWTH CONE MOTILITY. *Educ:* State Univ NY-Albany, BS, 74, Case Western Reserve Univ, PhD(anat), 80. *Prof Exp:* Assoc, Univ Iowa, 80-86; ASST PROF NEUROANAT, STATE UNIV NY, BUFFALO, 86- *Mem:* Soc Neurosci; Am Asn Anatomists; AAAS. *Res:* Mechanisms that regulate neurite outgrowth; the formation of connections in regenerating neurons; direct experimental manipulation and quantitative measurements of growth cone motility and its regulation by electrical activity; neurotransmitters and calcium. *Mailing Add:* Dept Anat Sci SUNY Farber Hall Buffalo NY 14214

COHART, EDWARD MAURICE, b New York, NY, Dec 8, 09; m 33; c 2. PUBLIC HEALTH. *Educ:* Columbia Univ, AB, 28, MD, 33, MPH, 47; Am Bd Prev Med & Pub Health, dipl. *Hon Degrees:* MA, Yale Univ, 56. *Prof Exp:* From assoc prof to prof pub health, 48-78, chmn dept epidemiol & pub health, 66-68, EMER PROF PUB HEALTH, SCH MED, YALE UNIV, 78- *Concurrent Pos:* Cancer control consult, USPHS, 47-48; dep comnr, Dept Health, NY, 55-56, mem adv coun, Nat Inst Environ Health Sci, 69-73. *Mem:* AAAS; AMA; Am Cancer Asn; Am Col Prev Med (vpres pub health, 73-74); Am Pub Health Asn. *Res:* Epidemiology of chronic disease; public health practice. *Mailing Add:* 625 Ellsworth Ave New Haven CT 06511

COHEE, GEORGE VINCENT, b Indianapolis, Ind, Feb 4, 07; m 30. PETROLEUM GEOLOGY. *Educ:* Univ Ill, BS, 31, MS, 32, PhD(geol), 37. *Prof Exp:* Asst geologist, Oil & Gas Div, State Geol Surv, Ill, 36-42; asst state geologist, State Geol Surv, Ind, 42-43; petrol analyst, Petrol Admin for War, 43; geologist, Fuels Br, US Geol Surv, 43-47, sr geologist, 47-51, chmn geol names comt, 52-77. *Concurrent Pos:* Chmn geol dept, Univ Ark, 51-52. *Mem:* Fel Geol Soc Am; Soc Econ Paleontologists & Mineralogists; hon mem Am Asn Petrol Geologists (secy-treas, 60-62); Sigma Xi. *Res:* Stratigraphy and petroleum geology. *Mailing Add:* 5508 Namakagan Rd Bethesda MD 20816

COHEN, AARON, b Corsicana, Tex, Jan 5, 31; m; c 3. SPACE & LIFE SCIENCE RESEARCH. *Educ:* Tex A&M Univ, BS, 52; Stevens Inst Technol, MS, 58. *Hon Degrees:* DSc, Stevens Inst Technol, 82; LHD, Univ Houston, 89. *Prof Exp:* Staff, 62-69, mgr, command & serv modules, Apollo Spacecraft Prog Off, 69-72, mgr, Space Shuttle Orbiter Proj, 72-82, dir, res & engr, 82-86, DIR, LYNDON B JOHNSON SPACE CTR, NASA, 86- *Honors & Awards:* W Randolph Lovelace II Award, Am Astronaut Soc; Von Karman Lectr, Am Inst Aeronaut & Astronaut; Am Soc Mech Engrs Medal, 84; Goddard Mem Trophy, 89; Gold Knight of Mgt Award, Nat Mgt Asn, 89. *Mem:* Nat Acad Eng; fel Am Inst Aeronaut & Astronaut; Am Soc Mech Engr; Int Asn Astronaut; fel Am Astronaut Soc. *Res:* Author of many articles for scientific and technical journals. *Mailing Add:* NASA Johnson Space Ctr Houston TX 77058

COHEN, ABRAHAM BERNARD, b Philadelphia, Pa, July 19, 22; m 53; c 2. PHOTOCHEMISTRY. *Educ:* Temple Univ, AB, 48; Cornell Univ, PhD(chem), 52. *Prof Exp:* Asst, Temple Univ, 47-48 & Cornell Univ, 48-50; sr res chemist, Photo Prod Dept, E I du Pont de Nemours & Co, Inc, 51-55, res supvr, 55-65, res fel, 65-66, res mgr, New Prod Develop, 69-70, mgr photopolymer systs, 70-72, dir res, 72-80, dir res electronics, 81-84, dir res advan technol, Photosysts & Electronic Prod Dept, 65-87, chief adv, Adv Technol Electronics Dept, 88-89; TECHNOL MGT CONSULT, 89- *Concurrent Pos:* Vis lectr, Soc Imaging Sci & Technol, 90. *Honors & Awards:* Indust Res Mag IR 100 Award, 69, 74, 75, 77 & 85; Kosar Mem Award, Soc Photog Sci & Eng, 79; Com Develop Asn Honor Award, 81; Gold Medal Res Leadership, Indust Res Inst, 89. *Mem:* Am Chem Soc; Soc Photog Sci & Eng. *Res:* Correlation of structure and properties of polymers; mechanism of polymer reactions; photopolymerization; nonconventional photographic systems; dimensionally stable film bases and coatings; photoresist films and equipment; photopolymer printing plates, color proofing systems and graphic arts systems; new venture management; electronics packaging systems laser imaging systems; 3D solid model imaging systems. *Mailing Add:* 283 Sayre Dr Princeton NJ 08540

COHEN, ADOLPH IRVIN, b New York, NY, Apr 7, 24; m 55; c 2. BIOLOGY, ANIMAL PHYSIOLOGY. *Educ:* City Col NY, BS, 48; Columbia Univ, MA, 50, PhD(zool), 54. *Hon Degrees:* DSc, Penn Col Optom. *Prof Exp:* from instr to asst prof anat, Wash Univ, 55-64, res assoc prof ophthal, 64-70, chmn univ comt neurobiol, 70-74, PROF ANAT & NEUROBIOL, SCH MED, WASH UNIV, 70-, PROF OPHTHAL & VISUAL SCI, 70- *Concurrent Pos:* USPHS fel, Univ Calif, Berkeley, 53-54 & Wash Univ, 54-55; mem bd trustees, Asn Res Vision & Ophthal, 73-78; chmn bd sci counrs, Nat Eye Inst, 79-80. *Honors & Awards:* Francis I Proctor Medal, 84. *Mem:* AAAS; Soc Neurosci; Am Soc Cell Biol; Asn Res Vision & Ophthal; Am Asn Anat. *Res:* Vision; cell biology; receptor physiology. *Mailing Add:* Dept Ophthal Box 8096 Sch Med Wash Univ 660 S Euclid Ave St Louis MO 63110

COHEN, ALAN MATHEW, b Chicago, Ill, Mar 22, 43. BIOLOGY, EMBRYOLOGY. *Educ:* Univ Ill, BSc, 64; Univ Va, PhD(biol), 69. *Prof Exp:* Asst prof anat, Sch Med, Johns Hopkins Univ, 71-; PROG OFFICER, ROBERT WOOD JOHNSON FOUND, VPRES. *Concurrent Pos:* USPHS fel anat, Harvard Med Sch, 69-71 & grant, 74-77; Nat Found March Dimes Basil O'Conner grant, 74-76, Nat Found March Dimes res grant, 77-80; USPHS grant, 78-81. *Mem:* AAAS; Am Soc Zoologists; Soc Develop Biol. *Res:* Regulation of cellular events during embryogenesis; neural crest as a model system for the study of development. *Mailing Add:* Asbury Rd Bloomsbury NJ 08804

COHEN, ALAN SEYMOUR, b Boston, Mass, Apr 9, 26; m 54; c 3. MEDICINE. *Educ:* Harvard Univ, AB, 47; Boston Univ, MD, 52. *Prof Exp:* Instr, Harvard Med Sch, 58-60; from asst prof to prof, 60-72, CONRAD WESSELHOEFT PROF MED, SCH MED, BOSTON UNIV, 72-, DIR ARTHRITIS & CONNECTIVE TISSUE DIS SECT, UNIV HOSP, 60-; CHIEF MED, HOSP & DIR THORNDIKE MEM LAB, BOSTON CITY HOSP, 73- *Concurrent Pos:* Fel med, Harvard Med Sch, 53, res fel, 56-58; consult, USPHS, 66-70; Bernadine Becker Mem lectr, 69; consult, Food & Drug Admin, 70-; mem gen med study sect A, Nat Inst Arthritis, Metab & Digestive Dis, 72-76; Wallace-Graham Mem lectr, Queen's Univ, Ont, 73; chmn med & sci comt, Mass Arthritis Found (pres, 81-82); mem House Deleg & bd trustees, Arthritis Found, 76-, mem budget & finance comt, 77-78; Tyndale vis prof, Univ Utah Sch Med, 78. *Honors & Awards:* Maimonides Award, Boston Med Soc, 52. *Mem:* Am Soc Clin Invest; Am Fedn Clin Res; Asn Am Physicians; Am Soc Cell Biol; Am Soc Exp Path. *Res:* Internal medicine; rheumatology; electron microscopy. *Mailing Add:* Thorndike Mem Lab Boston City Hosp Thorndike 314 818 Harrison Ave Boston MA 02118

COHEN, ALEX, b New York, NY, Feb 7, 31. ORGANIC CHEMISTRY, CLINICAL CHEMISTRY. *Educ:* Brooklyn Col, 54. *Prof Exp:* ORG CHEMIST, NAT BUR STANDARDS, 57- *Mem:* Am Chem Soc. *Res:* Clinical standard characterization-cholesterol; anticancer agents; stable isotope-labeled clinical compounds; uric acid; synthesis. *Mailing Add:* Chem A361 Nat Inst Standards & Technol Gaithersburg MD 20899

COHEN, ALFRED M, SURGICAL ONCOLOGY, COLON-RECTAL CANCER. *Educ:* Jonhs Hopkins Univ, MD, 67. *Prof Exp:* CHIEF, COLON-RECTAL SERV, DEPT SURG, MEM-SLOAN KETTERING CANCER CTR, 86- *Mailing Add:* Dept Surg Mem-Sloan Kettering Cancer Ctr 1275 York Ave New York NY 10021

COHEN, ALLEN BARRY, b Ft Wayne, Ind, July 14, 39; m 60; c 2. PULMONARY MEDICINE & DISEASES, ENPHYSEMA. *Educ:* George Washington Univ, BA, 60, MD, 63; Univ Calif, San Francisco, PhD(microbiol), 72. *Prof Exp:* Res fel, Sch Med, Univ Calif, San Francisco, 69-71, asst prof-in-residence, 71-75, assoc mem, Cardiovasc Res Inst, 73-75; chief pulmonary div, Temple Univ, Philadelphia, 75-83, prof med, 77-83, prof physiol, 78-83; EXEC ASSOC DIR, UNIV TEX HEALTH CTR, TYLER, 83 - *Concurrent Pos:* Dir respiratory intensive care unit, San Francisco Gen Hosp, 71-72, consult, 73-75; res career develop award, Nat Heart, Lung & Blood Inst, NIH, 73-78; Macy fac scholar, Temple Univ, 79-80. *Honors & Awards:* Cecil Mayer Award, Am Col Chest Physicians, 72. *Mem:* Am Fedn Clin Researchers; Am Physiol Soc; Am Soc Biol Chem; Am Soc Clin Invest; Am Thoracic Soc; Am Soc Cell Biol. *Res:* Pathogenesis of pulmonary emphysema; movements of neutrophils and the discharge of neutrophil enzymes into the lungs and methods of curtailing these events. *Mailing Add:* Univ Tex Health Ctr PO Box 2003 Tyler TX 75710

COHEN, ALLEN IRVING, b New York, NY, May 26, 32; m 54; c 3. ANALYTICAL CHEMISTRY, STRUCTURAL CHEMISTRY. *Educ:* City Col NY, BS, 54; Syracuse Univ, PhD(chem), 58. *Prof Exp:* Asst instr, Syracuse Univ, 54, AEC asst, 54-57; res chemist analytical chem, Gulf Res & Develop Co, 57-59; sr res chemist, 59-66, res assoc, 66-69, res group leader, 69-72, sect head molecular spectros, 72-82, sr res group leader, 82-83, ASST DIR, ANALYTICAL RES & DEVELOP, SQUIBB INST MED RES, 83- *Mem:* Am Chem Soc; Am Soc Mass Spectrometry; Am Pharmaceut Asn. *Res:* Structure determination of organic compounds and natural products by physicochemical methods; nuclear magnetic resonance and ultraviolet spectroscopy; organic polarography; radio chemistry, coprecipitation phenomena; developmental analytical methods; mass spectrometry and development of computer data acquisition programs; quantitative determination of drugs in body fluids; material sciences. *Mailing Add:* E R Squibb Bldg 101 PO Box 191 Rte One Col Farm Rd New Brunswick NJ 08903

COHEN, ALONZO CLIFFORD, JR, b Stone Co, Miss, Sept 4, 11; m 34; c 3. ESTIMATION, RELIABILITY. *Educ:* Ala Polytech Univ, BS, 32, MS, 33; Univ Mich, MA, 40, PhD(statist), 41. *Prof Exp:* Student engr, Westinghouse Elec & Mfg Co, 33-34; instr math, Ala Polytech Univ, 34-40; from instr to asst prof, Mich State Univ, 40-47; assoc prof statist, 47-52, prof math, 52-64, dir, Inst Statist, 59-65, prof, 64-78, EMER PROF STATIST, UNIV GA, 78- *Concurrent Pos:* Consult, Opers Analysis Off Hq, USAF, 50-72; consult, US Army Res Off, 50-75. *Honors & Awards:* Michael Res Award, Univ Ga, 54. *Mem:* Int Statist Inst; fel Am Soc Qual Control; Opers Res Soc Am; fel Am Statist Asn; Math Assoc Am. *Res:* Truncated frequency distributions; statistical methods of quality control; mathematical statistics; censored sampling; estimate in life-span distributions. *Mailing Add:* Dept Statist Univ Ga Athens GA 30602

COHEN, ALVIN JEROME, b Louisville, Ky, July 21, 18; m 43, 69; c 4. GEOCHEMISTRY. *Educ:* Univ Fla, BS, 40; Univ Ill, PhD(inorg chem), 49. *Prof Exp:* Analytical chemist, Tenn Valley Authority, Wilson Dam, Ala, 41; physicist, closed bomb ballistics, Ind Ord Works, 42; chemist war alcohol prod, Joseph Seagram & Sons, 43; asst, Purdue Univ, 46-47 & Univ Ill, 47-49; fel, Calif Inst Technol, 49-50; chemist phys chem, Naval Ord Test Sta, China Lake, 50-53; from fel to sr fel, Mellon Inst, 53-62; prof geochem, 63-88, EMER PROF GEOCHEM, UNIV PITTSBURGH, 88- *Concurrent Pos:*

Distinguished vis prof, Am Univ, Cairo, Egypt, 79. *Honors & Awards:* IR-100 Award, 62. *Mem:* Fel Am Mineral Soc; Geochem Soc; fel Meteoritical Soc; fel AAAS; fel Int Asn Geochem Cosmochem; Mineral Soc Gt Brit & Ireland. *Res:* Radiation effects in silicate minerals and glasses; geochemistry of meteorites; vacuum UV spectra of minerals and meteorites; color in gem minerals, especially quartz, topaz and spodumene; radiation coloration of planetary surfaces especially IO. *Mailing Add:* Dept Geol & Planetary Sci Univ Pittsburgh Pittsburgh PA 15260

COHEN, ANDREW SCOTT, b Trenton, NJ, Oct 25, 54; m 86. LIMNOLOGY, EVOLUTIONARY BIOLOGY. *Educ:* Middlebury Col, BA, 76; Univ Calif, Davis, PhD (geol), 82. *Prof Exp:* Asst prof geol, Colo Col, 82-86; ASST PROF GEOL, UNIV ARIZ, 86- *Mem:* Geol Soc Am; Paleont Soc; Soc Econ Paleontologists & Mineralogists; Soc Study Evolution; Soc Conserv Biologists. *Res:* Limnology, sedimentology and evolutionary biology of large lakes, particularly the Rift Lakes of Africa; generation of models based on modern lake systems which are applicable to the fossils and strata of ancient lakes. *Mailing Add:* Dept Geosci Univ Ariz Tucson AZ 85721

COHEN, ANNA (FONER), b Pittsburgh, Pa, July 10, 24; m 50; c 4. EDUCATION, ART. *Educ:* Univ Pittsburgh, BS, 45, PhD(educ), 71, MS 77; Carnegie-Mellon Univ, MS, 47, PhD(physics), 50. *Prof Exp:* Physicist, Oak Ridge Nat Lab, 51-58; instr, Carnegie-Mellon Univ, 58-70; prof physics, Point-Park Col, 71-74; res assoc, Univ Pittsburgh, 77-79; instr physics, Shady Side Acad, 79-81; PHYSICIST, US BUR MINES, 81- *Mem:* Am Phys Soc. *Res:* Safety in coal mines from fire and explosions; detection of explosive gases in mines; mine ventilation properties. *Mailing Add:* 5414 Albemarle Ave Pittsburgh PA 15217

COHEN, ANNE CAROLYN CONSTANT, b Durham, NC, Mar 1, 35; m 55; c 2. CRUSTACEA, SYSTEMATICS. *Educ:* Stanford Univ, BA, 56; Univ Md, MS, 72; George Washington Univ, PhD(biol), 87. *Prof Exp:* Res asst, Behav Sci Ctr, Stanford Univ, 56; mus technician, Oceanog Ctr, Smithsonian Inst, 63-66, mollusks, 68-69 & crustacea, 73-76, mus specialist, 76-82; teaching asst zool, Univ Md, 69-70; proj dir, Natural Hist Mus Los Angeles, 87; postdoctoral res fel biol, 87-91, ASST RESEARCHER, UNIV CALIF, LOS ANGELES, 91- *Mem:* Soc Women Geographers; Crustacean Soc; Western Soc Naturalists. *Res:* Systematics, phylogeny, morphology and behavior of OstraCoda, particularly bioluminescent signaling taxa; crustacean phylogeny. *Mailing Add:* Dept Life Sci Natural Hist Mus 900 Exposition Blvd Los Angeles CA 90007

COHEN, ARNOLD A, b Duluth, Minn, Aug 1, 14; m 42, 89; c 2. COMPUTER SCIENCE, PHYSICS. *Educ:* Univ Minn, BEE, 35, MS, 38, PhD(physics), 47. *Prof Exp:* Develop engr electron tubes, RCA Corp, 42-46; various tech mgt positions comput develop, Sperry Univac, 46-71; asst dean, Inst Technol, 71-81, SR FEL, CHARLES BABBAGE INST, UNIV MINN, 81- *Concurrent Pos:* Consult & mem sci adv bd, Dept Defense, 60-74; mem adv bd, Chem Abstr Serv, 70-73; dir & mem bd trustees, Charles Babbage Found, 78- *Honors & Awards:* Valuable Invention Citation, Am & Minn Patent Law Asns, 61; Centennial Medal, Inst Elec & Electronics Engrs, 84. *Mem:* Fel Inst Elec & Electronics Engrs. *Res:* History of computing; gaseous conduction devices; mass spectrometry; digital computer systems. *Mailing Add:* 6051 Laurel Ave Apt 217 Minneapolis MN 55416-1063

COHEN, ARTHUR DAVID, b Wilmington, Del, Feb 26, 42; m 70; c 2. COAL & PEAT PETROLOGY, PALYNOLOGY. *Educ:* Univ Del, BS, 64; Pa State Univ, PhD(geol), 68. *Prof Exp:* Asst prof geol, Univ Ga, 68 & Southern Ill Univ, Carbondale, 69-74; geologist, Coal Resources Br, US Geol Surv, 74-75; from assoc prof to prof geol, Univ SC, 75-82; staff mem, Los Alamos Nat Lab, 82-88; PROF GEOL, UNIV SC, 88- *Concurrent Pos:* NSF grants, 69, 71, 73, 75, 77, 79, 82 & 89, Dept Energy, 79, 80 & 81, Environ Protection Agency, 89 & 90; chmn, Coal Div, Geol Soc Am, 76; consult, projs dealing with peat & coal & extraction of pollutants from groundwater. *Mem:* AAAS; Sigma Xi; Am Asn Stratig Palynologists; Am Asn Petrol Geologists; Soc Org Petrol (pres, 89); fel Geol Soc Am. *Res:* Petrologic investigation of the peats of southern Florida with special reference to the origin of coal; coal petrology and geologic history of the Okefenokee Swamp from study of its peat sediments; economic characteristics of North Carolina, South Carolina and Georgia coastal peats; peat deposits of Central and South America; extraction of hazardous pollutants from wastewater using peat; over 150 technical publications. *Mailing Add:* Dept Geol Univ SC Columbia SC 29208

COHEN, ARTHUR LEROY, b Newport News, Va, Jan 22, 16; m 43; c 4. ELECTRON MICROSCOPY, RESEARCH ADMINISTRATION & CONSULTING. *Educ:* Stanford Univ, AB, 37; Harvard Univ, MA, 39, PhD(bot), 40. *Prof Exp:* Sheldon traveling fel from Harvard Univ, Hopkins Marine Sta, 40-41; res fel, Calif Inst Technol, 42-47; prof biol, Oglethorpe Univ, 47-62; assoc prof biol & dir, Electron Micros Ctr, 62-79, prof bot & biol sci, 68-79, EMER PROF BOT & BIOL SCI, WASH STATE UNIV, 80- *Concurrent Pos:* Res assoc, Cedars of Lebanon Hosp, Los Angeles, 45-47; Guggenheim fel, Delft Univ Technol, 56-57; vis prof, Yale Sch Med, 71; Fulbright lectr, Univ Sri Lanka, Kandya, 78; consult & lectr, Chulalongkorn Univ, Bangkok, Thailand, 85. *Mem:* Fel Royal Micros Soc; Electron Micros Soc Am; Electron Micros Soc Can; Soc Cell Biologists; AAAS; Sigma Xi. *Res:* Experimental morphogenesis of Myxomycetes; general biology, ultrastructure; critical point drying; enzymatic and chemical subcellular dissection. *Mailing Add:* Rte 1 Box 468 Pullman WA 99163-9740

COHEN, AVIS HOPE, b Chicago, Ill, Nov 29, 41; m 61; c 2. NEUROPHYSIOLOGY, MOTOR PHYSIOLOGY. *Educ:* Univ Mich, BS, 64; Cornell Univ, PhD(neurobiol), 77. *Prof Exp:* Fel, Karolinska Inst, 77-79; res assoc, Wash Univ, 79-80; RES ASSOC, CORNELL UNIV, 80- *Mem:* Soc Neurosci. *Res:* Organization of systems of neurons which are responsible for generating temporally patterned activity; the spinal neurons of lampreys which generate swimming. *Mailing Add:* 143 Graham Rd Ithaca NY 14850

COHEN, BARRY GEORGE, b New York, NY, June 2, 30; m 51; c 4. ELECTRICAL ENGINEERING, SEMICONDUCTOR PHYSICS. *Educ:* Brown Univ, ScB, 51; Johns Hopkins Univ, DEng, 59. *Prof Exp:* Mem tech staff, Bell Tel Labs, Inc, 59-68; Fulbright fel & vis prof elec eng, Israel Inst Technol, 68-69; PRES, RES DEVICES INC, 69- *Mem:* Am Phys Soc; Inst Elec & Electronics Engrs. *Res:* Infrared and optical properties of solids; semiconductor devices, including cryogenic and microwave devices. *Mailing Add:* Barron Am Inc 39 Cromwell Ct Berkeley Heights NJ 07922

COHEN, BENNETT J, b Brooklyn, NY, Aug 2, 25; m 52; c 2. LABORATORY ANIMAL MEDICINE, COMPARATIVE MEDICINE. *Educ:* Cornell Univ, DVM, 49; Northwestern Univ, MS, 51, PhD(physiol), 53; Am Col Lab Animal Med, dipl, 58. *Prof Exp:* Veterinarian, Northwestern Univ, 49-53; statewide veterinarian, Univ Calif, Berkeley, 53-57; veterinarian, Univ Calif, Los Angeles, 54-62, from instr to asst prof, 56-62; assoc prof physiol & dir animal care unit, 62-67, dir unit lab animal med, 67-85, PROF LAB ANIMAL MED, UNIV MICH, ANN ARBOR, 67-, RES SCI, INST GERONT, 86- *Concurrent Pos:* Assoc ed, J Geront, 80; mem gov bd, Int Coun Lab Animal Sci, 83-; mem, Geriatrics & Geront Rev Comt, NIA, 86- *Honors & Awards:* Charles A Griffin Award, Am Asn Lab Animal Sci, 66; Charles River Prize, Am Vet Med Asn, 80. *Mem:* Am Vet Med Asn; Am Asn Lab Animal Sci (pres, 58-60); Am Physiol Soc; fel Geront Soc. *Res:* Pathology of aging in laboratory animals; diseases of laboratory animals; aging and diabetic mice as models of cutaneous wound repair. *Mailing Add:* Unit Lab Animal Med Univ Mich Animal Res Facil Box 614 Ann Arbor MI 48109

COHEN, BERNARD, inorganic chemistry, for more information see previous edition

COHEN, BERNARD, b Newark, NJ, Apr 30, 29; m 55; c 3. NEUROLOGY, NEUROPHYSIOLOGY. *Educ:* Middlebury Col, AB, 50; NY Univ, MD, 54; Am Bd Psychiat & Neurol, dipl, 61. *Prof Exp:* Asst attend neurologist, Mt Sinai Hosp, New York, 61-66; assoc prof neurol & physiol, Mt Sinai Sch Med, 66-69, prof neurol, 69-76, dir, Neurobiol Grad Prog, 80-90, MORRIS B BENDER PROF NEUROL, MT SINAI SCH MED, 76- *Concurrent Pos:* Res fel neurophysiol, Col Physicians & Surgeons, Columbia Univ, 60-61; trainee, Nat Inst Neurol Dis & Blindness, 60-62; Nat Inst Neurol Dis & Stroke career res develop award, 67-73; assoc attend neurologist, Mt Sinai Hosp, New York, 66-69, attend neurologist, 69-; attend neurol, Elmhurst Gen Hosp; mem neurol study sect A, NIH, 77-82, VISB, 86-, expert in neurol, Food & Drug Admin, 75-80. *Honors & Awards:* Nylen-Hallpike Award, Barany Soc, 87. *Mem:* Am Physiol Soc; Am Acad Neurol; Am Neurol Asn; Barany Soc; Asn Res Nervous & Mental Dis (secy-treas, 80, pres, 86-87). *Res:* Physiology of oculomotor and vestibular systems. *Mailing Add:* Dept Neurol Mt Sinai Sch Med New York NY 10029

COHEN, BERNARD ALLAN, b Fond du Lac, Wis, Oct 6, 46; m 69; c 4. BIOMEDICAL ENGINEERING, NEUROSCIENCE & NEUROPHYSIOLOGY. *Educ:* Milwaukee Sch Eng, BSEE, 71; Marquette Univ, MSEE, 73, MSBME, 73, PhD(biomed eng), 75, PhD(neurophys), 75. *Prof Exp:* Res asst neurosci, Vet Admin Ctr, Wood, Wis, 72-75; sr res investr rehab, Emory Univ Rehab Res & Training Ctr, 75-78, asst prof rehab med, Sch Med, 75-78; DIR BIOMED ENG SECT, NEUROL, VET ADMIN CTR, WOOD WIS, 78-; ASST PROF NEUROL, MED COL WIS, 78-, DIR, ELECTRODIAG LABS, 79- *Concurrent Pos:* Instr eng, Milwaukee Sch Eng, 74-75; clin fel rehab med, Vet Admin Hosp, Atlanta, 75-76; curric consult, DeKalb Community Col, Decatur, 75-78; adj sr res engr, Ga Inst Technol, 77-78, adj asst prof mech eng, 77-78; consult biomed eng, Tech Adv Serv for Attys, Ft Washington, 77-; asst prof biomed eng, Marquette Univ, 78-; asst prof elec eng, Milwaukee Sch Eng, 78- *Mem:* Inst Elec & Electronics Engrs; Asn Advan Med Instrumentation; Am Soc Eng Educ; Instrument Soc Am; Sigma Xi. *Res:* Biomedical engineering involved with data acquisition and analysis of brain and nervous system aimed at the target population of patients suffering from neurological diseases. *Mailing Add:* 3264 N 50th St Milwaukee WI 53216

COHEN, BERNARD LEONARD, b Pittsburgh, Pa, June 14, 24; m 50; c 4. RISK ANALYSIS, RADIATION HEALTH. *Educ:* Case Western Reserve Univ, BS, 44; Univ Pittsburgh, MS, 48; Carnegie Inst Technol, DSc(physics), 50. *Prof Exp:* Asst, Carnegie Inst Technol, 47-49; physicist & group leader, Oak Ridge Nat Lab, 50-58; assoc prof, 58-61, dir, Scaife Nuclear Physics Lab, 65-78, PROF PHYSICS, UNIV PITTSBURGH, 61- *Concurrent Pos:* Consult, Oak Ridge Nat Lab, 58-66, Nuclear Sci Eng Corp, 59-61, Gen Atomic, 59-60, NSF, 62, Inst Defense Anal, 62, Brookhaven Nat Lab, 65, Los Alamos Sci Lab, 68, World Publ Co, 69-70, Elec Power Res Inst, 75, Gen Accounting Off, 76-78, Pac Legal Fedn, 77-78, McGraw-Hill Energy Systs, 78-80 & various other co, 79-; mem nat coun, Am Asn Physics Teachers, 73-76; mem exec comt, Am Phys Soc, 71-73, chmn, Div Nuclear Physics, 74-75; vis staff, Inst Energy Analysis, 74-75, Argonne Nat Lab, 78-79; chmn, Div Environ Sci, Am Nuclear Soc, 80-81; chmn, Intersoc liaison comt, Health Physics Soc, 85-87, publ comt, 88-90. *Honors & Awards:* Bonner Prize, Am Phys Soc, 81; Pub Info Award, Am Nuclear Soc, 85. *Mem:* Fel Am Phys Soc; fel AAAS; Am Nuclear Soc; Health Physics Soc; Soc Risk Analysis. *Res:* Nuclear structure; nuclear reactions and scattering; applied nuclear physics; environmental impacts of nuclear power; health effects of radiation, risk analysis and radon problems; energy. *Mailing Add:* Phys Dept Univ Pittsburgh Pittsburgh PA 15260

COHEN, BERNICE HIRSCHHORN, b Baltimore, Md, Apr 25, 24; c 2. HUMAN GENETICS, EPIDEMIOLOGY. *Educ:* Goucher Col, AB, 44; Johns Hopkins Univ, PhD(human genetics), 58, MPH, 59. *Prof Exp:* Nat Heart Inst fel, Div Med Genetics, Sch Med, 59-60, from asst prof to assoc prof, 60-70, prof epidemiol, Dept Chronic Dis, Sch Hyg, 70-, prof, Dept Med, Sch Med & Dept Biol, 73-, dir human genetics, Genetic Epidemiol Prog, 73-, PROF BIOL SCH ART & SCI, JOHNS HOPKINS UNIV & PROF EPIDEMIOL, SCH HYG. *Concurrent Pos:* Nat Inst Gen Med Sci res career develop award, Dept Chronic Dis, Johns Hopkins Univ, 60-70; assoc, Univ Sem in Genetics & Evolution of Man, Columbia Univ, 64-70; consult,

Baltimore City Hosps, Md, 66-; asst prof, Sch Med, Johns Hopkins Univ, 70- *Mem:* AAAS; Am Soc Human Genetics; Am Pub Health Asn; Genetics Soc Am; fel Am Col Epidemiol. *Res:* Human epidemiological genetics, especially the role of genetic factors in chronic diseases, differential fertility, aging and mortality; congenital anomalies; genetically determined marker traits and disease; maternal-fetal blood group incompatibility. *Mailing Add:* One E Univ Pkwy-Apt 411 Baltimore MD 21218

COHEN, BEVERLY SINGER, b New York, NY, Jan 2, 33; m 55; c 3. AEROSOL SCIENCE, RADIOLOGICAL PHYSICS. *Educ:* Bryn Mawr Col, BA, 53; Cornell Univ, MS, 61; NY Univ, PhD(environ health sci), 79. *Prof Exp:* Res asst radiol physics, Col Physicians & Surgeons, Columbia Univ, 53-56; instr physics, Mt St Mary Col, Newburgh, NY, 58-63; asst res scientist environ health sci, Med Ctr, NY Univ, 76-80; assoc res scientist & res asst prof, 80-85, res assoc prof, 85-89, RES PROF ENVIRON HEALTH SCI, INST ENVIRON MED, NY, 89- *Concurrent Pos:* Mem, aerosol technol comt, Am Indust Hyg Asn, 83-, chair, 86-87; fac adv, Environ Health Sci Coun, NY Univ, 83-; ed, ERC-Outreach, 86-89; chmn, Air Sampling Inst Comt, Am Conf Gov Indust Hygienists, 89-; dir, Am Asn Aerosol Res, 90- *Honors & Awards:* Spec Emphasis Res Career Award, Nat Inst Occup Safety & Health, 84. *Mem:* Am Phys Soc; Health Physics Soc; Am Asn Aerosol Res; Am Indust Hyg Asn; Sigma Xi; Am Conf Govt Indust Hygienists. *Res:* Measurement and dosimetry of inhaled environmental and occupational pollutant gases and aerosols. *Mailing Add:* Inst Environ Med Longmeadow Rd Tuxedo NY 10987

COHEN, BRUCE IRA, b Los Angeles, Calif, Oct 26, 48; m 75; c 1. PLASMA PHYSICS. *Educ:* Harvey Mudd Col, BS, 70; Univ Calif, Berkeley, MA, 72, PhD(physics), 75. *Prof Exp:* Res assoc, Plasma Physics Lab, Princeton Univ, 75-76; PHYSICIST, LAWRENCE LIVERMORE NAT LAB, UNIV CALIF, 76-, DEP GROUP LEADER, 83- *Concurrent Pos:* Jr staff scientist, Phys Dynamics, Inc, Calif, 73-75; adj lectr, Univ Calif, Berkeley, 80-86; fel, Am Phys soc, 87- *Mem:* Am Phys Soc; fel Am Phys Soc. *Res:* Theoretical plasma physics and physical oceanography; computational plasma physics and fluid mechanics. *Mailing Add:* Lawrence Livermore Nat Lab L630 PO Box 808 Livermore CA 94550

COHEN, BURTON D, b Waterbury, Conn, Aug 10, 26; m 51; c 3. INTERNAL MEDICINE. *Educ:* Yale Univ, AB, 50; Columbia Univ, MD, 54; Am Bd Internal Med, dipl, 63. *Prof Exp:* From clin asst prof to assoc prof, 70-79, PROF MED, ALBERT EINSTEIN COL MED, 79; CHIEF METAB SECT, BRONX-LEBANON HOSP, 62- DIR, DEPT MED, 80- *Concurrent Pos:* Asst vis physician, Bellevue Hosp, 59-68; clin investr, Vet Admin, 59-62; career scientist, Health Res Coun New York, 65-73. *Mem:* Am Fedn Clin Res; Am Diabetes Asn; fel Am Col Physicians. *Res:* Metabolism and renal disease. *Mailing Add:* 1276 Fulton Ave Bronx NY 10456

COHEN, CARL, b Brooklyn, NY, Nov 15, 20; m 49; c 3. IMMUNOGENETICS. *Educ:* Ohio State Univ, BSc, 46, MSc, 48, PhD(bact), 51. *Prof Exp:* Asst bact, Ohio State Univ, 46-50; mem staff immunogenetics, Jackson Mem Lab, 51-57; proj leader, Battelle Mem Inst, 57-62; prof biol & assoc prof exp path, Case Western Reserve Univ, 62-70; prof genetics & dir, Ctr Genetics, Univ Ill Med Ctr, 70-77; chief, Genetics & Transplantation Biol Br, Nat Inst Allergy & Infectious Dis, NIH, 77-79; prof genetics & surg, chief div surg immunol, Dept Surg, Univ Ill Med Ctr, 79-85; RETIRED. *Mem:* Fel AAAS; Am Asn Immunol; Genetics Soc Am; Am Soc Human Genetics. *Res:* Immunology; mammalian genetics. *Mailing Add:* 15 Fern Glade Rd Asheville NC 28804-1007

COHEN, CARL M, b Boston, Mass, Mar 29, 46; m 70; c 2. CELL BIOLOGY. *Educ:* Harvard Univ, PhD(physics), 75. *Prof Exp:* BIOMED RESEARCHER, ST ELIZABETH'S HOSP, BOSTON, MASS, 78-, CHIEF, DIV CELLULAR & MOLECULAR BIOL, 82- *Concurrent Pos:* Ad hoc reviewer, NIH, 80-88. *Mem:* Am Soc Cell Biol; Biophys Soc; AAAS; Sigma Xi. *Res:* Biochemistry and molecular biology of cell membranes, especially red blood cell membranes. *Mailing Add:* 15 Magnolia Ave Newton MA 02158

COHEN, CAROLYN, b Long Island City, NY, June 18, 29. BIOPHYSICS. *Educ:* Bryn Mawr Col, AB, 50; Mass Inst Technol, PhD(biophys), 54. *Prof Exp:* Fulbright scholar, King's Col, Univ London, 54-55; res assoc, Children's Cancer Res Found, 55-56; instr biol, Mass Inst Technol, 57-58; res assoc path, Children's Cancer Res Found, Children's Hosp Med Ctr, 58-74; PROF BIOL & MEM ROSENSTIEL BASIC MED SCI RES CTR, BRANDEIS UNIV, 72- *Concurrent Pos:* Res assoc biol, Mass Inst Technol, 55-58; res assoc biochem, Harvard Med Sch, 58-64, Harvard Med Sch lectr biophys, Children's Hosp, 64-74; Guggenheim fel, 78-79. *Honors & Awards:* Merit Award, NIH, 90. *Mem:* Am Crystallog Asn; Am Soc Biol Chemists; AAAS; Biophys Soc; Sigma Xi; Protein Soc; fel Am Acad Arts & Sci. *Res:* Structure of protein assemblies in the cell as determined by x-ray diffraction and electron microscopy; muscle structure and the contractile mechanism; protein folding; structural aspects of blood coagulation. *Mailing Add:* Four Bay View Rd Wellesley MA 02181

COHEN, CLARENCE B(UDD), b Monticello, NY, Feb 7, 25; m 47; c 2. TECHNOLOGY MANAGEMENT. *Educ:* Rensselaer Polytech Inst, BAE, 45, MAE, 47; Princeton Univ, MA, 52, PhD, 54. *Prof Exp:* Aeronaut engr, Nat Adv Comt Aeronaut, Ohio, 47-56, head propulsion aerodyn sect, 53-55, assoc chief spec proj br, 56; head hypersonics res sect, Ramo-Wooldridge Corp, 57, asst dir, Aerosci Lab, Thompson-Ramo-Woolridge Space Technol Labs, 58-65, mgr, Aerosci Lab, 65-69, dir technol appln, TRW systs Energy, 70-81, dir technol, TRW Elec & Defense, 81-87; CONSULT, TECHNOL MGT, 87- *Concurrent Pos:* Guggenheim fel, Princeton Univ, 50-52; trustee, Northrup Univ, 91. *Mem:* Fel Am Inst Aeronaut & Astronaut; Sigma Xi; Indust Res Inst; Lic Exec Soc. *Res:* Viscous flow; heat transfer; gas dynamics; hypersonics; aerodynamics; ballistic missiles and space vehicles; technology administration. *Mailing Add:* 332 Via El Chico Redondo Beach CA 90277

COHEN, DANIEL, public health, epidemiology, for more information see previous edition

COHEN, DANIEL ISAAC ARYEH, b Reading, Pa, Dec 26, 46; m 78. ENUMERATION & ALGORITHMS, GRAPH THEORY. *Educ:* Princeton Univ, AB, 67; Harvard Univ, MA, 70, PhD(math), 75. *Hon Degrees:* JD, Columbia Univ, 87. *Prof Exp:* Asst prof math, Northeastern Univ, 75-78; vis asst prof, Rockefeller Univ, 78-82; PROF COMPUT SCI, CITY UNIV NY, 80-, GRAD CTR, CHMN DEPT, 86- *Concurrent Pos:* Adj prof biostatist psychiat, Cornell Med Sch, NY Hosp & Payne Whitney Clin, 80-; vis prof, Columbia Univ, 86-87; inter-acad exchange, Acad Sci USSR, 78. *Mem:* Math Asn Am; Am Math Soc; Asn Comput Mach; NY Acad Sci; Am Asn Univ Prof; Inst Elec & Electronics Engrs. *Res:* Mathematics: combinatorial theory, comoputer science; psychiatry: diagnosis, experimental design. *Mailing Add:* 420 E 54th St Apt 10B New York NY 10022

COHEN, DANIEL MORRIS, b Chicago, Ill, July 6, 30; m 55; c 2. MUSEUM ADMINISTRATION, SYSTEMATIC ICHTHYOLOGY & MARINE BIOLOGY. *Educ:* Stanford Univ, AB, 52, MA, 53, PhD, 58. *Prof Exp:* Asst gen biol, Stanford Univ, 53-55, actg instr, 55-57; asst prof, Univ Fla, 57-58; syst zoologist fishes, Nat Marine Fisheries Serv, 58-60, lab dir, Ichthyol Lab, 60-70, dir, Systs Lab, 70-81, sr fisheries scientist, 81-82; CHIEF CUR, LIFE SCI, NATURAL HIST MUS, LOS ANGELES, 82- *Concurrent Pos:* Vis researcher, Brit Mus Natural Hist, 64-65; mem, Nat Acad Sci Comn Ecol Res Interocean Canal, 69-70; res assoc, Smithsonian Inst, 69-85; Ed-in-chief, Fishes of West NAtlantic, pt 6; pres, Biol Soc Wash, 71-72; mem oceanog deleg, Nat Acad Sci, People's Repub China, 78; adj prof biol, Univ Southern Calif. *Mem:* Fel AAAS; Am Soc Ichthyologists & Herpetologists (vpres, 69-70, pres, 85-); Japanese Soc Ichthyologists; Soc Syst Zool; Soc Hist Nat Hist; Soc Francaise Ichthyologists; Challenger Soc; fel AAAS. *Res:* Biology of fishes, particularly systematics; deepsea fishes; general marine biology; museum administration; systematics and biology of deep benthic fishes and other coldwater fishes, especially gadiform, ophidiiform and argentinoid fishes. *Mailing Add:* Natural Hist Mus Los Angeles Co 900 Exposition Blvd Los Angeles CA 90007

COHEN, DAVID, b Winnipeg, Man; US citizen. MAGNETISM. *Educ:* Univ Man, BA, 48; Univ Calif, Berkeley, PhD(exp nuclear physics), 55. *Prof Exp:* Assoc physicist, Defence Res Bd Can, 55-57; res assoc physics, Univ Rochester, 57-58; assoc physicist, Argonne Nat Lab, 58-65; assoc prof physics, Univ Ill, Chicago Circle, 65-68; BIOMAGNETISM GROUP LEADER, FRANCIS BITTER NAT MAGNET LAB, MASS INST TECHNOL, 68- *Concurrent Pos:* Estab Investigatorship, Am Heart Asn, 69-74. *Mem:* AAAS; Am Phys Soc. *Res:* Measuring magnetic fields produced by organs of the human body, especially by the brain (magnetoencephalography). *Mailing Add:* Bldg NW-14 Mass Inst Technol Cambridge MA 02139

COHEN, DAVID HARRIS, b Springfield, Mass, Aug 26, 38; m 60, 81; c 5. NEUROBIOLOGY. *Educ:* Harvard Univ, AB, 60; Univ Calif, Berkeley, PhD(psychol), 63. *Prof Exp:* Asst prof physiol, Sch Med, Case Western Reserve Univ, 64-68; assoc prof physiol, Sch Med, Univ Va, 68-71, prof, 71-79, dir neurosci prog, 75-79; prof & chmn Neurobiol & Behav, State Univ NY, Stony Brook, 79-86, prof anat sci, 79-86 & prof psychol, 80-86; VPRES RES & DEAN OF GRAD SCH, NORTHWESTERN UNIV, 86-, PROF NEUROBIOL, 86-, PROF PHYSIOL, 86- *Concurrent Pos:* NSF fel neurophysiol, Med Sch, Univ Calif, Los Angeles, 63-64; Nat Heart & Lung Inst career develop award, 69-74; mem adv panel neurobiol, NSF, 72-75; mem neurol A study sect, NIH, 77-81 & chmn, 83-87; assoc ed, Exp Neurol, 77-82 & J Neurosci, 80-85; mem cent coun, Int Brain Res Orgn, 78-82 & US Nat Comt, 80-; mem, US Nat Comt, Int Union Physiol Sci, 78-84; chmn, Asn Am Med Cols, 89-90; mem life sci adv bd, Off Sci Res, US Air Force, 85-; bd gov, Argonne Nat Lab, 86-; bd overseers, Fermi Nat Lab, 86-; bd trustees, Univ Res Assoc. *Mem:* Sigma Xi; Am Asn Anat; Am Physiol Soc; Pavlovian Soc NAm (pres, 78-79); Soc Neurosci (secy, 75-80, pres, 81-82); Nat Soc Med Res (vpres, 84-85); Asn Neurosci Depts & Prog (pres, 81-82); Int Brain Res Orgn; Asn Am Med Cols; fel Soc Behav Med; fel Acad Behav Med. *Res:* Neural mechanisms of learning; neural control of the cardiovascular system; comparative neurology. *Mailing Add:* Northwestern Univ Crown 2-221 633 Clark St Evanston IL 60208

COHEN, DAVID WALTER, b Philadelphia, Pa, Dec 15, 26; m 48; c 3. PERIODONTICS. *Educ:* Univ Pa, DDS, 50; Am Bd Periodont, dipl, 56. *Hon Degrees:* DSc, Boston Univ, 75; PhD, Hebrew Univ, 77, Univ Athens, 79, Louis Pasteur Univ Strasbourg, 86; DHL, Univ Detoit, 89. *Prof Exp:* From asst instr to assoc oral med & oral path, Sch Dent, 51-55, asst prof periodont, Dent Sch & Grad Sch Med & vchmn dept, Grad Sch Med, 55-59, assoc prof periodont, Dent Sch & Grad Sch Med & chmn dept, Grad Sch Med, 59-64, prof periodont & chmn dept, 69-72, dean, 72-83, EMER DEAN, SCH DENT MED, UNIV PA, 83-; PROF DENT MED, 73-, PRES, MED COL PA, 86- *Concurrent Pos:* Res fel path & periodont, Beth Israel Hosp, Boston, 50-51; asst vis chief oral med, Philadelphia Gen Hosp, 51-; clin asst periodont, Albert Einstein Med Ctr, 51-; nat consult periodont, US Air Force, 65-69, nat emer consult, 69-; dir, Am Bd Periodont, 66-72, vchmn, 70, chmn, 71-72; vis prof, Col Dent, Univ Ill; consult, Vet Admin Hosp, Philadelphia, Ft Dix Army Base & Walter Reed Army Med Ctr; dir dent res, Med Col Pa, Philadelphia, 73-; mem, Nat Dent Adv Coun, NIH, 79-83; mem bd trustees, Grad Hosp Philadelphia, 83-86; chmn, Nat Acad Pract Dent, 84- *Honors & Awards:* Spec Citation Periodont, Boston Univ, 69; Israel Peace Award, 70; Gold Medal Award, Am Acad Periodont, 71; William J Gies Found Periodont Award, 75; Samuel Charles Miller Medal, Am Acad Oral Med, 74; Glickman Mem Lectr, Tufts Univ, 76; W Jones Mem Lectr, Univ Wash, 80; Marvin Goldstein Lectr, Med Col Ga, 82. *Mem:* Inst Med-Nat Acad Sci; fel Am Acad Oral Path; Int Asn Dent Res; fel AAAS; fel Am Acad Periodont (pres); Nat Acad Pract Dent; NY Acad Sci; Am Soc Periodontists (pres, 66-67); fel Am Col Dentists; fel Royal Soc Health. *Res:* Vascular plexus of oral tissues; periodontal disease; treatment planning in dentistry; periodontal therapy; author or co-author of 10 books. *Mailing Add:* 3300 Henry Ave Philadelphia PA 19129

COHEN, DAVID WARREN, b Hartford, Conn, Feb 28, 40; m 64; c 2. MATHEMATICS. *Educ:* Worcester Polytech Inst, BS, 62; Univ NH, MS, 64, PhD(math), 68. *Prof Exp:* Instr math, Exten Serv, Univ NH, 65-68 & Univ, 67-68; from asst prof to assoc prof, 68-85, PROF MATH, SMITH COL, 85- *Mem:* Math Asn Am. *Res:* Quantum theory; mathematical physics. *Mailing Add:* Dept Math Smith Col Northampton MA 01063

COHEN, DONALD, b Tom's River, NJ, Feb 9, 20; m 43; c 4. RADIOCHEMISTRY. *Educ:* Univ Buffalo, BS, 41; Purdue Univ, MS, 48, PhD(chem), 50. *Prof Exp:* Assoc chemist, 49-72, CHEMIST, ARGONNE NAT LAB, 72- *Mem:* Sigma Xi; Am Chem Soc. *Res:* Electrode potentials; chemistry of the transuranium elements; Xenon chemistry and chemistry of molten copper. *Mailing Add:* 5833 Fairmont Ave Downers Grove IL 60515

COHEN, DONALD J, b Chicago, Ill, Sept 5, 40. CHILD PSYCHIATRY. *Educ:* Brandeis Univ, BS, 61; Yale Univ, MD, 66. *Prof Exp:* Med intern, Children Hosp, Boston Med Ctr, 66-67, resident child psychiat, 69-70; resident psychiat, Mass Mental Health Ctr, 67-69; fel child psychiat, Hillcrest Children's Hosp, Children Hosp, Wash, 70-72; DIR, CHILD PSYCHIAT, YALE-NEW HAVEN HOSP, 72-, PROF PEDIAT & PSYCHIAT, YALE-CHILD STUDY CTR, 72- *Concurrent Pos:* Researcher child psychol, Judge Baker Guid Ctr, 69-70; clin fel psychiat, Harvard Univ, 69-70; US Pub Health Serv, 70-72; assoc adult psychiat br, Nat Inst Mental Health, 70-72; spec asst, Dir Off Child Develop, Dept Health Educ & Welfare, Wash, DC, 70-72. *Honors & Awards:* Strecker Award; Itlesen Award. *Mem:* Inst Med-Nat Acad Sci; Am Psychiat Asn; Am Acad Child Psychiat; Am Psychoanalysis Asn. *Res:* Child psychiatry. *Mailing Add:* Yale Child Study Ctr Yale Univ Sch Med 230 S Frontage Rd PO Box 3333 New Haven CT 06510

COHEN, DONALD SUSSMAN, b Providence, RI, Nov 30, 34; m 58; c 2. APPLIED MATHEMATICS. *Educ:* Brown Univ, ScB, 56; Cornell Univ, MS, 59; NY Univ, PhD(math), 62. *Prof Exp:* Preceptorship, Dept Eng Mech & Inst Flight Struct, Columbia Univ, 62-63; asst prof math, Rensselaer Polytech Inst, 63-65; asst prof math, 65-67, assoc prof appl math, 67-71, PROF APPL MATH, CALIF INST TECHNOL, 71- *Concurrent Pos:* Ed, Soc Indust & Appl Math Rev. *Mem:* AAAS; Am Math Soc; Soc Indust & Appl Math; Sigma Xi. *Res:* Wave propagation and vibration problems; partial differential equations; special functions; variational techniques; non-linear boundary value problems; bifurcation theory; perturbation and asymptotic methods; differential equations. *Mailing Add:* Dept Appl Math Calif Inst Technol Pasadena CA 91125

COHEN, E RICHARD, b Philadelphia, Pa, Dec 14, 22; m 53; c 1. MATHEMATICAL PHYSICS, REACTOR PHYSICS. *Educ:* Univ Pa, AB, 43; Calif Inst Technol, MS, 46, PhD(physics), 49. *Prof Exp:* Asst instr physics, Univ Pa, 43-44; jr physicist acoust-electronic res, Calif Inst Technol, 44-45; theoret physicist, NAm Rockwell Corp, 49-56, res adv, 56-61, assoc dir, Res Dept, 61-62 & Sci Ctr, 62-69, mem tech staff, 69-75, DISTINGUISHED FEL, SCI CTR, ROCKWELL INT CORP, 75- *Concurrent Pos:* Sr lectr, Calif Inst Technol, 62-63, res assoc, 63-72; comn, IUPAP Symbols, Units, Nomenclature, Atomic Masses & Fundamental constants, 78-87, AIUPAC comn symbols, units & terminology, corresp mem, 83- *Honors & Awards:* E O Lawrence Award, 68. *Mem:* Fel AAAS; fel Am Phys Soc; fel Am Nuclear Soc; Asn Comput Mach; Sigma Xi. *Res:* Evaluation of the fundamental physical constants; nuclear reactor theory; molecular spectroscopy; electromagnetics; theory of measurement. *Mailing Add:* 17735 Corinthian Dr Encino CA 91316

COHEN, EDGAR A, JR, b Charleston, SC, Aug 29, 38; m 75. STOCHASTIC PROCESSES, SIGNAL PROCESSING. *Educ:* Duke Univ, AB, 60; Univ Cincinnati, MS, 64, PhD(math), 68. *Prof Exp:* Teaching asst math, Univ Fla, 60-61; Taft fel math, Univ Cincinnati, 62-67; off-campus teacher math, 68-69, MATHEMATICIAN, NAVAL SURFACE WEAPONS CTR, SILVER SPRING, MD, 67- *Mem:* Soc Indust & Appl Math; Asn Comput Mach. *Res:* Applied probability and statistics as related to naval defense issues; minefield effectiveness and strategic planning; modelling target dynamics, localization and target identification. *Mailing Add:* 5454 Marsh Hawk Way Columbia MD 21045

COHEN, EDWARD, b Glastonbury, Conn, Jan 6, 21; m 48, 81; c 5. CIVIL ENGINEERING. *Educ:* Columbia Univ, BS, 45, MS, 54. *Prof Exp:* Eng aide, Conn Hwy Dept, 40-42; asst engr, Dept Pub Works, East Hartford, Conn, 42-44; struct engr, Hardesty & Hanover, NY, 45-47 & Sanderson & Porter, NY, 47-49; vpres, Ammann & Whitney Int, Ltd, 63-73, vpres, Ammann & Whitney, Inc, 63-74, exec vpres, 74-77, partner, Ammann & Whitney Consult Engrs, 63-74, sr partner, 74-77, MANAGING PARTNER, AMMANN & WHITNEY CONSULT ENGRS, 77-, CHMN & CHIEF EXEC OFFICER, AMMANN & WHITNEY, INC, 77-; CHMN & CHIEF EXEC OFFICER, SAFEGUARD CONSTRUCT MGT CORP, 77- *Concurrent Pos:* Lectr archit, Columbia Univ, 48-51, Stanton Walker, Univ Md, 73, Univ Austin, 86, Johns Hopkins Univ, 88; consult, Rand Corp, Santa Monica, 58-72, Dept Defense, 62-63 & Hudson Inst, NY, 67-71, World Bank, 84, Tenn Valley Authority, 87-, Nat Trust Historic Preserv, NC, 90; chmn, Am Soc Consult Engrs Coun, 68-88, Reinforced Concrete Res Coun, 80-89, Building Res Coun, Comn Eng & Tech Systs, Nat Res Coun, 89-; mem, Nat Radio Astron Observ Tech Assessment Panel, WVa, 89; vpres, Eng Sci Div, NY Acad Sci, 91-, mem bd gov, 91-, chmn, Sci Sect Activ, 77-79. *Honors & Awards:* Illig Medal Appl Sci, Columbia Univ, 46, Egleston Medal, 81; Robert Ridgeway Award, Am Soc Civil Engrs, 46, State-of-the-Art Civil Eng Award, 74, Raymond Reese Award, 76, Ernest E Howard Gold Medal, 83, Serv to People Award, 87; Wason Medal, Am Concrete Inst, 56, Delmar L Bloem Award, 73; Laskowitz Gold Medal Aerospace Award, NY Acad Sci, 70, Academy Award, 89; Stanton Walker Lectr, Univ Md, 73; Goethals Medal, Soc Am Mil Engrs, 85; Grand Award for Eng Excellence, Am Consult Engrs Coun, 86; Henry M Shaw Lectr, NC State Univ, 87; Outstanding Eng Achievement, Nat Soc Prof Engrs, 87; US Presidential Award for Design Excellence, 88; Nat Historic Preserv Award for Eng, US Dept Interior, 88; Prize Bridge Award, Am Inst Steel Construct, 89. *Mem:* Nat Acad Eng; hon mem Am Soc Civil Engrs; hon mem Am Concrete Inst; fel Inst Civil Engrs Gt Brit; hon mem NY Acad Sci. *Res:* Buildings; bridges; guyed towers; structural design; shell structures; seismic design; hardened design; wind forces; dynamic analysis of structures, ultimate strength and plastic design; tower analysis; published in American and foreign professional journals and government manuals. *Mailing Add:* Ammann & Whitney Inc Consult Engrs 96 Morton St New York NY 10014

COHEN, EDWARD DAVID, b Haverhill, Mass, Mar 12, 37; m 58; c 3. PHYSICAL CHEMISTRY, CHEMICAL ENGINEERING. *Educ:* Tufts Univ, BSChE, 58; Univ Del, PhD(phys chem), 64. *Prof Exp:* Jr engr, Elkton Div, Thiokol Chem Corp, 58-60; res chemist, 64-67, sr res chemist, 67-72, res assoc, 72-77, res fel, 77-84, SR RES FEL, PHOTOPROD DEPT, E I DU PONT DE NEMOURS & CO, INC, 84- *Concurrent Pos:* adj prof, Ctr Interfacial Eng, Univ Minn. *Mem:* Am Inst Chem Engrs; Am Chem Soc; Soc Photog Sci & Eng; Am Asn Artificial Intel. *Res:* Rheology; radiation and polymer chemistry; analytical chemistry; gelatin and bio-polymers; emulsion chemistry; expert systems. *Mailing Add:* Du Pont Imaging Systs Photosysts Res Lab Parlin NJ 08859

COHEN, EDWARD HIRSCH, b Seattle, Wash, Aug 28, 47. RECOMBINANT DNA. *Educ:* Univ Chicago, BS, 68; Yale Univ, MPhil, 70, PhD(biol), 73. *Prof Exp:* Asst prof biol, Princeton Univ, 74-81; asst staff scientist, Fred Hutchinson Cancer Res Ctr, Seattle, 81-83; staff scientist, Intergrated Genetics, 83-89; SR RES SCIENTIST, DEPT MOLECULAR BIOL, DIAG PROD CORP, 89- *Concurrent Pos:* Fel zool, Univ Wash, 72-74. *Mem:* Am Soc Cell Biol; Genetics Soc Am; Am Soc Microbiol. *Res:* Organization and expression of DNA sequences in eukaryotic cells. *Mailing Add:* Dept Molecular Biol Diag Prod Corp 5700 W 96th St Los Angeles CA 90045

COHEN, EDWARD MORTON, b New York, NY, May 12, 36; m 59; c 3. PHARMACEUTICAL CHEMISTRY, ANALYTICAL CHEMISTRY. *Educ:* Columbia Univ, BS, 57; Rutgers Univ, MS, 60, PhD(pharmaceut chem), 65. *Prof Exp:* Res chemist, Johnson & Johnson Res Ctr, 62-65; res assoc, Merck Sharp & Dohme Res Labs, West Point, 65-70, sr res fel analytical chem, 70-78, dir, 76-81, sr dir pharmaceut res, 81-85; dir, Pharmaceut Res & Develop, Squibb Inst Med Res, 85-89; VPRES SCI OPERS, DANBURY PHARMACEUT, 89- *Mem:* AAAS; Am Chem Soc; Am Pharmaceut Asn. *Res:* Development of analytical methods for pharmaceutical dosage forms; measurement of physical properties of compounds; polarography and other electroanalytical methods; thin layer chromatography; thermal analysis; pharmaceutical formulations; quality control and quality assurance. *Mailing Add:* Four Plumtrees Rd Newton CT 06470-1730

COHEN, EDWARD P, b Glen Ridge, NJ, Sept 28, 32; m 63; c 4. IMMUNOLOGY, ALLERGY. *Educ:* Wash Univ, MD, 57; Am Bd Allergy & Immunol, cert, 74. *Prof Exp:* Intern med, Univ Chicago, 57-58; USPHS fel, NIH, 58-60; res fel, Univ Colo, 60-63, asst prof microbiol, 63-65; assoc prof, Rutgers Univ, 65-68; assoc prof, Univ Chicago, 68-76, prof microbiol, 76-79; dean, Sch Basic Med Sci, 79-82, dir, Off Res & Dev, 82-83, PROF MICROBIOL & IMMUNOL, UNIV ILL, 79- *Mem:* Am Asn Immunol; Am Soc Cell Biol; Am Acad Allergy. *Res:* Immune resistance to cancer. *Mailing Add:* Dept Microbiol Rm 109 Univ Ill Med Ctr 1852 W Polk St Chicago IL 60612

COHEN, EDWIN, b Cairo, Egypt, July 3, 34; US citizen; m 60; c 2. ELECTRICAL ENGINEERING, APPLIED MATHEMATICS. *Educ:* Cairo Univ, Egypt, BS, 57; Newark Col Eng, MS, 64; Polytech Inst NY, PhD(systs sci, elec eng), 70. *Prof Exp:* Engr elec power, Sadelmi, Ltd, Milan, 57-60; engr res & develop, Weston Instruments, 60-61; engr thermistors, Victory Eng, 61-62; mem fac, 62-81, PROF ELEC ENG, NJ INST TECHNOL, 71- *Concurrent Pos:* Consult, Calculagraph Co, 66, Kahle Eng Co, 73-74, Vector Eng, 78-80 & Systematics, 80-81; fel NASA, 81-82 & Pub Serv Elec & Gas, 82- *Mem:* Inst Elec & Electronics Engrs; Sigma Xi; Conf Int Grands Res Elec. *Res:* Computer application to analysis of power systems; approximation theory applied to electric filter design; programmable logic controllers for manufacturing machines. *Mailing Add:* 145 Milltown Rd Springfield NJ 07081

COHEN, EDWIN, b New York, NY, Aug 26, 24; m 49, 58; c 3. HUMAN FACTORS ENGINEERING, TRAINING SIMULATORS. *Educ:* Cornell Univ, AB, 44; Univ Okla, MS, 49, PhD(psychol), 55. *Prof Exp:* Res assoc to prog dir, Psychol Res Assocs, Washington, DC, 53-55; training specialist, Rand Corp, 55-56; proj dir, Educ Res Corp, Cambridge, Mass, 56-58; staff scientist & mgr, Human Factors Dept, Link Div, Singer Co, 58-86; RETIRED. *Concurrent Pos:* Adj fac, Simmons Col, 57-58, State Univ, NY, Binghamton, 68-73; chmn subcomt age 3B, Soc Automotive Engrs, 68-70; mem, Comt Simulation of Driving Task, Hwy Res Bd, 69-74; mem, NY State Bd Psychol, 77-87. *Mem:* Am Psychol Asn; Human Factors Soc. *Res:* Training, training equipment and vision. *Mailing Add:* Five Crestmont Rd Binghamton NY 13905-4115

COHEN, ELAINE, b NJ, July 17, 46; m 74; c 2. MECHANICAL ENGINEERING. *Educ:* Vassar Col, BA, 68; Syracuse Univ, MA, 70, PhD(math), 74. *Prof Exp:* res asst prof, 74-85, ASSOC PROF COMPUTER SCI, UNIV UTAH, 85- *Concurrent Pos:* Adj asst prof, Univ Utah, 76-82, adj assoc prof math, 82- *Mem:* Am Math Soc; Asn Comput Mach; Inst Elec & Electronics Engrs; Soc Indust & Appl Math. *Res:* Developing mathematical models for representing geometric shape information in computer; computer graphics; experimental computer-aided design and manufacturing systems; scientific visualization. *Mailing Add:* Comput Sci Univ Utah Salt Lake City UT 84112

COHEN, ELIAS, b Baltimore, Md, Sept 17, 20; m 46; c 2. IMMUNOLOGY. *Educ:* Univ Md, BSc, 42; Johns Hopkins Univ, MA, 49; Rutgers Univ, PhD(immunol), 52. *Prof Exp:* Biologist, Sch Hyg & Pub Health, Johns Hopkins Univ, 46, jr instr biol & asst, 47-49; jr instr, Rutgers Univ, 49-50, asst genetics & physiol, 50-52; instr clin path & asst dir clin labs, Sch Med, Univ Okla, 52-54, res assoc clin path, 54-56; assoc cancer res scientist immunohemat, Roswell Park Mem Inst, 56-69, prin cancer res scientist, 69-72, assoc chief cancer res scientist, 72-88, EMER CANCER SCIENTIST, ROSWELL PARK MEM INST, 88-; RES ASSOC PROF MICROBIOL & IMMUNOL, SCH MED, STATE UNIV BUFFALO, 68-, CLIN ASSOC PROF PATH, 77- *Concurrent Pos:* Lectr, Okla City Univ, 53-54; sr biologist, S R Noble Found, Inc, 54-56; lab consult, Erie County Lab, 57-64; lectr med genetics, Sch Med, State Univ NY Buffalo, 57-72; lectr, Queen's Univ, 58; mem acute leukemia task force & prof dir, Platelet Transfusion Eval Prog, Nat Cancer Inst, 64-76; ed newsletter, J Blood Banks Asn, NY, 66-; immunohemat consult, Children's Hosp, Buffalo, NY, 67-; US-USSR exchange scientist, 69, 73, 75, 76 & 85; pres, NY Publ Health Lab Asn, 74-75; consult, Nat Heart Lung, Blood Resources Inst, 84-85; vis prof, Tuskeegee Inst, 84; consult dir histocompatibility, Erie County Med Ctr, Buffalo, NY, 85-90; dir immunogenetics, Genetics Ctr, Scottsdale, Ariz, 88-89 & Genetrix, 90- *Mem:* Am Asn Path; Am Soc Hemat; Am Asn Immunol; Int Soc Blood Transfusion; Int Soc Hemat; Am Soc Histocompatability & Immunogenetics; Clin Immunol Soc. *Res:* Immunobiology and immunogenetics; comparative and human immunohematology; biomedical applications of erythrocytes, serum proteins and lectins; clinical immunology; histocompatibility for organ and bone marrow transplantation, disease associations, paternity testing. *Mailing Add:* Roswell Park Mem Inst 666 Elm St Buffalo NY 14263

COHEN, ELLIOTT, b New York, NY, Apr 14, 30; m 58; c 3. MEDICINAL CHEMISTRY. *Educ:* Syracuse Univ, BA, 51; Columbia Univ, PhD(chem), 56. *Prof Exp:* Chemist, NY State Psychiat Inst, 56-57; res chemist, 57-61, group leader org chem, 61-74, dept head, Cent Nerv Syst Dis Ther Sect, 74-77, DEPT HEAD, CARDIOVASC-CNS RES SECT, MED RES DIV, LEDERLE LABS, DIV AM CYANAMID CO, 77- *Mem:* Am Chem Soc. *Res:* Organic synthetic work in heterocyclic and cardiovascular drugs, geriatric agents and central nervous systems agents; mechanism of action and structure activity relationships. *Mailing Add:* Lederle Labs Div Am Cyanamid Co Pearl River NY 10965

COHEN, ELLIS N, b Des Moines, Iowa, June 5, 19; m 47; c 3. ANESTHESIOLOGY. *Educ:* Univ Minn, BS, 41, MD, 43, MS, 49. *Prof Exp:* From clin instr to assoc clin prof anesthesiol, Univ Minn, 49-60; from assoc prof to prof, 60-79, EMER PROF ANESTHESIOL, STANFORD UNIV, 79- *Mem:* AMA; Am Soc Anesthesiol. *Res:* Clinical pharmacology; toxicology; uptake and distribution and metabolism of drugs; analytic methods. *Mailing Add:* Dept Anesthesia Stanford Univ Sch Med 816 Lathrop Dr Palo Alto CA 94305

COHEN, EZECHIEL GODERT DAVID, b Amsterdam, Neth, Jan 16, 23; m 50; c 2. PHYSICS. *Educ:* Univ Amsterdam, BS, 47, PhD(physics), 57. *Prof Exp:* First asst physics, Univ Amsterdam, 50-61, assoc prof, 59-63; res assoc, Univ Mich, 57-58 & John Hopkins Univ, 58-59; PROF, ROCKEFELLER UNIV, 63- *Concurrent Pos:* Netherlands Orgn Pure Sci Res scholar, Univ Mich, 57-58 & Johns Hopkins Univ, 58-59; Van der Waals prof, Univ Amsterdam, 69; vis prof, Col France, Paris, 69, 72, 79, 83 & 90, Univ Del, 79 & Australian Nat Univ, 82, 88; Lorentz prof, Univ Leiden, Neth, 79; consult, Nat Bureau Standards & Los Alamos Nat Lab; ed, Physica & Physics Rep. *Mem:* Fel Am Phys Soc; Royal Dutch Acad Sci. *Res:* Statistical mechanics, particularly applied to equilibrium and nonequilibrium properties of gases and liquids at normal and low temperatures. *Mailing Add:* Rockefeller Univ 1230 York Ave New York NY 10021-6399

COHEN, FLOSSIE, b Calcutta, Brit India, May 10, 25; US citizen; m 58; c 1. IMMUNOLOGY, PEDIATRICS. *Educ:* Med Col, Calcutta, MB, 45; Univ Buffalo, MD, 50. *Prof Exp:* Hematologist, 56-58, from asst pediatrician & assoc hematologist to assoc pediatrician, 58- 73, DIR CLIN IMMUNOL & RHEUMATOL, CHILDREN'S HOSP MICH, DETROIT, 69- *Concurrent Pos:* Res assoc, Child Res Ctr Mich, Detroit, 58-60, sr res assoc, 60-; univ assoc, Dept Affil-Pediat, Hutzel Hosp, Detroit, 64-; attend pediatrician, Dept Pediat Med, Children's Hosp Mich, 71-, attend pediatrician, Dept Lab Med, Immunol Sect, 74-; immunol consult, William Beaumont Hosp, Royal Oak, Mich, 73-; mem biomet & epidemiol contract rev comt, Nat Cancer Inst, 74-78, chmn biomet & epidemiol contract rev comt, 77-78. *Mem:* Am Pediat Soc; Am Soc Hemat; Am Soc Human Genetics; Soc Pediat Res; Am Asn Immunologists. *Res:* Immunodeficiency diseases; pediatric rheumatological diseases; Acquired Immunodeficiency Syndrome (AIDS) in children. *Mailing Add:* Pediat Dept Sch Med Wayne State Univ Children's Hosp 3901 Beaubien Blvd Detroit MI 48201

COHEN, FREDRIC SUMNER, b Boston, Mass, Dec 17, 35; m 81; c 2. POLYMER CHEMISTRY, OPTICS. *Educ:* Oberlin Col, AB, 57; Brandeis Univ, PhD, 63; Mass Inst Technol, MSM, 76. *Prof Exp:* Res chemist, US Indust Chem Co, 59-60; sr res chemist, Diamond Alkali Co, 63-65; res chemist, Stauffer Chem Co, 65-67, sr res chemist, 67-69; scientist, Polaroid Corp, 69-70, res group leader, 71-80, tech serv mgr, 80-84, tech mgr, 84-85; EXEC VPRES, POWERCARD CORP, 85- *Mem:* Am Chem Soc; Soc Plastics Eng; Soc Rheol. *Res:* Physical polymer chemistry, especially related to optical systems and light polarizers; correlation of molecular and gross properties of synthetic polymers; polymer orientation and dyeing; electrochemical systems and materials. *Mailing Add:* Powercard Corp 393 Totten Pond Rd Waltham MA 02154-2014

COHEN, GARY H, b Brooklyn, NY, May 7, 34; m 59; c 2. MICROBIOLOGY, VIROLOGY. *Educ:* Brooklyn Col, BS, 56; Univ Vt, PhD(microbiol), 64. *Prof Exp:* Assoc prof, 67-81, PROF MICROBIOL, SCH DENT MED, UNIV PA, 81-, CHMN MICROBIOL, 85- *Concurrent Pos:* USPHS fel virol, Univ Pa, 64-67; USPHS res career develop award, 69-74; vis scientist, Swiss Inst Exp Cancer Res, Lausanne, 76-77; vis prof, Univ Lausanne, Switz. *Honors & Awards:* Fogarty Sr Scientist Award, 90. *Mem:* AAAS; Am Soc Virol; Am Soc Microbiol. *Res:* Soluble antigens of vaccinia virus-infected mammalian cells; antigens of herpes simplex virus; DNA synthesis in herpes-infected mammalian cells; ribonucleotide reductase; herpes simplex glycoproteins synthesis, processing and function; herpes infection in synchronized human cells; glycoprotein subunit vaccines; mapping of antigenic and functional sites on herpes simplex virus glycoproteins; immune response to herpes glycoproteins. *Mailing Add:* Dept Microbiol Univ Pa Sch Dent Med Philadelphia PA 19104

COHEN, GEORGE LESTER, b Brooklyn, NY, Dec 24, 39; m 62; c 3. PHYSICAL CHEMISTRY. *Educ:* Clarkson Col Technol, BS, 61, MS, 63; Univ Md, PhD(phys chem), 67. *Prof Exp:* Chemist, US Naval Ord Lab, Md, 62-67; proj supvr, Gillette Res Inst, 67-69; head phys chem res dept, Prod Div, Bristol-Myers, Inc, 67-76, mgr chem sci, 76-90; CONSULT, 90- *Concurrent Pos:* Adj asst prof, Rutgers Univ, Newark, 70-75; co-res dir, Ctr Prof Advan, 78- *Mem:* Am Chem Soc; Soc Cosmetic Chemists; Inst Elec & Electronics Engrs. *Res:* Chemistry of silver oxides, pharmaceutical, cosmetic, polymer, surface and colloid chemistry; laboratory computers. *Mailing Add:* 40 Gregory Lane Warren NJ 07059-5031

COHEN, GEORGE S(OL), electrical engineering, for more information see previous edition

COHEN, GERALD, b New York, NY, Feb 1, 30. NEUROCHEMISTRY. *Educ:* City Col New York, BS, 50; Columbia Univ, MA, 52, PhD(chem), 55. *Prof Exp:* Assoc biochem, Col Physicians & Surgeons, Columbia Univ, 54-65, asst prof, 65-73; RES PROF NEUROL, MT SINAI SCH MED, 73- *Honors & Awards:* Claude Bernard Sci Jour Award, Nat Soc Med Res, 68. *Mem:* Am Soc Biol Chem; Am Soc Pharmacol & Exp Therapeut; Am Chem Soc. *Res:* Biochemical pharmacology. *Mailing Add:* Mt Sinai Sch Med 100 St/5th Ave New York NY 10029

COHEN, GERALD H(OWARD), b Milwaukee, Wis, Oct 11, 22; m 46; c 2. ELECTRICAL ENGINEERING. *Educ:* Univ Wis, BS, 48, MS, 49, PhD(elec eng), 50. *Prof Exp:* Mem staff, Radiation Lab, Mass Inst Technol, 43; asst instr, Univ Wis, 49; res engr, Taylor Instrument Co, 50-58; prof, 58-84, EMER PROF ELEC ENG, UNIV ROCHESTER, 84- *Concurrent Pos:* Consult, Gen Dynamics, Worthington Corp, Bausch & Lomb, Inc, Transmation, Inc, Rochester Appl Sci Asn & Friden Corp; prof, Dept Ophthal-Elec Eng, Univ Rochester, 80- *Mem:* AAAS; NY Acad Sci; Inst Elec & Electronics Engrs. *Res:* Non-linear circuit analysis; ultrasonic physics; industrial automatic controls; biomedical engineering. *Mailing Add:* 701 S Winton Rd Rochester NY 14618

COHEN, GERALD STANLEY, b New York, NY, Nov 29, 26; m 51; c 2. BIOMEDICAL ENGINEERING. *Educ:* City Col New York, BS, 50; Univ Md, MS, 67; Univ NC, PhD, 79. *Prof Exp:* Electronic engr, Philco-Ford Corp, 50-58; electronic engr, Army Ord, Harry Diamond Labs, 58-60; chief electronic & elec eng sect, Div Res Servs, 60-68, CHIEF, HEALTH INFO SYSTS & TECH ASSESS, NAT CTR HEALTH SERV, 68- *Mem:* Instrument Soc Am; Inst Elec & Electronics Eng; Opers Res Soc Am. *Res:* Design and development of electronic instrumentation for medical research; computer applications in the delivery of health services. *Mailing Add:* 2101 E Jefferson St Rm 650 Rockville MD 20852

COHEN, GERALDINE H, b Suffern, NY, July 19, 42. VIROLOGY. *Educ:* Univ Miami, Fla, PhD(biol), 68; Univ Tex Med Br, Galveston, MD, 79; diplomate of the Am Bd Pediat, 84. *Prof Exp:* ASST CLIN PROF MED, UNIV TEX MED BR, 83-; STAFF PEDIATRICIAN, KELSEY-SEYBOLD CLIN, PA, 74- *Mem:* Am Soc Cell Biol; AAAS; AMA. *Mailing Add:* Kelsey-Seybold Clin PA 4290 Cypress Hill Dr Spring TX 77388

COHEN, GERSON H, b Philadelphia, Pa, July 8, 39; m 63; c 3. PROTEIN CRYSTALLOGRAPHY, COMPUTER GRAPHICS. *Educ:* Temple Univ, AB, 61; Cornell Univ, PhD(phys chem), 65. *Prof Exp:* RES CHEMIST, NIH, 65- *Concurrent Pos:* Vis scientist, Weizmann Inst Sci, 72 & 84. *Mem:* AAAS; Am Chem Soc; Am Crystallog Asn. *Res:* Computational problems in protein crystallography; computer graphics. *Mailing Add:* NIH Bldg 2 Rm 312 Bethesda MD 20892

COHEN, GLENN MILTON, b Elizabeth, NJ, Sept 8, 43; m 68. NEUROSCIENCES, EMBRYOLOGY. *Educ:* Rutgers Univ, BA, 65; Fla State Univ, PhD(physiol), 70. *Prof Exp:* Asst res scientist electron micros, Inst Rehab Med, New York Univ, 70-72; res fel motion sickness, NASA, 76-77; MEM STAFF, FLA INST TECHNOL, 77- *Concurrent Pos:* Res fel, Naval Air Sta, Pensacola, 84. *Mem:* Asn Res Otolaryngol; Electron Micros Soc Am; Southeast Electron Micros Soc; Soc Neurosci; Sigma Xi. *Res:* Normal and abnormal development of the ear; age-related (senescent) hearing losses (presbycusis). *Mailing Add:* Dept Biol Sci Fla Inst Technol 150 W University Blvd Melbourne FL 32901

COHEN, GLORIA, b Leeds, Eng, Jan 2, 30; m 51; c 3. INFORMATION SCIENCE. *Educ:* Univ Birmingham, Eng, DDS, 52. *Prof Exp:* Dental res fel, London Hosp Med Col, 52-53, clin asst, 53-55; gen dent practr, London, 55-67; info scientist, Data & Info Ctr, Res & Develop Div, 68-69, supvr biomed doc, 69-70, from asst mgr to mgr, 70-72, dir, Info Serv Dept, 72-76, dir, Corp Bus Res & Eval Dept, 76-78, CONSULT, INFO SERV, G D SEARLE & CO, 78- *Concurrent Pos:* Prog evaluator, NSF, 74-; mem task force sci & technol, Indust Res Inst, 74-78; mem, Pres Adv Comt Fed Policy & Progs, 79-80. *Mem:* AAAS; Am Soc Info Sci; Drug Info Asn; Am Rec Mgt Asn; Int Asn Dent Res; Am Dental Asn. *Res:* Health care administrative services. *Mailing Add:* 200 Kilpatrick Ave Wilmette IL 60091

COHEN, GORDON MARK, b Chicago, Ill, Jan 7, 48; US & Can citizen; m 78; c 3. SYNTHETIC POLYMER CHEMISTRY, MONOMER SYNTHESIS. *Educ:* McGill Univ, BSc, 69; Harvard Univ, AM, 70, PhD(chem), 74. *Prof Exp:* RES CHEMIST, E I DU PONT DE NEMOURS & CO INC, 74- *Mem:*

Am Chem Soc. *Res:* Preparative polymer chemistry and organic chemistry in polymer systems, mechanistic and synthetic aspects; control of polymer architecture through use of group transfer polymerization and other techniques; emulsion polymerization; polymer modificaton. *Mailing Add:* 627 Greythorne Rd Wynnewood PA 19096

COHEN, HARLEY, b Winnipeg, Manitoba, May 12, 33; m 56; c 3. THEORETICAL & APPLIED MECHANICS. *Educ:* Univ Manitoba, BSc, 56; Brown Univ, ScM, 58; Univ Minn, PhD(theory of elastic surfaces), 64. *Prof Exp:* Res engr, Boeing Airplane Co, 58-60; sr develop engr, Aeronaut Div, Honeywell Inc, 60-63, sr scientist, Corp Res Ctr, 64-65; asst prof aeronaut & eng mech, Univ Minn, 65-66; from assoc prof to prof, 66-83, dept head, 84-89, DISTINGUISHED PROF CIVIL ENG, UNIV MAN, 83-, DEAN FAC SCI, 89- *Concurrent Pos:* Nat Res Coun Can res grant, 66-86; James L Record vis prof, Univ Minn, 79; Killam vis scholar, Univ Calgary, 82; fel Brit Sci & Eng Res Coun & Univ Strathclyde, UK, 85; vis prof, Univ Pisa, Italy, 87, Italian CNR, Univ Torino, 90. *Mem:* Am Acad Mech; Soc Eng Sci; Soc Natural Philos. *Res:* Non-linear theories of rods and shells; directed elastic continua and high order theories of elasticity; continuum mechanics in general; nonlinear wave propagation in rods, plates and shells; over 85 publications in referred scientific journals; co-author research tract, 89. *Mailing Add:* Fac Sci Univ Man Winnipeg MB R3D 2E4 Can

COHEN, HAROLD KARL, b Trenton, NJ, Mar 12, 15; m 60; c 2. VETERINARY PATHOLOGY. *Educ:* City Col New York, BS, 38; Kans State Col, DVM, 47; Univ Wis, MS, 52, PhD(path), 55. *Prof Exp:* Instr vet path, Univ Wis, 50-55; vet pathologist, Ralph M Parsons Co, 55; vet, Eli Lilly & Co, 55-73, vet pathologist, 73-77; RETIRED. *Mem:* Am Col Vet Path. *Res:* Virology. *Mailing Add:* 6302 Brookline Dr Indianapolis IN 46220

COHEN, HAROLD P, b Brooklyn, NY, Sept 6, 24; m 57; c 3. BIOLOGICAL CHEMISTRY. *Educ:* City Col New York, BS, 48; Univ Iowa, MS, 51, PhD(biochem), 53. *Prof Exp:* Res assoc, Albert Einstein Med Ctr, 52-53; res assoc neurol, Med Sch, 53-55, asst prof, 55-63, ASSOC PROF NEUROL, COL MED SCI, UNIV MINN, MINNEAPOLIS, 65- *Mem:* AAAS; Am Soc Neurochem; Int Soc Neurochem; Am Soc Biol Chem. *Res:* Central nervous system metabolism; amino acid metabolism; biogenic amine detection and metabolism in CNS. *Mailing Add:* 6251D Magda Dr Maple Grove MN 55369

COHEN, HARRY, b Chicago, Ill, May 17, 16; m 44; c 1. CHEMISTRY. *Educ:* Univ Ill, BS, 38, MS, 39; Univ Wis, PhD(org chem), 41. *Prof Exp:* Fel, Univ Wis, 42-43; res chemist, Upjohn Co, Mich, 43-44; lectr org chem, 46-52, from asst prof to assoc prof chem, 52-65, PROF CHEM, ROOSEVELT UNIV, 65- *Concurrent Pos:* Res chemist, Armour & Co, 44-49. *Mem:* Am Chem Soc. *Res:* Organic synthesis; reactions of ketene diethylacetal; formation of heterocyclic compounds; formation of unsaturated compounds. *Mailing Add:* 430 S Michigan Chicago IL 60605-1301

COHEN, HARVEY JAY, b Brooklyn, NY, Oct 21, 40; m 64; c 2. ONCOLOGY, GERIATRICS. *Educ:* Brooklyn Col, BS, 61; Downstate Med Col, Brooklyn, MD, 65. *Prof Exp:* Chief med serv, 76-82, assoc chief staff educ, 82-84, DIR, GERIAT RES, EDUC & CLIN CTR, VA MED CTR, DURHAM, NC, 84-; PROF MED, MED CTR, DUKE UNIV, 80-, DIR, CTR STUDY AGING & HUMAN DEVELOP, 82- *Concurrent Pos:* Sect ed geriat biosci, J Am Geriat Soc, 84-87; Physician's Recognition Award, AMA, 84-87; Ed, J Geront, Med Sci, 88- *Mem:* Fel Am Geriat Soc; Am Soc Clin Oncol; Am Soc Hemat; Am Asn Cancer Res. *Res:* Interaction of aging and neoplasia; the immune system; disorders of immune function; immunoproliferative disorders. *Mailing Add:* Ctr Study Aging Duke Univ Med Ctr Box 3003 Durham NC 27710

COHEN, HASKELL, b Omaha, Nebr, Sept 12, 20; m 45; c 3. MATHEMATICS. *Educ:* Univ Omaha, AB, 42; Univ Chicago, SM, 47; Tulane Univ, PhD(math), 52. *Prof Exp:* Instr math, Univ Ala, 46-50; asst, Off Naval Res contract, Tulane Univ, 51-52; instr, Univ Tenn, 52-55; from asst prof to prof, La State Univ, 55-67; PROF MATH, UNIV MASS, AMHERST, 67- *Concurrent Pos:* Partic, US Air Force res contract, La State Univ, 56-57; consult, US Naval Ord Testing Sta, Calif, 58; mem, Inst Advan Study, 62. *Mem:* Am Math Soc; Math Asn Am. *Res:* Topology; fixed point theorems; dimension theory; topological semigroups. *Mailing Add:* Dept Math Univ Mass Amherst MA 01002

COHEN, HERBERT DANIEL, b New York, NY, Apr 27, 37; m 59. PHYSICS. *Educ:* Antioch Col, BS, 59; Stanford Univ, PhD(physics), 66. *Prof Exp:* Res assoc low temperature physics, Stanford Univ, 66-67; asst prof physics, Brandeis Univ, 67-72; assoc prof physics, Univ Vt, 72-77; assoc prof physics & assoc dean, arts & sci, State Univ NY, Binghamton, 77-; MEM STAFF, UNIV REDLANDS. *Concurrent Pos:* NSF fel, 66-67. *Mem:* AAAS; Am Asn Physics Teachers; Am Phys Soc. *Res:* Low temperature physics. *Mailing Add:* Ind Univ 1700 Mishawaka Ave South Bend IN 46634

COHEN, HERMAN, b New York, NY, Mar 1, 15; m 42; c 2. BIOCHEMISTRY. *Educ:* City Col New York, BS, 40; NY Univ, MS, 42, PhD, 51. *Prof Exp:* Res assoc, E R Squibb & Sons, 46-53; dir prod develop, Carter-Wallace Div, Princeton Labs Inc, 53-71, vpres & dir res, 71-76, sr vpres, Wampole Labs, 76-90; RETIRED. *Mem:* Fel AAAS; Soc Exp Biol & Med; Am Chem Soc; fel NY Acad Sci; Am Physiol Soc. *Res:* Endocrinology; estrogenic hormones, pituitary hormones; enzymes. *Mailing Add:* 26 Oxford Circle Skillman NJ 08558

COHEN, HERMAN JACOB, b New York, NY, Sept 18, 22. MATHEMATICS NUMBER THEORY. *Educ:* City Col New York, BA, 43; Univ Wis, MA, 46, PhD(math), 49. *Prof Exp:* Jr physicist, Nat Bur Stand, 43-44; asst math, Univ Wis, 45-49; instr, Tulane Univ, 49-50; Fulbright scholar, Univ Paris, 50-51; from instr to assoc prof, 54-70, PROF MATH, CITY COL NY, 70-, Fulbright Scholar, Univ Paris, 50-51. *Mem:* Am Math Soc; Math Asn Am. *Res:* General topology; plane continua; uniform spaces; theory of numbers; combinatorics. *Mailing Add:* Math Dept City Col New York NY 10031

COHEN, HIRSH G, b St Paul, Minn, Oct 6, 25; m 52; c 3. APPLIED MATHEMATICS. *Educ:* Univ Wis, BS, 47; Brown Univ, MS, 48, PhD(appl math), 50. *Prof Exp:* Asst prof eng res, Pa State Univ, 50-51; res assoc aeronaut eng, Israel Inst Technol, 51-53; asst prof math, Carnegie Inst Technol, 53-55; sr engr, NAm Aviation Co, 55; from assoc prof to prof math, Rensselaer Polytech Inst, 55-59; mgr appl math, Int Bus Mach Corp, 61-65, from asst dir to dir math sci dept, Int Bus Mach Res Ctr, 65-68, asst dir res, Int Bus Mach Res Div, 69-72, consult, to dir, 72-74, chmn res review bd, 74-81, dir, Yorktown Lab Opers, 81-83, vpres div opers, 85-88, consult to dir, 89; MEM STAFF, ALFRED P SLOAN FOUND, 89- *Concurrent Pos:* Fulbright vis res prof, Delft Univ Technol, 58-59; assoc scientist, Sloan-Kettering Inst Cancer Res, 64-70; consult biostatistician, Mem Hosp, 64-70; vis prof biomath, Med Col, Cornell Univ, 65-70; vis assoc appl math, Calif Inst Technol, 68-69; vis prof, Hebrew Univ, Jerusalem, 73-74; actg dir, Int Bus Mach Res, Zurich, 74 & 81; mem, Energy Research Adv Bd, Dept of Energy, 85; mem, Bd Math Sci, Nat Res Coun, 85-87, mem, Math Sci Educ Bd, 86-87. *Mem:* AAAS; Am Math Soc; Biophys Soc; Soc Indust & Appl Math (pres, 83-84); Sigma Xi. *Res:* Applied mathematical investigations in acoustics, vibration theory, hydrodynamics, cavitation, nonlinear differential equations; superconductivity theory; mathematical biology. *Mailing Add:* Alfred P Sloan Found 630 Fifth Ave New York NY 10111

COHEN, HOWARD DAVID, b San Francisco, Calif, Jan 10, 40; m 62; c 2. CHEMICAL PHYSICS, ENERGY CONVERSION. *Educ:* Univ Calif, Berkeley, BS, 62; Univ Chicago, PhD(chem), 65. *Prof Exp:* Res scientist, Nat Bur Standards, 65-66; corp appointee chem, Harvard Univ, 66-67; mem tech staff, NAm Rockwell Corp, 67-69; res scientist, Systs Sci & Software Corp, 69-74; res scientist, STD Res Corp, 74-75; sr staff mem, Mass Inst Technol 75-78; sr appln specialist, Analogic Corp, Wakefield, Mass, 78-84; appln mgr, Numerix Corp, Newton, Mass, 84-87; Alliant Comput Systs, Littleton, Mass, 87-88; sr analyst, Multiflow Comput, Wellesley, Mass, 88-89; consult, Gen Elec, Lynn, Mass, 89-90; SR APPLICATIONS ENGR, ANALOGIC CDA, PEABODY, MASS, 90- *Concurrent Pos:* Instr chem, Univ Calif Exten, 68. *Mem:* AAAS; Am Phys Soc; Soc Explor Geophysicists; Inst Elec & Electronics Engrs. *Res:* Hartree-Fock wave functions; polarizabilities; photoionization phenomena; molecular collisions; plasmas; numerical analysis; group theory; finite element methods; computers; magnetohydrodynamics power systems; array processors. *Mailing Add:* Four Hastings Rd Lexington MA 02173

COHEN, HOWARD JOSEPH, b New York, NY, Jan 12, 28; m 52; c 2. CHEMISTRY. *Educ:* City Col New York, BS, 54; George Washington Univ, MSA, 75. *Prof Exp:* Technician powder metall & inorg synthesis, Sylvania Elec Prod Inc, 52-55; res chemist, Nat Lead Co, 55-56 & US Indust Chem Co, 56-61; proj leader corp res, Glidden Co, 61-64 & chem res ctr, 64, sr chemist pigments & color group, 64-69, mkt res, 69-71, prod, 72-73, SR CHEMIST RES & DEVELOP, SCM CHEMICALS, 73- *Concurrent Pos:* Ed, The Chesapeake Chemist, 73-76; consult organometallic compounds; chmn, Md Sect, Am Chem Soc, 79. *Mem:* Am Chem Soc. *Res:* Organometallics; polymers; inorganics; metal organics; catalysts; pigments; process and product development of silica gels as pigments, desiccants, catalysts and additives for food, cosmetics and pharmaceuticals; beer; insecticides. *Mailing Add:* 7412 Shirley Rd Baltimore MD 21207

COHEN, HOWARD LIONEL, b New York, NY, May 27, 40; m 62; c 2. ASTRONOMY, ASTROPHYSICS. *Educ:* Univ Mich, BS, 62; Ind Univ, AM, 64, PhD(astron), 68. *Prof Exp:* Res asst & vis astronr, Lowell Observ, 64 & 66; from asst prof to assoc prof phys sci & astron, Univ Fla, 68-79, dir, Astron Teaching Observ, 80-86, grad coordr astron, 88-90, ASSOC PROF ASTRON, UNIV FLA, 79- *Concurrent Pos:* Mem adv bd, Sci-Expo Corp, Tucson, Ariz, 84-86; consult mkt & tech, Meade Instruments Corp, Costa Mesa, Calif, 85-86, regional sales mgr, 86. *Mem:* AAAS; Am Astron Soc; Royal Astron Soc; Astron Soc Pac; Int Planetarium Soc; Sigma Xi. *Res:* Photoelectric and photographic photometry of stars; spectroscopic and eclipsing binaries, variable stars; computer applications in astronomy; planetarium education; star clusters. *Mailing Add:* Dept Astron Univ Fla Space Sci Res Bldg 211 Gainesville FL 32611

COHEN, HOWARD MELVIN, b Ft Wayne, Ind, May 22, 36; m 57; c 2. SOLID STATE CHEMISTRY. *Educ:* George Washington Univ, BS, 58; Pa State Univ, PhD(geochem), 62. *Prof Exp:* Mem tech staff, 62-64, SUPVR, BELL LABS, INC, 65- *Mem:* Inst Elec & Electronics Engrs. *Res:* Glass technology; thermodynamics; defects in solids; phase equilibrium; high temperature chemistry of solids; magnetic oxides; thick film technology; printed wiring board technology. *Mailing Add:* 44 Tremont Terr Livingston NJ 07039

COHEN, HYMAN L, b New York, NY, Apr 11, 19; m 49; c 1. ORGANIC CHEMISTRY. *Educ:* City Col New York, BS, 39; Brooklyn Col, MA, 48; Univ Toronto, PhD(chem), 52. *Prof Exp:* Chemist, Felton Chem Co, NY, 46-47; res assoc, Jewish Hosp, Brooklyn, 47-48; chief chemist, Bell Craig, Ltd, Can, 48-49; res chemist, Glidden Co, Ill, 52-55; sr res chemist, 55-62, res assoc, Res Labs, Eastman Kodak Co, 62-82; vis prof, Weitzman Inst Sci, Rehovat, Israel, 83-84; RETIRED. *Mem:* Am Chem Soc. *Res:* Organometallic compounds; reactions of polymers. *Mailing Add:* 453 Seneca Pkwy Rochester NY 14613

COHEN, I BERNARD, b New York, NY, Mar 1, 14; m 44; c 1. HISTORY OF SCIENCE. *Educ:* Harvard Univ, SB, 37, PhD(hist sci), 47, LLD, 64. *Prof Exp:* Librn, Eliot House, Harvard Univ, 37-42, instr physics, 42-46 & hist sci, 46-49, from asst prof to assoc prof, 49-59, PROF HIST SCI, HARVARD UNIV, 59-, VICTOR S THOMAS PROF, 77- *Concurrent Pos:* Guggenheim fel, 56; NSF sr fel, 60-61; vis fel, Clare Hall, Cambridge Univ, 65; spec lectr, Univ London, 59; chmn, US Nat Comt Hist & Philos Sci, 61-62; vpres, Int Union Hist & Philos Sci. *Honors & Awards:* Lowell Lectr, Boston Univ, 61. *Mem:* Fel AAAS; Int Acad Hist Sci; Asn Hist Med; Am Acad Arts & Sci; Am Hist Sci Soc (pres, 61-62); Sigma Xi. *Res:* Newtonian science; science in America; effects of science on society. *Mailing Add:* Five Stella Rd Belmont MA 02178

COHEN, I KELMAN, b Troy, NY, March 30, 35; m 59; c 2. WOUND HEALING, PLASTIC & RECONSTRUCTIVE SURGERY. *Educ:* Columbia Univ, BS, 59; Univ NC, MD, 63. *Hon Degrees:* PhD, Kenyon Col, 84. *Prof Exp:* Resident surg, Univ NC, 64-68; resident plastic surg, Johns Hopkins Univ, 68-70; spec fel, Environ Therapeut, NIH, 70-72; PROF & CHMN PLASTIC SURG, MED COL VA, 72- *Concurrent Pos:* Kazanjian vis prof, Harvard Med Sch, 85 & vis prof, Univ Chicago, 86; chmn, Plastic Surg Res Coun. *Mem:* Am Surg Asn; Soc Univ Surgeons. *Res:* Wound healing and clinical and reconstructive surgery problems including keloid; collagen metabolism; inflamation wound hormones; fetal healing. *Mailing Add:* Med Col Va Commonwealth Univ Box 154 MCV Station Richmond VA 23298

COHEN, IRA M, b Chicago, Ill, July 18, 37; m 60; c 2. FLUID MECHANICS, MECHANICAL ENGINEERING. *Educ:* Polytech Univ, BAeroEng, 58; Princeton Univ, MA, 61, PhD(aeronaut eng, plasma physics), 63. *Hon Degrees:* MA, Univ Pa, 71. *Prof Exp:* Asst prof eng, Brown Univ, 63 66; from asst prof to assoc prof mech eng, 66-76, PROF MECH ENG, UNIV PA, 76- *Concurrent Pos:* Fulbright travel grant, 66; guest ord prof, Technische Hochschule Aachen, WGer, 66; mem tech staff, Sandia Lab, Alburquerque, NMex, 71, 74 & 77; consult, Joseph Oat Corp, Nat Air Oil Burner Co, Sylvan Pools, attys, 78- *Honors & Awards:* Spec Serv Citation, Am Inst Aeronaut & Astronaut, 87. *Mem:* Assoc fel Am Inst Aeronaut & Astronaut; Am Phys Soc; Sigma Xi; Am Asn Univ Prof; Am Soc Mech Engrs; Int Soc Hybrid Microelectronics. *Res:* Fluid mechanics; heat transfer; ionized gases; magnetohydrodynamics; microelectronics manufacturing. *Mailing Add:* Dept Mech Eng & Appl Mech Univ Pa Philadelphia PA 19104-6315

COHEN, IRVING ALLAN, b Brooklyn, NY, Nov 23, 44; c 4. ALCOHOLISM & DRUG ABUSE, HEALTH POLICY. *Educ:* Southeastern Univ, Washington, DC, BS, 75; Am Univ Caribbean, MD, 80; Johns Hopkins Univ, MPH, 83; Am Bd Prev Med, cert prev med & pub health, 86; cert, Am Med Soc Alcoholism & Other Drug Dependence, 87. *Prof Exp:* Resident physician internal med, Franklin Sq Hosp, Baltimore, 80-82; res physician, Sch Hyg & Pub Health, Johns Hopkins Univ, 82-84, chief resident prev med, 84-85; dep dir, NY State Res Inst Alcoholism, 85-88, dir, Tri-County Chem Dependency Progs, 88-89; CLIN ASST PROF MED, STATE UNIV NY, BUFFALO, 87-; ATTEND PHYSICIAN, DOWNTOWN CLIN, ERIE COUNTY MED CTR, 90- *Concurrent Pos:* Physician, var hosps & clins, Baltimore area, 82-85; mem, health risk appraisal coord comt, Off Dis Prev & Health Prom, US Dept Health & Human Serv, 84-; mem, alcoholism disability criteria task force, Soc Security Admin, 85-; pvt pract, Buffalo, NY, 88- *Mem:* Am Col Prev Med; Am Pub Health Asn; Am Soc Addiction Med; Inst Elec & Electronics Engrs; AMA; Am Col Int Physicians. *Res:* Improvement of efficacy of the alcoholism and drug abuse treatment system; improvement of primary and secondary prevention in the United States health system. *Mailing Add:* 220 Deleware Ave Suite 525 Buffalo NY 14202

COHEN, IRVING DAVID, b Brooklyn, NY, May 12, 45; m 66; c 4. ENVIRONMENTAL ENGINEERING, CHEMICAL ENGINEERING. *Educ:* City Col New York, BE, 67; NY Univ, ME, 70. *Prof Exp:* Sr process engr process design, Crawford & Russell, Inc, 67-71; assoc chem engr process design, Hoffmann-LaRoche, 71-72; sr proj mgr environ consult, Woodward-Envicon, Inc, 72-75; PRES ENVIRON CONSULT, ENVIRO-SCI INC, 75-; PRES ATMOSPHERIC MONITORING, AERO-INSTRUMENTATION RESOURCES, INC, 78- *Concurrent Pos:* Pres, ECRA Lab Inc. *Mem:* Am Inst Chem Engrs; Am Indust Hyg Asn; Asn Environ Prof; Air Pollution Control Asn; Int Asn Pollution Control. *Res:* Atmospheric emission treatment for submicron hydrocarbons; activated carbon in wastewater treatment; the environmental impacts associated with energy development and processes; insitu treatment of hazardous wastes spill. *Mailing Add:* 111 Howard Blvd Suite 108 Mt Arlington NJ 07856

COHEN, IRWIN, b Cleveland, Ohio, Feb 28, 24; m 45; c 3. MOLECULAR ORBITAL THEORY. *Educ:* Western Reserve Univ, AB, 44, MS, 48, PhD(chem), 50. *Prof Exp:* From asst prof to assoc prof, 49-58, PROF CHEM, YOUNGSTOWN STATE UNIV, 58- *Mem:* Am Chem Soc; Sigma Xi. *Res:* Molecular structure and bond order through molecular orbital population analysis; computer assisted instruction. *Mailing Add:* Dept Chem Youngstown State Univ Youngstown OH 44555

COHEN, IRWIN A, b New York, NY, Apr 28, 39. INORGANIC CHEMISTRY, LABORATORY COMPUTERS. *Educ:* Boston Univ, BA, 60; Northwestern Univ, PhD(inorg chem), 64. *Prof Exp:* Fel, Med Sch, Johns Hopkins Univ, 64-66; from asst prof to assoc prof chem, Polytech Inst New York, 66-74; assoc prof, 74-76, PROF CHEM, BROOKLYN COL, 76- *Mem:* Am Chem Soc. *Res:* Biochemically significant reactions of inorganic metal complexes, metalloporphyrins, organometallics; mechanisms of inorganic and biochemical reactions; laboratory utilization of micro-computers. *Mailing Add:* Dept Chem Brooklyn Col Brooklyn NY 11210

COHEN, ISAAC, b Cairo, Egypt, Nov 26, 36; US citizen. PLATELET BIOLOGY. *Educ:* Montpellier Univ, France, PhD, 61. *Prof Exp:* PROF BIOCHEM & ASSOC DIR ARTERIOSCLEROSIS, NORTHWESTERN UNIV, 78- *Mem:* Int Soc Thrombosis & Haemotasis; Am Soc Hemat; Am Heart Asn. *Res:* Several aspects of platelet biochemistry; microtubule actin interaction; platelet-fibrin interaction; membrane receptors. *Mailing Add:* Northwestern Univ 303 E Chicago Chicago IL 60611

COHEN, J(EROME) B(ERNARD), b Brooklyn, NY, July 16, 32; m 57; c 2. MATERIALS SCIENCE, CRYSTALLOGRAPHY. *Educ:* Mass Inst Technol, BS, 54, ScD(metall), 57. *Prof Exp:* Asst & res engr metall, Mass Inst Technol, 54, asst, 57; Fulbright scholar, Univ Paris, 57-58, sr scientist mat, Avco Corp, 58-59; from asst prof to assoc prof mat sci, 59-65, PROF MAT SCI, 65-, DEAN, TECHNOL INST, NORTHWESTERN UNIV, 86- *Honors & Awards:* Hardy medal, Am Inst Mining Metall & Petrol Engrs, 60; Westinghouse Prize, Am Soc Eng Educ, 76; Henry Marion Howe Medal, Am Soc Metals, 81, Engelhart Chair Mat Sci, 74. *Mem:* Fel Am Inst Mining Metall & Petrol Engrs; Am Crystallog Asn; Am Ceramic Soc; Royal Inst Gr Brit; fel Am Soc Metals. *Res:* X-ray diffraction; thermodynamics; physical metallurgy; ceramics; physics of solids; ordering; plastic deformation; clustering; catalysis. *Mailing Add:* Dept Mat Sci & Eng Northwestern Univ Evanston IL 60208-3100

COHEN, J CRAIG, b Alexandria, La, Oct 25, 50. VIROLOGY, GENETICS. *Educ:* Univ Miss, PhD(microbiol), 76. *Prof Exp:* ASSOC PROF MED & BIOCHEM, MED CTR, LA STATE UNIV, 82- *Mem:* Am Soc Microbiol; Am Soc Biol Chemists. *Mailing Add:* 97 Burgundy St New Orleans LA 70116

COHEN, JACK, b New York, NY, Jan 31, 37; m 59; c 3. PHARMACY, ANALYTICAL CHEMISTRY. *Educ:* Columbia Univ, BS, 57; Univ Iowa, MS, 59, PhD(pharm, analytical chem), 61. *Prof Exp:* Sr res chemist res, Lakeside Lab, Inc, 61-62; res chemist, Chas Pfizer & Co, Inc, Conn, 62-65; asst dir, Inst Pharmaceut Sci, 73-78, head, Pharmaceut Analysis Dept, 65-78, dir, 78-81, VPRES, QUAL ASSURANCE, SYNTEX LABS, INC, 81- *Mem:* Am Pharmaceut Asn; Am Soc Qual Control. *Res:* Pharmaceutical analysis; drug stability; quality control systems for pharmaceutical products. *Mailing Add:* Syntex Labs & Co 3401 Hillview Ave Palo Alto CA 94303

COHEN, JACK SIDNEY, b London, Eng, Sept 6, 38; US citizen; m 61; c 2. PHYSICAL BIOCHEMISTRY. *Educ:* Univ London, BSc, 61; Cambridge Univ, PhD(chem), 64. *Prof Exp:* Sci Res Coun UK fel, Weizman Inst, Israel, 64-66; fel, Harvard Med Sch, 66-67; sr res chemist, Merck Inst, 67-69; sr staff fel, Phys Sci Lab, Dir Comput Res & Technol, Nat Cancer Inst, NIH, 69-73, sr investr, Nat Inst Child Health & Develop, 73-83, head biophys pharmacol sect, Med Br, 83-90; PROF, DEPT PHARMACOL, GEORGETOWN UNIV MED SCH, WASHINGTON, DC, 90- *Concurrent Pos:* Mem, Pub Affairs Comt, Am Soc Biol Chemists, 80-90. *Mem:* Am Soc Biol Chemists; Am Chem Soc; Biophys Soc. *Res:* Pharmacology of oligonucleotides; Biophysical applications of stable isotopes; nuclear magnetic resonance studies of cellular metabolism; nucleotide and phosphorus chemistry. *Mailing Add:* Georgetown Univ Labs Four Research Ct Rockville MD 20850

COHEN, JACOB ISAAC, b Boston, Mass, Sept 8, 41; m 63; c 1. PHOTOGRAPHIC CHEMISTRY. *Educ:* Harvard Univ, BA, 63; Brandeis Univ, MA, 65, PhD(org chem), 67. *Prof Exp:* NIH fel, Dept Chem, Univ Chicago, 67-69; RES CHEMIST, EASTMAN KODAK CO, 69- *Mem:* Am Chem Soc; Soc Photog Sci & Eng. *Res:* Photographic research and development. *Mailing Add:* 69 Eastland Ave Rochester NY 14618-1026

COHEN, JAMES SAMUEL, b Houston, Tex, July 29, 46; m 68; c 2. ATOMIC & MOLECULAR PHYSICS, MUON PHYSICS. *Educ:* Rice Univ, BA, 68, MA, 70, PhD(physics), 73. *Prof Exp:* STAFF MEM, LOS ALAMOS NAT LAB, THEORET DIV, UNIV CALIF, 72- *Concurrent Pos:* Co-investr, New Res Initiative Prog, Los Alamos Sci Lab, 76-78; vis assoc prof, Physics Dept, Rice Univ, 79-80; vis scientist, Swiss Inst Nuclear Study, Villigen, Switz, 83, Ctr Nuclear Study, Saclay, France, 84, Joint Inst Nuclear Res, Dubna, USSR, 86; assoc ed, J Muon Catalyzed Fusion, 86-; deleg, US-USSR Fusion Exchange, Moscow, 87; prin investr, theory muon catalyzed fusion, Div Advan Energy Proj, US Dept Energy, 84-89. *Honors & Awards:* H A Wilson Award, Rice Univ, 73. *Mem:* Am Phys Soc; Sigma Xi. *Res:* Theoretical atomic and molecular physics, especially atom and molecule scattering, molecular structure, meson-molecule interactions, chemical kinetics; muon-catalyzed fusion. *Mailing Add:* 330 Valle del Sol Los Alamos NM 87544

COHEN, JAY O, b Jacksonville, Fla, Feb 17, 30; m 52; c 2. MEDICAL MICROBIOLOGY. *Educ:* Univ Fla, BS, 52, MS, 55; Purdue Univ, PhD(bact), 59. *Prof Exp:* MICROBIOLOGIST, CTR DIS CONTROL, USPHS, 59- *Mem:* Sigma Xi; Am Soc Microbiol; fel Am Acad Microbiol. *Res:* Coaglutination-identification of the O-antigens of Salmonella; immunogenicity of M protein of group A streptococci; natural antibodies for staphylococci in non-immunized animals; serological relationships of strains of staphylococci; epidemiology of staphyloccus disease; scientific writing and editing. *Mailing Add:* USPHS Ctr Dis Control Bldg 1 Rm 3237 Atlanta GA 30333

COHEN, JEFFREY M, b Elizabeth, NJ, Aug 30, 40; m 64; c 3. ASTROPHYSICS, THEORETICAL PHYSICS. *Educ:* Newark Col Eng, BS, 62; Yale Univ, MS, 63, PhD(physics), 66. *Prof Exp:* Res staff physicist, Yale Univ, 65-66, vis fel physics, 66-67; resident res assoc gravitation & astrophys, Inst Space Studies, New York, 67-69; mem physics, Inst Advan Study, NJ, 69-71; PROF PHYSICS, UNIV PA, 71- *Concurrent Pos:* Fel, Inst Space Studies, 66 & US AEC, 66-67; assoc, Nat Acad Sci-Nat Res Coun, 67-69; NSF grant, 70-76 & 77-; vis scientist, Max Planck Inst, 71; consult, Naval Res Lab, 74-; Air Force Off Sci Res grant, 78-; sr assoc, Nat Acad Sci & Lab High Energy Astrophys, NASA, 72-73. *Mem:* Int Astron Union; fel Am Phys Soc; Am Astron Soc; fel NY Acad Sci; Int Soc Gen Relativity & Gravitation; Sigma Xi. *Res:* Theoretical astrophysics; rotating bodies in general relativity; neutron star models and pulsars; gravitational collapse; relativistic astrophysics; cosmology. *Mailing Add:* Physics Dept Univ Pa Philadelphia PA 19104

COHEN, JOEL EPHRAIM, b Washington, DC, Feb 10, 44; m 70; c 2. POPULATION BIOLOGY, APPLIED MATHEMATICS. *Educ:* Harvard Univ, BA, 65, MA, 67, PhD(appl math), 70, MPH, 70, PhD, 73;. *Hon Degrees:* MA, Univ Cambridge, Eng, 74. *Prof Exp:* Asst & assoc prof biol, Harvard Univ, 71-75; PROF POP, ROCKEFELLER UNIV, 75- *Concurrent Pos:* Consult, Dept Math, Rand Corp, 68-74 & Computer Sci & Eng Bd, Nat Acad Sci, 69-73; mem, comt conserv nonhuman primates, Nat Res Coun, 71-74, comt Nat Statist, 82-85; lectr pop sci, Harvard Sch Pub Health, 71-73; chmn bd dirs, Soc Indust & Appl Math, 88-; fel, King's Col, Cambridge, Eng, 74-75; fel, Ctr Advan Study Behav Sci, 81-82; John Simon Guggenheim fel, 81-82; MacArthur fel, 81-86; vis prof statist, Stanford Univ, 82; mem, comt prob & policy, Social Sci Res Count, NY, 85-, educ adv bd, J S Guggenheim Found, 85-, comt selection, 90-; trustee, Russell Sage Found, 90-, Black Rock Forest Preserve, 90. *Honors & Awards:* Mercer Award, Ecol Soc Am, 72.

Mem: Fel AAAS; Am Soc Naturalists; Math Asn Am; Am Statist Asn; Soc Indust & Appl Math; Int Union Sci Study Pop. *Res:* Populations, epidemiology, demography, and ecology; applied mathematics especially linear algebra, stochastic processes, combinatorics, statistics and computing. *Mailing Add:* 500 E 63rd Apt 18C New York NY 10021

COHEN, JOEL M, b Worcester, Mass, Sept 27, 41; m 80. ALGEBRA. *Educ:* Brown Univ, ScB, 63; Mass Inst Technol, PhD(math), 66. *Prof Exp:* Instr math, Univ Chicago, 66-68; asst prof, Univ Pa, 68-75; assoc prof, 75-78, PROF MATH, UNIV MD, 78- *Concurrent Pos:* Vis prof, Univ Perugia, Italy, 77, 78 & 81 & Univ Rome, 80; prof algebra, Univ di Bari, Italy, 84-89. *Mem:* Am Math Soc; Math Asn Am; Ital Math Union. *Res:* Algebraic topology, chiefly low dimensional complexes and group theory; algebra and harmonic analysis, chiefly analysis on free groups and trees. *Mailing Add:* Dept Math Univ Md College Park MD 20742-4015

COHEN, JOEL RALPH, b Chelsea, Mass, Oct 20, 26; m 47; c 3. CLINICAL MICROBIOLOGY. *Educ:* Univ Mass, BS, 49, MS, 50, PhD(microbiol), 75. *Prof Exp:* Microbiologist & supvr labs, Springfield Hosp, 50-64, chief clin labs, 64-68; from assoc prof to prof biosci, 68-83, PROF BIOL & HEALTH SCI, SPRINGFIELD COL, 83-, COORDR HEALTH-RELATED PROGS, 87- *Concurrent Pos:* Col, US Army Res, 43-56, ret col, 86-, lectr microbiol, Sch Nursing, 51-53; vis lectr, Univ Mass, 52-62 & Springfield Col, 54-68; consult, Ludlow Hosp, 52-68, Noble Hosp, 52-74, Baystate Med Ctr, 77-, Wesson Maternity Hosp, 56-68, Springfield Munic Hosp, 60-, Vet Admin Hosp, Northampton, 63-85, Springfield Health Dept, 57-84 & Vet Admin, Washington, DC, 83-; regst & specialist microbiologist, Am Bd Microbiol. *Mem:* Fel AAAS; Sigma Xi; fel Am Pub Health Asn; Am Soc Microbiol; NY Acad Sci; fel Am Acad Microbiol; fel Royal Soc Health; Am Asn Blood Banks; Am Soc Allied Health Profs; Soc Armed Forces Lab Scientists. *Res:* Methods in clinical microbiology in area of incidence of intrahospital infections and microbial susceptibility to antibiotics; rapid identification of bacterial agents; history of public health laboratories and methodology in North America. *Mailing Add:* Allied Health Sci Ctr Springfield Col 263 Alden St Springfield MA 01109-3797

COHEN, JOEL SEYMOUR, b Baltimore, Md, Aug 27, 41. MATHEMATICS. *Educ:* Univ Md, BSEE, 64, PhD(math), 70. *Prof Exp:* ASST PROF MATH, UNIV DENVER, 69- *Mem:* Am Math Soc; Math Asn Am. *Res:* Functional analysis; Banach spaces; absolutely p-summing operators. *Mailing Add:* Dept Math Univ Denver Denver CO 80208

COHEN, JOHN DAVID, b Morristown, NJ, June 30, 46; m 81. AMORPHOUS SEMICONDUCTORS, MAGNETISM METALS. *Educ:* Univ Wash, BS, 68; Princeton Univ, MA, 73, PhD(physics), 76. *Prof Exp:* Res assoc, Univ Ill, Champaign-Urbana, 76-78; mem tech staff, AT&T Bell Labs, Murray Hill, NJ, 78-81; from asst prof to assoc prof, 81-89, PROF PHYSICS, UNIV ORE, EUGENE, 89- *Mem:* Am Phys Soc; Mat Res Soc. *Res:* Junction capacitance methods together with electron spin resonance to study the electronic properties of defect states in hydrogenated amorphous silicon and related alloys; influence of impurities on mobility gap states, on light-induced metastable defects and on defect distributions near interfaces. *Mailing Add:* Dept Physics Univ Ore Eugene OR 97403

COHEN, JONATHAN, b New York, NY, Feb 26, 15; m; c 4. PATHOLOGY. *Educ:* New York Univ, BS, 33; St Louis Univ, MD, 38. *Prof Exp:* Milton fel, Harvard Univ, 46, asst prof orthop surg, 50-70; prof, 70-88, EMER PROF ORTHOP SURG, TUFTS UNIV, 89- *Concurrent Pos:* Orthop pathologist, Children's Hosp Med Ctr, Boston, 50-; dep ed, J Bone & Joint Surg, 55-; guest dept metall, Radioactivity Ctr, Mass Inst Technol, 55-80. *Mem:* Am Acad Orthop Surg; Am Orthop Asn. *Res:* Interaction of tissues and metals; metabolism of bone seeking radio-isotopes. *Mailing Add:* 24 Grozier Rd Cambridge MA 02138-3315

COHEN, JONATHAN BREWER, b Akron, Ohio, Dec 17, 44. MOLECULAR PHARMACOLOGY. *Educ:* Harvard Col, BA, 66; Harvard Univ, MA, 67, PhD(chem), 72. *Prof Exp:* Res assoc neurobiol, Pasteur Inst, Paris, 71-74; asst prof, 75-80, ASSOC PROF PHARMACOL, HARVARD MED SCH, 80- *Concurrent Pos:* Lectr neurobiol, Ecole Normale Superieure, Paris, 73-74. *Mem:* Am Soc Pharmacol & Exp Therapeut; Biophys Soc; AAAS. *Res:* Molecular basis of synaptic function; mechanism of permeability control by acetycholine receptors; mode of action of drugs acting at cholinergic synapses; structural and functional properties of synaptic membranes. *Mailing Add:* 24 Grozier Rd Cambridge MA 02138-3315

COHEN, JORDAN J, b St Louis, Mo, June 18, 34; m 56; c 3. INTERNAL MEDICINE, NEPHROLOGY. *Educ:* Yale Univ, BA, 56; Harvard Univ, MD, 60. *Prof Exp:* Assoc physician, assoc div med res & dir div renal dis, RI Hosp, Providence, 65-71; PROF MED, MED SCH, TUFTS UNIV & DIR RENAL DIV, TUFTS-NEW ENG MED CTR, 71- *Concurrent Pos:* From asst prof to assoc prof, Brown Univ, 65-71; instr med, Harvard Med Sch, 67-71. *Mem:* Fedn Clin Res; Am Soc Nephrology; Am Soc Internal Med. *Res:* Renal mechanisms involved in acid-base homeostasis; application of computer techniques to simulation analyses of body fluid; electrolyte physiology. *Mailing Add:* SUNY Stony Brook Health Sci Ctr Stony Brook NY 11794-8430

COHEN, JUDITH GAMORA, b NY, May 5, 46; m 73. ASTROPHYSICS. *Educ:* Radcliffe Col, BA, 67; Calif Inst Technol, MS, 69, PhD(astron), 71. *Prof Exp:* Miller fel astron, Univ Calif, Berkeley, 71-73; asst astronomer, Kitt Peak Nat Observ, 73-77, assoc astronomer, 77-79; assoc prof, 79-87, PROF ASTRON, CALIF INST TECHNOL, 88- *Mem:* Am Astron Soc; Int Astron Union. *Res:* Nucleosynthesis, interstellar medium, globular clusters. *Mailing Add:* Calif Inst Technol Mail Code 105-24 Pasadena CA 91125

COHEN, JULES, b Brooklyn, NY, Aug 26, 31; m 56; c 3. INTERNAL MEDICINE, CARDIOLOGY. *Educ:* Univ Rochester, AB, 53, MD, 57. *Prof Exp:* Intern, Beth Israel Hosp, 57-58; resident, Strong Mem Hosp, 58-59; res assoc, NIH, 60-62; res asst, Royal Postgrad Md Sch London, 62-63; sr instr med, 64-65, from asst prof to assoc prof, 66-73, PROF MED, UNIV ROCHESTER, 73- *Concurrent Pos:* USPHS trainee med, Med Ctr, Univ Rochester, 59-60, res fel, 64-65, sr assoc dean, Sch Med, 82-; USPHS res grants, 63-69 & 75-78; Am Heart Asn res grant, 70-72; physician-in-chief, Rochester Gen Hosp, 76-82. *Mem:* Am Physiol Soc; Am Fedn Clin Res; Royal Soc Med; Am Heart Asn; Am Col Physicians; Am Col Cardiol. *Res:* Cardiac hypertrophy; hemoglobin function and tissue oxygenation; cardiomyopathies; medical education. *Mailing Add:* Strong Mem Hosp Box 601 Rochester NY 14642

COHEN, JULES BERNARD, b New York, NY, June 12, 33; m 56; c 2. ENVIRONMENTAL HEALTH ENGINEERING. *Educ:* City Col New York, BCE, 55; Univ Colo, MS, 58; Calif Inst Technol, PhD(environ health eng), 65. *Prof Exp:* Lectr civil eng, City Col New York, 55-56; staff engr, USPHS, Colo, 56-59 & Ohio, 59-62, sr staff engr, 65-66; sect chief environ eng, Arctic Health Res Ctr, 66-71, br chief environ sci, 71-73; tech coordr, Nat Enforcement Invests Ctr, Environ Protection Agency, 73-78, dep asst dir, 78-79; lab dir & vpres, Environ Lab, Sverdrup Technol Inc, 79-83, vpres Environ Div, Sverdrup Corp, 83-89; vpres eng, Hall Kimbrell Environ Serv, 89, VPRES PSI/HALL KIMBRELL DIV, 90- *Concurrent Pos:* Assoc prof, Univ Alaska, 71-73; consult, Environ Protection Agency Sci Adv Bd, Environ Eng Rev Panel; mem, Tenn Air Pollution Control Bd; dipl, Am Acad Environ Engrs. *Honors & Awards:* Bronze Medal, US Environ Protection Agency, 78. *Mem:* Am Soc Civil Engrs; Water Pollution Control Fedn; Air & Waste Mgt Asn; Sigma Xi. *Res:* Water and air pollution control; stream sanitation; environmental health engineering aspects of the adaptation of man to life in the Arctic and sub-Arctic. *Mailing Add:* 1601 Inverness Dr Lawrence KS 66074

COHEN, JULIUS, b Brooklyn, NY, Apr 16, 26. PHYSICS. *Educ:* NY Univ, AB, 50; Syracuse Univ, MS, 53. *Prof Exp:* Res assoc thin films, Syracuse Univ, 51-52; jr engr transistor physics, Gen Tel & Electronics Labs, 52-53, adv res engr emission & thin films, 54-67, eng specialist, 67-69; PHYSICIST, NAT BUR STANDARDS, 69- *Concurrent Pos:* Assoc physicist thermistor bolometers, Bulova Res & Develop Labs, 53-54; Japanese Ministry Educ guest scholar tunnel emission, Japan & Osaka Univs, 64-65. *Mem:* Am Phys Soc; Sigma Xi. *Res:* Thin films; semiconductors; electrical conductivity; emission; vacuum techniques; photoconductivity; piezoelectricity and pyroelectricity in polymers; systems analysis; optical instrumentation; physics engineering; infared; noise; thermography; Fourier spectroscopy. *Mailing Add:* MET B306 Nat Bur Standards Gaithersburg MD 20899

COHEN, JULIUS JAY, b Newark, NJ, Apr 26, 23; m 52; c 2. PHYSIOLOGY. *Educ:* Rutgers Univ, BS, 45; NY Univ, MD, 48. *Prof Exp:* Intern, Cincinnati Gen Hosp, 48-49; from jr to sr resident, Dept Med, Col Med, Univ Cincinnati, 49-51, res fel dept clin physiol, 51 & dept physiol, 53-54, instr physiol, 54-55, asst prof physiol & instr med, 55-59; assoc prof physiol, 59-66, actg chmn dept physiol, 67-68, PROF PHYSIOL, SCH MED & DENT, UNIV ROCHESTER, 66- *Concurrent Pos:* Markle Found med sci scholar, 55-60; mem physiol study sect, NIH, 65-70. *Mem:* AAAS; Soc Exp Biol & Med; Am Heart Asn; Am Physiol Soc. *Res:* Renal physiology and metabolism; relationships between intermediary metabolism of the kidney and its excretory function; comparative physiology; pathological physiology; isotopes. *Mailing Add:* Dept Med Box 601 Rochester Sch Med & Dent 601 Elmwood Ave Rochester NY 14642

COHEN, KARL (PALEY), b New York, NY, Feb 5, 13; m 38; c 3. NUCLEAR SCIENCE. *Educ:* Columbia Univ, AB, 33, AM, 34, PhD(phys chem), 37. *Prof Exp:* From asst to prof, Columbia Univ, 38-40, dir, Theoret Div SAM Labs, 40-44; head theoret physics group, Standard Oil Develop Co, 44-48; tech dir, H K Ferguson Co, 48-52; vpres, Walter Kidde Nuclear Labs, Inc, 52-55; mgr adv eng, Atomic Power Equip Dept, Gen Elec Co, 55-65, gen mgr, Breeder Reactor Dept, 65-71, mgr, Oper Planning, 71-73, chief scientist, Nuclear Energy Div, 73-78; CONSULT NUCLEAR ENERGY, 78- *Concurrent Pos:* Regents lectr, Univ Calif, Berkeley, 70; dir, US Nat Comt, World Energy Conf, 72-78; mem adv comt, Energy Proj, Int Inst Appl Systs Anal, Vienna, 76-81; mem adv coun, Ctr Theoret Studies, Univ Miami, 77-82; consult prof, Stanford Univ, 78-81. *Honors & Awards:* Krupp Prize for Energy Res, 77; Chem Pioneer Award, 79. *Mem:* Nat Acad Eng; fel AAAS; Am Phys Soc; fel Am Nuclear Soc (treas, 55-57, pres, 68-69); Inst Elec & Electronics Engrs. *Res:* Applied nuclear energy; isotope separation; gaseous diffusion; gas centrifuges; fast breeder reactors. *Mailing Add:* 928 N California Ave Palo Alto CA 94303-3405

COHEN, KENNETH SAMUEL, b Los Angeles, Calif, May 30, 37; m 61; c 4. INDUSTRIAL HYGIENE, TOXICOLOGY. *Educ:* San Diego State Univ, BS, 65; Calif Western Univ, PhD(occup health), 76. *Prof Exp:* Lab mgr, clin path, NIH Primate Res Colony, 63-65; mgr safety & qual anal, Biol Assocs, Inc, 67-72; dir indust hyg, Micronomics Int, Inc, 72-76; acting head indust hyg, Naval Regional Med Ctr, 76-78; TOXICOLOGIST & LECTR, OCCUP HEALTH, CONSULT HEALTH SERVS, 78- *Concurrent Pos:* Mem Toxicol & Textbook Comts, Am Indust Hyg Asn, 76-; mem educ comt, Am Soc Safety Engrs, 76-; mem Am Col Toxicol, 78- *Mem:* Am Pub Health Asn; Soc Occup Environ Health; Nat Fire Protection Asn; Am Conf Govt Indust Hygienists. *Res:* Toxic vapor simulation of industrial ventilation systems using SF_6 as an inert tracer gas to model emissions which are very toxic or difficult to measure. *Mailing Add:* PO Box 1625 El Cajon CA 92022

COHEN, LARRY WILLIAM, b Winnipeg, Man, Dec 24, 36; US citizen; m 59; c 2. GENETICS, MOLECULAR BIOLOGY. *Educ:* Univ Calif, Los Angeles, BA, 60, MA, 62, PhD(zool), 64. *Prof Exp:* NIH fel, Univ Calif, San Diego & Med Sch Univ Mich, 63-65; asst prof, Dept Biol Sci, Douglass Col, Rutgers Univ, 65-67; from asst prof to prof biol, Pomona Col, 67-89, chmn dept, 73-76, chmn dept biol, 81-83; PROF BIOL, CALIF STATE UNIV,

SAN MARCOS, 89- *Concurrent Pos:* Vis prof, Biol Dept, Univ Calif, 67 & 77; NIH spec fel, Univ Calif, Riverside, 71; vis prof, Dept Molecular Virol, Hadassa Med Sch, Hebrew Univ, Jerusalem, 74; vis scientist, City of Hope Nat Med Ctr, 78 & 81; vis prof, Rijksuniversiteit Groningen, Netherlands, 87. *Mem:* AAAS; Am Soc Microbiol; Genetics Soc Am; Protein Soc. *Res:* Microbial genetics; cell to cell agglutination phenomena; lysis in salmonella phage P22; recombinant DNA - directed mutagenesis for protein engineering. *Mailing Add:* Col Arts & Sci Calif State Univ 820 W Los Vallectios San Marcos CA 92096

COHEN, LAWRENCE, b Leeds, Eng, Nov 23, 26; m 51; c 3. ORAL MEDICINE, ORAL PATHOLOGY. *Educ:* Univ Leeds, BChD, 49, Univ London, MD, 56, PhD(histochem), 66. *Prof Exp:* Resident oral surg, Middlesex Hosp, London, Eng, 56-59; sr resident, Plastic & Jaw Unit, Stoke Mandeville Hosp, Buckinghamshire, 60-61; sr resident, Univ Col Hosp Dent Sch, 61-62, lectr oral path, Inst Dent Surg, London, 62-63, sr lectr oral med, 63-67; prof oral diag & head dept, Univ Ill Med Ctr, 67-76; chmn dept dent, Ill Masonic Med Ctr, 76-86; PRES, MED EMERGENCY SERV, BUFFALO GROVE, ILL, 86- *Concurrent Pos:* Consult, West Side Vet Admin Hosp, Chicago, Ill, 68-74 & Fed Trade Comn, 84-; vis prof med, Univ Chicago, 74-75; chmn panel rev of oral cavity drug preparations, Food & Drug Admin, 74-80. *Honors & Awards:* Samuel Charles Miller Award, Am Acad Oral Med, 80. *Mem:* Fel Royal Soc Med; Brit Bone & Tooth Soc; Int Asn Dent Res; Am Acad Oral Med (vpres, 82-83 & pres, 85-86). *Res:* Diseases of the oral mucosa. *Mailing Add:* 200 Kilpatrick Wilmette IL 60091

COHEN, LAWRENCE BARUCH, b Indianapolis, Ind, June 18, 39; m; c 2. NEUROPHYSIOLOGY. *Educ:* Univ Chicago, BS, 61; Columbia Univ, PhD(zool), 65. *Prof Exp:* From asst prof to assoc prof, 68-78, PROF PHYSIOL, SCH MED, YALE UNIV, 78- *Concurrent Pos:* NSF fel, Agr Res Coun Inst Physiol, Cambridge, Eng, 66-68. *Honors & Awards:* McMaster Award, Columbia Univ, 65; Elizabeth Roberts Cole Prize, Biophys Soc, 87. *Mem:* Soc Neurosci; Soc Gen Physiol; Biophys Soc. *Res:* Optical methods for measuring neuron activity in central nervous systems. *Mailing Add:* Dept Physiol Yale Univ Sch Med New Haven CT 06510

COHEN, LAWRENCE MARK, b New York, NY. OPTICAL DIAGNOSTICS, FLOWFIELD IMAGING. *Educ:* State Univ NY Buffalo, BS, 78; Mass Inst Technol, MS, 80; Stanford Univ, PhD(mech eng), 91. *Prof Exp:* Analytical engr, Pratt & Whitney Aircraft Co, 78; mem tech staff, TRW Inc, Space & Defense Sector, 80-84; res asst, Stanford Univ, 84-90; SR ENGR, AEROJET PROPULSION DIV, 91- *Mem:* Assoc mem Am Soc Mech Engrs; Optical Soc Am; sr mem Am Inst Aeronaut & Astronaut. *Res:* Nonintrusive measurements of scalor and vector properties of complex flows using optical diagnostics; experimental and analytical studies of combustion, supersonic flows and reacting environments. *Mailing Add:* 102 Soliday Court Folsom CA 95630

COHEN, LAWRENCE SOREL, b New York, NY, Mar 27, 33; m 61; c 2. INTERNAL MEDICINE, CARDIOLOGY. *Educ:* Harvard Col, AB, 54; NY Univ, MD, 58. *Hon Degrees:* MA, Yale Univ, 70. *Prof Exp:* Fel cardiol, Harvard Univ, 62-64; sr investr cardiol, NIH, 65-68; assoc prof med, Univ Tex Southwestern Med Sch Dallas, 68-70; prof, 70-81, EBENEZER K HUNT PROF MED, SCH MED, YALE UNIV, 81- *Concurrent Pos:* Attend physician, Yale-New Haven Med Ctr, 70- *Mem:* Am Fedn Clin Res; fel Am Col Cardiol; fel Am Col Physicians; Am Heart Asn. *Res:* Coronary artery disease; hemodynamics; radionuclide myocardial perfusion. *Mailing Add:* Dept Med Yale Univ Sch Med New Haven CT 06510

COHEN, LEONARD A, b New York, NY, Mar 21, 39; m 67; c 2. CELL BIOLOGY. *Educ:* Univ Wis-Madison, BS, 60; City Univ New York, PhD(biol), 72. *Prof Exp:* Res asst microbiol, Kingsbrooke Jewish Med Ctr, 60-62; res asst cell biol, Albert Einstein Col Med, 63-66; instr biol, City Univ New York, 67-72; res assoc cell biol, 73-78, & Div Nutrit, 78-84, SECT HEAD NUTRIT ENDOCRINOL, AM HEALTH FOUND, 84- *Mem:* AAAS; Sigma Xi; Am Asn Cancer Res; Tissue Cult Asn; Am Oil Chem Soc; Int Asn Breast Cancer Res. *Res:* The influence of dietary fat on mammary tumor development in rodent models mechanisms of action; analysis of intracellular response systems in cultured and neoplastic mammary epithelial cells. *Mailing Add:* 279 King St Chappaqua NY 10514

COHEN, LEONARD ARLIN, physiology; deceased, see previous edition for last biography

COHEN, LEONARD DAVID, b Philadelphia, Pa, Aug 7, 32; m 57; c 3. NUCLEAR PHYSICS, REACTOR PHYSICS. *Educ:* Univ Pa, BA, 54, MS, 56, PhD(physics), 59. *Prof Exp:* Reactor physicist, Knolls Atomic Power Lab, Gen Elec Co, 59-62, geophysicist, Space Sci Lab, 62-64; asst prof, 64-73, ASSOC PROF PHYSICS, DREXEL UNIV, 73- *Mem:* Am Phys Soc; Am Nuclear Soc; Am Geophys Union. *Res:* Space radiation physics; atmospheric physics using radon gas as a tracer of atmospheric turbulence; light scattering by non spherical atmospheric aerosols. *Mailing Add:* Dept Physics Drexel Univ Philadelphia PA 19104

COHEN, LEONARD GEORGE, b Brooklyn, NY, Feb 28, 41. COMMUNICATIONS, PLASMA PHYSICS. *Educ:* City Col New York, BEE, 62; Brown Univ, ScM, 64, PhD(plasma physics), 68. *Prof Exp:* Res asst eng, Brown Univ, 64-66, plasma physics, 66-68; MEM TECH STAFF, GUIDED WAVE RES LAB, BELL TEL LABS, 68- *Mem:* Sr mem Inst Elec & Electronics Engrs; Optical Soc Am; Sigma Xi. *Res:* Experimental and theoretical studies of the interaction between collisionless plasmas and electromagnetic fields; measured attenuation and depolarization of light transmitted along glass fibers; optical communications. *Mailing Add:* 69 Highland Circle Berkeley Heights NJ 07922

COHEN, LEONARD HARVEY, b Winnipeg, Man, Mar 19, 25; m 49; c 2. BIOCHEMISTRY. *Educ:* Univ Man, BSc, 48, MSc, 51; Univ Toronto, PhD(biochem), 54. *Prof Exp:* Asst pharmacol, Yale Univ, 54-55; res scientist, Roswell Mem Inst, 55-59; from asst prof pharmacol to prof biochem, Univ Man, 59-65; SR MEM, INST CANCER RES, 65- *Concurrent Pos:* Assoc prof phys biochem, Univ Pa, 65-81, adj prof, 81- *Mem:* Am Soc Biol Chemists; Am Soc Cell Biol; AAAS; Am Soc Develop Biol. *Res:* Development, differentiation; enzymology; chromosomal proteins; histones, gene regulation; cell cycle control. *Mailing Add:* Fox Chase Cancer Ctr Inst Cancer Res Philadelphia PA 19111

COHEN, LESLIE, b Baltimore, Md, Jan 14, 23; m 67; c 1. NUCLEAR PHYSICS. *Educ:* Johns Hopkins Univ, BA, 44, PhD(physics), 52. *Prof Exp:* Physicist, Bur Standards, 44-45; instr physics, Loyola Col, 47-48; jr instr, Johns Hopkins Univ, 48-51, asst, 51-52; res assoc, Knolls Atomic Power Lab, Gen Elec Co, 53-55; res assoc, Nuclear Physics Div, 55-81, Naval Res Lab, 55-85, exp phys, Condensed Matter & Radiation Sci Div, 81-85; RES STAFF MEM, SCI & TECHNOL DIV, INST DEFENSE ANALYSIS, 85- *Concurrent Pos:* Vis scientist, Optical Sci Ctr, Univ Ariz, 75-76. *Mem:* Am Phys Soc. *Res:* Nuclear spectroscopy; nuclear reactors; photonuclear and charged particle reactions; lasers. *Mailing Add:* 8801 Mansion Fram Place Alexandria VA 22309

COHEN, LEWIS H, b Dallas, Tex, Jan 2, 37. GEOCHEMISTRY. *Educ:* Mass Inst Technol, BS, 58; Univ Calif, Berkeley, MS, 61; Univ Calif, San Diego, PhD(earth sci), 65. *Prof Exp:* Asst prof geophys eng, Univ Calif, Berkeley, 65-66; from asst prof to assoc prof, 69-76, PROF GEOL, UNIV CALIF, RIVERSIDE, 76- *Mem:* Am Geophys Union; Geochem Soc; Mineral Soc Am. *Res:* Chemistry and physics at elevated temperatures and pressures; physical properties of rocks; radioactive waste disposal. *Mailing Add:* Dept Earth Sci Univ Calif Riverside CA 92521

COHEN, LOIS K, DENTAL RESEARCH. *Prof Exp:* DIR, DENT RES EXTRAMURAL PROG, NAT INST DENT RES, NIH, 89- *Mailing Add:* NIH Nat Inst Dent Res Extramural Prog Westwood Bldg Rm 503 5333 Westbard Ave Bethesda MD 20892

COHEN, LOUIS, b Chicago, Ill, Dec 5, 28; m 52; c 3. INTERNAL MEDICINE, CARDIOLOGY. *Educ:* Univ Chicago, BS, 48, MD, 53. *Prof Exp:* Res asst, Univ Chicago, 49-53, from intern to resident med, 53-55 & 57-58, from instr to assoc prof, 59-75 PROF MED, UNIV CHICAGO, 75- ATTEND PHYSICIAN & CONSULT CARDIOL & DIR AMBULATORY CARDIOL, UNIV CHICAGO HOSPS & CLINS, 61- *Concurrent Pos:* Nat Heart Inst trainee, 55; Am Heart Asn res fel, 58-60, advan res fel, 60-62; vis prof, Shaare Zedek Hosp, Jerusalem, Israel, 66, Sacred Heart Hosp, Eugene, Ore, 68, Univ Hawaii, 68, Pahlavi Univ, Iran, 77 & Hasharon Hosp, 78 & 80 & 83. *Mem:* Fel Am Col Physicians; fel Am Col Cardiol; fel Am Col Clin Pharmacol; Am Soc Pharmacol Exp Therapeut; Am Heart Asn; Sigma Xi. *Res:* Lipoproteins structure; experimental and clinical atherosclerosis; diagnostic enzymology; tissue culture; treatment and basic mechanisms in muscular dystrophy; myocardial infarction and salvage; regression of arteriosclerosis; mechanism of action of cardiac drugs; mechanism of sudden death. *Mailing Add:* Sch Med Univ Chicago Box 401 950 E 59th St Chicago IL 60637

COHEN, LOUIS ARTHUR, b Boston, Mass, July 12, 26; m 55; c 1. PHYSICAL ORGANIC CHEMISTRY, BIO-ORGANIC CHEMISTRY. *Educ:* Northeastern Univ, BS, 49; Mass Inst Technol, PhD(chem), 52. *Prof Exp:* Instr biochem, Med Sch, Yale Univ, 52-54; asst scientist, 54-57, from scientist to sr scientist, 57-64, CHIEF CHEM, SECT BIOCHEM MECHANISMS, NAT INST DIABETES & DIGESTIVE DIS, NIH, 65- *Concurrent Pos:* USPHS scientist, 54- *Mem:* Am Chem Soc. *Res:* Steroid synthesis; veratrum alkaloids; peptide synthesis; phosphorylation; oxidation of nitrogen compounds; amino acid metabolism and interconversion; protein structure; reaction kinetics and mechanism; nuclear magnetic resonance spectroscopy; biochemical mechanisms; drug design; antiviral antibiotics. *Mailing Add:* 9814 Inglemere Dr Bethesda MD 20817

COHEN, MAIMON MOSES, b Baltimore, Md, Jan 24, 35; m 55; c 3. CYTOGENETICS. *Educ:* Johns Hopkins Univ, AB, 55; Univ Md, MS, 59, PhD(agron), 63. *Prof Exp:* Jr asst health serv officer, NIH, 60-62; NSF fel human genetics, Univ Mich, 62-64, instr, 64-65; from asst prof to assoc prof pediat & assoc res prof microbiol, Med Sch, State Univ NY Buffalo, 68-72; prof human genetics & chmn dept, Hadassah-Hewbrew Univ Med Ctr, 72-78; prof pediat & assoc chief genetics, Children's Mem Hosp, Northwestern Univ Med Sch, 78-82; prof obstet, pediat & biol chem & chief, DIR DIV HUMAN GENETICS, CTR HUMAN GENETICS, 82-, PROF OBSTET, GYNEC, SCH MED, UNIV MD, 82- *Concurrent Pos:* Dir cytogenetics, Buffalo Children's Hosp, 67-72; Investr Israel Ministry Health, 76-79. *Honors & Awards:* Morris J Kaplun Int Prize, 75. *Mem:* AAAS; Genetics Soc Am; Soc Pediat Res; Am Soc Human Genetics; Tissue Cult Asn; Environ Mutagenic Soc; Europ Genetic Human Soc; Am Asn Dental Res; NY Acad Sci; Genetic Soc Israel. *Res:* Structure and function of human chromosomes; effect of various mutagenic agents on chromosomes of cultured tissues; chromosome instability syndromes. *Mailing Add:* Dept Genetics Univ Md Sch Med 655 W Baltimore St Baltimore MD 21201

COHEN, MARC SINGMAN, b Charleston, WVa, Feb 22, 50; m 75; c 3. UROLOGY. *Educ:* Univ SFla, BA, 72; Univ Miami, MD, 75. *Prof Exp:* Resident surg, Boston Univ Affil Hosp, 75-76; resident urol, Univ Tex Med Br, 76-80, instr, 80-81, from asst prof to assoc prof urol, microbiol, 81-89; ASSOC PROF SURG UROL, COL MED, UNIV FLA, 89- *Concurrent Pos:* Res award, Mead Johnson Nat Student Res Forum, 79; Res award, Am Soc Pediat Nephrologists, 79; Res fel, AUA Res Scholar, 80-82. *Mem:* AMA; Sigma Xi; fel Am Col Surg; Am Urol Asn. *Res:* Investigating the etiologies of urinary calculi, particularly the role of bacteria and bacterial byproducts; bacterial-urothelial interactions; bacterial association with carcinogenesis. *Mailing Add:* Div Urol Univ Fla Col Med Box J247 JUMUC Gainesville FL 32610

COHEN, MARGO NITA PANUSH, b Detroit, Mich, Oct 28, 40; m 61; c 3. INTERNAL MEDICINE, ENDOCRINOLOGY. *Educ:* Univ Mich, BS, 60, MD, 64; Univ Buenos Aires, PhD(biochem), 70; Am Bd Internal Med & Subspecialty Bd Endocrinol & Metab, Dipl. *Prof Exp:* Intern, Sinai Hosp, Detroit, 64-65; resident internal med, Henry Ford Hosp, 65-66; instr physiol, Univ Buenos Aires, 68-70; investr, Arg Nat Res Coun, 70-71; from asst prof to prof med, Sch Med, Wayne State Univ, 71-82, prof med, Univ Med & Dent NJ, 82-87; DIR, INST MED RES, UNIV CITY SCI CTR, PRES, EXOCELL, INC, 87- *Concurrent Pos:* USPHS diabetes trainee, Wayne State Univ & Sinai Hosp, Detroit, 66-68; NIH spec fel, Wayne State Univ, 68-69; vis scientist, Univ Manchester, 79; mem bd sci coun, Nat Inst Dent Res, NIH, 76-80, 86-; mem coun, Midwest Am Fedn Clin Res, 79-81; Fulbright scholar, 87; vis prof, Univ Tel Aviv, 86-87. *Honors & Awards:* Probus Award, 78; Burroughs Wellcome Award, 80. *Mem:* Am Fedn Clin Res; Am Physiol Soc; Am Diabetes Asn; Endocrine Soc; fel Am Col Physicians; Am Soc Clin Invest; Am Soc Biol Chem. *Res:* Diabetes, complications of; glomerular disease; basment membrane metabolism; protein glycation. *Mailing Add:* IMR/Exocell 3508 Market Philadelphia PA 19104

COHEN, MARLENE LOIS, b New Haven, Conn, May 5, 45; m 76; c 2. PHARMACOLOGY. *Educ:* Univ Conn, Storrs, BS, 68; Univ Calif, San Francisco, PhD(pharmacol), 73. *Prof Exp:* Fel, Roche Inst Molecular Biol, 73-75; sr pharmacologist, 75-80, res scientist, 80-85, sr res sci, 85-89, RES ADV, LILLY RES LABS, ELI LILLY & CO, 89- *Concurrent Pos:* Adj asst prof, Dept Pharmacol, Ind Univ Sch Med, 76-81, adj assoc prof, 81-87, adj prof, 87- *Mem:* Soc Exp Biol & Med; Am Soc Pharmacol & Exp Therapeut; Subcomt Women Pharmacol (chair, 84-). *Res:* Pharmacology, physiology and biochemistry of vascular and other smooth muscle. *Mailing Add:* Lilly Res Labs Div Eli Lilly & Co Indianapolis IN 46285

COHEN, MARSHALL HARRIS, b Manchester, NH, July 5, 26; m 48; c 3. EXTRAGALACTIC ASTRONOMY. *Educ:* Ohio State Univ, BEE, 48, MSc, 49, PhD(physics), 52. *Prof Exp:* Res assoc, Antenna Lab, Ohio State Univ, 52-54; from asst prof to assoc prof, Sch Elec Eng, Cornell Univ, 54-64, assoc prof astron, 64-66; prof, Dept Appl Electrophys, Univ Calif, San Diego, 66-68; exec officer astron, 81-84, PROF ASTRON, CALIF INST TECHNOL, 68- *Concurrent Pos:* Guggenheim fel, 60-61 & 80-81; assoc prof, Univ Paris VI, 89. *Honors & Awards:* Rumford Medal, Am Acad Arts & Sci, 71. *Mem:* Nat Acad Sci; Am Astron Soc; Int Union Radio Sci; Int Astron Union. *Res:* Extragalactic astronomy. *Mailing Add:* Dept Astron Calif Inst Technol Pasadena CA 91125

COHEN, MARTIN GILBERT, b Brooklyn, NY, Jan 13, 38; m; c 3. LASERS. *Educ:* Columbia Col, AB, 57; Harvard Univ, MA, 58, PhD(appl physics), 64. *Prof Exp:* Mem tech staff, Bell Tel Labs, 64-69; dir appl res prod mgr, 69-81, vpres res, 84-90, VPRES TECHNOL, QUANTRONIX CORP, 90- *Mem:* Am Phys Soc; Optical Soc Am; Inst Elec & Electronics Engr. *Res:* Optical modulation and deflection; interaction of materials and laser light, solid state lasers, laser Q-switching and mode-locking; laser-tissue interactions; medical laser systems. *Mailing Add:* Quantronix Corp PO Box 9014 49 Wireless Blvd Smithtown NY 11787-9014

COHEN, MARTIN JOSEPH, b Brooklyn, NY, May 6, 21; m 49; c 3. ELECTRONIC PHYSICS. *Educ:* Brooklyn Col, BA, 42; Princeton Univ, MA, 48, PhD(physics), 51. *Prof Exp:* Staff mem, Radiation Lab, Mass Inst Technol, 42-45; res assoc electronics, Princeton Univ, 45-49; staff physicist, Princeton Lab, Radio Corp Am, 49-52; sr physicist, Radiation Res Corp, 52-55; consult physicist, 55-57; vpres, Franklin Systs, Inc, 57-63, vpres, Franklin GNO Corp, 63-74; PRES, PCP, INC, 75- *Mem:* Am Phys Soc; Inst Elec & Electronics Engrs; Am Chem Soc; Sigma Xi. *Res:* Solid state electronics; ion dynamics applications in gaseous chemistry; chemical analysis by ion mobility spectrometry. *Mailing Add:* PCP Inc 2155 Indian Rd West Palm Beach FL 33409

COHEN, MARTIN O, b New York, NY, Jan 15, 40; m 63; c 2. PHYSICS, COMPUTER SCIENCE. *Educ:* Cooper Union, BEE, 60; Columbia Univ, MSNE, 61, DEngSc(nuclear sci & eng), 65. *Prof Exp:* Proj scientist, United Nuclear Corp, 65-68; asst mgr, Math Appln Group, Inc, 68-76, mgr res & develop, 76-85; dir software develop, Matrix Corp, 85-87; vpres, Interact, 87-89; CO-OWNER, GENESIS COMPUTER CONSULT, 89- *Mem:* Am Nuclear Soc; Mil Opers Res Soc. *Res:* Development of sophisticated computer software packages to solve radiation transport, computer generated imagery, mass properties, and other scientific problems; software solutions for interfacing dissimilar hardware, primarily in the medical field. *Mailing Add:* 32 Morris Rd Tappan NY 10983

COHEN, MARTIN WILLIAM, b Brooklyn, NY, Feb 18, 35; m 62. IMMUNOLOGY, HUMAN PATHOLOGY. *Educ:* Columbia Col, AB, 56; State Univ NY, MD, 60. *Prof Exp:* Intern, NY Hosp, 60-61; instr, Albert Einstein Col Med, 67-68, asst prof, 69-70; assoc attend pathologist, Beth Israel Med Ctr, 70-78; STAFF PATHOLOGIST, BROOKLYN VET ADMIN HOSP, 78- *Concurrent Pos:* Fel, NY Univ Med Ctr, 61-65; spec fel, Nat Inst Allergy & Infectious Dis, 65; asst attend pathologist, Bronx Munic Hosp Ctr, 67-70; asst prof, Mt Sinai Med Sch, 70-78; clin assoc prof, NY Univ Med Sch, 76- *Mem:* Int Acad Path. *Res:* Cellular aspects of hormonal immunity; immunopathology. *Mailing Add:* NY Univ Dental Ctr Dept Path 345 E 24th St New York NY 10010

COHEN, MARVIN LOU, b Montreal, Que, Mar 3, 35; US citizen; m 58; c 2. PHYSICS. *Educ:* Univ Calif, Berkeley, AB, 57; Univ Chicago, MS, 58, PhD(theoret solid state physics), 64. *Prof Exp:* Mem tech staff, Bell Tel Labs, 63-64; from asst prof to assoc prof, 64-69, PROF PHYSICS, UNIV CALIF, BERKELEY, 69- *Concurrent Pos:* A P Sloan fel, Cambridge Univ, Eng & Univ Calif, Berkeley, 65-67; prof, Miller Inst Basic Res Sci, Univ Calif, Berkeley, 69-70 & 76-77, chmn, 77-81 & 88; vis prof, Cambridge Univ, Eng, 66, Univ Paris, 68-88, Univ Hawaii, 78-79 & Technion, Israel, 87-88; Guggenheim fel, 90-91, Univ Hawaii, 78-79; US rep, Semiconductor Comn, Int Union Pure & Appl Physics, 75-81; mem selection comt, Presidential Young Investr Awards, 83; chmn, condensed matter physics search-screening somt, Nat Acad Sci, 81-82, chmn, 17th Int Conf Physics Semiconductors,84; mem, Comt Nat Synchrotron Radiation Facil, 83-84, vchmn, Govt-Univ-Indust Res Roundtable, 84-; mem Briefing Panel Implications Mechanisms Support in Res & Brief Panel Night Temperature Superconductivivity, Nat Res Coun, 87; mem, Oliver E Buckley Prize Comt, 80, chmn comt, Nat Acad Sci, 87; chmn, Div Solid State Phys, Am Phys Soc, 77-78; mem, div coun, Lawrence Berkeley Lab, 81-85, computer & staff comts, 81-82, rev bd, Ctr Advan Mat, 86-87; mem, adv bd, Int J Modern Physics B & Modern Physics Lett B, 87- & Tex Ctr Superconductivity, 88-; assoc ed, Mat Sci & Eng, 87-; Solid State Sci Rev Panel, Exxon Res Labs, 81 & chmn, Comt Rev Theoret Physics, Chem & Math, 84; mem, US deleg Bilateral Dialog Res & Develop US & Japan, Nat Res Coun, 89; mem, Int Orgn Comt, Inst de la Vie, France, 88- & sci policy bd, Stanford Synchrotron Radiation Lab, 90- *Honors & Awards:* Oliver E Buckley Prize Solid State Physics, Am Phys Soc, 79; Award Outstanding Accomplishment Solid State Physics, US Dept Energy, 81, Award Sustained Outstanding Res, 90. *Mem:* Nat Acad Sci; fel Am Phys Soc. *Res:* Theoretical condensed matter physics; superconductivity; semiconductors; optical bonding and electronic properties of solids; surfaces of solids; clusters; high pressure science. *Mailing Add:* Lawrence Berkeley Lab Mat Sci Div Univ Calif Dept Physics Berkeley CA 94720

COHEN, MARVIN MORRIS, b New York, NY, Apr 24, 40; m 61; c 4. SOLID STATE PHYSICS. *Educ:* Brooklyn Col, BA, 62; Am Univ, MS, 65, PhD(solid state physics), 67. *Prof Exp:* Res physicist, Harry Diamond Labs, 62-74; PHYSICIST, ENERGY RES & DEVELOP ADMIN, 74- *Concurrent Pos:* Adj prof, Am Univ, 67- *Mem:* AAAS; Am Phys Soc. *Res:* Impurity band conduction in solids; tunneling in solids; radiation damage in a fusion environment. *Mailing Add:* 5704 Dimes Rd Rockville MD 20855

COHEN, MAYNARD, b Regina, Sask, May 17, 20; US citizen; m 45; c 2. NEUROSCIENCES. *Educ:* Univ Mich, AB, 41; Wayne Univ, MD, 44; Univ Minn, PhD(path), 53. *Prof Exp:* Res assoc, Riks Hosp, Oslo, Norway, 51-52; from asst prof to prof neurol, Univ Minn, 53-63; prof & head div neurol & prof pharmacol, Col Med, Univ Ill, 63-71; chmn, Dept Neurol Sci, Presby-St. Luke's Hosp 63-83; Jean Schweppe Armoor prof, 83-87, PROF BIOCHEM, RUSH MED COL, 71-, CHMN EMER, 87- *Concurrent Pos:* NIH spec fel, Univ London, 57-58; consult, Nat Inst Neurol Dis & Blindness, 59-63 & NIMH, 68-72; mem prof adv bd, Epilepsy Founds; Fulbright lectr, Univ Oslo, 77; Crown Princess Marte fel, Am Scand Found, 86. *Honors & Awards:* Distinguished Serv Award, Am Acad Neurol, 89. *Mem:* Biochem Soc; Am Acad Neurol (past vpres, pres, 81-83); Am Asn Neuropath; Int Soc Neurochem; Asn Univ Prof Neurol (past pres); Norweg Acad Sci & Letters, 82. *Res:* Nervous and mental diseases; biochemistry of the nervous system; cerebrovascular disease; neurotoxic agents; phosphorylated compounds in the brain; amino acid and carbohydrate interrelationships in the brain; nuclear magnetic resonance studies of the brain. *Mailing Add:* Dept Neurol Sci Rush Med Col Chicago IL 60612

COHEN, MELVIN JOSEPH, b Los Angeles, Calif, Sept 28, 28; m 63; c 4. NEUROBIOLOGY. *Educ:* Univ Calif, Los Angeles, BA, 49, MA, 52, PhD(zool), 54. *Prof Exp:* NSF fel, Royal Vet Inst, Stockholm, Sweden, 54-55; instr biol, Harvard Univ, 55-57; from asst prof to prof, Univ Ore, 57-69; PROF BIOL, YALE UNIV, 69-, DIR, UNDERGRAD STUDIES, DEPT BIOL, 86- *Concurrent Pos:* Lalor Found fel, Marine Biol Labs, Woods Hole, Mass, 56; Guggenheim fel, Dept Zool, Oxford Univ, Eng, 64-65; USPHS fel, Dept Zool, Oxford Univ & Inst Marine Biol, Univ Bordeaux, Arcachon, France, 65; USPHS career develop award, 67-69; mem staff, Dept Zool, Glasgow Univ, Scotland, 72, Stanford Univ, Palo Alto, Calif, 83 & Pasteur Inst, Paris, 86. *Mem:* Nat Acad Sci; Am Soc Zoologists; Soc Gen Physiologists (pres 76-77); Soc Neurosci; Int Brain Res Orgn; fel AAAS; Sigma Xi. *Res:* Comparative neurophysiology; growth and regeneration of nerve cells. *Mailing Add:* Dept Biol Yale Univ PO Box 6666 New Haven CT 06511-8112

COHEN, MERRILL, b Boston, Mass, Feb 5, 26; m 50; c 3. POLYMERS, GEOTHERMAL ENERGY. *Educ:* Boston Univ, BA, 48; Univ Chicago, MS, 49, PhD(chem), 51. *Prof Exp:* Res assoc, Res Lab, Gen Elec Co, 51-52, specialist org chem, Thomson Lab, 52-55, mgr chem & insulation eng, Medium Steam Turbine Generator Dept, 55-70, mgr mat & processes lab, Medium Steam Turbine Generator Dept, 70-77; CONSULT, CHEMCO CONSULT INC, 77- *Concurrent Pos:* Jr chemist, Ionics Inc, 49, 50; mem tech adv comt, Nat Geothermal Info Resource Proj, Lawrence Berkeley Lab, Univ Calif, 74- *Mem:* Am Chem Soc; fel Am Inst Chem; Geothermal Res Coun. *Res:* Organic resin and polymer chemistry; epoxy, polyester, silicone resins; electrical insulation; laminated plastics; adhesives; protective coatings, high temperature synthetic lubricants; gas chromatography, air, water pollution; geothermal energy and steam turbine materials; fire resistant lubricants and control fluids, NOx removal technology, steam chemistry, supercritical oxidation. *Mailing Add:* Eight May St Marblehead MA 01945

COHEN, MICHAEL, b New York, NY, May 9, 30; m 58; c 3. THEORETICAL PHYSICS. *Educ:* Cornell Univ, AB, 51; Calif Inst Technol, PhD(physics), 56. *Prof Exp:* Res fel theoret physics, Calif Inst Technol, 55-57; mem, Inst Advan Study, NJ, 57-58; from asst prof to assoc prof, 58-75, assoc chmn grad affairs, 68-71, PROF PHYSICS, UNIV PA, 75- *Concurrent Pos:* Consult, Los Alamos Sci Lab; vpres, treas & hon trustee, Aspen Ctr Physics. *Mem:* Am Phys Soc. *Res:* Quantum and statistical mechanics; theory of liquid helium. *Mailing Add:* Dept Physics Univ Pa Philadelphia PA 19104-6396

COHEN, MICHAEL ALAN, b New Haven, Conn, Feb 20, 46; m 69, 91. DIAGNOSTIC MICROBIOLOGY. *Educ:* Clark Univ, AB, 68; Univ Fla, MS, 72; State Univ NY, Buffalo, PhD(med microbiol-immunol), 79. *Prof Exp:* Epidemiologist, Ctr Dis Control, 68-70; microbiol supvr, Long Island Jewish-Hillside Med Ctr, 76-77; microbiol chief, St Francis Hosp, 77-81; dir, microbiol-serol, Coney Island Hosp, 81-82; HEAD CLIN MICROBIOL, WARNER-LAMBERT-PARKE DAVIS PHARM RES, 82- *Concurrent Pos:* Clin adj, clin microbiol, Long Island Univ, 77-81; adj prof antibiotics, Eastern Mich Univ, 85- *Mem:* Am Soc Microbiol. *Res:* Development of antimicrobial agents. *Mailing Add:* 4021 Woodland Dr Ann Arbor MI 48103

COHEN, MICHAEL I, PEDIATRICS. *Prof Exp:* PROF & CHMN, DEPT PEDIAT, ALBERT EINSTEIN COL MED, 80- *Mem:* Inst Med-Nat Acad Sci. *Mailing Add:* Dept Pediat Montefiore Med Ctr Albert Einstein Col Med 110 E 210th St Bronx NY 10467

COHEN, MICHAEL PAUL, b San Mateo, Calif, July 8, 47. STATISTICAL SURVEY RESEARCH. *Educ:* Univ Calif, San Diego, BS, 69, Los Angeles, MS, 71, PhD(appl math), 78. *Prof Exp:* MATH STATISTICIAN, PRICE STATIST METHODS DIV, BUR LABOR STATIST, 79- *Concurrent Pos:* Teaching assoc dept math, Univ Calif, Los Angeles, 75-77, res asst, 76-77. *Mem:* Am Statist Asn; Int Asn Surv Statist; Am Math Soc; Math Asn Am; Inst Math Statist; Soc Indust & Appl Math. *Res:* Statistical and survey design aspects of the US Consumer Price Index and Consumer Expenditure; variance and composite estimation; statistical decision theory and nonparametric estimation; composite estimation; statistical decision theory; nonparametric estimation. *Mailing Add:* Nat Ctr Educ Statist Educ Dept Rm 408 555 New Jersey Ave NW Washington DC 20208-5654

COHEN, MITCHELL S(IMMONS), b Schenectady, NY, Nov 8, 30. ENGINEERING PHYSICS. *Educ:* Rensselaer Polytech Inst, BS, 52; Cornell Univ, PhD(eng physics), 58. *Prof Exp:* Staff mem, Lincoln Lab, Mass Inst Technol, 58-69; staff scientist, Micro-Bit Corp, 69-74; RES STAFF MEM, WATSON RES CTR, INT BUS MACH CORP, 74- *Mem:* Am Phys Soc; Inst Elec & Electronics Engrs. *Res:* Magnetic bubbles; electron optics. *Mailing Add:* IBM Corp T J Watson Res Ctr Div PO Box 218 Yorktown Heights NY 10598

COHEN, MONROE W, b Montreal, Que, May 3, 40; m 65; c 2. SYNAPTOGENESIS. *Educ:* McGill Univ, BSc, 61, PhD(neurophys), 65. *Prof Exp:* Res fel neurobiol, Harvard Med Sch, 65-68; from asst prof to assoc prof, 68-83, PROF PHYSIOL, MCGILL UNIV, 84- *Concurrent Pos:* NATO sci fel, Nat Res Coun Can, 66-68; Med Res Coun Can scholar, 71-76; ed, J Physiol, 75-77; Sci Res Coun Que res fel, 76-82; mem prog comt, Soc Neuroscience, 84-87. *Mem:* Can Physiol Soc; Soc Neurosci; Physiol Soc. *Res:* Formation and development of nerve-muscle synapses; regulation of the distribution of acetylcholine receptors in skeletal muscle. *Mailing Add:* Dept Physiol McGill Univ 3655 Drummond Montreal PQ H3G 1Y6 Can

COHEN, MONTAGUE, b London, Eng, July 24, 25; Can & Brit citizen; m 47; c 3. MEDICAL PHYSICS, RADIATION SAFETY. *Educ:* Univ London, BSc, 46, PhD(physics), 58; ARCS, 46; FCCPM, 80. *Prof Exp:* Physicist, Royal Aircraft Estab, 46-47 & Gen Elec Co Res Labs, 47-48; physicist med physics, London Hosp, 48-61, chief physicist, 66-75; prof officer, Int Atomic Energy Agency, 61-66; PROF RADIO PHYSICS, MCGILL UNIV, 75-, PROF MED PHYSICS, 79-; CUR, RUTHERFORD MUS, MCGILL UNIV, 85- *Concurrent Pos:* Consult radiother physics, Int Comn Radiation Units & Measurements, 64- & WHO, 70-; consult med physics, Int Atomic Energy Agency, 66-; dep ed, Brit J Radiol, 70-74, assoc ed, 79-85; consult, radiation protection, Ont Govt, 84- *Honors & Awards:* Roentgen Prize, Brit Inst Radiol, 74; Stanley Melville Mem Award, Brit Inst Radiol, 60; Cert Merit, Can Asn Radiol, 83. *Mem:* Inst Physics; Brit Inst Radiol (secy, 73-75); Hosp Physicists Asn; Am Asn Physicists Med; Can Asn Physicists; Can Col Physicists Med. *Res:* Dosimetry in radiotherapy with small sealed sources; quality control in diagnostic radiology; risk factors in radiation protection; history of science. *Mailing Add:* 12 Redpath Pl Montreal PQ H3G 1E1 Can

COHEN, MORREL HERMAN, b Boston, Mass, Sept 10, 27; m 50; c 4. THEORETICAL PHYSICS. *Educ:* Worcester Polytech Inst, BS, 47; Dartmouth Col, MA, 48; Univ Calif, Berkeley, PhD(physics), 52. *Hon Degrees:* DSc, Worcester Polytech Inst, 73. *Prof Exp:* Res assoc physics, Dartmouth Col, 47-48; from instr to assoc prof theoret physics, Univ Chicago, 52-60, prof, Dept Physics, Univ Chicago & James Franck Inst, 60-72, actg dir, 65-66, dir, James Franck Inst, 68-71, prof theoret biol, 68-72, Louis Block prof physics & theoret biol, 72-82, dir, Mat Res Lab, NSF, 77-81; SR SCI ADV & SCI AREA LEADER, CORP RES LABS, EXXON RES & ENG CO, 81- *Concurrent Pos:* Guggenheim fel, Cambridge Univ, 57-58; vis scientist, Nat Res Coun Can, 60; NSF sr fel, Univ Rome, 64-65; consult, Westinghouse Res Labs, 53, Chicago Midway Labs, 54, 55, Gen Elec Co Res Labs, 57-65, Argonne Nat Lab, 59-81, Boeing Sci Res Labs, 60, Hughes Res Lab, 60 & 62, Basic Sci Ctr, Nam Aviation Co, 62, Energy Conversion Devices Inc, 67-71 & 74-81, Monsanto Co, 72-78, Union Carbide Co, 76-79, Xerox Corp, 77 & 78 & Schlumberger Technol Corp, 78-81; NASA mem, Adv Panel Electrophysics, 62-66; mem adv comt, Nat Magnet Lab, 63-66; mem, Rev Comt Solid State Sci & Metall Div, Argonne Nat Lab, 64-66, chmn, 66, mem, Sci & Tech Adv Comt, 83-90, Rev Comt, Mat Sci Div, 84-; chmn, Gordon Conf Chem & Physics of Solids, NH, 68 & chmn, Fourth Int Conf Amorphous & Liquid Semiconductors, 71; assoc ed, J Chem Physics, 60-63; publ bd, Univ Chicago, 69-70 & bd eds, J Statist Physics, 70-75; NIH spec fel, 72-73; vis fel, Cambridge Univ, 72-73, assoc, 73-; vis prof, Univ Chicago, 73-74, Univ Va, 76, Res Inst Fundamental Physics, Kyoto Univ, Japan, 79, Brooklyn Col, 79; vis fel & assoc, Univ Cambridge Clare Hall, 72-; vchmn, Comn C3 Statist Mech, Int Union Pure & Appl Physics, 87-93; Asn Universities Inc distinguished lectr, Brookhaven Nat Labs, 91; Van der Waals prof, Dept Physics & Astron, Univ Amsterdam, 91-92. *Honors & Awards:* Shrum lectr, Simon Fraser Univ, 73. *Mem:* Nat Acad Sci; AAAS; Am Inst Physics; fel Am Phys Soc; Sigma Xi; NY Acad Sci. *Res:* Theoretical physics of condensed matter; developmental biology; quantum theory of solids; general physics of solids; theoretical biology; control of development in primitive organisms; author of 271 scientific publications. *Mailing Add:* Exxon Res & Eng Co Rte 22 E Annandale NJ 08801

COHEN, MORRIS, b Santa Ana, Calif, July 20, 21. CELL BIOLOGY, MEDICAL RESEARCH. *Educ:* Univ Calif, BS, 47, PhD(plant path), 51. *Prof Exp:* Asst plant path, Univ Calif, Berkeley, 48-51; res assoc bot, Univ Calif, Los Angeles, 51-54, asst res botanist, 54-57; electron microscopist, St Joseph Hosp, Burbank, Calif, 57-62, res assoc, 62; res biologist, 62-81, RES AFFIL, WADSWORTH VET ADMIN HOSP, 81- *Concurrent Pos:* Am Cancer Soc fel, 51-54. *Mem:* AAAS; Am Soc Cell Biol; Electron Micros Soc Am; NY Acad Sci; Am Soc Microbiol. *Res:* Submicroscopic morphology, pathogenesis of inflammatory diseases; bacteria; fungi; adipose tissue fine structure and function. *Mailing Add:* Electron Micros Res Lab Rm 330 Wadsworth Vet Admin Hosp Bldg 114 Los Angeles CA 90073

COHEN, MORRIS, b Chelsea, Mass, Nov 27, 11; m 37; c 1. MATERIALS SCIENCE & ENGINEERING, METALLURGY. *Educ:* Mass Inst Technol, BS, 33, ScD, 36. *Hon Degrees:* DTech, Royal Inst Technol, Sweden, 77; DSc Tech, Israel Inst Technol, 79; DEng, Colo Sch Mines, 85; DSc, Northeastern Univ, 89. *Prof Exp:* Asst phys metall, Mass Inst Technol, 34-36, from instr to prof, 36-62, Ford prof mat sci & eng, 62-74, inst prof, 74-82, EMER INST PROF METALL & MAT SCI, MASS INST TECHNOL, 82- *Concurrent Pos:* Official investr, Off Sci Res & Develop, 42-43; assoc dir, Manhattan Proj, Mass Inst Technol, 43-45. *Honors & Awards:* Howe Medal, Am Soc Metals, 45, 49, Albert Sauveur Achievement Award, 47, 77, Gold Medal, 68; Campbell Mem lectr, 48; Kamani Medal, Indian Inst Metals, 53; Inst Metals lectr, 57; Burgess Mem lectr, Carnegie Mem lectr & Sauveur lectr, 58; Woodside lectr, 59; Coleman lectr, Franklin Inst, 60; Houdremont Mem lectr, Int Inst Welding, 61; Howe Mem lectr, Metall Soc, 67, Leadership Award, 87; Hatfield Mem lectr, Brit Iron & Steel Inst, 62; Rockwell Mem lectr, 65; Gold Medal, Japan Inst Metals, 70; Robert S Williams lectr, Mass Inst Technol, 70; Pierre Chevenard Medal, Fr Soc Metall, 71; Procter Prize, Sigma Xi. *Mem:* Nat Acad Sci; Nat Acad Eng; AAAS; hon mem Am Soc Metals; fel NY Acad Sci; fel Am Acad Arts & Sci; hon mem Am Inst Mining Metall & Petrol Engrs; hon mem Japanese Iron & Steel Inst; hon mem Japanese Inst Metals; hon mem Brit Inst Metals; hon mem Indian Inst Metals; hon mem Korean Inst Metals. *Res:* Materials science and engineering; materials policy; physical metallurgy; phase transformations; strengthening mechanisms; mechanical behavior of metals. *Mailing Add:* 491 Puritan Rd Swampscott MA 01907-2819

COHEN, MORTON IRVING, b New York, NY, July 11, 23; m 58; c 2. NEUROPHYSIOLOGY. *Educ:* City Col New York, BS, 42; Columbia Univ, AM, 50, PhD(physiol), 57. *Prof Exp:* From instr to assoc prof, 57-73, prof physiol, 73-76, PROF NEUROSCI, ALBERT EINSTEIN COL MED, 76- *Mem:* Am Physiol Soc; Soc Neuroscience; Europ Neuroscience Asn. *Res:* Neural regulation of respiration and circulation; spontaneous activity of neurons in central nervous system; patterns of synaptic excitation and inhibition; computer analysis of neuroelectric data. *Mailing Add:* Dept Physiol Albert Einstein Col Med 1300 Morris Park Ave Bronx NY 10461

COHEN, MOSES E, US citizen. APPLIED MATHEMATICS. *Educ:* Univ London, BSc, 63; Univ Wales, PhD(theoret physics, appl math), 67. *Prof Exp:* Res fel astrophys, French Atomic Energy Comn, 68; asst prof math, Mich Technol Univ, 68-69; from asst prof to assoc prof, 69-74, PROF MATH, CALIF STATE UNIV, FRESNO, 74- *Mem:* Am Math Soc; assoc fel Brit Inst Math & Appln; assoc Brit Inst Physics & Phys Soc. *Res:* Astrophysics, especially cosmic radiation; applied mathematics, especially generating functions, combinatorial identities; solid state physics, especially solar cells. *Mailing Add:* Dept Math Calif State Univ Fresno CA 93740

COHEN, MURRAY SAMUEL, b Brooklyn, NY, May 19, 25; m 48; c 2. ORGANIC CHEMISTRY. *Educ:* Univ Mo, BS, 48, MA, 50, PhD(chem), 52. *Prof Exp:* Res chemist, Schenley Labs, 52-53; res chemist reaction motors div, Thiokol Chem Corp, 53-54, proj leader synthetic org & inorg res, 54-56, chief propellant synthesis sect, 56-62, mgr chem dept, 62-67; staff adv advan planning, Esso Res & Eng Co, 67-68, dir new ventures & dir fuel additives labs, 68-73; vpres res, Weston Chem Div & tech dir chem, Borg Warner Corp, 73-78; dir govt liason, Apollo Technol, subsid Econ Labs, Inc, 78-80, dir res & develop, 80-84; PRES, EPOLIN & ACCORT LABS, 84- *Mem:* Am Chem Soc; Soc Plastics Indust. *Res:* Solid and liquid propellants; light weight metal hydrides and derivatives; organometallic polymers; additive research, antioxidants, heat and ultra violet stabilizers, copper deactivators; pollution control, fuel additives; coatings and adhesives; polymeric additives; ultraviolet curing. *Mailing Add:* Symor Dr Convent St Morristown NJ 07960

COHEN, MYRON LESLIE, b New York, NY, March 7, 34; m 55; c 3. MEDICAL DEVICE DESIGN. *Educ:* Purdue Univ, BSME, 55; Univ Ala, MSE, 58; Polytech Inst Brooklyn, PhD(mech eng), 66. *Prof Exp:* Res engr, Allegany Ballistics Lab, Hercules Powder Co, 55-56; sr thermodynamics engr, Repub Aviation Corp, 58-60; instr mech eng, Polytech Inst Brooklyn, 60-66; asst prof mech eng, Stevens Inst Technol, 66-69, assoc prof mech eng, 69-77, dir, Med Eng Lab, 75-78, prof mech eng, Dept Mech Eng, 77-78; dir res & develop, hosp prod div, Chesebrough-Ponds, Inc, 78-83; pres, 83-90, CHMN BD & EXEC VPRES, CAS MED SYSTS, INC, 90- *Concurrent Pos:* Prof & sr US scientist, Inst Bioeng & Biomed Tech, Univ Karlsruhe, WGermany, 74-75; pres, CAS Inc, NJ, 75-78; vpres & ed dir, Freshet Press, 70-78; consult, US Navy, Johnson & Johnson, Vitro Corp Am; adj assoc prof surg, Sect Orthop Surg, Col Med & Dent NJ, 78- *Honors & Awards:* Humboldt Prize, 74. *Mem:* Am Soc Mech Engrs; Sigma Xi; Cardiovasc Systs Dynamics Soc; Soc Biomat; Am Inst Advan Med Instrumentation. *Res:* The use of engineering technologies in the design and development of products to be used in clinical medicine including wound management, blood pressure measurement and medical device development. *Mailing Add:* CAS Med Systs Inc 29 Business Park Dr Branford CT 06405

COHEN, MYRON S, b Chicago, Ill, May 7, 50; c 2. CELL BIOLOGY. *Educ:* Rush Med Col, MD, 74. *Prof Exp:* Asst prof, 80-85, ASSOC PROF MED, MICROBIOL & IMMUNOL, SCH MED, UNIV NC, 85- *Mem:* Infectious Dis Soc Am; Am Asn Immunol; Am Soc Microbiol; Am Col Physicians; Am Soc Clin Investr. *Res:* Phagocytic cells, bacterial pathogenesis. *Mailing Add:* Dept Med Sci Bldg 229 H Univ NC Sch Med 547 Burnett-Womack Clin Chapel Hill NC 27514

COHEN, NADINE DALE, b Yonkers, NY, Sept 15, 49; m 68; c 2. BIOCHEMISTRY. *Educ:* Rensselaer Polytech Inst, 70; Univ Rochester, 74. *Prof Exp:* Res assoc biochem, Case Western Reserve Univ, 73-76, sr res assoc, 76-79; asst prof biochem, Wright State Univ, 79-84; ASST DIR, MASS PUB

HEALTH BIOLOGIC LABS, 84- *Mem:* Sigma Xi; AAAS; Am Chem Soc; Am Soc Microbiol. *Res:* Biochemical characterization and clinical studies of acellular pertussis vaccines. *Mailing Add:* 93 Centre St Apt Two Brookline MA 02146-2801

COHEN, NATALIE SHULMAN, b New York, NY, Jan 16, 38; m 58; c 2. CELL PHYSIOLOGY, MITOCHONDRIAL BIOCHEMISTRY. *Educ:* Cornell Univ, BA, 59; NY Univ, MS, 61, PhD(biol), 65. *Prof Exp:* Res fel biol, Calif Inst Technol, 66-70; from res assoc to sr res assoc, 70-89, ASST PROF RES BIOCHEM, SCH MED, UNIV SOUTHERN CALIF, 89- *Concurrent Pos:* Res assoc biochem, Col Med, Univ Ariz, 73-74. *Mem:* AAAS; Am Soc Biochem & Molecular Biol; Sigma Xi. *Res:* Behavior of enzymes in the mitochondria; kinetics and the regulation of enzymes of the urea cycle; physiological and biochemical aspects of cellular functions; relation to cellular structure. *Mailing Add:* 1725 Homet Rd Pasadena CA 91106

COHEN, NATHAN WOLF, b Richmond, Va, Oct 3, 19; m 46; c 2. HERPETOLOGY. *Educ:* Univ Calif, Los Angeles, AB, 44; Univ Calif, Berkeley, MA, 50; Ore State Univ, PhD(zool), 55. *Prof Exp:* Asst physiol & zool, Univ Calif, 48-50, res zoologist, San Joaquin Exp Range, 50-51; instr zool, Fresno State Col, 51-52; instr physiol & biol, Modesto Jr Col, 55-63; sci coordr, Lib Arts Dept, Univ Calif Exten, Berkeley, 63, head, Letters & Sci Exten, 64-70 & Continuing Educ Sci & Math, 70-72, dir curric develop sci, 72-79; CONSULT, 79- *Concurrent Pos:* Res assoc herpet, emer res assoc, Los Angeles, County Mus Natural Hist & Mus Vert Zool, Univ Calif, Berkeley, 79- *Mem:* Fel AAAS; Am Soc Ichthyologists & Herpetologists; fel Herpet League; Sigma Xi. *Res:* Environmental physiology of terrestrial cold-blooded vertebrates; color photography of amphibians and reptiles; environmental physiology and behavior of amphibians and reptiles. *Mailing Add:* 1324 Devonshire Ct El Cerrito CA 94530

COHEN, NICHOLAS, b New York, NY, Nov 20, 38; m 74; c 4. IMMUNOLOGY, DEVELOPMENTAL BIOLOGY. *Educ:* Princeton Univ, AB, 59; Univ Rochester, PhD(biol), 66. *Prof Exp:* USPHS scholar med microbiol & immunol, Univ Calif, Los Angeles, 65-67; from asst prof to prof microbiol & immunol, 67-82, DIR, DIV IMMUNOL, SCH MED & DENT, UNIV ROCHESTER, 79-, PROF PSYCHIAT, 82- *Concurrent Pos:* Mem, Basel Inst Immunol, Switz, 75-76; USPHS career develop award, 73-77; mem, Immunobiol Study Sect, NIH, 76-80; vis prof & Fulbright Scholar, Agr Univ, Wageningen, Neth, 82-83; assoc ed, Brain, Behav & Immunity, Develop & Comp Immunol; chair, Div Comp Immunol, Am Soc Zool, 77-79. *Mem:* AAAS; Transplantation Soc; Am Soc Zool; Am Asn Immunol; Int Soc Develop Comp Immunol; Brit Soc Immunol. *Res:* Comparative and developmental immunology; transplantation and psychoneuroimmunology. *Mailing Add:* Dept Microbiol & Immunol Sch Med & Dent Univ Rochester Box 672 Rochester NY 14642

COHEN, NOAL, b Rochester, NY, Dec 29, 37; m 60; c 2. MEDICINAL CHEMISTRY. *Educ:* Univ Rochester, BS, 59; Northwestern Univ, PhD(org chem), 65. *Prof Exp:* Chemist, Eastman Kodak Co, 59-61; sr chemist, Dept Chem Res, 67-75, res fel, 75-80, res group chief, 80-84, res sect chief, 84-85, RES LEADER, HOFFMAN-LA ROCHE, INC, 85- *Concurrent Pos:* NSF fel, Stanford Univ, 65-67. *Mem:* Am Chem Soc; Sigma Xi; AAAS; NY Acad Sci; Int Soc Heterocyclic Chem. *Res:* Synthesis of natural products and other organic compounds possessing biological activity; medicinal chemistry. *Mailing Add:* 19 Euclid Pl Montclair NJ 07042

COHEN, NOEL LEE, b New York, NY, Sept 20, 30; m 57; c 1. OTOLARYNGOLOGY. *Educ:* NY Univ, BA, 51; State Univ Utrecht, MD, 57. *Prof Exp:* From instr to assoc prof, 62-72, prof clin otolaryngol, 72-80, PROF & ACTG CHMN OTOLARRYNGOL, SCH MED, NY UNIV, 80- *Concurrent Pos:* Consult, Manhattan Vet Admin Hosp, 74-; assoc attend physician, Univ Hosp, 68- & Bellevue Hosp, NY, 70- *Mem:* Am Neurotology Soc; Am Acad Ophthal & Otolaryngol; Am Col Surg; Soc Univ Otolaryngol; Am Laryngol, Rhinol & Otol Soc. *Res:* Neurotology. *Mailing Add:* Dept Otorhinolaryngol NY Univ Sch Med 550 First Ave New York NY 10016

COHEN, NORMAN, b New York, NY, Dec 13, 36; m 59, 87; c 3. CHEMICAL KINETICS, THERMOCHEMISTRY. *Educ:* Reed Col, AB, 58; Univ Calif, Berkeley, MA, 60, PhD(chem), 63. *Prof Exp:* Mem tech staff phys chem, 63-69, staff scientist, 69-73, head chem kinetics dept, 73-84, SR SCIENTIST, AEROSPACE CORP, 84- *Concurrent Pos:* Asst ed, Int J Chem Kinetics, 76-83. *Mem:* Am Chem Soc; Am Phys Soc; Sigma Xi. *Res:* Gas phase chemical kinetics and photochemistry; reactions of free radicals and atoms; kinetics of chemical lasers; decomposition of hydrogen halides; vibrational energy transfer; kinetics and thermochemistry of combustion and oxidation; evaluation of chemical kinetic and thermochemical data. *Mailing Add:* Aerospace Corp PO Box 92957 Los Angeles CA 90009-2957

COHEN, NORMAN, b Brooklyn, NY, Nov 6, 38; m 62; c 1. RADIOLOGICAL HEALTH, ENVIRONMENTAL RADIOACTIVITY. *Educ:* Brooklyn Col, BS, 60; NY Univ, MS, 65, PhD(environ sci), 70. *Prof Exp:* Chemist, Columbia Presby Hosp, NY, 60-61; res assoc radiobiol & radiochem, 66-74, assoc prof, 74-83, PROF ENVIRON MED, MED CTR, NY UNIV, 83-, DIR LAB RADIATION RES, 84- *Mem:* AAAS; Am Nuclear Soc; Health Physics Soc; Radiation Res Soc; Sigma Xi; NY Acad Sci; Soc Risk Analysis. *Res:* Radiobiological research of the metabolism of various radionuclides and elements in man and other primates; research and evaluation of toxicological properties of heavy metals in man and non-human primates. *Mailing Add:* Inst Environ Med A J Lanza Lab NY Univ Med Ctr Tuxedo NY 10987

COHEN, PAUL, b New York, NY, Aug 11, 12; m 37; c 1. NUCLEAR ENGINEERING. *Educ:* City Col, New York, BS, 34; Carnegie Inst Technol, MS, 41. *Prof Exp:* Fuel engr heat transfer rheology slags, US Bur Mines, Dept Interior, 36-49; mgr chem develop water coolant technol, Bettis Atomic Power Lab, Westinghouse Elec Co, 49-59; mgr chem develop, Westinghouse Atomic Power Dept, 59-67; consult, Westinghouse Advan Reactors Div, 67-77; CONSULT WATER TECHNOL, 77- *Mem:* Fel Am Soc Mech Eng; fel Am Nuclear Soc; Am Chem Soc; Nat Asn Corrosion Eng. *Mailing Add:* 4601 5th Ave Pittsburgh PA 15213-3666

COHEN, PAUL JOSEPH, b Long Branch, NJ, Apr 2, 34; m 63; c 2. MATHEMATICS. *Educ:* Univ Chicago, MS, 54, PhD(math), 58. *Hon Degrees:* DSc, Univ Ill, Urbana-Champaign, 71. *Prof Exp:* Instr math, Univ Rochester, 57-58 & Mass Inst Technol, 58-59; fel, Inst Advan Study, 59-61; from asst prof to assoc prof, 61-64, PROF MATH, STANFORD UNIV, 64- *Honors & Awards:* Bocher Prize & Res Corp Award, 64; Fields Medal, Int Math Union, 66; Nat Medal Sci, 67. *Mem:* Nat Acad Sci; Am Math Soc. *Res:* Axiomatic set theory; harmonic analysis; partial differential equations. *Mailing Add:* Dept Math Stanford Univ Stanford CA 94305-2125

COHEN, PAUL S(IDNEY), b Boston, Mass, Jan 20, 39. MOLECULAR BIOLOGY. *Educ:* Brandeis Univ, AB, 60; Boston Univ, AM, 62, PhD(genetics), 64. *Prof Exp:* USPHS trainee, St Jude Hosp, 64-66, asst prof microbiol, 66-69; assoc prof, 69-75, PROF MICROBIOL, UNIV RI, 75- *Mem:* Am Soc Biol Chemists; Am Soc Microbiol. *Res:* Molecular basis of E coli colonization. *Mailing Add:* Dept Microbiol Univ RI Kingston RI 02881-0812

COHEN, PHILIP, b New York, NY, Dec 13, 31; m 54; c 1. HYDROGEOLOGY, GROUNDWATER GEOLOGY. *Educ:* Univ Rochester, MS, 56. *Prof Exp:* Asst, Univ Rochester, 54-56; geologist, 56-67, res hydrologist, 67-68, hydrologist in charge LI prog, 68-72, staff scientist, Off of the Dir, 72-74, actg chief res & tech coord, 74-76, assoc chief res & tech coord, Land Info & Analysis Off, 76-79, ASST CHIEF HYDROLOGIST, SCIENTIFIC PUBLICATIONS & DATA MGT, & CHIEF HYDROLOGIST, WATER RESOURCES DIV, US GEOL SURV, 79- *Honors & Awards:* Ward Medal, 54; Meritorious Serv Award, Dept Interior, 75. *Mem:* Fel Geol Soc Am; Am Water Resources Asn; Am Inst Prof Geologists; Sigma Xi; Am Geol Union; Int Asn Hydrogeologists. *Res:* Artificial groundwater recharge; seawater encroachment; land-use planning implications of earth sciences. *Mailing Add:* 8705 Southern Pines Ct Vienna VA 22180

COHEN, PHILIP IRA, b Baltimore, Md, Oct 27, 48. SURFACE PHYSICS. *Educ:* Johns Hopkins Univ, BA, 69; Univ Wis-Madison, PhD(physics), 75. *Prof Exp:* Teaching asst, Univ Wis, 70, res asst, 70-75; res assoc physics & chem, Ctr Mat Res, Univ Md, College Park, 76-78; ASST PROF ELEC ENG, UNIV MINN, MINNEAPOLIS, 78- *Concurrent Pos:* Guest worker, Nat Bur Standards, 76- *Mem:* Sigma Xi; Am Inst Physics. *Res:* Surface physics and chemistry, surface crystallography, electron diffraction, and molecular beam epitaxy. *Mailing Add:* Dept Elec Eng Univ Minn Minneapolis MN 55455

COHEN, PHILIP PACY, b Derry, NH, Sept 26, 08; m 35; c 4. PHYSIOLOGICAL CHEMISTRY, ENZYMOLOGY. *Educ:* Tufts Col, BS, 30; Univ Wis, PhD(physiol chem), 37, MD, 38. *Hon Degrees:* DSc, Univ Mex, 79. *Prof Exp:* From asst prof to assoc prof, 43-47, prof physiol chem & chmn dept, 48-75, H C Bradley prof physiol chem, 68-79, EMER PROF, UNIV WIS-MADISON, 79- *Concurrent Pos:* Nat Res Coun fel, Univ Sheffield, 38-39 & Yale Univ, 39-40; Commonwealth Fund fel, Oxford Univ, 58; mem comt growth, Nat Res Coun, 54-56; mem bd sci coun, Nat Cancer Inst, 57-61, mem adv cancer coun, 63-67; mem adv comt biol & med, US AEC, 63-71; mem adv coun to dir, NIH, 66-70, mem adv coun arthritis, metab & digestive dis, 70-74; mem adv comt med res, Pan-Am Health Orgn, 67-75; mem, Nat Comn Res, 78-80; vis prof, Univ Calif, Los Angeles, 76 & Univ Mexico, 81. *Honors & Awards:* Hon mem fac, Univ Chile, 66. *Mem:* Nat Acad Sci; Am Soc Biol Chem (treas, 51-55); Brit Biochem Soc; hon mem Harvey Soc; hon mem Mex Nat Acad Med; hon mem Arg Biochem Soc; hon mem Japanese Biochem Soc; hon mem Chiba Med Soc Japan. *Res:* Intermediary nitrogen metabolism; action of thyroxine; differentiation and development; comparative biochemistry. *Mailing Add:* 529 Med Sci Bldg Univ Wis Madison WI 53706

COHEN, PINYA, b Burlington, Vt, Dec 23, 35; m 82; c 1. BIOCHEMISTRY, IMMUNOLOGY. *Educ:* Del Valley Col, BS, 57; Univ Ga, MS, 59; Purdue Univ, PhD(microbiol), 64. *Prof Exp:* Res microbiologist, NIH, 64-68, chief plasma derivatives sect, Lab Blood & Blood Prod, 68-72; dir plasma derivatives br, Bur Biologics, Food & Drug Admin, 72-76; dir regulatory affairs, Merieux Inst, 76-78, vpres qual control & regulatory affairs, 79-90; VPRES REGULATORY AFFAIRS, CONNAUGHT LABS, 90- *Concurrent Pos:* Mem adv panel, Comt Vaccine Innovation, Nat Acad Sci, 83-84. *Mem:* AAAS; Am Soc Microbiol; Am Asn Immunologists; Int Soc Blood Transfusion; NY Acad Sci. *Res:* Immunology of plasma proteins; immunology of viral and bacterial vaccines. *Mailing Add:* Connaught Labs Rte 611 PO Box 187 Swiftwater PA 18370-0187

COHEN, RAYMOND, b St Louis, Mo, Nov 30, 23; m 48, 86; c 3. MECHANICAL ENGINEERING. *Educ:* Purdue Univ, BSME, 47, MSME, 50, PhD, 55. *Prof Exp:* From instr to assoc prof, 47-60, asst dir, R W Herrick Labs, 70-71; PROF MECH ENG, PURDUE UNIV, 60-; DIR, R W HERRICK LABS, 71- *Concurrent Pos:* Consult, Gen Elec Co, 56-60 & Bendix Corp, 62-72; dept ed, Encycl Britannica, 57-62; consult, Gen Motors Res Labs, 74-78, Whirlpool Corp, 85-, Carrier Corp, 88-; fel sr scientist, NATO, 70-71; actg head mech eng, R W Herrick Labs, 88-89. *Honors & Awards:* E K Campbell Award, Am Soc Heating Refrig & Air Conditioning Engrs, 82, Int Activ Award, 86; Wilbur T Pentzer Achievement & Leadership Award, US Nat Comt for Int Inst Refrig, 90. *Mem:* Am Soc Mech Engrs; Am Soc Eng Educ; Nat Soc Prof Engrs; Am Soc Heating, Refrig & Air Conditioning Engrs; Int Inst Refrig; Sigma Xi; Inst Noise Control Eng. *Res:* Compressor technology; noise and vibration control. *Mailing Add:* R W Herrick Labs Purdue Univ W Lafayette IN 47907

COHEN, RICHARD LAWRENCE, b Philadelphia, Pa, Oct 6, 22; m 50; c 2. CHILD PSYCHIATRY. *Educ:* Univ Pa, AB, 43, MD, 47; Am Bd Psychiat & Neurol, cert psychiat, 53, cert child psychiat, 60. *Prof Exp:* Clin dir psychiat, Embreeville State Hosp, 51-52; dir child psychiat, Oakburne Hosp, 57-62; dir training, Philadelphia Child Guid Clin, 62-64; assoc prof child psychiat, Col Med, Univ Nebr, 64-67; assoc prof, 67-70, PROF CHILD PSYCHIAT, SCH MED, UNIV PITTSBURGH, 70-, CHIEF CHILD

PSYCHIAT, DEPT PSYCHIAT, 72-; DIR CHILDREN'S SERV, WESTERN PSYCHIAT INST & CLIN, 72-; EXEC DIR, PITTSBURGH CHILD GUID CTR, 72- *Concurrent Pos:* Psychiat consult, Univ Settlement House, Jewish Family Serv Philadelphia, Asn Jewish Children, Nat Teacher Corps & Student Health Serv & Univ Nebr; mem deans comt psychiat residency, 48-51; exec dir, Pittsburgh Child Guid Ctr, 72-79. *Mem:* AAAS; Am Acad Child Psychiat; Am Orthopsychiat Asn; Am Psychiat Asn; Am Pub Health Asn. *Res:* Prenatal prevention of developmental disorders in children; operations research into systems of delivery of medical service. *Mailing Add:* Western Psychiat Inst & Clin 0322 3811 O'Hara St Pittsburgh PA 15213

COHEN, RICHARD LEWIS, b New York, NY, Sept 8, 36; m 59; c 1. SOLID STATE SCIENCE. *Educ:* Haverford Col, BS, 57; Calif Inst Technol, MS, 59, PhD(physics), 62. *Prof Exp:* Mem tech staff physics, Bell Labs, 62-86, supvr protection apparatus develop, 86-89; VPRES ENG, PANAMAX, 90- *Concurrent Pos:* Res fel, Inst Physics, Munich Tech Univ, 64-65; mem ed bd, Rev Sci Instruments, 75-78; vpres eng, Panamax. *Honors & Awards:* Gold Medal, Electroplaters Soc, 76. *Mem:* Fel AAAS; fel Am Phys Soc; Inst Elec & Electronics Engrs. *Res:* Mössbauer effect to study nuclear and solid state physics, especially with rare-earth isotopes; x-ray photoelectric spectroscopy in solids; fiber optics; colloidal catalysts; gold electrodeposits; pulsed annealing; non-destructive testing; lightning protection. *Mailing Add:* Panamax 150 Mitchell Blvd San Rafael CA 94903

COHEN, RICHARD M, b 1946. MODELLING OF SEMICONDUCTORS. *Educ:* Univ Utah, PhD(mat sci), 83. *Prof Exp:* ASST PROF, UNIV UTAH, 85- *Concurrent Pos:* Consult, 85- *Mem:* Am Phys Soc. *Res:* Growth of III-V Materials; modelling of semiconductors quantum well structure; growth of thin and diamond-like film. *Mailing Add:* Dept Mat Sci & Eng Univ Utah Salt Lake City UT 84112

COHEN, ROBERT, b Indianapolis, Ind, Oct 15, 24; m 63; c 2. ENERGY CONVERSION, ELECTRICAL ENGINEERING. *Educ:* Wayne Univ, BS, 47; Univ Mich, MS, 48; Cornell Univ, PhD(elec eng), 56. *Prof Exp:* Res asst, Univ Mich, 47-48, Purdue Univ, 48-51 & Cornell Univ, 51-56; physicist, Aeronomy Lab, Environ Res Labs, Nat Oceanic & Atmospheric Admin, 56-73, prog mgr ocean thermal energy conversion, NSF Res Appl to Nat Needs, 73-75; br chief ocean thermal energy conversion, Div Solar Energy, Energy Res & Develop Admin, 75-76; prog mgr, Ocean Systs Br, Div Cent Solar Technol, US Dept Energy, 77-81; energy consult, 81-85; Energy Eng Bd, Nat Acad Sci, 85-90; ENERGY CONSULT, 91- *Concurrent Pos:* mem, energy policy comt, Inst Elec & Electronics Engrs, 75- *Honors & Awards:* Boulder Scientist Award, 64; Compass Distinguished Achievement Award, Marine Technol Soc, 80. *Mem:* Sigma Xi; Inst Elec & Electronics Engrs; Am Geophys Union; Int Solar Energy Soc; fel AAAS. *Res:* Ionospheric radio-wave propagation; irregularities in the ionosphere; equatorial ionosphere; ionospheric modification; aeronomy; Program management of the US ocean thermal energy conversion program; senior program officer, studies on alternative energy research and development strategies, magnetic fusion, and space- based power. *Mailing Add:* Energy Consult 1410 Sunshine Canyon Dr Boulder CO 80302

COHEN, ROBERT EDWARD, b Oil City, Pa, Jan 21, 47; m 78; c 2. POLYMER SYNTHESIS, POLYMER MORPHOLOGY. *Educ:* Cornell Univ, BS, 68; Calif Inst Technol, MS, 70, PhD(chem eng),72. *Prof Exp:* Fel eng sci, Oxford Univ, 72-73; from asst prof to prof, 73-88, BAYER PROF CHEM ENG, MASS INST TECHNOL, 88-, ASSOC CHMN FAC, 89- *Concurrent Pos:* Harold & Esther Egerton asst prof, 75-77, dir prog polymer sci & technol, Mass Inst Technol, 84-87; vis prof, Sandia Nat Lab, 79, Inst Guido Donegani, Italy, 81-82. *Mem:* Am Chem Soc; Am Inst Chem Engrs; Soc Rheology; Mat Res Soc; Brit Soc Rheology; NY Acad Sci. *Res:* Developing an understanding of basic relationships between polymer structure (molecular & morphological) & polymer properties; emphasis on formulating well characterized polymer systems which will aid in the elucidation of these structure/property relationships. *Mailing Add:* Dept Chem Eng Bldg 66 Rm 554 Mass Inst Technol Cambridge MA 02139

COHEN, ROBERT ELLIOT, b Newark, NJ, May 19, 53. PROTEIN BIOCHEMISTRY, ENZYMOLOGY. *Educ:* Univ Calif, Berkeley, AB, 74, PhD(biochem), 80. *Prof Exp:* Postdoctoral fel, Dept Molecular Biol, Univ Calif, Berkeley, 80-85; ASST PROF CHEM & BIOCHEM, DEPT CHEM & BIOCHEM, UNIV CALIF, LOS ANGELES, 85- *Concurrent Pos:* Awardee, Du Pont Young Fac Award, 85 & 86. *Mem:* Am Soc Biochem & Molecular Biol; Am Chem Soc; Am Soc Microbiol; Protein Soc. *Res:* Control of intracellular protein degradation; biochemistry and molecular biology of ubiquitin; protein folding, stability and conformational equilibria; protein-protein interactions. *Mailing Add:* Dept Chem & Biochem Univ Calif Los Angeles CA 90024-1569

COHEN, ROBERT JAY, b Milwaukee, Wis, May 31, 42; m 68; c 3. BIOPHYSICS, BIOCHEMISTRY. *Educ:* Univ Wis, BS, 64; Yale Univ, PhD(biophys chem), 69. *Prof Exp:* NIH fel, Calif Inst Technol, 69-71; asst prof, 71-76, ASSOC PROF BIOCHEM, COL MED, UNIV FLA, 76- *Concurrent Pos:* Vis prof, Univ Freiburg, 80. *Mem:* Am Chem Soc; AAAS; Biophysics Soc. *Res:* Sensory and hormonal transduction in model system Phycomyces; visual biochemistry physical chemistry of nucleic acids; electrokinetics, biomathematics, receptor physiology. *Mailing Add:* Dept Biochem & Molecular Biol Univ Fla Col Med Gainesville FL 32610

COHEN, ROBERT MARTIN, b Brooklyn, NY, Nov 27, 46; m 82; c 2. BRAIN IMAGERY, NEUROPHARMACOLOGY. *Educ:* Brooklyn Col, BS, 66; Princeton Univ, PhD(biochem), 72; Washington Univ, St Louis, MD, 74. *Prof Exp:* Chief resident, dept psychiat, Univ Chicago, 76-77; chief, Unit Clin Psychopharmacol, 80-83, CHIEF SECT CLIN BRAIN IMAGING, NIMH, 83- *Honors & Awards:* AE Bennett Award, Neuropsychiat Foun, 82. *Mem:* AAAS. *Res:* New tracers and other approaches to the study of neurotransmitter function in normal and abnormal physiology; experimental therapeutics of Alzheimers disease; pathophysiology of neuropsychiatric disorders. *Mailing Add:* Clin Brain Imaging Sect NIMH Lab 9000 Rockville Pike Bldg 10 Rm 4N317 Bethesda MD 20205

COHEN, ROBERT ROY, b Duluth, Minn, June 3, 39. VERTEBRATE ZOOLOGY, ORNITHOLOGY. *Educ:* Univ Minn, Duluth, BA, 61; Univ Colo, PhD(zool), 65. *Prof Exp:* Asst prof biol, Univ Sask, 65-67 & NMex Inst Mining & Technol, 67-69; assoc prof, 69-74, PROF BIOL, METROP STATE COL, DENVER, 74- *Mem:* AAAS; Sigma Xi; Wilson Ornith Soc; Am Ornith Union; Animal Behav Soc; Cooper Ornith Soc. *Res:* Breeding biology and population dynamics of the tree swallow; avian erythrocyte and hemoglobin kinetics. *Mailing Add:* Dept Biol Campus Box 053 Metrop State Col PO Box 173362 Denver CO 80204

COHEN, ROBERT SONNE, b New York, NY, Feb 18, 23; m 44; c 3. PHILOSOPHY OF SCIENCE, THEORETICAL PHYSICS. *Educ:* Wesleyan Univ, AB, 43; Yale Univ, MS, 43, PhD(physics), 48. *Hon Degrees:* LHD, Wesleyan Univ, 86. *Prof Exp:* Instr physics, Yale Univ, 43-44; mem sci staff, Div War Res, Columbia Univ, 44-46; Am Coun Learned Socs fel philos of sci, Yale Univ, 48-49, instr philos, 49-51; asst prof physics & philos, Wesleyan Univ, 49-57; assoc prof physics, 57-59, chmn dept, 58-73, actg dean, Col Lib Arts, 71-73, PROF PHYSICS & PHILOS, BOSTON UNIV, 59- *Concurrent Pos:* Mem tech staff, Joint Commun Bd, US Joint Chiefs of Staff, 44-46; consult, Fund Advan Educ, 51-53 & Nat Woodrow Wilson Found, 59-64; Ford fac fel, 55-56; vis prof, Mass Inst Technol, 58-59, Brandeis Univ, 59-60 & Univ Calif, San Diego, 69; chmn, Ctr Philos & Hist of Sci, Boston Univ, 61-; vis lectr, Polish & Czech Acad Sci, 62, Yugoslav Philos Asn, 63 & Hungarian Acad Sci, 64; ed, Boston Studies Philos Sci, 63-, Vienna Circle Collection, 70- & Studies Hist Mod Sci, 76-; mem staff, Oak Ridge Conf Sci & Contemporary Social Probs, 64; chmn, Am Inst Marxist Studies, 64-80; trustee, Inst Unity of Sci, 66-72, Wesleyan Univ, 68- & Tufts Univ, 84-; chmn US nat comt, Int Union Hist & Philos Sci, 67-69; chmn, dept philos, 86-88. *Mem:* Am Phys Soc; Am Asn Physics Teachers; Hist Sci Soc; Am Philos Asn; Philos Sci Asn (vpres, 73-75, pres, 82-84); AAAS. *Res:* Concept and theory formation in physical sciences; science and the social order; logical empiricism and natural science; dialectical materialism and science; general education in science; history of scientific concepts; comparative historical sociology of science. *Mailing Add:* Dept Philos Boston Univ 745 Commonwealth Ave Boston MA 02215

COHEN, ROCHELLE SANDRA, b Brooklyn, NY, June 20, 45. ENDOCRINOLOGY, NEUROBIOLOGY. *Educ:* Rutgers Univ, AB, 67; Univ Conn, MS, 70, PhD(physiol, endocrinol), 73. *Prof Exp:* Res asst endocrinol, Univ Conn, 68, teaching asst biol, 68-72; asst prof, State Univ NY Col Purchase, 72-74; NIH fel cell biol, Rockefeller Univ, 74-76, res assoc cell biol, 76-77; RES ASSOC & ASSOC PROF, RUDOLF MAGNUS INST PHARMACOL, UNIV ILL MED CTR, 77- *Concurrent Pos:* Asst prof biol, Empire State Col, 73-74; res assoc biochem genetics, Rockefeller Univ, 73-74. *Mem:* Am Soc Cell Biol; NY Acad Sci; Soc Neurosci. *Res:* Characterization of the synaptic junction by the use of biochemical, ultra structural and immunological techniques to find the relationship between its structure and function. *Mailing Add:* Dept Anat & Cell Biol Univ Ill PO Box 6998 Chicago IL 60680

COHEN, ROGER D H, b Singapore, Jan 23, 38; Australian citizen; m 68; c 4. BEEF CATTLE PRODUCTION, RUMINANT NUTRITION. *Educ:* Univ New Eng, Australia, BRurSc, 67, PhD, 80. *Prof Exp:* Res officer, Dept Agr, New SWales, 67-80; asst prof, Univ BC, 80-82; res scientist, 82-84, PROF, UNIV SASK, 84- *Concurrent Pos:* Dir, Termuende Res Sta, Univ Sask, 82-91, Sask Beef Cattle Res Ctr & Rop Bull Test Sta, 84-91, Can Soc Animal Sci, 86-88. *Mem:* Am Soc Animal Sci; Can Soc Animal Sci; Soc Range Mgt; Australian Soc Animal Production. *Res:* Beef cattle production and management, cow-calf and feedlot; nutrition; physiology of digestion and reproduction; stress physiology; pasture and forage management; computer simulation modeling of beef production systems. *Mailing Add:* Dept Animal & Poultry Sci Univ Sask Saskatoon SK S7N 0W0 Can

COHEN, RONALD R H, b Philadelphia, Pa, Jan 14, 47; m 70. ESTUARINE ECOLOGY, PHYTOPLANKTON ECOLOGY. *Educ:* Temple Univ, BA, 71; Univ Va, PhD(environ sci), 78. *Prof Exp:* Instr ecol, Univ Va, 77; ECOLOGIST-HYDROLOGIST ESTUARINE ECOL, DEPT INTERIOR, US GEOL SURV, 78- *Mem:* Am Soc Limnol & Oceanog; Ecol Soc Am; AAAS; Am Soc Plant Physiol; Sigma Xi. *Res:* Interaction between nutrients, light and primary productivity; estuarine phytoplankton dynamics and modelling; effect of invertebrates on phytoplankton biomass; statistical techniques in phytoplankton research. *Mailing Add:* Colo Sch Mines Dept Environ Sci & Eng Golden CO 80401

COHEN, SAMUEL ALAN, b Brooklyn, NY, Feb 3, 47; m 73; c 2. PLASMA PHYSICS, SURFACE PHYSICS. *Educ:* Mass Inst Technol, BS, 68, PhD(physics), 72. *Prof Exp:* Teaching asst physics, Mass Inst Technol, 68-72; mem res staff, 73-78, res physicist, 78-81, PRIN RES PHYSICIST, PRINCETON UNIV, 81-, LECTR, DEPT PLASMA PHYSICS & ASTROPHYS SCI, PLASMA PHYSICS LAB, 85- *Concurrent Pos:* Consult, Res Lab Electronics, Mass Inst Technol, 73- & Dept Energy-ETM Task Group on Plasma-Wall Interactions, 76- *Honors & Awards:* Goodwin Medal. *Mem:* Fel Am Phys Soc; Am Vacuum Soc. *Res:* Experimental work on surface physics and plasma physics related to controlled thermonuclear fusion and semiconductor processing. *Mailing Add:* Plasma Physics Lab Princeton Univ Forrestal Campus Princeton NJ 08543

COHEN, SAMUEL H, b Boston, Mass, May 11, 38; m 60; c 2. ELECTRON MICROSCOPY. *Educ:* Boston Univ, AB, 60; Northeastern Univ, MS, 72. *Prof Exp:* RES BIOLOGIST ELECTRON MICROS, US ARMY NATICK LABS, 64- *Concurrent Pos:* Ed, Food Microstruct J, 82-; adv bd, Microscope Technol & News, 89- *Mem:* Fel Royal Micros Soc; Electron Micros Soc Am. *Res:* Chromosome analysis and study of the effects of ultraviolet on cockroaches; effects of lysosomal proteases on bovine myofibrils; microstructural analysis of carbonaceous adsorbents; microscopic analysis of ballistic fibers. *Mailing Add:* 43 Lohnes Rd Framingham MA 01701

COHEN, SAMUEL MONROE, b Milwaukee, Wis, Sept 24, 46; m 68; c 4. BIOCHEMISTRY. *Educ:* Univ Wis-Madison, BS, 67, MD & PhD(oncol), 72. *Prof Exp:* Instr path, Med Sch, Univ Mass, 74-77, assoc prof, 77-81; PROF RES, EPPLEY INST CANCER RES & PROF & VCHMN PATH & MICROBIOL, UNIV NEBR MED CTR, 81 - *Concurrent Pos:* Staff pathologist, St Vincent Hosp, 75-76 & 77-81; vis prof path, Nagoya City Univ Med Sch, 76-77; assoc ed, Cancer Res, 82-90, Food Chem Toxicol, 90; NIH study sections, Chem Path, 82-86, immunol, virol & path, 89-93. *Mem:* Am Asn Path; Am Asn Cancer Res; Japanese Cancer Asn; AAAS; Soc Toxicol; Soc Toxicol Pathologists. *Res:* Mechanisms of carcinogenesis, including carcinogen metabolism, molecular biology, evaluation of factors influencing the different stages and development of a generalized computer-based model of carcinogenesis; diagnostic application of electron microscopy; uropathology. *Mailing Add:* Dept Path & Microbiol Univ Nebr Med Ctr 600 S 42nd St Omaha NE 68198-3135

COHEN, SANFORD I, b New York, NY, Sept 5, 28; m 52; c 4. PSYCHIATRY, PSYCHOPHYSIOLOGY. *Educ:* NY Univ, AB, 48; Chicago Med Sch, MD, 52; Am Bd Psychiat, dipl, 59. *Prof Exp:* Resident psychiat, Med Ctr, Univ Colo, 54 & Med Ctr, Duke Univ, 55, instr, 56-58, assoc, 58-59, from asst prof to assoc prof, 59-61, head div psychophysiol res, 60-65, prof & chmn exec comt inter-dept res training prog nerv syst sci, 64-68, head div psychosom med & psychophysiol res, 65-68; prof psychiat & biobehav sci & chmn dept, Med Ctr, La State Univ, New Orleans, 68-70; prof psychiat & chmn div, Med Sch, Boston Univ, 70-86, psychiatrist-in-chief, Univ Hosp, 70-86; PROF PSYCHIAT, UNIV MIAMI, SCH MED, 88- *Concurrent Pos:* Markle scholar, 57-62; fel psychoanal, Found Fund Res Psychiat, 58-63; instr, Wash Psychoanal Inst, 64-; mem ment health small grants comt, NIMH, 63-66; supt, Dr Solomon Carter Fuller Ment Health Ctr, 70-75; vis scientist, Health & Behav Br Div Basic Sci, NIMH, 86-88. *Mem:* AAAS; fel Am Psychiat Asn; Am Psychosom Soc; Am Fedn Clin Res; Soc Biol Psychiat. *Res:* Psychosomatic medicine; psychophysiology of emotions and behavior; neuroendocrinology and physiology of conditioned reflexes; effects of altered sensory environments; perceptual mode, personality and Pavlovian typology; life stress, hopelessness, and voodoo; death. *Mailing Add:* Dept Psychiat Univ Miami Sch Med 1400 NW Tenth Ave Rm 301 Miami FL 33136

COHEN, SANFORD NED, b Bronx, NY, June 12, 35; m 58, 84; c 1. PHARMACOLOGY, DEVELOPMENTAL BIOLOGY. *Educ:* Johns Hopkins Univ, AB, 56, MD, 60. *Prof Exp:* Res physician, Walter Reed Army Inst Res, 63-65; from instr to asst prof pharmacol, Sch Med, NY Univ, 65-71, asst prof pediat, 68-71, assoc prof pharmacol & pediat, 71-74; pediatrician-in-chief, Children's Hosp Mich, 74-81; chmn dept, 74-81, asst dean, 81-82, assoc dean, 82-86, dir, Develop Disabilities Inst, 83-86, PROF PEDIAT, SCH MED, WAYNE STATE UNIV, 81-, SR VPRES ACAD AFFAIRS & PROVOST, 86- *Concurrent Pos:* Nat Inst Child Health & Human Develop spec fel, 63-65; Markle Found & John & Mary R Markle scholars acad med, 68; physician-in-chg nurseries & assoc dir pediat, Bellevue Hosp, 69-74; vis prof, Georgetown Univ, 80. *Honors & Awards:* John F Kennedy lectr, Georgetown Univ, 80. *Mem:* Am Pediat Soc; Am Acad Pediat; Soc Pediat Res (vpres, 80-81); Am Soc Pharmacol & Exp Therapeut. *Res:* Developmental pharmacology; how developmental phenomena alter the rate of metabolism or the biological effects of drugs in the immature animal. *Mailing Add:* Wayne State Univ 1170 Mackenzie Hall Detroit MI 48202

COHEN, SAUL G, b Boston, Mass, May 10, 16; m 41; c 2. ORGANIC CHEMISTRY, BIOCHEMISTRY. *Educ:* Harvard Univ, AB, 37, MA, 38, PhD(org chem), 40. *Hon Degrees:* ScD, Brandeis Univ, 86. *Prof Exp:* Pvt asst, Harvard Univ, 39-40, instr chem, 40-41; res assoc, Nat Defense Res Comt, 41, res fel, 41-43; Nat Res fel & lectr chem, Univ Calif, Los Angeles, 43-44; res chemist, Pittsburgh Plate Glass Co, 44-45; sr chemist, Polaroid Corp, 45-50; assoc prof chem, Brandeis Univ, 50-52, prof, 52-74, chmn dept, 59-72, dean fac, 55-58, univ prof chem, 74-86, EMER PROF CHEM, BRANDEIS UNIV, 86- *Concurrent Pos:* Fulbright sr scholar & Guggenheim fel, UK, 58-59; consult, Polaroid Corp; mem, bd of overseers, Harvard Univ. *Honors & Awards:* James F Norris Award, Am Chem Soc, 72. *Mem:* Am Chem Soc; Am Acad Arts & Sci; Royal Soc Chem; Am Soc Biol Chem. *Res:* Mechanisms of organic reactions; free radicals; polymerization; stereochemistry; photography; enzyme reactions; photochemistry. *Mailing Add:* Dept Chem Brandeis Univ Waltham MA 02254-9110

COHEN, SAUL ISRAEL, b Boston, Mass, Feb 15, 26; m 52; c 2. NUTRITION. *Educ:* Northeastern Univ, BS, 45; Boston Univ, MA, 46; Harvard Univ, MS, 51; Columbia Univ, PhD, 56. *Prof Exp:* Asst, Harvard Med Sch, 44-45; jr chemist & jr bacteriologist, State Dept Pub Health, Mass, 46-48; asst, Sch Pub Health, Harvard Univ, 49-52; instr, NY Med Col, 55; res assoc, Harvard Med Sch, 56-64; assoc prof, NEssex Community Col, 65-66; RES ASST, RETINA FOUND, 67- *Concurrent Pos:* Teacher high sch, Boston, 67-68. *Mem:* AAAS. *Res:* Chemical interactions of the protein components of blood plasma; amino acids and proteins in liver repair. *Mailing Add:* Dept Chem Brandeis Univ Waltham MA 02154

COHEN, SAUL LOUIS, b Regina, Sask, May 10, 13; m 38; c 3. BIOCHEMISTRY, ENDOCRINOLOGY. *Educ:* Brandon Col, BA, 32; Univ Toronto, PhD(biochem, physiol), 36. *Prof Exp:* Exhib of 1851 fel, Swiss Fed Inst Technol, 36-37; instr physiol, Med Sch, Ohio State Univ, 37-42; asst prof, Med Sch, Univ Mich, 42-46; from asst prof to assoc prof biochem, Univ Minn, 46-56; res asst, 62-67, asst prof, 67-76, ASSOC PROF PATH CHEM, OBSTET & GYNEC, MED SCH, UNIV TORONTO, 76- *Honors & Awards:* Ortho Award, Soc Obstetricians & Gynecologists Can, 67. *Mem:* AAAS; Am Soc Biol Chem; Endocrine Soc; Can Fedn Biol Sci. *Res:* Biochemical endocrinology; hydrolysis, assay, concentration, isolation identification, metabolism and significance of the sex steroids, both free and conjugated. *Mailing Add:* Rm 7263 Med Sci Bldg Univ Toronto Toronto ON M5S 1A1 Can

COHEN, SAUL MARK, b Springfield, Mass, Oct 6, 24; m 53; c 1. ORGANIC CHEMISTRY, COATINGS. *Educ:* Univ Mass, BS, 48; Univ Ill, MS, 49, PhD(org chem), 52. *Prof Exp:* Asst, Univ Ill, 50-52; res chemist, Eastman Kodak Co, 52-55; res chemist, Shawinigan Resins Corp, 55-60, res group leader, 60-65; res specialist, Monsanto Co, 65-69, sr res specialist, 69-85; RETIRED. *Concurrent Pos:* Aerial navigator, USAF, 43-45. *Honors & Awards:* Arthur K Doolittle Award, Div Org Coatings & Plastic Chem, Am Chem Soc, 72. *Mem:* AAAS; NY Acad Sci; Am Chem Soc; fel Am Inst Chemists; Sigma Xi. *Res:* Organic synthesis; polymerization; polymer structure and properties; mechanism studies; automotive safety glass windshields; coatings. *Mailing Add:* 15 Lindsay Rd Springfield MA 01128

COHEN, SEYMOUR STANLEY, b New York, NY, Apr 30, 17; m 40; c 2. BIOCHEMISTRY, HISTORY OF SCIENCE. *Educ:* City Col New York, BS, 36; Columbia Univ, PhD(biochem), 41. *Hon Degrees:* DhC, Univ Louvain, 72, Univ Kuopio, 82. *Prof Exp:* Res assoc biochem, Columbia Univ, 42-43; instr pediat, Univ Pa, 45-47, from assoc prof to prof physiol chem, Dept Pediat, 47-57, Am Cancer Soc Charles Hayden prof biochem, 57-71, Hartzell prof therapeut res & chmn dept, 63-71; prof, Sch Med, Univ Colo, 71-72, Am Cancer Soc prof microbiol, 71-76; distinguished prof & Am Cancer Soc prof, 76-85, EMER PROF PHARMACOL SCI, STATE UNIV NY STONY BROOK, 85-; VIS PRES SCHOLAR, UNIV CALIF, SAN FRANCISCO, 88- *Concurrent Pos:* Abbott Lab fel, Columbia Univ, 40-41; Nat Res Found fel plant viruses, Rockefeller Inst, 41-42; Johnson Found fel, Univ Pa, 43-45; Guggenheim fel, Pasteur Inst, Paris, 47-48 & 82-83; Lalor fel, Marine Biol Lab, 50-51; Fogarty scholar, Nat Cancer Inst, 73-74; ed, Virol, 54-59, J Biol Chem, 60-65 & J Bact Rev, 69-73; vis prof, Radium Inst, 67, Col France, 70, Hadassah Med Sch, 74 & Univ Tokyo, 74; Smithsonian scholar, 73-74; trustee, Marine Biol Lab, 68-85, res assoc, 87; mem bd sci consult, Sloan-Kettering Inst; chmn, Coun for Anal & Proj, Am Cancer Soc, 71-75; fel, Nat Humanities Ctr, 82-83 & 85; Lady Davis fel, Hebrew Univ, 83. *Honors & Awards:* Eli Lilly Award, 51; Mead Johnson Award, 52; Cleveland Award, AAAS, 55; Borden Award, Am Asn Med Cols, 68; Passano Award, 74; French Soc Biol Chem Medal, 64; Forster Prize, Mainz Acad Sci & Letters, 78. *Mem:* Inst Med Nat Acad Sci; fel AAAS; fel Am Acad Arts & Sci; Soc Gen Physiol (pres, 68); hon mem Soc Française de Microbiologie. *Res:* Chemistry of viruses and nucleoproteins; metabolism of bacteria and virus infected cells; nucleic acids and phosphate compounds; polyamines; cancer research and infectious disease. *Mailing Add:* Ten Carrot Hill Rd Woods Hole MA 02543

COHEN, SHELDON A, b Detroit, Mich, Oct 11, 47. PSYCHOLOGY. *Educ:* NY Univ, PhD(psychol), 73. *Prof Exp:* From asst prof to assoc prof psychol, Univ Ore, 73-82; PROF PSYCHOL, CARNEGIE-MELLON UNIV, 82- *Honors & Awards:* Outstanding Contrib Health Psychol, Am Psychol Asn, 87. *Mem:* Am Psychol Asn; Soc Behav Med. *Res:* Influence of psychosocial and behavioralfactors. *Mailing Add:* Dept Psychol Carnegie Mellon Univ Pittsburgh PA 15213

COHEN, SHELDON GILBERT, b Pittston, Pa, Sept 21, 18. IMMUNOLOGY, ALLERGY. *Educ:* Ohio State Univ, BA, 40; NY Univ, MD, 43. *Hon Degrees:* DSc, Wilkes Col, 76. *Prof Exp:* Intern, Bellevue Hosp, 44; resident internal med, Vet Admin Hosp, Md, 47-48; resident allergy, Vet Admin Hosp & Univ Med Ctr, Univ Pittsburgh, 48-49; res fel, Addison H Gibson Lab Appl Physiol, Univ Pittsburgh, 49-50, res assoc, 50-51; res assoc immunol, Dept Biol, Wilkes Col, 51-56, from assoc prof to prof biol res, 57-68, prof exp biol, 68-72; consult, Nat Inst Allergy & Infectious Dis, NIH, 72-73, chief, Allergy & Immunol Br, Extramural Prog, 73-76, dir Immunol, Allergic & Immunol Dis Prog, 77-88; RETIRED. *Concurrent Pos:* Attend physician, Vet Admin Hosp, 51-60, consult internal med, 60-72, consult res, 61-72; chief allergy, Mercy Hosp, 51-72; reg med consult, Children's Asthma Res Inst & Hosp, Denver, Co, 69-72, mem bd dir, Lupus Found Am, 78-85, exec vpres, 81-85, med counr, 78-87; mem Medico adv bd, Care, 77-89; mem bd trustee, Marywood Col, Scranton, Pa, 83-89; mem sci adv bd, Terri Gotthelf Lupus Res Inst, 86-; adv, Immunol Unit, WHO, Geneva, Switzerland, 79, mem, Expert Adv Panel Immunol, 81-, Int Union Against Tuberc Respiratory Dis Comt Epidemiol Asthma, 81-83; mem sci adv bd, Allergy Inst, Int Life Sci Inst, 89-; consult, Ministry Pub Health, State Kuwait, 81-82; consult, Allergy/Clin Immunol Service, Walter Reed Army Med Ctr, Wash, DC, 82-; Solomon A Berson Med Alumni Achievement Award Health Sci, NY Univ, 88; vis prof, Dept Med, Northwestern Univ, 86; mem, Nat Inst Allergy & Infectious Dis; mem, Comt Standardization Allergens, 59-62. *Honors & Awards:* Distinguished Service Award, Am Acad Allergy, 71, Asthma & Allergy Found Am, 81; M Peshkin Award, Asn Care Asthma, 81; Clemens von Pirquet Award, Georgetown Univ, 81; Terri Gotthelf Lupus Res Inst Centennial Award, NIH, 87; M Murray Peshkin Lectr Award, Asn Care Asthma, 81; Mario Salazar Mallen Lectureship, Mex Soc Allergy & Immunol, Guadalajara, Mex, 82; Mathew Walzer Lectureship, VIII Latin-Am Allergy & Immunol Conf, Monterrey, Mex, 83; Michael J Bent Mem Lectureship, Meharry Med Col, 85; Bernard Halpern Lectr, V Int Symp Pediat Allergy, Mexico City, 87; First Francis H Chaffee Lectr, Rhode Island Soc Allergy, 87; Achievement Award, Int Asn Allerogol & Clin Immunol, 88; Spec Recognition Award, Am Acad Allergy & Immunol, 89. *Mem:* Am Asn Immunol; Soc Exp Biol & Med; fel Am Acad Allergy; fel Am Col Physicians; Asn Am Physicians; Am Rheumatism Asn. *Res:* Immunologic basis of hypersensitivity reactions; experimental eosinophilia; history of medicine. *Mailing Add:* Nat Inst Allergy & Infectious Dis NIH Control Data Bldg Rm 2075 Bethesda MD 20892

COHEN, SHELDON H, b Milwaukee, Wis, May 21, 34; m 62; c 3. INORGANIC CHEMISTRY. *Educ:* Univ Wis, BS, 56; Univ Kans, PhD(chem), 62. *Prof Exp:* From asst prof to assoc prof, 60-70, vpres acad affairs, 82-84, PROF CHEM & CHMN DEPT, WASHBURN UNIV, 70- *Concurrent Pos:* Feature ed, J Chem Educ. *Mem:* Am Chem Soc; AAAS. *Res:* Polarography of inorganic systems; preparation of compounds with unusual oxidation states; preparation and stability of inorganic complexes; chemical education. *Mailing Add:* Dept Chem Washburn Univ Topeka KS 66621

COHEN, SIDNEY, b Boston, Mass, June 3, 28; m 62; c 2. ORGANIC CHEMISTRY. *Educ:* Northeastern Univ, BS, 51; Tufts Univ, MS, 52; Univ Colo, PhD(org chem), 59. *Prof Exp:* Develop engr, Chem Div, Gen Elec Co, NY, 52-55; res assoc fluorine chem, Univ Colo, 60-61; from asst prof to assoc prof chem, Ft Lewis Col, 61-66; assoc prof, 66-69, PROF CHEM, STATE UNIV NY COL BUFFALO, 69- *Mem:* Am Chem Soc. *Res:* Small ring compounds; oxycarbons; fluorine chemistry; photochemistry. *Mailing Add:* NY State Col Buffalo NY 14222

COHEN, SIDNEY, b Malden, Mass, Jan 29, 13; m 54; c 2. MEDICINE, MICROBIOLOGY. *Educ:* Harvard Univ, AB, 33; Harvard Med Sch, MD, 37; Am Bd Internal Med, dipl. *Prof Exp:* Intern, Mt Sinai Hosp, 37-40; resident med, Beth Israel Hosp, 40-41; res assoc, 41-42, assoc path & med & in chg bact lab, 46-51, dir clin labs, 51-56, from assoc vis physician to vis physician, 49-58; prof, 71-82, EMER PROF MED, PRITZKER SCH MED, UNIV CHICAGO, 82- *Concurrent Pos:* Res fel bact, Harvard Med Sch, 40; asst bact & med, Harvard Med Sch, 41-42, instr med, 45-51, assoc, 51-58; dir microbiol lab, Michael Reese Hosp, 58-85. *Mem:* Infectious Dis Soc Am; Am Soc Microbiol; Am Col Physicians. *Res:* Medical microbiology with relation to staphylococcal infection; action of antibiotics and mechanism of microbial resistance to antibiotics. *Mailing Add:* Microbiol Lab Michael Reese Hosp 2900 S Ellis Ave Chicago IL 60616

COHEN, STANLEY, b Brooklyn, NY, Nov 17, 22; m 51, 80; c 3. BIOCHEMISTRY. *Educ:* Brooklyn Col, BA, 43; Oberlin Col, MA, 45; Univ Mich, PhD(biochem), 48. *Hon Degrees:* DSc, Univ Chicago, 85, Brooklyn Col, 87, Oberlin Col, 89. *Prof Exp:* Teaching fel, Dept Biochem, Univ Mich, 46-48; instr, Dept Biochem & Dept Pediat, Sch Med, Univ Colo, 48-52; Am Cancer Soc postdoctoral fel, Dept Radiol, Wash Univ, St Louis, Mo, 52-53, assoc prof, Dept Zool, 53-59; from asst prof to prof, 59-86, DISTINGUISHED PROF BIOCHEM, SCH MED, VANDERBILT UNIV, 86- *Concurrent Pos:* Res career develop award, NIH, 59-69; Am Cancer Soc res prof biochem, 76; Alfred P Sloan award, Gen Motors Cancer Res Found, 82; Lila Gruber Mem cancer res award, Am Acad Dermat, 83; Charles B Smith, vis res prof, Sloan Kettering, 84; Albert Lasker basic med res award, Albert & Mary Lasker Found, 86; mem, Bd Sci Counselors, Nat Inst Child Health & Human Develop, 86-; consult, Minority Res Ctr Excellence Meharry. *Honors & Awards:* Nobel Prize Physiol or Med, 86; Earl Sutherland Res Prize; Albion O Bernstein, M D Award, Med Soc NY; William Thomson Wakeman Award, Nat Paraplegia Found, 74; H P Robertson Mem Award, Nat Acad Sci, 81; Gairdner Found Int Award, 85; Feodor Lynen Lectr, Univ Miami, 86; Nat Medal of Sci, 86; Steenbock Lectr, Univ Wis, 86; Fred Conrad Kock Award, Endocrine Soc, H; Franklin Medal, Franklin Inst, 87; Albert A Michelson Award, Mus Sci & Indust, 87; Pincus Medal & Award, Worcester Found Exp Biol, 88. *Mem:* Nat Acad Sci; Am Soc Biol Chemists; Int Inst Embryol; Am Acad Arts & Sci. *Res:* Structure and function of growth factors; mechanism of hormone. *Mailing Add:* Dept Biochem Vanderbilt Univ Sch Med Nashville TN 37232-0146

COHEN, STANLEY, b Los Angeles, Calif, Feb 5, 27. THEORETICAL NUCLEAR PHYSICS, COMPUTER SOFTWARE. *Educ:* Univ Southern Calif, BS, 55; Cornell Univ, PhD(nuclear physics), 58. *Prof Exp:* Sr scientist, theoret group, Argonne Nat Lab, 60-78; PRES, SPEAKEASY COMPUT CORP, 78- *Mem:* Am Comput Mach Asn; Am Phys Soc. *Mailing Add:* 224 S Michigan Ave Chicago IL 60604

COHEN, STANLEY, b New York, NY, June 4, 37; c 3. IMMUNOPATHOLOGY. *Educ:* Columbia Univ, AB, 57, MD, 61. *Prof Exp:* Resident path, Mass Gen Hosp-Harvard, 62-64; fel path-immunol, Sch Med, NY Univ, 64-66; capt, Walter Reed Army Inst Res, 66-68; from assoc prof to prof path, State Univ NY, Buffalo, 68-74; prof path, Health Ctr, Univ Conn, 74-87; PROF & CHAIR PATH, HAHNEMANN UNIV, 87- *Concurrent Pos:* Assoc dir, Ctr Immunol, State Univ NY, Buffalo, 71-74, dir, 72-74, mem, Ctr Theoret Biol, 73-78; outstanding investr award, Nat Cancer Inst, 86; adj prof, Dept Path, Health Ctr, Univ Conn, 87-89. *Honors & Awards:* Kinne Award, Columbia Col, 54; Borden Award, Columbia Col Physicians & Surgeons, 61; Parke-Davis Award, Am Asn Pathologists, 77. *Mem:* Am Asn Immunologists; Am Asn Pathologists; Am Soc Clin Invest; Clin Immunol Soc; NY Acad Sci; Acad Path. *Res:* Mechanism of cytokine action of intracellular pathways that transduce signals for cell proliferation, movement and differentiation; novel cytosolic protein that induces DNA replication in cellfree systems. *Mailing Add:* Dept Path Hahnemann Univ Broad & Vine Sts Philadelphia PA 19102-1192

COHEN, STANLEY ALVIN, b Brooklyn, NY. ELECTRONICS ENGINEERING. *Educ:* City Col NY, BEE, 51; Polytechnic Inst Brooklyn, MEE, 54. *Prof Exp:* Electronic scientist, US Navy Appl Sci Lab, 51-55; engr, Sperry Gyroscope Co, 55-58; lectr elec eng, City Coll NY, 58-65; res engr, IIT Res Inst, 65-73; consult, Conic Res, 73-76; staff engr, E-Systs Inc, 76-77; proj leader, Space Div, Gen Elec Co, 77-78; SR ENGR, APPL PHYSICS LAB, JOHNS HOPKINS UNIV, 78- *Mem:* Inst Elec & Electronics Engrs. *Res:* Research, design, development and analysis of electronic equipment such as radar and communication systems; contributor of articles to professional journals on analysis of new concepts in electronic systems. *Mailing Add:* Appl Physics Lab Johns Hopkins Univ Johns Hopkins Rd Laurel MD 20723

COHEN, STANLEY NORMAN, b Perth Amboy, NJ, Feb 17, 35; m 61; c 2. MOLECULAR GENETICS, MICROBIOLOGY. *Educ:* Rutgers Univ, BS, 56; Sch Med, Univ Pa, MD, 60. *Prof Exp:* From asst prof to assoc prof med, 68-75, head, clin pharmacol div, 69-78, chmn genetics dept, 78-86, PROF MED, SCH MED, STANFORD UNIV, 75-, PROF GENETICS, 77- *Concurrent Pos:* Asst prof develop biol & cancer, Albert Einstein Col Med, 67-68; US Pub Health Serv grant, 69-74; Josiah Macy Found fac scholar, 75-76; mem, Comt Genetic Exp, Int Coun Sci Unions, 75; Guggenheim Found fel, 75; mem, comt genetic eng, Int Coun Sci Unions, 75; lectr, Harvey Soc, 79, Kinyon lectr, 81. *Honors & Awards:* Burroughs-Wellcome Scholar Award, 70; Mattia Award, Roche Inst Molecular Biol, 70; Harvey Soc lectr, 79; Lasker Award, 80; Marvin J Johnson Award, Am Chem Soc, 81; Wolf Prize, 81; Inst de la Vie Prize, 88; Cetus Award, Am Soc Microbiol, 88; Nat Medal Sci, 88; Nat Biotech Award, 88; Nat Medal Technol, 89. *Mem:* Nat Acad Sci; Inst Med-Nat Acad Sci; Am Soc Biol Chemists; Am Soc Microbiol; fel Am Acad Arts & Sci; Am Soc Clin Invest; Asn Am Physicians; Genetics Soc Am. *Res:* Molecular biology of bacterial plasmids and transposable elements; plasmid evolution and inheritance; control of gene expression. *Mailing Add:* Dept Genetics Sch Med Stanford Univ Stanford CA 94305

COHEN, STEPHEN ROBERT, b New York, NY, May 7, 28; m 54; c 2. NEUROCHEMISTRY, PHYSICAL CHEMISTRY. *Educ:* Cornell Univ, BChE, 51, PhD(phys chem), 56. *Prof Exp:* Asst phys chem, Cornell Univ, 51-55; instr chem, Brown Univ, 56-58; phys chemist, Itek Corp, 58-59; asst prof chem, Northeastern Univ, 59-61, City Col New York, 61-64; res assoc, Col Physicians & Surgeons, Columbia Univ, 64-66; sr res scientist, NY State Res Inst Neurochem & Drug Addiction, 66-70, assoc res scientist, 70-76; RES SCIENTIST V, NY STATE INST BASIC RES MENT RETARDATION, 76- *Concurrent Pos:* Adj prof chem, Long Island Univ, 86-, safety officer, radiation, 87-, sci, 89-, head, Off Lab Safety & Training, 90- *Mem:* Int Soc Neurochem; Am Chem Soc; Am Soc Neurochem; Royal Soc Chem; Fedn Am Socs Exp Biol. *Res:* Kinetics and reaction mechanisms; neurochemistry; active transport; extracellular spaces; data treatment. *Mailing Add:* NY State Inst Basic Res 1050 Forest Hill Rd Staten Island NY 10314

COHEN, STEVEN CHARLES, b New Kensington, Pa, Aug 27, 47; m 70; c 2. TECTONOPHYSICS, GEODYNAMICS. *Educ:* Drexel Univ, BS, 70; Univ Md, MS, 72, PhD(physics), 73. *Prof Exp:* GEOPHYSICIST/PHYSICIST, GODDARD SPACE FLIGHT CTR, NASA, 70- *Concurrent Pos:* Prin investr, Geoscience Laser Ranging Syst, 84-90. *Mem:* Am Geophys Union; Am Phys Soc. *Res:* Research activities on crustal deformation; continental collisions, mechanisms of earthquakes, geodynamic processes; application of space geodetic techniques to earth science studies. *Mailing Add:* Goddard Space Flight Ctr Code 921 Greenbelt MD 20771

COHEN, STEVEN DONALD, b Boston, Mass, Nov 22, 42; m 65; c 2. PHARMACOLOGY, BIOCHEMISTRY. *Educ:* Mass Col Pharm, BS, 65, MS, 67; Harvard Univ, DSc(toxicol), 70. *Prof Exp:* Res assoc toxicol, Harvard Univ, 70-72; from asst prof to assoc prof, 72-85, PROF TOXICOL, UNIV CONN, 85- *Concurrent Pos:* Nat Acad Sci Safe Drinking Water Comt, Conn Safe Drinking Water, Sci Adv Panel. *Mem:* AAAS; Soc Toxicol (secy, 87-89); Am Col Toxicol; Am Conf Govt Indust Hyg. *Res:* Biochemical mechanisms of toxicant actions and interactions; special emphasis on organophosphate insecticides and hepatotoxic chemicals. *Mailing Add:* Sect Pharmacol & Toxicol Univ Conn Sch Pharm Storrs CT 06268

COHEN, THEODORE, b Arlington, Mass, May 11, 29; m 54; c 2. ORGANIC CHEMISTRY. *Educ:* Tufts Univ, BS, 51; Univ Southern Calif, PhD(chem), 55. *Prof Exp:* Asst chem, Univ Southern Calif, 52-55; asst lectr, Glasgow Univ, 55-56; from instr to assoc prof, 56-66, PROF CHEM, UNIV PITTSBURGH, 66- *Concurrent Pos:* Fulbright grant & Ramsay fel, Glasgow Univ, 55-56. *Mem:* Am Chem Soc; Royal Soc Chem. *Res:* New synthetic methods; the synthesis of pheromones and other natural products. *Mailing Add:* Dept Chem Univ Pittsburgh Pittsburgh PA 15260

COHEN, WILLIAM, b Brooklyn, NY, Apr 19, 31; m 63. BIOCHEMISTRY. *Educ:* Long Island Univ, BS, 51; Purdue Univ, MS, 54; Fordham Univ, PhD(biochem), 58. *Prof Exp:* Asst, Purdue Univ, 51-54; assoc dept microbiol, Col Physicians & Surgeons, Columbia, 58-63; from asst prof to assoc prof, 63-74, PROF BIOCHEM, SCH MED, TULANE UNIV, 74- *Mem:* Am Soc Biol Chem. *Res:* Enzyme therapy. *Mailing Add:* Dept Biochem Tulane Univ 1430 Tulane Ave New Orleans LA 70112-2699

COHEN, WILLIAM C(HARLES), b Brooklyn, NY, May 30, 33; m 55; c 5. CHEMICAL ENGINEERING. *Educ:* Pratt Inst Technol, BChE, 54; Princeton Univ, MSE, 57, PhD(chem eng), 60. *Prof Exp:* From asst prof to assoc prof chem eng, Univ Pa, 58-72, dir continuing eng studies, 69-72; PROF CHEM ENG & ASSOC DEAN INDUST & ACAD AFFAIRS, NORTHWESTERN UNIV, 72-, DIR, BASIC INDUST RES INST, 89- *Concurrent Pos:* Consult, Systs Analysis Dept, Leeds & Northrup Co, 61-72; NSF fac fel sci, Univ Calif, Berkeley, 65-66, Amoco Oil Co, 86. *Mem:* Am Inst Chem Engrs; Sigma Xi. *Res:* Dynamics, control and optimization of process systems. *Mailing Add:* Tech Inst Rm 1387 Northwestern Univ Evanston IL 60208-3123

COHEN, WILLIAM DAVID, b Brooklyn, NY, Feb 24, 28; m 48; c 6. BIOCHEMISTRY. *Educ:* Univ Iowa, BS, 47; Univ Minn, MS, 50, PhD(biochem), 52, FRPath(C), 72. *Prof Exp:* Damon Runyan sr res fel, 52-53; chief biochemist, Mem Hosp, Worcester, Mass, 53-61; consult chemist, Minn State Dept Health, 62; asst prof obstet & gynec, Med Col, Univ Minn, 62-68; chief biochemist lab, St John's Gen Hosp, 68-81; RETIRED. *Concurrent Pos:* Assoc prof path, Mem Univ Nfld, 71-81, prov analyst, 68-81. *Mem:* Am Asn Clin Chem; Can Soc Clin Chemists. *Res:* Chemistry and metabolism of lipids and steroid hormones. *Mailing Add:* RR No 1 Orangedale NS B0E 2K0 Can

COHICK, A DOYLE, JR, b Nevada, Mo, July 14, 39; m 61; c 2. ECONOMIC ENTOMOLOGY, INFORMATION SCIENCE. *Educ:* Cent Mo State Col, BSEd, 61, MA, 64; Cornell Univ, PhD(entom), 68. *Prof Exp:* Teacher pub schs, Raytown, Mo, 61-65; asst entom, Cornell Univ, 65-68; res biologist, Chemagro Corp, 68-72, mgr data documentation, 72-81, MGR INSECTICIDE RES, AGR DIV, MOBAY CHEM CORP, 81- *Mem:* Entom Soc Am. *Res:* Odonate ethology; mammalian and insect ecology; stored products insect behavior and control; information storage and data processing systems; crop protection and pesticide development. *Mailing Add:* 532 E 28th Ave Kansas City MO 64116

COHLAN, SIDNEY QUEX, b New York, NY, July 31, 15; m 51; c 2. PEDIATRICS. *Educ:* Brooklyn Col, AB, 34; NY Med Col, MD, 39; Am Bd Pediat, dipl, 52. *Prof Exp:* From asst prof to assoc prof clin pediat, 53-70, PROF PEDIAT, MED SCH, NY UNIV-BELLEVUE MED CTR, 70- *Concurrent Pos:* Adj pediatrician, Beth Israel Hosp, 50-56, assoc pediatrician, 56-; vis physician, Children's Med Serv, Bellevue Hosp, 53-58, vis physician, Hosp, 58-; asst vis physician, NY Univ Hosp, 53-, pediat in chg, Dept Pediat, 58-; mem med bd, Irvington House, 59-; study sect child health & human develop, NIH. *Mem:* AAAS; Am Pediat Soc; Am Acad Pediat; Soc Pediat Res. *Res:* Vitamin A metabolism; developmental malformations. *Mailing Add:* NY Univ 530 First Ave New York NY 10016

COHLBERG, JEFFREY ALLAN, b Philadelphia, Pa, Feb 26, 43. PROTEIN CHEMISTRY, CELL BIOLOGY. *Educ:* Cornell Univ, AB, 66; Univ Calif, Berkeley, PhD(biochem), 72. *Prof Exp:* NIH fel, Inst Enzyme Res, Univ Wis, 72-75; ASSOC PROF CHEM, CALIF STATE UNIV, LONG BEACH, 75-; vis res prof, Ger Cancer Res Inst, Heidelberg, 82-83. *Mem:* Am Soc Cell Biol. *Res:* Structure and assembly of intermediate filaments. *Mailing Add:* Dept Chem Calif State Univ 1250 Bellflower Blvd Long Beach CA 90840

COHN, CHARLES ERWIN, b Chicago, Ill, Apr 25, 31; m 76. COMPUTER-AIDED EXPERIMENTATION. *Educ:* Univ Chicago, PhD(physics), 57. *Prof Exp:* Asst physicist, Argonne Nat Lab, 56-59, physicist, 59-85; sr res scientist, Ga Inst Technol, 89-90; CONSULT, 85-89 & 90- *Res:* Computer-aided experimentation. *Mailing Add:* 6311 Mark Trail Austell GA 30001-5126

COHN, DANIEL ROSS, b Berkeley, Calif, Nov 28, 43. LASERS, PLASMA PHYSICS. *Educ:* Univ Calif, Berkeley, AB, 66; Mass Inst Technol, PhD(physics), 71. *Prof Exp:* Staff physicist, Mass Inst Technol, 71-74, tech asst to dir, 74-75, assoc group leader quantum optics & plasma physics, 75-77, group leader plasma & laser systs, Francis Bitter Nat Magnet Lab, 77-80, HEAD FUSION SYSTS DIV, PLASMA FUSION CTR, MASS INST TECHNOL, 80- *Res:* Laser plasma interactions; infrared and submillimeter lasers; plasma diagnostics; controlled thermonuclear fusion; controlled thermonuclear fusion. *Mailing Add:* 26 Walnut Hill Rd Chestnut Hill MA 02167

COHN, DAVID L(ESLIE), b Highland Park, Ill, Apr 20, 43; m 65; c 1. ELECTRICAL ENGINEERING. *Educ:* Mass Inst Technol, SB, 66, SM, 66, PhD(elec eng), 70. *Prof Exp:* Consult, Sylvania, Farrel Lines, 66-70; asst prof elec eng, Southern Methodist Univ, 70-76; ASSOC PROF ELEC ENG, UNIV NOTRE DAME, 76- *Mem:* Inst Elec & Electronics Engrs. *Res:* Information theory; communication theory and coding. *Mailing Add:* Dept Elec Eng Univ Notre Dame Notre Dame IN 46556

COHN, DAVID VALOR, b New York, NY, Nov 8, 26; m 47; c 2. BIOCHEMISTRY, ENDOCRINOLOGY. *Educ:* City Col New York, BS, 48; Duke Univ, PhD, 52. *Prof Exp:* USPHS fel, Western Reserve Univ, 52-53; prin scientist & actg chief radioisotope serv, Vet Admin Hosp, 53-68, assoc chief staff & dir, Calcium Res Lab, 68-81; from instr to prof biochem, 53-81, assoc dean, Col Health Sci, Univ Kans Med Ctr, Kansas City, 74-81; vpres & dir res & develop, Immunonuclear Corp, 81-82; PROF BIOCHEM, DEPT ORAL HEALTH, SCH DENT, UNIV LOUISVILLE, 82-, CHAIR, 90- *Concurrent Pos:* From asst prof to prof, Sch Dent, Univ Mo, 62-81; mem gen med B study sect, USPHS, 71-75; chmn, Gordon Res Conf Chem, Physiol & Structure Bones & Teeth, 74; mem exec comt, Int Parathyroid Conf, 74-; pres, Int Conf Calcium Regulating Hormones, Inc; mem bd sci coun, Nat Inst Dent Res, NIH, 81- *Mem:* AAAS; Am Chem Soc; Am Soc Biol Chem; Endocrine Soc; Am Soc Bone Mineral Res. *Res:* Chemistry and physiology of peptide hormones including parathyroid hormone and calcitonin; biology of bone growth and resorption; calcium binding proteins; radioimmunoassay technology; secretory protein I/chromogranin A in secretory processes. *Mailing Add:* Dept Oral Health Health Sci Ctr Univ Louisville Louisville KY 40292-0001

COHN, DEIRDRE ARLINE, b New York, NY. ANDROGENS, IMMUNE RESPONSE. *Educ:* Hunter Col, BA, 48, MA, 65; State Univ NY, Downstate Med Ctr, PhD(anat), 71. *Prof Exp:* Lectr physiol, Hunter Col, 65-66; instr anat & cell Biol, State Univ NY, Downstate Med Ctr, 70-76, asst prof, 76-77; assoc prof, 77-87, PROF BIOL, YORK COL, CITY UNIV NEW YORK, 88- *Concurrent Pos:* Prin investr, NIH Androgen Sensitivity as a Factor in Immunocompetence, 77-; consult & site visitor, Minority Access to Res Careers Prog, NIH, 78-81; prog dir, Minority Biomed Support Prog, York Col, NIH, 79-85; adj prof immunol, City Univ New York Med Sch, 85- *Mem:* Am Asn Immunologists; Reticuloendothelial Soc; Asn Women Sci; NY Acad Sci; AAAS; Sigma Xi. *Res:* The influence of differential sensitivity to androgen; splenectomy on host variability in the immune response. *Mailing Add:* Dept Biol York Col City Univ NY 9400 Guy R Brewer Rd Jamaica NY 11451

COHN, ELLEN RASSAS, b Long Branch, NJ, Jan 12, 53; m 82; c 2. SPEECH PATHOLOGY. *Educ:* Douglass Col, Rutgers Univ, BA, 74; Vanderbilt Univ, MS, 76; Univ Pittsburgh, PhD(speech path), 81. *Prof Exp:* Clin asst prof, Dept Speech Path, 81-85, RES ASSOC, CLEFT PALATE CTR, UNIV PITTSBURGH, 80-, INSTR, COMMUN DEPT, 86-, ASST PROF, PHARM & THERAPEUT, SCH PHARM, 89- *Concurrent Pos:* Prin investr, Fac Develop Grant & Teaching Develop Grants, Univ Pittsburgh, 81 & Nat Inst Dent Res Grants; mem bd dirs, Montefiore Univ Hosp, Pittsburgh, Pa. *Mem:* Am Cleft Palate Asn; Am Speech & Hearing Asn; Speech Commun Asn. *Res:* Cleft palate and craniofacial disorders; velopharyngeal inadequacy; non-verbal communication; speech in relationship to orthodontic problems; video fluoroscopy and ultrasound. *Mailing Add:* 5436 Northumberland St Pittsburgh PA 15217

COHN, ERNST M, b Mainz, Ger, Mar 31, 20; nat US; m 81. PHYSICAL CHEMISTRY. *Educ:* Univ Pittsburgh, BS, 42, MS, 52. *Prof Exp:* Chemist minimum ignition energies gas mixtures, US Bur Mines, 42-44, chemist Fischer-Tropsch synthesis & magnetochem, 46-53; chemist, Bituminous Coal Res Br, Dept Interior, 53-60; phys chemist, US Army Res Off, 60-62; mgr solar & chem power, NASA, 62-76; CONSULT & COLUMNIST, 76- *Mem:* Am Chem Soc. *Res:* Kinetics of reactions in solids; carbides of iron, cobalt and nickel; Fischer-Tropsch synthesis; magnetochemistry; chemistry and physics of coal; batteries; fuel cells; bioelectrochemistry; electrocatalysis; photovoltaics and solar-cell arrays. *Mailing Add:* 1138 Appian Way Dothan AL 36303

COHN, GEORGE I(RVING), b Lake Forest, Ill, Jan 29, 21; m 53; c 4. ELECTRICAL ENGINEERING. *Educ:* Calif Inst Technol, BS, 42; Ill Inst Technol, MS, 47, PhD(elec eng), 51. *Prof Exp:* Instr elec eng, Ill Inst Technol, 47-49, from asst prof to prof, 49-61; chief scientist fluid physics div & mgr exploding mat lab, Electro-Optical Systs, Inc, 61-62; mgr penetration aids prog & electromagnetics group, Nat Eng Sci Co, 62-63; consult, 64-68; PROF ENG, CALIF STATE UNIV FULLERTON, 68- *Concurrent Pos:* Consult, Cook Res Co, Bendix Prod Div, Space Tech Labs, Inc, Motorola, Southwest Res Found, Armour Res Found, Jet Propulsion Lab, Autonetics Div, NAm Aviation, Inc & Magnavox Co. *Mem:* AAAS; Am Math Soc; Am Phys Soc; Am Soc Eng Educ; Optical Soc Am. *Res:* Electromagnetic theory; propagation; antennas; microwaves; tubes; accelerators; network theory; applied analysis; transform techniques; nonlinear differential equations; integration; electronics; teaching techniques; worktexts; lasers; plasmas; scattering. *Mailing Add:* 915 Monarch Dr La Canada CA 91011

COHN, GERALD EDWARD, b Buffalo, NY, June 10, 43; m 69; c 3. OPTICS, ENGINEERING PHYSICS. *Educ:* Columbia Univ, AB, 65; Univ Wis-Madison, MA, 68, PhD(physics), 72. *Prof Exp:* Trainee biophys, Pa State Univ, 71-72, scholar, 73; asst prof physics, Ill Inst Technol, 73-79; instrument scientist, 79-89, SR RES INSTRUMENT SCIENTIST, ABBOTT LABS, 89- *Concurrent Pos:* Cottrell Res Corp res grant, 74 & 76; fac res grant, Ill Inst Technol, 75 & 77; short course instr, Soc Photo-Optical Instrumentation Engrs, 90 & 91. *Mem:* Am Phys Soc; Am Asn Physics Teachers; Soc Photo-Optical Instrumentation Eng. *Res:* Applied holography; holographic interferometry; optical instrumen; electron spin resonance studies of the photobiology of natural and model membrane systems; photodynamic damage in yeast. *Mailing Add:* Abbott Labs Dept 93F Bldg AP-20 One Abbott Park Rd Abbott Park IL 60064

COHN, HANS OTTO, b Berlin, Ger, Dec 27, 27; nat US; m 60; c 3. PHYSICS, ELEMENTARY PARTICLE PHYSICS. *Educ:* Ind Univ, BS, 49, MS, 50, PhD(physics), 54. *Prof Exp:* Asst physics, Ind Univ, 49-51, res asst, 52-54; PHYSICIST, OAK RIDGE NAT LAB, 54-; RES PROF, UNIV TENN, 84- *Mem:* Fel Am Phys Soc. *Res:* Element particle physics; nuclear and high energy physics. *Mailing Add:* Oak Ridge Nat Lab PO Box 2008 Oak Ridge TN 37831-6374

COHN, HARVEY, b New York, NY, Dec 27, 23; m 51; c 2. MATHEMATICS. *Educ:* City Col New York, BS, 42; NY Univ, MS, 43; Harvard Univ, PhD(math), 48. *Prof Exp:* Asst prof math, Wayne Univ, 48-54; vis assoc prof, Stanford Univ, 54-55; assoc prof, Wayne Univ, 55-56; assoc prof, Wash Univ, 56-57, prof math & dir comput ctr, 57-58; prof math, Univ Ariz, 58-71, head dept, 58-67; DISTINGUISHED PROF MATH, CITY COL NEW YORK, 71- *Concurrent Pos:* Consult, Gen Motors Corp, 53, AEC, NY, 54, Nat Bur Standards, 56, IBM Corp, 57 & Argonne Nat Lab, 58-68; mem comt regional develop for math, Nat Res Coun, 62-65; mem adv comt, Autonomous Univ Guadalajara, 63-65; vis mem, Inst Advan Study, Princeton, 70-71; vis lectr, Univ Copenhagen, 76-77. *Honors & Awards:* Putnam Prize, Math, 42. *Mem:* Am Math Soc; Asn Comput Mach; Math Asn Am. *Res:* Number theory and modular functions, particularly use of computer techniques. *Mailing Add:* Dept Math City Col New York 138th St & Convent Ave New York NY 10031

COHN, ISIDORE, JR, b New Orleans, La, Sept 25, 21; m 75; c 2. SURGERY. *Educ:* Tulane Univ, BS, 42; Univ Pa, MD, 45, MSc, 52, DSc(med), 55; Am Bd Surg, dipl, 53. *Prof Exp:* From instr to assoc prof, 52-59, chmn dept, 62-89, PROF SURG, SCH MED, LA STATE UNIV MED CTR, NEW ORLEANS, 59- *Concurrent Pos:* Surgeon-in-chief, La State Univ Serv, Charity Hosp, 62-89; mem Am Bd Surg, 69-75; chmn, clin invest adv comt, Am Cancer Soc, 69-73; mem ed staff, Am Surg, Rev Surg, Am J Surg & Surg Digest; dir, Nat Pancreatic Cancer Proj, 75-85; Isidore Cohn Jr Prof Surg, La State Univ Sch Med, 87. *Mem:* Am Surg Asn; Am Col Surgeons; Am Gastroenterol Asn; Soc Univ Surgeons; Int Soc Surg; Sigma Xi. *Res:* Gastrointestinal surgery, strangulation intestinal obstruction and secondary interest in problems involving biliary, pancreatic and tumor problems associated with antibacterial agents; surgical research in germ-free animals. *Mailing Add:* 1542 Tulane New Orleans LA 70112

COHN, J GUNTHER, b Berlin, Ger, Mar 6, 11; nat US; m 40; c 1. CHEMISTRY. *Educ:* Univ Berlin, PhD(chem), 34. *Prof Exp:* Asst, Nobel Inst, Sweden, 34-36; asst & instr, Chalmers Tech Univ, Sweden, 37-41; Carnegie fel, Univ Minn, 41-42, Welsh fel, 43; res chemist, 43-63, vpres & dir, Res & Develop, 63-72, vpres res & technol, Engelhard Indust Div, Engelhard Minerals & Chem Corp, 72-76; CONSULT, 76- *Concurrent Pos:* Lectr, Royal Tech Univ, Stockholm, Sweden, 77- *Honors & Awards:* Distinguished Achievement Award, Int Precious Metals Inst, 87. *Mem:* Am Chem Soc; Electrochem Soc. *Res:* Corrosion of metals and reactions in solid state; catalytic properties of solids; photochemistry of solids; powder metallurgy; instrumentation of chemical processes; precious metals; electrochemistry. *Mailing Add:* 20 Lincoln Ave West Orange NJ 07052

COHN, JACK, b Rock Island, Ill, May 6, 32; m 53; c 4. THEORETICAL PHYSICS, ELECTRODYNAMICS. *Educ:* Univ Iowa, BA, 53, MS, 56, PhD(statist mech), 62. *Prof Exp:* From asst prof to assoc prof, 60-71, PROF PHYSICS, UNIV OKLA, 71- *Res:* Statistical mechanics; general relativity. *Mailing Add:* Dept Physics Univ Okla 660 Parrington Oval Norman OK 73019

COHN, JAY BINSWANGER, b Pelham, NY, Feb 22, 22; m 45; c 2. PSYCHIATRY. *Educ:* Amherst Col, BA, 42; Yale Univ, MD, 45; Univ Calif, Irvine, PhD(psychol), 74; Am Col Law, Anaheim, Calif, JD, 82. *Prof Exp:* Intern, St Elizabeth's Hosp, Washington, DC, 45-46; resident, Crile Hosp, Cleveland, Ohio, 48-49 & Cleveland Receiving Hosp, 51-53; lectr psychiat, Western Reserve Univ, 54-59; instr, Univ Calif, Los Angeles, 59-65; asst clin prof, Univ Southern Calif, 65-68, assoc prof, 68-71; prof psychiat & soc sci, Univ Calif, Irvine, 71-83; PROF DEPT PSYCHIAT & HUMAN BEHAV COL MED, NEUROPSYCHIAT INST, UNIV CALIF, LOS ANGELES, 83- *Concurrent Pos:* Consult, res projs, State Dept Ment Hyg, Ohio, 53-58, Vet Admin, Calif, 59-65, Indust Accident Comn, 60- & US Fed Court, Panel, 61- *Mem:* AAAS; fel Am Psychiat Asn; Am Psychol Asn; Am Acad Neurol; AMA; Acad Psychoanal. *Res:* Counter-transference and pharmacology; electrical measurement of counter-transference as measured in electrocardiogram. *Mailing Add:* Dept Psychiat Alcohol Res Unit Univ Calif 760 Westwood Blvd Los Angeles CA 90024

COHN, JAY NORMAN, b Schenectady, NY, July 6, 30; m 53; c 3. CARDIOVASCULAR DISEASES. *Educ:* Union Univ, NY, BS, 52; Cornell Univ, MD, 56; Am Bd Internal Med, dipl. *Prof Exp:* Intern med, Beth Israel Hosp, Boston, Mass, 56-57, asst resident, 57-58; chief resident, Vet Admin Hosp, 61-62, clin investr cardiovasc, 62-65; from instr to prof med, Georgetown Univ, 62-74; PROF MED & HEAD CARDIOVASC DIV, MED SCH, UNIV MINN, MINNEAPOLIS, 74- *Concurrent Pos:* Res fel cardiovasc, Georgetown Univ Hosp, 60-61; chief hypertension & clin hemodynamics, Vet Admin Hosp, DC, 65-74; chmn, Coun Circulation, Am Heart Asn, 79-81, chmn, Sci Sessions Prog Comt, 79-82; pres, Am Soc Hypertension, 90-92. *Honors & Awards:* Arthur S Flemming Award, 69. *Mem:* Am Col Cardiol; Am Heart Asn; Asn Am Physicians; Am Soc Clin Invest; Am Soc Pharmacol & Exp Therapeut; Am Phys Soc. *Res:* Hemodynamics of myocardial infarction; pathophysiology and treatment of hypertension and congestive heart failure in man; hemodynamic factors in clinical hypotension and shock; dynamics of regional blood flow in man. *Mailing Add:* Med Sch Univ Minn Minneapolis MN 55455

COHN, JONA, electrical engineering, for more information see previous edition

COHN, KIM, b New York, NY, Jan 25, 39; m 62, 82; c 1. INORGANIC CHEMISTRY. *Educ:* Queens Col, NY, BS, 60; Univ Mich, PhD(chem), 67. *Prof Exp:* Asst prof chem, Mich State Univ, 67-72; assoc prof, 72-80, PROF CHEM, CALIF STATE COL, BAKERSFIELD, 80- *Mem:* Am Chem Soc; Royal Soc Chem; Sigma Xi. *Res:* Synthesis, structure and bonding of non-transition elements. *Mailing Add:* Chem Dept Calif State Col 9001 Stockdale Hwy Bakersfield CA 93309

COHN, LAWRENCE H, b San Francisco, Calif, Mar 11, 37; m 60; c 2. CARDIAC SURGERY. *Educ:* Univ Calif-Berkeley, BA, 58; Stanford Univ, MD, 62. *Prof Exp:* PROF SURG, HARVARD MED SCH, 80-; CHIEF, DIV CARDIAC SURG, BRIGHAM & WOMENS HOSP, BOSTON, 87- *Concurrent Pos:* Coun Acad Affairs, Am Surg Asn. *Mem:* Am Surg Asn; Am Col Cardiol; Am Col Chest Physicians (pres, 86-87). *Res:* Cardiac valve surgery; myocardial infarction research. *Mailing Add:* Brigham-Womens Hosp 75 Francis St Boston MA 02115

COHN, LESLIE, b Philadelphia, Pa, Feb 3, 43. MATHEMATICS. *Educ:* Univ Pa, BA, 65; Univ Chicago, MS, 67, PhD(math), 69. *Prof Exp:* Asst prof, Univ Ill, Chicago Circle, 69-70; vis mem, Inst Advan Study, 70-71; asst prof, 71-77, ASSOC PROF MATH, JOHNS HOPKINS UNIV, 77- *Mem:* Am Math Soc. *Res:* Number theory; automorphic forms; representations of live groups. *Mailing Add:* The Citadel Charleston SC 29409

COHN, MAJOR LLOYD, b New York, NY, Oct 29, 27; m 58; c 2. MANAGEMENT OF PAIN. *Educ:* City Univ NY, BS, 50; Univ Geneva, Switz, MD, 56; Univ Minn, MS, 66; Univ Pittsburgh, PhD(biochem), 69. *Prof Exp:* Assoc attend, Montefiore Hosp, 61-62; asst prof anesthesiol, Univ Pittsburgh Sch Med, 69-79; dir anesthesia, 79-86, ASSOC PROF CHEM ANESTHESIOL, DREW UNIV MED & SCI, 79- *Concurrent Pos:* Adj prof pharmaceut, Duquesne Univ Sch Pharm, 77-79; dir, Pain Treat Ctr, King-Drew Med Ctr, 79- & Smart Diag Lab, 86- *Mem:* AAAS; Am Soc Internal Med; Soc Neurosci; Am Soc Pharmacol & Exp Therapeut; Am Pain Soc; Am Acad Pain Med. *Res:* High performance liquid chromatography analysis of adenine and guanosine nucleotides, nucleosides, and bases in rat brain; objective assessment of low-back pain therapies using low-frequency magnetic field technique. *Mailing Add:* Surf Med Ctr 3655 Lomita Blvd Suite 410 Torrance CA 90505

COHN, MILDRED, b New York, NY, July 12, 13; m 38; c 3. BIOCHEMISTRY, BIOPHYSICS. *Educ:* Hunter Col, BA, 31; Columbia Univ, PhD(phys chem), 38. *Hon Degrees:* ScD, Women's Med Col Pa, 66, Radcliffe Col, 78, Wash Univ, 81, Brandeis Univ, Hunter Col & Univ Pa, 84, Univ NC, 85; PhD, Weigmann Inst Sci, Israel, 88, Univ Miami, 90. *Prof Exp:* Res assoc biochem, George Washington Univ, 37-38, Cornell Univ Med Sch, 38-46 & Harvard Med Sch, 50-51; res assoc, Wash Univ Med Sch, 46-58, assoc prof, 58-60; from assoc prof to prof, 60-78, Benjamin Bush prof, 78-82, EMER PROF BIOCHEM & BIOPHYS, UNIV PA MED SCH, 82- *Concurrent Pos:* Career investr, Am Heart Asn, 64-78; vis prof biochem, Inst Biol Phys Chem, Paris, 66-67; chmn, Div Biol Chem, Am Chem Soc, 73-74; chancellor's distinguished vis prof, Univ Calif, Berkeley, 81; vis prof biochem, John Hopkins Med Sch, 85-; sr mem Inst Cancer Res, Fox Chase, 82-85. *Honors & Awards:* Garvan Medal, Am Chem Soc, 63; Cresson Medal, Franklin Inst, 75; Biochemists Award, Int Orgn Women, 79; Nat Medal Sci, 82; Distinguished Award, Col Physicians, 87; Remsen Award, Am Chem Soc, Md, 88. *Mem:* Nat Acad Sci; Am Philos Soc; Am Acad Arts & Sci; Am Soc Biol Chem (pres, 78-79); Am Chem Soc; Biophys Soc. *Res:* Metabolic studies with stable isotopes; mechanisms of enzymatic reactions; electron spin and nuclear magnetic resonance. *Mailing Add:* Dept Biochem &R Biophys Univ Pa Sch Med Philadelphia PA 19104-6089

COHN, NATHAN, electrical engineering; deceased, see previous edition for last biography

COHN, NORMAN STANLEY, b Philadelphia, Pa, June 26, 30; m 56; c 3. CYTOLOGY, CYTOCHEMISTRY. *Educ:* Univ Pa, AB, 52; Univ Ky, MS, 53; Yale Univ, PhD(bot), 57. *Prof Exp:* Asst biol, Yale Univ, 53-56; NSF res fel, Johns Hopkins Univ, 57-59; from asst prof to assoc prof, 59-68, chmn dept, 69-70, dean, Grad Col & dir res, 70-79, DISTINGUISHED PROF BOT, OHIO UNIV, 68- *Concurrent Pos:* Fulbright res scholar, State Univ Leiden, 65-66 & 68-69; Nat Acad Sci vis lectr, Czech & Yugoslavia, 68 & 73; head cellular & physiol biosci, NSF, 79-81. *Mem:* AAAS; Sigma Xi; fel Royal Micros Soc. *Res:* Radiation cytology; chromosome breakage and chemistry; immunocytochemistry; developmental plant biology; cytophotometry. *Mailing Add:* Dept Bot Ohio Univ Athens OH 45701

COHN, PAUL DANIEL, b Portland, Ore, July 14, 36. NUCLEAR ENGINEERING, ECONOMICS. *Educ:* Ore State Univ, BS, 58, MS, 59. *Prof Exp:* Proj mgr, space systs, NAm Rockwell Corp, 62-65; mgr, fast flux test facility core & fuels design, Pac Northwest Div, Battelle Mem Inst, 65-70; dep mgr nuclear proj, Battelle Northwest, 70-72, mgr nuclear projs, 72-73, mgr safety res, 73-74; consult, Energy Eng Assocs, 74-82; RETIRED. *Concurrent Pos:* Consult, reactor safety anal, United Nuclear Industs, 76-82, energy econ, Pac Northwest Div, Battelle Mem Inst, 77-82, prod eval, Exxon Nuclear, 77-82 & fuel cycle econ, Westinghouse-Hanford, 78-82. *Res:* Nuclear safety; fuel design; thermal hydraulics; economic modeling and analysis; production analysis. *Mailing Add:* 208 Ridge Tr Rd Bozeman MT 59715-9253

COHN, PETER FRANK, b New York, NY, Jan 19, 39; m 68; c 2. CARDIOLOGY. *Educ:* Columbia Univ, BA, 58, MD, 62. *Prof Exp:* From instr to assoc prof med, Med Sch, Harvard Univ, 69-81; PROF MED & CHIEF CARDIOL, HEALTH SCI CTR, STATE UNIV NY, STONY BROOK, 81- *Concurrent Pos:* Fel, Coun Clin Cardiol, Am Heart Asn. *Mem:* Fel Am Col Cardiol; Am Heart Asn; Asn Univ Cardiologists. *Res:* Pathophysiology, detection, prognosis and treatment of silent myocardial ischemia. *Mailing Add:* Dept Med Div Health Sci Ctr State Univ NY Rm T-17-020 Stony Brook NY 11794

COHN, RICHARD MOSES, b New York, NY, Sept 2, 19. ALGEBRA. *Educ:* Columbia Univ, BA, 39, MA, 41, PhD(math), 47. *Prof Exp:* Engr, Inspection Agency, Signal Corps, US Army, 42-44; lectr elem math, Columbia Univ, 46-47; from instr to assoc prof, 47-59, PROF MATH, RUTGERS UNIV, 59- *Mem:* Am Math Soc; Sigma Xi. *Res:* Difference algebra; differential algebra; electrical networks. *Mailing Add:* Dept Math Rutgers Univ New Brunswick NJ 08903

COHN, ROBERT, b Washington, DC, Feb 25, 09; m 37. NEUROLOGY. *Educ:* George Washington Univ, BS, 32, MD, 36. *Prof Exp:* Fel neurophysiol, St Elizabeth's Hosp, Washington, DC, 36-38; mem staff neurol & EEG, St Elizabeth's Hosp, 38-43; dir neurol res, US Naval Hosp, Bethesda, 46-71; PROF NEUROL, HOWARD UNIV, 71- *Concurrent Pos:* Clin prof, Howard Univ, 60-65; prof lectr, Mt Sinai Med Sch, NY, 69-; vis prof, Boston Univ, 70- *Mem:* Am Neurol Asn; Am EEG Soc; Asn Res Nerv & Ment Dis. *Res:* Clinical and experimental neurology and electroencephalography; physiology of sensation. *Mailing Add:* 7221 Pyle Rd Bethesda MD 20817

COHN, ROBERT M, b New York, NY, Mar 10, 41. BIOCHEMISTRY, INBORN ERRORS OF METABOLISM. *Educ:* Yale Univ, MD, 65. *Prof Exp:* Asst dir, 78-81, ASSOC DIR, CLIN LAB, CHILDREN'S HOSP, PHILADELPHIA, 81-; ASSOC PROF PEDIAT, SCH MED, UNIV PA, 81- *Mem:* Am Inst Nutrit; Am Soc Clin Nutrit. *Mailing Add:* Sch Med Univ Pa 34th & Civic Ctr Blvd Philadelphia PA 19104

COHN, SEYMOUR B(ERNARD), b Stamford, Conn, Oct 21, 20; m 48; c 3. ENGINEERING. *Educ:* Yale Univ, BE, 42; Harvard Univ, MS, 46, PhD(electromagnetic theory), 48. *Prof Exp:* Spec res assoc, Radio Res Lab, Harvard Univ, 42-45; Nat Res Coun fel, 46-48; res engr, Sperry Gyroscope Co, 48-53; lab mgr, Stanford Res Inst, 53-60; vpres & tech dir, Rantec Corp, 60-67; MICROWAVE CONSULT, S B COHN ASSOCS, 67- *Mem:* Nat Acad Eng; Sigma Xi; fel Inst Elec & Electronics Engrs. *Res:* Ultra high frequency and microwave transmission apparatus. *Mailing Add:* 300 S Glenroy Ave Los Angeles CA 90049

COHN, SIDNEY ARTHUR, b Toronto, Ont, May 8, 18; US citizen; m 46; c 2. BIOLOGY. *Educ:* Univ Conn, BS, 40, MS, 48; Brown Univ, PhD(biol), 51. *Prof Exp:* From instr to assoc prof, 51-66, PROF ANAT, UNIV TENN, MEMPHIS, 66-, ACTG CHMN DEPT, 80- *Mem:* Am Asn Anat; Int Asn Dent Res; Sigma Xi; Am Asn Dent Sch. *Res:* Developmental and histological studies of teeth; long-term effects of non-function on supportive tissues of teeth; temporomandibular joint; attachment of fibers of periodontal ligament; study of the supportive tissues of the teeth. *Mailing Add:* 5350 Denwood Ave Memphis TN 38119

COHN, STANLEY HOWARD, b Toronto, Ont, Aug 23, 26; m 49; c 5. INFORMATION SYSTEMS, INDUSTRIAL ENGINEERING. *Educ:* Univ Toronto, BA, 48, MA, 50. *Prof Exp:* Instr appl math, Fournier Inst Technol, 52-55; specialist, Avro Aircraft Ltd, Ont, 55-59; head digital comput, Space Task Group, NASA, 59-62; mgr sci comput, IBM Can, Ltd, 62-68; assoc prof indust eng, 68-89, EMER PROF, UNIV TORONTO, 89- *Concurrent Pos:* Nat Res Coun Can grant-in-aid res, 68-81; vis fel, Gen Pract Res Unit, UK, 77-78; vis prof, Technion, Israel, 86-87. *Mem:* Asn Comput Mach; Can Info Processing Soc. *Res:* Information systems engineering in primary health care; organizational information systems. *Mailing Add:* 37 Shakespeare Ave PO Box 1455 Niagara-on-the-Lake ON L0S 1J0 Can

COHN, STANTON HARRY, b Chicago, Ill, Aug 25, 20; m 49; c 5. PHYSIOLOGY, RADIOBIOLOGY. *Educ:* Univ Chicago, SB, 46, SM, 49; Univ Calif, Berkeley, PhD(physiol & radiobiol), 52. *Prof Exp:* Chemist, Kankakee Ord Works, 41-42; chemist, Sherwin Williams, 42-43; jr scientist biochem, Argonne Nat Lab, Univ Chicago, 46-49; asst radiobiol, Crocker Radiation Lab, Univ Calif, 49-50; head internal toxicity br, Biomed Div, US

Naval Radiation Lab, 50-58; scientist, Brookhaven Nat Lab, 58-70, sr scientist & head, Med Physics Div, Med Res Ctr, 70-86; RETIRED. *Concurrent Pos:* Mem subcomt inhalation hazards, Path Effects Atomic Radiation Comt, Nat Acad Sci; mem, Nat Coun Radiation Protection subcomt II, Int Radiation Dose, 61-; prof med, Med Sch, State Univ NY, Stony Brook. *Mem:* Radiation Res Soc; Am Physiol Soc. *Res:* Chemical dynamics of the mineral metabolism of bone; distribution and biological effects of internally deposited radioisotopes; whole-body neutron activation analysis and whole-body counting; application of nuclear technology to medical problems. *Mailing Add:* 844 W California Way Woodside CA 94062

COHN, THEODORE E, b Highland Park, Ill, Sept 5, 41; m 75; c 3. SENSORY INTELLIGENCE. *Educ:* Mass Inst Technol, SB, 63; Univ Mich, MS, 65, MA, 66, PhD(bioeng), 69. *Prof Exp:* From asst prof to assoc prof, 70-85, fac asst to vchancellor res, 84-85, PROF PHYSIOL-OPTICS, UNIV CALIF, BERKELEY, 85- *Concurrent Pos:* Vis fel, John Curtin Sch Med Res, Australian Nat Univ, 76; vis scholar, Univ Calif, San Diego, 89-90. *Mem:* Optical Soc Am; Asn Res Vision & Ophthal; sr mem Inst Elec & Electronics Engrs; Neurosci Soc; AAAS. *Res:* Visual effects of environmental and internal noise and target uncertainty; influence of a visual depth illusion on escalator disorientation and falls. *Mailing Add:* 283 Lake Dr Kensington CA 94708

COHN, VICTOR HUGO, b Reading, Pa, July 9, 30; m 53; c 3. DRUG ABUSE, DRUG METABOLISM. *Educ:* Lehigh Univ, BS, 52; Harvard Univ, AM, 54; George Washington Univ, PhD(biochem), 61. *Prof Exp:* Pharmacologist, Army Chem Ctr, 55-56; neuropharmacologist, Vet Admin Res Labs, 56-57; biochem pharmacologist, Nat Heart Inst, 57-61; from asst prof to assoc prof, 61-71, actg chmn dept, 70-71, 78-79 & 86, PROF PHARMACOL, MED CTR, GEORGE WASHINGTON UNIV, 71- *Concurrent Pos:* Vis investr, Jackson Labs, 66; drug policy adv, The White House, 79-80; mem, Bd Toxicol & Environ Health Hazards, Nat Res Coun-Nat Acad Sci, 80-84; vis scientist, Nat Inst Drug Abuse, 86-87. *Mem:* Fel AAAS; Am Soc Pharmacol & Exp Therapeut; Sigma Xi; Int Soc Biochem Pharmacol; Int Soc Study Xenobiotics. *Res:* Drug metabolism; drug abuse; fluorometric methods of analysis of biochemicals; histamine metabolism. *Mailing Add:* Dept Pharmacol George Washington Univ Med Ctr Washington DC 20037

COHN, WALDO E, b San Francisco, Calif, June 28, 10; m 38, 43; c 2. BIOCHEMISTRY. *Educ:* Univ Calif, BS, 31, MS, 32, PhD(biochem), 38. *Prof Exp:* Asst biochem, Huntington Labs, Harvard Med Sch, 39-42; tutor biochem sci, Harvard Univ, 39-42; biochem group leader, Plutonium Proj, Univ Chicago, 42-43; group leader, Manhattan Proj, Oak Ridge, 43-47; sr biochemist, Oak Ridge Nat Lab, 47-75; CONSULT, 75- *Concurrent Pos:* Fulbright scholar, Cambridge Univ, 55-56; Guggenheim fel, 55-56, 62-63; secy, Comn Biochem Nomenclature, Int Union Pure & Appl Chem-Int Union Biochem, 65-76; dir, Off Biochem Nomenclature, Nat Acad Sci-Nat Res Coun, 65-76. *Honors & Awards:* Chromatography Award, Am Chem Soc, 63. *Mem:* Fel AAAS; fel Am Acad Arts & Sci; Am Chem Soc; Am Soc Biochem & Molecular Biol; Am Soc Biol chemists (treas, 59-64). *Res:* Ion-exchange separations of rare earth elements and fission products, also of nucleic acid constituents and related biochemical substances; chemistry and structure of nucleic acids; biochemical nomenclature and editing. *Mailing Add:* Biol Div Oak Ridge Nat Lab Box 2009 Oak Ridge TN 37831-8077

COHN, ZANVIL A, b New York, NY, Nov 16, 26; m; c 2. MEDICINE. *Educ:* Bates Col, BS, 49; Harvard Univ, MD, 53. *Hon Degrees:* DSc, Bates Col, 87; MA, Oxford Univ, 88; Dr, Rijksuniversiteit, Leiden, Neth, 90. *Prof Exp:* Co-dir, joint MD-PhD prog, Rockefeller-Cornell Univs, 73-78; MEM STAFF, ROCKEFELLER UNIV, 58-, PROF & SR PHYSICIAN, 66-; ADJ PROF MED, CORNELL UNIV MED COL, 77-, HENRY G KUNKEL PROF IMMUNOL, 86- *Concurrent Pos:* Newton-Abraham vis prof & fel, Lincoln Col, UK, 88; Mem immunobiol study sect, NIH; consult, Nat Cancer Inst; mem Comn Radiation & Infection, Armed Forces Epidemiol Bd; ed, J Exp Med; adv-assoc ed, J Clin Invest, Am J Epidemiol, Cellular Immunol, Blood, Infection & Immunity & J Immunol. *Honors & Awards:* Squibb Award, Am Soc Infectious Dis, 72; Harvey Soc Lectr, 82; Joseph E Smadel Award, Infectious Dis Soc Am, 90. *Mem:* Nat Acad Sci. *Mailing Add:* Rockefeller Univ 1230 York Ave New York NY 10021

COHN-SFETCU, SORIN, b Romania, Aug 16, 44; Can citizen; m 68; c 1. DIGITAL SIGNAL PROCESSING. *Educ:* Polytech Inst Bucharest, Dipl Eng, 67; Univ Calgary, Can, MSc, 73; McMaster Univ, Can, PhD(elec eng), 76. *Prof Exp:* Res engr radio spectros, Romanian Inst Atomic Physics, 67-70; teaching asst physics & elec eng, Univ Calgary, 71-74; teaching asst elec eng, McMaster Univ, 74-76; mem sci staff digital switching, 76-78, mgr signal processing, 79-81, MGR BEHAV TECHNOL LABS, BELL-NORTHERN RES CAN, 81- *Concurrent Pos:* Adj prof elec eng, Univ Ottawa, 78- *Mem:* Sr mem Inst Elec & Electronics Engrs; Can Soc Prof Engrs. *Res:* Digital signal processing applications to microwave spectroscopy telecommunications systems; integrated voice-data-image workstations and office systems; man-machine interactions and interface technology; man-machine technology and applications. *Mailing Add:* Northern Telecom PO Box 3511 Sta C Ottawa ON K1Y 4H7 Can

COICO, RICHARD F, MEDICAL MICROBIOLOGY, PHYSIOLOGY. *Educ:* NY Univ, PhD(immunol), 82. *Prof Exp:* ASST PROF, MED SCH, NY UNIV, 82- *Mailing Add:* Dept Microbiol City Univ New York Med Sch 183th at Convent Ave New York NY 10031

COIL, WILLIAM HERSCHELL, b Ft Wayne, Ind, July 6, 25; m 46; c 3. PARASITOLOGY. *Educ:* Purdue Univ, BS, 48, MS, 49; Ohio State Univ, PhD(zool), 53. *Prof Exp:* Asst zool, Univ Tenn, 49-50; instr, Purdue Univ, 53-54; asst prof, Univ Nebr, 56-64; assoc prof, 64-69, assoc chmn, Dept Systs & Ecol, 68-72, PROF ZOOL, UNIV KANS, 69- *Concurrent Pos:* Mem exped, Oaxaca & Chiapas, Mex, 54 & 55; Muellhaupt scholar, Ohio State Univ, 54-56; NSF fel, Ore Inst Marine Biol, 57; res fel, Biol Sta, Univ Okla, 58; fac fel, Univ Nebr, 59; res assoc, Duke Marine Lab, 62-65; vis prof, Southern Ill Univ, 69, Kans State Univ, 75-76, Univ New Eng, New South Wales, Australia, 85. *Mem:* Am Soc Zool; Soc Syst Zool; Am Micros Soc; Am Soc Parasitol. *Res:* Histochemistry and electron microscopy of cestodes and trematodes, fascioliasis, life histories and bionomics. *Mailing Add:* Dept Syst & Ecol Univ Kans Lawrence KS 66045

COILE, RUSSELL CLEVEN, b Washington, DC, Mar 11, 17; m 51; c 3. OPERATIONS RESEARCH, INFORMATION SCIENCE. *Educ:* Mass Inst Technol, SB, 38, SM, 39, EE, 50; City Univ, London, PhD(info sci), 78. *Prof Exp:* Asst, Mass Inst Technol, 38-39; magnetician, Carnegie Inst, 39-42; engr, Colton & Foss, Inc, 46-47; opers analyst, Opers Eval Group, Mass Inst Technol, 47-62; dir Marine Corps Opers anal group, Ctr Naval Anal, Franklin Inst, 62-67; opers analyst, Univ Rochester, 67-78; opers analyst, Ketron, Inc, 78-81; consult engr, 81-82; dep exec dir, sci support Serv Co, 82-84, CHIEF SCIENTIST, PRC SCI SUPPORT SERV, 85- *Concurrent Pos:* Dir res, Opers Res Group, Off Naval Res, 53-54 & 56-57; Am del, Int Fedn Oper Res Socs Conf, Oslo, Norway, 63; mem, small arms adv comt, Advan Res Proj Agency, Dept Defense, 68-70, bd dirs, Int Test & Eval Asn, 85-; fel, Brit Fel Oper Res, 73; consult, Purdue Univ & NSF, 81-82. *Mem:* Opers Res Soc Am; Am Soc Info Sci; Int Test & Eval Asn. *Res:* Acoustics; electronics, radar; nomography; documentation; information retrieval; operational research; information science. *Mailing Add:* Eval Technol Inc 2150 Garden Rd Suite B3 Monterey CA 93950-2406

COISH, HAROLD ROY, theoretical physics; deceased, see previous edition for last biography

COKE, C(HAUNCEY) EUGENE, b Toronto, Ont; m 41. POLYMER CHEMISTRY. *Educ:* Univ Man, BSc, MSc; Univ Toronto, MA; Univ Leeds, PhD(polymer chem), 38. *Prof Exp:* Dir res, Courtaulds Can, Ltd, 39-42, several exec res & develop posts, 48-59; dir res & develop & mem exec comt, Hartford Fibres Co Div 59-62; tech dir, Textiles Div, Drew Chem Corp, 62-63; dir new prod, Fibers Div, Am Cyanamid Co, NJ, 63-70; pres, 70-78, CHMN BD, COKE & ASSOC CONSULT, 79-; VIS RES PROF, STETSON UNIV, 79- *Concurrent Pos:* Asst org chem, Yale Univ; asst phys chem, Univ Toronto; res fel, Ont Res Found; lectr phys chem, McMaster Univ; guest lectr, Sir George Williams Col, 48-59; chmn. Can Adv Comt on Int Orgn for Stand, 57-59, mem Can Nat Comt, 59; dir, Textile Technol Fedn Can, 58-59; pres, Aqua Vista Corp, Inc, 72-74. *Honors & Awards:* Bronze Medal, Can Asn Textile Colorists & Chem, 63; Bronze Medal, Am Asn Textile Technol, 71. *Mem:* Fel AAAS; NY Acad Sci; Am Asn Textile Technol (vpres, 61-62, pres, 63-65, secy, 69-70); fel Royal Soc Chem, Gt Brit; Can Asn Textile Colorists & Chem (vpres, 54, pres, 57-59); fel Inst Textile Sci (pres); fel Textile Inst, Gt Brit; fel Am Inst Chemists; fel Soc Dyers & Colorists, Gt Brit. *Res:* Basic organic and physical chemistry; equilibrium; manufacture of cellulosic and acrylic fibers; performance and end-use properties; physical characteristics of man-made fibers. *Mailing Add:* 26 Aqua Vista Dr Ormond Beach FL 32074

COKE, JAMES LOGAN, b Brownwood, Tex, Nov 1, 33; m 58; c 3. ORGANIC CHEMISTRY, BIOCHEMISTRY. *Educ:* Wash State Univ, BS, 56; Wayne State Univ, PhD(chem), 61. *Prof Exp:* Res assoc, Univ Wis, 60-61; from instr to asst prof, 61-67, assoc prof, 67-77, PROF ORG CHEM, UNIV NC, CHAPEL HILL, 77- *Concurrent Pos:* Res assoc, Univ Minn, 64. *Mem:* Am Chem Soc. *Res:* Physical organic chemistry; organic chemistry of natural products; plant biochemistry. *Mailing Add:* Dept Chem Univ NC Chapel Hill NC 27514

COKELET, EDWARD DAVIS, b Bremerton, Wash, Aug 9, 47; m 69; c 2. GEOPHYSICAL FLUID DYNAMICS. *Educ:* Univ Wash, BSc, 70, MSc, 71; Cambridge Univ, PhD(appl math), 76. *Prof Exp:* Oceanogr, Inst Oceanog Sci, UK, 75-78; OCEANOGR, PAC MARINE ENVIRON LAB, NAT OCEANIC & ATMOSPHERIC ADMIN, 78- *Mem:* Challenger Soc; Am Geophys Union; Estuarine Res Fedn. *Res:* Dynamics of steep and breaking water waves; estuarine circulation. *Mailing Add:* Pac Marine Environ Lab/NOAA Nat Oceanic & Atmospheric Admin 7600 Sandpoint Way Seattle WA 98115-0070

COKELET, GILES R(OY), b New York, NY, Jan 7, 32; m 63; c 2. CHEMICAL ENGINEERING, MICROCIRCULATION BIOPHYSICS. *Educ:* Calif Inst Technol, BS, 57, MS, 58; Mass Inst Technol, ScD(blood rheol), 63. *Prof Exp:* Chem engr, Dow Chem Co, 58-60; asst prof chem eng, Mass Inst Technol, 63-64 & Calif Inst Technol, 64-68; from assoc prof to prof, Mont State Univ, 68-78; PROF, SCH MED & DENT, UNIV ROCHESTER, 78- *Concurrent Pos:* Res engr, Dow Chem Co, 58-60. *Honors & Awards:* US Sr Scientist Award, Alexander von Humboldt-Stiftung, Ger, 81-82, 88. *Mem:* AAAS; Microcirculatory Soc; Soc Rheol; Int Biorheol Soc; Europ Microcirculatory Soc; NAm Soc Biorheol. *Res:* Rheology of disperse systems; applications of chemical engineering to medical problems. *Mailing Add:* Dept Biophys Med Sch Univ Rochester Rochester NY 14642

COKER, EARL HOWARD, JR, b Cottonwood, Calif, May 4, 34; m 60; c 4. PHYSICAL CHEMISTRY. *Educ:* Ore State Univ, PhD(phys chem), 63. *Prof Exp:* From asst prof to assoc prof, 61-71, PROF PHYS CHEM, UNIV SDAK, 71-, DIR, OFF RES, 83- *Mem:* Am Chem Soc; Am Sci Affil; AAAS. *Res:* Spectroscopy of complex ions in dilute solid solution; internal forces in ionic solids; point defects in ionic solids. *Mailing Add:* Dept of Chem Univ of SDak Vermillion SD 57069

COKER, SAMUEL TERRY, b Evergreen, Ala, Nov 29, 26; m 54; c 4. PHARMACOLOGY. *Educ:* Auburn Univ, BS, 51; Purdue Univ, MS, 53, PhD(pharmacol), 55. *Prof Exp:* Instr pharmacol, Univ Pittsburgh, 53-54; assoc prof, Univ Miss, 55-56 & Univ Mo, Kansas City, 56-59; dean, Sch Pharm, 59-73, PROF PHARMACOL, AUBURN UNIV, 73- *Mem:* Am Pharmaceut Asn; Acad Pharmaceut Sci; fel Am Found Pharmaceut Educ. *Res:* Toxicology, acute and chronic toxicity testing. *Mailing Add:* Sch Pharm Auburn Univ Auburn AL 36830-9323

COKER, WILLIAM RORY, b Athens, Ga, Dec 20, 39; m. EXPERIMENTAL NUCLEAR PHYSICS. *Educ:* Univ Ga, BS, 61, MS, 64, PhD(physics), 66. *Prof Exp:* Res fel nuclear physics, Ctr Nuclear Studies, 66-68, from asst prof to assoc prof, 68-80, PROF PHYSICS, UNIV TEX, AUSTIN, 80- *Mem:* Am Phys Soc. *Res:* Mechanisms of direct nuclear reactions; reactions to particle-unstable states; low and medium energy nuclear physics. *Mailing Add:* Dept Physics Univ Tex Austin TX 78712-1081

COKINOS, DIMITRIOS, US citizen. NUCLEAR ENGINEERING. *Educ:* Cent State Col, BS, 59; NC State Univ, MS, 60; Columbia Univ, PhD(nuclear sci & eng), 69. *Prof Exp:* Physics lectr, City Col of City Univ New York, 61-64; asst prof nuclear eng, Stevens Inst Technol, 61-70; lectr physics, City Col NY, 69-70; adj assoc prof physics, Hunter Col, 70-73; group leader, Gen Elec Co, 73-78; NUCLEAR ENGR, BROOKHAVEN NAT LAB, 78- *Concurrent Pos:* US AEC fel, Columbia Univ, 64-67, res scientist, 69-70. *Mem:* Am Nuclear Soc; Sigma Xi. *Res:* Nuclear and reactor physics; neutron physics; neutron thermalization; pulsed neutron source experiments; neutron cross sections; nuclear fuel cycles, core management; reactor dynamics and control, transient behavior; nuclear safety. *Mailing Add:* Brookhaven Nat Lab Bldg 475B Upton NY 11973

COLAHAN, PATRICK TIMOTHY, b Klamath Falls, Ore, May 31, 48; m 73. LARGE ANIMAL SURGERY. *Educ:* Univ Calif, Davis, BS, 70, DVM, 74; Am Col Vet Surgeons, dipl, 82. *Prof Exp:* Lectr large animal surg, Univ Calif, Davis, 77-78; asst prof, 78-84, chief, large animal surg, Large Animal Hosp, 84-86, CHIEF LARGE ANIMAL SURG, VET MED TEACHING HOSP, UNIV FLA, GAINESVILLE, 79-84 & 89-, ASSOC PROF LARGE ANIMAL SURG, 84- *Mem:* Am Vet Med Asn; Am Asn Equine Practr; Am Col Vet Surgeons; Orthop Res Soc. *Res:* Investigations of the diagnosis and-or treatment of disease of the respiratory, musculo-skeletal or digestive systems of horses; investigation of the mechanics of equine locomotion. *Mailing Add:* 7716 SW 53rd Pl Gainsville FL 32608

COLAIZZI, JOHN LOUIS, b Pittsburgh, Pa, May 10, 38; m 67; c 3. PHARMACEUTICS, PHARMACY. *Educ:* Univ Pittsburgh, BS, 60; Purdue Univ, MS, 62, PhD(pharm), 65. *Prof Exp:* Asst prof pharm, WVa Univ, 64-65; asst prof pharm, 65-68, assoc prof pharmaceut, 68-72, prof pharmaceut & chmn dept, Sch Pharm, Univ Pittsburgh, 72-78; DEAN COL PHARM, RUTGERS, STATE UNIV NJ, 78- *Concurrent Pos:* Mem comt revision, US Pharmacopeial Convention, 75- *Mem:* Am Pharmaceut Asn; Acad Pharmaceut Sci; Am Soc Hosp Pharmacists; Am Asn Cols Pharm. *Res:* Bioequivalency and bioavailability of drugs; pharmacy education; drug formulation factors related to pharmacological response. *Mailing Add:* Col Pharm Rutgers Univ Piscataway NJ 08855-0789

COLÁS, ANTONIO E, b Muel, Spain, June 22, 28; US citizen; m 55; c 4. BIOCHEMISTRY, REPRODUCTIVE PHYSIOLOGY. *Educ:* Univ Zaragoza, Lic, 51; Univ Madrid, MD, 53; Univ Edinburgh, PhD(biochem), 55. *Prof Exp:* Prof in chg physiol & biochem, Lit Univ Salamanca, 55-57; prof biochem, Univ Valle, Colombia, 57-62, dir grad div, 60-62; assoc prof biochem, obstet & gynec, Med Sch, Univ Ore, 62-66, prof obstet & gynec, 66-68, prof biochem, 68; PROF OBSTET & GYNEC & PHYSIOL CHEM, SCH MED, UNIV WIS-MADISON, 68- *Concurrent Pos:* Rockefeller Found grants, 57-62; Catedrático Extraordinario, Univ Zaragoza, Spain, 78-; NIH grants, 62-; mem, Biochem Endocinol Study Sect, NIH, 83-87; consult, NCI, 83- *Mem:* AAAS; Biochem Soc; Endocrine Soc; Am Soc Biochem & Molecular Biol. *Res:* Biochemistry and metabolism of steroid hormones; in vitro and in vivo studies during pregnancy; steroid hydroxylases; steroid hormone receptors. *Mailing Add:* Dept Obstet & Gynec Sch Med Univ Wis 1300 Univ Ave Madison WI 53706

COLASANTI, BRENDA KAREN, b Charleston, WVa, Dec 5, 45; m 68. NEUROPHARMACOLOGY. *Educ:* WVa Univ, BS, 66, PhD(pharmacol), 70. *Prof Exp:* NIMH fel neurol & psychopharmacol, Mt Sinai Sch Med, 70-72; from asst prof to assoc prof, 72-80, PROF PHARMACOL & SURG OPHTHAL, WVA UNIV, 80- *Concurrent Pos:* Acad investr award, NIH, 77-80; consult, Pharmacol Sci Rev Comt, NIH, 84-88. *Mem:* Sigma Xi; Asn Psychophys Study Sleep; Am Soc Pharmacol & Exp Therapeut; Asn Res Vision & Ophthal; Am Soc Neurochem; AAAS. *Res:* Mechanisms underlying the effects of autonomic drugs on intraocular pressure; central nervous system and ocular effects of marihuana cannabinoids. *Mailing Add:* Dept Pharmacol WVa Univ Med Ctr Morgantown WV 26506

COLBECK, SAMUEL C, b Pittsburgh, Pa, Oct 31, 40; div; c 3. PROPERTIES OF SNOW & ICE. *Educ:* Univ Pittsburgh, BS, 62, MS, 65; Univ Washington, PhD(geophysics), 70. *Prof Exp:* GEOPHYSICIST, US ARMY COLD REGIONS RES & ENG LAB, 70- *Concurrent Pos:* Vis asst prof earth sci, Dartmouth Col, 74, adj prof, 75-77, adj prof eng, 81-; assoc ed, J Water Resources Res, 78-84; sci ed, J Glaciol, 84-87. *Honors & Awards:* Horton Award, Am Geophys Union, 80. *Mem:* Int Glaciol Soc (pres, 87-90); Am Geophys Union; Am Asn Avalanche Prof. *Res:* The properties of snow, including mechanics, electromagnetics, metamorphism, water and heat flow; growth of ice crystals from both melt and vapor; snow friction and ski research. *Mailing Add:* US Army Cold Regions Res Eng Lab PO Box 1290 Hanover NH 03755-1290

COLBERT, CHARLES, b Minneapolis, Minn, Feb 19, 19; m 45; c 5. ELECTRICAL ENGINEERING, DIAGNOSTIC RADIOL. *Educ:* Univ Minn, BEE, 40; Union Grad Sch, Cincinnati, PhD, 74. *Prof Exp:* Jr engr equip develop, Gen Develop Lab, Signal Corps, US Army, 41, field engr, West Indies & SAm, 41-43, engr, Aircraft Signal Agency, Ohio, 44; proj engr, Mil TV Aerial Reconnaissance Lab, Wright Air Develop Ctr, 46-48, sect chief res & develop, 48-54; consult engr & pres, Westgate Lab, Inc, 54-62; asst to dean faculty for sponsored res, Antioch Col, 64-70, assoc prof eng, 66-70, sr investr, Fels Res Inst, 66-70, mem res staff, 64-66; dir radiol res lab, Res Inst & adj assoc prof eng, 71-77, ASSOC CLIN PROF RADIOL SCI, SCH MED, WRIGHT STATE UNIV, 79- *Concurrent Pos:* Consult engr, 54-66; exec dir, Ohio Res & Develop Found; consult, Goodyear Aircraft Corp, Martin Co, Appl Physics Lab, Johns Hopkins Univ, Autometric Corp, Aero Serv Corp, Philco Corp, Motorola, Inc, Harris-Intertype Corp, SINTRA, France Westgate Lab, Inc & Miami Valley Consortium of Col; dir, Radiol Res Lab, Greene Mem Hosp, 77-81; dir, Clin Radiol Testing Lab, Miami Valley Hosp, Dayton, 81-84; pres & chief exec officer, Skeletal Assessment Serv Co, Clin Radiol Testing Lab, Yellow Springs, Ohio, 84-86; pres, Personnel Identification & Entry Access Control, Inc, 85-; dir, Found Skeletal Health Res, 88- *Mem:* AAAS; Am Soc Artificial Internal Organs; Inst Elec & Electronics Engrs; emer mem Am Asn Physics in Med; fel Royal Numis Soc; Am Numis Soc. *Res:* Radar mapping; ancient numismatics; ancient languages; archeology; computer analysis of normal and diseased bone and forensic analyses from radiographic films; biometrics for entry control security. *Mailing Add:* Personnel Identification & Entry Access Control Inc PO Box 561 Yellow Springs OH 45387

COLBERT, EDWIN HARRIS, b Clarinda, Iowa, Sept 28, 05; m 33; c 5. VERTEBRATE PALEONTOLOGY. *Educ:* Univ Nebr, AB, 28; Columbia Univ, AM, 30, PhD(vert paleont), 35. *Hon Degrees:* DSc, Univ Nebr, 73, Univ Ariz, 76 & Wilmington Col, Ohio, 84. *Prof Exp:* Asst, Univ Nebr, 26-29; asst, Am Mus Natural Hist, 30-33, asst cur paleont, 33-42, cur fossil reptiles & amphibians, 43-70; prof, 45-69, EMER PROF VERT PALEONT, COLUMBIA UNIV, 70-; EMER CUR FOSSIL REPTILES & AMPHIBIANS, AM MUS NATURAL HIST, 70-; HON CUR VERT PALEONT, MUS NORTHERN ARIZ, 70- *Concurrent Pos:* Assoc cur, Acad Natural Sci, Philadelphia, 37-40; lectr, Bryn Mawr Col, 39-42 & Univ Calif, 45. *Honors & Awards:* Elliot Medal, Nat Acad Sci, 35; Am Mus Natural Hist Medal, 69; Antarctic Serv Medal, NSF, 77; Clinton Award, Buffalo Soc Nat Sci, 87; Romer-Simpson Medal, Soc Vert Paleont, 90. *Mem:* Nat Acad Sci; fel Geol Soc Am; fel Am Paleont Soc (vpres, 63); Soc Vert Paleont (secy & treas, 44, pres, 46-47); Soc Study Evolution (pres, 58). *Res:* Evolution of fossil vertebrates, particularly fossil amphibians, reptiles and mammals; fossil reptiles of North and South America and Asia; fossil mammals of North America and Asia; past distribution and intercontinental migrations of land-living vertebrates; paleoecology as based upon study of fossil vertebrates; fossil amphibians and reptiles of Antarctica. *Mailing Add:* Mus Northern Ariz Rte 4 Box 720 Flagstaff AZ 86001

COLBERT, MARVIN J, b Spokane, Wash, Nov 6, 23; m 51; c 3. OCCUPATIONAL MEDICINE. *Educ:* Yale Univ, BS, 46; Boston Univ, MD, 49; Am Bd Internal Med, dipl, 58. *Prof Exp:* Intern, Presby Hosp, Chicago, Ill, 49-50, asst resident internal med, 50; indust physician, Pub Serv Co Northern Ill, 52; asst resident internal med, Vet Admin Hosp, Boston, Mass, 53-54; resident, Univ Ill Res & Educ Hosps, Chicago, 54-55, instr, Dept Med, 56-58; physician, Steele Mem Clin, Belmond, Iowa, 55-56; from instr to assoc prof, Univ Ill Med Ctr, 56-65, dir health serv, 59-78, prof med, Col Med, 65-78; dir employee Health Serv, Evangel Hosp Asn, Oakbrook, 78-85; assoc med dir, Occupational Health, West Suburban Hosp, Oak Park, Ill, 86-89; RETIRED. *Concurrent Pos:* Vis prof, Chiengmai Med Sch & Hosp, Thailand, 65-66; lectr, Ill Acad Gen Pract, 67-68 & 77-78; attend physician, West Side Vet Admin Hosp & Univ Ill Hosp, Chicago. *Mem:* Am Fed Clin Res; Am Asn Automotive Med; Am Col Physicians; Sigma Xi. *Res:* Medical aspects of automotive safety; smallpox vaccination during pregnancy; nutritional studies of Eskimoes. *Mailing Add:* 5600 Plymouth Ct Downers Grove IL 60516

COLBORN, GENE LOUIS, b Springfield, Ill, Nov 23, 35; m 56, 77; c 5. ANATOMY, CLINICAL ANATOMY. *Educ:* Ky Christian Col, BA, 57; Milligan Col, BS, 62; Bowman Gray Sch Med, MS, 64, PhD(anat), 67. *Prof Exp:* From asst prof to assoc prof anat, Univ Tex Health Sci Ctr, San Antonio, 71-75; assoc prof, 75-88, DIR GROSS ANAT, MED COL GA, 75-, PROF ANAT, 88-, DIR, CTR CLIN ANAT, 88- *Concurrent Pos:* Fel, Dept Anat, Sch Med, Univ NMex, 67-68; grants, Med Res Found Tex, 68-69 & Am Heart Asn, 70-; consult dept surg, Ft Sam Houston, 69-75, Ft Gordon, 90-; secy-treas, State Anat Bd Ga, 75-80, pres, 80-; consult surg, Med Col Ga. *Mem:* Am Asn Anat; Am Asn Clin Anatomists. *Res:* Morphology of atrioventricular conduction system and autonomic ganglia; author of medical gross anatomy texts; texts of surgical anatomy; publications of clinical anatomy, hernia, Gastrointestinal system; inquinal anatomy; surgical vascular anatomy; surgical and clinical anatomy. *Mailing Add:* Dept Anat Med Col Ga Augusta GA 30912-2000

COLBORNE, WILLIAM GEORGE, b Carleton Place, Ont, Feb 18, 26; nat US; m 55; c 4. MECHANICAL ENGINEERING. *Educ:* Queen's Univ, Ont, BSc, 48, MSc, 51. *Prof Exp:* Researcher thermodyn, Nat Res Coun Can, 50; researcher combustion, Queen's Univ, Ont, 50-55, asst prof thermodyn, 54-58; assoc prof eng, 58-66, head, Dept Mech Eng, 59-78, PROF MECH ENG, UNIV WINDSOR, 66- *Mem:* Fel Am Soc Heat, Refrig & Air-Conditioning Engrs; Am Soc Mech Engrs. *Res:* Bistable fluidic amplifiers; computer aided design; computer simulation of buildings; energy conservation. *Mailing Add:* Dept Mech Eng Univ Windsor Windsor ON N9B 3P4 Can

COLBOURN, CHARLES JOSEPH, b Toronto, Can, Oct 24, 53; m 85; c 1. COMBINATORIAL DESIGNS, NETWORKS. *Educ:* Univ Toronto, BSc, 76, PhD(computer sci), 80; Univ Waterloo, MMath, 78. *Prof Exp:* Asst prof computer sci, Univ Sask, 80-83; assoc prof, 84-89, PROF COMBINATORICS, UNIV WATERLOO, 89- *Concurrent Pos:* Vis assoc prof, Univ Toronto, Can, 85; sr fel, Univ Auckland, NZ, 86, Inst Math & Applications, Minneapolis, 88, Simon Fraser Univ, Burnaby, Can, 89 & Curtin Univ, Perth, Australia, 90; vis prof, Auburn Univ, Ala, 89; adj prof, Carleton Univ, Ottawa, Can, 90-; found fel, Inst Combinatorics & Applications, 90- *Mem:* Soc Indust & Appl Math; Opers Res Soc Am; Inst Elec & Electronics Engrs. *Res:* Combinatorial design theory and its relation to computer science; efficient analysis of network reliability and performance. *Mailing Add:* Dept Combinatorics & Opt Univ Waterloo Waterloo ON N2L 3G1

COLBOW, KONRAD, b Bremen, Ger, May 23, 35; Can citizen; m 60; c 2. SOLID STATE PHYSICS. *Educ:* McMaster Univ, BSc, 59, MSc, 60; Univ BC, PhD(physics), 63. *Prof Exp:* Mem tech staff, Bell Tel Labs, 63-65; assoc prof, 65-75, PROF PHYSICS, SIMON FRASER UNIV, 75- *Concurrent Pos:* Humboldt Fel, Ger, 72-73. *Mem:* Am Phys Soc; Can Asn Physics. *Res:* Light interaction with matter; physics of biological membranes. *Mailing Add:* Dept Physics Simon Fraser Univ Burnaby BC V5A 1S6 Can

COLBURN, CHARLES BUFORD, physical inorganic chemistry; deceased, see previous edition for last biography

COLBURN, IVAN PAUL, b San Diego, Calif, June 5, 27; m 58; c 3. STRUCTURAL GEOLOGY, STRATIGRAPHY. *Educ:* Pomona Col, BA, 51; Claremont Grad Sch, MA, 53; Stanford Univ, PhD(geol), 61. *Prof Exp:* Exploration engr, Shell Oil Co, 53-56; Am Asn Petrol Geol res grant, 58-59; from asst prof to assoc prof geol, Calif State Col, Hayward, 61-64; assoc prof, 64-70, PROF GEOL, CALIF STATE UNIV, LOS ANGELES, 70- *Concurrent Pos:* NSF res grant, 63-65; Nat Sci grant, 69-71. *Mem:* Am Asn Petrol Geol; Soc Econ Paleont & Mineral; Geol Soc Am; Nat Asn Geol Teachers. *Res:* California Coast Range structure and stratigraphy; sedimentation and paleocurrent analysis of Jurassic-Cretaceous sediments in California Coast Ranges; statistical analysis of clastic rock fabrics. *Mailing Add:* Dept Geol Calif State Univ Los Angeles CA 90032

COLBURN, NANCY HALL, b Wilmington, Del, May 15, 41; m 81; c 2. BIOCHEMISTRY, CELL BIOLOGY. *Educ:* Swarthmore Col, BA, 63; Univ Wis, PhD(oncol), 67. *Prof Exp:* Asst prof molecular biol, Univ Del, 68-72; spec res fel dermat/carcinogenesis, Univ Mich, 72-74, asst prof biochem, Depts Biol Chem & Dermat, 74-75, vis scientist carcinogenesis, Dept Environ & Indust Health, 75-76; expert chem carcinogenesis in vitro, Lab Exp Path, 76-79, CHIEF, CELL BIOL SECT, LAB VIRAL CARCINOGENESIS, NAT CANCER INST, 79- *Concurrent Pos:* NIH spec res fel, 72-74. *Mem:* AAAS; Am Asn Cancer Res; NY Acad Sci; Sigma Xi. *Res:* Molecular and cellular mechanism of chemical carcinogenesis; use of epithelial cell culture model systems to test the somatic mutation theory and to study tumor promotion and preneoplastic progression. *Mailing Add:* 4511 Willowtree Dr Middletown MD 21769

COLBURN, THEODORE R, b Washington, DC, Oct 31, 39; m; c 2. MENTAL HEALTH. *Educ:* Univ Md, BS, 62, MS, 66, PhD(biomed eng), 69. *Prof Exp:* Control systs engr, Goddard Space Flight Ctr, NASA, 62-66; instr elec eng, Univ Md, 66-69; electronics engr, Sect Tech Develop, NIMH, NIH, 69-73, chief, Sect Tech Develop, NIMH-Nat Inst Neurol & Commun Dis & Stroke, 73-79, chief, Res Serv Br, Intramural Res Prog, NIMH, 79-90, assoc dir opers, 83-88, assoc dir info & technol, 88-90, DEP SCI DIR, NAT INST ALCOHOL ABUSE & ALCOHOLISM, NIH, 90- *Mem:* Sigma Xi; Inst Elec & Electronics Engrs. *Mailing Add:* Nat Inst Alcohol Abuse & Alcoholism NIH Bldg 31 Rm 1B-54 Bethesda MD 20892

COLBURN, WAYNE ALAN, b Allegan, Mich, Sept 10, 47. RESEARCH MANAGEMENT, PHARMACODYNAMICS. *Educ:* Albion Col, BA, 69; State Univ NY, Buffalo, PhD(pharmaceut), 77. *Prof Exp:* Res biochemist, UpJohn Co, 69-75; res asst, State Univ NY, Buffalo, 75-77; staff fel, Nat Inst Environ Health Sci, 77-78; res mgr, Hoffman-La Roche Inc, 78-85; sect dir, Parke-Davis Pharmaceut, 85-90; VPRES CLIN DEVELOP, HARRIS LABS, INC, 90- *Concurrent Pos:* Adj prof, Univ NC, 77-78; consult pharmacol, Cornell Univ Med Col, 81-, Univ Pittsburgh, 84 & Med Col Va, 86-; ed, J Clin Pharmacol, 87- *Mem:* Am Asn Pharmaceut Sci; Acad Pharmaceut Res & Sci; Am Col Clin Pharmacol; Am Soc Clin Pharmacol Ther; Int Soc Study Xenobiotics. *Res:* Role of absorption, distribution metabolism and excretion on the pharmacodynamics of therapeutic and toxic substances; factors that influence pharmacokinetics and pharmacodynamics of therapeutic agents through drug delivery systems. *Mailing Add:* Harris Labs Inc 7432 E Stetson Dr Scottsdale AZ 85251

COLBY, DAVID ANTHONY, b Brookline, Mass, Feb 21, 46; m 68; c 1. LIQUID CHROMATOGRAPHY. *Educ:* Va Polytech Inst, BA, 68, PhD(chem), 78; Clemson Univ, BS, 69. *Prof Exp:* Chemist, Dept Agr Chem Serv, Clemson Univ, 69-70; teaching asst, Chem Dept, Va Polytech Inst, 71-77; tech mkt chemist, Autolab Div, Spectra Physics Inc, 74-76, mgr tech support group, 77-78; sr res & develop chemist, R J Reynolds Tobacco Co, 78-82; sect head separation, Celanese, 82-83; sect head, Flavor Isolation & Identification, Brown & Williamson Tobacco Co, 83-85, sect head methods develop, 85-88, sr scientist anal res, 88-90, LIMS SYST MGR, BROWN & WILLIAMSON TOBACCO CO, 90- *Concurrent Pos:* Lectr, Am Chem Soc, 72-; admin tech asst, Process Control Div, Honeywell Inc, 74. *Mem:* Am Chem Soc; Am Soc Testing & Mat. *Res:* High performance liquid chromatography; advancing the use and understanding of this powerful separation technique in pure and applied research area including instrument design and automation. *Mailing Add:* 7311 Nottoway Circle Louisville KY 40214-3211

COLBY, EDWARD EUGENE, b Aberdeen, SDak, June 8, 29; m 53; c 4. PACKAGING FOOD PRODUCTS, CHEMICAL ENGINEERING. *Educ:* Princeton Univ, BS, 51. *Prof Exp:* Engr, Procter & Gamble Co, 54-56, group leader process develop, 56-65, sect head, 65-66, sect head food tech packaging, 66-77, assoc dir food packaging & tech serv, 77-88; RETIRED. *Concurrent Pos:* Mem indust adv bd, Sch Packaging, Rochester Inst Technol; mem, Packaging Coun. *Mem:* Gravure Tech Asn. *Res:* New packages and package materials. *Mailing Add:* 6255 Hamilton Ave Cincinnati OH 45224

COLBY, FRANK GERHARDT, b Muhlhausen, Ger, Apr 10, 15; m 52; c 2. CHEMISTRY. *Educ:* Univ Geneva, ChemEng, 39, DSc, 41. *Prof Exp:* Consult chemist, Havana, Cuba, 42-46; res chemist, Indust Tape, NJ, 46-47; chem lit specialist, Com Solvents Corp, 47-51; dir res info, R J Reynolds Tobacco Co, 51-70, mgr sci info, 70-79, assoc dir sci issues, R J Reynolds Industs, Inc, 79-83; CONSULT, 83- *Concurrent Pos:* Sci consult prod liability. *Mem:* Fel AAAS; Am Chem Soc. *Res:* Chemical, bioscience and technological literature; tobacco; research analysis. *Mailing Add:* 186 Riverside Dr New York NY 10024

COLBY, GEORGE VINCENT, JR, b Montpelier, Vt, Sept 4, 31; m 55; c 3. RADAR SYSTEM DESIGN, TECHNICAL PROGRAM MANAGEMENT. *Educ:* Mass Inst Technol, SB & SM, 54. *Prof Exp:* Tech engr, test equip, Measurements Lab, Gen Elec Co, 54-55; proj engr, US Army Signal Corps, 55-57; eng mgr, LFE Corp, 57-60; mem tech staff radar & commun, Lincoln Lab, Mass Inst Technol, 70-80; prin engr, missile electronics, Textron Defense Systs, 80-86; PROG DIR, LASER RADAR, AVCO RES LAB, 86- *Mem:* Inst Elec & Electronics Engrs. *Res:* Military voice and digital data communications systems; air traffic control radar and radar transponders; digital and analog circuit design; electronic missle guidance systems; laser radar. *Mailing Add:* Avco Res Lab 2385 Revere Beach Pkwy Everett MA 02149

COLBY, HOWARD DAVID, b New York, NY, June 16, 44; m 70. ENDOCRINOLOGY. *Educ:* City Col New York, BS, 65; State Univ NY Buffalo, PhD(endocrinol), 70. *Prof Exp:* NIH fel physiol, Med Sch, Univ Va, 70-72; from asst prof to assoc prof, 72-78, PROF PHYSIOL, MED SCH, WVA UNIV, 78- *Mem:* Am Physiol Soc; Endocrine Soc; Am Soc Pharmacol & Exp Therapeut; AAAS; Soc Toxicol. *Res:* Hormonal regulation of adrenocortical secretion; hormonal control of hepatic steroid and drug metabolism; endocrine toxicology. *Mailing Add:* Dept Pharmacol & Toxicol Philadelphia Col Pharmacol & Sci 43rd St & Woodland Ave Philadelphia PA 19104

COLBY, PETER J, b Grand Rapids, Mich, Mar 26, 33; c 3. AQUATIC BIOLOGY, FISHERIES. *Educ:* Mich State Univ, BS, 55, MS, 58; Univ Minn, PhD(fishery biol), 66. *Prof Exp:* Food technologist, Gen Foods Corp, 57-62; res asst fishery biol, Univ Minn, 62-66; aquatic biologist & proj leader, US Bur Com Fisheries, 66-70, US Bur Sport Fisheries, 70-71; RES SCIENTIST, ONT MINISTRY NATURAL RESOURCES, 71- *Concurrent Pos:* Res assoc environ & indust health, Sch Pub Health, Univ Mich, 70-71. *Mem:* Am Soc Zool; Am Fisheries Soc; Int Asn Gt Lakes Res; fel Am Inst Fisheries Res Biologists. *Res:* Physiological and behavioral responses of fish to environmental stress; response of fish communities to perturbations. *Mailing Add:* Ministry Natural Resources Fisheries Sect Box 5000 Thunder Bay ON P7C 5G6 Can

COLBY, RICHARD H, b Washington, DC, Nov 7, 39. CELL CULTURE, LIGHT MICROSCOPY. *Educ:* Mass Inst Technol, SB, 61; Univ Calif, Berkeley, PhD(biophys), 68. *Prof Exp:* Asst prof biol & physics, Paine Col, Augusta, Ga, 68-69; Leverhulme fel biol, Univ Nottingham, Eng, 69-70; guest prof zool, Univ Tübingen, Ger, 70-71; asst prof, 71-73, ASSOC PROF CELL BIOL, STOCKTON STATE COL, 73- *Mem:* Am Soc Cell Biol; Tissue Cult Asn. *Res:* Effects of atmosphere derived pollutants on forests; differentiation in cell culture; peroxidation of lipids. *Mailing Add:* NAMS Stockton State Col Pomona NJ 08240-9988

COLCLASER, ROBERT GERALD, b Wilkinsburg, Pa, Sept 21, 33; m 58; c 3. ELECTRICAL ENGINEERING, POWER ENGINEERING. *Educ:* Univ Cincinnati, BS, 56; Univ Pittsburgh, MS, 61, DSc(elec eng), 68. *Prof Exp:* Design engr circuit breakers, Westinghouse Elec Co, 56-63, test engr, 63-66, supvr eng training, 66-67, adv engr power systs, 67-70; from asst prof to assoc prof, 70-80, assoc dean, 74-80, PROF ELEC ENG & CHMN, DEPT ELEC ENG, UNIV PITTSBURGH, 80- *Concurrent Pos:* Mid Atlantic Power Res Comt res grant, 71-; consult, Gould-Brown Boverri, 75-80, Nuclear Regulatory Comn, 78- & Westinghouse Elec, Sharon, Pa, 80- *Mem:* Fel Inst Elec & Electronics Engrs; Am Soc Eng Educ. *Res:* Transient recovery voltage; arc interruption; system overvoltages; insulation coordination. *Mailing Add:* Dept Elec Eng Univ Pittsburgh Pittsburgh PA 15260

COLCORD, J E, b Portland, Maine, June 15, 22; m 47; c 1. PHOTOGRAMMETRY, SURVEYING. *Educ:* Univ Maine, BS, 47; Univ Minn, MS, 49; Ohio State Univ, PhC, 60. *Prof Exp:* Instr & surveyor summer camp, Maine, 47; instr surv, Univ Minn, 47-49; from instr to prof surv, 49-82, prof & assoc chmn civil eng, 83-87, PROF CIVIL ENGR, UNIV WASH, 87- *Concurrent Pos:* Vis prof, Univ Calif, Berkeley, 64 & Univ Ill; sci fac fel, NSF, 59-60; Am Heritage Asn fel, London, 76. *Honors & Awards:* Ford Bartlett Award, Am Soc Photogram. *Mem:* Am Soc Civil Engrs; fel Am Cong Surv & Mapping; Am Soc Photogram. *Res:* Surveying engineering; land surveying systems; remote sensing; multi-purpose cadastre. *Mailing Add:* Dept Civil Eng 121 More Hall FX-10 Univ Wash Seattle WA 98195

COLDREN, CLARKE L(INCOLN), b Uniontown, Pa, Jan 24, 26; m 49; c 4. CHEMICAL ENGINEERING. *Educ:* Pa State Univ, BS, 48; Univ Ill, MS, 50, PhD(chem eng), 54. *Prof Exp:* Engr chem & mech eng, Shell Develop Co, 52-54, 55-59, Koninklijke Shell Lab, Delft, 54-55, mgr tech develop lab, Shell Pipe Line Corp, 59-64, asst to gen mgr transp & supplies, Shell Oil Co, 64-68, mgr planning & anal unconventional raw mat, 68-69, mgr econ & financial res, Shell Chem Co, 69-72, mgr, Resins Bus Ctr, 72-78, strategic planner, 78-79, mgr transp support & planning, Shell Oil Co, 79-80, int logistics develop, 81-84; RETIRED. *Concurrent Pos:* Bus consult. *Mem:* Am Chem Soc; Am Inst Chem Engrs. *Res:* Fluid mechanics; non-Newtonian and multiphase flow; atomization; research administration; business economics; business administration; futurity; transportation. *Mailing Add:* 119 Litchfield Lane Houston TX 77024

COLDWELL, ROBERT LYNN, b Woodland, Wash, May 27, 41; m 62; c 2. STATISTICAL MECHANICS, COMPUTATIONAL PHYSICS. *Educ:* Univ Wash, BS, 63, PhD(physics), 69. *Prof Exp:* Fel & teacher physics, Washington & Lee Univ, 69-71; researcher, Northwestern Univ, 71-72; RESEARCHER PHYSICS, UNIV FLA, 72- *Mem:* Am Phys Soc. *Res:* Developing Monte Carlo and nonlinear optimization procedures for finding accurate ground and excited state eigen functions for systems of nuclei and electrons. *Mailing Add:* Dept Physics Univ Fla Gainesville FL 32611

COLE, AVEAN WAYNE, b Smithville, Miss, June 23, 34; m 60; c 2. WEED SCIENCE. *Educ:* Miss State Univ, BS, 62; Iowa State Univ, PhD(hort & plant physiol), 66. *Prof Exp:* Asst prof hort, Univ Wis, 66-68; PROF WEED SCI, MISS STATE UNIV, 68- *Mem:* Weed Sci Soc Am. *Res:* Weed control and related responses of both weed and crop plants to herbicides. *Mailing Add:* Drawer Pg Plant Path Miss State Univ Mississippi State MS 39762

COLE, BARBARA RUTH, b Hope, Kans, May 14, 41. PEDIATRIC NEPHROLOGY. *Educ:* Doane Col, AB, 63; Univ Kans, MD, 67. *Prof Exp:* From instr to asst prof, 72-81, ASSOC PROF PEDIAT, SCH MED, WASH UNIV, 81-, DIR PEDIAT RENAL DIV, 88- *Concurrent Pos:* Spec fel, NIH, 71-73; asst prof pediat, Univ Calif, San Diego, 77; exec comt, Network Coord Coun 9, 78-81; clin investr award, NIH, 79-82; asst pediatrician, St Louis Children's Hosp, 72- *Mem:* Am Soc Nephrology; Am Soc Clin Res; Am Soc Pediat Nephrology; Int Soc Nephrology; Polycystic Kidney Res Found; Am Physiol Soc. *Res:* Tubular diseases; study of kidney tubular function, transport and metabolism in adult and developing kidneys, employing quantitative histochemical techniques; pediatric dialysis techniques, atrial peptides. *Mailing Add:* Pediat Dept St Louis Children's Hosp 400 S Kingshighway St Louis MO 63110

COLE, BARRY CHARLES, IMMUNOGENETICS, MYCOPLASMOLOGY. *Educ:* Univ Birmingham, Eng, PhD(microbiol), 64. *Prof Exp:* PROF MYCOPLASMOLOGY, DEPT MED, UNIV UTAH, 65- *Mailing Add:* 8845 S1205 E Sandy UT 84070

COLE, BENJAMIN THEODORE, b New Brunswick, NJ, May 24, 21; m 43; c 2. PHYSIOLOGY. *Educ:* Duke Univ, BS, 49, MA, 51, PhD(physiol), 54. *Prof Exp:* Instr physiol, Sch Med, Duke Univ, 53-54; from asst prof to assoc prof zool, La State Univ, 54-58; res partic cell physiol & biol, Oak Ridge Nat Lab, 59-60; assoc prof, 60-63, head dept, 64-73, prof, 63-87, EMER DISTINGUISHED PROF BIOL, UNIV SC, 87- *Concurrent Pos:* Consult, Cell Physiol Sect, Biol Div, Oak Ridge Nat Lab. *Mem:* Fel AAAS; Soc Exp Biol & Med; Am Physiol Soc. *Res:* Digestion; absorption and metabolism of unsaturated fatty acids; in vitro autoxidation of unsaturated fatty acids; effects of sodium fluoroacetate on carbohydrate metabolism in cold-blooded animals. *Mailing Add:* Dept Biol Univ SC Columbia SC 29208

COLE, BRIAN J, b Seattle, Wash, Aug 25, 43. FUNCTIONAL ANALYSIS. *Educ:* Reed Col, BA, 64; Yale Univ, MA, 66, PhD(math), 68. *Prof Exp:* Asst prof math, Univ Calif, Los Angeles, 68-69; from asst prof to assoc prof, 69-83, PROF MATH, BROWN UNIV, 83- *Concurrent Pos:* Miller fel, Univ Calif, Berkeley, 72-74; vis prof math, Univ Calif, Los Angeles, 83; fac res grant, NSF. *Mem:* Am Math Soc. *Mailing Add:* Dept Math Brown Univ Providence RI 02912

COLE, C VERNON, b Wenatchee, Wash, Nov 12, 22. SOIL SCIENCE, ECOLOGY. *Educ:* Univ Mass, BS, 47, MS, 48; PhD(soil sci), 50. *Prof Exp:* RES SOIL SCIENTIST, AGR RES SERV, USDA, 50- *Mem:* Am Soc Agron; Soil Sci Soc Am; Ecol Soc Am. *Res:* Soil physical chemistry; agroecosystems. *Mailing Add:* Nat Res Ecol Lab Colo State Univ Ft Collins CO 80523

COLE, CHARLES F(REDERICK), JR, b Shidler, Okla, July 31, 26; m 55; c 2. ELECTRICAL ENGINEERING. *Educ:* Okla State Univ, BS, 50, MS, 56. *Prof Exp:* Radio-activity engr, Lane-Wells Co, Okla & Venezuela, 50- 52; field engr & systs engr, Western Elec Co, NY, 53-55; asst prof elec eng, La Polytech Inst, 56-60; radio engr & instr elec eng, Univ Tex, 60-61; reliability engr, Radio Corp Am Serv Co, NJ & Alaska, 61-63; res scientist, 63-70, SR RES ENGR & ELECTRONIC SPECIALIST, CONOCO INC, 70- *Mem:* Inst Elec & Electronics Engrs; AAAS. *Res:* Application of microwaves; solid state physics; nuclear, microwave systems; electronic instrumentation; physics. *Mailing Add:* 400 N Irving Ponca City OK 74601

COLE, CHARLES FRANKLYN, b Beaver Falls, Pa, Aug 3, 28; m 52; c 3. FISH BIOLOGY. *Educ:* Cornell Univ, BA, 50, PhD(vert zool), 57. *Prof Exp:* Instr zool, Univ Ark, 57-60; asst prof biol, Univ SFla, 60-62, assoc prof zool & chmn prog, 62-64; from assoc prof to prof fishery biol, Univ Mass, Amherst, 64-80; prof & chmn, Div Fish & Wildlife, 80-84, ASST DIR, SCH NATURAL RESOURCES, OHIO STATE UNIV, 86- *Mem:* AAAS; Am Soc Ichthyol & Herpet; Am Fisheries Soc; Ecol Soc Am; Am Soc Limnol & Oceanog. *Res:* Percid and sciaenid biology; ecology of estuarine fishes. *Mailing Add:* 210 Kottman Hall Ohio State Univ 2021 Coffey Rd Columbus OH 43210-1085

COLE, CHARLES N, b New York, NY, Oct 28, 46; m 69; c 4. MOLECULAR VIROLOGY. *Educ:* Oberlin Col, AB, 68; Mass Inst Technol, PhD(cell biol), 72. *Prof Exp:* Instr virol, Mass Inst Technol, 72-73; fel biochem, Stanford Univ, 74-77; asst prof human genetics, Sch Med, Yale Univ, 78-83; assoc prof, 83-88, PROF BIOCHEM, DARTMOUTH MED SCH, 88-, DIR, MOLECULAR GENETICS CTR. *Concurrent Pos:* Fel virol, Mass Inst Technol, 72-73. *Mem:* Am Soc Microbiol; AAAS; Union Concerned Scientists; Am Soc Virol. *Res:* Molecular biology with emphasis on regulation of eucaryotic gene expression and molecular virology; molecular biology of the DNA tumor virus, SV40, and the regulation of transcription termination and RNA processing in animal cells; recombinant DNA technology; oncogene and anti-oncogene action; RNA transport in yeast. *Mailing Add:* Dept Biochem Dartmouth Med Sch Hanover NH 03756

COLE, CHARLES RALPH, b Canfield, Ohio, Aug 31, 42; c 2. HYDROLOGY, COMPUTER SCIENCE. *Educ:* Kent State Univ, BS, 64; Wash State Univ, MS, 75. *Prof Exp:* Res scientist, Monsanto Res Corp, 64-66; STAFF SCIENTIST & TECH LEADER, PAC NORTHWEST DIV, BATTELLE MEM INST, 66- *Mem:* Am Geophys Union. *Res:* Surface and groundwater hydrology; development of a compatible set of research and management simulation models for surface and groundwater which are used to study and manage water resources. *Mailing Add:* Battelle Northwest PO Box 999 Richland WA 99352

COLE, CLARENCE RUSSELL, b Crestline, Ohio, Nov 20, 18; m 45; c 3. VETERINARY PATHOLOGY. *Educ:* Ohio State Univ, DVM, 43, MSc, 44, PhD(comp path), 47. *Prof Exp:* Instr vet path, 44-46, prof, 46-71, chmn dept, 47-67, from asst dean to dean col vet med, 60-72, REGENTS PROF VET PATH, OHIO STATE UNIV, 71- *Concurrent Pos:* Assoc vet res, Ohio Agr Res & Develop Ctr, 47-; consult, USAF, 43-, Walter Reed Med Ctr, 51-, Armed Forces Inst Path, 53- & USPHS, 53-; mem spec bd, Am Col Vet Pathologists, 59 & Am Col Toxicol, 78; lectr, Auburn Univ, 53. *Mem:* AAAS; Am Vet Med Asn; Am Asn Lab Animal Sci; Am Col Vet Path (vpres, 56, pres, 57); Int Acad Path. *Res:* Toxicologic pathology; lab animal diseases; animal infectious diseases; animal neoplasms; metabolic diseases, comparative pathology. *Mailing Add:* Col Vet Med Ohio State Univ 1925 Coffey Rd Columbus OH 43210

COLE, DALE WARREN, b Everett, Wash, May 28, 31; m 56; c 4. FOREST SOILS. *Educ:* Univ Wash, BSF, 55, PhD(forest soils), 63; Univ Wis, MS, 57. *Prof Exp:* Res assoc, Univ Wash, 60-64, from asst prof to assoc prof, 65-73, dir Ctr Ecosyst Studies, 72-82, PROF FOREST SOILS, COL FOREST RESOURCES, UNIV WASH, 74-, ASSOC DEAN, 82- *Mem:* AAAS; Soil Sci Soc Am; Am Geophys Union; Int Union Forest Res Orgn; Ecol Soc Am; Sigma Xi; Soc Am Foresters; Pac Sci Asn. *Res:* Mineral cycling in a forest ecosystem; processes of elemental leaching in forest soils; forest ecosystem response to fertilizers and municipal wastewater and sludge; soils and land-use planning. *Mailing Add:* Col Forest Resources AR-10 Univ Wash Seattle WA 98195

COLE, DAVID EDWARD, b Detroit, Mich, July 20, 37; m 65; c 2. MECHANICAL ENGINEERING. *Educ:* Univ Mich, BS(mech eng) & BS(math), 60, MS, 61, PhD(mech eng), 66. *Prof Exp:* From instr to asst prof, 65-71, ASSOC PROF MECH ENG, UNIV MICH, 71-, DIR CTR STUDY AUTOMOTIVE TRANSP, 78- *Concurrent Pos:* Consult, Ford Motor Co, 65-78, Outboard Marine Corp, 67-70, Gen Motors Corp, 69-70, Alcoa, Exxon, ICS, Dow Chem, Dept Transp, 78-, Aluminum ASN, Environ Protection Agency, Aisin Seiki, Bendix, NASA & Panhandle Eastern, 81-; chmn bd, Environ Dynamics Inc, 70-; pres, Appl Theory, Inc, 80- *Honors & Awards:* Ralph Teetor Educ Award, 67. *Mem:* Soc Automotive Engrs; Sigma Xi. *Res:* Exhaust emission and fuel consumption research on the 2 and 4-stroke spark ignited reciprocating and rotating engines; investigation of mixture motion in combustion bombs and reciprocating engines; advanced power plants and total vehicle design; future automotive technology trends; strategic planning in automotive industry. *Mailing Add:* 3946 Waldenwood Ann Arbor MI 48105

COLE, DAVID F, b Childress, Tex, Mar 3, 33. PHYSICAL CHEMISTRY, SURFACE CHEMISTRY. *Educ:* Univ Tex, Austin, BS, 57, PhD(phys chem), 64; Southern Methodist Univ, MBA, 78. *Prof Exp:* Jr chemist, Oak Ridge Nat Lab, 57; vis asst prof chem, La State Univ, 63-64; mem tech staff, Cent Res Labs, Tex Instruments, 65-89; CONSULT, 89- *Concurrent Pos:* Brookhaven Col, 90-91. *Mem:* Electrochem Soc; Am Chem Soc. *Res:* Advanced processes and yield improvement studies for the manufacture of semiconductor and other electronic devices; fuel cells; surface properties of solids. *Mailing Add:* PO Box 831544 Richardson TX 75083

COLE, DAVID LE ROY, b Preston, Idaho, Aug 6, 39; m 65; c 2. PHYSICAL CHEMISTRY, CHEMICAL KINETICS. *Educ:* Univ Utah, BS, 66, PhD(phys chem), 70. *Prof Exp:* SR RES CHEMIST, RES LABS, EASTMAN KODAK CO, 69- *Mem:* Am Chem Soc. *Res:* Kinetic studies of group III metal ion hydrolysis using E-jump perturbation techniques; investigation of fundamental reaction kinetics between metal ions and organic and inorganic ligands. *Mailing Add:* Bldg 59 Res Lab Eastman Kodak Co Kodak Park Rochester NY 14650

COLE, DOUGLAS L, b Great Bend, Kans, Mar 5, 47; m 76; c 3. PROCESS RESEARCH & DEVELOPMENT, MEDICINAL CHEMISTRY. *Educ:* Ft Hays State Univ, BSc, 69; Univ Ill, Champaign-Urbana, PhD(org chem), 75. *Prof Exp:* Sr res chemist, Merck Sharp & Dohme, 74-78, assoc dir, 83-85; group leader, Lederle Labs, 78-83; DIR, MARION LABS, 85. *Mem:* Am Chem Soc; Asn Off Anal Chemists; Drug Info Asn. *Res:* Natural product chemistry; numerous human and animal health product; analytical chemistry development and synthetic organic chemistry research. *Mailing Add:* Res & Develop Lab Marion Merrell Dow Inc 10236 Marion Park Dr Kansas City MO 64137

COLE, EDMOND RAY, b Huntington, WVa, Dec 17, 28; m 55; c 3. BIOCHEMISTRY. *Educ:* WVa Univ, BS, 52, MS, 55; Purdue Univ, PhD(biochem), 61. *Prof Exp:* NIH grants, Purdue Univ, 60-61 & Wayne State Univ, 61-63; res assoc coagulation, Rush-Presby-St Luke's Med Ctr, 63-69, asst dir coagulation, 69-73; assoc prof, 73-84, PROF BIOCHEM, RUSH MED COL, 84-, DIR COAGULATION, RUSH-PRESBY-ST LUKE'S MED CTR, 73- *Concurrent Pos:* Assoc scientist, Presby-St Luke's Hosp, 73-; Thrombosis Coun, Am Heart Asn. *Mem:* AAAS; Int Soc Thrombosis & Haemostasis; Am Soc Hemat; Sigma Xi. *Res:* Coagulation; fibrinolysis; tissue activators of plasminogen; enzymology. *Mailing Add:* 1617 Highland Ave Wilmette IL 60091

COLE, EDWARD ANTHONY, b Boston, Mass, Oct 16, 32; m 64; c 3. MICROBIOLOGY, PHYSIOLOGY. *Educ:* Univ Notre Dame, BS, 56, PhD(microbiol), 67. *Prof Exp:* Teacher, St Charles High Sch, Wis, 54-56, Vincentian Inst, NY, 56-57, Boysville High Sch, Mich, 57-62 & Bradley High Sch, NH, 62-63; teaching asst biol, Univ Notre Dame, 63-64; from asst prof to assoc prof, 66-72, PROF BIOL, ANNA MARIA COL WOMEN, 72- *Mem:* AAAS; NY Acad Sci. *Res:* Experimental hematology; transplantation. *Mailing Add:* Dept Biol Anna Maria Col Paxton MA 01612

COLE, EDWARD ISSAC, JR, b Raleigh, NC, May 9, 59; m 81; c 2. INTEGRATED CIRCUIT FAILURE ANALYSIS, ELECTRON BEAM ANALYSIS. *Educ:* Univ NC, Chapel Hill, BA, 81, MS, 85, PhD(physics), 87. *Prof Exp:* SR MEM TECH STAFF, SANDIA NAT LABS, 87- *Concurrent Pos:* Mem tech comt, Int Symp Testing & Failure Anal, 90-; concern leader, Task Force on Failure Anal, Joint Electron Device Eng Coun, 91- *Mem:* Am Phys Soc; Int Soc Hybrid Microelectronics. *Res:* New integrated circuit failure analysis techniques for benign device examination, with an emphasis on electron beam techniques, scanned probe methods, integrated circuit light emission analysis and digital image acquisition and processing. *Mailing Add:* 2116 White Cloud St NE Albuquerque NM 87112

COLE, EVELYN, b Aberdeen, Miss, June 10, 10. ZOOLOGY. *Educ:* Miss Univ Women, AB, 32; Duke Univ, MA, 43; Vanderbilt Univ, PhD(biol), 66. *Prof Exp:* From instr to asst prof biol, Greensboro Col, 45-57; from asst prof to prof, 60-75, EMER PROF BIOL, MURRAY STATE UNIV, 75- *Res:* Taxonomy and ecology of freshwater Ostracoda; taxonomy of invertebrates. *Mailing Add:* 1703 Ryan Ave Murray KY 42071

COLE, FRANCIS TALMAGE, b Lynbrook, NY, Oct 6, 25; wid; c 4. PHYSICS. *Educ:* Oberlin Col, AB, 47; Cornell Univ, PhD(physics), 53. *Prof Exp:* Asst physics, Cornell Univ, 47-51; from instr to assoc prof, Univ Iowa, 51-64; physicist, Lawrence Radiation Lab, 64-67; physicist, Fermi Nat Accelerator Lab, 67-88; CONSULT, 88- *Concurrent Pos:* Physicist, Midwestern Univs Res Asn, 55-59, head theory sect, 59-60, head physics div, 60-64. *Mem:* AAAS; Am Phys Soc. *Res:* Particle accelerators. *Mailing Add:* PO Box 500 Batavia IL 60510

COLE, FRANKLIN RUGGLES, b Newton, Mass, Aug 16, 25; m 47; c 2. PHARMACOGNOSY. *Educ:* Mass Col Pharm, BS, 51, MS, 53; Univ Utah, PhD(pharmacog), 56. *Prof Exp:* From asst prof to assoc prof, 56-65, PROF PHARMACOG, IDAHO STATE UNIV, 65- *Concurrent Pos:* Fulbright lectr, Cairo Univ, 65. *Mem:* AAAS; Am Soc Pharmacog; Am Pharmaceut Asn; Sigma Xi. *Res:* Re-evaluation of ethnobotanical drugs of the Bannock-Shoshone Indians. *Mailing Add:* 100 Ranch Dr Pocatello ID 83201

COLE, G(EORGE) ROLLAND, b Lincoln, Kans, Dec 12, 25; m 50; c 4. SCIENCE EDUCATION. *Educ:* Univ Kans, BS, 49, MA, 54, PhD(physics), 57. *Prof Exp:* Asst instr physics, Univ Kans, 51-56; res engr, Savannah River Lab, E I du Pont de Nemours & Co, Inc, 56-65, res physicist, Electrochem Dept, Wilmington, Del, 65-69, qual control supvr, Electronic Prod Div, Niagara Falls, 69-73, sr res physicist, Photo Prod Dept, 73-83, res assoc, 84-85; PROF, UNIV DEL, 85- *Mem:* Sigma Xi; Am Phys Soc. *Res:* Color centers in alkali halides; properties and irradiation behavior of uranium oxide; solid actinide compounds; thick film electronic components; photographic science; electron spectroscopy; chromium dioxide magnetic tape. *Mailing Add:* Seven Lantern Ct RD 3 Hockessin DE 19707

COLE, GARRY THOMAS, b Toronto, Ont, June 26, 41; m 64; c 2. MEDICAL MYCOLOGY, ELECTRON MICROSCOPY. *Educ:* Carleton Univ, Ottawa, BA, 63; Univ Waterloo, Ont, PhD(biol), 69. *Prof Exp:* Teacher sci, Clarke High Sch, Newcastle, Ont, 64-65; teaching fel biol, Univ Waterloo, Ont, 66-69; fel mycol, Univ Fla, Gainesville, 70-71; fac assoc, 70-71, asst prof, 71-75, assoc prof, 75-82, PROF BOT, UNIV TEX, AUSTIN, 82- *Concurrent Pos:* Vis asst prof, Lehrstühl für Zellenlehre, Univ Heidelburg, 74, vis assoc prof, 81; Humboldt fel, Inst für Meeres forschung, Bremerhaven, WGer, 75; NSF fel, Biochem Dept, Gifu Sch Med, Japan, 77-78. *Mem:* Am Soc Microbiol; Bot Soc Am; Mycol Soc Am; Med Mycol Soc Am; Electron Micros Soc Am. *Res:* Taxonomy and biology of conidial fungi, including morphogenetic, ultrastructural, cytological, host-parasite interaction, wall chemistry and immunological studies. *Mailing Add:* Dept Bot Univ Tex Austin TX 78712

COLE, GEORGE CHRISTOPHER, b Brooklyn, NY, Oct 12, 29; m 55; c 5. CELL BIOCHEMISTRY. *Educ:* St John's Univ, NY, BS, 50, MS, 52; Univ Wis, PhD(bact), 57. *Prof Exp:* Lab asst parasitol, NY Univ-Bellevue Med Ctr, 51-52; asst virol, Univ Wis, 54-57; sr res scientist microbiol, E R Squibb & Sons, 57-66; sr res microbiologist, Parke, Davis & Co, Detroit, 66-80; res assoc, Sch Med, Wayne State Univ, 81-83; lab supvr, Univ Mich & Univ Med Affil, PC, 83-86; CONSULT, 86- *Res:* Chemotherapy of infectious diseases; malaria cultivation in vitro; viral interference; viral and bacterial vaccines, serology and immunology; alpha and gamma interferons; cultivation and radioimmunoassay of biochemical functions of liver cells, monoclonal antibody. *Mailing Add:* 3834 Quarton Rd Bloomfield Hills MI 48013

COLE, GEORGE DAVID, b Minden, La, June 23, 25; m 47; c 3. CHEMICAL PHYSICS. *Educ:* Northwestern State Col, La, BS, 50; Univ Ala, PhD(physics), 63. *Prof Exp:* Asst prof physics & math, Nicholls State Col, 54-60; from asst prof to assoc prof, 64-72, from actg head to head dept, 68-72, prof physics & chmn, Dept Physics & Astron, 72-82, actg assoc vpres acad affairs, 80-83, ASSOC ACAD VPRES, UNIV ALA, 83- *Concurrent Pos:* Physics consult, Insts Int Educ, E & W Pakistan, 67-70. *Mem:* Am Phys Soc; Am Asn Physics Teachers. *Res:* Positron annihilation in liquid crystalline compounds; partial L- and M-shell fluorescence yields; polymorphic and mesomorphic behavior in organic compounds. *Mailing Add:* 13 Hickory Hill 1628-24 St E Tuscaloosa AL 35404

COLE, GEORGE WILLIAM, b Ridgewood, NJ, Oct 20, 33; m 55; c 4. INTERNAL MEDICINE, HEMATOLOGY. *Educ:* Univ Pa, BA, 54; Univ Miami, MD(med), 58. *Prof Exp:* Fel hemat, Dept Hemat, Jackson Mem Hosp, Miami, 61-63; instr hemat & internal med, Univ Miami, 63-66; from asst prof to assoc prof clin pathol, 66-73, ASST PROF INTERNAL MED, UNIV ALA, BIRMINGHAM, 66-, PROF PATH, 73-, PROF & MED DIR CLIN LAB SCI, SCH COMMUNITY & ALLIED HEALTH, 76-, VCHMN, DEPT PATH, 78- *Concurrent Pos:* Consult, Vet Admin Med Ctr, Birmingham, 66-; dir, Hemat Sect, Dept Clin Path, Univ Ala & Vet Admin Hosp Clin Labs, 66-76; dir, Lab Opers, CLin Labs, Univ Ala Hosp, 67-70, Outpatient Oper, 70-72, Gen Hemat & Chem, 74-; lab inspector, Col Am Pathologists, 70-74, chmn, guidelines for appropiate utilization of lab procedures, 74-79, adv, Clin Path Comt, 80-81; bd trustees, Am Path Found, 83-86. *Mem:* Acad Clin Lab Physicians & Scientists; AMA; Am Soc Clin Pathologists; Col Am Pathologists; Nat Comt Clin Lab Standards; Am Path Found. *Res:* Medical laboratory system engineering; construction of testing panels; analysis of test results; test reporting systems in health care systems. *Mailing Add:* Dept Path Med Ctr Univ Ala Birmingham AL 35294

COLE, GERALD AINSWORTH, b Hartford, Conn, Dec 25, 17; m 44; c 5. LIMNOLOGY. *Educ:* Middlebury Col, AB, 39; St Lawrence Univ, MS, 41; Univ Minn, PhD(zool), 49. *Prof Exp:* Teaching fel biol, St Lawrence Univ, 40-41; teacher chem, Milton Acad & instr biol, Phillips Acad, 42; asst zool, Univ Minn, 46-49; from asst prof to assoc prof biol, Univ Louisville, 49-58; from lectr to prof, 58-80, EMER PROF ZOOL, ARIZ STATE UNIV, 80- *Concurrent Pos:* Res, Douglas Lake State Biol Sta, Mich, 50; mem teaching staff, Lake Itasca Biol Sta, Minn, 64-66, 68 & 70; res, Coahuila, Mex, 67 & 68. *Mem:* Am Soc Limnol & Oceanog; Int Asn Theoret & Appl Limnol; Crustacean Soc. *Res:* Microcrustacea, amphipoda and regional limnology. *Mailing Add:* Rte 4 Box 892 Flagstaff AZ 86001

COLE, GERALD ALAN, b West New York, NJ, June 11, 31; m 58; c 3. VIROLOGY, IMMUNOLOGY. *Educ:* Wilson Teachers Col, BS, 52; Univ Md, PhD(microbiol), 66. *Prof Exp:* Med bacteriologist, Walter Reed Army Inst Res, 55-59; from asst prof to prof epidemiol, Sch Hyg & Pub Health, Johns Hopkins Univ, 67-81; virologist, 59-60, res assoc microbiol, 66-67, PROF MICROBIOL, SCH MED, UNIV MD, 81- *Concurrent Pos:* USPHS res grant, Sch Hyg & Pub Health, Johns Hopkins Univ, 70-80, res career develop award, 71-76; Josiah Macy Jr Found fac scholar award, John Curtin Sch Med Res, Australian Nat Univ, 74-75; mem ad hoc study group virus & rickettsial dis, US Army Res & Develop Command, 73-77; consult to Surgeon Gen, 73-77; res grant, Univ Md Sch Med, 80-85. *Mem:* AAAS; Am Soc Microbiol; Am Asn Immunol; Am Soc Trop Med & Hyg; Am Soc Virol. *Res:* Role of the immune response in the outcome of experimental viral infections. *Mailing Add:* Dept Microbiol & Immunol Sch Med Univ Md 665 W Baltimore St Baltimore MD 21201

COLE, GERALD SIDNEY, b Toronto, Ont, Mar 29, 36; m 69; c 1. TECHNICAL MANAGEMENT. *Educ:* Univ Toronto, BASc, 59, MA, 60, PhD(metall), 63. *Prof Exp:* Res scientist, Res Labs, Ford Motor Co, 64-67; French Atomic Energy Comn res fel, Grenoble Nuclear Res Ctr, 67-68; PRIN STAFF ENGR, SCI RES STAFF, FORD MOTOR CO, 68- *Concurrent Pos:* Foundry expert, UN Indust Develop Orgn, Mex, 75-76. *Mem:* Int Metall Soc; Am Inst Mining, Metall & Petrol Engrs; Am Soc Metals; Am Foundrymens Soc; Soc Mfg Engrs. *Res:* Foundry technology; manufacturing, solidification; ferrous metallurgy; aluminum components; automotive component design and production; metal matrix composites; 70 publications and 8 patents. *Mailing Add:* Ford Motor Co Box 2053 S-2016 Dearborn MI 48121-2053

COLE, HAROLD S, b Brooklyn, NY, Apr 20, 16. MEDICINE. *Educ:* Univ Md, BS, 37; NY Univ, MD, 42; Am Bd Pediat, dipl, 49. *Prof Exp:* Assoc prof, 48-74, prof pediat & chief, Sect Metab, Dept Pediat, 74-87, EMER PROF PEDIAT & COMMUNITY & PREV MED, NY MED COL, 87- *Concurrent Pos:* Assoc attend pediatrician, Flower & Fifth Ave Hosps, NY, 48-74, attend pediatrician, 74-80; assoc vis pediatrician, Metrop Hosp, 48-73, vis pediatrician, 73-87, head, pediat diabetes clin; med dir, Flower Hosp, 78-80; med dir, Bronx Develop Ctr, 83- *Mem:* Fel Am Acad Pediat; Lawson Wilkins Pediat Endocrine Soc; Am Pediat Soc; Am Diabetes Asn. *Res:* Adolescent medicine; diabetes mellitus in children; the infant of the diabetic mother; developmental disabilities. *Mailing Add:* 185 E 85th St New York NY 10028

COLE, HARVEY M, b North Weymouth, Mass, June 1, 20; m 46; c 4. CHEMICAL ENGINEERING. *Educ:* Northeastern Univ, BSChE, 43. *Prof Exp:* Chem engr, Godfrey L Cabot, Inc, 46-51, leader anal group, 51-57, head anal res & serv sect, 57-62, assoc dir res & develop, anal res & serv, 62-68, MGR CARBON BLACK RES SECT, CABOT CORP, 68- *Mem:* Am Chem Soc; Soc Appl Spectros. *Res:* Combustion; carbon formation in flames; radical reactions. *Mailing Add:* 162 North St Walpole MA 02081-2959

COLE, HENDERSON, b Wilmington, NC, Oct 2, 24; m 50; c 4. PHYSICS. *Educ:* Mass Inst Technol, BS, 50, PhD(physics), 52. *Prof Exp:* Asst physics, Mass Inst Technol, 49-52, instr, 53-55; Fulbright fel, Col France, 52-53; staff physicist, Res Ctr, Brooklyn Heights, NY, 55-78, SR SCIENTIST, IBM CORP, 78- *Concurrent Pos:* Vis scientist, IBM Corp, Venezula, 78- *Mem:* Am Phys Soc; Sigma Xi; Am Crystallog Asn. *Res:* Supervising service work; solid state physics; x-ray diffraction; computer control of instruments. *Mailing Add:* IBM Corp 472 Wheelers Farms Rd Milford CT 06460

COLE, HERBERT, JR, b Long Island, NY, Mar 29, 33; c 3. PLANT PATHOLOGY, AGRICULTURAL CHEMISTRY. *Educ:* Pa State Univ, BS, 54, MS, 55, PhD(plant path & agr biochem), 57. *Prof Exp:* From asst prof to assoc prof plant path, Pa State Univ, 57-66, agr chem coord, Col Agr, 64-66, from assoc prof to prof plant path & chem pesticides, 66-86, PROF PLANT PATH, PA STATE UNIV, 70-, HEAD DEPT, 86- *Concurrent Pos:* Consult pesticide litigation, Agr Chem Indust. *Mem:* Am Phytopath Soc; Can Phytopath Soc; Am Soc Testing & Mat. *Res:* Side effects of pesticides; control of plant diseases. *Mailing Add:* Dept Plant Path Pa State Univ University Park PA 16802

COLE, JACK ROBERT, b Milwaukee, Wis, Dec 28, 29; m 52; c 3. MEDICINAL CHEMISTRY. *Educ:* Univ Ariz, BS, 53; Univ Minn, PhD(pharmaceut chem), 57. *Prof Exp:* From asst prof to assoc prof, 57-62, head dept pharmaceut sci, 75-79, dean Col Pharm, 77-89, PROF MED CHEM, UNIV ARIZ, 62-, SR VPRES ACAD AFFAIRS & PROVOST, 90- *Concurrent Pos:* Actg vprovost, 87-88, actg provost, acad affairs, Univ Ariz, 88-89. *Res:* Chemistry of natural medicinal products; chromatography; synthetic organic medicinals. *Mailing Add:* Col Pharm Univ Ariz Tucson AZ 85721

COLE, JACK WESTLEY, b Portland, Ore, Aug 28, 20; m 43; c 4. EXPERIMENTAL SURGERY. *Educ:* Univ Ore, AB, 41; Wash Univ, MD, 44; Am Bd Surg, dipl, 53. *Prof Exp:* From jr instr to sr instr surg, Western Reserve Univ, 52-54, asst surgeon, Univ Hosps, 54-56, from assoc prof to prof surg, Sch Med, 56-63; prof & chmn dept, Hahnemann Med Col & Hosp, 63-66; dir, Div Oncol & Yale Comprehensive Cancer Ctr, 75-84; chmn dept surg, 66-74, ENSIGN PROF, YALE UNIV SCH MED, 66- *Concurrent Pos:* Chief, Hahnemann Serv, Div B, Dept Surg, Philadelphia Gen Hosp, 63-66; consult, Vet Admin Hosp, Philadelphia, 63-66; chief surg, Yale New Haven 66-74, attend surgeon, 74-; consult, West Haven Vet Hosp, 66-; vis fel, Woodrow Wilson Found, 81- *Mem:* Soc Exp Biol & Med; Am Fedn Clin Res; fel Am Col Surg; Soc Univ Surg; Am Surg Asn. *Res:* Cellular kinetics of gastrointestinal epithelium, normal and neoplastic. *Mailing Add:* Yale Comprehensive Cancer Ctr 333 Cedar St New Haven CT 06510

COLE, JAMES A, b Albany, NY, Nov 18, 39; m 63; c 2. SYSTEMS ENGINEERING. *Educ:* Union Col, NY, BS, 61; Johns Hopkins Univ, PhD(physics), 66. *Prof Exp:* Instr physics, Johns Hopkins Univ, 65-66; res assoc, State Univ NY, Stony Brook, 66-67, asst prof, 67-71; mgr systs eng, Elsytec Inc, NY, 71-74; mgr systs eng, 74-80, dir, 80-88, VPRES RES & DEVELOP, MEGADATA INC, BOHEMIA, NY, 88- *Concurrent Pos:* Guest res assoc, Brookhaven Nat Lab, 63-71. *Mem:* AAAS; Am Phys Soc; Inst Elec & Electronics Engrs. *Res:* Passive radar navigation systems; elementary particle physics; sound and vibration analysis. *Mailing Add:* 20 Upper Sheep Pasture Rd East Setauket NY 11733

COLE, JAMES CHANNING, b Oakland, Calif, May 22, 48; m 70. REGIONAL GEOLOGY, PETROLOGY. *Educ:* Univ Calif, Santa Barbara, BA, 69; Univ Colo, Boulder, PhD(geol), 77. *Prof Exp:* Geologist, Br Cent Environ Geol, 78-80, geologist, 80-81, chief geologist, Saudi Arabian Mission, 81-84, CHIEF ENVIRON, RESTORATION PROJ, US GEOL SURV, 84- *Concurrent Pos:* Vis lectr, Dept Geol Sci, Univ Colo, Boulder, 78-79. *Mem:* Geol Soc Am. *Res:* Structure and petrology of Precambrian rocks of the central United States; emplacement mechanisms of igneous bodies; uranium occurrences in igneous and metamorphic rocks; regional geology, petrology and structure of the Arabian Shield; metallogenesis of stockwork tungsten deposit. *Mailing Add:* Mail Stop 913 Box 25046 Fed Ctr US Geol Surv Denver CO 80225

COLE, JAMES EDWARD, b Detroit, Mich, Sept 10, 40; m 60; c 2. ETHOLOGY, VERTEBRATE ZOOLOGY. *Educ:* Western Mich Univ, BA, 62, MA, 63; Ill State Univ, PhD(zool), 68. *Prof Exp:* Instr biol, Highland Park Col, 63-64; fac asst zool, Ill State Univ, 65-67; assoc prof, Bloomsburg Univ, 68-71, prog coordr health sci, 73-79, chmn, Dept Biol, 79-86, dir, Scholars Prog, 86-87, interim assoc dean arts & sci, 87-88, PROF BIOL, BLOOMSBURG UNIV, 71-, PROG COORDR ALLIED HEALTH SERV, 73- *Concurrent Pos:* Mem, Galapagos Exped, 91. *Mem:* AAAS; Am Soc Zool; Animal Behav Soc; Am Inst Biol Sci. *Res:* Parent-young interactions of Cichlid fishes; behavior of lower vertebrates; allied health sciences. *Mailing Add:* Dept Biol & Allied Health Sci Bloomsburg Univ Bloomsburg PA 17815

COLE, JAMES WEBB, JR, b Norfolk, Va, July 22, 10; m 36; c 2. CHEMISTRY. *Educ:* Univ Va, BS, 32, MS, 34, PhD(phys chem), 36. *Prof Exp:* Res chemist, Exp Sta, E I du Pont de Nemours & Co, 36-37; from asst prof to prof, 37-75, dean, 58-75, EMER PROF CHEM, UNIV VA, 75- *Concurrent Pos:* Instr, USN, 44; prog dir, NSF, 52-53, mem adv panel, 53-; consult, Off Sci Res & Develop; investr, Nat Defense Res Comt. *Mem:* AAAS; fel Am Inst Chem; Am Chem Soc. *Res:* Thermal measurements; adsorption of gases on solids; vapor phase catalytic oxidation and hydrolysis; new analytical methods; gas analysis; synthesis of complex inorganic compounds and of organo metallic compounds; spectrophotometer used in analysis and in problems of structure; mechanisms of antioxidants and anticorrosion agents. *Mailing Add:* 900 Rosser Lane Charlottesville VA 22903

COLE, JEROME FOSTER, b Cincinnati, Ohio, Aug 8, 40; m 63; c 2. METALS TOXICOLOGY. *Educ:* Univ Cincinnati, BS, 62, MS, 66, ScD(environ health), 68. *Prof Exp:* Corp indust hygenist, Procter & Gamble, 68-69; mgr environ health, 69-72, vpres, 72-84, PRES INT LEAD ZINC RES ORGN, INC, 84- *Concurrent Pos:* Dir environ health, Lead Industs Asn, 71-83. *Mem:* Am Indust Hyg Asn; Soc Toxicol; Air & Waste Mgt Asn; Soc Environ Geochem Health. *Res:* Environmental health impact of lead, zinc and cadmium; product research on metallurgy, chemistry, and electro chemistry. *Mailing Add:* 908 Pinehurst Dr Chapel Hill NC 27514

COLE, JERRY JOE, b Kansas City, Mo, May 22, 38; m 61; c 2. ANALYTICAL CHEMISTRY. *Educ:* Univ Kansas City, BS, 60; Univ Iowa, MS, 63, PhD(anal chem), 65. *Prof Exp:* Asst prof chem, Ft Hays Kans State Col, 64-67; prof chem, Ashland Col, 74-88, chmn dept, 80-88; VPRES ACAD AFFAIRS, KANS COL TECHNOL, 88- *Mem:* Am Chem Soc. *Res:* Complexing of metal ions with organic and inorganic moieties; organic precipitants as a means of gravimetric estimation; amperometric titrations of inorganic ions. *Mailing Add:* Vpres Acad Affairs Kans Col Technol Salina KS 67401

COLE, JOHN E(MERY), III, b Upper Darby, Pa, May 20, 42; m 68; c 2. ACOUSTICS, FLUID MECHANICS. *Educ:* Drexel Univ, BSME, 65; Brown Univ, ScM, 67, PhD(fluid mech), 70. *Prof Exp:* Asst prof mech eng, Tufts Univ, 70-77; scientist, 77-80, PRIN SCIENTIST, CAMBRIDGE ACOUST ASSOCS, INC, 80- *Mem:* Am Soc Mech Engrs; Acoust Soc Am; Sigma Xi. *Res:* Vibrations; applied mathematics; structural dynamics, acoustics. *Mailing Add:* 62 Liberty Ave Lexington MA 02173

COLE, JOHN OLIVER, b Jamestown, NY, Apr 15, 15; m 44; c 2. CHEMISTRY. *Educ:* Bethany Col, Kans, BS, 39; Univ Colo, MS, 41, PhD(org chem), 43. *Prof Exp:* Asst chem, Univ Colo, 40-43, asst chem eng, 43; res chemist, Goodyear Tire & Rubber Co, 43-57, head anal sect, 57-65, mgr anal chem, 65-77, mgr spec projs, Org Chem Dept, 77-81; RETIRED. *Mem:* Am Chem Soc. *Res:* Organic and analytical chemistry; reaction of phenyl glyoxal with aliphatic amidines; autoxidation of elastomers; rubber chemicals; kinetics of autoxidation of hydrocarbons. *Mailing Add:* 200 Laurel Lake Dr No 147 Hudson OH 44236-2132

COLE, JOHN RUFUS, JR, b Baconton, Ga, Oct 27, 38; m 64; c 2. VETERINARY MICROBIOLOGY. *Educ:* Univ Ga, BS, 60, MS, 63, PhD(microbiol), 66. *Prof Exp:* Res asst microbiol, Poultry Dis Res Ctr, 61-63, microbiologist, Diag & Res Labs, 66-85, PROF COL VET MED, UNIV GA, 80- *Mem:* Am Soc Microbiol; Am Leptospiroses Res Conf (pres, 79); US Am Health Asn; Am Asn Vet Lab Diag. *Res:* Blood chemistry of chicks infected or endointoxicated with Escherichia coli; fat absorption from the small intestine of gnotobiotic chicks; efficacy of Pasteurella multocida bacterins in turkeys; fluorescent antibody tests for animal diseases; Leptospirosis; Mycobacteriosis; Salmonellosis (Porcine). *Mailing Add:* 2010 N Park Ave Tifton GA 31794

COLE, JON A(RTHUR), b Chicago, Ill, Apr 14, 39; m 60; c 4. SANITARY ENGINEERING. *Educ:* Ill Inst Technol, BSCE, 61; Univ Wis, Madison, MS, 64, PhD(civil eng), 70. *Prof Exp:* Consult civil engr, Harold F Steinbrecher, Ill, 60-63; from asst prof to assoc prof, 64-77, PROF ENG, WALLA WALLA COL, 77- *Mem:* Am Soc Civil Engrs; Water Pollution Control Fedn; Am Water Works Asn. *Res:* Physical processes in sanitary engineering; gravity thickening of compressible slurries. *Mailing Add:* Rte 1 Mojonnier Rd Walla Walla WA 99362

COLE, JONATHAN JAY, b New York, NY, Jan 14, 53; m 80; c 1. BIOGEOCHEMISTRY, PHYSIOLOGY. *Educ:* Amherst Col, BA, 76; Cornell Univ, PhD(ecol), 82. *Prof Exp:* res fel geol, Woods Hole Oceanog Inst, 81-82; res fel aquatic ecol, Marine Biol Lab, 82-83; ASST SCIENTIST AQUATIC MICROBIOL, INST ECOSYSTS STUDIES, NY BOT GARDEN, 83- *Concurrent Pos:* Assoc site coordr, Hubbard Brook Ecosyst Study, 83- *Mem:* Sigma Xi; Am Soc Limnol & Oceanog; Am Soc Microbiol. *Res:* Aquatic (marine and freshwater) microbiology and biogeochemistry; bacterial production; nutrient limitation in phytoplankton; biogeochemistry of molybdenum. *Mailing Add:* Inst Ecosysts Studies NY Bot Garden Box AB Millbrook NY 12545

COLE, JULIAN D(AVID), b Brooklyn, NY, Apr 2, 25; m 49; c 5. AERODYNAMICS. *Educ:* Cornell Univ, BME, 44; Calif Inst Technol, MS & AE, 46, PhD(aeronaut & math), 49. *Prof Exp:* Asst prof aeronaut & appl mech, Calif Inst Technol, 51-59, prof, 59-69; chmn, dept appl sci & eng, Univ Calif, Los Angeles, 69-76, prof, 69-82, prof math 76-82; PROF APPL MATH, RENSSELAER POLYTECH INST, 82- *Honors & Awards:* Von Karman Prize, Soc Indust & Appl Math, 84. *Mem:* Nat Acad Sci; Nat Acad Eng; Am Inst Aeronaut & Astronaut; Am Acad Arts & Sci. *Res:* Aeronautics and applied mathematics. *Mailing Add:* Dept Math Rensselaer Polytech Inst Troy NY 12180

COLE, LARRY KING, b Grundy, Va, July 11, 39; c 2. TOXICOLOGY, ENTOMOLOGY. *Educ:* WVa Univ, BA, 61; Univ Utah, MS, 68; Univ Ga, PhD(entom), 73. *Prof Exp:* Instr molecular biol, Juniata Col, 68-70; vis lectr entom, Univ Ill, Urbana, 73-74, res coordr, environ toxicol, 74-77; ASST PROF ENTOM, UNIV MASS, AMHERST, 77- *Concurrent Pos:* Res assoc, Ill Natural Hist Surv, 74-77; US rep, NATO Workshop Ecotoxicol, 77. *Mem:* Entom Soc Am; Am Chem Soc; AAAS. *Res:* Environmental toxicology; pesticide toxicology; model ecosystems. *Mailing Add:* 401 Stanley St Beckley WV 25801

COLE, LARRY LEE, b Houston, Tex, Nov 1, 41. MOLECULAR STRUCTURE OF COAL. *Educ:* Tex Southern Univ, BS, 67; Univ Houston, PhD(chem), 72. *Prof Exp:* From asst prof to assoc prof, 72-84, interim head dept, 89-90, TENURED ASSOC PROF CHEM, PRAIRIE VIEW A&M UNIV, 84-, DIR RES, 73- *Concurrent Pos:* Res partic, Pittsburgh Energy Technol Ctr, 86 & Oak Ridge Nat Lab, 87, 88, 89 & 91. *Mem:* Am Chem Soc; Nat Orgn Advan Black Chemists & Chem Engrs. *Res:* Molecular structure of coal; methods of hazardous waste remediation such as bioremediation and in situ vitrification; organic chemical analysis via thermal laser desorption; chemical kinetic studies to observe heats of solvation on transition states. *Mailing Add:* PO Box 155 Prairie View TX 77446

COLE, LARRY S, b Logan, Utah, Aug 31, 06; m 29; c 2. ELECTRICAL ENGINEERING. *Educ:* Univ Utah, BS, 40; Utah State Univ, MS, 45; Stanford Univ, DEng, 50. *Prof Exp:* Instr, 39-50, head dept, 50-65, prof, 50-76, actg dean col eng, 65-66, asst dean, 66-69, assoc dean col eng, 69-71, EMER PROF ELEC ENG, UTAH STATE UNIV, 76-; CONSULT ENGR, 79- *Concurrent Pos:* Consult, Cole & Clark, 47-49. *Mem:* Sr mem Inst Elec & Electronics Engrs; Nat Soc Prof Engrs. *Res:* Electronic ckt design; electroacoustics; electro-fishing systems; parasitic antennas. *Mailing Add:* 65 Heritage Cove Logan UT 84321

COLE, LEE ARTHUR, b Pittsburgh, Pa, May 2, 53; m 85. SOFTWARE SYSTEMS, MECHANICAL ENGINEERING. *Educ:* Ind Univ Pa, BS, 75; Dartmouth Col, PhD(theoret physics), 79. *Prof Exp:* Fel solid state physics, Tulane Univ, 79-81; staff scientist & proj mgr GaAs solar cells, Solar Energy Res Inst, US Dept Energy, 81-85; prog mgr, Solid State Electronics Div, Honeywell, 85-87; RES & DEVELOP MGR, GRAFTEK, INC, 87- *Concurrent Pos:* Vis prof, Dartmouth Col, 80. *Mem:* Am Phys Soc; Int Asn Math Modelling. *Res:* Computer aided design, manufacturing, and engineering. *Mailing Add:* 565 Mohawk Dr Boulder CO 80303

COLE, MADISON BROOKS, JR, b Worcester, Mass, Aug 30, 40; m 67; c 1. CELL BIOLOGY, ORTHOPEDICS. *Educ:* Colgate Univ, AB, 62; Univ Tenn, PhD(zool), 69. *Prof Exp:* Res asst biol, Brown Univ, 62-63 & Univ Tenn, 64-68; from instr to asst prof biol sci, Oakland Univ, 68-74; spec trainee biophys, NY Univ, 74-75; asst prof, 75-84, ADJ ASST PROF ORTHOP SURG, LOYOLA UNIV MED CTR, 84- *Concurrent Pos:* Prin investr, Hines VA Hosp, 84-; consult, SPI Supplies, West Chester, Pa, 81- *Mem:* AAAS; Orthop Res Soc; Am Soc Cell Biol; Am Asn Anat; Sigma Xi. *Res:* Electron microscopy; cytochemistry; growth and differentiation of bone and cartilage; electro-magnetic effects on living systems. *Mailing Add:* 213 W Oakley Dr SW No 160 Westmont IL 60559

COLE, MICHAEL ALLEN, b Denver, Colo, Dec 15, 43. MICROBIAL GENETICS, SOIL MICROBIOLOGY. *Educ:* Cornell Univ, BS, 67; NC State Univ, Raleigh, MS, 71, PhD(microbiol), 72. *Prof Exp:* Asst prof microbiol, Southern Ill Univ, Edwardsville, 72-73; ASST PROF SOIL MICROBIOL, UNIV ILL, URBANA, 74- *Mem:* Sigma Xi; Am Soc Microbiol. *Res:* Genetics of Rhizobium, particularly plasmid genetics; effects of agricultural chemicals and pollutants on soil micro-organisms. *Mailing Add:* Dept Agron Univ Ill 1102 S Goodwin Ave Urbana IL 61801

COLE, MILTON WALTER, b Washington, DC, Dec 14, 42; m 83; c 1. SOLID STATE PHYSICS, LOW TEMPERATURE PHYSICS. *Educ:* Johns Hopkins Univ, BA, 64; Univ Chicago, MS, 65, PhD(physics), 70. *Prof Exp:* Fel physics, Univ Toronto, 70-72; res assoc, Univ Wash, 72-74; from asst prof to assoc prof, 74-81, PROF PHYSICS, PA STATE UNIV, 81- *Concurrent Pos:* Consult, Jet Propulsion Lab, Calif Inst Technol, 74-; vis assoc, Calif Inst Technol, 81-82, prof, 82-83; vis prof physics & chem, Brown Univ, 82, Univ Marseille & Padua, 85; assoc prof physics, Brooklyn Col, 75-76; Fulbright scholar, Univ Oxford, 90. *Mem:* Sigma Xi; fel Am Phys Soc. *Res:* Surface physics, especially thin films, liquid helium and electronic properties; inhomogenous quantum systems; adsorption statistical mechanics and kinetics; surface science. *Mailing Add:* 104 Davey Lab Pa State Univ University Park PA 16802

COLE, MORTON S, b Chicago, Ill, June 16, 29; m 52; c 3. FOOD TECHNOLOGY. *Educ:* Univ Ill, BS, 51, MS, 53; Iowa State Univ, PhD(food technol), 61. *Prof Exp:* Scientist grocery prod develop, Pillsbury Co, 59-61, dried prod develop, 61-63; food technologist, Archer-Daniels-Midland Co, 63-64, group leader edible prod, 64-68; dir res, Paniplus Co, Div ITT Continental Baking Co, 68-73; assoc dir res, 74-76, dir res, Archer-Daniels-Midland Co, 76-81; PRES, COLE TECHNOL, INC, 85- *Mem:* Inst Food Technologists; Am Asn Cereal Chem; Am Soc Bakery Engrs. *Res:* Porphyrin pigments in fresh and processed meats; lipid oxidation; pectic enzymes; vegetable dehydration; edible oxygen and moisture barriers; fabricated foods from vegetable proteins; bakery ingredients; chemical surfactants. *Mailing Add:* 2506 Ivy Lane Decatur IL 62521

COLE, NANCY, b Boston, Mass, Oct 15, 02. MATHEMATICS, GEOMETRY. *Educ:* Vassar Col, AB, 24; Radcliffe Col, AM, 29, PhD(math), 34. *Prof Exp:* Instr, Oxford Sch, 24-26 & Vassar Col, 27-28; tutor, Radcliffe Col, 28-29; instr, Wells Col, 31-32; instr math, Sweet Briar Col, 33-42; from vis asst prof to vis assoc prof, Kenyon Col, 43-44; asst prof, Conn Col, 44-47; from asst prof to assoc prof, 47-71, EMER ASSOC PROF MATH, SYRACUSE UNIV, 71- *Concurrent Pos:* Actg head dept, Sweet Briar Col, 34-35 & 41-42; mem, People-to-People Univ Math Educ Deleg to People's Repub China, 83. *Mem:* Am Math Soc; Math Asn Am; Am Women Math. *Res:* Calculus of variations; index form associated with an extremaloid. *Mailing Add:* 212 Sandwich St Apt 4 Plymouth MA 02360-2438

COLE, NOEL ANDY, b Pampa, Tex, Feb 16, 49. RUMINANT NUTRITION, ANIMAL SCIENCE. *Educ:* W Tex State Univ, BS, 71; Okla State Univ, MS, 73, PhD(animal nutrit), 75. *Prof Exp:* Vis asst prof, Dept Animal Sci, Tex Tech Univ, 75-76; RESEARCHER ANIMAL NUTRIT, CONSERV & PROD RES LAB, AGR RES SERV, USDA, 76- *Concurrent Pos:* Adj prof animal sci, Tex Tech Univ. *Mem:* Am Soc Animal Sci; Coun Agr Sci & Technol; Am Inst Nutrit. *Res:* Beef cattle nutrition, especially as it relates to reducing environmental and physiological stresses occurring during the marketing and transporting of calves. *Mailing Add:* Conserv & Prod Lab Agr Res Serv USDA PO Drawer 10 Bushland TX 79012

COLE, NYLA J, b Wasco, Calif, Dec 5, 25; m 55. PSYCHIATRY. *Educ:* Univ Calif, Berkeley, BA, 47; Univ Rochester, MD, 51. *Prof Exp:* Intern, Univ Utah, 52, resident psychiat, 55, instr, 56-60, dir outpatient div, 56-62, asst prof, 60-68, dir adult psychiat, 62-65, assoc prof, 68-87, dir psychiat outpatient dept, 70-80, EMER PROF PSYCHIAT, COL MED, UNIV UTAH, 87- *Concurrent Pos:* Lectr, Dept Social Work, 61-68; mem, President's Comt Employ Handicapped, 67-72; chmn med subcomt, Gov Comt Employ Handicapped, 68-72. *Mem:* Fel Am Psychiat Asn; AMA. *Res:* Natural history of psychiatric disease; mental health care delivery systems; social survival of discharged patients. *Mailing Add:* 12230 Highway 20 Port Townsend WA 98368

COLE, PATRICIA ELLEN, b New York, NY, July 27, 44. BIOPHYSICAL CHEMISTRY. *Educ:* Brown Univ, AB, 66; Yale Univ, MPhil, 69, PhD(biophys chem), 72; Albert Einstein Col Med, MD, 84. *Prof Exp:* Fel biochem, Inst Cancer Res, Columbia Univ, 72-73, asst prof chem, 74-78; Res assoc biochem, Dept Molecular Biophys & Biochem, Yale Univ, 78-80; Am Cancer Soc spec fel, Albert Einstein Col Med, 82-84, internship pediat, Albert Einstein Col Med/Montefiore Med Ctr, 84-85; radiol residency, NY Hosp Cornell Med Ctr, 85-89, fel cardiovasc intervention radiol, 89-90; ASST ATTEND, MEM SLOAN KETTERING CANCER CTR, 90- *Concurrent Pos:* Damon Runyon fel, Damon Runyon Mem Fund Cancer Res, 72-73; NIH career develop award, 76-80. *Mem:* Am Chem Soc; Biophys Soc. *Res:* Biophysical studies of gene expression using fast reaction kinetics, NMR, fluorescence; promoter recognition and RNA transcription; molecular mechanisms of RNA replication by RNA viral replicases; RNA conformational changes; magnetic resonance in vivo spectroscopy. *Mailing Add:* 504 E 63rd St Apt 24R New York NY 10021

COLE, RALPH I(TTLESON), b St Louis, Mo, Aug 17, 05; m 31; c 2. ENGINEERING. *Educ:* Wash Univ, St Louis, BS, 27; Rutgers Univ, MS, 36. *Prof Exp:* Proj engr, Sig Corps Labs, Ft Monmouth, NJ, 29-31, asst chief radio sect, 31-38, engr in charge radio receiver work, 32-35, radio receiver, installation & direction finding, 35-38, chief radio sect, 38-40; chief radio direction finding sect, Eatontown Sig Lab, 40-42; chief Eng Div, Watson Labs, NJ, 45-47, chief engr, 47-52; mgr mil projs planning, Melpar, Inc, 52-63; mgr & tech consult, 63-73; prof lectr & adj prof, 63-69, dir insts & spec projs, 64-69, assoc prof eng & dir grad res & develop prog, 69-72, EMER ASSOC PROF ENG, AM UNIV, WASHINGTON, DC, 85- *Concurrent Pos:* Guest lectr, Air Univ, 47-49; mgr, WABCO Govt Serv, Washington, DC, 61-63; mem, Eng Manpower Comn, 64-69; resident prof, Univ Southern Calif, Patuxent River Naval Base, 80; bd mem, Asn Sci Technol & Innovation; consult, independent res & develop mgt, 73-; adj prof eng, Am Univ, Washington, DC, 73- *Mem:* Fel AAAS; Am Inst Aeronaut & Astronaut; fel Inst Elec & Electronics Engrs; Eng Mgt Soc. *Res:* Systems effectiveness; education research and development management; obsolescence and the engineer; technology transfer. *Mailing Add:* 3705 S Geo Mason Dr Falls Church VA 22041

COLE, RANDAL HUDIE, mathematics; deceased, see previous edition for last biography

COLE, RANDALL KNIGHT, b Putnam, Conn, Sept 21, 12; m 39; c 3. ANIMAL GENETICS. *Educ:* Mass State Col, BS, 34; Cornell Univ, MS, 37, PhD(animal breeding), 39. *Prof Exp:* Asst animal dis, Univ Conn, 34-35; instr poultry husb, 35-40, asst prof poultry sci & animal genetics, 40-48, from assoc prof to prof, 48-73, EMER PROF POULTRY SCI & ANIMAL GENETICS, CORNELL UNIV, 73- *Concurrent Pos:* Animal geneticist, Exp Sta, NY State Col Agr, Cornell Univ, 40-73; consult, Shaver Poultry Breeding Farms, Ltd, Cambridge, Ont, 56-; hon mem, First World Cong Genetics Appl to Livestock Prod, Madrid, 74. *Honors & Awards:* Poultry Sci Asn Res Award, 49; Tom Newman Mem Int Award, 69. *Mem:* AAAS; Am Genetic Asn; Poultry Sci Asn; World Poultry Sci Asn; Am Inst Biol Sci; Sigma Xi. *Res:* Genetics of disease resistance in poultry, especially to neoplasms; avian genetics; animal models of specific diseases. *Mailing Add:* 1154 Dryden Rd Ithaca NY 14850

COLE, RICHARD, b New York, NY, Apr 16, 24; m 47; c 2. PHYSICAL CHEMISTRY, ENVIRONMENTAL MANAGEMENT. *Educ:* City Col New York, BChE, 44; Univ Ill, MS, 48, PhD(phys chem), 52. *Prof Exp:* Asst phys chem, Univ Ill, 47-49 & 50-51; radiol chemist, US Naval Radiol Defense Lab, 52-61, head countermeasures eval br, 61-64, chem tech div, 64-66 & nuclear tech div, 66-69; sr res assoc, 69-74, VPRES, ENVIRON SCI ASSOC, SAN FRANCISCO, 74- *Mem:* Am Chem Soc; Opers Res Soc Am; Asn Environ Prof; Air Pollution Control Asn. *Res:* Formation, transport, deposition and removal of radioactive and other contamination; on-site inspection for clandestine nuclear operations; radioactivity in the oceans; environmental impact analysis and mitigation; airborne pollutants; hazardous materials and hazardous waste. *Mailing Add:* 1431 Tarrytown St San Mateo CA 94402

COLE, RICHARD ALLEN, b Suffern, NY, Oct 27, 42. AQUATIC ECOLOGY. *Educ:* State Univ NY Col Forestry, Syracuse Univ, BS, 64, MS, 66; Pa State Univ, PhD(zool), 69. *Prof Exp:* Res assoc, Mich State Univ, 69-73, asst prof aquatic ecol, 73-78; asst prof, 78-82, ASSOC PROF FISHERY SCI, NMEX STATE UNIV, 82- *Mem:* Am Soc Limnologists & Oceanogr; Am Fisheries Soc; Int Soc Limnol; Ecol Soc Am; NAm Benthological Soc. *Res:* Aquatic community ecology, particularly as influenced by eutrophication, thermal discharge and other watershed disturbances; modeling aquatic ecosystems. *Mailing Add:* Dept Fisheries & Wildlife Sci NMex State Univ PO Box 4901 Las Cruces NM 88003

COLE, RICHARD H, b Woodstock, Maine, Mar 2, 30; m 53; c 4. AGRONOMY. *Educ:* Univ Maine, BS, 52; Pa State Univ, PhD, 60. *Prof Exp:* Assoc county agent crops, Mass Exten Serv, 54-55; asst prof agron & agr eng, Univ Del, 60-65, chmn dept, 65-68, assoc prof plant sci, 68-70; assoc prof int agron, 70-76, POTATO PROD EXTEN SPECIALIST, PA STATE UNIV, 76- *Mem:* Potato Asn Am; Am Soc Agron; Crop Sci Soc Am; Weed Sci Soc Am. *Res:* Crop management; seed quality; weed control. *Mailing Add:* Dept Agron Pa State Univ 102 Tyson Bldg University Park PA 16802

COLE, ROBERT, b Bridgeport, Conn, Dec 31, 28; m 55; c 3. DROP DYNAMICS, HEAT TRANSFER. *Educ:* Clarkson Col Technol, BChE, 54, MChE, 59, PhD(chem eng), 66. *Prof Exp:* Res engr heat transfer, Nat Adv Comt Aeronaut, 54-56; from asst to assoc prof, 56-77, PROF CHEM ENG, CLARKSON COL TECHNOL, 77- *Concurrent Pos:* Vis prof, Technische Hogeschool Eindhoven, Nederland, 71-72 & 78-79 & Oxford Univ, Eng, 85-86; consult, AERE Harwell Lab, Eng, 85-86. *Mem:* Am Inst Chem Engrs; Am Soc Mech Engrs. *Res:* Adsorption on to dry ion exchange resins; liquid film flows; boiling heat transfer; nucleation, growth, and detachment; glass processing in space; bubble migration phenomena; bubble dynamics; glassmelt volatilization; holographic interferometry; speckle photography. *Mailing Add:* Dept Chem Eng Clarkson Univ Potsdam NY 13699-5705

COLE, ROBERT HUGH, chemical physics; deceased, see previous edition for last biography

COLE, ROBERT KLEIV, b San Francisco, Calif, Dec 19, 28; m 53; c 4. PHYSICS. *Educ:* Univ Calif, Berkeley, AB, 52; Univ Wash, PhD(physics), 59. *Prof Exp:* Physicist, Hanford Works, Gen Elec Corp, Wash, 52-54; vis asst prof, Univ Calif, Los Angeles, 59-62; asst prof, 62-66, ASSOC PROF PHYSICS, UNIV SOUTHERN CALIF, 66- *Mem:* Am Phys Soc. *Res:* Nuclear scattering; nuclear reactions studies; nuclear physics instrumentation. *Mailing Add:* 7120 Bianca Ave Van Nuys CA 91406

COLE, ROBERT STEPHEN, b Los Angeles, Calif, Apr 10, 43; m 65; c 2. SOLAR PHYSICS, QUANTUM PHYSICS. *Educ:* Univ Calif, Berkeley, AB, 65; Univ Wash, MS, 67; Mich State Univ, PhD(physics), 72. *Prof Exp:* Instr physics, St Martins Col, 67-69; asst prof physics, Univ NC, Asheville, 72-77, assoc prof, 77-; AT DEPT PHYSICS, EVERGREEN STATE COL, WASH. *Concurrent Pos:* Consult solar energy, Asheville Orthop Hosp & Rehab Ctr, 75- *Mem:* Am Asn Physics Teachers; Nat Asn Environ Educ; Int Solar Energy Soc. *Res:* Solar energy research, especially passive systems for residential heating and cooling. *Mailing Add:* MS Lab I Evergreen State Col 3095 Yerba Buena Rd Olympia WA 98505

COLE, ROGER DAVID, b Berkeley, Calif, Nov 17, 24; m 44; c 3. PROTEIN CHEMISTRY. *Educ:* Univ Calif, BS, 48, PhD(biochem), 54. *Prof Exp:* Asst phys chem, Atomic Res Inst, Univ Iowa, 48-49; chemist, Tidewater Oil Co, Calif, 49-51; jr res biochemist, Univ Calif, 54-55; fel, Nat Inst Med Res, London & Nat Found Infantile Paralysis, 55-56; res assoc biochem, Rockefeller Inst, 56-58; from asst prof to assoc prof, 58-65, chmn dept, 68-73, dir electron micros lab, 79-84, PROF BIOCHEM, UNIV CALIF, BERKELEY, 65-, ASST DEAN, 90- *Concurrent Pos:* Guggenheim fel, Cambridge Univ, 66-67; vis fel Brasenose Col, Oxford. *Honors & Awards:* Newmark Mem lectr. *Mem:* AAAS; Am Soc Biol Chem; Am Chem Soc. *Res:* Protein and peptide isolation and structural determination; relation of structure and biological activity of enzymes and hormones; histones; microtubules. *Mailing Add:* Dept Molecular & Cell Biol Univ Calif Berkeley CA 94720

COLE, ROGER M, b Akron, Ohio, Nov 5, 34; m 58; c 2. INORGANIC CHEMISTRY, RESEARCH MANAGEMENT. *Educ:* Kent State Univ, BS, 56; Univ Minn, MS, 58. *Prof Exp:* From res chemist to sr res chemist, 58-69, head Processing Res Lab, 69-73, asst dir, black & white photog res, 73-78, asst to dir res, 78-80, asst dir, res admin div, 80-86, DIR RES MGT, EASTMAN KODAK CO, 86- *Concurrent Pos:* Secy bd dir, Coun Chem Res, 80-81; mem Coun Chem Res 80- *Mem:* Am Chem Soc. *Res:* Mechanism of photographic development; photographic chemistry of silver ion complexes; mechanism and application of physical development; methods to simplify photographic processing; history of chemistry. *Mailing Add:* 225 Imperial Circle Rochester NY 14617

COLE, STEPHEN H(ERVEY), b New Brunswick, NJ, Dec 16, 42; m 70; c 2. METALLURGY, CHEMICAL ENGINEERING. *Educ:* Univ Del, BChE, 64; Alfred Univ, MS, 66; Columbia Univ, DEng, 71- *Prof Exp:* Scientist, Metal Mining Div, Kennecott Res Ctr, 71-76; res engr, Texasgulf, Inc, 77-81, gen mgr eng & construct, 81-; MEM STAFF, DURACELL CO. *Mem:* Am Inst Mining, Metall & Petrol Engrs; Am Inst Chem Engrs. *Res:* Hydrometallurgy; determination of rate-limiting factors in oxidation of copper sulfide through electrochemical techniques; air pollution abatement. *Mailing Add:* 97 Gary Rd Stamford CT 06903

COLE, TERRY, b Albion, NY, Mar 28, 31; m 55; c 3. CHEMICAL PHYSICS. *Educ:* Univ Minn, BS, 54; Calif Inst Technol, PhD(chem), 58. *Prof Exp:* Asst, Calif Inst Technol, 54-57; res scientist, Ford Motor Co, 59-71, mgr chem dept, 71-75, mgr chem eng dept, 75-76, sr staff scientist res staff, 76-80; CHIEF TECHNOLOGIST, JET PROPULSION LAB, CALIF INST TECHNOL, 80- *Concurrent Pos:* Fairchild scholar, Calif Inst Technol, 76-77; sr res assoc, Calif Inst Technol, 80- *Mem:* Am Phys Soc; Am Chem Soc; Am Inst Aeronaut & Astronaut. *Res:* Electron spin and nuclear magnetic resonance; radiation damage; free radical chemistry; quasielastic light scattering; thermoelectric energy conversion; solar energy. *Mailing Add:* Div Chem & Chem Eng Calif Inst Technol 164-30 Pasadena CA 91125

COLE, THOMAS A, b Harrisburg, Ill, Jan 9, 36; div; c 2. BIOCHEMISTRY, GENETICS. *Educ:* Wabash Col, BA, 58; Calif Inst Technol, PhD(biochem), 63. *Prof Exp:* From asst prof to assoc prof, 62-76, chmn dept, 68-79, NORMAN E TREVES PROF BIOL, WABASH COL, 76-; CONSULT-EXAMR, NCENT ASN COLS & SCHS, 70- *Concurrent Pos:* Comnr, Comn Undergrad Educ Biol Sci, 68-71; panelist, Educ Directorate, NSF, 75- *Mem:* Fel AAAS; NY Acad Sci; Sigma Xi. *Res:* Biochemistry of metamorphosing Drosophila; nutrition of paramecia; centrifugation techniques; identification of hydrolytic enzymes in substrate-included gels for electrophoresis; liposome. *Mailing Add:* Dept Biol Wabash Col Crawfordsville IN 47933

COLE, THOMAS EARLE, b Winter Park, Fla, Dec 13, 22; m 44; c 3. NUCLEAR ENGINEERING, PHYSICS. *Educ:* Rollins Col, BS, 46. *Prof Exp:* Physicist, 46-69, mgr, light water reactor technol prog, 80-84, sr develop staff mem nuclear reactors, Oak Ridge Nat Lab, 69-87; CONSULT, 88- *Concurrent Pos:* Consult, var co and countries, 56-62, Repub SAfrica, 62-65, US Energy Res & Develop Admin, 73-74. *Mem:* Am Phys Soc; Am Nuclear Soc; Sigma Xi. *Res:* Nuclear research reactors - all aspects; reactor safety and control; probabilistic reactor safety studies; nuclear energy centers, energy parks; electric utility industry. *Mailing Add:* 103 Disston Rd Oak Ridge TN 37830

COLE, THOMAS ERNEST, b Ft Wayne, Ind, Aug 17, 51; m 82. CHEMISTRY. *Educ:* Purdue Univ, BS, 74; Univ Tex, Austin, PhD(chem), 80. *Prof Exp:* Res asst, Purdue Univ, 80-86; ASSOC PROF, SAN DIEGO STATE UNIV, 86- *Mem:* Am Chem Soc. *Res:* Organometallics and organoboranes as intermediates for organic synthesis; transmetallation of organic groups between metals and boron compounds. *Mailing Add:* Dept Chem San Diego State Univ San Diego CA 92182

COLE, THOMAS WINSTON, JR, b Vernon, Tex, Jan 11, 41; m 64. ORGANIC CHEMISTRY. *Educ:* Wiley Col, BS, 61; Univ Chicago, PhD(chem), 66. *Prof Exp:* Asst prof chem, Atlanta Univ, 66-69, Fuller E Callaway prof, 69-82, chmn dept, 71-82; pres, WVa State Col, 82-86; PRES, CLARK ATLANTA UNIV, 86- *Concurrent Pos:* Vis prof, Dept Chem, Mass Inst Technol, 73-74. *Mem:* AAAS; Am Chem Soc; Nat Inst Sci. *Res:* Chemistry of cubane; small ring compounds; photochemistry; application of gas chromatography-mass spectrometry to problems in clinical chemistry; bio-organic chemistry. *Mailing Add:* James P Brawley Dr at Fair St Atlanta GA 30314

COLE, VERSA V, BIOCHEMISTRY. *Educ:* Univ Chicago, PhD(biochem), 31, MD, 38. *Prof Exp:* Physician, Saginaw County Home For Aged, 71-73; RETIRED. *Mailing Add:* c/o Grandview Home Aged 1728 S Peninsula Rd East Jordan MI 49727

COLE, W STORRS, geology, for more information see previous edition

COLE, WALTER EARL, organic chemistry; deceased, see previous edition for last biography

COLE, WALTER ECKLE, b Muskogee, Okla, Sept 2, 28; m 55; c 3. ENTOMOLOGY, MATHEMATICAL BIOLOGY. *Educ:* Colo State Univ, BSc, 50, MSc, 55; NC State Univ, PhD, 72. *Prof Exp:* Entomologist forest insect res, 54-60, actg proj leader pop dynamics res, 60-65, proj leader Pop Dynamics Res, US Forest Serv, 65-84, prin entomologist, 70-84; RETIRED. *Concurrent Pos:* Counr, Western Forest Insect Work Conf, 70-73; chmn working party, Int Union Forest Res Orgns. *Mem:* Am Statist Asn; Biomet Soc; Entom Soc Am; Japanese Soc Pop Ecol. *Res:* Population dynamics of forest insects; mensurational aspects of behavioral sampling; analysis and modeling of populations. *Mailing Add:* 944 Chatelain Rd Ogden UT 84403

COLE, WAYNE, b Indianapolis, Ind, Nov 5, 13; m 36; c 5. NATURAL PRODUCTS CHEMISTRY. *Educ:* DePauw Univ, AB, 35; Univ Ill, AM, 36, PhD(org chem), 38. *Prof Exp:* Asst chem, Univ Mich, 38-39; res chemist, Glidden Co, 39-46, from asst dir to dir res, 46-58; proj leader steroid res, Abbott Labs, 58-72, res fel, 72-79; CONSULT, 79- *Mem:* Fel AAAS; Am Chem Soc; Am Oil Chem Soc; Swiss Chem Soc. *Res:* Synthesis of carcinogenic hydrocarbons, protein derivatives; sterols and hormones; steroid and lipid research; peptide synthesis. *Mailing Add:* 1224 Norman Lane Deerfield IL 60015

COLE, WILBUR VOSE, b Waterville, Maine, Jan 19, 13; m 36; c 3. NEUROANATOMY. *Educ:* Univ NH, BS, 35; Kirksville Col Osteop, DO, 43; Northeast Mo State Univ, MA, 54. *Hon Degrees:* DSc, Nat Acad Pract. *Prof Exp:* Lab asst histol, Kirksville Col Osteop, 43-44, instr histol & embryol, Dept Anat, 44, asst prof histol & neuroanat, 44-51; assoc prof anat, Kansas City Col Osteop Med, 51-79, prof clin neurol, 69-79, dean, 73-79; RETIRED. *Mem:* Am Osteop Asn; Photog Biol Asn; Nat Asn Biol Teachers; Sigma Xi; Am Micros Soc. *Res:* Histopathology; polarized light in photomicroscopy; gallocyanin as a nuclear stain; comparative anatomy and physiology of motor and sensory endings in straited muscle. *Mailing Add:* Box 6 HCR 60 Prospect Harbor ME 04669

COLEBROOK, LAWRENCE DAVID, b Helensville, NZ, Dec 29, 30; m 59. PHYSICAL ORGANIC CHEMISTRY. *Educ:* Univ NZ, BSc, 54, MSc, 55, PhD(chem), 61. *Prof Exp:* From jr lectr to lectr chem, Univ Auckland, 57-60; fel, Univ Rochester, 61-63, from asst prof to assoc prof, 69-74, PROF CHEM, SIR GEORGE WILLIAMS CAMPUS, CONCORDIA UNIV, 74- *Concurrent Pos:* Vis prof, Univ BC, 77-78. *Mem:* Brit Chem Soc; assoc NZ Inst Chem; Am Chem Soc; Chem Inst Can. *Res:* Nuclear magnetic resonance spectroscopy; computers in chemistry; natural products. *Mailing Add:* Dept Chem Concordia Univ 1455 Maisonneuve Blvd W Montreal PQ H3G 1M8 Can

COLEGROVE, FORREST DONALD, b Madeira, Ohio, Nov 21, 29; m 56; c 2. APPLIED PHYSICS. *Educ:* Purdue Univ, BS, 51; Univ Mich, MS, 54, PhD(physics), 60. *Prof Exp:* mem tech staff, 59-80, SR MEM TECH STAFF PHYSICS, TEX INSTRUMENTS, INC, 80- *Concurrent Pos:* Vis scientist, Southwest Ctr Advan Studies, 65-69. *Mem:* Am Phys Soc. *Res:* Magnetic sensors; magnetic resonance; infrared systems; physics of the upper atmosphere. *Mailing Add:* North Lakes Dr Dallas TX 75240-7620

COLELLA, DONALD FRANCIS, b Utica, NY; m 69; c 1. PHARMACOLOGY. *Educ:* Rensselaer Polytech Inst, BS, 61; Col St Rose, MS, 68; Drexel Univ, MBA, 75. *Prof Exp:* Assoc res biologist pharmacol, Sterling-Winthrop Res Inst, 61-68; pharmacologist, Smith Kline & French Labs, 68-76, int US mkt mgr, 76-78, US prod mgr, 78-80, mgr new prod develop, 80-82, mgr, small new bus develop-licensing, 83-88, dir, worldwide bus develop-licensing, Smith Kline Beecham, 88-89; dir bus develop-pharmaceut, Alfred Benzon Inc, 90-91; VPRES PHARMACEUT, BTG USA, 90- *Mem:* NY Acad Sci; Am Chem Soc. *Res:* Cardiovascular and respiratory pharmacology; pharmacology of cardiac and smooth muscle; adrenergic mechanisms; theory of drug-receptor interactions; structure-activity relationships of medicinal agents. *Mailing Add:* Brit Technol Group USA Renaissance Business Park 2200 Renaissance Blvd Gulph Mills PA 19406

COLELLA, ROBERTO, b Milano, Italy, May 22, 35; m 60; c 3. SOLID STATE PHYSICS. *Educ:* Univ Milano, Laurea(physics), 58. *Prof Exp:* Trainee physics, Nat Nuclear Energy Comt, Casaccia, Rome, Italy, 60-61; staff scientist, Europe Atomic Energy Comn, Common Ctr Res, Ispra, Italy, 61-67; res assoc physics, Cornell Univ, 67-70 & Cath Univ, Washington, DC, 70-71; from asst prof to assoc prof, 71-75, PROF PHYSICS, PURDUE UNIV, WEST LAFAYETTE, 77- *Concurrent Pos:* USA co ed, Acta Crystallographica, 80-89. *Mem:* Am Phys Soc; Am Crystallog Asn; Italian Phys Soc. *Res:* Diffraction physics in perfect and imperfect crystals; phonons, charge densities, phase problem. *Mailing Add:* Dept Physics Purdue Univ West Lafayette IN 47907

COLEMAN, ALBERT JOHN, b Toronto, Ont, May 20, 18; m 53; c 2. APPLIED MATHEMATICS. *Educ:* Univ Toronto, BA, 39, PhD, 43; Princeton Univ, MA, 43. *Prof Exp:* Lectr math, Queen's Univ, Can, 43-45; travelling secy, World's Student Christian Fedn, 45-49; from lectr to assoc prof, Univ Toronto, 49-60; head dept, 60-80, prof, 80-83, EMER PROF & ADJ PROF MATH, QUEEN'S UNIV, ONT, 83- *Concurrent Pos:* Mem, Sci Coun Can, 73-77, Lambeth Conf, 78 & Natural Sci & Eng Res Coun Can. *Mem:* Am Math Soc; Math Asn Am. *Res:* Eddington's fundamental theory; group theory; quantum mechanics. *Mailing Add:* Dept Math & Statist Queens Univ Kingston ON K7L 3N6 Can

COLEMAN, ANNA M, b New Concord, Ohio, Jan 5, 13. INFORMATION SCIENCE. *Educ:* Geneva Col, BS, 33; Univ Pa, MS, 34; Univ Pittsburgh, PhD(chem), 58. *Prof Exp:* Teacher high sch, Pa, 34-42; Koppers Co res asst, Mellon Inst, 42-44, jr fel, 44-50; chem librn, Dow Corning Corp, 50-60, supvr res info serv, 60-67, mgr tech info serv, 67-78; RETIRED. *Mem:* Am Chem Soc; Sigma Xi. *Res:* Chemical documentation; dipole moments. *Mailing Add:* 2344 Perrysville Ave Pittsburgh PA 15214-3597

COLEMAN, ANNETTE WILBOIS, b Des Moines, Iowa, Feb 28, 34; m 58; c 3. CELL BIOLOGY. *Educ:* Barnard Col, AB, 55; Univ Ind, PhD(bot), 58. *Prof Exp:* NSF res fel, Johns Hopkins Univ, 58-61; res assoc, Univ Conn, 62-63; res assoc, 63-72, from asst prof to assoc prof, 72-84, PROF BIOL, BROWN UNIV, 84- *Concurrent Pos:* Guggenheim fel. *Honors & Awards:* Provasoli Award; Darbaker Award. *Mem:* Bot Soc Am; Soc Protozool; Phycol Soc Am; Soc Gen Physiol; NY Acad Sci. *Res:* Physiological control of mating in algae, inheritance of mating type; geographical distribution and speciation in algae; genetics; structure of chloroplast desoxyribonucleic acid; problems of cell fusion. *Mailing Add:* Div Biol & Med Sci Brown Univ Dept Biol Providence RI 02912

COLEMAN, BERNARD DAVID, b New York, NY, July 5, 30; m 65; c 2. CONTINUUM MECHANICS, MATHEMATICAL ANALYSIS. *Educ:* Ind Univ, BS, 51; Yale Univ, MS, 53, PhD(chem), 54. *Prof Exp:* Res chemist, E I du Pont de Nemours & Co, 54-57; sr fel, Mellon Inst, 57-88, prof math, biol & chem, Carnegie-Mellon Univ, 67-88; J WILLARD GIBBS PROF THERMOMECH & MATH, RUTGERS UNIV, 88-, DIR GRAD PROG & MECH, 91- *Concurrent Pos:* Visitor, Inst Math, Univ Bologna, 60-61; vis prof, Johns Hopkins Univ, 62-63, Univ Wash, 64, 67, 68 & 72 & Nat Bur Standards, Washington, DC, 81 & 82; adj prof, Univ Pittsburgh, 64-65 & 77; ed-in-chief, Springer Tracts in Natural Philos, 67-73; lectr, Univ Pisa, 66, 68, 69, 74 & 78, Scuola Normale Superiore, Italy, 69 & 70, Int Ctr Mech Sci, Udine, Italy, 71, 73 & 83 & Ravello, 88; chmn, Soc Natural Philos, 71-72; consult, fed, Rio De Janeiro, Brazil, Inst Math & Appln, Univ Minn, 83 & 89, Fed Univ Rio de Janiero, 84, Cornell Univ, 85 & 86. *Honors & Awards:* Bingham Medal Award, Soc Rheol, 84; Fourth Ann Aris Phillips Mem Lectr, Yale Univ, 91. *Mem:* Soc Natural Philos (treas, 67-68); Int Soc Interaction Mech & Math. *Res:* Viscoelasticity; liquid crystal physic foundations of thermodynamics; functional analysis; differential equations; mathematical biology; continuum physics. *Mailing Add:* Grad Prog Mech Rutgers Univ Box 909 Piscataway NJ 08855-0809

COLEMAN, BERNELL, b Lorman, Miss, Apr 26, 29; m 62; c 2. PHYSIOLOGY. *Educ:* Alcorn Agr & Mech Col, BS, 52; Loyola Univ, Ill, PhD(physiol), 64. *Prof Exp:* Asst path, Med Ctr, Univ Kans, 52-53; asst biochem, Univ Chicago, 56-57; asst cancer res, Hines Vet Admin Hosp, Ill, 57-59; instr physiol, Sch Med St Louis Univ, 63-65, asst prof, 65-69; assoc prof physiol & cardiovasc res, Chicago Med Sch, 69-76, prof physiol, 76; actg chmn physiol, 79-80, PROF PHYSIOL, COL MED, HOWARD UNIV, 76- *Concurrent Pos:* Mem coun basic sci, Am Heart Asn; mem Clin Sci Rev Group B, NIH, 81- *Mem:* AAAS; Am Heart Asn; Am Physiol Soc; Sigma Xi. *Res:* Myocardial catecholamines in hemorrhagic hypotension; electrolytes and water metabolism of the heart in hemorrhagic shock; effects of norepinephrine on electrolytes and water content of cardiac muscle; cardiodynamics in irreversible hemorrhagic shock; cardiodynamic and circulatory responses to heat; angiotensin II and cardiac function; carotid sinus reflex control of the heart; systolic stress and strain relation in normal and hypertrophied hearts. *Mailing Add:* Dept Physiol & Biophys Col Med Howard Univ 520 W St NW Washington DC 20059

COLEMAN, CHARLES CLYDE, b York, Eng, July 31, 37; US citizen; m 76; c 2. SOLID STATE PHYSICS. *Educ:* Univ Calif, Los Angeles, BA, 59, MA, 61, PhD(physics), 68. *Prof Exp:* Assoc prof physics, Calif State Univ, Los Angeles, 68-77, exec dir, Appl Physics Inst, 81-84, dir accelerator facil, 81-84, dir trustee fund, 81-87; proj specialist, Chinese Provincial Univ Develop Prog of the World Bank, 87-89; PROF PHYSICS, CALIF STATE UNIV, LOS ANGELES, 77- *Concurrent Pos:* Sr res fel, Darwin Col, Cambridge Univ, 75-76, NATO sr res fel, Cavendish Lab, 83-84; NSF res grant, 76-79, res corp grant, 87-; vis prof, Tech Univ Istanbul, 69, Univ Bucharest, 72, Arya Mar Univ, Teheran, 76, Univ Natal, Durban, 76, Univ Calif, Los Angeles, 90- *Mem:* Am Phys Soc; fel Brit Interplanetary Soc. *Res:* Semiconductors; crystal growth; thin films; interfaces; stimulated bioluminescence; modulation spectroscopy; ion implantation; intercalation. *Mailing Add:* Dept Physics Calif State Univ Los Angeles CA 90032

COLEMAN, CHARLES FRANKLIN, b Burley, Idaho, Dec 30, 17; m 52; c 3. SEPARATIONS CHEMISTRY, PHASE EQUILIBRIA. *Educ:* Univ Utah, BS, 41; Purdue Univ, MS, 43, PhD(phys chem), 48. *Prof Exp:* Asst chem, Univ Utah, 39-41 & Purdue Univ, 42-44; chemist, Substitute Alloy Mat Labs, Columbia Univ, 44, Tenn Eastman Corp & Clinton Eng Works, Oak Ridge, 44-46; asst chem, Purdue Univ, 46-47; chemist, Y-12 Plant, Carbide & Carbon Chem Corp, 48-51, chemist, Union Carbide Corp, 51-67, asst sect chief, 67-76, mgr chem develop, Separations Chem Progs, Union Carbide Corp, Oak Ridge Nat Lab, 76-84; RETIRED. *Concurrent Pos:* Consult, 85- *Mem:* AAAS; Am Chem Soc; Sigma Xi. *Res:* Calorimetry; phase equilibria; solution chemistry; separations chemistry; solvent extraction reagents, equilibria, kinetics, applications; actinide-lanthanide chemistry. *Mailing Add:* 106 Elliott Circle Oak Ridge TN 37830

COLEMAN, CHARLES MOSBY, b New York, NY, Oct 14, 25; m 51; c 5. CLINICAL CHEMISTRY, MICROBIOLOGY. *Educ:* Univ Mich, BS, 49; Univ Chicago, MS, 54; Univ Colo, PhD(microbiol), 56. *Prof Exp:* Asst med bact & virol, Univ Chicago, 52; biochemist, Nat Jewish Hosp, Denver, Colo, 56-58, chief clin labs, 58-60, res biochemist, 60-62; biochemist, Warner-Lambert Res Inst NJ, 63-64; chief clin chem, Vet Admin Hosp, Pittsburgh, 64-69; INVENTOR CLIN DIAGNOSTICS, 69-; PRES, COLEMAN LABS CORP, 77- *Mem:* AAAS; Am Chem Soc; Am Asn Clin Chem; NY Acad Sci. *Res:* Clinical laboratory diagnostic devices; analytical and organic chemistry; instrumentation and automation; biochemistry of trace metals and chelates; chemistry of mycobacteria. *Mailing Add:* 958 Washington Rd Pittsburgh PA 15228-2029

COLEMAN, COURTNEY (STAFFORD), b Ventura, Calif, July 19, 30; m 54; c 3. MATHEMATICAL ANALYSIS. *Educ:* Univ Calif, BA, 51; Princeton Univ, MA, 53, PhD, 55. *Prof Exp:* Instr math, Princeton Univ, 54-55; from instr to asst prof, Wesleyan Univ, 55-59; from asst prof to assoc prof, 59-66, PROF MATH, HARVEY MUDD COL, 66- *Concurrent Pos:* Vis scientist, Res Inst Advan Studies, 58-59 & 63-64; mem fac, Claremont Grad Sch, 68-; Vis math, Math Inst, Oxford Univ, 79-80, 86-87. *Mem:* Am Math Soc; Soc Indust & Appl Math; Math Asn Am. *Res:* Ordinary differential equations. *Mailing Add:* 675 Northwestern Dr Claremont CA 91711

COLEMAN, DAVID COWAN, b Bennington, Vt, Nov 7, 38; m 65; c 2. ECOLOGY, SOILS SCIENCE. *Educ:* Reed Col, BA, 60; Univ Ore, MA, 63, PhD(biol), 64. *Prof Exp:* Demonstr, Univ Col Swansea, Wales, 64-65; res assoc & fel, Savannah River Ecol Lab, Univ Ga, 65-72, asst prof zool, 67-72; res assoc agron, Natural Res Ecol Lab, Colo State Univ, 72-75, from asst prof to prof zool, 75-85; RES PROF ENTOM & ECOL, UNIV GA, 85- *Concurrent Pos:* Sr fel, Nat Res Adv Comt, Dept Sci & Indust Res, Soil Bur & Inst Nuclear Sci, Lower Hutt, NZ, 79-80; invited lectr, Brit Ecol Soc, 84. *Mem:* AAAS; Ecol Soc Am; Am Soc Microbiol; Brit Ecol Soc; Sigma Xi; Soil Sci Soc Am; Soc Nematol. *Res:* Decomposition and nutrient cycles; terrestrial ecosystems; flora-fauna interactions in the soil; nutrient cycling in forest floors. *Mailing Add:* Dept Entom Univ Ga Athens GA 30602

COLEMAN, DAVID MANLEY, b Duluth, Minn, Mar 24, 48; m 69, 90; c 3. ANALYTICAL CHEMISTRY, PHYSICAL CHEMISTRY. *Educ:* Southern Ill Univ, BA, 70; Univ Wis, MS, 72, PhD(anal chem), 76. *Prof Exp:* Proj assoc chem, Univ Wis, 76-77; asst prof, 77-87, ASSOC PROF CHEM, WAYNE STATE UNIV, 87- *Concurrent Pos:* Postdoctoral, Univ Wis, 45; Wayne State fac res fel, 78-79; nat mtg chmn, Fedn Analytical Chem & Spectros Soc, 87; fel Japan Soc Promotion Sci, Osaka Univ, 88-89; consult, Detroit Edison, 85-; gov bd chair, Fed Anal Chem & Spectros Soc, 91. *Honors & Awards:* J Lee Barrett Award, 86. *Mem:* Am Chem Soc; Soc Appl Spectros; Sigma Xi; Asn Anal Chemists (pres, 86); Fedn Anal Chem & Spectros Soc. *Res:* Optical emission spectroscopy; analytical spectroscopy; excitation mechanisms in atmospheric pressure; plasmas; excitation sources; optics; instrumentation. *Mailing Add:* Dept Chem Chem Bldg 171 Wayne State Univ Detroit MI 48202

COLEMAN, DONALD BROOKS, b Russellville, Ky, June 18, 34; m 54; c 2. MATHEMATICS. *Educ:* Union Univ, BA, 56; Purdue Univ, MS, 58, PhD(math), 61. *Prof Exp:* Asst instr math, Mich State Univ, 60-61; from asst prof to assoc prof, Vanderbilt Univ, 61-66; assoc prof, 66-75, PROF MATH, UNIV KY, 75- *Mem:* Am Math Soc; Math Asn Am. *Res:* Algebra. *Mailing Add:* Dept Math Univ Ky Lexington KY 40506

COLEMAN, GEOFFRY N, b Dover, Ohio, Oct 12, 46. SPECTROCHEMISTRY, TRACE ANALYSIS. *Educ:* De Pauw Univ, BA, 70, MA, 72; Colo State Univ, PhD(chem), 76. *Prof Exp:* Fel, Colo State Univ, 76; asst prof chem, Univ Ga, 76-; AT DEPT CHEM, UNIV ALA, UNIV. *Concurrent Pos:* Instr, Ctr Prof Advan, 81 & 82. *Mem:* Am Chem Soc; Soc Appl Spectros. *Res:* Trace analysis primarily by spectrochemical means; atomic emission, atomic absoprtion and atomic fluorescence; development and application of photoacoustic spectroscopy; applications of computers in chemistry. *Mailing Add:* Leeman Labs Inc 55 Technology Dr Lowell MA 01851-2729

COLEMAN, GEORGE HUNT, b San Gabriel, Calif, Oct 15, 28; m 53; c 3. NUCLEAR CHEMISTRY. *Educ:* Univ Calif, Berkeley, AB, 50; Univ Calif, Los Angeles, PhD(chem), 58. *Prof Exp:* Chemist, Calif Res & Develop Co, Livermore, 51-53; sr chemist, Lawrence Livermore Nat Lab, 57-69; assoc prof, 69-77, head dept, 76-79, PROF CHEM, NEBR WESLEYAN UNIV, 77-, CHAIR, DEPT CHEM, 89- *Concurrent Pos:* Vis scientist, Lawrence Livermore Nat Lab, 73, 83-84, Harwell Lab, 79-80. *Mem:* Am Chem Soc; Sigma Xi. *Res:* Radiocarbon dating; activation analysis. *Mailing Add:* Dept Chem Nebr Wesleyan Univ Lincoln NE 68504

COLEMAN, GEORGE W(ILLIAM), b Watertown, Mass, July 4, 00; m 44; c 1. CHEMISTRY, METALLURGY. *Educ:* Tufts Col, BS, 23; Mass Inst Technol, MS, 37; Temple Bar, ScD(metall), 39. *Prof Exp:* Res chemist & metallurgist, Waltham Watch Co, 23-33; res assoc, Mass Inst Technol, 35-37; chief metallurgist & chemist, Parker Mfg Co, 38-57, dir res, 57-75; CONSULT, 75- *Mem:* AAAS; Am Chem Soc; fel Am Inst Chem; Am Inst Mining, Metall & Petrol Engrs; NY Acad Sci. *Res:* Hair springs; forgings; electroplating; heat treating of steel; electropolishing. *Mailing Add:* Ten Meadow Lane PO Box 284 Southborough MA 01772-0003

COLEMAN, HOWARD S, b Everett, Pa, Jan 10, 17; m 41; c 4. ELECTROOPTICS, RENEWABLE ENERGY. *Educ:* Pa State Univ, BS, 38, MS, 39, PhD(physics), 42. *Prof Exp:* Asst, Pa State Univ, 39-40, instr phys sci, 40-42; dir optical inspection lab, 42-47; assoc prof physics & tech dir optical res lab, Univ Tex, 47-51; dir, Sci Bur, Bausch & Lomb Optical Co, NY, 51-56, vpres in charge res & eng, 56-62; head physics res & tech asst to vpres for res, Melpar, Inc, Va, 62-64; prof elec eng & dean, Col Eng, Univ Ariz, 64-68; res engr, Schellinger Res Labs, Univ Tex, El Paso, 68-69, dir spec projs, 69-75; dir, Div Cent Solar Technol, 76-81, prin dep asst secy conserv & renewable energy, 81-83, DIR SOLAR THERMAL TECHNOL, DEPT ENERGY, 83-, DIR GRANTS MGT, 90-; DIR, HOWARD S COLEMAN & ASSOCS, 75- *Concurrent Pos:* Consult, Xerox Corp, Burr-Brown Co, Kollsman Instrument Co, Melpar, Inc, Singer Co, NSF, NASA & Univ Calif, San Diego. *Honors & Awards:* Serv Tech Award, Depts Army & Navy-Am Acad Motion Picture Arts & Sci, 42. *Mem:* Nat Soc Prof Eng; Am Asn Physics Teachers; Am Soc Metals; Am Soc Eng Educ; fel Optical Soc Am. *Res:* Properties of optical instruments; guided missiles; electro-opticsysics; adoptive optics; lasers; hydrology; optical radiometry; night vision; countermeasures; camouflage; topography; simulation and consulting assistance in test and evaluation of weapon systems. *Mailing Add:* PO Box 26368 El Paso TX 79926

COLEMAN, JAMES ANDREW, b Niagara Falls, NY, Mar 11, 21; m 47; c 2. THEORETICAL PHYSICS. *Educ:* NY Univ, BA, 46; Columbia Univ, MA, 47. *Prof Exp:* Assoc physicist, Appl Physics Lab, Johns Hopkins Univ, 47-50; instr physics & astron, Conn Col Women, 50-57; PROF PHYSICS & CHMN DEPT, AM INT COL, 57- *Concurrent Pos:* Mem guided missile subcomt, Res & Develop Bd, 48-50; consult, USN, 51, 52 & 54; NSF fac fel, 58- *Mem:* AAAS; Am Phys Soc; Am Astron Soc; Am Asn Physics Teachers; Nat Asn Sci Writers. *Res:* Astronomy; relativity; cosmology. *Mailing Add:* Dept Physics Am Inst Col 1000 State St Springfield MA 01109

COLEMAN, JAMES EDWARD, b Newport, Ark, Oct 6, 28; m 54; c 3. PHYSICAL CHEMISTRY. *Educ:* La State Univ, BS, 50, MS, 52; Ohio State Univ, PhD(chem), 59. *Prof Exp:* Lab asst, La State Univ, 50-52; asst boron chem, Ohio State Univ, 52-58; chemist long range fuels, Esso Stand Oil Co, La, 59-60 & spec proj unit, Esso Res & Eng Co, NJ, 60-63, sr chemist, 63-64, sr chemist, Process Res Div, 64-68; assoc prof, 68-72, PROF CHEM, FAIRMONT STATE COL, 72- *Mem:* Am Chem Soc. *Res:* Boron chemistry; rocket propellants; synthesis and thermal stability of high energy fuels. *Mailing Add:* Dept Chem Fairmont State Col Fairmont WV 26554-2489

COLEMAN, JAMES J, b Chicago, Ill, May 15, 50; m 84; c 3. SEMICONDUCTOR LASERS, SEMICONDUCTOR EPITAXIAL GROWTH. *Educ:* Univ Ill, BS, 72, MS, 73, PhD(elec eng), 75. *Prof Exp:* Res assoc, Univ Ill, Urbana, 72-76; mem tech staff, Bell Labs, Murray Hill, NJ, 76-78, Rockwell Int, Anaheim, Calif, 78-82; PROF ELEC ENG, UNIV ILL, URBANA, 82- *Concurrent Pos:* Assoc ed, Inst Elec & Electronics Engrs Trans Electron Devices, 90- *Mem:* Inst Elec & Electronics Engrs; Am Phys Soc; Optical Soc Am. *Res:* Optical processs and electronic transport in various structures such as quantum well heterostructures; superlattices; real space transferred electron devices and low threshold high power index guided laser arrays. *Mailing Add:* 268 Microelectronics Lab Univ Ill 208 N Wright St Urbana IL 61801

COLEMAN, JAMES MALCOLM, b Vinton, La, Nov 19, 35; m 58; c 2. GEOMORPHOLOGY. *Educ:* La State Univ, BS, 58, MS, 62, PhD(geol), 66. *Prof Exp:* Assoc researcher geol, La State Univ, Baton Rouge, 60-66, asst dir, Coastal Studies Inst, 70-75, from asst prof to prof sedimentol, 66-80, DIR, COASTAL STUDIES INST, LA STATE UNIV, BATON ROUGE, 75-, BOYD PROF, 80-, EXEC VCHANCELLOR, 89-, DIR, SCH GEOSCI. *Concurrent Pos:* Leader numerous clastic sandstone seminars for indust; dir, recent & ancient deltaic deposits sem, NSF grants, 67, 69 & 71; mem, Sedimentary Processes Panel, Gulf Univs Res Consortium, 67; mem ad hoc comt on EPakistan, Nat Acad Sci; mem, Prog Develop Coun, Gulf Univs Res Consortium, 80. *Honors & Awards:* A I Levorsen Award, Am Asn Petrol Geologists, 74; Shepard Award, Soc Econ Paleontologists & Mineralogists, 80. *Mem:* Sigma Xi; Am Asn Petrol Geologists; fel Geol Soc Am. *Res:* Relationships between process and form and sedimentary characteristics of recent environments, especially in deltaic and offshore regions. *Mailing Add:* 155 Thomas Boyd Hall La State Univ Baton Rouge LA 70803

COLEMAN, JAMES R, b New York, NY, Nov 24, 37; m 59; c 2. ANALYTICAL CHEMISTRY. *Educ:* St Peter's Col, BS, 59; NY Univ, MS, 61; Duke Univ, PhD(cytol), 64. *Prof Exp:* Asst cytol, Inst Muscle Dis, 60; asst zool, Duke Univ, 61-62, res assoc cell biol, Dept Anat, 64; NIH fel, 64-65; from asst prof to assoc prof, Dept Radiation Biol & Biophys, Sch Med & Dent, Univ Rochester, 65-81; SR SCIENTIST, EASTMAN KODAK CO, 81- *Concurrent Pos:* Assoc dir educ, Univ Rochester. *Honors & Awards:* Nat lectr, Microbeam Soc, 79. *Mem:* Am Soc Cell Biol; Histochem Soc; Sigma Xi; Microbean Anal Soc; AAAS. *Res:* Cytology; histochemistry; light and electron microscopy; electron probe analysis. *Mailing Add:* 405 E 72nd St New York NY 10021

COLEMAN, JAMES ROLAND, b Kansas City, Mo, Oct 12, 46; m 71. NEUROSCIENCE, PHYSIOLOGICAL PSYCHOLOGY. *Educ:* Univ Calif, Los Angeles, BA, 69, MA, 71, PhD(psychol), 74. *Prof Exp:* MHTP trainee, Brain Res Inst, Univ Calif, Los Angeles, 69-73, res asst dept psychol, 73-74; fel neurosci, Duke Univ, 74-76; from asst prof to assoc prof psychol, Univ SC, 81-89. *Concurrent Pos:* Adj asst prof physiol, Sch Med, Univ SC, 78-81, adj assoc prof, 81-89. *Mem:* AAAS; Am Asn Anatomists; Asn Res Otolaryngol; Soc Neurosci. *Res:* Developmental mechanisms in central auditory structures; organization of central auditory and visual pathways. *Mailing Add:* 853 Gardendale Dr Columbia SC 29210

COLEMAN, JAMES STAFFORD, b CleElum, Wash, May 8, 28; m 64; c 6. ENERGY RESEARCH. *Educ:* Wash State Univ, BS, 50; Mass Inst Technol, PhD(physical chem), 53. *Prof Exp:* Mem staff, Los Alamos Nat Lab, 53-67; chemist, Div Res, AEC Comm, 67-69, tech adv, Off Gen Mgr, 69-75; asst div dir, Div Physical Res, Energy Res & Develop Admin, 75-77; DIV DIR, OFF ENERGY RES, DEPT ENERGY, 77- *Mem:* Sigma Xi; AAAS. *Res:* Engineering; mathematics; computer sciences; earth sciences. *Mailing Add:* 16112 Barnsville Rd Boyds MD 20841

COLEMAN, JOHN DEE, b Dozier, Tex, Oct 2, 32; m 54; c 3. ANIMAL NUTRITION, EDUCATIONAL ADMINISTRATION. *Educ:* Tex A&M Univ, DVM, 56; Auburn Univ, MS, 62; W Texas St Univ, MBA, 85. *Prof Exp:* Vet prev med, US Army, 60-62; sr res scientist feedlot dis, Abbott Labs, 64-68; pvt pract vet med, 68-70; assoc prof dis res, Tex A&M Univ, 70-76; asst vpres bacterin prod, Franklin Labs, Div Am Home Prod Corp, 76-80; DEAN, ROSS UNIV SCH VET MED, 87- *Mem:* Am Vet Med Asn; Am Soc Animal Sci; Am Asn Vet Nutritionists (secy-treas, 67-69); Am Asn Bovine Practitioners. *Res:* Diseases of feedlot cattle; nutrition of feedlot cattle; immunology of clostridial bacterin-toxoids. *Mailing Add:* 705 15th St Canyon TX 79015

COLEMAN, JOHN FRANKLIN, b Akron, Ohio, July 15, 39; m 61; c 3. ORGANIC CHEMISTRY, POLYMER CHEMISTRY. *Educ:* Univ Akron, BS, 61; Univ Ill, MS, 63, PhD(org chem), 66. *Prof Exp:* Asst gen chemist inorg chem, Univ Ill, 61-62, res asst org chem, 62-65; polymer res chemist, B F Goodrich Chem Group, 65-67, 69-70, sr adhesives res chemist, 70-73, group leader, 73-74, sect leader mat, 74-77, res & develop mgr, New Plastics, B F Goodrich Res & Develop Ctr, 77-79, gen mgr, 79-84; GEN MGR, CATALYST SYSTS DIV, 84-, DIR TECHNOL, US CHEM & PLASTICS, 87- *Mem:* Am Chem Soc. *Res:* Polymerization of vinyl monomers; polymer modifications; condensation polymerization; adhesives; fire retardant additives; thermally stable polymers; materials technology; organic peroxides. *Mailing Add:* 2465 Olentangy Dr Akron OH 44333-2831

COLEMAN, JOHN HOWARD, b Danville, Va, Aug 21, 25; m 64; c 5. PHYSICAL ELECTRONICS, LASER CHEMISTRY. *Educ:* Univ Va, BEE, 46; Princeton Univ, GS, 52. *Prof Exp:* Res physicist electronics, RCA Labs, NJ, 46-50; pres physics, Radiation Res Corp, 50-67; PRES PHYSICS, PLASMA PHYSICS CORP, 67- *Mem:* Am Phys Soc; AAAS; NY Acad Sci; Inst Elec & Electronics Engrs; Am Inst Metall Engrs. *Res:* Plasma deposition of semiconductor films and laser isotope separation; holds basic patents on amorphous silicon solar cells and photoreceptor drums. *Mailing Add:* Plasma Physics Corp PO Box 548 Locust Valley NY 11560

COLEMAN, JOHN RUSSELL, b Medford, Ore, Nov 4, 33; m 58; c 3. DEVELOPMENTAL BIOLOGY, CELL DIFFERENTIATION. *Educ:* Univ Minn, AB, 55; Ind Univ, MA, 57; Johns Hopkins Univ, PhD(biol), 61. *Prof Exp:* Res assoc develop biol, Univ Conn, 62-63; from asst prof to assoc prof, Brown Univ, 63-78, chmn develop biol sect, 80-85, chmn molecular cell & develop biol, 85-90, PROF BIOL, BROWN UNIV, 78- *Concurrent Pos:* NIH spec res fel, Univ Calif, San Diego, 69-70; mem cell biol study sect, NIH, 74-78; Fogarty int fel, Karolinska Inst, Stockholm, Sweden, 76-77; vis foreign investr, Nat Inst Basic Biol, Okazaki, Japan, 86; vis prof, Univ Mass Med Ctr, 90-91. *Mem:* AAAS; Am Soc Cell Biol; Am Soc Zool; Int Soc Develop Biologists (secy, 81-85 & 89-94); Int Soc Differentiation; Am Soc Develop Biol. *Res:* Differentiation of avian and mammalian cells in culture-myogenesis, regulation of gene expression. *Mailing Add:* Div Biol & Med Sci Brown Univ Providence RI 02912

COLEMAN, JOSEPH EMORY, b Iowa City, Iowa, Oct 11, 30; m 61; c 1. BIOCHEMISTRY, BIOPHYSICS. *Educ:* Univ Va, BA, 53, MD, 57; Mass Inst Technol, PhD(biophys), 63. *Prof Exp:* Intern med, Peter Bent Brigham Hosp, Harvard Sch Med, 57-58, Nat Acad Sci fel biophys, 48-59, NIH fel, 59-62, univ res fel, Biophys Res Lab, 58-63, sr resident med, Peter Bent Brigham Hosp, 63-64; from asst prof to assoc prof biochem, 64-74, PROF MOLECULAR BIOPHYS & BIOCHEM, YALE UNIV, 74- *Mem:* Am Chem Soc; Am Soc Biol Chemists. *Res:* Physical chemistry of proteins; mechanisms of enzyme action; metalloenzymes. *Mailing Add:* Dept Molecular Biophys & Biochem Yale Univ 333 Cedar St New Haven CT 06510

COLEMAN, JULES VICTOR, b Brooklyn, NY, Nov 2, 07; m 32; c 2. PSYCHIATRY. *Educ:* Cornell Univ, AB, 28; Univ Vienna, MD, 34; Am Bd Psychiat & Neurol, dipl, 46. *Prof Exp:* Dir, East Harlem Unit, Delinquency Proj, Bur Child Guid, 41-42; head ment hyg div, Med Ctr, Univ Colo, 46-50; psychiatrist, Dept Univ Health, Sch Med, Yale Univ, 50-52, clin prof psychiat, Sch Med, 52-62, chief, Ment Health Sect, Dept Epidemiol & Pub Health, 62-74, clin prof pub health & psychiat, 62-77, dir, Social & Community Psychiat Training Prog, 70-73; PVT PRACT, PSYCHIAT & PSYCHOANAL, 77- *Concurrent Pos:* Lectr, Sch Social Work, Univ Denver, 46-50; consult, Community Serv Comn, USPHS, 47-49, State Dept Health, Colo, 47-50 & Health Dept, NY, 49; mem nat tech fact finding comt, Mid-Century White House Conf Children & Youth, 50; assoc dir, Bur Ment Hyg, State Dept Health, Conn, 50-52; physician-in-chief, Psychiat Clin, New Haven Hosp, 52-56; consult, Vet Admin Hosp, West Haven, 53-; mem psychiat training rev comt, NIH, 65-69, chmn, 67-69; mem, Conn State Bd Ment Health, 65-71, chmn, 69-71; chief ment health & psychiat, Community Health Care Ctr Plan, New Haven, 71-78. *Mem:* Am Psychiat Asn; Am Orthopsychiat Asn; Asn Psychiat Clins Children (pres, 49-51); Am Psychoanal Asn. *Res:* Psychotherapy; health management organization psychiatry. *Mailing Add:* 255 Bradley St New Haven CT 06510

COLEMAN, LAMAR WILLIAM, b Philadelphia, Pa, Feb 19, 34; m 62; c 2. INERTIAL CONFINEMENT FUSION, DIAGNOSTICS. *Educ:* Va Mil Inst, 55; Ore State Univ, MS, 58, PhD(physics), 63. *Prof Exp:* Res assoc, nuclear physics, Ore State Univ, 63-64; physicist, 64-69, group leader, 69-79, dep assoc prog leader, 79-81, assoc prog leader fusion exp, 81-83, ASST PROG LEADER INERTIAL CONFINEMENT FUSION, LAWRENCE LIVERMORE NAT LAB, 83- *Mem:* Am Phys Soc. *Res:* Inertial confinement fusion and high resolution diagnostics development and applications; high energy density physics and pulsed power systems. *Mailing Add:* 947 Via Del Paz Livermore CA 94550

COLEMAN, LAWRENCE BRUCE, b Baltimore, Md, Feb 29, 48; m 71. SPECTROSCOPY & SPECTROMETRY. *Educ:* Johns Hopkins Univ, BA, 70; Univ Pa, PhD(physics), 75. *Prof Exp:* Fel physics, Univ Pa, 75-76; assoc prof, 76-91, PROF PHYSICS, UNIV CALIF, DAVIS, 91- *Concurrent Pos:* Regent's jr fac fel, Univ Calif, Davis, 77-78. *Mem:* Am Phys Soc; AAAS. *Res:* Experimental solid state physics; far infrared and infrared spectroscopy of quasi one dimensional solids, supersonic conductors, ultrathin inorganic, organic and polymeric films, Langmoir Blodgett films. *Mailing Add:* Dept Physics Univ Calif Davis CA 95616

COLEMAN, LESLIE CHARLES, b Toronto, Ont, Oct 22, 26; m 52; c 2. GEOLOGY. *Educ:* Queen's Univ, Can, BA, 50, MA, 52; Princeton Univ, PhD(geol), 55. *Prof Exp:* Instr geol, Tulane Univ, 55-56; vis asst prof, Lafayette Col, 56-57; asst prof mineral, Ohio State Univ, 57-60; from asst prof to assoc prof, 60-70, PROF GEOL, UNIV SASK, 70- *Mem:* Geol Soc Am; Mineral Soc Am; Geochem Soc; Geol Asn Can; Mineral Asn Can. *Res:* Distribution of trace metals in bedrock and relationship to the geology of the Hanson Lake area in Saskatchewan; mineralogy, petrology and geochemistry of meteorites and volcanic rocks. *Mailing Add:* Dept Geol Sci Univ Sask Saskatoon SK S7H 0W0 Can

COLEMAN, LESTER EARL, (JR), b Akron, Ohio, Nov 6, 30; m 51, 88; c 2. ORGANIC CHEMISTRY, POLYMER CHEMISTRY. *Educ:* Univ Akron, BS, 52; Univ Ill, MS, 53, PhD(chem), 55. *Prof Exp:* Chemist, Polymer Res Div, Goodyear Tire & Rubber Co, 51-52; asst gen chem, Univ Ill, 52-53, chemist res, Govt Synthetic Rubber Prog, 53-55; proj engr, Polymer Sect, Mat Lab, Wright Air Develop Ctr, US Air Force, 55-57; proj leader additive res dept, 59-64, dir org res, 64-68, asst div head res & develop, 68-72, asst to pres, 72-73, exec vpres, 74-76, pres, 76-82, DIR, LUBRIZOL CORP, 74-, CHIEF EXEC OFFICER, 78-, CHMN, 82- *Concurrent Pos:* Dir, Ferro Corp, 76-81, Soc Corp & Soc Nat Bank, 78-85, Norfolk Southern Corp, 80-, S C Johnson & Son, Inc, 81-, Harris Corp, 85- , Gen Corp, 87-89; chemist, Lubrizol Corp, 55. *Mem:* Am Chem Soc; Chem Mfrs Asn; Am Petrol Inst. *Res:* Synthesis and polymerization of vinyl monomers; synthetic organic chemistry; lubricant additives. *Mailing Add:* Lubrizol Corp 29400 Lakeland Blvd Wickliffe OH 44092

COLEMAN, LESTER F, b Burrwood, La, Jan 30, 22; m 53; c 3. CHEMICAL & NUCLEAR ENGINEERING. *Educ:* Tulane Univ, BE, 43. *Prof Exp:* Foreman & tech supvr, Y12 Plant, Tenn Eastman Corp, 43-45; from asst chem engr to assoc chem engr, 46-68, CHEM ENGR, ARGONNE NAT LAB, 68- *Mem:* Instrument Soc Am; Sigma Xi. *Res:* Nuclear fuel processing; radiation shielding design; process instrumentation; glovebox and gas purification system design; safety engineering; criticality hazards control. *Mailing Add:* 917 Sunset Rd Wheaton IL 60187

COLEMAN, MARCIA LEPRI, b New Haven, Conn. CHEMICAL PHYSICS, POLYMER PHYSICS. *Educ:* Mt Holyoke Col, BA, 69; Mass Inst Technol, PhD(chem physics), 73. *Prof Exp:* res chemist polymer physics & chem, Elastomer Chem Dept, E I du Pont de Nemours & Co Inc, 73-79, sr res supvr, Polymer Prod Dept, 79-83, res mgr, 83-85, prod mgr, Polymer Prod Dept, 85-86, planning mgr, Corp Plans Dept, 86-88, lab mgr, 88-90, TECHNOL DIR, DUPONT POLYMERS, E I DU PONT DE NEMOURS & CO INC, 90- *Mem:* AAAS; Am Chem Soc; Am Inst Physics. *Res:* Polymer chemistry and physics. *Mailing Add:* Polymers Exp Sta E I du Pont de Nemours & Co PO Box 80-0323 Wilmington DE 19880-0323

COLEMAN, MARILYN A, b Lancaster, SC, Mar 27, 46; m 68; c 2. GROWTH PHYSIOLOGY, REPRODUCTIVE PHYSIOLOGY. *Educ:* Univ SC, BS, 68; Auburn Univ, PhD(physiol), 76. *Prof Exp:* Teaching asst hist, Univ SC, 67-68; instr biol, Brunswick Co Schs, 68-69; teaching asst biol, Va Polytech Inst & State Univ, 70-72; res asst physiol, Auburn Univ, 73-76; asst prof poultry sci, physiol & mgt, 77-82, ADJ PROF PHARMACOL, OHIO STATE UNIV, 82-; PRES POULTRY PHYSIOL & MGT, MAC ASSOCS, 82- *Concurrent Pos:* Consult, poultry co allied indust, 77-; guest lectr, Regional Poultry Hatchability Schs, 77- *Mem:* Am Physiol Soc; Poultry Sci Asn; World Poultry Sci Asn; Sigma Xi. *Res:* Effects of environment and management on reproductive performance and hatchability of poultry; instrumental in development of computerized incubation system, software and other equipment for poultry incubation and reproductive management. *Mailing Add:* MAC Assocs 2532 Zollinger Rd Columbus OH 43221

COLEMAN, MARY SUE, b Richmond, Ky, Oct 2, 43; m 65; c 1. ENZYMOLOGY, PROTEIN STRUCTURE. *Educ:* Grinnell Col, BA, 65; Univ NC, PhD(biochem), 69. *Prof Exp:* Fel, Univ NC, 69-70 & Univ Tex, 70-71; fel, Univ Ky, 71-72, from instr to prof biochem, 85-90; PROF BIOCHEM, UNIV NC, CHAPEL HILL, 90- *Concurrent Pos:* Actg dir basic sci, McDowell Cancer Ctr, Univ Ky, 80-; PBC Study Sect, NIH, 80-84; Clin Cancer Prog Proj Subcomt, Nat Cancer Inst; assoc dir res, Markey Cancer Ctr, UK; assoc provost & dean res, Univ NC; res career develop award, NIH, 78-83. *Mem:* Am Chem Soc; AAAS; Am Soc Biol Chemists; Am Asn Cancer Res. *Res:* Purified human enzymes, their structure-function relationships and regulation of expression of these enzymes at genetic level. *Mailing Add:* Dept Biochem Univ NC Chapel Hill NC 27599-7260

COLEMAN, MICHAEL MURRAY, b Herne Bay, Eng, Jan 24, 38. POLYMER SCIENCE. *Educ:* Borough Polytech, Eng, BSc, 68; Case Western Reserve Univ, MS, 71, PhD(polymer sci), 73. *Prof Exp:* Assayer chem, Rhokana Corp Ltd, Zambia, 55-61; anal chemist, Johnson Mathey Ltd, Eng, 63-64; res chemist polymers, Revertex Ltd, Eng, 68-69 & E I du Pont de Nemours & Co, 73-75; from asst prof to assoc prof, 75-82, PROF POLYMERS, PA STATE UNIV, UNIVERSITY PARK, 82-, HEAD DEPT, MAT SCI & ENG, 83- *Mem:* Am Chem Soc; Am Phys Soc. *Res:* Polymer physical chemistry; polymer characterization; infrared, Raman and NMR spectroscopy as applied to polymers; vulcanization elastomers; polymer blends. *Mailing Add:* Dept Mat Sci Pa State Univ University Park PA 16802

COLEMAN, MORTON, b Norfolk, Va, Sept 15, 39; m 68; c 3. HEMATOLOGY, ONCOLOGY. *Educ:* Johns Hopkins Univ, BA, 59; Med Col Va, MD, 63. *Prof Exp:* Asst prof, 68-75, ASSOC PROF MED, DIV HEMAT-ONCOL, CORNELL UNIV, 75- *Concurrent Pos:* Asst attend physician, NY Hosp, 68-75, assoc attend physician, 75-; assoc dir oncol serv, NY Hosp-Cornell Med Ctr, 68-; consult, Doctors Hosp, NY, 69- & Manhattan Eye, Ear, Nose & Throat Hosp, 70-, New Rochelle Hosp, 80-; assoc dir clin chemother prog cancer control, Nat Cancer Inst-New York Hosp, 74-; chmn new agents comt, Cancer & Leukemia Group B, 75- *Mem:* Int Soc Hemat; Am Soc Hemat; Am Soc Clin Oncol; Soc Study Blood; Harvey Soc. *Res:* Clinical research in new chemotherapeutic agents for blood and lymphatic malignancies. *Mailing Add:* Div Hemat-Oncol Cornell Univ Med Col 525 E 68th St New York NY 10021

COLEMAN, NANCY PEES, b Hutchinson, Kans, Oct 5, 55; m 85. RISK ASSESSMENT, AIR POLLUTION. *Educ:* Old Dominion Univ, Norfolk, Va, BS, 76; Univ Okla Health Sci Ctr, MPH, 78, PhD(environ health), 85. *Prof Exp:* Sanitarian, Comn Officer Student Training & Exten Prog, USPHS, 77; environ toxicologist, Environ Consult Lab, 78-85; TOXICOLOGIST, OKLA STATE DEPT HEALTH, 85- *Concurrent Pos:* Asst prof, Univ Okla Health Sci Ctr, 85-; bd dirs, Am Acad Sanitarians, 90-92. *Mem:* Am Acad Sanitarians; Am Indust Hyg Asn; Am Conf Govt Indust Hygienists; Soc Environ Health & Geochem; Nat Environ Heatlh Asn; AAAS. *Res:* Environmental toxicology; non-carcinogenic risk assessment; lead and cadmium; air toxics; smelter emissions and related health effects. *Mailing Add:* Air Qual Serv Okla State Dept Health PO Box 53551 Oklahoma City OK 73152

COLEMAN, NEIL LLOYD, b Belvidere, Ill, Sept 3, 30; m 52. GEOLOGY, FLUID MECHANICS. *Educ:* Cornell Col, BA, 52; Univ Chicago, MS, 57, PhD(geol), 60. *Prof Exp:* Lab dir, 81-88, GEOLOGIST, SEDIMENTATION LAB, AGR RES SERV, USDA, 59-, GEOLOGIST, 88- *Concurrent Pos:* Assoc prof civil eng, Univ Miss, 61- *Honors & Awards:* Harold J Schoemaker Award, Int Asn Hydraul Res, 83. *Mem:* Am Geophys Union; Int Asn Hydraul Res; Sigma Xi. *Res:* Soil erosion; sediment transportation and deposition; mechanics of flow in natural and artificial streams or channels. *Mailing Add:* USDA-Agr Res Serv-NSL PO Box 1157 Oxford MS 38655

COLEMAN, NORMAN P, JR, b Richmond, Va, Mar 13, 42. APPLIED MATHEMATICS, ELECTRICAL ENGINEERING. *Educ:* Univ Va, BA, 65; Vanderbilt Univ, MA, 68, PhD(math), 69. *Prof Exp:* Mathematician, E I du Pont de Nemours & Co, Inc, 65; instr math, Vanderbilt Univ, 68-69; mathematician, Hq, US Army Weapons Command, 71-77, mathematician, 77-82, CHIEF AUTOMATION & ROBOTICS GROUP, US ARMY ARMAMENT RES & DEVELOP COMMAND, DOVER, NJ, 82- *Concurrent Pos:* Adj prof, Col Eng, Univ Iowa, 71-77; Byran Sch Bd, 82-85. *Mem:* Soc Indust & Appl Math; Am Math Soc; Sigma Xi. *Res:* Function algebras; necessary conditions for existence of complemented subspaces; operator theory; application of the theory of perturbation for linear operations to development of algorithms in optimal design and optimal control theory; development of microprocessor based pointing and tracking systems; robotics; general engineering; artificial intelligence. *Mailing Add:* US Army Armament Res & Develop Command Picatinny Arsenal NJ 07806-5000

COLEMAN, OTTO HARVEY, b Denver, Colo, June 26, 05; m 35; c 2. AGRONOMY. *Educ:* Colo State Univ, BS, 34, MS, 37. *Prof Exp:* Asst agronomist, State Agr Exp Sta, Univ Colo, 35-42; asst agronomist, Sugar Plant Field Sta, USDA, Miss, 42-44 & Fla, 45-46, assoc agronomist in charge sorgo breeding, 47-54, sta supt & res agronomist, Miss, 54-70, collabr sugar crops field sta, Sci & Educ Admin-Agr Res, 70-85; RETIRED. *Concurrent Pos:* Consult sugar prod, Sugarcane in Ghana, 75. *Mem:* AAAS; Am Soc Agron. *Res:* Genetics of barley and of HCN in Sudan grass; design of agricultural experiments; sugar cane breeding for syrup; breeding sorgo for syrup and sugar; sorghum genetics. *Mailing Add:* 6600 PSD Apt B204 Meridian MS 39305

COLEMAN, P(AUL) D(ARE), b Stoystown, Pa, June 4, 18; m 42; c 2. ELECTRICAL ENGINEERING. *Educ:* Susquehanna Univ, BA, 40; Pa State Univ, MS, 42; Mass Inst Technol, PhD(physics), 51. *Hon Degrees:* DSc, Susquehanna Univ, 78. *Prof Exp:* Asst physics, Pa State Univ, 40-42; physicist, Wright Air Develop Command, 42-46; res assoc physics, res lab electronics, Mass Inst Technol & physicist, Cambridge Air Res Ctr, 46-51; assoc prof elec eng, 51-57, dir, microwave lab, 54-64, PROF ELEC ENG, UNIV ILL, CHAMPAIGN-URBANA, 57-, DIR ELECTROPHYS LAB, 54- *Concurrent Pos:* Consult, Ramo-Wooldridge Corp & High Voltage Corp, 58, FXR, Inc, 59-65, McDonnell-Douglas Corp, Argonne Nat Lab, 65- & Northrop Corp, 78-; mem, Nat Acad Sci-Nat Bur Standards Panel 272 & Int Sci Radio Union Comn I; treas, Electron Device Res Conf, 66-; assoc ed, J Quantum Electronics. *Mem:* Fel Am Phys Soc; fel Inst Elec & Electronics Engrs; Sigma Xi; fel Optical Soc Am. *Res:* Quantum electronics; chemical and molecular lasers; far infrared physics; non-linear optics; millimeter and sub-millimeter waves; megavolt electronics; solid state heterojunction oscillators and detectors. *Mailing Add:* Dept Elec Engr 200 Elec Engr Res Lab Univ Ill Urbana IL 61801

COLEMAN, PATRICK LOUIS, PROTEIN CHEMISTRY, PROTEASE BIOCHEMISTRY. *Educ:* Purdue Univ, PhD(biochem), 72. *Prof Exp:* SR BIOCHEMIST, BIOSCI LABS, 84- *Mailing Add:* Bldg 270-35-06 3M Co 3M Ctr St Paul MN 55144

COLEMAN, PAUL DAVID, b New York, NY, Dec 2, 27; m 55; c 2. NEUROBIOLOGY. *Educ:* Tufts Univ, AB, 48; Univ Rochester, PhD(physiol psychol), 53. *Prof Exp:* Asst auditory physiol, Univ Rochester, 48-51, asst statist, 51-52; res psychologist, Army Med Res Lab, 54-56; asst prof & res assoc, Inst Appl Exp Psychol, Tufts Univ, 56-59; Nat Inst Neurol Dis & Stroke fel, Dept Anat, Johns Hopkins Univ, 59-62; assoc prof physiol, Sch Med, Univ Md, Baltimore County, 62-67; PROF ANAT, SCH MED, UNIV ROCHESTER, 67- *Concurrent Pos:* Assoc, Computer Ctr, Mass Inst Technol, 57-59. *Honors & Awards:* Lead Award, NIH. *Mem:* AAAS; Soc Neurosci; Am Asn Anat. *Res:* Neuroanatomy; brain-behavior relations; effects of early environment on quantitative aspects of brain; central nervous system aging. *Mailing Add:* Dept Anat Univ Rochester Med Ctr Rochester NY 14620

COLEMAN, PAUL JEROME, JR, b Evanston, Ill, Mar 7, 32; m 64; c 2. SPACE PHYSICS. *Educ:* Univ Mich, BS(eng math) & BS(eng physics), 54, MS, 58; Univ Calif, Los Angeles, PhD(space physics), 66. *Prof Exp:* Mem tech staff, Space Tech Labs, Inc, Calif, 58-61; head interplanetary sci prog, NASA, DC, 61-62; asst dir, Los Alamos Nat Lab, 81-85; res scientist, 62-66, assoc prof, 66-71, PROF GEOPHYS & SPACE PHYSICS, UNIV CALIF, LOS ANGELES, 71-, DIR, INST GEOPHYS & PLANETARY PHYSICS, 89- *Concurrent Pos:* Mem fel, John Simon Guggenheim, 75-76; Nat Coun Space, appointment by the Pres of US, 85-87; vis scholar, Univ Paris, France, 75-76; vis scientist, Lab Aeronomy, Nat Ctr Sci Res, Verrieres le Buisson, France, 75-76; pres, Univ Space Res Asn, 81-; adv bd, Inst Geophys & Planetary Physics, Los Alamos Nat Lab, 87-; consult, Indust, govt & non-profit, 63-; mem bd dirs, various corp, pub & pvt. *Honors & Awards:* Except Sci Achivement Medal for contribution to the exploration of the solar system, Nat Aeronaut & Space Admin, 70; Except Sci Achievement Medalfor contribution to explor of the moon, Nat Aeronaut & Space Admin, 72. *Mem:* AAAS; Am Geophys Union; Am Phys Soc; Am Inst Aeronaut & Astronaut; mem, Int Acad Astronaut. *Res:* Experimental space physics, studies of particles and magnetic fields in the solar wind and the earth's magnetosphere, magnetic fields of the earth, moon and planets, and interactions of the solar wind with planetary bodies. *Mailing Add:* Dept Earth & Space Sci Univ Calif Los Angeles CA 90024-1567

COLEMAN, PETER STEPHEN, b New York, NY, Feb 10, 38; m 69; c 1. BIOCHEMISTRY, BIOPHYSICS. *Educ:* Columbia Univ, AB, 59, PhD(biophys, biol), 66. *Prof Exp:* Res fel mechanochem, Polymer Dept, Weizmann Inst Sci, Israel, 67-68; res fel biochem & biophys, Yale Univ, 68-70, Nat Cancer Inst fel, 68-69; from asst prof to assoc prof biochem, NY Univ, 70-84, assoc prof basic med sci, 77-84, dir grad studies, dept biol, 78-79, PROF BIOCHEM, NY UNIV, 84- *Concurrent Pos:* Wellcome fel, MRC Lab Molecular Biol, Cambridge, 83. *Mem:* AAAS; Am Chem Soc; Am Soc Cell Biol; Biophys Soc; Am Soc Biochem & Molecular Biol. *Res:* Oxidative phosphorylation, mechanisms of coupling; cellular bioenergetics; tumor cell metabolism; membrane biochemistry; enzyme mechanism. *Mailing Add:* Dept Biol NY Univ New York NY 10003

COLEMAN, PHILIP HOXIE, b Fredericksburg, Va, May 11, 33; m 53; c 2. VETERINARY MEDICINE. *Educ:* Univ Ga, DVM, 56; Univ Wis, MS, 57, PhD(vet microbiol), 59. *Prof Exp:* Asst chief southeast rabies lab, Nat Commun Dis Ctr, USPHS, 50-60, asst chief zoonosis res unit, 60-61, in chg arbovirus lab, 61-66, asst chief biol reagents sect, 66-68, chief arbovirus infectious unit, 68-69; asst dean, Sch Basic Health Sci, 74-90, PROF MICROBIOL, MED COL VA, VA COMMONWEALTH UNIV, 69- *Mem:* Am Vet Med Asn; Sci Res Soc Am; Am Soc Microbiol; Soc Exp Biol & Med; Sigma Xi. *Res:* Pathogenesis of zoonotic viruses. *Mailing Add:* Microbiol Dept Med Col Va Sta Box 678 Richmond VA 23298-0678

COLEMAN, PHILIP LYNN, b Denver, Colo, Dec 25, 44; m 68; c 2. APPLIED PHYSICS. *Educ:* Calif Inst Technol, BS, 66; Univ Wis-Madison, PhD(physics), 71. *Prof Exp:* Res assoc space physics, Univ Wis-Madison, 71-72 & Rice Univ, 72-73; SCI STAFF MEM SHOCK PHYSICS, S-CUBED, 73- *Mem:* AAAS; Am Astron Soc. *Res:* Shock-wave physics; scientific data analysis software; instrumentation for transient, high pressure shocks; applied physics; electronics and electromagnetism. *Mailing Add:* S-Cubed PO Box 1620 La Jolla CA 92038-1620

COLEMAN, RALPH EDWARD, b Otwell, Ind, Jan 2, 43; m 67; c 3. CARDIOLOGY. *Educ:* Univ Evansville, BA, 65; Wash Univ, MD, 68. *Prof Exp:* Instr radiol, Wash Univ Sch Med, St Louis, 74-75, asst prof, 75-76; from asst prof radiol to assoc prof radiol, 76-79; PROF RADIOL, DUKE UNIV MED CTR, 79- *Concurrent Pos:* Trustee, Soc Nuclear Med, 83-, chmn, Residency Rev Comt Nuclear Med, 84-88 & Sci Affairs & Res Comt, 87; mem, Am Col Radiol Comt Nuclear Radiol, 82-, Am Bd Nuclear Med, 89-; pres, Inst Clin Res, 90- *Mem:* Soc Nuclear Med; AMA; Am Col Radiol; Am Col Nuclear Physicians; Radiol Soc NAm. *Res:* The use of the tracer techniques in evaluating disease processes; the quantitative distributions of labeled tracers is determined in these studies; disorders of the brain and heart and tumors in various parts of the body. *Mailing Add:* Dept Radiol Div Nuclear Med Duke Univ Med Ctr Box 3808 Durham NC 27710

COLEMAN, RALPH ORVAL, JR, b Corvallis, Ore, Dec 9, 31; m 64; c 2. SPEECH PATHOLOGY, AUDIOLOGY. *Educ:* Ore State Univ, BS, 54; Univ Ore, MS, 60; Northwestern Univ, PhD(speech path), 63. *Prof Exp:* Asst prof speech path, Univ Nebr, 63-65; res speech pathologist, Lancaster Cleft Palate Clin, 65-66; ASSOC PROF SPEECH PATH, ORE HEALTH SCI UNIV, PORTLAND, 66- *Concurrent Pos:* Mem summer fac, Eastern Ore Col, 69 & 71; guest researcher, Speech Transmission Lab, Royal Inst Technol, Stockholm, Sweden, 73; adj assoc prof, Univ Ore, 75- *Mem:* Am Asn Univ Professors; fel Am Speech & Hearing Asn; Am Asn Phonetic Sci; Int Asn Phonetic Sci. *Res:* Disorders of human communication; the development of language in humans; child development specifically relating to the development of communicative function; developmental disability in children. *Mailing Add:* 2923 SE Tolman Portland OR 97202

COLEMAN, RICHARD WALTER, b San Francisco, Calif, Sept 10, 22; div; c 1. ECOLOGY, BIOLOGY. *Educ:* Univ Calif, Berkeley, BA, 45, PhD(parasitol), 51. *Prof Exp:* Asst med entom, Univ Calif, Berkeley, 46, res asst med entom & helminth, 46-47 & 49-50; pvt supported res study, 51-61; prof biol & chmn dept, Curry Col, 61-63; chmn sci & math div, Monticello Col & Preparatory Sch, Ill, 63-64; vis prof biol, Wilberforce, 64-65; head, Dept Biol, 75-81, prof sci, 65-89, EMER PROF SCI, UPPER IOWA UNIV, 89- *Concurrent Pos:* Collabr nat hist div, Nat Park Serv, 52-53; spec consult, Arctic Health Res Ctr, USPHS Alaska, 54-62; appointed explorer by Comnr of Northwest Territories, Can, 66. *Mem:* AAAS; Am Bryol & Lichenol Soc; Am Soc Limnol & Oceanog; Sigma Xi; Am Inst Biol Sci; Nat Sci Teachers Asn; Nat Asn Biol Teachers; Ecol Soc Am; Am Malacol Union; Arctic Inst NAm. *Res:* Zoology; botany. *Mailing Add:* Dept Biol Upper Iowa Univ Fayette IA 52142-1857

COLEMAN, ROBERT E, b Trenton, NJ, Apr 22, 21; m 44; c 5. PLANT PHYSIOLOGY. *Educ:* Swarthmore Col, AB, 43; Univ Pa, MS, 53; Univ Hawaii, PhD(bot), 56. *Prof Exp:* Plant physiologist, Field Crops Res Br, Agr Res Serv, USDA, La, 48-54, Co-op Proj Exp Sta, Hawaiian Sugar Planters Asn, Hawaii, 54-56, Field Crops Res Br, Agr Res Serv, USDA, 56-60 & Co-op Proj Exp Sta, Hawaiian Sugar Planters Asn, 60-67; invest leader sugar cane & sweet sorghum invests, Tobacco & Sugar Res Br, Plant Sci Res Div, 67-72, staff scientist, Nat Prog Staff, Sci & Educ Admin-Agr Res, USDA, 72-80; RETIRED. *Mem:* Fel AAAS; Am Soc Plant Physiol; Soc Sugar Cane Technol; Sigma Xi; Am Inst Biol Sci. *Res:* Sugar cane physiology; post harvest deterioration; freeze injury; germination; flowering. *Mailing Add:* 4483 Foothill Dr Doylestown PA 18901

COLEMAN, ROBERT GRIFFIN, b Twin Falls, Idaho, Jan 5, 23; m 48; c 3. GEOLOGY. *Educ:* Ore State Col, BS, 48, MS, 50; Stanford Univ, PhD(geol), 57. *Prof Exp:* Instr geol, Ore State Col, 49 & La State Univ, 51-52; mineralogist, US AEC, 52-55; br chief isotope geol, US Geol Surv, 64-68, geologist, 55-81, br chief field geochem & petrol, 76-79; PROF GEOL, STANFORD UNIV, 81- *Concurrent Pos:* Vis lectr, Stanford Univ, 60; vis geologist, NZ Geol Surv, 62; tech adv, Saudi Arabia, 70-71; consult geol, Sultanate Oman, 73-74; res geologist, China, 85; vis prof, Sultan Qaboos Univ, Oman, 88, 90. *Mem:* Nat Acad Sci; fel Mineral Soc Am; fel Geol Soc Am; fel Am Geophys Union. *Res:* Mineralogy and geochemistry of uranium ore deposits; geology and mineralogy of silicate and sulfide minerals as related to their origin; glaucophane schists and ultramafic rocks of California, Oregon, New Zealand and China; new plate tectonic theories as applied to formation of ophiolites and glaucophane schists; evolution of Red Sea. *Mailing Add:* Geol Dept Stanford Univ Stanford CA 94305

COLEMAN, ROBERT J(OSEPH), b Pensacola, Fla, Jan 17, 41. ELECTRICAL ENGINEERING. *Educ:* Auburn Univ, BEE, 63, MSEE, 65, PhD(elec eng), 70. *Prof Exp:* Res asst, Auburn Res Found, 64-68; instr, Auburn Univ, 68-70; asst prof, 70-77, ASSOC PROF, UNIV NC, CHARLOTTE, 77- *Mem:* Am Inst Elec & Electronics Engrs; Am Soc Eng Educ. *Res:* Electromagnetic theory; microwave antennas; antenna research; design of circular and matrix antenna arrays; transverse electromagnetic-line arrays and elements; microwave theory and devices. *Mailing Add:* Col Eng Univ NC Charlotte NC 28223

COLEMAN, ROBERT MARSHALL, b Bridgton, Maine, Sept 27, 25; m 47; c 2. PARASITOLOGY, IMMUNOLOGY. *Educ:* Bates Col, BS, 50; Univ NH, MS, 51; Univ Notre Dame, PhD(parasitol), 54. *Prof Exp:* From instr to assoc prof microbiol, Russell Sage Col, 54-62; assoc prof, Boston Col, 62-68; head, Dept Biol Sci, 68-80, PROF IMMUNOL, UNIV LOWELL, 80- *Concurrent Pos:* Consult, AID, India, 65 & 68, Smithonian, Egypt, 76, WHO, 77, NSF, 80, Biotech Inst, Tufts, 83 & Cath Univ, 85, 86, 87, & 89, Harvard, 86. *Mem:* NY Acad Sci; Am Soc Microbiol; Am Soc Trop Med & Hyg; AAAS; Asn Med Lab Immunol. *Res:* Immunology of animal parasites; helminth antigens; malaria immuno-pathology; immune complexes; cytotoxic systems malaria; autoimmunity; author of one publication. *Mailing Add:* Dept Biol Sci Univ Lowell Lowell MA 01854

COLEMAN, ROBERT VINCENT, b Iowa City, Iowa, Oct 11, 30. PHYSICS. *Educ:* Univ Va, BA, 53, PhD(physics), 56. *Prof Exp:* Mem tech staff physics, Res Labs, Gen Elec Co, 55; sr res physicist, Res Labs, Gen Motors Corp, 56-58; asst prof physics, Univ Ill, 58-60; from assoc prof to prof, 60-76, COMMONWEALTH PROF PHYSICS, UNIV VA, 76-, CHMN DEPT, 78- *Mem:* Fel Am Phys Soc. *Res:* Solid state physics; growth and properties of crystals; magnetism and low temperature physics; metals. *Mailing Add:* Dept Physics Physics Bldg Univ Va McCormick Rd Charlottesville VA 22903

COLEMAN, RONALD LEON, b Wellington, Tex, Aug 20, 34; m 85; c 3. ENVIRONMENTAL TOXICOLOGY. *Educ:* Abilene Christian Col, BS, 56; Univ Okla, PhD(biochem), 63. *Prof Exp:* From instr to assoc prof biochem, 63-75, PROF & CHMN ENVIRON HEALTH, COL PUB HEALTH, UNIV OKLA, 75- *Mem:* Am Chem Soc; Am Pub Health Asn; Am Indust Hyg Asn; Am Conf Govt Indust Hygienists; Am Col Toxicologists. *Res:* Trace metal metabolism; toxicology of carbon monoxide, cadmium, nickel, chromium, silver and mercury; biochemical function of zinc, manganese, magnesium, iron, cadmium, nickel and chromium. *Mailing Add:* Dept Environ Health Univ Okla Health Sci Ctr Oklahoma City OK 73190

COLEMAN, SIDNEY RICHARD, b Chicago, Ill, Mar 7, 37; m. THEORETICAL PHYSICS. *Educ:* Ill Inst Technol, BS, 57; Calif Inst Technol, PhD(physics), 62. *Prof Exp:* Corning Glass Works res fel, 61-63, from asst prof to assoc prof, 63-69, PROF PHYSICS, HARVARD UNIV, 69- *Concurrent Pos:* Sloan Found fel, 64-66. *Honors & Awards:* Boris Pregel Award, NY Acad Sci; J Murray Luck Award for Sci Reviewing, Nat Acad Sci. *Mem:* Nat Acad Sci; Am Acad Arts & Sci; Am Phys Soc. *Res:* Theoretical high-energy physics; symmetry principles. *Mailing Add:* Lyman Lab Dept Physics Harvard Univ Cambridge MA 02138

COLEMAN, SYLVIA ETHEL, b Gainesville, Fla, Mar 23, 33; m 78. ULTRASTRUCTURE. *Educ:* Univ Fla, Gainesville, BS, 55, MS, 56, PhD(microbiol), 72. *Prof Exp:* Teaching asst, dept bact, Univ Fla, 55-56, dept microbiol, 67-69 & 70-72, NDEA fel, 69-70, electron microscopist, div biol sci, 72; med bacteriologist, Bact Lab, Ira Evans MD, St Petersburg, Fla, 59-60; med bact researcher, Vet Admin Med Ctr, Bay Pines, Fla, 60-63, res microbiologist, 63- 67, res biologist, 72-87, ADJ POSTDOCTORAL RES FEL, VET ADMIN MED CTR, GAINESVILLE, FLA, 87- *Concurrent Pos:* Adj asst prof microbiol, Univ Fla, 76-, adj res fel bot, 79. *Honors & Awards:* Pres Award, Am Soc Microbiol, 61; Grad Res Award, Sigma Xi, 72. *Mem:* Am Soc Microbiol; Am Soc Cell Biol; Am Asn Pathologists; Electron Micros Soc Am; Histochem Soc; Soc Biomat. *Res:* Microbial ultrastructure; marine bacteria; wood degradation by methane bacteria. *Mailing Add:* 122 NW 28th Terr Gainesville FL 32607

COLEMAN, THEO HOUGHTON, b Millport, Ala, Oct 25, 21; m 49. POULTRY SCIENCE. *Educ:* Ala Polytech Inst, BS, 43, MS, 48; Ohio State Univ, PhD(poultry genetics), 53. *Prof Exp:* Instr, Dept Poultry Sci, Ohio State Univ, 51-52; with com poultry farm, 53-54; from asst prof to prof poultry sci, 55-80, PROF ANIMAL SCI, MICH STATE UNIV, 80- *Mem:* AAAS; Genetics Soc Am; Am Genetic Asn; fel Poultry Sci Asn; World Poultry Sci Asn. *Res:* Fertility and hatchability in turkeys; effective use of visual aids in teaching; genetics, reproduction and behavior of quail. *Mailing Add:* Dept Animal Sci Mich State Univ East Lansing MI 48824

COLEMAN, THOMAS GEORGE, HYPERTENSION, COMPUTER ANALYSIS. *Educ:* Univ Miss, PhD(biomed eng), 67. *Prof Exp:* PROF PHYSIOL & BIOPHYS, MED CTR, UNIV MISS, 73- *Mailing Add:* Dept Physiol & Biophys Med Ctr Univ Miss 2500 N State St Jackson MS 39216

COLEMAN, WARREN KENT, b Stratford, Ont, Dec 30, 45; m 71. DROUGHT STRESS, COLD HARDINESS. *Educ:* Univ Western Ont, BA, 68, PhD(plant sci), 76. *Prof Exp:* Res assoc plant physiol, Dept Biol, Univ Calgary, 75-78; res dir, Agr Res & Develop, Golden West Res, 78; RES SCIENTIST, PLANT PHYSIOL, AGR CAN RES STA, FREDERICTON, 79- *Concurrent Pos:* Hon res assoc, Dept Biol, Univ NB, 80-; adjunct prof, Dept Plant Sci, NS Agr Col, 85-; assoc ed, Can J Plant Sci, 88- *Mem:* Agr Inst Can; Am Soc Hort Sci; Sigma Xi; Am Soc Plant Physiologists. *Res:* Responses to drought stress in Solanum species and cultivars; dormancy of potato tubers; cold hardiness responses in apple cultivars. *Mailing Add:* Agr Can Res Sta PO Box 20280 Fredericton NB E3B 4Z7 Can

COLEMAN, WILLIAM FLETCHER, b Montgomery, WVa, Sept 15, 44; m 65. PHYSICAL INORGANIC CHEMISTRY, MOLECULAR SPECTROSCOPY. *Educ:* Eckerd Col, BS, 66; Ind Univ, PhD(chem), 70. *Prof Exp:* Fel, Univ Ariz, 70-71; asst prof, 71-75, assoc prof chem, Univ NMex, 75-82; PROF, WELLESLEY COL, 82- *Concurrent Pos:* Vis prof, Stanford Univ, 78, 88. *Honors & Awards:* Catalyst Award, Chem Mfrs Asn, 86. *Mem:* Am Chem Soc; AAAS; Optical Soc Am; Soc Appl Spectros; Laser Inst Am; Sigma Xi; NY Acad Sci. *Res:* Photochemistry and spectroscopy of metal complexes; laser chemistry; photoelectrochemistry; inorganic complexes of biological interest; excited state electron spin resonance spectroscopy; energy transfer and nonradiative transitions. *Mailing Add:* Dept Chem Wellesley Col Wellesley MA 02181

COLEMAN, WILLIAM GILMORE, JR, b Birmingham, Ala, May 6, 42. MOLECULAR BIOLOGY. *Educ:* Talladega Col, BS, 64; Atlanta Univ, MS, 70; Purdue Univ, PhD(molecular biol), 73. *Prof Exp:* Lab instr physiol & cytol, Atlanta Univ, 64-66; biol teacher, David T Howard Community Adult Sch & elem sch teacher, Whiteford Elem Sch, 66-69; teaching asst cell biol, Purdue Univ, 70-73, vis prof, 73-74; staff fel, 74-77, sr staff fel, 77-79, RES MICROBIOLOGIST, LAB BIOCHEM PHARMACOL, NAT INST ARTHRITIS, METAB & DIGESTIVE DIS, NIH, 79-, AT LAB BIOCHEM PHARMACOL, NAT INST ARTHRITIS, DIABETES & DIGESTIVE & KIDNEY DIS, NIH; ASST PROF, HOWARD UNIV, WASHINGTON, DC, 79- *Concurrent Pos:* Lectr, Howard Univ, Wash, DC, 77-79; mem, Nat Inst Arthritis, Metab & Digestive Dis Equal Employ Opportunity Adv Comt, 74-, chmn employee rel subcomt, 74- *Mem:* Am Soc Microbiol. *Res:* Mode of assembly and function of the outer membrane of Escherichia coli; biosynthesis of L-glycero-D-manno heptose, a typical component of lipopolysaccharides. *Mailing Add:* Lab Biochem Pharmacol Bldg 8 Rm 2A02 Nat Inst Arthritis Diabetes & Digestive & Kidney Dis Bethesda MD 20894

COLEMAN, WILLIAM H, b Chestertown, Md, Mar 6, 37. MICROBIOLOGY. *Educ:* Wash Col, BS, 59; Univ Chicago, MS, 62, PhD(microbiol), 67. *Prof Exp:* Instr biol, Univ Chicago, 65-66, 67-68, sr res technologist microbiol, 66-67; fel, Univ Colo, 68-71; asst prof, 71-81, ASSOC PROF BIOL, UNIV HARTFORD, 81- *Concurrent Pos:* Lilly Found fel, Yale Univ, 75; sabbatical, Dept Microbiol & Immunol, Univ Colo Med Ctr, Denver, 79, 81, Vet Admin, Newington, Conn, 87. *Mem:* AAAS; Am Soc Microbiol; Sigma Xi. *Res:* RNA, protein purifications; protein synthesis in mammalian cells; interactions of fungal proteins and mammolian cell. *Mailing Add:* 40 Hill Farm Rd Bloomfield CT 06002

COLEN, ALAN HUGH, b Brooklyn, NY, Jan 29, 39; m 60; c 3. BIOPHYSICAL CHEMISTRY. *Educ:* Cornell Univ, BA, 60; Univ Wis, PhD(phys chem), 67. *Prof Exp:* Great Lakes Cols Asn-Kettering Found teaching intern chem, Kalamazoo Col, 66-67, asst prof, 67-70; res biochemist, Med Sch, Univ Kans, 70-72; RES CHEMIST, VET ADMIN HOSP, KANSAS CITY, MO, 72- *Concurrent Pos:* Consult, Upjohn Co, Mich, 67-68; Res Corp grant, 70; adj asst prof, Med Sch, Univ Kans, 77-81, adj assoc prof, 81- *Mem:* Am Chem Soc; Am Soc Biol Chem; Biophys Soc; Sigma Xi. *Res:* Kinetics of fast reactions in solution; relaxation kinetics; theory of fluids; critical phenomena; enzyme kinetics. *Mailing Add:* 8929 Westbrooke Overland Park KS 66212

COLER, MYRON ABRAHAM, b New York, NY, Mar 30, 13; m 42; c 2. ENGINEERING. *Educ:* Columbia Univ, AB, 33, BS, 34, ChE, 35, PhD(chem eng), 37, NY State Univ, PE, 40. *Prof Exp:* Res scientist, Manhattan Proj, US Govt, Columbia Univ, 43-45; asst prof chem eng, NY Univ, 45-46; founder, pres, chmn bd & dir, Markite Co, Markite Develop Co, Markite Corp, Markite Eng Co, 48-67; FOUNDER & DIR, COLER ENG CO, 67-, MARCUS & BERTHA COLER FUND, 69- *Concurrent Pos:* Adj prof & prof, dir surface technol prog, dir creative sci prog, NY Univ, 41-75; consult & comt mem, govt agencies, Naval Ordnance Lab, Oak Ridge Nat Lab, Brookhaven Nat Lab, Nat Inventors Coun, State Tech Serv Act Comt, NBS-Dept Com, Div Cult Studies, UNESCO-Dept State, NSF, indust & acad consult, 48-; hon prof, Polytech Inst NY, 76; hon adv, Family Issues & Substance Abuse Prog, San Francisco Univ, 85- *Honors & Awards:* Silver Insignia, US Govt, Manhattan Proj, 45. *Mem:* AAAS; Am Ceramic Soc; Am Chem Soc; Am Math Soc; Electrochem Soc. *Res:* Research and development of high technology new materials and devices for the energy and control fields; research methodology for science-based entrepreneurship and innovation. *Mailing Add:* 56 Secor Rd Scarsdale NY 10583

COLER, ROBERT A, b Hartford, Conn, July 23, 28; m 53; c 5. LIMNOLOGY. *Educ:* Champlain Col, BA, 52; Col Educ, Albany, MA, 54; State Univ NY Col Forestry, Syracuse, PhD(biol), 61. *Prof Exp:* Asst prof physics, NY State Col Educ, Cortland, 55-56; res biologist, Univ Md, 60-61; asst prof comp anat, Mass State Col, Bridgewater, 61-67; fel, 67-69, sr res assoc, 69-70, lectr, 70-71, dir environ technol training prog, 70-77, ASST PROF ENVIRON SCI, UNIV MASS, 71- *Concurrent Pos:* Dir, NSF res grant, 63-65. *Mem:* AAAS; Am Soc Zoologists; Sigma Xi. *Res:* Pollution biology and ecology; rhizosphere population responses to stress. *Mailing Add:* Div Pub Health Skinner Hall Univ Mass Amherst MA 01003

COLES, ANNA BAILEY, b Kansas City, Kans, Jan 16, 25; m 53; c 3. NURSING EDUCATION, NURSING ADMINISTRATION. *Educ:* Avila Col, BSN, 58; Cath Univ Am, MSN, 60, PhD(higher educ), 67. *Prof Exp:* Asst dir, Freedmen's Hosp, 60-63, admin asst, 63-67, dir nursing, 67-69; prof nursing & dean, Howard Univ, 68-86, emer dean, Col Nursing, 86-88; RETIRED. *Concurrent Pos:* Consult, Gen Res Adv Comn, NIH, 72-76, Vet Admin, 76-78 & Inst Med, 77; mem bd, Nat League Nursing, 77-81. *Mem:* Nat Inst Med; Nat League Nursing; Am Nurse's Asn; Am Asn Col Nursing. *Res:* Health manpower; nursing education. *Mailing Add:* 6841 Garfield Dr Kansas City KS 06102

COLES, DONALD (EARL), b Minn, Feb 8, 24; m 47; c 4. FLUIDS, MECHANICS. *Educ:* Univ Minn, BAeroE, 47; Calif Inst Technol, PhD(aeronaut), 53. *Prof Exp:* Res engr, jet propulsion lab, 50-53, res fel, 53-56, from asst prof to assoc prof, 56-64, PROF AERONAUT, CALIF INST TECHNOL, 64- *Honors & Awards:* Sperry Award, Am Inst Aeronaut & Astronaut, 54, Dryden Medal, 85. *Mem:* Nat Acad Eng, 84; fel Am Phys Soc; fel Am Inst Aeronaut & Astronaut. *Res:* Shear flows; rotating fluids; turbulence. *Mailing Add:* 1033 Alta Pine Dr Altadena CA 91001

COLES, EMBERT HARVEY, JR, b Garden City, Kans, Oct 12, 23; m 46; c 2. CLINICAL PATHOLOGY, MICROBIOLOGY. *Educ:* Kans State Univ, DVM, 45, PhD(bact, path), 58; Iowa State Univ, MS, 46. *Prof Exp:* Instr vet hyg, Iowa State Univ, 46-48; pvt pract, Kans, 48-54; from asst prof to prof path & head dept, 54-68, prof lab med & head dept, 68-82, EMER PROF CLIN PATH, KANS STATE UNIV, 85-, VET MED LAB CONSULT. *Concurrent Pos:* Dean fac vet med, Ahmadu Bello Univ, Zaria, Nigeria, 70-72; chief of party, Kans State Univ-Agency Int Develop Proj, Abu, Zaria, Nigeria, 70-72; consult, Develop Planning & Res Assoc, Ethiopia, 75, Food & Agr Orgn Conference Vet Educ, 78. *Mem:* Am Vet Med Asn; Conf Res Workers Animal Dis; Am Soc Vet Clin Path (pres, 65-66); Am Soc Microbiologists. *Mailing Add:* 1612 Denholm Dr Manhattan KS 66502

COLES, JAMES STACY, b Mansfield, Pa, June 3, 13; m 38, 81; c 3. OCEANOGRAPHY, PHYSICAL CHEMISTRY. *Educ:* Mansfield Univ, BS, 34; Columbia Univ, AB, 36, AM, 39, PhD(phys chem), 41. *Hon Degrees:* LLD, Brown Univ, 55, Univ Maine, 56, Colby Col, 59, Middlebury Col & Columbia Univ, 62 & Bowdoin Col, 68; DSc, Univ NB, 58; ScD, Merrimack Col, 64. *Prof Exp:* Instr chem, City Col, 36-41; from instr to asst prof, Middlebury Col, 41-43; res group leader, Underwater Explosives Res Lab, Woods Hole Oceanog Inst, Mass, 43-45, res supvr, 45-46, res consult, 46-68; from asst prof to assoc prof chem, Brown Univ, 46-52, exec off dept chem, 48-52; pres, Bowdoin Col, 52-68; dir, 58-68, pres, Res Corp, 68-82; RETIRED. *Concurrent Pos:* Mem adv comt, NSF, 53-55; trustee, Woods Hole Oceanog Inst, 53-88 & Am Savings Bank, 70-80; civilian aide to Secy US Dept Army, 54-57; mem adv comt educ, Int Geophys Year, 58; dir, Coun Libr Resources, 60-; Am Coun Educ, 62-65; Chem Fund, Inc, 68-; chmn mine adv comt, Nat Res Coun-Nat Acad Sci, 68-72; dir, Res-Cottrell, Inc, 68-79; mem adv comt, US Coast Guard Acad, 69-72; trustee-at-large & treas, Independent Col Funds of Am, 70-88; dir, Pennwalt Corp, 71-75, EDO Corp, 71-, Med Care Develop, Inc, Maine, 77-81 & Va Chem, Inc, 78-81; mem nat adv bd, Desert Res Inst, 77-81; emer pres, Bowdoin Col. *Mem:* Fel AAAS; Am Chem Soc; fel Am Acad Arts & Sci; NY Acad Sci; hon mem Am Inst Chem. *Res:* Physical properties of natural high polymers; ultracentrifuge; underwater explosives and measurement of shock waves; oceanography. *Mailing Add:* 45 Sutton Pl S No 17-I New York NY 10022

COLES, RICHARD WARREN, b Philadelphia, Pa, Sept 16, 39; m 62; c 2. PHYSIOLOGICAL ECOLOGY, ANIMAL BEHAVIOR. *Educ:* Swarthmore Col, BA, 61; Harvard Univ, MA & PhD(biol), 67. *Prof Exp:* Asst prof biol, Claremont Cols, 66-70; from adj asst prof to adj prof, 70- DIR, TYSON RES CTR, WASH UNIV, 70- *Concurrent Pos:* Consult, Ealing Corp, Mass, 65, Harland, Bartholomew & Assoc, 79-80, Nat Geog Soc & NSF; Claremont Cols grant, 67-69; dir res field exped, Colo Rockies, 67-70; dir NSF undergrad instrnl equip grant, Claremont Cols, 69-71; prin investr, NIH Animal Care Facil improvement grant, 72-73; ed consult, Nat Geog Mag, 73-90; secy-treas, Orgn Biol Field Sta, 76-79; mem bd, Open Space Coun, St Louis, Mo Nature Conservancy & Meramec River Recreation Asn, & St Louis chap, Audobon Soc; secy-treas, Orgn Biol Rield Sta, 76-; grants, Joyce Found, St Louis Audubon Soc, Webster Groves Nature Study Soc, NSF, Gaylord Found; vis instr, The Nature Pl, Florissant, Colo, 80-; nature tour leader, Holbrook Travel, 90-; consult, Weldon Springs Task Force, Univ Mo. *Mem:* AAAS; Am Inst Biol Sci; Wilderness Soc; Nat Audubon Soc; Ecol Soc Am; Explorer's Club; Am Soc Mammologists; Am Behav Soc; Am Soc Zoologists; Am Ornit Union; Am Birding Asn. *Res:* Thermoregulation of the beaver with special emphasis on the role of the beaver's expanded tail in thermoregulation; diving behavior and physiology of the dipper, or water ouzel; bird censusing in oak hickory forest winter and breeding season. *Mailing Add:* Tyson Res Ctr Wash Univ PO Box 258 Eureka MO 63025

COLES, ROBERT, b Boston, Mass, Oct 12, 29; m 60; c 3. PSYCHIATRY. *Educ:* Harvard Univ, AB, 50; Columbia Univ, MD, 54. *Hon Degrees:* Numerous from var US cols & univs. *Prof Exp:* Resident psychiat, Mass Gen Hosp, 55-56, mem staff, 60-62; clin asst, psychiat, 60-62, PROF PSYCHIAT & MED HUMANITIES, SCH MED, HARVARD UNIV, 78-, RES PSYCHIATRIST, HARVARD HEALTH SERVS, 63-, LECTR GEN EDUC, 66- *Concurrent Pos:* Nat Adv Comt Farm Labor, 65-; consult, Ford Found, 69-; mem adv coun, for Children Relief, 72-; author of numerous books and articles; mem, Field Found, Am Freedom from Hunger Found, Nat Rural Housing Coalition. *Honors & Awards:* Nat Educ TV award, 66; Inst Medal, Nat Acad Sci, 73; Pulitzer Prize, 73; Joseph Hale Award, 86. *Mem:* Am Acad Arts & Sci; Am Orthopsychiat Asn; Group Advan Psychiat; Am Psychiat Asn. *Res:* How children in various troubled nations develop a sense of national identity, of ideological commitment; how children deal with crisis; how children acquire religious and spiritual values; the effects of poverty and racial discrimination. *Mailing Add:* Harvard Univ Health Servs 75 Mt Auburn St Cambridge MA 02138

COLES, STEPHEN LEE, b Ogden, Utah, Jan 1, 44; c 2. CORAL REEF ECOLOGY, CORAL PHYSIOLOGY. *Educ:* Dartmouth Col, BA, 66; Univ Ga, MS, 68; Univ Hawaii, PhD(zool), 73. *Prof Exp:* Res asst, marine biol, Ga Marine Inst, Univ Ga, 65-66, teaching asst zool, 66-67, res asst, 67-68; teaching asst, Univ Hawaii, 68-69, res asst marine biol, Hawaii Inst Marine Biol, 69-73; marine biologist, Environ Dept, Hawaiian Elec Co, 73-85; res scientist, Univ Petrol & Minerals Res Inst, Saudi Arabia, 85-90; RES ASSOC, AECOS INC, 90-; RES ASSOC, HAWAII INST MARINE BIOL, 90- *Concurrent Pos:* Lectr, Windward Community Col, Univ Hawaii, 74-78. *Mem:* Sigma Xi; Ecol Soc Am; Am Soc Limnol Oceanog. *Res:* Coral reef ecology; environmental effects of power stations; physiology of reef corals; multivariate statistical analysis; biology of symbiosis; stability of marine biotic communities; oceanography and marine ecology of the Arabian Gulf. *Mailing Add:* Aecos Inc 970 N Kalaheo Ave Kailua HI 96734

COLES, WILLIAM JEFFREY, b Marquette, Mich, Oct 31, 29; m 55; c 3. MATHEMATICS. *Educ:* Northern Mich Col, BA, 50; Duke Univ, MA, 52, PhD(math), 54. *Prof Exp:* Asst, Duke Univ, 50-51; instr math, Univ Wis, 54-55; analyst, Dept Defense, Washington, DC, 55-56; from asst prof to assoc prof math, Univ Utah, 56-63; vis assoc prof, Math Res Ctr, US Army, Univ Wis, 63-64; PROF MATH, UNIV UTAH, 64- *Mem:* Am Math Soc; Math Asn Am; Soc Indust & Appl Math. *Res:* Differential equations. *Mailing Add:* Dept Math Univ Utah Salt Lake City UT 84112

COLEY, RONALD FRANK, b Chicago, Ill, Dec 27, 41; m 62; c 4. WASTE MANAGEMENT, NUCLEAR INSTRUMENTATION. *Educ:* St Procopius Col, 63; Iowa State Univ, PhD(inorg & phys chem), 69. *Prof Exp:* Asst chem, Iowa State Univ, 63-69; assoc, Biol & Med Res Div, Argonne Nat Lab, 69-70; chemist, Commonwealth Edison Co, 70-85; CONSULT, RLD CONSULT, 85-; SCIENTIST, ARGONNE NAT LAB, 90- *Concurrent Pos:* Res assoc, Biol & Med Res Div, Argonne Nat Lab, 70-79; mem, subcomt radioactive ref standards, Am Nat Standards Inst, 74-; consult nuclear instrumentation, 78-; adj fac, Dept Chem & Biochem, Ill Benedictine Col, 79- *Mem:* Am Chem Soc; Am Nuclear Soc. *Res:* Neutron therapy of cancer; Monte Carlo computations of neutron and gamma interactions with matter; gamma and neutron spectrometry; chemical and radionuclide analysis of nuclear power plant systems and effluents; computerized analytical instrumentation systems and networks. *Mailing Add:* Argonne Nat Lab EID/900 9700 S Cass Ave Argonne IL 60439

COLGATE, SAMUEL ORAN, b Amarillo, Tex, Oct 5, 33; m 55; c 2. PHYSICAL CHEMISTRY. *Educ:* WTex State Univ, BS, 55; Okla State Univ, MS, 56; Mass Inst Technol, PhD(phys chem), 59. *Prof Exp:* Asst prof, 59-66, ASSOC PROF CHEM, UNIV FLA, 66- *Concurrent Pos:* Vis prof, Harvard & Boston Univs, 69-70. *Mem:* Am Phys Soc; Sigma Xi. *Res:* Derivation of the intermolecular potential from scattering of molecular beams. *Mailing Add:* 4132 NW 38th St Gainesville FL 32606

COLGATE, STIRLING AUCHINCLOSS, b New York, NY, Nov 14, 25; m 47; c 3. PHYSICS. *Educ:* Cornell Univ, BA, 48, PhD(physics), 52. *Prof Exp:* Physicist, Radiation Lab, Univ Calif, 51-64; pres, 65-74, ADJ PROF PHYSICS, NMEX INST MINING & TECHNOL, 75-; PHYSICIST, LOS ALAMOS NAT LAB, 76-, SR FEL, 80- *Concurrent Pos:* Mem sci adv bd, USAF, 58-61, gas centrifuge comt, AEC, 61-69 & fluid dynamics comt, NASA, 61-63; consult, Conf Cessation Nuclear Weapons Tests, US State Dept, 59; lectr, Univ Calif, 59-64; ed, Nuclear Fusion; trustee-at-large, Assoc Univs, Inc, 70-73; chmn subfield surv plasma physics & physics of fluids panel, Nat Acad Sci Physics Surv Comt, 70-; trustee-at-large, Assoc Univs Res Astron, Inc, 73-78; mem, Space Sci Bd, 75-78; chmn panel, Physics Sun, Space Sci Bd, 79-82; mem Space & Earth Sci Advan Comt, NASA, 86-88. *Honors & Awards:* Rossi Prize, Cosmic Rays & Supernova, Am Astron Soc, 90. *Mem:* Nat Acad Sci; Am Acad Sci; Am Astron Soc; NY Acad Sci; Am Geophys Union; fel Am Phys Soc. *Res:* Accelerators; nuclear weapon physics; controlled thermonuclear fusion; plasma physics; atmospheric physics; astrophysics. *Mailing Add:* Theoret Div MS 275 BT-6 Los Alamos Nat Lab Los Alamos NM 87545

COLGLAZIER, MERLE LEE, b Holyoke, Colo, Aug 6, 20; m 48; c 3. VETERINARY PARASITOLOGY. *Educ:* Univ Colo, BA, 48. *Prof Exp:* From parasitologist to sr parasitologist, Animal Parasitol Inst, Agr Res Serv, Beltsville Agr Res Ctr, USDA, 48-68, zoologist vet chemother, 68-78; RETIRED. *Mem:* Am Soc Parasitologists. *Res:* Antiparasitic investigations dealing with chemotherapy and chemical control of helminthic diseases and parasites that affect domestic animals; poultry and fur-bearing animals raised in captivity. *Mailing Add:* 2712 Philben Dr Adelphi MD 20783

COLGROVE, STEVEN GRAY, b Lancaster, Pa, Nov 5, 53. GAS CHROMATOGRAPHY, MASS SPECTROMETRY. *Educ:* Lafayette Col, BS, 75; Iowa State Univ, PhD(anal chem), 80. *Prof Exp:* RES CHEMIST, EXXON RES & ENG CO, 80- *Mem:* Am Chem Soc; Sigma Xi; Am Soc Mass Spectrometry. *Res:* Analysis of complex mixtures of organic compounds; analysis of coal liquids, shale oil, and petroleum, using gas chromatography, mass spectrometry and liquid chromatography. *Mailing Add:* 1042 Masterson Dr Baton Rouge LA 70810-4729

COLI, G(UIDO) J(OHN), JR, b Richmond, Va, Sept 12, 21; m 47; c 4. CHEMICAL ENGINEERING. *Educ:* Va Polytech Inst, BS, 41, PhD(chem eng), 49. *Prof Exp:* Asst engr, Bur Indust Hyg, State Dept Health, Va, 41; assoc chemist, Naval Res Lab, 42-46; instr chem eng, Va Polytech Inst, 47-48; chem engr, Socony-Vacuum Oil Co, 49-50; mem staff, Allied Chem Corp, 50-74, group vpres, 68-74, dir, 70-74; exec vpres opers, 74-79, PRES, AM ENKA CO, 79-; DIR, AKZONA INC, 79- *Mem:* Am Chem Soc; fel Am Inst Chemists; Inst Mech Eng. *Res:* Molecular distillation; heat transfer. *Mailing Add:* 314 Town Mon Rd Asheville NC 28804-3821

COLICELLI, ELENA JEANMARIE, b Newark, NJ, Jan 27, 50. MOLECULAR CHEMISTRY. *Educ:* Col St Elizabeth, BS, 72; Univ Md, PhD(phys org chem), 77. *Prof Exp:* From instr to assoc prof, 76-88, PROF CHEM, COL ST ELIZABETH, 88- *Concurrent Pos:* Res assoc, Univ Md, College Park, 79; res chemist, Nat Inst Sci & Technol, 81 & Laramie Energy Technol Ctr, 83; alt counr, Am Chem Soc, 88-93; vis prof & Fulbright sr lectr, Univ Zimbabue & Fulbright Nat Univ Sci & Technol, 90-91. *Mem:* Am Chem Soc. *Res:* Spectroscopic structure determination of organic molecules, molecular configurations and binding constants. *Mailing Add:* Dept Chem Col St Elizabeth Two Convent Rd Morristown NJ 07960-0476

COLIGAN, JOHN E, b Canonsburg, Pa, July 3, 44. IMMUNOLOGY. *Educ:* Wabash Col, BA, 66; Ind Univ, MS, 68, PhD(microbiol), 71. *Prof Exp:* Pete Vaughn Scholarship, Wabash Col, 62; from asst res scientist to assoc res scientist, City Hope Med Ctr, 71-75; asst prof immunol, Rockefeller Univ, 75-77; res chemist, 77-80, sect head, 80-87, CHIEF, BIOL RESOURCES BR, NAT INST ALLERGY & INFECTIOUS DIS, 87-, SR EXEC SERV, 89- *Concurrent Pos:* Assoc ed, J Immunol, 83-87; ed, Immunol Res, 85- *Mem:* Am Asn Immunologists; Am Soc Biochem & Molecular Biol; Am Asn Cancer Res; Soc Complex Carbohydrates; Fedn Am Socs Exp Biol. *Res:* Molecular mechanisms that control the immune response with emphasis in the interaction of lymphocyte membrance proteins. *Mailing Add:* NIH Bldg 4 Rm 413 Bethesda MD 20892

COLILLA, WILLIAM, b Shanghai, China, Sept 1, 38; US citizen; m 65; c 1. BIOCHEMISTRY. *Educ:* Univ Ill, BS, 62; Southern Ill Univ, PhD(biochem), 70. *Prof Exp:* Res scientist plastic coatings, US Gypsum, 64-65; res assoc carbohydrate metab, Sch Med, Univ NDak, 70-74; SCIENTIST ENZYMOL PROCESSES, CLINTON CORN PROCESSING CO, STANDARD BRANDS, INC, 74- *Mem:* Am Chem Soc; Sigma Xi. *Res:* Plant biochemistry; carbohydrates and carbohydrate metabolism; enzymology and enzyme kinetics. *Mailing Add:* Res Div Clinton Corn Processing Co 1251 Beaver Channel Pkwy Clinton IA 52732

COLIN, LAWRENCE, b New York, NY, Jan 19, 31; m 53; c 2. ELECTRICAL ENGINEERING. *Educ:* Polytech Inst Brooklyn, BEE, 52; Syracuse Univ, MEE, 60; Stanford Univ, PhD(elec eng), 64. *Prof Exp:* Electronic engr, Rome Air Develop Ctr, US Air Force, 52-64; AEROSPACE TECHNOLOGIST, AMES RES CTR, NASA, 64- *Concurrent Pos:* Mem comn II, US Int Sci Radio Union, 58-; mem working group, Int Satellites Ionospheric Studies, Nat Aeronaut & Space Admin, 64- *Mem:* Inst Elec & Electronics Engrs; Am Geophys Union; Sigma Xi. *Res:* Radiowave propagation; upper atmospheric physics; space physics. *Mailing Add:* 142 Garth Scarsdale NY 10583

COLINGSWORTH, DONALD RUDOLPH, b Beaver Dam, Wis, June 20, 12; m 38; c 2. CHEMISTRY, BACTERIOLOGY. *Educ:* Univ Wis, BS, 34, MS, 36, PhD(bact), 38. *Prof Exp:* Biochemist, Red Star Yeast & Prod Co, 38-43; res chemist, Heyden Chem Corp, NJ, 43-44; sr res scientist, Upjohn Co, 44-56, mgr fermentation res & develop, 56-67, group mgr fermentation prod, 67-78; RETIRED. *Mem:* Am Chem Soc; Am Soc Microbiol. *Res:* Fermentation chemistry; antibiotics. *Mailing Add:* 1215 Miles Ave Kalamazoo MI 49001

COLINVAUX, PAUL ALFRED, b St Albans, Eng, Sept 22, 30; m 60; c 2. PALEOECOLOGY, PALEOCLIMATOLOGY. *Educ:* Cambridge Univ, BA, 56, MA, 60; Duke Univ, PhD(zool), 62. *Prof Exp:* Res officer pedology, Can Dept Agr, NB, 56-59; NATO fel biol, Queen's Univ, N Ireland, 62-63; res biologist, Yale Univ, 63-64; from asst prof to prof zool & anthropology, 64-91, mem, Byrd Polar Res Ctr, 64-91, STAFF SCIENTIST, SMITHSONIAN TROP RES INST, OHIO STATE UNIV, 91-,. *Concurrent Pos:* Guggenheim fel, 71-72; counr for Galapagos Islands, Charles Darwin Found, 75-; chmn, Sponsoring Inst, Inst Ecol, 78-80; mem, US Nat Comn Int Union Quaternary Res, Nat Acad Sci, 78-87, Adv Subcomt Ecol, NSF, 79-82; Nuffield fel, Green Col, Oxford, 86. *Honors & Awards:* Tansley Lectr, Brit Ecol Soc, 81. *Mem:* Am Soc Limnol & Oceanog; fel Arctic Inst NAm; Ecol Soc Am (treas); Am Soc Naturalists; fel Explorers Club; Brit Ecol Soc. *Res:* Environmental history of Bering land bridge and of equatorial South America and Panama; Galapagos limnology and ecology; pollen analysis of Galapagos, Andean, Amazonian, Panamanian and Arctic vegetation histories; chronology of Quaternary; ecological models of human history. *Mailing Add:* Dept Zool Ohio State Univ 1735 Neil Ave Columbus OH 43210

COLL, DAVID C, b Montreal, Que, June 17, 33; m 55; c 3. TELECOMMUNICATIONS, COMPUTERS. *Educ:* McGill Univ, BE, 55, ME, 56; Carleton Univ, Can, PhD(elec eng), 66. *Prof Exp:* Defence sci serv officer commun, Defence Res Bd, Can, 56-57; assoc prof, 67-73, chmn, Dept Systs Eng & Comput Sci, 75-78, PROF ELEC ENG, CARLETON UNIV, 73- *Concurrent Pos:* Sr consult, Lapp-Hancock Assocs; ed, Commun Soc, Inst Elec & Electronics Engrs, 82-90; pres, DCC Informatics Inc, 84-; chmn, Acad Requirements Comt, Asn Prof Engrs Ont, 85-88. *Mem:* Inst Elec & Electronics Engrs; Armed Forces Commun & Electronics Asn; Soc Motion Picture & TV Engrs. *Res:* Information technology: communications, computers, signal processing; television signals, systems and services; image communications. *Mailing Add:* Fac Eng Carleton Univ Ottawa ON K1S 5B6 Can

COLL, HANS, b Graz, Austria, June 8, 29. CHEMISTRY. *Educ:* La State Univ, MS, 55, PhD(chem), 57. *Prof Exp:* Res assoc phys chem, Cornell Univ, 57-60; res assoc, Mass Inst Technol, 60-62; res chemist, Shell Develop Co, Calif, 62-75; RES CHEMIST, RES LABS, EASTMAN KODAK CO, 75- *Concurrent Pos:* Grants, Off of Naval Res, 57-60 & NIH, 60-62. *Mem:* Am Chem Soc. *Res:* Chemistry of metal complexes; physical chemistry of detergents and high polymers. *Mailing Add:* 2800 Oakview Dr Rochester NY 14617-3210

COLLARD, HAROLD RIETH, b Paterson, NJ, July 29, 32; m 67; c 1. AIRBORNE INFRARED ASTRONOMY. *Educ:* Harvard Col, BA, 54; Stanford Univ, MS, 58, PhD(physics), 66. *Prof Exp:* Res scientist, Stanford Res Inst, 58-59, res asst, Hansen Lab Physics, 59-67; RES SCIENTIST, AMES RES CTR, NASA, 67- *Mem:* Am Geophys Union. *Res:* Electron scattering studies of structure of helium and tritium; measurement of heliocentric gradient of solar wind properties and interaction of solar wind with planetary magnetic fields; infrared astronomy instrumentation. *Mailing Add:* 496 Mariposa Ave Mountain View CA 94041-1704

COLLAT, JUSTIN WHITE, b New York, NY, Sept 29, 28; m 60; c 3. ANALYTICAL CHEMISTRY. *Educ:* Harvard Univ, AB, 49, AM, 51, PhD(chem), 53. *Prof Exp:* Fel chem, Mass Inst Technol, 54; from asst prof to assoc prof chem, Ohio State Univ, 54-66; asst prog adminr, Am Chem Soc, 66-70, prog adminr, Petrol Res Fund, 70-72, head dept res grants & awards, 72-81, dir mem div, 82-88, SECY, AM CHEM SOC, 89- *Mem:* Am Chem Soc; fel AAAS. *Res:* Electroanalytical chemistry. *Mailing Add:* Am Chem Soc 1155 16th St NW Washington DC 20036

COLLE, RONALD, b Milwaukee, Wis, Feb 11, 46. RADIOCHEMISTRY. *Educ:* Ga Inst Technol, BS, 69; Rensselaer Polytech Inst, PhD(nuclear & radiochem), 72; George Washington Univ, MSA, 79. *Prof Exp:* Res assoc nuclear & radiochem, Dept Chem, Brookhaven Nat Lab, 72-73; fac res assoc nuclear & radiochem, Univ Md, 73-74; consult & res chemist, Radioactiv Sect, Nat Bur Standards, 76-77, RES CHEMIST RADIATION METROL, PHYSICS LAB, NAT INST STANDARDS & TECHNOL, 77- *Concurrent Pos:* Vis res fel, Ctr Nuclear & Radiation Studies, State Univ NY, Albany,

71-72; adj fac mem & consult, Empire State Col & Albany Learning Ctr, State Univ NY, 73-75. *Honors & Awards:* Bronze Medal Award, US Dept Com, 81. *Mem:* Am Phys Soc; Am Chem Soc. *Res:* Nuclear physics; atomic physics and analytical chemistry; radiation metrology, standards development, measurement assurance for nuclear medicine, such as radiopharmaceuticals, and environmental radioactivity; research administration, environmental and technology assessment; measurement theory; radiation metrology. *Mailing Add:* Physics Lab Nat Inst Standards & Technol Gaithersburg MD 20899

COLLEN, MORRIS F, b St Paul, Minn, Nov 12, 13; m 37; c 4. MEDICINE. *Educ:* Univ Minn, BEE, 34, MB, 38, MD, 39; Am Bd Internal Med, dipl, 46. *Prof Exp:* Intern, Michael Reese Hosp, Chicago, 38-40; resident internal med, Los Angeles County Hosp, 40-42; chief med serv, Kaiser Found Hosp, Oakland, 42-52, med dir, 52-53; med dir, West Bay Div & chmn exec comt, 53-73, dir med methods res, 61-79, dir technol assessment, 79-83, CONSULT, PERMANENTE MED GROUP, 84- *Concurrent Pos:* Chief staff, Kaiser Found Hosp, San Francisco, 53-61; lectr, Sch Pub Health, Univ Calif, Berkeley, 65- & Med Sch, Univ Calif, San Francisco, 70-; consult, USPHS, 65-, mem adv comt on demonstration grants, 66-69; chmn health care systs study sect, US Dept Health, Educ & Welfare, 69-72; consult, WHO, Europe & Pan-Am Health Orgn & Tri-Serv Med Info Syst Prog, US Cong Off Technol Assessment Health Div, Dept Defense; prog chmn, 3rd Int Conf Med Informatics, Tokyo; Centennial scholar, Johns Hopkins Univ, 76; fel, Col Advan Studies Behav Sci, Stanford, 85-86; chmn, bd sci couns, Nat Library Med, 85-87; mem, Nat Acad Pract Med (chair, 83-88); pres, Col Med Info, 87-88; scholar-in-residence, Nat Lab Med, 87-92. *Mem:* Inst Med Nat Acad Sci; fel Am Col Physicians; fel Am Col Cardiol; fel Am Col Chest Physicians; Am Pub Health Asn; fel, Am Col Med Informatics; Nat Acad Pract Med. *Res:* Medical research and administration; internal medicine. *Mailing Add:* Permanente Med Group 3451 Piedmont Ave Oakland CA 94611

COLLER, BARRY SPENCER, b Brooklyn, NY, Nov 11, 45. HEMATOLOGY, PATHOLOGY. *Educ:* Columbia Col, BA, 66; NY Univ Med Sch, MD, 70. *Prof Exp:* Clin instr med, Georgetown Univ, Sch Med, 72-76; clin assoc, Clin Path Dept, NIH, 72-74, staff physician, 74-76; from asst prof to assoc prof med, 76-82, PROF MED & PATH, STATE UNIV NY, STONY BROOK, 82-, HEAD DIV HEMAT, DEPT MED, 84- *Concurrent Pos:* Clin chief, Hemat Lab, Univ Hosp, Stony Brook, NY, 76-; Guggenheim fel, 82; prin investr, NIH res grant, 84-; ed, Prog Hemostasis & Thrombosis, 85-; Roon lectr, Scripps Clin & Res Found, 86; consult, Centocor In, 86- *Honors & Awards:* Jane Nugent Cochems Prize, 77. *Mem:* Am Soc Clin Investrs; Am Soc Hemat (treas, 83-87); fel Am Col Physicians; Int Soc Hemostasis & Thrombosis; Am Fed Clin Res; Am Heart Asn. *Res:* The basic mechanisms of the adhesion of blood platelets to damaged blood vessels and aggregate with one another; monoclonal antibodies. *Mailing Add:* Div Hemat State Univ NY Stony Brook NY 11794

COLLETT, EDWARD, b Bronx, NY, Sept 15, 34; m 64; c 2. PHYSICAL OPTICS, LASERS. *Educ:* City Col NY, BS, 56; NMex State Univ, MS, 61; Cath Univ Am, PhD(physics), 68. *Prof Exp:* Radar systs engr, 56-67; optical physicist, US Army Res & Develop Command, Ft Monmouth, NJ, 67-80; pres, Measurement Concepts, Inc, Colts Neck, NJ, 80-90. *Honors & Awards:* Sam Stiber Commendation for Tech Excellence, Electronic Warfare Lab, Ft Monmouth, NJ, 78; Bronze Medallion for Sci Achievement, Dept of Army, 78 Army Sci Conf, 78. *Mem:* Optical Soc Am. *Res:* Physical optics with emphasis on classical optics and theoretical and experimental investigations of polarized light; flourescence polarization. *Mailing Add:* One Howard Ct Lincroft NJ 07738

COLLETTE, BRUCE BADEN, b Brooklyn, NY, Mar 14, 34; m 56; c 3. ICHTHYOLOGY, SYSTEMATICS. *Educ:* Cornell Univ, BS, 56, PhD(vert zool), 60. *Prof Exp:* asst lab dir, 63-82, SYST ZOOLOGIST, NAT MARINE FISHERIES SERV SYST LAB, NAT MUS NATURAL HIST, 60-, RES ASSOC, DEPT VERT ZOOL, SMITHSONIAN INST, 67-, LAB DIR, 82- *Concurrent Pos:* Ichthyol ed, Copeia, Am Soc Ichthyologists & Herpetologists, 64-69; sci ed, Nat Marine Fisheries Serv, 74-77; vis prof, Marine Sci Inst, Northeastern Univ, 67-85, adj prof, 80-; res assoc, Mus Comp Zool, Harvard Univ, 77-, instr, Bermuda Biol Sta, 84-; mem Bd Gov, Am Soc Ichthyologists & Herpetologists, 62-; mem coun, Am ASN Zool Nomenclature, 83-85, Soc Syst Zool, 86-88, Willi Henning Soc, 88-90. *Honors & Awards:* Robert H Gibbs Jr Mem Award, Am Soc Ichthyologists & Herpetologists, 89. *Mem:* fel AAAS; Am Soc Ichthyologists & Herpetologists (secy, 74-78, pres, 81); Soc Syst Zool; fel Am Inst Res Biol; Am Asn Zool Nomenclature (pres, 87); fel Willi Henning Soc. *Res:* Systematics, distribution and evolution of fishes, especially Scombridae, Hemiramphidae, Belonidae, Batrachoididae, and Percidae. *Mailing Add:* Nat Marine Fish Serv Syst Lab Nat Mus Natural Hist Washington DC 20560

COLLETTE, JOHN WILFRED, b Calgary, Alta, July 20, 33; nat US; m 52; c 3. ORGANIC CHEMISTRY , BIOMATERIALS. *Educ:* Univ Alta, BSc, 55; Univ Calif, PhD(chem), 58. *Prof Exp:* Res chemist, E I du Pont de Nemours & Co, Inc, 58-64, develop chemist, 64-66, res supvr, 66-69, supvr, Cent Res Dept, 69-80, asst lab dir, Fabric & Finishes Dept, 80-83, res mgr polymer sci, 83-87, ASST TO VPRES, CENT RES DEPT, E I DU PONT DE NEMOURS & CO, INC, 87- *Mem:* Am Chem Soc; The Chem Soc; AAAS. *Res:* Olefin polymerization; chemistry of organo metallic compounds; thermoplastic elastomers; high solids coatings. *Mailing Add:* 309 Brockton Rd Wilmington DE 19803

COLLEY, DANIEL GEORGE, b Buffalo, NY, Jan 21, 43; m 65; c 1. IMMUNOLOGY, MICROBIOLOGY. *Educ:* Cent Col Ky, BA, 64; Tulane Univ, PhD(microbiol), 68. *Prof Exp:* Fel, Yale Univ, 68-70; vis prof immunol, Fed Univ Pernambuco, 70-71; from asst prof to prof, Vanderbilt Univ, 71-79; res immunologist, 71-79, RES CAREER SCIENTIST, VET ADMIN MED CTR, 79-; PROF IMMUNOL, VANDERBILT UNIV, 79- *Concurrent Pos:* Vis prof immunol, Fed Univ Minas Gerais, 74-; assoc ed, J Immunol, 74-80; mem, Trop Med-Parasitol Study Sect, NIH, 74-78; vis scientist, Res & Control Dept, Ministry Health, St Lucia, W Indies, 75, 76, 77 & 78; mem, US-Japan Coop Med Sci Prog, Panel Parasitic Dis, Nat Inst Health, 78-88, chmn, 85-88; counr, Am Soc Trop Med & Hyg, 78-; mem, Microbiol-Infectious Dis Adv Comt, NIH, 80-83; consult, Ministry Health, Egypt, 81-87; chmn, Gordon Conf Parasitism, 85; assoc ed, J Exp Zool, 85- *Honors & Awards:* Henry Baldwin Ward Medal, Am Soc Parasitologists, 81; Bailey K Ashford Award, Am Soc Trop Med & Hyg, 89; Merit Award, Nat Inst Allergy & Infectious Dis, 90. *Mem:* AAAS; Am Asn Immunologists; Royal Soc Trop Med & Hyg; Am Soc Trop Med & Hyg; Am Soc Microbiol. *Res:* Immunobiology with specific activity in the areas of immuno-regulation of cell-mediated phenomena and lymphoid-eosinophil interactions, immunopathology and resistance with specific reference to schistosomiasis; parasitology and Chagas' disease. *Mailing Add:* Vet Admin Med Ctr Rm 324 1310 24th Ave S Nashville TN 37212

COLLIAS, ELSIE COLE, b Tiffin, Ohio, Mar 24, 20; m 48; c 1. ZOOLOGY. *Educ:* Heidelberg Col, BA, 42; Univ Wis, MS, 44, PhD(zool), 48. *Prof Exp:* Asst zool, Univ Wis, 42-46, asst econ entom, 47-48; asst prof biol, Heidelberg Col, 48-49; instr, Univ Wis, 50; assoc prof, Ill Col, 53-57; res assoc entom, Univ Calif, Los Angeles, 59-62; RES ASSOC, LOS ANGELES COUNTY MUS, 62-; RES ASSOC ZOOL, UNIV CALIF, 64- *Concurrent Pos:* Entomologist, USPHS, Ga, 46-47. *Honors & Awards:* Elliott Coues Award, Am Ornithol Union, 80. *Mem:* Animal Behav Soc; Am Ornithologists Union. *Res:* Bird behavior; invertebrate zoology; insect behavior. *Mailing Add:* Dept Biol Univ Calif Los Angeles CA 90024-1606

COLLIAS, EUGENE EVANS, b Cumberland, Wash, Feb 3, 25; c 2. OCEANOGRAPHY. *Educ:* Univ Wash, MS, 51 & Scripps Inst, Univ Calif, 59. *Prof Exp:* Res instr oceanog, Univ Wash, 48-56, sr oceanogr, 59-70, prin oceanogr, 70-80; NW consult, Oceanographers Inc, 71-89; RETIRED. *Mem:* Marine Technol Soc; Am Geophys Union; Sigma Xi. *Res:* Descriptive and chemical oceanography; field methods of oceanography. *Mailing Add:* 4318 First NE Seattle WA 98105

COLLIAS, NICHOLAS ELIAS, b Chicago Heights, Ill, July 19, 14; m 48; c 1. ANIMAL BEHAVIOR, ORNITHOLOGY. *Educ:* Univ Chicago, BS, 37, PhD(zool), 42. *Prof Exp:* Asst zool, Univ Chicago, 37-42; instr biol, Chicago City Jr Col, 46 & Amherst Col, 46-47; instr zool, Univ Wis, 47-51; conserv biologist, State Conserv Dept, Wis, 51-52; USPHS spec res fel, Cornell Univ, 52-53; prof biol, Ill Col, 53-58; from asst prof to prof, 58-85, EMER PROF ZOOL, UNIV CALIF, LOS ANGELES, 85- *Concurrent Pos:* Guggenheim fel, 62-63; hon res assoc, Los Angeles County Mus, 62-; hon res assoc, Percy Fitzpatrick Inst African Ornith, Univ Capetown, 69-70 & Nat Mus Kenya, Nairobi, 73-78. *Honors & Awards:* Elliott Coues Award, Am Ornith Union, 80. *Mem:* Am Soc Zool; fel Am Ornith Union; hon mem Cooper Ornith Soc; fel Animal Behav Soc; fel AAAS. *Res:* Animal sociology and ecology; field studies of behavior and populations in birds and mammals; hormones and behavior in birds; analysis of vocal communication in animals; nest building in birds; ornithology; behavioral energetics of birds. *Mailing Add:* Dept Biol Univ Calif Los Angeles CA 90024-1606

COLLIER, BOYD DAVID, b Sacramento, Calif, Aug 14, 37; m 58; c 2. POPULATION ECOLOGY. *Educ:* Univ Calif, Berkeley, BA, 60; Cornell Univ, MST, 64, PhD(evolutionary biol), 66. *Prof Exp:* From asst prof to assoc prof, 66-72, chmn dept, 78-84, PROF BIOL, SAN DIEGO STATE UNIV, 72- *Mem:* Ecol Soc Am; Brit Ecol Soc; Entom Soc Am. *Res:* Population ecology of terrestrial animals, particularly insects. *Mailing Add:* Dept Biol San Diego State Univ San Diego CA 92182

COLLIER, BRIAN, b York, Eng, June 8, 40; m 65; c 3. PHARMACOLOGY, PHYSIOLOGY. *Educ:* Univ Leeds, BSc, 62, PhD(pharmacol), 65. *Prof Exp:* Brit Med Res Coun fel, Cambridge Univ, 65-66; lectr physiol, 66-67, asst prof, 67-72, assoc prof, 72-78, PROF PHARMACOL, McGILL UNIV, 78- *Concurrent Pos:* Med Res Coun Can scholar, 68-73. *Mem:* Am Soc Pharmacol & Exp Therapeut; Can Physiol Soc; Pharmacol Soc Can; Soc Neurosci; Int Soc Neurochem; Am Soc Neurochem. *Res:* Identity, synthesis, storage, release and fate of neurotransmitter substances in the central and peripheral nervous system of mammals; physiology and pharmacology of cholinergic synapses. *Mailing Add:* McIntyre Bldg 3655 Drummond St Montreal PQ H3G 1Y6 Can

COLLIER, CLARENCE ROBERT, b Freeport, Ill, Mar 25, 19; m 42; c 3. PHYSIOLOGY. *Educ:* Andrews Univ, BA, 40; Loma Linda Univ, MD, 49. *Prof Exp:* Instr sci, Hylandale Acad, 40-41; from instr to asst prof med, Loma Linda Univ, 52-57, from assoc prof physiol to prof physiol & biophys & chmn dept, 57-70; assoc prof med,70-71, prof med & physiol, 71-83, EMER PROF, MED & PHYSIOL, UNIV SOUTHERN CALIF, 83- *Concurrent Pos:* Nat Found Infantile Paralysis res fel, Sch Pub Health, Harvard Univ, 55-56; sr res fel, NIH, 59-62; res assoc, Rancho Los Amigos Hosp, 52-55, dir res, 56-57, chief med sci serv, 62-64, consult, 58-; consult, Rand Corp, 63-65; consult physiol, Christian Med Col, Vellore, India, 72-; mem ed bd, J Appl Physiol, 65-70; subcomt Res Am Lung Asn Calif, 80-81, chmn 82-84, res screening comt, Calif Air Resources Bd, 84-85, chmn 85-; vis prof physiol, Sch Med, Univ De Monte Morelos, Mex, 87- *Mem:* AAAS; Am Physiol Soc; Am Fedn Clin Res; Sigma Xi; fel Am Col Physicians. *Res:* Pulmonary physiology; pulmonary exchange and circulation; mechanics of breathing; biological models. *Mailing Add:* 6930 Casa Contenta Dr Somerset CA 95684

COLLIER, DONALD W(ALTER), b Washington, DC, June 5, 20; div; c 2. CHEMICAL ENGINEERING, TECHNOLOGY STRATEGY. *Educ:* Cath Univ Am, BACh, 41; Princeton Univ, AM, 43, ChE & PhD(phys chem), 44. *Prof Exp:* Asst chem, Princeton Univ, 41-43, res assoc, Rubber Reserve Co Proj, 43-44; res chemist, Manhattan Proj contract, Sharples Corp, 44-46, res chem engr, 46-51; dir res, Thomas A Edison, Inc, 51-57, vpres, 55-57, vpres & dir res, McGraw-Edison Co Div, 57-59, pres, Res Lab, 59-60; vpres res, Borg-Warner Corp, 60-75, vpres technol, 75-78, sr vpres corp strategy, 78-83; CHMN GUARDIAN INTERLOCK DIV, GUARDIAN TECHNOL, 86- *Concurrent Pos:* Spec lectr, Princeton Univ, 43-44; mem, adv bd nucleonics,

Chem & Electronics Shares, Inc, 58-62; trustee, Rittenhouse Fund, 58-60; dir, Atomic Instruments Co, Mass, 53-56, Baird-Atomic, Inc, 56-57 & Sci and Nuclear Fund, 54-58; pres, Indust Res Inst, Inc, 66-67; dir, Alcohol Countermeasures, Inc, 76-83 & tech adv bd, US Dept Com, 73-81; Independent Consult, 83-; mem, Joint-US-Indian Coun, PACT(US AID), 86-; chmn bd, Innovisions Technol Inc, 88- *Mem:* Fel AAAS (vpres, 68); Am Chem Soc; fel Am Inst Chem Engrs; Asn Res Dirs (pres, 58-59). *Res:* Heterogeneous catalytic kinetics; catalytic reactor design; thermodynamics and distillation in hydrocarbon rubber systems; sedimentation and classification of fine particles; solvent extraction and leaching of solids; crystal formation and growth; electrochemistry; sound recording; research administration; automotive and industrial power transmission; environment control; science policy, electronic breath analysis. *Mailing Add:* 307 N Michigan Ave Suite 1008 Chicago IL 60601

COLLIER, FRANCIS NASH, JR, b New York, NY, Feb 11, 17; m 47; c 3. INORGANIC CHEMISTRY. *Educ:* Howard Col, BS, 42; Ohio State Univ, MS, 49, PhD, 57. *Prof Exp:* Asst prof chem & physics, Howard Col, 43-46, assoc prof chem, 49-53, assoc prof chem & physics & chmn dept physics, 56-57; assoc prof, 57-70, PROF CHEM, UNIV NC, 70- *Mem:* Am Chem Soc. *Res:* Kinetic studies of halomines in liquid ammonia; molecular addition compounds; anhydrous metalhalides; isotope exchange studies. *Mailing Add:* Dept Chem Univ NC Chapel Hill NC 27515

COLLIER, GERALD, b Monterey Park, Calif, Nov 16, 30; c 1. VERTEBRATE BIOLOGY, ANIMAL BEHAVIOR. *Educ:* Univ Calif, Los Angeles, BA, 53, MA, 58, PhD, 68. *Prof Exp:* From asst prof to assoc prof, 61-77, PROF BIOL, SAN DIEGO STATE UNIV, 69- *Concurrent Pos:* Grants, Mex & Panama, 62-65; Res Found grant, Mex & Costa Rica, 66 & 69; Sigma Xi grant, Costa Rica, 70-73; Nat Geog grant, Spain, 74; San Diego State Univ grant, Baja, CA, 80; Army Corps Engrs; environ consult, endangered species, State Calif, 85- *Mem:* Cooper Ornith Soc; Am Ornith Union; Wilson Ornith Soc. *Res:* Avian behavior and ecology; conservation and endangered species. *Mailing Add:* Dept Biol San Diego State Univ San Diego CA 92182-0057

COLLIER, HERBERT BRUCE, b Toronto, Ont, Oct 10, 05; m 30; c 2. BIOCHEMISTRY. *Educ:* Univ Toronto, BA, 27, MA, 29, PhD(biochem), 30. *Prof Exp:* From asst prof to assoc prof biochem, Col Med & Dent, WChina Union, 32-39; biochemist, Inst Parasitol, Macdonald Col, McGill Univ, 39-42; from asst prof to assoc prof biochem, Dalhousie Univ, 42-46; prof & head dept, Univ Sask, 46-49; head dept biochem, 49-61, prof, 49-64, prof clin biochem, 64-71, EMER PROF PATHOL, UNIV ALTA, 71- *Mem:* Am Chem Soc; Royal Soc Can; Sigma Xi. *Res:* Clinical biochemistry; chemistry of enzymes; drug action; biochemistry of erythrocytes. *Mailing Add:* 11808-100 Ave Apt 1306 Edmonton AB T5K 0K4 Can

COLLIER, HERMAN EDWARD, JR, b St Louis, Mo, Aug 8, 27; m 48; c 3. ANALYTICAL CHEMISTRY, INORGANIC CHEMISTRY. *Educ:* Randolph-Macon Col, BS, 50; Lehigh Univ, MS, 52, PhD(anal chem), 55. *Hon Degrees:* LLD, Lehigh Univ, LittD, Col Charleston, 76; ScD, Randolph-Macon Col, 77; LHD, Muhlenberg Col, 86, Moravian Col, 87. *Prof Exp:* Prof chem & chmn dept, Moravian Col, 55-57; res chemist, E I du Pont de Nemours & Co, 57-63; prof chmn & chmn, Div Natural Sci & Math Moravian Col, 63-69, pres, 69-86, emer pres, 86; DIR, BETHLEHEM STEEL CORP, 87-; PRES & DIR, I&I PLANNING ASSOCS INC, 88- *Concurrent Pos:* Mem sci adv bd, Environ Protection Agency, 79-85; chmn, High Level Radioactive Waste Disposal Ctr, 82-84; dir bd govs, Horizon Health Systs, Inc, 85-86; interim pres, Salem Acad & Col, 91- *Mem:* Am Chem Soc; Sigma Xi. *Res:* New developments in flame photometry; the hydrogenfluorine flame; quantitative analytical infrared analysis; determination of enol content; differential reaction rates as an analytical tool. *Mailing Add:* PO Box 212 Point Harbor NC 27964-0212

COLLIER, JACK REED, b Louisville, Ky, Aug 19, 26. EMBRYOLOGY. *Educ:* Univ Ky, BS, 48, MS, 50, PhD(zool), Univ NC, 54. *Prof Exp:* USPHS fel, Tokyo Metrop Univ, 54-55 & Calif Inst Technol, 55-56; instr biol, Univ Vt, 56-57; asst prof, La State Univ, 57-59; independent investr, Marine Biol Lab, Woods Hole, 59-63; assoc prof, Rensselaer Polytech Inst, 63-66; PROF BIOL, BROOKLYN COL, 66- *Concurrent Pos:* Mem corp, Marine Biol Lab; managing ed, Invert Reproduction & Develop. *Mem:* Am Soc Zool; Soc Develop Biol; Biophys Soc; Soc Gen Physiol; Am Soc Cell Biol; Int Soc Develop Biol. *Res:* Invertebrate embryology. *Mailing Add:* Dept Biol Brooklyn Col New York NY 11210

COLLIER, JAMES BRYAN, b Portland, Maine, Apr 9, 44; m 72. GEOMETRY. *Educ:* Carleton Col, BA, 66; Univ Wash, PhD(math), 72. *Prof Exp:* Fel math, Dalhousie Univ, 72-73; asst prof math, Univ Southern Calif, 73-80; MEM TECH STAFF, JET PROPULSION LAB, CALIF INST TECHNOL, LOS ANGELES, 77- *Mem:* Am Math Soc. *Res:* Convex functions; convex sets; polytopes; graphs. *Mailing Add:* 2217 Via Saldivar Glendale CA 91208

COLLIER, JESSE WILTON, b Killeen, Tex, Dec 20, 14; m 40; c 2. AGRONOMY, GENETICS. *Educ:* Tex A&M Univ, BS, 38, MS, 52; Rutgers Univ, PhD(agron), 57. *Prof Exp:* Jr soil surveyor, Soil Conserv Serv, USDA, 38-41; instr agr, Tex A&M Univ, 41-43; asst agr, Tex Agr Exp Sta, 43-53; res assoc, Rutgers Univ, 53-55; asst agronomist, 55-57, assoc prof, 57-72, PROF AGRON, TEX AGR EXP STA, 72-; EMER PROF AGRON, TEX A&M UNIV, 79- *Concurrent Pos:* From assoc prof to prof Agron, Tex A&M Univ, 64-79. *Mem:* Am Soc Agron; Crop Sci Soc Am; Sigma Xi. *Res:* Plant breeding methods in corn and grain sorghum; foundation seed production of several crops including small grains, corn, grain sorghum and vegetables. *Mailing Add:* 3919 Hilltop Dr Bryan TX 77801

COLLIER, JOHN ROBERT, b Indianapolis, Ind, Oct 4, 39; m 61; c 1. CHEMICAL ENGINEERING, POLYMER SCIENCE. *Educ:* SDak Sch Mines & Tech, BS, 61; Univ Ill, Urbana, MS, 62; Case Inst Technol, PhD(polymer sci & eng), 66. *Prof Exp:* Asst chem eng, Univ Ill, Urbana, 61-62; res asst polymer sci & eng, Case Inst Technol, 62-66; from asst prof to prof chem eng, Ohio Univ, 66-88, assoc dean, Grad Col, 72-78; PROF & CHMN, DEPT CHEM ENG, LA STATE UNIV, 88- *Concurrent Pos:* Res engr plastics dept, E I du Pont de Nemours & Co, Inc, WVa, 66, res & ed consult, 67, retained res consult, 68-; mem bd dirs, Wittenberg Univ, 77- *Mem:* Soc Plastics Engrs; Am Inst Chem Engrs; Am Chem Soc. *Res:* Interrelationships between processing conditions, morphology and resultant properties of semicrystalline polymers; tailoring of properties of polymers. *Mailing Add:* 6012 Riverbend Blvd Baton Rouge LA 70820

COLLIER, MANNING GARY, b Augusta, Ga, June 14, 51. GENERALIZED DISTRIBUTION THEORY. *Educ:* Furman Univ, BS, 73; Vanderbilt Univ, MS, 77, PhD(math), 80. *Prof Exp:* Asst prof math, The Citadel, 80-82; ASSOC PROF MATH, MARY WASHINGTON COL, 82- *Concurrent Pos:* Chairperson math, Mary Washington Col, 87-90. *Mem:* Am Math Soc; Math Asn Am; Sigma Xi. *Res:* Beurling generalization of Schwartz distribution (functional) theory and its applications to complex analysis. *Mailing Add:* Dept Math Mary Washington Col Fredericksburg VA 22401

COLLIER, MARJORIE MCCANN, EARLY DEVELOPMENT INVERTEBRATES, OOGENESIS. *Educ:* La State Univ, BS, 58; City Univ NY, PhD(biol), 75. *Prof Exp:* From asst prof to assoc prof, 76-90, DEPT CHAIR BIOL, ST PETER'S COL, 86-, PROF BIOL, 90- *Concurrent Pos:* Dir, Col Writing to Learn Prog, St Peter's Col, 90- *Mem:* Am Soc Zool; Sigma Xi; Marine Biol Lab. *Res:* Origin of cytoplasmic localization of the Ileyanassa egg during oogenesis and the nature and localization of the cytoplasmic determinants of the polar lobe. *Mailing Add:* Dept Biol St Peter's Col Kennedy Blvd Jersey City NJ 07306

COLLIER, MELVIN LOWELL, fluid mechanics, mechanical engineering; deceased, see previous edition for last biography

COLLIER, ROBERT JACOB, b Springfield, Mass, May 27, 26; m 56; c 2. ELECTRON OPTICS. *Educ:* Yale Univ, BS, 50, MS, 51, PhD(physics), 54. *Prof Exp:* Mem tech staff res & develop, 54-59, supvr, 59-71, SUPVR ELECTRON OPTICS GROUP, A&T BELL LABS, 71- *Mem:* Am Phys Soc; fel Optical Soc Am. *Res:* High power traveling wave tube amplifiers; optical memory devices; holography; scanning electron beam devices; design of electron optical systems for electron lithography. *Mailing Add:* 76 Whitman Dr New Providence NJ 07974

COLLIER, ROBERT JOHN, b Wichita Falls, Tex, Aug 6, 38; m 62; c 3. PATHOBIOLOGY, BIOCHEMISTRY. *Educ:* Rice Univ, BA, 59; Harvard Univ, MS, 61, PhD(biol), 64. *Prof Exp:* NIH fel Harvard Univ, 64; NSF fel, Inst Molecular Biol, Geneva, Switz, 64-66; from asst prof to prof bact, Univ Calif, Los Angeles, 66-84; PROF DEPT MICROBIOL & MOLECULAR GENETICS, HARVARD MED SCH, 84-, FAC DEAN GRAD EDUC & CHMN DIV MED SCI, 88- *Concurrent Pos:* Guggenheim Found fel, Pasteur Inst, Paris, France, 73-74. *Honors & Awards:* Eli Lilly & Co Award Microbiol & Immunol, Am Soc Microbiol, 72; Paul Ehrlich Award, 90. *Mem:* Nat Acad Sci; Am Soc Microbiol; Am Soc Biol Chem; AAAS. *Res:* Structure and activity of bacterial toxins; selective targeting of toxin action. *Mailing Add:* Dept Microbiol & Molecular Genetics Harvard Med Sch 260 Longwood Ave Boston MA 02115

COLLIER, ROBERT JOSEPH, b Madison, Ind, Feb 15, 47; c 2. PHYSIOLOGY, ENDOCRINOLGY. *Educ:* Eastern Ill Univ, BS, 69, MS, 74; Univ Ill, PhD(physiol), 76. *Prof Exp:* Res asst dairy sci, Univ Ill, 73-76; asst prof, 76-81, assoc prof dairy sci, Univ Fla, 81-85; RES ASSOC, MICH STATE UNIV, 76-; SR FEL & DIR DAIRY RES, MONSANTO CO, 88- *Concurrent Pos:* Fel, Monsanto Co, 85-88. *Mem:* Am Dairy Sci Asn; Am Soc Animal Sci; Sigma Xi; AAAS; Endocrinol Soc; Am Inst Nutrit. *Res:* Physiology of lactation and environmental effects on dairy cattle; environmental effects on endocrine system of dairy cattle; nutritional physiology. *Mailing Add:* Monsanto Co BB3F 700 Chesterfield Pkwy St Louis MO 63198

COLLIER, SUSAN S, b Washington, DC, Nov 5, 39. PHOTOCHEMISTRY, PHOTOGRAPHIC CHEMISTRY. *Educ:* Cornell Univ, BA, 61; Univ Rochester, PhD(spectros), 66. *Prof Exp:* Fel, Univ Rochester, 66; D J Wilson vis res assoc, Ohio State Univ, 66-69, J G Calvert spec fel air pollution, 69; sr chemist, 69-74, sr staff, Res Labs, 74-89, SR STAFF PROF PROD DIV, EASTMAN KODAK CO, 89- *Concurrent Pos:* Assoc ed, J Imaging Sci, 80-89. *Honors & Awards:* Am Chem Soc Award, 85. *Mem:* Am Chem Soc; fel Soc Photog Sci & Eng. *Res:* Chemical sensitization of photographic systems; spectral sensitization. *Mailing Add:* 7330 Selden Rd LeRoy NY 14482

COLLIGAN, GEORGE AUSTIN, b Far Rockaway, NY, Sept 10, 28; m 51; c 6. METALLURGICAL & CERAMIC ENGINEERING. *Educ:* Rensselaer Polytech Inst, BMetE, 50; Univ Mich, MS, 57, PhD(metall eng), 59. *Prof Exp:* Asst foundry metallurgist, Farrel-Birmingham Co, Conn, 50-52; foundry metallurgist, Turbine Div, Gen Elec Co, 52-54, proj engr, Res Lab, 54-55; instr metall eng, Col Eng & proj engr, Eng Res Inst, Univ Mich, 55-59; sr res scientist, Res Labs, United Aircraft Corp, 59-62; assoc prof, 62-66, assoc dean, 67-77, PROF ENG, THAYER SCH ENG, DARTMOUTH COL, 66- *Concurrent Pos:* Adj asst prof, Hartford Grad Div, Rensselaer Polytech Inst, 59-62; consult, United Aircraft Corp, Mass Mat Res Corp & Howmet Corp, 62- *Honors & Awards:* Thomas W Pangborn Gold Medal Award, Am Foundrymens Soc, 70. *Mem:* Am Ceramic Soc; Am Soc Metals; Am Inst Mining, Metall & Petrol Engrs; Am Foundrymens Soc; NY Acad Sci. *Res:* Solidification of metals; metal ceramic reaction studies; x-ray diffraction and fluorescence studies. *Mailing Add:* 33 Rayton Rd Hanover NH 03755

COLLIGAN, JOHN JOSEPH, b Watertown, NY, Feb 8, 37; m 58; c 5. SCIENCE POLICY, COMPUTER SCIENCE EDUCATION. *Educ:* Le Moyne Col, NY, BS, 58; Univ Notre Dame, MS, 60; State Univ NY Buffalo, PhD(policy sci), 72; Regis Col, Colo, MA, 84. *Prof Exp:* Jr engr, IBM Corp, 60-61, from assoc engr to sr assoc engr, 61-64; asst prof physics, Broome Tech Commun Col, 64-67; admin asst, Sch Advan Tech, State Univ NY, Binghamton, 67-68, asst dean, 68-69, assoc dean, 69-76, dir state tech serv, 67-72, dean, 76-84; dir, Pastoral & Matrimonial Renewal Ctr, 84-90; PRES, JOHN 17 CTR, 90- *Concurrent Pos:* Consult, Comn Col Physics Teachers, 69- *Res:* Family systems. *Mailing Add:* 400 Corey Ave Endwell NY 13760

COLLIN, PIERRE-PAUL, b Montreal, Que, July 23, 20; m 49; c 5. PEDIATRIC SURGERY. *Educ:* Univ Montreal, BA, 41, MD, 48. *Prof Exp:* From asst prof to prof surg, Fac Med, Univ Montreal, 66-89; DIR SURG, STE JUSTINE HOSP, 66-, SURGEON. *Concurrent Pos:* Consult, Hosp Marie-Enfant, 70- *Mem:* Fel Am Col Surg; fel Int Col Surg; fel Am Acad Pediat; fel Royal Col Surgeons Can; Am Pediat Surg Asn. *Mailing Add:* Dept Surg Univ Montreal Montreal PQ H3C 3J7 Can

COLLIN, ROBERT E(MANUEL), b Donalda, Alta, Oct 24, 28; nat US; m 52; c 3. ELECTRICAL ENGINEERING. *Educ:* Univ Sask, BSc, 51; Univ London, DIC, 53, PhD(microwaves), 54. *Prof Exp:* Sci officer, Can Armament Res & Develop Estab, 54-58; from asst prof to assoc prof elec eng, Case Western Reserve Univ, 58-65, chmn dept elec eng & appl physics, 78-82, dean eng, 87-89, PROF ELEC ENG, CASE WESTERN RESERVE UNIV, 65- *Concurrent Pos:* Mem, US Comn 6, Int Sci Radio Union. *Mem:* Nat Acad Eng; fel Inst Elec & Electronics Engrs. *Res:* Microwaves; antennas; electromagnetic diffraction; plasma physics. *Mailing Add:* Elec Eng & Appl Physics Dept Case Western Reserve Univ Cleveland OH 44106

COLLIN, WILLIAM KENT, b Los Angeles, Calif, Mar 31, 38; m 64; c 2. MEDICAL MICROBIOLOGY. *Educ:* Univ Calif, Davis, BA, 61; Univ Calif, Los Angeles, PhD(med microbiol & immunol), 68. *Prof Exp:* Fel electron micros virol, parasitol, Sch Pub Health, Univ Calif, Los Angeles, 68-70, from asst res virologist to asst res pediatrician, Dept Pediat, Sch Med, 70-74; biol res coordr dermatol, Redken Labs Inc, 74-75; lectr microbiol, 75-77, assoc, 78-80, PROF BIOL, CALIF STATE UNIV, FRESNO, 80-, PROF & CHMN BIOL. *Concurrent Pos:* NIH fel, 68. *Mem:* Am Soc Parasitologists; AAAS; Sigma Xi; Am Soc Microbiol. *Res:* Cytodifferentiation and histogenesis of cestode larval forms; lymphocytic responsiveness in viropathology; virol & parasitologic epidemology. *Mailing Add:* Dept Biol Calif State Univ Fresno CA 93740

COLLINGS, EDWARD WILLIAM, b New Plymouth, NZ, Jan 22, 30; m 53; c 2. EXPERIMENTAL SOLID STATE PHYSICS. *Educ:* Univ NZ, BSc, 51, MSc, 52, PhD(physics), 58; Victoria Univ Wellington, DSc(physics), 70. *Prof Exp:* mem fac physics, Victoria Univ Wellington, 52-61; sr res physicist, Franklin Inst Res Labs, Philadelphia, Pa, 62-66; sr res physicist, 66-87, RES LEADER, COLUMBUS LABS, BATTELLE MEM INST, 87- *Concurrent Pos:* Nat Res Coun Can fel, 58-60. *Mem:* AAAS; Am Phys Soc; Metall Soc; Am Inst Mining, Metall & Petrol Engrs; Mat Res Soc. *Res:* Electronic, magnetic, and low-temperature calorimetric properties of alloys (particularly titanium alloys and stainless steels) and intermetallic compounds; applied superconductivity; relationships between electronic and mechanical properties of metals; properties of rapidly quenched alloys and metallic glasses. other electronic properties; rapidly quenched and glassy metals; superconducting materials-properties, fabrication and applications. *Mailing Add:* Battelle Mem Inst 505 King Ave Columbus OH 43201

COLLINGS, PETER JOHN, b Jamaica, NY, Jan 29, 47; m 69; c 2. SOLID STATE CHEMICAL PHYSICS, OPTICS. *Educ:* Amherst Col, BA, 68; Yale Univ, MPh, 73, PhD(physics), 76. *Prof Exp:* Teaching fel physics, Yale Univ, 72-75, res asst, 73-75, actg instr, 75-76; asst prof, Kenyon Col, 76-83, chmn dept, 79-82, from assoc prof to prof physics, 83-90; PROF PHYSICS, SWARTHMORE COL, 90- *Concurrent Pos:* Res assoc, Liquid Crystal Inst, Kent State Univ, 77, 84; NSF grant, 77-78 & 85-88 & 90-93; Res Corp Cotrell Col Sci grant, 78-81 & 85-88 & 88-91; Alexander Von Humboldt fel, Univ Paderborn, Ger, 83-84; treas, Int Liquid Crystal Conf, 86; Coun Undergrad Res, 87- *Mem:* Am Asn Physics Teachers; Am Phys Soc; AAAS; Sigma Xi. *Res:* Physics and chemistry of liquid crystals; order parameter measurements; thermodynamic measurements; optical measurements. *Mailing Add:* Dept Physics Astron Swarthmore Col Swarthmore PA 19081

COLLINS, ALIKI KARIPIDOU, b Imathia, Greece, July 27, 58; US citizen; m 83; c 3. CHEMICAL VAPOR DEPOSITION, OPTICAL PROPERTIES. *Educ:* Tech Univ WBerlin, dipl physics, 82; Mass Inst Technol, PhD(mat sci), 87. *Prof Exp:* Res asst, Tech Univ WBerlin, 78-82; res asst, Mass Inst Technol, 83-87, postdoctoral, 87-88; PRIN RES SCIENTIST, MORTON INT, 88- *Concurrent Pos:* Sci secy & ed proc, NATO ASI, Int Sch Atomic & Molecular Spectros, 83-84; lectr, Simmons Col, 85, Worcester Polytechnic Inst, 87. *Mem:* Am Phys Soc; Inst Elec & Electronics Engrs; Mat Res Soc. *Res:* Optical and mechanical properties of chemical vapor deposited materials such as II-VI, SIC diamond; theoretical quantum mechanical cluster calculations; magnetic properties of rapidly solidified alloys. *Mailing Add:* 215 Grove St Newton MA 02166

COLLINS, ALLAN CLIFFORD, b Milwaukee, Wis, June 14, 42; m 62; c 2. BIOCHEMICAL PHARMACOLOGY, BEHAVIORAL GENETICS. *Educ:* Univ Wis-Madison, BS, 65, MS, 67, PhD(pharmacol), 69. *Prof Exp:* Instr pharmacol, Univ Wis-Milwaukee, 68-69; NIH fel, Med Sch, Univ Colo, Denver, 69-71; res pharmacologist, Vet Admin Hosp, Houston, Tex, 71-72; asst prof, 72-78, ASSOC PROF PHARMACOL, UNIV COLO, BOULDER, 78-, FEL, INST BEHAV GENETICS, 73- *Mem:* Am Soc Neurochem. *Res:* Biochemical bases of tolerance to and dependence upon alcohol, barbiturates and nicotine; use of behavior genetic techniques to test hypotheses concerning mechanisms by which these drugs exert their behavioral effects. *Mailing Add:* Inst Behav Genetics Campus Box 447 Univ Colo Boulder CO 80309

COLLINS, ALLAN MEAKIN, b Orange, NJ, Aug 7, 37; m 63; c 2. COMPUTERS & EDUCATION, COGNITIVE SCIENCE. *Educ:* Univ Mich, BBA, 59, MA, 62, PhD(psychol), 70. *Prof Exp:* Lectr cognitive sci, Harvard Univ, 76-77; sr scientist, 67-82, PRIN SCIENTIST, BOLT, BERANEK & NEWMAN INC, 82-; PROF EDUC & SOCIAL POLICY, COMPUTERS & EDUC, NORTHWESTERN UNIV, 89- *Concurrent Pos:* Guggenheim fel, 74 & Sloan fel, Univ Calif, Berkeley, 80; ed, Cognitive Sci, 76-80; chmn, Cognitive Sci Soc, 79-80, mem bd, 79-87; lectr, Chinese Acad Sci, 83; panel mem, Math, Sci & Technol Educ, Nat Acad Sci, 85-88, Cognitive Sci & Higher-Order Skills, 86-87 & Info Technol Pre-Col Educ, 87-88; co-dir, Ctr Technol Educ, Bankstreet Col, 91- *Mem:* Cognitive Sci Soc; fel Am Asn Artificial Intel; Am Educ Res Asn; AAAS. *Res:* Study new ways to teach and assess science using video and computer technology; development of intelligent tutoring systems computational modelling of human semantic processing. *Mailing Add:* 135 Cedar St Lexington MA 02173

COLLINS, ALVA LEROY, JR, b Sanford, Fla, May 15, 40; m 67. INORGANIC CHEMISTRY, ENERGY PROGRAM MANAGEMENT. *Educ:* Oberlin Col, AB, 62; Duke Univ, MA, 66, PhD(chem), 67; Wharton Col, MBA, 77. *Prof Exp:* Res chemist, Am Cyanamid Co, 66-67; res assoc inorg chem, Ind Univ, 67-68; asst prof chem, Sam Houston State Univ, 68-75; ASSOC, RESOURCE PLANNING ASSOCS, 77- *Mem:* Am Chem Soc. *Res:* Analysis of energy supply and demand; management planning of energy programs. *Mailing Add:* 5404 Ridgefield Rd Bethesda MD 20816-3336

COLLINS, ANITA MARGUERITE, b Allentown, Pa, Nov 8, 47. GENETICS, ANIMAL BEHAVIOR. *Educ:* Pa State Univ, BS, 69; Ohio State Univ, MSc, 72, PhD(genetics), 76. *Prof Exp:* Instr biol, Mercyhurst Col, 75-76; res geneticist, Honey Bee Breeding Genetics & Physiol Lab, 76-88, RES LEADER, HONEY BEE RES LAB, AGR RES SERV, USDA, 88- *Concurrent Pos:* Adj prof, La State Univ, 78-88. *Mem:* Asn Women Sci; Genetics Soc Am; Int Union Study Social Insects; Animal Behav Soc; Entom Soc Am; Am Genetic Soc. *Res:* Population genetics and behavior of honey bees; defensive behavior and response to alarm pheromone. *Mailing Add:* Honey Bee Res Lab 2413 E Hwy 83 Weslaco TX 78596-8344

COLLINS, ARLEE GENE, b Forest City, Iowa, Dec 20, 27; m 53; c 2. GEOCHEMISTRY. *Educ:* Kletzing Col, BA, 51; Kans State Col Pittsburg, MS, 55; Univ Tulsa, MS, 72. *Prof Exp:* Chemist, H C Maffitt, Consult Chemist, Iowa, 51-53 & Spencer Chem Co, Kans, 53-56; asst chief chemist, Consumers Coop Refinery, 56; proj leader geochem of petrol reservoirs, US Bur Mines, 56-75, proj leader, Bartlesville Energy Res Ctr, Energy Res & Develop Admin, 75-77; proj leader, Bartlesville Energy Technol Ctr, US Dept Energy, 77-82; PROJ LEADER, NAT INST PETROL & ENERGY RES, ENVIRON RES, 82- *Concurrent Pos:* Mem task group, Fed Adv Comt on Water Data; mem, Adv Counc, Petrol Data Syst; chmn, symp ground water contamination, Am Soc Test Mat. *Mem:* Am Chem Soc; Am Soc Testing & Mat; Int Asn Geochem & Cosmochem; Soc Petrol Engrs. *Res:* Geochemistry of oil and gas reservoirs for the characterization of reservoirs for enhanced recovery; environmental research. *Mailing Add:* 2280 SE Roselawn Ave Bartlesville OK 74006

COLLINS, ARLENE RYCOMBEL, b Buffalo, NY, Jan 2, 40; m 66; c 3. VIROLOGY, MICROBIOLOGY. *Educ:* D'Youville Col, BA, 61; State Univ NY, Buffalo, MA, 64, PhD(microbiol, virol), 67. *Prof Exp:* Instr microbiol, Med Col Wis, 69; asst prof, 71-78, ASSOC PROF MICROBIOL, STATE UNIV NY, BUFFALO, 78- *Concurrent Pos:* Fel, dept Microbiol, State Univ NY, Buffalo, 67-68; fel Med Col Wis, 69-70; vis investr, dept Immunopath, Scripps Clin & Res Found, La Jolla, Calif, 81. *Mem:* Am Soc Microbiol; Soc Gen Microbiol; Am Women Sci; Am Soc Virol. *Res:* Virology; human coronaviruses; persistent virus infections; monoclonal antibodies, viral antigens. *Mailing Add:* Virus Lab 203 Sherman Hall State Univ NY 3435 Main St Buffalo NY 14214

COLLINS, BARBARA JANE, b Passaic, NJ, Apr 29, 29; m 55; c 5. BOTANY, MICROBIOLOGY. *Educ:* Bates Col, BS, 51; Smith Col, MS, 53; Univ Ill, PhD(geol), 55, MS, 59. *Prof Exp:* Instr geol, Univ Ill, 57, teaching asst bot, 57-59; assoc prof biol, 63-76, PROF BIOL SCI, CALIF LUTHERAN UNIV, 77- *Concurrent Pos:* Consult, environ impact studies incl water anal. *Mem:* Am Soc Microbiol; Sigma Xi. *Res:* Taxonomy of gymnosperms and angiosperms and flora of Southern California; keys to the flora of regional areas; electron microscopy of clay minerals; endemic, rare and endangered species, their location and environmental factors controlling distribution. *Mailing Add:* Dept Biol Calif Lutheran Univ Thousand Oaks CA 91360

COLLINS, CARL BAXTER, JR, b San Antonio, Tex, Mar 4, 40; m 60, 73; c 4. LASERS. *Educ:* Univ Tex, BS, 60, MA, 61, PhD, 63. *Prof Exp:* Instr physics, Univ Tex, 62-64; from asst prof to assoc prof, Southwest Ctr Advan Studies, 64-69, assoc prof, Univ, 69-74, head grad physics prog, 72-75, PROF PHYSICS, UNIV TEX, DALLAS, 74-, DIR, CTR QUANTUM ELECTRONICS, 75- *Concurrent Pos:* Res asst, Univ Tex, 63-64; secy, C-17 Comn Int Union Pure & Appl Physics, 91-93. *Mem:* Fel Am Phys Soc; Inst Elec & Electronics Engrs; Sigma Xi. *Res:* Ion-electron recombination processes; gaseous electronics; low energy plasmas; atomic and molecular collision processes; high energy lasers; multiphoton spectroscopy; isotope separation; ultrashort wavelength lasers. *Mailing Add:* Dept Physics Univ Tex Dallas PO Box 830688 Richardson TX 75080-0688

COLLINS, CAROL DESORMEAU, b Schenectady, NY, Dec 9, 54; m 78; c 2. ALGAL ECOLOGY, MATHEMATICAL MODELING. *Educ:* Univ Vt, BS, 76; Rensselear Polytech Inst, MS, 78, PhD(biol), 80. *Prof Exp:* Teaching assoc, Rensselear Polytech Inst, 80-81; SR SCIENTIST, NY STATE SCI SERV, 81- *Concurrent Pos:* Consult, Tennessee Valley Authority, 80 & Waterways Exp Sta, US Army Corps Engrs, 80-81; bd trustees, Lake George Asn Res Fund, 82-; lectr, Rensselear Polytech Inst, 80-82, adj asst prof, 81-82. *Mem:* Am Soc Limnol & Oceanog; Phycol Soc Am; Ecol Soc Am; Am Soc

Microbiol. *Res:* Laboratory and field studies on the effects of environmental factors and their interactions on algal and macrophyte ecology, used in the development of mathematical simulation models of aquatic ecosystems; lake restoration. *Mailing Add:* 29 Schuyler Hills Rd Loudonville NY 12211

COLLINS, CAROL HOLLINGWORTH, b Lowell, Mass, Mar 21, 31; m 55; c 1. RADIOANALYTICAL CHEMISTRY, CHROMATOGRAPHY. *Educ:* Bates Col, BS, 52; Iowa State Univ, PhD(phys & org chem), 58. *Prof Exp:* Res scientist, Brookhaven Nat Lab, 62-63; sr res scientist, Western New York Nuclear Res Ctr, 64-67; asst prof org chem, State Univ NY Buffalo, 67-68; res scientist, Starks Assocs, 68-72; cancer res scientist, Roswell Park Mem Inst, 72-74; PROF ANALYTICAL CHEM, UNIV ESTADUAL DE CAMPINAS, 74- *Mem:* Soc Brasileira Progresso Ciencia; Soc Brasileira Quimica. *Res:* Chromatography applied to analyses of trace quantities of components generally in the presence of large quantities of similar species, using radio analytical determinations. *Mailing Add:* UNICAMP-Quimica-CP 6154 Campinas SP 13081 Brazil

COLLINS, CAROLYN JANE, b White Plains, NY, Sept 18, 42. TUMOR VIRUSES, CELL TRANSFORMATION. *Educ:* Skidmore Col, BA, 64; Duke Univ, PhD(biochem), 70. *Prof Exp:* DNA tumor viruses, Ger Ctr Cancer Res, 70-72; fel, Dept Microbiol, Univ Mich, Ann Arbor, 73-76; asst prof microbiol, Univ Va, 76-82; staff scientist, Dana Farber Cancer Inst, 82-88; CONSULT, 88- *Concurrent Pos:* F G Novy fel, Univ Mich, Ann Arbor, 73-74, USPHS fel, 75-76; prin invest-grant, Nat Cancer Inst, 78-81. *Mem:* Am Soc Microbiol; Sigma Xi; NY Acad Sci; AAAS. *Res:* Molecular biology of tumor viruses; mechanism of viral integration and host cell transformation; molecular basis of gene regulation and control of gene expression. *Mailing Add:* 20 Prescott St No 41 Cambridge MA 02138

COLLINS, CARTER COMPTON, b San Francisco, Calif, Aug 3, 25; m 61; c 2. BIOPHYSICS. *Educ:* Univ Calif, BS, 49, MS, 53, PhD(biophys), 66. *Prof Exp:* Consult biomech group, Med Ctr, Univ Calif, San Francisco, 50-54, assoc engr, Res & Develop Labs, 55, dir, 59-63; res mem, Inst Visual Sci, Pac Med Ctr, San Francisco, 63-69; assoc prof visual sci, Univ Pac, 69-71; SR SCIENTIST, SMITH-KETTLEWELL INST VISUAL SCI, INST MED SCI, 63-; SR SCIENTIST, SMITH-KETTLEWELL EYE RES INST, 88- *Concurrent Pos:* Res & develop engr, Donner Sci Co, 53; pres, Sutter Instruments, Inc, 53-; Consult, Neurosurg Inst, Mt Zion Hosp, 54 & Dept Physiol, Univ Pa, 57-58; lectr, Cardiovasc Res Inst, 59; mem armed forces comt vision, Nat Acad Sci-Nat Res Coun; Wm A Kettlewell Res Chair Opthal, 80-; mem Sci Adv Comt, Nat Inst Neurol & Commun Disorders & Stroke; assoc dir, Smith-Kettlewell Inst Visual Sci, Inst Med Sci, 75-; Fleischman Found Grantee, NIH & NSF, 69-84; Erskine fel, 77. *Honors & Awards:* Hektoen Silver Medal Award, AMA, 72, Hektoen Bronze Medal Award, 76. *Mem:* AAAS; NY Acad Sci; Asn Res Ophthal; Inst Elec & Electronics Eng; Sigma Xi; Am Asn Artificial Intel. *Res:* Conception and design of instrumentation for extracting information from physical and biological systems; basic correlates of recognition; neural information transfer codes; control of eye movements; vision substitution by tactile TV image projection. *Mailing Add:* 2232 Webster St San Francisco CA 94115

COLLINS, CHARLES THOMPSON, b Long Branch, NJ, Mar 9, 38; m. ZOOLOGY, ORNITHOLOGY. *Educ:* Amherst Col, AB, 60; Univ Mich, MS, 62; Univ Fla, PhD(zool), 66. *Prof Exp:* Am Mus Natural Hist Chapman res fel, 66-67; asst prof biol, Fairleigh Dickinson Univ, Florham-Madison Campus, 67-68; from asst prof to assoc prof, 68-77, PROF BIOL, CALIF STATE UNIV, LONG BEACH, 77- *Concurrent Pos:* Fulbright res scholar, India, 74-75. *Mem:* Fel Am Ornith Union; Wilson Ornith Soc; Cooper Ornith Soc; Am Inst Biol Sci; Asn Trop Biol; Sigma Xi. *Res:* Ornithology, particularly biology and ecology of swifts. *Mailing Add:* Dept Biol Calif State Univ Long Beach CA 90840

COLLINS, CLAIR JOSEPH, physical organic chemistry; deceased, see previous edition for last biography

COLLINS, CLIFFORD B, b Can, Nov 12, 16; nat US; m 42; c 3. PHYSICS. *Educ:* Univ Western Ont, BSc, 47; McMaster Univ, MA, 48; Univ Toronto, PhD(physics), 51. *Prof Exp:* Res assoc physics, Univ Toronto, 51-52; res assoc, Res Labs, 52-56, physicist lamp develop, 56-66, GROUP LEADER LAMP DEPT, GEN ELEC CO, 66- *Mem:* Am Phys Soc; Electrochem Soc. *Res:* Lamp development; materials liaison; technical planning. *Mailing Add:* RR 8 No 300 Mocksville NC 27028

COLLINS, CURTIS ALLAN, b Des Moines, Iowa, Sept, 40; m 62; c 2. PHYSICAL OCEANOGRAPHY. *Educ:* US Merchant Marine Acad, BS, 62; Ore State Univ, PhD(oceanog), 67. *Prof Exp:* Res scientist, Pac Oceanog Group, Nanaimo, BC, 68-70; sr tech adv oceanog, Cities Serv Oil Co, Okla, 70-72; prog mgr environ forecasting, Off Int Decade Ocean Explor, NSF, 72-80, prog mgr ocean dynamics, Ocean Sci, 80-87; PROF & CHAIR, DEPT OCEANOG, NAVAL POSTGRAD SCH, 87- *Mem:* Am Geophys Union; Oceanog Soc; Am Meteorol Soc. *Res:* Descriptive physical oceanography of the Pacific Ocean. *Mailing Add:* Code OC/CO Naval Postgrad Sch Monterey CA 93943-5100

COLLINS, DAVID ALBERT CHARLES, b Ottawa, Ont, Feb 11, 52; m 77; c 2. OIL RESERVOIR SIMULATION, NUMERICAL METHODS. *Educ:* Univ NB, BSc, 73; McGill Univ, MSc, 77, PhD(appl math), 83. *Prof Exp:* Res scientist, Computer Modelling Group, 82-84, simulation scientist, 84-86, sr simulation scientist, 86-88, coordr-black oil simulation, 86-90, STAFF SCIENTIST, COMPUTER MODELLING GROUP, 88- *Mem:* Soc Petrol Engrs; Soc Indust & Appl Math; Can Inst Mining. *Res:* Numerical techniques for oil reservoir simulation, including modelling of horizontal wells, multigrad and domain decomposition methods, hybrid grids, local grid refinement and adaptive implicit methods. *Mailing Add:* Computer Modelling Group 3512 33rd St NW Calgary AB T2L 2A6 Can

COLLINS, DEAN ROBERT, b Ft Sill, Okla, Dec 8, 35; m 62; c 3. NEUVAL NETWORKS, OPTICAL PROCESSING. *Educ:* Mass Inst Technol, SB & SM, 59; Univ Ill, Urbana, PhD(elec eng), 67. *Prof Exp:* Coop eng studies, Int Bus Mach Corp, 56-58; mem tech staff, Bell Tel Labs, 59-60 & 62-63; asst prof elec eng, Va Polytech Inst, 61-62; res assoc, Univ Ill, 63-66; mem tech staff, Semiconductor Res & Develop Lab, 66-69, Cent Res Labs, 69-73, br mgr, CCD Technol Br, 73-77, dir, CCD Technol Lab, 77-79, dir, Interface Technol Lab, 79-81, dr, Advan Technol Assessment, 81-84, DIR, SYST COMPONENTS LAB, TEX INSTRUMENTS, INC, 84- *Mem:* Sr mem Inst Elec & Electronics Engrs; Am Phys Soc; Electrochem Soc; Soc Info Display; Am Asn Artificial Intel; Int Neural Network Soc. *Res:* Bucket brigade shift registers; current gain degradation; effect of gold on the silicon-silicon dioxide interface; insulating thin films; charge coupled device imagers; memory and signal processing; liquid crystal displays, electroluminescent displays; printers; expert systems; gallium arsenide technology; optical signal processing; neural networks. *Mailing Add:* MS 134 Syst Components Lab Tex Instruments Dallas TX 75265

COLLINS, DELWOOD C, b Cairo, Ga, Oct 7, 37; div; c 3. BIOCHEMISTRY, ENDOCRINOLOGY. *Educ:* Emory Univ, AB, 59; Univ Ga, MS, 63, PhD(endocrinol, physiol), 66. *Prof Exp:* Res scientist, Div Pharmacol, Food & Drug Directorate, Ottawa, Ont, 67-68; res assoc steroid biochem, Univ Ottawa, 68-69; asst prof med & instr biochem, 69-72, assoc prof med, 69-77, assoc dir, training prog endocrinol & metab, 78-85, PROF MED, EMORY UNIV, 78-, ASSOC PROF BIOCHEM, 79-; COLLAB SCIENTIST, YERKES PRIMATE CTR, 71- *Concurrent Pos:* Squibb Ayrest traveling fel, 68; Nat Inst Arthritis, Metab & Digestive Dis career develop award, 72-77; assoc career scientist, Vet Admin, 73-; res consult, ORTHO Pharmaceut, 74-; mem Reproductive Biol Study Sect, NIH, 77-79; adj prof biol, Md Sch Emory, 82-; collab scientist, Yerks Primate Ctr, 71- *Mem:* AAAS; Soc Study Reproduction; Endocrine Soc; Can Biochem Soc; Am Chem Soc; Am Physiol Soc; Am Soc Primatology; Am Asn Clin Chem; NY Acad Sci; Am Asn Univ Prof; Sigma Xi; Am Fertil Soc; Soc Exp Biol & Med. *Res:* Metabolism and conjugation of steroids; radioimmunoassay of steroids; steroidogenesis in the ovary and testis; reproductive physiology of primates; physiological effects of estrogen; hormonal aspects of behavior; estrogens in disease; mechanisms for control of ovulation and spermatogenesis. *Mailing Add:* Med Res Serv Vet Admin Med Ctr Emory Univ Med Sch 1670 Clairmont Ave Decatur GA 30033

COLLINS, DESMOND H, b Daylesford, Australia, July 15, 38; m 64; c 3. PALEONTOLOGY. *Educ:* Univ Western Australia, BSc, 60; Univ Iowa, PhD(geol), 66. *Prof Exp:* Tech officer, Geol Surv Can, 64-65; sr res fel paleont, Brit Mus, London, 66-68; CUR INVERT PALEONT, ROYAL ONT MUS, 68- *Concurrent Pos:* Assoc prof zool, Univ Toronto, 70-; res assoc, McMaster Univ, 71-85. *Mem:* Am Paleont Soc; Brit Palaeont Asn; Int Paleont Asn; Geol Asn Can. *Res:* Paleozoic nautiloid cephalopod systematics and shell function; burgess shale fossils. *Mailing Add:* Invert Palaeont Royal Ont Mus 100 Queen's Park Toronto ON M5S 2C6 Can

COLLINS, DON DESMOND, b Cardwell, Mont, Jan 10, 34; m 72; c 3. PLANT ECOLOGY, PLANT GENETICS. *Educ:* Mont State Univ, BS, 61, PhD(genetics), 65. *Prof Exp:* From asst prof to assoc prof, 65-78, PROF ECOL, MONT STATE UNIV, 78- *Concurrent Pos:* Nat Park Serv grant, 67-68; NSF grant, 68-73, 78; Int Biol Prog grant, 69-72; consult, Grassland Info Synthesis Proj, Int Biol Prog, Grassland Biome, 68-72; ecologist, pvt indust, 80- *Mem:* Ecol Soc Am; Bot Soc Am; Nat Parks Asn. *Res:* Ecological research on effects of weather modification and vegetational response to campground usage; mountain grasslands; impact of pipeline systems. *Mailing Add:* Dept Bot Mont State Univ Bozeman MT 59717

COLLINS, EDWARD A, b Winnipeg, Man, May 22, 28; US citizen; m 52; c 3. PHYSICAL CHEMISTRY, POLYMER CHEMISTRY. *Educ:* Univ Man, BS, 50, MS, 52, PhD(phys chem), 67. *Prof Exp:* Lectr phys chem, Royal Mil Col, Ont, 51-52; develop chemist, 52-54, assoc develop scientist, 56-59, develop scientist, 59-62, sr scientist, 62-65, develop consult, B F Goodrich Chem Co, 65-78; assoc dir corp res, Diamond Shamrock Corp, 78-; AT MITECH CORP. *Concurrent Pos:* Adj prof, Rensselaer Polytech Inst, 68-78, Cleveland State Univ, 69-79 & Case Western Reserve Univ, 75-82. *Mem:* Am Chem Soc; Soc Rheol; Brit Soc Rheol. *Res:* Polymer characterization and relation of molecular structure to physical and mechanical properties; rheology. *Mailing Add:* 255 Glenview St Avon Lake OH 44012-1195

COLLINS, EDWIN BRUCE, b Conway, SC, Aug 5, 21; m 53. FOOD MICROBIOLOGY. *Educ:* Clemson Col, BS, 43; Iowa State Col, MS, 48, PhD(dairy bact), 49. *Prof Exp:* From instr dairy indust & jr dairy bacteriologist to assoc prof dairy indust & assoc dairy bacteriologist, 49-64, PROF FOOD SCI, UNIV CALIF, DAVIS & DAIRY BACTERIOLOGIST, EXP STA, 64- *Mem:* Am Soc Microbiol; Am Dairy Sci Asn; Sigma Xi. *Mailing Add:* Dept Food Sci & Technol Univ Calif Davis CA 95616

COLLINS, ELLIOTT JOEL, b New York, NY, June 12, 19; m 47; c 2. ENDOCRINOLOGY. *Educ:* Col Charleston, BS, 49; Princeton Univ, MA, 51, PhD(biol), 52. *Prof Exp:* Res scientist, Upjohn Co, 52-62; mgr regulatory affairs, Schering Corp, 62-84; RETIRED. *Mem:* Soc Exp Biol & Med; Endocrine Soc; Sigma Xi. *Res:* Endocrine physiology; metabolic bone disease; connective tissue disease. *Mailing Add:* 155 Rte 24 Mendham NJ 07945

COLLINS, F(RED) R(OBERT), b US, July 19, 26; m 50; c 5. ELECTRICAL ENGINEERING. *Educ:* Brown Univ, BS, 47. *Prof Exp:* Sect head, Alcoa Labs, Aluminum Co Am, Alcoa Conductor Prods Co, 58-62, asst div, chief, 62-67, chief elec eng div, 67-78, dir res & eng, 75-78, vpres tech, Div Alcoa, 78-84; RETIRED. *Mem:* Inst Elec & Electronics Engrs; Am Welding Soc. *Res:* Aluminum reduction; welding and brazing; soldering aluminum; aluminum transmission and distribution conductors; connectors and other products used by the electrical power industry. *Mailing Add:* 2481 S Avenida Loma Linda 510 One Allegheny Sq Green Valley AZ 85614

COLLINS, FRANCIS ALLEN, b Wichita, Kans, July 24, 31; m; c 3. AIRBORNE RADAR SYSTEMS. *Educ:* Univ Tex, BS, 54, MA, 57; Harvard Univ, PhD(appl physics), 64. *Prof Exp:* Physicist, Defense Res Labs, Univ Tex, 53-58; asst prof optics, Univ Rochester, 64-69; mem staff, Appl Res Labs, Univ Tex Austin, 69-77 & Hughes Aircraft Co, 77-80; MEM STAFF, TEX INSTRUMENTS, INC, 80- *Mem:* Am Phys Soc; Inst Elec & Electronics Engrs. *Res:* Airborne radar system conceptual design from user requirements and specifications. *Mailing Add:* 4512 Lone Grove Lane Plano TX 75093

COLLINS, FRANCIS SELLERS, b Staunton, Va, Apr 14, 50; m; c 2. MOLECULAR GENETICS. *Educ:* Univ Va, BS, 70; Yale Univ, MS, 72, PhD, 74; Univ NC, MD, 77. *Hon Degrees:* DSc, Emory Univ, 90; LHD, Mary Baldwin Col, 91. *Prof Exp:* Fel genetics & biochem, Sch Med, Univ NC, 76-77; intern med, NC Mem Hosp, Chapel Hill, 77-78, asst resident, 78-80, chief resident, 80-81; fel human genetics & pediat, Sch Med, Yale Univ, 81-84; from asst prof to assoc prof internal med & human genetics, Med Sch, Univ Mich, 84-91, chief, Div Med Genetics, Dept Internal Med, 87-91; asst investr, 87-88, assoc investr, 88-91, INVESTR, HOWARD HUGHES MED INST, 91-; PROF INTERNAL MED & HUMAN GENETICS, MED SCH, UNIV MICH, ANN ARBOR, 91- *Concurrent Pos:* Charles E Culpepper Found fel, 83-84; Cooley's Anemia Found fel, 83-84; consult molecular genetics, Genelabs Inc, Redwood City, Calif, 84-; Anthony Renda res grant, 84-85; Hartford Found fel, 85-87; NIH grants, 85-; assoc ed, Am J Human Genetics, 86-89, Genomics, 86-, Technique, 88-, Genes, Chromosomes & Cancer, 89- & Somatic Cell & Molecular Genetics, 89-; mem sci adv bd, Hereditary Dis Found, 87-, Nat Neurofibromatosis Found, 88-; dir, Neurofibromatosis Ctr, Univ Mich Med Ctr, 87-; vchmn, Gordon Conf on Molecular Genetics, 89, chmn, 91; co-dir, Ctr Molecular Genetics, 90-91. *Honors & Awards:* Jerome Conn Res Award, 86; Paul di Sant'Agnese Award, Cystic Fibrosis Found, 89; James A Shannon Lectr, Mass Gen Hosp, 90; Gairdner Found Int Award, 90; Von Recklinghausen Award, Nat Neurofibromatosis Found, 90; Young Investr Award, Am Fedn Clin Res, 91; Doris Tulcin Award for Cystic Fibrosis Res, 91. *Mem:* Inst Med-Nat Acad Sci; AAAS; Am Soc Human Genetics; Am Fedn Clin Res; Am Soc Microbiol; Am Soc Hemat; Am Soc Clin Invest; Human Genome Orgn. *Res:* Hereditary diseases; human genetics. *Mailing Add:* Howard Hughes Med Inst 1150 W Medical Center Dr 4570 MSRB II Ann Arbor MI 48109-0650

COLLINS, FRANK CHARLES, b Marton, NZ, Sept 18, 11; nat US; m 41, 87; c 1. PHYSICAL CHEMISTRY. *Educ:* Univ Calif, AB, 39; Columbia Univ, AM, 47, PhD(chem), 49. *Prof Exp:* Chemist, Shell Develop Co, Calif, 39-46; lectr anal chem, Columbia Univ, 47; asst prof, Polytech Inst NY, 48-51, prof phys & environ chem, 51-77, adj prof, 77-80; RETIRED. *Concurrent Pos:* Consult, Oil, Chem & Atomic Workers Int Union, 75-81 & Jour, 82- *Mem:* Am Chem Soc; AAAS; Fedn Am Scientists. *Res:* Energy and public policy; nuclear policy problems; occupational health and safety. *Mailing Add:* 15108 Alaska Rd Woodbridge VA 22205-0224

COLLINS, FRANK GIBSON, b Chicago, Ill, Feb 20, 38; m 60; c 2. FLUID MECHANICS, AERODYNAMICS. *Educ:* Northwestern Univ, BS, 61; Univ Calif, Berkeley, PhD(mech eng), 68. *Prof Exp:* Res asst aeronaut sci, Univ Calif, Berkeley, 62-68; asst prof Dept Aeronaut Eng & Eng Mech, Univ Tex, Austin, 68-74; assoc prof, 74-81, PROF AERONAUT & MECH ENG, UNIV TENN SPACE CTR, 81- *Concurrent Pos:* NSF res grant, 74-75; USAF/ASEE grant, 77; consult, Eng Res Corp, Inc, & FWG Assoc, Inc, Tullahoma, Tenn, 78-; fac, USAF-ASEE Res Prof, 83, NASA Summer Fac Res prog, 83, 85 & 86; mem, Am Inst Aeronaut & Astronaut processing Tech Comt, 84- *Mem:* Sigma Xi; Am Inst Aeronaut & Astronaut; Am Phys Soc; Am Soc Mech Engrs; assoc fel, Am Inst Aeronaut & Astronaut. *Res:* Boundary layer stability, material processing in space, transonic viscous interactions, stall suppression, kinetic theory, development of gas diagnostic techniques, heat transfer. *Mailing Add:* Space Inst Univ Tenn Tullahoma TN 37388

COLLINS, FRANK MILES, b Adelaide, S Australia, Mar 30, 28; US citizen; m 52; c 2. MICROBIOLOGY, IMMUNOLOGY. *Educ:* Univ Adelaide, S Australia, BSc, 49, PhD(immunochem), 61, DSc, 75. *Prof Exp:* Asst prof bacteriol, 54-60, asst prof microbiol, Univ Adelaide, 61-64; assoc mem, 65-68, MEM, TRUDEAU INST INC, 68- *Concurrent Pos:* Counr-at-large, Am Thoracic Soc, 71-74; mem bacteriol & mycol study sect, NIH, 75-79 & 87-91; mem, US Tuberc Panel, 80-86, Chmn, 87- *Mem:* Am Soc Microbiol; Am Thoracic Soc; fel Am Acad Microbiol; Soc Gen Microbiol; Am Asn Immunol; hon mem Leukocyte Biol Soc. *Res:* Mechanisms of immune response to infections caused by microbial intracellular parasites in mouse models of human disease; cellular immunity to mycobacteria, salmonellae and listeria infections in mice. *Mailing Add:* Trudeau Inst Inc PO Box 59 Saranac Lake NY 12983

COLLINS, FRANK WILLIAM, JR, b Toronto, Ont, Mar 31, 45; div; c 2. PLANT BIOCHEMISTRY, BOTANY. *Educ:* Univ Toronto, BSc, 66, PhD(bot), 72. *Prof Exp:* Fel, 72-73, res assoc, Dept Bot, Univ BC, 73-75; res assoc biochem, Inst Environ Studies, Univ Toronto, 75-77, lectr Dept Bot, 76-77; RES SCIENTIST BOT, 77-, RES SCIENTIST FOOD BIOCHEM, RES BR AGR, CANADA, 82-,. *Concurrent Pos:* Consult, NAm Contact Dermatitis Group, 73-75, Working Group Environ Aspects Heavy Metals, Nat Res Coun, 76-77. *Mem:* N Am Phytochem Soc. *Res:* Biochemistry of food processing with respect to phenolics, proteins, enzymes, terperoids, carotenoids, lipids; natural product chemistry of cereal grains wheat rye barley oats. *Mailing Add:* Food Res Inst Res Br Agr Can Cent Exp Farm Ottawa ON K1A 0C6 Can

COLLINS, FRANKLYN, b Toronto, Ont, Mar 13, 29; nat US; m 51; c 3. SOLID STATE PHYSICS. *Educ:* Mich State Col, BS, 50; Univ Buffalo, PhD(physics), 58. *Prof Exp:* mgr electronics res, res & develop labs, Airco Speer, 57-; PRES, OHMTEK. *Mem:* Am Phys Soc; Inst Elec & Electronics Engrs. *Res:* Carbon and graphite research; electronic components. *Mailing Add:* Ohmtek 2160 Liberty Dr Niagara Falls NY 14304

COLLINS, FREDERICK CLINTON, b Prairie Grove, Ark, May 31, 41; m 61; c 2. PLANT BREEDING, PLANT GENETICS. *Educ:* Univ Ark, Fayetteville, BSA, 63, MS, 65; Purdue Univ, PhD(plant breeding, genetics), 69. *Prof Exp:* Res asst agron, Univ Ark, 63-64; geneticist sugarbeet invests, Agr Res Serv, USDA, 64-66; from asst prof to prof agron, Univ Ark, Fayetteville, 69-83; WHEAT BREEDER/RES CTR MGR, NORTHRUP KING CO, BAY, ARK, 83- *Mem:* Am Soc Agron; Crop Sci Soc Am. *Res:* Plant breeding; genetic control and physiology of yield and chemical characteristics, particularly those involved with nutritional quality of plants. *Mailing Add:* Northrup King Co Box 729 Bay AR 72411

COLLINS, GALEN FRANKLIN, b Winona Lake, Ind, Dec 29, 27; m 56; c 4. CLINICAL DIAGNOSTICS, PHARMACEUTICAL DEVELOPMENT. *Educ:* Purdue Univ, BS, 49, MS, 52, PhD(pharm chem), 54. *Prof Exp:* Asst pharm chem, Purdue Univ, 49-52; pharm chemist, Miles Lab, Inc, 53-58, asst to dir, Miles-Ames Pharm Res Labs, 58-59, sect head, Ames Prods, 59-60; chief prod dir, Norwich Pharmacol Co, 60-63; mgr res, S E Massengill Co, 63-67, dir res, 67-71; vpres res & develop, Am Dade Div-Am Hosp Supply Corp, 71-75, vpres sci div, 75-82; MGR RES & DEVELOP, C B FLEET CO, INC, 82- *Mem:* Am Chem Soc; Am Pharmaceut Asn; Soc Cosmetic Chemists; Am Inst Chemists; AAAS; Sigma Xi. *Mailing Add:* 1431 Club Dr Lynchburg VA 24503

COLLINS, GARY BRENT, b Clare, Mich, Sept 23, 40; m 62; c 2. PHYCOLOGY, AQUATIC ECOLOGY. *Educ:* Cent Mich Univ, BA, 62, MA, 64; Iowa State Univ, PhD(bot), 68. *Prof Exp:* Asst prof bot, Ohio State Univ, 68-72; RES AQUATIC BIOLOGIST, US ENVIRON PROTECTION AGENCY, 72- *Concurrent Pos:* Adj prof biol, Univ Cincinnati, 74- *Mem:* AAAS; Am Micros Soc; Int Phycol Soc; Phycol Soc Am; NAm Benthological Soc. *Res:* Freshwater diatom ecology and taxonomy; plankton and periphyton methods development; algal ecology; quality assurance of biological methods. *Mailing Add:* 8109 Woodruff Rd Cincinnati OH 45255

COLLINS, GARY SCOTT, b Plainfield, NJ, Dec 15, 44; m 78; c 1. LOCAL ATOMIC & ELECTRONIC STRUCTURE OF SOLIDS. *Educ:* Rutgers Col, BA(physics), 66; Rutgers Univ, PhD(physics), 76. *Prof Exp:* Instr, Dept Physics, Rutgers Univ, 76-77; res assoc & instr, Dept Physics, Clark Univ, 77-79, vis res asst prof, 79-85; ASSOC PROF PHYSICS, WASH STATE UNIV, 85- *Concurrent Pos:* Prin investr, NSF Grants, 80-86, 87-90, 90-93, & 90-92; co-prin investr, NSF grants, 80-83 & 83-86; assoc investr, NSF grant, 90-93. *Mem:* AAAS; Am Phys Soc; Mat Res Soc; Metall Soc. *Res:* Local atomic and electronic structure of solids studied using techniques of nuclear solid state physics; perturbed gamma-gamma; angular correlations; Moessbauer effect and positron; annihilation spectroscopies; defects in medals and alloys; nanoclusters. *Mailing Add:* Dept Physics Wash State Univ Pullman WA 99164-2814

COLLINS, GEORGE BRIGGS, b Washington, DC, Jan 3, 06; m 34; c 3. PHYSICS. *Educ:* Johns Hopkins Univ, PhD(physics), 31. *Prof Exp:* Instr physics, Johns Hopkins Univ, 27-30; instr, Univ Notre Dame, 33-41; mem radiation lab, Mass Inst Technol, 41-46; prof physics & chmn dept, Univ Rochester, 46-50; chmn cosmotron dept, Brookhaven Nat Lab, 50-62, sr physicist, 62-71; prof physics & sr physicist, Va Polytech Inst & State Univ, 71-76, Univ Distinguished Emer Prof Physics, 77-81; RETIRED. *Concurrent Pos:* Coun mem, State Univ NY Stony Brook, 53-70; Fulbright fel, Belg, 57; with Deutsches Elektronen-Synchrotron, 65-66; chmn, Comt on Lib Educ & Professions, Va Polytech Inst & State Univ, 80-81. *Mem:* Fel Am Phys Soc. *Res:* Raman effect; hyperfine structure of iodine; far ultraviolet spectroscope; excitation of nuclei by electrons; nuclear physics; high energy electrons and protons; particle physics; development of wire chamber spectrometers; high energy multiparticle production. *Mailing Add:* 1380 Locust Ave Blacksburg VA 24060

COLLINS, GEORGE EDWIN, b Stuart, Iowa, Jan 10, 28; m 54; c 3. COMPUTER ALGEBRA, ANALYSIS OF ALGORITHMS. *Educ:* Univ Iowa, BA, 51, MS, 52; Cornell Univ, PhD(math), 55. *Prof Exp:* Res mathematician, Int Bus Mach Corp, 55-66; assoc prof, Univ Wis-Madison, 66-68, chmn dept, 70-72, prof comput sci, 68-86; PROF COMPUT SCI & MATH, OHIO STATE UNIV, 86- *Concurrent Pos:* Res fel math, Calif Inst Technol, 63-64; vis prof, Stanford Univ, 72-73, Univ Kaiserslautern, 74-75 & Univ Karlsruhe, 78; ed, Soc Indust & Appl Math J on Comput, 77-81; vis res fel, Gen Elec Res Ctr, 81-82; vis distinguished prof, Univ Del, 85-86. *Mem:* Asn Comput Mach; Am Math Soc; Math Asn Am; Soc Indust & Appl Math. *Res:* Algebraic algorithms; computer algebra systems; polynomial factorization; exact polynomial root calculation; quantifier elimination for the theory of real closed fields; cylindrical algebraic decomposition; resultants and subresultants. *Mailing Add:* Dept Comput Sci Ohio State Univ 2036 Neil Ave Mall Columbus OH 43210

COLLINS, GEORGE H, b Albany, NY, Sept 11, 27; m 51; c 6. PATHOLOGY, NEUROLOGY. *Educ:* Univ Vt, AB, 49, MD, 53. *Prof Exp:* From asst prof to prof path & med, Col Med, Univ Fla, 62-73; PROF PATH, COL MED, STATE UNIV NY UPSTATE MED CTR, 73- *Concurrent Pos:* Teaching fel neuropath, Harvard Med Sch, 58-59, res fel, 59-62; clin & res fel, Mass Gen Hosp, 59-62; asst resident path, Mass Gen Hosp, 58-59. *Mem:* AAAS; Am Asn Neuropath; Am Asn Pathol; Soc Neurosci. *Res:* Degeneration and regeneration in the central nervous system; glial cell function. *Mailing Add:* Dept Path State Univ NY Health Sci Ctr Syracuse NY 13210

COLLINS, GEORGE W, II, b Waukegan, Ill, July 18, 37; m 61; c 2. ASTRONOMY. *Educ:* Princeton Univ, AB, 59; Univ Wis, PhD(astron), 62. *Prof Exp:* Asst prof numerical anal, Univ Wis, 62-63; from asst prof to assoc prof, 63-69, PROF ASTRON, OHIO STATE UNIV, 69- *Concurrent Pos:* Res fel, Univ Sussex, 82, 89 & Univ Glasgow, 89. *Mem:* Fel AAAS; Am Astron Soc; fel Royal Astron Soc; Int Astron Union. *Res:* Eclipsing binary stars; stellar atmospheres; general problems in radiative transfer; numerical analysis. *Mailing Add:* 6143 Middlebury Dr W Worthington OH 43085

COLLINS, GLENN BURTON, b Folsom, Ky, Aug 7, 39; m 59; c 2. PLANT GENETICS, PLANT TISSUE & CELL CULTURE. *Educ:* Univ Ky, BS, 61, MS, 63; NC State Univ, PhD(genetics), 67. *Prof Exp:* Res asst plant genetics, Univ Ky, 61-63; plant cytogenetics, NC State Univ, 63-66; from asst prof to assoc prof, 66-70, assoc dean res, 85-87, PROF SOMATIC CELL GENETICS & CYTOGENETICS, UNIV KY, 75- *Concurrent Pos:* Sabbatical, John Innes Inst, Norwich, Eng, 73. *Honors & Awards:* Cooper Res Award, Col Agr, Univ Ky, 75; Univ Ky Res Found Award, 76; Philip Morris Res Award, Philip Morris Co, 80; Res Award, Crop Sci Soc, 85; Philip Morris Res Proj, 81 & 88. *Mem:* Fel Am Soc Agron; Am Genetics Soc; Tissue Culture Asn Am; Int Asn Plant Tissue & Cell Cult; Int Plant Molecular Biol Asn; fel Crop Sci Soc Am. *Res:* Genetics and breeding of tobacco; development and use of haploid procedures for breeding and genetic studies; use of tissue and cell culture methods in plant improvement; wide hybridization and gene transfer in crop plants; cellular and molecular transformation systems for crop plants. *Mailing Add:* N-222 Agr Sci BLD-N Univ Ky Lexington KY 40546-0091

COLLINS, H DOUGLAS, b Caribou, Maine, Jan 19, 28; m 50, 87; c 3. MEDICINE. *Educ:* Univ Maine, BA, 49; Harvard Univ, MD, 52; Am Bd Internal Med, dipl, 61. *Prof Exp:* Intern, Mass Gen Hosp, 52-53, asst resident med, 53-54, resident, 54-55; sr asst surg, USPHS, 55-57; asst resident med, Mass Gen Hosp, 72-73, staff, Dir Patient Care, 74-87; RETIRED. *Concurrent Pos:* Pvt pract gen internal med, Caribou, Maine, 57-72, 73-75 & 80-84, group pract, 84-87; dir, Central Maine Family Pract Residency, 75-79, Maine-Dartmouth Family Pract Residency, 79-80; gov, Am Bd Internal Med, 88-91. *Mem:* Inst Med-Nat Acad Sci; AMA; Am Col Physicians. *Mailing Add:* RFD Box 2179 Kingfield ME 04947

COLLINS, HENRY A, b Machipongo, Va, Sept 21, 32; m 62; c 3. AGRONOMY, PLANT PHYSIOLOGY. *Educ:* Md State Col, BS, 55; Rutgers Univ, MS, 57, PhD(farm crops), 62. *Prof Exp:* Asst farm crops, Rutgers Univ, 59-62; from asst prof to assoc prof biol, Tuskegee Inst, 62-68; res specialist, 68-73, mkt planning specialist, 73-77, RES & SR RES SPECIALIST, AGR DIV, CIBA-GEIGY CORP, 77- *Mem:* Am Soc Agron; Crop Sci Soc Am; Weed Sci Soc Am; Aquatic Plant Mgt Soc; Southern Weed Sci Soc. *Res:* Selective absorption of strontium by plants; factors affecting weed seed germination. *Mailing Add:* 3904 Hickory Tree Lane Greensboro NC 27405

COLLINS, HERON SHERWOOD, b Charlotte, NC, Nov 17, 22; m 52; c 4. MATHEMATICS. *Educ:* Wofford Col, BS, 48; Tulane Univ, MS, 50, PhD, 52. *Prof Exp:* Asst prof math, Univ Md, 52-53; asst prof, Univ SC, 53-54; from asst prof to assoc prof, 54-62, PROF MATH, LA STATE UNIV, 62- *Mem:* Am Math Soc. *Res:* Banach spaces and algebras; measure and integration theory; analytic function theory. *Mailing Add:* Dept Math La State Univ Univ Sta Baton Rouge LA 70803

COLLINS, HOLLIE L, b Laona, Wis, May 20, 38; m 57; c 3. TEACHING NATURAL HISTORY, RESEARCH FISHES & LEECHES. *Educ:* Wis State Univ, BS, 60; Mich State Univ, MS, 62, PhD(zool), 65. *Prof Exp:* PROF BIOL, UNIV MINN, DULUTH, 64-, DIR GRAD STUDIES BIOL, 70- *Mem:* Animal Behavior Soc; Am Fisheries Soc; Sigma Xi. *Res:* Behavior of aquatic organisms, especially social behavior of fishes; culture and biology of leeches. *Mailing Add:* Dept Biol Univ Minn Duluth MN 55812

COLLINS, HORACE RUTTER, b Shawnee, Okla, Feb 4, 30. GEOLOGY. *Educ:* Ohio Univ, BS, 54; WVa Univ, MS, 59. *Prof Exp:* Instr geol, Ohio Univ, 58-59; geologist, Coal Sect, Ohio Geol Surv, 59-60, head, 60-63, head regional sect, 63-66, asst state geologist, 66-68, asst chief Ohio Div, 67-68, state geologist & chief Ohio Div, 68-88; CONSULT GEOLOGIST, 88- *Concurrent Pos:* Vchmn, Coal div, Geol Soc Am, 71-72, chmn, 72-73; vpres Ohio sect, Am Inst Prof Geologists, 75, pres, 76; mem, Ohio Hazardous Waste Facil Approval Bd, 81-88, Ohio Low Level Radioactive Waste Adv Comt, 84-88. *Mem:* Geol Soc Am; Am Inst Prof Geologists; Asn Am State Geol; Am Inst Mining, Metall & Petrol Engrs; Am Asn Petrol Geologists. *Res:* Geology of Ohio; Pennsylvania stratigraphy; paleobotany; coal geology. *Mailing Add:* 1954 Milden Rd Columbus OH 43221

COLLINS, JACK A(DAM), b Columbus, Ohio, Nov 23, 29; m 58; c 4. MECHANICAL ENGINEERING. *Educ:* Ohio State Univ, BME, 52, MSc, 54, PhD(mech eng), 63. *Prof Exp:* From res asst to res assoc mech eng, Ohio State Univ, 52-63; assoc prof, Ariz State Univ, 63-72; assoc prof, 72-74, PROF MECH ENG, OHIO STATE UNIV, 74-, CHMN MECH DESIGN SECT OF MECH ENG DEPT, 75- *Concurrent Pos:* Consult, Babcock & Wilcox Res Ctr, Gen Elec Co, AiResearch Mfg Co, Worthington Industs & Owens/Corning Fiberglas. *Mem:* Am Soc Mech Engrs; Am Soc Eng Educ; Am Soc Testing & Mat; Soc Exp Stress Anal. *Res:* Experimental and analytical stress and deflection analysis; experimental and analytical failure analysis, including fatigue, creep, wear and fretting. *Mailing Add:* Dept Mech Eng Ohio State Univ 206 W 18th Ave Columbus OH 43210

COLLINS, JAMES FRANCIS, b Baltimore, Md, Jan 26, 42; m 69; c 2. ENVIRONMENTAL HEALTH. *Educ:* Loyola Col, Md, BS, 63; Univ NC, Chapel Hill, PhD(genetics), 68. *Prof Exp:* Fel, Nat Inst Arthritis, Metab & Digestive Dis, NIH, 68-70, staff fel, 70-72, sr staff fel, Nat Heart & Lung Inst, 73-75; res chemist, Audie L Murphy Vet Admin Hosp & asst prof med & biochem, Univ Tex Health Sci Ctr, San Antonio, 75-86, lectr life sci, Univ Tex, 83-85; STAFF TOXICOLOGIST, CALIF DEPT HEALTH SERV, BERKELEY, 86- *Concurrent Pos:* Consult, Biol & Chem Sci Degree Prog, Upward Mobility Col, Fed City Col Exp Prog, Washington, DC, 72; res preceptor, minority biol sci res prog, United Col San Antonio, 76-81, mem, adv bd, 76-78; instr, Univ Calif, Berkeley, Exten, 87- *Mem:* Am Soc Biochem & Molecular Biol. *Res:* Health effects of air pollution. *Mailing Add:* Calif Dept Health Serv ATES 2151 Berkeley Way Annex 11 Berkeley CA 94704

COLLINS, JAMES IAN, civil engineering, physical oceanography; deceased, see previous edition for last biography

COLLINS, JAMES JOSEPH, b Rochester, NY, Sept 5, 47; m 78; c 2. PUBLIC HEALTH & EPIDEMIOLOGY, DEMOGRAPHY. *Educ:* Loras Col, BA, 70; Univ Mo, Kansas City, MA, 76; Univ Ill, PhD(demog), 80. *Prof Exp:* Consult epidemiol, Argonne Nat Lab, 78-80, asst scientist, 80-82; MGR EPIDEMIOL, AM CYANAMID CO, 82- *Mem:* Pop Asn Am; Soc Epidemiol Res. *Res:* Development and evaluation of mathematical models for analysis of trends in disease incidence and mortality; epidemiology studies in occupational settings. *Mailing Add:* American Cyanamid/Lederle Labs Middletown Rd Bldg 28 Rm 102 Pearl River NY 10965

COLLINS, JAMES MALCOLM, b Atlanta, Ga, Mar 21, 38; m 59; c 2. BIOCHEMISTRY, MOLECULAR BIOLOGY. *Educ:* Univ Southern Miss, BS, 62; Univ Tenn, PhD(biochem), 68. *Prof Exp:* USPHS fel develop biol, Oak Ridge Nat Lab, 68-69; from asst prof to assoc prof, 69-79, PROF BIOCHEM, MED COL VA CAMPUS, VA COMMONWEALTH UNIV, 79- *Mem:* AAAS; Am Soc Biol Chemists; Am Chem Soc. *Res:* DNA synthesis; cell cycle; cancer. *Mailing Add:* 1607 Cloister Dr Richmond VA 23233

COLLINS, JAMES PAUL, b New York, NY, July 3, 47; m 70; c 2. ECOLOGY. *Educ:* Manhattan Col, BS, 69; Univ Mich, MS, 71, PhD(zool), 75. *Prof Exp:* From asst prof to assoc prof, 75-90, PROF ZOOL, ARIZ STATE UNIV, 90- *Concurrent Pos:* Chair, Dept Zool, 89- *Mem:* Ecol Soc Am; Am Soc Study Evolution; Am Soc Ichthyologists & Herpetologists; Sigma Xi; fel AAAS. *Res:* Investigation of the selective advantage of life history characters, especially in relation to the predictability of the organism's breeding habitat; the effect of competition and predation on life history characters. *Mailing Add:* Dept Zool Ariz State Univ Tempe AZ 85287-1501

COLLINS, JAMES R, MATHEMATICS. *Educ:* Univ Colo, Boulder, PhD(educ), 70. *Prof Exp:* Instr, Arapahoe Community Col, Littleton, Colo, 67-68; asst prof sch educ, Univ Mo, Kansas City, 72-74; from asst prof to assoc prof, 70-79, PROF COL EDUC, UNIV WYO, 79- *Mailing Add:* Dept Math Univ Wyo Col Educ Laramie WY 82071

COLLINS, JANET VALERIE, cell biology, insect physiology, for more information see previous edition

COLLINS, JEFFERY ALLEN, b Oakland, Calif, Feb 16, 44. INDUSTRIAL CHEMICALS. *Educ:* Purdue Univ, DVM, 67; Univ Houston, MBA, 80. *Prof Exp:* Vet lab animal med, Shell Develop Co, Div Shell Oil Co, Calif, 67-68; prod develop technologist, Shell Chem Co, New York, 68-70; staff vet anthelmintic prod, 70-72, supvr vet therapeut, Shell Develop Co, 72-76, BUS MGR INDUST CHEM, SHELL CHEM CO, DIV SHELL OIL CO, 76- *Concurrent Pos:* Consult parasitol, Am Asn Zoo Animal Vet, 73-; mkt res technologist, Shell Int Chem Co, Ltd, London, 73-74. *Mem:* Am Vet Med Asn; Am Chem Soc; Indust Vet Asn. *Res:* Veterinary parasitology; administrative role re all aspects of parasiticide research and product development; industrial chemicals-manufacturing, marketing, and technical support. *Mailing Add:* PO Box 1064 Kodiak AK 99615

COLLINS, JEFFREY JAY, b Monroe, La, May 12, 45; m 80. TOXICOLOGY, TUMOR BIOLOGY. *Educ:* Cornell Univ, Ithaca, NY, BS, 66; Harvard Univ, PhD(microbiol & molecular genetics), 72. *Prof Exp:* fel tumor virol, Imperial Cancer Res Fund Lab, Lincoln's Inn Fields, London, 72-74, vis scientist, 74; from asst prof to assoc prof exp surg, Duke Univ Med Ctr, 74-84, from asst prof to assoc prof microbiol & immunol, 75-84; proj officer & toxicologist, Nat Toxicol Prog, Nat Inst Environ Health Sci, 84-87; ASSOC DIR CLIN RES & ANTI-INFECTIVES, GLAXO, INC, 87- *Concurrent Pos:* Fac mem tumor virol, Duke Comprehensive Cancer Ctr, Durham, NC, 74-84; prin investr, Am Cancer Soc Res Grant, 77-84; mem & chmn, Am Cancer Soc Adv Comt Immunol & Immunol Therapy, 79-83; NIH fel; Procter & Gamble Univ Explor Res Award. *Mem:* Am Asn Immunologists; Am Soc Virol. *Res:* Use of viral structural proteins as specific antigenic targets for the passive serum therapy of virus-induced leukemias and sarcomas. *Mailing Add:* Glaxo Inc Five Moore Dr Research Triangle Park NC 27709

COLLINS, JERRY C, b Nashville, Tenn, Mar 19, 41. BIOMEDICAL ENGINEERING. *Educ:* Vanderbilt Univ, BE, 62; Purdue Univ, MS, 65; Duke Univ, PhD(elec eng), 70. *Prof Exp:* RES ASSOC PROF MED, MED CTR N, VANDERBILT UNIV, 87-; CONSULT, 89- *Concurrent Pos:* Mem, Basic Sci Coun, Am Heart Asn. *Mem:* Biomed Eng Soc; Am Heart Asn; Inst Elec & Electronics Engrs; Am Physiol Soc; Eng Med & Biol Soc; Am Math Soc; Sigma Xi. *Mailing Add:* Dept Med B-1318 Med Ctr N Vanderbilt Univ Nashville TN 37232

COLLINS, JERRY DALE, b Greeneville, Tenn, June 3, 43; m 61; c 2. ORGANIC CHEMISTRY, POLYMER CHEMISTRY. *Educ:* Tusculum Col, BS, 65; Univ Ark, Fayetteville, PhD(carbene chem), 70. *Prof Exp:* Res chemist, 70-71, spec process engr, 71-73, tech supvr nylon develop, Am Enka Div, Akzona Corp, Lowland, 73-77; specialist lexan prod develop, Lexan Div, Gen Elec Co, 77-; dir res, Develop & Eng, LST Corp, 79-80; pres, Jim Walter plastics div, Jim Walter Corp, 80-83; PRES, DIV N, RECTICEL FOAM CORP, 83- *Mem:* Am Chem Soc; Soc Plastics Engrs. *Res:* Steric effects associated with insertion of carbenes into carbon-hydrogen bonds; determination of insertion reaction mechanism; modification of fiber surface by carbene insertion reactions; new process development concerning nylon textile yarns. *Mailing Add:* Rte 5 Box 292 Greeneville TN 37743-9232

COLLINS, JIMMIE LEE, b Vicksburg, Miss, Nov 24, 34; m 59; c 3. FOOD SCIENCE. *Educ:* La State Univ, BS, 61; Univ Md, MS, 63, PhD(food tech), 65. *Prof Exp:* from asst prof to assoc prof, 65-77, PROF FOOD TECHNOL, UNIV TENN, KNOXVILLE, 77- *Concurrent Pos:* Prof activ, Thailand & Brazil. *Mem:* Am Inst Food Technol. *Res:* Textural characteristics of fruits and vegetables; enzymes of fresh and processed vegetables. *Mailing Add:* Dept Food Technol Univ Tenn 114C McLeod Food Bldg Knoxville TN 37996

COLLINS, JIMMY HAROLD, b Gaffney, SC, Dec 9, 48; m 79; c 4. CELL BIOLOGY, CELL MOTILITY. *Educ:* Duke Univ, Bs, 71; Univ Tex, Austin, PhD(chem), 77. *Prof Exp:* STAFF FEL, LAB CELL BIOL, NAT HEART & LUNG INST, NIH, 77- *Mem:* Am Soc Biol Chemist; Am Soc Cell Biol; Biophys Soc; Am Chem Soc. *Res:* Biochemistry of cytoskeletal and contractile proteins, including identification, isolation, characterization and interactions of these proteins; regulation of nonmuscle myosin adenosine triphosphatase activity and assembly, especially by calcium and phosphorylation. *Mailing Add:* 4832 Summerdale Ave Philadelphia PA 15261

COLLINS, JOHN A(DDISON), b Midway, Pa, Jan 6, 29; m 53; c 3. ELECTRICAL ENGINEERING, ELECTRONICS ENGINEERING. *Educ:* Wash & Jefferson Col, AB, 51; Mass Inst Technol, BSEE, 57. *Prof Exp:* Res engr instrumentation, Jet Propulsion Lab, Calif Inst Technol, 57-59, group supvr space photog, 61-64; staff engr, Instrumentation Lab, Mass Inst Technol, 59-61; mgr electronic systs photo-digital recorder, IBM Corp, 64-67; mgr, Electronics Div, Analog Technol Corp, Pasadena, 67-72; asst mgr, 73-81, sr scientist, Hughes Aircraft Co, 81-88; POWER CONVERSION CONSULT, 88- *Mem:* Inst Elec & Electronics Engrs; Am Inst of Aeronaut and Astronaut. *Res:* Space qualified power conversion systems; design, development and manufacture of traveling wave tube amplifiers for satellite and satellite earth stations, electronic power conversion subsystems, high voltage power supply design and analysis. *Mailing Add:* Cow Creek Enterprises 14671 Upper Cow Creek Rd Azalea OR 97410

COLLINS, JOHN BARRETT, b Cleveland, Ohio, Jan 25, 49; m 80; c 3. THEORETICAL ORGANIC CHEMISTRY. *Educ:* Holy Cross Col, BA, 71; Princeton Univ, MA, 73, PhD(chem), 76; NY Univ, MS, 83. *Prof Exp:* Res chemist drug design, Lederle Labs, 77-81, systs analyst, Stamford Res Labs, Am Cyanamid Co, 81-83; sr engr, 83-84, sr staff engr, 84-87, asst engr mgr, 87-89, ENGR MGR, SOFTWARE, PERKIN-ELMER CORP, 89- *Mem:* Am Chem Soc; AAAS; Quantum Chem Prog Exchange; Inst Elec & Electronics Engrs. *Res:* Molecular orbital theory, chemistry by computer, quantitative drug design; computer software development. *Mailing Add:* Five Arrowhead Rd Westport CT 06880

COLLINS, JOHN CLEMENTS, b Colchester, UK, Dec 8, 49. QUANTUM FIELD THEORY, RENORMALIZATION GROUP. *Educ:* Univ Cambridge, BA, 71, PhD(theoret physics), 75. *Prof Exp:* Res assoc, Princeton Univ, 75-76, asst prof, 76-80; from asst prof to prof physics, Ill Inst Technol, 80-90; PROF PHYSICS, PA STATE UNIV, 90- *Concurrent Pos:* Prin investr, Dept Energy grants; mem, Inst Adv Study,. *Res:* Elementary particle theory; quantum field theory. *Mailing Add:* Physics Dept Pa State Univ University Park PA 16802

COLLINS, JOHN HENRY, b Peabody, Mass, Sept 6, 42; m 67; c 2. PROTEIN SEQUENCING, LIQUID CHROMATOGRAPHY. *Educ:* Northeastern Univ, AB, 65; Boston Univ, PhD(biochem), 70. *Prof Exp:* Res fel muscle res, Boston Biomed Res Inst, 69-72, res assoc, 72-74, staff scientist, 74-76; asst prof cell biophysics, Baylor Col Med, 76-77; asst prof pharmacol & cell biophysics, Col Med, Univ Cincinnati, 77-79, assoc prof, 79-; AT DEPT BIOL, CLARKSON UNIV. *Concurrent Pos:* Res assoc, Harvard Col, 74-76; prin investr grants, Am Heart Asn, 75-76 & 81-84, Nat Heart Lung & Blood Inst, 76-83, Nat Inst Arthritis, Metab & Digestive Dis, 77-86. *Mem:* Am Soc Biol Chemists; Biophys Soc; NY Acad Sci; AAAS. *Res:* Protein structure and function; calcium binding, contractile and membrane proteins; chromatography; high performance liquid chromatography; amino acids; peptide isolation and characterization; sodium potassium adenosine triphatase; proteolipids; sarcoplasmic reticulum; actin, myosin and troponin; protein evolution. *Mailing Add:* Dept Biochem Univ Md Rm 237A Howard Hall Baltimore MD 21201

COLLINS, JOHNNIE B, b Grena Co, Miss, Sept 27, 43; m 74; c 3. SOILS & LAND MANAGEMENT. *Educ:* Alcorn State Univ, BS, 65; Mich State Univ, MS, 67, PhD(soil sci), &71. *Prof Exp:* Soil Scientist, Mich Agr Exp Sta, 65-71, Tex Agr Exp Sta, 71-79; grad asst soil sci, Mich State Univ, 65-71; assoc prof, Prairie View Agr & Mech Univ, 71-79; ASSOC PROF SOIL SCI & RES DIR, ALCORN STATE UNIV, 79- *Concurrent Pos:* Soil scientist, Agency Intl Devel 211(d) Tropical Soils, 71-79; supt exp sta, Miss Agr & Foresrty Exp Sta, 79- *Mem:* Am Soc Agron; Soil Sci Soc Am; AAAS; Int Soc Soil Sci; Soil Conserv Soc Am. *Res:* Agricultural development. *Mailing Add:* Alcorn State Univ PO Box 844 Lorman MS 39096

COLLINS, JON DAVID, b Flint, Mich, Mar 14, 35; m 57; c 4. AERONAUTICAL & ASTRONAUTICAL ENGINEERING, APPLIED MATHEMATICS. *Educ:* Univ Mich, BS, 57; Univ Colo, MS, 59; Univ Calif, Los Angeles, PhD(eng), 67. *Prof Exp:* Engr, Martin Co, 57-59; mem tech staff, Titan Prog Off, TRW Systs Group, TRW Inc, 59-63, mem tech staff, Dynamics Dept, 63-66, head, Systs Dynamics Sect, 66-67, head, Anal Dynamics Sect, 67-69; vpres, J H Wiggins Co, 69-81; PRES, ACTA INC, 82- *Concurrent Pos:* Res engr, Univ Calif, Los Angeles, 66-67; adj prof, Sch Bus Mgt, Northrop Univ, 75-83; lectr, Univ Southern Calif, 71-74, Calif State Univ, Long Beach, 83. *Honors & Awards:* New Technol Award, NASA, 76. *Mem:* Assoc fel Am Inst Aeronaut & Astronaut; Soc Risk Anal; Am Soc Mech Engrs; Am Soc Civil Engrs; Opers Res Soc Am; sr mem Systs Safety Soc. *Res:* Risk and safety analysis of structures, missile and space vehicles and nuclear power plants; development of systems analysis models for survivability of military systems; application of statistics to problems in engineering mechanics; author of over thirty technical papers in journals. *Mailing Add:* 27817 Longhill Dr Rancho Palos Verdes CA 90274

COLLINS, JOSEPH CHARLES, JR, b Pontiac, Mich, May 7, 31; m 53; c 4. ORGANIC CHEMISTRY. *Educ:* Wayne State Univ, BS, 53; Univ Wis, PhD(chem), 58. *Prof Exp:* Asst, Gen Motors Res, 53 & Chem Dept, E I du Pont de Nemours & Co, 56; res assoc, Univ Wis, 58 & Sterling-Winthrop Res Inst, 58-62; assoc prof chem, Ill Wesleyan Univ, 62-67; res assoc, Sterling-Winthrop Res Inst, 67-68, assoc dir chem div, 68-69, dir chem div, 69-76, vpres chem, 76-82, vpres, Tech Affairs, 82-87; PRES, MOLECULAR DESIGN ASSOC, 87- *Honors & Awards:* Caher Award, 53; Green Medallion Award, 66. *Mem:* AAAS; Am Chem Soc; The Chem Soc. *Res:* Organic synthesis and conformational analysis of biomolecular systems. *Mailing Add:* 28 Highland Dr East Greenfush NY 12061

COLLINS, KENNETH ELMER, b Newcastle, Wyo, July 10, 26; m 55; c 1. RADIOCHEMISTRY, RADIATION CHEMISTRY. *Educ:* San Jose State Univ, BA, 50; Iowa State Univ, MS, 55; Univ Wis, PhD(phys chem), 62. *Prof Exp:* Res scientist, Vet Med Res Inst, Ames, 55-58, Brookhaven Nat Lab, 61-63, Centre de Physique Nucleaire, Louvain, 63-64; asst prof anal & radio chem, State Univ NY at Buffalo, 64-68; res scientist, Western NY Nuclear Res Ctr, 69-70; PROF ANALYSIS & PHYS CHEM, UNIV ESTADUAL DE CAMPINAS, 74- *Concurrent Pos:* Adv, Int Atomic Energy Agency, 70-73; vis scientist, Centre de Recherches Nucleaires, Strasbourg, 79. *Mem:* Am Chem Soc; AAAS; Royal Soc Chem. *Res:* Nuclear and radio chemistry with emphasis on solid state recoil studies and liquid phase radiolytic reactions; high performance liquid chromatographic analysis of radio labelled species. *Mailing Add:* Instituto de Quimica CP 6154 Univ Estadual de Campinas Campinas SP 13081 Brazil

COLLINS, LORENCE GENE, b Vernon, Kans, Nov 19, 31; m 55; c 5. PHOTOGEOLOGY. *Educ:* Univ Ill, BS, 53, MS, 57, PhD(geol), 59. *Prof Exp:* Instr phys sci, Univ Ill, 58-59; from asst prof to prof geol, San Fernando Valley State Col, 59-74, fac res coordr, 70-78; PROF GEOSCI CALIF STATE UNIV, 74- *Mem:* Sigma Xi; Geol Soc Am; NAGT. *Res:* Mineral deposits in metamorphic terrains, particularly magnetite in granitic gneisses; refractive index studies of ferromagnesian silicates; studies of origin of myrmekite. *Mailing Add:* Dept Geol Calif State Univ Northridge CA 91330

COLLINS, MALCOLM FRANK, b Crewe, Eng, Dec 15, 35; m 61; c 3. PHYSICS. *Educ:* Cambridge Univ, BA, 57, MA, 61, PhD(physics), 62. *Prof Exp:* Physicist, Solid State Physics Div, Atomic Energy Res Estab, Harwell, Eng, 61-69; assoc prof, McMaster Univ, 69-73, assoc chmn dept, 74-76, chmn dept, 76-83, PROF PHYSICS, MCMASTER UNIV, 78-, DIR, MCMASTER NUCLEAR REACTOR, 87- *Concurrent Pos:* Guest, Brookhaven Nat Lab, 67-68, 76; fel, Alfred P Sloan Found, 70-72. *Mem:* Fel Can Asn Physicists; Can Inst Neutron Scattering. *Res:* Various aspects of slow neutron scattering, especially applications to magnetism, critical phenomena, transition metals, and crystallography; theory of paramagnetism and order-disorder phase transitions. *Mailing Add:* Dept Physics McMaster Univ Hamilton ON L8S 4M1 Can

COLLINS, MARY ELIZABETH, b Jersey City, NJ, July 19, 53; m. SOIL GENESIS, MORPHOLOGY & CLASSIFICATION. *Educ:* Cornell Univ, BS, 75; Iowa State Univ, MS, 77, PhD(soil genesis & classification), 80. *Prof Exp:* Soil scientist, Soil Conserv Serv, USDA, 72-80; res asst, Iowa State Univ, 75-77, res assoc, 77-80, teaching fel, 80-81; asst prof, 81-86, ASSOC PROF SOIL GENESIS, UNIV FLA, 86- *Concurrent Pos:* Lectr, Cornell Univ, 79-80; coordr, Soil Sci Inst; vis scientist, Nat Acad Sci, China, Spain & Ecuador; mem, People to People China Comt studying grasslands in China, 86; chair, Am Soc Agron Student Photo slide presentation contest, 87-; res in US VI, Honduras, Ecuador, China & Spain, 86-; vis prof, Univ Extremadura, Spain, 88-89. *Mem:* Am Soc Agron; Sigma Xi; Soil Sci Soc Am; Int Soil Sci Soc. *Res:* Formation of soil; classifying soils; explaining soil variability; interpreting soil potentials; archaeology. *Mailing Add:* G-2171 McCarty Hall Univ Fla Gainesville FL 32611

COLLINS, MARY JANE, b Miami, Fla, Oct 17, 40. ACOUSTICS. *Educ:* Vanderbilt Univ, BA, 61, MS, 63; Univ Iowa, PhD(hearing sci), 70. *Prof Exp:* Audiologist, Bristol Speech & Hearing Ctr, 64-65; res asst audiol, Vet Admin Hosp, Coral Gables, Fla, 65-66; audiologist, Vet Admin Hosp, Nashville, 70-72; asst prof audiol, Univ Wis, 72-73; assoc prof audiol, Tenn State Univ, 73-77; res assoc, Speech & Hearing Sci Dept, City Univ New York, 77-80; mem fac, dept speech path & audiol, Univ Iowa, 80-85; ASSOC PROF, DIV COMMUN DISORDERS, LA STATE UNIV, 85- *Concurrent Pos:* Asst prof audiol, Vanderbilt Univ, 70-72; adj asst prof, Kresge Labs, La State Univ Med Ctr, 76- *Mem:* Acoust Soc Am; Am Speech-Lang-Hearing Asn. *Res:* Psychophysical phenomena in the normal and pathological human auditory system. *Mailing Add:* Commun Dis La State Univ Baton Rouge LA 70803-2606

COLLINS, MARY LYNNE PERILLE, b Evanston, Ill, May 15, 49; m 74; c 1. MICROBIAL BIOCHEMISTRY. *Educ:* Emmanuel Col, BA, 71; Rutgers Univ, PhD(microbiol), 76. *Prof Exp:* NIH fel, NY Univ Med Ctr, 76-78, res asst prof microbiol, 78-80; asst prof, 80-86, ASSOC PROF BIOL SCI, UNIV WIS-MILWAUKEE, 86- *Mem:* AAAS; Am Soc Microbiol; Sigma Xi. *Res:* Microbial biochemistry and physiology, membrane biogenesis and differentiation, functional organization of membranes, bacterial antigenic structure; photosynthetic bacteria; methanotrophic bacteria. *Mailing Add:* Dept Biol Sci Univ Wis PO Box 413 Milwaukee WI 53201

COLLINS, MICHAEL, b Dayton, Ohio, Sept 18, 51; m 70; c 2. CROP SCIENCE. *Educ:* Berea Col, BS, 73; WVa Univ, MS, 75; Univ Ky, PhD(crop sci), 78. *Prof Exp:* Forestry tech strip mining, USDA Forest Serv, 72; res asst agron, WVa Univ, 73-75 & Univ Ky, 75-78; asst prof agron, Univ Wis, Madison, 78-84; PROF AGRON, UNIV KY, 84- *Mem:* Sigma Xi; Am Soc Agron; Crop Sci Soc Am. *Res:* Influence of management and fertility on forage chemical composition and nutritive value; chemical estimation of forage nutritive value. *Mailing Add:* Dept Agron Univ Ky Agr ASC-N Bldg N-122 Lexington KY 40546

COLLINS, MICHAEL ALBERT, b Oak Park, Ill, July 6, 42; div; c 1. HYDRODYNAMICS, HYDROLOGY. *Educ:* Ga Inst Technol, BS, 65, MS, 69; Mass Inst Technol, PhD(hydrodynamics), 70. *Prof Exp:* Teaching asst fluid mech, Ga Inst Technol, 66-67; engr hydraulics, Camp, Dresser & McKee Consult Engrs, 67-68; from asst prof to assoc prof, Sch Eng & Appl Sci, Southern Methodist Univ, 70-85, prof civil eng, 84-89, dir, Ctr Urban Water

Studies, 84-89; pres, Eaumac, Inc, 84-89; SR PROJ ENGR, WOODWARD-CLYDE CONSULTS, 90- *Concurrent Pos:* NSF traineeship, Ga Inst Technol, 65-66, Mass Inst Technol, 66-67; res asst hydrodynamics, Mass Inst Technol, 67-80; consult to govt, pvt bus & indust; exp witness hydrol. *Honors & Awards:* Res Award, Sigma Xi, 82. *Mem:* Am Geophys Union; Am Soc Civil Engrs; Am Water Resources Asn; Int Asn Hydraulic Res; AAAS; Am Inst Hydrol; Nat Water Well Asn. *Res:* Hydrodynamics evaluation of mixing and transport processes; mathematics of networks in physical applications with emphasis to role of mathematical programming techniques; hydrologic processes, including groundwater flow and nonpoint source pollution; water policy and water resources management. *Mailing Add:* Woodward-Clyde Consults PO Box 66317 Baton Rouge LA 70896

COLLINS, NICHOLAS CLARK, b Ogden, Utah, Mar 2, 46; m; c 4. POPULATION ECOLOGY, FISH BENTHOS INTERACTIONS. *Educ:* Pomona Col, BA, 68; Univ Ga, PhD(zool), 72. *Prof Exp:* Fel acarology, Ohio State Univ, 72-73; asst prof, 73-79, ASSOC PROF ZOOL, ERINDALE COL, UNIV TORONTO, 79- *Mem:* Ecol Soc Am; Am Soc Naturalists; Can Soc Zoologists; NAm Benthological Soc. *Res:* Mechanisms limiting production of benthivorous fishes in freshwater littoral zones; comparisons of feeding capabilities of visually-feeding fishes with tactile/olfactory benthivores; nighttime non-intrusive underwater video recording. *Mailing Add:* Dept Zool Erindale Col 3359 Mississauga Rd Mississauga ON L5L 1C6 Can

COLLINS, NORMAN EDWARD, JR, b Wilmington, Del, Mar 28, 40; m 63; c 2. AGRICULTURAL ENGINEERING, ENERGY MANAGEMENT. *Educ:* Univ Del, BS, 62; Univ Md, MS, 65; Univ Pa, PhD(civil eng), 75. *Prof Exp:* Instr, Univ Md, 64-65; from asst prof to assof prof, 65-86, DEPT CHMN, 81-, PROF, AGR ENG, UNIV DEL, 86- *Mem:* Am Soc Agr Engrs. *Res:* Energy use in crop and broiler production; computer modeling of on-farm broiler production. *Mailing Add:* Dept Agr Eng Agr Hall Univ Del Newark DE 19711

COLLINS, O'NEIL RAY, b Opelousas, La, Mar 9, 31; m 59; c 2. MYCOLOGY. *Educ:* Southern Univ, BS, 57; Univ Iowa, MS, 59, PhD(bot), 61. *Prof Exp:* Instr biol, Queens Col, 61-63; assoc prof, Southern Univ, 63-65; assoc prof, Wayne State Univ, 65-69; assoc prof, 69-73, PROF BOT, UNIV CALIF, BERKELEY, 73- *Mem:* Bot Soc Am; Mycol Soc Am; Sigma Xi. *Res:* Mating types in the Myxomycetes. *Mailing Add:* Dept Bot Univ of Calif Berkeley CA 94720

COLLINS, PAUL EVERETT, b White Rock, Minn, Feb 22, 17; m 48; c 2. SILVICULTURE. *Educ:* Gustavus Adolphus Col, BA, 39; Univ Minn, BS, 48, MS, 49, PhD, 67. *Prof Exp:* Asst, Univ Minn, 48-49; asst prof & exten forester, Kans State Col, 49-51; from asst prof & asst forester to assoc prof & assoc forester, SDak State Univ, 52-74, prof hort & forestry, 74-82; RETIRED. *Mem:* Soc Am Foresters; Sigma Xi. *Res:* Plains windbreak and shelter belt research on matters of cultural practices, design and tree breeding. *Mailing Add:* 1711 Olwien St SDak State Univ Brookings SD 57007

COLLINS, PAUL WADDELL, b Greenville, SC, Feb 26, 40; m 63; c 3. MEDICINAL CHEMISTRY. *Educ:* Univ SC, BS, 62; Med Col Va, PhD(med chem), 66. *Prof Exp:* Res fel org synthesis, Univ Va, 66-67; res scientist, 67-81, sr res scientist, 81-85, SECT HEAD, GASTROINTESTINAL DIS RES DEPT, G D SEARLE & CO, 85-, SR FEL, 89- *Mem:* Am Chem Soc. *Res:* Synthesis of heterocyclic spiro compounds; synthesis of cyclopropylogs of naturally occurring amines and amino acids; synthesis of prostaglandins, especially 16-hydroxy prostaglandins. *Mailing Add:* Gastrointestinal Dis Res Dept G D Searle & Co 4901 Searle Pkwy Skokie IL 60077

COLLINS, RALPH PORTER, b Alpena, Mich, Nov 26, 27; m 55; c 3. BOTANY. *Educ:* Mich State Univ, BA, 50, MS, 52, PhD(bot), 57. *Prof Exp:* Asst bot, Mich State Univ, 52-54 & 55-57; from instr to assoc prof, 57-69, head bot dept, 80-85, PROF BOT, UNIV CONN, 69-,. *Concurrent Pos:* NIH spec res fel, 64-65; Smithsonian Inst sr res fel, 72. *Mem:* AAAS; Mycol Soc Am; Am Chem Soc; Am Soc Pharmacog. *Res:* Fungal physiology; natural products chemistry. *Mailing Add:* Dept Biol U-42 Univ Conn Storrs CT 06268

COLLINS, RICHARD ANDREW, b Norristown, Pa, Oct 27, 24; m 55; c 3. PATHOLOGY. *Educ:* Pa State Univ, BS, 48; Univ Wis, MS, 50, PhD(biochem), 52; Marquette Univ, MD, 62. *Prof Exp:* Asst biochem, Marquette Univ, 58-60; intern, Mary Fletcher Hosp, Burlington, Vt, 62-63; Nat Heart Inst spec fel path, Univ Vt, 63-65; asst prof path, Med Col Wis, 65-67; pathologist, Regional Med Labs, 72-81; PATHOLOGIST, ST LUKE'S HOSP, MILWAUKEE, 81- *Mem:* Am Asn Pathologists; Am Asn Clin Chem; Am Soc Clin Path; Col Am Path; Int Acad Path. *Mailing Add:* St Luke's Med Ctr 2900 W Okla Ave Milwaukee WI 53215

COLLINS, RICHARD ARLEN, b Shaw, Miss, Aug 4, 30; c 5. FISHERIES BIOLOGY. *Educ:* Delta State Col, BSE, 53; Univ Southern Miss, MA, 57; Univ Southern Ill, PhD(zool), 68. *Prof Exp:* Marine biologist, Gulf Coast Res Lab, Miss, 57-59; assoc prof biol, State Col Ark, 59-65; instr zool, Univ Southern Ill, 65-68; PROF BIOL, UNIV CENT ARK, 68- *Mem:* AAAS; Am Fisheries Soc. *Res:* Fish culture; aquatic ecology. *Mailing Add:* Dept Biol Univ Cent Ark Conway AR 72032

COLLINS, RICHARD CORNELIUS, b Phoenix, Ariz, Feb 16, 41; m 61; c 1. PARASITOLOGY, ENTOMOLOGY. *Educ:* Ariz State Univ, BS, 63; Univ Ariz, MS, 66, PhD(entom, parasitol), 70. *Prof Exp:* Res asst entom, Dept Animal Path, Univ Ariz, 64-66, parasitol, 68-70; self-employed, 66-68; fel entom & parasitol, Dept Parasitol, Tulane Univ, 70-74; res entomologist parasitol, Ctr Dis Control, Bur Trop Dis, USPHS, 74-82; CONSULT, UNIV ARIZ, 82- *Mem:* Am Soc Parasitol; Entom Soc Am; Am Soc Trop Med & Hyg; AAAS. *Res:* Bionomics of parasites and insects; chemotherapy of parasitic diseases; transmission dynamics of vector-borne diseases of man and animals. *Mailing Add:* 6540 N Camino Libby Tucson AZ 85718

COLLINS, RICHARD FRANCIS, b St Paul, Minn, Jan 22, 38; m 60; c 3. CLINICAL MICROBIOLOGY. *Educ:* Shepherd Col, AB, 62; Wake Forest Univ, MA, 68; Univ Okla, PhD(parasitol & lab pract), 73. *Prof Exp:* Teacher biol, Alexandria, Va Pub Schs, 62-66; grad asst biol, Wake Forest Univ, 67-68; res asst microbiol, Univ Okla Health Sci Ctr, 70-72, instr, 72-73; asst prof microbiol, Rockord Sch Med, Univ Ill, 73-80, lab dir, 74-80; PROF MICROBIOL & DEPT HEAD, UNIV OSTEOP MED & HEALTH SCI, 80- *Concurrent Pos:* Consult, US Environ Protection Agency, 75-81 & Nat Bd Podiatry Examiners, 83-91. *Mem:* Sigma Xi; Am Soc Trop Med & Hyg; Am Soc Microbiol. *Res:* Clinical microbiology; serology of parasitic diseases; in vitro drug susceptibility of bacterial isolates. *Mailing Add:* Univ Osteop Med & Health Sci 3200 Grand Ave Des Moines IA 50312

COLLINS, RICHARD LAPOINTE, b New York, NY, May 10, 38; m 65. INVERTEBRATE ZOOLOGY, PROTOZOOLOGY. *Educ:* Boston Univ, AB, 61, MA, 62; Univ Calif, Berkeley, PhD(zool), 69. *Prof Exp:* Actg asst prof zool, Univ Calif, Berkeley, 68-69; ASST PROF ZOOL & PHYSIOL, LA STATE UNIV, BATON ROUGE, 69- *Mem:* AAAS; Am Inst Biol Sci; Soc Protozool; Am Micros Soc. *Mailing Add:* 2460 Brightside Lane Baton Rouge LA 70820

COLLINS, ROBERT C, b Los Angeles, Calif, Mar 21, 42; m 65; c 2. NEUROLOGY, NEUROCHEMISTRY. *Educ:* Univ Calif, Berkely, BA, 64; Cornell Univ, MD, 69. *Prof Exp:* Intern med, Mass Gen Hosp, 69-70; res assoc neurochem, Nat Inst Neurol & Commun Dis & Stroke, 70-72; resident neurol, New York Hosp, Cornell Univ, 72-75, asst prof, 75-76; from asst prof to prof neurol, Sch Med, Wash Univ, 76-87; PROF & CHMN, DEPT NEUROL, SCH MED, UNIV CALIF, LOS ANGELES, 87- *Concurrent Pos:* Vis physician neurol, Rockefeller Univ, 75-76; neurologist, Barnes Hosp & Jewish Hosp, 79-; ed, J Cerebral Blood Flow Metab, 83- *Mem:* Am Acad Neurol; Am Epilepsy Soc; Am Neurol Soc; Soc Neurosci; Am Soc Neurochem; Int Soc Cerebral Blood Flow & Metab. *Res:* Basic mechanisms of cerebral metabolism and blood flow and their role in epilepsy. *Mailing Add:* Dept Neurol Med Sch Univ Calif 710 Westwood Plaza Los Angeles CA 90024-1769

COLLINS, ROBERT JAMES, b Hazel Park, Mich, July 15, 28; m 52; c 2. PHYSIOLOGY, PHARMACOLOGY. *Educ:* Alma Col, BS, 51; Mich State Univ, 54, PhD, 58. *Prof Exp:* Bacteriologist, Mich Dept Health, 51-52; lab technician, E W Sparrow Hosp, 54-56; res head, Upjohn Co, 58-90; RETIRED. *Mem:* AAAS; Am Pharmacol Soc; Soc Exp Biol & Med. *Res:* Central nervous system. *Mailing Add:* CNS Res Upjohn Co 7000 Fortage Rd Kalamazoo MI 49001

COLLINS, ROBERT JOSEPH, b Philadelphia, Pa, July 23, 23; m 45; c 4. PHYSICS. *Educ:* Univ Mich, AB, 47, MS, 48; Purdue Univ, PhD(physics), 53. *Prof Exp:* Asst prof physics, Rose Polytech Inst, 49-50; asst, Purdue Univ, 50-53; mem staff, Bell Tel Lab, 53-62; Inst Defense Anal, DC, 62-63; head dept, 64-70, PROF ELEC ENG, UNIV MINN, MINNEAPOLIS, 63- *Mem:* Am Phys Soc; Optical Soc Am; fel Inst Elec & Electronics Engrs. *Res:* Optical properties of solid state; radiation effects in solids; quantum electronics. *Mailing Add:* Dept Elec Eng Univ Minn Elec Eng Bldg Rm 139 123 Church St SE Minneapolis MN 55455

COLLINS, RONALD WILLIAM, b Dayton, Ohio, Feb 5, 36; m 60; c 2. INORGANIC CHEMISTRY. *Educ:* Univ Dayton, BS, 57; Ind Univ, PhD(inorg chem), 62. *Prof Exp:* Inorg res chemist, Wyandotte Chem Corp, 62-65; from asst prof to assoc prof, Eastern Mich Univ, 65-71, dept head, 77-80, assoc vpres, Acad Affairs, 80-83, PROF CHEM, EASTERN MICH UNIV, 71-, PROVOST & VPRES ACAD AFFAIRS, 83- *Concurrent Pos:* Vis prof, Mich State Univ, 70-71. *Mem:* Am Chem Soc; Royal Soc Chem; Am Crystallog Asn; AAAS; Sigma Xi. *Res:* Inorganic compounds of group IV metals; x-ray crystallography; instructional uses of computers. *Mailing Add:* Dept Chem Eastern Mich Univ Ypsilanti MI 48197

COLLINS, ROYAL EUGENE, b Corsicana, Tex, Feb 25, 25; div; c 2. PHYSICS, PETROLEUM ENGINEERING. *Educ:* Univ Houston, BS, 49; Tex A&M Univ, MS, 50, PhD(physics), 54. *Prof Exp:* Res Mobil Field Res Lab, 50-52 & engr, Stanolind Oil Co, 54-55; sr res engr, Humble Oil Co, 55-59; assoc prof, Univ Houston, 59-71, prof, 71-79; Frank W Jessen prof petrol eng, Univ Tex, Austin, 79-90; PRES & RES & ENG CONSULT, 82- *Concurrent Pos:* Consult, Vet Admin Hosp, Houston, 58-62, Exxon Prod Res Corp, 59-68, M D Anderson Tumor Inst, 62, Col Med, Baylor Univ, 61-78, US Bur Mines, 61-63 & Subsurface Disposal Corp, 69-79. *Mem:* Am Phys Soc; Am Asn Physics Teachers; Soc Petrol Engrs; Sigma Xi; AAAS. *Res:* Fluid flow in porous media; petroleum production operations; enhanced oil recovery; quantum theory and related topics. *Mailing Add:* 5445 DTC Parkway Suite 640 Englewood CO 80111

COLLINS, RUSSELL LEWIS, b Coffeyville, Kans, Sept 21, 28; m 79; c 2. CHEMICAL PHYSICS. *Educ:* Univ Tulsa, BS, 48; Univ Okla, MS, 50, PhD(physics), 53. *Prof Exp:* Res physicist, Phillips Petrol Co, Okla, 53-58, group leader, 58-62; from asst prof to assoc prof physics, Univ Tex, Austin, 62-84; RETIRED. *Concurrent Pos:* Petrol Res Fund grant, 63-66; founder & pres, Austin Sci Assocs, Inc, 64-; Robert A Welch Found grant, 65-84; Off Naval Res contract, 69-70; US Army Mobile Equip Res & Develop Lab contract, 70-75. *Mem:* Am Phys Soc; Weed Sci Soc Am; Am Chem Soc. *Res:* Mossbauer effect as applied to iron organometallic complexes and stresses in ferrous metals; water-degradable polymers for controlled release of pesticides and medicines; thermodynamics of kinetic temperature. *Mailing Add:* HC01 Box 106-C Rockport TX 78382

COLLINS, STEPHEN, b Chicago, Ill, May 14, 27; m 54; c 3. ECOLOGY. *Educ:* Cornell Univ, BS, 49; Rutgers Univ, PhD(ecol), 56. *Prof Exp:* Resident naturalist, Palisades Nature Asn, NJ, 49-51; asst forester, Conn Agr Exp Sta, 57-62; from asst prof to prof biol, Southern Conn State Col, 61-89; RETIRED. *Concurrent Pos:* Consult, McGraw-Hill Bk Co, Inc, 57. *Mem:* Ecol Soc Am; Am Soc Mammal; Wildlife Soc; Wilderness Soc; Am Nature

Study Soc. *Res:* Relation of land use, climatic, biotic and fire factors on biotic communities and their successions; silvics, vertebrate ecology, biogeography and general natural history; environmental problems; natural areas; pesticides; biological photography. *Mailing Add:* 339 Brooks Rd Bethany CT 06525

COLLINS, STEVE MICHAEL, b Bloomington, Ind, Sept, 19, 47; c 2. BIOMEDICAL IMAGE PROCESSING. *Educ:* Univ Ill at Chicago Circle, BS, 71, MS, 74, PhD(elec eng), 77. *Prof Exp:* Consult, G D Searle Lab, 72-74; asst prof, Univ Iowa, 77-81, asst res scientist cardiol, 79, radiol, 80-81, assoc prof elec & comput, 81-87, assoc prof radiology, 82-87, PROF RADIOLOGY, UNIV IOWA, 87-, PROF ELECT & COMPUT ENG. *Concurrent Pos:* Pres, Univ Iowa Facul Senate, 90-91. *Mem:* Am Heart Asn; Inst Elec & Electronic Engrs; Asn Comput Mach; Soc Magnetic Resonance Med. *Res:* Development and evaluation of computer techniques for deriving quantitative information from images of the heart. *Mailing Add:* Elec & Comput Eng Univ Iowa Iowa City IA 52242

COLLINS, SYLVA HEGHINIAN, b Aleppo, Syria, Oct 9, 48; US citizen; m 80; c 3. STATISTICAL DESIGN, ANALYSIS OF CLINICAL TRIALS. *Educ:* Am Univ Beirut, BS, 71; Boston Univ, MA, 73, PhD(statist), 77; New York Univ, MS, 82. *Prof Exp:* Res statistician, Lederle Labs, 77-79; asst dir, 79-81, assoc dir biostatist, Ayerst Labs, 82-84; assoc dir, 85-86, DIR STATIST & DATA SYSTS, MILES PHARMACEUT, 86- *Mem:* Am Statist Asn; Biomet Soc; Soc Clin Trials; Drug Info Asn. *Res:* Statistical design and analysis of clinical trials; statistical modeling of computer systems performance. *Mailing Add:* 5 Arrowhead Rd Westport CT 06880

COLLINS, TERRENCE JAMES, b Auckland, NZ, Oct 12, 52; m 75; c 2. SYNTHETIC INORGANIC & ORGANOMETALLIC CHEMISTRY. *Educ:* Univ Auckland, BSc, 74, MSc, 75, PhD(chem), 78. *Prof Exp:* Res assoc, Stanford Univ, 78-80; ASSOC PROF CHEM, CARNEGIE MELLON UNIV, 87- *Concurrent Pos:* Dreyfus Teacher Scholar, 86-; Alfred P Sloan Res Fel, 87- *Mem:* Am Chem Soc; NZ Inst Chem. *Res:* Synthetic inorganic and organometallic chemistry; synthesis and reactivity of later transition metal complexes in high oxidation states; development of selective oxidizing agents. *Mailing Add:* Dept Chemistry Carnegie Mellon Univ Pittsburgh PA 15213

COLLINS, THOMAS C, b Atlanta, Ga, June 12, 36. SOLIDS STATE PHYSICS, THEORETICAL PHYSICS. *Educ:* Univ Ga, BS, 58, MAS, 60; Univ Fla, PhD(physics), 64, Univ George Washington, MAS, 79. *Prof Exp:* Assoc prof, Univ Mo, 79-81; VPROVOST RES, UNIV TENN, 85- *Honors & Awards:* Humboldt Award. *Mem:* Fel Am Phys Soc; Sigma Xi. *Res:* Issues of the twenty first century as well as semi-conductors and super conductors. *Mailing Add:* 404 Andy Holt Tower Univ Tennessee Knoxville TN 37996

COLLINS, TUCKER, PATHOLOGY, VASCULAR BIOLOGY. *Educ:* Univ Rochester, PhD(microbiol) & MD, 81. *Prof Exp:* ASST PROF PATH, MED SCH, HARVARD UNIV, 86- *Mailing Add:* Dept Path Brigham-Womens Hosp 75 Francis St Boston MA 02115

COLLINS, VERNON KIRKPATRICK, b Advocate, NS, May 14, 17; m 39; c 2. BIOCHEMISTRY. *Educ:* Acadia Univ, BSc, 37; McGill Univ, MS, 46. *Prof Exp:* Chemist, Can Packers, Montreal, 41-43; res chemist, Ogilvie Flour Mills, 43-46; chief chemist, Vio Bin Corp, Monticello, 46-77, RETIRED. *Mem:* Am Chem Soc; Am Oil Chem Soc. *Res:* Food spoilage; fat deterioration. *Mailing Add:* 201 S New St Champaign IL 61820-4620

COLLINS, VINCENT J, b Haverstraw, NY, Nov 24, 14; m 44; c 8. MEDICINE, ANESTHESIOLOGY. *Educ:* Marietta Col, BS, 36; Brown Univ, MS, 38; Yale Univ, MD, 42. *Prof Exp:* Assoc, Doctors Hosp, New York, 46-49; dir anesthesiol, St Vincent's Hosp, 49-57; asst dir, Bellevue Hosp, 57-61; assoc prof surg, Col Med, NY Univ, 61-66; dir, dept anesthesiol, Cook County Hosp, Chicago, 61-81; prof, 66-80, EMER PROF ANESTHESIOL, SCH MED, NORTHWESTERN UNIV, 80-; prof, 79-87, EMER PROF ANESTHESIOL, COL MED, UNIV ILL, 88- *Concurrent Pos:* Army US Med Corp, 82-86; consult, US Army Surg Gen, Medico-Legal; prof anesthesiol, Col Med, Univ Ill, 79-98. *Mem:* Am Soc Anesthesiol; AMA (secy, 57-62),; Sigma Xi; AAAS; Int Anesthesiol Res Soc; Asn Anesthesiol Gt Brit & Ireland. *Res:* Investigation of drugs, shock and endocrine problems. *Mailing Add:* Dept Anesthesiol Chicago Univ Ill 1740 W Taylor St Chicago IL 60612

COLLINS, VINCENT PETER, b Dublin, Ireland, Dec 3, 47. NEUROONCOLOGY, CLINICAL MORBID ANATOMY & HISTOLOGY. *Educ:* Nat Univ Ireland, MB, BCh, BAO, 71; Karolinska Inst, Stockholm, Swed, MD, 78; Royal Col Path, London, Eng, MRCPath, 88; FRCPI, 89. *Prof Exp:* Registrar path, Karolinska Hosp, Stockholm, Sweden 76-82, dir studies path, Dept Path, 83-84, consult pathologist path, 84- 86, sr consult pathologist path, 86-90; ASSOC PROF PATH, KAROLINKSA INST, STOCKHOLM, SWEDEN, 79-; SR RES FEL & LAB HEAD, LUDWIG INST CANCER RES, CLIN GROUP, STOCKHOLM BR, STOCKHOLM, SWEDEN, 86- *Concurrent Pos:* Mem, Sci Bd, Swed Can Soc, Stockholm, Sweden, 87- & Rev Comt, Swed Med Res Coun, Stockholm, Sweden, 91-; prof & sr consult path, Univ Gothenburg, Gothenburg, Sweden, 90-; hon sr consult pathologist, Karolinska Hosp, Stockholm, Sweden, 91- *Honors & Awards:* W O Russell Lectr in Anat Path, M D Anderson Hosp & Tumor Inst, Univ Tex, 90. *Mem:* Am Cell Biol Asn; Int Soc Neuropath; NY Acad Sci. *Res:* Molecular genetic analysis and phenotypical characterization of human brain tumors in vivo to document which tumor supressor genes and proto-oncogenes are involved in the oncogenesis of these tumor forms. *Mailing Add:* Dept Path I Sahlgrenska Hosp S-41345 Goteborg Sweden

COLLINS, WALTER MARSHALL, b Enfield, Conn, Nov 6, 17; m 43; c 4. ANIMAL GENETICS. *Educ:* Univ Conn, BS, 40, MS, 49; Iowa State Univ, PhD(poultry breeding), 60. *Prof Exp:* Instr poultry sci, Univ Conn, 47-49; from asst prof to assoc prof, 51-63, chmn genetics prog, 65-66, PROF ANIMAL & GENETICS SCI, UNIV NH, 63- *Concurrent Pos:* NIH spec fel, Univ Calif, Davis, 64-65. *Mem:* AAAS; Am Poultry Sci Asn; Genetics Soc Am; Am Genetic Asn; Int Soc Animal Blood Group Res. *Res:* Quantitative genetics; tumor biology. *Mailing Add:* Animal Sci Dept Kendall Hall Univ NH Durham NH 03824

COLLINS, WARREN EUGENE, b Memphis, Tenn, Jan 26, 47; m 71; c 2. NUCLEAR PHYSICS. *Educ:* Christian Bros Col, BS, 68; Vanderbilt Univ, MS, 70, PhD(physics), 72. *Prof Exp:* Assoc prof physics, Southern Univ, Baton Rouge, 72-73; asst prof, 73-74, assoc prof physics, Fisk Univ, 74; res assoc, Vanderbilt Univ, 75-77; prof physics, Southern Univ, Baton Rouge, 77-91, assoc, Vanderbilt Univ, 75-77; PROF PHYSICS, MEMPHIS STATE UNIV, 87- *Mem:* Am Asn Physics Teachers; Am Phys Soc. *Res:* Level structure studies of selenium and arsenic isotopes with mass numbers between 68 and 76, using various methods; singles analysis, gamma-gamma angular correlation, gamma-gamma coincidences, and life-time studies. *Mailing Add:* Dept Physics Memphis State Univ Memphis TN 38152

COLLINS, WILLIAM BECK, b Port Williams, NS, Dec 5, 26; m 50; c 1. PLANT PHYSIOLOGY, HORTICULTURE. *Educ:* McGill Univ, BSc, 48, MSc, 54; Rutgers Univ, PhD(hort, plant physiol), 61. *Prof Exp:* Res scientist, 48-73, potato physiologist, Res Sta, Can Dept Agr, 73-80, PROG SPECIALIST, ATLANTIC REGION RES BR, AGR CAN, 80- *Mem:* Can Soc Hort Sci; Potato Asn Am. *Res:* Relationships of growth regulators, endogenous and applied, to growth and development in potato; growth analysis and management studies with potato. *Mailing Add:* 42 Southill Dr Halifax NS B3M 2Y1 Can

COLLINS, WILLIAM EDGAR, b Terra Alta, WVa, Mar 31, 35; m 59; c 2. ENDOCRINOLOGY. *Educ:* WVa Univ BS, 57; Univ Wis, MS, 61, PhD(endocrinol), 65. *Prof Exp:* Prog specialist, Reprod Physiol for Ford Found, Inst Agr, Gujarat, India, 65-67; from asst prof to assoc prof, 67-74, actg dean, 74-75, prof biol, 74-, dean Col Arts & Sci, 75-, VPRES ACAD AFFAIRS & RES, WVA UNIV. *Mem:* Brit Soc Study Fertil; Endocrine Soc. *Res:* Function of corpus luteum and control of the estrous cycle. *Mailing Add:* Eight Alicia Ave-WO Morgantown WV 26056

COLLINS, WILLIAM EDWARD, b Brooklyn, NY, May 16, 32; m 70; c 1. MEDICAL SCIENCES. *Educ:* St Peter's Col, Jersey City, NY, BS, 54; Fordham Univ, Bronx, NY, MA, 56, PhD(psychol), 59. *Prof Exp:* Grad instr psychol, Fordham Univ, Bronx, NY, 58-59; res psychologist, US Army Med Res Lab, Ft Knox, 59-61; res psychologist, Fed Aviation Admin, Civil Aeromed Inst, 61-63, chief, 63-65, supvr, 65-86, mgr, 86-88, asst mgr, 88-89, DIR, FED AVIATION ADMIN, CIVIL AEROMED INST, 89- *Concurrent Pos:* Adj assoc prof, Dept Psychol, Univ Okla, 63-70, Health Sci Ctr, Dept Psychiat, 65-71, adj prof, Dept Psychol, 70- & Health Sci Ctr, Dept Psychiat, 71-; assoc ed, Aviation, Space & Environ Med J, 80- *Honors & Awards:* Longacre Award, Aerospace Med Asn, 71; Silver Medal, US Dept Transp, 86. *Mem:* Fel Aerospace Med Asn; fel Am Psychol Asn; fel AAAS; fel Am Psychol Soc; fel NY Acad Sci; Asn Aviation Psychologists (pres, 74-75). *Res:* Sensory psychology and physiology; animal and human vestibular function; visual vestibular interactions; effects on human performance of vestibular stimulation; performance limitations of color vision defectives; effects on performance of alcohol, sleep loss, and altitude in both static and dynamic environments. *Mailing Add:* AAM-3 FAA-CAMI PO Box 25082 Oklahoma City OK 73125-5060

COLLINS, WILLIAM ERLE, b Lansing, Mich, July 9, 29; m 56; c 2. MEDICAL ENTOMOLOGY, ECONOMIC ENTOMOLOGY. *Educ:* Mich State Univ, BS, 51, MS, 52; Rutgers Univ, PhD(entom), 54. *Prof Exp:* Asst entom, Rutgers Univ, 53-54; entomologist, Diamond Alkali Chem Co, 54; med entomologist, Biol Warfare Labs, Ft Detrick, Md, 55-58; exten specialist entom, Rutgers Univ, 58-59; med biologist, Lab Parasitic Dis, 59-74 & Bur Trop Dis, Ctr Dis Control, 74-81, MED BIOLOGIST, DIV PARASITIC DIS, CTR INFECTIOUS DIS, USPHS, 81- *Honors & Awards:* Joseph A LaPrince Medal for Malariology, Am Soc Trop Med & Hyg, 85. *Mem:* Am Soc Trop Med & Hyg; Am Mosquito Control Asn. *Res:* Medical entomology in field of virus and malaria transmission by arthropod vectors. *Mailing Add:* Div Parasitic Dis USPHS 1600 Clifton Rd Atlanta GA 30333

COLLINS, WILLIAM F, b Laceyville, Pa, May 9, 18; m 45; c 3. FOOD SCIENCE. *Educ:* Pa State Col, BS, 42, MS, 48, PhD(dairy sci), 49. *Prof Exp:* Dairy res chemist, Swift & Co, Chicago, 49-53, head, Ice Cream & Stabilizer Res Div, Res Labs, 53-56, mem staff, Mkt Develop, Hammond, Ind, 56-57, tech sales rep US & Can, Gen Gelatin Dept, Kearny, NJ, 57-67, mgr tech sci, Swift Chem Co, Oak Brook, Ill, 67-68; pres & gen mgr, Nutriprod Ltd & Topping Co, Can, 68-73; assoc prof food sci & technol & exten specialist, 73-84, EMER ASSOC PROF, VA POLYTECH INST & STATE UNIV, 84- *Mem:* Am Dairy Sci Asn; Inst Food Technologists; Int Asn Milk Food & Environ Sanitarians; Am Asn Candy Technologists; AAAS. *Res:* Milk quality and flavor as affected by microbial flora and ultra high pasteurization temperatures; asceptic packaging practices. *Mailing Add:* Dept Food Sci Va Polytech Inst & State Univ Blacksburg VA 24060

COLLINS, WILLIAM FRANCIS, JR, b New Haven, Conn, Jan 20, 24; m 51; c 3. NEUROSURGERY. *Educ:* Yale Univ, BS, 44, MD, 47; Am Bd Neurol Surg, dipl, 56. *Prof Exp:* Resident neurosurg, Barnes Hosp, St Louis, Mo, 47-49, from asst res neurosurgeon to res neurosurgeon, 51-53; from instr to assoc prof neurosurg, Sch Med, Western Reserve Univ, 54-63; prof neurol surg, dir div & neurosurgeon-in-chief, Med Col Va, 63-67; prof neurol surg & chief sect, 67-84, CUSHING PROF SURG, SCH MED, YALE UNIV, 70-, CHMN DEPT, 84- *Concurrent Pos:* Nat Found Infantile Paralysis fel, Wash Univ, 53-54; consult, West Haven Vet Admin Hosp, 67-; neurosurgeon in chief, Yale New Haven Med Ctr, 67-84, surgeon in chief, 84- *Mem:* Acad Neurol Surg; Am Col Surg; Asn Res Nerv & Ment Dis; Am Asn Neurol Surg; Neurosurg Soc Am; Am Surg Asn. *Res:* Neurophysiology of afferent pathway systems. *Mailing Add:* Yale Univ Sch Med 333 Cedar St New Haven CT 06510

COLLINS, WILLIAM HENRY, b Baltimore, Md, Mar 1, 30; m 51; c 2. CHEMICAL ENGINEERING, ENVIRONMENTAL SCIENCES. *Educ:* Loyola Col, BS, 52; Alexander Hamilton Inst, MBA, 65. *Prof Exp:* Chief chemist, chem mfg, John C Stalfort & Sons, 54-57; component develop supvr eng, Catalyst Res Corp, 57-60; chief engr, Miller Res Corp, 60-67; prog mgr, AAI Corp, 67-70; dir, Franklin Inst Res Labs, 70-74; br chief installation restoration, Chem Systs Lab, 74-84, DEP DIR RES, CHEM RES, DEVELOP & ENG CTR, ABERDEEN PROVING GROUND, 85- *Mem:* Sigma Xi; Am Defense Preparedness Asn; Am Chem Soc; AAAS. *Res:* Assessment of ecological stress resulting from industrial and military operations at various government installations; assessment of potential for contaminant migration; cataclysmic stress corrosion crack of steel; military ordnance design and development; chemical systems. *Mailing Add:* 1913 Forest Guard Ct Jarrettsville MD 21084

COLLINS, WILLIAM JOHN, b Mt Union, Pa, July 5, 34; m 58; c 3. INSECTICIDES, ENVIRONMENTAL TOXICOLOGY. *Educ:* Juniata Col, BS, 56; Rutgers Univ, PhD(entom), 65. *Prof Exp:* Asst prof entom, Rutgers Univ, 64-66; from asst dean to assoc dean, Col Biol Sci, 80-86, PROF ENTOM, OHIO STATE UNIV, 67-, ASSOC DEAN, GRAD SCH, 86- *Mem:* Entom Soc Am; Am Chem Soc; Soc Environ Toxicol Chem. *Res:* Toxicology; insecticide resistance, especially in cockroaches; mechanisms of resistance; environmental toxicology; aquatic invertebrate toxicology. *Mailing Add:* Dept Entom Ohio State Univ Columbus OH 43210-1220

COLLINS, WILLIAM KERR, b Vance Co, NC, June 21, 31; m 54; c 3. PLANT BREEDING, GENETICS. *Educ:* NC State Col, BS, 54, MS, 61; Iowa State Univ, PhD(plant breeding), 63. *Prof Exp:* Res instr, Crops Dept, NC State Col, 56-60; asst agron, Iowa State Univ, 60-63; agronomist, Res Dept, R J Reynolds Tobacco Co, 63-66; tobacco exten specialist, NC State Univ, 66-78, assoc prof, 66-70, Philip Morris exten specialist, 78-86, PROF CROP SCI, NC STATE UNIV, 70-, ASSOC HEAD, 86-, SPEC IN CHARGE, 86- *Honors & Awards:* Exten Educ Award, Am Soc Agron, 81. *Mem:* Am Soc Agron. *Res:* Fluecured tobacco variety evaluation for agronomic, chemical, physical and pathological properties; tobacco herbicides and sucker control chemicals; tobacco fertility; tobacco production in Greece, Iran, Afghanistan, British Honduras, Costa Rica, Uruguay, Italy and Venezuela. *Mailing Add:* Dept Crop Sci NC State Univ Box 7620 Raleigh NC 27695-7620

COLLINS, WILLIAM THOMAS, b Feb 21, 22; m; c 4. ANATOMIC PATHOLOGY. *Educ:* Univ Mich, MD, 44. *Prof Exp:* Dir lab, Dept Path, Lima Mem Hosp, 58-88; RETIRED. *Concurrent Pos:* Assoc clin prof path, Med Col Ohio. *Honors & Awards:* St George Medal Contrib Control Cancer, Am Cancer Soc, 89. *Mailing Add:* 4030 S Wapak Rd Lima OH 45806

COLLINSON, CHARLES WILLIAM, b Wichita, Kans, Dec 15, 23; m 78; c 2. GEOLOGY. *Educ:* Augustana Col, AB, 49; Univ Iowa, MS, 50-51, PhD(geol), 52. *Prof Exp:* Asst, Univ Iowa, 48 & 51-52; from asst geologist to prin geologist, 52-86, HEAD, LAKE MICH COAST & BASIN STUDIES, STATE GEOL SURV, ILL, 86- *Concurrent Pos:* Lectr, Univ Ill, 56-57 & Stanford Univ, 60; ed, Jour Paleont, 58-64; Guggenheim fel, Gt Brit & Ger, 63-64; mem Paleont Res Inst; prof geol, Univ Ill, Urbana-Champaign, 68-77. *Honors & Awards:* Dedicated Serv Award, Am Asn Petrol Geologists, 66; Distinguished Achievement Award, Soc Econ Paleontologists & Mineralogists, 86. *Mem:* Paleont Soc; fel Geol Soc Am; hon mem Soc Econ Paleontologists & Mineralogists (pres, 73-74). *Res:* Mississippian stratigraphy of the Mississippi Valley; Devonian-Mississippian conodonts; sedimentation and paleolimnology of Lake Michigan. *Mailing Add:* 616 W Hill Champaign IL 61820

COLLINSON, JAMES W, b Moline, Ill, June 24, 38; m 61; c 2. GEOLOGY. *Educ:* Augustana Col, Ill, AB, 60; Stanford Univ, PhD(geol), 66. *Prof Exp:* From asst prof to assoc prof, 66-82, PROF GEOL, OHIO STATE UNIV, 82-, ASSOC DEAN, COL MATH & PHYS SCI, 89- *Honors & Awards:* Fulbright Res Award, 80. *Mem:* Geol Soc Am; Paleont Soc; Am Asn Petrol Geol; Soc Econ Paleont & Mineral; Int Asn Sedimentologists. *Res:* Permian and Triassic stratigraphy and paleontology; Antarctic geology; central Utah overthrust belt. *Mailing Add:* Dept Geol Ohio State Univ Columbus OH 43210

COLLINS-WILLIAMS, CECIL, b Toronto, Ont, Dec 31, 18; m 44; c 2. PEDIATRICS, CLINICAL IMMUNOLOGY. *Educ:* Univ Toronto, BA, 41, MD, 44. *Prof Exp:* Sr staff physician, Hosp for Sick Children, 65-84; from assoc prof to prof, 68-84, EMER PROF PEDIAT, UNIV TORONTO, 84- *Concurrent Pos:* Consult, Hugh MacMillan Med Ctr, Toronto, 66- *Mem:* Am Col Allergists; Am Acad Allergy; Can Soc Allergy & Clin Immunol (pres, 62); Can Soc Immunol; Can Pediat Soc; Sigma Xi. *Mailing Add:* One Aberfoyle Cresent Suite SPH 5 Etobicoke ON M8X 2X8 Can

COLLIPP, BRUCE GARFIELD, b Niagara Falls, NY, Nov 7, 29; m 54; c 2. OCEAN ENGINEERING, PETROLEUM ENGINEERING. *Educ:* Mass Inst Technol, SB, 52, SM, 54. *Prof Exp:* Eng officer, Lykes Bros Steamship Co, 52-53; teaching asst naval archit, Mass Inst Technol, 53-54; mem tech staff, Shell Oil, 54-87; CONSULT OCEAN ENG, 87- *Concurrent Pos:* Vis comt mem, Mass Inst Technol, 71-80; lectr offshore eng, Univ Tex, 76- *Honors & Awards:* Holley Medal, Am Soc Mech Engrs; Gibbs Brothers Medal & Prize, Nat Acad Sci, 91; Cert of Merit, Soc Naval Architects & Marine Engrs, 91. *Mem:* Nat Acad Eng; Marine Technol Soc; Soc Naval Architects & Marine Engrs. *Res:* Semisubmersible, first rous; deepwater moorings; ship design and construction; fixed and floating offshore structures; dynamic positioning. *Mailing Add:* 511 Kickerillo Dr Houston TX 77079

COLLIPP, PLATON JACK, b Niagara Falls, NY, Nov 4, 32; m 56; c 3. PEDIATRICS. *Educ:* Univ Rochester, AB, 54, MD, 57. *Prof Exp:* UPSHS trainee biochem, Univ Wash, 57-59; from intern to resident pediat, Univ Southern Calif, 59-61, asst prof, 63-65; clin prof, State Univ NY Downstate Med Ctr, 65-69; PROF PEDIAT, HEALTH SCI CTR, STATE UNIV NY STONYBROOK, 69-, CHMN PEDIAT. *Concurrent Pos:* Assoc chmn pediat, Maimonides Med Ctr, 65-67; chmn pediat, Nassau County Med Ctr, 67- *Mem:* Am Acad Pediat; Am Physiol Soc; Soc Pediat Res. *Res:* Pediatric endocrinology and metabolism, especially growth hormone, growth disorders and childhood obesity. *Mailing Add:* 176 Memorial Dr Jesup GA 31545

COLLIS, RONALD THOMAS, b London, Eng, July 22, 20; m 51; c 1. ATMOSPHERIC PHYSICS. *Educ:* Oxford Univ, MA, 51. *Prof Exp:* Meteorologist, Royal Navy, UK, 50-55, Decca Radar Ltd, Eng, 55-58; head, Radar Aerophys Group, SRI Int, 58-67, dir, Atmospheric Sci Lab, 67-78, sr dir, Eng Res Group, Europe, 78-88; CONSULT, 88- *Concurrent Pos:* Consult, NASA, 74- & NSF, 70-; adj prof, San Jose State Univ, 85-; cert consult meteorologist, Am Meteorol Soc. *Mem:* Assoc fel Am Inst Aeronaut & Astronaut; fel Am Meteorol Soc; Am Geophys Union; fel Royal Meteorol Soc. *Res:* Atmospheric factors in propagation of electromagnetic energy; weather radar; lidar; instrumental and data processing aspects of air pollution, aviation, space applications and general meteorology. *Mailing Add:* 553 Cuesta Dr Aptos CA 95003

COLLISON, CLARENCE H, b Battle Creek, Mich, Oct 22, 45; m 68; c 3. ENTOMOLOGY, APICULTURE. *Educ:* Mich State Univ, BS, 68, MS, 73, PhD(entom), 76. *Prof Exp:* Instr entom, Mich State Univ, 75-76; from asst prof to assoc prof entom, Pa State Univ, 76-89; HEAD & PROF ENTOM, MISS STATE UNIV, 89- *Honors & Awards:* Driesbach Mem Award, Mich State Univ, 75. *Mem:* Entom Soc Am; Am Beekeeping Fedn; Sigma Xi. *Res:* Pollination and nectar secretion of buckwheat, birds-foot trefoil, and lima beans; biology and control of manure breeding flies, poultry litter beetles and Northern fowl mite; livestock ectoparasites; effects of pesticides on honeybee colonies; factors affecting drone honeybee production. *Mailing Add:* Dept Entomology Miss State Univ PO Drawer EM Mississippi State MS 39762

COLLISSON, ELLEN WHITED, b Alton, Ill, Aug 13, 46; div; c 3. VIRAL IMMUNOLOGY, MOLECULAR VIROLOGY. *Educ:* Univ Ill, BS, 68; Univ Ala, MS, 78, PhD(microbiol), 80. *Prof Exp:* Fel researcher, Univ Ala, 80-81, instr biol, Jefferson State Jr Col, 78-81; res assoc, Agr Res Serv, USDA, 81-88; DEPT VET MICROBIOL, TEX A&M UNIV, 88- *Mem:* Am Soc Microbiologists; Am Asn Avian Path; Am Soc Virol. *Res:* Antibody-mediated fluctuations of the immune system; fingerprint analysis of the oligonucleotides from the ten dsRNA segments of Bluetongue Virus; molecular biology and virology; Cache Valley virus, infectious bronchitis virus and feline immuno defficiency virus. *Mailing Add:* Dept Vet Microbiol Tex A&M Univ College Station TX 77843

COLLISTER, EARL HAROLD, b Galva, Ill, Mar 25, 23; m 45; c 3. PLANT PHYSIOLOGY, STATISTICS. *Educ:* Purdue Univ, BS, 47, MS, 48, PhD(agron & plant genetics), 50. *Prof Exp:* Assoc agronomist oilseed res, Tex Res Found, 50-51, agronomist & chmn field crops dept, 51-56, from sr agronomist to prin agronomist & chmn plant sci dept, 56-59; from asst dir & chief agronomist to dir & chief agronomist, High Plains Res Found, 59-67; exec vpres res sales mkt in US & overseas, World Seeds, Inc, 67-68; PRES, INT GRAIN, INC, 68-; PRES, TRANSERA RES, INC, 71- *Concurrent Pos:* Consult, Francisco Sugar Co, Cuba, 56-58; hon trustee, Int Sesanum Found; owner & pres, C Bar Ranch Properties. *Mem:* Am Soc Agron; Am Genetic Asn; AAAS; Sigma Xi. *Res:* Red clover; sesame; sunflowers; soybeans; triticale; wheat; safflower; corn; grain sorghum; millet. *Mailing Add:* Rte No 2 Box 449-A Fairfield Bay AR 72088

COLLMAN, JAMES PADDOCK, b Beatrice, Nebr, Oct 31, 32; m 55. MOLECULAR CONDUCTORS, BIOMINETIC CATALYSTS. *Educ:* Univ Nebr, BSc, 54, MS, 56; Univ Ill, PhD, 58. *Prof Exp:* From instr to prof chem, Univ NC, 58-67; PROF CHEM, STANFORD UNIV, 67-, DAUBERT PROF CHEM, 80- *Concurrent Pos:* A P Sloan Found fel, 63-66; Frontiers in Chem lectr, 64; Guggenheim fel, 77-78 & 85-86; Erskine fel, Univ Canterbury, New Zealand, 72. *Honors & Awards:* Calif Sect Award, Am Chem Soc, 72, Inorg Chem Award, 75, Pauling Award, 90, Distinguished Serv Award, 91. *Mem:* Nat Acad Sci; Am Chem Soc; Royal Soc Chem; AAAS. *Res:* Synthesis and electron transport properties of metal-metal bonds, reactions of coordinated dioxygen, homogeneous catalysis; reactions of coordinated ligands and homogeneous catalysis; mixed-functions oxygenase models. *Mailing Add:* Dept Chem Stanford Univ Stanford CA 94305

COLLOTON, JOHN W, b Mason City, Iowa, Feb 20, 31; m 60; c 3. PUBLIC HEALTH. *Educ:* Loras Col, BA, 53; Univ Iowa, MA, 57. *Prof Exp:* Hosp rels rep, Hosp Serv, Inc, Iowa, 57-58; bus mgr & admin asst, Univ Iowa Hosps & clins, 58-61, asst dir, 63-69, assoc dir, 69-71, DIR & ASST TO UNIV PRES, STATEWIDE HEALTH SERVS, UNIV IOWA HOSPS & CLINS, 71- *Concurrent Pos:* Mem, Coun Finance, Am Hosp Asn, 77, Comt Nominations, 83-87, Comt Med Educ, 84-; chmn, Coun Teaching Hosps, Asn Am Med Cols, 79; trustee, Nat Dermat Found, 80-82; mem bd dirs, Am Asn Hosp Planning, 82, Baxter Int, Inc, 89-, Nat Med Waste, Inc, Nashville, Tenn, 90; mem, Panel Nat Health Care Surv, Inst Med & Nat Res Coun, 90-; numerous other comts. *Honors & Awards:* Distinguished Serv Award, Am Hosp Asn, 90. *Mem:* Inst Med-Nat Acad Sci; fel Am Col Healthcare Execs; Am Hosp Asn; Asn Am Med Cols. *Mailing Add:* Univ Iowa Hosp & Clin Newton Rd Iowa City IA 52242

COLLUM, DAVID BOSHART, b Syracuse, NY, Apr 25, 55. CHEMISTRY. *Educ:* Cornell Univ, BS, 77; Columbia Univ, PhD(chem), 80. *Prof Exp:* ASST PROF CHEM, CORNELL UNIV, 80- *Mem:* Am Chem Soc. *Res:* Natural products synthesis; transition metal chemistry as applied to organic synthesis. *Mailing Add:* Dept Chem Baker Lab Cornell Univ Ithaca NY 14853

COLLURA, JOHN, b Waltham, Mass, Nov 11, 48; m 70; c 2. TRANSPORTATION. *Educ:* Merrimack Col, BS, 70; Villanova Univ, MS, 71; NC State Univ, PhD(civil eng), 76. *Prof Exp:* Design eng, Murphy Assocs, 69-70; field & off engr, J F White Construct & Contracting, 70; transport engr, Valley Forge Labs, 70-71; field & off engr, J F White Construct & Contracting, 71-72; teaching asst civil eng, NC State Univ, 72-73; transp analyst, NC Dept Transp, 73-76; asst prof transp, 76- ASSOC PROF CIVIL ENG, AT UNIV

MASS, AMHERST. *Concurrent Pos:* Transp consult, Mass Cent RR, 77, Champagne & Assocs, 80, Cape Cod Regional Transit Authority, 80-81 & Summit Land Trust, Fall River, Mass, 81; prin investr, Cape Cod Regional Transit Authority, 78, Mass Exec Off Transp & Contruct, 78, US Dept Transp, 78-81, Nat Sci Found, 78-79, Mass Dept Pub Works, 80-81, US Dept Transp, 81-83. *Mem:* Am Soc Civil Engrs; Transp Res Bd. *Res:* Transportation systems planning; evaluation and public transportation design and operations. *Mailing Add:* Dept Civil Eng Univ Mass Marston 214C Amherst MA 01003

COLLVER, MICHAEL MOORE, solid state physics, for more information see previous edition

COLMAN, BRIAN, b Stockport, Eng, Oct 19, 33; m 64; c 3. PLANT PHYSIOLOGY. *Educ:* Univ Keele, BA, 58; Univ Wales, PhD(plant physiol), 61. *Prof Exp:* Nat Res Coun Can fel biol, Queen's Univ, Ont, 61-62; res assoc, Univ Rochester, 62-65; assoc dean sci, 82-85, from asst prof to assoc prof, 65-78, PROF BIOL, YORK UNIV, ONT, 78- *Concurrent Pos:* Assoc ed, Can J Bot, 88- *Mem:* Am Soc Plant Physiologists; Phycol Soc Am; Can Soc Plant Physiologists (treas, 82-84, pres, 87-88). *Res:* Photosynthesis and photorespiration in algae and higher plants; inorganic carbon uptake and transport, and carbon metabolism, in algae and cyanobacteria. *Mailing Add:* Dept Biol York Univ 4700 Keele St Downsview ON M3J 1P3 Can

COLMAN, DAVID RUSSELL, b New York, NY, Jan 4, 49. CELL BIOLOGY, NEUROBIOLOGY. *Educ:* State Univ NY, PhD(microbiol), 78. *Prof Exp:* Prof researcher, 78-81, res asst prof, 81-84, ASST PROF CELL BIOL, SCH MED, NY UNIV MED CTR, 84- *Concurrent Pos:* Hirschl award, Irma T Hirschl Trust, 85-90. *Mem:* Am Soc Neurochem; Am Soc Cell Biol; NY Acad Sci; AAAS. *Res:* Molecular mechanisms underlying the synthesis and assembly of polypeptide components into the developing myclin sheath in vertebrate nervous systems. *Mailing Add:* Dept Cell Biol NY Univ Sch Med 550 First Ave New York NY 10016

COLMAN, MARTIN, b Johannesburg, SAfrica, Oct 28, 41; US citizen; m 64; c 2. RADIATION ONCOLOGY, NUCLEAR MEDICINE. *Educ:* Univ Witwatersrand, Johannesburg, MB, BCH, 64, MMed, 71; Royal Col Physicians & Royal Col Surgeons, DMRT, 71. *Prof Exp:* Instr nuclear med, Sch Med & radiologist, Div Radiother, Hosp, Johns Hopkins Univ, 72-73; assoc dir radiation oncol, Michael Reese Hosp, 73-77; asst prof, Univ Chicago, 77-81; asst prof, 77-81, CHIEF RADIATION ONCOL, UNIV CALIF, IRVINE, 77-, ASSOC PROF RADIOL SCI & OBSTET & GYNEC, 81- *Concurrent Pos:* Attend physician, Div Nuclear Med, Michael Reese Hosp, 73-77; consult, Radiation Therapy, Long Beach Vet Admin Hosp, 78-; asst prof, Dept Obstet & Gynec, Univ Calif, Irvine, 79-, assoc prof, 81- *Mem:* Am Soc Therapeut Radiol & Oncol; Radiol Soc NAm; Am Col Radiol; Radiation Res Soc; Royal Col Radiol. *Res:* Radiation Carcinogenesis; drug-radiation interactions; modification of radiation effects; optimization of radiation therapy of cancer. *Mailing Add:* Div Radiation Oncol Dept Radiol Sci Univ Calif Irvine CA 92717

COLMAN, NEVILLE, b Johannesburg, SAfrica, July 30, 45; US citizen; m 76; c 2. HEMATOLOGY & BLOODBANK, NUTRITION. *Educ:* Univ Witwatersrand, MB, BCh, 69, PhD(hemat), 74. *Prof Exp:* Intern med & obstet, Johannesburg Gen Hosp, 70; registr hemat, Sch Path, Univ Witwatersrand & SAfrican Inst Med Res, 71-74; registr med, Edenvale Hosp, 71-73; fel med, Vet Admin Hosp, 74-76, res assoc hemat, 76-80, dir, Blood Bank & Hemat Lab, Vet Admin Med Ctr, Bronx, 80-89; asst prof med, State Univ NY, Downstate Med Ctr Brooklyn, 76-80; fel med, Vet Admin Hosp, 74-76, res assoc hemat, 76-80, DIR, BLOOD BANK & HEMAT LAB, VET ADMIN MED CTR, BRONX, 80-; ASSOC PROF PATH, MT SINAI SCH MED, 80-, DIR CLIN LAB, 89- *Concurrent Pos:* NIH grant, Nat Cancer Inst, 74-76. *Honors & Awards:* Watkins-Pitchford Mem Prize, SAfrican Inst Med Res, 74. *Mem:* Am Soc Hemat; Am Soc Clin Nutrit; Am Inst Nutrit; Am Fedn Clin Res; Am Asn Blood Banks; Am Soc Clin Path; Col Am Pathologists. *Res:* Hematology, nutritional anemias, folate and vitamin B12 metabolism; folate analogues in chemotherapy; transfusion medicine; human immunodeficiency virus infection. *Mailing Add:* 670 W End Ave New York NY 10025

COLMAN, ROBERT W, b New York, NY, June 7, 35; m 57; c 2. HEMATOLOGY, HEMOSTASIS. *Educ:* Harvard Univ, AB, 56, MD, 60. *Hon Degrees:* MA, Univ Pa. *Prof Exp:* Assoc med, Harvard Med Sch, 67-69, from asst prof to assoc prof med, 69-73; from assoc prof to prof med & path, Univ Pa, 69-76; PROF MED & CHIEF HEMAT DIV, TEMPLE UNIV SCH MED, 78-, PROF THROMBOSIS RES, 81- *Concurrent Pos:* Dir thrombosis, Temple Univ Sch Med, 79-, prof path, 87-; mem, Coun Thrombosis, Am Heart Asn & Microcirculatory Thrombosis Task Group, Nat Heart Lung & Blood Inst; chmn, Int Comt Haemostasis & Thrombosis, 84-86. *Mem:* Am Soc Hemat. *Res:* Platelet receptors and plasma proteases; human factor V; high molecular weight kininogen and prekallikrein in coagulation; fibrinolysis and kinin formation; stimulus-response coupling in platelets as mediated by cAMP; characterization of a recently identified ADP receptor in platelets. *Mailing Add:* Thrombosis Res Ctr Temple Univ Sch Med 3400 N Broad St Philadelphia PA 19140

COLMAN, ROBERTA F, b New York, NY, July 21, 38; m 57; c 2. BIOCHEMISTRY, PROTEIN CHEMISTRY. *Educ:* Radcliffe Col, AB, 59, AM, 60, PhD(biochem), 62. *Prof Exp:* NIH fel enzym, 62-64; USPHS fel, Sch Med, Wash Univ, 64-66, asst prof enzym & protein chem, 66-67; assoc enzym, Harvard Med Sch, 67-69, from asst prof to assoc prof biochem, 69-73; PROF BIOCHEM, UNIV DEL, 73- *Concurrent Pos:* USPHS career develop award & res grant, Sch Med, Wash Univ, 66-67; Med Found Boston fel, Harvard Med Sch, 67-68, USPHS res grant, 67-, career develop award, 68-, Med Found res grant, 70-; Am Soc Biol Chem travel award, Int Cong Biochem, 67, 70 & 73; assoc ed, J Protein Chem; mem, Biochem Study Sect, NIH, 74-78; mem, Cellular & Molecular Basis Dis Rev Comt, Nat Inst Gen Med Sci, 79; mem, Study Sect, Am Heart Asn, 84-88 & Am Cancer Soc, 87-90; exed ed, Archives Biochem Biophysics, 84-; Travel Award, Int Cong Biochem, 85, 88; mem, Phys Biochem Study Sect, NIH, 89- *Honors & Awards:* Sci Achievement Award, Del Sect, Am Chem Soc, 90. *Mem:* AAAS; Am Soc Biol Chemists (treas, 81); Biophys Soc; Am Chem Soc. *Res:* Mechanism of enzyme action; active sites of dehydrogenases; chemical basis of regulation of allosteric enzymes; affinity labeling of purine nucleotide sites in proteins; investigation of specific enzymes such as glutathione reductase, glutamate dehydrogenase, isocitrate dehydrogenase, pyruvate kinase. *Mailing Add:* Dept Biochem Univ Del Newark DE 19716

COLMAN, STEVEN MICHAEL, b New Kensington, Pa, Apr 1, 49; m 72; c 2. QUATERNARY GEOLOGY, GEOMORPHOLOGY. *Educ:* Notre Dame Univ, BS, 71; Pa State Univ, MS, 74; Univ Colo, PhD(geol), 77. *Prof Exp:* RES GEOLOGIST, US GEOL SURV, COLO, 76- *Honors & Awards:* Outstanding Paper Award, Geol Soc Am, 76; Kirk Bryan Award, Geol Soc Am, 84. *Mem:* Geol Soc Am; Am Quaternary Asn. *Res:* Quaternary stratigraphy; Quaternary dating techniques; paleoclimatology; marine geology. *Mailing Add:* US Geol Surv Quissett Campus Woods Hole MA 02543

COLMANO, GERMILLE, b Pola, Italy, Aug 22, 21; nat US; m 47; c 3. VETERINARY PHYSIOLOGY, MEDICAL BIOPHYSICS. *Educ:* Univ Bologna, DVM, 49, PhD(physiol, biochem), 50. *Prof Exp:* Asst physiol, Univ Bologna, 47-49, from instr to asst prof, 49-51; asst vet, Phillips Vet Hosp, Colo, 51-52; asst vet sci, Univ Wis, 52-53; proj asst, Inst Enzyme Res, Wis, 54-56; scientist biophys, Res Inst Advan Study, 56-61; Stoner fel biophys res lab, Eye & Ear Hosp, Sch Med, Univ Pittsburgh, 61-62; prof physiol/biophys, Dept Vet Sci, Col Agr & Life Sci, Va Polytech Inst & State Univ, 62-78, dir vet biol & clin sci, Dept Vet Biosci, Vet Admin Maryland Region Col Vet Med, 78-86 Dept Vet Biosci, 86-89, DIR VET BIOL & CLIN SCI, DEPT BIOMED SCI, VA REGIONAL COL VET MED, VA POLYTECH INST & STATE UNIV, 89- *Concurrent Pos:* Horsley res award, Va Acad Sci, 84. *Mem:* NY Acad Sci; Am Soc Vet Physiol & Pharmacol; Biophys Soc; fel Royal Soc Health. *Res:* Lamellar function-structure; monomolecular films; absorption spectra of chromophores; trace elements; steroids and stress in health and disease; bacteriostasis of silver in intramedullary orthopedic pins; early cancer, myocandial infarction, trace and toxic minerals detection by computer analysis of spectrophotometric absorbance of body fluids; nerve-nerve mending, and nerve-prostheses connections with monolayers of metal(s)-stearate. *Mailing Add:* Dept Biomed Sci Va Regional Col Vet Med Va Polytech Inst & State Univ Blacksburg VA 24061-0442

COLMENARES, CARLOS ADOLFO, b Ocana, Colombia, June 17, 32; US citizen; m 56; c 2. PHYSICAL CHEMISTRY, CHEMICAL ENGINEERING. *Educ:* Univ Calif, Berkeley, BS, 53; Wash State Univ, MS, 56; Rensselaer Polytech Inst, PhD(chem eng), 60. *Prof Exp:* Actg instr chem eng, Wash State Univ, 54-56; instr, Rensselaer Polytech Inst, 56-60; CHEMIST, LAWRENCE LIVERMORE LAB, 60- *Mem:* Am Vacuum Soc. *Res:* Gas-solid interactions; surface chemistry and physics; radiation chemistry; catalysis. *Mailing Add:* 2211 Granite Dr Alamo CA 94507

COLODNY, PAUL CHARLES, b Springfield, Mass, Feb 17, 30. PHYSICAL CHEMISTRY, POLYMER CHEMISTRY. *Educ:* Univ Mass, BS, 51; Princeton Univ, MA, 53, PhD(chem), 57. *Prof Exp:* Fiber physicist, Dow Chem Co, 57-60; from res chemist to sr res chemist, Aerojet-Gen Corp, 60-63, res chem specialist, 63-65; sr res chemist, Gen Tire & Rubber Co, Ohio, 65-66; sr staff mem, Raychem Corp, 66-70; res scientist, Lockheed Missile & Space Co, 70-73; group leader, MCA Disco-Vision, 74-76; chief chemist, Reed Irrig Systs, 76-78; sr chem specialist, 81-84, assoc scientist, Aerojet Strategic Propulsion Co, 84-88; CONSULT, 88- *Concurrent Pos:* Consult, Raychem Corp, 71-73 & Carson-Alexion Corp, 76-78. *Mem:* Am Chem Soc; Soc Rheol; AAAS. *Res:* Fiber physics; textile fiber spinning; elastomers; solid rocket propellants; crystalline polymers; graphite, photo-polymerization; composite materials. *Mailing Add:* 10531 Mills Tower Dr No 41 Rancho Cordova CA 95670

COLOMBANI, PAUL MICHAEL, b Salem, Mass, May 7, 51; m 74. PEDIATRIC SURGERY, TRANSPLANTATION SURGERY. *Educ:* St Anselms Col, BA, 72; Univ Ky Col Med, MD, 76. *Prof Exp:* Resident surgery, George Wash Univ Sch Med, 76-80, chief resident surgery, 80-81; resident, 81-82, chief resident pediat surgery, 83-84, asst prof surgery & pediat & oncol, 84-87, ASSOC PROF SURGERY, PEDIAT & ONCOL, JOHNS HOPKINS SCH MED, 87- *Concurrent Pos:* Attending physician, John Hopkins Hosp, 83-, post doctorate fel, Transplantation Biol Lab, 83-84; Andrew Mellon Found Scholar, Johns Hopkins Univ Sch Med, 84-85. *Mem:* Am Col Surgeons fel; fel Am Acad Pediat; Soc Univ Surgeons; Transplantation Soc; Am Pediat Surgical Asn; Am Soc Transplant Surgeons. *Res:* Mechanism of action of the immunosuppresive drug cyclosporine at the subcellular level using biochemical techniques; immunopharmacology of other drugs that affect the calcium dependent problems during T-lymphocyte activation. *Mailing Add:* CMSC 7-116 Div Pediat Surgery Johns Hopkins Hosp, 600 N Broadway Baltimore MD 21205

COLOMBANT, DENIS GEORGES, b Bourg, France, Feb 12, 42; US citizen; m 65; c 3. PLASMA PHYSICS. *Educ:* Nat Sch Higher Arts & Trade, Paris, Ing, 63; Mass Inst Technol, MS, 66, PhD(nuclear eng), 69. *Prof Exp:* Res Scientist plasma physics, Atomic Energy Comn, Saclay, France, 69-70 & Limeil, France, 70-73; res scientist, SAI, McLean, Va, 73-76; RES SCIENTIST PLASMA PHYSICS, NAVAL RES LAB, WASHINGTON, DC, 76- *Mem:* Am Phys Soc; Sigma Xi. *Res:* Inertial confinement fusion. *Mailing Add:* 4439 Davenport St N W Washington DC 20016

COLOMBINI, MARCO, b Modena, Italy, June 28, 48; m 77; c 2. MEMBRANE BIOPHYSICS, MOLECULAR BIOLOGY OF PROTEINS. *Educ:* McGill Univ, BS, 70, PhD(biochem), 74. *Prof Exp:* Postdoctoral physiol, Albert Einstein Col Med, Bronx, NY, 74-76, asst prof neurosci, 76-79, asst prof physiol, 77-79; from asst prof to assoc prof, 79-89, PROF ZOOL, UNIV MD, COLLEGE PARK, 89- *Concurrent Pos:* Speaker biophys, Ettore

Majorana Ctr, Int Sch Biophys, Erice, Italy, 88. *Mem:* Biophys Soc; Am Soc Cell Biol; World Fedn Scientists. *Res:* Molecular mode of action of membrane transport systems, especially channel-forming proteins, including: voltage-gating, selectivity and modulation by other molecules; mitochondrial channel called VDAC. *Mailing Add:* Dept Zool Univ Md College Park MD 20742

COLOMBINI, VICTOR DOMENIC, b Boston, Mass, Feb 2, 24. GEOLOGY. *Educ:* Boston Univ, AB, 50, AM, 52 & 53, PhD, 61. *Prof Exp:* Geologist, USAEC, 55; asst prof geol, La Polytech Inst, 56-61; actg asst prof, Univ Miss, 61; assoc prof earth sci & head dept, Findlay Col, 61-66; mem fac, 66-68, asst prof, 68-73, ASSOC PROF GEOG, OHIO STATE UNIV, LIMA BR, 73- *Mem:* Geol Soc Am; Mineral Soc Am; Nat Asn Geol Teachers; Asn Am Geog; Nat Coun Geog Educ. *Res:* Physical geography; introductory economic geography; economic geology; crystallography; mineralogy; geomorphology. *Mailing Add:* 2276 E Elm St Lima OH 45804

COLON, FRAZIER PAIGE, b Athol, Mass, Mar 28, 34; m 60; c 3. MECHANICAL ENGINEERING, FLUID MECHANICS. *Educ:* Norwich Univ, BSME, 56; Worcester Polytech Inst, MSME, 68. *Prof Exp:* Asst engr, Allis-Chalmers Mfg Co, 56-61; proj engr, Alden Res Labs, Worcester Polytech Inst, 61-64; res observer, Nat Eng Lab, E Kilbride, Scotland, 64-66; instr fluid mech, Alden Res Lab, Worcester Polytech Inst, 66-74; proj mgr, Riley Stoker Corp, 74-76; sr mech eng, Chas T Main, Inc, 76-77; mgr eng, CE-KSB Pump Co, Inc, 77-85; GEN MGR, JVS CORP, 85- *Mem:* Am Soc Mech Engrs; Nat Soc Prof Eng. *Res:* Turbomachinery; design of new pumps; testing performance characteristics; optimizing design and increasing efficency; thermodynamics techniques for measuring pump efficiency. *Mailing Add:* PO Box 594 Durham NH 03824

COLON, JULIO ISMAEL, b Coamo, PR, June 19, 28; US citizen; m 55; c 2. VIROLOGY. *Educ:* Univ PR, BS, 50; Univ Chicago, PhD(microbiol), 59. *Prof Exp:* Fel, Nat Acad Sci-Nat Res Coun, 59-60; microbiologist, Ft Detrick, Md, 60-63; microbiologist, US Trop Res Med Lab, 63-65; ASSOC PROF VIROL, SCH MED, UNIV PR, SAN JUAN, 64- *Concurrent Pos:* Dir grad studies, Univ PR, San Juan, 72-74. *Mem:* Am Soc Microbiol; fel Am Acad Microbiologist; Tissue Culture Asn; Sigma Xi. *Res:* Intermediary metabolism; biochemistry and genetics of psittacosis group of microorganism; biochemistry and genetics of the arthropod-borne viruses; origin and source of viral infectious nucleic acid; radiobiology of viruses; viral and tumor immunology. *Mailing Add:* Micro Dept Univ PR Med Sci Campus-GPO 5067 San Juan PR 00936

COLONIAS, JOHN S, b Athens, Greece, June 8, 28; US citizen; m 59; c 2. COMMUNICATIONS NETWORKS, NUMERICAL SIMULATIONS. *Educ:* Ore State Univ, BS, 59; Univ Uppsala, Sweden, PhD(physics), 79. *Prof Exp:* Staff scientist res & develop/physics, Lawrence Berkeley Lab, Univ Calif, 64-85; chief scientist res & develop, Inst Technol Develop, 85-87; CHIEF SCIENTIST COMMUN, MOBILE TELECOMMUN TECHNOL, 87- *Concurrent Pos:* Consult, 70-; lectr, Univ Calif-Berkeley, 74-85, Mendez Educ Found, PR, 86-89; adj prof computer sci, Jackson State Univ, 86- *Mem:* Inst Elec & Electronic Engrs; Sigma Xi. *Res:* Numerical models in electromagnetic fields; communications networks and their design; numerical methods and computer simulations of various physical processes. *Mailing Add:* 27 Redbud Lane Madison MS 39110

COLONNIER, MARC, b Quebec, Que, May 12, 30; m 59; c 1. ANATOMY. *Educ:* Univ Ottawa, BA, 51, MD, 59, MSc, 60; Univ London, PhD(neurobiol), 63. *Prof Exp:* Med Res Coun Can fel, Univ Ottawa & Univ London, 59-63; asst prof anat, Univ Ottawa, 63-65; from asst prof to assoc prof neuroanat, Neurol Sci Lab, Univ Montreal, 65-69; prof anat & head dept, Univ Ottawa, 69-76; PROF ANAT, DEPT ANAT, UNIV LAVAL, 76- *Honors & Awards:* Lederle Med Fac Award, 66; Charles Judson Herrick Award, Am Asn Anat, 67. *Mem:* Can Asn Anat; Am Asn Anat; fel Royal Soc Can; Soc Neurosci. *Res:* Cerebral cortex; synapses; visual system. *Mailing Add:* Dept Anat Univ Laval Fac Med Quebec PQ C1K 7P4 Can

COLONY-COKELY, PAMELA, b Boston, Mass, Apr 18, 47; m 86; c 1. DEVELOPMENTAL GASTROENTEROLOGY, CELLULAR DIFFERENTIATION. *Educ:* Wellesley Col, BA, 69; Boston Univ, PhD(anat), 76. *Prof Exp:* Res asst, Boston Univ Sch Med, 69-71; res asst, Univ Hosp, 71-73; res asst, Div Gastroenterol, Peter Bent Brigham Hosp, 73-75; res asst, dept med, Harvard Med Sch, 75-77; assoc staff, dept med, Peter Bent Brigham Hosp, 76-79; instr, dept med, Harvard Med Sch, 77-79, sr fel & instr, 79-81; asst prof, Dept Anat, Pa State Univ Col Med, 81-82, asst prof, depts anat & med, 82-88, ADJ ASSOC PROF, DEPT SURG, PA STATE UNIV COL MED, 88-; RES ASSOC PROF BIOL, FRANKLIN & MARSHALL COL, 88- *Concurrent Pos:* Ger Exchange Serv grant, Res Proj, Fed Repub Ger, 78; sr fel award, Nat Found Ileitis & Colitus Inc, 79-81; Cancer Res Ctr instl grant award, Milton S Hershey Med Ctr, 82-83; PHS grant, Struct & Function Develop Rat Colon, 82-85, 87-91, R04 Ext Funding, 85-86; Prehealth adv, Pre-healing Arts Off, Franklin & Marshall Col, 88-90; independent assessor, Nat Health & Med Res Coun, Commonwealth Australia, 85-; reviewer, Prog Proj Grant, Maternal & Child Health Res Comt, NIH, 86, Nat Cancer Inst, 86. *Mem:* NY Acad Sci; AAAS; Am Soc Cell Biol; Am Gastroenterol Asn; Nat Asn Advisors Health Professions, 88-90. *Res:* Defining the morphological and biochemical changes associated with cellular differentiation in normal developing colon and in disease states; composition and cellular localization of different plasma membrane and secretory glycoproteins, particularly the mucins. *Mailing Add:* Biol Dept Franklin & Marshall Col PO Box 3003 Lancaster PA 17604-3003

COLOVOS, GEORGE, b Thessaloniki, Greece, Oct 15, 32; m 70; c 1. ANALYTICAL CHEMISTRY, ATMOSPHERIC CHEMISTRY. *Educ:* Univ Thessaloniki, Greece, BS, 57, PhD(chem), 66. *Prof Exp:* From instr to lectr, Univ Thessaloniki, Greece, 61-69; res assoc, Univ Ariz, 67-69 & 71-73 & Utah State Univ, 69-71; mem tech staff chem, Rockwell Int, 73-74, mgr chem lab, 75-76, mgr chem, 76-77, MGR TECH OPERS, ENVIRON MONITORING & SERV CTR, ROCKWELL INT, 77- *Concurrent Pos:* Mem sci comn, Nat Inst Occupational Safety & Health, 72-73. *Mem:* Am Chem Soc. *Res:* Development of analytical methods for determination of pollutants. *Mailing Add:* 1924 Ferndale Pl Thousand Oaks CA 91360-1873

COLP, JOHN LEWIS, b Carterville, Ill; c 3. GEOLOGICAL ENGINEERING, GEOTHERMAL ENGINEERING. *Educ:* Univ Ill, BS, 49; Univ NM, MS, 66; Tex A&M Univ, PhD(civil eng), 72. *Prof Exp:* Proj engr, El Segundo Div, Douglas Aircraft Co, 42-45; transmission engr, Allison Div, Gen Motors Corp, 48-50; plant engr, Inland Container Corp, Indianapolis, 50-53; chief engr, Charles Dowd Box Co, 53-55; mem tech staff, Sandia Labs, Albuquerque, 55-69; eng res assoc, Tex Eng Exp Sta, 69-72; mem tech staff, Sandia Nat Labs, Albuquerque, 72-82; VPRES, ISLAND SCI, INC, ALBUQUERQUE, 83- *Mem:* Am Geophys Union; Sigma Xi. *Res:* Magma energy utilization; geothermal energy; geology and engineering interface; marine sediment instrumentation; anchor pullout forces; earth penetration by projectiles. *Mailing Add:* 1023 Washington SE Albuquerque NM 87108

COLPA, JOHANNES PIETER, b Arnhem, Netherlands, Jan 26, 26; m 51; c 1. THEORETICAL CHEMISTRY, MAGNETIC RESONANCE. *Educ:* Univ Amsterdam, PhD(chem), 57. *Prof Exp:* Res chemist, Shell Res Labs, Amsterdam, 57-66; assoc prof chem, Univ Amsterdam, 63-69; PROF CHEM, QUEEN'S UNIV, ONT, 69- *Concurrent Pos:* Fel, Cambridge Univ, 61-62; consult, Max Planck Inst, Ger, 67-69; assoc ed, Molecular Physics, 64-69. *Mem:* Am Phys Soc; Netherlands Phys Soc; NY Acad Sci; Hist Sci Soc. *Res:* High pressure spectroscopy; pressure induced spectroscopic transitions; molecular orbital theory and magnetic resonance; theory of energy differences; reinterpretation of Hund's rules; theory of optical nuclear polarization; optical detection of magnetic resonance in paracyclophases; nuclear magnetic resonance and nuclear quadrupole resonance in excited stakes of molecules; phosphorescence; photochemical reactions in single crystals; tunnel effects. *Mailing Add:* Dept Chem Queen's Univ Kingston ON K7L 3N6 Can

COLQUHOUN, DONALD JOHN, b Toronto, Ont, Mar 29, 32; m 56; c 2. GEOLOGY. *Educ:* Univ Toronto, BA, 53, MA, 56; Univ Ill, PhD(geol), 60. *Prof Exp:* Asst geol, Univ Ill, 58-59; from asst prof to assoc prof, 60-64, chmn dept, 70-73, PROF GEOL & MARINE BIOL, UNIV SC, 67- *Concurrent Pos:* Proj geologist, SC State Develop Bd, 60-; grants, NSF, Am Geol Inst, US Geol Surv, 63- *Honors & Awards:* Russell Res Award Sci & Eng, 69. *Mem:* Geol Soc Am; Am Asn Petrol Geol. *Res:* Application of sedimentalogical studies and techniques to geomorphology and stratigraphy; interpretation of coastal plain terraces; regional stratigraphy. *Mailing Add:* Dept Geol & Marine Sci Univ SC Columbia SC 29208

COLQUITT, LANDON AUGUSTUS, b Ft Worth, Tex, Jan 25, 19; m 54; c 2. MATHEMATICS. *Educ:* Tex Christian Univ, BA, 39; Ohio State Univ, MA, 41, PhD(math), 48. *Prof Exp:* Instr math, Ohio State Univ, 46-48; from asst prof to prof, 48-49, EMER PROF MATH, TEX CHRISTIAN UNIV, 89- *Mem:* AAAS; Am Math Soc; Math Asn Am; Soc Indust & Appl Math. *Res:* Mathematical analysis; applied mathematics. *Mailing Add:* 2601 McPherson Ft Worth TX 76109

COLSON, ELIZABETH F, b Hewitt, Minn, June 15, 17. SOCIAL CHANGE. *Educ:* Univ Minn, BA, 38, MA, 40; Radcliffe Col, MA, 41, PhD(social anthrop), 45. *Hon Degrees:* DSc, Brown Univ, 79, Univ Rochester, 85,. *Prof Exp:* Dir, Rhodes-Livingston Inst Social Res, Northern Rhodesia, 47-51; sr lectr, Manchester Univ, 51-53; assoc prof African studies, Boston Univ, 55-59; prof anthrop, Brandeis Univ, 59-63; vis prof, Nortwestern Univ, 63-64; prof, 64-84, EMER PROF ANTHROP, UNIV CALIF, BERKELEY, 84- *Concurrent Pos:* Simon sr fel, Manchester Univ, 51; fel, Ctr Adv Studies Behav Sci, 67-68; Fairchild fel, Caltech, 75-76. *Honors & Awards:* Lewis Henry Morgan Lectr, Univ Rochester, 73; Rivers Mem Medal, Royal Anthrop Inst, 82; Malinowski Lectr, Soc Appl Anthrop, 85. *Mem:* Nat Acad Sci; hon fel Royal Anthrop Inst; fel Am Anthrop Asn; fel AAAS; Am Asn African Studies; Am Acad Arts & Sci; Asn Soc Anthropologists. *Res:* Longitudinal study, Gwembe Tonga and Zambia; development strategies; migration and resettlement. *Mailing Add:* 840 Arlington Blvd El Cerrito CA 94530

COLSON, STEVEN DOUGLAS, b Idaho Falls, Idaho, Aug 16, 41; m 62; c 7. CHEMICAL PHYSICS. *Educ:* Utah State Univ, BS, 63; Calif Inst Technol, PhD(chem), 68. *Prof Exp:* Asst prof, 68-73, assoc prof, 73-80, PROF CHEM, YALE UNIV, 80- *Concurrent Pos:* Jr fac fel, Yale Univ, 72-73; mem, Nat Res Coun Adv Bd to US Army Res Off, 72-75; assoc ed, J Chem Physics, 80-82; chmn, Gordon Res Conf Visible/UV MPI & Dissociation Processes; mem, Petrol Res Fund Adv Bd, 85-87. *Res:* Laser, mass spectrometric and photoelectron spectroscopic techniques; intermolecular interactions and energy transfer in molecular crystals, multiphoton gas and condensed phase spectroscopy. *Mailing Add:* Battele Pacific NW Lab K2-18 PO Box 999 Richland WA 99352-0999

COLTEN, HARVEY RADIN, b Houston, Tex, Jan 11, 39; m 59; c 3. IMMUNOLOGY, PEDIATRICS. *Educ:* Cornell Univ, BA, 59; Western Reserve Univ, MD, 63; Am Bd Allergy & Clin Immunol, dipl; Am Bd Pediat, dipl; Am Bd Allergy & Clin Immunol, dipl. *Hon Degrees:* MA, Harvard Univ. *Prof Exp:* From intern to resident pediat, Univ Cleveland Hosps, 63-65; res assoc immunol, NIH, 65-67, sr scientist, 67-69, head molecular separations unit, 69-70; from asst prof to prof pediat, Harvard Med Sch, 70-86, chief, Div Allergy, 73-76; sr assoc med, Children's Hosp, Boston, 76-86, chief, div cell biol, 76-86; HARRIET B SPOEHRER PROF PEDIAT, PROF MOLECULAR MICROBIOL & CHMN, DEPT PEDIAT, SCH MED, WASH UNIV, ST LOUIS, MO, 86- *Concurrent Pos:* asst prof pediat, George Wash Univ Sch Med, Washington DC, 69-70; assoc med, Children's Hosp Med Ctr, Boston, 70-73, chief, Allergy Div, 73-76, dir Cystic Fibrosis Prog, 76-86; assoc ed, J Immunol, 71-74, J Allergy & Clin Immunol 77-80, New Eng J Med, 78-81, J Immunol, 71-74, J Allergy & Clin Immunol, 77-80, J Immunochem, 72-75, J Clin Invest, 82-85; pediatrician-in-chief, Children's Hosp & Barnes Hosp, 86-, Jewish Hosp, 87-89; mem nominating comt, Inst Med. *Honors & Awards:* E Mead Johnson Award Pediat Res, 79. *Mem:* Inst Med-Nat Acad Sci; Am Asn Immunologists; Am Soc Clin Invest; Asn Am Physicians; Am Pediat Soc; AAAS; NY Acad Sci; fel Am Acad Allergy & Immunol; Am Thoracic Soc; fel Am Acad Pediat. *Res:* Immunology; allergy; pulmonary disease. *Mailing Add:* Dept Pediat Wash Univ Sch Med 400 S Kings Hwy St Louis MO 63110

COLTEN, OSCAR A(ARON), b Detroit, Mich, Apr 24, 12; m 35; c 2. CHEMICAL ENGINEERING. *Educ:* Wayne State Univ, BS, 32; Univ Mich, MS, 34, ScD(phys chem), 36. *Prof Exp:* Jr technologist, Shell Petrol Co, Mo, 36-38; technologist, Shell Oil Co, Tex, 38-41, sr technologist, NY, 41-46; tech asst, Shell Develop Co, Calif, 46-49; asst mgr econ res dept, Shell Chem Corp, 49-52, asst to gen mgr mfg, 52-54, sect leader mfg develop dept, 54-58, econ & eval, Indust Chem Div, 59-67, SR STAFF ENGR, SHELL CHEM CO, 67- *Concurrent Pos:* Dep br chief, Chem Div, Nat Prod Authority, Washington, DC, 51-52. *Mem:* AAAS; Am Chem Soc; Am Inst Chem Engrs. *Res:* Reaction kinetics; petroleum refining; petrochemicals; chemical economics. *Mailing Add:* 1693 West Belt Dr S Houston TX 77042-3107

COLTER, JOHN SPARBY, b Bawlf, Alta, July 23, 22; m 50; c 2. VIROLOGY. *Educ:* Univ Alta, BS, 45; McGill Univ, PhD(biochem), 51. *Prof Exp:* Mem staff, Virus & Rickettsial Res Div, Lederle Labs, 51-57 & Wistar Inst, Pa, 57-61; prof biochem & head dept, Univ Alta, 61-87; RETIRED. *Concurrent Pos:* Assoc ed, J Cellular Physiol, 61-76, Virol, 60-; mem, Res Adv Group, Nat Cancer Inst Can, 76-81, Sci Adv Coun, Alta Heritage Found Med Res, 82-; vis prof, Univ Brazil, 64; vis scientist, Swiss Trust Exp Cancer Res, 70, Dept Cell Biol, Univ Auchland, Med Res Coun Can, 77; sci adv to pres, Alta Heritage Found Med Res, 87-90; spec adv to assoc dean res, Fac Med, Univ Alta, 91- *Mem:* Am Soc Biol Chem; Am Soc Cell Biol; Am Asn Cancer Res; Can Biochem Soc; fel Royal Soc Can. *Res:* Biochemistry of virus infection; mechanisms of viral replication and oncogenesis. *Mailing Add:* 7920 119th St Edmonton AB T6G 2L4 Can

COLTHARP, FORREST LEE, b Caney, Kans, Oct 30, 33; m 52; c 3. MATHEMATICS EDUCATION. *Educ:* Okla State Univ, BS, 57, MS, 60, EdD(higher educ), 68. *Prof Exp:* Teacher & math consult, pub schs, Okla, 57-64; asst prof math educ, Kans State Col, 64-66; instr, Okla State Univ, 66-67; PROF MATH EDUC, PITTSBURG STATE UNIV, 67- *Concurrent Pos:* Chmn, Conv & Conf Comt, Nat Coun Teachers Math. *Mem:* Nat Coun Teachers Math. *Mailing Add:* 1402 S Homer Pittsburg KS 66762

COLTHARP, GEORGE B, b Maringouin, La, Nov 28, 28; m 53; c 2. FOREST HYDROLOGY, WATER POLLUTION. *Educ:* La State Univ, BS, 51; Colo State Univ, MS, 55; Mich State Univ, PhD(forest hydrol), 58. *Prof Exp:* Proj leader watershed mgt, Rocky Mountain Forest & Range Exp Sta, US Forest Serv, NMex, 58-61; asst mgr, Coltharp's Livestock Mkt, La, 61-64; from asst prof to assoc prof range watershed mgt, Utah State Univ, 64-74; ASSOC PROF FORESTRY, UNIV KY, 75- *Mem:* Am Water Resources Asn; Soc Am Foresters; Am Soc Range Mgt; Soil Conserv Soc Am; Sigma Xi. *Res:* Forest hydrology studies; wildland water quality and watershed management. *Mailing Add:* Dept Forestry Univ Ky Lexington KY 40506

COLTHUP, NORMAN BERTRAM, b Paris, France, July 6, 24; nat US; div; c 3. SPECTROSCOPY. *Educ:* Antioch Col, BS, 49. *Hon Degrees:* DSc, Fisk Univ, 74. *Prof Exp:* Infrared spectroscopist, Am Cyanamid Co, 44-86; RETIRED. *Concurrent Pos:* Infrared course lectr, Educ Dept, Am Chem Soc & Fisk Univ, I R Inst, 59-91. *Honors & Awards:* Williams-Wright Award, Coblentz Soc, 79. *Mem:* Coblentz Soc. *Res:* Vibrational spectroscopy; molecular structure studies using infrared spectroscopy; infrared spectra-structure correlations. *Mailing Add:* 71 Strawberry Hill Ave Apt 704 Stamford CT 06902

COLTMAN, CHARLES ARTHUR, JR, b Pittsburgh, Pa, Nov 7, 30; m 51; c 4. HEMATOLOGY, ONCOLOGY. *Educ:* Univ Pittsburgh, BS, 52, MD, 56; Ohio State Univ, MMS, 63; Am Bd Internal Med, dipl, 63, cert hemat, 72, cert med oncol, 73. *Prof Exp:* From intern to resident path, Del Hosp, Wilmington, 56-57; Med Corps, US Air Force, 57-77, flight surgeon, Walker AFB, NMex, 57-59, resident, Univ Hosp, Ohio State Univ, 59-63, staff hematologist, 63-66, chief hemat-oncol serv, Wilford Hall Air Force Med Ctr, 66-77, chmn dept med, 75-76; assoc prof, 77-78, PROF MED & DIR, CLIN MED ONCOL SECT, HEALTH SCI CTR, UNIV TEX, SAN ANTONIO, 78- *Concurrent Pos:* Asst med, Ohio State Univ Hosp, 59-61, from jr asst resident to sr asst resident, 59-62, asst instr med, 61-62, chief med resident & demonstr med, 62-63, attend physician & instr med, 63; mil consult to Surg Gen, US Air Force, 64-77; clin assoc prof physiol & med, Health Sci Ctr, Univ Tex, San Antonio, 70-75; med dir, Cancer Therapy & Res Ctr, San Antonio, 77-; chief, Med Oncol Sect, Med Serv, Audie Murphy Vet Admin Hosp, San Antonio, Tex, 77-; chmn, Southwest Oncol Group, 81-; pres, Am Soc Clin Oncol (ASCO), 88- *Honors & Awards:* Mederi Award, Aerospace Med Div, 69; Stitt Award, Asn Mil Surg US, 70; Harold Brown Award, US Air Force, 71; Legion of Merit, 78. *Mem:* AAAS; fel Am Col Physicians; Am Asn Cancer Res; Am Fedn Clin Res; Am Soc Hemat. *Res:* Research in clinical cancer chemotherapy. *Mailing Add:* 4450 Medical Dr San Antonio TX 78229

COLTMAN, JOHN WESLEY, b Cleveland, Ohio, July 19, 15; m 41; c 2. MUSICAL ACOUSTICS, ELECTRON OPTICS. *Educ:* Case Inst Technol, BS, 37; Univ Ill, MS, 39, PhD(physics), 41. *Prof Exp:* Res engr, Res Labs, Westinghouse Elec Corp, 41-44, sect mgr, 44-49, mgr, Electronics & Nuclear Physics Dept, 49-60, assoc dir, 60-64, dir math & radiation, 64-69, res dir indust & defense prod, 69-73, dir res planning, 73-80; RETIRED. *Concurrent Pos:* Mem adv group electron devices, US Dept Defense, 62-66; mem adv comt, NASA, 64-66; mem numerical data adv bd, Nat Acad Sci, 68-71; mem, Comn on Human Resources, Nat Res Coun, 76- *Honors & Awards:* Longstreth Medal, Franklin Inst, 60; Westinghouse Order of Merit, 68; Roentgen Medal, 70; Gold Medal, Radiol Soc NAm, 82. *Mem:* Nat Acad Eng; fel Am Phys Soc; fel Inst Elec & Electronics Engrs. *Res:* Slow neutrons; microwave tubes; x-ray fluorescence; scintillation counters; image amplifier tubes; energy conversion; control mechanisms; lasers; physics of wind instruments; acoustical properties of musical wind instruments, espcially the flute and organ pipe. *Mailing Add:* 3319 Scathelocke Dr Pittsburgh PA 15235

COLTMAN, RALPH READ, JR, b Pittsburgh, Pa, Nov 15, 24; m 44; c 3. SOLID STATE PHYSICS. *Educ:* Carnegie Inst Technol, BS, 50. *Prof Exp:* Physicist, 50-86, SCI CONSULT, OAK RIDGE NAT LAB, 86- *Honors & Awards:* Mat Sci Competition-Significant Implications for Energy, US Dept Energy, 82. *Mem:* Am Phys Soc; Sigma Xi. *Res:* Irradiation damage in metals; cryogenics. *Mailing Add:* 7905 Wiebelo Dr Knoxville TN 37931

COLTON, CLARK KENNETH, b New York, NY, July 20, 41; m 65; c 4. CHEMICAL ENGINEERING. *Educ:* Cornell Univ, BChE, 64; Mass Inst Technol, PhD(chem eng), 69. *Prof Exp:* Ford fel engr, Mass Inst Technol, 69-70, from asst prof to prof chem eng, 69-80, dep head, Chem Eng Dept, 77-78, Bayer prof, 80-85; CONSULT, 85- *Concurrent Pos:* Consult artificial kidney-chronic uremia prog, Nat Inst Arthritis & Metab Dis; mem adv bd on mil personnel supplies, Nat Res Coun; assoc ed, Am Soc Artificial Internal Organs J, 78-84. *Honors & Awards:* Allan P Colburn Award, Am Inst Chem Engrs, 77; Curtis W McGraw Res Award, Am Soc Eng Educ, 80; Gambro Award, Int Soc Blood Purification, 86. *Mem:* AAAS; Am Inst Chem Engrs; Am Chem Soc; Am Soc Artificial Internal Organs; NY Acad Sci; Am Diabetes Asn; Biomed Eng Soc; Int Soc Artificial Organs; Am Heart Asn. *Res:* Heat and mass transfer; biomedical engineering; artificial organs; membrane transport phenomena and separations; enzyme applications; quantitative physiology; biotechnology; applied immunology. *Mailing Add:* Rm 66-452 Dept Chem Eng Mass Inst Technol Cambridge MA 02139

COLTON, DAVID L, b San Francisco, Calif, Mar 14, 43; m 68; c 2. MATHEMATICS. *Educ:* Calif Inst Technol, BS, 64; Univ Wis, MS, 65; Univ Edinburgh, PhD(math), 67, DSc(math), 77. *Prof Exp:* From asst prof to assoc prof math, Ind Univ, Bloomington, 67-75; prof math, Univ Strathclyde, 75-78; PROF MATH, UNIV DEL, 78- *Concurrent Pos:* Asst prof, McGill Univ, 68-69; vis res fel, Univ Glasgow, 71-72; guest prof, Univ Konstanz, 74-75; assoc ed, Applicable Anal, Proc Edinburgh Math Soc, Inverse Probs. *Mem:* Soc Indust & Applied Math. *Res:* partial differential equations; integral equations; scattering theory. *Mailing Add:* Dept Math Univ Del Newark DE 19711

COLTON, ERVIN, b Omaha, Nebr, June 25, 27; m 55; c 4. INORGANIC CHEMISTRY. *Educ:* Ga Inst Technol, BS, 50; Univ Kans, MS, 52; Univ Ill, PhD(inorg chem), 54. *Prof Exp:* Asst prof chem, Ga Inst Technol, 54-56; prin res chemist, Int Minerals & Chem Corp, 56-58; res chemist, Allis-Chalmers Mfg Co, 58-62, mgr, Cerac Sect, New Prod Dept, 62-64, PRES & CHMN, CERAC, INC, 64- *Concurrent Pos:* Pres & chmn, Hot-Pressing, Inc, 67-88. *Mem:* Am Chem Soc; Am Ceramic Soc. *Res:* Liquid ammonia and inorganic fluorine chemistry; hydrazine chemistry and synthesis; high temperature refractories; potassium chemicals. *Mailing Add:* Box 1178 Milwaukee WI 53201

COLTON, FRANK BENJAMIN, b Poland, Mar 3, 23; c 4. CHEMISTRY. *Educ:* Northwestern Univ, BS, 45, MS, 46; Univ Chicago, PhD(chem), 49. *Prof Exp:* Fel biochem, Mayo Clin, 49-51; sr res chemist, G D Searle & Co, 51-61, asst dir chem rse, 61-70, res adv, 70-86; RETIRED. *Honors & Awards:* Nat Inventors Hall of Fame, 88. *Mem:* Am Chem Soc; Royal Soc Chem. *Res:* Organic chemistry; biochemistry; medicinal chemistry. *Mailing Add:* 3901 Lyons St Evanston IL 60203

COLTON, JAMES DALE, Owatonna, Minn, Sept 3, 45; m 70; c 2. SCALE MODELING, STRESS WAVE PROPAGATION. *Educ:* Univ Minn, BME, 67; Stanford Univ, MS, 68, PhD(appl math), 73. *Prof Exp:* Res eng, 68-74, asst dir eng mech, 74-81, DIR ENG MECH, POULTER LAB, SRI INT, 81- *Mem:* Am Soc Mech Eng; Sigma Xi. *Res:* Stress wave propagation in structures; scale modeling structural response; simulation of dynamic load environments. *Mailing Add:* 670 Georgia Ave Palo Alto CA 94306

COLTON, JONATHAN STUART, b Boston, Mass, Feb 8, 59; m 87. COMPOSITES & POLYMER PROCESSING, INTELLIGENT DESIGN SYSTEMS. *Educ:* Mass Inst Technol, BS, 81, MS, 82, PhD(mech eng), 86. *Prof Exp:* Asst prof, 85-91, ASSOC PROF MECH ENG, GA INST TECHNOL, 91- *Concurrent Pos:* Fac assoc, Lockheed Aeronaut Systs Co, Ga, 86; NSF presidential young investr award, 89. *Mem:* Sigma Xi; Nat Res Soc; Am Soc Mech Engrs; Soc Mfg Engrs; Soc Plastics Engrs; Soc Advan Mat & Process Eng. *Res:* High performance composite and polymer processing; filament winding, resin transfer and injection molding; intelligent design systems; design theories and methodologies for conceptual design. *Mailing Add:* Sch Mech Eng Ga Inst Technol Atlanta GA 30332-0405

COLTON, RICHARD J, b South Fork, Pa, Dec 12, 50; m 78; c 2. SURFACE CHEMISTRY, PHYSICAL CHEMISTRY. *Educ:* Univ Pittsburgh, BS, 72, PhD(phys chem), 76. *Prof Exp:* Teaching fel chem, Univ Pittsburgh, 72-76; Nat Res Coun res assoc, Nat Acad Sci, 76-77, RES CHEMIST, US NAVAL RES LAB, 77- *Concurrent Pos:* Chmn, E42 Subcomt Sec Ion Mass Spectrometry, Am Soc Testing & Mat, 84-86; vis scientist, Calif Inst Technol, 86-87; chmn, Appl Surf Sci Div, Am Vacuum Soc, 87-88. *Mem:* Am Chem Soc; Am Vacuum Soc; Am Soc Mass Spectrometry; Am Soc Testing & Mat. *Res:* Surface analysis by x-ray photoelectron and auger electron spectroscopy and secondary ion mass spectrometry; corrosion and contamination; thin films; organic overlayers; emission mechanisms polyatomic and molecular ions; new surface analytical tools; surface structure by scanning tunneling microscopy and atomic microscopy. *Mailing Add:* Chem Div Code 6170 US Naval Res Lab Washington DC 20375-5000

COLTON, ROGER BURNHAM, b Windsor Locks, Conn, Jan 1, 24; m 47, 73; c 5. ENVIRONMENTAL GEOLOGY. *Educ:* Yale Univ, BS, 47, MS, 49. *Prof Exp:* GEOLOGIST, US GEOL SURV, 49- *Mem:* Fel Geol Soc Am; Am Asn Petrol Geol; Am Inst Prof Geologists; Asn Eng Geologists. *Res:* Landslides, geomorphology and photogeology. *Mailing Add:* US Geol Surv Mail Stop 972 Box 25046 Fed Ctr Denver CO 80225

COLUCCI, ANTHONY VITO, b Chicago, Ill, Sept 24, 38. ENVIRONMENTAL HEALTH, TOXICOLOGY. *Educ:* Loyola Univ, Ill, BS, 61; Johns Hopkins Univ, ScD(pathobiol), 66. *Prof Exp:* Res asst bact res, Lutheran Gen Hosp, Park Ridge, Ill, 60-61; res assoc parasitol res, Loyola Univ, Ill, 61; NSF fel biol res, Univ Wis, 66-67; chemist III, R J Reynolds Tobacco Co, 67-70; chief biochem & physiol br, Human Studies Lab, US Environ Protection Agency, 70-74; vpres health progs, Greenfield, Attaway & Tyler Inc, 74-80; dir, Law Dept, R J Reynolds Tobacco Co, 80-87; CONSULT, 87- *Concurrent Pos:* Mem, adv comt coord res coun, subcomt health effect mobile source emissions, 70-; NSF consult, dir joint US Environ Protection Agency & NSF Span Am Mercury Health Effects Prog, 70-74. *Mem:* Soc Toxicol; Sigma Xi; Am Col Toxicol. *Res:* Air toxics; fate of smoke constituents in animals; intermediary metabolism of xenobiotics; risk assessment. *Mailing Add:* 7621 Lasater Rd Clemmons NC 27012-8436

COLUCCI, JOSEPH M(ICHAEL), b Brooklyn, NY, Aug 12, 37; m 57; c 3. MECHANICAL ENGINEERING. *Educ:* Mich State Univ, BS, 58; Calif Inst Technol, MS, 59. *Prof Exp:* Res engr, 59-63, assoc sr res engr, 63-67, sr res engr, 67-70, asst dept head, 70-72, DEPT HEAD, FUELS & LUBRICANTS DEPT, GEN MOTORS RES LABS, 72- *Mem:* Soc Automotive Engrs; AAAS; Sigma Xi; Fel, Soc Automotive Engrs. *Res:* Current and future fuels and lubricants for automotive applications; effects of fuels and lubricants on vehicle emissions; fuel economy, durability, performance; petroleum based and non-petroleum based materials. *Mailing Add:* Fuels & Lubricants Dept Gen Motors Res Labs Warren MI 48090

COLVARD, DEAN WALLACE, b Ashe Co, NC, July 10, 13; m 39; c 3. ANIMAL SCIENCE, ANIMAL ECONOMICS. *Educ:* Berea Col, BS, 35; Univ Mo, MA, 38; Purdue Univ, PhD, 50. *Hon Degrees:* Dr, Purdue Univ, 60; Belmont Abbey Col, 79; Univ NC, Charlotte, 80. *Prof Exp:* Instr agr, Brevard Col, 35-37; res asst, Univ Mo, 37-38; supt, NC Agr Res Sta, 38-46; prof animal sci, NC State Col, 47-53, head dept, 48-53, dean agr, 53-60; pres, Miss State Univ, 60-66; chancellor, 66-78, EMER CHANCELLOR, UNIV NC, CHARLOTTE, 79- *Concurrent Pos:* Trustee, Berea Col, 56-76, & St Andrews Presby Col, 69-76; managing consult, Sci Mus Charlotte, 80-81; chmn bd trustees, NC Sch Sci & Math, 78- *Res:* Research, teaching and writing; animal physiology; economic geography and education; science education and educational administration. *Mailing Add:* Off Emer Chancellor Univ NC Charlotte NC 28223

COLVER, C(HARLES) PHILLIP, b Coffeyville, Kans, Oct 21, 35; m 86; c 3. CHEMICAL ENGINEERING, FORENSIC ENGINEERING. *Educ:* Univ Kans, BS, 58, MS, 60; Univ Mich, PhD(chem eng), 63. *Prof Exp:* Engr, Gulf Oil Corp, 58-59; res assoc, Univ Mich, 61-63; from asst prof to prof chem eng, Univ Okla, 63-76, dir, Sch Chem Eng & Mat Sci, 70-74, assoc dean eng, 74-75; CONSULT ENGR, 76- *Concurrent Pos:* Vis prof chem eng, Univ Colo, 77-78. *Mem:* Nat Fire Protection Asn; Int Asn Arson Invest. *Res:* Explosions; fires; accident reconstruction; chemical and mechanical design; flammability of materials. *Mailing Add:* 0855 Mountain Laurel Dr Aspen CO 81611

COLVIN, BURTON HOUSTON, b West Warwick, RI, July 12, 16; m 47; c 3. MATHEMATICS. *Educ:* Brown Univ, AB, 38, AM, 39; Univ Wis, PhD(math), 43. *Prof Exp:* Instr math, Univ Wis, 40-43 & 46, asst prof, 47-51; tech aide appl math panel, Nat Defense Res Comt & Off Sci Res & Develop, 44-45; mathematician, Phys Res Staff, Boeing Co, 51-55, supvr math anal, 55-58, assoc head math res lab, Boeing Sci Res Labs, 58-59, head, 59-70, head math & Info Sci Lab, 70-72; chief, Appl Math Div, Nat Bur Standards, 72-78, dir Ctr Appl Math, 78-86, DIR, ACAD AFFAIRS, NAT INST STANDARDS & TECHNOL, 86- *Concurrent Pos:* Vis sci lectr, Soc Indust & Appl Math, 59-60 & Math Asn Am, 63-65; mem adv bd, Sch Math Study Group, 63-71, chmn, 65-66, Comn Sci Soc Pres, 75-78; chmn, Conf Bd Math Scis, 75-76; consult, NSF, 76-78; mem, Coun, AAAS, 65-67, vchmn, Comt on Sci Educ, 68-72, chmn, task force Tech Educ, 68-69; mem, Soc Indust & Appl Math, NEA, vis scientist lectr, 62-63, trustee, 62-65, 67-70, 78-80 & pres, 71-72; vis lectr, Math Asn Am, 63-65; mem bd dirs, Nat Phys Sci Consortium, 88- *Honors & Awards:* Silver Medal Award, Dept Com, 78, Gold Medal, 81. *Mem:* Fel AAAS; Soc Indust & Appl Math; Am Math Soc; Math Asn Am; Inst Math Statist; Nat Coun Teachers Math. *Res:* Differential equations; applied mathematics. *Mailing Add:* Acad Affairs Rm A515 Admin Bldg Nat Inst Standards & Technol Gaithersburg MD 20899

COLVIN, CLAIR IVAN, b Clyde, Ohio, Sept 16, 27. PHYSICAL CHEMISTRY. *Educ:* Ohio Univ, BS, 49; Univ Miami, MS, 61, PhD(phys chem), 63. *Prof Exp:* Anal chemist, Aluminum & Magnesium, Inc, 51-52 & Nat Carbon Co, 52-53; instr chem, Racine Exten Ctr, Univ Wis, 53-56; teacher high sch, 56-57; univ fel chem & USPHS grant, Sch Med, Univ Miami, 63-64; from asst prof phys chem to assoc prof chem, 64-70, PROF CHEM & CHMN DEPT, GA SOUTHERN COL, 70- *Mem:* Am Chem Soc; The Royal Soc Chem. *Res:* Kinetics of the thermal decomposition of ammonium nitrate in the presence of catalysts; molecular orbital calculations for conjugated organic compounds of biological interest. *Mailing Add:* Dept Chem Ga Southern Col Statesboro GA 30458

COLVIN, DALLAS VERNE, b Westport, Ore, May 30, 37; m 64. ANIMAL BEHAVIOR, ANIMAL ECOLOGY. *Educ:* Portland State Univ, BS, 63; Univ Colo, PhD(zool), 70. *Prof Exp:* Teaching asst biol, Univ Colo, Boulder, 67-68; from asst prof to assoc prof biol, Calif State Col, Dominquez Hills, 70-81; RETIRED. *Res:* Behavioral studies of small mammals; analysis and biological bases of ultrasounds used by Microtus as a form of communication between neonates and adults. *Mailing Add:* 1224 Bayberry Rd Lake Oswego OR 97034

COLVIN, HARRY WALTER, JR, b Schellsburg, Pa, Dec 5, 21; m 50; c 2. PHYSIOLOGY, ANIMAL SCIENCE. *Educ:* Pa State Univ, BS, 50; Univ Calif, Davis, PhD(comp physiol), 57. *Prof Exp:* Instr physiol, Okla State Univ, 56-57; from asst prof to assoc prof, Univ Ark, 57-65; from asst prof to assoc prof, 65-75, PROF PHYSIOL, UNIV CALIF, DAVIS, 75- *Concurrent Pos:* Coun Int Exchange Scholars Fulbright-Hays Award, 72-73, Yugoslavia, 86, Argentina; assoc ed, Hilgardia. *Mem:* Sigma Xi; AAAS; Am Dairy Sci Asn; Am Soc Animal Sci. *Res:* Rumen physiology; carbohydrate metabolism in calves; blood coagulation; animal nutrition. *Mailing Add:* Dept Animal Physiol Univ Calif Davis CA 95616

COLVIN, HOWARD ALLEN, b Houston, Tex, June 25, 53; m 78; c 3. CHEMISTRY. *Educ:* Univ Houston, BS, 75; Georgia Inst Technol, PhD(chem), 79. *Prof Exp:* sr res chemist, Goodyear Tire & Rubber, 79-83, prin chemist, 83-88, res & develop assoc, 88-90, SECT HEAD, NEW POLYMER, RES & DEVELOP, GOODYEAR TIRE & RUBBER CO, 91-, SECT HEAD, NEW SOLUTION POLYMERS & WINGTACK RES & DEVELOP, 91- *Mem:* Am Chem Soc. *Res:* Polymer blends and alloys; tackifying resin research; new monomer and polymer synthesis. *Mailing Add:* 534 Elm Ave Tallmadge OH 44278

COLVIN, ROBERT B, b Columbus, Ohio, May 7, 42; m 66; c 2. IMMUNOPATHOLOGY, NEPHROPATHOLOGY. *Educ:* Mass Inst Technol, BS, 64; Harvard Med Sch, MD, 68. *Prof Exp:* Intern surg, Mass Gen Hosp, 68-69, resident path, 69-72, NIH res fel path, 71-72; major USAMC res, Walter Reed Army Inst Path, 72-75; asst prof, 75-78, ASSOC PROF PATH, MASS GEN HOSP, HARVARD MED SCH, 78-, DIR IMMUNOL UNIT, 80- *Mem:* Am Asn Path; Am Asn Immunol; Int Acad Path; Transplantation Soc; Am Soc Nephrol; Asn Acad Pathologists. *Mailing Add:* Path Dept Harvard Med Sch Mass Gen Hosp Boston MA 02114

COLVIN, THOMAS STUART, b Columbia, Mo, July 17, 47. MACHINERY MANAGEMENT, CONSERVATION TILLAGE. *Educ:* Iowa State Univ, BS, 70, MS, 74, PhD(agr eng), 77. *Prof Exp:* Res asst, Iowa State Univ, 72-77; AGR ENGR, AGR RES SERV, USDA, 77- *Concurrent Pos:* Owner-mgr, Oxford Farms, 72- *Mem:* Am Soc Agr Eng; Am Soc Agron; Soil Conserv Soc Am; Sigma Xi. *Res:* Research on the physical aspects of soil and decision support systems for sustainable agriculture. *Mailing Add:* Nat Soil Tilth Lab 2150 Pammel Dr Ames IA 50011

COLVIS, JOHN PARIS, b St Louis, MO, June 30, 46; m 76; c 4. MATHEMATICS. *Educ:* Wash Univ, BS, 77. *Prof Exp:* assoc engr, 78-81, SR ENGR, MARTIN MARIETTA ASTRONAUTICS GROUP SPACE LAUNCH SYSTEMS CO, 81- *Mem:* Math Asn Am; AAAS. *Res:* Quantum postulate. *Mailing Add:* 4978 S Hoyt St Littleton CO 80123-1988

COLWELL, GENE THOMAS, b Chattanooga, Tenn, Aug 3, 37; m 73. THERMODYNAMICS, HEAT TRANSFER. *Educ:* Univ Tenn, BS, 59, MS, 62, PhD(eng sci), 66. *Prof Exp:* Res engr, Oak Ridge Nat Lab, 59-62, design specialist, 65- 66; instr mech eng, Univ Tenn, 62-65; from asst prof to assoc prof, 66-76, assoc dir dept, 76-78, PROF MECH ENG, GA INST TECHNOL, 76- *Concurrent Pos:* Eng consult. *Mem:* Am Soc Mech Engrs; Sigma Xi. *Res:* Energy engineering; aeronautical and astronautical engineering; gas turbines; turbomachinery. *Mailing Add:* Dept Mech Eng Ga Inst Technol Atlanta GA 30332

COLWELL, JACK HAROLD, b Wooster, Ohio, Dec 29, 31; m 56; c 2. EXPERIMENTAL THERMAL PHYSICS. *Educ:* Mt Union Col, BS, 53; Purdue Univ, MS, 58; Univ Wash, PhD(chem), 61. *Prof Exp:* Nat Res Coun Can fel, 61-63; chemist, Cryophysics Sect, 63-73, res chemist, Thermophysics Div, 73-81, PHYSICIST, TEMPERATURE & PRESSURE DIV, NAT BUR STANDARDS, 81- *Honors & Awards:* Dept Com Award, 72. *Mem:* Sigma Xi; Am Phys Soc. *Res:* Low temperature calorimetry; properties of molecular crystals, superconductors and magnetic insulators at low temperatures; temperature measurement and temperature scales at very low temperatures. *Mailing Add:* 5010 River Hill Rd Bethesda MD 20816

COLWELL, JOHN AMORY, b Boston, Mass, Nov 4, 28; m 54; c 4. INTERNAL MEDICINE, PHYSIOLOGY. *Educ:* Princeton Univ, AB, 50; Northwestern Univ, MD, 54, MS, 57, PhD(physiol), 68; Am Bd Internal Med, dipl, 62. *Prof Exp:* Intern med, Med Sch, Western Reserve Univ, 54-55; resident, Northwestern Univ, 55-57 & 59-60; instr, Med Sch, 60-62, assoc, 62-65, from asst prof to assoc prof, 65-71; PROF MED, MED UNIV SC, 71-, DIR ENDOCRINOL & METAB, NUTRIT DIV, 72-, UNIV RES COORDR, 73- *Concurrent Pos:* NIH fel, Northwestern Univ, 56-57, Am Diabetes Asn & univ fels, 60-61, NIH res grant, 62-71; clin investr, Vet Admin Res Hosp, Chicago, 61-63, chief sect, 63-71; assoc chief staff res & develop, Vet Admin Hosp, Charleston, 71- *Mem:* AAAS; fel Am Col Physicians; Am Diabetes Asn; Am Physiol Soc; Endocrine Soc. *Res:* Insulin secretion, degradation and action in animals and man; selected clinical studies in subjects with disorder of metabolism and endocrinology; platelet function in diabetes. *Mailing Add:* Dept Med Med Univ SC 171 Ashley Ave Charleston SC 29425

COLWELL, JOSEPH F, b Brush, Colo, Mar 16, 29; m 53; c 3. SOLID STATE PHYSICS. *Educ:* Colo State Univ, BS, 51; Cornell Univ, PhD(physics), 60. *Prof Exp:* Physicist, Navy Electronics Lab, 51-53; asst physics, Cornell Univ, 53-54 & 56-60; res staff mem, Gen Atomic Div, Gen Dynamics Corp, 60-67; staff mem, Gulf Radiation Technol Div, Gulf Energy & Environ Systs, Inc, 67-74; missile & aircraft tests, US Air Force, 78-85; TACTICAL TESTS & INFO DISTRIB, COMPUT SCI CORP, 85- *Mem:* AAAS; Am Phys Soc; Soc Explor Geophys. *Res:* Solid state theory, especially as related to direct energy conversion devices and radiation damage in solids; seismic methods of geophysical exploration; computer simulation of aircraft/missile missions. *Mailing Add:* 2030 Karren Lane Carlsbad CA 92008

COLWELL, RITA R, b Beverly, Mass, Nov 23, 34; m 56; c 2. MICROBIOLOGY. *Educ:* Purdue Univ, BS, 56, MS, 58; Univ Wash, PhD(marine microbiol), 61. *Hon Degrees:* DSc, Heriot-Watt Univ, Edinburgh, Scotland, 87. *Prof Exp:* Res asst prof, Univ Wash, 61-64; vis asst prof, Georgetown Univ, 63-64, from asst prof to assoc prof biol, 64-72; PROF MICROBIOL, UNIV MD, 72-; DIR, MD BIOTECHNOL INST, 87- *Concurrent Pos:* Guest scientist, Nat Res Coun Can, 61-63; mem classification res group, London; mem bd trustees, Am Type Cult Collection, chmn, 80; consult, Bur Higher Educ, Dept Health, Educ & Welfare, 68-78; consult div res grants, NIH, 70; consult adv comt sci educ & biolog oceanog, NSF, 70-81; consult, Environ Protection Agency, 75-, dir, Md Sea Grant Col, 78-83, vpres acad affairs, 83-87; mem res adv comt, AID, 82, Nat Sci Bd, 83-90; hon prof, Univ Queensland, Australia, 88; chmn bd gov, Am Acad

Microbiol, 89-94. *Honors & Awards:* Phi Sigma Serv Award, Am Chem Soc, 75; Fisher Award, Am Soc Microbiol, 85; Gold Medal, Int Biotechnol Inst, 90. *Mem:* Fel AAAS; Am Soc Microbiol (pres, 84-85); Soc Invert Path; fel Soc Indust Microbiol; fel Am Acad Microbiol; Sigma Xi (pres elect, 90). *Res:* Marine microbiology; numerical taxonomy; uses of computers in biology and medicine; microbial ecology and systematics; marine biotechnology. *Mailing Add:* Md Biotechnol Inst Univ Md College Park MD 20742

COLWELL, ROBERT KNIGHT, b Denver, Colo, Oct 9, 43; div; c 1. ECOLOGY, EVOLUTION. *Educ:* Harvard Univ, AB, 65; Univ Mich, Ann Arbor, PhD(zool), 69. *Prof Exp:* Asst to cur econ & ethnobot, Bot Mus, Harvard Univ, 66; Ford Found fel math biol, Univ Chicago, 69-70; asst prof, 70-76, ASSOC PROF ZOOL, UNIV CALIF, BERKELEY, 76- *Concurrent Pos:* Coordr, Org Trop Studies grad course in trop biol, 71. *Mem:* AAAS; Am Soc Naturalists; Asn Trop Biol; Ecol Soc Am; Soc Study Evolution. *Res:* Ecology and evolution of biological communities; tropical biology; behavioral and theoretical ecology. *Mailing Add:* Dept Zool Univ Calif 4079 Life Sci Bldg Berkeley CA 94720

COLWELL, ROBERT NEIL, b Star, Idaho, Feb 4, 18; m 42; c 4. FOREST MENSURATION. *Educ:* Univ Calif, BS, 38, PhD(plant physiol), 42. *Prof Exp:* Asst bot, Univ Calif, Berkeley, 38-42, from asst prof to prof, forestry, 47-69, assoc dir space sci lab, 69-85, dir Berkeley Off, Earth Satellite Corp, 70-85; RETIRED. *Concurrent Pos:* Chmn comt crop geog & veg anal, Nat Res Coun, 53-54. *Honors & Awards:* Abrams Award, 54; Fairchild Photogram Award, 57; Photo Interpretation Award, Am Soc Photogram, 64. *Mem:* Soc Am Foresters; Am Soc Photogram (vpres, 54); Int Soc Photogram. *Res:* Identification and mapping of vegetation types from aerial and space photographs; use of radioactive tracers in biological studies; applications of remote sensing to the space sciences; aerospace and earth sciences. *Mailing Add:* Waunita Dr Walnut Creek CA 94596

COLWELL, WILLIAM MAXWELL, b Blairsville, Ga, May 28, 31; m 56; c 4. VETERINARY MICROBIOLOGY. *Educ:* Berry Col, BS, 52; Univ Ga, DVM, 59, MS, 68, PhD(microbiol), 69; Am Col Vet Microbiologists, dipl. *Prof Exp:* Vet, Vanderbilt Vet Hosp, Durham, NC, 59-60 & diag lab, Ga Poultry Lab, Oakwood, 60-64; area vet, Elanco Prod Co, Ind, 64-66; Campbell fel, Avian Dis Res Ctr, Univ Ga, 66-69; asst prof vet sci, Univ Fla, 69-70; assoc prof poultry sci, NC State Univ, 70-74, prof vet sci, 74-78; dir vet, Res & Serv Lab, Holly Farms Foods, Inc, 78-89; MGR, TECH SERV LAB, TYSON FOODS INC, 89- *Mem:* Am Asn Avian Pathologists; Am Vet Med Asn. *Res:* Avian disease research; epidemiology of avian tumor viruses; avian respiratory viruses; oncogenic viruses of poultry; organ culture techniques; bioassay of aflatoxins. *Mailing Add:* Tyson Foods Inc 1600 River Rd Wilkesboro NC 28697

COLWELL, WILLIAM TRACY, b Joliet, Ill, Oct 18, 34; c 2. PHARMACEUTICAL CHEMISTRY, SYNTHETIC ORGANIC CHEMISTRY. *Educ:* Occidental Col, BA, 56; Univ Calif, Los Angeles, PhD(org chem), 62. *Prof Exp:* SR ORG CHEMIST PHARMACEUT CHEM, SRI INT, 62- *Mem:* Am Chem Soc. *Res:* Synthesis of pteridines and related heterocycles as antifolates; antiparasitic compounds; design, synthesis, and assay development of immunodulators. *Mailing Add:* SRI Int Menlo Park CA 94025

COLWICK, REX FLOYD, b Clifton, Tex, Mar 19, 22; m 43; c 1. AGRICULTURAL ENGINEERING. *Educ:* Tex A&M Univ, BS, 47, MS, 48. *Prof Exp:* Agr engr, USDA & Tex A&M Univ, 48-50, agr eng coordr, 51-81; invests leader crop prod systs res, USDA & prof agr & biol eng, Miss State Univ, 61-78, lab chief & location leader, Agr Res Serv, 78-80; RETIRED. *Concurrent Pos:* Consult, Cotton Res Adv Comt & Task Force, USDA, 55-78; consult agr eng, 81-88, IESC, Brazil, 88. *Mem:* Am Soc Agr Engrs. *Res:* Agricultural engineering research in crop production and harvesting equipment. *Mailing Add:* 1006 S Montgomery Starkville MS 39759

COLWILL, JACK M, b Cleveland, Ohio, June 15, 32; m 54; c 3. INTERNAL MEDICINE, FAMILY PRACTICE. *Educ:* Oberlin Col, BA, 55; Univ Rochester, MD, 57; Am Bd Internal Med, dipl, 64; Am Bd Family Practice, dipl, 77. *Prof Exp:* Intern, Barnes Hosp, Washington Univ, 57-58; res, Univ Wash Affiliated Hosps, 58-60, chief res, Univ Hosp, 60-61; from instr to sr instr med & dir med outpatient dept, Med Sch, Univ Rochester, 61-64; asst prof, 64-70, asst dean, Sch Med, 64-67, assoc dean, 67-69, assoc dean acad affairs, 69-76, actg chmn, Dept Family & Community Med, 76-77, ASSOC PROF MED, SCH MED, UNIV MO-COLUMBIA, 70-, PROF FAMILY & COMMUNITY MED, 76-, CHMN DEPT, 77-, DIR FAMILY MED RESIDENCY PROG, MED CTR, 74- *Mem:* Nat Acad Sci-Inst Med; AMA; Asn Am Med Cols. *Mailing Add:* Dept Family & Community Med Sch Med Univ Mo Columbia MO 65212

COLWIN, ARTHUR LENTZ, b Sydney, Australia, Jan 26, 11; nat US; m 40. ZOOLOGY. *Educ:* McGill Univ, BSc, 33, MSc, 34, PhD(embryol), 36. *Prof Exp:* Moyse traveling fel, Sir William Dunn Inst Biochem & Dept Exp Zool, Univ Cambridge, 34-35; Seessel fel, Osborn Zool Lab, Yale Univ, 36-37, Royal Soc Can fel, 37-38; instr biol, NY Univ, 38-39; from instr to prof biol, Queens Col, NY, 40-73, EMER PROF BIOL, 73-; ADJ PROF, ROSENSTIEL SCH MARINE & ATMOSPHERIC SCI, UNIV MIAMI, 73- *Concurrent Pos:* Instr embryol, Marine Biol Lab, Woods Hole, 48-50; Fulbright res scholar, Misaki Marine Biol Sta, Tokyo, 53-54; mem co, Marine Biol Lab, Woods Hole & trustee, 62-75, emer trustee, 75- *Mem:* Am Soc Zoologists; fel NY Acad Sci; Am Soc Cell Biol; Soc Develop Biol; Int Soc Develop Biol. *Res:* Normal and experimental embryology; cell division and differentiation; fertilization; sperm-egg association; egg cortical changes. *Mailing Add:* 320 Woodcrest Rd Key Biscayne FL 33149

COLWIN, LAURA HUNTER, b Philadelphia, Pa, July 5, 11; m 40. ZOOLOGY. *Educ:* Bryn Mawr Col, AB, 32, Univ Pa, MA, 34, PhD(protozool), 38. *Prof Exp:* Instr biol, Chatham Col, 36-37, asst prof, 37-40; instr zool, Vassar Col, 40-43; instr biol, Chatham Col, 45-46; lectr biol, 47-66, prof, 66-73, EMER PROF BIOL, QUEENS COL, NY, 73-; ADJ PROF, ROSENSTIEL SCH MARINE & ATMOSPHERIC SCI, UNIV MIAMI, 73- *Concurrent Pos:* Morrison fel, Am Asn Univ Women, Misaki Marine Biol Sta, Univ Toyko, 53-54; mem co & trustee, Marine Biol Lab, Woods Hole, 71-75, emer trustee, 75- *Mem:* Am Soc Zoologists; Am Soc Cell Biol; Soc Develop Biol; Int Soc Develop Biol; Int Soc Cell Biol. *Res:* Normal and experimental embryology; cell division and differentiation; fertilization; sperm-egg association; egg cortical changes. *Mailing Add:* 320 Woodcrest Rd Key Biscayne FL 33149

COMAI-FUERHERM, KAREN, b Detroit, Mich, June 20, 46; m 75; c 1. METABOLISM, ATHERSCLEROSIS. *Educ:* Wayne State Univ, BS, 69; Cornell Univ, PhD(biochem), 73. *Prof Exp:* Res asst, Med Ctr, NY Univ, 73-75; vis scientist, Hoffman La Roche, Nutley, NJ, 75-77, sr scientist, 77-81, res fel, 81-85, dir cardiovasc res, 85-87; CONSULT, 87- *Mem:* AAAS; Asn Women Sci; NY Acad Sci; Am Chem Soc; Am Oil Chemists Soc; Am Soc Pharmacol & Exp Therapeut. *Res:* Lipid metabolism primarily atherosclerosis and cardiovascular diseases; obesity therapy. *Mailing Add:* 151 Rutgers Pl Nutley NJ 07110-1810

COMAN, DALE REX, b Hartford, Conn, Feb 22, 06; m 37; c 2. PATHOLOGY. *Educ:* Univ Mich, AB, 28; McGill Univ, MD, 33. *Prof Exp:* Asst, Inst Path, McGill Univ, 33-34; resident, Univ Pa Hosp, 34-35; resident, Mass State Tumor Hosp, Pondville, 35-36; instr, Sch Med, NY Univ, 36-37; instr, 37-41, assoc, 41-42, from asst prof to prof exp path, 42-54, prof path, 54-72, chmn dept, 54-67, EMER PROF PATH, SCH MED, UNIV PA, 72-; RES ASSOC, JACKSON LAB, 77- *Honors & Awards:* Paget-Ewing Lectr Award, Metastasis Res Soc, 87. *Mem:* AAAS; NY Acad Sci; Am Soc Cell Biol; Int Soc Cell Biol; Soc Exp Path & Med. *Res:* Cancer. *Mailing Add:* Sand Point Rd Bar Harbor ME 04609

COMBA, PAUL GUSTAVO, b Tunis, Tunisia, Mar 6, 26; nat US. COMPUTER SCIENCE. *Educ:* Bluffton Col, AB, 47; Calif Inst Technol, PhD(math), 52. *Prof Exp:* Asst math, Calif Inst Technol, 47-51; from asst prof to assoc prof, Univ Hawaii, 51-60; math systs analyst, 60-63, mgr advan comput technol, 63, mgr prog lang eval, 63-64, SCI STAFF MEM, IBM CORP, 65- *Concurrent Pos:* Adj prof, NY Univ, 70 & Boston Univ, 77-79; lectr, Princeton Univ, 80-81. *Mem:* Fel AAAS; Soc Indust & Appl Math; Math Asn Am. *Res:* Computer programming languages; computer graphics; computer aided design; computer applications development; design and management of data bases; cryptography. *Mailing Add:* 126 Clark Rd Brookline MA 02146

COMBES, BURTON, b New York, NY, June 30, 27; m 48; c 3. INTERNAL MEDICINE, LIVER DISEASE. *Educ:* Columbia Univ, AB, 47, MD, 51; Am Bd Internal Med, dipl, 59. *Prof Exp:* Intern med, Columbia-Presby Med Ctr, 51-52, asst resident, 52-53, asst physician, 53-56; from instr to assoc prof, 57-67, PROF INTERNAL MED, UNIV TEX HEALTH SCI CTR DALLAS, 67- *Concurrent Pos:* Res fel, Col Physicians & Surgeons, Columbia Univ, 53-55, Am Heart Asn res fel, 55-56; Am Heart Asn res fel, Univ Col Hosp, Med Sch, Univ London, 56-57; USPHS res career develop award, 62-72; estab investr, Am Heart Asn, 57-62; consult, Dallas Vet Admin Hosp, Tex, 65-; mem adv coun, Nat Inst Arthritis, Metab & Digestive Dis, 75-79; chmn bd, Am Liver Found, 76-80; liver ed, Gastroenterol, Am Gastroenterol Asn, 77-81; mem, Nat Digestive Dis Adv Bd, 81- *Mem:* Am Fedn Clin Res; Asn Am Physicians; Am Soc Clin Invest; Am Asn Study Liver Dis (pres, 71); Am Gastroenterol Asn; Int Asn Study Liver; Soc Exp Biol Med. *Res:* Hepatic excretory function; liver immunology; liver function during pregnancy. *Mailing Add:* Dept Internal Med Tex SW Med Sch Dallas TX 75235

COMBIE, JOAN D, b New London, NH, June 30, 46. MICROORGANISMS FROM EXTREME ENVIRONMENTS, THERMALLY STABLE ENZYMES. *Educ:* Colby-Sawyer Col, BS, 68; Univ Ky, MS, 78, PhD(toxicol & pharmacol), 82. *Prof Exp:* Dir res, Alltech, 83-84; res scientist, Renewable Technologies, Inc, 84-87; DIR RES & VPRES, J K RES, 87- *Mem:* Am Soc Microbiol; Soc Indust Microbiol. *Res:* Microorganisms from extreme environments as a source of thermally stable enzymes, for use in coal desulfurization and for biohydrometallurgy. *Mailing Add:* J K Res 210 S Wallace Bozeman MT 59715

COMBS, ALAN B, b Boulder, Colo, July 4, 39; m 61; c 4. PHARMACOLOGY. *Educ:* Univ of the Pac, BSc, 62, MSc, 64; Univ Calif, Davis, PhD(comp pharmacol), 70. *Prof Exp:* From asst prof to assoc prof, 76-86, PROF PHARMACOL, SCH PHARM, UNIV TEX, AUSTIN, 86- *Mem:* AAAS; Am Asn Col Pharm; Soc Toxicol; Am Soc Pharmacol Exp Therapeut. *Res:* Cardiovascular pharmacology and toxicology. *Mailing Add:* Dept Pharmacol Univ Tex Sch Pharm Austin TX 78712

COMBS, CLARENCE MURPHY, anatomy; deceased, see previous edition for last biography

COMBS, GEORGE ERNEST, b Arcadia, Fla, Feb 21, 27; m 48; c 3. ANIMAL NUTRITION. *Educ:* Univ Fla, BSA, 51, MSA, 53; Iowa State Univ, PhD(animal nutrit), 55. *Prof Exp:* Asst animal husb, Univ Fla, 51-52, instr, 52-53; asst animal nutrit, Iowa State Univ, 53-55; from asst prof to assoc prof, 55-67, PROF ANIMAL NUTRIT, UNIV FLA, 67- *Mem:* Am Inst Nutrit; Am Soc Animal Sci. *Res:* Mineral, energy and amino acid metabolism with swine. *Mailing Add:* Dept Animal Sci Univ Fla Gainesville FL 32611

COMBS, GERALD FUSON, b Olney, Ill, Feb 23, 20; m 43; c 4. ANIMAL NUTRITION. *Educ:* Univ Ill, BS, 40; Cornell Univ, PhD(animal nutrit), 48. *Prof Exp:* Asst animal nutrit, Cornell Univ, 40-41 & 46-48; nutrit officer, US Army, 41-46; prof poultry nutrit, Univ Md, College Park, 48-69; dep chief nutrit prog, US Dept Health, Educ & Welfare, 69-71; nutrit & food safety coordr, USDA, 71-73; prof foods & nutrit & head dept, Univ Ga, 73-75; nutrit prog dir, Nat Inst Arthritis, Diabetes & Digestive & Kidney Dis, NIH, 75-83; ASST HUMAN NUTRIT, AGR RES SERV, USDA, BELTSVILLE, MD, 83- *Honors & Awards:* Poultry Nutrit Res Award, Am Feed Mfg, 53; Meterious Nutritionist Award, Distillers Res Inst, 70. *Mem:* AAAS; Poultry Sci Asn; Soc Exp Biol & Med; fel Am Inst Nutrit; Brit Nutrit Soc; Sigma Xi. *Res:* Poultry nutrition; factors concerned in bone formation; unidentified factors; energy-protein balance; antibiotics; amino acid requirements; human and international nutrition; basic and clinical human and animal nutrition research. *Mailing Add:* Nat Prog Staff Agr Res Serv USDA Rm 132 BARC-W Beltsville MD 20705

COMBS, GERALD FUSON, JR, b Ithaca, NY, June 10, 47; m 69; c 3. NUTRITION. *Educ:* Univ Md, College Park, BS, 69; Cornell Univ, MS, 71, PhD(nutrit), 74. *Prof Exp:* Asst prof biochem & nutrit, Auburn Univ, 74-75; asst prof, 75-80, ASSOC PROF NUTRIT, CORNELL UNIV, 80- *Honors & Awards:* Res Award, Poultry Sci Asn, 79. *Mem:* AAAS; Poultry Sci Asn; Am Inst Nutrit; Soc Exp Biol & Med; NY Acad Sci; Sigma Xi. *Res:* Nutrient interrelationships and mechanisms of action of selenium and vitamin E; influences of foreign compounds on selenium function. *Mailing Add:* Dept Poultry Sci Rice Hall Cornell Univ Ithaca NY 14853

COMBS, LEON LAMAR, III, b Meridian, Miss, Sept 19, 38; m 62; c 1. CHEMICAL PHYSICS. *Educ:* Miss State Univ, BS, 61; La State Univ, PhD(chem physics), 68. *Prof Exp:* Res chemist, Devoe & Reynolds Co, Inc, Ky, 61-64; from asst prof to assoc prof, 67-75, head Dept chem, 81-89, PROF CHEM & PHYSICS, MISS STATE UNIV, 75- *Concurrent Pos:* Vis prof quantum chem, Univ Uppsala, Sweden, 77-78. *Honors & Awards:* Outstanding Chemist Award, Am Chem Soc, Miss Sect, 81. *Mem:* Am Phys Soc; Am Chem Soc; Sigma Xi; fel Am Inst Chemists. *Res:* Quantum chemistry of small molecules; application of statistical mechanics to study phase transitions; theoretical conformational analysis; quantum mechanical studies in pharmacology. *Mailing Add:* Box CH Miss State Univ Mississippi State MS 39762

COMBS, ROBERT GLADE, b Maysville, Mo, Nov 17, 30; m 58; c 2. ELECTRICAL ENGINEERING, BIOENGINEERING. *Educ:* Univ Mo, BS, 56, MS, 59; Univ Fla, PhD(elec eng), 65. *Prof Exp:* Instr elec eng, Univ Mo, 56-58; instr, NC State Col, 58-59; asst prof, Univ Nebr, 59-65; from asst prof to assoc prof, 65-78, dir grad studies, 81-88, PROF ELEC ENG, UNIV MO, COLUMBIA, 78-, UNDERGRAD PROG DIR, 88- *Concurrent Pos:* Teaching consult, Training Prog, Hallam Nuclear Reactor Facil, Consumers Pub Power Dist, 60-61. *Mem:* AAAS; Nat Soc Prof Engrs; Inst Elec & Electronics Engrs; Am Soc Eng Educ. *Res:* Coaching effects on problem solving; competency-based instructional models; grades and grading. *Mailing Add:* 213 Elec Eng Bldg Univ Mo Columbia MO 65211

COMBS, ROBERT L, JR, b Fayetteville, Ark, Nov 6, 28; m 50; c 3. ENTOMOLOGY. *Educ:* Univ Ark, BS, 61, MS, 63; Miss State Univ, PhD(entom), 67. *Prof Exp:* Asst entom, Univ Ark, 63-64; asst entomologist, 64-70, assoc entomologist, 70-80, ENTOMOLOGIST, MISS STATE UNIV, 80- *Mem:* Entom Soc Am; Am Registry Prof Entomologists. *Res:* Veterinary entomology; applied and basic entomological problems. *Mailing Add:* Dept Entom Drawer EM Miss State Univ Mississippi State MS 39762

COMEAU, ANDRE I, b Richmond, Que, Oct 8, 45; m 71; c 3. GENETICS, PLANT BIOTECHNOLOGY. *Educ:* Sherbrooke Univ, BA, 64, BS, 67; Cornell Univ, PhD(entom), 71. *Prof Exp:* RES SCIENTIST, RES STA, AGR CAN, 71-; ASSOC PROF, LAVAL UNIV, 88- *Honors & Awards:* Leon Provancher Award Entom, Que Entom Soc, 83. *Mem:* Can Phytopath Soc; Int Asn Plant Tissue Cult; Asn Appl Biol UK. *Res:* Barley yellow dwarf virus and vector aphids on oats, triticale barley and wheat; anther culture; plant breeding; interspecific hybridation; plant tissue culture and biotechnology. *Mailing Add:* Gulf 1V2J3 Res Sta Agr Can 2560 Boul Hochelaga Ste-Foy PQ G1V 2J3 Can

COMEAU, ROGER WILLIAM, b Quincy, Mass, Apr 22, 33; div; c 3. PROBLEM-BASED MEDICAL EDUCATION. *Educ:* Boston Univ, AB, 55; State Univ NY Buffalo, PhD(physiol), 67. *Prof Exp:* Mem staff, Arthur D Little, Inc, 59-61; teaching asst physiol, State Univ NY Buffalo, 61-63, from asst instr to asst prof, 63-68; assoc dir sci info & regulatory affairs, Mead Johnson & Co, 68-69; assoc prof, Mid Ga Col, 70-75, prof biol, 75-79, chymn dept biol sci, 73-79; dir admissions, dir learning resources, 80-81, dir admin & student affairs, 81-82, asst dean admin & student affairs, 82-84, PROF PHYSIOL, 79-, ASSOC DEAN ADMIN & STUDENT AFFAIRS, MERCER UNIV SCH MED, 88- *Mem:* AAAS; Assoc Am Physiol Soc; Asn Am Med Col. *Res:* Membrane transport; teaching mammalian physiology; pharmacology and toxicology of cancer chemotherapeutic agents; problem-based medical education. *Mailing Add:* Sch Med Mercer Univ Macon GA 31207

COMEFORD, JOHN J, b Schenectady, NY, Apr 30, 28; m; c 2. ANALYTICAL CHEMISTRY. *Educ:* Colo State Univ, BS, 50; Wash State Univ, MS, 53; Georgetown Univ, PhD(molecular spectros), 66. *Prof Exp:* Phys chemist, Nat Bur Standards, 57-58, res chemist, 58-82; chemist, Dugway Proving Ground, 83-90; CONSULT, 91- *Concurrent Pos:* Secy comt on spectral absorption data, Nat Res Coun-Nat Acad Sci, 59-62; vis assoc prof, Dept Mat Sci & Eng, Univ Utah, 75. *Mem:* Am Chem Soc; Sigma Xi. *Res:* Low temperature matrix-isolation spectroscopy; development of test procedures for recycled petroleum products; infrared spectra of unstable molecules; mass spectrometry of combustion products; evaluation of lubricating oils; infrared lidar. *Mailing Add:* 4211 Shanna St Salt Lake City UT 84124

COMEFORD, LORRIE LYNN, b Redbank, NJ, Sept 30, 62; m 84. PHYSICAL ORGANIC CHEMISTRY. *Educ:* Worcester Polytech Inst, BS, 84; Brandeis Univ, PhD(phys org chem), 89. *Prof Exp:* Chemist, Alliance Technologies Corp, 89-90; ASST PROF CHEM, SALEM STATE COL, 90- *Mem:* Am Chem Soc. *Res:* Three dimensional structure of some ion pairs in solution using dipole moment determinations; uv-vis spectra of some laser dyes in aqueous solutions at pressures up to six kbar. *Mailing Add:* Chem & Physics Dept Salem State Col Salem MA 01970

COMEFORO, JAY E(UGENE), b Staten Island, NY, Aug 3, 22; m 45; c 2. CERAMICS SCIENCE & TECHNOLOGY, MINERAL SYNTHESIS. *Educ:* Rutgers Univ, BS, 44, MS, 46; Univ Ill, PhD(ceramics), 49; Rider Col, MBA, 72. *Prof Exp:* Res assoc, Rutgers Univ, 45-46; spec res assoc, Univ Ill, 46-48; ceramic engr, Electro Tech Lab, US Bur Mines, 49-53, head synthetic minerals sect, 53-54; engr in charge, Ceramics Lab, Sylvania Elec Prod, Inc, 54-55; mgr mkt & eng, Frenchtown Porcelain Co, 55-59; consult, 59-61; pres, Consol Ceramics & Metalizing Corp, 61-77; chief exec officer, Accuratus Ceramic Corp, 77-83; RETIRED. *Concurrent Pos:* Consult, Ceramic Consult & Manufacture, 77-85. *Mem:* Fel Am Ceramic Soc; Mineral Soc Am. *Res:* Synthesis of minerals, especially micas and asbestiform; machinable and oxide ceramics; metalizing ceramics, ceramic to metal seals. *Mailing Add:* 28 Fox Grape Rd Flemington NJ 08822

COMENETZ, GEORGE, engineering, for more information see previous edition

COMER, DAVID J, b Tuolumne, Calif, Jan 10, 39; m 58; c 6. ELECTRICAL ENGINEERING. *Educ:* San Jose State Col, BS, 61; Univ Calif, MS, 62; Wash State Univ, PhD(elec eng), 66. *Prof Exp:* Assoc elec eng, Int Bus Mach Corp, 59-64; asst prof elec eng, Univ Idaho, 64-66; assoc prof, Univ Calgary, 66-69; dean eng, 69-74, prof elec eng, Calif State Univ, Chico, 69-81; PROF, DEPT ELEC ENG, BRIGHAM YOUNG UNIV, 81-, DEPT CHAIR, 90- *Concurrent Pos:* Asst prof, San Jose State Col, 64; consult, Int Bus Mach Corp, 66 & Mobility Systs Inc, 67-68, Video Comput Syst, 79-81; mem bd dir, Comput Controls, Inc, 69-72. *Mem:* Inst Elec & Electronics Engrs. *Res:* Electronic circuit design; machine recognition of human speech; computer controlled machinery; mobile robotics. *Mailing Add:* Dept Elec Eng Brigham Young Univ Provo UT 84602

COMER, JAMES PIERPONT, b East Chicago, Ind, Sept 25, 34; m 59; c 2. CHILD PSYCHIATRY, PUBLIC HEALTH ADMINISTRATION. *Educ:* Ind Univ, AB, 56; Howard Univ, MD, 60; Univ Mich, MPH, 64. *Hon Degrees:* Univ New Haven, 77; Calumet Col, 78; Bank Str Col, 87; Albertus Magnus Col, 89; Quinnipiac Col & Depauw Univ, 90. *Prof Exp:* Intern, St Catherine's Hosp, 60-61; staff physician, NIMH, 67-68; asst prof, 68-70, assoc prof, 70-75, prof, 75-76, MAURICE FALK PROF CHILD PSYCHIAT, CHILD STUDY CTR, YALE UNIV, 76- ASSOC DEAN STUDENT AFFAIRS, MED SCH, 69- *Concurrent Pos:* Fel psychiat, Med Sch, Yale Univ, 64-66 & Child Study Ctr, 66-67; NIMH fel, Hillcrest Children's Ctr, Washington, DC, 67-68; Markle scholar, 69-74; adv & consult, Children's TV Workshop, 70-86; mem prof adv coun, Nat Asn Ment Health, 71-; mem comn, Joint Inst Judicial Admin-Am Bar Asn Juv Justice Standards Proj, 73-75; mem, Nat Adv Mental Health Coun, HEW, 76; Henry J Kaiser sr fel, Ctr Advan Studies in Behav Sci, 76-77; mem, Pub Comt Mental Health, 78-; mem, Assembly Behavioral Social Sci, Nat Res Coun, 80; hon chair, Lit Volunteers Greater New Haven, Inc, 90. *Honors & Awards:* Abe & Ethel Lapides Meritorious Pub Serv Award, 80; Rockefeller Pub Serv Award, 80; Agnes Purcell McGavin Award, Am Psychiat Asn, 85, Solomon Carter Fuller Award, 90 & Spec Presidential Commendation, 90; Lela Rowland Award in Prev, Nat Ment Health Asn, 89; Vera Paster Award, Am Orthopsychiat Asn, 90; Milton J E Senn Award, Am Acad Pediat, 90. *Mem:* Am Psychiat Asn; Am Orthopsychiat Asn; Am Acad Child Psychiat. *Res:* Race relations; elementary school education and mental health. *Mailing Add:* Child Study Ctr Med Sch Yale Univ PO Box 3333 New Haven CT 06510-8009

COMER, JOSEPH JOHN, b Brooklyn, NY, Dec 8, 20; m 47; c 4. INORGANIC CHEMISTRY, ELECTRON MICROSCOPY. *Educ:* Pa State Univ, BS, 44, MS, 47. *Prof Exp:* Chemist, Naval Res Lab, 44-45; electron microscopist, Cent Res Lab, Gen Aniline & Film Corp, 46-52; res assoc, Col Mineral Indust, Pa State Univ, 52-55, from asst prof to assoc prof mineral sci, 55-62, head mineral const labs, 57-62; scientist, Res Ctr, Sperry Rand Corp, 62-67; res chemist, Dept Electronic Mat, Rome Air Develop Ctr, US Air Force, 67-84, RETIRED. *Concurrent Pos:* Consult, US Air Force, Rome Air Develop Ctr, Hanscom AFB, 84-88. *Mem:* AAAS; Am Chem Soc; Electron Micros Soc Am (pres, 67). *Res:* Electron microscope studies of electronic, electrooptic materials and thin films. *Mailing Add:* 346 Powder Mill Rd Concord MA 01742

COMER, M MARGARET, b Jacksonville, Fla, Sept 11, 42. MOLECULAR GENETICS, REGULATION. *Educ:* Harvard Univ, AB, 64; Purdue Univ, PhD(biol sci), 72. *Prof Exp:* Res assoc bacteriol, Univ Wis-Madison, 72-75; res assoc microbiol, Univ Regensburg, WGer, 75-76; asst prof, 76-85, ASSOC PROF BIOL, CLARK UNIV, 85-, DIR, BIOCHEM & MOLECULAR BIOL PROG, 86- *Concurrent Pos:* NIH fel, Univ Wis-Madison, 74-75; adj assoc prof chem, Clark Univ, 86- *Mem:* AAAS; Am Soc Microbiol. *Res:* Molecular genetics of transfer RNAs and aminoacyl- tRNA synthetases in bacteria. *Mailing Add:* Dept Biol Clark Univ Worcester MA 01610

COMER, STEPHEN DANIEL, b Covington, Ky, May 2, 41; m 63; c 1. MATHEMATICAL LOGIC, ALGEBRA. *Educ:* Ohio State Univ, BSc, 62; Univ Calif, Berkeley, MA, 64; Univ Colo, Boulder, PhD(math), 67. *Prof Exp:* Asst prof math, Vanderbilt Univ, 67-74; vis asst prof, Clemson Univ, 74-75; assoc prof, 75-82, PROF MATH, THE CITADEL, 82- *Concurrent Pos:* Vis assoc prof, Univ Hawaii, 77-78 & Oxford Univ, 80-81; vis scholar, Iowa State Univ, 85-86; vis prof, Univ Udine, Italy, 87. *Mem:* Am Math Soc; Asn Symbolic Logic; Nat Coun Teachers Math; Asn Comput Mach; Soc Indust & Appl Math; London Math Soc; Math Asn Am. *Res:* Algebra and logic; algebraic logic; universal algebra; model theory; decision problems; sheaf theory; multivalued algebraic systems; algebraic combinatorics; lattic theory; programming language semantics; databases. *Mailing Add:* Dept Math & Comput Sci The Citadel Charleston SC 29409

COMER, WILLIAM TIMMEY, b Ottumwa, Iowa, Jan 11, 36; m 63; c 2. PHARMACEUTICAL RESEARCH, MEDICINAL CHEMISTRY. *Educ:* Carleton Col, BA, 57; Univ Iowa, PhD(org chem), 62. *Prof Exp:* Sr scientist, Mead Johnson & Co, 61-67, res group leader, 67-68, sect leader chem res, 68-70, from prin investr, to sr prin investr,70-74, dir, Pharmaceut Res, 75-77; vpres, 77-82, vpres, Pharmacodynamics, 82-85, SR VPRES, SCI & TECHNOL, BRISTOL MYERS, 85- *Concurrent Pos:* Sect ed, Ann Reports Med Chem. *Mem:* AAAS; Am Chem Soc; Sigma Xi; NY Acad Sci. *Res:* Medicinal chemistry; adrenergic agents; antihypertensive agents; serotonergics; medium ring heterocycles; dopamine antagonists; cardiovascular agents; antiasthmatics. *Mailing Add:* Bristol-Myers Pharm R-D Div 345 Park Ave Rm 5-330 New York NY 10154

COMERFORD, LEO P, JR, b Philadelphia, Pa, Apr 7, 47; m 73; c 3. GROUP THEORY. *Educ:* Villanova Univ, BS, 68; Univ Ill, Urbana, MS, 69, PhD(math), 73. *Prof Exp:* Teaching asst math, Univ Ill, Urbana, 68-73, Univ fel, 72-73; res assoc, Mich State Univ, 73-75; asst prof math, 75-81, assoc prof math, Univ Wis-Parkside, 81-88; assoc prof, 88-90, PROF MATH, EASTERN ILL UNIV, 90- *Concurrent Pos:* Prin investr, NSF res grant, 77-79. *Mem:* Am Math Soc; Math Asn Am. *Res:* Combinatorial group theory, especially equations over groups and small cancellation theory for groups. *Mailing Add:* Dept Math Eastern Ill Univ Charleston IL 61920

COMERFORD, MATTHIAS F(RANCIS), b Boston, Mass, Nov 6, 25; m 56; c 5. MATERIALS SCIENCE, PHYSICS. *Educ:* Mass Inst Technol, SB, 52, SM, 57, ScD(metall), 63. *Prof Exp:* Staff mem metall, Div Sponsored Res, Mass Inst Technol, 52-56; proj supvr, Alloyd Corp, 57-60; metallurgist, Smithsonian Astrophys Observ, 62-69; prin mem tech staff, RCA Info Systs Div, 69-70, leader mat lab, RCA Systs Develop Div, 70-72; independent consult, 72-77; asst res & develop mgr, 77-78, MGR RES & DEVELOP, HOLLIS ENG, INC, 78- *Concurrent Pos:* Res assoc, Mass Inst Technol, 62-63; assoc, Harvard Col Observ, 65-69; lectr, Fitchburg State Col, 72-78; mem, Int Electrotech Comn, 78- *Mem:* Am Soc Metals; Sigma Xi; AAAS; Am Soc Testing & Mat; Nat Asn Corrosion Engrs. *Res:* Mechanical and physical properties of materials; relation of properties to microstructure and environment; metallurgy of meteorites; materials and processes in computer manufacture; materials and processes in automatic soldering, cleaning and testing of printed circuit assemblies. *Mailing Add:* Hollis Eng Inc PO Box 1189 Charron Ave Nashua NH 03060

COMES, RICHARD DURWARD, b Nisland, SDak, Nov 16, 31; m 54; c 3. WEED SCIENCE, AQUATIC PLANT MANAGEMENT. *Educ:* Univ Wyo, BS, 58, MS, 60; Ore State Univ, PhD(weed sci), 71. *Prof Exp:* Res agronomist, 60-65, plant physiologist, 65-89, CONSULT, AGR RES, USDA, 90- *Concurrent Pos:* Coop Scientist, PL-480, India, Pakistan. *Honors & Awards:* Int Hon Award, Off Int Coop & Develop, USDA. *Mem:* Aquatic Plant Mgt Soc; Weed Sci Soc Am; Int Weed Sci Soc; Indian Soc Weed Sci. *Res:* Management of vegetation in aquatic and marginal areas; biology and ecology of aquatic and ditchbank vegetation; fate of herbicides in water; effect of herbicides in irrigation water on crops. *Mailing Add:* 946 Parkside Dr Prosser WA 99350

COMFORT, ALEXANDER, b London, Eng, Feb 10, 20. HUMAN BIOLOGY, GERONTOLOGY. *Educ:* Cambridge Univ, Eng, MA, MB, BCh, 44; Univ London, MRCS & LRCP, 45, DCh, 46, PhD(biochem), 49, DSc(geront), 62. *Prof Exp:* Lectr physiol, London Hosp Med Ctr, 48-51; Nuffield res fel geront, Univ Col, Univ London, 52-65, dir res & head MRC group on aging, 65-73; sr fel, Ctr Study Democratic Insts, Santa Barbara, Calif, 74-75; prof path, Univ Calif, Irvine, 76-78; fel, Inst Higher Studies, 75-82; consult geriat psychiat, Brentwood Vet Admin Hosp, Los Angeles, 77-84; ADJ PROF PSYCHIAT, UNIV CALIF, LOS ANGELES, 77-; CONSULT, UK NAT HEALTH SERV, 86- *Concurrent Pos:* Clin lectr psychiat, Stanford Univ, 74-80; adj prof, Dept Psychiat, Univ Calif, Los Angeles, 77-; ed & founder, Exp Gerontol; consult, UK NHS, 86- *Honors & Awards:* Chem Indust Basle Found Prize, 65; Dr Heinz Karger Mem Found Prize Geront, 69. *Mem:* Royal Soc Med; AMA; Brit Med Asn. *Res:* Epidemiology of Alzheimer's disease; implications of quantum theory for biology and psychology. *Mailing Add:* The Windmill House The Hill Cranbrook TN17 3AH England

COMFORT, JOSEPH ROBERT, b Fayetteville, Ark, July 18, 40; m 83; c 3. NUCLEAR PHYSICS. *Educ:* Ripon Col, AB, 62; Yale Univ, MS, 63, PhD(nuclear physics), 68. *Prof Exp:* Res physicist, Nuclear Struct Lab, Yale Univ, 67-68; fel nuclear physics, Argonne Nat Lab, 68-70; instr physics, Princeton Univ, 70-72; asst prof, Ohio Univ, Athens, 72-76; res assoc prof physics, Univ Pittsburgh, 76-81; assoc prof, 81-84, PROF PHYSICS, ARIZ STATE UNIV, 84- *Concurrent Pos:* Vis scientist, Univ Groningen, Neth, 74-75 & Ind Univ, 76; Physicist, Lawrence Livermore Nat Lab, Calif, 81. *Mem:* Am Phys Soc; Am Asn Physics Teachers. *Res:* Nuclear structure physics; nuclear reaction mechanisms; medium energy nuclear physics. *Mailing Add:* Dept Physics Ariz State Univ Tempe AZ 85287

COMFORT, WILLIAM WISTAR, b Bryn Mawr, Pa, Apr 19, 33; m 57; c 2. POINT-SET TOPOLOGY, TOPOLOGICAL GROUPS. *Educ:* Haverford Col, BA, 54; Univ Wash, MSc, 57, PhD(math), 58. *Hon Degrees:* MA, Wesleyan Univ, 69. *Prof Exp:* Asst math, Univ Wash, 56-58; Benjamin Peirce instr, Harvard Univ, 58-61; asst prof, Univ Rochester, 61-65; assoc prof, Univ Mass, Amherst, 65-67; chmn dept, 69-70 & 80-82, prof, 67-82, EDWARD BURR VAN VLECK PROF MATH, WESLEYAN UNIV, 82- *Concurrent Pos:* Managing ed proc, Am Math Soc, 74-75, assoc secy (Eastern Region), 82-; assoc ed, Am Math Monthly, 83-86. *Mem:* Am Math Soc; Math Asn Am; Asn Symbolic Logic; NY Acad Sci. *Res:* General topology; topological analysis; Stone-Cech compactification; the theory of ultrafilters; topological groups. *Mailing Add:* Dept Math Wesleyan Univ Middletown CT 06457

COMINGS, DAVID EDWARD, b Beacon, NY, Mar 8, 35; m 58; c 3. MEDICAL GENETICS, CELL BIOLOGY. *Educ:* Northwestern Univ, BS, 55, MD, 58; Am Bd Internal Med, dipl. *Prof Exp:* From intern to resident internal med, Cook County Hosp, Chicago, 58-61, fel hemat, 61-62; chief hemat, Madigan Gen Hosp, Tacoma, Wash, 62-64; DIR DEPT MED GENETICS, CITY OF HOPE MED CTR, 66- *Concurrent Pos:* fel med genetics, Univ Wash, 64-66; bd dirs, Am Soc Human Genetics, 74-76; genetics study sect, NIH, 75-78; co-chairman, ICN Univ Calif, Los Angeles Winter Conf Human Molecular Cytogenetics, 77; sci adv bd, Hereditary Dis Found, 75-82; chmn, Symp Molecular Cytogenetics, Uruguay, 77; sci adv bd, Nat Found, March Dimes, 77-; chmn work group, Genetics, Immunol, Virol & Presymptomatic Detection Huntington's Dis Comn, 76-77; ed, Am J Human Genetics, 78-86; pres med staff, City Hope Nat Med Ctr, 78-79; NIH Task Force Inborn Errors of Metab, 78-79; sci adv bd, Joseph's Dis Found, 81-, Neurofibromatosis Found, 84-, Nat Inst Alcoholism & Drug Abuse, 85; mem, Genetics Study Conf, Radiation Effects Res Found, Hiroshima, Japan, 84; pres, Am Soc Human Genetics, 88. *Mem:* AAAS; Am Soc Human Genetics; Am Soc Cell Biol; Am Fedn Clin Res; Am Soc Clin Invest. *Res:* Human genetics; biochemistry and physiology of chromosomes; Tonrette syndrome; human neurogenetics; molecular genetics and human brain. *Mailing Add:* Dept Med Genetics City of Hope Med Ctr Duarte CA 91010

COMINGS, EDWARD WALTER, b Phillipsburg, NJ, Feb 24, 08; m 31; c 3. CHEMICAL ENGINEERING, ENGINEERING ADMINISTRATION. *Educ:* Univ Ill, BS, 30; Mass Inst Technol, ScD(chem eng), 34. *Prof Exp:* Asst chem engr, Mass Inst Technol, 31-33; chem engr, Tex Co, NY, 33-35; asst prof chem eng, NC State Univ, 35-36; from asst prof to prof, Ill Univ, 36-51; prof chem eng & metall & head dept, Purdue Univ, 51-59; prof eng & dean col, 59-73, EMER PROF ENG, UNIV DEL, 73- *Concurrent Pos:* Off Sci Res & Develop official investr & assoc dir munitions develop lab, Univ Ill, 40-45; consult chem corp & other co; Fulbright lectr, Delft Technol Inst, Neth, 57; Guggenheim fel, 57; chmn, Eng Joint Coun Nominating Comt for Nat Medal of Sci; mem bd trustees, Del Tech & Community Col, 66-74; mem, Gov Coun Sci & Technol, 69-72; chmn high pressure res, Gordon Res Conf, 56, mem, Coun Gordon Res Conf; prof chem eng, Univ Petrol & Minerals, Saudia Arabia, 74-78, chmn dept, 75-77. *Honors & Awards:* Naval Ord Develop Award; Army-Navy Cert Appreciation; Walker Award, Am Inst Chem Engrs, 56. *Mem:* fel Am Inst Chem Engrs; Am Soc Eng Educ; Sigma Xi; fel AAAS. *Res:* Drying, extraction, thickening, jet mixing; author of fifty technical articles. *Mailing Add:* 509 Windsor Dr Newark DE 19711-7428

COMINS, NEIL FRANCIS, b New York, NY, May 11, 51; m 82; c 1. ASTROPHYSICS. *Educ:* Cornell Univ, BS, 72; Univ Md, MS, 74; Univ Col, Cardiff, Wales, PhD(astrophysics), 78. *Prof Exp:* From asst prof to assoc prof, 78-90, PROF ASTRONOMY, UNIV MAINE, ORONO, 90- *Concurrent Pos:* Summer fac fel, NASA/ASEE, Ames Res Ctr, 80-81 & 87-88. *Mem:* Fel Royal Astron Soc; Am Astron Soc; Can Astron Soc; Int Astron Union. *Res:* Numerical simulation of the formation, structure and stability of galaxies, stability of rotating stars, data analysis of noise limited experiments; observations of galaxies at Kitt Peak National Observatory and the Very Large Array; general relativity. *Mailing Add:* Dept Physics Univ Maine Bennett Hall Orono ME 04469

COMINSKY, LYNN RUTH, b Buffalo, NY, Nov 19, 53; m 80. X-RAY ASTRONOMY, HIGH ENERGY ASTROPHYSICS. *Educ:* Brandeis Univ, BA, 75; Mass Inst Technol, PhD(physics), 81. *Prof Exp:* Res physicist, Space Sci Lab, Univ Calif, Berkeley, 81-86; ASSOC PROF, DEPT PHYSICS & ASTRON, SONOMA STATE UNIV, 86- *Mem:* Am Astron Soc; Am Phys Soc; Sigma Xi; Asn Women Sci; Am Asn Physics Teachers. *Res:* Optical and x-ray properties of strong galactic x-ray sources, usually believed to be neutron stars in close binary systems; gamma-ray burst sources; reprocessing of x-rays and gamma-rays into optical radiation. *Mailing Add:* 2312 Humboldt Ave El Cerrito CA 94530

COMINSKY, NELL CATHERINE, b Las Animas, Colo, May 11, 20; m 49; c 3. HISTOLOGY. *Educ:* Univ Colo, BA, 42, MA, 44, PhD(zool), 46. *Prof Exp:* Asst prof comp anat, 46-69, assoc prof, 49-59, PROF HISTOL, UNIV HOUSTON, 59- *Honors & Awards:* Piper Award, Minnie Stephens Piper Found, San Antonio, 76. *Mem:* Sigma Xi. *Res:* Demonstration of monoamines in the brain of caiman sclerops; phylogenetic approach to basic sleep mechanisms. *Mailing Add:* 4836 Rockwood Houston TX 77004

COMIS, ROBERT LEO, b Troy, NY, July 16, 45; m 81; c 2. MEDICINE. *Educ:* Fordham Univ, BS, 67; State Univ NY, Upstate Med Ctr, 71. *Prof Exp:* Staff assoc oncol, Nat Cancer Inst, NIH, 72-74; oncol fel, Signey Farber Cancer Inst, Harvard, 75-76; asst med, Peter Bent Brigham Hosp, 75-76; asst prof & chief solid tumor oncol, 76-78, ASSOC PROF & CHIEF MED ONCOL, STATE UNIV NY, UPSTATE MED CTR, 78-; MED DIR, FOX CHASE CANCER CTR, 84- *Concurrent Pos:* Chmn respiratory comn, Cancer & Leukemia Group B, 79-; coordr cancer res, Barbara Kopp Res Ctr, 81- *Mem:* Am Soc Clin Oncol. *Res:* Clinical pharmacology of antineoplastic agents; innovative therapies for small cell anaplastic lung cancer. *Mailing Add:* Fox Chase Cancer Ctr 7701 Burholme Ave Philadelphia PA 19111

COMISO, JOSEFINO CACAS, b Narvacan, Ilocos Sur, Philippines, Sept 21, 40; US citizen; m 70; c 3. REMOTE SENSING, POLAR OCEANOGRAPHY. *Educ:* Univ Philippines, BS, 62; Fla State Univ, MS, 66; Univ Calif, Los Angeles, PhD(physics), 72. *Prof Exp:* Scientist, Philippine Atomic Res Ctr, 62-63; instr physics, Univ Philippines, 63-64; asst physicist, Univ Calif, Los Angeles, 72-73; res assoc, Univ Va, 73-77; mem sr staff, Comput Sci Corp, 77-79; PHYS SCIENTIST, NASA-GODDARD SPACE FLIGHT CTR, 79- *Concurrent Pos:* Prin investr, NASA-Goddard Space Flight Ctr, 82-, chief scientist, NASA Arctic Exp, 87. *Honors & Awards:* Achievement Award, NASA, 83, special award, 87. *Mem:* Am Geophys Union; Am Phys Soc; Int Glaciological Soc; Philippine-Am Acad Sci. *Res:* Satellite remote sensing of geophysical earth parameters in microwave and infrared region; polar oceanography; sea ice cover; radiative transfer modeling; use of artificial intelligence in parameter retrievals. *Mailing Add:* Lab Oceans Code 971 NASA-Goddard Space Flight Ctr Greenbelt MD 20771

COMIZZOLI, ROBERT BENEDICT, b Union City, NJ, Apr 22, 40; m 65; c 2. SEMICONDUCTORS, SOLID STATE ELECTRONICS. *Educ:* Boston Col, BS, 62; Princeton Univ, MA, 64, PhD(physics), 67. *Prof Exp:* mem tech staff res, RCA Labs, RCA Corp, 66-79, mgr qual assurance, RCA Solid State Tech Ctr, 79-81; mem tech staff, 81-88, DISTINGUISHED MEM TECH STAFF & SUPVR, AT&T BELL LABS, 88- *Concurrent Pos:* Chair, Dielectrics & Insulation Div Electrochem Soc. *Honors & Awards:* Callinan Award, Electrochem Soc, 88. *Mem:* Inst Elec & Electronics Engrs; fel Electrochem Soc. *Res:* Semiconductor devices; integrated circuits; power devices; reliability; passivation; semiconductor processing; electrophotography; environmental effects on electronic systems. *Mailing Add:* 95 Knickerbocker Dr Belle Mead NJ 08502

COMLEY, PETER NIGEL, b London, Eng, March 12, 51; US citizen; m 75; c 1. SUPERPLASTICITY. *Educ:* London Univ, BSc, 72; Imperial Col, ACGI, 72. *Prof Exp:* Res engr, Joseph Lucas Inc, 72-74; sr technologist, Int Nickel Corp, 74-82; DIR, SPF TECHNOL, MURDOCK INC, 82- *Mem:* Am Soc Metals; Soc Mining Engrs. *Res:* Development of superplastic forming of titanium, aluminum and stainless steels; production of spf parts for aircraft industry. *Mailing Add:* Murdock Inc 15800 S Avalon Blvd Compton CA 90220-4249

COMLY, HUNTER HALL, b Denver, Colo, July 21, 19; m 41; c 5. PSYCHIATRY. *Educ:* Yale Univ, BS, 41, MD, 43. *Prof Exp:* Intern pediat, Mass Gen Hosp, Boston, 44; asst resident, Children's Hosp, Iowa City, Iowa, 44-46, resident psychiat, Psychiat Hosp, 46-47; asst prof pediat in psychiat, Univ Iowa, 48-56; staff psychiatrist, Children's Div, Lafayette Clin, Detroit, Mich, 56-58; dir, Children's Ctr Wayne County, Detroit, 58-67; assoc prof child psychiat, Univ Iowa, 67-71, prof child psychiat & head div, 71-76; head, Children's Div, Ment Health Ctr Linn County, 76-86; CLINICIAN, CEDAR CTR PSYCHIAT CLIN, 86- *Concurrent Pos:* Fel child psychiat, Univ Minn Hosps, Minneapolis, 47-48; fac mem, Continuation Courses Child Psychiat, Univ Minn, 48, 54 & 61; ed, Presch Study Course, Nat Parent-Teacher Mag, 49-51; consult, Iowa Child Welfare Res Sta, 50-51; psychiatrist, Univ Iowa, 51-52, child psychiatrist, 52-56; workshop chmn, Psychopharmacol in Children's Learning & Behav Disorders, 63-66 & 69-70; pvt pract. *Mem:* AMA; fel Am Psychiat Asn; fel Am Orthopsychiat Asn; Am Acad Child Psychiat. *Res:* Learning and behavior disorders of children; effects of psychoactive drugs on learning and behavior. *Mailing Add:* Quail Creek 2-E North Liberty IA 52317

COMLY, JAMES B, b New York, NY, Nov 28, 36; m 59; c 1. INTELLIGENT SYSTEMS. *Educ:* Cornell Univ, BEE, 59; Harvard Univ, MA, 60, PhD(appl physics), 65. *Prof Exp:* NSF fel, Atomic Energy Res Estab, Eng, 65-66; res physicist, Gen Elec Res & Develop Ctr, 66-69, mgr planning & resources, 69-72, mgr, Thermal Br, 72-77, mgr, Energy Sci Br, 77-79, mgr, Combustion & Fuel Sci Br, 79-80, mgr, Thermal & Fuel Sci Br, 80-82, mgr, Comun & Syst Anal Br, 83-86, PHYSICIST, ARTIFICIAL INTEL, GEN ELEC RES & DEVELOP CTR, 87- *Concurrent Pos:* Chmn rev comt, Twin Rivers Energy Proj, Princeton Univ, 73-78; assoc ed, Int J Energy, 75-80; chmn energetics div, Am Soc Mech Engrs, 78-79; mem, Nat Res Coun eval panel, Nat Bur Standards EnCon Progs, 76-80, chmn, 80, exec bd, Spec Intrest Group & Oper Res Soc Am, 85-87; chmn, Nat Res Coun eval panel, Nat Bur Standards Ctr Chem Eng, 81-82; mem, Nat Res Coun Eval panel, Nat Inst Standards & Technol, Ctr Fire Res, 91- *Mem:* Am Phys Soc; Inst Elec & Electronics Engrs; AAAS; Am Asn Artificial Intel; Asn Comput Mach; Sigma Xi. *Res:* Artificial intelligence for automated reasoning, including reasoning with uncertainty and knowledge based planning; their application to situation assessment in military systems and financial systems; intelligent automated systems. *Mailing Add:* 1455 Dean St Schenectady NY 12309

COMMARATO, MICHAEL A, b Montclair, NJ, Apr 13, 40; m 67; c 1. PHARMACOLOGY. *Educ:* Rutgers Univ, BS, 62; Marquette Univ, PhD(pharmacol), 68. *Prof Exp:* USPHS fel pharmacol, Mich State Univ, 68-69; sr pharmacologist, William H Rorer, Inc, 69-71; sr scientist pharmacol, Warner Lambert Res Inst, 71-76; PHARMACOLOGIST, DIV CARDIO-RENAL DRUB PROD, FOOD & DRUG ADMIN, 76- *Mem:* Am Soc Pharmacol & Exp Therapeut; Am Heart Asn. *Res:* Review and evaluation of results of preclinical pharmacological and toxicological studies submitted in support of New Drug Applications. *Mailing Add:* Res Training & Develop Br Fed Bldg Div Heart & Vascular Dis NHLBI NIH 7550 Wisconsin Ave Rm 3C04 Bethesda MD 20892

COMMERFORD, JOHN D, b Deadwood, SDak, Aug 23, 29; m 53; c 5. AEROSOL PESTICIDES, REGULATORY AFFAIRS. *Educ:* Carroll Col, Mont, AB, 50; St Louis Univ, PhD(chem), 55. *Prof Exp:* Res chemist, Callery Chem Co, 54-57; sr res scientist, Anheuser-Busch, 57-67, mgr com develop, 67-69; dir tech develop, Corn Refiners Asn, Inc, 69-77; VPRES, SHIRLO, INC, MEMPHIS, 77- *Mem:* Southern Aerosol Tech Asn; Am Asn Cereal Chem; Sigma Xi. *Res:* Carbohydrates; corn products; boron hydrides; medicinal chemistry; aerosol technology. *Mailing Add:* 2640 Cedar Ridge Dr Germantown TN 38138

COMMINS, EUGENE DAVID, b New York, NY, July 1, 32; m 58; c 2. PHYSICS. *Educ:* Swarthmore Col, BA, 53; Columbia Univ, PhD(physics), 58. *Prof Exp:* Instr physics, Columbia Univ, 58-60; from asst prof to assoc prof, 60-69, PROF PHYSICS, UNIV CALIF, 69- *Concurrent Pos:* A P Sloan Found fel, Univ Calif, Berkeley, 62-66; Guggenheim Found fel, 67-68. *Mem:* Nat Acad Sci; fel Am Acad Arts & Sci; fel AAAS; fel Am Phys Soc. *Res:* Atomic spectroscopy; weak interactions. *Mailing Add:* Dept Physics Univ Calif Berkeley CA 94720

COMMISSIONG, JOHN WESLEY, b St Vincent, WI, Apr, 2, 44; Can citizen; m 69; c 4. NEUROPHYSIOLOGY, NEUROPHARMACOLOGY. *Educ:* Univ WI, BSc, 67; Univ Southampton, MSc, 71, PhD(neurophysiol), 74. *Prof Exp:* Asst prof physiol, Univ WI, 74-76; vis scientist pharmacol, NIH, 76-79; from asst prof to assoc prof physiol, McGill Univ, 79-89; CHIEF, UNIT NEURAL TRANSPLANTATION, NIH-NAT INST NEUROL DIS & STROKE, 90- *Mem:* Brit Pharmacol Soc; Am Soc Pharmacol & Exp Therapuet; Am Soc Neurochem; Soc Neurosci; Can Physiol Soc; NY Acad Sci. *Res:* Effect of descending monoaminergic projections on control of somatic motoneurons in mammalian spinal cord. *Mailing Add:* NIH-Nat Inst Neurol Dis & Stroke-CNB Bldg 10/5N214 9000 Rockville Pike Bethesda MD 20892

COMMITO, JOHN ANGELO, b Everett, Mass, Apr 23, 49; c 2. MARINE ECOLOGY, BENTHIC ECOLOGY. *Educ:* Cornell Univ, AB, 71; Duke Univ, PhD(zool), 76. *Prof Exp:* Teaching asst introd biol, Marine Lab, Duke Univ, 71-73, teaching asst marine sci, 72-74, teaching asst man & marine environ, 74, teaching asst invert zool, 75; asst prof biol, Univ Maine, 76-80; ASST PROF BIOL & DIR ENVIRON STUDIES PROG, HOOD COL, 80- *Concurrent Pos:* Mem search comt, Marine Sci Ctr, Dir Univ Maine, 77-78; mem selection comt marine res, Marine State Legis, 77-80; mem Washington county regional planning comn, Water Qual Rev Comt, 77-80; prin investr, Maine Sea grant, 78-80. *Mem:* AAAS; Am Soc Limnol & Oceanog; Ecol Soc Am; Am Inst Biol Sci; Atlantic Estuarine Res Soc. *Res:* Predation, competition and life history strategies of marine benthic polychaetes and bivalves; regulation of estuarine soft-bottom community structure; population biology. *Mailing Add:* Dept Biol Hood Col Frederick MD 21701

COMMON, ROBERT HADDON, b Larne, Northern Ireland, Feb 25, 07; m 35; c 6. AGRICULTURAL CHEMISTRY. *Educ:* Queen's Univ Belfast, BSc, 28, BAgr, 29, MAgr, 31, DSc, 57; Univ London, BSc, 30, PhD(biochem), 35, DSc, 44. *Hon Degrees:* LLD, Queen's Univ Belfast, 74. *Prof Exp:* Asst, Chem Res Div, Ministry Agr & instr agr chem, Queen's Univ Belfast, 29-47; prof, 47-74, chmn dept, 47-72, EMER PROF AGR CHEM, MACDONALD COL, MCGILL UNIV, 75- *Honors & Awards:* E W McHenry Award, Nutrit Soc Can, 75. *Mem:* Fel Royal Soc Can; fel Can Inst Chem; fel Agr Inst Can; fel Royal Soc Chem Gt Brit. *Res:* Mineral metabolism in the domestic fowl; biochemical effects of gonadal hormones in the fowl; composition and digestibility of feedstuffs; metabolism of estrogens in the fowl. *Mailing Add:* 12 Maple Ave Ste Anne de Bellevue PQ H9X 2E4 Can

COMMONER, BARRY, b New York, NY, May 28, 17; c 2. BIOLOGY. *Educ:* Columbia Univ, AB, 37; Harvard Univ, MA, 38, PhD(biol), 41. *Hon Degrees:* DSc, Hahnemann Med Col, 63, Colgate Univ, 72, Clark Univ, 74, Grinnell Col, 68, Lehigh Univ, 69, Williams Col, 70 & Ripon Col, 71, Cleveland State Univ, 80; LLD, Univ Calif, 67, Grinnell Col, 81. *Prof Exp:* Asst biol, Harvard Univ, 38-40; assoc ed, Sci Illustrated, NY, 46-47; from assoc prof to prof plant physiol, Wash Univ, 47-76, dept bot, 65-69, univ prof environ sci, 76-81, dir ctr biol natural systs, 65-81; instr, 40-42, DIR CTR BIOL NATURAL SYST, QUEENS COL, CITY UNIV NY, 81- *Concurrent Pos:* Naval liaison off, US Senate Comt Mil Affairs, 46; mem bd dirs, Scientists Inst Pub Info, 63-, co-chmn bd, 67-69, chmn, 69-78, chmn exec comt, 78; pres, St Louis Comt Nuclear Info, 65-66; mem bd dirs & exec comt sci div, St Louis Comt Environ Info, 66-; mem spec study group on sonic boom, US Dept Interior, 67-68; mem bd consult experts, Rachel Carson Trust for Living Environ, 67- & law ctr comn, Univ Okla, 69-70; mem bd, Univs Nat Anti-War Fund; mem adv comt, Coalition for Health of Communities, bd sponsors, In These Times, 76-; Secy's Adv Coun, US Dept Com, 76; adv comt, Ctr Develop Policy, 78; adv coun, Fund for Peace, 78. *Honors & Awards:* Newcomb Cleveland Prize, AAAS, 53; First Int Humanist Award, Int Humanist & Ethical Union, 70; Int Prize Safeguarding Environ, Cervia, Italy, 73; Comdr Order of Merit Repub Italy, 77; Premio Iglesias, Sardinia, Italy, 78, 82. *Mem:* Fel AAAS; Soc Gen Physiol; Am Inst Biol Sci; Am Chem Soc; Am Soc Plant Physiol. *Res:* Alterations in the environment in relation to modern technology; current status of the nitrogen cycle; roles of free radicals in biological processes; the origins and significance of the environmental and energy crises; environmental carcinogenesis; development of strategies to reduce the vulnerability of United States agriculture to disruptions from energy shortages; discovering origin of dioxins and furans in mass-burn incinerators; development of alternative trash-disposal systems. *Mailing Add:* 52 Remson St Brooklyn NY 11201

COMNINOU, MARIA, b Athens, Greece, Aug 12, 47. SOLID MECHANICS. *Educ:* Nat Tech Univ Athens, BS, 70; Northwestern Univ, MS, 71, PhD(theoret & appl mech), 73. *Prof Exp:* Instr math, Mass Inst Technol, 73-74; from asst prof to assoc prof, 74-85, PROF MECH ENG & APPL MECH, UNIV MICH, 85- *Honors & Awards:* Alfred Noble Prize, Am Soc Civil Engrs, Am Soc Mech Engrs, Inst Elect & Electronics Engrs, Western Soc Eng & Am Inst Mining Metall & Petrol Engrs; Henry Hess Award, Am Soc Mech Engrs. *Mem:* Am Soc Mech Engrs; Am Soc Civil Engrs; Am Acad Mech; Soc Women Engrs. *Res:* Wave propagation; fracture; dislocations; contact problems; elasticity. *Mailing Add:* Mech Engineering Dept Univ Mich Ann Arbor MI 48109-2125

COMP, PHILIP CINNAMON, b Kewanee, Ill, Feb 28, 45; m 74; c 3. MEDICINE, THROMBOTIC DISEASE. *Educ:* Reed Col, BA, 67; Univ Wash, MD, 71; Univ Okla, PhD(biochem), 76; Am Bd Internal Med, dipl, 74; Am Bd Allergy & Immunol, dipl, 83. *Prof Exp:* PROF MED, UNIV OKLA, 82-; DIR THROMBOSIS & COAGULATION LAB, STATE OKLA TEACHING HOSP, 82-; DIR ADULT SECT, STATE OKLA HEMOPHILIA PROG, 82- *Concurrent Pos:* Dir, Okla Ctr Molecular Med, 90- *Mem:* Am Soc Hemat; Am Col Physicians; Am Soc Clin Invest. *Res:* Etiology and diagnosis of hereditary and acquired thrombotic disease; prevention of thrombosis. *Mailing Add:* Okla Mem Hosp Rm EB 400 Oklahoma City OK 73126

COMPAAN, ALVIN DELL, b Hull, NDak, June 11, 43; m 69; c 4. SPECTROSCOPY, SPECTROMETRY. *Educ:* Calvin Col, AB, 65; Univ Chicago, MS, 66, PhD(physics), 71. *Prof Exp:* Res assoc physics, NY Univ, 71-73; from asst prof to prof physics, Kans State Univ, 73-87; PROF PHYSICS, UNIV TOLEDO, 87- *Concurrent Pos:* Traineeship, NSF, Univ Chicago, 69-71; Von Humboldt fel, Max Plank Inst Solid State Res, Stuttgart, FRG, 82-83; prin investr, grants/contracts, NSF, Off Naval Res,

Air Force Off Sci Res, Army Res Off, Solar Energy Res Inst. *Mem:* Am Phys Soc; AAAS; Sigma Xi; Mat Res Soc; Optical Soc Am; Union Concerned Scientists. *Res:* Raman scattering and photoluminescence studies of semiconductors; electron-phonon interactions in semiconductors; ion implantation effects in semiconductors; coherent anti-Stokes Raman scattering; laser annealing in semiconductors; pulsed laser deposition; growth and characterization of photovoltaci materials. *Mailing Add:* Physics Dept Univ Toledo 2801 W Bancroft St Toledo OH 43606-3390

COMPANION, AUDREY (LEE), b Tarentum, Pa, Aug 19, 32. QUANTUM CHEMISTRY. *Educ:* Carnegie Inst Technol, BS, 54, MS, 56, PhD(phys chem), 58. *Prof Exp:* From instr to assoc prof chem, Ill Inst Technol, 58-75; assoc prof, 75-76, PROF CHEM, UNIV KY, 76- *Honors & Awards:* Stone Award, Am Chem Soc. *Mem:* Am Chem Soc; Am Phys Soc. *Res:* Molecular orbital theories; electronic spectroscopy; crystal field theory; theories of chemisorption; theoretical studies of the stability, geometry and reactivity of small metal clusters; thermodynamics and henetics of reversible hydraction of mutals and alloys; vibrational analysis of molecules chemisorbed on metals. *Mailing Add:* Dept Chem Univ Ky Lexington KY 40506-0055

COMPANS, RICHARD W, b Syracuse, NY, Sept 15, 40; m 65. VIROLOGY. *Educ:* Kalamazoo Col, BA, 63; Rockefeller Univ, PhD(virol), 68. *Prof Exp:* Guest investr electron micros, Inst Sci Res Cancer, Villejuif, France, 68; Am Cancer Soc hon fel microbiol, John Curtin Sch Med Res, Australian Nat Univ, 68-69; from asst prof to assoc prof virol, Rockefeller Univ, 69-75; PROF MICROBIOL & SR SCIENTIST, CANCER RES CTR, UNIV ALA, BIRMINGHAM, 75-, PROF BIOCHEM, 85-; ASSOC DIR, CTR AIDS RES, 88- *Concurrent Pos:* Vis investr, Scripps Clin & Res Found, 81; mem, Virol Study Sect, USPHS, 76-80; vis prof, Univ Geneva, 88-89; chmn, RNA Virus Div, Am Soc Microbiol, 87-88. *Honors & Awards:* Gardner Award, 88. *Mem:* Am Soc Cell Biol; Am Soc Biol Chemists; Am Asn Immunol; Am Chem Soc; Am Soc Microbiol; Am Soc Virol. *Res:* Cell biology; biochemistry; structure and assembly of viruses; protein traffic in virus-infected cells; vaccine development. *Mailing Add:* Dept Microbiol UAB Sta Birmingham AL 35294

COMPARIN, ROBERT A(NTON), b Hurley, Wis, July 25, 28; m 56; c 5. MECHANICAL ENGINEERING. *Educ:* Purdue Univ, BSME, 54, MSME, 58, PhD(mech eng), 60. *Prof Exp:* Instr mech eng, Purdue Univ, 54-60; staff engr, Int Bus Mach Corp, 60-62; asst prof, Univ Maine, 62-64; from assoc prof to prof, Va Polytech Inst & State Univ, 64-74; chmn, Mech Eng Dept, NJ Inst Technol, 74-77; dean Fenn Col Eng, Cleveland State Univ, 77-83; AT DEPT MECH ENG, VA POLYTECH INST & STATE UNIV, 83- *Mem:* Am Soc Mech Engrs; Am Soc Eng Educ; Sigma Xi. *Res:* Fluid mechanics. *Mailing Add:* Dept Mech Eng Va Polytech Inst & State Univ Blacksburg VA 24061

COMPERE, EDGAR LATTIMORE, b Hamburg, Ark, Jan 23, 17; m 45; c 3. PHYSICAL CHEMISTRY. *Educ:* Ouachita Col, AB, 38; La State Univ, MS, 40, PhD(phys chem), 43. *Prof Exp:* Chemist, La Div, Stand Oil Co, NJ, 42-46; from asst prof to assoc prof chem, La State Univ, 46-51; sr chemist, Chem Div, 51-56, group leader corrosion sect, Reactor Exp Eng Div, 53-55, asst sect chief, 55-58, chief slurry mat sect, 58-61, sr chemist, Reactor Chem Div, 61-73, SR CHEMIST, CHEM TECHNOL DIV, OAK RIDGE NAT LAB, 73- *Concurrent Pos:* Sect ed, Nuclear Safety, 78- *Mem:* Fel AAAS; Am Chem Soc; Sigma Xi; Am Nuclear Soc; fel Am Inst Chemists. *Res:* Reaction kinetics; tritium; fractional liquid extraction; nuclear reactor chemistry; corrosion; reactor materials; fission product transport; leachability of radioactive solids. *Mailing Add:* 7144 Cheshire Dr Knoxville TN 37919

COMPERE, EDWARD L, JR, b Detroit, Mich, June 22, 27; m 54; c 3. ORGANIC CHEMISTRY. *Educ:* Beloit Col, BS, 50; Univ Chicago, MS, 54; Univ Md, PhD, 58. *Prof Exp:* Asst prof chem, Univ WVa, 58-59 & Kans State Teachers Col, 59-60; from asst prof to assoc prof, Mich Tech Univ, 60-64; assoc prof, 64-67, dir state tech serv prog, 66-67, PROF CHEM, EASTERN MICH UNIV, 67- *Mem:* AAAS; Am Chem Soc; Sigma Xi. *Res:* Physical-organic chemistry; inorganic chemistry; chemistry in aquatic biology; lattice-salt structure. *Mailing Add:* 237 Hillcrest Blvd Ypsilanti MI 48197

COMPHER, MARVIN KEEN, JR, b Clifton Forge, Va, May 17, 42; m. DEVELOPMENTAL BIOLOGY, ENDOCRINOLOGY. *Educ:* Wake Forest Univ, BS, 64; Univ Va, PhD(biol), 68. *Prof Exp:* Asst prof biol, Col Wooster, 68-72; from asst prof to assoc prof biol, 72-88, PROF BIOL, CHATHAM COL, 88- *Res:* Regulation of the newt thyroid gland; effects of hypothalamic lesions and pituitary autotransplantations on thyroid activity; Graves' Orbitopathy. *Mailing Add:* Dept Biol Chatham Col Woodland Rd Pittsburgh PA 15232

COMPTON, DALE L(EONARD), b Pasadena, Calif, June 18, 35; m 59; c 2. AERONAUTICAL ENGINEERING. *Educ:* Stanford Univ, BS, 57, MS, 58, PhD, 69; Mass Inst Technol, MS, 75. *Prof Exp:* Aerospace scientist, 57-72, tech asst to dir, 72-73, dep dir, Astronaut, 73-74, chief, Space Sci Div, 74-80 & 80-81, mgr, IRAS Telescope Proj, 81-83, dep dir, Projs Astronaut, 83-85, dir, Engr & Computer Systs, 85-88, CTR DEP DIR, AMES RES CTR, NASA, 88- *Mem:* Am Inst Aeronaut & Astronaut; AAAS; Sigma Xi. *Res:* Vehicle aerodynamics; re-entry radiative and convective heating; ablation. *Mailing Add:* 10131 Phar Lap Dr Cupertino CA 95014

COMPTON, ELL DEE, b Wilmington, Ohio, Mar 16, 16; m 44; c 1. ORGANIC CHEMISTRY, ENVIRONMENTAL MANAGEMENT. *Educ:* Univ Cincinnati, ChE, 39, MS, 40, PhD(tanning res), 42. *Prof Exp:* Monsanto Chem Co fel tanned calf skin, Univ Cincinnati, 42-43; chemist, Merrimac Div, Monsanto Chem Co, 43-46, group leader, 47-52; dir res, Eagle-Ottawa Leather Co, 52-60; res group leader, Maumee Chem Co, 60-61, appl res dir, 61-63, chem res dir, 63-69; lab dir, Sherwin-Williams Chem Div, 69-73, group dir, 73-78, dir, 78-81, CONSULT, REGULATORY AFFAIRS, SHERWIN-WILLIAMS, CO, 81- *Mem:* AAAS; Am Chem Soc. *Res:* Applications of organic chemicals; statistics. *Mailing Add:* 8457 Whitewood Rd Brecksville OH 44141-1529

COMPTON, JOHN S, b Palo Alto, Calif, Oct 16, 58; m. LOW-TEMPERATURE INORGANIC GEOCHEMISTRY. *Educ:* Univ Calif, San Diego, BA, 81; Harvard Univ, PhD(geol), 86. *Prof Exp:* ASST PROF MARINE GEOL, UNIV SFLA, 86- *Mem:* Geol Soc Am; Geochem Soc; Soc Econ Paleonotologists & Mineralogists. *Res:* Low-temperature geochemistry of marine sediments, especially the formation of authigenic minerals in organic-rich rocks; global carbon budget as it relates to organic carbon burial and the formation of organogenic minerals such as phosphorite, dolomite and pyrite. *Mailing Add:* Dept Marine Sci Univ SFla St Petersburg FL 33701-5016

COMPTON, LESLIE ELLWYN, b San Diego, Calif, Mar 24, 43; m 69; c 2. PHYSICAL CHEMISTRY, SYNTHETIC FUELS. *Educ:* Stanford Univ, BS, 66; Univ Calif, Santa Barbara, PhD(phys chem), 70. *Prof Exp:* Sr res chemist, Garrett Res & Develop Co, Occidental Petrol, 70-71; staff scientist, Sci Applications Inc, 72-75; sr res chemist, Occidental Res Corp, Occidental Petrol, 75-78; MEM TECH STAFF, JET PROPULSION LAB, CALIF INST TECHNOL, 79 - *Concurrent Pos:* Fel, Univ Calif, Santa Barbara, 71-72; consult, Radiation & Environ Mat, Inc, 71-72. *Mem:* Am Phys Soc; Am Chem Soc. *Res:* Gas phase and heterogeneous reactions of chemically high energy ions, cluster ions, atoms and small molecules; gas phase kinetics; synthetic fuels from oil shale, coal, tar sands; thermochemical hydrogen; instrumental analysis and instrumentation development. *Mailing Add:* Jet Propulsion Lab 4800 Oak Grove Dr Bldg 125 Rm 224 Pasadena CA 91109

COMPTON, OLIVER CECIL, b Seattle, Wash, Mar 1, 03; m 50. POMOLOGY. *Educ:* Univ Calif, BS, 31, MS, 32; Cornell Univ, PhD(pomol), 47. *Prof Exp:* Assoc, Exp Sta, Univ Calif, 32-40; from asst to instr pomol, Cornell Univ, 40-47; from asst horticulturist to assoc horticulturist, 48-60, from assoc prof to prof, 49-72, EMER PROF HORT, AGR EXP STA, ORE STATE UNIV, 72- *Mem:* Am Soc Hort Sci; Am Soc Plant Physiol. *Res:* Use of water by citrus and avocado trees; effect of aeration on absorption of nutritients by apple trees; physiological effects of fluorine on plants; response of tree and fruit to climate and nutrient level. *Mailing Add:* 1330 SW 35th St Corvallis OR 97333-1133

COMPTON, RALPH THEODORE, JR, b St Louis, Mo, July 26, 35; m 57; c 3. ELECTRICAL ENGINEERING. *Educ:* Mass Inst Technol, SB, 58; Ohio State Univ, MSc, 61, PhD(elec eng), 64. *Prof Exp:* Jr engr, Deco Electronics, Inc, 58-59; sr engr, Battell Mem Inst, 59-62; asst supvr, Antenna Lab, Ohio State Univ, 62-65, asst prof elec eng, 64-65; asst prof eng, Case Inst Technol, 65-67; NSF fel, Munich Tech Univ, 67-68; assoc prof elec eng, 68-78, PROF ELEC ENG, OHIO STATE UNIV, 78- *Concurrent Pos:* Battelle Mem Inst Grad fel, 61-62; postdoctoral fel, NSF, 67-68; vis engr, Naval Res Lab, 83-84. *Honors & Awards:* Sr Res Award, Eng Col, Ohio State Univ, 83. *Mem:* Fel Inst Elec & Electronics Engrs. *Res:* Electromagnetic theory; antennas; automatic control theory; radar; communications; adaptive antennas. *Mailing Add:* Dept Elec Eng Ohio State Univ 2015 Neil Ave Columbus OH 43210

COMPTON, ROBERT NORMAN, b Metropolis, Ill, Nov 28, 38; m 61; c 4. ATOMIC PHYSICS, MOLECULAR PHYSICS. *Educ:* Berea Col, BA, 60; Univ Fla, MS, 62; Univ Tenn, PhD(physics), 66. *Prof Exp:* Consult, 63-64, PHYSICIST, OAK RIDGE NAT LAB, 66-; PROF CHEM, UNIV TENN, 85- *Concurrent Pos:* Ford Found fel, Univ Tenn, 68, lectr, 69-, adj prof chem; sr vis scientist, FOM Inst Atomic & Molecular Physics, Amsterdam, 78. *Mem:* Fel Am Phys Soc; Optical Soc Am; Am Chem Soc. *Res:* Laser multiphoton ionization of atoms and molecules; interaction of electrons with atoms and molecules. *Mailing Add:* Oak Ridge Nat Lab PO Box X Oak Ridge TN 37830

COMPTON, ROBERT ROSS, b Los Angeles, Calif, July 21, 22; m 48; c 5. GEOLOGY. *Educ:* Stanford Univ, BA, 43, PhD(geol), 49. *Prof Exp:* Geologist, P-1, US Geol Surv, 43-44; from instr to assoc prof, 47-61, Prof Geol, 61-, EMER PROF GEOL, STANFORD UNIV. *Concurrent Pos:* Geologist, US Geol Surv, 51-52, 68-; NSF fel, 55-56; Guggenheim fel, 63-64. *Mem:* Geol Soc Am. *Res:* Igneous and metamorphic petrology and structure. *Mailing Add:* Dept Geol Stanford Univ Stanford CA 94305

COMPTON, THOMAS LEE, b Grafton, WVa, July 23, 42; m 65; c 2. BIOLOGY, ECOLOGY. *Educ:* Calif State Univ, San Jose, BA, 67; Univ Alaska, MS, 69; Univ Wyo, PhD(zool), 74. *Prof Exp:* Teaching asst biol, Univ Alaska, 67-69; instr zool, Pepperdine Univ, 69-70; teaching asst biol, Univ Wyo, 70-71; res assoc ecol, Wyo Game & Fish Comn, 71-74; chmn, Div Natural Sci, Letourneau Col, 76-81, prof biol, 74-81; DIR SCI CAMP ROCKIES & VIS PROF BIOL, FT LEWIS COL, DURANGO, COLO, 81- *Concurrent Pos:* Res grants, Alaska Dept Agr, 68-69, Wyo Game & Fish Comn, 71-74 & Forage Unlimited Inc, 76-77; dir, Sci Camp Rockies, 77-; Danforth Assoc. *Mem:* Am Sci Affil; Sci Res Soc NAm; Nat Asn Biol Teachers; Nat Asn Sci Teachers. *Res:* Conservation and stewardship of natural resources. *Mailing Add:* c/o Lemon Ranch 2694 CR 222 Durango CO 81301

COMPTON, WALTER DALE, b Chrisman, Ill, Jan 7, 29; m 51; c 3. PHYSICS. *Educ:* Wabash Col, BA, 49; Univ Okla, MS, 51; Univ Ill, PhD, 55. *Hon Degrees:* DSc, Mich Tech Univ, 76. *Prof Exp:* Physicist, US Naval Ord Test Sta, Inyokern, 51-52; physicist, US Naval Res Lab, 55-61; prof physics, Univ Ill, Urbana, 61-70, dir coord sci lab, 65-70; dir chem & phys sci, Ford Motor Co, 70-73, vpres res, 73-86; sr fel, Nat Acad Eng, 86-88; LILLIAN M GILBRETH DISTINGUISHED PROF INDUST ENG, PURDUE UNIV, 88- *Concurrent Pos:* Fel, US Naval Ord Test Sta, 55-56; US Naval Res Lab award, 71-73; mem adv bd, Naval Weapons Ctr; bd gov, Argonne Nat Lab; vis comn, Nat Bur Standards, 75-80, chmn, 80. *Mem:* Nat Acad Eng; Sigma Xi; fel Am Phys Soc; fel Soc Automotive Engrs; fel AAAS. *Res:* Solid state physics; radiation effects in solids; color centers in insulating crystals; luminescence; metal semiconductor junction; automotive technology; management of technology. *Mailing Add:* Sch Indust Eng Purdue Univ Grissom Hall West Lafayette IN 47907

COMPTON, WILLIAM A, b Richmond, Va, Aug 2, 27; m 79; c 3. QUANTITATIVE GENETICS, CORN BREEDING. *Educ:* NC State Col, BS, 58, MS, 60; Univ Nebr, PhD(agron), 63. *Prof Exp:* Res asst genetics, NC State Col, 58-60; res asst, Univ Nebr, 60-62, consult statistician, 62-63; asst prof, NC State Univ, 63-65; asst prof agron, Univ Minn, 65-67; assoc prof, 67-75, PROF AGRON, UNIV NEBR, LINCOLN, 75- *Concurrent Pos:* Consult, Agrarian Univ, Peru, 63- *Mem:* Am Soc Agron; Crop Sci Soc Am; AAAS. *Res:* Corn breeding; applied and basic quantitative genetics research with corn. *Mailing Add:* Dept Agron Univ Nebr Lincoln NE 68583-0915

COMSTOCK, DALE ROBERT, b Frederic, Wis, Jan 18, 34; m 56; c 2. MATHEMATICS. *Educ:* Cent Wash State Col, BA, 55; Ore State Univ, MS, 62, PhD(algebra), 66. *Prof Exp:* Instr math, Columbia Basin Col, 56-57 & 59-60; teaching asst, Ore State Univ, 61-64; from asst prof to assoc prof, 64-70, dean grad studies, 70- 90, PROF MATH, CENT WASH UNIV, 70- *Concurrent Pos:* Presidential Interchange Exec, US ERDA, 76-77; dean in residence, Coun Grad Schs US, 84-85; consult, NSF, US Info Agency, Dept Energy & Argonne Nat Lab. *Mem:* Math Asn Am; Am Math Soc; Soc Indust & Appl Math; Asn Comput Mach. *Res:* Algebra; computability; energy policy; graduate education. *Mailing Add:* Dept Math Cent Wash Univ Ellensburg WA 98926

COMSTOCK, GEORGE WILLS, b Niagara Falls, NY, Jan 7, 15; m 39; c 3. EPIDEMIOLOGY. *Educ:* Antioch Col, BS, 37; Harvard Univ, MD, 41; Univ Mich, MPH, 51; Johns Hopkins Univ, DrPH, 56. *Prof Exp:* Dir, Muscogee County Tuberc Study, USPHS, 46-55; chief epidemiol studies, Tuberc Prog, 56-62; assoc prof, 62-65, PROF EPIDEMIOL, JOHNS HOPKINS UNIV, 65- *Concurrent Pos:* Consult tuberc prog, USPHS, 62-; dir, Training Ctr for Pub Health Res, Hagerstown, Md, 63-; ed-in-chief, Am J Epidemiol, 79-89. *Mem:* Am Pub Health Asn; Am Thoracic Soc; Am Epidemiol Soc; Soc Epidemiol Res. *Res:* Epidemiology of chronic diseases, especially tuberculosis and cardio-respiratory diseases. *Mailing Add:* Johns Hopkins Res Ctr Box 2067 Hagerstown MD 21742-2067

COMSTOCK, JACK CHARLES, b Detroit, Mich, June 13, 43. PLANT PATHOLOGY. *Educ:* Mich State Univ, BS, 65, PhD(plant path), 71. *Prof Exp:* Res assoc corn path, Iowa State Univ, 71-74; asst pathologist, Hawaiian Sugar Planters Asn, 74-75, assoc pathologist, 75-89; RES PLANT PATHOLOGIST, AGR RES SERV, USDA, 89- *Concurrent Pos:* Affil fac mem bot sci, Univ Hawaii, 75-89. *Mem:* Am Phytopath Soc; AAAS; Sigma Xi. *Res:* Sugarcane pathology; disease control, screening resistance; sugar cane research. *Mailing Add:* Agr Res Serv SAA USDA Sugarcane Field Sta Star Rte Box 8 Canal Point FL 33438

COMSTOCK, VERNE EDWARD, b Kildeer, NDak, July 18, 19; m 41; c 6. PLANT BREEDING, PLANT GENETICS. *Educ:* State Col Wash, BS, 41, MS, 47; Univ Minn, PhD, 59. *Prof Exp:* Asst agronomist forage invest, Div Forage Crops & Dis, Bur Plant Indust, USDA, State Col Wash, 47-50, res agronomist flax breeding & invest, Southwestern Irrig Field Sta, Brawley, Calif, 50-53, flax qual invests, cereal crop sect, Crops Res Div, Agr Res Serv, Univ Minn, St Paul, 53-57, leader seedflax invests, Indust Crop Sect, 57-73; prof agron & plant genetics, EMERITUS PROF AGRON & PLANT GENTICS, UNIV MINN, ST PAUL, 74- *Mem:* Am Soc Agron. *Res:* Range grass breeding; flax cultural studies under irrigation and quality investigations, flax breeding for disease resistance. *Mailing Add:* Dept Agron Univ Minn St Paul MN 55108

CONAN, ROBERT JAMES, JR, b Syracuse, NY, Oct 30, 24. THERMODYNAMICS & MATERIAL PROPERTIES. *Educ:* Syracuse Univ, BS, 45, MS, 47; Fordham Univ, PhD(phys chem), 50. *Prof Exp:* Instr gen chem, Fordham Univ, 48; from asst prof to prof, 49-89, chmn dept chem, 59-67 & 73-84, EMER PROF PHYS CHEM, LE MOYNE COL, NY, 90- *Concurrent Pos:* Researcher, Stockholm, Sweden, 53 & Res Lab Phys Chem, Swiss Fed Inst Technol, 66-67, Univ SFla, 88; consult, Carrier Corp, 57-75, Owl Wire & Cable Co, 58-80, Thomas A Edison Re-recording Lab, 75-85; vis prof res, Univ SFla, Tampa, 90- *Mem:* Am Chem Soc. *Res:* Theory of liquids and solutions; surface phenomena; thermodynamics; education in chemistry. *Mailing Add:* 263 Robineau Rd Syracuse NY 13207

CONANT, CURTIS TERRY, b Boston, Mass; c 1. MECHANICAL & ELECTRO MECHANISMS, RELIABILITY & LIFE. *Educ:* Univ Calif Long Beach, BS, 69. *Prof Exp:* Nat mgr syst & serv terminals & systs optical, Data Source Corp Subsid Hercules Inc, 69-76; sr systs analyst online systs, TRW-Computer Systs & Serv, 76-78; VPRES ENG MAGNETIC READING SYSTS, AM MAGNETICS CORP, 78- *Concurrent Pos:* Mem-expert, working groups-magnetics, Am Nat Standards Inst, 86-; mem-magnetics, Tech Comts, Automotive Identification Mfrs, 87-; mem-expert, working groups-magnetics, Int Standards Orgn, 88- *Mem:* Soc Plastics Engrs; Am Soc Mech Engrs; Inst Elec & Electronics Engrs. *Res:* Optical character recognition readers; telecommunications multiple drop terminal systems; high and low density magnetics research and development encoding and decoding; high coercivity magnetics. *Mailing Add:* 3231 Carolina St San Pedro CA 90731

CONANT, DALE HOLDREGE, b Casper, Wyo, July 5, 39. PHOTOGRAPHIC CHEMISTRY. *Educ:* Col Idaho, BS, 61; Ore State Univ, MS, 64; Ohio State Univ, PhD(phys chem), 69. *Prof Exp:* Engr, Kaiser Refractories, 63-65; sr chemist, Rochester, 69-74, res assoc, Photog Res Div, 74-75, develop engr, Plate Mfg Div, 75-83, supvr, Lithoplate Prod Control, 84-85, SUPVR QUALITY SERV DIV, EASTMAN KODAK CO, 85- *Mem:* Am Chem Soc. *Res:* Interaction of chlorophyll and its derivatives with II-aromatic electron acceptor systems involving visible spectrum measurements, fluorescence quenching measurements and nuclear magnetic resonance measurements; research and development of photolithography. *Mailing Add:* 508 Canadian Parkway Ft Collins CO 80521

CONANT, FLOYD SANFORD, b Leroy, WVa, Nov 27, 14; m 38; c 2. POLYMER PHYSICS. *Educ:* Morris Harvey Col, BS, 34; WVa Univ, MS, 35. *Prof Exp:* Teacher high sch, 35-42; sr res scientist, Firestone Tire & Rubber Co, 42-75, res assoc, 75-80; mem staff, Standards Testing Labs, 81-89; RETIRED. *Concurrent Pos:* Mem US deleg, Tech Comt 45, Int Standards Orgn, 75-79, convenor working group 10, 77-79; assoc ed, Tire Sci & Technol, 82-90. *Mem:* Am Chem Soc; Am Soc Testing & Mat; Soc Automotive Engrs; Tire Soc. *Res:* Low temperature properties of elastomers; vibration properties of pneumatic tires; coefficient of friction of rubber; tire dynamics. *Mailing Add:* 143 Gunarh Dr Akron OH 44319-3715

CONANT, FRANCIS PAINE, b New York, NY, Feb 27, 26. HUMAN ECOLOGY. *Educ:* Cornell Univ, BA, 50; Columbia Univ, PhD(anthrop), 60. *Prof Exp:* Asst prof anthrop, Univ Mass, 60-61; assoc prof, 62-67, PROF ANTHROP, HUNTER COL, CITY UNIV N Y, 67- *Concurrent Pos:* Consult, Ministry Transp, Govt Kenya, 80-82. *Mem:* Am Anthrop Asn; AAAS; Royal Anthrop Inst; NY Acad Sci. *Res:* Human ecology; sex and gender roles; remote sensing and satellite data interpretation; African herding and farming systems; information systems; resource utilization. *Mailing Add:* Hunter Col City Univ New York Dept Anthrop 695 Park Ave New York NY 10021

CONANT, LOUIS COWLES, b Orford, NH, Sept 14, 02; m 30; c 2. GEOLOGY, GENERAL EARTH SCIENCE. *Educ:* Dartmouth Col, AB, 26; Cornell Univ, AM, 29, PhD(geol), 34. *Prof Exp:* Instr geol, Dartmouth Col, 26-27; asst, Cornell Univ, 27-28, instr, 28-29, 30-37; field geologist, NRhodesia, 29-30; asst geologist, Miss Geol Surv, 37-38, 40-42; supvr mineral surv, Works Progress Admin, 38-39; lectr, Smith Col, 39-40; geologist, US Geol Surv, 42-72, RETIRED. *Concurrent Pos:* Assoc prof, Univ Miss, 37-39, 40-42; geol map compiler, Agency Int Develop, Libya, 60-62. *Mem:* Fel AAAS; fel Geol Soc Am; Am Asn Petrol Geol. *Res:* Non-metallic economic geology; stratigraphy; Chattanooga Shale; east Gulf Coastal Plain; fuels; geology of Libya. *Mailing Add:* 10450 Lottsford Rd Apt 330 Mitchellville MD 20721

CONANT, ROBERT HENRY, b Rockland, Mass, Oct 5, 16; m 46; c 5. PHOTOGRAPHIC CHEMISTRY. *Educ:* Loyola Col, Md, BS, 37; Georgetown Univ, MS, 39, PhD(biochem), 42. *Prof Exp:* Instr chem, Georgetown Univ, 37-42; res chemist, Photo Repro Div, Gen Aniline & Film Corp, 45-47, sr res chemist & res group leader, 47-51, tech asst to film plant mgr, 51-54, sr opers supvr film emulsions dept, 54-56, sr emulsion specialist, 56-60, mgr film qual control dept, 60-63 & sensitometry dept, 63-64, res chemist, 64-67, qual control specialist gelatin, 67, prod qual specialist gelatin, 67-74, supvr emulsion mfg, Photo-Repro Div, GAF Corp, 74-80; RETIRED. *Res:* Photographic emulsion; gelatin; tests for sugars; S-amino acids in proteins; cystine and methionine distribution in proteins of egg whites. *Mailing Add:* 44-S Andrea Dr Vestal NY 13850-2241

CONANT, ROGER, b Mamaroneck, NY, May 6, 09; m 47, 79; c 2. HERPETOLOGY & ZOOGEOGRAPHY. *Hon Degrees:* ScD, Univ Colo, 71. *Prof Exp:* Cur reptiles, Toledo Zoo, Ohio, 29-33, educ dir, 31-33, cur, 33-35; cur reptiles, Philadelphia Zool Garden, 35-73, dir, 67-73; ADJ PROF BIOL, UNIV NMEX, 73- *Concurrent Pos:* Res assoc, Am Mus Natural Hist & Mus Southwestern Biol; consult, Am Philos Soc Proj for Adult Educ & Partic in Sci, 40-42 & Nat Res Coun, 59-62. *Mem:* Am Soc Ichthyol & Herpet (first vpres, 46, 56, secy, 58-60, pres, 62); Soc Study Amphibians & Reptiles; Zool Soc London; Herpetologists League; Soc Europ Herpet. *Res:* Distribution, natural history and speciation in reptiles and amphibians of the United States and Mexico; Asian and North American pit vipers of the genus Agkistrodon and its allies. *Mailing Add:* 6900 Las Animas NE Albuquerque NM 87110-3527

CONANT, ROGER C, b Milwaukee, Wis, Apr 12, 38; m 65; c 1. ELECTRICAL ENGINEERING, CYBERNETICS. *Educ:* Purdue Univ, BS, 61; Univ Ill, Urbana, MS, 63, PhD(info transfer), 68. *Prof Exp:* Asst prof, 68-77, ASSOC PROF ELEC ENG & COMPUTER SCI, UNIV ILL, CHICAGO, CIRCLE, 77- *Mem:* AAAS; Am Soc Cybernet; Inst Elec & Electronics Engrs. *Res:* Information transfer in complex systems; technical aids for the handicapped. *Mailing Add:* Dept Elec Eng & Computer Sci Univ Ill Box 4348 Chicago IL 60680

CONARD, GEORGE P(OWELL), b Brooklyn, NY, Sept 10, 19; m 46; c 3. PHYSICAL METALLURGY, MATERIALS SCIENCE. *Educ:* Brown Univ, BS, 41; Stevens Inst Technol, MS, 48; Mass Inst Technol, ScD, 52. *Prof Exp:* Instr math, Stevens Inst Technol, 46-48; instr metall, Mass Inst Technol, 48-50, asst prof, 50-52; from asst prof to prof metall, Lehigh Univ, 52-84, asst dir magnetics proj, 52-56, dir, Magnetic Mat Lab, 56-76, chmn dept metall & mat sci, 69-80, EMER PROF MAT SCI & ENG, LEHIGH UNIV, 84- *Concurrent Pos:* Vis prof, Yonsei Univ, Seoul, Korea, 75 & 82. *Mem:* Am Soc Eng Educ; Am Soc Metals; Metall Soc; Brit Metals Soc. *Res:* Deformation and annealing of metals; magnetics. *Mailing Add:* PO Box 52 Hellertown PA 18055

CONARD, GORDON JOSEPH, b Milwaukee, Wis, Sept 22, 39; m 85; c 3. BIOCHEMICAL PHARMACOLOGY, DRUG METABOLISM. *Educ:* Univ Wis-Madison, BS, 61, MS, 67, PhD(pharmacol), 69. *Prof Exp:* Instr, Sch Pharm, Univ Wis-Milwaukee, 68; res assoc biochem, Res Inst, Am Dent Asn, 68-72; sr biochem pharmacologist, Riker Lab, Inc, 3M Co, 73-74, from res specialist to sr res specialist, 75-83,; MGR DRUG METAB, 3M PHARMACEUTICALS, 83- *Concurrent Pos:* Reviewer, J Pharmaceut Sci, 72-; asst prof, Col Pharm, Univ Minn, Minneapolis, 73-80. *Mem:* Am Asn Pharmaceut Sci; Sigma Xi. *Res:* Metabolic disposition of new drug molecules in laboratory animals and in man with emphasis on relationships to pharmacologic and toxicologic activity. *Mailing Add:* 3M Pharmaceuticals Bldg 270-3S-05 3M Ctr St Paul MN 55144-1000

CONARD, ROBERT ALLEN, b Jacksonville, Fla, July 29, 13; m 48; c 4. MEDICAL RESEARCH. *Educ:* Univ SC, BS, 36, MD, 41. *Prof Exp:* Med officer, US Navy, 41-47, proj officer, Radiol Defense Lab, 47-50, with med res dist, 50-56; scientist & chief, Marshall Island Med Survs & sr scientist, Brookhaven Nat Lab, 56-79; RETIRED. *Concurrent Pos:* Prof path, State Univ NY Stony Brook. *Res:* Radiation effects; medical surveys of Marshallese people exposed to radioactive fallout; effects of radiation on animals. *Mailing Add:* 32 Ivy Lane Setauket NY 11733

CONAWAY, CHARLES WILLIAM, b Anniston, Ala, July 11, 43; m 73; c 2. INFORMATION SCIENCE. *Educ:* Jacksonville State Univ, AB, 64; Fla State Univ, MS, 65; Rutgers Univ, PhD(info sci), 74; Fla State Univ, cert(geront), 88. *Prof Exp:* Head ref librn, Fla Atlantic Univ, 66-68; res fel, Rutgers Univ, 68-71; asst prof, Sch Info & Libr Studies, State Univ NY, Buffalo, 71-77; ASSOC PROF INFO SCI, SCH LIBR & INFO STUDIES, FLA STATE UNIV, 77- *Concurrent Pos:* Fulbright lectr, prog libr sci, Univ Iceland, 75-76; prin investr, index usability proj, State Univ NY, Buffalo, 75-76; vis prof, Univ PR, 85; consult, GLARP, Bogota, Colombia, 86-87; consult, Occup Aspiration Scale, Montevided, Uruguay, 87; consult, Univ Zagreb, Yugoslavia, 87-; prin investr, mkt res proj, World Book Encyclopedia, 87-88; Carnegie Whitney grant, 87-88; prin investr, State Libr Fla, 88-89. *Mem:* Am Soc Info Sci; Special Libr Asn; Am Libr Asn & Res Roundtable; Asn Libr & Info Sci Educ. *Res:* Index usability; automatic and machine-aided indexing; on-line information retrieval systems; library systems analysis and automation; computerized bibliographic techniques; question clarification and information gathering behavior; information science education; microcomputers in libraries; expert systems, computational linguistics; gerontological information systems. *Mailing Add:* Sch Libr & Info Studies Fla State Univ Tallahassee FL 32306-2048

CONAWAY, HOWARD HERSCHEL, b Fairmont, WVa, Oct 2, 40; m 69; c 2. PHYSIOLOGY. *Educ:* Fairmont State Col, BS, 63; WVa Univ, MS, 67; Univ Mo-Columbia, PhD, 70. *Prof Exp:* NIH fel, Dept Pediat, Univ Mo-Columbia, 70-71; asst prof, 71-76, ASSOC PROF PHYSIOL & BIOPHYSICS, SCH MED, UNIV ARK, LITTLE ROCK, 76- *Mem:* AAAS; Endocrine Soc; Sigma Xi; Am Soc Bone & Mineral Res. *Res:* Endocrinology and metabolism; calcium metabolism. *Mailing Add:* Dept Physiol & Biophysics Sch Med Univ Ark Little Rock AR 72205

CONCA, ROMEO JOHN, b New Haven, Conn, May 11, 26; m 89; c 2. ORGANIC CHEMISTRY. *Educ:* Yale Univ, BS, 49, PhD(org chem), 53. *Prof Exp:* Asst, Princeton Univ, 52-53; res chemist, G D Searle & Co, 53-55; res chemist, ITT Rayonier, Inc, 55-59, group leader, 59-62, sect leader, 62-64, res supvr, 64-78, mgr tech support, Olympic Res Div, 78-81; OWNER, LOST MOUNTAIN WINERY, 81- *Mem:* Tech Asn Pulp & Paper Indust; Am Chem Soc. *Res:* Cellulose and wood chemistry; carbohydrates. *Mailing Add:* 730 Lost Mountain Rd Sequim WA 98382-9229

CONCANNON, JOSEPH N, b New York, NY, Sept 25, 20; m 64; c 3. PARASITOLOGY, RADIOBIOLOGY. *Educ:* Univ Dayton, BS, 42; Ohio State Univ, MS, 51; St John's Univ, MS, 56, PhD(parasitol), 59. *Prof Exp:* Teacher high sch, 42-51; instr biol, Cath Univ PR, 52-54; asst prof, Univ Dayton, 59-62; asst prof, 62-64, actg chairperson, Dept Biol Sci, 74-77, ASSOC PROF BIOL, ST JOHN'S UNIV, NY, 64- *Concurrent Pos:* NSF res fel, Univ Rochester, 62. *Mem:* AAAS; Soc Protozool; Am Soc Parasitol; Nat Sci Teachers Asn. *Res:* Protozoan parasitology, especially the trichomonads; thyroid physiology and insect control. *Mailing Add:* Dept Biol Sci St John's Univ Grand Central & Utopia Pwky Jamaica NY 11439

CONCIATORI, ANTHONY BERNARD, b New York, NY, Mar 4, 16; m 51; 2. ORGANIC CHEMISTRY. *Educ:* Fordham Univ, BSc, 38; Univ Cincinnati, PhD(chem), 49. *Prof Exp:* Res chemist, Interchem Res Labs, 39-44; sr res chemist, 49-60, res assoc, 60-68, sect head, 68-71, MGR, CELANESE CORP, 71- *Mem:* Am Chem Soc; Sigma Xi (pres, Sci Res Soc Am, 63-64). *Res:* Polymers; catalysis; coatings; fibers. *Mailing Add:* 27 Orchard St Chatham NJ 07928-2324

CONCORDIA, CHARLES, b Schenectady, NY, June 20, 08; m 48. ENGINEERING. *Hon Degrees:* DSc, Union Col, 71. *Prof Exp:* Lab asst, Gen Elec Co, 26-31, engr, 31-36, appl engr, 36-51, mgr gen anal eng, 51-64, consult eng, 64-73; CONSULT, 73- *Concurrent Pos:* Mem, Int Conf Large Elec Systs (CIGRÉ); chmn, Int Comt Power Syst Planning & Oper. *Honors & Awards:* Coffin Award, 42,; Lamme Medal, Inst Elec & Electronics Engrs; Steinmetz Award, 73; Centennial Award, Inst Elec & Electronics Engrs, 84; Philip Sporn Award, CIGRÉ, 89. *Mem:* Nat Acad Eng; fel Inst Elec & Electronics Engrs; fel Am Soc Mech Engrs; Nat Soc Prof Engrs; Asn Comput Mach (treas); fel AAAS; Cigré. *Res:* Air compressor design; governors; regulators; control systems; electric power system design, planning and operation; computing and electrical machinery; system dynamics. *Mailing Add:* 702 Bird Bay Dr W Venice FL 34292-4030

CONCUS, PAUL, b Los Angeles, Calif, June 18, 33; m 59; c 2. APPLIED MATHEMATICS, NUMERICAL ANALYSIS. *Educ:* Calif Inst Technol, BS, 54; Harvard Univ, AM, 55, PhD(appl math), 59. *Prof Exp:* Appl mathematician, Int Bus Mach Corp, 59-60; STAFF SR SCIENTIST, LAWRENCE BERKELEY LAB, 60-, PROF MATH, UNIV CALIF, BERKELEY, 78- *Concurrent Pos:* Lectr, Univ Calif, Berkeley, 63-65; sr vis fel, Sci Res Coun Gt Brit, 70-71; vis fel, Nat Res Coun, Univ Trento, Italy, 77 & 78; consult, var indust & govt concerns; assoc ed Soc Indust & Appl Math J Sci, Stat Comp, 80- *Mem:* Soc Indust & Appl Math. *Res:* Capillary fluid mechanics; numerical solution of partial differential equations, iterative methods. *Mailing Add:* Lawrence Berkeley Lab Univ Calif Berkeley CA 94720

CONDELL, WILLIAM JOHN, JR, b Melrose, Mass, Mar 29, 27; m 52; c 2. RESEARCH ADMINISTRATION, OPTICAL PHYSICS. *Educ:* Cath Univ Am, BChemE, 49, MS, 52, PhD(physics), 59. *Prof Exp:* Physicist, Naval Ord Lab, 51-52, Eng Res & Develop Lab, US Army, 52-58 & Lab Phys Sci, 58-66; physicist, Off Naval Res, 66-74, dir physics progs, 74-80, Chief Scientist, London, 80-81, leader physics div, 81-86; office mgr, Am Inst Physics, Washington, DC, 86-87; RETIRED. *Concurrent Pos:* Asst prof lectr, George Washington Univ, 57-66. *Mem:* Am Phys Soc; fel Optical Soc Am; Am Asn Physics Teachers. *Res:* Atomic spectroscopy; optics; lasers. *Mailing Add:* 4511 Gretna St Bethesda MD 20814

CONDELL, YVONNE C, b Quitman, Ga, Aug 29, 31; m 52. ENVIRONMENTAL SCIENCE. *Educ:* Fla Agr & Mech Col, BS, 52; Univ Conn, MA, 58, PhD(cellular biol), 65. *Prof Exp:* Teacher high sch, Fla, 55-57 & Minn, 58-60; instr biol, Fergus Falls Jr Col, 60-65; from asst prof to prof, 65-80, PROF MULTIDISCIPLINARY STUDIES & BIOL, MOORHEAD STATE UNIV, 80- *Concurrent Pos:* Lectr, Univ Conn, 63; Yvonne C Condell Am Fel, Am Asn Univ Women. *Mem:* AAAS; Soc Study Social Biol; Am Inst Biol Sci; Nat Asn Biol Teachers. *Res:* Hazardous waste disposal; biology education; science and society. *Mailing Add:* Dept Biol Moorhead State Univ Moorhead MN 56560

CONDER, GEORGE ANTHONY, b Albuquerque, NMex, Oct 7, 50; m 52; c 2. CHEMOTHERAPY, IMMUNOPARASITOLOGY. *Educ:* Pomona Col, BA, 72; Univ NMex, MS, 75; Brigham Young Univ PhD(zool), 79. *Prof Exp:* Res & teaching asst, Univ NMex, 73-75 & Brigham Young Univ, 75-78; res assoc, Brigham Young Univ, NIH, 78-79; fel, Michigan State Univ, NIH, 79-81; MEM STAFF RES & DEVELOP, UPJOHN CO, 81- *Concurrent Pos:* Res grant, Argonne Nat Lab, 75. *Mem:* Am Soc Parasitologists; Soc Protozoologists; Wildlife Dis Asn; Am Soc Trop Med & Hyg. *Res:* Immunological reactions at the host-parasite interface; serodiagnosis of parasitic diseases; development of radiation-attenated vaccines against parasites; chemotherapy of parasitic diseases. *Mailing Add:* 6835 East F Ave Richland MI 49083

CONDER, HAROLD LEE, b Salem, Ohio, Nov 26, 45; m 85; c 1. ORGANOMETALLIC CHEMISTRY. *Educ:* Youngstown Univ, BS, 67; Purdue Univ, PhD(inorg chem), 71. *Prof Exp:* Res asst chem, Tulane Univ, 71-73; ASSOC PROF CHEM, GROVE CITY COL, 73- *Concurrent Pos:* Vis asst prof, Purdue Univ, 81-82. *Mem:* Am Chem Soc. *Res:* Photochemical substitution reactions of transition metal phosphites and phosphines. *Mailing Add:* 14866 Kibler Rd New Springfiled OH 44443-9750

CONDIE, KENT CARL, b Salt Lake City, Utah, Nov 28, 36; m 63; c 1. PETROLOGY, GEOCHEMISTRY. *Educ:* Univ Utah, BS, 59, MA, 62; Univ Calif, San Diego, PhD(geochem), 65. *Prof Exp:* From asst prof to assoc prof geochem & petrol, Wash Univ, 64-70; assoc prof, 70-77, PROF GEOCHEM, NMEX INST MINING & TECHNOL, 77- *Mem:* Geol Soc Am; Geochem Soc; Am Geophys Union. *Res:* Trace element geochemistry; origin and growth of continents. *Mailing Add:* Dept Geosci NMex Inst Mining & Technol Socorro NM 87801

CONDIKE, GEORGE FRANCIS, b Brockton, Mass, Dec 1, 16; m 41; c 2. INORGANIC CHEMISTRY. *Educ:* DePauw Univ, AB, 40; Cornell Univ, PhD(inorg chem), 43. *Prof Exp:* Fel, Mellon Inst Indust Res, 43-44; sr engr, Sylvania Elec Prod, Inc, 44; tech rep, Rohm and Haas Co, 44-47; assoc prof chem, 47-53, prof, 56-82, dean col, 54-56, EMER PROF CHEM, FITCHBURG STATE COL, 82- *Concurrent Pos:* NSF grant, 64-66, mem, NSF Equip Comt, 68-69. *Mem:* Am Chem Soc. *Res:* Chelate compounds; donor-acceptor bonding. *Mailing Add:* 36 Ross Fitchburg MA 01420

CONDIT, PAUL BRAINARD, b Berkeley, Calif, Mar 12, 43; m 66; c 2. ORGANIC CHEMISTRY. *Educ:* Univ Calif, Riverside, BA, 65; Univ Mich, PhD(org chem), 70. *Prof Exp:* Res fel, Calif Inst Technol, 70-71; res chemist, Eastman Kodak Co, 71-76, lab head, 76-79; estimating supvr, Distrib Div, Ekco, 79-80, sr sales forecaster, Photog Div, 80-82; dir, Strategic Info, 82-85 & Mkt Res, Europe, 85-88, DIR MKT RES, HEALTH SCI, EASTMAN KODAK CO, 88- *Mem:* AAAS; Am Chem Soc; Soc Photog Sci & Eng; Sigma Xi. *Res:* Stereochemistry and mechanism of organic reactions; organic chemistry of color photography. *Mailing Add:* Eastman Kodak Co 343 State St NJ160 Rochester NY 14650

CONDIT, PAUL CARR, b Cleveland, Ohio, Sept 19, 14; m 39; c 2. PETROLEUM CHEMISTRY. *Educ:* Yale Univ, BS, 36, PhD(org chem), 39. *Prof Exp:* Res chemist, Calif Res Corp, 39-46, sr res chemist, 46-52, sect supvr, 52-67; mgr polymer div, Chevron Res Co, 67-69, sr res assoc, Patent Dept, 69-77; RETIRED. *Mem:* Am Chem Soc. *Res:* Petrochemical research and development; polymers; patents. *Mailing Add:* 720 Butterfield Rd San Anselmo CA 94960

CONDIT, PHILIP M, b Berkeley, Calif, Aug 2, 41; c 2. ENGINEERING. *Educ:* Univ Calif, Berkeley, BS, 63; Princeton Univ, MS, 65; Mass Inst Technol, MS, 75. *Prof Exp:* Dir eng, Boeing Com Airplanes, 72-83, vpres & gen mgr, Renton Div, 83-84, vpres sales & mkt, 84-86, exec vpres, 86-89, EXEC VPRES & GEN MGR, 777 DIV, BOEING COM AIRPLANES, 89- *Honors & Awards:* E C Well Award, Am Inst Aeronaut & Astronaut, 82 & Aircraft Design Award, 84. *Mem:* Nat Acad Eng; fel Am Inst Aeronaut & Astronaut; Royal Aeronaut Soc; Soc Automotive Engrs. *Mailing Add:* Boeing Com Airplane Group MS 75-26 PO Box 3707 Seattle WA 98124-2207

CONDIT, RALPH HOWELL, b Hollywood, Calif, May 12, 29; m 66; c 2. SOLID STATE CHEMISTRY. *Educ:* Princeton Univ, BA, 51, PhD(chem), 60. *Prof Exp:* Res adminr, Air Force Off Sci Res, 58-60; chemist, Lawrence Livermore Lab, Univ Calif, 60-90; CONSULT, 90- *Concurrent Pos:* Consult, Air Force Off Sci Res; dir, Geos Corp, Calif. *Mem:* Am Chem Soc; Am Phys Soc; Am Nuclear Soc. *Res:* Tracer techniques and diffusion in solids; hydrogen fuel economy; ceramics for turbines; laser isotope separation; materials problems in the fusion energy program; diffusion in minerals; plutonium. *Mailing Add:* 4669 Almond Circle Livermore CA 94550

CONDLIFFE, PETER GEORGE, b Christchurch, NZ, June 30, 22; nat US; m 42, 80; c 3. BIOCHEMISTRY, RESEARCH ADMINISTRATION. *Educ:* Univ Calif, BA, 47, PhD(biochem), 52. *Prof Exp:* Res asst, Inst Exp Biol, Univ Calif, 50-52; res assoc, Med Col, Cornell Univ, 52-54; chemist, Nat Inst Arthritis & Metab Dis, NIH, 54-66, chief, Europ Off, Paris, 66-68 & Conf & Sem Prog Br, John E Fogarty Int Ctr Advan Study Health Sci, 68-73, res biochemist & hormone distrib officer, Lab Nutrit & Endocrinol, Nat Inst Arthritis, Metab & Digestive Dis, 73-75, chief, Scholars & Fels Prog Br, Fogarty Int Ctr, 75-80 & Scholars in Residence Prog Br, Fogarty Int Ctr, 80-88, EMER SCIENTIST, LAB CELLULAR & DEVELOP BIOL, NAT INST DIABETES & DIGESTIVE & KIDNEY DIS, NIH, 88- *Concurrent Pos:* Lectr, USDA Grad Sch, Washington, DC, 55-60 & Found Advan Educ Sci, NIH, 60-75; fel, Nat Found Carlsberg Lab, Copenhagen, 59-60; sr attache, comp physiol lab, Mus Natural Hist, Paris, 66-68; vis prof, Japan Soc Prom Sci, Inst Endocrinol, Gunma Univ, Japan, 74; vis lectr, Inst Gen Path, 2nd Fac Med, Univ Naples, Italy, 83-85 & 89; consult, Int Biomed Rels, 88- *Mem:* AAAS; Am Soc Biol Chemists; Endocrine Soc. *Res:* Biochemical endocrinology; pituitary biochemistry; chemistry of pituitary hormones; reproductive biology; social implications of biomedical research; ethical issues in biology and medicine; international aspects of research in biomedicine. *Mailing Add:* Lab Cellular & Develop Biol Bldg 6 Rm B1-28 Nat Inst Diabetes & Digestive & Kidney Dis NIH Bethesda MD 20892

CONDO, ALBERT CARMAN, JR, b Hackensack, NJ, May 25, 24; m 47, 73; c 4. ENVIRONMENTAL SCIENCE ENGINEERING, MATERIALS SCIENCE ENGINEERING. *Educ:* Cornell Univ, BS, 49, MS, 51; Pa State Univ, Dipl, 80. *Prof Exp:* Instr anal chem, Cornell Univ, 49-51; from res chemist to prin chemist, Atlantic Richfield Co, 51-64, dir plastics develop, 64-69, mgr protective eng systs, 69-72, mgr arctic civil & erosion control/restoration Alyeska Pipeline Serv Co, 72-77, proj mgr com develop, 77-79, bus mgr geochem prod, Arco Chem, 81, specialty chem, 81-85; RETIRED. *Concurrent Pos:* From instr to asst prof, eve col, Drexel Univ, 54-72; pres ac4 chem consult, 85-91. *Mem:* Am Chem Soc; Soc Plastics Eng; Nat Asn Corrosion Eng; Sigma Xi; Soil Conserv Soc Am; Am Soc Civil Engrs; fel Am Inst Chemist. *Res:* Petrochemicals; polymerization; plastics; coatings and corrosion control; protective environmental and geotechnical systems; insulated roads on permafrost; hydraulic and thermal erosion control; restoration; revegetation; arctic thermal regimes; oil spill cleanup and remediation; specialty chemicals. *Mailing Add:* 3424 Ivy Lane Newtown Square PA 19073

CONDO, GEORGE T, b East St Louis, Ill, May 14, 34; m 58; c 2. HIGH ENERGY PHYSICS. *Educ:* Univ Ill, BS, 56, MS, 57, PhD(physics), 62. *Prof Exp:* Res assoc physics, Univ Ill, 62-63; from asst prof to assoc prof physics, 63-76, PROF PHYSICS & ASTRON, UNIV TENN, 76- *Concurrent Pos:* Consult, Oak Ridge Nat Lab, 63- *Mem:* Am Phys Soc. *Res:* Elementary particles using nuclear research emulsions and bubble chamber technique. *Mailing Add:* Dept Physics Univ Tenn Knoxville TN 37996

CONDON, FRANCIS EDWARD, b Abington, Mass, Oct 12, 19; m 43; c 7. CHEMISTRY. *Educ:* Harvard Univ, AB, 41, AM, 43, PhD(org chem), 44. *Prof Exp:* Res chemist, Phillips Petrol Co, 44-52; from asst prof to assoc prof, 52-67, PROF CHEM, CITY COL NEW YORK, 67- *Concurrent Pos:* NSF fac fel, Univ Southern Calif, 64-65. *Res:* Hydrocarbon chemistry; structure-reactivity correlations; hydration and base strength; hydrazines; electrophilic aromatic substitution. *Mailing Add:* 471 Larch Ave Bogota NJ 07603

CONDON, JAMES BENTON, b Buffalo, NY, Aug 20, 40; c 1. PHYSICAL CHEMISTRY, SURFACE CHEMISTRY. *Educ:* State Univ NY Binghamton, AB, 62; Iowa State Univ, PhD(phys chem), 68. *Prof Exp:* Develop chemist, Martin Marietta Energy Systs Inc, 68-88; ASSOC PROF CHEM, ROANE STATE COMMUNITY COL, 88- *Concurrent Pos:* Vis res prof, Forschungszentrum KFA Julich, 88- *Mem:* Am Chem Soc; AAAS; Sigma Xi; Am Electrochem Soc. *Res:* Catalysis; electrochemistry; corrosion; solid state chemistry; electrocrystallization; gas adsorption; solubility and diffusivity of gases in solids. *Mailing Add:* 511 Robertsville Rd Oak Ridge TN 37830

CONDON, JAMES JUSTIN, b New Orleans, La, Apr 15, 45; m 79. ASTRONOMY. *Educ:* Cornell Univ, BS, 66, PhD(astron), 72. *Prof Exp:* Fel, Arecibo Observ, 72; res assoc, Nat Radio Astron Observ, 72-74, asst scientist, 77, assoc scientist, 79-80; asst prof, 74-79, ASSOC PROF PHYS, VA POLYTECH INST & STATE UNIV, 79-; SCIENTIST, NAT RADIO ASTRON OBSERV, 80- *Concurrent Pos:* Mem, Users' Comt, Nat Radio Astron Observ, 75-79; mem sci adv comt, Nat Astron & Ionosphere Ctr, 77-79; fel Alfred P Sloan Found, 77-78. *Mem:* Am Astron Soc; Int Astron Union; Int Sci Radio Union. *Res:* Extragalactic radio astronomy; radio spectra, optical identifications, structure, variability and evolution of compact sources; radio sources in nearby galaxies; high-redshift 21 cm absorption lines; low-frequency variable sources. *Mailing Add:* Nat Radio Astron Observ Edgemont Rd Charlottesville VA 22903-2475

CONDON, MICHAEL EDWARD, b Providence, RI, Jan 17, 45; m 68; c 2. ORGANIC CHEMISTRY. *Educ:* Providence Col, BS, 66; Yale Univ, PhD(org chem), 70. *Prof Exp:* Res staff molecular biophysicist chem, Yale Univ, 70-71, res staff chemist, 71-72; fel, Squibb Inst for Med Res, 72-73, res chemist, 73-77, sr res chemist, 77-81; RES ASSOC AGR CHEM GROUP, FMC CORP, 81- *Mem:* Am Chem Soc. *Res:* Design and synthesis of new pesticides. *Mailing Add:* Am Cyanamid Co Agr Res Div PO Box 400 Princeton NJ 08540

CONDON, ROBERT EDWARD, b Albany, NY, Aug 13, 29; m 51; c 2. SURGERY, PHYSIOLOGY. *Educ:* Univ Rochester, AB, 51, MD, 57; Univ Wash, MS, 65; Am Bd Surg, dipl, 66. *Prof Exp:* Asst resident & intern, Univ Wash Hosps, Seattle, 57-63, chief resident surg, 63-65, res assoc, Sch Med, Univ Wash, 59-61, instr, 61-65; asst prof, Col Med, Baylor Univ, 65-67; from assoc prof to prof, Col Med, Univ Ill, Chicago, 67-71; prof surg & head dept, Col Med, Univ Iowa, 71-72; PROF, MED COL WIS, 79-, CHMN, DEPT SURG, 79-, AUSMAN FOUND PROF, 83- *Concurrent Pos:* Guggenheim fel, 63; hon clin asst, Royal Free Hosp, London, 63-64; asst chief surg, Vet Admin Hosp, Houston, Tex, 65-67; attend surgeon, Univ Ill Hosp, 67-71; chief surg serv, Univ Iowa Hosp, 71-72 & Vet Admin Hosp, Wood, Wis, 72-80; dir surg, Milwaukee County Med Complex, 79-; chief of surg, Froedtert Mem Lutheran Hosp, 80- *Mem:* Am Col Surg; Soc Univ Surg; Am Surg Asn; Soc Surg Alimentary Tract; Soc Clin Surg; Surg Infection Soc. *Res:* Gastric and intestinal physiology; hernia; infection. *Mailing Add:* Div Surg 8700 W Wisconsin Ave Milwaukee WI 53226

CONDOURIS, GEORGE ANTHONY, b Passaic, NJ, Dec 9, 25; m 49; c 5. PHARMACOLOGY. *Educ:* Rutgers Univ, BS, 49; Yale Univ, MS, 53; Cornell Univ, PhD(pharmacol), 55. *Prof Exp:* Instr, Med Col, Cornell Univ, 56-57; vis investr, Rockefeller Inst, 57; from asst prof to assoc prof, 57-67, PROF PHARMACOL, COL MED & DENT NJ, 67-, CHMN DEPT, 72- *Concurrent Pos:* NSF fel, Med Col, Cornell Univ, 54-55, res fel, 55-56. *Mem:* AAAS; Am Soc Pharmacol & Exp Therapeut. *Res:* Pharmacology of the nervous system; biometrics. *Mailing Add:* Dept Pharmacol & Toxicol UMDNJ-NJ Med Sch 185 S Orange Ave Newark NJ 07103-2757

CONDRATE, ROBERT ADAM, b Worcester, Mass, Jan 19, 38; m 60; c 3. SOLID STATE CHEMISTRY. *Educ:* Worcester Polytech Inst, BS, 60; Ill Inst Technol, PhD(chem), 66. *Prof Exp:* NSF fel & res assoc chem, Univ Ariz, 66-67; from asst prof to assoc prof, 67-78, PROF SPECTROS, ALFRED UNIV, 78- *Concurrent Pos:* Finger Lakes grant-in-aid, 69; vis prof, Los Alamos Sci Lab, 72-73; Corning Glass Works Found grant-in-aid, 75-76; Brockway Glass grant-in-aid, 76-77; ERDA Oil Shale grant, 76-77; Danforth Found Asn, 76-86; vis prof, GTE Sylvania, 80; US Bur Mines grant, 80; NSF grant, 83-86 & 87, Glass Inst & CACT grants, 87-90; consult spectroscopy, Statue of Liberty Found, 84-86; fac exchange scholar, State Univ NY, 88; lectr, Korea Inst Sci & Technol, 89. *Honors & Awards:* Soc Appl Spectros Award, 64; Samuel Scholes Award Res, 72. *Mem:* Am Chem Soc; Soc Appl Spectros; Am Phys Soc; fel Royal Soc Chem; fel Am Inst Chem; Am Ceramics Soc; Can Ceramic Soc. *Res:* Application of spectroscopy to elucidate the structures of crystalline and vitreous solids; to analyze the elemental content of materials. *Mailing Add:* Div Ceramic Eng & Sci Alfred Univ Alfred NY 14802

CONDRAY, BEN ROGERS, b Waco, Tex, July 4, 25; m 51; c 3. ORGANIC CHEMISTRY. *Educ:* Baylor Univ, BS, 48, PhD(org chem), 64; Purdue Univ, MS, 50. *Prof Exp:* From asst prof to assoc prof chem, ETex Baptist Col, 50-55; instr, Baylor Univ, 55-57; assoc prof, 58-65, PROF CHEM, E TEX BAPTIST COL, 65- *Mem:* AAAS; Am Chem Soc. *Res:* Synthesis of organosilicon compounds; separation and identification of natural products; organic-ozone chemistry and stabilities of organic ozonides; charge-transfer complexes; science education. *Mailing Add:* 2201 George Gregg Marshall TX 75670

CONE, CLARENCE DONALD, JR, b Savannah, Ga, Apr 17, 31; m 54; c 2. CELL BIOLOGY, ONCOLOGY. *Educ:* Ga Inst Technol, BChE, 54; Univ Va, MAeronE, 59; Med Col Va, PhD(biophys), 65, PhD(cell & molecular biol), 76. *Prof Exp:* Control engr, Buckeye Cellulose Corp, 54-55; res chemist, Herty Found Lab, 55-57; res aerodynamicist, Nat Adv Comt Aeronaut, 57-59; head hypersonic res, NASA, 59-61, head subsonic theory res, 61-64, res biophysicist, 64-65, head molecular biophys res, 65-72; dir, Molecular Biol Lab, Eastern Va Med Authority, 72-74; dir, cellular & Molecular Biol Lab, Vet Admin, 75-79; dir, Glenn Fedn Med Res, 80-85; CHIEF EXEC OFFICER, THERAPEUT SYSTS CORP, 80-; CHMN, APOLLO THERAPEUT LTD, 80- *Concurrent Pos:* Res assoc, Va Inst Marine Sci, 62-68 & Smithsonian Inst, 64-66; mem, Third Int Cong, Int Tech & Sci Orgn Soaring Flight, 63; consult, Biomed Res, 64-80; dir, Aging Res Proj, 78-83. *Honors & Awards:* Medal, Except Sci Achievement, NASA, 69. *Mem:* AAAS; Sigma Xi; Tissue Cult Asn; Biophys Soc; Am Soc Microbiol; Cancer Control Soc. *Res:* Molecular mechanisms of mitogenesis regulation; cytogenetic regulation; carcinogenesis mechanisms; cancer therapy; clinical pharmacology. *Mailing Add:* 104 Harbour Dr Yorktown VA 23690

CONE, CONRAD, b Wellington, NZ, Dec 3, 39; US citizen. ANALYTICAL CHEMISTRY, ORGANIC CHEMISTRY. *Educ:* Bristol Univ, BSc, 62, PhD(org chem), 66. *Prof Exp:* Fel, Mass Inst Technol, 65-68; fac assoc, Univ Tex, Austin, 68-70, asst prof, 70-73, res scientist org chem, 73-77; res chemist, Am Cyanamid Co, 78-79; SR RES CHEMIST, RES & DEVELOP, INTERNATIONAL FLAVORS & FRAGRANCES, 79- *Mem:* Am Chem Soc; Am Soc Mass Spectrometry. *Res:* Mass spectrometry of organic compounds; scientific data bases; laboratory computers. *Mailing Add:* Int Flavors & Fragrances Res & Develop 1515 Hwy 36 Union Beach NJ 07735-3500

CONE, DONALD R(OY), b Berkeley, Calif, Nov 3, 21; m 43; c 3. ELECTRONICS ENGINEERING. *Educ:* Univ Calif, Berkeley, BS, 43, MS, 51. *Prof Exp:* Engr res & develop, Lawrence Radiation Lab, Univ Calif, Berkeley, 46-52; tech dir, Chromatic TV Labs, 52-56; asst mgr, Litton Indust Elec Display Lab, 57-61; sr res engr, SRI Int, Menlo Park, Calif, 61-84; RETIRED. *Mem:* Inst Elec & Electronics Engrs; Sci Res Soc Am; Soc Info Display. *Res:* Electromechanical and electron optics; data processing and display. *Mailing Add:* 624 Santa Lucia Ave Los Osos CA 93402-1128

CONE, EDWARD JACKSON, b Mobile, Ala, Sept 17, 42; m 63; c 2. DRUG METABOLISM, DRUG TESTING. *Educ:* Mobile Col, BS, 67; Univ Ala, PhD(org chem), 71. *Prof Exp:* Lab instr chem, Mobile Col, 65-69; chemist, Shell Oil Co, 69-71; instr chem, Univ Ala, 71; fel tobacco chem, Univ Ky, 71-72; chemist, 72-78, CHIEF, LAB CHEM & DRUG METAB, NAT INST DRUG ABUSE ADDICTION RES CTR, 78- *Concurrent Pos:* Assoc prof, Sch Pharm, Univ Ky, 75-84; consult, NIMH, 78; adj prof, toxicol, Univ Md, Baltimore, 90- *Honors & Awards:* Outstanding Serv Medal, US Dept Health & Human Serv. *Mem:* Int Soc Study Xenobiotics; Am Soc Pharmacol & Exp Therapeut; Am Chem Soc; Soc Forensic Toxicol. *Res:* Studies on biotransformation of drugs of abuse in man and other animal species; development of analytical procedures useful for detection, isolation and quantification of drugs and their metabolites; mechanisms of action of drugs at molecular level. *Mailing Add:* NIDA Addiction Res Ctr PO Box 5180 Baltimore MD 21224

CONE, MICHAEL MCKAY, b Washington, DC, Oct 14, 47; m 71. PHYSICAL ORGANIC CHEMISTRY, ORGANIC CHEMISTRY. *Educ:* Princeton Univ, AB, 69; Yale Univ, MPhil, 74, PhD(org chem), 76. *Prof Exp:* Postdoctoral assoc, Cornell Univ, 76-77; res chemist, 77-81, asst div supt, 81-84, sr financial specialist, 84-85, sr tech specialist, 85-86, RES SUPVR, E I DU PONT DE NEMOURS & CO INC, 86- *Mem:* Am Chem Soc; AAAS; Sigma Xi. *Res:* Organic chemistry: kinetics and mechanism; homogeneous and heterogeneous catalysis; photopolymer chemistry. *Mailing Add:* PO Box 4240 Wilmington DE 19807-0240

CONE, RICHARD ALLEN, b St Paul, Minn, May 23, 36; m; c 2. BIOPHYSICS. *Educ:* Mass Inst Technol, SB, 58; Univ Chicago, SM, 59, PhD(physics), 63. *Prof Exp:* Res assoc biophys, Univ Chicago, 63-64; from instr to asst prof biol, Harvard Univ, 64-69; assoc prof, 69-73, PROF BIOPHYS, JOHNS HOPKINS UNIV, 73- *Concurrent Pos:* USPHS fel, Univ Chicago, 64. *Honors & Awards:* Cole Award, Biophys Soc, 79. *Mem:* Biophys Soc; Soc Study Reprod. *Res:* Mechanism of visual excitation; membrane structure and function; sperm motility and antisperm contraceptives; topical methods for preventing pregnancy as wellas sexually transmitted diseases; photoreceptor physiology. *Mailing Add:* Dept Biophys Johns Hopkins Univ 34th & Charles St Baltimore MD 21218

CONE, ROBERT EDWARD, b Brooklyn, NY, Aug 18, 43; m 66; c 2. IMMUNOLOGY. *Educ:* Brooklyn Col, BS, 64; Fla State Univ, MS, 67; Univ Mich, Ann Arbor, PhD(microbiol), 70. *Prof Exp:* Fel immunol, Walter & Eliza Hall Inst Med Res, Melbourne, Australia, 71-73; mem, Basel Inst Immunol, Switz, 73-74; from asst prof to assoc prof path & surg, Sch Med, 74-85, PROF PATH, UNIV CONN HEALTH CTR, FARMINGTON, 85- *Concurrent Pos:* Assoc prof path & surg, Yale Unv, 74-85. *Honors & Awards:* Damon Runyon Cancer Fund Fel, 70; Upjohn Co Res Award, 80, 81. *Mem:* Sigma Xi; AAAS; Am Asn Immunologists; Int Cell Res Orgn. *Res:* Structural and physiological properties of lymphocyte membranes; relationship of membrane structure to lymphocyte function. *Mailing Add:* Dept Path Univ Conn Health Ctr Farmington CT 06032

CONE, THOMAS E, JR, b Brooklyn, NY, Aug 15, 15; m 39; c 3. PEDIATRICS. *Educ:* Columbia Univ, BA, 36, MD, 39. *Prof Exp:* Chief pediat serv, US Naval Hosp, Philadelphia, Pa, 49-53; chief pediat serv, Nat Naval Med Ctr, Bethesda, Md, 53-63; chief med out-patient serv & asst dir adolescent unit, Children's Hosp Med Ctr, 63-73, chief med ambulatory serv & sr assoc med, 67-73, emer sr assoc med & clin genetics, 71-83; assoc clin prof pediat, Harvard Med Sch, 63-66, clin prof, 67-83, emer clin prof, 83-; RETIRED. *Concurrent Pos:* Sr consult, Children's Hosp, Washington, DC, 53-63; consult, Nat Inst Neurol Dis & Blindness, 58-63; mem nat adv coun, Nat Inst Child Health & Human Develop, 63-64; prof pediat, Southwest Med Sch, Univ Tex, 66-67; assoc ed, Pediatrics, 65-74. *Honors & Awards:* Officer, Order Naval Merit, Repub Brazil. *Mem:* AMA; fel Am Acad Pediat; Am Pediat Soc; NY Acad Sci; Am Asn Hist Med; Irish & Am Pediat Socs. *Res:* Physical growth and development of children; history of pediatrics; clinical pediatrics. *Mailing Add:* 300 Longwood Ave Boston MA 02115

CONE, WYATT WAYNE, b Plains, Mont, Mar 16, 34; m 56; c 3. ENTOMOLOGY. *Educ:* San Diego State Col, AB, 56; Wash State Univ, PhD(entom), 62. *Prof Exp:* From jr entomologist to assoc entomologist, 61-72, ENTOMOLOGIST, WASH STATE UNIV, 72- *Concurrent Pos:* Gov bd mem, Entom Soc Am, 84-86. *Mem:* Entom Soc Am; Entom Soc Can. *Res:* Insecticide effect on predator-prey relationship in mite populations; biology and control of otiorhynchus weevils on grapes; ecology of arthropods in native grasslands; biology and control of two-spotted spider mites, two-spotted spider mite sex attractant and pheromones. *Mailing Add:* Irrigated Agr Res & Exten Ctr Prosser WA 99350

CONEY, PETER JAMES, b Methewun, Mass. GEOLOGY. *Educ:* Cobly Col, BA, 51; Univ Maine, MS, 53; L'Ecole Nat Superierre Petrole, Paris, Eng, 55; Univ NMex, PhD(geol), 64. *Prof Exp:* Prof & chem geol, Middlebury Col, 64-75; PROF GEOL, UNIV ARIZ, 76-, PROF GEOSCI. *Concurrent Pos:* Vis prof, Dept Geophysics, Stanford Univ, 70-71; Fulbright prof, Escuela Polytecnica Naz, Ecuador, 73; geologist, US Geol Surv, 75-86. *Res:* Regional tectonics of mountain systems, particularly the North American Cordillera, in the framework of plate tectonics. *Mailing Add:* Dept Geosci Univ Ariz Tucson AZ 85721

CONFALONE, PAT N, b Fountain Hill, Pa, Oct 6, 45; m 78; c 2. BIOORGANIC CHEMISTRY, NATURAL PRODUCTS CHEMISTRY. *Educ:* Mass Inst Technol, BS, 67; Harvard Univ, MS, 68, PhD(org chem), 70, Post-Doctorate (R B Woodward), 71. *Prof Exp:* Fel, Harvard Univ, 70-71; res fel, Hoffmann-La Roche, 71-81; adj prof, Rutgers Univ, 80; RES DIR, E I DU PONT DE NEMOURS & CO, INC, 81- *Concurrent Pos:* Educ counr, Mass Inst Technol, 78-; chmn, Org Div Am Chem Soc. *Mem:* Am Chem Soc; Sigma Xi. *Res:* The total synthesis of natural products; synthesis of biologically active compounds for development as pharmaceuticals; areas of interest include computer molecular modeling, anti-cancer drugs, vitamins; anti-infectives, central nervous system agents, anti- inflammatories and cardiovascular compounds; enzyme inhibitors; bioorganic chemistry. *Mailing Add:* 722 Hertford Rd Wilmington DE 19803-1618

CONFER, ANTHONY WAYNE, b Hot Springs, Ark, July 29, 47; m 70; c 4. PATHOLOGY, IMMUNOLOGY. *Educ:* Okla State Univ, DVM, 72; Am Col Vet Pathologist, dipl, 77; Ohio State Univ, MS, 74; Univ Mo, Columbia, PhD(immunol), 78. *Prof Exp:* NIH trainee fel, Ohio State Univ, 72-74; pathologist, Armed Forces Inst Path, 74-76; Nat Cancer Inst fel, Univ Mo, 76-78; assoc prof vet path, La State Univ, 78-81; assoc prof, 81-85, prof vet path, 85-86, PROF & HEAD VET PATH, OKLA STATE UNIV, 86- *Concurrent Pos:* Vis prof, Univ BC, 90. *Honors & Awards:* Beecham Award Res Excellence, 85- *Mem:* Am Col Vet Path; Conf Res Workers Animal Dis; Am Vet Med Asn; Sigma Xi. *Res:* Activites in pathogenesis and immune respones in infectious diseases of domestic animals; bovine pneumonic pasteurellosis; bovine brucellosis; immune response to Pasteurella haemolytica and Brucella abortus, with major emphasis on immunities to specific antigens; molecular cloning of Pasteurella haemolytica genes. *Mailing Add:* Dept Vet Path Okla State Univ Stillwater OK 74078

CONFER, GREGORY LEE, b Erie, Pa, Nov 12, 54; m 87; c 1. DATA ACQUISITION & INSTRUMENTATION, SOFTWARE DEVELOPMENT & APPLICATION. *Educ:* Gannon Univ, Erie, Pa, BME, 77, MS, 80. *Prof Exp:* Sr engr, Gould Ocean Systs Div, 82-84; staff engr, 77-82, SR ENGR, GEN ELEC TRANSP SYSTS, 84- *Res:* Prototype development of a coal water slurry fired diesel electric locomotive. *Mailing Add:* 408 Joliette Ave Erie PA 16511

CONFER, JOHN L, b Dayton, Ohio, Sept 15, 40. AQUATIC ECOLOGY, POPULATION ECOLOGY. *Educ:* Earlham Col, BA, 62; Wash State Univ, MS, 64; Univ Toronto, PhD(zool), 69. *Prof Exp:* Asst prof zool, Univ Fla, 69-70; asst prof, 70-80, ASSOC PROF, DEPT BIOL, ITHACA COL, 80- *Res:* Population regulation, especially birds and zooplankton; models of foraging. *Mailing Add:* Dept Biol Ithaca Col Danby Rd Ithaca NY 14850

CONFORTI, EVANDRO, b SJ Rio Preto-Sao Paulo, Brazil, Aug 3, 47; c 3. FIBER OPTICS COMMUNICATION, COHERENT OPTICAL COMMUNICATION. *Educ:* Tech Inst Aeronaut, BSc, 70; Univ Toronto, MASc, 78; State Univ Campinas, PhD(antennas), 83. *Prof Exp:* Engr microelectronics, Univ Sao Paulo, 71-73; teaching asst microwaves, Univ Paraiba, 74-76, asst prof, 79-80; master student wave sci, Univ Toronto, 76-78; asst prof microwaves, 81-83, assoc prof fiber optics, 83-87, ADJ PROF FIBER OPTICS, STATE UNIV CAMPINAS, 87- *Concurrent Pos:* Consult, Telecommun Res Ctr, CPaD, 83- & IBM, Brazil, 89-; res prize, State Univ Campinas, 84, head, Dept Elec Eng, 84-86, dean fac elec eng, 86-87, prin investr, Lab Optical Commun, 87-, head, Dept Microwaves & Optics, 89- *Mem:* Inst Elec & Electronics Engrs. *Res:* Coherent optical communication; development and analysis of tunnable lasers; development of phase insensitive coherent optical receiver; optical switching using coherent techniques. *Mailing Add:* DMO-FEE State Univ Campinas CP 6101 Brazil

CONGDON, CHARLES C, b Dunkirk, NY, Dec 13, 20; m 47; c 5. PATHOLOGY. *Educ:* Univ Mich, AB, 42, MD, 44, MS, 50. *Prof Exp:* Instr path, Univ Mich, 49-51; vis scientist, Nat Cancer Inst, NIH, 51-52, med officer path, 52-55; sr biologist, Biol Div, Oak Ridge Nat Lab, 55-73; prof, Univ Tenn, Knoxville, 66-73, res prof & asst dir, 73-78, assoc dir, 78-81, prof med biol, Mem Res Ctr, 78-83; CONSULT, 83- *Mem:* AAAS; Am Asn Path; Am Asn Cancer Res; Int Soc Hemat; Soc Exp Biol & Med; Int Soc Exp Hemat. *Res:* Experimental Pathology. *Mailing Add:* 103 Westwind Dr Oak Ridge TN 37830-8619

CONGDON, JUSTIN DANIEL, b Kingston, Pa, Jan 5, 41; m 77. PHYSIOLOGICAL ECOLOGY, EVOLUTIONARY ECOLOGY. *Educ:* Calif State Polytech Univ, BS, 68, MS, 71; Ariz State Univ, PhD(zool), 77. *Prof Exp:* res scholar, Mus Zool, Univ Mich, 76-80; res assoc, 80-84, ASSOC ECOL, SAVANNAH RIVER, ECOL LAB, 85- *Concurrent Pos:* Adj res investr, Mus Zool, Univ Mich. *Mem:* Ecol Soc Am; Soc Study Amphibians & Reptiles; Herpetologists League; Sigma Xi. *Res:* Underlying physiological and ecological processes and adaptations that shape an animal's reproductive strategy and life history; how organisms allocate energy to the compartments of reproduction, growth, storage and maintenance. *Mailing Add:* 160 Old Nail Rd Beech Island SC 29841

CONGEL, FRANK JOSEPH, b Syracuse, NY, Mar 6, 43; m 65; c 3. NUCLEAR ENGINEERING. *Educ:* LeMoyne Col, Syracuse, NY, BS, 64; Clarkson Univ, Potsdam, NY, MS, 67, PhD(physics), 69. *Prof Exp:* Postdoctoral assoc, Argonne Nat Lab, 68-69; asst prof physics, Macalester Col, St Paul, Minn, 69-72; radiation physicist, US AEC, Washington, DC, 72-74; radiation physicist, 75-82, TECH MGR, US NAT RES COUN, 82- *Concurrent Pos:* Mem, Comt Radionuclide Contamination, Nat Coun Radiation Protection & Measurements, 91-93. *Honors & Awards:* Presidential Meritorious Rank Award, US Govt, 89. *Mem:* Health Physics Soc. *Mailing Add:* 7400 Cutty Sark Way Gaithersburg MD 20882-4301

CONGER, ALAN DOUGLAS, b Muskegon, Mich, Mar 23, 17; m 44; c 4. RADIOBIOLOGY, BIOPHYSICS. *Educ:* Harvard Univ, AB, 40, MA & PhD(biol), 47. *Prof Exp:* Sr res scientist biol, Oak Ridge Nat Lab, USAEC, 47-58; res prof radiobiol, Univ Fla, 58-65; prof radiobiol & head dept, Sch Med, Temple Univ, 65-82; prof radiobiol & head dept, Univ Pa Hosp & Med Sch, 84-86; RETIRED. *Concurrent Pos:* Fulbright sr res scholar, Radiobiol Res Unit, Med Res Coun, London, Eng, 52-53; USAEC grant nuclear sci, Univ Fla, 58-65; Nat Cancer Inst fel radiobiol, Sch Med, Temple Univ, 65-; sci proj officer, Oper Greenhouse Atomic Bomb Tests, Marshall Islands, 51-52; assoc ed, Radiation Res, 58-62, Radiation Bot, 61- & Mutation Res, 62-81; consult biol, Oak Ridge & Brookhaven Nat Labs, USAEC, 58-; mem radiation study sect, NIH, 63-66; chmn panel low dose report, Nat Acad Sci-Nat Res Coun, 72-73, mem subcomt radiobiol, 72- *Mem:* Genetics Soc Am; Am Soc Nat; Radiation Res Soc (vpres, pres, 71-73). *Res:* Effects of radiation on cells; genetic effects of radiation. *Mailing Add:* 506 Old Gulph Rd Narberth PA 19172

CONGER, BOB VERNON, b Greeley, Colo, July 2, 38; m 60; c 4. PLANT GENETICS. *Educ:* Colo State Univ, BS, 63; Wash State Univ, PhD(genetics), 67. *Prof Exp:* Asst prof genetics, Wash State Univ, 67-68; from asst prof to prof, dept energy, Comp Animal Res Lab & dept plant & soil sci, 68-86, CLYDE B AUSTIN DISTINGUISHED PROF PLANT & SOIL SCI, UNIV TENN, KNOXVILLE, 86- *Concurrent Pos:* Ed, Critical Rev in Plant Sci, 81-; spec serv agreement intern, Atomic Energy Agency, Vienna, Austria, 86-87. *Mem:* AAAS; fel Am Soc Agron; Am Genetic Asn; Tissue Cult Asn; fel Crop Sci Soc Am; Soc Plant Molecular Biol; Assoc Plant Tissue Culture. *Res:* Cell and tissue cult, cytogenetics and breeding cool season forage grasses. *Mailing Add:* Dept Plant & Soil Sci Univ Tenn PO Box 1071 Knoxville TN 37901-1071

CONGER, HARRY M, b Seattle, Wash, July 22, 30. MINING ENGINEERING. *Educ:* Colo Sch Mines, BS. *Prof Exp:* Mine engr & shift boss, ASARCO, Ariz, 55-64; mgr, Kaiser's Mine, Kaiser Steel Corp, 64-70, vpres & gen mgr, Kaiser Resources, Ltd, Fernie, BC, 70-72; vpres & gen mgr,

Midwestern Div, Consol Coal Co, Pinckneyville, Ill, 73-75; vpres & gen mgr, Base Metals Div, 75-77, pres & mem bd dirs, 77-86, CHIEF EXEC OFFICER, HOMESTAKE MINING CO, 78-, CHMN BD, 82- *Concurrent Pos:* Dir, ASA Ltd, Pac Gas & Elec Co, San Francisco, Baker Hughes Inc, Houston & Calmat Co, Los Angeles; chmn, Am Mining Cong, 86-89; mem bd gov, Nat Mining Hall Fame & Mus. *Mem:* Nat Acad Eng; Am Inst Mining Engrs; Mining & Metall Soc Am. *Mailing Add:* Homestake Mining Co 650 California St San Francisco CA 94108-2788

CONGER, JOHN JANEWAY, b New Brunswick, NJ, Feb 27, 21; m 44; c 2. DEVELOPMENTAL PSYCHOLOGY, ADOLESCENCE. *Educ:* Amherst Col, BA, 43; Yale Univ, MS, 47, PhD(psychol), 49. *Hon Degrees:* DSc, Ohio Univ, 81, Amherst Col, 83, Univ Colo, 89. *Prof Exp:* Asst prof psychol, Grad Sch, Ind Univ, 49-53, dir psychol res, dept psychiat, Sch Med, 50-53; chief psychologist, Vet Admin Hosp, Indianapolis, 52-53; head div clin psychol, Dept Psychiat, Sch Med, Univ Colo, 53-63, assoc dean, 61-63, vpres med affairs & dean, 63-70, prof psychiat & interim chmn dept, 83-84, actg chancellor, Health Sci Ctr, 84-85, prof clin psychol, 53-88, EMER PROF CLIN PSYCHOL & PSYCHIAT, SCH MED, UNIV COLO, 88- *Concurrent Pos:* Vpres, John D & Catherine T MacArthur Found, 80-83. *Honors & Awards:* Outstanding Contrib to Health Psychol, Am Psychol Asn, 83, Award for Distinguished Contrib to Psychol in the Pub Interest, 86. *Mem:* Inst Med-Nat Acad Sci; Acad Behav Med Res; fel Am Psychol Asn (pres, 81); fel AAAS; Soc Res Child Develop. *Res:* Textbooks and research volumes and articles in child and adolescent development, behavioral medicine and health psychology; mental health and social policy. *Mailing Add:* Univ Colo Health Sci Ctr Box C257 Denver CO 80262

CONGER, KYRIL BAILEY, b Berlin, Ger, Apr 11, 13; US citizen; m 46; c 4. SURGERY. *Educ:* Univ Mich, AB, 33, MD, 36; Am Bd Urol, dipl, 47. *Prof Exp:* Consult urol, US Army Mid Pac Area & Tripler Gen Hosp, Honolulu, 46; PROF UROL & HEAD DEPT, HOSP & MED SCH, TEMPLE UNIV, 47- *Concurrent Pos:* Consult, Vet Admin Hosp, Philadelphia & Mid-Atlantic Area, Vet Admin; attend urologist, St Christopher's Hosp for Children. *Mem:* Am Urol Asn; fel Am Col Surg. *Res:* Urological and prostatic surgery. *Mailing Add:* 3401 N Broad St Philadelphia PA 19140

CONGER, ROBERT PERRIGO, b Youngstown, Ohio, Dec 1, 22; m 48; c 2. PLASTICS CHEMISTRY. *Educ:* Cornell Univ, AB, 43, PhD(chem), 50. *Prof Exp:* Lab asst, Cornell Univ, 43-44, 46-50; instr chem, Col Med, Univ Tenn, 50-51; res chemist, Gen Labs, US Rubber Co, 51-61; mgr res, Congoleum Industs, Inc, 61-78, PRES, CONGER ASSOCS, CONSULTS, 78- *Mem:* Am Chem Soc; Soc Plastics Eng. *Res:* Foamed plastics, especially polyvinyl chloride; chemical embossing; clear films of elastomers such as polyvinyl chloride and polyurethane; compounding and modification of various elastomers; water based inks for rotogravure and silk screen printing; special plastisol and water based inks. *Mailing Add:* 87 Oak Ave Park Ridge NJ 07656-1282

CONGER, WILLIAM LAWRENCE, b Danville, Pa, June 19, 37; m 61, 79; c 3. EDUCATION ADMINISTRATION. *Educ:* Univ Louisville, BChE, 60; Univ Pa, PhD(chem eng), 65. *Prof Exp:* Sr res engr, Esso Corp, 65-66; from asst prof to assoc prof chem eng, Univ Ky, 66-83, dir summer progs, 80-83, actg dean univ exten progs, 82-83; PROF CHEM ENG & HEAD DEPT, VA POLYTECH INST & STATE UNIV, 83- *Mem:* Am Inst Chem Engrs; Am Soc Eng Educ. *Res:* Evaluation of closed thermodynamic cycles for the production of hydrogen (as an energy carrier) from water. *Mailing Add:* Dept Chem Eng Va Polytech Inst & State Univ Blacksburg VA 24061

CONGLETON, JAMES LEE, b Lexington, Ky, Dec 20, 42; m 68; c 2. PHYSIOLOGICAL ECOLOGY, FISHERIES. *Educ:* Univ Ky, BS, 64; Univ Calif, San Diego, PhD(marine biol), 70. *Prof Exp:* NIH fel, Univ Wash, 70-71; biologist aquaculture, Kramer, Chin & Mayo, Inc, 71-75; asst prof fisheries, Univ Wash, 75-80, spec unit asst, Coop Fishery Res Unit, 75-80; ASST LEADER, COOP FISHERY RES UNIT, UNIV IDAHO, 80- *Mem:* Am Fisheries Soc; Int Soc Develop Comp Immunol. *Res:* Disease resistance of fish. *Mailing Add:* Coop Fish & Wildlife Res Unit Dept Fish Resources Univ Idaho Moscow ID 83843

CONIBEAR, SHIRLEY ANN, b Amboy, Ill, Aug 20, 46; m 75; c 3. OCCUPATIONAL MEDICINE, EPIDEMIOLOGY. *Educ:* Shimer Col, BA, 68; Univ Ill, MD, 73, MPH, 76. *Prof Exp:* Intern, Conemaugh Valley Mem Hosp, 73-74; resident, Cook County Hosp, 74-76, dir, health hazards eval, 76-78; dir, occup med residency, 78-82, ASSOC PROF, SECT PREV MED, COL MED, UNIV ILL, 82-; ASSOC PROF, UNIV ILL SCH PUB HEALTH, 77- *Concurrent Pos:* Dir progs in med, Great Lakes Educ Resource Ctr, Univ Ill Col Med, Chicago, 77-82; epidemiologist, Argonne Nat Lab, 79-80; consult occup med, Carnow, Conibear & Assoc Ltd, 75- *Mem:* Am Col Prev Med; Am Pub Health Asn; AAAS; Am Occup Med Asn; Am Med Women's Asn; Am Indust Hyg Asn. *Res:* Clinical epidemiologic studies on groups of workers exposed to toxic materials in the workplace; diagnosis and treatment of patients exposed to toxic materials in the workplace or environment. *Mailing Add:* 333 W Wacker Dr Suite 1400 Chicago IL 60606

CONIGLIARO, PETER JAMES, b Milwaukee, Wis, Jan 27, 42; m 65; c 3. ORGANIC CHEMISTRY, CHEMICAL INSTRUMENTATION. *Educ:* Marquette Univ, BS, 63, MS, 65; Ohio State Univ, PhD(org chem), 67. *Prof Exp:* Res chemist, 68-70, sr res chemist, 70-80, RES MGR, S C JOHNSON & SON, 80- *Mem:* Am Chem Soc. *Res:* Specialized organic analysis; analysis of chemical specialty consumer products. *Mailing Add:* 8731 W Glenwood Dr Glendale WI 53129

CONIGLIO, JOHN GIGLIO, b Tampa, Fla, July 21, 19; m 42; c 3. BIOCHEMISTRY. *Educ:* Furman Univ, BS, 40; Vanderbilt Univ, PhD(biochem), 49. *Prof Exp:* From instr to prof, 51-90, EMER PROF BIOCHEM, SCH MED, VANDERBILT UNIV, 90- *Concurrent Pos:* AEC fel biophys, Colo Med Ctr, 49-50; AEC fel biochem, Sch Med, Vanderbilt Univ, 50-51; assoc ed, Lipids, 69-; contrib ed, Nutrit Rev, 84-; guest prof, Justus-Liebig Univ, Giessen, Germany, 85. *Honors & Awards:* Thomas Jefferson Award, Vanderbilt Univ, 78. *Mem:* AAAS; Am Inst Nutrit; Am Oil Chem Soc; Am Soc Biol Chem; Soc Exp Biol & Med. *Res:* Fat absorption and distribution; acetate utilization and fatty acid synthesis; effects of x-irradiation on fat absorption and on fatty acid synthesis; determination and metabolism of essential fatty acids; use of tracers in metabolism; interconversion of fatty acids; lipids in reproductive tissue. *Mailing Add:* Dept Biochem Vanderbilt Med Sch Nashville TN 37232-0146

CONINE, JAMES WILLIAM, b Newton, Iowa, May 13, 26; m 56; c 4. PHARMACEUTICAL CHEMISTRY. *Educ:* Univ Iowa, BS, 50, MS, 52, PhD(pharmaceut chem), 54. *Prof Exp:* Instr pharm, Univ Iowa, 52-54; pharmaceut chemist, Eli Lilly & Co, 54-87; RETIRED. *Mem:* AAAS; Am Chem Soc; Am Pharmaceut Asn. *Res:* Stability of pharmaceutical products; product development of tablets and capsules. *Mailing Add:* 5271 Channing Rd Indianapolis IN 46206

CONKIE, WILLIAM R, b Ayr, Scotland, Jan 10, 32; m 54; c 2. PHYSICS. *Educ:* Univ Toronto, BASc, 53; McGill Univ, MSc, 54; Univ Sask, PhD, 56. *Prof Exp:* Asst res officer physics, Atomic Energy Can, Ltd, 56-60; PROF PHYSICS, QUEEN'S UNIV, ONT, 60- *Mem:* Am Phys Soc; Can Asn Physicists. *Res:* Theoretical physics. *Mailing Add:* Dept Physics Queen's Univ Kingston ON K7L 3N6 Can

CONKIN, JAMES E, b Glasgow, Ky, Oct 14, 24; m 51; c 4. GEOLOGY, PALEONTOLOGY. *Educ:* Univ Ky, BS, 50; Univ Kans, MS, 53; Univ Cincinnati, PhD(geol), 60. *Prof Exp:* Paleontologist, Union Producing Co Div, United Gas Corp, 53-56; instr geol, Univ Cincinnati, 56-57; from instr to asst prof natural sci, 57-63, assoc prof geol, 63-67, chmn dept, 63-76, PROF GEOL, UNIV LOUISVILLE, 67- *Concurrent Pos:* Fulbright res fel micropaleont, Tasmania, 64-65; mem, Paleont Res Inst. *Mem:* Paleont Soc; fel Geol Soc Am. *Res:* Paleozoic Foraminifera; paleozoic stratigraphy, the Mississippian system and Devonian bone beds. *Mailing Add:* Dept Geol Univ Louisville Louisville KY 40292

CONKIN, ROBERT A, b Green City, Mo, Dec 31, 20; m 46; c 1. ANALYTICAL CHEMISTRY, ORGANIC CHEMISTRY. *Educ:* Northeast Mo State Teachers Col, BS, 42, AB, 46. *Prof Exp:* Analyst, Monsanto Co, Ill, 46-47, lab supvr, 47-50, chief chemist, Ala, 50-53, res chemist, Mo, 53-62, res group leader, 62-72, regist mgr, 72-81, mgr int regist, Agr Prod, 81-85; RETIRED. *Mem:* Am Chem Soc. *Res:* Development and application of micro-analytical methods and techniques in the field of agricultural pesticide residues in or on farm produce and products; data review and petition preparation for submission to Environmental Protection Agency and other regulatory agencies worldwide. *Mailing Add:* 845 Winesap Lane Kirkwood MO 63122-2245

CONKLIN, DOUGLAS EDGAR, b Waukegan, Ill, June 4, 42; m; c 2. AQUACULTURE, NUTRITION. *Educ:* Colo State Univ, BS, 64, MS, 66; NY Univ, PhD(biol), 73. *Prof Exp:* Res asst marine nutrit, Haskins Labs, Yale Univ, 66-71; Nat Oceanic & Atmospheric Admin, Southwest Fisheries Ctr, La Jolla, Calif, 71-73; res nutritionist, 73-78, ASSOC DIR AQUACULT, UNIV CALIF, DAVIS, 78-, LECTR ANIMAL SCI, 80- *Concurrent Pos:* Chmn Nutrit Task Force, World Maricult Soc, 78. *Mem:* Sigma Xi; World Maricult Soc. *Res:* Crustacean nutrition of primarily the lobster, Homarus americanus and nutritional requirements of Moina macrocopa. *Mailing Add:* Bodege Marine Labs Univ Calif PO Box 247 Bodega Bay CA 94923

CONKLIN, GLENN ERNEST, b Lyndon, Kans, June 2, 29; m 57; c 4. PHYSICS. *Educ:* Univ Wichita, BA, 51, MS, 53; Univ Kans, PhD(physics), 63. *Prof Exp:* Mem tech staff physics, Bell Tel Labs, 60-66; res scientist, Singer-Gen Precision, Inc, 66-70; RADIATION PHYSICIST, OFF DEVICE EVAL, CTR DEVICES & RADIOL HEALTH, FOOD & DRUG ADMIN, 70- *Mem:* Asn Advan Med Inst; Regulatory Affairs Prof Soc. *Res:* Dielectric behavior of plastics at millimeter wavelengths; application of optically pumped nuclear magnetic resonance to communications; nonionizing radiation. *Mailing Add:* Ctr Devices & Radiol Health 1390 Picard Dr Rockville MD 20850

CONKLIN, JAMES BYRON, JR, b Charlotte, NC, July 29, 37; m 62; c 2. COMPUTING CENTER ADMINISTRATION, COMPUTER SCIENCE. *Educ:* Mass Inst Technol, SB, 59, SM, 61, ScD(solid state physics), 64. *Prof Exp:* Asst prof, Univ Fla, 64-71, assoc prof physics, 71-81, dir ctr instrnl & res comput activ, 75-81; dir Comput Fac, Smithsonian Astrophys Observ, 81-88; DIR BTNET NETWORK INFO CTR, 88- *Concurrent Pos:* Assoc dir, Northeast Regional Data Ctr, State Univ Syst Fla, 73-74. *Mem:* AAAS; Asn Comput Mach; Inst Elec & Electronics Engrs Computer Soc. *Res:* Computers as instructional tools; numerical mathematics. *Mailing Add:* 721 Dale Dr Silver Spring MD 20910

CONKLIN, JOHN DOUGLAS, SR, b Middletown, NY, Mar 1, 33; m 59; c 2. BIOPHARMACEUTICS. *Educ:* Col Holy Cross, BSc, 56. *Prof Exp:* From jr res biochemist to res biochemist, Norwich-Eaton Pharmaceut, 59-65, sr res, Drug Distrib Unit, Biol Res Div, 65-80, unit leader, Biopharmaceut Unit, 80-85, head, Biopharmaceut Sect, Drug Safety Dept, 85-90, HEAD BIOPHARMACEUT SECT, REGULATORY & MED AFFAIRS DEPT, 90- *Mem:* Acad Pharmaceut Sci; Am Pharmaceut Asn; Am Asn Pharmaceut Sci. *Res:* The study of the bioavailability of drugs in man and animals by investigating drug absorption, distribution and excretion to optimize drug pharmacologic or therapeutic activity for clinical application. *Mailing Add:* Biopharmaceut Sect Regulatory & Med Affairs Dept Norwich-Eaton Pharmaceuticals Norwich NY 13815-1709

CONKLIN, R(OGER) N(ORTON), b Brooklyn, NY, Mar 18, 21; m 47; c 2. CHEMICAL ENGINEERING. *Educ:* Rensselaer Polytech Inst, BChE, 42, MChE, 47. *Prof Exp:* Chemist rubber technol, US Rubber Co, 42-45; instr chem & chem eng, Rensselaer Polytech Inst, 45-47; chemist, E I Du Pont De

Nemours & Co, Inc, 47-55, tech rep, 55-61, tech sales rep, Switz, 61-64, tech rep, 64-71, tech coordr Int Rubber Technol, 71-80, sr tech rep, 80-82; RETIRED. *Res:* Sales development work in rubber field. *Mailing Add:* 3810 Valley Brook Dr Oakwood Hills DE 19808

CONKLIN, RICHARD LOUIS, b Rockford, Ill, Dec 9, 23; m 50; c 4. PHYSICAL OPTICS. *Educ:* Univ Ill, BS, 44, MS, 48; Univ Colo, PhD(solid state physics), 57. *Prof Exp:* Jr scientist, Los Alamos Lab, Univ Calif, 44-46; asst physics, Univ Ill, 46-49; asst prof, Huron Col, 49-53; instr, Univ Colo, 53-57; assoc prof, 57-58, PROF PHYSICS, HANOVER COL, 58- *Concurrent Pos:* Vis colleague, Univ Hawaii, 67-68. *Mem:* Am Asn Physics Teachers; Am Phys Soc; Sigma Xi. *Res:* Electron accelerators and their application to nuclear physics; luminescence of solids. *Mailing Add:* PO Box 24 Hanover IN 46243

CONLAN, JAMES, b San Francisco, Calif, Apr 15, 23; m 46. MATHEMATICS. *Educ:* Univ Calif, BA, 45; Univ Md, PhD(math), 58. *Prof Exp:* Mathematician, Aberdeen Proving Ground, 50-53 & Naval Ord Lab, 53-67; assoc prof math, Howard Univ, 67-68; PROF MATH, UNIV REGINA, 68- *Mem:* Am Math Soc. *Res:* Partial differential equations; fluid mechanics. *Mailing Add:* Dept Math Univ Regina Regina SK S4S 0A2 Can

CONLEE, ROBERT KEITH, b Los Angeles, Calif, July 28, 42; m 63; c 6. EXERCISE PHYSIOLOGY, EXERCISE SCIENCE. *Educ:* Brigham Young Univ, BS, 69, MS, 70; Univ Iowa, PhD(phys educ), 75. *Prof Exp:* Res fel, Sch Med, Washington Univ, 75-77; from asst prof to assoc prof, 77-87, PROF PHYS EDUC, BRIGHAM YOUNG UNIV, 87- *Mem:* Am Col Sports Med; Am Physiol Soc; Sigma Xi. *Mailing Add:* Brigham Young Univ 118 RB Provo UT 84602

CONLEY, CARROLL LOCKARD, medicine, for more information see previous edition

CONLEY, FRANCIS RAYMOND, b Donora, Pa, Feb 11, 16; m 44; c 5. INORGANIC CHEMISTRY. *Educ:* St Bonaventure Univ, BS, 41. *Prof Exp:* Chemist, Ryder Scott Co, 41-43, lab foreman, 43, waterflood engr, 44-47, dir res, 47-51; res engr, Continental Oil Co, 51-55, res group leader, 55, supvr res chemist, 55-59, asst dir prod res, 59-65, dir prod res, 65-67, mgr prod res, 67-80; assoc gen mgr, Res & Develop Dept, Conoco Inc, 80-81; PETROL RECOVERY CONSULT, 81- *Mem:* Am Inst Mining, Metall & Petrol Engrs; Am Petrol Inst. *Res:* Petroleum secondary recovery; petrophysics reservoir mechanics; fluid flow in porous media; electric and radiation logging; coring; water flooding; production engineering and offshore technology; reservoir geology. *Mailing Add:* 1405 Meadowbrooks Ponca City OK 74604

CONLEY, HARRY LEE, JR, b Somerset, Ky, June 8, 35; m 58; c 3. PHYSICAL CHEMISTRY. *Educ:* Univ Ky, BS, 57; Univ Calif, Berkeley, MS, 60; Univ Va, PhD(chem), 64. *Prof Exp:* Sr chemist, Res Ctr, Sprague Elec Co, 63-68; from asst prof to assoc prof chem, 68-77, PROF CHEM, MURRAY STATE UNIV, 77- *Mem:* AAAS; Am Chem Soc. *Res:* Kinetics; electrochemistry; organophosphates; thermodynamics. *Mailing Add:* Dept Chem Murray State Univ Murray KY 42071-3311

CONLEY, JACK MICHAEL, b Wichita, Kans, Aug 9, 43; m 75; c 2. ANALYTICAL CHEMISTRY. *Educ:* Colo Sch Mines, MEC, 69; Univ Ill, MS, 72, PhD(anal chem), 75. *Prof Exp:* Teaching asst chem, Univ Ill, 69-72, res asst, 72-74; res scientist chem, Res & Develop Div, Union Camp Corp, 74-79; SR RES SCIENTIST, AM CYANAMID, 79- *Mem:* Am Chem Soc. *Res:* Chromatographic separations for pesticide development and manufacturing. *Mailing Add:* Am Cyanamid PO Box 0400 Princeton NJ 08540

CONLEY, JAMES FRANKLIN, b Forest City, NC, Dec 28, 31; m 54; c 3. GEOLOGY. *Educ:* Berea Col, AB, 54; Ohio State Univ, MS, 56; Univ S Carolina, PhD, 82. *Prof Exp:* Geologist, NC Div Mineral Resources, 56-65; GEOLOGIST, VA DIV MINERAL RESOURCES, 65- *Mem:* Fel Geol Soc Am; Am Inst Mining, Metall & Petrol Engrs; Am Inst Prof Geologists. *Res:* Geologic mapping in central and southern Virginia Piedmont. *Mailing Add:* Rte 1 Box 487-1 Troy VA 22974

CONLEY, ROBERT T, b Summit, NJ, Dec 27, 31; m 55; c 3. ORGANIC CHEMISTRY, POLYMER CHEMISTRY. *Educ:* Seton Hall Univ, BS, 53; Princeton Univ, MA, 55, PhD(chem), 57. *Prof Exp:* Asst prof chem, Canisius Col, 56-61; from asst prof to prof, Seton Hall Univ, 61-67; chmn dept chem, Wright State Univ, 67-68, dean col sci & eng, 68-77, prof chem, 67-77, vpres planning & develop, 74-77; pres, Seton Hall Univ, 77-79; PRES, CMG MGT GROUP, 80- *Concurrent Pos:* Consult, Carborundum Co. *Mem:* AAAS; Am Chem Soc; Sigma Xi. *Res:* Thermal stability of polymers; molecular rearrangements; reaction mechanisms; synthesis of compounds of pharmacological activity; synthetic methods in alicyclic systems; infrared spectroscopy. *Mailing Add:* 3784 Fallen Tree Lane Cincinnati OH 45236

CONLEY, VERONICA LUCEY, b Taunton, Mass, July 13, 19; m 45; c 4. PUBLIC HEALTH. *Educ:* Boston Univ, AB, 40; Yale Univ, MN, 43; Univ Chicago, MA, 53, PhD(sci educ), 59. *Prof Exp:* Instr nursing, Ohio State Univ, 46-48; secy comt cosmetics, AMA, 48-60, dir dept nursing, 60-62; exec dir, Nat Asn Practical Nurse Educ & Serv, Inc, 62-65; chief Allied Health Sect, Div Regional Med Progs, Pub Health Serv, Nat Cancer Inst, 67-73, chief Off of Comt & Rev Activities, 73- 79, chief, Community Spec Projs Br, 79-81, actg chief, Occup Cancer Br, 81-85, prog dir, Early Detection Br, 85-90; RETIRED. *Concurrent Pos:* Ed monthly series, Today's Health Mag, 52-58; exec ed, J Practical Nursing, 62-80. *Mem:* AAAS; affil mem AMA. *Res:* Educational and behavioral aspects of occupational cancer prevention. *Mailing Add:* 1941 Old Annapolis Blvd Annapolis MD 21401

CONLEY, WELD E, b Naples, NY, Aug 22, 11; m 43; c 3. FOOD SCIENCE & TECHNOLOGY, MECHANICAL ENGINEERING. *Educ:* Univ Rochester, BS, 33, MS, 34. *Prof Exp:* Chief engr, Thermal Eng Corp, 40-41; engr, Naval Ord Lab, 41-43, prin ord engr, 43-45; res engr, Rex Chainbelt Co, 45-53, chief develop engr, 53-55, mgr, Res & Develop Ctr, 55-61; tech dir, Votator Div, Chemetron Corp, 61-62, dir res, 62-63, mgr drying & sterilizing equip, 63-76; CONSULT ENGR, FOOD PROCESSING INDUST, 76- *Concurrent Pos:* Dir, Res & Develop Assocs, 54-62, vpres, 57-59, pres, 59-60, vchmn bd, 60-61, chmn bd, 61-62; tech consult, Food Processing Indust, 76-; fel, Inst Food Technol, 84. *Mem:* Fel Inst Food Technologists; Nat Soc Prof Engrs; Am Soc Heating, Refrig & Airconditioning Engrs; Am Soc Mech Engrs. *Res:* Development of processes and equipment for food canning and dehydration; development of underwater ordnance. *Mailing Add:* 1804 Corona Ct Louisville KY 40222

CONLIN, BERNARD JOSEPH, b Columbus, Wis, Mar 15, 35; m 65; c 1. DAIRY SCIENCE, ANIMAL BREEDING. *Educ:* Univ Wis, BS, 57, MS, 63; Univ Minn, PhD(dairy sci), 66. *Prof Exp:* 4-H Club exten agent, Agr Exten Serv, Univ Wis, 57-58; prom dir, E Cent Breeders Coop, 58-60; res asst dairy sci, Univ Wis, 60-61, exten dairy specialist, 61-62, res asst dairy sci, 62; res asst dairy husb, 62-66, from asst prof to assoc prof, 66-71, exten dairyman, 66-78, PROF ANIMAL SCI, UNIV MINN, ST PAUL, 71-, PROG LEADER AGR, 78-, DIR COMPUT APPLICATIONS, 80- *Concurrent Pos:* Consult, US Agency Int Develop to Tunisia, 78. *Mem:* Am Dairy Sci Soc; Holstein-Friesian Asn Am. *Res:* Relative merits of inbred and non-inbred sires for use in artificial insemination and uniformity of their sire families. *Mailing Add:* 4850 Lakeview Dr St Paul MN 55126

CONLY, JOHN F, b Ridley Park, Pa, Sept 11, 33. ENGINEERING. *Educ:* Univ Pa, BS, 56, MS, 58; Columbia Univ, PhD(eng mech), 62. *Prof Exp:* Res asst eng mech, Columbia Univ, 60-62; chmn dept, 71-84, from asst prof to assoc prof, 62-69, PROF AEROSPACE ENG, SAN DIEGO STATE UNIV, 69- *Concurrent Pos:* NSF res grant, 63- *Mem:* Am Inst Aeronaut & Astronaut; Am Soc Eng Educ. *Res:* Fluid and engineering mechanics. *Mailing Add:* Dept Aerospace Eng & Eng Mech San Diego State Univ San Diego CA 92182

CONN, ARTHUR L(EONARD), b New York, NY, Apr 5, 13; m 37, 72; c 3. CHEMICAL ENGINEERING, SCIENCE & TECHNOLOGY. *Educ:* Mass Inst Technol, SB, 34, SM, 35. *Prof Exp:* Asst dir Boston Sta, Sch Chem Eng Practice, Mass Inst Technol, 35-36; asst to dir res, Blaw-Knox Co, Pittsburgh, 36; exp chemist, Alco Prods Div, Am Locomotive Co, NY, 36-39; chem engr, Standard Oil Co, Ind, 39-43, group leader in charge Manhattan Proj work, 43-45, sect leader, 45-50, dir pilot plant div, 50-55, dir process div, 55-59, supt tech serv, 59-60, dir process develop, Amoco Oil Co, 60-62, coordr processes eng & admin, 62-64, sr consult engr, 64-67, dir govt contracts, 67-78; PRES, ARTHUR L CONN & ASSOCS LTD, 78- *Concurrent Pos:* Consult, Off Coal Res, US Dept Interior, 69-75, Energy Res & Develop Admin & Dept Energy, 75-78; consult & chmn adv comt on coal res, City Col, City Univ New York, 72-; chmn panel refining of coal & shale liquids, Nat Acad Eng, 77-80. *Honors & Awards:* Founders Award, Am Inst Chem Engrs. *Mem:* Fel AAAS; Am Chem Soc; fel Am Inst Chem Engrs (vpres, 69, pres, 70). *Res:* Petroleum process development; pilot plant operation; process design and economics; plant startup; fluid catalytic cracking, hydrocracking and synthetic fuels from oil shale and coal; ultraforming; process development in petroleum field; scale up of pilot plant results to commercial; science and technology of fossil fuels; technology of coal gasification and combined cycle power plants. *Mailing Add:* 1469 East Park Pl Chicago IL 60637-1835

CONN, ERIC EDWARD, b Berthoud, Colo, Jan 6, 23; m 59; c 2. BIOCHEMISTRY. *Educ:* Univ Colo, AB, 44; Univ Chicago, PhD(biochem), 50. *Prof Exp:* Chemist, Manhattan Dist, 44-46; instr biochem, Univ Chicago, 50-52; from instr to asst prof soils & plant nutrit, Univ Calif, Berkeley, 52-54, asst prof biochem, 54-58, assoc prof, 58-63,; assoc prof & assoc biochemist, 58-63, PROF BIOCHEM, UNIV CALIF, DAVIS, 63- *Concurrent Pos:* NIH sr res fel, 60, mem, USPHS Fel Rev Panel for Biochem Nutrit, 63-66; Fulbright scholar, NZ, 65-66; mem, US Nat Comt, Int Union Biochem, 65-68; asst ed, Plant Physiol, 68-72; ed, Arch Biochem & Biophys, 72-, Plant Physiol, 80 & Phytochem, 81- *Mem:* Nat Acad Sci; Am Soc Biol Chem; Am Soc Plant Physiol; Brit Biochem Soc; hon mem Phytochem Soc NAm; Am Soc Pharmacog; Sigma Xi. *Res:* Plant enzymes; intermediary metabolism of secondary plant products; cyanogenic glycosides; biosynthesis of aromatic amino acids; author of 170 technical publications. *Mailing Add:* Dept Biochem & Biophys Univ Calif Davis CA 95616

CONN, HADLEY LEWIS, JR, b Danville, Ind, May 6, 21; m 46; c 5. INTERNAL MEDICINE, CARDIOLOGY. *Educ:* Indiana Univ, BA, 42, MD, 44. *Hon Degrees:* MS, Univ Pa, 71. *Prof Exp:* Resident med, dept med, Univ Pa, 48-51, cardiol fel, 51-53, from asst prof to assoc prof, Sch Med, 56-72, dir, Clin Res Ctr, 70-72; assoc mem physiol, dept med, Brookhaven Nat Lab, 53-55; chmn dept, Presbyterian-Univ Pa Med Ctr, 64-69; chmn, dept med, 72-83, DIR CARDIOVASC INST, UNIV MED & DENT NJ SCH MED, RUTGERS UNIV, 83- *Concurrent Pos:* Estab investr, Am Heart Asn, 54-59; chmn, Am Fedn Clin Res, 61; secy, Nat Bd Med Examr, 62-65; vis prof med, Am Univ Beirut, Lebanon, 69-70; comt & coun mem, NIH, 70-76; pres, Detweiler Res Found, 72-85. *Mem:* Am Clin & Clinicians Asn; fel Am Col Physicians; Asn Prof Med; fel Am Col Cardiol; fel Am Heart Asn; Soc Clin Invest. *Res:* Basic science and clinical investigation in cardiology. *Mailing Add:* Dept Medi Univ Maryland New Jersey Robert Wood Johnson Med Sch One Robert Wood Johnson Pl CN19 New Brunswick NJ 08901-0019

CONN, HAROLD O, b Newark, NJ, Nov 16, 30; m 51. INTERNAL MEDICINE. *Educ:* Univ Mich, BS, 46, MD, 50; Am Bd Internal Med, dipl, cert gastroenterol. *Prof Exp:* Intern, Johns Hopkins Hosp, 50-51; asst resident physician, Grace New Haven Community Hosp, 51-52, chief resident physician, 55-56; from instr to assoc prof, 55-71, PROF INTERNAL MED, SCH MED, YALE UNIV, 71- *Concurrent Pos:* Browne res fel, Sch Med, Yale Univ, 52-53; dir med ed, Middlesex Mem Hosp, 56-57; clin investr, Vet

Admin Hosp, West Haven, 57-60, actg chief med serv, 59-60, chief, Hepatic Res Lab, 60-; counr, Am Asn Study Liver Dis, 67-, vpres elect, 71, vpres, 72, pres, 73; vis prof, Sch Med, Wash Univ, 68-69; mem med sch coun, Yale Univ Sch Med, 70; assoc ed, Gastroenterol, 73-77; dir, Am Gastroenterol Asn-NIH Workshop Diag Tech Hepatobiliary Dis, 78. *Honors & Awards:* William Beaumont Award Excellence in Clin Res, 74. *Mem:* Am Fedn Clin Res; Am Soc Clin Invest; Asn Am Physicians; Sydenham Soc (secy, 65-); Am Gastroenterol Asn. *Res:* Clinical management of liver disease; treatment of hepatic coma, esophageal varices and ammonia metabolism; abnormalities of protein metabolism. *Mailing Add:* Yale Univ Sch Med 333 Cedar St 05191336X CT 06510

CONN, JAMES FREDERICK, b Osborne, Kans, July 2, 24; m 48; c 2. CEREAL CHEMISTRY. *Educ:* Kans State Univ, BS, 48, MS, 49. *Prof Exp:* Wheat qual chemist, Int Multifoods Co, 49-57; sr res chemist, chem leavening cheese emulsification res, Monsanto Co, 57-66, res specialist, 66-78, sr res specialist, 78- 83, prin food technologist, 84-85; PRES, J F CONSULT INC, 86- *Concurrent Pos:* Expert, Teltech Inc, 89- *Mem:* fel Am Asn Cereal Chemists; Inst Food Technologists; Sigma Xi; Coun Agr Sci & Technol. *Res:* Determining the functions of ortho-, pyro- and polyphosphates in chemical leavening, process cheese and processed meat with particular interest in the interactions with other ingredients and the rheological effects. *Mailing Add:* 52 Chieftain Dr St Louis MO 63146

CONN, JEROME W, b New York, NY, Sept 24, 07; c 2. MEDICINE. *Educ:* Univ Mich, MD, 32. *Hon Degrees:* DSc, Rutgers Univ, 64; Univ Turin, MD, 75. *Prof Exp:* From instr to prof internal med, 35-68, Louis Harry Newburgh univ prof med, 68-74, consult clin invest, Dept Internal Med, 74-76, EMER DISTINGUISHED UNIV PROF INTERNAL MED, MED SCH, UNIV MICH, ANN ARBOR, 74- *Concurrent Pos:* Distinguished physician, Vet Admin, 73-76. *Honors & Awards:* Mod Med Mag Award, 57; Bernard Medal, Univ Montreal, 57; Banting Medal, Am Diabetes Asn, 58 & Banting Mem Award, 63; Henry Russell Lectr Award, Univ Mich, 61; Wilson Medal, Am Clin & Climat Asn, 62; Gairdner Found Int Prize, 65; Phillips Mem Award, Am Col Physicians, 65; Howard Taylor Award, Am Therapeut Soc, 67; Ruth Gray Mem Medal, Evanston Hosp, 68; Stouffer Int Prize, 69; Gold Medal, Int Soc Progress Internal Med, 69; Heath Med Award & Medal, Univ Tex, Houston, 71; Award, Am Col Nutrit, 73. *Mem:* Nat Acad Sci; Inst Med Nat Acad Sci; Asn Am Physicians; Am Diabetes Asn (pres, 62-63, first vpres, 61, 2nd vpres, 60); hon fel Am Col Surgeons. *Res:* Human nutrition; normal metabolism; disorders of metabolism; endocrinology; discoverer of primary aldorsteronism also known as Conn's Syndrome. *Mailing Add:* 2369 Gulf Shore Blvd N Naples FL 33940

CONN, P MICHAEL, b Oil City, Pa, May 12, 49; m. ENDOCRINOLOGY, CELL BIOLOGY. *Educ:* Univ Mich, BS & CEd, 71; NC State Univ, MS, 73; Baylor Col Med, PhD(cell biol), 76. *Prof Exp:* Fel endocrinol, NIH, 76-78; asst prof pharmacol, 78-82, ASSOC PROF, DUKE UNIV MED CTR, 82-; PROF & CHMN, DEPT PHARMACOL, COL MED, UNIV IOWA. *Concurrent Pos:* Sr fel aging, Ctr Study Aging & Human Develop, 78; res career develop award, NIH, 80; ed, J Clin Endocrinol Metab, 83; ed-in-chief, Endocrinol, 87, Methods in Neurosci & Receptors; vol ed, Methods in Enzymol. *Honors & Awards:* Weizmann Award, Endocrine Soc, 85; Abel Award, Am Soc Pharmacol & Exp Therapeut, 85; Miguel Alaman Award, Mex, 88. *Mem:* Endocrine Soc; Am Soc Cell Biol; Soc Study Reprod. *Res:* Mechanism of action of gonadotropin releasing hormone; mechanism of hormone action; molecular and cell biology of endocrine cells. *Mailing Add:* Dept Pharmacol Bowen Sci Bldg Col Med Univ Iowa 2511-BSB Iowa City IA 52242-1109

CONN, PAUL JOSEPH, b Kalispell, Mont, May 17, 40; m 61; c 3. PHYSICAL CHEMISTRY. *Educ:* Univ Calif, Davis, BS, 62; Univ Ore, MS, 64, PhD(phys chem), 66. *Prof Exp:* Staff res chemist, Shell Develop Co, 64-80, sr staff res chemist, 80-83; sr staff res scientist, Western Res Inst, 84-86; CONSULT, 86- *Concurrent Pos:* Exchange scientist, Koninklijke/Shell Lab, Amsterdam, 77-78; adj prof, Mont State Univ, 87- *Mem:* Am Chem Soc; Catalysis Soc. *Res:* Heterogeneous catalysis, preparation and characterization of catalysts, catalytic kinetics and mechanism; geochemistry; microbial surface processes-corrosion, leaching. *Mailing Add:* 514 S Third Ave Bozeman MT 59715-5253

CONN, PAUL KOHLER, b Akron, Ohio, July 25, 29; m 54; c 3. MATERIALS SCIENCE. *Educ:* Kenyon Col, AB, 51; Kans State Univ, MS, 53, PhD(phys chem), 56. *Prof Exp:* Sr engr & prin engr, Aircraft Nuclear Propulsion Dept, Gen Elec Co, 55-61, supvr chem res, Nuclear Mat & Propulsion Oper, 61-64, unit mgr, Solid State Chem Res, 64-69; prin scientist & assoc dir advan mat res, 69-78, staff prin scientist, High Energy Lasers, 77-80, CHIEF ENGR MAT & CHEM, BELL AEROSPACE CO, TEXTRON INC, 80-, DIR, ENG LAB & TEST, 80- *Concurrent Pos:* Mem, Environ Mgt Coun, Niagara County, NY, 82- *Mem:* Am Chem Soc; Nat Asn Corrosion Engrs; Sigma Xi. *Res:* High temperature materials and coatings; solid state chemistry, including kinetics, diffusion, fission product transport processes; gas-solid reactions; hot atom chemistry; radiation effects in materials; viscoelastic composite materials; organic coatings. *Mailing Add:* 4682 W Park Dr Lewiston NY 14092

CONN, REX BOLAND, b Marengo, Iowa, Aug 3, 27; m 50; c 3. MEDICINE, CLINICAL PATHOLOGY. *Educ:* Iowa State Univ, BS, 49; Yale Univ, MD, 53; Oxford Univ, BSc, 55; Univ Minn, Minneapolis, MS, 60. *Prof Exp:* Instr lab med, Univ Minn, Minneapolis, 60; asst prof med, WVa Univ, 60-61, from asst prof to prof, 61-68; prof lab med & dir subdept lab med, Johns Hopkins Univ, 68-77; prof path & lab med, Emory Univ & dir clin labs, Emory Univ Hosp, 77-87; PROF & VCHMN, DEPT PATH, THOMAS JEFFERSON UNIV, PHILADELPHIA, 87- *Concurrent Pos:* Mem path study sect, NIH, 68-72, mem, Path Training Comt, 72-73; mem Comt Chem Path, Am Bd Path, 74-80; mem bd dirs, Am Soc Clin Path, 75-81. *Mem:* Fel Am Soc Clin Path; fel Col Am Path; Am Asn Pathologists; Am Fedn Clin Res; Acad Clin Lab Physicians & Scientists (pres, 72-73). *Res:* Creatine metabolism; methods in clinical chemistry. *Mailing Add:* Dept Path Jefferson Med Col Thomas Jefferson Univ Hosp 125 S 11th St Philadelphia PA 19107

CONN, RICHARD LESLIE, b Watseka, Ill, Mar 14, 47; m 68; c 2. WEED SCIENCE, AGRONOMY. *Educ:* Univ Ill, BS, 69, MS, 70. *Prof Exp:* Regulatory specialist, 70-76, SR REGULATORY SPECIALIST PESTICIDE REGULATIONS, AGR DIV, CIBA-GEIGY CORP, 77-, STEWART ASSOC. *Mem:* Weed Sci Soc Am. *Res:* Herbicides; evaluation of chemicals affecting food production. *Mailing Add:* 2001 Jefferson-Davis Hwy Suite 603 Arlington VA 22202

CONN, ROBERT WILLIAM, b Brooklyn, NY, Dec 1, 42; m 63; c 2. FUSION ENGINEERING, PLASMA & MATERIALS TECHNOLOGY. *Educ:* Pratt Inst, BS, 64; Calif Inst Technol, MS, 65, PhD(eng sci), 68. *Prof Exp:* Res assoc, Brookhaven Nat Lab, 69-70; tac mem, Univ Wis-Madison, 70-80, vis assoc prof nuclear eng, 70-72, prof, 74-80, dir fusion technol, 74-79; co-dir, Ctr Plasma Physics & Fusion Eng, 82-86, PROF MECH & NUCLEAR ENG, UNIV CALIF, LOS ANGELES, 79-; DIR INST PLASMA & FUSION RES, 87- *Concurrent Pos:* NSF Fel, Euratom Community Ctr for Res, Ispra, Italy, 68-69; vis scientist, Argonne Nat Lab, 72; Romnes fac prof, Univ Wis, 77; guest scientist, Max-Planck Inst Plasma Physics, 77; guest scientist, Nuclear Res Ctr, Juelich, WGer, 83. *Honors & Awards:* Curtis W McGraw Award, Am Asn Eng Educ, 82; Outstanding Achievement Award, Am Nuclear Soc, 84; E O Lawrence Mem Award, US Dept Energy on behalf of President. *Mem:* Nat Acad Eng; fel Am Phys Soc; fel Am Nuclear Soc. *Res:* Theoretical and experimental work in plasma-materials interactions, thin films and coatings; plasma physics in tokamaks; fusion reactor engineering and reactor design. *Mailing Add:* Dept Mech & Nuclear Eng & Inst Plasma & Fusion Res Univ Calif 6291 44-139 Engr IV Los Angeles CA 90024-1597

CONNAMACHER, ROBERT HENLE, b Newark, NJ, Dec 20, 33; m 66; c 4. PHARMACOLOGY. *Educ:* Oberlin Col, AB, 55; NY Univ, MS, 59; George Washington Univ, PhD(pharmacol), 66. *Prof Exp:* From instr to asst prof, 67-73, ASSOC PROF PHARMACOL, SCH MED, UNIV PITTSBURGH, 73- *Concurrent Pos:* Fel molecular biol, Sch Med, Univ Pittsburgh, 66-67. *Mem:* Am Soc Microbiol; Biophys Soc; AAAS; NY Acad Sci; Fedn Am Sci. *Res:* Mechanism of action of antibiotics and biochemical mechanisms of adaptation; fatty alcohols as antimicrobial agents. *Mailing Add:* Dept Pharmacol 518 Scaife Hall Univ Pittsburgh Sch Med Pittsburgh PA 15261

CONNAR, RICHARD GRIGSBY, thoracic surgery, cardiovascular surgery; deceased, see previous edition for last biography

CONNELL, ALASTAIR MCCRAE, b Glasgow, Scotland, Dec 21, 29. GASTROENTEROLOGY. *Educ:* Univ Glasgow, BSc, 51, MB, ChB, 54, MD, 69; FRCP, 72; FACP, 73. *Prof Exp:* House physician, Stobhill Gen Hosp & house surgeon, Univ Dept Surg, Western Infirmary, Glasgow, 54-55; clin & res asst, Cent Middlesex Hosp, London, 57-60; mem sci staff, Med Res Coun Gastroenterol, Cent Middlesex Hosp & St Mark's Hosp London, 61-64; sr lectr clin sci, Queen's Univ, Belfast, 64-70; assoc dean, Col Med, Univ Cincinnati, 75-77, Mark Brown prof med & physiol & dir, Div Digestive Dis, 70-79; prof internal med & dean Col Med, Med Ctr, Univ Nebr, 79-84; prof internal med & vpres health sci, Med Col Va, Va Commonwealth Univ, 84-88; VCHANCELLOR & SPEC PROJ OFFICER, EAST CAROLINA UNIV, 90- *Concurrent Pos:* Consult physician, Northern Ireland Hosp Authority, Royal Victoria Hosp & SBelfast Hosp, Belfast, 64-70; attend physician, Cincinnati Gen Hosp, Ohio, 70-79; chief clinician, Div Digestive Dis, Med Ctr, Univ Cincinnati, 70-79; consult, Vet Admin Hosp, Cincinnati, 70-79 & Jewish Hosp, Cincinnati, 73-79. *Mem:* Brit Soc Gastroenterol; Am Gastroenterol Asn; Am Fedn Clin Res. *Res:* Motility of gastrointestinal tract; role of gastrointestinal hormones in control of motor activity; assessment of therapy of gastrointestinal disease; nutritional factors in pathogenesis of gastrointestinal disease. *Mailing Add:* 1501 Crystal Dr No 725 Arlington VA 22202

CONNELL, CAROLYN JOANNE, b Indianapolis, Ind; c 2. ENDOCRINOLOGY, CELL BIOLOGY. *Educ:* Ind Univ, MS, 68, PhD(zool), 70. *Prof Exp:* Fel anat, Stanford Univ Med Sch, 70-71 & Temple Univ Med Sch, 71-72; staff res scientist cell biol, Stanford Res Inst, 72-75; asst res reproductive endocrinol, Univ Calif Med Sch, 75-78, asst prof, 78-79; ASSOC PROF ANAT, COLO STATE UNIV, 79- *Concurrent Pos:* Prin investr, NIH grant, 75-83. *Mem:* Am Soc Cell Biologists; Asn Women Sci; AAAS; Am Asn Anatomists; Am Soc Androl. *Res:* Role that peptide and steroid hormones play as provocateurs of differentiation of the interstitial and tubular cells of the testis as reflected by intramembranous and intercellular specializations; author or coauthor of numerous publications. *Mailing Add:* 1412 Grizzly Peak Blvd Berkeley CA 94708

CONNELL, GEORGE EDWARD, b Saskatoon, Sask, June 20, 30; m 55; c 4. BIOCHEMISTRY. *Educ:* Univ Toronto, BA, 51, PhD(biochem), 55. *Hon Degrees:* LLD, Trent Univ, 84, Univ Western Ont, 85 & McGill Univ, 87. *Prof Exp:* Nat Res Coun Can fel, 55-56; Nat Acad Sci-Nat Res Coun fel biochem, Sch Med, NY Univ, 56-57; from asst prof to assoc prof biochem, Univ Toronto, 57-65, prof & chmn dept, 65-70, assoc dean fac med, 72-74, vpres, 74-77; pres, Univ Western Ont, 77-84; PRES, UNIV TORONTO, 84- *Concurrent Pos:* Mem, Med Res Coun Can, 66-70; mem, Ont Coun Health, 78-84; chmn, Coun Ont Univs, 81-83; mem bd gov, Upper Can Col, 82-88, Corp-Higher Educ Forum, 83-90, bd trustees, Nat Inst Nutrit & Royal Ontario Mus, 84-90, bd dir, Southam Inc, 85- *Honors & Awards:* Order Can, 87. *Mem:* Can Biochem Soc (pres, 73-74); Am Soc Biol Chemists; Can Soc Immunol; Chem Inst Can (chmn biochem div, 69-70); Sigma Xi; fel Royal Soc Can. *Res:* Protein chemistry; enzymology; immunochemistry; chemistry of human plasma proteins. *Mailing Add:* 240 Walmer Rd Toronto ON M5R 3R7 Can

CONNELL, JAMES ROGER, atmospheric science, for more information see previous edition

CONNELL, JOSEPH H, b Gary, Ind, Oct 5, 23; m 54; c 4. POPULATION BIOLOGY. *Educ:* Univ Chicago, BS, 46; Univ Calif, MA, 53; Glasgow Univ, PhD(zool), 56. *Prof Exp:* Res assoc marine biol, Woods Hole Oceanog Inst, 55-56; PROF ZOOL, UNIV CALIF, SANTA BARBARA, 56- *Concurrent Pos:* Guggenheim fel, 62-63, 71-72; vis fel, Res Sch Biol Sci, Australian Nat Univ, 80, 85-86, 88-89 & 91. *Honors & Awards:* Mercer Award, Ecol Soc Am, 63; Eminent Ecol Award, Ecol Soc Am, 85. *Mem:* AAAS; Am Soc Limnol & Oceanog; Brit Ecol Soc; Ecol Soc Am; Am Soc Naturalists. *Res:* Community ecology, especially succession and species diversity of tropical rain forests and coral reefs; population and community ecology of marine intertidal organisms, particularly predation, competition and spatial distribution. *Mailing Add:* Dept Biol Sci Univ Calif Santa Barbara CA 93106

CONNELL, RICHARD ALLEN, b Lincoln, Nebr, Oct 31, 29; m 56; c 2. PHYSICS. *Educ:* Nebr Wesleyan Univ, BA, 51; Northwestern Univ, PhD, 57. *Prof Exp:* Asst gaseous discharges, Northwestern Univ, 51-54, low temperature solid state, 54-57; assoc physicist superconductivity, Res Lab, Int Bus Mach Corp, 57-59, proj physicist, 59-62, prof engr, 62-63; sr physicist, Midwest Res Lab, 62-67; sr physicist, 67-70, MGR MAT SCI GROUP, PITNEY-BOWES, INC, 70- *Mem:* Am Phys Soc; Sigma Xi. *Res:* Superconductivity; thin film physics; biophysics; photoconductivity; electrophotography. *Mailing Add:* Pitney-Bowes Inc World Hq Loc 26-000 Stamford CT 06926-0700

CONNELL, WALTER FORD, b Kingston, Ont, Aug 24, 06; m 33; c 3. INTERNAL MEDICINE, CARDIOLOGY. *Educ:* Queen's Univ, Ont, MD & CM, 29; FRCP; FRCPS(C). *Hon Degrees:* LLD, Queen's Univ, Ont, 73. *Prof Exp:* Dir cardiol, Kingston Gen Hosp, 33-68; prof, 43-76, EMER PROF MED, QUEEN'S UNIV, ONT, 76- *Concurrent Pos:* Mem coun arteriosclerosis, Am Heart Asn. *Mem:* Am Heart Asn; Can Heart Asn. *Res:* Coronary heart disease. *Mailing Add:* 11 Arch St Kingston ON K7L 3N6 Can

CONNELL-TATUM, ELIZABETH BISHOP, b Springfield, Mass, Oct 17, 25; m 80; c 6. OBSTETRICS & GYNECOLOGY. *Educ:* Univ Pa, AB, 47, MD, 51; Am Bd Obstet & Gynec, dipl, 65. *Prof Exp:* Intern, Lankenau Hosp, Philadelphia, Pa, 51-52, resident path & anesthesia, 52-53; gen pract & anesthetist, Maine, 53-58; resident gynec, Grad Hosp, Univ Pa, 58-60; resident obstet, Mt Sinai Hosp, New York, 60-61; assoc prof obstet & gynec, NY Med Col, 62-69; assoc prof obstet & gynec, Col Physicians & Surgeons, Columbia Univ, 70-73, dir res & develop, Family Planning Serv, Int Inst Study Human Reproduction, 70-73; assoc dir health sci, Rockefeller Found, 73-78; assoc prof obstet & gynec, Northwestern Univ, 78-81; PROF GYNEC & OBSTET, EMORY UNIV, 81- *Concurrent Pos:* Am Cancer fel, Kings County Hosp, State Univ NY, 61-62; dir, Family Planning Ctr, NY Med Col-Metrop Hosp Med Ctr, 64-69; mem med adv bd, Planned Parenthood, New York, 64-; mem nat adv coun, Alan Guttmacher Inst, Planned Parenthood World Pop, 68-, chmn nat med comt, 74-; mem exec comt, Comt Med & Pub Health Asn Voluntary Sterilization, Inc; mem obstet & gynec adv comt, Food & Drug Admin, chmn over-the-counter rev panel; consult, Family Planning, New York Dept Health; family planning proj consult, Human Resources Admin; mem res adv comt, Agency Int Develop. *Mem:* Am Col Obstet & Gynec; Am Col Surg; Am Pub Health Asn; Am Fertil Soc; AMA. *Res:* Medicine; contraception. *Mailing Add:* Dept Gynec & Obstet Emory Univ Sch Med 69 Butler St Atlanta GA 30305

CONNELLY, CAROLYN THOMAS, b Brownville, Maine, Feb 25, 41; m 84. ANALYTICAL CHEMISTRY. *Educ:* Univ Maine, BA, 63; Northeastern Univ, PhD(anal chem), 68. *Prof Exp:* Teaching fel anal chem, Northeastern Univ, 63-67, lectr, Lincoln Col, 67-68; res chemist, US Army Natick Labs, 67-68; from asst prof to assoc prof, 68-77, PROF ANALYTICAL CHEM, WAYNESBURG COL, 77- *Mem:* Am Chem Soc; Sigma Xi. *Res:* Molten salts; electrochemistry; stability of foods; atomic absorption and emission spectroscopy applied to water pollution. *Mailing Add:* Dept Chem Waynesburg Col Waynesburg PA 15370

CONNELLY, CLARENCE MORLEY, b Jamestown, NY, Nov 4, 16; m 46; c 1. BIOPHYSICS. *Educ:* Cornell Univ, AB, 38; Univ Pa, PhD(biophys), 49. *Prof Exp:* Asst physics, Cornell Univ, 38-42; staff mem, Radiation Lab, Mass Inst Technol, 42-46; asst biophys, Univ Pa, 46-49; from instr to asst prof, Johns Hopkins Univ, 49-54; from asst prof to assoc prof, 54-84, assoc dean grad studies, 62-80, actg dean, 80-81, dean grad studies, 81-84, EMER PROF BIOPHYS, ROCKEFELLER UNIV, 84- *Concurrent Pos:* Ed, J Gen Physiol, 61-64. *Mem:* Soc Gen Physiol; Am Phys Soc; Biophys Soc; Sigma Xi. *Res:* Physiology of nerve and muscle. *Mailing Add:* 11612 Polaris Dr Grass Valley CA 95949-7609

CONNELLY, DONALD PATRICK, b Breckenridge, Minn, Oct 8, 42; m 67; c 3. LABORATORY MEDICINE, MEDICAL INFORMATICS. *Educ:* NDak State Univ, BSEE, 64, MSEE, 65; Univ Minn, MD, 71, PhD(health info syst), 77. *Prof Exp:* Engr logic design, Int Bus Mach Corp, 65-66; engr med appln, Mayo Clin, 67; asst prof, 74-82, ASSOC PROF LAB MED & PATH, UNIV MINN, 82- *Mem:* Acad Clin Lab Physicians & Scientists; Am Asn Med Systs & Informatics; Sigma Xi; Am Soc Clin Pathologists; Inst Elec & Electronics Engrs; Soc Med Decision Making. *Res:* Medical decision making; medical informatics. *Mailing Add:* UMHC Univ Minn Box 511 Minneapolis MN 55455

CONNELLY, J(OSEPH) ALVIN, b Knoxville, Tenn, Nov 9, 42; m 64; c 4. ELECTRICAL ENGINEERING. *Educ:* Univ Tenn, BS, 64, MS, 65, PhD(elec eng), 68. *Prof Exp:* From asst prof to assoc prof, 68-79, PROF ELEC ENG, GA INST TECHNOL, 79- *Concurrent Pos:* Instr solid state circuits course, ETenn State Univ, 69-75; consult, Harris Semiconductor, 72-75, Signetics, 77, Scientific Atlanta, 78, Honeywell, 79-85 & GEC Avionics, 84-; prog evaluator, ABET, 87-91. *Honors & Awards:* Centennial Award, Inst Elec & Electronics Engrs. *Mem:* Inst Elec & Electronics Engrs; Sigma Xi; Am Asn Univ Professors. *Res:* Solid state device and circuit theory; macromodeling of analog and digital circuits; integrated circuits and applications; phase locked loop communication systems; switched capacitor circuits; low noise design. *Mailing Add:* Sch Elec Eng Ga Inst Technol Atlanta GA 30332-0250

CONNELLY, JOHN JOSEPH, JR, b Syracuse, NY, Apr 14, 25; m 47; c 3. EXPERIMENTAL SOLID STATE PHYSICS. *Educ:* Rensselaer Polytech Inst, BAeroE, 45; Univ Va, MS, 55, PhD(physics), 56. *Prof Exp:* Sci adv to chief aircraft nuclear propulsion off & mat technologist, US AEC, 56-60; prog officer energy conversion, Power Br, Off Naval Res, 60-64; from assoc prof to prof physics, State Univ NY Col Fredonia, 64-85; RETIRED. *Concurrent Pos:* Mem comt elec eng systs, NASA; gast prof, ETH Zurich, 72-85. *Mem:* Assoc fel Am Inst Aeronaut & Astronaut; Am Phys Soc. *Res:* Very high speed ultra-centrifuges; high-temperature materials; advance type propulsion systems; direct energy conversion; crystal growth of halides, calcite and tellurates. *Mailing Add:* RD 1 Bernard Rd Cassadaga NY 14718

CONNELLY, THOMAS GEORGE, b Oak Park, Ill, Sept 22, 45; m 66; c 2. ANATOMY, ZOOLOGY. *Educ:* Monmouth Col, Ill, AB, 66; Mich State Univ, PhD(zool), 70. *Prof Exp:* asst prof, 72-79, asst res scientist, 75-79, ASSOC PROF ANAT, MED SCH, UNIV MICH, ANN ARBOR, 79-, ASSOC RES SCIENTIST, CTR HUMAN GROWTH & DEVELOP, 79- *Concurrent Pos:* NIH res fel, Biol Div, Oak Ridge Nat Lab, 70-72. *Mem:* Am Asn Anat; Int Soc Develop Biol; Am Soc Zool; Soc Develop Biol. *Res:* Amphibian lens and limb regeneration; computer assisted image analysis. *Mailing Add:* Dept Anat & Cell Biol 4643 Med Sci 2 Univ Mich Med Sch 1301 Catherine Rd Ann Arbor MI 48109

CONNER, ALBERT Z, b Philadelphia, Pa, Dec 21, 21. ANALYTICAL CHEMISTRY. *Educ:* Drexel Inst Technol, BS, 47; Univ Del, MS, 52. *Prof Exp:* Chemist, Publicker Industs, 46-47; res chemist, Hercules Powder Co, 47-65, sr res chemist, 65-70, res scientist, 70-76, RES ASSOC, HERCULES INC, 76- *Mem:* Am Chem Soc; Am Inst Chemists. *Res:* Chromatography; organic analysis; analysis of rocket propellants. *Mailing Add:* 2118 Swinnen Dr Wilmington DE 19810-4118

CONNER, BRENDA JEAN, b Savannah, Ga. BIOLOGY, MOLECULAR BIOLOGY. *Educ:* Emory Univ, BS, 67, MS, 69, PhD(molecular biol), 75. *Prof Exp:* Lab tech asst biol, Emory Univ, 67-70; NIH fel molecular biol, Univ Ga, 70-73; fel, 73-80, RES ASSOC CYTOGENETICS, CYTOLOGY & MOLECULAR BIOL, DEPT MED GENETICS, CITY OF HOPE MED CTR, 80-, RES ASSOC & ASST RES SCIENTIST, 85- *Mem:* AAAS; Am Soc Cell Biol; Am Soc Human Genetics. *Res:* Molecular genetics; gene structure and function; prenatal diagnosis of hemoglobin disorders and other genetic diseases; autoimmune disease; systemic lupus erythematosus. *Mailing Add:* Dept Molecular-Biochem Deckman Res Ctr City of Hope Hosp 1450 E Duarte Rd Duarte CA 91010

CONNER, GEORGE WILLIAM, b Jackson, Miss, Nov 19, 35. INSECT & HUMAN GENETICS. *Educ:* Vanderbilt Univ, BA, 57; Emory Univ, MS, 59; Univ Ariz, PhD(zool), 65. *Prof Exp:* Instr biol, Univ Southwestern La, 59-61; asst prof zool, Univ SDak, 64-68; asst prof biol, NTex State Teachers Col, 68-72; fel human genetics, Univ Edinburgh, Scotland, 72-74; asst prof biol, Univ NB, 74-75; PROF BIOL, LOYOLA COL, 75- *Concurrent Pos:* Fel, Muscular Dystrophy Asn Am, 73-74. *Mem:* Am Genetic Asn. *Res:* Genetics of complex loci in the hymenoptera; cytoplasmic inheritance and cytogenetics; human genetics. *Mailing Add:* Dept Biol Loyola Col Baltimore MD 21210

CONNER, HOWARD EMMETT, b Madison, Wis, Sept 26, 30; m 54; c 3. MATHEMATICS. *Educ:* Univ Wis, BS, 56; Mass Inst Technol, PhD(math), 61. *Prof Exp:* Staff assoc math, Lincoln Lab, Mass Inst Technol, 57-61; mem staff, US Army Math Res Ctr, 61-62, from asst prof to assoc prof, 62-73, PROF MATH, UNIV WIS-MADISON, 73- *Concurrent Pos:* Vis assoc prof, Rockefeller Univ, 67-68. *Mem:* Am Math Soc. *Res:* Systematic study of the properties of the solutions of the integral-differential equations used in gas dynamics, plasma theory and kinetic theory. *Mailing Add:* Dept Math Univ Wis 213 Van Vleck Hall 480 Lincoln Dr Madison WI 53706

CONNER, JACK MICHAEL, b Jackson, Miss, Nov 2, 35. INORGANIC CHEMISTRY, PHYSICAL CHEMISTRY. *Educ:* Millsaps Col, BS, 56; Univ Wyo, PhD(chem), 66. *Prof Exp:* Chemist, Baxter Labs, Inc, Miss, 57-58; res chemist, Sch Med, Univ Miss, 58-59; res assoc inorg chem, Univ Kans, 66-67; asst prof chem, 67-75, ASSOC PROF CHEM, REGIS COL, 75- *Mem:* Am Inst Chem; Am Chem Soc. *Res:* Preparation and chemistry of transition-metal coordination compounds and oxides; aqueous solution equilibria. *Mailing Add:* 1250 S Inca St Denver CO 80223

CONNER, JERRY POWER, b Sherman, Tex, Mar 20, 27; m 47; c 3. PHYSICS. *Educ:* Rice Inst, PhD(physics), 52. *Prof Exp:* MEM STAFF & PHYSICIST, LOS ALAMOS NAT LAB, 52- *Mem:* Am Phys Soc; Am Geophys Union; Am Astron Soc. *Res:* Space radiations. *Mailing Add:* 245 Rio Bravo Los Alamos NM 87544

CONNER, PIERRE EUCLIDE, JR, b Houston, Tex, June 27, 32; m 58; c 2. MATHEMATICS. *Educ:* Tulane Univ, BS, 52, MS, 53; Princeton Univ, PhD(math), 55. *Prof Exp:* Vis fel math, Inst Advan Study, 55-57; asst prof, Univ Mich, 57-58; from asst prof to prof, Univ Va, 48-71; NICHOLSON PROF MATH, LA STATE UNIV, 71- *Concurrent Pos:* Sloan fel, 60-64; vis mem, Inst Advan Study, 61-62; mem comt sci confs, DN Math, Nat Acad Sci, Nat Res Coun, 64-67; NICHOLSON PROF MATH, LA STATE UNIV, 71- Concurrent. *Mem:* Am Math Soc; AAAS. *Res:* Applications of the methods of differential and algebraic topology to the study of transformation groups, especially periodic maps. *Mailing Add:* Dept Math La State Univ Baton Rouge LA 70803

CONNER, ROBERT LOUIS, zoology; deceased, see previous edition for last biography

CONNERNEY, JOHN E P, b Boston, Mass, Sept 25, 50. PLANETARY MAGNETOSPHERES. *Educ:* Cornell Univ, BS, 72, PhD(appl physics), 79. *Prof Exp:* Resident res assoc, Nat Res Coun & Nat Acad Sci, 79-80; SPACE SCIENTIST, LAB EXTRATERRESTRIAL PHYSICS, GODDARD

SPACE FLIGHT CTR, NASA, 80- *Honors & Awards:* Medal Except Sci Achievement, NASA. *Mem:* Am Geophys Union. *Res:* Planetary and interplanetary magnetic fields and plasmas; interpretation of spacecraft observations (Voyager, Pioneer, Magsat) in context of magnetospheric theory (Earth, Jupiter, Saturn, Uranus, Neptune). *Mailing Add:* 5706 Berwyn Rd Berwyn Heights MD 20770

CONNERS, CARMEN KEITH, b Utah, May 20, 33; m 73. CHILD PSYCHOPHARMACOLOGY. *Educ:* Univ Chicago, BA, 53; Oxford Univ, MA, 55; Harvard Univ, PhD(psychol), 60. *Prof Exp:* Instr, Sch Med, Johns Hopkins Univ, 61-63, asst prof ped & med psychol, 63-67; from asst prof to assoc prof psychol, Mass Gen Hosp, 67-74; prof psychiat & behav sci, Sch Med & Health Sci, George Washington Univ, 77-79; res prof neurol & dir psychiat res, Childrens Hosp Nat Med Ctr, 79-89; PROF, DEPT PSYCHOL, DUKE UNIV, 89- *Mem:* Am Asn Rhodes Scholars; Am Psychol Asn; AAAS; NY Acad Sci; Soc Psychophysiol Res; Am Col Neuropsychopharmacol. *Res:* Rating scales for psychopharmalogic research and screening; psychostimulants in children; food additive treatments in children; nutritional and hormonal deficiencies in hyperactive children. *Mailing Add:* Dept Psychol Duke Univ Med Ctr Duke Univ Box 3320 Durham NC 27710

CONNERS, GARY HAMILTON, b Rochester, NY, Feb 15, 36; m 59; c 4. APPLIED MECHANICS. *Educ:* St Lawrence Univ, BS, 57; Mich State Univ, PhD(appl mech), 63. *Prof Exp:* Physicist, Delco Appliance Div Gen Motors Corp, 57-59, sr physicist, 62-64; instr appl mech, Mich State Univ, 61-62; assoc lectr mech eng, Univ Rochester, 63-64, asst prof mech & aerospace sci, 64-67; res supvr, Apparatus Div, 67-70, supvr appl math res, Hawk-Eye Works, 70-75, prog mgr res & eng, 75-79, prog dir strategic planning, Kodak Off, 79-81, DIR RES & ENG, KODAK APPARATUS DIV, EASTMAN KODAK CO, 81- *Concurrent Pos:* Assoc prof mech & aerospace sci, Univ Rochester; chmn, Appl Math, Univ Col, Univ Rochester, 78-81. *Mem:* Am Phys Soc; Am Soc Mech Eng; Optical Soc Am; Sigma Xi. *Res:* Solid mechanics including elasticity, plasticity and thermal mechanics; optical manufacturing; optical systems design; optical and mechanical properties of continuous media; liquid crystals; solid mechanics and mechanics of structures. *Mailing Add:* Five Old Acre Lane Rochester NY 14618

CONNETT, WILLIAM C, b Mexico City, Mex, Feb 22, 39; m 69. MATHEMATICAL ANALYSIS. *Educ:* Georgetown Univ, BS, 61; Univ Chicago, MS, 63, PhD(math), 69. *Prof Exp:* Asst prof, 69-77, ASSOC PROF MATH, UNIV MO, ST LOUIS, 77- *Mem:* Am Math Soc. *Res:* Multiple Fourier series; multiplier theory; singular integrals. *Mailing Add:* Dept Math Univ Mo 8001 Natural Bridge St Louis MO 63121

CONNEY, ALLAN HOWARD, b Chicago, Ill, Mar 23, 30; m 54; c 2. BIOCHEMISTRY, PHARMACOLOGY. *Educ:* Univ Wis, BS, 52, MS, 54, PhD(oncol), 56. *Prof Exp:* Res asst, Univ Wis, 52-56; pharmacologist, Nat Heart Inst, 57-60; head biochem pharmacol sect, Burroughs Wellcome & Co, 60-70; dir, dept biochem & drug metab, Hoffmann-La Roche Inc, 71-83, assoc dir, Exp Therapeut, 79-83, dir, Dept Exp Carinogenesis & metab, 83-85, head lab Exp Carcinogenesis & Metab, Roche Inst Molecular Biol, 85-87; PROF PHARMACOL & CHMN, DEPT CHEM BIOL & PHARMACOL, COL PHARM, RUTGERS STATE UNIV NJ, 87- *Concurrent Pos:* Ed bd, Clin Pharmacol & Therapeut, 69-; adj prof, Rockefeller Univ, 73-; assoc ed, Cancer Res, 78-86, 87-; adv bd Alcohol Res Ctr, Bronx VA Med Ctr & Mt Sinai Sch Med, 78-; exec comt, Div Drug Metab, Am Soc Pharmacol & Exp Therapeut, 87-; comt, Pub Affairs, Am Soc Pharmacol & Exp Therapeut, 88-; lectr, George Washington Univ, 88- *Honors & Awards:* E M Pepper Award Lectr, Columbia Uiv, 85-; First Bernard B Brodie Lectr, Pa State Univ, 85; Paul K Smith Lectr, George Washington Univ, 88-; Oustanding Investr Award, Nat Cancer Inst, 90. *Mem:* Nat Acad Sci; Am Soc Pharmacol & Exp Therapeut; Am Soc Biol Chem; Am Asn Cancer Res; Am Soc Toxicol; Acad Pharmaceut Sci; fel AAAS; fel NY Acad Sci. *Res:* Induced enzyme synthesis in mammals; metabolism of drugs, carcinogens, and steroid hormones; mechanism of drug action; ascorbic acid biosynthesis; carcinogenesis; published numerous articles in various journals. *Mailing Add:* Dept Chem Biol & Pharmacog Rutgers Univ Col Pharm Piscataway NJ 08855-0789

CONNICK, ROBERT ELWELL, b Eureka, Calif, July 29, 17; m 52; c 6. MECHANISMS OF INORGANIC REACTIONS. *Educ:* Univ Calif, BS, 39, PhD(phys chem), 42. *Prof Exp:* Asst chem, Univ Calif, Berkeley, 39-42, instr, 42-43, researcher, Manhattan Proj, 43-46, from asst prof to prof chem, 45-88, chmn dept, 58-60, dean, Col Chem, 60-65 & 87-88, vchancellor acad affairs, 65-67, vchancellor, 69-71, EMER PROF CHEM, UNIV CALIF, BERKELEY, 88- *Concurrent Pos:* Guggenheim fels, 49 & 59. *Mem:* Nat Acad Sci; Am Chem Soc; Sigma Xi; AAAS. *Res:* Radio chemistry; mechanisms of reactions; complex ions; aqueous solution chemistry of chromium and ruthenium; nuclear magnetic resonance studies of inorganic systems; sulfur chemistry. *Mailing Add:* Dept Chem Univ Calif Berkeley CA 94720

CONNICK, WILLIAM JOSEPH, JR, b New Orleans, La, Oct 16, 39; m 62; c 3. FORMULATION CHEMISTRY. *Educ:* Loyola Univ, New Orleans, BS, 61; Tulane Univ, MS, 63. *Prof Exp:* Chemist, Southern Regional Res Ctr, USDA, 63-64 & Chem Res & Develop Lab, US Army, 64-66; RES CHEMIST, SOUTHERN REGIONAL RES CTR, USDA, 66- *Mem:* Am Chem Soc; Weed Sci Soc Am; Controlled Release Soc. *Res:* Flame warfare formulation; textile finishing including oil and water repellants and flameproofing; weed science including chemical and biological herbicide formulation, allelopathy and seed germination stimulants. *Mailing Add:* Southern Regional Res Ctr Agr Res Serv USDA PO Box 19687 New Orleans LA 70179

CONNOLLY, DENIS JOSEPH, b Detroit, Mich, Nov 8, 38; m 87; c 4. APPLIED PHYSICS. *Educ:* Univ Detroit, BEE, 61; Case Western Reserve Univ, MSEE, 66, PhD(elec eng & appl physics), 71. *Prof Exp:* res engr, 61-79, sect head, 79-85, BR CHIEF & DEP DIV CHIEF, RES & DEVELOP, NASA LEWIS RES CTR, 85- *Mem:* Inst Elec & Electronics Engrs; Am Inst Aeronaut & Astronaut. *Res:* Solid State Electronics; microwave electronics. *Mailing Add:* NASA Lewis Res Ctr MS 54-1 21000 Brookpark Rd Cleveland OH 44135

CONNOLLY, JAMES D, b Chicago, Ill, Apr 25, 43; m 65; c 2. POLYMER CHEMISTRY. *Educ:* Roosevelt Univ, BS, 77. *Prof Exp:* SR CONSULT & DIV MGR, EHA DIV, WISS, JANNEY, ELSTNER ASSOCS, 84- *Mem:* Am Chem Soc; Am Soc Testing Mat. *Mailing Add:* 1404 Indigo Dr Mt Prospect IL 60056-1708

CONNOLLY, JAMES L, b Lynn, Mass, Apr 8, 48. PATHOLOGY. *Educ:* Vanderbilt Univ, MD, 74. *Prof Exp:* ASSOC PATHOLOGIST, BETH ISRAEL HOSP, 74-; ASST PROF PATH, HARVARD MED SCH, 82- *Mailing Add:* Dept Path Harvard Med Sch Beth Israel Hosp 330 Brookline Ave Boston MA 02215

CONNOLLY, JOHN CHARLES, b Jackson Heights, NY, June 19, 53; m 77; c 2. MATERIALS SCIENCE ENGINEERING. *Educ:* Rutgers Univ, NJ, BS, 77; Polytech Inst NY, MS, 83. *Prof Exp:* Res assoc, Optical Info Systs, Exxon Enterprises, 77-79; MEM TECH STAFF, DAVID SARNOFF RES CTR, 79- *Honors & Awards:* Cert Recognition, Nat Aeronaut & Space Admin, 84, 89, 90. *Mem:* Inst Elec & Electronics Engrs; Am Asn Crystal Growth. *Res:* Metalorganic chemical vapor deposition and liquid phase epitaxial growth techniques for multiple-layer heterojunction optoelectronic structures in the III-IV semiconductor material system; granted five patents. *Mailing Add:* David Sarnoff Res Ctr CN-5300 Princeton NJ 08543

CONNOLLY, JOHN E, b Omaha, Nebr, May 21, 23. SURGERY. *Educ:* Harvard Univ, AB, 45, MD, 48. *Prof Exp:* Giannini Found fel, Stanford Univ, 57, Markle scholar, 57-62, from instr to assoc prof surg, 57-65; PROF SURG & CHMN DEPT, UNIV CALIF, IRVINE, 65- , PROF THORACIC SURG, COL MED. *Concurrent Pos:* Chief consult, Vet Admin Hosp, Long Beach, 65-; chief surg, Med Ctr, Univ Calif, Irvine; staff mem, St Joseph's Hosp & Children's Hosp, Orange. *Mem:* Fel Am Col Surgeons; Am Surg Asn; Soc Univ Surgeons; Am Asn Thoracic Surg; Int Cardiovasc Soc. *Res:* Cardiovascular surgery. *Mailing Add:* Dept Surg Univ Calif Irvine CA 92717

CONNOLLY, JOHN FRANCIS, b Teaneck, NJ, Jan 22, 36; m 63; c 6. ORTHOPEDICS. *Educ:* St Peter's Col, NJ, BA, 57; NJ Col Med, MD, 61. *Prof Exp:* Asst prof orthop surg, Vanderbilt Univ, 68-73, asst prof biomed eng, 72-73; PROF ORTHOP SURG & REHAB, MED CTR, UNIV NEBR, OMAHA, 74- *Concurrent Pos:* Dir amputee clin & dir cerebral palsy clin, Med Ctr, Vanderbilt Univ, 68-73; chief orthop surg, Nashville Vet Admin Hosp, 68-73, Vet Admin Fund grant biomech fractures, 69-73, wear studies in arthritis, 76 & epiphyseal fractures, 77; chief orthop surg, Omaha Vet Admin Hosp, 74-; Vet Admin Fund grant elec osteogenesis, 80- *Mem:* Orthop Res Soc. *Res:* Pathophysiology and biomechanics of trauma and fracture healing, including nonunions and epiphyseal fractures; electrical stimulation of nonunions. *Mailing Add:* Dept Orthop Univ Nebr Med Ctr 42nd & Dewey Omaha NE 68105

CONNOLLY, JOHN FRANCIS, b Boston, Mass, Aug 15, 48; c 1. PSYCHOPHYSIOLOGY, EXPERIMENTAL PSYCHOPATHOLOGY. *Educ:* Holy Cross Col, AB, 70; Univ Sask, MA, 73; Univ London, PhD (psychol), 77. *Prof Exp:* Sr res fel psychol, Charing Cross Hosp Med Sch, Univ London, 76-82; sr res assoc psychol, Univ BC, 83-84; asst prof, 85-88, ASSOC PROF PSYCHOL & PSYCHIAT, DALHOUSIE UNIV, 88- *Concurrent Pos:* Consult ed, Psychophysiol, Brit J Psychiat, J Psychiat Res, Biol Psychol, 83-; referee, Med Res Coun, Can & UK, 83-; spec consult, NIMH, Bethesda, Md; BC Health Care Res Found; mem bd dirs, Int J Psychophysiol & Int Orgn Psychophysiol. *Mem:* Soc Neurosci; Soc Psychophysiol Res; Int Orgn Psychophysiol; Brit Psychophysiol Soc (secy, 80-83); Can Psychol Asn; NY Acad Sci. *Res:* Determining the role of input factors on performance of complex auditory tasks with particular reference to brain stem function; investigating the relationship between neurophysiological and neuroanatomical abnormalities in schizophrenia. *Mailing Add:* Dept Psychol Life Sci Ctr Dalhousie Univ Halifax NS B3H 4J1 Can

CONNOLLY, JOHN IRVING, JR, b Boston, Mass, June 23, 36. LOW TEMPERATURE PHYSICS, OPTICS. *Educ:* Mass Inst Technol, BS, 58; Univ Ill, Urbana, MS, 59, PhD(physics), 65; Northeastern Univ, MS, 71. *Prof Exp:* Fulbright lectr physics, Coun Sci Invests, 65-66; tech staff mem, Mitre Corp, 66-72; mem staff, 72-76, VPRES, SCI APPLNS, INC, 76- *Mem:* Am Phys Soc; Soc Indust & Appl Math; Inst Elec & Electronics Engrs; Sigma Xi. *Res:* Hypervelocity gems and power supplies. *Mailing Add:* 9906 Manet Rd Burke VA 22015

CONNOLLY, JOHN STEPHEN, b Butte, Mont, Nov 23, 36; m 63; c 3. PHYSICAL CHEMISTRY, PHOTOCHEMISTRY. *Educ:* Carroll Col, Mont, AB, 58; Univ Minn, MS, 60; Brandeis Univ, PhD(phys chem), 69. *Prof Exp:* Nuclear engr, Naval Reactors, USAEC, 60-61; res, US Naval Radiol Defense Lab, 61-63; fel, C F Kettering Res Lab, 68-69, staff scientist, 69-72; vis res chemist chem lasers, Aerospace Res Labs, Wright-Patterson AFB, 72-73; sr NAS/NRC res assoc photochem, US Army Natick Labs, Mass, 73-74; res dir laser hazards, Life Sci Div, Technol Inc, San Antonio, 74-76; consult comput & software, San Antonio, 76-77; sr scientist, Biochem Conversion Br, 77-79, sr scientist & task leader, 79-83, PRIN SCIENTIST & GROUP LEADER PHOTOCHEM, PHOTOCONVERSION RES BR, SOLAR ENERGY RES INST, GOLDEN, COLO, 83- *Concurrent Pos:* Lectr chem, Wright State Univ, 71; consult, US Army Natick Labs, 74-77; adj assoc prof bioeng, Univ Tex Health Sci Ctr, San Antonio, 75-78; chmn, Third Int Conf Photochem Conversion & Storage of Solar Energy & Int Orgn Comt, 78-80; vis fel, Dept Chem, Univ Colo, 85-; res prof, Dept Chem, Univ Denver, 89-; NIH postdoctoral fel, 64-66; Fulbright scholar, Fed Repub Ger, 89. *Mem:* AAAS; Am Chem Soc; Am Soc Photobiol; Inter-Am Photochem Soc; Europ Photochem Asn; Sigma Xi. *Res:* Applications of photochemistry to solar energy conversion and storage; photochemistry and photophysics of large molecules in condensed phase; interaction of laser radiation with molecules of biological interest; computerized data acquisition and analysis of kinetic and spectroscopic data. *Mailing Add:* Solar Energy Res Inst Photoconversion Res Br 1617 Cole Blvd Golden CO 80401

CONNOLLY, JOHN W, b Cincinnati, Ohio, Apr 4, 36; m 60; c 2. INORGANIC CHEMISTRY, BIOLOGICAL CHEMISTRY. *Educ:* Xavier Univ, Ohio, BS, 58; Purdue Univ, PhD(chem), 63. *Prof Exp:* Asst chem, Yale Univ, 62-64; asst prof, Marietta Col, 64-65; from asst prof to assoc prof, 65-77, PROF CHEM, UNIV MO, KANSAS CITY, 77- *Mem:* AAAS; Am Chem Soc; Sigma Xi. *Res:* Organometallic chemistry and enzyme model chemistry. *Mailing Add:* Dept Chem Univ Mo Kansas City MO 64110

CONNOLLY, JOHN WILLIAM DOMVILLE, b South Porcupine, Ont, July 18, 38; m 62, 77; c 3. THEORETICAL PHYSICS, SOLID STATE PHYSICS. *Educ:* Univ Toronto, BA, 60; Univ Fla, PhD(physics), 66. *Prof Exp:* Sr res assoc, Pratt & Whitney Aircraft Div, United Aircraft Corp, Conn, 67-70; assoc prof physics & mem staff, Quantum Theory Proj, Univ Fla, 70-76, prof, 76-78; assoc prog dir, Solid State Physics, NSF, 76-78, prog dir, Condensed Matter Theory, 78-84, dir, Off Adv Sci Comput, 84-87; DIR, CTR COMPUTATIONAL SCI, UNIV KY, LEXINGTON, 87- *Concurrent Pos:* Res affil, Mass Inst Technol, 69; res scientist, IBM, 82-83. *Mem:* AAAS; fel Am Phys Soc. *Res:* Electronic structure of solids and molecules. *Mailing Add:* Ctr Comput Sci Univ Ky 325 McVey Lexington KY 40506-0045

CONNOLLY, JOSEPH ANTHONY, muscle cell differentiation, membrane cytoskeleton interaction; deceased, see previous edition for last biography

CONNOLLY, KEVIN MICHAEL, b Baltimore, Md, Jan 10, 52. RHEUMATIC DISEASES. *Educ:* Va Polytech Inst & State Univ, PhD(microbiol), 79. *Prof Exp:* SR SCIENTIST, STERLING-WINTHROP RES INST, 82- *Mem:* Am Asn Immunologists; AAAS; Inflammation Res Asn; Sigma Xi. *Mailing Add:* Glaxo Inc Dept Chemother Five Moore Dr Research Triangle Park NC 27709

CONNOLLY, WALTER CURTIS, b Marysville, Ohio, May 1, 22; m 44; c 2. PHYSICS. *Educ:* Miami Univ, AB, 44; Univ Ill, MS, 46; Cath Univ Am, PhD, 54. *Prof Exp:* Asst physics, Univ Ill, 44-46; assoc prof, US Naval Acad, 46-55; sr scientist, Westinghouse Atomic Power, 55-56; assoc prof physics, Ala Polytech Inst, 56-58; sr physicist, Res Lab Eng Sci, Univ Va, 58-63; PROF PHYSICS, APPALACHIAN STATE UNIV, 63- *Mem:* Am Phys Soc; Am Asn Physics Teachers. *Res:* Energy sources; physics demonstration apparatus. *Mailing Add:* Dept Physics & Astron Appalachian State Univ Boone NC 28608

CONNOR, DANIEL HENRY, b Aylmer, Ont, Mar 26, 28; US citizen; m 53; c 3. PATHOLOGY. *Educ:* Queen's Univ, MD, CM, 53. *Prof Exp:* From intern med to jr resident path, Emergency Hosp, Washington, DC, 53-55; from resident to chief resident path, Med Sch Hosp, George Washington Univ, 55-57; chief lab serv, Irwin Army Hosp, Ft Riley, Kans, 57-59; assoc pathologist, Liver & Pediat Br, Armed Forces Inst Path, 59-60, assoc pathologist, Skin & Gastrointestinal Br, 60-61; prin investr study path of endomyocardial fibrosis, WHO, 62-64; assoc pathologist, infectious dis br, Armed Forces Inst Path, 64-67, chief infectious dis br, 67-79, chmn dept infectious & parasitic dis, 70-87; VIS PROF DEPT PATH, SCH MED, GEORGETOWN UNIV, 87- *Concurrent Pos:* Dir path, US Med Res & Develop Proj, Kampala, Uganda, 62-64; hon lectr path, Makerere Col Med Sch, Kampala, 62-64; adv African cardiopathies, WHO, 64, mem sci working group epidemiol, 76-81, mem expert adv panel parasitic dis, 83-, prin investr, Collaborating Ctr, Histopath Filarial Infections in Man, 77-86; assoc mem, Comn Parasitic Dis, Armed Forces Epidemiol Bd, 69; consult med parasitol, Am Bd Path, 72, mem test comt med microbiol-med parasitol, 74-81; consult onchocerciasis, WHO, 72, mem sci adv panel, Onchocerciasis Control Prog, Volta River Basin, 74-79, consult, Spec Comt Parasitic Dis, 81-86; team leader, Onchocerciasis Res Surv, Paul Carlson Med Prog, 77; clin prof uniformed serv, Univ Health Sci, 78; vis prof, Pro Tem, Cleveland Clin, 81; mem, US Nat Comt, Int Coun Socs Path, Nat Res Coun-Assembly Life Sci, 82-87, chmn, 84-87; chmn, subcomt African Path, US Nat Comt, ICSP, Nat Res Coun-Assembly Life Sci, 82-; mem bd dirs, Nat Found Infectious Dis, Washington, DC Area Chap, 83-85; vis prof, Mahidol Univ Bangkok, 79 & Univ Natal, Durban, 85; vis prof & external examr, Dept Parasitol, Univ Malaysia, Kuala Lumpur, 90. *Honors & Awards:* Decoration Meritorious Civilian Serv, Dept Army, 71. *Mem:* Int Acad Path; Am Asn Pathologists & Bacteriologists; Am Soc Exp Path; Col Am Pathologists; Am Soc Trop Med & Hyg; Ugandan Med Asn; AAAS; Int Soc Trop Dermat; Am Soc Dermatopath; Nat Found Infectious Dis. *Res:* The pathology and pathogenesis of tropical and exotic infectious diseases, especially those of tropical Africa, including onchocerciasis, Mycobacterium ulcerans infection, endomyocardial fibrosis, streptocerciasis and others. *Mailing Add:* 7209 Maple Ave Chevy Chase MD 20815

CONNOR, DANIEL S, b Cleveland, Ohio, Feb 25, 38. SYNTHETIC ORGANIC CHEMISTRY, APPLIED CHEMISTRY. *Educ:* Brown Univ, ScB, 60; Yale Univ, PhD(chem), 65. *Prof Exp:* RES CHEMIST, MIAMI VALLEY LABS RES DIV, PROCTER & GAMBLE CO, 65- *Mem:* Am Chem Soc; Sigma Xi. *Res:* Process chemistry. *Mailing Add:* PO Box 326 Ross OH 45061

CONNOR, DAVID THOMAS, b Batley, Eng, Nov 6, 39; m 67; c 2. ORGANIC CHEMISTRY. *Educ:* Univ Manchester, BSc, 62, MSc, 64, PhD(org chem), 65. *Prof Exp:* Res assoc org chem, Univ Chicago, 65-68; scientist, 69-74, sr scientist, 74-78, res assoc, 78-82, DIR, ALLERGY & INFLAMMATION CHEM, WARNER-LAMBERT/PARKE DAVIS PHARMACEUT RES DIV, 83-. *Mem:* Am Chem Soc. *Res:* Organic reaction mechanisms; synthetic methods for heterocyclic chemistry; natural product chemistry; antimicrobial agents; antiallergy, antisecretory and anti-inflammatory agents; arachidonic acid chemistry and biology; cytokines. *Mailing Add:* Warner-Lambert/Parke Davis 2800 Plymouth Rd Ann Arbor MI 48106

CONNOR, DONALD W, b Chicago, Ill, Jan 2, 23; m 43; c 4. SOLID STATE PHYSICS. *Educ:* Univ Chicago, SB, 43, SM, 48, PhD, 60. *Prof Exp:* Jr physicist, Argonne Nat Lab, 46-49; elec engr, Univ Chicago, 49-50; assoc physicist, Brookhaven Nat Lab, 51; scientist, Chicago Midway Labs, 51-52; assoc physicist, Argonne Nat Lab, 52-69, physicist, Solid State Div, 69-74, sr physicist, environ statement proj, 74-88. *Mem:* AAAS; Inst Elec & Electronics Eng; Am Phys Soc; Am Nuclear Soc; Sigma Xi. *Res:* Nuclear moments; lattice vibrations; neutron scattering. *Mailing Add:* 138 Warwick St Park Forest IL 60466

CONNOR, FRANK FIELD, b Chicago, Ill, June 15, 32; m 58; c 2. MATHEMATICS. *Educ:* Ill Inst Technol, BS, 54, MS, 56, PhD(math), 59. *Prof Exp:* Instr math, Ill Inst Technol, 58-60; res instr, La State Univ, 60-61, from asst prof to assoc prof, 61-70; chmn dept, 70-75, PROF MATH, N TEX STATE UNIV, 70- *Mem:* Am Math Soc; Math Asn Am. *Res:* Measure and integration; linear operators. *Mailing Add:* 1813 Stonegate Dr NTex State Univ Denton TX 76205

CONNOR, JAMES D, b Colleton, SC, Nov 19, 26; m; c 3. PEDIATRICS, MICROBIOLOGY. *Educ:* Clemson Col, BS, 53; Med Col SC, MD, 53; Am Bd Pediat, dipl, 59. *Prof Exp:* Intern, Wayne County Gen Hosp, Eloise, Mich, 53-54; resident pediat, Children's Hosp, Mich, 54-55; instr, Sch Med, Wayne State Univ, 55-57; instr, Sch Med, Univ Miami, 57-59, clin asst prof, 59-60, from asst prof to assoc prof, 61-69; assoc prof, 70-74, PROF PEDIAT, UNIV CALIF, SAN DIEGO, 74- *Concurrent Pos:* Chief resident pediat, Children's Hosp, Mich, 55-57; med coordr, Variety Children's Hosp, Miami, 58-60, res assoc virol, Variety Children's Res Found, 60-; USPHS fel virol, 60-62; res asst prof microbiol, Univ Miami, 61-70. *Mem:* AAAS; AMA; Am Acad Pediat; Am Pub Health Asn; Infectious Dis Soc Am. *Res:* Virology and pharmacology of antivirals, therapeutics and bacterial infection; pertussis. *Mailing Add:* Dept Pediat 0609-E Univ Calif San Diego 9500 Gilman Dr La Jolla CA 92093-0609

CONNOR, JAMES EDWARD, JR, b New Haven, Conn, Feb 14, 24; m 51; c 6. PHYSICAL ORGANIC CHEMISTRY. *Educ:* Harvard Univ, BA, 44, MS, 48, PhD(phys org chem), 49. *Prof Exp:* Sr res chemist, Atlantic Richfield Co, Glenolden, 49-54, supvr chemist, 54-60, asst mgr basic res div, 60-61, mgr chem res, 61-62, mgr res div, 62-71, mgr res & develop div, 62-71, mgr res & develop div, Res & Eng Dept, ARCO Chem Co Div, Glenolden, 71-79, vpres chem res & develop, 79-81, vpres res, 81-85, vpres res & develop, Arco Chem Co, Atlantic Richfield Co, Newton Square, Pa, 85-86; RETIRED. *Mem:* Am Chem Soc; NY Acad Sci; Soc Chem Indust. *Res:* Petroleum refining methods, especially hydrocarbon reactions and catalytic reactions; petrochemicals; chemical products. *Mailing Add:* 1421 Hillside Rd Wynnewood PA 19096

CONNOR, JOHN ARTHUR, b Kansas City, Mo, June 18, 40; c 3. NEURONAL PLASTICITY, MEMBRANE CONDUCTANCES. *Educ:* Univ Mo, Columbia, BS, 63; Northwestern Univ, MS, 64, PhD(elec eng), 67. *Prof Exp:* Postdoctoral physiol, Univ Wash, 67-69; from asst prof to prof physiol & biophys, Univ Ill, Urbana, 69-83; mem tech staff, Bell Labs, Murray Hill, 81-86, distinguished mem tech staff, 86-89; HEAD, DEPT NEUROSCI, ROCHE INST MOLECULAR BIOL, 89- *Honors & Awards:* Golgi Prize, Fidia Found, 91. *Mem:* Soc Neurosci; Biophys Soc. *Res:* Physiological basis of electrical pacemaking activity in neurons; control of developmental processes and excitability in neurons by second messengers such as calcium and cyclic nucleotides; high resolution fluorometric measurements of calcium in neurons of the central nervous system. *Mailing Add:* Rochester Inst Molecular Biol Hoffman-LaRoche 340 Kingsland Nutley NJ 07110

CONNOR, JOHN D, b Coatesville, Pa, Jan 15, 33; m 63; c 2. PHARMACOLOGY, NEUROSCIENCES. *Educ:* Philadelphia Col Pharm, BS, 60, MS, 62, PhD(pharmacol), 66. *Prof Exp:* Lab asst zool, Philadelphia Col Pharm, 60-62, lab asst pharmacol, 62-63; res assoc neuropharmacol, East Pa Psychiat Inst, 63-66; from asst prof to assoc prof, 69-75, PROF PHARMACOL, HERSHEY MED CTR, PA STATE UNIV, 75- *Concurrent Pos:* NIMH lab fel, 66-69; Borden Found Award; Fogarty fel, NIH, 78-79, Fulbright fel, 88-89. *Mem:* Am Soc Pharmacol & Exp Therapeut; Soc Neurosci; Sigma Xi. *Res:* Pharmacology and physiology of involuntary motor activity; temperature regulation; synaptic transmission in the central nervous system. *Mailing Add:* Dept Pharmacol Hershey Med Ctr Pa State Univ Hershey PA 17033-0950

CONNOR, JOHN MICHAEL, b Los Angeles, Calif, June 24, 38; m 58; c 2. ANIMAL SCIENCE & NUTRITION, RANGE SCIENCE & MANAGEMENT. *Educ:* Univ Nev, Reno, BS, 60, MS, 62. *Prof Exp:* Cattle rancher, Connor Ranch, 64-82; supt, Newlands Field Lab, Univ Nev, 78-83; SUPT, SIERRA FOOTHILL RANGE FIELD STA, UNIV CALIF, 83- *Mem:* Am Soc Animal Sci; Soc Range Mgt. *Res:* Range beef cattle management; range & pasture management. *Mailing Add:* Sierra Foothill Range Field Sta PO Box 28 Browns Valley CA 95918

CONNOR, JON JAMES, b Columbus, Ohio, Dec 12, 32; m 63. GEOLOGY. *Educ:* Ohio State Univ, BSc, 55; Univ Colo, PhD(geol), 63. *Prof Exp:* Geologist, US Geol Surv, 55-72, chief, Br Regional Geochem, 72-78, geologist, 78-91; RETIRED. *Concurrent Pos:* Liaison mem, US Nat Comt Geochem, Nat Acad Sci-Nat Res Coun, 72-75. *Honors & Awards:* Meritorious Serv Medal, Dept Interior. *Mem:* Soc Environ Geochem & Health; Geol Soc Am; Int Asn Math Geol; Soc Econ Geologists. *Res:* Geology of Colorado Plateau and Black Hills uranium deposits; groundwater investigations around Carlsbad, New Mexico; regional geochemistry of sedimentary rocks; environmental geochemistry and its relation to human and animal health; geochemistry of ore deposits in Beltage Rocks, Montana and Idaho. *Mailing Add:* 9780 W 36th Ave Wheat Ridge CO 80033

CONNOR, JOSEPH GERARD, JR, b West Chester, Pa, Aug 15, 36; m 61; c 3. ENGINEERING PHYSICS, SCIENCE EDUCATION. *Educ:* Georgetown Univ, BS, 57; Pa State Univ, MS, 61, PhD(physics), 63. *Prof Exp:* RES PHYSICIST, NAVAL SURFACE WARFARE CTR, WHITE OAK, 63- *Concurrent Pos:* Lectr, Montgomery Jr Col, 64-65, Trinity Col, DC, 65-72 & Montgomery Col, 81-86; adj prof, Montgomery Col, 86- *Mem:* AAAS; Am Asn Physics Teachers; Am Phys Soc. *Res:* Structural response; explosion effects in air and underwater. *Mailing Add:* 17805 Dominion Dr Sandy Spring MD 20860

CONNOR, LAWRENCE JOHN, b Kalamazoo, Mich, Aug 15, 45; m 68; c 2. ENTOMOLOGY, APICULTURE. *Educ:* Mich State Univ, BS, 67, MS, 69, PhD(entom), 72. *Prof Exp:* Asst prof entom, Ohio State Univ, 72-76; asst prof entom, Ohio Agr Res & Develop Ctr, 75-76; pres, Genetic Systs, Inc, 76-80; CONSULT & OWNER, BEEKEEPING EDUC SERV, 80-; OWNER, WICWAS PRESS, 88- *Concurrent Pos:* Exten entomologist, Ohio Coop Exten Serv, 72-76. *Mem:* Sigma Xi; Bee Res Asn; Assoc Registered Prof Entomologists. *Res:* Honey bee breeding and genetics; crop pollination requirements and mechanisms. *Mailing Add:* PO Box 817 Cheshire CT 06410-0817

CONNOR, RALPH (ALEXANDER), b Newton, Ill, July 12, 07; m 31; c 1. CHEMISTRY. *Educ:* Univ Ill, BS, 29; Univ Wis, PhD(org chem), 32. *Hon Degrees:* DSc, Phila Col Pharm, 54, Univ Pa, 59, Polytech Inst Brooklyn, 67; LLD, Lehigh Univ, 66. *Prof Exp:* Asst chem, Univ Wis, 29-31; instr org chem, Cornell Univ, 32-35; from asst prof to assoc prof, Univ Pa, 35-44; assoc dir res, Rohm and Haas Co & Resinous Prod & Chem Co, 45-48, vpres res, 48-70, dir & mem exect comt, 49-73, chmn bd, 60-70, vpres & chmn exec comt, 70-73; RETIRED. *Concurrent Pos:* Res chemist, du Pont Co, 34; tech aide, sect chief & div chief, Nat Defense Res Comn, 41-46; mem, div chem & chem tech, Nat Res Coun, 53-58; mem, Tech Adv Panel Biol & Chem Warfare, 54-60; chmn, US Nat Comt on Int Union Pure & Appl Chem; bd dirs, Am Chem Soc, 54-66, chmn, 56-58, mem various comts; bd dirs, Ursinus Col, 71- *Honors & Awards:* Naval Ord Develop Award; Gold Medal, Am Inst Chemists, 63, Chem Pioneer Award, 68; Chem Indust Medal, Soc Chem Indust, 65; Priestley Medal, Am Chem Soc, 67. *Mem:* Am Chem Soc; Am Inst Chemists. *Res:* Organic chemistry; catalysis; synthesis; explosives; mechanisms. *Mailing Add:* Royal Oaks GH No 1007 10035 Royal Oak Rd Sun City AZ 85351-3162

CONNOR, ROBERT DICKSON, b Edinburgh, Scotland, May 15, 22; m 48, 84; c 2. NUCLEAR PHYSICS. *Educ:* Univ Edinburgh, BSc, 42, PhD, 49. *Prof Exp:* From asst lectr to lectr physics, Univ Edinburgh, 48-57; from assoc prof to prof physics, Univ Man, 57-79, assoc dean arts & sci, 63-70, dean sci, 70-79, prof fac educ, 79-85, PROF PHYSICS, UNIV MAN, 85-, EMER DEAN SCI, 85-, SR SCHOLAR PHYSICS, 88- *Concurrent Pos:* Sci mem Fisheries Res Bd Can, 66-76; mem univ grants comn, Prov Man, 68-78. *Mem:* Sigma Xi. *Res:* Alpha, beta and gamma ray spectroscopy; history and philosophy of science; history of weights and measures. *Mailing Add:* Dept Physics Univ Man Winnipeg MB R3T 2N2 Can

CONNOR, SIDNEY G, b Joplin, Mo, Oct 24, 47; m 69; c 2. ELECTRONICS, INDUSTRIAL TECHNOLOGY. *Educ:* Wichita State Univ, BA, 73, MEd, 76; Kans State Univ, PhD(adult & occup educ), 86. *Prof Exp:* Lectr power & energy, 78-81, instr, 82-87, ASST PROF TECHNOL & CHAIR DEPT, WICHITA STATE UNIV, 88- *Concurrent Pos:* Region V chmn, Soc Mfg Engrs, 90-91; region IV dir, Nat Asn Indust Technol, 90- *Honors & Awards:* Int Award of Merit, Soc Mfg Engrs, 89. *Mem:* Soc Mfg Engrs; Nat Asn Indust Technol; Asn Supv & Curric Develop. *Res:* Technology education curricula for middle school and high school science and technology; principles of technology. *Mailing Add:* 8630 Longlake Wichita KS 67207

CONNOR, THOMAS BYRNE, b Baltimore, Md, Dec, 21, 21; m 57; c 2. MEDICINE, ENDOCRINOLOGY & METABOLISM. *Educ:* Loyola Col, Md, BA, 43; Univ Md, MD, 46. *Prof Exp:* Intern, Mercy Hosp, Baltimore, 46-47, resident, 49-51; from asst prof to assoc prof, 56-67, dir Clin Res Ctr, 61-80, PROF MED, SCH MED, UNIV MD, BALTIMORE, 67-, DIR DIV ENDOCRINOL & METAB, 56- *Concurrent Pos:* Fel endocrine & metab dis, Johns Hopkins Hosp, 51-56; asst physician outpatient dept, Diabetic Clin & Endocrine Clin, Johns Hopkins Hosp, 51-59; staff physician, Univ Md Hosp, 56-; consult med, Mercy Hosp, Baltimore, 60- & Baltimore Vet Admin Hosp, 65-; trustee, Endowment Bd, Good Samaritan Hosp, 79-; mem, bd trustees, Stella Mavis Hospice Operating Corp, 82- *Mem:* AAAS; Endocrine Soc; Am Clin & Climat Asn; Am Soc Bone & Mineral Res; fel Am Col Physicians. *Res:* Clinical research in calcium and bone metabolism and parathyroid disorders; endocrinology and metabolism. *Mailing Add:* Dept Med Sch Med Univ Md 655 W Baltimore St Baltimore MD 21201

CONNOR, WILLIAM ELLIOTT, b Pittsburgh, Pa, Sept 14, 21; m 46; c 5. INTERNAL MEDICINE. *Educ:* Univ Iowa, BA, 42, MD, 50; Am Bd Internal Med, dipl, 57; Am Bd Nutrit, dipl, 67. *Prof Exp:* Intern, USPHS Hosp, Calif, 50-51; asst med resident, San Joaquin Gen Hosp, 51-52; mem med staff, Enloe Hosp, 52-54; med resident, Vet Admin Hosp, Iowa, 54-56; from instr to prof internal med, Univ Iowa, 56-75, dir clin res ctr, 68-75; PROF CARDIOL & METAB-NUTRIT, DIR, LIPID-ATHEROSCLEROSIS LAB & ASSOC DIR, CLIN RES CTR, ORE HEALTH SCI UNIV, PORTLAND, 75- *Concurrent Pos:* Am Heart Asn res fel, Univ Iowa, 56-58; Am Col Physicians traveling fel, Oxford Univ, 60; vis prof, Med Cols & Basic Sci Med Inst, Karachi, Pakistan, 61; ed, J Lab & Clin Med, 70-73; mem & chmn, Coun Arteriosclerosis, Am Heart Asn; vis fel clin sci, Australian Nat Univ, Canberra, 70; chmn heart & lung prog, Proj Comn, Nat Heart & Lung Inst, NIH, 74-75; mem gen clin, Res Centers Comn, 76- *Mem:* AAAS; Am Inst Nutrit (pres, 78-79); Am Soc Clin Invest; Asn Am Physicians; Am Fedn Clin Res. *Res:* Atheriosclerosis; lipid metabolism; nutrition. *Mailing Add:* Dept Med L465 Ore Health Sci Univ 3181 SW Sam Jackson Park Rd Portland OR 97201

CONNOR, WILLIAM GORDEN, b El Paso, Tex, Nov 1, 36. MEDICAL PHYSICS, HEALTH PHYSICS. *Educ:* Tex Western Col, BSc, 62; Vanderbilt Univ, MSc, 64; Univ Calif, Los Angeles, PhD(med physics), 70. *Prof Exp:* Physicist, Michael Reese Hosp, Chicago, Ill, 64-66; asst prof radiol, Univ Wis-Madison, 70-72; from asst prof to assoc prof radiol, Univ Ariz, 72-89; RETIRED. *Mem:* AAAS; Am Asn Physicists Med; Health Physics Soc; NY Acad Sci; Am Col Med Physics. *Res:* Application of ionizing radiations to the treatment of malignant diseases; cellular repair of damage due to ionizing radiations; use of hyperthermia as an adjunct to radiation therapy; high linear-energy transfer radiation therapy. *Mailing Add:* 3002 E Loretta Dr Tucson AZ 85716

CONNOR, WILLIAM KEITH, b Houston, Tex, Dec 19, 31; m 59; c 3. ACOUSTICS. *Educ:* Rice Inst, BA, 55; Southern Methodist Univ, MS, 61. *Prof Exp:* Engr trainee, Gen Motors Proving Ground, 55-56; proj engr, Electro-Mech Labs, White Sands Proving Ground, 57-58, Gen Motors Proving Ground, Noise & Vibration Lab, 58-59 & 61-62; consult acoust, Rudmose Assocs Inc, 62-63; consult acoust, Tracor Inc, 63-69, dir acoust res dept, 69-84, prin consult, 84-91; INDEPENDENT CONSULT, ACOUST, 91- *Concurrent Pos:* Sr lectr, Sch Archit, Univ Tex, Austin, 73-80. *Mem:* Inst Noise Control Eng; Sigma Xi; fel Acoust Soc Am; Audio Eng Soc; Soc Automotive Engrs. *Res:* Community noise, psychoacoustics; architectural acoustics; noise and vibration control; electroacoustics. *Mailing Add:* 608 Broohaven Trail Austin TX 78746

CONNORS, ALFRED F, JR, b Brooklyn, NY, May 14, 50; m 78; c 1. CRITICAL CARE, PULMONARY DISEASE. *Educ:* St Louis Univ, BA, 71; Med Col Ohio, MD, 74. *Prof Exp:* Intern, Cleveland Metrop Gen Hosp, 74-75, med resident, 75-77; pulmonary fel, Univ Okla Health Sci Ctr, 78-79, res fel, 79-81; ASST PROF, CASE WESTERN RESERVE UNIV, 81- *Concurrent Pos:* Prin investr prediction hemodynamic status in critical-ill, Am Heart Asn, 86-88, prog on care of criticall-ill adults, Robert Wood Johnson Found, 88-; med dir, Cleveland Metrop Gen Hosp, 81- *Mem:* Am Thoracic Soc; Am Lung Asn; Soc Med Decision Making. *Res:* Clinical prediction roles for assessing cardiac function in the critically ill; investigation of ventilatory control during mechanical ventilation; study of physician decision-making in the critically ill. *Mailing Add:* 14201 Larchmere Blvd Shaker Heights OH 44120

CONNORS, DONALD R, b Saginaw, Mich, Oct 31, 28; m 50; c 4. NUCLEAR PHYSICS, LAW. *Educ:* Univ Notre Dame, BS 52, PhD(nuclear physics), 57; Duquesne Univ, JD, 68. *Prof Exp:* Sr engr exp physics, 56-60, mgr exp physics & anal, 60-65, mgr physics, reactor design & exp, 65-69, mgr reactor opers, 69-74, mgr light water breeder reactor proj design & opers, 74-77, MGR OPERS TRAINING, BETTIS ATOMIC POWER LAB, 77- *Concurrent Pos:* Chmn & mem Adv Safeguards Comt, Bettis Atomic Power Lab, 60- *Honors & Awards:* Order of Merit, Westinghouse Elec Corp. *Mem:* Am Nuclear Soc. *Res:* Nuclear reactor design and development; training of naval nuclear reactor operations personnel; reactor safety and environmental effects, public relations, legal aspects of reactor operation. *Mailing Add:* Bettis Atomic Power Lab PO Box 79 West Mifflin PA 15122

CONNORS, JOHN MICHAEL, ENDOCRINOLOGY. *Educ:* Univ Ill, PhD(physiol), 77. *Prof Exp:* ASST PROF PHYSIOL, SCH MED, WVA UNIV, 77- *Res:* The thyroid and pituitary glands. *Mailing Add:* Dept Physiol WVa Sch Med Morgantown WV 26506

CONNORS, KENNETH A, b Torrington, Conn, Feb 19, 32. PHARMACEUTICS, ANALYTICAL CHEMISTRY. *Educ:* Univ Conn, BS, 54; Univ Wis, MS, 57, PhD(pharm), 59. *Prof Exp:* Res assoc phys-org chem, Ill Inst Technol, 59-60 & Northwestern Univ, 60-62; from asst prof to assoc prof pharmaceut anal, 62-70, asst dean grad studies, 68-72, PROF PHARMACEUT ANALYSIS, UNIV WIS-MADISON, 70-, ACTG DEAN PHARM, 91- *Concurrent Pos:* NIH fel, 60-62. *Honors & Awards:* Justin L Powers Award Res Achievement, 80. *Mem:* Fel AAAS; Am Chem Soc; Am Pharmaceut Asn; Am Asn Pharmaceut Scientists. *Res:* Pharmaceutical analysis; mechanisms of organic reactions; molecular complexes especially of cyclodextrins; solvent effects. *Mailing Add:* Sch Pharm Univ Wis Madison WI 53706

CONNORS, NATALIE ANN, b St Louis, Mo. HISTOLOGY, HISTOCHEMISTRY. *Educ:* St Louis Univ, BS, 50, PhD(anat), 68; Univ Ill, Urbana-Champaign, MS, 52. *Prof Exp:* Technician, Monsanto Chem Co, Ill, 52-56; res assoc anat, 56-61, teaching asst, 61-68, instr, 68-70, asst prof, 70-85, ASSOC PROF ANAT, SCH MED, ST LOUIS UNIV, 85- *Mem:* Sigma Xi; Am Asn Anatomists; AAAS; Soc Neurosci. *Res:* Histology and histochemistry. *Mailing Add:* 620 N 66th St Belleville IL 62223

CONNORS, PHILIP IRVING, b Norfolk, Va, Oct 7, 37; m 59; c 3. ACADEMIC ADMINISTRATION, EXPERIMENTAL NUCLEAR PHYSICS. *Educ:* Univ Notre Dame, BS, 59; Pa State Univ, MS, 62, PhD(physics), 66. *Prof Exp:* Asst physics, Pa State Univ, 59-63; jr res assoc, Brookhaven Nat Lab, 63-65; res assoc, Dept Physics & Astron, Univ Md, 65-69, asst prof physics & dir col sci improv prog, 69-75; assoc prof & chmn, Div Environ & Natural Sci, Northern Va Community Col, Woodbridge Campus, 75-77; prof sci & acad dean, Cent New England Col, 77-79; PROF MARINE ENG, MASS MARITIME ACAD, 79- *Concurrent Pos:* Dir, Chesapeake Physics Asn, 69-75; Field Ctr Coordr, NSF Chautauqua Short Courses, 71-75; vis lectr comput sci & physics, Worcester State Col, 81- & Math, Bridgewater State Col, 82. *Mem:* Nat Sci Teachers Asn; Am Phys Soc; Am Asn Physics Teachers; Fel AAAS; Am Nuclear Soc. *Res:* Education and teacher training in physics on all levels; low energy nuclear experimental physics. *Mailing Add:* Mass Maritime Acad PO Box D Buzzards Bay MA 02532

CONNORS, ROBERT EDWARD, b Wellesley, Mass, Oct 1, 45. PHYSICAL CHEMISTRY, MOLECULAR SPECTROSCOPY. *Educ:* Univ Mass, BS, 67; Northeastern Univ, PhD(chem), 72. *Prof Exp:* Res assoc, Boston Univ, 72-76; asst prof, 76-81, L P Kinnicutt asst prof, 77-80, ASSOC PROF CHEM,

WORCESTER POLYTECH INST, 81- *Mem:* Am Chem Soc; AAAS. *Res:* Electronic spectroscopy and optical detection of magnetic resonance studies of organic and biological systems. *Mailing Add:* 56 Brownell St Worcester MA 01602

CONNORS, WILLIAM MATTHEW, b Canandaigua, NY, Sept 16, 21; m 90. BIOCHEMISTRY, INDUSTRIAL CHEMISTRY. *Educ:* St Bonaventure Col, BS, 42; Univ Southern Calif, MS, 47; James Martin Col, PhD, 71. *Hon Degrees:* LHD, James Martin Col, 69. *Prof Exp:* Chemist, Pillsbury Mills, 45; group leader, Nat Dairy Res Labs, Inc, 49-56; anal supvr, Gen Cigar Res & Develop Ctr, 56-73; MGR, CONNORS RES ASSOCS, 73- *Concurrent Pos:* Mem Manhattan Proj, 44-45. *Honors & Awards:* Meritorious Serv Award, Am Inst Chemists, 72; E F Rebman Plaque Environ Sci, 83. *Mem:* Fel AAAS; Am Chem Soc; fel Am Inst Chemists; NY Acad Sci; Am Soc Microbiol. *Res:* Biochemistry and analytical chemistry of tobacco products; commercial production of enzymes; nutrition of dairy products; energetics of ATPase; pharmacology and toxicology of uranium compounds; industrial engineering and chemical consulting; research and development. *Mailing Add:* PO Box 398 Bausman PA 17504

CONOLLY, JOHN R, b Sydney, Australia, July 23, 36; m 70; c 5. GEOLOGY, NATURAL RESOURCES. *Educ:* Univ Sydney, BSc, 58; Univ New South Wales, MSc, 60, PhD(geol), 63. *Prof Exp:* Sr demonstr geol, Univ NSW, 60-63; Ford Found fel, Lamont Geol Observ, NY, 63-65; vis prof, La State Univ, 65-66; Queen Elizabeth fel, Univ Sydney, 66-68; from assoc prof to prof, Univ SC, 69-72; explor geologist, BP Alaska Explor Inc, 72-74; Pres, Era NAm Inc, 75-78; managing dir, Sydney Oil Co Ltd, 79-88; INDEPENDENT CONSULT, 80- *Concurrent Pos:* Fulbright travel award, 63-66; consult, Scripps Inst Oceanog, 66; consult geologist, Univ Sydney, 68; consult, Oceanog Off, US Navy, 69-; consult, John R Conolly & Assocs Inc, 74-; adj assoc prof, Columbia Univ, 73-75; adj prof, City Col New York, 75 & C W Post Col, Long Island Univ, 75-76. *Honors & Awards:* Olle Prize, Royal Soc NSW, 67. *Mem:* Fel Geol Soc Am; Am Asn Petrol Geol; Geol Soc Australia; Australian Inst Mining & Metall; Petrol Explor Soc NY (vpres, 75-76, pres, 76-77). *Res:* Sedimentology of recent and ancient rocks; marine geology; petrology; glacial marine geology; origin of continental margins and geosyndines; exploration geology. *Mailing Add:* Six Balfour Rd Rose Bay New South Wales Australia

CONOMOS, TASSO JOHN, b New Kensington, Pa, Sept 11, 38; m 69; c 3. OCEANOGRAPHY, GEOCHEMISTRY. *Educ:* San Jose State Univ, BS, 61, MS, 63; Univ Wash, PhD(oceanog), 68. *Prof Exp:* Tech asst geol, San Jose State Univ, 60-62; phys sci technician paleont & stratig, US Geol Surv, 62-63; res assoc oceanog, Univ Wash, 63-64 & 68-69, teaching assoc, 64-65; intern, Smithsonian Inst, 66-68; fel, 69-70, res oceanogr, 70-81, regional res hydrologist, 81-85, REGIONAL HYDROLOGIST, WATER RESOURCES DIV, US GEOL SURV, 85- *Mem:* AAAS; Soc Econ Paleontologists & Mineralogists; Am Soc Limnol & Oceanog; Estuarine Res Fedn. *Res:* Geochemistry and distribution of suspended particulate matter in river-ocean mixing systems; descriptive chemical oceanography of near-shore and in-shore waters; sedimentological-geochemical studies of biogenic sediments; effects of man on estuarine processes. *Mailing Add:* 1260 Cotton St Menlo Park CA 94025

CONOMY, JOHN PAUL, b Cleveland, Ohio, July 31, 38; m 63; c 3. NEUROLOGY. *Educ:* John Carroll Univ, BS, 60; St Louis Univ, MD, 64. *Prof Exp:* Intern med, St Louis Univ Hosps, 64-65; resident neurol, Univ Hosps Cleveland, Case Western Reserve Univ, 65-68, fel neuropath, Cleveland Metrop Gen Hosp, 68; neurologist, US Air Force, 69 & 70; res fel neuroanat, Univ Pa, 70-71; asst prof med, Med Sch, Case Western Reserve Univ, 72-75, assoc prof med, 75-81; chmn dept Neurol, Cleveland Clin Found, 75-90; CONSULT, 90- *Concurrent Pos:* Career teaching fel award, Case Western Reserve Univ, 70; consult, Vet Admin Hosps, 72-; grants in aid, Vet Admin, 72 & 74, Mary B Lee Fund, 74, Mellon Fund & Reinberger Found, 76-81 & NIH, 78-81. *Mem:* Asn Res Nerv & Ment Dis; Soc Neurosci; fel Am Col Physicians; fel Am Acad Neurol; Asn Univ Profs Neurol. *Res:* Behavioral aspects of neurology, especially correlative studies of unit peripheral nerve and neuronal activity and behavior in animals and man; neurophysiologic action of central neurotransmitters; cerebrovascular disease. *Mailing Add:* Cleveland Clin Found 9500 Euclid Ave Cleveland OH 44106

CONOVER, CHARLES ALBERT, b Elizabeth, NJ, Apr 22, 34; m 57; c 1. ORNAMENTAL HORTICULTURE. *Educ:* Univ Fla, BSA, 62, MSA, 63; Univ Ga, PhD(plant sci), 70. *Prof Exp:* Asst ornamental horticulturist, Agr Exten Serv, 63-70, ORNAMENTAL HORTICULTURIST & CTR DIR, CENT FLA RES & EDUC CTR, UNIV FLA, 71- *Concurrent Pos:* Ornamental hort consult, United Brands Co, 69-74, Rainbird, 74-76 & Western Publ Co, 77- *Mem:* Int Soc Hort Sci; fel Am Soc Hort Sci. *Res:* Tropical ornamental plant nutrition; acclimatization of tropical ornamental plants for interior use; development of synthetic soil media; propagation of tropical ornamentals; commercial production of tropical ornamental foliage crops. *Mailing Add:* Cent Fla Res & Educ Ctr 2807 Binion Rd Apopka FL 32703

CONOVER, CLYDE S(TUART), b Springfield, Ill, June 25, 16; m 43; c 2. CIVIL ENGINEERING. *Educ:* Univ NMex, BS, 38. *Prof Exp:* Hydraul engr, Ground Water Invests, US Geol Surv, 38-48, asst dist engr, 48-51, dist engr, NMex, 51-57, asst chief, Ground Water Br, DC, 57-62, dist engr, Fla, 62-65, dist chief, Water Resources Div, 65-80, sr hydrologist, 80; RETIRED. *Mem:* Am Geophys Union; Am Soc Civil Engrs; Am Water Works Asn; Am Water Resources Asn; Nat Water Well Asn (vpres, 64-65). *Res:* Ground water hydraulics, management and quantitative evaluation of resource. *Mailing Add:* 411 Vinnedge Dr Tallahassee FL 32303

CONOVER, JAMES H, b Amsterdam, NY, Dec 22, 42; m 75; c 5. PHARMACOLOGY. *Educ:* Siena Col, BS, 65; NY Univ, MS, 67, PhD(biol), 69. *Prof Exp:* Asst pediat & human genetics, Mt Sinai Sch Med, City Univ New York, 69-70, instr, 70-71, assoc, 71-72, asst prof, 72-75; asst prof, Albert Einstein Sch Med, 75-77; asst prof, NY Univ Sch Med, 77-79; clin study analyst, Lederle Lab, Am Cyanamid Co, 79-81, mgr clin res assoc, 81-83, asst dir clin develop, 83-89; DIR REG AFFAIRS, PARKE-DAVIS, 89- *Concurrent Pos:* Career scientist award, Health Res Coun, New York City, 74; res scholar award, Nat Cystic Fibrosis Found, 76; prin investr grants, Nat Found March Dimes, 74-79, Health Res Coun, New York City, 74-75, Nat Cystic Fibrosis Found, 76-79. *Mem:* Soc Pediat Res; Am Soc Human Genetics. *Res:* Clinical safety analysis; new drug applications to Food and Drug Administration. *Mailing Add:* Parke-Davis 2800 Plymouth Rd Ann Arbor MI 48105

CONOVER, JOHN HOAGLAND, b McKeesport, Pa, Oct 26, 16; m 40; c 2. METEOROLOGY. *Prof Exp:* Weather observer, Blue Hill Meteorol Observ, Harvard Univ, 36-40, chief observer, 40-47, res asst & tech mgr, 47-52, meteorologist, 52-59, tech mgr, 52-57, actg dir, 57-58; meteorologist, Air Force Geophysics Lab, 59-77; RES ASSOC, MT WASH OBSERV, 69- *Concurrent Pos:* Lab instr, Harvard Univ, 41-43; instr, US Weather Bur, 44; consult meteorologist, 47-56 & 77-81. *Mem:* Am Meteorol Soc. *Res:* Climatic change; microclimates in the arctic; cloud studies; satellite meteorology; author of one book. *Mailing Add:* 15 Nobel Rd Dedham MA 02026

CONOVER, LLOYD HILLYARD, b Orange, NJ, June 13, 23; m 44, 90; c 4. MEDICINAL CHEMISTRY, ANIMAL DRUG RESEARCH. *Educ:* Amherst Col, AB, 47; Univ Rochester, PhD(chem), 50. *Prof Exp:* Res chemist, Pfizer, Inc, 50-58, res supvr, 58-61, res mgr, 61-68, dir chem res-chemother, 68-71, res dir, Pfizer Ltd, Eng, 71-75, vpres agr prod res & develop, 75-84; res consult, 84-88; RETIRED. *Honors & Awards:* Eli Whitney Award, 83; Third Century Award, 90. *Mem:* Am Chem Soc; Royal Soc Chem; Royal Soc Arts; Sigma Xi. *Res:* Synthesis of heterocycles; hydrogenolysis of oxygen functions; structure and synthesis of tetracycline antibiotics; synthesis of antiparasitic agents; drugs of microbiological origin; animal health agents. *Mailing Add:* One Earls Ave Flat Six Falkestone Kent CT20 2EX England

CONOVER, MICHAEL ROBERT, b Miami, Fla, Feb 16, 51; m 75; c 3. ANIMAL BEHAVIOR. *Educ:* Eckerd Col, BS, 73; Wash State Univ, MS, 75, MS, 78, PhD(zool), 78. *Prof Exp:* Asst prof animal behav, Ball State Univ, 78-79; assoc scientist, Conn Agr Exp Sta, 79-90; ASSOC PROF, UTAH STATE UNIV, 91- *Concurrent Pos:* Nat Sci Found fel, Univ Calif, Irvine, 80-81. *Mem:* Wildlife Soc; Am Ornith Union; Cooper Ornith Soc; Animal Behav Soc. *Res:* Applied ethology; avian communication; behavioral ecology of wildlife species; wildlife damage control. *Mailing Add:* Dept Fisheries & Wildlife Utah State Univ Logan UT 84322-5210

CONOVER, THOMAS ELLSWORTH, b Plainfield, NJ, Nov 20, 31; m 66; c 3. BIOCHEMISTRY. *Educ:* Oberlin Col, BA, 53; Univ Rochester, PhD(biochem), 59. *Prof Exp:* Res assoc, Johnson Found, Univ Pa, 62-64; asst mem, Inst Muscle Dis, 64-69; assoc prof, 70-81, PROF BIOL CHEM, HAHNEMANN UNIV, 81- *Concurrent Pos:* Nat Found fel, Wenner-Gren Inst, Univ Stockholm, 58-60; USPHS fel, Pub Health Res Inst, New York, 60-62; Muscular Dystrophy Asn fel, Inst Gen Path, Univ Padua, 69-70. *Mem:* Am Soc Biol Chem; AAAS. *Res:* Oxidative phosphorylation; mitochondrial structure and function; respiration and phosphorylation in cell nuclei; metabolic significance of cellular structures. *Mailing Add:* Dept Biol Chem Hahnemann Univ & Hosp Philadelphia PA 19102

CONOVER, WILLIAM JAY, b Hays, Kans, Dec 6, 36; m 60; c 5. STATISTICS. *Educ:* Iowa State Univ, BS, 58; Cath Univ, MA, 62, PhD(math statist), 64. *Prof Exp:* from asst prof statist to assoc prof statist & comput sci, Kans State Univ, 64-73; prof math & statist, Tex Tech Univ, 73-78, prof statist & coordr info systs & quant sci, 78-81, Horn prof statist & assoc dean grad progs & res, 81-84, Horn prof & coordr info systs & quant sci, 84-88, HORN PROF STATIST, TEX TECH UNIV, 88- *Concurrent Pos:* Consult, Water Resources Div, US Geol Surv, 62-68; NSF res grant, 67-68 & 78; NIH career develop award, 69-73; prof, Univ Zurich, 70-71 & Univ Calif, Davis, 76-77; consult, Upjohn Co, 75, Schering, 75, Sandia Labs, 77- & Vick Co, 78-; vis staff mem, Los Alamos Sci Labs, 75- *Honors & Awards:* Youden Prize; Owen Award. *Mem:* Inst Math Statist; Biomet Soc; fel Am Statist Asn. *Res:* Nonparametric statistics; stochastic models in hydrology and hydraulics. *Mailing Add:* Dept Info Systs & Quant Sci Tex Tech Univ Lubbock TX 79409-4320

CONOVER, WOODROW WILSON, b Terre Haute, Ind, July 30, 47; div. PHYSICAL BIOCHEMISTRY. *Educ:* Rose Hulman Polytech Inst, BS, 69; Ind Univ, Bloomington, PhD(chem), 73. *Prof Exp:* Fel biochem, Univ Chicago, 73-75; sr res assoc biochem & opers mgr, Stanford Magnetic Resonance Lab, Stanford Univ, 75-78; appln chemist, Nicolet Technol Corp, 78-83; GEN MGR INSTRUMENTS, NICOLET MAGNETICS CORP, 83- *Mem:* Am Chem Soc. *Res:* Exploitation of nuclear magnetic resonance instrumentation and techniques for the study of fundamental biochemical mechanisms. *Mailing Add:* 47647 Hoyt St Fremont CA 94539-7571

CONQUEST, LOVEDAY LOYCE, b Hilo, Hawaii, Jan 22, 48; m 83. STATISTICS. *Educ:* Pomona Col, BA, 70; Stanford Univ, MS, 72; Univ Wash, PhD(biostatist), 75. *Prof Exp:* Asst prof biostatist, Univ Hawaii, 75-76; from vis lectr to vis asst prof, Ctr Quant Sci, Col Fisheries, 76-77, vis asst prof, Dept Finance, Bus Econ & Quant Methods, Grad Sch Bus Admin, 77-78, asst prof, 78-83, ASSOC PROF STATIST, CTR QUANT SCI, COL FISHERIES, UNIV WASH, 83- *Concurrent Pos:* Statist consult, Tavolek, Inc, 76-80, Dept Surg Diabetics Vascular Study, Univ Wash, 78-79, Bierly & Assocs, Waterway & Natural Resource Consults, 79-80, Oceanog Inst Wash, 79-81, JRB Assocs, Sci Applications, Inc, 81-83, Municiplity Metrop Seattle, 81-, King County, 87-88. *Mem:* Am Statist Asn; Biomet Soc; Sigma Xi; Asn Women Math; Int Statist Inst. *Res:* Statistical methods for analysis of water quality data, including water variables and ecological measures; statistical analysis for environmental monitoring; quantitative fisheries management; streamside studies; environmental statistics; spacial statistics; landscape ecology. *Mailing Add:* Ctr Quant Sci HR-20 Univ Wash Seattle WA 98195

CONRAD, ALBERT G(ODFREY), electrical engineering; deceased, see previous edition for last biography

CONRAD, BRUCE, b Ann Arbor, Mich, July 2, 43; m 64. PARTIAL DIFFERENTAL EQUATIONS. *Educ:* Harvey Mudd Col, BS, 64; Univ Calif, Berkeley, PhD(math), 69. *Prof Exp:* Asst prof, 69-74, ASSOC PROF MATH, TEMPLE UNIV, 74- *Mem:* AAAS; Am Math Soc; Math Asn Am; Soc Indust & Appl Math. *Res:* Homology of groups; algebraic K-theory; hyperbolic systems of conservation laws. *Mailing Add:* Dept Math Temple Univ Philadelphia PA 19122

CONRAD, DANIEL HARPER, REGULATION OF IGE SYNTHESIS, FC RECEPTORS. *Educ:* WVa Univ, PhD(biochem), 73. *Prof Exp:* ASSOC PROF IMMUNOL, SUBDEPT IMMUNOL, SCH MED, JOHNS HOPKINS UNIV, 81- *Res:* Hypersensitivity. *Mailing Add:* 12 Hollis Ct Timonium MD 21093

CONRAD, EDWARD EZRA, b Richmond, Calif, June 11, 27; m 51; c 3. SOLID STATE PHYSICS, RADIATION CHEMISTRY. *Educ:* Univ Calif, BA, 50; Univ Md, MS, 55, PhD(radiation chem), 70. *Prof Exp:* Physicist solid state, Nat Bur Standards, 51-52; physicist, Harry Diamond Labs, 52-70, chief nuclear radiation effects lab, 70-76, actg assoc dir, 75-76; asst dep dir, 76-79, DEP DIR, DEFENSE NUCLEAR AGENCY, 79- *Concurrent Pos:* Secy Army res & develop fel, 59; lectr, Univ Md, 70-; rep, US Nat Comt Int Electrotech Comn, 58- *Mem:* Fel Inst Elec & Electronics Engrs; Am Phys Soc. *Res:* Semiconductors; radiation effects; magnetics and dielectric measurements; radiation chemistry; pulse radiolysis. *Mailing Add:* 7500 Marbury Rd Bethesda MD 20817

CONRAD, EUGENE ANTHONY, b Clinton, Mass, Aug 15, 27; m 49; c 2. PHARMACOLOGY. *Educ:* Col of the Holy Cross, BS, 50; Univ NH, MS, 52; Vanderbilt Univ, PhD(pharmacol), 56; NY Med Col, MPH, 89. *Prof Exp:* Res bacteriologist, Charles Pfizer & Co, 52; asst pharmacol, Vanderbilt Univ, 55-56; instr physiol & pharmacol, Bowman Gray Sch Med, 56-58; res assoc pharmacol, Sterling-Winthrop Res Inst, 58-60, asst dir coord sect, 60-62, clin pharmacologist, 62-63; admin asst dept drugs, AMA, Ill, 63-65, dir drug doc sect, 64-66; dir res admin, Wampole Lab, 66-67, dir clin res, 67-70; assoc dir med res, 70-84, gov liaison, 78-84, SR DIR CLIN RES, PURDUE FREDERICK CO, NORWALK, 84- *Concurrent Pos:* Vis lectr, Albany Med Col & Bowman Gray Sch Med, 59; mem behav pharmacol comt, NIMH, 65-66; res adv, Stamford Health Dept, 89- *Mem:* Am Soc Microbiol; AMA; Drug Info Asn; Soc Pharmacol & Exp Therapeut; fel Col Allergy Immunol; Acad Allergy & Immunol; Am Thoracic Soc; Am Pub Health Asn; Int Soc Pharmacoepidemiol. *Res:* Drug research; public health. *Mailing Add:* Dept Med Purdue Frederick Co 100 Connecticut Ave Norwalk CT 06856

CONRAD, FRANKLIN, b Smithville, Ohio, Sept 27, 21; m 49; c 4. INDUSTRIAL CHEMISTRY. *Educ:* Col Wooster, BA, 43; Ohio State Univ, MS, 48, PhD(chem), 52. *Prof Exp:* Chemist, 52-55, supvr, 55-58, proj mgr chlorinated hydrocarbons, 58-60, supvr, 60-63, asst dir contract res, 63-66, dir indust chem res, 66-82, dir specialty chem res, 82-86, dir electronic chem res, DIR ANIMAL NUTRIT-COM DEVELOP, ETHYL CORP, 86- *Mem:* Am Chem Soc. *Res:* Organometallics; metal hydrides; propellant chemicals; chlorinated hydrocarbons; all aspects of industrial chemicals research and development. *Mailing Add:* 1881 Madras Dr Baton Rouge LA 70815-4829

CONRAD, GARY WARREN, b Amsterdam, NY, Mar 24, 41. DEVELOPMENTAL BIOLOGY, CELL BIOLOGY. *Educ:* Union Col, BS, 63; Yale Univ, MS, 65, PhD(biol), 68. *Prof Exp:* NIH fel polysaccharide biochem, Univ Chicago, 68-70; from asst prof to assoc prof,71-80, PROF DEVELOP BIOL, KANS STATE UNIV, 80- *Concurrent Pos:* vis prof, Mt Desert Island Biol Lab, 71-90; sabbatical, Max Planck Inst Biochem, Munich, 77-78, Physiol Lab, Cambridge, England, 84-85; sr int fel, Fogarty Ctr, NIH, 84-85. *Mem:* Am Soc Cell Biol; Int Soc Develop Biologists; Soc Develop Biol; AAAS. *Res:* Differentiation of connective tissue; fibroblasts; control of synthesis and polymerization of extracellular matrices, especially in the cornea; mechanisms of cytokinesis; mechanisms of cell movement, cell adhesion and cell shape change. *Mailing Add:* Div Biol/Ackert Hall Kans State Univ Manhattan KS 66506

CONRAD, HANS, b Konradstahl, Ger, Apr 19, 22; nat US; m 44; c 3. MATERIALS SCIENCE. *Educ:* Carnegie Inst Technol, BS, 43; Yale Univ, MEng, 51, DEng(metall eng), 56. *Prof Exp:* Res metallurgist, Aluminum Co Am, 43-45; chief chem engr, Napier Co, 45-46; dir res, R Wallace & Sons, 46-53; res metallurgist, Chase Brass & Copper Co, 53-55; supvry metallurgist, Res Labs, Westinghouse Elec Corp, 55-59; sr tech specialist, Atomics Int Div, NAm Aviation, Inc, 59-61; head physics dept, Aerospace Corp, 61-64; dir mat sci & eng div, Franklin Inst, 64-67; prof mat sci & chmn dept metall eng & mat sci, Univ Ky, 67-81; PROF MAT SCI & ENG DEPT & DIR ELEC & MAGNETIC FIELDS EFFECTS LAB, NC STATE UNIV, 81- *Mem:* Fel Am Soc Metals; Am Inst Mining, Metall & Petrol Engrs; Am Soc Testing & Mat; Am Ceramic Soc. *Res:* Mechanical properties; superconductivity; crystal growths and defects; casting, working, forming, fabrication and finishing of metals; alloy development; electroplasticity; electrorheology. *Mailing Add:* Dept Mat Sci & Eng NC State Univ Raleigh NC 27695-7907

CONRAD, HARRY EDWARD, b Washington, DC, Jan 21, 29; m 52; c 2. BIOCHEMISTRY. *Educ:* La State Univ, BS, 49; Purdue Univ, MS, 52, PhD(biochem), 54. *Prof Exp:* Res chemist, Mead Johnson & Co, 54-58; res assoc, 58-60, prof biochem, Univ Ill, Urbana-Champaign, 72-89, from instr to assoc prof chem, 60-72, CHIEF SCIENTIST BIOCHEM, GLYCOMED INC 89- *Mem:* Am Chem Soc; Am Soc Biol Chemists. *Res:* Chemistry and biochemistry of mucopolysaccharides; changes in metabolism of mucopolysaccharides and complex cell surface carbohydrates during embryonic development. *Mailing Add:* Glycomed Inc 860 Atlantic Ave Alameda CA 94501-2200

CONRAD, HARRY RUSSELL, b Burlington, Ky, Oct 3, 25; c 2. NUTRITION. *Educ:* Univ Ky, BSc, 48; Ohio State Univ, MSc, 49, PhD(dairy sci), 52. *Prof Exp:* From instr to assoc prof, 52-64, PROF DAIRY SCI, OHIO AGR RES & DEVELOP CTR, 64- *Mem:* Am Dairy Sci Asn; Sigma Xi. *Res:* Rumen physiology; digestion; nitrogen metabolism and growth in cattle. *Mailing Add:* Dept of Dairy Sci Ohio Agr Res & Develop Ctr Wooster OH 44691

CONRAD, HERBERT M, b New York, NY, Feb 20, 27; m 51; c 3. BIOCHEMISTRY, NUTRITION. *Educ:* Cornell Univ, BS, 49; Univ Southern Calif, MS, 59, PhD(biochem), 65. *Prof Exp:* Chemist, Calif Grape Prod Corp, 50-52; chemist, Star-Kist Foods, Inc, 52-54; lab dir, Long Beach Water Dept, 54-59; proj engr, NAm Aviation, Inc, 62-67; dir biochem, RPC Corp, 67-71; PRES, ECOL SYSTS CORP, 71- *Concurrent Pos:* Consult munic water dist. *Mem:* Fel AAAS; Am Chem Soc; NY Acad Sci. *Res:* The effect of weightlessness upon physiological processes; mechanisms of plant hormones; biodegradation of hazardous materials. *Mailing Add:* Ecol Systs Corp 8517 Washington Blvd Culver City CA 90230

CONRAD, JOHN RUDOLPH, b San Antonio, Tex, Mar 21, 47; m 71. PLASMA PHYSICS. *Educ:* St Mary's Univ, Tex, BS, 68; Dartmouth Col, PhD(physics), 73. *Prof Exp:* Res assoc plasma physics, Inst Fluid Dynamics & Appl Math, Univ Md, 73-75; PROF NUCLEAR ENG, UNIV WIS, MADISON, 75- *Mem:* Am Phys Soc; Sigma Xi. *Res:* Experimental research in the areas of beam-plasma interaction; plasma transport properties, ion source and neutral beam-technology. *Mailing Add:* Dept Nuclear Eng Univ Wis 1500 Johnson Madison WI 53706

CONRAD, JOSEPH H, b Cass Co, Ind, Dec 7, 26; m 50; c 4. ANIMAL NUTRITION, BIOCHEMISTRY. *Educ:* Purdue Univ, BSA, 50, MS, 54, PhD(animal nutrit), 58. *Prof Exp:* From instr to prof animal sci, Purdue Univ, 53-71; PROF ANIMAL NUTRIT & COORDR TROP ANIMAL SCI PROGS, UNIV FLA, 71-, PROF ANIMAL SCI. *Concurrent Pos:* Animal nutritionist from Purdue Univ, Brazil Tech Asst Prog, Fed Univ Vicosa Minas Gerais, USAID, 61-65, hon prof, 65. *Honors & Awards:* Distinguished Nutrit Award, Distillers Feed Res Coun, 64; Int Animal Agr Award, Am Soc Animal Sci, 85, Gustav Bohstedt Award, Mineral & Trace Mineral Res, 87; Moorman Award, Nat Feed Ingredients Asn, 89. *Mem:* Am Soc Animal Sci; Latin Am Soc Animal Sci; Brazilian Soc Animal Sci; World Asn Animal Prod (vpres, 86). *Res:* Swine and cattle nutrition; amino acids; feed additives; livestock in small farm systems; tropical forage utilization; phosphorus and trace element deficiencies under tropical conditions. *Mailing Add:* Dept Animal Sci Univ Fla Gainesville FL 32611-0691

CONRAD, MALCOLM ALVIN, mineralogy, crystallography; deceased, see previous edition for last biography

CONRAD, MARCEL E, b New York, NY, Aug 15, 28; c 5. HEMATOLOGY, ONCOLOGY. *Educ:* Georgetown Univ, BS, 49, MD, 53; Nat Bd Internal Med, dipl, 61. *Prof Exp:* Intern med, Med Corps, US Army, 53-54, resident internal med, 55-58, chief resident, 58-59, asst chief hemat, 58-60 & 61-65, chief hemat, Walter Reed Army Med Ctr, 65-74, dir, Div Med, 69-71 & Clin Invest Serv, 71-74; prof med & dir, div hemat & oncol, Univ Ala, Birmingham, 74-83; PROF MED, DIR, DIV HEMAT & ONCOL & USA CANCER CTR, UNIV S ALA, MOBILE, 83- *Concurrent Pos:* From clin asst prof to clin assoc prof med, Sch Med, Georgetown Univ, 64-74. *Honors & Awards:* William Beaumont Award, 68. *Mem:* Am Physiol Soc; fel Am Col Physicians; Am Soc Clin Invest; fel Int Soc Hemat; Asn Am Physicians. *Res:* Gastroenterology; iron metabolism; hemolytic disorders; intestinal transport; hepatitis; oncology. *Mailing Add:* USA Cancer Ctr Univ SAla Mobile AL 36688

CONRAD, MARGARET C, physiology, for more information see previous edition

CONRAD, MICHAEL, b New York, NY, Apr 30, 41; m; c 1. BIOLOGICAL INFORMATION PROCESSING, MOLECULAR COMPUTING. *Educ:* Harvard Univ, AB, 63; Stanford Univ, PhD(biophys), 69. *Prof Exp:* Fel biophys, Ctr Theoret Studies, Univ Miami, 69-70; fel math, Univ Calif, Berkeley, 72; asst prof, Inst Info Sci, Univ Tübingen, 72-74; assoc prof biol, City Col, 74-75; assoc prof comput & commun sci, Univ Mich, Ann Arbor, 75-79; PROF COMPUT SCI, WAYNE STATE UNIV, 79- *Concurrent Pos:* Res assoc, Inst Info Sci, Univ Tübingen, 74-76; adj prof biol sci, Wayne State Univ, 79-88, distinguished univ res fel, 86-88; spec vis prof, Technol Univ Nagaoka, Japan, 87, adv, Advan Technol Inst, Tokyo, 87-; comput soc tutorial prog speaker, Inst Elect & Electronics Engrs, 88-; chmn Foreign Appl Assessment Ctr, Panel on Soviet Bloc Molecular Electronics Res, 88-89; distinguished lectr, Inst Elec & Electronic Engrs Computer Soc, 90-91; assoc managing ed, Biosysts, 83-; bd mem, Soc Math Biol, 83-86, secy & newsletter ed, 85- *Mem:* Biophys Soc; Soc Math Biol; Asn Comput Mach. *Res:* Computer modeling of biological systems; biophysics of information processing; brain models and intelligence; ecological and evolutionary problems; adaptability theory; molecular computing. *Mailing Add:* Dept Comput Sci Wayne State Univ Detroit MI 48202

CONRAD, PAUL, b Hempstead, NY, Oct 7, 21; m 43; c 1. MATHEMATICS. *Educ:* Univ Ill, PhD(math), 51. *Prof Exp:* From asst prof to prof math, Newcomb Col, Tulane Univ, 51-70; PROF MATH, UNIV KANS, 70- *Concurrent Pos:* NSF sr fel, Australian Nat Univ, 64-65; vis prof, Univ Paris, 67. *Honors & Awards:* Fulbright Lectr, Univ Ceylon, 56-57. *Mem:* Am Math Soc. *Res:* Ordered algebraic systems; group theory. *Mailing Add:* Dept Math Univ Kans Lawrence KS 66044

CONRAD, ROBERT DEAN, b El Reno, Okla, Sept 20, 23; m 47; c 3. LABORATORY ANIMAL MEDICINE. *Educ:* Univ Okla, BS, 49; Okla State Univ, MS & DVM, 53; Univ Calif, Davis, PhD(comp path), 70. *Prof Exp:* Asst prof vet microbiol, Wash State Univ, 53-59; pathologist, Heisdorf & Nelson Farms, Inc, 59-65; NIH fel epidemiol & prev med, Univ Calif,

Davis, 65-67, specialist, Div Exp Animal Resources, Sch Vet Med, 67-71; assoc prof med microbiol, Col Med & dir animal resource facilities, 71-76, VETERINARIAN & ASSOC PROF MED MICROBIOL, EPPLEY INST RES CANCER, UNIV NEBR, OMAHA, 76- *Concurrent Pos:* Dr Salsbury fel, 57-58. *Mem:* Am Vet Med Asn; Am Asn Avian Path; Am Soc Microbiol; Sigma Xi. *Res:* Bacterial and viral diseases of domestic and wild fowl. *Mailing Add:* 7115 N 57th St Omaha NE 68152

CONRAD, WALTER EDMUND, b Forward, Pa, Nov 16, 20; m 49, 82; c 2. ORGANIC CHEMISTRY. *Educ:* Wayne Univ, BS, 44, MS, 46; Univ Kans, PhD(org chem), 51. *Prof Exp:* Chemist, Armour Labs, 45-47; chemist, Sterling-Winthrop Res Inst, 47-48; res assoc, Med Sch, Tufts Univ, 51-53; res chemist, Celanese Corp, 55-57, assoc prof & chmn dept, Ohio Northern Univ, 57-59; prof chem, Southeast Mass Univ, 59-77, RETIRED. *Mem:* Am Chem Soc. *Res:* Schmidt reaction; effect of ultraviolet irradiation on pyrimidines; catalytic debenzylation; lubricants; chemistry of phosphoranes. *Mailing Add:* 836 Avon St Land-O-Lakes FL 34639

CONRADI, MARK STEPHEN, b St Louis, Mo, Jan 25, 52. MAGNETIC RESONANCE. *Educ:* Wash Univ, BS, 73, PhD(physics), 77. *Prof Exp:* Res asst, Physics Dept, Washington Univ, 73-77; staff scientist & Wigner fel, Chem Div, Oak Ridge Nat Lab, 77-79; assoc prof, Physics Dept, Col William & Mary, 79-85; ASSOC PROF, PHYSICS, WASH UNIV, ST LOUIS, 85- *Concurrent Pos:* Sloan fel, 83-85. *Mem:* Am Phys Soc. *Res:* Measurement of molecular motion in solids and liquids principally with nuclear magnetic relaxation; concentrating on glasses and disordered solids and solids at high pressure. *Mailing Add:* Dept Physics Wash Univ St Louis MO 63130

CONRATH, BARNEY JAY, b Quincy, Ill, June 23, 35; m 62; c 3. DYNAMICS OF PLANETARY ATMOSPHERES, RADIATIVE TRANSFER. *Educ:* Culver-Stockton Col, BA, 57; Univ Iowa, MA, 59; Univ NH, PhD(physics), 66. *Prof Exp:* SPACE SCIENTIST, GODDARD SPACE FLIGHT CTR, NASA, 60- *Concurrent Pos:* Vis prof astron, Cornell Univ, 84 & 89. *Honors & Awards:* Medal Except Sci Achievement, NASA. *Mem:* Am Geophys Union; Am Astron Soc. *Res:* Structure and dynamics of planetary atmospheres by means of remote sensing from spacecraft. *Mailing Add:* Code 6932 Goddard Space Flight Ctr Greenbelt MD 20771

CONREY, BERT L, b Glendale, Calif, Sept 9, 20; m 47. GEOLOGY. *Educ:* Univ Calif, Berkeley, AB, 47, MA, 48; Univ Southern Calif, PhD(geol), 59. *Prof Exp:* Field party chief petrol geol, Stanolind Oil & Gas Co, 48-50, div photogeologist, 50-51; from asst prof to assoc prof, 55-63, chmn dept, 60-64, PROF GEOL, CALIF STATE UNIV, LONG BEACH, 63- *Concurrent Pos:* NSF res grant, 63-64. *Mem:* Am Asn Petrol Geol; Geol Soc Am; Soc Econ Paleontologists & Mineralogists; Int Asn Sedimentol. *Res:* Marine geology and sedimentology. *Mailing Add:* Dept Geol Calif State Univ 1250 Bellflower Blvd Long Beach CA 90840

CONROW, KENNETH, b Philadelphia, Pa, Jan 22, 33; m 55; c 3. COMPUTER SCIENCE. *Educ:* Swarthmore Col, BA, 54; Univ Ill, PhD(org chem), 57. *Prof Exp:* From instr to asst prof chem, Univ Calif, Los Angeles, 57-61; from asst prof to assoc prof org chem, 61-71, asst dir, Comput Ctr, 74-76, assoc prof comput sci, 71-84, ASSOC DIR, COMPUT CTR, KAN STATE UNIV, 76-, MGR USER SERVS, 80- *Mem:* Am Chem Soc; Asn Comput Mach. *Res:* Programming language preprocessors; non-numeric programming. *Mailing Add:* Comput Ctr Cardwell Hall Kans State Univ Manhattan KS 66506

CONROY, CHARLES WILLIAM, periodontics, oral pathology; deceased, see previous edition for last biography

CONROY, JAMES D, b Dayton, Ohio, Dec 15, 33; m 62; c 3. VETERINARY PATHOLOGY & DERMATOLOGY. *Educ:* Ohio State Univ, DVM, 60; Univ Ill, Urbana-Champaign, PhD(vet path), 68. *Prof Exp:* Fel clin res internal med, Animal Med Ctr, NY, 60-61, fel comp dermat, 61-62, staff vet, 62-64, sr resident vet path, 64; USPHS fel vet med sci, 64-68, from instr to assoc prof vet path & hyg, Col Vet Med, Univ Ill, Urbana-Champaign, 68-77; PROF VET PATH, COL VET MED, MISS STATE UNIV, MISSISSIPPI STATE, 77-, DIR PATH, DVM PATH LABS, 81- *Honors & Awards:* Clin Proficiency Award, Upjohn Co, Kalamazoo, Mich, 60. *Mem:* Am Vet Med Asn; Am Col Vet Path; Int Acad Path; Am Acad Vet Dermat (secy-treas, 64-68, vpres, 68-70, pres, 71-73); Am Col Vet Internal Med; Am Col Vet Dermat. *Res:* Development of cutaneous prenatal pigmentation; melanocytic tumors in domestic animals; cytopathology of infectious diseases; comparative dermatology; veterinary pathology. *Mailing Add:* Dept Vet Med Path Lab PO Box 5204 Mississippi State MS 39762

CONROY, JAMES STRICKLER, b Philadelphia, Pa, Aug 24, 31; m 79; c 3. FUEL CHEMISTRY. *Educ:* Univ Pa, AB, 53; Pa State Univ, MS, 56, PhD(fuel tech), 59. *Prof Exp:* Asst prof, 57-62, chmn dept sci, 73-78, ASSOC PROF CHEM, WIDENER UNIV, 62- *Mem:* Am Chem Soc. *Res:* Condensed aromatic hydrocarbons; fuels and combustion. *Mailing Add:* 21 Karrens Way Downingtown PA 19335-1703

CONROY, LAWRENCE EDWARD, inorganic chemistry; deceased, see previous edition for last biography

CONRY, THOMAS FRANCIS, b West Hempstead, NY, Mar 7, 42; m 67; c 3. TRIBOLOGY, MECHANICAL SYSTEMS DYNAMICS. *Educ:* Pa State Univ, BS, 63; Univ Wis-Madison, MS, 67, PhD(mech eng), 70. *Prof Exp:* Prod engr, Gen Motors Corp, 63-66, res engr, 69-71; asst prof gen eng, 71-75, assoc prof gen & mech eng, 75-81, PROF GEN & MECH ENG, UNIV ILL, URBANA, 81-, HEAD, GEN ENG DEPT, 87- *Concurrent Pos:* Summer fac fel, Lewis Res Ctr, NASA, 74 & 75; staff consult, Sargent & Lundy, Engrs, 77 & 79; sr vis, Univ Cambridge, Eng, 78-79; tech ed, J Vibration Acoust Stress & Reliability Design, Am Soc Mech Engrs, 84-89; chair, Publ Comt & mem, Bd Commun, Am Soc Mech Engrs, 90-92. *Mem:* Fel Am Soc Mech Engrs; Sigma Xi. *Res:* Tribology; numerical methods for elasto hydrodynamic lubrication; surface topography and wear, failure of railroad roller bearings; rotor dynamics; mechanical systems dynamics; design optimization. *Mailing Add:* Dept Gen Eng Univ Ill 104 S Mathews Ave Urbana IL 61801-2996

CONS, JEAN MARIE ABELE, b Lancaster, Pa, 1934; m; c 2. DEVELOPMENTAL PHYSIOLOGY. *Educ:* Calif State Univ, San Francisco, BA, 60; Univ Calif, San Francisco, MS, 63, PhD(endocrinol), 72. *Prof Exp:* Lab technician develop anat, Univ Calif, San Francisco, 60-64, res anat, 64-68; consult, Med Sci Prog, Univ Calif, Berkeley, 72-73, fel, 72-74, asst res develop physiol, 74-75; asst prof, Col Notre Dame, Calif, 75-77; INSTR ANAT & PHYSIOL, COL SAN MATEO, 77- *Mem:* Inst Anat Sci. *Res:* Developmental patterns of pituitary and plasma glycoproteins and the actions of these hormones during development. *Mailing Add:* Col San Mateo 1700 W Hillsdale Blvd San Mateo CA 94402

CONSIGLI, RICHARD ALBERT, b Brooklyn, NY, Mar 2, 31; m 60; c 3. VIROLOGY, CANCER. *Educ:* Brooklyn Col, BS, 54; Univ Kans, MA, 56, PhD(bact), 60. *Prof Exp:* Asst bact, Univ Kans, 54-59, instr, 59-60; USPHS fel virol, Univ Pa, 60-62; from asst prof to prof bact, 62-85, DISTINGUISHED PROF BIOL, KANSAS STATE UNIV, 85-, SR SCIENTIST MID-AM CANCER CTR, UNIV, MED CTR, 75- *Concurrent Pos:* Pa Plan scholar, 61-62; NIH res grants, 62-88; career develop award, USPHS, 67-; consult, NSF; Cancer Res Manpower Rev Comt, NIH. *Honors & Awards:* Case Silver Medal, 85 & 86. *Mem:* Am Soc Microbiol; Tissue Cult Asn; Soc Exp Biol & Med; NY Acad Sci; Am Acad Microbiol; Am Soc Virol. *Res:* Investigation of the biochemical events during the animal virus infection of tissue culture cells; investigation of the host-parasite interrelationship during rickettsial infections; cancer research, biochemistry of tumor virus infection of cultured cells. *Mailing Add:* Sect Virol & Oncol Div Biol Kans State Univ Manhattan KS 66506

CONSROE, PAUL F, b Cortland, NY, Oct 18, 42. PHARMACOLOGY. *Educ:* Albany Col Pharm, BS, 66; Univ Tenn, Memphis, MS, 69, PhD(pharmacol), 71. *Prof Exp:* Assoc prof, 71-81, PROF PHARMACOL, COL PHARM, UNIV ARIZ, 81- *Concurrent Pos:* Nat Inst Drug Abuse grants, Univ Ariz, 72-; consult, Ariz Poison Control Inform Serv, 71- *Mem:* Soc Neurosci; Am Soc Pharmacol & Exp Therapeut. *Res:* Neuropsychopharmacological investigations of hallucinogens, marijuana and other psychotropic drugs. *Mailing Add:* Dept Pharmacol & Toxicol Univ Ariz Col Pharm Tucson AZ 85721

CONSTABLE, JAMES HARRIS, b Dayton, Ohio, Mar 9, 42; m 68; c 2. SOLID STATE PHYSICS. *Educ:* Ohio State Univ, BSc, 66, MSc, 67, PhD(physics), 69. *Prof Exp:* Res assoc physics, Ohio State Univ, 69-72, vis asst prof, 72-74; asst prof physics, State Univ NY, Binghamton, 74-76, dir tech studies div, 76-83, prof & chmn dept eng technol, 83-90, PROF ELEC ENG, STATE UNIV NY BINGHAMTON, 90- *Mem:* Am Phys Soc; Inst Elec & Electronics Engrs; Soc Mech Eng; Am Soc Eng Educ. *Res:* Experimental studies of the electrical, magnetic and thermal properties of the hydrogen solids; electrical noise measurements. *Mailing Add:* Watson Sch State Univ NY Binghamton NY 13902-6000

CONSTABLE, ROBERT L, b Detroit, Mich, Jan 20, 42; m 64; c 2. COMPUTER SCIENCE, MATHEMATICS. *Educ:* Princeton Univ, AB, 64; Univ Wis, MA, 65, PhD(math), 68. *Prof Exp:* Instr comput sci, Univ Wis, 68; from asst prof to assoc prof, 68-78, PROF COMPUT SCI, CORNELL UNIV, 78- *Mem:* Asn Symbolic Logic; Asn Comput Mach; Soc Indust & Appl Math. *Res:* Theory of computation, especially program verification, computational complexity and constructive type theory. *Mailing Add:* Dept Comput Sci 405 Upson Hall Cornell Univ Main Campus Ithaca NY 14853

CONSTANCE, LINCOLN, b Eugene, Ore, Feb 16, 09; m 36; c 1. BOTANY. *Educ:* Univ Calif, AM, 32, PhD(bot), 34. *Prof Exp:* Instr bot & cur herbarium, State Col Wash, 34-36, asst prof, 36-37; from asst prof to prof bot, 37-76, cur seed plant collections, 47-63, chmn dept bot, 54-55, dean col letters & sci, 55-62, vchancellor, 62-65, dir herbarium, 63-75, EMER PROF BOT, UNIV CALIF, BERKELEY, 76- *Concurrent Pos:* Vis lectr & acting dir, Gray Herbarium, Harvard Univ, 47-48; Guggenheim fel, 53-54. *Honors & Awards:* Fellows Medal, Cal Acad Sci, 85; Fel, AAAS; Asa Gray Award, Am Soc Plant Taxonomists, 86. *Mem:* Bot Soc Am (pres, 70); Torrey Bot Club; Am Soc Plant Taxonomists (pres, 53); fel Am Acad Arts & Sci; Linnean Soc London; Royal Swed Acad Sci; Bot Soc Arg. *Res:* Systematic botany of Umbelliferae; cytotaxonomy of Hydrophyllaceae. *Mailing Add:* Univ Herbarium & Dept Integrative Biol Univ Calif Berkeley CA 94720

CONSTANT, CLINTON, b Nelson, BC, Can, Mar 20, 12; US citizen; m 65. INORGANIC CHEMISTRY, CHEMICAL ENGINEERING. *Educ:* Univ Alta, BSc, 35; Western Reserve Univ, PhD, 41. *Prof Exp:* Develop engr, 36-38, foreman & engr, Acid Plant, Harshaw Chem Co, Ohio, 38-43; plant supt, 43-47, chief develop engr, Nyotex Chem, Inc, Tex, 47-48; sr chem engr, Harshaw Chem Co, Ohio, 48-50; mgr eng, Ferro Chem Corp, 50-52; tech asst mfg dept, Plant Develop & Eng, 52-61, mgr Fla Res Div, 61-63, proj mgr eng & design & mgr spec proj, Ga, Armour Agr Chem Co, Fla, 63-70; chem engr, Robert & Co Assoc, 70-79; chief engr, Almon Assoc Inc, 79-80; proj mgr, Eng Serv Assoc, Inc, 80-81; vpres eng, ACI Inc, Calif, 81-83, sr vpres ACI/MTI Inc, Calif, 83-86; AIR QUAL ENGR, SAN BERNARDINO COUNTY, CALIF, 86- *Concurrent Pos:* Reg Prof Engr, Calif, Wis, Dual Certifee Chem & Chem Eng, Patentee. *Mem:* AAAS; fel Am Inst Chem; assoc fel Am Inst Aeronaut & Astronaut; fel Am Inst Chem Eng; fel NY Acad Sci. *Res:* Chemical engineering design; phosphates; chemistry and production of anhydrous hydrofluoric acid; slide rule for complex chemical formulation; napalm and automated production. *Mailing Add:* PO Box 2529 Victorville CA 92393-2529

CONSTANT, FRANK WOODBRIDGE, b Minneapolis, Minn, June 1, 04; m 40; c 3. ENVIRONMENTAL PHYSICS. *Educ:* Princeton Univ, BS, 25; Yale Univ, PhD(physics), 28. *Prof Exp:* Nat res fel physics, Calif Inst Technol, 28-30; from instr to assoc prof, Duke Univ, 30-46; JARVIS PROF PHYSICS, TRINITY COL, CONN, 46- *Concurrent Pos:* Mem & off investr, Nat Defense Res Comt, Duke Univ, 42-46; instr, Dorr-Loomis Pre-Col Sci Ctr, 57, dir, 58 & 59. *Mem:* Fel Am Phys Soc; Am Asn Physics Teachers. *Res:* Ferromagnetism; theoretical physics; mechanics; electromagnetism; fundamental laws of physics. *Mailing Add:* 20 River Rd Essex CT 06426

CONSTANT, MARC DUNCAN, b Aledo, Ill, July 17, 41. ANALYTICAL CHEMISTRY. *Educ:* Monmouth Col, BA, 63; Southern Ill Univ, MA, 65; Kans State Univ, PhD(anal chem), 70. *Prof Exp:* supvr anal chem, Am Maize Prod Co, 69-80; GROUP LEADER METHODS, MILLER BREWING CO, 80- *Mem:* Am Chem Soc; Am Soc Brewing Chem. *Res:* Applications of instrumental analysis to the food processing industries, including gas and liquid chromatography, infrared, and general applications of automated wet methods. *Mailing Add:* Miller Brewing Co 3939 W Highland Blvd Milwaukee WI 53208-2866

CONSTANT, PAUL C, JR, b Kansas City, Mo, Sept 3, 22; m 49; c 4. ELECTRICAL ENGINEERING, MATHEMATICS. *Educ:* Univ Minn, BEE, 43; Univ Mo, Kansas City, MA, 56. *Prof Exp:* Asst instr elec eng, Univ Minn, 46-47; design engr, Ry Radiotel & Tel Co, Inc, 47-48; sr engr eng dept, Midwest Res Inst, 48-66, head electronics & elec eng sect & mgr tech utilization, 66-67, asst dir eng, 67-72, head environ measurements sect & mgr, field progs, 72-79, mgr spec projs, 77-79, admin mgr, 79-83, head detection & prod analysis sect, 83-85, ASST TO VPRES, MIDWEST RES INST, 85- *Concurrent Pos:* Lectr, Univ Mo-Kansas City, 56-; mem, Sect Comt on RF radiation hazards, US Stand Inst, 60-, vchmn, 60-69 & Int Oceanog Found; instr, Metrop Jr Col, Kansas City, Mo, 73; mem, Six subcomts Sect Comt C-95, Am Standards Asn, 60-73, Sect Comt C-95 Radio Frequency Hazards, 60-73, vchmn, 60-67; gen bus mgr, Local oscillator, 58-59; mem, tech prog, Mid-Am Electronics Conf, 56-57, chmn, 57, publ chmn, 57, mem bd dirs, 57-60, chmn 57-58, Conf Gen Chmn, 59; vchmn, meetings & papers, Kans City Sect, Inst Radio Engrs, 52-53, chmn, 53-54, treas, Kansas City Sect, 55-56, vchmn, 56-57, chmn, 57-58, publ chmn, 58-59, tech conf chmn, 59-60, mem,Nat Comt Electronic Computers, Inst Radio Engrs, 59-63; pres-elect, Kans City Chap, Sigma Xi, 75-76, pres, 76-77. *Mem:* Sigma Xi; Asn Comput Mach. *Res:* Mathematical and engineering analysis; design of electronic instrumentation and circuitry; electronic computers; bioengineering; system design; experimental design; research management, sampling and analysis of hazardous and toxic substances, test and evaluation. *Mailing Add:* 1212 W 113th St Kansas City MO 64114

CONSTANTIN, ROYSELL JOSEPH, b Rayne, La, Oct 21, 39; m 62; c 2. HORTICULTURE. *Educ:* Univ Southwestern La, BS, 61; La State Univ, MS, 62, PhD(hort), 64. *Prof Exp:* Asst prof hort, Fruit & Truck Sta, 64-65, from asst prof to prof, dept hort, 65-80, PROF & RES DIR, HAMMOND RES STA, LA STATE UNIV, 80- *Honors & Awards:* Nat Canners Award, Am Soc Hort Sci, 76, L M Ware Res Award, 78. *Mem:* Am Soc Hort Sci; Nat Sweet Potato Collabr; Azalea Soc Am; Sigma Xi. *Res:* Quality and production of fruit and vegetable products. *Mailing Add:* 5925 Old Covington Hwy Hammond LA 70403

CONSTANTINE, DENNY G, b San Jose, Calif, May 5, 25; m 52; c 2. VETERINARY MEDICINE, EPIDEMIOLOGY. *Educ:* Univ Calif, Davis, BS, 53, DVM, 55; Univ Calif, Berkeley, MPH, 65. *Prof Exp:* Head wildlife ecol care prog, Arctic Health Res Ctr, Alaska, 50-51, chief wildlife rabies res, Communicable Dis Ctr, Ga, 55-56, chief rabies field unit, NMex, 56-58, chief southwest rabies invest sta, 58-66, chief naval biomed res lab, Ctr Dis Control Activities, USPHS, 66-76; PUB HEALTH VET, CALIF DEPT HEALTH SERV, 76- *Concurrent Pos:* Asst mammalogist, Los Angeles County Mus, 42-46; pvt res, 46-50; zool mus cur, Univ Calif, Davis, 51-55, mammalogist & lab technician, Sch Vet Med, 54-55; consult, Armed Forces Bat Bomb Proj, 43-44. *Mem:* Wildlife Dis Asn; Am Vet Med Asn; Am Soc Mammal. *Res:* Public health, veterinary medicine; ecology, virology; physiology; mammalogy; wildlife diseases. *Mailing Add:* 1899 Olmo Way Walnut Creek CA 94598

CONSTANTINE, GEORGE HARMON, JR, b San Francisco, Calif, Sept 30, 36; m 62; c 3. PHARMACY. *Educ:* Univ Utah, BS, 60, MS, 62, PhD(pharmacog), 66. *Prof Exp:* Asst chem, 62-63, asst pharm, 63-64, asst pharmacog, 64-66, asst prof, 66-71, assoc prof, 71-80, PROF PHARMACOG, ORE STATE UNIV, 80-, ASST DEAN & HEAD ADV PHARMACOL, 80- *Mem:* Am Pharmaceut Asn; Acad Pharmaceut Sci; Am Soc Pharmacog. *Res:* Natural products; steroid and triterpenoid constituents of higher plants; plant alkaloids; marine biomedicinals. *Mailing Add:* 1520 NW 14th Pl Corvallis OR 97330

CONSTANTINE, HERBERT PATRICK, b Buffalo, NY, May 10, 29; m 54; c 2. PHYSIOLOGY, MEDICINE. *Educ:* Univ Buffalo, MD, 53. *Hon Degrees:* MA, Brown Univ, 67. *Prof Exp:* Nat Tuberc Asn fel, 57-59; Am Heart Asn res fel & instr med, Univ Rochester, 59-60; instr physiol, Univ Pa, 60-63; asst prof med, Boston Univ, 63-66; ASSOC PROF MED SCI, BROWN UNIV, 66- *Concurrent Pos:* NIMH grants, 64-68; consult, Vet Admin Hosp, Providence, RI, 66-; dir ambulatory care, RI Hosp, 70- *Mem:* Am Col Physicians; Am Fedn Clin Res; Am Thoracic Soc; Am Physiol Soc; Fedn Am Socs Exp Biol. *Res:* Respiratory physiology and mechanics; carbon dioxide reaction rates; laboratory automation; delivery of medical care. *Mailing Add:* Div Biol & Med Brown Univ Box G Providence RI 02912

CONSTANTINE, JAY WINFRED, b New York, NY, Feb 8, 26; m 52; c 2. PHYSIOLOGY. *Educ:* McGill Univ, BS, 51; Ohio State Univ, PhD(physiol), 59. *Prof Exp:* From asst to instr physiol, Ohio State Univ, 56-59; asst prof, NDak State Univ, 59-61; sr pharmacologist, 61-67, mgr gen pharmacol, 67-72, ASST DIR DEPT PHARMACOL, MED RES LABS, PFIZER, INC, 72- *Mem:* Am Soc Pharmacol & Exp Therapeut; Am Physiol Soc; Int Soc Hypertension. *Res:* Cardiovascular; platelet aggregation; general pharmacology. *Mailing Add:* Dept Pharmacol Pfizer Inc Med Res Labs Groton CT 06340

CONSTANTINE-PATON, MARTHA, b New York City, NY, July 16, 47; m 71; c 2. DEVELOPMENTAL NEUROBIOLOGY. *Educ:* Jackson Col, Tufts Univ, BS, 69; Cornell Univ, PhD(neurobiol & behav), 76. *Prof Exp:* Fel electron micros, Sect Neurobiol, Behav & Appl Physics, Cornell Univ, 75-76; asst prof biol, 76-81, ASSOC PROF DEVELOP NEUROBIOL, PRINCETON UNIV, 81- *Concurrent Pos:* Prin investr res grants, Nat Eye Inst, 77-, NSF, 81-; participant, Cent Nervous Syst Workshop, Cold Spring Harbor, 78; vis assoc prof, Dept Neurobiol, Harvard Med Sch, 82-83; mem, Visual Sci Study Sect B, NIH, 82- *Mem:* AAAS; Soc Neurosci; Soc Cell Biol; Sigma Xi; Soc Develop Biol; Fedn Am Socs Exp Biol. *Res:* Cellular interactions involved in sterotyped axon growth and patterned synaptogenesis during the early development of vertebrate nervous system. *Mailing Add:* Yale Univ Dept Biol Box 6666 New Haven CT 06511

CONSTANTINIDES, CHRISTOS T(HEODOROU), b Nicosia, Cyprus, Mar 22, 31; US citizen; m 62; c 2. ELECTRICAL ENGINEERING. *Educ:* Univ Manchester, BSc, 54, MSc, 55; Univ Kans, PhD(elec eng). *Prof Exp:* Res engr, Marconi's Wireless Tel Co Ltd, Eng, 55-58; asst prof elec eng, Univ Okla, 63-67; assoc prof, 67-75, PROF ELEC ENG, UNIV WYO, 75- *Mem:* Inst Elec & Electronics Engrs; Am Soc Eng Educ. *Res:* Non linear and optimum feedback control systems; distributed parameter systems. *Mailing Add.* Dept Elec Eng Box 3295 Univ Wyo Laramie WY 82071

CONSTANTINIDES, PANAYIOTIS PERICLEOUS, b Cyprus, July 5, 53. LIPOSOMES, ELECTRON PARAMAGNETIC RESONANCE SPECTROSCOPY. *Educ:* Univ Athens, Greece, BSc, 77; Brown Univ, PhD(biochem), 83. *Prof Exp:* Teaching asst inorg chem, Univ Athens, Greece, 76-77; teaching & res asst chem & biochem, Brown Univ, 77-82; fel, Yale Univ, 83-85, assoc res scientist, Dept Pharmacol, Sch Med, 86-; proj leader, Lipogen, Inc, 89; GROUP LEADER, DRUG DELIVERY DEPT, SMITHKLINE BEECHAM PHARMACEUT, 90- *Mem:* Am Biophys Soc. *Res:* Application of spectroscopic and other physical methods to characterize drug-membrane interactions; mechanism of action of anthracycline-antibiotics widely used in cancer chemotherapy. *Mailing Add:* Drug Delivery Dept Smithkline Beecham Pharmaceut PO Box 1539 King of Prussia PA 19406-0939

CONSTANTINIDES, PARIS, b Smyrna, Asia Minor, Dec 21, 19; Can citizen; m 50; c 1. ANATOMY, ELECTRON MICROSCOPY. *Educ:* Univ Vienna, MD, 43; Univ Montreal, PhD(exp med), 53. *Prof Exp:* Asst exp med, Inst Exp Med, Univ Montreal, 47-50; from asst prof to prof anat, Med Sch, Univ BC, 50-64, hon prof path, 64-65, prof anat, 65-67, prof path, 67-77; PROF PATH, MED SCH, LA STATE UNIV, 77- *Concurrent Pos:* Vis prof, Wash Univ, 63-64, 66-67 & Graz Univ, 76; ed, J Atherosclerosis Res, 61, chief ed, Am Hemisphere, 76-77; mem adv comt artificial heart, USPHS, 66; ed, Can J Physiol & Pharmacol, 69; consult, Nat Heart & Lung Inst & Med Res Coun Can Nat Sci Found. *Mem:* Am Asn Pathol; Am Heart Asn; Am Asn Anatomists. *Res:* Experimental pathology; aging degenerative diseases; endocrine. *Mailing Add:* Dept Pathol Med Sch La State Univ PO Box 33932 Shreveport LA 71130

CONSTANTINIDES, SPIROS MINAS, b Thessaloniki, Greece, Nov 4, 32; m 59; c 2. FOOD SCIENCE, BIOCHEMISTRY. *Educ:* Univ Thessaloniki, BS, 57; Mich State Univ, MS, 63, PhD(food sci), 66. *Prof Exp:* Res assoc food technol, Univ Thessaloniki, 57-61; NIH fel biochem, Mich State Univ, 66-68; from asst prof to assoc prof, 68-74, PROF FOOD & NUTRIT SCI & BIOCHEM, UNIV RI, 74- *Concurrent Pos:* Vis prof, Cath Univ Valparaiso, Chile, 72 & Univ Campinas, Brazil, 75-76; proj dir, food qual & inspection, Saudi Arabia, 78-86; dep dir, Intern Ctr Marine Resource Develop, 88- *Mem:* Am Chem Soc; Inst Food Technologists; Am Soc Biol Chemists. *Res:* Structure and function of enzymes; multiple molecular forms and control mechanisms of enzymes; lysosomes and proteolytic enzymes in marine animals; biochemical aspects of preservation of marine foods; utilization of unconventional food resources; food and nutritional science for developing nations; food quality control; transfer of technology in developing countries. *Mailing Add:* Dept Food & Nutrit Sci Woodward Hall Univ RI Kingston RI 02881

CONSTANTINOU, ANDREAS I, b Nicosia, Cyprus, Sept 26, 51; US citizen; m 75; c 2. CANCER BIOLOGY & CHEMOPREVENTION, EXPERIMENTAL THERAPEUTICS. *Educ:* Nat Univ Athens, Greece, BS, 75; Univ Toledo, Ohio, MS, 80; Med Col Ohio, PhD(biochem), 85. *Prof Exp:* Postdoctoral res, Med Sch, Northwestern Univ, 84-86 & Argonne Nat Lab, 86-90; RES BIOLOGIST, ITT RES INST, 90- *Concurrent Pos:* Invited partic, NATO Advan Study Inst, Spetsai, Greece, 86; instr, Col Dupage, 89-90; guest scientist, Argonne Nat Lab, 90- *Mem:* NY Acad Sci; Am Asn Cancer Res; AAAS; Sigma Xi. *Res:* Cancer chemoprevention; identifying compounds which may prevent chemically-induced carcinogenesis in animal models; identification of the compounds mechanism of action at the molecular level. *Mailing Add:* ITT Res Inst Ten W 35th St Chicago IL 60616

CONSTANTINOU, MICHALAKIS, b Kyrenia, Cyprus, July 16, 55; m 80; c 2. EARTHQUAKE ENGINEERING, SEISMIC BASE ISOLATION. *Educ:* Univ Patras, Greece, BS, 80; Rennselaer Polytech Inst, MS, 81, PhD(civil eng), 84. *Prof Exp:* Res asst, Rensselaer Polytech Inst, 80-84; engr, B Schwartz & Assoc, 84; asst prof, Drexel Univ, Philadelphia, 84-87; asst prof, 87-89, ASSOC PROF, STATE UNIV NY, BUFFALO, 89- *Concurrent Pos:* Consult, Watson Bowman Acme Corp, Gerb Vibration, Aeroflex Int & var eng firms, 87-; NSF presidential young investr award, 88-; dir grad studies, Dept Civil Eng, State Univ NY, Buffalo, 87- *Mem:* Am Soc Civil Engrs; Earthquake Eng Res Inst. *Res:* Seismic isolation of buildings and bridges; seismic isolation of equipment; shock and vibration isolation of equipment; large scale and prototype testing; mathematical modeling. *Mailing Add:* 243 Ketter Hall Dept Civil Eng State Univ NY Buffalo NY 14260

CONSTANTOPOULOS, GEORGE, b Greece, Feb 21, 23; US citizen; m 54; c 1. BIOCHEMISTRY. *Educ:* Univ Athens, BSc, 48; Wayne State Univ, PhD(biochem), 62. *Prof Exp:* Clin chemist, Hotel Dieu Hosp, Windsor, Ont, 55-58; res assoc biochem, Wayne State Univ, 62-63; res biochemist, 66-78, SR RES BIOCHEMIST, NAT INST NEUROL COMMUN DIS & STROKE, 78- *Concurrent Pos:* Univ res fel chem, Harvard Univ, 63-66, NIH fel, 64-65. *Mem:* AAAS; Am Chem Soc; Am Soc Biol Chemists; Am Soc Neurochem; NY Acad Sci; Am Soc Human Genetics. *Res:* Pathogenesis of heritable neurological disorders; glycosaminoglycans; lipids; neurochemistry; brain tumors. *Mailing Add:* 6406 Crane Terr Bethesda MD 20817

CONSTANTZ, GEORGE DORAN, b Washington DC, Sept 29, 47; m 81; c 1. POPULATION ECOLOGY, ANIMAL BEHAVIOR. *Educ:* Univ Mo-St Louis, BA, 69; Ariz State Univ, MS, 72, PhD(zool), 76. *Prof Exp:* Asst cur fish biol, Acad Natural Sci, 76-82; DIR, PINE CABIN RUN ECOL LAB, 82- *Concurrent Pos:* Adj prof, Univ Pa, 77-82; lectr, Shepherd Col, 82-83. *Mem:* Ecol Soc Am; Soc Study Evolution; AAAS. *Res:* Population and behavioral ecology of stream fishes; ecology of rivers. *Mailing Add:* Pine Cabin Run Ecol Lab High View WV 26808

CONSUL, PREM CHANDRA, b Meerut, India, Aug 10, 23; m 51; c 5. STATISTICS, MATHEMATICS. *Educ:* Agra Univ, BSc, 43, MSc, 46 & 47, PhD(math, statist), 57. *Prof Exp:* Lectr math, SD Col, Muzaffarnagar, India, 46-47; sr lectr, SM Col, Chandausi, 47-48; asst prof, NREC Col, Khurja, 48-50, assoc prof, 50-53, prof math & statist & head dept, 53-57; head dept statist & prin, MS Col, Saharanpur, 57-61; prof math & statist, Libya, Tripoli, 61-67; assoc prof, 67-71, PROF MATH & STATIST, UNIV CALGARY, 71- *Concurrent Pos:* Mem, Indian Statist Inst, 54-; convenor bd studies in statist, mem acad coun & fac sci, Agra Univ, 55-58, senate, 58-61, panel univ inspectros, 59-61. *Mem:* Int Statist Inst; Inst Math Statist; Am Statist Asn. *Res:* Probability distribution theory; statistical estimation and inference. *Mailing Add:* Dept Math & Statist Univ Calgary 2500 Univ Dr NW Calgary AB T2N 1N4 Can

CONTA, BARBARA SAUNDERS, IMMUNOGENETICS, LYMPHOKINES. *Educ:* Univ Rochester, PhD(microbiol), 75. *Prof Exp:* AT SCH MED, YALE UNIV, 80- *Mailing Add:* Dept Molecular Biol Abbott Labs Bldg AP9A Abbott Park IL 60064

CONTA, BART(HOLOMEW) J(OSEPH), b Rochester, NY, Mar 29, 14; m 37; c 3. MECHANICAL ENGINEERING. *Educ:* Univ Rochester, BS, 36; Cornell Univ, MS, 37. *Prof Exp:* Inst mech eng, Cornell Univ, 37-40; res engr, Texas Co, NY, 40-42; from instr to assoc prof heat-power eng, Cornell Univ, 42-47, in charge lab instr, Navy Steam Sch, 44-46; prof mech eng, Col Appl Sci, Syracuse Univ, 47-51; PROF MECH ENG, CORNELL UNIV, 51- *Concurrent Pos:* Ford Found vis prof, Univ Valle, Colombia, 64-65; NSF sci fac fel, Univ Calif, Berkeley, 66-67. *Mem:* Soc Hist Technol; AAAS; Am Asn Univ Prof. *Res:* Thermodynamics; energy conversion; history of science and technology. *Mailing Add:* Dept Mech & Aerospace Eng Cornell Univ Ithaca NY 14853

CONTANDRIOPOULOS, ANDRE-PIERRE, b France, July 1, 43; Can & French citizen; m 79; c 2. HEALTH ECONOMICS. *Educ:* Univ Aix-Marseille, Licence, 67; Univ Montreal, DSc, 68, PhD(econ), 76. *Prof Exp:* Asst prof epidemiol, Univ McGill, 72-75; dir, Interdisciplinary Res Group, Sante, 77-91, PhD Prog Community Health, 78-85; from asst prof to assoc prof, 75-85, PROF HEALTH ADMIN, UNIV MONTREAL, 85- *Concurrent Pos:* Mem, Eval Comt, Nat Health Res & Develop Prog, 72-, Working Group Res Into Health Care Orgn, 84-86; chmn, Eval Comt Res Group, FRSQ, 85-86, Can Health Econ Res Asn, 86; prin investr on over 40 res projs, 85-; vchmn bd, Montreal Chest Hosp, 87-; assoc, Can Inst Advan Res, 88- *Mem:* Am Pub Health Asn; Can Pub Health Asn; Am Eval Asn; Can Eval Soc; Can Health Econ Res Asn; Asn Univ Progs Health Admin. *Res:* Medical care organization; evaluation; health manpower planning; determinant of health; health economics. *Mailing Add:* Gris Univ Montreal CP 6128 Succ A Montreal PQ H3C 3J7 Can

CONTARIO, JOHN JOSEPH, b Detroit, Mich, Nov 23, 44; m 66; c 3. ANALYTICAL CHEMISTRY, CHROMATOGRAPHY. *Educ:* Eastern Mich Univ, BS, 66; Iowa State Univ, MS, 68, PhD(anal chem), 71. *Prof Exp:* Group leader chem phys testing, Chem Div, Abbott Labs, 70-72, sr isotope analyst radiopharmaceuts, Diag Div, 72-73, group leader antibiotic & instrumental anal, Chem Div, 73-74, sect head anal serv & methods develop, Chem-Agr Prod Div, 74-80; sect head anal chem, Merrell Dow Pharmaceut, 80-87; dir & gen mgr, Anal Serv Div, Hill Top Biol Labs, 87-89; MGR ANALYTICAL DEVELOP, CONSUMER HEALTH CARE DIV, MILES, INC, 89- *Mem:* Am Chem Soc; Sigma Xi; Am Soc Qual Control. *Res:* Analytical support and new test methods for pharmaceutical (new drug and dosage form) development. *Mailing Add:* 53387 Monticola Lane Bristol IN 46507-9694

CONTE, FRANK PHILIP, b South Gate, Calif, Feb 2, 29; m 54; c 3. COMPARATIVE PHYSIOLOGY, CELL PHYSIOLOGY. *Educ:* Univ Calif, Berkeley, AB, 51, PhD(physiol, biochem), 61. *Prof Exp:* Asst physiol, Univ Calif, Berkeley, 51-52 & aero-med lab, Wright Air Develop Ctr, Ohio, 52-53; asst biol div, Oak Ridge Nat Lab, Tenn, 53-56; asst chem, Wash State Univ, 56-57; asst prof, Cent Ore Col, 57-59; asst radiation physiol, Donner Lab, Univ Calif, Berkeley, 59-60; from asst prof to assoc prof, 61-71, PROF ZOOL, ORE STATE UNIV, 71- *Concurrent Pos:* Sr fel, NIH, 68-69; vis prof, Dept Zool, Duke Univ, 68-69; prog dir regulatory biol, NSF, 72-73. *Mem:* AAAS; Am Soc Zoologists; Am Physiol Soc; Soc Gen Physiologists; Am Zool Soc; Fedn Am Socs Exp Biol. *Res:* Cellular regeneration in aquatic vertebrate; regulation of internal body fluids in aquatic vertebrates; biogenesis of cell membranes; exocrine glands in invertebrates. *Mailing Add:* Dept Zool Ore State Univ Corvallis OR 97331-2914

CONTE, JOHN SALVATORE, b Philadelphia, Pa, June 12, 32; m 60. ORGANIC CHEMISTRY. *Educ:* La Salle Col, AB, 54; La State Univ, MS, 56; Univ Pa, PhD(chem), 59. *Prof Exp:* Asst instr gen chem, La State Univ, 54-56; asst instr org chem, Univ Pa, 56-57; res group leader, Org Synthetic Sect, Scott Paper Co, 59-69; PROD MGR, QUAKER CHEM CO, 75- *Mem:* Am Chem Soc; The Chem Soc. *Res:* Kinetics and mechanism of organic reactions; mechanism of polymer reactions; resin synthesis and free radical polymerization; specialty chemicals for the pulp and paper industries. *Mailing Add:* 908 Pierce Rd Norristown PA 19403

CONTE, SAMUEL D, b Lackawanna, NY, June 5, 17; m 48; c 5. NUMERICAL ANALYSIS, SOFTWARE METRICS. *Educ:* Buffalo State Teachers Col, BS, 39; Univ Buffalo, MS, 43; Univ Mich, MA, 48, PhD(math), 50. *Prof Exp:* From instr to assoc prof math, Wayne Univ, 46-56; mgr math anal dept, Space Tech Labs, 56-61; mgr math dept, Aerospace Corp, 61-62; dir comput sci ctr, 62-66, head comput sci dept, 62-79, PROF COMPUT SCI, PURDUE UNIV, 79- *Concurrent Pos:* Vis prog dir, NSF, 89-90. *Mem:* Soc Indust & Appl Math; Math Asn Am; Asn Comput Mach. *Res:* Numerical analysis and computation; computer science education; software metrics and software engineering. *Mailing Add:* 3746 Capilano Dr West Lafayette IN 47906

CONTENTO, ISOBEL, b Wuwei, China, Sept 15, 40; US citizen; div. IMMUNOLOGY, NUTRITION. *Educ:* Univ Edinburgh, BSc, 62; Univ Calif, Berkeley, MA, 64, PhD(immunochem), 69. *Prof Exp:* Asst prof biol, Merritt Col, 64-65; fac fel, Johnston Col, Univ Redlands, 69-76; ASSOC PROF NUTRIT, TEACHERS COL, COLUMBIA UNIV, NY, 77- *Mem:* Soc Nutrit Educ; Nat Asn Res Sci Teaching. *Res:* Psycho-social determinants of eating behavior; developmental influences on nutrition and knowledge and behavior in children and adolescents; piagetian analysis of the role of reasoning skills in consumer nutrition literacy. *Mailing Add:* Teachers Col Nutrit Res Ctr Box 188 Columbia Univ New York NY 10027

CONTI, JAMES J(OSEPH), b Coraopolis, Pa, Nov 2, 30; m 61; c 2. CHEMICAL ENGINEERING. *Educ:* Polytech Inst Brooklyn, BChE, 54, MChE, 56, DChE, 59. *Prof Exp:* Sr engr, Bettis Atomic Power Div, Westinghouse Elec Corp, 58-59; from asst prof to assoc prof, 59-64, head dept, 64-70, provost, 70-78, PROF CHEM ENG, POLYTECH INST BROOKLYN, 65-, VPRES EDUC DEVELOP & DIR, LONG ISLAND CAMPUS, 78- *Concurrent Pos:* Vis chem engr, Brookhaven Nat Lab, 60 & Res Lab, Diamond Alkali Co, 61; consult, Mobil Oil Co, 62-64, Nat Inst Gen Med Sci, NIH, HEW, 68- *Mem:* AAAS; Am Inst Chem Engrs; Am Soc Eng Educ. *Res:* Transport phenomena, including effects of induced pulsations on heat and mass transfer; chemical reaction kinetics; optimal design of multistage operations; phase equilibrium thermodynamics. *Mailing Add:* Long Island Ctr Polytech Univ Rt 110 Farmingdale NY 11735

CONTI, PETER SELBY, b NY, Sept 5, 34; m 61; c 3. ASTROPHYSICS, ASTRONOMY. *Educ:* Rensselaer Polytech Inst, BS, 56; Univ Calif, Berkeley, PhD(astron), 63. *Prof Exp:* Fel & res assoc astron, Calif Inst Technol & Hale Observ, 63-66; asst prof astron & astronr, Univ Calif, Santa Cruz, 66-71; PROF ASTROPHYS, UNIV COLO, BOULDER, 71- *Concurrent Pos:* Fulbright vis prof, Univ Utrecht, Neth, 69-70; dir, Assoc Univ Res Astron Inc, 77-86, chmn bd, 83-86; chmn, Dept Astrophys, Planetary & Atmospheric Sci, 80-86; counr, Am Astron Soc, 83-86. *Honors & Awards:* Gold Medal, Univ Liege, 75. *Mem:* Am Astron Soc; Int Astron Union; fel AAAS. *Res:* Spectroscopy of very hot and luminous stars; analysis of stellar winds. *Mailing Add:* Joint Inst Lab Astrophys Box 440 Univ Colo Boulder CO 80309-0440

CONTI, SAMUEL F, b Dec 24, 31; m; c 3. ELECTRON MICROSCOPY, MICROBIOL. *Educ:* Cornell Univ, PhD(microbiol), 59. *Prof Exp:* VCHANCELLOR RES & DEAN GRAD SCH, UNIV MASS, 80- *Mailing Add:* Grad Res Ctr Univ Mass A217 Amherst MA 01003

CONTI, VINCENT R, b New York, NY, Dec 6, 43; m 71. PHYSIOLOGY, CARDIOLOGY. *Educ:* Col of the Holy Cross, AB, 65; Tufts Univ, MD,69. *Prof Exp:* Resident, gen surg, Univ Calif, Los Angeles, 69-74; resident, cardiothoracic surg, Univ Ala, 76-79; PROF SURG, UNIV TEX MED BR, GALVESTON, 86- *Mem:* Soc Univ Surgeons; Am Asn Thoracic Surgeons; Am Col Cardiol; Am Heart Asn; Am Col Chest Physicians. *Res:* Physiology of myocardiol ischemia as it relates to surgical treatment of head disease; surgical treatment of heart disease. *Mailing Add:* Univ Tex Cardiovasc Surg Res Lab Univ Tex Med Br Rm 15 Galveston TX 77550

CONTOGOURIS, ANDREAS P, b Athens, Greece, Oct 25, 31; m 57; c 2. THEORETICAL PHYSICS. *Educ:* Nat Tech Univ Athens, dipl elec eng, 54; Cornell Univ, PhD(physics), 61. *Prof Exp:* Res assoc theoret particle physics, Cornell Univ, 61-62; prof theoret physics, Democritus Nuclear Res Ctr, Greece, 62-64; res assoc theoret particle physics, Europ Orgn Nuclear Res, Geneva, Switz, 64-66; assoc lectr theoret physics, Univ Paris, Orsay, 66-68; ASSOC PROF THEORET PARTICLE PHYSICS, McGILL UNIV, 68- *Concurrent Pos:* Greek rep to gov coun, Europ Orgn Nuclear Res, 65-66; Nat Res Coun Can grantee, 68- *Mem:* Am Phys Soc. *Res:* Theoretical problems and phenomenological applications related to the interactions of elementary particles at high energies. *Mailing Add:* Rutherford Phys Bldg McGill Univ Montreal PQ H3A 2T8 Can

CONTOIS, DAVID ELY, microbiology; deceased, see previous edition for last biography

CONTOPOULOS, GEORGE, b Egion, Greece, Oct 3, 28; m 64; c 4. DYNAMICAL ASTRONOMY, RELATIVITY & COSMOLOGY. *Educ:* Univ Athens, BSc, 50, PhD(astron), 53. *Prof Exp:* Prof astron, Univ Thessaloniki, Greece, 57-75; PROF ASTRON, UNIV ATHENS, GREECE, 75-; DIR, NAT OBSERV GREECE, 91- *Concurrent Pos:* Vis prof, Yale Univ, 62, Harvard Univ, 68, Mass Inst Technol, 69, Univ Chicago, 69 & 81, Univ Md, 71, 74 & 78, Cornell Univ, 82 & Univ Fla, 85, 86, 87, 89 & 90; res assoc, Yerkes Observ, 63, Inst Advan Study, Princeton Univ, 63, Inst Space Studies, NASA, 63, 65, 66 & 67 & Goddard Space Flight Ctr, 71, Europ Southern Observ, 76, 77, 79, 80, 82-88. *Honors & Awards:* Brouwer Prize, Am Astron Soc, 82. *Mem:* Int Astron Union (secy, 73-76). *Res:* Author of 10 books on astronomy and 110 papers on dynamical astronomy and relativity. *Mailing Add:* Dept Astron Univ Athens Athens 15783 Greece

CONTRACTOR, DINSHAW N, b Bangalore, India, Apr 23, 33; US citizen; m 58; c 5. HYDRAULIC MODELING, WATER QUALITY MODELING. *Educ:* Univ Baroda, BE, 57; State Univ Iowa, MS, 60; Univ Mich, Ann Arbor, PhD(civil eng), 63. *Prof Exp:* Res scientist, Hydronautics, Inc, Laurel, Md, 63-68; asst prof hydraulics, Va Polytech Inst & State Univ, 68-75, assoc prof, 75-81; vis prof hydraul, Water & Energy Res Inst, Univ Guam, 80-81; PROF, DEPT CIVIL ENG & ENG MECH, UNIV ARIZ, TUCSON, 81- *Mem:* Am Soc Civil Engrs; Am Soc Mech Engrs; Am Geophys Union. *Res:* Hydraulic modeling; optimization of hydraulic systems; simulation of groundwater systems; saltwater intrusion into aquifers; water quality modeling. *Mailing Add:* Dept Civil Eng & Eng Mech Univ Ariz Tucson AZ 85721

CONTRERA, JOSEPH FABIAN, b New York, NY, Nov 18, 38; m 62; c 1. ENDOCRINE PHYSIOLOGY, NEUROPHARMACOLOGY. *Educ:* NY Univ, BA, 60, MS, 61, PhD(endocrine physiol), 66. *Prof Exp:* Res asst neurochem, Sch Med, NY Univ, 60-62, res assoc physiol, Lab Exp Hemat, 63-66; asst prof, 67-70, assoc prof physiol, Univ Md, College Park, 70-77; neuropharmacologist, 77-81, SUPVRY NEUROPHARMACOLOGIST, OFF DRUG RES & REV, DEPT HEALTH & HUMAN SERV, FOOD & DRUG ADMIN, 81- *Concurrent Pos:* Lectr biol, Hunter Col, 63-64; post-doctoral trainee, neuropharmacol, Dept Pharmacol, Sch Med, Yale Univ, 66-67; vis instr, Dept Pharmacol & Exp Therapeut, Johns Hopkins Univ Sch Med, 75; adj assoc prof, neuroncol prog, Dept Radiation Ther, Sch Med, Univ Md, 79-80. *Mem:* AAAS; Soc Exp Biol & Med; Am Physiolsoc; Soc Neuro Sci; Sigma Xi. *Res:* Physiology and pharmacology of the autonomic and central nervous system; experimental hematology; endocrinology; drug development and risk assessment; evaluation of animal pharmacology and toxicology data supporting the safety of new experimental CNS drugs. *Mailing Add:* 17605 Cashell Rd Rockville MD 20853

CONTRERAS, PATRICIA CRISTINA, b Nov 25, 56. PHENCYLICIDINE, OPIOIDS. *Educ:* Univ Minn, PhD(pharmacol), 83. *Prof Exp:* RES SCIENTIST, G D SEARLE PHARMACEUT, 86- *Mem:* Soc Neurosci. *Res:* Mechanism of action of phencyclidine, and opiods using behavioral and biochemical methods is the primary focus of my lab. *Mailing Add:* G D Searle Pharmaceut 4901 Searle Pkwy Skokie IL 60077

CONTRERAS, THOMAS JOSE, b Morenci, Ariz, Mar 20, 45; m 65; c 2. BIOCHEMISTRY, PHYSIOLOGY. *Educ:* Northern Ariz Univ, BS, 67; Univ Utah, MS, 70. *Prof Exp:* Head spec chem clin serv, Long Beach Naval Hosp, 70; res chemist lipid chem, Naval Med Res Inst, 70-71; chief exp res div cryobiol, Naval Blood Res Lab, Boston, 71-76; res biochemist physiol, Armed Forces Radiobiology Res Inst, 76-82; diving physiologist, Hyperbaric Med Ctr, Naval Med Res Inst, Bethesda, 82-85; fleet health care mgr, Naval Med Res & Develop Command, Bethesda, 85-87; TECHNOL AREA MGR BIOMED/CBR, OFF NAVAL TECHNOL, ARLINGTON, 87- *Concurrent Pos:* Consult, Naval Blood Res Lab, 76-, Ctr Blood Res 76- & Blood Bank, Naval Hosp, Bethesda, 77- *Mem:* AAAS; Am Asn Blood Banks. *Res:* Granulocyte isolation and preservation for postirradiation treatment, in vitro and in vivo evaluation; exploring the gas exchange mechanisms underlying decompression sickness. *Mailing Add:* Technol Area Mgr Biomed Chem Biol & Radiol Off Naval Technol 800 N Quincy St Arlington VA 22217-5000

CONTROULIS, JOHN, b Chicago, Ill, Mar 31, 19; m 52; c 2. ORGANIC CHEMISTRY. *Educ:* Transylvania Col, AB, 40; Univ Cincinnati, MA, 41; Univ Mich, PhD(chem), 50. *Prof Exp:* Sr res chemist, 41-47, patent chemist, 49-50, asst dir prods develop, 50-62, mgr mkt res & develop, 62-71, supt capsule develop div, Parke Davis & Co, 71-78; MGR NEW PROD EVAL & PLANNING, WARNER-LAMBERT CO, 78- *Mem:* Am Chem Soc; AAAS; Sigma Xi. *Res:* Synthesis of organic chemicals for medicinal uses; organometallic compounds; nitrogen heterocyclic compounds; chloromycetin and intermediates. *Mailing Add:* Seven Indian Head Rd Morristown NJ 07960

CONVERSE, ALVIN O, b Ridley Park, Pa, Nov 6, 32; m 54; c 3. CHEMICAL ENGINEERING. *Educ:* Lehigh Univ, BS, 54; Univ Del, MChE, 58, PhD(chem eng), 61. *Prof Exp:* Tech serv engr, Sun Oil Co, 54-57; asst prof chem eng, Carnegie Inst Technol, 60-63; assoc prof, 63-69, PROF ENG, DARTMOUTH COL, 69- *Mem:* Am Chem Soc; Am Inst Chem Engrs. *Res:* Bioengineering; biochemical engineering. *Mailing Add:* Thayer Sch Eng Dartmouth Col Hanover NH 03755

CONVERSE, JAMES CLARENCE, b Brainerd, Minn, Apr 2, 42; m 65; c 4. WASTE MANAGEMENT, AGRICULTURAL ENERGY. *Educ:* NDak State Univ, BS, 64, MS, 66; Univ Ill, PhD(agr eng), 70. *Prof Exp:* From asst prof to assoc prof, 70-80, PROF AGR ENG, UNIV WIS- MADISON, 80, CHMN, 88- *Concurrent Pos:* Prin investr, Univ Wis-Madison, 70- *Mem:* Am Soc Agr Eng. *Res:* Domestic waste management with primary emphasis on mound development; methane production from animal and poultry manure; alcohol production from agricultural products. *Mailing Add:* Univ Wis Agr Eng Dept 460 Henry Mall Madison WI 53706

CONVERSE, RICHARD HUGO, b Greenwich, Conn, Sept 18, 25; m 47; c 3. PHYTOPATHOLOGY. *Educ:* Univ Calif, BS, 47, MS, 48, PhD, 51. *Prof Exp:* Asst prof plant path, SDak State Col, 50-52; asst prof bot & plant path, Okla State Univ & asst plant pathologist & agent, USDA, 52-57; plant pathologist, Plant Indust Sta, USDA, 57-67; plant pathologist, USDA, 67-72, res leader hort crops, Agr Res Serv, 72-82, res plant pathologist, 82-90; prof bot, Ore State Univ, 67-90; RETIRED. *Mem:* Am Soc Hort Sci; Am Phytopath Soc; fel Microbiol Soc Am. *Res:* Small fruit diseases, particularly virus and virus-like diseases of Rubus, Fragaria, and Vaccinium. *Mailing Add:* USDA ARS HCRL 3420 NW Orchard Ave Corvallis OR 97330-5088

CONVERTINO, VICTOR ANTHONY, b Troy, NY, Apr 26, 49; m 78; c 3. EXERCISE PHYSIOLOGY, ENVIRONMENTAL PHYSIOLOGY. *Educ:* San Jose State Univ, BA, 71, BS, 72; Univ Calif, Davis, MA, 74, PhD(physiol), 81. *Prof Exp:* Teaching asst exercise physiol, Univ Calif, Davis, 72, 74, res asst, 74-76, res assoc, 77-78, supvr stress test, 78-79; res assoc exercise physiol, Sch Med, Stanford Univ, 79-81; asst prof, Dept Phys Educ, Univ Ariz, 82-85; RES PHYSIOLOGIST, BIOMED RES LAB, NASA, KENNEDY SPACE CTR, 85- *Concurrent Pos:* Guest lectr, Dept Phys Educ, Univ Calif, Davis, 77-; test proctor, Calif Highway Patrol, Calif State Personnel Bd, 79; res assoc exercise physiol, NASA, 79-82; lectr, Dept Human Performance, San Jose State Univ, 81-82, & Dept Aeronaut & Astronaut, Stanford Univ, 82-83. *Honors & Awards:* New Investr Award, Am Col Sports Med, 82; Ellingson Literary Award, Aerospace Med Asn, 85; Vis Scholar Award, Am Col Sports Med, 86- *Mem:* Fel Am Col Sports Med; fel Aerospace Med Asn; AAAS; Am Physiol Soc; NY Acad Sci. *Res:* Investigation into the effects of exercise training and microgravity exposure on chronic cardiovascular and fluid-electrolyte-endocrine adaptations; aerospace medicine; cardiac rehabilitation applications. *Mailing Add:* NASA Life Sci Res Off Mail Code MD-RES-P Kennedy Space Center FL 32899

CONVERY, F RICHARD, b Olympia, Wash, June 12, 32; m 55; c 3. ORTHOPEDIC SURGERY. *Educ:* Univ Wash, BA, 54, MD, 58; Am Bd Orthop Surg, dipl, 69. *Prof Exp:* Resident, 61-66, from instr to assoc prof orthop surg, Sch Med, Univ Wash, 67-72; assoc prof & dir rehab, Div Orthop & Rehab, 72-77, PROF ORTHOP SURG, SCH MED, UNIV CALIF, SAN DIEGO, DIR REHAB, DIV ORTHOP & REHAB, SCH MED, 77- *Concurrent Pos:* Sr fel orthop surg, Sch Med, Univ Wash, 63-64; Southern Calif Arthritis Found clin fel, Ranchos Los Amigos Hosp, 66-67. *Mem:* Am Rheumatism Asn; Am Acad Orthop surg; Am Orthop Asn. *Res:* Degeneration and repair of articular cartilage as related to joint reconstruction; surgical management of rheumatoid arthritis; ongoing investigation of implant fixation to bone via application of external pressure and the use of cement composites. *Mailing Add:* Div Orthop & Rehab Univ Calif Sch Med 225 W Dickenson St San Diego CA 92103

CONVERY, ROBERT JAMES, b Philadelphia, Pa, Jan 17, 31; m 52; c 4. ORGANIC CHEMISTRY. *Educ:* St Joseph's Col, Philadelphia, BS, 52; Univ Notre Dame, MS, 56; Univ Pa, PhD(chem), 59. *Prof Exp:* Res chemist, Rohm and Haas Co, Pa, 54, Atlantic Refining Co, 55 & E I du Pont de Nemours & Co, 56-59; res chemist, Res & Develop Div, Sun Oil Co, 59-62 & Res Group, 62-63, asst sect chief, 63-64; from asst prof to assoc prof chem, 64-70, actg head dept, 64-66, assoc dean sci, 71-74, head dept, 66-71 & 74- 78, dean fac, 78-86, dean studies, 86-87, PROF CHEM, UNIV STEUBENVILLE, 70-, CHMN DEPT CHEM, 87- *Concurrent Pos:* Vis lectr, St Mary's Col, Ind, 54-55; instr, Evening Div, St Joseph's Col, Pa, 60-62. *Mem:* Sigma Xi. *Res:* Organic mechanisms and synthesis; free radicals. *Mailing Add:* Dept Chem Franciscan Univ Steubenville Steubenville OH 43952

CONVEY, EDWARD MICHAEL, b Hicksville, NY, Oct 12, 39; m 62; c 2. PHYSIOLOGY, ENDOCRINOLOGY. *Educ:* Mich State Univ, BS, 63, MS, 65; Rutgers Univ, PhD(physiol), 68. *Prof Exp:* From asst prof to assoc prof dairy physiol, Mich State Univ, 68-78, prof physiol, 78-83; exec dir, Basic Animal Res, 83-89, EXEC DIR, REGULATORY AFFAIRS ANIMAL RES, MERCK SHARPE & DOHME, INC, 89- *Mem:* Am Dairy Sci Asn; Am Soc Animal Sci; Endocrine Soc; AAAS; Soc Study Reproduction. *Res:* Lactational and reproduction physiology; anterior pituitary. *Mailing Add:* Regulatory Affairs Merck Sharpe & Dohme PO Box 2000 Rahway NJ 07065

CONVEY, JOHN, b Durham, Eng, Mar 29, 10; Can citizen; m 39; c 4. RESEARCH MANAGEMENT. *Educ:* Univ Alta, BSc, 33, MSc, 36; Univ Toronot, PhD(atomic physics), 39. *Hon Degrees:* DSc, McMaster Univ, Univ Windsor, Notre Dame Univ. *Prof Exp:* Lectr physics, Univ Alta, 35-36; prof, Univ Toronto, 46-48; dir mat sci, Dept Energy Mines & Resources, 48-75; CONSULT, CAN MINING INDUST, 75- *Concurrent Pos:* Consult, tech adv, Atomic Energy Can, 79-91. *Mem:* Can Inst Mining & Metall; fel Am Soc Metals Int; Am Inst Mining & Metall. *Res:* Atomic structure; applied atomic energy; study of stress relationships in rock-metal and coal mining. *Mailing Add:* Box 70 265 Shore Lane Wasaga Beach ON L0L 2P0 Can

CONWAY, BRIAN EVANS, b London, Eng, Jan 26, 27; m 54; c 1. PHYSICAL CHEMISTRY. *Educ:* Imp Col, London, BSc, 46, PhD & dipl, 49, DSc, 61. *Prof Exp:* Res assoc chem, Inst Cancer Res, London, 49-54; asst prof, Univ Pa, 54-56; from asst prof to assoc prof, 56-60, chmn dept, 66-69, PROF CHEM, UNIV OTTAWA, 60-, CHMN DEPT CHEM. *Concurrent Pos:* Consult tech ctr, Gen Motors; Commonwealth vis prof, Univs Southampton & Newcastle, 69-70. *Honors & Awards:* Noranda lectr award, Chem Inst Can, 64. *Mem:* The Chem Soc; Electrochem Soc; fel Brit Chem Soc; fel Royal Inst Chem; fel Royal Soc Can. *Res:* Electrochemistry; kinetics of electrode processes; adsorption at electrodes; isotopic effects in electrode reactions; polyelectrolytes; thermodynamics of polymer solutions. *Mailing Add:* Dept Chem Univ Ottawa 365 Nicholas St Ottawa ON K1N 6N5 Can

CONWAY, DWIGHT COLBUR, b Long Beach, Calif, Nov 14, 30; m 62; c 4. MASS SPECTROSCOPY. *Educ:* Univ Calif, BS, 52; Univ Chicago, PhD(chem), 56. *Prof Exp:* From instr to asst prof chem, Purdue Univ, 56-63; assoc prof, 63-67, PROF CHEM, TEX A&M UNIV, 67- *Mem:* Am Chem Soc; Am Phys Soc; Am Soc Mass Spectros; Sigma Xi. *Res:* Ion-molecule reactions; analysis of bacteria and polymers by pyrolysis mass spectroscopy. *Mailing Add:* Dept Chem Tex A&M Univ College Station TX 77843

CONWAY, EDWARD DAIRE, III, b New Orleans, La, Feb 7, 37; m 61; c 2. MATHEMATICAL ANALYSIS. *Educ:* Loyola Univ, BS, 59; Ind Univ, MA, 63, PhD(math), 64. *Prof Exp:* Vis mem, Courant Inst Math Sci, NY Univ, 64-65; asst prof math, Univ Calif, San Diego, 65-67; from asst prof to assoc prof, 67-74, PROF MATH, TULANE UNIV, 74- *Mem:* Am Math Soc; Soc Indust & Appl Math; Math Asn Am; AAAS. *Res:* Nonlinear partial differential equations. *Mailing Add:* 7725 Plum St New Orleans LA 70118

CONWAY, GENE FARRIS, b Cynthiana, Ky, Aug 24, 28; m 50; c 3. INTERNAL MEDICINE, CARDIOVASCULAR DISEASES. *Educ:* Univ Ky, BS, 49; Univ Cincinnati, MD, 52. *Prof Exp:* Intern med, Philadelphia Gen Hosp, 52-53; resident internal med, Louisville Gen Hosp, 53-54; chief med serv, USAF Hosp, Topeka, Kans, 54-56; from resident to chief resident internal med, Cincinnati Gen Hosp, 56-59, USPHS res fel cardiol, 59-61, from asst prof to assoc prof, 61-70, asst dir dept internal med, 73-75, assoc dean clin & house staff affairs, 75-78, PROF MED, COL MED, UNIV CINCINNATI, 70-, ASSOC DIR DEPT INTERNAL MED, 88- *Concurrent Pos:* Clin investr, Vet Admin Hosp, Cincinnati, 61-63, chief of cardiol, 63-, assoc chief of staff for res, 72-74; fel, Coun Clin Cardiol, Am Heart Asn, 69-; chief med serv, Vet Admin Hosp, Cincinnati, 72-75. prog adv cardiovasc res, Res Serv, Vet Admin, 74-76. *Mem:* Am Col Cardiol; Am Heart Asn; Ctr Soc Clin Res. *Res:* Chemistry of myocardial contractile proteins; myocardial biology; congestive heart failure; arrhythmias; ischemic heart disease. *Mailing Add:* MSB 6065 Univ of Cincinnati Med Ctr Cincinnati OH 45267-0557

CONWAY, H(ARRY) D(ONALD), b Chatham, Eng, Dec 3, 17; c 2. MECHANICS. *Educ:* Univ London, BSc, 42, PhD(mech), 45, DSc(mech), 49; Cambridge Univ, MA, 46, ScD, 71. *Prof Exp:* Eng apprentice, Dockyard, Eng, 34-38; stress analyst, Short Bros, 38-39; sci officer, Nat Phys Lab, 42-45; assoc prof, 47-48, PROF MECH, CORNELL UNIV, 48- *Concurrent Pos:* Guggenheim fel & vis prof, Imp Col, London, 53-54; Stone vis prof, Ohio State Univ, 58-59; NSF sr fel, 61-62. *Honors & Awards:* Baden-Powell Prize, Royal Aeronaut Soc. *Mem:* Am Soc Mech Engrs. *Res:* Vibrations; elasticity; plates and shells; lubrication; thin films adhesion. *Mailing Add:* 206 Thurston Hall Appl Mech Cornell Univ Main Campus Ithaca NY 14853

CONWAY, HERTSELL S, b Peoria, Ill, Sept 30, 14; wid; c 2. INFORMATION SCIENCE, CHEMICAL LITERATURE. *Educ:* Univ Chicago, SB, 32, PhD(org chem), 37. *Prof Exp:* Asst, Emulsol Corp, Ill, 37-39; chemist, Food & Drug Admin, USDA, 39-41 & Lambert Pharmacal Co, Mo, 41-46; org chemist, US Govt Rubber Labs, Ohio, 46-47; chemist, Standard Oil Co, Ind, 47-49, group leader, 49-55, sect leader, 55-60; res supvr, Am Oil Co, 60-65; sr info scientist, Standard Oil Co, Ind, 65-80; RETIRED. *Mem:* Am Chem Soc; Am Translators Asn. *Res:* Analysis of petroleum products and petrochemicals; indexing and searching; technical writing and editing. *Mailing Add:* 435 Old Stone Rd Munster IN 46321

CONWAY, JAMES JOSEPH, b Arlington, Mass, Feb 20, 29; m 52; c 5. APPLIED MATHEMATICS. *Educ:* Mass Maritime Acad, BS, 49; Boston Col Grad Sch, MEd, 58, MBA, 63; Tex Tech Univ, PhD(bus admin), 70. *Prof Exp:* Assoc prof math, Salem State Col, Mass, 60-70; PROF BUS STATIST & QUANT ANALYSIS, UNIV NEBR, OMAHA, 70- *Res:* Statistical and quantitative analysis of business data; corporate equity valuation models and their reliability and usefulness in the management planning process. *Mailing Add:* 4916 N 103rd St Omaha NE 68134

CONWAY, JOHN BELL, b Madison, Wis, April 5, 36; m 61; c 2. WATER QUALITY. *Educ:* San Diego State Univ, BS, 64, MS, 67; Univ Minn, MPH, 70, PhD(environ biol), 73. *Prof Exp:* Marine technician, Scripps Inst Oceanog, Univ Calif, 58-59; pub health sanitarian, Div Sanit, San Diego County Health Dept, 64-65; pub health biologist, Dept Natural Resources, State Wis, 67-69; res asst pub health, Sch Pub Health, Univ Minn, 70-72; asst prof, Dept Biol Sci, Wright State Univ, Ohio, 72-76; assoc prof, Bacteriol & Pub Health, Wash State Univ, 76-81; head, Div Occup & Environ Health, 84-86, assoc dir, Grad Sch Pub Health, 86-88, PROF ENVIRON HEALTH, SAN DIEGO STATE UNIV, 81- *Concurrent Pos:* Adj asst prof, Sch Pub Health, Univ Minn, 74-76; lectr, Ctr Lake Superior Environ Studies, Univ Wis, 78; disaster consult, Greene Co Health Dept, Ohio, 74-75; consult, Diver Health Risk Assessment, City of San Diego, 85-87. *Mem:* Sigma Xi; Nat Environ Health Asn; Am Pub Health Asn. *Res:* Water quality research, emphasizing chemical, biological and bacteriological indicators of waste and potable water; investigations of the water pollutants involved in diseases of man. *Mailing Add:* Grad Sch Pub Health San Diego State Univ San Diego CA 92182

CONWAY, JOHN BLIGH, b New Orleans, La, Sept 22, 39; m 64. MATHEMATICAL ANALYSIS. *Educ:* Loyola Univ, BS, 61; La State Univ, PhD(math), 65. *Prof Exp:* From asst prof to assoc prof, 65-77, PROF MATH, IND UNIV, BLOOMINGTON, 77- *Mem:* Am Math Soc; Math Asn Am. *Res:* Functional analysis. *Mailing Add:* Univ Tenn Knoxville TN 37996-1300

CONWAY, JOHN GEORGE, JR, b Pittsburgh, Pa, May 16, 22; m 47; c 7. ATOMIC SPECTROSCOPY. *Educ:* Univ Pittsburgh, BS, 44. *Prof Exp:* Chemist, Los Alamos Sci Lab, NMex, 44-46; asst physics, Univ Pittsburgh, 46; physicist, Lawrence Berkeley Lab, Univ Calif, Berkeley, 46-86; RETIRED. *Concurrent Pos:* Mem comt, Line Spectra of the Elements, Nat Acad Sci, 67; res assoc, Nat Ctr Sci Res, Orsay, France, 73-74 & 79-80; chmn comt line spectra of elements, Nat Res Coun, 74-75. *Honors & Awards:* William F Meggers Medal, Optical Soc Am, 80. *Mem:* Soc Appl Spectros; assoc Am Phys Soc; fel Optical Soc Am. *Res:* Spectroscopy; chemical analysis and absorption and emission of transuranium elements; the spectra of higher ionized atoms. *Mailing Add:* 1153 King Dr El Cerrito CA 94530

CONWAY, JOHN RICHARD, b Cincinnati, Ohio, Feb 28, 43; m 69; c 1. ENTOMOLOGY, HUMAN ANATOMY. *Educ:* Ohio State Univ, BS, 65; Univ Colo, MA & PhD(biol), 75. *Prof Exp:* Asst prof, Marycrest Col, 76-78; asst prof human anat & physiol, Elmhurst Col, 78-85; CONSULT, 85- *Mem:* Entom Soc Am; AAAS. *Res:* Biology of the honey ant, Myrmecocystus mexicanus. *Mailing Add:* Dept Biol Univ Scranton Scranton PA 18510

CONWAY, JOSEPH C, JR, b Wilkes-Barre, Pa, Mar 11, 39; m 65; c 3. TRIBOLOGY, MATERIALS PROCESSING. *Educ:* Pa State Univ, BS, 61, MS, 63, PhD(eng mech), 68. *Prof Exp:* Proj engr, Ord Res Lab, 64-76, from asst prof to assoc prof, 68-82, PROF ENG MECH, PA STATE UNIV, 82- *Concurrent Pos:* Consult failure anal. *Mem:* Am Soc Eng Educ; Am Soc Mech Engrs; Soc Exp Mech; Sigma Xi. *Res:* Numerical analysis of crack propagation problems; frictional wear of ceramics and ceramic composites; test methodology of ceramics and ceramic composites; non destructive testing. *Mailing Add:* 129 W Prospect Ave State College PA 16801

CONWAY, KENNETH EDWARD, b Philadelphia, Pa, June 7, 43; m 68; c 3. PLANT PATHOLOGY, MYCOLOGY. *Educ:* State Univ NY Potsdam, BS, 66; State Univ NY Col of Forestry, Syracuse, MS, 68; Univ Fla, PhD(bot), 73. *Prof Exp:* Teacher biol & ecol, Alachua County Bd Pub Instr, 70-73; res asst, 73-75, asst res scientist plant path, Univ Fla, 75-78; asst prof, 78-82, assoc prof, 82-87, PROF PLANT PATH, OKLA STATE UNIV, 87- *Concurrent Pos:* Secy-treas, Southern Div, Am Phytopath Soc, 83-86, vpres, 86, pres, 87 & counr, 89-92; pres-elect, Okla Acad Sci, 87-88 & pres, 88-90. *Mem:* Mycol Soc Am; Am Photopath Soc; Sigma Xi. *Res:* Etiology and control of soilborne and foliar diseases of horticultural crops (vegetables, fruits and ornamentals) and forest and shade trees; biological control of soilborne diseases. *Mailing Add:* Dept Plant Path 110NRC Okla State Univ Stillwater OK 74078-9947

CONWAY, LYNN ANN, b Mt Vernon, NY, Jan 2, 38. COMPUTER SCIENCE, ELECTRICAL ENGINEERING. *Educ:* Columbia Univ, BS, 62, MSEE, 63. *Prof Exp:* Mem res staff, IBM Corp, 64-69; sr staff engr, Memorex Corp, 69-73; mem res staff, Xerox Palo Alto Res Ctr, 73-78, res fel & mgr VLSI Syst Design Area & Knowledge Systs Area, 78-83; chief scientist & asst dir strategic comput, Defense Adv Res Projs Agency, 83-85; PROF ELEC ENG & COMPUT SCI & ASSOC DEAN COL ENG, UNIV MICH, ANN ARBOR, 85- *Concurrent Pos:* Consult, Syst Industs, 73-74; vis assoc prof elec eng & comput sci, Mass Inst Technol, 78-79; ed, Addison-Wesley VLSI Systs Series, 83-; USAF Adv Bd, 87-90; sr fel, Univ Mich Soc Fels, 87- *Honors & Awards:* Harold Pender Award, Univ Pa, 84; John Price Wetherill Medal, Franklin Inst, 85; Meritorious Civilian Serv Award, Secy Defense, 85; Nat Achievement Award, Soc Women Engrs, 90. *Mem:* Nat Acad Eng; fel Inst Elec & Electronics Engrs; AAAS; Am Asn Artificial Intel. *Res:* VLSI system architecture and design; computer architecture; machine intelligence technology; design methodology; artificial intelligence; collaboration technology. *Mailing Add:* 2402 EECS Bldg Univ Mich Ann Arbor MI 48109-2116

CONWAY, PAUL GARY, b Monson, Mass, July 31, 52; m 76. ANTIPSYCHOTIC DRUG RESEARCH & DEVELOPMENT. *Educ:* Ohio Northern Univ, BS, 75; Univ Toledo, MS, 78; Ohio State Univ, PhD(pharmacol), 82. *Prof Exp:* Instr pharmacol, Ohio Northern Univ, Ada, 75-76; postdoctoral fel pharmacol, Univ Pa, Philadelphia, 82-84; sr res scientist ophthal, 84-86, res assoc neuropharmacol, 86-88, GROUP LEADER NEUROPHARMACOL, HOECHST-ROUSSEL PHARMACEUT INC, 88- *Concurrent Pos:* Adj asst prof, Fairleigh-Dickinson Univ, 86-87. *Mem:* Am Soc Pharmacol & Exp Therapeut; Soc Neurosci; NY Acad Sci; Sigma Xi; Int Brain Res Orgn. *Res:* Biochemical basis of disease states: schizophrenia, depression and congestive heart failure; molecular pharmacology; catecholamine turnover; drug receptor mechanisms; cyclic nucleotides; cardiovascular biochemistry. *Mailing Add:* Dept Biol Res Hoechst-Roussel Pharmaceut Inc Somerville NJ 08876

CONWAY, RICHARD A, b Weymouth, Mass, Nov 10, 31; m 54; c 2. CIVIL ENGINEERING. *Educ:* Univ Mass, BS, 53; Mass Inst Technol, MS, 57. *Prof Exp:* Asst preventive med officer, US Army, 54-55; develop engr, group leader, develop assoc & corp fel, 57-87, SR CORP FEL, UNION CARBIDE CORP, 88- *Concurrent Pos:* Mem, Bd Environ Studies & Toxicol, 86-89, Bd Water Sci & Technol; sci adv bd, Environ Protection Agency, 82. *Honors & Awards:* Hering Medal, Am Soc Civil Engr, 74; Gascoigne Medal, Water Pollution Control Fedn, 67, Rudolfs Medal, 74, 83; Dudley Medal, Am Soc Testing Mat, 84. *Mem:* Nat Acad Eng; fel Am Soc Civil Engrs; Sigma Xi; Int Asn Water Pollution Res & Control; Am Soc Testing & Mat; Asn Environ Engr Profs; Soc Environ Toxicol & Chem; Asn Ground Water Scientist & Engrs; Am Acad Environ Engrs. *Res:* Synthetic organic chemicals in the environment; hazardous waste disposal; risk analysis related to chemicals in the environment; disposal of industrial waste waters. *Mailing Add:* Union Carbide Corp PO Box 8361 South Charleston WV 25303

CONWAY, RICHARD WALTER, b Milwaukee, Wis, Dec 12, 31; m 53; c 3. OPERATIONS RESEARCH. *Educ:* Cornell Univ, BME, 54, PhD(opers res), 58. *Prof Exp:* Assoc prof indust eng & opers res, 58-65, prof comput sci, 65-84, PROF INFO SYSTS, CORNELL UNIV, 85- *Concurrent Pos:* Consult, Gen Motors, Int Bus Mach Corp & Hewlett Parkard. *Mem:* Asn Comput Mach. *Res:* Production control; computer sciences. *Mailing Add:* Malott Hall Cornell Univ Ithaca NY 14853

CONWAY, THOMAS WILLIAM, b Aberdeen, SDak, June 6, 31; m 57; c 2. BIOCHEMISTRY. *Educ:* Col St Thomas, BS, 53; Univ Tex, MA, 55, PhD(biochem), 62. *Prof Exp:* Fel biochem, Rockefeller Inst, 62-64, res assoc, 64; from asst prof to assoc prof, 64-73, PROF BIOCHEM, UNIV IOWA, 73- *Concurrent Pos:* Vis prof, Univ Chile, 68; Am Cancer Soc scholar, London, 80-81. *Mem:* AAAS; Am Chem Soc; Am Soc Microbiol; Am Soc Biol Chem; hon mem Biol Soc Chile; Sigma Xi. *Res:* Mechanism and control of protein biosynthesis in interferon treated cells. *Mailing Add:* Dept Biochem Univ Iowa Iowa City IA 52240

CONWAY, WALTER DONALD, b Troy, NY, Feb 4, 31; m 58; c 4. SEPARATION METHODS, DRUG METABOLISM. *Educ:* Rensselaer Polytech Inst, BS, 52; Univ Rochester, PhD(org chem), 56. *Prof Exp:* Res chemist, Esso Res & Eng Co, 56-57, Nat Cancer Inst, HEW, 57-62 & Sterling-Winthrop Res Inst, 62-65; res assoc, Lab Chem Pharmacol, Nat Heart Inst, 65-67; from asst prof to assoc prof, 67-75, ASSOC PROF PHARMACEUT & MED CHEM, SCH PHARM, STATE UNIV NY BUFFALO, 75- *Concurrent Pos:* Mem US pharmacopeia comt rev, US Pharmacopeial Conv, 75-85. *Mem:* Am Chem Soc; Am Pharmaceut Asn; Acad Pharmaceut Sci; Am Soc Pharmacol & Exp Therapeut. *Res:* Analytical methodology; identification of drug metabolites; effect of species differences and route of administration on the metabolic fate of drugs; countercurrent chromatography. *Mailing Add:* Sch Pharm State Univ NY Amherst NY 14260

CONWAY, WILLIAM SCOTT, b Phillipsburg, Pa, Aug 12, 43; m 79; c 2. PHYTOPATHOLOGY. *Educ:* Pa State Univ, BS, 65; Univ NH, MS, 74, PhD(path), 76. *Prof Exp:* Asst prof hort, Delhi Agr & Tech Col, 76-79; RES PLANT PATHOLOGIST, BELTSVILLE AGR RES CTR, 79- *Mem:* Am Phytopath Soc; Am Soc Hort Sci. *Res:* Postharvest plant pathology; investigations into preventing loss due to pathogens of fruits in storage. *Mailing Add:* Hort Crops Qual Lab HSI ARS-USDA Agr Res Ctr Beltsville MD 20705

CONWAY DE MACARIO, EVERLY, b Buenos Aires, Arg, Apr 20, 39; US citizen; m 63; c 2. MOLECULAR GENETICS. *Educ:* Nat Univ Buenos Aires, PhD(pharm), 60, PhD(biochem), 62. *Prof Exp:* Res fel immunol, Karolinska Inst, Stockholm, 69-71; sr res scientist, Lab Cell Biol, Rome, 71-73; vis scientist, WHO, Lyons, France, 73-74 & Brown Univ, 74-76; RES SCIENTIST IMMUNOL, NY STATE DEPT HEALTH, 76-, PROF, GRAD SCH PUB HEALTH SCI, 85- *Concurrent Pos:* Reviewer sci, NATO, 80- & NY State Sci & Technol Found, 85-; prin investr, Dept Energy grant, 81-; lectr, USA, Ger, Italy, Holland, Belg, Peru & China. *Honors & Awards:* Winifred Cullis Award, Int Fedn Univ Women, 70; Gold Medal, Arg Soc Biochem, 80; Gold Medal, Found Microbiol, 82. *Mem:* Scand Soc Immunol; Ital Asn Immunologists; French Soc Immunol; Am Asn Immunologists; Am Soc Microbiol. *Res:* Regulation of gene expression during differentiation; evolutionary and ecological immunochemistry of archaebacteria using monoclonal antibodies; antigenicity and immunogenicity of methane producing bacteria; author, 100 articles in scientific refereed journals; editor of two books. *Mailing Add:* Wadsworth Ctr Labs & Res NY State Dept Health Albany NY 12201

CONWELL, ESTHER MARLY, b New York, NY, May 23, 22; m 45; c 1. SOLID STATE PHYSICS. *Educ:* Brooklyn Col, BA, 42; Univ Rochester, MS, 45; Univ Chicago, PhD(physics), 48. *Prof Exp:* Instr physics, Brooklyn Col, 45-51; mem tech staff, Bell Tel Labs, 51-52; eng specialist, 52-63, mgr physics dept, GTE Labs, 63-72; prin scientist, 72-80, RES FEL, XEROX, 80- *Concurrent Pos:* Vis prof, Univ Paris, 62-63; Abbie Rockefeller Mauze prof, Mass Inst Technol, 72. *Honors & Awards:* Annual Award, Soc Women Eng, 60. *Mem:* Nat Acad Sci; Nat Acad Eng; fel Am Phys Soc; Inst Elec & ElectronicsR Engrs. *Res:* Quasi i-d conductors; polymers. *Mailing Add:* Xerox Webster Res Ctr W114/40D 800 Phillips Rd Webster NY 14580

CONYERS, EMERY SWINFORD, b Cynthiana, Ky, Aug 16, 39; m 64; c 2. SOIL SCIENCE. *Educ:* Univ Ky, BS, 61; Ohio State Univ, MS, 63, PhD(soil chem), 66; St Marys Col, Calif; MBA. *Prof Exp:* Res asst soils, Ohio State Univ, 61-66, teaching assoc, 66; chemist, US Army Aviation Mat Labs, 66-68; res specialist, Dow Chem Co, 68-74, proj mgr construct mat, 74-75, group leader membrane systs, 75-80, sales mgr, 80-81, mgr govt rel, 81-87, dir govt pub affairs, 87-90, chem co, 81-90; DIR GOVT & PUB AFFAIRS, DOWELANCO, INDIANAPOLIS, IND, 90- *Mem:* Am Soc Agron; Soil Sci Soc Am; Sigma Xi. *Res:* Fixation and release of potassium by soils and clays, metal ion-clay interactions, chemical control of soil erosion, land treatment of wastewater and membrane systems for water treatment. *Mailing Add:* 230 Bentley Dr Zionsville IN 46077

CONYNE, RICHARD FRANCIS, b Canandaigua, NY, June 26, 19; m 47; c 4. POLYMER CHEMISTRY, PLASTICS. *Educ:* Univ Rochester, AB, 41. *Prof Exp:* Chemist & group leader, Rohm & Haas Co, 41-56, head lab, 56-68, res supvr, 68-73, int plastics mgr, 73-76, res proj leader, 76-79; PLASTICS CONSULT, 79- *Mem:* Am Chem Soc; AAAS; Soc Plastics Engrs. *Res:* Plastics; plasticizers; coatings; adhesives. *Mailing Add:* Wilkinson Rd Rushland PA 18956

CONZETT, HOMER EUGENE, b Dubuque, Iowa, Oct 16, 20; m 60; c 2. EXPERIMENTAL NUCLEAR PHYSICS. *Educ:* Univ Dubuque, BS, 42; Univ Calif, PhD(physics), 56. *Prof Exp:* Degaussing physicist, Bur Ord, Dept Navy, 42-44; res physicist, Radiation Lab, Univ Calif, 56-57; vis Fulbright lectr, Univ Tokyo, 57-58; res physicist, 58-64, dir 88-inch cyclotron, 64-71, SR RES PHYSICIST, LAWRENCE BERKELEY LAB, UNIV CALIF, 64- *Concurrent Pos:* Vis res physicist, Inst Nuclear Sci, Univ Grenoble, 66-67, Univ Marseille, 85-86. *Mem:* Fel Am Phys Soc. *Res:* Nuclear reactions and scattering below 100 mev; spin-polarization phenomena in nuclear physics. *Mailing Add:* Lawrence Berkeley Lab Bldg 88 Univ Calif Berkeley CA 94720

COOCH, FREDERICK GRAHAM, b Winnipeg, Man, May 4, 28; m 58; c 3. ECOLOGY, ORNITHOLOGY. *Educ:* Queen's Univ, Ont, BA, 51; Cornell Univ, MS, 53, PhD(wildlife mgt), 58. *Prof Exp:* Arctic ornithologist, Dept Environ, Can Wildlife Serv, 54-62, head biocide invests, 62-64, staff specialist, Migratory Birds Invests, head migratory bird pop sect, head Can banding off & migratory bird coordr, 64-79, sr scientist migratory birds, 79-89; ADJ PROF & COL PROF, DEPT FISHERIES & WILDLIFE, NMEX STATE UNIV, 89- *Mem:* Fel AAAS; fel Arctic Inst NAm; Am Ornith Union; Wildlife Soc. *Res:* Wildlife ecology, especially Arctic; insecticides; vertebrate systematics. *Mailing Add:* 665 Windsmill Ct Los Cruces NM 88003

COODLEY, EUGENE LEON, b Los Angeles, Calif, Jan 14, 20; m 47; c 3. INTERNAL MEDICINE, CARDIOLOGY. *Educ:* Univ Calif, Berkeley, BA, 40; Univ Calif, San Francisco, MD, 43. *Prof Exp:* Consult cardiac dis & rehab, Calif Dept Rehab, 55-61; dir dept med, Lidcombe Hosp, Australia, 61-62; dir dept med, Kern County Hosp, 65-67; dir, Dept Med, Philadelphia Gen Hosp, 67-; prof med, Hahnemann Med Col, 67-75, assoc chmn dept, dir, Div Internal Med & dir, Residency Training Prog, 74-75; PHYSICIAN, VET ADMIN MED CTR, 75- *Concurrent Pos:* Guest lectr, Europ Cong Rheumatism, 59 & Int Cong Surg, 65; consult, Dept Rehab, Sydney, Australia, 61-62. *Mem:* Fel Am Col Physicians; Am Col Angiol; fel Am Col Cardiol; fel Am Col Gastroenterol. *Res:* Enzyme research, development of new enzyme procedures for diagnosis in medicine. *Mailing Add:* Vet Admin Med Ctr Serv III 5901 E 7th St Long Beach CA 90822

COOGAN, ALAN H, b Brooklyn, NY, Dec 19, 29. PALEONTOLOGY, GEOLOGY. *Educ:* Univ Calif, Berkeley, BA, 56, MA, 57; Univ Ill, PhD, 62; Univ Akron, JD, 77. *Prof Exp:* Instr geol, Cornell Col, 57-60; geologist, Humble Oil & Refining Co, 62-65, geologist, Esso Prod Res Co, 65-67; assoc dean res, 69-77, PROF GEOL, KENT STATE UNIV, 67- *Mem:* Am Asn Petrol Geologists; Am Paleont Soc; Soc Econ Paleontologists & Mineralogists. *Res:* Paleontology and stratigraphy; carbonate petrology; environmental geology. *Mailing Add:* Dept Geol Kent State Univ Main Campus Kent OH 44242

COOGAN, CHARLES H(ENRY), JR, mechanical engineering; deceased, see previous edition for last biography

COOGAN, PHILIP SHIELDS, b Peoria, Ill, Feb 13, 38; m 87; c 4. PATHOLOGY. *Educ:* St Louis Univ, MD, 62. *Prof Exp:* USPHS res trainee path, Presby-St Lukes Hosp, 62-64; instr path, Univ Ill, 64-68; Blake fel path, Presby-St Lukes Hosp, 69-72; from asst prof to assoc prof path, Med Col, Rush Univ, 72-74; assoc prof, Northwestern Univ, 74-78; PROF & CHMN, DEPT PATH, COL MED, EAST TENN STATE UNIV, 78- *Concurrent Pos:* Assoc attending pathologist & chmn, Tissue Comt, Presby-St Luke's Hosp, 72-74; dir, Res Path, Med Col, Rush Univ, 72-74, mem res & educ comt, 74; assoc attending pathologist & dir anat path, Northwestern Mem Hosp, 74-78, mem, Instnl Review Bd, 76-78. *Honors & Awards:* Hektoen Award, Chicago Path Soc, 69. *Mem:* Am Soc Pathologists & Bacteriologists; Int Acad Path; AMA; AAAS; Am Soc Exp Path. *Res:* Interaction of sex steroids as carcinogens with 2-acetyl aminofluorene, a carcinogen capable of inducing carcinomas of the endometrium and urinary bladder in rabbits. *Mailing Add:* Dept Path Quillen Col Med ETenn State Univ Box 19540A Johnson City TN 37614-0002

COOHILL, THOMAS PATRICK, b Brooklyn, NY, Aug 25, 41; m 62; c 3. BIOPHYSICS. *Educ:* Univ Toronto, BSc, 62; Univ Toledo, MSc, 64; Pa State Univ, PhD(biophys), 68. *Prof Exp:* Asst prof cell biol, Med Sch, Univ Pittsburgh, 68-72; res physicist, Vet Admin Hosp, Leech Farm, 68-72; assoc prof, 72-80, PROF BIOPHYS, 72-, HEAD, DEPT PHYSICS & ASTRON, WESTERN KY UNIV, 87- *Concurrent Pos:* Nat lectr, Sigma Xi, 91. *Mem:* AAAS; Biophys Soc; Am Soc Photobiol (pres, 89-90); Sigma Xi. *Res:* Effects of ultraviolet light on fungi, on mammalian tissue culture cells, on human viruses and crustaceans; human tumor visuses; activation and development; effects of radiation on nematodes for space shuttle research; ozone depletion as it affects life on earth. *Mailing Add:* Physics Dept Western Ky Univ Bowling Green KY 42101

COOIL, BRUCE JAMES, b Colfax, Wash, Aug 21, 14; m 44; c 2. PLANT PHYSIOLOGY. *Educ:* State Col Wash, BS, 36; Univ Hawaii, MS, 39; Univ Calif, PhD(plant physiol), 47. *Prof Exp:* Asst physiologist guayule res proj, Bur Plant Indust, Soils & Agr Eng, USDA, Calif, 42-45; asst plant physiologist, US Regional Salinity Lab, 45-47; assoc plant physiologist, 47-54, prof bot & plant physiologist, 54-83, EMER PLANT PHYSIOLOGIST, AGR EXP STA, UNIV HAWAII, 84- *Mem:* Bot Soc Am; Am Soc Plant Physiol; Sigma Xi. *Res:* Translocation of organic materials in plants; mineral nutrition of plants; salt absorbtion and transport in plant roots. *Mailing Add:* St John Lab Univ Hawaii 3190 Maile Way Honolulu HI 96822

COOK, ADDISON GILBERT, b Caracas, Venezuela, Apr 1, 33; US citizen; m 56; c 3. ORGANIC CHEMISTRY. *Educ:* Wheaton Col, BS, 55; Univ Ill, PhD(org chem), 59. *Prof Exp:* Asst chem, Univ Ill, 55-56; fel, Cornell Univ, 59-60; from asst prof to assoc prof, 60-70, PROF CHEM & CHMN DEPT, VALPARAISO UNIV, 70- *Concurrent Pos:* Consult, Argonne Nat Labs, 62-73. *Mem:* Am Chem Soc. *Res:* Amines, heterocyclic compounds, bicyclic compounds, organophosphorus compounds and small ring compounds. *Mailing Add:* Dept Chem Valparaiso Univ Valparaiso IN 46383

COOK, ALAN FREDERICK, b Harrow, Eng, July 15, 39; US citizen; m 64; c 2. CHEMISTRY. *Educ:* Univ Birmingham, BSc, 61; Univ London, PhD(chem), 64. *Prof Exp:* Fel, Syntex Corp, Calif, 64-66; chemist, Hoffmann La Roche, 66-85, Enzo Biochem, 85-87, Lifecodes Corp, 87-90; CHEMIST, PHARMAGENICS INC, 90- *Mem:* Am Chem Soc. *Res:* Chemistry of nucleosides, nucleotides oligonucleotides and nucleic acids. *Mailing Add:* 19 Hillcrest Rd Cedar Grove NJ 07009

COOK, ALBERT MOORE, b Denver, Colo, June 24, 43; m 65; c 3. BIOMEDICAL ENGINEERING. *Educ:* Univ Colo, BS, 66; Univ Wyo, MS, 68, PhD(bioeng), 70. *Prof Exp:* PROF BIOMED ENGR, CALIF STATE UNIV, SACRAMENTO, 70-, CO-DIR ASSISTIVE DEVICE CTR, 77- *Concurrent Pos:* Consult, Glenrose Hosp Edmonton, Alta, 83-, & Lawrence Livermore Lab, 74-79; fac res award, Calif State Univ, Sacramento, 75; prin investr, Instrnl Equip Prog, NSF, 76-78, co-prin investr, Develop Action Module Prog for Speech Pathologists, 84-; co-prin investr, Oper Eng, commun enhancement grant, 79-82; mem adv comt on res to aid handicapped, NSF. *Honors & Awards:* Outstanding Biomed Engr Educ Award, Am Soc Engr Educ, 85- *Mem:* Inst Elec & Electronics Engrs; Am Soc Eng Educ; Rehab Soc NAm; Int Soc Alternative Augmentative Communication. *Res:* Development and effective application of assistive devices for the disabled; microprocessor applications in medicine and biology; augmentative communication system application and development. *Mailing Add:* Biomed Eng Prog 6000 Jay St Sacramento CA 95819

COOK, ALBERT WILLIAM, b Brooklyn, NY, July 23, 22; m 47; c 2. NEUROSURGERY. *Educ:* Dartmouth Col, AB, 44; Long Island Col Med, MD, 46; Am Bd Neurol Surg, dipl, 56. *Prof Exp:* Assoc prof surg & head div neurosurg, State Univ NY Downstate Med Ctr, 59-71, chmn dept, 71-76, prof neurosurg, 71-87; dir dept neurosci, Long Island Col Hosp, 77-87; dir, NY Pain Ctr, 87-90; DIR, SPINET TESTING CTR, 90- *Concurrent Pos:* Consult regional hosps, Brooklyn, NY & Vet Admin Hosps, 59- *Mem:* AMA; Am Col Surgeons; Am Asn Neurol Surgeons; Soc Neurosci; Asn Res Nerv & Ment Dis; Sigma Xi. *Res:* Craniocerebral trauma; intracranial blood clots; cerebrovascular disease; surgical treatment of vascular anomalies; cerebral hemodynamics; respiratory and metabolic aspects of cerebral lesions; cerebral neoplasms; pain control; electrical stimulation of the spinal cord; multiple sclerosis. *Mailing Add:* 950 Franklin Ave Garden City Long Island NY 11530

COOK, ALLAN FAIRCHILD, II, b New York, NY, May 9, 22; m 59; c 3. ASTROPHYSICS. *Educ:* Princeton Univ, BSE, 47, AM, 50, PhD(astron), 52. *Prof Exp:* Asst astron, Princeton Univ, 48-49; instr astron & physics, Carleton Col, 50-51; asst, Col Observ, Harvard Univ, 51-57, res assoc, 57-59, res fel, 59-61, res assoc, 61-62, lectr, 61-74; physicist, 61-64, ASTROPHYSICIST, SMITHSONIAN ASTROPHYS OBSERV, 64- *Concurrent Pos:* Res officer, Herzberg Inst Astrophys, Ottawa, Can, 82-85; mem, Voyager Imaging Sci Team. *Mem:* Am Astron Soc; fel Meteoritical Soc; fel Royal Astron Soc. *Res:* Asteroids; meteors; meteor spectra; planetary rings; planetary satellites. *Mailing Add:* Ctr Astrophys 60 Garden St Cambridge MA 02138

COOK, ANCEL EUGENE, b Sadieville, Ky, Sept 15, 09; m 37; c 3. PHYSICAL OPTICS, MICROELECTRONICS. *Educ:* Georgetown Col, BA, 35; Univ Ky, MS, 48. *Prof Exp:* Teacher pub sch, Ky, 35-42; physicist, Indust Mgr Off, US Navy, La, 42-43; asst prof physics, Georgetown Col, 47-49; instr, Univ Ky, 49-51; physicist, Bur Ships, Navy Dept, 54-59; physicist, Off Naval Res, 59-74; RETIRED. *Concurrent Pos:* Sci secy laser res & explor develop panel, US Navy, 63-74, mem microelectronics panel, 66-74, coordr adv groups, industry & independent res & develop progs; US Navy mem, Dept Defense Adv Group Electron Devices; asst to US Navy mem, Armed Serv Res Specialist Comt; secy, Govt Microelectronics Applns Confs. *Mem:* Am Phys Soc; Am Soc Naval Eng; sr mem Inst Elec & Electronics Engrs. *Res:* Lasers and fluidic research and exploratory development; physical sciences. *Mailing Add:* 214 Tahoma Rd Lexington KY 40503

COOK, BARNETT C, b Chicago, Ill, Nov 25, 23. PHYSICS. *Educ:* Northwestern Univ, BS, 46; Univ Chicago, PhD(physics), 56. *Prof Exp:* Physicist, Inst Nuclear Res, Univ Chicago, 52-56; asst prof physics, Univ Pa, 56-59; physicist, Midwest Univs Res Asn, 59-60; physicist, 60-61, asst prof physics, 61-75, ASSOC PHYSICIST, AMES LAB, IOWA STATE UNIV, 61-, ASSOC PROF PHYSICS, 74- *Mem:* AAAS; Am Phys Soc; Sigma Xi. *Res:* Photonuclear reactions; accelerator design. *Mailing Add:* Dept Physics Iowa State Univ Ames IA 50012

COOK, BENJAMIN JACOB, b Upper Darby, Pa, Sept 26, 30; m 52; c 2. INSECT PHYSIOLOGY, BIOCHEMISTRY. *Educ:* Providence Col, AB, 58; Rutgers Univ, MS, 61, PhD(insect physiol), 63. *Prof Exp:* Asst prof biol, Col Holy Cross, 62-64; adj prof zool & asst prof entom, NDak State Univ & insect physiologist, Metab & Radiation Lab, USDA, 64-72; insect physiologist, Western Cotton Res Lab, USDA, Phoenix, Ariz, 72-76; INSECT PHYSIOLOGIST, VET ENTOM RES UNITE, USDA, 76- *Concurrent Pos:* Assoc ed, Archives of Insect Biochem & Physiol, 83-84. *Mem:* Soc Gen Physiologists; Entom Soc Am. *Res:* Insect neurophysiology and neurochemistry; peripheral and central mechanisms of nerve impulse transmission; the isolation and pharmacodynamics of natural neurotransmitters and neurohormones that affect visceral muscle. *Mailing Add:* Food Animal Protection Res Lab F & B Rd Rte 5 Box 810 College Station TX 77845

COOK, BILLY DEAN, b Oklahoma City, Okla, July 28, 35; m 60, 84; c 2. PHYSICS. *Educ:* Okla State Univ, BS, 57; Mich State Univ, MS, 59, PhD(physics), 62. *Prof Exp:* Res instr, Mich State Univ, 62-63, asst prof res physics, 65-68; res assoc, 68-69, assoc prof, 69-76, PROF MECH & ELEC ENG, UNIV HOUSTON, 76- *Concurrent Pos:* Vis scientist, Lawrence Livermore Nat Lab, 83-84. *Mem:* Acoust Soc Am; Inst Elec & Electronics Engrs; Am Soc Nondestructive Testing. *Res:* Ultrasonic light diffraction; nonlinear propagation of acoustical waves; optics and noise control; nondestructive testing. *Mailing Add:* Dept Elec Eng Univ Houston Houston TX 77204-4792

COOK, CHARLES DAVENPORT, b Minneapolis, Minn, Nov 30, 19; m 45, 76; c 5. MEDICINE. *Educ:* Princeton Univ, AB, 41; Harvard Univ, MD, 44. *Hon Degrees:* MA, Yale Univ, 64. *Prof Exp:* Asst prof pediat, Harvard Med Sch, 57-63, assoc clin prof, 63-64; prof pediat & chmn dept, Yale Univ, 64-75; prof pediat & chmn dept, State Univ NY Downstate Med Ctr, 75-; dept pediat, Univ Rochester, 82-90; RETIRED. *Concurrent Pos:* Physician, Boston Children's Hosp, 58-64; chmn, Joint Coun Nat Pediat Socs, 70-73. *Mem:* Soc Pediat Res; Am Acad Pediat; Am Pediat Soc (secy-treas, 64-75); Am Physiol Soc. *Res:* Medical education; health care, delivery; quality assessment and assurance; ethics in medicine. *Mailing Add:* 9117 Dugway Rd Holcomb NY 14469

COOK, CHARLES EMERSON, b New York, NY, Oct 27, 26; m 48; c 3. ELECTRICAL ENGINEERING. *Educ:* Harvard Univ, SB, 49; Polytech Inst Brooklyn, MEE, 54. *Prof Exp:* Engr, Melpar, Inc, 49-51; from engr to res engr, Sperry Gyroscope Co, 51-63, res sect head, 63-64, sr res sect head, 64-67, res dept head, Sperry Rand Res Ctr, 67-71; sr tech staff, Mitre Corp, Bedford, MA, 71-88; SR RES SCIENTIST, ANRO ENG INC, LEXINGTON, MA, 88. *Honors & Awards:* Aerospace & Electron Systs Soc Pioneer Award, Inst Elec & Electronics Engrs, 88. *Mem:* Fel Inst Elec & Electronics Engrs; Sigma Xi. *Res:* Radar system analysis and synthesis; research and development of signal processing techniques to improve the detection and resolution performance of radar and sonar systems; communication systems analysis; computer modelling of interference environments and interference rejection techniques. *Mailing Add:* 288 Stearns St Carlisle MA 01741

COOK, CHARLES F(OSTER), JR, b Peacock, Tex, Jan 23, 32; m 56; c 2. ELECTRON BEAM LITHOGRAPHY, MATERIALS SCIENCE. *Educ:* Tex Tech Univ, BS, 54; Rutgers Univ, MS, 76. *Prof Exp:* Phys scientist & electron microscopist, US Army Electronic Technol & Devices Lab, 56-60, res phys scientist, 60-90; RETIRED. *Honors & Awards:* Di Paul Siple Medallion & Award, US Army Sci Conf, 70. *Mem:* Electron Micros Soc Am; Microbeam Anal Soc. *Res:* Electron beam lithography research and development; materials science research, including semiconductors, thin films, ferroelectrics, ferromagnetics and solid state materials. *Mailing Add:* 3432 Willowood Bartlesville OK 74003

COOK, CHARLES FALK, b Jonesboro, Ark, Dec 21, 28; m 49; c 2. NUCLEAR PHYSICS, CHEMICAL PHYSICS. *Educ:* Tex Christian Univ, BA, 48, MA, 50; Rice Inst, PhD(physics), 53. *Prof Exp:* Fel physics, Rice Inst, 53-54; asst prof, Univ Fla, 54-55; sr nuclear eng, Convair Div, Gen Dynamics Corp, Ft Worth, Tex, 55-56, nuclear test lab engr, 56-58; mgr, Nuclear Physics Sect, Res Div, Phillips Petrol Co, 58- 62, Physics Br mgr, 62-70, Physics & Anal Br mgr, 77-80, vpres res & develop, 80-90; RETIRED. *Mem:* Am Phys Soc; Sigma Xi. *Res:* Nuclear reaction cross sections; electron spin resonance; molecular energy levels in hydrocarbon molecules using spectroscopic techniques; analytical chemistry; molecular structure. *Mailing Add:* 3432 Willowood Bartlesville OK 74003

COOK, CHARLES GARLAND, b Waco, Tex, July 6, 59. KENAF & COTTON IMPROVEMENT, MULTI-ADVERSITY RESISTANCE BREEDING. *Educ:* Abilene Christian Univ, BS, 82; Tex A&M Univ, MS, 85, PhD(plant breeding), 89. *Prof Exp:* Res asst, Tex Agr Exp Sta, 84-85, res assoc, 86-89; RES GENETICIST, AGR RES SERV, USDA, 89- *Concurrent Pos:* Leader, Kenaf Agr Res Coord Group Steering Comt, 91; mem, New Crops Adv Comt, Kenaf & Roselle, 91. *Mem:* Am Soc Agron; Crop Sci Soc Am; Am Phytopath Soc. *Res:* Genetic improvement of Kenaf and cotton for increased yield, fiber and seed quality and resistance to diseases, nematodes, insects and environmental stresses; breeding and development of hybrid cotton utilizing cytoplasmic male sterility and pollen fertility restoration systems. *Mailing Add:* Agr Res Serv USDA 2413 E Hwy 83 Weslaco TX 78596

COOK, CHARLES J, b West Point, Nebr, Oct 2, 23; m 45; c 2. INDUSTRIAL PLANNING. *Educ:* Univ Nebr, BS, 48, MA, 50, PhD(physics), 53. *Prof Exp:* Res assoc, Univ Nebr, 53-54; physicist, SRI Int, 54-56, mgr, Molecular Physics Sect, 56-62, dir, Chem Physics Div, 62-69, exec dir phys sci, 69-76, vpres off res opers, 76-77, sr vpres res opers, 77-81; mgr int planning, Bechtel Group, Inc, 81-83, mgr automation, 83-87; founder & pres, Cam Cubed, 87-90; CONSULT, 90- *Concurrent Pos:* Instr, San Jose City Col, 57-58 & Foothill Col, 59-62; sr res assoc, Queen's Univ, Belfast, 62-63; partic, Advan Mgt Prog, Harvard Grad Sch Bus, 68; mem bd dirs, Robot Industs Asn. *Mem:* Am Asn Eng Socs; Soc Mfg Engrs; Am Inst Aeronaut & Astronaut; Comput & Automation Systs Asn; Am Soc Testing & Mat. *Res:* Ionic and atomic impact phenomena; technology utilization and transfer; international planning. *Mailing Add:* 47 Solana Dr Los Altos CA 94022

COOK, CHARLES S, b Rochester, NY, Feb 19, 38; m 64; c 2. MECHANICAL ENGINEERING. *Educ:* Cornell Univ, BME, 60; Univ Rochester, MS, 64, PhD(mech & aerospace sci), 66. *Prof Exp:* Res engr pulsed plasma accelerator diag, Gen Elec Co, 66-67, light detection & ranging systs, 67-71, magnetohydrodynamics power generation, 71-76, mgr magnetohydrodynamics & coal combustion component develop, Energy Systs Progs Dept, 76-81; vpres eng, Farrier, Inc, 81-84; MGR, INTEGRATED GASIFICATION COMBINED CYCLE TECHNOL, GEN ELEC ENVIRON SERV INC, 84- *Mem:* Sigma Xi; Am Soc Mech Engrs. *Res:* Plasma diagnostics; atmospheric probing with light detection & ranging; mie scattering; optical transmissivity of smoke plumes; magnetohydrodynamics power generation; high temperature heat exchangers; coal combustion; open and closed magnetohydrodynamics cycle systems; high temperature heat exchange, coal fired indirect gas turbine cycles; coal gasification combined cycles; air quality control systems. *Mailing Add:* Gen Elec Environ Serv 200 N Seventh St Lebanon PA 17042

COOK, CHARLES WAYNE, b Gove, Kans, Oct 20, 14; m 40; c 1. RANGE SCIENCE. *Educ:* Ft Hays Kans State Univ, BS, 40; Utah State Univ, MS, 42; Agr & Mech Univ, Tex, PhD, 50. *Prof Exp:* Range conservationist, Soil Conserv Serv, USDA, 42-43; res prof range mgt, Utah State Univ, 43-46, prof range sci, 46-67; prof range sci & head dept, Colo State Univ, 67-89; CONSULT, 89- *Honors & Awards:* Hoblitzelle Award, 53. *Mem:* AAAS; Am Inst Biol Sci; assoc Am Soc Animal Sci; assoc Soc Range Mgt; Soil Conserv Soc Am. *Res:* Range seeding and forage nutrition; bioecology of cactus on the central great plains; utilization of range plants and forage by herbivores; physiological responses of plants; energy flow in the range ecosystem. *Mailing Add:* 1180 Daleview Dr McLean VA 22102

COOK, CHARLES WILLIAM, b Yankton, SDak, Sept 27, 27; m 50; c 3. SPACE SYSTEMS & SCIENCE, EXPERIMENTAL NUCLEAR PHYSICS. *Educ:* Univ SDak, AB, 51; Calif Inst Technol, MS, 54, PhD(physics), 57. *Prof Exp:* Sr res engr, Convair, 57-58, design specialist & head nuclear physics, 58-60; chief res ballistic missile defense, Inst Defense Anal, 60-61; adv res proj agency, Dept Defense, 61-62; res & develop specialist, NAm Aviation, Inc, Calif, 62-63, corp dir electronics, 63-67; independent consult, 67-71; asst dir, Dept Res & Eng, Dept Defense, 71-74; dep under secy, Air Force Space Systs, 74-79, dep asst secy, Air Force Space Plans & Policy, Washington, DC, 79-88; CONSULT, INST DEFENSE ANALYSIS, ANSER, DEFENSE SCI BD, SYSTS PLANNING CORP & GLOBAL OUTPOST, INC, 88- *Concurrent Pos:* Consult, Inst Defense Anal, 62-63 & McGraw-Hill Book Co, Inc, 62-80. *Honors & Awards:* Meritorious Civil Serv Award, Secy Defense, 74, Distinguished Serv Award, 76; Except Civilian Serv Award, USAF, 81, 82, 87, 88; Group Achievement Award, NASA, 85, Distinguished Serv Medal, 88; Cert of Appreciation, Intel Res & Develop Coun, 87; Nat Intel Medal of Achievement, 88. *Mem:* Am Phys Soc; Am Inst Phys; Am Inst Aeronaut & Astronaut; Inst Elec & Electronics Engrs; Nat Space Club. *Res:* Energy generation and element synthesis reactions occurring in stellar interiors; electronics research and development. *Mailing Add:* 1180 Daleview Dr McLean VA 22102

COOK, CLARENCE EDGAR, b Jefferson City, Tenn, Apr 27, 36; m 57; c 3. BIOORGANIC CHEMISTRY. *Educ:* Carson-Newman Col, BS, 57; Univ NC, PhD(org chem), 61. *Prof Exp:* Am Chem Soc Petrol Res Fund fel, Cambridge Univ, 61-62; sr chemist, 62-68, group leader, 68-71, asst dir, Chem & Life Sci Div, 71-75, dir life sci, 75-80, dir bioorg chem, 75-85, RES VPRES, CHEM & LIFE SCI, RES TRIANGLE INST, 83- *Concurrent Pos:* Vis scholar, Univ Liverpool, 79; adj prof, Sch Pharm, Univ NC, 85-; mem, Drug

Abuse Biomed Res Review Comt, Nat Inst Drug Abuse, 85-89, ADAMHA Reviewers Reserve, 90- *Mem:* AAAS; Am Chem Soc; Int Soc Study Xenobiotics; fel NY Acad Sci; Am Soc Pharmacol & Exp Therapeut. *Res:* Drug metabolism; contraceptive drugs; synthesis of medicinal compounds; oxygen and nitrogen heterocycles; steroid chemistry; natural products; agricultural chemistry; development and application of antibodies. *Mailing Add:* Chem & Life Sci Unit Res Triangle Inst PO Box 12194 Research Triangle Park NC 27709

COOK, CLARENCE HARLAN, b Winthrop, Iowa, Jan 16, 25; m 45; c 4. MATHEMATICS. *Educ:* Univ Iowa, BA, 48, MS, 50; Univ Colo, PhD(math), 62. *Prof Exp:* Asst prof math, Western State Col, Colo, 50-52; asst, Univ Tex, 52-53; engr, Convair, Tex, 53-54; sr mathematician, Martin Co, Md, 54-55, prin engr, Colo, 55-59, sragg engr opers res, 60-62; asst prof math, Univ Okla, 62-65; from asst prof to assoc prof, 65-76, PROF MATH, UNIV MD, COLLEGE PARK, 76- *Mem:* Am Math Soc; Math Soc Belg; Sigma Xi. *Res:* Functional analysis; topology. *Mailing Add:* Dept Math Univ Md College Park MD 20742

COOK, CLARENCE SHARP, b St Louis Crossing, Ind, Aug 18, 18; m 43; c 2. PHYSICS. *Educ:* DePauw Univ, AB, 40; Ind Univ, MA, 42, PhD(physics), 48. *Prof Exp:* Asst physics, Ind Univ, 40-42 & 46-48; asst prof, Wash Univ, St Louis, 48-53; head nuclear radiation br, US Naval Radiol Defense Lab, 53-59, head radiation effects br, 59-60, head nucleonics div, 60-61, physics consult to sci dir, 62-65, head radiation physics div, 65-69; lectr, Univ Santa Clara, 69-70; chmn dept, 70-72 & 80-83, prof 70-85, EMER PROF PHYSICS, UNIV TEX, EL PASO, 85- *Concurrent Pos:* Mem bd exam scientists & engrs, US Civil Serv Comn, 55-58, chmn bd, 57-58, mem prof coun scientists & engrs, Calif-Nev area, 67-69, mem, Nat Res Coun Comt to Rev Methods Assigning Dose to Nuclear Vet, 84-85; Fulbright res scholar, Aarhus Univ, 61-62. *Mem:* Am Phys Soc; Am Asn Physics Teachers; Am Geophys Union; Health Physics Soc. *Res:* Energy education; effects of ionizing radiations. *Mailing Add:* Univ Tex Box 204 El Paso TX 79968

COOK, CLELAND V, b Milltown, Wis, Nov 18, 33. MATHEMATICS. *Educ:* Univ Wis, BS, 55, MA, 61; Univ SDak, MA, 62, Ed D, 69. *Prof Exp:* From instr to assoc prof, 62-77, PROF MATH, UNIV SDAK, 77- *Concurrent Pos:* Gen Elec fel, Nat Educ Asn, 59. *Mem:* Nat Coun Teachers Math; Math Asn Am. *Mailing Add:* RR2 Box 148 Vermillion SD 57069

COOK, DAVID ALASTAIR, b Haslemere, Surrey, Eng, May 19, 42; div. SMOOTH MUSCLE PHARMACOLOGY, MEDICAL EDUCATION. *Educ:* Oxford Univ, MA, 67, PhD(pharmacol), 74. *Prof Exp:* Fel, Univ Alta, 67-72, acad staff, 72-81, prof & chmn, Dept Pharmacol, 81-91, DIR, DIV STUDIES MED EDUC, UNIV ALTA, 90- *Concurrent Pos:* Res assoc, div Neurol, Univ Alta, 85-; lectr, Alta Alcoholism and Drug Abuse Comn, 80-; chmn, Hypertension Sect, Can Heart Found, 88-91. *Mem:* Pharmacol Soc Can (counr, 83-86); Western Pharmacol Soc (pres, 89); Soc Toxicol Can; Brit Pharmacol Soc; Can Asn Med Educ. *Res:* Studies of smooth muscle contractility, with special interest in signal transduction; cerebrovascular spasm secondary to subarachnoid hemorrhage; medical education. *Mailing Add:* Dept Pharmacol Univ Alta Edmonton AB T6G 2H7 Can

COOK, DAVID ALLAN, b Colby, Wis, Mar 11, 40; m 62; c 4. NUTRITIONAL BIOCHEMISTRY. *Educ:* Wis State Univ-River Falls, BS, 62; Iowa State Univ, PhD(animal nutrit), 67. *Prof Exp:* NIH res fel biochem, Univ Minn, St Paul, 67-68; sr investr, Dept Nutrit Res, 69-73, prin investr, 73-74, clin investr, Dept Clin Invest, 74-75, nutrit investr, 75-76, dir, Dept Nutrit Sci, Nutrit Div, 76-86, VPRES FOOD RES & DEVELOP, NUTRIT GROUP, MEAD JOHNSON, 87- *Mem:* AAAS; Brit Nutrit Soc; Am Inst Nutrit; Sigma Xi; Nutrit Today Soc; Am Soc Clin Nutrit. *Res:* Ruminant fatty acid metabolism and absorption; oxalic acid formation from aromatic amino acids; infant nutrition; bile acid metabolism; atherosclerosis; mineral metabolism and bioavailability; clinical nutrition; gastroenterology. *Mailing Add:* Food Res & Develop (R-10) Mead Johnson Nutrit Group Evansville IN 47721-0001

COOK, DAVID EDGAR, b Corpus Christi, Tex, Dec 13, 40; m 63; c 2. BIOCHEMICAL PHARMACOLOGY. *Educ:* Southwest Tex State Col, BS, 62; Okla State Univ, MS, 65; Univ Tex, Austin, PhD(chem), 70. *Prof Exp:* From instr to asst prof chem, Northeastern State Col, Okla, 64-70; NIH res fel & res assoc biochem, Inst Enzyme Res, Univ Wis-Madison, 70-73; from asst prof to assoc prof biochem, Col Med, Univ Nebr Med Ctr, Omaha, 78-85; head Div Sci & Math & prof chem, 85-86, vpres & prof chem, Dakota State Col, 86-89, ACAD VPRES & PROVOST, PROF CHEM, DAKOTA STATE UNIV, 89- *Concurrent Pos:* NSF sci fac fel, Univ Tex, Austin, 68-69; prin investr NIH grants, 75-79, 77-81, 84-85, consult, 87- *Mem:* Am Chem Soc; AAAS; Sigma Xi; Am Soc Biochem & Molecular Biol; Am Soc Pharmacol & Exp Therapeut. *Res:* Biochemical pharmacology; metabolism. *Mailing Add:* Dakota State Univ Madison SD 57042-1799

COOK, DAVID EDWIN, b Houston, Tex, Dec 3, 35; m 58; c 2. TOPOLOGY. *Educ:* Univ Tex, BA, 58, MA, 60, PhD(math), 67. *Prof Exp:* Spec instr math, Univ Tex, 61-64, teaching assoc, 64-67; asst prof, 67-69, ASSOC PROF MATH, UNIV MISS, 69- *Mem:* Am Math Soc; Math Asn Am; Nat Coun Math Teachers. *Res:* Point set topology. *Mailing Add:* Dept Math Univ Miss University MS 38677

COOK, DAVID GREENFIELD, b Birmingham, Eng, May 27, 41; Can citizen; m; c 2. INVERTEBRATE ZOOLOGY, SYSTEMATICS. *Educ:* Univ Liverpool, BSc, 64, PhD(zool), 67. *Prof Exp:* Fel systs, Marine Biol Lab, Woods Hole Mass, 67-69; fel zool, Nat Mus Natural Sci, Ottawa, Ont, 69-71; benthic biologist, Great Lakes Biolimnol Lab Can Ctr Inland Waters, Burlington Ont, 73-75; asst ed, Sci Info & Publ Br, Fisheries & Oceans Can, 75-88; ED, CAN J FISH AQUATIC SCI, 88- *Concurrent Pos:* Res assoc, Royal Ont Mus, Toronto. *Mem:* Freshwater Biol Asn. *Res:* Taxonomy, morphology and ecology of marine and freshwater microdrile oligochaetes; freshwater benthic ecology; scientific editing. *Mailing Add:* Sci Publ Br Fisheries & Oceans Can Ottawa ON K1A 0E6 Can

COOK, DAVID MARSDEN, b Troy, NY, Apr 3, 38; m 65; c 2. THEORETICAL PHYSICS, INSTRUCTIONAL COMPUTING. *Educ:* Rensselaer Polytech Inst, BS, 59; Harvard Univ, AM, 60, PhD(physics), 65. *Prof Exp:* From asst prof to assoc prof, 65-79, PROF PHYSICS, LAWRENCE UNIV, 79- *Concurrent Pos:* NSF sci fac fel, Dartmouth Col, 71-72. *Mem:* Am Phys Soc; Am Asn Physics Teachers. *Res:* chaos theory; applied mathematics; computers in physics education. *Mailing Add:* Dept Physics Lawrence Univ Box 599 Appleton WI 54912-0599

COOK, DAVID ROBERT, b Brockton, Mass, Sept 2, 52; m 80; c 1. MICROMETEOROLOGY, CLIMATOLOGY. *Educ:* Pa State Univ, BS, 74, MS, 77. *Prof Exp:* Res asst, Environ Res & Technol, Inc, 73 & 74; grad res asst, Inst Res Land & Water Resources, State Col Pa, 75-76; res asst, 77-89, RES ASSOC, ARGONNE NAT LAB, 89- *Mem:* Am Meteorol Soc; Sigma Xi. *Res:* Experimental micrometeorology; regional and local climatology; nitrogen oxides production by lightning; characterization of response, noise and calibration of specialized meteorological and air quality instrumentation. *Mailing Add:* 417 Bayview Ave Naperville IL 60565

COOK, DAVID RUSSELL, b Hastings, Mich, Aug 9, 22; m 52; c 2. INVERTEBRATE ZOOLOGY, ENTOMOLOGY. *Educ:* Univ Mich, BS, 48, MA, 51, PhD, 52. *Prof Exp:* USDA entomologist, US Nat Mus, 52-53; from asst prof to assoc prof, 53-65, PROF BIOL, WAYNE STATE UNIV, 65- *Concurrent Pos:* Entomologist & malariologist, Int, Co-op Admin Malaria Control Prog, US Opers Mission to Liberia, 56-58; Fulbright res fel, India, 62-63; Fulbright lectr, Argentina, 75. *Mem:* Sigma Xi. *Res:* Acarina; Hydracarina. *Mailing Add:* Dept Biol Wayne State Univ Detroit MI 48202

COOK, DAVID WILSON, b Wilkinson County, Miss, Nov 3, 39; m 61; c 2. MICROBIOLOGY. *Educ:* Miss State Univ, BS, 61, MS, 63, PhD(microbiol), 66. *Prof Exp:* HEAD MICROBIOL SECT, GULF COAST RES LAB, 66-, ASST DIR ACAD & SERV, 72- *Concurrent Pos:* Assoc mem grad fac, Miss State Univ, 68-; asst prof biol, Univ Miss, 71-; registr, Gulf Coast Res Lab, 71- *Mem:* Am Soc Microbiol; Inst Food Technologists; Sigma Xi. *Res:* Microbiology of the estuarine environment including pollution; nutrient turnover; diseases of fish and shellfish and spoilage of seafoods. *Mailing Add:* Gulf Coast Res Lab PO Box 7000 Ocean Springs MS 39564-7000

COOK, DESMOND C, b Geelong, Victoria, Australia, Oct 19, 49; m 87. MOSSBAUER SPECTROSCOPY, SURFACE PHYSICS & MAGNETISM. *Educ:* Monash Univ, Australia, BSc Hons, 72, PhD(physics), 78. *Prof Exp:* Asst prof, 81-87, ASSOC PROF PHYSICS, OLD DOMINION UNIV, NORFOLK, VA, 87- *Concurrent Pos:* Consult, Dreadnought Marine Inc, Norfolk, Va, 88-, Bethlehem Steel Corp, Pa, 89- *Honors & Awards:* Ayrton Premium Award, Inst Elec Engrs, London, 85. *Mem:* Am Inst Physics; Sigma Xi. *Res:* Magnetic and crystalline properties of materials, especially steel coatings and RAM. *Mailing Add:* Dept Physics Res E Old Dominion Univ Norfolk VA 23529

COOK, DONALD BOWKER, b Easthampton, Mass, Jan 14, 17; m 43; c 6. CRYOPHYSICS. *Educ:* Princeton Univ, AB, 38; Columbia Univ, MA, 39, PhD(physics), 50. *Prof Exp:* Physicist, Manhattan Proj, Columbia Univ & Carbon & Carbide Chem Corp, NY, 42-44, sect leader, 44-45, group leader, 45-46, physicist, Div Govt Aid Res, Columbia Univ, 46-50; physicist, Nevis Cyclotron Lab, 47; res physicist, E I du Pont de Nemours & Co, Inc, 50-57, sr res physicist, 57-86; RETIRED. *Mem:* Am Phys Soc; Am Chem Soc. *Res:* Multilayer films; gas diffusion; cryogenics; structure of polyfibers. *Mailing Add:* 16779 Timmons Rd Spring Run PA 17262

COOK, DONALD J, b Astoria, Ore, Feb 14, 20; m 44; c 2. MINERAL ENGINEERING. *Educ:* Univ Alaska, BS, 47, ME, 52; Pa State Univ, MS, 58, PhD(mineral beneficiation), 61. *Prof Exp:* Off engr, US Smelting, Refining & Mining Co, 47-50, mining engr, 54-57; assayer & engr, Alaska Div Mines & Minerals, 50-52; instr & asst, Univ Alaska, 52-54; asst mineral beneficiation, Pa State Univ, 57-59; from asst prof to prof mineral & beneficiation, Univ Alaska, 59-90, dir, Mineral Indust Res Lab Fac dean, Sch Mineral Eng, 83-90; DIR, ALASKA STATE OFF, TAIPEI, TAIWAN, 91- *Concurrent Pos:* NSF res grant, 60-64; vis prof, Cheng Kung Univ, Taiwan, 71, 81-82; vpres, Bd Engrs & Archit Exam, State of Alaska. *Mem:* Am Inst Mining, Metall & Petrol Engrs; Nat Soc Prof Engrs. *Res:* Mineral beneficiation; magnetic susceptibility and dielectric constants of rocks and minerals. *Mailing Add:* PO Box 80093 College AK 99708

COOK, DONALD JACK, b Rock Island, Ill, Feb 12, 15; m 39; c 2. ORGANIC CHEMISTRY. *Educ:* Augustana Col, AB, 37; Univ Ill, MA, 38; Ind Univ, PhD(org chem), 44. *Prof Exp:* City chemist, Rock Island, Ill, 38-39; Am Container Corp, 39-40; instr chem, Augustana Col, 40-41; chemist, Tex Co, NY, 41-42; asst sci, Ind Univ, 42-44; res chemist, Lubri-Zol Corp, 44-45; from asst prof to prof, 45-80, head dept, 64-80, EMER PROF, DEPAUW UNIV, 80- *Concurrent Pos:* Assoc prof dir, Div Sci Personnel & Ed, NSF, 61-62. *Mem:* Am Chem Soc. *Res:* N-substituted carbostyrils; preparation and properties of substituted lepidones; SeO2 oxidations. *Mailing Add:* 625 E Wash St Greencastle IN 46135

COOK, DONALD LATIMER, b Arena, Wis, July 31, 16; m 43; c 2. PHARMACOLOGY. *Educ:* Univ Wis, BS, 38, PhD(pharmaceut chem), 43. *Prof Exp:* Instr pharmacol & physiol, Sch Pharm, Western Reserve Univ, 42-43; pharmacologist, G D Searle & Co, 46-70, head dept pharmacol, 70-74, assoc dir dept biol res, Searle Labs, 74-83; RETIRED. *Mem:* Am Soc Pharmacol & Exp Therapeut; Soc Exp Biol & Med. *Res:* Phytochemical study of the leaves of Celastrus scandens Linne. *Mailing Add:* 818 S Knight Ave Park Ridge IL 60068

COOK, EARL FERGUSON, resource geography, volcanic geology; deceased, see previous edition for last biography

COOK, EDWARD HOOPES, JR, b Harrisburg, Pa, May 21, 29; m 51; c 2. PHYSICAL CHEMISTRY, INORGANIC CHEMISTRY. *Educ:* Elizabethtown Col, BS, 50; Pa State Col, PhD(chem), 53; Canisius Col, MBA, 84. *Prof Exp:* MGR, OCCIDENTAL CHEM CORP, 63- *Mem:* Am Chem Soc; Electrochem Soc. *Res:* Electrochemical development; industrial electrolytic; theoretical electrochemistry; mechanism of corrosion processes; phosphorous products. *Mailing Add:* 8424 Carol Ct Niagara Falls NY 14304

COOK, EDWIN FRANCIS, b San Francisco, Calif, Sept 11, 18; m 49; c 5. ENTOMOLOGY. *Educ:* Stanford Univ, AB, 43, AM, 44, PhD(biol sci), 48. *Prof Exp:* Actg instr gen biol, Stanford Univ, 46-48, actg asst prof entom, 48-49; prof entom, Univ Minn, St Paul, 49; RETIRED. *Mem:* AAAS; Am Soc Syst Zool; Entom Soc Am. *Res:* Systematic entomology. *Mailing Add:* 1727 Lindig St St Paul MN 55113

COOK, ELIZABETH ANNE, b Colorado Springs, Colo, Sept 19, 26. MICROBIOLOGY. *Educ:* Univ Colo, BA, 48; Ind Univ, MA, 56, PhD(bact), 58. *Prof Exp:* Instr bact, Mankato State Col, 57-58; from instr to asst prof, 62-70, assoc prof bact, 70-80, PROF BIOL SCI, DOUGLASS COL, RUTGERS UNIV, 80- *Mem:* AAAS; Am Soc Microbiol. *Res:* Microbiological assays for vitamin B12 and folic acid; bacterial nutrition. *Mailing Add:* Dept Biol Sci Douglass Col Rutgers the State Univ New Brunswick NJ 08903

COOK, ELLSWORTH BARRETT, b Springfield, Mass, Jan 29, 16; m 41. PHARMACOLOGY. *Educ:* Springfield Col, BS, 38; Tufts Univ, PhD(med sci), 51. *Prof Exp:* Res assoc, Fatigue Labs, Harvard, 41-45; head visual screening & statist facil, Med Res Lab Submarine Base, Conn, 45-49; environ physiologist, Cold Injury Res Team, Korea, 51-52; consult biometrician, Army Med Res Lab, Ft Knox, Ky, 52; res pharmacologist, Nat Naval Med Res Inst, 52-57; head dept exp biol, Naval Med Field Res Lab, Camp Lejeune, NC, 57-61; exec officer, Am Soc Pharmacol & Exp Therapeut, 61-77; RETIRED. *Concurrent Pos:* Lectr, Med Sch, Howard Univ, 62-77. *Mem:* AAAS; Am Soc Pharmacol & Exp Therapeut; Am Inst Biol Sci; Sigma Xi. *Res:* Acetylcholine analogues; standardization of bioassay techniques; biometrics and psychometrics. *Mailing Add:* 15500 Prince Frederick Way Silver Spring MD 20906

COOK, ELTON STRAUS, medicinal chemistry; deceased, see previous edition for last biography

COOK, ERNEST EWART, b Stratton St Margaret, Eng, Mar 23, 26; c 1. EXPLORATION GEOPHYSICS. *Educ:* Cambridge Univ, BA, 46, MA, 50. *Prof Exp:* Geophysicist, Cia Shell de Venezuela, 47-56, chief geophysicist, Pakistan Shell Oil Co, 56-57; chief geophysicist, Signal Oil & Gas Co, Venezuela, 57-60, geophysicist, Tex, 60-62, chief geophysicist, 62-65, asst mgr int explor, Calif, 65-68; vpres, Seismic Comput Corp, 68-71; pres, Invent Inc, 71-78, Zenith Explor Co, Inc, 78-82 & Jasmine Energy Inc, 82-87; PRES ENJAY ENTERPRISES INC, 76- *Mem:* Am Asn Petrol Geologists; Soc Explor Geophys; Am Geophys Union; Geol Soc London; Europ Asn Explor Geophys. *Res:* Seismic refraction methods and interpretation techniques; application of seismic velocities to geologic interpretation. *Mailing Add:* Enjay Enterprises Inc 12200 Northwest Freeway Suite 682 Houston TX 77092

COOK, EVERETT L, b Mounds, Okla, Sept 9, 27; m 48; c 4. AERONAUTICAL & STRUCTURAL ENGINEERING. *Educ:* Wichita State Univ, BS, 54, MS, 58; Okla State Univ, PhD(mech eng), 67. *Prof Exp:* Designer, Tech Training Aids, Inc, Okla, 50; instr, Spartan Col Aeronaut Eng, 51; stress analyst, Beech Aircraft Corp, Kans, 51-53; from instr to assoc prof aeronaut eng, Wichita State Univ, 53-67, dir digital comput, 64-67; assoc prof mech eng, Okla State Univ, 67-69; assoc prof aeronaut eng, Wichita State Univ, 69-80; at Gates Learjet Corp, 80-86; at Omac Inc, 86-89; at Bromon Corp, PR, 89-90; AT BEECH AIRCRAFT CORP, 90- *Concurrent Pos:* Consult, Beech Aircraft Corp, 68-70 & Gates Learjet Corp, 75-; Nat Res Coun sr res assoc, NASA Marshall Space Flight Ctr, 74-75. *Mem:* Am Inst Aeronaut & Astronaut; Am Soc Eng Educ. *Res:* Discrete element methods of analysis and design of flight vehicle structures. *Mailing Add:* 110 S Prescott Wichita KS 67209

COOK, EVIN LEE, b Waco, Tex, July 11, 18; m 41; c 1. PHYSICAL CHEMISTRY. *Educ:* Baylor Univ, AB, 39; Rice Inst, MA, 41; Univ Tex, PhD(phys chem), 49. *Prof Exp:* Res chemist petrol refining, Humble Oil & Refining Co, 41-43 & Pan Am Ref Corp, 44-47; mgr oil recovey res, Mobil Res & Develop Corp, 49-79, sr prod consult, 80-88; RETIRED. *Mem:* Am Chem Soc; Am Inst Mining, Metall & Petrol Engrs; Sigma Xi. *Res:* Hydrous oxides; aviation gasoline research; unit processes; surface chemistry; fluid flow through porous media. *Mailing Add:* 6417 McCommas Dallas TX 75214

COOK, FRANKLAND SHAW, b Toronto, Nov 30, 21; m 45; c 3. BOTANY. *Educ:* Univ Toronto, BA, 50, PhD(bot), 56. *Prof Exp:* From asst prof to prof, 52-87, EMER PROF BOT, UNIV WESTERN ONT, 87- *Mem:* Can Bot Asn; Am Bryol & Lichenological Soc; Can Soc Plant Physiol. *Res:* Growth and physiology of pollen and spores; physiology of mosses; emzyme kinetics. *Mailing Add:* Dept Bot One Larkspur Cr London ON N6H 3R1 Can

COOK, FRED D, b Ottawa, Ont, Oct 30, 21; m 48; c 3. SOIL MICROBIOLOGY. *Educ:* Univ BC, BSA, 45, MSA, 47; Univ Edinburgh, PhD(microbiol), 60. *Prof Exp:* Res officer soil microbiol, Exp Farm, Swift Current, Sask, 50-57; officer soil microbiol, Microbiol Res Inst, Ottawa, 57-64; mem fac, 64-69, prof, 69-84, EMER PROF SOIL SCI, UNIV ALTA, 84- *Mem:* Am Soc Microbiol; Can Soc Microbiol. *Res:* Petroleum microbiology; soil microbiology; microbial ecology. *Mailing Add:* 12812 66 Ave Edmonton AB T6H 1Y6 Can

COOK, FREDERICK AHRENS, b Chicago, Ill, June 15, 50; m 75; c 1. SEISMOLOGY, TECTONICS. *Educ:* Univ Wyo, BS, 73, MS, 75; Cornell Univ, PhD(geophysics), 81. *Prof Exp:* Geophysicist, Continental Oil Co, 75-77; fel, Cornell Univ, 80-; AT DEPT GEOL & GEOPHYS DEPT, UNIV CALGARY. *Mem:* Am Geophys Union; Geol Soc Am; Soc Explor Geophysicists; Am Asn Petrol Geologists. *Res:* Application of geophysical techniques to delineating the structure and evolution of the continental crust; computer modelling of seismic wave propagation. *Mailing Add:* 3037 29th St SW Calgary AB T3E 2K9 Can

COOK, FREDERICK LEE, b Baltimore, Md, Mar 15, 40; m 62; c 2. MATHEMATICS. *Educ:* Ga Inst Technol, BS, 61, MS, 63, PhD(math), 67. *Prof Exp:* Res mathematician, George C Marshall Space Flight Ctr, NASA, 67-69; from asst prof to assoc prof, Univ Ala, Huntsville, 69-88, chmn, Dept Math, 72-74, asst to pres, 74-75, CHMN DEPT MATH, UNIV ALA, HUNTSVILLE, 77-, PROF MATH & ASSOC DEAN, COL SCI, 88- *Mem:* Am Math Soc; Soc Indust & Appl Math; Math Asn Am; Sigma Xi. *Res:* Characterizations and applications of recursively generated Sturm-Liouville polynomial sequences; solutions of countably infinite systems of differential equations; determination of weight functions for given polynomial sequences; mathematical modeling in medicine and biology. *Mailing Add:* Col Sci Univ Ala Sci Bldg 110 Huntsville AL 35899

COOK, GEORGE A, b Anniston, Ala, Oct 25, 44; m; c 3. REGULATION OF FATTY ACID METABOLISM. *Educ:* Auburn Univ, PhD(biochem), 74. *Prof Exp:* Asst prof biochem, Sch Med, Univ Ind, 81-83; asst prof pharmacol, 83-85, ASSOC PROF PHARMACOL, COL MED, UNIV TENN, 85- *Mem:* Am Soc Biochem & Molecular Biol; Int Soc Heart Res; Sigma Xi; Am Soc Pharmacol & Exp Therapeut; Fel Coun Arteriosclerosis Am Heart Asn. *Res:* Regulation of Carnitine Palmitoyltransfusions. *Mailing Add:* Dept Pharmacol Col Med Univ Tenn 874 Union Ave Memphis TN 38163

COOK, GEORGE EDWARD, b Memphis, Tenn, Apr 4, 38; m 65. ELECTRICAL ENGINEERING. *Educ:* Vanderbilt Univ, BE, 60, PhD(elec eng), 65; Univ Tenn, MS, 61. *Prof Exp:* Lab instr elec eng, Univ Tenn, 60-61; asst eng res, 61-62, asst elec eng, 62-63, from instr to assoc prof, 63-76, PROF ELEC ENG, VANDERBILT UNIV, 76-, ASSOC DEAN ENG SCH, 87- *Concurrent Pos:* Consult & dir elec design, Price-Bass Co, 63-64; consult, vpres & dir eng, Merrick Eng, Inc, 64-69, 74-81, Adv Control Eng, 72-74 & CRC Welding Systs, 81-84; tech dir, Airtronics Indust Electronics Lab, 69- *Honors & Awards:* Gold Award, James F Lincoln Found, 81; NASA Space Act Award. *Mem:* Fel Inst Elec & Electronics Engrs; Am Welding Soc; Sigma Xi; Am Soc Metals; Nat Soc Prof Engrs; Am Soc Eng Ed; AAAS; Robotics Inst Am. *Res:* Automation and robotics; welding automation. *Mailing Add:* 1203 Hood Dr Rte 6 Brentwood TN 37027

COOK, GERALD, b Hazard, Ky, Oct 31, 37; m 62; c 2. ELECTRICAL ENGINEERING, AUTOMATIC CONTROL SYSTEMS. *Educ:* Va Polytech Inst, BS, 61; Mass Inst Technol, MS, 62, EE, 63, ScD(elec eng), 65. *Prof Exp:* Mem staff Apollo guid, Instrumentation Lab, Mass Inst Technol, 63-65; res engr automatic control, F J Seiler Res Lab, USAF Acad, 65-68; assoc prof elec eng, Univ Va, 68-73, prof, 73-81; prof & chmn, dept elec & biomed eng, Vanderbilt Univ, 81-85; EARLE C WILLIAMS PROF ELEC ENG, GEORGE MASON UNIV, 85- *Concurrent Pos:* Res asst, Electronic Systs Lab, Mass Inst Technol, 64-65; lectr, Univ Colo, 66-68; consult, Kaman Corp, 66-68, Phillip Morris Tobacco Co & Melpar Co, 69-, Babcock & Wilcox Corp, 71-, Sverdrup Corp, 81-84, Anal Sci Corp, 85-; assoc ed, Trans on Indust Electronics & Control Instrumentation, 73-83, ed, 83-90. *Honors & Awards:* Off Aerospace Res Tech Achievement Award, 68; Outstanding Res Award, Southeastern Sect, Am Soc Eng Educ, 71; Centennial Award, Inst Elec & Electronics Engrs, 84, Mittelmann Achievement Award, 89. *Mem:* Fel Inst Elec & Electronics Engrs (secy, 73-79); Indust Electronics & Control Instrumentation Soc (secy, 74-79, vpres, 79-81, pres, 82-83). *Res:* Theory of optimization and suboptimal control; approximation techniques; trajectory optimation in pursuit-evasion tactics; numerical optimization and use of computer to achieve optimum design; biomedical engineering; eye movement behavior; industrial automation; computer control; robotics. *Mailing Add:* Dept Elec & Computer Eng George Mason Univ Fairfax VA 22030

COOK, GLENN MELVIN, b Los Angeles, Calif, Sept 26, 35; m 60; c 2. PHYSICAL CHEMISTRY, TECHNICAL MANAGEMENT & CHEMICAL ENGINEERING. *Educ:* Univ Calif, Berkeley, BS, 57; Univ Ill, PhD(phys chem), 61. *Prof Exp:* Res asst, Radiation Lab, Univ Calif, 57; teaching asst, Univ Ill, 57-60; from assoc res scientist to res scientist, Lockheed Missiles & Space Co, Calif, 61-66; mem tech staff, Sprague Elec Co, 66-70; sr electro chemist, Ledgemont Lab, Kennecott Copper Co, 70-73, sr process scientist, 73-77; mgr & group leader electrochem res, Argonne Nat Lab, 77-85; OWNER, LECTRO PROS, 85- *Concurrent Pos:* Mem Bd Rev Metall Trans, 77- *Mem:* AAAS; Am Chem Soc; Am Phys Soc; Electrochem Soc; Am Inst Mining, Metall & Petrol Engrs. *Res:* Visible, fluorescent and phosphorescent spectra; molecular and ionic interactions in non-aqueous electrolytes; aqueous and non-aqueous batteries, fuel cells and cell design; electroplating and corrosion; electrochemical process; porous electrode; dilute solution hydrometallurgy; laboratory safety. *Mailing Add:* Elec Power Res Inst 3412 Hillview Ave PO Box 10412 Palo Alto CA 94303

COOK, GORDON SMITH, b Newark, NY, Mar 5, 14; m 39; c 3. ORGANIC CHEMISTRY. *Educ:* Hope Col, AB, 37; Syracuse Univ, MS, 39. *Prof Exp:* Asst gen chem, Syracuse Univ, 37-39; control & develop chemist, Chambers Works, E I du Pont de Nemours & Co, 40-49, tech rep org chem dept, Rubber Chem Div, 50-53, head div prod control & specifications, Elastomers Dept, 53-57, head foam-elastomers chem dept, 57-58, head prod control & specifications, 58-70, tech develop chemist, 70-79; RETIRED. *Mem:* Am Chem Soc; Am Soc Qual Control. *Res:* Synthetic elastomers; cellulose chemistry. *Mailing Add:* 21 Briar Rd Wilmington DE 19803

COOK, HAROLD ANDREW, b Wheeling, WVa, July 10, 41; m 64; c 1. AGRICULTURAL MICROBIOLOGY. *Educ:* West Liberty State Col, BS, 64; WVa Univ, MS, 66, PhD(microbiol), 69. *Prof Exp:* From asst prof to assoc prof, 69-77, PROF BIOL, WEST LIBERTY STATE COL, 77- CHMN DEPT MED TECHNOL, 70- *Concurrent Pos:* Reviewer, Am Biol Teacher Today, 74-; adj course dir, Chautauqua-Type Short Course Prog, NSF, 75-76;

consult, AAAS. *Mem:* Am Soc Microbiol. *Res:* Pollution; sanitary landfills; acid mine drainage; developing biology laboratory manual for non-science majors. *Mailing Add:* Dept Biol & Chem West Liberty State Col PO Box 296 West Liberty WV 26074

COOK, HARRY E, III, b Fresno, Calif, June 11, 35; div; c 3. GEOLOGY. *Educ:* Univ Calif, Santa Barbara, BS, 61, Univ Calif, Berkeley, PhD(geol), 66. *Prof Exp:* Consult petrog, Hales Labs, Calif, 63-65; res geologist, Denver Res Ctr, Marathon Oil Co, 65-70; assoc prof, Univ Calif, Riverside, 70-74; RES GEOLOGIST, US GEOL SURV, 74- *Concurrent Pos:* Asst prof, San Francisco State Univ, 64-65; adj prof geol, Univ Nev, 85-; leader field sem, Am Asn Petrol Geologists; lectr short courses, Soc Econ Paleontologist & Mineralogists; sedimentary counr & chmn mem comt, Soc Econ Paleontologists & Mineralogists, 85- *Honors & Awards:* George C Matson Award, Am Asn Petrol Geologists, 72. *Mem:* Int Asn Sedimentologists; Geol Soc Am; Am Asn Petrol Geologists; Soc Econ Paleontologists & Mineralogists. *Res:* Petroleum geology of carbonates; marine geology; sedimentary carbonate bank and reef facies; continental slope, submarine fans and aprons; diagenesis; petroleum reservoirs; debris flows and turbidity currents; petroleum geology of Western Canada, Nevada, Utah and Texas. *Mailing Add:* US Geol Surv Br Sedimentary Processing MS 999 345 Middlefield Rd Menlo Park CA 94025

COOK, HARRY EDGAR, b Americus, Ga, Feb 14, 39; m 61; c 2. MATERIALS SCIENCE. *Educ:* Case Inst Technol, BS, 60, MS, 62; Northwestern Univ, PhD(mat sci), 66. *Prof Exp:* Sr res scientist, Ford Motor Co, 67-69, sr engr, 69-70, prin engr chassis eng, 70-71, supvr, 71-72; from assoc prof to prof, metall & mech eng, Univ Ill, Champaign-Urbana, 72-77; sr res scientist metall, Ford Motor Co, 77-78, mgr mat eng, 78-79, body component eng, 79-81, & Metall Dept, 81-85; dir auto res, Chrysler Motors, 85-90; GRAYCE WICALL GAUTHIER PROF, DEPT MECH & INDUST ENG, UNIV ILL, URBANA, 90- *Honors & Awards:* Robert Lansing Hardy Medal, Am Inst Mining & Metall Engrs, 68; Teetor Award, Soc Automotive Engrs, 77. *Mem:* Nat Acad Eng; fel Am Soc Metals; Soc Automotive Engrs. *Res:* Phase transformations; friction materials; leadtime; competitiveness. *Mailing Add:* Dept Mech & Indust Eng Univ Ill Urbana IL 61801

COOK, HOWARD, b Spartanburg, SC, June 13, 33; m 60; c 3. TOPOLOGY. *Educ:* Clemson Col, BS, 56; Univ Tex, PhD(math), 62. *Prof Exp:* Spec instr math, Univ Tex, 58-62; asst prof, Auburn Univ, 62-64, res asst prof, Univ NC, Chapel Hill, 64; asst prof, 64-66; assoc prof, 66-71, PROF MATH, UNIV HOUSTON, 71- *Concurrent Pos:* Vis prof pure math, Univ Tasmania, 72-73. *Res:* Point set theory. *Mailing Add:* Dept Math Univ Houston 4800 Calhoun Rd Houston TX 77204

COOK, JACK E, b Ind, Feb 3, 31; m 54; c 2. ORGANIC POLYMER CHEMISTRY, PROOFING FILMS. *Educ:* DePauw Univ, BA, 53; Northwestern Univ, PhD, 57. *Prof Exp:* Res chemist exploratory plastics, Phillips Petrol Co, 57-64; SR RES SPECIALIST, MINN MINING & MFG CO, 64- *Res:* Acrylics; electroplating; elastomers; polymer characterization; injection molding; surface topography; polymer synthesis; personal identification systems; retroreflective products; adhesion; security systems; document security; antireflection technology; prepress proofing; polymeric films. *Mailing Add:* 56 Michael St St Paul MN 55119

COOK, JAMES ARTHUR, b Adairsville, Ga, May 19, 20; m 54; c 5. PLANT NUTRITION, SOIL FERTILITY. *Educ:* Cornell Univ, PhD(pomol), 51. *Prof Exp:* From asst plant physiologist to assoc plant physiologist, USDA, 48-52; asst viticult, Univ Calif, Davis, 53-58, assoc viticult & assoc prof viticult, 58-64, chmn dept viticult & enol, 62-66, viticult, Exp Sta & prof, 64-84, EMER PROF VITICULT & ENCOL, UNIV CALIF, 84- *Mem:* Am Soc Plant Physiol; Am Soc Hort Sci; Am Soc Enol & Viticulture. *Res:* Fertilizer requirements of dates, apples and grapes; foliage nutrient sprays; plant nutrition; mineral nutrition of grapevines; deficiencies as well as toxicities, as determined by visual symptoms, tissue analysis and field trial responses. *Mailing Add:* 1608 Astoria St Davis CA 95616

COOK, JAMES ARTHUR, CIRCULATORY SHOCK, ELEOSANOICS. *Educ:* Tulane Med Univ, PhD(physiol), 75. *Prof Exp:* PROF CARDIOVASC PHYSIOL, MED UNIV SC, 78- *Mailing Add:* Dept Physiol 171 Ashley Ave Med Univ SC Charleston SC 29425

COOK, JAMES DENNIS, b Tillsonburg, Ont, July 7, 36; US citizen. HEMATOLOGY. *Educ:* Queen's Univ, Kingston, Ont, Can, MD, 60, CM, 60, MSc, 63. *Prof Exp:* Instr med, Sch Med, Univ Wash, 67-69, asst prof, 69-73, assoc prof, 73-75; PHILLIPS PROF MED & DIR, HEMAT DIV, UNIV KANS MED CTR, 75- *Concurrent Pos:* Mem, Int Nutrit Anemia Consult Group, 75-; mem, Panel Malnutrit, US-Japan Coop Med Sci Prog, 75-; prin investr, Control Iron Deficiency grant, US Agency Int Develop, 78-; chmn, Iron Panel, Int Comt Standardization Hemat, 80- *Mem:* fel Am Col Physicians; Am Soc Hemat; Am Fed Clin Res; Am Soc Clin Nutrit; Soc Exp Biol & Med; Fedn Am Socs Exp Biol. *Res:* Iron deficiency; iron absorption in humans; serum ferritin as a measure of iron status. *Mailing Add:* Div Hemat Univ Kans Med Ctr 39th & Rainbow Blvd Kansas City KS 66103

COOK, JAMES ELLSWORTH, b Eureka, Kans, Oct 20, 23; m 50; c 4. VETERINARY PATHOLOGY, COMPARATIVE PATHOLOGY. *Educ:* Okla State Univ, DVM, 51; Am Col Vet Path, dipl, 56; Kans State Univ, PhD(path), 70. *Prof Exp:* Asst prof path, Okla State Univ, 51-52; vet pathologist, Jensen Salsburys Labs, 52-53; resident comp path, Armed Forces Inst Path, 53-56, assoc pathologist, Aerospace Path Br, 56-57, pathologist, Vet Path Div, 65-69; chief vivarium br primate med, 6571st Aerospace Med Lab, Holloman AFB, NMex, 57-63; instr path, Col Vet Med, Kans State Univ, 69-70, actg head dept, 72-75, head dept, 75-77 & 83-87, dir animal resource facil, 74-83, PROF PATH, KANS STATE UNIV, 70- *Mem:* Am Vet Med Asn; Am Soc Vet Clin Path. *Res:* Infectious and neoplastic diseases of primates and other laboratory animals; leptospirosis in domestic animals. *Mailing Add:* 301 N 15th St Manhattan KS 66502

COOK, JAMES H, JR, b Anderson, SC, Oct 28, 37; m; c 3. ELECTRICAL ENGINEERING. *Educ:* Ga Inst Technol, BEE, 61, MSEE, 70. *Prof Exp:* engr, Bendix Radio Div, Bendix Corp, 61-62, asst proj engr, 62-64; antenna engr, Sci-Atlanta Inc, 64-66, sr engr, 64-68, mgr antenna & microwave prod, 68-72, prod line mgr telecommun prod, 73-75, prod line mgr antenna & microwave, 75-78, prin engr, 78- 85, TECH DIR TELECOMMUN GROUP, SCI ATLANTA INC, 80-, PRIN ENGR, ELECTRONIC SYSTS GROUP. *Concurrent Pos:* Consult, Ga Tech Res Inst, 84-; chmn, antenna working group, adv comt reduced satellite spacing, Fed Commun Comn. *Mem:* Sr mem Inst Elec & Electronics Engrs; Am Inst Aeronaut & Astronaut. *Res:* Communications systems analysis; electromagnetics; antenna design and analysis. *Mailing Add:* Sci-Atlanta Inc 3845 Pleasantdale Rd Atlanta GA 30340

COOK, JAMES L, b Wabash, In, Oct 29, 46. VIROLOGY. *Educ:* Baylor Col Med, MD, 72. *Prof Exp:* STAFF PHYSICIAN, NAT JEWISH CTR IMMUNOL & RESPIRATORY MED, 80-; ASSOC PROF MED & MICROBIOL IMMUNOL, HEALTH SCI CTR, UNIV COLO, 85- *Res:* Tumor immunology. *Mailing Add:* Nat Jewish Ctr Microbiol & Immunol Dept 1400 Jackson St Denver CO 80206

COOK, JAMES MARION, b Franklin, Ky, Aug 16, 41; m 64; c 2. THEORETICAL PHYSICS, MATHEMATICAL ANALYSIS. *Educ:* Western Ky Univ, BS, 62; Vanderbilt Univ, PhD(physics), 67. *Prof Exp:* assoc prof, 66-77, PROF PHYSICS & CHEM, MID TENN STATE UNIV, 77- *Mem:* Am Asn Physics Teachers. *Res:* Differential equations. *Mailing Add:* Dept Chem & Physics Mid Tenn State Univ Murfreesboro TN 37132

COOK, JAMES MINTON, b Bluefield, WVa, Aug 6, 45; c 2. NATURAL PRODUCTS CHEMISTRY, MEDICINAL CHEMISTRY. *Educ:* WVa Univ, BS, 67; Univ Mich, PhD(org chem), 71. *Prof Exp:* NIH fel natural prod, Univ BC, 72-73; asst prof, 73-79, ASSOC PROF CHEM, UNIV WIS-MILWAUKEE, 79- *Mem:* Am Chem Soc; The Chem Soc; Am Soc Pharmacog. *Res:* Synthesis of beta adrenergic antagonists for antihypertensive drug studies; synthesis of B-carboline alkaloids and diazepam antagonists; construction of potential agents for treatment of Narcolepsey; synthesis of potential antimalarial agents; studies on the reactions of dicarbonyl compounds with dimethyl-3-ketoglutarate; general approach for the sythesis of polyquinanes; preparation of planar tetracordinate carbon atom. *Mailing Add:* Dept Chem Univ Wis Milwaukee WI 53201

COOK, JAMES RICHARD, b Maben, WVa, Nov 22, 29; m 59; c 2. CELL PHYSIOLOGY. *Educ:* Concord Col, BS, 50; WVa Univ, MS, 55; Univ Calif, Los Angeles, PhD(zool), 60. *Prof Exp:* NIH fel zool, Misaki Marine Biol Sta, Japan, 60-61; asst biophysicist, Lab Nuclear Med & Radiation Biol, Univ Calif, Los Angeles, 61-63; from asst prof to assoc prof zool, 63-74, PROF ZOOL & BOT, UNIV MAINE, ORONO, 74-, SAFETY OFFICER, RADIATION LAB, 81- *Concurrent Pos:* NIH res grant, 64-67, res career develop award, 68-78. *Mem:* Am Soc Cell Biol; Soc Protozool; Am Soc Plant Physiol; Fedn Am Socs Exp Biol. *Res:* Cell growth and division; adaptations of cells; chloroplast inheritance. *Mailing Add:* Dept Zool Murray Hall Univ Maine Orono ME 04469

COOK, JAMES ROBERT, b Hamilton, Ont, Aug 21, 41; m 63; c 5. PHYSICS, ENGINEERING. *Educ:* McMaster Univ, BEng, 64, PhD(physics), 69. *Prof Exp:* Programmer refinery simulation, Shell Oil, Can, 62-63; programmer-customer engr info processing, IBM, Can, 63-64; PRIN RES ENGR, ARMCO STEEL CORP, 69- *Concurrent Pos:* AISI sensor task chmn, Int Temperature Distrib Solid/Solidifying Bodies. *Mem:* Inst Elec & Electronics Engrs. *Res:* Solid state physics; mechanical and electrical engineering; modelling and optimization; real-time control of metallurgical processing; thermal process simulation; sensor based computer control of processes; sensor development; laser processing. *Mailing Add:* 5788 Trenton-Franklin Rd Curtiss St Middletown OH 45042

COOK, JOHN CAREY, b Ft Collins, Colo, Jan 13, 17; m 42; c 2. CIVIL ENGINEERING. *Educ:* Univ Idaho, BS, 47, CE, 66. *Prof Exp:* Testing engr, Idaho State Hwy Dept, 47-48, asst mat engr, 48-50; mat engr, AEC, Idaho, 50-53; owner-pres, Wash Testing Lab, 53-64; mgr, Pittsburgh Testing Lab, Wash, 64-66; res engr, Hwy Res Sect, Wash State Univ, 66-69, actg head, 69-70, head, 70-82, prof civil eng, 77-82; RETIRED. *Concurrent Pos:* Mem, Hwy Res Bd, Nat Acad Sci-Nat Res Coun, 67-, mem comt mineral aggregates & comt nuclear principles & appln, 69. *Mem:* Am Soc Testing & Mat. *Res:* Construction materials engineering testing and research, highways, airports, dams and buildings; environmental impact of highways on urban areas; nuclear construction application. *Mailing Add:* 206 E Weile No 2 Spokane WA 99208-5440

COOK, JOHN P(HILIP), b Washington, DC, Aug 24, 24; m 52; c 8. CIVIL ENGINEERING. *Educ:* Catholic Univ, BCE, 51, BArchE, 52; Rensselaer Polytech Inst, MCE, 55, DEngSc, 63. *Prof Exp:* Bridge designer, McEnteer Assoc, WVa, 55-58; from instr to assoc prof civil eng, Rensselaer Polytech Inst, 58-67; assoc prof, 67-70, JACOB LICHTER PROF ENG CONSTRUCT, UNIV CINCINNATI, 70- *Concurrent Pos:* Mem, Sealants Comt, Hwy Rds Bd, 63-68, 68-; consult, Thiokol Corp, NJ, 64-68, Watson-Bowman Assoc, NY, 70-74, SAfrican Bur Standards, 72-, Japan Archit Waterproofing Res Asn, 73 & Procter & Gamble, Ohio, 73-; mem, Bldg Res Inst, 65-70; mem, Adhesives Comn, 68- *Mem:* Am Soc Civil Engrs; Am Soc Eng Educ; Soc Plastics Indust; Am Soc Testing & Mat. *Res:* Behavior of elastomeric joint sealants for buildings and highway pavements; behavior of bonded-aggregate composite beams using polyester adhesives; field performance of highway pavements; corrosion resistance of concrete; impact yielding sign supports; composite behavior of metal roof decks. *Mailing Add:* Dept Civil Eng 71 Univ Cincinnati Main Campus Cincinnati OH 45221

COOK, JOHN SAMUEL, b Wilmington, Del, Sept 7, 27; m 65. CELL PHYSIOLOGY. *Educ:* Princeton Univ, AB, 50, MA, 53, PhD(biol), 55. *Prof Exp:* US Pub Health fel, Bern Univ, 55-56; from instr to assoc prof physiol, Sch Med, NY Univ, 56-66; prof biol, Univ Tenn-Oak Ridge Grad Sch Biomed Sci, 67-70, assoc dir sch, 68-70; STAFF MEM BIOL DIV, OAK RIDGE NAT LAB, 70- *Concurrent Pos:* Mem, US Nat Comt Int Union Psychol Sci, 76-82; assoc ed, Am J Physiol, 81-89 & News Physiol Sci, 86-; prog dir cell biol, NSF, 82-83; chmn managing bd, Nat Inst Pub Serv, 89- *Mem:* Fel AAAS; Soc Gen Physiol (pres, 79-80); Biophys Soc; Am Physiol Soc; Sigma Xi; Am Soc Cell Biol. *Res:* Membrane physiology; biogenesis and turnover of cell membranes; regulation of membrane transport systems. *Mailing Add:* Biol Div Oak Ridge Nat Lab PO Box 2009 Oak Ridge TN 37831-8077

COOK, JOHN WILLIAM, b Selma, Ala, Oct 25, 46. SOLAR PHYSICS. *Educ:* Mass Inst Technol, BS, 67; City Col, City Univ NY, MA, 70; Dartmouth Col, PhD(physics), 77. *Prof Exp:* Postdoctoral fel, Skylab Solar Workshop, High Altitude Observ, Boulder, Colo, 76-78; RES ASTROPHYSICIST, SOLAR PHYSICS BR, SPACE SCI DIV, NAVAL RES LAB, WASHINGTON, DC, 78- *Concurrent Pos:* Mem nominating comt, Solar Physics Div, Am Astron Soc, 80; mem, Space Physics Data Syst Steering Comt, NASA, 90-; mem co-investr team, sounding rocket, space shuttle & satellite sci experiments. *Mem:* Am Phys Soc; Am Astron Soc; Int Astron Union. *Res:* Physics of the solar atmosphere; analysis of high spatial and spectral resolution ultraviolet spectra; solar ultraviolet irradiance variability; atmospheres of cool stars. *Mailing Add:* Naval Res Lab Code 4163 4555 Overlook Ave Washington DC 20375-5000

COOK, JOSEPH MARION, b Oak Park, Ill, Feb 18, 24; m 56; c 3. MATHEMATICAL PHYSICS, SYSTEMS ANALYSIS. *Educ:* Univ Ill, BS, 47, MS, 48; Univ Chicago, PhD(math), 51. *Prof Exp:* Instr math, Johns Hopkins Univ, 51-52; NSF fel quantum mech, Harvard Univ, 52-53; assoc math, Argonne Nat Lab, 53-60; vis assoc prof, Univ Calif, Berkeley, 60-61; SR MATHEMATICIAN APPL MATH, ARGONNE NAT LAB, 61- *Mem:* Soc Indust & Appl Math; Am Math Soc; Am Phys Soc; Math Asn Am. *Res:* Applied mathematics, especially particle accelerator theory, nuclear reactor control. *Mailing Add:* Argonne Nat Lab Bldg 360 9700 S Cass Ave Argonne IL 60439

COOK, KELSEY DONALD, b Denver, Colo, Mar 16, 52. MASS SPECTROMETRY, SURFACTANT CHEMISTRY. *Educ:* Colo Col, BA, 74; Univ Wis-Madison, PhD(chem), 78. *Prof Exp:* Res fel, Univ Wis-Madison, 74-76, teaching asst, 75, res asst, 76-78; asst prof anal chem, Univ Ill, Urbana-Champaign, 78-84; asst prof, 84-86, ASSOC PROF ANALYTICAL CHEM, UNIV TENN, KNOXVILLE, 86- *Concurrent Pos:* Tech consult & phrasor sci, Oak Ridge Nat Labs, vis scientist, 91; Beckman fel, Univ Ill Ctr Advan Study, 82-83; Sci Alliance Res Incentive Award, Univ Tenn, 85-91; assoc ed, J Am Soc Mass Spectros. *Mem:* Am Chem Soc; Am Soc Mass Spectrometry (treas, 86-88); Soc Appl Spectros; Sigma Xi. *Res:* Development of electrohydrodynamic ionization mass spectrometry and characterization of polymer and related solutions by EHMS; fundamentals of desorption ionization. *Mailing Add:* Dept Chem Univ Tenn Knoxville TN 37996-1600

COOK, KENNETH EMERY, b Nebr, June 23, 28; m 52; c 3. ORGANIC CHEMISTRY. *Educ:* Hastings Col, BA, 53; Univ Nebr, MSc, 55, PhD(chem), 57. *Prof Exp:* From asst prof to assoc prof, 57-68, chmn dept, 62-74, PROF CHEM, ANDERSON COL, 68- *Mem:* Am Chem Soc; AAAS; Sigma Xi. *Res:* Synthetic organic chemistry; heterocyclic compounds. *Mailing Add:* 705 Maplewood Ave Anderson IN 46012

COOK, KENNETH LORIMER, b Middleton, NH, June 8, 15; m 46; c 3. GEOPHYSICS. *Educ:* Mass Inst Technol, BS, 39; Univ Chicago, PhD(geol & physics), 43. *Prof Exp:* Part-time instr phys sci, Univ Chicago, 41-43, geophysicist, US Bur Mines, Reno, Nev, 43-46; geophysicist, US Geol Surv, Nev, 46-49, Utah, 49-56; head dept, Univ Utah, 52-68, dir univ seismog stas, 52-76, prof geophys, 55-82, EMER PROF GEOPHYS, UNIV UTAH, 82- *Concurrent Pos:* Ed, Soc Explor Geophys. *Mem:* AAAS; Am Inst Mining, Metall & Petrol Eng; Soc Explor Geophys; Geol Soc Am; Am Geophys Union; Sigma Xi; Seismol Soc Am; Soc Explor Geophys (vpres, 67-68). *Res:* Mass spectroscopy; magnetic, gravitational and electrical geophysical interpretation; relative abundance of isotopes of potassium in Pacific kelps and rocks; vertical magnetic intensity over veins; resitivity data over filled sinks; regional magnetic and gravity surveys; gravity and magnetics of Utah; earth crustal and upper mantle studies; seismic research incident to large explosions. *Mailing Add:* Dept Geol & Geophys Univ Utah Salt Lake City UT 84112

COOK, KENNETH MARLIN, b Braddock, Pa, Aug 5, 20; m 44; c 2. ZOOLOGY. *Educ:* Univ Pittsburgh, BS, 43, MS, 48, PhD(biol sci), 53. *Prof Exp:* Asst instr biol, Univ Pittsburgh, 46-50; res fel, Mellon Inst, 50-52; res assoc, Grad Sch Pub Health, Univ Pittsburgh, 52-54; from asst prof to assoc prof, 54-67, PROF BIOL, COE COL, 67-, HEAD DEPT, 76-, HEINS-JOHNSON PROF BIOL, 78- *Honors & Awards:* Co-winner of award, Indust Med Asn, 55. *Mem:* Am Soc Zool; Sigma Xi; AAAS; Aerospace Med Asn; Am Inst Biol Sci. *Res:* Retention of particule matter in the human lung; measurement of pulmonary functional capacity; effects of antithyroid drugs on reproduction; oxygen consumption of small mammals; effects of centrifugation on physiological factors. *Mailing Add:* 2038 Westview Ave Waterloo IA 50701

COOK, KENNETH R, electrical engineering; deceased, see previous edition for last biography

COOK, L SCOTT, b Searcy, Ark, Dec 6, 48. CARDIOVASCULAR PHARMACOLOGY. *Educ:* Univ Tenn, PhD(pharmacol), 78. *Prof Exp:* SURG RESIDENT, UNIV OKLA HEALTH SCI CTR, 81- *Mailing Add:* Carle Clin Asn 60 W University Ave Urbana IL 61801

COOK, LAWRENCE C, b July 5, 25; US citizen; m 47; c 4. ORGANIC CHEMISTRY. *Educ:* ETenn State Univ, BS, 57. *Prof Exp:* Clin chemist, Mem Hosp, Johnson City, Tenn, 53-57; res org chemist, R J Reynolds Tobacco Co, 57-67, develop chemist, 67-82, res & develop mgr, 70-82, res & develop dir, 82-89; RETIRED. *Mem:* Am Chem Soc. *Res:* Clinical chemistry; isolation and identification of naturally occurring compounds. *Mailing Add:* 3045 Brookhill Dr Winston-Salem NC 27127

COOK, LEONARD, b Newark, NJ, June 27, 24; m 46; c 3. PHARMACOLOGY. *Educ:* Rutgers Univ, BA, 48; Yale Univ, PhD, 51. *Prof Exp:* Sr pharmacologist neuropharmacol, Smith, Kline & French Labs, 51-56, dir psychopharmacol res, 56-61, asst head pharmacol in charge res, 58-61, head psychopharmacol sect, 61-67, assoc dir pharmacol, 67-69; assoc dir pharmacol, Hoffmann-Laroche, Inc, 69-75, dir psychotherapeut res, 69-, dir pharmacol, 75-; CNS DIS RES, DEPT BIOMED PROD, E I DU PONT DE NEMOURS & CO. *Concurrent Pos:* Lectr, Woman's Med Col Pa, 59-; comt mem, Psychopharmacol Serv Ctr, NIH; Adj prof, Dept Psychiat, Med Sch, Rutgers Univ; pres, Psychopharmacol Div, Am Psychol Asn, 73. *Honors & Awards:* AMA Sci Exhibit Awards, 54. *Mem:* Am Soc Pharmacol & Exp Therapeut; Biomet Soc; Am Psychol Asn; fel Am Col Neuropsychopharmacol; fel Int Col Neuropsychopharmacol. *Res:* Neuropharmacology; central nervous system stimulants and depressants; drug potentiators; psycho-pharmacology; operant and classical conditioning techniques; neurobiochemistry; physiological conditioning; analgesics; gastrointestinal drugs; drugs affecting memory learning. *Mailing Add:* Dept Biomed Prod E I du Pont de Nemours & Co Exp Sta E-400 Rm 4414 PO Box 80400 Wilmington DE 19880-0400

COOK, LEROY FRANKLIN, JR, b Ashland, Ky, Dec 12, 31; m 57; c 3. PHYSICS. *Educ:* Univ Calif, Berkeley, AB, 53, MA, 57, PhD(physics), 59. *Prof Exp:* From instr to asst prof physics, Princeton Univ, 59-65; assoc prof, 65-68, actg head dept, 69-71, head dept physics & astron, 71-75 & 79-85, PROF PHYSICS, UNIV MASS, AMHERST, 68- *Concurrent Pos:* Consult, Inst Defense Anal, Arlington, Va, 63-67; vis fel, Clare Hall, Cambridge Univ, 71-72; chmn, Exec Comt New Eng Sect, Am Phys Soc, 90-91. *Mem:* Fel Am Phys Soc; Sigma Xi; AAAS. *Res:* Applications of gauge theories to weak, electromagnetic and strong interactions. *Mailing Add:* Dept Physics LGRTC Univ Mass Amherst MA 01003-0017

COOK, LESLIE G(LADSTONE), b Paris, Ont, July 12, 14; m 40; c 3. RADIO CHEMISTRY, PHYSICAL METALLURGY. *Educ:* Univ Toronto, BA, 36; Univ Berlin, PhD(chem), 39. *Prof Exp:* Chemist, Aluminum Co Can, 40-41; phys metallurgist, Aluminum Labs Ltd, 41-44; head, Chem Br, Atomic Energy Can, 44-53, dir, Chem & Metall Div, 53-56; proj analyst, Res Lab, Gen Elec Co, 56-59, mgr, Proj Anal Sect, 59-68, gen deleg policy & planning, Nat Res Coun Can, 68-69; mgr proj & prog planning, Corp Res Lab, ESSO Res & Eng, 69-77; pres, L G Cook Assocs Inc Energy Consults, 77-88; CONSULT, 88- *Mem:* Am Chem Soc; Am Nuclear Soc; Chem Inst Can. *Res:* Light alloy metallurgy; physical chemistry; research planning and management. *Mailing Add:* 26 Sunset Rd Newark DE 19711

COOK, MARIE MILDRED, b Bridgeport, Conn, Nov 22, 39. ZOOLOGY, DEVELOPMENTAL BIOLOGY. *Educ:* Georgian Ct Col, AB, 64; Rutgers Univ, MS, 70, PhD(zool), 74. *Prof Exp:* Teacher, Camden Cath High Sch, 66-69; instr & asst, 64-66, asst prof, 66-69, PROF BIOL & CHMN DEPT, GEORGIAN CT COL, 80- *Mem:* Am Soc Zoologists; Am Inst Biol Sci; Nat Asn Biol Teachers; Nat Sci Teachers Asn. *Res:* Developmental patterns and mechanisms of expression of esterase isozymes in hybrid species of the teleost genus Brachydanio. *Mailing Add:* Dept Acad Affairs Georgia Ct Col 900 Lakewood Ave Lakewood NJ 08701

COOK, MARK ERIC, b Houma, La, Apr 12, 56; m 81; c 3. POULTRY NUTRITION, ENVIRONMENTAL TOXICOLOGY. *Educ:* La State Univ, BS, 78, MS, 80, PhD(nutrit & immunol), 82. *Prof Exp:* Lectr, 82-83, assoc prof, poultry nutrit, 83-88, ASSOC PROF, ENVIRON TOXICOL, UNIV WIS-MADISON 88-, AFFIL PROF NUTRIT SCI, 89- *Concurrent Pos:* Assoc ed, Poultry Sci Asn, 88- *Mem:* Am Asn Avian Pathologists; AAAS; Am Chem Soc. *Res:* The immune modulatory effects of toxicants and nutrients in birds. *Mailing Add:* 260 Animal Sci Bldg Univ Wis Madison WI 53706

COOK, MARY ROZELLA, b Ardmore, Okla, Sept 30, 36. PSYCHOPHYSIOLOGY, BEHAVIORAL MEDICINE. *Educ:* Univ Okla, BA, 61, PhD(biol psychol), 70. *Prof Exp:* Res assoc, Inst Pa Hosp, 70-74; sr psychophysiologist, 74-77, prin psychophysiologist, 77-78, SECT HEAD BIOBEHAV SCI, MIDWEST RES INST, 78- *Concurrent Pos:* Instr, Dept Psychiat, Univ Pa, 70-74; ed, Biofeedback & Self-Regulation, 84-91. *Mem:* Soc Psychophysiol Res; Sigma Xi; Asn Appl Psychophysiology & Biofeedback. *Res:* Basic and applied studies in psychophysiology; effects of various stressors (environment, fatigue, work load, drugs) on human physiology and behavior; behavioral medicine including biofeedback; reversal theory. *Mailing Add:* Midwest Res Inst 425 Volker Blvd Kansas City MO 64110

COOK, MAURICE GAYLE, b Frankfort, Ky, Dec 26, 32; m 66; c 1. SOIL SCIENCE, AGRONOMY. *Educ:* Univ Ky, BS, 57, MS, 59; Va Polytech Inst & State Univ, PhD(agron), 61. *Prof Exp:* From asst prof to assoc prof, 61-70, PROF SOIL SCI, NC STATE UNIV, 70- *Concurrent Pos:* Vis prof agr chem & soils, Univ Agr Sci, Bangalore, India, 75-76; dir NC Div Soil & Water Conserv, Raleigh, 82-84. *Mem:* Fel Soil & Water Conserv Soc; fel Soil Sci Soc Am; Sigma Xi; fel Nat Asn Col & Teachers Agr; fel Am Soc Agron. *Res:* Soil mineralogy and its applications to soil genesis, morphology and classification; chemical and clay mineralogical interactions in soils. *Mailing Add:* Dept Soil Sci NC State Univ PO Box 7619 Raleigh NC 27695-7619

COOK, MELVIN ALONZO, b Swan Creek, Utah, Oct 10, 11; m 35; c 5. PHYSICAL CHEMISTRY, EXPLOSIVES. *Educ:* Univ Utah, BA, 33, MA, 34; Yale Univ, PhD(phys chem), 37. *Prof Exp:* Res chemist, E I du Pont de Nemours & Co, Inc, 37-47; prof metall, Univ Utah, 47-70; pres, IRECO Chem, 62-72, chmn, 62-74; CHMN, COOK ASSOCS, INC, 73- *Honors & Awards:* Frederick William Reynolds lectr, Univ Utah, 52; E V Murphree Award, Am Chem Soc, 68; Nitro Nobel Medallian, Swed Acad, Stockholm, 68; Dirs Award, Int Soc Explosives Specialists, 72; Chem Pioneer Award, Am Inst Chemists, 73; Distinguished Serv Award, Soc Explosives Engrs, 91. *Mem:* Am Chem Soc; Am Inst Chemists. *Res:* Commerical explosives, particularly slurry blasting agents; mechanism of detonation and rock blasting behavior. *Mailing Add:* Cook Assocs Inc 2026 Beneficial Life Tower Salt Lake City UT 84111

COOK, MICHAEL ARNOLD, b London, Eng, Dec 22, 44; m 68; c 3. PHARMACOLOGY, GASTROENTEROLOGY. *Educ:* Univ London, BSc, 67; Univ BC, PhD(physiol), 72. *Prof Exp:* Biochemist, Renal Unit, Royal Free Hosp, London, 67-68; Med Res Coun Can fel pharmacol, Univ Alta, 72-74; asst prof, 74-79, ASSOC PROF PHARMACOL, UNIV WESTERN ONT, 79 - *Concurrent Pos:* Assoc ed, Can J Physiol & Pharmacol, 80-87. *Mem:* Can Physiol Soc; NY Acad Sci; Pharm Soc Can; Soc Neurosci. *Res:* Studies on the neural and hormonal control of gastrointestinal motor activity; pharmacology of the gastrointestinal polypeptide hormones. *Mailing Add:* Dept Pharmacol & Toxicol Univ Western Ont London ON N6A 5C1 Can

COOK, MICHAEL MILLER, b Pittsburgh, Pa, Oct 10, 45; m 67. ORGANIC CHEMISTRY, CHEMISTRY. *Educ:* Carnegie-Mellon Univ, BS, 67; Stanford Univ, PhD(org chem), 73. *Prof Exp:* Res chemist water treat, Calgon Corp, Merck & Co, Inc, 72-75; res chemist org chem purification, 75-76, GROUP LEADER APPL RES & TECH SERV, DIV VENTRON CORP, MORTON THIOKOL INC, 76- *Mem:* Am Chem Soc; Am Oil Chemists Soc; Soc Photog Scientists & Engrs. *Res:* Water soluble polymers for scale corrosion control; liquid and solid separation; dithionite bleaching of pulp, and clay leaching; borohydride chemistry-application to organic chemical purification, metal recovery. *Mailing Add:* Morton Int 150 Andover St Danvers MA 01923-1428

COOK, NATHAN HENRY, b Ridgewood, NJ, Mar 17, 25; m 47; c 4. INSTRUMENTATION, COMPUTERS. *Educ:* Mass Inst Technol, SB, 50, SM, 51, ME, 54, ScD(mech eng), 55. *Prof Exp:* From asst prof to assoc prof, 53-65, prof mech eng, Mass Inst Technol, 65-85; RETIRED. *Concurrent Pos:* Consult, 53-85. *Honors & Awards:* Blackall Award, Am Soc Mech Engrs; Educ Award, Soc Mfg Engrs. *Mem:* Fel Am Soc Mech Engrs; Soc Mfg Engrs; Int Inst Prod Eng Res. *Res:* Materials; materials processing; friction; lubrication and wear; bearings; computerized manufacturing systems; computerized bio-medical data acquisition. *Mailing Add:* Box 95 North Eastham MA 02651-0095

COOK, NATHAN HOWARD, b Winston-Salem, NC, Apr 26, 39; m 61; c 2. CYTOLOGY, ZOOLOGY. *Educ:* NC Cent Univ, BS, 61, MA, 63; Okla State Univ, PhD(zool), 72. *Prof Exp:* Asst prof biol, Barber-Scotia Col, 62-68; teaching asst zool, Okla State Univ, 68-69; PROF BIOL, LINCOLN UNIV, 71-, HEAD DEPT, 74-, HEAD DEPT NATURAL SCI & MATH. *Concurrent Pos:* Proj dir, NSF Sci Equip Prog, 72-74 & Minority Biomed Support Prog Biol, NIH, Dept Health, Educ & Welfare, 72-; chmn, Mo Sickle Cell Anemia Adv Comt, 72- *Mem:* AAAS; Tissue Cult Asn; Sigma Xi. *Res:* In vitro effects of certain chemical carcinogens on the growth and chromosomes of mammalian cells. *Mailing Add:* Dept Natural Sci & Math Lincoln Univ Jefferson City MO 65101

COOK, NEVILLE G W, MINERAL ENGINEERING. *Prof Exp:* DONALD MCLAUGHLIN PROF MINERAL ENG, UNIV CALIF, 88- *Mem:* Nat Acad Eng. *Mailing Add:* Dept Mat Sci & Mineral Eng Univ Calif Hearst Mining Bldg Berkeley CA 94720

COOK, PAUL FABYAN, b Ware, Mass, Aug 2, 46; m 69; c 1. ENZYMOLOGY. *Educ:* Our Lady Lake Col, San Antonio, Tex, BA, 72; Univ Calif, Riverside, PhD(biochem), 76. *Prof Exp:* NIH fel biochem, Univ Wis, Madison, 76-80; asst prof biochem, La State Univ Med Ctr, 80-82; from asst prof to prof biochem, NTex State Univ, 82-88; PROF & CHMN, DEPT MICROBIOL & IMMUNOL, TEX COL OSTEOP MED, 88- *Concurrent Pos:* Ad hoc reviewer, NIH Study Sect Res Grants, 81-86, mem Biochem Study Sect, 87-; reviewer, biochem, J Am Chem Soc; Res Career Develop Award, NIH. *Mem:* Am Soc Microbiol; AAAS; Am Chem Soc; Biophys Soc; Am Soc Biol Chemists; NY Acad Sci; Sigma Xi. *Res:* Enzymology, specificlly mechanism of enzyme action which includes kinetic and chemical mechanism; extension of the kinetic theory as it applies to isotope effects in enzyme-catalyzed reactions and determination of mechanism. *Mailing Add:* Microbiol & Immunol Tex Col Osteop Med 3500 Camp Bowie Blvd Ft Worth TX 76107

COOK, PAUL LAVERNE, b Holland, Mich, Mar 2, 25; m 51; c 4. ORGANIC CHEMISTRY. *Educ:* Hope Col, BS, 50; Univ Ill, MS, 52, PhD, 54. *Prof Exp:* From instr to assoc prof, 54-64, prof, 65-90, EMER CHEM, ALBION COL, 90- *Mem:* Am Chem Soc; Sigma Xi. *Res:* Organic synthesis; products of pharmaceutical interest. *Mailing Add:* Dept Chem Albion Col Albion MI 49224

COOK, PAUL M, b Ridgewood, NJ, 1924; m; c 1. CHEMICAL ENGINEERING. *Educ:* Mass Inst Technol, BS, 47. *Prof Exp:* Mem staff, Stanford Res Inst, 49-52, head, Radiation Chem Lab, 52-54; chief exec officer, 57-90, FOUNDER CHMN, RAYCHEM CORP, 57- *Honors & Awards:* Winthrop Sears Award, Chem Indust Asn. *Mem:* Nat Acad Eng. *Res:* Commercial applications of radiations chemistry. *Mailing Add:* Bldg I Suite 185 300 Sand Hill Rd Menlo Park CA 94025

COOK, PAUL PAKES, JR, ecology, evolution; deceased, see previous edition for last biography

COOK, PHILIP W, b Underhill, Vt, Oct 6, 36. BOTANY. *Educ:* Univ Vt, BSc, 57, MSc, 59; Ind Univ, PhD(bot), 62. *Prof Exp:* NSF fel, 62-63; asst prof, 63-67, ASSOC PROF BOT, UNIV VT, 67- *Mem:* Bot Soc Am; Phycological Soc Am (pres, 71-72); Mycological Soc Am. *Res:* Fresh water algae; fungal parasites of algae. *Mailing Add:* Bot Dept Univ Vt Burlington VT 05401

COOK, RAY LEWIS, soils; deceased, see previous edition for last biography

COOK, RICHARD ALFRED, b Portland, Maine, Aug 9, 42; div. HUMAN NUTRITION. *Educ:* Univ Maine, Orono, BS, 65, MS, 68, PhD(nutrit), 73. *Prof Exp:* Res asst nutritionist, 65-74, asst prof, 74-80, ASSOC PROF NUTRIT, SCH HUMAN DEVELOP, UNIV MAINE, 80-, DIR SCH HUMAN DEVELOP, 85- *Mem:* Am Dietetic Asn; Sigma Xi; Soc Nutrit Educ; Am Pub Health Asn; Am Home Econ Asn. *Res:* Community nutrition; nutritional status assessment, monitoring, and surveillance; food consumption patterns. *Mailing Add:* 121 Fourth St Bangor ME 04401

COOK, RICHARD JAMES, b Alpena, Mich, Oct 20, 47; m 73. PHYSICAL ORGANIC CHEMISTRY. *Educ:* Univ Mich, BS, 69; Princeton Univ, MA, 71, PhD(chem), 73. *Prof Exp:* Res fel chem, Princeton Univ, 70-73; ASST PROF CHEM, KALAMAZOO COL, 73- *Concurrent Pos:* Res grant, Res Corp, 75-77. *Mem:* Am Chem Soc. *Res:* Electronic effects upon the barriers to inversion of substituted imines. *Mailing Add:* Dept Chem Kalamazoo Col Kalamazoo MI 49007-3295

COOK, RICHARD KAUFMAN, b Chicago, Ill, June 30, 10; m 38, 88; c 1. ACOUSTICS, PHYSICS. *Educ:* Univ Ill, BS, 31, MS, 32, PhD(physics), 35. *Prof Exp:* Asst physics, Univ Ill, 30-35; physicist, Nat Bur Standards, 35-42, chief sound sect, 42-66; chief geoacoustics group, Nat Oceanic & Atmospheric Admin, 66-71; spec asst acoust, Nat Bur Standards, 71-76; CONSULT PHYSICIST, 77- *Concurrent Pos:* Mem tech staff, Bell Tel Labs, 55-56; adj prof elec eng, Brooklyn Polytech Inst, 56; lectr mech eng, Cath Univ Am, 79 & 81. *Honors & Awards:* Gold Medal, US Dept Com, 64; Gold Medal, Acoust Soc Am, 88. *Mem:* Fel AAAS; fel Am Phys Soc; fel Acoust Soc Am (pres, 57-58); Am Geophys Union. *Res:* Solid state physics; applied mathematics; geophysics; physical acoustics; atmospheric sound propagation; acoustical measurements; mathematical acoustics. *Mailing Add:* 4111 Bel Pre Rd Rockville MD 20853

COOK, RICHARD SHERRARD, b Philadelphia, Pa, Apr 11, 21. ORGANIC CHEMISTRY. *Educ:* Philadelphia Col Pharm, BS, 43; Temple Univ, MA, 56. *Prof Exp:* Control chemist, Barrett Div, Allied Chem & Dye Corp, 43-45; chemist, 47-56, GROUP LEADER CHEM, ROHM AND HAAS CO, 56- *Mem:* Am Chem Soc; fel Am Inst Chemists; Sigma Xi. *Res:* Synthesis of new organic agricultural pesticides, including herbicides, fungicides and insecticides. *Mailing Add:* 107 S Lansdowne Ave Apt 204 Lansdowne PA 19050

COOK, ROBERT BIGHAM, JR, b Macon, Ga, Dec 4, 44; m 79; c 1. EXPLORATION GEOCHEMISTRY, ECONOMIC ANALYSIS OF MINERAL DEPOSITS. *Educ:* Colo Sch Mines, EM, 66; Univ Ga, MS, 68, PhD(geol), 71. *Prof Exp:* Geologist, Lindgren Explor Co, 70-72; from asst prof to assoc prof, 72-83, PROF ECON GEOL & HEAD DEPT, AUBURN UNIV, 83- *Mem:* Geol Soc Am; Am Inst Mining Engrs. *Res:* Exploration geochemistry for gold, silver, uranium, tin and tantalum; ground water quality and monitoring; coal geology and reserve estimation; economic modeling of potentially exploitable mineral deposits; ore mineralogy. *Mailing Add:* Dept Geol 210 Petrie Hall Auburn Univ Auburn AL 36849

COOK, ROBERT CROSSLAND, b New Haven, Conn, June 5, 47; div; c 2. CHEMICAL PHYSICS, POLYMER PHYSICS. *Educ:* Lafayette Col, BS, 69; Yale Univ, MPh, 71, PhD(theoret chem), 73. *Prof Exp:* asst prof chem, Lafayette Col, 73-81; RES SCIENTIST, LAWRENCE LIVERMORE NAT LAB, 81- *Concurrent Pos:* Vis prof chem, Dartmouth Col, 77, 78 & 79 & Colo State Univ, 80. *Mem:* Am Phys Soc; Am Chem Soc; Sigma Xi. *Res:* Physics and chemistry of polymeric materials; computer simulation of polymeric materials; deformation and failure of polymeric materials; statistical mechanics. *Mailing Add:* Lawrence Livermore Nat Lab L-482 Livermore CA 94550

COOK, ROBERT D(AVIS), b St Louis, Mo, Dec 20, 36; m 61; c 3. ENGINEERING MECHANICS. *Educ:* Univ Ill, BS, 58, MS, 60, PhD(appl mech), 63. *Prof Exp:* From asst prof to assoc prof, 63-74, PROF ENG MECH, UNIV WIS-MADISON, 74- *Concurrent Pos:* Consult. *Mem:* Am Soc Mech Engrs; Am Acad Mech. *Res:* Solid body mechanics; numerical methods of stress analysis; structural mechanics. *Mailing Add:* Dept Eng Mech Univ Wis Madison WI 53706

COOK, ROBERT DOUGLAS, b Montreal, Que, June 22, 41; m 70; c 2. ORGANOPHOSPHORUS CHEMISTRY, CHEMICAL EDUCATION IN THE DEVELOPING WORLD. *Educ:* Loyola Col, Montreal, BS, 62; Univ Calif Los Angeles, PhD(chem), 67. *Prof Exp:* Res chemist, Union Carbide Corp, 67-68; prof teaching, Am Univ Beirut, 68-86; PROF TEACHING, BISHOP'S UNIV, 86- *Concurrent Pos:* Vis prof, Mem Univ, Nfld, 76-77, Univ Toronto, 84-86. *Mem:* Am Chem Soc; Chem Inst Can. *Res:* Reactivity of organophosphorus compounds mainly derivatives of phosphinic acids; chemical education. *Mailing Add:* Dept Chem Bishop's Univ Lennoxville PQ J1M 1Z7 Can

COOK, ROBERT EDWARD, b Providence, RI, Sept 26, 46. POPULATION BIOLOGY, PLANT DEVELOPMENT. *Educ:* Harvard Univ, BA, 68; Yale Univ, PhD(biol), 73. *Prof Exp:* Instr biol, Yale Univ, 73-74; Cabot fel, 74-75, asst prof, 75-80, ASSOC PROF ECOL, DEPT BIOL, HARVARD UNIV, 80-, DIR, HERBARIUM, 89- *Concurrent Pos:* vis assoc prof, Cornell Univ, 81-82. *Res:* Plant population biology; plant development; history of ecology. *Mailing Add:* Dept Biol Harvard Univ Cambridge MA 02138

COOK, ROBERT EDWARD, b Springhill, WVa, Aug 26, 27; m 50; c 2. GENETICS. *Educ:* WVa Univ, BS, 49, MS, 56; NC State Col, PhD(genetics), 58. *Prof Exp:* Instr poultry, WVa Univ, 54-56; asst prof, Univ Fla, 58-61; coordr genetics, Agr Res Serv, USDA, 61-64, leader, Genetics Invests, 64-65; head dept poultry sci, Ohio State Univ, 65-69; HEAD DEPT POULTRY SCI, NC STATE UNIV, 69- *Mem:* Fel Poultry Sci Asn; World Poultry Sci Asn; Am Genetics Asn; Sigma Xi. *Res:* Basic genetics of the domestic fowl and systems of breeding for the improvement of poultry. *Mailing Add:* 3105 Cartwright Dr Raleigh NC 27612

COOK, ROBERT EDWARD, b 1946. ECOLOGY, SCIENCE EDUCATION. *Educ:* Harvard Col, AB, 68; Yale Univ, PhD(biol), 73. *Prof Exp:* Teaching asst, Yale Col, 70-71, actg instr, 72-73, instr, 73-74; Cabot fel, Harvard Univ, 74-75, from asst prof to assoc prof, dept biol, 75-82; prog dir pop biol & physiol ecol, NSF, 82-83; sr res assoc, 82-83, ASSOC PROF ECOL & SYSTS & DIR CORNELL PLANTATIONS, CORNELL UNIV, 83- *Concurrent Pos:* Mem Herbicide Assessment Comn, AAAS, 70; vis assoc prof ecol & systs, Cornell Univ, 81-82. *Mem:* Ecol Soc Am; Bot Soc Am; AAAS. *Res:* Plant population biology; plant development; history of ecology. *Mailing Add:* Cornell Plantations Cornell Univ One Plantation Rd Ithaca NY 14850

COOK, ROBERT HARRY, b Montreal, Que, Sept 30, 41; m 64; c 2. AQUACULTURE, GROWTH PHYSIOLOGY. *Educ:* McGill Univ, Montreal, BSc, 63; Dalhousie Univ, Halifax, MSc, 65, PhD, 68. *Prof Exp:* Biologist, Resource Develop Br, Dept Fisheries & Forestry, Maritime Region, Halifax, NS, 69-72; mgr, Environ Sci Br, Environ Protection Serv, Atlantic Region, Halifax, NS, 72-77; DIR, ST ANDREWS BIOL STA, DEPT FISHERIES & OCEANS, 77- *Concurrent Pos:* Mem, Huntsman Marine Sci Ctr, St Andrews, NB, 77-, bd, Int Joint Comn St Croix River, 78-; sci adv, Can parliamentary Comt Fisheries, Aquacult Can, 87-88; deleg, Maricult Comt, Int Coun Explor Sea, 88- *Mem:* Can Aquacult Asn; World Aquacult Soc. *Res:* Fisheries resource assessments; finfish and shellfish aquaculture development; marine ecology. *Mailing Add:* Can Dept Fisheries & Oceans St Andrews Biol Sta St Andrews NB E0G 2X0 Can

COOK, ROBERT JAMES, b Moorhead, Minn, Jan 14, 37; m 58; c 4. PLANT PATHOLOGY. *Educ:* NDak State Univ, BS, 58, MS, 61; Univ Calif, Berkeley, PhD(phytopath), 64. *Prof Exp:* NATO fel, Waite Inst, Australia, 64-65; asst, 65-68, PROJ LEADER, ROOT DIS & BIOL CONTROL RES UNIT, AGR RES SERV, USDA, WASH STATE UNIV, 68- *Concurrent Pos:* Guggenheim Mem Found fel, 73-74; adj prof, Path Univ Wash. *Honors & Awards:* Arthur Flemming Award, 75. *Mem:* Am Phytopath Soc; British Soc Plant Path; fel Am Phytopath Soc; Int Soc Plant Pathol. *Res:* Biological control of soil born plant pathogens; water relations of soil microorganisms; cereal root rots. *Mailing Add:* SW 910 Mies Pullman WA 99164-1030

COOK, ROBERT JAMES, b London, Eng, Mar 5, 50; m 80; c 1. METABOLIC REGULATION, PROTEIN STRUCTURE & FUNCTION. *Educ:* Univ Southampton, Eng, BSc, 71, PhD(biochem), 75. *Prof Exp:* Res assoc biochem, Univ Aston, 75-78, Vanderbilt Univ Med Sch, 78-80; res assoc pharmacol, Univ Nebr Med Ctr, 80-81; res instr, 81-85, RES ASST PROF BIOCHEM, VANDERBILT UNIV MED SCH, 85- *Mem:* Am Soc Biochem and Molecular Biol; Biochem Soc London. *Res:* Regulation of amino acid and vitamin metabolism in mammals; structure and function relationships of folate and flavin requiring enzymes. *Mailing Add:* Dept Biochem Vanderbilt Univ Sch Med Nashville TN 37232

COOK, ROBERT LEE, b Hollywood, Fla, Aug 30, 36; m 65; c 2. MOLECULAR SPECTROSCOPY. *Educ:* Univ Miami, BS, 58, MS, 60; Univ Notre Dame, PhD(phys chem), 63. *Prof Exp:* Res assoc microwave spectros, Duke Univ, 63-65, from instr to asst prof physics, 65-71; assoc prof physics, 71-74, PROF PHYSICS & CHEM, MISS STATE UNIV, 74-, DEP DIR, MHD ENERGY CTR, 85- *Mem:* Am Phys Soc; Am Asn Physics Teachers; Am Chem Soc; Sigma Xi; Am Inst Aeronaut & Astronaut. *Res:* Microwave spectroscopy; centrifugal distortion effects in asymmetric rotors; determination of molecular force constants; molecular structure; hyperfine interactions; ring conformations; spectrochemical analysis; development of optical techniques to characterize the gas streams of coal burning power plants, (magnetohydrodynamics, coal gasification). *Mailing Add:* Dept Physics Miss State Univ Mississippi State MS 39762

COOK, ROBERT MEROLD, b Bethany, Ill, Aug 5, 30; m 49; c 5. ANIMAL NUTRITION, BIOCHEMISTRY. *Educ:* Univ Ill, BS, 57, MS, 61, PhD(dairy sci), 62. *Prof Exp:* Res asst dairy sci, Univ Ill, 57-62; asst prof, Univ Idaho, 62-66; PROF ANIMAL SCI, MICH STATE UNIV, 66- *Mem:* Am Chem Soc; Am Dairy Sci Asn; Am Inst Nutrit; Am Soc Animal Sci. *Res:* Ruminant nutrition, control mechanisms regulating volatile fatty acid metabolism in ruminants; protein metabolism in rumen; xenobiotic metabolism in ruminants; international animal agriculture. *Mailing Add:* Dept Animal Sci Mich State Univ East Lansing MI 48823

COOK, ROBERT NEAL, b Marlette, Mich, Dec 10, 40; m 67; c 2. COMPUTER SCIENCE, APPLIED MATHEMATICS. *Educ:* Gen Motors Inst, BEE, 65; Univ Mich, MSE, 65; Univ Western Ont, MSc, 68, PhD(appl math), 76. *Prof Exp:* Sr engr, Gen Motors Corp, 59-68; res asst, Univ Western Ont, 68-71; asst prof comput sci, Frostburg State Col, 71-74 & Shippensburg State Col, 74-76; asst prof comput sci, Cent Mich Univ, 76-81; assoc prof, 81-85, PROF COMPUTER SCI TECHNOL, UNIV SOUTHERN COLO, 85-; VPRES, NETWORK ENG CORP, 87- *Mem:* Asn Computer Mach; Simulation Coun Inc; Int Asn Math & Computer Simulation; Inst Elec & Electronics Engrs Computer Soc. *Mailing Add:* Computer Sci Dept Univ Southern Colo 2200 Bonforte Blvd Pueblo CO 81001

COOK, ROBERT PATTERSON, b Aug 30, 47. DISTRIBUTED PROGRAMMING, OPERATING SYSTEMS. *Educ:* Vanderbilt Univ, BS, 69, MS, 71, PhD(comput sci), 78. *Prof Exp:* Syst analyst, Vanderbilt Comput Ctr, 67-71; lectr, Univ Fla, 72-76; asst prof comput sci, Univ Wis-Madison, 78-83; ASSOC PROF COMPUT SCI, UNIV VA, 83- *Concurrent Pos:* Consult, Rank Xerox, 71-72, Xerox Corp & Evergreen Assoc, 71-76, Nicolet Instruments, 80-82, Gen Dynamics, 81-82; prin investr grants, NSF, 78-87, Consortium Int Studies Educ IIP, 81-83 & 90-95, Off Naval Res, 86- & ARO, 87-89. *Mem:* Asn Comput Mach; Inst Elec & Electronics Engrs. *Res:* Computer architecture; operating systems; software reuse. *Mailing Add:* Comput Sci Dept Univ Va Charlottesville VA 22903

COOK, ROBERT SEWELL, b Unity, Wis, Nov 25, 29; m 53; c 2. VERTEBRATE ECOLOGY. *Educ:* Univ Wis-Stevens Point, BS, 51; Univ Wis-Madison, MS, 58, PhD(vet sci), 66. *Prof Exp:* Res asst wildlife ecol, Univ Wis-Madison, 54-57; biologist, Wis Conserv Dept, 57-59; gen secy, Appleton YMCA, 59-61; biol instr high sch, Wis, 61-63; res asst vet sci, Univ Wis-Madison, 63-66; res participation grant, 66-69, asst prof physiol, 66-70, Alumni Res Found grant, 67-68, asst prof environ control, 70-71, assoc prof environ control, Univ Wis-Green Bay, 71-77; dep dir, US Fish & Wildlife Serv, 77-81; DEPT HEAD, FISHERY & WILDLIFE BIOL DEPT, COLO STATE UNIV, 81- *Concurrent Pos:* Mem, US Fish & Wildlife Serv Missions, India, 79, 80, 82 & 86. *Mem:* AAAS; Wildlife Soc; Wildlife Dis Asn; Nature Conservancy Int Asn Fish & Wildlife Agencies. *Res:* Ecology of diseases in populations of wildlife that are transmissible to domestic animals and man; biological aspects of land-use planning; environmental impact analysis; communication and planning for fish and wildlife management. *Mailing Add:* Dept Fishery & Wildlife Biol Wagar Bldg Rm 135 Colo State Univ Ft Collins CO 80523

COOK, ROBERT THOMAS, b Nebraska City, Nebr, Apr 27, 37; m 61; c 3. PATHOLOGY, BIOCHEMISTRY. *Educ:* Univ Kans, AB, 58, MD, 62, PhD(biochem), 67. *Prof Exp:* Assoc pathologist, Walter Reed Army Inst Res, 67-69; asst prof path, Inst Path, Case Western Reserve Univ, 69-76; ASSOC PROF PATH, UNIV IOWA, 77- *Mem:* Am Asn Pathologists; Am Chem Soc. *Res:* Cell control mechanisms in neoplasia; alcohol effects on cell growth. *Mailing Add:* Dept Path Col Med Univ Iowa Iowa City IA 52240

COOK, RONALD FRANK, b Buffalo, NY, July 22, 39; m 64; c 3. ANALYTICAL CHEMISTRY, RESEARCH ADMINISTRATION. *Educ:* Univ Buffalo, BA, 61. *Prof Exp:* From chemist to res chemist, 62-72, supvr, Residue Lab, Agr Chem Group, 72-81, MGR, RESIDUE CHEMISTRY, FMC CORP, 81- *Concurrent Pos:* Assoc referee, Asn Off Anal Chemists, 65-75. *Mem:* Am Chem Soc. *Res:* Development and application of analytical methods to the determination of pesticide content of formulations and the residue content of agricultural commodities; management of residue laboratory function; interaction with state and federal agencies relative to regulating compliance. *Mailing Add:* 1224 Dickinson Dr Yardley PA 19067-2917

COOK, SHIRL ELDON, b Paris, Idaho, Mar 15, 18; m 44; c 1. CHEMISTRY. *Educ:* Brigham Young Univ, BS, 39; La State Univ, MS, 41. *Prof Exp:* Chemist, US Army, La, 41-42 & La Ord Plant, 42-45; chemist, Ethyl Corp, 45-76, supvr anal serv, 76-83; RETIRED. *Mem:* Am Chem Soc. *Res:* Alkyl metal compounds; organometallics. *Mailing Add:* 5734 Hyacinth Ave Baton Rouge LA 70808

COOK, STANTON ARNOLD, b Oakland, Calif, Dec 10, 29; m 59; c 2. TERRESTIAL ECOLOGY. *Educ:* Harvard Univ, AB, 51; Univ Calif, Berkeley, PhD(bot), 61. *Prof Exp:* From asst prof to assoc prof, 60-80, PROF BIOL, UNIV ORE, 80- *Res:* Interactions of plants; pathogens and herbivores in forests and range of western United States; terrestrial community ecology. *Mailing Add:* Dept Biol Univ Ore Eugene OR 97403

COOK, STEPHEN ARTHUR, b Buffalo, NY, Dec 14, 39; m 68; c 2. COMPUTATIONAL COMPLEXITY. *Educ:* Univ Mich, BS, 61; Harvard Univ, SM, 62, PhD(math), 66. *Prof Exp:* Asst prof math & comput sci, Univ Calif, Berkeley, 66-70; assoc prof, 70-75, PROF COMPUT SCI, UNIV TORONTO, 75-, UNIV PROF, 85- *Concurrent Pos:* Steacie mem fel, Nat Sci & Eng Res Coun Can, 77-78, Killiam res fel, 82-83. *Honors & Awards:* A M Turing Award, Asn Comput Mach, 82. *Mem:* Nat Acad Sci; fel Royal Soc Can; Asn Symbolic Logic; Am Acad Arts & Sci. *Res:* Computational complexity (the study of the intrinsic computational difficulty of problems); theory of large scale parallel computation. *Mailing Add:* Dept Comput Sci Univ Toronto Toronto ON M5S 1A4 Can

COOK, STUART D, b Boston, Mass, Oct 23, 36; m 60; c 3. NEUROLOGY, NEUROSCIENCES. *Educ:* Brandeis Univ, AB, 57; Univ Vt, MS, 59, MD, 62. *Prof Exp:* Intern med & surg, State Univ NY Upstate Med Ctr, 62-63; resident neurol, Albert Einstein Col Med, 65-68, instr, 68-69; asst prof, Col Physicians & Surgeons, Columbia Univ, 69-71; PROF MED, NJ MED SCH, UNIV MED & DENT, 71-, PROF NEUROSCI & CHMN, 72-, ACTG DEAN, 87- *Honors & Awards:* S Weir Mitchell Award, Am Acad Neurol, 69. *Mem:* Am Acad Neurol; Am Asn Neuropath; Am Fedn Clin Res; Harvey Soc; Reticuloendothelial Soc; Am Neurol Asn. *Res:* Neuroimmunology; demyelinating diseases. *Mailing Add:* NJ Med Sch 1855 Orange Ave Newark NJ 07103

COOK, THEODORE DAVIS, b Kentfield, Calif, Jan 23, 24; m 48; c 4. GEOLOGY. *Educ:* Univ Utah, BS, 48; Univ Calif, MA, 50. *Prof Exp:* Sr staff geologist, Shell Oil Co, mgr, stratagraphic serv, 76-89; RETIRED. *Mem:* Fel Geol Soc Am. *Res:* Tertiary micropaleontology and stratigraphy; North and Central America stratigraphy. *Mailing Add:* Seven Alpine Bellaire TX 77401

COOK, THOMAS BRATTON, JR, b Rich Pond, Ky, Aug 28, 26; m 47; c 2. SYSTEMS DESIGN & SYSTEMS SCIENCE, PHYSICS. *Educ:* Western Ky Univ, BS, 47; Vanderbilt Univ, MS, 49, PhD(physics), 51. *Prof Exp:* Mem staff, Weapons Effects Dept, Sandia Lab, 51-55, supvr, Vulnerability Studies Sect, 55-56, supvr nuclear burst studies div, 56-59, mgr nuclear burst dept, 59-62, dir nuclear burst physics & math res, 62-67, vpres, 67-82, exec vpres, Sandia Labs 82-86; RETIRED. *Concurrent Pos:* Mem, Sci Adv Group on Effects, Dept Defense, 61-72, Air Force Sci Adv Bd, 64-75, chmn,

Vulnerability Task Force, Defense Sci Bd, 66-70 & sci adv group, Joint Strategic Target Planning Agency, Dept Defense, 68-72. *Honors & Awards:* E O Lawrence Award, Dept Energy, 71, Distinguished Assoc Award, 86. *Mem:* Nat Acad Eng; fel Am Phys Soc; Sigma Xi. *Res:* Nuclear weapons effects, nuclear safety and control. *Mailing Add:* 80 Castlewood Dr Pleasanton CA 94566

COOK, THOMAS M, b Miami, Fla, Mar 12, 31; m 50; c 3. MICROBIOLOGY. *Educ:* Univ Md, BS, 55, MS, 57; Rutgers Univ, PhD(bact), 63. *Prof Exp:* Microbiologist, Merck Sharp & Dohme Res Labs, 57-61; assoc res biologist, Sterling-Winthrop Res Inst, 63-66; asst prof, 66-70, assoc prof, 70-77, PROF MICROBIOL, UNIV MD, COL PARK, 77- *Mem:* Am Soc Microbiol. *Res:* Microbial physiology and biochemistry; oxidative metabolism; fermentations; action of antimicrobial agents. *Mailing Add:* Dept Microbiol Univ Md College Park MD 20742-7611

COOK, THURLOW ADREAN, b Utica, NY, June 2, 39; m 61; c 2. MATHEMATICS. *Educ:* Univ Rochester, BA, 61; State Univ NY Buffalo, MA, 65; Fla State Univ, PhD(math), 67. *Prof Exp:* From asst prof to assoc prof, 67-88, PROF MATH, UNIV MASS, AMHERST, 88- *Concurrent Pos:* Dir, Freiburg Univ Exchange Prog, Freiburg, FRG, 78. *Mem:* Am Math Soc; Math Asn Am. *Res:* Functional analysis; foundations of quantum mechanics and empirical logics. *Mailing Add:* Dept Math Univ Mass Amherst MA 01003

COOK, VICTOR, b Palenville, NY, July 13, 29; m 57; c 2. PHYSICS. *Educ:* Univ Calif, Berkeley, AB, 57, PhD(physics), 62. *Prof Exp:* Res physicist, Lawrence Radiation Lab, Univ Calif, 62-63; from asst prof to assoc prof, 63-77, PROF PHYSICS, UNIV WASH, 77-; MEM STAFF, PHYS DIV, NSF. *Mem:* Am Phys Soc; Am Asn Physics Teachers; Union Concerned Scientists; AAAS. *Res:* Properties and interactions of elementary particles. *Mailing Add:* Phys Dept Univ Wash Seattle WA 98195

COOK, VIRGINIA A, algebra, number theory, for more information see previous edition

COOK, WARREN AYER, b Conway, Mass, July 22, 00; m 28; c 1. INDUSTRIAL HYGIENE. *Educ:* Dartmouth Col, AB, 23. *Prof Exp:* Head chem unit, Eng & Inspection Div, Travelers Ins Co, 25-28; chief indust hygienist, Bur Indust Hyg, State Dept Health, Conn, 28-37; dir, Div Indust Hyg & Eng Res, Zurich-Am Ins Co, 37-53; res assoc & from assoc prof to prof, Inst Indust Health, Univ Mich, 53-71, emer prof indust health, 71-; RETIRED. *Concurrent Pos:* Adj prof indust health, Univ NC, 71-; chmn, Comt Standards Anal Atmospheric Contamitants, Am Pub Health Asn, 35-38, Comt Eng Control Prev Silicosis, US Dept Labor, 36-37; mem, Med Adv Comt Healthful Working Conditions, Nat Asn Mfrs, 41-46, Prev Eng Comt, Indust Hyg Found Am, 37-45, Comt Chem & Toxicol, 46-54, Comt Acceptable Concentrations Dusts, Gases & Vapors, Am Nat Standards Inst, 47-73, Spec Hazards Comt, Am Ins Asn, 42-53, Comt Air Sampling & Anal, Am Soc Testing & Mat, 56-59 & 67-73, Subcomt Oxidants & Nitrogen Compounds Intersoc, Comt Manual Methods Air Sampling & Anal, Am Pub Health Asn, 67-74, Occup Health Standards Comt, Am Indust Hyg Asn, 83-; assoc ed, J Indust Hyg & Toxicol, 47-49; chmn, Occup Health Sect, Am Pub Health Asn, 53-54. *Honors & Awards:* Cummings Award, Am Indust Hyg Asn, 53; Bordon Found Award, Am Indust Hyg Asn, 79. *Mem:* Hon mem Am Indust Hyg Asn (pres, 40); hon mem Am Acad Occup Med; emer mem Am Soc Safety Eng; Am Acad Indust Hyg; Am Chem Soc. *Res:* Methods of determination of atmospheric contaminants; administrative phases of industrial hygiene; educational and training programs for occupational health personnel; study of relation of auto accident causation to carbon monoxide exposure and carboxy hemoglobin concentration of driver; causation and prevention of acroosteolysis of workers in polyvinychloride production plants in USA and Canada; control measures for occupational health hazard; author of several articles. *Mailing Add:* 6376 Plankton Dr Columbus OH 43213

COOK, WENDELL SHERWOOD, organic chemistry; deceased, see previous edition for last biography

COOK, WILLIAM BOYD, b Dallas, Tex, July 20, 18; m 42; c 1. ORGANIC CHEMISTRY. *Educ:* Univ Tex, AB, 40; Univ Colo, MS, 42; Univ Wyo, PhD(chem), 50. *Prof Exp:* Asst, Univ Colo, 40-42; chemist-analyst, Monsanto Chem Co, 42-43, res assoc, 43-47; from instr to asst prof chem, Univ Wyo, 47-53; assoc prof, Baylor Univ, 53-57; prof & head dept, Mont State Col, 57-65; vis prof, Stanford Univ, 65-67; dean col sci & arts, 67-68, PROF CHEM, COLO STATE UNIV, 67-, DEAN COL NATURAL SCI, 68- *Concurrent Pos:* Fund Adv Educ fel, 52-53; NSF grant, Cambridge Univ, 62-63; exec dir, Adv Coun Chem, 65-67, mem, 66-69; mem adv comt grants of Res Corp, 69-75; US rep comt teaching chem, Int Union Pure & Appl Chem, 77-; chmn bd publ, J Chem Educ, 75- *Honors & Awards:* Gold Medal, Am Chem Soc, 73. *Mem:* Fel AAAS; fel Am Inst Chemists; Am Chem Soc. *Res:* Isolation and structural studies of alkaloids; nitrogen heterocyclic compounds; science education curricula. *Mailing Add:* 1809 Cottonwood Point Ft Collins CO 80524-1747

COOK, WILLIAM JOHN, b Des Moines, Iowa, Apr 12, 29; m 53; c 2. FLUID DYNAMICS, THERMODYNAMICS. *Educ:* Iowa State Univ, BS, 57, MS, 59, PhD(mech eng), 64. *Prof Exp:* From instr to assoc prof, 59-76, PROF MECH ENG, IOWA STATE UNIV, 76- *Concurrent Pos:* Prin investr, unsteady viscous flows, Eng Res Inst, Iowa State Univ, 79-; vis prof gas dynamics, Nat Defense Acad, 84. *Mem:* Am Soc Eng; Am Soc Mech Engrs; Am Inst Aeronaut & Astronaut; Sigma Xi. *Res:* Experimental and analytical studies of time-dependent boundary layer flows; shock tube gas dynamic studies; instrumentation for measurements in time-dependent flows; thermodynamics and fluid dynamics of inviscid and viscous compressible flows. *Mailing Add:* 2636 Kellogg Ames IA 50010

COOK, WILLIAM JOSEPH, Tuscaloosa, Ala, July 18, 49; m 69; c 2. SURGICAL PATHOLOGY, AUTOPSY PATHOLOGY. *Educ:* Univ Ala, BS, 71; Univ Ala, Birmingham, MD, 74, PhD(biochem), 76. *Prof Exp:* Resident path, NIH, 76-78; res assoc, 71-74, fel, 74-75, investr, Inst Dent Res, 78-82, scientist, 82-85, assoc scientist, Comprehensive Cancer Ctr, 78-82, scientist, 82-85, from asst prof to assoc prof path, 78-85, SR SCIENTIST, INST DENT RES & COMPREHENSIVE CANCER CTR, UNIV ALA, BIRMINGHAM, 85-, PROF PATH, 85- *Mem:* Am Crystallog Asn; Am Asn Pathologists; Int Acad Path; NY Acad Sci; Am Soc Biol Chemists. *Res:* Structural studies of compounds of biological interest using x-ray crystallography; primarily proteins, but also small polypeptides and nucleic acid components. *Mailing Add:* Path Dept Univ Ala Univ Sta Birmingham AL 35294

COOK, WILLIAM R, JR, b Boston, Mass, Nov 28, 27; m 50; c 4. SOLID STATE CHEMISTRY, MINERALOGY. *Educ:* Oberlin Col, BA, 49; Columbia Univ, MA, 50; Case Western Reserve Univ, PhD, 71. *Prof Exp:* Crystallogr, Brush Develop Co, 51-53; head crystallog sect, Electronic Res Div, Clevite Corp, Gould, Inc, 53-74; SECY, CLEVELAND CRYSTALS, INC, 73-; CONSULT, 74- *Concurrent Pos:* Adj cur, Cleveland Mus Natural Hist. *Mem:* Am Crystallog Asn; Mineral Soc Am; fel Am Ceramic Soc; Am Chem Soc. *Res:* Ferroelectricity and piezoelectricity; nonlinear optical materials; mineral chemistry. *Mailing Add:* 684 Quilliams Rd Cleveland OH 44121

COOK, WILLIAM ROBERT, b Birmingham, UK, Dec 8, 30; m 65; c 3. VETERINARY MEDICINE. *Educ:* Royal Col Vet Surgeons, MRCVS, 52, FRCVS, 66; Univ Cambridge, PhD(vet sci), 76. *Prof Exp:* Practr vet med, Gen & Equine Pract, 52-59; house surgeon, Sch Vet Med, Univ Cambridge, 59-60; lectr surg, Royal Vet Col, Univ London, 60-65; sr lectr, Sch Vet Med, Univ Glasgow, 65-69; sci officer I equine res, Animal Health Trust, London, 69-77; prof equine med & surg, Sch Vet Med, Univ Ill, 77-80; mem fac, 80-, PROF SURG, SCH VET MET, TUFTS UNIV, 80- *Concurrent Pos:* External examr MVSc degree, Univ Glasgow, 75-77; external examr vet surg, Sch Vet, Univ Bristol, 75-77. *Honors & Awards:* Sir Frederick Hobday Mem Prize, Brit Equine Vet Asn, 76; Steele-Bodger Mem Scholar, Brit Vet Asn, 77. *Mem:* Royal Soc Med; Brit Vet Asn; Brit Equine Vet Asn (vpres, 76, pres, 77); Am Vet Med Asn; Am Asn Equine Practr. *Res:* Clinical research into diseases of the ear, nose and throat in the horse; hereditary diseases of the horse. *Mailing Add:* 81 Rice Ave Northboro MA 01532

COOKE, ANSON RICHARD, b Lawrence, Mass, Jan 12, 26; m 48; c 4. PLANT GROWTH REGULATION, HERBICIDES. *Educ:* Univ Mass, BS, 49, MS, 50; Univ Mich, PhD(bot), 53. *Prof Exp:* Asst bot, Univ Mich, 50-53; asst prof plant biochem, Univ Hawaii, 53-54; asst prof plant physiol, Okla State Univ, 55-56; res plant physiologist, E I du Pont de Nemours & Co, 56-63; dir biol res, Amchem Prod, Inc, 63-77; assoc dir, Union Carbide Agr Prod Co, 86-87; AGR CONSULT, 88- *Mem:* Am Chem Soc; Am Soc Plant Physiol; Weed Sci Soc Am; Plant Growth Regulator Soc Am. *Res:* Plant growth regulators; herbicides; flowering; stress physiology. *Mailing Add:* 1210 Huntsman Dr Durham NC 27713-2364

COOKE, CHARLES C, b Huntsville, Ohio, Sept 13, 16; m 44; c 4. PHYSICS, ENGINEERING. *Educ:* Ohio State Univ, BS, 38. *Prof Exp:* Lab asst physics, 39-41, jr engr, 43-45, sr proj engr glass forming, 45-56, res engr comput, 57-66, chief process anal, Control & Instrumentation, 66-73, CHIEF COMPUT TECHNOL, OWENS ILLINOIS INC, 73- *Concurrent Pos:* Vis prof, Physics Dept, Appalachian State Univ, 79-80. *Mem:* Instrument Soc Am; Asn Comput Mach; Sigma Xi. *Res:* Glass forming problems; glass manufacturing process control; technical computer applications. *Mailing Add:* 1666 Glenfield Lane Toledo OH 43614

COOKE, CHARLES ROBERT, b Oak Hill, WVa, June 12, 29; m 51; c 4. MEDICINE. *Educ:* WVa Univ, AB, 50, BS, 52; Johns Hopkins Univ, MD, 54. *Prof Exp:* From instr to asst prof, Johns Hopkins Univ, 63-70, assoc prof med, Sch Med, 70-85; PROF MED, UNIV TENN, MEMPHIS, CHIEF NEPHROLOGY SECT, VED ADMIN MED CTR, 85- *Concurrent Pos:* Am Heart Asn fel, 59-60; USPHS fel, 60-61; asst chief med, Baltimore City Hosps, 63-67; consult, Vet Admin, 67- *Mem:* Fel Am Col Physicians; Am Fedn Clin Res; Am Soc Nephrology. *Res:* Renal physiology and electrolyte metabolism. *Mailing Add:* 1030 Jefferson Ave Memphis TN 38104

COOKE, DAVID WAYNE, b Hopkinsville, Ky, Mar 16, 47; m 68. SOLID STATE PHYSICS. *Educ:* Western Ky Univ, BS, 69, MS, 70; Univ Ala, PhD(physics), 77. *Prof Exp:* Med physicist, Med Ctr, WVa Univ, 71-72; fel, Univ Ala, 77; ASST PROF PHYSICS, MEMPHIS STATE UNIV, 78- *Concurrent Pos:* Grad coun res fel, Univ Ala, 76-77, Lockheed Aircraft fel, 77. *Honors & Awards:* Res Award, Sigma Xi, 80. *Mem:* Sigma Xi; Am Phys Soc; Health Phys Soc; AAAS; NY Acad Sci. *Res:* Low temperature solid state luminescence; thermoluminescent dosimetry; muon spin rotation. *Mailing Add:* 44 Timber Ridge Los Alamos NM 87544

COOKE, DEAN WILLIAM, b Uniontown, Pa, Mar 12, 31; m 56; c 4. INORGANIC CHEMISTRY. *Educ:* Ohio State Univ, BS, 55, PhD(chem), 59. *Prof Exp:* From instr to assoc prof, 59-72, PROF CHEM, WESTERN MICH UNIV, 72- *Mem:* Am Chem Soc; Sigma Xi; AAAS. *Res:* Coordination chemistry; stereochemistry; reactions of coordinated ligands; homogeneous catalysis; optical activity; mechanisms of substitution reactions. *Mailing Add:* Dept Chem Western Mich Univ Kalamazoo MI 49008

COOKE, DERRY DOUGLAS, b Schenectady, NY, Jan 29, 37. PHYSICAL CHEMISTRY. *Educ:* Parsons Col, BS, 63; Clarkson Col Technol, PhD(phys chem), 69. *Prof Exp:* Fel phys chem, 68-73, RES ASST PROF CHEM, CLARKSON COL TECHNOL, 73- *Mem:* Am Chem Soc. *Res:* Aerosols; light scattering; submicron cylinders. *Mailing Add:* 29 1/2 Grant St Potsdam NY 13676

COOKE, FRANCIS W, b Jersey City, NJ, Nov 3, 34; m 57; c 4. BIOMATERIALS, SURGICAL IMPLANTS. *Educ:* Univ Notre Dame, BS, 57; Rensselaer Polytech Inst, PhD(mat eng), 65. *Prof Exp:* Metallurgist, Oak Ridge Nat Lab, 57-61; res asst mat eng, Rensselaer Polytech Inst, 61-65; sr res metallurgist, Franklin Inst Res Labs, 65-66, mgr metall lab, 66-71, mgr bioeng lab, 69-71; assoc prof, 71-78, head, Dept Interdisciplinary Studies, 72-82, PROF MAT ENG & BIOENG, CLEMSON UNIV, 78- *Concurrent Pos:* NIH consult, 77-; consult, Vet Admin, 83-; vis scientist, US Army Inst Dent Res, 85. *Mem:* Soc Biomat (pres, 77-78); Ortho Res Soc; Bioeng Soc; Am Soc Metals. *Res:* Development of new materials and devices for surgical (orthopaedic, oral, maxillofacial) implantation, structure and properties of bone, analysis of implant preformance and failure; bioengineering ethics. *Mailing Add:* Orthop Res Inst St Francis Regional Med Ctr 929 N St Francis Wichita KS 67214

COOKE, FRED, b Darlington, Eng, Oct 13, 36; m 63, 87; c 2. GENETICS. *Educ:* Cambridge Univ, BA, 60, MA, 63, PhD(biol), 65. *Prof Exp:* Asst prof, 64-69, assoc prof biol, 69-78, PROF BIOL, QUEEN'S UNIV, ONT, 78- *Concurrent Pos:* Killam fel, 85-86. *Honors & Awards:* Brewster Award, Am Ornithologists Union, 90. *Mem:* Genetics Soc Am; Genetics Soc Can; Am Ornithologists Union; Wilson Ornith Soc; Cooper Ornith Soc. *Res:* Population biology of a breeding colony of lesser snow geese using genetic, ecological, and behavioral approaches; population biology of the snow goose. *Mailing Add:* Dept Biol Queen's Univ Kingston ON K7L 3N6 Can

COOKE, GEORGE DENNIS, b Ravenna, Ohio, June 29, 37; div; c 3. AQUATIC ECOLOGY, LAKE MANAGEMENT AND RESTORATION. *Educ:* Kent State Univ, BS, 59; Univ Iowa, MS, 63, PhD(zool), 65. *Prof Exp:* Asst biol, Kent State Univ, 60; asst zool & ecol, Univ Iowa, 60-65; USPHS fel ecol, Univ Ga, 65-67; asst prof, 67-71, assoc prof, 71-76, res assoc, Ctr Urban Regionalism & Environ Systs, 72-80, PROF, DEPT BIOL SCI, KENT STATE UNIV, 76- *Concurrent Pos:* Prin invest grants, US Environ Protection Agency, 70-79, mem staff, Corvallis Lab, 79, consult, Clean Lakes Prog, 79-81; consult, Polish govt, 77-78. *Mem:* Am Water Resources Asn; Am Soc Limnol & Oceanog; Int Asn Theoret & Appl Limnol; N Am Lake Mgt Soc (pres, (80-81). *Res:* Eutrophication; lake restoration and mangement; experimental limnology. *Mailing Add:* Dept Biol Sci Kent State Univ Kent OH 44242

COOKE, HELEN JOAN, b Greenfield, Mass, May 21, 43; c 2. INTESTINAL PHYSIOLOGY, INTESTINAL DEVELOPMENT. *Educ:* Univ Mass, BS, 65; Univ Calif, Los Angeles, MS, 67; Univ Sydney, Australia, PhD(physiol), 71. *Prof Exp:* Instr renal physiol, Sch Med, Univ Iowa, 71-73, asst prof, 73-76; asst prof renal & gastro-intestine, Univ Kans Med Ctr, 76-80; from asst prof to assoc prof gastro-intestine, Sch Med, Univ Nev, 80-85; PROF, DEPT PHYSIOL, OHIO STATE UNIV, 85- *Concurrent Pos:* Vis prof, Sch Med, Univ Nev, 79-80; res career develop award, Arthritis, Metab & Digestive Dis, NIH, 82. *Mem:* Am Fedn Clin Res; Am Physiol Soc; Am Women Sci; Am Gastroenterol Asn; Sigma Xi; Soc Neurosci. *Res:* Development of intestinal transport processes; control of intestinal transport by nerves and hormones. *Mailing Add:* Dept Physiol Ohio St Univ 333 W Tenth Ave Columbus OH 43210

COOKE, HENRY CHARLES, b Poughkeepsie, NY, June 24, 13; m 39; c 2. MATHEMATICS. *Educ:* NC State Col, BS, 37, MS, 51. *Prof Exp:* Teacher high sch, NC, 37-40; from instr to asst prof, NC State Univ, 40-50, assoc prof math, 51-79, TV instr, 55-79; RETIRED. *Mem:* Sigma Xi. *Res:* Methods and techniques of audiovisual television presentation of instructional material. *Mailing Add:* 3350 Hampton Rd Raleigh NC 27607

COOKE, HERBERT BASIL SUTTON, b Johannesburg, SAfrica, Oct 17, 15; Can citizen; m 43; c 2. GEOLOGY. *Educ:* Cambridge Univ, BA, 36, MA, 41; Univ Witwatersrand, MSc, 41, DSc(geol), 47. *Hon Degrees:* LLD, Dalhousie Univ, 82. *Prof Exp:* Geologist, Cent Mining & Investment Co, Ltd, SAfrica, 36-38; lectr geol, Univ Witwatersrand, 38-47, sr lectr, 53-58, reader, 58-61; private consult, 47-52; from assoc prof to prof geol, 61-81, dean arts & sci, 63-68, EMER PROF GEOL, DALHOUSIE UNIV, 81- *Concurrent Pos:* Ed jour, SAfrican Asn Advan Sci, 45-57; Nuffield Found bursary, 55-56; vis res assoc, Univ Calif, Berkeley, 57-58; chmn, Bernard Price Inst Palaeont Res, 58-61; consult geol, 81-87; vis lectr, Inst Vertebrate Paleontol, Beijing, 84; pres, Royal Commonwealth Soc, Mainland BC, 86-91. *Honors & Awards:* Du Toit Mem Lectr, 57; Raymond Dart Lectr, 83. *Mem:* Fel Geol Soc Am; fel Royal Soc SAfrica; SAfrica Archaeol Soc (pres, 51); SAfrican Geog Soc (pres, 46); SAfrican Asn Advan Sci (vpres, 60); fel Geol Soc London; fel Royal Meteorol Soc; fel Geol Soc SAfrica; Nova Scotia Inst Sci (pres, 67-68). *Res:* Later Cenozoic geology and fossil mammals, particularly African. *Mailing Add:* 2133 154th St White Rock BC V4A 4S5 Can

COOKE, HERMON RICHARD, JR, mining geology, for more information see previous edition

COOKE, IAN MCLEAN, b Honolulu, Hawaii, Feb 6, 33; m 59; c 5. NEUROPHYSIOLOGY, COMPARATIVE PHYSIOLOGY. *Educ:* Harvard Univ, AB, 55, AM, 59, PhD(biol), 62. *Prof Exp:* Instr biol, Harvard Univ, 62; res assoc biophys, Univ Col London, 62-63; from instr to asst prof biol, Harvard Univ, 63-70; res assoc, Lab Cellular Neurophysiol, Nat Ctr Sci Res, Paris, 70-72; prog dir, Bekesy Lab Neurobiol, 75-90, PROF ZOOL, UNIV HAWAII, MANOA, 72- *Concurrent Pos:* Vis researcher, INSERM, Bordeaux, France, 80-81; exchange visitor, Lab de Neurogiologie, Ecole Normole Superieure, Paris, France, 91. *Mem:* Am Soc Zool; Soc Gen Physiol; Am Physiol Soc; Soc Neurosci; Biophys Soc. *Res:* Control and mechanisms of release of neurosecretory material; membrane channels of peptidergic neurons in culture; cellular neurophysiology. *Mailing Add:* Bekesy Lab Neurobiol Univ Hawaii-Manoa Honolulu HI 96822

COOKE, J DAVID, b Moncton, New Brunswick, Sept 23, 39; m 66; c 2. MOTOR CONTROL, NEUROPHYSIOLOGY. *Educ:* Mount Allison Univ, BSc, 60; Dalhousie Univ, MSc, 66, PhD(physiol), 70. *Prof Exp:* assoc prof, 71-88, PROF PHYSIOL, UNIV WESTERN ONT, 88- *Mem:* Can Physiol Soc; Soc Neurosci. *Res:* Control of arm movements in humans, including factors as limb mechanics, sensory information, reflex systems and central influences. *Mailing Add:* Dept Physiol Health Sci Bldg Univ Western Ont London ON N6A 5C1 Can

COOKE, JAMES BARRY, b London, Eng, Apr 28, 15; US citizen. DAMS. *Educ:* Univ Calif, BS, 39. *Prof Exp:* Engr, Pac Gas & Elec Co, 39-52, assoc consult, 52-61; CONSULT CIVIL ENGR, J BARRY COOKE, INC, 61- *Honors & Awards:* Rickey Medal, Am Soc Civil Engrs, 60, James Laurie Prize, 61, Thomas A Middlebrooks Award, 61, Terzaghi Lectr, 82. *Mem:* Nat Acad Eng; Am Soc Civil Engrs; Sigma Xi. *Res:* Dams, tunnels and hydroelectric projects. *Mailing Add:* 1050 Northgate Dr Suite 500 San Rafael CA 94903

COOKE, JAMES HORTON, theoretical physics, for more information see previous edition

COOKE, JAMES LOUIS, b Canyon, Tex, Sept 20, 29; m 50; c 2. ELECTRICAL ENGINEERING. *Educ:* Tex Tech Col, BS, 51; Univ Tex, MS, 52; Northwestern Univ, PhD(elec eng), 60. *Prof Exp:* Elec engr, Southwestern Pub Serv Co, 52-53, 55-56; asst prof, 56-58, from assoc prof to prof, 60-71, dir grad studies, 78-80, REGENT'S PROF ELEC ENG, LAMAR UNIV, 71- *Concurrent Pos:* Consult, Gulf States Utilities Co, 56-58, 60-65, 71-, J&J Mfg Co, 60-71 & Texaco Res, 66-70. *Mem:* Inst Elec & Electronics Engrs. *Res:* Systems analysis and control; power systems engineering, protection and reliability; on-line and off-line computer applications in these areas. *Mailing Add:* Dept Elec Eng Lamar Univ Box 10029 Beaumont TX 77710

COOKE, JAMES ROBERT, b Mooresville, NC, Apr 14, 39; m 61; c 3. AGRICULTURAL ENGINEERING. *Educ:* NC State Univ, BS, 61, MS, 65, PhD(biol & agr eng), 66. *Prof Exp:* Asst prof, 66-71, ASSOC PROF AGR ENG, CORNELL UNIV, 71- *Mem:* Am Soc Agr Engrs; Sigma Xi. *Res:* Engineering properties of biological materials and systems; biological engineering analysis. *Mailing Add:* 104 Riley-Robb Agr Eng Cornell Univ Main Campus Ithaca NY 14853

COOKE, JOHN COOPER, b Lawrence, Mass, May 12, 39; m 63; c 1. MYCOLOGY. *Educ:* Univ Mass, BS, 61, MA, 63; Univ Ga, PhD(mycol), 67. *Prof Exp:* Teacher high sch, RI, 64; asst prof biol, Elizabethtown Col, 67-69; res assoc, 69-70, from asst prof to assoc prof, 70-85, PROF BIOL, UNIV CONN, 85- *Mem:* Mycol Soc Am; British Mycol Soc; Am Inst Biol Sci; Torrey Bot Club; NAm Mycolog Asn. *Res:* Morphology of fungi; ecology of soil fungi. *Mailing Add:* Dept Ecol & Evol Biol Univ Conn Avery Pt 1084 Schennecossett Rd Groton CT 06340

COOKE, KENNETH LLOYD, b Kansas City, Mo, Aug 13, 25; m 50; c 3. DIFFERENTIAL EQUATIONS. *Educ:* Pomona Col, BA, 47; Stanford Univ, MS, 49, PhD(math), 52. *Prof Exp:* From instr to asst prof math, State Col Wash, 50-57; asst prof, 57-62, chmn dept, 61-71, PROF MATH, POMONA COL, 62-, WILLIAM B KECK DISTINGUISHED SERV PROF, 85- *Concurrent Pos:* Consult, Rand Corp, Calif, 56-65; assoc ed, J Math Anal & Appln, Utilitas Mathematica, 71- & J Computational & Appl Math, 74-81; researcher, Res Inst Advan Study, 63-64 & Differential & Integral Equations, 88-; NSF sci fac fel, Stanford Univ, 66-67; Fulbright res scholar, Univ Florence, 72; assoc, Ctr Study Dem Inst, 74-75; vis prof, Brown Univ, 78-79, Inst Math & Applications, Univ Minn, 83, Univ Graz, 83 & Cornell Univ, 87; prin investr res grant, NSF. *Mem:* Am Math Soc; Math Asn Am; Soc Indust & Appl Math; Soc Math Biol; Ital Math Union; Sigma Xi; Resource Modeling Asn. *Res:* Ordinary, partial and functional differential equations; difference equations; mathematical models in the biological and social sciences. *Mailing Add:* Dept Math Pomona Col Claremont CA 91711

COOKE, LLOYD MILLER, b La Salle, Ill, June 7, 16; m 57. CHEMISTRY, SCIENCE EDUCATION. *Educ:* Univ Wis, BS, 37; McGill Univ, PhD(org chem), 41. *Hon Degrees:* LLD, Col Ganado. *Prof Exp:* Lectr org chem, McGill Univ, 41-42; res chemist & sect leader, Corn Prod Refining Co, Ill, 42-46; group leader res, Food Prod Div, Union Carbide Corp, 46-49, mgr cellulose & casing res dept, 50-54, asst to mgr tech div, 54-59, asst dir res, 59-64, mgr mkt res, 64-67, mgr planning, 65-70, dir urban affairs, 70-73, corp dir univ rels, 73-76, dir community affairs, 76-78; sr consult pub affairs, 78-81; PRES, NAT ACTION COUN MINORITIES IN ENG, 81-84; PRES, LMC ASSOC, 86- *Concurrent Pos:* Trustee, Chicago Chem Libr Found; mem, Nat Sci Bd, 70-81; trustee, Carver Res Found, Tuskegee Inst, 71-78 & McCormick Theol Sem, Chicago, 73-79; consult, Off Technol Assessment, US Cong, 74-78; consult pre-col math & sci, 84-; sr adv to chancellor, NY pub schs, 84-87; trustee, NY Found Sci & Technol, 85- *Honors & Awards:* Proctor Prize in Sci, Sci Res Asn Am, 70; Honor Scroll Award, Am Inst Chemists, 70. *Mem:* NY Acad Sci; Am Inst Chem; Am Chem Soc; AAAS. *Res:* Structure of lignin; starch modifications and derivatives; cellulose derivatives; viscose chemistry; carbohydrate and polymer chemistry; secondary mathematics and science education. *Mailing Add:* One Beaufort St White Plains NY 10607-1336

COOKE, MANNING PATRICK, JR, b Suffolk, Va, July 24, 41; m 63; c 3. ORGANIC CHEMISTRY. *Educ:* Univ NC, AB, 63, MS, 66, PhD(chem), 67. *Prof Exp:* Fel org chem, Harvard Univ, 68-70 & Stanford Univ, 70-71; ASSOC PROF ORG CHEM, WASH STATE UNIV, 71- *Mem:* Am Chem Soc. *Res:* Synthesis and new synthetic methods in organic chemistry. *Mailing Add:* Dept Chem Wash State Univ Pullman WA 99164-4630

COOKE, NORMAN E(DWARD), b Vancouver, BC, Aug 30, 22; m 47, 77. CHEMICAL ENGINEERING. *Educ:* Univ BC, BASc, 45, MASc, 46; Mass Inst Technol, ScD(chem eng), 56. *Prof Exp:* Asst chem, Univ BC, 45-46; asst chemist, Pac Fisheries Exp Sta, Fisheries Res Bd, Can, 46-48, assoc chemist, 49-56; chem eng specialist, Eng Dept, Can Industs Ltd, 56-60, prin chem engr, 60-72, tech develop mgr, 72-74; chief process engr & surveyer, Nenniger & Chenevert Inc, 74-88, mgr tech develop, 78-88; ADJ PROF CHEM ENG, MCGILL UNIV, 88- *Concurrent Pos:* Spec lectr, McGill Univ, 63-; assoc comt water pollution, Nat Res Coun Can, 65-67, assoc comt sci criteria environ qual, 70-; Can Dept Nat Health & Welfare Environ Health Comt, Pub Health Res Adv Comt, 67-71; mem adv comt, Can Ctr Inland Waters, 69-; chmn subcomt metals & certain other elements, 70-72; chmn expert panel, Lead in Can Environ & Sulphur in Can Environ, 72-74; chmn task force environ protection, Can Coun Resource & Environ Ministers' Conf on Man & Resources, 73; auxilary prof chem eng, McGill Univ, 76-88. *Honors & Awards:* Indust Wastes Medal, Water Pollution Control Fedn, 60; Award in Indust Pract, Can Soc Chem Eng, 88; Serv Award, Can Asn Water Pollution Res & Control, 88. *Mem:* Fel Chem Inst Can; Can Soc Chem Eng (secy-treas, 64-67, vpres, 58 & 67-68, pres, 68-69); fel Am Inst Chem Engrs; fel Eng Inst Can; Can Soc Mech Engrs; Sigma Xi. *Res:* Air pollution control; water pollution control; mass transfer; process design; coal conversion. *Mailing Add:* 4444 Sherbrooke St W Apt 606 Westmont PQ H3Z 1E4 Can

COOKE, PETER HAYMAN, b Beverly, Mass, Feb 4, 43; c 2. CYTOLOGY. *Educ:* Springfield Col, BS, 64; Univ NH, PhD(zool), 67. *Prof Exp:* Res fel cell biol, Harvard Univ, 67-69; res fel muscle, Boston Biomed Res Inst, 69-71; asst prof physiol & cell biol, Univ Kans, 71-75; ASSOC PROF PHYSIOL, HEALTH CTR, UNIV CONN, 75- *Concurrent Pos:* Res fel, Muscular Dystrophy Asn Am, 69 & Am Heart Asn-Brit Heart Found, 74; vis res fel physics, The Open Univ, UK, 74-75; established investr, Am Heart Asn, 77-82. *Mem:* Am Soc Cell Biol; Sigma Xi; Fedn Am Socs Exp Biol. *Res:* Contractile mechanism of muscle. *Mailing Add:* EM Spec Dept Core Res Appl USDA ARS ERRC 600 E Mermaid Lane Philadelphia PA 19118

COOKE, RICHARD A, b Hartford, Conn, Jan 13, 54; m 83; c 2. THIN-FILM TECHNOLOGY, VACUUM SCIENCES. *Educ:* Univ Conn, BS, 76, MS, 78. *Prof Exp:* Res & develop scientist, Gillette Co, 78-88; SCIENTIST, SURMET CORP, 89- *Mem:* Am Vacuum Soc; Mat Res Soc. *Res:* Surface modification of metals, ceramics and polymers via vacuum and plasma techniques including sputtering, evaporation, chemical vapor deposition and plasma etching. *Mailing Add:* 33 B St Burlington MA 01803

COOKE, ROBERT CLARK, b Duluth, Minn; m 63; c 2. OCEANOGRAPHY. *Educ:* Randolph-Macon Col, BSc, 63; Dalhousie Univ, PhD(oceanog), 71. *Prof Exp:* Phys chemist, Hercules, Inc, 64-66; from asst to assoc oceanog, Dalhousie Univ, 71-87; RETIRED. *Mem:* Am Geophys Union; Geochem Soc; Can Meteorol & Oceanog Soc. *Res:* Oceanic processes at high pressure; effect of ion-pairing on environmental chemistry; physical effects at small dimensions; diffusion; gas transfer; organic solubilization and the phase rule. *Mailing Add:* 6301 Cobgurd Rd Halifax NS B3H 2A3 Can

COOKE, ROBERT E, b Attleboro, Mass, Nov 13, 20; m 42; c 5. PEDIATRICS. *Educ:* Sheffield Sci Sch, Bs, 41; Yale Univ Sch Med, MD, 44. *Hon Degrees:* ScD, Univ Miami, 71. *Prof Exp:* Intern, New Haven Hosp, 44-45, asst res, 45-46; NIH postdoctoral fel, Yale Univ, 48-50; resident, Grace-New Haven Hosp, 50-51, assoc pediatrician, 51-56; prof pediat, Johns Hopkins Hosp, 56-73, Univ Wis Sch Med, 73-77 & Med Col Pa, 77-80; consult, Div Off Mental Health & Develop Disability, Dept Ment Health, Commonwealth Mass, 80-82; A Conger Goodyear prof pediat, State Univ NY, Buffalo, 82-89, chmn, Sch Med, 85-89; chief pediatrician, Children's Hosp Buffalo, 85-89; RETIRED. *Concurrent Pos:* John & Mary R Markle scholar, 51-55; from asst prof to assoc prof pediat & physiol, Yale Univ, 51-56; chief pediatrician, Johns Hopkins Hosp, 56-73; Grover Powers prof pediat, Nat Asn Retarded Children, 57-59; pres, Panel Mental Retardation, 61-62; Given Found prof pediat, Johns Hopkins Univ Sch Med, 62-73; White House Adv Comt Mental Retardation, 63-65; chmn, joint comt pediat res, educ & pediat, Am Acad Pediat, 63-65; vis prof, dept sociol & prev med, Harvard Med Sch, 72-73 & St Louis Univ Sch Med, 81-82; bd dirs, Wis Regional Med Prog, 73-74, Univ City Sci Ctr, Pa, 78-79, Elwyn Inst, 77-, Int Spec Olympics & Erie County Asn Retarde Children, 83-; adv bd, Inst Med- Nat Acad Sci & Nat Asn Retarded Children; mem, spec comt life preserv, Dept Health & Mental Hyg, Md, 72-74, steering comt, Nat Acad Sci, Int Med Study Impact Legalized Abortion, 74-75, Nat Comn Protection Human Subj Biomed & Behav Res, 74-78; mem health manpower training assistance rev comt, Vet Admin, Washington, DC 74-76 & other comt, 78-79; consult to various hosp & clin inst, 57-; chmn sci adv bd, Joseph P Kennedy Found; chief med officer, Spec Olympics. *Honors & Awards:* E Mead Johnson Award, 54; Kennedy Int Award, 68; St Colleta Award, Caritas Soc, 67. *Mem:* Am Asn Mental Deficiency; AAAS; AMA; Am Asn Med Cols; Am Fedn Clin Res; Am Pub Health Asn; Am Pediat Soc; fel Am Soc Clin Invest; fel Asn Acad Health Ctr; Am Acad Pediat. *Res:* Author of over 120 publications in pediatrics; water and electrolyte physiology. *Mailing Add:* 865 Painted Bunting Lane Vero Beach FL 32963

COOKE, ROBERT SANDERSON, b Philadelphia, Pa, Nov 10, 44; m 74; c 2. ORGANIC CHEMISTRY, POLYMER CHEMISTRY. *Educ:* Wesleyan Univ, AB, 66; Calif Inst Technol, PhD(chem), 70. *Prof Exp:* Asst prof chem, Univ Ore, 70-75; res chemist, Allied Corp, 75-79, res assoc, 79-90, SR RES SCIENTIST, ALLIED-SIGNAL INC, 90- *Mem:* Am Chem Soc. *Res:* Kinetics and mechanistic studies related to organic processes, polymerization reactions, and polymer modification. *Mailing Add:* Allied-Signal Inc 101 Columbia Turnpike PO Box 1021 Morristown NJ 07962-1021

COOKE, ROGER, b Ann Arbor, Mich, Feb 22, 40; c 1. BIOPHYSICS. *Educ:* Mass Inst Technol, BS, 62; Univ Ill, MS, 64, PhD(physics), 68. *Prof Exp:* Res assoc, 70-71, from asst prof to assoc prof, 71-85, PROF BIOPHYS, UNIV CALIF, SAN FRANCISCO, 85- *Concurrent Pos:* USPHS fel, Univ Calif, San Francisco, 68-70; estab investr, Am Heart Asn, 71-76. *Mem:* Biophys Soc; Am Heart Asn. *Res:* Muscle biochemistry and biophysics; protein interactions; fluorescence; electron paramagnetic resonance. *Mailing Add:* Dept Biochem & Biophys Univ Calif PO Box 0048 San Francisco CA 94143

COOKE, ROGER LEE, b Alton, Ill, July 31, 42; m 68; c 3. MATHEMATICS. *Educ:* Northwestern Univ, BA, 63; Princeton Univ, MA & PhD(math), 66. *Prof Exp:* Asst prof math, Vanderbilt Univ, 66-68; from asst prof to assoc prof, 68-77, PROF MATH, UNIV VT, 77- *Concurrent Pos:* NSF res grant, 69-73, Int res & exchanges bd grants, 81-82, 88-89, Fulbright grant, 88-89. *Mem:* Am Math Soc; Math Asn Am; Hist Sci Soc. *Res:* History of 19th century mathematics. *Mailing Add:* Dept Math Univ Vt Burlington VT 05405

COOKE, RON CHARLES, b Chico, Calif, Dec 31, 47; m; c 1. SYNTHETIC ORGANIC CHEMISTRY, NATURAL PRODUCTS CHEMISTRY. *Educ:* Calif State Univ, Chico, BS, 70; Univ of the Pacific, MS, 73. *Prof Exp:* Chemist org chem, Calif State Univ, Chico, 70-71; instr pharmaceut sci, Univ Pac, 73-75; lab dir plant tissue cult, Bailey's Nursery, Inc, 75-77; mem tech staff, Flow Labs, Inc, 77-81; INSTR CHEM, CALIF STATE UNIV, CHICO, 81- *Mem:* Am Soc Pharmacog; Electron Micros Soc; Tissue Cult Asn; AAAS. *Res:* Plant morphogenesis and asexual propagation of plants by tissue culture; production of drugs by plants in tissue culture. *Mailing Add:* Dept Chem Calif State Univ Chico CA 95929-0210

COOKE, SAMUEL LEONARD, b Atlanta, Ga, Nov 30, 31; m 54; c 3. SYSTEMS SIMULATION, SYSTEMS ANALYSIS. *Educ:* Univ Richmond, BS, 52, MS, 54; Baylor Univ, PhD(phys chem), 57. *Prof Exp:* Res chemist, E I du Pont de Nemours & Co, 57-58; instrument designer, Intersci, Inc, 58-61; assoc prof chem, Ala Col, 61-63; from asst prof to prof chem, Univ Louisville, 63-85; FIELD PROF SYST SCI, UNIV SOUTHERN CALIF, 85- *Mem:* Am Chem Soc; Asn Comput Mach; Sigma Xi; Int Soc Syst Sci; Armed Forces Commun & Electronics Asn. *Res:* Information and communication system simulation and analysis; system science and system education; analytical chemistry; chemical application of computers. *Mailing Add:* 3015 Sherbrooke Rd Louisville KY 40205-2917

COOKE, T DEREK V, b Poona, India, Oct 11, 38; Can citizen. ORTHOPAEDIC SURGERY, CLINICAL MECHANICS. *Educ:* Cambridge Univ, BA, 60, BChir, 63, MB, 64, MA, 64; FRCPS(C), 69. *Prof Exp:* Head orthop, Mc Master Univ Med Ctr, Hamilton, 73; from asst prof to assoc prof, 73-84, PROF SURG & CHMN, QUEEN'S UNIV, 84-, DIR RES, 80- *Concurrent Pos:* Consult, Hotel Dieu Hosp, St Mary's of Lake Hosp & Kingston Gen Hosp, 73- *Mem:* Can Orthop Asn (pres); Orthop Res Soc; Knee Soc; Am Rheumatism Soc. *Res:* Etiology and pathogenesis of rheumatic diseases; mechanism for chronic inflammation in rheumatoid arthritis; mode of destruction of cartilage of rheumatoid and osteoarthritis; biology of chondrocytes; biomechanics of joints and factors predisposing to degenerative joint diseases; advancement of design of prostheses and surgical instrumentation. *Mailing Add:* Apps Med Res Ctr Queens Univ Fac Med Kingston Gen Hosp 76 Stuart St Kingston ON K7L 3N6 Can

COOKE, THEODORE FREDERIC, b Pittsfield, Mass, Jan 28, 13; m 40, 73; c 4. TEXTILE CHEMISTRY, POLYMER SCIENCE. *Educ:* Univ Mass, BS, 34; Yale Univ, PhD(phys chem), 37. *Prof Exp:* Res chemist, Standard Oil Develop Co, NJ, 37-40; res chemist, Org Chem Div, Am Cyanamid Co, 40-42, asst dir phys chem res, 45-48, asst dir appln res dept, 48-52, mgr textile resin lab, 52-54, asst to mgr textile resin dept, 54-58, mgr commercial develop, 58-60, dir chem res, 60-62, asst dir res & develop, 62-72, dir, Sci Serv Dept, Chem Res Div, 72-78; dir, Liaison Progs, Yale Univ, 78-85; RES ASSOC, TEXTILE RES INST, PRINCETON UNIV, 78-; VPRES RES & DEVELOP, CHEMTEK INC, 87- *Concurrent Pos:* Consult, Southern Regional Res Labs, USDA, 54-60; chmn, Gordon Res Conf Textiles; chmn comt textile finishing, Nat Res Coun Adv Bd Qm Res & Develop; consult, 78- *Honors & Awards:* Olney Medal, Am Asn Textile Chemists & Colorists. *Mem:* Am Chem Soc; fel Am Inst Chem; Asn Res Dirs. *Res:* Textile chemistry; fiber science; textile finishing, including durable press, flame resistant, water repellent, soil release and oil repellent finishes; cellulose chemistry; treatment of wood, including rotproofing. *Mailing Add:* 287 Weed Ave Stamford CT 06902

COOKE, WILLIAM BRIDGE, b Foster, Ohio, July 16, 08; m 42. MYCOLOGY. *Educ:* Univ Cincinnati, BA, 37; Ore State Col, MS, 39; State Col Wash, PhD(bot), 50. *Prof Exp:* Mycologist, Trop Deterioration Res Lab, US Qm Corps, 45-46; res assoc, dept plant path, State Col Wash, 50-51; mycologist, Bact Sect, Environ Health Ctr, USPHS, 52-53, prin mycologist, Robert A Taft Sanit Eng Ctr, 53-56, sr mycologist, Microbiol Activit, Cincinnati Water Res Lab, Fed Water Pollution Control Admin, Dept Health, Educ & Welfare, 56-66; mycologist biol treatment activities, Advan Waste Treatment Prog, US Dept Interior, 66-69; res assoc, dept bot, Miami Univ, 69-70; SR RES ASSOC, DEPT BIOL SCI, UNIV CINCINNATI, 70- *Mem:* Fel AAAS; Mycol Soc Am; Soc Indust Microbiol; Bot Soc Am. *Res:* Fungi of polluted water and sewage; taxonomy of Polyporaceae; flora and fungi of Mt Shasta and fungi of national parks; fungi of Ohio. *Mailing Add:* 1135 Wilshire Ct Cincinnati OH 45230

COOKE, WILLIAM DONALD, b Philadelphia, Pa, May 15, 18; m 46; c 6. ANALYTICAL CHEMISTRY. *Educ:* St Joseph's Col, Philadelphia, BS, 40; Univ Pa, MS & PhD, 49. *Prof Exp:* Chemist, Harshaw Chem Co, Pa, 40-42; Nat Res Coun fel, Princeton Univ, 49-51; from asst prof to assoc prof, 51-59, assoc dean col arts & sci, 62-64, dean grad sch, 64-69, vpres res, 69-83, PROF CHEM, CORNELL UNIV, 59- *Concurrent Pos:* Pres, Asn Grad Schs, 70-71, LeMoyne Col, 76-82; mem bd trustees, Fordham Univ, 70- & Assoc Univs, Inc, 74-83; mem, Nat Bd Grad Educ, 72-75. *Mem:* Am Chem Soc. *Res:* Electrochemical methods; absorption spectra; flame spectroscopy; nuclear magnetic resonance; gas chromatography. *Mailing Add:* Baker Lab Cornell Univ Ithaca NY 14853

COOKE, WILLIAM PEYTON, JR, b Hobart, Okla, Jan 4, 34; m 61; c 3. MATHEMATICS, STATISTICS. *Educ:* WTex State Univ, BS, 59; Tex Tech Col, MS, 61; Tex A&M Univ, PhD(statist), 68. *Prof Exp:* Instr math, Amarillo Col, 60-61; instr, Tex Tech Col, 61-62; from asst prof to assoc prof, WTex State Univ, 64-69; assoc prof, Univ Wyo, 69-77, prof statist, 77-85; PROF MATH, WTEX STATE UNIV, 85- *Mem:* Math Asn Am; Am Statist Asn. *Res:* Mathematical programming; statistical reliability. *Mailing Add:* Math & Phy Sci WTex State Univ Canyon TX 79016

COOK-IOANNIDIS, LESLIE PAMELA, b Kingston, Ont, Aug 23, 46; US citizen; m 72; c 2. APPLIED MATHEMATICS. *Educ:* Univ Rochester, BA, 67; Cornell Univ, MS, 69, PhD(appl math), 71. *Prof Exp:* NATO fel appl math, 71-72; instr & res assoc, Dept Theoret & Appl Mech & Dept Math, Cornell Univ, 72-73; adj prof math, Univ Calif, Los Angeles, 73-75, from asst prof to assoc prof, 75-84; assoc prof, 83-89, PROF, DEPT MATH, UNIV DEL, 89- *Concurrent Pos:* Vis assoc prof, Univ Maryland, College Park, Md, 87-88; NSF, vis prof for women, 87-88. *Mem:* Sigma Xi; Soc Indust & Appl Math; Am Phys Soc; Am Acad Mech; Am Math Soc. *Res:* Transonic aerodynamics; viscoelastic fluid flow. *Mailing Add:* Dept Math Univ Del Newark DE 19716

COOKSON, ALAN HOWARD, b London, Eng, July 3, 39; m 68; c 2. HIGH VOLTAGE ENGINEERING, DIELECTRICS & ELECTRICAL INSULATION. *Educ:* London Univ, Queen Mary Col, BSc, 61 & PhD (elec eng), 65. *Prof Exp:* Res fel high voltage breakdown, Queen Mary Col, 64-65; res officer high voltage insulation, Cent Elec Res Lab, 65-68; fel engr high voltage insulation, Westinghouse Res & Demonstration Ctr, 68-75, mgr, gas insulated cable res & demonstration, Power Circuit Breaker Div, Westinghouse Elec, 75-80, MGR POLYMERS, DIELECTRICS & ADVAN BATTERIES DEPT, SCI & TECHNOL CTR, WESTINGHOUSE ELEC CORP, 80. *Concurrent Pos:* Convenor, Working Group Compressed Gas Insulated Cables, Conf Int Des Grand Reseaux Elec a Haute Tension, 78-88; US rep, comt elec insulation, Conf Int Des Grand Reseaux Elec a Haute Tension, 82-88; ed, Dielectrics & Elec Insulation Soc Newslett, Inst Elec & Electronics Engrs, 83-88; mem adv comt high voltage, Miss State Univ, 83-88; vpres admin, Inst Elec & Electronics Engrs Dielectrics & Elec Insulation Soc. *Mem:* fel Inst Elec & Electronics Engrs; Inst Physics; Am Phys Soc. *Res:* Electrical breakdown of high pressure gases, including effects of contamination; development of ultra high voltage compressed gas insulated cables and equipment; application of electrical insulation in pulsed power systems. *Mailing Add:* Sci & Technol Ctr Westinghouse Elec Corp 1310 Beulah Rd Pittsburgh PA 15235

COOKSON, FRANCIS BERNARD, b Preston, Eng, Oct 30, 28; m 53; c 2. NEUROANATOMY, HISTOLOGY. *Educ:* Univ Manchester, BSc, 53, MB & ChB, 56; Royal Col Obstetricians & Gynecologists, Eng, dipl obstet, 57. *Prof Exp:* Demonstr anat, Univ Manchester, 57-58; pvt pract med, Eng, 58-64; asst prof anat, Univ Sask, 64-66; assoc prof, 66-71, asst dean med, 74-80, PROF ANAT, UNIV ALTA, 71-, HON LECTR MED, 66-, DIR HEALTH SERV, 77-, ASSOC DEAN MED, 80- *Concurrent Pos:* Med Res Coun res grants, 65-69; Alta Heart Found res grant, 67-68. *Mem:* Fel Am Heart Asn; Anat Soc Gt Brit & Ireland; Can Med Asn; Brit Med Asn. *Res:* Histopathology; experimental atherosclerosis, etiology pathogenesis and preventions; hypertension incidence in university students; medical education. *Mailing Add:* Univ Health Serv Univ Alta 10011th St 88th Ave Edmonton AB T6G 2R1 Can

COOKSON, JOHN T(HOMAS), JR, b East St Louis, Ill, July 7, 39; m 61; c 3. ENVIRONMENTAL ENGINEERING. *Educ:* Wash Univ, BS, 61, MS, 62; Calif Inst Technol, PhD(environ health eng), 66. *Prof Exp:* From asst prof to assoc prof civil eng, Univ Md, 65-75; pres, JTC Environ Consults, Inc, 73-89; VPRES ENVIRON SERVS, GEN PHYSICS CORP, 89- *Concurrent Pos:* US Dept Interior grants, 66-75; dir environ health traineeships, USPHS, 68-71; tech reviewer & adv, Consumers' Union & Environ Defense Fund, 74-; mem comt safe drinking water, Nat Acad Sci-Nat Res Coun-Nat Acad Eng, 78-; environ consult, US Dept Labor; prin consult, Environ Defense Fund, Can Broadcasting Co, Consumers Union & other govt & consumer orgn. *Mem:* Am Soc Civil Engrs; Am Water Works Asn; Water Pollution Control Fedn. *Res:* Removal of viruses from water; adsorption of organics on activated carbon; theory of filtration; surface chemistry of activated carbon; biological processes; design of activated carbon adsorption beds; drinking water quality; toxic chemical control; hazardous waste treatment. *Mailing Add:* Gen Physics Corp 6700 Alexander Bell Dr Columbia MD 21046

COOL, BINGHAM MERCUR, b Marion, Ill, Dec 21, 18; m 43; c 3. FORESTRY. *Educ:* La State Univ, BS, 40; Iowa State Univ, MS, 41; Mich State Univ, PhD(forestry), 57. *Prof Exp:* Asst, Iowa State Univ, 40-41; asst agr aide, Soil Conserv Serv, 41; asst forestry aide, Tenn Valley Authority, 42; asst county agent forestry, Ala Exten Serv, 45-47; timber mkt specialist, Miss Exten Serv, 47-48, state forest prods mrk specialist, 48-49; asst prof forestry, Ala Polytech Inst, 49-54, 56-58; asst, Mich State Univ, 54-56; assoc prof, 58-66, PROF FORESTRY, CLEMSON UNIV, 66- *Mem:* Soc Am Foresters. *Res:* Siviculture. *Mailing Add:* Dept Forestry Clemson Univ 201 Sikes Hall Clemson SC 29634

COOL, RAYMOND DEAN, b Winchester, Va, Mar 14, 02. ANALYTICAL & PHYSICAL CHEMISTRY, CHEMICAL MICROSCOPY. *Educ:* Bridgewater Col, BS, 22; Univ Va, MS, 26, PhD(chem), 28. *Prof Exp:* Sci instr, Salem High Sch, Va, 22-24; instr chem, Univ Nev, 28-29; instr, Univ Ore, 29-30; res assoc chem, Dept Pharmacol, Sch Med, Univ Pa, 30-34; from instr to asst prof, Univ Akron, 34-41; asst prof, Univ Okla, 41-46; asst prof, WVa Univ, 46; prof, Madison Col, 46-72, EMER PROF CHEM, JAMES MADISON UNIV, 72- *Concurrent Pos:* Chemist, US Naval Ord Lab, 49-52 & 54-57. *Mem:* AAAS; Am Chem Soc; Am Microchem Soc. *Res:* Microanalytical methods for iodides and rarer elements; volumetric methods for nitrites; polymorphic transitions; distribution coefficients in gas-liquid systems; chemical microscopy; metallic complexes of diketones and picolines; analytical chemistry. *Mailing Add:* 405 E College St Bridgewater VA 22812-1512

COOL, TERRILL A, b Boulder, Colo, Aug 18, 36; m 59; c 3. APPLIED PHYSICS, CHEMICAL PHYSICS. *Educ:* Univ Calif, Los Angeles, BS, 61; Calif Inst Technol, MS, 62, PhD(eng), 65. *Prof Exp:* From asst prof to assoc prof thermal eng, 65-73, assoc prof appl physics, Sch Appl & Eng Physics, 73-75, PROF APPL PHYSICS, SCH APPL & ENG PHYSICS, CORNELL UNIV, 75- *Mem:* Fel Am Phys Soc. *Res:* Chemical lasers; molecular energy transfer; laser spectroscopy. *Mailing Add:* Sch of Appl & Eng Physics 228 Clark Hall Cornell Univ Ithaca NY 14853

COOLBAUGH, RONALD CHARLES, b Missoula, Mont, Jan 21, 44; m 63; c 2. PLANT PHYSIOLOGY, PLANT BIOCHEMISTRY. *Educ:* Eastern Wash State Col, BA, 66; Ore State Univ, PhD(plant physiol), 70. *Prof Exp:* From asst prof to prof biol, Western Ore State Col, 70-80, dean arts & sci, 78-80; prof & chmn bot, Iowa State Univ, 80-87; PROF & HEAD BOT & PLANT PATH, PURDUE UNIV, 87- *Concurrent Pos:* Res assoc, Univ Calif, Los Angeles, 75-76. *Mem:* Am Soc Plant Physiologists; Sigma Xi; Am Inst Biol Sci; Plant Growth Reg Soc Am. *Res:* Biochemistry and physiology of plant growth substances; biosynthesis of Gibberellins and abscisic acid. *Mailing Add:* Dept Bot & Plant Path Lilly Hall Purdue Univ West Lafayette IN 47907

COOLER, FREDERICK WILLIAM, b Knoxville, Tenn, Dec 7, 30; m 52. FOOD TECHNOLOGY. *Educ:* Univ Tenn, BS, 52, MS, 58; Univ Md, PhD(food technol), 62. *Prof Exp:* Asst horticulturist, Univ Tenn, 56-58; asst food technol, Univ Md, 58-62; ASSOC PROF FOOD SCI & TECHNOL, VA POLYTECH INST & STATE UNIV, 62- *Mem:* Inst Food Technologists; Am Soc Hort Sci. *Res:* Rheological properties of foods; quality evaluation and control of foods; transportation, storage and quality maintenance of fruits and vegetables; food product development. *Mailing Add:* Dept Food Sci & Technol Va Polytech Inst Blacksburg VA 24061

COOLEY, ADRIAN B, JR, b Amelia, Tex, Sept 28, 28; m 48; c 3. PHYSICS, MATHEMATICS. *Educ:* Sam Houston State Univ, BS, 50, MA, 57; Univ Tex, Austin, PhD(sci educ), 70. *Prof Exp:* Teacher high sch, Tex, 50-56; asst prof physics, Sam Houston State Univ, 56-66; asst prof, Southwestern Univ, 66-67; teaching assoc educ, Univ Tex, Austin, 67-68; from asst prof to prof physics, Sam Houston State Univ, 68-90; RETIRED. *Mem:* Am Asn Physics Teachers; Nat Sci Teachers Asn. *Res:* Physical sciences. *Mailing Add:* 112 Sunset Lake Huntsville TX 77340

COOLEY, ALBERT MARVIN, chemical engineering; deceased, see previous edition for last biography

COOLEY, DENTON ARTHUR, b Houston, Tex, Aug 22, 20; m 49; c 5. SURGERY. *Educ:* Univ Tex, BA, 41; Johns Hopkins Univ, MD, 44. *Prof Exp:* Resident surg, Johns Hopkins Hosp, 44-50; sr surg registr, Brompton Hosp Chest Dis, London, 50-51; from assoc prof to prof surg, Baylor Col Med, 51-69; SURGEON-IN-CHIEF, TEX HEART INST, 69-; PROF CLIN SURG, UNIV TEX MED SCH, HOUSTON, 75- *Concurrent Pos:* Consult surg serv, Tex Children's & St Lukes Episcopal Hosps, Houston, 63- *Mem:* AMA; Am Asn Thoracic Surg; Soc Vascular Surg; Am Col Surg; Soc Clin Surg. *Res:* Cardiovascular surgery and diseases of the chest. *Mailing Add:* PO Box 20345 Tex Heart Inst Houston TX 77225

COOLEY, DUANE STUART, b Batavia, NY, May 9, 23; m 49; c 2. METEOROLOGY. *Educ:* Mass Inst Technol, BS, 48, MS, 49, PhD(meteorol), 59. *Prof Exp:* Instr meteorol, Weather Sch, Chanute AFB, Dept Air Force, 49-51, meteorologist, Air Force Cambridge Res Ctr, 51-57, supvry atmospheric physicist, 57-60, supvry physicist, Electronic Systs Div, 496L Syst Proj Off, Air Force Syst Command, 60-61; sr res scientist, Travelers Res Ctr, Inc, 61-69; actg exec scientist, US Comt Global Atmospheric Res Prog, Nat Acad Sci, 69-70; chief, Tech Procedures Br, Nat Oceanic & Atmospheric Admin, 70-85; SR ENG SPECIALIST, STANFORD TELECOMMUN, INC, 85- *Mem:* Fel Am Meteorol Soc; Am Geophys Union. *Res:* General atmospheric circulation; statistical and dynamical weather forecasting; atmospheric radiation; analysis and interpretation of meteorological satellite data. *Mailing Add:* 4503 Libbey Dr Fairfax VA 22032

COOLEY, JAMES HOLLIS, b New York, NY, March 25, 30; m 55; c 2. ORGANIC CHEMISTRY. *Educ:* Middlebury Col, AB, 52, MS, 54; Univ Minn, PhD(org chem), 58. *Prof Exp:* Asst, Univ Minn, 54-57; from asst prof to assoc prof, 57-68, PROF CHEM, UNIV IDAHO, 68- *Concurrent Pos:* Res assoc, Columbia Univ, 63-66. *Mem:* Am Chem Soc. *Res:* Chemistry of the hydroxylamine compounds. *Mailing Add:* Dept Chem Univ Idaho Moscow ID 83843-4199

COOLEY, JAMES WILLIAM, b New York, NY, Sept 18, 26; m 57; c 3. APPLIED MATHEMATICS. *Educ:* Manhattan Col, BA, 49; Columbia Univ, MA, 51, PhD(math), 61. *Prof Exp:* Programmer, Inst Advan Study, Princeton, 53-56; res asst math, Courant Inst, NY Univ, 56-62; RES STAFF MATH, IBM WATSON RES CTR, 62- *Concurrent Pos:* Prof comput sci, Royal Inst Technol, Sweden, 73-74. *Mem:* Fel Inst Elec & Electronics Engrs; Sigma Xi. *Res:* Numerical methods; solution of partial and ordinary differential equations; digital signal processing; discrete Fourier methods; mathematical modeling of nerve membranes. *Mailing Add:* IBM Watson Res Ctr Box 218 Yorktown Heights NY 10598

COOLEY, NELSON REEDE, b Mobile, Ala, Nov 30, 20; m 51; c 2. PROTOZOOLOGY. *Educ:* Spring Hill Col, BS, 42; Univ Ala, MS, 47; Univ Ill, PhD(zool), 54. *Prof Exp:* From asst to instr biol, Univ Ala, 46-48; instr anat, Druid City Hosp, Ala, 47; asst zool, Univ Ill, 48-53; instr zool & physiol, Okla Agr & Mech Col, 53-54, asst prof zool, 54-56; fishery res biologist, US Fish & Wildlife Serv, 56-70; res biologist, Gulf Breeze Lab, 70-74, microbiologist, 74-76; RES MICROBIOLOGIST, US ENVIRON PROTECTION AGENCY, 76-, QUAL ASSURANCE OFFICER, 80- *Honors & Awards:* Bronze Medal, US Environ Protection Agency, 75 & 81. *Mem:* Soc Protozoologists. *Res:* Physiology of ciliate protozoa; blood protozoa of bats; biological control of oyster predators; estuarine faunal biology; pesticides vs ciliate protozoan population growth; pesticide bioaccumulation by ciliates. *Mailing Add:* 3550 Rothschild Dr Pensacola FL 32503

COOLEY, RICHARD LEWIS, b Akron, Colo, Jan 11, 40; m 60; c 1. HYDROLOGY, GEOMORPHOLOGY. *Educ:* Ariz State Univ, BS, 62; Pa State Univ, PhD(geol), 68. *Prof Exp:* Hydrologist, Hydrol Eng Ctr, US Corps Engrs, 68-70; res assoc, 70-72, assoc prof, Ctr Water Resources Res, 72-77,

ADJ ASSOC PROF GEOL & GEOG, DESERT RES INST, UNIV NEV SYST, RENO, 77- *Concurrent Pos:* Inst Res Land & Water Resources scholar appointee, Pa State Univ, 68. *Mem:* Soc Econ Paleontologists & Mineralogists; Geol Soc Am; Am Geophys Union; AAAS; Sigma Xi. *Res:* Physics of water movement through porous media and its application in inferring the influence of geological conditions on ground water movement and recharge; analysis of ground water flow systems. *Mailing Add:* 575 S Coor Ct Lakewood CO 80228

COOLEY, ROBERT LEE, b Birmingham, Ala, Feb 20, 27; m 52; c 4. MATHEMATICS. *Educ:* Univ Ala, BS, 48; Univ Va, LLB, 51; Purdue Univ, MS, 57, PhD(math), 64. *Prof Exp:* From asst to assoc prof math, 67-80, PROF MATH, WABASH COL, 80- *Concurrent Pos:* Lilley endowment fel, Cambridge, Eng. *Mem:* Am Bar Asn; Math Asn Am; Nat Coun Teachers Math. *Res:* Topological algebra. *Mailing Add:* 1314 W Market St Crawfordsville IN 47933

COOLEY, STONE DEAVOURS, b Laurel, Miss, Jan 13, 22; m 48; c 3. PHYSICAL CHEMISTRY. *Educ:* Univ Tex, BSChE, 49, PhD(chem), 53. *Prof Exp:* Res chemist, Celanese Corp Am, 53-55, group leader, 55-59, sect head, 59-62, asst to tech dir, Celanese Chem Co, 62-63, facil mgr, Summit Res Lab, 64-68; asst dir res, Petro-Tex Chem Corp, 68-74, mgr tech serv, 74-77; mgr technol, Denka Chem Corp, 77-88; MGR TECHNOL, MOBAY SYNTHETICS CORP, 88- *Mem:* Am Chem Soc. *Res:* Hydrocarbon oxidation; synthetic rubber. *Mailing Add:* Mobay Synthetics Corp 8701 Park Place Blvd Houston TX 77017

COOLEY, WILLIAM C, b Lakeland, Fla, Dec 19, 24; m 49; c 4. AEROSPACE TECHNOLOGY, MECHANICAL ENGINEERING. *Educ:* Mass Inst Technol, SB, 44, ScD(mech eng), 51; Calif Inst Technol, MS, 47. *Prof Exp:* Instr mech eng, Mass Inst Technol, 49-50; res engr, Atomic Energy Res Dept, NAm Aviation, Inc, 51-53; prin engr, Gen Elec Co, 53-55, supvr mech eng, 55-56, mgr systs anal, 56-58; asst prog engr, Rocketdyne Div, NAm Aviation, Inc, 58-59; chief adv tech, NASA Hq, 59-61; from vpres to pres, Exotech, Inc, 61-69; pres, Terraspace Inc, 69-85; ASSOC PROF, ELEC & COMPUT ENG DEPT, GEORGE MASON UNIV, 85- *Concurrent Pos:* Consult, Oak Ridge Nat Lab, 50 & Gen Elec Co, 61-62. *Honors & Awards:* Pioneer Award, Water Jet Technol Asn, 85. *Mem:* Am Inst Aeronaut & Astronaut; Am Soc Mech Engrs. *Res:* Radiation effects; advanced energy conversion; shock hydrodynamics; spacecraft sterilization; planetary quarantine; high pressure liquid jet technology; computer-aided design. *Mailing Add:* Dept Elec & Computer Eng George Mason Univ 4400 Univ Dr Fairfax VA 22030

COOLEY, WILLIAM EDWARD, b St Louis, Mo, Mar 7, 30; m 52; c 4. DENTAL RESEARCH. *Educ:* Cent Methodist Col, Mo, AB, 51; Univ Ill, PhD(chem), 54. *Prof Exp:* Res chemist, Regulatory & Consumer Safety Serv, Procter & Gamble Co, 54-71, sect head, Prod Develop, 71-75, mgr, Regulatory Serv Toilet Goods Div, 75-81, mgr safety & regulatory serv, 80-82, mgr Health & Personal Care Div, 82-90, MGR REGULATORY & CLIN DEVELOP DIV, REGULATORY & CONSUMER SAFETY SERV, PROCTER & GAMBLE CO, 90- *Mem:* Am Chem Soc; Int Asn Dent Res; Am Asn Dent Res. *Res:* Chemistry of dental systems; fluorides; dentifrice abrasives; professional and regulatory affairs in oral health. *Mailing Add:* 531 Chisholm Trail Wyoming OH 45215

COOLEY, WILS LAHUGH, b Pittsburgh, Pa, Aug 23, 42; m 66; c 2. ELECTRICAL SAFETY. *Educ:* Carnegie-Mellon Univ, BS, 64, MS, 65, PhD(elec eng), 68. *Prof Exp:* Asst prof elec eng & biotech, Carnegie-Mellon Univ, 68-72; from asst prof to assoc prof, 72-80, PROF & ASSOC DEPT CHMN, WVA UNIV, 81- *Concurrent Pos:* NIH grant thoracic imped- ance measurements, 69-70; Pa Sci & Eng Fund grant seeding med instrumentation indust in Western Pa, 69-71; res assoc, Univ Calif, 71; US Dept Interior-Bur Mines res grants, 74, 78, 79, 80 & 81; eng consult, Pa State Univ, 75-76, Bendix Corp, 76-77, several law firms & mining co, 79-; NSF grant, eng design educ, 88-90. *Mem:* Inst Elec & Electronics Engrs; Am Soc Eng Educ; Sigma Xi; Nat Soc Prof Engrs. *Res:* Engineering education in design; electrical safety; transient protection of grounded equipment; mine power systems. *Mailing Add:* Rte 3 Box 506 Morgantown WV 26505

COOLIDGE, ARDATH ANDERS, b Chicago, Ill, July 22, 19; m 49; c 5. NUTRITION. *Educ:* Earlham Col, AB, 41; Iowa State Col, PhD(nutrit), 46. *Prof Exp:* Teacher pub sch, Ill, 41-42; asst prof foods & nutrit, Western Reserve Univ, 46-47; asst prof, Berea Col, 47-49; home economist, Sensory Testing Food Res, Armour & Co, 58-62; lit scientist, Nutrit Res Div, Nat Dairy Coun, 62-66; from asst prof to prof home econ, 66-82, EMER PROF FOODS & NUTRIT, PURDUE UNIV, CALUMET CAMPUS, 82- *Mem:* Am Dietetic Asn; Soc Nutrit Educ. *Res:* Biological utilization of ascorbic acid in apples. *Mailing Add:* 2612 Leer St South Bend IN 46614

COOLIDGE, EDWIN CHANNING, b Gambier, Ohio, Jan 30, 25; m 53; c 1. ANALYTICAL CHEMISTRY, ORGANIC CHEMISTRY. *Educ:* Kenyon Col, AB, 44; Johns Hopkins Univ, PhD(chem), 49. *Prof Exp:* Res org chemist, Procter & Gamble Co, 49-50 & 53-54; asst prof, Univ Utah, 51-52; res org chemist, Dugway Proving Ground, 52-53; asst prof chem, Hamilton Col, 54-58; asst prof, NMex Inst Mining & Technol, 58-61; assoc prof, 61-64, PROF CHEM, STETSON UNIV, 64- *Concurrent Pos:* Dir, Assoc Mid-Fla Cols Year Abroad Prog, 68, Freiburg, WGer, 69-70; consult, Tech Adv Serv for Attorneys, 75; sr Fulbright Lectr, Freiburg, WGer, 82-83. *Mem:* Am Chem Soc; Royal Soc Chem. *Res:* Organometallic chemistry; pyrrole and porphyrin synthesis; organo-phosphorus compounds; metal chelate stability and structure; ecological analysis. *Mailing Add:* Dept Chem Stetson Univ DeLand FL 32720

COOLIDGE, THOMAS B, BIOCHEMISTRY. *Educ:* Columbia Univ, PhD(biochem), 35. *Prof Exp:* Prof biochem, Univ Chicago, 66-76; RETIRED. *Mailing Add:* Univ Chicago 5755 S Dorchester St Chicago IL 60637

COOMBES, CHARLES ALLAN, b Nevada City, Calif, Feb 25, 34; m 56; c 3. PHYSICS. *Educ:* Univ Calif, AB, 55, PhD(physics), 60. *Prof Exp:* Asst prof physics, Idaho State Univ, 59-62 & San Jose State Col, 62-63; asst prof, 63-66, actg vdean fac arts & sci, 70-72, ASSOC PROF PHYSICS, UNIV CALGARY, 66-, ASST TO HEAD DEPT, 75- *Mem:* AAAS; Am Asn Physics Teachers; Am Physics Soc. *Res:* Dynamics of the lower atmosphere; foundations of quantum mechanics. *Mailing Add:* Dept Physics Univ Calgary Calgary AB T2N 1N4 Can

COOMBS, HOWARD A, hazards related to groundwater, potential earthquake hazards; deceased, see previous edition for last biography

COOMBS, MARGERY CHALIFOUX, b Nashua, NH, Aug 12, 45; m 69; c 2. VERTEBRATE PALEONTOLOGY. *Educ:* Oberlin Col, BA, 67; Columbia Univ, MA, 68, PhD(biol sci), 73. *Prof Exp:* Asst prof, 73-80, ASSOC PROF ZOOL, UNIV MASS, AMHERST, 80- *Mem:* Soc Vert Paleont; Paleont Soc. *Res:* Chalicothere systematics and function; early Miocene biostratigraphy; clawed herbivores. *Mailing Add:* Dept Zool Univ Mass Amherst MA 01003

COOMBS, ROBERT VICTOR, b Brighton, Eng, June 24, 37; m 63. REGULATORY AFFAIRS, ORGANIC CHEMISTRY. *Educ:* Univ London, BSc, 58, PhD(org chem), 61. *Prof Exp:* Fel, Univ Wis, 61-62 & 63-64; fel, Columbia Univ, 62-63; res chemist, Brit Drug Houses, Ltd, Eng, 64-66; fel, Synvar Res Inst, Calif, 66-67; sr scientist, Chem Dept, Sandoz-Warner, Inc, 67-71, group leader, 71-77, mem sr sci staff, 77-80, assoc sr sci staff, 80-88, MGR DOCUMENTATION & COMPLIANCE, SANDOZ RES INST, 88- *Mem:* Am Chem Soc; Royal Soc Chem. *Res:* Synthetic organic and medicinal chemistry. *Mailing Add:* Sandoz Res Inst 59 Rte 10 East Hanover NJ 07936

COOMBS, WILLIAM, JR, b Brooklyn, NY, Apr 30, 24; m 45; c 2. BIOMEDICAL ENGINEERING. *Educ:* Mass Inst Technol, BS, 47, MS, 48; Univ Rochester, MS, 56. *Prof Exp:* Res assoc biol, Mass Inst Technol, 47-48; develop engr, Eastman Kodak Co, 48-51; chief elec engr, Dept Physics, Univ Rochester, 51-59, head dept electronics res & develop, 59-63, dir electronics & biophys res & develop, 63-67, dir cent res labs, 67-68; gen mgr soflens contact lens, Bausch & Lomb, Inc, 68-71, vpres soflens contact lens div, 71-; RETIRED. *Concurrent Pos:* Consult, Xerox Corp & Taylor Instrument Co, 54-59; lectr univ sch, Univ Rochester, 55-60; indust rep, Ophthalmic Prosthetic Devices Subcomt, Food & Drug Admin, 76. *Mem:* Inst Elec & Electronics Eng. *Res:* Biomedical instruments and ophthalmic prosthetics. *Mailing Add:* 247 Gabilan Ave Sunnyvale CA 94086

COOMES, EDWARD ARTHUR, b Louisville, Ky, June 27, 09; m 40; c 5. PHYSICS. *Educ:* Univ Notre Dame, BS, 31, MS, 33; Mass Inst Technol, ScD(physics), 38. *Prof Exp:* From instr to assoc prof math & physics, 33-42, prof, 45-74, EMER PROF PHYSICS, UNIV NOTRE DAME, 74- *Concurrent Pos:* Mem staff radiation lab, Mass Inst Technol, 42-45; consult various industs & labs, 52-; adj prof physics & astronomy, St Mary's Col, 76-81; lectr, Forever Learning Inst, 78-90. *Mem:* Fel Am Phys Soc; sr mem Inst Elec & Electronics Engrs. *Res:* Solid state electronics; oxide cathodes; surface physics and chemistry; thermionics; energy conversion j. *Mailing Add:* 1546 Marigold Way South Bend IN 46617

COOMES, MARGUERITE WILTON, b Hertford, Eng; US citizen. BIOCHEMISTRY. *Educ:* NTex State Univ, BS, 73, MS, 75; Univ Tex Health Sci Ctr, Dallas, PhD(biochem), 80. *Prof Exp:* Fel biochem, Nat Inst Environ Health Sci, NIH, 80-82; instr, 82-83, ASST PROF BIOCHEM, COL MED, HOWARD UNIV, 84- *Mem:* Sigma Xi; Am Soc Biol Chemists. *Res:* Regulation of protein degradation in differentiating cells and in bacteria. *Mailing Add:* 3411 Murdock Rd Kensington MD 20895

COOMES, RICHARD MERRIL, b Provo, Utah, Oct 5, 39; m 63; c 2. ORGANIC CHEMISTRY, TOXICOLOGY. *Educ:* Utah State Univ, BS, 66, MS, 67; Colo State Univ, PhD(org chem), 69. *Prof Exp:* Fel chem, Ariz State Univ, 69-71; instr, Univ Va, 71-72; natural prod chemist, Gates Rubber Co, 72-73; group leader org res, Tosco Corp, 73-77, environ health coordr, 77-79, mgr environ health, 79-83; toxicologist, Getty Oil Co, 83-84; dir environ sci, Western Res Inst, 84-87; MGR, HAZARDOUS WASTE DIV, ERT, INC, 87- *Mem:* Am Chem Soc; Am Mgt Asn; Sigma Xi; Southwestern Asn Toxicologists. *Res:* Investigation of potential health effects of synthetic fuel processing, emphasizing toxicity, carcinogenicity, and chemistry of oil shale products and by-products: chemistry centers on polycyclic aromatic hydrocarbon analyses. *Mailing Add:* Stanford Place 3 Suite 1000 4582 South Ulster St Pkwy Denver CO 80237-2632

COON, CRAIG NELSON, b Big Springs, Tex, May 17, 44; m 62; c 2. NUTRITION, METABOLISM. *Educ:* Tex A&M Univ, BS, 66, MS, 70, PhD(biochem & nutrit), 73. *Prof Exp:* Asst prof nutrit & poultry sci, Univ Md, College Park, 73-75; from asst prof to assoc prof animal sci & nutrit, Wash State Univ, Pullman, 75-83; assoc prof, 83-85, PROF ANIMAL SCI & NUTRIT, UNIV MINN, ST PAUL, 91- *Concurrent Pos:* Quarter leave, Univ Reading, UK, 87 & Roslin Agr Res Ctr, Edinburgh, UK, 91; consult, Asia & Western Europe, Am Soybean Asn. *Mem:* AAAS; Poultry Sci Asn; Am Inst Nutrit. *Res:* Regulation of nitrogen metabolism; hormone-enzyme relationships as influenced by nutrition; amino acid availability, protein quality and energy utilization of feedstuffs; influence of environment of nutrition. *Mailing Add:* Dept Animal Sci Univ Minn 1404 Gortner Ave St Paul MN 55108

COON, DARRYL DOUGLAS, semiconductors; deceased, see previous edition for last biography

COON, JAMES HUNTINGTON, b Liberty, Mo, Nov 9, 14; m 55; c 3. SPACE PHYSICS. *Educ:* Ind Univ, AB, 37; Univ Chicago, PhD(physics), 42. *Prof Exp:* Res staff, Metall Lab, Univ Chicago, 42-43 & Los Alamos Sci Labs, 43-46; res assoc, Univ Wis, 47; res staff, 48-50, group leader, 50-77,

CONSULT, LOS ALAMOS NAT LABS, 77- *Mem:* Fel AAAS; fel Am Phys Soc; Am Astron Soc. *Res:* Thermal neutron diffusion and capture; nuclear interactions between light particles; scattering of fast neutrons; space physics. *Mailing Add:* Los Alamos Nat Labs MS D436 Los Alamos NM 87545

COON, JULIAN BARHAM, b Jackson, Miss, Nov 16, 39; m 61; c 2. RESEARCH ADMINISTRATION. *Educ:* Tex A&M, BS, 61; La State Univ, PhD(physics), 66. *Prof Exp:* Asst prof physics, Univ Houston, 68-73; mgr mining res, 78-81, mgr explor res, 81-84, VPRES RES & DEVELOP, CONOCO INC, E I DU PONT DE NEMOURS, 84- *Mem:* Soc Explor Geophysicists; Am Asn Petrol Geologists; Soc Petrol Engrs. *Res:* Research administration. *Mailing Add:* Res & Develop Dept PO Drawer 1267 E I du Pont de Nemours Ponca City OK 74603

COON, JULIUS MOSHER, b Liberty, Mo, Oct 29, 10; m 47, 75; c 3. PHARMACOLOGY, TOXICOLOGY. *Educ:* Ind Univ, AB, 32; Univ Chicago, PhD(pharmacol), 38; Univ Ill, MD, 45. *Prof Exp:* Asst pharmacol, Univ Chicago, 35-39, instr, 39-45, pharmacologist, Toxicity Lab, 41-45; pharmacologist, US Food & Drug Admin, Washington, DC, 46; from asst prof to assoc prof pharmacol, Univ Chicago, 46-53, dir, Toxicity Lab, 48-51, dir, USAF Radiation Lab, 51-53; prof pharmacol & chmn dept, 53-76, EMER PROF PHARMACOL, THOMAS JEFFERSON UNIV, 76- *Concurrent Pos:* Mem pharmacol test comt, Nat Bd Med Exam, 54-57; mem food protection comt, Nat Acad Sci-Nat Res Coun, 54-72, mem comt radiation preservation of food, 69-74, chmn, 76-79; mem toxicol study sect, NIH, 58-62, chmn sect, 62-64; mem pharmacol adv comt, Walter Reed Army Inst Res, 66-70; mem adv comt protocols for safety eval, Food & Drug Admin, 66-70, mem panel rev of internal analgesic agents, 72-77; mem expert adv panel food additives, WHO, 66-76; chmn panel food safety, White House Conf Food, Nutrit & Health, 69; mem nominating comt gen comt revision, US Pharmacopoeia, 70; mem comt admissions, Nat Formulary, 70-75; mem subcomt interpretation of relevant human experience versus newly acquired exp data, Citizens' Comn Sci, Law & Food Supply, 73-75; mem expert panel food safety & nutrit, Inst Food Technol, 74-85; mem select comt flavor evaluation criteria, Life Sci Res Off, Fedn Am Socs Exp Biol, 75-76; mem, Expert Panel on Cosmetic Ingredient Rev, Cosmetic, Toiletry & Fragrance Asn, Inc, 77-84; consult, Franklin Inst, Philadelphia, 78-83; mem bd sci adv, Am Coun on Sci & Health, 80- *Honors & Awards:* Merit Award, Soc Toxicol, 78, Educ Award, 83. *Mem:* Am Soc Pharmacol & Exp Therapeut (treas, 64-66); fel Soc Toxicol; Inst Food Technol; Soc Exp Biol & Med; fel NY Acad Sci. *Res:* Toxicology of insecticides; food toxicology; autonomic pharmacology. *Mailing Add:* Dept Pharmacol Thomas Jefferson Univ Philadelphia PA 19107

COON, LEWIS HULBERT, b Oklahoma City, Okla, Feb 26, 25; m 48; c 3. MATHEMATICS. *Educ:* Okla Agr & Mech Col, BS, 50; Ind Univ, MS, 51; Okla State Univ, MS, 58, EdD(math), 63. *Prof Exp:* Teacher Okla City Pub Schs, 50-57; asst prof math, Southwestern State Col, Okla, 58-61; staff asst, Okla State Univ, 62; asst prof educ res, Ohio State Univ, 63-65; PROF MATH, EASTERN ILL UNIV, 65- *Concurrent Pos:* Mem, NSF Col Conf, Carleton Col, 63. *Mem:* Math Asn Am; Am Math Asn; Nat Coun Teachers Math; Am Nat Metric Coun. *Res:* Hawthorne effect in mathematics education. *Mailing Add:* 1803 Meadowlake Dr Charleston IL 61920

COON, MINOR J, b Englewood, Colo, July 29, 21; m 48; c 2. MOLECULAR PHARMACOLOGY. *Educ:* Univ Colo, BA, 43; Univ Ill, PhD(biochem), 46. *Hon Degrees:* DSc, Northwestern Univ, 83 & Northeastern Univ, 87; MD, Karolinska Inst, 91. *Prof Exp:* Res asst, Univ Ill, Urbana, 46-47; from instr to assoc prof physiol chem, Univ Pa, 47-55; prof, 55-83, chmn, Dept Chem, 70-90, VICTOR C VAUGHAN DISTINGUISHED PROF BIOCHEM, UNIV MICH, ANN ARBOR, 83- *Concurrent Pos:* USPHS spec fel, NY Univ, 52-53; res fel, Swiss Fed Polytech Inst, Zurich, 61-62; NSF travel award, 69; mem, Biochem Study Sect, NIH, 63-66; mem res career award comt, Nat Inst Gen Med Sci, 66-70; mem, Int Adv Comt, 7th Int Conf Biochem & Biophys Cytochrome P-450, Moscow, 89-91 & fedn bd, Fedn Am Socs Exp Biol, 90-93. *Honors & Awards:* Paul Lewis Award, Am Chem Soc, 59; William C Rose Biochem Award, 78; Bernard B Brodie Award, Drug Metab, 80; Henry Russel Lectr, Univ Mich, 91. *Mem:* Nat Acad Sci; sr mem Inst Med Nat Acad Sci; Am Soc Biochem & Molecular Biol (secy, 81-83, pres elect, 90-91); Am Soc Pharmacol & Exp Therapeut; Am Soc Microbiol; Biophys Soc; fel Am Acad Arts & Sci; Am Chem Soc; AAAS; NY Acad Sci. *Res:* Enzyme reaction mechanisms; amino acid and lipid metabolism; cytochrome P-450; drug metabolism; detoxication. *Mailing Add:* Dept Biochem Univ Mich Med Sch 5416 Med Sci I Ann Arbor MI 48109-0606

COON, ROBERT L, CONTROL OF BREATHING. *Educ:* Marquette Univ, PhD(physiol), 72. *Prof Exp:* RES PHYSIOLOGIST, VET ADMIN MED CTR, WIS, 75-; ASST PROF PHYSIOL & ASSOC PROF ANESTHESIOL, MED COL WIS, 81- *Mailing Add:* Dept Anesthesiol & Physiol Med Col Wis Vet Admin Med Ctr Milwaukee WI 53295

COON, ROBERT WILLIAM, b Billings, Mont, July 13, 20; m 47; c 3. PATHOLOGY. *Educ:* NDak Agr Col, BS, 42; Univ Rochester, MD, 44; Am Bd Path, dipl, 51. *Hon Degrees:* DSc, Univ Vt, 81; DSc, NDak State Univ, 82; LHD, Marshall Univ, 85. *Prof Exp:* Intern path, Stong Mem Hosp, 44-45; asst, Sch Med, Emory Univ, 45-46; lab officer, US Naval Hosp, RI, 46-47; fel path, Univ Rochester, 48-49; from assoc to assoc prof, Col Physicians & Surgeons, Columbia Univ, 49-55; prof path & chmn dept, Col Med, Univ Vt, 55-73, assoc dean div health sci, 68-73; asst chancellor, Univ Maine Bd Trustees, 74-76; vchancellor health sci, WVa Bd Regents, Charlestown, WVa, 76; vpres & dean, Sch Med, Marshall Univ, 76-85; RETIRED. *Concurrent Pos:* Trustee, Am Bd Path, 60-72. *Mem:* Am Soc Clin Path (pres, 63-64); Am Soc Exp Path; Am Asn Pathologists & Bacteriologists; Col Am Pathologists; Int Acad Path; Fedn Am Socs Exp Biol. *Res:* Coagulation of blood. *Mailing Add:* 630 Hinesburg Rd Unit 7 South Burlington VT 05403

COON, SIDNEY ALAN, b Wash, Iowa, Apr 10, 42; m 69; c 2. FEW BODY THEORY, NUCLEAR INTERACTIONS. *Educ:* Univ Iowa, BS, 63; Univ Calif San Diego, MS; 66; Univ Md, PhD(physics), 72. *Prof Exp:* Vol physics & math, US Peace Corps, 66-68; res assoc, Univ Ariz, 72-74 & 76-79, Tech Univ Hanover, 75-76; lectr, Univ Ariz, 79-82, lectr physics, 84-89; ASSOC PROF, NMEX STATE UNIV, 89- *Concurrent Pos:* Vis assoc prof, Univ Wash, 82-83; Fulbright sr scholar, Univ Lisbon, Portugal, 83; prin investr, Nat Sci Found, 80-; physics adv comt, Ind Univ Cyclotron Facil, 89-92; steering comt, APS Few-Body Topical Group. *Mem:* Fel Am Phys Soc. *Res:* Nuclear theory and particle theory which include many-body theory, few-body dynamics, many-body forces and quark models of hadron and nuclei. *Mailing Add:* Physics Dept NMex State Univ Las Cruces NM 88003

COON, WILLIAM WARNER, b Saginaw, Mich, Aug 10, 25; m 49; c 2. SURGERY. *Educ:* Johns Hopkins Univ, MD, 49; Am Bd Surg, dipl, 57. *Prof Exp:* Dir blood bank, Univ Hosp, 56-64, from instr to assoc prof, 56-67, PROF SURG, UNIV MICH, ANN ARBOR, 67- *Concurrent Pos:* Attend surgeon, Ann Arbor Vet admin Hosp, 56-62, consult surgeon, 62- *Mem:* AAAS; Soc Univ Surg; Soc Exp Biol & Med; Sigma Xi; Am Surg Asn. *Res:* Blood coagulation; thromboembolism; metabolism. *Mailing Add:* 1406 Warrington Dr Ann Arbor MI 48103

COONCE, HARRY B, b Independence, Mo, Mar 19, 38; m 65; c 2. MATHEMATICS. *Educ:* Iowa State Univ, BS, 59. Univ Del, PhD(math), 69. *Prof Exp:* Asst prof math, Wichita State Univ, 64-66, Del State Col, 66-67 & US Naval Acad, 67-69; from asst prof to assoc prof, 69-77, PROF MATH, MANKATO STATE COL, 77- *Concurrent Pos:* Consult, Boeing Co, Kans, 65-67. *Mem:* Am Math Soc; Math Asn Am. *Res:* Geometric function theory; univalent functions; functions with bounded boundary rotation; variational methods. *Mailing Add:* Dept Math Mankato State Col Mankato MN 56001

COONEY, CHARLES LELAND, b Philadelphia, Pa, Nov 9, 44; m 78; c 2. BIOCHEMICAL ENGINEERING. *Educ:* Univ Pa, BS, 66; Mass Inst Technol, MS, 67, PhD(biochem eng), 70. *Prof Exp:* From instr to asst prof, 70-75, assoc prof, 75-81, PROF BIOCHEM ENG, MASS INST TECHNOL, 81- *Mem:* AAAS; Am Chem Soc; Am Inst Chem Engrs; Am Soc Microbiol; Inst Food Technol. *Res:* Fermentation technology; continuous microbial culture, production, isolation and application of enzymes; microbial protein production; engineering of microbial processes; biological processes for fuels and chemicals; recovery of biochemical product; computer control; computer-aided design. *Mailing Add:* Rm 66-472 Mass Inst Technol Cambridge MA 02139

COONEY, DAVID OGDEN, b Boston, Mass, Dec 19, 39; m 66; c 2. CHEMICAL ENGINEERING. *Educ:* Yale Univ, BE, 61; Univ Wis, Madison, MS, 63, PhD(chem eng), 66. *Prof Exp:* Fel chem eng, Univ Wis, Madison, 66; res engr, Chevron Res Co, Calif, 66-69; from asst prof to prof chem eng, Clarkson Univ, 69-80; prog dir, NSF, 80-81; dept head, 83-90, PROF CHEM ENG, UNIV WYO, 81- *Mem:* Am Inst Chem Engrs; Am Soc Eng Educ. *Res:* Mass transfer processes in general; adsorption; air and water pollution. *Mailing Add:* Dept Chem Eng Univ Wyo Box 3295 Univ Sta Laramie WY 82071

COONEY, JOHN ANTHONY, b Jersey City, NJ, Oct 31, 22; m 50; c 3. ATMOSPHERIC PHYSICS. *Educ:* Fordham Univ, BS, 49, MS, 50; NY Univ, PhD(physics), 67. *Prof Exp:* Res engr syst eng, Vitro Labs, Pullman Corp, 50-54; res engr microwave eng, Sperry Corp, 55-60; res scientist plasma & atmospherics, RCA Labs, 60-70; assoc prof physics, 70-80, PROF PHYSICS & ATMOSPHERIC SCI, DREXEL UNIV, 80- *Res:* Laser radar probing of the atmosphere. *Mailing Add:* 161 Greene St Philadelphia PA 19104

COONEY, JOHN LEO, b Washington, DC, June 26, 28; m 56; c 1. PHYSICAL CHEMISTRY. *Educ:* Loyola Col, Md, BS, 52; Fordham Univ, MS, 54, PhD(phys chem), 57. *Prof Exp:* Lectr phys chem, Notre Dame Col, Staten Island, 54-55; SR PHYS CHEMIST, E I DU PONT DE NEMOURS & CO, INC, 57- *Mem:* Am Phys Soc; Am Chem Soc. *Res:* Physical properties of polymers; thermodynamics of irreversible processes; polymer morphology; vapor phase reactions. *Mailing Add:* 1919 Kynwyd Dr Wilmington DE 19810-3841

COONEY, JOSEPH JUDE, b Syracuse, NY, Jan 16, 34; m 57; c 4. MICROBIOLOGY. *Educ:* Le Moyne Col, NY, BS, 56; Syracuse Univ, MS, 58, PhD(microbiol), 61. *Prof Exp:* From asst prof to assoc prof bact, Loyola Univ, 61-65; from assoc prof to prof biol, Univ Dayton, 65-77; prof & head, Chesapeake Biol Lab, Univ Md, 76-82; dir, 82-88, PROF, DEPT ENVIRON SCI, UNIV MASS, BOSTON HARBOR CAMPUS, 82- *Concurrent Pos:* Res grants, NIH, 63-64 & La Div, Am Cancer Asn, 64-; sabbatical leave, dept biochem, Med Sch, Tufts Univ, 71; grants, Firestone Coated Fabrics Div, Monsanto Corp, US Dept Interior, Md Dept Natural Resources, Md sea grant, Mass Inst Technol sea grant & Burroughs Welcome Foundation; sabbatical leave, dept microbiol, Univ Col, Galway, Ireland, 89. *Mem:* AAAS; Am Soc Microbiol; Soc Gen Microbiol; fel Soc Indust Microbiol; fel Am Acad Microbiol; Soc Environ Toxicol & Chem. *Res:* Microbial carotenoid pigments; metabolism of hydrocarbons; ecology and physiology of hydrocarbon-using organisms; oil pollution; microbiol interactions with heavy metals and organo metals. *Mailing Add:* Dept Environ Sci Univ Mass Boston Harbor Campus Boston MA 02125

COONEY, MARION KATHLEEN, microbiology; deceased, see previous edition for last biography

COONEY, MIRIAM PATRICK, b South Bend, Ind, May 6, 25. MATHEMATICS. *Educ:* St Mary's Col, Ind, BS, 51; Univ Notre Dame, MS, 53; Univ Chicago, SM, 63, PhD(math), 69. *Prof Exp:* Assoc prof, 63-72, PROF MATH, ST MARY'S COL, IND, 72- *Concurrent Pos:* Vis lectr, St Patrick's Col, Marynooth, Ireland, 80 & Northeast Normal Univ Changchun, Jilin Prov, People's Repub China, 85; mem, Nat Coun Teacher Math. *Mem:* Am Math Soc; Math Asn Am; Asn Women Math. *Res:* Algebra, especially finite groups; mathematical education on all levels. *Mailing Add:* Dept Math St Mary's Col Notre Dame IN 46556

COONS, FRED F(LEMING), b Woodland, Calif, Oct 14, 23; m 49; c 3. CHEMICAL ENGINEERING. *Educ:* Univ Calif, BSc, 44; Golden Gate Univ, MBA, 84. *Prof Exp:* Asst chem, Univ Calif, 43-44; chem engr, Spreckels Sugar Co, 44-53, chief engr, 53-58; chief proj engr, Am Sugar Refining Co, NY, 58-60; chief engr, 60-76, vpres, 76-81, exec vpres, 81-82, pres, Spreckels Sugar Div, 82-86, SR VPRES, AMSTAR CORP, 86- *Mem:* Am Chem Soc; Am Inst Chem Engrs. *Res:* Heat and heat transfer as connected with the sugar industry; filtration; settling; evaporation; diffusion; absorption; crystallization. *Mailing Add:* 44 Bret Harte Lane San Rafael CA 94901-5245

COONS, LEWIS BENNION, b Salt Lake City, Utah, July 28, 38; m 60; c 4. BIOLOGY. *Educ:* Utah State Univ, BS, 64, MS, 66; NC State Univ, PhD, 70. *Prof Exp:* Fel electron micros, Molecular Toxicol Prog, NC State Univ, 70-73; head, Electron Micros Ctr, Miss State Univ, 73-76; head ctr, 76-80, DIR ELECTRON MICROS, MEMPHIS STATE UNIV, 80- *Mem:* AAAS; Electron Micros Soc Am. *Res:* Scanning and transmission electron microscopy of animal cell and organ systems. *Mailing Add:* Dept Biol Memphis State Univ Memphis TN 38152

COOPE, JOHN ARTHUR ROBERT, b Liverpool, Eng, June 9, 31; Can citizen; m 54. MOLECULAR PHYSICS. *Educ:* Univ BC, BA, 50, MSc, 52; Univ Oxford, DPhil(theoret chem), 56. *Prof Exp:* Fel molecular physics, 56-57, lectr, 57-59, from instr to assoc prof, 59-73, PROF CHEM, UNIV BC, 73- *Mem:* Am Phys Soc. *Res:* Quantum theory; physics of atoms and small molecules; theoretical chemistry; solid state physics. *Mailing Add:* Dept Chem Univ BC 2075 Westbrook Pl Vancouver BC V6T 1W5 Can

COOPER, AARON DAVID, b Phildelphia, Pa, Nov 17, 28; m 56; c 2. REGULATORY, QUALITY ASSURANCE. *Educ:* Temple Univ, BS, 50; Univ Wis, MS, 52, PhD(pharmaceut), 54. *Prof Exp:* Res scientist, Anal Chem, Upjohn Co, 54-55; sr asst scientist, NIH, 55-57; res chemist, Atlas Powder Co, 57-61; head anal chem sect, 61-69, dir anal chem, 69-76, dir qual assurance, Vick Div Res & Develop, Richardson-Vick, Inc, 76-87; CONSULT, 87. *Concurrent Pos:* Mem gen comn of revision US Pharmacopeia, 70-75. *Mem:* Am Chem Soc. *Res:* Chromatography; pharmaceutical analyses; regulatory/FDA; quality control; biological analyses. *Mailing Add:* 11 Belmont St White Plains NY 10605

COOPER, ALFRED J(OSEPH), b New Orleans, La, Feb 2, 13; m 36; c 2. WATER RESOURCES. *Educ:* Tulane Univ, BE, 34; Univ Tenn, MSCE, 51. *Prof Exp:* Engr, R P Farnsworth, Inc, La, 34; draftsman, Tenn Valley Auth, 34-35, jr engr, 35-37, jr hydraul engr, 37-38, asst hydraul engr, 38-42, assoc hydraul engr, 42-46, hydraul engr, 46-52, head hydraul engr, 52-57, asst chief river control engr, 57-60, chief river control engr, 60-74; RETIRED. *Concurrent Pos:* Instr, Knoxville Adult Evening High Sch, 46-48; lectr, Univ Tenn, 48-49; consult, 78-87. *Mem:* Fel Am Soc Civil Engrs. *Res:* Correlation of rainfall and runoff; unit surface water runoff hydrographs; streamflow forecasting; analysis watershed hydrology; water control operations; river forecasting; reservoir operations. *Mailing Add:* 3843 Kenilworth Dr SW Knoxville TN 37919

COOPER, ALFRED R, JR, b New York, NY, Jan 1, 24; m 48; c 5. CERAMICS. *Educ:* Alfred Univ, BS, 48; Mass Inst Technol, ScD(ceramics), 60. *Prof Exp:* Instr ceramics, Mass Inst Technol, 58-59, from asst prof to assoc prof, 59-65; assoc prof metall, 65-69, PROF CERAMICS, CASE WESTERN RESERVE UNIV, 69- *Concurrent Pos:* Vis prof, Univ Sheffield, 64-; consult, Am Optical Co, 58-71, Pittsburgh Plate Glass Co, 61-, Dow Corning, 68-71 & Gen Elec Co, 70-; chmn, Gordon Conf Ceramics, 69, chmn, Gordon Conf Glass, 70. *Honors & Awards:* Raytheon Award, 66; Morey Award, Am Ceramic Soc, 86. *Mem:* Am Ceramic Soc; Brit Soc Glass Technol; Faraday Soc. *Res:* Diffusion in ceramic and glassy systems; multicomponent diffusion; process kinetics in ceramics; application of mathematical methods to ceramics; x-ray diffraction. *Mailing Add:* Dept Mat Sci & Eng Case Western Reserve Univ Cleveland OH 44106

COOPER, ALFRED WILLIAM MADISON, b Dublin, Ireland, June 12, 32; US citizen; m 61; c 3. ATMOSPHERIC OPTICS, LASER PHYSICS. *Educ:* Trinity Col, Dublin, BA, 55, MA, 58; Queen's Univ, Northern Ireland, PhD(physics), 61. *Prof Exp:* Res asst physics, Queen's Univ, Belfast, 55-56, asst lectr, 56-57; asst prof, 57-64, assoc prof, 64-76, PROF PHYSICS, NAVAL POSTGRAD SCH, 76-; DIR, NAVAL ACAD CTR, INFRARED TECHNOL, 85- *Concurrent Pos:* Mem tech staff, Aerospace Corp, 64, consult, 64-67. *Mem:* Optical Soc Am; Sigma Xi; Am Phys Soc; Asn Old Crows; Am Asn Univ Profs. *Res:* Infrared electro-optic and laser system design and analysis; atmospheric optics; system performance modeling. *Mailing Add:* Dept Physics Naval Postgrad Sch Monterey CA 93943

COOPER, ALLEN D, b New York, NY, Oct 18, 42. EXPERIMENTAL BIOLOGY. *Educ:* NY Univ, BA, 63, State Univ NY, MD, 67; Am Bd Internal Med, dipl, 71. *Prof Exp:* Intern straight med, V & VI Med Serv, Boston City Hosp, 67-68, resident internal med, 68-69; resident internal med, Med Ctr, Univ Calif, San Francisco, 69-70, chief resident, 70-71, NIH fel gastroenterol, 70-72; clin asst prof med, Univ Tex Med Sch, San Antonio, 72-74; from asst prof to assoc prof med gastroenterol, 74-90, assoc prof physiol, 87-89, PROF MOLECULAR & CELLULAR PHYSIOL, STANFORD UNIV, 90-, PROF MED, 90-; DIR, RES INST, PALO ALTO MED FOUND, 86-, VPRES RES, 88- *Concurrent Pos:* Prin investr, USAF Clin Invest, 72-74; assoc prog dir gastroenterol grant, NIH, 75-85, clin investr, 75-78, res career develop award, 78-83, prog dir gastroenterol, 85-86; Andrew W Mellon Found fel, 77-79; guest lectr atherosclerosis, Univ Calif, 79, pharmacol, Univ Cincinnati, 82; mem, Metab Study Sect, NIH, 81-85, Physiol Chem Study Sect, 88-91; vis prof med, Univ Iowa, 82, vis res prof, Univ Calif, San Diego, 84, vis prof med & pharmacol, Univ Fla, 85; mem, Res Comt, Am Asn Study Liver Dis, 83-86; mem prog comt, Coun Atherosclerosis, Am Heart Asn, 89-91; vis lectr, Okla Med Res Found, 89. *Mem:* Am Soc Biochem & Molecular Biol; fel Am Col Physicians; Am Soc Clin Invest; Am Gastroenterol Asn; Am Asn Study Liver Dis; AAAS; Am Fedn Clin Res. *Res:* Arteriosclerosis; lipoproteins; gall stone disease. *Mailing Add:* Res Inst Palo Alto Med Found 860 Bryant St Palo Alto CA 94301

COOPER, ARTHUR JOSEPH L, b London, Eng, April 19, 46; m 72; c 3. NEUROLOGY, ENZYMOLOGY. *Educ:* London Univ, BSc, 67, MSc, 69; Cornell Univ Med Col, PhD(biochem), 74. *Hon Degrees:* DSc, London Univ, 85. *Prof Exp:* Asst prof biochem in neurol, 77-81, asst prof dept biochem, 77-88, assoc prof dept neurol, 81-89, ASSOC PROF DEPT BIOCHEM, MED COL, CORNELL UNIV, 88-, RES PROF DEPT BIOCHEM & NEUROL, 89- *Concurrent Pos:* Vis investr, Dept Neuro, Mem Sloan Kettering Cancer Ctr, 80- 89. *Mem:* Am Chem Soc; Am Soc Biol Chemists; Fedn Am Soc Exp Biol; NY Acad Sci; AAAS; Am Soc Neurochem; Am Inst Chemist. *Res:* Amonia, amino acid and keto acid metabolism in brain and liver; study of enzymes and enzyme inhibitors; use of positron-emitters isotopes as tracers for brain biochemistry. *Mailing Add:* Dept Biochem Cornell Univ Med Col 1300 York Ave New York NY 10021

COOPER, ARTHUR WELLS, b Washington, DC, Aug 15, 31; m 53; c 2. ECOLOGY. *Educ:* Colgate Univ, BA, 53, MA, 55; Univ Mich, PhD(bot), 58. *Prof Exp:* Preceptor, Colgate Univ, 53-55, instr biol, 54-55; from asst prof to prof bot, NC State Univ, 58-71; dep dir, NC Dept Conserv & Develop 71, dir, 72-74; asst secy, NC Dept Natural & Econ Resources, 71-76; prof bot & forestry, 71-76, PROF FORESTRY, NC STATE UNIV, 76-, HEAD DEPT, 79- *Concurrent Pos:* chmn bd trustees, Inst Ecol, 82-84. *Honors & Awards:* Conservation Award, Am Motors Corp, 73; Sol Feinstone Award, 82. *Mem:* AAAS; Soc Am Forestry; Ecol Soc Am (vpres, 74, pres, 80). *Res:* Plant ecology; general plant sociology; microenvironments; resource management; forest productivity. *Mailing Add:* 719 Runnymede Rd Raleigh NC 27607

COOPER, BENJAMIN STUBBS, b Schenectady, NY, Apr 12, 41; m 62; c 4. NUCLEAR PHYSICS. *Educ:* Swarthmore Col, BA, 63; Univ Va, PhD(physics), 68. *Prof Exp:* Asst prof physics, Iowa State Univ, 67-73; cong sci fel, 73-74, MEM PROF STAFF, SENATE COMT ENERGY & NATURAL RESOURCES, 74- *Mem:* Am Phys Soc; AAAS. *Res:* Theoretical nuclear physics; science and public policy. *Mailing Add:* 1110 South Carolina Ave SE Washington DC 20003-2204

COOPER, BERNARD A, b Plainfield, NJ, July 2, 28; Can & US citizen; m 55; c 3. HEMATOLOGY. *Educ:* McGill Univ, BSc, 49, MD, CM, 53; FRCPS(C), 58. *Prof Exp:* Demonstr med, 60-62, lectr, 62-63, asst prof, 63-65, assoc prof med & clin med, 65-70, prof exp med, 70-76, PROF MED & PHYSIOL, MCGILL UNIV, 76-; DIR HEMAT DIV, ROYAL VICTORIA HOSP, 68-, LOUIS LOWENSTEIN PROF HEMATOL & ONCOL, 90- *Concurrent Pos:* Am Col Physicians res fel med, Harvard Univ, 56-57, Markle scholar med sci, 57-58; Med Res Coun Can career investr, 60-; asst physician, Royal Victoria Hosp, 63-68, physician, 69- *Mem:* Am Fedn Clin Res; Am Physiol Soc; Am Soc Clin Invest; fel Am Col Physicians; Can Soc Clin Invest. *Res:* Transport of vitamin B-12 across the intestine and into other cells, and its interrelationship with folate in human metabolism. *Mailing Add:* Royal Victoria Hosp 687 Pine Ave W Montreal PQ H3A 1A1 Can

COOPER, BERNARD RICHARD, b Everett, Mass, Apr 15, 36; m 62; c 3. SOLID STATE PHYSICS. *Educ:* Mass Inst Technol, BS, 57; Univ Calif, Berkeley, PhD(physics), 61. *Prof Exp:* Res assoc theoret physics, UK Atomic Energy Res Estab, Harwell, 62-63; res fel theoret solid state physics, Harvard Univ, 63-64; physicist, Gen Elec Res Lab, 64-68, Gen Elec Res & Develop Ctr, 68-74; CLAUDE WORTHINGTON BENEDUM PROF PHYSICS, WVA UNIV, 74- *Mem:* Fel Am Phys Soc; sr mem Inst Elec & Electronics Engrs; Am Vacuum Soc; Mat Res Soc; Sigma Xi. *Res:* Theory of magnetic properties of solids; physics of rare earth metals and compounds; theory of electronic and optical properties of solids; theory of electronic properties of surface and interfaces; theory of structural and mechanical properties of solids, alloy theory. *Mailing Add:* Dept Physics WVa Univ Morgantown WV 26506

COOPER, CARL (MAJOR), b Pa, Aug 18, 19; m 62; c 2. CHEMICAL ENGINEERING. *Educ:* Univ Okla, BS, 36; Mass Inst Technol, ScD(chem eng), 49. *Prof Exp:* Chem engr, Dept Res & tech asst to vpres, Phillips Petrol Co, 36-42; proj engr & head process anal group, Vulcan Copper & Supply Co, 42-48; from assoc prof to prof, 48-81, EMER PROF CHEM ENG, MICH STATE UNIV, 81- *Concurrent Pos:* Consult chem engr, 48-; mem, Develop Dept, Weyerhaeuser Timber Co, 51; tech dir, Vulcan-Cincinnati, 56. *Mem:* Am Chem Soc; Am Inst Chem Engrs; Am Soc Eng Educ. *Res:* Distillation; extraction; drying; heat transfer; hydrodynamics; thermodynamics; catalysis; corrosion; instrumentation; pulp and paper technology; petroleum technology; petrochemicals; industrial organic syntheses. *Mailing Add:* 4631 Ottawa Dr Okemos MI 48864

COOPER, CARY WAYNE, b Camden, Maine, Sept 1, 39; m 62; c 2. NEUROSCIENCES. *Educ:* Bowdoin Col, AB, 61; Rice Univ, PhD(biol), 65. *Prof Exp:* From instr to assoc prof, Sch Med, Univ NC, Chapel Hill, 68-77, prof pharmacol, Sch Med, 77-82; MED BR, UNIV TEX, GALVESTON, 82- *Concurrent Pos:* Nat Inst Dent Res fel pharmacol, Sch Dent Med, Harvard Univ, 65-66; Nat Inst Dent Res fel, Sch Med, Univ NC, Chapel Hill, 66-67, Nat Inst Arthritis & Metab Dis spec fel, 67-69, career develop award, 72-76, res grant, 74-90; Merck Co Found grant fac develop, 69-70; Wellcome res travel grant, 80, Int Sci Exchange travel grant, Can, 87. *Mem:* Endocrine Soc; Am Soc Pharmacol & Exp Therapeut; AAAS; Soc Exp Biol Med; Am Physiol Soc; Sigma Xi; Am Soc Bone & Mineral Res. *Res:* Endocrine physiology and pharmacology, especially calcium and bone metabolism and gastrointestinal hormones; hormonal control of calcium homeostasis and bone metabolism; parathyroid hormone and calcitonin; gastrin, CCK and somatostatin; neural peptides, radioimmunoassay; peptide hormones. *Mailing Add:* Med Br Univ Tex Galveston TX 77550-2774

COOPER, CECIL, b Philadelphia, Pa, Dec 24, 22; m 46; c 2. BIOCHEMISTRY. *Educ:* George Washington Univ, BS, 49; Univ Pa, PhD(biochem), 54. *Prof Exp:* Instr physiol chem, Johns Hopkins Univ, 55-56; from asst prof to assoc prof biochem, 56-68, PROF BIOCHEM, CASE WESTERN RESERVE UNIV, 68- *Mem:* Am Soc Biol Chemists; Am Chem Soc; Brit Biochem Soc; Fedn Am Soc Exp Biol. *Res:* Oxidative phosphorylation; metal-nucleotide complexes; whole body lipid metabolism; effect of ethanol on liver. *Mailing Add:* Dept Biochem 2109 Adelbert Rd Cleveland OH 44106-2624

COOPER, CHARLES BURLEIGH, b Parkesburg, Pa, Apr 18, 20; m 46; c 3. PHYSICS. *Educ:* Franklin & Marshall Col, BS, 41; Cornell Univ, MS, 43; Univ Md, PhD(physics), 51. *Prof Exp:* Asst physics, Cornell Univ, 41-44; tech supvr, Tenn Eastman Corp, Tenn, 44-46; instr physics, Univ Del, 46-49; asst physics, Univ Md, 49-51, asst prof, 51-52; vpres, Tagcraft Corp, Pa, 52-58; assoc prof, 58-65, PROF PHYSICS, UNIV DEL, 65- *Concurrent Pos:* Actg chmn, Physics Dept, Univ Del, 80-82. *Mem:* Am Phys Soc; Am Asn Physics Teachers. *Res:* Mass spectroscopy; particle-solid interaction; surface physics; sputtering; high temperature specific heat measurements; optical properties; calutron. *Mailing Add:* Dept Physics Univ Del Newark DE 19711

COOPER, CHARLES DEWEY, b Whittier, NC, Jan 11, 24; m 46; c 3. PHYSICS. *Educ:* Berry Col, BS, 44; Duke Univ, AM, 48, PhD(physics), 50. *Prof Exp:* From asst prof to prof physics, 50-89, dir arts & sci self study, 70-71, EMER PROF PHYSICS, UNIV GA, 89- *Concurrent Pos:* Res fel, Harvard Univ, 54-55; consult, Oak Ridge Nat Lab, 66-85. *Honors & Awards:* IR 100 Award, 83. *Mem:* Am Phys Soc. *Res:* Visible and ultraviolet spectroscopy; atmospheric physics; electron and negative ion physics; multiphoton ionization; third harmonic generation in rare gases. *Mailing Add:* Physics Dept Univ Ga Athens GA 30602

COOPER, CHARLES F, b Kenosha, Wis, Sept 26, 24. ECOLOGY. *Educ:* Univ Minn, BS, 50; Univ Ariz, MS, 57; Duke Univ, PhD(bot), 58. *Prof Exp:* Forester, Bur Land Mgt, US Dept Interior, 50-55 & Watershed Prog, Ariz, 56; asst prof natural resources, Humboldt State Col, 58-60; ecologist, Agr Res Serv, USDA, Idaho, 60-64; lectr, Sch Natural Resources, Univ Mich, 64-65, from assoc prof to prof natural resources ecol, 65-71; dir, Ctr Regional Environ Studies, 71-83, PROF BIOL, SAN DIEGO STATE UNIV, 71- *Concurrent Pos:* Fulbright res fel, Australia, 62-63; hydrologist, USPHS, 64-65; prog dir, Ecosystem Studies, NSF, 69-71; mem US deleg, Man & the Biosphere, Intergovt Coord Coun, UNESCO, 71; chmn US deleg, US-Taiwan Coop Sci Prog Sem Forest Ecol, 72; dir, San Diego County Water Authority, 73-91; mem exec comt, Inst Ecol, 77-81; Woodrow Wilson fel, Smithsonian Inst, 77; mgt authority, Tijuana River Nat Estuarine Res Reserve, 82- *Mem:* Fel AAAS; Soc Am Foresters; Ecol Soc Am; Am Geophys Union; Sigma Xi; Soc Conserv Biol. *Res:* Ecological implications of climate change; systems ecology; ecology and public policy. *Mailing Add:* Biol Dept San Diego State Univ San Diego CA 92182

COOPER, DALE A, b Smithville, Mo, Nov 13, 58; m 83; c 1. NUTRITIONAL SAFETY ASSESSMENT. *Educ:* Cent Mo State Univ, BS, 81; Iowa State Univ, PhD(biochem), 86. *Prof Exp:* Res assoc biochem, Cornell Univ, 87-89; TECH RES MGR NUTRIT, FOOD & BEVERAGE DIV, RES & CLIN DEVELOP, PROCTER & GAMBLE CO, 89- *Mem:* Am Inst Nutrit. *Res:* Nutritional safety assessment of food additives; Food and Drug Admin food regulatory issues; vitamin and mineral metabolism. *Mailing Add:* Procter & Gamble Co 6071 Center Hill Rd Cincinnati OH 45224

COOPER, DAVID B, b New York, NY, Jan 12, 33; m 60; c 2. ELECTRICAL ENGINEERING, APPLIED MATHEMATICS. *Educ:* Mass Inst Technol, BS, 54, MS, 57; Columbia Univ, PhD(appl math), 66. *Prof Exp:* Staff mem radar & detection systs, Sylvania Elec Prod, Inc, Calif, 57-59; sr res scientist & commun syst anal, Raytheon Co, Mass, 59-66; from asst prof to assoc prof, 66-78, PROF ELEC ENG, BROWN UNIV, 78- *Mem:* Asn Comput Mach; Inst Elec & Electronics Engrs. *Res:* Applied stochastic processes; computer vision; computer pattern recognition. *Mailing Add:* Div Eng Brown Univ Providence RI 02912

COOPER, DAVID GORDON, b Toronto, Ontario, Jan 9, 47. SURFACE SCIENCE, APPLIED MICROBIOLOGY. *Educ:* Univ Toronto, BSc, 70, PhD(chem), 76. *Prof Exp:* ASSOC PROF, DEPT CHEM ENG, MCGILL UNIV, 81- *Mem:* Chem Inst Can; Can Soc Chem Eng; Am Soc Microbiol; Am Orchid Soc; Can Geog Soc; Am Chem Soc. *Res:* Production, identification and application of surface active biological materials; development of computer controlled fermentations; reformation of waste polymers; biopolymer research. *Mailing Add:* Dept Chem Eng McGill Univ 3480 University St Montreal PQ H3A 2A7 Can

COOPER, DAVID YOUNG, b Henderson, NC, Aug 14, 24; m 55; c 2. MEDICINE. *Educ:* Univ Pa, MD, 48. *Prof Exp:* Intern, 48-49, asst instr pharmacol, 49-50, asst resident, 50, asst instr surg, 53-57, assoc, 57-59, from asst prof to assoc prof surg res, 59-68, PROF SURG RES, HOSP UNIV PA, 68- *Concurrent Pos:* Fel surg, Harrison Dept Surg Res, Univ Pa, 53-56, Kirby fel, 56-57, Finley fel, Col Surgeons, 57-60; estab investr, Am Heart Asn, 60- *Mem:* Am Physiol Soc; Endocrine Soc; Am Fedn Clin Res; Am Soc Biol Chem; Sigma Xi. *Res:* Adrenal physiology; relation of adrenal to hypertension; pulmonary physiology; oxygenases. *Mailing Add:* 424 Colebrook Lane Bryn Mawr PA 19010

COOPER, DERMOT M F, SECOND MESSENGERS, ADENYLATE CYCLASE. *Educ:* Univ Wales, Bangor, PhD(biochem), 76. *Prof Exp:* ASST PROF PHARMACOL & BIOCHEM, HEALTH & SCI CTR, UNIV COLO, 82- *Res:* Inositol phosphates. *Mailing Add:* Dept Pharmacol Campus Box C236 Univ Colo Health & Sci Ctr 4200 E 9th Ave Denver CO 80262

COOPER, DONALD EDWARD, b Denton, Md, Nov 8, 52; m 87; c 2. INFRARED FOCAL PLANE ARRAYS. *Educ:* Swarthmore Col, BA, 74; Stanford Univ, PhD(phys chem), 79. *Prof Exp:* Mem tech staff, Aerospace Corp, 80-86; MEM TECH STAFF, ROCKWELL INT SCI CTR, 86- *Mem:* Am Phys Soc. *Res:* Research and development of large-area infrared focal plane arrays, including detector physics, small-band gap semiconductor materials and processing. *Mailing Add:* Rockwell Sci Ctr PO Box 1085 MS A7A Thousand Oaks CA 91358

COOPER, DONALD RUSSELL, b Kalamazoo, Mich, Sept 8, 17; m 42; c 2. SURGERY. *Educ:* Univ Mich, AB, 39, MD, 42. *Prof Exp:* Intern, Univ Mich Hosp, 42-43, resident surg & instr, 46-50; assoc, 50-51, from asst prof to assoc prof, 51-59, PROF SURG, MED COL PA, 59- *Concurrent Pos:* Asst instr, Grad Sch Med, Univ Pa, 50-51, assoc, 51-52, from asst prof to assoc prof, 52-65; consult surgeon, Vet Admin Hosp, 57- *Mem:* Am Col Surg; AMA; Pan-Pac Surg Soc; Int Soc Surg; Soc Surg Alimentary Tract. *Res:* Gastrointestinal surgery; surgical research in gastrointestinal and vascular fields. *Mailing Add:* Dept Surg Med Col Pa 3300 Henry Ave Philadelphia PA 19129

COOPER, DOUGLAS ELHOFF, b New Boston, Ohio, May 21, 12; m 59. ORGANIC CHEMISTRY. *Educ:* Eastern Ky Univ, BS, 39; Univ Tenn, MS, 40; Purdue Univ, PhD(org chem), 43. *Prof Exp:* Asst chem, Univ Tenn, 39-40; res fel, Purdue Univ, 40-43; res chemist, Bristol Labs, 43-52; res chemist, Res Labs, Ethyl Corp, Detroit, 52-66, res assoc, 66-69 & mkt anal, 69-74; res assoc toxicol support, Ethyl Corp, Baton Rouge, 74-77; CONSULT CHEMIST, OCCUP ENVIRON, 77- *Mem:* Health Physics Soc; Am Asn Consult Chemists & Chem Engrs; Am Indust Hyg Asn; Am Soc Safety Engrs; Am Chem Soc. *Res:* Organic synthesis; antimalarial synthesis; penicillin isolation, purification and production; dosage forms for penicillin; derivatives of 4-chloroquinoline; chemical aspects of internal combustion engine problems; radioisotopic tracer applications; steroids; chemical and engineering support of toxicology and medical activities. *Mailing Add:* 1732 Avondale Arden Commons Sacramento CA 95825-1347

COOPER, DOUGLAS W, b Manhattan, NY, Dec 21, 42; m 84; c 2. AEROSOL PHYSICS. *Educ:* Cornell Univ, BA, 64; Penn State Univ, MS, 69; Harvard Univ, PhD(appl physics), 74. *Prof Exp:* From asst to assoc prof physics, Harvard Univ, 76-83; RES STAFF MEM, IBM, 83- *Concurrent Pos:* Bd dir, Am Asn Aersol Res, 90-92. *Honors & Awards:* W J Whitfield Award, Inst Environ Sci. *Mem:* AAAS; Am Inst Physics; Am Asn Aersol Res. *Res:* Interested in generation deposition; removal of particles. *Mailing Add:* Mfg Res 1-203 IBM TJ Watson Res Ctr Box 218 Yorktown Heights NY 10598

COOPER, DUANE H(ERBERT), b Ill, Aug 21, 23; m 49; c 2. PHYSICS, ELECTRICAL ENGINEERING. *Educ:* Calif Inst Technol, BS, 50, PhD(physics), 55. *Prof Exp:* ASSOC PROF ELEC ENG & RES ASSOC PROF PHYSICS, COORD SCI LAB, UNIV ILL, URBANA-CHAMPAIGN, 54- *Concurrent Pos:* Consult, Consumers Union, NY, 63, Shure Brothers Inc, Ill, 64 & Nippon Columbia Co, Ltd, Japan, 71-76; proponent mem, Nation Quadraphonic Radio Comt, Electronic Industs Asn, 72-77. *Honors & Awards:* Emile Berliner Award, Audio Eng Soc, 68. *Mem:* Am Phys Soc; hon mem Audio Eng Soc (pres, 75-76); Acoust Soc Am; sr mem Inst Elec & Electronics Engrs; Sigma Xi. *Res:* Computer technology; information theory; signal analysis; acoustics. *Mailing Add:* 918 W Daniel St Champaign IL 61821

COOPER, EARL DANA, b Washington, DC, Apr 16, 26; m 49; c 2. ELECTRONICS, MANAGEMENT. *Educ:* George Washington Univ, BEE, 48; Nova Univ, DPA, 81. *Prof Exp:* Elec engr, Navy Bur Ships, 48-54, ord engr, Navy Bur Ord, 55-57, supvry guided missile design engr, 57-59; sr systs engr air-to-surface weapon systs, Bur Naval Weapons, 59-62, tech dir air launched weapons systs, 62-66, tech dir advan systs concepts, 66-71, tech dir advan systs, 71-76, dep prog mgr, Vertical Short Take-Off & Landing Aircraft Prog, 77-78, tech dir advan systs, 78-79, tech dir res & technol, Naval Air Systs Command, 79-85; PROG DIR, FLA INST TECHNOL, 85- *Concurrent Pos:* Chmn, Naval Aviation Exec Inst, 74-75, mem exec bd, 75- *Honors & Awards:* Sr Exec Award, Secy Navy, 75. *Mem:* Am Defense Preparedness Asn; World Future Soc. *Res:* Aircraft and weapon system design; missile guidance; infrared detection; engineering administration; advanced aircraft systems and air launched weapons systems; executive development; public administration; space & military technology; counter-terrorism concepts/technologies; tourism innovation; automation and robotic technologies; international business planning and development; global market analysis and projections/forecast; media innovation and planning; business intelligence; strategic visioneering. *Mailing Add:* Fla Inst Technol 4875 Eisenhower Ave Alexandria VA 22304

COOPER, EDWIN LOWELL, b Oakland, Tex, Dec 23, 36; m 69. IMMUNOLOGY, BIOLOGY. *Educ:* Tex Southern Univ, BS, 57; Atlanta Univ, MS, 59; Brown Univ, PhD(biol), 63. *Prof Exp:* From asst prof to assoc prof, 64-73, PROF ANAT, UNIV CALIF, LOS ANGELES, 73- *Concurrent Pos:* Nat Cancer Inst fel, Univ Calif, Los Angeles, 62-64; Guggenheim fel & hon Fulbright scholar, Bact Inst, Karolinska Inst, Sweden, 70-71; vis asst prof, Nat Polytech Inst, Mex, 66; mem, Pan Am Cong Anat; mem adv comt, Nat Res Coun, 72-73, mem comn human resources, 73-74; mem bd sci counr, Nat Inst Dent Res, 74-78; founder, Int J Develop & Comparative Immunol, 76; Eleanor Roosevelt fel, Int Union Against Cancer, Ludwig Inst Cancer Res, Lausane Switz, 77-78. *Mem:* Fel AAAS; Am Soc Zool; Am Asn Immunol; Soc Invert Path; Am Asn Anat; Fedn Am Socs Exp Biol. *Res:* Transplantation immunology; developmental and comparative immuno-biology. *Mailing Add:* Dept Anat Univ Calif Sch Med Los Angeles CA 90024

COOPER, ELMER JAMES, b Milwaukee, Wis, Mar 31, 20; m 43; c 2. CEREAL CHEMISTRY. *Educ:* Marquette Univ, BS, 41, MS, 49. *Prof Exp:* Res chemist, Carnation Milk Co, 49-51 & Milwaukee County Hosp, 51; tech mgr cereal technol, Universal Foods Corp, 51-84; RETIRED. *Mem:* AAAS; Am Chem Soc; Am Asn Cereal Chemists; Am Soc Bakery Engrs. *Res:* Fermentation; baking; cereal technology. *Mailing Add:* 1633 N 57th St Milwaukee WI 53208-2131

COOPER, EMERSON AMENHOTEP, b Panama, Jan 15, 24; US citizen; m 49; c 3. ORGANIC CHEMISTRY. *Educ:* Oakwood Col, BA, 49; Polytech Inst Brooklyn, MS, 54; Mich State Univ, PhD(org chem), 59. *Prof Exp:* Instr sci, Oakwood Acad, 48-50; chemist, Metric Chem Co, 50-51; from asst prof to assoc prof, 51-59, PROF CHEM, OAKWOOD COL, 59-, CHMN DEPT, 77- *Mem:* Am Chem Soc; Sigma Xi. *Res:* Cyanine dyes; synthesis of ninhydrin analogs. *Mailing Add:* Dept Chem Oakwood Col Huntsville AL 35806

COOPER, EUGENE PERRY, b Somerville, Mass, Aug 15, 15; m 42; c 4. PHYSICS. *Educ:* Mass Inst Technol, BS, 37; Univ Calif, PhD(theoret physics), 42. *Prof Exp:* Asst prof physics, Univ NC, 41-43; res physicist, Franklin Inst, Philadelphia, 43-45 & US Naval Ord Test Sta, Inyokern, Calif, 45-47; assoc prof physics, Univ Ore, 47-48; res physicist, US Naval Ord Test Sta, Calif, 48-51; assoc sci dir, US Naval Radiol Defense Lab, 52-60, sci dir, 60-70, consult to tech dir, Naval Undersea Ctr, 70-77; independent res/explor develop dir, 77-84, PROG DIR RES, NAVAL OCEAN SYSTS CTR, 84- *Concurrent Pos:* Instr math & theoret physics, Univ Exten, Univ Calif; consult physicist, Off Sci Res & Develop, 44. *Mem:* AAAS; fel Am Phys Soc. *Res:* Theoretical atomic and nuclear physics; interactions of matter and radiation; beta disintegration; nuclear isomerism; hydrodynamics; electromechanical instrument theory; radioactivity; nuclear weapon effects; physical and biological radiation effects. *Mailing Add:* 8579 Prestwick Dr La Jolla CA 92037

COOPER, FRANKLIN SEANEY, b Robinson, Ill, Apr 29, 08; m 35; c 2. SPEECH, COMMUNICATION SCIENCE. *Educ:* Univ Ill, BS, 31; Mass Inst Technol, PhD(physics), 36. *Hon Degrees:* DSc, Yale Univ, 76. *Prof Exp:* Res engr, Gen Elec Res Labs, 36-39; assoc res dir, 39-55 & 75-88, pres & res dir, 55-75, EMER RES DIR, HASKINS LABS, 88- *Concurrent Pos:* Sci consult, Atomic Energy Comn Group, UN Secretariat, 46-47; consult, Off Secy of Defense, 49-50; mem vis comt, Modern Language Dept, Mass Inst Technol, 49-65; mem adv comt, Res Div, Col Eng, NY Univ, 49-65; adj prof phonetics, Columbia Univ, 55-65; mem bd dirs, Ctr Appl Linguistics, 68-74; adj prof linguistics, Univ Conn, 69-; chmn, Communicative Sci Interdisciplinary Cluster, President's Biomed Res Panel, 75; sr res assoc linguistics, Yale Univ, 70-76; mem, Nat Adv Neurol & Commun Dis Coun, NIH, 78-81 & Adv Bd, Nat Deafness & Other Commun Dis, 89-91. *Honors & Awards:* Pioneer Award in Speech Communication, Inst Elec & Electronics Engrs, 72; Warren Medal, Soc Exp Psychol, 75; Silver Medal in Speech Communication, Acoust Soc Am, 75; Fletcher-Stevens Award, Brigham Young Univ, 77. *Mem:* Nat Acad Eng; AAAS; Am Phys Soc; fel Acoust Soc Am; fel Inst Elec & Electronics Engrs. *Res:* Perception and production of speech; voice communications systems; prosthetic aids for the blind. *Mailing Add:* Haskins Labs 270 Crown St New Haven CT 06511-6695

COOPER, FREDERICK MICHAEL, b New York, NY, Apr 1, 44; m 72, 86; c 2. ELEMENTARY PARTICLE PHYSICS. *Educ:* City Col New York, BS, 64; Harvard Univ, MA, 65, PhD(physics), 68. *Prof Exp:* Instr physics, Cornell Univ, 68-70; asst prof, Belfer Grad Sch Sci, Yeshiva Univ, 70-75; STAFF MEM PHYSICS, LOS ALAMOS NAT LAB, 75-, DEP GROUP LEADER, ELEM PARTICLE PHYSICS GROUP T-8, 81- *Concurrent Pos:* Frederick Cottrell Res Corp res grant, 71; vis prof, Boston Univ, 85-86 & Brown Univ, 88-89. *Mem:* Am Phys Soc. *Res:* Develop bound state perturbation theory and strong coupling perturbation theory; initial value problems in field theory as applied to the time evolution of early universe and heavy ion collisions; supersymmetry and quantum mechanics; non perturbative methods in field theory and non-linear phenomena. *Mailing Add:* T-8 MS B285 Los Alamos Nat Lab PO Box 1663 Los Alamos NM 87545

COOPER, GARRETT, dermatology; deceased, see previous edition for last biography

COOPER, GARY PETTUS, b York, Ala, Aug 30, 33; m 61; c 2. NEUROPHYSIOLOGY, NEUROTOXICOLOGY. *Educ:* Univ Ala, BA, 56; Tulane Univ, MS, 59, PhD(physiol), 63. *Prof Exp:* Res physiologist, US Naval Radiol Defense Lab, 63-66; asst prof, 66-72, assoc prof, 72-77, PROF ENVIRON HEALTH, COL MED, UNIV CINCINNATI, 77- *Mem:* AAAS; Am Physiol Soc; Soc Neurosci; Radiation Res Soc; Fedn Am Socs Exp Biol. *Res:* Synaptic transmission; heavy metal toxicology; effects of toxins on sensory systems. *Mailing Add:* Dept Environ Health Univ Cincinnati Col Med 3223 Eden Ave Cincinnati OH 45267

COOPER, GEOFFREY KENNETH, b Yonkers, NY, Oct 31, 49; m 73; c 2. ORGANIC CHEMISTRY, BIOLOGICAL CHEMISTRY. *Educ:* Univ Calif, Berkeley, BSc, 71; Univ Ore, Eugene, PhD(chem), 77. *Prof Exp:* Res fel alkaloid chem, Univ Calif, Berkeley, 77-78; res chemist, Olympic Res Div, ITT Rayonier Inc, 79-83; Sr scientist, Arco Bioengineering Ctr, Dublin, Calif, 83-84; lab leader & sr scientist, Arco Plant Cell Res Inst, 84-87; prin scientist, Plant Cell Res Inst Inc, Dublin, Calif, 87- 90; PRIN RES CHEMIST, ICI AM WESTERN RES CTR, RICHMOND, CALIF, 90- *Mem:* Am Chem Soc; AAAS; fel Am Inst Chemists. *Res:* Synthetic organic chemistry; bio-organic chemistry and radiochemistry; analytical organic chemistry and spectroscopy. *Mailing Add:* 7534 Flagstone Dr Pleasanton CA 94588

COOPER, GEOFFREY MITCHELL, b Los Angeles, Calif, June 16, 48; m. MOLECULAR CARCINOGENESIS. *Educ:* Mass Inst Technol, BS, 69; Univ Miami, PhD(biochem), 73. *Prof Exp:* Fel virol, McArdle Lab Cancer Res, Univ Wis, 73-75; from asst prof to assoc prof, 75-84, PROF PATH, DANA-FARBER CANCER INST, HARVARD MED SCH, 84- *Honors & Awards:* US Steel Award, 84. *Mem:* Nat Acad Sci; AAAS; Am Asn Cancer Res. *Res:* Molecular basis of oncogenic transformation; roles of proto-oncogenes in signal transduction and early development. *Mailing Add:* Dept Pathology Dana-Farber Cancer Inst 44 Binney St No D710 Boston MA 02115

COOPER, GEORGE, IV, b Charlottesville, Va, June 25, 42; m; c 4. MEDICAL & PHYSIOLOGY. *Educ:* Williams Col, BA, 64; Cornell Univ Med Col, MD, 68, Am Bd Internal Med, 72. *Prof Exp:* Res investr, Div Biophys, Naval Med Res Inst, 75-76, head, Div Appl Physiol, 76-77; asst prof internal med, Col Med, Univ Iowa, 77-81; assoc prof internal med & physiol, dir basic cardiovasc res, Sch Med, Temple Univ, 81-85; PROF INTERNAL MED & PHYSIOL, MED UNIV SC, 85-, DIR CARDIOVASC RES, 85-, DIR CARDIOL, VET ADMIN MED CTR, 85- *Concurrent Pos:* Young Investr Competition Award, Am Col Cardiol, 73; numerous grants, NIH & Vet Admin, 78-95; consult, Nat Heart Lung & Blood Inst, 81; adj prof bioeng, Clemson Univ, 87- *Honors & Awards:* Charles L Horn Award, Cornell Med Col, 68; Louis N Katz basic sci res prize, 75. *Mem:* Am Fedn Clin Res; Am Heart Asn; Cardiac Muscle Soc; Am Physiol Soc; Int Soc Heart Res; Biophys Soc; AAAS; NY Acad Sci; Protein Soc; Soc Adv Drug Res Heart & Vascular Dis. *Res:* Cardiovascular physiology; molecular biology; biochemistry; pharmacology; anatomy; numerous publications. *Mailing Add:* Vet Admin Med Ctr 109 Bee St Charleston SC 29403

COOPER, GEORGE EMERY, b Burley, Idaho, May 17, 16; m 41; c 4. AEROSPACE ENGINEERING, HUMAN FACTORS ENGINEERING. *Educ:* Univ Calif, BS, 40. *Prof Exp:* Eng test pilot, Ames Aeronaut Lab, Nat Adv Comt Aeronaut, 45-51, chief res pilot & asst chief flight, 51-57, chief flight opers br, Ames Res Ctr, NASA, 57-73, actg asst dir safety, NASA Hq, 69-70; consult, G E Cooper Assocs, 74-80; PRES, ERGODYNAMICS INC, 80- *Concurrent Pos:* Mem, NASA Res & Technol Subcomt on Aircraft Operating Probs. *Honors & Awards.* Octave Chanute Award, Inst Aerospace Sci, Am Inst Aeronaut & Astronaut, 54, Wright Bros Lectureship in Aeronaut, 84; Adm Luis de Florez Award, Flight Safety Found, 66, Richard Hansford Burroughs Award, 71. *Mem:* Fel Soc Exp Test Pilots; fel Am Inst Aeronaut & Astronaut. *Res:* Flight and simulator research of advanced flight vehicles; handling qualities of aircraft and spacecraft; human factors; aviation safety; ergonomics; Cooper Harper Rating Scale. *Mailing Add:* 22701 Mt Eden Rd Saratoga CA 95070

COOPER, GEORGE EVERETT, b Tallahassee, Fla, June 29, 45; m 68; c 2. ANIMAL SCIENCE, AGRICULTURE. *Educ:* Fla A&M Univ, BS, 67; Tuskegee Inst, MS, 69; Univ Ill, Urbana, PhD(animal sci), 72. *Prof Exp:* Teaching asst animal nutrit, Tuskegee Inst, 67-69; fel, Univ Ill, Urbana, 69-72; asst prof, Tuskegee Inst, 72-76; animal nutritionist res & develop, Winrock Int Livestock Res & Training Ctr, Morrilton, Ark, 76-78; dean sch appl sci, Tuskegee Inst, 78-85; vpres acad affairs, Ala A&M Univ, Normal, 85-; PRIN ANIMAL NUTRITIONIST, USDA. *Mem:* Am Soc Animal Sci. *Res:* Animal nutrition; international development; nutritive value of agricultural wastes. *Mailing Add:* USDA/CSRS/PAS 901 D St SW Aerospace Ctr Rm 303 Washington DC 20250-2200

COOPER, GEORGE R, electrical engineering, for more information see previous edition

COOPER, GEORGE RAYMOND, b Denver, Colo, May 3, 16; m 43; c 1. PLANT ECOLOGY, PLANT PHYSIOLOGY. *Educ:* Univ Northern Colo, BS, 42; Iowa State Univ, MS, 48, PhD(plant ecol), 50. *Prof Exp:* Assoc prof, 50-59, PROF BOT, PLANT PHYSIOL & ECOL, UNIV MAINE, 59- *Mem:* AAAS; Am Soc Plant Physiol: Ecol Soc Am. *Res:* Remote sensing of the environment by aerial infrared photography; physiological changes in higher plants caused by fungicidal sprays. *Mailing Add:* Dept Bot Deering Hall Univ Maine Orono ME 04473

COOPER, GEORGE S, crop science, soil science; deceased, see previous edition for last biography

COOPER, GEORGE WILLIAM, b Mt Vernon, NY, Dec 16, 28; m 56; c 2. PHYSIOLOGY, HEMATOLOGY. *Educ:* NY Univ, AB, 49, PhD(physiol), 60; Columbia Univ, MA, 51. *Prof Exp:* Res asst radiobiol, Sloan-Kettering Inst, 57-59; USPHS fel physiol, Columbia Univ, 60-62; res assoc biol, NY Univ, 62-68; lectr, 62, from instr to asst prof, 62-72, ASSOC PROF BIOL, CITY COL NEW YORK, 73- *Mem:* AAAS; Am Soc Hemat; Am Soc Physiol; Soc Exp Biol & Med; Harvey Soc; Int Soc Exp Hematol. *Res:* Endocrine regulation of blood cell formation and release; effects of ionizing radiations on hematopoietic system and fetal development; action of hormones on enzyme systems; blood coagulation. *Mailing Add:* Dept Biol City Col New York Convent Ave New York NY 10031

COOPER, GERALD RICE, b Scranton, SC, Nov 19, 14; m 46; c 3. PHYSICAL CHEMISTRY, MEDICINE. *Educ:* Duke Univ, AB, 36, AM, 38, PhD(phys chem), 39, MD, 50. *Prof Exp:* Res assoc, Med Sch, Duke Univ, 39-50; intern & resident, US Vet Admin Hosp, Atlanta, 50-52; chief phys chem lab, 52-54, chief hemat & biochem sect, 54-63, chief med lab sect, 63-70, chief clin chem & hemat br, 70-72, chief metab biochem br, 72-78, RES MED OFFICER, CTR DIS CONTROL, USPHS, 78- *Concurrent Pos:* Fel, Med Sch, Duke Univ, 39-50. *Honors & Awards:* Hektoen Silver Medal, AMA, 54, Billings Silver Medal, 56; Commendation Medal, USPHS, 64; Am Asn Clin Chem Fisher Award, 75; Except Achievement Award, USPHS, 84; William C Watson Medal of Excellence, 89. *Mem:* AAAS; Am Asn Clin Chem (pres, 84); Am Soc Clin Path; Am Chem Soc; Soc Exp Biol & Med. *Res:* Electrophoresis; ultracentrifuge; diffusion; viscosity and electron microscopy of proteins and viruses; liver diseases and protein metabolism; lipids; standardization. *Mailing Add:* Div Lab Sci Ctr Environ Health Ctr Dis Control MS F20 Atlanta GA 30333

COOPER, GLENN ADAIR, JR, b Chicago, Ill, Aug 17, 31; m 56; c 3. WOOD SCIENCE & TECHNOLOGY, FORESTRY. *Educ:* Iowa State Univ, BS, 53, MS, 59; Univ Minn, PhD, 70. *Prof Exp:* Prin wood scientist & proj leader, Forestry Sci Lab, Southern Ill Univ, Carbondale, 59-74; staff res forest prods technologist, 74-76, asst to dep chief, US Forest Serv, Washington, DC, 76-79, DEP DIR, PACIFIC NW FOREST & RANGE EXP STA, US FOREST SERV, PORTLAND, ORE, 79- *Concurrent Pos:* Mem, Mt St Helens Sci Adv Bd, 82- *Mem:* Forest Prod Res Soc; Soc Am Foresters; Sigma Xi. *Res:* Wood-moisture relations; physical properties of wood; treatments and processes which alter wood properties; utilization of hardwoods; wood residues for energy. *Mailing Add:* 44755 NW Elk Mountain Rd Banks OR 97106

COOPER, GUSTAV ARTHUR, b College Point, NY, Feb 9, 02; m 30; c 2. PALEOBIOLOGY. *Educ:* Colgate Univ, BS, 24, MS, 26, DSc(geol), 53; Yale Univ, PhD(geol), 29. *Hon Degrees:* DSc, Colgate Univ, 53, Queens Col City Univ NY, 83. *Prof Exp:* Asst, Peabody Mus, Yale Univ, 28-29, res assoc, 29-30; from asst cur to cur invert fossils, US Nat Mus, 30-56, head cur dept geol, 56-63, chmn dept paleobiol, 63-67, sr paleobiologist, 67-72, emer paleobiologist, US Nat Mus Natural Hist, Smithsonian Inst, 72-87; RETIRED. *Honors & Awards:* Mary Clark Thompson Medal, 57, Nat Acad Sci, Daniel Giraud Elliott Medal, 79; Paleont Soc Medal, 64; Raymond C Moore Medal, Soc Econ Paleontogists & Mineralogists, 81; Penrose Medal, Geol Soc Am, 83; James Hall Medal, NY Geol Surv, 85. *Mem:* Fel Geol Soc Am; Paleont Soc (pres, 56-57). *Res:* Stratigraphy; paleontology; stratigraphy of the Hamilton group of New York; invertebrate fossils; investigations on modern and fossil Brachiopoda. *Mailing Add:* 3557 Springmoor Circle Raleigh NC 27615

COOPER, HENRY FRANKLYN, JR, b Augusta, Ga, Nov 8, 36; m 58; c 3. NUCLEAR WEAPONS EFFECTS, STRATEGIC SYSTEMS ANALYSIS. *Educ:* Clemson Univ, BS, 58, MS, 60; NY Univ, PhD(mech eng), 64. *Prof Exp:* Instr eng mech, Clemson Univ, 58-60; mem tech staff, Eng Mech Dept, Bell Tel Labs, Inc, NY, 60-64; proj officer eng mech, Air Force Weapons Lab, 64-67, sci adv, 67-72; mem sr tech staff & prog mgr, R & D Assocs, 72-79, dep div mgr, 82-83; dep strategic & space systs, Off Asst Secy, Air Force Res & Develop, 79-82; asst dir, 83-85, dep negotiation, 85-87, CHIEF NEGOTIATOR, DEFENSAND SPACE ARMS, US ARMS CONTROL & DISARMAMENT AGENCY, 88- *Concurrent Pos:* Mem var nat comts, Defense Sci Bd & Air Force Sci Adv Bd task forces; mem bd adv, Col Eng, Clemson Univ. *Mem:* Am Soc Mech Engrs; AAAS; Am Soc Civil Engrs; Am Inst Aeronaut & Astronaut; NY Acad Sci; US Strategic Inst. *Res:* Nuclear weapons effects; blast and shock phenomena; protective structures; wave propagation; structural dynamics; heat transfer; infrared transmission through the atmosphere; numerical methods; applied mathematics; operations research; technical management; arms control; negotiator for space and defense systems. *Mailing Add:* 7103 Holyrood Dr McLean VA 22101

COOPER, HERBERT A, b Grand Junction, Colo, Feb 21, 38; m 63; c 2. EXPERIMENTAL PATHOLOGY, PEDIATRICS. *Educ:* Univ Kans, BA, 60, MD, 64. *Prof Exp:* Intern, Charles T Miller Hosp, St Paul, Minn, 64-65; resident pediat, Mayo Grad Sch Med, 65-67, assoc consult, Mayo Clin, 67, resident pediat hemat, Mayo Grad Sch Med, 69-71; fel exp path, 71-74, from asst prof to assoc prof, 74-78, PROF PEDIAT & PATH, SCH MED, & DIR DIV PEDIAT HEMAT ONCOL UNIV NC, CHAPEL HILL, 85- *Concurrent Pos:* NIH res career develop award, Nat Heart, Lung & Blood Inst; mem coun thrombosis, Am Heart Asn; vis res prof, Theodor Kocher Inst & adj prof hemat, Inselspital, Univ Bern, Switz, 77-78; Am Bd Pediat, Inc; prin investr, Children's Cancer Study Group. *Mem:* Am Soc Hemat; Acad Clin Lab Physicians & Scientists; Am Soc Exp Pathologists; Int Soc Hemostasis & Thrombosis; Soc Pediat Res. *Res:* Hemostasis and thrombosis; the biochemistry of factor VIII and von Willebrand factor; interaction of platelets with von Willebrand and characterization of the platelet receptor for von Willebrand factor. *Mailing Add:* Dept Pediat CB No 7220 Burnette-Womack Bldg Chapel Hill NC 27599

COOPER, HOWARD GORDON, b Joliet, Ill, Feb 16, 27; m 53; c 2. RESEARCH ADMINISTRATION, ELECTROOPTICS. *Educ:* Univ Ill, BS, 49, MS, 50, PhD(physics), 54. *Prof Exp:* Mem tech staff, Bell Tel Labs, Inc, 54-70; DIR RES, RECOGNITION EQUIP INC, 70- *Mem:* Am Phys Soc; Optical Soc Am; Inst Elec & Electronics Eng; Pattern Recognition Soc; Sigma Xi; World Future Soc. *Res:* Technology forecasting; optical scanning; pattern recognition; image processing and information processing systems. *Mailing Add:* Recognition Equip Inc PO Box 660204 Dallas TX 75266-0204

COOPER, HOWARD K, b Chicago, Ill, Jan 26, 34; m 55; c 3. BIOMEDICAL ENGINEERING, TECHNICAL MANAGEMENT. *Educ:* Chicago Tech Col, BSEE, 55; Pac States Univ, MSEE, 65. *Hon Degrees:* ScD, Pac States Univ, 91. *Prof Exp:* Pres, Nucleonic Prod, 59-73 & Pacesetter, Inc, 84-87; group exec, Mica Corp, 73-76; chief exec officer, Ailtech Div, Eaton Corp, 76-84; pres & chief exec officer, Trimedyne, Inc, 87-91; CHMN, CHIEF EXEC OFFICER & PRES, CARDIAC SCI, INC, 91- *Concurrent Pos:* Assoc prof, Pierce Col, 63-66, & Pac States Univ, 80-84. *Mem:* Inst Radio Engrs; Inst Elec & Electronics Engrs; Am Asn Med Instrumentation; Asn Old Crows. *Res:* Life sciences and development; entrepreneurial organization and development of companies to develop and supply therapeutic medical devices on a worldwide basis. *Mailing Add:* 9975 Toledo Irvine CA 92718

COOPER, JACK LORING, polymer chemistry; deceased, see previous edition for last biography

COOPER, JACK ROSS, b Ottawa, Can, July 26, 24; nat US; m 51; c 3. PHARMACOLOGY. *Educ:* Queen's Univ, Ont, BA, 48; George Washington Univ, MA, 52, PhD(biochem), 54. *Prof Exp:* Asst, NY Univ Res Serv, Goldwater Mem Hosp, 48-50; from instr to assoc prof, 56-71, PROF PHARMACOL, YALE UNIV, 71- *Concurrent Pos:* Fel, Pub Health Res Inst, NY, 54-56; Wellcome travel grant, Eng, 59; NIH spec fel & vis scientist, Maudsley Hosp, London, 65-66 & Nat Inst Med Res, 78. *Mem:* Am Soc Pharmacol & Exp Therapeut; Int Soc Neurochem; Am Soc Neurochem; Soc Neurosci. *Res:* Acetylcholine; neurochemistry; neuropharmacology. *Mailing Add:* Dept Pharmacol Yale Univ Sch Med New Haven CT 06510

COOPER, JAMES ALFRED, b Twin Falls, Idaho, Dec 20, 42; c 2. AVIAN ECOLOGY. *Educ:* Univ Wash, BS, 66; Univ Mass, Amherst, MS, 69, PhD(biol), 73. *Prof Exp:* asst prof, 72-79, ASSOC PROF WILDLIFE ECOL, UNIV MINN, 79- *Mem:* Am Ornith Union; Wilson Ornith Soc; Wildlife Soc; Sigma Xi; AAAS. *Res:* Avian ecology, bioenergetics, incubation and nesting behavior, anatid reproduction and survival. *Mailing Add:* Dept Fisheries & Wildlife Univ Minn St Paul MN 55108

COOPER, JAMES BURGESS, b New Haven, Conn, Nov 1, 54; m 79. MOLECULAR & CELL BIOLOGY. *Educ:* Lebanon Valley Col, BS, 77; Wash Univ, PhD(biol), 82. *Prof Exp:* Fel, Arco Plant Cell Res Inst, 83-86; res assoc, Dept Biol Sci, Stanford Univ, 86-88; ASST PROF, DEPT BIOL SCI, UNIV CALIF, SANTA BARBARA, 88- *Concurrent Pos:* Postdoctoral fel, Arco Plant Cell Res Inst, 83-86; NSF presidential young investr award. *Mem:* Am Soc Plant Physiologists; Am Soc Cell Biol; Int Soc Plant Molecular Biol. *Res:* Formation of symbiotic root modules is used to investigate cellular and molecular mechanisms regulating plant growth development and plant microbe interactions particular role of extracellular matrix and photohormones. *Mailing Add:* Dept Biol Sci Univ Calif Santa Barbara CA 94106

COOPER, JAMES ERWIN, b Waxahachie, Tex, Aug 30, 33; m 54; c 2. ORGANIC GEOCHEMISTRY. *Educ:* NTex State Univ, BS, 54, MS, 55; Rice Univ, PhD(chem), 59. *Prof Exp:* Chemist, Core Labs, Inc, 55-56; sr res technologist, Field Res Labs, Socony Mobil Oil Co, Inc, 59-68, res assoc explor-prod res div, Mobil Res & Develop Corp, 68-73; ADJ ASSOC PROF GEOL, UNIV TEX, ARLINGTON, 74- *Concurrent Pos:* Lectr, Dallas Baptist Col, 69, 74-75. *Mem:* Am Chem Soc; Sigma Xi. *Res:* Organic reaction mechanisms; organic geochemistry and synthesis. *Mailing Add:* 2423 Bonnywood Lane Dallas TX 75233

COOPER, JAMES WILLIAM, b Buffalo, NY, Feb 7, 43; m 69; c 2. CHEMICAL INSTRUMENTATION, ORGANOMETALLIC CHEMISTRY. *Educ:* Oberlin Col, AB, 64; Ohio State Univ, MS, 67, PhD(chem), 69. *Prof Exp:* Instr chem, State Univ NY, Buffalo, 69-70; Nmr appln programmer, Digital Equip Co, 70-71; anal appln mgr, Nicolet Instrument Corp, 71-74; asst prof chem, Tufts Univ, 74-80; vpres software develop, 84-87, res staff mem, 88-89, VPRES SOFTWARE DEVELOP, BRUKER INSTRUMENTS, 80-, MGR LAB AUTOMATION, IBM, T J WATSON RES CTR, YORKTOWN HEIGHTS, NY, 89- *Mem:* Am Chem Soc. *Res:* Fourier transform Nmr; computers in chemistry; organometallic chemistry; non-benzenoid aromatic and pseudoaromatic systems. *Mailing Add:* 48 Old Driftway Wilton CT 06897

COOPER, JANE ELIZABETH, b Bethlehem, Pa, June 2, 37. GENETICS. *Educ:* Lindenwood Col, BA, 59; Univ Pa, PhD(zool), 65. *Prof Exp:* From instr to asst prof biol, Drexel Inst, 64-67; asst prof, 67-73, ASSOC PROF BIOL, PA STATE UNIV, 73- *Mem:* AAAS; Genetics Soc Am; Asn Adv Health Professions; Sigma Xi. *Res:* Swimming rates in Paramecium aurelia; control of protein system in Paramecium aurelia; host-endosymbiont interactions in Paramecium biaurelia; ethical issues in genetic counselling and genetic engineering. *Mailing Add:* Dept Biol Pa State Univ Media PA 19063

COOPER, JOHN (HANWELL), b Tynemouth, Eng, Mar 15, 22; Can citizen; m 53; c 2. HUMAN PATHOLOGY, HISTOCHEMISTRY. *Educ:* Glasgow Univ, MB & ChB, 45; FRCPath, 69. *Prof Exp:* Registr path, Victoria Infirmary, Scotland, 49-56; assoc, Gen Hosp, St John's Nfld, 56-57; dir, Glace Bay Hosps, NS, 57-62; assoc, Path Inst, Dept Pub Health, Prov NS, 62-63, dir anat path, 63-65; assoc prof, 62-66, PROF PATH, DALHOUSIE UNIV, 66-; DEP HEAD, DEPT PATH, VICTORIA GEN HOSP, 75- *Concurrent Pos:* Med dir, Path Inst, Dept Pub Health, Prov NS, 65-74. *Mem:* AAAS; Histochem Soc; Int Acad Path; Can Med Asn; Can Asn Path. *Res:* Connective tissue histochemistry and pathology, especially elastic sheath-elastofibril system, amyloid, splenic follicular hyaline, actinic elastosis and protein histochemistry. *Mailing Add:* 3196 Mayfield Ave Halifax NS B3L 4B4 Can

COOPER, JOHN (JINX), b Norwich, Eng, Nov 30, 37; m 62; c 3. ATOMIC PHYSICS, PLASMA PHYSICS. *Educ:* Cambridge Univ, BA, 59, MA, 63; Imp Col, London, dipl, 61, PhD(physics), 62. *Prof Exp:* Asst lectr physics, Imp Col, London, 62-63, lectr, 63-65; from asst prof to assoc prof, 65-70, mem, 65-67, fel, Joint Inst Lab Astrophys, 67-80, PROF PHYSICS & ASTROPHYS, UNIV COLO, BOULDER, 70- *Concurrent Pos:* Consult, Atomic Energy Auth, UK, 63-65 & Radio Stands Lab, Nat Bur Stands, Colo, 68- *Mem:* Fel Am Inst Physics; Am Phys Soc. *Res:* Experimental and theoretical interests in the radiation from hot gases of laboratory and astrophysical importance; line broadening, scattering of radiation and other aspects of plasma and laser spectroscopy. *Mailing Add:* Joint Inst Lab Astrophys Box 390 Univ Colo Boulder CO 80309

COOPER, JOHN ALLEN DICKS, b El Paso, Tex, Dec 22, 18; m 44; c 4. HOSPITAL ADMINISTRATION. *Educ:* NMex State Univ, BS, 39; Northwestern Univ, PhD(biochem), 43, MB, 50, MD, 51. *Hon Degrees:* Dr, Univ Brazil, 58, George Washington Univ, 83, Jefferson Med Col, 84; LLD, NMex State Univ, 71; DSc, Northwestern Univ, 72, Duke Univ, 73, Med Col Ohio, 74, Med Col Wis, 78, NY Med Col, 81, Georgetown Univ, 86; DMedSc, Med Col Pa, 73; DSc, Wake Forest Univ, 85; LHD Bayler Med Col, 87. *Prof Exp:* From instr to prof biochem & dean sci, Med Sch, Northwestern Univ, 43-69; pres, 69-86, EMER PRES, ASN AM MED COLS, 86- *Concurrent Pos:* Consult, Vet Admin Res Hosp, Chicago, 54-69, dir radioisotope serv, 54-65; vis prof, Univ Brazil, 56 & Univ Buenos Aires, 58; mem comt licensure, US AEC, 56-59, mem adv comt educ & training, Div Biol & Med, 57-63; mem policy adv bd, Argonne Nat Lab, 57-63, chmn rev comt, Div Biol & Med Res & Radiol Physics, 58-62; mem med adv comt, PR Nuclear Ctr, 59-60; mem bd pub health adv, State of Ill, 62-69, mem, Ill Legis Comn on Atomic Energy, 63-69, mem bd higher educ, 64-69, mem, Gov Sci Adv Coun, 65-69, chmn, Sci Adv Coun of Ill, 67-69; ed, J Med Educ, 62-71; mem adv comt on investigational drugs, Food & Drug Admin, 63-65, consult to adminr, 65-70; mem coun, Assoc Midwest Univs, 63-68, vpres bd dirs, 64-65, pres, 65-66; mem med adv comt, W K Kellogg Found, 63-68, mem Latin Am adv comt, 64-68; treas admin comt, Pan Am Fedn Asns Med Schs, 63-76; mem adv comt on health sci, Eng & Biotechnol & Int Fel Rev Panel, NIH, 64, adv coun health res facil, 65-69, spec consult to dir, 68-70, consult to div physician & health professions educ, Bur Health Manpower Educ, 70-73; chmn extramural educ surv comt, Mayo Found, 65-66; vpres, Argonne Univs Asn, 65-68, mem bd trustees, 65-69; adv to adminr int health manpower, Agency Int Develop, 66-71; chmn adv comt on study of training progs in gen med sci, Nat Acad Sci-Nat Res Coun, 66-71, mem adv comt off sci personnel, 67-70; mem adv comt for instnl rels, NSF, 67-71; mem res strengthening group, Spec Prog Res & Training in Tropical Diseases, WHO, 77-80; mem spec med adv comt, Veterans Admin, 81-86, distinguished physician, 88-; distinguished vis prof, Baylor Med Col, 87- *Honors & Awards:* Abraham Flexner Award, Asn Am Med Cols, 85; Honor Award, Dept Med Surg Vet Admin, 85. *Mem:* Nat Inst Med-Nat Acad Sci; hon mem Acad Med Inst

Chile; hon mem Am Hosp Asn; Am Soc Biochem & Molecular Biol; Asn Am Med Cols; hon fel Am Col Hosp Am; fel AAAS. *Res:* Educational administration; radiobiology. *Mailing Add:* 4118 N River St Arlington VA 22207

COOPER, JOHN C, JR, b Fullerton, Calif, Jan 16, 36; m 58; c 2. AUDIOLOGY. *Educ:* Auburn Univ, BS, 57; Wayne State Univ, MA, 65, PhD(audiol), 68. *Prof Exp:* Res asst audiol, Wayne State Univ, 65-67; asst prof, Vanderbilt Univ, 67-69; assoc prof, 69-80, PROF AUDIOL, MED SCH UNIV TEX, SAN ANTONIO, 80- *Concurrent Pos:* Assoc prof, Trinity Univ, 71-83. *Mem:* Am Acad Audiol; Am Speech & Hearing Asn; Acad Rehab Audiol (pres, 84); Southern Audiol Soc (pres, 79); Am Auditory Soc. *Res:* Hearing in the elderly; hearing screening; impedance audiometry. *Mailing Add:* 3619 Running Springs San Antonio TX 78261-9534

COOPER, JOHN NEALE, b San Antonio, Tex, May 25, 38; m 60, 78; c 2. PHYSICAL CHEMISTRY, INORGANIC CHEMISTRY. *Educ:* Calif Inst Technol, BS, 60; Univ Calif, Berkeley, PhD(chem), 64. *Prof Exp:* Lectr inorg chem, Makerere Univ Col, Uganda, 64-66; asst prof chem, Carleton Col, 66-67; from asst prof to assoc prof, 67-83, PROF CHEM, BUCKNELL UNIV, 83- *Concurrent Pos:* Vis asst prof, Univ Ill, Urbana, 71-72; fac res partic, Argonne Nat Lab, 75-76, Univ Cincinnati, 83-84. *Mem:* Am Chem Soc; Sigma Xi. *Res:* Kinetics and mechanisms of inorganic reactions. *Mailing Add:* Dept Chem Bucknell Univ Lewisburg PA 17837

COOPER, JOHN NIESSINK, b Kalamazoo, Mich, Feb 4, 14; m 36. PHYSICS. *Educ:* Kalamazoo Col, AB, 35; Cornell Univ, PhD(physics), 40. *Prof Exp:* Asst physics, Cornell Univ, 35-40; instr, Univ Southern Calif, 40-43; asst prof, Univ Okla, 43-46; from asst prof to prof, Ohio State Univ, 46-56; prof physics, Naval Postgrad Sch, 56-83; RETIRED. *Concurrent Pos:* Res physicist, Radiation Lab, Univ Calif, 44-45; staff mem, Sandia Corp, 51-54; consult, Ramo-Woolridge Corp, 55-58, Space Tech Labs, 58-66 & Kaman Nuclear, 65. *Mem:* Fel Am Phys Soc; Am Asn Physics Teachers; sr mem Inst Elec & Electronics Engrs. *Res:* X-rays; nuclear spectroscopy; stopping of protons; superconductivity. *Mailing Add:* 1716 Wateredge Dr Naples FL 33963

COOPER, JOHN RAYMOND, b Lafayette, Ala, Jan 2, 31; m 52; c 2. PHYSICS. *Educ:* Auburn Univ, BS, 52, PhD, 70; Ohio State Univ, MS, 55. *Prof Exp:* Sr physicist, Aircraft Nuclear Propulsion Dept, Gen Elec Co, 58-60; res physicist, Southern Res Inst, 60-65, sr physicist, 65-66; mem staff dept physics, 66-71, DIR NUCLEAR SCI CTR & ASST PROF PHYSICS, AUBURN UNIV, 71- *Mem:* Am Asn Physics Teachers. *Res:* Neutron scattering; scattering; neutron induced reactions; nuclear instrumentation; reactor physics; electrooptics. *Mailing Add:* Dept Physics Auburn Univ Main Campus Auburn AL 36849

COOPER, JOHN WESLEY, b Delta, Colo, Sept 7, 46; m 73. EXPERIMENTAL HIGH ENERGY PHYSICS. *Educ:* Univ Colo, Boulder, BA, 68; Univ Mich, Ann Arbor, MA, 69, PhD(physics), 75. *Prof Exp:* Res asst, Bubble Chamber Group, Univ Mich, 71-75; res assoc exp high energy physics, Univ Ill, 75-77, vis res asst prof, 77-80; asst prof physics, Univ Pa, 80-85; assoc scientist, 85-87, scientist I, 87-90, SCIENTIST II, FERMILAB, ILL, 90- *Mem:* Sigma Xi; Am Phys Soc. *Res:* Experimental elementary particle research using the Collider Detector at Fermilab and the Solenoid Detector at the Superconducting Super Collider Laboratory. *Mailing Add:* Fermilab MS 318 PO Box 500 Batavia IL 60510

COOPER, JOSEPH E, b Philadelphia, Pa, May 14, 21; m 54; c 5. BIOCHEMISTRY, SANITARY ENGINEERING. *Educ:* Lincoln Univ, Pa, AB, 49. *Prof Exp:* Res chem pollution, City Philadelphia, 54-65, sanitary eng water resources, 65-66; res chem tobacco, 66-71, RES BIOCHEM HIDES & LEATHER, SEA SCI & EDUC ADMIN, USDA, 71-, PROJ LEADER TANNERY POLLUTION, 71- *Mem:* Prof Black Chemists & Chem Engrs; Am Leather Chemists Asn; Nat Tech Asn. *Res:* Determination of carcinogenic fractions in cigarette smoke; recovery of by-products from tannery and industrial wastes for conversion to marketable items; pilot plant design and scale-up to plant size industrial waste treatment. *Mailing Add:* Sea Sci & Educ Admin USDA 600 E Mermaid Lane Wyndmoor PA 19118

COOPER, KEITH EDWARD, b Frome, Eng, Aug 7, 22; m 46; c 2. PHYSIOLOGY. *Educ:* Univ London, MB, BS, 45, BSc, 48, MSc, 50; Oxford Univ, MA, 60, DSc(physiol), 70. *Prof Exp:* Resident, St Mary's Hosp, Univ London, 45-46, lectr physiol, 46-48; mem div human physiol, Med Res Coun Labs, Eng, 50-54; from founding mem to dir, Med Res Coun Body Temperature Res Unit, London & Oxford, 54-69; head div, 69-78, assoc acad vpres, 78-79, prof & head med physiol, Fac Med, 69-78, vpres res, 79-85, prof med physiol, 85-88, EMER PROF MED PHYSIOL, FAC MED, UNIV CALGARY, 88- *Concurrent Pos:* Vis lectr, Vet Admin Hosp, Cincinnati, Ohio, 66; vis lectr clin invest, Yale Univ, 66; vis lectr biophys, Univ Western Ont & Rutgers Univ, 66; consult, Cowley Rd Hosp, Oxford, Eng, 66; examr, Oxford Univ, Univ WI & Queen's Univ, Belfast, 66; mem panel arctic med & climatic physiol, Can Defense Res Bd, 70-78; mem bd dirs, Alberta Res Coun, 79-85; vis prof, Univ Ife, Nigeria, 76 & Univ Giessen, Fed Repub Ger, 85. *Honors & Awards:* Sarazzin Lectr, Can Physiol Soc, 86. *Mem:* Brit Physiol Soc; Can Physiol Soc (vpres, 74, pres, 75-76); Am Physiol Soc; Brit Med Res Soc; Can Fedn Biol Sci (vchmn, 78-79, pres, 79-80); Sigma Xi. *Res:* Body temperature regulation; mechanism of fever; published numerous articles in various journals. *Mailing Add:* Med Physiol Univ Calgary 2500 University Dr NW Calgary AB T2N 1N4 Can

COOPER, KEITH RAYMOND, b Portsmouth, Va, Dec 7, 51; m 73; c 2. AQUATIC TOXICOLOGY, COMPARATIVE TOXICITY & PATHOLOGY. *Educ:* Col William & Mary, BS, 73; Tex A&M, MS, 76; Univ RI, PhD(animal path), 79; Thomas Jefferson Univ, MS, 81. *Prof Exp:* Nat Inst Environ Health Sci fel res, Thomas Jefferson Univ, 79-81; asst prof, 81-86, ASSOC PROF TOXICOL, RUTGERS STATE UNIV, 86- *Mem:* Soc Toxicol; Soc Environ Toxicol & Chem; Soc Invert Path; Nat Shellfisherics Soc; Sigma Xi. *Res:* Aquatic toxicology; comparative toxicity and pathology; development of animal models to study the mechanism of chemical agents; xenobiotic metabolism in aquatic animals. *Mailing Add:* Dept Toxicol & Pharmacol Col Pharm Rutgers State Univ Piscataway NJ 08855

COOPER, KENNETH WILLARD, b Flushing, NY, Nov 29, 12; m 37; c 2. CYTOGENETICS, ENTOMOLOGY. *Educ:* Columbia Univ, AB, 34, AM, 35, PhD(cytol), 39. *Hon Degrees:* MA, Dartmouth Col, 62. *Prof Exp:* Asst zool, Columbia Univ, 34-37; instr, Univ Rochester, 38-39; from instr to assoc prof biol, Princeton Univ, 39-53; prof & chmn dept, Univ Rochester, 53-57; grad res prof, Univ Fla, 57-59; prof cytol, Dartmouth Med Sch, 59-62, cytol & genetics, 62-67; prof biol, 67-82, EMER PROF BIOL, UNIV CALIF, RIVERSIDE, 82- *Concurrent Pos:* Guggenheim fel, Calif Inst Technol, 44, 45; vis lectr, Univ Colo, 50; mem vis comt, Brookhaven Nat Lab, 58-60, chmn vis comt, 60-61; mem gen training panel, NIH, 58-66, health res facilities, 66-69; mem FCP Coun, Smithsonian Inst, 72-76; consult, Energy Res & Develop Agency, 76-78. *Mem:* Fel Am Acad Arts & Sci; Int Soc Hymenopterists; Genetics Soc Am; Soc Study Evolution; Pac Coast Entomol Soc. *Res:* Experimental ecology; insect communication; interactions of bees and flowers; insect population biology; chromosome structure and segregation; interchromosomal effects; non-gametic functions of gametes; mechanisms of mitosis and meiosis; evolution; aculeate hymenoptera. *Mailing Add:* Dept Biol Univ Calif Riverside CA 92521

COOPER, LARRY RUSSELL, b Los Angeles, Calif, Sept 19, 34; m 65; c 5. ELECTRONICS, SOLID STATE PHYSICS. *Educ:* Univ Calif, Los Angeles, BS, 56; Ore State Univ, PhD(physics), 67. *Prof Exp:* Jr scientist physics, Edwards AFB, USAF, 56-57; res assoc, Ore State Univ, 63-67; Nat Acad Sci-Nat Res Coun res assoc, Cyclotron Lab, Naval Res Lab, 67-69; physicist, Nuclear Physics Br, 69-74, PHYSICIST, NUCLEAR PHYSICS PROG & ELECTRONICS & SOLID STATE SCI PROG, OFF NAVAL RES, 74- *Mem:* Am Phys Soc. *Res:* Electronics and solid state sciences; physics of electron and electrooptic devices; solid state surfaces and interfaces; radiation effects in electronic materials; surface analysis techniques; solid state physics. *Mailing Add:* Elec & Solid State Sci Prog Off Naval Res Code 1114SS Arlington VA 22217

COOPER, LEON N, b New York, NY, Feb 28, 30; m 69; c 2. NEURO-NETWORKS, LEARNING SYSTEMS. *Educ:* Columbia Univ, AB, 51, AM, 53, PhD(physics), 54. *Hon Degrees:* DSc, Columbia Univ, 73, Univ Sussex, 73, Univ Ill, 74, Brown Univ, 74, Gustavus Adolphus Col, 75, Ohio State Univ, 76 & Univ Marie Curie, Paris, 77. *Prof Exp:* NSF fel, Inst Advan Study, 54-55; res assoc physics, Univ Ill, 55-57; asst prof, Ohio State Univ, 57-58; from assoc prof to prof physics, 58-66, Henry Ledyard Goddard univ prof, 66-74, THOMAS J WATSON, SR PROF SCI, BROWN UNIV, 74- *Concurrent Pos:* Alfred P Sloan res fel, 59-66; John Simon Guggenheim Mem Found fel, 65-66; vis prof, var univs & summer schs; consult, var govt agencies, indust & educ orgns. *Honors & Awards:* Noble Prize, 72; Comstock Prize, Nat Acad Sci, 68. *Mem:* Nat Acad Sci; fel Am Phys Soc; fel Am Acad Arts & Sci; Am Philos Soc; Fedn Am Scientists; Soc Neurosci; AAAS; Sigma Xi. *Res:* Nuclear, low temperature and elementary particle physics; field theory; superconductivity; many body problems. *Mailing Add:* Dept Physics Brown Univ Providence RI 02912

COOPER, LOUIS ZUCKER, b Albany, Ga, Dec 25, 31; m 76; c 4. INFECTIOUS DISEASES, PEDIATRICS. *Educ:* Yale Univ, BS, 54, MD, 57. *Prof Exp:* USPHS fel, 61-63; instr med, Sch Med, Tufts Univ, 63-64; from instr to assoc prof pediat, Sch Med, NY Univ, 64-73; PROF PEDIAT, COL PHYSICIANS & SURGEONS, COLUMBIA UNIV, 73-; DIR PEDIAT SERV, ST LUKES-ROOSEVELT HOSP CTR, 73- *Concurrent Pos:* Career scientist, Health Res Coun, NY, 67-73; dir, Rubella Proj; consult, Bur Educ Handicapped, US Off Educ, 68-72; consult, President's Comt Ment Retardation, 70-; mem, NY State Comt Children, 70-74; mem & vchmn, Nat Adv Coun Develop Disabilities, HEW, 73-76. *Mem:* AAAS; fel Am Acad Pediat; Infectious Dis Soc Am; Am Pub Health Asn; Soc Pediat Res; Am Pediat Soc; Am Soc Virol. *Res:* Rubella; handicapped children; viral immunology and vaccines; chemotherapy of infectious diseases. *Mailing Add:* St Lukes-Roosevelt Hosp Pediat Serv Amsterdam Ave & 114th St New York NY 10023

COOPER, MARGARET HARDESTY, b St Louis, Mo, June 7, 44. ANATOMY, OTOLARYNGOLOGY. *Educ:* Drury Col, AB, 66; St Louis Univ, MS, 69, PhD(anat), 71. *Prof Exp:* Asst prof anat, Sch Med, Wayne State Univ, 71-78; ASSOC PROF OTOLARYNGOL & ANAT, ST LOUIS UNIV SCH MED, 78- *Mem:* Asn Res Otolaryngol; Am Asn Anat; Soc Neurosci; Cagal Club; Sigma Xi. *Res:* Neuroanatomy; light and electron microscopic studies of central nervous system nuclei; CT of themporal bone; CT of infratemporal fossa; reconstruction of tongue. *Mailing Add:* Dept Anat St Louis Univ Sch Med 1402 S Grand St Louis MO 63104

COOPER, MARTIN, b Chicago, Ill, Dec 26, 28; m; c 2. CELLULAR RADIO, RESEARCH & DEVELOPMENT MANAGEMENT. *Educ:* Ill Inst Technol, BS, 50, MS, 57. *Prof Exp:* Res engr data commun, Teletype Corp, 53-54; eng group leader commun, Motorola, Inc, 54-58, eng mgr mobile commun, 58-65, prod mgr portable commun, 65-67, opers mgr, 67-69, vpres & dir portable opers, 69-72, vpres & gen mgr systs div, 72-78, vpres & corp dir res & develop, 78-82; chmn, Cellular Bus Systs Inc, 82-86; CHMN, DYNA INC, 86- *Concurrent Pos:* Comt chmn, var tech comts, Electronics Industs Asn, 59-72; mem bd telecommun-comput appln, Nat Res Coun, 77-81, mem comt telecommun policy, Inst Elec Engrs, 78-82; mem Adv Bd, Univ Ill; vpres, IIT Alumni Bd. *Honors & Awards:* Centennial Medal, Inst Elec & Electronic Engrs, 84. *Mem:* Fel Inst Elec & Electronics Engrs (pres vehicular technol soc, 73 & 74); fel Radio Club Am. *Res:* Mobile and portable two-way radio communications; radio spectrum efficiency; quartz crystal material and resonator technology; data communication; management of technology. *Mailing Add:* 2704 Ocean Front Del Mar CA 92014

COOPER, MARTIN DAVID, b Los Angeles, Calif, Sept 8, 45; m 69; c 3. WEAK INTERACTION PHYSICS, PION NUCLEAR PHYSICS. *Educ:* Calif Inst Technol, BS, 67; Univ Md, PhD(physics), 71. *Prof Exp:* NSF fel, Univ Md, 70-71; res assoc, Univ Wash, 71-74; res assoc, 74-75, STAFF SCIENTIST, LOS ALAMOS NAT LAB, 75- *Concurrent Pos:* Vis scientist, CERN & Univ Basal, 83-84. *Mem:* Am Phys Soc. *Res:* Investigation of nuclear properties with pions; measurement of nuclear radii; observations of pion double charge exchange; discovery of the isovector monopole mode in nuclei; development of high resolution pi-zero spectrometer; best limits on neutrinoless muon decay; spokesman of MEGA experiment. *Mailing Add:* Los Alamos Nat Lab MS 846 Group MP-4 Los Alamos NM 87545

COOPER, MARTIN JACOB, b Detroit, Mich, June 27, 39; m 65. TECHNICAL RESEARCH & DEVELOPMENT MANAGEMENT. *Educ:* Univ Mich, BSE, 61, MS, 63; Brandeis Univ, PhD(physics), 67. *Prof Exp:* Nat Res Coun res assoc, Nat Bur Standards, 66-68, res assoc statist physics, 68-72, prog analyst, 72-74; Presidential interchange exec, Corp Res & Develop, Gen Elec Co, 74-75; Physicist & adv, Energy Res & Develop Admin, 75-77; dir strategic planning & anal, NSF, 77-79; mgr strategic planning, Occidental Petrol Res, 79-82; prin consult, Martin J Cooper & Assocs, 82-; TECH DIR, ROHRBACK TECH CORP. *Mem:* AAAS; Am Phys Soc; Am Chem Soc. *Res:* Many-body physics, phase transitions; science policy; research and development planning. *Mailing Add:* 6212 39th Ave NE Seattle WA 98115

COOPER, MARY WEIS, b St Louis Co, Mo, Aug 10, 42; m 68; c 2. OPERATIONS RESEARCH, APPLIED MATHEMATICS. *Educ:* Wash Univ, BA, 61, MS, 68, DSc(opers res), 70. *Prof Exp:* Analyst libr info retrieval, Monsanto Co, 59-61; sci programmer eng, McDonnell Automation Ctr, 62-64; res scientist, Advan Technol Ctr, 70-72; vis prof, 73, asst prof, 76-81, ASSOC PROF OPERS RES, SOUTHERN METHODIST UNIV, 81-, AT DEPT ENG MGT. *Concurrent Pos:* Consult, Degolyer-McNaughton Co, 70-71 & Advan Technol Ctr, 72-73. *Mem:* Opers Res Soc Am; Inst Mgt Sci; Am Women Sci; Sigma Xi. *Res:* Linear, non-linear and discrete programming. *Mailing Add:* Dept Math & Statist Southern Ill Univ Box 1653 Edwardsville IL 62026

COOPER, MAURICE ZEALOT, b Brooklyn, NY, Feb 19, 08; m 41; c 1. MEDICINE. *Educ:* Long Island Col Med, MD, 31. *Prof Exp:* Intern, Bethel Hosp, Brooklyn, NY, 31-32; physician, Health Dept, 33-36; physician, Vet Admin, 36-40, chief outpatient & reception serv, Vet Admin Hosp, Togus, Maine, 40-42, chief med officer, Vet Admin Regional Off, RI, 42-46, chief outpatient serv, New Eng Br, 46-49, chief plans & policy develop, Med Serv, Washington, DC, 49-56, dir med criteria ed bd, 56-59, asst dir, Med Serv, 59-62, chief-of-staff, Vet Admin Hosp, Seattle, Wash, 62-68, dir, Vet Admin Hosp, Tex, 68-69, dir, Domiciliary, Los Angeles Ctr, 69-70, dir, Vet Admin Outpatient Clin, 70-77; RETIRED. *Mem:* AMA; Am Col Physicians. *Res:* Residuals and evaluations of medical diseases; hospital and medical system administration, effects of malnutrition and other hardships on the morality and morbidity of former United States prisoners of war and civilian internees of World War II. *Mailing Add:* 11854 Darlington Ave 49771 Los Angeles CA 90049

COOPER, MAX DALE, b Hazlehurst, Miss, Aug 31, 33; m 60; c 4. PEDIATRICS, IMMUNOLOGY. *Educ:* Tulane Univ, MD, 57; Am Bd Pediat, dipl, 62. *Prof Exp:* Intern med, Saginaw Gen Hosp, 57-58; resident, Sch Med, Tulane Univ, 58-60, instr, 62-63; house officer, Hosp Sick Children, London, Eng, 60, res asst neurophysiol, 61; from instr to asst prof, Univ Minn, Minneapolis, 64-67; assoc prof microbiol, Univ Ala, Birmingham, 67-71, dir res, Rehab Res & Training Ctr, 68-70, dir, Cellular Immunobiol Unit, Tumor Inst, 76-87, DIR, DIV IMMUNOL, DEPT PEDIAT, UNIV ALA, BIRMINGHAM, 67-, PROF PEDIAT, 67-, PROF MICROBIOL, 71-, PROF, DEPT MED, 87- *Concurrent Pos:* Fel pediat allergy, Univ Calif, San Francisco, 61-62; Nat Tuberc Asn teaching traineeship award, Univ Minn, Minneapolis, 63-64; USPHS spec res fel, 64-66; sr scientist, Comprehensive Cancer Ctr, Univ Ala, Birmingham, 71-, Multipurpose Arthritis Ctr, 79-, Cystic Fibrosis Res Ctr, 81, dir, Cell Identification Lab, 87-, Ctr Interdisciplinary Res Immunol Dis, 87, Div Develop/Clin Immunol, 87-; assoc ed, J Immunol, 72-76, Arthritis & Rheumatism, 85-90, J Clin Immunol, 79-; vis scientist, Tumor Immunol Unit, Dept Zool, Univ Col London, 73-74, Inst Embryol, Nogent-Sur-Marne & Inst Pasteur, Paris, France, 84-85; mem, Comt Classification of Primary Immunodeficiencies, WHO, 73-, Comt Comp Aspects of Reprod Processes in Different Species, 73, Comt Eval Methods Human T & B Cell Regulation, 74; coun mem, Soc Pediat Res, 74-77, Am Asn Immunologists, 83-; mem, Immunobiol Study Sect, NIH, 74-78; mem, US-Japan Panel Immunol, 80-; ed, Current Topics Microbiol & Immunol, 81-; mem, Exec Cancer Comt, Med & Dent Staff Univ Ala Hosps, 82-; counr, Int Union Immunol Socs, 86-, Am Asn Immunologists, 83-86; investr, Howard Hughes Med Inst, Univ Ala, Birmingham, 88- *Honors & Awards:* Samuel J Meltzer Founder's Award, Soc Exp Biol & Med, 66; 3M Life Sci Award, 90; Sandoz Prize for Immunol, 90. *Mem:* Nat Acad Sci; Soc Pediat Res (vpres, 78); Am Soc Exp Path; Am Asn Immunol (pres, 88-89); Am Soc Clin Invest; AAAS; Sigma Xi; Am Asn Cancer Res; Am Pediat Soc; Fedn Am Scientists. *Res:* Developmental immunobiology with emphasis on B-cell and T-cell differentiation; clinical immunology with emphasis on immunodeficiency diseases and lymphoid malignancies. *Mailing Add:* Div Develop & Clin Immunol Univ Ala Med Ctr Rm 263TI UAB Sta Birmingham AL 35294

COOPER, MELVIN WAYNE, b Clovis, NMex, Jan 17, 43; m 69; c 3. PHILOSOPHY OF SCIENCE. *Educ:* Univ Tex, BA, 65, MD, 69; Tex Tech, MA, 89. *Prof Exp:* Asst prof, med, Southern Ill Univ, 77-78; assoc prof med, Tex Tech Univ, 81-89; prof med, Mich State Univ, 89-91; PROF MED, UNIV TEX, 91- *Concurrent Pos:* Consult, Inst Cardiol, Warsaw, Poland, 84-88; teching asst, Dept Philos, Tex Tech Univ, Lubbock, 88-89; assoc, Ctr Ethics & Humanities, Lansing, Mich, 89-91; dir element electrophysiol, Univ Tex Health Ctr, Tyler. *Mem:* Am Col Cardiol; Am Heart Asn; NY Acad Sci; AAAS; Am Col Chest Physicians; Am Philos Asn. *Res:* Fast-phase force frequency response; philosophic aspects of medical science; medical artificial intelligence. *Mailing Add:* PO Box 2003 Univ Tex Health Ctr Tyler TX 75710

COOPER, MILES ROBERT, b Elizabeth City, NC, Oct 21, 33; m 55; c 2. ONCOLOGY, HEMATOLOGY. *Educ:* Univ NC, Raleigh, BS, 55; Bowman Gray Sch Med, MD, 62. *Prof Exp:* From instr to assoc prof med, 67-75, PROF MED, BOWMAN GRAY SCH MED, 75- *Mem:* Am Col Physicians; Am Soc Hemat; Am Fedn Clin Res; Soc Clin Oncol. *Res:* Laboratory and clinical studies of the pathophysiology of leukocytes and platelets; evaluation of supportive therapies consisting of leukocyte and platelet transfusions and definitive therapy with various therapeutic programs in neoplastic disease. *Mailing Add:* Bowman Gray Sch Med 300 S Hawthorne Rd Winston-Salem NC 27103

COOPER, MORRIS DAVIDSON, b Norton, Va, Oct 27, 43; m 67; c 2. MEDICAL MICROBIOLOGY, IMMUNOLOGY. *Educ:* King Col, BA, 65; Tenn Technol Univ, MS, 67; Univ Ga, PhD(microbiol), 71. *Prof Exp:* Prof biol, Bluefield Col, 67-69; res fel microbiol, Harvard Univ, 71-73; asst prof, 73-79, ASSOC PROF MED MICROBIOL, SCH MED, SOUTHERN ILL UNIV, 80- *Concurrent Pos:* NIH fel, Sch Pub Health, Harvard Univ, 71-73; consult microbiol, Univ Ore, 74-76; vis assoc prof, Dept Med, Vanderbilt Univ, 81-82. *Mem:* Am Soc Microbiol; NY Acad Sci; AAAS; Sigma Xi; Fedn Am Soc Exp Biol. *Res:* Immune response to microbial antigens particularly Neisseria gonorrhoease; purification of microbial antigens; immune response to enzymes. *Mailing Add:* Dept Med Microbiol & Immunol Southern Ill Univ Sch Med 801 N Rutledge Box 19230 Springfield IL 62794-9230

COOPER, NEIL R, b Warrensburg, Mo, May 15, 34; m 61; c 4. BIOMEDICAL RESEARCH, VIROLOGY. *Educ:* Yale Univ, BA, 56, MD, 60. *Prof Exp:* PROF IMMUNOL, SCRIPPS CLIN & RES FOUND, 74- *Concurrent Pos:* Assoc ed, 71-75, 81-83, J Immunol, 83-87, ed complement, 83-87, ed in chief, 88- *Mem:* Am Fed Clin Res; Soc Clin Invest; Asn Am Physician; Am Asn Adv Sci; Am Soc Biochem & Molecular Biol; Am Asn Immunol; Am Soc Virol; Am Asn Pathol. *Res:* Immunobiology of the complement system; viral immunology; molecular events in infection of human cells by herpes viruses. *Mailing Add:* Dept Immunol Scripps Clin & Res Found 10666 N Torrey Pines Rd La Jolla CA 92037

COOPER, NORMAN JOHN, b Lurgan, Northern Ireland, Sept 18, 50; Brit citizen. SYNTHETIC INORGANIC & ORGANOMETALLIC CHEMISTRY. *Educ:* Balliol Col, Oxford Univ, BA, 73, DPhil, 76. *Prof Exp:* Res fel, 76-78, from asst prof to assoc prof chem, 78-83, LOEB ASSOC PROF NATURAL SCI, HARVARD UNIV, 83- *Concurrent Pos:* Fel, Alfred P Sloan Found, 82-86. *Mem:* Am Chem Soc; Royal Soc Chem; NY Acad Sci. *Res:* Synthetic and mechanistic organotransition metal chemistry; catalytic applications of transition metal complexes. *Mailing Add:* Chem Dept Univ Pittsburgh Pittsburgh PA 15260-0001

COOPER, NORMAN S, b Brooklyn, NY, Dec 23, 20; m 45; c 1. RHEUMATOLOGIC PATHOLOGY, IMMUNOPATHOLOGY. *Educ:* Columbia Univ, AB, 40; Univ Rochester, MD, 43; Am Bd Path, dipl, 52. *Prof Exp:* Asst path, Med Col, Cornell Univ, 45-46; asst microbiol, 49-50, instr microbiol, 50-51, from instr to assoc prof path, 51-67, PROF PATH, SCH MED & MEM GRAD FAC ARTS & SCI, NY UNIV, 67-; CHIEF LAB SERV, NY VET ADMIN MED CTR, 67- *Concurrent Pos:* Hoskins fel, 51-53; Polachek Found med res fel, 54-59; intern, NY Hosp, 44 & 48-49, asst res, 44-45, res pathologist, 45-46; consult pathologist, Bellevue Hosp, 67-; attend pathologist, Univ Hosp, 67- *Mem:* Am Asn Pathologists; Am Soc Cell Biol; Am Rheumatism Asn; Am Asn Immunol; Int Acad Path. *Res:* Metabolic idiosyncrasies of murine plasmacytoma; pathology of rheumatic diseases. *Mailing Add:* Dept Path NY Univ Vet Admin Med Ctr 408 First Ave New York NY 10010

COOPER, PAUL DAVID, pharmacology, organic chemistry, for more information see previous edition

COOPER, PAUL W, b New York, NY, Feb 25, 29. PATTERN RECOGNITION, TELECOMMUNICATIONS. *Educ:* Mass Inst Technol, BS, 50, MS, 51. *Prof Exp:* Mem tech staff, Bell Tel Labs, 53-56; sr engr, Appl Sci Div, Melpar, Inc, 58-60, staff consult, 60-62; eng specialist, Info Processing, Sylvania Electronic Syst, Gen Tel & Electronics Corp, 62-63, sr eng specialist, 63-66; prin staff engr, Lockheed Corp, 82-87; MEM TECH STAFF, AT&T BELL LABS, 87- *Concurrent Pos:* Res asst, Stanford Univ, 56-58; consult, 66-82; lectr, Prof State-of-the-Arts Prog, 70-82. *Mem:* Inst Elec & Electronics Engrs; Sigma Xi; Soc Mfg Engrs. *Res:* Adaptive systems; statistical pattern recognition; signal detection; information theory; communications systems; transistor circuitry and computers; pattern recognition applications; robotics; applied statistical techniques; design of experiments; telecommunications; machine vision; intelligent networks. *Mailing Add:* 138 Bodman Pl No 4 Red Bank NJ 07701-1017

COOPER, PETER B(RUCE), b Manchester, Conn, Mar 30, 36; m 60; c 2. CIVIL & STRUCTURAL ENGINEERING. *Educ:* Lehigh Univ, BS, 57, MS, 60, PhD(civil eng), 65. *Prof Exp:* Engr, Elec Boat Div, Gen Dynamics Corp, Conn, 57-58; res asst indust testing, Lehigh Univ, 58-60, res instr civil eng, 60-65, res asst prof, 65-66; from asst prof to assoc prof, 66-76, PROF CIVIL ENG, KANS STATE UNIV, 76- *Mem:* Am Soc Civil Engrs; Am Soc Eng Educ. *Res:* Experimental and analytical research on the behavior and load-carrying capacity of steel structures. *Mailing Add:* Dept Civil Eng Seaton Hall Kans State Univ Manhattan KS 66506

COOPER, PHILIP HARLAN, radiation biology, radiolgoical physics, for more information see previous edition

COOPER, R(OBERT) S(HANKLIN), b Kansas City, Mo, Feb 8, 32; m 56; c 2. ELECTRICAL ENGINEERING. *Educ:* Univ Iowa, BSEE, 54; Ohio State Univ, MSEE, 58; Mass Inst Technol, ScD(elec eng), 63. *Prof Exp:* From instr to asst prof elec eng, Mass Inst Technol, 59-72; asst dir defense res & eng, Dept Defense, 72-75; dep dir, NASA/Goddard Space Flight Ctr, 75-76, dir, 76-80; mem tech staff, Bell Tel Labs, Inc, 80-; DIR, DEFENSE ADVAN RES PROJ AGENCY. *Concurrent Pos:* Ford Found fel, 63-65; consult,

Lincoln Labs, Mass Inst Technol, 64-66, mem staff, 66-70, group leader, 70-72; mem bd gov, Nat Space Club, 76-; mem bd adv, Electronic & Aerospace Systs Conf, 77- *Mem:* Am Inst Aeronaut & Astronaut; Am Astronaut Soc; Inst Elec & Electronics Engrs. *Res:* Plasma physics; magnetohydrodynamics; systems engineering; electrical network synthesis; laser physics and systems. *Mailing Add:* Atlanic Areospace Elec Corp 6404 Ivy Lane Suite 300 Greenbelt MD 20770

COOPER, RALPH SHERMAN, b Newark, NJ, June 25, 31; m 56; c 2. THEORETICAL PHYSICS. *Educ:* Cooper Union, BChE, 53; Univ Ill, MS, 55, PhD(physics), 57. *Prof Exp:* Mem staff theoret physics, Los Alamos Sci Lab, Univ Calif, 57-65; chief scientist, Nuclear Lab, Donald W Douglas Labs, Douglas Aircraft Co, 65-69; staff mem, Los Alamos Sci Lab, Univ Calif, 69-70, alt group leader, 70-71, group leader 71-72, asst laser div leader, 72-75; dep dir, Radiation Physics Div, Physics Int Co, 75-80 & Res & Develop Div, 81-82; dir systs & appln, Inesco, Inc, 82-84; PRES, APOGEE RES CORP, 84-; ASSOC DEAN, SCH ENGR, CALIF STATE UNIV, 87- *Concurrent Pos:* Consult, Inst Defense Anal, Washington, DC, 62-65 & Douglas Aircraft Co, Calif, 63-65. *Mem:* Am Phys Soc; Am Nuclear Soc; Sigma Xi; Am Inst Aeronaut & Astronaut; Am Asn Artificial Intel. *Res:* Magnetic fusion, laser fusion and laser isotope enrichment; solid state physics; reactor physics; nuclear rocket propulsion; mission analysis; radiation effects; ion exchange column theory; integrated circuit fabrication. *Mailing Add:* 76 Santa Ana Ave Long Beach San Diego CA 90803

COOPER, RAYMOND DAVID, b Kansas City, Kans, Dec 13, 27; m 54; c 3. ENERGY, ENVIRONMENT. *Educ:* Univ Ill, BS, 51; Iowa State Col, MS, 54; Mass Inst Technol, PhD, 67. *Prof Exp:* Asst physics, Ames Lab, AEC, 53-54; nuclear physicist, Pioneering Res Div, Quartermaster Res & Eng Ctr, 54-62; chief linear accelerator sect, US Army Natick Labs, 62-70; radiation physicist, USAEC, 70-74; physicist, US Energy Res & Develop Admin, 74-75; asst dir integrated assessment, 75-78; consult, energy & environ policy, 78-79; res prof policy sci, Univ Md, Baltimore County, 79-83; sr staff officer, Nat Res Coun, Nat Acad Sci, 85-91; RETIRED. *Concurrent Pos:* Lectr, Tufts Univ, 56-61. *Mem:* AAAS; Sigma Xi. *Res:* Physical mechanisms in radiation biology; dosimetry; accelerators; radiation interactions; energy and environmental systems analysis; risk analysis. *Mailing Add:* 5220 Brittany Dr Apt 1507 St Petersburg FL 33715

COOPER, REGINALD RUDYARD, b Elkins, WVa, Jan 6, 32; m 54; c 4. ORTHOPEDIC SURGERY. *Educ:* Univ WVa, BA, 52, BS, 53; Med Col Va, MD, 55; Univ Iowa, MS, 60. *Prof Exp:* Resident surg, 56-57, resident orthop, 57-60, assoc, 62-65, from asst prof to assoc prof, 65-71, PROF ORTHOP SURG, UNIV IOWA, 71-, CHMN DEPT ORTHOP, 73- *Concurrent Pos:* Res fel orthop surg & anat, Johns Hopkins Hosp, 64-65; Am, Brit & Can Orthop Asns exchange fel, 69. *Honors & Awards:* Outstanding Orthop Res Award, Kappa Delta, 71. *Mem:* AMA; fel Am Col Surg; Am Acad Orthop Surg; Orthop Res Soc; Am Asn Mil Surg. *Res:* Electron microscopy of bone and skeletal muscle as related to disuse atrophy and regeneration. *Mailing Add:* Dept Orthop Surg Univ Iowa Iowa City IA 52240

COOPER, REID F, b Wash, DC, Oct 3, 55; m 82; c 2. CERAMIC MATERIALS SCIENCE, EXPERIMENTAL PETROLOGY & GEOPHYSICS. *Educ:* George Washington Univ, BS, 77; Cornell Univ, MS, 80, PhD(mat sci & eng), 83. *Prof Exp:* Sr res scientist, Corning Glass Works, 83-85; asst prof mat sci, 86-90, ASSOC PROF MAT SCI & GEOPHYS, UNIV WIS-MADISON, 90- *Concurrent Pos:* NSF presidential young investr, 87-92. *Mem:* Am Ceramic Soc; Am Geophys Union; Am Soc Metals Int; Mineral Soc Am. *Res:* Thermodynamic, kinetic and mechanical properties of ceramic, glass-ceramic and ceramic composite materials; physics of geological materials. *Mailing Add:* Dept Met Sci & Eng Univ Wis 1509 University Ave Madison WI 53706

COOPER, RICHARD GRANT, b New York, NY, Mar 8, 34. PHARMACOLOGY, PHYSIOLOGY. *Educ:* Univ Ky, BS, 56, MS, 60; Univ Tex, PhD(physiol), 64; Col Osteop Med & Surg, DO, 76. *Prof Exp:* From instr to asst prof physiol & agr chem, Univ Mo, 64-71, res assoc space sci res ctr, 65-71; assoc prof physiol, Col Osteop Med & Surg, 72-75; assoc prof clin sci, Okla Col Osteop Med, 77-83; PVT PRACT, OSTEOP MED, 83- *Concurrent Pos:* Res assoc, Mo Regional Med Program, 67-68. *Res:* Mammalian physiology; blood coagulation and fibrinolysis; hemorrhagic diseases; hemodynamics; depressed metabolism. *Mailing Add:* 7424 E 53rd Pl Tulsa OK 74145

COOPER, RICHARD KENT, b Detroit, Mich, Apr 13, 37; m 57, 86; c 3. ELECTROMAGNETICS, THEORETICAL PHYSICS. *Educ:* Calif Inst Technol, BS, 58, MS, 59; Univ Ariz, MS, 62, PhD(physics), 64. *Prof Exp:* Fulbright Scholar, Niels Bohr Inst, 64; prof physics & chmn dept, Calif State Univ, Hayward, 65-78; alt group leader, Storage Ring Technol Group, 76-82, GROUP LEADER, ACCELERATOR THEORY AND SIMULATION GROUP, LOS ALAMOS NAT LAB, 82- *Concurrent Pos:* Vis staff mem, Los Alamos Sci Lab, Univ Calif, 73-74, consult, 75-76; physicist, Lawrence Livermore Lab, Univ Calif, 74-75, consult, 75-76. *Mem:* Sigma Xi; Am Phys Soc. *Res:* Electromagnetic phenomena in charged particle beams; accelerator theory. *Mailing Add:* MS H829 Los Alamos Nat Lab PO Box 1663 Los Alamos NM 87545

COOPER, RICHARD LEE, b Rensselear, Ind, Feb 28, 32; m 52; c 4. PLANT BREEDING, PLANT GENETICS. *Educ:* Purdue Univ, BS, 57; Mich State Univ, MS, 58, PhD(plant breeding & genetics), 62. *Prof Exp:* Res assoc soybean breeding & genetics, Dept Agron & Plant Genetics, Univ Minn, 61-67; from assoc prof to prof agron, Univ Ill, Urbana-Champaign, 69-77, res leader, US Regional Soybean Lab, 67-76, prof plant breeding, 77; RES LEADER, OHIO AGR RES & DEVELOP CTR, US DEPT AGR SOYBEAN INVEST, WOOSTER, 77-; PROF PLANT BREEDING, OHIO STATE UNIV, COLUMBUS, 77- *Concurrent Pos:* Adv Coun, Potash & Phosphate Inst, 81-83. *Honors & Awards:* Soybean Res Recognition Award, Am Soybean Asn, 81. *Mem:* Am Soc Agron; Crop Sci Soc Am; Am Soybean Asn. *Res:* Solid seeding of soybeans with increases of 10-30% in yield; development of semidwarf soybean cultivars; maximum yield research in soybeans; breeding methodology (early generation testing); development of abiotic stress tolerant cultivars. *Mailing Add:* Agronomy 202 Kottman Hall Ohio State Univ Main Campus Columbus OH 43210

COOPER, ROBERT ARTHUR, JR, b St Paul, Minn, Aug 27, 32; m 59, 87; c 3. MEDICINE, ONCOLOGY. *Educ:* Univ Pa, AB, 54; Jefferson Med Col, MD, 58; Am Bd Path, cert path anat, 63. *Prof Exp:* From instr to prof path, Sch Med, Univ Ore, 62-69; from assoc prof to prof, 69-72, assoc dean curricular affairs, 69-72, head, Div Surg Path, 71-75, PROF ONCOL IN PATH & DIR, CANCER CTR, SCH MED, UNIV ROCHESTER, 74- *Concurrent Pos:* Am Cancer Soc fel, 60-61; teaching fel path, Harvard Med Sch, 62-63; Cancer Res Ctr grant, 75-; mem treatment comt, Breast Cancer Task Force, Nat Cancer Inst, 74-77, mem, Cancer Ctr Support Grant Rev Comt, 78-82, chmn, 81-82; mem jury, Albert & Mary Lasker Found Awards, 77; mem bd trustees, James P Wilmot Found, 81-; mem bd sci counselors, Div Cancer Prev & Control, Nat Cancer Inst, 83-85. *Mem:* Int Acad Path; AAAS; NY Acad Sci; Sigma Xi. *Res:* Experimental radiation pathology and toxicology; radiation carcinogenesis. *Mailing Add:* 99 S Main St Pittsford NY 14534

COOPER, ROBERT CHAUNCEY, b San Francisco, Calif, July 4, 28; m 56; c 4. INFECTIOUS DISEASE, WATER SUPPLY. *Educ:* Univ Calif, Berkeley, BS, 52; Mich State Univ, MS, 53, PhD(microbiol), 58. *Prof Exp:* from asst prof to assoc prof pub health, Sch Pub Health, 58-74, PROF, ENVIRON HEALTH SCI & DIR, SANIT ENG & ENVIRON HEALTH RES LAB, UNIV CALIF, BERKELEY, 80- *Mem:* Am Soc Microbiol. *Res:* Microbiological aspects of water quality; water quality and human health. *Mailing Add:* Sch Pub Health Univ Calif Berkeley CA 94720

COOPER, ROBERT MICHAEL, b Oakland, Calif, Oct 21, 39. PHARMACY. *Educ:* Univ Calif, San Francisco, PharmD. *Prof Exp:* Mem staff pharmaceut develop serv, Clin Ctr, NIH, 64-66; clin pharmacist, Long Beach Mem Hosp, Calif, 66; mem staff dose formulation unit, Cancer Chemother Serv Ctr, Nat Cancer Inst, 66-67; asst prof pharm, 67-71, actg chmn dept, 71-72, asst dean, 72-77 & 81-85, chmn dept, 78-82, ASSOC PROF PHARM, SCH PHARM, STATE UNIV NEW YORK BUFFALO, 71-, ASSOC DEAN, 85- *Mem:* Am Pharmaceut Asn; Am Soc Hosp Pharmacists; Am Soc Pharm Law. *Res:* Clinical pharmacy; sterile manufacturing. *Mailing Add:* Dept Pharm Cooke Hall Rm C126 State Univ NY Sch Pharm Buffalo NY 14260

COOPER, ROBIN D G, b Eastbourne, Eng, Sept 26, 38; m 70. ORGANIC CHEMISTRY, RESEARCH ADMINISTRATION. *Educ:* Univ London, BSc, 59, PhD(org chem), 62, DIC, 60; FRIC, DSc, 84. *Prof Exp:* Glaxo res fel org chem, Imp Col, Univ London, 62-63; Nat Acad Sci vis res assoc, US Army Natick Labs, 63-65; sr res chemist, 65-70, res scientist, 70-75, res assoc, 75-84, RES ADV, ELI LILLY & CO, 84- *Mem:* Am Chem Soc; fel Royal Soc Chem. *Res:* Structural determination and synthesis of antibiotics, especially penicillin and cephalosporins; B-lactam biosynthesis; author of numerous publications. *Mailing Add:* Dept MC 930 Lilly Corp Ctr Eli Lilly Co Bldg 88-2 Indianapolis IN 46285

COOPER, RONDA FERN, b Schenectady, NY, June 20, 43. MEDICAL MICROBIOLOGY, HEALING & PARAPSYCHOLOGY. *Educ:* Okla State Univ, BS, 64; Kans State Univ, MS, 66, PhD(microbiol), 71. *Prof Exp:* Res assoc & instr path, Kans State Univ, 67-73; asst prof biol, Univ NMex, 73-74; asst prof microbiol, La State Univ, Baton Rouge, 74-77; from asst prof to assoc prof microbiol, New Eng Col Osteop Med, 78-87; asst prof microbiol, Am Univ Caribbean, 88-90; MEM STAFF, RES & DEVELOP, CENT ARIZ VET LAB, 90- *Concurrent Pos:* Lalor Found fel, 75-; mem, Conf Pub Health Lab Dirs; consult, 87-88. *Mem:* Am Soc Microbiol; Inst Noetic Sci; Sigma Xi. *Res:* Interaction of microbiol toxins with host tissues and other aspects of host-parasite relationships; psychic and spiritual effects of bioplasmis energies on microbiol growth. *Mailing Add:* 4805 Westway Trail Amarillo TX 79109-6343

COOPER, SHELDON MARK, b New York, NY, Dec 5, 42; m 66; c 1. RHEUMATOLOGY, CLINICAL IMMUNOLOGY. *Educ:* Hobart Col, BS, 63; New York Univ, MD, 67; Am Bd Internal Med, dipl, 73; Subspecialty Bd Rheumatol, dipl, 76. *Prof Exp:* Teaching fel rheumatol & immunol, Med Ctr, New york Univ, 70-72; from asst prof to assoc prof, Dept Med, Sch med, Univ Southern Calif, 74-82; assoc prof med, 82-86, DIR RHEUMATOL & CLIN IMMUNOL UNIT, DEPT MED, COL MED, UNIV VT, 82-, PROF MED, 86- *Concurrent Pos:* Attend med, Los Angeles County-Univ Southern Calif Med Ctr, 74-82 & Med Ctr Hosp Vt, 82- *Mem:* Am Asn Immunologists; Am Col Rheumatism; Reticuloendothelial Soc; Am Fedn Clin Res. *Res:* Investigation of the immunologic mechanisms that are involved in the pathogenesis of autoimmune diseases such as rheumatoid arthritis and systemic lupus erythematosus. *Mailing Add:* Rheumatol & Immunol Unit Dept Med Univ Vt Given D-301 Burlington VT 05405

COOPER, STEPHEN, b Brooklyn, NY, Aug 6, 37; m 60; c 2. MICROBIOLOGY, GENETICS. *Educ:* Union Col, NY, BA, 59; Rockefeller Inst, PhD(microbiol), 63. *Prof Exp:* NSF res fels, Univ Inst Microbiol, Copenhagen Univ, 63-64 & Med Res Coun Microbial Genetics Res Unit, Univ London, 64-65; res assoc biol chem, Med Sch, Tufts Univ, 65-66; asst res prof pediat, State Univ NY, Buffalo, 66-70, asst res prof biochem, 67-70, lectr biol, 69-70; assoc prof, 70-78, PROF MICROBIOL, SCH MED, UNIV MICH, 78- *Concurrent Pos:* Fogarty Int fel, Imp Cancer Res Fund, London, 76. *Mem:* AAAS; Am Soc Microbiol. *Res:* Biochemistry and genetics of viruses; protein synthesis; control of DNA replication and cell division in bacteria; microbial genetics; development of dictyostelium. *Mailing Add:* Dept Microbiol Sch Med Univ Mich 1301 Catherine Rd Ann Arbor MI 48109-0620

COOPER, STEPHEN ALLEN, b Philadelphia, Pa, Apr 11, 46; m 71; c 2. ANALGESIOLOGY, ANXIETY AND PAIN CONTROL. *Educ:* Adelphi Univ, BS, 68; Univ Pa, DMD, 71; Georgetown Univ, PhD(pharmacol), 75. *Prof Exp:* Asst prof pharmacol, Georgetown Univ, 75-78; assoc dean res, Dent Sch, Univ Med & Dent, NJ, 78-85; assoc prof, oral surg & pharmacol, Sch Dent Med, Univ Pa, 86-90, dir, div pharmacol & therapeut, 87-90; ASSOC DEAN RES, ADVAN EDUC & CONTINUING EDUC, SCH DENT, TEMPLE UNIV, 90- *Concurrent Pos:* Vis prof, Georgetown Univ, 78-, Sch Dent Med, Univ Pa, 91-; clin prof, dept med, Med Col Pa; adv panel, USP. *Mem:* Am Soc Clin Pharmacol; Am Pain Soc; Am Asn Dent Res; Sigma Xi; Am Dent Asn. *Res:* Clinical evaluation of analgesic drugs; clinical evaluation of drugs used for intravenous sedation. *Mailing Add:* Sch Dent Temple Univ 3223 N Broad St Philadelphia PA 19140

COOPER, STUART L, b New York, NY, Aug 28, 41; m 65; c 2. BIOMATERIALS. *Educ:* Mass Inst Technol, BS, 63; Princeton Univ, MA, 65, PhD(chem eng), 67. *Prof Exp:* From asst prof to assoc prof, 67-74, chmn chem eng, 83-89, PROF, UNIV WIS-MADISON, 74- *Concurrent Pos:* Vis assoc prof, Univ Calif, Berkeley, 74; bd trustees, Argonne Univ Asn, 75-81; vis prof, Technion, Israel, 77. *Honors & Awards:* Clemson Award Basic Res, Soc Biomat, 87; Mat Eng & Sci Div Award, Am Inst Chem Engrs, 87. *Mem:* Am Chem Soc; fel Am Phys Soc; Am Inst Chem Engrs; Soc Plastic Engrs; Am Soc Artificial Internal Organs; Soc Biomat. *Res:* Structure-property relations of polymers including polyurethanes, ionomers and related block polymer systems; application of light and x-ray scattering, rheo-optical and thermal methods in polymer characterization; studies of protein and thrombus deposition on polymers used in biomedical applications; polyurethane for biomedical applications. *Mailing Add:* Dept Chem Eng Univ Wis-Madison 1415 Johnson Dr Madison WI 53706

COOPER, TERENCE ALFRED, b Oxford, Eng, Feb 8, 41; m 66; c 2. POLYMER CHEMISTRY. *Educ:* Oxford Univ, BA, 62, BSc, 64, DPhil(phys org chem) & MA, 66; Drexel Univ, MBA, 73. *Prof Exp:* Res staff chemist, Sterling Chem Lab, Yale Univ, 66-68; res chemist, Elastomers Res Lab, 68-80, RES ASSOC, POLYMER PROD DEPT, EXP STA, E I DU PONT DE NEMOURS & CO, INC, 80- *Mem:* Am Chem Soc; The Chem Soc; Am Inst Chem Engrs. *Res:* Materials science and polymer engineering; colloid chemistry; adhesives, sealant and coating. *Mailing Add:* 15 Stage Rd Newark DE 19711

COOPER, TERRANCE G, b Wyandotte, Mich, Aug 14, 42; c 2. IMMUNOLOGY, MICROBIOLOGY. *Educ:* Wayne State Univ, BSc, 65, MS, 67; Purdue Univ, PhD(biol), 69. *Prof Exp:* Fel biol, Mass Inst Technol, 69-71; asst prof biochem, Univ Pittsburg, 71-75, assoc prof biochem & genetics, 74-81, Audrey Avinoff prof biol, 81-84; VAN VLEET PROF MICROBIOL & IMMUNOL & CHMN DEPT, HEALTH SCI CTR, UNIV TENN, MEMPHIS, 85-, DIR, MOLECULAR RESOURCE CTR, 85- *Concurrent Pos:* Vis prof, Microbiol Lab, Univ Libre de Bruxelles, Belg, 75; vis scholar, Dept Biol, Univ Calif, Santa Barabara, 79-80; distinguished lectr, Concordia Univ, 82; vis heritage prof, Univ Alta, 86. *Honors & Awards:* Brown Hazen Award, Res Corp, 72. *Mem:* Am Soc Biol Chemists; Genetics Soc Am; Am Soc Microbiol; AAAS. *Res:* Regulation of gene expression; identification of the protein-nucleic interactions that mediate the control of transcription. *Mailing Add:* Dept Microbiol & Immunol Univ Tenn Ctr Health Sci 858 Madison Ave Memphis TN 38163

COOPER, THEODORE, b Trenton, NJ, Dec 28, 28; m 56; c 4. PHYSIOLOGY, PHARMACOLOGY. *Educ:* Georgetown Univ, BS, 49; St Louis Univ, MD, 54, PhD(physiol), 56. *Hon Degrees:* DSc, Col Osteopath Med & Surg, 76, Univ NMex 77, Univ Rochester, 79, Sch Med, Georgetown Univ, 77 & 81, Hahnemann, 83; LHD, Loyola Univ Chicago, 79. *Prof Exp:* Intern, St Louis Univ Hosps, 54-55; sr asst surgeon, Nat Heart Inst, 56-58, asst res cardiovascular surg, 58-59, mem staff clin surg, 59-60; from asst prof to prof surg, St Louis Univ, 60-66, mem bd grad studies, 62-66; prof pharmacol & surg & chmn dept pharmacol, Sch Med, Univ NMex, 66-68; dir, Nat Heart & Lung Inst, 68-74; dep asst secy health, 74-75, asst secy health, HEW, 75-77; dean med col & provost med affairs, Cornell Univ, 77-80; exec vpres, 80-84, vchmn bd, 84-87, CHMN BD & CHIEF EXEC OFFICER, UPJOHN CO, 87- *Concurrent Pos:* mem pharmacol & exp therapeut study sect, USPHS, 64-77; bd dir, Nat Health Coun, Inc, 76-85; expert adv panel cardiovasc dis, World Health Orgn, 77-83; med & sci adv coun, Nat Hemophilia Found, 77; vis fel, Woodrow Wilson Nat Fel Found, 78; bd trustees, Milton Helpern Libr Legal Med, 78; sci adv comt, Gen Motors Corp, 79-85; tech bd, Milbank Mem Fund,79-82; med adv comt, White House Conf on aging, 81-82. *Honors & Awards:* Harvey W Wiley Medal, Food & Drug Admin, Pub Health Serv, Silver Springs, Md, 76; Tom A Spies Mem Award, Int Acad Preventive Med, 77; Albert Lasker Award, 78; Am Col Prev Med Award, 84; Milton Helpern Libr Legal Med Award, 84. *Mem:* Inst Med-Nat Acad Sci; Am Physiol Soc; Am Col Cardiol; Am Fedn Clin Res. *Res:* Experimental and clinical cardiovascular physiology and pharmacology. *Mailing Add:* Chmn Bd UpJohn Co Kalamazoo MI 49001-0199

COOPER, THOMAS D(AVID), b Dayton, Ohio, Apr 7, 32; m 53; c 3. PHYSICAL METALLURGY. *Educ:* Univ Cincinnati, MetE, 55; Ohio State Univ, MS, 64. *Prof Exp:* Metall engr, Westinghouse Elec Corp, 55-56; proj engr light metals, USAF Mat Lab, 56-59, proj engr high temperature metals, 59-60, tech mgr, 60-64, high strength metals, 64-66, chief, Processing & Nondestructive Testing Br, 66-72, chief, Nondestructive Testing & Mech Br, 72-73, chief, Aeronaut Systs Br, 73-77, CHIEF MAT INTEGRITY BR, USAF MAT LAB, 77- *Mem:* Fel Am Soc Metals Int; assoc fel Am Inst Aeronaut & Astronaut; Am Inst Mining, Metall & Petrol Engrs; Sigma Xi; fel Am Soc Nondestructive Testing; Soc Automotive Engrs. *Res:* Promoting optimum applications of materials during development of aeronautical systems; operational organizations, including failure analysis, corrosion control, NDE, materials selection, specification and standards. *Mailing Add:* Mats Directorate Wright Lab/MLSA Wright-Patterson AFB OH 45433-6533

COOPER, THOMAS EDWARD, b Lindsay, Calif, May 31, 43; m 68; c 1. MECHANICAL ENGINEERING, BIOENGINEERING. *Educ:* Univ Calif, Berkeley, BS, 66, MS, 67, PhD(mech eng), 70. *Prof Exp:* Actg asst prof mech eng, Univ Calif, Berkeley, 70; asst prof, 70-76, ASSOC PROF MECH ENG, NAVAL POSTGRAD SCH, 76- *Concurrent Pos:* Consult, Lawrence Radiation Lab, 70- *Res:* Thermal modeling of heat transfer processes occurring in tissue surrounding cryosurgical and radio-frequency probes; biological thermal property determinations. *Mailing Add:* 1230 Buenua Vista Pacific Grove CA 93950

COOPER, TOMMYE, b Bandana, Ky, May 17, 38; m 61; c 1. AGRICULTURE, STATISTICS. *Educ:* Murray State Univ, BS, 60; Univ Ky, MS, 62, PhD(dairy sci, statist), 66. *Prof Exp:* Anal statistician, 66-67, chief methods & opers unit, Comput Lab, 67-68, actg dir, Data Systs Appln Div, 68-70, dir, 70-73, DIR, FT COLLINS COMPUT CTR, US DEPT AGR, 73- *Mem:* Am Dairy Sci Asn; Am Soc Animal Sci; Am Asn Comput Mach. *Res:* Dairy cattle reproductive and genetic research using digital computer and statistical techniques to achieve these goals. *Mailing Add:* 5109 Belle Isle Dr Dayton OH 45439-3225

COOPER, W(ILLIAM) E(UGENE), b Erie, Pa, Jan 11, 24; m 46; c 5. PRESSURE VESSELS, NUCLEAR SYSTEMS. *Educ:* Ore State Col, BS, 47, MS, 48; Purdue Univ, PhD(eng mech), 51. *Prof Exp:* Instr eng mech, Purdue Univ, 48-52; consult engr, Knolls Atomic Power Lab, Gen Elec Co, 52-63; eng mgr, Lessells & Assocs, Inc, 63-68; sr vpres & tech dir, Teledyne Mat Res Co, 68-76, CONSULT ENGR, TELEDYNE ENG SERV, 76- *Concurrent Pos:* Sr vpres codes & standards, Am Soc Mech Engrs, 81-84. *Honors & Awards:* William M Murray lectr, Soc Exp Stress Anal, 77; B F Langer Nuclear Codes & Standards Award, Am Soc Mech Engrs, 78; Pressure Vessel & Piping Award, Am Soc Mech Engrs, 83, Codes & Standards Medal, 86. *Mem:* Hon mem Am Soc Mech Engrs; Soc Exp Mech; Nat Acad Engrs. *Res:* Structural analysis and evaluation; application of material properties to design; pressure vessel and pressure piping codes and standards. *Mailing Add:* Teledyne Eng Serv 130 Second Ave Waltham MA 02254

COOPER, WALTER, b Clairton, Pa, July 18, 28; m 53; c 2. PHYSICAL CHEMISTRY. *Educ:* Washington & Jefferson Col, BA, 50; Univ Rochester, PhD(phys chem), 57. *Prof Exp:* From res chemist to sr res chemist, 56-66, RES ASSOC, EASTMAN KODAK CO, 66- *Mem:* AAAS; Am Chem Soc; Am Phys Soc; Sigma Xi. *Res:* Gas-phase kinetics; photographic theory; solid state chemistry of silver halides; luminescence properties of dyes. *Mailing Add:* 68 Skyview Lane Rochester NY 14625

COOPER, WILLIAM ANDERSON, b Archer City, Tex, Feb 4, 27; m 52; c 4. ZOOLOGY, PHYSIOLOGY. *Educ:* NTex State Univ, BS, 48, MS, 50; Tex A&M Univ, PhD(zool), 57. *Prof Exp:* Instr natural sci, Paris Jr Col, 53-54; asst biol, Tex A&M Univ, 54-55, instr, 55-57; assoc prof, 57-65, PROF BIOL, WTEX STATE UNIV, 65- *Concurrent Pos:* Res assoc, Agr & Mech Res Found, NIH, 55-57; consult water qual control, Can River Munic Water Authority, Nat Parks Serv, Ecology Audits, Inc, Landlocked Fisheries, Inc & Tex Parks & Wildlife, 71-74. *Mem:* Am Micros Soc; Am Soc Zoologists; Sigma Xi. *Res:* Effects of vitamin deficiencies on embryo development in white rats; limnology; water quality and fisheries biology research. *Mailing Add:* WTex State Univ Box 346 Canyon TX 79015

COOPER, WILLIAM CECIL, b Salisbury, Md, Apr 6, 09; m 32; c 2. PLANT PHYSIOLOGY. *Educ:* Univ Md, BS, 29; Calif Inst Technol, MS, 36, PhD, 38. *Prof Exp:* Jr pomologist, Bur Plant Indust, 29-38, assoc plant physiologist, Subtrop Fruit Prod, 38-42, plant physiologist, Trop Tree Crops Propagation, Office For Agr Relations, 43-44, citrus rootstock invest, Bur Plant Indust, 44-54, sr plant physiologist, Hort Crops Br, 55-59, leader citrus invests, Plant Sci Res Div, 59-77, COLLABR, SCI & EDUC ADMIN-AGR RES, USDA, 77- *Mem:* Am Soc Plant Physiol; Am Soc Hort Sci. *Res:* Plant hormones and root and flower formation; salt tolerance; cold hardiness; citrus rootstocks and citrus pheonology; fruit abscission; ethylene physiology. *Mailing Add:* 443 Lakewood Dr Winter Park FL 32789

COOPER, WILLIAM CLARK, b Manila, Philippines, June 22, 12; m 37; c 4. MEDICINE. *Educ:* Univ Va, MD, 34; Harvard Univ, MPH, 58; Am Bd Internal Med, Am Bd Prev Med & Am Bd Indust Hyg, dipl. *Prof Exp:* Intern & asst resident, Univ Hosp, Cleveland, 34-37; instr bact, Sch Hyg & Pub Health, Johns Hopkins Univ, 40; from asst surgeon to surgeon, NIH, 41-51, chief & med dir occup health field hqs, 52-57, chief epidemiol serv, Occup Health Prog, Bur State Serv, 57-61, dep chief, 61-62, chief div occup health, USPHS, 62-63; res physician, Sch Pub Health, Univ Calif, Berkeley, 63-65, prof in residence occup health, 65-72; vpres, Equitable Environ Health, Inc, 73-77; CONSULT OCCUP HEALTH, 78- *Concurrent Pos:* Med consult, AEC, 64-72. *Mem:* Fel AAAS; fel Am Col Chest Physicians; fel Am Acad Occup Med; fel Am Acad Indust Hyg; Cosmos Club; Int Comn Occup Health. *Res:* Nutrition; malaria; occupational health; epidemiology. *Mailing Add:* 2150 Shattuck Ave Suite 811 Berkeley CA 94704

COOPER, WILLIAM E, b Orono, Maine, May 8, 38; m 60; c 1. ZOOLOGY. *Educ:* Mich State Univ, BS, 60; Univ Mich, MS, 62, PhD(zool), 64. *Prof Exp:* Asst prof zool, Univ Mass, 64-65; from asst prof to assoc prof, 65-72, PROF ZOOL, MICH STATE UNIV, 72- *Mem:* Ecol Soc Am; Am Soc Zoologists. *Res:* Population dynamics and regulation of fresh-water invertebrate populations. *Mailing Add:* Dept Zool Mich State Univ Col Natural Sci East Lansing MI 48824

COOPER, WILLIAM EDWARD, b Akron, Ohio, Aug 26, 42; m 67; c 4. CRYOGENIC ENGINEERING, APPLIED SUPERCONDUCTIVITY. *Educ:* Oberlin Col, BA, 64; Harvard Univ, MA, 66, PhD(physics), 75. *Prof Exp:* Res assoc physics, Univ Mich, 71-73, Univ Pittsburgh, 75-79; PHYSICIST, FERMILAB, 79- *Mem:* Am Phys Soc. *Res:* Measurement of quench and field properties of superconducting magnets; helium liquification and refrigeration; experimental elementary particle physics; DZO experiment. *Mailing Add:* 452 S Raddant Batavia IL 60510

COOPER, WILLIAM JAMES, b Rochester, NY, Dec 1, 45; c 2. ORGANIC GEOCHEMISTRY, OXIDANT CHEMISTRY. *Educ:* Allegheny Col, BS, 69; Pa State Univ, MS, 71. *Prof Exp:* Res chemist, Environ Protection Res Div, US Army Med Bioeng Res & Develop Lab, 72-80; PROF RES, DRINKING WATER RES CTR, FLA INT UNIV, 80- *Mem:* Am Chem Soc; Am Water Works Asn; Am Soc Mass Spectrometry. *Res:* Chlorine-organic reactions in drinking water and ground water; analytical measurement of chlorine. *Mailing Add:* Drinking Water Res Ctr Fla Int Univ Miami FL 33199

COOPER, WILLIAM S, b Winnipeg, Man, Nov 7, 35; m 64; c 3. INFORMATION SCIENCE. *Educ:* Principia Col, BA, 56; Mass Inst Technol, MSc, 59; Univ Calif, Berkeley, PhD(logic & methodology), 64. *Prof Exp:* Alexander von Humboldt scholar, Univ Erlangen, 64-65; asst prof info sci, Univ Chicago, 66-70; actg assoc prof info sci & actg dir inst libr res, 70-71, assoc prof, 71-76, PROF INFO SCI, UNIV CALIF, BERKELEY, 76- *Concurrent Pos:* Miller prof, Miller Inst, Berkeley, Calif, 75-76; hon res fel, Univ Col, London, 77-78. *Mem:* AAAS; Am Soc Info Sci; Asn Comput Mach. *Res:* Foundation of language; information storage and retrieval; theory of indexing; question and answering system; evolutionary theory. *Mailing Add:* Sch Library & Info Studies Univ Calif Berkeley CA 94720

COOPER, WILLIAM WAILES, b Salisbury, Md, June 24, 41; m 63; c 4. CHEMICAL ENGINEERING, TECHNICAL MANAGEMENT. *Educ:* Mass Inst Technol, SB, 63, SM, 64, ScD(chem eng), 67. *Prof Exp:* Res engr, Abcor, Inc, Wilmington, 66-71, sr res engr, Cambridge, 71-73, mgr develop biomed prods, 73-75, mgr membrane mfg, 75-80, dir opers, 80-81, vpres opers, 81-89; VPRES OPERS & SPEC PROJS, KOCH MEMBRANE SYSTS, 89- *Mem:* Am Inst Chem Engrs; Am Chem Soc; Am Soc Artificial Internal Organs; Sigma Xi. *Res:* Biomedical engineering; membrane separation processes; plasma chemistry; ultrafiltration; reverse osmosis; microfiltration. *Mailing Add:* 11 Cedar Creek Rd Sudbury MA 01776

COOPER, WILSON WAYNE, b Checotah, Okla, July 28, 42; m 63; c 3. INDUSTRIAL CHEMISTRY, CHEMICAL ENGINEERING. *Educ:* Northeastern State Col, BS, 64; Univ Ark, Fayetteville, MS, 67, PhD(chem), 69. *Prof Exp:* Instr chem, Northwestern State Col, 66-67 & 69-70; res chem engr, 70-78, sr res engr, Pilot Plant Ctr, US Borax Res Corp, 78-88, SIR ENVIRON SCIENTIST, US BORAX & CHEM CO, 88- *Mem:* Am Chem Soc; Sigma Xi; Nat Asn Environ Professionals; Water Pollution Control Fedn. *Res:* Upcoming legislation, active regulations, applications and communicate to field operations. *Mailing Add:* US Borax 3075 Wilshire Blvd Los Angeles CA 90010-1294

COOPERMAN, BARRY S, b New York, NY, Dec 11, 41; m 63; c 1. BIOPHYSICS, ORGANIC CHEMISTRY. *Educ:* Columbia Univ, BA, 62; Harvard Univ, PhD(chem), 68. *Hon Degrees:* MA Univ Pa, 73. *Prof Exp:* NATO fel biochem, Pasteur Inst, 67-68; from asst prof to assoc prof, 68-77, PROF BIOORG CHEM, UNIV PA, 77- *Concurrent Pos:* Vice provost Res, Univ Pa 82- *Mem:* Am Soc Biol Chemists; Am Chem Soc. *Res:* Mechanism of phosphoryl transfer enxymes; photoaffinity labels for ribosome and adenylic acid receptor sites; peptidase inhibitor interaction with peptidases, mechanism of HIV-reverse transcriptase; structure and function of ribosomes. *Mailing Add:* Dept Chem 358 Chem Univ Pa Philadelphia PA 19104

COOPERMAN, JACK M, b New York, NY, Jan 13, 21; m 49; c 1. NUTRITIONAL BIOCHEMISTRY. *Educ:* City Col New York, BS, 41; Univ Wis, MS, 43, PhD(biochem), 45. *Prof Exp:* Sr res biochemist, Hoffman-La Roche Inc, 46-56; from asst prof to assoc prof, 57-71, PROF PEDIAT, NY MED COL, 71-, PROF COMMUNITY & PREV MED, 75-, DIR NUTRIT EDUC, 78- *Concurrent Pos:* Nat Found Infantile Paralysis fel, Univ Wis, 45-46; mem med adv bd, Cooley's Anemia & Res Found Children, Inc; staff mem, Off Sci Res & Develop; vis prof, Univ WI, Jamaica, 60 & All India Inst Med Sci, New Delhi, 78; dir nutrit biochem, Touro Col, 76-81; nutrit lectr, Cath Med Col Korea, Seoul, 89. *Honors & Awards:* Lederle Lectr Nutrit, Am Col Gen Practr Osteop Med & Surg, 85, Israeli Dietetic Asn, Tel Aviv, 88; John P McGrath Lectr Nutrit, CW Post Col, 87. *Mem:* Fel AAAS; Am Chem Soc; Soc Exp Biol & Med; fel NY Acad Sci; Am Inst Nutrit; Am Soc Biol Chem; Am Soc Clin Nutrit; Biochem Soc. *Res:* Nutrition of hamster; nutrition of monkey; relation of nutrition of prevention of infectious diseases; microbiological vitamin assays; vitamins; unknown growth factors for animals and bacteria; enzymes; amino acids and anemias; vitamin B-12, folic acid and riboflavin metabolism; placental transfer; purine and pyrimidine metabolism and synthesis. *Mailing Add:* Dept Community & Prev Med NY Med Col Munger Pavillion Valhalla NY 10595

COOPERRIDER, DONALD ELMER, b Thornville, Ohio, Sept 21, 14; m 36. VETERINARY MEDICINE. *Educ:* Ohio State Univ, DVM, 36, MS, 42. *Prof Exp:* Jr veterinarian, USDA, 36-40; veterinarian, Civilian Conserv Corps, 40-41; asst vet parasitol, Ohio State Univ, 41-42; vet diagnostician, State Dept Agr, Ohio, 46; assoc veterinarian vet parasitol, Okla Agr & Mech Col, 46-48; assoc parasitologist, Univ Tenn, 48-49; assoc prof vet parasitol & path, Univ Ga, 49-54; chief diag labs, State Dept Agr, NC, 54-59; parasitologist, 59-68, chief, Bur Diag Labs, Fla Dept Agr, 68-76; vet consult, UN-Pan Am Health Orgn, 76-77; prof vet parasitol, Ross U Vet Sch, 89-90. *Mem:* Am Vet Med Asn; US Animal Health Asn; Poultry Sci Asn; Am Asn Vet Parasitol (secy-treas, 61-69, pres, 71-73); Am Asn Vet Lab Diagnosticians (pres, 71-73). *Res:* Veterinary parasitology. *Mailing Add:* 1001 East Gulf Dr Sanibel FL 33957

COOPERRIDER, NEIL KENNETH, mechanical engineering, for more information see previous edition

COOPERRIDER, TOM SMITH, b Newark, Ohio, Apr 15, 27; m 53; c 2. PLANT TAXONOMY. *Educ:* Denison Univ, BA, 50; Univ Iowa, MS, 55, PhD(bot), 58. *Prof Exp:* Asst bot, Univ Iowa, 53-57; NSF fel, 57-58; from instr to asst prof biol sci, Kent State Univ, 58-62; asst prof bot, Univ Hawaii, 62-63; from asst prof to assoc prof, 63-69, PROF BIOL SCI, KENT STATE UNIV, 69-, CUR HERBARIUM, 66-, DIR, BOT GARDENS, 72- *Concurrent Pos:* Consult, US Fish & Wildlife Serv, Dept Interior, 76-83 & Davey Tree Expert Co, 79-85. *Mem:* Bot Soc Am; Am Soc Plant Taxon; Int Asn Plant Taxon; fel AAAS; Sigma Xi. *Res:* Angiosperm taxonomy; floristics; interspecific hybridization; biosystematics. *Mailing Add:* Dept Biol Sci Kent State Univ Kent OH 44242

COOPERSMITH, MICHAEL HENRY, b Brooklyn, NY, Aug 11, 36; m 59; c 3. THEORETICAL PHYSICS. *Educ:* Swarthmore Col, BA, 57; Cornell Univ, PhD(physics), 62. *Prof Exp:* NSF fel, Univ Paris, 61-62; res assoc physics, Univ Chicago, 62-64; asst prof, Case Inst Technol, 64-69; ASSOC PROF PHYSICS, UNIV VA, 69- *Res:* Statistical mechanics; phase transitions; homogeneity properties of thermodynamic systems; many-body theory; mobility of ions in helium; irreversible statistical mechanics. *Mailing Add:* Dept Physics Univ Va McCormick Rd Charlottesville VA 22903

COOPERSTEIN, RAYMOND, b New York, NY, Nov 19, 24; m 52; c 3. INORGANIC CHEMISTRY. *Educ:* City Col New York, BS, 47; Syracuse Univ, MS, 49; Pa State Univ, PhD(inorg chem), 52. *Prof Exp:* Res assoc ceramics, Exp Sta, Sch Mineral Indust, Pa State Univ, 52-53; sr chemist inorg chem, Navy Ord Div, Eastman Kodak Co, 53-57; prin engr, Dept Aircraft Nuclear propulsion, Gen Elec Co, 57-59; sect chief, Beryllium Corp, 59-61; chemist refractory mat, Lawrence Radiation Lab, Univ Calif, Berkeley, 61-65; mgr new prod develop, Wood Ridge Chem Corp, 65; sr engr, Nuclear Reactor Dept, Gen Elec Co, 65-67; sr engr, Douglas United Nuclear, Inc, 67-72; mem staff, JRB Assocs, McLean, Va, 73-75; chem engr, US Nuclear Reg Comn, 75-80; environ specialist, 80-86, INDUST SPECIALIST & PHYSICAL SCIENTIST, US DEPT ENERGY, 86- *Concurrent Pos:* Sr scientist, Nat Phys Lab, Hebrew Univ, Jerusalem, 72-73. *Mem:* Am Chem Soc; Am Ceramic Soc; Am Soc Testing & Mat; Sigma Xi. *Res:* Coordination compounds; ferrites; photoconductors; nuclear ceramics; inorganic synthesis; silicate and molten salt chemistry; nuclear facility licensing; radioactive waste management; nuclear safety; environmental protection; nuclear materials production. *Mailing Add:* 10935 Deborah Dr Potomac MD 20854

COOPERSTEIN, SHERWIN JEROME, b New York, NY, Sept 14, 23; m 47; c 2. CELL PHYSIOLOGY, SECRETION. *Educ:* City Col NY, BS, 43; NY Univ, DDS, 48; Western Reserve Univ PhD(anat),51. *Prof Exp:* Instr biol, City Col NY, 43 & 46-48; instr anat, Western Reserve Univ, 48-49, sr instr, 51-52, from asst prof to assoc prof, 52-64, asst dean, 57-64; PROF ANAT & HEAD DEPT, SCHS MED & DENT MED, UNIV CONN HEALTH CTR, 64- *Concurrent Pos:* Res assoc physiol, Col Dent, NY Univ, 46-48; fel anat, Western Reserve Univ, 49-51; mem adv panel med student res, NSF, 60-61; mem anat sci training comt, Nat Inst Gen Med Sci, 66-70; ed adv diabetes lit index, Nat Inst Arthritis, 66-79; mem spec study sect diabetes ctrs, Nat Inst Arthritis, Metab & Digestive Dis, 73-75; mem study sect, systs & integrative biol, Instnl Nat Res Serv Awards applns, NIH, 77; mem adv panel on res personnel needs in basic biomed sci, Nat Acad Sci, 76-83. *Mem:* AAAS; Am Diabetes Asn; Am Chem Soc; Am Asn Anat; Am Soc Biol Chem; Marine Biol Lab; Asn Am Med Col. *Res:* Metabolism of the islets of Langerhans; insulin secretion; mechanism of the diabetogenic action of alloxan. *Mailing Add:* Dept Anat Univ Conn Sch Med Farmington CT 06032

COOPERSTOCK, FRED ISAAC, b Winnipeg, Man, Aug 20, 40; m 62; c 2. THEORETICAL PHYSICS. *Educ:* Univ Man, BSc, 62; Brown Univ, PhD(physics), 66. *Prof Exp:* Res scholar theoret physics, Dublin Inst Adv Studies, 66-67; from asst prof to assoc prof, 71-78, PROF PHYSICS, UNIV VICTORIA, 78- *Concurrent Pos:* Can-France sci exchange visitor, Inst Henri Poincare, Paris, 73-74 & 80-81; Lady Davis vis prof, Technion-Israel Inst Technol, Haifa, Israel, 87-88. *Mem:* Am Phys Soc; Int Comt Gen Relativity & Gravitation. *Res:* General relativity; gravitational waves; relativistic astrophysics and cosmology; two-body problem; static and stationary solutions of Einstein-Maxwell equations; elementary particles. *Mailing Add:* Dept Physics Univ Victoria Victoria BC V8W 2Y2 Can

COOR, THOMAS, b Houston, Tex, Nov 21, 22; c 2. APPLIED PHYSICS, INSTRUMENTATION. *Educ:* Rice Univ, BA, 43; Princeton Univ, PhD(physics), 48. *Prof Exp:* Res physicist, Brookhaven Nat Lab, 50-53; sr res assoc plasma physics, Princeton Univ, 53-64; chief scientist & founder, Princeton Appl Res Corp, 62-80; pres & found, Biotech Int, Inc, 81-87; RETIRED. *Concurrent Pos:* Sci secy, UNO, 57-58; protein eng, Aaston Biochem; mem bd dir advan magnetics, York Ltd. *Mem:* Am Phys Soc. *Res:* Instrumentation physics. *Mailing Add:* Nine Hemlock Rd Cambridge MA 02138

COORTS, GERALD DUANE, b Emden, Ill, Feb 3, 32; m 57; c 3. FLORICULTURE, GREENHOUSE MANAGEMENT. *Educ:* Univ Mo, BS, 54, MS, 58; Univ Ill, PhD(hort), 64. *Prof Exp:* Asst prof hort, Univ RI, 64-68; from assoc prof to prof plant & soil sci, Southern Ill Univ, Carbondale, 68-85, chmn dept, 73-85; PROF AGR, COL AGR & HOME ECON & DEAN, TENN TECH UNIV, COOKEVILLE, 85-,. *Concurrent Pos:* Floricult mkt specialist, 59-61; consult hort ther, 78-85; pres, Am Assoc State Col Agr & Renew Res, 87-90, Tenn Coun Chapters, Soil & Water Cons, 89-90; mem, Coun Agr Sci & Technol. *Mem:* Am Hort Soc; Am Soc Hort Sci; Plant Growth Regulator Soc; Soc Am Florists; Am Soc Agron. *Res:* Plant growth regulators; growing ornamental plants in artificial media; grades and standards on cut roses; post-harvest physiology of roses; mineral nutrition of ornamental crops. *Mailing Add:* Col Agr & Home Econ Tenn Tech Univ Box 5165 Cookeville TN 38505

COOTE, DENIS RICHARD, b 1945; Can citizen; m 67; c 3. SOIL DEGRADATION, AGRICULTURE. *Educ:* WScotland Agr Col, N Dip Agr Engr, 65; Cornell Univ, MS, 69, PhD(soils), 73. *Prof Exp:* Teacher agr eng, Jamaica Sch Agr, 65-66; engr soil & water conserv, Ministry Agr & Fisheries, Jamaica, 69-70; contract scientist agr & water qual, Great Lakes Studies, 73-77; PHYS SCIENTIST LAND & ENVIRON DEGRADATION, LAND RESOURCE RES CTR, AGR CAN, 77- *Mem:* Can Soc Soil Sci; Can Soc Agr Eng; Soil & Water Conserv Soc; Int Soc Soil Sci; Am Soc Agr Eng. *Res:* Soil degradation and associated environmental impacts; soil erosion. *Mailing Add:* K W Neatby Bldg Land Resource Res Ctr Agr Can Ottawa ON K1A 0C6 Can

COOTS, ALONZO FREEMAN, b Little Rock, Ark, May 6, 27; m 47. PHYSICAL CHEMISTRY. *Educ:* Vanderbilt Univ, BE, 49, PhD(chem), 54. *Prof Exp:* Instr phys chem, Fisk Univ, 52-53; res chemist, Anal Sta, Del, 56-58; ASSOC PROF CHEM, NC STATE UNIV, 58- *Mem:* Am Chem Soc. *Res:* Radioisotope principles and techniques; general and physical chemistry; radiochemistry. *Mailing Add:* Dept Chem NC State Univ Raleigh NC 27650

COOTS, ROBERT HERMAN, b Kansas City, Mo, Feb 24, 28; m 57; c 1. BIOCHEMISTRY, TOXICOLOGY. *Educ:* Univ Mo, BS, 54; Univ Wis, MS, 56, PhD(biochem), 58. *Prof Exp:* Res biochemist, Miami Valley Labs, Procter & Gamble Co, 58-65, head, Physiol Chem Sect, 65-67, Dent Res Sect, 67-69, Dent & Toxicol Res Sect, 69-70, Pharmacol & Metab Res Sect, 70-71, assoc dir, Res & Develop Dept, 71-75, Human & Environ Safety Div, 75-89; RETIRED. *Mem:* Am Soc Biol Chem; Sigma Xi; Am Col Toxicol. *Res:* Intermediary metabolism; lipid metabolism; pharmacology and toxicology. *Mailing Add:* 1119 Meredith Dr Cincinnati OH 45231

COOVER, HARRY WESLEY, JR, b Newark, Del, Mar 6, 19; m 41; c 3. ORGANOMETALLIC CHEMISTRY, POLYMER CHEMISTRY. *Educ:* Hobart Col, BS, 41; Cornell Univ, MS, 42, PhD(org chem), 44. *Prof Exp:* Res chemist, Eastman Kodak Co, 44-49; sr res chemist, Tenn Eastman Co, 49-54, res assoc, 54-63, div head polymers res div, 63-65, dir res, 65-73, vpres, 70-73, exec vpres, Tenn Eastman Co, 73-81, vpres, Eastman Kodak Co, 81-84; pres, New Bus Develop Group, Loctite Corp, 85-88; RETIRED. *Concurrent Pos:* Consult Int Mgt, 84- *Honors & Awards:* Southern Chemist Award, Am Chem Soc, 60, Earl B Barnes Award, 85; Indust Res Inst Medal, 84. *Mem:* Nat Acad Eng; Am Chem Soc; Textile Res Inst; Am Asn Textile Technologist; Indust Res Inst; NY Acad Sci. *Res:* Adhesives; insecticides; fungicides; high polymer chemistry; organophosphorus chemistry; synthetic fiber research. *Mailing Add:* PO Box 3866 1201 Eastman Rd Kingsport TN 37664

COPE, CHARLES S(AMUEL), b Philadelphia, Pa, Mar 17, 28; m 70; c 1. CHEMICAL ENGINEERING. *Educ:* Cornell Univ, BChE, 49; Yale Univ, PhD(chem eng), 56. *Prof Exp:* sr res assoc, Polymer Prod Dept, E I du Pont de Nemours & Co, 56-86; RETIRED. *Mem:* Am Inst Chem Engrs; Am Chem Soc; Sigma Xi. *Res:* Synthesis and behavior of fluorocarbon polymers. *Mailing Add:* 218 N Hills Dr Parkersburg WV 26101-9230

COPE, DAVID FRANKLIN, b Crumpler, WVa, June 28, 12; m 36; c 2. NUCLEAR SCIENCE, ENERGY CONVERSION. *Educ:* WVa Univ, AB, 33, MS, 34; Univ Va, PhD(physics), 52. *Prof Exp:* Instr phsyics, WVa Univ, 35-36; instr math, Agr & Mech Col, Texas, 37-38, physics, 46-47; asst prof, NMex Agr & Mech Col, 47-52; physicist, AEC, 52, chief res br, 53-56, dep dir, Res & Develop Div, Oak Ridge Opers, 56-59, dir, Reactor Div, 59-66, sr site rep reactor develop, 66-74, ENERGY CONSULT, OAK RIDGE NAT LAB, 74-; NUCLEAR ENERGY EXPERT & ENERGY EXPERT, ENERGY RES & DEVELOP OFF, OAK RIDGE 74- *Concurrent Pos:* US team leader, US-Mexico Study of Dual Purpose Nuclear Plants, 68-; chmn, tech subcomt, NSF Energy Fac Sifting Comn, 74-; spec assignment, Fed Energy Off, Washington, DC, 74; consult, Fed Energy Admin, 76; energy consult, 75- *Mem:* AAAS; Am Nuclear Soc; Am Phys Soc; Sigma Xi. *Res:* Magnetic properties of iron and nickel at high temperatures; analysis of rocket trajectories; nuclear science and technology, evaluation of nuclear energy centers; the nation's future energy requirement and assessmant of various options for meeting these needs; evaluative studies on ethical problems and nuclear energy. *Mailing Add:* 113 Orange Lane Oak Ridge TN 37830

COPE, FREDERICK OLIVER, b York, Pa, Aug 8, 46; m 69; c 1. ONCOLOGY, CELL BIOLOGY. *Educ:* Delaware Valley Col Sci & Agr, BSc, 69; Millersville Univ, Pa, MS, 76; Univ Conn-Storrs, PhD(nutrit biochem), 82. *Prof Exp:* Res div, Thin Films & Polymer Chem Dept, Armstrong Cork Co, Inc, 71-72; head, Vaccine, Sera & Enzyme Labs, Wyeth Labs, Inc, 72-78; fel nutrit biochem, Univ Conn, 78-82; Nat Cancer Inst fel, McArdle Lab Cancer Res, Univ Wis-Madison, 82-84; head cancer chemoprev, Southern Res Inst, Ala, 84-88; HEAD, CANCER PATIENT SUPPORT RES & DEVELOP, ROSS LABS, COLUMBUS, OHIO, 88- *Concurrent Pos:* Grad admin bd fel awards, med oncol & oncol, Univ Wis-Madison, 83-85; vis scientist in residence, Lab Cancer Prev, Nat Cancer Ctr Res Inst, Tokyo, 87; provisional fac mem, Dept Oncol, Comprehensive Cancer Ctr, Ohio State Univ, 88-; mem bd trustees, Leukemia Soc Am, Ohio & Hy-Gene Inc, Calif, 88- *Mem:* Am Asn Cancer Res; Soc Exp Biol & Med; Am Inst Nutrit; AAAS; Sigma Xi; NY Acad Sci; Am Soc Parenteral & Enteral Nutrit; Int Asn Vitamin & Nutrit Oncol; Comt Sci Invest of Claims of the Paranormal. *Mailing Add:* Cancer Patient Support Res & Develop Med Dept Ross Lab 625 Cleveland Ave Columbus OH 43215

COPE, JOHN THOMAS, JR, b Akron, Ohio, July 24, 21; m 45; c 3. SOIL FERTILITY. *Educ:* Auburn Univ, BS, 42, MS, 46; Cornell Univ, PhD(soil sci), 50. *Prof Exp:* Farm mgr, Fla, 46-47; assoc soil chemist, Nitrogen Res, Auburn Univ, 50-52, assoc agronomist in charge exp fields, 52-59, agronomist, 59-66, soil tester, 66-78, prof agron & soils, 78-85; RETIRED. *Mem:* Am Soc Agron; Soil Sci Soc Am. *Res:* Soil fertility; nitrogen fertilization; soil organic matter; crop rotations; soil and crop management; soil test fertilizer recommendations and computer programs; soil test calibration; fertility index. *Mailing Add:* 244 Virginia Ave Auburn AL 36830

COPE, OLIVER, b Germantown, Pa, Aug 15, 02; m 32; c 2. SURGERY. *Educ:* Harvard Univ, AB, 23, MD, 28; Am Bd Surg, dipl. *Hon Degrees:* Dr, Univ Toulouse, 50. *Prof Exp:* Traveling fel, Harvard Univ, 33, from instr to asst prof, 34-38, from assoc prof to prof, 48-69, actg head dept, 68-69, EMER PROF SURG, MED SCH, HARVARD UNIV, 69- *Concurrent Pos:* From asst to assoc surgeon, Mass Gen Hosp, 34-46, vis surgeon, 46-69, actg chief surg serv, 68-69, mem bd consults, 69-, sr surgeon, 72-89, hon surgeon, 89-; investr, Off Sci Res & Develop, 42-45; mem subcomt burns, Nat Res Coun, 43-45; dir res under contract with Off Naval Res, 47-52; chief of staff, Boston Unit, Shriners Burn Inst, 64-69, emer chief of staff, 69- *Mem:* Inst of Med of Nat Acad Sci; fel Am Col Surgeons; Am Surg Asn (pres, 62-63); Int Soc Surg; AAAS. *Mailing Add:* Mass Gen Hosp 32 Fruit St Boston MA 02114

COPE, OLIVER BREWERN, b San Francisco, Calif, June 16, 16; m 42; c 2. FISHERY BIOLOGY. *Educ:* Stanford Univ, AB, 38, AM, 40, PhD(entomol), 42. *Prof Exp:* Agent, Bur Entomol & Plant Quarantine, US Dept Agr, Calif, 39; asst, Stanford Univ, 39-42; aquatic biologist, US Fish & Wildlife Serv, 46-50, chief cent valley invest, 50-52, chief Rocky Mt invests, 52-59, fish-pesticide res lab, 59-69, chief br fish husb res, US Bur Sport Fisheries & Wildlife, 69-71; phys scientist, Off of Water Resources Res, 71-74; consult-ed, 74-85; RETIRED. *Concurrent Pos:* Jr quarantine inspector, Calif State Dept Agr, 41. *Mem:* AAAS; Coun Biol Ed; Am Fisheries Soc; Wildlife Soc; Am Inst Fisheries Res Biol. *Res:* Insect morphology; fresh water fisheries; economic poisons and fish; fish husbandry. *Mailing Add:* 15 Adamswood Rd Asheville NC 28803

COPE, OSWALD JAMES, b Hanley, Eng, June 8, 34; US citizen; m 83; c 3. INFORMATION STORAGE MEDIA, SOLVENT COATING TECHNOLOGY. *Educ:* Univ London, BSc, 55; Univ Alta, PhD(org chem), 62. *Prof Exp:* Asst lectr chem, Western Col, 56-57; chemist, Paint Res Labs, Can Indust Ltd, 57-58; asst chem, Univ Alta, 58-62; res assoc org chem, Purdue Univ, 62-63; chemist, Plastics Dept, Exp Sta, E I du Pont de Nemours & Co, Del, 63-69; sr chemist, Memorex Corp, Calif, 69-70; mgr res & develop, Xidex Corp, 70-74, vpres, 74-85; RETIRED. *Concurrent Pos:* Post doctorate, Purdue Univ, 62-63; consult, 85- *Mem:* Am Chem Soc; Soc Photog Sci & Eng; Sigma Xi. *Res:* Polymer synthesis and structure property relationships; non-silver imaging processes; vesicular and diazo film; optical recording of digital information; solvent coating technology. *Mailing Add:* 429 High St Santa Cruz CA 95060

COPE, VIRGIL W, b Storm Lake, Iowa, Feb 4, 43; m 65; c 2. INORGANIC CHEMISTRY. *Educ:* Iowa State Teachers Col, BA, 65; Univ Kans, PhD(inorg chem), 68. *Prof Exp:* From asst prof to assoc prof, 68-87, PROF CHEM, UNIV MICH-FLINT, 87- *Concurrent Pos:* Vis prof chem, Boston Univ, 74-75; assoc prof, Wash Univ, 81-82. *Mem:* Am Chem Soc. *Res:* Mechanisms of reactions of coordination compounds, especially electron transfer reaction, photosubstitution and photo reduction; reactions of primary air pollutants with polycyclic aromatic hydrocarbons. *Mailing Add:* Dept Chem Univ Mich Flint Col 1321 E Court St Flint MI 48503-6207

COPE, WILL ALLEN, b Inverness, Ala, June 23, 22; m 50; c 1. GENETICS, PLANT BREEDING. *Educ:* Ala Polytech Inst, BS, 48, MS, 49; NC State Col, PhD(field crops), 56. *Prof Exp:* Asst prof field crops, 55-64, res assoc prof crop sci, 64-69, assoc prof crop sci & genetics, 69-71, PROF CROP SCI, NC STATE UNIV, 71-, PROF GENETICS, 76- *Concurrent Pos:* Res agronomist, sci & educ admin-agr res, USDA, 55- *Mem:* Am Soc Agron; Genetics Soc Am. *Res:* Breeding and genetics of Trifolium repens. *Mailing Add:* 3710 Eakley Ct Raleigh NC 27606

COPELAND, ARTHUR HERBERT, JR, b Columbus, Tex, June 11, 26; m 47; c 3. MATHEMATICS. *Educ:* Univ Mich, BS, 49, MA, 50; Mass Inst Technol, PhD(math), 54. *Prof Exp:* Instr, Mass Inst Technol, 51-54; from instr to assoc prof math, Purdue Univ, 54-61; assoc prof, Northwestern Univ, 61-68; PROF MATH, UNIV NH, 68- *Mem:* Am Math Soc; Soc Indust Appl Math. *Res:* Algebraic topology; homotopy theory; Hopf spaces. *Mailing Add:* Dept Math Univ NH Durham NH 03824

COPELAND, BILLY JOE, b Mannsville, Okla, Nov 20, 36; m 63; c 3. ECOLOGY. *Educ:* Okla State Univ, BS, 59, MS, 61, PhD(zool), 63. *Prof Exp:* Asst limnol, Okla State Univ, 59-62; res scientist marine ecol, Inst Marine Sci, Univ Tex, Austin, 62-65, from asst prof to assoc prof, 65-70; assoc prof zool & bot & dir, Pamlico Marine Lab, 70-73, PROF ZOOL & BOT & DIR, NC SEA GRANT COL PROG, NC STATE UNIV, 73-,. *Concurrent Pos:* Grants estuarine ecol & pollution control; consult, indust; pres, Sea Grant Asn. *Mem:* Am Soc Limnol & Oceanog; Ecol Soc Am; Water Pollution Control Fedn. *Res:* Estuarine ecology; effects of water resources development and pollution on fresh and saltwater ecology; systems analysis of estuarine ecosystems. *Mailing Add:* 105 1911 Bldg Box 8605 NC State Univ Raleigh NC 27695-8605

COPELAND, BRADLEY ELLSWORTH, b Wilkinsburg, Pa, Sept 8, 21; m 43; c 2. MEDICINE, CLINICAL PATHOLOGY. *Educ:* Dartmouth Col, AB, 43; Univ Pa, MD, 45. *Prof Exp:* Assoc path, New Eng Deaconess Hosp & New Eng Baptist Hosp, 50-59, clin pathologist, 59-72, chmn dept path, 72-75, chief, Div Clin Path, 75-79; CHIEF LAB SERV, VET ADMIN MED CTR, CINCINNATI, 79-; PROF PATH, MED SCH, UNIV CINCINNATI, 79- *Concurrent Pos:* Chmn comt world stands, World Asn Socs Anat & Clin Path, 60-81; chmn subcomt documentation, Comt Path, Nat Acad Sci-Nat Res Coun, 69-71; chmn, Cholesterol NBS-CDC Traceability Prog VA Path Sect. *Mem:* Am Chem Soc; AMA; Am Soc Clin Path; Col Am Path. *Res:* Electrolytes; magnesium; hemoglobin; statistics and quality control; instrumentation. *Mailing Add:* Lab Serv Vet Admin Med Ctr 3200 Vine St Cincinnati OH 45220

COPELAND, CHARLES WESLEY, JR, b Hueytown, Ala, Oct 1, 32; m 57; c 2. STRATIGRAPHY, INVERTEBRATE PALEONTOLOGY. *Educ:* Birmingham-Southern Col, BS, 54; Univ NC, Chapel Hill, MS, 61. *Prof Exp:* Paleontologist & core libr supvr, Geol Surv Ala, 61-68, geol mapping supvr, 64-68, chief paleont & stratig div, 68-74, chief geol div, 74-83, ASST STATE GEOLOGIST, GEOL SURV ALA, 83- *Mem:* Geol Soc Am; Sigma Xi; Brit Paleont Soc; Soc Econ Paleontologists & Mineralogists; Am Asn Petrol Geologists. *Res:* Subsurface stratigraphy, paleontology, and stratigraphy; coastal plain faults; solution collapse phenomena; ground water aquifer studies. *Mailing Add:* 203 32nd Place E Tuscaloosa AL 35405

COPELAND, DAVID ANTHONY, b Jasper, Ala, Dec 4, 42; m 67; c 2. CONSTRUCTION MANAGEMENT SYSTEMS, CHEMICAL PHYSICS. *Educ:* Univ Ala, BSChem, 65, MA, 66; La State Univ, PhD(chem physics), 70. *Prof Exp:* Res assoc inorg chem, La State Univ, 69-70; asst prof, 70-74, assoc prof inorg phys chem, Univ Tenn, Martin, 74-77; contruct systs specialist, Ala Power Co, 77-80, sr construct systs specialist, 80-81, supvr construct systs, 81-88, SUPVR POWER GENERATION SYSTS, ALA POWER CO, BIRMINGHAM, 88- *Mem:* IEEE Computer Soc; Sigma Xi. *Res:* Theoretical electronic structures of transition metal tetrahedral complexes and the theories of electrons in polar solvents; developing project management data processing systems; developing knowledge based expert systems. *Mailing Add:* Power Generation Systs Dept Ala Power Co PO Box 2641 Birmingham AL 35291-0371

COPELAND, DONALD EUGENE, b Mendon, Ohio, Feb 6, 12; m 41; c 3. BIOLOGY. *Educ:* Univ Rochester, AB, 35; Amherst Col, MA, 37; Harvard Univ, PhD(zool), 41. *Prof Exp:* Instr biol, Univ NC, 41-42; from asst prof to assoc prof, Brown Univ, 46-54; prof assoc, Med Sci Div, Nat Res Coun, 53-55; exec secy, Div Res Grants, NIH, 56-59; prof zool, 59-70, prof biol, 70-77, EMER PROF BIOL, TULANE UNIV, 77- *Concurrent Pos:* Dir Advan Physiol Prog, Surgeon Gen Off, US Air Force, 51-54; mem physiol panel, Res & Develop Bd, US Dept Defense, 52-53; trustee emeritus, Marine Biol Lab, Woods Hole, 68-, res investr, 77- *Mem:* Am Soc Zool; Am Asn Anat; Am Soc Cell Biol; Am Physiol Soc. *Res:* Electron microscopy of inorganic ion transport; fine structure of oxygen elevating tissue in fish eye and swim bladder; fine structure of teleost eye function and luminescent organs. *Mailing Add:* 41 Fern Lane Woods Hole MA 02543

COPELAND, EDMUND SARGENT, b Lancaster, Pa, Mar 2, 36; m 65; c 2. BIOPHYSICAL CHEMISTRY, MAGNETIC RESONANCE. *Educ:* Cornell Univ, AB, 58; Univ Rochester, MS, 61, PhD(radiation biol), 64. *Prof Exp:* Fel biophys, Roswell Park Mem Inst, 64-65; USPHS fel biochem, Norsk Hydro's Inst Cancer Res, 65-67; chemist, Walter Reed Army Inst Res, 67-76; exec secy, Path B (AHR) Study Sect, 76-79, EXEC SECY, CHEM PATH STUDY SECT, DIV RES GRANTS, NIH, 79- *Concurrent Pos:* Vis scientist, Clin Neuropharmacol Br, Nat Inst Ment Health, NIH, 76-88; lectr, grad sch, NIH. *Mem:* Am Chem Soc; Biophys Soc. *Res:* Mechanism of action of radioprotective drugs; chemical protection against phototoxicity; the opiate receptor, mechanism of action of antidepressants; science administration. *Mailing Add:* Div Res Grants Westwood Bldg Rm 322 NIH 5333 Westbard Ave Bethesda MD 20892

COPELAND, FREDERICK CLEVELAND, b Brunswick, Maine, Oct 9, 12; m 39; c 3. BIOLOGY. *Educ:* Williams Col, AB, 35; Harvard Univ, AM, 37, PhD(biol), 40. *Prof Exp:* Asst biol, Harvard Univ, 37-40; instr, Trinity Col, Conn, 40-46, dir admis, 45-46; from asst prof to prof, 46-78, dean admis, 46-78, EMER PROF BIOL, WILLIAMS COL, 78-, EMER DEAN ADMIS, 78- *Concurrent Pos:* Trustee, Hotchkiss Sch, Lenox Sch, Educ Rec Bur & Trinity Sch, New York, Helene W Toolan Inst Med Res. *Honors & Awards:* Rogerson Cup Award, 69. *Mem:* Genetics Soc Am. *Res:* Cytogenetics; plant growth rates. *Mailing Add:* 1571 Oblong Rd Williamstown MA 01267

COPELAND, GARY EARL, b Maud, Okla, Aug 15, 40. CHEMICAL PHYSICS. *Educ:* Univ Okla, BS, 64, MS, 65, PhD(physics), 70. *Prof Exp:* Engr plasma physics, Tex Instruments, 64-65; instr physics, Univ Okla, 69-70; fel laser physics, Langley Res Ctr, NASA, 70-71; asst prof physics, Old Dominion Univ, 71-80, asst prof geophys sci, 76-80, assoc prof physics & geophys, 80-82, PROF PHYSICS, OLD DOMINION UNIV, 82- *Concurrent Pos:* Consult, Va State Air Pollution Control, 72- & Earth Resources Observ Bd, Goddard Space Flight Ctr, NASA, 74. *Mem:* Am Phys Soc; Am Geophys Union; Sigma Xi. *Res:* Chemical physics; laser physics; atmospheric physics; computer modeling. *Mailing Add:* Dept Physics Old Dominion Univ Norfolk VA 23529

COPELAND, JAMES CLINTON, b Chicago, Ill, Nov 15, 37; m 60; c 5. MICROBIAL GENETICS, MOLECULAR GENETICS. *Educ:* Univ Ill, Urbana-Champaign, BS, 59; Univ Tenn, MS, 61; Rutgers Univ, PhD(microbiol), 65; Univ Chicago, MBA, 81. *Prof Exp:* Am Cancer Soc fel molecular genetics, Albert Einstein Col Med, 65-67; from asst geneticist to assoc geneticist, Argonne Nat Lab, 67-72; assoc prof microbiol, Ohio State Univ, 72-77; dir res & develop, Moffett Tech Ctr, CPC Int, 77-81; pres, Enzyme Technol Corp, 81-86; PRES, TECHNOL INVESTMENTS, 86- & OXYRASE, INC, 87- *Concurrent Pos:* Vis lectr, Northern Ill Univ, 70-71, adj assoc prof, 71-72; NIH res career develop award, 72-77; ed, Microbial Genetics Bull, 74-77. *Mem:* AAAS; Soc Ind Microbiol. *Res:* Structure and function of microbial chromosomes, especially their organization, regulation, replication, evolution, information storage and retrieval; genetic engineering; commercialization of biotechnology. *Mailing Add:* 298 N Countryside Dr Ashland OH 44805

COPELAND, JAMES LEWIS, b Champaign, Ill, Apr 7, 31; m 55; c 2. PHYSICAL CHEMISTRY. *Educ:* Univ Ill, BS, 52; Ind Univ, PhD(phys chem), 62. *Prof Exp:* Res assoc & fel phys chem, Inst Atomic Res, Iowa State Univ, 61-62; from asst prof to assoc prof chem, 62-68, PROF CHEM, KANS STATE UNIV, 74-, ASSOC HEAD, DEPT CHEM, 81- *Concurrent Pos:* Grants, Bur of Gen Res, 63, Res Coord Coun, 64-65 & NSF, 65-74. *Mem:* Am Chem Soc; Am Phys Soc; Sigma Xi. *Res:* Physical chemistry of fused salt systems, including properties, electrochemistry, and theories of behavior; physical chemistry of high-temperature systems in general; high temperature reaction kinetics. *Mailing Add:* Dept Chem Kans State Univ Willard Hall Manhattan KS 66506

COPELAND, JOHN ALEXANDER, b Atlanta, Ga Feb 6, 41; m 60; c 2. SOLID STATE PHYSICS. *Educ:* Ga Inst Technol, BS, 62, MS, 63, PhD(physics), 65. *Prof Exp:* Res physicist, Ga Inst Technol, 65; mem tech staff, Bell Labs, 65-, HEAD, REPEATER RES DEPT, 76-; VPRES ENG TECHNOL, SANGAMO WESTERN SCHLUMBERGER. *Concurrent Pos:* Ed, Inst Elec & Electronics Engrs Transitions on Electron Devices, 71-73. *Honors & Awards:* Morris N Liebmann, Inst Elec & Electronics Engrs, 70. *Mem:* Inst Elec & Electronics Engrs; Am Phys Soc. *Res:* Solid state devices for lightwave communications and high-speed logic applications. *Mailing Add:* Hayes Microcomput Prod PO Box 105203 Atlanta GA 30348

COPELAND, MILES ALEXANDER, b Quebec City, Que, May 5, 34; m 62; c 2. ELECTRICAL ENGINEERING. *Educ:* Univ Man, BSc, 57; Univ Toronto, MASc, 62, PhD(elec eng), 65. *Prof Exp:* From asst prof to assoc prof, 65-77, PROF ELEC ENG, CARLETON UNIV, CAN, 77- *Concurrent Pos:* Consult, Northern Elec Co, Ont, 65-91, Gen Elec, 79-85. *Mem:* Fel Inst Elec & Electronics Engrs. *Res:* Electrical engineering related to VLSI design, especially mixed analog/digital; communications applications; semi-conductors; solid state electronics. *Mailing Add:* Dept Electronics Carleton Univ Ottawa ON K1S 5B6 Can

COPELAND, MURRAY JOHN, b Toronto, Ont, Apr 23, 28; m 54; c 1. MICROPALEONTOLOGY, INVERTEBRATE PALEONTOLOGY. *Educ:* Univ Toronto, BA, 49, MA, 51; Univ Mich, PhD(geol), 55. *Prof Exp:* Field officer geol, 49-55, geologist, 55-65, RES SCIENTIST, GEOL SURV, CAN, 65- *Concurrent Pos:* Pres, Int Res Group, Paleozoic Ostracoda, Int Paleont Asn. *Mem:* Geol Asn Can. *Res:* Micropaleontology; Paleozoic Ostracoda; Paleozoic and Mesozoic Arthropoda. *Mailing Add:* Geol Surv Can Bldg 601 Booth St Ottawa ON K1A 0E8 Can

COPELAND, RICHARD FRANKLIN, b Tyler, Tex, Oct 10, 38; m 68; c 1. CRYSTALLOGRAPHY, CHEMICAL INFORMATION. *Educ:* Tex A&M Univ, BS, 61, MS, 63, PhD(chem), 65; Univ Cent Fla, MS, 86. *Prof Exp:* Robert A Welch fels, Tex A&M Univ, 65-66 & Univ Tex, Austin, 66-67; asst prof chem, Ball State Univ, 67-69 & Univ Mich, 69-71; assoc prof, 71-78, PROF CHEM & COMPUT SCI, BETHUNE-COOKMAN COL, 78- *Concurrent Pos:* NSF grants, 68-69, 74- & NIH, 74-82. *Mem:* Am Chem Soc; Am Crystallog Asn; Asn Comput Mach; Inst Elec & Electronics Engrs Comput Soc. *Res:* Chemical information retrieval; information science; educational uses of computers. *Mailing Add:* 1131 Lakewood Park Dr Daytona Beach FL 32017-3940

COPELAND, ROBERT B, b Arab, Ala, Jan 24, 38; m 59; c 2. MEDICINE. *Educ:* Auburn Univ, BS, 60; Med Col Univ Ala, MD, 63; Am Bd Internal Med, cert, 69. *Prof Exp:* Intern, Mass Gen Hosp, Boston, 63-64, asst resident med, 64-65, clin & res fel cardiol, 65-66, sr resident med, 66-67; assoc, Clark-Holder Clin, LaGrange, Ga, 67-77; PHYSICIAN & PRES, SOUTHERN CARDIOVASC ASSOC, LA GRANGE, GA, 77-; MED DIR DEPT CARIOVASC MED, W GA MED CTR, 76- *Concurrent Pos:* Vis fel med, Royal Free Hosp, London, 67; founder, Ga Heart Clin, 71; chief med, W Ga Med Ctr, 73-78 & 80; from clin asst prof to clin assoc prof, Emory Univ, 68-80, clin prof med, 80-; lectr med, LaGrange Col, Ga, 77-80; clin assoc prof med-cardiol, Univ Ala, Birmingham, 77-80, clin prof, 80-, external rev comt, dept med, 82- *Mem:* Am Col Physicians; Am Col Cardiol; Am Col Chest Physicians; Am Heart Asn; Am Med Asn. *Res:* Identification of immunogenetic risk factors in early age onset coronary artery disease. *Mailing Add:* 1550 Doctors Dr LaGrange GA 30240

COPELAND, THOMPSON PRESTON, b Ark, March 11, 21; m 50; c 2. ENTOMOLOGY. *Educ:* Ouachita Baptist Col, BS, 47; George Peabody Col, MA, 50; Univ Tenn, PhD, 62. *Prof Exp:* Instr biol, Ouachita Baptist Col, 47-49; asst prof, Union Univ, Tenn, 51-52; assoc prof, 54-64, chmn dept biol, 64-80, PROF BIOL, E TENN STATE UNIV, 64- *Concurrent Pos:* Danforth assoc, 64. *Mem:* AAAS; Am Entom Soc. *Res:* Taxonomy and ecology of Protura, Collembola and Zoraptera. *Mailing Add:* One Tallapoosa Rd Johnson City TN 37604

COPELAND, WILLIAM D, b Colorado Springs, Colo, Mar 16, 34; m 57; c 3. PHYSICAL METALLURGY. *Educ:* Carleton Col, BA, 56; Univ Minn, PhD(metall), 66. *Prof Exp:* PROF METALL, COLO SCH MINES, 66-, DEAN GRAD SCH, 72-, AT METALL ENG DEPT. *Concurrent Pos:* Am Coun Educ admin fel, 70-71. *Mem:* Am Soc Metals; Metall Soc; Am Inst Mining, Metall & Petrol Engrs; Sigma Xi. *Res:* Transport processes in high temperature compounds; oxidation-corrosion mechanisms. *Mailing Add:* Dept Metall Eng Colo Sch Mines Golden CO 80401

COPELIN, EDWARD CASIMERE, geochemistry, for more information see previous edition

COPELIN, HARRY B, b Staten Island, NY, Aug 9, 18; m 43; c 3. ORGANIC CHEMISTRY. *Educ:* Cornell Univ, BS, 40, MS, 41. *Prof Exp:* Chemist, E I du Pont de Nemours & Co, Inc, 41-58, tech assoc, 58-63, res assoc, 63-71, res fel, 71-82; RETIRED. *Res:* Organic chlorine compounds; coordination chemistry; catalysis; heterocyclic compounds; hydroformylation processes; hydrogen peroxide process; poly tetrahydrofuran process; supercritical gas extraction. *Mailing Add:* 980 Ridge Rd Lewiston NY 14092

COPEMAN, ROBERT JAMES, b Regina, Sask, Can, Dec 18, 42; m 64; c 3. PHYTOBACTERIOLOGY. *Educ:* McGill Univ, BSc, 65; Wis Univ, PhD(plant path), 70. *Prof Exp:* asst prof, 69-78, ASSOC PROF PLANT SCI, PLANT SCI DEPT, UNIV BC, 78- *Mem:* Can Phytopath Soc (secy treas, 76-80, treas, 80-82); Am Phytopath Soc. *Res:* Identification and detection of phytopathogenic bacteria in epiphytological studies and crop certification programs. *Mailing Add:* Plant Sci Dept 248-2357 Main Mall Univ BC 2075 Westbrook Pl Vancouver BC V6T 1W5 Can

COPENHAVER, JOHN HARRISON, JR, b Ralston, Nebr, Dec 21, 22; m 46; c 5. BIOCHEMISTRY. *Educ:* Dartmouth Col, BA, 46; Univ Wis, MS, 49, PhD(zool), 50. *Hon Degrees:* MA, Dartmouth Col. *Prof Exp:* Nat Heart Inst fel enzyme res, Univ Wis, 50-51; asst prof pharmacol, Southwestern Med Sch, Univ Tex, 51-53; from asst prof to assoc prof zool, 53-60, PROF BIOL SCI, DARTMOUTH COL, 60- *Mem:* Am Soc Biol Chem. *Res:* Oxidative phosphorylation; enzyme chemistry of renal function; cell transport mechanisms. *Mailing Add:* Ten Woodcock HC61 Box 237 Etna NH 03750

COPENHAVER, THOMAS WESLEY, b Sioux City, Iowa, Oct 17, 45; m 78. BIOSTATISTICS. *Educ:* Univ Nebr, BS, 68; Univ Ark, MS, 70; Colo State Univ, PhD(statist), 76. *Prof Exp:* MGR PRECLIN STATIST, WYETH LABS, AM HOME PROD, 74- *Mem:* Biomet Soc; Am Statist Asn; Sigma Xi. *Res:* Statistical methods for analyzing dose-response curves. *Mailing Add:* Wyeth Labs PO Box 8299 Philadelphia PA 19101

COPES, FREDERICK ALBERT, b Tomahawk, Wis, Dec 25, 37; m 59; c 5. ECOLOGY, FISHERIES. *Educ:* Wis State Univ, Stevens Point, BS, 60; Univ NDak, MS, 65; Univ Wyo, PhD(zool), 70. *Prof Exp:* Teacher high sch, Wis, 60-63; from instr to asst prof, 64-67, assoc prof biol & zool, 70-81, PROF BIOL, ZOOL & FISHERIES, UNIV WIS-STEVENS POINT, 78-, ECOL SPECIALIST, MUS NATURAL HIST. *Concurrent Pos:* Secy, Wis Environ Pract Comn, 74-78; pres, NAm Native Fish Asn; mem, Green Bay Rap comt; mem Habitat Adv Bd, Great Lakes Fishery Comn. *Mem:* Am Fisheries Soc; Sigma Xi; NAm Native Fish Asn (vpres, 77-80); Am Inst Fish Res Biol. *Res:* Fishes of the Red River tributaries of North Dakota; ecology of the native fishes of east Wyoming; ecology of fishes; fishery production; vital statistics of the Lake Michigan Whitefish Fishery; fisheries habitat improvement; Great Lakes fisheries. *Mailing Add:* Dept Biol Univ Wis Stevens Point WI 54481

COPES, PARZIVAL, b Nakusp, BC, Jan 22, 24; m 46; c 3. FISHERIES ECONOMICS, FISHERIES RESOURCE MANAGEMENT. *Educ:* Univ BC, BA, 49, MA, 50; London Sch Econ, PhD(econ), 56. *Prof Exp:* Economist & statistician, Dom Bur Statist, Ottawa, 53-57; from assoc prof to prof econ, Mem Univ Nfld, 57-64, dir econ res, Inst Social & Econ Res, 61-64, head, Dept Econ, 61-64; head, Dept Econ & Com, Simon Fraser Univ, 64-69, chmn, 72-75, dir, Ctr Can Studies, 78-86, PROF ECON, SIMON FRASER UNIV, 64-, DIR, INST FISHERIES ANALYSIS, 80- *Concurrent Pos:* Dir & prin investr, Can-Foreign Arrangements Proj, 76; dir, Social Sci Fedn Can, 79-81; mem coun, Western Regional Sci Asn, 79-83, Pac Regional Sci Conf Orgn, 85-, pres, 77-85; mem exec comt, Int Inst Fisheries Econ & Trade, 82-86. *Mem:* Social Sci Fedn Can (vpres, 81-83); Can Regional Sci Asn (pres, 83-85); Can Econ Asn (vpres, 72-73). *Res:* Academic articles, book contributions and consulting reports in areas of fisheries economics, fisheries resource management and fisheries policy. *Mailing Add:* 2341 Lawson Ave West Vancouver BC V7V 2E5 Can

COPLAN, MICHAEL ALAN, b Cleveland, Ohio, Apr 26, 38; c 2. CHEMICAL PHYSICS, SPACE PHYSICS. *Educ:* Williams Col, BA, 60; Yale Univ, MS, 61, PhD(electrolytic conductance), 63. *Prof Exp:* NIH res fel electrochem, Univ Paris, 63-64, NATO fel, 64-65; res assoc, Univ Chicago, 65-67; from res asst prof to res assoc prof, 67-81, RES PROF INST PHYS SCI & TECHNOL, UNIV MD, 81-, DIR CHEM PHYSICS PROG, 85- *Concurrent Pos:* Vis prof, Univ Bern, Switz, 79-80. *Honors & Awards:* Pub Serv Group Achievement Award, NASA, 79. *Mem:* Am Phys Soc; Sigma Xi. *Res:* Chemical physics; atomic and molecular collisions; charged particle optics; space physics. *Mailing Add:* Inst Phys Sci & Technol Univ Md Comput & Space Sci Bldg College Park MD 20742

COPLAN, MYRON J, b Chicago, Ill, Jan 5, 22; m 52; c 5. MEMBRANE SYNTHESIS & OPERATIONS. *Educ:* Brooklyn Col, BA, 43. *Prof Exp:* Prod supt, Montrose Chem Co, 43-48; head chem eng, Inst Textile Technol, 48-51; sr res asst, Fabric Res Labs Inc, 51-61, dir, 61-65, vpres, 65-72; dir, Albany Int Co, 72-79, sr corp scientist, 79-83, vpres, 83-87; PRES, INTELLEQUITY, 87- *Concurrent Pos:* Lectr, Northeastern Univ, 53-55; rev ed, Indust & Eng Chem, 65-68. *Honors & Awards:* IR-100. *Mem:* Am Chem Soc; Am Inst Chem Engrs; Fiber Soc; Textile Inst; NY Acad Sci. *Res:* Polymer physical and organic chemistry; fiber technologies; rheology of spinning; morphology and function of polymer systems; membrane structure, product and application in fluid separations such as reverse osmosis, dialysis, gas purification. *Mailing Add:* 47 Speen St Natick MA 01760-4114

COPLEY, ALFRED LEWIN, b Germany, June 19, 10; nat US; wid; c 1. PHYSIOLOGY, EXPERIMENTAL MEDICINE. *Educ:* Univ Heidelberg, MD, 35; Univ Basel, MD, 36. *Hon Degrees:* Dr Med, Univ Heidelberg, 72. *Prof Exp:* Res assoc med, Univ Basel, 36-37; intern, Trinity Lutheran Hosp, Mo, 39; res assoc, Hixon Lab Med Res, Univ Kans, 40-42; asst resident med, Goldwater Mem Hosp, New York, 42-43; head lab exp surg, Univ Va, 43-44, res assoc prev med & bact, 44-45; res assoc lab cellular physiol, NY Univ, 45-49, res fel hemat, Mt Sinai Hosp, 48-49; res assoc med & asst clin prof, NY Med Col, 49-52; sr researcher & head lab blood & vascular physiol, Res Labs, Int Children's Ctr, Paris, 52-55; sr researcher, Nat Inst Hyg & head res lab microcirculation, Nat Ctr Blood Transfusion, 55-57; dir exp res vascular dis, Head Med Res Labs, Charing Cross Hosp, London, 57-59; assoc prof physiol, 60-61, head hemorrheol lab, 60-65, prof physiol, 62-64, res prof pharmacol, New York Med Col, 65-74, res prof med, 72-74; RES PROF LIFE SCI & BIOENG & DIR LAB BIORHEOL, POLYTECH UNIV, 74- *Concurrent Pos:* Celler fel, Univ Kans, 40; fel surg, Univ Va, 43-44; res lab cellular physiol, Grad Sch Arts & Sci, NY Univ, 45- 49; gen lectr, Int Rheol Cong, Neth, 48; dir, AEC Proj, 49-52; chief investr, Off Naval Res Proj, 50-52; mem med adv bd, Nat Blood Res Found, 54-; mem, Marine Biol Lab Corp, Woods Hole, 54-; gen lectr, Int Cong Blood Transfusion, Japan, 60; co-ed-in-chief & co-founder, Biorheol, 62-; co-ed, Rheologica Acta; assoc chief of staff res & educ, Vet Admin Hosp, East Orange, NJ, 65-67, chief hemorrhage & thrombosis res labs, 65-71; chmn sci organizing comt & conf chmn, Int Conf Hemorheol, Iceland, 66; adj res prof bioeng, NJ Inst Technol, 67-; invited lectr, Int Cong Rheol, Kyoto, Japan, 68, St Louis, France, 72; co-ed-in-chief & founder, Thrombosis Res, 72-; co-founder & adv, Acupuncture & Electrotherapeutic Res, 75-; co-ed-in-chief & founder, Clin Hemorheology, 81-; vis prof path, Royal Col Surgeons Eng, 59. *Honors & Awards:* Poiseuille Gold Medal, Int Cong Biorheol, 72. *Mem:* Fel AAAS; Am Physiol Soc; Soc Study Blood; Int Soc Hemorheol (pres, 66-69); Int Soc Biorheol (pres, 69-72, past pres, 72-78). *Res:* Blood clotting; mechanisms of thrombosis, hemorrhage and hemostasis; blood vessel wall; surface hems rheology and adsorption of proteins; comparative hematology; blood platelets; physiology of the spleen; immunology; brain and liver metabolism; cholinesterase; radiobiology; biorheology of hair; experimental tuberculosis; microcirculation; hemorheology; survey and studies of snake venoms; biorheology; endoendothelial fibrin ogenin lining theory; rostulation of vessel-blood organ. *Mailing Add:* Lab Biorheol Polytech Univ 333 Jay St Brooklyn NY 11201

COPLEY, LAWRENCE GORDON, b Reading, Eng, July 17, 39. ENGINEERING MECHANICS. *Educ:* Univ Queensland, BMechEng, 60; Harvard Univ, SM, 62, PhD(appl physics), 65. *Prof Exp:* Mech engr, Queensland Railways, Australia, 60-61, res engr appl physics, 65-66; sr scientist, Cambridge Acoust Assocs, Inc, 66-70; CONSULT, 70- *Mem:* Acoust Soc Am; Inst Noise Control Eng. *Res:* Environmental acoustics and vibration. *Mailing Add:* PO Box 479 Needham MA 02192

COPLIEN, JAMES O, b Monroe, Wis, Jan 7, 54; m 78; c 3. OBJECT-ORIENTED DESIGN, DEVELOPMENT PROCESS. *Educ:* Univ Wis-Madison, BS, 77, MS, 79. *Prof Exp:* Mem tech staff develop, 79-82, appl res, 82-90, MEM TECH STAFF RES, AT&T BELL LABS, 90-; INSTR, TECHNOL EXCHANGE, ADDISON-WESLEY, 90- *Concurrent Pos:* Consult, Sun Microsysts, 89-90. *Res:* Advanced uses of the Ctt programming language and object- oriented design; large scale system development process dynamics. *Mailing Add:* AT&T Bell Labs Rm IHC 1G341 1000 E Warrenville Rd Naperville IL 60566-7050

COPLIN, DAVID LOUIS, b Albuquerque, NMex, July 7, 45; m 68; c 2. PLANT PATHOLOGY. *Educ:* Univ Calif, Davis, BS, 67; Univ Wis, Madison, MS, 71, PhD(plant path & bact), 72. *Prof Exp:* Res assoc plant path, Univ Nebr, 72-74; asst prof, 74-80, ASSOC PROF PLANT PATH, OHIO STATE UNIV, 80- *Concurrent Pos:* Assoc ed, Phytopath, 78-80 & 86-88 & Plant Dis, 81-83; co-mgr, USDA Competitive Grants Prog, 82. *Mem:* Am Phytopath Soc; Am Soc Microbiol; Sigma Xi. *Res:* Molecular genetics and physiology of plant pathogenic bacteria. *Mailing Add:* Dept Plant Path Ohio State Univ Columbus OH 43210

COPP, ALBERT NILS, b Aurora, Ill, Feb 22, 37; m 60; c 3. CERAMICS ENGINEERING. *Educ:* Univ Mo Rolla, BS, 62; Pa State Univ, MS, 65, PhD(ceramic sci), 69. *Prof Exp:* Develop engr, Western Elec Co, 65-66; sr scientist, Basic Incorp, 69-74; dir res, C-E Basic, Div Combustion Eng, 74-80, dir res, C-E Refractories, 80-89, EXEC VPRES, BASIC INC, 89- *Mem:* Am Soc Testing & Mat; Am Ceramic Soc; Am Inst Mining, Metall & Petrol Engrs. *Res:* Development of granular and shaped basic refractories used by the metals refining industry; development and improvement of methods to extract and process refractory raw materials. *Mailing Add:* Basic Inc PO Box 600 Betsville OH 44815

COPP, DOUGLAS HAROLD, b Toronto, Ont, Jan 16, 15; m 39; c 3. PHYSIOLOGY. *Educ:* Univ Toronto, BA, 36, MD, 39; Univ Calif, PhD(biochem), 43. *Hon Degrees:* LLD, Univ Toronto, 70, Queen's Univ, Ont, 70; DSc, Univ Ottawa, 73, Acadia Univ, NS, 75, Univ BC, 80; FRCP(C), 74. *Prof Exp:* Lectr biochem, Univ Calif, 42-43, from instr to asst prof physiol, 44-50; head dept, 50-80, PROF PHYSIOL, UNIV BC, 50- *Concurrent Pos:* Consult, Nat Res Coun, DC, 48-49; mem subcomt human applns, Adv Comt Isotope Distribution, US AEC, 49-50; mem, Panel Radiation Protection, Defense Res Bd Can, 51-59, chmn, 57-59; mem assoc comt dent res, Nat Res Coun Can, 53-59, chmn, 57-59, mem adv comt med res, 54-57; chmn, Gordon Res Conf Bones & Teeth, 57; mem sci secretariat, UN Int Conf Peaceful Uses Atomic Energy, 58; pres, Nat Cancer Inst Can, 68-70; Beaumont lectr, 70; mem, Coun Royal Soc Can, 74-84; coordr health sci, Univ BC, 75-77; Mem, Sci Coun of B C, 81-88. *Honors & Awards:* Gairdner Found Award, 67; Nicolas Andry Award, Asn Bone & Joint Surgeons, 68; Officer, Order of Can, 71, Companion, 80; Jacobaeus Mem lectr, Gothenbourg, 71 & Helsinki, 80; Flavelle Medal, Royal Soc Can, 72; Steindler Award, Orthop Res Soc, 74; Gold Medal, Sci Coun BC, 80; William F Neuman Award, Am Soc for Bone And Mineral Res, 1983. *Mem:* Am Physiol Soc; Soc Exp Biol & Med; Endocrine Soc; Can Physiol Soc (pres, 63-64); Royal Soc Can; Royal Soc London. *Res:* Iron and bone metabolism; fission product metabolism; severe phosphorus deficiency; regulation of blood calcium; parathyroid function; calcitonin discovery; bone blood flow; ultimobranchial function; corpuscles of stannius. *Mailing Add:* 4755 Belmont Ave Vancouver BC V6T 1A8 Can

COPPA, NICHOLAS V, b Upper Darby, Pa, July 29, 55; m 87; c 3. HIGH TEMPERATURE SUPERCONDUCTIVITY, DIFFUSIONLESS SOLID STATE CHEMISTRY. *Educ:* Syracuse Univ, BA, 77, MS, 81; Temple Univ, PhD(chem), 90. *Prof Exp:* Teaching asst chem, Syracuse Univ, 78-80; scientist, Ayerst Labs Res Inc, Subsid Am Home Prod, 85-86; teaching asst phys chem, Temple Univ, 87-89; STAFF MEM, LOS ALAMOS NAT LAB, 90- *Concurrent Pos:* Res asst, Syracuse Univ, 79-80; scientist, Diabetes Res Ctr, Univ Pa, 82-85; postdoctoral fel, Los Alamos Nat Lab, 90- *Mem:* Mat Res Soc; Am Chem Soc; AAAS. *Res:* High temperature superconductors and precursors; exploration in non-oxide high temperature superconductors; processing kinetics and thermodynamics. *Mailing Add:* Explor Res & Develop MS K763 Los Alamos Nat Lab Los Alamos NM 87545

COPPAGE, WILLIAM EUGENE, b Geary, Okla, Nov 24, 34; div; c 2. ALGEBRA. *Educ:* Tex A&M Univ, BA, 55, MS, 56; Ohio State Univ, PhD(math), 63. *Prof Exp:* Assoc prof math, Ind State Col, 63-64; asst prof, 64-67, coord dept math, 65-66, assoc prof, 67-88, PROF MATH INCARNATE WORD COL, WRIGHT STATE UNIV, 88- *Concurrent Pos:* Consult, Inst Defense Anal, 63-69; consult, Aerospace Med Res Labs, 66-71. *Mem:* Am Math Soc; Math Asn Am; Nat Coun Teachers Math. *Res:* Non-associative algebra and applications to geometry. *Mailing Add:* Incarnate Wood Col 4301 Broadway San Antonio TX 78209

COPPEL, CLAUDE PETER, b Zweibrucken, Ger, Aug 14, 32; nat US; m 60; c 2. PHYSICAL CHEMISTRY, PETROLEUM ENGINEERING. *Educ:* Univ Denver, BS, 54; Univ Calif, PhD(chem), 58. *Prof Exp:* Asst, Univ Calif, 54-55, asst radiation lab, 55-58; res scientist, Marathon Oil Res Ctr, Colo, 58-63; res chemist, Arthur D Little, Inc, Mass, 63-66; from res chemist to sr res chemist, 66-74, SR RES ASSOC, CHEVRON OIL FIELD RES CO, 74- *Mem:* AAAS; Am Chem Soc; Soc Petrol Engrs; Sigma Xi. *Res:* Colloidal and surface chemistry; high temperature physical measurements; fluid flow in porous media; well stimulation technology. *Mailing Add:* Chevron Oil Field Res Co PO Box 446 La Habra CA 90631

COPPEL, HARRY CHARLES, b Galt, Ont, Jan 2, 18; nat US; m 50; c 2. ENTOMOLOGY. *Educ:* Ont Agr Col, BSA, 43; Univ Wis, MSc, 46; NY State Col Forestry, Syracuse Univ, PhD(forest entom), 49. *Prof Exp:* Agr res officer, Entom Lab Sci Serv, Can Dept Agr, 43-57; from asst prof to assoc prof, 57-65, PROF ENTOM, UNIV WIS-MADISON, 65- *Mem:* Entom Soc Am; Entom Soc Can; Int Orgn Biol Control. *Res:* Biological control of forest insect pests. *Mailing Add:* Dept Entom Univ Wis 337 Russell Lab Madison WI 53706

COPPENGER, CLAUDE JACKSON, b Beaumont, Tex, Oct 12, 27. PHYSIOLOGY. *Educ:* Stephen F Austin State Col, BS, 51; Tex A&M Univ, MS, 53, PhD(physiol), 64. *Prof Exp:* Asst physiol, Tex A&M Univ, 51-53, biochem, 54-55; instr, Pub Sch, Tex, 57-58; assoc prof zool, Amarillo Jr Col, 58-61; res assoc radiation biol, Tex A&M Univ, 61-63; from asst prof to assoc prof physiol, 63-74, chmn dept physiol & behav biol, 68-78, PROF BIOL, SAN FRANCISCO STATE UNIV, 74- *Mem:* AAAS; Am Zool Soc; Sigma Xi. *Res:* Mammalian physiology; physiological and behavioral sexual differentiation of mammals. *Mailing Add:* Dept Biol San Francisco State Univ San Francisco CA 94132

COPPENS, ALAN BERCHARD, b Hollywood, Calif, June 26, 36; c 2. PHYSICS, ACOUSTICS. *Educ:* Cornell Univ, BEngPhys, 59; Brown Univ, MS, 62, PhD(physics), 65. *Prof Exp:* Asst prof, 64-69, ASSOC PROF PHYSICS, US NAVAL POST GRAD SCH, 69- *Mem:* Acoust Soc Am; fel Scientist's Inst Pub Info. *Res:* Finite-amplitude acoustic processes; properties of liquids; ocean acoustics. *Mailing Add:* Dept Physics Naval Postgrad Sch Monterey CA 93940

COPPENS, PHILIP, b Amersfoort, Holland, Oct 24, 30; m 56; c 3. CRYSTALLOGRAPHY, CHEMISTRY. *Educ:* Univ Amsterdam, Drs, 57, PhD(crystallog), 60. *Prof Exp:* Res asst, Weizmann Inst, 57-60, res assoc, 62-65; res assoc, Brookhaven Nat Lab, 60-62, scientist, 65-68; assoc prof, 68-71, PROF CHEM, STATE UNIV NY BUFFALO, 71- *Concurrent Pos:* Vis prof, Fordham Univ, 66-67; res grants, NSF, Petrol Res Fund, AEC & NIH, 69; mem Nat Comt Crystallog, Nat Res Coun-Nat Acad Sci, 72-75; adj prof, Univ Grenoble, 74-75, dept appl & eng sci, Cornell Univ, 83-; mem, Comn Charge Spin & Momentum Densities, Int Union Crystallog, 72-81 & Comn Neutron Diffraction, 75-78; mem comt facil, NSF, 79-80; mem exec comt, Int Union Crystallog, 87- *Mem:* AAAS; Am Crystallog Asn (pres, 78); Am Chem Soc; corresp mem, Royal Dutch Acad Sci. *Res:* Crystal structure determination; crystallographic computing; neutron diffraction; electron density determination by accurate diffraction methods; crystallography at liquid helium temperatures; crystallographic applications of synchrotron radiation. *Mailing Add:* 565 N Forest Rd Williamsville NY 14221

COPPER, JOHN A(LAN), b Minneapolis, Minn, Sept 5, 34; m 57; c 2. GAS DYNAMICS. *Educ:* Univ Minn, BAE & BBA, 57; Univ Southern Calif, MS, 59; Calif Inst Technol, AE, 61. *Prof Exp:* Assoc develop engr, Coop Wind Tunnel, Calif Inst Technol, 57-59, develop engr, 59-60; mem tech staff, Nat Eng Sci Co, 60-61; res engr, Douglas Aircraft Co, Inc, 61-63, sect chief hypervelocity facil, 63-68, br chief, 68-76, CHIEF ENGR, McDONNELL DOUGLAS ASTRONAUT CO, McDONNELL DOUGLAS CORP, 76- *Mem:* Am Inst Aeronaut & Astronaut; Sigma Xi. *Res:* Missile flight mechanics. *Mailing Add:* 1212 Via Coronel Palos Verdes Estates CA 90274

COPPER, PAUL, b Surabaya, Indonesia, May 6, 40; Can citizen; div; c 2. PALEONTOLOGY. *Educ:* Univ Sask, BA, 60, MA, 62; Univ London, DIC & PhD(paleontol), 65. *Prof Exp:* Nat Res Coun fel paleontol, Queen's Univ, Ont, 65-67; asst prof geol, 67-70, assoc prof, 70-75, chmn dept, 74-77, PROF GEOL, LAURENTIAN UNIV, 75- *Honors & Awards:* Thomas Huxley Prize, London, 68. *Mem:* Geol Asn Can; Paleont Soc; Paleont Asn. *Res:* Brachiopod morphology, ecology and evolution, especially Atrypida; evolution and paleoecology of reef ecosystems; mass extinction and recovery of ecosystems. *Mailing Add:* Dept Geol Laurentian Univ Sudbury ON P3E 2C6 Can

COPPERSMITH, DON, b St Marys, Penn, July 4, 50; m 80; c 2. CRYPTOGRAPHY. *Educ:* Mass Inst Technol, BS, 72; Harvard Univ, MS, 75, PhD(Math), 77. *Prof Exp:* RES STAFF MEM, IBM, 77- *Concurrent Pos:* Assoc ed, Trans Info Theory, Inst Elec & Electronics Engrs. *Mem:* Int Assoc Cryptol Res. *Res:* Cryptography, conventional and public key; algebraic complexity of computation; combinatorics and coding theory. *Mailing Add:* IBM Corp PO Box 218 Yorktown Heights NY 10598

COPPERSMITH, FREDERICK MARTIN, b New York, NY, June 7, 47; m 74; c 2. RESEARCH ADMINISTRATION. *Educ:* Columbia Univ, BS, 68; NY Univ, ME, 72. *Prof Exp:* Prod engr, proj engr, sr engr & siting engr, 68-85, mgr res & develop, 85-90, DIR RES & DEVELOP PLANNING & OPERS, CONSOL EDISON CO, 90- *Mem:* Am Inst Chem Engrs; Am Soc Metals. *Res:* Energy conversion, distribution and transmission; emissions control and environmental science. *Mailing Add:* Consol Edison NY Rm 1428 Four Irving Pl New York NY 10003

COPPI, BRUNO, b Gonzaga, Italy, Nov 19, 35; m 63; c 3. PHYSICS. *Educ:* Milan Polytech Inst, Dr, 59. *Prof Exp:* Res leader & lectr, Milan Polytech Inst & Univ Milan, 60-61; mem plasma physics lab, Princeton Univ, 61-63; asst prof & res asst physics, Univ Calif, San Diego, 64-67; mem, Inst Advan Study & Princeton Univ, 67-69; PROF PHYSICS, MASS INST TECHNOL, 69- *Concurrent Pos:* Ital Acad Sci grant, 62; Ital Acad Sci fel, 62, hon guest, 72; mem, Int Ctr Theoret Physics, Trieste, 66; consult, Princeton Univ & Am Sci & Eng, Cambridge, Mass, 69- & Naval Res Lab, Washington, DC; sci counr to bd dirs, Nat Comt Nuclear Energy Italy, 71; prof on leave, Advan Norm Sch, Pisa, 73; hon mem, Int Sch Plasma Physics, 75; vis scientist, Princeton Univ, 62-63. *Honors & Awards:* Gold Medal, Milan Polytech Inst, 60; US Cert Contrib to Fusion Res, 77. *Mem:* Fel Am Acad Arts & Sci; fel Am Phys Soc; Ital Physics Soc. *Res:* Basic plasma physics; controlled thermonuclear fusion research; astrophysics; space physics; neutron transport theory; Alcator, Frascati Torus, and Ignitor experiments. *Mailing Add:* Dept Physics Mass Inst Technol Rm 26-217 Cambridge MA 02139

COPPIN, CHARLES ARTHUR, b Belton, Tex, July 18, 41; m 64; c 2. MATHEMATICAL ANALYSIS. *Educ:* Southwestern Univ, Tex, BS, 63; Univ Tex, Austin, MS, 65, PhD(math), 68. *Prof Exp:* Instr math, Univ Tex, Austin, 68; asst prof, 68-75, ASSOC PROF MATH, UNIV DALLAS, 75-, CHMN DEPT, 73- *Mem:* Am Math Soc; Math Asn Am. *Res:* Integration theory; primitive dispersion sets; relations on topological spaces. *Mailing Add:* 3213 Konet St Irving TX 75060

COPPINGER, RAYMOND PARKE, b Boston, Mass, Feb 7, 37; m 58; c 2. CANINE BEHAVIOR, AGRICULTURE STUDIES. *Educ:* Boston Univ, AB, 59; Univ Mass, Amherst, MA, 64; Amherst Col via Univ Mass, PhD(biol), 68. *Prof Exp:* PROF BIOL, HAMPSHIRE COL, 69- *Concurrent Pos:* Res assoc, Dept Biol, Amherst Col, 68-70; dir, Livestock Dog Proj, 76-; field work, Eng, Italy, Yugoslavia, Turkey, China, Spain, Venequela & Arg. *Honors & Awards:* Chevron Conservation Award, 90. *Mem:* Am Soc Mammalogists; Sigma Xi. *Res:* Comparative biology and developmental behavior of canids; evolutionary and adaptive significance of acceptance or rejection of novelty; feeding behavior of birds exposed to novel food items. *Mailing Add:* 401 Chestnut Hill Rd Montague MA 01351

COPPOC, GORDON LLOYD, b Larned, Kans, Nov 11, 39; m 62; c 2. PHARMACOLOGY, VETERINARY PHARMACOLOGY. *Educ:* Kans State Univ, BS, 61, DVM, 63; Harvard Univ PhD(pharmacol), 68. *Prof Exp:* Fel, Harvard Univ, 63-66; instr pharmacol, Sch Med, Univ NC, 66-67; res pharmacologist, US Air Force Sch Aerospace Med, 67-69; res fel, Ben May Lab Cancer Res, Univ Chicago, 69-71; asst prof, 71-73, assoc prof pharmacol, Sch Vet Sci & Med, 73-77, PROF VET PHARMACOL, SCH VET MED, PURDUE UNIV, 77-, HEAD, DEPT VET PHARMACOL & PHYSIOL, 79- *Concurrent Pos:* Vis scholar, Dept Pharmacol, Univ Wash, 90. *Mem:* AAAS; Am Vet Med Asn; Am Acad Vet Pharmacol & Therapeut (pres-elect, 89-91, pres, 91-93); Asn Am Vet Med Col; Am Asn Vet Physiol & Pharmacol; Sigma Xi; Am Soc Pharmacol Exp Therapeut. *Res:* Biochemical pharmacology; Pharmacokinetics; chemotherapy; computer applications to education; interactive education. *Mailing Add:* Sch Vet Med Purdue Univ West Lafayette IN 47907

COPPOC, WILLIAM JOSEPH, b Cumberland, Iowa, July 14, 13; m 39; c 2. CHEMISTRY. *Educ:* Ottawa Univ, Kans, BSc, 35; Rice Univ, MA, 37, PhD(chem), 39. *Hon Degrees:* DSc, Ottawa Univ, 55. *Prof Exp:* Chemist, Port Arthur, Tex, 39-44, asst to asst chief chemist, 44-47, 48-49, actg asst supvr, Grease Res, Beacon, NY, 47-48, asst dir, Res, 49-51, assoc dir, New York, 51-53, dir, 53-54, mgr res, Beacon, 54-57, res & develop, 57-60, sci planning & Info, 60-65, gen mgr res & tech dept, 65-68, vpres res & tech dept, 68-71, vpres environ protection, 71-78, CONSULT, TEXACO, INC, 78- *Concurrent Pos:* Mem, Gordon Res Conf Coun, 49-56, chmn, 55-56; mem environ studies bd, Nat Res Coun-Nat Acad Sci, 74-77; Woodrow Wilson vis fel, 81-84. *Honors & Awards:* Environ Consult Distinguished Serv, Am Soc Mining Mineral & Petrol Eng, 83. *Mem:* AAAS; Am Chem Soc; fel Am Inst Chem; Soc Automotive Eng; Sigma Xi. *Res:* Coagulation of colloidal solutions; composition of hydrates; development of block type greases; ball and roller bearing greases; various industrial lubricants; soluble and non-soluble cutting oils; research administration; atmospheric chemistry; fuel technology; petroleum engineering. *Mailing Add:* 4650 54th Ave S 301 W College Harbor St Petersburg FL 33711

COPPOCK, CARL EDWARD, b Dayton, Ohio, Dec 1, 32; m 59; c 3. ANIMAL NUTRITION. *Educ:* Ohio State Univ, BS, 54; Tex A&M Univ, MS, 55; Univ Md, PhD(dairy sci), 64. *Prof Exp:* Dairy husbandman, Agr Res Serv, USDA, 58-64; asst prof dairy cattle nutrit, Cornell Univ, 64-69, assoc prof animal sci, 69-77; PROF ANIMAL SCI, TEX A&M UNIV, 77- *Concurrent Pos:* Animal husbandman, Int Voluntary Serv, Laos, 56-58; vis prof, Purdue Univ, 70-71. *Mem:* Am Dairy Sci Asn; Am Soc Animal Sci; Am Inst Nutrit; Am Forage & Grassland Soc; Fedn Am Socs Exp Biol. *Res:* Energy metabolism of lactating cows; nonprotein nitrogen utilization by lactating cows; glucose availability to high producing cows; chloride-bicarbonate relationships in lactating cows. *Mailing Add:* Dept Animal Sci Rm 218-B Kleberg Ctr Tex A&M Univ College Station TX 77843

COPPOCK, DONALD LESLIE, b Washington, DC, Sept 25, 47. REGULATION OF CELL PROLIFERATION. *Educ:* Swathmore Col, BS, 69; Univ Calif, PhD(biol), 82. *Prof Exp:* Postdoctoral fel cancer biol, Dana Farber Cancer Inst, 82-85; res assoc biol, Brandeis Univ, 85-88; DIR ONCOL RES LAB, CANCER BIOL, WINTROP UNIV HOSP, 88-; ASST PROF, DEPT MED, MED SCH, STATE UNIV NY, STONY BROOK, 89- *Mem:* Am Soc Cell Biol; Am Asn Cancer Res. *Res:* Investigation of the molecular basis for aberrant cell proliferation in cancer cells with emphasis on regulation of the quiescent state. *Mailing Add:* 222 Station Plaza N No 300 Mineola NY 11501

COPPOCK, GLENN E(DGAR), b Cullman, Ala, May 18, 24; m 54; c 3. AGRICULTURAL ENGINEERING. *Educ:* Auburn Univ, BS, 49; Okla State Univ, MS, 55. *Prof Exp:* Res agr engr, tillage mach lab, 49-52, Okla State Univ, 53-56, cotton field sta, Calif, 56-57, res agr engr, USDA, Fla Dept Citrus, Univ Fla, 57-86; RETIRED. *Mem:* Am Soc Agr Engrs. *Res:* Harvesting equipment and methods for peanuts, castor beans and citrus; mechanization of citrus harvest. *Mailing Add:* 970 14th St NE Winter Haven FL 33881

COPPOCK, ROBERT WALTER, b Battle Creek, Mich, Aug 21, 42; m 65; c 2. PHARMACOLOGY, VETERINARY MEDICINE. *Educ:* Andrews Univ, BSc, 65; Mich State Univ, DVM, 69; Am Bd Vet Toxicol, dipl, 78; Okla State Univ, MS, 81; Univ Ill PhD, 84, Am Bd Toxicol, 87. *Prof Exp:* Clin vet pharmacologist, Parke Davis & Co, 70-74; vet med, Inter Ridge Bet Clin, 75-77; vet toxicologist & pharmacologist, dept physiol sci, Col Vet Med, Okla State Univ, 77-79; vet toxicologists, Univ Ill, Urbana, 80-84; HEAD, CLIN INVEST BR, ALTA ENVIRON CTR, 84- *Concurrent Pos:* Legal toxicol, field invests. *Mem:* Vet Med Asn BC; Am Vet Med Asn; Can Vet Med Asn; Sigma Xi; Soc Toxicol; Am Acad Vet & Comp Toxicol; Am Cad Vet Disaster Med. *Res:* Oil field toxicology; mycotoxins, heavy metals. *Mailing Add:* Clin Invest Br Bag 4000 Vegreville AB T0B 4L0 Can

COPPOLA, ELIA DOMENICO, b S Salvatore Telesino, Italy, Aug 2, 41; m 72; c 3. ANALYTICAL CHEMISTRY. *Educ:* Southern Conn State Col, BA, 67; Rensselaer Polytech Inst, PhD(anal chem), 72. *Prof Exp:* Chem technician, Upjohn Co, Conn, 64-67; res chemist anal chem, Conn Agr Exp Sta, 71-76; sr res chemist, 72-76 GROUP LEADER ANALYSIS, OCEAN SPRAY CRANBERRIES, MIDDLEBORO, 76- *Mem:* Am Chem Soc; Inst Food Technologists; Asn Official Anal Chemists. *Res:* The development of new analytical methods in food analysis; high-pressure liquid chromatography and thin-layer chromatography of food ingredients and contaminants; gas chromatography and high performance liquid chromatography of pesticide residues. *Mailing Add:* Ocean Spray Cranberries One Ocean Spray Dr Middleboro MA 02349-0001

COPPOLA, JOHN ANTHONY, endocrinology, pharmacology, for more information see previous edition

COPPOLA, PATRICK PAUL, b Buffalo, NY, June 30, 17; m 44; c 5. PHYSICAL CHEMISTRY, ELECTRONICS. *Educ:* Canisius Col, BS, 41, MS, 51. *Prof Exp:* Chemist, Anal Lab, Union Carbide & Carbon Corp, 41-42, chemist group leader, Metals Lab, 42-43, chemist, Res Lab, Manhattan Dist Contract, 43-46; res chemist, Philips Res Lab, 46-55; sr res engr, Radio Corp Am, 55-56; physicist & leader res group, Gen Elec Co, 56-60, mgr chem & phys electronics adv develop, 60-66, mgr mat develop, 66-68, mgr eng, Visual Commun Prod Dept, 68-72, mgr eng, Video Display Equip Oper, 72-80; RETIRED. *Mem:* Am Chem Soc; NY Acad Sci; fel Am Inst Chem. *Res:* Electronic emission; vacuum physics; semiconductors; materials and processes; new business planning and organization. *Mailing Add:* 301 Fayette Dr Fayetteville NY 13066

COPPOLILLO, HENRY P, b Cervicati, Italy, July 27, 26; US citizen; m 62; c 3. PSYCHIATRY. *Educ:* Univ Rome, MD, 55. *Prof Exp:* Intern, Cook County Hosp, Ill, 55-56; res psychiat, Univ Chicago, 56-59; asst attend physician, Michael Reese Hosp, 59-61, pediat liaison psychiatrist, 61-65, assoc physician, 61-66, asst chief child psychiat, 63-65; asst prof, Med Sch, Univ Mich, 66-68, dir day care serv, Children's Psychiat Hosp, 66-68, assoc prof psychiat, Univ, 68-71, prof, 71, asst to chmn dept, 68-69, assoc chmn, 69-70, actg chmn, 70; prof psychiat & dir div child psychiat, Vanderbilt Univ, 71-76; PROF PSYCHIAT & DIR DIV CHILD PSYCHIAT, UNIV COLO MED CTR, 76- *Concurrent Pos:* Fel, Michael Reese Hosp, Chicago, 61-63; res asst psychiat, Northwestern Univ, 59-61, instr, 61-63; child psychiat consult, McLean County Health Clin, Ill, 59-66 & Med Sch, Univ Wis, 64; lectr, Col Lit, Sci & Arts, Univ Mich, 67; fac asst, Mich Psychoanal Inst, 69-70, lectr, 70-71. *Mem:* Fel Am Psychiat Asn; Am Acad Child Psychiat; fel Am Col Psychiat; Sigma Xi. *Res:* Child and adult psychiatry and psychoanalysis. *Mailing Add:* 5600 S Greenwood Piz Blvd Englewood CO 80111

COPSON, DAVID ARTHUR, b Boston, Mass, June 16, 18; c 7. BIOPHYSICS, MICROBIOLOGY. *Educ:* Univ Mass, BS, 40; Mass Inst Technol, PhD(tech), 53. *Prof Exp:* Group mgr res div, Raytheon Co, 53-58, consult microwave heating, 58-60; prof biol, Univ PR, Mayaguez 60-84; RETIRED. *Concurrent Pos:* Consult, Matrix Publ Co, IBM Corp, Whirlpool Corp, Sunbeam, FMC Corp & Am Can Co, 58- *Mem:* AAAS; Int Microwave Power Inst; Biophys Soc. *Res:* Microwave heating and biophysics; athermic microwave effects; radiation and theoretical biology; information theory and biological communication; radiation hazards; automation of telecommunications analysis; electromagnetic spectrum characteristics, documentation and classification; integration of new technology in science education; informational bioelectromagnetics; microwave heating; food and nutrition; radiation and environment; United State corporate objectives. *Mailing Add:* Dept Biol Univ PR Mayaguez PR 00709

COPSON, HARRY ROLLASON, corrosion; deceased, see previous edition for last biography

COPULSKY, WILLIAM, b Zhitomir, Russia, Apr 4, 22; US citizen; m 48; c 3. INDUSTRIAL CHEMISTRY. *Educ:* NY Univ, BA, 42, PhD(econ), 57; City Col New York, MBA, 48. *Prof Exp:* Asst res dir, J J Beruner & Staff, 46-48; asst to pres, R S Aries & Assocs, 47-51; mgr com develop, W R Grace & Co, 51-69, dir electronucleonics labs, 69-77, vpres oper serv group, 74-86; ASSOC PROF CHEM, BARUCH COL, 87- *Concurrent Pos:* Adj assoc prof, Baruch Col, 67-77. *Mem:* Am Chem Soc; Am Statist Asn; Am Mkt Asn. *Res:* Long range planning; forecasting of new technology; correlation techniques in forecasting; industrial chemical development. *Mailing Add:* 23-35 Bell Blvd Bayside NY 11360

CORAK, WILLIAM SYDNEY, b Philadelphia, Pa, Mar 10, 22; m 46; c 5. MATERIALS DEVELOPMENT, LOW TEMPERATURE PHYSICS. *Educ:* Univ Pa, BS, 43; Ohio State Univ, MS, 47; Univ Pittsburgh, PhD, 54. *Prof Exp:* Chem control engr, Davison Chem Corp, 43-44; res assoc, Johns Hopkins Univ, 44-45 & Ohio State Univ, 45-47; res engr, Westinghouse Res Labs, 47-55, mgr, Westinghouse Semi-Conductor Div, 55-64, mgr solid state technol, Sci & Technol Dept, Systs Develop Div, Westinghouse Defense Ctr, 64-80, sr adv scientist, Adv Technol Div, Westinghouse Electronics Systs Ctr, 80-88; CONSULT SOLID STATE TECHNOL, CTR NONDESTRUCTIVE EVALUATION, JOHNS HOPKINS UNIV, 88- *Mem:* Am Phys Soc; Inst Elec & Electronics Engrs; Sigma Xi; Am Soc Metals Int. *Res:* Low temperature gas thermodynamics; low temperature techniques; superconductivity; low temperature specific heats; materials and process development; integrated circuit technology; solid state physics. *Mailing Add:* 695 Carlisle Dr Arnold MD 21012

CORAM, DONALD SIDNEY, b Philadelphia, Pa, July 6, 45; m 66; c 2. GEOMETRIC TOPOLOGY, BATTLE MANAGEMENT. *Educ:* Univ Del, Newark, BA, 66; Univ Wis-Madison, MA, 67, PhD(math), 71. *Prof Exp:* Prof math, Okla State Univ, 71-85; MGR, BDM, INC, 85- *Concurrent Pos:* Sr lectr, US Fulbright Exchange Prog, Univ Zagrels, Yugoslavia, 83-84. *Mem:* Am Math Soc; Soc Indust & Appl Math; Armed Forces Commun & Electronics Asn; Inst Elec & Electronics Engrs. *Res:* Geometric topology; battle management, command, control and communications. *Mailing Add:* BDM Inc 7915 Jones Branch Dr McLean VA 22102-3396

CORAN, ARNOLD GERALD, b Boston, Mass, Apr 16, 38; m; c 3. PEDIATRIC SURGERY. *Educ:* Harvard Univ, AB, 59, MD, 63. *Prof Exp:* Intern surg, Peter Bent Brigham Hosp, Boston, Mass, 63-64, resident, 64-68; instr surg, Harvard Med Sch, 67-69; asst clin prof surg, George Washington Univ, 70-72; from asst prof to assoc prof, Univ Southern Calif Sch Med, 72-74, head physician pediat surg, Med Ctr, 72-74; PROF SURG & HEAD SECT PEDIAT SURG, UNIV MICH MED SCH, 74-, PROF PEDIAT, 83-; SURGEON-IN-CHIEF, C S MOTT CHILDREN'S HOSP, ANN ARBOR, MICH, 81- *Concurrent Pos:* Asst surg resident, Children's Hosp Med Ctr, Boston, Mass, 65-66, sr surg resident, 66, chief surg resident, 68, Peter Bent Brigham Hosp, 69; NIH fel, 65-66, grant, 86-87, 87-88, 88-89 & 89-90; clin fel, Am Cancer Soc, 67-68; res fel neonatal metab, Univ Hosp, Oslo, Norway, 69; ed, Pediat Rounds & Pediat Surg Int; prin investr, Sch Med, Univ Southern Calif, Los Angeles, 72-74, Sch Med, Univ Mich, 74-76, Towsley Found, 76-77, Upjohn Co, 77-79, Cutter Labs Nutrit Res, 78-79, Abbott Labs Nutrit Res, 79-80, Univ Mich Clin Res Ctr grant, 85- & 88-; examr, Am Bd Surg, 80-, consult, 89-; mem, Pre- & Postoperative Care Comt, Am Col Surgeons; Dozer vis prof, Ben Gurion Univ, Beer Sheba Israel, 90; coun mem, Sect Pediat Surg, PanAm Med Asn; chmn, Educ Comt, Am Pediat Surg Asn, 90-, gov, 91-, Am Col Surgeons, 91-; vis prof & guest lectr numerous orgns. *Honors & Awards:* Bronze Medal, Scand Soc Pediat Surg, 90. *Mem:* Am Acad Pediat; Am Col Chest Physicians; Am Col Surgeons; Am Fedn Clin Res; Am Geriat Soc; NY Acad Sci; Am Inst Nutrit; Am Soc Clin Nutrit; Am Surg Asn; Am Trauma Soc. *Res:* Pediatric trauma. *Mailing Add:* Pediat Surg F7516 Mott Childrens Hosp Univ Mich Med Ctr 1500 E Med Center Dr Box 0245 Ann Arbor MI 48109-0245

CORAN, AUBERT Y, b St Louis, Mo, Mar 24, 32; m 58, 84; c 3. PHARMACY, ORGANIC CHEMISTRY. *Educ:* St Louis Col Pharm, BS, 53, MS, 55. *Prof Exp:* Instr chem, St Louis Col Pharm, 53-55; res chemist, Mo, 55-62, res group leader, WVa, 62-64, scientist, Mo, 64-70, sect mgr, Rubber Chem Res, Ohio, 70-75, DISTINGUISHED SCI FEL, MONSANTO CO, 75- *Concurrent Pos:* Ed, Rubber Chem & Technol J, 78-83. *Honors & Awards:* Thomas Midgley Award, Am Chem Soc, 80,; Edgar M Queeny Award, Monsanto Co, 83; Colwyn Medal, Rubber & Plastic Inst, 84; Award Tech Excellence, Rubber Div, Am Chem Soc, 83, Thermoplastic Elastomers Award, 91. *Mem:* AAAS; Am Chem Soc; Sigma Xi. *Res:* Polymer and rubber science and technology; vulcanization kinetics and chemistry; thermoplastic elastomers; elastomer filler interactions. *Mailing Add:* Monsanto Co Akron OH 44313

CORAOR, GEORGE ROBERT, b Jacksonville, Ill, May 10, 24; m 47; c 3. ORGANIC CHEMISTRY. *Educ:* Ill Col, AB, 47; Univ Ill, PhD(org chem), 50. *Prof Exp:* Res assoc org chem, Mass Inst Technol, 50-51; sr chemist, 51-53, group leader, 53-55, sect head, 55-60, div head org chem, 60-76, SR CONSULT, E I DU PONT DE NEMOURS & CO, INC, 76- *Mem:* AAAS; Am Chem Soc; Sigma Xi. *Res:* Free radical reaction; coordination chemistry; photochemistry. *Mailing Add:* 1035 Springhouse Rd Allentown PA 18104

CORBASCIO, ALDO NICOLA, pharmacology, for more information see previous edition

CORBATO, CHARLES EDWARD, b Los Angeles, Calif, July 12, 32; m 57; c 3. GEOLOGY, GEOPHYSICS. *Educ:* Univ Calif, Los Angeles, BA, 54, PhD(geol), 60. *Prof Exp:* Instr geol, Univ Calif, Riverside, 59; from instr to asst prof, Univ Calif, Los Angeles, 59-66; assoc prof, 66-69, chmn, Dept Geol & Mineral, 72-80, PROF GEOL SCI, 69-, ASSOC PROVOST ACAD AFFAIRS, OHIO STATE UNIV, 87- *Mem:* Geol Soc Am; Am Geophys Union; Int Asn Math Geol; Am Inst Prof Geologists. *Res:* computer applications to geological problems; structural geology. *Mailing Add:* Off Acad Affairs Ohio State Univ 190 N Oval Mall Columbus OH 43210-1358

CORBATÓ, FERNANDO JOSE, b Oakland, Calif, July 1, 26; m 62, 75. PHYSICS, SOFTWARE SYSTEMS. *Educ:* Calif Inst Technol, BS, 50; Mass Inst Technol, PhD(physics), 56. *Prof Exp:* Res assoc, Mass Inst Technol, 56-59, asst dir prog res, Comput Ctr, 59-60, assoc dir, 60-63, dep dir, 63-66, group leader, Comput Systs Res Group, Lab Comput Sci, 63-72, co-head, Systs Res Div, 72-74, co-head, Automatic Prog Div, 72-74, assoc head comput sci & eng, 74-78, assoc prof, 62-75, Cecil H Green prof, 78-80, dir, comput & telecommun resources, 80-83, PROF COMPUT SCI & ENG, MASS INST TECHNOL, 65-, ASSOC HEAD COMPUT SCI & ENG, 83- *Concurrent Pos:* Mem comput sci & eng bd, Nat Acad Sci, 71-73. *Honors & Awards:* W W McDowell Award, Inst Elec & Electronic Engrs, 66; Harry Goode Memorial Award, Am Fedn Info Processing Socs, 80; Comput Pioneer Award, Inst Elec & Electronics Engrs Comput Soc, 82; A M Turing Award, Asn Comput Mach, Inc, 90. *Mem:* Nat Acad Eng; Asn Comput Mach; fel Inst Elec & Electronic Engrs; fel Am Acad Arts & Sci; Am Phys Soc; fel AAAS. *Res:* Computer operating systems; time-sharing systems; automatic programming and knowledge-based application systems. *Mailing Add:* Dept Comput Sci Mass Inst Technol 545 Technol Sq Rm 514 Cambridge MA 02139

CORBEELS, ROGER, b Kessel-Loo, Belgium, Apr 16, 36; m 62. FUEL SCIENCE, HIGH TEMPERATURE CHEMISTRY. *Educ:* Cath Univ Louvain, BS, 56, MS, 58, PhD(chem), 60. *Prof Exp:* Vis res assoc flame kinetics, Aerospace Res Labs, Wright-Patterson Air Force Base, Ohio, 62-64; sr chemist, 64-70, res chemist, 70-73, SR RES CHEMIST, TEXACO INC, 73- *Mem:* Am Chem Soc; Combustion Inst. *Res:* Kinetics of combustion reactions; coal chemistry. *Mailing Add:* Pine Ridge Dr Wappingers Falls NY 12590

CORBEIL, LYNETTE B, b Vancouver, BC, May 19, 38; m; c 2. PATHOLOGY. *Educ:* Univ Guelph, MSc, 65; Cornell Univ, PhD(immunol & pathogenic bacteriol), 74. *Prof Exp:* NIH Fel immunol, NY State Vet Col, Cornell Univ, 74-75; assoc prof immunol, Col Vet Med, Kans State Univ, 78-79; from assoc prof to prof vet microbiol, Wash State Univ, 82-87; NIH fel infectious dis, Sch Med, Univ Calif, San Diego, 75-78, assoc prof path & dir, Lab Animal Diag Lab, 79-82, ADJ PROF PATH, UNIV CALIF, SAN

DIEGO, 87-, COURTESY APPT, DEPT VET MICROBIOL & IMMUNOL, DAVIS, 89- *Concurrent Pos:* Prin investr, USDA grants; chair, Educ Comt, Am Asn Vet Immunol, 83-87; bd mem, Conf Res Workers in Animal Dis, 85-90, pres, 90; vis prof, Univ Calif, San Diego, 85-86. *Mem:* Can Vet Med Asn; Am Vet Med Asn; Am Soc Microbiol; Am Asn Immunol; Fedn Am Socs Exp Biol; Am Asn Vet Immunologists. *Res:* Haemophilus somnus; Tritrichomonas foetus. *Mailing Add:* Dept Path Med Ctr 8416 Univ Calif 225 Dickinson St San Diego CA 92103-8416

CORBEN, HERBERT CHARLES, b Dorset, Eng, Apr 18, 14; US citizen; m 41, 57; c 3. BIOMATHEMATICS, HISTORY & PHILOSOPHY OF SCIENCE. *Educ:* Univ Melbourne, MA & MSc, 36; Cambridge Univ, PhD(theoret physics), 39. *Prof Exp:* Lectr math & physics, Univ Col Armidale, NSW, 41; dean, Trinity Col & lectr math & physics, Univ Melbourne, 42-44, sr lectr, 45-46; from assoc prof to prof physics, Carnegie Inst Technol, 46-56; mem tech staff, Ramo-Wooldridge Corp, 54-55, 56-57, assoc dir res lab, 57-61; dir quantum physics lab, TRW, Inc, 61-68, chief scientist, Phys Res Ctr, 66-68; prof physics, Cleveland State Univ, 68-72, dean faculties & grad studies, 68-70, vpres acad affairs, 70-72; chmn, Phys Sci Div, Scarborough Col, Univ Toronto, 72-77, prof physics, 72-78 & 80-82; sr prof & scholar residence, Harvey Mudd Col, Claremont, 82-88; RETIRED. *Concurrent Pos:* Rouse Ball res student, Trinity Col, Cambridge Univ, 36-39; Fulbright vis prof, Univs Genoa, Milan & Bologna, 51-53; consult, Ramo-Wooldridge Corp, 55-56; vis prof, Harvey Mudd Col, Claremont, Calif, 78-80. *Mem:* Fel Am Phys Soc. *Res:* Quantum theory; relativity theory; theory of nuclear forces; electromagnetic propagation; nuclear reactor theory; theory of elementary particles; insect bird and machine flight; history and philosophy of science; author of one book. *Mailing Add:* 4304 OLeary St Pascagoula MS 39581

CORBETT, GAIL RUSHFORD, b Rapid River, Mich, May 23, 36; m 59; c 2. PLANT ECOLOGY, TAXONOMIC BOTANY. *Educ:* Univ Mich, BA, 58, MS, 60, PhD(bot), 67. *Prof Exp:* Instr bot, WVa Univ, 63, res consult flora, 68-76, instr bot, Univ Akron, 77; CONSULT FLORA, 77- *Concurrent Pos:* Consult, eng firm, 80, drilling firm, 82; vol in park at Cuyahoga Valley Nat Recreation Area, lead wildflower walks, 86; taught advan placement biol at Western Reserve Acad, Hudson, Ohio, 87-88; taught conserv of natural resources, Ill State Univ, 90, adj appoinment, Biol Dept, 90-91. *Honors & Awards:* Bradley Moore Davis Award, 58. *Mem:* Sigma Xi; Ecol Soc Am; Am Soc Plant Taxonomists; AAAS; Am Forestry Asn; Nature Conservancy. *Res:* Phytosociological study of disturbed habitats. *Mailing Add:* 504 Wellesley Dr Normal IL 61761

CORBETT, JAMES JOHN, b Chicago, Ill, July 2, 40; m 62; c 3. NEURO-OPHTHALMOLOGY. *Educ:* Brown Univ, BA, 62; Chicago Med Sch, MD, 66. *Prof Exp:* Instr neurol, Jefferson Med Col, 73-75, clin asst prof, 75-77; from asst prof to assoc prof neurol, Univ Iowa, 77-85, prof neurol & ophthal, 85-90; CHMN & PROF NEUROL, UNIV MISS, 91- *Concurrent Pos:* Silversides lectr, Toronto Western Gen Hosp, 82; Leon Louis Tureen vis prof, St Louis Univ, 80-85. *Mem:* Am Acad Neurol; Am Neurol Asn. *Res:* Afferent visual system; increased intracranial pressure; migraine; the incidence of clinical and surgical management of pseudotumor cercbri (idopathic intracranial hypertension). *Mailing Add:* Dept Neurol Univ Miss Jackson MS 39216-4505

CORBETT, JAMES MURRAY, b Welland, Ont, Can, Jan 29, 38; m 58; c 3. PHYSICS. *Educ:* Univ Toronto, BASc, 60; Univ Waterloo, MSc, 61, PhD(physics), 66. *Prof Exp:* Lectr, 62-66, asst prof, 66-74, ASSOC PROF PHYSICS, UNIV WATERLOO, 74- *Concurrent Pos:* Nat Res Coun Can res fel, Dept Metall, Oxford Univ, 67-68. *Mem:* Electron Micros Soc Am; Can Asn Physicists. *Res:* Nucleation, growth and structure of vacuum-deposited films; electron microscopy of thin crystals. *Mailing Add:* Dept Physics Univ Waterloo Waterloo ON N2L 3G1 Can

CORBETT, JAMES WILLIAM, b New York, NY, Aug 25, 28; m 54, 72; c 2. SOLID STATE PHYSICS. *Educ:* Univ Mo, BS, 51, MA, 52; Yale Univ, PhD(physics), 55. *Hon Degrees:* DSc, King Mem Col, 77, Univ Mo-Columbia, 89, Univ Novosibirsk, 90. *Prof Exp:* Res assoc chem, Yale Univ, 55; res assoc, Gen Elec Res Lab, 55-68; prof physics, 68-81, chmn, physics dept, 69-70, DIR, INST STUDY DEFECTS IN SOLIDS, STATE UNIV NY, ALBANY, 74-, DISTINGUISHED SERV PROF, 81-, DIR, JOINT LABS ADV MAT, 87- *Concurrent Pos:* Adj prof, Rensselaer Polytech Inst, 64-68; chmn, Int Conf Radiation Effects in Semiconductors, 70 & Int Conf Radiation-Induced Voids in Metals, 71; exchange scholar, US-USSR Exchange Prog, Nat Acad Sci, 73; Guggenheim fel, 75; lectr oenology, State Univ NY Albany, 72-; vis prof, Am Univ Cairo, 73, Ecole Normale Superieure & Univ Paris VII, 76, Tbilisi State Univ, 79 & Univ Linköping, 84 & 85, Beijing, Fudan & Chengdu, 86. *Honors & Awards:* Ivane Javakishvili Medal, 77; Alexander I Shokin Medal, 90. *Mem:* Fel Am Phys Soc; Am Asn Physics Teachers; sr mem Inst Elec & Electronic Engrs; fel NY Acad Sci. *Res:* Point defects in semiconductors and metals; radiation damage; reaction kinetics; nucleation theory. *Mailing Add:* Dept Physics State Univ NY Albany NY 12222

CORBETT, JOHN DUDLEY, b Yakima, Wash, Mar 23, 26; m 48; c 3. INORGANIC CHEMISTRY. *Educ:* Univ Wash, BS, 48, PhD(chem), 52; Oxford Univ, MPI, 79; Justus Liebig Univ, MPI, 86. *Prof Exp:* From assoc chemist to sr chemist, Ames Lab, US AEC, 52-63, div chief, 68-73; from asst prof to prof chem, 52-63, chmn dept, 68-73, prog dir, Ames Lab, Dept Energy, 74-78; PROF CHEM, IOWA STATE UNIV, 63-, DISTINGUISHED PROF, COL SCI & HUMANITIES, 83- *Concurrent Pos:* vis scientist, Max-Planck Inst fur Festkoerperforschung, Stuttgart, Tech Univ Denmark & Oxford Univ, 79. *Honors & Awards:* Inorg Chem Award, Am Chem Soc, 86; Sr US Scientist Award, Alexander von Humboldt Found, Bonn, WGer, 85. *Mem:* Am Chem Soc; Mat Res Soc; Am Asn Univ Prof. *Res:* Inorganic and physical chemistry; unfamiliar oxidation states; solid state chemistry. *Mailing Add:* Dept Chem Iowa State Univ Ames IA 50011

CORBETT, JOHN FRANK, b Doncaster, Eng, May 8, 35; c 3. ORGANIC CHEMISTRY, COSMETIC CHEMISTRY. *Educ:* Royal Inst Chem, London, ARIC, 57; Univ Reading, PhD(org chem), 61. *Prof Exp:* Res scientist chem, Gillette Res Labs, Eng, 61-62, head dept org chem, 63-69; sr res scientist chem, Gillette Co, Chicago, 70-71; mgr special proj, Gillette Res Labs, Eng, 72; dir res chem, Clairol Res Lab, 72-74, vpres res, 74-76, vpres tech develop, 76-87, VPRES TECHNOL, CLAIROL RES LAB, STAMFORD, CONN, 87- *Honors & Awards:* Lit Award, Soc Cosmetic Chemists, 72; CIBS Award, 83. *Mem:* Fel Royal Inst Chem; The Chem Soc; Soc Cosmetic Chemists; Fel Soc Dyers & Colorists. *Res:* Synthesis and properties of dyestuffs, mechanism of oxidative dyeing processes. *Mailing Add:* Clairol Res Lab Two Blachley Rd Stamford CT 06922-0001

CORBETT, JULES JOHN, b Natrona, Pa, Apr 12, 19; m 50; c 3. MICROBIOLOGY. *Educ:* Univ Chicago, BS, 50; Ill Inst Technol, MS, 57. *Prof Exp:* Instr microbiol & chem, Sch Nursing & bacteriologist, Englewood Hosp, 50-54; dir labs, Beverly Med Arts Bldg, 54-55; from instr to prof, 56-84, chmn biol dept, 74-79, EMER PROF BIOL, ROOSEVELT UNIV, 84- *Concurrent Pos:* Bacteriologist, Borden Co, Ill, 55-64. *Mem:* AAAS; Am Soc Microbiol; Sigma Xi; NY Acad Sci. *Res:* Immunology and pathogenicity of chromobacterium violaceum. *Mailing Add:* 8318 S Komensky Ave Chicago IL 60652

CORBETT, M KENNETH, b Port Lorne, NS, Sept 12, 27; m 50; c 3. PLANT PATHOLOGY. *Educ:* McGill Univ, BSc, 50; Cornell Univ, PhD(plant path), 54. *Prof Exp:* Asst plant path, Cornell Univ, 50-54; asst plant pathologist & asst prof, Univ Fla, 54-60, assoc virologist, Agr Exp Sta, 60-66; assoc prof, 66-69, PROF BOT, UNIV MD, 69- *Concurrent Pos:* Guggenheim fel, Netherlands, 64-65. *Mem:* Am Phytopath Soc. *Res:* Virology; biochemistry; plant physiology; plant breeding and pathology. *Mailing Add:* Dept Bot Univ Md Col Agr College Park MD 20740

CORBETT, MICHAEL DENNIS, b Great Bend, Kans, Feb 23, 44; m 80; c 5. GENERAL ENVIRONMENTAL SCIENCES, PATHOLOGY. *Educ:* Univ Kans, BS, 66, PhD(med chem), 70. *Prof Exp:* Asst prof pharmacog, Univ Miss, 70-73, assoc prof, 73-74; sr chemist, Midwest Res Inst, 75-76; assoc prof marine chem, Univ Miami, 76-81, prof, 81; PROF TOXICOL, UNIV FLA, 81- *Concurrent Pos:* Res career develop award, Nat Inst Environ Health Sci, 78-83; adj prof pharmacol, Univ Fla, 81-; ed adv, Chem Res Toxicol, ACS J. *Mem:* Am Chem Soc; AAAS; Sigma Xi; fel, World Life Res Inst; Soc Toxicol; Int Soc Study Xenobiotics. *Res:* Bioorganic chemistry of agricultural and other industrial organic chemicals; metabolism and toxicology of aromatic amine and nitro compounds; biochemical activation and nucleic acid binding of toxic materials in leukocytes and hepatocytes; discovered the "nitroso-glyoxylate" reaction, a new organic chemical reaction. *Mailing Add:* Food Sci & Human Nutrit Dept Univ Fla Gainesville FL 32611

CORBETT, ROBERT B(ARNHART), b Akron, Ohio, May 15, 17; m 41, 55; c 3. METALLURGICAL ENGINEERING. *Educ:* Carnegie Inst Technol, BS, 39; Univ Pittsburgh, MS, 42, PhD(metall eng), 44. *Prof Exp:* Student engr, Jones & Laughlin Steel Corp, 39-40; instr ferrous & non-ferrous metall, Univ Pittsburgh, 40-42; metallurgist, US Bur Mines, 42-45; res metallurgist, Heppenstall Co, Pa, 45-51, dir res & asst to pres, 51-56, dir metall, Midvale-Heppenstall Co, 56-59, consult, 59-62; pres, Corbett Assocs, Inc, 64-69; tech dir, Roll Mfg Inst, 77-81; mgr res & develop, Mackintosh-Hemphill Div, Gulf & Western Mfg Co, 81-85; PRES, CORBETT ASSOCS, 83- *Concurrent Pos:* Vpres, Satec Corp, 52-64; Calsicat Corp, 58-64. *Mem:* Am Inst Mining, Metall & Petrol Engrs; Am Soc Testing & Mat; Am Soc Metals; Asn Iron & Steel Engrs; Am Steel & Foundry Asn. *Res:* Metallurgy; testing machines; design and consult. *Mailing Add:* 111 Meadow Lane Bakerstown PA 15007

CORBETT, ROBERT G, b Chicago, Ill, Mar 13, 35; m 59; c 2. ECONOMIC GEOLOGY, GEOCHEMISTRY. *Educ:* Univ Mich, BS, 58, MS, 59, PhD(geol), 64. *Prof Exp:* From asst prof to assoc prof geol, WVa Univ, 62-69, prin investr, Water Res Inst, 67-69; from assoc prof to prof geol, Univ Akron, 69-89, coordr res, 72-83, head dept, 83-89, EMER PROF GEOL, UNIV AKRON, 89-; PROF GEOL, ILL STATE UNIV, 89-, CHAIRPERSON DEPT, 89- *Concurrent Pos:* Consult, Argonne Nat Labs, 76-79, Ohio Dept Nat Resources, 84-, Several state & Local govt agencies, 84- & three Fortune 500 Corps, 86-; spec ed, Ohio J Sci, 88. *Mem:* Mineral Soc Am; Geochem Soc; Nat Asn Geol Teachers; Geol Soc Am; Inst Prof Geologists; Asn Ground Water Scientists & Engrs. *Res:* Formation of hydroxylapatite; geology and mineralogy of uranium deposits; chemical characteristics of natural waters; mineral deposits; chemical characteristics of brines; applied hydrogeology. *Mailing Add:* Dept Geog & Geol Ill State Univ Normal IL 61617

CORBETT, THOMAS HUGHES, b Appelton, Wis, Oct 11, 38; m 61; c 2. CANCER CHEMOTHERAPY. *Educ:* Univ Wis, BS, 64, PhD(oncol), 70. *Prof Exp:* Technician cardiovasc, Sch Med, Univ Wis, 63; fel carcinogenesis, Oak Ridge Nat Lab, 70-72; head soli tumor biol & treat div cancer res, Southern Res Inst, 72-; AT MICH CANCER FOUND, DETROIT; DEPT INTERN MED & HARPERH, WAYNE STATE UNIV, DETROIT, MICH. *Mem:* Am Asn Cancer Res. *Res:* Cancer chemotherapy of transplantable tumors in mice; chemical carcinogenesis. *Mailing Add:* Wayne State Univ Sch Med 3800 Woodward 540 E Canfield Detroit MI 48201

CORBIN, ALAN, b New York, NY, Sept 3, 34; m 59; c 1. NEUROENDOCRINOLOGY, REPRODUCTIVE PHYSIOLOGY. *Educ:* City Col New York, BS, 56; Univ Iowa, MS, 60, PhD(physiol), 61. *Prof Exp:* Instr anat, Albert Einstein Col Med, 61-63; instr, Inst Pharmacol, Milan, Italy, 63-64; sr investr endocrinol, Sci Div, Abbott Labs, 64-66; sr res scientist, Squibb Inst Med Res, 66-71; head endocrinol, Wyeth Labs, 71-80, assoc dir biol res exp ther, 80-87; DIR, REPRODUCTIVE ENDOCRINOL DIV, WYETH-AYERST RES, 87- *Concurrent Pos:* Fel neuroanat, Albert Einstein Col Med, 61-63 & Inst Pharmacol, Milan, Italy, 63-64; NIH fel, 61-64; adj prof physiol, Rutgers Univ, New Brunswick; mem res grant review adv panel, NSF; mem, finance comt, Endocrine Soc. *Mem:* Endocrine Soc; Int Soc Res Reproduction; Int Soc Neuroendocrinol; Am Physiol Soc; Am

Col Obstetricians & Gynecologists. *Res:* Neuroendocrinology of mammalian reproduction; contraception; fertility regulation; general endocrinology; neuroanatomy; pharmacology of hypothalamic releasing factors; neuropharmacology; osteoporosis; metabolic disorders (diabetes, atherogenesis); female health care; menopause. *Mailing Add:* Develop Dept Wyeth-Ayerst Res Box 8299 Philadelphia PA 19101

CORBIN, FREDERICK THOMAS, b Franklin, NC, Dec 2, 29; m 52; c 2. WEED SCIENCE, PLANT PHYSIOLOGY. *Educ:* Wake Forest Col, BS, 51; Univ NC, MEd, 56; NC State Univ, PhD(physiol, microbiol), 65. *Prof Exp:* Teacher high sch, NC, 53-56; instr chem & physics, Mars Hill Col, 56-60; asst prof physics, ECarolina Col, 60-62; res asst, 62-65, asst prof weed sci, 65-80, PROF CROP SCI, NC STATE UNIV, 80- *Concurrent Pos:* Agr Res Serv grant, 65-68. *Mem:* AAAS; Weed Sci Soc Am; Am Phys Soc. *Res:* Interactions between major classes of chemical pesticides; biotransformation of herbicides in plants and cell suspension cultures. *Mailing Add:* Dept Crop Sci NC State Univ 4205 Williams Hall Raleigh NC 27695-7620

CORBIN, JACK DAVID, b Franklin, NC, Feb 8, 41; m 65; c 1. PHYSIOLOGY, BIOCHEMISTRY. *Educ:* Tenn Technol Univ, BS, 63; Vanderbilt Univ, PhD(physiol), 68. *Prof Exp:* Assoc prof, 71-, PROF PHYSIOL, SCH MED, VANDERBILT UNIV. *Concurrent Pos:* NIH fel, Univ Calif, Davis, 68-70; Am Diabetes Asn fel, Univ Calif, Davis & Vanderbilt Univ, 70-71 & Vanderbilt Univ, 71-72; NIH res grant, 72, Diabetes Ctr grant, 74. *Mem:* AAAS; Am Soc Biol Chemists. *Res:* Molecular endocrinology; hormone regulation; cyclic nucleotide regulation; protein kinase. *Mailing Add:* Dept Physiol Howard Hughes Med Inst Vanderbilt Univ Sch Med Nashville TN 37232

CORBIN, JAMES EDWARD, b Providence, Ky, July 14, 21; m 50; c 4. ANIMAL NUTRITION. *Educ:* Univ Ky, BS, 43, MS, 47; Univ Ill, PhD(animal nutrit), 51. *Prof Exp:* Dir res, Nat Oats Co, 51-54; mgr special chows res, Ralston Purina Co, 54-59, mgr dog res, 59-67; dir pet care ctr, 67-73; PROF ANIMAL SCI, UNIV ILL, URBANA, 73- *Concurrent Pos:* Chmn dog nutrit subcomt, Nat Res Coun, 68-; chmn dog & cat standards, Inst Lab Animal Resources, Nat Acad Sci. *Mem:* Am Soc Animal Sci; Am Asn Lab Animal Sci (pres, 72-73); Am Inst Nutrit; Am Asn Lab Animal Sci; Brit Small Animal Vet Med Asn. *Res:* Nutrition of dogs and laboratory animals. *Mailing Add:* Dept Animal Sci Univ Ill 1207 W Gregory Dr 160 Animal Sci Bldg Urbana IL 61801

CORBIN, JAMES LEE, b Coshocton, Ohio, Oct 20, 35; m 60. ORGANIC CHEMISTRY, ANALYTICAL CHEMISTRY. *Educ:* Bowling Green State Univ, BA, 57; Mich State Univ, PhD, 62. *Prof Exp:* Proj chemist, Am Oil Co, Ind, 62-63; Investr, Kettering Res Lab, 63-84, prin res scientist, Battelle-Kettering Lab, 84-87; RETIRED. *Mem:* Am Chem Soc; The Chem Soc. *Res:* Structure of natural products; synthesis; design and synthesis of novel ligand systems; metal complexes; siderophores; HPLC analysis of plant and bacterial products; methods development. *Mailing Add:* 165 Miami Dr Yellow Springs OH 45387

CORBIN, KENDALL BROOKS, b Oak Park, Ill, Dec 31, 07; m 32; c 2. NEUROLOGY, MEDICAL ADMINISTRATION. *Educ:* Stanford Univ, AB, 31, MD, 35. *Prof Exp:* Instr anat, Stanford Univ, 34-38; from assoc prof to prof, Col Med, Univ Tenn, 38-46, chief div, 41-46, in charge neurol, 43-46; prof neuroanat, 46-54, prof neurol, 54-72, EMER PROF NEUROL, MAYO GRAD SCH, 72- *Concurrent Pos:* Nat Res Coun fel, Neurol Inst, Northwestern Univ, 37-38; assoc dir, Mayo Found, 50-54, pres staff, 68, chmn bd develop, 69-72, head sect neurol, Mayo Clin, 57-63, sr consult, 63-72, emer consult, Mayo Clin & Mayo Found, 72-; pres, Friends of Gardens, Marie Selby Bot Gardens, Fla, 75-78. *Honors & Awards:* AOA Award for Res, Stanford, 34. *Mem:* Am Physiol Soc; Soc Exp Biol & Med; fel Am Neurol Asn; Am Asn Anat; fel Am Acad Neurol. *Res:* Neuroanatomy; neurophysiology; clinical neurology; medical administration. *Mailing Add:* 211 W Second St Apt 1817 Rochester MN 55901

CORBIN, KENDALL WALLACE, b Memphis, Tenn, Apr 5, 39; m 61, 81; c 2. EVOLUTIONARY BIOLOGY, POPULATION ECOLOGY. *Educ:* Carleton Col, BA, 61; Cornell Univ, PhD(vert biol), 65. *Prof Exp:* NIH res fel, 65-67; res assoc & lectr biol, Yale Univ, 67-70; from asst prof to assoc prof ecol & behav biol, 70-82, cur systs, 70-83, chmn prog evolutionary & syst biol, 74-82, PROF ECOL & BEHAV BIOL, UNIV MINN, MINNEAPOLIS, 82-, DIR GRAD STUDIES, GRAD PROG BIOL, 82-, DIR GRAD PROG ECOL, 84- *Concurrent Pos:* Dir grad studies, Grad Prog Biol, Univ Minn, Minneapolis, 82-; mem, Test Develop Comt, Biol Achievement Tests, Col Bd & Educ Testing Serv, 83-86, chmn, 85-86. *Mem:* AAAS; Soc Study Evolution; fel Am Ornith Union (secy, 77-81); Cooper Ornith Soc; Am Soc Naturalists; Wilson Ornith Soc. *Res:* Evolutionary relationships among bird species of North and South America using comparative biochemical data on the structure of protein molecules; studies of protein polymorphisms and genetic structure of natural populations; gene flow and natural selection in animal populations; scientific and educational software written for personal computers. *Mailing Add:* Dept Ecol & Behav Biol Univ Minn 318 Church St SE Minneapolis MN 55455

CORBIN, MICHAEL H, b 1949, US citizen; m 74; c 2. ENGINEERING. *Educ:* Univ Va, BS, 70; Mass Inst Technol, MS, 72. *Prof Exp:* Res asst, Mass Inst Technol, 70-72; engr, County of Fairfax, Va, 72-76; mgr, 76-86, TECH DIR, WESTON, 86- *Mem:* Dipl Am Acad Environ Engrs; Sigma Xi. *Res:* Solid waste and hazardous waste management including disposal, collection, storage, treatment and recycle-recovery; industrial waste disposal and sludge management disposal facility design and permitting; site characterization and site remediation; technology development. *Mailing Add:* Hawk Hill Rd RD 2 Downington PA 19335

CORBIN, THOMAS ELBERT, b Orange, NJ, Sept 6, 40; m 66; c 1. ASTRONOMY. *Educ:* Harvard Univ, AB, 62; Georgetown Univ, MA, 69; Univ Va, PhD(astron), 77. *Prof Exp:* ASTRONOMER, US NAVAL OBSERV, 64- *Mem:* Am Astron Soc; Int Astron Union; Sigma Xi. *Res:* Positional astronomy; fundamental astrometry; proper motion systems. *Mailing Add:* US Naval Observ Astrometry Dept Washington DC 20392-5100

CORBITT, MAURICE R(AY), b St Matthews, SC, Jan 28, 31. CHEMICAL ENGINEERING. *Educ:* Clemson Univ, BS, 53. *Prof Exp:* Tech engr, viscose rayon, E I du Pont de Nemours & Co, Tenn, 53-54, cellulose sponges, 56-59; res assoc, Sonoco Prod Co, 60-85, tech group leader, 80-88; RETIRED. *Mem:* Am Soc Testing & Mat; Tech Asn Pulp & Paper Indust. *Res:* Viscose chemistry of rayon and cellulose sponges; engineering and physical testing of superior paper and plastic products for textile, paper, film, packaging and construction industries. *Mailing Add:* Rte One Box 238 St Matthews SC 29135

CORBO, VINCENT JAMES, b Port Chester, NY, Apr 28, 43; m 65; c 2. PROCESS DEVELOPMENT. *Educ:* Manhattan Col, BChemE, 65; Princeton Univ, MA, 67, PhD(chem eng), 70. *Prof Exp:* Res engr, 69-74, sr res engr, 74-75, res supvr, 75-77, res mgr chem eng, 77, MGR, ENG SCI DIV, DEVELOP DEPT, HERCULES INC, 77- *Mem:* Am Chem Soc; Am Inst Chem Engrs. *Res:* Optimal control theory; computer applications; numerical analysis. *Mailing Add:* Five Southview Path Chadds Ford PA 19317

CORBY, DONALD G, b Jamestown, NDak, Jan 13, 34; m 59; c 6. PEDIATRICS, HEMATOLOGY. *Educ:* Univ NDak, BSc, 57; Northwestern Univ, MD, 59; Am Bd Pediat, dipl, 65 & 74. *Prof Exp:* Intern, Evanston Hosp Asn, Ill, 59-60; resident pediat, Children's Mem Hosp, Chicago, 60-61; resident pediat, US Army, 61, chief pediat, 225th Sta Hosp, Europe, 61-63, resident pediat, Brooke Gen Hosp, Ft Sam Houston, Tex, 63-65, asst chief pediat, William Beaumont Gen Hosp, El Paso, 65-68, dir spec educ & clin res pediat, 70-71; dir, Div Med, Walter Reed Army Inst Res, 87-88, assoc dir, plans & overseas opers, 88-89; CLIN ASSOC PROF PEDS, UNIFORMED SERV UNIV HEALTH SCI, 88-; COMDR, LETTERMAN ARMY INST RES, 89- *Concurrent Pos:* Fel pediat hemat, Univ Ill, 68-70; assoc clin prof pediat, Univ Colo Med Sch, 70; chief, Dept Clin Invest, Fitzsimmons Army Med Ctr, 71-87; consult res & develop, US Army Surgeon Gen, 89-91. *Honors & Awards:* Cert of Achievement & Army Commendation Medal, William Beaumont Gen Hosp, 68. *Mem:* fel Am Acad Pediat; Am Acad Clin Toxicol; Am Soc Hemat; Soc Pediat Res; Am Pediat Soc. *Res:* Toxicology; specifically prevention and treatment of accidental poisoning in childhood; coagulation physiology. *Mailing Add:* Letterman Inst Res Med San Francisco CA 94129-6800

CORCORAN, EUGENE FRANCIS, b Arthur, NDak, Nov 28, 16; m 40; c 2. OCEANOGRAPHY. *Educ:* NDak Agr Col, BS, 40; Univ Calif, PhD(oceanog), 58. *Prof Exp:* Instr math, San Diego State Col, 46-49; asst marine biochem, Scripps Inst, Univ Calif, 49-50, res biochemist, 52-57; from asst prof to assoc prof, 57-72, prof biol & living resources, Dorothy H & Lewis Rosenstiel Sch Marine & Atmospheric Sci, 72-82, EMER PROF, UNIV MIAMI, 82- *Concurrent Pos:* Assoc prog dir facils & spec progs, NSF, 67-68. *Mem:* AAAS; Am Chem Soc; Geochem Soc; Am Soc Limnol & Oceanog; Am Inst Chemists; Sigma Xi. *Res:* Organic productivity of sea water; photosynthesis in marine plants; organic constituents and trace metals of sea water and marine sediments. *Mailing Add:* 5990 SW 85th St Miami FL 33143

CORCORAN, GEORGE BARTLETT, III, b Madison, Wis, May 14, 48; m 73; c 2. BIOCHEMICAL TOXICOLOGY, DRUG METABOLISM. *Educ:* Ithaca Col, BA, 70; Bucknell Univ, MSc, 73; George Washington Univ, PhD(pharmacol), 80. *Prof Exp:* Chemist, Sun Oil Co, Pa, 70; teaching asst chem, Bucknell Univ, Pa, 70-72; guest worker, NIH, 72-73, civil servant, 73-77; res asst pharmacol, Baylor Col Med, 77-80, teaching fel, 80-81; asst prof, Dept Pharmaceut, State Univ NY, Buffalo, 81-88; ASSOC PROF, PHARMACOL & TOXICOL, COL PHARM & ADJ ASSOC PROF, BIOCHEM, SCH MED, UNIV NMEX, ALBUQUERQUE, 88- *Concurrent Pos:* Prin investr, NIH grants, 84-; chair, toxicol prog, Univ New Mex, Albuquerque. *Mem:* Am Soc Pharmacol & Exper Therapeut; Soc Toxicol; Am Asn Pharmaceut Sci; AAAS; Am Asn Col Pharm. *Res:* Study of drug biotransformation and toxicity, mechanisms of cellular injury, chemically reactive drug metabolites, mechanisms of action of clinically important antidotes, risk factor identification in controlling toxic drug reactions, influence of obesity and nutritional status on drug metobolism and drug toxicity. *Mailing Add:* Col Pharmacy Univ NMex Albuquerque NM 87131-1066

CORCORAN, JOHN W, b Dayton, Ohio, Sept 23, 24; m 49; c 2. PEDODONTICS. *Educ:* Miami Univ, BA, 48; Western Reserve Univ, DDS, 53; Am Bd Pedodontics, dipl, 62. *Prof Exp:* Intern, Univ Hosps, Cleveland, Ohio, 53-54, resident, 54-55; pvt pract pedodontics, 55-65; from asst prof to assoc prof pedodontics, Case Western Reserve Univ, 65-68; PROF PEDIAT DENT & CHMN DEPT, MED UNIV SC, 68-, STAFF MEM, DEPT DENT, UNIV HOSP, 69- *Concurrent Pos:* Supvr, Dent Fillings Clin, City Cleveland Babies & Children's Hosp, 58-65; assoc dent surgeon & dent surgeon, Univ Hosps, Cleveland, 65-68; dir dent serv, Coastal Habilitation Ctr, 68- *Mem:* Am Dent Asn; Am Acad Pedodontics; Am Soc Dent for Children. *Res:* Clinical research for the handicapped child. *Mailing Add:* 746 Willowlake Rd James Island SC 29412

CORCORAN, JOHN WILLIAM, b Des Moines, Iowa, June 12, 27; m 76. BIOCHEMISTRY, CHEMISTRY. *Educ:* Iowa State Univ, BS, 49; Western Reserve Univ, PhD, 55. *Prof Exp:* Instr biochem, Columbia Univ, 56-57; from asst prof to assoc prof, Western Reserve Univ, 57-68; prof biochem & chmn dept, Med Sch, 68-78, prof biochem & chem, 78-90, EMER PROF BIOCHEM, NORTHWESTERN UNIV, CHICAGO, 90- *Concurrent Pos:* Vis fel, Columbia Univ, 55-56; Am Heart Asn res fel, 55-58; USPHS career develop award, 64-; estab investr, Am Heart Asn, 58-63; acad guest, Lab Org

Chem, Swiss Fed Inst Technol, 64-65; pharmaceut consult, 68-; vis prof, Univ London, 78 & Univ Colo, Boulder, 85- *Mem:* Am Soc Biol Chem; Am Soc Microbiol; Am Chem Soc; Sigma Xi; Am Soc Pharmacog. *Res:* Mechanisms of sensitivity and resistance to antibiotics; antibiotic chemistry; natural product chemistry; carbohydrate chemistry; biosynthesis of natural products; carbohydrate metabolism; actinomycete metabolism. *Mailing Add:* PO Box 2557 Estes Park CO 80517

CORCORAN, MARJORIE, b Dayton, Ohio, July 21, 50; m 72; c 1. PARTICLE PHYSICS. *Educ:* Univ Dayton, BS, 72; Ind Univ, PhD(physics), 77. *Prof Exp:* Res assoc physics, Univ Wis, 77-79; vis asst prof, 80-81, ASST PROF PHYSICS, RICE UNIV, 81- *Mem:* Am Phys Soc. *Res:* Hadronic reactions resulting in high transverse momentum secondary particles; such interactions are believed to arise from the hard scattering of the hadrons constituents; spin dependence of hadronic interactions. *Mailing Add.* T W Bonner Nuclear Labs Rice Univ PO Box 1892 Houston TX 77251

CORCORAN, MARY RITZEL, b Los Angeles, Calif, July 3, 28; m 57; c 2. GENETICS, PLANT PHYSIOLOGY. *Educ:* Univ Calif, Los Angeles, BA, 53, PhD(bot), 59. *Prof Exp:* Asst bot, Univ Calif, Los Angeles, 54-59, jr res botanist, 59-62; from asst prof to assoc prof, 62-71, PROF BIOL, CALIF STATE UNIV, NORTHRIDGE, 71- *Mem:* AAAS; Bot Soc Am; Am Soc Plant Physiol. *Res:* Plant growth inhibitors; plant growth mutants. *Mailing Add:* Calif State Univ 18111 Nordhoff Northridge CA 91330

CORCORAN, VINCENT JOHN, b Chicago, Ill, Oct 7, 34; wid; c 5. LASERS. *Educ:* Univ Notre Dame, BSEE, 57; Univ Ill, MSEE, 58; Univ Fla, PhD(elec eng), 68. *Prof Exp:* Staff mem infrared, Lab Appl Sci, Univ Chicago, 58-62; vpres lasers, Astromarine Prod Corp, 62-63; sr res scientist, Martin Marietta Aerospace, 63-73; res staff mem lasers, Inst Defense Anal, 73-80; VPRES & CHIEF SCIENTIST, POTOMAC SYNERGETICS, INC, 80- *Mem:* Optical Soc Am; Inst Elec & Electronics Engrs. *Res:* Laser research and development on Nd; Yag Laser oscillator; amplifiers, new nonlinear geometries, phase conjunction, stimulated Kaman scattering and optical phased arrays. *Mailing Add:* 2034 Freedom Lane Falls Church VA 22043

CORCORAN, WILLIAM P, b Assumption, Minn, Apr 14, 22; m 48; c 7. ENERGY ENGINEERING, ENERGY CONSERVATION. *Educ:* Univ Minn, BEE, 48. *Prof Exp:* Jr engr, Am & Foreign Power, 48-56; sr engr & supvr, Atomics Int Div NAm Aviation, 56-61 & eval mgr, Hallam Nuclear Power Facil, 61-62; br chief, Allison Div, Gen Motors Corp, 62-76; sr engr & assoc dir, Indianapolis Ctr Advan Res, 76-78 & 80-83; sr analyst, Solar Energy Res Inst, 78-80; PRES, ADVAN ENERGY SYSTS, 83- *Concurrent Pos:* Lectr, Univ Calif Los Angeles, 58-61; instr, Purdue Univ, 81-82. *Res:* Energy conservation; energy utilization; alternate sources of energy. *Mailing Add:* 7042 E Second St Tucson AZ 85710-1224

CORCOS, ALAIN FRANCOIS, b Paris, France, June 7, 25; US citizen; m 50; c 2. GENETICS. *Educ:* Mich State Univ, BS, 51, MS, 52, PhD(plant breeding), 60. *Prof Exp:* Plant breeder, Grant Merrill Orchards, Inc, 58-60; assoc biol, Univ Calif, Santa Barbara, 61-63; instr, Ore Col Educ, 63-64; res assoc virol, Inst Cancer Res, Philadelphia, 64-65; from asst prof to prof natural sci, 65-89, prof bot & plant path, 89-90, EMER PROF BOT & PLANT PATH, MICH STATE UNIV, 91- *Mem:* Hist Sci Soc. *Res:* Genetics and molecular biology of Arabidopsis thaliana; history of biology and genetics; Mendel. *Mailing Add:* 530 Marshall East Lansing MI 48823

CORCOS, GILLES M, b Paris, France, Sept 10, 26; US citizen; m 78; c 3. FLUID MECHANICS. *Educ:* Univ Mich, BS, 49, MS, 50, PhD(aero eng), 52;. *Hon Degrees:* Doctorat d'Etat, Univ Grenoble, 80. *Prof Exp:* Fel fluid mech, Johns Hopkins Univ, 52-54; aerodyn engr, Douglas Aircraft Co, 54-58; from asst prof to assoc prof, 58-67, PROF AERONAUT SCI, UNIV CALIF, BERKELEY, 67- *Concurrent Pos:* Liaison scientist, Off Naval Res, London, Eng, 64-65; assoc forestry dept, Johns Hopkins Univ, 61- *Mem:* AAAS; Am Phys Soc. *Res:* Turbulent boundary layers; airfoil theory; aerodynamic noise; geophysical flows; blood flow; ocean currents; stratified flow turbulence; mantle convection; fluid dynamics of and transport in turbulent shear flows; deterministic models of turbulent flows. *Mailing Add:* Dept Mech Eng Univ Calif Berkeley CA 94720

CORDANO, ANGEL, PEDIATRIC NUTRITION, NUTRITIONAL REQUIREMENTS. *Educ:* San Marcos Univ, Lima, Peru, MD, 54. *Prof Exp:* DIR PEDIAT NUTRIT, BRISTOL MYERS-MEAD JOHNSON NUTRIT GROUP, 77- *Res:* Infant formulas. *Mailing Add:* Med Dept Bristol Myers-Mead Johnson Nutrit Group 2404 Pennsylvania Ave Evansville IN 47721

CORDEA, JAMES NICHOLAS, b Akron, Ohio, July 15, 37; m 59; c 3. PHYSICAL METALLURGY. *Educ:* Case Inst Technol, BS, 59; Ohio State Univ, MS, 62, PhD(metall eng), 65. *Prof Exp:* Welding engr, Battelle Mem Inst, 59-62; res engr, Armco Steel Inc, 64-68, sr res metallurgist, 68-70, supvry res metallurgist, 70- 77, sr staff metallurgist, 77-81, res mgr, 81-84, prin res engr, 85-89, RES MGR, ARMCO STEEL INC, 89- *Concurrent Pos:* Chmn ship mat, fabrication & inspection adv group, Ship Res Comt, Nat Res Coun, 78-79; Vis prof, Metall Eng Dept, Purdue Univ, 74-75. *Mem:* Fel Am Soc Metals; Am Inst Mining, Metall & Petrol Engrs; Nat Asn Corrosion Engrs. *Res:* Welding processes; steel or ferrous metallurgy; basic science; oil field materials; advanced coating processes. *Mailing Add:* Res Ctr Armco Steel Inc 705 Curtis St Middletown OH 45043

CORDELL, BRUCE MONTEITH, b Shelby, Mich, Sept 10, 49; div. SPACE UTILIZATION, PLANETARY SCIENCE. *Educ:* Mich State Univ, BS, 71; Univ Calif, Los Angeles, MS, 73; Univ Ariz, PhD(planetary sci), 77. *Prof Exp:* Weizmann res fel, Calif Inst Technol, 77-78; asst prof earth sci, Cent Conn State Univ, 78-80; asst prof geophys & physics, Calif State Univ, Bakersfield, 80-84; sr engr, 84-88, ENGR SPECIALIST & PROG MGR, GEN DYNAMICS, SPACE SYST DIV, SAN DIEGO, 88- *Concurrent Pos:* Consult, Reuben H Fleet Space Theater, San Diego, 82- *Mem:* Am Geophys Union; Am Astron Soc; Am Inst Aeronaut & Astronaut; British Interplanetary Soc. *Res:* Advanced space mission system analysis; extraterrestrial resources; planetary physics; manned planetary exploration and utilization; space program policy and management. *Mailing Add:* PO Box 1271 Solana Beach CA 92075

CORDELL, FRANCIS MERRITT, b South Pittsburgh, Tenn, Sept 11, 32; m 50; c 1. ASTRONOMICAL OBSERVATORIES & TELESCOPE MOUNTINGS. *Educ:* US Armed Forces Inst, Dept Army, External Studies, Hamilton Col, BLit, 66, Del, PhD(physics), 73. *Prof Exp:* Low speed code operator, Dept Army, 49-52; mat tester, Tenn Valley Authority, 52-53, instrument mech, 53-57, sr instrument mech, 57-80, instrumentation supvr, 80-87; RETIRED. *Concurrent Pos:* Prin restorer, Telescope & Observ Restoration Proj, Univ of South, 82. *Mem:* Instrument Soc Am; AAAS; Astron Soc Pac. *Res:* Portable astrophotographic observatory design--research and development of high pressure gas hypersensitization technique for photographic emulsions; restoration of astronomical instruments. *Mailing Add:* Medius Lodge-Dogwood Trail South Pittsburgh TN 37380

CORDELL, GEOFFREY ALAN, b London, Eng, Sept 1, 46; m 81; c 1. PHARMACOGNOSY, SPECTROSCOPY. *Educ:* Manchester Univ, BSc, 67, MSc, 68, PhD(org chem), 70. *Prof Exp:* Fel chem, Mass Inst Technol, 70-72; res assoc, Univ Ill, 72-74, from asst prof to assoc prof pharmacog, 74-80, assoc dean res, 80-86, asst vchancellor, 86-87, PROF PHARMACOG, COL PHARM, UNIV ILL, CHICAGO, 80-, DEPT HEAD MED CHEM & PHARMACOG, 88- *Concurrent Pos:* Consult, NIH; Alexander von Humboldt fel, 81-82; consult, pharm indust. *Mem:* Am Chem Soc; fel Royal Soc Chem; Am Soc Pharmacog (pres, 85-86); fel Am Asn Pharmaceut Scientists; Am Asn Col Pharm; Phytochem Soc Europ. *Res:* Isolation, structure elucidation, synthesis and biosynthesis of biologically active nature products, particularly anticancer and antimalarial agents; expertise in the use of nmr spectroscopy. *Mailing Add:* 9447 Hamlin Ave Evanston IL 60203

CORDELL, RICHARD WILLIAM, b Brooklyn, NY, Oct 4, 39; m 66; c 2. ANALYTICAL CHEMISTRY. *Educ:* Villanova Univ, BS, 61; Ohio Univ, PhD(anal chem), 66. *Prof Exp:* from asst prof to assoc prof, 75-79, PROF CHEM, HEIDELBERG COL, 80- *Concurrent Pos:* Reader, Advan Placement Exam, 69-86; consult, Educ Testing Exam, 75, Col Bd, 84-; mem, Am Chem Soc Exam Comt, 84-; asst examr, Int Baccalaureate, 87- *Mem:* Am Chem Soc; Sigma Xi. *Res:* Platinum group metals; organic analytical reagents. *Mailing Add:* Dept Chem Heidelberg Col Tiffin OH 44883

CORDELL, ROBERT JAMES, b Quincy, Ill, Jan 7, 17; m 42; c 3. PETROLEUM GEOLOGY. *Educ:* Univ Ill, BS, 39, MS, 40; Univ Mo, PhD(geol), 49. *Prof Exp:* Asst, Univ Ill, 39-40; asst instr geol, Univ Mo, 41-42, instr 46-47; instr Colgate Univ, 47-51; res geologist, Sun Oil Co, 51-53, dir res, Abilene Labs, 53-55, mgr geol res, 55-63, supvr basic res, 63-66, sr sect mgr basic res, 66-68 & paleontology, 68-70, res scientist, 70-75, sr res scientist, 75-76, sr prof scientist, 76-77; PRES, CORDELL REPORTS, INC, 77- *Concurrent Pos:* Mem, US Potential Gas Comt, 75-83. *Honors & Awards:* Spec Serv Award, Am Asn Petrol Geologists. *Mem:* Fel AAAS; Soc Econ Paleontologists & Mineralogists; Am Asn Petrol Geologists; fel Geol Soc Am. *Res:* Origin, migration and accumulation of oil and natural gas; determination of geological parameters-categories preferentially associated with petroleum; prediction of amount and distribution of undiscovered oil and gas; determination of favorable areas for petroleum exploration. *Mailing Add:* 305 W Shore Dr Richardson TX 75080

CORDEN, MALCOLM ERNEST, b Portland, Ore, Nov 8, 27; m 48; c 6. PLANT PATHOLOGY. *Educ:* Ore State Univ, BS, 52, PhD(plant path), 55. *Prof Exp:* Asst plant pathologist, Crop Protection Inst, Conn Agr Exp Sta, 55-58; from asst prof to prof, 58-85, EMER PROF PLANT PATH, ORE STATE UNIV, 85- *Mem:* Am Phytopath Soc. *Res:* Physiology of parasitism in fungal diseases of plants and the mechanism of fungicidal action in wood. *Mailing Add:* Dept Bot & Plant Path Ore State Univ Corvallis OR 97331

CORDEN, PIERCE STEPHEN, b Chattanooga, Tenn, June 26, 41; m 75; c 2. PHYSICS. *Educ:* Georgetown Univ, BS, 63; Univ Pa, MS, 66, PhD(physics), 71. *Prof Exp:* phys sci officer weapons technol & arms control, US Arms Control & Disarmament Agency, Wash, DC, 71-82, exec secy, US Deleg Conf Disarmament, Geneva, 83-89, CHIEF EUROP SECURITY NEGOTIATIONS DIV, US DELEG CONF DISARMAMENT, GENEVA, 89- *Concurrent Pos:* Res fel sci & soc, Tech Univ Twente, Netherlands, 77-78; legis asst, US Senate, 81. *Honors & Awards:* ACDA Meritorious Hon Award, 80. *Mem:* Am Phys Soc; Arms Control Asn. *Res:* Science and technology for arms control and disarmament, especially verification of nuclear testing and aspects of environmental warfare and chemical weapons limitations; conventional weapons control. *Mailing Add:* US Arms Control & Disarmament Agency Washington DC 20451

CORDER, CLINTON NICHOLAS, b Oberlin, Kans, Aug 1, 41; m 61; c 5. PHARMACOLOGY, MEDICINE. *Educ:* Univ Kans, BS, 64; Marquette Univ, PhD(pharmacol), 68; Wash Univ, MD, 71. *Prof Exp:* Res asst pharmacol, Sch Med, Wash Univ, 71-72; asst prof pharmacol & med, Sch Med, Univ Pittsburgh, 72-78, assoc prof, 78-79; prof & chmn dept pharmacol & prof med, Oral Roberts Univ, Tulsa, Okla, 79-84; dir res & educ, St Anthonys Hosp, Okla Cardiovasc Inst, 84-87; HAED CLIN PHARMACOL DIV, OKLA MED RES FOUND, OKLA CITY, 86- *Concurrent Pos:* Nat Inst Neurol Dis & Blindness fel, 68-69; Am Cancer Soc scholar, 69-71; NIH trainee clin pharm, Univ Pittsburgh, 72-74; rep US pharmacopeal Conv, 75, 80, 85 & 90; bd dir, Okla affil, Am Heart Asn, 85-89; chmn, Okla Med Asn. *Mem:* Am Soc Pharmacol & Exp Therapeut; Am Pharmaceut Asn; Am Chem Soc; AMA; Sigma Xi; Am Soc Clin Pharm & Therapeut; Asn Med Sch Pharmacol. *Res:* Quantitative microbiochemical analysis, and drug biotrand formation with emphasis on kidney, heart and hepatic tissues; clinical pharmacology; cardiovascular, endocrine and general pharmacology. *Mailing Add:* Okla Med Res Found 825 NE 13th St Oklahoma City OK 73104-5046

CORDERO, JULIO, b San Jose, Costa Rica, Jan 10, 23; US citizen; m 63; c 1. MAGNETOHYDRODYNAMICS, LASERS. *Educ:* Wayne State Univ, BS, 48; Univ Minn, MS, 51. *Prof Exp:* Meteorologist, Pan Am World Airways, 44; assoc scientist, Univ Minn, 51-53; scientist, Fluidyne Eng Corp, 53-55, proj engr, 55-60; sr scientist, Avco Res & Adv Develop Div, 60-64, sr staff scientist, 64-69; mem tech staff, Anal Serv, Inc, 69-75; mem res staff magnetohydrodynamics, 75-76, CHIEF ENGR, MAGNETOHYDRODYN RES FACIL, MASS INST TECHNOL, 76-; AT AVCO-ASD. *Mem:* Am Phys Soc; Am Astron Soc (secy, 69-70); Am Meteorol Soc; Am Inst Aeronaut & Astronaut; Sigma Xi. *Res:* Shock wave phenomena; supersonic and hypersonic aerodynamics; high temperature gas dynamics; radiation damage; plasma devices; rarefied gas flow; viscous and nonstationary flow; thermodynamics; ballistic and flight mechanics; impact phenomena; systems analysis; energy conversion; lasers. *Mailing Add:* 23 Mohawk St Danvers MA 01923

CORDERY, ROBERT ARTHUR, b London, Eng, Sept 25, 52. STATISTICAL PHYSICS, PHASE TRANSITIONS. *Educ:* Univ Toronto, BSc, 75, MSc, 76, PhD(physics), 80. *Prof Exp:* Fel physics, Rutgers Univ, 80-82; asst prof physics, Northeastern Univ, 82-84; STAFF, PITNEY BOWES INC, 84-, SR MEM TECH STAFF, 89- *Mem:* Am Phys Soc; Inst Elec & Electronics Engrs; Soc Indust & Appl Math. *Mailing Add:* Pitney Bowes Inc 26-07 Walter Wheeler Dr Stamford CT 06926

CORDES, ARTHUR WALLACE, b Freeport, Ill, June 29, 34; m 56; c 4. INORGANIC CHEMISTRY. *Educ:* Northern Ill Univ, BS, 56; Univ Ill, MS, 58, PhD(inorg chem), 60. *Prof Exp:* From asst prof to assoc prof, 59-66, PROF INORG CHEM, UNIV ARK, FAYETTEVILLE, 66- *Mem:* Am Chem Soc; Am Crystallog Asn. *Res:* Inorganic chemistry of the elements phosphorus, nitrogen, sulfur and arsenic; crystal structure determinations. *Mailing Add:* Dept Chem Univ Ark Fayetteville AR 72701

CORDES, CARROLL LLOYD, b Eagle Pass, Tex, Nov 29, 38; m 71. COASTAL & ESTUARINE RESOURCES MANAGEMENT MODELS. *Educ:* Stephen F Austin State Univ, BS, 61; NC State Univ, MS, 65, PhD(zool), 71. *Prof Exp:* Res biologist virology, Baylor Univ, 61-63; res assoc, Univ Tex, 65-66; from asst prof ecol to assoc prof, Univ Southwestern La, 69-77; WILDLIFE ECOLOGIST RESOURCE ECOL, FISH & WILDLIFE SERV, 77-, CHIEF, WINTERING WATERFALL RES BR, US DEPT INTERIOR. *Concurrent Pos:* Consult, Acadiana Planning & Evangeline Econ Develop Dist, 75-77; tech advr, St Tammany Parish Mosquito Abatement Dist, 77-79, dist comnr, 79-; res consult, La State Bd Regents, 81- *Mem:* Int Asn Impact Assessment; Ecol Soc Am; Wildlife Soc; Coastal Soc. *Res:* Development of a national environmental data base and management evaluation system for fish and wildlife resources in coastal ecosystems; development of information and technology transfer methods. *Mailing Add:* Nat Wetlands Res Ctr US Fish & Wildlife Serv 1010 Gause Blvd Slidell LA 70548

CORDES, CRAIG M, b Long Beach, Calif, July 27, 44. ALGEBRAIC THEORY. *Educ:* Stanford Univ, BS, 65; Univ Md, PhD(math), 70. *Prof Exp:* From asst prof to assoc prof, 70-82, PROF MATH, LA STATE UNIV, 82-, ASSOC DEAN ARTS & SCI, 83- *Mem:* Am Math Soc; Math Asn Am. *Res:* Algebraic theory of quadratic forms. *Mailing Add:* Dept Math La State Univ Baton Rouge LA 70803

CORDES, DONALD ORMOND, b Auckland, NZ, Feb 9, 30; m 59; c 2. VETERINARY MEDICINE, PATHOLOGY. *Educ:* Univ Sydney, BVSc, 57; Wash State Univ, MS, 64; Am Col Vet Pathologists, dipl. *Prof Exp:* Vet, Te Awamutu Vet Club, NZ, 57-60; vet res officer path, Ruakura Animal Res Ctr, NZ, 60-65, pathologist, Ruakura Animal Health Lab, 65-69, supt, 69-79; vis assoc prof, Col Vet Med, Cornell Univ, 79-80; supt, Cent Animal Health Lab, NZ, 80-81; CHMN, DEPT PATHOBIOL, COL VET MED, VA TECH, 81- *Concurrent Pos:* Consult, dept animal dis, Univ Conn, 64-65 & Brit Ministry Agr, Fisheries & Food, 72-73; external examr, Univ Zimbabwe, 85; consult, Food & Agr Orgn, UN, Syria, 88, Malawi, 89. *Mem:* Australian Col Vet Scientists; Am Vet Med Asn; Australian Soc Vet Path; Int Soc Vet Epidemiol; Am Soc Tropical Vet Med; US Animal Health Asn. *Res:* Pathology of animal diseases with emphasis on reproductive diseases and infectious diseases. *Mailing Add:* Col Vet Med Va Tech Blacksburg VA 24061

CORDES, EUGENE H, b York, Nebr, Apr 7, 36; m 57; c 2. BIOCHEMISTRY, ORGANIC CHEMISTRY. *Educ:* Calif Inst Tech, BS, 58; Brandeis Univ, PhD(biochem), 62. *Prof Exp:* From instr to prof chem, Ind Univ, Bloomington, 62-78, chmn dept, 72-78; exec dir, 78- PRES BIOCHEM, MERCK, SHARP & DOHME RES LABS. *Concurrent Pos:* Grants, NSF, 62-78 & NIH, 64-78. *Mem:* Am Chem Soc; Am Soc Biol Chem. *Res:* Mechanism of enzyme-catalyzed reactions; mechanism and catalysis of carbonyl addition reactions; lipoprotein chemistry. *Mailing Add:* 867 Leslie Rd Villanova PA 19085-1117

CORDES, HERMAN FREDRICK, b Upland, Calif, May 30, 27. PHYSICAL CHEMISTRY. *Educ:* Pomona Col, BA, 50; Stanford Univ, PhD(chem), 54. *Prof Exp:* res chemist, Naval Weapons Ctr, 54-81; RETIRED. *Mem:* AAAS; Am Chem Soc; Sigma Xi. *Res:* Chemical kinetics; physical chemistry of propellant and explosive ingredients; mass spectrometry. *Mailing Add:* 1028 N Peg St Ridgecrest CA 93555

CORDES, WILLIAM CHARLES, b St Louis, Mo, Aug 17, 29; m 57; c 6. PLANT PHYSIOLOGY, CYTOCHEMISTRY. *Educ:* Univ Mo, BS, 55, MA, 57, PhD(plant physiol), 60. *Prof Exp:* From instr to asst prof biol, Creighton Univ, 60-65; actg chmn dept, 70-72, ASST PROF BIOL, LOYOLA UNIV CHICAGO, 65- *Mem:* AAAS; Bot Soc Am; Japanese Soc Plant Physiol. *Res:* Physiology of differentiation in plants; appearance of enzyme systems in time; plant wound enzymes. *Mailing Add:* Dept Biol Loyola Univ Lake Shore Campus 6525 N Sheridan Rd Chicago IL 60626

CORDINER, JAMES B(EATTIE), JR, b Spokane, Wash, Nov 15, 15; m 47, 79. CHEMICAL ENGINEERING. *Educ:* Univ Wash, Seattle, BS, 37, MS, 38, PhD(chem eng, chem), 41. *Prof Exp:* Chem engr, US Bur Mines, 46-50, asst supv engr, 51-52; tech report ed, cent res dept, Food Mach & Chem Corp, 53-54; sr chem engr, Kaiser Aluminum & Chem Corp, 55-57; from asst prof to assoc prof chem eng, 58-69, prof chem eng, 69-81, EMER PROF CHEM ENG, LA STATE UNIV, 81- *Mem:* Am Soc Eng Educ; Am Chem Soc; Soc Am Mil Engrs; Am Inst Chem Engrs. *Res:* Coal gasification; combustion; physical properties of materials; unit operations; sugar technology; technical report editing. *Mailing Add:* 3155 Fritchie Dr Baton Rouge LA 70809-1504

CORDING, EDWARD J, CIVIL ENGINEERING. *Prof Exp:* PROF CIVIL ENG, UNIV ILL, 67- *Mem:* Nat Acad Eng. *Mailing Add:* 2221 Newmark Chem Eng Lab MC 250 Univ Ill 205 N Mathews Urbana IL 61801

CORDINGLY, RICHARD HENRY, b Denver, Colo, Aug 9, 31; m 56; c 2. PULP & PAPER TECHNOLOGY. *Educ:* Univ Colo, BS, 53; Inst Paper Chem, MS, 55, PhD(chem), 58. *Prof Exp:* Proj chemist papermaking, 58-60, asst tech dir, 61-62, res engr, 63-70, MGR RES & DEVELOP, WEYERHAEUSER CO, 71- *Mem:* Tech Asn Pulp & Paper Indust. *Res:* Wood pulping and bleaching; pulp refining; paper forming, pressing and drying; surface treating; paper product technology and new paper product development. *Mailing Add:* 7505 90th Ave SW Tacoma WA 98498

CORDON, MARTIN, b West New York, NJ, Aug 10, 28; m 53; c 3. ORGANIC CHEMISTRY. *Educ:* Rutgers Univ, BS, 50; Univ Calif, Los Angeles, PhD(chem), 55. *Prof Exp:* Sr res chemist, 55-70, res assoc, 70-79, SR RES ASSOC, COLGATE-PALMOLIVE CO, 79- *Mem:* Am Chem Soc. *Res:* Biochemistry of saliva; chemistry of hair; organic synthesis; cleaning and polishing agents for oral products. *Mailing Add:* 55 Grant Ave Highland Park NJ 08904

CORDON-CARDO, CARLOS, b Feb 25, 57; m. IMMUNO-PATHOLOGY, CELL BIOLOGY. *Educ:* Univ Barcelona, MD, 80; Cornell Univ, PhD(cell biol & genetics), 85. *Prof Exp:* Spec fel, 81-84, ASST ATTEND PATHOLOGIST & RES ASSOC CELL BIOL & GENETICS, SLOAN-KETTERING CANCER INST, 84- *Concurrent Pos:* Asst mem, Sloan-Kettering Cancer Inst, 84- *Mailing Add:* Dept Path Sloan-Kettering Cancer Ctr 1275 York Ave New York NY 10021

CÓRDOVA, FRANCE ANNE-DOMINIC, b Paris, France, Aug 5, 47; US citizen; m 85; c 2. OBSERVATIONAL ASTRONOMY, HIGH ENERGY ASTROPHYSICS. *Educ:* Stanford Univ, BA, 69; Calif Inst Technol, PhD(physics), 79. *Prof Exp:* Group leader educ, Contemp Univ, Ford Found Proj, 69; teacher physics trigonometry fil, Damien High Sch, 70-71; res asst astrophys, Calif Inst Technol, 75-79, fel, 79; staff mem & proj leader astrophys, Los Alamos Nat Lab, 79-89; PROF & HEAD, DEPT ASTRON & ASTROPHYS, PA STATE UNIV, 89- *Concurrent Pos:* Consult, Univ Calif, Berkeley, 69; guest prin investr satellite observ, 76-; mem, adv coun, NSF, 81-83 & Int Users Comt, Roentgen Satellite, WGer, 85-90; NATO postdoctoral fel, Mullard Space Sci Lab, UK, 83; mem, Extreme Ultraviolet Satellite Working Group, NASA, 89-, ext adv comt, Particle Astrophys Ctr, NSF, 89-, astron adv comt, NSF, 90-, Space Telescope Sci Inst, 90- & US Nat Comn, Int Astron Union, 90-; vchmn, High Energy Astrophys Div, Am Astron Soc, 89, chmn, 90; bd dirs, Asn Univs for Res Astron, 90- *Mem:* Am Astron Soc; Int Astron Union. *Res:* Nature of astrophysical sources of high energy radiation; close binary systems containing an accreting black hole, neutron star or white dwarf; use of space and ground based observing facilities that cover the entire electromagnetic spectrum from radio to gamma ray frequencies; space flight instrumentation for astronomical research; software development for astrophysics. *Mailing Add:* Dept Astron & Astrophys Pa State Univ 525 Davey Lab University Park PA 16802

CORDOVA, VINCENT FRANK, b Philadelphia, Pa, Feb 24, 36; m 65; c 4. CRIMINALISTICS, TOXICOLOGY. *Educ:* Edison Col, BA, 77. *Prof Exp:* Biochemist toxicol, Med Examr Philadelphia, 57-65; police chemist criminalistics, Philadelphia Police Dept, 65-68, lab dir, 68-76; DIR CRIMINALISTICS, NAT MED SERV, 76- *Concurrent Pos:* Toxicol consult, Coroner's Off, Del County, Pa, 63-68; instr, Temple Community Col, 66-67, Lehigh Community Col, 69-73 & Breathalyzer Courses, Philadelphia Police Lab, 69-76; tech dir, Law Enforcement Assistance Admin grant, 72-73; forensic consult, Toxicon Ltd, 76- *Mem:* Am Chem Soc; fel Am Acad Forensic Sci; Soc Appl Spectros; Am Soc Testing & Mat. *Res:* Toxicology; criminalistics; environmental pollution analyses, gas chromatography, mass spectrometry, data systems. *Mailing Add:* 501 Lafayette Lane Warminster PA 18974-2231

CORDOVA-SALINAS, MARIA ASUNCION, ORAL ELECTROPHYSIOLOGY. *Educ:* Med Univ, DMD, 86. *Prof Exp:* Asst prof physiol, Med Univ-SC, 80-86; DENTIST, 87- *Mailing Add:* 159 Wentworth St Charleston SC 29401

CORDOVI, MARCEL A(VERY), b Sofia, Bulgaria, Nov 17, 15; nat US; m 41; c 1. METALLURGICAL ENGINEERING, CORROSION. *Educ:* Am Col, Bulgaria, Dipl, 35; Polytech Inst, Czech, IngC, 38; Polytech Inst Brooklyn, BME, 41, MME, 42, ME, 47, PE, 77. *Prof Exp:* Asst, Welding Res Coun, 43-44; supvr res & develop, Babcock & Wilcox Co, 44-49, staff engr, 49-53, mgr nuclear mat res & develop, 53-58; supvr power appln, 58-66, group leader, 66-68, mgr appln eng, 68-75, asst to pres, Inco US, Inc, 75-77, DIR SPEC PROJS, INCO LTD, 77-, VPRES, 80- *Concurrent Pos:* Adj prof, Polytech Inst Brooklyn, 50-70 & NY Univ, 58-60; consult, Brookhaven Nat Lab, 50-59, US del, Int Inst Welding, 50-60; lectr, City Col New York, 58-60. *Honors & Awards:* Award, Am Soc Metals, 57; award, Am Soc Mech Engrs, 68. *Mem:* Am Nuclear Soc; Am Soc Testing & Mat; Am Soc Metals; Am Soc Mech Engrs; Nat Asn Corrosion Engrs; Sigma Xi. *Res:* Metals sciences; solid state reactions; nuclear technology. *Mailing Add:* 12510 Queens Blvd Kew Gardens NY 11415

CORDS, CARL ERNEST, JR, b South Bend, Ind, Aug 24, 33; m 58; c 2. VIROLOGY, MEDICAL MICROBIOLOGY. *Educ:* Ariz State Univ, BS, 58; Univ Wash, PhD(microbiol), 64. *Prof Exp:* NIH fel, 64-65, from instr to asst prof microbiol, 65-72, ASSOC PROF MICROBIOL, SCH MED, UNIV NMEX, 72- *Mem:* AAAS; Am Soc Virol; Am Soc Microbiol. *Res:* Genetics and mechanisms of replication of animal viruses; tissue culture. *Mailing Add:* Dept Microbiol Univ NMex Sch Med Albuquerque NM 87131

CORDS, DONALD PHILIP, b Evanston, Ill, Sept 18, 40; m 64; c 2. ORGANIC CHEMISTRY. *Educ:* Northwestern Univ, BA, 62; Ind Univ, PhD(org chem), 66. *Prof Exp:* Res chemist, 66-73, sr supvr, 73-79, tech serv supvr, 79-80, res supvr, 80-84, res & tech serv mgr, 84-85, bus develop mgr, 85, CUSTOMER SERV & MFG MGR, E I DU PONT DE NEMOURS & CO INC, 85- *Mem:* Am Chem Soc. *Res:* Applications of organometallics in organic synthesis; free radical rearrangement and elimination reactions; organic reaction mechanisms. *Mailing Add:* Specialty Div E I du Pont de Nemours & Co Wilmington DE 19898

CORDTS, RICHARD HENRY, JR, b Teaneck, NJ, May 1, 34; m 66; c 2. NUTRITION. *Educ:* Rutgers Univ, BS, 56, MS, 61, PhD(nutrit), 64. *Prof Exp:* Asst, Rutgers Univ, 61-64; dir nutrit res, Whitmoyer Labs, Inc, Rohm and Haas Co, 64-65, dir nutrit, 65-68; assoc dir res & develop, Nat Molasses Co, 68-70; nutritionist, 70-73, assoc mgr animal health tech serv, 73-75, mgr, 75-78, dir agr mkt res & planning, 79-80, dir chem res planning & admin, 80-85, res investr, 85-86, REGULATORY AFFAIRS PROJ MGR, HOFFMANN-LAROCHE, 86- *Mem:* Poultry Sci Asn; Animal Nutrit Res Coun; Am Asn Dairy Sci; Am Animal Sci Asn. *Res:* Roughage utilization by ruminants; protein metabolism in monogastrics. *Mailing Add:* One Sylvan Pl Nutley NJ 07110

CORDUNEANU, CONSTANTIN C, b Iasi, Romania, July 26, 28; m 49. STABILITY THEORY, PERIODIC FUNCTIONS. *Educ:* Univ Iasi, Romania, MS, 51, PhD(math), 56. *Prof Exp:* Instr math, Univ Iasi, Romania, 49-50, asst, 50-55, lectr, 55-62,from assoc prof to prof, 62-77; PROF MATH, UNIV TEX, ARLINGTON, 79- *Concurrent Pos:* Vis prof, Univ RI, 67-78 & Univ Tenn, Knoxville, 78-79; assoc ed, Nonlinear Anal, Oxford, 77-; ed, Libertas Math, Am Romanian Acad Arts & Sci, 81- *Honors & Awards:* State Dept Educ Prize, Romania, 63. *Mem:* Am Math Soc; Soc Indust & Appl Math; Math Asn Am; Am Romanian Acad Arts & Sci. *Res:* Qualitative theory of various classes of differential, integral and related functional equations with main emphasis on stability and oscillations. *Mailing Add:* Dept Math Univ Tex Arlington TX 76019

CORDY, DONALD R, b Fall River, Wis, Feb 17, 13. VETERINARY PATHOLOGY. *Educ:* Univ Calif, Los Angeles, BA, 34; Iowa State Col, DVM, 37; Cornell Univ, MS, 38, PhD(vet path), 40. *Prof Exp:* Instr vet path, State Col Wash, 40-42, assoc prof, 46-50; from asst to prof, 50-83, head dept, 58-69, EMER PROF VET PATH, UNIV CALIF, DAVIS, 83- *Mem:* Am Col Vet Pathologists; Am Vet Med Asn; Am Asn Pathologists. *Res:* Pathology of communicable animal diseases; animal neuropathology. *Mailing Add:* Dept Vet Path Univ Calif Davis CA 95616

CORE, EARL LEMLEY, botany; deceased, see previous edition for last biography

CORE, HAROLD ADDISON, b Cassville, WVa, Nov 4, 20; m 43; c 2. FORESTRY. *Educ:* WVa Univ, BSF, 42; State Univ NY, MS, 49, PhD, 62. *Prof Exp:* From asst prof to assoc prof wood prod eng, State Univ NY Col Forestry, Syracuse, 46-66; PROF FORESTRY, COL AGR, UNIV TENN, KNOXVILLE, 66- *Mem:* Soc Am Foresters; Forest Prod Res Soc. *Res:* Wood and fiber anatomy; foreign woods. *Mailing Add:* 224 Clearfield Rd Knoxville TN 37922

CORE, JESSE F, b Ford City, Pa, Mar 6,13. MINING ENGINEERING. *Educ:* Pa State Univ, BS, 37. *Prof Exp:* vpres, coal opers, US Steel, 51-76; ADJ PROF MINING ENG, PA STATE UNIV, 77- *Mem:* Nat Acad Eng; Am Inst Mining, Metall & Petrol Engrs; hon mem Am Inst Mech Engrs. *Mailing Add:* 420 S Corl St No 5 State College PA 16801-4153

CORELL, ROBERT WALDEN, b Detroit, Mich, Nov 4, 34; m 85; c 4. ENGINEERING, OCEAN ENGINEERING. *Educ:* Case Inst Technol, BSME, 56, PhD(mech eng), 64; Mass Inst Technol, MSME, 59. *Prof Exp:* Res engr, Gen Elec Corp, 55-57; instr mech eng, Univ NH, 57-58, asst prof, 59-60; res asst, Case Inst Technol, 60-64; assoc prof, 64-66, chmn dept, 64-72, PROF MECH ENG, UNIV NH, 66-; DIR, MARINE SYST ENG LAB, 77- *Concurrent Pos:* Eng assoc, Huggins Hosp, 58-60; instrumentation consult, Highland View Hosp, 61-64; res assoc, Scripps Inst Oceanog, Univ Calif, San Diego, 71-72; consult, Off Naval Res & var industs & bus; dir, Sea Grant Prog & Marine Prog, 75-87; vis prof, Univ Wash, 85; asst & dir geosci, NSF, 87; sr sci assoc geosci, NSF, 87. *Mem:* AAAS; Am Geophys Union. *Res:* Research in ocean sciences, geosciences, ocean engineering, and arctic science and engineering. *Mailing Add:* NSF 1800 G St NW Rm 510 Washington DC 20550

CORELLI, JOHN C, b Providence, RI, Aug 6, 30; m 59; c 2. RADIATION TECHNOLOGY, ION SOURCES. *Educ:* Providence Col, BS, 52; Brown Univ, MS, 54; Purdue Univ, PhD(physics), 58. *Prof Exp:* Physicist, Knolls Atomic Power Lab, Gen Elec Co, 58-61; PROF NUCLEAR ENG & ENG PHYSICS, RENSSELAER POLYTECH INST, 62- *Concurrent Pos:* Spec fel, NIH, 71. *Mem:* Am Phys Soc; Am Nuclear Soc. *Res:* Use of ion beams, neutrons and electrons to alter the properties of solids; liquid metal ion sources for use in focused ion beam columns to produce microelectronic devices and perform materials analysis. *Mailing Add:* Ctr Integrated Electronics Rensselaer Polytech Inst Troy NY 12180

CORELLI, JOHN CHARLES, b Providence, RI, Aug 6, 30; m 59; c 2. MATERIALS SCIENCE ENGINEERING, SOLID STATE SCIENCE. *Educ:* Providence Col, BSc, 52; Brown Univ, MSc, 54; Purdue Univ, PhD(physics), 58. *Prof Exp:* Physicist, Knolls Atomic Power Lab, Gen Elec Co, 58-62; assoc prof, 62-65, PROF, DEPT NUCLEAR ENG & SCI, RENSSELAER POLYTECH INST, 65- *Concurrent Pos:* NIH fel, Univ Rochester, 71. *Mem:* Am Phys Soc; Am Nuclear Soc; Electrochem Soc. *Res:* Radiation damage studies in silicon and ceramic materials; radiation alteration of polymer materials; microeleotronic materials; ion beam modification of materials. *Mailing Add:* 1A Salem Ct Albany NY 12203

COREN, RICHARD L, b New York, NY, Sept 1, 32; m 55; c 3. PHYSICS, ELECTRICAL ENGINEERING. *Educ:* City Col New York, BS, 54; Polytech Inst Brooklyn, MS, 56, PhD(physics), 60. *Prof Exp:* Teaching fel physics, Polytech Inst Brooklyn, 54-56, res assoc, 56-60; res specialist magnetic films lab, Philco Corp, Pa, 60-64; prin physicist mat develop lab, Univac Div, Sperry Rand Corp, 64-65; from asst prof to assoc prof elec eng, Drexel Univ, 65-69, chmn fac coun, 69-71, asst head, dept grad affairs, 80-89, PROF ELEC & COMPUTER ENG, DREXEL UNIV, 69-, DIR, DEPT OFF CAMPUS PROGS, 83- *Concurrent Pos:* Adj instr, Cooper Union Col, 55-59; lectr, La Salle Col, 62-64; consult, Rockefeller Inst, 61-62; partner, C&M Assocs, Tech Consults, 67- *Mem:* AAAS; Inst Elec & Electronics Engrs; Am Phys Soc; Am Soc Eng Educ. *Res:* Electrical, magnetic, and optical studies of magnetic material; physiology of reproduction; antennathy. *Mailing Add:* Dept Elec & Computer Eng Drexel Univ Philadelphia PA 19104

CORET, IRVING ALLEN, b Salt Lake City, Utah, Apr 28, 20; m 46; c 1. PHARMACOLOGY. *Educ:* Emory Univ, AB, 40, MD, 43. *Prof Exp:* Intern, Piedmont Hosp, Ga, 44; instr pharmacol, Col Physicians & Surgeons, Columbia Univ, 47-48; from instr to assoc prof, 50-73, PROF PHARMACOL, SCH MED, ST LOUIS UNIV, 73- *Concurrent Pos:* Rockefeller fel, Col Physicians & Surgeons, Columbia Univ, 46-47; NIH fel, Univ Pa, 48-50. *Mem:* AAAS; Am Soc Pharmacol & Exp Therapeut; Sigma Xi. *Res:* Antibiotics; autonomic and cellular pharmacology. *Mailing Add:* 1402 S Grand Blvd St Louis MO 63104

COREY, ALBERT EUGENE, b Gardner, Mass, July 4, 28; m 50; c 2. ORGANIC CHEMISTRY. *Educ:* Rensselaer Polytech, BS, 50. *Prof Exp:* Assoc chemist, Allied Chem & Dye Corp, 51-54; res chemist, Plymouth Cordage Co, 54-55; res specialist, Shawinigan Resins Corp, 55-69 & Plastics Prod & Resins Div, Monsanto Co, 69-73, sr res specialist, Monsanto Plastics & Resins Co, 73-87; PRIN TECHNOLOGIST, ADHESIVES, MONSANTO CHEM CO, 87- *Mem:* Am Chem Soc. *Res:* Glycerol; hydroxylation; non-electrolytic hydrogen peroxide; polymers; emulsion polymers; paper and surface coatings; textile applications; pressure sensitive adhesive applications. *Mailing Add:* 730 Worcester St Indian Orchard MA 01151-9998

COREY, ARTHUR THOMAS, b Anne Arundel County, Md, Apr 8, 19; m 43; c 2. AGRICULTURAL & GROUNDWATER ENGINEERING. *Educ:* Univ Md, BS, 47; Colo State Univ, MS, 49; Rutgers Univ, PhD(soil physics), 52. *Prof Exp:* Physicist, Gulf Res & Develop Co, Gulf Oil Corp, 52-56; res civil engr, 56-61, from assoc prof to prof & res agr engr, 58-77, EMER PROF AGR ENG, EXP STA, COLO STATE UNIV, 77- *Concurrent Pos:* Consult, petrol indust, 56-59; res grants, NSF, 58-59, 61-63, 65-67, 68-71 & 80-82 & NIH, 62-64; prof hydraul eng from Colo State Univ to Seato Grad Sch Eng, Bangkok, USAID, 59-60; prof agr eng, Ore State Univ, 77-79; consult, 77. *Mem:* Am Soc Agr Engrs; Soil Sci Soc Am; Am Soc Civil Engrs. *Res:* Flow of immiscible fluid mixtures through porous media; irrigation and drainage engineering; water conservation. *Mailing Add:* 1309 Kirkwood Dr Apt 605 Ft Collins CO 80525

COREY, BRIAN E, b Akron, Ohio, Dec 8, 47; m; c 1. RADIO ASTRONOMY, GEODESY. *Educ:* Oberlin Col, BA, 69; Princeton Univ, PhD(physics), 78. *Prof Exp:* Vis scholar, Dept Earth & Planetary Sci, Mass Inst Technol, 76-78, res assoc, 78-80; RES STAFF, HAYSTACK OBSERV, 80- *Mem:* Am Phys Soc; Am Astron Soc; Am Geophys Union. *Res:* Geodetic measurements of tectonic plate motions and earth orientation by very-long-baseline interferometry; radio astrometry and astrophysics; experimental relativity; measurements of cosmic microwave background; radio astronomical instrumentation. *Mailing Add:* Haystack Obs Westford MA 01886

COREY, CLARK L(AWRENCE), b Mont, Sept 17, 21; m 44. METALLURGY, MATERIALS. *Educ:* Univ Mich, BS, 48, MS, 49, PhD, 51. *Prof Exp:* Res metallurgist, Babcock & Wilcox Co, 51-53; res engr, Univ Mich, 53-56; from assoc prof to PROF ENG, WAYNE STATE UNIV, 56- *Concurrent Pos:* Consult, Atomic Power Develop Asn, 57- *Mem:* Soc Metals; NY Acad Sci; Am Inst Mining, Metall & Petrol Engrs. *Res:* Physics of materials; x-ray crystallography; theory of alloying; theory of deformation. *Mailing Add:* 28183 Wildwood Farmington Hills MI 48018

COREY, ELIAS JAMES, b Methuen, Mass, July 12, 28; m 61; c 3. ORGANIC CHEMISTRY. *Educ:* Mass Inst Technol, BS, 48, PhD(chem), 51; Univ Helsinki, PhD, 90. *Hon Degrees:* DSc, Univ Chicago, 68, Hofstra Univ, 74, Colby Col, 77, Oxford Univ, 82, Univ Liege, 85, Univ Ill, Urbana, 85, Kenyon Col, 89 & Merrimack Col, 90; LLD, Hokkaido Univ, 90. *Prof Exp:* Res chemist, A D Little Co, Inc, 48; from instr to prof chem, Univ Ill, 51-59; SHELVON EMERY PROF CHEM, HARVARD UNIV, 59- *Concurrent Pos:* Consult, Chas Pfizer Co. *Honors & Awards:* Nobel Prize in Chem, 90; Pure Chem Award, Am Chem Soc, 60, Fritzsche Award, 68, Synthetic Chem & Harrison Howe Awards, 70, Linus Pauling Award, 73, Remsen Award, 74; Ciba Found Medal & Evans Award, Ohio State Univ, 72; Dickson Prize Sci, Carnegie Mellon Univ, 73; George Ledlie Prize Sci, Harvard Univ, 73; Arthur C Cope Award, 76; Nichols Medal, 77; Franklin Medal Sci, 78; J G Kirkwood Award, Yale Univ, 80, Silliman Award, 86; C S Hamilton Award, Univ Nebr, 80; Chem Pioneer Award, Am Inst Chemists,

81, Gold Medal Award, 89; Paul Karrer Award, Univ Zurich, 82; Paracelsus Award, Swiss Chem Soc, 84; Madison Marshall Award, Am Chem Soc, 85; V D Mattia Award, Roche Inst Molecular Biol, 85. *Mem:* Nat Acad Sci; Franklin Inst; hon mem Royal Soc Chem; hon foreign mem Chem Soc Finland; Am Chem Soc; hon mem Pharmaceut Soc Japan; hon mem Chem Soc Japan. *Res:* Stereochemistry; structural, synthetic and theoretical organic chemistry. *Mailing Add:* Dept Chem Harvard Univ Cambridge MA 02138

COREY, EUGENE R, b Oregon City, Ore, Nov 18, 35; m 62. INORGANIC CHEMISTRY, CRYSTALLOGRAPHY. *Educ:* Willamette Univ, BS, 58; Univ Wis, PhD(inorg chem), 63. *Prof Exp:* Instr chem, Univ Wis, 63-64; assoc prof, Univ Cincinnati, 64-69; assoc prof, 69-81, PROF CHEM, UNIV MO-ST LOUIS, 81- *Mem:* Am Chem Soc; Am Crystallog Asn; The Chem Soc; NY Acad Sci. *Res:* Chemical crystallography; single crystal structure determinations of organometallic compounds by the method of x-ray diffraction. *Mailing Add:* Dept Chem Univ Mo 8001 Natural Bridge Rd St Louis MO 63121

COREY, GILBERT, b Olathe, Colo, Feb 7, 23; m 49; c 3. AGRICULTURAL ENGINEERING. *Educ:* Colo State Univ, BS, 48, MS, 49, PhD(agr eng), 65. *Prof Exp:* Asst irrigationist, Univ Idaho, 49-54; irrig engr, Agr Res Serv, USDA, 54-55; irrig specialist & consult, Eduador Proj, Univ Idaho, 55-57, assoc prof agr eng, 57-72, chmn dept, 66-72, prof agr eng & agr engr, 72-77; WATER MGT SPECIALIST, OFF AGR, USDA, 76- *Mem:* Am Soc Agr Engrs; Am Soc Eng Educ; Nat Soc Prof Engrs; Sigma Xi. *Res:* Irrigation engineering, principally fundamentals governing the hydraulics of flow of water into, over and through a soil profile. *Mailing Add:* 1521 Beulah Rd Vienna VA 22180

COREY, HAROLD SCOTT, b Bridgeport, Conn, May 14, 19; m 47; c 5. MECHANICAL ENGINEERING. *Educ:* Fitchburg State Col, BS, 47, EdM, 49. *Prof Exp:* Instr mach tool lab, Univ Mass, 48-49; instr graphics & descriptive geom, Syracuse Univ, 49-51; asst prof mech eng, 51-64, assoc prof, 64- EMER ASSOC PROF MECH ENG WORCESTER POLYTECH INST. *Mem:* Am Soc Eng Educ. *Res:* Design; materials processing. *Mailing Add:* 169 Greenwood St Worcester MA 01609

COREY, JOHN CHARLES, b Toronto, Ont, May 7, 38; US citizen; m 66; c 3. SOIL PHYSICS. *Educ:* Univ Toronto, BSA, 60; Univ Calif, MS, 62; Iowa State Univ, PhD(soil physics), 66. *Prof Exp:* Res physicist, 66-73, tech supvr, 73-78, res supvr, 78-82, RES MGR, SAVANNAH RIVER LAB, E I DU PONT DE NEMOURS & CO, INC, 82- *Mem:* Am Soc Agron; Int Soc Soil Sci. *Res:* Management of radioactive and industrial waste; remote sensing; sewage sludge application to forested land; waste site remediation; bioreclamation; technical transfer. *Mailing Add:* 212 Lakeside Dr Aiken SC 29801

COREY, JOYCE YAGLA, b Waverly, Iowa, May 26, 38; m 62. INORGANIC CHEMISTRY. *Educ:* Univ NDak, BS, 60, MS, 61; Univ Wis, PhD(inorg chem), 64. *Prof Exp:* From instr to asst prof chem, Villa Madonna Col, 64-68; from asst prof to assoc prof chem, 68-79, PROF CHEM, UNIV MO-ST LOUIS, 80- *Concurrent Pos:* Vis prof chem, Univ Wis, 81 & 84. *Mem:* Am Chem Soc; Royal Soc Chem; Sigma Xi. *Res:* Synthetic organometallic chemistry of group IV, specifically analogs of antidepressants. *Mailing Add:* Dept Chem Univ Mo St Louis MO 63121

COREY, KENNETH EDWARD, b Cincinnati, Ohio, Nov 11, 38; m 61; c 2. URBAN GEOGRAPHY. *Educ:* Univ Cincinnati, PhD(geog), 69. *Prof Exp:* PROF & CHMN URBAN PLANNING GEOG, UNIV MARYLAND, 79-, DIR, URBAN STUDIES, 79- *Concurrent Pos:* Vis prof, Peking Univ, 86. *Mem:* Royal Geographical Soc; Asn Am Geographers. *Res:* Urban and regional planning. *Mailing Add:* Michigan State Univ 205 Berkey Hall East Lansing MI 48824

COREY, MARION WILLSON, b Jackson, Miss, Feb 11, 32; m 56; c 3. CIVIL & STRUCTURAL ENGINEERING. *Educ:* Auburn Univ, BS, 54; Miss State Univ, MS, 60; Ga Inst Technol, PhD(civil eng), 68. *Prof Exp:* From instr to asst prof civil eng, Miss State Univ, 58-61; instr civil technol, Southern Tech Inst, Ga Inst Technol, 61-63; assoc prof civil eng, 63-70, PROF CIVIL ENG, MISS STATE UNIV, 70- *Concurrent Pos:* Engr, Mitchell Eng Co, Miss, 62. *Mem:* Am Soc Civil Engrs; Am Soc Eng Educ. *Res:* Electronic computing; computer-aided design. *Mailing Add:* Dept Civil Eng Miss State Univ Mississippi State MS 39762

COREY, PAUL FREDERICK, b Lakewood, NJ, Apr 9, 50; m 72; c 1. CLINICAL CHEMISTRY. *Educ:* Va Polytech Inst & State Univ, BS, 72; State Univ NY, Buffalo, PhD(medicinal chem), 77. *Prof Exp:* Res assoc pharm, Univ Wis, Madison, 77-79; SR STAFF SCIENTIST, MILES LABS, INC, ELKHART, IND, 79- *Mem:* Am Chem Soc; AAAS. *Res:* Design and development of biologically active compounds; design and synthesis of novel indicators and reagents useful in medical diagnostics, especially colorimetric enzyme substrates for immunoassays. *Mailing Add:* 1529 Middlebury St Elkhart IN 46516

COREY, RICHARD BOARDMAN, b Wisconsin Rapids, Wis, Dec 25, 27. SOIL CHEMISTRY, SOIL FERTILITY. *Educ:* Univ Wis-Madison, BS, 49, MS, 51, PhD(soil chem), 53. *Prof Exp:* From asst prof to assoc prof, 54-65, PROF SOIL SCI, UNIV WIS-MADISON, 65- *Concurrent Pos:* Vis prof, Postgrad Col, Nat Sch Agr, Mex, 64; consult, Univ Wis Proj, US AID, Porto Alegre, Brazil, 65, Embrapa, 76; Prof, Univ Ife, Nigeria, 67-71, dean fac agr, 70-71. *Mem:* AAAS; Am Soc Agron; Soil Sci Soc Am; Int Soc Soil Sci. *Res:* Reactions of phosphorus and potassium in soils; development of methods for determining available nutrients in soils; land application of waste materials; reactions of heavy metals in soils. *Mailing Add:* 5017 Tomahawk Trail Madison WI 53705

COREY, ROBERT ARDEN, entomology, for more information see previous edition

COREY, SHARON EVA, b Princeton, NJ, Sept 12, 45; m 70; c 2. PHARMACOLOGY, MORPHOLOGY. *Educ:* Grove City Col, BS, 67; WVa Univ, PhD(pharmacol), 71. *Prof Exp:* Fel cell biol, WVa Univ, 71-73, res assoc electron microscopy, 73-74; ASST PROF PHARMACOL, UNIV PITTSBURGH, 76- *Mem:* AAAS; Am Asn Col Pharm. *Res:* Autonomic pharmacology; morphology-pharmacology; effects of neurotoxins on the peripheral and central nervous systems, especially vinblastine and vincristine. *Mailing Add:* Dept Pharmacol Salk Hall 0809 Univ Pittsburgh 4200 Fifth Ave Pittsburgh PA 15260

COREY, VICTOR BREWER, b Bynumville, Mo, Feb 9, 15; m 42; c 2. PHYSICS. *Educ:* Cent Col, Mo, AB, 37; Univ Iowa, MS, 39, PhD(physics), 42. *Prof Exp:* Asst physics, Univ Iowa, 37-42; res physicist, Sylvania Elec Prods, Inc, Pa & NY, 42-45; head electro-acoustics sect, Curtiss-Wright-Cornell Res Lab, 45-46; head electro-acoustics dept & actg head nuclear physics dept, Fredric Flader, Inc, 46-48, res coordr, 48-49, mgr eng physics div, 49-50, dir res,50-51; exec engr, Electronics Div, Willys Motors, Inc, Toledo, 51-53; tech dir, Donner Sci Co, 53-59; pres, Palomar Sci Corp, Calif, 60-63; vpres & bd dir, United Control Corp, 63-67, gen mgr transducer div, 66, int vpres, 66-71, VPRES NEW BUS DEVELOP, SUNDSTRAND DATA CONTROL, INC, 71- *Mem:* AAAS; Am Phys Soc; Acoust Soc Am; Inst Elec & Electronics Engrs; Am Inst Aeronaut & Astronaut. *Res:* Simulators and analog computers; servomechanisms; telemetering; missile guidance, control components; trainers; servo accelerometers, electromechanical amplifiers, integrators; applied physics instrumentation; electronic test and measurement equipment; digital transducers; analog-digital converters; microelectronics; radio altimeters; ultrasonic acoustics; inertial sensors. *Mailing Add:* 270 Cagney Lane Apt 111 Newport Beach CA 92663

CORFIELD, PETER WILLIAM REGINALD, b Manchester, Eng, Sept 14, 37; m 63; c 4. INORGANIC CHEMISTRY, X-RAY CRYSTALLOGRAPHY. *Educ:* Univ Durham, BSc, 59, PhD(x-ray), 63. *Prof Exp:* Res assoc, Crystallog Lab, Univ Pittsburgh, 63-65; res instr chem, Northwestern Univ, 65-66, instr, 66-67; asst prof, Ohio State Univ, 67-73; assoc prof, 73-77, PROF CHEM, KING'S COL, 77- *Concurrent Pos:* Sabbatical year, Univ Cambridge, UK, 85-86. *Mem:* AAAS; Am Crystallog Asn; Am Chem Soc; Royal Soc Chem. *Res:* Crystal and molecular structure by x-ray methods of inorganic substances and of proteins; development of computer programs in crystallography. *Mailing Add:* Dept Chem The Kings Col Briarcliff Manor NY 10510-1200

CORFMAN, PHILIP ALBERT, b Berea, Ohio, July 19, 26; m 50; c 4. OBSTETRICS & GYNECOLOGY. *Educ:* Oberlin Col, BA, 50; Harvard Univ, MD, 54; Am Bd Obstet & Gynec, dipl. *Prof Exp:* Mem staff clin obstet & gynec, Rip Van Winkle Clin, Hudson, NY, 59-63; prog assoc population res, 65-67, asst to dir population res, 67-68, dir ctr population res, Nat Inst Child Health & Human Develop, 68-84; officer res & development, Spl Programme Human Develop, 84-86; SUPVRY MED OFFICER, FERTILITY & MATERNAL HEALTH DRUGS, ROCKVILLE, MD, 87- *Concurrent Pos:* Josiah Macy, Jr Found res fel cervical carcinogenesis & population res, Col Physicians & Surgeons, Columbia Univ, 63-65; adv, WHO. *Mem:* Am Col Obstet & Gynec. *Res:* Population research, including biological studies in animals and humans; research administration of this field. *Mailing Add:* NIH Landow Bldg Rm A721 Bethesda MD 20205

CORIA, JOSE CONRADO, b Cuerhavaca, Mex, Nov 27, 52; m 82; c 1. CHEMICAL ANALYSIS SILICON WAFERS, ION CHROMATOGRAPHY. *Educ:* Universidad de Morelos, BS, 81; Univ Ariz, MS, 85; Univ NMex, PhD(anal chem), 90. *Prof Exp:* Res asst, Especialidades Quimicas, S A, 80-81, Inst Invest Elec, 81-82; teaching asst quant anal, Univ Ariz, 82-85; res asst, Univ NMex, 85-89; researcher H, Inst Invest Elec, 90-91; ANALYSIS SCIENTIST, MEMC ELECTRONIC MAT, INC, 91- *Mem:* Am Chem Soc; Soc Appl Spectros. *Res:* Chemical analysis of metal contaminants present on silicon water surfaces; fluorescence of ions in concentrated acids; fiber optic sensors; metal-ligand complexation equilibria; ion chromatography; ICP-MS; raman spectrometry. *Mailing Add:* PO Box 8 St Peters MO 63376

CORIELL, KATHLEEN PATRICIA, b Cumberland, Md, Feb 19, 35; m 63. COMPUTER SCIENCES. *Educ:* Univ Md, BS, 59, MS, 66; Howard Univ, PhD(physics), 69. *Prof Exp:* Physicist, Nat Bur Standards, 59-63; asst prof physics, Hood Col, 69-70; instr, Montgomery Col, Md, 72; programmer analyst, Greenwich Data Systs, 73; mem tech staff, Comput Sci Corp, 74-76; STAFF, COMPUT SYSTS ANALYST, FED GOVT, 76- *Mem:* Sigma Xi. *Res:* Computer programming and analysis. *Mailing Add:* 15208 Spring Meadows Dr Germantown MD 20874-3438

CORIELL, LEWIS L, b Sciotoville, Ohio, June 19, 11; m 36; c 3. BACTERIOLOGY. *Educ:* Univ Mont, BA, 34; Univ Kans, MA, 36, PhD(bact), 40, MD, 42. *Prof Exp:* Asst instr bot, Univ Mont, 34; from asst instr to instr bact, Univ Kans, 34-40; instr pediat, 46-49, assoc prof immunol pediat, 49-63, PROF PEDIAT, UNIV PA, 63-; DIR, INST MED RES, 55- *Concurrent Pos:* Nat Res sr fel virus dis, Inst Med Res, 47-49; med dir, Camden Munic Hosp for Contagious Dis, 49-61; pediatrist, Cooper Hosp, 49-; sr physician, Children's Hosp, 54-; consult, Philadelphia Naval Hosp, 56-66. *Mem:* AAAS; assoc AMA; assoc Am Soc Microbiol; assoc Asn Mil Surg US; assoc Soc Pediat Res. *Res:* Preservation of bacteria lymphilization; natural immunity and streptomycin therapy of tularemia; botulism; herpes simplex; herpes zoster; natural immunity of cats; poliomyelitis; antibiotics; tissue culture; cancer; pediatrics. *Mailing Add:* Inst for Med Res Copewood & Davis Sts Camden NJ 08103

CORIELL, SAM RAY, b Greenfield, Ohio, Dec 21, 35; m 63. MATERIALS SCIENCE, PHYSICAL METALLURGY. *Educ:* Ohio State Univ, BSc, 56, PhD(phys chem), 61. *Prof Exp:* Phys chemist, Statist Physics Sect, 61-63, PHYS CHEMIST, METALL DIV, NAT BUR STANDARDS, 63- *Concurrent Pos:* Nat Res Coun-Nat Bur Standards res assoc, 61-63. *Mem:* AAAS; Am Chem Soc; Am Phys Soc; Am Assoc Crystal Growth; Metall Soc. *Res:* Crystal growth; solidification; heat flow, diffusion, and fluid flow. *Mailing Add:* Nat Inst Standards & Technol Gaithersburg MD 20899

CORINALDESI, ERNESTO, b Italy, Aug 20, 23. QUANTUM MECHANICS. *Educ:* Univ Rome, BS, 44; Univ Manchester, PhD(theoret physics), 51. *Prof Exp:* Asst, Univ Rome, 45-47; res fel, Nat Res Coun Can, 51-52; Higgins vis fel & instr, Palmer Phys Lab, Princeton Univ, 52-53; asst, Dublin Inst Adv Studies, 53-55; Imp Chem Indust res fel, Univ Glasgow, 55-57; lectr math, Univ Col North Staffordshire, 57-58; dir grad sch nuclear studies & in charge of theoret physics course, Nat Inst Nuclear Physics, Pisa, 59-61; mem, Inst Adv Study, 61-62; assoc vis prof, Univ Iowa, 62-63; vis prof, Univ Toronto, 63-64; vis prof, Boston Univ, 65; fel scientist, Westinghouse Res & Develop Ctr, Pa, 65-66; PROF PHYSICS, BOSTON UNIV, 66- *Concurrent Pos:* Prof ruolo, Univ Calabria, 76-77. *Mem:* Italian Phys Soc. *Res:* Quantum mechanics. *Mailing Add:* Dept Physics Boston Univ 590 Commonwealth Ave Boston MA 02215

CORK, BRUCE, b Peck, Mich, Oct 21, 15; m 46; c 4. PARTICLE PHYSICS. *Educ:* Univ Mich, BS, 37; Polytech Inst New York, MS, 41; Univ Calif, Berkeley, PhD, 60. *Prof Exp:* Physicist, Tube Lab, Radio Corp Am, 37-40; physicist, Radiation Lab, Mass Inst Technol, 41-45 & Dept Physics, 46; physicist, Los Alamos Sci Lab, Univ Calif, 45-46 & Lawrence Radiation Lab, 46-68; assoc lab dir high energy physics, Argonne Nat Lab, 68-73; PHYSICIST, LAWRENCE BERKELEY LAB, UNIV CALIF, BERKELEY, 73- *Concurrent Pos:* Chmn Los Alamos Meson Physics Facil policy comt, Los Alamos Sci Lab, 73-75. *Honors & Awards:* Sigma Xi. *Mem:* Am Phys Soc. *Res:* Nuclear scattering; antiproton; antineutron; parity nonconservation; resonant states of nucleons; electron-positron annihilations. *Mailing Add:* 2724 Las Aromas Piedmont CA 94611

CORK, DOUGLAS J, b St Paul, Minn, May 5, 50; m 75; c 1. GAS BIOENGINEERING, AUTOTROPHIC BIOENGINEERING. *Educ:* Univ Ariz, BS, 72, MS, 74, PhD(agr biochem), 78. *Prof Exp:* Res group leader indust microbiol, Aquaterra Biochem, 78-79; asst prof biochem eng, Univ Miss, 79-80; asst prof biochem eng, Ill Inst Technol, 80-; AT DEPT BIOL, ILL INST TECHNOL. *Concurrent Pos:* Consult, Nat Chemsearch Corp, 79, Nat Distillers Chem Corp, 80, Schaeffer & Roland Engrs, 80, Chicago Metrop Sanit Dist, 80, Genex Corp, 81-, WR Grace & Co, 81-, Inst Gas Technol, 81- *Mem:* Am Inst Chem Engrs; Am Chem Soc; Soc Indust Microbiol; Am Soc Microbiol. *Res:* Continuous gypsum sulfate reduction and sulfur production by biochemical and genetic engineering technology; process engineering of microbial systems of fuels and chemicals from waste gases and liquids. *Mailing Add:* Dept Biol Ill Inst Technol Chicago IL 60616

CORK, LINDA K COLLINS, b Texarkana, Tex, Dec 14, 36; div; c 2. VETERINARY PATHOLOGY. *Educ:* Tex A&M Univ, BS, 69, DVM, 70; Wash State Univ, PhD(exp path), 74; Am Col Vet Pathologists, dipl, 75. *Prof Exp:* Asst prof vet path, Univ Ga, 74-76; asst prof, 76-89, PROF, DIV COMP MED & DEPT PATHOL, SCH MED, JOHNS HOPKINS UNIV, 90- *Concurrent Pos:* NIH fel, Wash State Univ, 70-74; mem ed bd, Vet Path, 77-; mem, comt animal models & genetic stocks, 77-, mem comt Animals as Models of Environ Hazards, Nat Acad Sci, 87-; mem, adv coun, Div Res Resources, NIH, 85; assoc dir res, Alzheimer's Dis Res Ctr, Sch Med, Johns Hopkins Univ. *Mem:* Am Col Vet Pathologists; Am Asn Neuropathologist; Am Asn Path; Am Vet Med Asn; AAAS; Inst Med-Nat Acad Sci. *Res:* Comparative neuropathology; neurovirology; animal models of neurologic disease. *Mailing Add:* Div Comp Med Sch Med Johns Hopkins Univ 720 Rutland Ave Baltimore MD 21205

CORKE, CHARLES THOMAS, b Stratford, Ont, Mar 19, 21; m 45; c 4. SOIL MICROBIOLOGY. *Educ:* Univ Western Ont, BSc, 50, MSc, 51; Rutgers Univ, PhD, 54. *Prof Exp:* From assoc prof to prof microbiol, Univ Guelph, 54-81; AT DEPT ENVIRON BIOL, UNIV GUELPH. *Mem:* Agr Inst Can. *Res:* Pesticides degradation; microbial transformations of substituted anilines. *Mailing Add:* 37 Maple Ave New Glasgow NS B2H 2B1 Can

CORKERN, WALTER HAROLD, b Washington Parish, La, Mar 28, 39; m 59; c 3. ORGANIC CHEMISTRY. *Educ:* La State Univ, BS, 61; Univ Ark, PhD(org chem), 66. *Prof Exp:* Res chemist, E I du Pont de Nemours & Co, Tex, 65-66; asst prof, 66-72, assoc prof, 72-77, PROF CHEM, SOUTHEASTERN LA UNIV, 77- *Mem:* AAAS; Am Chem Soc. *Res:* Acid catalized ketone rearrangements; mechanistic studies; organic reaction mechanisms. *Mailing Add:* Dept Chem Southeastern La Univ Box 725 Univ Sta Hammond LA 70402

CORKIN, SUZANNE HAMMOND, b Hartford, Conn, May 18, 37; div; c 3. BEHAVIORAL NEUROBIOLOGY. *Educ:* Smith Col, BA, 59; McGill Univ, MSc, 61, PhD(psychol), 64. *Prof Exp:* Res & teaching asst neuropsychol, Dept Psychol, McGill Univ, 61-64; fel, Montreal Neuro Inst, 61-64; res assoc, Dept Psychol, 64-77, res assoc, Clin Res Ctr, 64-79, lectr neuropsychol, Dept Psychol, 77-79, prin res scientist, 79-81, assoc prof, 81-87, SR INVESTR, CLIN RES CTR, MASS INST TECHNOL, 79-, PROF BEHAV NEUROSCI, DEPT BRAIN & COGNITIVE SCI, 87- *Concurrent Pos:* Consult psychol (neurosurg), Mass Gen Hosp, 75- *Mem:* Am Psychol Asn; AAAS; Soc Neurosci; Sigma Xi. *Res:* Brain-behavior relationships in man; somatosensory system; disorders of memory function; neuroplasticity. *Mailing Add:* Dept Psychol E10-003A Mass Inst Technol 77 Massachusetts Ave Cambridge MA 02139

CORKUM, KENNETH C, b Aurora, Ill, Aug 9, 30. PARASITOLOGY, INVERTEBRATE ZOOLOGY. *Educ:* Aurora Col BS, 58; La State Univ, MS, 60, PhD(zool), 63. *Prof Exp:* NIH fel parasitol, Sch Med, Tulane Univ, 63-65; from asst prof to assoc prof zool & physiol, 65-77, PROF ZOOL & PHYSIOL, LA STATE UNIV, BATON ROUGE, 77- *Mem:* Am Soc Parasitol; Am Micros Soc; Soc Syst Zool; Am Soc Zool. *Res:* Marine trematode taxonomy; cestode life cycles. *Mailing Add:* Dept Zool & Physiol La State Univ Baton Rouge LA 70803

CORLESS, JOSEPH MICHAEL JAMES, b Orlando, Fla, July 28, 44; m 74; c 1. BIOPHYSICS, VISION. *Educ:* Georgetown Univ, BS, 66; Duke Univ, PhD(anat), 71, MD, 72. *Prof Exp:* Teaching asst bot & zool, Georgetown Univ, 65-66; instr micros anat, 68 & 69, teaching asst phys anthrop, 69, instr micros anat, 70, assoc anat, 72-73, asst prof, 74-80, ASSOC PROF ANAT & ASST PROF OPHTHAL, MED CTR, DUKE UNIV, 80- *Concurrent Pos:* Fel, Med Res Coun Lab Molecular Biol, Cambridge, Eng, 73-74. *Mem:* Biophys Soc; Sigma Xi; Soc Photochem & Photobiol; Asn Res in Vision & Ophthal. *Res:* Structure of visual photoreceptors; biomembrane structure. *Mailing Add:* Dept Cell Biol & Ophthal Box 3011 Duke Univ Med Ctr Durham NC 27710

CORLETT, MABEL ISOBEL, b Noranda, Que, Feb 7, 39. MINERALOGY. *Educ:* Queen's Univ, Ont, BSc, 60; Univ Chicago, SM, 62, PhD(mineral), 64. *Prof Exp:* Res asst mineral, Inst Crystallog & Petrog, Swiss Fed Inst Technol, 65-69; res assoc, Queen's Univ, Ont, 69-71, from asst prof geol sci to assoc prof geol sci, 71-86,; RETAILER, 86- *Mem:* Mineral Soc Am; Mineral Asn Can; Swiss Soc Mineral & Petrog. *Res:* Electron probe microanalysis; mineral chemistry. *Mailing Add:* 313 University Ave Kingston ON K7L 3R3 Can

CORLETT, MICHAEL PHILIP, b Toronto, Ont, Apr 28, 37. MYCOLOGY. *Educ:* Univ Toronto, BA, 59, MA, 62, PhD(mycol), 65. *Prof Exp:* RES MYCOLOGIST, BIOSYSTEMATICS RES INST, AGR CAN, 65- *Mem:* Mycol Soc Am; Can Bot Asn; Can Phytopath Soc. *Res:* Histology, morphology and development of fungi; taxonomy of the ascomycetes and fungi imperfecti. *Mailing Add:* 57 Quinpool Crescent Nepean ON K2H 6H9 Can

CORLEY, CHARLES CALHOUN, JR, b Charlotte, NC, June 30, 27; m 52; c 4. HEMATOLOGY, INTERNAL MEDICINE. *Educ:* Clemson Univ, BS, 53; Emory Univ, MD, 53. *Prof Exp:* From instr to assoc prof, 56-70, PROF HEMAT, SCH MED, EMORY UNIV, 70-, PROF MED. *Concurrent Pos:* USPHS fel hemat, Emory Univ, 56-57. *Mem:* Am Col Physicians; Am Soc Hemat; Am Fed Clin Res; AMA. *Res:* Chemotherapy of hematologic malignancies. *Mailing Add:* Dept Med Emory Univ Sch Med Atlanta GA 30322

CORLEY, GLYN JACKSON, b Carson, La, Jan 23, 16; m 39; c 5. ACTUARIAL MATHEMATICS. *Educ:* Northwestern State Univ, BA, 38; Columbia Univ, MA, 40; George Peabody Col, PhD(math), 59. *Prof Exp:* Instr math & physics, Springfield Col, 40; from instr to assoc prof math, Northwestern State Univ, 41-62; prof, ETex State Univ, 62-67, head dept, 66-67; prof math & head dept, 67-80, EMER PROF & EMER CHMN, LA STATE UNIV, SHREVEPORT, 80- *Concurrent Pos:* Instr, Univ Tex, 46-47 & Vanderbuilt Univ, 54-55; head actuary, Werntz & Assocs, 80-85. *Mem:* Math Asn Am. *Res:* Analysis; statistics; actuarial mathematics. *Mailing Add:* 6117 Gaylyn Dr Shreveport LA 71105-4823

CORLEY, JOHN BRYSON, b Calgary, Alta, Aug 29, 13; m 47; c 2. FAMILY MEDICINE. *Educ:* Univ Alta, BA, 36, MD, 42. *Prof Exp:* Sr partner, Chinook Med Clin, Calgary, 58-73; prof family pract, Med Univ SC, 73-83; RETIRED. *Concurrent Pos:* Prin investr, Nat Health Res Grant, Develop Grad Training Family Physicians, 66-69; vis prof, dept community med, Univ Conn, 72; consult, educ adv comt, Col Family Physicians Can, 73-76; assoc ed self-assessment, Continuing Educ for Family Physician, 73-77; mem, Clin Prob Solving Skills Comt, Nat Bd Med Examr, 74-75. *Mem:* Royal Col Med; fel Am Soc Clin Hypnosis; fel Can Col Family Physicians; Int Soc Res Med Educ; Am Educ Res Asn. *Res:* Development of prototype models of formative evaluation of post-graduate programs of specialty medical education. *Mailing Add:* 588 Oyster Rake Dr Johns Island SC 29455

CORLEY, RONALD BRUCE, b Durham, NC, Oct 22, 48; m 80; c 4. MOLECULAR IMMUNOLOGY, IMMUNOGENETICS. *Educ:* Duke Univ, BS, 70, PhD(microbiol), 75. *Prof Exp:* Mem scientist immunol, Basel Inst Immunol, Switzerland, 75-77; ASSOC PROF MICROBIOL & IMMUNOL, MED CTR, DUKE UNIV, 77- *Concurrent Pos:* Vis scientist, Basel Inst Immunol, Switzerland, 77, 79 & 80; mem fac, Comprehensive Cancer Ctr, Duke Univ, 78-; scholar award, Leukemia Soc Am, 79-84; prin investr res grant, NIH, 80- *Mem:* Am Asn Immunologists; Am Soc Histocompatibility & Immunogenetics. *Res:* Gene expression in B lymphocytes; diversity of immunoglobin V genes in autoimmune B lymphocytes; helper T cells in B cells activation and diversification. *Mailing Add:* Div Immunol Box 3010 Med Ctr Duke Univ Durham NC 27710-3010

CORLEY, TOM EDWARD, b Kellyton, Ala, Apr 25, 21; m 48; c 2. AGRICULTURAL ENGINEERING. *Educ:* Ala Polytech Inst, BS, 43, MS, 49. *Prof Exp:* Asst agr eng, Auburn Univ, 46-48, asst agr engr, 48-53, assoc, 53-63, prof, 63-66, asst dir, Agr Exp Sta, 66-83, assoc dir, 83-84, EMER ASSOC DIR AGR EXP STA, AUBURN UNIV, 84- *Mem:* Am Soc Agr Engrs; Sigma Xi. *Res:* Farm machinery development, design and testing. *Mailing Add:* 519 Heard Ave Auburn AL 36830

CORLEY, WILLIAM GENE, b Shelbyville, Ill, Dec 19, 35; m 59; c 3. STRUCTURAL ENGINEERING. *Educ:* Univ Ill, BS, 58, MS, 60, PhD(struct eng), 61. *Prof Exp:* Res asst, Univ Ill, 59-61; develop engr, Portland Cement Asn, Skokie, 64-66, mgr struct res sect, 66-74, dir, 74-78, div dir, Eng Develop Div, Res & Develop Labs, 78-87; VPRES, CONSTRUCTION TECHNOL LABS, INC, 87- *Concurrent Pos:* First lt, US Army Corps of Engrs, 61-64. *Honors & Awards:* Martin P Korn Award & Wason Res Medal, Am Concrete Inst, 70, Bloem Award, 78, Reese Struct Res Award, 86; T Y Lin Award, Am Soc Civil Engrs; Arthur J Boase Award, Reinforced Concrete Res Coun, 86. *Mem:* Am Concrete Inst; Am Soc Civil Engrs; Nat Soc Prof Engrs; Earthquake Eng Res Inst. *Res:* Structural concrete for earthquake resistant construction; uses of structural concrete in buildings and bridges. *Mailing Add:* 744 Glenayre Dr Glenview IL 60025

CORLISS, CHARLES HOWARD, b Medford, Mass, Oct 30, 19; m 43; c 2. ATOMIC SPECTROSCOPY, LINE INTENSITY. *Educ:* Mass Inst Technol, BS, 41. *Prof Exp:* physicist, Nat Bur Standards, 42-81; PHYSICIST, FOREST HILLS LAB, 81- *Honors & Awards:* Silver Medal, US Dept Com, 63. *Mem:* AAAS; Optical Soc Am; Royal Astron Soc. *Res:* Description and analysis of atomic spectra; measurement of spectral intensity; transition probabilities; light sources; astronomical spectroscopy; compilation of energy levels. *Mailing Add:* Forest Hills Lab 2955 Albemarle St NW Washington DC 20008

CORLISS, CLARK EDWARD, b Coats, Kans, Nov 11, 19; m 50; c 3. ANATOMY. *Educ:* Univ Vt, BS, 42; Univ Mass, MS, 49; Brown Univ, PhD(biol), 52. *Prof Exp:* Asst prof microanat, Med Sch, Dalhousie Univ, 51-53; from instr to prof, 63-85, EMER PROF ANAT, CTR HEALTH SCI, UNIV TENN, MEMPHIS, 85- *Concurrent Pos:* Emb ed, Stedman's Med Dict, 76-90; assoc ed, Anat Rec, 83-88. *Mem:* Am Asn Anat; Soc Develop Biol. *Res:* Study of normal and abnormal development of the central nervous system in vertebrate embryos. *Mailing Add:* 1293 Dogwood Dr Memphis TN 38111

CORLISS, EDITH LOU ROVNER, b Cleveland, Ohio, Sept 8, 20; m 43; c 2. PHYSICS. *Educ:* Mass Inst Technol, BS & MS, 41. *Prof Exp:* Jr physicist, Nat Bur Standards, 41-42 physicist, 44-80 & US Weather Br, 42-43, physicist, Nat Bur Standards, 44-80; jr astronomer, US Naval Observ, 43-44; physicist, Nat Bur Standards, 44-80; MEM STAFF, FOREST HILLS LAB, 81- *Concurrent Pos:* Res fel, Elec Eng Dept, Imperial Col, London, 62-63. *Mem:* Am Phys Soc; Acoust Soc Am; Am Hort Soc; Inst Elec & Electronics Eng; Am Speech Lang Hearing Asn. *Res:* Analysis of transients; speech communication; physical problems in measurement of hearing; musical acoustics. *Mailing Add:* Forest Hills Lab 2955 Albemarle St NW Washington DC 20008-2135

CORLISS, JOHN BURT, b Pasadena, Calif, Apr 16, 36; m 61; c 3. ORIGIN OF LIFE, SUBMARINE HOT SPRINGS. *Educ:* Ariz State, BS, 58; Scripps Inst Oceanog, PhD(oceanog), 70. *Prof Exp:* ASSOC PROF OCEANOG & RES SCIENTIST, ORE STATE UNIV, 70- *Concurrent Pos:* Vis prof, Yale Univ, 71-72; mem, Gen Evolution Res Group. *Res:* Submarine hydrothermal systems: chemistry, geology and biology; dynamical systems theory and the origin of life on earth; simulation of evolutionary systems on massively parallel processor arrays. *Mailing Add:* Computer Systs Res Facil Code 935 NASA-Goddard Space Flight Ctr Greenbelt MD 20771

CORLISS, JOHN OZRO, b Coats, Kans, Feb 23, 22; wid; c 5. PROTOZOOLOGY. *Educ:* Univ Chicago, BS, 44; Univ Vt, AB, 47; NY Univ, PhD(biol), 51. *Hon Degrees:* Univ Clermont, Clermont-Ferrand, France, Docteur Honoris Causa, 73. *Prof Exp:* US Atomic Energy Comn res fel biol sci, Col France, 51-52; instr zool, Yale Univ, 52-54; from asst prof to prof, 54-64, head dept biol sci, Univ Ill, 64-69; dir syst biol, NSF, 69-70; prof & head dept, 70-87, prof, 87-89, EMER PROF ZOOL, UNIV MD, COLLEGE PARK, 89- *Concurrent Pos:* Vis prof, Univs London & Exeter, 60-62; ed, Trans Am Micros Soc, 66-80; chmn US Nat comt, Int Union Biol Sci, 71-73; mem & comnr, Int Comn Zool Nomenclature, 72-, counr, 87-; mem, Corp of Marine Biol Lab, Woods Hole, Mass, 74-; ed, J Protozool, 80-83; vis hon prof, East China Normal Univ, Shanghai, Repub China, 80, 86; vis res prof, Univ Geneva, Switzerland, 80; adj prof biol, Univ NM, Albuquerque, 88-; Sigma Xi fac award, Univ Md, 85. *Mem:* Soc Protozool (secy, 58-61, pres, 64-65); Am Soc Zool (pres, 71-72); Soc Syst Zool (pres, 69-70); Am Micros Soc (pres, 65-66); Coun Biol Eds; Int Soc Eval Protistol (pres, 79-80); Int Phycol Soc. *Res:* Comparative morphology, systematics, evolution, and phylogeny of ciliate protozoa; anatomy of the infraciliature; morphogenesis; nomenclature; international collection of ciliate type-specimens; evolution, phylogeny and classification of protists history of protistology. *Mailing Add:* PO Box 53008 Albuquerque NM 87153

CORLISS, LESTER MYRON, b NJ, Mar 29, 19; m 41; c 2. SOLID STATE PHYSICS. *Educ:* City Col New York, BS, 40; Harvard Univ, MA, 48, PhD(chem physics), 49. *Prof Exp:* Chemist, Manhattan Proj, 43-46; sr chemist, Brookhaven Nat Lab, 49-85; RETIRED. *Concurrent Pos:* NSF sr fel, Univ Grenoble, 59-60; mem comn neutron diffraction, Int Union Crystallog, 66-75, chmn 69-75. *Mem:* Fel Am Phys Soc. *Res:* Neutron diffraction; magnetism; phase transformations; critical phenomena. *Mailing Add:* Fearrington Post 252 Pittsboro NC 27312

CORMACK, ALLAN MACLEOD, b Johannesburg, SAfrica, Feb 23, 24; m 50; c 3. NUCLEAR PHYSICS, COMPUTED TOMOGRAPHY. *Educ:* Univ Cape Town, BSc, 44, MSc, 45. *Hon Degrees:* DSc, Tufts Univ, 80. *Prof Exp:* From jr lectr to lectr, Univ Cape Town, 46-56; res fel, Harvard Univ, 56-57; From asst prof to prof, 57-80, UNIV PROF PHYSICS, TUFTS UNIV, 80- *Concurrent Pos:* Assoc ed, J Computed Tomography, 77-; Adv comt, Inverse Probs, 85. *Honors & Awards:* Nobel Prize in Med, 79; Ballou Medal, Tufts Univ, 78; Mike Hogg Medal, Univ Tex Cancer Syst, 81; Nat Medal Sci, 90. *Mem:* Nat Acad Sci; Fel Am Phys Soc; Fel Am Acad Arts & Sci; hon mem Swedish Radiol Soc; hon mem SAfrica Inst Physics; Foreign Asn, Royal Soc SAfrica. *Res:* Mathematics. *Mailing Add:* Physics Dept Tufts Univ Medford MA 02155

CORMACK, GEORGE DOUGLAS, b Killam, Alta, Sept 11, 33; m 59; c 3. COMMUNICATIONS ENGINEERING. *Educ:* Univ BC, BASc, 55, MSc, 60, PhD(physics), 62. *Prof Exp:* Instrumentation engr, Comput Devices Can, Ltd, 57-59; NATO overseas fel, 62-64; from asst prof to assoc prof, Carleton Univ, 64-76; systs consult, Dept Commun, Govt Can, 76-80; vpres eng, Can Cable TV Asn, 80-82; PROF ELEC ENG, UNIV ALBERTA, 82- *Concurrent Pos:* Consult, Nat Res Coun Can, 64-65, Comput Devices Can, Ltd, 65-68 & Northern Elec Co, 66-76; sr mem, Bell Tel Labs, 70-71. *Mem:* Inst Elec & Electronics Engrs. *Res:* Computer techniques; electromagnetics; communication systems; CATV; demographics; economics of rural communications. *Mailing Add:* Dept Elec Eng Univ Alberta Edmonton AB T6G 2M7 Can

CORMACK, JAMES FREDERICK, b Portland, Ore, Mar 13, 27; m 52; c 5. ENVIRONMENTAL CHEMISTRY. *Educ:* Reed Col, BA, 48; Ore State Col, PhD(biochem), 53. *Prof Exp:* Asst, Ore State Col, 48-49; res chemist, Crown-Zellerbach, 53-59, supvr water progs, 59-84, tech specialist, Environ Serv Div, 84-86; MGR WATER PROGS, CORP ENVIRON SERV, JAMES RIVER, 86- *Mem:* AAAS; Am Chem Soc; Tech Asn Pulp & Paper Indust. *Res:* Stream improvement; waste treatment; hazardous waste disposal. *Mailing Add:* Corp Environ Serv James River Corp Camas WA 98607

CORMACK, ROBERT GEORGE HALL, b Cedar Rapids, Iowa, Feb 2, 04; m 39; c 2. BOTANY. *Educ:* Univ Toronto, BA, 29, MA, 31, PhD(biol), 34. *Prof Exp:* Demonstr bot, Univ Toronto, 29-36; lectr, 36-45, from asst prof to prof, 45-69, EMER PROF BOT, UNIV ALTA, 69- *Concurrent Pos:* Bot consult, Northwest Proj Study Group, MacKenzie Valley, NWT, Can, 70-71. *Mem:* Can Inst Forestry; Royal Soc Can. *Res:* Developmental anatomy; forest and wildlife conservation. *Mailing Add:* 9835-113th St Apt 1106 Edmonton AB T5K 1N4 Can

CORMAN, EMMETT GARY, b Kansas City, Kans, Aug 2, 30; m 57; c 4. THEORETICAL PHYSICS, NUCLEAR PHYSICS. *Educ:* Univ Kans, BS, 52, MS, 54, PhD(physics), 60. *Prof Exp:* Assoc physicist, Vitro Corp Am, 52-53; physicist, Los Alamos Sci Lab, 54-56; RES PHYSICIST, LAWRENCE RADIATION LAB, UNIV CALIF, LIVERMORE, 60- *Res:* Optical model analysis of high energy nucleons; Monte Carlo photon-matter interactions; physics of nuclear weapons; numerical solutions on high speed computers; physics of plasma-magnetic field interactions; multi-group and Monte Carlo ion transport; Monte Carlo transport of electrons and positrons from Gamma Annihilation. *Mailing Add:* Comp Physics Div Bldg 311 L-298 Lawrence Livermore Lab PO Box 808 Livermore CA 94550

CORMIER, ALAN DENNIS, b Lynn, Mass, Feb 16, 45; m 68; c 2. ANALYTICAL CHEMISTRY, PHYSICAL CHEMISTRY. *Educ:* Northeastern Univ, BA, 68; Univ NH, PhD(spectros), 72. *Prof Exp:* Fel spectros, Marquette Univ, 72-74; instr phys chem, Boston Univ, 74-76; PROF LEADER ELECTROANAL CHEM, INSTRUMENTATION LAB INC, 76- *Concurrent Pos:* NDEA res fel, Univ NH, 69-72; consult/sci programmer, US Forest Serv, 70-74. *Mem:* Am Chem Soc; Sigma Xi. *Res:* Sensors, techniques and instrumental fluidic design for quantitative analytical determinations in biological fluids; potentiometric and polargraphic electrodes and selected spectrophotometric methods. *Mailing Add:* 36 Charles St Newbury Port MA 01950

CORMIER, BRUNO M, b Laurierville, Que, Nov 14, 19; m. PSYCHIATRY. *Educ:* Univ Montreal, BA, 42, MD, 47; McGill Univ, dipl psychiat, 52. *Prof Exp:* From asst prof to assoc prof psychiat, 53-67, dir criminal res, 67-74, lectr psychiat, 63-74, assoc prof psychiat, 74-78, PROF FORENSIC PSYCHIAT, MCGILL UNIV, 78- *Concurrent Pos:* Clin asst psychiat, Royal Victoria Hosp, Montreal, 53-55, assoc psychiatrist, 56-63; 55-70; consult, NY State Correctional Serv, 66-72. *Mem:* Am Psychiat Asn; Can Psychiat Asn; Can Corrections Asn; Can Psychoanal Soc. *Res:* Clinical criminology; psychopathology of deprivation of liberty; persistent criminality. *Mailing Add:* 5088 Cote Saint Antoine Montreal PQ H4A 1N5 Can

CORMIER, MILTON JOSEPH, b DeRidder, La, Nov 29, 26; m 51; c 3. BIOCHEMISTRY. *Educ:* Southwestern La Inst, BS, 48; Univ Tex, MA, 51; Univ Tenn, PhD(microbiol), 56. *Prof Exp:* Assoc biochemist, Biol Div, Oak Ridge Nat Lab, 56-58; from asst prof to prof bioluminescence, 58-66, RES PROF BIOLUMINESCENCE, UNIV GA, 66- *Mem:* Am Chem Soc; Am Soc Microbiol; Am Soc Biol Chem. *Res:* Mechanisms of bioluminescent reactions. *Mailing Add:* Dept Biochem Univ Ga Boyd G SRC Athens GA 30602

CORMIER, RANDAL, b Truro, NS, Mar 9, 30; m 55; c 6. GEOLOGY. *Educ:* St Francis Xavier Univ, Can. BSc, 51; Mass Inst Technol, PhD(geol), 56. *Prof Exp:* Assoc prof, 57-74, PROF GEOL, ST FRANCIS XAVIER UNIV, 74- *Mem:* Can Geol Soc; Can Inst Min & Metall. *Res:* Geochronology; geochemical prospecting. *Mailing Add:* Dept Geol St Francis Xavier Univ Antigonish NS B2G 1C0 Can

CORMIER, REGINALD ALBERT, b New York, NY, Aug 11, 1930; m 57; c 6. MICROWAVE TRANSMITTERS. *Educ:* City Col NY, BEE, 57. *Prof Exp:* Prin engr, REL/Reeves, 71-72; transmitter dept mgr, Comtech Telecommun, 73-78; MEM TECH STAFF, JET PROPULSION LAB, CALIF INST TECHNOL, 79- *Concurrent Pos:* Consult, Comcrylt Ltd, 78- *Res:* Radar and communication transmitters for operation in Nat Aeronautics and Space Administrations deep space network; klystron development; use of gyrotron oscillators as microwave amplifiers; electromanetic breakdown of gases; applied propositional calculus to the development of expert systems. *Mailing Add:* 5033 Mount Royal Dr Los Angeles CA 90041

CORMIER, ROMAE JOSEPH, b New York, NY, May 17, 28; m 54; c 4. MATHEMATICS. *Educ:* Univ Chattanooga, BS, 51; Univ Tenn, MA, 56 & Univ Mo, 63. *Prof Exp:* Mathematician ballistics, Vitro Corp Am, 55; ASST PROF MATH, NORTHERN ILL UNIV, 56- *Concurrent Pos:* Consult, Gen Elec Co, 65, NAm Mineral Explor, 69- & Commonwealth Edison, 85; Solutions ed, J Recreational Math, 77-; consult, Commonwealth Edison, 85. *Mem:* Am Math Soc; Int Tensor Soc; Math Asn Am; Austrian Math Soc; Ital Math Union; Planetary Soc. *Res:* Steiner symmetrization; differential equations; combinatorics; number theory; logic. *Mailing Add:* Dept Math Northern Ill Univ De Kalb IL 60115

CORMIER, THOMAS MICHAEL, b Waltham, Mass, Nov 15, 47. NUCLEAR STRUCTURE, NUCLEAR REACTIONS. *Educ:* Mass Inst Technol, SB, 71, PhD(physics), 74. *Prof Exp:* Res assoc, Mass Inst Technol, 74-75, State Univ NY, Stony Brook, 75-77; vis scientist, Max Planck Inst Nuclear Physics, Heidelberg, 77-78; prof, Univ Rochester, 78-88; PROF, TEX A&M UNIV, 88- *Concurrent Pos:* Alfred P Sloan fel, 81-83. *Mem:* Am Phys Soc; AAAS. *Mailing Add:* Cyclotron Inst Tex A&M Univ College Station TX 77843

CORN, HERMAN, b Philadelphia, Pa, Oct 7, 21; c 3. PERIODONTOLOGY. *Educ:* Temple Univ, DDS, 44; Am Bd Periodont, dipl, 63. *Prof Exp:* Instr periodont, Grad Sch Med, 58-63, assoc, Sch Dent Med & Grad Sch Med, 63-65, asst prof, Sch Dent Med, 65-68, assoc prof, 68-74, PROF PERIODONT, SCH DENT MED, UNIV PA, 74- *Concurrent Pos:* Oral surgeon, Lower Bucks County Hosp, 54-; lectr, Dent Schs, Wash Univ & Univ Ky, 68, Temple Univ & Case Western Reserve Univ, 69 & Boston Univ, 69-; lectr post grad educ, Sch Dent Med, Univ Pa; clin assoc prof & dir prev dent, Med Col Pa, 74-, clin prof, Div Dent Med, 86- *Honors & Awards:* Master Clinician Award, Am Acad Periodont, 87. *Mem:* Fel Am Col Dentists; Am Dent Asn; fel Am Acad Periodont; Am Soc Prev Dent (pres, 73-74). *Res:* Clinical periodontics; periodontal therapy; mucogingival surgery; adult orthodontics. *Mailing Add:* 565 Sanctuary Dr No A202 Longboat Key FL 34228

CORN, MORTON, b New York, NY, Oct 18, 33; m 55; c 2. SANITARY ENGINEERING. *Educ:* Cooper Union, BChE, 55; Harvard Univ, MS, 56, PhD(sanit eng), 61. *Prof Exp:* Asst sanit engr, R A Taft Sanit Eng Ctr, Ohio, 56-58; res assoc, Sch Pub Health, Harvard Univ, 60-61; from asst prof to assoc prof environ health, Grad Sch Pub Health, Univ Pittsburgh, 62-67, prof indust health & air eng & prof chem eng, 67-79; PROF & DIV DIR, ENVIRON HEALTH ENG, JOHNS HOPKINS UNIV, 80- *Concurrent Pos:* NSF fel, Sch Hyg, Univ London, 61-62; consult, Health Physics Div, Oak Ridge Nat Lab, 63-, Div Biol & Med, US Atomic Energy Comn, 63-, Los Alamos Sci Lab, 65- & Nat Acad Sci-Nat Res Coun Comt Biol Effects Air Pollution, 70-; consult nat air pollution control admin, USPHS, 63-, chmn air pollution res grants adv comt, 69-71; US rep to panel of experts on aerosols, Int Atomic Energy Agency, Vienna, 67; mem, Air Pollution Adv Comt, Allegheny County, Pa; adv ed, Atmospheric Environment, Environ Letters, J Air Pollution Control Asn, Ann Occcup Hyg (London); asst secy labor, Occup Safety & Health Act, 75-77; mem, panel experts occup health, WHO, 75-, sci adv bd, US Environ Protection Agency, 78-81, nat adv comt health & vital statist, Dept Health & Human Serv, 79-81, comt on risk assessment, Nat Acad Sci, 81-82, Mine Health Res Adv Comn of NIOSH, 86- & GM/UAW JT Health Adv Bd, 88 -; trustee, Assoc Univs, Inc, Wash, DC, 91- *Honors & Awards:* Cummings Award, Am Indust Hyg Asn, 86. *Mem:* AAAS; Am Chem Soc; Am Indust Hyg Asn; Am Inst Chem Engrs; Air Pollution Control Asn. *Res:* Aerosol physics; air pollution, particularly engineering control and analytical assessment; industrial hygiene and ventilation; risk assessment. *Mailing Add:* Dept Environ Health Sci Johns Hopkins Univ 615 N Wolfe St Baltimore MD 21205

CORNACCHIO, JOSEPH V(INCENT), b New York, NY, Dec 27, 34; m 60; c 2. COMPUTER SCIENCE, ENGINEERING. *Educ:* Pa State Univ, BS, 56; Syracuse Univ, MS, 59, PhD(elec eng), 62. *Prof Exp:* Instr elec eng, Syracuse Univ, 59-62; mem staff, Int Bus Mach Corp, 62-72; assoc prof, 69-72, prof, 72-77, chmn, Dept Comput Sci, 77-78 & 81-84, PROF, DEPT COMPUT SCI, WATSON SCH ENG, STATE UNIV NY, BINGHAMTON, 77- *Concurrent Pos:* Vis scholar, Univ Mich, 76-77. *Mem:* Asn Comput Mach; Inst Elec & Electronics Engrs. *Res:* Distributed computing; operating systems. *Mailing Add:* Dept Comput Sci State Univ NY Binghamton NY 13901

CORNATZER, WILLIAM EUGENE, b Mocksville, NC, Sept 28, 18; m 46; c 2. BIOCHEMISTRY. *Educ:* Wake Forest Col, BS, 39; Bowman Gray Sch Med, MD, 51; Univ NC, MS, 41, PhD(biochem), 44; Am Bd Clin Chem, dipl. *Prof Exp:* Asst zool, Wake Forest Col, 37-38, asst phys chem, 38-39; asst biol chem, Univ NC, 39-41; asst prof biochem, Bowman Gray Sch Med, 46-51; prof & head dept, 51-83, Chester Fritz distinguished prof, 73-83, EMER PROF BIOCHEM, SCH MED, UNIV NDAK, 83-; DIR, IRELAND RES LAB, 53- *Concurrent Pos:* NSF travel award, Int Cong Biochem, Paris, 52, Tokyo, 67; USAEC travel award, Int Cancer Cong, London, 58; Int Union Physiol Sci travel award, Int Cong Pharmacol, Stockholm, 61; Am Inst Nutrit travel award, Int Cong Nutrit, Prague, 69 & Mexico City, 72; consult med div, Oak Ridge Inst Nuclear Studies, 51-; mem, Am Bd Clin Chem; mem biochem test comt, Nat Bd Med Exam. *Mem:* AAAS; Am Chem Soc; Am Soc Biol Chem; Am Fedn Clin Res; Soc Exp Biol & Med; Am Inst Nutrit. *Res:* Properties of proteins; quinine metabolism and absorption; antimalarial testing; phospholipid metabolism; liver function tests and disease; radiation of effects and toxicity of isotopes; lipotropic agents; biosynthesis of membrane lipids in health and disease. *Mailing Add:* Dept Biochem Univ NDak Med Sch Grand Forks ND 58201

CORNBLATH, MARVIN, b St Louis, Mo, June 18, 25; m 48; c 3. PEDIATRICS, BIOCHEMISTRY. *Educ:* Wash Univ, MD, 47. *Prof Exp:* Asst pediat, Wash Univ, 49-50; from instr to asst prof, Johns Hopkins Univ, 53-59; from asst prof to assoc prof, Northwestern Univ, 59-61; from assoc prof to prof, Univ Ill, 61-68; chmn dept pediat, 68-78, PROF PEDIAT, SCH MED, UNIV MD, BALTIMORE, 68-; LECTR PEDIAT, JOHNS HOPKINS UNIV, 75- *Concurrent Pos:* USPHS fel biochem, Wash Univ, 50-51; res assoc, Sinai Hosp, Md, 53-59; asst chmn div pediat, Michael Reese Hosp, Ill, 59-61; spec asst to sci dir & actg br chief, perinatal & pediat med, Nat Inst Child Health & Human Develop, NIH, Bethesda, 78-82. *Mem:* Soc Pediat Res; Am Soc Biol Chemists; Am Physiol Soc; Brit Biochem Soc; Am Pediat Soc; Endocrine Soc; Sigma Xi. *Res:* Physiological and biochemical maturation of newborn and premature infants; carbohydrate metabolism and enzymology. *Mailing Add:* 3809 St Paul St Baltimore MD 21218

CORNEIL, ERNEST RAY, b Nov 11, 32; m 56; c 4. SYSTEMS & CONTROL ENGINEERING. *Educ:* Queen's Univ, Ont, BSc, 55; Univ London, PhD & DIC, 60. *Prof Exp:* Lectr, 58-61, asst prof, 61-64, dir, Comput Ctr, 64-65, assoc prof, 64-70, head dept, 78-81, PROF MECH ENG, QUEEN'S UNIV, ONT, 70- *Concurrent Pos:* Consult info systs & comput appln; eng consult, Transport Can, 75-76, 82-83; Ford Found indust residency, 67-68. *Mem:* Am Soc Mech Engrs; Am Soc Heating, Refrig & Air-Conditioning Engrs. *Res:* Control systems using nonlinear feedback and adaptive systems; heat pump design and application; North American railway electrification. *Mailing Add:* Dept Mech Eng Queen's Univ Kingston ON K7L 3N6 Can

CORNELISSEN GUILLAUME, GERMAINE G, b Schaerbeek, Belgium, Nov 22, 49; m 75. CHRONOBIOLOGY, SIGNAL PROCESSING. *Educ:* Univ Brussels, Belgium, BA, 69, MS, 71, Med, 71, PhD(physics), 71. *Prof Exp:* Fel Physics, Inst Sci Res Indust & Agr, Brussels, Belgium, 74-76; res fel chronobiol, 76-82, res assoc, 82-87, ASST PROF & ASST DIR CHRONOBIOL, CHRONOBIOL LABS, UNIV MINN, 87- *Concurrent Pos:* Bd mem, Int Soc Chronobiol, 85-; secy Nam br & mem Int Soc Res Civilization Dis & Environ, 87-; NIH grant, 88. *Honors & Awards:* Chronobiol Award, 83. *Mem:* Int Soc Chronobiol; Am Phys Soc; Soc Indust & Appl Math; Biomet Soc; Am Statist Asn; AAAS. *Res:* Biological rhythms in health and disease with emphasis on the chronobiology of human blood pressure. *Mailing Add:* Univ Minn Chronobiol Labs 420 Wash Ave SE 5-187 Lyon Labs Minneapolis MN 55455

CORNELIUS, ARCHIE J(UNIOR), b Westmoreland, Kans, Oct 16, 31; m 53; c 4. MECHANICAL ENGINEERING. *Educ:* Kans State Univ, BS, 58; Okla State Univ, MS, 63, PhD(mech eng), 65. *Prof Exp:* Res engr, Jersey Prod Res Co, Okla, 58-64; res engr, Phillips Petrol Co, 65-80, eng dir, 80-86; ENG CONSULT, 86- *Mem:* Am Inst Mining, Metall & Petrol Engrs. *Res:* Petroleum recovery by thermal methods; heat transfer to supercritical fluids; improved oil recovery methods. *Mailing Add:* 2212 SE Pkwy Dr Bartlesville OK 74006

CORNELIUS, BILLY DEAN, b Dublin, Tex, Nov 6, 39; m 67; c 1. STERILIZATION OF FOOD PRODUCTS, STERILIZATION OF PHARMACEUTICAL LIQUID PRODUCTS & DEVICES. *Educ:* Tri-State Col, BS, 63; Ill Inst Technol, MS, 72. *Prof Exp:* Sterilization engr res, Libby, McNeill & Libby Res & Develop, 63-72, process engr develop & design, Libby, McNeill & Libby Corp, 72-73; sterilization engr, Abbott Labs, Hosp Prod Div, 73-78, sterilization mgr, 78-85, sr res scientist, Ross Div, 85-87, sect head, 87-90, ASSOC RES FEL, ROSS DIV, ABBOTT LABS, 90- *Mem:* Inst Thermal Processing Specialists; Inst Food Technologists. *Res:* Sterilization processes; steam, dry heat and chemical sterilization in batch and continuous systems. *Mailing Add:* 221 Raccoon Run Powell OH 43065

CORNELIUS, CHARLES EDWARD, b Walnut Park, Calif, Dec 19, 27; m 48; c 4. PHYSIOLOGY. *Educ:* Univ Calif, BS, 49 & 51, DVM, 53, PhD(comp path), 58. *Hon Degrees:* DSc, Univ Fla, 82. *Prof Exp:* Lectr clin path, Univ Calif, Davis, 54-58, from asst prof to assoc prof, 58-66, assoc dean, 62-64; prof physiol & dean col vet med, Kans State Univ, 66-71; dean, Col Vet Med, Univ Fla, 71-81, prof path, 74-81, chmn, Dept Vet Sci, 76-78; dir, Calif Primate Res Ctr, 81-87, PROF PHYSIOL, SCH VET MED, UNIV CALIF, DAVIS, 87- *Concurrent Pos:* Mem res grants study sect gen med, NIH, 65-69, Nat Adv Comt health res facilities, 69-71; chmn, Comt Animal Health, Nat Acad Sci, 72-73; mem, Biomed Res Develop grant panel, NIH, 79-80, Inst Lab Animal Resources Coun, NAS, 83-85; co-chmn, Orgn Comt for Vet Sch in Israel, 82- *Honors & Awards:* Fel, Hebrew Univ, Israel, 88. *Mem:* Am Physiol Soc; Am Gastroenterol Asn; Am Vet Med Asn; Am Asn Liver Dis. *Res:* Physiologic mechanisms responsible for jaundice; biochemistry of bile pigments in health and disease in animals. *Mailing Add:* Dept Physiol Sci Univ Calif Davis CA 95618

CORNELIUS, E(DWARD) B(ERNARD), b Laredo, Tex, Nov 6, 18; m 46; c 2. CHEMICAL ENGINEERING. *Educ:* Princeton Univ, BS, 40, ChE, 41. *Prof Exp:* Jr develop engr, Houdry Process & Chem Co Div, Air Prod & Chem, Inc, 41-46, asst res chemist, 46-50, proj dir, 50-57, sect chief, 57-70, sr res chemist & admin asst, 71-74; sr res chemist, Matthey Bishop Inc, 74-78; sr res chemist, Ashland Petrol Co Div, Ashland Oil Co, 78-80, sr res group leader, 80-86; RETIRED. *Concurrent Pos:* Mem Int Cong Catalysis, 56-86. *Mem:* Am Chem Soc; Am Inst Chem Engrs; Clay Minerals Soc; Nat Catalysis Soc; Sigma Xi. *Res:* Preparation, characterization and performance of hydrocarbon conversion catalysts useful in petroleum and chemical industry; 30 US patents and 40 foreign patents. *Mailing Add:* 6543 Waterford Circle Sarasota FL 34238-2637

CORNELIUS, LARRY MAX, b Washington, Ind, Apr 30, 43; m 65; c 1. VETERINARY MEDICINE. *Educ:* Purdue Univ, DVM, 67; Univ Mo-Columbia, PhD(clin path), 71. *Prof Exp:* Intern, Angell Mem Animal Hosp, Boston, Mass, 67-68; res assoc vet med, Univ Mo-Columbia, 68-74; mem fac, 74-80, assoc prof animal med, 80-, PROF SMALL ANIMAL MED, SCH VET MED, UNIV GA. *Concurrent Pos:* Speaker, Technicon Int Cong, Chicago, 69; Nat Inst Arthritis & Metab Dis fel, 69-72. *Mem:* Am Vet Med Asn; Comp Gastroenterol Soc. *Res:* Fluid and electrolyte balance in veterinary medicine. *Mailing Add:* Dept Small Animal Med Univ Ga Sch Vet Med Athens GA 30601

CORNELIUS, RICHARD DEAN, b Chicago, Ill, Sept 18, 47; m 70; c 3. INORGANIC CHEMISTRY, BIOINORGANIC CHEMISTRY. *Educ:* Carleton Col, BA, 69; Univ Iowa, PhD(inorg chem), 74. *Prof Exp:* Res assoc biochem, Univ Wis-Madison, 74-77; from asst prof to assoc prof chem, Wichita State Univ, 77-85; PROF CHEM & CHMN, LEBANON VALLEY COL, 85- *Mem:* Am Chem Soc. *Res:* Coordination chemistry; solution kinetics; science education. *Mailing Add:* Dept Chem Lebanon Valley Col Annville PA 17003

CORNELIUS, STEVEN GREGORY, b Alton, Ill, Mar 17, 51; m 72; c 3. SYSTEMS ANALYSIS & MODELING. *Educ:* Univ Ill, Urbana, BS, 72, PhD(animal sci & nutrit), 76. *Prof Exp:* Res chemist, US Meat Animal Res Ctr, 76-78; asst prof, 78-84, ASSOC PROF, NUTRIT & ANIMAL SWINE, DEPT ANIMAL SCI, UNIV MINN, 84- *Mem:* Am Soc Animal Sci; Animal Nutrit Res Coun; Biomet Soc. *Res:* Factors influencing the digestion and absorption of nutrients by the pig; application of systems analysis techniques to animal agriculture problems; swine growth and the influence of environment on growth; general swine nutrition. *Mailing Add:* Dept Animal Sci 122 Peters Hall Univ Minn 1404 Gortner Ave St Paul MN 55108

CORNELIUSSEN, ROGER DUWAYNE, b Fargo, NDak, Aug 1, 31; m 54; c 4. POLYMER MORPHOLOGY, FAILURE MECHANISMS. *Educ:* Concordia Col, Moorhead, Minn, BA, 53; Univ Chicago, MS, 60, PhD(phys chem), 62. *Prof Exp:* Chemist, Am Oil Co, 58-60; asst prof chem, Luther Col, 61-64; chemist polymer chem, Res Traingle Inst, 64-67; sr chemist, NStar Res Inst, 67-68; pres polymer physics, Res Serv, Inc, 69-70; assoc prof, 70-80, PROF MAT ENG, DREXEL UNIV, 80- *Concurrent Pos:* Dir, Drexel Polymer Prog, 71-, Ctr Insulation Technol, 80-; vis prof, Ford Motor Co & Bell Telephone Labs, Columbus, Ohio, 79; lectr, Soc Plastics Engrs, 79-; ed, Drexel Polymer Notes, 84- *Mem:* Am Chem Soc; Soc Plastics Engrs; Am Soc Testing Mat; Am Soc Eng Educ; Am Phys Soc. *Res:* Polymer morphology; structure properly relationships; mechanical properties; failure analysis; molecular composites; blends and alloys. *Mailing Add:* Dept Mat Engr 3-261 Drexel Univ Philadelphia PA 19104

CORNELL, ALAN, b Fall River, Mass, May 4, 29; m 52; c 3. FOOD SCIENCE. *Educ:* Univ Mass, BS, 51; Mass Inst Technol, PhD(food sci), 60. *Prof Exp:* Res chemist, Res Ctr, Philip Morris, Inc, 60-65, sr scientist, 65-67; mgr food sci div, Cent Labs, Ralston Purina Co, 67-69; asst vpres res & develop, 69-73, VPRES RES & PROD DEVELOP, CONSOL CIGAR CORP, 73- *Concurrent Pos:* Consult, Tech Mgt & Res Productivity, Prod Safety. *Mem:* Am Chem Soc; Inst Food Technol; NY Acad Sci. *Res:* Subjective effects on constituents in tobacco and resultant smoke; subjective evaluation; product development of consumer products; research management; tobacco and smoke chemistry; quality control; product safety and health. *Mailing Add:* Ten Arnold Dr Bloomfield CT 06002

CORNELL, C(ARL) ALLIN, b Mobridge, SDak, Sept 19, 38; m 59, 81; c 5. STRUCTURAL ENGINEERING, PROBABILISTIC METHODS. *Educ:* Stanford Univ, AB, 60, MS, 61, PhD(struct eng), 64. *Prof Exp:* Actg asst prof struct eng, Stanford Univ, 63-64; from asst prof to prof civil eng, Mass Inst Technol, 64-82; PROF CIVIL ENG, STANFORD UNIV, 82- *Concurrent Pos:* Ford Found eng fel, 64-66; vis prof, Univ Calif, Berkeley, 70-71; consult, various co, 77-; Fulbright-Hayes Advan Res grant; Guggenheim fel. *Honors & Awards:* Huber Res Prize; Moisseiff Award, Am Soc Civil Engrs, Norman Medal & Fruedenthal Medal. *Mem:* Nat Acad Eng; Seismol Soc Am (pres, 86-87); Am Soc Civil Engrs. *Res:* Application of probability theory and statistics in civil and structural engineering. *Mailing Add:* Dept Civil Eng Stanford Univ Stanford CA 94305-4020

CORNELL, CREIGHTON N, b Rolfe, Iowa, Mar 20, 33; m 58; c 2. PHYSIOLOGY, BIOCHEMISTRY. *Educ:* Univ Mo, BS, 54, DVM, 62, MS, 69. *Prof Exp:* Instr agr chem & vet physiol, 62-63, instr agr chem, 63-65, asst prof agr chem, 65-75, ASST PROF BIOCHEM, UNIV MO-COLUMBIA, 75- *Res:* Biochemistry and physiology of the intoxication occurring in cattle feeding on toxic fescue grass pastures or hay, causing gangrene and sloughing of the distal portions of the extremities. *Mailing Add:* Dept Biochem Univ Mo Columbia Med Sch Columbia MO 65212

CORNELL, DAVID, b Elmira, NY, Aug 19, 25; m 48; c 4. CHEMICAL ENGINEERING. *Educ:* Univ Mich, BSChE, 47, MSChE, 49, PhD(chem eng), 52. *Prof Exp:* Asst res & develop labs, Blaw-Knox Div, Blaw-Knox Co, 48-50; asst prof chem eng, Univ Tex, 52-55; res engr, Monsanto Chem Co, 55-58; assoc prof petrol eng, Okla State Univ, 58-61; prof, 61-90, EMER PROF CHEM ENG, MISS STATE UNIV, 90- *Concurrent Pos:* Res engr, Tex Petrol Res Comt, Univ Tex, 52-53. *Mem:* Am Inst Chem Engrs; Am Chem Soc; Am Soc Petrol Engrs; Am Soc Eng Educ; Sigma Xi. *Res:* Unsteady state flow of natural gas in natural gas reservoirs; flow through porous media. *Mailing Add:* 208 Windsor Rd Starkville MS 39759

CORNELL, DAVID ALLAN, b St Paul, Minn, Dec 29, 37; m 62; c 2. PHYSICS. *Educ:* Principia Col, BS, 59; Univ Calif, Berkeley, PhD(physics), 64. *Prof Exp:* From instr to assoc prof, 64-78, prof, 78-80, ASSOC PROF PHYSICS, PRINCIPIA COL, 80- *Mem:* Am Asn Physics Teachers. *Res:* Nuclear magnetic resonance phenomena in metals and alloys; hydrogen diffusion in metallic crystals. *Mailing Add:* Dept Physics Principia Col Elsah IL 62028

CORNELL, DONALD GILMORE, b St Augustine, Fla, Dec 20, 31; m 81. SURFACE CHEMISTRY. *Educ:* Univ Fla, BS, 54; George Washington Univ, PhD(phys chem), 76. *Prof Exp:* Chemist, E I Du Pont de Nemours & Co, Kinston, NC, 60-63; res chemist, USDA, Washington, DC, 63-74, RES CHEMIST, AGR RES SERV, USDA, PHILADELPHIA, PA, 74- *Mem:* Am Chem Soc; AAAS. *Res:* Surface and colloid chemistry; spectroscopy and reactions in lipid-protein monolayers; infrared and circular dichroism spectroscopy of lipids and protein lipid complexes. *Mailing Add:* Agr Res Serv USDA 600 E Mermaid Lane Philadelphia PA 19118

CORNELL, HOWARD VERNON, b Berwyn, Ill, Apr 13, 47; m 71; c 1. BIOGEOGRAPHY, ECOLOGY. *Educ:* Tufts Univ, BS, 69; Cornell Univ, PhD(ecol), 75. *Prof Exp:* From asst prof to assoc prof, 75-90, PROF BIOL SCI, UNIV DEL, 90- *Mem:* AAAS; Ecol Soc Am; Am Soc Naturalists. *Res:* Ecology of host plant-herbivore systems. *Mailing Add:* Sch Life & Health Sci 117 Wolf Hall Univ Del Newark DE 19716

CORNELL, JAMES MORRIS, b Bismarck, NDak, Sept 8, 37; m 57; c 2. ANIMAL BEHAVIOR. *Educ:* Univ Wash, BS, 60, MS, 62, PhD(psychol), 63. *Prof Exp:* Asst prof psychol, Mont State Col, 63-64; asst prof, 64-68, ASSOC PROF PSYCHOL, UNIV WATERLOO, 68- *Concurrent Pos:* Nat Res Coun Can res grants, 64- *Mem:* AAAS; Am Psychol Asn; Animal Behavior Soc; Sigma Xi. *Res:* Conidtioning in animals; imprinting; memory. *Mailing Add:* Dept Psycol Univ Waterloo Waterloo ON N2L 3I6 Can

CORNELL, JAMES S, b Harrisburg, Pa, Apr 9, 47; m 88; c 3. BIOLOGICAL CHEMISTRY, ENDOCRINOLOGY. *Educ:* Mich State Univ, BS, 69; Univ Calif, Los Angles, PhD(biol chem), 73; Cornell Univ Med Col, MD, 88. *Prof Exp:* Fel, Univ Calif, Los Angeles, 73; res assoc, 73-75, ASSOC PROF BIOCHEM & ASST DIR, LAB CLIN BIOCHEM, MED COL, CORNELL UNIV, 75- *Mem:* Endocrine Soc; Am Chem Soc; AAAS; NY Acad Sci. *Res:* Reproductive biochemistry, the protein chemistry of the placenta; factors influencing the induction of labor. *Mailing Add:* 784 Flint Pl Paramus NJ 07652-3744

CORNELL, JOHN ALSTON, b New York, NY, Oct 24, 22; c 2. PRODUCT TESTING-DENTAL. *Educ:* Mass Inst Technol, BS & MS, 48. *Prof Exp:* Dir res, H D Justi & Son, 50-57 & Saatomer, 57-76; EXEC DIR, WESTWOOD RES LAB, 76- *Concurrent Pos:* Consult, Princeton Polymer Lab, 85-88. *Mem:* Am Chem Soc; Int Dent Res Asn; AAAS. *Res:* Dental research; rubber chemistry; author of 25 publications; awarded 27 patents. *Mailing Add:* Westwood Res Lab Inc Box 3126 West Chester PA 19381

CORNELL, JOHN ANDREW, b Westerly, RI, Apr 29, 41; m 63; c 2. APPLIED STATISTICS. *Educ:* Univ Fla, BSEd, 62, MStat, 66; Va Polytech Inst, PhD(statist), 69. *Prof Exp:* Appl statistician, Tenn Eastman Co, 65-66; asst prof statist, Univ Fla, 68-72; lectr statist, Birkbeck Col, Univ London, 72-73; from assoc prof to prof statist, 73-87, PROF STATIST, UNIV FLA, 88- *Concurrent Pos:* Statist consult, Tenn Eastman Co, 65-66 & Agr Exp Sta, Univ Fla, 68-; vis prof statist, Colo State Univ, 87-88. *Honors & Awards:* J Youden prize, 74; Shewell Award, 81. *Mem:* Fel Am Statist Asn; Am Soc Qual Control; Int Statist Inst. *Res:* Design and analysis of statistical experiments in agricultural, social, physical and biological sciences, design and analysis of experiments with mixtures. *Mailing Add:* Dept Statist 411 Rolfs Hall Univ Fla Gainesville FL 32611

CORNELL, NEAL WILLIAM, b Savannah, Ga, July 29, 37; m 67. BIOCHEMISTRY. *Educ:* Univ Redlands, BS, 60; Univ Calif, Los Angeles, PhD(biochem), 64. *Prof Exp:* Res fel biol chem, Harvard Med Sch, 63-66; from asst prof to assoc prof chem, Pomona Col, 66-72; Med Res Coun investr, Oxford Univ, 72-74; RES CHEMIST, LAB METAB, NAT INST ALCOHOL ABUSE & ALCOHOLISM, 74- *Concurrent Pos:* Investr, Marine Biol Lab, Woods Hole, 70, 71 & 72, mem corp, 71-; vis prof, dept biochem, Univ Copenhagen, 83; guest investr, Woods Hole Oceanog Inst, 83-85. *Mem:* Brit Biochem Soc; Am Soc Biol Chem; Am Inst Nutrit; Am Soc Cell Biol; Biophys Soc. *Res:* Metabolic control; cell biology; enzyme rates in living cells; distribution and translocation of cellular materials; comparative biochemistry of regulatory enzymes; analytical biochemistry. *Mailing Add:* Dept Cell Biochem Marine Biol Lab Woods Hole MA 02543

CORNELL, RICHARD GARTH, b Cleveland, Ohio, Nov 18, 30; m 61; c 3. BIOSTATISTICS. *Educ:* Univ Rochester, AB, 52; Va Polytech Inst, MS, 54, PhD(statist), 56. *Prof Exp:* Asst, Va Polytech Inst, 52-54, Oak Ridge Inst Nuclear Studies fel, 55-56; statistician, Commun Dis Ctr, USPHS, Ga, 56-58, chief lab & field sta, Statist Unit, 58-60; from assoc prof to prof statist, Fla State Univ, 60-71; chmn biostatist dept, 71-84, PROF BIOSTATIST, UNIV MICH, ANN ARBOR, 71- *Mem:* Fel Am Statist Asn; Biomet Soc. *Res:* Biometry. *Mailing Add:* Dept Biostatist Univ Mich Ann Arbor MI 48109

CORNELL, ROBERT JOSEPH, b Westerly, RI, Oct 1, 40; m 67; c 2. POLYMER CHEMISTRY. *Educ:* Clarkson Col Technol, BS, 62, MS, 64; Worcester Polytech Inst, PhD(org chem), 67. *Prof Exp:* Res chemist polymer chem, 67-70, res scientist, Polymer Res Group, 70-73, sr group leader paracril res, 73-77, res & develop mgr spec chem, Chem Div, 78-81, RES & DEVELOP MGR CHEM, CHEM DIV, UNIROYAL INC, 81- *Mem:* Am Chem Soc; Sigma Xi. *Res:* Plastic additives; petroleum additives, chemical blowing agents and rubber chemicals. *Mailing Add:* 69 Ramsey Ave Naugatuck CT 06770

CORNELL, SAMUEL DOUGLAS, b Buffalo, NY, Apr 16, 15; m 39, 69; c 4. SCIENCE ADMINISTRATION, ACADEMIC ADMINISTRATION. *Educ:* Yale Univ, BA, 35, PhD(physics), 38. *Prof Exp:* Develop physicist, Eastman Kodak Co, NY, 38-42; sci warfare adv, Res & Develop Bd, Washington, DC, 46-52; exec officer, Nat Acad Sci-Nat Res Coun, 52-65; pres, Mackinac Col, 65-70; consult sci admin, Nat Acad Sci, 70-72, asst to the pres, 72-81; RETIRED. *Mem:* AAAS; Am Phys Soc. *Res:* Molecular spectroscopy; design of motion picture equipment. *Mailing Add:* 1807 Sun Mountain Dr Santa Fe NM 87505

CORNELL, STEPHEN WATSON, b Chicago, Ill, May 21, 42; m 69; c 2. POLYMER ENGINEERING, CHEMICAL ENGINEERING. *Educ:* Rensselaer Polytech Inst, BChE, 64; Case Western Reserve Univ, MS, 66, PhD(polymer sci), 69. *Prof Exp:* Proj scientist res & develop, Chem & Plastics Div, Union Carbide Corp, 68-74; mgr mat res, Corp Res & Develop Div, Continental Can Co, 74-76; mgr mat & prod develop, Continental Plastics Indust, 76-79, dir spec proj, Continental Packaging Co, Continental Group, Inc, 80-85; dept mgr, package develop, Am Nat Can Co, 85-88, dir, plastic packaging technol, 88-90; VPRES TECHNOL, SILGAN PLASTICS CORP, 90- *Mem:* Am Chem Soc; Am Phys Soc; Soc Plastics Engrs. *Res:* New product development; new packaging development; food packaging development; polymer-filler interactions; surface modification; degradable polyolefin compositions; multilayer packaging; plastic blow molded bottles; plastic thermoformed containers; flexible packaging; plastics recycling; blow molded bottle technology. *Mailing Add:* 1816 Wakeman St Wheaton IN 60187

CORNELL, W(ARREN) A(LVAN), b Attleboro, Mass, July 2, 21; m 43; c 4. COMMUNICATIONS NETWORKS. *Educ:* Rensselaer Polytech Inst, BEE, 42. *Prof Exp:* Dir digital commun syst eng, Bell Labs, 66-67, dir data commun eng, 67-70, dir opers res, 70-81, dir network anal, 81-82; RETIRED. *Mem:* Inst Elec & Electronics Engrs; Opers Res Soc Am. *Res:* Electronic telephone switching; digital computers; military surveillance; communication systems; operations research. *Mailing Add:* 18 Crest Dr Little Silver NJ 07739

CORNELL, WILLIAM CROWNINSHIELD, b Attleboro, Mass, May 5, 41; m 69; c 2. MICROPALEONTOLOGY, PALYNOLOGY. *Educ:* Univ RI, BS, 65, MS, 67; Univ Calif, Los Angeles, PhD(geol), 72. *Prof Exp:* Asst prof, 71-75, ASSOC PROF GEOL, UNIV TEX, EL PASO, 75- *Concurrent Pos:* Actg chmn dept geol sci, Univ Tex, El Paso, 75-77 & 78-79, asst dean, Col

Sci, 81- *Mem:* Soc Econ Paleontologists & Mineralogists; AAAS; Am Asn Stratig Palynologists; Sigma Xi; NAm Micropaleontological Soc; Paleontol Soc. *Res:* Mesozoic and Cenozoic siliceous; organic walled phytoplankton. *Mailing Add:* Dept Geol Sci Univ Tex El Paso TX 79968-0555

CORNELL-BELL, ANN HALL, CELL BIOLOGY, IMMUNE REACTION. *Educ:* Northeastern Univ, PhD(gastric physiol), 84. *Prof Exp:* RES ASSOC, FEL EYE RES INST & DEPT OPHTHAL, HARVARD UNIV, 84- *Mailing Add:* Molecular Neurobiol Yale Sch Med 333 Cedar St New Haven CT 06510

CORNELY, PAUL BERTAU, b French West Indies, Mar 9, 06; nat US; m 34; c 1. PUBLIC HEALTH. *Educ:* Univ Mich, AB, 28, MD, 31, DrPH, 34; Am Bd Prev Med, dipl, 50. *Hon Degrees:* DSc, Univ Mich, 68; DPS, Univ of the Pac, 73. *Prof Exp:* From asst prof to assoc prof, 34-47, dir health serv, 37-47, head dept community health pract, 55-70, prof, 47-73, EMER PROF PREV MED & PUB HEALTH, COL MED, HOWARD UNIV, 73- *Concurrent Pos:* Med dir, Freedmen's Hosp, 47-58; consult, Nat Urban League, 44-47 & Am Health Educ African Develop, US AID; pres, Community Group Health Found, 68-74; mem, President's Comn Population Growth & Am Future, 70-72; Mem staff, Health Servs Eval Systs Sci, Inc, 73-81. *Mem:* Fel Am Pub Health Asn (pres, 69-70); fel Am Col Prev Med; hon fel Am Col Health Care Execs. *Res:* Medical education; distribution and supply of professional personnel; Negro health problems; student health program; health motivation among low income families. *Mailing Add:* 1220 East West Hwy Apt 622 Silver Spring MD 20910

CORNER, JAMES OLIVER, b Toronto, Ont, July 19, 17; nat US; m 45; c 3. ORGANIC POLYMER CHEMISTRY. *Educ:* Dartmouth Col, AB, 39; Univ Ill, AM, 40, PhD(org chem), 42. *Prof Exp:* Chemist, Exp Sta, E I Du Pont De Nemours & Co, Inc, 42-52, res supvr, Exp Sta, 52-57, sr supvr, 57-61, res mgr, 61, res mgr, Textile Res Lab, 62-63, lab dir, Dacron Res Lab, 63-64, actg tech supt, Kinston Plant, 64-65, res dir, Dacron Tech Div, 65-68, Nylon Tech Div, 68-69 & Indust Fibers Div, 70-71, tech mgr, Nylon Tech Div, 71-76, lab dir, Textile Res Lab, 76-78, mgr mkt res div, Mkt Commun Dept, 78-82; RETIRED. *Mem:* Am Chem Soc. *Res:* Synthetic organic chemistry; polymer chemistry; textile fibers. *Mailing Add:* 655 Horseshoe Hill Rd Hockessin DE 19707-9570

CORNER, THOMAS RICHARD, b Waterloo, NY, Jan 6, 40; m 61; c 3. MICROBIOLOGY. *Educ:* Cornell Univ, BS, 62, MS, 64; Univ Rochester, PhD(microbiol), 68. *Prof Exp:* NIH fel, 68-70, asst prof, 70-85, ASSOC PROF MICROBIOL, ASSOC CHAIR & DIR UNDERGRAD STUDIES, MICH STATE UNIV 87-; MICH STATE UNIV 85- *Concurrent Pos:* Res biologist, Lab Molecular Biol, Nat Inst Neurol Dis & Stroke, NIH, 77-78. *Mem:* Am Soc Microbiol; Soc Gen Microbiol. *Res:* Structure and function of bacterial membranes and cell walls; action of organochlorine compounds and other ecotoxicants on bacteria; biomechanical and biorheological properties of cells; bacterial spore heat resistance mechanisms. *Mailing Add:* Dept Microbiol Mich State Univ East Lansing MI 48823

CORNET, I(SRAEL) I(SAAC), b New York, NY, Dec 13, 12; m 40; c 2. MECHANICAL ENGINEERING. *Educ:* Univ Calif, Los Angeles, AB, 33; Univ Calif, PhD(soil sci), 42. *Prof Exp:* Reader & lab asst chem, Univ Calif, Los Angeles, 32-33; analyst & technician, agron div, col agr, Univ Calif, 34-36; jr soil surveyor, soil conserv serv, USDA, Ore, 38-40 & 41-42; asst civil engr, US War Dept, 42; res engr, Off Sci Res & Develop & Nat Defense Res Comt projs, Calif, 42-43, instr eng, sci & mgt war training, 43; sr metal engr, Permanente Metals Corp, Calif, 43-46; engr, Nat Adv Comt Aeronaut proj, 46,; from asst prof to assoc prof, 46-57, prof 57-78 EMER PROF MECH UNIV CALIF, BERKELEY, 78- *Concurrent Pos:* Guggenheim fel, 56-57; res prof, Miller Inst Basic Res Sci, 60-61. *Mem:* AAAS; Am Chem Soc; Am Inst Mining, Metall & Petrol Engrs; Am Soc Eng Educ; Nat Asn Corrosion Engrs. *Res:* Corrosion science and engineering; metal failure. *Mailing Add:* Dept Mech Eng Univ Calif Berkeley CA 94720

CORNETET, WENDELL HILLIS, JR, b Huntington, WVa, Oct 20, 23; m 67; c 2. ELECTRICAL ENGINEERING. *Educ:* WVa Univ, BSEE, 48, MSEE, 51; Ohio State Univ, PhD(elec eng), 58. *Prof Exp:* Instr electronics, WVa Univ, 48-51; res assoc, 51-56, from instr to assoc prof, 56-66, PROF ELECTRONICS, OHIO STATE UNIV, 66- *Concurrent Pos:* Consult, Indust Nucleonics Corp, 62-68 & Energystics Inc, 77-78; trustee, Nat Electronics Conf, 69-71. *Mem:* Sigma Xi; Inst Elec & Electronics Engrs. *Res:* Electron devices; microwave electronics. *Mailing Add:* Dept Elec Eng Ohio State Univ 2015 Neil Ave Columbus OH 43210

CORNETT, LAWRENCE EUGENE, b San Francisco, Calif, Aug 27, 51; m 74; c 2. ENDOCRINOLOGY. *Educ:* Univ Calif, Riverside, BS, 73; Univ Calif, Davis, PhD(physiol), 78. *Prof Exp:* Teaching fel reproductive endocrinol, Univ Calif, San Francisco, 78-80; from asst prof to assoc prof, 80-90, PROF PHYSIOL & BIOPHYS, UNIV ARK MED SCI, 90- *Concurrent Pos:* Prin investr, NIH grants, 83- & NSF grant, 85 - *Mem:* Endocrine Soc; Am Physiol Soc; Soc Study Reproduction. *Res:* Mechanism of hormone action; isolation and biochemical characterization of membrane receptors for the neurohypophyseal hormone vasopressin and the catecholamines, norepinephrine and epinephrine. *Mailing Add:* Dept Physiol & Biophys Slot 505 Univ Ark Med Sci 4301 W Markham St Little Rock AR 72205-7199

CORNETT, RICHARD ORIN, b Driftwood, Okla, Nov 14, 13; m 43; c 3. PHYSICS, COMMUNICATIONS SCIENCE. *Educ:* Okla Baptist Univ, BS, 34; Okla Univ, MS, 37; Univ Tex, PhD(physics), 40. *Hon Degrees:* DSc, Hardin Simmons Univ, 54; LittD, Jacksonville Univ, 64; LLD, Belknap Col, 67. *Prof Exp:* From instr to assoc prof physics, Okla Baptist Univ, 35-41; asst supvr physics, eng, sci & mgt defense training prog, Pa State Univ, 41-42; lectr electronics, Harvard Univ, 42-45; prof physics & asst to the pres, Okla Baptist Univ, 45-46, head dept physics & vpres, 46-47, exec vpres, 47-51; exec secy educ comn, Southern Baptist Convention & ed, Southern Baptist Educr, 51-58; specialist col & univ orgn, Off Educ, Dept Health, Educ & Welfare, Washington, DC, 59, exec asst to dir div higher educ, 60-61, actg asst comnr & dir div, 61-63, dir div educ admin, 64-65; vpres, 65-75, chmn, Ctr Studies in Language & Commun, 81-82, prof audiol, 82-84, RES PROF & DIR CUED SPEECH PROGS, GALLUDET UNIV, 75-, EMER PROF AUDIOL, 85- *Concurrent Pos:* Originator of Cued Speech, method of commun for deaf, 66; consult, Am Optometric Asn, 68-72; prin investr, Autocuer Field Test, 73-75, 80-81, 88-89. *Honors & Awards:* Nitchie Award in Human Communication, New York League for Hard of Hearing, 88. *Mem:* Alexander Graham Bell Soc for the Deaf. *Res:* Acoustics; theory of hearing; diplacusis; communication for the hearing impaired; development of electronic lipreading aid for the deaf; design of hearing aids for improved auditory processing. *Mailing Add:* 8702 Royal Ridge Lane Laurel MD 20708

CORNETTE, JAMES L, b Bowling Green, Ky, May 8, 35; m 62; c 2. MATHEMATICS, DATA APPLICATIONS. *Educ:* WTex State Col, BS, 55; Univ Tex, MA, 59, PhD(math), 62. *Prof Exp:* From asst prof to assoc prof, 62-70, PROF MATH, IOWA STATE UNIV, 70- *Concurrent Pos:* Fulbright lectr, Nat Univ Malaysia, 73-74; res lab, Nat Cancer Inst Health, 85-87. *Mem:* Am Math Soc; Math Asn Am; Soc Indust & Appl Math. *Res:* Biomathematics; point set topology; mathematical models in population genetics. *Mailing Add:* Dept Math Iowa State Univ Ames IA 50011

CORNFELD, DAVID, b Philadelphia, Pa, Apr 5, 26; m 56; c 3. PEDIATRICS. *Educ:* Univ Pa, MD, 48, MSc, 66. *Prof Exp:* Asst instr, Univ Pa, 51-52, assoc, 54-57, from asst prof to assoc prof, 57-72, actg chmn pediat, 72-80, PROF PEDIAT, SCH MED, UNIV PA, 72-, ASSOC CHMN DEPT, 80- *Concurrent Pos:* Asst chief univ serv, Philadelphia Gen Hosp, 55-68; dir pediat clin, Hosp Univ Pa, 56-62; dir outpatient dept & sr physician, Children's Hosp Philadelphia, 62-, co-dir nephrology serv, 63-72, dir gen pediat, 78-87, dep physician-in-chief, 74-; ed-in-chief, Clin Pediat, 78-80. *Mem:* Am Acad Pediat; Am Pediat Soc. *Res:* Nephrology; patient care. *Mailing Add:* Dept Pediat Univ Pa Philadelphia PA 19104

CORNFORD, EAIN M, b Wellington, NZ, July 2, 42; m 65; c 3. BLOOD BRAIN BARRIER, SCHISTOSOMAISIS. *Educ:* Drake Univ, BA, 64, MA, 67; Iowa State Univ, PhD(zool), 72. *Prof Exp:* CHIEF NEUROPHARMACOL, VET ADMIN-WADSWORTH MED CTR, 77-; ASSOC PROF NEUROL, SCH MED, UNIV CALIF, LOS ANGELES, 85- *Concurrent Pos:* Inst Zool, Iowa State Univ, 68-72. *Mem:* Am Physiol Soc; Am Epilepsy Soc; Soc Neurosci; Am Soc Parasitol. *Res:* Blood-brain barrier and brain exchange of nutrients and drugs; schistosome parasites with emphasis on glucose influx and glucose utilization. *Mailing Add:* Dept Neurol Univ Calif Los Angeles Sch Med Los Angeles CA 90024-1769

CORNFORTH, CLARENCE MICHAEL, b Kansas City, Mo, Aug 9, 40; m 78; c 3. OPERATIONS RESEARCH. *Educ:* US Naval Acad, BS, 62; Iowa State Univ, PhD(physics), 69. *Prof Exp:* USN, 62-74, weapons officer, 72-74, oper res analyst eng, Naval Ship Weapons Eng Sta, 74-77 & Ctr Naval Anal, 77-80, rep to comdr, Third Fleet, Ctr Naval Anal, 80-82, rep to comdr-in-chief, Pac Fleet, 82-83, rep to Tactical Training Group, Pac Fleet, 83-86, chief, Third Fleet Anal Team, 87, sr analyst, NOSC, 87-90, CHIEF, WARGAMING BR & COMDR-IN-CHIEF, PAC STAFF, USN, 90- *Concurrent Pos:* Sci analyst surface warfare to dep chief naval opers, 77-80. *Mem:* AAAS; US Naval Inst; Am Phys Soc; Mil Opers Res Soc. *Res:* Military operations research. *Mailing Add:* US Comndr-in-Chief Pac Staff Box 15-J553 H M Smith Camp HI 96861-5025

CORNFORTH, JOHN WARCUP, chemical enzymology, for more information see previous edition

CORNGOLD, NOEL ROBERT DAVID, b New York, NY, Jan 20, 29; m 52; c 2. REACTOR PHYSICS, STATISTICAL MECHANICS. *Educ:* Columbia Univ, AB, 49; Harvard Univ, AM, 50, PhD(physics), 54. *Prof Exp:* Asst phys sci, Harvard Univ, 50-51; res assoc, Brookhaven Nat Labs, 51-54, from assoc physicist to physicist, 54-66; PROF APPL PHYSICS, DIV ENG & APPL SCI, CALIF INST TECHNOL, 66- *Concurrent Pos:* Consult var nat labs, 67-; chmn steering comt appl physics, Calif Inst Technol, 82- *Honors & Awards:* Osborn Lectr, Univ Mich, 90. *Mem:* Fel Am Nuclear Soc; Am Phys Soc. *Res:* Theory of neutron scattering and transport; reactor physics; statistical physics; plasma physics. *Mailing Add:* Eng & Applied Sci 128-95 Calif Inst Technol Pasadena CA 91125

CORNHILL, JOHN FREDRICK, b Chatham, Ont, July 21, 49; m 72; c 2. CARDIOVASCULAR PHYSIOLOGY, EXPERIMENTAL PATHOLOGY. *Educ:* Univ Western Ont, BSc, 72; Univ Oxford, PhD(eng), 76. *Prof Exp:* Lectr biophysics, Univ Western Ont, 75-76; ASST PROF SURG, OHIO STATE UNIV, 76- *Concurrent Pos:* Ont scholar, Univ Western Ont, 68; overseas sci scholar, Royal Comn Exhib 1851, London, Eng, 72-75. *Honors & Awards:* Edward Chapman Res Prize, Magdalen Col, Univ Oxford, 77. *Mem:* Am Physiol Soc; Can Physiol Soc; Biophys Soc; Can Cardiovasc Soc; Am Soc Artifical Internal Organs. *Res:* Arterial wall mass transport and atherosclerosis; cardiovascular hemodynamics and its relationship to atherosclerosis; dynamics of prosthetic heart valves; coronary hemodynamics; environmental factor and heart disease and cancer. *Mailing Add:* Biomed Eng Ctr Ohio State Univ 270 Bevis Hall 1080 Carmack Rd Columbus OH 43210-1091

CORNIE, JAMES ALLEN, b Buhl, Idaho, Mar 7, 37; m 58; c 2. PHYSICAL METALLURGY. *Educ:* Univ Idaho, BS, 60 & 61; Rensselaer Polytech Inst, MS, 65; Univ Pittsburgh, PhD(metall eng), 71. *Prof Exp:* Engr casting, Kennecott Copper Co, 61-63; engr phys metall, Pratt & Whitney Aircraft, 63-65; sr engr alloy develop, Astronuclear Lab, Westinghouse Elec Co, 65-70, sr scientist phys metall, Res Lab, 70-77; MGR METAL MATRIX MAT, SPECIALTY MAT DIV, AVCO CORP, 77-; MAT PROCESSING CTR LAB, MASS INST TECHNOL. *Concurrent Pos:* Westinghouse fel, Univ Pittsburgh, 69-70; NSF fel, 77-78. *Mem:* Metall Soc; Am Inst Mining, Metall & Petrol Engrs; Am Soc Metals. *Res:* Interfaces in composites; composites technology; metal matrix composites; scanning electron microscopy; transmission electron microscopy; x-ray diffraction; Auger spectroscopy; phase relationships; reaction kinetics. *Mailing Add:* 118 Chelmsford St Apt 8 Cambridge MA 02138

CORNILSEN, BAHNE CARL, b Savanna, Ill, Apr 3, 45; m 77; c 1. SOLID STATE CHEMISTRY, CERAMIC SCIENCE. *Educ:* Ill Inst Technol, BS, 68; Marquette Univ, MS, 71; NY State Col Ceramics, PhD(ceramics), 75. *Prof Exp:* Res chemist, Globe-Union Inc, 68-69; fel, Ore Grad Ctr Study & Res, 75-77; asst prof, 77-82, ASSOC PROF CHEM, MICH TECHNOL UNIV, 82- *Concurrent Pos:* Vis prof, NY State Col Ceramics, 86. *Mem:* Am Chem Soc; Am Ceramic Soc; Soc Appl Spectros; Sigma Xi; Electrochem Soc; Ceramic Sci Hon Soc. *Res:* Preparation and structure of solid state compounds, including the point defect structure of nonstoichiometric materials; laser Raman vibrational spectroscopy; EXAFS spectroscopy used to study the structure of solids, nonstoichiometry; solid state phase transitions; interfacial structure. *Mailing Add:* Dept Chem & Chem Eng Mich Technol Univ Houghton MI 49931

CORNING, MARY ELIZABETH, b Norwich, Conn, Oct 19, 25. INTERNATIONAL BIOMEDICAL COMMUNICATIONS, PHYSICAL CHEMISTRY. *Educ:* Conn Col, BA, 47; Mt Holyoke Col, MA, 49. *Hon Degrees:* DSc, Mt Holyoke Col, 82; dipl, Paulista Sch Med, Brazil, 83. *Prof Exp:* Asst chemist, Mt Holyoke Col, 47-49; chemist, Nat Bur Standards, 49-58; spec asst to sci adv, US Dept State, 58-60; proj dir, Ed & Int Activities Planning Group, NSF, 60-61, spec asst to head, Off Int Sci Activities, 61-62, assoc prom dir, 62-64; chief, publ & trans div, Nat Libr Med, Dept Health & Human Serv, 64-66, spec asst to dep dir, 66-67, spec asst to dir, 67-72, asst dir int progs, 72-83; CONSULT, BIOMED COMMUN NAT & INT BODIES & FOREIGN GOVT, 84- *Concurrent Pos:* Asst ed, J Optical Soc Am, 50-60; mem chem panel, US Civil Serv Bd Examr, 53-60, tech ed & writer panel, 55-60; US Nat Liaison Officer to Orgn Econ Coop & Develop, 62; mem, US Nat Comt, Int Comn Optics, 61-67, Int Fedn Doc, 64-65 & Int Coun Sci Union Abstr Bd, 73-83; Fed Exec fel, Brookings Inst, 77; mem bd dirs & exec comt, Gorgas Mem Inst Trop & Prev Med, Inc, 72-83, secy, 74-83, emer dir, 84-91. *Honors & Awards:* Silver Medal, Pan Am Health Orgn, 83. *Mem:* Fel AAAS; Med Libr Asn; Am Chem Soc; fel Optical Soc Am. *Res:* Far ultraviolet absorption spectra of organic compounds; color and chemical constitution; spectrophotometry; organization of science; science information and documentation; international biomedical research and communications. *Mailing Add:* 90 Peck St Norwich CT 06360

CORNISH, KURTIS GEORGE, b Salt Lake City, Utah, Dec 1, 42; m 66; c 7. NEURO CONTROL CARDIOVASCULAR SYSTEM. *Educ:* Wake Forest Univ, PhD(physiol), 77. *Prof Exp:* ASSOC PROF PHYSIOL & BIOPHYS, COL MED, UNIV NEBR, 83- *Concurrent Pos:* Mem, Coun Circulation, Am Heart Asn. *Mem:* Am Physiolog Soc; fel Am Soc Primatologists. *Res:* Cardiovascular reflex control systems in the primate. *Mailing Add:* Dept Physiol Col Med Univ Nebr Omaha NE 68198-4575

CORNMAN, IVOR, b Cleveland, Ohio, May 22, 14; m 47. ZOOLOGY. *Educ:* Oberlin Col, AB, 36; NY Univ, MS, 39; Univ Mich, PhD, 49. *Prof Exp:* Asst zool, Univ Mich, 41-42; fel cytol, Sloan-Kettering Inst, 46-49; asst prof anat, Med Sch, George Washington Univ, 49-55; head dept cellular physiol, Hazleton Labs, Va, 55-59, asst dir res, 59-64; independent biol consult, 64-68; vpres, Environ Develop, Inc, 68-69; BIOL CONSULT, 69- *Concurrent Pos:* Mem corp, Marine Biol Lab, Woods Hole, 47-; guest, Univ West Indies, 64-78. *Mem:* Am Soc Zool. *Res:* Cancer chemotherapy; carcinogenesis; experimental alteration of cell division in normal and malignant plant and animal cells; marine biology; biological systems adapted to development of new drugs. *Mailing Add:* 21 Strawberry Hill Rd Woods Hole MA 02543

CORNONI-HUNTLEY, JOAN CLAIRE, b Melrose, Mass, May 17, 31; m 76. AGING RESEARCH. *Educ:* Mary Washington Col, Univ Va, AB, 53; Univ NC, MPH, 62, PhD(epidemiol), 70. *Prof Exp:* From asst prof to assoc prof epidemiol, Sch Pub Health, Univ NC, 70-76; spec asst-to-assoc dir, office stat res, Nat Ctr Health Statists, 76-78; sr res epidemiologist, 78-81, dep assoc dir, 81-88, EPIDEMIOL, DEMOG & BIOMET PROG, DEPT HEALTH & HUMAN SERVS, NIH, NAT INST AGING, 85- *Concurrent Pos:* Adj prof, Sch Pub Health, Univ NC, 78- *Honors & Awards:* Director's Award, NIH, 85. *Mem:* Am Col Epidemiol; Biomet Soc; Am Statist Asn; Am Pub Health Asn; Soc Epidemiol Res; Int Epidemiol Asn. *Res:* Epidemiologic studies of factors associated with hypertension, physical disability and health consequences of weight and body mass. *Mailing Add:* 7550 Wisconsin Ave Fed Bldg Rm 612 Bethesda MD 20892

CORNSWEET, TOM NORMAN, b Cleveland, Ohio, Apr 29, 29. VISION, OPHTHALMIC INSTRUMENTATION. *Educ:* Cornell Univ, AB, 51; Brown Univ, MSc, 53, PhD(exp psychol), 55. *Prof Exp:* From instr to asst prof psychol, Yale Univ, 55-59; from asst prof to prof, Univ Calif, Berkeley, 59-65; staff scientist biomed res, Stanford Res Inst, 65-71; chief scientist, Acuity Systs, Inc, Va, 71-73; assoc prof ophthal, Baylor Col Med, 74-76; PROF PSYCHOL, UNIV CALIF, IRVINE, 76- *Concurrent Pos:* NIH grants, Yale Univ, Univ Calif, Berkeley & Stanford Res Inst, 56-71; NASA grants, Stanford Res Inst, 66-71; Retina Res Found grant, Baylor Col Med, 74- *Mem:* AAAS; fel Optical Soc Am; Asn Res Vision & Ophthal. *Res:* Glaucoma; oculomotor system; ophthalmic instrumentation; visual aids for visually handicapped. *Mailing Add:* Dept Ophthal Univ Calif Irvine Col Med Irvine CA 92717

CORNWALL, JOHN MICHAEL, b Denver, Colo, Aug 19, 34; m 65. THEORETICAL PHYSICS. *Educ:* Harvard Univ, AB, 56; Univ Denver, MS, 59; Univ Calif, Berkeley, PhD(physics), 62. *Prof Exp:* NSF fel physics, Calif Inst Technol, 62-63; mem, Inst Advan Study, 63-65; from asst prof to assoc prof, 65-74, PROF PHYSICS, UNIV CALIF, LOS ANGELES, 74- *Concurrent Pos:* Consult, Aerospace Corp, El Segundo, 62-, NASA, 75, MITRE Corp, Los Alamos Nat Lab; Alfred P Sloan Found fel, 67-69; mem adv bd, Nat Inst Theoret Physics, 79-, chmn bd, 81-82; mem adv bd, High-Energy Physics Div, Lawrence Berkeley Lab, 85-90 & AT Div, Los Alamos Nat Lab, 87- *Mem:* Am Phys Soc; Am Geophys Union; AAAS; NY Acad Sci. *Res:* Theoretical elementary particle and high energy physics; emphasis on quantum chromodynamics; theoretical space and plasma physics; Van Allen belt studies. *Mailing Add:* Dept Physics Univ Calif Los Angeles CA 90024

CORNWELL, CHARLES DANIEL, b Williamsport, Pa, Dec 27, 24; m 51. MOLECULAR SPECTROSCOPY. *Educ:* Cornell Univ, AB, 47; Harvard Univ, MS, 49, PhD(chem physics), 51. *Prof Exp:* Res assoc chem, Univ Iowa, 50-52; from instr to assoc prof, 52-62, PROF CHEM, UNIV WIS-MADISON, 62 - *Mem:* AAAS; Am Phys Soc; Am Chem Soc. *Res:* Nuclear magnetic resonance, nuclear quadrupole resonance and microwave molecular rotational spectra; relationship of spectroscopic parameters to molecular structure and chemical binding; spin relaxation in gases. *Mailing Add:* Dept Chem Univ Wis Madison WI 53706

CORNWELL, DAVID GEORGE, b San Rafael, Calif, Oct 8, 27; m 59; c 2. BIOCHEMISTRY. *Educ:* Col of Wooster, BA, 50; Ohio State Univ, MA, 52; Stanford Univ, PhD(chem), 55. *Prof Exp:* Asst physiol chem, Ohio State Univ, 50-52; asst chem, Stanford Univ, 52-53; fel, Nat Res Coun, Harvard Univ, 54-56; from asst prof to assoc prof, 56-63, PROF PHYSIOL CHEM, OHIO STATE UNIV, 63-, CHMN DEPT, 65-, ASSOC DEAN ACAD AFFAIRS. *Concurrent Pos:* Mem nutrit study sect, NIH, 66-70, nutrit sci training comt, 70-73. *Mem:* Am Chem Soc; Biophys Soc; Am Soc Biol Chem; Am Inst Nutrit; Am Oil Chem Soc. *Res:* Chemistry and metabolism of lipids, lipoproteins, fat soluble vitamins and membranes. *Mailing Add:* Dept Physiol Chem 260 Meiling Hall Ohio State Univ 370 W Ninth Ave Columbus OH 43210

CORNWELL, JOHN CALHOUN, b Chester, SC, Dec 21, 44. EQUINE PHYSIOLOGY. *Educ:* Clemson Univ, BS, 70; La State Univ, MS, 72, PhD(animal sci), 76. *Prof Exp:* Grad asst instr equine reproductive physiol, Animal Sci Dept, La State Univ, 70-76, instr, Sch Vet Med, 76-78; ASSOC PROF HORSE SCI, ANIMAL SCI DEPT, NC STATE UNIV, 78- *Mem:* Am Soc Animal Sci; Am Asn Vet Anatomists; Sigma Xi. *Res:* Hormone levels in periparturient mares and newborn foals; induction of estrus in mares and control of ovalation; seasonal variation in stallion semen; freezing stallion semen; puberty in the colt. *Mailing Add:* Dept Animal Sci NC State Univ Box 7621 Raleigh NC 27695-7621

CORNWELL, LARRY WILMER, b Quincy, Ill, July 29, 41; m 63; c 4. OPERATIONS RESEARCH, COMPUTER SCIENCE. *Educ:* Culver-Stockton Col, BA, 63; Southern Ill Univ, MS, 65; Univ Mo, Rolla, PhD(math), 72. *Prof Exp:* Teacher math, Murphysboro High Sch, 64-65; instr, Forest Park Community Col, 65-66; asst prof, Culver-Stockton Col, 66-69; prof math, Western Ill Univ, 71-80; PROF MATH, BRADLEY UNIV, 80- *Concurrent Pos:* Resident assoc, Argonne Nat Lab, 76-77, consult, 77-78. *Mem:* Opers Res Soc Am; Inst Mgt Sci; Asn Comput Mach; Am Prod & Inventory Control Soc; Am Asn Univ Prof. *Res:* Nonlinear programming; nonlinear least squares; unconstrained and constrained optimization; generation of the distribution of the product of two normal random variables; heuristic programming; mathematical modeling; applications of mathematics; business applications of microcomputers. *Mailing Add:* Bus Mgt & Admin Bradley Univ Peoria IL 61625

CORNYN, JOHN JOSEPH, b Pittsburgh, Pa, Dec 20, 44. COMPUTER SCIENCE, UNDERWATER ACOUSTICS. *Educ:* Carnegie-Mellon Univ, BS, 67, MS, 69; Univ Md, MS, 74. *Prof Exp:* Physicist underwater acoust, Naval Res Lab, 68-72, physicist signal processing, 72-76; physicist underwater acoust, Naval Ocean Res & Develop Activ, 76-81; comput consult & pres, Third Wave Data Systs, Inc, 81-86; SR STAFF ENGR PHYSICS & UNDERWATER ACOUST, MARTIN MARIETTA CORP, BALTIMORE, 86- *Mem:* Acoust Soc Am; Inst Elec & Electronics Engrs Comput Soc. *Mailing Add:* PO Box 72489 Baltimore MD 21237

CORONITI, FERDINAND VINCENT, b Boston, Mass, June 14, 43; m 69; c 2. SPACE PHYSICS, PLASMA PHYSICS. *Educ:* Harvard Univ, BA, 65; Univ Calif, Berkeley, PhD(physics), 69. *Prof Exp:* Res asst physics, Univ Calif, Berkeley, 65-69; asst res physicist, 69-70, from asst prof to assoc prof physics & space physics, 70-78, PROF PHYSICS & ASTRON, UNIV CALIF, LOS ANGELES, 78- *Concurrent Pos:* Consult, TRW Syst, 69- & Los Alamos Sci Lab, 73-74. *Mem:* Am Geophys Union; Int Union Radio Sci; Am Astron Soc; Am Phys Soc. *Res:* Magnetospheric dynamics; Jupiter radiation belts and magnetospheric structure; magnetic field reconnection; nonlinear plasma theory; plasma astrophysics. *Mailing Add:* Dept Physics Univ Calif Los Angeles CA 90024

COROTIS, ROSS BARRY, b Woodbury, NJ, Jan 15, 45. STRUCTURAL ENGINEERING. *Educ:* Mass Inst Technol, SB, 67, SM, 68, PhD(civil eng), 71. *Prof Exp:* Asst prof civil eng, Northwestern Univ, 71-74, assoc prof, 75-79, prof, 79-81; chmn dept, 83-90, PROF CIVIL ENG, JOHNS HOPKINS UNIV, 81-, HACKERMAN PROF, 82-, ASSOC DEAN ENG, 90- *Concurrent Pos:* Live loads chmn, Bldg Code Comt, Am Nat Standards Inst, 78-84; comt chmn safety bldgs, 84-87, Am Soc Civ Engrs & structural safety, Am Concrete Inst, 86-88; mem, bldg res bd, Int Technol Coun & Nat Res Coun; pres, Md Sect, Am Soc Civil Engrs, 88-89. *Honors & Awards:* Huber Res Prize, Am Soc Civil Engrs, 84. *Mem:* Am Concrete Inst; Am Soc Civil Engrs; Am Soc Eng Educ. *Res:* Structural safety; earthquake engineering; wind energy. *Mailing Add:* Dept Civil Eng Johns Hopkins Univ Baltimore MD 21218-2699

CORPE, WILLIAM ALBERT, b Walworth, Wis, Jan 11, 24; m 49; c 2. MICROBIAL ATTACHMENT TO SURFACES. *Educ:* Univ Wis, Madison, BS, 48, MS, 50; Pa State Univ, PhD(microbiol), 56. *Prof Exp:* Teaching asst bacteriol, Univ Wis, 48-49, res asst, 49-50; asst prof, Western Ky State Col, 50-53; res asst microbiol, Pa State Univ, 53-56; asst prof, Dept Biol, Columbia Univ & Barnard Col, NY, 56-60; from assoc prof to prof, 60-89, EMER BROF BIOL SCI, COLUMBIA UNIV, 89- *Concurrent Pos:* Res fel, Microbiol Dept, US Pub Health Serv, Univ New South Wales, Australia, 63-64; res collabr, Brookhaven Nat Lab, 65-71. *Mem:* Am Soc Microbiol; Can Soc Microbiol; NY Acad Sci; Soc Indust Microbiol; AAAS. *Res:* Bacterial physiology and ecology; cyanide metabolism; hydrolytic enzyme secretion; exopolymer production; mechanisms of attachment to and colonization of solid surfaces by microbial species; ecology and ultra structure of methylotrophs; freshwater picoplankton. *Mailing Add:* Dept Biologic Sci Columbia Univ 1110 Schermerhorn Hall New York NY 10027

CORRADINO, ROBERT ANTHONY, b Lancaster, Pa, Aug 6, 38; m 63; c 1. CELL PHYSIOLOGY, ENDOCRINOLOGY. *Educ:* Millersville State Col, BS, 60; Purdue Univ, MS, 62; Cornell Univ, PhD(physiol), 66. *Prof Exp:* Res assoc, Wyeth Labs, Inc, Radnor, Pa, 62-64; res assoc, 66-72, sr res assoc, 72-80, ASSOC PROF, DEPT PHYSIOL, CORNELL UNIV, 80- *Concurrent Pos:* Prin investr res grants, Nat Inst Arthritis, Metab & Digestive Dis & NIH, 71-; Res Career Develop Award, NIH, 75-80; grant reviewer, NIH, NSF, & USDA, 76- *Mem:* Am Physiol Soc; Am Inst Nutrit; Soc Exp Biol & Med; Tissue Cult Asn; Sigma Xi. *Res:* Hormone mechanisms at the cellular and molecular levels with specific emphasis on the mode of action of cholecalciferol (vitamin D-3). *Mailing Add:* Dept Physiol NY State Col Vet Med Cornell Univ Ithaca NY 14853

CORREIA, JOHN ARTHUR, b Brookline, Mass, June 8, 45; m 67; c 1. MEDICAL PHYSICS. *Educ:* Lowell Technol Inst, BS, 67, PhD(nuclear physics), 73. *Prof Exp:* Res fel physics, 72-74, asst physicist, 74-79, ASSOC PHYSICIST, MASS GEN HOSP, 79- *Concurrent Pos:* Res fel, Sch Med, Harvard Univ, 72-74, res assoc, 74-79, asst prof radiol, 79-83, assoc prof radiol, 83- *Mem:* Inst ELec & Electronics Engrs; AAAS; Soc Nuclear Med; AAAS; NY Acad Sci. *Res:* Application of computers to nuclear medicine; development of instrumentation and computational schemes for the measurement of cerebral blood flow and for three dimensional reconstruction using radioisotopes; cyclotrons in medicine; positron imaging. *Mailing Add:* 306 Salem St Wakefield MA 01880

CORREIA, JOHN SIDNEY, b Berkeley, Calif, Mar 21, 37; m 89. PHYSICAL ORGANIC. *Educ:* St Mary's Col Calif, BS, 59; Univ Calif, Berkeley, PhD(chem), 62. *Prof Exp:* PROF CHEM, ST MARY'S COL, 62- *Mem:* Am Chem Soc. *Res:* Mechanisms and methods of phase transfer oxidation reactions; 1,6-additions to conjugated systems. *Mailing Add:* St Mary's Col PO Box 4527 Moraga CA 94575

CORREIA, MARIA ALMIRA, b Goa, India, Jan 4, 46; US citizen. PHARMACOLOGY. *Educ:* Univ Bombay, BPharm, 67; Univ Minn, PhD(pharmacol), 73. *Prof Exp:* Instr, Goa Col Pharm, India, 67; res asst, Dept Pharmacol, Univ Minn, 67-72; postdoctoral res fel, Dept Med, Univ Calif San Francisco, 73-74, postgrad res pharmacologist, 74-75, from asst prof to assoc prof, 75-87, PROF, DEPTS PHARMACOL & MED, UNIV CALIF SAN FRANCISCO, 87-, PROF, DEPT PHARM, SCH PHARM, 90- *Concurrent Pos:* Grants, var insts & corp, 75-93; mem, Pharmacol Toxicol Rev Comt, NIH-NIGMS, 80, Pharmacol Study Sect, 81 & 82-86, Gen Med B Study Sect, 90, Toxicol Study Sect, 91-95; grad student adv, Dept Pharmacol, Univ Calif San Francisco, 81-83 & 87-, dir, Pharmacol Grad Prog, 89-; vis prof, Queens Univ, Ont, 88, Vanderbilt Univ, 89; consult, G D Searle, Ill & NIH-HLBI. *Honors & Awards:* Am Porphyria Found Award, 90. *Mem:* Sigma Xi; Am Soc Pharmacol & Exp Therapeut; AAAS; Soc Toxicol; Am Soc Biol Chemists. *Res:* Pharmacology; toxicology; drug metabolism. *Mailing Add:* Dept Pharmacol Univ Calif San Francisco Box 0450 S-1210 San Francisco CA 94143

CORRIERE, JOSEPH N, JR, b Easton, Pa, Apr 3, 37; m 87; c 4. UROLOGY. *Educ:* Univ Pa, BA, 59; Seton Hall Col Med, MD, 63. *Prof Exp:* Resident urol, Hosp Univ Pa, 64-69; asst prof, Univ Pa, 71-74; PROF UROL & DIR DIV, UNIV TEX MED SCH HOUSTON, 74- *Concurrent Pos:* USPHS res trainee urol, Univ Pa, 67-68. *Mem:* Asn Acad Surg; Soc Univ Surg; Acad Pediat; Am Col Surg; Am Urol Asn; Am Asn Surg Trauma. *Res:* Urinary tract infection and the use of isotopes in the urinary tract. *Mailing Add:* Dept Urol Univ Tex Med Sch-Houston 6431 Fannin Suite 6018 MSMB Houston TX 77030

CORRIGAN, DENNIS ARTHUR, b Hartford, Conn, June 25, 51; m 85; c 1. ELECTRODE KINETICS, SPECTROELECTROCHEMISTRY. *Educ:* Purdue Univ, BS, 74; Univ Wis, PhD(anal chem), 79. *Prof Exp:* STAFF RES SCIENTIST, GEN MOTORS RES LABS, 79- *Mem:* Electrochem Soc; Am Chem Soc. *Res:* Batteries; electrocatalysis; electroplating; electrode rate measurements; cyclic voltammetry. *Mailing Add:* Phys Chem Dept Gen Motors Res Labs Warren MI 48090

CORRIGAN, JAMES JOHN, JR, b Pittsburgh, Pa, Aug 28, 35; m 60; c 2. HEMATOLOGY, PEDIATRICS. *Educ:* Juniata Col, BS, 57; Univ Pittsburgh, MD, 61. *Prof Exp:* Intern med, Univ Colo, 61-62, resident pediat, 62-64; assoc, Sch Med, Emory Univ, 66-67, asst prof, 67-70; assoc prof, 71-74, PROF PEDIAT, COL MED, UNIV ARIZ, 74-, PROF INTERNAL MED, 87-; DIR, MOUNTAIN STATES HEMOPHILIA CTR, 76- *Concurrent Pos:* Fel pediat hemat, Col Med, Univ Ill, 64-66; NIH res grants, 67-; Ga Heart Inc res grant, 68-70; chief of staff, Univ Med Ctr, 84-87; pres, Pima County Med Soc, 88. *Honors & Awards:* Ross Award, Pediat Res, 75. *Mem:* Am Soc Pediat; fel Am Acad Pediat; Am Soc Hemat; Soc Pediat Res; Int Soc Thrombosis & Haemostasis. *Res:* Disorders of blood coagulation mechanisms, particularly those conditions associated with disseminated intravascular coagulation and hyperfibrinolysis. *Mailing Add:* Dept Pediat Univ Ariz Col Med Tucson AZ 85724

CORRIGAN, JOHN JOSEPH, b Chicago, Ill, Jan 17, 29; m 54; c 4. BIOCHEMISTRY. *Educ:* Carleton Col, AB, 50; Univ Ill, MS, 57, PhD(entom), 59. *Prof Exp:* Physiologist, Baxter Labs, 50-52; asst entom, Ill Natural Hist Surv, 52-53; res fel biochem, Sch Med, Tufts Univ, 59-62, sr instr, 62-63, asst prof, 63-70; assoc dean col arts & sci, 70-86, assoc prof, 70-75, PROF LIFE SCI & ADJ PROF BIOCHEM, IND STATE UNIV, 75- *Concurrent Pos:* Lectures annually on nitrogen metab to first yr students at the Terre Haute Ctr for Med Educ, Ind Univ Sch Med. *Mem:* AAAS; Am Chem Soc; Am Soc Zoologists. *Res:* Metabolism of amino acids and proteins in insects; biochemistry of insect metamorphosis; silk fibroin biosynthesis in lepidoptera; biochemical basis of insecticide action; stereospecific metabolism of amino acids; biochemistry of development. *Mailing Add:* Dept Life Sci Ind State Univ Terre Haute IN 47809

CORRIGAN, JOHN RAYMOND, b Fargo, NDak, Apr 28, 19; wid; c 4. ORGANIC CHEMISTRY. *Educ:* Univ Portland, BS, 40; Univ Notre Dame, MS, 42, PhD(chem), 49. *Prof Exp:* Asst, Univ Notre Dame, 41-43; res assoc, Frederick Stearns & Co, 43-47; res fel, Sterling-Winthrop Res Inst, 47-49, res assoc, 49-51; res assoc, Sharp & Dohme, 51-55; sr chemist, Mead Johnson & Co, 55-57, group leader, 57-59, from asst dir to dir dept chem develop, 59-68, dir process eng, 68-78; owner, small bus, 78-84; RETIRED. *Mem:* Emer mem, Am Chem Soc. *Res:* Synthetic organic medicinals. *Mailing Add:* 905 Conway No 39 Las Cruces NM 88005

CORRIPIO, ARMANDO BENITO, b Mantua, Cuba, Mar 6, 41; US citizen; m 62; c 4. CHEMICAL ENGINEERING, AUTOMATIC CONTROL SYSTEMS. *Educ:* La State Univ, Baton Rouge, BSc, 63, MS, 67, PhD(chem eng), 70. *Prof Exp:* Control systs engr, Dow Chem Co, La, 63-68; from instr to assoc prof, 68-81, PROF CHEM ENG, LA STATE UNIV, BATON ROUGE, 81- *Concurrent Pos:* Consult, Dow Chem Co, 68-69 & La Div, 74-; investr, Proj Themis, Off Sci Res, USAF, 69-; distinguished fac fel, La State Univ Found, 74; vis engr, Mass Inst Technol, 78-79. *Honors & Awards:* Charles E Coater Award, 90. *Mem:* Am Inst Chem Engrs; Instrument Soc Am. *Res:* Mathematical modeling and dynamic simulation of chemical processes; automatic control systems design, especially application of digital computers to process control; practical application of modern control theory to the control of non-linear systems; computer-aided process design. *Mailing Add:* Dept Chem Eng La State Univ Baton Rouge LA 70803

CORRIVEAU, ARMAND GERARD, genetics, forest science; deceased, see previous edition for last biography

CORRUCCINI, LINTON REID, b Corvallis, Ore, Jan 1, 44; m 72; c 2. LOW TEMPERATURE PHYSICS, SOLID STATE PHYSICS. *Educ:* Swarthmore Col, BA, 66; Cornell Univ, PhD(physics), 72. *Prof Exp:* NSF fel physics, Cornell Univ, 67-71; mem tech staff, Aerospace Corp, 71-73; from asst prof to assoc prof, 73-84, PROF PHYSICS, UNIV CALIF, DAVIS, 84- *Mem:* Am Phys Soc. *Res:* Liquid helium three-helium four solutions; superfluid helium 3; nuclear magnetic resonance; magnetic order at low temperatures. *Mailing Add:* Dept Physics Univ Calif Davis CA 95616

CORRUCCINI, ROBERT SPENCER, b Takoma Park, Md, May 21, 49; m 82; c 2. ODONTOLOGY, ANTHROPOLOGICAL EPIDEMIOLOGY. *Educ:* Colo Univ, BA, 71; Univ Calif, Berkeley, PhD(anthrop), 75. *Prof Exp:* Res assoc, Smithsonian Inst, 76-78; from asst prof to assoc prof, 78-86, PROF ANTHROP, SOUTHERN ILL UNIV, 86- *Concurrent Pos:* NATO fel, Univ Florence, Italy, 80-81; prof, Univ Pisa, Italy, 87. *Mem:* Am Asn Physical Anthropologists; Soc Syst Zool; fel Human Biol Coun; Am Antrhop Asn. *Res:* Dental anthropology; hominid evolution; cross-cultural approaches to epidemiology; quantitative methods; morphometrics; relative growth in primate structures. *Mailing Add:* Dept Anthrop Southern Ill Univ Carbondale IL 62901-4502

CORRY, ANDREW F, b Lynn, Mass, 1922; m; c 3. ELECTRIC POWER. *Educ:* Mass Inst Technol, BS, 44. *Prof Exp:* Mem staff, Boston Edison Co, 47-75, vpres elec opers & eng, 75-80, sr vpres & corp res officer, 79-83; CONSULT, 83- *Honors & Awards:* Habirshaw Medal, Inst Elec & Electronics Engrs Centennial Medal. *Mem:* Nat Acad Eng; fel Inst Elec & Electronics Engrs. *Res:* Research and development of electric transmission technology. *Mailing Add:* PO Box 310 West Hyannisport MA 02672-0310

CORRY, THOMAS M, b Pittsburgh, Pa, June 9, 26; m 51; c 4. ELECTRICAL ENGINEERING. *Educ:* Carnegie Inst Technol, BS, 51. *Prof Exp:* Develop engr radar, A B Dumont Labs, 51-52; proj engr control systs, Femco, Inc, 52-54; supvr engr, Westinghouse Res Labs, 54-63 & Defense Res Labs, Gen Motors Corp, 63-75; PRES, ENERGY SCI CORP, 76- *Mem:* Inst Elec & Electronics Engrs. *Res:* Aircraft electrical systems; supervisory control systems; semiconductor power switching circuits; thermoelectric generators; high energy physics; experimental machines. *Mailing Add:* Energy Sci Corp PO Box 152 Goleta CA 93017

CORSARO, ROBERT DOMINIC, b Elizabeth, NJ, Nov 5, 44; m 67; c 2. PHYSICAL CHEMISTRY, ACOUSTICS. *Educ:* Lebanon Valley Col, BS, 66; Univ Md, PhD(chem), 71. *Prof Exp:* RES CHEMIST ACOUST, NAVAL RES LAB, 70- *Honors & Awards:* Res Publ Award, Naval Res Lab. *Mem:* Am Chem Soc; Acoust Soc Am; Sigma Xi. *Res:* Acoustic propagation in materials; acoustic techniques in material science; ultrasonics. *Mailing Add:* Code 5135 Naval Res Lab Washington DC 20375-5000

CORSE, JOSEPH WALTERS, b Denver, Colo, Sept 7, 13; m 36; c 4. PHYTOCHEMISTRY. *Educ:* Univ Calif, Los Angeles, BA, 36; Univ Ill, PhD(org chem), 40. *Prof Exp:* Asst, Univ Ill, 37-39; res chemist, Eli Lilly Co, 40-46; asst, Calif Inst Technol, 46-47; res assoc, Univ Calif, Los Angeles, 47-48; CHEMIST, WESTERN REGIONAL RES LAB, AGR RES SERV, US DEPT AGR, 48- *Mem:* Am Chem Soc; Phytochem Soc NAm. *Res:* biosynthesis of penicillin; enzyme reactions and synthesis of metabolites; chemical interactions in plant growth. *Mailing Add:* Western Regional Res Ctr USDA 800 Buchanan St Albany CA 94710

CORSINI, A, b Hamilton, Ont, Apr 28, 34; m 61; c 1. ANALYTICAL CHEMISTRY. *Educ:* McMaster Univ, BSc, 56, PhD(anal chem), 61. *Prof Exp:* Atomic Energy Comn fel chelate chem, Univ Ariz, 61-63; from asst prof to assoc prof anal & inorg chem, 63-72, PROF CHEM, MCMASTER UNIV, 72- *Concurrent Pos:* Nat Res Coun Can grant, 63- *Mem:* Chem Inst Can. *Res:* Instrumental analysis; chemistry of metal chelates, heteropolytungstates, metalloporphyrins and analytical applications. *Mailing Add:* Dept Chem McMaster Univ Hamilton ON L8S 4L8 Can

CORSINI, DENNIS LEE, b Los Angeles, Calif, Nov 8, 42; m 63; c 3. AGRICULTURAL BIOCHEMISTRY. *Educ:* Univ Calif, Los Angeles, BA, 65; Univ Idaho, PhD(agr biochem), 71. *Prof Exp:* Fel plant path, Univ Calif, Riverside, 71-72; res assoc plant sci, 73-76, RES PATHOLOGIST, SCI &

EDUC ADMIN, USDA, UNIV IDAHO, 76- *Mem:* Potato Asn Am; Am Phytopathological Soc; Am Chem Soc. *Res:* Develop new potato cultivars with multiple disease resistance. *Mailing Add:* Univ Idaho Res & Exten Ctr Aberdeen ID 83210

CORSON, DALE RAYMOND, b Pittsburg, Kans, Apr 5, 14; m 38; c 4. NUCLEAR PHYSICS. *Educ:* Col Emporia, AB, 34; Univ Kans, AM, 35; Univ Calif, PhD(physics), 38. *Hon Degrees:* LHD, Emporia Col, 70; LLD, Columbia Univ, 72 & Hamilton Col, 73; DSc, Univ Rochester, 75, Elmira Col, 77 & Wilkes Col, 85. *Prof Exp:* Asst, Univ Calif, 36-38, res fel, 38-39, instr physics, 39-40; from asst prof to assoc prof, Univ Mo, 40-45; staff mem, Los Alamos Sci Lab, 45-46; from asst prof to prof physics, Cornell Univ, 46-63, chmn dept, 56-59, dean, Col Eng, 59-63, provost, 63-69, pres, 69-77, chancellor, 77-79, EMER PRES, CORNELL UNIV, 79- *Concurrent Pos:* Mem staff, Radiation Lab, Mass Inst Technol, 41-43; radar consult, US Army Air Force, 43-45. *Honors & Awards:* Pub Welfare Medal, Nat Acad Sci; Arthur M Bueche Award, Nat Acad Eng. *Mem:* Nat Acad Eng; fel Am Acad Arts & Sci; Am Phys Soc. *Res:* Cosmic rays; nuclear physics; engineering. *Mailing Add:* 615 Clark Hall Cornell Univ Ithaca NY 14853

CORSON, GEORGE EDWIN, JR, plant morphogenesis; deceased, see previous edition for last biography

CORSON, HARRY HERBERT, b Nashville, Tenn, Feb 2, 31; m 53; c 2. MATHEMATICS. *Educ:* Vanderbilt Univ, AB, 52; Duke Univ, MA, 54, PhD(math), 57. *Prof Exp:* Res instr math, Tulane Univ, 56-58; res asst prof, 58-62, assoc prof, 62-65, PROF MATH, UNIV WASH, 65- *Concurrent Pos:* Fel, Off Naval Res, 58-59. *Mem:* Am Math Soc. *Res:* Topology; linear spaces. *Mailing Add:* Dept Math Univ Wash Seattle WA 98195

CORSON, JOSEPH MACKIE, b Mar 8, 24; m; c 3. SURGICAL PATHOLOGY, IMMUNOHISTOCHEMISTRY. *Educ:* Jefferson Med Col, MD, 47. *Hon Degrees:* MA, Harvard Univ, 90. *Prof Exp:* CHIEF, SURG PATH DIV, BRIGHAM & WOMEN'S HOSP, 57-; PROF PATH, SCH MED, HARVARD UNIV, 84- *Concurrent Pos:* Consult staf, Dana Farber Cancer Inst, 82- *Mem:* Am Asn Pathologists; Am Med Asn; AAAS. *Res:* Pathology of mesotheliomas; immunohistochemistry of mesotheliomas and soft tissue sarcomas. *Mailing Add:* Dept Path Brigham & Women's Hosp 75 Francis St Boston MA 02115

CORSON, SAMUEL ABRAHAM, b Odessa, Russia, Dec 31, 09; nat US; m 47; c 3. PHYSIOLOGY, PHARMACOLOGY. *Educ:* NY Univ, BS, 30; Univ Pa, MS, 31; Univ Tex, PhD, 42. *Prof Exp:* Asst physiol, Sch Med, Univ Pa, 32-35; res assoc cell physiol, NY Univ, 35-37; instr physiol, Div Gen Educ, NY, 37-39; consult physiologist, NY, 38-40; instr zool & physiol, Univ Tex, 40-42; asst prof physiol, Sch Med, Univ Okla, 42-43; instr pharmacol, Sch Med, Georgetown Univ, 43-44; from instr to asst prof physiol, Med Col, Univ Minn, 44-47, instr sci Russian & coordr contemp Russian, Exten Div, 45-47; assoc prof, Col Med, Howard Univ, 47-50; chief, Dept Physiol, Toledo Hosp Inst Med Res, 50-51; prof pharmacol & head dept, Kirksville Col Osteop & Surg, 51-54; assoc prof pharmacol & physiol, Sch Med, Univ Ark, 54-59; res assoc hist med, Sch Med, Yale Univ, 59-60; from assoc prof to prof psychiat, 60-80, prof biophys, Col Biol Sci, 69-80, dir Lab Cerebrovisceral Physiol, 60-80, EMER PROF PSYCHIAT (PSYCHOPHYSIOL), COL MED, OHIO STATE UNIV, 80- *Concurrent Pos:* Consult hosp staff, Univ Ark, 58-59; ed in chief, Int J Psychobiol, 70-73; mem consult roster, Behav Med Br, Div Heart & Vascular Dis, Nat Heart, Lung & Blood Inst, NIH, 79-; Book rev ed, Pavlovian J Biol Sci, 81- *Honors & Awards:* Anokhin Mem Medal, P K Anokhin Inst, USSR Acad Sci, 79. *Mem:* Fel AAAS; Am Physiol Soc; fel Am Col Cardiol; Am Psychosom Soc; Sigma Xi. *Res:* Cerebrovisceral and renal physiology; psychopharmacology; cybernetics and systems approach in psychophysiology; individual differences in reactions to psychologic stress; Pavlovian and operant conditioned reflexes; interaction of pharmacologic and psychosocial factors in the control of aggression and hyperkinesis; minimal brain dysfunction; paradoxical effects of amphetamines; psychobiologic integrative mechanisms; physiologic basis of psychosomatic medicine; pet-facilitated therapy; psychiatry, psychobiology and psychology in the USSR and East European countries; comparative studies on Pavlovian and Freudian Contributions to psychiatry; gerontology; stress coping mechanisms; psychophysiologic substrates of type A and B behavior; cross-cultural, international, and interpolitical communication. *Mailing Add:* Dept Psychiat Upham Hall Ohio State Univ Col Med 473 W 12 Ave Columbus OH 43210

CORSTVET, RICHARD E, b Big Bend, Wis, Oct 12, 28; m 54; c 3. VETERINARY MICROBIOLOGY, VETERINARY PATHOLOGY. *Educ:* Univ Wis, BS, 51, MS, 55; Univ Calif, Davis, PhD(comp path), 65. *Prof Exp:* Lab technician avian med, Sch Vet Med, Univ Calif, Davis, 55-59, pub health, 59-65; asst res microbiologist, 65; from asst prof to assoc prof microbiol, Col Vet Med, Okla State Univ, 65-72, actg head dept, 77-80, prof microbiol & pub health, 72-; AT SCH VET MED, LA STATE UNIV, BATON ROUGE. *Mem:* Poultry Sci Asn; US Animal Health Asn; NY Acad Sci; Am Asn Avian Path; Sigma Xi. *Res:* Infection and immunity in Newcastle disease; relationship of various disease syndromes to the wholesomeness of market poultry; carrier states in animals; multiple infections in animals. *Mailing Add:* Sch Vet Med La State Univ Baton Rouge LA 70803

CORT, SUSANNAH, b Prague, Czech, Aug 8, 57; US citizen. GENERAL MEDICAL SCIENCES, PUBLIC HEALTH & EPIDEMIOLOGY. *Educ:* Charles Univ, MD, 82. *Prof Exp:* Resident med, Univ Hosp, Czech, 82-84; res assoc clin res, Univ Ariz, 84-86, Biomed Res Consults, 86-88; assoc dir epidemiol res, Frosted, 88-90; asst med dir clin res, 90-91, DIR RES, COMMUNITY RES INITIATIVE, 91- *Concurrent Pos:* Consult, Biomed Res Consults, Ltd, 88- *Res:* Hypertension and diabetes mellitus; clinical studies in AIDS research; epidemiologic studies on HIV seroprevalence; new pharmaceutical drugs in AIDS; writing clinical research protocols. *Mailing Add:* Community Res Initiative 31 W 26th St New York NY 10010

CORT, WINIFRED MITCHELL, b Cleveland, Ohio; m 47; c 2. MICROBIAL BIOCHEMISTRY, ENZYMES & NUTRIENTS & ANTIOXIDANTS. *Educ:* Univ Ill, BS, 41, MS, 43, PhD(bact, biochem), 46. *Prof Exp:* Res microbiologist, Commercial Solvents Corp, 43-45; res biochemist, Chas Pfizer & Co, 45-47; res microbiologist, Nat Dairy Res Labs, Inc, 51-54; sect leader, Res & Develop Div, 54-58, sect leader enzymes, Evans Res & Develop Corp, 58-64; group leader, Res & Develop, Beechnut Life Savers, 64-65; sr biochemist, 65-73, group leader, 73-81, assoc dir prod develop, 81-85, CONSULT, HOFFMANN-LA ROCHE INC, 85- *Concurrent Pos:* Lectr, Adelphi Univ, 59-62; consult, Evans Res & Develop Corp, 61-62 & Food & Drug Res, Inc, 62. *Mem:* Am Soc Microbiol; Am Chem Soc; Am Pharmacol Asn; AAAS; Inst Food Technol; Am Oil Chem Soc; Sigma Xi. *Res:* Antioxidants, antibiotics, vitamin stabilization; fermentation technology, enzymology and drug delivery systems. *Mailing Add:* 4395 Brandywine Dr Sarasota FL 34241

CORTH, RICHARD, b New York, NY, Apr 14, 25; m 44; c 3. PHYSICAL CHEMISTRY, ANALYTICAL CHEMISTRY. *Educ:* Brooklyn Col, BS, 48; Polytech Inst Brooklyn, PhD(radiation chem), 63. *Prof Exp:* Res engr, Westinghouse Elec Corp, 55-56, from sr res scientist to fel res scientist, 56-85; RETIRED. *Mem:* Am Soc Photobiol. *Res:* Visual perception; photobiology; radiochemistry; radiation chemistry; solid state diffusion; gas-metal reactions; gas chromatography; activation analysis. *Mailing Add:* 1388 Mecklenburg Rd Ithaca NY 14850

CORTNER, JEAN A, b Nashville, Tenn, Nov 10, 30; c 3. BIOCHEMICAL GENETICS. *Educ:* Vanderbilt Univ, BA, 52, MD, 55. *Prof Exp:* Guest investr human genetics & asst physician, Rockefeller Inst, 62-63; chief dept pediat, Roswell Park Mem Inst, 63-67; prof pediat & chmn dept, State Univ NY, Buffalo, 67-74, physician-in-chief, Children's Hosp, 67-74; physician-in-chief, Children's Hosp Philadelphia, 74-86, chmn dept, 74-86, PROF SCH MED, UNIV PA, 74-, DIR LIPID-HEART RES CTR, 80-, DIR NUTRIT CTR, CHILDREN'S HOSP, 85- *Concurrent Pos:* NIH vis fel dept pediat & biochem, Babies Hosp & Columbia Univ, 61-63; NIH fel, Galton Lab Human Genetics, Univ London, 72-73. *Mem:* AAAS; Am Soc Human Genetics; Am Fedn Clin Res; Soc Pediat Res; Am Acad Pediat. *Res:* Human genetics; pediatrics; the study of lipoprotein metabolism in adults with dominantly inherited hypolipidemins associated with premature coronary artery disease and in their children who are at risk. *Mailing Add:* Dept Pediat Children's Hosp Univ Pa Philadelphia PA 19104

CORTRIGHT, EDGAR MAURICE, b Hastings, Pa, July 29, 23; m 45; c 2. AEROSPACE ENGINEERING. *Educ:* Rensselaer Polytech Inst, BS, 47, MS, 49. *Hon Degrees:* DEng, Rensselaer Polytech Inst, 75; DSc, George Washington Univ, 73. *Prof Exp:* Aeronaut res scientist aerodyn, Lewis Flight Propulsion Ctr, NASA, Cleveland, 48-58; chief advan tech progs, NASA, Washington, 58-60, asst dir lunar and planetary progs, 60-61, dep assoc adminr, Off Space sci & Appln, 61-67 & Off Manned Space Flight, 67-68; dir, Langley Res Ctr, Va, 68-75; vpres tech dir, Owens-Ill, Inc, Toledo, 75-78; pres, Lockheed Calif Co, 79-83; PRES, E M CORTRIGHT & ASSOCS, 83- *Honors & Awards:* NASA medals, 66, 67 & 77; Space Flight award, Am Astronaut Soc, 71; Spirit St Louis Award, Am Soc Mech. *Mem:* Nat Acad Eng; Fel Am Astronaut Soc; Am Inst Aerospace & Astronaut (pres, 76); AAAS; Sigma Xi. *Res:* Author of two novels on space for US Govt. *Mailing Add:* 1010 Wormley Creek Dr Yorktown VA 23690

CORTY, CLAUDE, b Bochum, Ger, May 16, 24; nat US; m 48; c 2. CHEMICAL ENGINEERING. *Educ:* Mass Inst Technol, BS, 44; Univ Mich, MS, 48, PhD(chem eng), 51. *Prof Exp:* Chemist, Atlas Powder Co, 46-47; res engr, E I Du Pont de Nemours & Co Inc, 51-57, res supvr, Indust & Biochem Dept, 57-81, res mgr, Biochem Dept, 81-90; RETIRED. *Mem:* Am Chem Soc; Am Inst Chem Engrs; Sigma Xi. *Res:* Heat transfer; agricultural formulations. *Mailing Add:* 1511 Emory Rd Wilmington DE 19803

CORUM, JAMES FREDERIC, b Natick, Mass, Aug 15, 43; m 70; c 4. ELECTRICAL ENGINEERING. *Educ:* Univ Lowell, BS, 65; Ohio State Univ, MSc, 67, PhD(elec eng), 74. *Prof Exp:* Electronic engr antennas, Nat Security Agency, 65-66; res asst radio astron, Ohio State Univ, 67-70; prof math & physics, Ohio Inst Technol, 70-74; asst prof elec eng, WVa Univ, 74-79; pres, Radiation Dynamics Inc, 79-80; assoc prof elec eng, W VA Univ, 81-87; DIR ENG RES, PINZONE COMMUNS PRODS,INC, 87- *Concurrent Pos:* Consult, AM/FM Radio Stations, Satellite Television Earth Stations & US Govt Agencies. *Mem:* Inst Elec & Electronics Engrs; Am Asn Physics; Sigma Xi; AAAS; Am Soc Eng Educ; Am Geophys Union. *Res:* Relativistic electrodynamics; differential geometry; general relativity; antennas and practical radiating systems; communications; wave propagation; satellite communications; electronic location finder. *Mailing Add:* 2982 Barclay Sq Columbus OH 43209

CORWIN, GILBERT, b Dayton, Ohio, Feb 10, 21; m 49, 77; c 2. GEOLOGY & RESOURCES, WESTERN PACIFIC ISLANDS. *Educ:* Harvard Col, SB, 43; Univ Minn, PhD(petrol), 52. *Prof Exp:* Field lab geologist, US Geol Surv, 46-55, admin chief geologist, Pac Geol Surv Sect, 55-59, res geologist, Geol Div, 59-63 & staff geologist, 63-82; CONSULT GEOLOGIST, JOHNSTON ASSOC, 82- *Concurrent Pos:* Chmn, Geol Panel, US Geol Surv Bd Expert Examrs for US Civil Serv Comn, 62-68; mem, interagency comt oceanog & other various comts, 63-66; chief, off Marine Geol & Hydrol, US Geol Surv, 64-66; vchmn, Marine Geol Panel, US-Japan Marine Resources & Eng Comt, 69-82; staff rep, interagency comt Marine Sci & eng, Dept Interior, 72-77, staff observer, Nat Adv Comt Oceans & Atmosphere, 73-77; asst to exec secy, Int Geol Correlation Prog, US Nat Comt, 72-82. *Mem:* fel Geol Soc Am; AAAS; Am Geophys Union; Marine Technol Soc. *Res:* Field geologic mapping; western Pacific islands; petrologic studies of volcanic rock samples from the islands. *Mailing Add:* 2362 N Nelson St Arlington VA 22207-5152

CORWIN, H(AROLD) E(ARL), b New York, NY, Feb 8, 19; m 49; c 3. CHEMICAL ENGINEERING. *Educ:* City Col New York, BChE, 40; Syracuse Univ, MChE, 51, PhD(chem eng), 52; Univ Conn, MBA, 72. *Prof Exp:* Prin engr, draftsman & jr naval architect, Philadelphia Navy Yard, 42-44; designer & chem engr, Solvay Process Div, Allied Chem Co, 46-49; prin chem engr, Battelle Mem Inst, 51-53; proj coordr & lectr chem eng, Syracuse Univ, 54-56; dir process & qual control, Molded-Packaging Div, 56-63, asst dir res, 63-69, VPRES & TECH DIR, FIBER PROD DIV, DIAMOND INT CORP, 69- *Mem:* Am Chem Soc; Am Soc Qual Control; Am Inst Chem Engrs; Tech Asn Pulp & Paper Indust. *Res:* Pulp and paper; packaging; deinking; resinous cements; color; process development and instrumentation; quality control. *Mailing Add:* 1392 Riverbank Rd Stamford CT 06903-2017

CORWIN, HAROLD G, JR, b Pasadena, Calif, June 5, 43; m 77. ASTRONOMICAL CATALOGUING, EXTRAGALACTIC ASTRONOMY. *Educ:* Univ Kans, BA, 65, MA, 67; Univ Edinburgh, UK PhD(astron), 81. *Prof Exp:* Res scientist assoc, Dept Astron, Univ Tex, Austin, 71-76, res scientist, McDonald Observ, 81-91; res assoc, Royal Observ, Edinburgh, 76-81; SCIENCE SUPPORT STAFF, IPAC, CALIF INST TECHNOL, 91- *Concurrent Pos:* Consult, McDonald Observ, Univ Tex, Austin, 76-81, lectr, Dept Astron, 81-91; consult, Dept Astron, Univ Calif, Los Angeles, 77-80, Willmann-Bell, Inc, 88- *Mem:* Int Astron Union; Am Astron Soc; Royal Astron Soc; Astron Soc Pac; Sigma Xi. *Res:* Galaxy catalogues, printed and on-line; galaxy photometry and spectroscopy; large scale structure of the universe; night sky photometry; photometric standard stars; history of observational cosmology. *Mailing Add:* Ipac MS 100-22 Calif Inst Technol Pasadena CA 91125

CORWIN, HARRY O, b Los Angeles, Calif, Apr 30, 38; m 61; c 1. GENETICS. *Educ:* Univ Calif, Santa Barbara, BA, 61; Univ Calif, Los Angeles, MA, 62, PhD(zool), 66. *Prof Exp:* Teaching asst biol, embryol, anat & genetics, Univ Calif, Los Angeles, 62-66; asst prof genetics, 66-71, assoc prof biol, 71-80, PROF BIOL, UNIV PITTSBURGH, 80- *Mem:* AAAS. *Res:* Genetic analysis of mutagens effect on Drosophila melanogaster. *Mailing Add:* Dept Biol Sci A-234 Langley Hall Univ Pittsburgh 4200 Fifth Ave Pittsburgh PA 15260

CORWIN, JEFFREY TODD, b Riverhead, NY, Oct 15, 51. NEUROBIOLOGY, SENSORY PHYSIOLOGY. *Educ:* Cornell Univ, BS, 73; Univ Hawaii, MS, 75; Univ Calif, San Diego, PHD(neurosci), 80. *Prof Exp:* Vis fel neurophysiol, Plymouth Lab, Marine Biol Asn UK, Plymouth, Eng, 80-81; from asst prof to assoc prof zool, Univ Hawaii, 81-88; ASSOC PROF OTOLARYNGOL & NEUROSCI, UNIV VA SCH MED, 88- *Concurrent Pos:* Mem, Bekesy Lab Neurobiol & Pac Biomed Res Ctr, Univ Hawaii, 81-88; Grass Found fel neurophysics, Marine Biol Lab, Woods Hole, Mass, 83, assoc dir Grass Found Fel Prog, 84 & 85; NIH res career develop award, 86-91. *Mem:* Soc Neurosci; Am Soc Cell Biol; Am Soc Zoologists; Asn Res Otlaryngol. *Res:* Comparative neurophysiology and neuroanatomy, especially regeneration in the auditory system; sensory biology of fishes and amphibians; development of auditory and lateral line sensory systems. *Mailing Add:* Dept Otolaryngol Univ Va Sch Med Charlottesville VA 22908

CORWIN, LAURENCE MARTIN, b Rochester, NY, Aug 26, 29; m 52; c 3. BIOCHEMISTRY. *Educ:* Univ Chicago, PhB, 48; Syracuse Univ, BA, 50; Wayne State Univ, PhD(biochem), 56. *Prof Exp:* Res fel biol, Calif Inst Technol, 57-58; biochemist, NIH, 58-61; biochemist, Walter Reed Army Inst Res, 61-67; assoc prof, 67-73, PROF MICROBIOL & NUTRIT SCI, SCH MED, BOSTON UNIV, 73- *Mem:* Am Soc Biol Chemists; Am Inst Nutrit; Am Soc Microbiol. *Res:* Biochemical function of vitamin E; biochemical genetics in microorganisms; effect of vitamin E in immunology and cancer. *Mailing Add:* Dept Microbiol Boston Univ Sch Med Boston MA 02118

CORWIN, LAWRENCE JAY, b E Orange, NJ, Jan 20, 43; m 78; c 1. MATHEMATICAL ANALYSIS. *Educ:* Harvard Univ, BA, 64, AM, 65, PhD(math), 68. *Prof Exp:* C L E Moore Instr, Mass Inst Technol, 68-70; vis mem, Courant Inst Math Sci, NY Univ, 70-71; asst prof, Yale Univ, 71-75; assoc prof, 75-80, PROF MATH,RUTGERS UNIV, NEW BRUNSWICK, 80-, ASSOC PROVOST MATH & PHYS SCI, 88- *Concurrent Pos:* Sloan Found fel, 72-74; mem, Inst Advance Sci Stud, 74, 78. *Mem:* Am Math Soc; Math Asn Am. *Res:* Harmonic analysis on nilpotent lie groups; representations of p-adic groups. *Mailing Add:* 117 N Fourth Ave Highland Park NJ 08904

CORWIN, THOMAS LEWIS, b Newburgh, NY, Oct 9, 47; m 71; c 1. MATHEMATICAL STATISTICS. *Educ:* Villanova Univ, BS, 69; Princeton Univ, MS, 71, PhD(statist), 73. *Prof Exp:* Sr assoc statist, Daniel H Wagner Assoc, 80-84; PRES, METRON INC, 84- *Mem:* Inst Math Statist; Am Statist Asn; Soc Appl & Indust Math. *Res:* Development of analytical tools to perform real-time, operational analysis of naval search problems; developing Bayesian information processing techniques for use in search problems. *Mailing Add:* Metron Inc 11911 Freedom Dr Suite 800 Reston VA 22090

CORY, JOSEPH G, b Tampa, Fla, Jan 27, 37; m 63; c 3. BIOCHEMISTRY. *Educ:* Univ Tampa, BS, 58; Fla State Univ, PhD(chem), 63. *Prof Exp:* Fel chem, Fla State Univ, 63; fel biochem, Albert Einstein Med Ctr, 63-64, asst mem, 64-65; asst prof, Fla State Univ, 65-69; assoc prof biochem, 69-73, prof med microbiol, 73-74, prof biochem & chmn dept, Col Med, 74-86, PROF, DEPT INTERNAL MED, UNIV SFLA, 86- *Concurrent Pos:* Research Career Development Award, USPHS, 69-73. *Mem:* Am Chem Soc; Am Soc Biol Chem; Am Asn Cancer Res; Soc Exp Biol & Med. *Res:* Enzymology of nucleotide interconversions and nucleic acid synthesis; cancer biochemistry. *Mailing Add:* Dept Biochem East Carolina Univ Sch Med Brody Med Sci Bldg Greenville NC 27858

CORY, MICHAEL, b New York, NY, Dec 5, 41; m 66; c 2. MEDICINAL CHEMISTRY. *Educ:* San Jose State Col, BS, 64; Univ Calif, Santa Barbara, PhD(org chem), 71. *Prof Exp:* Chemist org chem, Stanford Res Inst, 64-68 & 71-77; res scientist org chem, 77-85, group leader, 85-88, ASST DIV DIR, WELLCOME RES LABS, BURROUGHS WELLCOME CO, 88- *Mem:* Am Chem Soc; Am Asn Cancer Res. *Res:* Drug design; enzyme inhibitor design; interaction of drugs with macromolecules. *Mailing Add:* 55 Cedar St Chapel Hill NC 27514

CORY, ROBERT MACKENZIE, b Washington, DC, Feb 22, 43; m 66; c 1. ORGANIC, SYNTHETIC & NATURAL PRODUCTS CHEMISTRY. *Educ:* Harvey Mudd Col, BS, 65; Univ Wis, PhD(org chem), 71. *Prof Exp:* Fel, Univ Colo, 71-72 & Rice Univ, 72-73; asst prof chem, Ohio Univ, 73-74; asst prof, 74-81, ASSOC PROF CHEM, UNIV WESTERN ONT, 81- *Mem:* Am Chem Soc; Chem Inst Can; Royal Soc Chem; Int Union Pure & Appl Chem. *Res:* Conjugated macrocyclic systems including cyclacenes and cyclic polyparaphenylenes, including cyclophanes, molecular recognition, molecular electronics. *Mailing Add:* Dept Chem Univ Western Ont London ON N6A 5B7 Can

CORY, WILLIAM EUGENE, b Dallas, Tex, Apr 5, 27; m 47; c 2. ELECTRICAL ENGINEERING. *Educ:* Tex A&M Univ, BS, 50; Univ Calif, Los Angeles, MS, 59. *Prof Exp:* Elec engr, USAF Security Serv, 50-51 & 52-57; electronic systs engr, Lockheed Aircraft Corp, 57-59, aircraft develop engr, 59; sr res engr, Southwest Res Inst, 59-61, mgr commun, 61-65, dir, Electronic Systs Res Dept, 65-72, vpres, Electronic Systs & Geosci Div, 72-89; PRES, CORY CONSULT, 89- *Honors & Awards:* Lawrence G Cumming Award, Electromagnetic Compatibility Soc, Inst Elec & Electronics Engrs, 83. *Mem:* Fel Inst Elec & Electronics Engrs; Nat Soc Prof Engrs; Bioelectromagnetics Soc. *Res:* Electromagnetic compatibility, interference and propagation; signal detection, recognition, extraction, processing, storage and retrieval; communications, navigation and identification systems engineering applied to defense, urban and ocean problems; research and development management. *Mailing Add:* Cory Consult 4135 High Sierra San Antonio TX 78228

CORYELL, MARGARET E, b Great Falls, Mont, Jan 25, 13; m 41; c 3. BIOLOGICAL CHEMISTRY. *Educ:* Univ Colo, AB, 35; Univ Mich, MA, 37, PhD, 41. *Prof Exp:* Asst, Childrens Fund Mich, 40-48; instr cell biol, Med Col Ga, 56-76, asst res prof, Cell & Molecular Biol, 76-80; RETIRED. *Mem:* Sigma Xi; Am Chem Soc. *Res:* Biochemical defects in mentally retarded children. *Mailing Add:* Metab Eval-Res Lab Gracewood State Sch & Hosp Gracewood GA 30812

CORY-SLECHTA, DEBORAH ANN, b St Paul, Minn, Jan 31, 50; m 75; c 2. BEHAVIORAL TOXICOLOGY & PHARMACOLOGY. *Educ:* Western Mich Univ, BS, 71, MA, 72; Univ Minn, PhD(psychol), 77. *Prof Exp:* Jr staff fel, Nat Ctr Toxicol Res, 77-79; res fel, 79-82, res assoc, 82-84, ASST PROF, SCH MED & DENT, UNIV ROCHESTER, 84- *Concurrent Pos:* Mem, Environ Health Sci Rev Comt, Nat Inst Environ Health, 85- *Mem:* Soc Toxicol; Am Psychol Asn; Behav Toxicol Soc (secy-treas, 83-); Behav Pharmacol Soc; Behav Teratology Soc; Asn Behav Anal. *Res:* Determination of the relationships between the behavioral and biological effects produced by chronic low-level exposure to heavy metals, especially lead, in young and aged organisms. *Mailing Add:* Sch Med & Dent Univ Rochester PO Box RBB Rochester NY 14642

COSBY, LYNWOOD ANTHONY, b Richmond, Va, June 11, 28; m 51; c 7. ELECTRONICS. *Educ:* Univ Richmond, BS, 49; Va Polytech Inst, MS, 51. *Prof Exp:* Instr physics, Va Mil Inst, 50-51; from physicist to br head advan tech, US Naval Res Lab, 51-71, head, div electronic warfare, 71-, chmn, Naval Electronic Warfare Adv Group, 73-; AT TELEDYNE MEC. *Concurrent Pos:* Mem threat environ steering comt, Airborne Warning & Control Syst, 75- *Mem:* Fel Inst Elec & Electronics Engrs; Sigma Xi; Am Soc Naval Engrs; Asn Old Crows (pres, 81-). *Res:* Administrative and technical leadership for advancement of technology and systems for all Navy electronic warfare applications; development of circuits, electron devices, antennas and the design of special research development technical and engineering facilities. *Mailing Add:* 2111 Wilson Blvd No 1100 Arlington VA 22201-3068

COSCARELLI, WALDIMERO, b Brooklyn, NY, Nov 5, 26; m 48; c 3. MICROBIOLOGY, BIOCHEMISTRY. *Educ:* Wagner Col, BS, 56; Rutgers Univ, PhD(microbiol), 61. *Prof Exp:* Bacteriologist, Chase Chem Co, NJ, 56; microbiologist, Schering Corp, 56-58; mem tech staff microbiol deterioration, Bell Tel Labs, Inc, 61-68; sr scientist, 68-69, mgr microbiol, Div Cent Res, Shulton, Inc, 69-73, mgr, Clin Res Serv, 73-76, assoc dir prod safety and govt regulatory affairs, 76-77, dir prod & regulatory affairs, consumer prod div, Shulton Inc, 77-79,; DIR PROD & RES DIV, AM CYANAMID CO, 79- *Mem:* Am Soc Microbiol; Soc Cosmetic Chemists; NY Acad Sci. *Res:* Effects of microorganisms on plastics, including bacteria and fungi; isolation of organisms to study their metabolic requirements from plastics; transformations of steroid compound by microorganisms. *Mailing Add:* 245 Lake Rd Basking Ridge NJ 07900

COSCIA, ANTHONY THOMAS, b New York, NY, Nov 1, 28; m 51; c 4. POLYMER CHEMISTRY, ENVIRONMENTAL SCIENCES. *Educ:* Fordham Col, BS, 50; NY Univ, PhD(chem), 56. *Prof Exp:* Asst, NY Univ, 50-52; res chemist, Fleischmann Labs, 52-55; res chemist, group leader org & polymer chem, 60-66, mgr org & polymer res, Indust Chem Div, 66-71, mgr flocculants, Indust Chem & Plastics Div, 71-72, mgr polymer res, Chem Res Div, 72-80, consult technol, Indust Prod Res, 80-83, GROUP LEADER, TEXTILE CHEM, AM CYANAMID SOC, 83- *Mem:* Am Chem Soc. *Res:* Polysaccharide isolation and identification; synthesis of nitrogen heterocycles; synthesis and polymerization of new monomers; water soluble polymers; resins and coatings for paper; new product research on flocculants, water treating chemicals, mining, enhanced oil recovery polymers and textile chemicals. *Mailing Add:* 17 Woodchuck Ct Norwalk CT 06854-3308

COSCIA, CARMINE JAMES, b New York, NY, July 26, 35; m 56; c 3. BIOCHEMISTRY. *Educ:* Manhattan Col, BS, 57; Fordham Univ, MS, 60, PhD(org chem), 62. *Prof Exp:* NATO res fel org chem, Swiss Fed Inst Technol, 62-63, USPHS fel, 63-64; res fel biochem, Univ Pittsburgh, 64-65; from asst prof to assoc prof, 65-73, PROF BIOCHEM, SCH MED, ST LOUIS UNIV, 73- *Concurrent Pos:* Vis prof, Unit Molecular Biol, Med Res Coun, Cambridge, Eng, 85-86. *Mem:* Am Soc Biol Chemists; Am Chem Soc. *Res:* Secondary metabolism in plants and animals. *Mailing Add:* Dept Biochem St Louis Univ Sch Med 1402 S Grand Blvd St Louis MO 63104-1082

COSCINA, DONALD VICTOR, b New Britain, Conn, Oct 11, 43; m 66; c 2. BIOPSYCHOLOGY, NEUROCHEMISTRY. *Educ:* Univ Vt, BA, 65; Bucknell Univ, MA, 67; Univ Chicago, PhD(biopsychol), 71. *Prof Exp:* Vis scientist neurochem, 70-71, jr res scientist, 71-73, sr res scientist, 73-77, HEAD DEPT BIOPSYCHOL, CLARKE INST PSYCHIAT, 77-; PROF PSYCHIAT & PSYCHOL, UNIV TORONTO, 86- *Concurrent Pos:* Ed, J Pharmacol, Biochem & Behav, 73-, Neurosci & Biobehav Rev, 77- & J Nutrit & Behav, 81-87; asst prof psychiat, Univ Toronto, 75-80; assoc prof psychol, Univ Toronto, 77-86, assoc prof, 80-86; grant reviewer, Nat Inst Ment Health, NIH, 82-84; res assoc, Toronto Gen Hosp, 85- *Honors & Awards:* Res Award, Clarke Inst Psychiat, 74. *Mem:* Can Asn Neurosci; Can Col Neuropsychopharmacol; Soc Neurosci; NAm Asn Study Obesity; Soc Study Ingestion Behav. *Res:* Determination of neuroanatomical and neurochemical controls in the brain of food intake and body weight regulation, particularly in anorexia nervosa and obesity. *Mailing Add:* Sect Biopsychol Clarke Inst Psychiat 250 College St Toronto ON M5T 1R8 Can

COSENS, KENNETH W, b Fairgrove, Mich, June 22, 15; m 36; c 4. SANITARY & ENVIRONMENTAL ENGINEERING. *Educ:* Mich State Univ, BS, 38, MS, 46, CE, 48. *Prof Exp:* Asst city engr, Pontiac, Mich, 38-41; from instr to asst prof civil eng, Mich State Univ, 41-47; aircraft struct engr, Fisher Body Div, Gen Motors Corp, 44-45, struct engr, Olds-Mobile Div, 45; asst prof civil eng, Univ Tex, 47-49; assoc prof civil & sanit eng, Ohio State Univ, 49-70, acting chmn dept, 53-55; sanit engr, Alden E Stilson & Assoc, 70-74, head pub works div & partner, 74-80; supply adminr, Div Water, Columbus, Ohio, 81-84; RETIRED. *Concurrent Pos:* Consult, 43-64; Dipl, Am Sanit Eng Intersoc Bd, 56. *Honors & Awards:* George W Fuller Award, Am Water Works Asn, 80. *Mem:* Fel Am Soc Civil Engrs; Nat Soc Prof Engrs; Am Water Works Asn; Water Pollution Control Fedn. *Res:* Water quality, especially taste and odor studies; waste water quality including aerobic and anaerobic disposal methods for small installations; garbage disposal with household grinders; lime recovery from water softening sludge. *Mailing Add:* 2620 Chester Rd Columbus OH 43221

COSGAREA, ANDREW, JR, b Highland Park, Mich, Aug 8, 34; m 58; c 3. METALLURGICAL & CHEMICAL ENGINEERING. *Educ:* Univ Mich, BSChE, 55, MSNucE, 56, PhD(chem, metall eng), 59. *Prof Exp:* Res asst liquid metals, Res Inst, Univ Mich, 56-59; from asst prof to assoc prof eng, Univ Okla, 59-65, dir nuclear reactor lab, 60-64; assoc chem engr & group leader, Chem Eng Div, Argonne Nat Lab, 65-66; sect chief, Avco Systs Div, 66-68, proj mgr, 68-69, prin consult scientist, 69-72; DIR, RES & DEVELOP, FRANKLIN MINT CORP, 72- *Concurrent Pos:* NSF & Mil Serv res grants, 60-72; mem bd dirs & pres, Int Precious Metals Inst, 76- *Mem:* AAAS; Am Inst Metall Engrs; Audio Eng Soc. *Res:* Physical and chemical properties of precious metals; thermal and chemical response of systems; thermodynamics; energetics. *Mailing Add:* Five Goodman St Rochester NY 14607

COSGRIFF, THOMAS MICHAEL, b Minneapolis, Minn, June 5, 45; m 82. HEMATOLOGY, MEDICAL ONCOLOGY. *Educ:* Univ Minn, BA, 66, BS, 67, MD, 70. *Prof Exp:* Intern internal med, State Univ NY, Buffalo, 70-71; ward officer, US Army Drug Treatment Ctr, Long Binh, Vietnam, 71-72; chief, Outpatient Clin, McDonald Army Hosp, Ft Eustis, Va, 72-73; resident internal med, Brooke Army Med Ctr, Ft Sam Houston, Tex, 73-75; chief, Dept Med, Patterson Army Hosp, Ft Monmouth, NJ, 75-76; fel hemat-oncol, Univ Utah Med Ctr, 76-78; clin investr, Div Exp Therapeut, Walter Reed Army Inst Res, 78-89; CHIEF, HEMAT-ONCOL SERV, DEPT MED, FITZSIMONS ARMY MED CTR, 89-; ASSOC CLIN PROF MED, UNIV COLO HEALTH SCI CTR, 90- *Concurrent Pos:* Dir, Coagulation Lab, US Army Med Res Inst Infectious Dis, 78-89, chief, Med Div, 85-89; staff physician, Dept Med, Walter Reed Army Med Ctr, 78-89; clin asst prof med, Uniformed Serv Univ Health Sci, 80-87, clin assoc prof, 87-89. *Mem:* Fel Am Col Physicians; Am Soc Hemat; Am Soc Clin Oncol; AAAS; Am Soc Trop Med & Hyg; Int Soc Thrombosis & Haemostasis. *Res:* Development of new antimalarials; resistance to antimalarials, mechanisms of hemorrhage in viral infections; pathogenetic mechanisms in hemorrhagic fever with renal syndrome; chemotherapy of cancer. *Mailing Add:* Hemat-Oncol Serv Dept Med Fitzsimons Army Med Ctr Aurora CO 80045-5001

COSGROVE, CLIFFORD JAMES, b Torrington, Conn, Jan 31, 27; m 52; c 3. DAIRY SCIENCE, FOOD SCIENCE. *Educ:* Univ Conn, BS, 51; Southern Conn State Col, BS, 53; Univ RI, MS, 57. *Prof Exp:* Assoc prof animal sci, 53-76, PROF FOOD SCI & NUTRIT, UNIV RI, 76- *Concurrent Pos:* Expert dairy technol, Food & Agr Orgn of UN, 76-77, Chile & Bolivia, 78-79. *Res:* Food technology, especially dairy products and their substitution by imitations or synthetics. *Mailing Add:* Dept Food Sci & Nutrit 15 Woodward Hall Univ RI Kingston RI 02881-0804

COSGROVE, DANIEL JOSEPH, b Westover AFB, Mass, Sept 15, 52; m 79. WATER RELATIONS OF PLANTS, CELL WALL MECHANICS & BIOCHEMISTRY. *Educ:* Univ Mass, BA, 74; Stanford Univ, PhD(biol sci), 80. *Prof Exp:* Vis res scientist membrane res, Kernforschungsanlage, Jülich, Germany, 80-81; res assoc bot, Univ Wash, 81-83; asst prof, 83-87, ASSOC PROF BIOL, PA STATE UNIV, 87- *Concurrent Pos:* J S Guggenheim fel, 89. *Honors & Awards:* Presidential Young Investr Award, NSF, 84; McKnight Found Award, 86. *Mem:* AAAS; Am Soc Plant Physiologists; Am Soc Gravitational & Space Biol; Am Soc Photobiol; Am Inst Biol Scis. *Res:* Biophysical and cellular control of plant cell growth; investigation of how mechanical and biochemical properties of plant cell walls regulate growth and how water is absorbed by expanding cells. *Mailing Add:* 208 Mueller Lab Pa State Univ University Park PA 16802

COSGROVE, GERALD EDWARD, b Dubuque, Iowa, July 13, 20; m 43; c 7. PATHOLOGY. *Educ:* Univ Notre Dame, BS, 45; Univ Mich, MD, 44; Am Bd Path, dipl, 50. *Prof Exp:* Hosp pathologist, Klamath Falls, Ore, 50-52; hosp pathologist, Rapid City, SDak, 52-55; hosp pathologist, Gorgas Hosp, Ancon, CZ, 55-57; biologist, Biol Div, Oak Ridge Nat Lab, 57-77; pathologist, Zool Soc San Diego 77-; DIR, SILVERWOOD WILDLIFE SANCTUARY, SAN DIEGO, 85- *Mem:* Radiation Res Soc; Am Soc Exp Path; Am Soc Parasitol; Am Soc Ichthyologists & Herpetologists; Wildlife Dis Asn. *Res:* Comparative pathology; parasitology; delayed effects of radiation; relation of organisms to their environment. *Mailing Add:* Zool Soc San Diego PO Box 551 San Diego CA 92112

COSGROVE, JAMES FRANCIS, b Bridgeport, Conn, Aug 3, 29; m 53; c 8. MATERIALS SCIENCE, ANALYTICAL CHEMISTRY. *Educ:* Col of Holy Cross, BS, 51, MS, 52. *Prof Exp:* Res asst chem, Col of Holy Cross, 51-52; res chemist, Sylvania Elec Prod, NY, 52-60; sect head radiochem, 60-68, mgr mat anal, 68-72, res mgr mat eval, 72-78, DIR PRECISION MAT TECHNOL LAB, GEN TEL & ELECTRONICS LABS, INC, 78- *Mem:* Am Chem Soc; AAAS; Sigma Xi. *Res:* Development and application of trace methods of analysis; development and engineering in luminescent materials, ceramics, refractory and hard metals, precious metals; development of process monitoring techniques. *Mailing Add:* Vpres-Eng Lighting Products Div 100 Endicott St Danvers MA 01923

COSGROVE, STANLEY LEONARD, b Dorset, Eng, Mar 10, 26; US citizen; m 51; c 5. ORGANIC CHEMISTRY, ENGINEERING. *Educ:* Oxford Univ, BA, 48, BSc, 49, DPhil(org chem), 50. *Prof Exp:* Fel org chem, Univ Chicago, 50-51; res technician petrol chem, Shell Oil Co, 51-55; asst div chief org chem, Battelle Mem Inst, 55-61; res mgr appl phys, Columbia Broadcasting Syst, 61-62; from asst dean to assoc dean eng, 76-85, assoc prof chem eng, 62-85, EMER ASSOC DEAN ENG, UNIV CINCINNATI, 85- *Concurrent Pos:* Vis prof, McGill Univ, 69-70; consult, Nuclear Eng Co, 73-76. *Mem:* Am Chem Soc; Am Soc Eng Educ. *Res:* Applied organic chemistry; polymer chemistry; radioactive waste and toxic waste immobilization. *Mailing Add:* 3853 Middleton Ave Cincinnati OH 45220

COSGROVE, WILLIAM BURNHAM, b New York, NY, June 11, 20; m 49; c 2. CELL BIOLOGY, COMPARATIVE PHYSIOLOGY. *Educ:* Cornell Univ, AB, 41; NY Univ, MS, 47, PhD(zool), 49. *Prof Exp:* Assoc zool, Univ Iowa, 49-51, from asst prof to assoc prof, 51-57; from assoc prof to prof, Univ Ga, 57-69, head dept, 64-72, alumni found prof, 69-90, EMER PROF ZOOL, UNIV GA, 90- *Honors & Awards:* Michael Award, 67. *Mem:* AAAS; Am Soc Cell Biol; Am Soc Zool; Soc Protozool (pres, 70-71). *Res:* Comparative physiology of respiratory pigments; cell biology of trypanosomatids; physiology of protozoa; membrane transport. *Mailing Add:* 537 Fearrington Post Pittsboro NC 27312

COSMAN, BARD CLIFFORD, b New York, NY, Mar 1, 63; m 89. GENERAL SURGERY. *Educ:* Harvard Univ, AB, 83; Columbia Univ, MPH & MD, 87. *Prof Exp:* RESIDENT GEN SURG, STANFORD UNIV HOSP, 87- *Concurrent Pos:* Staff physician, Spinal Cord Injury Serv, Palo Alto Vet Admin Med Ctr, 89-91; postdoctoral fel surg, Dept Surg, Stanford Univ, 89-91. *Res:* Gastrointestinal complications of chronic spinal cord injury. *Mailing Add:* Dept Surg Rm S-067 Stanford Univ 300 Pasteur Dr Stanford CA 94305-5101

COSMAN, DAVID JOHN, b Notingham, Eng, Apr 4, 54; British citizen. MOLECULAR BIOLOGY, MAMMALIAN EXPRESSION VECTORS. *Educ:* Cambridge Univ, BA, 76; Pa State, PhD(microbiol), 80. *Prof Exp:* Fogarty fel, Lab Molecular Virology, 80-82; staff sci, 83-84, VPRES & DIR, MOLECULAR BIOL & SR STAFF SCIENTIST, IMMUNEX RES & DEVELOP CORP, 84- *Mem:* AAAS. *Res:* Molecular biology of cytokine and cytokine receptors involved in immune regulation, including cloning recombinant express and mutagenesis to study structure/function relationships. *Mailing Add:* Immunex Corp 51 University St Seattle WA 98101

COSMIDES, GEORGE JAMES, b Pittsburgh, Pa, July 23, 26; m 48; c 1. TOXICOLOGY, ENVIRONMENTAL CHEMICALS. *Educ:* Univ Pittsburgh, BS, 52; Purdue Univ, MS, 54, PhD(pharmacol), 56. *Prof Exp:* Pharmacist, Pittsburgh, 52 & Purdue Univ, 52-54; sr scientist, Smith, Kline & French Labs, 56-57; asst prof pharmacol, Univ RI, 57-59; sr res pharmacologist, Psychopharmacol Serv Ctr, NIMH, 59-63, prog adminr, Pharmacol & Behav Sci Training Prog, Nat Inst Gen Med Sci, 63-64, exec secy pharmacol-toxicol prog comt, 64-74, dir, pharmacol-toxicol prog, pharmacol res assoc training prog & mem, pharmacol res assoc traing comt, NIH, 64-74, DEP DIR TOXICOL INFO PROG & DEP ASSOC DIR SPECIALIZED INFO SERV, NAT LIBR MED, NIH, 74- *Concurrent Pos:* Consult, Sci Criminal Invest, RI, 57-59; mem reference panel, Am Hosp Formulary Serv, 63-76; adj prof pharmacol, Univ Pittsburgh, 63-73; mem comt probs of drug safety, Nat Acad Sci-Nat Res Coun, 64-69 & US Nat Comt for CODATA, Numerical Data Adv Bd, 77-; consult pharmacol & toxicol, WHO, 65; consult biomed commun for radio & TV, NIH, 70-74, Surgeon Gen, USPHS, 71-, mem rev panel disposal lethal chem. *Honors & Awards:* Vavro Mem Award & Bristol Award, 51-52; Distinguished Scientist Award Pharmaceut Sci, AAAS, 71. *Mem:* AAAS; Am Soc Info Sci; Am Soc Pharmacol & Exp Therapeut; Environ Mutagen Soc; Soc Toxicol; Int Union Pharmacol, 77- *Res:* Psychopharmacology; effects of drugs on human behavior; drug metabolism; environmental chemicals; environmental toxicology; toxicology information systems; computer-based data retrieval systems; drug use, misuse and abuse; drug safety and efficacy; rational pharmacotherapy. *Mailing Add:* 639 Crocus Dr Rockville MD 20850-2046

COSMOS, ETHEL, b Springfield, Mass, Aug 1, 23. DEVELOPMENTAL NEUROBIOLOGY. *Educ:* Univ Rochester, PhD(physiol), 57. *Hon Degrees:* DSc, Am Int Col, 85. *Prof Exp:* PROF NEUROSCI, HEALTH SCI CTR, MCMASTER UNIV, 74- *Mailing Add:* Dept Biomed Sci Health Sci Ctr McMaster Univ 1200 Main St W Hamilton ON L8N 3Z5 Can

COSPER, DAVID RUSSELL, b Ypsilanti, Mich, Oct 3, 42; m 65; c 3. ORGANIC CHEMISTRY, POLYMER CHEMISTRY. *Educ:* Purdue Univ, BS, 64; Univ Wis, Madison, PhD(chem), 69. *Prof Exp:* Chemist, 69-73, group leader, 73-80, res scientist, 80-84, RES ASSOC, NALCO CHEM CO, 84- *Mem:* Am Chem Soc; Sigma Xi; Tech Asn Pulp & Paper Indust; Am Ceramic Soc. *Res:* Synthesis of organic compounds and polymers for water treatment applications; design of chemical additives for pulp and paper manufacturing. *Mailing Add:* 6824 Valley View Dr Downers Grove IL 60516

COSPER, PAULA, b Birmingham, Ala, Mar 4, 47; m. MEDICAL GENETICS. *Educ:* Birmingham-Southern Col, BS, 69; Univ Ala, Birmingham, MS, 72, PhD(physiol biophys), 74. *Prof Exp:* Fel med genetics, 74-75, res assoc, 75-76, instr, 76-78, asst prof physiol & med genetics, 78-83, asst prof genetics, 83-85, ASSOC PROF GENETICS, DEPT OBSTET-GYNEC, UNIV ALA, BIRMINGHAM, 85- *Mem:* Am Soc Human Genetics; Tissue Cult Asn; Sigma Xi. *Res:* Human cytogenetics; drug effects on human chromosomes; prenatal cytogenetic studies. *Mailing Add:* 3609 Rockhill Rd Birmingham AL 35223

COSPER, SAMMIE WAYNE, b Greggton, Tex, Oct 8, 33; m 54; c 3. NUCLEAR PHYSICS. *Educ:* Univ of Southwestern La, BS, 60; Purdue Univ, PhD(nuclear physics), 65. *Prof Exp:* Res appointee nuclear chem, Lawrence Radiation Lab, Univ Calif, 65-67; assoc prof, 67-70, head dept, 67-72, prof physics, 70-72, dean, Col Liberal Arts, 72-74, ACAD VPRES, UNIV SOUTHWESTERN LA, 74- *Mem:* Am Asn Physics Teachers; Am Phys Soc; Sigma Xi. *Res:* Low-energy charged particle nuclear reactions and scattering; charged particle identification techniques; charged particles from spontaneous fission; nuclear physics; instrumentation. *Mailing Add:* Univ Southwestern La PO Box 41810 Lafayette LA 70504

COSPOLICH, JAMES D, b New Orleans, La, Dec 19, 44; m 67; c 3. ELECTRICAL ENGINEERING DESIGN OF PETROCHEMICAL FACILITIES & ONSHORE-OFFSHORE OIL & GAS PRODUCTION FACILITIES. *Educ:* La State Univ, BS, 67, MS, 72. *Prof Exp:* Elec engr, Waldemar S Nelson & Co Inc, 67-74, mgr elec engr, 74-85, asst vpres, 75-80, vpres, 80-85, sr vpres, 85-91, EXEC VPRES, WALDEMAR S NELSON & CO INC, 91- *Mem:* Inst Elec & Electronic Engrs; Nat Soc Prof Engrs; Instrument Soc Am; Gas Processors Asn; Illum Eng Soc; Am Asn Eng Soc Inc; Am Petrol Inst. *Res:* Specialize in application of electrical power, controls, and instrumentation systems in hazardous areas in onshore and offshore production facilities and petro-chemical plants. *Mailing Add:* 61 Rosedown Dr Destrehan LA 70047

COSS, RONALD ALLEN, b Long Beach, Calif, Apr 24, 47; m 79; c 2. CELL BIOLOGY, RADIATION BIOLOGY. *Educ:* Univ Calif, Riverside, BA, 69; Univ Colo, PhD(biol), 74. *Prof Exp:* NIH teaching & res asst, Dept Molecular, Cellular & Develop Biol, Univ Colo, Boulder, 69-74; NIH fel cell biol, Rockefeller Univ, 74-76; fel, Dept Radiol & Radiation Biol, Colo State Univ, 76-81, asst prof, Dept Radiol & Radiation Biol, 81-82; asst prof, dept radiation ther & nuclear med, 82-87, ASSOC PROF, DEPT RADIATION ONCOL & NUCLEAR MED, THOMAS JEFFERSON UNIV, 87- *Mem:* Am Soc Cell Biol; Radiation Res Soc; AAAS; Sigma Xi; Am Radium Soc. *Res:* Mechanisms of mitosis and cytokinesis; cell cycle related changes in the cytoskeleton and plasma membrane; mechanisms of action of hyperthermia and radiosensitization by hyperthermia; cell cycle phase specific proteins. *Mailing Add:* Exp Radiation Oncol Lab-Radiation Oncol & Nuclear Med Thomas Jefferson Univ Tenth & Walnut St Philadelphia PA 19107

COSSACK, ZAFRALLAH TAHA, b Damascus, Syria, Nov 20, 48; US citizen; m 80; c 3. HUMAN & CLINICAL NUTRITION, MEDICAL RESEARCH. *Educ:* Univ Alexandria, Egypt, BSc, 71, MSc, 74; Univ Ariz, PhD(nutrit), 80. *Prof Exp:* Res asst nutrit, Univ Alexandria, Egypt, 72-74; res tech nutrit, Iowa State Univ, 76; from res asst to res assoc nutrit, Univ Ariz, 78-80; res assoc clin nutrit, Wayne State Univ, Mich, 81-84; ASSOC PROF & DIR CLIN NUTRIT, ODENSE UNIV, DENMARK, 84- *Concurrent Pos:* Coordr clin res, Vet Admin Hosp, Mich, 81-84; consult endocrinol, Diabetes & Endocrinol Clin Lab, Mich, 82-84; lect chem, Schoolcraft Col, Mich, 83; sr investr Med Res, Nuclear Reactor & Children Hosp, Delft & Rotterdam, Neth, 84; vis prof, Odense Univ, Denmark, 84-, lectr, 84-, dir, Human & Clin Nutrit, 84-; consult, Nutrit Int, Neth, 85-, Cow & Gate Res Div, Eng, 85- *Mem:* Am Soc Clin Nutrit; Am Inst Nutrit; Am Fedn Clin Res; Am Col Nutrit; Brit Soc Nutrit; Int Soc Trace Elements Res in Humans; Europ Soc Pediat Res; Europ Soc Parenteral & Enteral Nutrit. *Res:* Biochemical, physiological and clinical importance of zinc, copper, iron and selenium in human subjects; nutrition therapy particularly for patients with cancer; role of trace elements in human health and disease. *Mailing Add:* Vet Admin Med Ctr Med Res 151 Allen Park MI 48101

COSSAIRT, JACK DONALD, b Indianapolis, Ind, Oct 11, 48; m 73; c 2. ENVIRONMENTAL HEALTH, ELEMENTARY PARTICLE PHYSICS. *Educ:* Ind Cent Col, BA, 70; Ind Univ, Bloomington, MS, 72, PhD(physics); Am Bd Health Physics, dipl, 82. *Prof Exp:* Res assoc, Cyclotron Inst, Tex A&M Univ, 75-78; assoc scientist, 78-83, appl scientist I, 85-89, APPL SCIENTIST II, FERMI NAT ACCELERATOR LAB, 89- *Concurrent Pos:* Assoc ed, Healthphysics, 86- *Mem:* Am Phys Soc; Health Physics Soc. *Res:* Mass measurements of nuclei far from stability using charged particle transfer reactions: nuclear structure in alpha particle pickup reactions and polariation measurements; accelerator shielding for hadrons and muons; high energy spin physics. *Mailing Add:* Fermi Nat Accelerator Lab PO Box 500 Batavia IL 60510

COSSINS, EDWIN ALBERT, b Romford, Eng, Feb 28, 37; m 62; c 2. PLANT PHYSIOLOGY, BIOCHEMISTRY. *Educ:* Univ London, BSc, 58, PhD(plant biochem), 61, DSc(plant biochem), 81. *Prof Exp:* Res assoc biol sci, Purdue Univ, 61-62; from asst prof to assoc prof, Univ Alta, 62-69, actg head dept, 65-66, coordr, Introd Biol Prog, 74-77, McCalla res prof, 82-83, assoc dean sci, 83-89, PROF BOT, UNIV ALTA, 69- *Concurrent Pos:* Vis prof, Univ Geneva, Switz, 72-73. *Honors & Awards:* Centennial Medal, Govt Can, 67. *Mem:* Am Soc Plant Physiol; Can Soc Plant Physiol (pres, 76-77); Soc Plant Physiologists Japan; fel Royal Soc Can. *Res:* Pteroylglutamates in plants; amino acid biosynthesis; biochemistry of germinating tissues; intermediary metabolism of plant tissues. *Mailing Add:* Dept Bot Univ Alta Edmonton AB T6G 2E9 Can

COST, J(OE) L(EWIS), b Sidney, Ohio, Feb 26, 20; m 43; c 1. CHEMICAL ENGINEERING. *Educ:* Ohio State Univ, BChE, 43; Univ Del, MS, 49. *Prof Exp:* Process engr chem eng, Elliott Co, 44-49; chief process engr low temperature eng, Stacey Bros, 49-53; chief process eng, Air Prods, Inc, 53-57, mgr eng, 57-64, tech dir, Air Prod, Ltd, London, Eng, 64-67, mgr process develop, 67-69, mgr eng chem, 69-76, assoc dir eng, 76-80, TECH ADV CHEM, AIR PROD & CHEM, 80- *Mem:* Am Chem Soc; Am Inst Chem Engrs. *Res:* Engineering of complete low temperature liquefaction and separation facilities; air separation, helium recovery, hydrogen purification and methane; hydrogen and helium liquefaction plants; engineering of company owned chemical plants. *Mailing Add:* 800 Hausman Rd Unit 227 Allentown PA 18104

COST, JAMES R(ICHARD), b Oak Park, Ill, Sept 5, 28; m 53; c 4. PHYSICAL METALLURGY. *Educ:* Univ Wis, BS, 51; Univ Ill, MS, 59, PhD(metall eng), 62. *Prof Exp:* Res metallurgist, Ford Motor Co, 61-65; assoc prof mat sci & metall eng, Purdue Univ, 65-70, prof, 70-78; STAFF MEM, LOS ALAMOS NAT LAB, 78- *Concurrent Pos:* Adj lectr, Univ Mich, 62-63; vis scientist, Centre de Etudes Nucleaires, Grenoble, France, 75-76; staff, US Dept Energy, Off Basic Energy Sci, 76-77. *Mem:* Sigma Xi; Am Inst Mining, Metall & Petrol Engrs; Am Soc Metals; Mat Res Soc. *Res:* Metallic glasses, irradiation effects; composite materials; metal-gas equilibrium; point defects; internal friction; diffusion; permanent magnets. *Mailing Add:* MS-G730 Los Alamos Nat Lab Los Alamos NM 87545

COST, THOMAS LEE, b Birmingham, Ala, Dec 24, 37; m 59; c 2. AEROSPACE ENGINEERING, STRUCTURAL ENGINEERING. *Educ:* Univ Ala, BS, 60, PhD(eng mech), 69; Univ Ill, MS, 62. *Prof Exp:* Res engr, Boeing Co, Wash, 61-62; head appl mech team, Rohm & Haas Co, 62-69; PROF AEROSPACE ENG, UNIV ALA, 69- *Concurrent Pos:* Consult, Army Res Off, 70-, Hercules Inc, 70-, US Army Missile Res & Develop Command, 70- & Olin Co, 77-; pres, Athena Eng Co, 73- *Mem:* Am Inst Aeronaut & Astronaut; Am Acad Mech; Am Soc Eng Educ; Am Soc Civil Eng; Am Soc Mech Eng. *Res:* Structural analysis; finite element methods; viscoelasticity; thermal stress; dynamic structural response. *Mailing Add:* Box 2901 University AL 35486

COSTA, DANIEL LOUIS, b Fall River, Mass, Apr 29, 48; m 75; c 5. TOXICOLOGY, OCCUPATIONAL HEALTH. *Educ:* Providence Col, BS, 70; Rutgers Univ, MS, 73; Harvard Univ, MS, 73, ScD(physiol), 77. *Prof Exp:* Res assoc toxicol, Assoc Univs Inc, 77-79, asst scientist, 79-80, assoc scientist, 80-85; INSTR PHYSIOL, STATE UNIV NY STONY BROOK, 79-; RES PHYSIOLOGIST, ENVIRON PROTECTION AGENCY, 85- *Concurrent Pos:* Consult occup health, 76-; chief toxicol br, Environ Protection Agency, 87- *Mem:* AAAS; Am Thoracic Soc; NY Acad Sci; Sigma Xi; Dipl Am Bd Toxicol. *Res:* Development of animal models of lung disease; assessment of physiological and chemical interaction of gaseous and particulate substances on the lungs of animals. *Mailing Add:* 311 Avalon Ct Chapel Hill NC 27514

COSTA, DANIEL PAUL, b Van Nuys, Calif, Jan 23, 52. ZOOLOGY. *Educ:* Univ Calif, Los Angeles, BA, 74, Santa Cruz, PhD(biol), 78. *Prof Exp:* Physiologist, Scripps Inst Oceanog, San Diego, Calif, 78-; AT LONG MARINE LAB, UNIV CALIF, SANTA CRUZ. *Mem:* AAAS; Ecol Soc Am; Am Soc Mammalogists. *Res:* Physiological ecological of marine mammals and birds with an emphasis on material and energy flux; water balance and reproductive energetics of marine mammals. *Mailing Add:* Long Marine Lab 100 Shaffer Rd Univ Calif Santa Cruz CA 95060

COSTA, ERMINIO, b Cagliari, Italy, Mar 9, 24; nat US; m 50; c 3. NEUROSCIENCE. *Educ:* Univ Cagliari, Sardinia, MD, 47. *Prof Exp:* From asst prof to prof pharmacol, dept pharmacol, Med Sch, Univ Cagliari, Sardinia, 48-56; Fulbright res fel, Dept Physiol & Pharmacol, Chicago Med Sch, 55; med res assoc, Thudichum Psychiat Res Lab, Galesburg, Ill, 56-60; vis scientist, Lab Chem Pharmacol, Nat Heart Inst, NIH, 60-61, dep chief, 60-61, head sect clin pharmacol, 63-65; assoc prof pharmacol, Col Physicians & Surgeons, Columbia Univ, 65-68; chief, Lab Preclin Pharmacol, Div Intramural Progs, NIMH, St Elizabeth's Hosp, 68-85; PROF PHARMACOL & CELL BIOL & DIR, FIDIA-GEORGETOWN INST NEUROSCI, 85- *Concurrent Pos:* Mem sci rev panel, Drug Interactions Eval Prog, Am Pharmaceut Asn, 74-75; prof lectr pharmacol, George Washington Univ, 74-; Wellcome vis prof, 82-83; invited lectr, Distinguished Lect Series Neurosci, Univ Calif, Irvine, 84-85; mem, Nat Adv Bd, World Cong Biol Psychiat, 85; mem, sci comt, Lorenzini Found, 85. *Honors & Awards:* Benet Found Award, 60; Res Achievement Award, Acad Pharmaceut Sci, 75; Gold Medal, Soc Biol Psychiat, 76; Troy C Daniels Lectr, 83; Williams Lectr, Emory Univ, 84. *Mem:* Nat Acad Sci; fel AAAS; Am Soc Pharmacol & Exp Therapeut; Am Acad Neurol; fel Am Col Neuropsychopharmacol (vpres); Am Physiol Soc; NY Acad Sci; Int Brain Res Orgn; Int Soc Psychoneuroendocrinol; Soc Exp Biol & Med; Am Soc Neurochem. *Res:* Antirheumatic drugs; anticoagulants; curari; cholinesterase inhibitors; chemotherapy of cancer; chemotherapy of tuberculosis; glucagon; effect of drugs and physiological control of the turnover rate of neuronal catecholamines, indolealkylamines, acetylcholine and cyclic adenosine monophosphate; mechanism of habituation to amphetamines and morphine; concept of supramolecular organization of transmitter receptors polytypic signaling in synaptic benzodiazepines; allosteric modulation and endogenous allosteric modulators in synaptic transmission; nonanalgesic function of opioid peptides; molecular biological techniques in measuring dynamic state of neuromodulatory peptides; tolerance of benzodiazepines and morphine; receptor-mediated modulation of gene expression; role of early inducible genes in transmitter-mediated DNA transcription; transmitter-mediated regulation of NGF biosynthesis in glial cells; GABA and neuropeptides in the modulation of chromaffin cell function; neuroleptics and opioid peptide interaction; turnover rate of neurotransmitters by stable isotopes. *Mailing Add:* Fidia-Georgetown Inst Neurosci 3900 Reservoir Rd Washington DC 20007

COSTA, JOHN EMIL, b Ithaca, NY, Aug 14, 47; m 73; c 2. GEOMORPHOLOGY, ENVIRONMENTAL GEOLOGY. *Educ:* State Univ NY Col Oneonta, BS, 69; Johns Hopkins Univ, PhD(geog & environ eng), 73. *Prof Exp:* Instr geog, Towson State Col, 71-72; geologist, Md Geol Surv, 70-73; from asst prof to assoc prof geog, Univ Denver, 73-83; res hydrologist, 83-86, CHIEF RES, CASCADES VOLCANO OBSERV, US GEOL SURV, 86- *Concurrent Pos:* Adj prof, Univ Wis, 78, Colo State Univ, 83-86 & Colo Sch Mines, 86; ed, Geol Soc Am Bull, 89- *Mem:* Geol Soc Am; Am Geophys Union; Asn Earth Sci Ed. *Res:* Fluvial processes; geomorphic responses to extreme events. *Mailing Add:* US Geol Surv 5400 MacArthur Blvd Vancouver WA 98661

COSTA, LORENZO F, b Genova, Italy, Feb 10, 31; m 66; c 4. ELECTRONIC SPECTROSCOPY, SPECTRO-ENGINEERING. *Educ:* Birmingham-Southern Col, BS, 61; State Univ NY Buffalo, MS, 68. *Prof Exp:* Cancer scientist, Roswell Park Mem Inst, 61-67; res chemist, 67-74, sr res chemist, 74-77, RES ASSOC, EASTMAN KODAK CO, 77- *Mem:* Soc Appl Spectros; Am Chem Soc; Soc Photog Scientists & Engrs; Am Optical Soc. *Res:* Photophysics, luminescence surface analysis; absolute luminescence methodology; optical spectroscopy; spectrophotometry of optically turbid media; photosensitivity mechanisms; molecular electronics; electromagnetic photosensing devices; solid state physics. *Mailing Add:* 120 Wood Hill Dr Rochester NY 14609

COSTA, MAX, b Cagliari, Italy, Jan 10, 52; US citizen; m 74. PHARMACOLOGY, BIOCHEMISTRY. *Educ:* Georgetown Univ, BS, 74; Univ Ariz, PhD(pharmacol), 76. *Prof Exp:* Res asst met health, NIMH, 70-72; res asst cancer, Nat Cancer Inst, 72-74; res asst, Dept Pharmacol, Univ Ariz, 74-76, res assoc, Dept Radiation Oncol, 76; asst prof, Sch Med, Univ Conn, 77-79; ASST PROF, DEPT MED PHARMACOL & TOXICOL, TEX A&M UNIV, 79- *Concurrent Pos:* Nat Cancer Inst & NIH fels, 77; lectr, Advan Study Inst, NATO, 77; consult, Amax Inc, 78-79. *Mem:* AAAS; Am Soc Pharmacol & Exp Therapeut; Am Soc Biol Chemists; Am Soc Cell Biol. *Res:* Metal carcinogenesis in tissue culture; regulation of ornithine decarboxylase by asparagine and cAMP; differences in enzyme regulation in normal and neoplastic cells. *Mailing Add:* NY Univ Med Ctr Long Meadow Rd Tuxedo NY 10987

COSTA, PAUL T, JR, b Franklin, NH, Sept 16, 42; m 64; c 3. GERONTOLOGY, BEHAVIORAL MEDICINE. *Educ:* Clark Univ, AB, 64; Univ Chicago, MA, 68, PhD(human develop), 70. *Prof Exp:* Chief stress & coping sect, 78-84, CHIEF, LAB PERSONALITY & COGNITION, GERONT RES CTR, NAT INST AGING, NIH, 85-; ASSOC PROF BEHAV BIOL, JOHNS HOPKINS SCH MED, 79- *Concurrent Pos:* Vis assoc prof med psychol, Duke Univ, 77-78, adj assoc prof, Med Ctr, 84-; consult ed, J Geront, 77-84, Int J Aging & Human Develop, 75-, J Aging & Health, 88; adj assoc prof psychol, Univ Md, 79-83, adj prof, 84-; assoc prof, Johns Hopkins Univ, 79-84. *Honors & Awards:* Ardie Lubin Mem lectr, Naval Health Res Ctr, 83. *Mem:* Fel Geront Soc Am; fel Am Psychol Asn; fel Soc Behav Med; Am Pub Health Asn; Am Psychosom Soc. *Res:* Directing and conducting research on the stressors, coping mechanisms and enduring personality dispositions on psychological and health outcomes; assessment of type A and coronary prone behavior; learned illness behavior; personality traits and cardiovascular physiology. *Mailing Add:* Geront Res Ctr Nat Inst Aging 4940 Eastern Ave Baltimore MD 21224

COSTA, RAYMOND LINCOLN, JR, b Baltimore, Md, Feb 12, 48; m 69; c 2. PHYSICAL ANTHROPOLOGY, ANATOMY. *Educ:* Univ Calif, Berkeley, BS, 69; Univ Pa, PhD(anthrop), 77. *Prof Exp:* Instr anthrop, Pa State Univ, 75-76 & Univ Pa, 76; anthropologist, Smithsonian Inst, 77-78; ASST PROF ORAL ANAT, COL DENT, UNIV ILL MED CTR, 78- *Mem:* Am Asn Phys Anthropologists; Am Anthrop Asn; Paleopaleopathol Asn; AAAS. *Res:* Comparative paleopathology of dental and oral diseases in dried skeletal material derived from archeological sites; Hominid and pre-Hominid paleontology. *Mailing Add:* Grad Col Health Sci Ctr Chicago IL 60612

COSTACHE, GEORGE IOAN, b Plolesti, Romania, Apr 28, 43; Can citizen; m 68; c 1. NUMERICAL TECHNIQUES IN ELECTROMAGNETICS, FINITE ELEMENT METHOD. *Educ:* Polytech Inst Bucharest, dipl eng, 66, DSc(electromagnetics), 74. *Prof Exp:* From asst prof to assoc prof elec eng, Polytech Inst Bucharest, 66-76; researcher, Univ Man, 76-77; scientist, Bell-Northern Res, 77-85; assoc prof, 85-88, PROF ELEC ENG, UNIV OTTAWA, 88- *Concurrent Pos:* Adj prof elec eng, Univ Ottawa, 79-85; consult, Bell-Northern Res, Ottawa, 85- *Mem:* Inst Elec & Electronics Engrs Electromagnetic Compatibility Soc; Inst Elec & Electronics Engrs Microwave Tech Soc; Inst Elec & Electronics Engrs Antennas & Propagation Soc; Inst Elec & Electronic Engrs Magnetics Soc. *Res:* Numerical modeling of electromagnetics; electromagnetic interference and compatibility; finite element method; moments method; skin-effect and proximity effect on frequency and time-domain. *Mailing Add:* Elec Eng Dept Univ Ottawa Ottawa ON K1N 6N5 Can

COSTAIN, CECIL CLIFFORD, b Ponoka, Alta, June 16, 22; m 49; c 2. PRIMARY CESIUM CLOCKS, SATELLITE TIME TRANSFER. *Educ:* Univ Sask, BA, 41 & 46, MA, 47; Cambridge Univ, PhD(physics), 51. *Prof Exp:* Res officer microwave spectros, 51-71, head, time sect, 72-84, PRIN RES OFFICER, ELEC & TIME SECT, DIV PHYSICS, NAT RES COUN, CAN, 84- *Concurrent Pos:* Mem, Comn 31, Int Astron Union, 72- *Mem:* Can Asn Physicists (secy-treas, 70-74, vpres, 79, pres, 80); fel Inst Elec & Electronics Engrs; fel Royal Soc Can. *Res:* Microwave spectroscopy; determination of structure of polar molecules and dimers; time and frequency standards, time dissemination and satellite time tranfer. *Mailing Add:* 49 Cedar Rd Ottawa ON K1J 6L5 Can

COSTAIN, JOHN KENDALL, b Boston, Mass, Nov 18, 29; m 56; c 3. GEOPHYSICS. *Educ:* Boston Univ, BA, 51; Univ Utah, PhD(geol), 60. *Prof Exp:* Geophysicist, Socony Vacuum Oil Co Venezuela, 51-54; asst prof physics, San Jose State Col, 59-60; from asst prof to assoc prof geophys, Univ Utah, 60-67; assoc prof, 67-69, PROF GEOPHYS, VA POLYTECH INST & STATE UNIV, 69- *Concurrent Pos:* NSF grants, 64-; Dept Energy contracts, 75- *Mem:* Am Geophys Union; Am Asn Petrol Geol; Geol Soc Am; Soc Explor Geophys. *Res:* Exploration seismology; terrestrial heat flow; geothermal energy. *Mailing Add:* Dept Geol Sci Va Polytech Inst Blacksburg VA 24061

COSTANTINO, MARC SHAW, b Oakland, Calif, Nov 8, 45; m 66; c 3. HIGH PRESSURE PHYSICS, GEOPHYSICS. *Educ:* Rensselaer Polytech Inst, BS, 67; Princeton Univ, PhD(solid state physics), 72. *Prof Exp:* Physicist, US Mil Acad, West Point, 71-74; PHYSICIST, LAWRENCE LIVERMORE LAB, 74- *Mem:* Am Phys Soc. *Res:* High pressure thermodynamics; physics of geological materials; mechanical equation of state of energetic materials; high pressure spectroscopy. *Mailing Add:* Lawrence Livermore Nat Lab L-369 Livermore CA 94550

COSTANTINO, ROBERT FRANCIS, b Everett, Mass, Mar 21, 41; m 63; c 5. POPULATION BIOLOGY. *Educ:* Univ NH, BS, 63; Purdue Univ, MS, 65, PhD(genetics), 67. *Prof Exp:* Asst prof genetics, Pa State Univ, 67-72; PROF ZOOL, UNIV RI, 72- *Concurrent Pos:* Vis Prof, Purdue Univ, 79 & Univ Calif, Davis, 86. *Mem:* Genetics Soc Am; AAAS; Sigma Xi; Am Soc Naturalists. *Res:* Population genetics-demography of tribolium; genetics of competing species. *Mailing Add:* Dept Zool Univ RI Kingston RI 02881

COSTANZA, ALBERT JAMES, b Pittsburgh, Pa, Dec 3, 17; m 43; c 3. POLYMER CHEMISTRY, RUBBER CHEMISTRY. *Educ:* Univ Pittsburgh, BS, 40; Univ Akron, MS, 50. *Prof Exp:* Jr chemist, US Govt, 40-43; proj chemist, Govt Synthetic Rubber Lab, Univ Akron, 46-50; sr res chemist, 50-71, sect head specialty rubbers, 71-74, RES SCIENTIST, GOODYEAR TIRE & RUBBER CO, 74- *Mem:* Am Chem Soc. *Res:* Synthetic latices and polymers; emulsion and emulsifier free polymerizations; liquid polymers; functional and group polymers; reactive copolymers; specialty rubbers. *Mailing Add:* 924 Mem Pkwy No 103 Akron OH 44303-1274

COSTANZA, MARY E, b Quincy, Mass, Feb 21, 37. MEDICAL ONCOLOGY. *Educ:* Radcliffe Col, BA, 58; Univ Calif, Berkeley, MA, 63; Univ Rochester, MD, 68. *Prof Exp:* From instr med & oncol to asst prof oncol, Sch Med, Tufts Univ & New Eng Med Ctr, 72-79; PROF MED, SCH MED, UNIV MASS, 79-, DIR, DIV ONCOL, 80- *Concurrent Pos:* Mem, Breast Cancer Task Force, Nat Cancer Inst, 76-80, grant rev br, 82-85. *Mem:* Am Asn Cancer Res; Am Soc Clin Oncologists; Am Asn Cancer Educ; Soc Study Breast Dis. *Res:* Cancer therapeutics; breast cancer research. *Mailing Add:* Sch Med Univ Mass 55 Lake Ave Worcester MA 01605

COSTANZA, MICHAEL CHARLES, b New York, NY. COMMUNITY INTERVENTION STUDIES, BREAST CANCER SCREENING. *Educ:* Univ Calif, Los Angeles, AB, 70, MS, 73, PhD(biostatist), 77. *Prof Exp:* Res asst prof, Dept Epidemiol & Environ Health, Univ Vt, 77-79, from asst prof to assoc prof, 80-90, dir, 85-90, PROF BIOSTATIST & STATIST, DEPT MATH & STATIST, MED BIOSTATIST/BIOMET FAC & STATIST PROG, UNIV VT, 91- *Concurrent Pos:* Sci reviewer, Nat Cancer Inst, 88-, Nat Heart, Lung, Blood Inst, 86- & Nat Inst Alcohol & Alcohol Abuse, 91-; consult, Vt Alcohol Res Ctr, 89- & Wood Heating Alliance, 90-; dir, Breast Screening Prog Proj: Biometrycore, Nat Cancer Inst, 90- & Vt Regional Cancer Ctr: Biometrycore, 86-; secy, Subsect Teaching Statist Health Sci, Am Statist Asn, 91-93. *Mem:* Am Statist Asn; Inst Math Statist; Biomet Soc; Am Pub Health Asn; AAAS. *Res:* Designing community intervention studies (eg cancer prevention/control, smoking/alcohol prevention/cessation); variable selection in discriminant analysis; sequential and multistage estimation and testing; applications of biostatistics in collaborate medical/public health research. *Mailing Add:* Med Biostatist 24C Hills Bldg Univ Vt Burlington VT 05405

COSTANZA, ROBERT, b Pittsburgh, Pa, Sept 14, 50. ENERGY ANALYSIS, ECOLOGICAL ECONOMICS. *Educ:* Univ Fla, BA, 73, MA, 74, PhD(syst ecol), 79. *Prof Exp:* Res assoc, Ctr Wetlands, Univ Fla, 74-75, grad res asst, 75-79; energy analyst, Fla Gov Energy Off, 79; res asst, 79-81, asst prof, 81-84, assoc prof, Ctr Wetland Resources, LA State Univ, 84-88; ASSOC PROF, CHESAPEAKE BIOL LAB, UNIV MD, 88- *Concurrent Pos:* Consult, US Environ Protection Angency, 80-84; Kellogg Nat Fel, 82-85; chief ed, ecol econ. *Honors & Awards:* Nat Wildlife Fed Outstanding Pub, 85-86. *Mem:* AAAS; Ecol Soc Am; Am Soc Naturalists; Am Inst Biol Sci; Int Soc Ecol Modeling. *Res:* Research in systems ecology, resource management and energy analysis with special emphasis on modeling human/environment interactions; estimation of environmental costs of coastal wetland modifications, net energy analysis of energy sources, and integrated analysis of the coastal zone. *Mailing Add:* Chesapeake Biol Lab Univ MD Box 38 Solomons MD 20688

COSTANZI, JOHN J, b Old Forge, Pa, Apr 25, 36; m 81; c 2. MEDICAL ONCOLOGY, INTERNAL MEDICINE. *Educ:* Univ Scranton, BS, 57; Georgetown Univ, MD, 61. *Prof Exp:* Intern, Walter Reed Gen Hosp, 61-62; resident, Wilford Hall USAF Med Ctr, 62-65, fel hemat-oncol, 65-66, asst chief, 66-72; dir intern educ, Univ Tex Med Sch, San Antonio, 69-72; prog dir clin cancer ctr planning grant, 72-75, from asst prof to assoc prof, 72-79, prof med, Univ Tex Med Br, Galveston, 79-87, prog dir, cancer ctr, 75-87, Am Cancer Soc Prof Clin Oncol, 85-87; MED DIR, THOMPSON CANCER SURVIVAL CTR, 88-; PROF MED, UNIV TENN, KNOXVILLE, 88- *Concurrent Pos:* Prin investr southwest oncol group, Univ Tex Med Br, Galveston, 75-87, clin cancer educ grant, 79-87; mem, prev oncol rev comt, Nat Cancer Inst, NIH, 82-87, mem, cancer control sci prog rev comt, 83-87, chmn, clin investrs award comt, 84-87; mem, acquired immune deficiency syndrome task force, Tex Dept Health, 83-87; mem, Tex Legis Task Force on Career, 84-87. *Mem:* Cent Soc Clin Res; fel Am Col Physicians; Am Soc Clin Oncol; Am Soc Cancer Res; Am Soc Hemat. *Res:* Cancer chemotherapy; interferon; cell kinetics and its role in cancer treatment; malignant melanoma; multiple myeloma; lung cancer; breast cancer. *Mailing Add:* Thompson Cancer Survival Ctr 1915 White Ave Knoxville TN 37916

COSTANZO, LINDA SCHUPPER, b 1947; m 71; c 2. RENAL PHYSIOLOGY, CALCIUM METABOLISM. *Educ:* State Univ NY, PhD(pharmacol), 73. *Prof Exp:* ASSOC PROF PHYSIOL & BIOPHYSICS, MED COL VA, VA COMMONWEALTH UNIV, 85- *Concurrent Pos:* Mem, Physiol Test Comt, Nat Bd Med Examrs; sr investr, NY Heart Asn & estab investr, Am Heart Asn. *Mem:* Am Physiol Soc; Am Soc Nephrology. *Res:* Renal hormone action. *Mailing Add:* Dept Physiol & Biophysics Med Col Va Box 551 MCV Sta Richmond VA 23298

COSTANZO, RICHARD MICHAEL, b Brooklyn, NY, July 18, 47; c 2. NEUROPHYSIOLOGY, OLFACTION & TASTE. *Educ:* State Univ NY, Stony Brook, BS, 69; State Univ NY, Upstate Med Col, PhD(physiol), 75. *Prof Exp:* NIH res fel, Rockefeller Univ, 75-77; res assoc physiol, Sch Med, New York Univ, 77-78, instr, 78-79; asst prof, 79-85, ASSOC PROF PHYSIOL, MED COL VA, 85- *Concurrent Pos:* Prin investr, Nat Inst Neurol & Commun Dis & Stroke res grant, NIH, 80-, Res Career Develop Award, 84-89. *Mem:* Am Physiol Soc; Soc Neurosci; Asn Chemoreception Sci; NY Acad Sci; Int Brain Res Orgn; Sigma Xi. *Res:* Physiology of olfaction and taste; olfactory nerve cell regeneration; brain transplants; diagnosis and measurement of smell and taste disorders. *Mailing Add:* Dept Physiol Med Col Va Box 551 MCV Sta Richmond VA 23298-0551

COSTEA, NICOLAS V, b Bucharest, Romania, Nov 10, 27; US citizen. IMMUNOHEMATOLOGY. *Educ:* Univ Paris, MD, 56. *Prof Exp:* Instr med, Sch Med, Tufts Univ, 59-63; Lederle fac award, 63-66; from asst prof to prof med, Sch Med, Univ Ill, Chicago Circle, 70-72; chief hemat-oncol div, Univ Calif, Los Angeles-San Fernando Valley Prog, 72; PROF MED, SCH MED, UNIV CALIF, LOS ANGELES, 72-, COORDR, HEMAT-ONCOL PROG, 72- *Concurrent Pos:* Dir hemat, Cook County Hosp, 70-72; NIH grants. *Mem:* Am Soc Hemat; Am Fedn Clin Res; Am Asn Immunologists; Am Rheumatism Asn. *Res:* Hematology. *Mailing Add:* Hemat Dept Vet Admin Sepulveda Hosp Los Angeles CA 91343

COSTELLO, CATHERINE E, b Medford, Mass, May 11, 43. MASS SPECTROMETRY, ORGANIC BIOCHEMISTRY. *Educ:* Emmanuel Col, AB, 64; Georgetown Univ, MS, 67, PhD(org chem), 70. *Prof Exp:* Chemist, Div Food Chem, US Food & Drug Admin, Dept Health Educ & Welfare, 66-67 & USDA, 69; res assoc mass spectrometry, 70-79, prin res scientist, 79-88, SR RES SCIENTIST, MASS INST TECHNOL, 88- *Concurrent Pos:* Assoc dir, Mass Spectrometry Facil, Mass Inst Technol, 76-; lectr, Northeastern Univ, 80-85; NIH Biotechnology Res Rev Comt, 85-89; consult, Eunice Kennedy Shriver Ctr for Ment Retardation, Watham, Mass, Nat Res Coun Can; vis prof, Zhongshen Univ, Guangzhon, China, 85. *Mem:* Am Chem Soc; Am Soc Mass Spectrometry; AAAS; Asn Women Sci. *Res:* Applications of advanced methods of organic chemical analysis, particularly mass spectrometry to biochemistry and to biomedical problems including metabolic disorders, drug metabolism and toxicology. *Mailing Add:* Rm 56-029 Mass Inst Technol Cambridge MA 02139

COSTELLO, CHRISTOPHER HOLLET, b St John's, Nfld, Jan 19, 13; nat US; m 46; c 5. PHARMACOLOGY. *Educ:* Mass Col Pharm, BS, 36, MS, 41, PhD(chem & pharmacol), 50. *Prof Exp:* Pharmacist, Mass Gen Hosp, 35-36; chief chemist, Pinkham Co, 38-41; instr chem, Mass Col Pharm, 46-49; vpres & sci dir, Columbus Pharmacol Co, 49-60; dir res, Pharmaceut Labs, Colgate-Palmolive Co, 60-62, dir res biol prod, 62-69, assoc dir corp res, 69-75, adminr regulatory affairs, 76-79; consult, 78-81; RETIRED. *Concurrent Pos:* Drug indust liaison, Rev Panel Oral Cavity Drugs, Food & Drug Admin, 74-80; consult, Proprietary Asn, 70. *Mem:* Fel AAAS; Am Chem Soc; Am Pharm Asn; fel Am Inst Chemists; Soc Clin Pharmacol & Therapeut. *Res:* Pharmaceutical and cosmetic chemistry; chemistry and pharmacology of botanical drugs; isolated estrone from licorice; sustained release medication; antilipemic agents; proprietary drugs; dermatological, dental and hair products. *Mailing Add:* Lower Rd Putnam Station NY 12861

COSTELLO, DAVID FRANCIS, environmental biology, for more information see previous edition

COSTELLO, DONALD F, b New York, NY, Mar 12, 34; m 61; c 7. STATISTICS, COMPUTER SCIENCE. *Educ:* Manhattan Col, BS, 54; Univ Notre Dame, MS, 59. *Prof Exp:* Instr math, Univ Alaska, 56-57 & La Salle Col, 59-60; lect, Univ Nebr, 60-63; asst prof & dir comput ctr, Wis State Univ, 63-64; res assoc, Univ Wis, 64-65; asst dir comput ctr, 65-70, dir comput ctr, 72-80, dir acad comput, Univ Comput Network, 76-80, ASSOC PROF COMPUT SCI, UNIV NEBR, 70-, LECTR MGT, 76- *Concurrent Pos:* Asst prof, Colo State Univ, 67-68; proj dir, Mo Valley Planning Info Ctr; pres, Costello & Assocs, Mgt Consults. *Mem:* Asn Comput Mach; Inst Math Statist. *Res:* Analytic techniques useful in managing information systems environments. *Mailing Add:* Univ Notre Dame Notre Dame IN 46556

COSTELLO, ERNEST F, JR, b Fall River, Mass, June 9, 23; m 48; c 5. PHYSICS. *Educ:* Boston Univ, AB, 49; Lehigh Univ, MS, 51, PhD(physics), 59. *Prof Exp:* Asst physics, Lehigh Univ, 49-52, instr, 52-59; assoc prof, 59-67, chmn dept, 68-69, dir div lib arts & sci, 69-74, asst dir, 63-69, dean sci & eng, 74-77, health prof adv, 77-81, PROF PHYSICS, MERRIMACK COL, 67-,. *Concurrent Pos:* Sr engr physics, Raytheon Co, 62-65; vis scientist, Lincoln Labs, 60-61. *Mem:* Am Phys Soc; Am Asn Physics Teachers; Sigma Xi. *Res:* Magnet thin films; magnetic powders; thin film circuits. *Mailing Add:* Dept Physics Merrimack Col North Andover MA 01845

COSTELLO, RICHARD GRAY, b New York, NY, Apr 16, 38; m 59; c 3. COMPUTER HARDWARE, CONTROL SYSTEMS DESIGN. *Educ:* Cooper Union, BS, 59; NY Univ, MS, 60; Univ Wis, PhD(manual control), 66. *Prof Exp:* Engr, Autonetics Div, N Am Aviation, Inc, 60-61; assoc engr, Cornell Aeronaut Labs, 66-67; assoc prof, 67-80, chmn elec eng curric group, 77-81, PROF ELEC ENG, COOPER UNION, 80- *Concurrent Pos:* Consult, US Consumer Prod Safety Comn, 76, Con Ed, NY, 85, NY Telephone, 85, US Justice Dept, 87, United High Technol, 88 & for various lawyers & inventors, NY. *Honors & Awards:* Rossi Prize res writing, 69. *Mem:* Inst Elec & Electronics Engrs. *Res:* Control theory; man-machine systems; guidance and manual control of vehicular systems; micro computers in bioengineering and system control; medical electronics; transducers; Czochralski YAG laser crystal growth; heat pump control and data acquisition. *Mailing Add:* Dept Elec Eng Cooper Square New York NY 10003

COSTELLO, WALTER JAMES, b San Antonio, Tex, Nov 5, 45; m 81; c 2. NEUROBIOLOGY. *Educ:* Trinity Univ, BA, 66, MS, 69; Boston Univ, PhD(neurobiol), 78. *Prof Exp:* Res asst physiol, Trinity Col, 66-69; teaching fel biol, Boston Univ, 72-74; res asst neurobiol, Marine Prog-Woods Hole, Boston Univ, 74-77; fel neurogenetics, Yale Univ, 77-81; asst prof, 81-86, ASSOC PROF BIOMED SCI, OHIO UNIV, 86- *Concurrent Pos:* Fel, NIH, 78 & Muscular Dystrophy Asn Am, 78-; researcher, Bermuda Biol Sta, 78 & Marine Biol Lab, Woods Hole, 79 85 & 84-; ed, Fine Sci Points, 87- *Mem:* Am Soc Zoologists; AAAS; Soc Neurosci; Soc Cell Biol; Crustacean Soc. *Res:* Developmental neurobiology; neurogenetics; cell signalling. *Mailing Add:* Dept Zool & Biomed Sci Ohio Univ Athens OH 45701

COSTELLO, WILLIAM JAMES, b Cavalier, NDak, Aug 2, 32; m 58; c 2. ANIMAL SCIENCE, MEAT SCIENCE. *Educ:* NDak State Univ, BS, 54; Okla State Univ, MS, 60, PhD(meat sci), 63. *Prof Exp:* Res nutritionist, Res Lab, John Morrell & Co, 62-65; asst prof, 65-75, assoc prof animal sci, 75-86, exten meat specialist, 82-87, PROF ANIMAL SCI, SDAK STATE UNIV, 86- *Concurrent Pos:* Co-auth, The Meat We Eat, 85. *Mem:* Am Meat Sci Asn; Am Soc Animal Sci; Inst Food Technol; Coun Agr Sci & Technol. *Res:* Quality studies of beef and pork; quantity aspects of beef, pork and lamb carcasses; fast freezing and packaging beef, fresh meat processing (restructuring); tranquilizers associated with beef cattle marketing. *Mailing Add:* Dept Animal & Range Sci SDak State Univ Box 2170 Brookings SD 57007

COSTER, JOSEPH CONSTANT, b Diekirch, Luxembourg; US citizen. PHYSICS. *Educ:* Mass Inst Technol, SB, 57; Univ Calif, Berkeley, PhD(physics), 67. *Prof Exp:* PROF PHYSICS, WESTERN ILL UNIV, 67- *Mem:* Am Asn Physics Teachers. *Res:* S-matrix theory. *Mailing Add:* Dept Physics Western Ill Univ Adams St Macomb IL 61455

COSTERTON, J WILLIAM F, b Vernon, BC, July 21, 34; m 55; c 4. MICROBIOLOGY. *Educ:* Univ BC, BA, 55, MA, 56; Univ Western Ont, PhD(microbiol), 60. *Prof Exp:* Prof biol, Baring Union Christian Col, Punjab, India, 60-62, dean sci, 63-64; fel bot, Cambridge Univ, 65; prof assoc microbiol, McGill Univ, 66-67, asst prof, 68-70; assoc prof, 70-75, PROF MICROBIOL, UNIV CALGARY, 75- *Mem:* Can Soc Microbiol; Am Soc Microbiol. *Res:* Architecture of bacterial cell walls, including extracellular carbohydrate coats, especially as it relates to physiological processes, to the presence of periplasmic enzymes and to adhesion to inert or cellular surfaces. *Mailing Add:* Dept Biol Univ Calgary 2500 Univ Dr NW Calgary AB T2N 1N4 Can

COSTES, NICHOLAS CONSTANTINE, b Athens, Greece, Sept 20, 26; US citizen; m 58; c 3. GEOTECHNICAL ENGINEERING, RESEARCH ADMINISTRATION. *Educ:* Darmouth Col, NH, AB, 50; Darmouth Col-Thayer Sch Eng, MSCE, 51 & NC State Univ, 55; Harvard Univ, AM & ME, 61; NC State Univ, PhD(eng), 65. *Prof Exp:* Teaching fel soil mech, Dept Civil Eng, NC State Univ, 51-53; mat engr geotech eng, NC State Hwy & Pub Works Co, 53-56; contract scientist snow & ice mech, US Snow Ice & Permafrost Res Estab, 56-59; res civil eng, US Cold Regions Res & Eng Lab, 59-62; instr & Ford Found fel soil mech & viscoelasticity, Dept Civil Eng, NC State Univ, 62-64; SR RES SCIENTIST SPACE SCI LAB, GEORGE C MARSHALL SPACE FLIGHT CTR, NASA, 65- *Concurrent Pos:* Mem numerous ad hoc comt & tech mgt teams, NASA, 65-; team leader, Apollo 11 Soil Mech Invest Sci Team, 67-70; co-investr, Lunar Geol Exp, Apollo Missions 12 & 13, 69-71, Apollo Soil Mech Exp, Apollo Mission 14-17, 71-74; lectr, Dept Eng Mech, Univ Ala, Huntsville, 72-; invited lectr, Acad Inst, Govt Agencies & Pvt Orgns. *Honors & Awards:* Norman Medal, Am Soc Civil Engrs, 72; Martin Schilling Award, Am Inst Aeronaut & Astronaut, 79. *Mem:* Sigma Xi; AAAS; Am Geophys Union; fel Am Soc Civil Engrs; assoc fel Am Inst Aeronaut & Astronaut. *Res:* Soil mechanics and geotechnical engineering; mechanical behavior of snow and ice as engineering materials; spacelab experiment definition on soil behavior; interior flow process modeling as related to space shuttle main engine; author or coauthor of over 50 publications. *Mailing Add:* 4216 Huntington Rd SE Huntsville AL 35802

COSTICH, EMMETT RAND, b Rochester, NY, July 15, 21; m 45; c 6. ORAL SURGERY. *Educ:* Colgate Univ, BA, 43; Univ Pa, DDS, 45; Univ Rochester, MS, 49, PhD(path), 54. *Prof Exp:* Instr dent surg, Sch Med & Dent, Univ Rochester, 47-55; from asst prof to assoc prof dent, Sch Dent, Univ Mich, 55-62; chmn dept oral surg, Col Dent, Univ Ky, 62-69, prof, 62-80, assoc dean, Col Dent, 69-72, asst for extramural educ prog coordr to vpres Med Ctr, 72-78, prof oral & maxillofacial surg, 80-90, dean, 86-87, EMER PROF ORAL & MAXILLOFACIAL SURG, COL DENT, UNIV KY, 90- *Concurrent Pos:* Res assoc & supvr periodont & oral path, Eastman Dent Dispensary, 49-55; consult, US Vet Hosp, Lexington, Ky, 62-90 & NIMH Hosp, 63-73. *Mem:* Am Dent Asn; Int Asn Dent Res; Am Asn Dent Sch. *Res:* Transplantation; bone physiology. *Mailing Add:* Col Dent Rm D510 Univ Ky Lexington KY 40536-0084

COSTILL, DAVID LEE, b Feb 7, 36; US citizen; m 60; c 2. HUMAN PHYSIOLOGY. *Educ:* Ohio Univ, BSEd, 59; Miami Univ, MA, 61; Ohio State Univ, PhD(physiol), 65. *Prof Exp:* Instr physiol, Ohio State Univ, 63-64; asst prof, State Univ NY Cortland, 64-66; PROF PHYS ED & BIOL, HUMAN PERFORMANCE LAB, BALL STATE UNIV, 66-, DIR. *Concurrent Pos:* Vis fac, Desert Res Inst, 68; hon lectr, Univ Sulford, Eng, 72; res assoc, Gymnastik-och idrottshogskolan, Stockholm, 72-73. *Mem:* Am Physiol Soc; Am Soc Zoologists. *Res:* Alterations in skeletal muscle water and electrolytes following prolonged exercise and dehydration in man; glycolytic-oxidative enzyme and fiber composition in human skeletal muscle. *Mailing Add:* 3907 W Ethel Muncie IN 47304

COSTIN, ANATOL, b Bucharest, Rumania, Aug 26, 26; m 55; c 2. NEUROSCIENCES. *Educ:* Inst Med & Pharm, Bucharest, 52; Rumanian Acad Sci, PhD(physiol, physiopath), 56. *Prof Exp:* Asst prof physiol, Inst Med & Pharm, Bucharest, 49-57; sr res fel, Inst Physiol, Rumanian Acad Sci, 56-61; sr lectr pharmacol, Hadassah Med Sch, Hebrew Univ, Israel, 61-70; res assoc, 70-72, assoc prof anat, 72-76, ASSOC CLIN PROF ANAT, UNIV CALIF, LOS ANGELES, 84-; ASSOC PROF, DREW UNIV, 84- *Concurrent Pos:* Res asst anat, Univ Calif, Los Angeles, 67-68. *Mem:* Am Physiol Soc; Soc Neurosci; Am Asn Anatomists; Royal Soc Med; Fedn Am Socs Exp Biol. *Res:* Neurophysiology; neurochemistry. *Mailing Add:* Dept Anat Univ Calif Los Angeles CA 90024

COSTLEY, GARY E, b Caldwell, Iowa, Oct 26, 43. SCIENCE ADMINISTRATION. *Educ:* Ore State Univ, PhD(nutrit & biochem). *Prof Exp:* Sr vpres sci & qual, 81-85, EXEC VPRES SCI & TECHNOL, KELLOGG CO, 85- *Mailing Add:* Kellogg Co 235 Porter St Battle Creek MI 49016

COSTLOW, JOHN DEFOREST, b Brookville, Pa, Jan 28, 27; m 52; c 2. MARINE INVERTEBRATE ZOOLOGY. *Educ:* Western Md Col, BS, 50; Duke Univ, PhD(zool), 55. *Prof Exp:* Asst biol, Western Md Col, 50; asst zool, 50-51, 53-54, marine zool, Off Naval Res, 51-53, res assoc, NSF contracts, 54- 59, from asst prof to assoc prof, 59-68, dir Marine Lab, 68-90, PROF ZOOL, DUKE UNIV, 68- *Concurrent Pos:* Chmn, NC Marine Fisheries Comn, 85-88. *Mem:* Am Soc Limnol & Oceanog; Am Soc Zool; Brit Marine Biol Asn. *Res:* Organogenesis; larval development; molting, growth and physiology of Cirripedia; larval development of Crustacea in relation to environmental factors; endocrine mechanisms in larvae of marine Crustacea; environmental factors affecting development, survival and morphological abnormalties in developing invertebrate larvae; including natural and anthropogenic factors and hormonal and physiological factors regulating barnacle development. *Mailing Add:* Marine Lab Duke Univ Beaufort NC 28516

COSTLOW, MARK ENOCH, b Bridgeport, Conn, June 14, 42; div; c 1. ENDOCRINOLOGY, ONCOLOGY. *Educ:* Univ Conn, BA, 65; Univ Kans, PhD(microbiol), 72. *Prof Exp:* Fel, Temple Med Sch, 71-72; fel, Med Sch, Univ Tex, 72-76; asst mem, 76-82, ASSOC MEM, ST JUDE CHILDREN'S RES HOSP, 82- *Concurrent Pos:* Adj asst prof biochem, Ctr Health Sci, Univ Tenn, 77-82, adj assoc prof, 82- *Mem:* Tissue Culture Asn; Am Physiol Soc; Am Soc Cell Biol; Endocrine Soc; Am Asn Cancer Res. *Res:* Hormone receptor regulation in mammary carcinoma; mechanism of prolactin action; growth and differentiation in mammary glands; glucocorticord receptors in childhood leukemia. *Mailing Add:* Dept Skin Biol Res Schering-Plough Inc 3030 Jackson Ave Memphis TN 38151

COSTLOW, RICHARD DALE, b Johnstown, Pa, July 19, 25; m 45; c 4. MICROBIOLOGY, BIOCHEMISTRY. *Educ:* Pa State Univ, BS, 49, MS, 52, PhD(bact), 54. *Prof Exp:* Res microbiologist, US Army Biolabs, Ft Detrick, 54-69, chief virus & rickettsia div, 69-71; dir cancer chemother dept, Microbiol Assocs, Inc, 71-75; CHIEF, CANCER DETECTION BR, DIV CANCER PREV & CONTROL, NAT CANCER INST, 75- *Mem:* Am Soc Microbiol; AAAS; NIH, Sigma Xi; NY Acad Sci. *Res:* Bacterial physiology; enzymology; microbiological chemistry; virology; research administration. *Mailing Add:* 607 Schley Ave Frederick MD 21701

COSTOFF, ALLEN, b Milwaukee, Wis, Sept 26, 35. ZOOLOGY, ENDOCRINOLOGY. *Educ:* Marquette Univ, BS, 57, MS, 59; Univ Wis-Madison, PhD(zool, biochem), 69. *Prof Exp:* Res technician life sci, Marquette Univ, 59-60; head biol res, Aldrich Chem Co, 60-61; teaching asst zool, Univ Wis-Madison, 61-63, NIH-NSF res asst, 63-65, NIH-Ford Found trainee, 65-69, trainee reproductive physiol, 69-70; instr, 70-73, asst prof endocrinol, 73-77, ASSOC PROF ENDOCRINOL, MED COL GA, 77- *Concurrent Pos:* Primate reproduction, 70. *Mem:* AAAS; Endocrine Soc; Am Asn Anatomists; Am Soc Cell Biol; Sigma Xi; Am Physiol Soc; NY Acad Sci. *Res:* Ultrastructure of anterior pituitary gland; fractionation and bioassay of pituitary secretory granules; hypothalamo-pituitary-ovarian axis in aging; morphometry of gonadotropes and mammotropes in rat adenohypophyses; antigonadotropin effect of prolactin; control of gonadotropins and prolactin. *Mailing Add:* Dept Physiol & Endocrinol Med Col Ga Augusta GA 30912

COSTON, TULLOS OSWELL, ophthalmology; deceased, see previous edition for last biography

COSTRELL, LOUIS, b Bangor, Maine, June 26, 15; m 42; c 3. NUCLEAR PHYSICS, NUCLEAR SCIENCE. *Educ:* Univ Maine, BS, 39; Univ Md, MS, 49. *Prof Exp:* Elec designer, Westinghouse Elec Corp, Pa, 40-41; elec engr, Bur Ships, US Dept Navy, 41-46; PHYSICIST, NAT BUR STANDARDS, 46- *Concurrent Pos:* Chmn, comt N42 on nuclear instrumentation, Am Nat Standards Inst, 62-; tech adv to US Nat Comt of Int Electrotech Comn, 62- *Honors & Awards:* Gold Medal Award, US Dept Com, 67, Edward Bennett Rosa Award,79; Harry Diamond Mem Award, Inst Elec & Electronics Engrs, Centennial Medal, 84. *Mem:* Am Phys Soc; fel Inst Elec & Electronics Engrs. *Res:* Nucleonic instrumentation; electronics. *Mailing Add:* Nat Bur Standards Gaithersburg MD 20899

COSULICH, DONNA BERNICE, b Albuquerque, NMex, Dec 2, 18. DRUG METABOLISM. *Educ:* Univ Ariz, BS, 39, MS, 40; Stanford Univ, PhD(org chem), 43. *Prof Exp:* Res chemist, Calco Chem Div, 43-55, RES CHEMIST, LEDERLE LABS, AM CYANAMID CO, 55- *Honors & Awards:* Sr Res Award, Am Cyanamid Co, Geneva, 58; Sci Achievement Award, Am Cyanamid Co, 84; Bruce Cain Mem Award, 87. *Mem:* Am Chem Soc; Am Soc Mass Spectros. *Res:* Identification of drug metabolites and analyses in biological fluids by mass spectrometry; crystal and molecular structures of pharmaceutical compounds by x-ray crystallographic analysis; pteroylglutamic acid and related compounds; proof of structure work by degradation of hemicellulose, echinocystic acid and antibiotics; pharmaceutical chemistry; synthetic organic chemistry. *Mailing Add:* 200 Braunsdorf Rd Dept A-3 Pearl River NY 10965

COSWAY, HARRY F(RANCIS), b Toledo, Ohio, Jan 24, 32; m 64; c 4. CHEMICAL ENGINEERING. *Educ:* Univ Toledo, BS, 53; Northwestern Univ, MS, 54; Univ Mich, PhD(chem eng), 58. *Prof Exp:* Res scientist, Cent Res Div, Stamford Res Lab, Am Cyanamid Co, 58-62 & Plastics & Resins Div, 62-66; sr chem engr, Polaroid Corp, 66-74; SR RES ENGR, FORMICA CORP, 75- *Mem:* Am Inst Chem Engrs; Sigma Xi. *Res:* Low temperature vapor-liquid equilibria; distillation and heat transfer; chemical kinetics; rheology; non-Newtonian fluid flow; fluid metering and dispensing systems. *Mailing Add:* 685 La Placita Greenwood IN 46142

COTA, HAROLD MAURICE, b San Diego, Calif, Apr 16, 36; m 59; c 3. ENVIRONMENTAL & CHEMICAL ENGINEERING. *Educ:* Univ Calif, Berkeley, BS, 59; Northwestern Univ, MS, 61; Univ Okla, PhD(chem eng), 66. *Prof Exp:* Res engr, Lockheed Missiles & Space Co, Calif, 60-62; res asst chem eng, Univ Okla, 62-65; assoc prof, 66-73, PROF ENVIRON ENG, CALIF POLYTECH STATE UNIV, SAN LUIS OBISPO, 73- *Concurrent Pos:* Mem, Cent Coast Region, Calif State Water Qual Control Bd, 70-85, chmn, 76-77; consult noise & air pollution control, 77-; dir, Cal Poly Environ Protection Agency Air Pollution Control Areawide Training Ctr, 83-; mem, Calif Resource Bd Res Screening Comt, 84-; chmn, W Coast Sect, Air Pollution Control Asn, 86-88. *Honors & Awards:* Lyman Ripperton Award, Air Pollution Control Asn, 84. *Mem:* Air Pollution Control Asn; Am Inst Chem Engrs; Am Indust Hygiene Asn; Inst Noise Control Eng; Am Acad Environ Eng. *Res:* Solid-gas interactions in the atmosphere; electrochemistry; thermodynamics; noise control; environmental modeling. *Mailing Add:* Dept Environ Eng Calif Polytech State Univ San Luis Obispo CA 93407

COTABISH, HARRY N(ELSON), b Lakewood, Ohio, Apr 20, 16; m 42; c 4. CHEMICAL ENGINEERING, PHYSICAL CHEMISTRY. *Educ:* Case Inst Technol, BS, 38, MS, 39. *Prof Exp:* Chemist, Parker Appliance Co, 38; res dept, Mine Safety Appliances Co, 39-42, res chemist, 42-50, sr chemist & group supvr, 50-53, prin chemist & supvr, 53-57, chief chemist, Tech Prod Div, 57-61, mgr develop & eng, safety prod group, 61-67, assoc dir, Res & Eng Div, 67-81; RETIRED. *Mem:* Am Chem Soc; Am Inst Chem Engrs; Am Indust Hyg Asn. *Res:* Catalytic removal of combustibles in air; handling and mixing of gases; design of cryogenic equipment; development of chemical analysis instrumentation and protective equipment for use in toxic environments. *Mailing Add:* 3007 Sturbridge Ct Allison Park PA 15101-1537

COTANCH, PATRICIA HOLLERAN, b Pittsburgh, Pa, Aug 26, 45; m 71; c 1. NURSING. *Educ:* Univ Pittsburgh, BSN, 69, Med, 74, PhD(higher educ & nursing), 79. *Prof Exp:* Staff nurse, Braddock Hosp, 66-68; instr, Montefiore Hosp, 69-74 & Univ Pittsburgh, 75-76; asst prof, 76-78, ASSOC PROF NURSING, DUKE UNIV, 79- *Concurrent Pos:* Res proj assoc, Sch Nursing, Duke Univ, 76-79; fac mem, Duke Univ Comprehensive Cancer Ctr, 77-, res consult, 79-; prin investr, Behav Med Sect, Nat Cancer Inst, 80- *Mem:* Soc Behav Med; Oncol Nurses Soc. *Res:* Clinical application of behavioral medicine techniques for symptom management for cancer patients receiving chemotherapy. *Mailing Add:* Dept Nursing/Baccalaureate Ga State Univ University Plaza Atlanta GA 30303

COTANCH, STEPHEN ROBERT, b Quincy, Ill, May 7, 47; div; c 1. THEORETICAL NUCLEAR & PARTICLE PHYSICS. *Educ:* Ind Univ, BS, 69; Princeton UNiv, 71-72; Fla State Univ, PhD (theoret physics), 73. *Prof Exp:* Res assoc physics, Univ Pittsburgh, 73-76; from asst prof to assoc prof, 76-86, PROF PHYSICS, NC STATE UNIV, 86- *Concurrent Pos:* Prin investr, Dept Energy grant & Res Corp; vis prof, Univ Melbourne, Australia, 85. *Mem:* Am Phys Soc. *Res:* Theoretical studies of hyperon and hypernuclear electromagnetic production; many-body reaction theory emphasizing fundamental symmetry constraints and applications to photonuclear processes. *Mailing Add:* Dept Physics NC State Univ Raleigh NC 27695-8202

COTA-ROBLES, EUGENE H, b Nogales, Ariz, July 13, 26; m 57; c 3. MICROBIOLOGY, CYTOLOGY. *Educ:* Univ Ariz, BS, 50; Univ Calif, MA, 54, PhD(microbiol), 56. *Prof Exp:* Instr bact, Univ Calif, Riverside, 56-58, from asst prof to prof microbiol, 58-70, asst dean col letters & sci, 68-69, spec asst to chancellor, 69-70; prof & head dept, Pa State Univ, 70-73; vchancellor acad admin, 73-75, acad vchancellor & dir affirmative action, 75-80, PROF BIOL, UNIV CALIF, SANTA CRUZ, 73-, PROVOST, CROWN COL. *Concurrent Pos:* USPHS fel, Sabbatsberg Hosp, Stockholm, Sweden, 57-58; fel, Biochem Inst, Univ Uppsala, 63-64; consult, Mex-Am & Puerto Rican predoctoral fel selection comt, Ford Found, 69-71. *Mem:* AAAS; Am Soc Microbiol; Electron Micros Soc Am. *Res:* Biochemical organization of microbial cells; chemical structure of microbial membranes; membrane morphogenesis in lipid containing bacterophages. *Mailing Add:* Acad Affairs Dept 359 University Hall Univ Calif Berkeley CA 94720

COTE, LOUIS J, b Detroit, Mich, July 18, 21; m 48; c 4. MATHEMATICAL STATISTICS. *Educ:* Univ Mich, AB, 43, AM, 47; Columbia Univ, PhD(math statist), 54. *Prof Exp:* Asst prof math & statist, Purdue Univ, 54-56; asst prof math, Syracuse Univ, 56-59; ASSOC PROF MATH & STATIST, PURDUE UNIV, 59- *Concurrent Pos:* Consult, Midwest Appl Sci Corp, 61- *Mem:* Am Math Soc; Inst Math Statist; Am Statist Soc; Math Asn Am; Sigma Xi. *Res:* Sums of random variables; statistical estimation. *Mailing Add:* 719 Bexley Rd West Lafayette IN 47906

COTE, LUCIEN JOSEPH, b Angers, Que, Jan 4, 28; m 60; c 1. BIOCHEMISTRY, NEUROLOGY. *Educ:* Univ Vt, BS, 51, MD, 54. *Prof Exp:* Physician, Buffalo Gen Hosp, NY, 54-56; resident neurol, Neurol Inst, New York, 58-61; NIH fel, NY State Psychiat Inst, 61-68, asst prof, 68-70, ASSOC PROF NEUROL, MED CTR, COLUMBIA UNIV, 70-, ASSOC PROF REHAB MED, 80- *Mem:* AAAS. *Res:* Clinical neurochemistry. *Mailing Add:* Dept Clin Neurol Columbia Univ 630 W 168th St New York NY 10032

COTE, PHILIP NORMAN, b Norwich, Conn, Oct 1, 42; m 67; c 3. ORGANIC CHEMISTRY, DYESTUFF CHEMISTRY. *Educ:* Univ Conn, BA, 64; Univ RI, PhD(chem), 70. *Prof Exp:* Spec instr chem, Univ RI, 70-71; chemist res & develop, Toms River Chem Corp, 72-76; res chemist, 76-78, GROUP LEADER, DISPERSE DYES, SODYECO DIV, MARTIN MARIETTA CHEMS, 78- *Concurrent Pos:* Res assoc, Univ RI, 71-72. *Mem:* Am Chem Soc; Am Asn Textile Chemists & Colorists; Sigma Xi. *Res:* Electronic effects in free radical rearrangement reactions; synthesis of textile dyestuffs and related intermediates. *Mailing Add:* 5936 Colchester Pl Charlotte NC 28210-5404

COTE, ROGER ALBERT, b Manchester, NH, Aug 28, 28; m 52; c 5. PATHOLOGY, COMPUTER SCIENCES. *Educ:* Assumption Col, Mass, BA, 50; Univ Montreal, MD, 55; Marquette Univ, MS, 64; FRCP(C). *Prof Exp:* Chief immunohemat sect, Vet Admin Hosp, Milwaukee, Wis, 60-61, asst chief lab serv, 61-62, actg chief, 62-63, chief anat path, 63-64, chief lab serv, Boston, 64-69, chief, Boston Area Reference Lab Syst, 67-69; chmn dept, 69-77, secy, Fac Med, 83-88, PROF PROF PATH, MED SCH, UNIV SHERBROOKE, 69- *Concurrent Pos:* Asst prof, Sch Med, Marquette Univ, 62-64; instr, Sch Med, Harvard Univ, 64-69; asst prof, Sch Med, Tufts Univ, 64-68, assoc prof, 68-69, lectr, Sch Dent Med, 67-69; co-dir, Lab Comput Proj, Boston Vet Admin Hosp, 67-69; mem path & lab med res eval comt, Vet Admin Cent Off, Washington, DC, 67-69; ed-in-chief, Systematized Nomenclature of Med, 77-79; Can rep sci adv bd, Institut de Recherche d'Informatique et d'Automatique; mem, comn world standards, World Asn of Socs Path & Salutatis Unitas, 74-; consult health div, Statist Can, 81-84; chmn, Working Group 6, Int Med Informatics Asn, 82- *Mem:* Int Acad Path; Am Soc Clin Path; fel Col Am Path; NY Acad Sci; Can Orgn Advan Comput Health; fel Am Col Med Info. *Res:* Pulmonary pathology, emphysema; clinical pathology, methodology; medical nomenclature; computer technology; laboratory medicine. *Mailing Add:* Dept Path Fac Med Univ Sherbrooke Sherbrooke PQ J1H 5N4 Can

COTE, WILFRED ARTHUR, JR, b Willimantic, Conn, May 27, 24; m 47; c 5. ELECTRON MICROSCOPY, WOOD SCIENCE. *Educ:* Univ Maine, BS, 49; Duke Univ, MF, 50; State Univ NY Col Forestry, Syracuse, PhD(wood prod eng), 58. *Prof Exp:* Instr wood technol, 50-58, from asst prof to assoc prof wood prod eng, 58-65, dean, Sch Environ & Resource Eng, 76-80, dir, Renewable Mat Inst, 80-83, PROF WOOD PROD ENG, STATE UNIV NY COL ENVIRON SCI & FORESTRY, 65-, DIR NC BROWN CTR ULTRASTRUCTURE STUDIES, 70- *Concurrent Pos:* Fulbright res fel, Univ Munich, 59-60; Walker-Ames vis prof, Univ Wash, 66; vis prof, Tech Univ Denmark, 72; guest scholar, Japan Soc Promotion Sci Sponsorship, Kyoto Univ, 78; distinguished serv prof, Col Environ Sci & Forestry, State Univ NY, 89-; Erskine fel, Univ Canterbury, Christchurch, NZ, 86. *Honors & Awards:* Distinguished Serv Award, Soc Wood Sci & Technol, 90. *Mem:* Electron Micros Soc Am; Soc Wood Sci & Technol; fel Int Acad Wood Sci(pres, 84-87); Int Asn Wood Anat (secy, 70-76). *Res:* Ultrastructure of the cells of wood as revealed by light and electron microscopy; interaction of wood ultrastructure with adhesives, coatings and processing chemicals. *Mailing Add:* 207 Brookford Rd Syracuse NY 13224

COTELLESSA, ROBERT F(RANCIS), b Passaic, NJ, June 7, 23; m 48; c 3. ELECTRICAL ENGINEERING. *Educ:* Stevens Inst Technol, ME, 44, MS, 49; Columbia Univ, PhD(physics), 62. *Prof Exp:* Instr math, Stevens Inst Technol, 46-48; intermediate engr, Allen B DuMont Labs, Inc, 49-51; tech coordr elec eng, Sch Eng & Appl Sci, NY Univ, 51-55, from asst prof to assoc prof, 55-62, prof elec eng & dir lab electrosci res, 62-68; prof elec & comput eng & chmn dept, Sch Eng, Clarkson Col Technol, 68-80; provost & prof elec eng & computer sci, Stevens Inst Technol, 80-83, sr vpres acad affairs & dean eng, 85-87, vpres acad & student affairs, 86-88, prof, 84-88, EMER PROF ELEC & COMPUTER ENG, CLARKSON UNIV, 88- *Concurrent Pos:* Consult, Radio Corp Am, 56, Bridgeport, Conn Eng Inst, US Naval Appl Sci Lab, 63-69, Naval Ship Res & Develop Lab, 69-76 & Sprague Elec Co, 64-65; mem, Eng Soc Libr Bd, 82-, chmn, 88-90; dir, Eng Info, Inc, 84-89; mem adv panel, elec sci sect, NSF, Washington, DC, 75-78; mem adv coun, Res Corp, Pub Serv Elec & Gas Co, 81-84; mem, consumer adv coun, Niagara Mohawk Power Corp, 90- *Honors & Awards:* Centennial Medal, Inst Elec & Electronics Engrs, Haraden Pratt Award, US Activ Bd Citation of Honor, Meritorious Serv Award. *Mem:* Fel AAAS; Sigma Xi; Am Soc Eng Educ; Am Phys Soc; Nat Soc Prof Engrs; fel Inst Elec & Electronics Engrs. *Res:* Solid state electronics; low temperature physics; physics of thin films. *Mailing Add:* Dept Elec & Computer Eng Clarkson Univ Potsdam NY 13699-5720

COTERA, AUGUSTUS S, JR, b Houston, Tex, Jan 2, 31; m 55; c 2. GEOLOGY. *Educ:* Univ Tex, BS, 52, MA, 56, PhD(geol), 62. *Prof Exp:* Res geologist, Tex Pac Coal & Oil Co, 52 & 54; explor geologist, Carter Oil Co, 56-57; from asst prof to assoc prof geol, Allegheny Col, 61-67; assoc prof, 67-70, vpres acad affairs, 80, asst to pres, 81, PROF GEOL, NORTHERN ARIZ UNIV, 70-, CHMN DEPT, 68-, CHMN FAC, 79- *Concurrent Pos:* NSF res grant, 64-68; NSF cause grant, 79-82. *Mem:* Geol Soc Am; Soc Econ Paleont & Mineral. *Res:* Sedimentary petrology, paleogeography and environments of deposition during the Mississippian, Pennsylvanian and Cretaceous Periods; eolian deposition on Navajo Reservation; field mapping in Sonora, Mexico. *Mailing Add:* Off Pres Northern Ariz Univ Flagstaff AZ 86011

COTHERN, CHARLES RICHARD, b Indianapolis, Ind, Sept 6, 37; m 76; c 2. ENVIRONMENTAL PHYSICS. *Educ:* Miami Univ, BA, 59; Yale Univ, MS, 60; Univ Man, PhD(physics), 65. *Prof Exp:* Lectr physics, Univ Man, 61-65; asst prof, 65-70, assoc prof physics, Univ Dayton, 70-78; PHYSICIST, ENVIRON PROTECTION AGENCY, 78- *Mem:* Am Asn Physics Teachers; Am Phys Soc; AAAS; Inst Environ Sci; Sigma Xi (pres, 75-76). *Res:* Health physics; low energy nuclear spectroscopy; environmental physics-heavy metals in air, water and sludge; surface physics-photoelectron and Auger spectroscopies. *Mailing Add:* 5304 Merivale Rd Chevy Chase MD 20815-3704

COTHRAN, WARREN RODERIC, insect ecology; deceased, see previous edition for last biography

COTMAN, CARL WAYNE, b Cleveland, Ohio, Apr 5, 40; m; c 4. BIOCHEMISTRY, CELL BIOLOGY. *Educ:* Wooster Col, BA, 62; Wesleyan Univ, MA, 64; Ind Univ, PhD(biochem), 68. *Prof Exp:* From asst prof to assoc prof, 68-74, PROF PSYCHOBIOL, UNIV CALIF, IRVINE, 74-, PROF NEUROCHEM & MOLECULAR PSYCHOL, 80- *Concurrent Pos:* Mem, Neurol Bd Study Sect, NIH, 72-77; prin investr, Nat Inst Neurol & Commun Dis & Stroke, 75, NIMH, 75 & Nat Inst Aging, 77; res sci grant, Nat Inst Drug Abuse, 76. *Mem:* AAAS; Am Soc Neurochem; Soc Neurosci; Am Soc Cell Biol. *Res:* Chemistry of the nervous system; formation and long-term modification of synaptic connections in the central nervous system; recovery after brain damage. *Mailing Add:* Dept Psychobiol Univ Calif Irvine CA 92717

COTNER, JAMES BRYAN, JR, b Springfield, Ohio, June 14, 59; m 86. LIMNOLOGY, MICROBIAL ECOLOGY. *Educ:* Wittenberg Univ, BA, 81; Kent State Univ, MSc, 84; Univ Mich, PhD(biol), 90. *Prof Exp:* Instr biol, Kent State Univ, 81-84; res asst, Mich State Univ, 84-86; res asst, Univ Mich, 87-90; POSTDOCTORAL FEL, NAT OCEANIC & ATMOSPHERIC ADMIN, GREAT LAKES ENVIRON RES LAB, 90- *Concurrent Pos:* Environ tech, Ohio Environ Protection Agency, 81-84; teaching asst, Mich State Univ, 84-86 & Univ Mich, 86-90; travel award, Am Soc Limnol & Oceanog, 89. *Mem:* Am Soc Limnol & Oceanog; Int Asn Great Lakes Res; Int Asn Theoret & Appl Limnol; Sigma Xi. *Res:* Study factors controlling microbial growth in freshwater and marine habitats; how bacterial growth and nutrient regeneration affect primary production. *Mailing Add:* Great Lakes Environ Res Lab 2205 Commonwealth Blvd Ann Arbor MI 48105

COTRAN, RAMZI S, b Haifa, Palestine, Dec 7, 32; m 56; c 4. MEDICINE, PATHOLOGY. *Educ:* Am Univ Beirut, BA, 52, MD, 56; Harvard Univ, MA, 56; Am Bd Path, dipl. *Hon Degrees:* MA, Harvard Univ, 72. *Prof Exp:* From intern to chief resident path, Mallory Inst, Boston, 56-59; from instr to assoc prof, 60-72, FRANK B MALLORY PROF PATH, HARVARD MED SCH, 72-; PATHOLOGIST-IN-CHIEF, BRIGHAM & WOMEN'S HOSP, 74- *Concurrent Pos:* Fel, Mem Ctr Cancer & Allied Dis, NY, 59-60; vis res fel, Sloan-Kettering Inst Cancer Res, NY, 59-60; assoc dir, Mallory Inst Path, 69-74; mem coun circ & renal sect, Am Heart Asn. *Honors & Awards:* Elected to Inst of Med, 86. *Mem:* Am Asn Path & Bact; Am Soc Nephrol; Am Soc Cell Biologists; Int Acad Path; Am Soc Exp Path. *Res:* Pathology and pathogenesis of renal disease; electron microscopy; pathogenesis of inflammation; endothelial injury. *Mailing Add:* Brigham & Women's Hosp 75 Francis St Boston MA 02115

COTRUVO, JOSEPH ALFRED, b Toledo, Ohio, Aug 3, 42; m 83; c 1. ORGANIC CHEMISTRY. *Educ:* Toledo Univ, BS, 63; Ohio State Univ, PhD(chem), 68; Univ Bologna, postdoc, 69. *Prof Exp:* Chemist, Chemsampco, Inc, 70-73; tech policy analyst, 73-75, dir, Criteria & Standards Div, Off Drinking Water, 76-90, SCI ADV, WATER SUPPLY OFF, US ENVIRON PROTECTION AGENCY, 75-, DIR, HEALTH & ENVIRON REV DIV, OFF TOXIC SUBSTANCES, 90- *Concurrent Pos:* Prof lectr, Prince Georges Community Col, 74-75; adj prof, Dept Chem, George Washington Univ, 81; dir drinking water pilot proj, Comt Challenges Mod Soc, NATO, 77-81. *Honors & Awards:* Environ Leadership Award, Nat Sanitation Found; Donald Boyd Award, Asn Met Water Agencies. *Mem:* Am Chem Soc; Am Water Works Asn. *Res:* Synthesis and application of insect sex attractants; heterocyclic chemistry; diazo compounds and electronic properties of carbenes; chemical products of chlorine and ozone disinfection of drinking water and waste water. *Mailing Add:* 5015 46th St NW Washington DC 20016

COTSONAS, NICHOLAS JOHN, JR, b Boston, Mass, Jan 28, 19; m 70; c 3. MEDICINE. *Educ:* Harvard Col, AB, 40; Georgetown Univ, MD, 43. *Prof Exp:* Rotating intern, Gallinger Munic Hosp, Washington, DC, 44, res clin dir, Tuberc Div, 46-47, asst med res, Med Div, 47-48, chief med res, 48-49, chief med officer, 51-53; from instr to prof, 49-70, PROF MED & DEAN, PEORIA SCH MED, UNIV ILL COL MED, 70-,. *Concurrent Pos:* Chief med serv, Res & Educ Hosp, Univ Ill Med Ctr, 68-70; fel coun clin cardiol, Am Heart Asn. *Mem:* AMA; fel Am Col Cardiol; Am Fedn Clin Res; Sigma Xi. *Res:* Internal medicine; cardiology; medical education. *Mailing Add:* 21 Spinning Wheel Rd Apt 6K Hinsdale IL 60521

COTT, JERRY MASON, b San Antonio, Tex, Oct 16, 46; m 73; c 3. PSYCHOPHARMACOLOGY, NEUROPHARMACOLOGY. *Educ:* Cent State Univ, BS, 69; Univ NC, Chapel Hill, PhD(pharmacol), 75. *Prof Exp:* Lectr pharmacol, Univ Ibadan, 76-78; res scientist psychopharmacol, Astra Lakemedel Ab, 78-80; rev pharmacologist, neuropharmacol div, Food & Drug Admin, Rockville, Md, 80-84; res scientist, CNS Clin Res, Bristol-Myers Co, 84-88; sr prin scientist, Boehringer Ingelheim, 88-91; CHIEF PSYCHOTHERAPEUT DRUG DEVELOP PROG, NIMH, 91- *Concurrent Pos:* Guest worker, Psychobiol Sect, NIMH, Bethesda, Md, 81-84. *Mem:* Soc Neurosci; Brit Brain Res Asn; Europ Brain & Behav Soc; AAAS; Am Soc Pharmacol Exp Ther; Drug Info Asn; Soc Toxicol. *Res:* Role of brain monoamines and peptides in the mediation of behaviors which may be correlated with human mental disease and movement disorders; mechanism of action of neuroleptics, antidepressants and anxiolytics. *Mailing Add:* ADAMHA-NIMH 5600 Fishers Lane Park Lawn Bldg Rm 11-105 Rockville MD 20857

COTTAM, GENE LARRY, b Coffeyville, Kans, Nov 3, 40; m 63; c 3. BIOCHEMISTRY. *Educ:* Univ Kans, BA, 62; Univ Mich, MS, 63 & 65, PhD(biochem), 67. *Prof Exp:* USPHS trainee biochem, Univ Tex & Vet Admin Hosp, Dallas, 67-68; PROF BIOCHEM, UNIV TEX HEALTH SCI CTR DALLAS, 68- *Mem:* Am Biochem Soc; Am Chem Soc. *Res:* Elucidation of mechanism of chemical reactions catalyzed by enzymes; active sites; metabolic control. *Mailing Add:* Dept Biochem Univ Tex Health Sci Ctr Dallas TX 75235-9038

COTTAM, GRANT, b Sandy, Utah, Aug 26, 18; m 42; c 5. ECOLOGY, ENVIRONMENTAL SCIENCES. *Educ:* Univ Utah, BA, 39; Univ Wis, PhD(bot), 48. *Prof Exp:* Teacher high sch, Utah, 39-40; asst prof bot, Univ Hawaii, 48-49; from asst prof to prof, 49-86, chmn dept, 70-73 & 79-81, chmn instrnl prog, Inst Environ Studies, 74-78, EMER PROF BOT, UNIV WIS-MADISON, 86- *Concurrent Pos:* Guggenheim fel, 54-55. *Mem:* Ecol Soc Am; Am Inst Biol Sci. *Res:* Methods of phytosociology; forest phytosociology; interdisciplinary environmental research. *Mailing Add:* Dept Bot Birge Hall Univ Wis Madison WI 53706

COTTELL, PHILIP L, b Ladysmith, BC, June 21, 41; m 65; c 2. FOREST HARVESTING & PRODUCTS, ERGONOMICS. *Educ:* Univ BC, BSF, 66, MF, 67; Yale Univ, PhD(forestry), 72. *Prof Exp:* Assoc res forester, Pulp & Paper Res Inst Can, Montreal, 67-69, res forester, Vancouver, 72-74; res dir, Forest Eng Res Inst Can, 75-77; assoc prof, Fac Forestry, Univ BC, 77-80; dir, Wood Harvesting Res, 80-85, DIR BLDG MAT DIV, MACMILLAN BLOEDEL LTD, 85- *Concurrent Pos:* Adj prof, Fac Forestry, Univ BC, 80- *Mem:* Can Inst Forestry; Asn BC Prof Foresters; Sigma Xi; Can Pulp & Paper Asn; Human Factors Asn Can. *Res:* Analysis of factors influencing productivity and economics of forest harvesting operations. *Mailing Add:* 3971 W 33 Ave Vancouver BC V6N 2H7 Can

COTTEN, GEORGE RICHARD, b Warsaw, Poland, Mar 21, 29; US citizen; m 57; c 3. POLYMER CHEMISTRY, PHYSICAL CHEMISTRY. *Educ:* Univ London, BS, 52, PhD(chem), 56. *Prof Exp:* Nat Res Coun Can fel, 56-57; res chemist, Visking Co Div, Union Carbide Corp, 58-60 & Fibers Div, Am Cyanamid Co, 60-63; group leader carbon black res dept, 63-69, from res assoc to sr res assoc, Res & Develop Div, 69-77, MEM TECH STAFF, CARBON BLACK RES & DEVELOP, CABOT CORP, 77- *Mem:* Am Chem Soc; assoc Brit Inst Rubber Indust. *Res:* Reaction kinetics; characterization and viscoelastic properties of polymers; rubber reinforcement; statistics; polymer characterization. *Mailing Add:* 37 Lawrence Lane Lexington MA 02173

COTTER, DAVID JAMES, b Glens Falls, NY, July 24, 32; m 53; c 3. PLANT ECOLOGY. *Educ:* Univ Ala, BS, 52, AB, 53, MS, 55; Emory Univ, PhD(biol), 58. *Prof Exp:* Res assoc, Atomic Energy Comn, Emory Univ, 57-58; from asst prof to assoc prof biol, Ala Col, 58-66; PROF BIOL, & CHMN DEPT, GA COL MILLEDGEVILLE, 66- *Concurrent Pos:* Res assoc, Atomic Energy Comn, Emory Univ, 63; consult, Bd Educ & dir, NSF Summer Insts, 63 & 64, Talladega, Shelby & Montgomery Counties, 64 & 65; NSF Undergrad Res Participation, 67, 70, 71, 73, 74, & 75; fel, Inst Radiation Ecol, Savannah River Plant, 65-67; staff biologist, Comn Undergrad Educ Biol Sci, Am Inst Biol Sci; consult, Outdoor Educ Inst, Ga Col, 69-; pres, Asn Southeastern Biologists, 73. *Mem:* Ecol Soc Am; Sigma Xi. *Res:* Radiation biology; effects of environmental stress on plant communities. *Mailing Add:* Dept Biol & Environ Sci Ga Col Milledgeville GA 31061

COTTER, DONALD JAMES, b Providence, RI, Jan 13, 30; m 52; c 3. HORTICULTURE, PLANT PHYSIOLOGY. *Educ:* Univ RI, BS, 52; Cornell Univ, MS, 54, PhD(veg crops), 56. *Prof Exp:* From asst prof to assoc prof hort, Univ Ky, 56-69; PROF HORT, NMEX STATE UNIV, 69- *Mem:* Am Soc Hort Sci; Sigma Xi; Int Soc Hort Sci. *Res:* Physiological and environmental studies on vegetable plants; greenhouse media for crop culture; water use on urban landscapes. *Mailing Add:* Dept Hort NMex State Univv Las Cruces NM 88001

COTTER, DOUGLAS ADRIAN, b Brockport, NY, Aug 15, 43; div; c 2. COMPUTER ENGINEERING, ELECTRICAL ENGINEERING. *Educ:* Duke Univ, BSEE, 65; NC State Univ, ME, 67, PhD(elec eng), 70. *Hon Degrees:* DMD, Harvard Bus Sch, 83. *Prof Exp:* Elec engr design, Elec Res Lab, Corning Glass Works, 66-70, sr res engr, Biomed Tech Ctr, 70-72, mgr comput syst res & develop, 73-76, mgr instrument develop, Sullivan Sci Park, 76-78, mgr bus develop, Corning Europe, 78-80, mgr, Med/Sci Res & Develop Portfolio, 80-83, dir, Clin Info Systs, Corning Med, 83-84; PRES, HEALTHCARE DECISIONS, INC, 84- *Concurrent Pos:* Adj asst prof elec eng, NC State Univ, 73-76. *Mem:* Sr mem Inst Elec & Electronics Engrs; Sigma Xi; Am Mgt Asn; Am Asn Clin Chem; Soc Med Decision Making. *Res:* Automatic design systems for digital computers; pattern recognition in medical applications; biomedical instrumentation. *Mailing Add:* Corning Med-Sci Portfol Res-Develop Div Medfield Indust Park Medfield MA 02052

COTTER, EDWARD, b Everett, Mass, May 30, 36. SEDIMENTOLOGY. *Educ:* Tufts Univ, BS, 58; Princeton Univ, MA, 61, PhD(geol), 63. *Prof Exp:* Res geologist, Jersey Prod Res Co, Okla, 62-64; asst prof geol, Tufts Univ, 64-65; asst prof, 65-76, ASSOC PROF GEOL, BUCKNELL UNIV, 76- *Mem:* Geol Soc Am; Soc Econ Paleont & Mineral; Int Asn Sedimentol. *Res:* Sedimentary petrology of carbonate and clastic rocks. *Mailing Add:* Dept Geol Bucknell Univ Lewisburg PA 17837

COTTER, EDWARD F, b Baltimore, Md, Feb 15, 10; m 45; c 1. INTERNAL MEDICINE. *Educ:* Univ Md, MD, 35. *Prof Exp:* Asst med, 39-40, instr path, 40-42, asst neurol, 40-47, instr, 47-48, from asst prof to assoc prof med, 47-77, EMER ASSOC PROF MED, SCH MED, UNIV MD, BALTIMORE CITY, 77- *Concurrent Pos:* Hitchcock fel neurosurg, Sch Med, Univ Md, 40-42; pvt pract, 46-63 & 69-; consult, Vet Admin, 50; dir demyelinating dis clin, Univ Md Hosp, 54-63; chief dept med, Md Gen Hosp, 52-63, dir educ, Dept Med, 63-69. *Mem:* AMA; Am Col Physicians. *Res:* Clinical medicine. *Mailing Add:* 110 Upnor Rd Baltimore MD 21212

COTTER, MARTHA ANN, b Granite City, Ill, Mar 18, 43; m 75. STATISTICAL MECHANICS, LIQUID CRYSTAL THEORY. *Educ:* Southern Ill Univ, Ba, 64; Georgetown Univ, PhD(chem), 69. *Prof Exp:* Assoc chem, Cornell Univ, 69-70, assoc chem physics, Bell Telephone Labs, 70-72; asst prof, 72-78, ASSOC PROF CHEM, RUTGERS UNIV, 78- *Concurrent Pos:* Res fel, Alfred P Sloan Found, 74-78; vis scientist, Bell Telephone Labs, Murray Hill, 75-76. *Res:* Equilibrium statistical mechanics; molecular theory of liquid crystalline and micellar phases; theory of liquids and liquid mixtures. *Mailing Add:* Dept Chem Rutgers Univ New Brunswick NJ 08903

COTTER, MAURICE JOSEPH, b New York, NY, Apr 20, 33. PHYSICS. *Educ:* Fordham Univ, AB, 54, MS, 59, PhD(physics), 62. *Prof Exp:* Mathematician, US Navy Bur Ships, Washington, DC, 54-55; jr res assoc, Brookhaven Nat Lab, 60-62; from asst prof to assoc prof, 62-79, PROF PHYSICS, QUEENS COL, NY, 79- *Concurrent Pos:* Res assoc, Chem Dept, Brookhaven Nat Lab, 69-; guest scientist, Dept Physics, State Univ NY Stony Brook, 75-76. *Mem:* Am Phys Soc; Sigma Xi. *Res:* Nuclear reactor physics; neutron spectroscopy; application of neutron activation analysis techniques to oil paintings. *Mailing Add:* Dept Physics Queens Col Flushing NY 11367

COTTER, RICHARD, b Brooklyn, NY, Dec 29, 43; m 69; c 2. BIOCHEMISTRY, CELL PHYSIOLOGY. *Educ:* St Johns Univ, BS, 67, PhD(cell physiol), 74; Adelphi Univ, MS, 69. *Prof Exp:* Biologist virol, State Dept Health, NY, 67; biol res asst biochem & pharmacol, US Army Res Inst Environ Med, 69-71; asst prof, New York Med Col, 76-77; mgr biochem nutrit sect, pharmacol dept, Baxter Travenol Labs, 85-87; assoc dir appl res, Clintec Nutrit Co, Baxter Health Care Corp, 87-89; CORP DIR RES & DEVELOP, GERBER PROD CO, 89- *Concurrent Pos:* Fel, NY Med Col, 74-76. *Mem:* AAAS; Soc Toxicol; Am Soc Parenteral & Enteral Nutrit; Am Physiol Soc; Am Inst Nutrit; Am Soc Clin Nutrit; Am Soc Qual Control. *Res:* Biochemistry and clinical nutrition; protein, carbohydrate, lipid and vitamin metabolism; isolation, characterization and physiological role of biological receptors; biochemical aspects of pharmacology and toxicology; development of new assay techniques; metabolism of injury and repair; organ preservation; cardiovascular disease prematurity; gerintology; immunochemistry, including monoclonal antibodies and immunoabsorption technology; infant and children care and nutrition; nutrition. *Mailing Add:* Res & Develop Dept Gerber Prod Co 445 State St Fremont MI 49412

COTTER, ROBERT JAMES, b Washington, DC, July 15, 43; m; c 1. ANALYTICAL CHEMISTRY. *Educ:* Col of Holy Cross, BS, 65; Johns Hopkins Univ, MS, 71, PhD(phys chem) 72. *Prof Exp:* Instr chem, Towson State Col, 72-74; asst prof anal chem, Gettysburg Col, 74-78; DIR MASS SPECTROS, NSF REGIONAL CTR & ASSOC PROF PHARMACOL, SCH MED, JOHNS HOPKINS UNIV, 78- *Concurrent Pos:* NASA fel, 65-68, fel, Witco Chem Corp, 68-69; Petrol Res Fund grant, 78; NSF grant, 78, 81, 82, 84 & 86; NIH grant, 85 & 86. *Mem:* Am Soc Mass Spectrometry; Am Chem Soc. *Res:* Mass spectroscopy; instrumentation and computer design applied to analytical chemistry; ionization methods for nonvolatile, high molecular weight biological compounds and laser desorption. *Mailing Add:* 314 Overhill Rd Baltimore MD 21210-2906

COTTER, ROBERT JAMES, b New Bedford, Mass, Apr 15, 30; m 59; c 3. ENGINEERING PLASTICS, COMPOSITES. *Educ:* Brown Univ, ScB, 51; Mass Inst Technol, PhD(org chem), 54. *Prof Exp:* Chemist, Res Dept, Bakelite Div, 54-56, group leader, Res & Develop Dept, Plastics Div, 56-61, sr group leader, Chem & Plastics Opers Div, 61-71, res assoc, 71-75, corp res fel, Union Carbide Corp, 75-86; sr res assoc, Amoco Performance Prod, 86-90; VPRES, DURHAM TECHNOL CONSULTS, 91- *Mem:* AAAS; Am Chem Soc; Sigma Xi; fel Asn Inst Chem. *Res:* Product and process development; polystyrenes; engineering plastics; organic chemicals; pollution control polymers; polyarylethers; matrix resins; emerging technologies, membranes, epoxy resins. *Mailing Add:* Four Surrey Lane Durham NH 03824

COTTER, SUSAN M, b Evanston, Ill, June 26, 43; m. VETERINARY MEDICINE, HOMATOLOGY & ONCOLOGY. *Educ:* Univ Ill, BS, 64, DMV, 66; Am Col Vet Int Med, dipl, 76. *Prof Exp:* Mem staff internal med, Angell Mem Animal Hosp, 66-81; assoc prof, 81-90, PROF MED, TUFTS UNIV SCH VET MED, 90- *Concurrent Pos:* Res assoc oncol, Sch Pub Health, Harvard Univ, 75-80, lectr cancer biol, 80- *Honors & Awards:* Woman Vet of Year, Women's Vet Med Asn, 78; Carnation Award, 83; Bucham Award, 85; Pumna Award, 89. *Mem:* Am Vet Med Asn; Vet Cancer Soc; Int Soc Res Leukemia & Related Dis. *Res:* Comparative oncology and hematology, primarily as it concerns feline leukemia virus; studies concerning epidemiology, immunology, and treatment of feline leukemia; comparative transfusion medicine. *Mailing Add:* Dept Med Tufts Univ Sch Vet Med 200 Westboro Rd North Grafton MA 01536

COTTER, WILLIAM BRYAN, JR, b Hartford, Conn, May 8, 26; m 48; c 6. GENETICS, ANATOMY. *Educ:* Wesleyan Univ, BA, 49, MA, 51; Yale Univ, PhD(zool), 56. *Prof Exp:* Asst prof biol, Col Charleston, 55-56; asst prof, Wesleyan Univ, 56-57; asst prof biol, Col Charleston, 57-59; teaching fel anat, Med Col SC, 59-60; from asst prof to assoc prof, 60-74, actg chmn dept, 63-64, PROF ANAT, MED CTR, UNIV KY, 74- *Mem:* AAAS; Am Soc Human Genetics; Soc Study Evolution; Genetics Soc Am; Am Asn Anat; Sigma Xi. *Res:* Physiological genetics and evolution; effects of genes on behavior. *Mailing Add:* Dept Anat Univ Ky Med Ctr Lexington KY 40536

COTTINGHAM, JAMES GARRY, b Salt Lake City, Utah, June 7, 27; m 46; c 2. ACCELERATOR ENGINEERING, SOLAR ENERGY. *Educ:* Univ Ill, BS, 49. *Prof Exp:* Assoc physicist accelerator develop, Brookhaven Nat Lab, 59- 49-58; sr engr radar, RCA Corp, 58-59; elec engr accelerator develop, 59-76, SR ENGR SOLAR & ACCELERATOR ENG, BROOKHAVEN NAT LAB, 76-, SR ELEC ENGR, 79- *Res:* Development of proton accelerators such as Cosmotron, AGS and SSC; solar heat pump, cooling and heliostat systems. *Mailing Add:* Ten Cedar Ctr Center Moriches NY 11934

COTTLE, MERVA KATHRYN WARREN, b Calgary, Alta, Oct 8, 28; m 50; c 3. PHYSIOLOGY, PHARMACOLOGY. *Educ:* Univ BC, BA, 49, MA, 51; Univ Wash, PhD(physiol), 56; Univ Alta, dipl(HSA), 82. *Prof Exp:* Lectr physiol, Univ Alta, 56-67; Wellcome res fel, Inst Physiol, Agr Res Coun, Cambridge, Eng, 67-68; res assoc & lectr pharmacol, Univ Alta, 69-76, sessional lectr, 76-77, asst prof pharm, 78-81; consult epidemiol, Alta Govt, 82-83; sr researcher, dept physiol, Laval Univ, 83-84; RES ASSOC OCCUP MED, UNIV ALTA, 85- *Honors & Awards:* Elected Mem, Sigma Xi, 54. *Mem:* Can Physiol Soc; Soc Toxicol Can. *Res:* Histological studies of central nervous system and peripheral nervous system, including afferent and efferent innervation of the heart; fetal maternal physiology; geriatic medicine and epidemiology, brown adipose tissue innervation and development; occupational medicine and epidemiology. *Mailing Add:* Dept Occup Med Health Serv Admin & Community Med Univ Alta Edmonton AB T6G 2G3 Can

COTTLE, RICHARD W, b Chicago, Ill, June 29, 34; m 59; c 2. MATHEMATICS, OPERATIONS RESEARCH. *Educ:* Harvard Univ, AB, 57, AM, 58; Univ Calif, Berkeley, PhD(math), 64. *Prof Exp:* Instr math, Middlesex Sch, Mass, 58-60; mem tech staff, Bell Tel Labs, Inc, NJ, 64-66; assoc prof, 69-73, PROF OPERS RES, STANFORD, UNIV, 73- *Concurrent Pos:* Alexander von Humboldt Found sr scientist award. *Mem:* Am Math Soc; Math Asn Am; Inst Mgt Sci; Math Prog Soc; Soc Indust & Appl Math. *Res:* Mathematical programming. *Mailing Add:* Dept Opers Res Stanford Univ Stanford CA 94305

COTTLE, WALTER HENRY, b Edmonton, June 9, 21; m 50; c 3. PHYSIOLOGY. *Educ:* Univ BC, BA, 49, MA, 51; Univ Wash, Seattle, PhD, 56. *Prof Exp:* From asst prof to assoc prof physiol, 56-76, prof phys educ, EMER PROF, UNIV ALTA, 76- *Mem:* Can Physiol Soc; Can Asn Sport Sci; Am Physiol Soc. *Res:* Thermoregulatory responses to environmental temperature and exercise. *Mailing Add:* 3951 W 20th Ave Vancouver BC V6S 1G3 Can

COTTOM, MELVIN C(LYDE), b Coffeyville, Kans, Oct 11, 24; m 68. ELECTRICAL ENGINEERING. *Educ:* Univ Kans, BSEE, 45, MS, 48. *Prof Exp:* Instr elec eng, Univ Kans, 45-50; elec design engr, Black & Veatch, Consult Engr, Mo, 50-55; ASST PROF ELEC ENG, KANS STATE UNIV, 55- *Mem:* Sr mem Inst Elec & Electronics Engrs. *Res:* Electrical noise generated at the sliding contact; a study of random noise generated at slip ring-carbon brush contacts. *Mailing Add:* Dept Elec & Comput Eng Kans State Univ Manhattan KS 66506

COTTON, FRANK ALBERT, b Philadelphia, Pa, Apr 9, 30; m 59; c 2. CHEMISTRY. *Educ:* Temple Univ, AB, 51; Harvard Univ, PhD(chem), 55. *Hon Degrees:* Numerous DSc from US & foreign univs, 63-89. *Prof Exp:* From instr to prof chem, Mass Inst Technol, 55-69; Dreyfus prof, 69-71, Robert A Welch prof, 72-73, ROBERT A WELCH DISTINGUISHED PROF CHEM, TEX A&M UNIV, 73-, DIR, LAB MOLECULAR STRUCT & BONDING, 82- *Concurrent Pos:* Guggenheim fel, 56 & 89; Alfred P Sloan Found fel, 61-65; vis prof, Nat Univ Buenos Aires, 65, Univ Strasbourg, 75; lectr, var US & Foreign Univs, 62-78; Todd Prof, Univ Cambridge, 85-86; hon fel, Indian Acad Sci, 85; var ed bd activities, 61-86. *Honors & Awards:* Am Chem Soc Award, 62, Award Advan Inorg Chem, 74, Nichols Medal & Harrison Howe Award, 75, Edgar Fahs Smith Award & Linus Pauling Medal, 76, Southwest Regional Award, 77, Will & Gibbs Medal, 80; Baekeland Medal, 63, FP Dwyer mem lectr & medallist, Univ New South Wales, 66; William Lloyd Evans Award, Ohio State Univ, 73; John G Kirkwood Medal, Yale Univ, 78; Nyhdm Medal, Royal Soc Chem, 82; Nat Medal Sci, 82; Award Phys & Math Sci, NY Acad Sci, 83; Chem Sci Award, Nat Acad Sci, 90. *Mem:* Nat Acad Sci; hon mem NY Acad Sci; Am Acad Arts & Sci; Am Chem Soc; hon mem Royal Soc Chem; Am Soc Biol Chemists; Ital Acad Sci; Indian Acad Sci; Indian Nat Sci Acad; Royal Danish Acad Sci & Lett; hon mem Ital Chem Soc. *Res:* Application of valence theory and physical and preparative studies to elucidate molecular structures and bonding in inorganic compounds; molecular structure of enzymes and proteins; author var articles on inorganic chemistry. *Mailing Add:* Dept Chem Tex A&M Univ College Station TX 77843

COTTON, FRANK ETHRIDGE, JR, b Corinth, Miss, Aug 14, 23; m 78; c 3. INDUSTRIAL ENGINEERING, PETROLEUM ECONOMICS. *Educ:* Miss State Univ, BS, 46 & 47; Univ Pittsburgh, MLitt, 51, PhD(econ), 62. *Prof Exp:* Indust engr, Westinghouse Elec Corp, Pa, 47-51; sr engr-economist, int hqs, Gulf Oil Corp, 51-58; dir eng exten serv, 58-68, assoc prof indust eng, 59-62, PROF & HEAD DEPT INDUST ENG, MISS STATE UNIV, 62- *Concurrent Pos:* Consult, indust, NASA & res orgns, 58-; rep dir & bd dirs, Engrs Coun Prof Develop, 74-82. *Mem:* Fel Inst Indust Engrs (vpres, 67-69, pres, 70-71); Am Soc Eng Educ; Soc Petrol Engrs. *Res:* Petroleum economics; engineering economic analysis; technical and management planning. *Mailing Add:* PO Box 6257 Mississippi State MS 39762

COTTON, IRA WALTER, b New York, NY, Dec 29, 45; m 73; c 2. COMPUTER SYSTEMS INTEGRATION, DATA COMMUNICATIONS. *Educ:* Brown Univ, BA, 67; Univ Pa, MSE, 70; George Washington Univ, DBA, 79. *Prof Exp:* Sr systs anal, Univac, Div Sperry Rand Corp, 67-71; mem tech staff, Mitre Corp, 71-72; supvry computer scientist, US Dept Com, Nat Bur Standards, 72-80; prin, Booz, Allen & Hamilton, Inc, 80-90; CHIEF ENG, MARTIN MARIETTA CORP, 90- *Concurrent Pos:* Partner, Computer Network Assocs, 74-; adj assoc prof, George Washington Univ, 80-82. *Mem:* Asn Comput Mach; Inst Elec & Electronics Computer Soc; Int Coun Computer Commun. *Res:* Computer systems integration; data communications especially local area networks; office automation; cost-benefit analysis and computer procurement and acquisition; automation requirements analysis. *Mailing Add:* 27 Lily Pond Ct Rockville MD 20852-4230

COTTON, JOHN EDWARD, b Minneapolis, Minn, Dec 21, 24; m 51; c 7. ANALYTICAL CHEMISTRY. *Educ:* San Francisco State Col, AB, 53; Univ Ore, MS, 55, PhD(phys chem), 59. *Prof Exp:* Asst, Univ Ore, 54-59; phys chemist, Boeing Co, 59-87; RETIRED. *Mem:* AAAS; Am Chem Soc; Sigma Xi; Am Soc Mass Spectrometry. *Res:* Transport properties of gases; polarography; gas chromatography; mass spectroscopy. *Mailing Add:* 2512 102nd Ave NE Bellevue WA 98004

COTTON, ROBERT HENRY, b Newton, Mass, Nov 17, 14; m 48; c 3. FOOD SCIENCE, NUTRITION. *Educ:* Bowdoin Col, BS, 37; Mass Inst Technol, SM, 39; Pa State Univ, PhD(plant nutrit), 44. *Prof Exp:* Chemist, Gen Elec Co, Mass, 39-40; asst plant nutrit, Pa State Univ, 40-43, instr & asst prof human nutrit res, 43-45; dir, Plymouth, Fla Div, Nat Res Corp, 45-47; prof & supvry chemist, Citrus Exp Sta, Univ Fla, 47-48; dir res, Holly Sugar Corp, 48-53 & Huron Milling Co, 54-58; dir res, ITT Continental Baking Co, Inc, div Int Tel & Tel Corp, 58-65, vpres, 65-79, chief scientist, ITT Food Group, 74-79; pres, Consulfood Inc, 81-89; RETIRED. *Concurrent Pos:* Mem indust adv comt, Sugar Res Found, 48-53 & sci adv comt, Am Inst Baking, 58-84; chmn tech liaison comt, Am Bakers Asn-US Dept Agr, 59-77; mem vis comt, Dept Nutrit & Food Sci, Mass Inst Technol, 65-69; mem panel V-3, White House Conf Food, Nutrit & Health, 69; chmn comt cereals & gen prod adv bd mil personnel supplies, Nat Res Coun-Nat Acad Sci, 69-74, adv comt nutrit guidelines for foods, 70, dir gen, Found Chile, Santiago, Chile, 75-78; consult nutrit, New Prod Develop & Mgt, 79-89. *Honors & Awards:* C M Frey Award, Am Asn Cereal Chem, 78; Babcock-Hart Award, Inst Food Technol & Nutrit Found, 79. *Mem:* Fel AAAS; Am Chem Soc; Inst Food Technol; Asn Res Dir (pres, 64-65); Am Asn Cereal Chem (pres, 65-66). *Res:* Plant nutrition; food technology; analytical chemistry; new techniques to increase storage life of orange juice powder; cattle nutrition; sugar beet technology; by-product development; application food science to combat malnutrition; cereal science; helped develop first successful frozen orange juice concentrate; directed development of high fiber, low calorie bread and various new frozen foods; baking chemistry. *Mailing Add:* 570 Hammock Ct Marco FL 33937

COTTON, THERESE MARIE, b Peru, Ill. RAMAN SPECTROSCOPY, ELECTROCHEMISTRY. *Educ:* Bradley Univ, AB, 61; Northwestern Univ, Ill, PhD(chem), 76. *Prof Exp:* Res asst, Northern Regional Lab, USDA, 61-63; res asst, Argonne Nat Lab, 66-71; fel, Northwestern Univ, 76-81; asst prof anal chem, Ill Inst Technol, 81-; Dept Chem, Univ Nebr, Lincoln; PROF, IOWA STATE UNIV. *Concurrent Pos:* Vis asst prof, Northwestern Univ, 81-89. *Mem:* Am Chem Soc; Am Soc Photobiol; Biophys Soc; Electrochem Soc; Asn Women Sci. *Res:* Electroanalytical methods for the study of electron transfer reactions in biomolecules, especially photosynthetic preparations; application of surface enhanced Raman spectroscopy to proteins adsorbed at electrodes and as a sensitive analytical method for detecting low concentrations of chromophoric species. *Mailing Add:* Iowa State Univ Ames IA 50011-0061

COTTON, WILLIAM REUBEN, b Little Falls, NY, Oct 27, 40; m; c 2. ATMOSPHERIC SCIENCE, CLOUD PHYSICS. *Educ:* State Univ NY Albany, BS, 64, MS, 66; Pa State Univ, PhD(meteorol), 70. *Prof Exp:* Res asst cloud physics, Atmospheric Sci Res Ctr, State Univ NY, 64-66 & Pa State Univ, 66-70; res meteorologist, Exp Meteorol Lab, Nat Oceanic & Atmospheric Admin, 70-74; asst prof, 75-78, assoc prof, 78-80, PROF ATMOSPHERIC SCI, COLO STATE UNIV, 80- *Mem:* Am Meteorol Soc; Sigma Xi. *Res:* Numerical modeling and observational analysis of the physics and dynamics of cumulus clouds and convective mesoscale systems; modification of convective systems. *Mailing Add:* 2721 Davis Ranch Rd PO Box 74 Belvue CO 80512

COTTON, WILLIAM ROBERT, b Miami, Fla, Nov 29, 31; m 74; c 4. MICROSCOPIC ANATOMY, ORAL MICROANATOMY. *Educ:* Univ Md, DDS, 55; Northwestern Univ, MS, 63; Roosevelt Univ, MA, 73; George Washington Univ, EdS, 80. *Prof Exp:* Asst dent officer, Naval Training Ctr, Bainbridge, Md & Camp Lejeune, NC, Naval Med Res Inst, 55-57 & Mobilization Team, Miami, Fla, 57-58, asst dent officer & clin supvr, Dent Detachment, Marine Corps Sch, Quantico, Va, 58-59, postgrad officer, Naval Dent Sch, Bethesda, Md, 59-60, asst dent officer, USS F D Roosevelt (CVA-42), Mayport, Fla, 60-61, head, Exp Path Div, Dent Res Dept, Naval Med Res Inst, Nat Naval Med Ctr, 63-67, dent officer, USS Fulton (AS-11), New London, Conn, 67-69, chief, Histopath Div, Naval Dent Res Inst, Great Lakes, Ill, 69-76, exec officer, 72-73, dep cmndg officer, 73-76, chmn, Dent Sci Dept, 76-79, dir, Casualty Care Res Prog Ctr, Naval Med Res Inst, 79-81; assoc prof & dir res, dept oper dent, Temple Univ, 81-83; PROF & CHMN, DEPT OPER DENT, GEORGETOWN UNIV, 83- *Concurrent Pos:* Mem sect, Working Group Five, Comn on Mat, Instruments, Equipment & Therapeut, Int Dent Fedn, 74-, consult, Intercomn Group Uniform Definition of Dent Terms, 77-; mem adv comt, Dent Lab Technol Prog, Sch Tech Careers, Southern Ill Univ, Carbondale, 74-84; lectr microbiol, Nat Naval Dent Ctr, 76-81; res assoc, Nat Res Coun, 77-81; ed rev, J Dent Res, 76-85 & J Oper Dent, 84-; consult, Naval Dent Res Inst, Great Lakes, Ill, 81-86; consult & lectr, Naval Dent Sch, Bethesda, Md, 83-; bd dir, Dist Columbia Dent Soc, 86- *Mem:* NY Acad Sci; Soc Exp Biol & Med; Am Dent Asn; Int Asn Dent Res; Am Asn Dent Res. *Res:* Biological and toxicity testing of dental materials in animals and human clinical trials; bone resorption research-introduced osteopetrotic (tl) rat which has deficient bone resorption; dental caries research. *Mailing Add:* 11816 Winterset Terr Potomac MD 20854

COTTON, WYATT DANIEL, b Mexia, Tex, Feb 2, 43; m 68; c 3. ORGANIC CHEMISTRY, FOOD PRODUCT DEVELOPMENT. *Educ:* Calif State Univ, Los Angeles, BS, 69; Univ Calif, Los Angeles, PhD(org chem), 74. *Prof Exp:* res chemist phys & synthetic org chem, 74-77, TECH BRAND MGR, FOOD PROD DEVELOP, PROCTER & GAMBLE CO, 77- *Mem:* Am Chem Soc. *Res:* Amino acid synthesis; sultone chemistry; synthetic sweeteners; fats and oil chemistry; food science; soaps and detergents. *Mailing Add:* 12070 Springdale Lake Dr Cincinnati OH 45246

COTTONE, JAMES ANTHONY, b Syracuse, NY, Apr 10, 47; div. INFECTION CONTROL, FORENSIC ODONTOLOGY. *Educ:* Univ Pa, BA, 68; Tufts Univ, DMD, 72; Ind Univ, MS, 77. *Prof Exp:* Fel oral diag & med, Ind Univ Sch Dent, 75-77; from asst prof to assoc prof, 77-85, PROF & HEAD ORAL MED, UNIV TEX HEALTH SCI CTR, SAN ANTONIO DENT SCH, 85- *Concurrent Pos:* Consult, Bexar County Med Examr Off, 77-, Coun Dent Therapeut, Am Dent Asn, 84-; consult, Nat High Blood

Pressure Prog, 80-; consult, Merck Sharp & Dohme, 82-, Mead Johnson Pharmaceut Div, 83-, Parke Davis Co, 84-86; ed, Oral Surg, Oral Path, Oral Med, 84- *Mem:* Am Dent Asn; Orgn Teachers Oral Diag, (pres, 83-84); Fedn Dent Diag Sci (secy, 83-84); Am Soc Forensic Odontol (pres, 86-87); Am Asn Dent Schs (vpres, 89-). *Res:* Hepatitis B, delta hepatitis, AIDS; infection control; oral diagnosis; forensic dentistry. *Mailing Add:* Dept Dent Diag Sci Univ Tex Dent Sch 7703 Floyd Curl Dr San Antonio TX 78284

COTTONY, HERMAN VLADIMIR, b Nizhni-Novgorod, Russia, Mar 27, 09; nat US; wid; c 2. ELECTRICAL ENGINEERING. *Educ:* Cooper Union, BS, 32, EE, 46; Columbia Univ, MS, 33. *Prof Exp:* Res engr, Sonotone Corp, 35-37; physicist, Nat Bur Standards, 37-41, electronic engr & chief antenna res sect, 46-65; prog leader antennas, Environ Sci Serv Admin, 65-70; consult, Off Telecommun, Inst Telecommun Sci, 70-73; CONSULT ENGR, 73- *Concurrent Pos:* Mem, Comns A and B, US Nat Comt, Int Union Radio Sci; ed, Trans Antennas & Propagation, 62-65. *Mem:* Fel Inst Elec & Electronics Engrs; Am Phys Soc; AAAS; Antennas & Propagation Soc; Microwave Theory & Tech Soc; Econ Soc. *Res:* Antennas; measurement of electromagnetic fields; radio wave propagation; antenna data processing; radar; systems engineering. *Mailing Add:* 5204 Wilson Lane Bethesda MD 20814-2408

COTTRELL, IAN WILLIAM, b York, Eng, June 18, 43; m 69. CARBOHYDRATE CHEMISTRY. *Educ:* Univ Edinburgh, BSc, 65, PhD(polysaccharide chem), 68. *Prof Exp:* Fel carbohydrate chem, Trent Univ, 68-70; res chem guar gum, Stein Hall & Co, Celanese Corp, 70-72; sr res chemist, Kelco Co, Merck, Inc, 72-73, sect head, 73-75; asst res dir, Kelco Div, Merck & Co, Inc, 76-79, dir, basic res & develop, 79-; TECH DIR, WATER SOLUBLE POLYMER DIV, RHONE-POULENC. *Mem:* Am Chem Soc; AAAS; Brit Biophys Soc. *Res:* Preparation of industrially useful polymers with special emphasis on polysaccharides. *Mailing Add:* Rhone-Poulenc 9800 Bluegrass Pkwy Jeffersontown KY 40299

COTTRELL, ROGER LESLIE ANDERTON, b Birmingham, Eng, Jan 20, 40; m 67; c 2. HIGH ENERGY PHYSICS, COMPUTING. *Educ:* Manchester Univ, Eng, BSc, 62, PhD(nuclear physics), 67. *Prof Exp:* Asst lectr physics, Manchester Univ, Eng, 65-66, res asst nuclear physics, 66-67; STAFF MEM, STANFORD LINEAR ACCELERATOR CTR, STANFORD UNIV, 67- *Concurrent Pos:* Vis scientist, Europ Orgn Nuclear Res, Switz, 72-73; mem software working group, Nuclear Instrumentation Module-Computer-Aided Measurement & Control, AEC, 72-74; vis scientist, IBM UK Kingdom Labs, Ltd, Hursley, 79-80; mem, Tech Coord Comt, High energy Physics Network, Dept Energy, 85- Energy Sci Comt, 88- *Mem:* Brit Comput Soc. *Res:* Local and wide area computer-to-computer networks; terminal-to-computer networks; distributed computing; one US patent. *Mailing Add:* Stanford Linear Accelerator Ctr PO Box 4349 Stanford CA 94309

COTTRELL, ROY, b Center, Mo, Nov 3, 39. APPLICATION DEVELOPMENT. *Educ:* Univ Mo, BA, 61. *Prof Exp:* Dir oper syst develop, 87-89, DIR RES, EVAL & IMPLEMENTATION TECHNOL, UNION PAC TECHNOLOGIES, 89- *Concurrent Pos:* Chmn, Asn Comput Mach, 74-76, Computer Personnel Res, 76-78 & Bus Data Processing Group, 81-83, mem, Spec Interest Group Bd, 82- *Mem:* Asn Comput Mach. *Mailing Add:* Union Pac Technologies 7930 Clayton Rd Rm 256 St Louis MO 63117-1368

COTTRELL, STEPHEN F, b Trenton, NJ, Apr 17, 43. BIOLOGY. *Educ:* Wilkes Col, BA, 65; Rutgers Univ, MS & PhD(cell biol), 70. *Prof Exp:* Res fel biochem, Univ Chicago, 70-72; res fel chem, Ind Univ, Bloomington, 72-73; res assoc planetary biol, NASA-Ames Res Ctr, Moffett Field, Calif, 73-75; from asst prof to assoc prof, 75-83, PROF BIOL, CITY UNIV NY, BROOKLYN COL, 83- *Concurrent Pos:* Prin investr, NASA, 76-77 & 82-84, NIH, 77-80, 78-82, 79-84 & 81-82, Dept Energy, 80-83; co-prin investr, USDA, 81-84; travel grant, Geront Soc Am, 81. *Res:* Mitochondrial involvement in the aging process; mitochondrial autonomy and biogenesis with special emphasis on control interactions between the nucleocytoplasmic and mitochondrial protein synthesizing and genetic systems; examination of mitochondrial structural and functional development during the yeast cell cycle; ethanol production from biomass. *Mailing Add:* Dept Biol Col Univ NY Brooklyn Col Bedford & Ave H Brooklyn NY 11210

COTTRELL, THOMAS S, b Chicago, Ill, Feb 2, 34; m 59; c 3. PATHOLOGY. *Educ:* Brown Univ, AB, 55; Columbia Univ, MD, 65. *Prof Exp:* Fel path, Columbia Univ-Presby Hosp, 65-67; fel path, Yale Univ, 67-68; instr, Columbia Univ, 68-69; assoc prof path, New York Med Col, 68-79; ASSOC PROF PATH & ASSOC DEAN, SCH MED, STATE UNIV NY STONY BROOK, 79- *Concurrent Pos:* Asst vis pathologist, Metrop Hosp, NY, 68-79; Markle Scholar, Acad Med, 69-74; assoc attend physician, West County Med Ctr Hosp, 78-79; attend pathologist, Univ Hosp, Stony Brook, NY, 80-; John & Mary R Markel scholar acad med, 69-74. *Mem:* Fel NY Acad Med; NY Acad Sci; AAAS. *Res:* Ultrastructural correlation of normal and abnormal cardio-respiratory physiology. *Mailing Add:* Sch Med State Univ NY Stony Brook NY 11794-8431

COTTRELL, WILLIAM BARBER, nuclear engineering; deceased, see previous edition for last biography

COTTS, ARTHUR C(LEMENT), b Kansas City, Mo, July 27, 22; m 51; c 2. ELECTRICAL ENGINEERING. *Educ:* Kans State Univ, BS, 49, MS, 50. *Prof Exp:* Sr instr radio & TV, Cent Radio & TV Sch, Mo, 46; asst appl mech, Kans State Univ, 49, asst elec eng, 49-50; engr, Eng Mech Div, Midwest Res Inst, 50-51; engr, Appl Physics Lab, Johns Hopkins Univ, 51-55; res engr & head analog comput appln group, Midwest Res Inst, 55-57; proj supvr, Polaris Submarine Navig, Appl Physics Lab, 57-71, MEM PRIN PROF STAFF, APPL PHYSICS LAB & PROG SUPVR, POLARIS SUBMARINE PATROL ANALYST, JOHNS HOPKINS UNIV, 71- *Concurrent Pos:* Lectr, Cath Univ Am, 51-53. *Mem:* AAAS; Sigma Xi; sr mem Inst Elec & Electronics Engrs. *Res:* Operation of reflex klystron oscillators; control systems; inertial navigation; system analysis and evaluation. *Mailing Add:* 11102 Nicholas Dr Silver Springs MD 20902

COTTS, DAVID BRYAN, b Washington, DC, July 26, 54; m 80. POLYMER PHYSICS, LIGHT SCATTERING. *Educ:* Eckerd Col, BS, 75; Carnegie-Mellon Univ, PhD(chem), 79. *Prof Exp:* POLYMER CHEMIST, SRI INT, 79- *Mem:* AAAS; Am Chem Soc. *Res:* Polymer physical chemistry, molecular characterization, morphology, rheology; conducting polymers; application of light scattering techniques to the study of solids and liquids. *Mailing Add:* 1400 Oak Creek Dr Apt 306 Palo Alto CA 94304-2065

COTTS, PATRICIA METZGER, b New York, NY, Sept 14, 53; m 79. LIGHT-SCATTERING, HIGH TEMPERATURE POLYMERS. *Educ:* Col William & Mary, BS, 75; Carnegie-Mellon Univ, PhD(chem), 79. *Prof Exp:* Fel, 80-81, RES STAFF MEM, IBM RES LAB, 81- *Mem:* Am Chem Soc; Am Phys Soc. *Res:* Structure/property relations in polymers with emphasis on new high performance polymers and characterization of these with light scattering techniques. *Mailing Add:* Dept K91-80 IBM Almaden Res Ctr 650 Harry Rd San Jose CA 95120-6099

COTTS, ROBERT MILO, b Green Bay, Wis, Aug 22, 27; m 50; c 4. NUCLEAR MAGNETIC RESONANCE. *Educ:* Univ Wis, BS, 50; Univ Calif, PhD(physics), 54. *Prof Exp:* Instr physics, Stanford Univ, 54-57; from asst prof to assoc prof, 57-68, PROF PHYSICS, CORNELL UNIV, 68- *Concurrent Pos:* Physicist, Nat Bur Standards, 63-64; vis prof physics, Univ BC, 70-71 & Univ Warwick, UK, 78, vis fel, Cambridge Univ, UK, 86. *Mem:* Am Phys Soc; Am Asn Physics Teachers. *Res:* Solid state physics. *Mailing Add:* Clark Hall Cornell Univ Ithaca NY 14853

COTTY, VAL FRANCIS, b New York, NY, July 11, 26; m 51; c 3. PHARMACOLOGY, TOXICOLOGY. *Educ:* St John's Col, BS, 48, MSc, 50; NY Univ, PhD(biol), 55. *Prof Exp:* From instr to asst prof biol, St John's Col, 50-55; staff mem, Boyce Thompson Inst, 55-57; head dept biochem, Bristol-Myers Co, 57-70, dir biol res, 70-82, dir res, 82-84, dir res & develop affairs, 84-86; pres, Proquest, Inc, 87-88; CONSULT, 89- *Concurrent Pos:* Res assoc, Sch Med, NY Univ, 61-65. *Honors & Awards:* Fel, Royal Soc Med. *Mem:* NY Acad Sci; Am Chem Soc; Am Soc Microbiol; Int Asn Dent Res; Soc Toxicol. *Res:* Pharmacology and toxicology of drugs. *Mailing Add:* 236 Avon Rd Westfield NJ 07090

COTY, WILLIAM ALLEN, b Los Angeles, Calif, Apr 10, 48; m 65; c 3. BIOCHEMISTRY, ENDOCRINOLOGY. *Educ:* Calif Inst Technol, BS, 69; Sch Med, Johns Hopkins Univ, PhD(biochem), 74. *Prof Exp:* Fel biochem endocrinol, Col Med, Baylor Univ, 74-76; ASST PROF BIOCHEM, SCH MED, UNIV CALIF, LOS ANGELES, 77- *Concurrent Pos:* Fel, Nat Inst Arthritis, Metab & Digestive Dis, 74-76. *Mem:* Am Soc Biol Chemists; Endocrine Soc. *Res:* Mechanism of steroid hormone action; mechanism and regulation of calcium transport. *Mailing Add:* Microgenics Corp 2380 A Bissola Concord CA 94520

COUCH, HOUSTON BROWN, b Estill Springs, Tenn, July 1, 24; m 45; c 5. PLANT PATHOLOGY. *Educ:* Tenn Polytech Inst, BS, 50; Univ Calif, PhD(plant path), 54. *Prof Exp:* From asst prof to assoc prof plant path, Pa State Univ, 54-65; head dept, 65-75, PROF PLANT PHYSIOL & PATH, VA POLYTECH INST & STATE UNIV, 65- *Mem:* Am Phytopath Soc; Am Agron Soc; Soil Sci Soc Am; Crop Sci Soc Am; Sigma Xi. *Res:* Diseases of turfgrasses and forage crops; physiology of parasitism; role of physical environment in plant disease development. *Mailing Add:* Dept Plant Physiol & Path Va Polytech Inst & State Univ Blacksburg VA 24061-0331

COUCH, JACK GARY, b Pocatello, Idaho, Apr 5, 36; m 55; c 8. NUCLEAR PHYSICS. *Educ:* Utah State Univ, BS, 58; Vanderbilt Univ, MS, 59; Tex A&M Univ, PhD(physics), 66. *Prof Exp:* Chmn, dept phys sci, Brigham Young Univ, Hawaii Campus, 59-61; asst prof physics, Wash State Univ, 66-67; chmn dept, Southern Ore State Col, 67-74, from assoc prof to prof physics, 67-80; radiation physics, Fermi Nat Lab, 80-85; health physics head, Plasma Physics Lab, Princeton Univ, 85-88; PROF PHYSICS & HEALTH PHYSICS, BLOOMSBURG UNIV, 88- *Concurrent Pos:* Vis assoc prof, Univ Ill, 74-75. *Mem:* Am Asn Physics Teachers; Am Phys Soc; Health Physics Soc. *Res:* Heavy ion reactions; neutron scattering; health physics; molecular infrared spectroscopy; physics education; radon; health physics. *Mailing Add:* Dept Physics Bloomsburg Univ Bloomsburg PA 17815

COUCH, JAMES RUSSELL, b Grandview, Tex, June 10, 09; m 34; c 4. BIOCHEMISTRY, NUTRITION. *Educ:* Agr & Mech Col Tex, BS, 31, MS, 34; Univ Wis, PhD(biochem), 48. *Prof Exp:* From asst poultry husbandman to poultry husbandman, Exp Sta, Univ Tex, 31-41; prof, 48-74, EMER PROF BIOCHEM & POULTRY SCI, TEX A&M UNIV, 74- *Honors & Awards:* Am Feed Mfrs Award, Poultry Sci Asn, 51. *Mem:* AAAS; fel Poultry Sci Asn; Am Inst Nutrit; Fedn Am Socs Exp Biol; Am Soc Biol Chem. *Res:* Poultry nutrition; embryology; physiology; nutritional significance and metabolic functions of vitamins and trace elements in the domestic fowl; vitamin B12; folic acid; antibiotics; unidentified growth factors; trace elements; proteins and amino acids; fats and fatty acids. *Mailing Add:* Dept Poultry Sci Tex A&M Univ College Station TX 77843

COUCH, JAMES RUSSELL, JR, b Bryan, Tex, Oct 25, 39; m 64; c 2. NEUROLOGY, NEUROPHARMACOLOGY. *Educ:* Tex A&M Univ, BS, 61; Baylor Col Med, MD, 65, PhD(physiol), 66. *Prof Exp:* Nat Heart Inst fel, Baylor Col Med, 65-66; NIH staff fel, Lab Neuropharmacol, NIMH, 67-69; Nat Inst Neurol Dis & Stroke spec trainee, Wash Univ, 69-72; from asst prof to assoc prof neurol, Univ Kansas Med Ctr, 72-79; PROF & CHIEF, DIV NEUROL, SOUTHERN ILL UNIV SCH MED, SPRINGFIELD, ILL, 79- *Concurrent Pos:* Consult, Kansas City Vet Admin Hosp, Kansas City Gen Hosp, Kansas, Marion Vet Admin Hosp; attend staff, Mem Med Ctr, St John's Hosp, Springfield & Lincoln Develop Ctr, Lincoln, Ill, 79- *Mem:* Fel Am Acad Neurol (asst secy-treas, 85-87, secy-treas, 87-); Soc Neurosci; Am Neurol Asn; Am Asn Study Headache. *Res:* Neuropharmacology and neurotransmitters; movement disorders; headache; stroke. *Mailing Add:* Chief-Div Neurol Southern Ill Univ Sch Med PO Box 19230 Springfield IL 62794-9230

COUCH, JOHN ALEXANDER, b Washington, DC, Feb 12, 38; m 63; c 2. PATHOBIOLOGY, PROTOZOOLOGY. *Educ:* Univ Ala, BS, 61; Fla State Univ, MS, 64, PhD(morphogenesis, cell biol, parasitol), 71. *Prof Exp:* Teaching asst zool, Fla State Univ, 61-64; parasitologist, Biol Lab, Nat Marine Fisheries Serv, Nat Oceanic & Atmospheric Admin, 64-71; chief, pathobiol br, 85-88, PATHOBIOLOGIST, BIOL LAB, US ENVIRON PROTECTION AGENCY, 71-, SR RES SCIENTIST, 87- *Concurrent Pos:* US Dept Interior training assignment, Fla State Univ, 67-68; scientist-aquanaut, Tektite II, 70; fac assoc, Univ WFla, 75-; mem grad sch comt, 80- *Honors & Awards:* Sci & Technol Achievement Award for Ecol, Toxicol & Pathol Res, US Environ Protection gency, 81, 82, 86, 87 & 89. *Mem:* Sigma Xi; Gulf Estuarine Res Soc; Soc Invert Path; Soc Protozoologists. *Res:* Evolution of commensal-host relationships and morphogenesis of marine protozoa; aquatic animal pathology; toxicological pathology, interaction of pollutants and natural disease; neoplasia; experimental carcinogenesis; viruses of invertebrates. *Mailing Add:* Biol Lab US Environ Protect Agency Sabine Island Gulf Breeze FL 32561

COUCH, JOHN NATHANIEL, botany; deceased, see previous edition for last biography

COUCH, LEON WORTHINGTON, II, b Durham, NC, July 6, 41; m 64; c 3. ELECTRICAL ENGINEERING. *Educ:* Duke Univ, BS, 63; Univ Fla, ME, 64, PhD(elec eng), 68. *Prof Exp:* Asst elec eng, 63-68, from asst prof to assoc prof, 68-84, PROF ELEC ENG, 84-, ASSOC CHMN, ELEC ENG DEPT, UNIV FLA, 90- *Honors & Awards:* Centennial Medal, Inst Elec & Electronics Engrs, 84. *Mem:* Sr mem Inst Elec & Electronics Engrs; Am Soc Eng Educ. *Res:* Performance of communication and radar systems and subsystems in the presence of noise; modulation systems; realistic systems, analog and digital. *Mailing Add:* 3524 NW 51st Ave Gainesville FL 32605

COUCH, MARGARET WHELAND, b Chicago, Ill, Aug 27, 41; m 64; c 3. ORGANIC CHEMISTRY. *Educ:* Duke Univ, BS, 63; Univ Fla, MS, 66, PhD(org chem), 69. *Prof Exp:* res asst, dept radiol, 69-70, ASST RES PROF, DEPT RADIATION CHEM, UNIV FLA, 71-; RES CHEMIST, VET ADMIN MED CTR, GAINESVILLE, 71- *Mem:* Am Chem Soc; Sigma Xi. *Res:* Identification by means of mass spectrometry-gas chromatography of aromatic acids and amines present in biological fluids of patients with neurological disorders; radiopharmaceuticals for receptor imaging. *Mailing Add:* 3524 NW 51st Ave Gainesville FL 32605

COUCH, RICHARD W, b Dayton, Ohio, June 9, 31; m 62; c 2. GEOPHYSICS, SEISMOLOGY. *Educ:* Mich State Univ, BS, 58; Ore State Univ, MS, 63, PhD(geophys), 69. *Prof Exp:* Electronics engr, Gen Dynamics/Electronics, NY, 58-60, mem res staff, 60-62; res asst marine geophys, 63-65, res fel geophys, 65-66, from instr to asst prof, 66-73, ASSOC PROF GEOPHYS, SCH OCEANOG, ORE STATE UNIV, 73- *Mem:* Soc Exploration Geophys; Seismol Soc Am. *Res:* Structure and tectonics of continental margins of the Eastern Pacific Ocean; geophysical exploration for geothermal resources; earthquake seismology and geological hazards. *Mailing Add:* Sch Oceanog Ore State Univ Corvallis OR 97331

COUCH, RICHARD WESLEY, b Pryor, Okla, Mar 30, 37; m 60; c 2. PLANT PHYSIOLOGY. *Educ:* Okla State Univ, BS, 59; Univ Tenn, MS, 61; Auburn Univ, PhD(bot biochem), 66. *Prof Exp:* Asst county agent, Exten Serv, Univ Tenn, 61-63; prof biol, Athens Col, 65-73, chmn dept, 66-73; assoc prof, 73-77, PROF BIOL, ORAL ROBERTS UNIV, 77- *Concurrent Pos:* Consult, US Corps Engrs, contract, 68- *Mem:* Nat Sci Teachers Asn. *Res:* Herbicidal plant physiology; aquatic biology; aquatic weed control; aquatic plants of Oklahoma. *Mailing Add:* Dept Biol Oral Roberts Univ 7777 S Lewis Ave Tulsa OK 74171

COUCH, ROBERT BARNARD, b Guntersville, Ala, Sept 25, 30; m 55; c 4. INTERNAL MEDICINE, INFECTIOUS DISEASES. *Educ:* Vanderbilt Univ, BA, 52, MD, 56. *Prof Exp:* From intern to chief resident, Vanderbilt Univ Hosp, 56-61; clin assoc surg, Nat Cancer Inst, 57-59, sr investr, Lab Clin Invest, Nat Inst Allergy & Infectious Dis, 61-65, head clin virol, 65-66; assoc prof, 66-71, PROF MICROBIOL, IMMUNOL & MED, BAYLOR COL MED, 66- *Mem:* Am Fedn Clin Res; Soc Exp Biol & Med; Am Soc Microbiol; Am Soc Clin Invest; Am Asn Physicians; Am Soc Epidemiol; Infectious Dis Soc. *Res:* Clinical and general virology; immunology. *Mailing Add:* Dept Microbiol Baylor Col Med Houston TX 77030

COUCH, TERRY LEE, b Middletown, Pa, Jan 8, 44; m 65; c 2. ENTOMOLOGY. *Educ:* Franklin & Marshall Col, AB, 65; Pa State Univ, MS, 68, PhD(entom), 70. *Prof Exp:* Res entomologist, Abbott Labs, 70-75, group leader entom res, chem & agr prod div, 75-76, assoc res fel, sci ladder, 76-; dir res & develop, Pet Chemicals Inc, 83-86; PRES, BECKER MICROBIAL PROD, INC, 86- *Mem:* Entom Soc Am; Am Inst Biol Sci; Sigma Xi; Soc Invert Path. *Res:* Discovery and development of microbial insecticides and narrow spectrum and chemical insecticides. *Mailing Add:* 9464 NW 11th St Plantation FL 33322

COUCHELL, GUS PERRY, b Henderson, NC, Apr 14, 39; m 68. NUCLEAR PHYSICS. *Educ:* NC State Univ, BS, 61, MS, 63; Columbia Univ, PhD(physics), 68. *Prof Exp:* Asst prof physics & appl physics, 68-73, assoc prof, 73-79, PROF PHYSICS, UNIV LOWELL, 79- *Concurrent Pos:* Prin investr grants, NSF, 80- *Mem:* Am Nuclear Soc. *Res:* Isobaric analog states; nuclear spectroscopy; fast neutron scattering; nuclear resonance fluorescence; energy spectra of delayed neutrons following neutron-induced fission. *Mailing Add:* Five Old Colony Dr No A Westford MA 01886-1074

COUCHMAN, JAMES C, b Cincinnati, Ohio, Aug 28, 29; m 48; c 3. PHYSICS. *Educ:* Cent Col, Iowa, BA, 53; Vanderbilt Univ, MA, 55; Tex Christian Univ, PhD(math physics), 65. *Prof Exp:* Radiation safety area rep, Argonne Nat Lab, 55-56; nuclear engr, Convair, Tex, 56-58; sr nuclear engr, Gen Dynamics/Ft Worth, 58-61, 63-65; sci specialist, Edgerton, Germeshausen & Grier, Calif, 65-67; PROJ NUCLEAR PHYSICIST, GEN DYNAMICS CORP, 67- *Mem:* Health Physics Soc; Am Nuclear Soc. *Res:* Nuclear weapons effects research; aerospace nuclear safeguards; health physics; use of digital computer techniques in studying nuclear reactor, criticality; evaluation of potential environmental hazards associated with the uses of nuclear energy. *Mailing Add:* Gen Dynamics Corp MZ-5998 Fort Worth TX 76101

COUCHMAN, PETER ROBERT, b London, Eng, Jan 5, 47; m 77; c 2. POLYMER THERMODYNAMICS, SURFACE TENSION. *Educ:* Univ Surrey, Eng, BSc, 69; Univ Va, MSc, 72, PhD(mats sci), 76. *Prof Exp:* Postdoctoral fel polymer sci, Univ Mass, 76-77; asst prof macromolecular sci, Case Western Reserve Univ, 78; asst prof, 78-81, assoc prof, 81-87, PROF MECH & MAT SCI, RUTGERS UNIV, 87- *Concurrent Pos:* Consult, AT&T Bell Labs, 84-85. *Honors & Awards:* Gwathmey Award, Univ Va, 76. *Mem:* Fel Am Phys Soc; fel NAm Thermal Anal Soc. *Res:* Transition and equation of state behavior of homopolymers, copolymers and blends; calculation of surface tension for simple and polymeric liquids. *Mailing Add:* Dept Mechs & Mat Sci Rutgers Univ PO Box 909 Piscataway NJ 08854

COUCOUVANIS, DIMITRI N, b Athens, Greece, Nov 20, 40; US citizen; m 63; c 3. INORGANIC CHEMISTRY, CRYSTALLOGRAPHY. *Educ:* Allegheny Col, BS, 63; Case Inst Technol, PhD(chem), 67. *Prof Exp:* Res assoc, Case Inst Technol, 67 & Columbia Univ, 67-68; assoc prof, 68-75, prof chem, Univ Iowa, 75-; DEPT CHEM, UNIV MICH, ANN ARBOR. *Mem:* Am Chem Soc. *Res:* Synthesis and structure of polynuclear coordination complexes and their use as models for metal containing enzymes. *Mailing Add:* Dept Chem Univ Mich Ann Arbor MI 48109

COUDRON, THOMAS A, b Marshall, Minn, 1951. ANALYTICAL BIOCHEMISTRY, INSECT BIOCHEMISTRY. *Educ:* St John's Univ, BS, 73; NDak State Univ, PhD(biochem), 78. *Prof Exp:* Res assoc biochem, Univ Chicago, 78-80; PROJ LEADER BIOCHEM, BIOL CONTROL INSECTS, USDA, 80- *Concurrent Pos:* USDA grant; adj fac, Univ Mo. *Mem:* AAAS; Sigma Xi; Am Chem Soc. *Res:* Hormonal control of insect development; effects of natural products on insects and regulation of insect development. *Mailing Add:* USDA-ARS Biol Control Res Park Rte K PO Box 7629 Columbia MO 65205-5001

COUEY, H MELVIN, b Shedd, Ore, May 22, 26; m 55; c 3. PLANT PHYSIOLOGY. *Educ:* Ore State Col, BS, 51, MS, 54; Iowa State Col, PhD(plant physiol), 56. *Prof Exp:* From assoc plant physiologist to plant physiologist, Sci & Educ Admin-Agr Res Serv, USDA, Fresno, Calif, 56-63, sr plant physiologist, 63-68, invests leader, Northwest Fruit Invests, Wenatchee, Wash, 68-73, location leader prod, harvesting & handling tree fruits, 73-77, res leader, Commodity Treat, Handling & Transp Unit, Hawaii, 77-87, res leader, Handling & Distrib Unit, 77-87; res leader, Nat Clonal Germplasm Repository, Agr Res Serv, USDA, Ore, 87-89; RETIRED. *Mem:* AAAS; Am Soc Hort Sci; Am Soc Plant Physiol; Am Phytopath Soc. *Res:* Post-harvest physiology of fruits; fruit storage and storage disorders; physiology of fungus spores. *Mailing Add:* 36483 Hwy 226 SE Albany OR 97321-9735

COUFAL, HANS-JÜRGEN, b Ruhla, Ger, Jan 17, 45; m 73; c 3. PHYSICS. *Educ:* TH Munich, BS,66, MS, 70;TU MUNICH, PhD(physics),75. *Prof Exp:* Asst prof physics, Munich, 75-77; vis prof physics, Berlin, 77-78; vis scientist, IBM Res Div, 78-79; asst prof physics, Munich, 79-80; RES STAFF, IBM RES DIV, ALMADEN RES CTR, 81- *Concurrent Pos:* ed, J Photoacoust, 80-83, Appl Physics, 86- *Mem:* Am Physical Soc; German Phys Soc. *Res:* Radiation induced thermal and acoustic transients, their use for the analysis of surfaces, absorbates and thin film and applications of these methods for spectroscopy, imaging and non-destructive evaluation. *Mailing Add:* IBM Res Div Almaden Res Ctr K33/802 650 Harry Rd San Jose CA 95120-6099

COUGER, J DANIEL, b Olney, Tex, Oct 20, 29; m 51; c 4. COMPUTER SCIENCE. *Educ:* Phillips Univ, BA, 51; Univ Mo, Kansas City, MA, 58; Univ Colo, DBA(mgt sci), 64. *Prof Exp:* Indust engr, Nat Gypsum Co, 53-54; supvr, Indust Eng Dept, Hallmark Cards, Inc, 54-58; sect chief, Comput Dept, Martin-Marietta, Inc, 58-65; prof mgt sci, 65-, prof computer sci, 80-, DISTINQUISHED PROF COMPUTER & MGT SCI, UNIV COLO, COLORADO SPRINGS. *Concurrent Pos:* Consult, Int Bus Mach Corp & Dow Chem Corp. *Honors & Awards:* Distinguished Serv Award, Asn Systs Mgt. *Mem:* Asn Comput Mach; Opers Res Soc Am; Soc Mgt Info Systs; Asn Systs Mgt. *Res:* Creativity/innovation in information systems; design of computer-based management information systems. *Mailing Add:* Sch Bus Univ Colo Austin Bluffs Pkwy Colorado Springs CO 80933-7150

COUGHANOWR, D(ONALD) R(AY), b Brazil, Ind, Mar 11, 28; m 55; c 3. CHEMICAL ENGINEERING. *Educ:* Rose Polytech Inst, BS, 49; Univ Pa, MS, 51; Univ Ill, PhD(chem eng), 56. *Prof Exp:* Chem engr, Standard Oil Co, Ind, 51-53; from asst prof to prof chem eng, Purdue Univ, 56-67; PROF CHEM ENG & HEAD DEPT, DREXEL UNIV, 67- *Concurrent Pos:* Mem staff, Case Inst Technol, 63-64 & Electronic Assocs, Inc, 64. *Mem:* Instrument Soc Am; Am Soc Eng Educ; Am Inst Chem Engrs. *Res:* Process dynamics and control; analog computation; mass transfer; aerosols. *Mailing Add:* Dept Chem Eng Drexel Univ 32nd & Chesnuts Sts Philadelphia PA 19104

COUGHENOUR, MICHAEL B, b Oak Park, Ill, Sept 22, 52. ECOLOGY. *Educ:* Univ Ill, BS, 73, MS, 74; Colo State Univ, PhD(ecol), 78. *Prof Exp:* Res assoc ecol, dept biol, Syracuse Univ, 78-; NATURAL RESOURCE ECOL LAB, COLO STATE UNIV, FT COLLINS. *Concurrent Pos:* NSF fel, 78-80. *Mem:* Ecol Soc Am; Am Soc Naturalists; AAAS; Sigma Xi. *Res:* Grassland ecology; systems ecology; nutrient cycling; grazing systems; plant-animal interactions; decomposition ecology; production ecology; applied ecology; ruminant ecology. *Mailing Add:* Natural Resource Ecol Lab Colo State Univ Ft Collins CO 80523

COUGHLIN, JAMES ROBERT, b Albany, NY, Sept 8, 46; div; c 3. RISK ASSESSMENT & COMMUNICATION, FOOD SAFETY. *Educ:* Siena Col, NY, BS, 68; Univ Calif, Davis, MS, 74, PhD(agr & environ chem), 79. *Prof Exp:* Food safety scientist, Armour Foods Res Ctr, 79-81; mgr, toxicol affairs, Gen Foods Corp, 81-87, dir, prod safety & external tech affairs, Gen Foods Worldwide, 87-89, dir, sci relations, Kraft Gen Foods, 90-91; PRIN, ENVIRON CORP, 91- *Concurrent Pos:* Fel trainee, Dept Environ Toxicol, Univ Calif, Davis, 79; chmn, sci adv group, Nat Coffee Asn, 85-88; chmn, Toxicol Safety Eval Div, Inst Food Technologists, 86, exec comt, Int Relations, 89-; pres, Int Coffee Sci Asn, Paris, 90- *Mem:* AAAS; Inst Food Technologists; Am Chem Soc; Toxicol Forum; Int Life Sci Inst Nutrit Found. *Res:* Safety assessment and toxicological evaluation of components in the food supply and environment, particularly affecting human health. *Mailing Add:* Environ Corp One Park Plaza Suite 700 Irvine CA 92714

COUGHLIN, RAYMOND FRANCIS, b Chicago, Ill, Oct 6, 43; m 70; c 2. ALGEBRA. *Educ:* Lewis Col, BA, 65; Loyola Univ Chicago, MA, 67; Ill Inst Technol, PhD(math), 69. *Prof Exp:* From instr to asst prof math, Loyola Univ, Chicago, 67-70; asst prof, 70-77, ASSOC PROF MATH, TEMPLE UNIV, 77- *Mem:* Am Math Soc; Math Asn Am. *Res:* Non-associative algebras satisfying the associo-symmetric identity and non-associative rings satisfying the m-associative ring identity. *Mailing Add:* Dept Math Temple Univ Broad & Montgomery Philadelphia PA 19122

COUGHLIN, ROBERT WILLIAM, b Brooklyn, NY, June 18, 34; m 60; c 2. CHEMICAL ENGINEERING, APPLIED CHEMISTRY. *Educ:* Fordham Univ, BS, 56; Cornell Univ, PhD(chem eng), 61. *Prof Exp:* Fulbright fel, Univ Heidelberg, 60-61; chem engr, Esso Res & Eng Co, NJ, 61-64; sr scientist, Isotopes Inc, 64, mgr eng sci & data processing, 64-65; from asst prof to prof chem eng, Lehigh Univ, 65-78, assoc dir, Ctr Marine & Environ Sci, 68-78; dept head, 77-83, PROF CHEM ENG, UNIV CONN, STORRS, 77- *Concurrent Pos:* Adj asst prof, NY Univ, 62-64; vis lectr, Stevens Inst Technol, 64-65; prof & dir inst environ studies & grad prog environ eng & sci, Drexel Univ, 71; prin investr various res grants & contracts from govt agencies, found & indust corp, consult to various indust corp, univs & govt agencies; dir fermentation & separations lab, Biotechnol Ctr, Univ Conn, 87-; dir & consult, Sym Biotech Inc, 87-; fel Am Inst Chem Engrs, 87-; dir & consult, 87-, chmn bd, Sym Biotech Inc, 90- *Honors & Awards:* Robinson Award, Lehigh Univ, 67. *Mem:* Am Chem Soc; Am Inst Chem Engrs; Catalysis Soc. *Res:* Surface chemistry; microbiology; biochemical engineering and biotechnology; catalysis; kinetics; fuel technology and petroleum engineering; chemical reactor engineering; electrochemistry; applied biochemistry. *Mailing Add:* Dept Chem Eng U-139 Univ Conn Storrs CT 06268

COUGHRAN, WILLIAM MARVIN, JR, b Fresno, Calif, Feb 13, 53; m 72; c 2. SCIENTIFIC COMPUTING, COMPUTATIONAL ELECTRONICS. *Educ:* Calif Inst Technol, BS & MS, 75; Stanford Univ, MS, 77, PhD(computer sci), 80. *Prof Exp:* MEM TECH STAFF, COMPUT SCI RES CTR, AT&T BELL LABS, 80- *Concurrent Pos:* Adj assoc prof, Dept Computer Sci, Duke Univ, 84-91; vis scientist, Integrated Systs Lab, Swiss Fed Inst Technol, Zurich, 89; lectr, Dept Computer Sci, Stanford Univ, 91. *Mem:* Soc Indust & Appl Math; Asn Comput Mach; Sigma Xi. *Res:* Numerical methods for coupled systems of advection-diffusion equations; preconditioners, iterative methods, time-integration schemes and error estimation and adaptive mesh strategies for semiconductor device modeling. *Mailing Add:* AT&T Bell Labs 600 Mountain Ave Murray Hill NJ 07974-2070

COUILLARD, PIERRE, b Montmagny, Que, Mar 19, 28; m 55; c 3. CELL PHYSIOLOGY. *Educ:* Laval Univ, BA, 47, BSc, 51; Univ Pa, PhD(zool), 55. *Prof Exp:* Fel, Belgium, 55-56; from asst prof to prof, 56-70, head dept, 63-67, PROF BIOL, UNIV MONTREAL, 70- *Mem:* AAAS; Soc Protozool; Fr Asn Physiol; Can Soc Cell Biol; Int Soc Evolution Protozool. *Res:* Physiology of amoeba. *Mailing Add:* 5770 Northmount Montreal PQ H3S 2H5 Can

COULL, BRUCE CHARLES, b New York, NY, Sept 16, 42; m 67; c 2. ECOLOGY, BIOLOGICAL OCEANOGRAPHY. *Educ:* Moravian Col, BS, 64; Lehigh Univ, MS, 66, PhD(biol), 68. *Prof Exp:* Res asst biol, Lehigh Univ, 64-68; NSF award biol oceanog, Marine Lab, Duke Univ, 68-70; asst prof zool, Clark Univ, 70-73; assoc prof, 73-78, PROF BIOL & MARINE SCI, UNIV SC, 78-, DIR, MARINE SCI, 82-, CAROLINA RES PROF, 86- *Concurrent Pos:* Prin investr, NSF grant, 72-92; Fulbright res scholar, Victoria Univ, NZ, 81. *Mem:* Am Micros Soc (pres, 88); Ecol Soc Am; Sigma Xi; Int Asn Meiobenthologists (chmn, 73-75); Am Soc Zool. *Res:* Meiobenthic ecology; harpacticoid copepod systematics; benthic ecology; population dynamics; zoogeography; ecotoxicology. *Mailing Add:* Marine Sci Prog Univ SC Columbia SC 29208

COULMAN, G(EORGE) A(LBERT), b Detroit, Mich, June 29, 30; m 56; c 2. CHEMICAL ENGINEERING. *Educ:* Case Inst Technol, BS, 52, PhD(chem eng), 62; Univ Mich, MS, 58. *Prof Exp:* Develop engr, pilot plant, Dow Corning Corp, 54-57; mgr develop, Am Metal Prod, 58-60; asst prof chem eng, Univ Waterloo, 61-64; from asst prof to prof chem eng & eng res, Mich State Univ, 64-76; chmn dept, 76-85, PROF CHEM ENG, CLEVELAND STATE UNIV, 76-, DEAN, FENN COL ENG, 88- *Concurrent Pos:* Nat Res Coun res grant, 62-64; NSF grants, 65-, numerous govt & indust res grants. *Mem:* Am Inst Chem Engrs; Am Chem Soc; Am Soc Eng Educ; Nat Soc Prof Engrs. *Res:* Systems engineering applied to chemical processing, particularly the analysis and modeling of process systems for use in system design methods; dynamic characteristics of chemical processes and use in improved operation and control. *Mailing Add:* Dept Chem Eng 1960 E 24th St Cleveland OH 44115

COULOMBE, HARRY N, b Long Beach, Calif, Oct 7, 39; m 75; c 1. VERTEBRATE BIOLOGY, ENVIRONMENTAL MANAGEMENT. *Educ:* Univ Calif, Los Angeles, 62, MA, 65, PhD(zool), 68. *Prof Exp:* Asst prof zoophysiol, Inst Arctic Biol, Univ Alaska, 68-69; asst prof ecol, San Diego State Col, 69-74, dir bur ecol & mem exec comt, Ctr Regional Environ Studies, 70-74, proj mgr, 74-75; oil shale res mgr, Western Energy & Land Use Team, 75-77, leader, Nat Habitat Assessment Group, 77-79, asst team leader, Western Energy & Land Use Team, 79-84, environ contaminants coord, 84-87, wildlife res coordr, Res & Develop, 87-89, CHIEF, ENVIRON CONTAMINANTS RES BR, PATUXENT WILDLIFE RES CTR, US FISH & WILDLIFE SERV, 90- *Concurrent Pos:* Analytical & modeling coordr, Tundra Biome, Anal of Ecosyst Sect, US Int Biol Prog, 68-; chmn ad hoc comt rabies control, County of San Diego, 70-; adj prof, Col State Univ, 75-84. *Mem:* AAAS; Ecol Soc Am; Am Soc Mammal; Cooper Ornith Soc; Wildlife Soc. *Res:* Project development and management in regional environmental planning; ecosystem modeling and simulation; predator population ecology; ecology of cetaceans; adaptive physiology of vertebrates. *Mailing Add:* ECR Br Patuxent Wildlife Res Ctr Laurel MD 20708

COULOMBE, LOUIS JOSEPH, b Lac-St-Jean, Que, Dec 16, 20; m 47; c 7. AGRICULTURE. *Educ:* Laval Univ, BA, 43, BSA, 47; McGill Univ, MSc, 49, PhD, 56. *Prof Exp:* Res scientist, Can Dept Agr, 48-61; mem tech dept & sales, Niagara Brand Chem Co, 61-64; res scientist, Can Dept Agr, 64-85; RETIRED. *Mem:* Can Phytopath Soc. *Res:* Creation of apple cultivars resistant to scab, mildew and fire blight; ecological aspect of fruit pesticides; new programs for scab and mildew control. *Mailing Add:* 1415 DuVallon Beloeil PQ J3G 3Y4 Can

COULSON, DALE ROBERT, b Monessen, Pa, Oct 26, 38; m 67; c 2. HETEROGENEOUS CATALYSIS, ORGANOFLUORINE CHEMISTRY. *Educ:* Carnegie Inst Technol, BS, 60; Columbia Univ, MA, 61, PhD(chem), 64. *Prof Exp:* NSF fel photochem, Univ Chicago, 64-65, NIH fel, 65-66; res chemist, 66-74, res supvr, 74-83, lab adminr, 83-86, RES CHEMIST, E I DU PONT DE NEMOURS & CO INC, 86- *Mem:* Am Chem Soc. *Res:* Heterogeneous catalysis. *Mailing Add:* 2314 Empire Dr Wilmington DE 19810

COULSON, JACK RICHARD, b Manhattan, Kans, Jan 31, 31; m 64; c 2. ENTOMOLOGY. *Educ:* Iowa State Univ, BS, 52. *Prof Exp:* Biologist, Insect Identification & Parasite Introd Res Br, Washington, DC, 56-61, biologist, Plant Indust Sta, Md, 61-63, entomologist, 63-64, entomologist, Introduced Beneficial Insects Lab, NJ, 64-65 & Europ Parasite Lab, Paris, France, 65-67, asst to br chief taxon & biol control, Plant Indust Sta, Beltsville, Md, 67-72, chief, Beneficial Insect Introd Lab, Insect Identification & Beneficial Insect Introd Inst, 72-85, HEAD, ARS BIOL CONTROL DOC CTR, INSECT BIOCONTROL LAB, AGR RES SERV, USDA, BELTSVILLE, MD, 85- *Mem:* Entom Soc Am; Int Orgn Biol Control (pres, 88-). *Res:* Taxonomic entomology, especially bibliographic; biological control of insect pests. *Mailing Add:* Beltsville Agr Res Ctr E Bldg 476 USDA Agr Res Serv Beltsville MD 20705

COULSON, KINSELL LEROY, b Hatfield, Mo, Oct 7, 16; m 47. METEOROLOGY, ATMOSPHERIC PHYSICS. *Educ:* Northwest Mo State Col, BS, 42; Univ Calif, Los Angeles, MA, 52, PhD(meteorol), 59. *Prof Exp:* Meteorologist, Univ Chicago, 42-43 & US Naval Ord Test Sta, Calif, 49-51; res meteorologist, Univ Calif, Los Angeles, 51-59 & Stanford Res Inst, 59-60; mgr geophys, Space Sci Lab, Gen Elec Co, 60-65; prof agr eng, Univ Calif, Davis, 65-66, prof meteorol, 67-79; dir, Mauna Loa Observ, Hilo, Hawaii, 79-84; EMER PROF METEOROL, UNIV CALIF, DAVIS, 84- *Concurrent Pos:* USSR exchange fel, Nat Acad Sci, 72 & 80; consult, Johnson Space Ctr, 84- & Ames Res Ctr, NASA, 86- *Mem:* AAAS; fel Am Meteorol Soc; Am Geophys Union; Solar Energy Soc; Sigma Xi; Soc Photo-Optical Instrumentation Engrs. *Res:* Atmospheric radiation, especially solar radiation regime; molecular and aerosol scattering of radiation in planetary atmospheres; reflection from planetary surfaces; planetary albedo; space environment. *Mailing Add:* 119 Bryce Way Vacaville CA 95687

COULSON, LARRY VERNON, b LaFollette, Tenn, Oct 15, 43; m 66; c 3. PARTICLE PHYSICS, RADIATION PHYSICS. *Educ:* Kans State Univ, BS, 65; Univ Va, PhD(physics), 70. *Prof Exp:* Fel particle physics, Rice Univ, 70-72; asst head, safety sect, Fermi Nat Accelerator Lab, 78-83, dep head, safety sect, 83-85, head safety sect, 85-89, SR RADIATION SAFETY OFFICER, FERMI NAT ACCELERATOR LAB, 76-; ASST DIR, ENVIRON, SAFETY & HEALTH, SUPERCONDUCTING SUPER COLLIDER LAB, 89- *Res:* Radiation related problems and dosimetry of accelerator produced radiation; isotope production cross sections at high energies. *Mailing Add:* Superconducting Super Collider Lab 2550 Beckleymeade Ave Dallas TX 75237

COULSON, PATRICIA BUNKER, b Kankakee, Ill, Apr 27, 42; c 2. REPRODUCTIVE ENDOCRINOLOGY, CELL PHYSIOLOGY. *Educ:* Univ Ill, BS, 64, MS, 65, PhD(reproductive endocrinol), 70. *Prof Exp:* Lab asst reproductive endocrinol, Univ Ill, 65-66; res reproduction, 70-72, asst prof endocrinol, Univ Tenn, Knoxville, 72-77, asst res prof, Mem Res Ctr, 77-78; assoc prof physiol, Col Med, East Tenn State Univ, Johnson City, 78-81; assoc prof obstet gynec & assoc prof med biol, Univ Tenn Ctr Health Sci, Knoxville, 81-85; LAB DIR, TENN ENDOCRINE REFERENCE LAB, 85- *Concurrent Pos:* Assoc prof obstet & gynec, ETenn State Univ-COM, Johnson City, 87-90; dir, Blout Mem Hosp Endocrine Lab, 86-90. *Mem:* AAAS; Sigma Xi; Soc Study Reproduction; Am Tissue Cult Asn; Fedn Am Soc Exp Biol; Endocrine Soc. *Res:* Hormone action; endocrine control of the female reproductive tract emphasizing modulation of receptors for estrogen, progesterone and the gonadotropic hormones in the uterus, vagina, pituitary, hypothalamus, ovary and mammary cells; hormonal control of cancer cells; clinical assays for reproductive hormones; flow cytometry; clinical chemistry. *Mailing Add:* Tenn Endocrine Ref Lab 1810 Ailor Lab Knoxville TN 37921

COULSON, RICHARD, KIDNEY ADENOSINE RECEPTORS. *Educ:* Univ London, Eng, PhD(biochem), 69. *Prof Exp:* Assoc prof renal endocrinol & res chemist, State Univ NY Health Sci Ctr & Vet Admin Med Ctrs, 82-89; ASSOC PROF, INTERNAL MED, JAMES A HALEY VET HOSP, UNIV SFLA, 89- *Mailing Add:* James A Haley Vet Hosp 13000 Bruce B Downs Blvd Res 151 Tampa FL 33612

COULSON, RICHARD L, b Turnervalley, Alta, Can, Dec 8, 43; c 2. CARDIA MUSCLE, MEDICAL EDUCATION. *Educ:* Univ Toronto, PhD(physiol), 71. *Prof Exp:* Assoc prof, 78-85, PROF PHYSIOL, SOUTHERN ILL UNIV, 86- *Concurrent Pos:* Asst prof cardiol, Temple Univ, 73-78; fel Med Res Coun. *Mem:* Am Physiol Soc; Biophys Soc; Coun Circulation; AERA. *Res:* NMR muscle energetics; cognitive science in medicine. *Mailing Add:* Dept Physiol Southern Ill Univ Carbondale IL 62901

COULSON, ROBERT N, b Dallas, Tex, Mar 1, 43. INSECT ECOLOGY, FOREST ENTOMOLOGY. *Educ:* Furman Univ, BS, 65; Univ Ga, MS, 67, PhD(entom), 69. *Prof Exp:* Prin entomologist pest control sect, Tex Forest Serv, 69-73; from asst prof to assoc prof, 73-79, PROF ENTOM, TEX A&M UNIV, 79- *Concurrent Pos:* Res assoc entom, Univ Ga, 67-70. *Mem:* Entom Soc Am; Entom Soc Can; Ecol Soc Am. *Res:* Forest insect community and population ecology in relation to pest management. *Mailing Add:* Dept Entom Tex A&M Univ College Station TX 77843

COULSON, ROLAND ARMSTRONG, b Rolla, Kans, Dec 20, 15; m 44; c 2. BIOCHEMISTRY. *Educ:* Univ Wichita, BA, 37; La State Univ, MS, 39; Univ London, PhD(biochem), 44. *Prof Exp:* From instr to assoc prof, 44-53, PROF BIOCHEM, SCH MED, LA STATE UNIV, NEW ORLEANS, 53- *Mem:* Soc Exp Biol & Med; Am Soc Biol Chem; Am Asn Clin Chem; Am Phys Soc; Am Inst Nutrit; Sigma Xi. *Res:* Nutrition; biochemical studies on Alligator mississippiensis; theory of metabolic rate and anaerobic glycolysis; protein digestion, absorption, and amino acid metabolism. *Mailing Add:* Dept Biochem Sch Med La State Univ New Orleans LA 70112

COULSON, WALTER F, b Harrogate, Eng, Dec 17, 26; nat US; c 5. SURGICAL PATHOLOGY, CYTOLOGY. *Educ:* Univ Edinburgh, MB, ChB, 49, BSc, 54, MD, 67; FRCPath, 72. *Prof Exp:* Resident med, Royal Infirmary, Edinburgh, Scotland, 49-50; registr path, Western Gen Hosp, 54-55; lectr, Univ Edinburgh, 55-60; asst prof, Univ Utah, 60-64, assoc prof, 64-68; assoc prof, 68-70, vchmn dept, 71-82, PROF PATH, CTR HEALTH SCI, SCH MED, UNIV CALIF, LOS ANGELES, 70- *Concurrent Pos:* Brown res fel, Yale Univ, 57-58; consult, Salt Lake Gen Hosp, 60-68; chief lab serv, Vet Admin Hosp, 62-68; res assoc, Univ Col, Univ London, 66-67; head div surg path, Univ Calif, Los Angeles, 68-86; consult, Dept Path, Radiation Effects Res Found, Japan, 80-81; vis sr lectr, Dept Path, Ninewells Hosp & Med Sch, Scotland, 81; vis pathologist, Royal Brisbane Hosp, Queensland, Australia, 87; vis sr lectr, Dept Path, Univ Otago Med Sch, Dunedin, New Zealand, 88. *Mem:* Am Asn Pathologists; Path Soc Gt Brit & Ireland; Int Acad Path; Europ Soc Path; Int Asn Study Lung Cancer. *Res:* Morphology and biochemistry of connective tissue, including bone, particularly mechanical properties and the changes induced by copper deficiency. *Mailing Add:* Dept Path Ctr Health Sci Univ Calif Los Angeles CA 90024-1732

COULSTON, MARY LOU, b Tarrytown, NY, June 16, 38; m 82. WATER QUALITY MONITORING. *Educ:* State Univ NY Albany, BS, 62; Univ PR, Rio Piedras, MS, 68; Univ Kans, Lawrence, PhD(ecol), 72. *Prof Exp:* Asst prof biol & marine ecol, Univ Calif, Berkeley, 72-76; sci coordr, Nat Undersea Res Ctr, Nat Oceanic & Atmospheric Admin, WI Lab, Fairleigh Dickinson Univ, 83-88; assoc prof, Univ VI, 88-90; exten specialist, VI Marine Adv Serv Sea Grant, Univ VI, 90-91; LAB DIR, OCEAN SYSTS RES, INC, 88- *Concurrent Pos:* Researcher, Island Resources Found, Inc, 88- *Mem:* Sigma Xi; Am Water Works Asn. *Res:* Laboratory analysis of drinking water; environmental testing and evaluation of marine ecosystems; water quality monitoring programs as mitigation for construction impacts. *Mailing Add:* 128 Gallows Bay St Croix VI 00820

COULTER, BYRON LEONARD, b Phenix City, Ala, Aug 16, 41; m 62; c 1. THEORETICAL PHYSICS, SOLAR ENERGY. *Educ:* Univ Ala, BS, 62, PhD(physics), 66. *Prof Exp:* From asst prof to assoc prof, 66-77, PROF PHYSICS, ECAROLINA UNIV, 77- *Mem:* Am Phys Soc; Am Asn Physics Teachers; Int Solar Energy Soc; Sigma Xi. *Res:* Computer simulation of physics problems; available work from solar radiation; computer simulations. *Mailing Add:* Dept Physics ECarolina Univ Greenville NC 27834

COULTER, CHARLES L, b Akron, Ohio, Jan 10, 33; m 55; c 3. STRUCTURAL CHEMISTRY, BIOPHYSICAL CHEMISTRY. *Educ:* Miami Univ, AB, 54, MA, 56; Univ Calif, Los Angeles, PhD(phys chem), 60. *Prof Exp:* USPHS fel, Med Res Coun Unit for Molecular Biol, Cambridge, Eng, 60-62; fel, Lab Molecular Biol, NIH, 62-64, prog dir anal biochem, Div Res Facil & Resources, 65-66; from asst prof to assoc prof anat, Univ Chicago, 66-76; head biol struct sect, 76-88, DIR RES FAC IMPROV PROG, DIV RES RESOURCES, NIH, 88- *Mem:* AAAS; Am Crystallog Asn. *Res:* Protein crystallography; crystal structures of biologically important compounds; structural biochemistry. *Mailing Add:* Res Facil Improv Prog Nat Ctr Res In Resources Bethesda MD 20892

COULTER, CLAUDE ALTON, b Phenix City, Ala, Mar 30, 36; m 60; c 2. SYSTEMS ANALYSIS & MODELING. *Educ:* Samford Univ, BA, 56; Univ Ala, MS, 59; Harvard Univ, MA, 63, PhD(physics), 64. *Prof Exp:* Asst prof physics, Univ Ala, 63-66; from asst prof to assoc prof, Clark Univ, 66-71; from assoc prof to prof physics, Univ Ala, Tuscaloosa, 71-77; STAFF MEM, LOS ALAMOS NAT LAB, 82- *Concurrent Pos:* Consult, Phys Sci Directorate, US Army Missile Command, Redstone Arsenal, Ala, 63-; vis staff mem, Los Alamos Sci Lab, 74-81. *Mem:* Am Phys Soc; Inst Nuclear Mat Mgt; Sigma xi. *Res:* Systems studies for nuclear materials safeguards. *Mailing Add:* 182 Loma Del Escolar Los Alamos NM 87544

COULTER, DWIGHT BERNARD, b Iowa City, Iowa, Jan 8, 35; m 61; c 2. VETERINARY PHYSIOLOGY. *Educ:* Iowa State Univ, DVM, 60, MS, 65, PhD(physiol), 69. *Prof Exp:* From instr to assoc prof physiol, Col Vet Med, Iowa State Univ, 62-72; assoc prof physiol, 72-79, actg asst vpres for res, 85-88, PROF PHYSIOL, COL VET MED, UNIV GA, 79-, ASSOC DEAN, 89- *Concurrent Pos:* Adj prof, Col Vet Med, Miss State Univ, 79-80. *Mem:* Am Vet Med Asn; Am Physiol Soc; Conf Res Workers Animal Dis; Sigma Xi. *Res:* Comparative cardiovascular physiology; comparative neurophysiology. *Mailing Add:* Dept Physiol Univ Ga Col Vet Med Athens GA 30602

COULTER, ELIZABETH JACKSON, b Baltimore, Md, Nov 2, 19; m 51; c 1. BIOSTATISTICS, HEALTH ECONOMICS. *Educ:* Swarthmore Col, Pa, AB, 41; Radcliffe Col, AM, 46, PhD(econ), 48. *Prof Exp:* Asst dir health study, Bur Labor Statist, San Juan, PR, 46; res asst, Milbank Mem Fund, New York, 48-51; economist, Off Defense Prod, Washington, DC, 51-52; res analyst, Children's Bur, US Dept Health, Educ & Welfare, Wash, DC, 52-53; statistician, Ohio Dept Health, Columbus, 54-55, chief statistician, 55-65; assoc prof, Sch Pub Health, Univ NC, Chapel Hill, 65-72, prof biostatist, 72-90, assoc dean undergrad studies, 79-86, EMER PROF BIOSTATIST, SCH PUB HEALTH, UNIV NC, CHAPEL HILL, 90- *Concurrent Pos:* Lectr in econ, Ohio State Univ, Columbus, 54-55, clin asst prof prev med, 63-65; asst clin prof biostatist, Sch Pub Health, Univ Pittsburgh, Pa, 58-62; assoc prof econ, Univ NC, Chapel Hill, 65-78; adj assoc prof hosp admin, Duke Univ, 72-79. *Mem:* Fel Am Pub Health Asn; Am Econ Asn; Am Statist Asn; Am Acad Polit & Social Sci; AAAS; Sigma Xi; Am Eval Asn; Biometric Soc; Asn Health Serv Res. *Res:* Socio-economic factors in natality, mortality and utilization of health services; quantitative approaches in health planning and evaluation; application of economics in the health field. *Mailing Add:* 1825 N Lake Shore Dr Chapel Hill NC 27514

COULTER, GLENN HARTMAN, b Orangeville, Ont, Jan 28, 47; m 80; c 2. REPRODUCTIVE PHYSIOLOGY, ANIMAL SCIENCE. *Educ:* Univ Guelph, BSc, 69; Cornell Univ, PhD(reproductive physiol), 73. *Prof Exp:* Res specialist reproductive physiol, Cornell Univ, 72-73, res assoc, 73-74; RES SCIENTIST REPRODUCTIVE PHYSIOL, LETHBRIDGE RES STA, AGR CAN, 74-, HEAD, LIVESTOCK SCI SECT, 88- *Concurrent Pos:* Adj prof, Western Col Vet Med, Univ Sask, Saskatoon, Can; vis animal scientist, dept animal sci, Wash State Univ, Pullman. *Mem:* Am Soc Animal Sci; Soc Stud Fertility; Am Dairy Sci Asn; Am Soc Andrology; Sigma Xi. *Res:* Reproductive physiology of the male with emphasis on the testicular development, function and temperature regulation; general breeding soundness and management factors effecting reproduction in young beef bulls. *Mailing Add:* Res Sta Agr Can PO Box 3000 Main Lethbridge AB T1J 4B1 Can

COULTER, HENRY W, b Gastonia, NC, Nov 11, 27. GEOLOGICAL ENGINEERING. *Educ:* NC State Univ, BA, 53. *Prof Exp:* DIR LAND & RESOURCES, NC RESOURCES & COMMUNITY DEVELOP, 56- *Mem:* Fel Geol Soc Am; Soc Mining Eng. *Mailing Add:* 2243-49 St NW Washington DC 20007

COULTER, HERBERT DAVID, JR, b Enid, Okla, Dec 31, 39; m 64; c 2. ANATOMY. *Educ:* Westminster Col, Mo, BA, 61; Univ Tenn, PhD(anat), 68. *Prof Exp:* Instr, 68-70, ASST PROF ANAT, SCH MED, UNIV MINN, MINNEAPOLIS, 70- *Mem:* AAAS; Am Soc Cell Biol; Electron Micros Soc Am; Soc Neurosci; Am Asn Anatomists. *Res:* Immunocytochemistry; neurocytology. *Mailing Add:* Himalayan Inst Honesdale PA 18431

COULTER, JAMES B, ENGINEERING. *Prof Exp:* RETIRED. *Mem:* Nat Acad Eng. *Mailing Add:* 1069 Double Gate Rd Davidsonville MD 21035

COULTER, JOE DAN, b Victoria, Tex, July 25, 44; m 67; c 3. NEUROPHYSIOLOGY, NEUROANATOMY. *Educ:* Univ Okla, BA, 66, PhD(psychiat & behav sci), 71. *Prof Exp:* NIH fel, Marine Biomed Inst, Univ Tex Med Br Galveston, 71-73; Found fund res psychiat fel, Inst Physiol, Univ Pisa, 73-74 & Univ Edinburgh, 74-75; from asst prof to prof physiol, biophys, psychiat & behav sci, Marine Biomed Inst, Univ Tex Med Br, Galveston, 75-85; PROF ANAT & HEAD DEPT, COL MED, UNIV IOWA, IOWA CITY, 85- *Concurrent Pos:* Assoc dir, Sensor Physiol & Perception Prog, NSF, 80-81; mem, Neurol B Study Sect, NIH, 81-85, bd mem, Coun Acad Soc, Asn Am Med Cols, 85- *Mem:* AAAS; Am Asn Anatomists; Am Physiol Soc; Soc Neurosci (treas, 85-88); Am Soc Cell Biol. *Res:* Monoaminergic and peptidergic transmitters; neuroanatomy; sensory motor control; neuronal proteins and development. *Mailing Add:* Dept Anat Univ Iowa Med Col 1-472 BSB Iowa City IA 52242

COULTER, LOWELL VERNON, b Marion, Ohio, July 3, 13; m 37; c 2. PHYSICAL CHEMISTRY, SOLID STATE CHEMISTRY. *Educ:* Heidelberg Col, BS, 35; Colo Col, AM, 37; Univ Calif, Berkeley, PhD(chem), 40. *Prof Exp:* Instr chem, Colo Col, 35-37; asst, Univ Calif, 37-40; instr, Univ Idaho, 40-42; instr, Boston Univ, 42-44, from asst prof to assoc prof chem, 46-55, prof, 55-79, chmn dept, 61-72; RETIRED. *Concurrent Pos:* Group leader, Manhattan Proj, Monsanto Chem Co, Ohio, 44-45. *Mem:* Fel AAAS; Am Chem Soc; Am Phys Soc; Sigma Xi. *Res:* Application of the third law of thermodynamics; low temperature calorimetry; solution calorimetry; properties of liquid ammonia solutions of metals; thermodynamic properties of clathrates. *Mailing Add:* 14 Foxhill St Westwood MA 02090-1118

COULTER, MALCOLM WILFORD, b Suffield, Conn, Dec 30, 20; m 48; c 5. WILDLIFE ECOLOGY. *Educ:* Univ Conn, BS, 42; Univ Maine, MS, 48; Syracuse Univ, PhD, 66. *Prof Exp:* Proj leader furbearers, Vt State Fish & Game Serv, 48-49; instr wildlife resources & asst leader res unit, Dept Wildlife Resources, Univ Maine, Orono, 49-52, from asst prof to prof wildlife resources, 52-82, assoc dir, Sch Forest Resources, 68-80, chmn dept, 81-82; RETIRED. *Concurrent Pos:* Mem, Maine Land Use Regulation Comn, 75-78, chmn, 76-78. *Mem:* AAAS; hon mem Wildlife Soc; Am Soc Mammal. *Res:* Ecology and behavior of furbearing animals; waterfowl breeding biology, ecology, behavior and population dynamics; marsh ecology and management. *Mailing Add:* RD 1 Box 150 E Holden East Eddington ME 04429

COULTER, MURRAY W, b El Dorado, Ark, May 2, 32; m 59; c 2. PLANT GROWTH REGULATION, PHYSIOLOGY OF FLOWERING. *Educ:* Emory Univ, BA, 54; Univ Ariz, MS, 56; Univ Calif, Los Angeles, PhD(bot & plant biochem), 63. *Prof Exp:* Asst prof biol, div sci & math, Calif State Univ, Northridge, 59-62; res scientist, Space Sci Ctr, Inst Geophys Planetary Physics, Univ Calif, Los Angeles, 62-63, res fel plant physiol, dept bot & plant biochem, 63-64; asst prof, 64-67, ASSOC PROF BOT & PLANT PHYSIOL, DEPT BIOL SCI, TEX TECH UNIV, 67- *Concurrent Pos:* Consult, Space

& Info Systs Div, NAm Aviation, Inc, 62-64, Food & Agr Orgn, UN, 85- *Mem:* Fel AAAS; Am Soc Plant Physiologists; Japanese Soc Plant Physiol; Indian Soc Plant Physiol; Am Inst Biol Sci; Bot Soc Am. *Res:* Molecular basis for the control of plant growth and development using genetic mutants and photoperiodic plants in which development can be manipulated to allow study of the resulting molecular changes; Gibberellin studies with microorganisms and genetic mutants of maize; photoperiod and endogenous rhythms as related to biological clocks; environmental control of plant growth and development; hormones and plant growth regulators; biochemistry. *Mailing Add:* Dept Biol Sci Tex Tech Univ Lubbock TX 79409

COULTER, NEAL STANLEY, b Columbus, Ga, May 3, 44; m 66; c 2. COMPUTER SCIENCE, INFORMATION SCIENCE. *Educ:* Univ Ala, BS, 65, MA, 66; Ga Inst Technol, MS, 72, PhD(info & comput sci), 74. *Prof Exp:* Assoc res engr appl math, Boeing Co, Ala, 66-67; asst prof math & comput sci, Columbus Col, 67-75; assoc prof & dir comput systs, 75-81, from assoc prof coput & info systs, to prof comput & info syst, 81-87, PROF & CHMN, DEPT COMPUT SCI, FLA ATLANTIC UNIV, 87- *Concurrent Pos:* Fac assoc, IBM, 81; long term vis staff mem, Los Alamos Nat Lab, 84-85. *Mem:* Asn Comput Mach; Inst Elec & Electronics Engrs; Comput Soc; Sigma Xi. *Res:* Software engineering, software metrics; cognitive processes and computer programming; software metrics; computer science education. *Mailing Add:* Dept of Comput Sci Fla Atlantic Univ Boca Raton FL 33431

COULTER, NORMAN ARTHUR, JR, b Atlanta, Ga, Jan 9, 20; m 51; c 1. PSYCHIATRY, BIOMATHEMATICS. *Educ:* Va Polytech Inst, BS, 41; Harvard Univ, MD, 50. *Prof Exp:* Instr math, Va Polytech Inst, 46; Nat Res Coun fel, Johns Hopkins Univ, 50-52; asst prof physiol, Ohio State Univ, 52-55, from asst prof to assoc prof physiol & biophys, 55-65; assoc prof, Univ NC, Chapel Hill, 65-67, sci dir, AF Fortune Biomed Comput Ctr, 70-78, chmn Biomed Eng & Math Curric, 69- 82, prof bioeng & biomath, 67-90; RETIRED. *Concurrent Pos:* Consult, NIH; bd dir, Physicians for Social Responsibility. *Mem:* Biophys Soc; Am Physiol Soc; Biomed Eng Soc; Inst Elec & Electronic Engr; Soc Gen Systs Res; Physicians for Social Responsibility. *Res:* Hemodynamics; teleogenic system theory; biological cybernetics; biomathematics; biomedical computing; study of cooperative phenomena in physical, biological, and social systems, a new field called synergetics (haken). *Mailing Add:* Dept Surg Univ NC 152 Macnider 202 H Chapel Hill NC 27514

COULTER, PAUL DAVID, b Dayton, Ohio, Apr 4, 38; m 59; c 3. TECHNICAL MANAGEMENT. *Educ:* Univ Dayton, BS, 60; Univ Kans, PhD(chem), 65. *Prof Exp:* Prog mgr, Control Line Facil, Parma, Ohio, 73-76, plant mgr, Greenville, Sc, 76-78 & Clarksburg, WVa, 78-80, pres, Union Carbide Graphito, Yabucoa, 80-83, dir technol, Parma, Ohio, 83-86, VPRES TECHNOL, UNION CARBIDE CORP, 86- *Mem:* Am Chem Soc; AAAS. *Mailing Add:* Union Carbide Corp PO Box 6116 Cleveland OH 44101

COULTER, PHILIP W, b Phenix City, Ala, Apr 19, 38; m 86; c 3. THEORETICAL ELEMENTARY PARTICLE PHYSICS. *Educ:* Univ Ala, BS, 59, MS, 61; Stanford Univ, PhD(physics), 65. *Prof Exp:* Res assoc physics, Univ Mich, 65-67; asst res physicist, Univ Calif, Irvine, 67-68, asst prof physics, 67-71; assoc prof, 71-76, PROF PHYSICS, UNIV ALA, 76-, CHMN DEPT, 81- *Mem:* Am Phys Soc; Sigma Xi. *Res:* Quantum chromodynamics and string theory. *Mailing Add:* Dept Physics & Astron Univ Ala Tuscaloosa AL 35487

COULTER, RICHARD LINCOLN, b Pittsburgh, Pa, Feb 12, 45; m 74. ATMOSPHERIC PHYSICS. *Educ:* Kalamazoo Col, BA, 67; Rutgers Univ, MS, 69; Pa State Univ, PhD(meteorol), 76. *Prof Exp:* Sr proj assoc meteorol, Pa State Univ, 76-77; asst meteorologist, 77-81, METEOROLOGIST, ARGONNE NAT LAB, 81- *Concurrent Pos:* Consult. *Mem:* Sigma Xi; Am Meteorol Soc. *Res:* Remote sensing of the atmosphere; acoustic sounding in the atmosphere; micrometeorology. *Mailing Add:* 257 Woodstock Ave Clarendon Hills IL 60514

COULTER, SAMUEL TODD, dairy manufacturing, for more information see previous edition

COULTER, WALLACE H, b Little Rock, Ark. TECHNICAL MANAGEMENT. *Educ:* DSc, Westminster Col, Fulton, Miss, 75; Clarkson Col, Potsdam, NY, 79; DEng, Univ Miami, Fla, 79. *Prof Exp:* CHMN, CORP VPRES & DIR ENG, COULTER CORP, HIALEAH, FLA, 47- *Honors & Awards:* Morris E Leeds Award, Inst Elec & Electronics Engrs. *Mem:* Inst Elec & Electronics Engrs; AAAS; Sigma Xi. *Res:* Research and development of new methods and instrumentation for automatically identifying, measuring and determining the characteristics of biological cells and other particles. *Mailing Add:* Coulter Elect Inc 600 W 20 St Hilaleah FL 33010

COULTHARD, THOMAS LIONEL, agricultural & sanitary engineering, for more information see previous edition

COUNCE, SHEILA JEAN, b Hayes Center, Nebr, Mar 18, 27; m 60. GENETICS, EMBRYOLOGY. *Educ:* Univ Colo, BA, 48, MA, 50; Univ Edinburgh, PhD(genetics), 54. *Prof Exp:* Lab asst biol, embryol & genetics, Univ Colo, 48-50, instr, 50; demonstr genetics, Univ Edinburgh, 51; lab assoc, Jackson Mem Lab, 54-55; Macauley fel, Univ Edinburgh, 55-56; NSF fel, Zurich, Switz, 56-57; from res asst to res assoc biol, Yale Univ, 57-65; assoc anat, 67-68, res assoc zool, 65-78, from asst prof to assoc prof, 69-78, PROF ANAT, SCH MED, DUKE UNIV, 78- *Concurrent Pos:* Vis prof, Max Planck Inst, Tabingen, Ger, 72-73; vis prof zool, Univ Siena, 77. *Mem:* Fel AAAS; Soc Develop Biol; Am Soc Zool; Am Soc Naturalists; Genetics Soc Am. *Res:* Developmental genetics; experimental embryology, especially with insects; history of experimental biology in late 19th century Germany and US. *Mailing Add:* Dept Cell Biol Box 3011 Duke Univ Med Ctr Durham NC 27710

COUNCIL, MARION EARL, b Palmetto, Fla, May 20, 29; m 53; c 3. ELECTRICAL ENGINEERING. *Educ:* Univ Fla, BSEE, 57; La State Univ, MSEE, 60; Okla State Univ, PhD(eng), 65. *Prof Exp:* Transmission & distrib engr, Gulf State Utilities Co, 57-60; instr elec eng, La State Univ, 60-62 & Okla State Univ, 62-65; prof, La State Univ, 65-73; dir, Sch Elec Eng & Comput Sci, Univ Okla, 78-80, Okla Gas & Elec prof eng, 73-; PIPES PROF ELEC ENERGY & POWER, LA TECH UNIV, RUSTON. *Concurrent Pos:* Ed-in-chief, Elec Power Systs J, 77-; prin investr, elec distrib syst proj, US Dept Energy & mandatory lighting stand proj, Okla Dept Energy, 78-79; head, Dept Electronics & Instruments, Barbay Engrs, Inc, 80-81. *Honors & Awards:* Halliburton Teaching Excellence Award; Western Elec Award for Excellence in Teaching. *Mem:* Inst Elec & Electronics Engrs; Am Soc Eng Educ. *Res:* Electric power systems, specifically distribution and planning; energy conservation; electrical components. *Mailing Add:* Dept Elec Eng La Tech Univ PO Box 3161 Tech Sta Ruston LA 71272

COUNCILL, RICHARD J, b Greenville, SC, May 26, 23; m 49; c 5. PETROGRAPHY, SEDIMENTOLOGY. *Educ:* Univ NC, BS, 48, MS, 56. *Prof Exp:* Geologist, US Geol Surv, 48-51; econ geologist, NC Dept Conserv & Develop, 52-55; indust geologist, Atlantic Coast Line RR, 55-60, gen indust geologist, 60-67; gen indust geologist, Seaboard Coast Line RR, 67-73, MGR INDUST DEVELOP & CHIEF GEOLOGIST, SEABOARD COAST LINE RR & LOUISVILLE & NASHVILLE RR, 73-; DIR, INT RESOURCES DEVELOP CORP, 68- *Mem:* Am Inst Prof Geol; Am Inst Mining & Metall Eng; fel Geol Soc Am. *Res:* Resources development. *Mailing Add:* 3001 Paddle Boat Lane Jacksonville FL 32223

COUNSELL, RAYMOND ERNEST, b Vancouver, BC, Aug 20, 30; US citizen; m 57; c 3. PHARMACOLOGY, MEDICINAL CHEMISTRY. *Educ:* Univ BC, BSP, 53; Univ Minn, PhD(pharmaceut & org chem), 57. *Prof Exp:* Lectr, Univ BC, 53-54; sr res chemist, G D Searle & Co, 57-64; assoc prof, 64-69, PROF MED CHEM, UNIV MICH, ANN ARBOR, 69-, PROF PHARMACOL, 72- *Concurrent Pos:* Res assoc, Am Cancer Soc, 64-71; mem med chem study sect A, NIH, 68-72; E Roosevelt Inst fel cancer res, Univ Milan & Univ Uppsala, 72-73; mem prog comt pharmacol & toxicol, Nat Inst Gen Med Sci, 75-79; consult, Nat Inst Health, Govt Agencies & Pharmaceut Indust; sect ed, Ann Reports Med Chem; Fogarty Sr Int fel, 86; assoc chmn pharmacol, 88-90, interim chmn pharmacol, Univ Mich, Ann Arbor, 90- *Honors & Awards:* Czerniak Prize, Ahavat Zion Found, Israel, 74; T O Soine Mem Award, 81. *Mem:* Fel AAAS; Am Chem Soc; Am Soc Pharmacol & Exp Therapeut; fel Am Asn Pharmaceut Sci; Soc Nuclear Med. *Res:* Synthesis and molecular mode of action of chemical regulators of biological processes, especially adrenal hormone biogenesis; organ imaging agents for diagnosis of disease, especially cancer. *Mailing Add:* Dept Pharmacol Univ Mich Med Sch Ann Arbor MI 48109-0626

COUNSELMAN, CHARLES CLAUDE, III, b Baltimore, Md, Apr 27, 43; m 66; c 2. GEODESY, PLANETARY SCIENCES. *Educ:* Mass Inst Technol, BSEE, 64, MSEE, 65, PhD(instrumentation), 69. *Prof Exp:* from asst prof to assoc prof, 69-82, PROF PLANETARY SCI, MASS INST TECHNOL, 82- *Concurrent Pos:* Satellite geol inventor/consult, 78-; vpres, Macrometrics Inc, 81-84; consult, Western Geophys Co, 84- *Honors & Awards:* Except Sci Achievement Medal, NASA, 80; Carl Pulfrich Prize, 83. *Mem:* Int Astron Union; Int Sci Radio Union; Am Geophys Union (pres, geod sect, 88-90); Inst Elec & Electronics Engrs; fel Am Geophys Union. *Res:* Space geodesy. *Mailing Add:* Dept Earth & Planetary Sci Mass Inst Technol Cambridge MA 02139

COUNTER, FREDERICK T, JR, b Lowell, Mass, Dec 16, 34; m 63. VIROLOGY. *Educ:* Mass Col Pharm, BS, 56, MS, 58; Univ Mass, PhD(microbiol), 63. *Prof Exp:* Sr bacteriologist, 63-68, mgr biol develop, 68-69, HEAD IMMUNIZATION BIOL RES & DEVELOP, ELI LILLY & CO, 69- *Mem:* AAAS; Am Soc Microbiol; NY Acad Sci. *Res:* Biological research and development in bacterial and virus vaccines. *Mailing Add:* 550 E Steele Ford Greenfield IN 46140

COUNTRYMAN, DAVID WAYNE, b Ottumwa, Iowa, May 21, 43; m 65; c 2. FOREST MANAGEMENT. *Educ:* Iowa State Univ, BS, 66, MF, 68; Univ Mich, PhD(forest mgt & planning), 73. *Prof Exp:* Actg exten forester, Iowa State Univ Exten Serv, USDA, 67-68, forester, Poplar Bluff Ranger Dist, Forest Serv, 68-70, Region 9 Off, 70-73 & Washington, DC Off, 73-74; assoc prof, 75-80, PROF FOREST MGT, IOWA STATE UNIV, 80- *Mem:* Soc Am Foresters; Sigma Xi. *Res:* Forest administration and planning systems and processes; policy analysis and land use planning. *Mailing Add:* 251 Bessey Hall Iowa State Univ Ames IA 50011

COUNTS, DAVID FRANCIS, b Houston, Tex, May 29, 48. CONNECTIVE TISSUE PHARMACOLOGY, MOLECULAR REGULATION OF REPAIR PROCESSES. *Educ:* Tulane Univ, BS, 70; Med Col Ga, MS, 73, PhD(cell & molecular biol), 75. *Prof Exp:* Fel biochem, Molecular Biol, Roche Inst, 74-76; asst prof vet microbiol & path, Col Vet Med, Wash State Univ, 76-80; res assoc prof biochem, Col Med, Univ Vt, 80-83; res assoc skin biol, Gillette Res Inst, 83-84; sr biochemist skin biol, Lilly Res Lab, 84-88; team leader wound care, Marion Labs, 88-90, SR SCIENTIST SPONSORED RES, MARION MERRELL DOW INC, 90- *Mem:* Sigma Xi; Am Soc Pharmacol & Exp Therapeut; AAAS; NY Acad Sci. *Res:* Disease in which the primary response leads to an alteration of connective tissue metabolism; identify specific intervention sites for the prevention of connective tissue destruction on the inappropriate synthesis of connective tissue; numerous publications. *Mailing Add:* Marion Merrell Dow Inc PO Box 9627 Kansas City MO 64134

COUNTS, GEORGE W, b Idabel, Okla, June 14, 35. CLINICAL RESEARCH. *Educ:* Univ Okla, BS, 57, MS, 60; Univ Iowa, MD, 65; Am Bd Internal Med, dipl, 70. *Prof Exp:* Intern internal med, Ohio State Univ Hosp, Columbus, 65-66, resident, 66-68; postdoctoral fel, Dept Med, Infectious Dis Div, Univ Wash, Seattle, 68-70, from assoc prof to prof med, 75-89; asst prof, Dept Epidemiol & Pub Health, Univ Miami, Fla, 70-72, Dept Med, 70-75,

Dept Path, 72-75; CHIEF, CLIN RES MGT BR, DIV AIDS, NAT INST ALLERGY & INFECTIOUS DIS, NIH, BETHESDA, MD, 89- *Concurrent Pos:* Attend physician, Jackson Mem Hosp, Miami, Fla, 71-75, dir, Clin Microbiol Sect, Div Clin Path, 72-75, dir, Infection Control Dept, 72-75; consult, Pac Med Ctr, Seattle, Wash, 77-85; head, Infectious Dis Clin, Harborview Med Ctr, Seattle, Wash, 84-85; active staff, Swed Hosp Med Ctr, Seattle, Wash, 86-89; mem, Anti-infective Drugs Adv Comt, Food & Drug Admin, 90-; mem, Community Based Clin Trials Adv Comt, Am Found AIDS Res, 90- *Mem:* Fel Am Col Physicians; Am Fedn Clin Res; fel Infectious Dis Soc Am; Int Soc Antiviral Res. *Res:* Author or co-author of over 95 publications. *Mailing Add:* Clin Res Mgt Br Nat Inst Allergy & Infectious Dis NIH Control Data Bldg Rm 207P 6003 Executive Blvd Rockville MD 20892

COUNTS, JON MILTON, b Richlands, Va, July 31, 37. PUBLIC HEALTH ADMINSTRATION. *Educ:* Univ Ariz, BS, 59; Tulane Univ, MPH, 63; Univ NC, DrPH(lab pract), 66. *Prof Exp:* Bacteriologist pub health, 60-62, asst dir, 66-73, dir lab licensure, 70-75, actg chief, 73-75, CHIEF BUR EPIDEMIOL & LAB HEALTH, SERV, STATE LAB, ARIZ DEPT HEALTH SERV, 75- *Concurrent Pos:* Chmn comt lab mgt, Nat Commun Dis Ctr Task Force, 71-73; qual assurance coordr, Adv Comt, Region IX, Environ Protection Agency, 74- *Mem:* Am Pub Health Asn; Am Soc Microbiologists; Conf State & Prov Pub Health Lab Dirs. *Res:* Laboratory management; development of clinical laboratories in quality assurance; disease control, immunization and epidemiology. *Mailing Add:* 1520 W Adams Phoenix AZ 85007

COUNTS, WAYNE BOYD, b Prosperity, SC, Oct 27, 36; m 62; c 3. ORGANIC CHEMISTRY. *Educ:* Furman Univ, BS, 58; Univ NC, PhD(org chem), 64. *Prof Exp:* Staff fel, Nat Cancer Inst, 63-66; prof chem, Lincoln Mem Univ, 66-69, head dept, 68-69; assoc prof, 69-73, coordr dept, 80-85, PROF CHEM, GA SOUTHWESTERN COL, 73-, DEPT HEAD, 86- *Mem:* Am Chem Soc; Sigma Xi. *Res:* Organic synthesis; heterocyclic chemistry. *Mailing Add:* Dept Chem Ga Southwestern Col Americus GA 31709

COUPER, JAMES R(ILEY), b St Louis, Mo, Dec 10, 25; m 53; c 2. CHEMICAL ENGINEERING, PROCESS ECONOMICS. *Educ:* Washington Univ, St Louis, BS, 49, MS, 50, DSc(chem eng), 57. *Prof Exp:* Res chemist, Presstite Eng Co, Mo, 50; res engr, Mo Portland Cement Co, 50-51; eng trainee, Monsanto Chem Co, 52-53, chem engr, 53-56, sr engr, 56-58, prod supvr, 58-59; assoc prof chem eng, 59-65, actg head dept, 68-69, admin asst res coordr, 65-68, head dept chem eng, 69-79, prof, 65-89, EMER PROF CHEM ENG, UNIV ARK, FAYETTEVILLE, 89- *Mem:* Fel Am Inst Chem; Am Inst Chem Engrs; Am Chem Soc; Am Soc Eng Educ; Am Asn Cost Engrs; Sigma Xi. *Res:* Heat transfer and thermal properties of two-phase systems; rheology of complex materials; fluidization; plant design and economics. *Mailing Add:* 2506 Sweetbriar Ave Fayetteville AR 72703-6537

COUPERUS, MOLLEURUS, b Essen, Ger, Jan 27, 06; nat US; m 39; c 4. DERMATOLOGY. *Educ:* Andrews Univ, BA, 27; Loma Linda Univ, MD, 34. *Prof Exp:* Prof dermat & syphilol, Loma Linda Univ, 47-80, chmn dept dermat, 64-77, prof phys anthrop, 75-80. *Mem:* Am Acad Dermat; Soc Invest Dermat. *Res:* Tumors of the skin; physical anthropology. *Mailing Add:* PO Box 6 Angwin CA 94508

COUPET, JOSEPH, b Port-Au-Prince, Haiti, Sept 7, 37; US citizen; m 63; c 2. BIOCHEMICAL PHARMACOLOGY. *Educ:* NY Univ, BS, 62, MS, 69, PhD(biochem pharmacol), 72. *Prof Exp:* Res asst, Hillside Psychiat Hosp, 62-64; res assoc, Med Ctr, NY Univ, 64-73; SR RES SCIENTIST, BIOCHEM PHARMACOL, LEDERLE LABS, 73- *Concurrent Pos:* Instr, NY Univ Med Ctr, 72-73. *Mem:* Am Chem Soc; AAAS. *Res:* Cyclic nucleotide system; kinetics of receptor binding pertaining to the mechanism of action of antipsychotic and anxiolytic agents. *Mailing Add:* 28 Haller Crescent Spring Valley NY 10977

COUPLAND, ROBERT THOMAS, b Winnipeg, Man, Jan 24, 20; m 45; c 1. GRASSLAND ECOLOGY, SYSTEMS ECOLOGY. *Educ:* Univ Man, BSA, 46; Univ Nebr, PhD(bot), 49. *Prof Exp:* Student asst range studies, Dom Dept Agr, Sask, 41-46; officer in chg, Dom Forest Serv, Man, 46; asst bot, Univ Nebr, 46-47; from asst prof to assoc prof, 48-57, head dept, 48-82, prof plant ecol, 57-85, EMER PROF, UNIV SASK, 85- *Concurrent Pos:* Dir Matador Proj, Int Ctr Grasslands Studies, Int Biol Prog, 67-77, chmn int coord comt grasslands & mem productivity terrestrial sect comt; ed, Natural Grasslands in Ecosysts of World Series (Vol 8, Part A & B). *Honors & Awards:* Can Centennial Medal. *Mem:* Fel AAAS; Am Inst Biol Sci; Brit Ecol Soc; Int Asn Ecol; Ecol Soc Am. *Res:* Nutrient balance and energy flow in cropland and natural grassland ecosystems; classification of grasslands; autecology of native plants, weeds and crop plants; studies of succession resulting from grazing and land abandonment. *Mailing Add:* Dept Crop Sci & Plant Ecol Univ Sask Saskatoon SK S7H 0W0 Can

COURANT, ERNEST DAVID, b Goettingen, Ger, Mar 26, 20; nat US; m 44; c 2. THEORETICAL PHYSICS. *Educ:* Swarthmore Col, BA, 40; Univ Rochester, MS, 42, PhD(physics), 43. *Hon Degrees:* DSc, Swarthmore Col, 88. *Prof Exp:* Asst physics, Univ Rochester, 40-43; sci officer, Nat Res Coun Can, 43-46; res assoc theoret physics, Cornell Univ, 46-48; consult, Brookhaven Nat Lab, 47-48, from assoc physicist to physicist, 48-60, sr physicist, 60-89, EMER AVI DISTINGUISHED SCIENTIST, BROOKHAVEN NAT LAB, 90- *Concurrent Pos:* Vis asst prof, Princeton Univ, 50-51; Fulbright res grant, Cambridge Univ, 56; consult, Gen Atomic Div, Gen Dynamics Corp, 58; vis prof, Yale Univ, 61-62, Brookhaven prof, 62-67; prof, State Univ NY, Stony Brook, 67-85; vis physicist, Nat Accelerator Lab, 68-69; vis scientist, Europ Orgn Nuclear Res, 74. *Honors & Awards:* Pregel Prize, NY Acad Sci, 79; Fermi Award, US Dept Energy, 86; R R Wilson Prize, Am Phys Soc, 87. *Mem:* Nat Acad Sci; fel AAAS; Am Phys Soc; NY Acad Sci. *Res:* Theory of solids; chain reactors, particle accelerators and nuclear reactions. *Mailing Add:* 109 Bay Ave Bayport NY 11705

COURANT, HANS WOLFGANG JULIUS, b Ger, Oct 30, 24; nat US; div; c 3. HIGH ENERGY PHYSICS. *Educ:* Mass Inst Technol, BS, 49, PhD(physics), 54. *Prof Exp:* Asst, Mass Inst Technol, 49-54, res assoc physics, 54; Fulbright grant, Polytech Sch, Paris, 54-55; asst res physicist, Univ Calif, 55-56; from instr to asst prof, Yale Univ, 56-61; Ford Found fel, 61-62; NSF sr fel, Europ Orgn Nuclear Res, Geneva, 62-63; assoc prof, 63-68, PROF PHYSICS, UNIV MINN, MINNEAPOLIS, 68- *Concurrent Pos:* Vis assoc physicist, Brookhaven Nat Lab, 56-58, 59-, guest physicist, 57, 58-; physicist, Radiation Lab, Univ Calif, 57; consult, Argonne Nat Lab, 64-70; vis prof, Univ Heidelberg, 69-70; vis res physicist LAPP, Annecy France, 78-79. *Mem:* Europ Phys Soc. *Res:* Experimental high energy physics; cosmic rays. *Mailing Add:* Sch Physics Univ Minn Physics Rm 316 Minneapolis MN 55455

COURCHENE, WILLIAM LEON, b Springfield, Mass, Nov 15, 26, m 48; c 3. PHYSICAL CHEMISTRY. *Educ:* Univ Mass, BS, 48; Cornell Univ, PhD(chem), 52. *Prof Exp:* Asst, Cornell Univ, 48-52; res chemist, Procter & Gamble, 52-67, sect head, 67-74, assoc dir, 74-88; RETIRED. *Mem:* Am Chem Soc; AAAS; Sigma Xi; NY Acad Sci. *Res:* Infrared and mass spectroscopy; solution thermodynamics; proteins; perfumes & flavors. *Mailing Add:* 968 Wittshire Lane Cincinnati OH 45255-5702

COURCHESNE, ERIC, b Berkeley, Calif, Apr 3, 49; m 70; c 1. DEVELOPMENTAL NEUROPSYCHOLOGY, DEVELOPMENTAL NEUROPHYSIOLOGY. *Educ:* Univ Calif, Berkeley, BA, 70; Univ Calif, San Diego, PhD(neurosci), 75. *Prof Exp:* Res neuroscientist, Sch Med, Univ Calif, San Diego, 75; scholar psychiat, Sch Med, Stanford Univ, 76-77, res assoc psychol, 77-78; asst prof neurosci, 78-86, ASSOC PROF NEUROSCI, SCH MED, UNIV CALIF, SAN DIEGO, 86- *Mem:* AAAS; Nat Soc Autistic Children; NY Acad Sci. *Res:* Relationships between the maturation of the central nervous system and the development of cognitive, attentional and memory abilities in normal and neurologically impaired individuals; developmental neuroradiology. *Mailing Add:* Dept Neurosci M-0008 Univ Calif San Diego Med Sch La Jolla CA 92093

COURI, DANIEL, b Streator, Ill, Oct 4, 30; m 59; c 4. TOXICOLOGY, BIOCHEMICAL PHARMACOLOGY. *Educ:* Queens Col, BS, 54; New York Univ, MS, 58; State Univ NY, Brooklyn, PHD(pharmacol), 65. *Prof Exp:* PROF PHARMACOL, DEPT PHARMACOL, OHIO STATE UNIV, 65-, DIR, DIV TOXICOL, 67- *Mem:* Am Soc Pharmacol & Exp Therapeut; Am Indust Hyg Asn; Am Acad Forensic Sci; Am Conf Govt Indust Hygienists; Soc Toxicol; Sigma Xi. *Res:* Mechanism of action of chemical agents on living systems resulting in beneficial or deleterious effects, especially the enzymatic and molecular aspects. *Mailing Add:* Dept Pharm Med Sch Ohio State Univ 370 W Ninth Ave Columbus OH 43210

COURSEN, BRADNER WOOD, b Roselle Park, NJ, Feb 10, 29; m 54; c 3. CELLULAR AGING. *Educ:* Drew Univ, BA, 52; Univ Md, MS, 57, PhD(fungus physiol), 59. *Prof Exp:* Jr chemist, Res Dept, Am Can Co, 52-55; instr biol & plant physiol, Lawrence Univ, 59-61, from asst prof to assoc prof biol, 61-68; assoc prof, 68-69, PROF BIOL, COL WILLIAM & MARY, 69- *Concurrent Pos:* Pres, Wis State Bd Exam in Basic Sci, 65-69. *Mem:* Bot Soc Am. *Res:* Mechanism of action of antibiotics; fungus physiology and metabolism; biology and biochemistry of cellular aging. *Mailing Add:* Dept Biol Col William & Mary Williamsburg VA 23185

COURSEN, DAVID LINN, b Newark, NJ, May 7, 23; m 45; c 4. ROCK MECHANICS. *Educ:* Columbia Univ, BA, 43; Cornell Univ, PhD(chem), 51. *Prof Exp:* Res chemist, E I du Pont de Nemours & Co, Inc, 50-58, sr res chemist, 58-59, tech assoc, 59-61, res assoc, 61-66, prod mgr, 66-67, res & develop mgr, 67-70, res fel, 70-88; res fel, ETI, 88-89; PRES, DYNATEC EXPLOSIVES CONSULTS, 89- *Concurrent Pos:* Assoc, Woods Hole Oceanog Inst. *Mem:* Am Chem Soc; Int Soc Rock Mech; Soc Explor Geophysicists; Sigma Xi; Soc Explosive Engrs. *Res:* Crystal structure; x-ray diffraction; explosives; radiography; underwater acoustics; mining and quarrying oil and gas well stimulation; working of mineral deposits in place. *Mailing Add:* PO Box 2952 Sedona AZ 86336

COURSEY, BERT MARCEL, b Birmingham, Ala, Mar 27, 42; m 65; c 3. RADIOCHEMISTRY, NUCLEAR MEDICINE. *Educ:* Univ Ga, BS Chem, 65, PhD(phys chem), 70. *Prof Exp:* Rubber chemist mfg, SEastern Rubber Mfg Co, 64-67; teaching asst, Chem Dept, Undergrad Lab, Univ Ga, 67-68; mem staff, Army Off Nuclear Power, US Army Engr Reactors Group, 69-71; res chemist radioactivity, US Nat Bur Standards, 72-87; GROUP LEADER RADIATION INTERACTIONS & DOSIMETRY, NAT INST STANDARDS & TECHNOL, 88- *Concurrent Pos:* Ed, Int J Appl Radiation & Isotopes, 77-, Int J Nuclear Med & Biol, 85-, Radioactivity & Radiochem, 90- *Honors & Awards:* Bronze Medal, Dept Com. *Mem:* Am Chem Soc. *Res:* Methods for the accurate measurement of radiation and radioisotopes. *Mailing Add:* Radiation Interactions & Dosimetry Group Rm C214 Bldg 245 Nat Bur Standards Gaithersburg MD 20899

COURT, ANITA, b Chicago, Ill, Aug 15, 30. REACTOR PHYSICS, PHYSICAL INORGANIC CHEMISTRY. *Educ:* Ill Inst Technol, BS, 52; Mich State Univ, PhD(chem), 56. *Prof Exp:* Asst chemist, M W Kellogg Co Div, Pullman, Inc, 56-59; chemist, Brookhaven Nat Lab, 59-70; scientist, Westinghouse Elec Corp, 70-75, prin engr, 75-76, adv engr, Advan Reactors Div, 76-82; RETIRED. *Mem:* Am Chem Soc; Am Nuclear Soc; Sigma Xi. *Res:* Shielding, safety analysis, core mechanics testing. *Mailing Add:* 3721 Beckwith Crete IL 60417

COURT, ARNOLD, b Seattle, Wash, June 20, 14; m 41, 88; c 3. CLIMATOLOGY. *Educ:* Univ Okla, BA, 34, Univ Wash, Seattle, MS, 49; Univ Calif, Berkeley, PhD(geog), 56. *Prof Exp:* Meteorologist, US Weather Bur, 38-43; climatologist & head climat unit, Res & Develop Br, Off Qm Gen, US Dept Army, 46-51; res meteorologist, Statist Lab, Univ Calif, Berkeley, 52-54, lectr meteorol & climat, 56-57 & 58; meteorologist, Pac Southwest Forest & Range Exp Sta, US Forest Serv, 56-60; chief appl climat br, Geophys

Res Directorate, Air Force Cambridge Res Labs, 60-62; res scientist, Lockheed-Calif Co, 62-64; chmn dept geog, 70-72, prof, 62-85, EMER PROF CLIMAT, CALIF STATE UNIV, NORTHRIDGE, 85- *Concurrent Pos:* Chief meteorologist, US Antarctic Serv, Little Am, 40-41; consult, Lockheed-Calif Co, 64-78. *Honors & Awards:* Spec Cong Medal for US Antartic Serv, 43. *Mem:* Fel AAAS; Am Geophys Union; Asn Am Geog; fel Am Meteorol Soc; Am Statist Asn; fel Royal Meteorol Soc UK. *Res:* History and principles of climatic diagrams; statistical analysis of climate; cloud seeding evaluation; wind energy computations. *Mailing Add:* 17168 Septo St Northridge CA 91325

COURT, WILLIAM ARTHUR, b Canmore, Alta, Feb 14, 43; m 66; c 2. ORGANIC CHEMISTRY, ANALYTICAL CHEMISTRY. *Educ:* Carleton Univ, BSc, 65, MSc, 68; Univ NB, PhD(chem), 70. *Prof Exp:* RES SCIENTIST, AGR CAN RES STA, 73- *Mem:* Am Chem Soc; Chem Inst Can. *Res:* Synthesis and structural elucidation of compounds of biological interest; chemical and biological studies of agricultural products; analytical methodology of plant constituents of biological interest; synthetic organic and natural products chemistry; agricultural and food chemistry. *Mailing Add:* Agr Can Chem Lab Res Sta Delhi ON N4B 2W9 Can

COURTENAY, WALTER ROWE, JR, b Neenah, Wis, Nov 6, 33; m 60; c 2. MARINE BIOLOGY, ICHTHYOLOGY. *Educ:* Vanderbilt Univ, BA, 56; Univ Miami, Fla, MS, 60, PhD(marine biol), 65. *Prof Exp:* Temp instr zool, Duke Univ, 63-64, vis asst prof, 64-65; asst prof biol, Boston Univ, 65-67; assoc ichthyol, Mus Comp Zool, Harvard Univ, 65-68; from asst prof to assoc prof, 67-72, PROF ZOOL, FLA ATLANTIC UNIV, 72- *Concurrent Pos:* NSF res grants, 65-69. *Mem:* Am Soc Ichthyol & Herpet; Am Fisheries Soc; Soc Syst Zool; Am Inst Fishery Res Biologists; Sigma Xi. *Res:* Distributions and impacts of introduced fishes; sonic mechanisms in fishes; systematics, functional morphology and ecology of fishes. *Mailing Add:* Dept Biol Sci Fla Atlantic Univ Boca Raton FL 33431

COURTER, R(OBERT) W(AYNE), b Denver, Colo, June 12, 35; m 56; c 3. FLUID MECHANICS, AERODYNAMICS. *Educ:* Univ Tex, BS, 57, MS, 58, PhD(aerospace eng), 65. *Prof Exp:* Assoc engr, McDonnell Aircraft Corp, Mo, 58-60; instr aerospace eng, Univ Tex, 60-64; asst prof mech eng, Univ Wyo, 64-68; ASSOC PROF MECH ENG, LA STATE UNIV, BATON ROUGE, 68-, ASST TO DEAN, 80- *Concurrent Pos:* Am Soc Eng Educ-Ford Found resident eng, Gen Dynamics Corp, Tex, 69-70. *Mem:* Am Inst Aeronaut & Astronaut; Am Soc Eng Educ. *Res:* Hypervelocity impact; ballistics; flight mechanics. *Mailing Add:* Dept Mech Eng La State Univ Baton Rouge LA 70803

COURTNEY, CHARLES HILL, b Baltimore, Md, Jan 1, 47; m 69; c 2. PARASITOLOGY. *Educ:* Clemson Univ, BS, 69; Univ Fla, MS, 73; Auburn Uni, DVM, 77; Ohio State Univ, PhD (vet parasitol), 82. *Prof Exp:* Instr parasitol, Ohio State Univ, 78-82; asst prof, 82-85, ASSOC PROF PARASITOL, UNIV FLA, 85-, CHMN, INFECTIOUS DIS, 90- *Concurrent Pos:* Consult, control bovine fascioliasis, Govt Cayman Islands. *Mem:* Am Vet Med Asn; Am Soc Parasitologists; Am Asn Vet Parasitologists; Am Soc Trop Vet Med; Am Heartworm Soc. *Res:* Epidemiology and control of livestock helminths in the tropics and subtropics with emphasis on ruminants; diagnosis and treatment of canine dirofilariasis. *Mailing Add:* Dept Infectious Dis Col Vet Med Univ Fla Bldg 471 Mowry Rd Gainesville FL 32611-0633

COURTNEY, GLADYS (ATKINS), b Erwin, Tenn, June 10, 30; m 67; c 2. MAMMALIAN PHYSIOLOGY, ENDOCRINOLOGY. *Educ:* La Col, BS, 56; La State Univ, MS, 58; Univ Ill Med Ctr, USPHS fel, 63-64, PhD(physiol), 64. *Prof Exp:* Gen duty nurse, Baptist Hosp, La, 52-56; teaching asst zool, La State Univ, 56-58; teaching asst physiol, Univ Ill, Urbana, 58-60; teaching asst & asst instr, Univ Ill Med Ctr, 60-65; asst prof biol, Malone Col, 65-68; assoc prof nursing & physiol, Univ Ill Med Ctr, 69-71, proj dir nurse scientist prog, Col Nursing, 69-76, prof nursing & physiol & head dept gen nursing, 71-76; prof & dean, Sch Nursing, Univ Mo-Columbia, 76-80; dean, 81-89, PROF, COL NURSING, MICH STATE UNIV, 81- *Mem:* Am Nurses' Asn; Nat League Nursing. *Res:* Relationship of adrenal cortex to ovarian function; prolongation of pseudopregnancy in hamsters; adrenal cortical function during environmental stress and functions of Academic Nursing Centers. *Mailing Add:* Col Nursing Mich State Univ East Lansing MI 48824

COURTNEY, JOHN CHARLES, b Washington, DC, June 11, 38; m 69. RADIATION PROTECTION. *Educ:* Cath Univ Am, BCE, 60, MNE & DEng, 62. *Prof Exp:* Asst engr, Allis-Chalmers Mfg Co, 62-63; nuclear res officer, McClellan AFB, 65-68; physics specialist, Aerojet Nuclear Systs Co, 68-71; PROF NUCLEAR ENG, LA STATE UNIV, 71- *Concurrent Pos:* Consult, Oak Ridge Nat Lab, 65 & pvt consult, 71-; vis prof, Argonne Nat Lab, Idaho, 75-86; counr/rep, La State Univ Syst to Oak Ridge Assoc Univ, 76- *Mem:* Am Nuclear Soc; Health Physics Soc. *Res:* Radiation protection, nuclear radiation shielding; radiological safety analysis; applied health physics; comparitive environmental impact assessments. *Mailing Add:* 10326 Hackberry Ct Baton Rouge LA 70809

COURTNEY, KENNETH OLIVER, medical research; deceased, see previous edition for last biography

COURTNEY, KENNETH RANDALL, b Snohomish, Wash, Dec 24, 44; m 88; c 3. PHYSIOLOGY, PHARMACOLOGY. *Educ:* Wash State Univ, BS, 67; Univ Wash, MS, 68, PhD(physiol biophys), 74. *Prof Exp:* Instr math, Centralia Col, 69-71; fel neurobiol, Univ Colo, 74-75; res assoc epilepsy, Med Ctr, Stanford Univ, 75-76 & res assoc membrane physiol, 76-77; cardiovasc pharmacologist, Stanford Res Inst, 76-78; SR SCIENTIST PHYSIOL & PHARMACOL, PALO ALTO MED FOUND, 78- *Concurrent Pos:* NIH fel, 67-68 & 71-74; prin investr, Heart, Lung & Blood Inst grant, NIH, 77-80; prin investr, NIH grants, 79-90; consult assoc prof anesthesia, Stanford Univ Med Ctr, 83-88. *Mem:* Biophys Soc; Am Heart Asn. *Res:* Stability of excitable tissues, including electrophysical characterization of anesthetic and; antiarrythmic drug actions; cardiac arrhythmias. *Mailing Add:* Palo Alto Med Found 860 Bryant St Palo Alto CA 94301

COURTNEY, RICHARD J, b Greenville, Pa, July 2, 41. BIOCHEMISTRY. *Educ:* Grove City Col, BS, 63; Syracuse Univ, MSc, 66, PhD(microbiol), 68. *Prof Exp:* Fel, Dept Virol, Baylor Col Med, 68-70, from asst prof to assoc prof, 70-77; from assoc prof to prof, Dept Microbiol, Univ Tenn, 78-85; prof & chmn, Dept Biochem & Molecular Biol, Sch Med, La State Univ, 85-91; PROF & CHMN, DEPT MICROBIOL & IMMUNOL, PA STATE UNIV, HERSHEY, PA, 91- *Concurrent Pos:* Mem, Am Cancer Soc Study Sect Microbiol & Virol & NIH Epidemiol & Dis Control Study Sect, 83-86, Virol Study Sect, 88-92, adv comt personnel res, 89-93. *Mem:* Am Soc Biochem & Molecular Biol; Am Soc Microbiol; Am Soc Virol; Sigma Xi. *Res:* Biochemistry; author of numerous scientific publications. *Mailing Add:* Dept Microbiol & Immunol Pa State Univ 500 University Dr PO Box 850 Hershey PA 17033

COURTNEY, RICHARD JAMES, b Greenville, Pa, July 2, 41; m 66; c 2. VIROLOGY. *Educ:* Grove City Col, BS, 63; Syracuse Univ, MS, 66, PhD(microbiol), 68. *Prof Exp:* NIH fel, Baylor Col Med, 68-70, from asst prof to assoc prof virol, 70-78; assoc prof, 78-81, PROF MICROBIOL, UNIV TENN, KNOXVILLE, 81- *Mem:* Am Soc Microbiol; Sigma Xi. *Res:* The biochemical and immunological characterization of the proteins and glycoproteins of herpes simplex viruses and to identify their functional role in virus-infected and transformed cells. *Mailing Add:* Dept Microbiol 2103 Chase Bend Shreveport LA 71118-4612

COURTNEY, THOMAS HUGH, b Dallas, Tex, Sept 26, 38; m 60; c 4. PHYSICAL METALLURGY, MATERIALS SCIENCE. *Educ:* Mass Inst Technol, SB, 60, ScD(phys metall), 64; Cornell Univ, MS, 62. *Prof Exp:* Res metallurgist, Res Ctr, Babcock & Wilcox Co, Ohio, 64-66; res assoc phys metall, Mass Inst Technol, 66-68; from asst prof to prof mech eng, Univ Austin Tex, 68-75; prof metall eng, Mich Technol Univ, 75-86, dean Grad Sch, 83-86; PROF & CHMN, MAT SCI, UNIV VA, 86- *Concurrent Pos:* Guest scientist, Powder Metall Labs, Max Planck Inst Metals Res, Stuttgart, WGermany; vis prof, Univ Calif, Santa Barbara. *Honors & Awards:* Adams Memorial Membership, Am Welding Soc, 73; Fel, Am Soc Metals Int, 87. *Mem:* Am Soc Metals; Am Inst Mining, Metall & Petrol Engrs; Mat Res Soc. *Res:* Fiber composite materials; mechanical behavior; powder metallurgy; stability of microstructure; materials processing. *Mailing Add:* Dept Mat Sci Univ Va Charlottesville VA 22903

COURTNEY, WELBY GILLETTE, b Hamilton, Ohio, Sept 17, 25; m 48; c 3. PHYSICAL CHEMISTRY. *Educ:* Oberlin Col, BA, 49; Iowa State Col, PhD(phys chem), 51. *Prof Exp:* Fel, Inst Atomic Res, Iowa, 51-52; sr chemist, Chem Construct Corp, NY, 52-55; res chemist, Freeport Sulphur Co, La, 55-56; sr scientist, Exp, Inc, 56-62; sect supvr heterogeneous combustion, Thiokol Chem Corp, 62-65, mgr physics dept, 65-68; supvry res chemist, US Bur Mines, 68-90; RETIRED. *Concurrent Pos:* Consult phase transformation, Switz, 58. *Res:* Kinetics and thermodynamics of phase transformation; coupled kinetics-gas dynamics; reduction of respirable and float dust in coal and noncoal mines; explosion phenomena in coal mines. *Mailing Add:* 154 Canterbury Lane McMurray PA 15317

COURTNEY, WILLIAM HENRY, III, b Lexington, Ky, Dec 27, 48; m 78; c 3. TOMATO BREEDING. *Educ:* Univ Ky, BS, 74; Univ Mo, MS, 76; Univ NH, PhD(plant sci), 79. *Prof Exp:* Asst prof, Univ BC, 80-82; res scientist, Ont Ministry Agr & Food, 82-85; ASSOC PROF, CLEMSON UNIV, 85- *Mem:* Am Soc Horticult Sci; Sigma Xi. *Res:* Tomato breeding and genetics; tissue culture. *Mailing Add:* Coastal Res & Educ Ctr 2865 Savannah Hwy Charleston SC 29414

COURTNEY-PRATT, JEOFRY STUART, b Hobart, Australia, Jan 31, 20; m 57. OPTICS, COMMUNICATIONS. *Educ:* Univ Tasmania, BE, 42; Cambridge Univ, PhD, 49, ScD, 58. *Prof Exp:* From asst res officer to res officer, Lubricants & Bearings iv, Coun Sci & Indust Res, Australia, 41-44; res officer ballistic instrumentation, Brit Admiralty, Eng, 44-45; asst dir res, Dept Phys Chem & Dept Physics, Cambridge Univ, 52-57; head, Dept Mech & Optics, Bell Tel Labs, 58-69, head, Appl Physics Dept, Bell Tel Labs, 69-82, Am Bell, Inc, 83 & AT&T Info Systs Labs, 84 & 85; PRES, PHYSICS INC, 85- *Concurrent Pos:* Consult, Tube Investments, Ltd, 53-55, Gen Elec Co, 55, Bell Tel Labs, 56-58 & Brit Ministry Supply; lectr, US, 52 & Ger Phys Soc, 55; independent consult, 85-; sr instr fiber optics, Learning Tree Int, 87-; consult partner, Solutions Group, 90- *Honors & Awards:* Stewart Prize, 46; Civic Medal, High Speed Photog Cong, Paris, 54; Boys Prize, 54; Gold Medal, Photog Soc Vienna, 61; Dupont Gold Medal, Soc Motion Picture & TV Eng, 61, Progress Medal, 69; Alan Gordon Mem Award, Soc Photo-Optical Instrument Engrs, 74. *Mem:* Hon mem Soc Motion Picture & TV Eng; fel Optical Soc Am; fel Royal Photog Soc Gt Brit; Brit Inst Mech Eng; Brit Inst Physics. *Res:* Applied physics; instrumentation; high speed photography; optics; ballistics; physics of contact of solids friction; electrical contacts; optical measurements on satellites; acoustics; electro-optics; recording systems; teleconferencing; graphics transmission systems; data communications; fiber optics. *Mailing Add:* Wigwam Rd Locust NJ 07760-2337

COURTRIGHT, JAMES BEN, b El Dorado, Kans, Dec 7, 41; m 67; c 2. BIOCHEMICAL GENETICS, BIOCHEMISTRY. *Educ:* Yale Univ, AB, 63; Johns Hopkins Univ, PhD(genetics), 67. *Prof Exp:* Sci asst, Max Planck Inst Biol, 67-69; fac assoc, Univ Tex, Austin, 69-70; asst prof, 70-75, PROF BIOL, MARQUETTE UNIV, 75- *Concurrent Pos:* Am Cancer Soc fel, 68-69; NSF grant, 69-70; Am Heart Asn grant, 71-73; Am Cancer Soc grant, 71-76; NASA grant, 78-80; NIH grant, 79-82 & 80-83. *Mem:* Am Genetic Asn; AAAS; Am Soc Microbiol; Genetics Soc Am. *Res:* Genetic control and regulation of gene expression in higher organisms; tissue specific gene expression in drosophila; inhibition of enzyme activity by inhibitors. *Mailing Add:* Dept Biol Marquette Univ Milwaukee WI 53233

COURVILLE, JACQUES, b Montreal, Que, Mar 17, 35; m 58; c 2. NEUROANATOMY. *Educ:* Univ Montreal, BA, 54, MD, 60, MSc, 62; Univ Oslo, Dr Med, 68. *Prof Exp:* Asst prof neurol & neurosurg, McGill Univ, 66-70; assoc prof, 70-76, PROF NEUROANAT, UNIV MONTREAL, 76-, HEAD DEPT NEUROSCI, 80- *Concurrent Pos:* Can Med Res Coun scholar, 67; mem res group neurol sci, Univ Montreal, 72. *Mem:* Am Asn Anatomists; Can Asn Anatomists; Soc Neurosci. *Res:* Experimental neuroanatomy with silver impregnation methods and injections of labelled amino acids; electron microscopy; red nucleus, facial nucleus, intracerebellar nuclei, cerebellar cortex, thalamic nuclei, inferior olive and pontine nuclei. *Mailing Add:* Dept Anat Univ Montreal CP 6128 Sta A Montreal PQ H3C 3J7 Can

COURY, ARTHUR JOSEPH, b Coaldale, Pa, Dec 5, 40; m 67; c 2. ORGANIC CHEMISTRY, POLYMER CHEMISTRY. *Educ:* Univ Del, BS, 62; Univ Minn, PhD(org chem), 65; Univ Minn, MBA, 80. *Prof Exp:* Sr res chemist II, Gen Mills Chem, Inc, 65-76; sr staff scientist & mgr polymer develop, 76-83, CORP RES FEL, MEDTRONIC, INC, 83-, DIR CORP PRES, 89- *Concurrent Pos:* chmn, Minn Sect, Am Chem Soc, 89-90; adj prof, Univ Minn, 89- *Honors & Awards:* IR-100 Award, Soc Prof Engrs, 84. *Mem:* Am Chem Soc; Electrochem Soc; Soc Biomat; Europ Soc Artificial Organs; Europ Soc Biomat. *Res:* Synthesis of monomers and polymers; preparation and uses of biomedical polymers. *Mailing Add:* 2225 Hillside Ave St Paul MN 55108

COUSER, RAYMOND DOWELL, b Oklahoma City, Okla, Mar 11, 31; m 54; c 2. ENTOMOLOGY, INVERTEBRATE ZOOLOGY. *Educ:* Northeastern State Col, BSEd, 56; Univ Okla, MS, 59, PhD(zool), 67. *Prof Exp:* Instr biol, Baker Univ, 58-61; assoc prof zool, 65-75, PROF BIOL, ARK TECH UNIV, 75- *Res:* Insect physiology; cold tolerance. *Mailing Add:* Dept Biol Ark Tech Univ Russellville AR 72801

COUSER, WILLIAM GRIFFITH, b Lebanon, NH, July 11, 39; m 89; c 2. NEPHROLOGY, IMMUNOPATHOLOGY. *Educ:* Harvard Col, Cambridge, BA, 61; Dartmouth Med Sch, BMS, 63; Hardvard Med Sch, Boston, MD, 65. *Prof Exp:* Asst prof med, Pritzker Sch Med, Univ Chicago, 71-72; from asst prof to assoc prof med, Sch Med, Boston Univ, 72-82; PROF MED, UNIV WASH, 82- *Concurrent Pos:* Head, Div Nephrology, Univ Wash, 82-; mem, Path A Study Sect, NIH, 85-89, chmn, 88-89; mem, Nephrology Bd, Am Bd Internal Med, 86-; counr, Western Asn Physicians, 89-93, Am Soc Nephrol, 90- & Am Asn Physicians, 90- *Mem:* Am Soc Clin Invest (vpres, 83-84); Am Asn Physicians; Am Soc Nephrol; Western Asn Physicians; Am Asn Immunologists; Am Asn Pathologists; Int Soc Nephrology. *Res:* Defining the immunologic basis for human glomerular disease with particular emphasis on mechanisms of immune complex formation in glomeruli and the factors which mediate immunologically induced renal injury. *Mailing Add:* Div Nephrol Rm-11 Seattle WA 98195

COUSIN, MARIBETH ANNE, b Beloit, Wis, Jan 26, 49. FOOD MICROBIOLOGY, DAIRY SCIENCE. *Educ:* Univ Wis, BS, 71, MS, 72, PhD(food sci), 76. *Prof Exp:* Res asst food microbiol, Univ Wis, 71-76; mgr dairy res & develop, Great Atlantic & Pac Tea Co, 77-78; PROF FOOD MICROBIOL, PURDUE UNIV, 78- *Concurrent Pos:* Sabbatical, Rijks Instituut voor Volksgezondheid en Milieu Hygiene, Bilthoven, Neth, 87; mem Indonesian Second Univ Develop Proj, 87; vis scientist, Agr Univ, Bogor, Indonesia, 88; elected or app various offices, Am Dairy Sci Asn, Inst Food Technologists & Am Soc Microbiol, 81- *Honors & Awards:* Richard M Hoyt Mem Award, Nat Milk Producers Fedn, Am Dairy Sci Asn, 77. *Mem:* Am Dairy Sci Asn; Inst Food Technologists; Soc Indust Microbiol; Nutrit Today Soc; Am Soc Microbiol. *Res:* Food fermentation; isolation and identification of fermentation microorganisms; immunological detection of mold in food products; microbiology of aseptically processed foods. *Mailing Add:* Dept Food Sci Purdue Univ West Lafayette IN 47907

COUSINEAU, GILLES H, b Montreal, Que, Sept 19, 32; m 59; c 2. MOLECULAR BIOLOGY, BIOCHEMISTRY. *Educ:* Col Ste-Marie de Montreal, BA, 54; Univ Montreal, BS, 58; NY Univ, MS, 61; Brown Univ, PhD(biochem), 64. *Prof Exp:* Res asst biol, Haskins Labs, 58-61; teaching fel, Brown Univ, 61-65; asst prof molecular biol, 65-71, ASSOC PROF BIOL SCI, UNIV MONTREAL, 71- *Concurrent Pos:* Res asst, Marine Biol Lab, Woods Hole, 58-63, independent investr, 66-; fel biol, Zool Sta, Naples, Italy, 64-65; grants, Nat Res Coun Can, 65-69, Montreal, 65-69, Damon Runyon Mem Fund, 66-68, Childs Mem Fund, 66-67, Donner Can Found, 66-67, Soc Sigma Xi, 66-67 & Defense Res Bd Can, 67-69. *Mem:* Electron Micros Soc Am; Can Soc Cell Biol; Am Soc Cell Biol; NY Acad Sci; Int Soc Develop Biol. *Res:* Synthesis of macromolecules in developing sea urchin eggs; cell division and differentiation. *Mailing Add:* Lab Molecular Biol Univ Montreal Montreal PQ H3C 3J7 Can

COUSINS, ROBERT JOHN, b New York, NY, Apr 5, 41; m 69; c 3. NUTRITIONAL BIOCHEMISTRY. *Educ:* Univ Vt, BA, 63; Univ Conn, MS, 65, PhD(nutrit biochem), 68. *Prof Exp:* Res assoc biochem, Univ Wis, 68-69, NIH fel, 69-70; asst prof, Rutgers Univ, 71-74, assoc prof nutrit, 74-77, dir grad prog nutrit, 76-82, prof nutrit biochem, 77-79, distinguished prof, 79-82; BOSTON FAMILY PROF HUMAN NUTRIT & BIOCHEM, UNIV FLA, 82-, DIR CTR NUTRIT SCI, 82- *Concurrent Pos:* Future Leader Grantee Award, Nutr Fedn Inc, 71-74; J L Prall vis prof, Va Polytech Inst & State Univ, 80; eminent scholar chair, Univ Fla, 82-; Wellcome vis prof, Auburn Univ, 86; mem bd dirs, Fedn Am Socs Exp Biol, 89-92, vpres, 90-91, pres, 91-92; ed, J Nutrit, 90- *Honors & Awards:* Mead Johnson Award, Am Inst Nutrit, 79; Osborne & Mendel Award, Outstanding Basic Res in Nutrit, 89; Mary E Shank Lectr, Univ Md, 89; James Waddell Mem Lectr, Univ Wisc, 89, Lectr Biochem & Molecular Biol, 89. *Mem:* Am Chem Soc; Am Inst Nutrit; Am Soc Biochem & Molecular Biol; Soc Exp Biol Med; Soc Toxicol; Biochem Soc. *Res:* Function of metals in mammalian systems, emphasizing nutrition and disease; hormonal regulation of zinc and copper metabolism, absorption, and transport; control of metalloprotein biosynthesis and degradation; nutritional regulation of gene expression. *Mailing Add:* Ctr Nutrit Sci Human Nutrit Bldg Univ Fla Gainesville FL 32611

COUSTEAU, JACQUES-YVES, b St Andre de Cubzac, France, June 11, 10; c 2. OCEANOGRAPHY. *Educ:* Stanislas Acad, Paris, BS, 27. *Hon Degrees:* Dsc, Univ Calif, Berkeley & Brandeis Univ, 70, Rensselaer Polytech Inst, 77, Harvard Univ, 79 & Univ Ghent, 83. *Prof Exp:* Founder, Group Study & Res Submarines, Toulon, France, 46; founder & pres, French Oceanog Co, 50 & Ctr for Marine Studies, 52; dir, Oceanog Mus, 57-88; PRIN, COUSTEAU SOC, 73- *Concurrent Pos:* Leader, Calypso Oceanog Exped; producer, World of Jacques Yves Cousteau, 66-68 & Undersea world of Jacques Costeau, 68-76. *Honors & Awards:* Potts Medal, Franklin Inst, 70; Gold Medal, Nat Geog Soc, 71; Int Environ Prize, UN, 77; Lindbergh Award, 82; Bruno H Schubert Found Prize, 83; US Presidential Medal of Freedom, 85; Founders Award, Nat Acad TV Arts & Sci. *Mem:* Foreign mem Nat Acad Sci; fel Soc Film & Arts; French Soc Writers. *Res:* Observing and recording the vitality of the living sea and man's relationship to it; author of more than fifty books in collaboration with various co-authors. *Mailing Add:* Cousteau Soc 425 E 52nd St New York NY 10022

COUTANT, CHARLES COE, b Jamestown, NY, Aug 2, 38; m 62; c 2. AQUATIC ECOLOGY, ZOOLOGY. *Educ:* Lehigh Univ, BA, 60, MS, 62, PhD(biol), 65. *Prof Exp:* Res scientist, Pac Northwest Div, Battelle Mem Inst, 65-70; prog mgr aquatic ecol, 70-78, SR RES ECOLOGIST, OAK RIDGE NAT LAB, 79- *Concurrent Pos:* Mem Comt Water Qual Criteria, Nat Acad Sci, 71-72; consult, Nat Comn Water Qual, 73-74; mem adv coun, Elec Power Res Inst, 73-80; adj assoc prof, Univ Tenn, Knoxville, 76- *Honors & Awards:* Darbaker Prize, Pa Acad Sci, 64; Battelle Excellence in Sci & Technol Award & Dir Award, Battelle Mem Inst, 68. *Mem:* AAAS; Sigma Xi; Am Fisheries Soc; Ecol Soc Am; Am Inst Fishery Res Biologists. *Res:* Effects of human alterations, particularly power plant cooling, on aquatic systems, including thermal effects, fish behavior, predator-prey relationships, population and community dynamics; energy-environment policy analysis; transport and fate modeling of chemicals in the environment. *Mailing Add:* 120 Miramar Circle Oak Ridge TN 37830

COUTCHIE, PAMELA ANN, b San Mateo, Calif. DATABASE DESIGN AND ACCESS. *Educ:* Univ Calif, Santa Barbara, BA, 71; Univ Calif, Davis, PhD(zool), 77; State Univ NY, Albany, MS, 88. *Prof Exp:* Res fel, Zool Dept, Univ Toronto, 77-80; asst prof, Biol Dept, Siena Col, 80-84; tech dir, DOHVM Environ, NY State Dept Health, 86-89; COMPUTER SYSTS ANALYST, HEALTH RES INC, 89- *Concurrent Pos:* Res assoc, Zool Dept, Univ Toronto, 81. *Mem:* Am Soc Zoologists; Sigma Xi; AAAS. *Res:* End user (both policy setters and epidemiologists) access to large databases; techniques to establish data quality in large data bases. *Mailing Add:* 440 New Scotland Ave Albany NY 12208-2727

COUTINHO, CLAUDE BERNARD, b Bombay, India, Aug 19, 31; m 57; c 5. BIOCHEMISTRY, EXPERIMENTAL MEDICINE. *Educ:* Univ Bombay, BS, 52; Univ Bristol, dipl, 53; Univ Belfast, BM, 55; Inst Divi Thomae, PhD(exp med, biochem), 63; Rutgers Univ, MBA, 75; Univ Auton de Cuidad Juarez, MD, 88. *Prof Exp:* Res microbiologist, Inst Laryngol & Otol, London, 55-57; asst chief chemother & microbiol, Aspro-Nicholas Pharmaceut Co, Ltd, Eng, 57-58, dept chief, 58-60; res scientist, Warner-Lambert Res Inst, 63-65; sr res assoc, Arthur D Little, Inc, 65-67; from sr scientist to sr res group chief, Hoffman-La Roche, 67-75, sr clin pharmacologist, 75-78; assoc dir clin pharmacol, Revlon Pharmaceut Corp, 78-79, dir clin pharmacol, 79-80; sr dir, Dept Clin Pharmacol, Ives Labs Div, Am Home Prod, 80-82; dir & clin res coordr, Farmitalia Carlo Erba, Milan, 82-84; CONSULT BASIC & CLIN RES & DEVELOP, INT PHARMACEUT INDUST, 84- *Concurrent Pos:* USPHS fel, 60-63. *Mem:* AAAS; Am Soc Microbiol; Am Soc Pharmacol & Exp Therapeut; Am Soc Toxicol; NY Acad Sci; Am Soc Clin Pharmacol & Therapeut. *Res:* Mechanism of pharmacological and toxicological drug action as related to biochemical changes and drug metabolism; pharmacokinetics; pharmacodynamics; tissue immunity; cancer; cellular metabolism; clinical pharmacology of therapeutic agents; pharmaceuticals and general chemotherapy. *Mailing Add:* Therapeutic Development 195 S Terrace Boonton NJ 07005

COUTINHO, JOHN, US citizen; m 42; c 2. ENGINEERING. *Educ:* NY Univ, MAeroE, 41; Berlin Tech Univ, Dr Eng, 70. *Prof Exp:* Engr, Grumman Aerospace Corp, 39-72; GEN ENGR, US ARMY MAT SYSTS ANALYTICAL ACTIV, 72- *Concurrent Pos:* Adj prof aircraft design, Polytech Inst NY, 51-60; proj mgr, battlefield damage assessment & repair, Depot Systs Command Corpus Christi Army Depot, Tex. *Honors & Awards:* Laskowitz Gold Medal, NY Acad Sci, 64; Brumbaugh Award, Am Soc Qual Control, 72; Centennial Medal, Am Soc Mech Engrs, 80; Systs Effectiveness & Safety Award, Am Inst Aeronaut & Astronaut, 83. *Mem:* Fel Am Soc Mech Engrs; fel Am Soc Qual Control; fel NY Acad Sci. *Res:* Reliability and maintainability for aircraft and space systems; stress analysis; structural design; control systems design; battelfield damage assessment and repair; combat resilience. *Mailing Add:* 602 Westgate Rd Aberdeen MD 21001-1625

COUTO, WALTER, b Montevideo, Uruguay, Jan 29, 32; m 62; c 2. SOIL FERTILITY, PASTURE DEVELOPMENT. *Educ:* Univ of Repub, Uruguay, Ing Agr, 61; Postgrad Col, Mex, MC, 71; Cornell Univ, PhD(soils), 76. *Prof Exp:* Adv soil fertility, Agromax SA, 62-63; res assoc, Ctr Invest Agr Alberto Boerger, La Estanzuela, 64-73; asst prof trop soils, NC State Univ, 76-77; sr scientist soils & pasture develop, Ctr Int Agr Trop, Cali, Colombia, 77-83; consult soil fertil & pasture develop, IICA, Brasilia, Brazil, 83-84; assoc prof soils, NC State Univ, 85-87; consult, soil fertil & pasture develop & res adminr, 87-90; EXPERT LAND EVAL & AGROECOL ZONING, FAO, BRASILIA, BRAZIL, 90- *Mem:* Am Soc Agron; Soil Sci Soc Am; Int Soc Soil Sci. *Res:* Soil fertility and soil chemistry in relation to major soil classification systems; soil fertility restrains for pasture development in the tropics; pasture establishment methods and pasture renovation methods; tropical pasture response to fertilizers. *Mailing Add:* c/o Representacao da Fao no Brazil CP 071058 Brasilia DS 70330 Brazil

COUTTS, JOHN WALLACE, b Neepawa, Man, Feb 2, 23; nat US; m 59; c 1. PHYSICAL CHEMISTRY. *Educ:* Univ Man, BSc, 45, MSc, 47; Purdue Univ, PhD(chem), 50. *Prof Exp:* Asst, Purdue Univ, 46-48; asst prof chem, Mt Union Col, 50-55; from assoc prof to prof, 55-86, chmn dept chem, 61-84, EMER PROF CHEM, LAKE FOREST COL, 86- *Concurrent Pos:* Fulbright lectr, Univ Peshawar, Pakistan, 58-59; NSF fac fel, Univ Calif, Berkeley, 67-68, vis prof, 68; vis prof, Rensselaer Polytech Inst, 74-75. *Mem:* AAAS; Am Chem Soc; Sigma Xi. *Res:* Molecular structure. *Mailing Add:* 106 E Sheridan Rd Lake Bluff IL 60044

COUTTS, RONALD THOMSON, b Glasgow, Scotland, June 19, 31; m 57; c 3. ORGANIC CHEMISTRY, MEDICINAL CHEMISTRY. *Educ:* Univ Glasgow, BSc, 55, PhD(pharmaceut chem), 60, DSc(med chem & drug metab), 76. *Prof Exp:* Asst lectr pharmaceut chem, Royal Col Sci & Technol, Univ Glasgow, 56-59; lectr, Sunderland Tech Col, Eng, 59-63; from asst prof to assoc prof, Univ Sask, 63-66; assoc prof, 66-69, PROF PHARMACEUT CHEM, UNIV ALTA, 69- *Concurrent Pos:* Asst sci ed, Can J Pharmaceut Sci, Can Pharmaceut Asn, 66-69, sci ed, 69-72; vis res prof, Dept Pharm, Chelsea Col, Univ London, Eng, 72-73. *Mem:* Fel Chem Inst Can; fel Brit Pharmaceut Soc; fel Royal Inst Chem; fel Royal Soc Chem; fel Royal Soc Can. *Res:* Synthesis, pharmacology, toxicology and metabolism of physiologically-active amines and analogs; mass spectrometry of medicinal compounds and metabolites; neurochemistry. *Mailing Add:* Fac Pharm Univ Alta Edmonton AB T6G 2N8 Can

COUTURE, JOSAPHAT MICHEL, b Montreal, Que, Can, Dec 3, 49; m 82. NUMERICAL METHODS, DIFFERENTIAL EQUATIONS. *Educ:* Sherbooke Univ, BSc, 71; McGill Univ, MSc, 75, PhD(physics), 81. *Prof Exp:* Res fel, 80-82, ASST RES OFFICER, ATOMIC ENERGY CAN LTD, 85- *Concurrent Pos:* Secy Subatomic Grant, Selction comt, Nat Sci & Eng Res Coun Can, 84- *Res:* Problem solving of electromagnetic theory by computer methodology. *Mailing Add:* Theoret Physics Br Chalk River Nuclear Labs Chalk River ON K0J 1P0 Can

COUTURE, ROGER, b Hull, PQ, May 7, 30; m 58; c 6. INTERNAL MEDICINE, NEPHROLOGY. *Educ:* Univ Ottawa, BA & BSc, 52; McGill Univ, MD, 57; FRCP(C), 64. *Prof Exp:* Lectr, 65-70, asst prof, 70-73, ASSOC PROF MED, FAC MED, UNIV OTTAWA, 73- *Mem:* Can Med Asn; Royal Col Physicians & Surgeons Can; Can Soc Nephrology. *Res:* Hemodialysis and renal transplantation. *Mailing Add:* Dept Med Univ Ottawa 451 Smyth Rd Ottawa ON K1H 8M5 Can

COUTURIER, GORDON W, b Sparta, Mich, Sept 14, 42; m 64; c 2. DATA-VOICE COMMUNICATIONS-NETWORKS, CASE TECHNOLOGY. *Educ:* Mich State Univ, BSEE, 64, MSEE, 65; Northwestern Univ, PhD, 71. *Prof Exp:* Mem tech staff, Bell Tel Labs, 65-72; eng proj leader, ITT Telecommun, 72-80; dir eng, GTE Subscriber Equip Group, 80-82 & Paradyne Corp, 82-87; CONSULT, C&C CONSULT, 87-; ASST PROF COMPUTER INFO SYSTS, UNIV TAMPA, 88- *Concurrent Pos:* Seminar leader, Systs Technol Forum, 87- *Mem:* Inst Elec & Electronics Engrs Computer Soc; Asn Comput Mach; Decision Sci Inst. *Res:* Application of expert systems in the business environment; developed a development project evaluation system which accessed a skills bank and recommended projects initiation or not based upon financial analysis of project's estimated costs/benefits. *Mailing Add:* 3127 Hillside Lane Safety Harbor FL 34695-5338

COUVILLION, JOHN LEE, b Jackson, Miss, Oct 20, 41; m 63; c 4. ORGANIC CHEMISTRY. *Educ:* La State Univ, New Orleans, BS, 63, Baton Rouge, MS, 65, PhD(chem), 67. *Prof Exp:* Teaching asst org chem, La State Univ, Baton Rouge, 63-65; res chemist, Tech Ctr, Celanese Chem Co, 67-72; res chemist, Food & Drug Admin, Washington, DC, 72-74; assoc prof chem, 74-79, prof chem & dir develop, 79-85, PROF CHEM & DEAN, STUDENT & INSTRNL SERV, LA STATE UNIV, EUNICE, 85- *Concurrent Pos:* Ed, Proc La Acad Sci, 76. *Mem:* Am Chem Soc. *Res:* Organic chemical synthesis; heterogeneous catalysis; toxicology; biochemical toxicology. *Mailing Add:* Dept Chem La State Univ PO Box 1129 Eunice LA 70535

COVA, DARIO R, b Benld, Ill, Feb 7, 28; m 53; c 3. CHEMICAL ENGINEERING. *Educ:* Univ Ill, BS, 51, MS, 53, PhD, 55. *Prof Exp:* Res engr, Monsanto Co, St Louis, 54-63, sr fel, 67-90; RETIRED. *Concurrent Pos:* Lectr, Wash Univ, 57-58. *Mem:* Am Inst Chem Engrs; Sigma Xi. *Res:* Distillation; interphase mass transfer; mixing; process development. *Mailing Add:* 723 Craig Dr St Louis MO 63122-5710

COVALT-DUNNING, DOROTHY, b Washington, DC, Jan 1, 37; m 60. ANIMAL PHYSIOLOGY, NEUROSCIENCES. *Educ:* Mt Holyoke Col, MA, 60; Tufts Univ, PhD(biol), 66. *Prof Exp:* Asst zool, Yale Univ, 60-61; asst biol, Harvard Univ, 61-62; fel, Max-Planck Inst Behav Physiol, 67; temp instr zool, Duke Univ, 68-69; asst prof, 69-74, ASSOC PROF BIOL, WVA UNIV, 74- *Mem:* AAAS; Am Soc Zool; Am Soc Mammal; Animal Behav Soc. *Res:* Sensory physiology and behavior of bats; prey-predator interactions between bats and moths; auditory neurophysiology and behavior in insects. *Mailing Add:* Dept Biol WVa Univ PO Box 6057 Morgantown WV 26506-6057

COVAN, JACK PHILLIP, b Cleveland, Ohio, Aug 20, 12; m 39; c 2. INDUSTRIAL ENGINEERING. *Educ:* Ohio State Univ, BSME & BSIE, 35; Univ Ill, MS, 42. *Prof Exp:* Draftsman, Jeffry Mfg Co, Ohio, 35-36; indust engr, Am Steel Wire Co, 36-37; instr mech eng, Univ Ill, 37-43; prod engr, Murphy Chair Co, Ky, 43-44; prod mgr, SPO Inc, Ohio, 44-46; assoc prof indust eng, Tex A&M Univ, 46-54; plant mgr, Orton Ceramic Found, Ohio, 54-55; prof indust eng, 55-74, EMER PROF INDUST ENG, TEX A&M UNIV, 74- *Concurrent Pos:* On loan to USAID prog, E Pakistan Univ Eng & Technol, 61-63. *Mem:* Am Inst Indust Engrs; Am Soc Eng Educ; Am Soc Tool & Mfg Engrs. *Res:* Tool design; manufacturing engineering; statistical quality control. *Mailing Add:* 747 S Rosemary Dr Bryan TX 77802

COVAULT, DONALD O, b Ft Wayne, Ind, Apr 19, 26; m 51; c 4. CIVIL ENGINEERING, TRAFFIC ENGINEERING. *Educ:* Purdue Univ, BSCE, 48, MSCE, 50, PhD, 59. *Prof Exp:* Instr civil eng, Purdue Univ, 48-50, res engr, 55-58; instr, Univ Colo, 50-51; hwy engr, Wis Hwy Dept, 51-55; PROF CIVIL ENG, GA INST TECHNOL, 58- *Concurrent Pos:* Consult, Ind State Hwy Dept, 57, Dept Traffic & Transp, Metrop Dade County, Fla, 59 & Ga State Hwy Dept, 59. *Mem:* Am Soc Civil Engrs; Inst Transp Engrs; Am Rd & Transp Builders Asn. *Res:* Highway engineering; traffic and planning of transportation facilities. *Mailing Add:* 2000 LaVista Circle Tucker GA 30084

COVEL, MITCHEL DALE, b Oakland, Calif, July 10, 17; m 71. MEDICINE. *Educ:* Univ Calif, AB, 39, MD, 42. *Prof Exp:* Assoc clin prof med, Loma Linda Univ, 51-60; from asst clin prof to assoc clin prof, Ctr Health Sci, 60-73, dep dir regional med progs, 67-73, CLIN PROF MED, MED CTR, UNIV CALIF, LOS ANGELES, 65-, DEP DIR REGIONAL MED PROGS, INST CHRONIC DIS & REHAB, 68-, ASSOC DEAN CLIN AFFAIRS, SCH MED, 79- *Concurrent Pos:* Chief outpatient serv, Los Angeles County Hosp, 49-52; mem staff, Hosp Good Samaritan, 51-; fel coun clin cardiol, Am Heart Asn; gov, Am Col Physicians, Southern Calif, Region I, 81-85. *Mem:* Am Heart Asn; fel Am Col Physicians; fel Am Col Cardiol; fel NY Acad Sci; Am Heart Asn. *Mailing Add:* 9730 Wilshire Blvd Beverly Hills CA 90212

COVELL, CHARLES VANORDEN, JR, b Washington, DC, Dec 10, 35; m 58; c 3. ENTOMOLOGY. *Educ:* Univ NC, BA, 58; Va Polytech Inst, MS, 62, PhD(entom), 65. *Prof Exp:* Teacher, Norfolk Acad Sch Boys, 58-60; asst entom, Va Polytech Inst, 60-64; from instr to assoc prof, 64-74, PROF BIOL, UNIV LOUISVILLE, 74- *Concurrent Pos:* Consult, Health Dept, Louisville-Jefferson County, Ky, 65- *Mem:* Entom Soc Am; Am Entom Soc; Lepidopterists Soc; Am Mosquito Control Asn. *Res:* Taxonomy, distribution, ecology and life history of lepidopterous insects, especially the moth family Geometridae; faunistics of Kentucky insects, especially Lepidoptera and mosquitoes. *Mailing Add:* Dept Biol Univ Louisville Louisville KY 40292

COVELL, DAVID GENE, b Toledo, Ohio, Oct 3, 49. BIOENGINEERING. *Educ:* Univ Mich, BSE, 71, MSE, 75, PhD(bioeng), 79. *Prof Exp:* Mich heart fel, Dept Physiol, Univ Mich, 79-80; staff fel math biol, Lab Theoret Biol, NIH, 80-; MICH HEART ASN FEL, DEPT PHYSIOL, UNIV MICH, ANN ARBOR. *Mem:* Inst Elec & Electronics Engrs; AAAS; Biomed Eng Soc. *Res:* Mathematical characterization of biological functions; mathematical models used to more efficiently analyze experimental results and design further testing as well as designing methods to better control biological processes. *Mailing Add:* PO Box B Bldg 430 Frederick MD 21702

COVELL, JAMES WACHOB, b San Francisco, Calif, Aug 13, 36; div; c 2. CARDIOVASCULAR PHYSIOLOGY, BIOMEDICAL ENGINEERING. *Educ:* Carleton Col, BA, 58; Univ Chicago, MD, 62. *Prof Exp:* Intern surg, Univ Chicago Hosps & Clin, 62-63, resident, 63-64; res assoc cardiol, Nat Heart & Lung Inst, 64-66, sr investr, 66-68; from asst prof to assoc prof, 68-76, PROF MED, SCH MED, UNIV CALIF, SAN DIEGO, 76- *Concurrent Pos:* USPHS fel, 63-64, career develop award, 70-75. *Mem:* Am Fedn Clin Res; Am Heart Asn; Am Physiol Soc; Fedn Am Soc Exp Biol; Cardiac Muscle Soc. *Res:* Factors influencing cardiac performance in the normal and diseased heart; mechanics of muscle contraction; on-line data analysis and control of hemodynamic parameters. *Mailing Add:* 2732 Inverness Dr La Jolla CA 92037

COVENEY, RAYMOND MARTIN, JR, b Marlboro, Mass, Oct 15, 42; m 65; c 3. GEOLOGY. *Educ:* Tufts Univ, BS, 64; Univ Mich, MS, 68, PhD(geol), 71. *Prof Exp:* Mine geologist, Dickey Explor Co, 69-71; from asst prof to assoc prof, 71-84, actg dept chair, 90-91, PROF GEOL, UNIV MO, KANSAS CITY, 84- *Concurrent Pos:* Prin investr, NSF, 83-85 & 89-91. *Honors & Awards:* Veatch Res Award. *Mem:* Fel Geol Soc Am; Mineral Soc Am; Sigma Xi; Am Geophys Union; Soc Econ Geologists; AAAS. *Res:* Geology, geochemistry, and mineralogy of ore deposits and black shales; fluid inclusion research; determinative mineralogy with emphasis on x-ray and optical techniques. *Mailing Add:* Dept Geosci Univ Mo Kansas City MO 64110-2499

COVENTRY, MARK BINGHAM, b Duluth, Minn, Mar 30, 13; m 37; c 3. ORTHOPEDIC SURGERY. *Educ:* Univ Mich, MD, 37; Univ Minn, MS, 42. *Prof Exp:* PROF ORTHOP SURG, MAYO MED SCH, 58-; CONSULT, SECT ORTHOP SURG, MAYO CLIN, 42- *Concurrent Pos:* Head dept orthop, Mayo Clin, 68-74. *Mem:* Am Acad Orthop Surg; Clin Orthop Soc; Am Orthop Asn; Am Col Surg; hon fel Brit Orthop Asn; Sigma Xi. *Mailing Add:* Mayo Clin Rochester MN 55905

COVER, HERBERT LEE, b Elkton, Va, Dec 28, 21; m 45; c 1. PHYSICAL CHEMISTRY. *Educ:* Univ Va, BS, 45, MS, 46, PhD(chem), 49. *Prof Exp:* From instr to prof qual & quant anal, 49-74, PROF CHEM, MARY WASHINGTON COL, 74- *Res:* Catalytic oxidation of carbon monoxide; catalytic hydration of ethylene; analytical chemistry and electronics. *Mailing Add:* 1607 Franklin St Fredericksburg VA 22401

COVER, RALPH A, b Boston, Mass, Dec 11, 43. FREE ELECTRON LASER PHYSICS, MANY-BODY PHYSICS. *Educ:* Providence Col, BS, 65; Brandeis Univ, PhD(physics), 72; Calif State Univ, MBA, 90. *Prof Exp:* Staff scientist, KMS Fusion, Inc, 71-78; staff scientist, R&D Assocs, 78-80; SR SCIENTIST, ROCKETDYNE DIV, ROCKWELL INT, 80- *Concurrent Pos:* Mem, NIH Sponsored Hons Sci Prog, Providence Col. *Res:* Free electron lasers, optics, charged particle interation with matter and plasma physics. *Mailing Add:* 20807 Roscoe Blvd Unit 12 Canoga Park CA 91306

COVER, RICHARD EDWARD, b Youngsville, Pa, Nov 7, 26; m 48; c 3. ANALYTICAL CHEMISTRY. *Educ:* Pa State Univ, BS, 47; Polytech Inst Brooklyn, MS, 60, PhD(anal chem), 62. *Prof Exp:* Chemist, US Steel Corp, 48-50 & Fairchild Camera & Instrument Co, 50-57; sr chemist, Socony Mobil Oil Co, 57-64; PROF ANALYTICAL CHEM, ST JOHN'S UNIV, NY, 64- *Concurrent Pos:* Mem, Simulation Coun, 65- *Mem:* Am Chem Soc; Soc Indust & Appl Math; Sigma Xi. *Res:* Electroanalytical chemistry; gas chromatography; chemical kinetics; computer applications to chemistry; applied mathematics. *Mailing Add:* 9130 98th St Woodhaven NY 11421

COVER, THOMAS MERRILL, b Pasadena, Calif, Aug 7, 38; m 68. INFORMATION THEORY, STATISTICS. *Educ:* Mass Inst Technol, BS, 60; Stanford Univ, MS, 61, PhD(elec eng), 64. *Prof Exp:* assoc prof elec eng, 64-72, assoc prof statist, 70-72, PROF ELEC ENG & STATIST, STANFORD UNIV, 72- *Concurrent Pos:* Consult, Rand Corp, 60-64, FMC Corp, 67-68, Stanford Res Inst, 63-65, Sylvania Elec Prod Inc, 68-72, SRI, 76- & Calif State Lottery, 88-; assoc ed, Inst Elec & Electronics Engrs Trans on Info Theory, 71-72; vis prof, Mass Inst Technol, 71; Vinton Hayes res fel, Harvard, 71; assoc ed, PAMI, Mgt Sci; bk rev ed, Inst Elec & Electronics Engrs Info Theory Soc, 90. *Honors & Awards:* Info Theory Prize Paper Award, Inst Elec & Electronics Engrs, 72, Shannon Lectr, 90. *Mem:* Fel Inst Elec & Electronics Engrs; Soc Indust & Appl Math; Am Math Soc; fel Inst Math Statist. *Res:* Applied statistics; pattern recognition and learning. *Mailing Add:* Dept Elec Eng Stanford Univ Stanford CA 94305

COVERT, EUGENE EDZARDS, b Rapid City, SDak, Feb 6, 26; m 46; c 4. AERONAUTICAL ENGINEERING, UNSTEADY FLUID MECHANICS. *Educ:* Univ Minn, BS, 46, MS, 48; Mass Inst Technol, ScD, 58. *Prof Exp:* Preliminary design aerodynamicist, Naval Air Design Ctr, Johnsville, Pa, 48-52; res engr to assoc dir, Aerophys Lab, Mass Inst Technol, 52-63, from assoc prof to prof aeronaut & astronaut, 63-85, dir, Gas Turbine & Plasma Dynamics Lab, 79-84, dept head, 85-90, PROF AERONAUT & ASTRONAUT, MASS INST TECHNOL, 90-, DIR, CTR AERODYN STUDIES, 84- *Concurrent Pos:* Consult, Hercules Inc, 63- & Sverdrup Technol Inc, 76-; chief scientist, USAF, 72-73; consult & dir, Megatech Inc, 71-85; vchmn, Sci Adv Bd, USAF, 77-81, chmn, 82-86; mem, Gas Turbine Technol Adv Comt, NASA, 79; chmn, Power & Energetics Panel, NATO, 80-83; consult, Lincoln Labs, Mass Inst Technol, 63; consult, Defense Sci Bd, 73; mem, bd dirs, Allied Signal Corp, Rohr Industs Inc & Phys Sci Inc; tech dir, Europ Off Res & Develop, 78-79; mem, Aerospace Am Adv Comt, 84- & Presidential Comn on Space Shuttle Challenger Accident, 86; consult, United Technol Corp, 87-90 & Defense Sci Bd, 87-; vchmn, Aeronaut & Space Eng Bd, 88-90, mem, 88- *Honors & Awards:* Ground Testing Award, Am Inst Aeronaut & Astronaut, 90; Agard von Karman Medal, 90. *Mem:* Nat Acad Eng; fel Am Inst Aeronaut & Astronaut; fel Royal Astron Soc; Sigma Xi; fel Royal Aeronaut Soc; NY Acad Sci; fel AAAS. *Res:* Unsteady aerodynamics of boundary layers, both in unseparating and separating state. *Mailing Add:* Mass Inst Technol Rm 33-207 77 Massachusetts Ave Cambridge MA 02139

COVERT, ROGER A(LLEN), b Muncie, Ind, May 9, 29; m; c 2. METALLURGY, CORROSION. *Educ:* Purdue Univ, BS, 51; Mass Inst Technol, SM, 54, ScD(metall), 57. *Prof Exp:* Asst, Mass Inst Technol, 56-57; serv engr, E I du Pont de Nemours & Co, 57-58; proj supvr & asst tech dir, Alloyd Electronics Corp, 58-62; res consult, Avesta Jernwerks, Sweden, 62-63; sr metallurgist, 63-67, corrosion engr, 67-68, in charge corrosion eng, Prod Develop, 67-70, sect mgr corrosion eng, Prod Res & Develop, Int Nickel Co, NY, 70-72, proj mgr mkt develop, 72-75, asst mgr distribr sales mkt, Int Nickel Co, Europe, London, UK, 75-77, mgr plating sales mkt, 77-80, gen sales mgr, 80-81, GEN SALES MGR PLATING & CHEM IND, INT NICKEL CO, NY, 81- *Mem:* AAAS; Electrochem Soc; Am Inst Mining, Metall & Petrol Engrs; Nat Asn Corrosion Engrs; Electroplating Soc. *Res:* Corrosion of metals; physical metallurgy; theoretical electrochemistry; electroplating. *Mailing Add:* 94 Deepdale Dr Middletown NJ 07748

COVEY, CURTIS CHARLES, b Stockton, Calif, Oct 30, 51; m 90. CLIMATE DYNAMICS, EXTRATERRESTRIAL ATMOSPHERES. *Educ:* Mass Inst Technol, SB, 73; Univ Calif, Santa Barbara, MA, 75; MA, 78; Univ Calif, Los Angeles, PhD(geophys & space physics), 82. *Prof Exp:* Res fel, Nat Ctr Atmospheric Res, 82-83; Asst prof div meteorol & phys oceanog, Rosentiel Sch Marine & Atmospheric Sci, Univ Miami, 84-87; PHYSICIST, LAWRENCE LIVERMORE NAT LAB, 87- *Mem:* Am Meteorol Soc; Am Geophys Union. *Res:* Numeric modeling of geophysical fluid dynamical systems; the earth's atmosphere, ocean and climate; extraterrestrial environments. *Mailing Add:* Physics/G Div Lawrence Livermore Nat Lab Livermore CA 94550

COVEY, DOUGLAS FLOYD, b Baltimore, Md, May 25, 45; m 72; c 2. ENDOCRINOLOGY, STEROID HORMONES. *Educ:* Loyola Col, BS, 67; Johns Hopkins Univ, MA, 69, PhD(chem), 73. *Prof Exp:* Fel chem, Johns Hopkins Univ, 73-74, pharmacol, 74-77; from asst prof to assoc prof pharmacol, 77-90, PROF MOLECULAR BIOL & PHARMACOL, SCH MED, WASH UNIV, 90- *Concurrent Pos:* NIH res career develop award, 82-87. *Mem:* Am Chem Soc; Endocrine Soc; Am Soc Pharmacol & Exp Therpeut; AAAS; Am Soc Biochem & Molecular Biol. *Res:* Design and evaluation of inhibitors of steroid biosynthesis, enzymology and endocrinology of steroid hormones; mechanism of action of anticonvulsant drugs; anticonvulsants. *Mailing Add:* Dept Molecular Biol & Pharmacol Sch Med Wash Univ 660 S Euclid Ave St Louis MO 63110

COVEY, IRENE MABEL, b Newark, NJ, Jan 1, 31; m 56; c 3. GENERAL CHEMISTRY. *Educ:* Douglass Col, BSc, 52; Univ Mich, PhD(chem), 57. *Prof Exp:* Asst prof, 71-77, ASSOC PROF CHEM, UNIV CONN, WATERBURY, 77- *Mem:* Am Chem Soc. *Mailing Add:* Univ Conn 32 Hillside Ave Waterbury CT 06710

COVEY, RONALD PERRIN, JR, b Jamestown, NY, Aug 19, 29; m 52; c 3. PLANT PATHOLOGY. *Educ:* Univ Minn, BS, 56, MS, 59, PhD(plant path), 62. *Prof Exp:* Asst plant pathologist, 62-73, ASSOC PLANT PATHOLOGIST, TREE FRUIT RES CTR, UNIV WASH, 73- *Mem:* Am Phytopath Soc. *Res:* Bacterial and fungal disease of tree fruits; phytophthora cactorum. *Mailing Add:* Fruit Exp Sta Wenatchee WA 98801

COVEY, RUPERT ALDEN, b Manchester, NH, July 24, 29; m 56; c 3. ORGANIC CHEMISTRY. *Educ:* Middlebury Col, AB, 51; Univ NH, MS, 53; Univ Mich, PhD(pharmaceut chem), 58. *Prof Exp:* Sr res chemist agr chem, Naugatuck Chem Div, US Rubber Co, 57-67; sr res chemist, 67-70, res scientist, 70-79, sr res scientist, 79-81, RES ASSOC, UNIROYAL CHEM CO, INC, 81- *Honors & Awards:* Maurice R Chamberland Award, Am Chem Soc, 85. *Mem:* Am Chem Soc; Sigma Xi. *Res:* Crop protection chemicals; synthesis of organic compounds as pesticides; organic sulfite esters; heterocyclic compounds. *Mailing Add:* Uniroyal Chem Co Inc World Hq 1-2A Middlebury CT 06749

COVEY, WILLIAM DANNY, b Sacramento, Calif, June 10, 40. PHYSICAL INORGANIC CHEMISTRY, CHEMICAL KINETICS. *Educ:* Univ Calif, Berkeley, BS, 67; Sacramento State Col, MS, 70; Univ Ill, Champaign-Urbana, PhD(phys inorg chem), 73. *Prof Exp:* Asst chemist qual control, Pac Gas & Elec Co, Calif, 62-65; chem tech, Consortium Electrochem Indust, Munich, Ger, 65-67; res assoc bio-inorg kinetics, Ohio State Univ, 72-74 & Occidental Col, 74-76; sr chemist anal chem serv, Avery Int, 77; asst prof chem, Occidental Col, 77-80; PROF CHEM, L A PIERCE COL, 80- *Concurrent Pos:* Sigma Xi grant-in-aid, 75. *Mem:* Am Chem Soc; AAAS; Sigma Xi. *Mailing Add:* 11007 Ayres Ave Los Angeles CA 90064

COVEY, WINTON GUY, JR, b Glen Daniel, WVa, Feb 3, 29; m 52; c 2. MICROMETEOROLOGY. *Educ:* Johns Hopkins Univ, AB, 49; Tex A&M Univ, MS, 59, PhD(soil physics), 65. *Prof Exp:* Micrometeorologist, Tex A&M Univ, 55-59; soil physicist, Tex Agr Exp Sta, 59-60; res soil scientist, Agr Res Serv, USDA, 60-65; assoc prof meteorol, Cornell Univ, 65-68; assoc prof, 68-74, PROF NATURAL SCI, CONCORD COL, 74- *Mem:* Geol Soc Am; Nat Speleol Soc. *Mailing Add:* Dept Phys Sci Concord Col Athens WV 24712

COVINO, BENJAMIN GENE, b Lawrence, Mass, Sept 12, 30; m 53; c 2. PHYSIOLOGY. *Educ:* Col of the Holy Cross, AB, 51; Boston Col, MS, 53; Boston Univ, PhD(physiol), 55; State Univ NY Buffalo, MD, 61. *Prof Exp:* Res scientist, Arctic Aeromed Lab, 55-57; asst prof pharmacol, Sch Med, Tufts Univ, 57-59; asst prof physiol, Sch Med, State Univ NY, Buffalo, 59-63; med dir, Astra Pharmaceut Prod, Inc, 63-77, vpres res & develop, 67-77, vpres & dir res, Astra Res Lab, 77-79; PROF ANESTHESIOL, MED SCH, UNIV MASS, WORCESTER, 75-; CHMN, DEPT ANESTHESIOL, BRIGHAM & WOMEN'S HOSP, 79-; PROF ANESTHESIOL, HARVARD MED SCH, 79- *Concurrent Pos:* Lectr, Harvard Med Sch, 78-; clin assoc anesthesiol, Mass Gen Hosp; consult physiologist & clin pharmacologist, St Vincent's Hosp, Worcester. *Mem:* Am Physiol Soc; Am Fedn Clin Res; Int Col Angiol; Am Col Clin Pharmacol & Chemother. *Res:* Cardiovascular physiology and pharmacology of local anesthetics; physiology of temperature regulation. *Mailing Add:* Brigham & Women's Hosp 75 Francis St Boston MA 02115

COVITCH, MICHAEL J, b Altoona, Pa, Aug 27, 49; m 72; c 4. PHYSICAL CHEMISTRY. *Educ:* Lehigh Univ, BS, 71; Univ Rochester, MS, 73; Case Western Reserve Univ, PhD(macromolecular sci), 77. *Prof Exp:* Dept mat sci, Univ Rochester, 72-73; electrochem & polymer res dept, Diamond Shamrock Corp, Painesville, Ohio, 77-83; DEPT STRATEGIC RES, LUBRIZOL CORP, WICKLIFFE, OHIO, 83- *Mem:* Soc Automotive Engrs. *Res:* Polymer physical chemistry; ion exchange membranes; rheology; lubricants; electrochemistry; microscopy; thermal analysis. *Mailing Add:* Lubrizol Corp 29400 Lakeland Blvd Wickliffe OH 44092

COWAN, ALAN, b Selkirk, Scotland, July 6, 42; m 68; c 2. PSYCHOPHARMACOLOGY. *Educ:* Univ Glasgow, BSc, 64, Univ Strathclyde, PhD(pharmacol), 68. *Prof Exp:* Sect leader, Reckitt & Colman, Eng, 68-76; from asst prof to assoc prof, 77-88, PROF, DEPT PHARMACOL, SCH MED, TEMPLE UNIV, 88- *Mem:* Am Soc Pharmacol & Exp Therapeut; Int Soc Psychoneuroendocrinol; Brit Pharmacol Soc. *Res:* Psychopharmacology of opioids and behaviorally active peptides; pharmacology of substance abuse; dermatopharmacology. *Mailing Add:* Dept Pharmacol Sch Med Temple Univ 3420 N Broad St Philadelphia PA 19140

COWAN, ARCHIBALD B, b Ambia, Ind, July 14, 15; m 42; c 3. WILDLIFE DISEASES, WILDLIFE MANAGEMENT. *Educ:* Univ Mich, BSF, 40, PhD(wildlife mgt), 54. *Prof Exp:* Asst leader pheasant res proj, State Natural Hist Surv, Ill, 46-49; wildlife res biologist, Patuxent Res Refuge, 54-56; from asst prof to assoc prof wildlife mgt, Univ Mich, Ann Arbor, 56-80; RETIRED. *Mem:* Wildlife Soc; Wildlife Dis Asn (treas, 58-63, vpres, 65-67, pres, 67-69). *Res:* Parasites and diseases of wildlife. *Mailing Add:* 1341 Wines Dr Ann Arbor MI 48103

COWAN, DANIEL FRANCIS, b Mineola, NY, Aug 7, 34; m 60; c 3. PATHOLOGY. *Educ:* Antioch Col, BA, 56; McGill Univ, MDCM, 60. *Prof Exp:* NIH spec fel comp path, 66-67; from asst prof to assoc prof path, Mich State Univ, 67-73; chmn dept path, Eastern Va Med Sch, 73-77; prof path, Univ Tex Med Sch, Houston, 77-83, Temple Univ Med Sch, 83-86 & State Univ NY, Buffalo, 86-90; PROF PATH, UNIV TEX MED BR, 90- *Mem:* AMA; fel Col Am Pathologists; fel Am Soc Clin Pathologist; Wildlife Dis Asn. *Res:* Research in anatomic pathology; comparative pathology; diseases of marine mammals; surgical pathology. *Mailing Add:* Dept Path Univ Tex Med Br 101 Keiller Bldg F09 Galveston TX 77850

COWAN, DARREL SIDNEY, b Los Angeles, Calif, Feb 26, 45. GEOLOGY. *Educ:* Stanford Univ, BS, 66, PhD(geol), 72. *Prof Exp:* Geologist, Shell Oil Co, 71-74; from asst prof to assoc prof, 74-85, PROF GEOL, UNIV WASH, 85-, CHMN, 89- *Concurrent Pos:* Assoc ed, Tectonics, Am J Sci, mem planning comt, Joint Oceanog Inst Deep Earth Sampling, 85- *Mem:* Geolog Soc Am; Am Geophys Union. *Res:* History of accretion and collision of major tectonostratigraphic terranes to the continental margin of western North America; deformation and structural evolution of ancient and modern subduction complexes at convergent plate boundaries; tectonics of convergent and collisional plate boundaries; origin of chaotic melanges. *Mailing Add:* Geolog Sci AJ-20 Univ Wash Seattle WA 98195

COWAN, DAVID J, b San Antonio, Tex, Aug 5, 36; m 65. PHYSICS. *Educ:* Univ Tex, BS, 58, MA, 60, PhD(physics), 65. *Prof Exp:* Assoc physicist, IBM Corp, 60-61; NIH traineeship, 64-65; asst prof, 65-74, chmn dept, 74-80, ASSOC PROF PHYSICS, GETTYSBURG COL, 74- *Mem:* Am Phys Soc; Am Asn Physics Teachers. *Res:* Investigation of biological ultrastructure by x-ray diffraction. *Mailing Add:* Dept Physics Gettysburg Col Gettysburg PA 17325

COWAN, DAVID LAWRENCE, b Havre, Mont, Oct 18, 36; m 60; c 1. SOLID STATE PHYSICS. *Educ:* Univ Wis, BS, 56, MS, 58, PhD(physics), 64. *Prof Exp:* Res engr, NAm Aviation, Inc, Calif, 58-60; res assoc physics, Cornell Univ, 64-67; mem tech staff, Sandia Corp, NMex, 67-68; asst prof, 68-74, ASSOC PROF PHYSICS, UNIV MO-COLUMBIA, 74- *Mem:* Am Phys Soc. *Res:* Ferromagnetic properties of metals; electron spin resonance; nuclear magnetic resonance; optical properties of solids. *Mailing Add:* Dept Physics Univ Mo Columbia MO 65211

COWAN, DAVID PRIME, b Aurora, Ill. BIOLOGY. *Educ:* Univ Mich, BS, 70, PhD(zool), 78. *Prof Exp:* ASSOC PROF BIOL, WESTERN MICH UNIV, 79- *Res:* Evolution and ecology of insect life histories and behavior. *Mailing Add:* Dept Biol Western Mich Univ Kalamazoo MI 49008

COWAN, DONALD D, b Toronto, Ont, Mar 11, 38; m; c 3. COMPUTER SCIENCE. *Educ:* Univ Toronto, BASc, 60; Univ Waterloo, MSc, 61, PhD(appl math), 65. *Prof Exp:* Fel math, 60-62, lectr, 61-65, asst prof, 65-67, chmn, Dept Appl Anal & Comput Sci, Fac Math, 66-72, assoc prof, 67-75, assoc dean grad studies, Fac Math, 74-78, PROF MATH, DEPT COMPUT SCI, UNIV WATERLOO, 75- *Concurrent Pos:* Nat Res Coun grant, 65-79; Defence Res Bd grant, 68-74. *Mem:* Asn Comput Mach; Inst Elec & Electronics Engrs; Soc Indust & Appl Math; Can Info Processing Soc. *Res:* Computer systems and communications. *Mailing Add:* Dept Comput Sci Fac Math Univ Waterloo Waterloo ON N2L 3G1 Can

COWAN, DWAINE O, b Fresno, Calif, Nov 25, 35; m 63. ORGANIC CHEMISTRY. *Educ:* Fresno State Col, BS, 58; Stanford Univ, PhD(org chem), 62. *Prof Exp:* Res fel chem, Calif Inst Technol, 62-63; from asst prof to assoc prof, 63-72, PROF CHEM, JOHNS HOPKINS UNIV, 72- *Concurrent Pos:* Sloan res fel, 68-70; Guggenheim fel, Phys Chem Inst, Univ Basel, 70-71. *Mem:* Am Chem Soc; Brit Chem Soc. *Res:* Physical organic chemistry; mixed valence organometallic compounds, electron transfer reactions, organic solid state chemistry, synthesis and study of organic metals, organic photochemistry; a charge transfer salt from tetrathiafulvalene and tetracyanoquinodimethane. *Mailing Add:* Dept Chem Johns Hopkins Univ Baltimore MD 21218

COWAN, ELLIOT PAUL, b Philadelphia, Pa, Aug 2, 55; m 83; c 1. NEUROIMMUNOLOGY, IMMUNOGENETICS. *Educ:* Williams Col, BA, 77; Wash Univ, PhD(biol & biomed sci), 83. *Prof Exp:* Staff fel, Nat Inst Allergy & Infectious Dis, 83-86, sr staff fel, Nat Inst Neurol Dis & Stroke, 86-90, SPEC EXPERT, NEUROIMMUNOL BR, NAT INST NEUROL DIS & STROKE, NIH, 90- *Mem:* Am Asn Immunologists. *Res:* Understanding of molecular mechanisms for tissue-specific regulation of major histocompatibility complex genes in the central nervous system; molecular mechanisms influencing susceptibility to multiple sclerosis. *Mailing Add:* Neuroimmunol Br NINDS NIH Bldg 10 Rm 5B-16 Bethesda MD 20892

COWAN, EUGENE WOODVILLE, b Ree Heights, SDak, Sept 30, 20; m 56; c 2. PHYSICS. *Educ:* Univ Mo, BS, 41; Mass Inst Technol, SM, 43; Calif Inst Technol, PhD(physics), 48. *Prof Exp:* Instr elec commun, Mass Inst Technol, 43-44, staff mem, Radiation Lab, 44-45; res fel, 48-50, from asst prof to assoc prof, 50-61, PROF PHYSICS, CALIF INST TECHNOL, 61- *Mem:* Am Phys Soc. *Res:* Radar; electrical circuits; electrical measurements; cosmic rays. *Mailing Add:* Dept Physics 103-33 Calif Inst Technol Pasadena CA 91125

COWAN, F BRIAN M, b Chatham, Ont, Apr 15, 38; m 62; c 2. ZOOLOGY, PHYSIOLOGY. *Educ:* Queen's Univ, BA, 62; Univ Toronto, MA & dipl electron micros, 65, PhD(zool), 70. *Prof Exp:* Instr zool, 69-70, asst prof, 70-77, ASSOC PROF BIOL, UNIV NB, 77- *Mem:* AAAS; Am Soc Zoologists; Int Soc Stereology; Can Soc Zoologists. *Res:* Activation of salt secreting epithelia during times of osmotic stress using euryhaline reptilian species as test system and studying ultrastrucure, stereology, histochemistry and biochemistry; electrical events in cardiac cycle in animals with apparent physiological alterations in plasma ion concentrations; effect of pesticides on blood cell structure. *Mailing Add:* Dept Biol Univ NB Col Hill Box 4400 Fredericton NB E3B 5A3 Can

COWAN, FREDERICK FLETCHER, JR, b Washington, DC, Jan 17, 33; m 60; c 1. PHARMACOLOGY. *Educ:* George Washington Univ, BS, 55; Georgetown Univ, PhD(pharmacol), 59. *Prof Exp:* Pharmacist, Bretler Pharm, 55-56; instr pharmacol, Schs Med & Dent, Georgetown Univ, 60-62, asst prof, 62-66; assoc prof, 66-73, PROF PHARMACOL & CHMN DEPT, DENT SCH, UNIV ORE, 73- *Concurrent Pos:* Fel, Nat Heart Inst, 59; prin co-investr, USPHS res grants, 61- & prin investr, 64-; prin investr, NSF res grant, 62. *Mem:* AAAS; Am Pharmaceut Asn; Am Soc Pharmacol & Exp Therapeut; Sigma Xi. *Res:* Pharmacology of the autonomic ganglia and the peripheral sympathetic nervous system; tachyphylaxis of sympathomimetic amines; pharmacology of carotid body chemoreceptors. *Mailing Add:* Health Sci Ctr Sch Dent Univ Ore 611 SW Campus Dr Portland OR 97201

COWAN, FREDERICK PIERCE, b Bar Harbor, Maine, July 3, 06; m 34. HEALTH PHYSICS. *Educ:* Bowdoin Col, AB, 28; Harvard Univ, AM, 31, PhD(physics), 35. *Prof Exp:* Instr physics, Bowdoin Col, 28-29; asst, Harvard Univ, 29-34, instr, 34-35; from instr to asst prof, Rensselaer Polytech Inst, 35-43; res assoc, Radio Res Lab, Harvard Univ, 43-45; res assoc, Eng Div, Chrysler Corp, Detroit, 45-47; head health physics div, Brookhaven Nat Lab, 47-71; mem Atomic Saftey & Licensing Bd, Nuclear Regulatory Comm, 72-85; RETIRED. *Concurrent Pos:* Asst, Radcliffe Col, 34-35; chmn, Am Bd Health Physics, 63-64; mem comt 3, Int Comn Radiation Protection, 65-73; mem, Int Comn Radiation Units & Measurements, 65-73; mem, Nat Coun Radiation Protection, 67-73, bd dirs, 69-73; mem adv comt radiobiol aspects of supersonic transport, Fed Aviation Admin, 67-74. *Mem:* AAAS; Am Phys Soc; Radiation Res Soc; Health Physics Soc (pres elect, 56-57, pres, 57-58). *Res:* Thermal measurements on electron tubes; electronic temperature control; radiation dosimetry. *Mailing Add:* 6152 N Verde Trail Apt B-125 Boca Raton FL 33433

COWAN, GARRY IAN MCTAGGART, b Victoria, BC, July 9, 40; m 77; c 2. SYSTEMATICS. *Educ:* Univ BC, BSc, 63, MSc, 66, PhD(zool), 68. *Prof Exp:* Asst prof, 68-71, ASSOC PROF BIOL, MEM UNIV NFLD, 71- *Concurrent Pos:* Nat Res Coun Can res grant in aid, 68-; res scientist, Marine Sci Res Lab, 70-87. *Mem:* Am Soc Ichthyol & Herpet; Soc Study Evolution; Soc Syst Zool. *Res:* Systematics of marine organisms based on their morphological and biochemical characteristics; ecology of salmonid fishes, acid tolerance of trout. *Mailing Add:* Dept Biol Mem Univ Nfld St John's NF A1C 5S7 Can

COWAN, GEORGE A, b Worcester, Mass, Feb 15, 20; m 46. RADIOCHEMISTRY. *Educ:* Worcester Polytech Inst, BS, 41; Carnegie Inst Technol, DSc(chem), 50. *Prof Exp:* Res scientist, Metall Lab, Univ Chicago, 42-45; res scientist, Los Alamos Nat Lab, Univ Calif, 45-46, mem staff radiochem, 49-81, group leader, 55-81, assoc div leader, Test Div, 56-70, div leader, 71-79, assoc dir, 79-81, sr fel, 81-88, EMER SR FEL, LOS ALAMOS NAT LAB, UNIV CALIF, 88-; PRES, SANTA FE INST, SANTA FE, NMEX, 88- *Concurrent Pos:* Mem, White House Sci Coun, 82-86; pres, Santa Fe Inst, Santa Fe, NMex, 88-91; chmn, Los Alamos Nat Bank. *Honors & Awards:* E O Lawrence Award, 65; Enrico Fermi Award, 90. *Mem:* Fel AAAS; Am Chem Soc; fel Am Phys Soc; Am Nuclear Soc; Sigma Xi. *Res:* Radiochemical diagnostics; nuclear reactions; solar neutrino physics. *Mailing Add:* 721 42nd St Los Alamos NM 87544

COWAN, IAN MCTAGGART, b Edinburgh, Scotland, June 25, 10; m 36; c 2. MAMMALOGY, WILDLIFE ECOLOGY. *Educ:* Univ BC, BA, 32; Univ Calif, PhD(vert zool), 35. *Hon Degrees:* LLD, Univ Alta, 71 & Simon Fraser Univ, 81; DEnviron Sci, Univ Waterloo, 75; DSc, Univ BC, 77, Univ Victoria, 84. *Prof Exp:* Insect pest investr, Dom Govt, Can, 29; field asst, Nat Mus Can, 30-31; asst biologist, BC Prov Mus, 35-38, asst dir, 38-40; from asst prof to prof zool, 40-75, head dept, 53-64, dean fac grad studies, 64-75, EMER PROF ZOOL, UNIV BC, 75- *Concurrent Pos:* Carnegie traveling fel, Am Mus & US Nat Mus, 37; Nuffield fel, 52; Am Inst Biol Sci foreign vis lectr, 63; Erskine fel, Univ Canterbury, 69; mem, Environ Protection Bd, 69-75; chmn, Acad Bd BC, 69-75; mem, Can Environ Adv Coun, 72-75, chmn, 75-79; chmn, Can Comt Whales & Whaling, 77- & Acad Coun Prov BC, 78-84; chancellor, Univ Victoria, 79-85; bd dir, Nature Trust of BC, chmn, Habitat Conserv Bd of BC. *Honors & Awards:* Order of Can, 72; Leopold Medal & Arthur Einarsen Award, Wildlife Soc, 70; Fry Medal, Can Soc Zool, 76. *Mem:* Am Ornith Union; Wildlife Soc (pres, 49); Am Soc Mammalogists; Soc Syst Zool; fel Royal Soc Can. *Res:* Ecology and environmental physiology of native ungulates; problems of insularity in vertebrates; vertebrate distribution and speciation in Canada; behavior of carnivorous mammals. *Mailing Add:* 3919 Woodhaven Terr Victoria BC V8N 1S7 Can

COWAN, JACK DAVID, b Leeds, Eng, Aug 24, 33; m 58; c 2. APPLIED MATHEMATICS, BIOMATHEMATICS. *Educ:* Univ Edinburgh, BSc, 55; Mass Inst Technol, SM, 60; Imp Col, dipl elec eng, 59, Univ London, PhD(elec eng), 67. *Prof Exp:* Res engr & mathematician, Instrument & Fire Control Sect, Ferranti Ltd, Scotland, 55-58; staff mem neurophysiol group, Res Lab Electronics, Mass Inst Technol, 60-62; acad vis math biol, Imp Col, 62-67; chmn dept & prof theoret biol, 67-73, prof biophys & theoret biol, 73-81, PROF MATH, UNIV CHICAGO, 81- *Concurrent Pos:* Res assoc & consult, Northeastern Univ & USAF Cambridge Res Lab, 60; vis prof, Univ Naples, 65; guest worker, Nat Phys Lab, Eng, 66-67; mem, NIH comn training grants in epidemiol & biomet, 69-73; chmn adv coun, Lab Cybernet, Nat Res Coun, Italy, 69-73; mem vis comt, Dept Psychol, Mass Inst Technol, 72-75; mem, Am Math Soc comt math in life sci, 70-77 & 88-; mem comn commun & control, Int Union Pure & Appl Biophys, 75-78. *Honors & Awards:* Alexander von Humboldt Sr Scientist, 77; Weingart Dist Lectr, CalTech, 82. *Mem:* AAAS; Am Math Soc; Am Phys Soc; Soc Am Neurosci; Sigma Xi; Soc Ind Appl Math; Soc Math Biol. *Res:* Theory of nervous activity; theoretical biology; vision research; applied mathematics. *Mailing Add:* Dept Math Univ Chicago 5734 S University Ave Chicago IL 60637-5315

COWAN, JAMES W, b Beaver Falls, Pa, Aug 23, 30; m 60. NUTRITION, BIOCHEMISTRY. *Educ:* Pa State Univ, BS, 55, MS, 59, PhD(biochem), 61. *Prof Exp:* Assoc prof food tech & nutrit, 61-73, assoc dean sch agr sci, 70-73, prof nutrit & dean sch agr sci, Am Univ Beirut, 73-77; DIR INT PROGS & STUDIES OFF, NAT ASN STATE UNIVS & LAND GRANT COLS, 77-, DIR FED RELA, AGR NAT RESERV & INTERNAL AFFAIRS, 77- *Mem:* AAAS; Am Nutrit Soc; Am Soc Clin Nutrit. *Res:* Iodine metabolism; iron utilization. *Mailing Add:* Nat Asn State Univs & Land Grant Cols One Du Pont Circle Washington DC 20036

COWAN, JOHN, b Toronto, Can, Aug 30, 43; m. PITUITARY ENDOCRINOLOGY, GROWTH HORMONE. *Educ:* Univ Toronto, PhD(physiol), 69. *Prof Exp:* Prof & head dept physiol, Sch Med, 79-86, VPRES RESOURCES & PLANNING, UNIV OTTAWA, 87- *Mem:* Can Physiol Soc (pres, 82); Am Physiol Soc; Can Fedn Biol Soc (pres, 80). *Mailing Add:* Univ Ottawa 550 Cumberland St Ottawa ON K1N 6N5 Can

COWAN, JOHN ARTHUR, b Winnipeg, Man, July 8, 21; m 46; c 4. FLUID PHYSICS. *Educ:* Univ Man, BSc, 43; Univ Toronto, MA, 47, PhD(physics), 50. *Prof Exp:* Lectr, Dept Math, Univ Man, 46; sci officer, Can Defense Res Bd, 50-57; chmn dept, 57-68, PROF PHYSICS, UNIV WATERLOO, 57- *Mem:* Can Acoust Asn; Can Asn Physicists. *Res:* Speech communication; experimental studies of transport properties of liquids. *Mailing Add:* 445 Blythwood Pl Waterloo ON N2L 4A4 Can

COWAN, JOHN C, b Danville, Ill, Oct 25, 11; m 38; c 2. ORGANIC CHEMISTRY. *Educ:* Univ Ill, AB, 34, PhD(chem), 38. *Prof Exp:* Asst, Univ Ill, 35-38, du Pont fel, 38-39; instr, DePauw Univ, 39-40; assoc chemist, Northern Regional Res Lab, Agr Res Serv, USDA, 40-41, from chemist to head oil & protein sect, Agr Res Serv, 53-57, chief oilseed crops lab, 57-73; CONSULT, 73- *Concurrent Pos:* Mem, Soybean Res Coun, 45-73; adj prof chem, Bradley Univ, 73-; mem adv panel, Soybean Utilization Res. *Honors & Awards:* Superior Serv Award, USDA, 48, Superior Serv Team Awards, 52, 63; A E Bailey Medal, Am Oil Chem Soc, 61; Meritorious Serv Award, Am

Soybean Asn, 70; Chevreul Medal, French Asn for Study of Oil Substances, 75. *Mem:* Am Chem Soc; Am Oil Chem Soc (pres, 68-69); Inst Food Technol; Am Asn Cereal Chemists. *Res:* Condensation and vinyl polymers; edible oil spreads; flavor stability of soybean oil; metal-inactivation agents; cyclic fatty acids; aldehydic esters; amino acid derivatives; uses for soybeans, linseed oil, and soybean protein products. *Mailing Add:* 262 Loma Entrada Santa Fe NM 87501

COWAN, JOHN D, JR, b Willshire, Ohio, Sept 7, 18; m 40; c 2. ELECTRICAL ENGINEERING, MECHANICAL ENGINEERING. *Educ:* Ohio State Univ, BSME, 47, MSEE, 55. *Prof Exp:* Consult, Dept Aviation Psychol, Ohio State Univ, 48-54, Shop-lab coordr, 48-55, from instr to prof, 50-83, EMER PROF ELEC ENG, OHIO STATE UNIV, 83- *Concurrent Pos:* Consult, NAm Aviation Co, 57-74 & Ohio State Fire Marshals Arson Crime Lab, 78-85; mem, Nat Comt Elec Eng Films, 62-68; consult, electrocution, explosives & elec fires. *Mem:* Sr mem Inst Elec & Electronics Engrs. *Res:* Electromechanical systems; circuit theory. *Mailing Add:* 233 Montrose Way Columbus OH 43214

COWAN, JOHN JAMES, b Washington, DC, Apr 3, 48; m 71. THEORETICAL ASTROPHYSICS, RADIO OBSERVATIONS. *Educ:* George Washington Univ, BA, 70; Case Inst Technol, MS, 72; Univ Md, College Park, PhD(astron), 76. *Prof Exp:* Fel, Harvard-Smithsonian Ctr Astrophys, Harvard Univ, 76-79; from asst prof to assoc prof, 79-89, PROF PHYSICS & ASTRON, UNIV OKLA, NORMAN, 89- *Concurrent Pos:* Prin investr, NSF, 82-; vis assoc, Smithsonian Ctr, Astrophys, Harvard Univ, 87-88; consult, Lawrence Livermore Nat Lab, 88-90. *Mem:* Sigma Xi; Int Astron Union; Am Astron Soc. *Res:* Theoretical studies in nuclear astrophysics; origin of the elements; age of the universe; radio observations of supernovae using very large array. *Mailing Add:* Dept Physics & Astron Univ Okla Norman OK 73019

COWAN, KEITH MORRIS, immunochemistry, for more information see previous edition

COWAN, MAYNARD, JR, b Independence, Mo, Dec 15, 25; m 45; c 3. PHYSICS. *Educ:* William Jewel Col, BA, 48; Univ NMex, MS, 51. *Prof Exp:* Mem staff, Weapons Effects Studies, 51-60, div supvr, Magneto Physics Res Div, 60-69, mgr, Shock Simulation Dept, 69-75, mgr, Simulation Res Dept, 75-80, MGR, ADVAN ENERGY CONVERSION SYSTS DEPT, SANDIA LABS, 80- *Mem:* Am Phys Soc. *Res:* Technical management; pulsed power engineering; plasma physics; nuclear burst effects. *Mailing Add:* 1107 Stagecoach SE Albuquerque NM 87112

COWAN, RAYMOND, b Marion, Ind, Nov 14, 14; m 40; c 2. PHYSICS. *Educ:* Butler Univ, BS, 54, MS, 55. *Prof Exp:* From asst prof to assoc prof, 55-67, chmn, Sci Div, 60-79, PROF PHYSICS, FRANKLIN COL, 67-, CHMN DEPT, 57- *Mem:* Am Asn Physics Teachers. *Res:* Electronics. *Mailing Add:* 1231 Adams St Franklin IN 46131

COWAN, RICHARD SUMNER, b Crawfordsville, Ind, Jan 23, 21; m 41, 86; c 4. PLANT TAXONOMY. *Educ:* Wabash Col, AB, 42; Univ Hawaii, MS, 48; Columbia Univ, PhD, 52. *Prof Exp:* Tech asst, NY Bot Garden, 48-52, asst cur, 52-57; assoc cur, Nat Mus Natural Hist, Smithsonian Inst, 57-62, asst dir, 62-65, dir, 65-73, sr botanist, dept bot, 73-85; RETIRED. *Concurrent Pos:* NSF fel, 52-53; temp sci officer, West Australian Herbarium, 87-91. *Mem:* AAAS; Am Inst Biol Sci; Am Soc Plant Taxon; Asn Trop Biol; Bot Soc Am; Int Asn Plant Taxon. *Res:* Phanerogamic taxonomy; Leguminosae of northern South America; Rutaceae of Guayana Highland; flora of Venezuela; botanical bibliography; Leguminosae of Western Australia. *Mailing Add:* Three Bass Close East Cannington WA 6107 Australia

COWAN, ROBERT DUANE, b Lincoln, Nebr, Nov 24, 19; m 44, 90; c 4. ATOMIC PHYSICS, SPECTROSCOPY. *Educ:* Friends Univ, AB, 42; Johns Hopkins Univ, PhD(physics), 46. *Hon Degrees:* PhD, Lund Univ, Sweden, 82. *Prof Exp:* Jr instr physics, Johns Hopkins Univ, 43-46, lab asst, 44-46; Nat Res fel, Univ Chicago, 46-47, res assoc spectros, 47-48; prof physics & head dept, Friends Univ, 48-51; physicist, 51-85, CONSULT, LOS ALAMOS NAT LAB, 85- *Concurrent Pos:* Fulbright lectr, Peru, 58-59; vis prof, Purdue Univ, 71; consult, Culham Lab, Eng, 77, Zeeman Lab, Amsterdam, 84, Lund Univ, Sweden, 88, Manne Siegbahn Inst, Stockholm, 89-90. *Honors & Awards:* William F Meggers Award, Optical Soc Am, 84. *Mem:* Fel Am Phys Soc; fel Optical Soc Am. *Res:* Visible and ultraviolet spectroscopy; selfabsorption of spectral lines; equations of state at extreme pressures and temperatures; theoretical atomic spectroscopy. *Mailing Add:* LANL MS212 T Div Los Alamos Nat Lab Los Alamos NM 87545

COWAN, ROBERT LEE, b Beaver County, Pa, Nov 20, 20; m 45; c 5. ANIMAL NUTRITION. *Educ:* Pa State Col, BS, 43, MS, 49, PhD, 52. *Prof Exp:* Chemist food processing, Gen Foods, Inc, 43-44; asst agr biol chem, 46-48, from instr to assoc prof, 48-65, prof, 65-83, EMER PROF ANIMAL NUTRIT, PA STATE UNIV, 83- *Concurrent Pos:* Fulbright res scholar, Massey Col, NZ, 56-57. *Mem:* Fel Am Soc Animal Sci; NY Acad Sci; Am Inst Nutrit; Wildlife Soc. *Res:* Nutritive values of forages; preservation of grass silage; nutrition of deer and mink; importation, propagation and nutritional studies with African blue duikers; miniature laboratory ruminants. *Mailing Add:* RD 1 Box 42 Julian PA 16844

COWAN, RUSSELL (WALTER), b Oakland, Calif, Feb 26, 12; m 48. APPLIED MATHEMATICS. *Educ:* Univ Calif, AB, 32, MA, 33, PhD(math), 35. *Prof Exp:* Instr math & astron, Col St Scholastica, 35-38; from instr to asst prof math, Univ Ala, 38-47; assoc prof, Univ Fla, 47-66; prof math, Lamar Univ, 66-82; RETIRED. *Mem:* Am Math Soc; Math Asn Am. *Res:* Analysis; differential equations; difference equations; solution of a linear difference equation of the second order with quadratic coefficients; differential equations, special functions and gamma functions. *Mailing Add:* 1425 23rd St Beaumont TX 77706

COWAN, W MAXWELL, b Johannesburg, SAfrica, Sept 27, 31; m 56; c 3. NEUROANATOMY. *Educ:* Univ Witwatersrand, BSc, 52; Oxford Univ, DPhil(neuroanat), 56, BM & BCh, 58. *Hon Degrees:* MA, Oxford Univ, 58. *Prof Exp:* Asst anat, Univ Witwatersrand, 51-53; dept demonstr, Oxford Univ, 53-58, lectr, 59-66, tutor, Pembroke Col, 56-66; assoc prof, Univ Wis-Madison, 66-68; prof & head dept, Sch Med, Wash Univ, 68-80; prof & vpres, Salk Inst Biol Studies, La Jolla, Calif, 80-86; prof, dept neurol & neurosurg & adj prof, dept biol, Wash Univ Sch Med, St Louis, Mo, 86-87; ADJ DISTINGUISHED PROF NEUROSCI, JOHNS HOPKINS UNIV SCH MED, 88-; VPRES & CHIEF SCI OFFICER, HOWARD HUGHES MED INST, BETHESDA, MD, 88- *Concurrent Pos:* Fel, Pembroke Col, 56-66, lectr, Balliol Col, 62-66; vis assoc prof, Wash Univ, 64-65; managing ed, J Comp Neurol, 69-80; ed, Ann Rev Neurosci, 75-; non-resident fel, Salk Inst, 78-80; ed-in-chief, J Neurosci, 80-87; mem, Neurosci Res Prog; provost & exec vchancellor, Wash Univ, St Louis, Mo, 86-87. *Honors & Awards:* Marion Hines lect, Emory Univ, 73, Robert W Woodruff lectr, 82; Clinton N Woolsey lectr, Univ Wis, 76; Tyler lectr, Calif Inst Technol, 77; Scott lectr, Mich State Univ, 80; Oliver Lowry lectr, Wash Univ Sch Med, 84; Karl Spencer Lashley Prize for Neurobiol; Jenkinson Mem lectr, Oxford, 86; Cecil & Ada Green distinguished lectr, Univ Tex Med Br, 87. *Mem:* Inst Med-Nat Acad Sci; foreign assoc US Nat Acad Sci; fel AAAS; Anat Soc Gt Brit & Ireland; fel Royal Micros Soc; Soc Neurosci (pres, 77-78); fe; Am Acad Arts & Sci; fel Royal Soc; foreign mem Norweg Acad Sci; hon mem Am Neurol Asn; Am Philos Soc; Sigma Xi; Am Asn Anatomists. *Res:* Neurobiology, especially the structure and development of the mammalian forebrain and the avian visual system. *Mailing Add:* Howard Hughes Med Inst 6701 Rockledge Dr Bethesda MD 20817

COWAN, W R, b Ont, Jan 15, 42. QUATERNARY GEOLOGY. *Educ:* Univ Colo, PhD(geol), 75. *Prof Exp:* Supvr, Ont Geol Surv, 68-79; mgr, environ, Northern Pipeline Agency, 79-81 & Tecman Eng Ltd, 81; vpres geol, Pallisser Consult, 83-87; MGR MINING, ONT MINISTRY NORTHERN DEVELOP & MINES, 87- *Mem:* Fel Geol Soc Am; Can Geol Soc; Can Geol Tech Soc. *Mailing Add:* 159 Cedar St 4th Floor Sudbury ON P3E 6A5 Can

COWAN, WILLIAM ALLEN, b Pittsfield, Mass, Oct 4, 20; m 46. ANIMAL BREEDING. *Educ:* Mass State Col, BS, 42; Univ Minn, MS, 48, PhD(animal breeding, dairy prod), 52. *Prof Exp:* Asst farm mgr, Grafton State Hosp, Mass, 42-43; instr & farm supt, RI State Col, 43-45; asst & assoc prof animal husb, Univ Mass, 46-52; head dept, 52-74, prof, Animal Industs, Univ Conn, 52-88, head dept, 80-88; RETIRED. *Mem:* Am Soc Animal Sci; Am Dairy Sci Asn; Am Genetic Asn. *Res:* Animal production; dairy herd and livestock management. *Mailing Add:* Animal Dept Sci Univ Conn Storrs CT 06268

COWARD, DAVID HAND, b Buffalo, NY, Nov 16, 34; m 60; c 3. ELEMENTARY PARTICLE PHYSICS. *Educ:* Cornell Univ, BEng, 57; Stanford Univ, MS, 58, PhD(physics), 63. *Prof Exp:* EXP PHYSICIST, STANFORD LINEAR ACCELERATOR CTR, STANFORD UNIV, 63- *Concurrent Pos:* Consult, MITRE Corp, McLean, Va, 76; sci assoc, Europ Orgn Nuclear Res, CERN, Geneva, 76-77 & 86-87; vis, CERN, 89-90. *Mem:* Fel Am Phys Soc. *Res:* Present research interests encompass the physics of heavy quark systems using experiments at electron-positron storage rings; physics of CP violation in the neutral kaon system. *Mailing Add:* Stanford Linear Accelerator Ctr PO Box 4349 Stanford CA 94309

COWARD, JAMES KENDERDINE, b Buffalo, NY, Oct 13, 38; m 75; c 1. BIO-ORGANIC CHEMISTRY, MEDICINAL CHEMISTRY. *Educ:* Middlebury Col, AB, 60; Duke Univ, MA, 64; State Univ NY Buffalo, PhD(med chem), 67. *Prof Exp:* NIH fel, Univ Calif, Santa Barbara, 66-68; asst prof, Dept Pharmacol, Sch Med, Yale Univ, 69-74, assoc prof, 75-79; assoc prof, 79-82, PROF CHEM, RENSSELAER POLYTECH INST, NY, 82- *Concurrent Pos:* Vis prof, Salk Inst Biol Sci, 77-78; mem exp therapeut study sect, NIH, 79-82. *Mem:* Am Chem Soc; Royal Soc Chem; Am Soc Biol Chem; AAAS. *Res:* Investigation of the mechanism of enzyme-catalyzed reactions as a basis for the design and synthesis of new mechanism-based enzyme inhibitors. *Mailing Add:* Dept Chem Six Haverhill Ct Ann Arbor MI 48105

COWARD, JOE EDWIN, b Searcy, Ark, Mar 27, 38; m 59; c 2. VIROLOGY, MICROBIOLOGY. *Educ:* State Col Ark, BSE, 59; Univ Ark, Fayetteville, MS, 62; Univ Miss, PhD(microbiol), 68. *Prof Exp:* Res assoc, Col Physicians & Surgeons, Columbia Univ, 68-70, asst prof microbiol, 70-75; assoc prof microbiol, 75-80, ASSOC PROF MICROBIOL & IMMUNOL, SCH MED, LA STATE UNIV MED CTR, NEW ORLEANS, 80- *Mem:* Am Soc Microbiol. *Res:* Viral replication as studied by electron microscopy; development of the slow viruses and their relationship to human disease. *Mailing Add:* Dept Microbiol Sch Med La State Univ Med Ctr 1542 Tulane Ave New Orleans LA 70112

COWARD, NATHAN A, b Belton, SC, Jan 7, 27; m 53; c 4. PHYSICAL CHEMISTRY. *Educ:* The Citadel, BS, 50; Univ Rochester, PhD(phys chem), 54. *Prof Exp:* Res chemist, F H Levey Co Div, Columbian Carbon Co, 54-57 & Am Viscose Corp, 57-59; from instr to assoc prof, 59-69, actg chmn dept, 64-65, 68-69, PROF CHEM, UNIV WIS-SUPERIOR, 69- *Concurrent Pos:* Vis prof, Univ Wis-Madison, 67-68. *Mem:* Am Chem Soc; Sigma Xi. *Res:* Fluorescence; photochemistry; printing inks; analytical methods and instrumentation; liquid phase weak complexes of lanthanide ions; trace metals in fuels and natural waters. *Mailing Add:* Rte 1 Box 910 South Range WI 54874-9784

COWARD, STUART JESS, b New York, NY, June 7, 36; div; c 2. DEVELOPMENTAL BIOLOGY. *Educ:* Univ Miami, Fla, BS, 58; Univ Iowa, MS, 61; Univ Calif, Davis, PhD(zool), 64. *Prof Exp:* Res assoc, State Univ NY, Buffalo, 64-65; asst prof, 65-69, chmn div interdisciplinary studies, 78-83, ASSOC PROF ZOOL, UNIV GA, 69- *Concurrent Pos:* NIH fel, 64-65; NIH spec fel, Harvard Univ Med Sch, 70-71; Ga Coun Sci Educ, 85-; chmn, SOTAB Prog, 79- *Mem:* AAAS; Soc Develop Biol; Am Soc Cell Biol. *Res:* Regeneration in planaria; biochemical aspects of amphibian development; Avian behavior. *Mailing Add:* Dept Zool Univ Ga Athens GA 30602

COWBURN, DAVID, b Sale, Cheshire, Eng, July 13, 45; m 77; c 2. PHYSICAL BIOCHEMISTRY, BIOPHYSICS. *Educ:* Univ Manchester, Eng, BSc Hons, 65; Univ London, PhD(biophys), 70. *Hon Degrees:* DSc, Univ London, 81. *Prof Exp:* Res assoc, Columbia Univ Col Physicians & Surgeons, 70-73; asst prof, 73-78, ASSOC PROF, ROCKEFELLER UNIV, 78- *Concurrent Pos:* Fel, Res Training Prog, Dept Psychiat, Columbia Univ Col Physicians & Surgeons, 71-73. *Mem:* Am Soc Biochem & Molecular Biol; NY Acad Sci; Am Chem Soc; Soc Magnetic Resonance Med; Protein Soc. *Res:* Structure-function relationships in biological chemistry; intracellular communication by hormones, neurotransmitters and trophic factors; nuclear magnetic resonance. *Mailing Add:* Rockefeller Univ Box 299 1230 York Ave New York NY 10021

COWDEN, RONALD REED, b Memphis, Tenn, July 9, 31; m 56. CYTOLOGY, EMBRYOLOGY. *Educ:* La State Univ, Baton Rouge, BS, 53; Univ Vienna, DrPhil(zool), 56. *Prof Exp:* USPHS fel biol, Oak Ridge Nat Lab, 56-57; asst prof biol, Johns Hopkins Univ, 57-60; asst mem cell biol, Inst Muscle Dis, 60-61; asst prof path, Col Med, Univ Fla, 61-66, USPHS career develop award, 62-66; assoc prof anat, La State Univ Med Ctr, New Orleans, 66-68; prof biol sci & chmn dept, Univ Denver, 68-71; prof anat & chmn dept, Albany Med Col, 72-75; assoc dean basic sci, 75-80, prof biophys & chmn dept, 80-86, EMER PROF BIOPHYS, COL MED, ETENN STATE UNIV, 86- *Mem:* Am Soc Cell Biologists; Soc Develop Biol; Am Asn Anatomists; Am Zool Soc; Royal Micros Soc. *Res:* Quantitative cytochemistry; cytochemistry of oogenesis and development; comparative hematology and immunity. *Mailing Add:* 74272 Military Rd Covington LA 70433

COWDREY, ERIC JOHN, b Ste Agathe des Monts, Que, Sept 15, 45; m 75. HEALTH PHYSICS, PHYSICS. *Educ:* Concordia Univ, BSc, 67, MSc, 70. *Prof Exp:* Res scientist textiles, Bobtex Corp, 70-71; res physicist paper, Pulmac Instruments Ltd, 72-74; res physicist explosives res, Can Industs Ltd, 74-78; health physicist nuclear res, Atomic Energy Can Ltd, Man, 78-80, head, Radiation & Indust Safety Br, 80-87; RADIATION PROTECTION OFFICER, MAN CANCER FOUND, 88- *Concurrent Pos:* mem, Conf Radiation Control Prog Dirs Inc. *Mem:* Health Physics Soc; Can Radiation Protection Asn. *Res:* Radiation safety. *Mailing Add:* Dept Med Physics Man Cancer Found 100 Olivia St Winnipeg MB R3E 0V9 Can

COWDRY, REX WILLIAM, b Des Moines, Iowa, Feb 12, 47; m 81; c 1. PERSONALITY DISORDERS RESEARCH. *Educ:* Yale Col, BA, 68; Harvard Univ, MD & MPH, 73. *Prof Exp:* Resident, Mass Ment Health Ctr, 73-76; staff psychiatrist, NIMH, 76-83, clin dir, Intramural Res Prog, 84-87, actg dep dir, Inst, 86-88, CHIEF EXEC OFFICER, NEUROPSYCHIAT RES HOSP, NIMH, 87- *Concurrent Pos:* Assoc clin prof, Dept Psychiat, Med Sch, Georgetown Univ, 83-; chmn med bd, Warren Grant Magnusen Clin Ctr, NIH, 86-87. *Honors & Awards:* Adminr Award, Alcohol Drug Abuse & Ment Health Asn Admin, 85 & 88. *Mem:* Am Psychiat Asn. *Res:* Biological factors in and psychopharmacologic treatment of personality disorders and mood disorders. *Mailing Add:* NIMH Neurosci Ctr St Elizabeth's Hosp Washington DC 20032

COWELL, BRUCE CRAIG, b Buffalo, NY, Oct 20, 37; m 65; c 2. LIMNOLOGY. *Educ:* Bowling Green State Univ, BA, 58, MA, 59; Cornell Univ, PhD(limnol ecol), 63. *Prof Exp:* Fishery res biologist limnol, NCent Reservoir Invest, US Fish & Wildlife Serv, 63-67; from asst prof to assoc prof zool, 67-85, PROF BIOL, UNIV SFLA, 85- *Concurrent Pos:* Assoc ed, Environ Pollution, 87-90. *Mem:* Am Soc Limnol & Oceanog; Ecol Soc Am; Int Soc Limnol; NAm Benthological Soc. *Res:* Production and population dynamics of plankton and benthic invertebrate communities; lake management and restoration. *Mailing Add:* Dept Biol Univ SFla Tampa FL 33620-5150

COWELL, JAMES LEO, b Wilkes-Barre, Pa, May 4, 44; m 69; c 1. VACCINE RESEARCH & DEVELOPMENT. *Educ:* King's Col, Pa, BS, 66; St John's Univ, NY, MS, 68; Univ Ill, Urbana, PhD(microbiol), 72. *Prof Exp:* Res asst biochem, Cornell Univ, 72-74; res asst microbiol, Sch Med, NY Univ, 74-78; res microbiologist, Div Bact Prod, Bur Biologics, Food & Drug Admin, 78-86; mgr bacterial res, 86-89, SCI DIR, LAB RES, PRAXIS BIOLOGICS INC, 89- *Mem:* Am Soc Microbiol; Nat Found Infectious Dis; NY Acad Sci. *Res:* Bacterial infections, pathogenesis and vaccine development. *Mailing Add:* 37 Sugarmills Circle Fairport NY 14450

COWELL, WAYNE RUSSELL, b Wakefield, Kans, June 27, 26; m 53; c 3. APPLIED MATHEMATICS. *Educ:* Kans State Univ, BS, 48, MS, 50; Univ Wis, PhD(math), 54. *Prof Exp:* Asst math, Kans State Univ, 48-50 & Univ Wis, 50-54; asst prof, Univ Mont, 54-59; mem tech staff, Bell Tel Labs, 59-61; assoc mathematician, 61-72, COMPUTER SCIENTIST, ARGONNE NAT LAB, 72- *Concurrent Pos:* Asst to pres, Argonne Univs Asn, 68-71. *Mem:* Asn Comput Mach; Soc Indust & Appl Math. *Res:* Mathematical software; software tools. *Mailing Add:* Math & Comput Sci Div Argonne Nat Lab Argonne IL 60439

COWEN, CARL CLAUDIUS, JR, b Madison, Ind, Nov 15, 45; m 70; c 2. OPERATOR THEORY, COMPLEX ANALYSIS. *Educ:* Ind Univ, Bloomington, AB, 67, AM, 71; Univ Calif, Berkeley, PhD(math), 76. *Prof Exp:* Vis lectr math, Univ Ill, Urbana, 76-78; from asst prof to assoc prof math, 78-89, PROF MATH, PURDUE UNIV, 89- *Mem:* Am Math Soc; Math Asn Am. *Res:* Toeplitz operators and composition operators on Hardy spaces and related complex function theory. *Mailing Add:* Dept Math Purdue Univ West Lafayette IN 47907

COWEN, DAVID, b New York, NY, July 29, 07. NEUROPATHOLOGY. *Educ:* Columbia Col, AB, 28; Columbia Univ, MD, 32. *Prof Exp:* Asst pathologist, Neurol Inst, NY, 37-46; instr neurol, 37-39, from instr to assoc prof, 39-63, prof, 63-76, EMER PROF NEUROPATH, COL PHYSICIANS & SURGEONS, COLUMBIA UNIV, 76-, SPEC LECTR, 77- *Concurrent Pos:* From asst attend neuropathologist to attend neuropathologist, Columbia-Presby Med Ctr, 45-73, consult neuropathologist, 73-; consult neuropathologist, Vet Admin Hosp, East Orange, NJ, 53-; emer consult neuropathologist, Lenox Hill Hosp, NY. *Honors & Awards:* Neuropath Contributions Award, Am Asn Neuropath, 79. *Mem:* Am Asn Neuropath (pres, 61-62); Am Neurol Asn; Harvey Soc; Am Asn Path; NY Acad Med. *Res:* Infections of the nervous system; neurological diseases of perinatal origin. *Mailing Add:* Columbia Univ Div Neuropath Col Physicians & Surgeons 630 W 168th St New York NY 10032

COWEN, JERRY ARNOLD, b Toledo, Ohio, July 17, 24; m 46; c 5. PHYSICS. *Educ:* Harvard Col, BS, 48; Mich State Col, MS, 50, PhD, 54. *Prof Exp:* Asst physics, Mich State Col, 49-53; asst prof, Colo Agr & Mech Col, 53-55; from asst prof to assoc prof, 55-69, PROF PHYSICS, COL NATURAL SCI, MICH STATE UNIV, 69- *Concurrent Pos:* NSF fel, Saclay Nuclear Res Ctr, France, 64-65; res physicist, Nat Bur Standards, 55 & Lockheed Res Lab, 60-61. *Mem:* Am Phys Soc; Am Asn Physics Teachers. *Res:* Microwave resonance. *Mailing Add:* Dept Physics Mich State Univ East Lansing MI 48824

COWEN, RICHARD, b Workington, Eng, Jan 24, 40; m 71; c 2. PALEONTOLOGY. *Educ:* Cambridge Univ, BA, 62, PhD(geol), 66. *Prof Exp:* Res asst geol, Cambridge Univ, 65-67; from asst prof to assoc prof, 67-79, chmn dept, 76-79, PROF GEOL, UNIV CALIF, DAVIS, 79- *Mem:* Paleontol Soc; AAAS. *Res:* Anatomical and functional studies in living and fossil invertebrates, especially brachiopods, their implications for evolution and paleobiogeography. *Mailing Add:* Dept Geol Univ Calif Davis CA 95616

COWEN, WILLIAM FRANK, b Oshkosh, Wis, Aug 20, 45; m 67; c 3. WATER CHEMISTRY. *Educ:* Univ Wis, Madison, BS, 67, MS, 69, PhD(water chem), 74. *Prof Exp:* Res chemist environ chem, US Army Bioeng Res & Develop Lab, Environ Protection Div, 73-78; SR ANALYTICAL CHEMIST & TECH SUPVR, MARCUS HOOK LAB, CATALYTIC INC, 78- *Mem:* Am Chem Soc; Water Pollution Control Fedn. *Res:* Methods for analysis of trace organic compounds in water supplies and in wastewaters. *Mailing Add:* 5530 Princeton Rd Macungie PA 18062-9053

COWETT, EVERETT R, b Ashland, Maine, Mar 6, 35; m 60; c 5. AGRONOMY, PLANT PHYSIOLOGY. *Educ:* Univ Maine, BS, 57, MS, 58; Rutgers Univ, PhD(farm crops), 61. *Prof Exp:* Asst prof agron, Univ NH, 61-63; rep field res plant protection, Geigy Chem Corp, 63-65, herbicide specialist, 65-67, field res mgr plant protection, 67-70, mgr plant sci res, Ciba-Geigy Ltd, Ardsley, 70-73, DIR TECH SERVS, AGR DIV, CIBA-GEIGY CORP, 73- *Mem:* Am Soc Agron; Weed Sci Soc Am; Entom Soc Am. *Res:* Plant physiology; forage crop management; weed control; insect and disease control; plant growth regulators. *Mailing Add:* PO Box 18300 Greensboro NC 27419

COWETT, RICHARD MICHAEL, b New York, NY, Sept 20, 42; m 66; c 3. PEDIATRICS, NEONATALOGY. *Educ:* Lawrence Col, Wis, BA, 64; Univ Cincinnati, MD, 68. *Hon Degrees:* MA, Brown Univ, 82. *Prof Exp:* Intern pediat, Cleveland Metrop Gen Hosp, 68-69, resident, 69-70; resident, Children's Hosp Med Ctr, Boston, 70-71; physician & chief, pediat serv, USAF Med Corps, Goose, AFB, Labrador, Can, 71-73; fel neonatology, Woman & Infants Hosp RI, Providence, 73-76; from asst prof to assoc prof, 76-88, PROF PEDIAT, DEPT PEDIAT, DIV BIOL & MED SCI, BROWN, 89- *Concurrent Pos:* Teaching fel, dept pediat, Harvard Med Sch, 70-71; teaching fel med sci, Div Biol & Med Sci, Brown Univ, 74-75, adj clin instr pediat, 75-76; physician-in-chg, Spec Care Nursery, Women & Infants Hosp RI, 81-87, staff neonatologist, 1988- *Mem:* Am Acad Pediat; Am Pediat Soc; Soc Pediat Res; Am Diabetes; Am Inst Nutrit; Perinatal Res Soc; Am Soc Clin Nutrit; Am Pediat Soc. *Res:* Metabolic adaptation of the pregnant woman and the newborn; kinetic analyses, using non-radioactive and radioactive isotopes. *Mailing Add:* Dept Pediat Women & Infants Hosp RI 101 Dudley St Providence RI 02905-2401

COWGER, MARILYN L, b Douglas, Nebr, June 7, 31. PEDIATRICS, BIOCHEMISTRY. *Educ:* Univ Omaha, BA, 53; Univ Nebr, MD, 56; Am Bd Pediat, dipl, 62. *Prof Exp:* Intern med, Bryan Mem Hosp, Lincoln, Nebr, 56-57; fel pediat, Mayo Clin, Minn, 57-60; from asst to assoc prof pediat, Sch Med, Univ Wash, 60-70; assoc prof, 70-76, PROF PEDIAT, ALBANY MED COL, 76- *Concurrent Pos:* NIH trainee, 60-63, career develop award, 65-74; Am Heart Asn advan res fel, 63-65; res assoc prof chem, State Univ NY Albany, 70-76, adj prof chem, 76-84. *Mem:* Am Acad Pediat; Soc Pediat Res; NY Acad Sci; AAAS; Western Soc Pediat Res; AOA. *Res:* Mechanism of bilirubin toxicity; electron transport inhibitors; clinical and biochemical studies of the porphyrias; mechanism of mental retardation in inborn errors of metabolism; natural collaborative studies for maternal phenysketonuria (PKU) and treatment of growth hormone deficiency. *Mailing Add:* Dept Pediat Albany Med Col 43 New Scotland Ave Albany NY 12208

COWGILL, ROBERT WARREN, b Topeka, Kans, Jan 31, 20; m 44; c 2. BIOCHEMISTRY. *Educ:* Univ Kans, BA, 41; Rensselaer Polytech Inst, MS, 42; Johns Hopkins Univ, PhD(biochem), 50. *Prof Exp:* Res chemist, Hercules Powder Co, 42-45; instr biochem, Wash Univ, 50-52; instr, Univ Calif, 53-56; asst prof, Univ Colo, 56-62; prof, 62-80, EMER PROF BIOCHEM, BOWMAN GRAY SCH MED, 80- *Mem:* Am Soc Biol Chemists; Am Chem Soc; Sigma Xi. *Res:* Mechanism of enzyme action; protein structure. *Mailing Add:* 183 Glen Abbey Rd Johns Island SC 29455

COWGILL, URSULA MOSER, b Bern, Switz, Nov 9, 27; US citizen; m 54. GEOCHEMISTRY. *Educ:* Hunter Col, AB, 48; Kans State Univ, MS, 52; Iowa State Univ, PhD(soil chem), 56. *Prof Exp:* Mem staff, Lincoln Lab, Mass Inst Technol, 57-58 & Doherty Charitable Found Inc, Guatemala, 58-60; res assoc ecol & anthrop geochem, Yale Univ, 60-68; prof biol sci, Univ Pittsburgh, 68-81, prof anthrop, 72-81; environ scientist, 81-84, ASSOC ENVIRON CONSULT, DOW CHEM CO, 84- *Concurrent Pos:* Mem sci adv bd, US Environ Protection Agency, 76-80, Int Joint Comn, 84-90. *Mem:* Mineral Soc UK; Clay Minerals Soc; Int Asn Theoret & Appl Limnol; Soc Environ Geochem & Health; Am Soc Limnol & Oceanog. *Res:* Ecological and anthropological geochemistry; primitive agriculture; phosphate mineralogy; history of lake basins; exotic demography; prosimian behavior; limnology; plankton nutrition; hydrological studies; geochemistry; aquatic toxicology. *Mailing Add:* Aquatic Toxicol Dow Chem Co 1702 Bldg Midland MI 48674

COWHERD, CHATTEN, JR, b Kansas City, Mo, May 16, 39; m 61; c 4. AIR POLLUTION. *Educ:* Rockhurst Col, BS, 60; Johns Hopkins Univ, PhD(chem eng), 64. *Prof Exp:* From asst prof to assoc prof eng sci, Rockhurst Col, 64-69; sr engr, 70-74, prin engr, 74-76, SECT HEAD, MIDWEST RES INST, 76- *Mem:* Air Pollution Control Asn; Sigma Xi. *Res:* Convective heat and mass transfer; aerosol sizing and flux measurement; air pollution source and ambient sampling; kinetics and mechanisms of atmospheric dust generation by wind erosion or materials handling; cost-effectiveness evaluation of pollution control. *Mailing Add:* Midwest Res Inst 425 Volker Blvd Kansas City MO 64110

COWIE, LENNOX LAUCHLAN, b Jedburgh, Scotland, Oct 18, 50. ASTROPHYSICS, ASTRONOMY. *Educ:* Edinburgh Univ, BSc, 69; Harvard Univ, PhD(theoret physics), 76. *Prof Exp:* Res assoc, Princeton Univ, 76-78, res staff mem astrophys, 78-81, res astron & assoc prof, 79-81; assoc prof physics, Mass Inst Technol, 80-83; astronr & chief acad affairs br, Space Telescope Sci Inst, 83-86; prof physics & astron, Johns Hopkins Univ, 84-86; ASSOC DIR, INST ASTRON, UNIV HAWAII, 86- *Concurrent Pos:* Fairchild Distinguished Scholar, Calif Inst Technol, 80. *Honors & Awards:* Bok Prize, Harvard Univ, 84; Warner Prize, Am Astron Soc, 85. *Mem:* Fel Royal Astron Soc; Am Astron Soc; fel Am Phys Soc; Int Astron Union. *Res:* Gas dynamics of the interstellar and intergalactic medium. *Mailing Add:* Univ Hawaii Manoa 2680 Woodlawn Dr Honolulu HI 96822

COWIE, MARTIN, b Aberdeen, Scotland, Feb 27, 47; Can citizen; m 69; c 3. INORGANIC CHEMISTRY, X-RAY CRYSTALLOGRAPHY. *Educ:* McMaster Univ, BSc, 69; Univ Alta, PhD(x-ray crystallog), 74. *Prof Exp:* From asst prof to assoc prof, 76-87, PROF CHEM, UNIV ALTA, 87- *Concurrent Pos:* NATO fel, Northwestern Univ, 74-76. *Mem:* Chem Inst Can; Am Chem Soc. *Res:* Metal-metal cooperativity in small molecule activation; chemistry of binuclear diphosphine-bridged, metal complexes; mixed-metal complexes. *Mailing Add:* Dept Chem Univ Alta Edmonton AB T6G 2G2 Can

COWIN, STEPHEN CORTEEN, b Elmira, NY, Oct 26, 34; m 56; c 2. APPLIED MECHANICS, BIOMEDICAL ENGINEERING. *Educ:* Johns Hopkins Univ, BS, 56, MS, 58; Pa State Univ, PhD(eng mech), 62. *Prof Exp:* Res asst, Johns Hopkins Univ, 56-58; sr engr, Aircraft Armaments, Inc, Md, 58-59; from instr to asst prof eng mech, Pa State Univ, 59-63; from asst prof to prof mech eng, Tulane Univ, 63-77, prof mech, Dept Biomed Eng, 77-85, Alden J Laborde prof eng, 85-88; DISTINGUISHED PROF MECH ENG, CITY UNIV NEW YORK, 88. *Concurrent Pos:* Instr, Loyola Col, Md, 58-59; US Army Res Off Durham grant, 64-67; consult, Cornell Aeronaut Lab, 66-68, Lawrence Livermore Nat Lab, 69-83, Off Energy Related Inventions, Nat Bur Standards, 77-78, Bendix Advan Technol Ctr, 81-83 & Exxon Prod Res Co, 82-; Edward G Schlieder Educ Found grant, 70-71 & 80-83; NSF grants, 71-85; sr vis fel, Sci Res Coun Gt Brit & Univ Strathclyde, 74, 80; prof-in-charge, Tulane Univ-Newcomb Jr Yr Abroad Prog, Gt Brit, 74-75; assoc ed, J Appl Mech, 74-82 & J Biomech Eng, 82-; chmn, Master Sci Appl Math Prog & Eng Curric Prog, 75-79; vis res prof, Inst Math Statist & Comput Sci, State Univ Campinas, Brazil, 78; Nat Inst Arthritis, Metab & Digestive Dis grant, NIH, 78-84; adj prof orthop, Sch Med, Tulane Univ, 78-; vis scientist, Cath Pontifical Univ, Rio de Janeiro, Div Int Progs, NSF, 84. *Mem:* Soc Rheology; Soc Natural Philos (treas, 77-79); Soc Eng Sci; Math Asn Am; Sigma Xi; fel Am Acad Mech; fel AAAS; Am Soc Biomech; fel Am Soc Mech Engrs; Soc Indust & Appl Math. *Res:* Continuum mechanics, rheology, mechanics of granular media, continuum theories representing the microstructure of materials; theory of constitutive relations; biomechanics. *Mailing Add:* 107 W 86th St New York NY 10024

COWLES, DAVID LYLE, b Forks, Wash, Sept 1, 55; m 80; c 2. ECOLOGICAL PHYSIOLOGY. *Educ:* Walla Walla Col, BS, 78, MS, 81; Univ Calif, Santa Barbara, PhD(biol), 87. *Prof Exp:* Res biologist, Univ Calif, Santa Barbara, 83-87; ASST PROF ECOL PHYSIOL & MARINE BIOL, LOMA LINDA UNIV, 87- *Mem:* Am Soc Zoologists; Am Soc Limnologists & Oceanographers; AAAS. *Res:* The ecological physiology of deep-sea species, especially respiration and energetics of swimming in pelagic crustaceans. *Mailing Add:* Dept Natural Sci Loma Linda Univ Loma Linda CA 92350

COWLES, EDWARD J, b Careywood, Idaho, July 15, 18; m 48; c 2. CHEMISTRY. *Educ:* Univ Wash, Seattle, BS, 40, PhD(chem), 53. *Prof Exp:* Jr chemist, Western Regional Res Lab, Bur Agr & Indust Chem, USDA, 41-42; shift chemist, Rayonier, Inc, 46; instr chem & Ger, Grays Harbor Col, 46-47; chemist, Philippine Fishery Prog, 47-48; assoc prof chem, Whitworth Col, 52-54; from asst prof to assoc prof, 54-69, PROF CHEM, UNIV MINN, DULUTH, 69- *Concurrent Pos:* Consult, Hilding Res Fund, US Dept Health, Educ & Welfare, 62-72. *Mem:* AAAS; Am Chem Soc; Sigma Xi. *Res:* Electrophilic substitution reactions of azulene; identification of the pigments of maize; determination of ammonia; spectra of azulene and related compounds; solubility of inorganic compounds; empirical applications of the computer to chemical problems. *Mailing Add:* Dept Chem Univ Minn 1011 W Arrowhead Rd Duluth MN 55811-2211

COWLES, HAROLD ANDREWS, b Flandreau, SDak, Apr 12, 24; m 53; c 3. ECONOMIC & INDUSTRIAL ENGINEERING. *Educ:* Iowa State Univ, BS, 49, MS, 53, PhD(eng valuation), 57. *Prof Exp:* From instr gen eng to assoc prof indust eng, 49-61, PROF INDUST ENG, IOWA STATE UNIV, 61- *Concurrent Pos:* Anson Marston distinguished prof eng, 84. *Mem:* Inst Indust Engrs; Opers Res Soc Am; Am Soc Eng Educ. *Res:* Industrial property depreciation, mortality and valuation; engineering economy; operations research-cost effectiveness studies. *Mailing Add:* 633 Agg Ave Ames IA 50010

COWLES, JOE RICHARD, b Edmonson County, Ky, Oct 29, 41; m 65; c 2. PLANT PHYSIOLOGY. *Educ:* Western Ky Univ, BS, 63; Univ Ky, MS, 65; Ore State Univ, PhD(plant physiol), 68. *Prof Exp:* Res assoc plant physiol, Purdue Univ, 68-69, Univ Ga, 69-70; from asst prof to prof biol, Univ Houston, 70-90, chmn biol, 81-90; PROF & HEAD, VA POLYTECH INST & STATE UNIV, 90- *Concurrent Pos:* Prin investr, 73- *Mem:* Am Soc Plant Physiol; AAAS; Sigma Xi; Am Soc Gravity & Space Biol. *Res:* Physiological and biochemical parameters associated with the establishment of symbiosis in leguminous plants, developmental changes associated with the regulation of aromatic biosynthesis in plants, and biomass conversion. *Mailing Add:* Dept Biol Va Polytech Inst & State Univ Blacksburg VA 24061

COWLES, JOHN RICHARD, b Berkeley, Calif, Aug 26, 45; div; c 2. AUTOMATED THEOREM PROVING. *Educ:* Univ Wyo, BA, 68; Univ Nebr, MA, 70; Pa State Univ, PhD(math), 75. *Prof Exp:* Grad asst math, Univ Nebr, 68-71 & Pa State Univ, 71-75; mem staff, Dept Math, Inst Advan Study, 75-76 & 78; lectr, Univ Va, 76-78; asst prof, 78-86, ASSOC PROF, UNIV WYO, 86- *Mem:* Am Math Soc; Asn Comput Mach; Asn Symbolic Logic; Math Asn Am; Asn Automated Reasoning. *Res:* Number theory and mathematical logic; automated theorem provers; theory of computation, program correctness and synthesis; number theory; mathematical logic. *Mailing Add:* Dept Comput Sci Univ Wyo PO Box 3682 Laramie WY 82071

COWLES, WILLIAM WARREN, b New York, NY, Feb 22, 34; m 56; c 3. ELECTRICAL & ELECTRONICS ENGINEERING. *Educ:* Princeton Univ, BSE, 55; Yale Univ, MEng, 57, DEng(elec eng), 63. *Prof Exp:* Instr elec eng, Yale Univ, 57-63; asst prof, Moore Sch Elec Eng, Univ Pa, 63-70; mem tech staff, Res & Develop Lab, SCM Corp, 70-73, prin electronics engr, 73-75, sr mem tech staff, 75-80; sr staff engr, Perkin-Elmer Corp, 80-86; TECH ADV, PITNEY-BOWES, 86- *Concurrent Pos:* Consult, Hamilton-Standard Div, United Aircraft Corp, 56-57 & McNeal Electronics Conn, 57-63; mem tech staff, Bell Tel Labs, Inc, summers 54 & 64; engr, Proj Matterhorn, Princeton Univ, 56-59; NSF grant, Yale Univ, 61 & 62; proj supvr, Moore Sch Elec Eng, Univ Pa, 63-70. *Mem:* Inst Elec & Electronics Engrs; Soc Info Display. *Res:* Image and word processing; microprocessors; video displays; impact and non-impact printers; incremental motion systems; stepping motor design. *Mailing Add:* Ten Musket Lane West Redding CT 06896-1524

COWLEY, ALAN H, b Manchester, Eng, Jan 29, 34; m 76; c 4. INORGANIC CHEMISTRY. *Educ:* Univ Manchester, BS, 55, MS, 56, PhD(chem), 58. *Prof Exp:* Fel chem, Univ Fla, 58-60; tech officer, Billingham Div, Imp Chem Industs Gt Brit, 60-62; from asst prof to assoc prof, 62-70, PROF CHEM, UNIV TEX, AUSTIN, 70- *Concurrent Pos:* Welch res grant, 64-; NIH res grant, 64-67; NSF grant, 69-; res panel, Army Res Off, 79. *Honors & Awards:* Jeremy I Musher lectr, Hebrew Univ, Jerusalem, Israel, 79; Main-Group Elem Chem Award, Royal Soc Chem, 80. *Mem:* Am Chem Soc; Royal Soc Chemists. *Res:* Chemistry of non-metals; inorganic free radicals; nuclear magnetic resonance; photoelectron spectroscopy. *Mailing Add:* Dept Chem Univ Tex Welch Hall 4 330 Austin TX 78712-1167

COWLEY, ALLEN WILSON, JR, b Harrisburg, Pa, Jan 21, 40; m 65. PHYSIOLOGY, CARDIOVASCULAR PHYSIOLOGY. *Educ:* Trinity Col, BA, 61; Hahnemann Med Col, MS, 65, PhD(physiol & biophys), 68. *Prof Exp:* NIH grant, Med Ctr, Univ Miss, 68-69, from instr to assoc prof, 68-74, prof physiol & biophys, 74-80, dir grad studies, 78-80; CHMN PHYSIOL, MED COL WIS, 80- *Concurrent Pos:* NIH res grant hypertension res, 71-85; Am Heart Asn estab investr, 73-78; mem adv bd, Coun High Blood Pressure Res, Am Heart Asn; mem, Study Sect, NIH, 80-83. *Mem:* Int Soc Hypertension; AAAS; Am Physiol Soc; Am Heart Asn. *Res:* Quantitative analysis of interacting neural and hormonal control systems in the overall regulation of arterial blood pressure. *Mailing Add:* Dept Physiol Med Col Wis 8701 Watertown Plank Rd Milwaukee WI 53226

COWLEY, ANNE PYNE, b Boston, Mass, Feb 25, 38; m 60; c 2. ASTRONOMY. *Educ:* Wellesley Col, BA, 59; Univ Mich, MA, 61, PhD(astron), 63. *Prof Exp:* Res assoc astron, Yerkes Observ, Univ Chicago, 63-68; res assoc astron, Univ Mich, Ann Arbor 68-73, assoc res scientist, 73-77, res scientist, 77-83, prof, 83-85; PROF PHYSICS & ASTRON, ARIZ STATE UNIV, 83- *Concurrent Pos:* Guest worker, Dom Astrophys Observ, 74-75 & 77-78. *Mem:* Am Astron Soc; Int Astron Union. *Res:* Stellar spectroscopy; x-ray sources. *Mailing Add:* Physics & Astron Dept Ariz State Univ Tempe AZ 85287

COWLEY, BENJAMIN D, JR, b Louisville, Ky, Oct 18, 56. EXPERIMENTAL BIOLOGY. *Educ:* Baylor Col Med, MD, 81; Am Bd Internal Med, dipl, 84, dipl nephrology, 88. *Prof Exp:* Engr, E I Du Pont de Nemours & Co, Inc, 74 & 75; res asst, Rice Univ, 76-77; intern & resident internal med, Univ Kans Med Ctr, Kansas City, 81-84, fel nephrology, 84-86, res fel, Dept Biochem, 86-87, res instr internal med, 86-87; guest scientist, Lab Kidney & Electrolyte Metab, Nat Heart, Lung & Blood Inst, 87-89; ASST PROF MED, UNIV KANS MED CTR, KANSAS CITY, 89- *Concurrent Pos:* Max Allen scholar, Dept Internal Med, Univ Kans Med Ctr, 86, adj asst prof biochem & molecular biol, 89-; Nat Kidney Found fel, 86-87; nat res serv award, Nat Heart, Lung & Blood Inst, 87-89. *Mem:* Am Soc Cell Biol; AAAS; Am Soc Nephrology; Am Fedn Clin Res. *Res:* Kidney disease. *Mailing Add:* Lab Kidney & Electrolyte Metab NHLBI NIH Bldg 10 Rm 6N323 Bethesda MD 20892

COWLEY, CHARLES RAMSAY, b Aguana, Guam, Sept 13, 34; US citizen; m 60; c 2. ASTRONOMY. *Educ:* Univ Va, BA, 55, MA, 58; Univ Mich, PhD(astron), 63. *Prof Exp:* From instr to asst prof astron, Univ Chicago, 63-67; from asst prof to assoc prof, 67-77, PROF ASTRON, UNIV MICH, 77- *Mem:* Am Astron Soc; Royal Astron Soc; Int Astron Union. *Res:* Stellar atmospheres. *Mailing Add:* Dept Astron Univ Mich Ann Arbor MI 48109

COWLEY, GERALD TAYLOR, b Barron, Wis, Aug 1, 31; m 57; c 5. MYCOLOGY, MICROBIOLOGY. *Educ:* Univ Wis, BS, 53, MS, 57 & PhD (bot), 62. *Prof Exp:* Teacher high sch, Wis, 55-56; asst prof biol, 62-67, asst dean col arts & sci, 69-72, asst vprovost, Div Lib & Cult Disciplines, 72-73, ASSOC PROF BIOL, UNIV SC, 67-, ASST HEAD DEPT, 73- *Mem:* Mycol Soc Am. *Res:* Ecology and physiology of soil and litter microfungi. *Mailing Add:* Dept Biol Sci Univ SC Columbia SC 29208

COWLEY, JOHN MAXWELL, b Peterborough, SAustralia, Feb 18, 23; m 51; c 2. PHYSICS, MATERIALS SCIENCE. *Educ:* Univ Adelaide, BSc, 42, MSc, 45, DSc(phys), 57; Mass Inst Technol, PhD(phys), 49. *Prof Exp:* Res officer, Commonwealth Sci & Indust Res Orgn, Australia, 47-62; prof phys, Univ Melbourne, Australia, 63-70; dir, High Resolution Electron Micros Facil, 83-90, GALVIN PROF PHYSICS, ARIZ STATE UNIV, 70-, REGENTS PROF, 88- *Concurrent Pos:* Res assoc, Mass Inst Technol, 62-63; mem Comn Electron Diffraction, Int Union Crystallog, 57-63 & 84-87, chmn, 87-93; Int Union Crystallog rep, Solid State Comn, Int Union Pure & Appl Physics, 69-78; mem US Nat Comt Crystallog, 73-78 & 84-87; co-ed, Acta Crystallog, 71-80. *Honors & Awards:* B E Warren Award, Am Crystallog Asn, 76, Ewald Prize, 87; Res Medal, Royal Soc Victoria, 62. *Mem:* Electron Micros Soc Am; Am Crystallog Asn; fel Inst Physics, London; fel Australian Inst Physics; fel Australian Acad Sci; fel Am Physics Soc; fel Royal Soc London. *Res:* Electron diffraction and electron microscopy; diffraction and imaging theory; disorder and imperfections in crystals and crystal surfaces. *Mailing Add:* Dept Physics Ariz State Univ Tempe AZ 85287-1504

COWLEY, THOMAS GLADMAN, b Clifton Springs, NY, Aug 20, 38; m 77; c 3. ANALYTICAL CHEMISTRY, SPECTROCHEMISTRY. *Educ:* Rochester Inst Technol, BS, 61; Iowa State Univ, PhD(anal chem), 66. *Prof Exp:* Assoc, AEC Ames Lab, 66-67; sr res scientist, 67-81, PROJ COORDR, CONOCO, INC, 81- *Concurrent Pos:* Chmn, Am Soc Testing & Mat. *Mem:* Am Chem Soc; Soc Appl Spectros; Am Soc Testing & Mat. *Res:* Chemical and physical behavior of spectroscopic sources and their uses in analytical chemistry; development of flame emission spectroscopy as an analytical tool; application of spectrochemistry to environmental sciences; analysis of petroleum and petroleum products; physical and chemical analysis of plastics. *Mailing Add:* 2316 El Camino Ponca City OK 74604

COWLING, ELLIS BREVIER, b Waukegan, Ill, Dec 11, 32; m 56; c 2. BIOGEOCHEMISTRY. *Educ:* State Univ NY Col Forestry, Syracuse, BS, 54, MS, 56; Univ Wis, PhD(plant path & biochem), 59; Univ Uppsala, FilDr(physiol bot), 70. *Prof Exp:* Chemist wood properties res, Dow Chem Co, 55-56; wood pathologist, Forest Prod Lab, USDA, 56-59; asst prof forest path, Sch Forestry, Yale Univ, 60-68; assoc prof plant path & forestry, 65-68, assoc dean res, Col Forest Resources, 78-91, PROF PLANT PATH, FORESTRY & WOOD & PAPER SCI, NC STATE UNIV, RALEIGH, 68-, UNIV PROF AT-LARGE & PROF FOREST RESOURCES, 89- *Concurrent Pos:* USPHS fel, Royal Pharmaceut Inst Stockholm, Sweden, 59-60; vis prof, Inst Physiol Bot, Univ Uppsala, Sweden, 70-71; assoc ed, Ann Rev Phytopath, Ann Revs, Inc, Palo Alto, Calif, 71-; mem bd agr, Nat Res Coun, 88-; US Forest Serv res grant, 89-90; Environ Protection Agency res grant, 91-93. *Honors & Awards:* Res Award, Sigma Xi, 68; O Max Gardner Award, 81. *Mem:* Nat Acad Sci; AAAS; fel Int Acad Wood Sci; fel Am Phytopath Soc. *Res:* Forest and wood products pathology; physiology of trees and of tree diseases; air-borne chemicals and their ecological effects. *Mailing Add:* Col Forest Resources NC State Univ Raleigh NC 27695

COWLING, VINCENT FREDERICK, b St Louis, Mo, Dec 15, 18; m 44; c 1. MATHEMATICS. *Educ:* Rice Inst, BA, 41, MA, 43, PhD(math), 44. *Prof Exp:* Instr math, Ohio State Univ, 45-46; asst prof, Lehigh Univ, 46-49; from assoc prof to prof, Univ Ky, 49-61; prof, Rutgers Univ, 61-67; dean, Col Sci & Math, State Univ NY, Albany, 72-80, prof math, 66-87; RETIRED. *Concurrent Pos:* Ford Found fel, Yale Univ, 52-53. *Mem:* Am Math Soc; Math Asn Am. *Res:* Complex variable theory; applications of functional analysis. *Mailing Add:* 106 Linda Lane Palm Beach Shores FL 33404

COWLISHAW, JOHN DAVID, b Grand Rapids, Mich, Sept 10, 38; m 82; c 3. BIOPHYSICS. *Educ:* Univ Mich, BS, 60, MS, 61; Pa State Univ, PhD(biophys), 68. *Prof Exp:* Asst engr, Air Arm Div, Westinghouse Elec Corp, 61-63; instr physics, Westminster Col, Pa, 63-65; asst prof, 68-74, ASSOC PROF BIOL SCI, OAKLAND UNIV, 74- *Concurrent Pos:* Actg assoc dean, Col Arts & Sci, Oakland Univ. *Mem:* Am Asn Univ Professors. *Res:* Theory of information and energy-processing by biological systems. *Mailing Add:* Dept Biol Sci Oakland Univ Rochester MI 48309-4401

COWMAN, RICHARD AMMON, b Brainerd, Minn, Apr 24, 38; m 64; c 2. MICROBIOLOGY, BIOCHEMISTRY. *Educ:* Univ Minn, BS, 61; NC State Univ, MS, 63, PhD(food sci), 66. *Prof Exp:* Instr food microbiol, NC State Univ, 66-67, asst prof, 67-70; asst prof microbiol & biochem, Inst Oral Biol, Univ Miami, 70-74; RES MICROBIOLOGIST, US VET ADMIN HOSP, 70-, RES ASSOC PROF MICROBIOL, UNIV MIAMI, 77- *Mem:* Am Chem Soc; Am Soc Microbiol; Soc Cryobiol; NY Acad Sci. *Res:* Freeze-damage of microorganisms; nutritional aspects of microorganisms involved in dental caries. *Mailing Add:* Vet Admin Hosp Dent Res Unit 151 1201 NW 16th St Miami FL 33156

COWPER, GEORGE, b Newcastle-upon-Tyne, Eng, Sept 6, 21; Can citizen; m 51; c 3. PHYSICS. *Educ:* Durham Univ, BSc, 43. *Prof Exp:* Sci officer, Telecommun Res Estab, Eng, 43-48; from jr res officer to assoc res officer, Atomic Energy Can, Ltd, 48-58, sr res officer & head radiation dosimetry br, 58-67, head Health Physics Br, 67-84; CONSULT, 84- *Concurrent Pos:* Chmn subcomt electronic mat res, Defence Res Bd Can, 57-63, mem, 63-65, mem panel radiation protection & treatment, 58-66; chmn panel basic req personnel monitoring, Int Atomic Energy Agency, 63-64, 77-78; WHO vis prof, Bhabha Atomic Res Ctr, Bombay, India, 71; news ed, Health Physics, Health Phys Soc, 71-75; health physics adv comt, Oak Ridge Nat Lab, 71-75; assoc comt sci criteria environ qual, Nat Res Coun Can, 75-; mem, Main Comn, Int Comn Radiation Units & Measurements, 77-85, sr adv, 85-89; ed bd, Radiation Protection Dosimetry, 85. *Mem:* Health Phys Soc (dir, 68-71); Int Comn Radiation Units & Measurements; Int Radiation Protection Asn (vpres, 80-84). *Res:* Instrumentation for health physics; radiation dosimetry; detection of ionizing radiation as applied to nuclear chemistry and geophysical prospecting. *Mailing Add:* Four Cartier Circle Deep River ON K0J 1P0 Can

COWPER, GEORGE RICHARD, b Galt, Ont, June 9, 30. APPLIED MATHEMATICS. *Educ:* Queen's Univ, BSc, 52, MSc, 54; Brown Univ, PhD(appl math), 59. *Prof Exp:* Res officer appl mech, Nat Aeronaut Estab, Nat Res Coun Can, 59-86; RETIRED. *Mem:* Am Soc Mech Engrs. *Res:* Finite element method; theory of elasticity. *Mailing Add:* 871 Pleasant Park Rd Ottawa ON K1G 1Z1 Can

COWPERTHWAITE, MICHAEL, b Keswick, Eng, June 18, 32; m 56; c 4. CHEMISTRY, THERMODYNAMICS. *Educ:* Manchester Univ, Eng, BSc, 54, MSc, 56, PhD(chem), 58. *Prof Exp:* Fel chem, Cornell Univ, 58-61; sr sci officer detonation, Explosives Res & Develop Estab, 61-64; SR CHEM PHYSICIST DETONATION, SRI INT, 64- *Mem:* Combustion Inst; Sigma Xi. *Res:* Shock wave physics and chemistry; detonation; thermodynamics. *Mailing Add:* SRI Int 333 Ravenswood Ave Menlo Park CA 94025

COWSAR, DONALD ROY, b Baton Rouge, La, Dec 12, 42; m 66; c 3. BIOMATERIALS, POLYMER CHEMISTRY. *Educ:* La State Univ, BS, 64; Rice Univ, PhD(org chem), 69. *Prof Exp:* Res chemist, Southern Res Inst, 68-70, sr chemist, 70-73, head biomat sect, 74-77, head, Biosysts Div, 77-80, assoc dir, 80-82, dir appl sci dept, 82-87, dir prog develop, 87-89; PRES, TRATECH INT, 89- *Concurrent Pos:* Consult, NIH, 73-89. *Mem:* Am Chem Soc; Controlled Release Soc. *Res:* New biomaterials and biomedical devices including controlled-release drug-delivery systems, microcapsules, liposomes, polymeric drugs, polymeric pesticides, biodegradable implants, hemoperfusion devices, prosthetic polymers, and hydrogels; in vivo evaluation of materials and devices. *Mailing Add:* 4657 Round Forest Dr Birmingham AL 35213

COWSER, KENNETH EMERY, b Chicago, Ill, Apr 12, 26; m 47; c 3. HEALTH PHYSICS, ENVIRONMENTAL ENGINEERING. *Educ:* Univ Ill, Champaign-Urbana, BS, 47; Univ Tenn, Knoxville, MS, 59; Oak Ridge Sch Reactor Technol, grad, 63. *Prof Exp:* Eng aide, Ill Dept Hwys, 46; dist engr, Ill Dept Pub Health, 47-51, actg regional engr, 51-53; engr leader res & develop, Oak Ridge Nat Lab, 53-67, sect chief res & develop, Radioactive Waste Disposal Sect, Health Physics Div, 67-71, asst to assoc dir biomed & environ sci, 71-75; asst to asst adminr environ & safety, Energy Res & Develop Admin, 75-77; sect chief med physics & internal dose sect, Health Physics Div, Oak Ridge Nat Lab, 77, mgr, Life Sci Synthetic Fuels Prog, 77-83; mgr prog develop, Off Environ, Safety, Health, Martin Marietta Energy Systs, 84-85; MGR, RES TECHNOL DIV, MAXIMA CORP, 85- *Concurrent Pos:* Mem subcomt radioactive waste disposal, Am Standards Asn, 60-69; mem panel, Internal AEC, 67-70 & 73-77; mem subcomt radioactive waste mgt, Am Nat Standards, 69-71; mem sci comt, task group krypton-85 & sci comt radiation nuclear power generation, Nat Coun Radiation Protection & Measurements; vis lectr health physics prog, Vanderbilt Univ. *Honors & Awards:* Award of Excellence in Books, Soc Tech Commun, 85. *Mem:* Health Physics Soc; Sigma Xi; AAAS. *Res:* Waste management; siting and operating nuclear and fossil energy facilities; plant and environmental monitoring and testing; environmental and health assessments. *Mailing Add:* 937 W Outer Dr Oak Ridge TN 37830

COWSIK, RAMANATH, b Nagpur Madgta Pradesh, India, Aug 29, 40; div; c 1. PARTICLE PHYSICS. *Educ:* Bangalore, India, BSc, 58; Bombay Univ, DPhil, 68. *Prof Exp:* Asst prof, Univ Calif, Berkeley, 70-73; vis scientist, Max Planck Inst Ext Physics, Munich, 73-74; assoc prof, 77-84, JR RES ASSOC, TATA INST FUNDAMENTAL RES, BOMBAY, 61-, READER, 75-, PROF, 84- *Concurrent Pos:* Vis prof, Wash Univ, St Louis, 87- *Honors & Awards:* Sarabhai Award in Space Sci, Phys Res Lab, Hari Om Soc, Ahmedabad, 81; Bhatnagar Award in Phys Sci, Indian Nat Sci Acad, New Delhi, 84; Group Achievement Award, NASA, 86. *Mem:* Fel Indian Acad Sci; fel Indian Nat Sci Acad; Am Phys Soc; Int Astron Union. *Res:* Weakly interacting particle relicts from the big bang as the constituents of dark matter and set the upper bound on the turn of their mass in particular as neutrinos; the cosmological significance of the hard x-ray extensive studies in high energy astrophysics of nonthermal emissions from quasars and supernova remnants and in the areas interfacing with particle physics; author of five journal publications. *Mailing Add:* McDonnell Ctr Space Sci Physics Dept Box 1105 Wash Univ One Brookings Dr St Louis MO 63130-4899

COX, AARON J, b Topeka, Kans, Apr 6, 41; m 70. PHYSICS. *Educ:* Univ NMex, BS, 63, MS, 65; Univ Ariz, PhD(physics), 70. *Prof Exp:* Res asst physics, Univ Ariz, 67-70; assoc prof, 70-80, PROF PHYSICS, UNIV REDLANDS, 80- *Concurrent Pos:* Res physicist chem, Univ Calif, Riverside, 76- *Mem:* Optical Soc Am; Sigma Xi. *Res:* Quantum optics and laser spectroscopy; picosecond laser spectroscopy; dye laser. *Mailing Add:* 1365 Pacific Rdls Redlands CA 92374

COX, ALVIN JOSEPH, JR, b Manila, PR, Mar 6, 07; m 46; c 3. PATHOLOGY. *Educ:* Stanford Univ, AB, 27, MD, 31. *Prof Exp:* From asst to prof path, 31-64, PROF DERMAT, SCH MED, STANFORD UNIV, 64-, EMER PROF PATH, 80- *Concurrent Pos:* Fel Harvard Univ, 57-58; exchange asst, Path Inst, Freiburg, 35-36; consult pathologist, Western Utilization Res Br, USDA, 37; consult pathologist, San Francisco Hosp, 41-59. *Mem:* Fel AMA; Am Asn Path & Bact; Soc Exp Biol & Med; Soc Invest Dermat; Am Soc Dermatopath (pres, 73). *Res:* Coccidioidal infection; pathology of stomach; arteriosclerosis; experimental tumor production; dermatopathology. *Mailing Add:* 300 Pasteur Dr Stanford CO 94305

COX, ANDREW CHADWICK, b Hattiesburg, Miss, July 20, 36; m 61; c 2. BIOCHEMISTRY, BIOPHYSICS. *Educ:* Univ Tex, BS, 59; Univ Houston, MS, 63; Duke Univ, PhD(biochem), 67. *Prof Exp:* Fel, Am Cancer Soc, 66-68; asst prof biochem, 69-72, ASSOC PROF BIOCHEM, SCH MED, UNIV OKLA, 72- *Mem:* Am Chem Soc; Am Heart Asn; Am Soc Biol Chemists. *Res:* Elucidation of the mechanisms of platelet activation and the contributions of platelets to regulation of hemostasis. *Mailing Add:* Dept Biochem Univ Okla Health Sci Ctr Oklahoma City OK 73190

COX, ANN BRUGER, CELLULAR RADIATION BIOLOGY. *Educ:* Univ Chicago, SB; Boston Univ, PhD(biol), 76. *Prof Exp:* Res assoc, Colo State Univ, 79-86; PHYSIOLOGIST, USAF ARMSTRONG LAB, 87- *Concurrent Pos:* Assoc ed, Advances Radiation Biol. *Mem:* Am Soc Cell Biol; Radiation Res Soc; AAAS; Sigma Xi; Aerospace Med Asn. *Res:* Space radiobiology and aging; studies of late effects of particulate radiations on normal mammalian tissues with emphases on the skin, the eye and the brain. *Mailing Add:* Directed Energy Div Det 4 USAF Armstrong Lab RZ Brooks AFB TX 78235-5301

COX, ARTHUR NELSON, b Van Nuys, Calif, Oct 12, 27; m 73; c 5. ASTROPHYSICS, HYDRODYNAMICS. *Educ:* Calif Inst Technol, BA, 48; Ind Univ, AM, 52, PhD(astron), 53. *Hon Degrees:* DSc, Ind Univ, 73. *Prof Exp:* Staff mem physics, 48-49, staff mem field testing, 53-57, group leader, 57-74, STAFF MEM THEORET DIV, LOS ALAMOS SCI LAB, 75- *Concurrent Pos:* Vis prof, Univ Calif, Los Angeles, 66; NATO sr fel sci, Univ Liege, Belg, 68, Fulbright res scholar, 68-69; NSF prog dir, 73-74; pres, Int Astron Union Comt 35, 82-85, lab fel, 83-, coordr, 86-88. *Mem:* AAAS; Am Astron Soc; Int Astron Union; Sigma Xi. *Res:* Calculations of stellar stability and pulsation; compilation of equations of state and opacities for astrophysics; studies of stellar atmosphere, interior structure and stellar evolution; hydrodynamical problems in astrophysics; total solar eclipses. *Mailing Add:* 1700 Camino Redondo Los Alamos NM 87544

COX, B(RIAN), b Liverpool, Eng, Aug 4, 31; m 55; c 2. MATERIALS SCIENCE, NUCLEAR REACTOR MATERIALS. *Educ:* Cambridge Univ, BA, 52, MA & PhD(chem), 55. *Prof Exp:* Sect leader reactor chem, Atomic Energy Res Estab, Harwell, 55-63; leader corrosion group, Atomic Energy Can, Ltd, 63-67, head, mat sci br, 67-85, dir, Chem & Mat Div, 85-86, prog consult, Reactor Mat Div, 86-88; PROF & CHMN, CTR NUCLEAR ENG, UNIV TORONTO, 89- *Concurrent Pos:* Consult, Atomic Energy Can, Ltd, Ont Hydro, Philadelphia Elec, Elec Power Res Inst, Nuclear Fuel Indust Res Group. *Honors & Awards:* A B Campbell Award, Nat Asn Corrosion Engrs, 61; Zirconium Medal, W J Kroll Inst, Colo Sch Mines, 76; W B Lewis Award, Can Nuclear Asn, 88. *Mem:* Can Nuclear Soc; Am Nuclear Soc; Am Soc Metals; Int Metallog Soc. *Res:* Oxidation, corrosion and stress corrosion cracking of zirconium alloys; effects of irradiation on these and the basic physical processes involved, especially diffusion, charge transport, point defect production and agglomeration; in reactor peformance asessments and failure analyses on zirconium alloy components in nuclear reactors and chemical plants. *Mailing Add:* Ctr Nuclear Eng Univ Toronto 184 College St Toronto ON M5S 1A4 Can

COX, BENJAMIN VINCENT, b Chicago, Ill, Jan 25, 34; m 59; c 2. COMMUNICATION THEORY, DATA COMPRESSION. *Educ:* Univ Utah, BS, 63, MES, 69, PhD(elec eng), 79. *Prof Exp:* Proj engr, Sperry, 63-72, prin engr, 74-78, mgr advan technol, 78-84; supvry engr, Environ Protection Data Base, Naval Civil Eng Lab, Port Hueneme, Calif, 72-74; DIR ADVAN TECHNOL, UNISYS, 84- *Concurrent Pos:* Mem, Tech Comt Commun Systs, Am Inst Aeronaut & Astronaut, 79-83, Tech Comt Digital Avionics Systs, 80-83, MTMWG Commun Task Force, 80-84; mem, Marine Traffic Mgt Working Group, Utah Am Inst Aeronaut & Astronaut Coun & Am Inst Aeronaut & Astronaut Educ Chmn, 80-84; adj prof, Univ Utah, 80-; Unisys fel, 88. *Honors & Awards:* Governor's Medal for Sci & Technol, 87. *Mem:* Inst Elec & Electronics Engrs; Am Inst Aeronaut & Astronaut; AAAS; Am Defense Preparedness Asn; Air Force Asn. *Res:* Communication systems; data compression applied to radar, speech, and image data; development of robust algorithms for speech detection; application of universal algorithms to sensor data compression; developed new signal processing algorithms for improving data link and spread spectrum detection. *Mailing Add:* 2760 E Blue Spruce Dr Salt Lake City UT 84117

COX, BEVERLEY LENORE, b Huntingdon, Pa, Jan 11, 29. ZOOLOGY, PHYSIOLOGY. *Educ:* Pa State Univ, BS, 51, MS, 53; Univ Okla, PhD(invert physiol), 60. *Prof Exp:* Instr physiol, Univ Okla, 59-60; res assoc & NIH fel, Univ Ore, 60-61; from asst prof to assoc prof, 61-70, PROF BIOL, CENT STATE UNIV, OKLA, 70- *Concurrent Pos:* NIH fel cardiovasc physiol, Sch Med, Univ Okla, 63-, vis assoc prof, 67, 68, 69 & 74, NSF fel isotope & nuclear reactor technol 63, 64, 65; secy-treas, Okla Soc Physiologists, 86-88, pres-elect, 88-89 & pres, 89-90. *Mem:* AAAS; Am Soc Zool; Sigma Xi. *Res:* Physiology of insect tarsal chemoreception; gull food-finding behavior; recorded gull calls as repellants and attractants; endocrinology of mammalian salivary glands, molting in arachnids and electrolyte balance in lower vertebrates; bat blood studies. *Mailing Add:* Rte 1 Box 184 Oklahoma City OK 73131-9801

COX, BRADLEY BURTON, b Danville, Ky, Oct 29, 41; div; c 1. ELEMENTARY PARTICLE PHYSICS. *Educ:* Duke Univ, PhD(physics), 67. *Prof Exp:* Res assoc high energy physics, Johns Hopkins Univ, 67, consult, 67-69, asst prof high energy physics, 69-73; group leader, Proton Lab, Fermilab, 73-74, assoc head, 74-76, head Proton Lab, 76-77, head superconducting magnet group, Proton Dept, 77-78, proj leader & design physicist, High Intensity Lab, 78-79, head, Res Serv Dept-Fermilab Nat Accelerator Lab, 81-83, dept chmn physics dept, 83-84, head DO calorimetry group, 84-85; MEM, HIGH ENERGY PHYSICS REV COMT, LAWRENCE BERKELEY LAB, 77-; PROF PHYSICS, UNIV VA, 88- *Concurrent Pos:* Mem high energy discussion group, Brookhaven Nat Lab, 69-72; consult, Nuclear Effects Lab, 69-70; mem, sci adv comt, Fermi Nat Accelerator Lab, 77-78, sci spokesman for Fermilab exp & res group leader, Fermilab Physics Dept, 72-, Super Conducting-Super Collider Detector Task Force, 85-86; vis prof, Shandong Univ, People's Repub China. *Mem:* AAAS; Am Phys Soc; fel Am Phys Soc. *Res:* Electromagnetic, weak and strong interactions of elementary particles; low energy fusion reactions and identical target-projectile heavy ion nuclear reactions; direct photon production in hadronic interactions; Drell Yan production of lepton pairs in antiproton interactions; production of charmonium in hodronic interactors; production of beauty hadrons in high energy interactions. *Mailing Add:* Univ Va Dept Physics J W Beans Lab McCormick Rd Charlottesville VA 22901

COX, BRIAN, b Liverpool, UK, Aug 4, 31; Can citizen; m 55; c 2. NUCLEAR REACTOR MATERIALS, CORROSION IN HIGH TEMPERATURE WATER. *Educ:* Univ Cambridge, UK, BA, 52, PhD(chem), 55, MA, 59. *Prof Exp:* Sr sci officer, UK Atomic Energy Authority, Harwell, 55-63; leader, Corrosion Group, Atomic Energy Can Ltd, Chalk River, 63-67, br head mat sci, 67-85, div dir chem & mat, 85-86, prog consult reactor mat, 86 89; PROF NUCLEAR ENG, MAT SCI, CTR NUCLEAR ENG, UNIV TORONTO, 89. *Concurrent Pos:* Consult, Int Atomic Energy Authority, Vienna, 88-, Elec Power Res Inst, Palo Alto, Calif, 88-, Philadelphia Elec Co, Pa, 89-; dir, Can Nuclear Asn, 91- *Honors & Awards:* A B Campbell Award, Nat Asn Corrosion Engrs, 61; W B Lewis Medal, Can Nuclear Asn, 88. *Mem:* Can Nuclear Soc; Can Nuclear Asn; Am Nuclear Soc; Am Soc Metals; Int Metall Soc. *Res:* Behavior of materials, especially zirconium alloys used for fuel cladding and pressure tubes in nuclear reactors; effects of irradiation on their properties; failure analyses of zirconium alloy components. *Mailing Add:* Ctr Nuclear Eng Univ Toronto 184 College St Toronto ON M5S 1A4 Can

COX, BRIAN MARTYN, b Sutton, Eng, Nov 3, 39; m 64; c 3. NEUROSCIENCE, DRUG ADDICTION. *Educ:* Coun Nat Acad Awards, London, UK, Dipl Technol, 62; Univ London, PhD(pharmacol), 65. *Prof Exp:* Lectr pharmacol, Chelsea Col, Univ London, 65-74; res assoc, Med Sch, Stanford Univ, 73-75; assoc dir, Addiction Res Found, Palo Alto, Calif, 75-81; PROF PHARMACOL, UNIFORMED SERV UNIV HEALTH SCI, 81- *Concurrent Pos:* Lectr & consult assoc prof, Dept Pharmacol, Med Sch, Stanford Univ, 75-78; mem, Exec Comt, Int Narcotics Res Conf, 81-; res scientist award, US Public Health Serv, Nat Inst Drug Abuse, 78-81; chmn, Biochem Subcomt, Nat Inst Drug Abuse Biomed Res Rev Comt, 81-85. *Mem:* Brit Pharmacol Soc; Am Soc Pharmacol & Exp Therapeut; Soc Neurosci; Int Narcotics Res Conf. *Res:* Structure, properties and functions of endogeneous opioid peptides and other related neuropeptides; mechanisms associated with the development of drug tolerance and dependence; pharmacology of drugs of abuse. *Mailing Add:* Dept Pharm Uniformed Serv Univ Health Sci 4301 Jones Bridge Rd Bethesda MD 20814-4799

COX, CATHLEEN RUTH, b Vallejo, Calif, Oct 20, 48; m. REPRODUCTIVE BEHAVIOR, SOCIOBIOLOGY. *Educ:* Univ Calif, San Diego, BA, 70; Stanford Univ, PhD(psychol), 76. *Prof Exp:* Fel animal behav, Am Mus Natural Hist, 76-78; res assoc psychol, Barnard Col, Columbia Univ, 78-79; fel biol, Univ Calif, Los Angeles, 79-82; asst prof psychol, Calif State Univ, Northridge, 80-84; DIR RES, LOS ANGELES ZOO, 81- *Concurrent Pos:* Vis prof psychol, Scripps Co, 79; res assoc biol, City Col, City Univ NY, 76-78; instr, Stanford Univ, 75; res assoc ornith, Los Angeles Mus Natural Hist, 79-; lectr biol, Univ Calif, Los Angeles, 80- *Honors & Awards:* WC Allee Award in Animal Behav, 76. *Mem:* Animal Behav Soc; Am Ornith Union; Asn Women Sci; Am Asn Zool Parks & Aquaria; Am Primatological Soc. *Res:* Animal social behavior, specifically aggressive, sexual and reproductive; reproductive success of endangered species housed in zoos. *Mailing Add:* Los Angeles Zoo 5333 Zoo Dr Los Angeles CA 90027

COX, CHARLES DONALD, b Danville, Ill, Sept 10, 18; m 42; c 2. MEDICAL MICROBIOLOGY. *Educ:* Univ Ill, BS, 40, MS, 41, PhD(bact), 47. *Prof Exp:* Asst bact, Univ Ill, 40-42, 46-47; asst prof, Med Col Va, 47-49; assoc prof, Pa State Univ, 49-51; prof microbiol & pub health & head dept, Sch Med, Univ SDak, 51-60; head microbiol br, Off Naval Res, Washington, DC, 60-62; head dept, 62-72, prof, 62-84, EMER PROF MICROBIOL, UNIV MASS, AMHERST, 84- *Concurrent Pos:* Mem ed bd, J Bact, 57-70; mem adv panel environ biol, NASA, 65-72; mem comt naval med res, Nat Res Coun, 66-73; mem exobiol panel, Space Sci Bd, Nat Acad Sci, 74-77. *Mem:* AAAS; Am Soc Microbiol; Am Acad Microbiol; NY Acad Sci. *Res:* Medical bacteriology and immunology; physiology and virulence of spirochetes. *Mailing Add:* 7558 Longmeadow Lane Athens OH 45701-9217

COX, CHARLES PHILIP, b Eng, Dec 15, 19. STATISTICS, BIOMETRICS. *Educ:* Oxford Univ, BA, 40, MA, 47. *Prof Exp:* Head sect 4a, Army Oper Res Group, Eng, 45-46; head statist, Nat Inst Res Dairying, 48-61; prof 61-90, EMER PROF STATIST, IOWA STATE UNIV, 90- *Concurrent Pos:* Ministry of Agr res scholar, Exp Sta, Rothamsted, 47-48. *Mem:* Am Statist Asn; Biomet Soc. *Res:* General statistical methodology; biological assay. *Mailing Add:* Statist Lab Iowa State Univ Ames IA 50011

COX, CHARLES SHIPLEY, b Hawaii, Sept 11, 22; m 51; c 5. GEOPHYSICS. *Educ:* Calif Inst Technol, BS, 44; Univ Calif, PhD, 54. *Prof Exp:* Asst oceanogr, 54-60, assoc prof, 60-66, chmn ocean res div, 73-76, PROF OCEANOG, SCRIPPS INST OCEANOG, UNIV CALIF, SAN DIEGO, 66- *Concurrent Pos:* NSF fel, 57; Fulbright res fel, 57-58; vis prof, Mass Inst Technol, 69-70. *Mem:* Fel Am Geophys Union; Royal Astron Soc; fel AAAS. *Res:* Physical oceanography; geophysics relating to oceanic microstructure; microseisms and electrical structure of lithosphere. *Mailing Add:* Scripps Inst Oceanog Univ Calif San Diego La Jolla CA 92093-0230

COX, CLAIR EDWARD, II, b Lawrence Co, Ill, Sept 2, 33; m 58; c 4. SURGERY, UROLOGY. *Educ:* Univ Mich, MD, 58; Am Bd Urol, dipl, 66. *Prof Exp:* Intern med, Med Ctr, Univ Colo, 58-59, resident surg, 59-60; resident urol, Med Ctr, Univ Calif, San Francisco, 60-63; from instr to prof, Bowman Gray Sch Med, 63-73; CHMN DEPT UROL, COL MED, UNIV TENN, MEMPHIS, 73- *Concurrent Pos:* Nat Cancer Inst res grant, 69- *Mem:* Am Urol Asn; Am Col Surg; Am Asn Genitourinary Surg; Infectious Dis Soc Am; Soc Univ Urol. *Res:* Epidemiology of urinary tract infection; urological cancer. *Mailing Add:* Dept Urol Health Sci Ctr Univ Tenn 800 Madison Ave Memphis TN 38163

COX, CYRUS W(ILLIAM), b West Terre Haute, Ind, Apr 20, 24; m 46; c 4. ELECTRICAL ENGINEERING. *Educ:* Rose Polytech Inst, BS, 49; Purdue Univ, MS, 51. *Prof Exp:* Elec engr, US Bur Reclamation, 49-50; from instr to assoc prof, 51-64, PROF ELEC ENG, S DAK SCH MINES & TECHNOL, 64- *Concurrent Pos:* Consult, Argonne Nat Lab, 65- & Detroit Edison Co, 78-79. *Mem:* Am Soc Eng Educ; Nat Soc Prof Engrs; Inst Elec & Electronics Engrs; Sigma Xi. *Res:* Undergraduate electric circuits and systems; nondestructive testing of nuclear fuels; energy efficiency in rotating machines; models for voltage and steam control in generating systems; real-time rating of power system components. *Mailing Add:* Dept Elec Eng S Dak Sch Mines & Technol Rapid City SD 57702

COX, DAVID BUCHTEL, b Denver, Colo, Jan 25, 27; m 53; c 2. RHEOLOGY, ORGANIC CHEMISTRY. *Educ:* DePauw Univ, AB, 48; Stanford Univ, MS, 50; Univ NMex, PhD(org chem), 53. *Prof Exp:* From technologist to sr technologist, Socony Mobil Oil Co, Inc, 53-58, res assoc, 58-67; sr res chemist, Battelle Mem Inst, 67-69, assoc chief lubrication mech div, 69-70; chief chemist, Chem-Trend, Inc, 70-73, tech dir, 73-76, vpres tech opers, 76-84, tech admin, 84-91; RETIRED. *Mem:* Am Chem Soc; Soc Rheology; AAAS; Soc Mfg Engrs. *Res:* Rheology of lubricants, fuels and other petroleum products; polymer solutions; colloid chemistry of lubricants; mold release agents for urethane foams; die-casting lubricants; fire-resistant hydraulic fluids; cutting and grinding fluids; mold release agents for rubber; thermoset plastics. *Mailing Add:* 1620 Sheridan Dr Ann Arbor MI 48104

COX, DAVID ERNEST, b Rochford, Eng, Dec 2, 34; m 57, 86; c 4. CRYSTALLOGRAPHY, SOLID STATE CHEMISTRY. *Educ:* Univ London, BSc, 55, PhD(inorg chem), 58. *Prof Exp:* Tech officer ceramics, Steatite & Porcelain Prod Div, Imp Chem Industs, Ltd, 58-59; from chemist to sr chemist, Res Labs, Westinghouse Elec Corp, 59-63; assoc physicist, 63-66, physicist, 66-89, SR PHYSICIST, BROOKHAVEN NAT LAB, 89- *Concurrent Pos:* Asst ed, J Physics & Chem of Solids, 69; mem, Neutron Diffraction Comn, Int Union Crystallog, 72-81, chmn 78-81 & mem, Comn Powder Diffraction, 90-; assoc ed, J Physics & Chem Solids, 76-; ed, J Physics & Chem Solids, 81- *Mem:* Fel Am Phys Soc; Am Chem Soc; Am Asn Crystal Growth; Am Crystallog Asn; Mat Res Soc; AAAS. *Res:* Neutron and x-ray synchrotron powder diffraction; magnetic and crystal structures; synthesis and characterization of inorganic materials; high technetium oxides, zeolites. *Mailing Add:* Physics Dept Brookhaven Nat Lab Upton NY 11973

COX, DAVID FRAME, b New York, NY, Feb 19, 31; m 54; c 1. AGRICULTURAL & FOREST SCIENCES. *Educ:* Cornell Univ, BS, 53; NC State Col, MS, 57; Iowa State Univ, PhD(animal breeding, genetics), 59. *Prof Exp:* Assoc prof animal sci, 59-66, PROF STATIST, IOWA STATE UNIV, 66- *Mem:* Biomet Soc; Am Soc Animal Sci. *Res:* Design and analysis of experiments. *Mailing Add:* Dept Statist Iowa State Univ Ames IA 50011

COX, DAVID JACKSON, b New York, NY, Dec 22, 34; m 58, 83; c 3. PHYSICAL BIOCHEMISTRY. *Educ:* Wesleyan Univ, BA, 56; Univ Pa, PhD(biochem), 60. *Prof Exp:* Instr biochem, Univ Wash, Seattle, 60-63; from asst prof to assoc prof chem, Univ Tex, Austin, 63-73; prof biochem & head dept, Kans State Univ, 73-89; PROF CHEM & DEAN ARTS & SCI, INDIANA PURDUE UNIV, FT WAYNE, 89- *Concurrent Pos:* Investr, Howard Hughes Med Inst, 60-63; NSF fel & vis prof, Univ Va, 70-71. *Mem:* AAAS; Am Chem Soc; Am Soc Biochem & Molecular Biol. *Res:* Physical chemistry of macromolecules; protein chemistry. *Mailing Add:* Sch Arts & Sci Ind Univ-Purdue Univ 2101 Coliseum Blvd Ft Wayne IN 46805

COX, DENNIS DEAN, b Denver, Colo, Apr 7, 50; m 70; c 1. NONPARAMETRIC REGRESSION, TIME SERIES ANALYSIS. *Educ:* Univ Colo, Boulder, BA, 72; Univ Denver, MS, 76; Univ Wash, PhD(math), 80. *Prof Exp:* Develop engr, Honeywell Marine Syst Ctr, 77-80; asst prof statist, Univ Wis, 80-85; PROF STATIST, UNIV ILL, 85- *Mem:* Inst Math Statist; Math Asn Am; Am Math Soc; Am Statist Asn; Royal Statist Soc; AAAS; Soc Indust Appl Math. *Res:* Statistical inference for curve estimation and time series; probability theory on abstract spaces; approximation theory; differential operators. *Mailing Add:* Univ Ill 725 S Wright St Champaign IL 61820

COX, DENNIS HENRY, b St Paul, Minn, Aug 12, 25; m 50; c 1. ANIMAL NUTRITION. *Educ:* Univ Minn, BS, 50, MS, 53; Univ Fla, PhD(animal nutrit), 55. *Prof Exp:* Assoc toxicologist, Ga Coastal Plain Exp Sta, 55-62; asst prof, Univ Iowa, 62-63 & Coker Col, 63-64; asst prof foods & nutrit, Pa State Univ, 64-68, assoc prof nutrit, 68-71; RES CHEMIST, DEPT HEALTH, EDUC & WELFARE, 71- *Mem:* Am Chem Soc; Am Inst Nutrit; Sigma Xi. *Res:* Various aspects of animal and human nutrition, especially trace and macro minerals and their various interrelationships; trace element methodology. *Mailing Add:* 38 Highland Falls Dr Ormond Beach FL 32174

COX, DENNIS PURVER, b Seattle, Wash, Sept 12, 29; wid; c 4. ECONOMIC GEOLOGY. *Educ:* Stanford Univ, BS, 51, MS, 54, PhD(geol), 56. *Prof Exp:* Geologist mineral deposits explor, Anaconda Co, 56-59; vis prof geol, Univ Bahia, Brazil, 59-61; geologist mil geol, 61-65, geologist mineral resources, 65-72, mgr mineral resource specialist prog, 76-78, copper resources specialist, US Geol Surv, 72-82; CONSULT, 83- *Mem:* Geol Soc Am; Soc Econ Geol. *Res:* Geology of porphyry copper deposits; mineral resource modeling and resource estimation. *Mailing Add:* 3918 Grove Ave Palo Alto CA 94303

COX, DIANE WILSON, b Belleville, Ont, May 18, 37; m 61; c 3. HUMAN GENETICS. *Educ:* Univ Western Ont, BSc, 59; Univ Toronto, MA, 61; McGill Univ, PhD(genetics), 68. *Prof Exp:* Res asst genetics, McGill Univ, 63-64; fel, Hosp for Sick Children, 67-70, investr, 71-77; asst prof, Depts Pediat & Med Genetics, 75-77, assoc prof, 77-85, PROF DEPTS PEDIAT, MED GENETICS, MED BIOPHYS & INST MED SCI, UNIV TORONTO, 85-; PROF, RES INST HOSP FOR SICK CHILDREN, 85- *Concurrent Pos:* Fel, Can Col Med Genetics, 76- *Mem:* Genetics Soc Can (pres, 85-); Am Soc Human Genetics; Soc Pediat Res. *Res:* Human DNA polymorphisms; serum protein polymorphism, particularly relating to disease; immunoglobelin heavy chain gene cluster; di-antitrypsin and other protease inhibitors; gene mapping; population studies; molecular evolution; apolipoprotein disorders. *Mailing Add:* Pediat Dept & Res Inst 555 University Ave Toronto ON M5G 1X8 Can

COX, DOAK CAREY, b Wailuku, Maui, Hawaii, Jan 16, 17; m 41; c 5. ENVIRONMENTAL MANAGEMENT, ENVIRONMENTAL GEOLOGY. *Educ:* Univ Hawaii, BS, 38; Harvard Univ, AM, 41, PhD, 65. *Prof Exp:* Geologist, US Geol Surv, 41-46; geophysicist, Exp Sta, Hawaiian Sugar Planters Asn, 46-60; geophysicist, Inst Geophys, Univ Hawaii, 60-64, dir, Water Resources Res Ctr, 64-70, prof geol, 60-85, dir Environ Ctr, 70-85. *Concurrent Pos:* Consult, Pac Islands, 46-64; hydrologist, Arno Exped, Pac Sci Bd, 50; secy, Tsunami Comt Int Union Geod & Geophys, 60-67; Tsunami adv, Hawaii State Civil Defense Div, 60-66; chmn oceanog panel, Comt on Alaska Earthquake, Nat Acad Sci-Nat Res Coun, 64-73; Hawaii State Water Comn, 78; mem, Earthquake Adv Comt, Hawaii State Civil Defense Div, 90- *Mem:* Fel AAAS; fel Geol Soc Am; Seismol Soc Am; Am Geophys Union. *Res:* Hawaii and Pacific geology and hydrology; tsunamis; natural hazard management. *Mailing Add:* 1929 Kakela Dr Honolulu HI 96822

COX, DONALD CLYDE, b Lincoln, Nebr, Nov 22, 37; m 61; c 2. ELECTRICAL ENGINEERING, RADIO PHYSICS. *Educ:* Univ Nebr, Lincoln, BS, 59, MS, 60; Stanford Univ, PhD(elec eng), 68. *Hon Degrees:* DSc, Univ Nebr, Lincoln, 83. *Prof Exp:* Res asst elec eng, Stanford Univ, 63-66, res assoc radio propagation, Stanford Electronics Labs, 67-68; mem tech staff, radio transmission res dept, 68-73, supvr satellite systs res dept, 73-83, head radio & satellite systs res dept, Bell Labs, 83, MGR, RADIO RES DIV, BELLCORE, BELL COMMUN RES, 84- *Concurrent Pos:* Mem Comns B, C & F, US Nat Comt, Int Union Radio Sci, 69-; assoc ed, Inst Elec & Electronics Engrs Trans Antennas & Propagation, 83-86; admin comt, Inst Elec & Electronics Engrs Antennas & Propagation Soc, 86-88. *Honors & Awards:* Guglielmo Marconi Prize, Inst Int Commun, Italy, 83; Morris E Leeds Award, Inst Elec & Electronics Engrs, 85. *Mem:* Fel Inst Elec & Electronics Engrs; Sigma Xi; fel AAAS. *Res:* Radio propagation; communications systems; satellite systems; electronics; atmospheric physics; radio measurement techniques. *Mailing Add:* 24 Aiden Lane Tinton Falls NJ 07724

COX, DONALD CODY, b Peoria, Ill, Mar 31, 36; m 63; c 2. VIROLOGY, MOLECULAR BIOLOGY. *Educ:* Northwestern Univ, BA, 58; Univ Mich, PhD(epidemiol sci), 65. *Prof Exp:* From asst prof to assoc prof microbiol, Univ Okla, 65-72, assoc prof bot & microbiol, 72-78; prof & chairperson, dept microbiol, Miami Univ, 78-89; CONSULT, 89- *Mem:* AAAS; Sigma Xi; Am Soc Microbiol; Am Soc Virol. *Res:* Studies concerning biochemical alterations occurring in virus-infected cells and virus therapy of neoplastic disease; cancer therapy. *Mailing Add:* Dept Microbiol Miami Univ Oxford OH 45050

COX, DONALD DAVID, b Maben, WVa, Aug 2, 26; m 46; c 3. PLANT SCIENCE. *Educ:* Marshall Univ, AB, 49, MA, 50; Syracuse Univ, PhD(plant sci), 58. *Prof Exp:* Teacher pub sch, WVa, 49-50; prof biol, Marshall Univ, 50-62; consult biol sci curric study, Univ Colo, Boulder, 62-63; actg chmn dept, 74-76, PROF BIOL, STATE UNIV NY COL OSWEGO, 63-, DIR, RICE CREEK BIOL FIELD STA, 81- *Concurrent Pos:* Chmn dept biol, Marshall Univ, 57-62; res botanist, Corps Engrs, US Army, Miss Waterways Exp Sta, 62; staff biologist, Comn Undergrad Educ Biol Sci, NSF, 70-71. *Mem:* Ecol Soc Am; Sigma Xi. *Res:* Post glacial forests in New York State as determined by the method of pollen analysis. *Mailing Add:* Dept Biol State Univ NY Col Oswego NY 13126

COX, DUDLEY, b Brooklyn, NY, Mar 3, 29; m 55; c 2. MICROBIOLOGY, CELL PHYSIOLOGY. *Educ:* Howard Univ, BS, 56; Long Island Univ, MS, 58; NY Univ, PhD(microbiol), 66. *Prof Exp:* RES ASSOC MICROBIOL, HASKINS LABS, 58-; PROF BIOL, PACE UNIV, 66- *Mem:* AAAS; Soc Protozool. *Res:* Nutrition and metabolic activity of microorganisms; conversion of cellulosic wastes to useful products; methanogenesis. *Mailing Add:* 582 Macon St Brooklyn NY 11233

COX, EDMOND RUDOLPH, JR, b Pascagoula, Miss, Nov 15, 32; m 59; c 1. PHYCOLOGY, MICROBIOLOGY. *Educ:* Univ Ala, BS, 57, MS, 60, PhD(bot), 67. *Prof Exp:* From instr to assoc prof biol, Middle Tenn State Univ, 61-68; chmn dept, 68-77, PROF BIOL, TENN WESLEYAN COL, 68- *Mem:* Phycol Soc Am; Int Phycol Soc. *Res:* Soil algae. *Mailing Add:* Dept Biol Tenn Wesleyan Col Townsend Hall College St Box 40 Athens TN 37303

COX, EDWARD CHARLES, b Alberni, Can, June 28, 37; US citizen; m 60; c 3. GENETICS & DEVELOPMENTAL BIOLOGY. *Educ:* Univ BC, BSc, 59; Univ Pa, PhD(biochem), 64. *Prof Exp:* Instr biochem, Univ Pa, 61-62; fel genetics, Stanford Univ, 64-67; asst prof biol & biochem, Princeton Univ, 67-72, assoc prof biol & assoc dean of col, 72-77, chmn dept, 77-87, PROF BIOL & MOLECULAR BIOL, PRINCETON UNIV, 77- *Mem:* Genetics Soc Am. *Res:* The genetic control of mutation rates; developmental genetics of the cellular slime molds. *Mailing Add:* Dept Molecular Biol Princeton Univ Princeton NJ 08544

COX, EDWIN, III, b Richmond, Va, Oct 31, 31; m; c 4. CHEMICAL ENGINEERING, PHYSICAL CHEMISTRY. *Educ:* Va Mil Inst, BS, 53; Univ Va, MChE, 60. *Prof Exp:* Asst instr chem, Univ Va, 57-59; assoc, Cox & Gillespie, 59-60, partner & chem engr, 60-66; CHEM ENGR, EDWIN COX ASSOCS, 66- *Concurrent Pos:* Secy-treas, Commonwealth Lab, Inc, 59-67, pres, 67-; mem personal staff, Gov Va, 60-68 & 72-90. *Mem:* Fel Am Inst Chem; Nat Soc Prof Engrs; Am Chem Soc; Am Soc Metals; Am Soc Testing & Mat. *Res:* Surface chemistry; phosphate rock; phosphorus production; phosphoric acid production; fertilizers; material science; catalysis; air and water pollution and abatement. *Mailing Add:* PO Box 8025 Richmond VA 23223

COX, ELENOR R, b Georgetown, Tex. PHYCOLOGY. *Educ:* Rice Univ, BA, 52; Univ Tex, Austin, MS, 61, PhD(bot), 66. *Prof Exp:* From asst prof to assoc prof, 67-81, PROF BIOL, TEX A&M UNIV, 81- *Mem:* Phycol Soc Am (secy, 75-78, vpres & pres-elect, 79, pres, 80); Sigma Xi. *Res:* Phylogeny of the photosynthetic dinoflagellates. *Mailing Add:* Dept Biol Tex A&M Univ College Station TX 77843

COX, FRED WARD, JR, b Atlanta, Ga, Dec 10, 14; m 39; c 3. INDUSTRIAL CHEMISTRY, RESEARCH ADMINISTRATION. *Educ:* Ga Inst Technol, BS, 36; Univ Wis, PhD(org chem), 39. *Prof Exp:* Res chemist, Rohm and Haas Co, Philadelphia, 38; res chemist, Goodyear Tire & Rubber Co, Akron, 39-42, group leader, 42-45; group leader, Southern Res Inst, 45-49, head appl chem div, 46-49; asst dir, Eng Exp Sta, Ga Inst Technol, 49-53; chief chemist, Deering Milliken Res Trust, 53-56; mgr, Atlas Res & Develop Lab, Tyler Corp, 56-79, dir res & develop, Atlas Powder Co, 79-80; RETIRED. *Concurrent Pos:* Mem, Tamaqua Area Sch Bd, 63-73, pres, 70-73; mem, Schuylkill County Bd Educ, 65-73 & Schuylkill County Area Voc Schs Oper Bd. *Mem:* Am Chem Soc; Am Soc Testing & Mat; Franklin Inst; Soc Rheology. *Res:* High polymer technology; organic process development; textiles; explosives; research administration; explosive slurries and emulsions; blasting initiators. *Mailing Add:* 13651 Meeker Blvd Sun City West AZ 85375-3725

COX, FREDERICK EUGENE, b Quincy, Mass, Nov 4, 38; m 67; c 2. PEDIATRICS, INFECTIOUS DISEASES. *Educ:* Boston Univ, BA, 60; Boston Univ, MD, 64. *Prof Exp:* Surg intern, Univ Hosp, Boston, 64-65; gen med officer, US Coast Guard Acad Hosp, USPHS, New London, Conn, 65-67; resident path, Boston City Hosp, Mass, 67-68; resident pediat, St Elizabeth's Hosp, Brighton, Mass, 68-69 & Boston Floating Hosp, 69-70; fel pediat infectious dis, Cleveland Metrop Gen Hosp, Ohio, 70-72; fel, St Jude Children's Res Hosp, Memphis, Tenn, 72-74, res assoc infectious dis serv, 74-77; ASSOC PROF PEDIAT & CHIEF INFECTIOUS DIS DIV, MED COL GA, 77- *Mem:* Fel Am Acad Pediat; Sigma Xi; Am Soc Microbiol; Pediat Infectious Dis Soc. *Res:* Bacterial and fungal adherence. *Mailing Add:* Dept Pediat Med Col Ga Augusta GA 30912

COX, FREDERICK RUSSELL, b Sutherland, Nebr, Mar 11, 32; m 59; c 2. SOIL SCIENCE. *Educ:* Univ Nebr, BS, 53, MS, 58; NC State Univ, PhD(soils), 61. *Prof Exp:* Assoc prof, 61-76, PROF SOIL SCI, NC STATE UNIV, 76- *Mem:* Am Soc Agron; Soil Sci Soc Am. *Res:* Soil fertility; micronutrient research. *Mailing Add:* Dept Soil Sci NC State Univ Box 7619 Raleigh NC 27695

COX, GENE SPRACHER, b Norton, Va, Mar 21, 21; m 46; c 2. FOREST SOILS. *Educ:* Duke Univ, BS, 47, MF, 48, PhD(forestry), 53. *Prof Exp:* Asst prof forestry, Stephen F Austin State Univ, 51-53; from asst prof to assoc prof, Univ Mont, 53-60; assoc prof, 60-63, PROF FORESTRY, UNIV MO, COLUMBIA, 63- *Mem:* Am Soc Agron; Soil Sci Soc Am; Soc Am Foresters; Ecol Soc Am; Sigma Xi. *Res:* Forest ecology. *Mailing Add:* Sch Forestry, Fish & Wildlife Univ Mo Columbia MO 65211

COX, GEORGE ELTON, b Ayden, NC, July 22, 31; m 56; c 2. EXPERIMENTAL & TOXICOLOGIC PATHOLOGY. *Educ:* Univ NC, BA, 51, MS, 54; Univ Ill, MD, 56. *Prof Exp:* Intern path, Presby Hosp, Chicago, 56-57; resident, Presby-St Luke's Hosp, Chicago, 57-60; dir exp path, Evanston Hosp, 61-62; res assoc biol chem, Col Med, Univ Ill, 62-64; res assoc & asst prof path, Med Units, Univ Tenn, 64-65; res assoc, May Inst Med Res, Cincinnati, 65-69; asst prof path, Col Med, Univ Cincinnati, 66-69; pathologist, Food & Drug Res Labs, Inc, 69-81; RES ASSOC & SR PATHOLOGIST, MOBIL OIL CORP, PRINCETON, NJ, 81- *Mem:* Am Asn Pathologists. *Res:* Experimental study of etiology of atherosclerosis; spontaneous neoplasia in laboratory animals; histopathologic effects of industrial chemicals. *Mailing Add:* Div Path & Immunotoxicol Mobil Oil Corp PO Box 1029 Princeton NJ 08558

COX, GEORGE STANLEY, b Roswell, NMex, Jan 26, 46; m 67; c 2. BIOCHEMISTRY, MOLECULAR BIOLOGY. *Educ:* NMex State Univ, BS, 68; Univ Iowa, PhD(biochem), 73. *Prof Exp:* Fel, Roche Inst Molecular Biol, 72-74; staff fel, Lab Molecular Biol, Nat Inst Arthritis, Metab & Digestive Dis, NIH, 74-76; asst prof biochem & biophys, Iowa State Univ, 76-83; ASSOC PROF BIOCHEM & BIOPHYS, UNIV NEBR, 83- *Concurrent Pos:* Ad hoc reviewer, GMA-1 study sect, NIH. *Honors & Awards:* Masua Hon lectr, Mid Am State Univ Asn, 85. *Mem:* Am Soc Microbiol; Sigma Xi; Am Asn Cancer Res; Am Soc Biochemists. *Res:* Regulation of gene expression in eukaryotes including transcription, translation, and nuclear-cytoplasmic transport of macromolecules; molecular oncology; regulation of gene expression in eukaryotic cells, including DNA methalation and trans-acting factors. *Mailing Add:* Dept Biochem Univ Nebr Med Ctr 600 S 42nd St Omaha NE 68198-4525

COX, GEORGE W, b Williamson, WVa, Feb 10, 35; m 57; c 2. ECOLOGY, ORNITHOLOGY. *Educ:* Ohio Wesleyan Univ, AB, 56; Univ Ill, MS, 58, PhD(zool), 60. *Prof Exp:* Asst prof biol, Univ Alaska, 60-61 & Calif Western Univ, 61-62; from asst prof to assoc prof, 62-69, PROF BIOL, SAN DIEGO STATE UNIV, 69- *Concurrent Pos:* Prog dir ecol, NSF, 78-79. *Mem:* AAAS; Ecol Soc Am; Am Ornith Union; Soc Conserv Biol. *Res:* Terrestrial population and community ecology; evolution of migration in birds; ecology of fossorial rodents; agricultural and conservation ecology. *Mailing Add:* Dept Biol San Diego State Univ San Diego CA 92182-0057

COX, GEORGE WARREN, b Bronx, NY, July 30, 60; m 87. IMMUNOPHARMACOLOGY. *Educ:* Ohio Wesleyan Univ, BA, 82; Ohio State Univ, PhD(pharmacol), 87. *Prof Exp:* Biotechnol training prog fel, 87-90, SR STAFF FEL, NAT CANCER INST, NIH, 90- *Mem:* Am Soc Pharmacol & Exp Therapeut; AAAS; Int Soc Immunopharmacol. *Res:* Immunological studies of cytokines, including monokines and lymphokines; immunology and molecular biology studies of macrophages and activation of macrophages for anti-tumor functions; regulation of macrophage gene expression and function. *Mailing Add:* 8265 Waterside Ct Frederick MD 21701

COX, GERALDINE ANNE VANG, b Philadelphia, Pa, Jan 10, 44; m 65. ENVIRONMENTAL SCIENCE, BIOLOGY. *Educ:* Drexel Univ, BS, 66, MS, 67, PhD(environ sci), 70. *Prof Exp:* Tech coordr environ progs, Raytheon Co, 70-76; White House fel, spec asst to secy, US Dept Labor, 76-77; environ scientist, Am Petrol Inst, 77-79; VPRES & TECH DIR, CHEM MFG ASN, 79- *Concurrent Pos:* Mem, prog comn, Water Pollution Control Fedn, 74-79; chmn, Marine Water Qual Comt, 75-80; mem, Environ Measurement Panel, Nat Bur Standards, Nat Acad Sci, 77-80; mem, Transp Adv Comn, USCG 80-; chmn, Marine Occup Safety & Health Comn; bd mem, Comn Environ Improv, Am Chem Soc, 80-88, assoc, 89-; bd mem, Fedn Org Prof Women, 80-82, pres, 82-84; mem, Hazardous Waste Comn, 85-87; mem, Eng Affairs Coun, Asn Am Eng Soc, 87-, chmn, 91-; lectr, Sch Nursing, Univ Pa; lab suprvr, Sanitary Landfill Proj & instr; food technologist, Keebler Biscuit Co, C Schmidt & Sons; chair, Marine Water Qual Control Comn. *Honors & Awards:* Outstanding Young Women Am, 75; Achievement Award, Soc of Women Engrs, 84. *Mem:* Am Soc Testing & Mat; Water Pollution Control Fedn; Fedn Orgn Prof Women; Am Chem Soc; Am Nat Standards Inst; Soc Women Engrs; Asn Am Eng Soc. *Res:* Oil pollution; marine pollution; fresh water pollution; ecological damage assessment; environmental health. *Mailing Add:* 2501 M St NW Washington DC 20037

COX, GORDON F N, b Montreal, Que, Nov 25, 48; US citizen; m 69. ARCTIC OFFSHORE ENGINEERING, SEA ICE PHYSICS. *Educ:* McGill Univ, Montreal, BSc Hons, 70; Dartmouth Col, NH, AM, 73, PhD(glaciol), 74. *Prof Exp:* Sr res engr, Amoco Prod Co, 74-78; arctic mgr consult, Oceanog Serv, Inc, 78-80; GEOPHYSICIST ARCTIC RES, COLD REGIONS RES & ENG LAB, 80- *Concurrent Pos:* Mem, Nat Acad Sci Sea Ice Mech Panel, 80-81 & Glaciol Panel, 80-83; consult, Oceanog Serv, Inc, 80- *Mem:* Int Glaciol Soc; Am Geophys Union; Arctic Inst NAm; Am Soc Mech Engrs. *Res:* Mechanical properties of sea ice considering ice salinity, structure, temperature, strain-rate, and confining pressure; ice stress sensor to measure ice loads on arctic offshore structures. *Mailing Add:* US Army Cold Regions Res & Eng Lab 72 Lyme Rd Hanover NH 03755-1290

COX, H C, b Melrose, NMex, May 18, 27; m 54; c 2. ENTOMOLOGY. *Educ:* Univ NMex, BS, 50; Iowa State Univ, MS, 52, PhD(entom), 55. *Hon Degrees:* LLD, NMex State Univ, 81. *Prof Exp:* Asst, Iowa State Univ, 50-53; res scientist, Entom Res Div, USDA, Iowa, 53-55, Utah, 55-58, Miss, 58-60, dir, Southern Grains Insects Res Lab, Ga, 60-67, asst dir, Entom Res Div, Beltsville, Md, 67-71, dir, 71-72, assoc admin, Southern Region, Agr Res Serv, New Orleans, La, 72-76, admin, Western Region, Oakland, Calif, 76-85; RETIRED. *Concurrent Pos:* Fel, Princeton Univ, 66-67. *Mem:* Entom Soc Am; Am Asn Prof Entomologists; Sigma Xi. *Mailing Add:* 316 Dover Dr Walnut Creek CA 94598

COX, HENRY MIOT, b Stephens Co, Ga, 07; m 35; c 1. MATHEMATICS, PSYCHOMETRICS. *Educ:* Emory Univ, BS, 28; Duke Univ, AM, 31. *Prof Exp:* Instr math & asst exam, Univ Syst Ga, 30-39; from asst dir to dir, 39-73, EMER DIR, BUR INSTR RES & EXAM SERV, UNIV NEBR, LINCOLN, 73- *Concurrent Pos:* Exec dir, Ann High Sch Math Exam, Math Asn Am, Soc Actuaries, Mu Alpha Theta & Nat Coun Teachers Math, 70-76. *Mem:* AAAS; Math Asn Am; Psychomet Soc; Nat Coun Measurement in Educ. *Res:* Measurement; instructional research; student guidance. *Mailing Add:* 1145 N 44th St Lincoln NE 68503

COX, HERALD REA, virology; deceased, see previous edition for last biography

COX, HIDEN TOY, b Greenville, SC, Mar 3, 17; m 43; c 1. PLANT MORPHOLOGY, PLANT ANATOMY. *Educ:* Furman Univ, BS, 36, BA, 37; Univ NC, MA, 39, PhD(bot), 47. *Prof Exp:* Asst prof biol, Howard Col, 41-46; assoc prof bot, Agnes Scott Col, 46-49; from assoc prof to prof, Va Polytech Inst, 49-55; dep exec dir, Am Inst Biol Sci, 53-54, exec dir, 55-63; coordr res, 63-67, dean sch letters & sci, 67-71, PROF BIOL, CALIF STATE UNIV, LONG BEACH, 71- *Concurrent Pos:* Asst adminr pub affairs, NASA, 61-62; Beattie lectr, 64; mem, Calif Curric Comn, 65-66, 67-69; chmn, Calif Sci Adv Comt, 66-67. *Honors & Awards:* Distinguished Serv Citation, NASA, 62. *Mem:* Fel AAAS; Sigma Xi; Asn Advan Biomed Educ. *Res:* Comparative wood anatomy; plant embryology. *Mailing Add:* Dept Biol Calif State Univ 1250 Bellflower Long Beach CA 90840

COX, HOLLACE LAWTON, JR, b Oak Park, Ill, Nov 17, 35; m 83; c 1. LASERS & ELECTROOPTICS, INTEGRATED OPTICS. *Educ:* Univ Rochester, AB, 59; Ind Univ, Bloomington, PhD(chem physics), 67. *Prof Exp:* Mem tech staff, Tex Instruments, Inc, 67-69; Robert A Welch res fel, Baylor Univ, 70-73; Robert A Welch res fel path, Univ Tex Syst Cancer Ctr, M D Anderson Hosp & Tumor Inst Houston, 73-75, res fel physics, 75-76; instr, Mallinckrodt Inst Radiol, Wash Univ Sch Med, 76-77; asst prof, Dept Radiation Oncol, Univ Kans Med Ctr, 77-80; assoc prof, Therapeut Radiol Dept, Sch Med, Univ Louisville, 80-83, lectr, 83-85, asst prof, Elec Eng Dept, 85-88, ASSOC PROF, ELEC ENG DEPT, UNIV LOUISVILLE, 88- *Mem:* Inst Elec & Electronics Engrs; Sigma Xi; Am Phys Soc; Optical Soc Am; Soc Photo-Optical Instrumentation Engrs. *Res:* Lasers and electrooptics; integrated optics, optical communications; fiber optics; nonlinear optical interactions in materials for use in integrated optical devices; laser spectroscopy; stopping power of alpha particles. *Mailing Add:* Dept Elec Eng Speed Sch Eng & Sci Univ Louisville Louisville KY 40292

COX, J E, b Waco, Tex, Aug 29, 35. MECHANICAL ENGINEERING. *Educ:* Southern Methodist Univ, BS, 58, MS, 60; Okla State Univ, PhD(eng), 63. *Prof Exp:* Engr, Astronaut Div, Chance Vought Corp, 56-59; instr, Okla State Univ, 61-63; prof mech eng, Univ Houston, 63-81; DIR GOVT AFFAIRS, AM SOC HEATING, REFRIG & AIR CONDITIONING ENGRS, 81- *Concurrent Pos:* Am Soc Mech Engrs cong fel with Sci & Technol Comt, US House Rep, 75. *Mem:* Fel Am Soc Mech Engrs; Am Soc Heating Refrig & Air Conditioning Engrs; Am Soc Asn Execs; Sigma Xi. *Res:* Heat transfer; fluid mechanics; thermodynamics. *Mailing Add:* PO Box 57104 Washington DC 20037

COX, JAMES ALLAN, b Chisholm, Minn, Sept 19, 41; m 65; c 2. ANALYTICAL CHEMISTRY, ELECTROCHEMISTRY. *Educ:* Univ Minn, BS, 63; Univ Ill, PhD(chem), 67. *Prof Exp:* Lectr chem, Univ Wis, 67-69; from asst to assoc prof, 69-78, prof chem Southern Ill Univ, Carbondale, 78-87; PROF & CHMN, DEPT CHEM, MIAMI UNIV, 87- *Mem:* Am Chem Soc. *Res:* Trace methods for anions; ion exchange membrane applications to trace analysis, modified electrodes, sensors. *Mailing Add:* Dept Chem Miami Univ Oxford OH 45056

COX, JAMES LEE, b Wayne Co, Ind, Oct 5, 38; m 71. ANIMAL NUTRITION. *Educ:* Purdue Univ, BS, 61; Univ Ill, MS, 63, PhD(animal nutrit), 67. *Prof Exp:* Sr res physiologist, 67-71, animal sci data analyst, 71-74, RES FEL, MERCK SHARP & DOHME RES LABS, 74- *Mem:* AAAS; Am Soc Animal Sci; Poultry Sci Asn; Sigma Xi; World Poultry Sci Asn. *Res:* Design and analysis of animal experiments; environmental effects on pregnant animals; mineral nutrition and physiological functions; influence of drugs on growth of animals. *Mailing Add:* Animal Sci Res Merck & Co Inc Box 2000 Rahway NJ 07065

COX, JAMES LESTER, JR, b Reidsville, NC, Nov 12, 42; m 63; c 2. PLASMA PHYSICS. *Educ:* NC State Univ, BS, 63, MS, 65, PhD(physics), 69. *Prof Exp:* From asst prof to assoc prof, 69-83, PROF PHYSICS, OLD DOMINION UNIV, 83- *Concurrent Pos:* Asst dean, Col Sci, Old Dominion Univ, 87-89, chmn, Dept Physics, 89- *Mem:* Sigma Xi; Am Phys Soc; AAAS. *Res:* Relativistic electron beams. *Mailing Add:* 905 Bentley Heath Common Virginia Beach VA 23452

COX, JAMES REED, JR, b Nashville, Tenn, Mar 25, 32. ORGANIC CHEMISTRY. *Educ:* Vanderbilt Univ, BA, 54, MA, 55; Harvard Univ, PhD(chem), 59. *Prof Exp:* NSF fel chem, Univ Munich, 58-59; from asst prof to assoc prof, Ga Inst Technol, 59-66; ASSOC PROF CHEM, UNIV HOUSTON, 66- *Mem:* Am Chem Soc. *Res:* Structure-reactivity relationships; mechanisms of reaction of organic and bio-organic chemical systems. *Mailing Add:* Dept Chem Univ Houston Houston TX 77004

COX, JEROME R(OCKHOLD), JR, b Washington, DC, May 24, 25; m 51; c 3. ELECTRICAL ENGINEERING. *Educ:* Mass Inst Technol, BS, 47, MS, 49, ScD(elec eng), 54. *Prof Exp:* Asst, Acoust Lab, Mass Inst Technol, 48-52; acoust engr, Liberty Mutual Life Ins Co, 52-55; from asst prof to assoc prof, 55-61, dir, Biomed Comput Lab, 64-75, PROF ELEC ENG, WASH UNIV, 61-, CHMN & PROF, DEPT COMPUT SCI, 75- *Concurrent Pos:* Res assoc, Cent Inst Deaf, 55-64; Nat Adv Res Resources Comt, NIH, 65-69; chmn comput labs, Wash Univ, 67-; co-chmn, Comput Cardiol Int Conf, 74-; pres, Biomed Res Panel, 75-76; cardiol adv comt, Nat Heart & Lung Inst, 75-78; chmn, Nat Inst Gen Med Sci Panel, 77-78, Div Comp Res & Technol Rev Comt, 83-84; prophet adv panel, NIH, 83-88; mem, Genome Data Base Sci Adv Comt, Howard Hughes Med Inst & Nat Neural Circuitry Database Comt, Inst Med, Nat Acad Sci, 89-, Nat Adv Coun, Human Genome Res, NIH, 90- *Mem:* sr mem Inst Med-Nat Acad Sci; fel Acoust Soc Am; fel Inst Elec & Electronics Engrs; Asn Comput Mach; fel Am Col Med Informatics. *Res:* Computer design; biomedical engineering; statistical communication theory; acoustics; medical imaging; computer networking. *Mailing Add:* Dept Comput Sci Box 1045 Wash Univ St Louis MO 63130

COX, JOHN DAVID, b San Juan, PR, Oct 5, 54; US citizen; m 80; c 2. ENGINEERING PHYSICS, IMAGING SYSTEMS. *Educ:* Univ Fla, BS, 78, ME, 79, PhD(eng physics), 82. *Prof Exp:* Res engr, RTS Labs, Inc, 81-85; CHMN, FUTURETECH INDUST, INC, 85-; VPRES ENG, GEN IMAGING SYSTS, 85- *Concurrent Pos:* Res scientist, nuclear eng dept, Univ Fla, 80- *Honors & Awards:* Tech Achievement Award, NASA, 84. *Mem:* Inst Elec & Electronics Engrs; Am Nuclear Soc; Am Soc Nondestructive Testing; Soc Photo-optical Instrumentation Engrs. *Res:* Solid-state x-ray imaging sensors and imaging systems. *Mailing Add:* PO Box 13455 Gainesville FL 32604

COX, JOHN E(DWARD), b Danville, Ill, Mar 28, 23; m 56; c 2. CERAMIC ENGINEERING. *Educ:* Univ Ill, BS, 50, MS, 51, PhD(eng), 53; Univ Hartford, BS, 68. *Prof Exp:* Res engr, NJ Zinc Co, 53-54; res fel, Mellon Inst, 54-60; dir res & develop, O Hommell Co, 60-62; sr res scientist, United Aircraft Corp, 62-68; dir res, Ceramic Dept, R T Vanderbilt Co, 68-73; midwest mgr, O Hommel Co, 73-75, mat mgr, 75-78; mgr coatings, Alpha Metals, 78-81; SPEC PROJ MGR, O HOMMEL CO, 81- *Mem:* Am Ceramic Soc; Am Soc Testing & Mat. *Res:* Porcelain enamel, glass and related ceramic materials; metal fibers. *Mailing Add:* 1500 Cochran Rd Unit 403 Pittsburgh PA 15243

COX, JOHN JAY, JR, b Pa, Apr 1, 29; m 53; c 2. METALLURGY. *Educ:* Carnegie Inst Technol, MS & ScD, 53. *Prof Exp:* Res supvr mat tech, Eng Dept, 53-64, res supvr chlorine prod sect, 64-66, develop mgr electronic prod, 66-70, mgr electronic packaging, 70-74, tech mgr, 74-78, prod mkt mgr, 78-81, dir sales, electronics prod dept, E I du Pont de Nemours & Co, Inc, 81-84; CONSULT, 86- *Concurrent Pos:* vpres, Opers Semiconductor res cord, 84-86. *Mem:* Sigma Xi; Am Inst Mining, Metall & Petrol Engrs; Int Soc Hybrid Microelectronics; Am Soc Metal; Inst Elec & Electronics Engrs; Int Electronic Packaging Soc. *Res:* Physical metallurgy; phase equilibria; precipitation from solids; x-ray diffractions; alloy development; electron microscopy; structure of crystalline polymers; thick film electronic materials. *Mailing Add:* Seven Tilbury Ct Chapel Hill NC 27514

COX, JOHN LAYTON, b Pendleton, Ore, Dec 30, 43; m 66; c 2. INORGANIC CHEMISTRY, PHYSICAL CHEMISTRY. *Educ:* Eastern Ore State Col, BS, 66; Univ Wyo, PhD(chem), 71. *Prof Exp:* Res scientist coal gasification, Natural Resources Res Inst, Univ Wyo, 71-72, dept mineral eng, 72-73; sr res engr fossil fuels, Battelle Northwest Labs, Battelle Mem Inst, 74-78; mgt basic res, Gas Res Inst, 78-85; PRES, SEPARATIONS TECHNOL ASSOCS, INC, 85- *Mem:* Am Chem Soc; Sigma Xi. *Res:* Synthetic fuels from fossil fuels; kinetics and catalysis; gas separations. *Mailing Add:* 2143 George Washington Way Richland WA 99352

COX, JOHN PAUL, theoretical astrophysics, for more information see previous edition

COX, JOHN THEODORE, b Fairborn, Ohio, June 27, 50. PROBABILITY THEORY, INTERACTING PARTICLE SYSTEMS. *Educ:* Harvey Mudd Col, BS, 72; Cornell Univ, MS, 75, PhD(math), 76. *Prof Exp:* Asst prof math, Ga Tech, 76-77; asst prof, Univ Southern Calif, 77-79; from asst prof to assoc prof, 79-86, PROF MATH, SYRACUSE UNIV, 86- *Concurrent Pos:* Vis prof math, Cornell Univ, 87-88. *Mem:* Am Math Soc; fel Inst Math Statist. *Mailing Add:* Dept Math Syracuse Univ Syracuse NY 13244-1150

COX, JOHN WILLIAM, b St Louis, Mo, Aug 31, 28; m 50; c 1. MEDICAL EDUCATION, CARDIOLOGY. *Educ:* St Louis Univ, MD, 51, PhD(physiol), 53; Am Bd Internal Med, dipl. *Prof Exp:* Chief res, Labs, Vet Admin Hosp, St Louis, 53-54; USN, 54-, chief cardiopulmonary lab, Dept Internal Med, US Naval Hosp, San Diego, 54-56, resident internal med, 56-59, staff supvr, Dept Internal Med, US Naval Hosp, San Diego, 59-61, chief med & dir clin serv, US Naval Hosp, Subic Bay, Philippines, 61-63, head cardio-respiratory dis br, US Naval Hosp, Philadelphia, 63-65, chief med & dir res, 65-69, head training & clin serv br, 69-72, dir med educ & training, Navy Dept, 72-77, cmndg officer, Naval Health Sci Educ & Training Command, 75-77, asst chief, Bur Med & Surg, 77-78, cmndg officer, Naval Regional Med Ctr, San Diego, 78-80, cmndg officer, Naval Health Sci Educ & Training Command, 75-78, cmdg officer, Naval Regional Med Ctr, 78-80, Surgeon Gen of the Navy, 80-83; assoc dir Grad Sch Pub Health, San Diego, 83-86; DIR HEALTH SERV, SAN DIEGO COUNTY, 87- *Concurrent Pos:* Assoc prof med, Jefferson Med Col, 64-72; House Deleg Am Med Asn, 72-84. *Honors & Awards:* Borden Award Med Res, 51. *Mem:* Fel Am Col Physicians; Am Col Cardiol (treas, 74-79); Am Heart Asn; Asn Mil Surg US. *Res:* Clinical investigations; cardiopulmonary physiology. *Mailing Add:* 5929 Del Cerro Blvd San Diego CA 08854

COX, JOSEPH ROBERT, b Lafayette, Ind, Mar 15, 34; m 56; c 3. PHYSICS. *Educ:* Harvard Univ, BA, 56; Ind Univ, MS, 57, PhD(physics), 62. *Prof Exp:* Asst prof physics, Univ Miami, 62-64; from asst prof to assoc prof, 64-76, PROF PHYSICS, FLA ATLANTIC UNIV, 76- *Concurrent Pos:* Res assoc, Yale Univ, 67-68. *Mem:* Am Phys Soc. *Res:* Scattering theory. *Mailing Add:* 1601 NW Seventh St Boca Raton FL 33486

COX, JULIUS GRADY, b Ayden, NC, Dec 6, 26; m 46; c 2. INDUSTRIAL ENGINEERING, OPERATIONS RESEARCH. *Educ:* Auburn Univ, BS, 48, MS, 50; Purdue Univ, PhD(indust eng), 64. *Prof Exp:* Instr math, Auburn Univ, 49-51; chief statistician, weapons sect, air proving ground ctr, Eglin AFB, Fla, 51-53, opers analyst, 55-56; head comput, math serv, Vitro Corp, 56-58; assoc prof mech eng, 58-62, head, Dept Indust Eng, 63-66, from asst dean eng to assoc dean eng, 66-68, dean eng, 69-72 & 79-80, PROF INDUST ENG, AUBURN UNIV, 71-, EXEC VPRES, 80- *Concurrent Pos:* Consult, Battelle Mem Inst, 77- *Honors & Awards:* Civil Serv Award, 59. *Mem:* Am Soc Eng Educ; Am Inst Indust Engrs; Opers Res Soc Am; Sigma Xi. *Res:* Application of techniques of mathematics, engineering and computer science to problems in military operations; applied decision theory; applied statistics. *Mailing Add:* 310 Dunstan Hall Auburn Univ Auburn AL 36849

COX, LAWRENCE EDWARD, b Salina, Kans, May 4, 44; div; c 2. SPECTROCHEMISTRY. *Educ:* Kans State Univ, BS, 66; Ind Univ, PhD(chem), 70. *Prof Exp:* Fel chem, Univ Ga, 70-72; STAFF MEM CHEM, LOS ALAMOS NAT LAB, 72- *Mem:* Am Chem Soc; Sigma Xi. *Res:* surface analysis utilizing x-ray photoelectron and auger spectroscopies. *Mailing Add:* 85 Joya Loop Los Alamos NM 87544

COX, LAWRENCE HENRY, b Yonkers, NY, Feb 19, 47; m 70; c 1. STATISTICAL CONFIDENTIALITY, STATISTICAL GRAPHICS. *Educ:* Manhattan Col, BSc, 68; Brown Univ, PhD(math), 73. *Prof Exp:* Asst prof math, Univ Md, 73-74; prin res math, Bur Census, 74-80, asst chief, Statist Res Div, 80-83, sr math statistician, 83-87; DIR, BD MATH SCI, NAT ACAD SCI, 87- *Concurrent Pos:* Adj prof, Univ Md, 79-86; chair, Comt Privacy & Confidentiality, Am Statist Asn, 82-85 & Sect Statist graphics, 86-88; mem bd dir, Am Statist Asn, 87-89 & Nat Comput Graphics Asn, 88-90; vis lectr, Prog in Statist, Comt Pres Statist Soc, 84-86. *Honors & Awards:* Math Medal, Manhattan Col, 68. *Mem:* AAAS; Sigma Xi; fel Am Statist Asn; Nat Comput Graphics Asn; Am Math Soc; Math Asn Am; Opers Res Soc Am. *Res:* Statistical confidentiality and other applications of mathematical theory to statistical problems; statistical graphics applications and policy; science policy. *Mailing Add:* Bd Math Sci Nat Acad Sci 2101 Constitution Ave NW Washington DC 20418

COX, LIONEL AUDLEY, b Winnipeg, Man, Sept 18, 16; m 42; c 3. ORGANIC CHEMISTRY, NUTRITION CHEMISTRY. *Educ:* Univ BC, BA, 41, MA, 43; McGill Univ, PhD(chem), 46, PEng, 75. *Prof Exp:* Teacher sci, Univ Sch, Victoria, BC, 35-40; chief chemist & consult, Sidney Roofing & Paper Co, BC, 41-44; res chemist, Am Viscose Corp, Pa, 46-51, sr res chemist, 51-53; vpres & dir res, Johnson & Johnson, Ltd, Can, 53-61, vpres & dir res & eng, Personal Prod Co, div Johnson & Johnson, NJ, 61-65; dir res, MacMillan Bloedel Ltd, Can, 65-73, dir technol assessment, 73-77; CONSULT, LIONEL A COX, INC, CAN, 77- *Concurrent Pos:* Lectr, Univ BC, 43-44; mem Sci Coun Can; mem bd dir, Educ Inst BC; mem, Nat Res Coun Can; mem bd trustees, Univ Hosp & George Derby Long-Term Care Soc, Vancouver, Can. *Mem:* Fel AAAS; The Chem Soc; Can Res Mgt Asn; Am Chem Soc; fel Chem Inst Can; fel Tech Asn Pulp & Paper Indust; Sigma Xi; emer mem, Indust Res Inst. *Res:* Natural and man-made fibers; pulp and paper; pollution control technology; research and development management; management and administration of research and development engineering; improving interactions between academics, government and industry; consumer and hospital care products. *Mailing Add:* Lionel A Cox Inc 4185 Yuculta Cres Vancouver BC V6N 4A9 Can

COX, MALCOLM, EPITHELIAL TRANSPORT. *Educ:* Harvard Univ, MD, 70. *Prof Exp:* ASSOC PROF MED, UNIV PA, 85-; ASSOC CHIEF RES STAFF, PHILADELPHIA VET ADMIN MED CTR, 86- *Mailing Add:* Vet Admin Ctr University & Woodland Aves Philadelphia PA 19104

COX, MARTHA, b Chappaqua, NY, Oct 23, 08. PHYSICS. *Educ:* Cornell Univ, AB, 29; Bryn Mawr Col, AM, 36, PhD(physics), 42. *Prof Exp:* Asst to res physicist, Taylor Instrument Co, 29-30; lectr physics, Huguenot Univ Col, SAfrica, 31-33; demonstr, Bryn Mawr Col, 34-36; teacher, Shipley Sch, 36-38; instr, Bryn Mawr Col, 39-43; asst prof, Newcomb Col, Tulane Univ, 43-44; res physicist, Lukas Harold Corp, Ind, 44-45; physicist, US Naval Avionics Facil, 45-46, from actg head to head physics div, 46-58, tech consult, Appl Res Dept, 58-73, mgr mat lab & consult div, 73-75; RETIRED. *Mem:* Am Phys Soc; Sigma Xi. *Res:* Design of avionics equipment; guidance and control; mathematical and numerical analysis. *Mailing Add:* Apt 1106 8140 Township Line Rd Indianapolis IN 46260-2423

COX, MARY E, b Detroit, Mich, Nov 11, 37; m 61; c 1. OPTICAL SENSORS, FIBER OPTICS. *Educ:* Albion Col, BA, 59; Univ Mich, MA, 61; Univ Calif, Los Angeles, PhD, 84. *Prof Exp:* Lectr physics, Univ NDak, 62-66; lectr, 66-71, from asst prof to assoc prof, 71-84, assoc dean, Col Arts & Sci, 80-81, actg dean, Col Arts & Sci, 81-82, CHMN DEPT, UNIV MICH, FLINT 87-, PROF PHYS & ENG, 88-; RES SCIENTIST, CRUMP INST MED ENG, UNIV CALIF, LOS ANGELES, 84- *Concurrent Pos:* Prin investr, Off Naval Res, 69-72 & NIH, 75-77; consult, Off Naval Res, 69-74, Alza Corp, 74-79 & Univ Basel, 79-81; fac develop grant, NSF, 77-78; res sci, Crump Inst Med Eng, Univ Calif, Los Angeles, 84-87; A E Mann Found, 87- *Honors & Awards:* Ralph & Marjorie Crump Prize in Biomed Eng, Univ Calif, Los Angeles, 84. *Mem:* Am Asn Physics Teachers; Optical Soc Am; Soc Photo-Optical Instrumentation Engrs. *Res:* Application of coherent optics to biomedical problems; devices for optical engineering; optical materials; fiber-optic devices for biomedical engineering. *Mailing Add:* Dept Phys & Eng Sci Univ Mich-Flint Flint MI 48502-2186

COX, MICHAEL, b Washington, DC, July 30, 41; m 72. GEOPHYSICAL FLUID DUNAMICS. *Educ:* George Washington Univ, BA, 63. *Prof Exp:* Res physicist, Geophys Fluid Dynamics Lab, Environ Servs Admin, Washington, DC, 63-68; RES PHYSICIST, GEOPHYS FLUID DYNAMICS LAB, NAT OCEANIC & ATMOSPHERIC ADMIN, PRINCETON UNIV, 68- *Mem:* Am Geophys Union. *Res:* Large-scale ocean circulation. *Mailing Add:* 19 Bay Harbor St Point Pleasant Beach NJ 08742

COX, MICHAEL MATTHEW, b Wilmington, Del, May 19, 52; m 86; c 2. GENETIC RECOMBINATION, ENZYMOLOGY. *Educ:* Univ Del, BA, 74; Brandeis Univ, PhD(biochem), 80. *Prof Exp:* Postdoctoral fel biochem, Stanford Univ Sch Med, 79-82; asst prof, 83-87, ASSOC PROF, BIOCHEM, DEPT BIOCHEM, UNIV WIS, MADISON, 87- *Concurrent Pos:* Res career develop award, NIH, 84; Dreyfus teacher scholar, 86. *Honors & Awards:* Eli Lilly Award, Am Chem Soc Div Biol Chem, 89. *Mem:* Am Soc Biochem & Molecular Biol; Am Chem Soc; AAAS. *Res:* Enzymology and molecular biology of genetic recombination; protein-DNA interactions; variations in DNA structure; use of site-specific recombination in chromosomal targeting of introduced DNA. *Mailing Add:* Dept Biochem Univ Wis 420 Henry Mall Madison WI 53706

COX, MILTON D, b Indianapolis, Ind, Jan 13, 39; m 66; c 3. FACULTY DEVELOPMENT, MATHEMATICS EDUCATION. *Educ:* DePauw Univ, BA, 61; Ind Univ, MA, 64, PhD(quasi-finite fields), 66. *Prof Exp:* ASSOC PROF MATH, MIAMI UNIV, 66-, ASSOC PROVOST, 83- *Concurrent Pos:* Pres, Ohio sect, Math Asn Am, 86-87. *Mem:* Am Math Soc; Math Asn Am. *Res:* Class field theory; quasi-finite fields. *Mailing Add:* Dept Math & Statist Miami Univ Main Campus Oxford OH 45056

COX, NEIL D, b Lawton, Okla, Aug 6, 32; m 65; c 5. CHEMICAL ENGINEERING, APPLIED STATISTICS. *Educ:* Univ Tex, BSChE, 55; Univ Wis, MSChE, 60, PhD(chem eng), 62. *Prof Exp:* Prod foreman soap mfg, Procter & Gamble Co, 55-56; prof chem eng, Univ Ariz, 62-73; pain scientist, EG&G Idaho, Inc, 74-87; SCI APPLNS INT CORP, 87- *Concurrent Pos:* Ford Found grant eng, 65-66. *Mem:* Am Inst Chem Engrs; Am Nuclear Soc. *Res:* Data analysis; reliability of processes; energy conservation; nuclear waste processing. *Mailing Add:* 3061 Conquista Ct Las Vegas NV 89121-3866

COX, NELSON ANTHONY, b New Orleans, La, Jan 6, 43; m 63; c 3. MICROBIOLOGY. *Educ:* La State Univ, Baton Rouge, BS, 66, MS, 68, PhD(poultry sci), 71. *Prof Exp:* Microbiol consult, Supreme Sugar Refinery, La, 69-70; MICROBIOLOGIST, RICHARD B RUSSELL AGR RES CTR, AGR RES SERV, USDA, 71- *Honors & Awards:* Ralston Purina Res Award, Southern Asn Agr Scientists, 72; Poultry & Egg Inst Am Award, 77. *Mem:* Am Soc Microbiologists; Inst Food Technologists; Poultry Sci Asn; Soc Appl Bact; World's Poultry Sci Asn. *Res:* Destruction of salmonellae on poultry carcasses; development of improved sampling and cultural methods for detection of salmonellae in poultry; microbiological evaluation of immersion versus air chilling of broilers. *Mailing Add:* PO Box 5677 Athens GA 30613

COX, PARKER GRAHAM, b Grosse Pointe Farms, Mich, June 3, 13; m 46; c 3. ENGINEERING. *Educ:* Univ Mich, BSE, 34, MSE, 36. *Prof Exp:* Engr, Zenith Carburetor Div, Bendix Aviation Corp, 36-43, priorities mgr, 40-45, termination mgr, 44-46; owner & engr, Multiple Prod Co, 46-47; engr, Genco Elec Co, 47-48; physicist, Bausch & Lomb Optical Co, 48-56; proj engr, Mishawaka Div, Bendix Corp, 56-58, chief engr, 58-67, mem doc staff, Launch Support Div, 67-74; eng/tech writer, Planning Res Corp, 74-82; ENG & TECH WRITER, APPL ENERGY SYSTS, 82- *Concurrent Pos:* Mem, Nat Aerospace Stand Comt, 62-67. *Res:* Fluid filtration and lubrication of pneumatic equipment. *Mailing Add:* 2755 Las Palmas Titusville FL 32780-5430

COX, PAUL ALAN, b Salt Lake City, Utah, Oct 10, 53. EVOLUTIONARY BIOLOGY, ETHNOBOTANY. *Educ:* Brigham Young Univ, BS, 76; Univ Wales, MSc, 78; Harvard Univ, AM, 78, PhD(biol), 81. *Prof Exp:* Teaching fel biol, Harvard Univ, 71-81; Miller res fel bot, Univ Calif, Berkeley, 81-83; from asst prof to assoc prof, 83-91, PROF BOT, BRIGHAM YOUNG UNIV, 91- *Concurrent Pos:* Presidential young investr, 86-91. *Honors & Awards:* Bowdoin Prize, Harvard Univ, 75. *Mem:* Asn Trop Biol; Bot Ecol Soc; Am Soc Naturalists; Soc Ethnobiol; Econ Bot Soc; Ecol Soc Am. *Res:* Plant breed system evolution pollination biology; tropical ecology ethnobotany. *Mailing Add:* Dept Bot & Range Sci Brigham Young Univ Provo UT 84602

COX, PRENTISS GWENDOLYN, b New Augusta, Miss, June 9, 32; m 53; c 4. DEVELOPMENTAL BIOLOGY. *Educ:* Southern Miss Univ, BS, 57; Univ Miss, MS, 61; Case Western Reserve Univ, PhD(regeneration), 68. *Prof Exp:* Instr sci, Clark Mem Col, 57-64; NIH res fel, Case Western Reserve Univ, 68-69; assoc prof, 69-77, PROF BIOL, MISS COL, 77-, HEAD DEPT, 81- *Concurrent Pos:* Vis prof microbiol, Univ Miss Sch Med, 78-84. *Mem:* AAAS; Sigma Xi; Am Inst Biol Sci; Am Soc Zool; Soc Develop Biol. *Res:* Vertebrate regeneration, especially lizard tail regeneration and in vitro culture of myogeneic cells from the regenerating tail; iron metabolism in heart muscle cultures. *Mailing Add:* Dept Biol Sci Miss Col Clinton MS 39058-4065

COX, RAY, b Donalsonville, Ga, Dec 2, 43; m 65; c 2. BIOCHEMISTRY. *Educ:* Berry Col, BS, 65; Auburn Univ, MS, 67, PhD(biochem), 70. *Prof Exp:* CLIN ASST PROF BIOCHEM, VET ADMIN HOSP & UNIV TENN, 72- *Concurrent Pos:* Fel, Fels Res Inst, Sch Med, Temple Univ, 71-72. *Mem:* Am Asn Cancer Res. *Res:* Biochemistry of chemical carcinogenesis. *Mailing Add:* Dept Biochem Univ Tenn Vet Admin Hosp 1030 Jefferson Ave Memphis TN 38104

COX, RAYMOND H, b Meadville, Pa, Mar 26, 36. MATHEMATICS. *Educ:* Allegheny Col, BS, 58; Univ NC, MA, 61, PhD(math), 63. *Prof Exp:* Asst prof, 63-69, ASSOC PROF MATH, UNIV KY, 69- *Mem:* Am Math Soc; Sigma Xi. *Res:* Hilbert space theory; linear space theory. *Mailing Add:* Dept Math Univ Ky Lexington KY 40506

COX, RICHARD HARVEY, b Oakland, Ky, May 21, 43; m 67; c 1. NUCLEAR MAGNETIC RESONANCE, ORGANIC CHEMISTRY. *Educ:* Univ Western Ky, BS, 63; Univ Ky, PhD(org chem), 66. *Prof Exp:* Res chemist, Northern Regional Lab, USDA, summer 63; res fel chem, Mellon Inst, 67; from asst prof to assoc prof chem, Univ Ga, 68-78; res chemist, Nat Inst Environ Health Sci, 78-81; sr scientist, Philip Morris USA, 81-84, mgr anal res div, 84-88, mgr qual assurance plants, 88-90, MGR FLAVOR TECHNOL DIV, PHILIP MORRIS USA, 91- *Mem:* Am Chem Soc; AAAS. *Res:* Applications of nuclear magnetic resonance in chemistry; conformational analysis; theoretical aspects of nuclear magnetic resonance spectroscopy. *Mailing Add:* Philip Morris USA Res Ctr PO Box 26583 Richmond VA 23261

COX, RICHARD HORTON, b Paia, Hawaii, Oct 10, 20; m 42; c 6. HYDROLOGY & WATER RESOURCES. *Educ:* Calif Inst Technol, BS, 42, MS, 46. *Prof Exp:* Range supvr, Calif Inst Technol, 42-46; civil engr, McBryde Sugar Co, Inc, Hawaii, 46-56; mgr real estate div, Alexander & Baldwin Inc, 56-71, vpres properties group, 71-74, vpres eng, 74-86; ENG CONSULT, 86- *Concurrent Pos:* Mem, Hawaii Comn Water Resource Mgt, 87- *Mem:* Am Soc Civil Engrs; assoc Am Geophys Union. *Res:* Irrigation water resource. *Mailing Add:* 1951 Kakela Dr Honolulu HI 96822

COX, RICHARD T(HRELKELD), physics, for more information see previous edition

COX, ROBERT HAROLD, b Philadelphia, Pa, Sept 10, 37; m 62; c 2. PHYSIOLOGY, BIOENGINEERING. *Educ:* Drexel Inst Technol, BS, 61, MS, 62; Univ Pa, PhD(biomed eng), 67. *Prof Exp:* Assoc physiol, 67-69, asst prof, 69-72, ASSOC PROF PHYSIOL, UNIV PA, 72-, ASSOC PROF BIOMECH, 73-, ASSOC DIR BOCKUS RES INST, 70-, PROF DEPT PHYSIOL. *Concurrent Pos:* Nat Heart & Lung Inst grant, Bockus Res Inst, Univ Pa, 75-78. *Mem:* Inst Elec & Electronics Engrs; AAAS; Sigma Xi; Am Physiol Soc. *Res:* Vascular smooth muscle mechanics; arterial wall physiology; hypertension; carotid sinus reflex. *Mailing Add:* Dept Physiol Univ Pa 415 S 19th St Philadelphia PA 19146

COX, ROBERT SAYRE, JR, b San Francisco, Calif, Jan 30, 25; c 7. PATHOLOGY. *Educ:* Stanford Univ, BS, 46, MS, 48, PhD(chem), 52; Univ Chicago, MD, 52. *Prof Exp:* Chief lab serv, US Army, Rodrigues US Army Hosp, 59-62; dir labs path, Santa Clare Valley Med Ctr, 62-80; PROF & CHMN PATH, CREIGHTON UNIV, ST JOSEPH HOSP, 80- *Mem:* Am Med Asn; Col Am Pathologists; Am Soc Clin Pathologists; Am Asn Blood Banks; AAAS. *Res:* Management and delivery of laboratory services. *Mailing Add:* 601 N 30th St Omaha NE 68131

COX, RODY POWELL, b New Brighton, Pa, June 24, 26; m 53; c 3. MEDICAL GENETICS. *Educ:* Univ Pa, MD, 52. *Prof Exp:* Asst clin instr med, Univ Mich, 53-54; instr, Univ Pa, 54-56, assoc, 56-68, asst prof med & res med, 58-60; res assoc genetics, Glasgow Univ, 60-61; from asst prof to assoc prof med, NY Univ, 61-71, assoc prof pharmacol, 70-71, prof med & pharmacol, 71-79; prof med, vchmn & chief med serv, Vet Admin Med Ctr, Case Western Univ, 79-88; dean, 88-89, PROF INTERNAL MED, UNIV TEX SOUTHWESTERN MED CTR, 89- *Concurrent Pos:* Fel, Arthritis & Rheumatism Found, 57-59; USPHS res fel, 60-61; career scientist, Health Res Coun, NY, 61-; dir summer res inst, Will Rogers Hosp, 62-66; dir coun, Asn Career Scientists, 66-68; dir med scientist training prog, NY Univ, 67-79; mem metab study sect, NIH, 70-73, chmn genetics study sect, 78-81; mem panel clin sci, Nat Res Coun, 76-, dir, Div Human Genetics, NY Univ, 73-79; co-dir, med scientist training prog, Case Western Reserve Univ, 80-88. *Mem:* Am Soc Clin Invest; Am Soc Human Genetics; Asn Am Physicians; Am Col Physicians; Am Clin Climat Asn. *Res:* Biochemical genetics; somatic cell genetics; mammalian cell regulatory mechanism; mechanisms of hormone action; pharmacology; tissue culture. *Mailing Add:* Univ Tex Southwestern Med Ctr 5323 Harry Hines Blvd Dallas TX 75235-8889

COX, RONALD BAKER, b Chattanooga, Tenn, Sept 27, 43. STRATEGIC PLANNING, ENGINEERING MANAGEMENT. *Educ:* Univ Tenn, BS, 65, MS, 68; Rice Univ, PhD(mech eng), 70; Vanderbilt Univ, MBA, 80. *Prof Exp:* Engr, Westinghouse Elec Corp, 65-66, Dupont, 66-67; dir eng, Indust Boiler corp, 67-70; chief opers officer, Prog Res, US Dept Transp, Washington, DC, 74-77; asst prof mech eng, 70-74, assoc prof mech eng, 77-79, DEAN ENG, UNIV TENN, CHATTANOOGA, 79- *Concurrent Pos:* Consult, Int Eng & Mfg, 70-; trustee, Tenn Eng Found, 87-; dir, Integrated Voice Solutions Inc, 87-, Univ Tenn Res Corp, 89- *Honors & Awards:* Serv Award, Am Soc Mech Engrs, 82. *Mem:* Sigma Xi; Nat Soc Prof Engrs; Am Soc Mech Engrs; Am Soc Engrs in Educ; Am Soc Eng Mgt; Am Soc Safety Engrs. *Res:* Energy conversion; thermal system design; vehicular accident reconstruction; engineering and technical management; strategic planning and technology policy as related to corporate and national issues. *Mailing Add:* 907 Glamis Circle Signal Mountain TN 37377

COX, STEPHEN KENT, b Galesburg, Ill, Sept 2, 40; m 61; c 3. ATMOSPHERIC PHYSICS. *Educ:* Knox Col, Ill, BA, 62; Univ Wis, Madison, MS, 64, PhD(meteorol), 67. *Prof Exp:* Res meteorologist, Atmospheric Physics & Chem Lab, Environ Sci Serv Admin, 64-66; scientist, Space Sci & Eng Ctr, Univ Wis, 66-69; from asst prof to assoc prof, 69-76, PROF ATMOSPHERIC SCI, COLO STATE UNIV, 72-, DEPT HEAD, 89- *Concurrent Pos:* Scientist, dept meteorol, Univ Wis, 67-69,; chmn flight facil adv panel, Nat Ctr Atmospheric Res, 70-73; Global Atmospheric Res Prog-Atlantic Trop Exp Radiation Subprog scientist, 74, mem adv panel, Nat Acad Sci Global Atmospheric Res Prog-Atlantic Trop Exp & Nat Acad Sci Monsoon Exp, 74-80, US Global Atmospheric Res Prog-Atlantic Trop Exp Radiation Subprog coordr, 75-79; chmn, NASA fire Sci Exp Team, 84-, adv panel, Res Aviation Flight Fac, 87- *Mem:* fel Am Meteorol Soc. *Res:* Atmospheric heat budget; radiation parameterization for numerical models; meteorological field experiments; radiative transfer. *Mailing Add:* Dept Atmospheric Sci Colo State Univ Ft Collins CO 80523

COX, THOMAS C, b Missoula, Mont, Nov 1, 50; m 78; c 2. CYTOLOGY, NEUROSCIENCES. *Educ:* Ariz State Univ, PhD(zool), 79. *Prof Exp:* Asst prof, 82-87, ASSOC PROF PHYSIOL, SCH MED, SOUTHERN ILL UNIV, 87- *Mem:* Am Physiol Soc; Biophys Soc. *Res:* Physiology and biophysics of ion transport; transepithelial sodium and potassium transport; development of amiloride sensitive sodium channels in epithelia; calcium channel in mammalian sperm cells. *Mailing Add:* Dept Physiol Sch Med Southern Ill Univ Carbondale IL 62901

COX, WILLIAM EDWARD, b Pulaski, Va, Feb 18, 44; m 65; c 1. WATER RESOURCES MANAGEMENT. *Educ:* Va Polytech Inst & State Univ, BS, 66, MS, 68, PhD(civil eng), 76. *Prof Exp:* Asst prof, Civil Eng Tech, Va Commonwealth Univ, 68-72; res assoc, Va Water Resources Res Ctr, 72-77, from asst prof to assoc prof, 77-86, PROF, CIVIL ENG DEPT, VA POLYTECH INST & STATE UNIV, 86- *Concurrent Pos:* Prin investr res grants, Off Water Res & Technol, 75-80, Va Water Resources Res Ctr, 75, 76, 83, 84, 85 & 88, Va Dept Conserv & Hist Resources, 85, US Geol Surv, 86 & 87, US Army Corps Engrs, Inst Water Resources, 91; mem, UN Educ Sci & Cult Org Working Group Importance Water Resources Socio-Econ Develop, 81-89; consult, GKY & Assocs, Inc, 81, Camp Dresser & McKee, 85, City of Newport News, Va, 89-91; lectr, Ferrum Col, 81, 83, 85, 87. *Mem:* Am Soc Civil Engrs; Am Water Resources Asn; Int Water Resources Asn; Sigma Xi; Water Pollution Control Fedn. *Res:* Water resources planning and management, with emphasis on public policy and institutional arrangements. *Mailing Add:* Civil Eng Dept Va Polytech Inst & State Univ Blacksburg VA 24061

COXETER, HAROLD SCOTT MACDONALD, b London, Eng, Feb 9, 07; m 36; c 2. MATHEMATICS. *Educ:* Cambridge Univ, BA, 29, PhD(geom), 31. *Hon Degrees:* LLD, Univ Alta, 57, Trent Univ, 73, Univ Toronto, 79; DMath, Waterloo Univ, 69; DSc, Acadia Univ, 71, Carleton, 84, McMaster, 88. *Prof Exp:* Fel, Trinity Col, Cambridge Univ, 31-35; from asst prof to assoc prof, 36-48, PROF MATH, UNIV TORONTO, 48- *Concurrent Pos:* Rockefeller Found fel, Princeton Univ, 32-33, Procter fel, 34-35; vis prof, Univ Notre Dame, 47, Columbia Univ, 49, Dartmouth Col, 64, Fla Atlantic Univ, 65, Univ Amsterdam, 66, Univ Edinburgh, 67, Univ E Anglia, 68, Australian Nat Univ, 69, Univ Sussex, 72 & Univ Warwick, 76, Calif Inst Technol, 77 & Univ Bologna, 78; ed in chief, Can J Math, 49-58; pres, Int Cong Mathematicians, 74. *Honors & Awards:* Tory Medal, 49. *Mem:* Am Math Soc; Math Asn Am; fel Royal Soc; fel Royal Soc Can; for mem Royal Neth Acad Arts & Sci; foreign hon mem Am Acad Arts & Sci. *Res:* Regular and semi-regular polytopes; abstract groups; non-Euclidean geometry; configurations. *Mailing Add:* 67 Roxborough Dr Toronto ON M4W 1X2 Can

COXON, JOHN ANTHONY, b Blackpool, Lancashire, Eng, July 15, 43; Brit & Can citizen; m 77; c 2. HIGH RESOLUTION LASER SPECTROSCOPY, ELECTRONIC STATES OF DIATOMIC MOLECULES. *Educ:* Cambridge Univ, BA, 64; Univ E Anglia, MSc, 65, PhD(phys chem), 67. *Prof Exp:* Fel chem, Royal Comn Exhib 1851, Queen Mary Col Univ, 67-69, Imp Chem Indust, 69-71, Sci Res Coun, 71-73; sr scientist combustion, Brit Gas Corp, 73-74; Killam res prof, 74-79, assoc prof, 79-80, PROF CHEM, DALHOUSIE UNIV, HALIFAX, NS, 80- *Concurrent Pos:* Vis fel, Australian Nat Univ, 81. *Mem:* Can Inst Chem; Can Asn Physicists. *Res:* Electronic spectra of small molecules by high resolution laser spectroscopy and chemiluminescent emission spectroscopy; least-squares fitting of spectral data to potential energy functions with account of Born-Oppenheimer breakdown. *Mailing Add:* Dept Chem Dalhousie Univ Halifax NS B3H 4J3 Can

COY, DANIEL CHARLES, b Ames, Iowa, July 25, 63. THERMODYNAMICS, COMPUTER GRAPHICS. *Educ:* Iowa State Univ, BS, 86. *Prof Exp:* Co-op student res & develop & prod, Dow Chem Co, 83-85; prod engr mfg, Proctor & Gamble, 86-87; RES ASST THERMODYN, CHEM ENG DEPT, IOWA STATE UNIV, 88- *Mem:* Am Inst Chem Engrs. *Res:* Computer-based visualization of the thermodynamic properties that govern phase equilibrium and thermodynamic stability in multiphase, multicomponent fluid systems; development of computational tools for general use in scientific visualization. *Mailing Add:* Dartmoor Rd Ames IA 50010

COY, DAVID HOWARD, b Manchester, Eng, Sept 15, 44; m 66; c 1. ENDOCRINOLOGY. *Educ:* Univ Manchester, BSc, 66, PhD(chem), 69. *Prof Exp:* Res assoc chem, Univ Toledo, 69-70; teaching assoc biochem, Med Col Ohio, 70-72; from asst prof to assoc prof, 72-81, PROF MED, SCH MED, TULANE UNIV, 81-, ADJ PROF BIOCHEM, 83- *Concurrent Pos:* Res assoc, Vet Admin Hosp, New Orleans, 72-82; assoc ed, Peptides, Peptide & Protein Rev, Brain Res Bull. *Mem:* Endocrine Soc; AAAS; Am Chem Soc; NY Acad Sci; Southern Soc Clin Invest. *Res:* Chemistry and biological properties of neuro-gastrointestinal peptide hormones and their analogs. *Mailing Add:* Dept Med Tulane Med Ctr 1430 Tulane Ave New Orleans LA 70112-2699

COY, RICHARD EUGENE, b New Kensington, Pa, Oct 28, 25; m 57; c 2. DENTISTRY. *Educ:* Univ Pittsburgh, BS, 49, DDS, 51, MS, 59. *Prof Exp:* Assoc prof prosthodontics, Univ Pittsburgh, 60-70; PROF, 70-87, asst dean clin affairs, 75-87, EMER PROF PROSTHODONTICS, SCH DENT MED, SOUTHERN ILL UNIV, EDWARDSVILLE, 88- *Concurrent Pos:* Consult, Vet Admin Hosps, 63-; assoc prof, Med Ctr, St Louis Univ, 71-; chmn continuing educ, Am Asn Dent Schs, 74, chmn prosthodont, 76. *Mem:* Fel Am Col Dentists; fel Am Col Prosthodontists; Int Asn Dent Res; Int Asn Dento-Facial Abnormalities; Am Equilibration Soc (pres, 86); hon fel Int Col Cranic Mandimolar Orthop; master fel Am Acad Implant Prosthodontics. *Res:* Study of craniofacial abnormalities; study of pain involved in temporomandibular dysfunction syndrome. *Mailing Add:* 17 Crestwood Dr Edwardsville IL 62025

COYE, ROBERT DUDLEY, b Los Angeles, Calif, Dec 17, 24; m 46; c 3. ANATOMIC PATHOLOGY. *Educ:* Williams Col, BA, 48; Univ Rochester, MD, 52. *Prof Exp:* Resident path, Univ Rochester, 52-55; from asst prof to prof, Univ Wis-Madison, 55-72; dean, Sch Med, Wayne State Univ, 72-80, prof path, 80-; RETIRED. *Res:* Pathology of kidney. *Mailing Add:* RR 2 Box 85 Lawrence Rd Suttons Bay MI 49682

COYER, JAMES A, b Eau Claire, Wis. MOLECULAR BIOLOGY. *Educ:* Univ Wis, BS, 70; Univ Southern Calif, PhD(marine biol), 79. *Prof Exp:* Oceanogr, Ocean Sci Dept, Interstate Electronics Corp, 80-81; chair & asst prof biol, Div Sci & Math, Mary Mount Col, 81-86; postdoctoral fel, Dept Microbiol, State Univ NY, 86-90; RES ASSOC, DEPT MOLECULAR GENETICS & CELL BIOL, UNIV CHICAGO, 90- *Concurrent Pos:* Lectr, Dept Ecol & Evolutionary Biol, Univ Calif, Irvine, 80; consult, Calif Dept Fish & Game, 80-81; vis asst prof, Div Biol Sci, Cornell Univ, 81- *Mem:* Western Soc Naturalists. *Res:* Environmental control of N2 fixing and luciferase gene expression in cultural and natural populations of marine bacteria and single-cell detection of nitrogenase and luciferase mRNA's using in situ hybridizations; role of sea urchins in determining community structure of California keep beds. *Mailing Add:* Hopkins Marine Sta Univ Chicago Pacific Grove CA 93950

COYER, PHILIP EXTON, b Harrisburg, Pa, May 20, 48; m 77; c 2. PHYSIOLOGY, NEUROPHYSIOLOGY. *Educ:* Franklin & Marshall Col, AB, 70; Col William & Mary, MA, 72; Univ Mass, Amherst, PhD(zool), 77. *Prof Exp:* Teaching asst biol, Col William & Mary, 70-71; teaching asst dept zool, Univ Mass, 72-76; res asst zool & neurophysiol, Univ Mass, 76-77; res assoc, Med Ctr, Univ Ala, 77-81, res asst prof neurol, 81-83; asst dir neurol lab, Pa Hosp, 83-87; asst mgr, 87-91, MGR SCI & PROF AFFAIRS, HOECHST-ROUSSEL PHARMACEUTICALS, 91- *Concurrent Pos:* NIH traineeship neurophysiol & cerebral metab dept neurol, Univ Ala, 78-; Grass fel neurophysiol, Marine Biol Lab, Woods Hole, Mass; grad fel, Col William & Mary. *Mem:* Soc Neurosci; Sigma Xi; Am Soc Zoologists; Int Soc Oxygen Transp Tissue; Am Physiol Soc; Am Col Clin Pharmacol. *Res:* Neurophysiology; neuronal basis underlying rhythmic respiratory patterns; single cell analysis of oxygen effects on neural functioning; oxygen transport in the nervous system; cerebral metabolism; medical liaison. *Mailing Add:* 558 Hickory Ridge Ct St Louis MO 63131

COYIER, DUANE L, b Aurora, Ill, Mar 14, 26; m 47; c 3. HORTICULTURE. *Educ:* Univ Wis, BS, 50, PhD(plant path), 61. *Prof Exp:* Plant pathologist, Tree Fruit Dis Invests, 61-73, PLANT PATHOLOGIST, FUNGUS & BACT DIS TREE FRUITS, FOLIAR DIS ORNAMENTAL PLANTS, SCI EDUC ADMIN-FED RES, USDA, 73- *Mem:* Am Phytopath Soc; Sigma Xi; Coun Agr Sci & Technol; Int Plant Propagators Soc. *Res:* Erysiphacae which attack ornamental plants with emphasis on rose powdery mildew. *Mailing Add:* USDA Agr Res Serv Western Region 3420 SW Orchard St Corvallis OR 97330

COYKENDALL, ALAN LITTLEFIELD, b Hartford, Conn, Jan 11, 37; m 59; c 1. ORAL MICROBIOLOGY. *Educ:* Bates Col, BS, 59; Tufts Univ, DMD, 63; George Washington Univ, MS, 70. *Prof Exp:* Fel microbiol, US Naval Dent Res Inst, 64-65, microbiologist, US Naval Med Res Inst, 67-71, microbiologist, US Naval Dent Res Inst, 71-72; res assoc dent, 72-75, clin investr, Vet Admin Hosp, Newington, 75-78; ASST PROF MICROBIOL, UNIV CONN SCH DENT MED, 78- *Mem:* AAAS; Int Asn Dent Res; Am Soc Microbiol; Sigma Xi. *Res:* Tooth replantation; taxonomy of oral bacteria; retention of sugar in the mouth; nucleic acids of cariogenic streptococci; guanine-cytosine content and homologies. *Mailing Add:* 15 Whispering Rod Rd Unionville CT 06085

COYLE, BERNARD ANDREW, b Pukekohe, NZ, May 2, 34; m 60; c 5. BIOCHEMISTRY, INORGANIC CHEMISTRY. *Educ:* Univ NZ, BSc, 55, MSc, 56; Northwestern Univ, PhD(inorg chem), 69. *Prof Exp:* Instr chem, City Col San Francisco, 60-66; asst prof, NCent Col, Ill, 69-71; prof chem, City Col San Francisco, 71-80; DEAN ACAD AFFAIRS, PALMER COL WEST, 80- *Concurrent Pos:* Guest scientist, Argonne Nat Lab, 67-73; vis chemist, Brookhaven Nat Lab, 70-71; lectr chem, Univ San Francisco, 74-77 & San Francisco State Univ, 74-81. *Mem:* Am Chem Soc; Pattern Recognition Soc; Soc Magnetic Resonance Imaging. *Res:* Computers in chemical education; educational administration; crystallography. *Mailing Add:* Palmer Col W 1095 Dunford Way Sunnyvale CA 94087-3765

COYLE, CATHERINE LOUISE, b New York, NY, May 16, 52; c 1. BIOINORGANIC CHEMISTRY, CATALYSIS. *Educ:* Hunter Col, City Univ New York, BS, 74; Calif Inst Technol, PhD(chem), 78. *Prof Exp:* Fel chem, Stanford Univ, 77-79; res chemist, 79-80, sr chemist, 80-83, STAFF CHEMIST, EXXON RES & ENGR, 83-, PROJ LEADER, 90- *Concurrent Pos:* Res award, Sigma Xi, 77; adj asst prof, Hunter Col, City Univ New York, 81-82. *Mem:* Am Chem Soc. *Res:* Bioinorganic chemistry with specific emphasis on the kinetics of metalloprotein reactions; preparation and characterization of transition metal-sulfur synthetic analogs of the active sties in proteins and catalysts. *Mailing Add:* Exxon Res & Eng Rte 22 East Annandale NJ 08801

COYLE, EDWARD JOHN, b Bryn Mawr, Penn, Apr 22, 56. ELECTRIC ENGINEERING. *Educ:* Univ Del, BS, 77; Princeton Univ, MS, 80, MA, 80, PhD(elec eng), 82. *Prof Exp:* Asst prof elec eng, 82-86, ASSOC PROF ELEC ENG, PURDUE UNIV, 86-; PRES, INFO SCI & SYST INC, 85- *Concurrent Pos:* Consult, Delco Electronics, 85-86, Gen Dynamics, 86-; prin investr, Res Grant, NSF, 82- *Mem:* Inst Elec & Electronics Engrs; Asn Comput Mach; Oper Res Soc Am. *Res:* Performance analysis and design of computer communication networks; modeling and analysis of the behavior of queveing networks; design and analysis of nonlinear filters; design of optional adaptive and nonadaptive filters under the mean absolute error criterion. *Mailing Add:* Sch Elec Eng Purdue Univ West Lafayette IN 47907

COYLE, FREDERICK ALEXANDER, b Port Jefferson, NY, May 31, 42; c 2. SYSTEMATICS, BEHAVIOR. *Educ:* Col Wooster, BA, 64; Harvard Univ, MA, 66, PhD(biol), 70. *Prof Exp:* PROF BIOL, WESTERN CAROLINA UNIV, 69- *Concurrent Pos:* Res assoc, Entom Dept, Am Mus Natural Hist & assoc invert zool, Mus Comp Zool, Harvard Univ. *Mem:* Am Arachnol Soc; Brit Arachnol Soc. *Res:* Systematics, behavior and ecology of spiders, especially mygalomorph spiders. *Mailing Add:* Dept Biol Western Carolina Univ Cullowhee NC 28723

COYLE, HARRY MICHAEL, b Johnstown, Pa, Jan 7, 27; m 55; c 4. CIVIL ENGINEERING. *Educ:* US Mil Acad, BS, 50; Mass Inst Technol, MS, 56; Univ Tex, Austin, PhD(civil eng), 65. *Prof Exp:* Instr civil eng & res engr, Univ Tex, 62-65; asst prof civil eng & res engr, 65-68, assoc prof, 68-72, PROF CIVIL ENG, TEX A&M UNIV, 72-, HEAD GEOENG GROUP & GEOTECH DIV, 73- *Concurrent Pos:* Consult, Exxon, Shell & McClelland Eng; pres, Tex Sect, Am Soc Civil Engrs, 82. *Mem:* Nat Soc Prof Engrs; Am Soc Civil Engrs; Am Soc Testing & Mat. *Res:* Soil-structure interaction; soil mechanics and foundation engineering; pile-soil interaction. *Mailing Add:* Dept Civil Eng Tex A&M Univ College Station TX 77843

COYLE, JOSEPH T, b Chicago, Ill, Oct 9, 43; m 68; c 3. PSYCHIATRY. *Educ:* Col Holy Cross, AB, 65; Johns Hopkins Univ, MD, 69; Am Bd Psychiat & Neurol, cert, 80. *Prof Exp:* Intern pediat, Johns Hopkins Hosp, 69-70, resident psychiat, 73-76; res assoc, Lab Clin Sci, NIMH, NIH, 70-73; asst prof pharmacol, Sch Med, Johns Hopkins Univ, 74-76, asst prof pharmacol & psychiat, 76-78, assoc prof, 78-80, prof neurosci, psychiat & pharmacol, 80-91; EBEN S DRAPER CHAIR, CONSOL DEPT PSYCHIAT & SR PSYCHIATRIST-IN-CHIEF, HARVARD MED SCH, 91- *Concurrent Pos:* Henry Strong Denison res scholar, 68-69; co-dir, Outpatient Pharmacother Clin, Henry Phipps Psychiat Clin, Johns Hopkins Hosp, 77-82; res career develop award, NIMH, NIH, 77078, chmn, Cellular Neurobiol & Psychopharmacol Res Rev Comt, 85-89, Nat Adv Ment Health Coun, 90-; dir, Div Child Psychiat, prof psychiat, neurosci, pharmacol & pediat, Sch Med, Johns Hopkins Univ, 82-91, distinguished serv prof child psychiat, 85-91; mem, sci adv bd, Hereditary Dis Found, 82-; Javits Neurosci investr award, Nat Adv Neurol & Commun Dis & Stroke Coun, NIH, 85-91; mem coun res, Am Psychiat Asn, 86-; mem sci adv bd, John F Merck Found, 90-; prof neurosci, Harvard Med Sch, 91- *Honors & Awards:* John Jacob Abel Award, Am Soc Pharmacol & Exp Therapeut, 79; Sato Int Mem Award, Japanese Pharmaceut Soc, 79; Daniel Efron Award, Am Col Neuropsychopharmacol, 82; Harold C Voris Mem Lectr in Neurosci, Univ Ill, 83; Grass Lectr, Univ Mo, 83, Univ SC, 87; Tarbox Distinguished Neuroscientist Lectr, Tex A&M Univ, 87; Halbert Robinson Distinguished Lectr, Univ NC, 89; Gold Medal Award, Soc Biol Psychiat, 91. *Mem:* Inst Med-Nat Acad Sci; Am Soc Pharmacol & Exp Therapeut; Soc Neurosci; Am Soc Neurochem; Int Soc Develop Psychobiol; Am Psychiat Asn; Sigma Xi; Am Col Neuropsychopharmacol; Psychiat Res Soc. *Res:* Neuropsychopharmacology; developmental neurobiology; biochemistry of the synthesis, storage and catabolism of neurotransmitters; author or co-author of 4 books and over 400 publications. *Mailing Add:* Sch Med Johns Hopkins Univ 500 Overhill Rd Baltimore MD 21210

COYLE, MARIE BRIDGET, b Chicago, Ill, May 13, 35. MEDICAL MICROBIOLOGY, MICROBIAL GENETICS. *Educ:* St Louis Univ, MS, 63; Kans State Univ, PhD(genetics), 65; Am Bd Med Microbiol, dipl, 77. *Prof Exp:* Instr sci, Columbus Hosp Sch Nursing, Chicago, 57-59; fel microbiol, Univ Chicago, 64-67, res assoc molecular genetics, 67-70; instr microbiol, Med Ctr, Univ Ill, Chicago Circle, 70-71; fel microbiol, Temple Univ, 71-73; asst prof, 73-80, ASSOC PROF LAB MED & MICROBIOL, SCH MED, UNIV WASH, 80- *Mem:* Am Soc Microbiol; Sigma Xi; Am Asn Univ Professors; Acad Clin Lab Physicians & Scientists; fel Am Acad Microbiol. *Res:* Mitotic recombination in Neurospora crassa; DNA repair mechanisms in mammalian cells after treatment with alkylating agents; genetics of bacterial virulence; antibiotic resistance and susceptibility testing; molecular epidemiology of pathogenic bacteria. *Mailing Add:* Dept Lab Med & Microbiol Sch Med Univ Wash Harborview Med Ctr Seattle WA 98104-2499

COYLE, PETER, b Hanover, NH, Mar 4, 39; m 67. CEREBRAL COLLATERAL CIRCULATION, NEUROANATOMY. *Educ:* Univ Vt, BA, 62; Univ Mich, MS, 64, PhD(anat), 67. *Prof Exp:* From instr to assoc prof, 67-88, PROF ANAT, MED SCH, UNIV MICH, ANN ARBOR, 88- *Concurrent Pos:* Rackham fac res grant, Med Sch, Univ Mich, Ann Arbor, 68-70, USPHS grants, 70-74 & 81-88; Mich Heart Asn grant, 77-78 & 79-81; Univ Mich Phoenix Proj grant, 77-78; Nat Res Serv Awards sr fel, 84-85; fel, High Blood Pressure coun. *Mem:* AAAS; Am Asn Anatomists; Neurosci Soc; Sigma Xi. *Res:* Experimental and descriptive studies of the cerebrovasculature of normotensive and genetically stroke-prone rats. *Mailing Add:* Dept Anat 2790 Danbury Lane Ann Arbor MI 48103

COYLE, THOMAS DAVIDSON, b Glen Cove, NY, Sept 25, 31; m 54. INORGANIC CHEMISTRY, MATERIALS SCIENCE. *Educ:* Univ Rochester, BS, 52; Harvard Univ, AM, 59, PhD(chem), 61. *Prof Exp:* NATO fel, 61-62; chemist, 62-64, chief inorg chem sect, 64-78, chief chem stability & corrosion div, 78-83, chief inorg mat div, 83-85, RES CHEMIST, NAT BUR STANDARDS, 85- *Mem:* AAAS; Am Chem Soc; Royal Soc Chem. *Res:* Inorganic and organometallic chemistry; boron chemistry; nuclear magnetic resonance and applications to inorganic chemistry; coordination chemistry of main group elements; inorganic halides; stability of inorganic materials; low-temperature synthesis of ceramic materials. *Mailing Add:* 11400 Game Preserve Gaithersburg MD 20878

COYNE, DERMOT P, b Dublin, Ireland, July 4, 29; US citizen; m 57; c 6. PLANT BREEDING. *Educ:* Univ Col, Dublin, BAgrSc, 53, MAgrSc, 54; Cornell Univ, PhD(plant breeding), 58. *Hon Degrees:* DSc, Nat Univ Ireland, 81. *Prof Exp:* Asst mgr agr develop, Campbell Soups, Ltd, Eng, 58-60; from asst prof to prof, 61-86, GEORGE HOLMES REGENTS PROF HORT, UNIV NEBR, LINCOLN, 86- *Concurrent Pos:* Chmn tech adv comt, Int Ctr Trop Agr, Cali, Colombia, 74-79; assoc ed, 74-78, Am Soc Hort Sci, vpres, Res Div, 79-80, pres, 84-85, chmn bd dir, 85-86; chmn, Bean Improv Coop, 68-77. *Honors & Awards:* Nat Canner's Asn Award, 67; Asgrow Award, 74; Campbell Award, 75; Bean Improv Coop Meritorious Award, 75; Marion W Meadows Award, Am Soc Hort Sci, 76 & 80; Outstanding Scientist Award, Sigma Xi, 88. *Mem:* Fel Am Soc Hort Sci; fel AAAS; Am Genetic Asn; Crop Sci Soc; Sigma Xi; Am Phytopath Soc; Am Asn Univ Professors; Am Soc Agron. *Res:* Germplasm identification; breeding and genetics of resistance to bacterial, rust, BCMV and white mold pathogens in beans; genetics and breeding investigation in the following areas in beans resistance to iron induced leaf chlorosis, photoperiodism, adaptation, plant architecture, physiological and morpho genetical components of yields and seed quality; breeding for fruit quality in winter squash. *Mailing Add:* Dept Hort Univ Nebr Lincoln NE 68503

COYNE, DONALD GERALD, b Hutchinson, Kans, Oct 28, 36; m 59; c 2. HIGH ENERGY PHYSICS. *Educ:* Univ Kans, BS, 58; Calif Inst Technol, PhD(physics), 67. *Prof Exp:* Teaching asst physics, Calif Inst Technol, 58-59, res asst, 59-66; fel, Lawrence Radiation Lab, 66-70; asst prof, Princeton Univ, 70-76, res physicist, 76-80, sr res physicist, 81-; AT STANFORD LINEAR ACCELERATOR CTR, CRYSTAL BALL GROUP. *Concurrent Pos:* Mem positron-electron proj policy comt, Stanford Univ/Univ Calif, 75-78; chmn, Stanford Linear Accelerator Ctr/Lawrence Berkeley Lab Orgn, 78-79. *Mem:* Am Phys Soc. *Res:* Experimental high energy particle physics; apparatus and analysis for experiments using electron/positron colliding beam machines. *Mailing Add:* 8717 Empire Grade Bonny Doon CA 95060

COYNE, EDWARD JAMES, JR, b Tacoma, Wash, Sept 27, 53; m 76; c 3. COMPOSITES, MATERIAL BEHAVIOR. *Educ:* Vanderbilt Univ, BE, 75; Ga Inst Technol, MS, 77, PhD(metall), 79; Ga State Univ, MBA, 83. *Prof Exp:* Res asst, Ga Inst Technol, 75-79; from scientist assoc to scientist, 79-87, RES GROUP ENGR, LOCKHEED AERONAUT SYSTS CO, GA, 87- *Mem:* Am Inst Aeronaut & Astronaut; Am Inst Mining, Metall & Petrol Engrs; Am Soc Metals; Soc Advan Mat & Process Eng; Southeastern Electron Micros Soc. *Res:* Development and assessment of metallic matrix and organic matrix materials for aerospace applications; ductility, fatigue resistance and fracture toughness of metals; advanced aluminum alloys, primarily aluminum-lithium alloys. *Mailing Add:* 4980 Carriage Lakes Dr Roswell GA 30075

COYNE, GEORGE VINCENT, b Baltimore, Md, Jan 19, 33. ASTRONOMY. *Educ:* Fordham Univ, AB, 57, PhilosL, 58; Georgetown Univ, PhD(astron), 62. *Prof Exp:* Res assoc astron & investr NASA grant, Res Inst Natural Sci, Woodstock Col, 63-70; asst prof, 70-76, assoc dir observ, 77-78, SR RES FEL, LUNAR & PLANETARY LAB & LECTR, DEPT ASTRON, UNIV ARIZ, 76-, ACTG DIR & HEAD OBSERV, 79; DIR, VATICAN OBSERV, 78- *Concurrent Pos:* Asst astronr, Vatican Observ, Italy, 70- *Mem:* Int Astron Union; Am Astron Soc; Astron Soc Pac; Am Phys Soc; Optical Soc Am. *Res:* Evolution in young stellar associations; polarimetry; interstellar material; stars with extended atmospheres. *Mailing Add:* Steward Observ Univ Ariz Tucson AZ 85721

COYNE, JAMES E, b Springfield, Mass, Sept 25, 25. METALLURGY, PLASTIC DEFORMATION METALS. *Educ:* Univ Notre Dame, BS, 53; Rensselaer Polytech Inst, 57. *Prof Exp:* VPRES & GEN MGR, EASTERN DIV, WYMAN-GORDON CO, 90- *Concurrent Pos:* Pres, Forging Indust Educ Res Found, 81-88. *Mem:* Fel Am Soc Metals; Am Inst Mining Metall & Petrol Engrs. *Mailing Add:* Wyman-Gordon Co 105 Madison St Worcester MA 01613

COYNE, MARY DOWNEY, b Lynn, Mass, Jan 17, 38; m 61; c 2. INTRACELLULAR MESSENGERS, ELECTROPHYSIOLOGY. *Educ:* Emmanuel Col, Mass, AB, 59; Wellesley Col, MA, 61; Univ Va, PhD(physiol), 64. *Prof Exp:* Instr physiol, Sch Med, Univ Va, 66-67, asst prof, 67-68; asst prof physiol, La State Univ Med Ctr, New Orleans, 68-70; asst prof, 70-74, chmn dept, 75-78, 82-84, assoc prof, 74-80, PROF BIOL SCI, WELLESLEY COL, 80- *Concurrent Pos:* USPHS fel, 64-66, res grant, 67-72, res corp grant, 80-82; vis lectr pharmacol, Harvard Med Sch, 78-79; NSF Fac Opportunity Award biochem, Brandeis Univ, 85-86; area grants, NIH Endocrine Study Sect, 89- *Mem:* Endocrine Soc; AAAS; Sigma Xi. *Res:* Cellular endocrinology of adrenal cortex; intracellular messengers; ion channels. *Mailing Add:* Dept Biol Sci Wellesley Col Wellesley MA 02181

COYNE, PATRICK IVAN, b Wichita, Kans, Feb 26, 44; m 64; c 2. PHYSIOLOGICAL ECOLOGY. *Educ:* Kans State Univ, BS, 66; Utah State Univ, PhD(range sci), 70. *Prof Exp:* Plant physiologist arctic tundra, US Army Cold Regions Res & Eng Lab, 70-72; asst prof forest tree physiol, Forest Soil Lab, Univ Alaska, 73-74; plant physiologist air pollution effects on plants, Lawrence Livermore Lab, Univ Calif, 75-79; res plant physiologist ecology range forage grasses, USDA & ARS Southern Plains Range Res Sta, 79-85; DIR, FT HAYS EXP STA, KANSAS STATE UNIV, 85- *Concurrent Pos:* Consult, US Army Cold Regions & Eng Lab, 73-74. *Mem:* AAAS; Soc Range Mgt; Am Soc Agron; Crop Sci Soc Am; Soil Sci Soc Am; Coun Agr Sci & Technol. *Res:* Environmental physiology of plants especially leaf photosynthesis and water relations as related to stress. *Mailing Add:* Ft Hays Exp Sta Kans State Univ Hays KS 67601

COYNER, EUGENE CASPER, b Conover, NC, Dec 25, 18; m 43; c 3. CHEMISTRY. *Educ:* Univ Ill, BS, 40; Univ Minn, PhD(org chem), 44. *Prof Exp:* From asst to instr chem, Univ Minn, 40-44; res chemist, chem dept, E I du Pont de Nemours & Co, 44-46; asst prof chem, Univ Tenn, 46-48; group leader, res dept, Mallinckrodt Chem Works, 49-54; res supvr, E I du Pont de Nemours & Co, 55, tech asst sales, 55-66, tech assoc, Freon Prod Div, 66-69, sr bus analyst, Org Chem Dept, 69-79, sr indust economist, 79-80; consult chem econ, SRI Int, 81-84; RETIRED. *Concurrent Pos:* Mem, Air Qual Comt, Sierra Club. *Mem:* AAAS; Am Chem Soc. *Res:* General organic chemistry; fluoro-organic compounds; product development. *Mailing Add:* 837 Seaview Dr El Cerrito CA 94530

COZAD, GEORGE CARMON, b Corning, Kans, Mar 5, 27; m 51; c 4. MEDICAL MYCOLOGY, IMMUNOLOGY. *Educ:* Univ Kans, AB, 50; Univ Okla, MS, 54; Duke Univ, PhD(microbiol), 57; Am Bd Med Microbiol, dipl. *Prof Exp:* From res assoc med mycol to assoc prof microbiol, 57-70, PROF MICROBIOL, UNIV OKLA, 70- *Concurrent Pos:* La State Univ trainee, Costa Rica & Colombia, 69; NSF fel, 78; WHO assignment, Tunisia, 81; army res prog partic, US Army Med Res Inst Infectious Dis, 85-90. *Mem:* Am Soc Microbiol; Mycol Soc Am; Int Soc Human & Animal Mycol; Am Thoracic Soc; Am Soc Trop Med & Hyg. *Res:* Pathogenic mechanisms of systemic fungal agents; immunology of systemic fungus diseases; effects of fungus infection on basic immune mechanisms of host. *Mailing Add:* Dept Bot & Microbiol Univ Okla Norman OK 73019

COZZARELLI, FRANCIS A(NTHONY), b Jersey City, NJ, Apr 8, 33; m 58; c 6. CREEP MECHANICS, STRUCTURAL DAMPING. *Educ:* Stevens Inst Technol, BS, 55, MS, 58; Polytech Inst Brooklyn, PhD(appl mech), 64. *Prof Exp:* Stress analyst, Gibbs & Cox, Inc, NY, 55-57; from instr to asst prof eng sci, Pratt Inst, 57-62; from asst prof to prof eng sci, 62-80, prof civil eng, 80-82, PROF MECH & AERONAUT ENG, STATE UNIV NY BUFFALO, 82- *Concurrent Pos:* Vis prof, Technol Univ Delft, 68-69 & Polytech Milano, 76; sci fac fel Award, NSF, 68-69; Fulbright res award, 76-77; vis scientist, Euratom, Ispra, Italy, 77. *Mem:* Am Soc Mech Engrs; Soc Eng Sci; Am Acad Mech; Sigma Xi. *Res:* Creep in plates and shells; compressibility effects; temperature effects; random parameters in creep; inelastic wave propagation; irradiation induced creep; creep rupture; effect of geothermal gradients on lithospheric flexure; earthquake engineering; creep damage; geodynamics. *Mailing Add:* 393 Starin Ave Buffalo NY 14216

COZZARELLI, NICHOLAS ROBERT, b Jersey City, NJ, Mar 26, 38; m 67; c 1. BIOCHEMISTRY. *Educ:* Princeton Univ, AB, 60; Harvard Univ, PhD(biochem), 66. *Prof Exp:* NSF fel biochem, Stanford Univ, 66-68; from asst prof to assoc prof, Univ Chicago, 68-77, prof biochem, 77-82; prof, Dept Molecular Biol, 82-89, chmn dept, 86-89, PROF, DEPT MOLECULAR & CELL BIOL, UNIV CALIF, BERKELEY, 89- *Concurrent Pos:* Mem, NIH Study Sect, 75-81; actg ed, Virol, 75; consult, Abbott Labs, 80-87; dir, Virus Lab, Univ Calif, Berkeley, 86- & Prog Math & Molecular Biol, 88- *Honors & Awards:* Ciba-Geigy-Drew Award in Biomed Res, 90. *Mem:* Nat Acad Sci; Am Soc Microbiol; Am Soc Biochem & Molecular Biol; AAAS. *Res:* Synthesis of DNA in vitro; mechanism of DNA transposition; author of 14 technical publications. *Mailing Add:* Dept Molecular & Cell Biol Univ Calif 401 Barker Hall Berkeley CA 94720

COZZENS, ROBERT F, b Alexandria, Va, Sept 6, 41. PHYSICAL CHEMISTRY, MATERIAL SCIENCE. *Educ:* Univ Va, BS, 63, PhD(chem), 66. *Prof Exp:* Nat Res Coun-Nat Acad Sci fel, US Naval Res Lab, 66-67; PROF CHEM, GEORGE MASON UNIV, 67- *Concurrent Pos:* Consult, chem div, US Naval Res Lab, 67-; chmn, Gordon Res Conf; consult, numerous co. *Mem:* AAAS; Am Chem Soc; Sigma Xi; AIAA; Am Defense Preparedness Asn; Mat Res Soc. *Res:* Photochemistry and energy transfer processes, polymer intramolecular energy transfer, polymer photoconductors; interactions of high energy laser and microwave radiation with materials. *Mailing Add:* Dept Chem 4400 Univ Dr George Mason Univ Fairfax VA 22030

CRABBÉ, PIERRE, chemistry; deceased, see previous edition for last biography

CRABILL, EDWARD VAUGHN, b Winamac, Ind, May 30, 30. ANATOMY. *Educ:* DePauw Univ, BA, 52; NY Univ, PhD(anat), 57. *Prof Exp:* From instr to assoc prof anat, Albany Med Col, 56-77; PROF ANAT, UNIV PITTSBURGH, 77- *Mem:* Sigma Xi. *Res:* Pituitary cytology; hair growth; skin physiology. *Mailing Add:* 2261 Clairmont Dr Pittsburgh PA 15241

CRABLE, GEORGE FRANCIS, b New Castle, Pa, June 10, 22; m 45; c 3. PHYSICS. *Educ:* Geneva Col, BS, 43; Univ Mich, MS, 47; Duke Univ, PhD(physics), 51. *Prof Exp:* Asst, Carnegie Inst Technol, 43-44; chemist, Koppers Co, 44-45; physicist, Gulf Res & Develop Co, Pa, 51-61; chmn dept physics, Geneva Col, 61-63; physicist, 63-70, SR RES PHYSICIST, DOW CHEM CO, 70- *Mem:* Am Phys Soc; Am Chem Soc; NY Acad Sci; Sigma Xi. *Res:* Microwave; mass and infrared spectroscopy; nuclear magnetic resonance; x-ray photoelectron spectroscopy. *Mailing Add:* 3212 Kingsley Dr Bloomington IN 47401-2338

CRABLE, JOHN VINCENT, b New Castle, Pa, Sept, 14, 23. INDUSTRIAL HYGIENE, OCCUPATIONAL HEALTH. *Educ:* Duquesne Univ, BS, 49. *Prof Exp:* Jr fel anal chem, Mellon Inst, 50-57; res chemist, Gulf Res & Develop Co, 57-62; res chemist, Phys & Chem Anal Br, 62-70 & chief, 70-76, chief, Measurements Res Br, 76-80, CHIEF ANAL CHEM, METHODS RES BR, NAT INST OCCUP SAFETY & HEALTH, 80- *Concurrent Pos:* Chmn, Int Orgn Standardization Workplace Atmospheres, 79-; co-ed, Methods Biol Monitoring, Am Pub Health Asn, 81-; chmn, Joint Task Group Lab Safety-Standard Methods for Water & Waste Water, 81-; consult, Inst Health, Beijing, Peoples Repub China, 84- *Mem:* Am Chem Soc; Am Pub Health Asn; Am Indust Hyg Asn. *Res:* Toxic substances and their products found in the workplace in environmental and biological samples. *Mailing Add:* 1146 Meriweather Ave Cincinnati OH 45208-2811

CRABTREE, DAVID MELVIN, b Upper Lake, Calif, Aug 29, 45; m 72; c 3. MARINE BIOLOGY, INVERTEBRATE ZOOLOGY. *Educ:* Pac Union Col, BS, 68, MA, 70; Loma Linda Univ, PhD(biol), 75. *Prof Exp:* CHMN DEPT BIOL, ANTILLIAN COL, 75- *Honors & Awards:* Aquanaut Cert, Nat Oceanic & Atmospheric Admin. *Mem:* AAAS. *Res:* Invertebrate growth lines, ecological and paleoecological applications; effects of environmental conditions on coral growth. *Mailing Add:* Dept Biol Antillian Col Box 118 Mayaguez PR 00708

CRABTREE, DOUGLAS EVERETT, b Boston, Mass, June 14, 38; m 59; c 3. MATHEMATICS. *Educ:* Bowdoin Col, BA, 60; Harvard Univ, MA, 61; Univ NC, PhD(math), 65. *Prof Exp:* Fel math, Univ NC, 61-64; asst prof, Univ Mass, 64-66; asst prof, Amherst Col, 66-72; INSTR MATH, PHILLIPS ACAD, MASS, 72- *Mem:* Am Math Soc; Math Asn Am. *Res:* Theory of matrices; theory of rings. *Mailing Add:* Dept Math Phillips Acad Andover MA 01810

CRABTREE, GARVIN (DUDLEY), b Eugene, Ore, Nov 29, 29; m 65; c 2. HORTICULTURE, WEED SCIENCE. *Educ:* Ore State Univ, BS, 51; Cornell Univ, MS, 55, PhD, 58. *Prof Exp:* From asst prof & asst horticulturist to assoc prof, 58-80, PROF HORT, ORE STATE UNIV, 80- *Concurrent Pos:* Res assoc, Mich State Univ, 75-76. *Mem:* Weed Sci Soc Am; Am Soc Hort Sci; Plant Growth Regulator Soc Am. *Res:* Weed control in horticultural crops and application of plant growth regulators. *Mailing Add:* Dept Hort Ore State Univ Corvallis OR 97331

CRABTREE, GEORGE WILLIAM, b Little Rock, Ark, Nov 28, 44; m 67. METALS PHYSICS, SUPERCONDUCTIVITY. *Educ:* Northwestern Univ, BS, 67; Univ Wash, MS, 68; Univ Ill, Chicago, PhD(physics), 74. *Prof Exp:* Asst scientist, 74-79, assoc scientist, 79-85, SR SCIENTIST SOLID STATE PHYSICS, ARGONNE NAT LAB, 85- *Concurrent Pos:* Vis prof, Univ Ore, 77 & Univ Grenoble, CRTBT, 86. *Honors & Awards:* Outstanding Res Award, Dept Energy, 82 & 85. *Mem:* Fel Am Phys Soc; AAAS. *Res:* Metals physics; Fermi surfaces of delocalized f electrons in rare earth and actinide compounds, superconductivity in magnetic rare earth compounds, actinide compounds, and organic conductors; normal state thermodynamic and transport properties of metals; high temperature superconductivity. *Mailing Add:* MSD 223 Argonne Nat Lab Argonne IL 60439

CRABTREE, GERALD WINSTON, b Manchester, Eng, June 29, 41; Can citizen; m 65; c 2. BIOCHEMISTRY, PHARMACOLOGY. *Educ:* Univ Guelph, BSA, 63, MS, 65; Univ Alta, PhD(purine metab), 70. *Prof Exp:* Res assoc purine metab, 70-72, instr, 72-74, asst prof purine metab, 74-76, asst prof biochem pharmacol, 76-80, ASSOC PROF MED SCI, BROWN UNIV, 80- *Mem:* AAAS; Can Biochem Soc. *Res:* Purine nucleotide metabolism in intact mammalian cells; effects of purine analogues on normal purine metabolic pathways; metabolism of purine analogues; purine metabolism of schistosomes. *Mailing Add:* Div Biol & Med Sci Brown Univ Providence RI 02912

CRABTREE, JAMES BRUCE, b Wichita, Kans, Dec 11, 18. MATHEMATICS. *Educ:* Univ Kans, AB, 41, MA, 42; Harvard Univ, PhD(math), 50. *Prof Exp:* Instr math, Univ Chicago, 47-48; from asst prof to assoc prof, Univ NH, 50-56; ASSOC PROF MATH, STEVENS INST TECHNOL, 56- *Mem:* Am Math Soc; Math Asn Am; Soc Indust & Appl Math. *Res:* Functional analysis; analytic functions on algebras. *Mailing Add:* Dept Math Stevens Inst Technol Hoboken NJ 07030

CRABTREE, KOBY TAKAYASHI, b Tokyo, Japan, Apr 29, 34; US citizen; m 58; c 2. MICROBIOLOGY, CIVIL ENGINEERING. *Educ:* Ohio Wesleyan Univ, BA, 58; Univ Wis, MS, 63, PhD(bact, civil eng), 65. *Prof Exp:* Fel civil eng, Univ Wis, 65-66; from asst prof to assoc prof bact, 66-73, PROF BACT, UNIV WIS, MARATHON CAMPUS, 73-; CHMN DEPT BIOL SCI, UNIV WIS CTR SYST, 75- *Concurrent Pos:* Wis State Dept Natural Resources grants, 67-70; consult, Foth & Van Dyke, Green Bay, Wis, 76-, Niagara of Wis Paper Co, 85-, Champion Paper Co, Lufkin, Tex, 88-, Neenah Paper-Whiting, 89-, Zimpro/Passavant, Wis, 89-, Chesapeake Corp, West Point, VA, 90-, Crompton & Knowles Corp, Reading, PA, 90-, Mosinee Paper Corp, 90- *Mem:* AAAS; Am Soc Microbiol; Am Chem Soc. *Res:* Microbial ecology; water and solid waste pollution; methane fermentation; ecology of hyphomicrobia. *Mailing Add:* Dept Bact Univ Wis 518 Seventh Ave Wausau WI 54401

CRABTREE, ROBERT H, b London, Eng, Apr 17, 48. ALKANE ACTIVATION, HYDROGEN & HYDRIDES. *Educ:* Oxford Univ, BA, 69; Sussex Univ, PhD(chem), 73. *Hon Degrees:* DSc, Sussex Univ, 83. *Prof Exp:* Res chem, National Center Scientific Res, France, 73-77; from asst prof to assoc prof, 77-85, PROF CHEM, YALE UNIV, 85- *Honors & Awards:* Corday-Morgan Medal, Royal Soc Chem, 84. *Mem:* Am Chem Soc; Royal Soc Chem, London. *Res:* Research into the behavior of metals in chemical systems some of which are of commercial or biological significance. *Mailing Add:* Yale Chem Dept 255 Prospect St New Haven CT 06511

CRABTREE, ROSS EDWARD, b Arkansas City, Kans, Mar 20, 32; m 54; c 2. DRUG METABOLISM, RADIOIMMUNOASSAYS. *Educ:* Southwestern State Col, Okla, BS, 54; Purdue Univ, MS, 56, PhD, 57. *Prof Exp:* Sr phys chemist, Lilly Clin, Eli Lilly & Co, 57-63, sr anal res chemist, 63-70, res scientist, 70-85; RETIRED. *Mem:* Health Phys Soc. *Mailing Add:* 5433 Fallwood Dr No 113 Indianapolis IN 46220

CRACRAFT, JOEL LESTER, b Wichita, Kans, July 31, 42. VERTEBRATE MORPHOLOGY, BIOSYSTEMATICS. *Educ:* Univ Okla, BS, 64; La State Univ, MS, 66; Columbia Univ, PhD(biol), 69. *Prof Exp:* Res fel, Am Mus Natural Hist, 69-70; asst prof, 70-76, ASSOC PROF ANAT, UNIV ILL MED CTR, 77- *Mem:* Am Soc Naturalists; Soc Study Evolution; Soc Syst Zool; Am Soc Zoologists; Am Ornithologists Union. *Res:* Functional morphology of birds; multivariate morphometric analysis of size and shape; avian evolution; systematic theory; vertebrate biogeography. *Mailing Add:* Dept Anat Col Med Univ Ill 1853 W Polk St Chicago IL 60612

CRADDOCK, ELYSSE MARGARET, b Sydney, Australia, Sept 7, 44; m 73; c 2. EVOLUTIONARY BIOLOGY, MOLECULAR EVOLUTION. *Educ:* Univ Sydney, BSc, 65, PhD(cytoevolution), 71. *Prof Exp:* Fel evolution, Univ Hawaii, 71-72 & Yale Univ, 72; res fel pop biol, Australian Nat Univ, 73; res scientist, dept biol, NY Univ, 74-80; ASST PROF BIOL, DIV NATURAL SCI, STATE UNIV NY, PURCHASE, 84- *Mem:* Genetics Soc Am; Soc Study Evolution; AAAS; NY Acad Sci; Soc Develop Biol. *Res:* Structure, organization, regulation and evolution of the vitellogenin gene cluster in Hawaiian Drosophilia; hormonal and behavioral correlates of female reproductive development; population differentiation and speciation of Hawaiian Drosophila. *Mailing Add:* Div Natural Sci State Univ NY Purchase NY 10577

CRADDOCK, GARNET ROY, b Chatham, Va, May 7, 26; m 49; c 2. AGRONOMY. *Educ:* Va Polytech Inst, BS, 52; Univ Wis, PhD(soils), 55. *Prof Exp:* From asst prof to assoc prof, 55-67, head dept agron & soils, 72-83, PROF AGRON, CLEMSON UNIV, 67- *Concurrent Pos:* Asst soil scientist, Exp Sta, Clemson Univ, 55-57, assoc soil scientist, 57-66; party chief, Seychelles Food Crop Proj, 83-84. *Mem:* Am Soc Agron; Soil Sci Soc Am; Soil Conserv Soc Am. *Res:* Potassium and its relationship to soil mineralogy; pediological investigations relative to southeastern soils. *Mailing Add:* Dept Agron & Soils Col Agr Sci Clemson Univ Clemson SC 29631

CRADDOCK, J(OHN) CAMPBELL, b Chicago, Ill, Apr 3, 30; m 53; c 3. STRUCTURAL GEOLOGY, TECTONICS. *Educ:* DePauw Univ, BA, 51; Columbia Univ, MA, 53, PhD(geol), 54. *Prof Exp:* Asst geol, Columbia Univ, 53-54; geologist, Shell Oil Co, 54-56; from asst prof to assoc prof geol, Univ Minn, Minneapolis, 56-67; chmn dept, 77-80, PROF GEOL, UNIV WIS, MADISON, 67- *Concurrent Pos:* Dir, Antarctic Res Expeds, 59-69, Alaskan Expeds, 68-81, Spitsbergen Expeds, 77-86; geologist, Minn Geol Surv, 59 & 64, Antarctic Res Expeds, 79-80; vis scientist, NZ Geol Surv, 62-63, co-chief scientist, Leg 35, Deep Sea Drilling Proj, 74; consult, Corps Engrs, US Army, NStar Res Inst, US State Dept, Phillips Petrol, Texaco; US mem, Working Group Geol, Sci Comt Antarctic Res, 68-82, chmn, 73-80; deleg, Sci Comt Antarctic Res, Int Union Geol Sci, 74-87; mem, Comn Struct Geol, Int Union Geol Sci, 68-76, comn tectonics, 76-84; vpres, Comn Geol Map of World; chmn, Antarctic panel, Circum-Pacific Map Proj, 78-89; chmn, N Cent Sect, Geol Soc Am, 82-83, Struct Geol & Tectonics, 83-84. *Honors & Awards:* US Antarctic Serv Medal, 68; Bellingshausen-Lazarev Medal, Soviet Acad Sci, 70; Distinguished Serv Award, Geol Soc Am, 88. *Mem:* Fel AAAS; fel Geol Soc Am; Am Asn Petrol Geol; Am Geophys Union; Seismol Soc Am. *Res:* Overthrusts; folds; transcurrent faults; Antarctic, Spitsbergen and Alaskan geology; gravity; Precambrian geology. *Mailing Add:* Dept Geol & Geophys Univ Wis Madison WI 53706

CRADDOCK, JOHN HARVEY, b Memphis, Tenn, May 30, 36; m 67; c 2. PRODUCT & ENVIRONMENTAL SAFETY. *Educ:* Memphis State Univ, BS, 58; Vanderbilt Univ, PhD(inorg chem), 61. *Prof Exp:* Res chem, M W Kellogg Co Div, Pullman, Inc, 61-65; res specialist, Cent Res Dept, Monsanto Co, 65-68, res group leader, 68-73, supvr com develop, Polymers & Petrochem Co, 73-74, mgr com develop, Indust Chem Co, 74-78, mgr prod & environ safety, 78-81, dir, environ safety, Indust Chem Co, 81-85, DIR ENVIRON SAFETY/REGULATORY AFFAIRS, MONSANTO CO, 85- *Concurrent Pos:* Res assoc, Princeton Univ, 64; mem, Comt on Disposal of Hazardous Indust Wastes, Nat Res Coun, Nat Acad Sci, 81-82; chmn, CMA PCB prog panel, 84; mem, Comt Alternatives to Chloroflurocarbons as Solvents, Nat Res Coun, NAS, 90-91. *Mem:* Am Chem Soc; Royal Soc Chem; Am Inst Chem; NY Acad Sci. *Res:* Homogeneous catalysis; petrochemical reactions and processes; coordination chemistry; chemical food preservatives and antimicrobials; polychlorinated biphenyls, chemistry, health and environmental effects and disposal. *Mailing Add:* Monsanto Co 800 N Lindbergh Blvd St Louis MO 63167

CRADDOCK, MICHAEL KEVIN, b Portsmouth, Eng, Apr 15, 36; div; c 1. NUCLEAR PHYSICS, PARTICLE PHYSICS. *Educ:* Oxford Univ, MA, 61, DPhil(nuclear physics), 64. *Prof Exp:* Sci officer nuclear physics, Rutherford High Energy Lab, Nat Inst Res Nuclear Sci, Chilton, Eng, 61-64; from asst prof to assoc prof, 64-77, PROF PHYSICS, UNIV BC, 77- *Concurrent Pos:* Group leader beam dynamics, Triumf, BC, 68-81, div head, Accelerator Res, 81-88; dep leader, KAON Factory, 88- *Mem:* Brit Inst Physics & Phys Soc (fel Phys Soc); Can Asn Physicists; fel Am Phys Soc. *Res:* Medium energy proton scattering; polarized ion sources; cyclotrons; high-intensity proton synchrotrons. *Mailing Add:* Dept Physics Univ BC 6224 Agriculture Rd Vancouver BC V6T 2A6 Can

CRADDUCK, TREVOR DAVID, b London, Eng; m 59; c 3. MEDICAL PHYSICS. *Educ:* Bristol Univ, Eng, BSc, 58; Univ Sask, MSc, 63, Dr Phil (physics), 66. *Prof Exp:* Apprentice elec eng, Gen Elec Co, Coventry, Eng, 58-60; res fel, Nat Cancer Inst Can, 61-65; physicist & tech dir nuclear med, Manitoba Cancer Found, Winnipeg, 65-67; consult physicist, Foothills Hosp, Calgary, Alta, 67-70; physicist, Toronto Gen Hosp, 70-76; from asst prof to assoc prof, 77-86, PROF FAC MED, UNIV WESTERN ONT, 86-, PHYSICIST, VICTORIA HOSP, DEPT NUCLEAR MED, LONDON, ONT, 77- *Concurrent Pos:* Assoc prof, Fac Med, Univ Calgary, 68-70; asst prof, Fac Med, Univ Toronto, 73-76. *Mem:* Soc Nuclear Med (secy, 74-77); Am Asn Physicists Med; Can Col Physicists Med; Am Bd Med Physics. *Res:* Physics in nuclear medicine; scintillation cameras; computers applied to nuclear medicine and radiology. *Mailing Add:* Dept Nuclear Med Univ Western Ont London ON N6A 4G5 Can

CRAFT, GEORGE ARTHUR, b Youngstown, Ohio, Nov 16, 16; m 46; c 1. MATHEMATICS. *Educ:* Miami Univ, BS, 39; Ind Univ, MA, 50; Ohio State Univ, PhD(math), 57. *Prof Exp:* Instr math, Mont State Univ, 50-53 & Ohio State Univ, 57-58; asst prof, Denison Univ, 58-61; asst prof, 61-63, ASSOC PROF MATH, HARPUR COL, STATE UNIV NY BINGHAMTON, 63- *Mem:* Am Math Soc. *Res:* Continuous transformations in Euclidean n-space; group spaces and topological vector spaces. *Mailing Add:* Dept Math Harpur Col State Univ NY Binghamton NY 13901

CRAFT, HAROLD DUMONT, JR, b Newark, NJ, May 28, 38; m 62, 88; c 2. RADIO ASTRONOMY, RADIOPHYSICS. *Educ:* Cornell Univ, BEE, 61; NY Univ, MEE, 63; Cornell Univ, PhD(radio astron), 70. *Prof Exp:* Staff mem tech commun syst, Bell Tel Lab, 61-65; mem tech radio propagation, COMSAT Lab, 69-71; tech coordr astron, Nat Astron & Ionosphere Ctr, Cornell Univ, 71-73; dir opers, Arecibo Observ, 73-81, ACTG DIR, NAT ASTRON & IONOSPHERE CTR, CORNELL UNIV, 81-, ASSOC VPRES FACIL & BUS OPERS. *Mem:* AAAS; Int Union Radio Sci; Inst Elec & Electronics Engrs; Am Astron Soc; Am Geophys Union. *Res:* Pulsar physics and radio emission mechanisms; radio propagation studies particularly with respect to atmospheric and ionospheric effects. *Mailing Add:* Cornell Univ Humphreys Serv Bldg Ithaca NY 14853

CRAFT, THOMAS JACOB, SR, b Monticello, Ky, Dec 27, 24; m 48; c 2. CELL BIOLOGY, DEVELOPMENTAL BIOLOGY. *Educ:* Cent State Univ, BS, 48; Kent State Univ, MA, 50; Ohio State Univ, PhD(develop biol), 63. *Prof Exp:* Lab asst biol, Cent State Col, Ohio, 47-48; asst lab instr vert anat, Kent State Univ, 49-50; from instr to assoc prof, Biol, Cent State Univ, 50-67, dir, 64-65, prof biol, 67-79; dir, Div Natural Sci & Math, Fla Mem Col, Miami, 81-82, dir energy prog, 80-81, dir of planning, 82-84, dean fac, 84-87; RETIRED. *Concurrent Pos:* Consult, NSF, US AID, India, 67-69; Eli Lilly grant; mem, Nat Adv Res Resources Coun, NIH; adv, Ohio Health Manpower Linkage Syst Proj, Ohio Dept Health; mem exec comt, Ohio Acad Sci; adj prof anat, Sch Med, Wright State Univ, 73-79; mem Beacon Coun Miami, Fla, 85-86; sr consult, Sci & Educ Consults; mem, adv panel, Elem Sci Proj, Educ Develop Ctr, 86-90 & health subcomt, Gov Comn Socially Disadvantaged Black Male, Ohio 88-90. *Mem:* Fel AAAS; Sigma Xi. *Res:* Experimental morphology and embryology; pigment cell biology; swimming in common brown bats; melanogenesis and the homograft reaction; effect of atmospheric pollutants; energy utilization among the low income segment of society. *Mailing Add:* 1280 Wenofred Dr PO Box 252 Wilberforce OH 45384

CRAFTON, PAUL A(RTHUR), b New York, NY, June 11, 23; m 53; c 2. ENGINEERING. *Educ:* City Col New York, BME, 44; Univ Md, MS, 50, PhD, 56. *Prof Exp:* Res engr, Naval Res Lab, 44-67; from assoc prof to prof eng admin, Washington Univ, 60-83, CONSULT, 67-; PROF BUS SCI & COMPUT SERV, OKLA UNIV, 85- *Concurrent Pos:* Pres, United Continental Securities Inc, 71-76; pres, Constellation Corp, 71-83, patent agt, 75-; consult, 56- *Mem:* NY Acad Sci; Sigma Xi; Nat Soc Prof Engrs; Inst Elec & Electronics Engrs. *Res:* Information systems, management and intellectual property; systems analysis. *Mailing Add:* PO Box 6076 Norman OK 73070

CRAFTS, ROGER CONANT, b Lewiston, Maine, Jan 26, 11; m 38; c 2. ANATOMY. *Educ:* Bates Col, BS, 33; Columbia Univ, PhD(anat), 41. *Prof Exp:* Instr, Col Physicians & Surgeons, Columbia Univ, 39-40; from instr to assoc prof anat, Sch Med, Boston Univ, 41-50; prof anat & chmn dept, 50-79, Francis Brunning prof, 79-81, EMER FRANCIS BRUNNING PROF ANAT, COL MED, UNIV CINCINNATI, 81- *Concurrent Pos:* Instr, Boston City Hosp, 42-49; consult, Div Fels, NIH, 60-63. *Mem:* Fel AAAS; Am Asn Anatomists. *Res:* Endocrinology; reproduction; pituitary gland; endocrines and hemopoiesis. *Mailing Add:* 1722 Larch Ave No 403 Cincinnati OH 45224-2955

CRAGG, GORDON MITCHELL, b Capetown, SAfrica, Sept 4, 36; US citizen; m 66. SYNTHETIC ORGANIC & NATURAL PRODUCTS CHEMISTRY. *Educ:* Rhodes Univ, SAfrica, BSc Hons, 56; Oxford Univ, Eng, DPhil, 63. *Prof Exp:* Postdoctoral chem, Univ Calif, Los Angeles, 63-64; sr lectr, Univ SAfrica, 65-72; sr lectr, Univ Cape Town, 73-79; res assoc & asst to dir, Cancer Res Inst, Ariz State Univ, 79-84; chemist, 85-88, CHIEF, NAT PROD BR, NAT CANCER INST, NIH, 89- *Mem:* Am Chem Soc; Am Soc Pharmacog; Soc Econ Bot. *Res:* Natural products chemistry; isolation and structural elucidation of novel anticancer and anti-HIV agents from natural sources. *Mailing Add:* 5117 Elsmere Ave Bethesda MD 20814

CRAGGS, ROBERT F, b South Charleston, Ohio, June 9, 37. TOPOLOGY. *Educ:* Ohio Univ, Athens, AB, 59; Univ Wis-Madison, MS, 60, PhD(math), 66. *Prof Exp:* Instr math, Ohio Univ, Athens, 61-63; fel math, Inst Adv Study, Princeton, 66-68; from asst prof to assoc prof, 68-86, PROF MATH, UNIV ILL, URBANA, 86- *Concurrent Pos:* Sabbatical vis, Sci Inst, Univ Iceland, 74-75. *Mem:* Am Math Soc. *Res:* Topology of 3- and 4-manifolds; Heegaard theory; Poincare conjecture. *Mailing Add:* Dept Math Univ Ill 1409 W Green Urbana IL 61801

CRAGLE, RAYMOND GEORGE, animal physiology, nutrition; deceased, see previous edition for last biography

CRAGO, SYLVIA S, MUCOSAL IMMUNITY. *Educ:* Univ Ala, Birmingham, PhD(microbiol), 79. *Prof Exp:* RES ASST PROF, MOLECULAR IMMUNOL, SCH MED, UNIV NMEX, 84- *Res:* Immuno-regulation of immunoglobin; immuno-phenotyping of lymphoid malignancies. *Mailing Add:* Dept Cell Biol Sch Med Univ NMex 900 Caminode Salud Albuquerque NM 87131

CRAGOE, EDWARD JETHRO, JR, b Tulsa, Okla, July 16, 17; m 42; c 3. MEDICINAL CHEMISTRY, ORGANIC CHEMISTRY. *Educ:* Baker Univ, BA, 39; Univ Nebr-Lincoln, MA, 41, PhD(org chem), 44. *Prof Exp:* Res assoc, Sharp & Dohme, Inc, 44-56; instr chem, Pa State Col, 47-48; res assoc, Merck Eco Inc, 56-60, from asst dir to assoc dir, 60-75, sr dir med chem, 75-79, distinguished sr scientist, Merck Sharp & Dohme Res Lab, 79-88; CONSULT, 88- *Concurrent Pos:* Consults, adv bds & high-tech drug co. *Mem:* Am Chem Soc; Sigma Xi; fel NY Acad Sci; fel Am Inst Chemists; Soc Chem Indust. *Res:* Synthetic drugs; chemotherapy; antidiabetic agents; antihypertensive agents; mental health drugs; gastrointestinal drugs; organic synthesis; heterocyclic compounds; agents for renal lithiasis; antiallergy drugs; antiasthmatic agents; hypolemic drugs; antiatherosclerosis drugs; antiallergy agents; ophthalmic drugs and drugs for brain injury; cation transport inhibitors; anion transport inhibitors; 550 scientific papers; 270 patents. *Mailing Add:* PO Box 631548 Nacogdoches TX 75963-1548

CRAGON, HARVEY GEORGE, b Ruston, La, Apr 21, 29. COMPUTER ENGINEERING. *Educ:* La Polytech Inst, BS, 50. *Prof Exp:* Engr, Southern Bell Telephone, 50-53 & Hughes Aircraft, 53-57; mem, Aro, Inc, 57-58; sr fel, Tex Instruments Inc, 58-84; CENTENNIAL CHMN ENG, UNIV TEX, 84-, PROF ELEC & COMPUT ENG, 86- *Honors & Awards:* Fel, Inst Elec & Electronics Engrs, 83, Emanuel R Piore Award, 84; Eckert-Mauchly Award, ACM-Inst Elec & Electronics Engrs, 86. *Mem:* Nat Acad Eng; Inst Elec & Electronics Engrs; Asn Comput Mach. *Res:* Computer architecture design methodology; high speed computers. *Mailing Add:* Dept Comput Eng Univ Tex Austin TX 78712-1084

CRAGWALL, J(OSEPH) S(AMUEL), JR, b Richmond, Va, Aug 3, 19; m 41; c 1. HYDROLOGY, CIVIL ENGINEERING. *Educ:* Univ Va, BCE, 40. *Prof Exp:* Engr, Solvay Process Co, 40-41; hydraul engr, US Geol Surv, 41-58, dist engr, Tenn, 58-62, dist chief, 62-67; staff asst planning, Water Resources Prog, US Dept Interior, 67-68; asst chief hydrol for opers, Water Resources Div, US Geol Surv, 68-73, asst dir progs, 73-74, chief hydrologist, 74-79, assoc dir, 79-80; hydrologist, 80-86; RETIRED. *Concurrent Pos:* Chmn, US Nat Comt Sci Hydrol, 75-79; mem, Va Water Control Bd, 82-86. *Honors & Awards:* Distinguished Serv Award, Dept Interior, 76; Julians Hinds Award, Am Soc Civil Engrs, 85. *Mem:* Am Soc Civil Engrs; Am Geophys Union; Nat Soc Prof Engrs; Am Water Resources Asn. *Res:* Hydrology and hydraulics of surface streams; measurement of open-channel flow; floods and low flow water resource appraisal. *Mailing Add:* 4901 English Dr Annandale VA 22003

CRAIB, JAMES F(REDERICK), b Buffalo, NY, June 25, 13; wid; c 2. ELECTRONICS. *Educ:* Rensselaer Polytech Inst, EE, 36. *Prof Exp:* Mem staff res & develop electronics, Stromberg-Carlson, 36-40; gen mgr, Technifinish Labs, 40-41; elec eng, develop mil electronics, Hazeltine Corp, 41-45; gen mgr electronic develop, Kryptar Corp, 46-48; elec eng, Airborne Instruments Lab Div, Cutler-Hammer Inc, 48-66, res scientist, 66-72, head, Cent Res Div, 72-75; RETIRED. *Mem:* Inst Elec & Electronic Engrs. *Res:* Electronic circuitry; radar. *Mailing Add:* Box 25 Miami AZ 85539-0025

CRAIG, ALAN DANIEL, b Hempstead, NY, Feb 11, 35; m 60; c 4. INDUSTRIAL CHEMISTRY, POLYMER CHEMISTRY. *Educ:* Hofstra Univ, BA, 56; Univ Pa, PhD(inorg chem), 61. *Prof Exp:* Res chemist res ctr, Hercules Inc, 61-66, mgr high energy res div, 67-70, sr venture analyst, New Enterprise Dept, 70- 73; vpres, Adria Labs, Inc, 74-78; mgr prod develop, Hercules, Inc, 78-82, dir appl technol, 82-83, dir res & develop, Hercules Specialty Chem, 84-85; dir aquisitions & planning, Hercules Engineered Polymers Co, 86-89; DIR BUS DEVELOP, HERCULES ADVAN MAT CO, 89- *Concurrent Pos:* dir, Texas Alkyls Inc, 82-, BHC Labs, 85-, Med Ct Del, 88- *Mem:* Am Chem Soc; Sigma Xi. *Res:* Nitrogen-fluorine chemistry; silicon and other light metal hydride chemistry; pharmaceutical development; polymer fabrication and modification; medical device development; composite materials. *Mailing Add:* Hercules Inc Hercules Plaza Wilmington DE 19894

CRAIG, ALAN KNOWLTON, b Ft Sill, Okla, Mar 7, 30; div. PALEOECOLOGY, MARINE RESOURCE MANAGEMENT. *Educ:* La State Univ, BS, 58, PhD(geog), 66. *Prof Exp:* Explor geologist, Gulf Oil Corp, 56-60; consult geol, 61-66; PROF GEOG, FLA ATLANTIC UNIV, 66- *Concurrent Pos:* Consult, Carribean Fisheries Mgt Coun, 78-80; vis prof, Univ del Norte, Antofagasta, Chile, 81; sr res fel, Fulbright, 81; adv mem S Atlantic Fisheries Mgt Coun, 81-; mem UNESCO - IGCP - Proj 200, sea level changes, 83-87. *Mem:* AAAS; fel Am Inst Fishery Res Biologists. *Res:* Paleoenvironmental studies of Andean South America; maritime ecology of the Carribean and western South America; fisheries management of the spiny lobster. *Mailing Add:* PO Box 55 Boca Raton FL 33432

CRAIG, ALBERT BURCHFIELD, JR, b Sewickley, Pa, April 19, 24; m 47; c 4. PHYSIOLOGY, MEDICINE. *Educ:* Cornell Univ, MD, 48. *Prof Exp:* Intern, Dept Med, 48-49, asst resident, 50-51, instr, 53-55, instr physiol & med, 55-59, from asst prof to assoc prof physiol, 59-72, PROF PHYSIOL, SCH MED & DENT, UNIV ROCHESTER, 72- *Concurrent Pos:* Fel med, Sch Med & Dent, Univ Rochester, 49-50, USPHS res fel, 53-55; estab investr, Am Heart Asn, 61-66. *Mem:* Am Col Sports Med; Am Physiol Soc. *Res:* Man in water; respiration. *Mailing Add:* Dept Physiol Univ Rochester Sch Med & Dent 285 Clover Hills Dr Rochester NY 14642

CRAIG, ALBERT MORRISON (MORRIE), b San Francisco, Calif, Oct 13, 43. TOXICOLOGY, BIOMEDICAL. *Educ:* Ore State Univ, BS, 65, PhD(biochem, biophys), 70. *Prof Exp:* Am Cancer Soc fel, Ore State Univ, 70; NIH fel, Univ Ore, 71-74, res assoc molecular biol, 74-75; researcher & instr biol, Univ Hawaii, Hilo, 75-76; res assoc, 76-78, ASST PROF & SR RESEARCHER, SCH VET MED, ORE STATE UNIV, 78- *Concurrent Pos:* Lectr, Univ London, 71, Gordon Conf, 74. *Honors & Awards:* Samuel Talbot Award, 69. *Mem:* Biophys Soc; Sigma Xi; AAAS; Am Col Clin Chem. *Res:* Immunotoxicity, clinical chemistry and subchronic toxicity of clinical carcinogens. *Mailing Add:* PO Box 1107 Corvallis OR 97339

CRAIG, ARNOLD CHARLES, b Johnstown, NY, Sept 5, 33; m 59; c 3. PHYSICAL ORGANIC CHEMISTRY. *Educ:* Syracuse Univ, BA, 54; Cornell Univ, PhD(org chem), 59. *Prof Exp:* Res chemist, Eastman Kodak Co, 59-65; from asst prof to assoc prof, 65-75, PROF CHEM, MONT STATE UNIV, 75- *Mem:* Am Chem Soc. *Res:* Synthesis of natural products and heterocycles; color and constitution relation of sensitizing dyes; effect of environment on radiative transitions. *Mailing Add:* Dept Chem Mont State Univ Bozeman MT 59717

CRAIG, BURTON MACKAY, b Vermilion, Alta, May 29, 18; m 45; c 2. AGRICULTURAL BIOCHEMISTRY. *Educ:* Univ Sask, BScA, 44, MSc, 46; Univ Minn, PhD, 50. *Prof Exp:* Lab asst, Prairie Regional Lab, Univ Sask, 41-44, asst, 44-46, from asst res officer to prin res officer, 50-69, assoc dir, 69-70, dir, Fats & Oils Lab, 70-83; RETIRED. *Concurrent Pos:* Protein, Oil & Starch Tech Auditor, 88- *Honors & Awards:* Canola Award, 81; W J Eva Award, 83. *Mem:* Am Oil Chem Soc; Chem Inst Can. *Res:* Nutrition of fats; biosynthesis of fatty acids; fatty acid composition of oils and fats; gas liquid chromatography in fats and oils. *Mailing Add:* 423 Lake Cres Nat Res Coun Saskatoon SK S7H 3A3 Can

CRAIG, CATHERINE LEE, b Oxnard, Calif, Feb 2, 51. ECOLOGY, BEHAVIOR ETHOLOGY. *Educ:* Stanford Univ, AB, 73; Univ Calif, Berkeley, MA, 76; Cornell Univ, PhD(ecol & evol), 85. *Prof Exp:* ASSOC PROF EVOL ECOL, YALE UNIV, 85- *Mem:* Ecol Soc Am; Soc Am Naturalists; Soc Study Evol; Asn Trop Biol; Am Arachnol Soc; AAAS. *Res:* Coevolutionary interactions between predators and prey, in particular, role of silk proteins in the foraging ecology of web weaving spiders and the implications of silk molecular structure to the evolution of the silk genome; effects of insect perception and learning on spider foraging modes. *Mailing Add:* Dept Biol Osborn Mem Lab PO Box 6666 New Haven CT 06511-7444

CRAIG, CHARLES ROBERT, b Buckhannon, WVa, Jan 24, 36; m 60; c 1. PHARMACOLOGY. *Educ:* Univ Wis, PhD(pharmacol), 64. *Prof Exp:* Sr investr neuropharmacol, G D Searle & Co, 64-66; from asst prof to assoc prof pharmacol, 66-75, PROF PHARMACOL, HEALTH SCI CTR, WVA UNIV, 75- *Mem:* Am Soc Pharmacol & Exp Therapeut; Am Soc Neurochem; Soc Neurosci. *Res:* Pharmacological changes in the chronically epileptic rat. *Mailing Add:* Dept Pharmacol WVa Univ Health Sci Ctr Morgantown WV 26506

CRAIG, DEXTER HILDRETH, b Pontiac, Mich, Feb 12, 24; m 49; c 2. GEOLOGY. *Educ:* Univ Mich, BS, 50; Univ Tex, MA, 52. *Prof Exp:* From asst geologist to geologist, Marathon Oil Co, Tex, 52-61, from geologist to sr res geologist, Colo, 61-85, adv sr geologist, Denver Res Ctr, 85-86; CONSULT GEOLOGIST, 86- *Mem:* AAAS; Am Asn Petrol Geol; Soc Econ Paleont & Mineral; Am Inst Mining, Metall & Petrol Eng; Sigma Xi; Geol Soc Am. *Res:* Geology of petroleum deposits; stratigraphy and sedimentology of carbonate rocks. *Mailing Add:* 6654 S Sycamore St Littleton CO 80120-3039

CRAIG, DONALD SPENCE, b Ridgetown, Ont, Aug 13, 23; m 52; c 3. NUCLEAR PHYSICS. *Educ:* Queen's Univ, Ont, BSc, 45; Univ Wis, PhD(physics), 52. *Prof Exp:* Jr res physicist, Nat Res Coun Can, 45-47; asst res physicist, Atomic Energy Ltd, Can, 52-57, from assoc res officer to sr res officer, 57-88; RETIRED. *Mem:* Am Nuclear Soc; Can Asn Physicists. *Res:* Reactor physics. *Mailing Add:* 18 Cabot Pl Deep River ON K0J 1P0 Can

CRAIG, DOUGLAS ABERCROMBIE M, b Nelson, NZ, Oct 24, 39; m 62; c 2. INSECT MORPHOLOGY, INVERTEBRATE LIMNOLOGY. *Educ:* Univ Canterbury, BSc, 62, PhD(zool), 66. *Prof Exp:* Asst lectr zool, Univ Canterbury, 62-66; session lectr, 66-67, lectr, 67-68, asst prof, 68-74, assoc prof entom, 74-79, PROF ENTOM, UNIV ALTA, 79- *Mem:* Entom Soc Can; Entom Soc Can; Micros Soc Can; AAAS. *Res:* Phylogeny of nematocera; filter-feeding of larval simuliidae; larval sensory organs; hydrodynamics of filter-feeding; extensive collecting of simuliidae in South Pacific; taxonomy of Pacific simuliidae. *Mailing Add:* Dept Entom Univ Alta Edmonton AB T6G 2E3 Can

CRAIG, EDWARD J(OSEPH), b Springfield, Mass, July 17, 24; m 47; c 5. ELECTRICAL ENGINEERING. *Educ:* Union Col, NY, BS, 48; Mass Inst Technol, ScD(elec eng), 54. *Prof Exp:* Instr math, Union Col, NY, 48-49; asst elec eng, Mass Inst Technol, 49-51, instr, 51-53; asst prof commun, Northeastern Univ, 53-55, assoc prof elec eng, 55-56; assoc prof, Union Col, 56-60, chmn dept, 73-74, 77-79, co-chmn, elec eng & comput sci, 77-83, dean eng, 84-88, PROF ELEC ENG, UNION COL, 60- *Concurrent Pos:* Consult, Tube Dept, Adv Tech Lab, Gen Elec Co, 56-70, 84-85; Fulbright prof elec eng, Univ Liberia, 79-80; vis prof elec eng, Univ PR, 89-90. *Mem:* Inst Elec & Electronic Engrs. *Res:* Electromagnetic field theory; microwave tubes; system theory; digital computers. *Mailing Add:* Div Eng & Appl Sci Union Col Schenectady NY 12308

CRAIG, ELIZABETH ANNE, b Peterborough, NH, Sept 12, 46. PHYSIOLOGY. *Educ:* Univ RI, BS, 68; Wash Sch Med, PhD(microbiol), 72. *Prof Exp:* Postdoctoral fel, Inst Molecular Biol, St Louis Univ & Dept Path, Wash Univ Sch Med, St Louis, 72-75; sr fel, Dept Biochem & Biophys, Univ Calif, San Francisco, 76-79; from asst prof to assoc prof, 79-88, PROF, DEPT PHYSIOL CHEM, UNIV WIS-MADISON, 88- *Concurrent Pos:* NIH fel, Inst Molecular Virol, St Louis Univ, 72-75, res career develp award, Dept Physiol Chem, Univ Wis-Madison, 81-86; Am Cancer Soc sr fel, Univ Calif, San Francisco, 76-78; ad hoc mem, Genetics Study Sect, NIH, 83 & 85, mem, 86-90; H I Romnes fac fel, Univ Wis, 85-90. *Mem:* Am Soc Biol Chemists; Am Soc Microbiol; Genetics Soc Am; Am Soc Cell Biol; AAAS. *Res:* Function of stress proteins; regulation of gene expression. *Mailing Add:* Dept Physiol Chem Univ Wis Med Sci Bldg 1300 University Ave Madison WI 53706

CRAIG, FRANCIS NORTHROP, b Englewood, NJ, June 2, 11; m 40; c 3. PHYSIOLOGY. *Educ:* Rutgers Univ, BS, 32; Harvard Univ, AM, 33, PhD(biol), 37. *Prof Exp:* Asst physiol, Harvard Univ, 33; asst, Radcliffe Col, 34-35; asst bot, Columbia Univ, 36; res asst, Harvard Med Sch, 36-39; instr physiol, Col Med, NY Univ, 43-46; physiologist, Edgewood Arsenal, 46-49, chief appl physiol br, 49-73, chief med physiol br, Biomed Lab, 73-78; RETIRED. *Concurrent Pos:* Teaching fel physiol, Harvard Med Sch, 38-39, res fel med, 39-41, res fel anesthesia, 41-43; vis scientist, Chem Defense Exp Estab, UK, 54. *Mem:* Fedn Am Soc Exp Biol; Am Physiol Soc; Soc Exp Biol & Med; Am Col Sports Med. *Res:* Cell metabolism; kidney physiology; environmental physiology; respiration; physiology of exercise. *Mailing Add:* 13801 York Rd G-5 Cockeysville MD 21030

CRAIG, GEORGE BLACK, b Toronto, Ont, Dec 4, 21; m 48; c 3. PHYSICAL METALLURGY. *Educ:* Univ Toronto, BASc, 49, MASc, 50, PhD(metall), 52. *Prof Exp:* Sr scientist, Dept Mines & Tech Surv, Govt Can, 52-56; assoc prof phys metall, 56-64, assoc dean, fac appl sci & eng, 70-75, chmn, dept metall & mat sci, 76-78, PROF PHYS METALL, UNIV TORONTO, 64- *Mem:* Fel Am Soc Metals. *Res:* Deformation. *Mailing Add:* Dept Metall & Mat Sci Rm 136 Univ Toronto Toronto ON M5S 1A4 Can

CRAIG, GEORGE BROWNLEE, JR, b Chicago, Ill, July 8, 30; m 54; c 4. ENTOMOLOGY. *Educ:* Ind Univ, BA, 51; Univ Ill, MS, 52, PhD(entom), 56. *Prof Exp:* Asst, dept entom, Univ Ill, 51-53; from asst prof to prof, 57-74, CLARK DISTINGUISHED PROF BIOL, UNIV NOTRE DAME, 74- *Concurrent Pos:* WHO travel fel, 60, consult, 63; entomologist, Prev Med Unit, Md, 54 & Chem Corps Med Labs, Army Chem Ctr, 54-57; dir, N D Vector Biol Lab, 60- & WHO Int Ref Ctr Aedes, 66-; res dir, Int Ctr Insect Physiol & Ecol, Nairobi, Kenya, 70-77; NIH study sect trop med, 69-74; consult to surg gen, US Army 80-88; gov bd, Entom Soc Am, 78-83; coun, AM Soc Trop Med & Hyg, 80-84; coun, Am Comt Med Entomol, 84-88; examr, Am Registry Prof Entomol. *Mem:* Fel NAS; fel AAAS; fel Entom Soc Am; Am Soc Trop Med & Hyg; Genetics Soc Am; fel Am Acad Arts & Sci; Am Mosquito Control Asn (pres, 86-88). *Res:* Culicidae; genetics, systematics, bionomics and physiology of Aedes; systematics of aedine eggs; evolutionary mechanisms; zoogeography; Arctic insects; vector genetics. *Mailing Add:* Dept Biol Univ Notre Dame Notre Dame IN 46556

CRAIG, HARMON, b New York, NY, Mar 15, 26; m 47; c 3. GEOCHEMISTRY, OCEANOGRAPHY. *Educ:* Univ Chicago, PhD(geol), 51. *Hon Degrees:* Dr, Univ Pierre et Marie Curie, Paris, 83. *Prof Exp:* Res assoc geochem, Enrico Fermi Inst Nuclear Studies, Univ Chicago, 51-55; res assoc geochem, 55-65, chmn, dept earth sci, Univ Calif, San Diego & Div Earth Sci, Scripps Inst Oceanog, 65-68; PROF GEOCHEM & OCEANOG, GRAD DEPT & GEOL RES DIV, SCRIPPS INST OCEANOG & REVELLE COL, UNIV CALIF, SAN DIEGO, 68- *Concurrent Pos:* Guggenheim fel, Univ Pisa, 62-63; mem oceanog expeds, Monsoon, 61, Zephyrus, 62, Carrousel, 64, Nova, 67, Scan, 70, Antipode, 71, Geochemical Ocean Sect Study expeds, 72-77, Vulcan, Pluto, 81, Benthic, 82 & 83, Alcyone, 85 & var diving expeds, E Pac, Tibet & China, 73-84; dir GEOSECS, 70. *Honors & Awards:* V M Goldschmidt Medal, Geochem Soc, 79; Arthur L Day Medal, Geol Soc Am, 83; Arthur L Day Prize & lectr, Nat Acad Sci, 87; Vetlesen Prize, 87. *Mem:* Nat Acad Sci; Am Geophys Union; Am Acad Arts & Sci; Geol Soc Am; Int Glaceological Soc; Geochem Soc. *Res:* Isotopic geochemistry; volcanic gases and geothermal areas; polar ice cores; oceanography; atmospheric chemistry; isotopic studies Greek marbles; Lake Tanganyika studies; African Rift Valley. *Mailing Add:* Isotope Lab A-20 Scripps Inst Oceanog Univ Calif San Diego La Jolla CA 92093

CRAIG, JAMES CLIFFORD, JR, b Connellsville, Pa, Aug 10, 36; m 61; c 3. FOOD ENGINEERING, OPTIMIZATION & SIMULATION. *Educ:* Johns Hopkins Univ, BES, 58. *Prof Exp:* Chem engr, 59-74, res leader, 74-81, chief, eng sci lab, 81-85, RES LEADER, ENG SCI RES UNIT, EASTERN REGIONAL RES CTR, AGR RES SERV, US DEPT AGR, 85- *Mem:* Am Inst Chem Engrs; Inst Food Technologists; Am Leather Chemists Asn; Am Soc Agr Engrs; Sigma Xi. *Res:* Food and agricultural products; process studies; engineering science phenomena; optimization and simulation. *Mailing Add:* 600 E Mermaid Lane Eastern Regional Res Ctr US Dept Agr Philadelphia PA 19118

CRAIG, JAMES I, b Philadelphia, Pa, Feb 13, 42; m 65; c 3. SOLID MECHANICS. *Educ:* Mass Inst Technol, BS, 64; Stanford Univ, MS, 66, PhD(aeronaut, astronaut), 69. *Prof Exp:* from asst prof to assoc prof, 68-80, PROF AEROSPACE ENG, 81-, CAE CAD LAB DIR, GA INST TECHNOL, 85- *Mem:* Am Inst Aeronaut & Astronaut; AAAS. *Res:* Experimental mechanics; aerospace structures; experimental stress analysis; building structural dynamics; systems design and simulation. *Mailing Add:* Dept Aerospace Eng Ga Inst Technol Atlanta GA 30332

CRAIG, JAMES MORRISON, b Drayton, NDak, June 21, 16; m 39; c 2. MICROBIOLOGY. *Educ:* San Jose State Col, AB, 38; Stanford Univ, MA, 48; Ore State Univ, PhD(microbiol), 68. *Prof Exp:* Chemist, B Cribari & Sons, 39-40; chemist & bacteriologist, Eng-Skell Co, 40-42; chief chemist & bacteriologist, Goldfield Consol Mines, 42-45; chief chemist & asst supt, L De-Martini Co, 45-48; instr zool & biol, 48-51, asst prof bact, 51-57, assoc prof, 57-68, PROF MICROBIOL, SAN JOSE STATE UNIV, 68- *Concurrent Pos:* NSF grant, Ore State Univ, 58 & Univ PR, 63; AEC grant, Univ Wash; NIH spec grant, Ore State Univ, 64. *Mem:* AAAS; Am Inst Biol Sci; Am Soc Microbiol; Soc Indust Microbiol; Nat Sci Teachers Asn. *Res:* General bacteriology; aquatic, sanitary and industrial microbiology. *Mailing Add:* 2201 Gundersen Dr San Jose CA 95125

CRAIG, JAMES PORTER, JR, b Mobile, Ala, Oct 4, 26; m 50; c 2. PHYSICAL CHEMISTRY. *Educ:* La State Univ, BS, 48, MS, 50, PhD(chem), 53. *Prof Exp:* Chemist, Bound Brook Labs, Am Cyanamid Co, 50-51; finish chemist, Nylon plant, Chemstrand Corp, 53-54, res chemist, 54-60, group leader phys chem, Res Ctr, NC, 60-70, group leader phys chem, 70-76, supvr acrilan develop, Tech Ctr, Monsanto Textile Div, 76-; RETIRED. *Mem:* AAAS; Am Chem Soc. *Res:* Colloid and surface chemistry of polymers; polymer physical chemistry. *Mailing Add:* 907 Daventry St Baton Rouge LA 70808-5830

CRAIG, JAMES ROLAND, b Philadelphia, Pa, Feb 16, 40; m 62; c 2. GEOCHEMISTRY. *Educ:* Univ Pa, BA, 62; Lehigh Univ, MS, 64, PhD(geol), 65. *Prof Exp:* Fel, Carnegie Inst, 65-67; asst prof geochem, Tex Tech Univ, 67-70; ASST PROF GEOCHEM & PROF GEOL, VA POLYTECH INST & STATE UNIV, 70-, CHMN GEO SCI, 90- *Concurrent Pos:* Assoc ed, Am Min, 74-78; counr, Mineral Soc Am, 82-85; mem Va Waste Mgt Bd, 86- *Mem:* Mineral Soc Am; Soc Econ Geologists; Mineral Asn Can. *Res:* Phase relations of ore minerals; ore deposits; thermochemistry and phase equilibria of ore minerals; evaporite minerals; gold mineralogy and deposits. *Mailing Add:* Dept Geol Sci Va Polytech Inst & State Univ Blacksburg VA 24061

CRAIG, JAMES VERNE, b Bonner Springs, Kans, Feb 7, 24; m 48; c 3. GENETICS. *Educ:* Univ Ill, BS, 48, MS, 49; Univ Wis, PhD(genetics), 52. *Prof Exp:* First asst animal sci, Univ Ill, 52-54, asst prof, 54-55; assoc prof poultry husb, 55-60, PROF ANIMAL SCI, KANS STATE UNIV, 60- *Concurrent Pos:* NIH spec res fel, Poultry Res Ctr, Scotland, 61-62; animal behav unit, Univ Queensland, Australia. *Honors & Awards:* Poultry Sci Res Award, 61. *Mem:* AAAS; fel Poultry Sci Asn; Am Soc Animal Sci. *Res:* Animal behavior; quantitative genetics. *Mailing Add:* Dept Animal Sci Kans State Univ Manhattan KS 66506

CRAIG, JAMES WILLIAM, b West Liberty, Ohio, Jan 23, 21; m 48, 72; c 8. MEDICINE. *Educ:* Case Western Reserve Univ, BS, 43, MD, 45. *Prof Exp:* From instr to assoc prof med, Sch Med, Case Western Reserve Univ, 52-72; assoc dean, 72-89, prof, 72-90, EMER PROF MED, SCH MED, UNIV VA, 91- *Mem:* Soc Exp Biol & Med; Am Fedn Clin Res; Am Diabetes Asn; Am Inst Nutrit. *Res:* Internal medicine; clinical research in intermediary metabolism, particularly carbohydrate metabolism and diabetes mellitus. *Mailing Add:* Sch Med Univ Va Charlottesville VA 22908

CRAIG, JOHN CYMERMAN, b Berlin, Ger, Jan 23, 20; m 45; c 2. ORGANIC CHEMISTRY. *Educ:* Univ London, BSc, 42, PhD(org chem), 45; Univ Sydney, DSc, 61. *Prof Exp:* Res chemist, Boots Pure Drug Co, Eng, 45-47; lectr org chem, Univ London, 47-48; from lectr to sr lectr, Univ Sydney, 48-60; from vchmn to chmn dept, 63-70, PROF CHEM & PHARMACEUT CHEM, UNIV CALIF, SAN FRANCISCO, 60-, ASSOC DEAN RES, 83- *Concurrent Pos:* Nuffield Found Dom traveling fel, Dyson Perrins Lab, Oxford Univ, 56-57; vis scientist, Lab Chem Natural Prod, NIH, 59; mem panel, psychopharmacol chem NIMH, 63-68 & mem Preclin psychopharmacol res rev comt, NIMH, 68-72; mem panel, Polycyclic Org Matter, Nat Acad Sci, 70-72 & Vapor-Phase Org Pollutants, 72-75; fac res lectr, Acad Senate, Univ Calif, San Francisco, 74-75. *Honors & Awards:* Res Achievement Award, Am Pharmaceut Asn Res Found, 67; Res Achievement Award, Am Asn Pharmaceut Scientists, 89. *Mem:* Am Chem Soc; Royal Soc Chem; fel Am Asn Pharmaceut Scientists; fel Acad Pharmaceut Sci; Am Col Neuropsychopharmacol. *Res:* Biomedical mass spectrometry; deuterium labeling in clinical research; acetylene chemistry and biosynthesis; mechanisms of biological reactions; chemistry of sulfur compounds. *Mailing Add:* Dept Pharmaceut Chem Univ Calif San Francisco CA 94143-0446

CRAIG, JOHN FRANK, b Cheltenham, Eng, Jan 22, 45; m 70; c 4. POPULATION DYNAMICS & ENERGETICS OF FISH. *Educ:* Univ London, BSc, 66, MPhil, 73; Univ Bristol, Cert, 67; Univ Lancaster, PhD(fish biol), 79. *Prof Exp:* Head biol, Chipping Norton Sch, 67-69; res asst, Northeast London Polytech, 69-72; prin teacher, Eyemouth Sch, 72; scientist & prin sci officer, Fresh Water Biol Asn, 73-82; head fish ecol, Alta Environ Ctr, 82-84; res scientist, Dept Fisheries & Oceans, Freshwater Inst, 84-90; INDEPENDENT CONSULT, 90- *Concurrent Pos:* Adj prof, Univ Manitoba; consult, Int Develop Res Ctr. *Mem:* Sigma Xi; Fisheries Soc Brit Isles; Inst Biol. *Res:* Population dynamics of fresh water fish in particular percids; life history strategies, senescence, energetics, and recruitment and how these effect yield and population dynamics. *Mailing Add:* Crawfordton Garden Cottage Moniaive Thornhill Dumfriesshire DG3 4HF Scotland

CRAIG, JOHN HORACE, b Macon, Ga, Dec 25, 42; m 68. ORGANIC CHEMISTRY. *Educ:* George Washington Univ, BS, 64; Georgetown Univ, PhD(org chem), 69. *Prof Exp:* NIH Fel chem, Univ Ill, Urbana-Champaign, 69-70, res assoc org chem, 70-71; asst prof, 71-75, ASSOC PROF ORG CHEM, CALIF STATE UNIV, SAN BERNARDINO, 75- *Mem:* Am Chem Soc. *Res:* Reaction mechanisms; strained sigma and pi bonds; bridged small and medium ring compounds; aliphatic nitrogen heterocycles; conformational analysis; stereo-chemistry; molecular rearrangements; physiologically active compounds. *Mailing Add:* Dept Chem Calif State Univ San Bernardino CA 92407

CRAIG, JOHN MERRILL, b Pasadena, Calif, Oct 14, 13; m 49; c 3. PATHOLOGY. *Educ:* Univ Calif, AB, 36, MA, 38; Harvard Univ, MD, 41. *Prof Exp:* Asst prof path, Children's Hosp, Harvard Med Sch, 55-59; prof, Sch Med, Univ Pittsburgh, 59-60; clin prof, Harvard Med Sch, 60-70, prof path, 70-79; pathologist-in-chief, Boston Hosp Women, 60-79; resident psychiatrist, Wes Ros Park Mental Health Clin, 79-82, psychiatrist, 82-90; ASSOC PSYCHIATRIST, BRIGHAM WOMEN'S HOSP, 81- *Concurrent Pos:* Res assoc, Children's Cancer Res Found, 50-59; pathologist, Children's Hosp, 50-59; dir labs, E S Magee Hosp, 59-60; mem path study sect, USPHS, 59-64; assoc ed, Am J Path, 64-70. *Mem:* Am Soc Exp Path; Am Asn Path & Bact; Soc Pediat Res; Int Acad Path; Am Psychiat Asn. *Res:* Experimental and morphological pediatric, gynecologic and obstetric pathology. *Mailing Add:* 41 Sargent-Beechwood Brookline MA 02146

CRAIG, JOHN PHILIP, b West Liberty, Ohio, Nov 29, 23. MICROBIOLOGY, INFECTIOUS DISEASES. *Educ:* Western Reserve Univ, MD, 47; Harvard Univ, MPH, 53. *Prof Exp:* Epidemiologist, 406th Med Gen Lab, Tokyo, Japan, 51-52; fel virol, Div Med & Pub Health, Rockefeller Found, 53-54; from asst prof to assoc prof, 54-72, PROF MICROBIOL & IMMUNOL, STATE UNIV NY DOWNSTATE MED CTR, 72- *Concurrent Pos:* NIH spec res fel, Lister Inst Prev Med, London, Eng, 58-59 & Pakistan-SEATO Cholera Res Lab, Dacca, EPakistan, 64; Nat

Inst Allergy & Infectious Dis res grants, 60-82; mem, NIH Cholera Adv Comt, 70-73; mem cholera panel, US-Japan Coop Med Sci Prog, 70-, chmn, 72-77; chmn steering comt bact enteric infections, Diarrheal Dis Control Prog, WHO, 79-83; mem, Bact & Mycol Study Sect, Nat Inst Allergy & Infectious Dis, NIH, 80-84; vis prof bact, Osaka Univ, Japan, 79-80. *Mem:* Fel Infectious Dis Soc Am; Am Asn Immunologists; Am Soc Trop Med & Hyg; fel Am Pub Health Asn; Am Acad Microbiol; Am Soc Microbiol. *Res:* Seroepidemiology of diphtheria antitoxic immunity; development of in vivo assays for cholera enterotoxin, antitoxin, and toxoid; biotypic variation in vibrios associated with disease; immunomodulation of in vivo response to toxoids. *Mailing Add:* Down State Med Ctr 450 Clarkson Ave Box 44 Brooklyn NY 11203

CRAIG, KENNETH DENTON, b Calgary, Alta, Nov 21, 37; m 71; c 3. CLINICAL PSYCHOLOGY, HEALTH PSYCHOLOGY. *Educ:* Sir George Williams Univ, BA 58; Univ BC, MA, 60; Purdue Univ, PhD(psychol), 64. *Prof Exp:* PROF PSYCHOL, UNIV BC, 63-, DIR GRAD PROG CLIN PSYCHOL, 87- *Concurrent Pos:* Assoc prof psychol, Univ Calgary, 69-71; dir grad prog clin psychol, Univ BC, 80-83; dir, Can Psychol Asn, 82-88; Killan sr fel, 89-90. *Honors & Awards:* Distinguished Contrib to Psychol, Can Psychol Asn, 91. *Mem:* Can Psychol Asn (pres, 86-87); Social Sci Fedn Can (treas, 88-92); Int Asn Study Pain; Acad Behav Med Res; Soc Behav Med. *Res:* Health psychology and the psychology of pain focusing on socialization and family factors in pain experience and behavior; non verbal communication of painful distress and pain in children. *Mailing Add:* Dept Psychol Univ BC Vancouver BC V6T 1Y7 Can

CRAIG, LOUIS ELWOOD, b Clifton Hill, Mo, Dec 10, 21; m 43; c 4. ORGANIC CHEMISTRY. *Educ:* Cent Col, Mo, BA, 43; Univ Rochester, PhD(org chem), 48. *Prof Exp:* Chemist, Am Cyanamid Co, Conn, 43-46 & Gen Aniline & Film Corp, 48-54; supv res chemist, Grand River Chem Div, Deere & Co, 54-56, dir res, 56-58, dir res & tech serv, John Deere Chem Co, 58-65; mgr mkt res & develop, Kerr-McGee Chem Corp, 65-68, mgr mfg, 68-69, vpres mfg, 69-70, vpres info ser, Kerr-McGee Corp, 70-72, vpres, Chem Mfg Div, 72-77, dir energy info div, Kerr-McGee Chem Corp, 77-79, DIR INFO DIV, KERR-MCGEE CORP, 79- *Mem:* AAAS; Am Chem Soc; Am Mgt Asn. *Res:* Polymerization; pharmaceuticals; heterocyclics; inorganic chemicals; fertilizers; marketing; manufacturing. *Mailing Add:* 4921 NW 32 Oklahoma City OK 73122-1109

CRAIG, NESSLY COILE, b Honolulu, Hawaii, Nov 20, 42; m 68; c 1. CELL BIOLOGY. *Educ:* Reed Col, BA, 63; Univ Pa, PhD(biol), 67. *Prof Exp:* Fel molecular biol, Inst Cancer Res, 67-70; asst prof, 70-75, ASSOC PROF BIOL SCI, UNIV MD BALTIMORE COUNTY, 75- *Concurrent Pos:* USPHS fel, NIH, 68-70. *Mem:* AAAS; Am Soc Cell Biol; Am Soc Microbiol. *Res:* Synthesis and processing of eukaryotic RNA; control of protein synthesis; control of growth and development in higher organisms. *Mailing Add:* Dept Biol Sci Univ Md Baltimore County Baltimore MD 21228

CRAIG, NORMAN CASTLEMAN, b Washington, DC, Nov 12, 31; m 55; c 3. SPECTROSCOPY. *Educ:* Oberlin Col, BA, 53; Harvard Univ, MA, 55, PhD(chem), 57. *Prof Exp:* From asst prof to assoc prof, 57-65, assoc dean, 67-68, chmn dept, 73-80, PROF CHEM, OBERLIN COL, 65- *Concurrent Pos:* Hon fel chem, Univ Minn, 63-64; NSF sci fac fel, Univ Calif, Berkeley, 70-71; AAAS Chataugua lectr, 71-72; vis prof chem, Princeton Univ, 74-75, NIH, 78-79; summer lectr, Taiyuan Univ Tech, Shanxi, China, 82; mem, PRF Adv Bd ACS, 79-81, Adv Comt SEE, Directorate NSF, 85-88, CPT, 87-; mem, PRF Adv Bd ACS, 79-81, Adv Comt SEE, Directorate NSF, 85-88, CPT, 87- *Honors & Awards:* CMA Catalyst Award, Chem Educ, 87. *Mem:* AAAS; Am Chem Soc; Sigma Xi. *Res:* Infrared and Raman spectroscopy; normal coordinate analysis; laser applications. *Mailing Add:* Dept Chem Oberlin Col Oberlin OH 44074

CRAIG, PAUL N, b Minneapolis, Minn, Jan 2, 21; m 43; c 4. QUANTITATIVE STRUCTURE-ACTIVITY RELATIONSHIP. *Educ:* Hamline Univ, BS, 41, Univ Minn, PhD(org chem), 48. *Prof Exp:* Chemist, US Naval Res Lab, 42-45; medicinal chemist, Smith Kline & French Labs, 48-57, group leader sci info, 57-63, asst sect head, 63-71; consult, Craig Chem Consult Corp, 71-73; dir, Toxicol Info Dept, Franklin Inst Res Lab, 73-79; spec expert toxicol info, Nat Toxicol Prog, 79-83, Nat Libr Med, 83-84; CONSULT, 80- *Concurrent Pos:* Mem, Comt Modern Methods Handling Chem Info, Nat Acad Sci-Nat Res Coun, 60-66, med-chem adv panel, Walter Reed Army Inst Res, 80-; Nat Inst Neurol & Commun Dis & Stroke, NIH, Epilepsy Prog Adv Bd, 89- *Mem:* Fel AAAS; Am Chem Soc. *Res:* Structure-activity relationships as applied to toxic organic chemicals; mathematical models for estimation of LD-50, carcinogenesis and mutagenesis; mathematical models of toxicity. *Mailing Add:* 4931 Mariners Dr Shady Side MD 20764

CRAIG, PAUL PALMER, b Reading, Pa, July 29, 33; m 85; c 2. PHYSICS, SCIENCE. *Educ:* Haverford Col, BS, 54; Calif Inst Technol, PhD(physics), 59. *Prof Exp:* Mem staff cryogenics, Los Alamos Sci Lab, 58-62; assoc physicist, Brookhaven Nat Lab, 62-66, group leader, 62-72, physicist, 66-72; PROF PHYSICS, UNIV CALIF, DAVIS, 77- *Concurrent Pos:* Guggenheim Found fel, 65-66; assoc prof, State Univ NY Stony Brook, 67-71; mem bd trustees, Environ Defense Fund, 67-71; mem staff, NSF, 71-74, dep dir & actg dir, Off Energy Res & Develop Policy, 74-75; dir, Energy & Resources Coun, Univ Calif, 75-78; scientist, Lawrence Berkely Lab, Univ Calif, 77- *Mem:* Fel Am Phys Soc; Sigma Xi; Sierra Club. *Res:* Energy policy; alternative energy strategies; energy conservation; cryogenics; nuclear arms race. *Mailing Add:* 2425 Elendil Lane Davis CA 95616

CRAIG, PETER HARRY, b Pittsburgh, Pa, Dec 25, 29; m 54, 87; c 4. TOXICOLOGY. *Educ:* Pa State Univ, BS, 52; Univ Pa, VMD, 55, MS, 58. *Prof Exp:* Instr vet path, Univ Pa, 55-58; mem staff, Armed Forces Inst Path, 58-59 & 60-61; mem staff, Aviation Med Acceleration Lab, 59-60; asst prof vet path, Univ Pa, 61-65; assoc prof path, State Univ NY Vet Col, Cornell Univ, 65-73; MGR, HEALTH RISK ASSESSMENT & TOXICOL, MOBIL OIL CORP, 89- *Mem:* Am Vet Med Asn; NY Acad Sci; Am Col Vet Path; Sigma Xi; Am Col Toxicol; Soc Risk Anal. *Res:* Experimental pathology, transplantation, cancer, radiation; laboratory animal, aviation, and veterinary pathology; toxicology; risk analysis. *Mailing Add:* Mobil Environ Health Box 1031 Princeton NJ 08543-1031

CRAIG, RAYMOND ALLEN, b Mansfield, Ohio, Apr 19, 20; m 48; c 3. CHEMISTRY. *Educ:* Muskingum Col, BS, 43; Ohio State Univ, MS & PhD(phys & org chem), 48. *Prof Exp:* Chemist, Am Petrol Inst, Ohio, 43-44; asst chem, Ohio State Univ, 44-47; res chemist nylon res lab, E I du Pont de Nemours & Co, Inc, 48-51, res supvr, 52-54, dacron res lab, 54-57, sr res supvr, 57-59, res mgr, 59-63, textile res lab, Del, 63-66, dir, Benger Lab, Va, 66-68, tech dir, Du Pont Int, SA, Switz, 68-74, tech dir, 74-85; RETIRED. *Mem:* Am Chem Soc. *Res:* High polymer chemistry and polymer chemistry; synthetic fibers. *Mailing Add:* 512 Kerfoot Farm Rd Woodbrook Wilmington DE 19803

CRAIG, RAYMOND S, b DeLand, Fla, Jan 11, 17; m 50; c 2. PHYSICAL CHEMISTRY. *Educ:* Stetson Univ, BS, 39; Univ Pittsburgh, PhD(chem), 44. *Prof Exp:* Res assoc, Allegany Ballistics Lab, 44-46; sr res fel, 46-49, from asst res prof to assoc res prof, 49-59, PROF CHEM, UNIV PITTSBURGH, 59- *Mem:* Am Chem Soc; Sigma Xi. *Res:* Magnetic and thermal properties of metals and intermetallic compounds; low temperature calorimetry. *Mailing Add:* Dept Chem Univ Pittsburgh 172 E Latan Dr Pittsburgh PA 15243

CRAIG, RICHARD, b Carnegie, Pa, July 14, 37; m 61; c 3. PLANT GENETICS, PLANT BREEDING. *Educ:* Pa State Univ, BS, 59, MS, 60, PhD(genetics & breeding), 63. *Prof Exp:* Asst hort, 59-61 & comput sci, 61-62, instr plant breeding & asst prof, 63-71, assoc prof, 71-81, PROF PLANT BREEDING, PA STATE UNIV, 81-, DIR INST ADV FLORICULT, 86- *Honors & Awards:* Spec Recommendation Hort Achievement, 64; Bronze Medal Res, African Violet Soc Am, 81. *Mem:* Fel Am Soc Hort Sci; Am Genetic Asn. *Res:* Genetics, cytology and breeding of floricultural plants, including geraniums, exacum, regal pelargoniums, African violets and primula; genetics, anatomy and biochemistry of pest resistance in geraniums. *Mailing Add:* Dept Hort Pa State Univ Main Campus University Park PA 16802

CRAIG, RICHARD ANDERSON, b New York, NY, Aug 9, 36. PHYSICS. *Educ:* Univ Ill, BS, 59, MS, 60, PhD(physics), 66. *Prof Exp:* Asst physics, Univ Ill, 59-66; res assoc physics, Univ Ore, 66-67; lecturer physics, Univ Calif, Riverside, 67-69; asst prof physics, Ore State Univ, 69-71; mem staff, Battelle Inst, 71-74, res scientist, Columbus Labs, 74-78, SR RES SCIENTIST, PAC NORTHWEST LAB, WASH STATE UNIV, 78- *Concurrent Pos:* Lectr, elect engrs, Tri-Cities Campus, Univ Wash, 88- *Mem:* Am Physical Soc; Sigma Xi. *Res:* Condensed matter physics; especially interaction of electromagnetic radiation with condensed matter on a macroscopic or microscopic basis. *Mailing Add:* 1831 Dove Lane West Richland WA 99352

CRAIG, RICHARD ANSEL, meteorology; deceased, see previous edition for last biography

CRAIG, RICHARD GARY, b Wilmington, Del, June 3, 49; m 82; c 1. COMPUTER SIMULATION, STATISTICS. *Educ:* Dickinson Col, BA, 71; Pa State Univ, MS, 76, PhD(geol), 79. *Prof Exp:* Asst prof, 78-81, ASSOC PROF GEOL, KENT STATE UNIV, 81- *Concurrent Pos:* Consult, Battelle Mem Inst, Pac Northwest Labs, 81-, Rockwell Int, 84-87; mem, adv panel, Oak Ridge Nat Labs, 84-87; mem, Shuttle Imaging Radar Sci Working Group, NASA, 84-85. *Mem:* Geol Soc Am; AAAS; Am Quaternary Asn; Int Asn Math Geol; Am Geophys Union. *Res:* Computer simulation of large geologic systems; models of local and global climate change over long time spans; long-term stability of repositions for high level nuclear waste. *Mailing Add:* Dept Geol Kent State Univ Kent OH 44242-0001

CRAIG, ROBERT BRUCE, b Washington, DC, Apr 22, 44; m 66; c 2. ANIMAL ECOLOGY. *Educ:* Univ Calif, Davis, BS, 70, MS, 72, PhD(ecol), 74. *Prof Exp:* Res assoc zool, Univ Calif, Davis; res staff mem ecol, Environ Sci Div, Oak Ridge Nat Lab, 74-78, prog mgr ecol, 78-81; DEPT MGR, HENNINGSON, DURHAM & RICHARDSON, 81- *Mem:* Ecol Soc Am; Am Ornithologists Union; Am Nuclear Soc. *Res:* Predation theory and predatory behavior of insectivorous birds; environmental impact analysis and environmental effects of developing energy technologies. *Mailing Add:* Gen Del Santa Barbara CA 93105

CRAIG, ROBERT GEORGE, b Charlevoix, Mich, Sept 8, 23; m 45; c 3. DENTAL MATERIALS RESEARCH. *Educ:* Univ Mich, BS, 44, MS, 51, PhD(chem), 55. *Hon Degrees:* Dr, Univ Geneva, 89. *Prof Exp:* Chemist, anal res, Linde Air Prods, 44-50 & friction & lubrication, Texaco, 54-55; assoc res chemist high polymers, Eng Res Inst, 55-57, from asst prof to assoc prof, 57-65, PROF DENT, UNIV MICH, 65-, PROF BIOL & MAT SCI, 87- *Concurrent Pos:* Chmn, dept dent mat, Univ Mich, 69-87. *Honors & Awards:* Thomas Young Award, 69; Wilmer Souder Award, Int Asn Dent Res, 75; Clemson Award, Soc Biomat, 78; Hollenback Prize, Acad Oper Dent, 91. *Mem:* Am Chem Soc; Am Asn Dent Schs; Int Asn Dent Res; Soc Biomat; Acad Sports Dent. *Res:* Colloid and surface chemistry; polymer chemistry; dental materials; bio-engineering; friction and wear; stress analysis; elastomers and polymer-ceramic composites used in dentistry; use of experimental and numerical stress analysis to determine the optimum design of dental restorations. *Mailing Add:* 1503 Wells St Ann Arbor MI 48104

CRAIG, ROY PHILLIP, b Durango, Colo, May 10, 24. ENVIRONMENTAL MANAGEMENT, SCIENCE WRITING. *Educ:* Univ Colo, BA, 48; Calif Inst Technol, MS, 50; Iowa State Col, PhD(chem), 52. *Prof Exp:* Asst, Iowa State Col, 49-52; prof chemist, Dow Chem Co, 52-56, group leader, 56-60; lectr phys chem, Univ Colo, 61-65, assoc prof phys chem & coordr phys sci, div integrated studies, 65-68; vis prof, Univ Hawaii, 69 & Colo Col, 79; pres, 74-81, DIR, FOUR CORNERS ENVIRON RES

INST, 74- *Concurrent Pos:* Tech & educ consult & sci writer, 69- *Mem:* AAAS. *Res:* Impact of science on society; relation between science and the humanities; physical sciences for nonscience majors; solution adsorption; environmental management; solar energy utilization. *Mailing Add:* PO Box 335 Ignacio CO 81137

CRAIG, STANLEY HAROLD, b New York, NY, Oct 24, 09. RADIOLOGY. *Educ:* NY Univ, BS, 30; Univ Basel, MD, 35. *Prof Exp:* Resident radiol, Beth Israel Hosp, NY, 35-38; PROF RADIOL, AM COL GASTROGENTEROL, NY MED COL, 48- *Concurrent Pos:* Assoc prof, Med Col, New York Univ, 47-54; consult radiol, Med Dept, NY Times, 48- *Mem:* Radiol Soc NAm; Int Skeletal Soc; AMA; fel Am Col Gastroenterol; fel Am Col Radiol. *Mailing Add:* 3800 Oaks Clubhouse Dr Pompano Beach FL 33060

CRAIG, SUSAN WALKER, b New York, NY; m 68. IMMUNOBIOLOGY. *Educ:* Univ Pa, BA, 67; Johns Hopkins Univ, PhD(biol), 73. *Prof Exp:* Fel molecular pharmacol, Dept Pharmacol, 73-75, ASST PROF PHYSIOL CHEM, SCH MED, JOHNS HOPKINS UNIV, 75- *Concurrent Pos:* Fel, Jane Coffin Childs Mem Fund Med Res, 73-75. *Res:* Cell surface control of lymphocyte physiology. *Mailing Add:* Dept Biolchem Sch Med Johns Hopkins Univ 725 N Wolfe St Baltimore MD 21205

CRAIG, SYNDEY POLLOCK, III, b Washington, DC, Mar 27, 45; c 3. NUCLEIC ACID BIOCHEMISTY, MOLECULAR EVOLUTION. *Educ:* Purdue Univ, BS, 66; Calif Inst Technol, Pasadena, PhD(develop biol & genetics), 70. *Prof Exp:* Staff fel, Nat Inst Child Health & Human Develop, NIH, 70-72; asst prof biol, Newcombe Col, Tulane Univ, 72-74; assoc prof, Univ SC, Columbia, 75-78; from assoc prof to prof, Univ Southwestern La, Lafayette, 79-82; CHMN BIOL DEPT, WARREN WILSON COL, 82- *Concurrent Pos:* Adj prof, dept parasitol & lab pract, Sch Pub Health, Univ NC, Chapel Hill, 83- *Mem:* Genetics Soc Am; Am Soc Microbiol; Am Soc Cell Biol; Sigma Xi. *Res:* Investigation of mitochondrial biogenesis in early embryogenesis; gene expression in chicken lens development; the sequence organization and evolution of DNA in Trypanosoma cruzi, the etiologic agent for Chagas' Disease. *Mailing Add:* Dept Biol Box 5036 Warren Wilson Col 701 Warren Wilson Colleg Rd Swannanoa NC 28778

CRAIG, WILFRED STUART, b New Hartford, Mo, Aug 11, 16; m 44; c 3. ENTOMOLOGY. *Educ:* Univ Mo, BS & AB, 47; Iowa State Col, PhD(entom), 53. *Prof Exp:* Asst state entomologist, Mo, 47-49; asst entomologist, Iowa State Col, 49-53, asst state entomologist, Iowa State Univ, 53-60, state entomologist, Univ & entomologist, Exp Sta, 60-65; exten entomologist, 65-85, EMER PROF, UNIV MO, COLUMBIA, 85- *Mem:* Entom Soc Am; Sigma Xi. *Res:* Biology and control of nursery insects; taxonomy of Anthicidae and other Coleoptera. *Mailing Add:* 2313 Braemore Rd Columbia MO 65203

CRAIG, WILLIAM WARREN, b Kansas City, Mo, Apr 1, 35; m 61; c 5. GEOLOGY. *Educ:* Univ Mo, Columbia, BA, 57, MA, 61; Univ Tex, Austin, PhD(geol), 68. *Prof Exp:* Assoc prof geol, Northeastern Mo State Col, 65-68; from asst prof to assoc prof, 68-77, dept chair, 77-81, PROF GEOL, UNIV NEW ORLEANS, 77-, DEPT CHAIR, 90- *Mem:* Geol Soc Am; Soc Econ Paleont & Mineral. *Res:* Cretaceous nonmarine Ostracodes; midcontinent Ordovician and Silurian stratigraphy and conodont biostratigraphy; stratigraphy and structure core area Ouachita Mountains, Arkansas. *Mailing Add:* Dept Geol Geophys Univ New Orleans Lakefront New Orleans LA 70148

CRAIG, WINSTON JOHN, b Melbourne, Australia, Sept 21, 47; m 71; c 1. NUTRITIONAL & LIPID BIOCHEMISTRY. *Educ:* Univ Newcastle, Australia, BSc Hons, 68; Univ Queensland, Australia, PhD(organ chem), 71; Loma Linda Univ, MPH, 75. *Prof Exp:* Fel marine chem, Australian Nat Univ, 71-72; res assoc insect chem, Col Environ Sci & Forestry, State Univ NY, 72, res assoc marine chem, sci & forestry, Univ Okla, 72-73; res asst nutrit, Loma Linda Univ, 74; instr sci, Kingsway Col, Can, 74-76; asst prof sci, chem & health, Adventist Col, WAfrica, 76-79; asst prof nutrit, Sch Health, Loma Linda Univ, 79-82, assoc prof, 82-84; ASSOC PROF HUMAN NUTRIT, UNIV MASS, AMHERST, 85- *Concurrent Pos:* Consult, Med Ctr, Loma Linda Univ, 81-84; lectr, Cent Am Master Sci Pub Health Prog, 79-83; guest lectr, Col Osteop Med, Pomona, Calif, 82-83; vis prof, Framingham State Col, 85. *Mem:* Am Dietetic Asn; Am Col Nutrit; Inst Food Technol; Am Inst Nutrit. *Res:* Effect of dietary factors (dietary fiber, type of fat, trace minerals and caffeine) upon a persons health and nutritional well-being; advantages of a vegetarian diet; effect of diet on serum lipids of children. *Mailing Add:* Dept Home Econ Andrews Univ Berrien Springs MI 49104

CRAIGE, ERNEST, b El Paso, Tex, 1918; m 46; c 4. MEDICINE. *Educ:* Univ NC, BA, 39; Harvard Univ, MD, 43. *Prof Exp:* From intern to resident med, Mass Gen Hosp, 43-48, resident cardiol, 49-50, asst med, 50-52; from asst prof to assoc prof, 52-62, PROF MED, SCH MED, UNIV NC, CHAPEL HILL, 62- *Concurrent Pos:* Clin & res fel med, Mass Gen Hosp, 49-S0; fel med, Harvard Med Sch, 49-50; chief cardiol, NC Mem Hosp, 52-78. *Mem:* Asn Univ Cardiologists; Am Heart Asn; fel Am Col Cardiol; Am Clin & Climat Asn. *Res:* Echo-phono-cardiography and other non-invasive methods of studying cardiac function. *Mailing Add:* 338 Clin Sci Bldg 229H Univ NC Sch Med Chapel Hill NC 27514

CRAIGHEAD, JOHN EDWARD, b Pittsburgh, Pa, Aug 14, 30; m 57; c 2. PATHOLOGY, VIROLOGY. *Educ:* Univ Utah, BS, 52, MD, 56. *Prof Exp:* Intern med, Barnes Hosp, St Louis, Mo, 56-57; jr asst resident, Peter Bent Brigham Hosp, 60-61, sr asst resident, 61-62, chief resident path, 62-63, assoc, 63-68; PROF PATH, COL MED, UNIV VT, 68-, CHMN DEPT, 74- *Concurrent Pos:* Teaching fel, Harvard Med Sch, 61-63; assoc, 63-66, asst prof, 66-68; assoc mem comn viral infections, Armed Forces Epidemiol Bd, 66; mem infectious dis adv comt, Nat Inst Allergy & Infectious Dis, 71-75. *Mem:* Am Soc Exp Path; Am Asn Path & Bact; Int Acad Path; Soc Exp Biol & Med; Am Soc Clin Path; Sigma Xi. *Res:* Biology, pathogenesis and pathology of virus disease in man and animals; diabetes; pulmonary disease. *Mailing Add:* Nine Iranistan Rd Burlington VT 05401

CRAIGHEAD, JOHN J, b Washington, DC, Aug 14, 16; m 44; c 3. ECOLOGY. *Educ:* Pa State Col, BA, 39; Univ Mich, MS, 40, PhD, 50. *Prof Exp:* Biologist, NY Zool Soc, 47-49; dir survival training for armed forces, US Dept Defense, 50-52; wildlife biologist & leader Mont Coop Wildlife Res Unit, Bur Sport Fisheries & Wildlife, 52-77; PROF ZOOL & FORESTRY, UNIV MONT, 52- *Concurrent Pos:* Grants, Wildlife Mgt Inst, raptor predation; NSF, ecol of grizzly bear & radiotracking grizzly bears; AEC, radiotracking & telemetering systs large western mammals; NASA, satellite tracking large mammals & habitat mapping; Bur Sport Fisheries & Wildlife, surv peregrine eyries & studies ecol grizzly bear; US Forest Serv & Mont Fish & Game Dept, eval grizzly bear habitat. *Honors & Awards:* Bur Sport Fisheries & Wildlife Superior Performance Award, 65; Citation for Organizing & Admin Navy's Land Survival Training Prog, Secy Defense; Arthur S Einarsen Award, Wildlife Soc, 77; Am Motors Conserv Award, 78; John Oliver LaGorce Gold Medal, Nat Geol Soc, 79. *Mem:* Wildlife Soc (vpres, 62-63); AAAS; Wilderness Alliance. *Res:* Raptor predation; waterfowl; population dynamics. *Mailing Add:* 5125 Orchard Lane Missoula MT 59801

CRAIGIE, EDWARD HORNE, comparative neuroanatomy; deceased, see previous edition for last biography

CRAIGMILES, JULIAN PRYOR, b Thomasville, Ga, Jan 17, 21; m 48; c 3. AGRONOMY. *Educ:* Univ Ga, BSA, 42, MSA, 48; Cornell Univ, PhD, 52. *Prof Exp:* Asst, Coker Pedigreed Seed Co, SC, 41; asst agronomist, Ga Crop Improv Asn, 46-48; agronomist, Ga Exp Sta, 48-64; PROF & RESIDENT DIR, AGR RES & EXTEN CTR, TEX A&M UNIV, 64- *Concurrent Pos:* Res award, Ga Plant Food Educ Soc, 59- *Honors & Awards:* Outstanding Researcher, Sears Roebuck & Co, 58. *Mem:* Am Soc Agron; Weed Sci Soc Am; Am Genetic Asn; Crop Sci Soc Am. *Res:* Soybean breeding; rice and forage management; seed production of rice, soybeans and forages; developed cytoplasmic sterile sudan grass, which led to development and release of hybrid sudan grass; developement of varieties of grasses and soybeans; management of rice, soybeans, forages and other crops. *Mailing Add:* 5925 Honeysuckle Beaumont TX 77706

CRAIK, CHARLES S, b East Liverpool, Ohio, June 5, 54. PROTEIN ENGINEERING, STRUCTURAL BIOLOGY. *Educ:* Allegheny Col, Meadville, Pa, BS, 76; Columbia Univ, New York, NY, PhD(chem), 81. *Prof Exp:* Grad student chem, Columbia Univ, 76-81; postdoctoral fel, 81-85, asst prof, 85-91, ASSOC PROF BIOCHEM, UNIV CALIF, SAN FRANCISCO, 91- *Concurrent Pos:* Assoc ed, Protein Eng, 88- *Mem:* Am Chem Soc; AAAS; Am Soc Biol Chemists; Protein Soc. *Res:* Analysis and use of structure-function relationships in proteins; protein engineering using genetic, recombinant and biophysical methods; structure based inhibitor and drug design. *Mailing Add:* Dept Biochem & Pharmaceut Chem Univ Calif Box 0446 S-926 San Francisco CA 94143-0446

CRAIK, EVA LEE, b Gatesville, Tex, Aug 12, 19; m 41; c 4. BIOLOGY, SCIENCE EDUCATION. *Educ:* Tex Women's Univ, BS, 40; Hardin-Simmons Univ, MEd, 60; NTex State Univ, EdD(biol), 66. *Prof Exp:* Teacher, Ranger High Sch & Jr Col, Tex, 40-43, pub sch, Kans, 43-44 & pub schs, Tex, 59-62; from instr to prof biol, Hardin-Simmons Univ, 62-84, coordr sci educ, 62-84; RETIRED. *Concurrent Pos:* Vol mission serv corp, Ricks Inst, Monrovia, Liberia, 84. *Mem:* Nat Sci Teachers Asn; Nat Asn Biol Teachers. *Res:* Relative effectiveness of inductive-deductive and deductive-descriptive methods of teaching college zoology. *Mailing Add:* 1802 N 11th Abilene TX 79603

CRAIN, CULLEN MALONE, b Goodnight, Tex, Sept 10, 20; m 43; c 2. ELECTRONICS. *Educ:* Univ Tex, BS, 42, MS, 47, PhD(elec eng), 52. *Prof Exp:* Engr, Philco Corp, Philadelphia, 42-43; instr elec eng, Univ Tex, 43-44; proj engr, Off Res & Inventions, US Dept Navy, Washington, DC, 45-46; asst prof elec eng & elec engr, Res Lab, Univ Tex, 46-52, assoc prof, 52-57; group leader, Electronic Dept, Rand Corp, 57-68, dep dept head, Eng Sci Dept, 68-76, assoc dept head, 76-86, dept head, Eng & Appl Sci Dept, 86-88; RETIRED. *Concurrent Pos:* Chmn, US Comn F, Internation Union Radio Sci, 63-66; mem frequency mgt adv coun, Off Telecommun Policy, Exec Off of the President, 65-87. *Mem:* Nat Acad Eng; fel Inst Elec & Electronics Engrs. *Res:* Propagation of radio waves; radio communication systems; radio physics; nuclear effects. *Mailing Add:* 463 17th Santa Monica CA 90402

CRAIN, DONALD LEE, b St Joseph, Mo, Apr 20, 33; m 53; c 3. ORGANIC CHEMISTRY. *Educ:* William Jewell Col, AB, 54; Purdue Univ, PhD(org chem), 58. *Prof Exp:* Res chemist, Res & Develop, 58-78, group leader, 62-65, sect suprv, 65-68, br mgr, 68-76, budget mgr chem group, 76-77, dir plastics & anal, 77-81, MGR, POLYMERS & MAT DIV, RES & DEVELOP, PHILLIPS PETROL CO, 81- *Mem:* Am Chem Soc; Soc Plastics Engrs. *Res:* Petrochemicals; polymers, composites and advanced materials. *Mailing Add:* 1324 SE Ridgewood Dr Bartlesville OK 74006

CRAIN, RICHARD CULLEN, b Portsmouth, Va, Feb 8, 51; m 72; c 2. LIPIDOLOGY. *Educ:* Dartmouth Col, BA, 73; Univ Rochester, PhD(biochem), 78. *Prof Exp:* NIH postdoctoral fel, Cornell Univ, 78-80; asst prof, 80-86, ASSOC PROF, UNIV CONN, 86- *Mem:* Am Soc Plant Physiologists; Sigma Xi; Am Soc Biochem & Molecular Biol. *Res:* Structure and function of biological membranes; signal transduction in photosynthetic organisms. *Mailing Add:* Molecular & Cell Biol U-125 Univ Conn Storrs CT 06269-3125

CRAIN, RICHARD WILLSON, JR, b Denver, Colo, July 2, 31; m 55; c 6. ENERGY SYSTEMS, APPLIED THERMODYNAMICS. *Educ:* Univ Wash, BS, 53, MS, 55; Univ Mich, PhD(mech eng), 65. *Prof Exp:* Assoc prof, 65-74, PROF MECH ENG, WASH STATE UNIV, 74- *Concurrent Pos:* Sr engr, Varian Assocs, Palo Alto, Calif, 87-88; chmn dept, Wash State Univ, 76-87, actg dean, 83-84. *Mem:* Am Soc Mech Engrs; Am Soc Eng Educ; Sigma Xi. *Res:* Evaluation of thermodynamic properties, especially at cryogenic temperatures; equations of state for gases; thermal pollution of water bodies; applied thermodynamics; thermal power systems; waste heat utilization. *Mailing Add:* Dept Mech Eng Wash State Univ Pullman WA 99164-2920

CRAIN, STANLEY M, b New York, NY, Feb 5, 23; m 46; c 2. NEUROPHYSIOLOGY, DEVELOPMENTAL NEUROBIOLOGY. *Educ:* Brooklyn Col, AB, 43; Columbia Univ, PhD(biophys), 54. *Prof Exp:* Group leader radiol res, Health Physics Div, Argonne Nat Lab, 45-47; res electrophysiologist, Dept Neurol, Columbia Univ, 50-57; res cell physiologist & sect head nerve tissue cult lab, Abbott Labs, Ill, 57-61; asst prof anat, Dept Neurol, Col Physicians & Surgeons, Columbia Univ, 61-65; from an assoc prof to prof physiol, 65-74, PROF NEUROSCI & PHYSIOL/BIOPHYS, ALBERT EINSTEIN COL MED, 69- *Concurrent Pos:* Grass res fel, Marine Biol Lab, Woods Hole, Mass, 57; NIH res career develop fel, 61-65; Kennedy scholar, Rose F Kennedy Ctr Ment Retardation & Human Develop, 65-75; assoc ed, J Neurobiol, 70-86, Develop Neurosci, 78- & Develop Brain Res, 81-; chmn med adv bd, Dysautonomia Found, 77-79 & 81-82. *Mem:* Am Pharmacol Exp Ther Soc; Am Asn Anatomists; Am Soc Cell Biologists; Soc Neurosci; Tissue Cult Asn; Int Narcotics Res Soc. *Res:* Electrophysiologic, pharmacologic and cytologic research on organotypic tissue cultures of fetal mammalian brain, spinal cord sensory ganglia and neuromuscular systems during maturation; in vitro model systems for analyses of cellular mechanisms underlying opioid addiction. *Mailing Add:* Dept Neurosci & Physiol/Biophys Albert Einstein Col Med 1300 Morris Park Ave Bronx NY 10461

CRAIN, WILLIAM RATHBONE, JR, b Cuero, Tex, Nov 30, 44; div; c 2. MOLECULAR BIOLOGY OF DEVELOPMENT. *Educ:* Univ Tex, Austin, BA, 67; Univ Houston, MS, 69; Grad Sch Biomed Sci, Univ Tex, PhD(molecular biol), 74. *Prof Exp:* Fel molecular biol, Calif Inst Technol, 74-76; staff scientist molecular biol, 76-84, SR SCIENTIST, WORCESTER FOUND EXP BIOL, 84-; ASSOC PROF, MOLECULAR GENETICS & MICROBIOL, UNIV MASS MED SCH, 84- *Concurrent Pos:* Instr embryol course, Marine Biol Lab, Woods Hole, MA, 79-85. *Mem:* AAAS; Am Soc Cell Biol; Soc Develop Biol. *Res:* Regulation of gene expression during embryonic development; structure and evolution of action genes. *Mailing Add:* Dept Cell Biol Worcester Found Exp Biol 222 Maple Ave Shrewsbury MA 01545

CRAINE, ELLIOTT MAURICE, b Burlington, Iowa, Oct 5, 24; m 50; c 2. ANALYTICAL BIOCHEMISTRY. *Educ:* Univ Ill, AB, 49, BS, 51, PhD(biochem), 52. *Prof Exp:* Chemist, Agr Res Serv, USDA, 54-58; res chemist, Upjohn Co, 58-63; res biochemist, Hess & Clark Div, Rhodia, Inc, 63-65, head biochem, 65-80; mgr metab & anal chem, 80-86, DIR CHEM, WIL RES LABS, 86- *Mem:* AAAS; Am Chem Soc. *Res:* Chemistry of seed proteins; isolation of antibiotics; drug metabolism; biochemistry of ruminants; metabolism of foreign compounds in animals, plants and soil; drugs and pesticides; analytical biochemistry; isolation of natural chemicals. *Mailing Add:* 827 Ridge Rd Ashland OH 44805-1006

CRAINE, LLOYD BERNARD, b St Paul, Minn, July 8, 21; m 45; c 4. ELECTRICAL ENGINEERING, ELECTRICAL MEASUREMENT. *Educ:* Ore State Col, BS, 47, MS, 50, EE, 53. *Prof Exp:* Instr elec eng, Ore State Col, 48-50; asst prof, Univ Idaho, 50-56; assoc prof, 56-64, PROF ELEC ENG, WASH STATE UNIV, 64- *Concurrent Pos:* Prog dir aeronomy, NSF, 65-66; Cong fel, Inst Elec & Electronic Engrs, 74-75; prof dir, Wash Energy Exten Serv, 77-78; Consult, 87- *Mem:* Inst Elec & Electronics Engrs; Am Soc Eng Educ; AAAS. *Res:* Electromagnetic wave propagation; electrical transmission distribution neutral/grounding problems; electrical measurements; energy technology and behavior modification. *Mailing Add:* Dept Elec & Comput Eng Wash State Univ Pullman WA 99164-2752

CRAKER, LYLE E, b Reedsburg, Wis, Feb 3, 41; m 63; c 3. PLANT PHYSIOLOGY. *Educ:* Univ Wis, Madison, BS, 63; Univ Minn, St Paul, PhD(agron), 67. *Prof Exp:* from asst prof to prof plant physiol, 69-89, actg assoc dean, Grad Sch, 78-80, DEPT HEAD, PLANT & SOIL SCI, UNIV MASS, 89- *Concurrent Pos:* Ed, Herb, Spice & Medicinal Plant Digest, 83- & J of Herbs, Spices & Medicinal Plants, 91-; co-ed-in-chief, Herbs, Spices & Medicinal Plants, Recent Adv in Bot Hort & Pharmacol, 85- *Honors & Awards:* Oberly Award, 85. *Mem:* Am Soc Plant Physiol; Am Soc Agron; Crop Sci Soc Am; Am Soc Hort Sci. *Res:* Utilization of lights and plant hormones for regulated crop production and storage; effects of air pollution on plant development; culture of herbs and aromatic plants. *Mailing Add:* Dept Plant & Soil Sci Univ Mass Amherst MA 01002

CRALL, JAMES MONROE, b Monongahela, Pa, July 13, 14; m 43; c 2. PLANT BREEDING, PLANT PATHOLOGY. *Educ:* Purdue Univ, BS, 39; Univ Mo, AM, 41, PhD(plant path), 48. *Prof Exp:* Tomato blight agent, Exp Sta, Purdue Univ & Bur Plant Indust, USDA, 38-39; asst bot, Univ Mo, 39-42, jr plant pathologist, Exp Sta, 46-48; res asst prof plant path, Iowa State Univ & assoc path, USDA, 48-52, dir, Agr Res Ctr, Leesburg, Agr Exp Sta, Inst Food & Agr Sci, 52-77, PROF PLANT PATH, UNIV FLA, 52- *Concurrent Pos:* Weather forecaster, US, India, China, US Army Air Force, 43-44. *Mem:* AAAS; Am Soc Hort Sci; Am Phytopath Soc; Coun Agr Sci & Technol. *Res:* Diseases of watermelon and grape; watermelon breeding; new watermelon varieties for commercial production in US and other areas; fruit yields, quality and disease resistance; inheritance studies in watermelon. *Mailing Add:* IFAS Cent Fla Res & Educ Ctr Univ Fla 5336 Univ Ave Leesburg FL 32748

CRALLEY, JOHN CLEMENT, b Carmi, Ill, Oct 16, 32; m 63; c 3. HUMAN ANATOMY, PHYSIOLOGY. *Educ:* Univ Ill, BS, 56, MS, 60, PhD(zool), 65. *Prof Exp:* Asst prof zool, 63-76, ASSOC PROF ANAT, ILL STATE UNIV, 76-, ASST TO CHMN & DIR TEACHING ASSTS, 88- *Mem:* Am Asn Anatomists. *Res:* Blood vascular systems; functional anatomy of the human foot. *Mailing Add:* Dept Biol Ill State Univ Normal IL 61761

CRAM, DONALD JAMES, b Chester, Vt, Apr 22, 19; m 69. ENZYME MIMICS, SYNTHETIC IONOFORES. *Educ:* Rollins Col, BS, 41; Univ Nebr, MS, 42; Harvard Univ, PhD(org chem), 47. *Hon Degrees:* DSc, Uppsala Univ, 77, Univ Southern Calif, 83, Rollins Col, 88, Univ Nebr & Univ Western Ont, 89. *Prof Exp:* Res chemist, Merck & Co, NJ, 42-45; instr org chem & Am Chem Soc fel mold metabolites, 47-48, from asst prof to prof chem, 48-85, SAUL WINSTEIN PROF ORG CHEM, UNIV CALIF, LOS ANGELES, 85- *Concurrent Pos:* Consult, Upjohn Co, 52-87 & Union Carbide, 61-81; Guggenheim fel, 54-55; mem, Adv Bd Org Reactions, Am Chem Soc, 70-; lectr, numerous US & foreign univs, 74- *Honors & Awards:* Nobel Prize chem, 87; Am Chem Soc Award, 53 & 65; Herbert Newby McCoy Award, 65 & 75; Soc Chem Mfg Asn Award, 65; Arthur C Cope Award, Am Chem Soc, 74, Roger Adams Award, 85; R C Fuson Lectr, Univ Nev, 79; Beckman Lectr, Calif Inst Technol, 88; Seaborg Medal, 89. *Mem:* Nat Acad Sci; Am Chem Soc; Swiss Chem Soc; Am Acad Arts & Sci. *Res:* Stereochemistry, especially of carbanions and substitutions at sulfur; macro ring chemistry; synthetic multiheteromacrocycles that model enzyme systems in complexation and catalysis; highly structured molecular complexes; host-guest chemistry. *Mailing Add:* Dept Chem & Biochem Univ Calif 405 Hilgard Ave Los Angeles CA 90024

CRAM, LEIGHTON SCOTT, b Emporia, Kans, Oct 21, 42; m 65. BIOPHYSICS. *Educ:* Kans State Teachers Col, BA, 64; Vanderbilt Univ, MS, 66; Pa State Univ, PhD(biophys), 69. *Prof Exp:* Fel, 69-71, DEP DIV LEADER, LIFE SCI DIV, LOS ALAMOS NAT LAB, 71- *Concurrent Pos:* Consult, Particle Technol Inc, 72-75 & Coulter Electronics, 75- *Mem:* AAAS; Biophys Soc; Int Soc Anal Cytol. *Res:* Photon emission from thin films bombarded by electrons and DNA conformation in bacterial viruses; development and application of high speed methods of obtaining data on cellular properties and intracellular components, and cytogenetics of tumorigenic cells. *Mailing Add:* 540 Navajo Los Alamos NM 87544

CRAM, SHELDON LEWIS, physics, for more information see previous edition

CRAM, WILLIAM THOMAS, b South Burnaby, BC, Oct 20, 27; m 51; c 2. ENTOMOLOGY. *Educ:* Univ BC, BSA, 50; Ore State Univ, MS, 55, PhD(entom), 64. *Prof Exp:* Scientist, res sta, Can Dept Agr, 50-84; RETIRED. *Mem:* Entom Soc Am; Entom Soc Can. *Res:* Ecology of root weevils as pests of berry crops; nutrition, host selection and fecundity of root weevils, especially Otiorhynchus sulcatus; pests of berry crops. *Mailing Add:* 12939-22A Ave White Rock BC V4A 7E4 Can

CRAMBLETT, HENRY G, b Scio, Ohio, Feb 8, 29; m 60; c 2. PEDIATRICS, VIROLOGY. *Educ:* Mt Union Col, BS, 50; Univ Cincinnati, MD, 53; Am Bd Pediat, dipl; Am Bd Med Microbiol, dipl. *Hon Degrees:* DSc Mt Union Col, 74. *Prof Exp:* Intern med, Harvard Med Serv, Boston City Hosp, 53-54; resident pediat, Children's Hosp, Cincinnati, Ohio, 54-55; res assoc, Clin Ctr, NIH, 55-57; chief resident & instr pediat, Univ Iowa, 57-58, asst prof, 58-60; dir virol lab, Bowman Gray Sch Med, 60-64, from assoc prof to prof pediat, Col Med, Ohio State Univ, 60-64; chmn dept med microbiol, 66-73, dean col med, 73-80, actg vpres med affairs, 74-80, vpres health sci, 80-82, PROF PEDIAT, OHIO STATE UNIV, 64-, PROF MED MICROBIOL, 66-, POMERENE CHMN MED, 82- *Concurrent Pos:* Leukemia chemother grant, 58-60; NIH career develop award, 62; Carey & Hofheimer scholastic awards, Univ Cincinnati; dir res & div microbiol, Children's Hosp, Columbus, 64-66, exec dir, Res Found, 66-73; dir, Med & Postgrad ed, King Faisal Specialist Hosp, Riyadh, Saudi Arabia, 83-84; assoc, int med affairs to vpres health serv, Ohio State Univ, 84-, exec dir, independent study prog, 85-89. *Mem:* Soc Pediat Res; Soc Exp Biol & Med; Am Acad Pediat; Infectious Dis Soc Am; fel Am Acad Microbiol. *Res:* Pediatric virology; infectious diseases; antibiotic research and immunology of the newborn and infant. *Mailing Add:* Col Med Ohio State Univ Col Med 370 W Ninth Ave Columbus OH 43210

CRAMER, ARCHIE BARRETT, b Winnipeg, Man, Oct 23, 09; m 35; c 2. FOOD CHEMISTRY. *Educ:* Univ Man, BSc, 34, MSc, 35; McGill Univ, PhD(wood chem), 39. *Prof Exp:* Asst, Carnegie Inst Technol, 39-41; res chemist, Miner Labs, 41-46; chief chemist, 46-63, VPRES RES & TECH DIR, F & F LABS, 63- *Honors & Awards:* Stroud Jordan Award, 63. *Mem:* AAAS; Am Chem Soc; Am Inst Chem; Am Asn Candy Technol; Nat Confectioners Asn US. *Res:* Wood and lignin chemistry; chemistry and preparation of lecithin; general carbohydrate chemistry; general chemistry of candy and food manufacture; preparation of certain oils and derivatives from lignin; preparation of glycerine derivatives; carbohydrate food flavor. *Mailing Add:* 11657 Chenault St - 108 Los Angeles CA 90049

CRAMER, CALVIN O, b Prairie du Sac, Wis, May 18, 26; m 52; c 2. AGRICULTURAL & CIVIL ENGINEERING. *Educ:* Univ Wis-Madison, BS(agr) & BS(civil eng), 52, MS, 54, PhD(civil eng), 67. *Prof Exp:* From instr to asst prof agr eng, Univ Wis-Madison, 54-64; res engr, US Forest Prod Lab, 64-66; assoc prof agr eng, 67-71, PROF AGR ENG, UNIV WIS-MADISON, 72- *Mem:* Am Soc Agr Engrs; Am Soc Civil Engrs. *Res:* Dairy cattle and swine housing; timber design; farm waste disposal. *Mailing Add:* Dept Agr Eng 460 Henry Mall Univ Wis Madison WI 53706

CRAMER, CARL FREDERICK, b Raton, NMex, June 7, 22; m 49; c 2. PHYSIOLOGY. *Educ:* Univ NMex, BS, 44, MS, 47; Univ Calif, PhD(physiol), 53. *Prof Exp:* Res assoc physiol, Med Sch, Tulane Univ, 54; ASSOC PROF PHYSIOL, UNIV BC, 55- *Mem:* Am Physiol Soc; Am Inst Nutrit; Can Physiol Soc; Sigma Xi. *Res:* Bone and gastrointestinal mineral physiology; turnover of radioactive tracers and removal of fission products. *Mailing Add:* Dept Physiol Univ BC 3506 W 36th Ave Vancouver BC V6N 2S2 Can

CRAMER, DAVID ALAN, b Ann Arbor, Mich, Aug 7, 25; m 45; c 3. ANIMAL NUTRITION. *Educ:* Colo State Univ, BS, 49, MS, 55; Ore State Univ, PhD(animal nutrit & biochem), 60. *Prof Exp:* Res chemist, Arapahoe Chem, Inc, 55-56; asst prof animal nutrit & vet sci, Calif State Polytech Col, 56-58; from asst prof to assoc prof biochem of animal prods, 60-68, prof, 68-80, PROF ANIMAL SCI, COLO STATE UNIV, 80- *Concurrent Pos:* Fulbright scholar, Dept Sci & Indust Res, NZ, 65-66. *Mem:* Am Soc Animal Sci; Am Meat Sci Asn; Am Oil Chem Soc; Inst Food Technol; Am Dairy Sci Asn. *Res:* Lipid metabolism in domestic animals, especially in the ruminant; meat flavor analysis; rumen metabolism. *Mailing Add:* Dept Animal Sci Colo State Univ Ft Collins CO 80523

CRAMER, DONALD V, b Long Beach, Calif, Oct 22, 41. BIOLOGY. *Educ:* Univ Calif, BS, 64, DVM, 66; Harvard Univ, PhD(exp path), 78; Am Col Vet Pathologists, cert, 71. *Prof Exp:* Res asst, Primate Res Ctr, Univ Calif, 64-65; intern, Angell Mem Animal Hosp, 66-67; res assoc path, Med Sch, Harvard Univ, 67-71; from instr to assoc prof path, Sch Med, Univ Pittsburgh, 71-89; ACTG DIR, SURG RES & DIR, TRANSPLANTATION BIOL RES, CEDARS-SINAI MED CTR, 89- *Concurrent Pos:* Clin asst, Presby-Univ Hosp, Pittsburgh, 71-89; res career develop award, NIH, 78-83; Alexander von Humboldt Found res fel, Fed Repub Ger, 80-81; mem, Pittsburgh Cancer Inst, 86-89. *Mem:* Am Col Vet Pathologists; Am Asn Pathologists & Bacteriologists; Am Asn Pathologists; Soc Toxicol Pathologists; Int Transplantation Soc; Am Soc Histocompatibility & Immunogenetics; Am Genetic Asn; Am Soc Transplant Physicians; Int Soc Heart & Lung Transplantation. *Res:* Genetics. *Mailing Add:* Transplantation Res Lab Cedars-Sinai Med Ctr 150 N Robertson Blvd Beverly Hills CA 90211

CRAMER, EVA BROWN, b New York, NY, June 6, 44; m 68; c 2. CELL BIOLOGY, IMMMUNOLOGY. *Educ:* Cornell Univ, BS, 65; Jefferson Med Col, MS, 67, PhD(anat), 69. *Prof Exp:* Instr, Col Physicians & Surgeons, Columbia Univ, 70-71; instr, Harvard Med Sch, 71-73; instr, 73-76, asst prof, 76-80, ASSOC PROF ANAT & CELL BIOL, STATE UNIV NY DOWNSTATE MED CTR, 80-; PROF ANAT & CELL BIOL, STATE UNIV NY HEALTH SCI CTR, BROOKLYN, 88-, ACTG ASSOC PROVOST SCI AFFAIRS, 88- *Concurrent Pos:* NIH fel, Col Physicians & Surgeons, Columbia Univ, 69-70; NH Heart Asn fel, Univ NH, 72-73; prin investr, Nat Inst Allergy & Infectious Dis-NIH, 80-; prin investr, Tobacco Res Coun, 84-87; NY Heart Asn, 88-89; adj assoc prof, Rockerfeller Univ, 83-86; fac marshal, Commencement Ceremony Med Sch Class, 84, 87-90. *Honors & Awards:* Charles W Labelle Prize, Jefferson Med Col. *Mem:* Am Asn Anatomists; Am Soc Cell Biol; NY Acad Sci; Sigma Xi; Harvey Soc. *Res:* leukocyte chemotaxis, diapedesis; immunology; inflammatory mediators; cell biology; inflammation, virus and leukocyte interaction; isolation of an epithelial permeability factor. *Mailing Add:* Dept Anat & Cell Biol SUNY Health Sci Ctr Brooklyn 450 Clarkson Ave Brooklyn NY 11203

CRAMER, GISELA TÜRCK, b Rohrbach, Ger, Mar 13, 34; m 67; c 3. BIOLOGICAL CHEMISTRY. *Educ:* Univ Frankfort, Dr phil nat(pharmaceut chem), 66. *Prof Exp:* Fel, Dept Pharmacol, Col Med, Univ Fla, 73; lab asst, Sch Med Sci, Univ Nev, Reno, 74-75, res asst, 75-81; MEM FAC, SCH MED, MERCER UNIV, GA, 81- *Concurrent Pos:* Anna Fuller Fund fel, Sch Med, Yale Univ, 66-67 & Col Med, Univ Fla, 67-68. *Res:* Cellular mechanisms of antitumor drugs; photodynamic action of dyes. *Mailing Add:* Sch Med Mercer Univ Macon GA 31027

CRAMER, HARRISON EMERY, b Johnstown, Pa, May 27, 19; m 42; c 4. AIR POLLUTION, METEOROLOGY. *Educ:* Amherst Col, AB, 41; Mass Inst Technol, SM, 43, ScD(meteorol), 48. *Prof Exp:* Instr meteorol, Mass Inst Technol, 42-44; meteorologist, Am Export Air Lines, 44; res assoc, Mass Inst Technol, 46-48, res meteorologist, 48-65; sr staff scientist, GCA Technol Div, 65-66, dir environ sci lab, 67-72, vpres, 69-72; pres, H E Cramer Co, Inc, 72-85; METEOROL & AIR POLLUTION CONSULT, 85- *Mem:* Fel AAAS; fel Am Meteorol Soc; Am Geophys Union; fel Royal Meteorol Soc; Air & Waste Mgt Asn. *Res:* Development and implementation of mathematical models for air pollution and toxic hazard applications, meteorological instrumentation and air pollution data systems; quantitative assessment of air pollution and toxic hazard problems; analysis of air pollution and meteorological measurements. *Mailing Add:* 1581 Millbrook Rd Salt Lake City UT 84106

CRAMER, HOWARD ROSS, b Chicago, Ill, Sept 17, 25; m 50, 82. STRATIGRAPHY, INVERTEBRATE PALEONTOLOGY. *Educ:* Univ Ill, BS, 49, MS, 50; Northwestern Univ, PhD(geol), 54. *Prof Exp:* From instr to asst prof geol, Franklin & Marshall Col, 53-58; from asst prof to prof, geol, Emory Univ, 58-87; RETIRED. *Concurrent Pos:* Consult, 53- *Mem:* Geol Soc Am; Am Asn Petrol Geol; Paleont Soc; Nat Asn Geol Teachers. *Res:* Bibliography; geology and environment; petroleum and stratigraphy of Georgia. *Mailing Add:* 2047 Deborah Dr Atlanta GA 30345

CRAMER, JAMES D, b Canton, Ohio, Aug 4, 37; m 57; c 2. NUCLEAR PHYSICS. *Educ:* Calif State Univ, BS, 60; Univ Ore, MS, 62; Univ NMex, PhD(physics), 69. *Prof Exp:* Staff mem, Los Alamos Sci Lab, 62-70; staff scientist, Sci Applns Inc, 70-75, vpres, 75-78, dir, 72-80, sr vpres, 78-80; PRES, SCIENCE & ENG ASSOCS INC, 80- *Concurrent Pos:* Consult, Los Alamos Sci Lab, 70-72. *Mem:* Am Phys Soc; Inst Elec & Electronics Engrs. *Res:* Physics of gases and nuclear physics. *Mailing Add:* 4152 Dietz Farm Circle NW Albuquerque NM 87107

CRAMER, JANE HARRIS, b Chicago, Ill, Dec 1, 42; m 66; c 2. BIOTECHNOLOGY, PLANT MOLECULAR BIOLOGY. *Educ:* Carleton Col, BA, 64; Northwestern Univ, PhD(microbiol), 70. *Prof Exp:* Fel, 70-74, asst scientist, 74-77, assoc scientist molecular biol, Univ Wis-Madison, 77-82; res scientist, 81-85, sr res scientist, 85-89, MGR, REGULATORY AFFAIRS, AGRIGENETICS CO, 89- *Mem:* Am Soc Microbiol; Int Soc Plant Molecular Biol; Sigma Xi; Asn Women in Sci. *Res:* Development of superior crop species through biotechnology; use of DNA restriction fragment length polymorphism mapping for agronomic crop improvement. *Mailing Add:* Agrigenetics Co 5649 E Buckeye Rd Madison WI 53716

CRAMER, JOHN ALLEN, b June 25, 43; US citizen; m 71; c 2. PHYSICS. *Educ:* Wheaton Col, Ill, BS, 65; Ohio Univ, MS, 68; Tex A&M Univ, PhD(physics), 75. *Prof Exp:* Instr physics, Wheaton Col, Ill, 68-71; asst prof physics, King's Col, NY, 76-80; asst prof, 80-83, ASSOC PROF PHYSICS, OGLETHORPE UNIV, GA, 83- *Mem:* Am Phys Soc; AAAS; Am Sci Affil. *Res:* Magnetic effects in gases and solids. *Mailing Add:* Dept Physics Oglethorpe Univ Atlanta GA 30319

CRAMER, JOHN GLEASON, b Houston, Tex, Oct 24, 34; m 61; c 3. NUCLEAR PHYSICS. *Educ:* Rice Univ, BA, 57, MA, 59, PhD(physics), 61. *Prof Exp:* Res assoc nuclear physics, Ind Univ, 61-63, asst prof physics, 63-64; from asst prof to assoc prof, 64-73, PROF PHYSICS, UNIV WASH, 73- *Concurrent Pos:* Guest prof, Univ Munich, 71-72 & Halm-Meitnor Inst, Berlin, 82-83. *Mem:* Fel AAAS; fel Am Phys Soc. *Res:* Heavy ion reactions and scattering; nuclear reactions and reaction mechanisms through measurements of reaction cross sections and angular correlations; ultra-relativistic heavy ion reactions; foundations of quantum mechanics. *Mailing Add:* Nuclear Physics Lab GL-10 Univ Wash Seattle WA 98195

CRAMER, JOHN WESLEY, b Freeport, Ill, Oct 15, 28; m 67; c 3. BIOCHEMISTRY. *Educ:* Beloit Col, BS, 50; Univ Wis, MS, 52, PhD(biochem), 55. *Prof Exp:* Instr oncol, Univ Wis, 59; instr pharmacol, Sch Med, Yale Univ, 59-61, asst prof, 61-67; assoc prof, Sch Med, Univ Fla, 67-73; prof pharmacol, Sch Med Sci, Univ Nev, Reno, 74-81; PROF PHARMACOL, SCH MED, MERCER UNIV, 81- *Concurrent Pos:* Fel biochem, Univ Wis, 55-57, fel oncol, 57-59; NIH & Am Cancer Soc grants. *Mem:* AAAS; Am Soc Pharmacol & Exp Therapeut; Am Asn Cancer Res; Am Soc Microbiol; Tissue Cult Asn; Sigma Xi. *Res:* Nutrition; vitamin D; cancer aromatic hydrocarbons; cell and viral nucleic acid synthesis and inhibition; antimetabolites. *Mailing Add:* Sch Med Mercer Univ Macon GA 31207

CRAMER, JOSEPH BENJAMIN, b Rochester, NY, Aug 24, 14; m 46; c 2. PSYCHIATRY, PSYCHOANALYSIS. *Educ:* Univ Rochester, BA, 36; NY Med Col, MD, 41; Inst Psychoanal, Chicago, cert psychoanal, 55. *Prof Exp:* Scottish Rite res fel schizophrenia, Sch Med, NY Univ, 47-49; assoc prof child psychiat, Univ Pittsburgh, 51-55; assoc prof & dir child psychiat, 55-68, prof psychiat & pediat, 68-85, EMER PROF, ALBERT EINSTEIN COL MED, 85- *Concurrent Pos:* Dir, Pittsburgh Guid Ctr, 51-53; consult, Scranton Guid Ctr, Pa, 55-57, Jewish Child Care Asn NY, 55-65 & Wiltwyck Sch, 73-74; mem steering comt, Conf Training in Child Psychiat, 63; mem, Conf Psychiat & Med Educ, 67. *Mem:* Am Acad Child Psychiat; Am Psychoanal Asn; Am Psychiat Asn; Am Orthopsychiat Asn. *Res:* Childhood schizophrenia; psychotherapy; medical education. *Mailing Add:* 201 E 77th St Apt 5D New York NY 10021

CRAMER, MICHAEL BROWN, b Dayton, Ohio, July 4, 38; m 62; c 3. CLINICAL RESEARCH. *Educ:* Purdue Univ, BS, 61, MS, 63, PhD(pharmacol), 65. *Prof Exp:* Prin investr, Armed Forces Radiobiol Res Inst, Defense Atomic Support Agency, US Army, 65-67; res pharmacologist, Procter & Gamble Co, 67-70; asst prof pharmcol, Col Pharm, Univ Houston, 70-75, assoc prof, 75-80; asst dir res supports, Merrell Nat Lab, Richardson-Merrell Inc, 80-81, sr clinic res scientist, 81-84, SR CLIN PROJ MGR, MARION MERRELL DOW INC, 84- *Concurrent Pos:* Zool instr, Montgomery Jr Col, 66-67; lectr pharm, Tex Tripartate Comt Continuing Educ, 73-80, Adv Course Pharmacol, Food & Drug Admin, Univ Houston, 74-79; fac adv, Poison Prevention Prog, Univ Houston, 72-80. *Mem:* AAAS; Am Soc Clin Pharmacol & Therapeut. *Res:* New drug development: preparation of clinical research protocols, monitoring, evaluation of data and preparation of sponsor's summaries; evaluation and interpretation of clinical and preclinical data in preparation of investigator's brochures, INDs and NDAs. *Mailing Add:* Marion Merrell Dow Res Inst 2110 E Galbraith Rd Cincinnati OH 45215-6300

CRAMER, RICHARD (DAVID), b Mifflin, Pa, Aug 12, 13; m 37; c 4. ORGANOMETALLIC CHEMISTRY. *Educ:* Juniata Col, BS, 35; Harvard Univ, MA, 36, PhD(chem), 40. *Prof Exp:* Asst prof org chem, Carnegie Inst Technol, 40-41; res chemist, E I Du Pont De Nemours & Co, Inc, 41-78. *Honors & Awards:* Del Sect Awards, Am Chem Soc, 65, 67. *Mem:* Am Chem Soc. *Res:* Synthesis of organics from radioactive carbon; chemistry of organic fluorine compounds; polymer synthesis; modified polyethylenes and synthesis of fluorine containing monomers; radioactive lactic acid in biological studies; chemistry of triphenylfuryl ketones; organo-transition metal chemistry. *Mailing Add:* 515 S Bank Rd Landenberg PA 19350-1012

CRAMER, ROGER EARL, b Findlay, Ohio, Sept 14, 43; m 67; c 2. INORGANIC CHEMISTRY. *Educ:* Bowling Green State Univ, BS, 65; Univ Ill, MS, 67, PhD(inorg chem), 69. *Prof Exp:* From asst prof to assoc prof, 69-80, assoc chmn dept, 80-84, PROF CHEM, UNIV HAWAII, 80-, DEPT CHMN, 86- *Concurrent Pos:* Vis prof, Northwestern Univ, 78. *Mem:* Am Chem Soc. *Res:* Nuclear magnetic resonance of paramagnetic molecules; coordination chemistry; metal-ion complexes in medicine; organouranium chemistry. *Mailing Add:* Dept of Chem 2524 The Mall Univ Hawaii Honolulu HI 96822

CRAMER, WILLIAM ANTHONY, b New York, NY, June 11, 38; m 64; c 4. BIOPHYSICS. *Educ:* Mass Inst Technol, BS, 59; Univ Chicago, MS, 60, PhD(biophys), 65. *Prof Exp:* NSF fel, Univ Calif, San Diego, 65-67, res assoc, 67-68; from asst prof to assoc prof, 68-78, PROF BIOL SCI, PURDUE UNIV, 78-, assoc head, Dept Biol Sci, 84-86. *Concurrent Pos:* Nat Inst Gen Med Sci res career develop award, 70-75; Europ Molecular Biol Org sr fel, 74-75; mem molecular biol panel, NSF, 78-80; panel, NAS-NRC predoctral fel, 79-81, chmn, 81; USDA panel, photosynthesis grants, 84-85; mem cell biochem panel, NSF, 88-90; chmn, bioenergetics subgroup, Biophys Soc, 89-92, Garden Conf Biochem & Photosynthesis, 90, Study sect, Phys Biochem, NIH, 91-95. *Honors & Awards:* H N McCoy Award, Sci Achievement, 88. *Mem:* Am Soc Biol Chem; Biophys Soc. *Res:* Photosynthetic electron transport and energy coupling; bioenergetics; mechanism of action of colicin. *Mailing Add:* Dept Biol Sci Purdue Univ West Lafayette IN 47907

CRAMER, WILLIAM HERBERT, physical chemistry, for more information see previous edition

CRAMER, WILLIAM SMITH, b Frederick, Md, Aug 25, 14; m 47; c 2. PHYSICS. *Educ:* Ursinus Col, BS, 37; Brown Univ, MS, 38, PhD(physics), 48. *Prof Exp:* Asst math, Univ Md, 38-39; teacher math & sci, Pikeville Col, 39-40; physicist, US Naval Ord Lab, 42-56 & Off Naval Res, 56-66, physicist, Naval Ship Res & Develop Ctr, Washington, DC, 66-76; physicist, Mar Assocs, Inc, Rockville, 76-83; RETIRED. *Mem:* Fel Acoust Soc Am (vpres, 75-76). *Res:* Acoustic properties of plastics; underwater and physical acoustics. *Mailing Add:* 11512 Colt Terr Silver Spring MD 20902

CRAMPTON, GEORGE H, b Spokane, Wash, Nov 20, 26; m 47; c 2. BIOCHEMISTRY, PHARMACOLOGY. *Educ:* Wash State Univ, BS, 49, MS, 50; Univ Rochester, PhD(psychol), 54. *Prof Exp:* Colonel, US Army, 44-71; prof psychol, 71-86, EMER PROF PSYCHOL, WRIGHT STATE UNIV, 87- *Mem:* Soc Neurosci; Aerospace Med Asn; Barany Soc. *Res:* Pharmacology and physiology related to motion sickness. *Mailing Add:* 1842 N Dawnview Terr Oak Harbor WA 98277-9208

CRAMPTON, JAMES MYLAN, b Mitchell, SDak, Nov 26, 23; m 50; c 6. PHARMACOLOGY. *Educ:* Creighton Univ, BS, 50; Univ Fla, MS, 51, PhD(pharmacol), 53. *Prof Exp:* Asst prof pharmacol, Xavier Univ, 55-58; asst prof biol sci, Sch Pharm, 58 & 67, prof, 67-70, dir dept biol sci, 58-71, dir clin studies, 71-74, assoc dean sch pharm, 75-78, PROF BIOPHARM, SCH PHARM & PROF PHYSIOL & PHARMACOL, SCH MED, CREIGHTON UNIV, 70- *Mem:* Am Pharmaceut Asn. *Res:* Castrix toxicity; salicyclamide. *Mailing Add:* Dept Pharm & Physiol Creighton Univ Calif at 24th Omaha NE 68178

CRAMPTON, JANET WERT, b Danville, Ill, Jan 19, 34; m 55. EARTH SCIENCES. *Educ:* Ind Univ, Bloomington, BS, 55; Mt Holyoke Col, MA, 64. *Prof Exp:* Teacher sci & math, West Point Schs, NY, 65-69; teacher sci, Christ Episcopal Sch, Rockville, Md, 73-75; res assoc geol, Earth Satellite Corp, Chevy Chase, Md, 76-77; mus technician, Sci Event Alert Network, Smithsonian Inst, Washington, DC, 81-85; sr ed, Geotimes, Am Geol Inst, Alexandria, Va, 85-87; managing ed, EOS, Am Geophys Union, 87-90; special projects mgr, Am Geophys Union, 90- *Concurrent Pos:* Pres, Md Div, Am Asn Univ Women, 83-85. *Mem:* Am Asn Univ Women; Asn Women Geoscientists; Nat Asn Geol Teachers; Nat Sci Teachers Asn; Asn Earth Sci Ed; Asn Women Sci. *Res:* Write and edit for the geoscience community. *Mailing Add:* Six Lakeside Overlook Rockville MD 20850-2730

CRAMPTON, STUART J B, b New York, NY, Nov 3, 36; m 61; c 3. ATOMIC PHYSICS. *Educ:* Williams Col, BA, 58; Oxford Univ, BA, 60; Harvard Univ, PhD(physics), 64. *Prof Exp:* From asst prof to assoc prof, 65-75, PROF PHYSICS, WILLIAMS COL, 75- *Concurrent Pos:* NSF fel, Harvard Univ, 64-65; Alfred P Sloan Found res fel, 67-69; NATO sr fel, 75; vis prof, MIT, 77-78, Univ Paris, 82-83, Univ Mass, 86-87. *Honors & Awards:* Award for Res Undergrad Inst, Am Phys Soc, 89. *Mem:* Fel Am Phys Soc; Am Asn Physics Teachers. *Res:* Atom-atom interactions, particularly electron spin exchange collisions and atom-surface interactions at low temperatures. *Mailing Add:* Dept Physics Williams Col Williamstown MA 01267

CRAMPTON, THEODORE HENRY MILLER, b Patchogue, NY, Apr 4, 26; m 55. MATHEMATICS, RADIOBIOLOGY. *Educ:* Hamilton Col, AB, 49; Ind Univ, MA, 54, PhD(math), 55. *Prof Exp:* Instr math, Mt Holyoke Col, 55-57, asst prof, 57-58; instr nuclear weapons, US Army Engr Sch, Va, 58-61; mathematician, Armed Forces Radiobiol Res Inst, Nat Naval Med Ctr, Md, 61-64; instr math, US Mil Acad, 64-66, asst prof, 66-69; chief weapons effects sect, Joint Strategic Target Planning Staff,, 69-72; systs analyst, Int Security Affairs, Energy Res & Develop Admin, 72-77; sr res fel & chief climate proj, res directorate, Nat Defense Univ, 77-81; chief, Arms Control & Policy Off, Defense Nuclear Agency, 81-87; RETIRED. *Mem:* Sigma Xi. *Res:* Class field theory; radiation transport theory; radiation dosimetry; international affairs analysis and climatic change; radiation shielding. *Mailing Add:* Crampton Six Lakeside Overlook Rockville MD 20850-2730

CRAMTON, THOMAS JAMES, b Chadron, Nebr, Oct 23, 38; m 60; c 2. MATHEMATICS. *Educ:* Harvard Univ, AB, 60; Clark Univ, AM, 62; Dartmouth Col, PhD(math), 70. *Prof Exp:* Instr math, US Naval Nuclear Power Sch, Md, 62-66, dir off dept, 63-66; asst prof, Purdue Univ, Ft Wayne, 69-76; actuary, Louis Behr Orgn, Inc, 77-84; mgr, Touche Ross & Co, 84-86; ACTUARY, EPACO, 86- *Res:* Inverse eigenvalue problems in dimension 2 or higher. *Mailing Add:* 96 N Kenilworth Ave Glen Ellyn IL 60137

CRANBERG, LAWRENCE, b New York, NY, July 4, 17; m 53; c 2. ACCURACY & INTEGRITY IN RESEARCH. *Educ:* Col City New York, BS & MS, 37; Harvard Univ, AM, 40; Univ Pa, PhD(physics), 49. *Prof Exp:* Sr physicist, Signal Corps Eng Labs, 40-50; staff mem, Los Alamos Sci Lab, 50-64; prof physics, Univ Va, 63-71; vpres res & develop, Accelerators, Inc, 71-73; PRES, TDN INC, 74-; PRES, TEX FIREFRAME CO, 75- *Concurrent Pos:* Fel, Guggenheim Found, 58; vis prof physics, Case Inst Technol, 58; vis staff mem, Atomic Energy Res Estab, Harwell, 61-62; prin investr, NSF, Univ Va, 63-69; consult physicist, 73-; ed, Nuclear & Chem Waste Mgt, Pergamon Press, 80- *Honors & Awards:* Edward G Budd Lectr, Franklin Inst, 66. *Mem:* Fel Am Phys Soc; Am Asn Physics Teachers; Am Asn Univ Prof. *Res:* Neutron physics; accelerator physics and applications to neutron therapy; combustion and domestic heating; ethical problems of scientists and applications to research administration; application of research to public policy. *Mailing Add:* 1205 Constant Springs Dr Austin TX 78746

CRANCH, EDMUND TITUS, b Brooklyn, NY, Nov 15, 22; m 46; c 3. MECHANICS. *Educ:* Cornell Univ, BME, 45, PhD(mech), 52. *Prof Exp:* Mem tech staff appl mech, Bell Tel Labs, 47-48; from asst prof to prof mech & mat, 51-62, head dept, 56-62, assoc dean eng, 67-77, prof mech, Cornell Univ, 62-78, dean, Col Eng, 72-78; pres, Worcester Polytech Inst, 78-85; pres, Wang Inst Grad Studies, 85-87; DISTINGUISHED PROF, UNIV NH, 87- *Concurrent Pos:* Consult, Lincoln Lab, Mass Inst Technol, 51-53, Cornell Aeronaut Lab, 55-61, Aerojet-Gen Corp, Calif, 57, Bausch & Lomb Corp, NY, 61, Int Bus Mach Corp, 63- & Battelle Mem Inst, 63-; NSF fac fel, Stanford Univ, 58-59, sr fel, Swiss Fed Inst Technol, 64-65. *Mem:* Fel Am Soc Mech Engrs; Am Soc Testing & Mat; Soc Exp Stress Anal; Am Soc Eng Educ; Sigma Xi. *Res:* Dynamics; vibration theory; elasticity; applied mathematics; materials; shells. *Mailing Add:* Six Gen Amerst Rd Amherst NH 03031

CRANDALL, ARTHUR JARED, b Syracuse, NY, Mar 14, 39; m 64; c 3. PHYSICS. *Educ:* St Lawrence Univ, BS, 61; Mich State Univ, MS, 64, PhD(physics), 67. *Prof Exp:* Asst prof, 67-74, ASSOC PROF PHYSICS, BOWLING GREEN STATE UNIV, 74- *Mem:* Acoust Soc Am. *Res:* Physics teaching; ultrasonics; diffraction effects and interaction of light with sound. *Mailing Add:* Dept Physics & Astron Bowling Green State Univ Bowling Green OH 43403

CRANDALL, DANA IRVING, b New York, NY, Sept 4, 15; m 42; c 3. BIOCHEMISTRY. *Educ:* Columbia Univ, AB, 36; Univ Pa, PhD(biochem), 45. *Prof Exp:* From instr to asst prof physiol chem, Univ Pa, 46-50; from assoc prof to prof, 50-80, EMER PROF BIOL CHEM, UNIV CINCINNATI, 80- *Mem:* Am Soc Biol Chemists. *Res:* Metabolism of tyrosine in mammalian tissues; properties and mechanism of action of homogentisate dioxygenase; role of oxygenases in mammalian metabolism. *Mailing Add:* 1231 Halpin Ave W Cincinnati Col Med Cincinnati OH 45208

CRANDALL, DAVID HUGH, b Chicago, Ill, July 9, 42; m 76; c 3. ELECTRON PHYSICS, RESEARCH ADMINISTRATION. *Educ:* Sioux Falls Col, BS, 64; Univ Nebr-Lincoln, MS, 67, PhD(physics), 70. *Prof Exp:* Vis asst prof physics, Univ Mo-Rolla, 70-71; res assoc, Joint Inst Lab Astrophys, Nat Bur Standards & Univ Colo, Boulder, 71-74; res physicist, Oak Ridge Nat Lab, 74-79, mgr, Atomic & Plasma Physics Res Group, 79-83; br chief, Exp Plasma Res Br, 83-87, DIR APL PLASMA PHYSICS DIV, OFF FUSION ENERGY, DEPT ENERGY, 87- *Mem:* Fel Am Phys Soc; Sigma Xi; AAAS. *Res:* Atomic collisions between ions and atoms or electrons using ion beams of a few keV energy for charge exchange, impact excitation and ionization; development of plasma diagnostics; plasma physics; science policy and education. *Mailing Add:* ER 54 G217 GTN US Dept Engery Washington DC 20545

CRANDALL, DAVID L, b Creston, Iowa, Sept 1, 52; m 82; c 2. CARDIOVASCULAR PHYSIOLOGY, ENDOCRINOLOGY. *Educ:* Tulane Univ, BS, 74; Iowa State Univ, MS, 77, PhD(physiol), 79. *Prof Exp:* NIH fel endocrinol, Sch Med, Emory Univ, 79-81; SR RES PHARMACOLOGIST, CARDIOVASC BIOL, LEDERLE LABS, AM CYANAMID, 81- *Concurrent Pos:* Lectr, NY Med Col, 85. *Mem:* Sigma Xi; NY Acad Sci; Am Physiol Soc; Fedn Am Soc Exp Biol; Harvey Soc. *Res:* Prevention of myocardial infarction and the relationship of obesity to cardiovascular disease. *Mailing Add:* Bldg 56 Rm 240 Am Cyanamid Co Pearl River NY 10965

CRANDALL, EDWARD D, b Brooklyn, NY, Aug 10, 38; m 63; c 2. PULMONARY & CRITICAL CARE MEDICINE, LUNG CELL BIOLOGY. *Educ:* Cooper Union, BChE, 60; Northwestern Univ, MS, 62, PhD(chem eng), 64; Univ Pa, MD, 72. *Prof Exp:* From asst prof to assoc prof chem eng, Univ Notre Dame, 64-68; lectr, Dept Physiol, Univ Pa, 68-74, from asst prof to assoc prof physiol, med & bioeng, 74-79; from assoc prof to prof med, Univ Calif, Los Angeles, 79-85; prof med & chief, Div Pulmonary Critical Care Med, Cornell Univ Med Col, 85-90; PROF MED & CHIEF, DIV PULMONARY CRITICAL CARE MED, UNIV SOUTHERN CALIF, 91- *Concurrent Pos:* USPHS spec fel, 68-70; med internship & residency, Hosp, Univ Pa, 72-74, pulmonary fel, Hosp, 74-76; Webster prof internal med, Cornell Univ Med Col, 86-90. *Honors & Awards:* Res Career Develop Award, NIH, 75 & Merit Award, 87. *Mem:* Am Col Physicians; Am Thoracic Soc; Am Soc Clin Invest; Am Physiol Soc; Am Inst Chem Eng. *Res:* Lung epithelial transport; pulmonary physiology and disease; alveolar epithelial cell biology; cell membrane structure and function. *Mailing Add:* Div Pulmonary Critical Care Med Univ Southern Calif GNH 11-900 2025 Zonal Ave Los Angeles CA 90033

CRANDALL, ELBERT WILLIAMS, b Normal, Ill, Nov 4, 20; m 51; c 4. ORGANIC CHEMISTRY. *Educ:* Ill State Norm Univ, BEd, 42; Univ Mo, MA, 48, PhD(org chem), 50. *Prof Exp:* Prof chem, Ky Wesleyan Col, 50-51; res & develop, US Rubber Co, 51-52; prof chem, Pittsburg State Univ, 52-85; RETIRED. *Mem:* Am Chem Soc. *Res:* Nitric acid oxidations; near infrared spectra of polymers; organic compounds in natural waters; conversion of waste cellulose to ethanol. *Mailing Add:* 16309 W 123rd St Olathe KS 66062

CRANDALL, IRA CARLTON, b South Amboy, NJ, Oct 30, 31; m 54; c 3. POWER DISTRIBUTION, COMMUNICATIONS. *Educ:* Ind Inst Technol, BS, 54, BS, 58; US Naval Postgrad Sch, BS, 62; Univ Sussex, PhD(eng mgt), 64. *Hon Degrees:* LLD, St Matthew Univ, 70; EDD, Mt Sinai Univ, 72. *Prof Exp:* Teacher, Madison Twp Sch Bd, 54-55; Naval officer, eng electronics, USN, 55-72; chief engr, Williamson Eng, Inc, 73-82; VPRES, GAYNER ENGRS, 82- *Concurrent Pos:* Pres, IC's Enterprises, 72-; prin, I C Crandall Consult Eng, 72-; chief exec officer, I C Crandall & Assocs, Inc, 72-82; vpres, Dickinson Enterprises, Inc, 73-76; res consult, Int Res Assocs, 82- *Mem:* Asn Energy Engrs; Inst Elec & Electronics Engrs; Soc Am Mil Engrs; Asn Naval Aviation. *Res:* Energy conservation; two-way cable television. *Mailing Add:* PO Box 3268 Walnut Creek CA 94598

CRANDALL, JACK KENNETH, b Fillmore, Calif, June 8, 37; div; c 1. ORGANIC CHEMISTRY. *Educ:* Univ Calif, Berkeley, BS, 60; Cornell Univ, PhD(org chem), 63. *Prof Exp:* NIH fel, 64; from instr to assoc prof, 64-72, PROF CHEM, IND UNIV, BLOOMINGTON, 72- *Concurrent Pos:* Alfred P Sloan res fel, 68-70; John Simon Guggenheim fel, 70-71; Fulbright res fel, 85-86. *Mem:* Am Chem Soc; The Chem Soc. *Res:* Synthetic organic chemistry; small ring compounds; epoxides; allenes; organometallic chemistry; C-13 nuclear magnetic resonance. *Mailing Add:* Dept Chem Ind Univ Bloomington IN 47401

CRANDALL, JOHN LOU, b Hart, Mich, Sept 18, 20; m 43, 82; c 1. ENVIRONMENTAL SCIENCES, CHEMISTRY. *Educ:* Mass Inst Technol, BS, 42, PhD(phys chem), 48. *Prof Exp:* Res chemist, Savannah River Lab, E I du Pont de Nemours & Co, Inc, 48-52, res mgr exp phys, 53-67, from asst dir to dir advan oper planning, Atomic Energy Div, 67-75, dir environ sci sect, 75-78, dir adv planning, 78-80, prog mgr, 80-85, CONSULT, SAVANNAH RIVER LAB, E I DU PONT DE NEMOURS & CO, INC, 85- *Mem:* Am Chem Soc; fel Am Nuclear Soc. *Res:* High polymers; photochemistry; reactor physics; operations analysis; environmental management; nuclear waste management. *Mailing Add:* Midland Valley Estates 204 Laurel Circle Graniteville SC 29829

CRANDALL, JOHN R, b Seattle, Wash, Aug 19, 14; m 48. PETROLEUM RESERVOIR ENGINEERING. *Educ:* Carnegie Inst Technol, BS, 35. *Prof Exp:* Labor oil drilling & prod, Humble Oil & Refining Co, 37-38, eng assignments from jr petrol engr to sr supv engr, 39-58, petrol econ engr on loan to Standard Oil NJ, 58-59, div staff engr, 59-61, staff coordr petrol econ & diversification studies, 61-63, res assoc, Prod Res Div, 63-64; res coordr, Esso Prod Res Co, 64-66; planning adv, Coal & Shale Oil Dept, Humble Oil & Refining Co, 66-69 & Carter Oil Co, 69-76; MANAGING TRUSTEE, RODERIC CRANDALL TRUST, 71-; PRES, CRANDALL & ASSOCS, INC, 79- *Concurrent Pos:* On loan, Non-hydrocarbon Mineral Activ, Standard Oil Co, NJ, 70-71. *Mem:* Soc Petrol Engrs. *Mailing Add:* Crandall & Assocs Inc 721 Ourlane Circle Houston TX 77024

CRANDALL, LEE W(ALTER), b Hartford, Wis, July 26, 13; m 38; c 1. CIVIL & STRUCTURAL ENGINEERING. *Educ:* Univ Wis, BS, 36, MS, 37; Stanford Univ, PhD(civil eng), 52. *Prof Exp:* Instr civil eng, Univ Colo, 37-38, asst prof, 43-46, assoc prof, 47-48; asst struct engr, Bur Reclamation, 39-42; from assoc prof to prof civil eng, 48-77, prof, 77-80, EMER PROF CIVIL & ENVIRON ENG & MEM GRAD FAC, UNIV WIS-MADISON, 80- *Concurrent Pos:* Fulbright res scholar, Finland, 53-54; consult, Food & Agr Orgn, UN Rome, at the Forest Prod Res Inst, Philippines, 61-62. *Mem:* Am Soc Civil Engrs; Am Soc Eng Educ; Nat Soc Prof Engrs; Nat Asn RR Passengers; Sigma Xi. *Res:* Economic selection; design of timber and steel structures; distress or failed structures; timber allowable loads; reports and expert witness testimony in courts of 12 states on unusual, distressed or failed structures, products or components for attorneys, casualty insurers, fabricators and owners. *Mailing Add:* 1008 Beloit Ct Madison WI 53705-2233

CRANDALL, MICHAEL GRAIN, b Baton Rouge, La, Nov 29, 40; m 62; c 3. MATHEMATICS. *Educ:* Univ Calif, Berkeley, BS, 62, MA, 64, PhD(math), 65. *Prof Exp:* Instr math, Univ Calif, Berkeley, 65-66; asst prof math, Stanford Univ, 66-69; from asst prof to prof, Univ Calif, Los Angeles, 69-74; prof math, Univ Wis-Madison, 74-90; PROF MATH, UNIV CALIF, SANTA BARBARA, 88- *Concurrent Pos:* Vis prof math, Math Res Ctr, Univ Wis-Madison, 71-72; assoc ed, Nonlinear Anal, 76-88, Annales Inst Fouries, 84-88 & Applicable Anal, 85-; mem coun, Am Math Soc, 84-87, chair, Western Sect Prog Comt, 91; dir, Nonlinear Anal Prog IAC, Univ Calif, Santa Barbara, 88-; managing ed, Commun Partial Differential Equations, 89-; progress math lectr, 90. *Mem:* Am Math Soc; Soc Indust Appl Math. *Res:* Nonlinear differential equations in both abstract and concrete cases; the abstract setting involves nonlinear evolution governed by accretive operators; concrete problems are nonlinear elliptic, parabolic and hyperbolic equations. *Mailing Add:* Dept Math Univ Calif Santa Barbara CA 93106

CRANDALL, PAUL HERBERT, b Essex, Vt, Feb 15, 23; m 51; c 4. MEDICINE. *Educ:* Univ Vt, BS, 43, MD, 47; Am Bd Neurol Surg, dipl, 56. *Prof Exp:* Asst chief neurosurg serv, Wadsworth Vet Admin Hosp, Calif, 54-56; from instr to assoc prof surg & neurosurg, 54-72, PROF NEUROSURG & NEUROL, SCH MED, UNIV CALIF, LOS ANGELES, 72- *Concurrent Pos:* Consult, Vet Admin Hosp & Wadsworth Gen Hosp, Calif. *Mem:* AMA; Am Asn Neurol Surg; NY Acad Sci; fel Am Col Surg; Sigma Xi. *Res:* Stereotaxic surgery; depth electrode studies in epilepsy; radioisotopic brain scanning. *Mailing Add:* 760 Westwood Plaza Los Angeles CA 90024

CRANDALL, PHILIP GLEN, b Manhattan, Kans, Nov 4, 48; m 69; c 2. HORTICULTURE. *Educ:* Kans State Univ, BS, 70; Purdue Univ, MS, 72, PhD(food sci), 75. *Prof Exp:* Asst prof, 75-80, ASSOC PROF CITRUS RES, INST FOOD & AGR SCI, AGR RES & EDUC CTR, UNIV FLA, LAKE ALFRED, 80-, FOOD SCIENTIST, 75- *Mem:* Inst Food Technologists; Int Soc Citricult. *Res:* Citrus processing; pectin. *Mailing Add:* Dept Food Sci Univ Ark 272 Young Ave Fayetteville AR 72701

CRANDALL, RICHARD B, b Greencastle, Ind, Sept 8, 28; m 58. PARASITOLOGY, IMMUNOLOGY. *Educ:* DePauw Univ, AB, 49; Univ Mass, MA, 53, Purdue Univ, PhD(zool), 59. *Prof Exp:* NIH trainee, Sch Pub Health, Univ NC, 59-60; instr med microbiol, 60-64, from asst prof to assoc prof, 65-73, PROF PARASITOL, COL MED, UNIV FLA, 74- *Mem:* Am Soc Parasitol; Am Soc Trop Med & Hyg; Am Asn Immunol. *Res:* Immunology of parasitic infections; medical parasitology and microbiology. *Mailing Add:* Dept Immunol & Med Microbiol Box J266 Univ Fla Col Med Gainesville FL 32610

CRANDALL, STEPHEN H, b Cebu, Philippines, Dec 2, 20. APPLIED MECHANICS. *Prof Exp:* FORD PROF ENG, MASS INST TECHNOL, 75- *Mem:* Nat Acad Eng. *Mailing Add:* Dept Mech Eng Rm 3-362 Mass Inst Technol Cambridge MA 02139

CRANDALL, WALTER ELLIS, b Norwich, Conn, Dec 18, 16; m 44; c 2. NUCLEAR PHYSICS. *Educ:* Worcester Polytech Inst, BS, 40; Univ Calif, PhD(physics), 52. *Prof Exp:* Physicist, US Naval Ord Lab, 41-42; physicist, Radiation Lab, Univ Calif, 48-52 & 54-62; physicist, Calif Res & Develop Co, 53; mgr, Sci & Technol Dept, Northrop Corp, 62-70, vpres & mgr, Northrop Corp Labs, 70-75, chief scientist, Northrop Res & Technol Ctr, 75-83; CONSULT, 83- *Concurrent Pos:* Consult, Boeing Airplane Co. *Mem:* Am Phys Soc; Am Inst Elec & Electronics Engrs. *Res:* High energy nuclear physics; neutral meson; nuclear radii; nuclear weapons; magnetics; electrooptics; biophysics. *Mailing Add:* 21930 Carbon Mesa Rd Malibu CA 90265

CRANDALL, WILLIAM B, b Andover, NY, Feb 24, 21; m 42; c 3. CERAMICS, MATERIALS SCIENCE. *Educ:* Alfred Univ, BS, 42, MS, 44. *Prof Exp:* Res assoc, Off Naval Res proj, Alfred Univ, 46-50, proj dir, 50-57; assoc prof ceramic eng, State Univ NY Col Ceramics, 57-63; consult mat res, Wireless Div, Matsushita Elec Co, Japan, 63; asst dir res, Pfaudler Co, Pfaudler-Permutit Corp, NY, 63, dir res, 63-64, vpres mat res, Pfaudler Co, Sybron Corp, NY, 65-70; dir ceramics res, IIT Res Inst, 70-74; Dir, Alfred Univ Res Found, 74-; AT DEPT CERAMIC ENG, ALFRED UNIV. *Concurrent Pos:* Co-founder, Materiadyne Corp, NY, 60-63; mem, Mat Adv Bd, Nat Acad Sci, 60-63; mem adv ceramic dept, Univ Ill, Urbana-Champaign, 69- *Mem:* Fel Am Ceramic Soc; Nat Inst Ceramic Engrs; Int Plansee Soc Powder Metall. *Res:* Ceramic materials, such as glass, crystalline oxides, carbides, cermets, coatings and composites. *Mailing Add:* Dept Ceramic Eng Alfred Univ - Main St Alfred NY 14802

CRANDALL-STOTLER, BARBARA JEAN, b Jamestown, NY, Mar 4, 42; m 69. BOTANY. *Educ:* Keuka Col, BA, 64; Univ Cincinnati, MS, 66, PhD(bot), 68. *Prof Exp:* Fel bot, Univ Tex, Austin, 68-69; from asst prof to assoc prof, 70-82, PROF BOT, SOUTHERN ILL UNIV, 82- *Honors & Awards:* Diamond Award, NSF & Am Bot Soc, 75. *Mem:* Am Bryol & Lichenology (secy-treas, 81-85); Bot Soc Am; Brit Bryol Soc; Int Asn Bryologists; Sigma Xi. *Res:* Morphogenesis and developmental anatomy of the mosses, liverworts, and hornworts. *Mailing Add:* Dept Bot Southern Ill Univ Carbondale IL 62901

CRANDELL, DWIGHT RAYMOND, b Galesburg, Ill, Jan 25, 23; m 43; c 2. GEOLOGY. *Educ:* Knox Col, BA, 46; Yale Univ, MS, 48, PhD(geol), 51. *Prof Exp:* Field asst, 47, GEOLOGIST, US GEOL SURV, 48- *Concurrent Pos:* Asst, Yale Univ, 48-51. *Honors & Awards:* Kirk Bryan Award, Geol Soc Am, 72. *Mem:* Fel Geol Soc Am. *Res:* Volcanic hazards. *Mailing Add:* Ign Geotherm Proc Br US Geol Surv MS 903 Fed Ctr Denver CO 80225

CRANDELL, GEORGE FRANK, b Astoria, Ore, Dec 5, 32; m 56; c 5. BIOLOGICAL OCEANOGRAPHY. *Educ:* Ore State Univ, BS, 60, MS, 63, PhD(oceanog), 66. *Prof Exp:* From asst prof to assoc prof, 66-75, PROF OCEANOG, HUMBOLDT STATE UNIV, 75- *Mem:* Am Soc Limnol & Oceanog. *Res:* Zooplankton and benthic ecology. *Mailing Add:* Dept Oceanog Humboldt State Univ Arcata CA 95521

CRANDELL, MERRELL EDWARD, b Clearfield, Pa, Mar 19, 38; m 60; c 2. PHYSICS. *Educ:* Hobart Col, BS, 59; Syracuse Univ, MS, 61, PhD(physics), 67. *Prof Exp:* Asst physics, Syracuse Univ, 59-63; instr physics, Le Moyne Col, NY, 63-67; asst prof, 67-70, ASSOC PROF PHYSICS, MUSKINGUM COL, 70- *Concurrent Pos:* Vis res fel, Kammerlingh Onnes Lab, State Univ Leiden, 74-75. *Mem:* Am Phys Soc; Am Asn Physics Teachers; Sigma Xi. *Res:* Optical and transport properties of metals and alloys. *Mailing Add:* 107 Thompson Ave New Concord OH 43762

CRANDELL, ROBERT ALLEN, b Three Rivers, Mich, July 30, 24; m 50; c 4. VETERINARY VIROLOGY. *Educ:* Mich State Col, BS, 47, DVM, 49; Univ Calif, MPH, 55; Am Col Vet Prevent Med, dipl; Am Col Vet Microbiologists, dipl. *Prof Exp:* Practitioner vet med, 49-51; base vet, Vet Corps, US Air Force, Selfridge AFB, Mich, 51-52, asst chief animal farm, Ft Detrick, Md, 52-54, lab officer, Naval Biol Lab, 54-56, asst chief virol br, Armed Forces Inst Path, 56-60, chief virol br, US Air Force Epidemiol Lab, Lackland AFB, Tex, 60-67, res epidemiologist, Pan Am Foot-and-Mouth Dis Ctr, Brazil, 67-69, chief biosci div, US Air Force Sch Aerospace Med, 69-70, chief epidemiol div, 70-71; sr microbiologist, Col Vet Med, Vet Diag Med, 71-87, prof, Dept Path & Hyg, Univ Ill, Urbana, 71-87; CONSULT, 87- *Concurrent Pos:* Mem, Animal Resources Adv Comt, NIH, 75-79; actg dir, Lab Vet Diag Med, Col Vet Med, Univ Ill, 76-78, dir, 78-81; head, diag microbiol, Tex Vet Med Diag Lab, 81-87; ed, J Vet Diag Invest. *Mem:* Am Vet Med Asn; US Animal Health Asn; Am Asn Vet Lab Daiagnosticians; Conf Res Workers Animal Dis; Am Asn Vet Lab Diagnostics (pres, 83-85). *Res:* Isolation and charcaterization of new viruses of the domestic animals; virologic studies and control of zoonoses and diseases of laboratory animals; basic studies with rabies virus; diagnostic virology; research administration. *Mailing Add:* 30 Forest Dr Bryan TX 77840

CRANDELL, WALTER BAIN, b New York, NY, July 26, 11; m 35; c 4. SURGERY, METABOLISM. *Educ:* Dartmouth Col, AB, 34; NY Univ, MD, 37. *Prof Exp:* Intern, Mary Hitchcock Mem Hosp, 37-38; clin asst chest surg, Hitchcock Clin, Hanover, NH, 39; asst resident surg, Mass Gen Hosp, Boston, 39-41; from asst resident to resident chest surg, Bellevue Hosp, New York, 41-42, asst vis consult surg & chest surg, 46-47; attend thoracic surg, Vet Admin Hosp, Bronx, NY, 47; chief surg serv, Vet Admin Hosp, White River Jct, 47-79; prof surg, Dartmouth Med Sch, 75-79; RETIRED. *Concurrent Pos:* Instr anat, Dartmouth Med Sch, 39; from asst clin prof to assoc clin prof surg, Dartmouth Med Sch, 47-68, clin prof surg, 68-75; adj surgeon, Lenox Hill Hosp, New York, 47; instr anat & oper surg, Postgrad Sch, NY Univ, 46, instr anat, Col Med, 46-47; consult surg, Mary Hitchcock Hosp, Hanover, NH, 70-79. *Mem:* Am Col Surg; Am Asn Thoracic Surg. *Res:* Metabolic disorders in surgical patients, especially disorders of acid-base and osmolality; pulmonary and renal insufficiency. *Mailing Add:* Three Meadow Lane Hanover NH 03755

CRANDLEMERE, ROBERT WAYNE, b South Weymouth, Mass, Mar 5, 47; m 66; c 2. APPLIED CHEMISTRY, FORENSIC SCIENCE. *Educ:* Suffolk Univ, BS, 70, MS, 75; Am Inst Chemists, cert, 76. *Prof Exp:* Assoc res scientist, Factory Mutual Res Corp, 67-70; chemist, NE Indust Chem Corp, 72-73; chief chemist & vpres, Briggs Eng & Testing Co, Inc, 73-; EXEC VPRES & CHIEF CHEMIST, CERT ENG & TESTING CO, INC. *Concurrent Pos:* Mem, Chem Week Adv Bd, 76. *Mem:* Am Chem Soc; Am

Inst Chemists; AAAS; Nat Fire Protection Asn. *Res:* Investigation of weathering and other effects on bituminous materials used in roofing and development of test procedures in analysis of cement concrete. *Mailing Add:* Cert Eng & Testing Co Inc 25 Mathewson Dr Weymouth MA 02189

CRANE, ANATOLE, b New Brunswick, NJ, Feb 23, 33; m 56; c 2. MICROBIOLOGY. *Educ:* Univ Ill, BS, 54, MS, 56; Univ Ind, PhD(bact), 60. *Prof Exp:* Proj leader microbiol, 59-68, MGR MICROBIOL, QUAKER OATS CO, 68- *Mem:* Am Soc Microbiol; Am Asn Cereal Chemists; Inst Food Technologists. *Res:* Food microbiology and preservation. *Mailing Add:* RR 3 Box 213G Algonquin IL 60102

CRANE, AUGUST REYNOLDS, medicine, pathology; deceased, see previous edition for last biography

CRANE, CHARLES RUSSELL, b Mangum, Okla, Jan 19, 28; m 54; c 4. TOXICITY OF SMOKE, HAZARDOUS MATERIAL. *Educ:* Univ Okla, BS, 51, MS, 52; Fla State Univ, PhD(biochem), 56. *Prof Exp:* Asst prof biochem, Okla State Univ, 56-61; chief biochem res, Civil Aeromed Inst, Fed Aviation Admin, Dept Transp, 61-88; SCI & TECH CONSULT, 85- *Mem:* AAAS; Am Chem Soc. *Res:* Aviation toxicology; drug metabolism; pesticides; enzymology; composition and inhalation toxicology of combustion/pyrolysis products. *Mailing Add:* 9104 S Hillcrest Dr Oklahoma City OK 73159

CRANE, FREDERICK LORING, b Mass, Dec 3, 25; m 50; c 4. BIOCHEMISTRY. *Educ:* Univ Mich, BS, 50, MS, 51, PhD(bot), 53. *Hon Degrees:* Dr Med, Karolinska Inst, Stockholm, 89. *Prof Exp:* Trainee, Inst Enzyme Res, Univ Wis, 53-57, asst prof, 57-59; asst prof chem, Univ Tex, 59-60; assoc prof, 60-62, PROF BIOL, PURDUE UNIV, WEST LAFAYETTE, 62- *Concurrent Pos:* Newcombe fel, Univ Mich, 52-53; NSF sr fel, Univ Stockholm, 63-64; NIH career investr, 64; Fulbright vis fel, Australian Nat Univ, 71-72 & 79-80; dir, NATO Advan Res Workshop Plasma Membrane Oxidoreductases in Control of Growth & Develop, Cordoba, Spain, 88. *Honors & Awards:* Eli Lilly Award, Am Chem Soc, 61. *Mem:* Am Soc Plant Physiol; Am Soc Biol Chem; Am Soc Cell Biol; Am Inst Biol Sci; Plant Growth Soc Am. *Res:* Vitamin biosynthesis in plants; fatty acid metabolism; biological oxidations and energy coupling; coenzyme Q and plastoquinones; ultrastructure of subcellular particles; hormone control of cell function and growth. *Mailing Add:* Dept Biol Sci Purdue Univ West Lafayette IN 47907

CRANE, GEORGE THOMAS, b Nephi, Utah, Nov 6, 28; m 54; c 4. MICROBIOLOGY, VIROLOGY. *Educ:* Utah State Univ, BS, 58; Colo State Univ, 62-63; Brigham Young Univ, MSc, 69. *Prof Exp:* Microbiologist, Dis Ecol Sect, USPHS, Colo, 58-66, Environ & Ecol Br, Dugway, 66-85; RETIRED. *Mem:* Emer mem Am Soc Trop Med & Hyg. *Res:* Arbovirus and rickettsial surveillance of the fauna of North America; arbovirus isolations in western United States. *Mailing Add:* 698 N Nelson Tooele UT 84074

CRANE, HEWITT DAVID, b Jersey City, NJ, Apr 27, 27; m 54; c 3. ELECTRICAL ENGINEERING. *Educ:* Columbia Univ, BS, 47; Stanford Univ, PhD(elec eng), 60. *Prof Exp:* Engr, Western Union Tel Co, 48-49; comput maintenance, Int Bus Mach Corp, 50-52; engr, Inst Advan Study, 52-55 & Radio Corp Am, 55-56; STAFF SCIENTIST, SRI INT, 56- *Concurrent Pos:* VChmn, Nat Joint Comput Conf, 62. *Honors & Awards:* Inst Elec & Electronics Engrs Award, 62; NASA Award for sci achievement, 70; Indust Res IR-100 Awards, 74, 76 & 90. *Mem:* Fel Inst Elec & Electronics Engrs; fel Optical Soc Am. *Res:* Digital device systems; neurophysiological and sensory devices and phenomena. *Mailing Add:* SRI Int 333 Ravenswood Ave Menlo Park CA 94025

CRANE, HORACE RICHARD, b Turlock, Calif, Nov 4, 07; m 34; c 2. PHYSICS, SCIENCE EDUCATION. *Educ:* Calif Inst Technol, BS, 30, PhD(physics), 34. *Prof Exp:* Res fel physics, Calif Inst Technol, 34-35; instr & res physicist, 35-38, from asst prof to prof, 38-46, chmn dept, 65-72, EMER PROF PHYSICS, UNIV MICH, ANN ARBOR, 78- *Concurrent Pos:* Res assoc, Mass Inst Technol, 40 & dept terrestrial magnetism, Carnegie Inst Technol, 41; dir proximity fuze proj, Univ Mich, 42-45; pres, Midwestern Univs Res Asn, 57-60; mem policy adv bd, Argonne Nat Lab, 57-67; mem, Comn Col Physics, 62-70, vpres, 68-70; mem standing comt controlled thermonuclear res, Atomic Energy Comn, 69-72; chmn bd gov, Am Inst Physics, 71-75; mem, Comn Human Resources, 77-80; mem, Coun Int Exchange Scholars, 77-80. *Honors & Awards:* Davisson-Germer Prize, Am Phys Soc, 68; Oersted Medal, Am Asn Physics Teachers, 76, Melba Newell Phillips Award, 88; Nat Medal Sci, 86. *Mem:* Nat Acad Sci; fel AAAS; fel Am Phys Soc; Am Asn Physics Teachers (pres, 65); fel Am Acad Arts & Sci. *Res:* Nuclear physics; high energy accelerators; g-factor of electron; physics teaching methods; geomagnetism. *Mailing Add:* 830 Avon Rd Ann Arbor MI 48104

CRANE, JOSEPH LELAND, b Wilmot, NH, Jan 9, 35. MYCOLOGY. *Educ:* Univ Maine, BS, 61; Univ Del, MS, 64; Univ Md, PhD(mycol), 67. *Prof Exp:* From asst mycologist to assoc mycologist, 67-79, MYCOLOGIST, ILL NATURAL HIST SURV, 79-; ASSOC PROF BOT & PLANT PATH, UNIV ILL, 76- *Mem:* Bot Soc Am; Mycol Soc Am; Am Inst Biol Sci; Brit Mycol Soc; Soc Mycol France. *Res:* Aquatic hyphomycetes; marine fungi; taxonomy of fungi imperfecti. *Mailing Add:* Ill Natural Hist Surv Natural Resource Bldg E Peabody Dr Champaign IL 61820

CRANE, JULES M, JR, b New York, NY, Sept 5, 28; m 55; c 4. ICHTHYOLOGY, MARINE BIOLOGY. *Educ:* NY Univ, AB, 54; Calif State Col Los Angeles, MA, 63. *Prof Exp:* Chmn dept, 64-70 & 74-78, PROF BIOL, CERRITOS COL, 62- *Concurrent Pos:* Lectr, Calif State Col Long Beach, 65-73. *Mem:* AAAS; Am Soc Ichthyologists & Herpetologists; Marine Biol Soc UK. *Res:* Behavioral and biochemical aspects of bioluminescence in fishes; evolution of deep sea fishes; marine ecology. *Mailing Add:* 165 Stanford Lane Seal Beach CA 90740

CRANE, JULIAN COBURN, b Morgantown, WVa, Mar 7, 18; m 42. POMOLOGY. *Educ:* Univ Md, BS, 39, PhD(hort), 42. *Prof Exp:* Asst, Univ Md, 39-42, asst horticulturist, 42; assoc agronomist, Off for Agr Rels, USDA, 43-45, horticulturist, 46; from asst prof to prof pomol, Col Agr, Univ Calif, Davis, 46-85; RETIRED. *Concurrent Pos:* NSF fel, 57. *Mem:* AAAS; Am Soc Plant Physiol; fel Am Soc Hort Sci; Am Inst Biol Sci. *Res:* Plant physiology; hormones in fruit set, growth and maturation; fig and pistachio culture. *Mailing Add:* 44166 Lakeview Dr El Macero CA 95618-1050

CRANE, LANGDON TEACHOUT, JR, b Detroit, Mich, Feb 23, 30; m; c 4. SOLID STATE PHYSICS, RESEARCH ADMINISTRATION. *Educ:* Amherst Col, BA, 52; Univ Md, PhD(physics), 60. *Prof Exp:* Res physicist, Sci Lab, Ford Motor Co, 59-63; from asst prog dir to assoc prog dir, Physics Sect, NSF, 63-68, prog dir atomic & molecular physics, 68-69; res prof & dir inst fluid dynamics & appl math, Univ Md, 69-74; specialist sci & technol & head mgt & Policy Sci Sect, Sci Policy Res Div, Cong Res Serv, Libr Cong, Wash, DC, 74-90; RETIRED. *Concurrent Pos:* Mem comt atomic & molecular physics, Nat Acad Sci-Nat Res Coun, 71-73; prof, Inst Fluid Dynamics & Appl Math, Univ Md, 74-76. *Mem:* Fel AAAS; fel Am Phys Soc; Sigma Xi. *Res:* Low temperature, atomic, molecular and plasma physics; areas of superconductivity; low temperature specific heats and magnetic properties of metals and alloys; nuclear magnetic resonance; analysis of national policies and legislation affecting or employing science and technology and effectiveness of federally sponsored research programs; technology transfer and innovation. *Mailing Add:* 4008 Rosemary St Chevy Chase MD 20015

CRANE, LAURA JANE, b Middletown, Ohio, Nov 2, 41; m 72. TECHNICAL MANAGEMENT. *Educ:* Carnegie-Mellon Univ, BS, 63; Harvard Univ, MA, 64; Rutgers Univ, PhD(biochem), 73. *Prof Exp:* Asst scientist clin biochem, Warner-Lambert Pharmaceut Co, 67-68; assoc scientist polymer chem, W R Grace, Inc, 68-69; from fel to res assoc biochem, Roche Inst Molecular Biol, Hoffmann-La Roche Inc, 72-75; scientist clin biochem, Warner-Lambert Co, 75-78, sr scientist & group leader clin coagulation & haemat, 78-79; mgr lab prod res & develop, 79-80, asst dir, 80-85, DIR LAB PROD RES & DEVELOP, J T BAKER CHEM CO, 85- *Mem:* AAAS; NY Acad Sci; Am Chem Soc; Am Soc Biochemists. *Res:* Development of bonded phases on silica gel for high performance liquid chromatography and other applications. *Mailing Add:* PO Box 160 Mill Rd Buttzville NJ 07829

CRANE, LEE STANLEY, b Cincinnati, Ohio, Sept 7, 15; m 76; c 2. ENGINEERING, INDUSTRIAL ENGINEERING. *Educ:* George Washington Univ, BSE, 38. *Hon Degrees:* DSc, Susquehanna Univ, 84. *Prof Exp:* Var lab & eng positions, Southern Rwy Co, 37-63; dir indust eng, Pa RR Co, 63-65; vpres eng & res, Southern Rwy Co, 65-70, exec vpres opers, 70-76, chief admin officer, 76-77, pres, 76-79, chief exec officer, 77-79, chmn, 79-80; chmn & chief exec officer, Consol Rail Corp, 81-88; RETIRED. *Concurrent Pos:* Dir, Am Security & Trust Co, Washington, DC & US Rwy Asn; trustee, George Washington Univ. *Honors & Awards:* Joseph C Scheleen Award, Am Soc Traffic & Transp, 82; Seley Award, Transp Asn Am, 82; Nat Defense Transp Award, 83. *Mem:* Nat Acad Eng; fel Am Soc Testing & Mat (past pres); fel Soc Automotive Engrs; Am Rwy Eng Asn; Am Soc Traffic & Transp; fel Am Soc Mech Engrs; US Rwy Asn. *Res:* Application of modern and creative engineering concepts to more productive railroad equipment and operations. *Mailing Add:* 1351 Monk Rd Gladwyne PA 19035

CRANE, MARK, in-vitro cultivation, parasite surface, for more information see previous edition

CRANE, PATRICK CONRAD, b Salt Lake City, Utah, Mar 17, 48. RADIO ASTRONOMY, PHYSICS. *Educ:* Mass Inst Technol, SB, 70, PhD(physics), 77. *Prof Exp:* Res assoc, 76-79, sci assoc II, 79-83, sci assoc I, 84, systs scientist, 85-88, FREQUENCY COORD, NAT RADIO ASTRON OBSERV, 84-, ASSOC SCIENTIST, 88- *Concurrent Pos:* Mem, Comt Radio Frequencies, Nat Acad Sci, 88- *Mem:* Am Astron Soc; Am Phys Soc; Am Asn Physics Teachers; Int Union Radio Sci; Int Astron Union; Astron Soc Pac. *Res:* Galaxies; galactic nuclei; radio interferometry and aperture synthesis; extragalactic radio sources; radio-frequency interference. *Mailing Add:* Nat Radio Astron Observ PO Box O Socorro NM 87801-0387

CRANE, PAUL LEVI, b Clayton, NMex, Oct 17, 25; m 51; c 3. PLANT BREEDING, AGRONOMY. *Educ:* NMex State Univ, BS, 50; Iowa State Univ, MS, 51; Purdue Univ, PhD, 56. *Prof Exp:* Asst agronomist, 51-60, assoc prof, 60-71, PROF AGRON, PURDUE UNIV, 71- *Mem:* Sigma Xi. *Res:* Genetics and breeding of maize. *Mailing Add:* Dept Agron Purdue Univ West Lafayette IN 47907

CRANE, PHILIPPE, b New Haven, Conn, July 18, 43; m 65; c 3. ASTROPHYSICS, PHYSICS. *Educ:* Yale Univ, BS, 65, MS, 66, PhD(physics), 69. *Prof Exp:* Res physicist, Yale Univ, 69-70; from instr to asst prof physics, Princeton Univ, 70-75; head, Image Processing, 81-86, ASTRONOMER & PHYSICIST, EUROP SOUTHERN OBSERV, 75- *Concurrent Pos:* US Mem ESA Faint Object Camera Instrument Sci Team, NASA & Syst Gen Corp, 78-83. *Mem:* Am Phys Soc; Am Astron Soc; Int Astron Union. *Res:* Extragalactic astronomy; radio sources, galaxies; instrument development related to research interests. *Mailing Add:* Karl-Schwarzschild-Strasse 2 Europ Southern Obs D-8046 Garching Bei Munich Germany

CRANE, ROBERT KELLOGG, b Palmyra, NJ, Dec 20, 19; m 41; c 2. BIOLOGICAL CHEMISTRY, PHYSIOLOGY. *Educ:* Wash Col, BS, 42; Harvard Univ, PhD(biol chem), 50. *Hon Degrees:* ScD, Wash Col, 82. *Prof Exp:* Res assoc, Reynolds Exp Lab, Atlas Powder Co, 42-43; instr chem, Northeast Mo State Teachers Col, 43-44; asst biochemist, Mass Gen Hosp, 49-50; from instr to assoc prof biol chem, Sch Med, Washington Univ, 50-61; prof biochem & head dept, Chicago Med Sch, 61-66; prof physiol & chmn dept, Univ Med & Dent NJ, Rutgers Med Sch, Piscataway, 66-86. *Mem:* Am Physiol Soc; Am Soc Cell Biologists; Biophys Soc; Am Chem Soc; Am Soc Biol Chemists. *Res:* Intermediary metabolism; transport of carbohydrates. *Mailing Add:* Crane's Lab Off Mill Rd Buttzville NJ 07829-0160

CRANE, ROBERT KENDALL, b Worcester, Mass, Dec 9, 35; m 57; c 4. RADIOPHYSICS, METEOROLOGY. *Educ:* Worcester Polytech Inst, BS, 57, MS, 59, PhD(elec eng), 70. *Prof Exp:* Mem tech staff, Mitre Corp, 59-64; mem tech staff, Lincoln Lab, Mass Inst Technol, 64-76; dep div mgr, Environ Res & Technol Inc, 76-81; RES PROF ENG, THAYER SCH ENG, DARTMOUTH COL, 81- *Concurrent Pos:* Mem US Study group 5; asst ed, Trans Antennas & Propagation, Inst Elec & Electronic Engrs, 72-74; mem, InterUnion Comn Radio Meteorol, 78-81; chmn, Subcomt Space & Terrestrial Sci, Comt Radio Frequencies, Nat Res Can, 84-88; adv comt, Antenna & Probe Soc, 85-87; consult meteorologist, Am Meterologist Soc, 78-; chmn Comn F, Int Sci Radio Union, 87-89. *Mem:* Fel Inst Elec & Electronics Engrs; Am Meteorol Soc; Am Geophys Union; Sigma Xi; Int Sci Radio Union. *Res:* Applied meteorology; radar meteorology; radio propagation. *Mailing Add:* Thayer Sch Eng Dartmouth Col Hanover NH 03755

CRANE, ROGER L, b Monroe, Iowa, June 27, 33; m 59; c 3. NUMERICAL ANALYSIS, OPTIMIZATION. *Educ:* Iowa State Univ, BS, 56, MS, 61, PhD(math), 62. *Prof Exp:* Engr, Univac Remington-Rand Corp, Minn, 56-59; instr math, Iowa State Univ, 62-63; mem tech staff, Radio Corp Am Labs, 63-72, FEL TECH STAFF, RCA CORP, 72- *Concurrent Pos:* Resident assoc, Appl Math Div, Argonne Nat Lab, 78-79. *Mem:* Soc Indust & Appl Math; Inst Elec & Electronics Eng; Asn Comput Mach; Math Prog Soc. *Res:* Numerical analysis and scientific applications of digital computers. *Mailing Add:* RCA Corp Davin Sarnoff Res Ctr Fel-Tech Staff Princeton NJ 08543-5300

CRANE, SARA W, b Tallahassee, Fla, Nov 24, 51; m 70. PHYSICS. *Educ:* Univ Ill at Chicago Circle, BS, 76, MS, 78, PhD(physics), 82. *Prof Exp:* Res physicist, Amoco Res Ctr, Standard Oil Co, 81-85; RETIRED. *Mem:* Am Phys Soc; Inst Elec & Electronics Engrs; AAAS. *Res:* Magnetic properties of alloys; materials development primarily concerned with ceramics. *Mailing Add:* 3110 Furman La Apt 202 Alexandria VA 22306

CRANE, SHELDON CYR, b Long Beach, Calif, Dec 22, 18; m 56; c 1. ANALYTICAL INSTRUMENTATION, ENVIRONMENTAL SCIENCES. *Educ:* Calif Inst Technol, BS, 40; Univ Calif, Los Angeles, PhD(phys biol sci), 49. *Prof Exp:* Res fel chem, Calif Inst Technol, 49-55; proj engr, Gyro Div, Giannini Controls Corp, 56-59; specialist engr inertial navig, Monterey Eng Lab, Dalmo Victor Corp, 59-60; res opers engr, Combat Develop Exp Ctr, Stanford Res Inst, 60-61; partner, Del Monte Tech Assocs, 61-62; opers res analyst, Combat Develop Exp Ctr, Stanford Res Inst, Ft Ord, 62-63, US Army Concept Team, Viet Nam, 63-64, combat develop exp ctr, 64, sr res engr marine tech, Systs Sci Dept, 64-65, bioengr med res, Life Sci, 65-72; sr res assoc, Haile Sellassie I Univ, Addis Ababa, Ethiopia, 72-74; head, Monitoring Div, Guam Environ Protection Agency, 74-76; head, Anal Lab, Environ Health & Safety, Univ Hawaii, 76-79, res assoc chem, 79-86; RETIRED. *Concurrent Pos:* Res asst, Linus Pauling Proj, Calif Inst Technol, 43-46; instr, Undergrad gas chromatography, Univ Hawaii, 86-90. *Mem:* Am Chem Soc; Sigma Xi. *Res:* Studies of respiratory function and effects of artificially induced emphysema in rats and monkeys; development of pilot manufacturing facility for concentrating the molluscicidal principle in the soap berry Endod for schistosomiasis control; instrumentation for air pollution monitoring; noise pollution studies; chemical analysis for air and water pollution; analog and digital systems in chemical instrumentation; gas chromatography. *Mailing Add:* 6616 Kii Pl Honolulu HI 96825

CRANEFIELD, PAUL FREDERIC, b Madison, Wis, Apr 28, 25. PHYSIOLOGY. *Educ:* Univ Wis, PhB, 46, PhD(physiol), 51; Albert Einstein Col Med, MD, 64. *Prof Exp:* From instr to to assoc prof, Dept Physiol, State Univ NY Downstate Med Ctr, 53-62; exec secy, Comt Pub & Med Info & ed bull, NY Acad Med, 63-66; assoc prof physiol, 66-75, PROF PHYSIOL, ROCKEFELLER UNIV, 75- *Concurrent Pos:* Rockefeller Found-Nat Res Coun fel, Dept Biophys, Johns Hopkins Univ, 51-53; sr res fel, Dept Psychiat, Albert Einstein Col Med, 60-64; assoc prof pharm, Col Physicians & Surgeons, Columbia Univ, 64-66, adj assoc prof, 66-75, adj prof, 75-; ed, J Gen Physiol, 66- *Honors & Awards:* Schumann Prize, 56; Einthoven Medal, Univ Leiden, 83; Medal, NY Acad Sci, 88. *Mem:* Am Physiol Soc; Biophys Soc; Soc Gen Physiologists; Am Asn Hist Med; fel Int Acad Hist Med. *Res:* Electrophysiology of single cardiac cells; electrophysiological basis of cardiac arrhythmias; antiarrhythmic drugs; history of 19th century physiology and mental retardation. *Mailing Add:* Rockefeller Univ New York NY 10021

CRANFORD, JACK ALLEN, b San Francisco, Calif, Dec 23, 39. MAMMALIAN ECOLOGY, PHYSIOLOGICAL ECOLOGY. *Educ:* San Francisco State Col, BA, 67, MA, 70; Univ Utah, PhD(ecol), 77. *Prof Exp:* Res asst ecol, Dr J G Hall, N Rim, Grand Canyon, Ariz, 70; park naturalist park ecol, Mt Rainier Nat Park, Longmire, Wash, 71-72; instr biol, Hartnell Col, Salinas, Calif, 73; asst prof, 77-83, ASSOC PROF ZOOL, VA POLYTECH INST & STATE UNIV, 83-, CUR MAMMALS, 77- *Concurrent Pos:* Acad fac, Corresp Study Sch, Univ Utah, 74-82; prin investr, Va Polytech Inst & State Univ res grants, 78-91, Dept Interior res grant, 80-83, Dept Energy, 86, Va Fish & Game, 86-87, Va Mus Nat Hist, 89-91. *Honors & Awards:* A Brazier Howell Award, Am Soc Mammalogists, 76. *Mem:* Am Soc Mammalogists; Am Soc Zoologist; Am Soc Naturalists; Am Ecol Soc; Am Inst Biol Sci; Sigma Xi; Brit Ecol Soc. *Res:* Ecology, behavior and physiology of mammals; with special emphasis on field ecology, circadian rhythms, cirannual rhythms and physiological regulation of hibernation in small mammals. *Mailing Add:* Dept Biol Va Polytech Inst & State Univ Blacksburg VA 24061

CRANFORD, JERRY L, neuropsychology, psychophysiology, for more information see previous edition

CRANFORD, ROBERT HENRY, b Columbia, Tenn, Sept 10, 35; m 60; c 3. MATHEMATICS. *Educ:* Mid Tenn State Col, BS, 57; La State Univ, MS, 59, PhD(algebra), 64. *Prof Exp:* Asst prof math, NTex State Univ, 64-73; assoc prof, 73-76, PROF MATH, TEX EASTERN UNIV, 76-, CHMN DEPT MATH & COMPUT SCI, 80- *Mem:* Math Asn Am; Nat Coun Teachers Math. *Res:* Ideal theory in commutative rings with unity. *Mailing Add:* Dept Math & Comput Sci Univ Tex 3900 Univ Blvd Tyler TX 75701

CRANFORD, WILLIAM B(RETT), b Chatham, Ont, Dec 20, 20; m 50; c 2. CHEMICAL ENGINEERING. *Educ:* Univ Toronto, BASc, 42, MASc, 47. *Prof Exp:* Res engr, Indust Cellulose Res Ltd, 47-60, chemist in charge chem eng & new prod lab, 60-63, mgr, Process Develop Div, Int Cellulose Res, Ltd, 63-70; VPRES & SECY, CIP RES LTD, 70- *Mem:* Sr mem Can Tech Asn Pulp & Paper Indust; sr mem Can Pulp & Paper Asn. *Res:* Cellulose and related substances and its application to industrial processes involving either pulping and bleaching of pulp and paper or the recovery of by-products. *Mailing Add:* CIP Res Ltd 179 Main St W Hawkesbury ON L8P 1J1 Can

CRANG, RICHARD FRANCIS EARL, b Clinton, Ill, Dec 2, 36; m 58; c 2. PLANT CYTOLOGY, ELECTRON MICROSCOPY. *Educ:* Eastern Ill Univ, BS, 58; Univ SDak, MA, 62; Univ Iowa, PhD(bot), 65. *Prof Exp:* Teacher high sch, Ill, 58-61; asst prof biol, Wittenberg Univ, 65-69; from assoc prof to prof biol sci, Bowling Green State Univ, 74-80; PROF PLANT BIOL & DIR, CTR ELECTRON MICROS, UNIV Ill, 80- *Concurrent Pos:* Mem educ comt, 72-78, chmn, Nat Cert Bd, 81-88, dir, Electron Microscopy Soc Am, 91-; adj prof anat, Med Col Ohio, 74-80. *Honors & Awards:* Dimond Award, Bot Soc Am, 74. *Mem:* AAAS; Am Soc Cell Biol; Bot Soc Am; Electron Micros Soc Am; Sigma Xi. *Res:* Plant ultrastructure, particularly fine structure of fungal cells as effected by environmental pollutants; analyses with transmission and scanning electron microscopy and x-ray microanalysis. *Mailing Add:* 905 S Goodwin Ave 74 A Bevier Hall Urbana IL 61801

CRANNELL, CAROL JO ARGUS, b Columbus, Ohio, Nov 15, 38; m 61; c 3. ASTROPHYSICS. *Educ:* Miami Univ, Ohio, BA, 60; Stanford Univ, PhD(physics), 67. *Prof Exp:* Fel exp nuclear physics, Cath Univ, 67-70; res assoc high energy cosmic ray physics, Fed City Col & Goddard Space Flight Ctr, NASA, 70-73; res scientist, Imp Col, Univ London, 73-74; vis assoc professor, Caltech, 90. *Concurrent Pos:* Asst prof, Howard Univ, 68-69; vis scientist, Mass Inst Technol, 74; vis assoc prof, Calif Inst Technol. *Honors & Awards:* Outstanding Achievement Award, Women in Aerospace, 90. *Mem:* Am Phys Soc; Am Astron Soc; AAAS. *Res:* Solar physics. *Mailing Add:* Code 682 NASA Goddard Space Flight Ctr Greenbelt MD 20771

CRANNELL, HALL L, b Berkeley, Calif, Feb 23, 36; m 61; c 3. NUCLEAR PHYSICS. *Educ:* Miami Univ, BA, 56, MA, 58; Stanford Univ, PhD(physics), 64. *Prof Exp:* Res assoc high energy physics, Stanford Univ, 64-67; assoc prof, 67-72, chmn dept, 79-81 & 86-89, PROF PHYSICS, CATH UNIV AM, 72- *Concurrent Pos:* Chmn, Bates Users Group, 72-73, mem bd dirs, 81; vis sr scientist & vis prof, Westfield Col, Univ London, 73-74; vis prof, Mass Inst Technol, 74; prog dir intermediate energy nuclear physics, NSF, 82-84; chmn, CEBAF Users Group, 84-85, mem exec comt, 84-86; prog adv comt, Los Alamos Meson Physics Fac, 85-88; vis prof, Caltech, 90. *Mem:* AAAS; fel Am Phys Soc; Am Asn Phys Teachers; Am Asn Univ Prof. *Res:* Electron scattering at medium energies; astrophysics; nuclear instrumentation. *Mailing Add:* 10000 Branch View Ct Silver Spring MD 20903

CRANO, JOHN CARL, b Akron, Ohio, Nov 16, 35; m 58; c 3. ORGANIC CHEMISTRY. *Educ:* Univ Notre Dame, BS, 57; Case Inst Technol, MS, 59, PhD(org chem), 62. *Prof Exp:* Sr res chemist, Chem Div, Pittsburgh Plate Glass Co, 61-63, res supvr, 63-65, res assoc, 65-78, sr res assoc, chem div, 78-81, sr res supvr, 81-84, mgr, prod res, Chem Group, 84-86, SCIENTIST, DISCOVERY RES, PPG INDUSTS INC, 86- *Mem:* AAAS; Am Chem Soc. *Res:* Mechanism of halocarbon pyrolyses; fluorine chemistry; halogen exchange reactions; polymerization; bromine chemistry; organic photochromic chemistry. *Mailing Add:* PPG Industs Inc PO Box 31 Barberton OH 44203

CRANSTON, FREDERICK PITKIN, JR, b Denver, Colo, Aug 28, 22; m 46; c 5. PHYSICS. *Educ:* Colgate Univ, BA, 43; Stanford Univ, MS, 50, PhD(physics), 59. *Prof Exp:* Asst physics, Stanford Univ, 47-53; mem staff nuclear physics, Los Alamos Sci Lab, 53-62; assoc prof, 62-66, PROF PHYSICS, HUMBOLDT STATE UNIV, 66- *Concurrent Pos:* Consult, Lawrence Livermore Lab, Univ Calif, 64-70. *Mem:* Am Phys Soc; Am Asn Physics Teachers; Fedn Am Scientists; Sigma Xi. *Res:* Nuclear spectroscopy; radiation hazards; nuclear weapons effects; x-ray fluorescence; forensic physics. *Mailing Add:* Dept Physics Humboldt State Univ Arcata CA 95521

CRANSTON, JOSEPH WILLIAM, JR, b Hartford, Conn, Oct 3, 44; m 69; c 2. INFECTIOUS DISEASE THERAPY, DRUG INFORMATION. *Educ:* Univ Conn, BSPharm, 67; Univ Mich, PhD(pharmacol), 72. *Prof Exp:* Res fel, Albert Einstein Col Med, 72-74; asst prof, Sch Med, Univ Louisville, 75-76; res scientist II, Frederick Cancer Res Ctr, 76-81; sr res scientist, 81-88, DIR, DEPT DRUGS, AMA, 88- *Concurrent Pos:* Consult, USP DI, 87-90. *Mem:* Am Soc Pharmacol & Exp Therapeut; Am Soc Microbiol; AAAS; Sigma Xi. *Res:* Evaluation of medical literature in the areas of infectious diseases and antimicrobial agents. *Mailing Add:* Dept Drugs AMA 515 N State St Chicago IL 60610

CRAPO, HENRY HOWLAND, b Detroit, Mich, Aug 12, 32; m 62; c 3. MATHEMATICS. *Educ:* Univ Mich, AB, 54; Mass Inst Technol, PhD(math), 64. *Prof Exp:* Prof math, Univ Waterloo, 65-77; prof math, Mem Univ Nfld, 77-81; researcher, Math Res Ctr, Univ de Montreal, 81-83; RESEARCHER, NAT INST RES INFORMATICS & AUTOMATION, 83- *Concurrent Pos:* Mem coun, Can Math Cong, 75-79; mem, Ctr Math Res, Montreal, assoc dir, Res Group Struct Topology; vis prof archit, Univ Montreal, 78-81. *Mem:* Am Math Soc; Math Asn Am; NY Acad Sci; Soc Math de France. *Res:* Combinatorial geometry; lattices and ordered sets; applications of projective and combinatorial geometry to architecture and structural engineering; scene analysis; robotics. *Mailing Add:* BP 6 Sergines Pont sur Yonne 89140 France

CRAPPER, DONALD RAYMOND, b Windsor, Ont, Can, Oct 4, 32; m 58; c 2. NEUROPHYSIOLOGY, NEUROLOGY. *Educ:* Univ Toronto, MD, 57; FRCP(C), 66. *Prof Exp:* HC Storrs fel ment retardation, Neuropath, Letchworth Village, NY, 58-60; fel neurophysiol, State Univ NY, Buffalo,

60-63; PROF PHYSIOL & PROF MED, UNIV TORONTO, 66- *Concurrent Pos:* Staff neurologist, Toronto Gen Hosp, 66-; consult neurologist, Clark Inst Psychiat, 76-81; dir, Centre Res Neurodegenerative Dis, Univ Toronto. *Honors & Awards:* Starr Medal, Order Can. *Mem:* Can Neurol Soc; Soc Neurosci. *Res:* Environmental factors and brain aging; chromatin structure and brain aging; neurophysiology of dementia. *Mailing Add:* Dept Physiol Univ Toronto Toronto ON M5S 1A1 Can

CRAPUCHETTES, PAUL W(YTHE), b San Francisco, Calif, Feb 12, 17; m 41; c 7. ANALYTICAL ENGINEERING. *Educ:* Univ Calif, BS, 40. *Prof Exp:* Test engr, Gen Elec Co, NY, 41, design engr, 42-43, 44-45; proj engr, Electronic Res Assocs, Calif, 46; vpres & tech dir, Litton Industs, 64-76, dir eng, Vacuum Tube Div, 46-64; RETIRED. *Concurrent Pos:* Mem bd dirs, Dirks Electronics, 69-70; mem bd trustees, Fuller Theol Sem, 69-70. *Mem:* Fel Inst Elec & Electronics Engrs. *Res:* Gaseous conductors; ultra-high frequency devices, particularly magnetron design; thyratron counter and multiple collector for counters; H-2 thyrathron for radar pulsing. *Mailing Add:* Box 13 Goodyear Rd Benicia CA 94510

CRARY, ALBERT PADDOCK, b Pierrepont, NY, July 25, 11; m 68; c 1. GEOPHYSICS. *Educ:* St Lawrence Univ, BS, 31; Lehigh Univ, MS, 33. *Hon Degrees:* PhD, St Lawrence Univ, 59. *Prof Exp:* Explor geophysicist, Independent Explor Co, 35-41; proj scientist, Woods Hole Oceanog Inst, 41-42; explor geophysicist, United Geophys Co, 42-46; proj scientist, US Air Force Cambridge Res Ctr, 46-60; chief scientist, Off Antarctic Prog, NSF, 60-67, dep dir, 67-69, dir, Div Environ Sci, 69-75 & dir, Div Earth Sci, 76; res assoc, Haverford Col, 79-82. *Concurrent Pos:* Dep chief scientist, US Nat Comt, Int Geophys Year, Antarctic Prog, Nat Acad Sci, 56-58. *Honors & Awards:* Vega Medal, Swedish Soc Anthrop & Geog; Cullum Medal, Am Geog Soc, 59; Patrons Medal, Royal Geog Soc, 63. *Mem:* AAAS; Am Geophys Union; Am Geog Soc; Sigma Xi. *Res:* Seismic exploration for oil; atmospheric acoustic studies; geophysical work in Arctic Ocean Basin and Antarctica. *Mailing Add:* 3010 New Mexico Ave Washington DC 20016-3519

CRARY, SELDEN BRONSON, b Schenectady, NY, May 10, 49; m 75; c 1. MICROSENSORS, DESIGN OF EXPERIMENTS. *Educ:* Brown Univ, ScB, 71; Univ Wash, MS, 73, PhD(physics), 78. *Prof Exp:* Asst prof physics, Amherst Col, 78-82; sr res scientist, Gen Motors Res Labs, 83-87, staff res scientist, 87-88; ASST RES SCIENTIST, DEPT ELEC ENG & COMPUT SCI, UNIV MICH, 88- *Concurrent Pos:* Vis asst prof physics, Univ Mass, Amherst, 81-83; pres, Mich Microsensor, Inc, 88- *Mem:* Am Phys Soc; Inst Elec & Electronic Engrs; AAAS. *Res:* Microsensors; optimal design of experiments. *Mailing Add:* Solid-State Electronics Lab Dept Elec Eng & Comput Sci Univ Mich 1301 Beal Ave Ann Arbor MI 48109-2122

CRAS, PATRICK, b Antwerp, Belg, June 19, 58; m 84; c 2. AMYLOIDOSIS. *Educ:* Univ Antwerp, Belg, MD, 83. *Prof Exp:* Resident, Path & Neuropath, Neurol Dept, Univ Antwerp, Belg, 83-89; FOGARTY FEL, INST PATH, DIV NEUROPATH, CASE WESTERN RESERVE UNIV, 89- *Concurrent Pos:* Fogarty fel, NIH, 89; NATO fel, NATO, 89. *Mem:* Am Asn Neuropathologists; Soc Neurosci. *Res:* Structural and biochemical abnormalities that are involved in the pathogenesis of senile plaques and neuropathology tangles of Alsheimer disease. *Mailing Add:* 2085 Adelbert Rd Cleveland OH 44106

CRASEMANN, BERND, b Hamburg, Ger, Jan 23, 22; nat US; m 52. ATOMIC PHYSICS. *Educ:* Univ Calif, Los Angeles, AB, 48; Univ Calif, Berkeley, PhD(physics), 53. *Prof Exp:* Asst prof, 53-63, head dept, 76-84, dir, Chem Physics Inst, 84-87, PROF, PHYSICS, UNIV ORE, 63- *Concurrent Pos:* Consult, Lawrence Radiation Lab, Berkeley, 54-68; guest assoc physicist, Brookhaven Nat Lab, 61-62; vis prof, physics dept, Univ Calif, Berkeley, 68-69; chmn ad hoc panel on accelerator-related atomic physics res, Comt Atomic & Molecular Physics, Nat Acad Sci-Nat Res Coun, 74-82, vchmn, Comt, 78-82, mem, Panel on Radiation Physics, 85-; vis scientist, Ames Res Ctr, NASA, 75-76; vis prof, Univ Paris VI & Lab Curie, 77; mem proposal rev panel, Stanford Synchrotron Radiation Lab, 78-87; mem physics, subcomt, adv comt for review of atomic, molecular & plasma physics, NSF, 81; chmn organizing comt, Int Conf on X-ray & Atomic Inner-Shell Physics, 82, 84 & 87; assoc ed, Atomic Data & Nuclear Data Tables, 82-; vis scholar, Physics Dept, Stanford Univ, 83; chmn, Advan Light Source Users Exec Comt, 84-87. *Mem:* Fel Am Phys Soc; Am Asn Physics Teachers; fel AAAS. *Res:* Atomic inner-shell processes; application of synchrotron radiation to atomic physics; interface of atomic and nuclear physics. *Mailing Add:* Dept Physics Univ Ore Eugene OR 97403-1274

CRASS, MAURICE FREDERICK, III, b Akron, Ohio, Nov 15, 34; m 60; c 3. VASCULAR CONTRACTILE FUNCTION, BLOOD FLOW REGULATION. *Educ:* Univ Md, BS, 57, MS, 59; Vanderbilt Univ, PhD(physiol), 65. *Prof Exp:* Instr physiol, Sch Med, Vanderbilt Univ, 65-66; asst prof med & physiol, Col Med, Univ Fla, 69-70; from asst prof to assoc prof biochem & med, Col Med, Univ Nebr, 70-73; assoc prof, 73-81, PROF PHYSIOL, SCH MED, TEX TECH UNIV, 81- *Concurrent Pos:* Res fel, Tenn Heart Asn, 65-66; res fel med, Col Med, Univ Fla, 66-69. *Mem:* AAAS; Am Physiol Soc; Soc Exp Biol & Med; Int Heart Soc; Western Pharmacol Soc. *Res:* Cardiovascular physiology, pharmacology and metabolism; regulation of vascular smooth muscle contractile function; cardiovascular effects of parathyroid hormone and related mechanisms. *Mailing Add:* Dept Physiol Sch Med Tex Tech Univ Sch Health Sci Ctr Lubbock TX 79430

CRAST, LEONARD BRUCE, JR, b Adams, NY, June 1, 36; m 55; c 4. ORGANIC CHEMISTRY, MEDICINAL CHEMISTRY. *Educ:* Syracuse Univ, AB, 58. *Prof Exp:* Jr chemist, 58-65, SR RES SCIENTIST CHEM, BRISTOL LABS, INC, 65- *Mem:* Am Chem Soc. *Res:* Semi-synthetic penicillins and cephalosporins, most notably cephapirin, cefadroxil, and cefatrizine. *Mailing Add:* 107 Black Walnut Dr PO Box 706 Durham CT 06422

CRASWELL, KEITH J, b Kent, Wash, Dec 17, 36; m 58; c 2. MATHEMATICS. *Educ:* Univ Wash, BS, 59, MSc, 61, PhD(math), 63. *Prof Exp:* Actg instr math, Univ Wash, 62-63; staff mem, Statist Res Div, Sandia Corp, 63-65; asst prof math & statist, Colo State Univ, 65-67; assoc prof, 67-80, PROF MATH, WESTERN WASH STATE UNIV, 80- *Mem:* Inst Math Statist; Am Math Soc; Math Asn Am. *Res:* Stochastic processes. *Mailing Add:* Dept of Math & Comput Sci Western Wash Univ 516 High St Bellingham WA 98225

CRATER, HORACE WILLIAM, b Washington, DC, May 26, 42; m 63; c 2. PARTICLE PHYSICS, QUANTUM MECHANICS. *Educ:* Col William & Mary, BS, 64; Yale Univ, MS, 65, MPhil & PhD(physics), 68. *Prof Exp:* Vis mem, Inst Advan Study, 68-70; asst prof physics, Vanderbilt Univ, 70-75; mem staff, 75-80, ASST PROF PHYSICS, SPACE INST, UNIV TENN, 80- *Mem:* Am Phys Soc. *Res:* Theoretical particle physics, current algebra; meson-meson scattering; phenomenological quantum field theory; non-perturbative techniques in quantum field theory; Pade approximants; two body relativistic quantum mechanic and quark models. *Mailing Add:* 5044 McLendon Dr Antioch TN 37013

CRATIN, PAUL DAVID, b Moline, Ill, Feb 26, 29; m 57; c 5. PHYSICAL CHEMISTRY. *Educ:* Spring Hill Col, BS, 51; St Louis Univ, MS, 54; Tex A&M Univ, PhD(thermodyn), 62. *Prof Exp:* Anal chemist, Chemstrand Corp, 55-58; instr chem, Ga Inst Technol, 58-59 & Tex A&M Univ, 59-62; res chemist, Jersey Prod Res Co, 62-64; assoc prof, Spring Hill Col, 64-67; res assoc, St Regis Paper Co, 67-68; assoc prof, Univ Miami, 68-70; chmn dept chem, 71-73, PROF CHEM, CENT MICH UNIV, 71- *Concurrent Pos:* Vis prof, Univ Strathclyde, UK, 88. *Mem:* Am Chem Soc. *Res:* Excess thermodynamic properties of binary liquid solutions; application of thermodynamics to surface and interfacial phenomena; kinetics and mechanisms of homogeneous chemical reactions; hyperbaric processes as related to oceanography. *Mailing Add:* Dept Chem Cent Mich Univ Mt Pleasant MI 48859

CRATTY, LELAND EARL, JR, b Oregon, Ill, June 3, 30; m 56; c 3. SOLID STATE CHEMISTRY, SURFACE CHEMISTRY. *Educ:* Beloit Col, BS, 52; Brown Univ, PhD(chem), 57. *Prof Exp:* Res chemist, Linde Co, 56-58; from asst prof to assoc prof, 58-73, PROF CHEM, HAMILTON COL, 73- *Concurrent Pos:* Vis fel, Mellon Inst, 64-65; chemist, Ames Lab, US Atomic Energy Comn, 69-70; vis scientist, Dept Mats Sci, Univ Pa, 82-83 & 87-88. *Mem:* AAAS; Am Chem Soc; Sigma Xi. *Res:* Super ionic conduction in solids; adsorption state in clean surfaces; surface chemistry and catalysis, particularly at metal and alloy surfaces; rates of adsorption at very low pressures. *Mailing Add:* RD Box 52 B Deansboro NY 13328-9708

CRAVALHO, ERNEST G, b San Mateo, Calif, Feb 25, 39. BIOENGINEERING & BIOMEDICAL ENGINEERING. *Educ:* Univ Calif, Berkeley, BS, 61, MS, 62, PhD(mech eng), 67. *Prof Exp:* Matsushita prof med mech eng, Mass Inst Technol, 75-86, EDWARD HOOD TAPLIN PROF MED ENG, DIV HEALTH SCI & TECHNOL, HARVARD UNIV-MASS INST TECHNOL, 86-; CHIEF BIOMED ENG, MASS GEN HOSP, 86- *Concurrent Pos:* Assoc prof, Dept Mech Eng, Mass Inst Technol, 70-76, prof, 76-, assoc dean acad progs, Sch Eng, 75-77, assoc dir, Whitaker Col Health Sci, Technol & Mgt, 79-83; adj prof, Northeastern Univ, 72-75; assoc dir med eng & med physics, Harvard-Mass Inst Technol, 77-, actg dir, Biomed Eng Ctr Clin Instrumentation, 79; assoc dir med eng & med physics, Harvard Med Sch, 79-85; mem bd dirs, Biomed Eng Soc, 83-86. *Mem:* Inst Med-Nat Acad Sci; Biomed Eng Soc; Soc Cryobiol; Am Soc Mech Engrs. *Mailing Add:* Dept Biomed Eng Mass Gen Hosp Fruit St Boston MA 02114

CRAVEN, BRYAN MAXWELL, b Wellington, NZ, Feb 12, 32. STRUCTURAL CHEMISTRY. *Educ:* Univ Auckland, BSc, 53, MSc, 54, PhD(chem), 58. *Prof Exp:* Jr lectr, Univ Auckland, 56; res assoc & instr, Univ Pittsburgh, 57-59, from asst res prof to assoc prof, 59-71, chmn dept crystallog, 74-77, PROF CHEM CRYSTALLOG, UNIV PITTSBURGH, 71-, CHMN DEPT, 85- *Concurrent Pos:* Rothmans sr fel, Univ Sydney, 62-64; co-ed, Acta Crystallog, 88- *Mem:* Am Crystallog Asn (pres, 89). *Res:* Crystal structure determination of biomolecules including lipids; charge density from x-ray & neutron diffraction; thermal vibrations in crystals. *Mailing Add:* Dept Crystallog Thaw Hall 304 Univ Pittsburgh 4200 5th Ave Pittsburgh PA 15260

CRAVEN, CLAUDE JACKSON, physics; deceased, see previous edition for last biography

CRAVEN, DONALD EDWARD, b Omaha, Nebr, Jan 13, 44; m 83; c 1. EPIDEMIOLOGY. *Educ:* Wesleyan Univ, BA, 66; Albany Med Col, MD, 70. *Prof Exp:* Resident internal med, McGill Univ, 71-74; fel infectious dis, Boston Univ, 74-76; res assoc, Bur Biologics, NIH, 76-79; from asst prof to assoc prof med, Boston City Hosp, 79-83, assoc prof biostatist & epidemiol, Sch Pub Health, 83-89, PROF MED & MICROBIOL, BOSTON UNIV, 89-, PROF EPIDEMIOL & BIOSTATIST, 89- *Mem:* Fel Infectious Dis Soc; fel Am Col Physicians; fel Royal Col Physicians; Am Soc Microbiol; sr mem Am Fedn Clin Res. *Res:* Nosocomial infections; gram-negative bacteremia; pathogenesis of meningococcal infections; hepatitis B infections; acquired immune deficiency syndrome. *Mailing Add:* 17 Ellery Sq Cambridge MA 02138

CRAVEN, JOHN P, b Brooklyn, NY, Oct 30, 24; m 51; c 2. ENGINEERING PHYSICS, OCEAN LAW. *Educ:* Cornell Univ, BS, 46; Calif Inst Technol, MS, 47; Univ Iowa, PhD, 51; George Washington Univ, JurDr, 58. *Prof Exp:* Assoc hydraul, Univ Iowa, 49-51; phys sci adminr, David Taylor Model Basin, 51-59; chief scientist, Special Proj Off, US Naval Bur Weapons, 59-70; dean marine progs, 70-80, prof, 70-85, EMER PROF OCEAN ENG, UNIV HAWAII, 85-, CHMN, NATURAL ENERGY LAB, 74-; PROF LAW, RICHARDSON SCH LAW, 84- *Concurrent Pos:* Vis prof, Dept Naval Archit & Marine Eng, Mass Inst Technol, 69-70; marine affairs coordr, State of Hawaii, 70-; consult, Hudson Inst, 70-76 & Pac Sierra Res Corp, 70-85; dir,

Law of Sea Inst, 76-; adj prof, Hubert H Humphrey Inst, 83-86; mem, Coun Foreign Rels, 85- *Honors & Awards:* Distinguished Civilian Serv Award, US Dept Defense; Meritorious Civilian Serv Award, US Navy; Lockheed Award, Marine Technol Soc, 82. *Mem:* Nat Acad Eng; Am Soc Mech Engrs; Am Phys Soc; Am Bar Asn; fel Am Geophys Union; fel Marine Technol Soc (pres, 70-71); World Acad Arts & Sci; Sigma Xi. *Res:* Ocean engineering and development. *Mailing Add:* Richardson Sch Law 2515 Dole St Honolulu HI 96822

CRAVEN, PATRICIA A, b New York, NY, Mar 21, 44; m 67; c 2. PROTEIN KINASE C, EICOSANOID METABOLISM. *Educ:* Columbia Univ, AB, 64; Univ Pittsburgh, PhD(biochem), 68. *Prof Exp:* Res assoc, Dept Biochem, 68-74, asst res prof, Dept Med & res chemist, Dept Vet Affairs, 74-85, ASSOC CAREER SCIENTIST, DEPT VET AFFAIRS & ASSOC RES PROF, DEPT MED, UNIV PITTSBURGH, 85- *Mem:* Am Fedn Clin Res; Am Soc Biochem & Molecular Biol. *Res:* Role of alterations in the inositol phospholipid protein Kinase C signalling system; synthesis of vasoactive eicosanoids in the pathogenesis of microvascular complications in the diabetic kidney; diabetic nephropathy; colonic growth and colonic cancer; cyclic nucleotides. *Mailing Add:* Gen Med Res Vet Admin Med Ctr University Dr C Pittsburgh PA 15240

CRAVENS, WILLIAM WINDSOR, b Daviess Co, Ky, Oct 24, 14; m 39; c 5. NUTRITION, RESEARCH ADMINISTRATION. *Educ:* Univ Ky, BS, 35; Iowa State Col, MS, 37; Univ Wis, PhD(biochem), 40. *Prof Exp:* Asst animal chem & nutrit, Iowa State Col, 35-37; from asst to instr poultry, Univ Wis, 37-41, from asst prof to prof poultry husb, 41-53; dir feed res, Cent Soya Co, 53-68, vpres res, 68-78; RETIRED. *Honors & Awards:* Am Feed Mfrs Award, 50. *Mem:* Am Chem Soc; Poultry Sci Asn; Soc Exp Biol & Med; Am Soc Animal Sci; Am Inst Nutrit. *Res:* Nutrition of livestock and poultry; animal production; feed formulation; soybean processing; food technology. *Mailing Add:* 11112 Carriage Pl Ft Wayne IN 46845-1402

CRAVER, CLARA DIDDLE (SMITH), b Portsmouth, Ohio, Dec 3, 24; m 46, 70; c 4. SPECTROCHEMISTRY, PETROLEUM CHEMISTRY. *Educ:* Ohio State Univ, BSc, 45. *Hon Degrees:* DSc, Fisk Univ, 74. *Prof Exp:* Tech man spectros, Esso Res Labs, 45-48; res engr, Battelle Mem Inst, 49-55, group leader, 55-57, consult, 57-58; SPECTROS CONSULT & OWNER CHEMIR LABS, 58- *Concurrent Pos:* ed, Coblentz Soc Spectral Data, 56-; instr chem, Ohio State Univ, 57-58; guest lectr, Infrared Inst, Fisk Univ, 59-69 & Fisk Infrared Inst, Sao Paulo, Brazil, 65 & fac mem, 65; guest lectr, Infrared Inst, Univ Minn, 61-62, 64-65 & 67-68 & Canisius Col, 61; consult, Nat Stand Ref Data Prog, 65-; consult, US Coast Guard, 75; chmn, Gordon Res Conf Vibrational Spectros, 74, past chmn, Joint Comt Atomic Molecular Phys Data, mem; dir, Fisk Infrared Inst, Vanderbilt Univ, 87-88, tech dir Polytech Labs Inc, 76-88; Am Inst Chemists fel, 80-92. *Honors & Awards:* A K Doolittle Award, Union Carbide & Am Chem Soc, 56; Award of Merit, Am Soc Testing & Mat, 82; Williams-Wright Award, Coblentz Soc Indust Spectros, 84. *Mem:* Am Chem Soc; Optical Soc Am; hon mem Coblentz Soc Indust Spectros; Soc Appl Spectros; Am Soc Test & Mat; Am Inst Chemist; Sigma Xi. *Res:* Application of chemical spectroscopy to research and analysis in polymers, coatings, marine environment, air pollution, petroleum products, asphalts, paper and cellulose chemistry; standard data compilations and computer retrieval of spectroscopic data; absorption spectroscopy; applications of infrared spectroscopy; polymer science; forensic science. *Mailing Add:* 761 W Kirkham Ave Glendale MO 63122

CRAVER, JOHN KENNETH, b Jonesboro, Ill, May 1, 15; m 39, 70; c 4. POLYMER CHEMISTRY, CORPORATE MANAGEMENT. *Educ:* Southern Ill Univ, BEd, 37; Syracuse Univ, MS, 38. *Prof Exp:* Res chemist, Monsanto Co, 38-46, coordr plasticizers, 46-51, develop mgr resin mat & functional fluids, 51-54; develop dir, Gen Mills, Inc, 55-56; develop mgr, Res & Eng Div, Monsanto Co, 56-61, res assoc org div, 61-67, sr res assoc, Cent Res Dept, 67-70, mgr, corp long range forecasting, 70-79; PRES, FUTURESEARCH & MGT CONSULT, 79- *Concurrent Pos:* Chmn, Div Org Coatings & Plastics Chem, Am Chem Soc, 57, Chem Mkt & Econ, 75, Gordon Res Conf, 66 & AAAS, 84. *Honors & Awards:* Gold C Award, Com Develop Asn. *Mem:* Fel AAAS; Am Chem Soc; Com Develop Asn; Commanderie de Bordeaux. *Res:* Long range planning; decision analysis; paper and polymer chemistry; applied polymer research; commercial development; strategic management. *Mailing Add:* 761 W Kirkham Glendale MO 63122

CRAVIOTO, HUMBERTO, b Pachuca, Mex, Oct 4, 24. NEUROPATHOLOGY, NEUROLOGY. *Educ:* Sci & Lit Inst, Mex, BS, 45; Nat Univ Mex, MD, 52; State Univ NY, MD, 64. *Prof Exp:* Jr & sr resident path, Univ Vt, 52-54; resident neurol, Bellevue Med Ctr, NY Univ, 54-56, instr neurol & neuropath, 56-58, asst prof neuropath, 58-64; assoc prof path & neurol, Sch Med, Univ Southern Calif, 64-68; assoc prof, 68-77, PROF NEUROPATH, SCH MED, NY UNIV, 77- *Concurrent Pos:* Alexander von Humboldt Soc res fel electron micros, Berlin, 61-62. *Honors & Awards:* Fel, NY Acad Sci, 77. *Mem:* Am Acad Neurol; Am Asn Neuropath (asst secy-treas, 61-64); Histochem Soc; NY Acad Sci; AAAS. *Res:* Electron microscopy; brain tumor immunology. *Mailing Add:* NY Univ Sch Med 550 First Ave New York NY 10016

CRAVITZ, LEO, bacteriology, immunology; deceased, see previous edition for last biography

CRAWFORD, BRYCE (LOW), JR, b New Orleans, La, Nov 27, 14; m 40; c 3. PHYSICAL CHEMISTRY, MOLECULAR SPECTROSCOPY. *Educ:* Stanford Univ, AB, 34, AM, 35, PhD(chem), 37. *Hon Degrees:* DSc, Hamline Univ. *Prof Exp:* Asst chem, Stanford Univ, 33-35; Nat Res Found fel, Harvard Univ, 37-39; instr chem, Yale Univ, 39-40; from asst prof to prof chem, Univ Minn, 40-85, chmn dept, 55-60, dean grad sch, 60-72, mem grad record exam bd, 68-72, regents prof, 82-85, EMER REGENTS PROF CHEM, UNIV MINN, MINNEAPOLIS, 85- *Concurrent Pos:* Guggenheim Found fel, Calif Inst Technol & Oxford Univ, 50- 51; Fulbright fel, Oxford Univ, 51; chmn, Coun Grad Schs US, 62-63; mem bd dirs, North Star Res Found, Midwest Res Inst, 63-; Fulbright prof, Univ Tokyo, 66; mem adv comt, Off Sci Personnel, Nat Res Coun, 67-71, mem gov bd, 75-78; ed, J Phys Chem, 70-80. *Honors & Awards:* Presidential Cert Merit; Minn Award, Am Chem Soc, 68; Pittsburgh Spectros Award, 77; Ellis Lippincott Award, 78; Priestley Medal, 82. *Mem:* Nat Acad Sci (home secy, 79-87); Am Chem Soc; Asn Grad Schs (pres, 70); Optical Soc Am; Am Philos Soc; Am Acad Arts & Sci. *Res:* Molecular spectra; statistical thermodynamics; molecular dynamics. *Mailing Add:* Dept Chem Five Smith Hall Univ Minn 207 Pleasant St SE Minneapolis MN 55455

CRAWFORD, CLIFFORD SMEED, b Beirut, Lebanon, July 30, 32; US citizen; m 58; c 3. DESERT INVERTEBRATES. *Educ:* Whitman Col, BA, 54; Wash State Univ, MS, 58, PhD(entom), 61. *Prof Exp:* Instr biol, Portland State Col, 61-64; from asst prof to assoc prof, 64-73, chmn dept, 75-78, PROF BIOL, UNIV NMEX, 73- *Mem:* AAAS; Entom Soc Am; Ecol Soc Am; Int Asn Ecol; Sigma Xi. *Res:* Ecology and physiology of arid-land invertebrates. *Mailing Add:* Dept Biol Univ NMex Albuquerque NM 87131

CRAWFORD, CRAYTON MCCANTS, b Greenville, SC, Sept 20, 26; m 55; c 3. PHYSICAL CHEMISTRY. *Educ:* Clemson Col, BS, 49; Univ NC, PhD(phys chem), 57. *Prof Exp:* Mem staff, Los Alamos Sci Lab, 55-59; asst prof, 59, ASSOC PROF PHYS CHEM, MISS STATE UNIV, 59- *Concurrent Pos:* Ed, J Miss Acad Sci, 68-76. *Mem:* Am Chem Soc; Sigma Xi. *Res:* Precision absorptiometry; thermodynamics; photochemical kinetics. *Mailing Add:* Dept Chem Box CH Miss State Univ Mississippi State MS 39762

CRAWFORD, DANIEL J, b Boston, Mass, Apr 17, 35. DEVELOPMENT OF MANNED REAL TIME FLIGHT SIMULATION INFRASTRUCTURE, TIME CRITICAL DATA COMMUNICATIONS. *Educ:* Univ Mass, Amherst, BS, 61; George Washington Univ, MSEE, 75. *Prof Exp:* Real-Time Simulation applications analyst, Langley Res Ctr, NASA, 62-76, evaluator sci supercomputers, 76-83, proj mgr, Advan Real-Time Simulation Syst Proj, 83-88, GROUP LEADER ADVAN SYSTS GROUP & SR RES ENGR, LANGLEY RES CTR, NASA, 88- *Concurrent Pos:* Chief systs engr, Diversified Int Sci Corp, 90- *Honors & Awards:* I-R 100 Award, 86. *Mem:* Assoc fel & sr mem Am Inst Aeronaut & Astronaut. *Res:* Project manager in development and integration of hardware and software systems primarily concerned with real-time man-in- the-loop flight simulation; presently working on developing improvements in terminal area; air traffic control; one US patent for real-time simulation clock. *Mailing Add:* 881 Cascade Dr Newport News VA 23602

CRAWFORD, DANIEL JOHN, b Columbus Junction, Iowa, May 27, 42; m 61; c 2. PLANT SYSTEMATICS. *Educ:* Univ Iowa, BA, 64, MS, 66, PhD(bot), 69. *Prof Exp:* From asst prof to assoc prof bot, Univ Wyo, 69-77; PROF BOT, OHIO STATE UNIV, 77- *Mem:* Bot Soc Am; Soc Study Evolution; Am Soc Plant Taxon; Am Soc Plant Taxonomist (pres, 88-90). *Res:* Systematics and evolution of flowering plants; use of allozymes and chloroplast DNA in elucidating plant relationships. *Mailing Add:* Dept Bot 108 Bot/Zool Bldg Ohio State Univ 1735 Neil Ave Columbus OH 43210-1293

CRAWFORD, DAVID LEE, b Hays, Kans, Nov 30, 35; m 57; c 2. FOOD SCIENCE, FOOD BIOCHEMISTRY. *Educ:* Ore State Col, BS, 58, Ore State Univ, MS, 61, PhD(food sci), 66. *Prof Exp:* Chemist I, 58-60, from asst to asst prof food sci, 64-70, assoc prof food sci & technol, 70-76, PROF FOOD SCI & TECHNOL, ORE STATE UNIV, 76-, PROG DIR SEAFOODS LAB, 66- *Concurrent Pos:* Grants, Sea Grant Prog, Bur Com Fisheries, 66- & Ore Dept Fish & Wildlife, 66- *Mem:* Inst Food Technologists. *Res:* Basic and applied research of food science, especially investigation of basic chemistry of biological systems as related to food preservation, quality and physiological response. *Mailing Add:* Seafoods Lab Ore State Univ 250 36th St Astoria OR 97103

CRAWFORD, DAVID LIVINGSTONE, b Tarentum, Pa, Mar 2, 31; m 63; c 3. ASTRONOMY. *Educ:* Univ Chicago, PhD(astron), 58. *Prof Exp:* Asst, Yerkes Observ, Univ Chicago, 53-57; asst prof physics & astron, Vanderbilt Univ, 58-60; assoc dir, 70-74, ASTRONR, KITT PEAK NAT OBSERV, 60- *Concurrent Pos:* Exec dir, Int Dark-Sky Asn, 88- *Mem:* Am Astron Soc; Astron Soc Pac; Int Astron Union. *Res:* Galactic structure; stellar photometry; observational instruments and techniques; light pollution and control. *Mailing Add:* Kitt Peak Nat Observ Box 26732 Tucson AZ 85726

CRAWFORD, DONALD LEE, b Santa Anna, Tex, Sept 28, 47; m 70; c 1. MICROBIOLOGY. *Educ:* Oklahoma City Univ, BA, 70; Univ Wis, MS, 72, PhD(bact), 73. *Prof Exp:* Asst prof biol, George Mason Univ, 73-76; from asst prof to assoc prof, 76-82, PROF BACT BIOCHEM, UNIV IDAHO, 82- *Mem:* Am Soc Microbiol; Sigma Xi; Soc Indust Microbiol. *Res:* Microbial physiology with emphasis on metabolism of organic matter and toxic chemicals by bacteria and on the bioremediation of chemically contaminated soils; genetic engineering of actinomycete bacteria. *Mailing Add:* Dept Bact Biochem Univ Idaho Moscow ID 83843

CRAWFORD, DONALD W, b St Louis, Mo, Mar 9, 28; m 57; c 3. CARDIOVASCULAR DISEASES. *Educ:* Washington Univ, BA, 50, MD, 54. *Prof Exp:* Intern, Grady Mem Hosp, Emory Univ, 54-55; resident med, Med Ctr, Stanford Univ, 58-60; NIH trainee clin cardiol, Univ Calif, San Francisco, 60-62, fel cardiopulmonary physiol, Cardiovasc Res Inst, 62-63; chief cardiol sect & cardiopulmonary lab, Long Beach Vet Admin Hosp, 63-66; asst clin prof med, Col Med, Univ Calif, 65-66; asst prof, 66-70, assoc prof, 70-80, PROF MED, SCH MED, UNIV SOUTHERN CALIF, 80- *Concurrent Pos:* Fel coun clin cardiol & coun arteriosclerosis, Am Heart Asn. *Honors & Awards:* Long Beach Vet Admin Hosp Award, 65. *Mem:* Am Fedn Clin Res; fel Am Col Cardiol. *Res:* Atherosclerosis and clinical physiology. *Mailing Add:* Dept Med Univ Southern Calif Med Ctr 2011 Zonal Ave HMR 810 Los Angeles CA 90033

CRAWFORD, DUANE AUSTIN, b Oblong, Ill, Feb 28, 29; m 52; c 2. PETROLEUM ENGINEERING. *Educ:* Eastern Ill State Univ, Dipl, 50; Mo Sch Mines, BS, 52; Pa State Univ, MSc, 59. *Prof Exp:* Jr petrol engr, Marathon Oil Co, 50, 54-55, assoc petrol engr, 55-56; res assoc, Pa State Univ, 56-58; ASSOC PROF PETROL ENG, TEX TECH UNIV, 58- *Concurrent Pos:* Reservoir engr, Gulf Oil, 61, Amoco Petrol Corp, 62, 66, 73-83 & 85 & Humble Oil & Ref Co, 69; consult engr, Amoco, 73-84 & 87. *Honors & Awards:* Slonneger Award, Soc Petrol Engrs. *Mem:* Soc Petrol Engrs; Soc Prof Well Log Analysts. *Res:* Reservoir mechanics and fluid flow problems of petroleum reservoirs and ground water hydrology; sweep efficiencies and playa lake ground water recharge; pressure fall-off analysis; pressure transient analysis; oil property evaluations; co-inventor of surgical device for prostate surgery and co-inventor of blood vessel ligator. *Mailing Add:* Petrol Eng Dept PO Box 4099 Tech Sta Lubbock TX 79409

CRAWFORD, EUGENE CARSON, JR, b Mt Gilead, NC, Nov 13, 31; m 55; c 2. PHYSIOLOGY, ZOOLOGY. *Educ:* ECarolina Col, AB, 57; Duke Univ, MA, 60, PhD(physiol), 65. *Prof Exp:* From asst prof to assoc prof, 65-75, PROF ZOOL, PHYSIOL & BIOPHYS, UNIV KY, 75-, DIR SCH BIOL SCIS, 86- *Concurrent Pos:* NIH fel zool, Duke Univ, 65; NSF res grants, 68-77; res fel, Max Planck Inst Exp Med, 73. *Mem:* AAAS; Am Soc Zool; Am Physiol Soc. *Res:* Comparative physiology; temperature regulation; respiration metabolism. *Mailing Add:* Sch Biol Sci Univ Ky Lexington KY 40506

CRAWFORD, FRANK STEVENS, JR, b Scranton, Pa, Oct 25, 23; m 62; c 2. PHYSICS. *Educ:* Univ Calif, AB, 48, PhD(physics), 53. *Prof Exp:* Res assoc, Radiation Lab, 53-58, from asst prof to assoc prof, 58-65, PROF PHYSICS, UNIV CALIF, BERKELEY, 65- *Mem:* Am Phys Soc; Am Asn Physics Teachers. *Res:* Experimental nuclear physics. *Mailing Add:* Lawrence Radiation Lab Univ Calif Berkeley CA 94720

CRAWFORD, FREDERICK WILLIAM, b Birmingham, Eng, July 28, 31; m 63; c 2. PLASMA PHYSICS. *Educ:* Univ London, BSc, 52, MSc, 58, DSc, 75; Univ Liverpool, PhD, 55, dipl ed, 56, DEng, 65. *Prof Exp:* Res trainee, J Lucas Ltd, Birmingham, Eng, 48-52; scientist, Mining Res Estab, Nat Coal Bd, Isleworth, 56-57; sr lectr elec eng, Col Advan Technol, Birmingham, 58-59; res assoc, microwave lab, Stanford Univ, 59-61; vis scientist, Fr Atomic Energy Comn, Saclay, 61-62; res physicist, Microwave Lab, Stanford Univ, 62-64, adj prof, 64-67, assoc prof, 67-69, prof elec eng & chmn, Inst Plasma Res, 69-82, consult prof, 83-84; VCHANCELLOR, ASTON UNIV, 80- *Concurrent Pos:* Consult, Compagnie Francaise Thomson-Houston, Paris, 61-62; chmn comn H, Int Sci Radio Union, 78-81. *Mem:* Fel Brit Inst Math & Appl; fel Inst Elec & Electronics Engrs; fel Am Phys Soc; fel Brit Inst Elec Engrs; fel Brit Inst Physics. *Res:* Wave propagation phenomena; diagnostics in laboratory and space plasmas. *Mailing Add:* VChancellor's Off Astron Univ Birmingham B4 7ET England

CRAWFORD, GEORGE WILLIAM, b Statesville, NC, Oct 21, 06; m 34. PHYSICS. *Educ:* Davidson Col, BS, 29; Univ NC, MS, 49; Ohio State Univ, PhD, 59. *Prof Exp:* Teacher, pub schs, NC, 29-40; instr physics, NC State Col, 46-50; asst prof, Davidson Col, 51-60; from assoc prof to prof, 60-73, EMER PROF PHYSICS, COL WILLIAM & MARY, 73- *Mem:* AAAS; Am Asn Physics Teachers; Optical Soc Am. *Mailing Add:* 205 John Wythe Pl Williamsburg VA 23185

CRAWFORD, GEORGE WOLF, b San Antonio, Tex, May 7, 22; m 46; c 3. ENVIRONMENTAL PHYSICS, NUCLEAR PHYSICS. *Educ:* Univ Tex, BS, 47, MA, 49, PhD(physics), 51. *Prof Exp:* Res physicist, Optical Res Lab, Univ Tex, 49-51; assoc prof physics, Clemson Col, 51-55; asst dir, Tex Petrol Res Comt & asst prof petrol eng, Univ Tex, 55-59; nuclear physicist & chief physics br, Sch Aerospace Med, Brooks AFB, 59-63; chmn dept, 75-81, PROF PHYSICS, SOUTHERN METHODIST UNIV, 63- *Concurrent Pos:* Res physicist, Deering Millikin Res Trust, 52-54 & Grad Res Ctr Southwest, 63-64. *Mem:* Am & Int Solar Energy Socs; Am Phys Soc; Am Asn Physics Teachers; Sigma Xi. *Res:* Solar and wind energy storage and conversion. *Mailing Add:* Dept Physics Southern Methodist Univ Dallas TX 75275

CRAWFORD, GERALD JAMES BROWNING, experimental physics; deceased, see previous edition for last biography

CRAWFORD, GLADYS P, b Krum, Tex, Apr 12, 27; m 53; c 3. BIOLOGY, MICROBIOLOGY. *Educ:* Univ NTex, BS, 46, MS, 49. *Prof Exp:* Instr biol & chem, Our Lady Victory Jr Col, 46-49; instr, 48-71, ASST PROF BIOL, UNIV NTEX, 71- *Mem:* Am Soc Microbiologists. *Mailing Add:* Univ NTex Box 5218 Denton TX 76201

CRAWFORD, HEWLETTE SPENCER, JR, b Syracuse, NY, June 4, 31; m 52; c 4. WILDLIFE ECOLOGY, FOREST ECOLOGY. *Educ:* Univ Mich, BS, 54, MS, 57; Univ Mo, Columbia, PhD(ecol, forest wildlife), 67. *Prof Exp:* Asst wildlife mgt, Univ Mich, 53-54 & 56-57; res forester, Southern Exp Sta, Forest Serv, USDA, 57-64, proj leader, Cent States Exp Sta, 64-68, prin wildlife ecologist & proj leader wildlife habitat res, Southeastern Forest Exp Sta, 68-74, prin ecologist, Northeastern Forest Exp Sta, 74-75, prin res wildlife ecologist, 75-91; CONSULT, CRAWFORD ASSOCS, 91- *Concurrent Pos:* Mem grad fac, Col Forest Resources, Univ Maine, 75- *Mem:* Wildlife Soc; Soc Am Foresters. *Res:* Wildlife habitat research, especially as influenced by forest management practices. *Mailing Add:* Rte 1 Box 170 East Holden ME 04429

CRAWFORD, IRVING POPE, microbial genetics; deceased, see previous edition for last biography

CRAWFORD, ISAAC LYLE, b Houston, Tex, Aug 22, 42; m 63; c 2. NEUROBIOLOGY. *Educ:* Univ Tex, Austin, BA, 64; Pa State Univ, PhD(pharmacol), 74. *Prof Exp:* Physical scientist chem, US Dept Interior, 65-67; physiologist, NIMH, 67-70; res asst pharmacol, Pa State Univ, 70-74; inst, Pa State Col Med, 74-75; ASST PROF NEUROL & PHARMACOL, UNIV TEX SOUTHWESTERN MED SCH, 75-; PHARMACOLOGIST NEUROPHARMACOLOGY, VET ADMIN MED CTR, DALLAS, 75- *Concurrent Pos:* Grad fel, Med Sch, Georgetown Univ, 68-70; fac lectr, Baylor Univ Med Ctr Dallas, 76; consult, Tex Dept Mental Health & Mental Retardation, 77; lectr, Tex A&M Univ, 77; vis scientist, NIMH, Washington, 78; consult, Working Conf on Circulation, Neurobiol & Behav, NIH, 81. *Mem:* Am Physiol Soc; Neurosci Soc; Epilepsy Asn Am; NY Acad Sci; Am Chem Soc. *Res:* Mechanism of drug action on neurons in limbic brain sites are studied using neurochemical and neurophysiologic approaches; epileptic seizure disorders, and the relationship of the central nervous system to hypertension. *Mailing Add:* Epilepsy Ctr 127 4800 Lancaster Rd Dallas TX 75216-7191

CRAWFORD, JAMES DALTON, b Clyde, Tex, Aug 2, 19; m 43; c 1. ORGANIC CHEMISTRY. *Educ:* Hardin-Simmons Univ, BA, 52, MA, 53. *Prof Exp:* Chemist, Cardinal Chem Inc, 53-54, dir res, 54-60; TECH DIR, CONTINENTAL PROD TEX, 60- *Mem:* AAAS; Am Chem Soc; Nat Asn Corrosion Engrs. *Res:* Oil field chemicals for prevention of corrosion in production of petroleum; water treatment; demulsifiers; surfactants. *Mailing Add:* Rte 2 Box 657 Abilene TX 79601

CRAWFORD, JAMES GORDON, b Alma, Mich, Sept 12, 29; m 53; c 3. MICROBIOLOGY, BIOCHEMISTRY. *Educ:* Alma Col, BS, 51; Univ Mich, MS, 53, PhD(bact), 55. *Prof Exp:* Sr asst scientist, Div Biol Standards, NIH, 55-58; head biol control labs, Pfizer, Inc, 58-61, proj dir infectious hepatitis res, 62-65, dir biol develop, 63-68, DIR FERMENTATION DEVELOP, PFIZER, INC, 68- *Concurrent Pos:* Vis prof, Ind State Col, 63- *Mem:* Am Soc Microbiol; NY Acad Sci; Fedn Am Scientist. *Res:* Etiology and prophylaxis of infectious hepatitis; development of human and veterinary vaccines. *Mailing Add:* 4351 E Dallas Rd Rte 24 Terre Haute IN 47802

CRAWFORD, JAMES JOSEPH L, b June 23, 31; US citizen; m 51; c 6. MICROBIOLOGY. *Educ:* Univ Mo, BA, 53, MA, 54; Univ NC, PhD(microbiol), 62. *Prof Exp:* Asst instr microbiol, Sch Med, Univ Mo, 53-54; teaching asst, Med Sch, Univ Minn, 54-56; from res asst to res assoc, Sch Med, 57-60, instr oral microbiol, Sch Dent, 60-65, trainee, Dept Bact, Sch Med, 62-63, from asst prof to assoc prof oral microbiol, Sch Dent, 65-74, assoc prof endodontics & lectr bacteriol, 74-80, PROF ENDODONTICS, RES & BACTERIOL, SCH MED, UNIV NC, CHAPEL HILL, 80- *Concurrent Pos:* Consult, Microbiol Labs, NC Mem Hosp, 59-; prin investr, USPHS, 65-68; consult, Womack Army Hosp, Ft Bragg, 67-77; coordr infection control, Dent Clin, Univ NC, 73-; consult, Am Dent Assoc & dent schs & indust. *Mem:* Am Soc Microbiol; Am Dent Asn; Am Soc Infection Control in Dent; Am Assoc Dent Schs. *Res:* Mechanisms of host resistance; clinical and oral microbiology. *Mailing Add:* 117 Warren Way Chapel Hill NC 27514-9370

CRAWFORD, JAMES WELDON, b Napoleon, Ohio, Oct 27, 27; m 55; c 1. PSYCHIATRY, EXPERIMENTAL PSYCHOLOGY. *Educ:* Oberlin Col, AB, 50; Univ Chicago, MD, 54, PhD, 61. *Prof Exp:* Clin instr psychiat, Chicago Med Sch, 62-63, clin assoc, 63-65, clin asst prof, 65-69, clin assoc prof & assoc dir undergrad educ, Dept Psychiat & Behav Sci, 69-70; chmn & organizer of Dept Psychiat, Ravenswood Hosp Med Ctr, 70-79; ASSOC PROF PSYCHIAT, PSYCHOL & SOCIAL SCI, RUSH MED COL, RUSH UNIV, CHICAGO, 84- *Concurrent Pos:* Nat Inst Neurol Dis & Blindness res grants, 55-59; staff psychiatrist, Field Clin, Chicago, Ill, 62-65, partner, 65-78; assoc staff, Mt Sinai Hosp, 66-; courtesy staff, Louis A Weiss Mem Hosp, 71-; attend staff, Fox River Hosp, 71-74; clin assoc prof, Dept Psychiat, Abraham Lincoln Sch Med, Univ Ill, 71-; pres, 78-80, chmn, J W Crawford Asn, Inc, 80-83; asst attend staff, Presby St Luke's Hosp, 84-; pvt pract, clin & occup psychiat. *Mem:* AAAS; fel Am Psychiat Asn; Asn Am Med Cols; Am Med Asn; Am Asn Univ Profs; Sigma Xi; fel Am Orthopsychiat Asn. *Res:* Changes in attitudes of medical students during psychiatric training differences between individual and couples or family therapy; visual behavior and thyroid function in cats; various organic therapies in psychiatry; empathy in psychotherapy; administration in psychiatry; work stress psychiatry in business and industry; developmental theory in psychiatry and psychoanalysis. *Mailing Add:* 2418 Lincoln St Evanston IL 60201-2151

CRAWFORD, JAMES WORTHINGTON, b Newport, RI, Feb 25, 44; m 65; c 1. ORGANIC CHEMISTRY, PHOTOCHEMISTRY. *Educ:* The Citadel, BS, 65; Univ SC, PhD(org photochem), 69. *Prof Exp:* From staff scientist to sr engr, Off Prod Div, IBM Corp, 69-79; PROG MGR, MEAD OFF SYSTS, 79- *Mem:* Am Chem Soc. *Res:* Development of organic photoconducting polymers and small molecule photoconductors for use in electrophotography; development of ink jet technology and ink jet printing machines. *Mailing Add:* 9823 Limerick Dr Dallas TX 75218

CRAWFORD, JEAN VEGHTE, b Buffalo, NY, Mar 13, 19. ORGANIC CHEMISTRY. *Educ:* Mt Holyoke Col, AB, 40; Oberlin Col, AM, 42; Univ Ill, PhD(chem), 50. *Prof Exp:* Instr chem, Mt Holyoke Col, 42-45; chemist, Eastman Kodak Co, 45-47; adj prof chem, Randolph-Macon Woman's Col, 50-51; from asst prof to prof, 51-74, CHARLES FITCH ROBERTS PROF CHEM, WELLESLEY COL, 74- *Mem:* Am Chem Soc. *Res:* Heterocyclic nitrogen compounds; mechanism of organic reactions. *Mailing Add:* 617 Washington Wellesley MA 02181

CRAWFORD, JOHN ARTHUR, b Fort Dodge, Iowa, July 24, 46; m 67. ANIMAL HABITAT RELATIONSHIPS. *Educ:* Creighton Univ, BS, 68; Univ Nebr, Omaha, MS, 71; Tex Tech Univ, PhD(range & wildlife mgt), 74. *Prof Exp:* Teaching asst biol, Univ Nebr, Omaha, 69-71; res asst wildlife, Tex Tech Univ, 71-74; asst prof, 79, ASSOC PROF WILDLIFE, ORE STATE UNIV, 79- *Mem:* Wildlife Soc; Am Ornithologists Union. *Res:* The habitat requirements of upland gamebirds and determination of animal habitat relationships. *Mailing Add:* Dept Fisheries Wildlife Ore State Univ Corvallis OR 97330

CRAWFORD, JOHN CLARK, b Liberty, Tex, Sept 27, 35; m 55; c 2. PHYSICS. *Educ:* Phillips Univ, BA, 57; Kans State Univ, MS, 59, PhD(physics), 62. *Prof Exp:* Mem staff physics, Sandia Nat Lab, 62-67, div supvr solid state electronics res, 67-71, mgr, Electron Tube Develop Dept, 71-77, dir, Components & Standards Systs, 77-87, VPRES LIVERMORE PROGS, SANDIA NAT LAB, 87- *Mem:* Sr mem Inst Elec & Electronics Engrs. *Res:* X-ray diffraction as applied to ferroelectric whiskers; pulsed high magnetic fields; thin films semiconductors; piezoelectric devices; neutron sources; technical management. *Mailing Add:* Sandia Nat Lab Orgn 8000 PO Box 969 Livermore CA 94550

CRAWFORD, JOHN DAVID, US citizen. BIFURCATION THEORY, HYDRODYNAMIC STABILITY. *Educ:* Princeton Univ, BA, 77; Univ Calif, Berkeley, 83. *Prof Exp:* Res physicist, Univ Calif, San Diego, 85-89; ASST PROF PHYSICS, UNIV PITTSBURGH, 90- *Mem:* Am Phys Soc; Am Math Soc; AAAS; Sigma Xi; Soc Indust & Appl Math. *Res:* Bifurcation theory; nonlinear dynamics; plasma physics; fluid dynamics; mathematical physics. *Mailing Add:* Dept Physics Pittsburgh PA 15260

CRAWFORD, JOHN DOUGLAS, b Boston, Mass, Apr 16, 20; m 49; c 3. ENDOCRINOLOGY, METABOLISM. *Educ:* Harvard Med Sch, MD, 44. *Prof Exp:* From asst prof to prof, 54-90, EMER PROF PEDIAT, HARVARD MED SCH, 90- *Concurrent Pos:* Chief pediat, Burns Inst, Boston Unit, Shriners Hosps Crippled Children; chief, Endocrine-Metab Unit, Children's Serv, Mass Gen Hosp, Boston. *Mem:* Am Pediat Soc; Soc Pediat Res; Am Soc Clin Invest; Endocrine Soc. *Res:* Clinical endocrinology; renal physiology. *Mailing Add:* Children's Serv Mass Gen Hosp Boston MA 02144

CRAWFORD, JOHN OKERSON, b New York, NY, April 6, 49; m 75; c 1. PRECISION OPTO-MECHANICS, COMPUTER CONTROL OF OPTICS. *Educ:* Newark Col Eng, BS, 72. *Prof Exp:* Engr jewelry, Crestmark Mfg, 72-73; eng supvr temperature, Thermo Elec, 73-79; eng mgr surg equip, US Surg Corp, 79-81; eng mgr optics, Macbeth Div Kollmorgen, 81-83; sr staff engr optics, Perkin-Elmer, 83-89; DIR ENG OPTICS, ORIEL CORP, 89- *Mem:* Optical Soc Am; Int Soc Optical Eng. *Res:* Design of optical research instruments, sensor systems and software, including spectrographs, radiometry test equipment, precision mechanics, motion control, robotics, mathematical systems modeling, and math methods. *Mailing Add:* 95 Settlers Farm Rd Monroe CT 06468

CRAWFORD, JOHN S, b Toronto, Ont, May 12, 21; m 44; c 3. INTERNAL MEDICINE, REHABILITATION MEDICINE. *Educ:* Univ Toronto, MD, 44; FRCP(C), 52. *Prof Exp:* J J McKenzie fel path, Banting Inst, Univ Toronto, 48-49, McLaughlin traveling fel med, Fac Med, 53-54, clin teacher med & rehab med, 54-60, assoc rehab med, 60-63, assoc med, 60-64, asst prof rehab med, 63-68, asst prof med, 64-68, assoc prof med & rehab med, 68-75, prof med, Fac Med, 68-80, prof rehab med & chmn dept, 73-85,; DIR REHAB MED, TORONTO WESTERN HOSP, 54- *Concurrent Pos:* Mem med adv bd, Rehab Found Disabled, 61-; med dir, Hillcrest Hosp, 64-75; chmn med adv bd, Toronto Rehab Ctr, 66-; mem div rehab, Coun Health, Prov Ont, 69- *Mem:* Can Asn Phys Med & Rehab (pres, 65). *Res:* Electrodiagnosis; electromyegraphy. *Mailing Add:* Dept Rehab Med Toronto Western Hosp Toronto ON M5S 1A1 Can

CRAWFORD, LESTER M, b Demopolis, Ala, Mar 13, 38; m 63; c 2. ANTIBIOTIC TOXICOLOGY. *Educ:* Auburn Univ, DVM, 63; Univ Ga, PhD(pharmacol), 69. *Hon Degrees:* MDV, Budapest Univ, Hungary, 87. *Prof Exp:* Asst dean, Col Vet Med, Univ Ga, 68-70, assoc dean, 70-75; pharmacologist, Bur Vet Med, Food & Drug Admin, 75-76; assoc prof pharmacol, Col Vet Med, Univ Ga, 76-78; dir, Bur Vet Med, Food & Drug Admin, 78-80; head, dept physiol-pharmacol, Col Vet Med, Univ Ga, 81-82; dir, Ctr Vet Med, Food & Drug Admin, 82-85; assoc adminr, 86-87, ADMINR, FOOD SAFETY & INSPECTION SERV, USDA, 87- *Honors & Awards:* A M Mills Award, Univ Ga, 79; K F Meyer Gold-Headed Cane Award, 80. *Mem:* AAAS; NY Acad Sci; Sigma Xi; Am Vet Med Asn; Am Col Vet Pharmacol & Ther; hon Fr Acad Vet Med, 84. *Res:* Antibiotic toxicology; veterinary medical history; veterinary educational methodology. *Mailing Add:* FSIS-USDA 331-E 14th & Independence Ave SE Washington DC 20250

CRAWFORD, LLOYD V, b Memphis, Tenn, July 23, 23. PEDIATRICS, IMMUNOLOGY. *Educ:* Univ Tenn Col Med, MD, 46. *Prof Exp:* Chief, Univ Tenn Ctr Health, 60-85, CO-DIR, ALLERGY & CLIN IMMUNOL, UNIV TENN CTR HEALTH, 86- *Mem:* Am Bd Pediat; Am Bd Allergy & Immunol; Am Acad Allergy & Immunol; Am Asn Cert Allergists; AMA; Am Col Allergists. *Mailing Add:* Le Bonheur Children's Hosp Univ Tenn Ctr Health Sci 848 Adams Ave Memphis TN 38103

CRAWFORD, LLOYD W(ILLIAM), b Prince Albert, Sask, Nov 20, 28; m 57; c 2. CHEMICAL ENGINEERING. *Educ:* Univ Sask, BSc, 50, MSc, 51; Univ Cincinnati, PhD(chem eng), 61. *Prof Exp:* Res assoc chem eng, Univ Sask, 52-56; res engr, Film Dept, Yerkes Res & Develop Lab, E I du Pont de Nemours & Co, NY, 61-64; assoc prof mech eng, Tenn Polytech Inst, 64-67; assoc prof chem eng, 67-76, PROF MECH ENG, SPACE INST, UNIV TENN, 76- *Mem:* Am Inst Chem Engrs; Am Chem Soc; Sigma Xi. *Res:* Heat transfer and thermodynamics; mass transfer and fluid flow. *Mailing Add:* PO Box 1655 Tullahoma TN 37388

CRAWFORD, MARIA LUISA BUSE, b Beverly, Mass, July 18, 39; m 63. METAMORPHIC PETROLOGY, MINERALOGY. *Educ:* Bryn Mawr Col, BA, 60; Univ Calif, Berkeley, PhD(geol), 65. *Prof Exp:* From asst prof to assoc prof, 65-79, chmn dept, 76-88, PROF GEOL, BRYN MAWR COL, 79-, WILLIAM R KENAN JR PROF, 85- *Concurrent Pos:* Counr, Geol Soc Am, 82-85 & Mineral Asn Can, 85-88; NSF vis professorship for women, 88-89; counr, Mineral Soc Am, 89-92. *Mem:* Geol Soc Am; Mineral Soc Am; Am Geophys Union; Am Women in Sci; Mineral Asn Can. *Res:* Petrology, mineralogy and geochemistry of metamorphic and igneous rocks; fluid inclusion studies; structure and tectonics of metamorphic terranes. *Mailing Add:* Dept Geol Bryn Mawr Col Bryn Mawr PA 19010-2899

CRAWFORD, MARTIN, b Lake City, Tenn, Mar 10, 34; m 68; c 2. MECHANICAL ENGINEERING. *Educ:* Univ Tenn, BS, 54, MS, 58; Ga Inst Technol, PhD(mech eng), 63. *Prof Exp:* Instr mech eng, Univ Tenn, 55-56; asst prof, Va Polytech Inst, 56-58; instr Ga Inst Technol, 58-59, asst prof, 59-62; assoc prof, Va Polytech Inst, 62-68; assoc prof, 68-70, PROF MECH ENG, UNIV ALA, BIRMINGHAM, 70- *Mem:* AAAS; Am Soc Eng Educ; Am Soc Mech Engrs. *Res:* Heat transfer; thermodynamics; fluid mechanics; air pollution control theory. *Mailing Add:* Sch Eng Univ Ala UAB Sta 3627 Redmont Rd Birmingham AL 35213

CRAWFORD, MICHAEL HOWARD, b Madison, Wis, July 10, 43; m 68; c 3. CARDIOVASCULAR DISEASES, INTERNAL MEDICINE. *Educ:* Univ Calif, Berkeley, BA, 65, Univ Calif, San Francisco, MD, 69. *Prof Exp:* Resident med, Univ Calif Hosp, San Francisco, 69-71; sr resident, Beth Israel Hosp, Boston, Mass, 71-72; fel cardiol, Sch Med, Univ Calif, San Diego, 72-74, from clin instr to asst prof med, 73-76; from asst prof to assoc prof, 76-82, PROF MED, UNIV TEX HEALTH SCI CTR, 82- *Concurrent Pos:* Teaching fel, Harvard Med Sch, 71-72; prin investr, Vet Admin, San Antonio, 76-; adj scientist, Southwest Found Biomed Res, 80-; mem bd dirs, Am Soc Echocardiography, 80-83. *Honors & Awards:* Paul Dudley White Award, Asn Mil Surgeons US, 81. *Mem:* Am Heart Asn; Am Col Cardiol; Am Soc Echocardiography; Am Col Physicians; Asn Univ Cardiologists. *Res:* Response of the cardiovascular system to physiologic stress, its modification by coronary artery and valvular disease and its alteration by pharmacologic therapy. *Mailing Add:* Univ NMex Sch Med Univ NMex Hosp Albuquerque NM 87131

CRAWFORD, MICHAEL KARL, b Bay Shore, NY, June 29, 54; m 87. SPECTROSCOPY OF SOLIDS, SUPERCONDUCTIVITY. *Educ:* State Univ NY, Stony Brook, BS, 76; Columbia Univ, PhD(phys chem), 81. *Prof Exp:* Postdoctoral fel phys chem, Chem Dept, Columbia Univ, 81-82 & infrared astron, Physics Dept, Univ Calif, Berkeley, 82-85; MEM RES STAFF, DUPONT CO, 85- *Mem:* Am Phys Soc; Mat Res Soc; Electrochem Soc. *Res:* High temperature superconductivity; infrared and raman spectroscopy of solids; synchrotron x-ray diffraction. *Mailing Add:* DuPont Exp Sta E356-209 Wilmington DE 19880-0356

CRAWFORD, MORRIS LEE JACKSON, b Ellijay, Ga, Jan 20, 33; m 60; c 1. PHYSIOLOGICAL PSYCHOLOGY. *Educ:* Univ Ga, BS, 59, MS, 60, PhD(psychol), 62. *Prof Exp:* Asst prof res & psychol, Univ Ga, 62-63; Nat Inst Neurol Dis & Blindness fel, 63-64; instr res, Sch Med, Univ Miss, 65; asst prof psychol, Col Med, Baylor Univ, 66-70; PROF NEURAL SCI, UNIV TEX HEALTH CTR HOUSTON, 78-, DEPT OPHTHAL, HEALTH & SCI CTR. *Concurrent Pos:* Prof physiol optics, Baylor Col Med, 82- *Mem:* AAAS; Soc Nuerosci; Asn Res Vision & Ophthal. *Res:* Behavioral science; central nervous system control of visual behavior; visual system development in primates. *Mailing Add:* Sensory Sci Ctr Health Sci Ctr Univ Tex Houston TX 77030

CRAWFORD, MYRON LLOYD, b Orem, Utah, Oct 29, 38; m 58; c 5. ELECTROMAGNETIC COMPATIBILITY METROLOGY. *Educ:* Brigham Young Univ, BS, 60; Univ Colo, MS, 68. *Prof Exp:* Appl mathematician, Nat Bur Standards, 60-63, electronic engr, 63-71, proj leader, 71-86; SUPVRY ELEC ENGR, NAT INST STANDARDS & TECHNOL, 87- *Concurrent Pos:* Chmn, EMI Standards Comt, Soc Automotive Engrs, 77-87; Mem, C-63 Electromagnet Compatibility Comt, Am Nat Standards Inst, 80-, Electromagnetic Compatibility Standards Comt, Inst Elec & Electronics Engrs, 83-86; mem, US tech adv group, Int Orgn Standardization, 80-, chmn, 82-86. *Honors & Awards:* Cert Achievement, Inst Elec & Electronics Engrs, Electromagnetic Compatibility Soc, 79; Recognition Award, Soc Automotive Engrs, 87. *Mem:* Fel Inst Elec & Electronic Engrs; fel Inst Elec & Electronics Engrs, Electromagnetic Compatibility Soc. *Res:* Electromagnetic compatibility problems for evaluation of electronic equipment and systems; improvements in electromagnetic compatibility standards and measurement technology. *Mailing Add:* 7235 Empire Dr Boulder CO 80303

CRAWFORD, NICHOLAS CHARLES, b Dayton, Ohio, June 2, 42; m 69; c 2. HYDROGEOLOGY, CLIMATOLOGY. *Educ:* Tenn Technol Univ, BS, 65; E Tenn State Univ, MS, 67; Clark Univ, PhD(geog), 78. *Prof Exp:* Instr geog, Austin Peay State Univ, 67-68; from instr to asst prof geog, Peabody Col, 68-76; asst prof, Vanderbilt Univ, 72-76; from asst prof to assoc prof, 76-83, PROF GEOG & GEOL, WESTERN KY UNIV, 83- *Mem:* Asn Am Geographers; Nat Speleol Soc; Sigma Xi; Geol Soc Am. *Res:* Hydrogeology, Karst hydrology, speleology. *Mailing Add:* Dept Geog & Geol Western Ky Univ Bowling Green KY 42101

CRAWFORD, NORMAN HOLMES, b Stettler, Alta, July 30, 35; US citizen; m 64; c 2. HYDROLOGY, HYDRAULICS. *Educ:* Univ Alta, BSc, 58; Stanford Univ, MS, 59, PhD(hydraul, hydrol), 62. *Prof Exp:* Res asst hydraul, Stanford Univ, 60-62, asst prof, 62-68; dir, 68-76, PRES, HYDROCOMP INT, 76- *Concurrent Pos:* Consult, United Nations & Govt India; lectr, Univ Padua, Univ New SWales. *Mem:* Am Geophys Union. *Res:* Digital computer simulation of river systems. *Mailing Add:* 2386 Branner Dr Menlo Pk CA 94025

CRAWFORD, OAKLEY H, b Bridgeton, NJ, Sept 29, 38; m 63; c 2. PARTICLE-SOLID INTERACTIONS. *Educ:* Carson-Newman Col, BS, 59; Univ Ill, PhD(phys chem), 66. *Prof Exp:* Sr res fel appl math, Queen's Univ, Belfast, 66-67; asst prof chem, Pa State Univ, 67-74; assoc prof chem, Barnard Col, Columbia Univ, 74-76; STAFF SCIENTIST, CHEM DIV, OAK RIDGE NAT LAB, 76- *Mem:* AAAS; Am Phys Soc; Am Chem Soc. *Res:* Reaction dynamics in crossed molecular beams; atomic and molecular scattering; calculation of chemical reaction rates from first principles; particle-solid interactions; molecular scattering and intramolecule forces; stopping power of swift ions; channeling; atomic physics of channeled ions. *Mailing Add:* Oak Ridge Nat Lab Bldg 4500-S MS-123 PO Box 2008 Oak Ridge TN 37831-6123

CRAWFORD, PAUL B(ERLOWITZ), b Stamford, Tex, July 28, 21; m 48; c 3. PETROLEUM ENGINEERING. *Educ:* Tex Tech Col, BS, 43; Univ Tex, MS, 46, PhD(chem eng), 49. *Prof Exp:* Res engr, Am Cyanamid & Chem Corp, 43-44; instr, Univ Tex, 46-47; res engr, Magnolia Petrol Co, 47-52; ASST DIR TEX PETROL RES COMT, TEX A&M UNIV, 52-, PROF COMN, PETROL ENG, 62- *Concurrent Pos:* Chmn long range planning comt, Sul Ross Sch; pres libr bd, Carnegie Libr; mem bd, United Fund; mem, Interstate Oil Compact Comn, chmn libr comt, vchmn res comt; mem, Water-for-Tex Comt. *Mem:* Am Inst Mining, Metall & Petrol Engrs. *Res:* Petroleum production; underground combustion; formation fracturing; catalytic cracking of gas-oils; heat transfer; natural gas; thermal oil recovery; nuclear bomb applications. *Mailing Add:* 1100 Edgewood Bryan TX 77803

CRAWFORD, PAUL VINCENT, b Concord, NH, Jan 9, 33; m 65; c 2. PHYSICAL GEOGRAPHY, CARTOGRAPHY. *Educ:* Univ Okla, BA, 56, MA, 58; Univ Kans, PhD(geog), 69. *Prof Exp:* Instr geog, RI Col, 60-65; asst prof, 69-71, assoc prof, 71-76, PROF GEOG, BOWLING GREEN STATE UNIV, 76- *Mem:* Asn Am Geog; Am Cong Surv & Mapping; Sigma Xi. *Res:* Cartographic perception; three dimensional mapping; interrelationships of soil genesis and geomorphic processes. *Mailing Add:* Dept Geog Bowling Green State Univ Bowling Green OH 43403

CRAWFORD, RAYMOND BERTRAM, medicine, for more information see previous edition

CRAWFORD, RAYMOND MAXWELL, JR, b Charleston, SC, July 28, 33; m 51; c 5. NUCLEAR & CHEMICAL ENGINEERING. *Educ:* Wayne State Univ, BSc, 58, MSc, 60; Univ Calif, Los Angeles, PhD(nuclear eng), 69. *Prof Exp:* Instr chem eng, Wayne State Univ, 60-63; asst prof eng, San Fernando Valley State Col, 63-66; mem tech staff reactor safety, Atomics Int, 69-71; assoc nuclear engr, Argonne Nat Lab, 71-74; asst head, Nuclear Safeguards & Assoc, Sargent & Lundy, 74-81; vpres, Sci Appl Inc, 81-83; exec dir, Nutech Engrs, 83-86; PRIN, ENG RES GROUP, 86- *Concurrent Pos:* Consult, Atomic Power Develop Assoc, Inc, 62-63. *Honors & Awards:* Analog Comput Educ Users Group Achievement Award, 65. *Mem:* AAAS; Am Inst Chem Engrs; Am Chem Soc; Am Nuclear Soc; Nat Soc Prof Engrs; Western Soc Engrs; NY Acad Sci. *Res:* Reactor safety analysis of light-water reactors and liquid-metal fastbreeder reactors; mathematical studies of control and stability of lumped-parameter and distributed-parameter systems; chemical catalysis on solid surfaces; high-vacuum technology. *Mailing Add:* 1005 E Kennebec Lane Naperville IL 60563

CRAWFORD, RICHARD BRADWAY, b Kalamazoo, Mich, Feb 16, 33; m 54; c 4. MOLECULAR BIOLOGY, EMBRYOLOGY. *Educ:* Kalamazoo Col, AB, 54; Univ Rochester, PhD(biochem), 59. *Prof Exp:* Fel biochem, Univ Rochester, 59; instr microbiol, Sch Dent, Univ Pa, 59-60, assoc, 60-61, from asst prof to assoc prof microbiol, 61-67, instr biochem, Sch Med, 59-61, assoc, 61-67; assoc prof, 67-74, chmn dept, 78-87, PROF BIOL, TRINITY COL, CONN, 74- *Concurrent Pos:* Sr investr, Mt Desert Island Biol Lab, Salisbury Cove, Maine, 62-82, mem corp, 62-, asst dir, 66-82, trustee, 65-71 & 76- 80; vis scientist, Scripps Inst Oceanog, 74, Jackson Lab, 88, Univ Warwick, UK, 88. *Mem:* AAAS; Am Soc Zoologists; Sigma Xi. *Res:* Role of lipids in bioenergetics; biochemistry of fertilization and embryogenesis; xenobiotic effects on development; comparative enzymology. *Mailing Add:* Dept Biol Trinity Col Hartford CT 06106

CRAWFORD, RICHARD DWIGHT, b Kirksville, Mo, Nov 16, 47; m 66; c 1. WILDLIFE ECOLOGY. *Educ:* Northeast Mo State Univ, BS, 68, MS, 69; Iowa State Univ, PhD(wildlife biol), 75. *Prof Exp:* Sec sch teacher math-biol, Adair County RII Sch Dist, 68-69; instr wildlife biol, Iowa State Univ, 73-75; asst prof wildlife ecol, 75-80, chmn, dept biol, 81-82, ASSOC PROF WILDLIFE ECOL, 80- *Concurrent Pos:* Assoc ed, J Wildlife Mgt, 80-82. *Mem:* Sigma Xi; Wildlife Soc; Am Ornithologists Union; Wilson Ornith Soc; Cooper Ornith Soc. *Res:* Population ecology; habitat management and conservation of waterfowl; upland game and nongame birds. *Mailing Add:* PO Box 8238 Grand Forks ND 58202

CRAWFORD, RICHARD H, b St Paul, Minn, Mar 21, 23; m 44; c 3. MECHANICAL ENGINEERING, MECHANICAL DESIGN. *Educ:* Univ Minn, BSME, 45; Univ Colo, MSME, 51. *Prof Exp:* Stress analyst design, Ryan Aeronaut Co, 44-45; plastics engr design & develop, Erie Resistor, 45; staff mem, Proj Lincoln, Mass Inst Technol, 54-56; plastics researcher, Eastman Kodak Co, 56-59; chief engr res & develop, Ball Bros Res Corp, 59-62; prof 47-78, EMER PROF MECH DESIGN & ECON EVAL, UNIV COLO, 78-; ENG CONSULT & OWNER, PONDEROSA ASSOCS, LTD, 74- *Concurrent Pos:* Consult, var firms, univs, and govt bureaus. *Honors & Awards:* Merit Award, Soc Mfg Engrs. *Mem:* Am Soc Mech Engrs; Soc Plastics Engrs (vpres & pres), Nat Soc Prof Engrs; Am Soc Eng Educ; Soc Mfg Engrs (vpres & pres); Am Soc Metals. *Res:* Engineering design; failure analysis; product review and evaluation. *Mailing Add:* Ponderosa Assocs Ltd 1300 Plaza Court N Lafayette CO 80026-1473

CRAWFORD, RICHARD HAYGOOD, JR, b Troy, Ala, Aug 12, 54; m 73; c 3. COMPUTER-AIDED ENGINEERING, COMPUTER GEOMETRY. *Educ:* La State Univ, BSME, 82; Purdue Univ, MSME, 85, PhD(mech eng), 89. *Prof Exp:* ASST PROF MECH ENG, UNIV TEX, AUSTIN, 90- *Mem:* Am Soc Mech Engrs; Soc Mfg Engrs; Soc Automotive Engrs; Am Asn Artificial Intel; Asn Comput Mach. *Res:* Mechanical design; computer-aided engineering; computational geometry; applications of artificial intelligence in engineering; solid freeform fabrication and rapid prototyping. *Mailing Add:* Dept Mech Eng Univ Tex Austin TX 78712-1063

CRAWFORD, RICHARD WHITTIER, b Modesto, Calif, Aug 18, 36; m 60; c 2. MASS SPECTROSCOPY. *Educ:* San Jose State Univ, BS, 58. *Prof Exp:* CHEMIST, LAWRENCE LIVERMORE NAT LAB, 58- *Mem:* Am Soc Mass Spectros; Am Chem Soc. *Res:* Principal developer of capillary column GC/MS/FT-IR; development of instrument database systems; development of techniques for the TQMS to determine S & N species from oil shale pyrolysis; computer automation of instruments; development of analytical techniques using charge exchange and specific addition products with the TQMS; management of a certified environmental analytical laboratory. *Mailing Add:* Lawrence Livermore Nat Lab PO Box 808 L-310 Livermore CA 94551

CRAWFORD, ROBERT FIELD, b Martinez, Calif, Feb 18, 30; m 55; c 2. AGRONOMY, SOIL SCIENCE. *Educ:* Calif State Polytech Col, BS, 56; Cornell Univ, MS, 58, PhD(agron), 60. *Prof Exp:* Lab asst, Union Oil Co, 52; agr res scientist, US Borax Res Corp, 60-64, mgr agr res & develop dept, 64-67; assoc prof chem & dean col, 67-80, VPRES, BUS & FINANCIAL AFFAIRS, BIOLA COL, 80- , ASSOC PROF CHEM, VPRES PLANNING & INFO SERV, BIOLA UNIV. *Mem:* Am Soc Agron; Weed Sci Soc Am. *Res:* Crop production and physiology; plant-animal relations; chemistry. *Mailing Add:* Biola Univ La Mirada CA 90639-0001

CRAWFORD, ROBERT JAMES, b Edmonton, Alta, July 8, 29; m 56; c 4. ORGANIC CHEMISTRY. *Educ:* Univ Alta, BSc, 52, MSc, 54; Univ Ill, PhD, 56. *Prof Exp:* From asst prof to assoc prof, 56-67, chmn dept, 79-84, PROF CHEM, UNIV ALTA, 67- *Concurrent Pos:* Mem bd Govs, Univ Alta. *Mem:* Am Chem Soc; Chem Inst Can; The Chem Soc. *Res:* Reaction mechanisms; azo and diazo chemistry; racemization of cycloalkanes and epoxides. *Mailing Add:* Dept Chem Univ Alta Edmonton AB T6G 2G2 Can

CRAWFORD, ROBERT R, b Detroit, Mich, Oct 9, 44. LOGICS PROGRAMMING, PROLOG. *Educ:* Univ Chicago, BS, 65; Ind Univ, MA, 69, PhD(math), 70. *Prof Exp:* Asst prof math, Western Ky Univ, 70-76; assoc prof math & comput sci, 76-85, PROF COMPUT SCI, WESTERN KY UNIV, 85- *Concurrent Pos:* Contrib ed, TugLines J. *Mem:* Asn Comput Mach; Math Asn Am. *Mailing Add:* Dept Computer Sci Western Kentucky Univ Bowling Green KY 42101

CRAWFORD, RONALD L, b Santa Anna, Tex, Sept 28, 47; m 67; c 1. HAZARDOUS WASTE MANAGEMENT, BIOTECHNOLOGY. *Educ:* Okla City Univ, BA, 70; Univ Wis, MS, 72, PhD(bact), 73. *Prof Exp:* Postdoctoral assoc bact, Dept Biochem, Univ Minn, 73-74, from asst prof to prof microbiol, Gray Freshwater Biol Inst, 75- 87; res scientist, NY State Dept Health, Albany, 74-75; dept head, 87-90, PROF BACT, DEPT BACT & BIOCHEM, UNIV IDAHO, 87-, DIR, INST MOLECULAR & AGR GEN ENG, 88-, CO-DIR, CTR HAZARDOUS WASTE REMEDIATION RES, 90- *Concurrent Pos:* Ed, Appl & Environ Microbiol, 82-92; sci adv, Biotrol, Inc, 87-90; chair, Appl & Environ Microbiol, Gordon Res Conf, 89; affil sr sci fel, Battelle Pac Northwest Labs, 89- *Mem:* Am Soc Microbiol; AAAS. *Res:* Microbial ecology, physiology, and genetics; elucidation of mechanisms of microbial degradation of man-made chemicals; use of microorganisms to restore chemically-contaminated environments to their pristine states. *Mailing Add:* Inst Molecular & Agr Genetic Eng Univ Idaho, Food Res Ctr 202 Moscow ID 83843

CRAWFORD, RONALD LYLE, b Santa Ana, Tex, Sept 28, 47; m 67; c 1. MICROBIOLOGY, ECOLOGY. *Educ:* Okla City Univ, BA, 70; Univ Wis, MS, 72, PhD(bact), 73. *Prof Exp:* Assoc, Univ Minn, St Paul, 73-74; res scientist, Div Labs & Res, NY State Health Dept, 74-75; from asst prof to assoc prof, 75-83, PROF MICROBIOL, FRESHWATER BIOL INST, UNIV MINN, TWIN CITIES, 83- *Concurrent Pos:* Ed, Appl & Environ Microbiol, 82-86. *Mem:* Sigma Xi; Am Soc Microbiol. *Res:* Degradation of aromatic compounds in natural environments; microbial ecology; lignin biodegradation; biological oxidations; hazardous waste detoxification. *Mailing Add:* Freshwater Biol Inst PO Box 100 Navarre MN 55392

CRAWFORD, ROY DOUGLAS, b Vancouver, BC, June 6, 33. POULTRY GENETICS. *Educ:* Univ Sask, BSA, 55; Cornell Univ, MS, 57; Univ Mass, PhD(poultry genetics), 63. *Prof Exp:* Poultry geneticist, Res Br, Can Dept Agr, Ottawa, 57-58, Charlottetown, 58-63 & Kentville, 63-64; from asst prof to assoc prof, 64-74, PROF POULTRY GENETICS, UNIV SASK, 74- *Mem:* Poultry Sci Asn; Am Genetic Asn; Agr Inst Can; World Poultry Sci Asn. *Res:* Physiological and behavioral genetics of domestic fowl; effects of environment and management on domestic fowl. *Mailing Add:* Dept Animal & Poultry Sci Univ Sask Saskatoon SK S7H 0W0 Can

CRAWFORD, ROY KENT, b Wilmington, NC, Oct 21, 41. MATERIALS SCIENCE, NEUTRON SCATTERING. *Educ:* Kans State Univ, BS, 63; Princeton Univ, PhD(physics), 68. *Prof Exp:* Sloan fel, Princeton Univ, 68-69, res assoc mat sci, 69-70, mem res staff & lectr, 70-71; asst prof physics, Univ Ill, Urbana, 71-75; res assoc, 75-77, PHYSICIST, ARGONNE NAT LAB, 77- *Mem:* Am Phys Soc; AAAS. *Res:* Neutron scattering instrumentation; neutron scattering studies of phase transition and surface adsorbed molecules. *Mailing Add:* IPNS Div Argonne Nat Lab Argonne IL 60439

CRAWFORD, STANLEY EVERETT, b Dallas, Tex, Nov 9, 24; m 48; c 3. PEDIATRICS, MEDICAL ADMINISTRATION. *Educ:* Univ Tex, BA, 45, MD, 48; Am Bd Pediat, dipl, 54. *Prof Exp:* Intern, Univ Chicago Clins & Albert Merritt Billings Hosp, 48-49; resident, Univ Tex, Galveston, 50-51; resident Univ Minn Hosps, 52-53; pvt pract, Children's Clin, Jackson, Tenn, 54-61; assoc prof pediat, LeBonheur Children's Hosp, Col Med, Univ Tenn, 63-68; prof pediat & chmn dept, Med Ctr, Univ Tex, San Antonio, 68-73, dean, 73-80; dean & vpres med affairs, Univ SAla, 80-85; CONSULT, 86- *Concurrent Pos:* Off examr, Am Bd Pediat, 67-, mem residency rev comt, 70; mem exec comt, Children's Hosp Found, San Antonio, 68, bd dirs, 68-; pediatrician-in-chief, Bexar County Hosp Dist Teaching Hosps, 68-73; civilian regional consult, Wilford Hall, USAF Hosp, 69-; mem adv comt, Foster Grandparent Proj, 70. *Mem:* AMA; fel Am Acad Pediat; Asn Med Schs; Sigma Xi. *Res:* Clinical pediatrics. *Mailing Add:* Oakwell Farms 80 Granburg Circle San Antonio TX 78218

CRAWFORD, SUSAN YOUNG, b Vancouver, BC; US citizen; m 56; c 1. CO-CITATION ANALYSIS, SOCIOLOGY OF SCIENCE. *Educ:* Univ BC, BA, 48; Univ Toronto, BLS, 50; Univ Chicago, MA, 56, PhD(library-info sci), 70. *Prof Exp:* Dir, Div Library & Archival Serv, AMA, 60-81; assoc prof library sci, Columbia Univ, NY, 72-75; PROF BIOMED COMMUN & DIR, SCH MED LIBR, WASH UNIV, ST LOUIS, 81- *Concurrent Pos:* Mem, bd regents, US Nat Libr Med, 71-75; assoc ed, J Am Soc Info Sci, 79-81; ed-in-chief, Bull Med Libr Asn, 82-88; chmn, Consult Ed Panel, Med Libr Asn, 82-88; mem ed bd, J Am Soc Info Sci, 84-87, Specialty Group Award, 88 & 89; prin investr, Current Contents Online, 85; bd overseers, Tufts Univ, 88-89; chair, Med Info Syst, Am Soc Info Sci, 87-88, ed, 88-90. *Honors & Awards:* Eliot Award, Med Libr Asn, 76, Spec Award Ed, 88; J Doc hon lectr, 83. *Mem:* Sigma Xi; AAAS; Am Library Asn; Spec Libraries Asn; Med Library Asn; Soc Social Studies Sci; Am Med Informatics Asn; Acad Health Info Profs. *Res:* Communication among scientists; biomedical communication; statistical surveys of health sciences libraries in the US; technology in health sciences libraries. *Mailing Add:* Box 8132 660 S Euclid Ave St Louis MO 63110

CRAWFORD, THOMAS CHARLES, b Muskegon, Mich, July 31, 45; m 72. ORGANIC CHEMISTRY. *Educ:* Kalamazoo Col, BA, 67; Univ Calif, Los Angeles, MS, 69, PhD(chem), 74. *Prof Exp:* RESEARCHER SYNTHETIC ORG CHEM, PFIZER, INC, 74- *Mem:* Am Chem Soc; The Chem Soc; Sigma Xi. *Res:* Synthesis of biologically important molecules, development of new synthetic methods and utilization of organometallics in synthesis. *Mailing Add:* Pfizer Inc Eastern Point Rd Groton CT 06340-5196

CRAWFORD, THOMAS H, b Oct 22, 31; US citizen; m 54; c 2. INORGANIC CHEMISTRY, ENVIRONMENTAL SCIENCES. *Educ:* Univ Louisville, BS, 58, PhD(phys chem), 61. *Prof Exp:* From asst prof to assoc prof, 61-70, chmn dept, 71-75, actg dean, Col Arts & Sci, 73-74, fac assoc, Off of Pres, 75-77, asst exec vpres admin, assoc vpres acad affairs, 79-83, act univ provost, 83-84, PROF CHEM, UNIV LOUSVILLE, 70-, ASSOC UNIV PROVOST, 84- *Concurrent Pos:* Vis assoc prof, Calif Inst Technol, 68-69. *Mem:* AAAS; Am Chem Soc; Sigma Xi; AAAS. *Res:* Preparation and characterization of transition metal coordination compounds and the study of molecular complexes of antimony trichloride and organic substrates. *Mailing Add:* Off Univ Provost Univ Louisville Louisville KY 40292

CRAWFORD, THOMAS MICHAEL, b Cleveland, Ohio, Aug 13, 28; m 51; c 3. FOOD TECHNOLOGY. *Educ:* Ohio State Univ, BA, 50, MSc, 54, PhD(food tech), 57. *Prof Exp:* Technician, Cleveland Clin Found, 50; chemist, Walter Reed Army Med Serv Grad Sch, 53; asst, Ohio Agr Exp Sta, 54-57; food technologist, Res & Develop Div, Nat Dairy Prod Corp, 57-62; sect chief res & develop ctr, Pet Milk Co, Ill, 62-66, group mgr, 66-68; tech mgr, 68-71, sr scientist, Food Technol Group, Res & Develop, Pillsbury Co, Minneapolis, 71-75; mgr res & develop dept, Cent Labs, Stokely-Van Camp, Inc, 75-77, asst dir res & develop & qual control, 77-78, vpres res & qual control, 78-83; tech consult, 83-86; PRES, O A LABS & RES INC, 86- *Mem:* Inst Food Technol; Am Soc Hort Sci; Am Chem Soc; Sigma Xi. *Res:* Product development and quality control of canned, frozen, refrigerated and shelf stable fruits, vegetables, meats, cereals and fabricated single and multicomponent food products. *Mailing Add:* 538 Hunters Dr No A Carmel IN 46032

CRAWFORD, TODD V, b Los Angeles, Calif, Aug 9, 31; m 59; c 3. GENERAL ENVIRONMENTAL SCIENCES. *Educ:* Calif Polytech State Col, BS, 53; Univ Calif, Los Angeles, MA, 58, PhD(meteorol), 65. *Prof Exp:* Res asst numerical meteorol, Univ Calif, Los Angeles, 57-58, asst agr engr & lectr, Davis, 58-65, physicist, Lawrence Livermore Lab, 65-72; res mgr & sect dir environ sci, E I du Pont de Nemours & Co, 72-89; CONSULT SCIENTIST, SAVANNAH RIVER LAB, WESTINGHOUSE SAVANNAH RIVER CO, 89- *Concurrent Pos:* Dept fel, E I Du Pont de Nemours & Co, 72-89. *Mem:* Am Meteorol Soc; fel Am Nuclear Soc. *Res:* Energy balance, turbulent properties of the lower atmosphere and atmospheric diffusion from small scale to continental scale; managing a multi disciplined environmental research group. *Mailing Add:* 1040 Woodland Dr New Ellenton SC 29809

CRAWFORD, VAN HALE, b Philadelphia, Pa, Aug 10, 46; m 76; c 2. TRANSITION METAL CHEMISTRY. *Educ:* Denison Univ, BS, 68; Univ NC, PhD(inorg chem), 76. *Prof Exp:* Vis asst prof chem, State Univ NY Col, Oswego, 75-77; asst prof, 77-84, chmn dept, 85-88, ASSOC PROF CHEM, MERCER UNIV, 84- *Mem:* Am Chem Soc. *Res:* Electronic and magnetic properties of condensed transition metal complexes; electrochemistry of transition metal macrocycles. *Mailing Add:* Dept Chem Mercer Univ Macon GA 31207

CRAWFORD, VERNON, b Amherst, NS, Feb 13, 19; nat US; m 43; c 2. PHYSICS. *Educ:* Mt Allison Univ, BA, 39; Dalhousie Univ, MSc, 44; Univ Va, PhD(physics), 49. *Hon Degrees:* LLD, Mt Allison Univ, 75. *Prof Exp:* Lectr physics, Dalhousie Univ, 44-47; assoc prof, 49-56, dir physics, 64-68, vpres acad affairs, 68-80, chancellor, chief admin officer, univ system & chief exec officer bd regents, 80-85, prof physics, GA Inst Technol, 56; RETIRED. *Mem:* Am Phys Soc. *Res:* Optics; electromagnetics. *Mailing Add:* 1526 Walthall Ct NW Atlanta GA 30318

CRAWFORD, WILLIAM ARTHUR, b Norman, Okla, Mar 25, 35; m 63. GEOCHEMISTRY. *Educ:* Kans State Univ, BS, 57; Univ Kans, MS, 60; Univ Calif, Berkeley, PhD(geol, geochem), 65. *Prof Exp:* Teaching asst geol, Univ Kans, 59-60; instr geol, Univ Calif, Berkeley, 65; from asst prof to assoc prof, 65-81, PROF GEOL, BRYN MAWR COL, 81- *Mem:* Mineral Soc Am; Nat Asn Geol Teachers; fel Geol Soc Am; Am Inst Prof Geologists; Asn Geoscientists Int Develop; Asn Women Geoscientists. *Res:* Chemical analysis of rocks and minerals; studying field relations and mapping of igneous and metamorphic rock bodies. *Mailing Add:* Dept Geol Bryn Mawr Col Bryn Mawr PA 19010

CRAWFORD, WILLIAM HOWARD, JR, b Montclair, NJ, Apr 14, 37; m 58; c 2. PATHOLOGY, ORAL PATHOLOGY. *Educ:* Univ Southern Calif, BA, 58, DDS, 62, MS, 64. *Prof Exp:* Asst prof, 66-68, asst dean, 69-71, assoc dean acad affairs, 71-77, interim dean sch dent, 72-75, ASSOC PROF PATH & ORAL PATH, SCH DENT, UNIV SOUTHERN CALIF, 68-, DEAN SCH DENT, 77- *Concurrent Pos:* Nat Inst Dent Res res fel, 62-64. *Mem:* AAAS; fel Am Acad Oral Path; fel Int Col Dent; fel Am Col Dent; Am Asn Dent Schs; Am Dent Asn. *Res:* Detailed histology and pathogenesis of keratinizing cysts of the oral cavity. *Mailing Add:* Sch Dent Univ Southern Calif 1337 Via Gabriel Palos Verdes Estates CA 90274

CRAWFORD, WILLIAM STANLEY HAYES, b St John, NB, Apr 17, 18; m 43; c 2. MATHEMATICS. *Educ:* Mt Allison Univ, BA, 39; Univ Minn, MA, 42, PhD(math), 50. *Hon Degrees:* DSc, Univ Nebr, 79. *Prof Exp:* Instr math, Univ Minn, 42-43; from asst prof to prof & head dept, 43-73, dean sci, 56-62, dean fac, 62-65, vpres, 62-69, pres, 75-80, OBED EDMUND SMITH PROF MATH, MT ALLISON UNIV, 46- *Concurrent Pos:* Mem, NB Higher Educ Comn, 67-74. *Mem:* AAAS; Am Math Soc; Math Asn Am; Can Math Soc (vpres, 67-69). *Res:* Analysis; integration in function space; history of mathematics. *Mailing Add:* PO Box 1050 Sackville NB E0A 3C0 Can

CRAWFORD-BROWN, DOUGLAS JOHN, b Ridgewood, NJ, Nov 18, 53; m 75. RISK ANALYSIS, RADIATION CARCINOGENESIS. *Educ:* Ga Inst Technol, BS, 75, MS, 77 & PhD(health physics), 80. *Prof Exp:* Res scientist, radiation physics, Oak Ridge Nat Lab, 77-82; PROF RADIATION PHYSICS, UNIV NC, CHAPEL HILL, 82- *Concurrent Pos:* Comt mem, Int Comn Radiol Protection, 86- *Honors & Awards:* IBM Fac Develop award, Int Bus Mach, 83- *Mem:* Radiation Res Soc. *Res:* Theories related to the metabolism, dosimetry and radiobiology of radionuclides; logical and ethical foundations of risk analysis. *Mailing Add:* Dept Environ Sci & Eng Univ NC Chapel Hill NC 27516-7402

CRAWLEY, EDWARD FRANCIS, b Boston, Mass, Sept 7, 54; m 90. AERONAUTICAL & ASTRONAUTICAL ENGINEERING. *Educ:* Mass Inst Technol, SB, 76, SM, 78, ScD(aeronaut & astronaut), 80. *Prof Exp:* Boeing asst prof, 80-84, assoc prof, 84-90, PROF, MASS INST TECHNOL, 90-, DIR, SPACE ENG RES CTR, 88- *Concurrent Pos:* Vis lectr, Beijing Inst Aeronaut & Astronaut, 84; dir eng educ, Int Space Univ; assoc ed, Handbk Astronaut, Am Inst Aeronaut & Astronaut; mem, Space Sta Adv Comt, NASA; presidential young investr, NSF, 85-90; chmn, Struct & Dynamics Comt, Gas Turbine Inst, Am Soc Mech Engrs, 86-88; adv, Comt Space Sta, Nat Res Coun, 87; vis scholar, Dept Aeronaut & Astronaut, Stanford Univ, 88. *Mem:* Am Inst Aeronaut & Astronaut; Sigma Xi; Am Soc Mech Engrs. *Res:* Design of large spacecraft and space systems; dynamics of multi-element and jointed structures; development of intelligent structures with embedded actuators, sensors and processors; aeroelastic response of turboprops, turbofans, and active lifting surfaces; dynamics of spacecraft with on-board fluids; author of over 30 journal publications. *Mailing Add:* Mass Inst Technol Bldg 37-341 Cambridge MA 02139-8999

CRAWLEY, GERARD MARCUS, b Airdrie, Scotland, Apr 10, 38; m 61; c 4. NUCLEAR PHYSICS. *Educ:* Univ Melbourne, BSc, 59, MSc, 61; Princeton Univ, PhD(physics), 65. *Prof Exp:* Res assoc, Cyclotron Lab, Mich State Univ, 65-66; Queen Elizabeth fel, Dept Nuclear Physics, Australian Nat Univ, 66-68; from asst prof to assoc prof, 68-74, PROF PHYSICS, MICH STATE UNIV, 74- *Concurrent Pos:* Fulbright scholar & Ford Int fel, 61; vis fel, Australian Nat Univ, 74-75; prog officer nuclear physics, NSF, 75- *Mem:* Am Phys Soc. *Res:* Nuclear reactions, particularly inelastic scattering and multinuclear transfer reactions. *Mailing Add:* Cyclotron Lab Mich State Univ East Lansing MI 48824

CRAWLEY, JACQUELINE N, b Philadelphia, Pa, June 14, 50; m; c 1. NEUROCHEMISTRY, ANIMAL BEHAVIOR. *Educ:* Univ Pa, BA, 71; Univ Md, MS, 73, PhD(zool), 76. *Prof Exp:* Instr biol, Univ Md, 72-76; res fel psychopharmacol, Sch Med, Yale Univ, 76-79; sr neurobiologist, E I Du Pont Co, 81-83; staff fel, Clin Psychobiol Br, 79-81, CHIEF, UNIT BEHAV NEUROPHARMACOL, NIMH, 83- *Concurrent Pos:* Panel mem, NSF; receiving ed, J Neuroendocrinol, Pharmacol Biochem & Behavior, Topics Psychopharmacol. *Mem:* Soc Neurosci; Animal Behav Soc; Sigma Xi; NY Acad Sci; Am Col Neuropsychopharmacol. *Res:* Neurochemical substrates of behavior; clinically-oriented basic research in neuropsychopharmacology. *Mailing Add:* 6203 Stratford Rd Chevy Chase MD 20815-5420

CRAWLEY, JAMES WINSTON, JR, b Louisville, Ky, Mar 17, 47; m 67; c 2. MATHEMATICS, COMPUTER SCIENCE. *Educ:* Carson-Newman Col, BS, 68; Univ Tenn, Knoxville, PhD(math), 76. *Prof Exp:* Asst prof, 76-80, assoc prof math & comput sci, 80-83, PROF MATH COMPUT SCI, SHIPPENSBURG UNIV, 83- *Concurrent Pos:* Consult, Info Systs Design & Implementation. *Mem:* Asn Comput Mach; Math Asn Am. *Mailing Add:* Dept Math & Comput Sci Shippensburg Univ Shippensburg PA 17257

CRAWLEY, LANTZ STEPHEN, b Martinsburg, WVa, Oct 18, 44; m 73; c 2. CHEMISTRY. *Educ:* Am Univ, BS, 67; Univ Pittsburgh, PhD(chem), 70. *Prof Exp:* Fel chem, Harvard Univ, 70-71; sr res chemist & group leader med res, Am Cyanamid Co, 72-77, mgr admin consumer prod, 77-78, sect mgr, 78-79, prod mgr, 79-80, bus mgr indust prod, 80-81, dir res consumer prod, 81-83, VPRES RES, AGR RES DIV, AM CYANAMID CO, 83- *Mem:* Am Chem Soc. *Res:* Discovery and development of new products for plant and animal agriculture and biotechnology. *Mailing Add:* One Arvida Dr Pennington NJ 08534

CRAWMER, DARYL E, b Columbus, Ohio, June 1, 49; m 68; c 4. ELECTRONICS ENGINEERING. *Prof Exp:* Prin res scientist thermal spray, Battelle Mem Inst, 72-90; DIR ENG THERMAL SPRAY, MILLER THERMAL, INC, 90- *Concurrent Pos:* Safety comt chmn, Thermal Spray Div, Am Soc Metals Int, 90-, mem, Thermal Spray Coun, 86- *Honors & Awards:* IR-100, Indust Res Mag, 76; Cert of Recognition, NASA, 88. *Mem:* Am Soc Metals Int. *Res:* Thermal spray equipment; coating systems; applications; process control; methodology and hardware; coating microstructural control; bond enhancement. *Mailing Add:* W5329 County Rd O Appleton WI 54915

CRAWSHAW, LARRY INGRAM, b Los Angeles, Calif, Nov 5, 42; m 68; c 1. COMPARATIVE PHYSIOLOGY. *Educ:* Univ Calif, Los Angeles, BA, 64; Univ Calif, Santa Barbara, PhD(physiol psychol), 70. *Prof Exp:* Res asst psychol, Univ Calif, Los Angeles, 63-64; res & teaching asst, Univ Calif, Santa Barbara, 68-70, NSF res assoc, 68-70; NSF res assoc physiol, Scripps Inst Oceanog, Univ Calif, San Diego, 70-71, NIMH fel, 71-72; asst fel, John B Pierce Found, 72-76; asst prof, Sch Med, Yale Univ, 73-76; assoc prof biol, Portland State Univ, 76-78; asst prof rehab med & pharmacol, Col Physicians & Surgeons, Columbia Univ, 78-81; ASSOC PROF, DEPT BIOL, PORTLAND STATE UNIV, 81- *Mailing Add:* Portland State Univ Dept Biol PO Box 751 Portland OR 97207

CRAWSHAW, RALPH SHELTON, b New York, NY, July 3, 21; m 48; c 2. PSYCHIATRY. *Educ:* Middlebury Col, AB, 43; NY Univ Col Med, MD, 47. *Prof Exp:* Pvt pract, Washington, DC, 54; staff psychiatrist, C F Menninger Mem Hosp, Topeka, Kans, 54-57; asst chief, Vet Admin Ment Hyg Clin, Topeka, Kans, 57-60; staff psychiatrist, Community Child Guidance Clin, Portland, Ore, 60-63; clin dir, Tualatin Valley Guidance Clin, Beaverton, Ore, 61-67; PVT PRACT, PORTLAND, ORE, 60- *Concurrent Pos:* Holladay Park Hosp, 61-85; lectr, dept child psych, Univ Ore Med Sch, 61-63, clin prof, dept psych, 76-, dept pub health & prev med, 84-; lectr, Sch Social Work, Portland State Univ, 64-67; trustee, Millicent Found, 64-67; consult, Bur Hearings & Appeals, HEW, 64-, Albina Child Develop Ctr, 65-75; mem, Gov Adv Comt Mental Health, 66-72, Gov Adv Comt Med Care to the Indigent, 76-80; pres, Benjamin Rush Found, 68-89; contrib ed, Pharos, Aerospace Psychol, Prism, 72-75, Western J Med; reviewer, JAMA. *Honors & Awards:* Ralph Crawshaw Ann Lectr, Ore Found Med Excel, 87; Benjamin Rush Award, AMA, 90. *Mem:* Nat Acad Sci Inst Med; AMA; fel Am Psychiat Asn; Nat Med Asn; Am Psychol Asn; Royal Soc Med; AAAS; Am Med Writers' Asn; Soc Psychol Study Social Issues. *Res:* Author of three chapters and numerous publications. *Mailing Add:* 2525 NW Lovejoy Portland OR 97210

CRAXTON, ROBERT STEPHEN, b London, Eng, May 14, 49; m 86; c 1. LASER FUSION, HYDRODYNAMIC SIMULATIONS. *Educ:* Univ Cambridge, BA, 70; Imperial Col, London, PhD(physics), 76. *Prof Exp:* Res asst, Imperial Col, 74-76; res assoc, 76-79, scientist, 79-84, SR SCIENTIST, LAB LASER ENERGETICS, UNIV ROCHESTER, 84- *Mem:* Am Phys Soc; Inst Physics. *Res:* Plasma physics, especially two-dimensional hydrodynamic simulations of laser-produced plasmas, optical ray-tracing and thermal self-focusing; glass laser physics, especially high efficiency frequency conversion. *Mailing Add:* Lab Laser Energetics Univ Rochester 250 E River Rd Rochester NY 14623-1299

CRAYTHORNE, N W BRIAN, b Belfast, Northern Ireland, Jan 1, 31; m 57, 81; c 2. ANESTHESIOLOGY. *Educ:* Queen's Univ, Belfast, MB, BCh, 54; Am Bd Anesthesiol, dipl, 63. *Prof Exp:* Asst instr anesthesiol, Univ Pa, 57-58, instr, 58-59; clin asst, Royal Victoria Hosp, 59-60; consult anesthetist, Queen Mary Vet Hosp, 60-61; prof anesthesiol & chmn dept, WVa Univ Med Ctr, 61-70; prof anesthesia & chmn dept, Univ Cincinnati Med Ctr, 70-74; PROF ANESTHESIOL & CHMN DEPT, SCH MED, UNIV MIAMI, 75- *Concurrent Pos:* Demonstr anesthesiol, McGill Univ, 60-61; anesthetist-in-chief, Cincinnati Gen Hosp, 70-75. *Mem:* Am Soc Anesthesiologists; Int Anesthesia Res Soc; Am Med Asn; Fel Am Col Anesthesiologists; fac fel Royal Col Surg Ireland. *Mailing Add:* Dept Anesthesiol Sch Med Univ Miami PO Box 016370 1600 NW Tenth Ave Miami FL 33101

CRAYTON, PHILIP HASTINGS, b Seneca Falls, NY, Jan 22, 28; m 51; c 2. CERAMIC ENGINEERING. *Educ:* Alfred Univ, BA, 49; Univ Buffalo, MA, 51, PhD(chem), 56. *Prof Exp:* Res chemist, Metals Res Labs, Union Carbide Metals Co, 52-59, sect leader, 59-61; sr res chemist res & develop div, Carborundum Co, 61-63; from asst prof to prof chem, 63-90, head div eng & sci, 74- 77, EMER PROF CHEM, NY STATE COL CERAMICS, ALFRED UNIV, 90- *Concurrent Pos:* Consult, Carborundum Co, 63-66; res assoc, Max Planck Inst Metall Res, Stuttgart, 70 & 78, Lawrence Livermore Labs, 86. *Mem:* Am Ceramic Soc. *Res:* SHS synthesis and properties of the transition metal carbides and borides; description and mechanism of ceramic hot pressing; chemistry of process metallurgy; synthesis of oxides, carbides, nitrides and borides. *Mailing Add:* Div Eng & Sci NY State Col Ceramic Alfred Univ Alfred NY 14802

CREAGAN, ROBERT JOSEPH, b Rockford, Ill, Aug 24, 19; m 48; c 4. APPLIED PHYSICS. *Educ:* Ill Inst Technol, BS, 42; Yale Univ, MS, 43, PhD(physics), 49. *Prof Exp:* Physicist, Argonne Nat Lab, 46-47; physicist, Atomic Power Div, Westinghouse Elec Co, 49-57; dir nuclear prog, Bendix Aviation Corp, 57-60; eng mgr, Atomic Power Div, Westinghouse Elec Co, 60-69, proj mgr liquid metal fast breeder reactor, Adv Reactor Div, 69-75, dir technol assessment, 75-78, consult scientist, 78-84; RETIRED. *Mem:* Nat Acad Eng; fel Am Nuclear Soc; Am Phys Soc. *Res:* Nuclear power reactors; instrumentation; energy. *Mailing Add:* 2305 Haymaker Rd Monroeville PA 15146

CREAGER, CHARLES BICKNELL, b Bicknell, Ind, Oct 5, 24; m 51; c 3. NUCLEAR PHYSICS. *Educ:* Western Reserve Univ, BS, 51, MS, 53; Ind Univ, PhD, 59. *Prof Exp:* From asst prof to assoc prof physics, Kans Wesleyan Univ, 53-69, prof physics & fac dir res & grants, 69-71; PROF PHYSICS & CHMN DIV PHYS SCI, EMPORIA STATE UNIV, 71- *Concurrent Pos:* Res assoc, Ind Univ, 56-59. *Mem:* AAAS; Am Phys Soc; Am Asn Physics Teachers; Sigma Xi. *Res:* Beta-gamma ray spectroscopy. *Mailing Add:* 1200 Commercial St Emporia KS 66801

CREAGER, JOAN GUYNN, b Austin, Ind, Dec 8, 32; m 52; c 4. EMBRYOLOGY. *Educ:* Trinity Univ, Tex, BS, 55, MS, 58; George Washington Univ, PhD(zool), 64. *Prof Exp:* Part-time res analyst, Bionetics Res Labs, Va, 63-67; res assoc, off sci personnel, Nat Acad Sci, Washington, DC, 67-69; assoc prof biol, Northern Va Community Col, Alexandria Campus, 69-72 & 74-80; PROF SCI, MARYMOUNT UNIV, ARLINGTON, 80- *Concurrent Pos:* Staff biologist, Comn Undergrad Educ Biol Sci, 69; consult, AAAS, 70-; ed, Am Biol Teacher, 74-80. *Res:* Personnel research, especially two-year college biologists; curriculum development in undergraduate education in biology. *Mailing Add:* Marymount Univ 2807 N Glebe Rd Arlington VA 22207-4299

CREAGER, JOE SCOTT, b Vernon, Tex, Aug 30, 29; m 51, 87; c 2. GEOLOGICAL OCEANOGRAPHY. *Educ:* Colo Col, BS, 51; Agr & Mech Col Tex, MS, 53, PhD(oceanog), 58. *Prof Exp:* Asst geol & eng oceanog, Agr & Mech Col, 51-52, 55-58; phys sci asst, US Army Beach Erosion Bd, 53-55; from asst prof to assoc prof, 58-66, prof geol oceanog & assoc dean, Col Arts & Sci, 66-81, PROF GEOL SCI, UNIV WASH, 81- *Concurrent Pos:* Prog dir oceanog, NSF, 65-66; chmn, Joides Planning Comt, 70-72 & 76-78; co-ed, Quaternary Res, 70-79; vis geol scientist, Am Geol Inst, 62-65; vis scientist prog, Am Geophys Union, 65-72; US nat coord, Int Indian Ocean Exped, 65-66; assoc ed, Jour Sedimentary Petrol, 63-76; secy & treas, Marine Technol Soc, 72-78. *Mem:* Fel AAAS; fel Geol Soc Am; Soc Econ Paleont & Mineral; Am Asn Quaternary Res; Am Geophys Union; Int Asn Sedimentol; Int Asn Math Geologists. *Res:* Submarine geology relating to continental shelves, slopes and shorelines; sedimentation and micropaleontology; bottom sediment transport. *Mailing Add:* Sch Oceanog WB-10 Univ Wash Seattle WA 98195

CREAGH-DEYTER, LINDA T, b Denton, Tex, May 25, 41; m 61, 87; c 2. PHYSICAL ORGANIC CHEMISTRY. *Educ:* Univ NTex, BS, 62, MS, 63, PhD(chem), 67. *Prof Exp:* Res technician org synthesis, NTex State Univ, 58-63, teaching fel, 62-66; mem tech staff, cent res & eng, Tex Instruments Inc, 66-68; asst prof chem, Tex Woman's Univ, 68-69; mem staff res & develop, Tex Instruments Inc, 69-73; sr scientist, Xerox Inc, 73-75, mgr, 75-82; consult, Creare Innovations, 82-85, dir ink develop & chem, 85-86; DIR INK DEVELOP & CHEM, SPECTRA INC, 86- *Mem:* Am Chem Soc; Soc Photog Scientist & Engrs; Soc Info Display. *Res:* Laser development and applications; photochemistry; liquid crystals chemistry, characterization for displays; development and evaluation of materials for marking technologies. *Mailing Add:* HC 63 Box 5A Lebanon NH 03766

CREAN, GERALDINE L, NEUROSCIENCE, DEVELOPMENTAL BIOLOGY. *Educ:* Univ Conn, PhD(physiol), 83. *Prof Exp:* RES SCIENTIST, PFIZER INC, 84- *Mailing Add:* Dept Immunol & Infect Dis Pfizer Cent Res Eastern Point Rd Groton CT 06340

CREASEY, SAVILL CYRUS, b Portland, Ore, July 17, 17; m 43; c 2. GEOLOGY. *Educ:* Univ Calif, Los Angeles, AB, 39, AM, 41, PhD(geol), 49. *Prof Exp:* GEOLOGIST, US GEOL SURV, 41- *Honors & Awards:* Meritorious Serv Award, US Geol Surv, 71. *Mem:* Fel Geol Soc Am; Soc Econ Geol. *Res:* Geology of base metal deposits, especially porphyry coppers; structural geology; petrology of igneous and metamorphic rocks. *Mailing Add:* 4179 Old Adobe Rd Palo Alto CA 94306

CREASEY, WILLIAM ALFRED, b London, Eng, May 12, 33; US citizen; c 1. BIOCHEMISTRY, PHARMACOLOGY. *Educ:* Oxford Univ, BA, 55, MA & DPhil(radiobiol), 59. *Prof Exp:* Asst tutor org chem, St Catherine's Col, Oxford Univ, 58-59; from res asst to assoc prof pharmacol, Yale Univ, 68-76; prof pharmacol & ped, Sch Med, Univ Pa, 76-82; assoc clin pharmacol dir & clin support & Anal Lab dir, Squibb Inst Med Res, 82-86; DIR, BIOMED INFO SERV, INFO VENTURES, INC, PHILADELPHIA, PA, 86- *Concurrent Pos:* USPHS fel, Yale Univ, 59-61; mem cancer clin invest rev comt, Nat Cancer Inst, 72-76; mem working cadre, Nat Bladder Cancer Proj, 74-77; mem develop therapeut comt, Nat Cancer Inst, 77-80. *Mem:* Am Col Clin Pharmacol; Am Soc Clin Pharmacol & Therapeut; Am Soc Pharmacol & Exp Therapeut; Am Soc Biochem & Molecular Biol; Am Asn Cancer Res; Am Soc Clin Oncol; NY Acad Sci. *Res:* Biochemical effects of ionizing radiations; influence of dietary pyrimidines on pyrimidine and lipid metabolism; enzymatic studies with agents that influence nucleotide metabolism; metabolic studies with antineoplastic and antimitotic agents; studies with plant derivatives; clinical pharmacology. *Mailing Add:* Information Ventures Inc 1500 Locust St Suite 3213 Philadelphia PA 19102

CREASIA, DONALD ANTHONY, b Milford, Mass, Mar 28, 37; m 63; c 2. RESPIRATORY TOXICOLOGY, IMMUNOTOXICOLOGY. *Educ:* Univ Vt, BA, 61; Harvard Univ, MS, 67; Univ Tenn, PhD(immunotoxicol), 71. *Prof Exp:* Biochemist, Mason Res Inst, 61-63; chemist, Sch Med, Harvard Univ, 63-65, res assoc, 67-70; toxicologist, Biol Div, Oak Ridge Nat Lab, 70-77; prog head, in vivo carcinogenesis prog, Frederick Cancer Res Facil, Nat Cancer Inst, 77-83; RES CHEMIST DEPT DEFENSE, US ARMY RES & DEVELOP COMMAND, FT DETRICK, FREDERICK, MD, 83- *Concurrent Pos:* Prin investr, Oak Ridge Nat Lab, Nat Cancer Inst, 70-77, Frederick Cancer Res Facil, 77-, lectr, 81-; student adv, Hood Col, 79-; consult, Enviro-Control, Inc, 79. *Mem:* Sigma Xi; AAAS; Air Pollution Control Asn; Nat Asn Environ Professionals; Am Col Toxicol; NY Acad Sci; Soc Toxicol; Govt Toxologist Asn. *Res:* Respiratory toxicology of inhaled atmospheric pollutants; physiological parameters studied include: respiratory deposition and clearance of inhaled particles; respiratory immunology; respiratory cancer and metastastic foci in the lungs; pulmonary function; respiratory physiology; respiratory toxicology of inhaled natural toxins; efficacy of response to inhaled peptides and mediators of physiological activity. *Mailing Add:* 6187 Viewsite Dr Frederick MD 21701

CREASY, LEROY L, b White Plains, NY, Feb 21, 38; m 60; c 2. PLANT PHYSIOLOGY. *Educ:* Cornell Univ, BS, 60, MS, 61; Univ Calif, Davis, PhD(plant physiol), 64. *Prof Exp:* NSF fel, Low Temperature Res Sta, Cambridge Univ, Eng, 64-65; from asst prof to assoc prof, 65-78, PROF POMOL, CORNELL UNIV, 78- *Mem:* Phytochemistry Soc NAm; Am Soc Hort Sci; Am Soc Plant Physiol. *Res:* Physiology and biochemistry of secondary plant products derived from phenylalanine. *Mailing Add:* 134-A Plant Sci Bldg Pomol Cornell Univ Main Campus Ithaca NY 14853

CREASY, WILLIAM RUSSEL, b Bloomsburg, Pa, Apr 27, 58. MASS SPECTROMETRY, LASER ABLATION. *Educ:* Franklin & Marshall Col, BA, 80; Univ Rochester, MS, 82, PhD(chem), 86. *Prof Exp:* Postdoctoral assoc, Naval Res Lab, Washington, DC, 86-87; res assoc, 87-88, STAFF CHEMIST, IBM, INC, ENDICOTT, 88- *Mem:* Am Phys Soc; Am Chem Soc; Am Soc Mass Spectrometry. *Res:* Laser ionization mass spectrometry in analytical and physical chemistry is used for identification of polymers and surface analysis. *Mailing Add:* 315 Garfield Ave Endicott NY 13760

CREAVEN, PATRICK JOSEPH, b London, Eng, Jan 31, 33; m 63; c 4. CLINICAL PHARMACOLOGY. *Educ:* Univ London, MB, BS, 56, PhD(biochem), 64. *Prof Exp:* Asst lectr biochem, St Mary's Hosp Med Sch, Univ London, 63-64, lectr, 64-66; chief biochem sect, Tex Res Inst Ment Sci, 66-69; chief oncol pharmacol, Med Oncol Br, Nat Cancer Inst, Vet Admin Hosp, 69-75; assoc chief cancer res clinician, Dept Exp Therapeut & Dept Med A, 75-79, CHIEF CANCER RES CLINICIAN, DIV CLIN PHARMACOL & THERAPEUT, ROSWELL PARK MEM INST, 79- *Concurrent Pos:* Asst prof, Col Med, Baylor Univ, 66-69; asst prof, Grad Sch Biomed Sci, Univ Tex, 67-69, assoc prof, 69; assoc res prof, Grad Sch, State Univ NY Buffalo, 75- *Mem:* Fel Royal Soc Health; fel Am Col Clin Pharmacol; Am Soc Pharmacol & Exp Therapeut; NY Acad Sci; Am Asn Cancer Res; Am Soc Clin Oncol. *Res:* Clinical pharmacology, pharmacokinetics, metabolism and biochemical pharmacology of anti-cancer agents; initial clinical testing of anticancer agents; induction of drug metabolizing enzymes. *Mailing Add:* Roswell Park Cancer Inst Elm & Carlton Sts Buffalo NY 14263

CREBBIN, KENNETH CLIVE, b Nelson, BC, June 5, 24; US citizen; m 55; c 3. ACCELERATOR PHYSICS. *Educ:* Univ Calif, Berkeley, AB, 49, MA, 51. *Prof Exp:* Physicist particle accelerators, Lawrence Berkeley Lab, Univ Calif, 51-87; RETIRED. *Mem:* Am Phys Soc. *Res:* Particle dynamics, beam diagnostics and control in particle accelerators, Bevatron/Bevalac. *Mailing Add:* 3070 Sun Ct Richmond CA 94803

CRECELIUS, ERIC A, b Chardon, Ohio, Mar 31, 45; m 66; c 2. CHEMICAL SPECIATION. *Educ:* Univ Wash, PhD(oceanog), 74. *Prof Exp:* STAFF RES SCIENTIST, BATTELLE NORTHWEST, 74- *Mem:* Am Chem Soc; Am Soc Limnol & Oceanog. *Res:* Concentrations, fate and effects of metals and organic contaminants in marine and fresh water ecosystems. *Mailing Add:* 439 W Sequim Bay Rd Sequim WA 98382

CRECELIUS, ROBERT LEE, b Volin, SDak, Dec 8, 22; m 45; c 4. CHEMISTRY. *Educ:* Mont State Col, BSChem E, 47, MS, 49; Univ Wyo, PhD(chem), 54. *Prof Exp:* Chem engr, US Bur Mines, 49-52, 53-54; sr staff res chemist, Shell Develop Co, 54-84; RETIRED. *Res:* Catalytic polyforming of shale oil; para-Claisen rearrangement; high pressure hydrogenation of shale oil; oil reaction processes; hydrotreating; hydrocracking; catalytic cracking; catalysis; catalyst formulation and development. *Mailing Add:* 12527 Blackstone Ct Houston TX 77077

CRECELY, ROGER WILLIAM, b Rochester, NY, Jan 12, 42; m 67. ANALYTICAL CHEMISTRY. *Educ:* Univ Rochester, BS, 64; Emory Univ, PhD(chem), 69. *Prof Exp:* LAB DIR, BRANDYWINE RES LAB, INC, 70- *Mem:* Am Chem Soc; Coblentz Soc. *Res:* Proton and carbon nuclear magnetic resonance studies; high pressure liquid chromatography; infrared spectroscopic analysis of polymers. *Mailing Add:* 21 Cornwall Dr Newark DE 19711-7730

CREDE, ROBERT H, b Chicago, Ill, Aug 11, 15; m 47; c 3. INTERNAL MEDICINE. *Educ:* Univ Calif, Berkeley, AB, 37; Univ Calif, San Francisco, MD, 41. *Prof Exp:* Commonwealth fel med & instr med, Col Med, Univ Cincinnati, 47-49; from asst prof to prof med, Univ Calif, San Francisco, 49-86, asst dean, Sch Med, 56-60, assoc dean acad affairs, Depts Med & Family & Community Med, 80-89, EMER PROF MED, SCH MED, UNIV CALIF, SAN FRANCISCO, 86- *Concurrent Pos:* Chmn, div Ambulatory & Community med, Univ Calif, San Franciso, 67-80. *Mem:* Am Psychosom Soc; Am Fedn Clin Res; Am Geriat Soc; Asn Teachers Prev Med; Soc Gen Internal Med. *Res:* Psychosomatic medicine; delivery of health services; community medicine. *Mailing Add:* 5224 Dean's Off Sch Med Univ Calif San Francisco CA 94143

CREE, ALLAN, b Congress, Ariz, July 10, 10; m 37; c 2. PETROLEUM GEOLOGY. *Prof Exp:* Tech asst, Lowell Observ, 28-33; teacher high sch, Ariz, 35-36; supv critic phys sci & math, Ohio Univ, Athens, 36-41; asst geol, Univ Colo, 41-43; geologist, Shell Oil Co, Inc, Wyo, 43-46; asst prof geol, Univ Nev, 46-48; geologist, Cities Serv Oil Co, 48-50, dist geologist, 50-53, actg div geologist, 53-54, chief geologist, Cities Serv Petrol, Inc, 54-59, Cities Serv Co, 59-63 & Int Cities Serv Oil Co, 63-67, mgr explor, Int Div, Cities Serv Oil Co, 67-70, explor coordr, Cities Serv Int, 70-72; petrol consult, Ulster Petrols Ltd, 72-85, mem, bd dirs, 74-85; consult, 78-85; RETIRED. *Concurrent Pos:* Dir OSEC Petrol AG, Munich, 72-73; mem bd dirs, OSEC Inc, Oklahoma City, 76-78. *Mem:* Am Asn Petrol Geol; fel Explorers Club. *Res:* Subsurface geology; structural and stratigraphic geology and photogeologic mapping; petrographic and petrologic study of igneous rocks; metasomatism and replacement phenomena resulting in igneous rocks; exploration for petroleum on concessions in the Middle East, Europe, Africa, South America, Indonesia, Canada, Australia, Southeast Asia and India. *Mailing Add:* Box 3945 West Sedona AZ 86340

CREECH, HUGH JOHN, b Exeter, Ont, June 27, 10; nat US; m 37; c 2. IMMUNOCHEMISTRY, CANCER. *Educ:* Univ Western Ont, BA, 33, MA, 35; Univ Toronto, PhD(biochem), 38. *Prof Exp:* Asst chem, Harvard Univ, 38-41; from asst prof to assoc prof, Univ Md, 41-45; immunochemist, Lankenau Hosp Res Inst, 45-47; chmn admin comt, Inst Cancer Res, 47-54, head dept chemother & immunochem, 47-57, chmn div chemother, 57-70, SR MEM, INST CANCER RES, 49- *Concurrent Pos:* Lectr, Bryn Mawr Col, 45-47; mem US comt, Int Union Against Cancer, 57-60 & 80-84; archivist, Am Asn Cancer Res, 83- *Mem:* Sigma Xi; hon mem Am Asn Cancer Res (secy-treas, 52-77, vpres, 77-78, pres, 78-79). *Res:* Synthetic organic chemistry; chemotherapy and immunology of polysaccharides and protein complexes in cancer research; chemoantigens; carcinogenesis; alkylating agents; antimalarials; fluorescent antibodies; frameshift mutagens. *Mailing Add:* AACR 702 Preston Rd Erdenheim PA 19118-1327

CREECH, RICHARD HEARNE, b Boston, Mass, Apr 6, 40; m 63; c 2. CLINICAL INVESTIGATION, MEDICAL ONCOLOGY. *Educ:* Johns Hopkins Univ, AB, 61; Univ Pa, MD, 65. *Prof Exp:* Intern resident internal med, Hosp Univ Pa, 65-67; clin assoc med oncol, Nat Cancer Inst, 67-70; fel immunol & hemat, Hosp Univ Pa, 70-71; chief med oncol, Pa Gen Hosp, Univ Pa Serv, 71-72; dir,Out-Patient Dept, 75-84 & Med Admin, 80-84, ASSOC PHYSICIAN MED ONCOL, AM ONCOL HOSP, FOX CHASE CANCER CTR, 72- *Concurrent Pos:* Assoc in med, Univ Pa, 71-73, asst prof clin med, 73-79, from adj asst prof to adj asoc prof med,79-85; prin investr, Eastern Coop Oncol Group, Am Oncol Hosp, 72-84; attend physician, Jeanes Hosp, 85-, Lower Bucks Hosp, 86- & Holy Redeemer Hosp, 87- *Mem:* Am Soc Clin Oncol; Am Asn Cancer Res; AMA. *Res:* Clinical investigation of maximally effective, but minimally toxic chemotherapy regimens for the treatment of breast cancer; clarification of the roles of chemotherapy and radiotherapy in the management of small cell carcinoma of the lung. *Mailing Add:* 7500 Cent Ave Suite 203 Philadelphia PA 19111

CREECH, ROY G, b Center, Tex, Jan 24, 35; m 57; c 3. HORTICULTURE, MOLECULAR BIOLOGY. *Educ:* Stephen F Austin State Univ, BS, 56; Purdue Univ, MS, 58, PhD(genetics), 60. *Prof Exp:* Asst genetics, Purdue Univ, 56-60; from asst prof to prof plant genetics & breeding, Pa State Univ, 60-72; PROF AGRON, MISS STATE UNIV, 72- *Concurrent Pos:* Vis prof, Univ Ill, 69-70; mem Meterol Study Panel, Agr Res Inst, 74-, chmn, 76-77; mem tech adv comt, US-Israel Binat Agr Res & Develop Fund, 79-; chmn, Sect O Agr AAAS, 84-86; bd gov, Agr Res Inst, 84-86; pres, Agron Sci Found, 86-87, bd dir, 83-89. *Mem:* Genetics Soc Am; fel Crop Sci Soc Am (pres, 79); fel Am Soc Agron; Am Soc Hort Sci; Nat Sweet Corn Breeders Asn (pres, 73); fel AAAS; Coun Agr Sci & Technol; Am Genetics Asn; Soil & Water Conserv Soc Am; Am Asn Cereal Chem. *Res:* Genetic regulation of metabolic pathways in plants, especially carbohydate metabolism in maize; plant germplasm. *Mailing Add:* Dept Agron Miss State Univ PO Box 5248 Mississippi State MS 39762

CREED, DAVID, b Colchester, Eng, Sept 22, 43; m 72; c 1. PHOTOCHEMISTRY. *Educ:* Univ Manchester, BS, 65, MS, 66, PhD(chem), 68. *Prof Exp:* Res assoc biol, Southwest Ctr, Advan Studies, 68-69; res assoc biol, Univ Tex, Dallas, 69-71; SRC res fel, Royal Inst, Univ London, 71-72; temp lectr, Univ Warwick, 72-73; Robert A Welch Found fel, Univ Tex, Dallas, 73-77; from asst prof to assoc prof, 77-87, actg chmn, 85-87, PROF CHEM & CHMN DEPT, SOUTHERN MISS, 87- *Mem:* Inter-Am Photochem Soc; Soc Photobiol; Chem Soc; Am Chem Soc. *Res:* Reaction mechanisms in organic photochemistry; the molecular basis of some biological effects of ultraviolet light; photochemical aspects of solar energy conversion; polymer photochemistry. *Mailing Add:* Univ Southern Miss Box 5043 Hattiesburg MS 39406-5043

CREEK, JEFFERSON LOUIS, b Oak Ridge, Tenn, Jan 7, 45; m 68; c 2. PHYSICAL CHEMISTRY, ANALYTICAL CHEMISTRY. *Educ:* Mid Tenn State Univ, BS, 67; Southern Ill Univ, MS, 70, PhD(phys chem) 76. *Prof Exp:* Fel phys chem, Univ Calif, Los Angeles, 75-77; RES CHEM PHYS CHEM, CHEVRON OIL FIELD RES CO, 77- *Mem:* Am Chem Soc; Sigma Xi; Soc Petrol Engrs of Am Inst Mining Engrs. *Res:* Solution thermodynamics and phase behavior of reservior fluid systems. *Mailing Add:* 1406 E Alto Ln Fullerton CA 92631-2006

CREEK, KIM E, CELLULAR DIFFERENTIATION. *Educ:* Purdue Univ, PhD(biol sci), 80. *Prof Exp:* SR STAFF FEL, LAB CELLULAR CARCINOGENESIS & TUMOR PROMOTION, NAT CANCER INST, NIH, 83- *Res:* Vitamin A. *Mailing Add:* Dept Chem Univ SC Phys Sci Bldg Columbia SC 29208

CREEK, ROBERT OMER, b Harrisburg, Ill, June 30, 28; m 58; c 2. ENDOCRINOLOGY. *Educ:* Univ Ill, BS, 50; Southern Ill Univ, MS, 55; Univ Ind, PhD(zool), 60. *Prof Exp:* NIH fel, Sloan-Kettering Inst, NY, 60-61; trainee endocrinol, Univ Wis, 61-64; from asst prof to assoc prof physiol, Sch Med, Creighton Univ, 64-72, prof, 72-; AT EASTERN KY UNIV, RICHMOND. *Mem:* Endocrine Soc; Sigma Xi. *Res:* Mechanisms of hormone action; effect of thyroid stimulating hormone on the thyroid. *Mailing Add:* Moore 235 Eastern Ky Univ Richmond KY 40475

CREEL, DONNELL JOSEPH, b Kansas City, Mo, June 17, 42. ELECTROPHYSIOLOGY. *Educ:* Univ Mo, Kansas City, BA, 64, MA, 66; Univ Utah, PhD(neuropsychol), 69. *Prof Exp:* Res assoc, Vet Admin Hosp, Kansas City, 69-71; asst prof psychol, Univ Mo-Kansas City, 71; chief psychol res, Vet Admin Hosp, Phoenix, 71-76; res neuropsychologist, 76-79, RES CAREER SCIENTIST, VET ADMIN HOSP, SALT LAKE CITY, 79-; DEPT OPHTHAL, UNIV UTAH, SALT LAKE CITY, 83- *Concurrent Pos:* Adj assoc prof, Ariz State Univ, 71-76 & Univ Utah, 76- *Mem:* Asn Res Vision & Ophthal; Sigma Xi; Int Pigmentation Cell Soc. *Res:* Functional anatomy of sensory systems; inherited anomalies of sensory systems specifically those correlated with genes controlling pigmentation, albinism; the scalp-recorded evoked potential, and ERG as diagnostic tools. *Mailing Add:* Dept Ophthal Univ Utah Med Ctr Salt Lake City UT 84132

CREEL, GORDON C, b Daisetta, Tex, Oct 14, 26; m 49; c 3. GENETICS. *Educ:* Howard Payne Col, BA, 49; Univ Tex, MA, 58; Mont State Univ, PhD(genetics), 64. *Prof Exp:* Teacher high sch, 50-57; asst prof biol, Wayland Baptist Col, 58-62; NSF res grant, 62-63; prof biol & chmn div, Howard Payne Col, 64-65; PROF BIOL, ANGELO STATE UNIV, 65- *Mem:* AAAS; Am Genetic Asn. *Res:* Genetics of vertebrates. *Mailing Add:* Dept Biol Angelo State Univ 2601 Weest Ave N San Angelo TX 76901

CREELY, ROBERT SCOTT, b Kentfield, Calif, Aug 29, 26; m 52; c 4. PETROLEUM EXPLORATION. *Educ:* Univ Calif, Berkeley, BS, 50, PhD(geol), 55. *Prof Exp:* Geologist, Wm Ross Cabeen & Assoc, Colo, 55-60; from instr to assoc prof geol, Colo State Univ, 60-68, chmn dept, 65-68; chmn dept, 68-76, PROF GEOL, SAN JOSE STATE UNIV, 68- *Mem:* AAAS; Geol Soc Am; Nat Asn Geol Teachers. *Res:* Structural geology; petrology. *Mailing Add:* Dept of Geol San Jose State Univ San Jose CA 95192

CREESE, IAN N, b Bristol, Eng, Apr 4, 49; m 72. PSYCHOPHARMACOLOGY, NEUROCHEMISTRY. *Educ:* Univ Cambridge, Eng, BA, 70, MA, 72, PhD(psychol), 73. *Prof Exp:* Fel pharmacol, Sch Med, Johns Hopkins Univ, 73-76, res assoc, 76-78; from asst prof to assoc prof, Sch Med, Univ Calif, San Diego, 78-84, dir grad prog neurosci, 85-87, prof neurosci, 84-87; PROF & CO-DIR, CTR MOLECULAR & BEHAV NEUROSCI, RUTGERS STATE UNIV NJ, 87- *Concurrent Pos:* Alfred P Sloan Found fel, 78-80; NIMH study sect preclin psychopharm, 78-82; Smith-Kline vis prof, Australia, 79. *Mem:* Soc Neurosci; Am Soc Neurochem; Am Soc Pharmacol & Exp Therapeut; Am Col Neuropsychopharmacol. *Res:* Mechanisms of central nervous system; neurotransmitter and drug receptors; receptor changes in psychiatric and neurological diseases. *Mailing Add:* Ctr Molecular & Behav Neurosci Rutgers State Univ NJ 195 University Ave Newark NJ 07102

CREESE, THOMAS MORTON, b New York, NY, June 19, 34; m 65; c 2. MATHEMATICS. *Educ:* Mass Inst Technol, BS, 56; Univ Calif, Berkeley, MA, 63, PhD(math), 64. *Prof Exp:* Res mathematician, Univ Calif, Berkeley, 64; asst prof, 64-79, ASSOC PROF MATH, UNIV KANS, 79- *Concurrent Pos:* Lectr, Univ Oslo, 67-68. *Mem:* Am Math Soc; Math Asn Am. *Res:* Complex and functional analysis; several complex variables. *Mailing Add:* Dept of Math Univ of Kans Lawrence KS 66045

CREGER, CLARENCE R, b Carona, Kans, June 8, 34. BIOCHEMISTRY. *Educ:* Kans State Univ, BS, 56, MS, 57; Tex A&M Univ, PhD(biochem, nutrit), 61. *Prof Exp:* Res asst, Kans State Univ, 56-57 & Tex A&M Univ, 58-61; res assoc, Allied Mills Inc, 61-62; from asst prof to prof poultry sci, biochem & nutrit, 62-77, PROF BIOCHEM & BIOPHYS, TEX A&M UNIV, 77-, PROF POULTRY SCI. *Mem:* Am Chem Soc; Am Inst Chem; Poultry Sci Asn. *Res:* Metabolism of various radio elements; intermediary metabolism in biosynthetic mechanism. *Mailing Add:* Dept Poultry Sci Tex A&M Univ 101 Kleberg Ctr College Station TX 77843-2472

CREGER, PAUL LEROY, b Cresco, Iowa, Oct 26, 30; m 60; c 3. ORGANIC CHEMISTRY, MEDICINAL CHEMISTRY. *Educ:* Univ Ill, BS, 52; Univ Nebr, MS, 55, PhD(chem), 57; Univ Mich, MBA, 76. *Prof Exp:* From assoc res chemist to res chemist, 58-71, SR RES CHEMIST, PARKE, DAVIS & CO, 71- *Concurrent Pos:* Fel, Columbia Univ, 57-58. *Honors & Awards:* Indust Res Award, Am Chem Soc, 71. *Mem:* Am Chem Soc. *Res:* Atherosclerosis; gemfibrozil; carbonions; metalated carboxylic acids; medicinal chemistry. *Mailing Add:* Parke-Davis Res Div Warner-Lambert Co 2800 Plymouth Rd Ann Arbor MI 48103

CREIGHTON, CHARLIE SCATTERGOOD, b Orlando, Fla, Aug 23, 26; m 52; c 3. ENTOMOLOGY. *Educ:* Clemson Univ, BS, 50. *Prof Exp:* Biol aide tobacco insect invests, Entom Res Serv, 52-56, entomologist, 56-61, RES ENTOMOLOGIST, VEG INSECT INVESTS, USDA, 61- *Mem:* Entom Soc Am. *Res:* Biology and control of insects affecting tobacco; investigation of the biology, ecology and control of insects affecting vegetable crops in the South. *Mailing Add:* 1019 Fort Sumter Dr Charleston SC 29412

CREIGHTON, DONALD JOHN, b Stockton, Calif, Jan 25, 46; m 69; c 2. ENZYMOLOGY, PROTEIN CHEMISTRY. *Educ:* Calif State Univ, Fresno, BS, 68; Univ Calif, Los Angeles, PhD(biochem), 72. *Prof Exp:* Fel biochem, Inst Cancer Res, Philadelphia, 72-75; asst prof, 75-80, ASSOC PROF CHEM, UNIV MD, 80- *Concurrent Pos:* Prin investr, Am Cancer Soc & NIH, 79- *Mem:* Am Chem Soc; Sigma Xi. *Res:* Mechanism of action of sulfhydryl proteases as well as on the mechanism of action of glutathione-dependent enzymes. *Mailing Add:* Chem Dept Univ Md Baltimore County 5401 Wilkens Ave Baltimore MD 21228

CREIGHTON, DONALD L, b Hays, Kans, Jan 3, 32; m 53; c 1. MECHANICAL ENGINEERING, MATERIALS SCIENCE. *Educ:* Univ Kans, BSME, 54, MSME, 61; Univ Ariz, PhD(mech eng), 64. *Prof Exp:* Res & develop engr, Aeronaut Div, Minneapolis-Honeywell Regulator Co, 57-58; test engr, Rocketdyne Div, NAm Aviation, Inc, 58-59; instr mech eng, Univ Kans, 59-61; res assoc mat & vacuum, Univ Ariz, 63-64; asst prof mech eng, 64-68, assoc prof, 68-78, PROF MECH & AEROSPACE ENG, UNIV MO-COLUMBIA, 78- *Concurrent Pos:* Consult, Fed Trade Comn, 80-82. *Mem:* Am Soc Metals; Am Soc Testing & Mat; Am Soc Eng Educ; Am Soc Biomat; Am Welding Soc; Am Soc Mech Engrs; Sigma Xi. *Res:* Materials; cryogenics; failure analysis; fatigue of materials; heat transfer; biomaterials; fracture mechanics; design; systems design. *Mailing Add:* Dept Mech & Aerospace Eng Univ Mo 1307 Woodhill Rd Columbia MO 65203

CREIGHTON, HARRIET BALDWIN, b Delavan, Ill, June 27, 09. BOTANY. *Educ:* Wellesley Col, AB, 29; Cornell Univ, PhD(bot), 33. *Prof Exp:* Asst bot, Cornell Univ, 29-32, instr cytol & microtechnique, 32-34; asst prof bot, Conn Col, 34-40; from assoc prof to prof, 40-74, EMER PROF BOT, WELLESLEY COL, 74- *Concurrent Pos:* Fulbright lectr, Australia, 52-53 & Peru, 59-60. *Mem:* Fel AAAS (secy, 60-63, vpres, 64); Genetics Soc Am; Am Soc Naturalists; Soc Develop Biol; Bot Soc Am (secy, 50-54, vpres, 55, pres, 56); Sigma Xi. *Res:* Cell physiology; plant growth hormones; plant morphogenesis. *Mailing Add:* Dept Biol Sci Wellesley Col Wellesley MA 02181

CREIGHTON, JOHN ROGERS, b San Mateo, Calif, Feb 23, 35; c 2. COMPUTER ENGINEERING. *Educ:* Univ Calif, Berkeley, BS, 58, Univ Calif, Davis, MS, 70, PhD(eng appl sci), 75. *Prof Exp:* Jr engr, electronics, Univ Calif, Berkeley, 59-63; staff physicist, Physics Int, 63-68; PHYSICIST & CHEMIST, LAWRENCE LIVERMORE NAT LAB, UNIV CALIF, BERKELEY, 69- *Concurrent Pos:* Vis lectr, Electric & Comput Eng, Univ Calif, Davis, 82-84. *Mem:* Am Phys Soc; Inst Elec & Electronic Engrs; Asn Comput Mach; Combustion Inst. *Res:* Software engineering; databases; parallel computer architectures; computational chemical kinetics; modeling and simulation. *Mailing Add:* Lawrence Livermore Nat Lab PO Box 808 Livermore CA 94550

CREIGHTON, PHILLIP DAVID, b Irwin, Pa, June 18, 45; m; c 2. ORNITHOLOGY, BEHAVIORAL ECOLOGY. *Educ:* Tarkio Col, Mo, BA, 66; Colo State Univ, MS, 70, PhD(zool), 74. *Prof Exp:* Jr biologist, Midwest Res Inst, 66-68; from asst prof to prof biol, Towson State Univ, 73-90; DEAN, HENSON SCH SCI & TECHNOL, SALISBURY STATE UNIV, MD, 90- *Concurrent Pos:* Assoc res consult, Ecol Consult, Inc, 70-73; co-dir, Inst Animal Behav, Towson State Univ, 79-90, assoc dean, 88-90. *Honors & Awards:* Francis A Roberts Award, Cooper Ornith Soc; Josselyn Van Tyne Award, Am Ornith Union. *Mem:* Am Ornith Union; Wilson Ornith Soc; Cooper Ornith Soc; Sigma Xi; Am Soc Zoologists; Asn Field Ornith. *Res:* Community inter-relationships as demonstrated by variations of resource utilization; niche segregation and behavioral time budgets; emphasis is placed on behavioral ecology; nestling energetics and parental behavior of birds. *Mailing Add:* Henson Sch Sci Salisbury State Univ Salisbury MD 21801

CREIGHTON, STEPHEN MARK, b Sask, Aug 2, 20; m 46; c 1. ORGANIC CHEMISTRY, PHYSICAL CHEMISTRY. *Educ:* Queen's Univ, Ont, BA, 48, MA, 49; Yale Univ, PhD(org chem), 54. *Prof Exp:* Chemist, Can Packers Ltd, 49-51; chemist, Hooker Chem Corp, 53-56, res assoc polyolefins, 56-59, res supvr polymers, 59-63; HEAD PROD RES & DEVELOP, RES COUN ALTA, 63- *Concurrent Pos:* Lectr & adv, Niagara Univ, 57-58. *Mem:* Am Chem Soc; fel Chem Inst Can. *Res:* Steroids; fat and fatty acids; chlorinated hydrocarbons; pesticides; light stability and fire-retardance of plastics; polyurethane foams; polyolefins; polythers; polyesters; expoxies; sugar and sugar cane; building products. *Mailing Add:* 9140 142nd St Edmonton AB T5R 0M5 Can

CREIGHTON, THOMAS EDWIN, b St Louis, Mo, Apr 20, 40; div; c 3. MOLECULAR BIOLOGY. *Educ:* Calif Inst Technol, BS, 62; Stanford Univ, PhD(biol), 66. *Prof Exp:* Air Force Off Sci Res-Nat Acad Sci res fel, Med Res Coun Lab Molecular Biol, Cambridge Univ, 66-67; asst prof biol, Yale Univ, 67-69; sr scientist, Med Res Coun Lab Molecular Biol, Cambridge, 69-90; SR SCIENTIST, EUROP MOLECULAR BIOL LAB, HEIDELBERG, 90- *Concurrent Pos:* Vis scientist, Weizmann Inst, Israel, 76. *Res:* Protein structure, folding, and chemistry. *Mailing Add:* Europ Molecular Biol Lab Meyerhofstrasse 1 Heidelberg Germany

CREININ, HOWARD LEE, b Chicago, Ill, Sept 5, 42; m 63; c 2. FOOD SCIENCE. *Educ:* Univ Ill, Urbana, BS, 63, PhD(food sci), 71. *Prof Exp:* Prod develop chemist food sci, Res & Develop Div, Lever Brothers Co, NJ, 70-74; prod develop scientist food sci, Marschall Div, Miles Labs, 74-77; sr proj leader protein chemist, Ross Labs, Columbus, 77-82; supvr prod utilization & mgr grocery prod res & develop, Tech Ctr, Kroger Co, 82-88; dir res & develop & qual assurance, Miami Magarine Co, Ohio, 88-91; DIR FAT & OIL RES, CENT SOYA CO, 91- *Mem:* Am Oil Chemists Soc; Inst Food Technologists. *Mailing Add:* Cent Soya Co 1946 W Cook Rd Ft Wayne IN 46818

CRELIN, EDMUND SLOCUM, b Red Bank, NJ, Apr 26, 23; m 48; c 4. HUMAN ANATOMY, HUMAN DEVELOPMENT. *Educ:* Cent Col, Iowa, BA, 47; Yale Univ, PhD(anat), 51. *Hon Degrees:* DSc, Cent Col, Iowa, 69. *Prof Exp:* From instr to assoc prof human anat, 51-68, PROF ANAT, SCH MED, YALE UNIV, 68-; CHMN HUMAN GROWTH & DEVELOP STUDY UNIT, YALE-NEW HAVEN MED CTR, 72- *Concurrent Pos:* Consult, Ciba-Geigy Pharmaceut Co Inc, 61-; assoc ed, Anat Record, 68-74. *Honors & Awards:* Award, Sch Med, Yale Univ, 61; Kappa Delta Res Award, Am Acad Orthop Surgeons, 76. *Mem:* AAAS; AMA; Sigma Xi; Am Asn Anatomists. *Res:* Structure and physiology of connective tissues; developmental biology; anthropology. *Mailing Add:* Dept Surg Yale Univ Sch Med New Haven CT 06510

CRELLING, JOHN CRAWFORD, b Philadelphia, Pa, June 13, 41; m 67; c 2. COAL PETROLOGY, COAL GEOLOGY. *Educ:* Univ Del, BA, 64; Pa State Univ, MS, 67, PhD(geol), 73. *Prof Exp:* Res geologist, Bethlehem Steel Corp, 72-77; from asst prof to assoc prof, 77-87, PROF GEOL, SOUTHERN ILL UNIV, CARBONDALE, 87- *Concurrent Pos:* Sr vis res fel, Univ Newcastle-Upon-Tyne, Eng, 81; vis assoc prof fuels eng, Univ Utah, 84; sr vis res fel, Univ Newcastle-Upon-Tyne, Eng, 86, 91. *Mem:* Geol Soc Am; Iron & Steel Soc - Asn Inst Mining Engrs; AAAS; Am Chem Soc; Soc Org Petrol; Sigma Xi. *Res:* Basic and applied coal petrology including fluorescence and photoacoustic microscopy and fluorescence spectral analysis, coal maceral separation, coal maceral properties, automated petrographic analysis, weathered coal and carbonization; dispersed organic analysis. *Mailing Add:* Dept Geol Southern Ill Univ Carbondale IL 62901

CREMENS, WALTER SAMUEL, b Tampa, Fla, Aug 14, 26; m 50; c 2. PHYSICAL METALLURGY, MATERIAL SCIENCE. *Educ:* Mass Inst Technol, BS, 49, MS, 54, ScD, 57, Ga State Univ, MBA, 74. *Prof Exp:* Engr, Schnectady Works Lab & Res Lab, Gen Elec Co, 49-50, Thomson Lab, 50-53; res asst metall, Mass Inst Technol, 53-57; civilian tech liaison off, Hq, US Air Forces, Europe, 57-60; res metallurgist, 60-62, head powder metall sect, Mat & Struct Div, Lewis Res Ctr, NASA, 62-67; staff scientist & mgr graphite composites task force, Lockheed Aeronaut Systs Co, 67-70, mat scientist adv struct dept, 70-81, prin struct & mat scientist, adv res orgn, 81-82, staff scientist, adv struct dept, 81-84, sr staff engr, adv mfg progs, Ga Div, 84-89; MAT & MFG ENGR, US ARMY MATERIEL COMMAND, 89- *Honors & Awards:* Wright Bros Medal, Soc Automotive Engrs, 81. *Res:* Fiber composites; powder metallurgy; fracture mechanics; high temperature alloys; research management; composite materials; advanced manufacturing concepts and automation. *Mailing Add:* 10210 Ridgemor Dr Silver Spring MD 20901

CREMER, NATALIE E, b Minot, NDak, Sept 13, 19. IMMUNOLOGY, VIROLOGY. *Educ:* Univ Minn, BS, 44, MS, 56, PhD, 60. *Prof Exp:* Res microbiologist, State of Calif Dept Health Serv, 62-66, res specialist immunol, 66-81, res scientist, 81-84; RETIRED. *Concurrent Pos:* NIH res fel immunochem, Calif Inst Technol, 60-62; lectr, Sch Pub Health, Univ Calif, 71-84; consult, Immunologic Devices Panel, NIH, 74-78, Food & Drug

Admin, 78-85; mem, proj rev B comt, neurol disorders prog, NIH, 82-86. *Mem:* AAAS; Am Soc Microbiol; Am Asn Immunologists; NY Acad Sci. *Res:* Immunologic factors in viral infection; viral antibodies and antigens; latent viral infection in chronic degenerative diseases; host-parasite relationships. *Mailing Add:* 110 Holly Oak Lane Alameda CA 94501

CREMER, SHELDON E, b Parkersburg, WVa, Oct 24, 35; m 65; c 2. ORGANIC CHEMISTRY. *Educ:* Carnegie Inst Technol, BS, 57; Univ Rochester, PhD(org chem), 61. *Prof Exp:* Asst prof chem, Ill Inst Technol, 63-69; assoc prof, 69-74, PROF CHEM, MARQUETTE UNIV, 74- *Concurrent Pos:* Fels, Ohio State Univ, 61-62 & Univ Ill, 62-63; sr res associateship, Nat Res Coun, Wright-Patterson Air Force Base, Dayton, Ohio, 74-75; Alfrd P Sloan Fel, 71-74. *Mem:* Fel Am Inst Chem; Brit Chem Soc; Am Chem Soc; Sigma Xi. *Res:* Stereochemistry; organo-phosphorus chemistry; organosilicon, organogermanium and organotin chemistry; organic synthesis. *Mailing Add:* Dept Chem Marquette Univ 535 N 14th St Milwaukee WI 53233

CREMERS, CLIFFORD J, b Minneapolis, Minn, Mar 27, 33; m 54; c 5. HEAT TRANSFER & FLUID MECHANICS. *Educ:* Univ Minn, BS, 57, MS, 61, PhD(mech eng), 64. *Prof Exp:* Res fel mech eng, Univ Minn, 59-61, instr, 61-64; asst prof, Ga Inst Technol, 64-66; assoc prof, 66-71, chmn dept, 75-84, PROF MECH ENG, UNIV KY, 71- *Concurrent Pos:* Lectr, Gen Elec A Course, 65-66; NSF grants, 65, 66, 69, 75 & 81; NASA res grant, 68-74; consult, UNESCO, 76-; Lady Davis vis prof, Technion, Israel Inst Technol; vis scholar, Imp Col Sci & Technol, London, 73. *Mem:* AAAS; fel, Am Soc Mech Engrs; assoc fel Am Inst Aeronaut & Astronaut; Am Inst Eng Educ; Sigma Xi. *Res:* Turbulent flow in non-circular ducts; plasma heat transfer; heat transfer in frost; direct energy conversion; lunar thermophysical properties and heat transfer; heat transfer in arc welding; heat transfer in electronic cooling. *Mailing Add:* Dept Mech Eng Univ Ky Lexington KY 40506-0046

CRENSHAW, DAVID BROOKS, b Columbia, Mo, May 15, 45; m 66; c 2. ANIMAL BREEDING. *Educ:* Univ Mo-Columbia, BS, 68, MS, 69, PhD(animal husb & cytogenetics), 72. *Prof Exp:* Asst prof, 72-75, assoc prof animal sci, Tex A&I Univ, 75-, assoc prof, Col Agr, 80-; AT DEPT AGR, ETEX STATE UNIV, COMMERCE. *Mem:* Am Soc Animal Sci. *Res:* Genetics of fertility in Texas Longhorn cattle; cytogenetics of early embryonic mortality in beef cattle; hormonal manipulation of postpartum estrus in range beef cattle. *Mailing Add:* Dept Agr ETex State Univ ETex Sta Commerce TX 75428

CRENSHAW, JOHN WALDEN, JR, b Atlanta, Ga, May 17, 23; m 46; c 2. POPULATION GENETICS. *Educ:* Emory Univ, AB, 48; Univ Ga, MS, 51; Univ Fla, PhD(zool), 55. *Prof Exp:* Instr zool, Univ Mo, 55-56; asst prof biol, Antioch Col, 56-60; asst prof zool, Southern Ill Univ, 60-62; assoc prof zool, Univ Md, 62-65 & prof, 65-67; prof zool, Univ RI, 67-72; dir, 72-81, prof, sch biol, 72-85, EMER PROF & DIR, SCH BIOL, GA INST TECHNOL, 85-; ADJ PROF, SKIDWAY MARINE EXTEN, SAVANNAH, 85- *Concurrent Pos:* Am Philos Soc res grant, 57-58; NSF res grants, 58-64 & sci fac fel, Univ Calif, Berkeley, 59-60; USPHS res grants, 62-67 & res contracts, 68-70 & 75-78; consult, NSF Int Sci Activities, 63-67, Biol Sci Curriculum Study, 60-68 & US AID, Latin Am, 69. *Mem:* Fel AAAS; Am Soc Nat; Genetics Soc Am; Am Genetic Asn; Am Inst Biol Sci; Sigma Xi. *Res:* Selection for rapid growth rate in the hard clam and the bay scallop. *Mailing Add:* Rte 1 Box 367B Godsey Rd Jackson GA 30233

CRENSHAW, MILES AUBREY, b Earlysville, Va, Mar 22, 32; div; c 3. COMPARATIVE PHYSIOLOGY. *Educ:* Univ Va, BA, 59; Duke Univ, MA, 62, PhD(zool), 64. *Prof Exp:* Asst physiol, Med Sch, Univ Va, 54-59; asst zool, Duke Univ, 60, instr, 64; instr physiol, Sch Dent Med, Harvard Univ & staff assoc histol, Forsyth Dent Ctr, 64-67; asst prof dent sci, 67-71, assoc prof oral biol, Dent Res Ctr, 71-79, adj assoc prof zool, 73-79, assoc prof marine sci, 76-79, PROF PEDONTICS & MARINE SCI, SCH DENT, UNIV NC CHAPEL HILL, 79- *Mem:* AAAS; Am Dent Asn; Am Soc Zoologists; Am Soc Zoologists; Int Asn Dent Res. *Res:* Comparative biology of mineralizing tissues. *Mailing Add:* Dent Res Ctr Univ NC Chapel Hill NC 27514

CRENSHAW, PAUL L, b Nacogdoches, Tex, May 19, 33; m 54; c 4. WELL COMPLETION, REMEDIAL WORK IN WELLS. *Educ:* Tex Tech Univ, BS, 54. *Prof Exp:* Var pos, Dowell Div, Dow Chem, USA, 57-65, region eng mgr, 65-73, mgr, mech res & develop, 73-76, mgr, treatment eng & develop, 76-81; mgr technol, Western Co NAm, 81-85; OWNER & CONSULT, ENERGY EFFICIENCY ENG, INC, 85- *Concurrent Pos:* Coordr, Deep Well Team, Dowell Div, Dow Chem, USA, 71-73, consult, Safety & Loss Prev Comt, 76-81; dir, Soc Petrol Engrs, 74-76, chmn, three nat comts, 74-81. *Honors & Awards:* Serv Award, Soc Petrol Engrs, 88. *Mem:* Soc Petrol Engrs. *Res:* Author & lecturer of industry seminars on oil and gas well cementing, completion and workover fields, and acidizing; training; oilfield safety. *Mailing Add:* PO Box 331509 Ft Worth TX 79163-1509

CRENSHAW, RONNIE RAY, b Earlington, Ky, Dec 24, 36; m 59; c 2. MEDICINAL CHEMISTRY. *Educ:* Vanderbilt Univ, BA, 58, PhD(org chem), 63. *Prof Exp:* Assoc chemist, Mead Johnson Labs, Ind, 58-59; sr chemist, 63-77, asst dir med chem, 77-80, ASSOC DIR MED CHEM, BRISTOL LABS, 80-, PROJ DIR, GASTROINTESTINAL AREA, 80- *Mem:* Am Chem Soc. *Res:* Synthetic organic chemistry; organic sulfur chemistry; pharmaceuticals. *Mailing Add:* Bristol-Myers Squibb Co Five Research Pkwy Rm 550-0 Wallingford CT 06492-7660

CRENTZ, WILLIAM LUTHER, b Baltimore, Md, May 1, 10; m 57. CHEMISTRY. *Educ:* Univ Md, BS, 32, MS, 33. *Prof Exp:* Statistician, Smokeless Coal Code Authority, Washington, DC, 34-35; researcher, US Procurement Div, 35-37; econ analyst, Coal Div, US Dept Interior, 37-43; coal technologist, US Bur Mines, 43-52, chem engr, 52-60, asst to chief, Div Bituminous Coal, 60-63, from asst dir to dir coal res, 63-70, asst dir energy, 70-73; consult, Off Secy, US Dept Interior, 74-75, Energy Res Develop Admin, 75-77, Dept Energy, 75-79; RETIRED. *Honors & Awards:* Chevalier de l'Ordre de la Courounne, Belg, 59; Hon Mining Medal, USSR, 70. *Res:* Coal preparation and utilization; petroleum and natural gas research; shale oil research; production and conservation of helium. *Mailing Add:* 1120 N Shore Dr NE St Petersburg FL 33701

CREPET, WILLIAM LOUIS, b New York, NY, Aug 10, 46; m 72, 80; c 2. PALEOBOTANY, EVOLUTIONARY BIOLOGY. *Educ:* State Univ NY Binghamton, BA, 69; Yale Univ, MPh, 71, PhD(biol), 73. *Prof Exp:* Lectr bot, Ind Univ, 73-75; from asst prof to prof biol, Univ Conn, 75-89, head dept, 85-89; PROF & CHAIR, BAILEY HORTORIUM, CORNELL UNIV, 89- *Mem:* Am Inst Biol Sci; AAAS; Bot Soc Am; Int Asn Angiosperm Paleobot; Explorers Club. *Res:* Evolution of the angiosperms with emphasis on the evolution of floral structure and the evolution of pollination mechanisms and pollen; cycadophyte evolution; systematics and evolution of Fagaceae; biodiversity. *Mailing Add:* Bailey Hortorium 462B Mann Libr Cornell Univ Ithaca NY 14853

CRERAR, DAVID ALEXANDER, b Toronto, Ont, July 23, 45; m 72; c 1. GEOCHEMISTRY. *Educ:* Univ Toronto, BSc, 67, MSc, 69; Pa State Univ, PhD(geochem), 74. *Prof Exp:* From asst prof to assoc prof, 74-82, PROF GEOCHEM, PRINCETON UNIV, 82- *Concurrent Pos:* Shell Distinguished Chair Geochem, Shell Co Found, 81-85. *Honors & Awards:* Lindgren Award, Soc Econ Geologists. *Mem:* Geochem Soc; Soc Econ Geologists. *Res:* Hydrothermal geochemistry; processes of ore deposition; chemistry of natural water; environmental geochemistry; thermodynamics. *Mailing Add:* Dept Geol Princeton Univ Guyot Hall Princeton NJ 08544

CRESCITELLI, FREDERICK, b Providence, RI, June 23, 09; m 41; c 3. VISUAL PHYSIOLOGY. *Educ:* Brown Univ, PhB, 30, MS, 32, PhD(physiol), 34. *Hon Degrees:* MD, Univ Linkoping, Sweden, 75. *Prof Exp:* Keen fel, Brown Univ, 34-35; instr, Colby Jr Col, 35-36; res assoc zool, Univ Iowa, 36-40; instr, Univ Wash, 40-42; actg asst prof, Stanford Univ, 42-43; vis asst prof, Univ Southern Calif, 43-44; physiologist, Chem Warfare Serv, US Army, Edgewood Arsenal, Md, 44-46; Mellon fel, Johns Hopkins Hosp, 46; assoc prof zool, 46-51, prof, 51-76, EMER PROF CELL BIOL, UNIV CALIF, LOS ANGELES, 76- *Concurrent Pos:* Nat Res Coun fel, 35-36; vis prof, Univ Cologne, 66-67; ed, Vision Res; mem adv bd, J Comparable Physiol; mem adv bd, Nat Eye Inst, 74-75. *Mem:* Am Physiol Soc; Soc Gen Physiol; fel Optical Soc Am. *Res:* Neurophysiology; effects of drugs; visual physiology; aviation physiology; physiology and biochemistry of retina; microspectrophotometry. *Mailing Add:* Dept Biol LS(B) Univ Calif 405 Hilgard Ave Los Angeles CA 90024

CRESPI, HENRY LEWIS, b Joliet, Ill, Mar 18, 26; m 55; c 8. PHYSICAL CHEMISTRY. *Educ:* Univ Ill, BS, 52, PhD(chem), 55. *Prof Exp:* Asst chemist, Argonne Nat Lab, 55-59, assoc chemist, 59-76, asst dir chem div, 76-83, chemist, 83-87; RETIRED. *Concurrent Pos:* Vis lectr, St Procopius Col, 57-62; spec term appointee, Argonne Nat Lab, 87- *Mem:* Am Chem Soc. *Res:* Biological effects of deuterium, with special reference to algae and other microorganisms; chemistry of proteins. *Mailing Add:* 8524 Evergreen Lane Darien IL 60559

CRESPO, JORGE H, b Santiago, Chile, May 19, 44; m 71; c 2. MEDICAL EDUCATION. *Educ:* San Simon Univ, Cochabamba, Bolivia, MD, 70. *Prof Exp:* Res assoc indust toxicol, Dept Environ Med, Med Col Wis, 72-73; ASST PROF MED, DEPT LATERAL MED, WRIGHT STATE UNIV, DAYTON, OHIO, 78-; ASSOC DIR, LATERAL MED RESIDENCY PROG, KETTERING MED CTR, OHIO, 78- *Mem:* Am Col Physicians; AMA. *Res:* Rapid diagnostic methods in infectious diseases. *Mailing Add:* Kettering Med Ctr 3535 Southern Blvd Kettering OH 45429

CRESS, ANNE ELIZABETH, b Ky, Nov 30, 52; m 80; c 2. DNA DAMAGE & REPAIR, DNA REPLICATION. *Educ:* Univ Ariz, PhD(biochem), 80. *Prof Exp:* ASSOC PROF CANCER BIOL & DNA REPLICATION, DEPT RADIATION & ONCOL, UNIV ARIZ, 84- *Mem:* Am Soc Cell Biol; Radiation Res Soc. *Res:* Regulation of eukaryotic DNA replication and repair. *Mailing Add:* Health Sci Ctr Tucson AZ 85724

CRESS, CHARLES EDWIN, b Rowan Co, NC, Aug 17, 34; m 57. STATISTICS, QUANTITATIVE GENETICS. *Educ:* NC State Univ, BS, 56, MS, 61; Iowa State Univ, PhD(statist), 65. *Prof Exp:* Asst prof statist, Rutgers Univ, 65-66; from asst prof to assoc prof, 66-75, PROF CROP & SOIL SCI, MICH STATE UNIV, 75- *Mem:* Biomet Soc; Am Statist Asn; Am Soc Agron. *Res:* Methods of statistical data analysis. *Mailing Add:* Dept Crop & Soil Sci Mich State Univ East Lansing MI 48824

CRESS, DANIEL HUGG, b Canon City, Colo, Sept 13, 44; m 66; c 2. SENSOR ANALYST. *Educ:* Univ Southern Colo, BS, 67; Tex Tech Univ, MS, 71; PhD(physics), 74. *Prof Exp:* Physicist, Waterways Exp Sta, 73-79, engr, Transp Test Ctr, 79-81, PHYSICIST, WATERWAYS EXP STA, US ARMY, 81- *Concurrent Pos:* Prin investr various grants, 78-81; secy, Res Study Group II, NATO, 79-80; post doc, Colo Sch Mines, 83. *Mem:* Sigma Xi; Am Inst Physics; Opt Soc Am. *Res:* Mechanical and electromagnetic wave propagation emphasizing polarimetic-based electromagnetic imaging and seismic/acoustic coupling and propagation in outdoor environment. *Mailing Add:* US Army Waterways Exp Sta PO Box 631 Vicksburg MS 39180

CRESSEY, ROGER F, b Stoughton, Mass, June 9, 30. PARASITOLOGY. *Educ:* Boston Univ, AB, 56, AM, 58, PhD(biol), 65. *Prof Exp:* Instr biol, Boston Univ, 64-65; CUR CRUSTACEA, SMITHSONIAN INST, 65-, CHMN DEPT INVERT ZOOL, 85- *Concurrent Pos:* Res assoc, Mote Marine Lab, Fla; adj lectr biol sci, George Washington Univ; ed, Biol Soc Wash; adj prof, Dunbarton Col, 74; panel mem, Food & Agr Orgn, UN, 74- *Mem:* Am Soc Ichthyologists & Herpetologists; Am Soc Zool; Soc Syst Zool. *Res:* Taxonomy, evolution and systematics of copepods parasitic on fishes; taxonomy of synodontid fishes. *Mailing Add:* Dept Invert Zool Smithsonian Inst Washington DC 20560

CRESSIE, NOEL A C, b Fremantle, Australia, Dec 1, 50. STATISTICS FOR SPATIAL DATA, ROBUST STATISTICS. *Educ:* Univ Western Australia, BSc, 72; Princeton Univ, MA, 73, PhD(statist), 75. *Prof Exp:* Sr lectr math, Flinders Univ S Australia, 76-83; PROF STATIST, IOWA STATE UNIV, 83- *Mem:* Am Statist Asn; Bernoulli Soc; Int Statist Inst; Royal Statist Soc; Statist Soc Australia. *Res:* Theory and applications of stochastic models for the analysis of spatial data. *Mailing Add:* Dept Statist 102C Snedecor Iowa State Univ Ames IA 50011

CRESSMAN, GEORGE PARMLEY, b West Chester, Pa, Oct 7, 19; m 75; c 4. METEOROLOGY. *Educ:* Univ Chicago, PhD(meteorol), 49. *Prof Exp:* Asst meteorol, Univ Chicago, 45-49; consult, Air Weather Serv, 49-54; dir, Joint Numerical Weather Prediction Unit, 54-58, Nat Meteorol Ctr, 58-65, Nat Weather Serv, 65-79, res meteorologist, 79-85; res meteorologist, Univ Md, 85-87; RETIRED. *Honors & Awards:* Losey Award, Am Inst Astronaut & Aeronaut, 67; Appl Meteorol Award, Am Meteorol Soc, 72 & Cleveland Abbe Award, 75; Int Meteorol Orgn Prize, 78. *Mem:* Am Meteorol Soc (pres, 78). *Res:* Synoptic meteorology; atmospheric dynamics. *Mailing Add:* 11 Old Stage Ct Rockville MD 20852

CRESSMAN, WILLIAM ARTHUR, b Philadelphia, Pa, Feb 1, 41; m 63; c 2. PHARMACOLOGY. *Educ:* Philadelphia Col Pharm, BSc, 63, MSc, 65, PhD(biopharmaceut), 67. *Prof Exp:* Group leader, 67-73, sr proj coordr, 73-76, dir new prod, 76-77, exec dir new prod & int mkt develop, 77-81, vpres new prod planning, 81-85, ASST VPRES LICENSING & BUS DEVELOP, WYETH-AYERST RES, 85- *Mem:* Am Pharmaceut Asn; Acad Pharmaceut Sci; Am Chem Soc; Am Soc Pharmacol & Exp Therapeut; Am Soc Clin Pharmacol; fel NIH; fel Am Found Pharmaceut Educ. *Res:* Physical pharmacy; drug dosage form design and evaluation; drug metabolism and kinetics; bioavailability and new drug development. *Mailing Add:* Develop Plan & Coord Bristol-Myers Squibb Co PO Box 4000 Princeton NJ 08543-4000

CRESSWELL, MICHAEL WILLIAM, b Gloucester, Eng, Apr 26, 37; m 63; c 2. ELECTRONICS. *Educ:* Univ London, BSc, 58; Pa State Univ, MS, 61, PhD(physics), 65; Univ Pittsburgh, MBA, 75. *Prof Exp:* Sr engr, Power Devices Lab, Westinghouse Res & Develop Ctr, 65-80, mgr, 80-89; CONSULT, 89- *Mem:* Am Phys Soc; Inst Elec & Electronics Engrs. *Res:* Nuclear physics, including observation and analysis of reactions with photographic emulsions; semiconductors, including influence of mechanical stress on electrical properties of semiconductor materials; computer process control in power semiconductor device manufacture; effects of nuclear radiation on silicon devices. *Mailing Add:* 691 Presque Isle Dr Pittsburgh PA 15239

CRESSWELL, PETER, b Mexborough, Eng, Mar 6, 45; m 69; c 2. IMMUNOLOGY. *Educ:* Univ Newcastle Upon Tyne, BSc, 66, MSc, 67; London Univ, PhD(biochem & immunol), 71. *Prof Exp:* Res assoc biochem & molecular biol, Harvard Univ, 71-73; assoc, 73-74, from asst prof to assoc prof, 74-85, PROF IMMUNOL, MED CTR, DUKE UNIV, 85- *Mem:* Am Asn Immunol; Am Soc Home Inspectors. *Res:* Structure and function of products of the human major histocompatibility complex; lymphocyte surface markers; lymphocyte activation and function; somatic cell hybridization; genetics of human lymphocyte antigens; cell surface lectins. *Mailing Add:* Box 3010 Div Immunol Med Ctr Duke Univ Durham NC 27710

CRETIN, SHAN, b New Orleans, La, Dec, 5, 46; m 76; c 3. HEALTH OPERATIONS RESEARCH. *Educ:* Mass Inst Technol, SB, 68, PhD(opers res), 75; Yale Univ, MPH, 70. *Prof Exp:* Res assoc public health, Dept Epidemiol & Pub Health, Yale Univ, 70-71; asst prof, biostatist, Harvard Sch Pub Health, 74-76; from asst prof to assoc prof, Health Serv, Sch Pub Health, Univ Calif, Los Angeles, 76-88, prof & chair, 88-90; PRES, SHANCRETIN & ASSOCS, 90- *Concurrent Pos:* Consult, Rand Corp, 76-, Nat Ctr Health Serv Res, 80-; assoc ed, Mgt Sci & Eval Rev, 81-; vis assoc prof, Mass Inst Technol & Harvard Med Sch, 82; vis prof, Shanghai Med Univ, 87. *Mem:* Opers Res Soc Am; Am Pub Health Asn; Sigma Xi; Am Soc Qual Control. *Res:* Design and evaluation of health insurance plans in rural China; applications of total quality management to health systems; applications of operations research to the evaluation of health services. *Mailing Add:* 402 15th St Santa Monica CA 90402

CREUTZ, CARL EUGENE, b Pittsburgh, Pa, Oct 15, 47; m 76; c 2. BIOCHEMISTRY, BIOPHYSICS. *Educ:* Stanford Univ, BS, 69; Univ Wis-Madison, MS, 70; Johns Hopkins Univ, PhD(biophys), 76. *Prof Exp:* Staff fel res, Nat Inst Health, 76-81; asst prof, 81-87, ASSOC PROF PHARMACOL, UNIV VA, 87- *Mem:* Am Soc Cell Biol; Biophys Soc; Am Soc Biochem & Molecular Biol. *Res:* Basic mechanisms underlying the release of hormones and neurotransmitters by exocytosis. *Mailing Add:* Dept Pharmacol Univ Va Charlottesville VA 22908

CREUTZ, CAROL, b Washington, DC, Oct 20, 44; m 65; c 1. INORGANIC CHEMISTRY. *Educ:* Univ Calif, Los Angeles, BS, 66; Stanford Univ, PhD(chem), 70. *Prof Exp:* Asst prof chem, Georgetown Univ, 70-72; res assoc, Brookhaven Nat Lab, 72-75, assoc chemist, 75-77, chemist, 77-89, SR CHEMIST, BROOKHAVEN NAT LAB, 90- *Mem:* Am Chem Soc. *Res:* Synthesis and properties of unusual transition metal complexes; dynamics of inorganic reactions in solution. *Mailing Add:* Bldg 555 Brookhaven Nat Lab Upton NY 11973

CREUTZ, EDWARD (CHESTER), b Beaver Dam, Wis, Jan 23, 13; m 37, 74; c 3. NUCLEAR PHYSICS. *Educ:* Univ Wis, BS, 36, PhD(physics), 39. *Prof Exp:* Instr physics, Princeton Univ, 39-42; group leader, Manhattan Proj, Univ Calif, Los Alamos, 42-46; assoc prof physics, Carnegie Inst Technol, 46-48, prof physics, head dept & dir nuclear res ctr, 48-55; vpres res, Gen Atomic Div, Gen Dynamics Corp, 55-70; asst dir res, NSF, 70-78; dir Bernice P Bishop Mus, 78-85; CONSULT, 84- *Concurrent Pos:* Consult, Manhattan Proj, 46, Oak Ridge Nat Lab, 46-58, Lawrence Radiation Lab, Univ Calif, 46, 56 & NSF, 60-70; mem coun exec bd, Argonne Nat Lab, 46-51; appointments comt, Am Inst Physics, 55-58, vis scientists prog comt, 58-64, col physics comt, 59, adv comt corp assocs, 64 & dir-at-large bd gov, 65-68; scientist-at-large, Proj Sherwood Div Res, US AEC, 55-56; dir, John Jay Hopkins Lab Pure & Appl Sci, 55-67; mem adv coun & seawater conversion tech adv comt, Water Resources Ctr, Univ Calif, 58-61; adv panel gen sci, US Dept Defense, 59-63; adv comt to off sci personnel, Nat Res Coun, 60-62; res adv comt electrophys, NASA, 64-70; mem, comt sr reviewers, Energy Res & Develop Admin, 73-78, fusion power coord comt, 74-78. *Mem:* Nat Acad Sci; fel Am Phys Soc; fel Am Nuclear Soc; NY Acad Sci; Am Soc Eng Educ; fel AAAS. *Res:* Proton-proton and proton-lithium scattering; artificial radioactivity; metallurgy of uranium and beryllium; deuteron-neutron reactions; neutron absorption in uranium; synchrocyclotron design; meson reactions; nuclear reactors; thermonuclear reactions; gas flow in porous media. *Mailing Add:* PO Box 2757 Rancho Santa Fe CA 92067

CREUTZ, MICHAEL JOHN, b Los Alamos, NMex, Nov 24, 44; m 66; c 1. THEORETICAL PHYSICS. *Educ:* Calif Inst Technol, BS, 66; Stanford Univ, MS, 68, PhD(physics), 70. *Prof Exp:* Res assoc, Stanford Linear Accelerator Ctr, 70; fel physics, Univ Md, 70-72; asst physicist, 72-74, assoc physicist, 74-76, physicist, 76-85, SR PHYSICIST, BROOKHAVEN NAT LAB, 85- *Concurrent Pos:* Mem, High Energy Adv Panel, 84- 88, Prog Adv Panel, Div Adv Sci Comput, NSF, 87-90. *Mem:* Am Phys Soc. *Res:* Theoretical elementary particle physics; computer simulations of quantum field theory; computational physics. *Mailing Add:* Dept Physics Brookhaven Nat Lab Upton NY 11973

CREVASSE, GARY A, b Cedar Key, Fla, Oct 16, 34; m 59. FOOD SCIENCE. *Educ:* Univ Fla, BS, 61, MS, 63; Mich State Univ, PhD(food sci), 67. *Prof Exp:* Asst prof food sci, Univ Ariz, 67-68; DIR RES COLLAGEN SAUSAGE CASINGS, BRECHTEEN CO DIV, HYGRADE FOOD PROD CORP, 68- *Mem:* Inst Food Technol; Sigma Xi. *Res:* Meat science, specifically collagen and its application to meat industry as a packaging medium; proteolytic enzyme effects on acid soluble collagen and isolation of the resulting components. *Mailing Add:* Brechteen Co 50750 E Russell Schmidt Mt Clemens MI 48045

CREVELING, CYRUS ROBBINS, b Washington, DC, May 30, 30; m 54; c 2. BIOCHEMISTRY, PHARMACOLOGY. *Educ:* George Washington Univ, BS, 53, MS, 55, PhD(pharmacol), 62. *Prof Exp:* Chemist, Naval Ord Res Lab, Washington, DC, 52-53; asst biochem, George Washington Univ, 54-55; chemist, Hunter Mem Labs, DC, 56-58; biochemist, Nat Heart Inst, 58-62; from asst to assoc biochem, Sch Med, Harvard Univ, 62-64; PHARMACOLOGIST, NAT INST DIABETES, DIGESTIVE & KIDNEY DIS, 64- *Concurrent Pos:* Mem staff, Pharmacol Study Sect, Dis Res Grants, NIH, 66-70; assoc, Med Sch, Howard Univ, 70-; chmn, Int Conf Transmethylation, 78-; mem task force environ cancer & heart & lung dis, Environ Protection Agency, 79-83; prof, Med Col Va, 84-; mem, US Army Med Res & Develop Command Rev Bd, 86- *Mem:* Am Soc Pharmacol & Exp Therapeut; Am Chem Soc; Soc Neurosci. *Res:* Biosynthesis and metabolism of biogenic amines; drug metabolism; central nervous system pharmacology; cyclic adenosine monophosphate; mechanism of action of local anesthetics; structure and function of catechol-o-methyl transfer. *Mailing Add:* Sect Pharmacodynamics Nat Inst Diabetes, Digestive & Kidney Dis Bethesda MD 20892

CREW, GEOFFREY B, b New York, NY, June 13, 56. SPACE PHYSICS, PLASMA PHYSICS. *Educ:* Dartmouth Col, AB, 78; Mass Inst Technol, PhD(physics), 83. *Prof Exp:* SPONSOR RES TECH STAFF, MASS INST TECHNOL, 83- *Mem:* Am Phys Soc; Am Geophys Union. *Mailing Add:* Ctr Space Res Mass Inst Technol 77 Mass Ave Rm 37 275 Cambridge MA 02139

CREW, JOHN EDWIN, b Chicago, Ill, July 10, 30; m 58; c 4. ELEMENTARY PARTICLES. *Educ:* Univ Chicago, BS, 52, MS, 53; Univ Ill, PhD(physics), 57. *Prof Exp:* Physicist, X-ray Sect, Radiation Physics Lab, Nat Bur Standards, 57-59; res asst prof physics, Univ Ill, 59-61; assoc prof, Millikin Univ, 61-63; assoc prof, 63-69, PROF PHYSICS, ILL STATE UNIV, 69- *Mem:* Am Asn Physics Teachers. *Res:* High energy nuclear physics; penetration of matter by fast electrons. *Mailing Add:* Dept Physics Ill State Univ Normal IL 61761

CREW, MALCOLM CHARLES, b Columbus, Ohio, May 11, 27; m 52; c 3. DRUG METABOLISM. *Educ:* Ohio Wesleyan Univ, BA, 48; Columbia Univ, MA, 50, PhD(chem), 54. *Prof Exp:* Asst chem, Columbia Univ, 49-51; pharmaceut res chemist, Wallace & Tiernan, Inc, 52-56; sr proj leader, Fleischmann Labs, Stand Brands, Inc, Conn, 56-65; scientist, Biochem Dept, Warner-Lambert Res Inst, 65-68, sr scientist, 68-71, sr res assoc, Dept Drug Metab, 71-77; SECT HEAD, RES & DEVELOP, NORWICH-EATON PHARMACEUT, 77- *Mem:* Am Chem Soc; Am Soc Pharmacol & Exp Therapeut; Int Soc Study Xenobiotics. *Res:* Toxicology (safety assessment); drug metabolism; bioanalytical methodology; analytical instrumentation. *Mailing Add:* Norwich Eaton Pharmaceut PO Box 191 Norwich NY 13815

CREWE, ALBERT VICTOR, b Bradford, Eng, Feb 18, 27; m 49; c 4. ELECTRON MICROSCOPY. *Educ:* Univ Liverpool, BSc, 47, PhD(cosmic rays), 50. *Prof Exp:* Lectr physics, Univ Liverpool, 50-55; res assoc, Univ Chicago, 55-56, from asst prof to prof physics, 56-77, dean, Phys Sci Div, 71-81, WILLIAM E WRATHER DISTINGUISHED SERV PROF PHYSICS, UNIV CHICAGO, 77- *Concurrent Pos:* Dir, Argonne Nat Lab, 61-67, dir particle accelerator div, 58-61; tech dir cyclotron, Univ Chicago, 56-58; William E Raether distinguished professorship, 77; pres, Orchid One Corp, 87-90. *Honors & Awards:* Michelson Medal, Franklin Inst, 77, Dudell Medal, Royal Micros Soc Eng, 82. *Mem:* Nat Acad Sci; Am Phys Soc; Electron Micros Soc Am; fel Royal Micros Soc; Am Acad Arts & Sci; fel Am Phys Soc. *Res:* High energy physics; electron microscopes. *Mailing Add:* Enrico Fermi Inst 5640 S Ellis Ave Chicago IL 60637

CREWS, ANITA L, b Memphis, Tenn, Jan 9, 52. STRATIGRAPHY. *Educ:* Univ Tex, Austin, BA, 73; Univ Houston, MS, 75; Univ Calif, Los Angeles, PHD(geol), 80. *Prof Exp:* Res geologist, Union Oil Co of Calif, 75-76; ASST PROF SEDIMENTOLOGY & STRATIG, UNIV MINN, 80- *Mem:* Soc Econ Paleontologists & Mineralogists; Am Asn Petrol Geologists; Geol Soc Am; Int Asn Sedimentologists. *Mailing Add:* Dept Earth Sci Univ Minn Minneapolis MN 55455

CREWS, DAVID PAFFORD, b Jacksonville, Fla, Apr 18, 47; m; c 1. PSYCHOBIOLOGY, BEHAVIORAL ENDOCRINOLOGY. *Educ:* Univ Md, Col Park, 69; Rutgers Univ, PhD(psychobiol), 73. *Prof Exp:* Trainee psychobiol, Inst Animal Behav, Rutgers Univ, 69-73; res assoc zool, Univ Calif, Berkeley, 73-75; assoc herpet, Mus Comp Zool, 75-76, lectr biol, 76-77, from asst prof to assoc prof biol & psychol, Harvard Univ, 77-82; PROF ZOOL & PSYCHOL, UNIV TEXAS, AUSTIN, 82- *Concurrent Pos:* Res scientist develop award, 77-87; Sloan fel basic neurosci, 78-80; res scientist award, 88- *Honors & Awards:* Early Career Award, APA, 79. *Mem:* Endocrine Soc; Soc Neurosci; Soc Study Reproduction; Animal Behav Soc; Am Soc Zoologists; fel AAAS. *Res:* Effects of external stimuli, both physical and social, on hormone secretion, and the consequent effects of these endocrine secretions on the animal's reproductive behavior and physiology; sex reversal. *Mailing Add:* Dept Zool Univ Tex Austin TX 78712

CREWS, FULTON T, b Raleigh, NC, July 2, 50; m 74; c 1. NEUROPHARMACOLOGY, IMMUNOPHARMACOLOGY. *Educ:* Syracuse Univ, BS, 71; Univ Mich, PhD(pharmacol), 78. *Prof Exp:* Res technician, Endo Pharmaceut, 71-73; staff fel, Nat Inst Mental Health, 78-81; ASSOC PROF PHARMACOL, MED SCH, UNIV FLA, 81- *Mem:* Neurosci Soc; NY Acad Sci; AAAS; Int Soc Neurochem; Am Soc Pharmacol & Exp Therapeut. *Res:* The interaction of receptors and membranes; the role of lipids in receptor activation and how changes in membranes during aging and chronic alcohol consumption alter receptor-mediated signal transduction; the activation of phosphatidylinositol turnover in rat brain and rat mast cells; the stimulation of muscarinic cholinergic and alpha-adrenergic receptors as they relate to the loss of cognitive function which occurs during Alzheimer's disease and alcohol-related dementias; mast cell studies dealing with mechanisms of immune cell activation; the actions of various drugs on receptor-stimulated mast cell activation; mechanisms of receptor activation; development of better anti-inflammatory and immunostimulant compounds; receptor function impairment during aging and in alcoholics; hormone, neurotransmitter and prostaglandin receptors; receptor stimulated changes in phospholipid metabolism and membrane fluidity; antidepressants and other psychoactive drugs and their actions on central nervous system receptors and membranes. *Mailing Add:* Box J-267 Dept Pharmacol & Exp Therapeut Med Sch Univ Fla Gainsville FL 32610

CREWS, PHILLIP, b Urbana, Ill, Aug 15, 43; m 67, 82; c 2. ORGANIC CHEMISTRY, MARINE NATURAL PRODUCTS. *Educ:* Univ Calif, Los Angeles, BS, 66; Univ Calif, Santa Barbara, PhD(org chem), 69. *Prof Exp:* NSF fel, Princeton Univ, 69-70; asst prof, 70-86, PROF CHEM, UNIV CALIF, SANTA CRUZ, 86- *Concurrent Pos:* Vis prof, Univ Hawaii, 74 & Univ Ariz, 81. *Mem:* Am Chem Soc; Royal Soc Chem. *Res:* Application of nuclear magnetic resonance to problems of organic structure and stereochemistry; marine natural products chemistry. *Mailing Add:* Dept Chem Univ Calif Santa Cruz CA 95064

CREWS, ROBERT WAYNE, b Pendleton, Ore, Feb 11, 19; m 45; c 4. PHYSICS. *Educ:* Ore State Univ, BS, 47, MA, 48, PhD(physics), 52. *Prof Exp:* Sr res physicist eng, Stanford Res Inst, 52-64; mem fac, Col San Mateo, 64-65; mem staff, 65-84, EMER PROF, CHABOT COL, 84- *Res:* Charged particle interactions; electron devices. *Mailing Add:* 2221 Louis Rd Palo Alto CA 94303

CRIBBEN, LARRY DEAN, b Jackson, Ohio, July 3, 40; m 72. PLANT ECOLOGY. *Educ:* Rio Grande Col, BS, 62; Univ Okla, MNS, 68; Ohio Univ, PhD(bot), 72. *Prof Exp:* ASST PROF BIOL, MONTCLAIR STATE COL, 72- *Mem:* Sigma Xi. *Res:* The effects of acid mine drainage on river bottom plant communities. *Mailing Add:* Dept Biol Montclair State Col Upper Montclair NJ 07043

CRIBBINS, PAUL DAY, b Jacksonville, Fla, June 24, 27; m 51; c 3. CIVIL ENGINEERING. *Educ:* US Merchant Marine Acad, BS, 48; Univ Ala, BS, 52; Purdue Univ, MS, 57, PhD(civil eng), 59. *Prof Exp:* Cost engr, E I du Pont de Nemours & Co, 52; instr civil eng, Purdue Univ, 52-53 & 55-59; from asst prof to assoc prof, 59-67, PROF CIVIL ENG, NC STATE UNIV, 67- *Concurrent Pos:* Mem adv panel, Nat Coop Res Prog, US Merchant Marine, 45-49; mem hwy engr econ comn, Hwy Res Bd, Nat Acad Sci-Nat Res Coun, 61-; consult transp projs, Venezuelan Ministry Pub Works & World Bank, 62- *Mem:* Am Soc Civil Engrs; Am Soc Eng Educ; Soc Naval Archit & Marine Engrs; Inst Transp Eng; Nat Soc Prof Engrs. *Res:* Transportation research related to traffic flow, economic impact and highway design; feasibility and design of inland waterways. *Mailing Add:* Civil Engr Box 7908 NC State Univ Raleigh Main Campus Raleigh NC 27695-7908

CRICHTON, DAVID, b Central Falls, RI, Mar 7, 31. ANALYTICAL CHEMISTRY. *Educ:* Hope Col, AB, 52; Purdue Univ, MS, 55; Univ Iowa, PhD(spectrophotom), 62. *Prof Exp:* From instr to assoc prof, 55-71, PROF CHEM, CENT COL, IOWA, 71- *Mem:* Am Chem Soc; Sigma Xi. *Res:* Spectrophotometry; stability constants. *Mailing Add:* 3214 College Ave Berkeley CA 94705

CRICK, FRANCIS HARRY COMPTON, b Northampton, Eng, June 8, 16. BIOLOGY. *Educ:* Univ Col, London, BSc, 37; Cambridge Univ, PhD, 54. *Prof Exp:* Scientist, Brit Admiralty, 40-47; med res coun student, Strangeways Lab, Cambridge, 47-49; Lab Molecular Biol, Med Res Coun, 49-77; J W KIECKHEFER DISTINGUISHED RES PROF, SALK INST BIOL STUDIES, SAN DIEGO, 77- *Concurrent Pos:* Protein struct proj, Brooklyn Polytech, 53-54; vis prof, chem dept, Harvard Univ & vis lectr, Rockefeller Inst, 59; fel, Churchill Col, Cambridge, 60-61; vis prof biophysics, Harvard Univ, 62; non resident fel, Salk Inst Biol Studies, 62-73; adj prof, chem & psychol dept, Univ Calif, San Diego, 77- *Honors & Awards:* Nobel Prize in physiol, 62; Royal Medal, Royal Soc London, 72; Copley Medal, Royal Soc London, 75; Albert Medal, Royal Soc Arts, 88; Korkes Mem lectr, Duke Univ, 60; Herter lectr, Johns Hopkins Sch Med, 60; Prix Charles Leopold Mayer, Fr Acad Sci, 61; Vanuxem lectr, Princeton Univ, 64; William T Sedgwick Mem lectr, Mass Inst Technol, 65; A J Carlson Mem lectr, Univ Chicago, 65; John Danz lectr, Univ Wash, Seattle, 66; Rickman Godlee lectr, Univ Col, London, 68; Shell lectr, Stanford Univ, 69; Eighth Sir Hans Krebs lectr & Medal, Fed Europ Biochem Soc, 77; Beatty Mem lect, Mc Gill Univ, 85 & other various awards & lect from various donors & univ, 66-88. *Mem:* Foreign assoc Nat Acad Sci; fel Royal Soc; foreign hon mem AAAS; hon mem Am Soc Biol; Ger Acad Sci; foreign mem Am Philos Soc; hon fel Indian Acad Sci; fel Indian Nat Sci Acad; assoc foreign mem Fr Acad Sci; hon mem Royal Irish Acad. *Mailing Add:* Salk Inst Biol Studies 10010 N Torrey Pines Rd La Jolla CA 92037

CRICK, REX EDWARD, Cunningham, Kans, Dec 21, 43. BIOGEOCHEMISTRY. *Educ:* Univ Kans, BA, 73, MSc, 76; Univ Rochester, Phd(geol), 78. *Prof Exp:* Vis asst prof paleont, Dept Geol, Univ Mich, 78-79; ASST PROF GEOL, UNIV TEX, ARLINGTON, 79- *Mem:* Am Asn Petrol Geologists; Geol Soc Am; Int Paleont Asn; Paleont Asn London; Paleont Soc Am; Sigma Xi. *Res:* Systematics, paleobiology, and paleobiogeography of Phanerozoic ammonoid and nautiloid cephalopods. *Mailing Add:* Dept Geol UTA Box 19049 Univ Tex Arlington TX 76019

CRIDDLE, RICHARD S, b Logan, Utah, Sept 20, 36. BIOCHEMISTRY. *Educ:* Utah State Univ, BS, 58; Univ Wis, MS, 60, PhD(biochem), 62. *Prof Exp:* Asst prof biophys, 62-73, PROF BIOPHYS & BIOCHEM & BIOPHYSICIST EXP STA, UNIV CALIF, DAVIS, 73- *Res:* Protein structure and protein-protein interactions in relation to enzyme activity; biosynthesis of organelles. *Mailing Add:* Dept Biochem Univ Calif Davis CA 95616

CRIDER, FRETWELL GOER, b Centerville, Ala, June 8, 23; m 47; c 4. PHYSICAL CHEMISTRY. *Educ:* Univ NC, BS, 45, PhD(phys chem), 53. *Prof Exp:* Instr chem, Armstrong Col, Ga, 47-48; asst, Univ NC, 48-52 & Off Naval Res, 52-53; sr res technologist, Field Res Labs, Socony Mobil Oil Co, Inc, 53-64; chmn dept chem & physics, Armstrong State Col, 64-72; dean, Gordon Jr Col, 72-73; DEAN ADMIN, MID GA COL, 73-, PROF CHEM, 77- *Concurrent Pos:* Exec dir, Armstrong Res Inst, 66-72. *Mem:* Am Chem Soc; Combustion Inst; fel Am Inst Chemists. *Res:* Kinetics of photochemical decomposition; combustion kinetics; oxidation of carbon and hydrocarbons; diffusion flames; mechanics and chemistry of geological formations; corrosion; pollution chemistry; water analysis. *Mailing Add:* Dean Admin Mid Ga Col Cochran GA 31014

CRIGLER, JOHN F, JR, b Charlotte, NC, Sept 11, 19; m 44; c 4. PEDIATRICS. *Educ:* Duke Univ, AB, 39; Johns Hopkins Univ, MD, 43. *Prof Exp:* Intern med, Mass Mem Hosp, 43-44; intern , Johns Hopkins Hosp, 46-47, asst resident, 47-48, physician-in-chg outpatient dept, 48-49, resident, 49-50; instr pediat, 55-56, assoc, 56-62, asst prof, 62-68, ASSOC PROF PEDIAT, HARVARD MED SCH, CHILDREN'S HOSP, BOSTON, 68- *Concurrent Pos:* Fel pediat endocrinol, Johns Hopkins Hosp, 50-51; Nat Found fel biol, Mass Inst Technol, 51-55; instr, Sch Med, Johns Hopkins Univ, 48-50; assoc physician, Children's Hosp Med Ctr, 56-61, physician, 61-62, sr assoc med, 62-, dir gen clin res ctr for children, 64-76, chief endocrine div, Dept Med, 66-88, emer chief, 89. *Mem:* Am Acad Pediat; Endocrine Soc; Soc Pediat Res; Am Pediat Soc; Am Fedn Clin Res; Sigma Xi. *Res:* Pediatric endocrinology and metabolism; hormonal aspects of physical growth and development; steroid hormone metabolism. *Mailing Add:* 15 Tappan Rd Wellesley MA 02181

CRILEY, BRUCE, b Chicago, Ill, Apr 13, 39; m 67; c 2. EXPERIMENTAL EMBRYOLOGY. *Educ:* Univ Ill, BS, 60, MS, 62, PhD(zool), 67. *Prof Exp:* Instr embryol, Univ Ill, 65-66; from asst prof to assoc prof embryol & chmn organismic biol div, Univ Colo, 66-71; prof, 73-80, GEORGE C & ELLA BEACH LEWIS PROF BIOL, ILL WESLEYAN UNIV, 80-, CHMN DEPT, 71- *Mem:* AAAS; Am Zool Soc. *Res:* Neuroembryology; vertebrate morphogenesis. *Mailing Add:* Dept Biol Ill Wesleyan Univ Bloomington IL 61702

CRILL, WAYNE ELMO, MECHANISMS OF EPILEPSY, CELLULAR BIOPHYSICS. *Educ:* Univ Wash, MD, 62. *Prof Exp:* PROF NEUROPHYSIOL, PHYSIOL & BIOPHYS, UNIV WASH, 77-, CHMN DEPT PHYSIOL & BIOPHYS, 83- *Mailing Add:* Dept Physiol & Biophys SJ40 Univ Wash Seattle WA 98195

CRIM, F(ORREST) FLEMING, b Waco, Tex, May 30, 47; m 69; c 1. MOLECULAR DYNAMICS. *Educ:* Southwestern Univ, BS, 69; Cornell Univ, PhD(chem), 74. *Prof Exp:* Mem res staff, Eng Res Ctr, Western Elec Co, 74-76; staff mem, Los Alamos Nat Lab, 76-77; from asst prof to assoc prof, 77-84, PROF CHEM, UNIV WIS, 84-, HELFAER PROF, DEPT CHEM, 85- *Concurrent Pos:* Res fel, Alfred P Sloan Found, 81; teacher-scholar, Camille & Henry Dreyfus Found, 82; Romnes Fac Fel, Univ Wis, 83. *Honors & Awards:* Alexander von Humbolt Sr US Scientist Award, 86. *Mem:* Am Chem Soc; fel Am Phys Soc; AAAS. *Res:* Chemical physics; molecular dynamics; state resolved unimolecular reaction studies; spectroscopy and dynamics of highly vibrationally excited molecules; molecular energy transfer. *Mailing Add:* 1101 University Ave Madison WI 53706

CRIM, GARY ALLEN, b Louisville, Ky, July 13, 49; m 84. BIOMATERIALS, OPERATIVE DENTISTRY. *Educ:* Univ Ky, DMD, 74; Ind Univ, MSD, 81. *Prof Exp:* ASSOC PROF RESTORATIVE DENT, SCH DENT, UNIV LOUISVILLE, 77- *Concurrent Pos:* Consult, Nat Dent Bd Exam, 85- *Mem:* Int Asn Dent Res; Acad Dent Mat; Am Asn Dent Schs; Int Asn Dent Studies; Acad Oper Dent. *Res:* Microleakage of composite restorative resins and dental adhesive systems in conjunction with cavity modifications. *Mailing Add:* Sch Dent Health Sci Ctr Univ Louisville Louisville KY 40292

CRIM, JOE WILLIAM, b Wichita Falls, Tex, Nov 19, 45; div; c 1. ENDOCRINOLOGY, COMPARATIVE ENDOCRINOLOGY. *Educ:* Univ Calif, Berkeley, AB, 68, MA, 72, PhD(zool), 74. *Prof Exp:* Teaching asst dept zool, Univ Calif, Berkeley, 70-71; res sci cancer res, Am Med Ctr Denver, Spivak, Colo, 74-75; res assoc zool, Univ Wash, 75-78; from asst prof to assoc prof, 78-90, PROF DEPT ZOOL, UNIV GA, 90- *Concurrent Pos:* Res fel Pub Health Serv, Nat Inst Arthritis, Metab & Digestive Dis, Univ Wash, 75-77. *Mem:* Am Soc Zool; Sigma Xi. *Res:* Comparative endocrinology of neuropeptides; receptor biology; endocrinology of insect gut; biology of gut peptides. *Mailing Add:* Dept Zool Univ Ga Athens GA 30602

CRIM, STERLING CROMWELL, b Corsicana, Tex, Jan 5, 27; m 54; c 2. MATHEMATICS. *Educ:* Baylor Univ, BS, 50; NTex Univ, MEd, 53; George Peabody Col, MA, 58; Univ Tex, Austin, PhD(math), 68. *Prof Exp:* Teacher high sch, Tex, 51-53 & jr high sch, 53-57; teacher math, WGa Col, 58-59; spec instr, Univ Tex, Austin, 59-64; PROF MATH, LAMAR UNIV, 64- *Mem:* Math Asn Am. *Res:* Integral transforms. *Mailing Add:* Dept Math Lamar Univ Lamar Univ Sta Beaumont TX 77710

CRIMINALE, WILLIAM OLIVER, JR, b Mobile, Ala, Nov 29, 33; m 62; c 2. FLUID MECHANICS, APPLIED MATHEMATICS. *Educ:* Univ Ala, BS, 55; Johns Hopkins Univ, PhD(aeronaut), 60. *Prof Exp:* Asst fluid mech & appl math, Johns Hopkins Univ, 56-60; asst prof, dept aerospace & mech sci, Princeton Univ, 62-68; assoc prof oceanog & geophys, 69-73, chmn dept appl math, 76-84, PROF OCEANOG & GEOPHYS, UNIV WASH, 73-, PROF APPL MATH, 76- *Concurrent Pos:* NATO fel, Inst Appl Math, Ger, 60-61; sr vis, Cambridge Univ, 61-; consult, Aerospace Corp, Calif, 63, 65; guest prof, Can Armament Res & Develop Estab, Que, 65, Inst Mech Statist Turbulence, France, 67-68, Royal Inst Technol, Stockholm & Inst Oceanog, Gothenburg, 73-74 & Univ St Andrews, Scotland, 85; consult adv group aeronaut res & develop, NATO, 67-68; Nat Acad Sci exchange scientist, USSR, 69; consult, Boeing Sci Res Labs, 69-70; sr res award, Alexander von Humboldt Found, Ger,73-74; guest prof, Stanford Univ, 91; Royal Soc fel, 91-92 (England, Scotland); consult, NASA, 91- *Mem:* AAAS; Am Geophys Union; Am Phys Soc; Soc Indust & Appl Math; Fed Am Scientists. *Res:* Non-linear mechanics; geophysical fluid dynamics, especially stability and turbulence. *Mailing Add:* Dept Appl Math FS-20 Univ Wash Seattle WA 98195

CRIMMINS, MICHAEL THOMAS, b E St Louis, Ill, Jan 3, 54; m 87; c 1. SYNTHETIC ORGANIC & NATURAL PRODUCTS CHEMISTRY. *Educ:* Hendrix Col, BA, 76; Duke Univ, PhD(chem), 80. *Prof Exp:* Nat Cancer Inst-NIH fel, Calif Inst Technol, 80-81; asst prof, 81-86, ASSOC PROF CHEM, UNIV NC, CHAPEL HILL, 86- *Mem:* Am Chem Soc. *Res:* Development of synthetic methods and applications to the total synthesis of biologically and topologically interesting natural products; intramolecular photocycloadditions as applied to synthesis. *Mailing Add:* Dept Chem Univ NC Chapel Hill NC 27599-3290

CRIMMINS, TIMOTHY FRANCIS, b Hempstead, NY, Oct 14, 39; m 66. ORGANIC CHEMISTRY. *Educ:* St Johns Univ, NY, BS, 61; Purdue Univ, MS, 63, PhD(org chem), 65. *Prof Exp:* Res assoc of Dr C R Hauser & dir res, Duke Univ, 65-66; assoc prof, 66-76, PROF CHEM, UNIV WIS, OSHKOSH, 76- *Mem:* Am Chem Soc. *Res:* Organosodium metalation reactions employing tetramethylenediamine activated pentasodium. *Mailing Add:* Dept Chem Univ Wis Oshkosh WI 54901-8645

CRINE, JEAN-PIERRE C, b St Angeau, France, Dec 13, 44; Can citizen; m 70; c 1. PROPERTIES OF MATERIALS, AGING OF SOLID & LIQUID DIELECTRICS. *Educ:* Univ Quebec, Montreal, BSc, 72, MSc, 74, Polytech Inst Montreal, PhD(eng phys), 78. *Prof Exp:* Asst prof nat sci, Univ Quebec, Montreal, 76-79; res scientist, 78-84, SR RES SCIENTIST, HYDRO-QUEBEC RES INST, 84- *Concurrent Pos:* Can mem, Tech Comt 10 C, Int Electrotechnical Comn, 86-; lectr, Eng Phys Dept, Polytech Inst, 88. *Mem:* Sigma Xi; Inst Elec & Electronics Engrs; Mat Res Soc. *Res:* Measurement & interpretation of properties of solid & liquid dielectrics; analysis of the influence of morphology, environment & aging on dielectric materials used in high power equipment. *Mailing Add:* CP 1000 IREQ Varennes PQ J3X 1S1 Can

CRIPPEN, GORDON MARVIN, b Cheyenne, Wyo, Apr 2, 45; m 70; c 1. BIOPHYSICAL CHEMISTRY, THEORETICAL CHEMISTRY. *Educ:* Univ Wash, BS, 67; Cornell Univ, PhD(biophys chem), 72. *Prof Exp:* Res chemist phys chem, Cardiovasc Res Inst, Univ Calif, San Francisco, 72-73; instr chem & physics, Bur Schs, Hamburg, Ger, 73-75; asst prof phys chem, Sch Pharm, Univ Calif, San Francisco, 75-80; asst prof phys chem, dept chem, Tex A&M Univ, College Station, 80-82; PROF PHARMACEUT CHEM, COL PHARM, UNIV MICH, ANN ARBOR, 82- *Mem:* Am Chem Soc. *Res:* Theoretical studies on the conformation of proteins, calculation of binding of ligands to proteins and related subjects. *Mailing Add:* 1921 Norway Ann Arbor MI 48104

CRIPPEN, RAYMOND CHARLES, b Brooklyn, NY, Mar 1, 17; m 41; c 2. ANALYTICAL CHEMISTRY, BIOCHEMISTRY. *Educ:* Iowa State Univ, BS, 39; Johns Hopkins Univ, MS, 48; St Thomas Inst Advan Studies, PhD(anal chem), 70. *Prof Exp:* Group leader chem coatings, E I du Pont de Nemours & Co Ltd, Ft Madison, Iowa, 40-45; res & develop mgr indust chem, Penniman & Browne, Baltimore, Md, 48-49; dir indust consult, Crippen Labs, Div Foster D Snell Inc, Baltimore, Md, 49-61; group leader chromatog, Res Ctr, Atlas Chem Indust, Wilmington, 61-66; sect head anal res & develop, Stauffer Chem Co, Adrian, Mich, 66-68; group leader anal methods develop, Richardson-Merrell Corp, Cincinnati, Ohio, 68-70; instr chem & geol, Northern Ky Univ, Highland Heights, 70-75; dir, pres & consult chem, Crippen Labs, Inc, 75-87, consult & tech dir, 87-89; RETIRED. *Concurrent Pos:* Lectr, forensic anal-testing, cert prof chem; mem cert comt, Am Inst Chemists, 86- *Mem:* Am Chem Soc; Soc Appl Spectros; Am Soc Testing & Mat; fel Am Inst Chemists. *Res:* Development of analytical methods for styrene in rubber polymers; isolation and identification of physiological active peptides in yeast extract; development of agent for reduction of chronic diseases in plants, animals and humans; reduction of EMR (ELF) frequencies. *Mailing Add:* Coffee Run Condo D-2G Hockessin DE 19707-9594

CRIPPS, DEREK J, b London, Eng, Sept 17, 28; m 63; c 4. DERMATOLOGY, PHOTOBIOLOGY. *Educ:* Univ London, MB & BS, 53, MD, 65; Univ Mich, MS, 61; Am Bd Dermat, dipl, 69. *Prof Exp:* Intern med, London Hosps, 53-54; sr registr, Inst Dermat, Eng, 62-65; asst prof med, 65-68, assoc prof dermat, 68-72, PROF DERMAT & CHMN DEPT, MED CTR, UNIV WIS-MADISON, 72- *Concurrent Pos:* NIH grant, 66-, EPA grant, 76-; mem, Study Comt, NIH, 69- *Mem:* Brit Dermat Asn; Am Acad Dermat; Am Fedn Clin Res; Soc Invest Dermat; fel Am Col Physicians. *Res:* Investigation of persons sensitive to sunlight and diseases of prophyrin metabolism. *Mailing Add:* Dept Med Univ Wis Madison WI 53792

CRIPPS, HARRY NORMAN, b Webster, NY, May 14, 25; m 48; c 4. ORGANIC CHEMISTRY. *Educ:* Ga Inst Technol, BS, 46; Univ Rochester, BS, 48; Univ Ill, PhD(chem), 51. *Prof Exp:* Chemist, Eastman Kodak Co, 48; CHEMIST, CENT RES & DEVELOP DEPT, E I DU PONT DE NEMOURS & CO, 51- *Mem:* Sigma Xi. *Res:* Organic synthesis; polymers. *Mailing Add:* 416 Snuff Mill Lane Hockessin DE 19707-9643

CRISCUOLO, DOMINIC, b New Haven, Conn, June 14, 08; m 46; c 2. BIOCHEMISTRY. *Educ:* Univ Pittsburgh, BS, 32; Trinity Univ, MS, 53. *Prof Exp:* Biochemist, Food Res Labs, 39-41, Sch Aviation Med, Randolph Field, 46-58 & Surg Res Lab, Lackland AFB Hosp, 58-68; chemist, Regional Environ Lab, Kelly AFB, 68-73; ADMINR, ST ANTHONY'S SPEC EDUC SCH, 73- *Mem:* Fel AAAS; Am Chem Soc; fel Am Inst Chemists; NY Acad Sci; Sigma Xi. *Res:* Enzymes; hematology; blood volumes; acclimatization to adverse environments; anesthesiology. *Mailing Add:* 321 Concord Ave San Antonio TX 78201

CRISLER, JOSEPH PRESLEY, b Hedley, Tex, Sept 12, 22; m 60; c 4. MICROCHEMISTRY. *Educ:* WTex State Univ, BS, 42, MA, 47; Univ Colo, PhD(chem micros), 62; Drexel Inst Technol, dipl elec micros, 63. *Prof Exp:* Head dept chem, Buena Vista Col, 47-54; instr inorg chem, Univ Colo, 54-55; asst prof anal chem, Tex Col Arts & Industs, 55-56; instr, Colo Sch Mines, 56-59; chemist micros, Univ Colo, 59-62; res chemist, US Naval Ord Sta, 62-69; CHIEF CHEM BR, DEPT HUMAN RESOURCES, GOVT DC, 69- *Mem:* Am Chem Soc; fel Royal Micros Soc. *Res:* Micro methods for the determination of trace elements, erythrocyte protoporphyrin and drugs of abuse in body fluids; chemical microscopy. *Mailing Add:* PO Box 251 Clinton MD 20735-0251

CRISLEY, FRANCIS DANIEL, b Braddock, Pa, Aug 19, 26; m 60; c 3. MICROBIOLOGY, BACTERIOLOGY. *Educ:* Univ Pittsburgh, BS, 50, MS, 52, PhD(microbiol), 59. *Prof Exp:* Asst bact, Univ Pittsburgh, 50-52; instr, Miami Univ, 52-54 & Univ Pittsburgh, 54-58, res assoc microbiol, Dept Biol Sci, 59-61; microbiologist, Robert A Taft Sanit Eng Ctr, Ohio, 61-67; chmn dept biol, 67-75, PROF BIOL, NORTHEASTERN UNIV, 67- *Concurrent Pos:* Vis scientist, Natick Army Res & Develop Command, Natick, Mass, 78. *Mem:* Am Soc Microbiol; Soc Indust Microbiol; Sigma Xi. *Res:* Bacterial physiology; microbial ecology; public health microbiology; microbial toxins; food and industrial microbiology. *Mailing Add:* Dept Biol 360 Huntington Ave Northeastern Univ Boston MA 02115

CRISP, CARL EUGENE, b Buhl, Idaho, Aug 1, 31; m 55. BIOCHEMISTRY, PLANT PHYSIOLOGY. *Educ:* Univ Idaho, BS, 55, MS, 59; Univ Calif, Davis, PhD(plant physiol), 65. *Prof Exp:* Jr flight test engr, Lockheed Missile Systs Div, 55; teaching asst plant physiol, Univ Calif, Davis, 59-64; plant physiologist, Insecticide Eval Proj, US Forest Serv, 67-81; SR RES BIOCHEMIST RES & DEVELOP, CHEVRON CHEM CO, RICHMOND, CALIF, 81- *Mem:* AAAS; Am Chem Soc; Am Inst Biol Sci; Am Soc Plant Physiol; Sigma Xi. *Res:* Investigations into the anatomy and chemical composition of plant surfaces, the biopolymer cutin, the design of systemic insecticides, the mechanisms of phloem transport in plants and biotechnology research and development. *Mailing Add:* Three Mira Lane Orinda CA 94563

CRISP, MICHAEL DENNIS, b Elmhurst, Ill, Apr 27, 42; m 65; c 3. QUANTUM OPTICS, LASERS. *Educ:* Bradley Univ, AB, 64; Washington Univ, MS, 66, PhD(physics), 68. *Prof Exp:* Res assoc physics, Columbia Univ, 68-70; scientist, Owens-Ill Inc, 70-72, sr scientist, Toledo, 72-77; legis asst for energy, off of US Sen Howard H Baker, Jr, 77-78; prof staff mem, Sen Comt on Com, Sci & Transp, 78-80; dir, Fed Govt Tech Liaison, Owens-Ill, Inc, Washington, DC, 80-83; EXEC ASST TO DIR ENERGY RES, US DEPT ENERGY, WASHINGTON, DC, 83- *Concurrent Pos:* Res assoc, Argonne Nat Lab, 65; instr physics, Univ Toledo, 72; Cong sci & eng fel, AAAS/Optical Soc Am, 76. *Mem:* Am Phys Soc; Optical Soc Am; Inst Elec & Electronics Engrs; Soc Info Display; AAAS. *Res:* Interaction of coherent light with matter, including laser physics, nonlinear optics, coherent pulse propagation and coherent spectroscopy; neoclassical radiation theory, laser-induced damage and gas-discharge display devices. *Mailing Add:* US Dept Energy 1000 Independence Ave Washington DC 20585

CRISP, ROBERT M, JR, b Ft Smith, Ark, Aug 20, 40; m 64, 83; c 3. ENGINEERING. *Educ:* Univ Ark, BSIE, 63, MSIE, 64; Univ Tex, PhD(mech eng), 67. *Prof Exp:* Assoc systs engr, Int Bus Mach Corp, 64-65; from assoc prof to prof indust eng, 67-85, PROF COMPUT SCI ENG, UNIV ARK, 85- *Honors & Awards:* Outstanding Res Award, Haliburton, 87 & 88. *Mem:* Inst Elec & Electronic Engrs Comput Soc. *Res:* microcomputer applications, interfacing, productivity software and parallel processing algorithms; application of operations research to production problems; energy systems analysis. *Mailing Add:* Dept Comput Sci Eng Univ Ark Fayetteville AR 72701

CRISP, THOMAS MITCHELL, JR, b San Antonio, Tex, Sept 29, 39; m 65; c 2. TOXICOLOGY, ENDOCRINOLOGY. *Educ:* Univ St Thomas, Houston, BA, 61; Rice Univ, MA, 64; Univ Tex, PhD(anat), 66. *Prof Exp:* Res asst endocrinol, Dent Br, Univ Tex, 60-61; teaching asst biol, Rice Univ, 61-63; from instr to assoc prof anat, Sch Med & Dent, Georgetown Univ, 66-90; REPRODUCTIVE BIOLOGIST, US ENVIRON PROTECTION AGENCY, WASHINGTON, DC, 90- *Concurrent Pos:* NIH teaching fel anat, Med Br, Univ Tex, 63-66; consult, Army Oral Biol Prog, 66-70 & Navy Dent Sch, Bethesda Naval Hosp, 67-69; hon res fel, Dept Clin Endocrinol, Women's Hosp, Univ Birmingham, 74-75; spec res fel, Rockefeller Found, 74-75; vis scientist, Dept Biochem, Univ Basel, Basel, Switz, 83 & 84. *Mem:* AAAS; Am Asn Anatomists; Soc Study Reprod; Endocrine Soc. *Res:* Electron microscopy of mammalian corpora lutea; ovarian tissue culture; human breast cancer. *Mailing Add:* US Environ Protection Agency Off Health & Environ Assessment Human Health Assessment Group RD-689 401 M Street SW Washington DC 20460

CRISPELL, KENNETH RAYMOND, b Ithaca, NY, Oct 30, 16; m 42; c 6. MEDICINE. *Educ:* Philadelphia Col Pharm & Sci, BS, 38; Univ Mich, MD, 43; Am Bd Internal Med, dipl, 50. *Prof Exp:* From instr to assoc prof internal med, Univ Va, 49-58; prof med & dir dept, NY Med Col, 58-60; asst dean, Sch Med, 62, actg dean, 62-64, dean, 64-71, vpres health sci, 71-76, prof, 76-87, EMER PROF MED & LAW, UNIV VA, 87- *Concurrent Pos:* Fel internal med, Ochsner Clin, La, 47-48; fel biophys, Tulane Univ, 48-49; Commonwealth fel, Univ Va, 49-51; physician, Univ Hosp, Va, 49-58 & 60. *Mem:* Endocrine Soc; Am Soc Clin Invest; Am Thyroid Asn; AMA; fel Am Col Physicians. *Res:* Endocrinology. *Mailing Add:* Med Ctr Box 423 Univ Va Charlottesville VA 22908

CRISPENS, CHARLES GANGLOFF, JR, b Bellevue, Pa, Aug 3, 30; m 53; c 1. HISTOLOGY, ONCOLOGY. *Educ:* Pa State Univ, BS, 53; Ohio State Univ, MS, 55; Wash State Univ, PhD(zool), 59. *Prof Exp:* Fel, Jackson Lab, Bar Harbor, Maine, 59-60; from instr to assoc prof anat, Sch Med, Univ Md, 60-68; fac med, Univ Sherbrooke, 68-69; PROF BIOL, UNIV ALA, BIRMINGHAM, 69- *Concurrent Pos:* Mem bd dirs, Southeastern Cancer Res Asn, 72-86; vis prof, Rockefeller Univ, 79. *Honors & Awards:* Lederle Med Fac Award, 64. *Mem:* AAAS; Am Asn Anat; Am Asn Cancer Res; Sigma Xi. *Res:* Lactate dehydrogenase virus and murine tumorigenesis, humoral immunity in mice in relation to age and tumorigenesis, copper complexes and Tween 80 as antineoplastic agents. *Mailing Add:* Dept Biol Univ Ala UAB Sta Birmingham AL 35294

CRISS, CECIL M, b Wheeling, WVa, Apr 22, 34; m 58; c 2. THERMODYNAMICS & MATERIAL PROPERTIES. *Educ:* Kenyon Col, AB, 56; Purdue Univ, PhD(phys chem), 61. *Prof Exp:* Asst prof chem, Univ Vt, 61-65; from asst prof to assoc prof, 65-76, PROF CHEM, UNIV MIAMI, 76-, CHMN DEPT, 84- *Mem:* AAAS; Am Chem Soc; The Chem Soc; Sigma Xi. *Res:* Thermodynamic properties of aqueous and nonaqueous ionic solutions at higher temperatures; ionic heat capacities; ionic entropies, ionic volumes and oxidation-reduction potentials in nonaqueous solutions. *Mailing Add:* Dept Chem Univ Miami Box 249118 Coral Gables FL 33124

CRISS, DARRELL E, b Terre Haute, Ind, Aug 25, 21; m 43; c 7. ELECTRICAL ENGINEERING, COMPUTER SCIENCE. *Educ:* Rose Polytech Inst, BS, 43; Univ Ill, MS, 50, PhD(elec eng), 59. *Prof Exp:* From instr to assoc prof elec eng, Rose-Hulman Inst Technol, 46-59, head dept, 61-65, assoc dean fac, 65-67, dean, 67-70, dir inst res, 70-75 & Computer Ctr, 62-75, prof elec eng, 59-88, head computer sci, 75-88; RETIRED. *Concurrent Pos:* Consult, Ind State Bd Regist for Engrs, 51-75, Nat Sch Aeronaut, 63- & Western Paper & Mfg Co, 64; consult elec eng, Peabody Coal Co, 76-77; consult comput sci, NSF, 77- *Res:* Control systems; nonlinear analysis; advanced computer logic; system simulation; computer education. *Mailing Add:* RR 25 Box 66 Terre Haute IN 47802

CRISS, JOHN W, b Clarksburg, WVa, Mar 24, 41; m 68; c 2. APPLIED X-RAY PHYSICS, X-RAY SPECTROMETRY. *Educ:* WVa Univ, BA, 64. *Prof Exp:* Mathematician, X-Ray Optics Br, Naval Res Lab, Washington, DC, 64-80; PRES, SCI APPLNS, SOFTWARE DEVELOP & MKT, CRISS SOFTWARE INC, 79- *Concurrent Pos:* Miscellanous consult & teaching. *Mem:* Am Phys Soc; Asn Comput Mach; Soc Indust & Appl Math; Sigma Xi. *Res:* Physical-mathematical modeling of x-ray generation, transport and measurement. *Mailing Add:* 12204 Blaketon St Largo MD 20772

CRISS, THOMAS BENJAMIN, b Clarksburg, WVa, June 30, 49; m 80; c 3. THEORETICAL PHYSICS, IMAGE PROCESSING. *Educ:* WVa Univ, BS, 71; Univ Tex, Austin, PhD(physics), 75. *Prof Exp:* PHYSICIST SUBMARINE NAVIG, JOHNS HOPKINS APPL PHYSICS LAB, 76-, PHYSICIST BIOMED IMAGE PROCESSING, 81- *Concurrent Pos:* Lectr, Grad Sch Engr, Johns Hopkins Univ, 82-84. *Res:* Relativity; astrophysics; image processing. *Mailing Add:* 13833 Dayton Meadows Ct Dayton MD 21036

CRISS, WAYNE ELDON, b Washington, Iowa, Mar 7, 40. CANCER, ENZYMOLOGY. *Educ:* William Penn Col, BS, 62; Univ Fla, MS, 64, PhD(biochem), 68. *Prof Exp:* Asst prof obstet & gynec, Med Sch, Univ Fla, 70-75, biochem, 74-75; dir cancer res, Comprehensive Cancer Ctr, Howard Univ, 75-, prof-doctor biochem & oncol, 80-; AT DEPT MED, SCH MED HACETEPPE UNIV, ANKARA, TURKEY. *Concurrent Pos:* Am Cancer Soc fel, Fels Inst, Temple Univ, 68-70; Nat Cancer Inst res career develop award, 74-79; Eleanor Roosevelt Int Award, Am Chem Soc, 75-76; consult, NSF, NIH, Cancer Res, Biochimica Biophysica Acta & Kanser. *Mem:* Endocrine Soc; Am Asn Univ Profs; Nat Tissue Cult Asn; Am Asn Cancer Res; Am Soc Biol Chemists; Union Internationale Contre le Cancer. *Res:* Cancer metabolism, modified regulatory control mechanisms in cancer tissues. *Mailing Add:* Sch Med Hacettepe Univ Ankara Turkey

CRISSMAN, HARRY ALLEN, b Lock Haven, Pa, Aug 16, 35; div; c 1. FLOW CYTOMETRY & CELL ANALYSIS. *Educ:* Lock Haven Univ, BS, 62; Pa State Univ, MS, 68, PhD(zool), 71. *Prof Exp:* Teacher biol & chem, Ovid Cent High Sch, 65-67; instr biol, Pa State Univ, 67-68; asst cell biol, 71-73, POST DOCTORAL FEL, STAFF CELL BIOL, LOS ALAMOS NAT LAB, 73- *Concurrent Pos:* Fulbright scholar, dept hemat, Univ Hosp, Nijmegen, Neth, 79-80, vis scientist, Neth Res Asn, 82; consult, flow cytometry div, Becton-Dickinson Co, 83; lectr, French-Int Flow Cytometry, 84, 85 & 88. *Mem:* Soc Anal Cytol; Cell Kinetics Soc; Histochem Soc; Am Soc Cell Biol. *Res:* Study of effects of chemotherapy and x-irradiation on cell metabolism (DNA, RNA, protein) and chromatin structure using flow cytometry techniques; development of cell preparative and flurochrome labeling methods for flow cytometry. *Mailing Add:* Cell Biol Group Los Alamos Nat Lab LS-4 MS M888 Los Alamos NM 87545

CRISSMAN, JACK KENNETH, JR, b Bellefonte, Pa, Jan 25, 44; m 66; c 2. CELLULAR & MOLECULAR BIOLOGY. *Educ:* Juniata Col, BS, 65; WVa Univ, PhD(biol), 72. *Prof Exp:* Instr biol, WVa Univ, 68-72, res assoc biochem, 72-73; asst prof biol, Wabash Col, 73-79; RES BIOCHEMIST, SUPELCO INC, 79- *Concurrent Pos:* Res fel, Am Cancer Soc, 71-72; res assoc, WVa Univ, 72-73; Grass Found res fel, 76-; film & book reviewer, AAAS, 77-; NSF prog reviewer, Div Sci Educ Develop & Res, 78- *Mem:* AAAS; Am Inst Biol Sci; Sigma Xi; Soc Study Reprod; Am Oil Chemists Soc. *Res:* Reproductive physiology-function and biochemistry of the mammalian cryptoorchid testis. *Mailing Add:* RD 5 Box 191 Bellefonte PA 16823

CRISSMAN, JOHN MATTHEWS, b Evanston, Ill, Oct 21, 35; m 68; c 2. POLYMER SCIENCE. *Educ:* Pa State Univ, PhD(physics), 63. *Prof Exp:* PHYSICIST, NAT BUR STANDARDS, 63- *Concurrent Pos:* Nat Acad Sci-Nat Res Coun physicist, Nat Bur Standards, 63-65. *Mem:* Am Phys Soc. *Res:* Mechanical and other bulk properties of polymeric materials. *Mailing Add:* Rm A209 Polymers Bldg Nat Bur Standards Gaithersburg MD 20899

CRISSMAN, JUDITH ANNE, b Clarion, Pa, July 11, 42. INORGANIC CHEMISTRY. *Educ:* Thiel Col, BA, 64; Univ NC, Chapel Hill, PhD(inorg chem), 70. *Prof Exp:* From asst prof to assoc prof chem, 68-80, PROF CHEM, MARY WASHINGTON COL, 80- *Mem:* Am Chem Soc. *Res:* Schiff base complexes of transition metals. *Mailing Add:* Dept Chem Mary Washington Col Fredericksburg VA 22401-5359

CRIST, BUCKLEY, JR, b Plainfield, NJ, Jan 12, 41; m 66; c 2. POLYMER SCIENCE. *Educ:* Williams Col, BA, 62; Duke Univ, PhD(chem), 66. *Prof Exp:* Phys chemist, Camille Dreyfus Lab, Res Triangle Inst, 66-73; from asst prof to assoc prof, 73-84, PROF MAT SCI & ENG & CHEM ENG, NORTHWESTERN UNIV, EVANSTON, 84- *Concurrent Pos:* Consult, 75- *Mem:* Am Chem Soc; Am Phys Soc; Mat Res Soc. *Res:* Morphology and mechanical properties of semicrystalline polymers; scattering of light, x-rays and neutrons by polymers; molecular motion and relaxation effects. *Mailing Add:* Dept Mat Sci & Eng Northwestern Univ Evanston IL 60208

CRIST, DELANSON ROSS, b New York, NY, July 16, 40; m 62; c 6. PHYSICAL ORGANIC CHEMISTRY. *Educ:* Swarthmore Col, AB, 62; Mass Inst Technol, PhD(org chem), 67. *Prof Exp:* NSF fel, 67-68, asst prof org chem, 68-74, NATO sr fel, 71, ASSOC PROF ORG CHEM, GEORGETOWN UNIV, 74- *Mem:* AAAS; Am Chem Soc. *Res:* Mechanisms of organic reactions of synthetic or biological importance; investigations concerning the stability and reactions of species containing positive-charged nitrogen atoms. *Mailing Add:* Dept Chem Georgetown Univ Washington DC 20057

CRIST, JOHN BENJAMIN, b Washington, DC, Jan 16, 41; m 83; c 3. WOOD SCIENCE & TECHNOLOGY, FORESTRY. *Educ:* Va Polytech Inst & State Univ, BS, 67, PhD(wood sci), 72. *Prof Exp:* Forest prod technologist, Duluth, Minn, 71-76, forest prod technologist wood sci, NCent Forest Exp Sta, 76-80, FIELD REP RESOURCE USE, FOREST SERV, USDA, 80- *Concurrent Pos:* Adj asst prof, Dept Forest Prod, Col Forestry, Univ Minn, 74-80, adj assoc prof, 80-91. *Mem:* Forest Prod Res Soc; Soc Wood Sci & Technol; Tech Asn Pulp & Paper Indust. *Res:* Anatomy and physical properties of woody materials grown under short rotation; intensive culture, and the properties of various wood products made from the above raw materials. *Mailing Add:* Rte 12 Box 203-A Morgantown WV 26505

CRISTINI, ANGELA, b Oct 2, 48; US citizen; m 83; c 1. BIOLOGY, MARINE ECOLOGY. *Educ:* Northeastern Univ, BA, 71; City Univ New York, PhD(biol), 77. *Prof Exp:* Fel biol, Univ WFal, 76-77; from asst prof to assoc prof, 77-88, Ramapo Col NJ, PROF BIOL, 88- *Concurrent Pos:* Theodore Roosevelt Mem fel, Am Mus Natural Hist, 75; asst dir environ res, NJ Dept Environ Protection, 87-89. *Mem:* AAAS; Am Soc Zoologists; Estuarine Res Fedn; Sigma Xi. *Res:* Physiological ecology of invertebrates; hydromineral balance in decapod crustaceans; molt physiology of crustaceans; effects of toxic substances on invertebrate physiology. *Mailing Add:* Sch Theoret & Appl Sci Ramapo Col NJ Mahwah NJ 07430

CRISTOFALO, VINCENT JOSEPH, b Philadelphia, Pa, Mar 19, 33; m 64; c 6. PHYSIOLOGY, BIOCHEMISTRY. *Educ:* St Joseph's Col, BS, 55; Temple Univ, MA, 58; Univ Del, PhD, 62. *Hon Degrees:* MA, Univ Pa, 82. *Prof Exp:* Asst instr gen biol, Temple Univ, 57-58, USPHS fel, 61-62, instr, dept chem, 62-72, res assoc, 63; res asst, Univ Del, 58-60, NSF fel, 60-61; from asst prof to prof, 67-90, EMER PROF BIOCHEM, DEPT ANIMAL BIOL, SCH VET MED, UNIV PA, 90-; AUDREY MEYER MARS PROF GERONT RES, MED COL PA, 90-, DIR CTR GERONT RES, 90-, PROF PHYSIOL/BIOCHEM, 90- *Concurrent Pos:* Res asst, Oak Ridge Nat Lab, 59; mem, Wistar Inst Anat & Biol, 63-69, assoc mem, 69-76; reviewer, NIH, Molecular Biol Study Sect, 74-75 & Molecular Cytol Study Sect, 75-79, aging rev comt, 83-86; actg dir, Ctr Study Aging, Univ Pa, 78-81, dir, 81-90; ed, J Geront, 88-; adj prof, Wistar Inst, 90-; actg chmn, Dept Physiol, Med Col Pa, 90-, dir, Div Physiol, 90- *Honors & Awards:* Kleemeier & Brookdale Awards, Geront Soc Am; Keston Lectr. *Mem:* AAAS; Tissue Culture Asn; Am Cell Biol Geront Soc; Soc Exp Biol & Med. *Res:* Effects of oxygen and radiation on development; intermediary metabolism; neoplastic tissues; tissue culture cells; aging in cell and tissue culture; cell proliferation. *Mailing Add:* 444 Haverford Ave Narberth PA 19072

CRISTOL, STANLEY JEROME, b Chicago, Ill, June 14, 16; m 57; c 2. CHEMISTRY. *Educ:* Northwestern Univ, BS, 37; Univ Calif, Los Angeles, MA, 39, PhD(org chem), 43. *Prof Exp:* Asst chem, Univ Calif, Los Angeles, 37-38; res chemist, Stand Oil Co, Calif, 38-41; from asst to instr chem, Univ

Calif, Los Angeles, 41-43; res fel, Univ Ill, 43-44; res chemist, USDA, Md, 44-46; from asst prof to prof, 46-79, chmn dept, 60-62, fac res lectr, 60, DISTINGUISHED PROF CHEM, UNIV COLO, BOULDER, 79- *Concurrent Pos:* Guggenheim fel, 55-56 & 81-82; assoc ed, Chem Rev, 57-59; vis prof, Stanford Univ, 60, Univ Geneva, Switz, 75 & Univ Lausanne, Switz, 81; mem, NSF Adv Comt, 57-63, 69-72 & NIH Adv Comt, 68-72; consult, various co; legal expert, Org Chem. *Honors & Awards:* Stearns Award, 71; James Flack Norris Award, Am Chem Soc, 72. *Mem:* Nat Acad Sci; AAAS; Am Chem Soc; Royal Soc Chem. *Res:* Organic chemistry; mechanisms of organic reactions; photochemistry. *Mailing Add:* Dept Chem & Biochem Univ Colo Boulder CO 80309-0215

CRISWELL, BENNIE SUE, b Huntsville, Tex, Nov 17, 42; m 64, 79; c 5. IMMUNOLOGY, INFECTIOUS DISEASES. *Educ:* NTex State Univ, BS, 64; Registry Med Technol, cert, 64; Baylor Col Med, MS, 68, PhD(immunol), 69. *Prof Exp:* Instr, 72-74, asst prof microbiol, Baylor Col Med, 74-80; assoc prof med technol, Univ Ariz, 80-91; dir clins, 91, DIR LAB, REPRODUCTIVE INST TUCSON, INC, 85-; DIR CLIN, REPRODUCTIVE INST, ATLANTA, GA, 91- *Concurrent Pos:* NASA fel, Manned Spacecraft Ctr, Houston, 69-72; vis scientist, NASA Johnson Space Ctr, 72- *Mem:* Sigma Xi; Am Fertil Soc; Am Asn Immunol. *Res:* Reproductive biology. *Mailing Add:* 8221 E Snyder Rd Tucson AZ 85715

CRISWELL, DAVID RUSSELL, b Ft Worth, Tex, July 17, 41; c 2. SPACE PHYSICS, AUTOMATION & ROBOTICS. *Educ:* NTex State Univ, BS, 63, MS, 64; Rice Univ, Phd(space physics), 68. *Prof Exp:* Prof staff eng, TRW Syst, Houston Opers, 68-70; staff scientist res admin, Lunar & Planetary Inst, 71-80; vis res physicist, Calif Space Inst, Univ Calif, San Diego, 80-82; exec dir, Consortium Space/Terrestrial Automation & Robotics, 82-88; DIR, INST SPACE SYSTS OPERS, UNIV HOUSTON, 90- *Concurrent Pos:* Exec dir, Lunar & Planetary Rev Panel, 71-78; prin investr, NASA grant, 74-77, mem staff, NASA Ames Res Ctr, Space Indust, 75, Sci Asn Inc, 76, prin investr, NASA grant, 77-78, dir automation & robotics, Nat Space Prog, 84-85; consult, aerospace, space physics, automation & robotics, 82- *Res:* Space industrialization; plasma physics as applied to moon solar wind interactions; physical descriptions of psychology; seismology; management and cybernetics; automation and robotics; space transportation; 22 patents. *Mailing Add:* Inst Space Systs Opers Univ Houston Houston TX 77204-5502

CRISWELL, MARVIN EUGENE, b Chappell, Nebr, Oct 31, 42; m 65; c 4. STRUCTURAL ENGINEERING, CONSTRUCTION MATERIALS. *Educ:* Univ Nebr, BS, 65; Univ Ill, MS, 66, PhD(civil eng), 70. *Prof Exp:* Res struct engr, Waterways Exp Sta, US Army Corps Engrs, 67-70; from asst prof to assoc prof, 70-84, PROF CIVIL ENG, COLO STATE UNIV, FT COLLINS, 84-, ASSOC DEPT HEAD, 89- *Concurrent Pos:* Dow award young fac, Am Soc Eng Educ, 78; prin, Eng Data Mgt, Inc, 82-85; bd dirs, Am Soc Eng Educ, 90-92. *Mem:* Am Soc Civil Engrs; Am Soc Eng Educ; Am Concrete Inst; Am Soc Testing & Mat; Forest Prod Res Soc. *Res:* Structural engineering design and behavior; reinforced concrete slab systems; wood joist floor systems; transmission line structures; roller compacted concrete dams; inflatible structures for a future moon base; wood and fiber reinforced concrete as construction materials; structural reliability; reliability-based design codes. *Mailing Add:* Dept Civil Eng Colo State Univ Ft Collins CO 80523

CRITCHFIELD, CHARLES LOUIS, b Shreve, Ohio, June 7, 10; m 35; c 4. MATHEMATICAL PHYSICS. *Educ:* George Washington Univ, BS, 34, MA, 36, PhD(physics), 39. *Prof Exp:* Instr, Univ Rochester, 39-40 & Harvard Univ, 41-42; physicist, Geophys Lab, 42-43 & Monsanto Chem Co, 46-47; assoc prof physics, George Washington Univ, 46; from assoc prof to prof, Univ Minn, 47-55; dir sci res, Convair Div, Gen Dynamics, 55-60; vpres res, Telecomput Corp, 60-61; physicist, Los Alamos Sci Lab, 43-46, assoc div & group leader nuclear physics, 61-77; RETIRED. *Concurrent Pos:* Mem planetology subcomt, NASA, 64-69. *Mem:* Fel Am Phys Soc. *Res:* Scalar potentials in the Dirac equation for nuclear and particle physics; elliptic functions. *Mailing Add:* 391 El Conejo Los Alamos NM 87544

CRITCHFIELD, WILLIAM BURKE, b Minneapolis, Minn, Nov 21, 23. FOREST GENETICS. *Educ:* Univ Calif, BS, 49, PhD(bot), 56. *Prof Exp:* Forest geneticist, Cabot Found, Harvard Univ, 56-59; Geneticist, Pac Southwest Forest & Range Exp Sta, US Forest Serv, 59-88; RETIRED. *Mem:* Bot Soc Am. *Res:* Geographic variation and evolution in forest trees; leaf variation and shoot development in trees. *Mailing Add:* Pac Southwest Forest & Range Exp Sta PO Box 245 Berkeley CA 94701

CRITCHLOW, BURTIS VAUGHN, b Hotchkiss, Colo, Mar 5, 27; m 48; c 4. ANATOMY. *Educ:* Occidental Col, BA, 51; Univ Calif, Los Angeles, PhD(neuroendocrinol), 57. *Prof Exp:* From instr to prof anat, Baylor Col Med, 57-72; prof anat & chmn dept, Ore Health Sci Univ, 72-82; DIR, ORE REGIONAL PRIMATE RES CTR, 82- *Concurrent Pos:* Sr res fel, 59-64; vis investr, Nobel Inst Neurophysiol, Karolinska Inst, Stockholm, 61-62; USPHS res career develop award, 64-69; mem, reprod biol study sect, NIH, 69-73, consult. *Mem:* Am Asn Anatomists; Endocrine Soc; Soc Neurosci; Int Brain Res Orgn; Int Soc Neuroendocrinol. *Res:* Brain and endocrine interrelations; neural control of pituitary functions. *Mailing Add:* Ore Regional Primate Ctr 505 NW 185th Ave Beaverton OR 97005

CRITCHLOW, D(ALE), b Harrisville, Pa, Jan 6, 32; m 55; c 3. ELECTRICAL ENGINEERING. *Educ:* Grove City Col, BS, 53; Carnegie Inst Technol, MS, 54, PhD(elec eng), 56. *Prof Exp:* Engr magnetic amplifiers develop, Magnetics, Inc, 52-53; proj engr magnetic devices res, Carnegie Inst Technol, 53-56, assoc prof elec eng, 56-58; staff engr, 58-59, adv engr, 59-64, res staff mem, Yorktown Heights, 64-80, sr engr, Res Ctr, IBM Corp, Hopewell Jct, 80-86, SR TECH STAFF, 86-, IBM FEL, IBM CORP GEN TECH DIV, ESSEX JCT, VT, 86- *Concurrent Pos:* Consult, Magnetic Amplifiers Inc, 56- & Johnstone Foundry, 56. *Mem:* Nat Acad Eng; Sigma Xi; Inst Elec & Electronic Engrs. *Res:* Mosfet logic circuits; dram technology. *Mailing Add:* IBM Corp Essex Jct VT 05452

CRITES, JOHN LEE, b Wilmington, Ohio, July 10, 23; m 46; c 2. HELMINTHOLOGY, WILDLIFE DISEASE. *Educ:* Univ Idaho, BS, 49, MSc, 51; Ohio State Univ, PhD(zool), 56. *Prof Exp:* Asst zool, Univ Idaho, 49-51; from instr to assoc prof, 55-67, PROF ZOOL, OHIO STATE UNIV, 67-, CHMN DEPT, 81- *Concurrent Pos:* Res investr, Stone Lab Hydrobiol, 56 & Marine Lab Duke Univ, 58; USAID-NSF consult, Aligarh Muslim Univ & Banaras Hindu Univ, India, 64; prof parasitol, Stone Lab Hydrobiol, 64-81, dir acad, 72-77 & res dir, 77-81. *Mem:* Am Soc Parasitologists; Wildlife Dis Asn; Int Soc Nematol; Am Micros Soc; AAAS; Int Great Lakes Res Soc. *Res:* Nematode parasites of animals; nematode parasites of plants; free-living nematodes, both marine and fresh water; biology, pathology, development, life histories and taxonomy of parasites of fish and wild animals. *Mailing Add:* Ohio State Univ Dept of Zool 1735 Neil Ave Columbus OH 43210

CRITOPH, E(UGENE), b Vancouver, BC, March 29, 29; m 52; c 4. ENGINEERING PHYSICS. *Educ:* Univ BC, BASc, 51, MASc, 57. *Prof Exp:* Sr res off, 53-67, head, Reactor Physics Br, 67-75, dir, Fuels & Mat Div, 75-76, dir, Advan Projs & Reactor Physics Div, 76-78, vpres & gen mgr, 79-86, VPRES STM, ATOMIC ENERGY CAN LTD, 86- *Concurrent Pos:* Chmn Europ Am Comt Reactor Physics, 68-69. *Honors & Awards:* W B Lewis Award, 86. *Mem:* Am Nuclear Soc; Can Asn Physicists; Can Nuclear Soc. *Res:* Reactor physics; power reactor design and the physics of heavy water-uranium lattices. *Mailing Add:* 1002-221 Lyon St N Ottawa ON K1R 7X5 Can

CRITS, GEORGE J(OHN), b Norristown, Pa, Feb 11, 22; m 48; c 2. CHEMICAL ENGINEERING, INDUSTRIAL WATER TREATMENT TECHNOLOGY. *Educ:* Pa State Univ, BS, 43; Columbia Univ, MS, 50. *Prof Exp:* With Kellex Corp, 43-45; with Harshaw Chem Co, 45; with Barrett Div, Allied Chem & Dye Corp, 45-46; design & chem engr, Los Alamos Sci Labs, 46-47; chem engr, Welding Engr, 48-49; physicist & chem engr, J Razek Labs, 49; mgr chem res, 50-67, tech consult, 67-68, tech dir, Cochrane Div, Crane Co, 68-87; TECH DIR, AQUA-ZEOLITE SCI INC, 88- *Concurrent Pos:* Mem, Franklin Inst; thirteen US patents & foreign patents. *Mem:* Am Chem Soc; Am Soc Testing & Mat; Nat Asn Corrosion Engrs. *Res:* Water and waste treatment research; ion exchange technology in equipment manufacturing and operation for industry. *Mailing Add:* 268 Glendale Rd Havertown PA 19083

CRITTENDEN, ALDEN LA RUE, b Wichita, Kans, Nov 27, 20; m 58. CHEMISTRY. *Educ:* Univ Ill, BS, 42, PhD(chem), 47. *Prof Exp:* From instr to asst prof, 47-60, ASSOC PROF CHEM, UNIV WASH, 60- *Mem:* Am Chem Soc. *Res:* Polarography. *Mailing Add:* 6233 37th NE Seattle WA 98115

CRITTENDEN, EUGENE CASSON, JR, b Washington, DC, Dec 25, 14; m 42; c 2. ELECTRO-OPTICS. *Educ:* Cornell Univ, Ithaca, AB, 34 & PhD(physics), 39. *Prof Exp:* Asst prof physics, Case Inst Technol, Case Western Reserve, 38-44; res physicist, Univ Calif Berkeley, 44-46; prof physics, Case Inst Technol, Case Western Reserve, 46-52; head solid state physics, Atomic Energy, N Am Aviation, 52-53; prof physics, Naval Postgrad Sch, 53-77, chmn, 64-67, distinguished prof, 77-86; CONSULT & RES CONTRACTOR, 87- *Mem:* Am Phys Soc; Optical Soc Am. *Res:* Nuclear physics; solid state physics; electro-optics. *Mailing Add:* 1176 Castro Rd Monterey CA 93940

CRITTENDEN, JOHN CHARLES, b Nov 12, 49; US citizen; m; c 3. ADSORPTION, PHYSICAL-CHEMICAL TREATMENT. *Educ:* Univ Mich-Ann Arbor, BSE, 71, MSE, 72, PhD(civil & environ eng), 76. *Prof Exp:* Sr vpres, Limno Tech, Inc, 75-77; asst prof civil & environ eng, Washington State Univ, 77-78 & Univ Ill-Urbana, 78-79; from asst prof to assoc prof civil & environ eng, 79-84, adj assoc prof chem eng, 81-84, PROF CIVIL & ENVIRON ENG & ADJ PROF CHEM ENG, MICH TECHNOL UNIV, 84- *Concurrent Pos:* Mem, bd dirs, Asn Environ Eng Professors, 83-85. *Honors & Awards:* Rudolph Hering Medal, Am Soc Civil Engrs, 80; Corecipient NALCO Chem Award, Asn Environ Eng Prof, 81. *Mem:* Am Soc Civil Engrs; Water Pollution Control Fedn; Int Soc Humic Substances; Asn Environ Eng Professors; Am Water Works Asn; Am Inst Chem Engrs. *Res:* Water and wastewater treatment using granular activated carbon and air stripping; fate of organic and inorganic pollutants in the subsurface. *Mailing Add:* Dept Civil Eng Mich Technol Univ Houghton MI 49931

CRITTENDEN, LYMAN BUTLER, b New Haven, Conn, May 27, 26; m 59; c 2. GENETICS. *Educ:* Calif Polytech Inst, BS, 51; Purdue Univ, MS, 55, PhD, 58. *Prof Exp:* Geneticist, Nedlar Farms, 51-53; instr genetics, Purdue Univ, 55-57; geneticist, Creighton Bros, 57-58 & Nat Cancer Inst, 58-61; geneticist, Regional Poultry Res Lab, 61-67 & Animal Physiol & Genetics Inst, Beltsville, Md, 67-75, RES GENETICIST, REGIONAL POULTRY RES LAB, MICH, USDA, 75- *Honors & Awards:* Merck Award Poultry Sci, 80; Tom Newman Int Award Poultry Res. *Mem:* AAAS; Poultry Sci Asn; Genetics Soc Am. *Res:* Quantitative genetics; genetics of disease resistance; virology. *Mailing Add:* 1747 Burkley Rd Williamston MI 48895

CRITTENDEN, RAY RYLAND, b Galesburg, Mich, Mar 19, 31; m 62; c 3. PHYSICS. *Educ:* Willamette Univ, BA, 54; Univ Wis, MS, 56, PhD(physics), 60. *Prof Exp:* Assoc scientist, Brookhaven Nat Lab, 60-63; from asst prof to assoc prof, 63-74, PROF PHYSICS, COL ARTS & SCI, GRAD SCH, IND UNIV, BLOOMINGTON, 74- *Mem:* Am Phys Soc. *Res:* High energy nuclear physics. *Mailing Add:* 8587 W Rice Rd Bloomington IN 47401

CRITTENDEN, REBECCA SLOVER, b Lake City, Tenn, July 10, 36; m 66. ALGEBRA. *Educ:* Georgetown Col, BS, 58; Univ NC, Chapel Hill, MA, 62, PhD(math), 63. *Prof Exp:* Asst prof math, Georgetown Col, 63-64, Va Polytech, 64-66 & Vanderbilt Univ, 66-67; asst prof, 67-70, ASSOC PROF MATH, VA POLYTECH INST & STATE UNIV, 70- *Concurrent Pos:* Woodrow Wilson fel. *Mem:* Am Math Soc; Math Asn Am. *Res:* Ring theory. *Mailing Add:* Dept Math Va Polytech Inst & State Univ Blacksburg VA 24061

CRITTENDEN, RICHARD JAMES, b Milwaukee, Wis, Feb 28, 30; m 53; c 3. MATHEMATICS. *Educ:* Williams Col, AB, 52; Oxford Univ, BA, 54; Mass Inst Technol, PhD(math), 60. *Prof Exp:* Asst prof math, Northwestern Univ, 60-64; assoc ed, Math Rev, Am Math Soc, 64-68, exec ed, 68-71; chmn dept, 71-74, PROF MATH, COL GEN STUDIES, UNIV ALA, BIRMINGHAM, 71- *Mem:* Am Math Soc. *Res:* Differential geometry; Riemannian geometry; G-structures. *Mailing Add:* Dept Math Col Gen Studies Univ Ala Birmingham AL 35294

CRITZ, JERRY B, b Boonville, Mo, Apr 9, 34; m 55; c 2. PHYSIOLOGY. *Educ:* Univ Mo, BS, 56, MA, 58, PhD(physiol), 61. *Prof Exp:* From asst prof to assoc prof physiol, Univ SDak, 61-67; assoc prof, Univ Western Ont, assoc prof physiol, Ctr Med Educ, Sch Med, Ind Univ, 73-75; prof physiol, Sch Med, Southern Ill Univ, Carbondale, 75-77; HEALTH SCIENTIST ADMINR, NAT HEART LUNG & BLOOD INST, CARDIAC FUNCTION BR, 77- *Mem:* AAAS; Soc Exp Biol Med. *Res:* Cardiovascular effects of exercise and training. *Mailing Add:* Nat Inst Health DRG Westwood Bldg Rm 339 5333 Westbard Ave Bethesda MD 20892

CRIVELLO, JAMES V, b Grand Rapids, Mich, July 30, 40. ORGANIC CHEMISTRY. *Educ:* Aquinas Col, BS, 62; Univ Notre Dame, PhD(org chem), 66. *Prof Exp:* RES CHEMIST, RES & DEVELOP CTR, GEN ELEC CO, 66- *Mem:* Am Chem Soc. *Res:* Polymer science; thermally stable polymers; synthesis and characterization; oxidation and nitration chemistry; organic photochemistry, photoinitiated cationic polymerizations. *Mailing Add:* Gen Elec Res & Develop Ctr Bldg K-1 PO Box 8 Schenectady NY 12301

CROASDALE, HANNAH THOMPSON, botany, for more information see previous edition

CROAT, JOHN JOSEPH, b St Marys, Iowa, May 23, 43; m; c 1. PHYSICAL METALLURGY, SOLID STATE PHYSICS. *Educ:* Simpson Col, BA, 65; Iowa State Univ, MS, 69, PhD(metall), 72. *Prof Exp:* Jr metallurgist, Ames Lab, AEC, 65-69; res asst, Iowa State Univ, 69-72; sr res scientist, 72-80, staff scientist, 80-84, asst chief engr, 84-85, eng mgr, 85-86, CHIEF ENGR, DELCO REMY DIV, GEN MOTORS, CORP, 87. *Honors & Awards:* Appln of Physics Prize, Am Inst Physics, 86; New Materials Prize, Am Physics Soc, 86. *Mem:* Inst Elec & Electronic Engrs; Am Phys Soc; Sigma Xi. *Res:* Magnetic and magnetoelastic properties of rare earth transition metal alloys; application of permanent magnets. *Mailing Add:* 206 Amhurst Circle Noblesville IN 46060

CROAT, THOMAS BERNARD, b St Mary's, Iowa, May 23, 38; m 65; c 2. BOTANY. *Educ:* Simpson Col, BA, 62; Univ Kans, MA, 66, PhD(bot), 67. *Prof Exp:* High sch teacher biol, Virgin Islands Govt- Knoxville, Iowa Pub Sch, 62-64; asst res botanist, 67-71, cur phanerograms, 74-77, P A SCHULZE CUR BOT, MO BOT GARDEN, 77- *Concurrent Pos:* Vis fel, Smithsonian Trop Res Inst, 68-71; cur, Summit Herbarium & Libr, 70-71; fac assoc biol, Wash Univ, 70-; mem, NSF Adv Comt Resources in Syst Bot, 72-; mem, Comn Orgn for Flora Neotropica, 75-; adj prof biol, Univ Mo & St Louis Univ, 74- *Mem:* Am Soc Plant Taxonomists; Inst Soc Plant Taxonomists; Asn Trop Biol; Bot Soc Am; Am Inst Biol Sci. *Res:* Systematic and evolutionary studies of the Araceae and Sapindaceae; floristics of the neotropics; phenological behavior of tropical floras. *Mailing Add:* Mo Bot Garden Box 299 St Louis MO 63166-0299

CROBER, DONALD CURTIS, b Morrisburg, Ont, July 20, 39; m 64; c 2. ANIMAL GENETICS, ANIMAL PHYSIOLOGY. *Educ:* McGill Univ, BSc, 61, MSc, 64; Univ BC, PhD(animal genetics), 71. *Prof Exp:* Biometrician, Can Wildlife Serv, Ottawa, 69-71; PROF ANIMAL SCI, NS AGR COL, 71- *Concurrent Pos:* Vis sr lectr dept animal sci, Makerere Univ, Kampala, Uganda, 72 & dept animal prod, Univ Sci & Technol, Kumasi, Ghana, 73-74. *Mem:* Poultry Sci Asn; Can Soc Animal Sci; World's Poultry Sci Asn. *Res:* Poultry breeding and physiology; poultry management. *Mailing Add:* Dept Animal Sci NS Agr Col Truro NS B2N 5E3 Can

CROCE, CARLO MARIA, b Milan, Italy, Dec 17, 44; nat US; c 1. MOLECULAR GENETICS. *Educ:* Univ Rome, MD, 69. *Prof Exp:* Assoc scientist, 70-71, res assoc, 71-74, assoc mem, 74-76, prof, 76-80, ASSOC DIR & INST PROF, WISTAR INST ANAT & BIOL, 80- *Concurrent Pos:* Vis scientist, Carnegie Inst, 78-79; mem, Mammalian Genetics Study Sect, NIH, Bethesda, 79-83, adv comt cell & develop biol, Am Cancer Soc, 83-86. *Res:* Study of the organization of the human genome; regulation of gene expression during mammalian embryogenesis; differentiation in vitro. *Mailing Add:* Wistar Inst Anat & Biol 36th St at Spruce Philadelphia PA 19104

CROCE, LOUIS J, b New York, NY, Sept 17, 21; m 48; c 2. PHYSICAL CHEMISTRY, ORGANIC CHEMISTRY. *Educ:* St John's Univ, NY, BS, 48; NY Univ, PhD(phys org chem), 52. *Prof Exp:* Group leader org res, Evans Res & Develop Co, 51-52; res chemist, Socony-Mobile Oil Co, Inc, 52-55; res chemist, Petro-Tex Chem Corp, 55-58, group leader, 58-60, res supvr, 60-61, res mgr, 61-66, assoc dir res, 66-74, dir res, 74-82; dir res, Tenneco Oil, 82-86; CONSULT, TENNECO OIL, 86- *Mem:* AAAS; Am Chem Soc; Sigma Xi; NY Acad Sci. *Res:* Petrochemicals, including monomer synthesis, catalytic dehydrogenation, dehydrocyclization, ammoxidation; reactions of olefins and dienes; polyolefins; petroleum research in hydrocracking, hydrocarbon separations, alkylation. *Mailing Add:* 135 Driftwood Dr Seabrook TX 77586

CROCHIERE, RONALD E, b Wausau, Wis, Sept 28, 45; m 69; c 1. DIGITAL SIGNAL PROCESSING, SPEECH CODING. *Educ:* Milwaukee Sch Eng, BS, 67; Mass Inst Technol, MS, 68, EE, 72, PhD(elec eng), 74. *Prof Exp:* Mem tech staff, Raytheon Co, 68-70; mem tech staff, Bell Labs, 74-83, HEAD DEPT, AT&T BELL LABS, 83- *Honors & Awards:* Centennial Medal, Inst Elec & Electronics Engrs, 84. *Mem:* Fel Inst Elec & Electronics Engrs. *Res:* Contributed to development of multirate digital signal processing concepts and digital filter bank theory; concepts of digital speech coding; inventor of sub-band speech coding. *Mailing Add:* AT&T Bell Labs Murray Hill NJ 07974

CROCKER, ALLEN CARROL, b Boston, Mass, Dec 25, 25; m 53; c 3. PEDIATRICS. *Educ:* Mass Inst Technol, 42-44; Harvard Univ, MD, 48. *Prof Exp:* Lab house officer, 48-49, jr asst resident med, 49-51, from asst physician to assoc physician, 56-62, res assoc med, 62-66, SR ASSOC MED, CHILDREN'S HOSP, 66-, DIR DEVELOP EVAL CTR, 67- *Concurrent Pos:* Fel path, Children's Hosp, 53-56; res assoc path, Harvard Med Sch, 56-60, res assoc pediat, 60-66, asst prof, 66-69, assoc prof, 69-, tutor med sci, 64-79; pres, Persons with Develop Disabilities, Am Asn Univ Affil, 82-83. *Mem:* Am Asn Ment Retardation (vpres med, 80-82); Am Asn Univ Affil Progs for Persons with Develop Disabilities (pres, 82-83); Nat Down Syndrome Cong (vpres, 84-85); Soc Behav Pediat (pres, 87-88). *Res:* Clinical investigation; pediatric metabolic diseases; biochemistry of the lipids; mental retardation. *Mailing Add:* Children's Hosp 300 Longwood Ave Boston MA 02115

CROCKER, BURTON B(LAIR), b Atlanta, Ga, Jan 1, 20; m 54, 84; c 2. CHEMICAL ENGINEERING. *Educ:* Ga Sch Technol, BS, 41; Mass Inst Technol, SM, 47. *Prof Exp:* Jr eng aide, Dept Chem Eng, Tenn Valley Authority, 39-41, opers mgr, Ala Ordnance Works, 41-42; jr chem engr, Phosphate Div, Monsanto Co, 45-46, chem engr, 47-50, sr engr, 50-52, resident engr, Res Dept, 52-53, aast chief chem eng, Eng Dept, Inorg Chem Div, 53-59, technologist, 59-66, advan technologist, Phosphorus Dept, 66-69, sr eng fel corp eng, 70-81, distinguished eng fel, corp eng dept, 81-85; DIR CHEM ENG, SITEX CORP, 86- *Concurrent Pos:* Lectr, Air Pollution Control, Univ Tulsa, 73 & 74, Am Chem Soc tour lectr, 71 & 72, Univ Liege, Belg, 77; adj prof environ eng, S Ill Univ, Edwardsville, 87; consult coal gasification process develop, Memphis, Tenn, 80-81; chmn tech comt, Particulate Solid Res, Inc, 79-85. *Honors & Awards:* Lawrence K.Cecil Environ Award, Am Inst Chem Engrs, 82. *Mem:* Fel Am Inst Chem Engrs; Air Pollution Control Asn (dir & vpres, 72-75). *Res:* Combustion of fuels; refractory design; phosphorus, phosphate salt process and equipment design; design of air pollution control equipment; air pollution studies, including effects on vegetation; meteorology; process systems; fluidization; gas-solids processing and separation, thermal processing and destruction; solid and hazardous waste processing, treatment and disposal. *Mailing Add:* 15716 Eldon Ridge Dr Chesterfield MO 63017

CROCKER, DENTON WINSLOW, b Salem, Mass, May 1, 19; m 46; c 4. INVERTEBRATE ZOOLOGY. *Educ:* Northeastern Univ, BA, 42; Cornell Univ, MA, 48, PhD(zool), 52. *Prof Exp:* Instr biol, Amherst Col, 51-53; from instr to assoc prof, Colby Col, 53-60; prof, 60-83, chmn dept, 60-77, EMER PROF BIOL, SKIDMORE COL, 83- *Mem:* AAAS; Am Inst Biol Sci; Sigma Xi. *Res:* Systematics and physiological ecology of crayfishes. *Mailing Add:* 118 White St Saratoga Springs NY 12866

CROCKER, DIANE WINSTON, b Cambridge, Mass, Nov 23, 26; m 49, 74; c 3. PATHOLOGY, DATA PROCESSING. *Educ:* Wellesley Col, BA, 46; Brown Univ, MS, 48; Boston Univ, MD, 52. *Prof Exp:* Asst path, Univ Southern Calif, 55-56; asst, Col Physicians & Surgeons, Columbia Univ, 57-58; instr, Harvard Med Sch, 58-65, assoc, 65-68, asst clin prof, 68-69; from asst prof to prof path & chief dept, Health Sci Ctr Hosp, Temple Univ, 69-73; chief anat path data processing, Los Angeles County-Univ Southern Calif Med Ctr & prof path, Univ Southern Calif, 73-; AT COOK COUNTY HOSP, CHICAGO. *Concurrent Pos:* Los Angeles County Heart Asn res grant, Los Angeles Children's Hosp, 55-56; Am Cancer Soc fel, Francis Delafield Hosp, New York, 56-57 & Presby Hosp, 58; surg pathologist & chief cytol, Peter Bent Brigham Hosp, Boston, 58-70, attend & consult, 70-71; attend, Vet Admin Hosp, West Roxbury, Mass, 60-70. *Mem:* AAAS; fel Royal Soc Med; Am Soc Nephrology; Int Acad Path; Int Soc Nephrology; Sigma Xi. *Res:* Renal disease and hypertension. *Mailing Add:* Cook County Hosp 627 S Wood St Chicago IL 60612

CROCKER, IAIN HAY, b Hamilton, Ont, July 28, 28; m 55; c 3. ANALYTICAL CHEMISTRY. *Educ:* McMaster Univ, BSc, 50. *Prof Exp:* Anal chemist, NAm Cyanamid Co, 50-52; chemist, NRX Reactor Opers, Atomic Energy, Can, Ltd, 52-53, chem process develop, 53-55, develop chem, 55-62, assoc res officer, 62-72, sect head, Mass Spectrometry & Fuel Anal, 72-79, br mgr, gen chem, Chalk River Nuclear Labs, 79-88; RETIRED. *Mem:* Fel Chem Inst Can; Am Soc Mass Spectrometry; Can Nuclear Soc. *Res:* Mass spectrometry-spark source and thermionic; reactor materials analyses; burnup; trace analyses; environmental analyses. *Mailing Add:* Box 496 Eight Mountain View Crescent Deep River ON K0J 1P0 Can

CROCKER, THOMAS TIMOTHY, b Barranquilla, Colombia, May 9, 20; div; c 4. INFECTIOUS DISEASES, VIROLOGY. *Educ:* Univ Calif, AB, 42, MD, 44. *Prof Exp:* Asst med, Hosp & Sch Med, Univ Calif, 45-46; asst med, Grace-New Haven Community Hosp, 48-49; from asst prof to prof med, Univ Calif, San Francisco, 50-71; PROF MED & CHMN DEPT COMMUN & ENVIRON MED, UNIV CALIF, IRVINE, 71-, PROF COMMUNITY & ENVIRON MED, SCH MED. *Concurrent Pos:* Nat Res Coun fel virol, Sch Med, Yale Univ, 49-50; Markle Found scholar, 50-55; Guggenheim fel, Clare Col, Cambridge Univ, 57-58; res assoc, Cancer Res Inst, Univ Calif, San Francisco, 57-71; consult, Calif State Dept Pub Health, Nat Cancer Inst, & Biol Div, Oak Ridge Nat Lab; mem extramural grants adv comt, Air Pollution Control Off, Environ Protection Agency, Agency; mem sci adv comn, Calif Air Resources Bd & Southern Calif Air Qual Mgt Dist; co-chmn panel on polycyclic org matter, Environ Protection Agency. *Mem:* AAAS; Am Asn Cancer Res; Am Soc Cell Biol; Am Thoracic Soc; Am Pub Health Asn. *Res:* Cellular physiology of mammalian respiratory epithelia and tumors; environmental toxicology including chemical carcinogenesis with correlation between rodent and primate susceptibility to epithelial metaplasia and transformation; health effects of air pollutants. *Mailing Add:* Dept Community & Environ Med Col Med Univ Calif Irvine CA 92717

CROCKET, DAVID SCOTT, b Cranford, NJ, Mar 6, 31; m 53; c 5. ENVIRONMENTAL CHEMISTRY. *Educ:* Colby Col, AB, 52; Univ NH, MS, 54, PhD(inorg chem), 60. *Prof Exp:* Trainee, Owens-Corning Fiberglass, 53-54; asst prof inorg chem, Lafayettte Col, 59-65, asst dean acad affairs, 65-

67, assoc dean col, 67-68, dean spec progs, 68-73, assoc, 73-80, dir res, 80-83, ASSOC PROF CHEM, LAFAYETTE COL, 65-, DIR ENVIRON SCI PROG, 90- *Concurrent Pos:* Dir, Univ City Sci Ctr, 72-80. *Mem:* Am Chem Soc; Sigma Xi. *Res:* Study of complex fluorides, particularly in the solid state by means of x-ray diffraction; infrared spectro-photometry; application of scanning electron microscopy to geochemical and environmental problems. *Mailing Add:* Chem Dept Lafayette Col Easton PA 18042

CROCKET, JAMES HARVIE, b Fredericton, NB, June 27, 32; m 58; c 2. GEOCHEMISTRY, GEOLOGY. *Educ:* Univ NB, BSc, 55; Oxford Univ, BSc, 57; Mass Inst Technol, PhD(geochem), 61. *Prof Exp:* From asst prof to assoc prof, 61-74, PROF GEOL, MCMASTER UNIV, 74- *Res:* Neutron activation studies of the geochemistry of precious metals in basic rocks; genesis of ore deposits. *Mailing Add:* Dept Geol McMaster Univ Hamilton ON L8S 4L8 Can

CROCKETT, ALLEN BRUCE, b Ft Worth, Tex, June 8, 44; m 82. ENVIRONMENTAL IMPACTS, ENVIRONMENTAL LAW. *Educ:* Univ Okla, BS, 67; Univ Colo, PhD(ecol), 75, Univ Denver, JSD, 81. *Prof Exp:* Lectr, Univ Colo Boulder, 74-75; ecologist & geologist, Environ Consult, 75-77; ecologist & geologist, CDM Environ Consult, 77-80, Western Resource Develop, 80-85; ecologist & geologist, Morrison-Knudsen Engrs, 85-90; ENVIRON CONSULT & NAT RESOURCE MGR S M STOLLER, 90- *Concurrent Pos:* Lectr, Univ Colo, Boulder, 86 & 87. *Mem:* Ecol Soc; Am Ornith Union; Cooper Ornith Soc; Wilson Ornith Soc; Geol Soc Am. *Res:* Avian behavioral ecology; woodpeckers; habitat utilization; restoration of wildlife habitat; sedimentary stratigraphy. *Mailing Add:* 3372 16th St Boulder CO 80304

CROCKETT, DAVID JAMES, b Warrensberg, MO, Nov 10, 42; Can citizen; div; c 1. NEUROPSYCHOLOGY. *Educ:* Whitman Col, BA, 64; Univ Brit Columbia, MA, 68; Univ Victoria, PhD(psychol), 72. *Prof Exp:* ASSOC PROF PSYCHOL, UNIV BRIT COLUMBIA, 79- *Concurrent Pos:* Psychologist, Dept Psychol, Univ Hosp, Univ Brit Columbia Site, 79- *Mem:* Can Psychol Assoc; Biofeedback Soc Am; Acad Aphasia; Int Neuropsychol Soc; Am Psychol Asn. *Res:* Investigation of neuropsychological basis of psychiatric disorders; description of changes in memory functioning associated with Alzheimer's disease; evaluation of treatment effectiveness on chronic pain conditions. *Mailing Add:* Univ Brit Columbia Dept Psychiat 2255 Wesbrook Mall Vancouver BC V6T 2B5 Can

CROCKETT, JERRY J, b Chickasha, Okla, May 25, 28; m 51; c 3. PLANT ECOLOGY, ENVIRONMENTAL PLANNING. *Educ:* Northwestern Okla State Univ, BS, 51; Ft Hays Kans State Univ, MS, 60; Univ Okla, PhD(bot & ecol), 62. *Prof Exp:* Chemist, Continental Oil Co, 51-55 & Tretolite Co, 55-59; from asst to assoc prof ecol, Okla State Univ, 62-67; assoc dean, Col Lett & Sci, Univ Idaho, 67-68; dir exten, Col Arts & Sci, 68-77, PROF ECOL & BIOL, OKLA STATE UNIV, 77- *Mem:* Ecol Soc Am; Am Soc Range Mgt; Grassland Res Found (secy, 62-64); Sigma Xi. *Res:* Mechanisms of secondary succession; productivity of grasslands; strip-mine reclamation; terrestrial pollution; reclamation of abandoned lands; efficiency of nitrogen fixation in native legumes. *Mailing Add:* Dept Bot Okla State Univ Stillwater OK 74074

CROCTOGINO, REINHOLD HERMANN, b Teplitz, Czechoslovakia, Aug 2, 42; m; c 2. PULP & PAPER TECHNOLOGY. *Educ:* Univ BC, BASc, 66; McGill Univ, PhD(chem eng), 71. *Prof Exp:* Res engr, Voith Gesellschaft Mit Beschraenkter Haftung, 71-76; SR SCIENTIST & DIV DIR, PULP & PAPER RES INST CAN, 76- *Concurrent Pos:* Auxiliary prof chem eng, McGill Univ, 76- *Honors & Awards:* H I Weldon Medal, Can Pulp & Paper Asn, 80. *Mem:* Can Pulp & Paper Asn; Tech Asn Pulp & Paper Indust. *Res:* Physical processes used in the manufacture of pulp and paper; fundamentals of calendering and drying of paper, pulp washing; developing new calendering and pulp washing techniques; process control; process simulation; sensor development. *Mailing Add:* Pulp & Paper Res Inst Can 570 St John's Blvd Pointe Claire PQ H9R 3J9 Can

CROFFORD, OSCAR BLEDSOE, b Chickasha, Okla, Mar 29, 30; m 57; c 3. MEDICAL RESEARCH, INTERNAL MEDICINE. *Educ:* Vanderbilt Univ, AB, 52, MD, 55. *Prof Exp:* Intern med, Hosp, Vanderbilt Univ, 55-56, asst resident, 56-57, USPHS res fel clin physiol, Univ, 59-62, resident med, Hosp, 62-63; USPHS fel clin biochem, Univ Geneva, 63-65; from asst prof to assoc prof, 65-74, PROF MED, SCH MED, VANDERBILT UNIV, 74-, ASSOC PROF PHYSIOL, 70- *Concurrent Pos:* Investr, Howard Hughes Med Inst, 65-71; mem, Metab Study Sect, NIH, 70-74, chmn, 72-74; Addison B Scoville, Jr Chair Diabetes & Metab, Sch Med, Vanderbilt Univ, 73-, div head diabetes & metab, Dept Med, 73-, dir, Diabetes-Endocrinol Ctr, 73-78; chmn, Nat Comn Diabetes, 75-76; dir, Diabetes Res & Training Ctr, 78- *Honors & Awards:* Lilly Award, Am Diabetes Asn, 70, Charles H Best Award, 76; Humanitarian Award, Juv Diabetes Found, 76. *Mem:* Am Diabetes Asn (pres, 81); Am Physiol Soc; Endocrine Soc; Am Soc Clin Invest; Asn Am Physicians. *Res:* Hormone control of metabolism in adipocytes; mechanism of action of insulin; sugar transport; pathophysiology and treatment of Diabetes Mellitus; pathophysiology and treatment of obesity. *Mailing Add:* Dept Med & Physiol Vanderbilt Univ Sch Med A-5119 Med Ctr Nashville TN 37232

CROFT, ALFRED RUSSELL, ecology; deceased, see previous edition for last biography

CROFT, BARBARA YODER, b Port Chester, NY, Aug 11, 40; m 77. NUCLEAR MEDICINE, PHYSICAL CHEMISTRY. *Educ:* Swarthmore Col, BA, 62; Johns Hopkins Univ, MA, 64, PhD(phys chem), 67. *Prof Exp:* Sr scientist, Johnston Labs, 67-68; programmer, Comput Sci Ctr, 68, from instr to asst prof, 69-87, ASSOC PROF RADIOL, SCH MED, UNIV VA, 87- *Concurrent Pos:* Res partic, Med Div, Oak Ridge Assoc Univs, Inc, 72; expert, Int Asn Educ Assessment, 84- *Mem:* Soc Nuclear Med; Am Chem Soc; Am Asn Phys Med; Am Col Nuclear Physicians; AAAS. *Res:* Computers in nuclear medicine; xenon ventilation and perfusion studies; spectroscopy. *Mailing Add:* Dept Radiol Box 170 Univ Va Sch Med Charlottesville VA 22908

CROFT, BRUCE, b Melbourne, Australia, 52. COMPUTER SCIENCE. *Educ:* Monash Univ, BSc, 73, MSc, 74; Univ Cambridge, PhD(computer sci), 79. *Prof Exp:* From asst prof to assoc prof, 79-91, PROF, COMPUTER INFO SCI DEPT, UNIV MASS, AMHERST, 91- *Concurrent Pos:* Prof computer sci, Univ Col, Dublin, 85-86; chair, SIG Info Retrieval, Asn Computer Mach, 87- *Mem:* Asn Computer Mach. *Mailing Add:* Dept Computer Sci Lederle Grad Res Ctr Univ Mass Amherst MA 01003

CROFT, CHARLES CLAYTON, b Washington, DC, June 16, 14; m 44; c 2. MICROBIOLOGY, PUBLIC HEALTH. *Educ:* Univ Md, BS, 36, MS, 37; Johns Hopkins Univ, ScD(bact), 49; Am Bd Microbiol, dipl. *Prof Exp:* Lab technician, Md State Health Dept, 37; sr bacteriologist, Ariz State Lab, 38-43; asst chief labs, Ohio Dept Health, 49-61, chief div labs, 61-81; RETIRED. *Concurrent Pos:* Consult, WHO, 60; chmn, Intersoc Comt Lab Serv Related to Health, 61-64; Medicare Lab Prog, Ohio Dept Health, 81- *Mem:* Am Soc Microbiol; Fel Am Pub Health Asn; Fel Am Acad Microbiol. *Res:* Public health laboratory administration; virology; syphilis serology; enteric bacteriology; hemolytic antibody. *Mailing Add:* 149 Leland Ave Columbus OH 43214-1511

CROFT, GEORGE THOMAS, b Washington, DC, Sept 29, 26; m 48; c 3. TECHNICAL MANAGEMENT, RESOURCE MANAGEMENT. *Educ:* Western Md Col, BS, 48; Univ Pa, PhD(physics), 53. *Prof Exp:* Physicist, Sound Div, Naval Res Lab, 47-48; physicist, Frankford Arsenal, 49-50, Frankford Arsenal, 49-50; res physicist, Edison Lab, McGraw Edison, Inc, 53-58; supvr appl res, Pitney-Bowes Inc, 58-61, mgr, Appl Res Lab, 61-68, dir corp res & develop, Addressograph Multigraph Corp, 71, vpres, dir res & develop, 72, vpres corp res & develop, Ohio, 75-76; vpres res & develop, Optical Prod Div, Am Optical Corp, 76-80; CONSULT, 80-; HEAD MATH & SCI DEPT, SAVANNAH TECH INST, GA, 88- *Concurrent Pos:* Mem, Res & Develop Coun, Am Mgt Asn, 75-80, Corp Assocs Adv Comt, Am Inst Physics, 77-80 & dean's adv coun, Eng Sch, Univ Mass, 77-83; pres, Technol Resources Mgt Group. *Mem:* Am Phys Soc; Inst Elec & Electronic Engrs; Electrochem Soc; Am Mgt Asn; Am Inst Physics; Sigma Xi. *Res:* Low temperature physics; semiconductors; mechanisms of electrochemical reactions; mathematical analysis; graphics communication systems; medical instruments; automated systems design; specialized systems controlled with personal computers and software. *Mailing Add:* 39 Acorn Lane Hilton Head Island SC 29928

CROFT, PAUL DOUGLAS, b Ft Erie, Ont, Oct 16, 37; US citizen; m 61; c 3. NUCLEAR CHEMISTRY. *Educ:* Univ Western Ont, BSc, 59; Univ Calif, Berkeley, PhD(chem), 64. *Prof Exp:* Res assoc & instr chem, State Univ NY Stony Brook, 64-66, lectr chem & dir chem labs, 66-72; dean for admin, Univ Mich, Flint, 72-75; EXEC OFFICER CHEM, UNIV CALIF, SAN DIEGO, 75- *Concurrent Pos:* Res collab, Brookhaven Nat Labs, 64-66. *Mem:* AAAS; Am Chem Soc; Chem Inst Can. *Mailing Add:* Chem Dept B 032 Univ Calif San Diego La Jolla CA 92093-0332

CROFT, THOMAS A(RTHUR), b Denver, Colo, Feb 15, 31; m 64; c 3. ELECTRICAL ENGINEERING. *Educ:* Dartmouth Col, BA, 53, MS, 54; Stanford Univ, PhD(elec eng), 64. *Prof Exp:* Elec engr, Convair/Astronaut, Gen Dynamics Corp, 57-59; adj prof, Radiosci Lab, Stanford Univ, 59-75; PRIN SCIENTIST, SRI INT, 75- *Concurrent Pos:* Consult, Stanford Res Inst, 65-75, Appl Tech Inc, 65-75, Barry Res Corp, 68-75 & Nat Acad Sci, 71;; Assoc ed, Radio Sci, US Nat Comt-Int Sci Radio Union Comn III. *Mem:* Am Geophys Union; AAAS Div Planetary Sci; Inst Elec & Electronics Engrs; Soc Motion Picture & TV Engrs; Inst Union Radio Sci. *Res:* Ionospheric physics; simulation of the performance of radar systems by means of digital-computer wave propagation analysis and synthesis; solar wind density and planetary atmospheric structure measurement by radio propagation to spacecraft. *Mailing Add:* 76 Moulton Dr Atherton CA 94027

CROFT, THOMAS STONE, b Marfa, Tex, May 9, 38. ORGANIC CHEMISTRY. *Educ:* Univ Fla, BS, 61; Univ Colo, PhD(org chem), 66. *Prof Exp:* Sr chemist, 66-71, res specialist, 71-77, PROD DEVELOP SPECIALIST, 3M CO, 77- *Mem:* Am Chem Soc. *Res:* Organic fluorine chemistry; heterocycles, chemical specialty formulations, including emulsions, polishes, abrasion resistant and fire retardant coatings; polymeric coatings and surfaces. *Mailing Add:* 11103 Centennial Trail Austin TX 78726-1411

CROFT, WALTER LAWRENCE, b Longview, Miss, June 28, 35; m 68; c 1. PHYSICS. *Educ:* Miss State Univ, BS, 57; Vanderbilt Univ, PhD(physics), 64. *Prof Exp:* From asst prof to assoc prof, 62-71, PROF PHYSICS, MISS STATE UNIV, 71- *Mem:* Int Solar Energy Soc; Am Asn Physics Teachers. *Res:* Nuclear spectroscopy; solar energy; microwave spectroscopy. *Mailing Add:* Dept Physics Miss State Univ Mississippi State MS 39762

CROFT, WILLIAM JOSEPH, b New York, NY, Nov 29, 26; m 49; c 2. CRYSTALLOGRAPHY. *Educ:* Columbia Univ, BS, 50, MA, 52, PhD(mineral, crystallog), 54. *Prof Exp:* Asst mineral, Columbia Univ, 52-54; instr, Hofstra Col, 54-55; staff mem crystallog, Lincoln Lab, Mass Inst Technol, 55-57; sr staff mem, Radio Corp Am, 57-61; staff mem, Sperry Rand Res Ctr, 61-69; RES CHEMIST, ARMY MAT & MECH RES CTR, 69- *Concurrent Pos:* Vis prof, Brown Univ, 70-72 & 80-86. *Mem:* Electron Micros Soc Am; fel Mineral Soc Am; Am Crystallog Asn; Mineral Soc Gt Brit & Ireland; Sigma Xi. *Res:* X-ray crystallographic studies of oxides; high and low temperature phase transformations; lattice distortion and crystallite size in multiple oxides; electron microscopy and diffraction. *Mailing Add:* Army Mat Technol Res Lab Watertown MA 02172

CROFTS, ANTONY R, b Harrow, Middlesex, Eng, Jan 26, 40; m; c 3. PHOTOSYNTHETIC ELECTRON TRANSPORT. *Educ:* Cambridge Univ, BA, 61, PhD(biochem), 65. *Prof Exp:* PROF BIOPHYS, UNIV ILL, URBANA-CHAMPAGNE, 78- *Concurrent Pos:* Lectr biochem, Univ Bristol, UK, 66-72, reader biochem, 72-77; Guggenheim fel, 85-86; univ scholar, Univ Ill Urbana-Champaign, 89-92. *Mem:* Am Soc Biochem &

Molecular Biol; Biophys Soc; fel AAAS. *Res:* Research on structure/function relationships in photosynthetic energy conversion and applications of biophysical methods to photosynthesis in intact plants: function and structure of UQH2: cyt c2 oxidoreductase of Rb sphaeroides; mechanism of the cyt b6/f complex of green plants; mechanism of the two-electron gate of photosystem II, and inhibition by herbicides; role of thioredoxin in activation and control of chloroplast ATP-ase and other enzymes; photosynthesis in intact plants, - a biophysical approach through novel instrumentation; molecular engineering of the enzymes of photosynthetic electron transport; control of photosynthetic electron transfer in the coupled steady-state; mechanism of UV-photo inhibition; prediction of structure in membrane proteins. *Mailing Add:* Biophys Div 156 Davenport Hall Univ Ill 607 S Mathews Urbana IL 61801

CROFTS, GEOFFREY, b Winnipeg, Man, May 12, 24; m 49; c 3. ACTUARIAL MATHEMATICS. *Educ:* Univ Man, BCom, 46. *Prof Exp:* Assoc prof actuarial math, Univ Man, 49-55 & Occidental Col, 55-59; actuarial training dir, Occidental Life Ins Co Calif, 59-64; dean & dir, Grad Sch Actuarial Sci, Northeastern Univ, 64-81; AT DEPT ACTUARIAL STUDIES, UNIV HARTFORD, CONN, 82- *Mem:* Fel Soc Actuaries; Casualty Actuarial Soc. *Res:* Actuarial science. *Mailing Add:* Dept Actuarial Studies Univ Hartford 200 Bloomfield Ave West Hartford CT 06117-1599

CROKE, EDWARD JOHN, b Oak Park, Ill, Nov 13, 35; m 65; c 2. COMPUTER APPLICATIONS, TECHNICAL MANAGEMENT. *Educ:* Univ Ill, BS, 57; Princeton Univ, MS, 60; Nogoya Univ, DEng, 77; Univ Chicago, MBA, 81. *Prof Exp:* Res asst aerospace eng, James Forrestal Res Ctr, Princeton Univ, 58-60; tech proj officer aerospace tech, US Air Force, 60-63; res engr nuclear reactor devices, Reactor Eng Div, 63-66, energy engr, Advanced Concepts Group, 66-67, sect mgr natural resources & environ studies, Environ Natrual Res Sect, 67-69, dep dir, Ctr Environ Studies, 69-71, DIV DIR, ENERGY, ENERGY & ENVIRON SYSTS DIV, ARGONNE NAT LAB, 71- *Concurrent Pos:* Adj prof, Dept Energy Eng, Univ Ill, 77-; mem tech adv comt, Gas Res Inst, 80-; consult, Ill Dept Energy & Natural Resources, 81- *Mem:* Sigma Xi. *Res:* Technical and economic assessment of energy supply, conservation technologies, environmental control technologies, and polices; micro and mainframe computer data base management, mathematical modeling and systems analysis. *Mailing Add:* 815 Forest Ave River Forest IL 60305

CROKER, BYRON P, b Pittsburg, Pa, Nov 23, 45. MEDICAL & HEALTH SCIENCES. *Educ:* Univ Pittsburgh, BS, 67; Duke Univ, MD, 71, PhD(path), 71. *Prof Exp:* Path internship, Duke Univ Med Ctr, 71-72; fel immunopath, Scripps Clin & Res Found, La Jolla, Calif, 72-76; path residency, Duke Univ Med Ctr, 76-77, asst prof, 77-82; ASSOC PROF, UNIV FLA, GAINESVILLE, 82-; CHIEF LAB SERV, VET ADMIN MED CTR, GAINESVILLE, 88- *Concurrent Pos:* Fel, Am Cancer Soc, 74-76; staff physician, Vet Admin Hosp, Durham, NC, 77-82; mem, Duke Comprehensive Cancer Ctr, 80-82. *Mem:* Int Acad Path; Am Asn Pathologists; Am Soc Nephrol; AAAS. *Res:* Pathology of human tumors; pathology of human transplantation. *Mailing Add:* Lab Serv 113 Vet Admin Med Ctr Gainesville FL 32608-1197

CROKER, ROBERT ARTHUR, b New York, NY, Sept 4, 32; m 72; c 2. ENVIRONMENTAL HISTORY, NATURAL RESOURCES. *Educ:* Adelphi Col, AB, 58; Univ Miami, MS, 60; Emory Univ, PhD(biol), 66. *Prof Exp:* Asst fisheries, Univ Miami, 58-59; marine biologist, Bur Sport Fisheries & Wildlife, US Dept Interior, 60-62 & Bur Commercial Fisheries, 62-64; res asst, Marine Inst, Univ Ga, 64; from asst to prof zool, 66-88, PROF ENVIRON HISTORY & CONSERV, UNIV NH, 88- *Concurrent Pos:* Res asst Oyster Lab, Rutgers Univ, 60; panel mem, Nat Res Coun, 73-77; vis prof, Univ Austral, Chile, 82. *Mem:* Fel AAAS; Ecol Soc Am; Sigma Xi; Am Soc Environ Hist. *Res:* History of American ecology and conservation. *Mailing Add:* Dept Natural Resources James Hall Univ NH Durham NH 03824

CROLEY, THOMAS EDGAR, b Gladewater, Tex, Apr 30, 40; m 64; c 2. ANATOMY, MICROBIOLOGY. *Educ:* ETex State Univ, BS, 63, MS, 64; Southwestern Univ, cytotechnol degree, 66; Baylor Univ, PhD(anat), 71. *Prof Exp:* Adj prof, NTex State Univ, 71-72; asst prof anat, La State Univ Med Ctr, 72-77; asst prof anat, Tex Col Osteop Med, 77-; AT DEPT PHYS THERAP, TEX WOMAN'S UNIV. *Concurrent Pos:* NASA-Manned Spacecraft Ctr fel, 71-72. *Mem:* Am Soc Clin Path; Int Asn Dent Res. *Res:* Calcified tissue research, particularly developing teeth; hormone relationship to epithelial development; placenta morphology. *Mailing Add:* 8194 Walnut Hill Lane Dallas TX 75231

CROLL, IAN MURRAY, b Regina, Sask, Dec 7, 29; m 55; c 2. PHYSICAL CHEMISTRY. *Educ:* Univ Man, BSc, 50, MSc, 58; Univ Calif, Los Angeles, PhD(chem), 58. *Prof Exp:* Chemist, Defence Res Bd Can, 51-52; chemist fed systs div, Int Bus Mach Corp, NY, 58-61, mem res staff, Watson Res Ctr, 61-63, mgr chem res, 63-67, mgr mem & storage res, 67-72, dir technol, 72-73, MGR MAT TECHNOL, IBM CORP, 73- *Mem:* Am Chem Soc; Inst Elec & Electronics Engrs; Electrochem Soc. *Res:* Thermodynamics; solution chemistry; electrochemistry; electrodeposition; structure and magnetic properties of metal films. *Mailing Add:* 1235 Ocean View Blvd Pacific Grove CA 93950-2789

CROMARTIE, THOMAS HOUSTON, b Raleigh, NC, Aug 9, 46. BIOCHEMISTRY. *Educ:* Duke Univ, BS, 68; Mass Inst Technol, PhD(chem), 73. *Prof Exp:* Res assoc, Mass Inst Technol, 73-75; asst prof chem, Univ Va, 75-81; supvr, Stauffer Chem Co, 81-87; MGR, ICI AMERICAS, 88- *Mem:* Am Chem Soc; Sigma Xi; AAAS; Photochem Soc N Am. *Res:* Reaction mechanisms of enzymes, especially flavoenzymes and isomerases; rational design of herbicides and insecticides; heavy atom and hydrogen kinetic isotope effects; pesticide metabolism. *Mailing Add:* Jealotts Hill Res Sta Berkshire RG 126 EY England

CROMARTIE, WILLIAM JAMES, b Garland, NC, May 19, 13; m 45; c 5. MICROBIOLOGY, INFECTIOUS DISEASES. *Educ:* Emory Univ, MD, 37; Am Bd Path, dipl, 48, Am Bd Internal Med, dipl, 51. *Prof Exp:* Intern, Grady Hosp, Emory Univ, 37-38; asst resident path, Vanderbilt Univ Hosp, 38-39, instr, Sch Med, 39-41; asst resident med, Bowman-Gray Sch Med, 42; dir res lab & asst chief med serv, Vet Admin Hosp, 46-49; assoc prof med & bact, Med Sch, Univ Minn, 49-51; from asoc prof to prof med & bact, 51-85, assoc dean clin sci, 69-82, EMER PROF, MED & BACT, SCH MED, UNIV NC, CHAPEL HILL, 85- *Concurrent Pos:* Instr, Med Col, Southwestern Univ, 47-49; mem, Adv Panel Microbiol, Off Naval Res, 50-55, Bact Test Comt, Nat Bd Med Examrs, 61-69; chief staff, NC Mem Hosp, 69-74; mem, Infectious Dis Adv Comt, Nat Inst Allergy & Infectious Dis, 71-75. *Mem:* Am Acad Microbiol; fel Am Col Physicians; Am Asn Path & Bact; Am Soc Exp Path; Am Soc Microbiol. *Res:* Pathogenesis of rheumatoid arthritis and rheumatic fever. *Mailing Add:* 204 Weaver Rd Glendale Chapel Hill NC 27514

CROMARTIE, WILLIAM JAMES, JR, b Dallas, Tex, Oct 14, 47. ECOLOGY, NATURAL HISTORY. *Educ:* St Johns Col, AB, 69; Cornell Univ, PhD(ecol), 74. *Prof Exp:* Asst, 74-80, dir, Ctr Environ Res, 79-91, ASSOC PROF ENVIRON STUDIES, STOCKTON STATE COL, 80- *Mem:* AAAS; Ecol Soc Am; Entom Soc Am; Brit Ecol Soc. *Res:* Biogeography and evolution of plant-arthropod associations; coastal plain biogeography, especially pine barrens and pocosins; insects associated with pines; endangered and threatened plants and animals. *Mailing Add:* Nams Dept Stockton State Col Jim Leeds Rd Pomona NJ 08240

CROMBIE, DOUGLASS D, b Alexandra, NZ, Sept 14, 24; US citizen; m 51. AERONAUTICS, RADIO WAVE PROPAGATION. *Educ:* Univ Otago, NZ, BS, 47, MS, 49. *Prof Exp:* Assoc adminr, Nat Telecommun & Info Admin, 71-80, chief scientist, 80-85; SR ENGR, AEROSPACE CORP, 85- *Concurrent Pos:* Ed, Radio Sci, 67. *Honors & Awards:* Gold Medal Award, US Dept Com, 70. *Mem:* Nat Acad Eng; Inst Elec & Electronic Engrs; Union Radio Sci Int. *Res:* Scattering of radio waves by the ocean surface; multi-mode and nonreciprocal propagation of very low frequency signals in the earth-ionosphere waveguide; microwave propagation through high altitude nuclear burst environments. *Mailing Add:* The Aerospace Corp PO Box 92957 M4-928 Los Angeles CA 90009

CROMER, ALAN H, b Chicago, Ill, Aug 15, 35; wid. SCIENCE WRITING, PHYSICS. *Educ:* Univ Wis, BS, 54; Cornell Univ, PHD(theoret physics), 60. *Prof Exp:* Res fel physics, Harvard Univ, 59-61; from asst prof to assoc prof, 61-70, PROF PHYSICS, NORTHEASTERN UNIV, 70- *Concurrent Pos:* Pres, Edutech, Inc, Newton, 80- *Mem:* AAAS; Am Phys Soc; Am Asn Physics Teachers. *Res:* Physics for the life sciences; theoretical mechanics; physics in science and industry; educational software. *Mailing Add:* 1173 Commonwealth Ave Allston MA 02134

CROMER, DON TIFFANY, b South Bend, Ind, July 27, 23; m 45; c 5. X-RAY CRYSTALLOGRAPHY. *Educ:* Univ Wis, BS, 47, PhD(chem), 51. *Prof Exp:* Chemist, Nat Lead Co, 50-52; mem staff, Los Alamos Nat Lab, 52-; RETIRED. *Concurrent Pos:* Vis scientist, Puerto Rico Nuclear Ctr, 65, Ctr Nat Sci Res, Grenoble, France, 72-73; prog mgr, Div Mat Sci, Dept Energy, 80-82; consult, Los Alamos Nat Lab, 85- *Honors & Awards:* E O Lawrence Award, Atomic Energy Comn, 69. *Mem:* Am Chem Soc; Am Crystallog Asn. *Res:* X-ray crystallography. *Mailing Add:* 11052 Academy Ridge Rd NE Albuquerque NM 87111

CROMER, JERRY HALTIWANGER, b Anderson, SC, Apr 4, 35; m 62. DEVELOPMENTAL BIOLOGY, PHYSIOLOGY. *Educ:* Wofford Col, BS, 57; Univ SC, MS, 65; Vanderbilt Univ, PhD(develop biol), 68. *Prof Exp:* Asst prof, 68-74, ASSOC PROF BIOL, CONVERSE COL, 74- *Concurrent Pos:* Fac res grant, 68-69. *Mem:* Sigma Xi. *Res:* Quantitative analysis of impulse transmission in the aortic depressor nerve of rabbits; development of hepatic xanthine dehydrogenase in hatching chick embryos; histochemistry; enzyme development; developmental genetics and physiology. *Mailing Add:* Dept Biol Converse Col Spartanburg SC 29301

CROMER, JOHN A, b Logansport, Ind, Nov 29, 38; m 58; c 4. PHYSIOLOGY. *Educ:* Taylor Univ, BA, 62; Ball State Univ, MS, 68; Univ NDak, PhD(physiol), 72. *Prof Exp:* Asst prof biol, Pasadena Col, 71-73; chmn dept biol, Point Loma Col, 73-74; asst dean & health prof adv, Duke Univ, 76-78; asst dean student affairs, Oral Roberts Univ, 78, asst dean med educ, 78-80; VPRES ACAD AFFAIRS & DEAN, ILL COL OPTOM, 80- *Concurrent Pos:* Partic, Human Genetics & Societal Probs, NSF Chautauqua Course, 72-73. *Mem:* Sigma Xi; AAAS; Undersea Med Soc. *Res:* Central nervous system during exposure to hyperbaric environments; described high pressure nervous syndrome from sterotaxically implanted electrodes in the basal ganglia to provide evidence of the origin of EEg changes. *Mailing Add:* 3241 S Mich Ave Ill Col Optom Chicago IL 60616

CROMPTON, ALFRED W, b Durban, SAfrica, Feb 21, 27; m 54; c 3. VERTEBRATE PALEONTOLOGY. *Educ:* Univ Stellenbosch, BSc, 47, MSc, 49, PhD(zool), 51; Cambridge Univ, PhD(paleont), 53. *Prof Exp:* Cur fossil vertebrates, Nat Mus, Bloemfontein, SAfrica, 54-56; dir, SAfrican Mus, Cape Town, 56-64; dir, Peabody Mus Natural Hist & prof biol & geol, Yale Univ, 64-70; dir, Mus Comp Zool, 70-83, FISHER PROF NATURAL HIST, HARVARD UNIV, 84- *Concurrent Pos:* Lectr, Univ Cape Town, 58-64. *Mem:* Fel Zool Soc London; fel Am Acad Arts & Sci; Soc Vertebrate Paleont; Am Soc Zool. *Res:* Evolution of mammals and dinosaurs; functional anatomy of feeding in mammals. *Mailing Add:* Mus Comp Zool Harvard Univ Cambridge MA 02138

CROMPTON, CHARLES EDWARD, b St George Island, Alaska, Oct 8, 22; m 45; c 2. PHYSICAL CHEMISTRY. *Educ:* Univ Calif, Berkeley, BS, 43; Univ Tenn, PhD(chem), 49. *Prof Exp:* Res chemist, Manhattan Proj, Berkeley, Calif & Oak Ridge, Tenn, 43-46; tech reviewer, Isotopes Div, USAEC, 49-51; dep dir, 53-56; dir radioactivity div, US Testing Co, 51-53;

assoc tech dir, Nat Lead Co Ohio, 56-60; dir nuclear chem, Martin-Nuclear Div, Martin Marietta Corp, 60-62; tech dir pigments div, Chem Group, Glidden Co, 62-63, dir res, Consol Inorg Res, 63-66; dir res inorg div, Chemetron Corp, 66-74, dir develop, Chem Group, 74-77, dir res & develop, Chem Prod Div, 77-80; PRES, CPI CONSULTS, INC, 80- *Concurrent Pos:* Consult, US Atomic Energy 51-52 & Am Med Asn, 54-55; mem sci adv bd, New Eng Nuclear Corp, 56-; consult, Martin Co, Md, 62-63; mem comt on depleted uranium, Nat Mat Adv Bd, 70-71. *Mem:* Fel Am Inst Chem; Am Chem Soc; Asn Res Dirs; Com Develop Asn; Sigma Xi. *Res:* Nuclear chemistry; homogeneous and heterogeneous catalysis; kinetics surface chemistry of inorganic solids, metals and ceramics; catalytic hydrogenation; physical chemistry and metallurgy of uranium and plutonium; pigments; fire retardants; plastic additives; phosgene and phosgene derivatives. *Mailing Add:* 923 Portsmouth Circle Maryville TN 37801-6794

CROMPTON, DAVID WILLIAM THOMASSON, b Bolton, Lancashire, UK, Dec 5, 37; m 62; c 3. EPIDEMIOLOGY, HELMINTHOLOGY. *Educ:* Univ Cambridge, BA, 61, MA, 64, PhD(parasitol), 64, ScD(parasitol), 77. *Prof Exp:* Fel biol sci, Sidney Sussex Col, Cambridge, 64-85; PROF ZOOL, UNIV GLASGOW, 85- *Concurrent Pos:* Lectr parasitol, Univ Cambridge, 68-85; co-ed, Parasitol, 72-82; adj prof nutrit sci, Cornell Univ, Ithaca, NY, 81-; mem, Expert Comt Parasitol, WHO, 81-, Aquatic Life Sci Comt, Nat Environ Res Coun, 81-84, Biol Sect Comt, Royal Soc Edinburgh, 90-; vis prof, Univ Nebr, Lincoln, 82; head, Collab Ctr Ascariasis, WHO, 89- *Honors & Awards:* Sci Medal, Zool Soc London, 77. *Mem:* Fel Royal Soc Edinburgh; Brit Soc Parasitol; Am Soc Parasitol; Royal Soc Trop Med & Hyg; Am Inst Nutrit; fel Zool Soc London. *Res:* Effects of intestinal helminth infections on human nutritional status with emphasis on Ascaris lumbricoides. *Mailing Add:* Dept Zool Univ Glasgow Glasgow G12 8QQ Scotland

CROMROY, HARVEY LEONARD, b Lawrence, Mass, Jan 5, 30; m 57, 70; c 4. RADIOBIOLOGY, ENTOMOLOGY. *Educ:* Northeastern Univ, BSc, 51; NC State Col, PhD(entom), 58. *Prof Exp:* Entomologist, NC State Bd Health, 56-57; assoc prof entom & biophys, Col Agr, Univ PR, 57-60; res scientist, Oak Ridge Nat Lab, 60-61; sci adminr radiol health, USPHS, 61-64; assoc prof nuclear sci, 64-70, PROF ENTOM & NEMATOL, UNIV FLA, 70- *Concurrent Pos:* Adv mem on agr, Southern Interstate Nuclear Bd, 73-81; consult, Int Atomic Energy Agency, Vienna, Austria, 76-85; guest prof, Inst Appl Zool, Univ Bonn, WGer, 78. *Mem:* Enom Soc Am; Am Acarological Soc. *Res:* Taxonomy of plant feeding mites; uses of isotopes in entomology; mites of honey bees. *Mailing Add:* Dept Entom & Nematol Bldg 970 Hull Rd Univ Fla Gainesville FL 32611-0740

CROMWELL, FLORENCE S, b Lewiston, Pa, May 14, 22. OCCUPATIONAL THERAPY. *Educ:* Miami Univ, BS, 43; Washington Univ, St Louis, MA, 49. *Prof Exp:* Mem staff therapist, La Co Gen Hosp, 49-53; occup therapist, Goodwill Indust, 54-55; res therapist, United Cerebral Palsy Asn, 56-60; dir occup ther, Orthopedic Hosp, La, 61-67; consult educ & prog develop, Los Angeles Job Corps Ctr, 76-78; CONSULT, 89- *Concurrent Pos:* Part-time instr occup ther, Univ Southern Calif, 52-61, assoc proj occup ther, 70-76, actg chmn, 73-76; emer ed, Occup Ther Healthcare, 84-89. *Mem:* Inst Med-Nat Acad Sci; fel Am Occup Ther Asn. *Mailing Add:* 1179 Yocum St Pasadena CA 91103

CROMWELL, GARY LEON, b Salina, Kans, Oct 6, 38; m 60; c 3. ANIMAL NUTRITION. *Educ:* Kans State Univ, BS, 60; Purdue Univ, MS, 65, PhD(animal nutrit), 67. *Prof Exp:* Teacher high sch, Kans, 60-64; res asst animal nutrit, Purdue Univ, 64-67; from asst prof to assoc prof, 67-76, PROF ANIMAL SCI, UNIV KY, 76- *Honors & Awards:* UK Res Award, 81; Animal Nutrit Res Award, Am Feed Indust Asn, 85. *Mem:* Am Soc Animal Sci; Brit Soc Animal Prod; Am Soc Prof Animal Scientists. *Res:* Mineral and amino acid nutrition of swine; efficacy and safety of feed additives. *Mailing Add:* Dept Animal Sci Univ Ky Lexington KY 40506

CROMWELL, LESLIE, b Manchester, Eng, Apr 2, 24; m 56; c 3. BIOMEDICAL ENGINEERING. *Educ:* Univ Manchester, BScTech, 43, MScTech, 61; Univ Calif, Los Angeles, MS, 51, & PhD, 67. *Prof Exp:* Res engr, Salford Elec Instruments, 43-44; design engr, Aircraft Elec Div, English Elec Co, 44-46; chief engr elec eng, B Cromwell & Co, Ltd, Eng, 46-48; lectr eng, Univ Calif, Los Angeles, 48-53; from asst prof to prof eng, Cailf State Univ, Los Angeles, 53-80, head dept, 55-64, chmn dept interdisciplinary eng, 68-86, dean sch eng, 77-80; RETIRED. *Concurrent Pos:* Consult, Ed Develop Prog, Univ Calif, Los Angeles, 62-64. *Mem:* Inst Elec & Electronics Engrs; Am Soc Eng Educ. *Res:* Aircraft electrical equipment design; industrial electronics; electric power equipment; electric machinery; control systems; biomedical instrumentation. *Mailing Add:* 27661 Via Graanados Mission Viejo CA 92692

CROMWELL, NORMAN HENRY, b Terre Haute, Ind, Nov 22, 13; m 55; c 2. ORGANIC CHEMISTRY, CHEMOTHERAPY. *Educ:* Rose Hulman Inst, BS, 35; Univ Minn, PhD(org chem), 39. *Hon Degrees:* DSc, Univ Nebr, 87. *Prof Exp:* Asst chem, Univ Minn, 35-39; from instr to prof org chem, 39-60, mem res coun, 47-49, mem grad coun, 51-60, Howard S Wilson prof chem, 60-70, chmn dept, 64-70, vpres grad studies & res, Univ Nebr Syst, 72-73; interim dir, Eppley Inst Cancer Res Med Ctr, 79-83; regents prof, 73-84, EMER PROF CHEM, UNIV NEBR-LINCOLN, 84- *Concurrent Pos:* Consult, Parke, Davis & Co, 44-45, Smith Kline & French, Pa, 46-49, USPHS, 53-, Philip Morris, Inc & Nat Cancer Inst, 54-65; Fulbright scholar & Guggenheim fel, 50 & 58; hon res assoc, Univ Col, Univ London, 50-51, 58-59 & Calif Inst Technol, 58; guest, Mass Inst Technol, 67; pres, Int Heterocyclic Chem Cong, 69, plenary lectr, 75; asst ed, Int J Heterocyclic Chem, 73-; exchange scientist, India, 77. *Honors & Awards:* Fortieth Midwest Award, Am Chem Soc, 84. *Mem:* Am Chem Soc; The Chem Soc; Sigma Xi; Int Soc Heterocyclic Chem. *Res:* Reaction mechanisms; aziridines and azetidines; amino ketones; beta chloro amines; benzacridines; synthetic and physical organic chemistry; small-ring compounds; heterocyclics of biological interest in carcinogenesis and carcinostasis; pyrrols; activated allylsystems. *Mailing Add:* 6600 Shamrock Rd Univ Nebr 837 Hamilton Hall Lincoln NE 68506

CRONAN, JOHN EMERSON, JR, b Long Beach, Calif, Dec 2, 42; m 73; c 2. BIOCHEMISTRY, MOLECULAR BIOLOGY. *Educ:* San Fernando Valley State Col, BA, 65; Univ Calif, Irvine, PhD(molecular biol), 68. *Prof Exp:* Instr biol chem, Sch Med, Washington Univ, 68-70; from asst prof to assoc prof molecular biophys & biochem, Yale Univ, 74-78; PROF MICROBIOL, UNIV ILL, 78- *Concurrent Pos:* NIH fel, Sch Med, Washington Univ, 68-70; NIH res grant & NSF res grant, Yale Univ, & Univ Ill, 71- NIH career development award, 72-77; mem, Metab Biol Panel, NSF, 75-76, Biochem Panel, NSF, 81-87; mem physiol chem study comt, Am Heart Asn, 81- *Mem:* Am Soc Microbiol; Am Soc Biol Chemists. *Res:* Microbial lipid metabolism and chemistry; biogenesis of biological membranes; bacteriophage infection physiology; biochemical genetics; gene synthesis. *Mailing Add:* Dept Microbiol Univ Ill 407 S Goodwin Urbana IL 61801

CRONAUER, DONALD (CHARLES), b Sewickley, Pa, Nov 4, 36; m 62; c 3. CHEMICAL ENGINEERING. *Educ:* Carnegie-Mellon Univ, BS, 58, MS, 59, PhD(chem eng), 62. *Prof Exp:* Proj chem engr, Res & Develop Dept, Amoco Chem Corp, 61-65; sr chem engr, Borg Warner Develop Div, Marbon Chem, 66-67, res assoc, 68-70; res engr, Gulf Res & Develop Co, 70-73, sr res engr, 73-76, staff engr, 76-85; SR RES ENGR, AMOCO OIL CO, 85- *Honors & Awards:* Richard A Glenn Award. *Mem:* Am Chem Soc; Am Inst Chem Engrs; Sigma Xi. *Res:* Process development with emphasis on kinetics and catalysis; mechanism of coal hydrogenation and liquefaction including catalysis and hydrogen transfer; fundamentals of conversion of shale and heavy fuels; upgrading of synthetic fuels. *Mailing Add:* Six S 180 Cape Rd Naperville IL 60540

CRONE, ANTHONY JOSEPH, b Boston, Mass, Jan 3, 47; m 74; c 1. EARTHQUAKE HAZARD ASSESSMENT. *Educ:* Clark Univ, BA, 69; Univ Colo, PhD(geol), 75. *Prof Exp:* Proj geologist oil & gas explor, Bond Explor Co, Inc, 74-76; consult, 76-78; GEOLOGIST, US GEOL SURV, 78- *Mem:* Fel Geol Soc Am; Am Geophys Union; Seismolog Soc Am. *Res:* Assessment of earthquake hazards in the midcontinent and northern Rocky Mountains; utilizing geologic and geophysical information to investigate causes and to determine the size and frequency of major earthquakes in study areas. *Mailing Add:* MS 966 US Geol Surv Box 25046 Denver Fed Ctr Denver CO 80225

CRONE, LAWRENCE JOHN, b Orangeville, Ill, May 18, 35. PLANT PATHOLOGY. *Educ:* Carthage Col, AB, 56; Rutgers Univ, PhD(plant path), 62. *Prof Exp:* Asst prof, 62-67, ASSOC PROF BIOL, UNIV WIS-WHITEWATER, 67- *Res:* Diseases of trees and ornamental plants. *Mailing Add:* Dept Biol Univ Wis 800 W Main Whitewater WI 53190

CRONEMEYER, DONALD CHARLES, b Chanute, Kans, Nov 10, 25; m 53; c 5. PHYSICS, GENERAL. *Educ:* Mass Inst Technol, ScD(physics), 51. *Prof Exp:* Res physicist, Gen Elec Co, 51-57 & Bendix Aviation Res Lab, Mich, 57-67; RES PHYSICIST, T J WATSON RES CTR, IBM CORP, 68- *Concurrent Pos:* Instr physics, Wayne State Univ, 58-67; assoc prof physics, Wheaton Col, 67-68. *Mem:* Creation Res Soc; Am Phys Soc; Inst Elec & Electronic Engrs. *Res:* Electrical and optical properties of semi-conductors; lasers; magnetic properties of materials; properties of superconductors. *Mailing Add:* 211 Barnes St Ossining NY 10562

CRONENBERG, AUGUST WILLIAM, b Teaneck, NJ, Apr 22, 44. ENGINEERING SCIENCE. *Educ:* Newark Col Eng, BS, 66; Northwestern Univ, MS, 69, PhD(eng sci), 71. *Prof Exp:* Assoc engr eng sci, Argonne Nat Lab, 71-74; asst prof chem & nuclear eng, Univ NMex, 74-77; sr scientist mat eng, EG&G Idaho, Inc, 77-79; CONSULT ENG SCI & ANAL, 79- *Concurrent Pos:* NSF fel, Argonne Nat Lab, 72-73. *Mem:* Am Nuclear Soc; Am Inst Chem Engrs; Am Soc Mech Engrs. *Res:* Material science; boiling heat transfer; nuclear fuel behavior; nuclear reactor safety; thermophysical properties. *Mailing Add:* 8100 Mountain Rd NE Suite 220 Albuquerque NM 87110

CRONENBERGER, JO HELEN, b LaGrange, Tex, Mar 17, 39. COMPUTER SCIENCES, BIOPHYSICS. *Educ:* Univ Tex, Austin, BA(microbiol) & BS(med technol), 62; Univ Houston, PhD(biophysics), 72. *Prof Exp:* Spec chemist clin chem, Methodist Hosp, Houston, 61-64; dir med technol, Bentaub Hosp, Baylor Col Med, 64-65, res assoc immunol, 65-68; Max-Planck Inst fel, 72-74; res fel, Ger Res Soc, 74-75; asst prof life sci, Univ Tex, San Antonio, 75-80, dir med technol, 75-80 & adj asst prof path, Univ Tex Health Sci Ctr, 76-80; dir Clin Lab Sci, dept MAHP, 80-86, ASSOC PROF PATH, MED SCH UNIV NC, CHAPEL HILL, 80-, ASSOC PROF BIOMED ENG, 90- *Concurrent Pos:* Facul fel Inst Acad Technol, Univ NC. *Honors & Awards:* Hematol Merit Award, Am Soc Med Technol, 63. *Mem:* Biomed Eng Soc; Am Soc Clin Path; Soc Appl Learning Tech. *Res:* High resolution color digitized images; their communication and use in clinical consultation, continuing medical education and medical education; their organization as database. *Mailing Add:* Dept MAHP Wing E CB 7145 Univ NC Med Sch Chapel Hill NC 27599-7145

CRONENWETT, WILLIAM TREADWELL, b Texarkana, Tex, Jan 3, 32; div; c 2. ELECTRICAL ENGINEERING. *Educ:* Tex Col Arts & Indust, BS, 54; Univ Tex, MS, 60, PhD, 66. *Prof Exp:* Res engr, Elec Eng Res Lab, Univ Tex, 55-59; res engr, Electro-Mech Co, Tex, 60-62; Welch Found fel physics, Univ Tex, 64-66; res fel, chem dept, Univ Leicester, 66-68; asst prof, 68-75, ASSOC PROF, DEPT ELEC ENG & COMPUT SCI, UNIV OKLA, 75-, ELEC & ELECTRONIC DESIGN CONSULT, 69- *Concurrent Pos:* Expert witness & expert invest, 81- *Mem:* Inst Elec & Electronics Engrs; Am Welding Soc. *Res:* Electric field theory; electronic conduction processes; electronic instrument design; medical instrument design. *Mailing Add:* Dept Elec Eng & Comput Sci Univ Okla 202 W Boyd St Norman OK 73019

CRONHEIM, GEORG ERICH, b Berlin, Ger, Jan 15, 06; nat US; m 32. PHARMACOLOGY. *Educ:* Univ Berlin, PhD(phys chem), 30. *Prof Exp:* Asst & instr, Inst Phys Chem, Univ Berlin, 30-32; head, Biochem Lab, Inst Phys Agron & res chemist, Inst Exp Med, Leningrad, 32-36; res chemist, Phys

Inst, Univ Stockholm, 37-39; res chemist, NY State Res Inst, 40; chief chemist, G F Harvey Co, NY, 41-45; res dir, S E Massengill Co, 46-52; dir biol sci, 52-67, tech liaison exec, 67-73, CONSULT, RIKER LABS, INC, 73- *Mem:* Am Soc Clin Pharmacol; Am Soc Pharmacol; NY Acad Sci. *Res:* Biochemistry; pharmacology and clinical pharmacology. *Mailing Add:* Riker Labs Inc Box 1 Northridge CA 91324

CRONHOLM, LOIS S, b St Louis, Mo, Aug 15, 30; div; c 2. MICROBIOLOGY. *Educ:* Univ Louisville, BA, 62, PhD(bot), 67. *Prof Exp:* Nat Inst Health fel, Dept microbiol, 66-79, from asst prof to assoc prof, 72-79, prof microbiol, 79-, res assoc, water resources lab, 73-, assoc, dept microbiol, Sch of Me, 75-, dean, Col Arts & Sci, 79- PROF BIOL AT UNIV LOUISVILLE; Dean Arts & Sci Temple Univ, Phila, Pa. *Concurrent Pos:* Assoc dean, Col Arts & Sci, Univ Louisville, 77-78, actg dean, 78-79. *Mem:* Am Soc Microbiol; AAAS; Am Water Resources Asn. *Res:* Bacterial endotoxins; histoplasma capsulatum and bird roosts; viruses in water and waste water; bacterial aerosols from waste water. *Mailing Add:* Dean's Off Col Arts & Sci Temple Univ Philadelphia PA 19122

CRONIN, JAMES LAWRENCE, JR, b St Louis, Mo, Mar 14, 19. APPLIED MECHANICS, CIVIL ENGINEERING. *Educ:* Washington Univ, St Louis, BS, 41, MS, 49, DSc(appl mech), 57. *Prof Exp:* Res engr, Res Found, Washington Univ, St Louis, 46-49, instr, 46-51; from asst prof to assoc prof eng, St Louis Univ, 51-61, prof civil eng, 61-68, prof elec eng, 68-71; chief elec sect, US Army Corps Engrs, St Louis Dist, 72-77, chief struct sect, 77-89; RETIRED. *Concurrent Pos:* Staff mem, Alvey Conveyor Mfg Co, 46-; indust consult, Emerson Elec Co, 53-68; vis prof mgt sci, Southern Ill Univ, Edwardsville, 72-85, adj lectr civil eng, 85- *Res:* Theoretical mechanics. *Mailing Add:* 1543 Windridge Ctr St Louis MO 63131

CRONIN, JAMES WATSON, b Chicago, Ill, Sept 29, 31; m 54; c 3. ELEMENTARY PARTICLE PHYSICS. *Educ:* Southern Methodist Univ, BS, 51; Univ Chicago, MS, 53, PhD(physics), 55. *Prof Exp:* Asst physicist, Brookhaven Nat Lab, 55-58; from asst prof to prof, Princeton Univ, 58-71, PROF PHYSICS, UNIV CHICAGO, 71- *Honors & Awards:* Nobel Prize, 80; E O Lawrence Award; Res Corp Award; Franklin Medal. *Mem:* Nat Acad Sci; Am Phys Soc; Am Acad Arts & Sci. *Res:* Elementary particles; development of improved detection techniques; C P violation; ultra-high-energy-x-ray astronomy. *Mailing Add:* Dept Physics Univ Chicago Chicago IL 60637

CRONIN, JANE SMILEY, b New York, NY, July 17, 22; div; c 4. BIOMATHEMATICS. *Educ:* Univ Mich, PhD(math), 49. *Prof Exp:* Mathematician, US Air Force Cambridge Res Ctr, 51-54; instr, Wheaton Col, 54-55; mathematician, Am Optical Co, 56; from asst prof to prof math, Polytech Inst Brooklyn, 57-65,; PROF MATH, RUTGERS UNIV, 65- *Concurrent Pos:* NSF vis professorship for women, Courant Inst, 84-85. *Mem:* Soc Indust & Appl Math; Am Math Soc; Int Soc Chronobiol. *Res:* Topological degree; functional analysis; ordinary differential equations; applications of analysis in medicine and biology. *Mailing Add:* 110 Valentine St Highland Park NJ 08904

CRONIN, JOHN READ, b Marietta, Ohio, Mar 5, 37; m 63; c 3. BIOCHEMISTRY. *Educ:* Col Wooster, BA, 59; Univ Colo, PhD(biochem), 64. *Prof Exp:* Res assoc biochem, Sch Med, Yale Univ, 64-66; from asst prof to assoc prof, 66-78, PROF, DEPT CHEM, ARIZ STATE UNIV, 78- *Concurrent Pos:* NIH fel, 64-65; temp staff mem org geochem, Carnegie Inst Wash, 74-75. *Mem:* AAAS; Am Chem Soc; Am Soc Biol Chem; Meteoritical Soc; Sigma Xi; Int Soc Study Origin Life. *Res:* Organic chemistry of carbonaceous meteorites; chemical evolution. *Mailing Add:* Dept Chem Ariz State Univ Tempe AZ 85287-1604

CRONIN, LEWIS EUGENE, b Aberdeen, Md, May 11, 17; m 45; c 3. MARINE ECOLOGY, ZOOLOGY. *Educ:* Western Md Col, AB, 38; Univ Md, MS, 42, PhD(zool), 46. *Hon Degrees:* DSc, Western Md Col, 66. *Prof Exp:* Teacher high sch, Md, 38-40; biologist, State Dept Res & Educ, Md, 43-50; dir marine lab & assoc prof biol sci, Univ Del, 50-55; dir & biologist, Dept Res & Educ, 55-61; dir, Natural Resources Inst & Chesapeake Biol Lab, 61-75, assoc dir res, Ctr Environ & Estuarine Studies, 75-77, prof Univ Md, 61-84, dir, Chesapeake Res Consortium, 77-84; CONSULT, 84- *Concurrent Pos:* Consult coastal ecol off chief engr, CEngr, US Army; consult pesticides, USDA & Environ Protection Agency & Mem Md Water Pollution Comn, 55-63; mem, Md Bd Nat Resources, 55-69, Chesapeake Res Coun, chmn, 66-67, Md Comn Pesticides, chmn, 67-69 & Md Comn Environ Educ, 70-; mem, Interstate Comn Potomac River Basin, 63-, Secys Comn Pesticides & Relationship Environ Health, Dept Health, Educ & Welfare, 69, Adv Comt Water Resources Sci Info Ctr, Dept Interior, 69-73 & Panel Oceanog & Marine Res, Comn Space Prog Earth Observation, Adv to Dept Interior, Nat Acad Sci; mem, Nat Marine Fish Adv Comn, Dept Com, 71-73; mem, Law of the Sea Adv Comn, US State Dept, 73-; mem, Mid-Atlantic Fisheries Mgt Coun, 76-77. *Honors & Awards:* Award, Oyster Inst NAm, Isaac Walton League & Chesapeake Bay Seafood Indust Asn; Am Motors consult Award, 71. *Mem:* AAAS; Am Soc Zool; Am Inst Fishery Res Biol; Am Fisheries Soc; Nat Shell Fisheries Asn (pres, 60); Estuarine Res Fedn (pres, 71-73). *Res:* Fisheries and pollution; resource management; environmental education; estuarine ecology and biology; research administration. *Mailing Add:* 12 Mayo Ave Bay Ridge Annapolis MD 21403

CRONIN, MICHAEL JOHN, b Los Angeles, Calif, Nov 19, 49; m 79. NEUROENDOCRINOLOGY. *Educ:* Loyola-Marymount Univ, Los Angeles, BS, 71; Univ Southern Calif, PhD(physiol), 76. *Prof Exp:* Fel neuroendocrinol, Univ Calif, San Francisco, 76-79; ASST PROF PHYSIOL, SCH MED, UNIV VA, 79- *Mem:* Endocrine Soc; Soc Neurosci; Am Physiol Soc; Int Soc Neuroendocrinol; Fedn Am Soc Exp Biol. *Res:* Study of hormones released by the brain to elucidate the subsequent mechanisms by which the anterior pituitary responds to these hormones in mammals and cell culture systems. *Mailing Add:* Dept Develop Biol 460 Pt San Bruno Blvd South San Francisco CA 94080

CRONIN, MICHAEL THOMAS IGNATIUS, b Glasgow, Scotland, Feb 1, 24; nat US; m 50; c 9. PATHOLOGY. *Educ:* Vet Col Ireland, MRCVS, 45; Univ Dublin, MSc, 46, PhD(path, bact), 48; Georgetown Univ, MD, 65. *Prof Exp:* Res officer, Equine Res Sta, Eng, 50-52; dir, Regional Diag Lab, Va, 52-53; bacteriologist, Dept Animal Path, Univ Ky, 53-55; assoc pathologist, Penrose Res Lab & asst prof vet path, Univ Pa, 55-57; head dept path & toxicol, Schering Corp, 57-61; pathologist, Woodard Res Corp, 61-65; intern, Grace-New Haven Community Hosp, Med Ctr, Yale Univ, 65-66; asst resident, Yale Univ-New Haven Hosp, 66-67 & Vet Admin Hosp, 67-68; Pathologist, 68-72, PATHOLOGIST, MEM HOSP, MERIDEN, 74- *Concurrent Pos:* Consult pathologist, Woodard Res Corp, 65-71; pathologist, Masonic Home & Hosp, Wallingford, 68-70, 72-84; consult ed, Am Scientist, 71-; assoc pathologist, Hosp of St Raphael, 72-76. *Mem:* Am Asn Path; emer mem Am Col Vet Path; Int Acad Path; Am Med Asn; Sigma Xi. *Mailing Add:* 67 Edgehill Rd New Haven CT 06511

CRONIN, ROBERT FRANCIS PATRICK, physiology, for more information see previous edition

CRONIN, THOMAS MARK, b Bronx, NY, July 20, 50; m 79; c 3. EARTH SCIENCES, PALEONTOLOGY. *Educ:* Colgate Univ, BA, 72; Harvard Univ, MA, 74, PhD(geol), 77. *Prof Exp:* Res assoc, 77-78, GEOLOGIST, US GEOL SURV, 78- *Mem:* Paleont Soc Am; Sigma Xi; AAAS; Am Quaternary Asn; Soc Econ Palentol Minerals; Soc Study Evolution. *Res:* Cenozoic ostracode evolution, biostratigraphy, paleoecology and taxonomy; Cenozoic paleoceanography and paleoclimatology; evolutionary paleontology. *Mailing Add:* Nat Ctr MS-970 US Geol Surv Reston VA 22092

CRONIN, THOMAS WELLS, b Baltimore, Md, May 19, 45; m 74; c 2. CRUSTACEAN VISUAL PHYSIOLOGY. *Educ:* Dickinson Col, BS, 67; Duke Univ, MA, 69, PhD(zool), 79. *Prof Exp:* Vol biol, US Peace Corp, 69-73; staff res biologist, 79-81, fel biol, Yale Univ, 81-; AT DEPT BIOL SCI, UNIV MARYLAND, BALTIMORE COUNTY. *Mem:* AAAS; Am Soc Zoologists. *Res:* Visual and physiological adaptation of invertebrates, especially crustaceans and their larvae, to marine and other aquatic environments; eye structure; visual physiology; phototaxis; endogenous rhythms. *Mailing Add:* Dept Biol Sci Univ Md Baltimore County 5401 Wilkens Ave Catonsville MD 21228-5398

CRONIN, TIMOTHY H, b Boston, Mass, June 4, 39; m 63; c 2. RESEARCH ADMINISTRATION. *Educ:* Boston Col, BS, 60; Mass Inst Technol, PhD(org chem), 64. *Prof Exp:* Res chemist, Pfizer Inc, 65-70, proj leader, 70-71, mgr, 71-75, dir med chem, Infectious Dis, 75-81, dir res, Agr Prod Res & Develop, 81-90, VPRES, PFIZER INC, 90- *Mem:* Am Chem Soc; Sigma Xi. *Res:* Intramolecular Diels-Alder reactions; synthetic organic chemistry. *Mailing Add:* 40 Laurel Hill Dr Niantic CT 06357-1506

CRONIN-GOLOMB, MARK, ELECTROOPTICS, NONLINEAR OPTICS. *Educ:* Univ Sydney, BSc, 79; Calif Inst Technol, PhD(physics), 83. *Prof Exp:* Staff scientist, Ortel Corp, 84-86; ASSOC PROF ELEC ENG, TUFTS UNIV, 86- *Concurrent Pos:* Consult, Foster Miller, Inc, 86-, Litton Itek Optical Systs, 91-; assoc ed, Optics Letters, 87-89. *Mem:* Optical Soc Am; Am Phys Soc; Inst Elec & Electronics Engrs. *Res:* Development of novel nonlinear optical devices; real time holographic distortion correction (phase conjugation); optical computing; nonlinear optical interactions in photorefractive crystals. *Mailing Add:* Electro-Optics Technol Ctr Four Colby St Tufts Univ Medford MA 01867

CRONK, ALFRED E(DWARD), b Hudson, Wis, July 1, 15; m 40; c 4. AERONAUTICAL ENGINEERING. *Educ:* Col St Thomas, BS, 37; Univ Minn, MS, 46. *Prof Exp:* From instr to assoc prof aeronaut eng, Univ Minn, 43-56; prof & head dept, 56-78, EMER PROF AERONAUT ENG, TEX A&M UNIV, 78- *Concurrent Pos:* Consult, Flui-Dyne Eng Corp, Gen Mills, Inc, Minneapolis Honeywell Regulator Co & Minn Mining & Mfg Co. *Mem:* Assoc fel Am Inst Aeronaut & Astronaut; Am Soc Eng Educ. *Res:* Stability and control of aircraft; experimental aerodynamics. *Mailing Add:* 727 N Rosemary Bryan TX 77802-4308

CRONK, CHRISTINE ELIZABETH, b Detroit, Mich, Oct 25, 44; m 68. PHYSICAL GROWTH. *Educ:* Western Mich Univ, BA, 66, MA, 75; Harvard Univ, MS, 77, ScD, 80. *Prof Exp:* Anthropologist, Children's Hosp Med Ctr, 72-80; res assoc, Fels Res Inst, 80-; AT NESBITT COL, DREXEL UNIV. *Mem:* Human Biol Coun; Am Asn Phys Anthrop; Am Pub Health Asn; Sigma Xi. *Res:* Description and control of growth; natural history of body fatness; effects of malnutrition on fetal and postnatal growth; ultrasound measurement of fetal growth. *Mailing Add:* Nesbitt Col Drexel Univ 32nd & Chestnut Philadelphia PA 19104

CRONK, TED CLIFFORD, b Ridgway, Pa, May 25, 46. FOOD SCIENCE. *Educ:* Pa State Univ, BS, 68, MS, 71; Cornell Univ, PhD(food sci), 75. *Prof Exp:* Asst prof food & nutrit, Univ Ga, 74-75; sr scientist, Pillsbury Co, 75-80, res mgr, Frozen Foods Div, 80-90; VPRES FOOD SECTOR, GRAND METROPOLITAN, 90- *Mem:* Mycol Soc Am; Inst Food Technologist. *Mailing Add:* 330 University Ave SE Minneapolis MN 55414

CRONKHITE, LEONARD WOOLSEY, JR, b Newton, Mass, May 4, 19; c 4. MEDICINE. *Educ:* Bowdoin Col, BS, 41; Harvard Med Sch, MD, 50. *Hon Degrees:* LLD, Northeastern Univ, 70; LHD, Curry Col, 77; LLD, Bowdoin Col, 79. *Prof Exp:* Pres, Children's Hosp Med Ctr, 61-77; pres, Med Col Wis, 77-84; RETIRED. *Concurrent Pos:* Trustee, Bowdoin Col, 73- *Mem:* Nat Inst Med-Nat Acad Sci; Asn Am Med Cols; Soc Med Adminr (pres, 70-71). *Mailing Add:* 11 Quarry Rd Brunswick ME 04011

CRONKITE, DONALD LEE, b Denver, Colo, Dec 10, 44; m 68; c 3. CELL BIOLOGY, GENETICS. *Educ:* Ind Univ, AB, 66, PhD(zool), 72. *Prof Exp:* Asst prof, Univ Redlands, 72-78; from asst prof to assoc prof, 81-88, PROF BIOL, HOPE COL, 88- *Concurrent Pos:* Exchange of Persons fel, Japan Soc Prom Sci, Tohoku Univ, 76; vis assoc prof, Zool Dept, College Park Univ,

Maryland, 86. *Mem:* Genetics Soc Am; Am Soc Cell Biol; Soc Protozoologists. *Res:* Genetics of membrane function in paramecium; mechanism of mating in paramecium; cellular water regulation; ciliate-algal symbiosis. *Mailing Add:* Dept Biol Hope Col Holland MI 49423

CRONKITE, EUGENE PITCHER, b Los Angeles, Calif, Dec 11, 14; m 40; c 1. MEDICINE. *Educ:* Stanford Univ, AB, 36, MD, 41. *Hon Degrees:* DSc, Univ Long Island, 62; MD, Univ Ulm WGer. *Prof Exp:* Intern, Stanford Univ Hosp, 40-41, asst resident, 41-42; hematologist, Naval Med Res Inst, 46-54; physician & chmn med dept, 67-79, HEMATOLOGIST, MED RES CTR, BROOKHAVEN NAT LAB, 54- *Concurrent Pos:* Mem, Naval Studies Bd, Nat Acad Sci; prof med & dean, Brookhaven Clin Campus, State Univ NY Stony Brook; hematologist, Atomic Bomb Tests, 46, 51-54; ed-in-chief, Exp Hemat, 84-88. *Honors & Awards:* Wellcome Prize, 48; Alfred Benzon Prize, Govt Denmark; Ludwig Heilmeyer Gold Medal, Govt WGer, 74; Semmelweis Medal, Govt Hungary, 75; Alexander von Humboldt Sr Scientist Award, WGer, 79-80.. *Mem:* Nat Acad Sci; Am Soc Hemat; Soc Exp Biol & Med; Am Physiol Soc; Am Soc Clin Invest; Am Asn Physicians; Int Soc Esp Hemat (pres, 80). *Res:* Control of hemopoiesis in health and disease. *Mailing Add:* Med Res Ctr Brookhaven Nat Lab Upton NY 11973

CRONKRIGHT, WALTER ALLYN, JR, b East Orange, NJ, July 23, 31; m 54; c 3. POLLUTION CHEMISTRY, ANALYTICAL CHEMISTRY. *Educ:* Rutgers Univ, BS, 52; Drexel Inst Technol, MS, 60; Polytech Inst Brooklyn, PhD(anal chem), 68. *Prof Exp:* Anal chemist, M W Kellogg Co, Lab, Pullman Inc, 59-61, res chemist, 61-65, supvr, 65-68, sect head, Pullman-Kellogg Div, Pullman Inc, 68-71, MGR ANALYTICAL SERVS, PULLMAN-KELLOGG DIV, PULLMAN INC, 71- *Mem:* Am Chem Soc; Am Soc Test & Mat; Sigma Xi. *Res:* Development of improved processes for removal of SO2 from waste gases; standardization of analytical methods for atmospheric pollutants. *Mailing Add:* 13503 Westport Houston TX 77079

CRONN, DAGMAR RAIS, b Vicksburg, Miss, Nov 9, 46; m 68. ATMOSPHERIC CHEMISTRY & PHYSICS. *Educ:* Univ Wash, BS, 69, MS, 72, PhD(atmospheric chem), 75. *Prof Exp:* From asst prof to assoc prof, Wash State Univ, 75-85; DEAN, COL SCI, UNIV MAINE, 85- *Concurrent Pos:* Prin investr, Environ Protection Agency, 75-82, Dept Energy, 77-82, Nat Atmospheric & Space Admin, 77-83 & NSF, 79-86; consult, Environ Protection Agency, 80-82 & Univ Calif, 81-82; W K Kellogg Found fel, 81-84. *Mem:* Am Chem Soc; Air Pollution Control Asn; Am Geophys Union; Am Meteorol Soc; Asn Women Sci. *Res:* Analysis of trace gases in the atmosphere; impact of anthropogenic and natural emissions on stratospheric ozone; antarctic meteorology and air chemistry; tropospheric/stratospheric exchange mechanisms; measurement technique development for atmospheric uses. *Mailing Add:* Off Dean Col Sci Univ Maine 263 Aubert Hall Orono ME 04469

CRONQUIST, ARTHUR JOHN, b San Jose, Calif, Mar 19, 19; m 40; c 2. SYSTEMATIC BOTANY. *Educ:* Utah State Col, BS, 38, MS, 40; Univ Minn, PhD(bot), 44. *Hon Degrees:* DSc(biol sci), Utah State Univ, 87. *Prof Exp:* Tech asst, NY Bot Garden, 43-44, asst cur, 44-46; asst prof bot, Univ Ga, 46-48; asst prof, State Col Wash, 48-51; assoc cur, NY Bot Garden, 52-57, cur, 57-65, sr cur, 65-71, dir bot, 71-74, SR SCIENTIST, NY BOT GARDEN, 74- *Concurrent Pos:* Tech consult, Econ Co-op Admin, Brussels, 51-52; adj prof, Columbia Univ, 64- & Lehman Col, City Univ New York, 69-; ed, Bot Rev, 69- *Honors & Awards:* Asa Gray Award, Am Soc Plant Taxonomists, 85; Leidy Medal, Acad Natural Sci, 70; Linnean Medal for Botany, Linnean Soc London, 86. *Mem:* Am Soc Plant Taxon (pres, 62); Bot Soc Am (pres, 73); Asn Trop Biol; Ecol Soc Am; Torrey Bot Club (pres, 76); Int Asn Plant Taxon. *Res:* Taxonomy of American species of Compositae; flora of temperate North America; systems of angiosperms; general system of plants. *Mailing Add:* New York Bot Garden Bronx NY 10458

CRONSHAW, JAMES, b Oswaldtwistle, Eng, Mar 11, 33; m 56; c 1. CELL BIOLOGY, PLANT PHYSIOLOGY. *Educ:* Univ Leeds, BSc, 54, PhD(bot), 57, DSc, 73. *Prof Exp:* Res officer, Commonwealth Sci & Indust Res Orgn, Australia, 57-62; asst prof biol, Yale Univ, 62-65; assoc prof, 65-70, PROF BIOL, UNIV CALIF, SANTA BARBARA, 70- *Mem:* Fel AAAS; Bot Soc Am; Electron Micros Soc Am; Am Soc Cell Biol; fel Royal Micros Soc. *Res:* Electron microscopy of phloem and xylem cells; structure and function of the avian adrenal gland; effects of petroleum on marine birds; endocrinology. *Mailing Add:* Dept Biol Sci Univ Calif Santa Barbara CA 93106

CRONSON, HARRY MARVIN, b Providence, RI, May 31, 37; m 62; c 2. SATELLITE COMMUNICATIONS, MONOLITHIC MICROWAVE INTEGRATED CIRCUITS. *Educ:* Brown Univ, ScB, 59, ScM, 61, PhD(elec eng), 63. *Prof Exp:* Corinna Borden Keen fel from Brown Univ, Oxford Univ, 63-64; asst prof electrophys, Grad Ctr, Polytech Inst Brooklyn, 64 -66; staff scientist, Avco Space Systs Div, 66-67, sr staff scientist, 67-68; sr scientist, Ikor, Inc, 68-71; mem tech staff, Sperry Res Ctr, 71-83; group leader, 83-90, PRIN ENGR, MITRE CORP, 90- *Concurrent Pos:* NSF res grant, 65-66. *Mem:* Fel Inst Elec & Electronics Engrs; Int Union Radio Sci. *Res:* Microwave and millimeter wave metrology; subnanosecond pulse technology; advanced technology for satellite terminals; monolithic microwave integrated circuits. *Mailing Add:* Mitre Corp MS R354 Burlington Rd Bedford MA 01730-0208

CRONVICH, JAMES A(NTHONY), b New Orleans, La, Oct 26, 14; m 51; c 5. ELECTRICAL ENGINEERING. *Educ:* Tulane Univ, BE, 35, MS, 37; Mass Inst Technol, SM(elec eng), 38. *Prof Exp:* Instr elec eng, Tulane Univ, 38-41; from asst to assoc elec engr, Panama Canal, 41-42; from asst prof to assoc prof elec eng, 42-49, assoc prof biophysics, Dept Med, 46-49, prof elec eng & biomed eng, 49-82, head dept elec eng, 56-82, EMER PROF, TULANE UNIV, 82- *Concurrent Pos:* Asst ed, Am Heart J, 78-82. *Mem:* Fel Inst Elec & Electronics Engrs; Am Soc Eng Educ; Sigma Xi. *Res:* Bioengineering in cardiovascular studies; electrical measurements. *Mailing Add:* 101 Colonial Club Dr Harahan LA 70123

CRONVICH, LESTER LOUIS, b New Orleans, La, Aug 21, 16; m 45; c 1. AERODYNAMICS, FLUID MECHANICS. *Educ:* Tulane Univ, BEng, 36, MS, 38; Univ Wis, PhD(appl math), 42. *Prof Exp:* Instr math, Northwestern Univ, 41-42; stress analyst aircraft struct, McDonnell Aircraft Corp, 42-45; asst proj supvr aerodyn, John Hopkins Univ, 51-55, proj supvr, 55-57, asst group supvr, 57-58, mathematician, Appl Physics Lab, 45-86, group supvr, 58-86, asst div supvr aeronaut, 74-86; RETIRED. *Concurrent Pos:* Mem, Navy Aeroballistics Comt, 58-77, Am Inst Aeronaut & Astronaut Tech Comt on Atmospheric Flight Mech, 67-70, Publ Comt, 71-74 & NASA Res & Technol Adv Coun Comt on Aerodyn & Configurations, 77, Aeronaut Adv Comt, 84-85. *Mem:* Am Inst Aeronaut & Astronaut. *Res:* Stability and control in aerodynamics. *Mailing Add:* 9322 Mellenbrook Rd Columbia MD 21045

CRONYN, MARSHALL WILLIAM, b Oakland, Calif, June 22, 19; m 42; c 3. ORGANIC CHEMISTRY. *Educ:* Reed Col, BA, 40; Univ Mich, PhD(org chem), 44. *Prof Exp:* Res assoc comt med res, Univ Mich, 44-46; lectr & Am Chem Soc fel, Univ Calif, 46-48, from instr to asst prof chem, 48-52; from asst prof to prof chem, Reed Col, 52-89, chmn dept, 66-73, vpres & provost, 82-89, EMER PROF CHEM, REED COL, 89- *Concurrent Pos:* NIH res fel, Cambridge Univ, 60-61; consult, USPHS, 61-66; mem bd, Ore Resonance & Technol Develop Corp, 90-91. *Mem:* AAAS; Sigma Xi. *Res:* Organic sulfur chemistry; antitumor agents; hydrogen carriers for automotive fuel. *Mailing Add:* 3232 NW Luray Terr Portland OR 97210

CROOK, JAMES RICHARD, infectious disease, epidemiology; deceased, see previous edition for last biography

CROOK, JOSEPH RAYMOND, b Reno, Nev, Oct 16, 36; m 58; c 4. INORGANIC CHEMISTRY. *Educ:* Univ Nev, BS, 58; Ill Inst Technol, PhD(chem), 63. *Prof Exp:* NSF fel, Univ Colo, 63-64; asst prof inorg chem, San Jose State Col, 64-66; from asst to assoc prof chem, Cleveland State Univ, 66-70; ASSOC PROF CHEM, WESTERN WASH UNIV, 70- *Mem:* Am Chem Soc; Sigma Xi. *Res:* Synthetic and mechanistic chemistry of transition elements. *Mailing Add:* Dept Chem Western Wash Univ Bellingham WA 98225

CROOK, PHILIP GEORGE, b Washington, DC, Oct 21, 25. MICROBIOLOGY. *Educ:* Univ Md, BS, 49; Univ NMex, MS, 51; Pa State Univ, PhD(bact), 55. *Prof Exp:* Asst instr bact, Pa State Univ, 53-55; assoc prof biol, Hope Col, 55-69, chmn dept, 62-66; Chas A Dana prof biol, 69-84, chmn dept, 69-76, emer dana prof biol, Colgate Univ, 84-; RETIRED. *Concurrent Pos:* NSF grant, Am Cancer Soc grant & Nat Cancer Inst grant, Hope Col, 58-60. *Mem:* AAAS; Am Soc Parasitol; Am Soc Microbiol; Soc Protozool. *Res:* Intermediate metabolism of carbohydrates in bacteria and protozoa; axenic culture of protozoa; biochemical action of hormones. *Mailing Add:* One Euclid Ave Cortland NY 13045

CROOKE, JAMES STRATTON, b New York, NY, June 4, 28; m 56; c 3. ENGINEERING MANAGEMENT, PROJECT FINANCING. *Educ:* US Merchant Marine Acad, Kings Pt, NY, BS, 50; Columbia Univ, MA, 56. *Prof Exp:* Engr, Union Carbide, 50-52, proj engr, 56-58; chief engr, USN, 52-54; traffic mgr, NY Tel Co, 54-56; indust officer, USAID, 58-62; asst sales mgr, M W Kellogg Co, 62-63; econ assoc, Mobil Oil Corp, 63-67; dir int develop, UN Indust Develop Orgn, 67-90; CONSULT ECON AFFAIRS, EMBASSY ST KITTS-NEVIS, BWI, 90- *Concurrent Pos:* Pres, Inst Indust Engrs, NY, 69-71 & UN Asn, 91; comnr, Eng Manpower Comt, 74-89; mem, Int Planning Comt, Inst Indust Engrs, 75; team deleg leader to USSR, 90 & 91. *Mem:* Inst Indust Engrs. *Res:* International science and industry. *Mailing Add:* One Windsor Lane Scarsdale NY 10583

CROOKE, PHILIP SCHUYLER, b Summit, NJ, Mar 10, 44; m 68. APPLIED MATHEMATICS. *Educ:* Stevens Inst Technol, BS, 66; Cornell Univ, PhD(applied math), 70. *Prof Exp:* From asst prof to assoc prof, 70-76, PROF MATH, VANDERBILT UNIV, 86- *Mem:* Soc Indust & Appl Math. *Res:* Partial differential equations; isoperimetric inequalities; dusty gas equations; mathematical modeling of fermentation processes; mathematical modeling of mechanic ventilation. *Mailing Add:* Box 6205 Sta B Vanderbilt Univ Nashville TN 37235

CROOKE, ROBERT C, ENGINEERING. *Prof Exp:* RETIRED. *Mem:* Nat Acad Eng. *Mailing Add:* 2135 Vineyard Dr Templeton CA 93465-9711

CROOKE, STANLEY T, b Indianapolis, Ind, Mar 28, 45; c 1. SIGNAL TRANSDUCTION, DRUG DISCOVERY & DEVELOPMENT. *Educ:* Baylor Col Med, PhD, 71, MD, 75. *Prof Exp:* Assoc dir res & develop, Bristol Labs, Syracuse, NY, 77-79, vpres assoc dir, 79-80; vpres res & develop, Smithkline & French Labs, King of Prussia, Pa, 80-82, dir molecular pharmacol, 81-88, pres res & develop, 82-88; vpres, Smithkline Beckman, Philadelphia, Pa, 83-88; chief exec officer, 89-91, CHMN BD & CHIEF EXEC OFFICER, ISIS PHARMACEUT, CARLSBAD, CALIF, 91- *Concurrent Pos:* Adj prof pharmacol, Baylor Col Med, Houston, Tex, 82, Univ Pa, Philadelphia, 82-89; ed, Anti-Cancer Drug Design, 84; distinguished prof, Univ Ky, 86; trustee, Franklin Inst, Philadelphia, Pa, 87-89; mem sci adv bd, BCM Technol, Houston, Tex, 89; bd dirs, Genetic Eng Systs, Houston, Tex, 89. *Mem:* AAAS; Am Asn Cancer Res; Am Soc Microbiol; Am Soc Pharmacol & Exp Therapeut; Am Soc Clin Pharmacol & Therapeut; Am Soc Clin Oncol. *Res:* RNA biochemistry and antisense technology; molecular pharmacology; signal transduction mechanisms particularly for leukotrienes and mechanisms of action of anticancer drugs; recipient of one patent. *Mailing Add:* 2280 Faraday Ave Carlsbad CA 92008

CROOKER, NANCY USS, b Chicago, Ill, Apr 1, 44; m; c 1. SPACE PHYSICS. *Educ:* Knox Col, Ill, AB, 66; Univ Calif, Los Angeles, MS, 68, PhD(meteorol), 72. *Prof Exp:* Engr scientist atmospheric sci, McDonnell Douglas Astronaut Co, 69; res assoc space physics, Cornell Univ, 73 & Mass Inst Technol, 73-75; from asst res scientist to assoc res scientist, 75-87, RES SCIENTIST, UNIV CALIF, LOS ANGELES, 87- *Concurrent Pos:* Consult, Physics Dept, Boston Col, 75-76, Aerospace Corp, 83-85. *Mem:* Am Geophys Union. *Res:* The coupling of the solar wind and the earth's magnetosphere. *Mailing Add:* Dept Atmospheric Sci Univ Calif Los Angeles CA 90025-1565

CROOKER, PETER PEIRCE, b Westerly, RI, Apr 4, 37; m 67; c 2. OPTICAL PHYSICS. *Educ:* Ore State Univ, BS, 59; Naval Postgrad Sch, PhD(physics), 67. *Prof Exp:* Proj officer, David Taylor Model Basin, 59-60; instr physics, Naval Postgrad Sch, 60-67; fel, Lincoln Lab, Mass Inst Technol, 68-70; asst prof physics, Calif State Polytech Col, 70; from asst prof to assoc prof, 70-82, PROF PHYSICS, UNIV HAWAII, 82- *Mem:* Am Phys Soc. *Res:* Electron paramagnetic resonance; acoustic paramagnetic resonance; raman spectroscopy; Rayleigh and Brillouin spectroscopy in liquid crystals; structures and phase transitions of liquid crystals. *Mailing Add:* Dept Physics & Astron Univ Hawaii 2505 Correa Rd Honolulu HI 96822

CROOKS, GEORGE CHAPMAN, b North Brookfield, Mass, Jan 21, 05; m 34; c 2. PHYSIOLOGICAL CHEMISTRY. *Educ:* Amherst Col, AB, 28; Mass State Col, PhD(biochem), 37. *Prof Exp:* Asst chem, Mass State Col, 28-30; instr, 30-35, 36-39,; assoc prof, 46-50, 52-69, prof, 69-71, EMER PROF CHEM, UNIV VT, 71- *Concurrent Pos:* Consult, Univ Vt, 71- *Mem:* Am Chem Soc; Sigma Xi. *Res:* Blood proteins; chemical and nutritive studies on fish muscle; carotene studies in blood, milk and plant materials. *Mailing Add:* 74 Spear St S Burlington VT 05403

CROOKS, MICHAEL JOHN CHAMBERLAIN, b Victoria, BC, Oct 25, 30; m 59; c 3. LOW TEMPERATURE PHYSICS. *Educ:* Reed Col, BA, 53; Univ BC, MA, 57; Yale Univ, PhD(physics), 63. *Prof Exp:* Res assoc physics, Duke Univ, 62-63; asst prof, 63-72, ASSOC PROF PHYSICS, UNIV BC, 72- *Concurrent Pos:* Sr vis res fel, Univ of Sussex, Eng, 71-72 & 80-81; leader Can team, Int Physics Olympiad, 86- *Mem:* Am Phys Soc; Am Asn Physics Teachers; Can Asn Physicists. *Res:* Properties of liquid and solid states of helium-three and helium-four; critical phenomena in super fluid helium. *Mailing Add:* Dept of Physics Univ of BC Vancouver BC V6T 2A6 Can

CROOKS, PETER ANTHONY, b Maltby, Yorkshire, UK, Oct 17, 42; m 72; c 4. DRUG DESIGN & METABOLISM. *Educ:* Univ Manchester, UK, BSc, 66 MSc, 67, PhD(pharm), 70. *Prof Exp:* Asst lectr pharm, Univ Manchester, UK, 68-70, lectr, 70-80, sr lectr, 80-81; assoc prof pharm, 81-91, ASSOC PROF TOXICOL, UNIV KY, 83-, PROF PHARM, 91- *Concurrent Pos:* Pres, Univ Manchester Pharmaceut Soc, 75-76; vis prof toxicol, Univ Naples, 76; res assoc pharmacol, Med Sch, Yale Univ, 76-78; mem, Conf Sci Comt Pharmaceut, Soc Great Brit, 80-82; regional ed, J Enzyme Inhibition, 87- *Mem:* Royal Pharmaceut Soc Great Brit; Am Chem Soc; Sigma Xi; AAAS; Am Asn Pharmaceut Scientists; Int Soc Study Xenobiotics. *Res:* Rational design and determination of the mechanism of action of drug molecules; enzyme mechanisms in drug and xenobiotic metabolism; molecular basis for the toxicology of xenobiotics. *Mailing Add:* Univ Ky Rose St Lexington KY 40506

CROOKS, STEPHEN LAWRENCE, b Sunderland, Eng, May 15, 57; US citizen; m 79; c 2. PULMONARY DISEASE, SIGNAL TRANSDUCTION. *Educ:* Hamline Univ, BSc, 81; Univ Minn, PhD(med chem), 85. *Prof Exp:* Asst res prof chem, Univ SC, 85-87; SR CHEMIST, 3M PHARMACEUTICALS LABS, ST PAUL, MINN, 87- *Mem:* Am Chem Soc. *Res:* Agents for pulmonary disease, especially asthma; drug design via computer modeling; synthesis of novel candidates for treatment of asthma. *Mailing Add:* 3M Pharmaceuticals Lab 270-2S-06 3M Ctr St Paul MN 55144

CROOKSHANK, HERMAN ROBERT, b Linneus, Mo, June 7, 16; m 45; c 3. CLINICAL CHEMISTRY. *Educ:* Northeast Mo State Teachers Col, BS, 38; Univ Iowa, MS, 40, PhD(biochem), 42. *Prof Exp:* Asst biochem, Univ Iowa, 38-42; instr, Med Col, Univ Ala, 46-49; asst secy, Am Inst Biol Sci, Washington, DC, 49; animal nutritionist, Animal Husb Res Div, USDA, 49-70, res chemist, Sci & Educ Admin, 70-79; RETIRED. *Concurrent Pos:* Assoc prof animal husb, Tex Technol Col, 54-59; asst exec secy, Div Biol & Agr, Nat Res Coun, 49-; dipl, Am Bd Nutrit & Am Bd Clin Chem. *Mem:* Am Chem Soc; fel Am Inst Chem; emer fel Am Soc Animal Sci. *Res:* Grass tetany and urinary caluli; pesticide residues in livestock; physiological and biochemical interactions of mineral in livestock. *Mailing Add:* 3812 Plainsman Lane Bryan TX 77802-3909

CROOKSTON, J(AMES) A(DAMSON), b Benld, Ill, Dec 8, 19; m 42; c 2. CERAMIC ENGINEERING. *Educ:* Mo Sch Mines, BS, 42, MS, 47; Univ Ill, PhD, 49. *Prof Exp:* Ceramic res engr, A C Spark Plug Div, 42-44; ceramic res engr, Harbison-Walker Refractories Co, 49-51, lab mgr, 51-54, asst to dir res, 54-55; from mgr new prod develop to res dir, A P Green Fire Brick Co, 55-61; dir res, A P Green Refractories Co, 61-85, vpres res, 66-85; RETIRED. *Concurrent Pos:* Chmn technol comt, Retractories Inst. *Mem:* Am Ceramic Soc; Nat Inst Ceramic Engrs; Sigma Xi. *Res:* Refractories; clay; base exchange phenomenon. *Mailing Add:* Two Country Club Mexico MO 65265

CROOKSTON, MARIE CUTBUSH, b Dean, Victoria, Australia, Aug 15, 20; wid; c 2. IMMUNOHEMATOLOGY. *Educ:* Univ Melbourne, BSc, 40. *Prof Exp:* Head lab clin path, St Andrew's Hosp, 41-46; res scientist, Blood Transfusion Res Unit, Med Res Coun, Eng, 47-57; res assoc immunohemat, Univ Tornto, 64-78, assoc prof path, 78-86; RETIRED. *Concurrent Pos:* Nuffield grant, India, 52. *Honors & Awards:* Landsteiner Award, Am Asn Blood Banks, 76. *Mem:* Brit Blood Transfusion Soc; Can Soc Immunol; Brit Soc Immunol; Int Soc Blood Transfusion; Can Soc Hemat; Can Soc Transfusion Med. *Res:* Hemolytic disease of the newborn; Duffy blood group system; survival of red cells after low temperature storage; effect of incompatible antibodies on red cell survival; blood-group chimeras; blood-group antigens in plasma; specificity and behavior of auto-antibodies. *Mailing Add:* 246 Russell Hill Rd Toronto ON M4V 2T2 Can

CROOKSTON, REID B, b Pittsburgh, Pa, Mar 21, 39; m 66; c 2. CHEMICAL & PETROLEUM ENGINEERING. *Educ:* Univ Pittsburgh, BS, 61; Carnegie-Mellon Univ, MS, 63, PhD(chem eng), 66. *Prof Exp:* Res engr, Gulf Sci & Technol Co, 66-68, sr res engr, 68-73, dir enhanced recovery processes, 73-81, tech consult, 81-85; chief reservoir engr, 85-90, RESERVOIR ENG ADV, CHEVRON USA, INC, 90- *Concurrent Pos:* Instr, Univ Pittsburgh, 77-80. *Mem:* Soc Petrol Engrs; Sigma Xi. *Res:* Chemical reactor engineering; mathematical modeling; optimization; process development; reservoir simulation; enhanced oil recovery processes. *Mailing Add:* 14600 Westdale Dr Bakersfield CA 93312-9121

CROOM, FREDERICK HAILEY, b Lumberton, NC, Aug 6, 41; m 63; c 2. MATHEMATICS. *Educ:* Univ NC, BS, 63, PhD(math), 67. *Prof Exp:* Asst prof math, Univ Ky, 67-71; from asst prof to assoc prof, 71-81, PROF MATH, UNIV SOUTH, 81-, ASSOC DEAN, COL ARTS & SCI, 84- *Concurrent Pos:* Vis assoc prof, La State Univ, 77-78; vis scholar, Vanderbilt Univ, 81. *Mem:* Am Math Soc; Math Asn Am; Sigma Xi. *Res:* Algebraic topology; point set topology. *Mailing Add:* Dept Math Univ South Sewanee TN 37375

CROOM, HENRIETTA BROWN, b Burlington, NC, Sept 23, 40; m 63; c 2. BIOCHEMISTRY. *Educ:* Univ NC, AB, 62, PhD(biochem), 68. *Prof Exp:* Res assoc anat, Univ Ky, 69-70; from asst prof to assoc prof, 72-88, PROF BIOL, UNIV SOUTH, 88- *Concurrent Pos:* Vis prof, dept microbiol, La State Univ, 77-78; vis scholar, Dept Molecular Biol, Vanderbilt Univ, 81-82 & Dept Zool, Univ Hawaii, Manoa, 89; Pew fel, Univ Ky, 88-89. *Mem:* Sigma Xi; Am Soc Cell Biol. *Res:* Interaction between guanine nucleotides and tubulin from bovine brain; mitochondrial DNA analysis of spider phylogeny. *Mailing Add:* Biol Dept Univ South Sewanee TN 37375

CROOM, WARREN JAMES, JR, b Aug 6, 50; c 2. DIGESTIVE FUNCTIONS. *Educ:* Univ Ill, PhD(nutrit sci), 78. *Prof Exp:* ASSOC PROF PHYSIOL, NC STATE UNIV, 84- *Mem:* Am Inst Nutrit; Am Soc Animal Sci. *Res:* Gastrointestinal physiology and drugs; research emphasizes the study of novel cholinergic agonist with high affinity for gastrointestinal track, special efforts are being made to use this chemical model to enhance digestive function indomestic livestock. *Mailing Add:* Dept Animal Sci NC State Univ Box 7621 Raleigh NC 27695-7621

CROPP, FREDERICK WILLIAM, III, B Wheeling, WVa, Dec 9, 32; m 55; c 8. GEOLOGY. *Educ:* Col Wooster, BA, 54; Univ Ill, MS, 56, PhD(geol), 58. *Prof Exp:* Asst, Univ Ill, 54-58, from instr to asst prof geol, 58-64; assoc prof & assoc dean, 64-68, dean & vpres acad affairs, 68-77, PROF GEOL, COL WOOSTER, 68-; PRES & CHIEF EXEC OFFICER, ENVIRON EXPERIENCES, INC, 90- *Concurrent Pos:* Ellis L Phillips Found intern acad admin, 62-63. *Mem:* Geol Soc Am; Nat Asn Geol Teachers; Am Asn Petrol Geol. *Mailing Add:* Dept Geol Col Wooster Wooster OH 44691

CROPP, GERD J A, b Delmenhorst, WGer, July 2, 30; US citizen; m 57, 88; c 3. PHYSIOLOGY, PULMONOLOGY. *Educ:* Univ Western Ont, MD, 58, PhD(biophys), 65. *Prof Exp:* From asst prof to assoc prof, Sch Med, Univ Colo, 65-72, assoc clin prof pediat, 72-76; prof pediat, State Univ NY & dir, Children's Lung & Cystic Fibrosis Ctr, Children's Hosp, Buffalo, 76-88; PROF PEDIAT, COLUMBIA PRESBY MED CTR, COLUMBIA UNIV, 88- *Concurrent Pos:* Dir, Dept Clin Physiol, Nat Asthma Ctr, Denver, Colo, 72-76. *Mem:* Soc Pediat Res; Am Physiol Soc; Am Thoracic Soc; Am Col Chest Physicians; Am Acad Pediat. *Res:* Control of respiration; exercise; pathophysiology of asthma; pathophysiology of cystic fibrosis; pulmonary disease. *Mailing Add:* Div Pediat Pulmonary Dis Columbia Univ Col Physicians & Surgeons 630 W 168th St New York NY 10032

CROPPER, WALTER V, instrumentation, research administration; deceased, see previous edition for last biography

CROPPER, WENDELL PARKER, b Gary, Ind, June 22, 51. ECOLOGY. *Educ:* Cornell Col, Iowa, BA, 73; Emory Univ, Georgia, MS, 77, PhD(biol), 80. *Prof Exp:* Res asst, Emory Univ, 74-77, teaching asst, 77-80, asst prof biol & environ sci, 81; res assoc, Sch Forest Resources & Conserv, Univ Fla, 81-84; ctr environ res, Cornell Univ, 85-86; CONSULT, 86- *Concurrent Pos:* Vis asst prof & asst res scientist, Dept Forestry, Univ Fla, 86- *Mem:* Ecol Soc Am; Asn Southeastern Biologists; AAAS; Am Inst Biol Sci; Int Soc Ecol Modelling. *Res:* Lead uptake and translocation in an urban tree species; carbon cycling below ground in managed forest ecosystems and on a computer model of carbon cycling in forests; analysis of the stability of ecosystem models of nutrient cycling. *Mailing Add:* Dept Forestry 118 NZ Hall Univ Fla Gainesville FL 32601

CROSBIE, EDWIN ALEXANDER, b Washington, Pa, July 23, 21; m 46; c 2. PHYSICS. *Educ:* Washington & Jefferson Col, BA, 42, MA, 48; Univ Pittsburgh, PhD(physics), 52. *Prof Exp:* ASSOC PHYSICIST, ARGONNE NAT LAB, 52-, SR SCIENTIST, 87- *Mem:* Am Phys Soc; Sigma Xi. *Res:* Theoretical nuclear physics; high energy particle physics, particle accelerators; field theory. *Mailing Add:* 305 Somonauk Park Forest IL 60466

CROSBY, ALAN HUBERT, b Columbia, Pa, Oct 17, 22; m 53; c 3. ORGANIC CHEMISTRY. *Educ:* Ursinus Col, BS, 43; Univ Va, MS, 50, PhD, 51. *Prof Exp:* From asst prof to prof chem, Northwestern State Col La, 50-69, head dept phys sci, 57-69; prof, 69-86, EMER PROF CHEM, DEPT CHEM & PHYSICS, LOCK HAVEN STATE COL, 86- *Mem:* Am Chem Soc. *Res:* Chemistry of quinones. *Mailing Add:* 89 Jay St Mill Hall PA 17731

CROSBY, DAVID S, b St George, Utah, June 4, 38; m 62; c 2. APPLIED STATISTICS, MATHEMATICAL STATISTICS. *Educ:* Am Univ, BA, 62; Univ Ariz, MA, 64, PhD(math), 66. *Prof Exp:* Mathematician, Harry Diamond Labs, 62-64; asst prof math & statist, 66-75, chmn dept, 75-77, PROF MATH, STATIST & COMPUT SCI, AM UNIV, 75- *Concurrent Pos:* Statist consult, NESPIS/NOAA, 68- *Mem:* Math Asn Am; Inst Math Statist; Am Statist Asn. *Res:* Multidimensional probability distributions; statistical problems of inversion techniques. *Mailing Add:* Dept Math & Statist Am Univ 4400 Massachusetts Ave NW Washington DC 20016

CROSBY, DONALD GIBSON, b Portland, Ore, Sept 11, 28; m 53; c 2. ENVIRONMENTAL CHEMISTRY, PESTICIDE CHEMISTRY. *Educ:* Pomona Col, BA, 50; Calif Inst Technol, PhD(chem), 54. *Prof Exp:* Chemist, Union Carbide Chem Co, 54-55, group leader biol chem, 56-61; assoc toxicologist, Exp Sta, 61-62, toxicologist & lectr food sci & technol, 62-69, chmn, Agr Toxicol & Residue Res Lab, 62-66 & Regional Res Proj W-45, 68-70, PROF ENVIRON TOXICOL, UNIV CALIF, DAVIS, 69-, TOXICOLOGIST, EXP STA, 63- *Concurrent Pos:* Mem comn pesticide chem, Inst Union Pure Appl Chem, 74-82; mem mat hazard adv comt,

Environ Protection Agency, 75-79; assoc ed, J Agr Food Chem, 79-83. *Mem:* Fel AAAS; fel Am Chem Soc; Oceanic Soc; fel Int Acad Environ Safety. *Res:* Chemistry of natural products; pesticide chemistry and metabolism; chemical ecology; environmental chemistry; marine environment toxicology. *Mailing Add:* Dept Environ Toxicol Univ Calif Davis CA 95616

CROSBY, EDWIN ANDREW, b Greenwich, Conn, May 18, 24; m 47, 72; c 4. AGRICULTURAL PROBLEMS. *Educ:* Univ Conn, Storrs, BS, 48; Univ Calif, Davis, MS, 50, PhD(plant physiol), 54. *Prof Exp:* Teaching asst pomol, dept pomol, Univ Calif, Davis, 48-49, res asst, 49-54; asst prof hort, Rutgers Univ, 54-56; asst dir, Raw Prod Res Bur, Nat Canners Asn, 56-66, dir, Agr Div, 66-73, dir, agr & environ affairs, 73-75, vpres, 75-78, sr vpres agr & environ affairs, Nat Food Processors Asn, 78-83; pres, Agr Res Inst, 83-86; RETIRED. *Concurrent Pos:* Mem agr bd subcomt, Nat Res Coun-Nat Acad Sci, 65-67; from secy to pres, Agr Res Inst, 68-71; group exec, Resources & Support Servs, Nat Food Processors Asn, 81-83. *Honors & Awards:* Diamond Pin Award, Nat Jr Hort Asn, 66. *Mem:* Fel Am Soc Hort Sci (pres, 76-77); AAAS; Coun Agr Sci & Technol. *Res:* Agricultural research promotion. *Mailing Add:* 7712 Savannah Dr Bethesda MD 20817

CROSBY, EMORY SPEAR, b Georgetown, SC, Jan 17, 28; m 53; c 2. FOREST PATHOLOGY, PLANT PHYSIOLOGY. *Educ:* Western Ky Univ, BS, 60, MA, 61; Clemson Univ, PhD(plant path), 65. *Prof Exp:* Assoc prof biol, Armstrong State Col, 66-68; assoc prof, 68-80, PROF BIOL, THE CITADEL, 80- *Res:* Fungus morphology and physiology. *Mailing Add:* Dept Biol Citadel Mil Col SC Charleston SC 29409

CROSBY, GARY WAYNE, structural geology, geophysics, for more information see previous edition

CROSBY, GAYLE MARCELLA, b Battle Creek, Mich. DEVELOPMENTAL BIOLOGY. *Educ:* Albion Col, BA, 55; Univ Mich, MS, 57; Brandeis Univ, PhD(develop biol), 67. *Prof Exp:* Asst prof biol, St Mary's Col, 69-73; res asst anat, Univ Wis, 73-75; ASST PROF ANAT, HOWARD UNIV, 75- *Mem:* Soc Develop Biol. *Res:* Determination and analysis of the morphogenetic factors responsible for normal development of the vertebrate limb. *Mailing Add:* 4614 Nottingham Dr Chevy Chase MD 20815

CROSBY, GLENN ARTHUR, b Hempfield Township, Pa, July 30, 28; m 50; c 3. PHYSICAL CHEMISTRY. *Educ:* Waynesburg Col, BS, 50; Univ Wash, PhD(phys chem), 54. *Prof Exp:* Res assoc chem, Fla State Univ, 55-57, vis asst prof physics, 57; from asst prof to assoc prof chem, Univ NMex, 57-67; chmn chem physics prog, 77-84, prof chem & chem physics, 67-90, PROF CHEM & MAT SCI, WASH STATE UNIV, 90- *Concurrent Pos:* Fulbright lectr, Univ T7o06ubingrn, WGer, 78-79; vis prof physics, Univ Canterbury, Christchurch, NZ, 74; consult, Indust & Govt Agencies; vis prof phys chem, Univ Hohenheim, WGer, 78-79; mem, Res Corp Cottrell Progs Adv Bd, 81-; US sr scientist Award, A von Humboldt Found, 78-79; mem comt educ, Am Chem Soc, 81-88, 90-92, vchair, 84-88, chair, 90-, subcomt precol educ, 81-87, chmn, Div Chem Educ, 82; vis res lectr, Japan, 86; vis res prof, Tohoku Univ, Sendai, Japan, 91; mem, Comn Life Sci, Nat Acad Sci, 90-; mem, steering comt, Coun Chief State School Officers, 90- *Mem:* Fel AAAS; Am Chem Soc; Am Phys Soc. *Res:* Perturbations of ions and molecules by chemical environments; luminescence of transition-metal and rare-earth complexes; optical, magnetic and electrical properties of complexes and solids with extended interactions; design of electro-optical materials. *Mailing Add:* NE 1825 Valley Rd Pullman WA 99163

CROSBY, GUY ALEXANDER, b Beverly, Mass, June 10, 42; m 66; c 2. SYNTHETIC ORGANIC CHEMISTRY, POLYMER CHEMISTRY. *Educ:* Univ NH, ScB, 64; Brown Univ, PhD(chem), 69. *Prof Exp:* Asst prof pharm, Col Pharm, Rutgers Univ, 46-49, assoc prof chem, 49-52, prof, 52-, emer prof pharmaceut chem & chmn dept; fel chem, Stanford Univ, 69-70; res scientist, Alza Corp, 70-72; dir chem synthesis, Dynapol Corp, 72-80; dir org chem, FMC Corp Agr Chem Group, 80-; RETIRED. *Concurrent Pos:* Prin scientist, Nat Inst Dent Res Contract, 76-79; consult asst prof, Stanford Univ, 75-78. *Mem:* Am Chem Soc; Soc Chem Indust (London). *Res:* Chemistry of taste; new synthetic sweeteners; polymeric reagents; natural products; food chemistry; agricultural chemistry. *Mailing Add:* 171 Main St Apt 11 Madison NJ 07940

CROSBY, LON OWEN, b Webster City, Iowa, Aug 6, 45; m 67; c 2. NUTRITION, NUTRITIONAL BIOCHEMISTRY. *Educ:* Iowa State Univ, BS, 67; Purdue Univ, PhD(nutrit), 71. *Prof Exp:* Res assoc biochem nutrit, Cornell Univ, 71-73; head monogastric nutrit, Syntex Corp, 73-75; sr prog scientist nutrit, Enviro Control, Inc, 75-79; res asst prof nutrit & exec dir, Clin Nutrit Ctr, Univ Pa, 79-83, dir, Cent Lab, VACS No 221, 83-86; PRES, NUMEDLOC, INC, 79- *Concurrent Pos:* Asst ed, Nutrit & Cancer: An Int J; consult, Philadelphia Vet Admin Med Ctr; adminr, Nutrit Support Serv, Univ Pa Hosp, 79-83; res collabr, Brookhaven Nat Lab, 88- *Mem:* Fel Am Col Nutrit; Am Soc Parenteral & Enteral Nutrit; Am Chem Soc; Inst Food Technologists. *Res:* Interrelationship between diet, nutrition and disease, including nutritional and biochemical interactions and the use of nutrition in disease prevention and treatment; medical instrumentation; human performance. *Mailing Add:* Numedloc Inc 430 Hollybush Rd Bryn Mawr PA 19010

CROSBY, MARSHALL ROBERT, b Jacksonville, Fla, June 3, 43; div; c 2. SYSTEMATICS OF MOSSES. *Educ:* Duke Univ, BS, 65, PhD(bot), 69. *Prof Exp:* Cur cryptogams, 68-74, chmn dept bot, 74-79, dir res, 77-86, dir, Bot Info Serv, 86-88, ASST DIR, MO BOT GARDEN, 88- *Concurrent Pos:* Res assoc, Wash Univ, 68-69, adj asst prof bot, 70-73, fac assoc, 74-79, adj prof biol, 80-87; adj prof biol, Univ Mo, St Louis, 83- *Mem:* Bot Soc Am; Am Soc Plant Taxonomists; Am Bryol & Lichenological Soc; Int Asn Plant Taxon; AAAS; Sigma Xi; British Bryological Soc; Nordic Bryological Soc. *Res:* Systematics of mosses with concentration on monographic studies of selected tropical families and genera and studies of the overall classification of the mosses at the familial and generic levels. *Mailing Add:* Mo Bot Garden P O Box 299 St Louis MO 63166-0299

CROSBY, ROBERT H(OWELL), JR, chemical engineering, for more information see previous edition

CROSBY, WARREN MELVILLE, b Topeka, Kans, Mar 19, 31; m 54; c 1. OBSTETRICS & GYNECOLOGY. *Educ:* Washburn Univ, Topeka, BS, 53; Univ Kans, MD, 57; Am Bd Obstet & Gynec, dipl, 65. *Prof Exp:* Intern, St Luke's Hosp, Kansas City, Mo, 57-58; resident obstet & gynec, Univ Calif, San Francisco, 58-62; from instr to assoc prof, 62-70, PROF OBSTET & GYNEC & VCHMN DEPT, UNIV OKLA, 70- *Mem:* AMA; fel Am Asn Obstet & Gynec; Soc Gynec Invest; fel Am Col Obstet & Gynec; Cent Asn Obstet & Gynec. *Res:* Erythroblastosis Fetalis; intrauterine fetal transfusion; studies of impact and decompression in pregnant women and animals; fetal malnutrition. *Mailing Add:* Dept Gynec & Obstet Univ Okla Health Sci Ctr Box 26901 Oklahoma City OK 73190

CROSBY, WILLIAM HOLMES, JR, b Wheeling, WVa, Dec 1, 14; m 40, 59, 86; c 7. HEMATOLOGY. *Educ:* Univ Pa, AB, 36, MD, 40. *Prof Exp:* Intern, Walter Reed Gen Hosp, US Army, 40-41, instr, Off Cand Sch, Med Field Serv, 41-43 & 45-46, regimental surgeon, 85th Infantry Div, Italy, 43-45, resident intern med, Brooke Gen Hosp, 46-48, res fel hematol, Pratt Diagnostic Hosp, 49-50, med specialist, Queen Alexandra Mil Hosp, London, 50-51, chief dept hematol, Walter Reed Army Inst Res, 51-65, chief hematol serv, Walter Reed Gen Hosp, 53-65, dir div med, 59-65, cancer chemother prog, 60-65; chief hemat, dir blood res & sr physician, NEng Med Ctr Hosps & prof med, Sch Med, Tufts Univ, 65-72; head, Div Hemat & Oncol & dir, L C Jacobson Blood Ctr, Scripps Clin & Res Found, 72-79; res hematologist, Dept Hemat, Walter Reed Army Inst Res, 79-83; prof med, Uniformed Serv Univ, 81-84; distinguished physician, Vet Admin, 83-86; DIR HEMAT, CHAPMAN CANCER CTR, JOPLIN, MO, 86- *Concurrent Pos:* Dir surg res team, Korea, 52-53; spec lectr hemat, Dept Med, George Washington Univ, 53-65; mem hematol test comt, Am Bd Internal Med, 70-76; adj prof med, Univ Calif, San Diego, 72-80; clin prof med, George Washington Univ, 80-86. *Honors & Awards:* Stitt Award, 64; McCollum Award, 70. *Mem:* Am Soc Clin Invest; fel Am Col Physicians; Am Soc Hematol; Asn Am Physicians; Asn Mil Surgeons; fel AAAS. *Res:* Clinical hematology; iron metabolism; marrow; spleen. *Mailing Add:* Chapman Cancer Ctr PO Box 2644 Joplin MO 64803

CROSLEY, DAVID RISDON, b Webster City, Iowa, Mar 4, 41; div; c 1. PHYSICAL CHEMISTRY, ATOMIC AND MOLECULAR PHYSICS. *Educ:* Iowa State Univ, BS, 62; Columbia Univ, MA, 63, PhD(phys chem), 66. *Prof Exp:* Res assoc physics, Joint Inst Lab Astrophys, 66-68; asst prof phys chem, Univ Wis-Madison, 68-75; proj leader, Ballistic Res Labs, Aberdeen Proving Ground, Md, 75-79; sr chem physicist, 79-80, prog mgr, 80-88, ASSOC DIR, MOLECULAR PHYSICS LAB, SRI INT, MENLO PARK, CALIF, 88- *Concurrent Pos:* Joint Inst Lab Astrophys fel, 66-68; vis prof, Ruhr Universität, Boehum, Fed Repub Ger, 88, Université de Paris, Orsay, France, 89. *Mem:* Fel Am Phys Soc; AAAS; Combustion Inst; Am Chem Soc. *Res:* Spectroscopy and chemical dynamics, reaction and energy transfer, of small state-selected molecules; laser-based diagnostics for combustion; atmospheric chemistry; materials processing. *Mailing Add:* Molecular Physics Dept SRI Int Menlo Park CA 94025

CROSS, ALEXANDER DENNIS, b Leicester, Eng, Mar 29, 32. ORGANIC CHEMISTRY. *Educ:* Univ Nottingham, BSc, 52, PhD(org chem), 55; FRIC. *Hon Degrees:* DSc, Univ Nottingham, 66. *Prof Exp:* Fulbright travel scholar & fel, Univ Rochester, 55-57; sr fel & res asst, Dept Sci & Indust Res, Imp Col, London, 57-58, from asst lectr to lectr, 58-60; sr chemist, Syntex SAm, 61-62, asst dir chem res, 62-64, from assoc dir to dir, Syntex Inst Steroid Chem, 64-67, vpres, Syntex Res, 66-67, vpres, Com Rels, 67-70 & Chem Group 67-72, pres, Syntex Sci Systs, 70-74, pres, Int Pharm Div, 74-78, sr vpres, Corp Econ & Strategic Planning, Syntex Corp, 78-79, exec vpres, 79-86, PRES, ZOECON CORP, 86- *Concurrent Pos:* Vis lectr, Stanford Univ, 67. *Mem:* Am Chem Soc; The Chem Soc. *Res:* Elucidation of structure and synthesis of natural products; synthesis of biologically active steroids; applications of spectroscopic methods to problems in stereochemistry and structure. *Mailing Add:* 286 Parklane Atherton CA 94025-5457

CROSS, CARROLL EDWARD, b Palo Alto, Calif, Nov 5, 35. TOXICOLOGY. *Educ:* Columbia Univ, MD, 61. *Prof Exp:* PROF MED PHYSIOL, SCH MED, UNIV CALIF, DAVIS, 76- *Mailing Add:* Div Pulmonary & Critical Care Med Dept Med & Physiol Univ Calif Davis Sch Med 4301 X St Sacramento CA 95817

CROSS, CHARLES KENNETH, b Toronto, Ont, June 25, 27; m 54; c 2. ANALYTICAL CHEMISTRY. *Educ:* Univ Toronto, BA, 52, MA, 53. *Prof Exp:* Res chemist, Can Packers Inc, 53-90; RETIRED. *Mem:* Chem Inst Can. *Res:* Trace contaminants in food, particularly nitrosamines in cured meats; analytical chemistry in the field of fats and oils. *Mailing Add:* Four Barford Rd Etobicoke ON M9W 4H2 Can

CROSS, CLARK IRWIN, b Olds, Alta, Sept 20, 13; US citizen; m 46; c 2. PHYSICAL GEOGRAPHY. *Educ:* Univ Wash, MS, 47, PhD(geog), 51. *Prof Exp:* Instr geog, Southern Methodist Univ, 46-47; from instr to prof, Univ Fla, 49-80, chmn dept geog & phys sci, 78-80, emer prof geog & phys sci, 80-87; RETIRED. *Concurrent Pos:* State climatologist Fla. *Mem:* Soc Am Foresters; Asn Am Geog; Sigma Xi. *Res:* Geography of USSR; air photo interpretation. *Mailing Add:* 5016 NW 20th Pl Gainesville FL 32611

CROSS, DAVID RALSTON, b Lawrence, Mass, Mar 16, 28; m 58; c 3. SOLAR ELECTROCHEMISTRY, APPLIED SOLAR ENERGY. *Educ:* Wesleyan Univ, AB, 49, MA, 51; Syracuse Univ, PhD(phys & org chem), 60. *Prof Exp:* Sr res chemist, Res Labs, Eastman Kodak Co, NY, 57-64; from asst prof to assoc prof phys chem, 64-74, PROF CHEM, WESTERN MD COL, 74- *Concurrent Pos:* NSF res grant, 68-; vis prof, Case Western Reserve Univ, 70-71; Gas Res Inst grant, Am Soc Electronics Engrs, 81- *Mem:* Solar Energy Soc; Electrochem Soc. *Res:* Chromatography; chemiluminescence; photochemistry; chemical and electrode kinetics; electrochemistry; solvated electrons; photochemistry of phytochrome. *Mailing Add:* 159 Leppo Rd Westminster MD 21157

CROSS, EARLE ALBRIGHT, JR, b Memphis, Tenn, Nov 23, 25; m 48; c 4. INSECT TAXONOMY, INSECT ECOLOGY. *Educ:* Utah State Univ, BS, 51; Univ Kans, MA, 55, PhD(entom), 62. *Prof Exp:* Instr entom, Purdue Univ, 57-58; vis instr, Univ Kans, 58-59, NIH res assoc, 59-60; from asst prof to prof biol sci, Northwestern State Col La, 60-70; assoc prof biol, 70-73, PROF BIOL, UNIV ALA, 73- *Concurrent Pos:* NIH res grant, 64-67; US Forest Serv coop res matching grant, 66-67 & 75-76; NSF teaching grant, 72 & res grant, 72-74; Ala Sch Mines energy develop grant, 78-81. *Mem:* Cent States Entom Soc; Entom Soc Am; Soc Vector Ecol. *Res:* Systematics and ecology of insects and acarines; systematics and ecology of mite family Pyemotidae; ecological energetics; strip mine succession. *Mailing Add:* Box 1927 University AL 35487

CROSS, EDWARD F, b Brooklyn, NY, Nov 16, 08; m 35; c 2. MECHANICAL & CIVIL ENGINEERING. *Educ:* Stevens Inst Technol, MechE, 29; Columbia Univ, MA, 38. *Hon Degrees:* DEng, Walla Walla Col, 74. *Prof Exp:* Pub utility engr, Columbia Eng & Mgt Corp, 29 32; construct engr, Godfrey M Weinstein Construct Co, 32-33; mach design engr, Multi-Needle Eng Corp, 33-35; teacher, High Schs, NY, 35-37, 38-41, NJ, 37-38 & Manhattan Sch Aviation Trades, NY, 41-47; from asst prof to prof mech & civil eng, Walla Walla Col, 47-59, head dept, 47-74, emer dean eng, 79-85; RETIRED. *Concurrent Pos:* Consult engr, Walla Walla Col, 47-88 & City of College Place, 52-66. *Honors & Awards:* Western Elec Fund Award, 65. *Mem:* Nat Soc Prof Engrs; Am Soc Eng Educ. *Res:* Thermodynamics; engineering mechanics; structural analysis and design; fluid mechanics; mechanical and electrical design of buildings. *Mailing Add:* 626 SE Fourth St College Place WA 99324

CROSS, ERNEST JAMES, JR, b Loganton, Pa, Aug 3, 30; m 52; c 6. AERODYNAMICS, FLIGHT MECHANICS. *Educ:* Pa State Univ, BS, 59; Univ Tex, Austin, MS, 65, PhD(aerospace eng), 68. *Prof Exp:* Aeronaut engr, Aircraft Lab, Wright-Patterson AFB, 59-62, res engr gas dynamics, Hypersonic Res Lab, Aerospace Res Lab, 64-69, chief, VSTOL Technol Div, Air Force Flight Dynamics Lab, 70-71 & Protype Div, 71-73; prof & dir, Aerospace Eng Dept, Raspet Flight Res Lab, Miss State Univ, 73-79; prof & head, Aerospace Eng Dept, Tex A&M Univ, 79-84; DEAN, COL ENG & TECHNOL, OLD DOMINION UNIV, 84- *Concurrent Pos:* Mem, Int Tripartite Tech Coord Panel, VSTOL, 71-72, USAF/NASA VSTOL Transp Technol Ad Hoc Working Group, 72-73, Flight Dynamics Adv Panel, Air Force Flight Dynamics Lab, 75; US rep, US/Fed Repub Ger, res & develop prog, VSTOL Corp & proj officer, US/France Mutual Weapons Develop Data Exchange Agreement, VSTOL Technol, 71-73. *Mem:* Assoc fel Am Inst Aeronaut & Astronaut; Am Soc Eng Educ; Sigma Xi. *Res:* Experimental investigation of the aerodynamics and performance of subsonic aircraft and ground vehicles including automobiles, trucks and rail systems; facilities utilization including wind tunnel and full-scale flight test. *Mailing Add:* Dean Col Eng & Technol Old Dominion Univ Norfolk VA 23529-0236

CROSS, FRANK BERNARD, b Kansas City, Mo, Sept 17, 25; m 54; c 3. ZOOLOGY. *Educ:* Okla Agr & Mech Col, BS, 47, MS, 49, PhD(zool), 51. *Prof Exp:* Instr biol, 51-53, asst prof zool, 53-59, assoc prof & assoc dir, State Biol Surv, 59-67, PROF SYSTS & ECOL & DIR STATE BIOL SURV, 67-, CUR FISHES, MUS NATURAL HIST, 67- *Mem:* AAAS; Am Fisheries Soc; Am Soc Ichthyologists & Herpetologists. *Res:* Ichthyology and fishery biology. *Mailing Add:* Dyche Hall Univ of Kans Lawrence KS 66045

CROSS, GEORGE ALAN MARTIN, b Cheshire, Eng, Sept 27, 42; m 86; c 1. MOLECULAR BIOLOGY, BIOCHEMISTRY. *Educ:* Univ Cambridge, Eng, BA, 64, PhD(microbiol), 68. *Prof Exp:* Scientist biochem parasitol, Med Res Coun, Eng, 69-77; Head, Dept Immunochem & Molecular Biol, Wellcome Found Res Labs, 77-82; PROF MOLECULAR PARASITOL, ROCKEFELLER UNIV, 82- *Concurrent Pos:* Instr biol, Marine Biol Lab, Woods Hole, 80-84; consult, Wellcome Found, 82-87, WHO, 83-87 & New Eng Biolabs, 86-; coed, Molecular & Biochem Parasitol, 86- *Honors & Awards:* Fleming Prize, Soc Gen Microbiol, 78;; Chalmers Med, Royal Soc Trop Med & Hyg, 83; Paul Ehrlich & Ludwig Darmstaedter Prize, 84. *Mem:* Am Soc Cell Biol; fel Royal Soc; Biochem Soc; Soc Gen Microbiol. *Res:* Cell biology and genetics of protozoal parasites, including trypanosomes and malaria; structure and genetic modulation of surface glycoproteins and receptors; posttranslational modification of proteins; tropical parasitology. *Mailing Add:* Dept Molecular Parasitol Rockefeller Univ 1230 York Ave New York NY 10021-6399

CROSS, GEORGE ELLIOT, b Auburndale, NS, Apr 17, 28; m 52; c 6. PURE MATHEMATICS. *Educ:* Dalhousie Univ, BA, 52, MA, 54; Univ BC, PhD(math), 58. *Prof Exp:* From instr to asst prof math, Victoria Col, BC, 56-59; from asst prof to assoc prof, Univ Western Ont, 59-63; assoc prof, 63-65, dean grad studies, 67-72, chmn dept pure math, 78-83, PROF MATH, UNIV WATERLOO, 65- *Mem:* Can Math Cong. *Res:* General theories of integration. *Mailing Add:* Dept Pure Math Univ Waterloo Waterloo ON N2L 3G1 Can

CROSS, GERALD HOWARD, b Marshall, Minn, Nov 15, 33; m 55; c 3. WILDLIFE MANAGEMENT. *Educ:* Univ Minn, Duluth, BA, 55; Ohio State Univ, MA, 64; NDak State Univ, PhD(zool), 73. *Prof Exp:* Exten specialist wildlife, 73-76, DEPT HEAD FISHERIES & WILDLIFE, VA POLYTECH INST & STATE UNIV, 76- *Mem:* Wildlife Soc; Am Fisheries Soc. *Res:* Wildlife land use interrelationships. *Mailing Add:* Dept Wildlife Mgt Va Polytech Inst & State Univ Blacksburg VA 24061

CROSS, HAROLD DICK, b Wellington, Kans, Apr 2, 30; m 47; c 4. MEDICINE. *Educ:* Colby Col, BA, 53; Yale Univ, MD, 57. *Prof Exp:* Physician, Eastern Maine Med Ctr & Pvt Pract, 58-91; DIR EMERGENCY SERV, NORTHERN MAINE MED CTR, 91- *Concurrent Pos:* Internship, Eastern Maine Med Ctr, 57-58; clin assoc prof med & family pract, Col Med, Univ Vt, 71-84; co-dir prob oriented med info serv, Augusta Gen Hosp, 69-70. *Mem:* Inst Med-Nat Acad Sci. *Res:* Delivery of health care; ambulatory; systems approach to health care. *Mailing Add:* 71 Dubley St Hampden ME 04444

CROSS, HIRAM RUSSELL, b Dade City, Fla, Feb 11, 44; m 63; c 2. MEATS & MUSCLE BIOLOGY. *Educ:* Univ Fla, BSA, 66, MS, 69; Tex A&M Univ, PhD(meat sci), 72. *Prof Exp:* Instr animal sci, Univ Fla, 67-69 & meat sci, Tex A&M Univ, 69-72; meat grader, Agr Mkt Serv, USDA, 66-67, mkt specialist, meat sci, 72-73, res scientist, 73-80, res leader, 80-83; SECT LEADER, MEAT SCI, TEX A&M UNIV, 83- *Concurrent Pos:* Mem, Agr Mkt Serv, USDA Task Force Develop US Feeder Grades, 72-74; proj leader, Beef Qual Study, USDA, 73-76, Beef Compos Study, 75-76, Instrument Grading Task Force, 79-81, OIG Task Force, 80-81. *Mem:* Am Meat Sci Asn (pres, 82-83); Am Soc Animal Sci; Inst Food Technologists. *Res:* The conversion of muscle to meat, including early postmortem events affecting muscle quality; animal growth and composition. *Mailing Add:* Dept Animal Sci Tex A&M Univ Kleberg Ctr College Station TX 77843

CROSS, JAMES THOMAS, b Ardmore, Tenn, Sept 2, 24; m 48. NUMBER THEORY, APPLIED MATHEMATICS. *Educ:* Brown Univ, AB, 51; Harvard Univ, MS, 56; Univ Tenn, PhD(math), 62. *Prof Exp:* From instr to assoc prof, 55-68, PROF MATH, UNIV SOUTH, 68- *Concurrent Pos:* Teacher math, Limestone County Bd Educ, 51-54. *Mem:* Math Asn Am; Am Asn Univ Prof. *Res:* Study the extension of elementary number theoretic results to algebraic integers. *Mailing Add:* Univ South SPO 1246 Sewanee TN 37375

CROSS, JOHN MILTON, b Little Falls, NJ, Jan 2, 15; m 45; c 4. PHARMACEUTICAL CHEMISTRY. *Educ:* Rutgers Univ, BS, 36; Univ Md, MS, 39, PhD(pharm), 43. *Prof Exp:* Res chemist, Merck & Co, Inc, NJ, 42-46; asst prof pharm, 46-49, assoc prof chem, 49-52, PROF, 52-, EMER PROF PHARMACEUT CHEM & CHMN DEPT, COL PHARM, RUTGERS UNIV; TECH DIR, FOOD & PHARM PROD DIV, FMC CORP, 91- *Concurrent Pos:* Dir org chem, FMC Corp, 80-91. *Mem:* Am Pharmaceut Asn; Am Chem Soc. *Res:* Preparation of iodine compounds for roentgenology; fats and oils; analysis of foods and drugs; synthesis of fungicides; photo-decomposition of foods and drugs. *Mailing Add:* FMC Corp Food & Pharm Prods Div PO Box 8 New Brunswick NJ 08903

CROSS, JOHN PARSON, b Bloomington, Ill, Mar 23, 50; m 74; c 2. ORGANIC CHEMISTRY. *Educ:* Mass Inst Technol, SB, 72, PhD(org chem), 76. *Prof Exp:* Res chemist, Milliken Res Corp, 76-78; develop chemist, Milliken Chem, Div Milliken & Co, 78-80; sr res chemist, 80-81, GROUP LEADER, NEW BUS DEVELOP, SPECIALTIES TECHNOL DIV, EXXON CHEM CO, 82- *Mem:* Am Chem Soc; Sigma Xi. *Res:* Specialty chemicals and process development; water soluble polymers; enhanced oil recovery surfactants. *Mailing Add:* Exxon Chem Co 8230 Stedman St Houston TX 77029-3999

CROSS, JOHN WILLIAM, b Memphis, Tenn, Feb 9, 47; m 73; c 2. PLANT BIOCHEMISTRY, PLANT CELL BIOLOGY. *Educ:* Vanderbilt Univ, BA, 69; Calif Inst Technol, PhD(cell biol & genetics), 76. *Prof Exp:* Carnegie fel, dept plant biol, Carnegie Inst of Wash, 75-78; staff scientist biochem, Plant Genetics Group, Pfizer Cent Res, 78-81, sr res scientist, 81-85; at Sogetal Inc, 85-90; SR RES BIOCHEMIST, UNIV CALIF, DAVIS, 90-; AT CALIF DEPT FOOD & AGR, 91- *Concurrent Pos:* Vis acad researcher, Univ Calif, Davis, 90- *Mem:* AAAS; Am Soc Plant Physiologists; Am Chem Soc; Crop Sci Soc Am. *Res:* Regulation of protein synthesis and of organelle development; mechanism of action and metabolism of plant growth regulators; function and organization of cellular organelles; developmental biochemistry of plants. *Mailing Add:* Calif Dept Food & Agr 1220 N St PO Box 942871 Sacramento CA 94271-0001

CROSS, JON BYRON, b New York, NY, July 16, 37; m 83; c 3. CHEMICAL PHYSICS, PHYSICAL CHEMISTRY. *Educ:* Univ Colo, BS, 60; Univ Ill, PhD(chem physics), 67. *Prof Exp:* MEM STAFF, LOS ALAMOS NAT LAB, UNIV CALIF, 66- *Concurrent Pos:* Adj prof chem, Univ NMex, 78- *Mem:* Am Phys Soc; Am Chem Soc. *Res:* Dynamics of chemical reactions; measurement of reactive scattering differential cross sections and product translational energy distributions using crossed molecular beam apparatus and mass spectrometer detector; photoionization of large organic molecules; photodissociation dynamics; gas-surface interactions using molecular beam techniques; atomic oxygen interactions with spacecraft; space station materials requirements. *Mailing Add:* Los Alamos Nat Lab MS G738 Los Alamos NM 87545

CROSS, LEE ALAN, b Flint, Mich, Sept 23, 34; m; c 3. LASERS. *Educ:* Univ Mich, BS, 57, MS, 58, PhD(physics), 66. *Prof Exp:* Consult lasers, Laser Systs Ctr, Lear-Seigler Inc, 66-67; prof physics, Western Ill Univ, 67-78; RES PHYSICIST & ASSOC PROF IMPACT PHYSICS, RES INST, UNIV DAYTON, 78- *Concurrent Pos:* Fel, Dept Chem, Univ Mich, 66-67; consult, Avionics Div, Lear-Seigler Inc, 67-68; vis scholar, Dept Chem, Univ Mich, 78- *Mem:* Am Phys Soc; Inst Elec & Electronics Engrs. *Res:* Pulsed and CW metal vapor lasers; laser Q-switching; molecular spectroscopy; impact physics. *Mailing Add:* 444 Volusia Ave Dayton OH 45409

CROSS, LESLIE ERIC, b Leeds, Eng, Aug 14, 23; m 50; c 6. ELECTRICAL ENGINEERING, PHYSICS. *Educ:* Univ Leeds, BSc, 48, PhD(physics), 52. *Prof Exp:* Exp off electronics, Brit Admiralty, 43-46; asst lectr physics, Univ Leeds, 49-51; Imp Chem Indust fel, 51-54; sr res assoc, Brit Elec Res Assoc, 54-61; sr res assoc, Pa State Univ, 61-63, assoc prof physics, 63-68, assoc dir mat res lab, 69-85, dir mat res lab, 85-89, PROF ELEC ENG, PA STATE UNIV, 68- *Mem:* Nat Acad Eng; fel Am Ceramic Soc; Phys Soc Japan; fel Am Inst Phys; fel Inst Elec & Electronics Engrs; fel Am Optical Soc. *Res:* Material science; ferroelectric and antiferroelectric properties of titanates and niobates; thermodynamics of ferroelectricity; high permittivity materials; dielectric measuring techniques; dielectric properties of glass systems. *Mailing Add:* Mat Res Lab Rm 167 Pa State Univ Univ Park PA 16802

CROSS, RALPH DONALD, b Quincy, Ill, Dec 31, 31; m 55; c 4. CLIMATOLOGY, WATER RESOURCES GEOGRAPHY. *Educ:* Eastern Mich Univ, AB, 60; Univ Okla, MA, 61; Mich State Univ, PhD(geog), 68. *Prof Exp:* Instr geog, Southeast Mo State Col, 61-63; asst prof, Okla State Univ, 66-68 & Boston Univ, 68-71; assoc prof, 71-79, PROF GEOG, UNIV SOUTHERN MISS, 79- *Concurrent Pos:* Boston Univ Grad Sch grant soil moisture, Payne County, Okla, 68-70; Univ Southern Miss grant, Miss State Atlas, 72-74; res consult, Miss Marina Resources Coun, 75, Miss Water Resources Inst, 75-76, US Fish & Wildlife Serv, 79-81 & Miss Dept Natural Resources, 81-82. *Mem:* Am Water Resources Asn; Air Pollution Control Asn; Asn Am Geog; Nat Weather Asn. *Res:* Hydroclimatology; water resources; a regional interest in the Union of Soviet Socialist Republics; air quality analysis; agricultural climatology; regional interest in Canadian studies. *Mailing Add:* Southern Sta Box 5051 Univ Southern Miss Hattiesburg MS 39406

CROSS, RALPH EMERSON, b Detroit, Mich, June 3, 10. MACHINE TOOLS. *Educ:* Mass Inst Technol, BS, 33. *Prof Exp:* Pres, chmn & chief exec officer, Cross Co, 67-79, chmn bd, Cross & Trecker Corp, 79-82, emer chmn, 82-86; RETIRED. *Mem:* Nat Acad Eng; Nat Mach Tool Builders Asn (pres, 75); hon mem Soc Mfg Engrs; Soc Automotive Engrs. *Mailing Add:* 4120 Shelldrake Lane Boynton Beach FL 33436

CROSS, RALPH HERBERT, III, b Oakland, Calif, Aug 17, 38; m 59; c 2. COASTAL ENGINEERING, OCEANOGRAPHY. *Educ:* Univ Calif, Berkeley, BS, 61, MS, 62, PhD(civil eng), 66. *Prof Exp:* Jr civil engr, Univ Calif, Berkeley, 61-66; asst prof civil eng, Mass Inst Technol, 66-71; asst vpres, Alpine Geophys Assoc, 71-74; ASST VPRES & SR COASTAL ENGR, WOODWARD-CLYDE CONSULTS, 74- *Mem:* Am Soc Civil Engrs. *Res:* Water waves; coastal structures; nearshore oceanography; coastal site surveys. *Mailing Add:* Woodward-Clyde Consults 500 12th St Suite 100 Oakland CA 94607-4014

CROSS, RICHARD JAMES, b New York, NY, Mar 31, 15; m 39; c 5. INTERNAL MEDICINE, COMMUNITY HEALTH. *Educ:* Yale Univ, BA, 37; Columbia Univ, MD, 41, ScD(med), 49. *Prof Exp:* From instr to asst prof med, Columbia Univ, 47-59, asst dean, 57-59; assoc dean, Univ Pittsburgh, 59-63; assoc prof, Sch Med, Temple Univ, 63-64; adminr, Asn Am Med Cols, 64-65; prof med & assoc dean, 65-70, chmn dept, 70-80, prof, 70-85, EMER PROF COMMUNITY MED, UNIV MED & DENT NJ-ROBERT WOOD JOHNSON RUTGERS MED SCH, PISCATAWAY, NJ, 85- *Concurrent Pos:* Dean, Fac Med, Univ Ghana, 63-64. *Mem:* AMA; Am Asn Sex Educr, Coun & Therapists; Soc Sci Study Sex; Sex Info & Educ Coun US. *Res:* Treatment of thrombo-embolic disease; aerobic phosphorylation; renal tubular transport in rabbit kidney slices; medical school administration and education; teaching human sexuality to health professionals. *Mailing Add:* 210 Elm Rd Princeton NJ 08540

CROSS, RICHARD LESTER, b Hoboken, NJ, Aug 26, 43; m 71; c 1. BIOCHEMISTRY, BIOENERGETICS. *Educ:* Hartwick Col, BA, 66; Yale Univ, PhD(chem), 70. *Prof Exp:* From asst prof to assoc prof, 73-82, PROF & CHMN BIOCHEM, STATE UNIV NY HEALTH SCI CTR, SYRACUSE, 82- *Concurrent Pos:* Jane Coffin Childs Fund fel, Univ Calif, Los Angeles, 70-73; mem res comt, Am Heart Asn, 78-81; vis prof biochem, Pub Health Inst, New York, 81-82; mem, Molecular Biol Panel, NSF, 82-85; co-chmn/chmn, Gordon Res Conf Bioenergetics, 89-91. *Mem:* Sigma Xi; Fedn Am Scientists; Am Soc Biol Chem. *Res:* Energy-transducing membrane systems; mechanism of Adenosine Triphosphate synthesis by oxidative phosphorylation; structure of nucleotide binding sites. *Mailing Add:* Dept Biochem State Univ NY Health Sci Ctr Syracuse NY 13210

CROSS, ROBERT EDWARD, b Toledo, Ohio, July 24, 42; m 67; c 3. CLINICAL BIOCHEMISTRY, TOXICOLOGY. *Educ:* Univ Toledo, BS, 65, MS, 67; Univ Fla, PhD(org chem), 71. *Prof Exp:* Assoc prof, 79-88, PROF, DEPT PATH, MED & BIOCHEM, UNIV NC SCH MED, 88-; DIR, TOXICOL & CLIN PHARMACOL LAB, UNIV NC HOSPS, 78- *Concurrent Pos:* Dir, Clin Chem Labs, Univ NC Hosps, 81-90. *Mem:* Am Soc Clin Pathologists; Am Asn Clin Chem; Acad Clin Lab Physicians & Scientists; AAAS; NY Acad Sci; Sigma Xi. *Res:* Development of improved instrumentation and methodology in clinical chemistry; interpretation of laboratory tests; clinical enzymology; toxicology and clinical pharmacology. *Mailing Add:* Clin Chem Labs Univ NC Hosps 1071 Patient Support Tower Chapel Hill NC 27514

CROSS, ROBERT FRANKLIN, b Columbus, Ohio, May 6, 24; m 46; c 4. VETERINARY PATHOLOGY. *Educ:* Ohio State Univ, DVM, 46, MSc, 50; Purdue Univ, PhD(vet path), 61. *Prof Exp:* Instr vet path, Ohio State Univ, 48-50; vet pathologist, Territory of Hawaii, 50-58; from instr to assoc prof, Purdue Univ, 58-66; VET PATHOLOGIST, OHIO AGR RES & DEVELOP CTR, 66- *Concurrent Pos:* Lectr, Univ Hawaii, 55-58. *Mem:* Am Vet Med Asn; Am Col Vet Path; Conf Res Workers Animal Dis; Am Soc Vet Clin Path. *Res:* Veterinary pathology. *Mailing Add:* 1655 Linwood Dr Wooster OH 44691

CROSS, RONALD ALLAN, microbial genetics; deceased, see previous edition for last biography

CROSS, STEPHEN P, b Los Angeles, Calif, Apr 10, 38; div; c 4. MAMMALOGY, WILDLIFE BIOLOGY. *Educ:* Calif State Polytech Col, BS, 60; Univ Ariz, MS, 62, PhD(zool), 69. *Prof Exp:* From instr to asst prof biol, 63-67, from asst prof to assoc prof, 69-77, PROF BIOL, SOUTHERN ORE STATE COL, 78- *Concurrent Pos:* Consult, Bur Reclamation, 73-74, Park Serv, 76-77 & US Forest Serv, 77-78; wildlife biologist, Bur Land Mgt, 79-82; vis prof, Northern Ariz Univ, 84-85. *Mem:* Wildlife Soc; Am Soc Mammal; Northwest Sci Asn. *Res:* Vertebrate utilization of special and unique habitats; behavioral ecology of small mammals, including bats, especially as related to mans manipulation of habitats. *Mailing Add:* Dept Biol Southern Ore State Col Ashland OR 97520

CROSS, TIMOTHY ALBERT, b Hyannis, Mass, Oct 16, 53; m 76; c 2. NUCLEAR MAGNETIC RESONANCE. *Educ:* Trinity Col, Hartford, Conn, BS, 76; Univ Pa, Philadelphia, PhD(chem), 81. *Prof Exp:* ASSOC PROF CHEM & DIR NUCLEAR MAGNETIC RESONANCE FACIL, FLA STATE UNIV, TALLAHASSEE, 84- *Concurrent Pos:* NSF Presidential Young investr award, 85-90; NIH ad hoe Molecular Cytol Study Sect, 88; Sloan fel, 88-90; NSF III Rev Panel, 89, Biophys Panel, 89- *Mem:* Am Chem Soc; NY Acad Sci; Biophys Soc; AAAS. *Res:* Development of a solid state nuclear magnetic resonance method for the elucidation of a structural and dynamic characterization of polypeptides and proteins with atomic resolution, with application to the linear polypeptide, Gramicidin A. *Mailing Add:* Dept Chem Fla State Univ 600 W Col Ave Tallahassee FL 32306-3006

CROSS, TIMOTHY AUREAL, b Pittsburgh, Pa, Jan 22, 46; m 70; c 3. SEDIMENTOLOGY, TECTONICS. *Educ:* Oberlin Col, BA, 67; Univ Mich, Ann Arbor, MS, 69; Univ Southern Calif, PhD(geol), 76. *Prof Exp:* Explor geologist, Texaco, Inc, 69-72; asst prof sedimentology & tectonics, Univ NDak, Grand Forks, 75-78; assoc prof, Purdue Univ, 78-79; res geologist, Exxon Prod Res Co, Houston, 79-84; ASSOC PROF GEOL, COLO SCH MINES, 84- *Concurrent Pos:* Vis prof, Univ Strasbourg, France, 90. *Mem:* Am Asn Petrol Geologists; Soc Econ Paleontologists & Mineralogists; Geol Soc Am; Int Asn Sedimentologists. *Res:* History of igneous activity, western United States with relation to plate tectonics; Carbonate and clastic sedimentology; petroleum and coal geology; foreland basin sedimentation and tectonics; sequence stratigraphy. *Mailing Add:* Dept Geol Colo Sch Mines Golden CO 80401

CROSS, VIRGINIA ROSE, b Portland, Ore, May 15, 50; m 74; c 3. SURFACE CHEMISTRY, ADHESION SCIENCE. *Educ:* Ore State Univ, BS, 72; Mass Inst Technol, PhD(phys chem), 76; Univ Houston, MBA, 86. *Prof Exp:* Res chemist phys chem, Celanese Plastics Co, 76-79 & Am Hoechst Co, 79-80; STAFF RES CHEMIST, EXXON CHEM CO, 80- *Mem:* Am Chem Soc; Sigma Xi. *Res:* Research and development in single-site catalyzed polyolefin and specialized polyolefins. *Mailing Add:* 3419 Ledgestone Dr Houston TX 77059

CROSS, WILLIAM HENLEY, entomology, ecology; deceased, see previous edition for last biography

CROSSAN, DONALD FRANKLIN, b Wilmington, Del, Apr 8, 26; m 48; c 3. PLANT PATHOLOGY. *Educ:* Univ Del, BS, 50; NC State Col, MS, 52, PhD(plant path), 54. *Prof Exp:* Res asst veg dis, NC State Col, 50-54; assoc prof, 54-66, asst dean col agr sci & asst dir agr exp sta, 66-69, assoc dean col agr sci, 69-76, assoc dir, Del Agr Exp Sta, 69-79; vpres, Univ Res, 72-76, PROF PLANT SCI, COL AGR SCI, UNIV DEL, 66-, DEAN, 76-, DIR, DEL AGR EXP STA, 79- *Mem:* AAAS; Am Phytopath Soc. *Res:* Vegetable plant diseases; disease resistance; fungus physiology. *Mailing Add:* Col Agr Sci Univ Del Newark DE 19717-1303

CROSSER, ORRIN KINGSBERY, b Akron, Ohio, Mar 2, 29; m 56; c 2. CHEMICAL ENGINEERING. *Educ:* Univ Mo, BS, 50, MS, 51; Rice Univ, PhD(chem eng), 55. *Prof Exp:* Instr chem eng, Univ Mo, 51-52; from asst prof to prof, Univ Okla, 55-66, actg dir dept, 65-66; PROF CHEM ENG, UNIV MO-ROLLA, 66- *Concurrent Pos:* Fulbright travel grant, 61-62. *Mem:* Am Inst Chem Engrs; Am Chem Soc; Nat Soc Prof Engrs; Am Soc Eng Educ. *Res:* Mass and heat transport phenomena; chemical kinetic technology. *Mailing Add:* Dept Chem Eng Univ Mo Rolla MO 65401

CROSSFIELD, A SCOTT, b Berkeley, Calif, Oct 2, 21; m 43; c 6. AERONAUTICAL ENGINEERING. *Educ:* Univ Wash, BS, 49, MS, 50. *Hon Degrees:* DSc, Fla Inst Technol, 82. *Prof Exp:* Prod expediter, Boeing Airplane Co, 41-42; design specialist, Seeger Aircraft Specialties, 46-49; aeronaut res pilot, Nat Adv Comt for Aero High Speed Flight Sta, Calif, 50-55; eng test pilot & design specialist, chief eng test pilot, N Am Aviation, Inc, 55-60, div dir test & qual assurance, Space & Info Systs Div, 60-66, tech dir res & develop, Space Div, 66-67; div vpres flight res & develop, Eastern Air Lines, 67-73; sr vpres, Hawker Siddeley Aviation, Inc, 74-75; TECH CONSULT, HOUSE REP COMT SCI & TECHNOL, 77- *Concurrent Pos:* Wind tunnel operator, Aeronaut Lab, Univ Wash, 46-50; aeronaut consult, 75-77; Ira C Eaker Hist fel, Air Force Asn Aerospace Educ Found, 82 & 83; Am Inst Aeronaut & Astronaut Lectr, 87-88. *Honors & Awards:* Lawrence Sperry Award, Inst Aeronaut Sci, 54, Octave Chanute Award, 58; Inst Aeronaut Sci Award, Inst Aeronaut & Astronaut, 59; Astronautics Award, Am Rocket Soc, 60; Kincheloe Award, Soc Exp Test Pilots, 60; Harmon Trophy, Clifford B Harmon Trust, 60; Collier Trophy, Nat Aeronaut Asn, 62; Montgomery Award, Nat Soc Aerospace Prof, 62; Outstanding Contrib Qual Control Award, Am Soc Qual Control, 67; Meritorious Serv Aviation Award, Nat Bus Aircraft Asn, 84. *Mem:* Fel Am Inst Aeronaut & Astronaut; fel Soc Exp Test Pilots; hon fel Aerospace Med Asn; hon mem Exp Aircraft Asn; hon mem Flying Physicians Asn; fel Inst Aerospace Sci; Sigma Xi. *Res:* Wind tunnel experiments and flight test of high speed research aircraft. *Mailing Add:* 12100 Thoroughbred Rd Herndon VA 22071

CROSSLEY, DAVID JOHN, b Salisbury, Eng, May 10, 44. GEOPHYSICS. *Educ:* Univ Newcastle-upon-Tyne, BSc, 66; Univ BC, MSc, 69, PhD(geophys), 73. *Prof Exp:* Res scientist, ICI Fibers Div, 66-67; fel, 73-74, res assoc physics, Mem Univ Nfld, 74-76; ASST PROF APPL GEOPHYSICS, McGILL UNIV, 76- *Mem:* Can Asn Physicists; Am Geophys Union. *Res:* Solid earth and planetary geophysics, especially dynamics of the earth's core and free oscillation theory; theory; inversion of geophysical data in areas of pure and applied geophysics. *Mailing Add:* Dept Geol Sci McGill Univ 853 Sherbrooke St W Montreal PQ H3A 2M5 Can

CROSSLEY, DERYEE ASHTON, JR, b Kingsville, Tex, Nov 6, 27; m 50, 61; c 1. ENTOMOLOGY. *Educ:* Tex Technol Col, BA, 49, MS, 51; Univ Kans, PhD(entom), 57. *Prof Exp:* Instr biol, Tex Technol Col, 49-51; asst, Univ Kans, 51-56; biologist, Oak Ridge Nat Lab, 56-67; prof entom, 67-84, RES

PROF, UNIV GA, 85- *Honors & Awards:* Cert Appreciation, US Forest Serv, 84, Partnership Award, 90. *Mem:* Ecol Soc Am; Entom Soc Am; Am Soc Naturalists. *Res:* Nutrient cycling in ecosystems; role of soil arthropods in ecosystems. *Mailing Add:* Dept Entom Univ Ga Athens GA 30601

CROSSLEY, F(RANCIS) R(ENDEL) ERSKINE, b Derby, Eng, July 21, 15; nat US; m 41; c 2. MECHANICAL ENGINEERING, MECHANISMS. *Educ:* Cambridge Univ, Eng, BA, 37, MA, 41; Yale Univ, DEng, 49. *Prof Exp:* Asst, Gen Motors Res Labs, 37-38; designer, Ford Motor Co, 39-42; instr mech drawing, Univ Detroit, 42-44; from asst prof to assoc prof, mech eng, Yale Univ, 44-65; vis fel, Inst Sci & Technol, Univ Manchester, England, 64; prof mech eng, Ga Inst Technol, 65-68; prof mech & aerospace eng, 69-78, prof civil eng, 78-80, EMER PROF, UNIV MASS, AMHERST, 80- *Concurrent Pos:* Guest lectr, Tech Univ Braunschweig, 59; ed, Proc Int Conf Teachers Mechanisms, 61; organizer, Int Conf Mechanisms, Yale Univ, 61; Fulbright sr lectr, Munich & Aachen Tech Univs, 62-63; invited lectr tour, seven German tech univs, 64; ed-in-chief, J Mechanisms, 66-71; mem bd dirs, Int Asn Exchange Students Tech Experience, US Div; chmn, Tenth Am Soc Mech Engrs Mechanisms Conf, 68; pres, US Coun Theory Mach & Mechanisms, Inc, 68-70; first vpres, Int Fedn Theory Mach & Mechanisms, 69-75; exec ed, Int Jour Mechanism & Mach Theory, 71-80; vis researcher, Tech Univ Warsaw, Poland, 75; vis prof, Rhein-Westphal Tech Hochschule Aachen, 75-76; Fulbright lectr, Bucharest, Rumania, 76; guest lectr, Acad Sci, USSR, Inst Mach Studies, 76; hon chmn, Fifth World Cong Theory Mach & Mechanisms, Montreal, 79; staff scientist, Conn State legis, Off Legis Res, Hartford, 81-83; adj prof, mech eng, Univ Fla, 88-91. *Honors & Awards:* von Humboldt Sr Scientist Award, Fed Repub Ger, 75; Mechanisms Award, Am Soc Mech Engrs, 76; Centennial Medal, Am Soc Mech Engrs, 80. *Mem:* Fel Am Soc Mech Engrs; Am Soc Eng Educ; Sigma Xi; hon mem Union German Engrs; hon mem Int Fedn Theory Mach & Mechanisms. *Res:* Dynamics and vibration of machines; kinematics and mechanisms; resource recovery and garbage incineration. *Mailing Add:* 282 Pine Orchard Rd Branford CT 06405

CROSSLEY, FRANK ALPHONSO, b Chicago, Ill, Feb 19, 25; m 50; c 1. TITANIUM SCIENCE, TITANIUM TECHNOLOGY. *Educ:* Ill Inst Technol, BS, 45, MS, 47, PhD(metall eng), 50. *Prof Exp:* Instr metall, Ill Inst Technol, 47-49; prof & dept head foundry eng, Tenn State Univ, Nashville, 50-52; sr scientist, Res Inst, Ill Inst Technol, 52-66; sr mem res lab, Lockheed Missiles & Space Co, 66-74, dept mgr, 74-79, consult engr, 79-86; res dir propulsion mat, Aerojet Propulsion Res Inst, Techsysts Co, Sacramento, Calif, 86-87; res dir mat eng & develop, 87-90, TECH PRIN, RES & ENG, GENCORP AEROJET PROPULSION DIV, SACRAMENTO, CALIF, 90- *Mem:* Fel Am Soc Metals; Minerals Metals & Mat Soc; Am Inst Aeronaut & Astronaut; Sigma Xi. *Res:* Titanium science and technology including: phase relationships, heat treatment, mechanical properties and alloy development; stress-corrosion of metal alloys; grain refinement of metals; high conductivity copper alloys. *Mailing Add:* 7575 Woodborough Dr Roseville CA 95661-9563

CROSSLEY, PETER ANTHONY, b Newcastle, Gt Brit, Nov 17, 39. IMAGE PROCESSING, COMPUTATIONAL VISION. *Educ:* Cambridge Univ, Eng, BA, 62, PhD(elec eng), 65. *Prof Exp:* Mem tech staff, David Sarnoff Res Ctr, RCA, 65-74; mem res staff, Fairchild Res Lab, 74-75, sr mem res staff, 78-87; consult mem res staff, Xerox Corp, 75-78; mem tech staff, Schlumberger Palo Alto Res, 87-89; RES SCIENTIST, SCHLUMBERGER LAB COMPUTER SCI, 89- *Mem:* Soc Photo-Optical Instrumentation Engrs. *Res:* Image processing and computational vision; semiconductor device physics and fabrication. *Mailing Add:* Schlumberger Lab Computer Sci 8311 N RR 620 Austin TX 78736

CROSSMAN, DAVID W, b Boston, Mass, May 23, 46. SOFTWARE ENGINEERING, FULL TEXT DATA RETRIEVAL. *Educ:* Univ Calif, Berkeley, BA, 78. *Prof Exp:* Consult, 80-84; CHIEF ENGRS, TIMEPLACE, WALTHAM, MASS, 85- *Mem:* Math Asn Am; Asn Comput Mach. *Mailing Add:* 61 Myrtle St Boston MA 02114

CROSSMAN, EDWIN JOHN, b Niagara Falls, Ont, Sept 21, 29; m 52; c 2. ICHTHYOLOGY. *Educ:* Queen's Univ, Can, BA, 52; Univ Toronto, MA, 54; Univ BC, PhD, 57. *Prof Exp:* Fishery biologist, Biol Sta, Queen's Univ, Can, 50; biologist, Toronto Anglers & Hunters Conserv Proj, 51-54; fishery biologist, BC Game Comn, 54-57; asst cur, Dept Fishes, 57-64, assoc cur, Dept Ichthyol & Herpet, 64-68, cur, 68-74, CUR-IN-CHG, DEPT ICHTHYOL & HERPET, ROYAL ONT MUS, 74-; PROF ZOOL, UNIV TORONTO, 68- *Mem:* Am Fisheries Soc Am; Soc Ichthyologists & Herpetologists; Soc Syst Zool; Am Inst Fishery Res Biol; Can Soc Zool. *Res:* Biology, distributions and systematics of freshwater fishes, particularly esocoid fishes. *Mailing Add:* 25 Anson Toronto ON M1M 1X2 Can

CROSSMON, GERMAIN CHARLES, b Prattsburg, NY, May 9, 05; m 31. MICROSCOPY. *Educ:* Alfred Univ, BS, 28. *Prof Exp:* Teacher high sch, NJ, 29-31; bacteriologist, Sch Med & Dent, Univ Rochester, 35-42; biol & chem microscopist & indust hyg chemist, Bausch & Lomb, Inc, 42-74; CONSULT INDUST HYG & MICROS, 74- *Mem:* AAAS; Am Indust Hyg Asn; Am Chem Soc; Am Micros Soc. *Res:* Biological and chemical microscopy; industrial hygiene chemistry including original publications on the identification of asbestos and beryllium compounds by dispersion staining microscopy. *Mailing Add:* 23 Esternay Lane Pittsford NY 14534

CROSSNER, KENNETH ALAN, b New York, NY, Oct 10, 46; m 72; c 1. POPULATION ECOLOGY, MAMMALOGY. *Educ:* City Col New York, BS, 69, MA, 72; Rutgers Univ, PhD(ecol), 76. *Prof Exp:* Asst prof biol, Univ Col, Rutgers Univ, 76-77; asst prof biol, Seton Hall Univ, 77-84; ASSOC PROF BIOL, PASS CITY COL, 87- *Concurrent Pos:* Admin, Rutgers Univ, 84-86. *Mem:* Am Soc Mammalogists; Ecol Soc Am; Wildlife Soc; Mammal Soc; NY Acad Sci; Sigma Xi. *Res:* Mammal population ecology and their responses to urbanization and other environmental perturbations; distribution of mammals in suburbia; ecological needs of mammal populations in urban reservations. *Mailing Add:* 101 Hollywood Ave Somerset NJ 08873-1532

CROSSON, ROBERT SCOTT, b Fairbanks, Alaska, Oct 19, 38; m 59; c 2. GEOPHYSICS, SEISMOLOGY. *Educ:* Univ Wash, BS, 61; Univ Utah, MS, 63; Stanford Univ, PhD(geophys), 66. *Prof Exp:* From asst prof to assoc prof, 66-78, PROF GEOPHYS & GEOL, UNIV WASH, 78- *Mem:* Seismol Soc Am; Soc Explor Geophys; Am Geophys Union; AAAS. *Res:* Physical properties and structure of earth's crust and upper mantle; characteristics and distribution of earthquakes; tectonic and hazard implications of earthquakes; seismology of volcanic regions. *Mailing Add:* Geophys Prog AK-50 Univ Wash Seattle WA 98195

CROSSWHITE, CAROL D, b Perth Amboy, NJ, Aug 1, 40; m 61; c 3. DESERT ECOLOGY. *Educ:* Univ Calif, Riverside, BS, 61; Univ Wis, Madison, MS, 64, PhD(entom), 68. *Prof Exp:* Instr entom, Univ Wis, Madison, 68-70; CUR & EDUC COORDR, BOYCE THOMPSON SOUTHWESTERN ARBORETUM, UNIV ARIZ, 71- *Concurrent Pos:* Trustee, Ariz-Sonora Desert Mus, 79-84. *Mem:* Ecol Soc Am; Soc Syst Zool. *Res:* Ecology of the Sonoran Desert; insect-plant relationships in desert regions; pollination ecology. *Mailing Add:* PO Box AB Superior AZ 85273

CROSSWHITE, F JOE, b Springfield, Mo, Oct 13, 29; m 49; c 3. MATHEMATICS EDUCATION. *Educ:* Univ Mo, BSEd, 53, MEd, 58; Ohio State Univ, PhD(math educ), 64. *Prof Exp:* Teacher math, Salem High Sch, Mo, 53-57; instr, Keokuk Community Col, Iowa, 57-61; from instr to prof math educ, Ohio State Univ, 62-84; RETIRED. *Concurrent Pos:* US Off Educ fel, Stanford Univ, 68-69; prog mgr, NSF, 75-76; chmn, Conf Bd Math Sci, 84- *Mem:* Nat Coun Teachers Math (pres, 84-); Math Asn Am; Am Educ Res Asn. *Res:* Teaching and learning of mathematics at the precollege level including the education of teachers for this level. *Mailing Add:* 2048 E Briar St Springfield MO 65804

CROSSWHITE, FRANK SAMUEL, b Atchison, Kans, Sept 23, 40; m 61; c 3. ECONOMIC BOTANY, DESERT PLANT SCIENCE. *Educ:* Ariz State Univ, BS, 62; Univ Wis, Madison, MS, 65, PhD, 71. *Prof Exp:* Res asst bot, Univ Wis, Madison, 69-70; asst prof, Univ Wis Ctr, Waukesha, 70-71; CUR BOT, BOYCE THOMPSON SOUTHWESTERN ARBORETUM, UNIV ARIZ, 71- *Concurrent Pos:* Ed, Desert Plants J, 79- *Mem:* Int Asn Plant Taxon. *Res:* Classification of Scrophulariaceae; pollination ecology; history of botany; geography of North American plants; life forms of desert plants; desert ethnobotany; ecology of the Sonoran Desert. *Mailing Add:* Univ Ariz PO Box AB Boyce Thompson Southwestern Arboretum Superior AZ 85273

CROSSWHITE, HENRY MILTON, JR, b Riverdale, Md, March 26, 19; m 44; c 3. PHYSICS. *Educ:* Western Md Col, AB, 40; Johns Hopkins Univ, PhD(physics), 46. *Prof Exp:* Asst, Nat Res Coun war contract, Johns Hopkins Univ, 42-46, instr physics, 46-47, asst prof, 47-50, res assoc, 50-58, res scientist, 58-63, prin res scientist, 63-65, adj prof, 65-75; sr chemist, chem div, Argonne Nat Lab, 75-84. *Mem:* Optical Soc Am. *Res:* Visible and ultraviolet spectroscopy. *Mailing Add:* 4308 Kenny St Beltsville MD 20705

CROSTHWAITE, JOHN LESLIE, b Winnipeg, Man, Dec 22, 40; m 65; c 2. MECHANICAL ENGINEERING, NUCLEAR FUEL WASTE CONTAINER DESIGN DEVELOPMENT. *Educ:* Univ Man, BSc, 62; Univ Birmingham, MSc, 64. *Prof Exp:* Lectr thermodyn, Univ Man, 65-66; ASSOC RES OFFICER NUCLEAR FUEL DEVELOP & WASTE MGT & HEAD CONTAINER DESIGN & TESTING SECT, AECL RES, 66- *Mem:* Can Soc Chem Eng. *Res:* Nuclear fuel development; heat transfer and thermodynamics in relation to nuclear energy development; nuclear waste management studies; development of durable containers for nuclear fuel waste disposal; spent fuel dry storage. *Mailing Add:* AECL Res Whiteshell Labs Pinawa MB R0E 1L0 Can

CROTEAU, RODNEY, b Springfield, Mass, Dec 6, 45; m 67; c 2. BIOCHEMISTRY. *Educ:* Univ Mass, Amherst, BS, 67, PhD(food sci), 70. *Prof Exp:* NIH res assoc biochem, Ore State Univ, 70-72; res assoc, 73-75, from asst agr chemist to assoc agr chemist, 75-80, assoc prof, 80-82, PROF CHEM & FEL, WASH STATE UNIV, 82- *Concurrent Pos:* Dir Inst, 86-89. *Mem:* Am Chem Soc; Am Soc Plant Physiologists; Phytochem Soc NAm; Am Soc Biol Chemists. *Res:* Biosynthesis and catabolism of terpenoids. *Mailing Add:* Inst Biol Chem Wash State Univ Pullman WA 99164-6340

CROTHERS, DONALD M, b Fatehgarh, India, Jan 28, 37; US citizen; m 60; c 2. BIOPHYSICS NUCLEIC ACIDS. *Educ:* Yale Univ, BS, 58; Cambridge Univ, BA, 60; Univ Calif, San Diego, PhD(chem), 63. *Prof Exp:* From asst prof to assoc prof, 64-71, chmn chem, 75-81, PROF, YALE UNIV, 71-, ALFRED P KEMP PROF CHEM, 85- *Concurrent Pos:* ed bd, J Molecular Biol, 71-75, Nucleic Acids Res, 73-82, Biochem, 75-79, Biopolymers, 77-79, 87-90, Annual Rev Phys Chem, 81-85; Biophys Chem Study Sect, NIH, 72-74, chmn, 74-76; Mellon fel, 58-60, Guggenheim fel, 78-79. *Honors & Awards:* Alexander von Humboldt US Sr Scientist Award, 81. *Mem:* Nat Acad Sci; fel Am Acad Arts & Sci. *Res:* Author of numerous articles and publications. *Mailing Add:* Yale Univ Chem 225 Prospect St New Haven CT 06511

CROTTY, WILLIAM JOSEPH, b NJ, Sept 25, 22; m 46; c 4. PLANT MORPHOGENESIS. *Educ:* City Col New York, BS, 48; Rutgers Univ, PhD(bot), 52. *Prof Exp:* Teaching fel bot, Rutgers Univ, 49-52; from asst prof to assoc prof, 52-73, chmn dept, 52-73, dir undergrad studies, 73-77, PROF BIOL, WASH SQ COL, NY UNIV, 73- *Mem:* Torrey Bot Club; Bot Soc Am; Am Soc Cell Biol; Fedn Am Soc Exp Biol. *Res:* Plant development. *Mailing Add:* Dept Biol Wash Sq Col NY Univ Rm 952 Brown New York NY 10003

CROUCH, BILLY G, b Port Lavaca, Tex, May 14, 30. CLINICAL PHARMACOLOGY, RADIATION BIOLOGY. *Educ:* Baylor Univ, BS, 54, MS, 55; Univ Tenn, PhD(clin physiol), 58. *Prof Exp:* Instr biol, Baylor Univ, 54-55; asst clin physiol, Col Med, Univ Tenn, 55-58, lectr, 56-58; radiation biologist, US Naval Radiol Defense Lab, Calif, 60-63; clin res pharmacologist, Sterling-Winthrop Res Inst, 63-64, sr clin pharmacologist, 64-65; dir med commun, Winthrop Labs, 65-68; dir drug regulatory affairs, Sterling Drug,

Inc, 68-76; exec vpres, 77-78, PRES & CHIEF EXEC OFFICER, SANOFI RES CO, INC, 78- *Concurrent Pos:* Nat Acad Sci-Nat Res Coun Donner fel, Radiobiol Sect, Nat Defense Res Coun, Neth, 58-59; Nat Heart Inst res fel, 59-60; vis lectr, Fac Med, State Univ Leiden, 59-60; vpres & mem bd dir, Choay Labs, Inc, 77-; mem nat comn clin use interferon, French Ministry Health, 80-81; mem bd dir, Girpi, Paris, 80-82; dir med affairs, Sanofi, SAm, Paris, 80-82; mem bd gov, Am Hosp Paris, 81-; vpres, Sanofi Pharmaceuticals, Inc, 81- *Mem:* Am Physiol Soc; Am Fedn Clin Res; Radiation Res Soc; Royal Soc Med; Neth Soc Radiobiol. *Res:* Clinical physiology; bone marrow transplantation in irradiation sickness; clinical evaluation of new drug substances. *Mailing Add:* 101 Park Ave 34th Fl New York NY 10178

CROUCH, EDUARD CLYDE, COLLAGEN BIOCHEMISTRY, BASEMENT MEMBRANES. *Educ:* Univ Wash, Seattle, PhD(biochem), 78, MD, 79. *Prof Exp:* ASSOC PATHOLOGIST, JEWISH HOSP, ST LOUIS, 83-; ASST PROF PATH, WASHINGTON UNIV, 83-; ASST PROF PATH, JEWISH HOSP, 83- *Res:* Mechanisms of fibrosis. *Mailing Add:* Dept Path Jewish Hosp 216 S Kings Hwy St Louis MO 63110

CROUCH, GLENN LEROY, b Los Angeles, Calif, Jan 14, 29; div; c 4. WILDLIFE ECOLOGY, FORESTRY. *Educ:* Univ Idaho, BS, 59; Colo State Univ, MS, 61; Ore State Univ, PhD(range mgt), 64. *Prof Exp:* Range scientist, 64-76, RES WILDLIFE BIOLOGIST, US FOREST SERV, 76- *Mem:* Wildlife Soc; Soc Range Mgt; Soc Am Foresters. *Res:* Forest wildlife relationships; protection of forest regeneration from animals; range science and management. *Mailing Add:* 620 Matthews St No 314 Ft Collins CO 80524

CROUCH, JAMES ENSIGN, b Urbana, Ill, Jan 28, 08; m 31; c 2. ORNITHOLOGY, ANATOMY. *Educ:* Cornell Univ, MS, 31; Univ Southern Calif, PhD(vert zool), 39. *Prof Exp:* Mem, Dept Zool, 32-42, prof, 42-73, chmn, Div Life Sci, 62- 69, EMER PROF ZOOL, SAN DIEGO STATE UNIV, 73- *Concurrent Pos:* Auth. *Mem:* AAAS. *Res:* Bird behavior; life history of Phainopepla nitens lepida. *Mailing Add:* 10430 Russell Rd La Mesa CA 91941

CROUCH, MADGE LOUISE, b Winston-Salem, NC, Sept 21, 19. PUBLIC HEALTH ADMINISTRATION. *Educ:* Methodist Hosp Sch Nursing, Brooklyn, dipl, 41; Columbia Univ, BS, 47; George Washington Univ, MA, 61. *Prof Exp:* Instr basic nursing & microbiol, Methodist Sch Nursing, Brooklyn, 41-43; nat nursing dir, Am Nat Red Cross, 48-65; asst to chief, Blood Bank Prod Lab, Div Biol Standards, NIH, 65-71; dir, Blood Bank Prod Br, Div Blood & Blood Prod, Ctr Biol Eval & Res, 72-77, dep dir, Div Biol Eval, Bur Biol, 77-81, dep, Off Biol Res & Rev, Food & Drug Admin, 81-88, dep dir, Off Compliance, 88-89; RETIRED. *Concurrent Pos:* Mem comt plasma & plasma substitutes, Div Med Sci, Nat Acad Sci, 69-70; mem secy task force nat blood policy, Dept of Health, Educ & Welfare, 72-74. *Mem:* Int Soc Blood Transfusion. *Res:* Development of improved blood banking procedures for prolonged storage of red blood cells; greater efficacy of components; training of personnel; techniques and equipment. *Mailing Add:* 5801 Rossmore Dr Bethesda MD 20814

CROUCH, MARSHALL FOX, b St Louis, Mo, Nov 22, 20; m 49; c 4. PHYSICS. *Educ:* Univ Mich, BS, 41; Univ Wash, PhD(physics), 50. *Prof Exp:* Res assoc, Nat Defense Res Comt, Univ Mich, 41-42; mem staff, Radiation Lab, Mass Inst Technol, 42-43 & Los Alamos Lab, 43; PROF PHYSICS, CASE WESTERN RESERVE UNIV, 50- *Concurrent Pos:* Fulbright res prof, Univ Tokyo, 56-57; dep sci attache, Am Embassy, Tokyo, 59-61. *Mem:* Am Phys Soc. *Res:* Neutrino physics; cosmic rays. *Mailing Add:* Dept Physics Case Western Reserve Univ Cleveland OH 44106-7079

CROUCH, MARTHA LOUISE, b East Lasing, Mich, Sept 15, 51. DEVELOPMENTAL BIOLOGY. *Educ:* Ore State Univ, BS, 74, Yale Univ, MPh, 78, PhD(biol), 79. *Prof Exp:* ASST PROF PLANT SCI, BIOL DEPT, IND UNIV, 79- *Concurrent Pos:* Res assoc biol dept, Univ Calif, Los Angeles, 80-81. *Mem:* Soc Develop Biol; Am Soc Plant Physiol; Botanical Soc Am. *Res:* Developmental biology of higher plants; switch from embryonic growth to germination in embryos of Brassica napus by studying the regulation of storage protein synthesis. *Mailing Add:* Dept Biol Ind Univ Bloomington IN 47401

CROUCH, NORMAN ALBERT, b Monroe, Wis, June 7, 40; m 63; c 3. VIROLOGY. *Educ:* Univ Wis, BS, 62, MS, 66, PhD(med microbiol), 69. *Prof Exp:* Fel, Col Med, Baylor Univ, 69-70 & Pa State Univ, 70-72; asst prof microbiol, Univ Iowa, 72-78; ASST PROF MICROBIOL, UNIV ILL, 78-, DIR, VIROLOGIC/MICROBIOLOGIC SERVICES, 82- *Mem:* Am Soc Microbiol. *Res:* Noncytocidal virus-induced alterations which affect the behavior and function of cells in the infected animal host. *Mailing Add:* Dept Biomed Sci Col Med Univ Ill 1601 Parkview Ave Rockford IL 61107

CROUCH, ROBERT J, b Dallas, Tex, July 30, 39. NUCLEIC ACID ENTYMOLOGY. *Educ:* Univ Ill, Champagne, PhD(chem), 68. *Prof Exp:* RES CHEMIST, NAT INST CHILD HEALTH & HUMAN DEVELOP, NIH, 71- *Mem:* Am Soc Biol Chemists. *Mailing Add:* Lab Molecular Genetics Bldg 6 Rm 331 Nat Inst Child Health & Human Develop NIH Bethesda MD 20892

CROUCH, ROGER KEITH, b Jamestown, Tenn, Sept 12, 40; div; c 3. SOLID STATE PHYSICS, CRYSTAL GROWTH. *Educ:* Tenn Technol Univ, BS, 62; Va Polytech Inst & State Univ, MS, 68, PhD(physics), 71. *Prof Exp:* Physicist semiconductors, 62-85, Floyd Thompson fel, 79-80, CHIEF SCIENTIST, MICROGRAVITY SCI & APPLICATIONS DIV, NASA HQ, 85- *Concurrent Pos:* Vis scientist, Mass Inst Technol, 79-80. *Mem:* Am Phys Soc; Am Asn Crystal Growth; AAAS; Metals Soc; Electrochem Soc. *Res:* Materials processing in space; semiconductor crystal growth; infrared detectors and arrays; electrical and optical properties of impurities in semiconductors; thermophysical property measurements. *Mailing Add:* Mail Code EN NASA Hq Washington DC 20546

CROUCH, STANLEY ROSS, b Turlock, Calif, Sept 23, 40; m 90. ANALYTICAL CHEMISTRY. *Educ:* Stanford Univ, MS, 63; Univ Ill, PhD(chem), 67. *Prof Exp:* Instr chem, Univ Ill, 67, vis asst prof, 67-68; from asst prof to assoc prof, 68-77, PROF CHEM, MICH STATE UNIV, 77- *Mem:* AAAS; Am Chem Soc; Optical Soc Am; Soc Appl Spectros. *Res:* Kinetics and mechanisms of analytical reactions; fast reaction kinetics; computers in chemistry; analytical spectroscopy. *Mailing Add:* Dept Chem Mich State Univ East Lansing MI 48824-1322

CROUNSE, NATHAN NORMAN, b Omaha, Nebr, May 25, 17; m 39; c 3. ORGANIC CHEMISTRY. *Educ:* Iowa State Col, BS, 38; Univ Iowa, PhD(chem), 42. *Prof Exp:* Chemist, C M Bundy Co, 42-43; sr chemist org res, Hilton-Davis Chem Co, 43-47 & Inst Med Res, Christ Hosp, 47-51; assoc dir res, Hilton-Davis Chem Co, 51-62, dir chem res, 62-75, sr scientist, 75-80; CONSULT, 80- *Mem:* Am Chem Soc. *Res:* pigments, fluorescent compounds; colorless duplicating papers; structure studies in dyes and drug metabolism; germicides; aliphatic synthesis; dyes. *Mailing Add:* 302 Club Circle Myrtle Beach SC 29572-4700

CROUNSE, ROBERT GRIFFITH, b Albany, NY, Mar 23, 31; m 55; c 2. DERMATOLOGY, BIOCHEMISTRY. *Educ:* Yale Univ, BS, 52, MD, 55; Am Bd Dermat, dipl, 63. *Prof Exp:* Intern, Grace-New Haven Hosp, Conn, 55-56; clin assoc, Nat Cancer Inst, 57-58; res asst prof, Sch Med, Univ Miami, 61, from asst prof to assoc prof, 63-64; assoc prof & chmn, sub-dept, Sch Med, Johns Hopkins Univ, 64-67; prof dermat & biochem & res dir, Med Col Ga, 67-73, assoc dean, Sch Med, & dir, Off Instrnl Systs, 71-73; prof dermat, 73-75, assoc dean, 73-80, chmn, dept med allied health prof, 78-80, prof exp surg, dept surg, 80-82, actg chmn, dept microbiol, 83-85, assoc dean basic sci, 83-87, ACTG CHMN, DEPT PREV MED & PUB HEALTH POLICY, EAST CAROLINA, UNIV GREENVILLE, NC, 87- *Concurrent Pos:* Fel dermat, Sch Med, Yale Univ, 56-57; Nat Cancer Inst spec res fel biochem, Sch Med, Univ Miami, 61-62; Nat Inst Arthritis & Metab Dis res career develop award, 62-64. *Mem:* Am Chem Soc; AMA; Soc Invest Dermat; Am Fedn Clin Res. *Res:* Biochemistry of epidermal keratinization; protein chemistry of epidermal structures; skin physiology and pharmacology; trace elements and nutrition; disorders of human hair. *Mailing Add:* Sch Med Brody Rm AD 55 E Carolina Univ Greenville NC 27834

CROUSE, DALE MCCLISH, b Los Angeles, Calif, Aug 27, 41; m 66; c 2. OTHER COMPUTER SCIENCES, TECHNICAL MANAGEMENT. *Educ:* Stanford Univ, BS, 64; Ore State Univ, MS, 67; Univ Del, PhD(org chem), 78. *Prof Exp:* Mfg asst, 78-79, St supvr, 80-86, RES CHEM, RES & DEVELOP, E I DU PONT DE NEMOURS & CO, 69-; COMPUT CONSULT, 87- *Mem:* Am Chem Soc; Sigma Xi. *Res:* Dialkyl carbene chemistry; Wittig reaction; organophosphorus; nuclear magnetic resonance spectroscopy; fluorocarbons; aromatic intermediates; specialty chemicals; local area networks; corporate computer architecture. *Mailing Add:* 125 Old Oak Rd Oaklands Newark DE 19711

CROUSE, DAVID AUSTIN, b Canton, Ill, Aug 29, 44; m 68; c 2. HEMATOLOGY, DEVELOPMENTAL IMMUNOLOGY. *Educ:* Western Ill Univ, BS, 66, MS, 68; Univ Iowa, PhD(radiobiol), 74. *Prof Exp:* NDEA IV fel radiobiol, Univ Iowa, 71-74; app, Argonne Nat Lab, 75-76; MEM FAC, DEPT ANAT, UNIV NEBR MED CTR, OMAHA, 76- *Mem:* Radiation Res Soc; Am Asn Immunol; Int Soc Exp Hematol; Soc Cell Kinetics. *Res:* hematopoietic transplantation; regulation of the hematopoietic stem cell compartment; radiation and drug effects on hematopoietic and lymphoid cells; microenviromental interactions between stem cell and stromal cell populations in the lymphohematopoietic system. *Mailing Add:* Dept Anat 42nd & Dewey Ave Omaha NE 68105

CROUSE, DAVID J, JR, b Johnstown, Pa, Jan 16, 20; m 54; c 2. CHEMICAL ENGINEERING. *Educ:* Pa State Univ, BS, 42. *Prof Exp:* Chem engr, Elec Storage Battery Co, 42-46 & Jaunty Fabric Corp, 47-48; ASST SECT CHIEF ATOMIC ENERGY RES, OAK RIDGE NAT LAB, 48-; AT TENN TECH UNIV. *Honors & Awards:* Indust Res-100 Award, 80. *Mem:* Am Nuclear Soc; Am Chem Soc; Sigma Xi. *Res:* Recovery of metals from ores by solvent extraction techniques; processing of atomic reactor fuels. *Mailing Add:* Tenn Tech Univ Box 5055 Cookville TN 38505

CROUSE, GAIL, b Connellsville, Pa, May 10, 23; m 51; c 3. ANATOMY. *Educ:* Heidelberg Col, BS, 50; Univ Mich, AM, 52, PhD(zool), 56. *Prof Exp:* From instr to asst prof anat, Hahnemann Med Col, 55-61; from asst prof to assoc prof, 61-83, PROF ANAT, SCH MED, TEMPLE UNIV, 83- *Mem:* Am Asn Anatomists. *Res:* Experimental mammalian tissue transplantation, especially differentiation of embryonic tissue after transplantation to the brain; reactivity to prostheses in animals; transplantation site; progressive development of the autonomic nervous system in human fetuses. *Mailing Add:* Dept Anat Temple Univ Sch Med Philadelphia PA 19122

CROUSE, GRAY F, b North Wilkesboro, NC, Feb 9, 49; m 83; c 2. MOLECULAR BIOLOGY. *Educ:* Duke Univ, BS, 70; Harvard Univ, PhD(chem), 76. *Prof Exp:* Postdoctoral scholar biol, Stanford Univ, 76-80; scientist, Lab Molecular Biol, Frederick Cancer Res Facil, Nat Cancer Inst, 80-84; ASSOC PROF BIOL, EMORY UNIV, 84- *Concurrent Pos:* Chmn, Dept Biol, Emory Univ, 87-90; mem, Sci Adv Comt Nucleic Acids & Protein Synthesis, Am Cancer Soc, 91- *Mem:* AAAS; Sigma Xi; Am Soc Microbiol. *Res:* Molecular and somatic cell genetics; regulation of eukaryotic gene expression; RNA processing; DNA repair, especially DNA mismatch repair. *Mailing Add:* Dept Biol Emory Univ Atlanta GA 30322

CROUSE, HELEN VIRGINIA, CHROMOSOME BEHAVIOR & SEX DETERMINATION. *Educ:* Univ Mo, PhD(zool), 40. *Prof Exp:* Res assoc & lectr cytol, Fla State Univ, 64-76; RETIRED. *Mailing Add:* Rte 3 Box 213 Hayesville NC 28904

CROUSE, PHILIP CHARLES, b Brooklyn, NY, July 14, 51; m 69. MECHANICAL ENGINEERING, MARINE SCIENCE. *Educ:* Univ Tex, Austin, BS, 73, MBA, 75. *Prof Exp:* Planning & eval analyst, Atlantic Richfield Co, 75-78; sr eval analyst, Champlin Petrol Co, 78; coordr, Sun Prod Co, 78-79 & 81; sr analyst, US Senate Comt Budget, 79-80; CHMN & PRES, PHILLIP C CROUSE & ASSOC, INC, 81- *Concurrent Pos:* Distinguished lectr comt chmn, Soc Petrol Engrs, 87-88, dir, 88-92. *Mem:* Soc Petrol Engrs. *Res:* International impacts of petroleum; finding, development and production of recovery processes, types of wells, and locations of wells; enhanced oil recovery processes; horizontal drilling. *Mailing Add:* Philip C Crouse Assocs Inc 10124 Solta Dallas TX 75218

CROUT, JOHN RICHARD, b Portland, Ore, Dec 30, 29; m 54; c 3. CLINICAL PHARMACOLOGY. *Educ:* Oberlin Col, AB, 51; Northwestern Univ, MD, 55, MS, 56; Am Bd Internal Med, dipl, 62. *Hon Degrees:* Dr, Uppsala Univ, Uppsala, Sweden. *Prof Exp:* Intern, Passavant Mem Hosp, Chicago, Ill, 55-56; asst resident med, Vet Admin Res Hosp, Chicago, 56-57; clin assoc, Nat Heart Inst, 57-60; asst resident, NY Univ-Bellevue Hosp Med Ctr, 60-61; instr pharmacol, Harvard Med Sch, 61-63; from asst prof to assoc prof pharmacol & internal med, Univ Tex Southwestern Med Sch, 63-70; prof pharmacol & med, Col Human Med, Mich State Univ, 70-71; dep dir, Bur Drugs, Food & Drug Admin, 71-72, dir, Off Sci Eval, 72-73, dir, 73-82; dir, Off Med Applications Res, NIH, 83-84; VPRES, MED & SCI AFFAIRS, BOEHRINGER MANNHEIM PHARMACEUTICALS CORP. *Concurrent Pos:* Res fel pharmacol, Harvard Med Sch, 61-62; USPHS fel, 61-63, Burroughs-Wellcome scholar clin pharmacol, 65-70; mem, Coun High Blood Pressure Res, Am Heart Asn, Anesthesiol Training Grant Comt, Nat Inst Gen Med Sci, 66-68, Pharmacol, Toxicol Prog Comt, 69-71, Comt Myocardial Info Study Ctrs, NIH, 67 & Ad Hoc Sci Adv Comt, Food & Drug Admin, 70-71; field ed, J Am Soc Pharmacol & Exp Therapeut, 68-71; consult, WHO, 74-84, mem bd, US Pharmacopeial Convention, 85- *Honors & Awards:* Distinguished Serv Award, USPHS. *Mem:* Am Fedn Clin Res; Am Soc Pharmacol & Exp Therapeut; Am Soc Clin Pharmacol & Therapeut; Am Soc Clin Invest; fel Am Col Physicians; Soc Clin Trials. *Res:* Catecholamine metabolism; pheochromocytoma; cardiovascular pharmacology; hypertension; clinical pharmacology; drug regulation policy. *Mailing Add:* Boehringer Mannheim Pharmaceut Corp 15204 Omega Dr Rockville MD 20850

CROUTHAMEL, CARL EUGENE, b Lansdale, Pa, Dec 25, 20; m 44; c 3. INORGANIC CHEMISTRY. *Educ:* Eastern Nazarene Col, BS, 42; Boston Univ, MA, 43; Iowa State Col, PhD(chem), 50. *Prof Exp:* Inorg chem res, Manhattan Proj, 43-44; res assoc, Inst Atomic Res, Iowa State Col, 46-50; sr chemist, Argonne Lab, 50-66, 67-73; sr engr, Exxon Nuclear Corp, 73-83; sr chemist, Argonne Lab, 83-86; RETIRED. *Concurrent Pos:* Consult, Argonne Lab, 86- *Mem:* AAAS; NY Acad Sci; Am Chem Soc; Sigma Xi; Am Nuclear Soc. *Res:* Inorganic chemistry of transuranium elements; rare earths; fused salt-liquid metals chemistry; development of high temperature fast breeder reactor fuel and materials; fuel performance of LWR reactor fuel. *Mailing Add:* 22112 SE 40th LN Issaquah WA 98027-7218

CROUTHAMEL, WILLIAM GUY, b Sellersville, Pa, Feb 11, 42; m 66; c 4. PHARMACY, PHARMACEUTICS. *Educ:* Philadelphia Col Pharm & Sci, BS, 65, MS, 67; Univ Ky, PhD(pharmaceut), 70. *Prof Exp:* From asst prof to assoc prof pharmaceut, WVa Univ, 70-75; assoc prof pharm, Univ Md, Baltimore City, 75-78; group chief, Hoffman-La Roche, 78-80, sect head, Biopharmaceut, 80-84, dir, 84-90, SR DIR, DEPT DRUG METAB, HOFFMANN-LA ROCHE, 90- *Mem:* Am Pharmaceut Asn; Sigma Xi; Am Asn Pharm Sci. *Res:* Effect of factors such as pH and intestinal blood flow on the kinetics of gastrointestinal drug absorption; effect of disease states on drug kinetics; dosage form effects on the absorption of drugs. *Mailing Add:* Dept Drug Metab Hoffmann-La Roche Inc Nutley NJ 07110

CROVETTI, ALDO JOSEPH, b Lake Forest, Ill, Apr 2, 30; m 59. ORGANIC CHEMISTRY. *Educ:* Lake Forest Col, BA, 51; Univ Ill, MS, 52, PhD(org chem), 55. *Prof Exp:* Res chemist, 57-65, group leader, 65-67, res org chemist, 67-70, sect head & dept mgr, Agr Prod Res & Develop, Agr & Indust Prod Res & Develop, 70-75, mgr agr chem res & develop, 75-78, mgr agr chem field res & develop, 78-83, MGR LICENSING, COMMERCIAL DEVELOP & UNIV RELATIONS, 83- *Concurrent Pos:* Mem bd dirs, Tech Transfer Conferences, 88- *Mem:* Am Chem Soc; Sigma Xi; Am Soc Hort Sci; Plant Growth Regulator Soc Am; Am Soc Plant Physiologist; Coun Agr Sci & Technol; Licensing Exec Soc; Am Phylopathological Soc; Entom Soc Am; Sigma Xi; Asn Univ Technol Managers; Soc Indust Microbiol. *Res:* Synthesis in the fields of heterocyclic compounds, mainly nitrogen; agricultural chemistry and biological research, including field research. *Mailing Add:* D-986 Abbott Labs 14th & Sheridan Rd North Chicago IL 60064

CROW, EDWIN LEE, b Clinton, Mo, Apr 26, 39; m 62; c 1. ORGANIC CHEMISTRY. *Educ:* Univ Mo, PhD(chem), 66. *Prof Exp:* Chemist, Electrochem Dept, Del, 65-68, develop chemist, Tex, 68-70, res supvr, Wilmington, Del, 70-74, tech supt, 74-78, prod supt, Gibbstown, NJ & Memphis, Tenn, 78-80, environ mgr, 80-81, res mgr chem develop, 80-87, ASSOC DIR PLANNING & ADMIN, DU PONT PHARMACEUT RES & DEVELOP, E I DU PONT DE NEMOURS & CO, WILMINGTON, DEL, 87- *Mem:* Am Chem Soc; Drug Info Soc; AAAS; Pharmaceut Mfrs Asn. *Res:* Catalytic reactions; organometallic chemistry; chemistry of carbenes; chemistry of olefins; organic processes; pharmaceutical chemistry. *Mailing Add:* 2604 W 17th St Wilmington DE 19806

CROW, EDWIN LOUIS, b Cadiz Twp, Wis, Sept 15, 16; m 42; c 2. STATISTICS, MATHEMATICS. *Educ:* Beloit Col, BS, 37; Univ Wis, PhM, 38, PhD(math), 41. *Prof Exp:* Instr math, Case Sch Appl Sci, 41-42; mathematician res & develop div, Bur Ord, US Navy Dept, 42-46; mathematician, US Naval Ord Test Sta, 46-54, head statist br, 50-54; consult statist, Boulder Labs, Nat Bur Stand, 54-65 & Environ Sci Serv Admin, 65-70; consult statist, Off Telecommun, 70-73, statistician, Nat Ctr Atmospheric Res, 75-82, STATISTICIAN, INST TELECOMMUN SCI, US DEPT COM, 74- *Concurrent Pos:* Instr exten div, Univ Calif, Los Angeles, 47-54 & Metrop State Col, 74; Govt Employees Training Act trainee, London, 61-62; adj prof appl math, Univ Colo, 63-81; assoc ed, J Am Statist Asn, 67-75, Commun Statist, 73-; contrib ed, Current Index Statist, 81- *Honors & Awards:* Bronze Medal, US Dept Com, 70; Outstanding Publ Award, Inst Telecomm Sci, 80 & 82; Ed Award, Am Meteorol Soc, 87. *Mem:* Fel AAAS; Soc Indust & Appl Math; Inst Math Statist; fel Am Statist Asn; Royal Statist Soc; Am Math Soc. *Res:* Expansion problems associated with ordinary differential equations; mathematical statistics with applications in ordnance, radio propagation, radio standards, communication systems and weather modification. *Mailing Add:* 605 20th St Boulder CO 80302

CROW, FRANK WARREN, b Madison, Wis, Feb 10, 51; m 78. ANALYTICAL CHEMISTRY, MASS SPECTROMETRY. *Educ:* Univ Wis-Stevens Point, BS, 73; Univ Va, PhD(chem), 78. *Prof Exp:* Res scientist anal chem, Columbus Labs, Battelle Mem Inst, 78-79; asst dir, dept chem, Midwest Ctr Mass Spectrometry, Univ Nebr, 79-84; RES SCIENTIST, THE UPJOHN CO, KALAMAZOO, MI, 84- *Mem:* Am Chem Soc; Am Soc Mass Spectrometry. *Res:* Gas chromatography; computer applications; negative ion chemical ionization mass spectrometry; chemical derivatization; mass spectrometry; fast atom bombardment. *Mailing Add:* Upjohn Co Unit 482-259-12 7000 Portage Rd Kalamazoo MI 49001-0102

CROW, FRANKLIN ROMIG, b Liverpool, Pa, Dec 28, 17; m 42; c 3. AGRICULTURAL ENGINEERING. *Educ:* Pa State Univ, BS, 40; Okla State Univ, MS, 52. *Prof Exp:* Agr engr, US Soil Conserv Serv, 40-43, 46-49; from instr to prof, 49-83, EMER PROF AGR ENG, OKLA STATE UNIV, 83- *Mem:* Am Soc Agr Engrs; Am Geophys Union. *Res:* Use of monomolecular films to reduce evaporation from water supply reservoirs; watershed hydrology. *Mailing Add:* 2011 W Tenth Stillwater OK 74074

CROW, GARRETT EUGENE, b Phoenix, Ariz, Dec 11, 42; m 72; c 3. SYSTEMATIC BOTANY, AQUATIC PLANTS. *Educ:* Taylor Univ, AB, 65; Mich State Univ, MS, 68, PhD(bot), 74. *Prof Exp:* Instr, Mich State Univ, 73-74, fel syst bot & man & biosphere, 74-75; asst prof, 75-81, ASSOC PROF SYST BOT & DIR, HERBARIUM, UNIV NH, 81- *Concurrent Pos:* US-USSR scientific exchange, 81; vis prof, Univ Nacional, Costa Rica, 84. *Mem:* Int Asn Plant Taxonomists; Am Soc Plant Taxonomists; Sigma Xi. *Res:* Plant systematics, particularly Caryophyllaceae Sagina and Lentibulariaceae Utricularia; scanning electron microscope studies in seed morphology in Caryophyllaceae, floristics, ecology and phytogeography of arctic, subantarctic, alpine plants, bogs and aquatic plants; flora and endangered species of plants of New Hampshire; aquatic plants of Costa Rica. *Mailing Add:* Dept Plant Biol Univ NH Durham NH 03824

CROW, JACK EMERSON, b Angola, NY, Aug 17, 39; m 60; c 3. SOLID STATE PHYSICS. *Educ:* Cleveland State Univ, BE, 62; Univ Rochester, PhD(physics), 67. *Prof Exp:* Fel, dept physics, Brookhaven Nat Lab, 67-68, from asst physicist to assoc physicist, 68-73; assoc prof, 73-77, PROF, DEPT PHYSICS, TEMPLE UNIV, 73-, DIR, CTR FOR MAT RES, 85- *Concurrent Pos:* Chmn, Int Conf Crystalline Elec Fields & Struct Effects in f-Electron Systs, 79, mem int adv comt, 4th Int Conf Crystal Field & Struct Effects f-Electron Systs, 81, mem prog comt, 6th Int Conf Crystal Fields Effects & Heavy Fermion Physics, 88, co-chmn, Int Conf Highly Correlated f-Electron Systs, 89; chmn, dept physics, Temple Univ, 79-82; dir, Solid States Prog, NSF, 84-86. *Mem:* Am Phys Soc; Mats Res Soc; Sigma Xi. *Res:* Solid state and low temperature physics; electronic and magnetic properties of f-electron materials; pure and applied superconductivity. *Mailing Add:* Phys Dept Temple Univ Philadelphia PA 19122

CROW, JAMES FRANKLIN, b Phoenixville, Pa, Jan 18, 16; m 41; c 3. POPULATION GENETICS. *Educ:* Friends Univ, AB, 37; Univ Tex, PhD(genetics), 41. *Prof Exp:* Tutor zool, Univ Tex, 37-40; from instr zool to asst prof zool & prev med, Med Sch, Dartmouth Col, 41-48; from asst prof to prof zool & genetics, 48-58, prof genetics, 58-86, actg dean, Med Sch, 63-65, chmn, Dept Genetics & Med Genetics, 65-70, EMER PROF MED GENETICS, UNIV WIS-MADISON, 86- *Mem:* Nat Acad Sci Inst Med; Genetics Soc Am (pres, 60); Am Soc Human Genetics (pres, 63); AAAS; Soc Study Evolution. *Res:* Genetics of Drosophila; population genetics. *Mailing Add:* Dept Genetics Univ Wis Madison WI 53706

CROW, JOHN H, b San Pedro, Calif, Nov 18, 42; m 66. PLANT ECOLOGY. *Educ:* Whittier Col, BA, 64; Wash State Univ, PhD(bot), 68. *Prof Exp:* Asst prof bot, 68-71, chmn dept, 72-78, ASSOC PROF BOT, RUTGERS UNIV, 71- *Concurrent Pos:* Ecol consult, State of Wash, 68-69, US Dept Interior, 74- & State of NJ Dept Pub Advocate, 75 & State Alaska Dept Fish & Game, 76-; mem panel productivity, Nat Wetlands Tech Coun, 78-79; solid waste coordr & dir, Dept Environ, Warrey Co, NJ, 81. *Honors & Awards:* A G Tausley Award, 81. *Mem:* AAAS; Am Inst Biol; Ecol Soc Am; Sigma Xi. *Res:* Ecological investigations of the salt marshes of Pacific Coastal Alaska; food habits of marsh, estuarine and marine organisms; secondary productivity; Atriplex salt balance physiological research; wetland vegetation of New Jersey; upland forest composition and environmental studies; salt marsh pollution; solid waste and resource recovery. *Mailing Add:* Dept Bot Rutgers Univ Newark NJ 07102

CROW, TERRY TOM, b Sapulpa, Okla, Sept 16, 31; m 54; c 3. ELECTROMAGNETISM. *Educ:* Miss State Univ, BS, 53; Vanderbilt Univ, MA, 57, PhD(physics), 60. *Prof Exp:* Asst prof physics, Miss State Univ, 60-62; physicist, Lawrence Radiation Lab, Univ Calif, 62-64; assoc prof, 64-67, PROF PHYSICS, MISS STATE UNIV, 67-, DEPT HEAD, 79- *Mem:* Am Phys Soc. *Res:* Electromagnetic theory. *Mailing Add:* 1111 Robinhood Rd Starkville MS 39759

CROWDER, GENE AUTREY, b Wichita Falls, Tex, Oct 25, 36; m 65; c 2. PHYSICAL CHEMISTRY. *Educ:* Cent State Col, Okla, BS, 58; Univ Fla, MS, 61; Okla State Univ, PhD(phys chem), 64. *Prof Exp:* Asst res chemist, Petrol Chems, Inc & Cities Serv-Continental Oil Co, 58-59; from asst prof to prof chem, WTex State Univ, 64-90; PROF CHEM & HEAD DEPT, LA TECH UNIV, 90- *Mem:* Am Chem Soc; Soc Appl Spectros; Sigma Xi. *Res:* Molecular spectroscopy; vibrational assignments and normal coordinate calculations; rotational isomerism. *Mailing Add:* Dept Chem La Tech Univ Ruston LA 71272-0001

CROWDER, LARRY A, b Mattoon, Ill, Mar 12, 42; m 64; c 2. INSECT PHYSIOLOGY, TOXICOLOGY. *Educ:* Eastern Ill Univ, BS, 64; Purdue Univ, MS, 66, PhD(entom), 70. *Prof Exp:* Assoc prof insect physiol, Univ Ariz, 69-80, prof, 80-; AT DEPT ENTOM, OKLA STATE UNIV, STILLWATER. *Mem:* AAAS; Entom Soc Am; Am Chem Soc; Am Registry Prof Entomologists; Sigma Xi. *Res:* Significance of 5-hydroxytryptamine in insects; mode of action of cyclodiene insecticides; regulation of insect diapause; insect resistance. *Mailing Add:* Dept Entom Okla State Univ 501 Life Sci W 139 Ag Hall Stillwater OK 74078

CROWDER, LARRY BRYANT, b Fresno, Calif, June 9, 50; m 73; c 2. FISHERIES ECOLOGY, SPECIES INTERACTIONS. *Educ:* Calif State Univ-Fresno, BA, 73; Mich State Univ, East Lansing, MS, 75, PhD(zool), 78. *Prof Exp:* Lectr ecol modeling, Calif State Univ-Fresno, 73; res asst zool, Mich State Univ, 73-77, teaching fel, 78; proj assoc, Lab Limnol, Univ Wis-Madison, 78-80, asst res scientist, 80-82, lectr limnol, 80; assoc prof ecol & fisheries, 82-87, ASSOC PROF ECOL & FISHERIES, NC STATE UNIV, 87- *Concurrent Pos:* Co-prin investr, NSF, 78-80, prin investr sea grants, Univ Wis, 82- & Univ NC, 83- & NSF, 85-; ed, Aquatic Ecol Newslett, Ecol Soc Am, 83-85; vis asst prof, Ctr Limnol, Univ Wis-Madison, 83-87. *Honors & Awards:* Res Award, Sigma Xi, 86. *Mem:* Am Fisheries Soc; Am Soc Limnol & Oceanog; Ecol Soc Am; Sigma Xi; Am Soc Fish Res Biol. *Res:* Species interactions and community ecology of aquatic systems; freshwater ponds, Great Lakes and marine systems; predation and competition; larval fish recruitment. *Mailing Add:* 1508 Banbury Rd Raleigh NC 27607

CROWE, ARLENE JOYCE, b Wakaw, Sask, Can, Oct 8, 31. BIOCHEMISTRY. *Educ:* Univ Alta, BSc, 50; McGill Univ, MSc, 56, PhD(biochem), 62. *Prof Exp:* Res asst liver dis, Hammersmith Hosp, Postgrad Sch, London, 56-57; chief technologist biochem, Montreal Children's Hosp, Que, 57-59; BIOCHEMIST, HOTEL DIEU HOSP, 62-; ASST PROF, BIOCHEM DEPT, QUEEN'S UNIV, ONT, 83- *Concurrent Pos:* Lectr, Queen's Univ, Ont, 62-; ed, Can Soc Clin Chem Newslett. *Honors & Awards:* Ames Award, Can Soc Clin Chem, 78. *Mem:* Can Soc Clin Chem; Am Asn Clin Chem; NY Acad Sci; Sigma Xi. *Mailing Add:* Dept Biochem Hotel Dieu Hosp Kingston ON K7L 3H6 Can

CROWE, CAMERON MACMILLAN, b Montreal, Que, Oct 6, 31; m 69. CHEMICAL ENGINEERING. *Educ:* McGill Univ, BEng, 53; Cambridge Univ, PhD(chem eng), 57. *Prof Exp:* Sr develop engr, Du Pont of Can, 57-59; from asst prof to assoc prof, 59-70, chmn dept, 71-74, PROF CHEM ENG, McMASTER UNIV, 70- *Concurrent Pos:* C D Howe Mem fel, Rice Univ, 67-68; assoc ed, Can J Chem Eng, 75-81; dir, Can Soc Chem Eng, 84-87. *Mem:* Am Inst Chem Engrs; fel Chem Inst Can; Can Soc Chem Eng (vpres, 90-91). *Res:* Design, optimization and simulation of reactors and chemical systems; applications of mathematics and numerical methods. *Mailing Add:* Dept Chem Eng McMaster Univ Hamilton ON L8S 4L7 Can

CROWE, CHRISTOPHER, b London, Eng, Dec 4, 28; m 52; c 3. GEOPHYSICAL EXPLORATION, PHYSICAL PROPERTIES OF EARTH. *Educ:* Western Ont Univ, BSc, 52, PhD(physics), 56. *Prof Exp:* Teacher pub schs, Ont, 46-47; physics demonstr, Western Ont Univ, 52-55; 1851 Exhib Overseas scholar, dept geod & geophys, Cambridge Univ, 56-58; asst prof geophys, Pa State Univ, 58-63; staff geophysicist, Tex Instruments, Inc, 64-66, scientist, 66-69; sr geophysicist, Sun Oil Co, 69-70, res scientist, 70-77; staff res scientist, Res Ctr, Amoco Prod Co, 77-89; CONSULT, RES & TECH, ALL SCHS & UNIV, 89- *Concurrent Pos:* Consult, Earth Sci Curric Proj, Am Geol Inst, 63-65; instr, Nat Educ Corp, Spartan Sch Aeronaut, 90- *Mem:* AAAS; Brit Inst Physics; Am Geophys Union; Soc Explor Geophys; Can Asn Physicists. *Res:* Heat flow and thermal conductivity probes; seismic, electrical and electromagnetic exploration; physical properties of rocks; seismic absorption and dispersion; direct hydrocarbon detection; new types 1 and 3 component geophones; seismic record processing; multi-component transducer. *Mailing Add:* 9211 S 69th East Ave Tulsa OK 74133

CROWE, DAVID BURNS, b New Brighton, Pa, Oct 6, 30; m 56; c 2. ZOOLOGY. *Educ:* Washington & Jefferson Col, BA, 52; Univ Mich, MS, 57; Univ Louisville, PhD(biol), 64. *Prof Exp:* Instr biol, State Univ NY, 58-61; PROF BIOL, UNIV WIS-EAU CLAIRE, 64- *Mem:* Am Soc Ichthyologists & Herpetologists. *Res:* Orientation of social piscine groups with emphasis on integrative role of the sense organs. *Mailing Add:* Dept Biol Univ Wis Eau Claire WI 54701

CROWE, DENNIS TIMOTHY, JR, b Milwaukee, Wis, Nov 21, 46; m 71; c 2. VETERINARY EMERGENCY & CRITICAL CARE, VETERINARY SURGERY. *Educ:* Iowa State Univ, DVM, 72; Colo State Univ, cert, 73, Ohio State Univ, cert, 76. *Prof Exp:* Surgeon & chief small animal surg, Westcott Hosp & Animal Emergency Room, 76-78; asst prof surg, Kans State Univ, 78-80; asst prof, 80-87, ASSOC PROF SMALL ANIMAL SURG, UNIV GA, ATHENS, 87-, DIR SHOCK-TRAUMA TEAM, 81-,CHIEF EMERGENCY & CRITICAL CARE, 88- *Concurrent Pos:* Chmn, Qual Assurance Comt Northeast Ga, EMS coun, 83-86, mem, 86- *Mem:* Vet Critical Care Soc (secy 80-82); Vet Emergency & Critical Care Soc (pres 84-86); Am Vet Med Asn; Am Animal Hosp Asn; Am Col Vet Surg; Am Col Vet Emergency & Critical Care; Soc Critical Care Med. *Res:* Clinical & experimental investigations of both surgical & emergency & critical care problems in small animal med & surgery; cardipulmonary resuscitation; pulmonary changes following severe shock; abdominal surgery; nutritional support in critically ill animals; thyroid function studies in critically ill animals. *Mailing Add:* 630 Sandstone Dr Athens GA 30605

CROWE, DONALD WARREN, b Lincoln, Nebr, Oct 28, 27; m 53, 79; c 5. MATHEMATICS. *Educ:* Univ Nebr, BS, 49, MA, 51; Univ Mich, PhD(math), 59. *Prof Exp:* Instr math, Univ BC, 55-57; lectr, Univ Toronto, 57-59 & Univ Col, Ibadan, 59-62; from asst prof to assoc prof, 62-68, PROF MATH, UNIV WIS, MADISON, 68- *Concurrent Pos:* Vis distinguished prof, Calif State Univ, Chico, 78 & 81. *Mem:* Math Asn Am. *Res:* Geometry, especially combinatorial problems and applications to archeology. *Mailing Add:* Dept Math Univ Wis Madison WI 53706

CROWE, GEORGE JOSEPH, b Brooklyn, NY, Oct 1, 21; m 52; c 11. SOLID STATE PHYSICS. *Educ:* Manhattan Col, BS, 43; Columbia Univ, MA, 47; Carnegie Inst Technol, MS, 61, PhD(physics), 66. *Prof Exp:* Instr physics, Manhattan Col, 46-48; assoc prof, Seton Hill Col, 48-64; proj physicist, Carnegie Inst Technol, 64-65, instr physics, 65; assoc prof, Manhattan Col, 65-74, chmn dept, 68-77, prof physics, 74-87. *Mem:* AAAS; Am Phys Soc; Am Asn Physics Teachers; Sigma Xi. *Res:* Point defects in alkali halides; electron irradiated cadmium sulfide x-ray spectroscopy. *Mailing Add:* 182 Ivy St Hempstead NY 11552

CROWE, JOHN H, b Columbia, SC, Apr 12, 43. COMPARATIVE PHYSIOLOGY. *Educ:* Wake Forest Univ, BS, 65, MA, 67; Univ Calif, Riverside, PhD(biol), 70. *Prof Exp:* From asst prof to assoc prof, 70-79, PROF ZOOL, UNIV CALIF, DAVIS, 79-, CHMN DEPT. *Concurrent Pos:* Consult, Review Panelist, NSF, 75-78; chmn, Div Comp Physiol & Biochem, Am Soc Zoologist, 80-81. *Mem:* Sigma Xi; Am Soc Zoologists; fel AAAS; Am Micros Soc; Biophys Soc Am. *Res:* Physiology of the induction of cryptobiotic states; intracellular water; water and membrane structure; cryobiology. *Mailing Add:* Dept Zool 2320 Storer Hall Univ Calif Davis CA 95616

CROWE, KENNETH MORSE, b Boston, Mass, Oct 6, 26; m 63; c 3. PHYSICS. *Educ:* Brown Univ, BS, 48, PhD(physics), 53. *Prof Exp:* Physicist, Radiation Lab, Univ Calif, 49-51; res assoc, High-Energy Physics Lab, Stanford Univ, 51-56; from asst prof to assoc prof, 58-69, PROF PHYSICS, UNIV CALIF, BERKELEY, 69-, PHYSICIST, LAWRENCE RADIATION LAB, 56- *Mem:* Am Phys Soc. *Res:* High-energy particle physics. *Mailing Add:* Radiation Lab Univ Calif Berkeley CA 94704

CROWE, RICHARD GODFREY, JR, b Dayton, Ohio, July 1, 44; m 78; c 2. TECHNICAL MANAGEMENT. *Educ:* Morehead State Univ, BA, 67, MBE, 70. *Prof Exp:* Assoc dir res, Morehead State Univ, 67-70; pres, Exec Serv, Lexington, 70-74; asst prof bus econ, Cumberland Col, 74-77; PROF BUS ECON, HAZARD COMMUNITY COL, 77- *Mem:* Am Arbitration Asn; Int Asn Financial Planners; Am Bus Law Asn. *Res:* Economic education. *Mailing Add:* PO Box 388 Ary KY 41712

CROWELL, CLARENCE ROBERT, b Sweetsburg, Que, July 29, 28. ELECTRICAL ENGINEERING, MATERIALS SCIENCE. *Educ:* McGill Univ, BA, 49, MSc, 51, PhD(physics), 55. *Prof Exp:* Jr res officer elec eng, Radio & Elec Eng Div, Nat Res Coun Can, 54-55; asst prof physics, McGill Univ, 55-60; mem tech staff semiconductor device res & develop, Bell Labs, 60-66; PROF ELEC ENG & MAT SCI, UNIV SOUTHERN CALIF, 66- *Concurrent Pos:* Consult, Chem Physics Dept, Hughes Res Lab, 69-; Humboldt fel, 74. *Mem:* Am Vacuum Soc; fel Inst Elec & Electronics Engrs; Sigma Xi. *Res:* Semiconductor device and interface physics; charge storage and charge transport in metal-semiconductor systems. *Mailing Add:* Dept Elec Eng PHE 604 Univ Southern Calif Los Angeles CA 90089-0271

CROWELL, EDWIN PATRICK, b Elizabeth, NJ, Feb 27, 34; m 58; c 6. CHEMISTRY. *Educ:* Seton Hall Univ, BS, 56; Univ Richmond, MS, 62. *Prof Exp:* Control anal chemist, Ethylene Oxide Unit, Gen Aniline & Film Corp, 56-58; assoc chemist, Anal Res, Phillip Morris, Inc, 58-62; res chemist, Agr Div, Am Cyanamid Co, 62-63; res scientist, 63-70, group leader, 70-74, sect leader, 74-81, ASST LAB DIR, RES & DEVELOP DIV, UNION CAMP CORP, 81- *Mem:* Am Chem Soc; Tech Asn Pulp & Paper Indust; Am Indust Hyg Asn. *Res:* Development of instrumental and chemical analytical procedures in support of research program in paper pulp and chemical byproducts; tobacco and agricultural chemicals. *Mailing Add:* 13 Blackfoot Rd Trenton NJ 08638-1022

CROWELL, HAMBLIN HOWES, b Portland, Ore, Aug 23, 13; m 39; c 1. ENTOMOLOGY. *Educ:* Ore State Col, BS, 35, MS, 37; Ohio State Univ, PhD(entom), 40. *Prof Exp:* Asst biol sci surv, Ore State Col, 35-37 & zool & entom, Ohio State Univ, 37-40; sanit inspector, Health Dept, CZ, 40-44; asst entomologist, Exp Sta, 46-51, assoc prof, 51-66, actg head entom dept, 58-59, prof, 66-77, EMER PROF ENTOM, ORE STATE UNIV, 77- *Mem:* AAAS; Entom Soc Am; Sigma Xi. *Res:* Economic entomology and malacology. *Mailing Add:* 2771 SW De Armond Dr Corvallis OR 97333

CROWELL, JOHN CHAMBERS, b State College, Pa, May 12, 17; m 46; c 1. GEOLOGY. *Educ:* Univ Tex, BS, 39; Univ Calif, Los Angeles, MA, 46, PhD(geol), 47. *Hon Degrees:* DSc, Cath Univ Louvain, 66. *Prof Exp:* Asst geologist, Shell Oil Co, Inc, 41-43; from instr to prof geol, Univ Calif, Los Angeles, 47-67, chmn dept, 57-60, 63-67; prof, 67-87, EMER PROF GEOL, UNIV CALIF, SANTA BARBARA, 87- *Concurrent Pos:* Guggenheim Found fel, 53-54; Fulbright res prof, Austria, 53-54; NSF sr res fel, Scotland, 60-61; distinguished lectr, Am Asn Petrol Geologists; nat lectr, Sigma Xi, 80-82. *Honors & Awards:* Chrestien Mica Gondwanaland Medal, Mining, Geol & Metall Inst India, 72. *Mem:* Nat Acad Sci; fel AAAS; Geol Soc Am; Am Asn Petrol Geol; Am Geophys Union; Soc Econ Paleont & Mineral; Sigma Xi. *Res:* Structural and general geology; tectonics; paleoclimatology of ancient ice ages. *Mailing Add:* Inst Crustal Studies Univ Calif Santa Barbara CA 93106

CROWELL, JOHN MARSHALL, b Mobile, Ala, June 30, 42. TECHNICAL MANAGEMENT. *Educ:* Ga Inst Technol, BS, 64; Johns Hopkins Univ, PhD(physics), 73. *Prof Exp:* physicist, Los Alamos Sci Lab, Univ Calif, 73-86; VPRES, MULTIWARE, INC, 87- *Mem:* Am Phys Soc; Inst Elec & Electronic Engrs; Asn Comput Mach. *Res:* Intelligent instrumentation; nuclear instrumentation; environmental instrumentation. *Mailing Add:* PO Box 128 Davis CA 95617-0128

CROWELL, JULIAN, b Shelbyville, Tenn, Jan 24, 34; m 58; c 3. ACTUARIAL SCIENCE. *Educ:* Univ Tenn, BS, 56; Vanderbilt Univ, MS, 59, PhD(physics), 66. *Prof Exp:* Lectr physics, Gordon Col, Rawalpindi, Pakistan, 58-61; asst prof physics, Roanoke Col, 65-67; assoc prof physics & math, St Andrews Presby Col, 67-69; assoc prof math, Bosphours Univ, Istanbul, Turkey, 69-78, chmn dept, 71-76; prof math & physics, Inst nat D'Electricite et d'Electronique, Boumerdes, Algeria, 78-82; asst prof math, Stockton State Col, 82-83; ACTUARIAL ASSOC, MASS MUTUAL, 83- *Concurrent Pos:* Consult- 66-70. *Mem:* Math Asn Am; Soc Actuaries; Am Acad Actuaries. *Res:* Positron annihilation in metals; collective phenomena in metals. *Mailing Add:* 48 Biltmore St Springfield MA 01108

CROWELL, KENNETH L, b Glen Ridge, NJ, July 19, 33; m 62; c 2. VERTEBRATE ECOLOGY, ZOOGEOGRAPHY. *Educ:* Yale Univ, BS, 55; Univ Pa, PhD(zool), 61. *Prof Exp:* Instr zool, Duke Univ, 61-62; mem fac biol, Marlboro Col, 62-66; fel zool, Calgary Univ, 66-67, instr, 67; from asst prof to assoc prof, 67-84, PROF BIOL, ST LAWRENCE UNIV, 84- *Concurrent Pos:* Chapman Mem Fund grant, Am Mus Natural Hist, 59-60, 66; Jessup Fund grant, Acad Natural Sci Philadelphia, 61; Res Soc grant, Soc Sigma Xi, 62; NSF grants, 62-72; dir, Planned Parenthood Northern NY, 69-71; trustee, Adirondack Conservancy Comt, 75-; ed, NY Fed Bird Clubs. *Mem:* Fel AAAS; Ecol Soc Am; elected mem Am Ornith Union; Am Soc Mammal; Nature Conservancy; fel Linnean Soc London. *Res:* Population dynamics and niche segregation through studies of competition and habitat selection in mammals of islands of Gulf of Maine and birds of Bermuda and West Indies, insular biogeography, rare species and range expansion in Adirondack region. *Mailing Add:* Dept Biol St Lawrence Univ Canton NY 13617

CROWELL, MERTON HOWARD, b Corry, Pa, June 5, 32. ELECTRO-OPTICS. *Educ:* Pa State Univ, BS, 56; New York Univ, MS, 60; Polytech Inst Brooklyn, PhD(electro-phys), 70. *Prof Exp:* Group supvr electron devices, Bell Tel Labs, 56-69; dir device res, 69-74, MGR, ELECTRO-OPTICAL LAB, PHILIPS LABS, 76- *Mem:* Fel Inst Elec & Electronics Engrs. *Res:* Simulation of electro-optics; author or coauthor of over 50 publications. *Mailing Add:* 10 Scarborough Dr Nashua NH 03063

CROWELL, RICHARD HENRY, b Northeast, Pa, Apr 6, 28; m 55; c 2. MATHEMATICS. *Educ:* Harvard Univ, AB, 49; Princeton Univ, MA, 53, PhD(math), 55. *Hon Degrees:* MA, Dartmouth Col, 67. *Prof Exp:* Asst anal res group, Forrestal Res Ctr, Princeton Univ, 55-57; lectr, Mass Inst Technol, 57-58; from asst prof to assoc prof, 58-67, chmn dept, 73-79, PROF MATH, DARTMOUTH COL, 67-, CHMN DEPT MATH & COMPUT SCI, 86- *Mem:* Am Math Soc; Math Asn Am. *Res:* Topology, knot theory and algebraic topology. *Mailing Add:* Dept Math & Comput Sci Dartmouth Col Hanover NH 03755

CROWELL, RICHARD LANE, b Springfield, Mo, Sept 27, 30; m 53; c 4. VIROLOGY. *Educ:* Univ Buffalo, BA, 52; Univ Minn, MS, 54, PhD(microbiol), 58; Am Bd Med Microbiol dipl. *Prof Exp:* Nat Cancer Inst trainee & instr microbiol, Univ Minn, 58-60; from asst prof to assoc prof, 60-71, PROF MICROBIOL, HAHNEMANN UNIV SCH MED, 71-, CHMN DEPT MICROBIOL & IMMUNOL, 79- *Concurrent Pos:* Nat Inst Allergy & Infectious Dis res career develop award, 62-72; consult, Virus-Tissue Resources Br, Nat Cancer Inst, 66-68; vis res scientist, Univ Uppsala, 69-70; consult, Smith Kline Corp, 75-77 & Lehn & Fink Prod, Inc, 76-80; vis res scientist, Merck Sharp & Dohme Res Labs, West Point, Pa, 84-85. *Mem:* Sigma Xi; Am Soc Microbiol (pres, 91-92); AAAS; Am Asn Immunologists; fel Am Acad Microbiol; Am Soc Virol. *Res:* Human enteroviruses in mammalian cell cultures, with emphasis on enzyme levels, chronic infection, viral interference, cell susceptibility, characterization of viral receptors of cells and viral proteins; relationship between cellular differentiation and factors controlling virus susceptibility as studied in tissue culture systems; role of coxsackie viruses in heart disease, muscle disease and diabetes. *Mailing Add:* Dept Microbiol & Immunol Hahneman Univ Broad & Vine Philadelphia PA 19102-1192

CROWELL, ROBERT MERRILL, b Sandusky, Ohio, May 29, 21; m 46; c 3. ZOOLOGY, ENTOMOLOGY. *Educ:* Bowling Green State Univ, AB, 45, MA, 47; Ohio State Univ, PhD(zool), 57. *Prof Exp:* Asst biol, Bowling Green State Univ, 45-47; instr, Kent State Univ, 47-48; asst zool, Duke Univ, 48-51; instr biol, Col Wooster, 51-55; from asst prof to assoc prof, 56-67, PROF BIOL, ST LAWRENCE UNIV, 67- *Concurrent Pos:* NSF res grants, 61-63 & 69-71; trainee, Ohio State Univ, 67-68; vis scholar, Univ Amsterdam, 76; guest investr, Freshwater Biol Asn, 77. *Mem:* Entom Soc Am; Am Soc Parasitol; Entom Soc Can. *Res:* Systematics; developmental cycles; host-parasite relationships of the Hydracarina. *Mailing Add:* St Lawrence Univ Canton NY 13617

CROWELL, (PRINCE) SEARS, JR, b Natick, Mass, May 2, 09; m 38; c 3. DEVELOPMENTAL BIOLOGY, INVERTEBRATE ZOOLOGY. *Educ:* Bowdoin Col, AB, 30; Harvard Univ, AM, 31, PhD(biol), 35. *Prof Exp:* Instr, Brooklyn Col, 35-36; from instr to assoc prof zool, Miami Univ, Ohio, 36-48; from asst prof to prof, 48-79, EMER PROF ZOOL, IND UNIV, BLOOMINGTON, 79- *Concurrent Pos:* Mem corp, Marine Biol Lab, Woods Hole, trustee, 58-66, 67-75, mem exec comt trustees, 63-66, 67-70 & secy bd trustees, 72-75; managing ed, Am Zoologist, Am Soc Zoologists, 61-65. *Mem:* AAAS; Am Soc Zoologists; Soc Develop Biol; Int Soc Develop Biol. *Res:* Morphogenesis and natural history of hydroids; regeneration in worms. *Mailing Add:* Dept Biol Indiana Univ Bloomington IN 47405

CROWELL, THOMAS IRVING, b Glen Ridge, NJ, July 9, 21; m 50; c 2. PHYSICAL ORGANIC CHEMISTRY. *Educ:* Harvard Univ, BS, 43; Columbia Univ, AM, 47, PhD(chem), 48. *Prof Exp:* Asst, Manhattan Proj, 43-45; asst chem, Columbia Univ, 45-46; instr math, Bard Col, 46-47; from asst prof to prof chem, Univ Va, 48-84, chmn dept, 57-62, prof chem, 61-84, EMER PROF CHEM, UNIV VA, 84- *Mem:* Am Chem Soc; Sigma Xi. *Res:* Kinetics of nucleophilic displacement and addition reactions in both aqueous and fused-salt solutions; pH-rate functions and kinetic solvent isotope effects of nitrostyrene hydrolysis and borate ester solvolysis. *Mailing Add:* Dept Chem Univ Va Charlottesville VA 22901

CROWELL, WAYNE ALLEN, b Sterling, Colo, Nov 25, 40; m 63; c 2. VETERINARY PATHOLOGY, VETERINARY MEDICINE. *Educ:* Colo State Univ, BS, 63, DVM, 64; Univ Ga, PhD, 73. *Prof Exp:* Post vet, US Army, 66-68; res vet animal sci, Com Solvents Corp, 68-70; res assoc to asst prof path, Univ Ga, 70-74; assoc prof path, La State Univ, 74-76; assoc prof path, 76-, PROF PATH, GRAD SCH & SCH VET MED, UNIV GA. *Honors & Awards:* Nat Award of Excellence, Am Vet Med Asn, 78. *Mem:* Am Vet Med Asn; Am Col Vet Path; Int Acad Path; Soc. Vet Urol; Am Asn Vet Med Col. *Res:* Renal disease of animals, amyliodosis, pathology of parasitic diseases in animals, educational methods. *Mailing Add:* Dept Path Univ Ga Athens GA 30602

CROWL, GEORGE HENRY, glacial geology, geomorphology, for more information see previous edition

CROWL, ROBERT HAROLD, b Wellsville, Ohio, Apr 17, 25; m 47; c 3. ANIMAL GENETICS. *Educ:* Harvard Univ, SB, 49; Miami Univ, MS, 50; Ohio State Univ, PhD, 64. *Prof Exp:* Instr zool, Miami Univ, 50-51; sales rep pharmaceut, Bowman Bros Drug Co, 51-54, Winthrop Labs, 54-59 & Columbus Pharmacal Co, 59-60; teaching asst zool, Ohio State Univ, 60-64; from asst prof to assoc prof biol, 64-71, ADJ PROF BIOL, PFEIFFER COL, 71- *Mem:* NY Acad Sci; AAAS. *Res:* Pysiological genetics. *Mailing Add:* 28 Delight Lane Misenheimer NC 28109

CROWLE, ALFRED JOHN, b Mexico, DF, Mex, Apr 15, 30; US citizen; m 54; c 2. IMMUNOLOGY, MICROBIOLOGY. *Educ:* San Jose State Univ, AB, 51; Stanford Univ, PhD(microbiol), 54. *Prof Exp:* Researcher, Webb-Waring Lung Inst, 56-59, from instr to assoc prof, Sch Med, 56-74, PROF MICROBIOL, SCH MED, UNIV COLO MED CTR, DENVER, 74-, HEAD DIV IMMUNOL, WEBB-WARING LUNG INST, 59- *Concurrent Pos:* Nat Tuberc Asn med res fel, Stanford Univ, 53-55, Nat Acad Sci-Nat Res Coun fel, 55-56; NY Tuberc & Health Asn fel, Sch Med, Univ Colo Med Ctr, Denver, 59-61. *Mem:* Reticuloendothelial Soc; Am Asn Immunol; Am Soc Microbiol; Soc Exp Biol & Med; Am Col Allergists. *Res:* Immunochemistry, immunodiffusion and applications of immunodiffusion techniques; cell-mediated immunity in myco-bacterial diseases and acquired immune deficiency syndrome. *Mailing Add:* Webb-Waring Inst Div Immunol 4200 E Ninth Ave Denver CO 80220

CROWLEY, JAMES M, b Baltimore, Md, Feb 19, 49. MATHEMATICAL CONTROL THEORY. *Educ:* Col Holy Cross, BA, 71; Va Polytech Inst, MS, 73; Brown Univ, PhD(appl math), 82. *Prof Exp:* Mathematician, Foreign Technol Div, Air Force Systs Command, 73-77; from instr to assoc prof math, US Air Force Academy, 77-86; dir math & info sci, US Air Force Off Sci Res, 86-88, ASST CHIEF SCIENTIST SYSTS COMMAND, US AIR FORCE, 88- *Mem:* Am Math Soc; Soc Indust & Appl Math; Inst Elec & Electronic Engrs. *Res:* Distributed controls and computational methods. *Mailing Add:* 8513 Wagon Wheel Rd Alexandria VA 22309

CROWLEY, JAMES PATRICK, b Birmingham, Eng, Oct 15, 43; US citizen; m 68; c 2. HEMATOLOGY & INFLAMMATION. *Educ:* Providence Col, AB, 65; Georgetown Univ, Washington, MD, 69. *Hon Degrees:* MA, Brown Univ, 82. *Prof Exp:* Clin asst med, Harvard Med Sch, 69-72; res med officer, Blood Res, US Navy, 72-75; asst prof, 76-81, ASSOC PROF MED, BROWN UNIV, PROVIDENCE, 81- *Concurrent Pos:* Consult, Armed Forces Radio Biol Res Inst, 74-75, Naval Blood Res Lab, Boston, 75-; adj asst prof med, Sch Med, Boston Univ, 80-; vis prof, Va Commonwealth Med Sch, 80. *Mem:* Am Fedn Clin Res; Soc Exp Biol & Med; Am Soc Hemat; Am Soc Microbiol; NY Acad Sci. *Res:* Plasma factors that modulate and facilitate the interaction of phagocytic cells with tumor cells and with bacteria. *Mailing Add:* Div Clin Hemat RI Hosp 593 Eddy St Providence RI 02902

CROWLEY, JOHN JAMES, b San Diego, Calif, Feb 20, 46; m 69; c 3. BIOSTATISTICS. *Educ:* Pomona Col, BA, 68; Univ Wash, MS, 70, PhD(biomath), 73. *Prof Exp:* Fel biostatist, Stanford Univ, 73-74; asst prof statist & human oncol, Univ Wis-Madison, 74-79, assoc prof, 79-81; assoc prof, 81-84, PROF BIOSTATIST, UNIV WASH, 84- *Concurrent Pos:* Mem, Fred Hutchinson Cancer Res Ctr, 81- *Mem:* Am Statist Asn; Biomet Soc; Inst Math Statist; AAAS. *Res:* Methods for analyzing censored survival data; biostatistical methods in cancer research. *Mailing Add:* Dept Biostatist Univ Wash MS SC-32 Seattle WA 98195

CROWLEY, JOSEPH MICHAEL, b Philadelphia, Pa, Sept 9, 40; m 63; c 5. ELECTRICAL ENGINEERING, FLUID MECHANICS. *Educ:* Mass Inst Technol, BS, 62, MS, 63, PhD(elec eng), 65. *Prof Exp:* Guest scientist, Max Planck Inst, Gottingen, 65-66; asst prof, 66-71, assoc prof, 71-79, PROF ELEC ENG, UNIV ILL, URBANA, 79-; PRES, ELECTROSTATIC APPLNS, 86- *Concurrent Pos:* NATO fel, 65-66; consult, 68-; pres, Joseph M Crowley, Inc, 81-88. *Mem:* Am Phys Soc; Inst Elec & Electronics Engrs; Electrostatic Soc Am; Soc Info Display. *Res:* Feedback control of wave motion in fluids and plasmas; electrohydrodynamics of jets and drops; instabilities in viscous two-phase flows; electrical effects in biological systems; electrostatics; ink jet printers; computer output devices. *Mailing Add:* Electrostatic Applications 16525 Jackson Oaks Dr Morgan Hill CA 95037

CROWLEY, LAWRENCE GRANDJEAN, b Newark, NJ, July 2, 19; m 45; c 3. SURGERY. *Educ:* Yale Univ, BA, 41, MD, 44. *Prof Exp:* From instr to asst prof surg, Sch Med, Yale Univ, 51-53; attend, Southern Calif Permanente Med Group, 53-55; clin asst prof, Sch Med, Univ Southern Calif, 55-64; from assoc prof to prof, Sch Med, Stanford Univ, 64-74, assoc dean, 72-74; prof surg & dean, Sch Med, Univ Wis-Madison, 74-77; dep dean, Sch Med, 77-80, PROF SURG, SCH MED, STANFORD UNIV, 77-, ACTG VPRES MED AFFAIRS & ACTG DEAN, SCH MED, 80-, MED DIR HOSP, 77- *Concurrent Pos:* Am Cancer Soc clin fel, Med Ctr, Yale Univ, 49-51; mem, Bd Dirs, Casa Colina Rehab Hosp, Pomona, Calif, 59-74, pres, 63-66; mem spec grants comt, Calif Div, Am Cancer Soc, 67-70; mem, Cancer Categorical Comt, Calif Regional Med Prog, 68-71; mem, Cancer Adv Coun, State Calif, 70-74. *Mem:* Am Col Surgeons; Sigma Xi. *Res:* Endocrine relationships with mammary carcinoma. *Mailing Add:* Stanford Univ Sch Med 900 Welch Rd Stanford CA 94305

CROWLEY, LEONARD VINCENT, b Binghamton, NY, Jan 12, 26; m 51; c 5. MEDICINE, PATHOLOGY. *Educ:* Univ Vt, MD, 49; Ohio State Univ, MSc, 56. *Prof Exp:* Instr path, Col Physicians & Surgeons, Columbia Univ, 52-54; instr, Med Sch, Ohio State Univ, 54-56; from asst prof to assoc prof, Col Med, Univ Vt, 56-60; ASST PROF PATH, MED COL, UNIV MINN, MINNEAPOLIS, 62- *Mem:* AMA; Am Fedn Clin Res; Am Soc Exp Path. *Res:* Immunohematology; clinical chemistry. *Mailing Add:* 5337 Kellogg Ave Minneapolis MN 55424

CROWLEY, MICHAEL SUMMERS, b Chicago, Ill, Dec 24, 28; m 50; c 2. CERAMICS ENGINEERING, MATERIALS SCIENCE. *Educ:* Iowa State Univ, BS, 53; Pa State Univ, PhD(geochem), 59. *Prof Exp:* Res engr mat, Standard Oil Co, Ind, 53-55; Am Petrol Inst fel geochem, Pa State Univ, 55-58; sr res engr, 58-80, SR RES ASSOC, ENGR RES & SERV DIV, AMOCO CORP, 80- *Concurrent Pos:* Consult, US Dept Energy, 75- & Argonne Nat Labs, 78; co-chmn, Metals Prop Coun Task Force on Mat for Coal Gasification, 72-77. *Mem:* Fel Am Ceramic Soc; Nat Inst Ceramics Engrs; Sigma Xi; Nat Soc Prof Engr; Am Soc Testing Mat. *Res:* Non-metallic materials of construction such as refractories, insulation, fire proofing, coatings and plastics. *Mailing Add:* 1270 Deer Run Trail Crete IL 60417

CROWLEY, PATRICK ARTHUR, b Titusville, Pa, June 6, 41; m 64; c 2. NUCLEAR PHYSICS & FLUID PHYSICS, OPERATIONS RESEARCH. *Educ:* Carnegie Inst Technol, BS, 63; Univ Pittsburgh, PhD(physics), 68. *Prof Exp:* Staff physicist, Columbia Res Corp, 68-70; physicist, US Army Foreign Sci & Technol Ctr, 70-78; proj officer, Defense Commun Agency, 87-91; PROJ OFFICER, DEFENSE NUCLEAR AGENCY, 78-87 & 91- *Concurrent Pos:* Consult, Armed Forces Commun & Electronics Asn. *Mem:* Am Inst Physics; Am Geophys Union. *Res:* Nuclear structures and reactions; ionospheric physics; radiation effects; atmospheric phenomena; nuclear weapons effects. *Mailing Add:* 12450 Wendell Holmes Rd Herndon VA 22071-2461

CROWLEY, PHILIP HANEY, b Ft Worth, Tex, Dec 8, 46; m 72; c 2. ECOLOGY, MATHEMATICAL BIOLOGY. *Educ:* Rice Univ, BA, 69, MS, 72; Mich State Univ, PhD(zool), 75. *Prof Exp:* Res assoc, Dept Elec Eng & Systs Sci, Mich State Univ, 75-76; from asst prof to assoc prof, 76-89, PROF ECOL, THOMAS HUNT MORGAN SCH BIOL SCI, UNIV KY, 89- *Concurrent Pos:* Fel, Brit Roy Soc, 91-92. *Mem:* Ecol Soc Am; Brit Ecol Soc; Am Soc Naturalists. *Res:* Predator-prey interactions; population regulation and ecosystem stability; population and community ecology of freshwater systems; mathematical modeling of biological systems. *Mailing Add:* Thomas Hunt Morgan Sch Biol Sci Univ Ky Lexington KY 40506

CROWLEY, THOMAS HENRY, b Bowling Green, Ohio, June 7, 24; m 47; c 14. COMPUTER SCIENCE. *Educ:* Ohio State Univ, BEE, 48, MA, 50, PhD(math), 54. *Prof Exp:* Res assoc antenna lab, Ohio State Univ, 48-54; mem tech staff, Bell Tel Labs, Inc, 54-64, head comput res dept, 64-65, dir comput sci res ctr, 65-68, exec dir safeguard design div, 68-75, exec dir, Bus Systs & Technol Div, 75-79, exec dir, comput technol & design, 79-; VPRES, SOFTWARE SYSTS, WESTERN ELEC CO. *Mem:* Inst Elec & Electronics Engrs. *Res:* Computers; switching and control. *Mailing Add:* 47 Woodland Ave No 108 Summit NJ 07901

CROWLEY, WILLIAM FRANCIS, JR, b Meriden, Conn, Dec 28, 43; m 68; c 4. REPRODUCTIVE ENDOCRINOLOGY, NEUROENDOCRINOLOGY. *Educ:* Holy Cross Col Worcester, Mass, BA, 65; Tufts Med Sch, MD, 69. *Prof Exp:* Internship internal med, Mass Gen Hosp & Harvard Med Sch, 69-70, asst residency, 70-71; gen med officer, Newport Naval Hosp, 71-73; sr residency internal med, Mass Gen Hosp & Harvard Med Sch, 73-74, Endocrinol fel, 74-77; ASSOC PROF MED, HARVARD MED SCH, 84-; CHIEF, REPRODUCTIVE ENDOCRINOL, MASS GEN HOSP, 84- *Concurrent Pos:* vis prof, Yale Univ, 82, Duke Univ, 83, George Washington Univ, 85; consult, Endocrine & Metabolism Div, Food & Drug Admin, 86-, Contraceptive Develop Br Adv Bd, Nat Inst Child Health & Human Develop, 82- *Mem:* Am Bd Internal Med; fel Am Philosophical Soc; Am Soc Clin Invest; Am Fedn Clin Res. *Res:* Clinical investigation in the neuroendocrine control of growth, sexual maturity and reproduction; hypotholaime control of GURH secretion and the physiology of gurradotroyon secretion. *Mailing Add:* Reproductive Endocrine Unit Mass Gen Hosp Boston MA 02114

CROWLEY, WILLIAM ROBERT, b Bristol, Conn, May 20, 48. NEUROPHARMACOLOGY, NEUROENDOCRINOLOGY. *Educ:* Univ Conn, BA, 70; Villanova Univ, MS, 72; Rutgers Univ, PhD(psychobiol), 76. *Prof Exp:* ASST PROF PHARMACOL, CTR HEALTH SCI, UNIV TENN, 78- *Concurrent Pos:* NIMH fel, 76-78. *Mem:* Soc Neurosci; Int Soc Psychoneuroendocrinol; AAAS. *Res:* Neurochemical mechanisms in control of reproductive processes. *Mailing Add:* Dept Pharmacol Crowe Bldg Rm 301 Ctr Health Sci 874 Union Ave Memphis TN 38163

CROWNOVER, RICHARD MCCRANIE, b Quincy, Fla, Oct 11, 36; m 58; c 3. MATHEMATICAL ANALYSIS. *Educ:* Ga Inst Technol, BS, 58, MS, 60; La State Univ, PhD(math), 64. *Prof Exp:* Instr math, Ga Inst Technol, 59-61 & La State Univ, 63-64; asst prof, 64-72, ASSOC PROF MATH, UNIV MO-COLUMBIA, 72- *Mem:* Am Math Soc; Math Asn Am; Sigma Xi; Soc Indust & Appl Math. *Res:* Spaces and algebras of functions and operators. *Mailing Add:* Dept Math Univ Mo Columbia MO 65211

CROWSON, CHARLES NEVILLE, b Ottawa, Ont, Sept 21, 19; m 51; c 2. PATHOLOGY. *Educ:* Queen's Univ, Ont, BA, 41, MA, 43; McGill Univ, MD, 49; Univ Edinburgh, PhD(path), 54; FRCPS(C). *Prof Exp:* Biochemist, Nat Res Coun, Can, 42-43; lectr path, Queen's Univ, Ont, 54-55; dir labs, Deer Lodge Hosp, 55-60; DIR LABS, MISERICORDIA GEN HOSP, 60- *Concurrent Pos:* Asst prof, Univ Man, 55-67; dir, Cent Med Lab, 59-; consult pathologist aviation crash path, Prov of Man, 55- & patologist, Can Forces Med Coun, 71- *Mem:* Am Soc Clin Path; Am Asn Path & Bact; Can Asn Pathologists; Int Acad Path. *Res:* Hepato-renal syndrome and renal pathology; pathogenesis of nephrotoxic chemical lesions. *Mailing Add:* Dept Path Misericordia Gen Hosp 99 Cornish Winnipeg MB R3C 1A2 Can

CROWSON, HENRY L, b Okeechobee, Fla, Apr 16, 27; m 51; c 3. MATHEMATICS. *Educ:* Univ Fla, BChE, 53, MS, 55, PhD(amth), 59. *Prof Exp:* Instr math, Univ Fla, 58-59. asst prof, 59-60; staff mathematician, 60-65, ADV MATHEMATICIAN, IBM CORP, 65- *Mem:* Am Math Soc; Math Asn Am; Soc Indust & Appl Math. *Res:* Classical mathematical analysis, especially ordinary differential equations; signal analysis. *Mailing Add:* PO Box 3038 Laredo TX 78044-3038

CROWTHER, C RICHARD, b Waterloo, Iowa, July 16, 24; m 49; c 3. FORESTRY. *Educ:* Iowa State Col, BS, 47, MS, 56; Univ Mich, PhD, 71. *Prof Exp:* Forester, Cent States Exp Sta, US Forest Serv, 47; assoc ed, Naval Stores Rev, 47-48; ed, Lake Mills Graphic, Iowa, 48-53; from instr to assoc prof, Mich Technol Univ, 56-72, prof forestry, 72-84; RETIRED. *Concurrent Pos:* Alumni Found res grant, 64; McIntire-Stennis grants, 72 & 80; Nat Park Serv grant, 80. *Mem:* Soc Am Foresters. *Res:* Forest recreation. *Mailing Add:* 1030 Ash St Hancock MI 49930

CROWTHER, ROBERT HAMBLETT, b Amherst, NH, Mar 17, 25; m 46; c 3. PETROLEUM PROCESS ENGINEERING ECONOMICS & RESEARCH, COMPUTERIZED PROCESS CONTROL & OPTIMIZATION. *Educ:* Fenn Col, Cleveland, Ohio, BChE, 50; Kans State Col, Manhattan, MS, 51. *Prof Exp:* Group leader, Standard Oil Co, Ind, 51-60; res proj mgr, Am Oil Co, 60-64; computer appln div mgr, Amoco Chem Corp, 64-69; com develop mgr, C W Nofsinger Co, 69-78; PRES, APPL SCI CORP, 78- *Concurrent Pos:* Res award, Sigma Xi, 52. *Mem:* Fel Am Inst Chem Engrs; Am Petrol Inst. *Res:* Hydrocarbon processing; catalytic cracking; catalytic reforming of gasoline; hydrotreating of petroleum products; computer control systems for refineries; enhanced oil recovery; combustion systems. *Mailing Add:* 4205 W 91st St Prairie Village KS 66207-2664

CROXALL, WILLARD (JOSEPH), b Aberdeen, Wash, Nov 11, 10; m 32, 84; c 5. CHEMISTRY. *Educ:* Univ Notre Dame, BS, 32, MS, 33, PhD(org chem), 35. *Prof Exp:* Garyin fel, Univ Notre Dame, 35; res chemist, Gen Chem Co, NY, 35-36; res chemist, Rohm & Haas Co, 36-46, res chemist & lab head hydrocarbon explor, 46-48, res chemist & lab head insecticides, 48-50, from asst dir res to dir res, Sumner Chem Co, Inc, 50-56, mem bd dirs, 53-36, gen mgr, 56-59, asst to mgr & coordr res & develop, Miles Chem Co, 59-62, dir, 62-64, vpres, 64-69, vpres & sr sci officer, Process Indust Group, 69-72; RETIRED. *Mem:* Am Chem Soc. *Res:* Emulsion polymerization; synthetic monomers; acetylene chemistry; organic insecticides and fungicides; organic chemical development. *Mailing Add:* 6800 Placida Rd No 225 Englewood FL 34224

CROXDALE, JUDITH GEROW, b Oakland, Calif, Aug 27, 41; div; c 1. BOTANY. *Educ:* Univ Calif, Berkeley, AB, 71, PhD(bot), 75. *Prof Exp:* Asst prof bot, Va Polytech Inst & State Univ, 75-79; asst prof, 79-81, ASSOC PROF BOT, UNIV WIS-MADISON, 81- *Concurrent Pos:* Vis prof, Wash Univ, 80, Fla State Univ, 81-82, Hebrew Univ, 87-88; vis res scientist, Nat Inst for Environ Sci, Tsukuba, Japan, 88, 89; Lady Davis vis prof. *Mem:* Bot Soc Am; Am Soc Plant Physiologists; Soc Develop Biol. *Res:* Structural and functional aspects of vegetative plant development; quantitative histochemistry; plant patterning. *Mailing Add:* Dept Bot Univ Wis Madison WI 53706

CROXTON, FRANK CUTSHAW, b Washington, DC, June 26, 07; m 30; c 2. CHEMISTRY, RESEARCH ADMINISTRATION. *Educ:* Ohio State Univ, AB, 27, AM, 28, PhD(phys chem), 30. *Hon Degrees:* DSc, Denison Univ, 65. *Prof Exp:* Res chemist, Standard Oil Co of Ind, 30-39; res chemist, Battelle Mem Inst, 39-42, from asst supvr to supvr, 42-47, asst dir, 47-53, tech dir, 53-64, asst dir, 64-72; CONSULT, 72- *Concurrent Pos:* Mem bd trustees, Denison Univ Res Found, 52-75 & Children's Hosp Res Found, 66-75. *Mem:* Fel Brit Inst Chem Engrs; Am Chem Soc; fel Am Inst Chem Eng; Com Develop Asn; Sigma Xi (pres, 71-72). *Res:* Petroleum technology; lubrication oil refining; lubricant additive; structure of aluminum carbide. *Mailing Add:* 1921 Collingswood Rd Columbus OH 43221-3739

CROY, LAVOY I, b Pauls Valley, Okla, Dec 4, 30; m 54; c 5. CROP PHYSIOLOGY, PLANT BIOCHEMISTRY. *Educ:* Okla State Univ, BS, 55, MS, 59; Univ Ill, Urbana, PhD(crop physiol), 67. *Prof Exp:* From instr to assoc prof, 55-78, PROF AGRON, OKLA STATE UNIV, 78- *Concurrent Pos:* Asst to head dept agron, Univ Ill, Urbana-Champaign, 64-66; assoc ed, Am Soc Agron, 77-81. *Mem:* AAAS; Am Soc Agron; Crop Sci Soc Am; Am Soc Plant Physiol. *Res:* Protein production in cereal grains; physiology of nitrate reduction in wheat; genetics of physiology and morphology of wheat. *Mailing Add:* Dept Agron Okla State Univ Main Campus Stillwater OK 74078

CROZAZ, GHISLAINE M, b Brussels, Belg, Aug 31, 39; m 73. GEOCHEMISTRY. *Educ:* Univ Brussels, BSc, 61, PhD(nuclear geochem), 67. *Prof Exp:* Res assoc, Univ Brussels, 67-71; vis assoc geochem, Calif Inst Technol, 72; from asst prof to assoc prof, 73-82, PROF EARTH & PLANETARY SCI, WASHINGTON UNIV, 82- *Concurrent Pos:* Res assoc, Wash Univ, 69-73. *Mem:* Meteoritical Soc; Am Geophys Union; Geochemical Soc; AAAS. *Res:* Geochemistry of extraterrestrial materials. *Mailing Add:* Dept of Earth & Planetary Sci Washington Univ St Louis MO 63130

CROZIER, EDGAR DARYL, b Montreal, Que, Apr 9, 39; m 64; c 2. SURFACES, HIGH PRESSURES. *Educ:* Univ Toronto, BSc, 61; Queen's Univ, Ont, PhD(phys chem), 65. *Prof Exp:* Nat Res Coun Can overseas fel, 65-67; from asst prof to assoc prof, 67-81, PROF PHYSICS, SIMON FRASER UNIV. 81- *Mem:* Can Asn Physicists. *Res:* Use of x-ray absorption fine structure spectroscopy to examine amorphous, liquid or solid metals and semiconductors versus temperature or pressure; study of surfaces. *Mailing Add:* Dept of Physics Simon Fraser Univ Burnaby BC V5A 1S6 Can

CRUCE, WILLIAM L R, b Galveston, Tex, Oct 2, 42; c 1. BRAIN EVOLUTION, RETICULAR FORMATION. *Educ:* Univ Chicago, BS, 64; Rockefeller Univ, PhD(physiol), 71. *Prof Exp:* Fel neuroanat, Univ Nijmegen, Netherlands, 71-72 & Univ Wis-Madison, 72-74; asst prof anat, Howard Univ, Washington, DC, 74-77; assoc prof, 77-84, PROF NEUROBIOL, NORTHEASTERN OHIO UNIV COL MED, 84-; PROF BIOL, GRAD SCH, KENT STATE UNIV, 84- *Concurrent Pos:* Vis investr neurobiol, Armed Forces Radiobiol Res Inst, Bethesda, 75-77; mem, Neurobiol Grants Rev, NSF, 77-80; ad hoc mem, Neurol B Grants Rev, NIH, 80; vis prof, neurosci dept, Univ Calif, San Diego, La Jolla, Calif, 85, 86 & 88; adj prof biomed & eng, Univ Akron, 84-; assoc prof biol, Grad Sch, Kent State Univ, 77-84. *Mem:* Neurosci Soc; Am Asn Anatomists; Am Physiol Soc; Asn Comput Mach; Am Zool Soc. *Res:* Evolution of the brain; motor systems; spinal cord; brainstem; neuroanatomy; neurophysiology; neuronol response to injury. *Mailing Add:* Neurobiol Dept Northeastern Ohio Univ Col Med Rootstown OH 44272

CRUDDACE, RAYMOND GIBSON, b Richmond, Eng, June 3, 36; US citizen. XRAY ASTRONOMY. *Educ:* Imp Col, Univ London, BSc, 58; Linacre Col, Oxford Univ, DPhil(physics), 68; Univ Calif, Berkeley, MA, 73. *Prof Exp:* Sci officer, Rocket Propulsion Estab, UK Ministry Aviation, 59-62; asst prof mech eng, Sacramento State Col, 65-66; physicist, Nuclear Rocket Opers, Aerojet-Gen Corp, 65-69; res physicist, Space Sci Lab, Univ Calif, Berkeley, 71-74; ASTROPHYSICIST, SPACE SCI DIV, NAVAL RES LAB, 74- *Mem:* Am Astron Soc; Brit Interplanetary Soc. *Mailing Add:* Code 4129-1 4555 Overlook Ave Sw Washington DC 20375-5000

CRUDEN, DAVID MILNE, b London, UK, Dec 4, 42; Can citizen; m 71; c 2. ROCK MECHANICS, LANDSLIDES. *Educ:* Oxford Univ, UK, BA, 64; Alberta Univ, Can, MS, 66; London Univ, PhD(geol), 69. *Prof Exp:* Asst prof geol, Univ Ottawa, 70-71; from asst prof to assoc prof 71-80, PROF GEOL, UNIV ALTA, 80- *Concurrent Pos:* Mem, Can Geosci Coun, 80-82, Can Nat Comt Geol, 80-82, Coord comt, Interunion Comn Lithosphere, 87-90; coun mem, Int Assoc Eng Geol, 80-82, chmn, Comn Landslides, 88-; secy, Tech Comt Landslides, 87-89, chmn, Int Soc Soil Mech & Found Eng 89- *Mem:* Geol Soc Am; Can Geotech Soc; Geol Assoc Can; Can Inst Mining; Am Geophysical Union. *Res:* Behavior of landforms and natural materials; prediction of catastrophic changes, landslides, rockfall-avalanches, ground subsidence and possible adaptions to them. *Mailing Add:* Dept Civil Eng Univ Alberta Edmonton AB T6G 2G7 Can

CRUDEN, ROBERT WILLIAM, b Cleveland, Ohio, Mar 18, 36; m 67; c 2. ECOLOGY, EVOLUTION. *Educ:* Hiram Col, BA, 58; Ohio State Univ, MSc, 60; Univ Calif, Berkeley, PhD(bot), 67. *Prof Exp:* From asst prof to assoc prof, 67-78, PROF BOT, UNIV IOWA, 78-; asst dir, 88-89, ACTING DIR, IOWA LAKESIDE LAB, 89- *Concurrent Pos:* Ed, Ecol Soc Am, 83-86. *Mem:* Ecol Soc Am; Brit Ecol Soc; Am Soc Plant Taxonomists; Am Bot Soc; Asn Trop Biol. *Res:* Breeding systems, pollination biology, other animal-plant interactions and other life history parameters of plants and their evolution; systematics. *Mailing Add:* Dept of Bot Univ of Iowa Iowa City IA 52242

CRUESS, RICHARD LEIGH, b Dec 17, 29; Can citizen; m 53; c 2. EDUCATIONAL ADMINISTRATION, SURGERY. *Educ:* Princeton Univ, BA, 51; Columbia Univ, MD, 55. *Prof Exp:* From asst prof to assoc prof orthop, 63-73, PROF SURG, MCGILL UNIV, 73-, DEAN, FAC MED, 81- *Honors & Awards:* Steindler Award, Am Acad Orthop Surg, 83; Malcolm Brown Award, Royal Col Physicians & Surgeons Can, 81. *Mem:* Royal Soc of Can; Can Orthop Asn (secy, 71-76, pres, 77-78); Am Orthop Res Soc (pres, 75-76); Can Orthop Res Soc (pres, 71-72); Royal Col Physicians & Surgeons Can. *Res:* Clinical subjects; osteonecrosis of bone; physiologic and biological research in skeletal tissue. *Mailing Add:* Fac Med McGill Univ 3655 Drummond ST Montreal PQ H3G 1Y6 Can

CRUICE, WILLIAM JAMES, b New York, NY, Aug 22, 37; m 65; c 2. FIRE & EXPLOSION,CHEMICALS & PROCESSES. *Educ:* St John's Univ, BS, 60, BA, 61, MS, 64. *Prof Exp:* From chemist to sr chemist aerospace, Reaction Motors Div, Thiokol Corp, 63-69; asst vpres res eng, US Banknote Corp, 69-70; VPRES & GEN MGR, HAZARDS RES CORP, 69- *Mem:* Am Inst Chem Engrs; Am Soc Testing & Mat; Nat Fire Protection Asn; Am Inst Chemists. *Res:* Fire and explosion hazards of chemicals and chemical processes. *Mailing Add:* Four Cambridge Rd Convent Station NJ 07961

CRUICKSHANK, ALEXANDER MIDDLETON, b Marlboro, NH, Dec 13, 19; m 45; c 2. INORGANIC CHEMISTRY, ANALYTICAL CHEMISTRY. *Educ:* RI State Col, BS, 43, MS, 45; Univ Mass, PhD(chem), 54. *Prof Exp:* Asst chem, RI State Col, 43-45, instr, 45-48; instr, Univ Mass, 48-53; from asst prof to assoc prof, 53-69, PROF CHEM, UNIV RI, 69- *Concurrent Pos:* Instr, Holyoke Jr Col, 50-53; asst to dir, Gordon Res Confs, AAAS, 47-68, dir, 68- *Mem:* Am Chem Soc; Am Asn Textile Chem & Colorists; Sigma Xi. *Res:* Polarography; metal complexes of amino and hydroxyl acids; alkyd resins; textiles. *Mailing Add:* Gordon Res Ctr Univ of RI Kingston RI 02881

CRUICKSHANK, MICHAEL JAMES, b Glasgow, Scotland, July 9, 29. MARINE MINING, DEEP SEABED MINING. *Educ:* Camborne Sch Mines, ACSM, 53; Colo Sch Mines, MSc, 62; Univ Wis, Madison, PhD(oecanog & limnol), 78. *Prof Exp:* Dist engr, Anglo-Greek Magnesite Co, Ltd, 53-55; min mgr, Muirshiel Barytes Co, Ltd, 56-61; res geologist, Scripps Inst Oceanog, 63-64; asst prof mining, Univ Alaska, 64-65; res specialist, Lockheed Missiles & Space Co, Ltd, 65-66; res supvr, US Bur Mines, 66-69; res coordr, Marine Mineral Technol Ctr, Nat Oceanic & Atmospheric Admin, 69-73; marine mining engr, US Geol Surv, 73-; TECH DIR, MARINE MINERALS TECH CTR, UNIV HAWAII. *Concurrent Pos:* Ed-in-chief, Draft Environ Statement Deep Sea Mining, 73-74; Sloan fel, Grad Sch Bus, Stanford Univ, 75-76; mem, Group Experts Sci Aspects Marine Pollution, UN, 77; chmn, Task Force Outer Continental Shelf Mining Policy, US Dept Interior, 78; consult, Offshore Exploration, UN, Thailand, 79-80; vis prof, Henry Krumb Sch Mines, Columbia Univ, 81-83. *Mem:* Soc Mining Engrs; Marine Technol Soc; World Dredging Asn. *Res:* International development of mineral resources and technology in marine and coastal areas. *Mailing Add:* 811 Olomehani St Honolulu HI 96813

CRUICKSHANK, P A, b Bremerton, Wash, Oct 11, 29; m 53; c 3. PESTICIDE CHEMISTRY. *Educ:* Univ Wash, BSc, 51; Mass Inst Technol, PhD(org chem), 55. *Prof Exp:* Chemist, E I du Pont de Nemours Co, 55-58; group leader peptides, Res Inst Med & Chem, 58-63; interdisciplinary scientist, 63-66, RES MGR, AGR CHEM GROUP, FMC CORP, 66- *Mem:* Am Chem Soc. *Res:* Synthesis of peptides and steroids; chromatographic methods of analysis; insect hormones; insecticides; nematicides; quantitative structure activity relationships; computer assisted molecular design. *Mailing Add:* 211 Dodds Lane Princeton NJ 08540

CRUIKSHANK, DALE PAUL, b Des Moines, Iowa, Aug 10, 39; m 64, 75; c 3. PLANETARY PHYSICS, GEOSCIENCE. *Educ:* Iowa State Univ, BS, 61; Univ Ariz, MS, 65 & PhD(planetary geol), 68. *Prof Exp:* Res asst, Lunar & Planetary Lab, Univ Ariz, 61-68, res assoc, 69-70; from asst astrom to astrom, Inst Astron, Univ Hawaii, 70-87; SPACE SCIENTIST, AMES RES CTR, NASA, 88- *Concurrent Pos:* Mem, Can-France-Hawaii Telescope Corp, 80-87; assoc ed, Icarus, Int J of Solar Syst, 82- *Mem:* Int Astron Union; Am Astron Soc; Am Geophys Union. *Res:* Physics and geology of the planets and their satellites, asteroids, and the comets as determined largely through astronomical techniques of spectroscopy and photometry; detection and study of atmospheres. *Mailing Add:* Ames Res Ctr NASA Mail Stop 245-6 Moffett Field CA 94035

CRUIKSHANK, DONALD BURGOYNE, JR, b Boise, Idaho, Aug 2, 39; m 65; c 1. ACOUSTICS, GENERAL PHYSICS. *Educ:* Kalamazoo Col, BA, 64; Univ Rochester, MS, 66, PhD(acoustics), 70. *Prof Exp:* Asst prof physics, Cornell Col, 69-71; from asst prof to assoc prof, 71-80, PROF PHYSICS, ANDERSON UNIV, 80- *Concurrent Pos:* Sr scientist, Bolt, Beranek & Newman, 76-77. *Honors & Awards:* Lilly Fac Open Fel, 76. *Mem:* Acoust Soc Am; Am Asn Physics Teachers; fel Res Found. *Res:* Finite-amplitude acoustic resonance oscillations in closed, rigid-walled tubes. *Mailing Add:* Dept of Physics Anderson Univ Anderson IN 46012

CRUISE, DONALD RICHARD, b Los Angeles, Calif, Feb 17, 34; m 65; c 2. APPLIED MATHEMATICS. *Educ:* Fresno State Col, AB, 56. *Prof Exp:* mathematician, Naval Weapons Ctr, China Lake, 56-89; STAFF ENGR, COMARCO, INC, 89- *Mem:* Sigma Xi. *Res:* Mathematical models for chemical equilibrium. *Mailing Add:* 2051 S Inyo St Ridgecrest CA 93555-9548

CRUISE, JAMES E, b Port Dover, Ont, June 26, 25. PLANT TAXONOMY. *Educ:* Univ Toronto, BA, 50; Cornell Univ, MS, 51, PhD, 54. *Prof Exp:* Lab instr, Cornell Univ, 51-56; from asst prof to prof biol, Trenton State Col, 56-63; assoc prof, Univ Toronto, 63-69, prof biol & cur, 69-75; DIR, ROYAL ONT MUS, 75- *Mem:* Am Soc Plant Taxon; Can Bot Asn; Int Soc Plant Taxon. *Res:* Taxonomy of vascular plants; flora of Ontario. *Mailing Add:* 100 Queens Park Toronto ON M5S 1A1 Can

CRUM, EDWARD HIBBERT, b South Charleston, WVa, Sept 21, 40; m 63; c 2. BIOCHEMICAL ENGINEERING. *Educ:* WVa Inst Technol, BS, 62; Mo Sch Mines, MS, 64; Univ Mo-Rolla, PhD(chem eng), 67. *Prof Exp:* From asst prof to assoc prof, 66-77, PROF CHEM ENG, WVA INST TECHNOL, 77-, CHMN DEPT, 80- *Concurrent Pos:* Res fel, NASA; NSF fel, Union Carbide Corp. *Mem:* Am Chem Soc; Am Inst Chem Engrs; Instrument Soc Am. *Res:* Thiobacillus thiooxidans cell walls; air and waste water treatment; technological forecasting. *Mailing Add:* PO Box 117 Charlton Heights WV 25040

CRUM, FLOYD M(AXILAS), b Baton Rouge, La, Apr 28, 22; m 43; c 5. ELECTRICAL ENGINEERING. *Educ:* La State Univ, BS, 47, MS, 51. *Prof Exp:* Assoc prof elec eng, La State Univ, 47-55; REGENTS PROF & HEAD ELEC ENG DEPT, LAMAR UNIV, 55- *Concurrent Pos:* Design engr, Stone & Webster Consult Engrs, 47; electronic scientist, Naval Res Lab, 51-53, 55 & 61-69 & Boeing Aircraft, 54; consult, Sun Oil Geophys Lab, 55-61; Naval res electronic consult engr, 62-76; prof, elec engr, Lamar Univ, 76. *Mem:* Am Soc Eng Educ; sr mem Inst Elec & Electronics Engrs; sr mem Instrument Soc Am; Sigma Xi. *Res:* Solid state electronics; network synthesis; information theory. *Mailing Add:* Rt 8 Box G55 PO Box 10029 Beaumont TX 77705

CRUM, GLEN F(RANCIS), b Humboldt, Ill, June 13, 25; m 49; c 2. CHEMICAL ENGINEERING. *Educ:* Univ Ill, BS, 48, MS, 58, PhD(chem eng), 61. *Prof Exp:* Res engr, Gen Elec Co, 48-52; res assoc, Upjohn Co, 52-56; res specialist, Monsanto Co, 60-67; res & develop engr, El Paso Prod Co, 67-76, res assoc, 76-83; RETIRED. *Concurrent Pos:* Adj prof chem, appl math & physics, Univ Tex. *Mem:* Am Chem Soc; Am Inst Chem Engrs. *Res:* Distillations; chemical reactions; catalysts. *Mailing Add:* 838 Eaglebrooke Dr Manchester MO 63021-7531

CRUM, HOWARD ALVIN, b Mishawaka, Ind, July 14, 22; m 60; c 2. BRYOLOGY. *Educ:* Western Mich Col, BS, 47; Univ Mich, MS, 49, PhD(bot), 51. *Prof Exp:* Instr, Western Mich Col, 46-47; teaching fel, Univ Mich, 48-49; res biol & actg asst prof, Stanford Univ, 51-53; asst prof, Univ Louisville, 53-54; biologist, Nat Mus Can, 54-65; PROF BOT & CUR BRYOPHYTES & LICHENS, UNIV MICH, 65- *Concurrent Pos:* Ed, Am Bryological & Lichenological Soc, 53-62, assoc ed, 62-77; assoc ed, Brit Bryological Soc, 72-77; pres, Mich Bot Club, 73-74, assoc ed, 75-76, ed, 77-84. *Honors & Awards:* Henry Allen Gleason Award, 81. *Mem:* Am Bryological & Lichenological Soc (pres, 62-63); Brit Bryological Soc. *Res:* Taxonomy of bryophytes; mosses; monographic studies in the genus Sphagnum; Lecanora esculenta (desert manna) and its potential use by grazing animals. *Mailing Add:* Herbarium 2002 N Univ Bldg Univ of Mich Ann Arbor MI 48109

CRUM, JAMES DAVIDSON, b Ironton, Ohio, July 29, 30; m 56; c 3. ORGANIC CHEMISTRY. *Educ:* Ohio State Univ, BSc, 52, PhD(org chem), 58; Marshall Univ, MSc, 53. *Prof Exp:* Asst, Marshall Univ, 52-53; from asst to asst instr, Ohio State Univ, 53-57; NIH res fel, Harvard Univ, 58-59; from instr to asst prof chem, Case Western Reserve Univ, 59-66; assoc prof, 66-69, chmn dept, 70-72, PROF CHEM, CALIF STATE UNIV, SAN BERNARDINO, 69-, DEAN, SCH NATURAL SCI, 72- *Concurrent Pos:* Consult, Nat Sci Found-US Agency Int Develop, 65-68. *Mem:* Fel AAAS; fel Am Inst Chem; Am Chem Soc; fel India Inst Chem; Royal Soc Chem; Sigma Xi. *Res:* Natural products; carbohydrates and steroids; reaction mechanisms; synthesis. *Mailing Add:* Chem Dept Calif State Univ 5500 Univ Pkwy San Bernardino CA 92407-2397

CRUM, JOHN KISTLER, b Brownsville, Tex, July 28, 36. INORGANIC CHEMISTRY, SCIENCE ADMINISTRATION. *Educ:* Univ Tex, BSChE, 60, PhD(chem), 64; Harvard Univ, Advan Mgt Prog dipl, 75. *Prof Exp:* From asst ed to assoc ed anal chem, Am Chem Soc, 64-68, managing ed publ, 68-70, group mgr jour, 70-71, dir books & jour div, 71-75, treas & chief financial officer, 75-84, dep exec dir & chief operating officer, 81-83, EXEC DIR, AM CHEM SOC, 83- *Concurrent Pos:* Consult, Joint Comt Atomic & Molecular Phys Data; mem, Coun Eng & Sci Soc Execs. *Mem:* Am Chem Soc; Royal Soc Chem; NY Acad Sci. *Res:* Chemistry of reactions in liquid ammonia; design of scientific publications; chemical literature; information retrieval. *Mailing Add:* Am Chem Soc 1155 16th St NW Washington DC 20036

CRUM, LAWRENCE ARTHUR, b Caldwell, Ohio, Oct 25, 41; m 63; c 4. ACOUSTICS. *Educ:* Ohio Univ, BS, 63, MS, 65, PhD(physics), 67. *Prof Exp:* Res fel, Harvard Univ, 67-68; asst prof, 68-72, ASSOC PROF PHYSICS, US NAVAL ACAD, 72-; assoc dir, 87-89, DIR, NAT CTR PHYS ACOUST, 90- *Concurrent Pos:* Consult, Planning Syst Inc, 72- & Naval Ord Sta, 73-; pres, Acad Assoc, Inc, 75-; sr vis scientist, Univ Col, Cardiff, Wales, 76-77; assoc prof physics, Univ Miss, 77-78, prof, 78-; Fulbright fel, Univ London, 85-86. *Mem:* Am Asn Physics Teachers; fel Acoust Soc Am; Am Inst Ultrasound Med. *Res:* Acoustic cavitation and effects of sound fields on bubbles. *Mailing Add:* Nat Ctr Phys Acoust PO Box 847 University MS 38677

CRUM, RALPH G, b Youngstown, Ohio, July 22, 30; m 59; c 3. APPLIED MECHANICS, CIVIL ENGINEERING. *Educ:* Carnegie Inst Technol, BS, 53, MS, 54, PhD, 56. *Prof Exp:* Asst civil eng, Carnegie Inst Technol, 56-61; sr scientist, Ford Motor Co, 61-62; prin scientist, NAm Aviation, 62-70; assoc prof, 70-75, PROF CIVIL ENG, YOUNGSTOWN STATE UNIV, 75- *Concurrent Pos:* Ford Found fel, 60; mat & mech consult, 61-68; environ consult, 69-70; NASA fel, 71-73; assoc forensic eng, Verna Eng, Inc, 78- *Mem:* Am Soc Civil Engrs; Am Soc Test & Mat; Soc Exp Stress Anal; Am Soc Eng Educ; Sigma Xi. *Res:* Materials, mechanics, transportation engineering; forensic engineering. *Mailing Add:* Eng Technol Youngstown State Univ Youngstown OH 44555

CRUMB, GLENN HOWARD, b Burlingame, Kans, Dec 21, 27; m 50; c 4. SCIENCE, MATH ADMINISTRATION. *Educ:* Kans State Teachers Col, BS, 51, MS, 56; Univ Nebr, PhD(sci educ), 64. *Prof Exp:* Teacher high schs, Kans, 51-56; instr phys sci, Kans State Teachers Col, 56-59; physics, Univ Nebr High Sch, 59-63; assoc prof phys sci, Kans State Teachers Col, 63-65, prof, 65-71, head dept, 64-71, dir res & grants ctr, 69-71; prof educ & dir grant & contract servs, Western Ky Univ, 71-83; dir, Ctr Math/Sci Environ Educ, 84-89. *Concurrent Pos:* Shell Oil Co Merit fac fel, 69; Dupont fel, Harvard Univ, 56. *Honors & Awards:* Christa McAuliffe Award Merit Appl Res, Am Asn Col Teachers. *Mem:* Am Asn Physics Teachers; Nat Sci Teachers Asn. *Res:* School networks using telecommunications to support computer aided instructions and computer managed instruction; science teaching strategies; reasoning in science. *Mailing Add:* 2251 Smallhouse Rd Bowling Green KY 42107

CRUMB, STEPHEN FRANKLIN, electrical engineering; deceased, see previous edition for last biography

CRUME, ENYEART CHARLES, JR, b Dayton, Ohio, Nov 14, 31; c 3. PLASMA PHYSICS, NUCLEAR ENGINEERING. *Educ:* Wabash Col, AB, 53; Wesleyan Univ, MA, 55; Univ Tenn, PhD(physics), 72. *Prof Exp:* Asst proj engr exp nuclear eng, CANEL Proj, Pratt & Whitney Aircraft, United Aircraft Corp, 56-64; physicist nuclear criticality safety, Y-12 Plant, Nuclear Div, Union Carbide Corp, 64-72; theoretical physicist plasma physics, Fusion Energy Div, 72- 87, exp physicist, 87-90, PHYSICIST NUCLEAR CRITICALITY SAFETY, OFF OPER READINESS & SAFETY, OAK RIDGE NAT LAB, MARTIN MARIETTA ENERGY SYSTS INC, 91- *Mem:* Am Phys Soc. *Res:* Theory and numerical simulation of transport processes in fusion energy related plasmas, especially transport of plasma impurities and its consequences; spectroscopic measurements of impurities in plasmas. *Mailing Add:* Oak Ridge Nat Lab PO Box 2008 Oak Ridge TN 37831-6012

CRUMMETT, WARREN B, b Moyers, WVa, Apr 4, 22; m 48; c 2. SPECTROSCOPY, QUALITY OF DATA. *Educ:* Bridgewater Col, BA, 43; Ohio State Univ, PhD(chem), 51. *Prof Exp:* Control chemist, Solvay Process Co, 43-46; anal chemist, Dow Chem Co, 51-55, group leader, 55-61, asst dir anal labs, 61-71, anal scientist, 71-76, res scientist, 76- 83, res fel, 83-88; RETIRED. *Concurrent Pos:* Mem adv bd, Anal Chem, 74-76; mem environ measurements adv comt, Sci Adv Bd, Environ Protection Agency, 76-78, consult, 78-80; consult, Sci Adv Bd, US Air Force, 81, Environ Adv Comt, Oak Ridge, 84- *Honors & Awards:* H H Dow Medal, 80, Am Chem Soc. *Mem:* Soc Chem Indust; Asn Off Anal Chem; Am Chem Soc; NY Acad Sci; Sigma Xi; Int Union Pure Appl Chem. *Res:* Ion exchange; platinum metals; ultraviolet and near infrared spectrophotometry; purity of organic compounds; liquid chromatography; analytical systems; environmental analysis; measurements near the limit of detection. *Mailing Add:* 808 Crescent Dr Midland MI 48640

CRUMMY, ANDREW B, b Newark, NJ, Jan 30, 30; m 58; c 3. RADIOLOGY. *Educ:* Bowdoin Col, AB, 51; Boston Univ, MD, 55. *Prof Exp:* Intern, Univ Wis Hosps, 55-56, resident radiol, 58-61; asst prof, Univ Colo, 63-64; from asst prof to assoc prof, 64-70, PROF RADIOL, UNIV WIS-MADISON, 70- *Concurrent Pos:* Fel radiol, Mt Auburn Hosp, Cambridge, Mass, 61-62; fel cardiovasc radiol, Yale Univ, 62-63. *Mem:* Fel Am Col Angiol; fel Am Col Radiol; fel Am Heart Asn; AMA; fel Soc Cardiovasc Radiol. *Res:* Diagnostic radiology; cardiovascular radiology. *Mailing Add:* Dept Radiol D4-346 CSC Univ Wis Sch Clin Ctr Madison WI 53792

CRUMMY, PRESSLEY LEE, b Glade Mills, Pa, Oct 1, 06; m 34. HISTOLOGY, HUMAN ANATOMY. *Educ:* Grove City Col, BS, 29; Univ Pittsburgh, MS, 32, PhD(zool), 34. *Hon Degrees:* DSc, Kirksville Col Osteop Med, 87. *Prof Exp:* Teacher rural sch, Pa, 25-26, high sch, 29-30; asst biol, Grove City Col, 27-29 & zool, Univ Pittsburgh, 30-34; head dept sci, High Sch, Pa, 34-35; from instr to asst prof biol, Juniata Col, 35-47, prof, 47-49, from actg registr to registr, 42-49; from assoc prof to prof, 49-77, EMER PROF ANAT, KIRKSVILLE COL OSTEOP MED, 77- *Res:* Histology; gross human anatomy; morphological anomalies; biological effects of x-rays. *Mailing Add:* 910 E Harrison St Kirksville MO 63501

CRUMP, JESSE FRANKLIN, b Pine Bluff, Ark, July 20, 27; m 59. BIOMEDICAL ENGINEERING, INTERNAL MEDICINE. *Educ:* Univ Nebr, BSEE, 50, MD, 56. *Prof Exp:* Intern, Methodist Hosp, Brooklyn, 56-57, asst resident, 57-58; resident internal med, Long Island Col Hosp, 61-62; from asst prof to assoc prof elec eng, 62-68, ASSOC PROF BIOENG, POLYTECH INST NEW YORK, 68-; RADIOLOGIST, BROOKLYN HOSP, 84- *Concurrent Pos:* NIH res fel psychosom med, State Univ NY, 58-59 & cardiol, Long Island Col Hosp, Brooklyn, 59-61; consult, Dept Psychol, Princeton Univ, 57-62, Biomed Eng Rehab Dept, Montefiore Hosp, New York, 65-77; vis res physiologist, Princeton Univ, 62-75; assoc attend, Long Island Col Hosp, 62-81; prin investr, USPHS grant, 66-68; radiol resident, Long Island Col Hosp, 81-84. *Mem:* AAAS; NY Acad Sci; Asn Advan Med Instrumentation; Inst Elec & Electronics Engrs. *Res:* Auditory, cardiology and cardiovascular physiology; biomedical instrumentation. *Mailing Add:* 19 Giletta Ct Closter NJ 07624

CRUMP, JOHN C, III, b Richmond, Va, Feb 21, 40; m. SOLID STATE PHYSICS. *Educ:* Hampden-Sydney Col, BS, 60; Univ Va, MS, 62, PhD(physics), 64. *Prof Exp:* Res assoc physics, Univ Va, 64-66; res physicist, Oak Ridge Nat Lab, 66-69; ASSOC SR SCIENTIST, PHILIP MORRIS INC, 69- *Mem:* Am Phys Soc; Electron Micros Soc Am; Sigma Xi. *Res:* Dislocation phenomena and irradiation effects in solid materials; aerosol and thermokinetic physics. *Mailing Add:* Philip Morris USA Res Ctr Po Box 26583 Richmond VA 23261

CRUMP, JOHN WILLIAM, b Santa Rosa, Calif, Jan 18, 32; m 55; c 4. ORGANIC CHEMISTRY. *Educ:* Univ Calif, BA, 53; Univ Ill, PhD(org chem), 57. *Prof Exp:* Res chemist, Dow Chem Co, 57-62; from asst prof to assoc prof, 62-69, chmn dept, 69-87, PROF CHEM, ALBION COL, 69- *Concurrent Pos:* NSF sci fac fel, Univ Leiden, Neth, 70-71; on leave, Lucknow Christian Col, India, 77-78. *Mem:* Am Chem Soc; Sigma Xi. *Res:* Aromatic rearrangements and ipso-substitutions. *Mailing Add:* 14685 E Michigan Albion MI 49224

CRUMP, KENNY SHERMAN, b Haynesville, La, Oct 13, 39; m 61; c 3. MATHEMATICS, STATISTICS. *Educ:* La Tech Univ, BS, 61; Univ Denver, MA, 63; Mont State Univ, PhD(math), 68. *Prof Exp:* Instr, Mont State Univ, 63-66; assoc prof math, La Tech Univ, 66-78, prof math & statist, 78-81; pres, Sci Res Systs, Inc, 81-85, pres, K S Crump & Co, Inc, 85-87; EXEC VPRES, CLEMENT INT CORP, 87- *Concurrent Pos:* Res assoc, State Univ NY Buffalo, 67-68; vis scientist, Nat Inst Environ Health Sci, 74-75. *Mem:* Am Statist Asn; Biometric Soc; Inst Math Statist; Soc Epidemiol Res; AAAS. *Res:* Applications of statistics and stochastic processes to problems related to biology and health. *Mailing Add:* Clement Int Corp 1201 Gaines St Ruston LA 71270

CRUMP, MALCOLM HART, b Culpeper, Va, Aug 10, 26; m 52; c 3. GASTROENTEROLOGY. *Educ:* Va Polytech Inst, BS, 51; Univ Ga, DVM, 58; Univ Wis, MS, 61, PhD(physiol), 65. *Prof Exp:* Res assoc physiol, Univ Wis, 60-65; asst prof, 64-66, ASSOC PROF PHYSIOL, IOWA STATE UNIV, 66- *Mem:* Am Vet Med Asn; assoc Am Physiol Soc. *Res:* Pharmacological action of mycotoxins; drug metabolism; comparative gastroenterology; control mechanisms in the large intestine. *Mailing Add:* Dept Physiol Iowa State Univ Col Vet Med Ames IA 50011

CRUMP, MARTHA LYNN, b Madison, Wis, Aug 23, 46. ECOLOGY, BIOLOGY. *Educ:* Univ Kans, BA, 68, MA, 71, PhD(ecol & systs), 74. *Prof Exp:* Fel, City Univ New York, 74-76; asst prof ecol, 76-80, ASSOC PROF ZOOL, UNIV FLA, 80- *Concurrent Pos:* NSF res grant, 75-76 & 78-79; res assoc, Am Mus Natural Hist, 76-; affil asst cur herpetol, Fla State Mus, 78- *Mem:* AAAS; Brit Ecol Soc; Soc Study Amphibians & Reptiles; Soc Study Evolution; Ecol Soc Am; Sigma Xi. *Res:* Population and community ecology of amphibians; life history strategies of tropical amphibians; energetics and behavior of amphibian reproduction. *Mailing Add:* Dept Zool 622 BRW Bldg Univ Fla Gainesville FL 32611

CRUMP, STUART FAULKNER, b Boston, Mass, Feb 20, 21; m 44; c 4. PHYSICS, RESEARCH ADMINISTRATION. *Educ:* Brown Univ, BA, 43. *Prof Exp:* Physicist, David W Taylor Model Basin, 43-45, 56-59, naval architect, 45-49, 52-56, hydraul engr, 49-52, phys sci administr, 59-60; physicist, Navy Bur Ships, David W Taylor Naval Ship Res & Develop Ctr, 61-62, phys sci adminstr, 62-67, contract res adminstr, 67-79; sr consult, Vector Res Co & Designers & Planners, Inc, 79-88; sr consult, Atlantic Res Prof Serv, 88-89; SR CONSULT, J J MCMULLEN ASSOC, INC & TECHNOL PLANNING ASSOC, 90- *Concurrent Pos:* Consult, 61-62. *Mem:* Am Soc Naval Engrs; Marine Technol Soc. *Res:* Hydromechanics of naval architecture; underwater acoustics; cavitation research; engineering physics; contract research administration; ship silencing. *Mailing Add:* 101 Evans St Rockville MD 20085

CRUMPACKER, DAVID WILSON, b Enid, Okla, Mar 29, 29; m 55; c 4. GENETICS, AGRONOMY. *Educ:* Okla Agr & Mech Col, BS, 51; Univ Calif, Davis, PhD(genetics), 59. *Prof Exp:* Asst agron, Univ Calif, Davis, 55-59; assoc prof, Colo State Univ, 59-70; chmn dept, 77-80, PROF ENVIRON, POP & ORGANISMIC BIOL, UNIV COLO, BOULDER, 70- *Concurrent Pos:* USPHS spec fel, Rockefeller Univ, 65-66. *Mem:* AAAS; Am Soc Agron; Crop Sci Soc Am; fel Am Soc Naturalists; Genetics Soc Am. *Res:* Evolution and population genetics of maize and Drosophila; biometrical genetics of wheat; maize breeding. *Mailing Add:* Dept Biol Box 334 Univ Colo Boulder CO 80309

CRUMPLER, THOMAS BIGELOW, b Louisa, Ky, July 20, 09; m 35; c 2. ANALYTICAL CHEMISTRY. *Educ:* Va Polytech Inst, BS, 31, MS, 32; Univ Va, PhD(chem), 36. *Prof Exp:* Fel, Univ Va, 36-37; from instr to assoc prof, 37-43, head dept, 42-62, prof, 43-74, EMER PROF CHEM, TULANE UNIV, 74- *Concurrent Pos:* Ford fac fel, 51-52. *Mem:* Fel Am Inst Chem; Am Chem Soc; Sigma Xi. *Res:* Photoelectric colorimetry and polarimetry; colorimetric reagents; paper chromatography. *Mailing Add:* PO Box 831 Highlands NC 28741

CRUMRINE, DAVID SHAFER, b Memphis, Tenn, Aug, 12, 44; m 67; c 2. PHYSICAL ORGANIC CHEMISTRY. *Educ:* Ashland Col, AB, 66; Univ Wis-Madison, PhD(org chem), 71. *Prof Exp:* Fel org chem, Mass Inst Technol & Ga Inst Technol, 71-72; asst prof, 72-76, ASSOC PROF ORG PHYS CHEM, LOYOLA UNIV, CHICAGO, 76- *Concurrent Pos:* Vis foreign scientist, Inst Molecular Sci, Okazaki, Japan, 80 & Dyson Perrins Lab, Oxford Univ, 81; vis prof, Univ Chicago, 88. *Mem:* Am Chem Soc; Sigma Xi. *Res:* S-33 NMR of sulfonates; mechanistic and exploratory organic photochemistry; synthesis of novel systems; dipolar cycloadditions; neuropeptide-dye conjugates. *Mailing Add:* Dept Chem Loyola Univ Chicago IL 60626

CRUSBERG, THEODORE CLIFFORD, b Meriden, Conn, Feb 23, 41; m 66; c 3. MICROBIOLOGY, WATER QUALITY. *Educ:* Univ Conn, BA, 63; Yale Univ, MS, 64; Clark Univ, PhD(chem), 68. *Prof Exp:* NIH trainee biochem, Sch Med, Tufts Univ, 68-69; ASSOC PROF BIOL & BIOTECHNOL, WORCESTER POLYTECH INST, 69- *Concurrent Pos:* Consult, water qual, biotechnol. *Mem:* AAAS; Sigma Xi; Soc Indust Microbiol. *Res:* Heavy metal removal and recovery from wastewaters by fungal mycelia. *Mailing Add:* Dept Biol & Biotechnol Worcester Polytech Inst Worcester MA 01609-2280

CRUSE, CARL MAX, b St Charles, Va, Aug 4, 36; m 58; c 2. CHEMICAL ENGINEERING. *Educ:* Univ Tenn, BS, 59. *Prof Exp:* Res chem engr, Monsanto Chem Co, 59-61; process develop engr, Holston Defense Corp, Eastman Kodak Co, 61-63; res engr, Monsanto Co, 63- 69, process specialist, 69-71, mfg supvr, 71-77, prin engr, 77-79, mfg supt, 79-83, total qual supt, 83-86, SR MFG SPECIALIST, STERLING CHEM INC, 86- *Concurrent Pos:* statist process control consult, 86-87. *Res:* Kinetic studies of the pyrolysis of hydrocarbon mixtures; low-pressure carbonylation of alcohols using noble metal catalysts. *Mailing Add:* Sterling Chem Inc PO Box 1311 Texas City TX 77592-1311

CRUSE, JULIUS MAJOR, JR, b New Albany, Miss, Feb 15, 37. IMMUNOLOGY, PATHOLOGY. *Educ:* Univ Miss, BA & BS, 58; Univ Tenn, MD, 64, PhD(path), 66. *Prof Exp:* USPHS res fel path, Univ Tenn, Med Ctr, Memphis, 64-67; prof immunol, dir Immuni Res Unit & prof biol, Grad Sch, Univ Miss, 67-74, asst prof microbiol & path, Univ Miss Med Ctr, 68-74, ASSOC PROF MICROBIOL, PROF PATH & DIR GRAD STUDIES PROG PATH, UNIV MISS MED CTR, 74-, DIR IMMUNOPATH SECT, 78- *Concurrent Pos:* Lectr path, Univ Tenn Med Units, 67-74, vis prof, 68-74; chmn Safety Adv Comt & mem Radiation Adv Comt, Univ Miss, 72-74; mem Cancer Res Adv Comt, Univ Miss Med Ctr, 72-, Grad-Med Sci Coun, 75-, Transfusion Med Educ Comt, 88-; mem res comt, Vet Admin Hosp, Jackson, Miss, 74-78; adj prof immunol, Miss Col, 77- *Honors & Awards:* Physician's Recognition Award in Continuing Med Educ, AMA, 69 & 75. *Mem:* Fel AAAS; Am Asn Path & Bact; Am Soc Exp Path; Am Chem Soc; Am Soc Microbiol; fel Am Acad Microbiol; AMA; NY Acad Sci; Sigma Xi; fel Am Acad Microbiol; Am Asn Immunologists; Am Soc Clin Path; Am Soc Histocompatibility & Immunogenetics; Transplantation Soc; Reticuloendothelial Soc. *Res:* Transplantation immunology; immunologic monitoring of transplant patients by flow cytometry and monoclonal antibody techniques; immunological of transplantable tumors; histocompatibility and disease association; immulologic classification of human leukemias and lymphomes by monoclonal antibodies; immunofluorescence and immunoperoxidase examinatory of renal biopsies; author of approximately 200 papers and abstracts and three books on immunology and immunopathology. *Mailing Add:* Dept Path Univ Miss Med Ctr 2500 N State St Jackson MS 39216

CRUSE, RICHARD M, b Waverly, Iowa, July 14, 50; m 69; c 2. SOIL SCIENCE, AGRONOMY. *Educ:* Iowa State Univ, BS, 72; Univ Minn, MS, 75, PhD(soil physics), 78. *Prof Exp:* Asst prof soil mgt, NC State Univ, 78-79; from asst prof to assoc prof, 83-88, PROF SOIL MGT, IOWA STATE UNIV, 88- *Mem:* Am Soc Agron; Soil Sci Soc Am; Soil Conserv Soc Am. *Res:* Impact of various soil management practices on soil environmental conditions and plant root growth. *Mailing Add:* Dept Agron Iowa State Univ Ames IA 50011

CRUSE, ROBERT RIDGELY, b Tucson, Ariz, Aug 20, 20; m 47; c 1. APPLIED CHEMISTRY. *Educ:* Antioch Col, BS, 42. *Prof Exp:* Res engr, Battelle Mem Inst, 42-47; anal chemist, US Bur Mines, Ariz, 53-55, extractive metallurgist, 55; assoc indust chemist, Southwest Res Inst, 55-61; res chemist, Nitrogen Div, Allied Chem Corp, 61-68 & Agr Prod Technol Res, USDA, 68-82; RETIRED. *Concurrent Pos:* Instr org chem, Trinity Univ, 56-57; assoc, Southwest Agr Inst, Tex, 59-60; consult, 82- *Mem:* Fel AAAS; fel Am Inst Chem; Am Chem Soc; Sigma Xi. *Res:* Chemurgy; organic synthesis; pharmaceuticals; fuels; insecticides; fertilizers; foods. *Mailing Add:* 1106 W Third St Weslaco TX 78596

CRUSER, STEPHEN ALAN, b Greensburg, Ind, Dec 12, 42; m 67; c 2. ANALYTICAL CHEMISTRY, PHYSICAL CHEMISTRY. *Educ:* Ind Univ, Bloomington, BS, 64; Univ Tex, Austin, PhD(chem), 68. *Prof Exp:* Chemist, 68-69, sr chemist, 70-72, res chemist, 72-77, sr res chemist, 78-81, asst supvr, 81-84, SUPVR, TEXACO INC, BELLAIRE, 84- *Mem:* Nat Asn Corrosion Engrs. *Res:* Electroanalytical chemistry; corrosion control; inhibitor mechanisms. *Mailing Add:* 12042 Corona Lane Houston TX 77072

CRUTCHER, HAROLD L, b Cheraw, Colo, Nov 18, 13; m 43; c 2. METEOROLOGY, CHEMISTRY. *Educ:* Southeastern State Col, BA, 33, BS, 34; NY Univ, MS, 51, PhD(meteorol), 60. *Prof Exp:* Teacher pub schs, Okla, 35-39; from jr observer to sr observer, Nat Oceanic & Atmospheric Admin, 39-42, meteorologist, 42-45, forecaster, 45-47, meteorologist, Opers Div, 47-50, Climat & Hydrol Serv Div, 51-52 Climat Serv Div, 52-53 & Nat Climatic Ctr, 53-63, res meteorologist, Nat Climatic, 63-77; ADJ PROF, UNIV NC, ASHEVILLE, NC, 54- *Concurrent Pos:* Mem opers anal standby unit, Univ NC, 59-69; consult, Meteorol, Qual Control, Chem. *Honors & Awards:* Silver Award, US Dept Com, 62, Gold Medal, 71. *Mem:* Am Meteorol Soc; Am Geophys Union; Am Soc Qual Control; Am Statist Asn; fel Am Inst Chem; Am Chem Soc. *Res:* Upper air climatology; stochastic and dynamic models; quality assurance; environmental pollution. *Mailing Add:* 35 Westall Ave Asheville NC 28804-3529

CRUTCHER, KEITH A, b Ft Lauderdale, Fla, May 29, 53; m 72; c 4. NEUROBIOLOGY, COMPARATIVE NEUROANATOMY. *Educ:* Pt Loma Col, BA, 74; Ohio State Univ, PhD(anat), 77. *Prof Exp:* Fel neurobiol, Med Ctr, Duke Univ, 77-80; from asst prof to assoc prof anat, Univ Utah, 80-87; ASSOC PROF NEUROSURG, UNIV CINCINNATI, 87- *Mem:* AAAS; Am Asn Anatomists; Soc Neurosci. *Res:* Neuronal plasticity with particular emphasis on transmitter-specific rearrangement which may underly the recovery of function following injury to the central nervous system. *Mailing Add:* Dept Neurosurg Col Med Unic Cincinnati 231 Bethesda Ave Cincinnati OH 45267-0515

CRUTCHER, RICHARD METCALF, b Lexington, Ky, April 18, 45; m 72; c 2. INTERSTELLAR MEDIUM. *Educ:* Univ Ky, BS, 67; Univ Calif, Los Angeles, MA, 69, PhD(astron), 72. *Prof Exp:* Res fel, Calif Inst Technol, 72-74; asst prof, 74-78, ASSOC PROF ASTRON, UNIV ILL, 78- *Concurrent Pos:* Res assoc, Radio Astron Lab, Univ Calif, Berkeley, 80-81. *Mem:* Am Astron Soc; Int Astron Union; Union Radio Sci Int. *Res:* Physics and chemistry of the interstellar medium by observations at ultraviolet, visual, infrared and radio wavelengths. *Mailing Add:* 341 Astron Bldg 1011 W Springfield Univ Ill Urbana IL 61801

CRUTCHFIELD, CHARLIE, b Norwood, Pa, Dec 29, 28; m 62; c 2. ANALYTICAL CHEMISTRY, INORGANIC CHEMISTRY. *Educ:* Univ Pa, BA, 50, MS, 53, PhD(chem), 60. *Prof Exp:* Anal chemist, Dalare Assocs, Pa, 50-55; instr gen & anal chem, Flint Jr Col, 59-61; electrochemist, Stanford Res Inst, 61-66, anal chemist, 66-69; tech dir, Truesdail Labs, Inc, 69-84; ALLIED - AEROSPACE, 84- *Mem:* Am Chem Soc; fel Royal Soc Chem; AAAS; fel Am Inst Chemists; Sigma Xi; Am Acad Clin Toxicol. *Res:* Complex compounds; corrosion. *Mailing Add:* 2708 Pinelawn Dr La Crescenta CA 91214

CRUTCHFIELD, FLOY LOVE, endocrinology, for more information see previous edition

CRUTCHFIELD, MARVIN MACK, b Oxford, NC, Sept 15, 34; m 56; c 2. PHYSICAL CHEMISTRY, DETERGENT CHEMICALS & PHOSPHATES. *Educ:* Duke Univ, BS, 56; Brown Univ, PhD(phys chem), 60. *Prof Exp:* Sr res chemist, 60-63, res specialist, 63-65, scientist, 65-70, sci fel, 70-83, SR FEL, MONSANTO CO, 84- *Mem:* AAAS; Am Chem Soc; Sigma Xi; Am Oil Chemist's Soc. *Res:* Peroxoanions; nuclear magnetic resonance; metal ion complexing; chemistry of phosphorus compounds; surface active agents; detergents; builders; polyelectrolytes; biodegradation; molecular structure-physical property relationships; calcium ion equilibria; physical chemistry of short fibers. *Mailing Add:* Monsanto Co 800 N Lindbergh Blvd St Louis MO 63167

CRUTHERS, LARRY RANDALL, b Kenosha, Wis, Mar 15, 45; m 67; c 2. VETERINARY PARASITOLOGY. *Educ:* Univ Wis, Stevens Point, BS, 67; Kans State Univ, MS, 71, PhD(parasitol), 73. *Prof Exp:* Instr biol, Kans State Univ, 70-73; res investr, E R Squibb & Sons, 74-77, sr res investr parasitol, 77-80; sr res parasitologist, 80-82, res assoc, Diamond Shamrock Corp, 82-83; group leader, SDS Biotech Corp, 83-84, assoc res dir, 84-85; SCI DIR, PROF LAB & RES SERV, INC, 89- *Concurrent Pos:* Sch adminr, 86-88. *Mem:* Am Asn Vet Parasitologists; Am Soc Parasitol. *Res:* Chemotherapy of helminth and protozoan parasites of economic and companion animals. *Mailing Add:* 852 Jo Anne Circle Chesapeake VA 23320

CRUZ, ALEXANDER, b New York, NY, July 12, 41; m 63; c 2. ECOLOGY, VERTEBRATE ZOOLOGY. *Educ:* City Univ New York, BS, 64; Univ Fla, PhD(ecol & zool), 73. *Prof Exp:* Microbiologist, New York Dept Health, Bur Labs, 64-68; asst prof, 73-80, ASSOC PROF BIOL, ENVIRON, POP & ORG, UNIV COLO, BOULDER, 80- *Concurrent Pos:* Consult, Biol & Health Sci Educ Opportunities Prog, Univ Colo, 73- & Ecol Analysts Inc, 74-; fac res initiation fel, Univ Colo, 74. *Mem:* Ecol Soc Am; Am Ornithologists Union; Wilson Ornith Soc; Sigma Xi. *Res:* Ecology and behavior of vertebrates with a special interest in avian ecology and behavior, ornithology, community analysis, tropical and insular biology. *Mailing Add:* Dept Environ Pop & Biol B334 Univ Colo Boulder CO 80302

CRUZ, ANATOLIO BENEDICTO, JR, b Marikina, Philippines, June 20, 33; US citizen; m 55; c 4. SURGERY, SURGICAL ONCOLOGY. *Educ:* Univ Philippines, AA, 52, MD, 57; Univ Minn, MS, 63. *Prof Exp:* Instr surg, Univ Minn, Minneapolis, 64 & Univ E Med Ctr, Quezon City, Philippines, 64-65; researcher cardiovasc surg, Univ Alta, 65-66; from asst prof to assoc prof surg, 66-74, PROF SURG & ANAT, UNIV TEX HEALTH SCI CTR, 74- *Concurrent Pos:* Co-prin investr, Nat Surg Adj Breast Proj, NIH grant, 66-; consult gen surg, US Air Force Med Ctr, San Antonio; consult thoracic surg, San Antonio State Chest Hosp, 67-; dir tumor clin & chief, head & neck, soft tissue tumor, surg oncol & burn serv, Bexar County Teaching Hosps, 69-; mem serv & rehab comt, Am Cancer Soc, Bexar County Unit, 72-; staff physician gen surg, Audie L Murphy Mem Vet Hosp, 73-; prin investr, Southwest Oncol Group, NIH grant, 76-; mem bd dirs, STex Health Educ Ctr, 78- *Mem:* Am Soc Clin Oncol; Am Soc Prev Oncol; Soc Surg Oncol Inc;

Sigma Xi; Am Col Surgeons. *Res:* Development of combination modalities in the treatment of cancer in addition to activities related to understanding the problems in cancer in general. *Mailing Add:* Univ Tex Med Sch 7703 Floyd Curl Dr San Antonio TX 78284-7842

CRUZ, CARLOS, b Aguadilla, PR, Dec 24, 40; US citizen; m 71; c 3. ENTOMOLOGY. *Educ:* Univ PR, Mayag ez, BSA, 63; Rutgers Univ, MS, 68, PhD(entom), 72. *Prof Exp:* Agr agt, Agr Exten Serv, 63-65; res asst sugar cane, Agr Exp Sta, PR, 65-66; res asst entom, Rutgers Univ, 66-68 & Agr Exp Sta, PR, 68-69; from asst entomologist to assoc entomologist, 72-82, ENTOMOLOGIST, AGR EXP STA, UNIV PR, 82-, PROF & DIR CROP TECTION DEPT, 82- *Mem:* Entom Soc Am; PR Soc Agr Sci; Sigma Xi; Caribbean Food Crop Soc; Soc Entomol PR. *Res:* Insect pest management on legumes and vegetables; study of all measures of insect control, particularly the use of resistant varieties, parasites, predators and selective insecticides. *Mailing Add:* Crop Protection Dept Univ PR Mayaquez PR 00708

CRUZ, JOSE B(EJAR), JR, b Bacolod City, Philippines, Sept 17, 32; US citizen; m 53, 90; c 5. CONTROL SYSTEMS, SYSTEMS ENGINEERING. *Educ:* Univ Philippines, BSEE, 53; Mass Inst Technol, SM, 56; Univ Ill, PhD, 59. *Prof Exp:* Instr elec eng, Univ Philippines, 53-54; res asst lab electronics, Mass Inst Technol, 54-56; from instr to pro elec eng, Univ Ill, Urbana, 56-86, assoc mem, Ctr Advan Studies, 67-68; PROF ELEC & COMPUTER ENG, UNIV CALIF, IRVINE, 86- *Concurrent Pos:* Vis assoc prof, Univ Calif, Berkeley, 64-65; ed, Inst Elec & Electronics Engrs Trans Automatic Control, 71-73; vis prof, Mass Inst Technol & Harvard Univ, 73; pres, Dynamic Systs, 78-; assoc ed, J Franklin Inst, 77-82, & J Optimization Theory & Applns, 81-; ed, Advances in Large Scale Systs, 82-; pres Control Syst Soc, 79, div dir, 80-81, vpres tech active, 82-83; mem, prof eng exam comt, State Ill, 82-86; vpres publ activ, Inst Elec & Electronic Engrs, 84-85. *Honors & Awards:* Halliburton Eng Educ Award, 81; Curtis W McGraw Res Award, Am Soc Eng Educ, 72; Emberson Award, Inst Elec & Electronic Engrs, 89. *Mem:* Nat Acad Eng; fel Inst Elec & Electronics Engrs; Nat Soc Prof Engrs; Am Soc Eng Educ; fel AAAS; Opers Res Soc Am. *Res:* Control of systems containing uncertainties; strategies for systems with multiple decision makers; sensitivity analysis of dynamic systems; manufacturing systems control. *Mailing Add:* Dept Elec & Computer Eng Univ Calif-Irvine Irvine CA 92717

CRUZ, JULIO C, b Sullana, Peru, Feb 5, 34; m; c 4. GAS EXCHANGE IN THE LUNG. *Educ:* Univ San Mascos, Lima, Peru, MD, 63. *Prof Exp:* ASSOC PROF PHYSIOL & ASST PROF ANESTHESIOL, MED COL OHIO, 83- *Concurrent Pos:* Sr res asst, Nat Res Coun- Nat Acad Sci, 74-75. *Mem:* NY Acad Sci; Am Soc Anesthesiologists; Am Physiol Soc; Am Med Asn; Peruvian Soc Physiol Sci (pres, 74-76). *Mailing Add:* Dept Anesthesiol Med Col Ohio 3000 Arlington Ave Toledo OH 43699

CRUZ, MAMERTO MANAHAN, JR, b Manila, Philippines, July 15, 18; US citizen; m 50; c 2. CHEMISTRY. *Educ:* Univ Philippines, BS, 39; Mass Inst Technol, SM, 41; State Univ NY Col Forestry, Syracuse, PhD(chem), 54. *Prof Exp:* Chemist, Am Viscose Corp, Pa, 42-47; lectr chem processes, Mapua Inst Technol, Philippines, 48; chem engr, Philippines Mission to Japan, 48, tech investr, 48-49; sr res chemist, Am Viscose Corp, Pa, 53-55; sect chief colloid chem, Olin Mathieson Chem Corp, Conn, 55-57; group leader pulp & bleaching, Ketchikan Res, Am Viscose Corp, 57-59, paper fibers, Com Develop Dept, 59-60, head avicel pilot opers, Corp Res Dept, 61-62, avicel plant mgr, 62, sect leader prod develop, 63, & spec prod, Am Viscose Div, 63, sr scientist new prod explor res, 63-64 & chem div, 64-66, mgr probing res & mgr new polymer prod, Cent Res Dept, 66-70, mgr prod & res develop, Avicon, Inc, FMC Corp, 71-79; PRES & TECH DIR, MORCA, INC, 79- *Mem:* Fel AAAS; Am Chem Soc; NY Acad Sci; Am Inst Chemists; Am Leather Chemists Asn. *Res:* Textile engineering; pulping and bleaching of wood pulps; rayon fiber and cellophane film technology; cellulose reactions; cellulose derivatives; synthetic polymeric coatings; wet strength paper and non-wovens; wood chemistry; collagen based medical products; microcrystalline cellulose. *Mailing Add:* 12 Mullray Ct Deptford NJ 08096-6713

CRUZAN, JOHN, b Bridgeton, NJ, Jan 6, 42; m 64; c 2. ECOLOGY. *Educ:* King's Col, BA, 65; Univ Colo, PhD(zool), 68. *Prof Exp:* From asst prof to assoc prof, 68-79, PROF BIOL, GENEVA COL, 79-, HEAD, BIOL DEPT, 82- *Concurrent Pos:* Asst to vpres Acad Affairs, Geneva Col, 89- *Mem:* Am Soc Mammal; Am Sci Affiliation; Ecol Soc Am; Am Inst Biol Sci; AAAS. *Res:* Ecology, behavior and geography of mammals and other vertebrates; soil community structure. *Mailing Add:* Dept Biol Geneva Col Beaver Falls PA 15010

CRYBERG, RICHARD LEE, b Los Angeles, Calif, Nov 2, 41; m 63; c 2. ORGANIC CHEMISTRY. *Educ:* Iowa State Univ, BS, 63; Ohio State Univ, PhD(org chem), 69. *Prof Exp:* Chief chemist, G F Smith Chem Co, 63-68; sr res chemist, T R Evans Res Ctr, Diamond Shamrock Corp, 69-74, group leader chem, 74-76, group leader polymers, 76-77, group leader anal, 77-78; mgr, Analytica, 79-81; mgr ag chem res, SDS Biotech, 81-86; mgr, 86-89, DIR ANAL RICERCA, INC, 89- *Mem:* Am Chem Soc. *Res:* Spectroscopy, x-ray and microscopy. *Mailing Add:* 9531 Robinson Rd Chardon OH 44024

CRYER, COLIN WALKER, b Leeds, Eng, Aug 28, 35; m 68; c 1. COMPUTER SCIENCE. *Educ:* Univ Pretoria, BS, 54, MS, 58; Cambridge Univ, PhD(comput sci), 62. *Prof Exp:* Asst res officer, Nat Phys Lab, SAfrica, 55-58; res officer, Nat Res Inst Math Sci SAfrica, 62-63; res fel comput sci, Calif Inst Technol, 63-65; mem res staff, Math Res Ctr, 65-66, from asst prof to assoc prof, 66-77, PROF COMPUT SCI, UNIV WIS-MADISON, 77- *Mem:* Am Math Soc; Asn Comput Mach. *Res:* Numerical solution of integral and partial differential equations. *Mailing Add:* Univ Munster Inst Numerische Math Einsteinstrasse 62 Munster Germany

CRYER, DENNIS ROBERT, b Dearborn, Mich, Mar 30, 44; m 86; c 1. ATHEROSCLEROSIS, HUMAN GENETICS. *Educ:* Johns Hopkins Univ, BA, 68; Albert Einstein Col Med, MD, 77. *Prof Exp:* Med scientist trainee biochem, Albert Einstein Col Med, 70-77; resident pediat, Children's Hosp Philadelphia, 77-79, asst chief resident, 80-81; clin fel human genetics, 81-83, clin asst prof, 83-84, asst prof pediat, 84-, DIR, CARDIOVASCULAR GROUP, CHILDREN'S HOSP PHILADELPHIA, UNIV PA. *Concurrent Pos:* Teaching fel molecular biol, Sch Med, Univ Pa, 79-80, instr pediat, 80-81; asst physician human genetics & investr, Lipid-Heart Res Ctr, Children's Hosp Philadelphia, 84-, clin assoc physician, Gen Clin Res Ctr, 85-; Fac Develop Award, Merck & Co-Univ Pa, 84; mem Coun Arteriosclerosis, Am Heart Asn. *Mem:* Am Heart Asn; Am Soc Human Genetics; NY Acad Sci. *Res:* Identifying predictors of atherosclerosis risks in childhood; adaptation of stable isotope methodology with gas chromatography-mass spectrometry analysis for the safe study of lipoprotein metabolism and specific hepatic secretory protein metabolism in man. *Mailing Add:* PO Box 4500 Princeton NJ 08543-4500

CRYER, JONATHAN D, b Toledo, Ohio, Feb 10, 39; m 61; c 3. STATISTICS. *Educ:* DePauw Univ, BA, 61; Univ NC, PhD(statist), 66. *Prof Exp:* Statistician, Res Triangle Inst, 63-66; asst prof statist, Univ Iowa, 66-70, dir statist consult ctr, 85-86, assoc chmn & dir grad studies, 87-89, ASSOC PROF STATIST, UNIV IOWA, 70-, STATISTICIAN. *Concurrent Pos:* Consult, Amana Refrig, Inc, 74-, Am Col Testing, 77- & Midwest Energy Co, 90- *Mem:* Inst Math Statist; Am Statist Asn. *Res:* Statistics time series analysis and statistical process improvement. *Mailing Add:* Dept Statist Univ Iowa Iowa City IA 52242

CRYER, PHILIP EUGENE, b ElPaso, Ill, Jan 5, 40; m 63; c 2. ENDOCRINOLOGY, METABOLISM. *Educ:* Northwestern Univ, BA, 62, MD, 65. *Prof Exp:* Resident, Barnes Hosp, St Louis, 65-67; fel metab, 67-68, instr med, 71-72, asst prof, 72-77, assoc prof, 77-81, PROF MED, SCH MED, WASHINGTON UNIV, 81- *Concurrent Pos:* Resident med, Barnes Hosp, St Louis, 68-69; dir, Clin Res Ctr, Sch Med, Washington Univ, 78-, Metabolism Div, 85- *Mem:* Am Fedn Clin Res; Am Diabetes Asn; Endocrine Soc; Cent Soc Clin Res; Am Soc Clin Invest; Assoc Am Physicians. *Res:* The sympathoadrenal system in human metabolic physiology and pathophysiology. *Mailing Add:* Dept Med Sch Med Washington Univ 660 S Euclid St Louis MO 63110

CRYNES, BILLY LEE, b Worthington, Ind, Mar 16, 38; m 57; c 3. CHEMICAL ENGINEERING. *Educ:* Rose Polytech Inst, BS, 63; Purdue Univ, MA, 66, PhD(chem eng), 68. *Prof Exp:* Technician, Lab, Commercial Solvents Corp, 61-63; design engr, Pilot Plant, E I du Pont de Nemours & Co, 63-64; from asst prof to prof, 67-87, head dept, 78-87 PROF CHEM ENG, OKLA STATE UNIV, 73-; DEAN ENG, UNIV OKLA, 87- *Mem:* Am Chem Soc; Am Inst Chem Engrs; Am Soc Eng Educ. *Res:* Pyrolysis of hydrocarbons; hydrodesulfurization catalysts; synthetic crudes from coal; reactor engineering. *Mailing Add:* Carson Eng Rm 107 Univ Okla 202 W Boyd Norman OK 73019

CRYSTAL, GEORGE JEFFREY, b Newark, NJ, Apr 5, 48. CARDIOVASCULAR PHYSIOLOGY, CORONARY PHYSIOLOGY. *Educ:* Rutgers Univ, New Brunswick, NJ, AB, 70, MS, 74, PhD(physiol), 77. *Prof Exp:* Teaching asst physiol, Rutgers Med Sch, Piscataway, NJ, 74-77; fel physiol, Univ Tex Health Sci Ctr, Dallas, 77-79, asst prof physiol, 79-; RES ASST PROF, DEPT ANESTHESIOL, UNIV ILL COL MED. *Concurrent Pos:* Mem, Am Heart Asn. *Mem:* Am Physiol Soc. *Res:* Reflex control of the cardiovascular system; metabolic regulation of coronary blood flow. *Mailing Add:* Dept Anesthesiol Univ Col Med Ill Masonic Med Ctr 836 W Wellington Ave Chicago IL 60657

CRYSTAL, MAXWELL MELVIN, b New York, NY, Oct 9, 24; m 49; c 1. MEDICAL & VETERINARY ENTOMOLOGY. *Educ:* Brooklyn Col, AB, 47; Ohio State Univ, MSc, 48; Univ Calif, PhD(parasitol), 56. *Prof Exp:* Asst helminth & med entom, Univ Calif, 51-55; asst prof biol, NY Teachers Col, New Paltz, 56-57; instr parasitol, Albert Einstein Col Med, 57-58; asst med physics, Montefiore Hosp, 59-60; res entomologist, Screwworm Res Lab, Agr Res Serv, USDA, 61-77, Biol Eval Chem Lab, Agr Environ Qual Inst, 77-79, Insect Reproduction Lab, 79-81, res entomologist, Livestock Insects Lab, Agr Res Serv, 81-87; RETIRED. *Concurrent Pos:* Nat Cancer Inst res grant, 58-59; copy ed, Am Fisheries Soc, 88-89; assoc entomologist, Environ Protection Agency, 90- *Mem:* AAAS; Sigma Xi; Inst Biol Sci; Entom Soc Am; Am Registry Prof Entomologists; Am Soc Parasitologists. *Res:* Research on behavior, physiology, and biology of northern fowl mites; programs for controlling northern fowl mites on poultry. *Mailing Add:* 10813 Bucknell Dr Silver Spring MD 20902

CRYSTAL, RONALD GEORGE, b Newark, NJ, Apr 23, 41. INTERNAL MEDICINE, BIOCHEMISTRY. *Educ:* Tufts Univ, BA, 62; Univ Pa, MS, 63, MD, 68. *Prof Exp:* Intern internal med, Mass Gen Hosp, Boston, 68-69, resident, 69-70; res assoc molecular hemat, 70-72, head pulmonary biochem sect, 72-75, CHIEF PULMONARY BR, NAT HEART, LUNG & BLOOD INST, 75- *Concurrent Pos:* Clin fel, Harvard Univ, 69-70 & Univ Calif, San Francisco, 73; adj prof genetics, George Washington Univ, 75- *Mem:* Am Soc Biol Chemists; Am Soc Clin Invest; Am Thoracic Soc; Am Fed Clin Res; Asn Am Physicians. *Res:* Lung structure and function in health and disease; genetic disorders of the lung; inflammation and immune processes in lung. *Mailing Add:* Nat Heart Lung & Blood Inst 6D03 Bldg 10 NIH Bethesda MD 20892

CSAKY, TIHAMER ZOLTAN, b Hungary, Aug 12, 15; nat US; m 53; c 2. PHYSIOLOGY, PHARMACOLOGY. *Educ:* Univ Budapest, MD, 39. *Prof Exp:* Asst prof physiol, Univ Budapest, 40-45; res adj, Hungarian Biol Res Inst, 46-47; res assoc, Duke Univ, 49-51; from asst prof to assoc prof pharmacol, Sch Med, Univ NC, 51-61; prof pharmacol, Sch Med, Univ Ky, Lexington, 61-79, chmn dept, 61-76; prof, 79-85, EMER PROF PHARMACOL, SCH MED, UNIV MO, COLUMBIA, 85- *Concurrent Pos:*

Res fel, Biochem Inst, Helsinki, 47-48 & Microbiol Inst, Uppsala, 48-49; vis prof & USPHS spec fel, Univ Milan, 68-69; A Von Humboldt sr scientist award, Univ Bochum & Saar Univ, Homburg, Germany. *Mem:* Int Soc Biochem Pharmacol; Soc Exp Biol & Med; Am Soc Pharmacol & Exp Therapeut; Am Physiol Soc; Soc Gen Physiol. *Res:* Biological transport. *Mailing Add:* 7930 Rte N Univ Mo Sch Med Columbia MO 65203

CSALLANY, AGNES SAARI, b Budapest, Hungary, Apr 20, 32; US citizen; m 54; c 1. ORGANIC CHEMISTRY. *Educ:* Budapest Tech Univ, BS, 54, MS, 55, ScD, Budapest Tech Univ, 70. *Prof Exp:* Head qual control lab, Duna Canning Co, Budapest, 55-56; from res asst to res assoc biochem, Dept Animal Sci, Univ Ill, Urbana, 57-65, res asst prof, 65-69, res asst prof, Dept Food Sci, 69-73; assoc prof food sci & nutrit, 73-80, PROF FOOD SCI & NUTRIT, UNIV MINN, ST PAUL, 80- *Concurrent Pos:* Deleg, Int Nutrit Cong, 60, 69, 75, 79 & 85 & Int Biochem Cong, 64; Am Inst Nutrit travel grants, 69, 75, 79 & 85. *Mem:* Am Chem Soc; Am Inst Nutrit; Sigma Xi; Am Oil Chem Soc; Int Soc Free Radical Res; Inst Food Technologits. *Res:* Food chemistry and biochemistry; vitamin E; oxidative degradation; antioxidant action; free radicals. *Mailing Add:* Dept Food Sci & Nutrit Univ Minn 1334 Eckles Ave St Paul MN 55108

CSANADY, GABRIEL TIBOR, b Budapest, Hungary, Dec 10, 25; m 45, 69; c 4. PHYSICAL OCEANOGRAPHY, FLUID MECHANICS. *Educ:* Munich Tech Univ, Dipl Ing, 48; Univ New South Wales, PhD(mech eng), 58. *Prof Exp:* Engr, Brown, Boveri Co, Ger, 48-49, Elec Comn New South Wales, Australia, 49-52 & Elec Comn Victoria, 52-54; lectr mech eng, Univ New South Wales, 54-61; assoc prof, Univ Windsor, 61-63; prof, Univ Waterloo, 63-73, chmn dept, 63-67; sr scientist, Woods Hole Oceanog Inst, 73-87; EMINENT PROF, OLD DOMINION UNIV, NORFOLK, VA. *Concurrent Pos:* Sr res fel aerodyn, Col Aeronaut Eng, 60-61; vis prof, Univ Wis, 69; Fairchild scholar, Calif Tech, 82. *Honors & Awards:* President's Prize, Can Meteorol Soc, 70; Ed Award, Am Meteorol Soc, 75; Chandler-Misener Award, Int Asn Great Lakes Res, 77. *Mem:* Am Meteorol Soc; Am Soc Mech Engrs; Am Geophys Union; Royal Meteorol Soc; Can Meteorol Soc. *Res:* Dynamics of shallow seas; continental shelves; coastal ocean; upper ocean dynamics; air-sea interaction; effluent disposal; ocean's role in climate. *Mailing Add:* Dept Oceanog Old Dominion Univ Norfolk VA 23529-0276

CSAVINSZKY, PETER JOHN, b Budapest, Hungary, July 10, 31; US citizen; m 76. THEORETICAL SOLID STATE PHYSICS. *Educ:* Tech Univ Budapest, diplom ing chem, 54; Univ Ottawa, PhD(theoret physics), 59. *Prof Exp:* Mem tech staff, Semiconductor Div, Hughes Aircraft Co, 59-60; sr physicist, Electronics Div, Gen Dynamics Corp, 60-62; mem tech staff, Tex Instruments Inc, 62-65 & TRW Systems Inc, 65-70; assoc prof, 70-75, PROF PHYSICS, UNIV MAINE, ORONO, 75- *Concurrent Pos:* Lectr, Univ Calif Berkely, 71, Univ Calif Los Angeles, 72-77; vis prof, Univ Southern Calif, 77-78; vpres, New Eng Sect, Am Phys Soc, 86-87, pres, 88-89, sect adv to coun, 89-90. *Mem:* Fel Am Phys Soc; AAAS; Sigma Xi. *Res:* Quantum theoretical studies of charge transport in solids; the structure of impurities and defects in solids; theory of impurity states and charge transprot in semiconductors; density-functional theory. *Mailing Add:* Dept Physics Univ Maine Orono ME 04469

CSEJKA, DAVID ANDREW, b Passaic, NJ, June 9, 35; m 65; c 5. PHYSICAL CHEMISTRY, ANALYTICAL CHEMISTRY. *Educ:* Fordham Univ, BS, 56; Iowa State Univ, PhD(phys chem), 61. *Prof Exp:* Sr res chemist, 61-74, res assoc, 74-76, group leader, 76-80, MGR CHEM SECT, OLIN CORP, 80- *Mem:* Sigma Xi; Am Chem Soc; NY Acad Sci. *Res:* Chemical and physical properties of aqueous and non-aqueous solutions; electroanalytical chemistry, polarography, coulometry, voltammetry. *Mailing Add:* Olin Chem Res Ctr 350 Knotter Dr Cheshire CT 06410-0586

CSEJTEY, BELA, JR, b Budapest, Hungary, Jan 26, 34; US citizen; div; c 1. STRUCTURAL GEOLOGY, PETROLOGY. *Educ:* Princeton Univ, PhD(geol), 63. *Prof Exp:* Res asst geophys, Dept Geol, Princeton Univ, 63; geologist petrol, Richfield Oil Corp, 63-66; GEOLOGIST, US GEOL SURV, 66- *Concurrent Pos:* Exchange scientist, Geochem Res Lab, Hungarian Acad Sci, 73. *Mem:* Fel Geol Soc Am; AAAS; Sigma Xi; Am Polar Soc. *Res:* Study of plutonic rocks and structural features of southern Alaska in order to decipher the plate-tectonic evolution of the region. *Mailing Add:* US Geol Surv MS 904 345 Middlefield Rd Menlo Park CA 94025

CSENDES, ERNEST, b Satu-Mare, Rumania, Mar 2, 26; nat US; m 53; c 2. ORGANIC CHEMISTRY. *Educ:* Col Hungary, BA, 44; Univ Heidelberg, BS, 48, MS & PhD, 51. *Prof Exp:* Asst, Univ Heidelberg, 51; res assoc biochem, Tulane Univ, 52; fel, Harvard Univ, 52-53; res chemist, Org Chem Dept, E I Du Pont de Nemours & Co, Del, 53-56 & Elastomer Chem Dept, 56-61; tech dir, Armour & Co, 61-62, dir res & develop, Agr Chem Div, 62-63; vpres corp develop, Occidental Petrol Corp, 63-64, exec vpres res eng & develop, 64-68, exec vpres & chief operating officer, Occident Res Eng Corp, 63-68, dir, Occidental Res & Eng, Ltd, UK, 64-68; pres & chief exec, Tex Repub Industs, 68-84; chmn & chief exec officer, Micronic Technol, Inc, 81-85; MANAGING PTNR, INTER-CONSULT LTD, 84- *Concurrent Pos:* Pres & chief exec, TRI Ltd, Bermuda, 71-84, chmn & dir, TRI Int Ltd, Bermuda, 71-84, managing dir, TRI Holdings SA, Luxembourg, 71-84 & TRI Capital NV, Neth, 71-, managing ptnr, Inter-Consult, Ltd, 84- *Honors & Awards:* Pro-Mundi Beneficio Gold Medal, Brazilian Acad Humanities. *Mem:* Fel AAAS; fel Am Inst Chemists; Am Chem Soc; Am Inst Chem Engrs; Am Inst Aeronaut & Astronaut; fel Royal Soc Chem; NY Acad Sci; Ger Chem Soc. *Res:* Regional development projects related to energy resources and agriculture; international finance related to industrial facilities; banking, trusts, insurance; administration of research, engineering and industrial development; technical, economic management and international consulting. *Mailing Add:* 514 Marquette St Pacific Palisade CA 90272-3314

CSERMELY, THOMAS J(OHN), b Szombathely, Hungary, June 25, 31; US citizen; m 61; c 1. COMPUTER ENGINEERING, BIOMEDICAL ENGINEERING. *Educ:* Tech Univ Budapest, Dipl, 53; Syracuse Univ, PhD(physics-sci teaching), 68. *Prof Exp:* Instr, Tech Univ Budapest, Inst Theoret Physics, 53-56; res engr, Res & Develop Div, Carrier Corp, NY, 57-67; res assoc biophys, Syracuse Univ, 67-68; asst prof physiol & biophys, State Univ NY Upstate Med Ctr, 68-76; asst prof physics dept, Le Moyne Col, 76-77; asst prof, 77-80, assoc prof elec & comput eng, 80-89, ASSOC PROF BIOENG, SYRACUSE UNIV, 89- *Concurrent Pos:* Consult, Design Bur for Power Stas, Budapest, 56. *Honors & Awards:* Wolverine Diamond Key Award, Soc Heat, Refrig & Air-Conditioning Engrs, 65. *Mem:* AAAS; Am Phys Soc; Am Asn Physics Teachers; Inst Elec & Electronics Engrs; Biophys Soc; NY Acad Sci. *Res:* Systems analysis and dynamics; mathematical modeling and computer simulation of systems; neural networks; medical computation and data systems. *Mailing Add:* Bioeng Dept 417 Link Hall Syracuse Univ Syracuse NY 13244-1240

CSERNA, EUGENE GEORGE, b Budapest, Hungary, Jan 2, 20; nat US; m 50; c 3. STRUCTURAL GEOLOGY. *Educ:* Univ Budapest, Dr Polit Sci, 43; Columbia Univ, MA, 50, PhD(geol), 56. *Prof Exp:* Asst geol, Columbia Univ, 50-51; geologist & explorer, Gulf Oil Corp, 51-54; instr geol, Hunter Col, 55; field geologist, US Geol Surv, 55-57; asst prof, Idaho State Col, 57-59; PROF GEOL, CALIF STATE UNIV, FRESNO, 59- *Concurrent Pos:* NSF fac fel, Swiss Fed Inst Technol, 66-67; Nat Acad Sci exchange fel, Hungary & Romania, 74; exchange deleg, Nat Acad Sci, Czech, 79. *Mem:* Asn Petrol Geologists; Sigma Xi; Geol Soc Am. *Res:* Structure and stratigraphy of Fra Cristobal Area, New Mexico, Poison Spider, Oil Mountain, Pine Mountain in Wyoming; revision of portions of the New Tectonic Map of the United States. *Mailing Add:* Dept Geol Calif State Univ Fresno CA 93740

CSERR, HELEN F, b Boston, Mass, June 23, 37; m 62; c 1. PHYSIOLOGY, IMMUNOLOGY. *Educ:* Middlebury Col, BA, 59; Harvard Univ, PhD(physiol), 65. *Prof Exp:* Instr neurol, Harvard Med Sch, 68-70; lectr, 70-71, from asst prof to assoc prof, 71-82, PROF MED SCI, BROWN UNIV, 82- *Concurrent Pos:* United Cerebral Palsy Res & Educ Found fel physiol, Harvard Med Sch, 65-68; Nat Inst Neurol Dis & Stroke career develop award, 73-78; trustee, Mt Desert Island Biol Lab, 71-77, 83-86, mem, Exec Comt, 75-77; mem, Physiol Study Sect, NIH, 75-79; Javits Neurosci Investr Award, Nat Inst Neurol & Communicative Dis & Stroke, 87-94; adv comt, RI Found, 90; trustee, Grass Found, 90- *Mem:* AAAS; Soc Gen Physiol; Royal Soc Med; Am Physiol Soc; Soc Neurosci; Physiol Soc. *Res:* Physiology of cerebrospinal fluid and brain interstitial fluid; neuroimmunology. *Mailing Add:* Div Biol & Med Brown Univ Box G-B318 Providence RI 02912

CSICSERY, SIGMUND MARIA, b Budapest, Hungary, Feb 3, 29; US citizen; m 56. SURFACE CHEMISTRY, PETROLEUM CHEMISTRY. *Educ:* Budapest Tech Univ, MS, 51; Northwestern Univ, PhD(org chem), 61. *Prof Exp:* Res chemist, Res & Eng Div, Monsanto Chem Co, Ohio, 57-59 & Calif Res Corp, 61-66; sr res chemist, Chevron Res Co, 66-70, sr res assoc, 70-86; CONSULT & CHIEF TECH ADV, UN INDUST DEVELOP ORGN/UN DEVELOP PROGRAMME, 86- *Concurrent Pos:* Exten instr, Univ Calif, Berkeley, 65-; pres, Calif Catalysis Soc, 73-74; dir, Catalysis Soc NAm, 76-80. *Mem:* Am Chem Soc. *Res:* Catalysis and chemistry of petroleum hydrocarbons; structure, activation, and preparation of acidic and metal-containing heterogeneous catalysts; application of molecular sieves as catalysts; shape selective catalysis. *Mailing Add:* PO Box 843 Lafayette CA 94549

CSONKA, PAUL L, b Aug 10, 38; US citizen. PARTICLE BEAMS, LASERS. *Educ:* Johns Hopkins Univ, PhD(theoret physics), 63. *Prof Exp:* Res assoc, Johns Hopkins Univ, 63-64; fel theoret physics, Lawrence Radiation Lab, Univ Calif, 64-66; NSF fel, Europ Orgn Nuclear Physics, Cern, 66-67; from asst prof to assoc prof, 68-76, dir, Inst Theoret Sci, 77-79, RES ASSOC, UNIV ORE, 68-, PROF THEORET PHYSICS, 76- *Concurrent Pos:* Alfred P Sloan fel, 70-72; vis scientist & prof, various univs & labs. *Mem:* Am Phys Soc. *Res:* Particle beams; invariance principles; causality; x-ray lasers. *Mailing Add:* Inst Theoret Sci Univ Ore Eugene OR 97403

CSORGO, MIKLOS, b Egerfarmos, Hungary, Mar 12, 32; Can citizen; m 57; c 2. MATHEMATICS. *Educ:* Sch Econ, Budapest, BA, 55; McGill Univ, MA, 61, PhD(math), 63. *Prof Exp:* Lectr statist, Sch Econ, Budapest, 55-56; instr math, Princeton Univ, 63-65; from asst prof to assoc prof, McGill Univ, 65-72; PROF MATH, CARLETON UNIV, 72- *Concurrent Pos:* Vis prof, Math Inst, Univ Vienna, 69-70. *Mem:* Can Math Soc; Am Math Soc; Inst Math Statist; Statist Soc Can; Int Statist Inst; Bernoulli Soc Math Statist & Probability. *Res:* Probability theory; mathematical statistics. *Mailing Add:* Dept Math & Statist Carleton Univ Colonel By Dr Ottawa ON K1S 5B6 Can

CUADRA, CARLOS A(LBERT), b San Francisco, Calif, Dec 21, 25; m 47; c 3. INFORMATION SCIENCE, ONLINE INFORMATION SYSTEMS. *Educ:* Univ Calif, AB, 49, PhD(psychol), 53. *Prof Exp:* Psychol res supvr, Vet Admin Hosp, Downey, Ill, 53-56; training specialist air defense, Syst Develop Div, Rand Corp, 56-57; head personnel planning & training group, Syst Develop Corp, 57-59, proj team, Air Force 466L Syst, 59-60, proj team, Intel Systs Br, 60-62, develop staff, Spec Develop Dept, 63-64, tech asst to head technol directorate, 65-66, head info systs technol staff, 66-67, mgr, Libr & Doc Systs Dept, 68-71, mgr, Educ & Libr Systs Dept, 71-74, gen mgr, SDC Search Serv, 74-78; FOUNDER & PRES, CUADRA ASSOC, INC, 78- *Concurrent Pos:* Ed, Ann Rev Info Sci & Technol, 64-75; consult, Comt Sci & Tech Commun, Nat Acad Sci, 68-69; mem, Nat Comn Libr & Info Sci, 71-84; adj prof, Grad Sch Libr & Info Sci, Univ Calif, Los Angeles, 79-80. *Honors & Awards:* Miles Conrad Award, Nat Fedn Abstr & Indexing Serv, 80. *Mem:* Am Soc Info Sci; Spec Libr Asn. *Res:* Behavioral aspects of professional communication; research methodology; use of technology in libraries; state-of-the-art reviews in information science; on-line database systems and services from scientific, technical and financial standpoint; professional communications; library and information science, including evaluation of current and prospective research. *Mailing Add:* 13213 Warren Ave Mar Vista CA 90066

CUANY, ROBIN LOUIS, b Glasgow, Scotland, Oct 17, 26; US citizen; m 51; c 6. PLANT GENETICS. *Educ:* Cambridge Univ, BA, 47, MA, 51; Iowa State Univ, PhD(crop breeding, genetics), 58. *Prof Exp:* Cytogeneticist cotton breeding sect, Empire Cotton Growing Corp & Sudan Govt, 48-53; asst dept agron, Iowa State Univ, 54-56; res assoc biol dept, Brookhaven Nat Lab, 56-58, asst geneticist, 58-59; geneticist nuclear energy prog, Inter-Am Inst Agr Sci, 59-62; asst prof bot, Univ Iowa, 62-68; from asst prof to assoc prof, 68-85, PROF AGRON & CROPS, COLO STATE UNIV, 85- *Concurrent Pos:* Chmn, High Altitude Reveg Comt, 74-82; vis scientist, Div Trop Crops & Pastures, Commonwealth Sci & Indust Res Orgn, 85-86. *Mem:* Am Soc Agron; Am Genetic Asn; Soc Range Mgt. *Res:* Genetics and breeding of maize; grasses for forage and revegetation; photoperiodism and climatic adaptation; seed dormancy and seedling establishment; use of marker genes in detecting outcrossing and somatic mutation; cytogenetics; forage and range feed supply for cattle in the Gambia. *Mailing Add:* Dept Agron Colo State Univ Ft Collins CO 80523

CUATRECASAS, JOSÉ, b Catalonia, Spain, Mar 19, 03; US citizen; m 33; c 3. THE COCOA PLANT & ITS ALLIES, TAXONOMY OF HUMIRIACEAE. *Educ:* Univ Barcelona, Spain, Lic Pha, 23; Univ Madrid Spain, Dr Pha(systematic bot), 28. *Prof Exp:* Asst prof bot, Univ Barcelona, 24-30; prof syst bot, Univ Madrid, 31-39; cur trop flora, Madrid Bot Garden, Spain, 32-39, cur syst bot, Univ Nat Bogota, Colombia, SAm, 39-42, cur Colombian Bot, Chicago National Hist Mus, 47-50; dir, Madrid Bot Garden, Spain, 32-37, Sch Trop Agr, Cali, Colombia, 42-43, dir Bot Comn & prof bot, Sch Trop Agr, Cali, Colombia, 43-47; RES ASSOC, SMITHSONIAN INST BOT, 55- *Concurrent Pos:* Guggenheim fel, 51-52; investr bot, 52-55, prin investr, NSF & res assoc bot, Smithsonian Inst, 55-78; sci dir, Orgn Flora, Neotropica, 62-71, pres, 72-75. *Honors & Awards:* H A Gleason Award, NY Bot Garden, 63; N Monturiol Medal, Govt Catalonia, Spain, 87. *Mem:* Am Soc Plant Taxonomists; Soc Study Evolution; Int Asn Plant Taxonomy; Asn Trop Biol; Ecol Soc Am; Soc Biogeog Paris; fel AAAS. *Res:* Investigation of flora and ecology of the neotropical regions through extensive explorations and collections and direct observations of the plant life; systematics of the compositae; plant distribution in Northern and South America. *Mailing Add:* Dept Bot NHB 166 Smithsonian Inst Mus Natural Hist Tenth & Constitution Ave NW Washington DC 20560

CUATRECASAS, PEDRO, b Madrid, Spain, Sept 27, 36; US citizen; m 59; c 4. BIOCHEMISTRY, MEDICINE. *Educ:* Wash Univ, BA, 58, MD, 62. *Hon Degrees:* Dr, Univ Barcelona, 84; DSc, Mt Sinai Sch Med, City Univ New York, 85. *Prof Exp:* Intern & resident internal med, Johns Hopkins Univ Hosp, 62-64; clin assoc clin endocrinol, Nat Inst Arthritis & Metab Dis, 64-66, med officer, Lab Chem Biol, 67-70; assoc prof med, pharmacol & exp therapeut, Burroughs Wellcome prof clin pharmacol & dir div, Sch Med, Johns Hopkins Univ, 70-72, prof pharmacol & exp therapeut & assoc prof med, Sch Med, Johns Hopkins Univ, 72-75; vpres res develop & med, Wellcome Res Labs & dir, Burroughs Wellcome Co, 75-86; sr vpres res & develop, Glaxo Inc, 86-89; PRES, PHARMACEUT RES DIV, WARNER-LAMBERT CO, 89- *Concurrent Pos:* USPHS spec fel, NIH Lab Chem Biol, 66-67; prof lectr biochem, George Washington Univ, 67-70; mem, Adv Comt Personnel for Res, Am Cancer Soc, 73-75; dir, Burroughs Wellcome Fund, 75-86; adj prof, Dept Med & Depts Pharmacol & Physiol, Duke Univ, 75-; adj prof, Dept Med & Dept Pharmacol, Univ NC, 75-; ed, J Solid-Phase Biochem, 75-80; mem, Adv Comt Cancer Res Prog, Univ NC, 75-; distinguished lectr med sci, Dept Med, Mayo Clin, 79; pharmacy alumni lectr, Univ Toronto, 81; mem, comn drugs rare dis, Pharmaceut Mfrs Asn, 82-85; Ernest Cotlove Mem Lectr, Acad Clin Lab Physicians & Scientists, 81, Edward E Smissman Lectr, Univ Kans, 81, Jeanette Piperno Mem Lectr, Temple Univ, 82, John C Krant Jr Lectr, Univ Md, 82, Chilton Lectr, Univ Tex, 83, John Allen Love Lectr, 83, Windsor C Cutting Lectr, Univ Hawaii, 84, Schueler Distinguished Lectr, Tulane Univ, 85, Harvey Lectr, 85; adj prof, Dept Med Chem, Col Pharm, Univ Mich Med Sch, 90-, Dept Pharmacol, 90- *Honors & Awards:* John Jacob Abel Prize Pharmacol, 72; Eli Lilly Award, Am Diabetes Asn, 75; Laude Prize, Pharmaceut World, 74 & 75; Beerman Lectr & Beerman Award, Soc Investigative Dermat, 81; Goodman & Gilman Award, Am Soc Pharmacol & Exp Therapeut, 82. *Mem:* Nat Acad Sci; Sigma Xi; Inst Med-Nat Acad Sci; Am Soc Pharmacol & Exp Therapeut; Am Soc Clin Invest; Am Chem Soc; Am Soc Biol Chemists; Am Fedn Clin Res. *Res:* Cell membranes; protein and glycolipid chemistry; mechanism of action of hormones; membrane receptors; cell growth; affinity chromatography; pharmacology. *Mailing Add:* Warner-Lambert Co 2800 Plymouth Rd Ann Arbor MI 48105

CUBBERLEY, ADRIAN H, b Westfield, NJ, Feb 1, 18; m 41; c 2. TECHNOLOGY ASSESSMENT, TECHNOLOGY TRANSFER. *Educ:* Columbia Univ, AB, 39. *Prof Exp:* Res chemist, Norda Essential Oil & Chem Co, 39-42; res chemist, Barrett Div, Allied Chem Corp, 42-54, admin asst res & develop, 54-58 & plastics div, 58-62, proj mgr com develop, 62-65, dep sci dir, Allied Chem SA, Belg, 66-69, sr technico-econ planner, Allied Chem Corp, NJ, 69-70; res planner, Corp Res Ctr, Int Paper Co, Tuxedo Park, NY, 70-74, mgr res intel, Corp Res & Develop Div, 74-77, mgr sci & tech studies, 77-79, mgr indust health & prod safety, 79-80, mgr health serv, 80-81, mgr spec studies, sci & tech, 81-85; consult, Washingtonville, NY, 85-87, CONSULT, CHAPEL HILL, NC, 87- *Mem:* Am Chem Soc; fel Am Inst Chem (treas, 85-86); Tech Asn Pulp & Paper Indust; Com Develop Asn; Am Soc Photogram & Remote Sensing. *Res:* Research management, strategy and planning; technical liaison; commercial development; technology assessment; remote sensing of forest environment. *Mailing Add:* 110 Charlesberry Lane Chapel Hill NC 27514

CUBBERLEY, VIRGINIA, b Trenton, NJ, Apr 30, 46; m 69; c 2. TOXICOLOGY. *Educ:* Fairleigh Dickinson Univ, BS, 68; Rutgers Univ, MS, 75; NY Univ, PhD(biol), 80. *Prof Exp:* Asst pharmacologist, Carter Wallace Inc, 68-71; asst biochemist, Hoffmann La Roche, 72-74; assoc res scientist, Dept Environ Med, NY Univ, 75-78; asst prof, Pace Univ, NY, 79-; res biologist, Am Cyanamid Co, 81-; CLIN PROG SPECIALIST, E I DU PONT DE NEMOURS, 90- *Concurrent Pos:* Consult, Am Cyanamid Co, 80-81. *Mem:* Am Soc Cell Biol. *Res:* Data collection and evaluation. *Mailing Add:* 1208 Delpa Dr Newark DE 19711

CUBICCIOTTI, DANIEL DAVID, b Philadelphia, Pa, June 28, 21; m 48; c 3. HIGH TEMPERATURE CHEMISTRY, PHYSICAL INORGANIC CHEMISTRY. *Educ:* Univ Calif, BS, 42, PhD(chem), 46. *Prof Exp:* Asst chem, Manhattan Proj, Univ Calif, 44-46; instr, NY Univ, 46-47 & Univ Calif, 47-48; res asst prof, Ill Inst Technol, 48-51; res chemist, Atomic Energy Res Lab, NAm Aviation, Inc, 51-55; res chemist, Stanford Res Inst, 55-63, sci fel, 63-72; tech specialist, Nuclear Energy Div, Gen Elec Co, 72-74; sr scientist, SRI Int, 74-80; SCI SPEC, NUCLEAR DIV, ELEC POWER RES INST, 80- *Concurrent Pos:* Mem, Nat Res Coun-Nat Acad Sci Comt High Temperature Sci & Technol, 67-74, chmn, 71-74; US rep, Int Union Pure & Appl Chem Comn High Temperatures & Refractories, 70-74; div ed jour, Electrochem Soc, 75-90; Plenary Lectr,ConfThermo-Nuclear Mat, IntUnionPure & Applied Chem, 84. *Honors & Awards:* Plenary Lectr, Conf Thermo Nuclear Mat, Int Union Pure & Applied Chem, 84. *Mem:* Electrochem Soc; Metall Soc; Am Nuclear Soc. *Res:* Reactions of metals with molten salts and with gases at high temperatures; evaporation of materials; chemistry of nuclear reactor fuels; thermodynamics of inorganic systems; corrosion by high temperature water; stress corrosion cracking. *Mailing Add:* Elec Power Res Inst PO Box 10412 Palo Alto CA 94303-0813

CUCCI, CESARE ELEUTERIO, b Italy, Dec 22, 25; US citizen; m 49, 66; c 3. MEDICINE. *Educ:* Univ Perugia, MD, 49; Univ Rome, dipl cardiol, 53; Am Bd Pediat, dipl, 58, cert cardiol, 63. *Prof Exp:* Assoc prof, 60-76, PROF CLIN PEDIAT, NY UNIV, 76-; CHIEF CHILDREN'S CARDIAC SERV, LENOX HILL HOSP, 63- *Concurrent Pos:* Vis pediatrician, Bellevue Hosp, New York, 65-; consult pediat cardiol, Booth Mem Hosp, 60- & Flushing Hosp & Methodist Hosp, 63- *Mem:* AAAS; fel Am Col Physicians; Am Acad Pediat; Am Col Chest Physicians; Am Col Cardiol. *Res:* Cardio-pulmonary physiology; pediatric cardiology. *Mailing Add:* 37 Montebello Rd Suffern NY 10901

CUCHENS, MARVIN A, b Ft Walton Beach, Fla, June 7, 48; m 87; c 2. CARCINOGENESIS OF LYMPHIOD CELLS, FLOW CYTOMETRY. *Educ:* Univ Fla, PhD(immunol), 77. *Prof Exp:* From asst prof to assoc prof, 79-89, PROF MICROBIOL, MED CTR, UNIV MISS, 89- *Concurrent Pos:* Sci dir, Flow Cytometry Facil. *Mem:* Am Asn Immunologists; Am Asn Cancer Res; Am Chem Soc; Soc Anal Cytol. *Res:* Role of the Peyer's patch in chemical carcinogenesis of lymphiod cells. *Mailing Add:* Dept Microbiol Univ Miss 2500 N State St Med Ctr Jackson MS 39216-4505

CUCULO, JOHN ANTHONY, b Providence, RI, June 23, 24; m 46; c 3. POLYMER CHEMISTRY, MATERIALS SCIENCE. *Educ:* Brown Univ, ScB, 46; Duke Univ, PhD(chem), 50. *Prof Exp:* Res chemist, Polychem Dept, E I du Pont de Nemours & Co, Inc, 50-60, sr res chemist, Textiles Fibers Dept, 60-68; res asst, 68-69, from assoc prof to prof fiber sci, 69-87, HOECUST CELANESE PROF FIBER & POLYMER SCI, NC STATE UNIV, 87- *Concurrent Pos:* Consult various indust firms, 72-; prin investr, NSF grant, 75- *Mem:* Am Chem Soc; Nat Asn Advan Sci; Sigma Xi; Fiber Soc. *Res:* Fiber and polymer science; high speed spinning-high performance fibers; solubility, liquid crystal formation and extrusion of cellulose solutions. *Mailing Add:* 1900 Rangecrest Rd Hidden Valley Raleigh NC 27612

CUDABACK, DAVID DILL, b Napa, Calif, Jan 18, 29; m 53; c 1. RADIO & INFRARED ASTRONOMY, TELESCOPE DEVELOPMENT. *Educ:* Univ Calif, Berkeley, BA, 51, PhD(astron), 62. *Prof Exp:* Physicist, Lawrence Radiation Lab, Univ Calif, 50-57; astronr, Radio Astron Inst, Stanford Univ, 58-62. *Concurrent Pos:* Physicist, Los Alamos Sci Lab, 53-54. *Mem:* AAAS; Am Astron Soc; Int Astron Union; Int Sci Radio Union; Sigma Xi. *Res:* Radio and infrared studies of interstellar material and star formation; instrumentation for same; high altitude observatory development for same. *Mailing Add:* Dept Astron Univ Calif Berkeley CA 94720

CUDE, JOE E, b Austin, Tex, Feb 23, 39; m 68. TOPOLOGY, ALGEBRA. *Educ:* Southwest Tex State Univ, BS, 60; Univ Tex, MA, 62, PhD(math), 66. *Prof Exp:* Asst prof math, Wash State Univ, 66-69; assoc prof & chmn dept, Dallas Baptist Col, 69-72; PROF MATH & HEAD DEPT, TARLETON STATE COL, 72- *Mem:* Am Math Soc; Math Asn Am. *Res:* Topological rings. *Mailing Add:* Dept Math Tarleton State Univ Tarleton Sta Stephenville TX 76402

CUDE, WILLIS AUGUSTUS, JR, b Luling, Tex, Jan 2, 22; m 44; c 4. INORGANIC CHEMISTRY, CRYSTALLOGRAPHY. *Educ:* Univ Tex, Austin, BS, 42, PhD(chem), 68; Ohio State Univ, MS, 53. *Prof Exp:* Meteorologist, US Army Air Corps, US, 43-44 & Okinawa, 45-49; instr meteorol, Chanute Air Force Base, Ill, 49-51; sci liaison nuclear sci, Los Alamos Sci Labs, 53-56; res scientist, Aeronaut Res Labs, Wright Patterson AFB, Ohio, 56-59; from instr to assoc prof chem, USAF Acad, 69-63; asst prof, 63-66, ASSOC PROF CHEM, SOUTHWEST TEX STATE UNIV, 67- *Concurrent Pos:* Partic, NSF Col Teacher Res Prog, Univ Ark, Fayetteville, 68-69. *Mem:* Am Crystallog Asn; Am Chem Soc; Sigma Xi. *Res:* Coordination compounds; structure of mu-bridged complexes. *Mailing Add:* 102 E Hillcrest San Marcos TX 78666

CUDERMAN, JERRY FERDINAND, b Crosby, Minn, Aug 1, 35; m 61; c 2. EXPERIMENTAL ATOMIC PHYSICS. *Educ:* Univ Minn, BME, 58; Ore State Univ, PhD(physics), 66. *Prof Exp:* Exp engr, United Aircraft Corp, 58-59; STAFF MEM, SANDIA LAB, 65- *Mem:* Am Phys Soc. *Res:* Atomic physics; fast alkali metal atom interactions with other atomic species and surfaces; laser plasma physics; x-ray spectroscopy; arc physics; propellant technology development for increasing natural gas production; systems research. *Mailing Add:* 4300 Andrew Dr NE Albuquerque NM 87109

CUDWORTH, ALLEN L, applied acoustics, for more information see previous edition

CUDWORTH, KYLE MCCABE, b Minneapolis, Minn, June 7, 47. ASTRONOMY. *Educ:* Univ Minn, BPhys, 69; Univ Calif, Santa Cruz, PhD(astron), 74. *Prof Exp:* Asst prof, 74-81, ASSOC PROF ASTRON, YERKES OBSERV, UNIV CHICAGO, 81- *Concurrent Pos:* Alfred P Sloan Found fel, 80-82. *Mem:* Am Astron Soc; Am Sci Affil; Astron Soc Pacific; Int Astron Union. *Res:* Measuring proper motions and parallaxes of stars; star clusters and planetary nebulae; stellar photometry. *Mailing Add:* Yerkes Observ Box 258 Williams Bay WI 53191-0258

CUE, BERKELEY WENDELL, JR, b Arlington, Mass, June 10, 47. ORGANIC CHEMISTRY, MEDICINAL CHEMISTRY. *Educ:* Univ Mass, Boston, BA, 69; Univ Ala, PhD(org chem), 74. *Prof Exp:* Fel org chem, Ohio State Univ, 73-74 & Univ Minn, 74-75; RES & DEVELOP SCIENTIST AGR PROD, PFIZER INC, 75- *Mem:* Am Chem Soc. *Res:* Heterocyclic chemistry; medicinal chemistry; total synthesis of natural products. *Mailing Add:* 266 Stoddads WF Rd Gales Ferry CT 06335-1130

CUE, NELSON, b Cavite City, Philippines, Aug 10, 41; US citizen; m; c 2. EXPERIMENTAL PHYSICS. *Educ:* Feati Univ, Philippines, BS, 61; Univ Wash, PhD, 67. *Prof Exp:* Res assoc, Univ Wash, 67 & State Univ NY, StonyBrook, 67-70; from asst prof to prof physics, State Univ NY, Albany, 82-90, chmn, 84-90; PROF PHYSICS, HONG UNIV SCI & TECHNOL, 90-, HEAD, 90- *Concurrent Pos:* Vis prof physics, Univ Lyon I, 78-79; co-dir, NATO Advan Study Inst, State Univ NY, Albany, 78; lectr, Int Winter Sch Nuclear Physics, Beijing, 80; NATO & Heineman sr scientist fel, 79-80; hon prof, Sichuan Univ, 84-; mem Int Comt, Int Conf Atomic Collisions in Solids, 85-; co-chmn, Eleventh Int Conf Atomic Collisions in Solids, Georgetown Univ, 85. *Mem:* Am Phys Soc; Am Asn Physics Teachers. *Res:* Interactions of relativistic electrons with solids; atomic collisions in solids; breakup of fast molecular ions; ion implantation; corrosion. *Mailing Add:* Dept Physics Hong Kong Univ Sci & Technol Seven Centon Rd Tsimsatsui Hong Kong

CUELLAR, ORLANDO, b San Antonio, Tex, Sept 6, 34; m 59; c 3. EVOLUTIONARY BIOLOGY, HERPETOLOGY. *Educ:* Univ Tex, Austin, BA, 64; Tex Tech Univ, MS, 65; Univ Colo, Boulder, PhD(biol), 69. *Prof Exp:* Res asst, Univ Tex, 61-63; teaching asst, Tex Tech Univ, 64-65; res asst, Univ Colo, 65-67; US Pub Health Serv fel, Univ Colo, Boulder, 67-69, Univ Mich, Ann Arbor, 69-71; ASSOC PROF BIOL, UNIV UTAH, 72- *Concurrent Pos:* Predoctoral fel, USPHS, Univ Colo, 67-69, Univ Mich, 69-72; panelist, Nat Res Coun NSF grad fels, 75-76, NSF nat needs res grants, 77- & NSF educ & human resources, 90; prin investr, NIH Gen Med Sci grant, 73-76; Nat Res Coun fel, Univ Hawaii, Hilo, 81-82; postdoctoral fel, Nat Res Coun-Ford Found, Univ Hawaii, Hilo, 81-82. *Honors & Awards:* NSF-NATO Res Award, Salamanca, Spain, 86; US-Soviet Exchange Res Award, Nat Acad Sci, 87. *Mem:* Am Soc Zool; AAAS; Am Asn Ichthyologists & Herpetologists; Am Soc Naturalists; Asn Study Animal Behav. *Res:* Ecology; evolution and cytogenetics of parthenogenetic animals; behavior and population biology of lizards. *Mailing Add:* Dept Biol Univ Utah Salt Lake City UT 84112

CUETO, CIPRIANO, JR, b Tampa, Fla, July 15, 23; m 41; c 3. PHARMACOLOGY, TOXICOLOGY. *Educ:* Univ Tampa, BS, 49; Emory Univ, MS, 57, PhD(pharmacol), 64. *Prof Exp:* Chemist, Bur Agr & Indust Chem, USDA, Fla, 50-51, chemist, Bur Entom & Plant Quarantine, Ga, 51-53; chemist, Tech Develop Labs, Dept Health, Educ & Welfare, USPHS, Ga, 53-58, chemist, Phoenix Field Sta, Ariz, 58-59, chemist, Toxicol Sect, Commun Dis Ctr, 59-66, supvry res pharmacologist & chief pharmacol sect, Fla, 66-69; chief staff officer, Human Safety Eval of Pesticides, USDA & Environ Protection Agency, Washington, DC, 69-71; chief chronic studies div, Nat Ctr Toxicol Res, Ark, 71-74; pharmacologist-toxicologist, Nat Cancer Inst, 74-78, chief, Toxicol Br, 78-79; assoc dir, Dept Toxicol, Litton Bionetics, 80; SR SCI ADVR, CLEMENT ASSOCS, VA, 81- *Mem:* Am Chem Soc; Fed Am Soc Exp Biol; Soc Toxicol; Am Col Toxicol. *Res:* Toxicology of pesticides; adrenocortical inhibitions and its effects on the response of the cardiovascular system to catecholamines; chemical carcinogen bioassay and research; data evaluation and risk analysis. *Mailing Add:* CTC/Cueto Toxicol Consults 205 Nelson St Rockville MD 20850

CUFF, DAVID J, b Edmonton, Alta, May 11, 33; m 61; c 1. CARTOGRAPHY, ENERGY RESOURCES. *Educ:* Univ Alta, BSc, 54; Pa State Univ, MSc, 68, PhD(geog), 72. *Prof Exp:* Geologist, Texaco Explor Co, Calgary, 54-56, Gallup, Buckland & Farney Co, 56-63 & Can Indust Gas Corp, 63-64; asst prof, 68-78, ASSOC PROF GEOG, TEMPLE UNIV, 78- *Mem:* Am Congress Surveying & Mapping; Asn Am Geogrs; Can Cartog Asn. *Res:* Map design; US energy resources. *Mailing Add:* Dept Geog Temple Univ 309 Gladfelter Bldg Philadelphia PA 19122

CUFFEY, ROGER JAMES, b Indianapolis, Ind, May 2, 39; div; c 2. INVERTEBRATE PALEONTOLOGY. *Educ:* Ind Univ, Bloomington, BA, 61, MA, 65, PhD(paleont), 66. *Prof Exp:* From asst prof to assoc prof, 67-79, PROF PALEONT, PA STATE UNIV, UNIV PARK, 79- *Mem:* AAAS; Paleont Soc; Soc Vert Paleont; Geol Soc Am; Soc Econ Paleont & Mineral; Int Bryozool Asn (pres, 86-89). *Res:* Taxonomy, morphology, evolution and paleoecology of fossil and living bryozoans; role of bryozoans in fossil and living reefs. *Mailing Add:* Dept Geosci Pa State Univ Univ Park PA 16802

CUFFIN, B(ENJAMIN) NEIL, b McKeesport, Pa, Apr 21, 41. ELECTROCARDIOGRAPHY, MAGNETOCARDIOGRAPHY. *Educ:* Pa State Univ, BS, 63, MS, 65, PhD(elec eng), 74. *Prof Exp:* Antenna engr, Radio Corp Am, 65-66; instr elec eng, Pa State Univ, 69-74; STAFF SCIENTIST, MASS INST TECHNOL, 74- *Concurrent Pos:* Nat Heart, Lung & Blood Inst fel, 76-78; prin investr, NIH grants, 79- *Res:* Computer modeling and simulation of forward and inverse problem of electroencephalography, magnetoencephalography, electrocardiography and magnetocardiography. *Mailing Add:* Magnet Lab Bldg NW-14 Mass Inst Technol Cambridge MA 02139

CUGELL, DAVID WOLF, b New Haven, Conn, Sept 19, 23; m. PULMONARY DISEASES. *Educ:* Yale Univ, BS, 45; LI Col Med, MD, 47. *Prof Exp:* Asst med, Albany Med Col, 49-50; assoc, 55-57, from asst prof to assoc prof, 57-69, BAZLEY PROF MED, SCH MED, NORTHWESTERN UNIV, 69- *Concurrent Pos:* Res fel, Harvard Med Sch, 50-51, 53-55, Am Heart Asn fel, 53-55; USPHS res career develop award, 62-68; dir pulmonary function labs, Northwestern Univ; attend physician, Vet Admin Res Hosp, Cook County Hosp, Northwestern Mem Hosp; mem pulmonary dis subspeciality bd, Am Bd Internal Med, 66-72. *Mem:* Am Col Physicians; Am Thoracic Soc; Am Col Chest Physicians; Am Fedn Clin Res. *Res:* Cardiopulmonary disease. *Mailing Add:* Northwestern Mem Hosp Rm 456 Chicago IL 60611

CUJEC, BIBIJANA DOBOVISEK, b Ljubljana, Yugoslavia, Dec 25, 26; Can citizen; m 56; c 4. NUCLEAR PHYSICS. *Educ:* Univ Ljubljana, MSc, 51, PhD(physics), 59. *Prof Exp:* Res asst physics, Inst J Stefan, Ljubljana, 50-55, res officer, 55-61; res assoc, Univ Pittsburgh, 61-63 & Univ Alta, 63-64; from asst prof to assoc prof, 64-70, PROF PHYSICS, LAVAL UNIV, 70- *Concurrent Pos:* Vis assoc, Calif Inst Technol, 71-72 & 78-79, Cern, Geneva, Switz, 85-86. *Mem:* Am Phys Soc; Can Asn Physicists. *Res:* Nuclear structure; nuclear reactions and scattering. *Mailing Add:* Dept Physics Laval Univ Quebec PQ G1K 7P4 Can

CUKIER, ROBERT ISAAC, b New York, NY, Oct 10, 44. THEORETICAL CHEMISTRY. *Educ:* Harpur Col, BA, 65; Princeton Univ, MS, 67, PhD(chem), 69. *Prof Exp:* NATO fel, Lorentz Inst Theoret Physics, 69-71; res chemist, Univ Calif, San Diego, 71-72; from asst prof to assoc prof chem, 72-80, PROF CHEM, MICH STATE UNIV, 80- *Mem:* Am Phys Soc. *Res:* Applications of nonequilibrium statistical mechanics to physical chemistry problems. *Mailing Add:* Dept Chem Mich State Univ East Lansing MI 48824

CUKOR, GEORGE, microbiology, virology, for more information see previous edition

CUKOR, PETER, b Szolnok, Hungary, Aug 29, 36; US citizen; m 64; c 3. ANALYTICAL CHEMISTRY, SCIENCE POLICY. *Educ:* City Col New York, BChEng, 61; St John's Univ, NY, MS, 63, PhD(phys & anal chem), 66; Boston Univ, MA, 91. *Prof Exp:* Res technologist anal chem, Customer Serv Dept, Mobil Oil Co, 64-66; eng specialist anal chem & metal-org, Gen Tel & Electronics Res Labs Inc, 67-69, head Absorption Spectros & Chromatog Sect, 69-78, Org Anal & Org Mat Sect, 78-79, tech mgr, Advan Technol Lab, 79-82, sr tech analyst, Strategic Technol Off, 82-85; DIR, RES ADMIN & SUPPORT, GTE LABS, INC, 85- *Concurrent Pos:* Adj asst prof polymer chem, Worcester Polytech Inst. *Mem:* AAAS; Am Chem Soc; NY Acad Sci; fel Am Inst Chemists; Sigma Xi. *Res:* Absorption and emission spectroscopy; thermal analysis; gas and liquid chromatography; conducting polymers; radioimmunoassay; photopolymers; photoresist technology; advanced ceramics; electroluminescence; industry/university interactions. *Mailing Add:* GTE Labs 40 Sylvan Rd Waltham MA 02254

CULBERSON, CHARLES HENRY, b Philadelphia, Pa, March 16, 43; m 79; c 1. OCEANOGRAPHY. *Educ:* Univ Wash, BS, 65; Ore State Univ, MS, 68, PhD(oceanog), 72. *Prof Exp:* Res assoc oceanog, Sch Oceanog, Ore State Univ, 72-75; res assoc chem, Dept Chem, Univ Fla, 75-77; ASST PROF OCEANOG, COL MARINE STUDIES, UNIV DEL, 77- *Res:* Physical chemistry of seawater, especially the acid-base system and carbonate chemistry; chemistry of estuaries. *Mailing Add:* Dept Marine Sci Univ Del Newark DE 19716

CULBERSON, CHICITA FRANCES, b Philadelphia, Pa, Nov 1, 31; m 53. ORGANIC CHEMISTRY, NATURAL PRODUCTS. *Educ:* Univ Cincinnati, BS, 53; Univ Wis, MS, 54; Duke Univ, PhD(chem), 59. *Prof Exp:* Res assoc org chem, 59-61, sr res assoc bot, 61-81, lectr, 71-81, ADJ PROF, DUKE UNIV, 81- *Mem:* AAAS; Am Chem Soc. *Res:* Secondary products of lichen-forming fungi. *Mailing Add:* Dept Bot Duke Univ Durham NC 27706

CULBERSON, JAMES LEE, b Pana, Ill, Sept 18, 41; m 62; c 3. ANATOMY, COMPARATIVE NEUROSCIENCE. *Educ:* Ill Wesleyan Univ, AB, 63; Tulane Univ, PhD(anat), 68. *Prof Exp:* Instr anat, Tulane Univ, 67-68; from instr to prof, 68-84, actg chmn, 87-89, PROF ANAT, MED CTR, WVA UNIV, 84- *Concurrent Pos:* Vis sr lectr, Univ Otago, Dunedin, NZ, 78; res fel, MRC molecular neurobiol unit, Cambridge, 90, Fogarty Int fel, 90, vis fel, Clare Hall, Cambridge, 90. *Mem:* Am Asn Anatomists; Soc Neuroscience. *Res:* Comparative neuroanatomy and neurophysiology. *Mailing Add:* Dept Anat WVa Univ Health Sci Ctr Morgantown WV 26506

CULBERSON, ORAN L(OUIS), b Houston, Tex, May 2, 21; m 49; c 3. CHEMICAL ENGINEERING. *Educ:* Agr & Mech Col, Tex, BS, 43; Univ Tex, MS, 48, PhD(chem eng), 50. *Prof Exp:* Chem engr, Gulf Res & Develop Co, 50-53; chem engr, Celanese Petrochem Res Lab, Celanese Corp Am, 53-57, mgr data processing syst, Celanese Chem Div, Tex, 57-61, mgr math serv dept, NC, 61-65; assoc prof chem eng, 65-69, PROF CHEM ENG, UNIV TENN, KNOXVILLE, 69- *Res:* Solubilities in liquid phase for systems water-methane and water-ethane at temperatures to 340 degrees Fahrenheit and pressures to 10,000 pounds per square inch; economic, design and operations research. *Mailing Add:* 12124 Warrior Trail Concord TN 37922

CULBERSON, WILLIAM LOUIS, b Indianapolis, Ind, Apr 5, 29; m 53. BOTANY. *Educ:* Univ Cincinnati, BS, 51; Université de Paris, France, dipl, 52; Univ Wis, PhD(bot), 54. *Prof Exp:* From instr to assoc prof, 55-70, prof, 70-84, HUGO L BLOMQUIST PROF BOT, DUKE UNIV, 84- *Concurrent Pos:* Grants, NSF & Lalor Found, 57-91; ed, Bryologist, Am Bryol & Lichenological Soc, 62-70; ed, Brittonia, Am Soc Plant Taxon, 75; ed, Syst Bot, Am Soc Plant Taxonomy, 76-77; vis prof res, Mus Nat d'Hist Naturelle, Paris, 80. *Mem:* Bot Soc Am (pres elect, 90-91); Am Soc Plant Taxon; Am Bryol & Lichenological Soc (pres, 88-89); Mycol Soc Am; Brit Lichen Soc. *Res:* Systematics, ecology, morphology and natural-product chemistry of lichens. *Mailing Add:* Dept Bot Duke Univ Durham NC 27706

CULBERT, JOHN ROBERT, b Rossville, Ill, Dec 18, 14; m 45; c 2. FLORICULTURE. *Educ:* Univ Ill, BS, 37; Ohio State Univ, MS, 39. *Prof Exp:* Instr floricult, Pa State Col, 39-43 & 45-46; from asst prof to prof, Univ Ill, Urbana-Champaign, 46-80, emer prof floricult, 80-86. *Concurrent Pos:* Consult, Framptons Nurseries, Eng, 59-60. *Honors & Awards:* Res award, Soc Am Florists, 56. *Mem:* Am Soc Hort Sci; Sigma Xi. *Res:* Breeding of chrysanthemums. *Mailing Add:* 401 W Main St Clinton IL 61727

CULBERTSON, BILLY MURIEL, b Hillside, Ky, Aug 23, 29; m 49; c 3. ORGANIC CHEMISTRY, POLYMER CHEMISTRY. *Educ:* Augustana Col, Ill, BS, 59; Univ Iowa, MS, 62, PhD(org chem), 63. *Prof Exp:* Sr res chemist, Archer Daniels Midland Co, Minn, 62-67; sr res chemist, Ashland Chem Co, 67-69, res assoc, 69-81, sr res assoc, 82-89; PROF, OHIO STATE UNIV, 89- *Concurrent Pos:* NSF vis scientist, Minn Acad Sci, 68-69; ed, Minn Chemist, Am Chem Soc, 68-70; vis scientist, Indust Res Inst & Mfg Chemists Asn, 74-76; chmn, Prog Comt, Polymer Chem Div, Am Chem Soc, 74-76, Div Polymer Chem, Inc, 77-79; ed, Polymer Preprints, Div Polymer Chem, Inc, 82-; chmn, Gordon Res Conf Polymers, Colby-Sawyer Col, 86. *Honors & Awards:* Columbus Sect Award, Am Chem Soc, 82, Polymer Chem Serv Award, 86. *Mem:* Am Chem Soc; Mfg Chemists Asn; Chem Soc; Sigma Xi; AAAS. *Res:* Organic syntheses, monomer syntheses and polymerization studies; polymerization mechanisms and kinetics studies; cyclopolymerization studies; synthesis of plastics for space applications; synthesis and physical studies of polymers having high thermal stability; amine acylimide monomers and polymers; synthesis of reaction injection molding materials; new composite resin syntesis; synthesis and characterization of new biomedical materials; author of four books and 80 publications in archival literature; 40 US patents. *Mailing Add:* Ohio State Univ Dent Col 305 W 12th Ave Columbus OH 43210

CULBERTSON, CLYDE GRAY, b Vevay, Ind, July 27, 06; m 31. PATHOLOGY. *Educ:* Ind Univ, BS, 28, MD, 31. *Hon Degrees:* DS, Ind Univ, 89. *Prof Exp:* From instr to asst prof path, Sch Med, Ind Univ, 31-42, prof clin path & chmn dept, 43-63; from asst dir to dir biol div, 46-63, res adv, 63-70, Lilly res consult, Lilly Res Labs, 70-71; prof path, Sch Med, Ind Univ, Indianapolis, 63-77; consult, Eli Lilly & Co, 71-73; EMER CLIN PROF ANAT PATH, SCH MED, IND UNIV, INDIANAPOLIS, 77- *Concurrent Pos:* Dir lab, Sch Med, Ind Univ, 31-46, dir labs, State Bd Health, Ind, 33-46. *Mem:* AMA; Am Asn Immunologists; Soc Protozoologists; Tissue Cult Asn; Am Soc Clin Path (secy-treas, 49-58, vpres, 59-68, pres, 70). *Res:* Clinical pathology; medical microbiology and parasitology. *Mailing Add:* 1901 Taylor Rd Columbus IN 47201

CULBERTSON, EDWIN CHARLES, b Charleroi, Pa, Jan 5, 50; m 71; c 2. POLYMER CHEMISTRY. *Educ:* Washington & Jefferson Col, BA, 72; Univ SC, PhD(inorg chem), 76. *Prof Exp:* Chemist, Celanese Plastics Co, 77-79; tech chemist, Mobay Chem Co, 79-80; RES CHEMIST, AM HOECHSTCELE CORP, 80- *Mem:* Am Chem Soc. *Res:* Development of new coatings for polyester pentaerythritol film which are used in the reprographics and microfilm areas. *Mailing Add:* 107 Saratoga Dr Greer SC 29650

CULBERTSON, GEORGE EDWARD, b Cranes Nest, Va, Oct 23, 37; m 59; c 3. MATHEMATICS. *Educ:* Va Polytech Inst, BS, 59, MS, 62, PhD(math), 70. *Prof Exp:* Engr, Radford Army Ammunition Plant, Hercules, Inc, Va, 62-63, sr engr, 63-64, asst area supvr, 64; asst prof math, US Naval Acad, 64-70; assoc prof, 70-71, REGISTR & ASST DEAN, CLINCH VALLEY COL, UNIV VA, 71-, PROF MATH, 75- *Concurrent Pos:* Pres, Systech Corp, Md, 69-70. *Mem:* Math Asn Am; Nat Coun Teachers Math; Sigma Xi. *Res:* A basic variable study of the propellant used in the Minuteman Missile; geometry of paths of particles traveling in force fields under certain constraints; an extension of Halphen's Theorem. *Mailing Add:* Dept Math Clinch Valley Col Univ Va Wise VA 24293

CULBERTSON, TOWNLEY PAYNE, b Durant, Okla, May 13, 29; m 69; c 1. CHEMISTRY. *Educ:* Univ Okla, BS, 51, MS, 52; Univ Iowa, PhD(org chem), 58. *Prof Exp:* Fel org chem, Univ Ill, 58-59; assoc res chemist, Warner-Lambert/Parke, 60-65; res chemist, David Pharmaceut, 65-70; SR RES CHEMIST, RES DIV, WARNER-LAMBERT CO, 70- *Mem:* Am Chem Soc. *Res:* Synthetic modifications of steroids; antiviral compounds; aminoglycoside antibiotics; diquins as antifungal agents; cephalosporins; nalidixic acid type compounds as antibacterial agents. *Mailing Add:* Chem Div Warner-Lambert Parke Davis Ann Arbor MI 48103

CULBERTSON, WILLIAM RICHARDSON, b Coeburn, Va, May 16, 16; m 50; c 2. MEDICINE. *Educ:* Transylvania Col, AB, 37; Vanderbilt Univ, MD, 41; Am Bd Surg, dipl, 55. *Prof Exp:* Intern, St Joseph's Hosp, Lexington, Ky, 41-42; asst resident surg, Cincinnati Gen Hosp, 46-53; from asst prof to assoc prof, 57-70, PROF SURG, UNIV CINCINNATI, 70- *Mem:* Fel Am Col Surg; Am Surg Asn; Soc Surg Alimentary Tract; AMA; Sigma Xi. *Res:* Bacteriology of surgical infections; hemorrhage shock; shock due to sepsis. *Mailing Add:* Dept Surg Cincinnati Gen Hosp Cincinnati OH 45229

CULBRETH, JUDITH ELIZABETH, b Greenville, SC, Apr 28, 43; m; c 1. SOFTWARE SYSTEMS. *Educ:* Furman Univ, BS, 65; Univ Md, College Park, PhD(inorg chem), 71. *Prof Exp:* Teaching asst chem, Univ Md, College Park, 65-68; instr, Northern Va Community Col, 68-69; from jr instr to instr, Univ Md, College Park, 69-72; clin chemist, Diag Labs, 73-80; SR ENGR DIGITAL SYSTS, GEN ELEC CO, 80- *Concurrent Pos:* Lectr, Univ NC, Wilmington, 81- *Mem:* Am Asn Clin Chem; Am Chem Soc; Sigma Xi. *Res:* Computer calculation of group electronegatives; correlation of computed charges to observed physical and chemical properties; development of clinical chemistry assays for drugs; heavy metals. *Mailing Add:* 121 Watauga Rd Wilmington NC 28412

CULICK, FRED E(LLSWORTH) C(LOW), b US, Oct 25, 33; m 60; c 3. AERONAUTICAL ENGINEERING. *Educ:* Mass Inst Technol, SB & SM, 57, PhD(aero eng), 61. *Prof Exp:* Res fel, 61-63, from asst prof to assoc prof, 63-71, PROF MECH ENG & JET PROPULSION, CALIF INST TECHNOL, 71- *Mem:* Am Phys Soc; Am Inst Aeronaut & Astronaut. *Res:* Fluid mechanics; acoustics; combustion and propulsion; robotics; applied aerodynamics and flight mechanics; aeronautical history. *Mailing Add:* 1375 E Hull Lane Altadena CA 91001

CULLATI, ARTHUR G, COMPUTER TECHNOLOGY. *Educ:* Bridgewater State Col, BS, 58, MEd, 61; Worcester Polytech Inst, MS, 66; Somerset Univ, Eng, PhD(computer sci eng), 88. *Prof Exp:* Res engr, mathematician, analyst, programmer & consult, Cray Res Inc, Control Data Corp, NAm Aviation, 66-69; group leader large scale systs, Cent Computer Facil, Ames Res Ctr, NASA, 79-86, chief numerical aerodyn simulation, Computations Br, 86-88; chief, Computer Technol Br, Nat Inst Environ Health Sci, 88-90; HEAD SCI COMPUT, ENVIRON PROTECTION AGENCY, 90- *Mailing Add:* Nat Data Processing Div Sci Comput Br-86 US Environ Protection Agency Research Triangle Park NC 27711

CULLEN, ABBEY BOYD, JR, b Oxford, Miss, Dec 25, 15; m 43, 84; c 1. APPLIED PHYSICS. *Educ:* Univ Miss, BA, 37, MS, 42; Univ Va, PhD(physics), 47. *Prof Exp:* Technician develop physiol res apparatus, Sch Med, Univ Miss, 37-40; engr & physicist, Naval Res Lab, 42-45; From assoc prof to prof physics, Univ Miss, 47-81, chmn dept physics & Astron, 57-76, dir radiation res, 76-81; RETIRED. *Mem:* Am Phys Soc. *Res:* Microwave electron accelerator; electronic instruments. *Mailing Add:* 812 Lincoln Ave Univ Miss Oxford MS 38655

CULLEN, ALEXANDER LAMB, b London, Eng, Apr 30, 20; m 40; c 3. MICROWAVES, MEASUREMENTS. *Educ:* Imp Col London, BSc, 41; Univ London, PhD(elec eng), 52, DSc, 67. *Hon Degrees:* DSc, Chinese Univ Hong Kong, 81; DEng, Univ Sheffield, 85; DSc, Univ Kent Canterbury, 86. *Prof Exp:* Scientist, Royal Aircraft Estab, Farnborough, 40-46; prof elec eng, Univ Sheffield, 55-67; lectr elec eng, Univ Col London, 46-55, Pender prof, 67-80, Sci & Eng Res Coun sr fel, 80-83, HON RES FEL ELEC ENG, UNIV COL LONDON, 83- *Concurrent Pos:* Consult, 80-; vpres, Int Union Radio Sci, 81-87 & treas, 81-84; mem, Independent Broadcasting Authority, 82-90; hon prof, Polytech Univ, Xian, China, 83. *Honors & Awards:* Faraday Medal, Inst Elec Engrs, UK, 84; Royal Medal, Royal Soc UK, 84. *Mem:* Fel Inst Physics UK; Int Union Radio Sci (pres, 87-90); hon fel Inst Elec Engrs UK; fel Inst Elec & Electronic Engrs; fel Royal Soc UK; fel Royal Acad Eng. *Res:* Microwave and millimetre measurements; antenna design; surface waves; material measurements. *Mailing Add:* Three Felden Dr Hemel Hempstead Herts HP3 0BD England

CULLEN, BRUCE F, b Iowa City, Iowa, May 6, 40; m 60; c 3. ANESTHESIOLOGY. *Educ:* Stanford Univ, BS, 62; Univ Calif, Los Angeles, MD, 66. *Prof Exp:* Staff anesthesiologist, NIH, 70-72; asst prof, 72-75, from asst prof to assoc prof, anesthesiol, Univ Wash, 72-76; prof anesthesiol & chmn dept, Univ Calif, Irvine, 76-84; PROF ANESTHESIOL, UNIV WASH, SEATTLE, 84- *Mem:* Am Soc Anesthesiologists; Int Anesthesia Res Soc; Asn Univ Anesthetists. *Res:* Immunologic and cellular effects of anesthesia. *Mailing Add:* Dept Anesthesiol Univ Wash 325 Ninth Ave Seattle WA 98104

CULLEN, CHARLES G, b Elmira, NY, Nov 6, 32; m 54; c 2. MATHEMATICS. *Educ:* State Univ NY Albany, BA, 54; Univ NH, MA, 56; Case Inst Technol, PhD(math), 62. *Prof Exp:* Instr math, Worcester Polytech Inst, 56-59; instr math, Case Inst Technol, 59-61; asst prof, 62-66, ASSOC PROF MATH, UNIV PITTSBURG, 66- *Mem:* Soc Indust & Appl Math; Math Asn Am. *Res:* Linear algebra; matrix analysis; functions of matrices and matrix algebras; numerical analysis. *Mailing Add:* Dept Math Univ Pittsburgh Pittsburgh PA 15260

CULLEN, DANIEL EDWARD, b Oak Park, Ill, Feb 16, 42; m 63; c 2. OPERATIONS RESEARCH. *Educ:* Stanford Univ, BS, 63; Univ Ill, MS, 64; Wash Univ, ScD(appl math), 67. *Prof Exp:* Mem tech staff appl math, Bell Tel Labs, 67-68; VPRES MGT & TECH STUDIES, MATHEMATICA, 68- *Mem:* Am Math Soc; Soc Indust & Appl Math; Opers Res Soc Am; Inst Mgt Sci; Am Asn Pub Opinion Res. *Res:* Development and use of analytic and scientific methods for the solution of business and government problems. *Mailing Add:* 980 Stuart Rd Princeton NJ 08540

CULLEN, GLENN WHERRY, b Nashville, Tenn, June 27, 31; m 56; c 1. SOLID STATE CHEMISTRY. *Educ:* Univ Cincinnati, BS, 53; Univ Ill, MS, 54, PhD(chem), 56. *Prof Exp:* HEAD MAT SYNTHESIS RES GROUP, DAVID SARNOFF LABS, RCA CORP, 58- *Concurrent Pos:* Assoc ed, J Crystal Growth, 74-; div ed jour, Electrochem Soc, 75-78; mem, Air Force Sci & Technol Adv Group, 80; mem comt high purity silicon, Nat Mat Adv Bd, 80; chmn, Electronic Div, The Electrochem Soc, 77-79; chmn, Gordon Conf Crystal Growth, 80, Int Conf Vapor Growth & Epitaxy, 84, Gordon Conf Inorg Films & Interfaces, 89; pres, Fed Mat Socs, 89-90. *Honors & Awards:* Electronics Div Award, Electrochem Soc, 82. *Mem:* Am Asn Crystal Growth; Am Chem Soc; hon mem & fel Electrochem Soc; Mat Res Soc. *Res:* Semiconductor materials chemistry; homoepitaxial and heteroepitaxial thin film silicon growth; bulk silicon growth and characterization; silicon on insulators; characterization of electronic materials. *Mailing Add:* 56 Adams Dr Princeton NJ 08540

CULLEN, HELEN FRANCES, b Boston, Mass, Jan 4, 19. MATHEMATICS. *Educ:* Radcliffe Col, AB, 40; Univ Mich, AM, 44, PhD(math), 50. *Prof Exp:* From asst prof to assoc prof, 49-71, PROF MATH, UNIV MASS, AMHERST, 71 - *Concurrent Pos:* Asst eng, Gen Elec, Lynn, Mass, 40-43; teaching fel, Univ Mich, 46-49. *Mem:* Am Math Soc; Math Asn Am; Sigma Xi; Am Phys Soc. *Res:* Topology; relativity. *Mailing Add:* Dept Math Univ Mass Sta Lederle Towers Amherst MA 01003

CULLEN, JAMES ROBERT, b Brooklyn, NY, Jan 28, 36; m; c 1. SOLID STATE PHYSICS. *Educ:* St John's Univ, NY, BS, 58; Univ Md, PhD(physics), 65. *Prof Exp:* Physicist, US Army Night Vision Lab, Ft Belvoir, Va, 65-66; PHYSICIST, US NAVAL ORD LAB, 66- *Concurrent Pos:* Lectr, George Washington Univ, 66; vis asst prof, Dept Physics, Univ Wis-Milwaukee, 68-69. *Mem:* Am Phys Soc. *Res:* Superconductivity, fluctuation phenomena, theory of ultrasonic waves in superconductors; semiconductors, interactions between impurities; magnetism, theory of electron-electron interactions and their effects on susceptibility and neutron scattering experiments; Hall effect in rare-earth metals; metal-insulator transition in transition metal-oxides, especially in magnetite. *Mailing Add:* Naval Surface R-45 Warfare Ctr Silver Spring MD 20903

CULLEN, JOHN KNOX, b Denver, Colo, Jan 2, 36; m 57; c 2. PHYSIOLOGY, ELECTRONICS ENGINEERING. *Educ:* Univ Md, BS, 60, MS, 69; La State Med Ctr, New Orleans, PhD, 75. *Prof Exp:* Engr, NIH, 60-62, res engr, Psychopharmacol Res Ctr, 62-64; res engr, Neurocommun Lab, Sch Med, Johns Hopkins Univ, 64-68; PROF, KRESGE HEARING LAB, LA STATE UNIV MED CTR, NEW ORLEANS, 68-, COMMUN DISORDERS, LA STATE UNIV, 86- *Concurrent Pos:* Consult, noise measurement. *Mem:* AAAS; Acoust Soc Am; Asn Res Otolaryngol. *Res:* Biomedical research; speech, hearing science. *Mailing Add:* Commun Disorders La State Univ Rm 163 M & D A Bldg Baton Rouge LA 70803

CULLEN, JOSEPH WARREN, cancer prevention & control; deceased, see previous edition for last biography

CULLEN, MARION PERMILLA, b Pittsburgh, Pa, Dec 3, 31. BIOCHEMISTRY, NUTRITION. *Educ:* Pa State Univ, BS, 54, MS, 56; Univ Calif, Berkeley, PhD(nutrit, biochem), 68. *Prof Exp:* Asst biochemist, Dept Animal Nutrit, Am Meat Inst Found, Chicago, 56-60; fel, Sch Pub Health, Harvard Univ, Boston, 60-62; res asst & res biochemist, Dept Nutrit Sci, Univ Calif, Berkeley, 62-68; fel, Finney Howell Cancer Res Lab, Dept Surg, Johns Hopkins Univ, 69-71; sr sci proj officer, Dept Pediat, Guy's Hosp Med Sci, Univ London, 81-85; consult, Health Sci Admin, 86-87; nutrit epidemiologist & biostatiscian, Ctr Nutrit, Meharry Med Col, 87-88; NUTRIT DATA LIAISON, TENN DEPT HEALTH, BUR HEALTH SERV, NASHVILLE, 88- *Concurrent Pos:* Teacher horse mgt, Pa State Univ, 55; lectr human biochem & nutrit, Albany Med Col, 72-75. *Mem:* NY Acad Sci; AAAS; Am Pub Health Asn; Am Chem Soc, Div Biol Chem; Am Chem Soc; fel Royal Soc Med. *Res:* Energy metabolism of animal by-products; correlation of iron deficiency anemia in adolescent population with food consumption; pantothenic acid deficiency serum protein isolation in vitro labeling of serum proteins in turnover rate studies; isolation of serum born virus antigen and antibody assay techniques; author of numerous reviews, textbooks and articles; designed and developed total system to assess renal pore size in premature infants; developed pediatric and prenatal tracking operation. *Mailing Add:* 865 Bellvue Rd G 17 Nashville TN 37221

CULLEN, MICHAEL JOSEPH, b Cleveland, Ohio, July 27, 45; m 68; c 2. NEUROANATOMY, NEUROCYTOLOGY. *Educ:* Case Western Reserve Univ, AB, 67, PhD(anat), 75. *Prof Exp:* Staff fel, Lab Neuropath & Neuroanat Sci, Nat Inst Neurol & Commun Disorders & Stroke, 75-78; asst prof anat, 78-83, ASSOC PROF ANAT & CELL BIOL, SCH MED, UNIV SOUTHERN CALIF, 84- *Mem:* Am Asn Anatomists; Soc Neurosci; Am Soc Cell Biol; Am Soc Neurochem. *Res:* Cellular mechanisms of myelin formation and maintenance; neuron-glial cell interactions; mechanisms of neuron connectivity and plasticity. *Mailing Add:* Dept Anat & Cell Biol 1333 San Pablo St Los Angeles CA 90033

CULLEN, SUSAN ELIZABETH, b New York, NY, Jan 12, 44; m 74; c 2. IMMUNOCHEMISTRY, IMMUNOGENETICS. *Educ:* Col Mt St Vincent, NY, BS, 65; Albert Einstein Col Med, NY, PhD(microbiol & immunol), 71. *Prof Exp:* Res assoc, Basel Inst Immunol, Switz, 71-72; res assoc, Albert Einstein Col Med, NY, 72-74; vis fel, Immunol Br, Div Cancer Biol & Diag, Nat Cancer Inst, NIH, 74-76; from asst prof to assoc prof, 76-85, PROF MICROBIOL & IMMUNOL, SCH MED, WASH UNIV, ST LOUIS, 85-, DIR SPONSORED PROJ SERV, RES OFF, 90- *Concurrent Pos:* Career develop award, NIH, 77-82; mem, Allergy & Immunol Study Sect, NIH, 80-84; prin investr, Am Cancer Soc, 76-81, NIH, 76-; assoc ed & sect head, Publ Comt, J Immunol, 78-90; adv ed, Molecular Immunol, 82-; mem, Nat Cancer Inst Div Cancer Biol & Diag Bd Sci Counr, 86-90. *Mem:* Am Asn Immunologists. *Res:* Biochemical approach correlating between structure and function of lymphoid and myeloid cell surface molecules involved in the regulation of immune responsiveness. *Mailing Add:* Sch Med Box 8013 Washington Univ 724 S Euclid St Louis MO 63110

CULLEN, THEODORE JOHN, b St Louis, Mo, Dec 19, 28; m 47; c 4. MATHEMATICS. *Educ:* DePaul Univ, BS, 55, MS, 56. *Prof Exp:* Asst math, Univ Ill, 56-57; asst prof, Ariz State Col, 57-59 & Los Angeles State Col, 59-68; assoc prof, 68-77, PROF MATH, CALIF STATE POLYTECH UNIV, POMONA, 77- *Concurrent Pos:* Engr, Jet Propulsion Lab, Pasadena, 62- *Mem:* Math Asn Am. *Res:* Abstract algebra; hilbert and banach spaces; set theory and point set toplogy; functional analysis; numerical analysis; computers. *Mailing Add:* Dept Math Calif State Polytech Univ 3801 W Temple Ave Pomona CA 91768

CULLEN, THOMAS M, b Birmingham, Ala, Dec 16, 35. METALLURGICAL ENGINEERING, HIGH TEMPERATURE MATERIALS FOSSIL POWER. *Educ:* Univ Ala, BS, 57; Univ Mich, MS, 59, PhD(metall eng), 63. *Prof Exp:* Mgr reliability, Dept Qual & Reliability, 83-90, MGR MAT & CHEM, SYSTS ENG DEPT, POTOMAC ELEC POWER CO, 91- *Mem:* Fel Am Soc Metals; Am Soc Mech Engrs. *Mailing Add:* Systs Eng Dept Potomac Elec Power Co 1900 Pennsylvania Ave NW Washington DC 20068

CULLEN, WILLIAM CHARLES, b Buffalo, NY, Nov 6, 19; m 52; c 3. ORGANIC CHEMISTRY. *Educ:* Canisius Col, BS, 48. *Prof Exp:* Chemist, Bldg Res Div, Inst Appl Tech, 48-67, sect chief, 67-73, dep chief struct, mat & safety div, 73-78, dep dir, Off Eng Standards, Nat Eng Lab, Nat Bur Standards, 78-81; CONSULT, 81- *Concurrent Pos:* US Dept Com sci fel, 65-66; res assoc, Nat Roofing Contracters Asn, 81- *Honors & Awards:* Silver Medal Award, Dept Com, 68; J A Piper Award, Nat Roofing Contractors Asn, 74; Edward Bennett Rosa Award, Nat Bur Standards, 75; Walter C Voss Award, Am Soc Testing & Mat, 78; Gold Medal Award, Dept Com, 80. *Mem:* Hon mem Am Soc Testing & Mat. *Res:* Mechanisms of chemical degradation and physical deterioration of organic roofing materials; methods to determine engineering properties of roof systems; durability and performance of building materials; application and impact of engineering standards. *Mailing Add:* 11718 Tifton Dr Potomac MD 20854

CULLEN, WILLIAM ROBERT, b Dunedin, NZ, May 4, 33; m 56, 79; c 3. ANALYTICAL CHEMISTRY. *Educ:* Univ Otago, NZ, BSc, 55, MSc, 57; Cambridge Univ, PhD(chem), 59. *Prof Exp:* From instr to assoc prof, 58-69, PROF CHEM, UNIV BC, 69- *Concurrent Pos:* Nat Res Coun Can fel, 66-67. *Mem:* Am Chem Soc; Chem Inst Can. *Res:* Organometallic chemistry of the main group elements; coordination chemistry of arsines and phosphines; conformational problems in coordination chemistry; ferrocene/ferrocyne chemistry; asymmetric catalysis; biological transformations of inorganic and organometallic compounds; analytical environmental chemistry. *Mailing Add:* Dept Chem Univ BC Vancouver BC V6T 1Y6 Can

CULLER, DONALD MERRILL, b Hicksville, Ohio, Nov 17, 29; m 50; c 3. ELECTRONICS ENGINEERING, SYSTEMS DESIGN. *Educ:* Carnegie Inst Technol, BS, 51; Univ Calif, MS, 53. *Prof Exp:* Dynamics engr, Gen Dynamics/Convair, 51-53; proj engr, Farnsworth Electronic Co, Div Int Tel & Tel Corp, 53-54, mgr, Systs Anal & Design Sect, 54-55, assoc head adv develop, 55-57, head, 57-58, assoc lab dir space electronics, ITT Fed Labs Div, 58-61, lab dir, 61-63, dir, Astrionics Ctr, 63-66; lab mgr, 66-71, adv systs mgr, 71-75, BUS AREA MGR, TRW DEFENSE SYSTS GROUP, 75- *Concurrent Pos:* Secy, Nat Telemetry Conf, 56. *Mem:* Inst Elec & Electronics Engrs; assoc fel Am Inst Aeronaut & Astronaut. *Res:* Missile and space systems analysis and design; guidance and control systems; satellite communication systems; optical and RF tracking equipment; computer and information systems; laser systems; systems sciences. *Mailing Add:* TRW Inc Space & Technol Group One Space Park Bldg RZ-1094 Redondo Beach CA 90278

CULLER, F(LOYD) L(EROY), JR, b Washington, DC, Jan 5, 23; m 46; c 1. ENERGY, ELECTRICAL ENERGY. *Educ:* John Hopkins Univ, BS, 43. *Prof Exp:* Design engr, Eastman Kodak Co, 43-44, develop engr, Tenn Eastman Co, 44-47; sr design engr, Radio-Chem Plant Design, Oak Ridge Nat Lab, Monsanto Chem Co, 47-48; dir chem tech div, Nuclear Div, Union Carbide Corp, 48-64, asst lab dir, 64-70; assoc dir, Oak Ridge Nat Lab, 63-71, dep dir, 71-77; pres, 78-88, EMER PRES, ELEC POWER RES INST, 88- *Concurrent Pos:* Mem sci adv comt, Int Atomic Energy Agency, 74-; adv bd, DOE: Energy Res, 82-88; bd dirs, Houston Light & Power. *Honors & Awards:* E O Lawrence Award, 65; Robert E Wilson Award, Am Inst Chem Engrs, 72; Atoms Peace Award, 68. *Mem:* Nat Acad Eng; fel Am Inst Chem; fel Am Nuclear Soc; fel Am Inst Chem Engrs; Am Chem Soc. *Res:* Nuclear energy, research and development; fuel reprocessing, reactor technology; economics; radioactive waste disposal and treatment; fuel cycle; radiochemical plant design, construction and operation; laboratory-scale chemical research and development; solvent extraction; electric energy, fossil, solar, wind, geothermal conservation, storage, energy policy, economics. *Mailing Add:* 1385 Corinne Lane Menlo Park CA 94025

CULLERS, ROBERT LEE, b North Manchester, Ind, May 19, 37; m 70; c 2. GEOCHEMISTRY. *Educ:* Ind Univ, Bloomington, BS, 59, MA, 62; Univ Wis-Madison, PhD(geochem), 71. *Prof Exp:* Teacher math, Vevay Town Schs, Ind, 60-61; teacher math-chem, Goshen City Schs, 61-63; teacher chem, Park Ridge, Ill, 64-67; res asst, Univ Wis-Madison, 67-69, NSF fel geochem, 69-71; from asst prof to assoc prof geochem, 71-81, PROF GEOCHEM & PETROLOGY, KANS STATE UNIV, 81- *Mem:* Geochem Soc Am; Nat Asn Geol Teachers; Geol Soc Am. *Res:* Trace element geochemistry; experimental igneous and metamorphic trace element partitioning; trace element geochemistry. *Mailing Add:* Dept Geol Thompson Hall Kans State Univ Manhattan KS 66506

CULLEY, BENJAMIN HAYS, mathematics; deceased, see previous edition for last biography

CULLEY, DUDLEY DEAN, JR, b Jackson, Miss, May 14, 37. AQUACULTURE, VITICULTURE. *Educ:* Millsaps Col, BS, 59; Univ Miss, MEd, 61; Miss State Univ, MS, 64, PhD(zool), 68. *Prof Exp:* Teacher, Univ High Sch, Miss, 59-61; teacher-counr pub schs, 61-63; asst prof fisheries, pollution biol & aquatic ecol, 68-74, assoc prof, 74-80, PROF FORESTRY & WILDLIFE MGT, LA STATE UNIV, BATON ROUGE, 80- *Concurrent Pos:* Consult aquacult & viticult. *Res:* Crustacean and amphibian aquaculture; reproductive biology; aquatic plant management and utilization; agricultural waste management; viticulture. *Mailing Add:* Sch Forestry Wildlife & Fisheries 249 Agr Ctr La State Univ Baton Rouge LA 70803

CULLIMORE, DENIS ROY, b Oxford, Eng, Apr 7, 36; m 62. MICROBIAL ECOLOGY, BACTERIOLOGY. *Educ:* Univ Nottingham, BSc, 59, PhD(agr microbiol), 62. *Prof Exp:* Lectr microbiol, Univ Surrey, 62-68; from asst prof to assoc prof, 68-74, PROF MICROBIOL, UNIV REGINA, 74-; DIR, REGINA WATER RES INST, 75- *Concurrent Pos:* Consult, var govt agencies & indust. *Mem:* Can Soc Microbiol; Brit Soc Appl Bact; Am Soc Microbiol; fel Royal Soc Arts. *Res:* Bioassay systems using algae; effects of herbicides on soil and water microflora; simplified classification of bacteria; novel uses of microorganisms; factors effecting the biofouling of wells and the influence of iron bacteria on these systems. *Mailing Add:* Dept Biol Univ Regina Regina SK S4S 0A2 Can

CULLINAN, HARRY T(HOMAS), JR, b New York, NY, May 27, 38; m 64; c 3. CHEMICAL ENGINEERING. *Educ:* Univ Detroit, BChE, 61; Carnegie Inst Technol, MS, 63, PhD(chem eng), 65. *Prof Exp:* Res engr, Westinghouse Res Labs, 63-64; from asst prof to prof chem eng, State Univ NY Buffalo, 64-76, chmn dept, 69-76; ACAD DEAN, INST PAPER CHEM, 76-, VPRES ACAD AFFAIRS, 77- *Mem:* Am Inst Chem Engrs; Tech Asn Pulp & Paper Indust. *Res:* Multicomponent diffusion; thermodynamic properties of multicomponent mixtures; pulp and paper process engineering. *Mailing Add:* Pescara Pl Donvale Victo 311 Australia

CULLINANE, THOMAS PAUL, b Brockton, Mass, Dec 24, 42; m 67; c 1. ENGINEERING. *Educ:* Boston Univ, BS, 66; Northeastern Univ, MSIE, 68; Va Polytech Inst & State Univ, PhD(eng), 72. *Prof Exp:* Instr eng, Northeastern Univ, 68-69; asst prof, Va Polytech & State Univ, 71-72, Univ Ala, Huntsville, 72-73 & Univ Notre Dame, 74-78; assoc prof, Univ Mass, 78-79, Univ Notre Dame, 79-81; ASSOC PROF ENG, NORTHEASTERN UNIV, 81-, AT DEPT INDUST ENG & INFO SYSTS. *Concurrent Pos:* Mem bd dir, Gainsboro Elec Mfg Co, 71-73. *Mem:* Am Inst Indust Engrs; Col Indust Coun Mat Handling Educ; Inst Mgt Sci; Soc Mfg Engrs; Am Soc Eng Educ. *Res:* Application of operations research methodologies for the optimization of production systems and economic decision making. *Mailing Add:* Dept Indust Eng & Info Systs Northeastern Univ 360 Huntington Ave Boston MA 02115

CULLIS, CHRISTOPHER ASHLEY, b Harrow, Eng, Nov 20, 45; m 71; c 6. PLANT MOLECULAR BIOLOGY. *Educ:* London Univ, BSc, 67; Univ E Anglia, MSc, 68, PhD(genetics), 72. *Prof Exp:* Higher sci officer, John Innes Inst, 71-73, sr sci officer, 73-80, prin sci officer, 80-85; PROF BIOL, GENETICS, MOLECULAR BIOL, CASE WESTERN RESERVE UNIV, 86-, DEAN MATH & NATURAL SCI, 89- *Concurrent Pos:* Hon lectr, Univ E Anglia, 78-85; vis prof, Stanford Univ, 82-83, Case Western Reserve Univ, 85-86; Nuffield & Lever Holme travelling fel, 82-83. *Mem:* Genetics Soc Am; AAAS; Int Soc Plant Molecular Biol. *Res:* Molecular events in response to environmental stress in higher plants; genetic mapping with forest trees using molecular markers. *Mailing Add:* 2915 Southington Rd Shaker Heights OH 44120

CULLIS, PIETER RUTTER, b Barnard Castle, Eng, May 12, 46; Can citizen; m; c 2. BIOCHEMISTRY. *Educ:* Univ BC, BSc, 67, PhD(physics), 72. *Prof Exp:* Fel biochem, Univ Oxford, 73-76 & State Univ Utrecht, 77; from asst prof to assoc prof, 78-85, PROF BIOCHEM, UNIV BC, 85- *Concurrent Pos:* Scholar, Med Res Coun, Can, 78-83, scientist, 83- *Honors & Awards:* Ayerst Award, 86; Killam Res Prize, 88- *Res:* Polymorphic phase behavior of lipids in relation to membrane function; modelling of the biological membrane; drug delivery in liposomol systems. *Mailing Add:* Dept Biochem Univ BC 2146 Health Sci Mal Vancouver BC V6T 1W5 Can

CULLISON, ARTHUR EDISON, b Lawrence Co, Ill, Oct 30, 14; m 39; c 3. ANIMAL NUTRITION. *Educ:* Univ Ill, BS, 36, MS, 37, PhD(animal husb), 48. *Prof Exp:* Asst animal husb, Univ Ill, 36-39; asst prof, Miss State Col, 39-43; prof, Ala Polytech Inst, 46-48; head dept animal husb, Univ Ga, prof, 58-78; RETIRED. *Mem:* Fel AAAS; Am Soc Animal Sci. *Res:* Ruminant nutrition; silage preservation; by-products for livestock feeding; recycling animal wastes. *Mailing Add:* 142 S Stratford Dr Athens GA 30605

CULLISON, WILLIAM LESTER, b Baltimore, Md, Aug 26, 31; m 53; c 1. ENGINEERING, BUSINESS. *Educ:* US Merchant Marine Acad, BS, 53; LaSalle Exten Univ, LLB, 68; Fla Atlantic Univ, MBA, 75. *Prof Exp:* Engr officer, Am Trading & Prod Co, 56-57; coordr petrol res, Am Petrol Inst, 57-61, asst dir sci & technol, 61-67 & Air & Water Conserv, 68; dir tech opers pulp & paper res, 68-75, TREAS EXEC DIR, TECH ASN PULP & PAPER INDUST, 75- *Concurrent Pos:* Mem, Decimalized Inch Comt, Am Standards Asn, 60-64, Calibration of Instruments Comt, 61-64 & Dimensional Metrol Comt, 61-63; mem, Pulp & Paper Comt, US Standards Inst, 69; pres, Collab Testing Servs, Inc, 77- *Honors & Awards:* Key Award, Am Soc Asn Execs, 85. *Mem:* Am Soc Mech Engrs; Soc Res Adminr; Tech Asn Pulp & Paper Indust; Am Soc Asn Execs. *Res:* Administration of chemical, geologic, medical and engineering research projects; liaison between scientists and management. *Mailing Add:* Tech Asn Pulp & Paper Indust Technology Park PO Box 105113 Atlanta GA 30348

CULLUM, JANE KEHOE, b Norfolk, Va; c 2. APPLIED MATHEMATICS. *Educ:* Va Polytech Inst, BS, 60, MS, 62; Univ Calif, Berkeley, PhD(appl math), 66. *Prof Exp:* Tech asst to dir res, 75, mgr numerical methods & appl anal, Math & Sci Dept, T J Watson Res Ctr, 79-82, RES STAFF MEM MATH SCI, IBM RES, 66-, SR MGR, APPLIED MATH, MATH SCI, 82- *Concurrent Pos:* Mem Adv Panel Elec Sci & Anal Sect Eng Div, NSF, 75-76; mem, bd gov, IEEE Control Systs Soc, 78-80, 82-87. *Honors & Awards:* Lilian Moller Gilbreth, Soc Women Engrs, 59. *Mem:* Soc Indust & Appl Math (secy, 72-76, vpres, 80-83); Inst Elec & Electronics Engrs Control Systs Soc. *Res:* Design, analysis and development of algorithms for solving various kinds of scientific and engineering problems, especially problems involving large eigenvalue computations, optimization and/or data analysis. *Mailing Add:* Hunters Brook Condominiums 14-2 Ridgeview Lane Yorktown Heights NY 10598

CULLUM, MALFORD EUGENE, b Guymon, Okla, Apr 14, 51. RETINOIDS CHEMISTRY. *Educ:* Okla Panhandle State Univ, BS, 73; Univ Okla, MS, 77, PhD(biochem), 79. *Prof Exp:* Sr technician, Okla Med Res Found, 77-78; res assoc biochem, Iowa State Univ, 79-81; res assoc, 81-87, ASST PROF FOOD SCI & HUMAN NUTRIT, MICH STATE UNIV, 88- *Concurrent Pos:* First Award, Nat Cancer Inst, 87- *Mem:* NY Acad Sci; Sigma Xi; Am Inst Nutrit. *Res:* Function of vitamin A at the molecular level; control of cell proliferation by nutrients. *Mailing Add:* Dept Food Sci & Human Nutrit 334 Food Sci Bldg East Lansing MI 48824

CULP, ARCHIE W, b St Joseph, Mo, Jan 20, 31; m 57; c 3. MECHANICAL & NUCLEAR ENGINEERING. *Educ:* Mo Sch Mines, BS, 52, MS, 54; Univ Mo-Columbia, PhD, 70. *Prof Exp:* Nuclear engr, Gen Dynamics/Convair, Tex, 55-56; nuclear engr, Astra, Inc, Conn, 56-58, dir nuclear eng, 58-61; asst prof mech eng, 61-64, ASSOC PROF MECH ENG, UNIV MO-ROLLA, 64- *Concurrent Pos:* NSF fac fel, 67-68. *Mem:* Am Nuclear Soc; Am Soc Eng Educ. *Res:* Design of nuclear reactors; principles of nuclear engineering. *Mailing Add:* Dept Mech & Aerospace Eng Univ Mo Box 249 Rolla MO 65401

CULP, FREDERICK LYNN, b Duquesne, Pa, May 12, 27; m 53; c 2. PHYSICS. *Educ:* Carnegie Inst Technol, BS, 49, MS, 60; Vanderbilt Univ, PhD, 66. *Prof Exp:* Asst, US Steel Res Lab, 50-51; electronics engr, Pratt & Whitney Air Craft Corp, 52-54; res physicist, Stand Piezo Co, Pa, 54-55 & Westinghouse Lab, 55-56; asst elec & magnetism, Carnegie Inst Technol, 57-59, asst shaped charges & hyperballistics, 57-59; from asst prof to assoc prof, 59-65, prof physics & chmn dept, 65-86, EMER PROF PHYSICS, TENN TECHNOL UNIV, 86- *Mem:* AAAS; Am Asn Physics Teachers; Am Phys Soc; Sigma Xi. *Res:* Physics of fluids; electronic instrumentation; piezoelectricity; photoconductivity; shaped charges and hyperballistics; exploding wires; relaxation times in gases. *Mailing Add:* Dept Physics Tenn Technol Univ Cookeville TN 38501

CULP, LLOYD ANTHONY, b Elkhart, Ind, Dec 23, 42; m 65; c 2. VIROLOGY, CELL BIOLOGY. *Educ:* Case Inst Technol, BS, 64; Mass Inst Technol, PhD(biochem), 69. *Prof Exp:* Fel virol, Harvard Med Sch & Mass Gen Hosp, 69-71; from asst prof to assoc prof microbiol, 72-83, PROF MOLECULAR BIOL & MICROBIOL, SCH MED, CASE WESTERN RESERVE UNIV, 83-, PROF ONCOL & GEN MED SCI, 87- *Concurrent Pos:* Pinney cancer scholar, Case Western Reserve Univ, 73-75; Nat Cancer Inst career develop award, 74-79; mem, Cellular Physiol Study Sect, NIH, 78-82; vis scientist, Univ Calif, San Diego, 78-79. *Mem:* AAAS; Am Soc Microbiol; Sigma Xi; Am Soc Cell Biol; NY Acad Sci; Soc Complex Carbohydrates. *Res:* Study of the molecular mechanism of substratum adhesion of normal growth-controlled mammalian cells and possible alteration after transformation to a malignant state by oncogenic viruses or specific oncogenes; study of neuronal cell adhesion to extracellular matrices and requirements for growth cone extension. *Mailing Add:* Dept Molecular Biol & Microbiol Sch Med Case Western Reserve Univ Cleveland OH 44106

CULP, ROBERT D(UDLEY), b McAlester, Okla, Feb 28, 38; m 60; c 2. CELESTIAL MECHANICS, ASTRONAUTICS. *Educ:* Univ Okla, BS, 60; Univ Colo, Boulder, MS, 63, PhD(aerospace eng), 66. *Prof Exp:* Engr, Martin Co, Colo, 60-62; res assoc orbital mech, 66-68,from asst prof to assoc prof, 68-80, PROF AEROSPACE ENG SCI, UNIV COLO, BOULDER, 80-, CHMN, 90- *Concurrent Pos:* Assoc dir, Colo Ctr Astrodynamics Res, 85- *Mem:* Assoc fel Am Inst Aeronaut & Astronaut; fel Am Astronaut Soc; Am Soc Eng Educ; Sigma Xi. *Res:* Orbital mechanics; optimal transfer; optimization techniques; artificial space debris; atmospheric entry and satellite decay; collision hazards for orbiting objects. *Mailing Add:* Dept Aerospace Eng Sci Univ Colo Boulder CO 80309-0429

CULPEPPER, ROY M, b Tallassee, Ala, Aug 30, 47; m 76; c 3. MEDICINE. *Educ:* Univ Ala, MD, 73. *Prof Exp:* Asst prof med, Univ Tex, Houston, 80-85; ASST PROF MED, MED COL VA, 85- *Honors & Awards:* NIH First Award. *Mem:* Fel Am Col Physicians; Am Physiol Soc; Am Soc Nephrol; Am Fed Clin Res. *Res:* Regulation of salt transport in mammalian renal tubules. *Mailing Add:* Dept Nephrol Med Col Va Box 160 Richmond VA 23298

CULVAHOUSE, JACK WAYNE, b Mt Park, Okla, Sept 15, 29; m 52; c 3. MAGNETISM, PHASE TRANSITION. *Educ:* Univ Okla, BS, 51, AM, 54; Harvard Univ, PhD(physics), 58. *Prof Exp:* Physicist, Gen Elec Co, 51-53; asst prof physics, Univ Okla, 57-58; from asst prof to assoc prof, 58-64, PROF PHYSICS, UNIV KANS, 64- *Concurrent Pos:* Consult, Hycon Eastern Inc, 58-59; vis scientist, Ames Lab, 76, Sandia Nat Lab, 87; Guggenheim fel, 68-69. *Mem:* Fel Am Phys Soc. *Res:* Computer simulation of condensed matter systems; microwave properties of high Tc superconductors. *Mailing Add:* Dept Physics & Astron Univ Kans Lawrence KS 66044

CULVER, CHARLES GEORGE, b Bethlehem, Pa, Dec 4, 37; m 64; c 1. CIVIL & STRUCTURAL ENGINEERING. *Educ:* Lehigh Univ, BS, 59, MS, 61, PhD(civil eng), 65. *Prof Exp:* Engr, Sandia Corp, 61; instr, Lehigh Univ, 61-63; proj engr, US Navy Marine Eng Lab, 63-66; from asst prof to assoc prof civil eng, Carnegie-Mellon Univ, 66-72; res mgr, Nat Bur Standards, 72-80, dep dir, Ctr Bldg Technol, 80-83, chief, Struct Div, 83-88; DIR, OFF CONSTRUCT & ENG, OCCUP SAFETY & HEALTH ADMIN, 88- *Concurrent Pos:* Commerce sci fel, Exec Off of President-Off Mgt & Budget, 78-79. *Honors & Awards:* Civil Eng Res Prize, Am Soc Civil Engrs, 73; Silver Medal, Dept Commerce, 77. *Mem:* Am Soc Civil Engrs; Am Arbit Asn; Nat Acad Forensic Eng; Nat Soc Prof Engrs; Struct Stability Res Coun; Am Soc Safety Engrs. *Res:* Structural vibrations; structural stability; earthquake engineering; building research; fire research. *Mailing Add:* Off Construct & Eng Rm N3306 OSHA 200 Constitution Ave NW Washington DC 20210

CULVER, DAVID ALAN, b Oak Ridge, Tenn, Feb 14, 45; m 67; c 2. AQUATIC ECOLOGY, ZOOLOGY. *Educ:* Cornell Univ, AB, 67; Univ Wash, MS, 69, PhD(zool), 73. *Prof Exp:* Asst prof biol, Queen's Univ, Ont, 73-75; asst prof, 75-81, ASSOC PROF ZOOL, OHIO STATE UNIV, COLUMBUS, 81- *Concurrent Pos:* Res grant, Off Water Resources Technol, US Dept Interior, 76-83, US Environ Protection Agency, 77-78, Sea Grant, US Dept Com, 80-92, US Fish Wildlife Serv, 83-84, 87-92 & USDA, 90-92; vis scientist, Dept Zool, Univ Adelaide, Australia, 84-85; training grants, US Dept Educ, 86-87, 88-89, Howard Hughes Med Inst, 89-94. *Mem:* Am Soc Limnol & Oceanog; Ecol Soc Am; Int Asn Theoret & Appl Limnol; Int Asn Great Lakes Res; Am Fisheries Soc; Sigma Xi. *Res:* Plankton ecology: primary and secondary productivity, cyclomorphosis, eutrophication; meromictic lakes: plankton, chemical cycles, sedimentary history, stratification; ecology of Australian Larval fish; limnology of Laurentian Great Lakes; aquaculture of larval fishes. *Mailing Add:* Dept Zool 1735 Neil Ave Columbus OH 43210

CULVER, DAVID CLAIR, b Waverly, Iowa, Sept 23, 44; m 68. ECOLOGY. *Educ:* Grinnell Col, BA, 66; Yale Univ, PhD(biol), 70. *Prof Exp:* Fel pop biol, Univ Chicago, 70-71; from asst prof to prof biol, 71-81, PROF & CHMN ECOL & EVOLUTIONARY BIOL, NORTHWESTERN UNIV, EVANSTON, 81- *Concurrent Pos:* Vis assoc prof human ecol, Harvard Sch Pub Health, 75. *Mem:* Nat Speleol Soc; Ecol Soc Am. *Res:* Cave biology; control of species diversity; theoretical ecology. *Mailing Add:* Dept Biol Sci Northwestern Univ 633 Clark St Evanston IL 60201

CULVER, JAMES F, b Macon, Ga, June 10, 21; m 47. OPHTHALMOLOGY. *Educ:* Univ Ga, MD, 45; Am Bd Ophthal, dipl, 53. *Prof Exp:* Clin asst, Med Sch, Northwestern Univ, 52; pvt pract, 53-59; chief, Ophthal Br, US Air Force Sch Aerospace Med, 59-66, asst dir res & develop, Aerospace Med Div, Brooks AFB, Tex, 66-69, chief med res group, Off Surgeon Gen, 69-73, brig gen, 81, dep surgeon gen, OPS, US AIR FORCE; PVT PRACT OPHTHAL, 84- *Concurrent Pos:* Fel ophthal, Wesley Mem & Passavant Mem Hosps, Chicago, 51-52; mem, Consult Group & Med Debriefing Team, Projs Mercury & Gemini, 61-, Exec Coun, Armed Forces-Nat Res Coun Comt Vision, 62- & Vision Comt, Adv Group Aerospace Res & Develop, Aerospace Med Panel, NATO, 64-77. *Honors & Awards:* Tuttle Award, Aerospace Med Asn, 66. *Mem:* AMA; Aerospace Med Asn; Am Acad Ophthal & Otolaryngol. *Res:* Medicine and surgery; aerospace medicine and ophthalmology. *Mailing Add:* 381 Magnolia Place DeBary FL 32713

CULVER, RICHARD S, b Berkeley, Calif, July 26, 37; m 62; c 4. MATERIALS SCIENCE. *Educ:* Vanderbilt Univ, BE, 59; Stanford Univ, MSc, 60; Cambridge Univ, PhD(mech of mat), 64. *Prof Exp:* Asst prof mech & actg dir mining res lab, Colo Sch Mines, 64-68; lectr mech eng, Ahmadu Bello Univ, Nigeria, 68-70; res assoc explosives appln, Univ Calgary, 70-72; dean students, 74-79, assoc prof, 72-79, PROF BASIC ENG, COLO SCH MINES, 79- *Concurrent Pos:* Consult, Union Carbide Nuclear Co, Tenn, 61-63. *Mem:* Am Soc Mech Engrs; Am Soc Eng Educ; Nat Asn Student Personnel Adminrs; Sigma Xi. *Res:* Explosive metal forming; dynamics of impact; dynamic behavior of materials; engineering education. *Mailing Add:* 16 Schubert St Binghamton NY 13901

CULVER, ROGER BRUCE, b Brigham City, Utah, Sept 6, 40; div; c 3. ASTRONOMY. *Educ:* Univ Calif, Riverside, BA, 62; Ohio State Univ, MSc, 66, PhD(astron), 71. *Prof Exp:* Instr astron & math, 66-70; PROF ASTRON, COLO STATE UNIV, 70- *Concurrent Pos:* Mem user's comt, Kitt Peak Nat Observ, 74-76. *Mem:* Am Astron Int Astron Union; Astron Soc Pac; Sigma Xi. *Res:* Physical properties of cool stars having unusual chemical compositions. *Mailing Add:* Dept Physics Colo State Univ Ft Collins CO 80523

CULVER, WILLIAM HOWARD, b Eau Claire, Wis, Feb 17, 27; m 59; c 3. PHYSICS. *Educ:* Mass Inst Technol, BS, 50; Univ Calif, Los Angeles, MS, 54, PhD, 61. *Prof Exp:* Res asst, Mass Inst Technol, 50-52 & Scripps Inst Oceanog, Univ Calif, 52-54; physicist, Rand Corp, 55-61 & Inst Defense Anal, 61-66; mgr, Quantum Electronics Dept, IBM Corp, 66-72, corp laser strategist, 67-69; CHMN BD, OPTELECOM INC, GAITHERSBURG, MD, 72- *Concurrent Pos:* Mem, Spec Group Optical Masers, Dept Defense, 62-66; mem interdept comt atmospheric sci, Dept Com, 64-65; mem, Nat Acad Sci adv comt to Nat Bur Standards Cent Radio Propagation Lab, 65; mem, NASA Res & Technol Adv Comt on Commun & Tracking, 67-69. *Mem:* Optical Soc Am; Am Phys Soc; Inst Elec & Electronics Engrs. *Res:* Quantum electronics; optical and microwave spectroscopy; applications of lasers to communication systems. *Mailing Add:* 15930 Luanne Dr Gaithersburg MD 20877

CUMBERBATCH, ELLIS, b Eng, Apr 19, 34; m 57; c 3. APPLIED MATHEMATICS. *Educ:* Univ Manchester, BSc, 55, PhD(appl math), 58. *Prof Exp:* Res fel appl mech, Calif Inst Technol, 58-60; res assoc, Courant Inst Math Sci, NY Univ, 60-61; lectr, Univ Leeds, 61-64; from assoc prof to prof math, Purdue Univ, West Lafayette, 64-81; PROF MATH, CLAREMONT GRAD SCH, 81- *Res:* Fluid dynamics; mathematics. *Mailing Add:* Dept Math Claremont Grad Sch Claremont CA 91711

CUMBERLAND, WILLIAM GLEN, b Can, Feb 7, 48; m 74; c 4. SAMPLING THEORY, DEMOGRAPHY. *Educ:* McGill Univ, BSc, 68; Johns Hopkins Univ, MA, 72, PhD(statist), 75. *Prof Exp:* Fel pop dynamics, Johns Hopkins Univ, 75-76; ASST PROF BIOSTATIST, UNIV CALIF, LOS ANGELES, 76- , ASSOC PROF. *Mem:* Am Statist Asn; Biometric Soc; Inst Math Statist. *Res:* Variance estimation in finite population sampling; stochastic models for population growth. *Mailing Add:* Div Biostatist Univ Calif Sch Pub Health Los Angeles CA 90024

CUMBIE, BILLY GLENN, b Dickens, Tex, Mar 21, 30; m 51; c 3. BOTANY. *Educ:* Tex Tech Col, BS, 51, MS, 52; Univ Tex, PhD(bot), 60. *Prof Exp:* Spec instr bot, Univ Tex, 57-58; from instr to asst prof, Tex Tech Col, 58-61; asst prof bot, 61-67, PROF BIOL SCI, UNIV MO-COLUMBIA, 71- *Mem:* Bot Soc Am; Int Asn Wood Anat. *Res:* Development of the vascular cambium and xylem in dicotyledons. *Mailing Add:* Div Biol Sci Univ Mo Columbia MO 65211

CUMMEROW, ROBERT LEGGETT, b Toledo, Ohio, Jan 7, 15; m 48; c 2. PHYSICS. *Educ:* Univ Toledo, BE, 38; Univ Pittsburgh, MS, 40, PhD(physics), 47. *Prof Exp:* Asst physics, Univ Pittsburgh, 38-42; spec res assoc underwater sound lab, Harvard Univ, 42-45; mem staff, US Navy Underwater Sound Lab, Conn, 45-46; res assoc, Knolls Atomic Power Lab, Gen Elec Co, 48-55; group leader solid state physics, Nat Carbon Res Labs, 55-63; group leader, Union Carbide Res Inst, 63-70, SR SCIENTIST, TARRYTOWN TECH CTR, UNION CARBIDE CORP, 70- *Mem:* Am Phys Soc; Sigma Xi. *Res:* Elastic and damping constants of metals and alloys; underwater sound detection equipment; paramagnetic resonance absorption in salts of the iron group; radiation-induced property changes in semiconductors and metals; photo-effects in semiconductors; plastic behavior of semiconductors and refractory materials; superconductivity. *Mailing Add:* 24 Lochwood Dr Clinton CT 06413

CUMMIN, ALFRED SAMUEL, b London, Eng, Sept 5, 24; nat US; m 45; c 1. PHYSICAL CHEMISTRY. *Educ:* Polytech Inst Brooklyn, BS, 43, PhD, 46; Univ Buffalo, MBA, 59. *Prof Exp:* Res chemist, Substitute Alloy Mats Lab, Manhattan Proj, Columbia Univ, 43-44; plant supvr & head res, Metal & Plastic Processing Co, 46-51; res chemist, Gen Chem Div, Allied Chem & Dye Corp, 51-53; sr chemist, Congoleum Nairn, 53-54; prof math & sci, US Merchant Marine Acad, 54; capacitor div, Gen Elec Co, 54-56, supvr dielectric adv develop, 54-56; mgr, Indust Prof Res Dept, Spencer Kellog & Sons, Inc, 56-59; mgr, Plastics Div, Trancoa Chem Corp, 59-62; assoc dir, Prod Develop & Serv Labs, Chem Div, Merck & Co, Inc, NJ, 62-69; dir prod develop, Chem Div, Borden Co, 69-72, tech dir, Borden Chem Div, Borden Inc, NY, 72-73, corp tech dir, 73-78, vpres prod safety & qual, 78-80, vpres sci & technol, 81-87, SR VPRES SCI & TECHNOL, BORDEN INC, 87- *Concurrent Pos:* Instr, Polytech Inst Brooklyn, 46-47; asst prof, Adelphi Col, 52-54; adj prof mgt, Mgt Inst, NY Univ, 68- *Mem:* Am Chem Soc; Inst Food Technol; Fedn Coatings Technol; Am Soc Test Mat; Nutrit Found, Int Life Sci Inst. *Res:* Polymers; electrochemistry; food packaging; colloid chemistry; preservatives; agricultural chemicals; dielectrics and insulating materials; paints; adhesives; surgical adhesives; nutrition; industrial hygiene; occupational health. *Mailing Add:* Borden Inc 960 Kingsmill Pkwy Columbus OH 43229

CUMMING, BRUCE GORDON, b London, Eng, Oct 12, 25; Can citizen; m 69. PLANT PHYSIOLOGY. *Educ:* Univ Reading, BS, 52; McGill Univ, PhD(plant physiol, agron), 56. *Prof Exp:* Res off, Plant Res Inst, Can Dept Agr, 57-65; prof bot, Western Ont Univ, 65-71; chmn dept, 71-74, PROF BIOL, UNIV NB, 71- *Concurrent Pos:* Del gen assembly, Int Union Biol Sci, 70 & 73. *Mem:* Can Soc Plant Physiol (pres, 69-70). *Res:* Physiology; photobiology; photoperiodism; endogenous rhythms and phytochrome, particularly in germination and flowering; tissue culture and morphogenesis. *Mailing Add:* Dept Biol Univ NB Col Hill Box 4400 Fredericton NB E3B 5A3 Can

CUMMING, GEORGE LESLIE, b Saskatoon, Saskatchewan, Feb 11, 30; m 54; c 3. GEOCHRONOLOGY, SEISMOLOGY. *Educ:* Univ Saskatchewan, BA, 51, MA, 53; Univ Toronto, PhD(geophysics), 55. *Prof Exp:* Geophysicist, Gulf Oil Co, 55-59; PROF GEOPHYSICS, UNIV ALTA, 59- *Concurrent Pos:* Chmn, physics dept, Univ Alta, 84-88. *Mem:* Am Geophys Union; Geol Soc Can; Can Geophys Union. *Res:* Geochronology with emphasis on lead isotope variations; crustal seismic exploration using reflection and refraction. *Mailing Add:* Dept Physics Univ Alta Edmonton AB T6G 2J1 Can

CUMMING, JAMES B, b Jamaica, NY, June 6, 28; m 53; c 3. NUCLEAR CHEMISTRY. *Educ:* Yale Univ, BS, 49; Columbia Univ, MA, 51, PhD(nuclear chem), 54. *Prof Exp:* Res assoc, 54-55, from assoc chemist to chemist, 55-69, SR CHEMIST, BROOKHAVEN NAT LAB, 69- *Concurrent Pos:* Vis physicist, Fermilab, 81-83. *Mem:* Am Phys Soc; Am Chem Soc; Sigma Xi. *Res:* Nuclear reactions induced by energetic light-ions and heavy-ions; nuclear decay schemes. *Mailing Add:* Chem Dept Bldg 555 Brookhaven Nat Lab Upton NY 11973

CUMMING, LESLIE MERRILL, b Joggins, NS, Nov 5, 25; m 58; c 3. GEOLOGY. *Educ:* Univ NB, BSc, 48, MSc, 51; Univ Wis, PhD(geol), 55. *Prof Exp:* Asst to prov geologist, NB Dept Mines, 44-46; field asst, 47-49, tech officer, 49-54, geologist, 55-67, RES SCIENTIST, GEOL SURV CAN, 67- *Mem:* Am Paleont Soc; fel Geol Asn Can; fel Royal Can Geog Soc; Brit Palaeontolograph Soc; Can Inst Mining & Metall. *Res:* Regional Paleozoic geology of the Appalachians; regional geology of the Hudson Bay lowlands and Arctic Canada; Paleozoic faunas; graptolite morphology. *Mailing Add:* Geol Surv Can 601 Booth St Ottawa ON K1A 0E8 Can

CUMMINGS, CHARLES ARNOLD, b Cincinnati, Ohio, Sept 23, 30; m 57; c 3. NEURAL NETWORKS-ARTIFICIAL INTELLIGENCE PRODUCT APPLICATION, INTERACTIVE TELEVISION. *Educ:* Univ Cincinnati, BS, 75. *Prof Exp:* Res engr advan electronic systs, NAm-Rockwell, 59-68; SR STAFF DEDICATED COMPUTER, ARTIFICIAL INTEL & NEVAAL NETWORKS, DIV HASBRO TOY CO, KENNER PROD CO, 68- *Mem:* Nat Soc Prof Engrs; Inst Elec & Electronic Engrs; Am Asn Artificial Intel; AAAS. *Res:* Granted several patents. *Mailing Add:* Kenner Prod Div Hasbro Corp 1014 Vine St Cincinnati OH 45202

CUMMINGS, DAVID, b New York, NY, Feb 11, 32. GEOLOGY, GEOPHYSICS. *Educ:* City Col New York, BS, 57; Univ Tenn, MS, 59; Mich State Univ, PhD(geol), 62. *Prof Exp:* Mineralogist & petrologist, Electrotech Res Lab, US Bur Mines, 59; geologist, US Geol Surv, 62-68; PROF GEOL, OCCIDENTAL COL, 68- *Concurrent Pos:* Consult, US Geol Surv, 68-75; independent consult, 70-; pres, Ryland-Cummings, Inc, 79- *Mem:* Am Geophys Union; Asn Eng Geologists; Geol Soc Am; Int Asn Eng Geologists; Am Asn Petrol Geologists. *Res:* Theoretical and applied mechanics used to solve structural geologic problems, engineering geologic problems; applied geophysics. *Mailing Add:* SAIC No 255 14062 Denver West Pkwy Golden CO 80401

CUMMINGS, DENNIS PAUL, b Yonkers, NY, Apr 19, 40. MICROBIOLOGY. *Educ:* Manhattan Col, BS, 61; St John's Univ, NY, MS, 63, PhD(microbiol), 68. *Prof Exp:* Asst microbiol, St John's Univ, NY, 62-63; res microbiologist, 68-72, sr res scientist, 72-78, SUPVR MICROBIOL DEVELOP LAB, MILES LABS, INC, 78- *Mem:* Am Soc Microbiol; Soc Indust Microbiol. *Res:* Medical microbiology; antifungal and antibacterial agents; analytical microbiology; microbiological aspects of pharmaceutical development. *Mailing Add:* 444 Painter Dr West Haven CT 06516

CUMMINGS, DONALD JOSEPH, b Staten Island, NY, Mar 4, 30; m 58; c 3. BIOPHYSICS. *Educ:* George Washington Univ, BS, 55; Univ Chicago, MS, 57, PhD(biophys), 59. *Prof Exp:* Biophysicist, NIH, 53-55; USPHS fel, Copenhagen Univ, 59-60; res physicist, Nat Inst Neurol Dis & Blindness, 60-64; from asst prof to assoc prof, 64-71, PROF MICROBIOL, MED SCH,

UNIV COLO, DENVER, 71- *Concurrent Pos:* Josiah Macy Jr fac scholar, 77-78; Fogarty int fel, 84-85. *Res:* Effect of canavanine on head morphogenesis in T-even bacteriophages of Entameba coli; replication and function of mitochondria from Paramecium and Podospora; biochemistry and morphology of aging in Paramecium and Podospora. *Mailing Add:* Dept Microbiol Univ Colo Med Sch 4200 E Ninth Ave Denver CO 80262

CUMMINGS, EDMUND GEORGE, b Albany, NY, Aug 2, 28; m 55; c 2. PHYSIOLOGY. *Educ:* Union Col, BS, 50; NC State Col, MS, 53, PhD(ecol), 55. *Prof Exp:* Asst zool, NC State Col, 51-55, instr, Exten Sch, 54; instr, Duke Univ, 55-56; physiologist & chief respiratory sect, Edgewood Aresenal, 56-78, physiologist, chem systs lab, 78-80, chief, Chem-Physiol Toxicol, Chem Systs Lab Res Div, 80-87; CHIEF, BIOTECHNOLOGY DIV, CRDEC, ABERDEEN PROVING GROUND, 87- *Concurrent Pos:* From asst prof to assoc prof, Hartford Col, 59-69; asst prof, Univ Md Exten Sch, 64- *Mem:* AAAS; Am Soc Zool; Am Physiol Soc; Sigma Xi. *Res:* Culture of slime molds; food habits of game birds; osmoregulation in fish; respiratory and exercise physiology; thermoregulation and anticholinergics; environment and skin penetration; evaluation of respiratory protective devices; toxicity of anticholinesterases. *Mailing Add:* 205 E Heather Rd Bel Air MD 21014

CUMMINGS, FRANK EDSON, b Berkeley, Calif, Feb 19, 40; m 67; c 2. PHYSICAL CHEMISTRY. *Educ:* Harvey Mudd Col, BS, 62; Harvard Univ, PhD(chem), 72. *Prof Exp:* From instr to assoc prof, 67-81, PROF CHEM, ATLANTA UNIV, 81-, CHMN DEPT, 79- *Concurrent Pos:* Chief investr, NIH grant, 72-76 & NSF grant, 76-77; proj dir, Dept Energy grant, 78-81. *Mem:* Am Phys Soc; Am Chem Soc. *Res:* Long range forces between atoms. *Mailing Add:* Dept Chem Atlanta Univ Atlanta GA 30314-4385

CUMMINGS, FREDERICK W, b New Orleans, La, Nov 21, 31; c 1. THEORETICAL PHYSICS, MATHEMATICS. *Educ:* La State Univ, BS, 55; Stanford Univ, PhD(physics), 60. *Prof Exp:* Res scientist theoret physics, Aeronutronic Div, Philco Corp, Calif, 60-63; from asst prof to assoc prof, 63-74, PROF PHYSICS, UNIV CALIF, RIVERSIDE, 74- *Concurrent Pos:* ed, Coop Phenomena. *Mem:* AAAS; Am Phys Soc; Am Sci Teachers Asn. *Res:* Coherence in radiation; solid state; many particles; biophysics; theoretical physics. *Mailing Add:* Dept Physics Univ Calif 900 University Ave Riverside CA 92521

CUMMINGS, GEORGE AUGUST, b Cortland, Ind, Dec 17, 27; m 53; c 3. SOIL SCIENCE. *Educ:* Purdue Univ, BS, 51, MS, 57, PhD(agron), 61. *Prof Exp:* Teacher high sch, Ind, 51-58; from asst prof to assoc prof, 61-78, PROF SOIL SCI, NC STATE UNIV, 78- *Mem:* Am Soc Hort Sci; Am Soc Agron; Soil Sci Soc Am. *Res:* Determination of effects of plant nutrition upon biochemical constituents of plants; influence of nutrition upon yield and quality of fruits. *Mailing Add:* Dept Soil Sci NC State Univ Raleigh NC 27650

CUMMINGS, GEORGE H(ERBERT), b Cape May, NJ, Dec 28, 13; m 39; c 4. CHEMICAL ENGINEERING. *Educ:* Pa State Univ, BS, 35, PhD(chem eng), 41; Mass Inst Technol, SM, 36. *Prof Exp:* Asst chem, Pa State Univ, 36-39, instr chem eng, 39-41; chem engr, Houston Plant, Rohm and Haas Co, 41-51, chief chem engr, 51-76; RETIRED. *Mem:* Am Chem Soc; Am Inst Chem Engrs. *Res:* Solvent extraction; distillation; petroleum refining; chemical engineering process design. *Mailing Add:* 5108 Doe Valley Lane Austin TX 78759-7128

CUMMINGS, JOHN ALBERT, b Evanston, Ill, May 3, 31; m 51; c 2. RADIOBIOLOGY, MARINE BIOLOGY. *Educ:* Wis State Univ-Whitewater, BS, 53; Univ Wis, MS, 59; Univ Northern Colo, EdD, 66. *Prof Exp:* Teacher high sch, Wis, 53-61; prof biol, Univ Wis-Whitewater, 61-90; CONSULT, 90- *Concurrent Pos:* Dir, NSF Inserv Inst Molecular Biol, 64-65 & 68-69, dir, NSF Inst Environ Sci, 71-72; AEC grant, Nuclear Sci Instrumentation for Radiation Biol Lab, Univ Wis-Whitewater, 64-65; acad dir, Univ Wis Syst, Pigeon Lake Field Sta, 76-80. *Mem:* Nat Asn Biol Teachers; Am Inst Biol Sci. *Res:* Concentration of various radioisotopes in Orconectes virilis in various intermolt stages; radioisotope accumulation by radioautographic methods in crustaceans and reptiles; crayfish population of Wisconsin. *Mailing Add:* 1264 Satinwood Lane Whitewater WI 53190

CUMMINGS, JOHN FRANCIS, b Newark, NJ, Sept 3, 36; m 61; c 5. VETERINARY ANATOMY, COMPARATIVE NEUROLOGY. *Educ:* Cornell Univ, BS, 58, DVM, 62, MS, 63, PhD(vet anat), 66. *Prof Exp:* Asst vet anat, 63-65, from asst prof to assoc prof,67-77, PROF ANAT, NY STATE COL VET MED, CORNELL UNIV, 77- *Concurrent Pos:* Consult, Div Neuropsychiat, Walter Reed Army Inst Res, 75-84. *Res:* Neuroanatomy and neuropathology. *Mailing Add:* Dept Anat NY State Col Vet Med Cornell Univ Ithaca NY 14853-6401

CUMMINGS, JOHN PATRICK, b Westfield, Mass, June 28, 33; c 7. ECOLOGY, PHYSICAL CHEMISTRY. *Educ:* St Michael's Col, BS(philos) & BS(chem), 55; Univ Tex, Austin, PhD(chem), 69; Univ Toledo, JD, 73, MSCE, 77. *Prof Exp:* Instr, Univ Tex, 63-66; dist mgr, ITT Hamilton Mgt Corp, 62-68; sr chemist, Owens-Ill Inc, 68-70, ecologist, 71-76, prod safety mgr, 76-78, legal coun, 78-79, mgr legis compliance,; consult, 84-88; VPRES, SCS ENGRS, 88- *Concurrent Pos:* Consult, Environ Protection Agency, Glass Packaging Inst, Nat Ctr Resource Recovery, 71-76. *Mem:* Sigma Xi; Royal Soc Chem; Am Ceramic Soc; Am Chem Soc; Am Crystallog Asn. *Res:* Ecology, physical chemistry, international law, resource recovery and legislative aspects. *Mailing Add:* 843 Barcelona Dr Fremont CA 94536

CUMMINGS, JOHN RHODES, b Detroit, Mich, Feb 4, 26; m 53; c 2. PHARMACOLOGY. *Educ:* Kalamazoo Col, BA, 50; Wayne State Univ, MS, 52, PhD(pharmacol), 54. *Prof Exp:* Asst prof pharmacol, Med Sch, Tufts Univ, 54-57; group leader cardiovasc pharmacol, Lederle Labs Div, Am Cyanamid Co, 57-69, head, Dept Cardiovasc-Renal Pharmacol, 69-73; dir, Dept Pharmacol, Ayerst Res Labs Div, Am Home Prod Corp, 73-91; RETIRED. *Mem:* Am Soc Pharmacol & Exp Therapeut; Soc Exp Biol & Med; Am Heart Asn; Am Chem Soc. *Res:* Cardiovascular pharmacology, particularly in the fields of arrhythmia, hypertension and diuretic research. *Mailing Add:* 65 St Paul St W No 409 Montreal PQ H2Y 1Z1 Can

CUMMINGS, LARRY JEAN, b Chicago, Ill, Oct 1, 37; m 63; c 2. ALGEBRA. *Educ:* Roosevelt Univ, BS, 61; DePaul Univ, MS, 63; Univ BC, PhD(math), 67. *Prof Exp:* Teaching asst math, Univ BC, 63-67; ASST PROF MATH, UNIV WATERLOO, 67-, AT DEPT COMBINATORICS. *Res:* Multilinear algebra; generalized matrix functions and matrix inequalities. *Mailing Add:* Dept Combinatorics Univ Waterloo Waterloo ON N2L 3G1 Can

CUMMINGS, MARTIN MARC, b Camden, NJ, Sept 7, 20; m 42; c 3. MEDICINE. *Educ:* Bucknell Univ, BS, 41; Duke Univ, MD, 44. *Hon Degrees:* DSc, Bucknell Univ, 68 & Duke Univ, 85; ScD, Univ Nebr, Emory Univ & Georgetown Univ, 71; MD, Karolinska Inst, Sweden, 72; hon Dr, Acad Med, Lodz, Poland, 77. *Prof Exp:* Med intern, Boston Marine Hosp, USPHS, 44, asst resident med, 45, med officer, Tuberc Div, 46-47, dir, Tuberc Eval Lab, Commun Dis Ctr, 47-49; dir, Tuberc Res Lab, Vet Admin Hosp, Atlanta, Ga, 49-53; dir res serv, US Vet Admin, 53-59; prof microbiol & chief dept, Univ Okla, 59-61; chief off int res, NIH, 61-63, assoc dir res grant, 63; dir, 64-84, EMER DIR, NAT LIBR MED, USPHS, 84- *Concurrent Pos:* Asst prof & assoc prof, Sch Med, Emory Univ, 51-53; prof lectr, Sch Med, George Washington Univ, 53-54; distinguished prof, Georgetown Univ, 85. *Honors & Awards:* John C Leonard Award, Asn Hosp Med Educ, 79; Abraham Horowitz Award, Pan Am Health Orgn, 83. *Mem:* Nat Inst Med-Nat Acad Sci; Am Soc Clin Invest; Am Clin & Climat Asn; Am Acad Microbiol; Am Col Med Informatics; Royal Col Physicians; fel Med Libr Asn. *Res:* Laboratory diagnosis of tuberculosis and experimental methods; epidemiology of sarcoidosis; library and information science. *Mailing Add:* NIH Med Libr 8600 Rockville Pike Bethesda MD 20814

CUMMINGS, MICHAEL R, b Chicago, Ill, July 7, 41; m 66; c 2. DEVELOPMENTAL GENETICS. *Educ:* St Mary's Col, Minn, BA, 63; Northwestern Univ, MS, 65, PhD(biol), 68. *Prof Exp:* Instr genetics, Northwestern Univ, 68-69; asst prof genetics, 69-72, ASSOC PROF BIOL SCI, UNIV ILL, CHICAGO, 72- *Concurrent Pos:* Res assoc prof, Inst Study Develop Disabilities, Univ Ill, Chicago, 78-; Assoc prof genetics, Univ Ill, Chicago, 79- *Mem:* Soc Develop Biol; Am Soc Cell Biol; Gen Soc Am; Am Soc Human Genetics. *Res:* Human genome organization; developmental genetics; Down Syndrome. *Mailing Add:* Dept Biol Sci Univ Ill Chicago IL 60680

CUMMINGS, NANCY BOUCOT, b Philadelphia, Pa, Feb 21, 27; div; c 3. NEPHROLOGY, INTERNAL MEDICINE. *Educ:* Oberlin Col, BA, 47; Univ Pa, MD, 51. *Prof Exp:* Rotating intern, Pa Hosp, 51-52; resident internal med, Univ Pa Hosp, 52-54; res & clin asst med, Royal Hosp St Bartholomew, London, 54-55; res & clin asst med, Manchester Royal Infirmary, Eng, 55; asst med, Peter Bent Brigham Hosp, 55-58; guest worker, Nat Inst Arthritis '& Metab Dis, 59-62; res med officer, Walter Reed Army Inst Res, 62-66; res med officer, Div Exp Med, Naval Med Res Inst & consult nephrol, Dept Med, Naval Hosp, Md, 66-72; prog officer, 72-73, spec asst to dir renal & urol dir, 73-74, actg assoc dir renal & urol dis, 74-76, assoc dir kidney, urol & hemat dis, 76-84, ASSOC DIR RES & ASSESSMENT, NAT INST ARTHRITIS, DIABETES, DIGESTIVE & KIDNEY DIS, NIH, 84- *Concurrent Pos:* Res fel med, Harvard Med Sch, 55-58, res fel biol chem, 58-59; Am Heart Asn fel, 55-57, advan res fel, 57-62; Nat Found res fel, Royal Hosp St Bartholomew & Manchester Royal Infirmary, 54-55; mem, Res Comt, Washington Heart Asn, 69-78, chmn res comt, Am Heart Asn, Nat Capitol Area, 76-78. & med adv bd, Washington Chap, Nat Kidney Found, 69-; from clin instr med to clin asst prof, Georgetown Univ, 60-81, clin assoc prof, 81-88; clin prof med, 88-, co-chmn, Adv Comt Epidemiol & Statist Kidney Dis, Nat Kidney Found, mem, Sci Adv Bd & trustee at large; consult, Coord Comt Coun Urol; mem, Res Comt, Am Urol Asn; chmn, Comm Med Ethics, Episcopal Diocese Wash. *Honors & Awards:* Robert Abbe lectr, Col Physicians Philadelphia, 75; Distinguished Serv Award, Nat Kidney Found; Jacob Ehrenzeller Award, 86. *Mem:* AAAS; Am Soc Hemat; Am Soc Nephrology; Cosmos Club; Am Fedn Clin Res; Am Soc Pediat Nephrol; Am Soc Transplant Physicians. *Res:* Epidemiology and statistics of renal disease; biochemistry of uremia; renal physiology and pathophysiology; ethical issues in end-stage renal disease; biomedical ethics. *Mailing Add:* Nat Inst Diabetes & Digestive & Kidney Dis 9000 Rockville Pike NIH Bldg 31 Rm 9A18 Bethesda MD 20892

CUMMINGS, NORMAN ALLEN, b New York, NY, Mar 26, 35; m 60; c 2. INTERNAL MEDICINE. *Educ:* NY Univ, AB, 55; State Univ NY, MD, 59; Am Bd Internal Med, dipl, 72. *Prof Exp:* Intern med, Jewish Hosp, Brooklyn, 59-60, resident, 61; resident med, Med Sch, Univ Mich, 61-62; res assoc protein chem, Nat Cancer Inst, 64-66; asst prof internal med, Col Med, Baylor Univ, 66-67; med officer & head connective tissue dis prog, Oral Med & Surg Br, Nat Inst Dent Res, 67-74; chief, Clin Immunol & Connective Tissue Dis Div & dir, 74-78, from assoc prof to prof med, 74-83, CLIN PROF MED, UNIV LOUISVILLE SCH MED, 83- *Concurrent Pos:* Vis fel rheumatic dis, NY Univ-Bellevue Hosp Med Ctr, 62; USPHS physician trainee grant rheumatic dis, 62-64; fel biophys & arthritis, Med Sch, Univ Mich, 62-64; Kayser Found sci grant, 66-67; asst attend physician, Ben Taub Gen Hosp & Vet Admin Hosp, Houston, 66-67; attend physician, Louisville Gen Hosp; consult, Louisville Vet Admin Hosp, Jewish Hosp & Nat Inst Dent Res, NIH, 67-74; consult, Norton's Children's Hosp & Louisville Baptist Hosp. *Mem:* Am Rheumatism Asn. *Res:* Protein-mucopolysaccharide interactions; metallo-proteins of serum; joint pH; protein solubility and conformation; cryoprecipitation of cryoglobulins; oral-mucosal manifestations of connective tissue diseases; immunochemistry and cellular immunology in rheumatic diseases. *Mailing Add:* 801 Barret Ave Suite 224 Jewish Hosp Prospect KY 40204

CUMMINGS, PETER THOMAS, b Wingham, New South Wales, Australia, Feb 10, 54; m 75; c 1. STATISTICAL THERMODYNAMICS, CHEMICAL PROCESS DESIGN. *Educ:* Univ Newcastle, New South Wales, Australia, BMath, 76; Univ Melbourne, Victoria, Australia, PhD(math), 80. *Prof Exp:* Researcher physics, Univ Guelph, Ont, 80-81; res assoc chem & mech eng, State Univ NY, Stony Brook, 81-83; from asst prof to assoc prof chem eng,

83-91, PROF CHEM ENG, UNIV VA, 91- *Concurrent Pos:* Vis fel, Res Sch Chem, Australian Nat Univ, Canberra, 87 & 90; distinguished vis scholar, Dept Chem Eng, Univ Mass, 90; fac res partic, Oak Ridge Nat Lab, 90. *Mem:* Am Inst Chem Engrs; Am Phys Soc; Am Chem Soc; Sigma Xi. *Res:* Statistical mechanics of dense molecular fluids, especially integral equations and molecular simulation; computer aided chemical process optimization and design; mathematical modeling of bacterial transport. *Mailing Add:* Dept Chem Eng Univ Va Charlottesville VA 22903-2442

CUMMINGS, RALPH WALDO, JR, b Ithaca, NY, July 20, 38; m 61; c 2. AGRICULTURAL ECONOMICS, AGRICULTURAL RESEARCH PLANNING. *Educ:* Univ NC, AB, 60; Univ Mich, PhD(econ), 65. *Prof Exp:* Asst prof econ, Univ Ill, 65-70; adv agr econ, Harvard Adv Group & Nat Develop Planning Agency Indonesia, 70-72; consult, Bur Near East & S Asia, AID, 70, agr economist, 72- 76, prog officer, Int Agr Develop Serv, 76-80, agr economist, Rockefeller Found, 72-82, AGR ECONOMIST, AID, 82-, COORDR, IARC STAFF, 91- *Concurrent Pos:* Asst dir, Midwestern Univs Consortium Int Activ, 66-67; chief agr econ div, Off Agr Develop, AID, India, 67-69. *Mem:* Am Agr Econ Asn. *Res:* Role of technology in agricultural development. *Mailing Add:* 1880 Massachusettes Ave McLean VA 22101

CUMMINGS, SUE CAROL, b Dayton, Ohio, Apr 24, 41; m 79. INORGANIC CHEMISTRY, BIOINORGANIC CHEMISTRY. *Educ:* Northwestern Univ, BA, 63; Ohio State Univ, MSc, 65, PhD(inorg chem), 68. *Prof Exp:* Vis res assoc inorg chem, Aerospace Res Labs, Wright-Patterson AFB, 68-69; from asst prof to assoc prof, 69-77, PROF CHEM, WRIGHT STATE UNIV, 77-, CHMN DEPT, 88- *Concurrent Pos:* Petrol Res Fund grant, 69-72; Cottrell res grant, 73; Nat Heart & Lung Inst res grant, 73-81; vis scientist, C F Kettering Res Labs, Ohio, 81-82; NSF grad fel panel, 89-91. *Mem:* AAAS; Am Chem Soc. *Res:* Coordination chemistry; synthesis, characterization, stereochemistry and reactions of metal complexes containing multidentate, macrocyclic or cage-type ligands; synthetic oxygen carriers; study of metal complexes as models for biologically important molecules; poly-nuclear metal complexes. *Mailing Add:* Dept Chem Wright State Univ Dayton OH 45435

CUMMINGS, THOMAS FULTON, b Taxila, India, Oct 25, 25; US citizen; m 48; c 4. PHYSICAL CHEMISTRY, ANALYTICAL CHEMISTRY. *Educ:* Mass Inst Technol, BSc, 47; Case Inst Technol, MSc, 52, PhD(chem), 55. *Prof Exp:* Chem engr, Res Ctr, B F Goodrich Co, 48-49; asst chem, Case Inst Technol, 49-52; instr, Westminster Col, 52-55; from asst prof to assoc prof, 55-67, PROF CHEM, BRADLEY UNIV, 67- *Concurrent Pos:* Vis prof chem, Anal Inst, Univ Vienna, 66-67, 87-88 & Univ Birmingham, Eng, 73-74 & 80-81; consult chem patent infringement; prin investr, 4 NSF grants & 2 Water Resources grants. *Mem:* Am Chem Soc; Am Sci Affiliation; Soc Appl Spectros; Sigma Xi. *Res:* Kinetics; instrumental analysis; water and air pollution analysis, government support. *Mailing Add:* Dept of Chem Bradley Univ Peoria IL 61625

CUMMINGS, WILLIAM CHARLES, b Boston, Mass, Apr 6, 32; m 55; c 2. UNDERWATER ACOUSTICS, MARINE BIOLOGY. *Educ:* Bates Col, BS, 54; Univ Miami, MS, 60, PhD(marine sci), 68. *Prof Exp:* Biol oceanogr, Univ RI, 58-60; instr bioacoustics, Univ Miami, 60-65; oceanogr, Naval Ocean Systs Ctr, 65-70, head, Underwater Bioacoustics Br, 70-77; chief scientist, San Diego Nat Hist Mus, 77-79; DIR, OCEANOGRAPHICS CONSULT, 79- *Concurrent Pos:* Consult, 63-; prof writer & ed, 68- *Mem:* Fel Acoust Soc Am; Oceanic Soc; Am Cetacean Soc; Sigma Xi. *Res:* Underwater acoustics; bioacoustics; research methodology; environmental science; marine mammalogy and polar research. *Mailing Add:* 5948 Eton Ct San Diego CA 92122

CUMMINS, ALVIN J, b Wheeling, WVa, Apr 26, 19; m 47; c 3. INTERNAL MEDICINE, GASTROENTEROLOGY. *Educ:* Georgetown Univ, BS, 41; Johns Hopkins Univ, MD, 44. *Prof Exp:* Asst instr med, Med Sch, Univ Pa, 51-52, instr, 52-53, assoc, 53-54, from asst prof to assoc prof, 57-63, prof med & chief sect gastroenterol, 63-71; RETIRED. *Concurrent Pos:* Fel med, Cornell Med Ctr, 50-51; hon consult, Blytheville AFB, Ark, 61; chmn, Gastroenterol Res Group, 61-62; consult, Vet Admin Hosp, Memphis, 62-71; pvt pract gastroenterol; clin prof med, Univ Tenn, 71- *Mem:* AMA; Am Col Physicians; Am Gastroenterol Asn; Am Fedn Clin Res; Sigma Xi. *Res:* Clinical and investigative gastroenterology; intestinal absorption; intestinal blood flow; pharmacology of gastrointestinal tract. *Mailing Add:* 4320 Chickasaw Cove Memphis TN 38117

CUMMINS, CECIL STRATFORD, b Monkstown, Ireland, Nov 20, 18; m 59; c 2. MICROBIOLOGY. *Educ:* Univ Dublin, BA, 41, MB, BCh, BAO, 43. *Hon Degrees:* ScD, Univ Dublin, 64. *Prof Exp:* House physician, Sir Patrick Dun's Hosp, Dublin, Ireland, 43; asst to prof bact, Trinity Col, Dublin, 44; lectr bact, London Hosp Med Col, London, 48-64, reader, 64-67; PROF MICROBIOL, ANAEROBE LAB, DIV BASIC SCI, COL AGR, VA POLYTECH INST & STATE UNIV, 67- *Mem:* Am Soc Microbiol; Brit Soc Gen Microbiol; Path Soc Gt Brit. *Res:* Chemical morphology; taxonomy. *Mailing Add:* Anaerobe Lab Div Basic Sci Va Polytech Inst & State Univ Blacksburg VA 24061

CUMMINS, DAVID GRAY, b Cookeville, Tenn, June 29, 36; m 59; c 3. AGRONOMY. *Educ:* Tenn Polytech Inst, BS, 57; Univ Tenn, MS, 59; Univ Ga, PhD(agron), 62. *Prof Exp:* Soil scientist, US Forest Serv, Ky, 62-63; from asst prof to assoc prof, 63-76, PROF DEPT AGRON, GA EXP STA, UNIV GA, 76- *Concurrent Pos:* Assoc planning dir, USAID grant, Peanut Collab Res Support Prog, Univ Ga, 80-82, prog dir, 82- *Mem:* Am Soc Agron; Crop Sci Soc Am; Am Peanut Res Educ Soc. *Res:* Determining more efficient practices for production of corn and sorghum for grain and silage. *Mailing Add:* Ga Sta Griffin GA 30223-1797

CUMMINS, ERNIE LEE, b Warrenton, Ore, July 13, 21; m 43; c 1. SCIENCE EDUCATION. *Educ:* Ore State Univ, BS, 43, MS, 52, EdD(sci educ), 60. *Prof Exp:* Chemist, Scott Paper Co, 47-48 & Evans Prod Co, 49-50; teacher gen sci, Jr High Sch, Ore, 51-53 & High Sch, 53-57; asst prof phys sci & sci educ, Ore Col Educ, Western Ore State Col, 57-61, Ore Col Educ, 57-61, assoc prof phys sci, 61-66, prof phys sci & sci educ, 66-85, EMER PROF PHYS SCI & SCI EDUC, 85- *Concurrent Pos:* Dir in-serv insts, NSF, 62-69. *Mem:* Fel AAAS; Nat Asn Res Sci Teaching; Am Asn Physics Teachers; Nat Sci Teachers Asn. *Res:* Physics; general science. *Mailing Add:* 210 Walnut Dr Monmouth OR 97361

CUMMINS, HERMAN Z, b Rochester, NY, Apr 23, 33; m 63. QUANTUM OPTICS, SOLID STATE PHYSICS. *Educ:* Ohio State Univ, BS & MS, 56; Columbia Univ, PhD(physics), 63. *Prof Exp:* Res physicist, Radiation Lab, Columbia Univ, 62-64; from asst prof to prof physics, Johns Hopkins Univ, 64-71; prof, NY Univ, 71-73; DISTINGUISHED PROF PHYSICS, CITY COL NEW YORK, 73- *Concurrent Pos:* Sloan Found fel, 69, Guggenheim fel, 84. *Mem:* Fel Am Phys Soc. *Res:* Phase transitions in liquids and crystals, ferroelectrics and light scattering spectroscopy; critical phenomena. *Mailing Add:* Dept Physics City Col NY 138th St Convent Ave New York NY 10031

CUMMINS, JACK D, b Shreveport, La, Dec 28, 39; m 60; c 2. INORGANIC CHEMISTRY. *Educ:* Western State Col Colo, BA, 61; Univ MNex, MS, 66, PhD(inorg chem), 67. *Prof Exp:* Asst prof, 66-74, PROF CHEM, METROP STATE COL, 74-, CHMN DEPT, 71- *Concurrent Pos:* Res grant, Educ Media Inst, Univ Colo, 67-68; fel, Univ Mo, St Louis, 69-70. *Mem:* Am Chem Soc; Sigma Xi. *Res:* Study of bonding by characterization and preparation of boron hydrides and boronium cations; x-ray crystallography of amalgams. *Mailing Add:* Dept of Chem Metrop State Col 1006 11th St Denver CO 80204

CUMMINS, JAMES NELSON, b Dix, Ill, Jan 22, 25; m 48; c 5. POMOLOGY, FRUIT VIROLOGY. *Educ:* Univ Ill, BS, 48; Southern Ill Univ, MS, 60, PhD(bot), 65; Univ Wis, MS, 61. *Prof Exp:* Exec secy, Ill Fruit Coun, 48-50; self employed orchardist, 53-55; instr, Anna-Jonesboro High Sch, Ill, 55-57 & Mt Vernon High Sch & Mt Vernon Community Col, 57-60; asst prof sci educ, Southern Ill Univ, 61-67; PROF POMOL, NY STATE AGR EXP STA, CORNELL UNIV, 67- *Concurrent Pos:* Fulbright scholar, Justin-Liebig Univ, WGer, 82-83; Consult, Brazil, 87, Spain, 90, Columbia, 90. *Honors & Awards:* Shepard Award, Am Pomol Soc, 80. *Mem:* Am Soc Hort Sci; Am Pomol Soc; Int Soc Hort Sci; Europ Union Plant Breeding. *Res:* Breeding, testing and development of rootstocks for deciduous fruit trees; rhizogenesis in stem tissues of woody plants; virus diseases of deciduous fruit trees. *Mailing Add:* Hedrick Hall New York State Agr Exp Sta Geneva NY 14456

CUMMINS, JOHN FRANCES, b Bronx, NY, July 24, 39; m 75; c 2. WHITE DWARF STARS, HISTORY OF SCIENCE. *Educ:* St Bonaventure Univ, BA, 61; Whitefriars Theol, MA, 66; Univ Notre Dame, MS, 70; Pa State Univ, MS, 81, PhD(sci technol & soc), 85. *Prof Exp:* Dean, St Albert's Col, 67-70; grad fel, interdisciplinary studies, Pa State Univ, 80-82; mem fac physics, 70-82, assoc prof sci & eng, 85-86, CHMN DEPT SCI & ENG, ORANGE COUNTY COMMUNITY COL, 75- *Concurrent Pos:* Vis assoc prof physics, Dartmouth Col, 84-86. *Mem:* AAAS; Am Asn Physics Teachers; Am Astron Soc. *Res:* Human aspects of scientific enterprise and investigation of the sense of responsibility currently felt by scientists towards future generations. *Mailing Add:* Dept Phys Sci Orange County Community Col 115 South St Middletown NY 10940

CUMMINS, JOSEPH E, b Whitefish, Mont, Feb 5, 33; m 62; c 1. GENETICS, CELL BIOLOGY. *Educ:* Washington State Univ, BS, 55; Univ Wis, PhD(bot), 62. *Prof Exp:* NIH res fel zool, Univ Edinburgh, 62-64; cancer researcher, McArdle Lab, Univ Wis, 64-66; asst prof biol sci, Rutgers Univ, 66-67; asst prof zool, Univ Wash, 67-72; asst prof, 72-76, ASSOC PROF PLANT SCI, UNIV WESTERN ONT, 72- *Concurrent Pos:* Fel, Karolinska Inst, Stockholm, Sweden, 69. *Mem:* Am Soc Cell Biol; Brit Biochem Soc; Genetics Soc Am; Environ Mutagens Soc; Can Genetics Soc. *Res:* Cell cycle; morphogenesis; environmental mutagenesis; environmental pollution. *Mailing Add:* 738 Wilkins London ON N6C 4Z9 Can

CUMMINS, KENNETH BURDETTE, b New Washington, Ohio, July 27, 11. MATHEMATICS. *Educ:* Ohio Wesleyan Univ, AB, 33; Bowling Green State Univ, MA, 39; Ohio State Univ, PhD(math educ), 58. *Prof Exp:* Teacher pub schs, Ohio, 33-40, 41-55, 56-57; from asst prof to prof, 57-81, chmn dept, 64-65, EMER PROF MATH, KENT STATE UNIV, 81- *Mem:* Am Math Asn. *Res:* Methodology in the teaching of collegiate and high school mathematics; an experiential approach to mathematics education. *Mailing Add:* 421 S Center St New Washington OH 44854

CUMMINS, KENNETH WILLIAM, b Chicago, Ill, Mar 28, 33; div; c 2. LIMNOLOGY. *Educ:* Lawrence Col, BA, 55; Univ Mich, MS, 57, PhD(zool), 61. *Prof Exp:* Instr zool, Univ Mich, 60-61; asst prof biol sci, Northwestern Univ, 61-62 & biol, Univ Pittsburgh, 62-68; prof biol, Kellogg Biol Lab, Mich State Univ, 68-78; Prof fisheries & wildlife, Ore State Univ, Corvallis, 78-84; Appalachian Environ Lab, Univ Md, Frostburg, 84-89; PROF & DIR, PYMATUNING LABS ECOL, UNIV PITTSBURGH, 89- *Concurrent Pos:* Prin investr, USPHS res grant, 63-, AEC res grant, 66- & NSF res grants, 70-; chmn water ecosysts, Inst Ecol, 73-; aquatics ed, Ecol Soc Am, 74- *Mem:* Ecol Soc Am; Am Soc Limnol & Oceanog; Brit Freshwater Biol Asn; Int Asn Theoret & Appl Limnol. *Res:* Structure and function of stream ecosystems. *Mailing Add:* Dept Biol Sci Rm 162 Univ Pittsburg Crawford Hall Pittsburgh PA 15260

CUMMINS, LARRY BILL, b Uvalde, Tex, Nov 18, 41; m 62; c 2. VETERINARY MEDICINE. *Educ:* Tex A&M Univ, BS, 70, DVM, 71. *Prof Exp:* Resident vet, Yerkes Regional Primate Res Ctr, 71-73; vet, Atlanta Zool Garden & Emory Univ, 73-74; assoc found vet, Dept Animal Resources, Southwest Found Res & Educ, San Antonio, Tex, 74-76, Dept Microbiol &

Infectious Dis, 76-81; dir, Vet Med, Bio-Systs Res, Inc, Salida, Colo, 81-84 & Sci Support Serv, NMex State Univ, Primate Res Inst, Holloman AFB, NMex, 81-84; DIR, WHITE SANDS RES CTR, ALAMOGORDO, NMEX, 84- *Concurrent Pos:* Vet, Grant Park Zoo, Atlanta, Ga, 73-74, Dept Physiol, Sch Med, Emory Univ, Atlanta, Ga, 73-74; vet consult, Lion Country Safari, Stockbridge, Ga, 72-74, Atlanta Zool Garden, 74-84, San Antonio Zool Soc, 74-84 & Western Med Serv, Inc, Devine, Tex, 79-84; vet & owner, Lakehills Vet Clin, Tex, 79-81; gen mgr, USA Partner, Beijing C&C Sci Res Corp, Beijing, China, 88-; vpres, Coulston Int Corp, Alamogorda, NMex, 87-, chief operating officer, 88- & bd dirs, 89-; pres, Nat Chimpanzee Task Force, 88-89. *Mem:* Am Vet Med Asn; Am Asn Lab Animal Sci; Am Asn Wildlife Vets; Asn Primate Vets (pres, 89-90). *Res:* Infectious diseases of nonhuman primates; clinical medicine of zoologic animals. *Mailing Add:* White Sands Res Ctr 1300 LaVelle Rd Alamogordo NM 88310

CUMMINS, RICHARD L, b Sullivan, Ill, Aug 1, 29; m 60; c 3. ELECTRICAL ENGINEERING. *Educ:* Univ Ill, BS, 53, MS, 59, PhD(elec eng), 62. *Prof Exp:* Jr elec engr, Res Lab, Cook Elec Co, 54-55, sr elec engr, 55-58; asst, Digital Comput Lab, Univ Ill, 58-62, res assoc, 62-63; asst prof, 63-80, ASSOC PROF ELECTRONIC ENG, OKLA STATE UNIV, 80- *Mem:* Inst Elec & Electronics Engrs. *Res:* Graph theory; network topology and synthesis; digital computer applications; minicomputers. *Mailing Add:* Sch Elec Comput Eng Okla State Univ Main Campus Stillwater OK 74078

CUMMINS, RICHARD WILLIAMSON, b Allen, Mich, May 10, 20; m 43; c 2. SYNTHETIC ORGANIC CHEMISTRY. *Educ:* Univ Mich, BS, 42; Polytech Inst Brooklyn, MS, 46, PhD(chem), 53. *Prof Exp:* Res chemist, Westvaco Mineral Prod Div, Food Mach & Chem Corp, 42-64, sr res chemist, 64-68, res assoc, 68-73, sr res assoc, 73-78, res assoc, FMC Corp, 78-82, internal consult, 81-82; RETIRED. *Mem:* Am Chem Soc. *Res:* Organic synthesis reaction mechanisms; organic phosphorus compounds; triazine chemistry; detergent builders; flame retardants functional fluids. *Mailing Add:* 14 Wynnewood Dr Cranbury NJ 08512-3199

CUMMINS, STEWART EDWARD, b Elkhart, Ind, July 7, 32; m 56; c 4. ELECTRONICS ENGINEERING, MATERIALS SCIENCE. *Educ:* Purdue Univ, BSEE, 56, MSEE, 57. *Prof Exp:* Proj engr, Electronic Technol Lab, USAF, 57-61, proj engr electronic mat, 61-67, sr res scientist ferroelec, 67-72, sr res scientist memory technol, 72-74, tech prog mgr magnetic bubble memory, Avionics Lab, 74-89; TECH PROG MGR MIMIC MAT DEVICE CORRELATION, WRIGHT STATE UNIV, 90- *Honors & Awards:* Res Award, Res Soc Am, 67; Burka Award, Air Force Avionics Lab, 72; Harrell V Noble Award, Inst Elec & Electronics Engrs Dayton Sect, 85. *Mem:* Inst Elec & Electronics Engrs; Sigma Xi. *Res:* Memory devices including ferroelectric and related materials; applied physics; optical devices; gallium arsenide material/device studies and high temperature superconductor materials and devices. *Mailing Add:* Univ Res Ctr Wright State Univ Dayton OH 45435

CUMMINS, W(ILLIAM) RAYMOND, b Hamilton, Ont, May 9, 44; m 67; c 2. STRESS PHYSIOLOGY, ARCTIC BIOLOGY. *Educ:* McMaster Univ, BSc, 67; Mich State Univ, PhD(bot & biochem), 72. *Prof Exp:* Res fel biophys, Stanford Univ, 71-73; asst prof biol, York Univ, Toronto, 73-74; asst prof, 74-79, ASSOC PROF BOT, ERINDALE COL, UNIV TORONTO, 79- *Concurrent Pos:* Res fel plant physiol, Plant Res Lab, Atomic Energy Comn; consult, Sci Adv Bd, Govt Northwest Territories, 81-82; chmn, Arctic Working Group, Univ Toronto, 81-86; mem, prog comt, Am Soc Plant Physiologists, 88-89. *Mem:* Can Soc Plant Physiologists; Am Soc Plant Physiologists; Can Asn Univ Teachers; Can Soc Plant Molecular Biol. *Res:* Responses to cold and drought stress particularly in arctic plants; investigate role & alternative pathway respiration; assess limits to agricultural production in arctic and subarctic Canada. *Mailing Add:* Erindale Col Univ Toronto Rm 3042 Mississauga ON L5L 1C6 Can

CUMMISKEY, CHARLES, b St Louis, Mo, Feb 12, 24. INORGANIC CHEMISTRY. *Educ:* Dayton Univ, BS, 43; Northwestern Univ, MS, 52; Univ Notre Dame, PhD(chem), 56. *Prof Exp:* Teacher sec schs, 43-52; res assoc, Univ Notre Dame, 53-55; from instr to assoc prof chem, 55-65, head dept, 57-66, 78-84, vpres & dean faculties, 66-75, PROF CHEM, ST MARY'S UNIV, TEX, 65- *Concurrent Pos:* Am Chem Soc vis scientist to sec schs, 62-66; Piper Found prof, 79. *Mem:* Am Chem Soc; Sigma Xi. *Res:* Chemical effects of radioactive decay; ion exchange methodology, especially as applied to inorganic analytical separations and complex ions. *Mailing Add:* Dept Chem St Mary's Univ San Antonio TX 78284

CUNDALL, DAVID LANGDON, b Altrincham, Cheshire, Eng, July 21, 45; Can citizen; m 69; c 1. VERTEBRATE MORPHOLOGY, SYSTEMATICS. *Educ:* McGill Univ, BSc, 67; Univ Ark, MS, 70, PhD(zool), 74. *Prof Exp:* Asst prof biol, Va Commonwealth Univ, 74-75; asst prof, 75-80, ASSOC PROF BIOL, LEHIGH UNIV, 80- *Mem:* Am Soc Zool; Am Soc Ichthyologists & Herpetologists; Animal Behav Soc; Soc Syst Zool. *Res:* Feeding mechanics in squamate reptiles; systematics of snakes; patterns of growth in the snake skull; habitat selection and resource partitioning in snakes. *Mailing Add:* Dept Biol Lehigh Univ Bethlehem PA 18015

CUNDIFF, LARRY VERL, b Abilene, Kans, Dec 9, 39; m 60; c 3. ANIMAL BREEDING, POPULATION GENETICS. *Educ:* Kans State Univ, BS, 61; Okla State Univ, MS, 64, PhD(animal breeding), 66. *Prof Exp:* Asst prof animal sci, Univ Ky, 65-67; REGIONAL COORDR & GENETICIST, US MEAT ANIMAL RES CTR, NCENT REGION, AGR RES SERV, USDA, 67-, RES LEADER GENETICS & BREEDING, 76- *Concurrent Pos:* Prof animal sci, Univ Nebr, Lincoln, 67- *Honors & Awards:* Animal Breeding & Genetics Res Award, Am Soc Animal Sci. *Mem:* Am Soc Animal Sci; Am Genetics Asn. *Res:* Beef cattle breeding. *Mailing Add:* US Meat Animal Res Ctr PO Box 166 Clay Center NE 68933

CUNDIFF, MILFORD MEL FIELDS, b Baker, Ore, Dec 7, 36; m 81; c 2. VERTEBRATE PHYSIOLOGY, ECOLOGY. *Educ:* Univ Colo, BS, 60, PhD(zool), 66. *Prof Exp:* Teaching assoc biol, Univ Colo, 61-62; from instr to assoc prof, Austin Col, 64-70; ASSOC PROF BIOL & COORDR BIOL SCI, 70-, DIR, NATURAL SCI PROG, UNIV COLO, BOULDER, 88- *Mem:* AAAS; Am Inst Biol Sci. *Res:* Altitude physiology of blood; nuclear activation analysis of trace metals in biological systems; ecology of the flat tops wilderness in Colorado; science education; coral reef ecology; Rocky Mountain elk ecology. *Mailing Add:* Natural Sci Prog Univ Colo Box 331 Boulder CO 80309

CUNDIFF, ROBERT HALL, b Winchester, Ky, Apr 10, 22; m 44; c 2. RESEARCH ADMINISTRATION, ANALYTICAL CHEMISTRY. *Educ:* Univ Ky, BS, 48, MS, 49. *Prof Exp:* Res chemist, Com Solvents Corp, Ind, 49-52; res chemist, R J Reynolds Tobacco Co, 52-66, head anal serv sect & blends & filter develop sect, 66-70, mgr tobacco prod develop div, 70-76, dir, Tobacco Develop Dept, 76-78, dir, blends & processes, 78-83; CONSULT, TOBACCO INDUST, 83- *Honors & Awards:* Philip Morris Award, 67. *Res:* Tobacco products development, including chewing and smoking tobacco, cigarettes and filter technology; tobacco processing, blending and flavoring; analytical chemistry of tobacco, tobacco additives and tobacco smoke; development administration. *Mailing Add:* 718 Chester Rd Winston-Salem NC 27104-1706

CUNDY, KENNETH RAYMOND, b Spearfish, SDak, Dec 22, 29; m 57. MEDICAL MICROBIOLOGY, CLINICAL MICROBIOLOGY. *Educ:* Stanford Univ, BA, 50; Univ Wash, MS, 53; Univ Calif, Davis, PhD(comp path), 65; Am Bd Med Microbiol, dipl, 69. *Prof Exp:* Res microbiologist, Univ Calif, Berkeley, 57-60; bacteriologist, Gerber Prod Co, Calif, 60-61; res microbiologist, Univ Calif, Davis, 61-65; from instr to assoc prof, 67-71, assoc prof microbiol & immunol, 71-77, PROF MICROBIOL & IMMUNOL, SCH MED, TEMPLE UNIV, 77-, DIR CLIN MICROBIOL LABS, UNIV HOSP & HEALTH SCI CTR, 70- *Concurrent Pos:* Fel microbiol, Sch Med, Temple Univ, 65-67 & Nat Cystic Fibrosis Res Found, 68-69; assoc dir microbiol, Lab, St Christopher's Hosp Children, 67-68, dir diag microbiol, 68-70. *Mem:* Fel Am Acad Microbiol; Sigma Xi; AAAS; Am Soc Microbiol; NY Acad Sci. *Res:* Mechanisms of pathogenicity as related to microorganisms, particularly with reference to Pseudomonas and anaerobes; infectious diseases and their laboratory diagnosis. *Mailing Add:* Dept Microbiol & Immunol Temple Univ Sch Med Philadelphia PA 19140

CUNHA, BURKE A, b Hartford, Conn, Mar 25, 42. INFECTIOUS DISEASES. *Educ:* Pa State Univ, MD, 72. *Prof Exp:* CHIEF INFECTIOUS DIS & VCHMN MED, WINTHROP-UNIV HOSP, 79-, DIR MED EDUC, 84-; ASSOC PROF MED, STATE UNIV NY, STONY BROOK, 83- *Mailing Add:* Winthrop-Univ Hosp 259 First St Mineola NY 11501

CUNHA, TONY JOSEPH, b Los Banos, Calif, Aug 22, 16; m 41; c 3. ANIMAL HUSBANDRY, NUTRITION. *Educ:* Utah State Agr Col, BS, 40, MS, 41; Univ Wis, PhD(animal nutrit), 44. *Prof Exp:* From instr to assoc prof animal husb, State Col Wash, 44-48; prof & head dept, Univ Fla, 48-75; dean, Sch Agr, 75-80, EMER DEAN, CALIF POLYTECH UNIV, 80-, UNIV CONSULT AGR, 80-; EMER PROF ANIMAL SCI, UNIV FLA, 75- *Concurrent Pos:* Am Soc Animal Sci res award, 68-; chmn, swine nutrient requirements comt, Nat Res Coun, 65-75 & animal nutrit comt, 72-76; mem, White House Conf on Food, Nutrit & Health, 69; consult, Univ Tenn-AEC, Oak Ridge & USDA Meat Animal Res Ctr, Nebr; mem, World Fod & Nutrit Study, 78- *Honors & Awards:* Int Award in Agr, 76. *Mem:* AAAS; Am Soc Animal Sci (vpres, 61, pres, 62); Soc Exp Biol & Med; Am Inst Nutrit. *Res:* Nutrition and feeding of swine, beef, cattle, sheep, horses and small animals. *Mailing Add:* PO Box 81410 Las Vegas NV 89180-1410

CUNIA, TIBERIUS, b Edessa, Greece, Jan 10, 26; US citizen; m 57; c 4. STATISTICS. *Educ:* McGill Univ, MSc, 57. *Prof Exp:* Forest mensuration, Can Int Paper Co, 51-52, forest opers, 52-54, comput appln, 54-58, forest statistician, 58-68; prof forest mensuration & statist, 68-70, PROF STATIST & OPERS RES, STATE UNIV NY COL ENVIRON SCI & FORESTRY, SYRACUSE, 70- *Honors & Awards:* Alexander von Humboldt Sr Scientist Award. *Mem:* Am Statist Asn; Biomet Soc; Can Inst Forestry; Soc Am Foresters. *Res:* Forest mensuration and biomass inventory; applications of the statistical methodology, operations research methods and computers to forestry problems. *Mailing Add:* State Univ NY Col Environ & Forestry Syracuse NY 13210

CUNICO, ROBERT FREDERICK, b Detroit, Mich, Feb 25, 41; m 66. ORGANIC CHEMISTRY. *Educ:* Univ Detroit, BS, 62; Purdue Univ, PhD(organosilicon chem), 66. *Prof Exp:* Res assoc with Dr Melvin S Newman, Ohio State Univ, 66-68; from asst prof to assoc prof, 68-81, PROF CHEM, NORTHERN ILL UNIV, 81- *Mem:* Am Chem Soc. *Res:* Organosilicon chemistry; synthetic methods in organic chemistry. *Mailing Add:* Dept Chem Northern Ill Univ De Kalb IL 60115

CUNNANE, STEPHEN C, b London, UK, Oct 29, 52; UK & Can citizen; m 77; c 2. LIPID METABOLISM, STABLE ISOTOPES. *Educ:* Bishop's Univ, BSc, 75, BEd, 76; McGill Univ, PhD(physiol), 80. *Prof Exp:* Asst prof, 86-90, ASSOC PROF NUTRIT SCI, UNIV TORONTO, 90- *Concurrent Pos:* Vis prof, Inserm U26, Paris, 86-, Univ London, 91-; assoc ed, J Nutrit Med, 90- *Mem:* Am Inst Nutrit; Can Soc Nutrit Sci; NY Acad Sci; Am Oil Chemists Soc. *Res:* Regulation of long chain fatty acid metabolism in relation to nutritional status and/or disease. *Mailing Add:* Dept Nutrit Sci Univ Toronto Toronto ON M5S 1A8 Can

CUNNEA, WILLIAM M, b Chicago, Ill, Oct 31, 27; m 71. MATHEMATICS, ALGEBRA. *Educ:* Univ Chicago, PhB, 46, MS, 56; Univ Calif, Berkeley, PhD(math), 62. *Prof Exp:* Asst prof, 61-66, ASSOC PROF MATH, WASH STATE UNIV, 66- *Concurrent Pos:* Vis scholar, Univ Victoria, 68-69. *Mem:* Math Asn Am; Am Math Soc. *Res:* Ideal and valuation theory; algebraic geometry; lattice theory; history of mathematics. *Mailing Add:* Dept Math Wash State Univ Pullman WA 99164-2930

CUNNIFF, PATRICIA A, b Washington, DC, Dec 6, 38; m 60; c 4. PHYSICAL CHEMISTRY. *Educ:* Dunbarton Col Holy Cross, BA, 59; Univ Md, College Park, MS, 62, PhD(phys chem), 72. *Prof Exp:* Ed asst, Anal Chem, Am Chem Soc, 61-62; from asst prof to assoc prof, 72-79, pres, fac orgn, 81-83, PROF CHEM, PRINCE GEORGE'S COMMUNITY COL, 79- *Concurrent Pos:* Div Chem, NSF, 86; Sci Policy Sabbatical Fel, Am Chem Soc, 87-88. *Mem:* AAAS; Am Chem Soc; Sigma Xi. *Res:* Academic education in chemistry; instrumental analysis; energy education; Federal Res & Develop Funding. *Mailing Add:* 7006 Partridge Pl Univ Park MD 20782

CUNNIFF, PATRICK F, b New York, NY, Oct 25, 33; m 60; c 4. ENGINEERING MECHANICS. *Educ:* Manhattan Col, BCE, 55; Va Polytech Inst, MS, 56, PhD(eng mech), 62. *Prof Exp:* Res engr, Mech Div, US Naval Res Lab, 60-63; from asst prof to assoc prof, 63-69, chmn dept, 75-82, PROF MECH ENG, UNIV MD, 69- *Concurrent Pos:* Assoc Dean, Grad Sch, 88-90. *Mem:* Am Soc Mech Engrs; Am Soc Eng Educ. *Res:* Structural response to shock and vibration excitations; nonlinear vibrations; acoustics and noise pollution problems. *Mailing Add:* Dept Mech Eng Univ Md College Park MD 20742

CUNNING, JOE DAVID, b Mt Ayr, Iowa, May 31, 36; m 63; c 2. CHEMICAL ENGINEERING. *Educ:* Iowa State Univ, BS, 58, MS, 62, PhD(chem eng), 65. *Prof Exp:* Engr, 58-60, res engr, Va, 65-67, supvr res & develop, 67-70, sr supvr, 70-73, supt, 73-78, mfg mgr, 78-82, RES DIR, E I DU PONT DE NEMOURS & CO, WILMINGTON, 82- *Mem:* Am Inst Chem Engrs; Am Chem Soc; Sigma Xi. *Res:* Textile fiber research and development; polymerization processes; product development; heat transfer on fiber processing; reaction kinetics; distillation; nylon and polyester fiber development. *Mailing Add:* 501 Andover Rd Wilmington DE 19803

CUNNINGHAM, ALICE JEANNE, b Walnut Ridge, Ark, Sept 23, 37. ELECTROCHEMISTRY, CHEMICAL EDUCATION. *Educ:* Univ Ark, BA, 59; Emory Univ, PhD(chem), 66. *Prof Exp:* Chemist, Layne Res, Tenn, 59; instr chem, Brenau Acad, 59-61 & Atlanta Pub Schs, 61-62; res assoc, Univ Tex, 67-68; vis asst prof, Agnes Scott Col, 66-67, from asst prof to assoc prof, 72-79, chmn dept, 78-90, WILLIAM R KENAN PROF CHEM, AGNES SCOTT COL, 79- *Concurrent Pos:* Distinguished vis scholar, Emory Univ, Atlanta, Ga, 84-85, vis prof, 85-86; mem, comt prof training, Am Chem Soc, 79-89, chmn, 86-89. *Mem:* AAAS; Am Chem Soc; Sigma Xi. *Res:* Electrochemistry and mechanisms of biological oxidation-reduction reactions; invivo voltammetry in the brain. *Mailing Add:* Dept Chem Agnes Scott Col Decatur GA 30030-3797

CUNNINGHAM, BRUCE ARTHUR, b Winnebago, Ill, Jan 18, 40; m 65; c 2. MOLECULAR BIOLOGY & EMBRYOLOGY. *Educ:* Univ Dubuque, BS, 62; Yale Univ, PhD(biochem), 66. *Prof Exp:* NSF fel biochem, from asst prof to assoc prof biochem, 68-78, PROF, DEVELOP & MOLECULAR BIOL, ROCKEFELLER UNIV, 78- *Concurrent Pos:* hon Sterling Fel, Yale Univ, 65-66; Teacher-Scholar grant, Camille & Henry Dreyfus Found, 70; career scientist, Irma T Hirschl Trust, 75-80. *Mem:* AAAS; Am Chem Soc; Am Asn Immunologists; Am Soc Biol Chemists; Harvey Soc; Sigma Xi; Protein Soc; Am Soc Cell Biol. *Res:* Structure and function of cell-surface proteins; structure of antibodies; primary structure of proteins. *Mailing Add:* Dept Develop & Molecular Biol Rockefeller Univ New York NY 10021

CUNNINGHAM, BRYCE A, b Brainerd, Minn, June 21, 32; m 56; c 4. BIOCHEMISTRY. *Educ:* Univ Minn, BA, 55, BS, 58, PhD(biochem), 63. *Prof Exp:* Asst prof, 63-72, ASSOC PROF BIOCHEM, KANS STATE UNIV, 72- *Mem:* AAAS; Am Chem Soc. *Res:* Enzyme chemistry; peroxidases. *Mailing Add:* Tech Innovation Ctr Univ Iowa Oakdale Campus Iowa City IA 52242

CUNNINGHAM, CAROL CLEM, b Woodland, Calif, July 14, 38; m 60; c 3. ALCOHOLISM RESEARCH, BIOENERGENETICS. *Educ:* Oklahoma State Univ, BS, 61, MS, 63; Univ Ill, PhD(biochem), 68. *Prof Exp:* From asst prof to assoc prof, 70-82, PROF BIOCHEM, WAKE FOREST UNIV SCH MED, 82- *Concurrent Pos:* Mem, study sect, Nat Inst Alcohol Abuse & Alcoholism, 87- *Honors & Awards:* Res Develop Award, Nat Inst Alcohol Abuse & Alcoholism, 79-84. *Mem:* Am Soc Biol Chemists & Molecular Biologists; Res Soc Alcoholism. *Res:* Interaction of the mitochondrial and synthase with phospholipids; investigations of the effects of chronic ethanol consumption on liver function, with an emphasis on characterizing alterations in energy metabolism. *Mailing Add:* Dept Biochem Bowman Gray Sch Med Wake Forest Univ Winston-Salem NC 27103

CUNNINGHAM, CHARLES G, b Springfield, Vt, Dec 5, 40; m 68; c 2. GEOCHEMISTRY. *Educ:* Amherst Col, Mass, BA, 67; Univ Colo, MS, 69, Stanford Univ, PhD(geol), 73. *Prof Exp:* Asst prof econ geol, Syracuse Univ, NY, 73-74; RES GEOLOGIST, US GEOL SURVEY, 74- *Concurrent Pos:* Postdoctoral res assoc, Lab Ctr Prog Comt, Nat Res Coun, 84-87; secy, Econ Geol Publ Co, 84-; counr, Soc Econ Geologists, 88-91. *Mem:* fel Soc Econ Geologists; fel Geol Soc Am. *Res:* Specialist in volcanic-hosted, epithermal precious metal deposits and fluid-inclusion geothermometry and geobarometry as applied to the evolution of hydrothermal; ore-forming systems; author of over 160 publications on subject. *Mailing Add:* US Geol Survey 959 Nat Ctr Reston VA 22092

CUNNINGHAM, CHARLES HENRY, b Washington, DC, Apr 12, 13; m 40; c 4. VETERINARY MICROBIOLOGY. *Educ:* Univ Md, BS, 34; Iowa State Univ, MS, 37, DVM, 38; Mich State Univ, PhD, 53; Am Bd Med Microbiol, dipl; Am Col Vet Microbiol, dipl. *Prof Exp:* Asst prof & vet inspector, Univ Md, 38-42; assoc prof poultry husb, Univ RI, 42-45; from assoc prof to prof, 45-78, EMER PROF MICROBIOL & PUB HEALTH, MICH STATE UNIV, 78- *Concurrent Pos:* Consult, vet microbiol, Univ Tenn Int Agr Prog & vet virol, Food Agr Orgn, 69; mem, Food Agr Orgn Prog, WHO, 71-78. *Mem:* Fel AAAS; Am Asn Avian Path; NY Acad Sci; Am Soc Microbiol; Am Vet Med Asn; Conf of Res Workers in Animal Dis; Am Acad of Microbiol; Am Col Vet Microbiologists. *Res:* Virology; veterinary virology; animal diseases. *Mailing Add:* 617 Snyder Rd East Lansing MI 48823-3420

CUNNINGHAM, CLARENCE MARION, b Cooper, Tex, July 24, 20; m 51; c 4. PHYSICAL CHEMISTRY. *Educ:* Agr & Mech Col Tex, BS, 42; Univ Calif, MS, 48; Ohio State Univ, PhD(chem), 54. *Prof Exp:* Instr chem, Calif State Polytech Col, 48-49; cryogenic engr, AEC Proj, 52; consult, Herbert L Johnston, Inc, 53; assoc prof, 54-85, EMER PROF CHEM, OKLA STATE UNIV, 85- *Mem:* AAAS; Am Chem Soc; Am Phys Soc; Am Asn Univ Prof. *Res:* Theoretical and experimental investigation of physical absorption of gases on solid surfaces at low temperatures; thermodynamic properties of nonaqueous solutions. *Mailing Add:* 924 Lakeridge Ave Stillwater OK 74075

CUNNINGHAM, DAVID A, b Toronto, Ont, Mar 18, 37; m 61; c 3. PHYSIOLOGY. *Educ:* Univ Western Ont, BA, 60; Univ Alta, MSc, 63; Univ Mich, PhD(phys educ, physiol), 66. *Prof Exp:* Res assoc epidemiol, Univ Mich, 66-69; assoc prof, 69-80, PROF PHYSIOL & PHYS EDUC, UNIV WESTERN ONT, 80- *Concurrent Pos:* Res assoc, Dept Nat Health & Welfare, Can, 69-72. *Mem:* Am Physiol Soc; Can Physiol Soc; Can Asn Sports Sci; Am Col Sports Med. *Res:* Physiology of exercise with special interest in the relationship of heart disease, aging, and physical activity. *Mailing Add:* Dept of Physiol Univ of Western Ont London ON N6A 3K7 Can

CUNNINGHAM, DENNIS DEAN, b Des Moines, Iowa, Aug 16, 39. CELL BIOLOGY, BIOCHEMISTRY. *Educ:* State Univ Iowa, BA, 61; Univ Chicago, PhD(biochem), 67. *Prof Exp:* From asst prof to prof, 70-87, PROF MICROBIOL & CHMN, COL MED, UNIV CALIF, IRVINE, 87- *Concurrent Pos:* NSF res fel, Princeton Univ, 67-70; USPHS study sect, 75-80; assoc ed, J Cell Physiol, 78-; mem, Comt Develop & Cell Biol, Am Cancer Soc, 82-86; NIH res career develop award, 75-80. *Res:* Control of proteases by protease nexins. *Mailing Add:* Dept Microbiol & Molecular Genetics Univ Calif Col Med Irvine CA 92717

CUNNINGHAM, DOROTHY J, b Jersey City, NJ, Nov 7, 27. PHYSIOLOGY, ENVIRONMENTAL HEALTH. *Educ:* Caldwell Col Women, BA, 49; Cath Univ, MS, 51; Yale Univ, PhD(physiol), 66. *Prof Exp:* Res asst physiol, Sch Med, Univ Pa, 57-58; asst prof biol sci, Montclair State Col, 58-62; fel epidemiol & pub health & environ physiol, Sch Med, Yale Univ, 66-67, lectr, 67-69, asst prof, 69-70, asst fel, John B Pierce Found Lab, 66-70; assoc prof, Sch Health Sci, 70-74, PROF PHYSIOL, SCH HEALTH SCI, HUNTER COL, 75- *Concurrent Pos:* Lectr dept community med, Div Environ Med, Mt Sinae Sch Med, 71-82; Yale Sci & Eng Asn (exec bd, 75-); adj prof environ med, Inst Environ Med, NY Univ Med Ctr, 81-; vis fel, John B Pierce Found Lab, 82-; res affil environ health, Yale Univ Sch of Med, 85- *Mem:* AAAS; Am Physiol Soc; fel NY Acad Sci; assoc fel NY Acad Med; Harvey Soc; Sigma Xi; Am Fedn for Clin Res. *Res:* Temperature regulation; environmental physiology. *Mailing Add:* 750 Park Ave Apt 15D New York NY 10021

CUNNINGHAM, EARLENE BROWN, b Cleveland, Ohio, Aug 27, 30; div. BIOCHEMISTRY, MOLECULAR BIOLOGY. *Educ:* Univ Ill, BS, 49; Univ Calif, Los Angeles, MS, 51; Univ Southern Calif, PhD(phys org chem), 54. *Prof Exp:* Asst clin path, Sch Med, Ind Univ, 54-56, res assoc med, 56-59; from asst prof to assoc prof, Col Med, Howard Univ, 59-64; res physiologist, Univ Calif, Berkeley, 64-68; res assoc chem, Univ SC, 68-69, lectr, 69-71; assoc prof biochem, Med Univ SC, 71-78; ASSOC PROF BIOCHEM, NJ MED SCH, 78- *Honors & Awards:* Lederle Med Fac Award, 61-64. *Mem:* Am Chem Soc; Am Soc Biochem & Molecular Biol. *Res:* Biochemical regulation involving the covalent modification of membrane-associated regulatory enzymes; role of inositol 1,4,5-trisphosphate, IP3, in the regulation of cell growth and development. *Mailing Add:* Dept Biochem UMDJN-NJ Med Sch Newark NJ 07103-2714

CUNNINGHAM, ELLEN M, b Chicago, Ill, June 20, 40. MATHEMATICAL LOGIC. *Educ:* St Mary of the Woods Col, BA, 63; Cath Univ Am, MA, 70; Univ Md, PhD(math), 74; Univ Evansville, MS, 85. *Prof Exp:* From asst prof to assoc prof, 74-90, PROF MATH, ST MARY OF THE WOODS COL, 90-, CHMN, DEPT SCI & MATH, 86- *Mem:* Math Asn Am; Comp Prof Social Responsibility; Asn Women in Math. *Res:* Chain models and infinite-quantifier languages. *Mailing Add:* St Mary of the Woods Col St Mary of the Woods IN 47876

CUNNINGHAM, FAY LAVERE, b Lansing, Mich, July 26, 22; m 48; c 2. CHEMICAL ENGINEERING, TECHNICAL MANAGEMENT. *Educ:* Mich State Univ, BS, 48. *Prof Exp:* Mech engr, Manhattan Dist, Mass Inst Technol, 44-46; chem engr, Upjohn Co, 48-58, res sect head, 58-70, res mgr, 70-72 & Chem Eng Develop, 72-83, dir specialty chem prod, 84-88; RETIRED. *Concurrent Pos:* Radiation monitor-opers crossroads, Atomic Energy Comn, 46; ltd consult, 88. *Honors & Awards:* Upjohn Award, 63. *Mem:* AAAS; Am Chem Soc; Am Inst Chem Engrs. *Res:* Chemical and fermentation process development; scale up of organic reactions; separation and purification operations; chemical production operations. *Mailing Add:* 4323 Sunset Dr Kalamazoo MI 49008

CUNNINGHAM, FLOYD MITCHELL, b Estherville, Iowa, Nov 13, 31; c 6. ENGINEERING MECHANICS, AGRICULTURAL ENGINEERING. *Educ:* Iowa State Univ, BS, 53, MS, 57, PhD(agr eng, eng mech), 60. *Prof Exp:* From asst prof to assoc prof agr eng, Va Polytech Inst, 60-66; asst prof eng mech, 66-68, ASSOC PROF ENG MECH, UNIV MO-ROLLA, 68- *Mem:* Am Soc Mech Engrs; Sigma Xi. *Res:* Design relationships for fertilizer distributors; vibration of shells and plates. *Mailing Add:* Dept Eng Mech Univ Mo Rolla MO 65401

CUNNINGHAM, FRANKLIN E, b Huntington, WVa, Mar 2, 27; m 51; c 5. FOOD SCIENCE, BIOCHEMISTRY. *Educ:* Kans State Univ, BS, 57; Univ Mo, MS, 59, PhD(poultry prod), 63. *Prof Exp:* Res chemist biochem, Western Regional Res Lab, USDA, 63-69; assoc prof, 69-80, PROF ANIMAL SCI, KANS STATE UNIV, 80- *Mem:* Poultry Sci Asn; Inst Food Technol. *Res:* Constituents of poultry and eggs; new and improved food items or processes. *Mailing Add:* Dept Animal Sci Call Hall Kans State Univ Manhattan KS 66502

CUNNINGHAM, FREDERIC, JR, b Cooperstown, NY, Sept 6, 21; m 47; c 3. MATHEMATICS. *Educ:* Harvard Univ, BS, 43, MA, 47, PhD(math), 53. *Prof Exp:* Teaching fel math, Harvard Univ, 47-50; from instr to asst prof, Univ NH, 51-56; lectr, Bryn Mawr Col, 56-57; asst prof, Wesleyan Univ, 57-59; assoc prof, 59-69, PROF MATH, BRYN MAWR COL, 69- *Mem:* AAAS; Am Math Soc; Math Asn Am; Math Soc France. *Res:* L-structure and related structures on Banach spaces. *Mailing Add:* Dept Math Bryn Mawr Col Bryn Mawr PA 19010

CUNNINGHAM, FREDERICK WILLIAM, physics; deceased, see previous edition for last biography

CUNNINGHAM, GEORGE LEWIS, JR, b New Orleans, La, Nov 20, 23; m 53; c 2. PHYSICAL CHEMISTRY. *Educ:* Tulane Univ, BE, 44; Univ Calif, Berkeley, MS, 47, PhD(phys chem), 50. *Prof Exp:* Asst prof chem, La State Univ, 50-53; from assoc prof to prof, Southwest State Col, Okla, 53-61; assoc prof, Eastern Ill Univ, 61-66, prof chem, 66-86; RETIRED. *Mem:* AAAS; Am Chem Soc. *Res:* Microwave spectra; molecular structure; digital computer applications to chemistry. *Mailing Add:* Dept Chem Eastern Ill Univ Charleston IL 61920

CUNNINGHAM, GLENN N, b Spring City, Tenn, Sept 13, 40; m 62; c 2. BIOCHEMISTRY, NUTRITION. *Educ:* Univ Tenn, Knoxville, BS, 61; NC State Univ, MS, 64, PhD(nutrit, biochem), 66. *Prof Exp:* Fel biochem, Clayton Found Biochem Inst, Univ Tex, Austin, 66-68; asst prof chem, Northeast La State Col, 68-69; from asst prof to assoc prof, 69-78, PROF CHEM, UNIV CENT FLA, 78- *Concurrent Pos:* Res Award, United Coop Farmers Found, 78. *Mem:* AAAS; Am Chem Soc; Am Inst Biol Sci. *Res:* Biochemical regulation; eucaryotic developmental biochemistry; mode of action of bioactive compounds. *Mailing Add:* PO Box 801 Oviedo FL 32765

CUNNINGHAM, GLENN R, b Wewoka, Okla, Feb 7, 40; m 62; c 2. ENDOCRINOLOGY, ANDROLOGY. *Educ:* Univ Okla, BA, 62, MD, 66. *Prof Exp:* Med intern, Duke Med Ctr, 66-67, med resident, 67-68, fel, 68-70, chief resident, 70-71; chief endocrinol, Brooke Gen Hosp, Tex, 71-73; dir, Endocrinol Lab, Methodist Hosp, Houston, 80-86; ASST PROF CELL BIOL, BAYLOR COL MED, HOUSTON, TEX, 75-, ASSOC PROF MED, 79- & ACTG CHIEF, DIV ENDOCRINOL & METAB, 90- *Concurrent Pos:* Asst prof med, Baylor Col Med, 73-79; chief, Endocrinol Sect, Med Serv, Va Med Ctr, Houston, Tex & dir, Core Hormone Radioimmunoassay Lab, Diabetes & Endocrinol Res Ctr, 90- *Mem:* Am Fertil Soc; Am Fed Res; Endocrine Soc; Am Soc Androl. *Res:* Basic and clinical studies of benign prostatic hyperplasia, impotence and androgen therapy. *Mailing Add:* VA Med Ctr 151 2002 Holcome Blvd Houston TX 77030

CUNNINGHAM, GORDON ROWE, b Oswego, Kans, May 11, 22; m 46; c 3. FORESTRY. *Educ:* Mich State Univ, BS, 48, PhD(forest econ), 66; Pa State Univ, MS, 50. *Prof Exp:* Asst exten forester, Univ Ill, 49-54; from asst prof to assoc prof forestry & exten forester, Cornell Univ, 54-61; exten forester, Univ Wis-Madison, 63-80, from asst prof to prof forestry, 63-80; RETIRED. *Honors & Awards:* M N Taylor Award, 90. *Mem:* Soc Am Foresters; Am Forestry Asn. *Res:* Management of small woodlands. *Mailing Add:* 4818 Woodburn Dr Madison WI 53711

CUNNINGHAM, HARRY N, JR, b Imperial, Pa, Mar 7, 35; wid; c 2. VERTEBRATE ECOLOGY. *Educ:* Univ Pittsburgh, BS, 55, MS, 60, PhD(ecol), 66. *Prof Exp:* Instr biol, Mt Union Col, 59-61; asst prof, Thiel Col, 63-67, chmn dept, 65-67; asst prof, 67-71, ASSOC PROF BIOL, BEHREND COL, PA STATE UNIV, 71- *Concurrent Pos:* NSF instnl grant, Pa State Univ, 67-; consult, Aquatic Ecol Assocs, 73-78, mammals, Presque Isle, Pa, Land Mgt. *Mem:* Am Soc Mammal; Sigma Xi. *Res:* Soil microcommunities; small mammal distributions. *Mailing Add:* Div Sci Eng Technol Behrend Col Pa State Univ Erie PA 16563

CUNNINGHAM, HOWARD, b Philadelphia, Pa, Aug 22, 42. ORGANIC CHEMISTRY. *Educ:* Univ Pa, BS, 64, PhD(org chem), 69. *Prof Exp:* Sr develop chemist, Res Ctr, Lever Bros Co, NJ, 69-71; tech serv rep, Chem Div, Pfizer, Inc, 71-74; mkt res analyst, 74-76, MGR PROD SAFETY, WITCO CHEM CORP, 76- *Mem:* Am Chem Soc. *Res:* Organic synthesis-pharmaceuticals; preparation and evaluation of new actives in detergent formulations; organic specialty chemicals; managing and directing compliance with federal, state and local regulations affecting chemical products. *Mailing Add:* 323 W Champlost Ave Philadelphia PA 19120-1826

CUNNINGHAM, HUGH MEREDITH, b Brandon, Man, Dec 28, 27; m 50; c 4. FOOD CHEMISTRY. *Educ:* Univ Man, BSA, 49, MSc, 50; Cornell Univ, PhD(animal nutrit), 53. *Prof Exp:* Res scientist, Can Dept Agr, 53-70; head food additives & contaminants, Food Div, Food & Drug Directorate, 70-79, HEAD TASK FORCE, REASSESSMENT CHEM SAFETY, DEPT NAT HEALTH & WELFARE, 79- *Concurrent Pos:* Res scientist, Inst Animal Physiol, Cambridge, Eng, 67-68. *Mem:* Am Soc Animal Sci; Nutrit Soc Can; Agr Inst Can; Prof Inst Pub Serv Can; Can Soc Animal Prod. *Res:* Lipid biochemistry; physiology; nutrition. *Mailing Add:* 1054 Castle Hill Crescent Ottawa ON K2C 2A8 Can

CUNNINGHAM, JAMES GORDON, b Pittsburgh, Pa, Dec 13, 40; c 3. NEUROPHYSIOLOGY, NEUROLOGY. *Educ:* Purdue Univ, DVM, 64; Univ Calif, Davis, PhD(physiol), 71. *Prof Exp:* Lectr physiol, Univ Ibadan, 64-66; ASSOC PROF PHYSIOL & NEUROL, MICH STATE UNIV, 72- *Concurrent Pos:* Consult, Peace Corps & Agency for Int Develop. *Mem:* AAAS; Am Vet Neurol Asn; Am Soc Vet Physiol & Pharmacol. *Res:* Neurophysiologic basis of seizures; animal models for epilepsy. *Mailing Add:* Dept Physiol 221 Giltner Hall Mich State Univ East Lansing MI 48824

CUNNINGHAM, JOEL L, b Mooresville, NC, Jan 1, 44; m 65; c 2. COMMUTATIVE RINGS, QUADRATIC FORMS. *Educ:* Univ Tenn, Chattanooga, BA, 65; Univ Ore, MA, 67, PhD(math), 69. *Prof Exp:* Asst prof math, Univ Ky, 69-74; from asst prof to assoc prof, 74-79, dean continuing educ, Univ Tenn, Chattanooga, 74-79; vpres academic affairs, 79-84, PROF MATH, SUSQUEHANNA UNIV, 79-, PRES, 84- *Concurrent Pos:* Prin investr, NSF Grant, 70-71; proj dir, NSF Summer & Inservice Inst, 71-73. *Mem:* Am Math Soc; Math Asn Am; Asn Comput Mach. *Res:* Commutative rings and quadratic forms. *Mailing Add:* 501 University Ave Selinsgrove PA 17870

CUNNINGHAM, JOHN (ROBERT), b Regina, Sask, Jan 5, 27; m 51; c 5. MEDICAL PHYSICS. *Educ:* Univ Sask, BEng, 50, MSc, 51; Univ Toronto, PhD(physics), 55. *Prof Exp:* Physicist, Grain Res Lab, 51-53 & Can Defense Res Bd, 55-58; physicist, Ont Cancer Inst, Toronto, 58-89; CONSULT MED PHYSICS, 89- *Concurrent Pos:* Tech adv to Govt of Ceylon, Int Atomic Energy Agency, 64-65; assoc ed, Physics in Med Biol, 80-87. *Honors & Awards:* Coolidge Award, Am Asm Physicists in Med, 88. *Mem:* Can Asn Physicists; Asn Med Physicists India; Am Asn Med Physicists; Can Col Phys Med; Hosp Physicist Asn. *Res:* Scattering absorption and measurement of x-rays, gamma rays; application to medical physics and radiobiology; computer applications in radiotherapy. *Mailing Add:* 960 Teron Rd Apt 910 Kanata ON K2K 2B6 Can

CUNNINGHAM, JOHN CASTEL, b Aberdeen, Scotland, Jan 28, 42; m 68, 81; c 2. INSECT PATHOLOGY. *Educ:* Glasgow Univ, BSc, 64; Oxford Univ, DPhil(insect virol), 67. *Prof Exp:* RES SCIENTIST, FOREST PEST MGT INST, 67- *Mem:* Soc Invert Path; Entom Soc Can; Can Inst Forestry. *Res:* Various aspects of insect virology, including use of viruses in forest insect control; safety testing and registration of viruses, formulation and application technology. *Mailing Add:* Forest Pest Mgt Inst Forestry Can Serv PO Box 490 Sault Ste Marie ON P6A 5M7 Can

CUNNINGHAM, JOHN E, b Malone, NY, Apr 18, 31; m 59; c 4. GEOLOGY. *Educ:* Dartmouth Col, AB, 53; Univ Ariz, PhD(geol), 65. *Prof Exp:* Asst geologist, Bear Creek Mining Co, 56, 57 & 68; asst prof geol & anthrop, Eastern NMex Univ, 62-63; geologist, Minerals Br, Superior Oil Co, 63; from asst prof to assoc prof, 64-79, PROF GEOL & ANTHROP, WESTERN NMEX UNIV, 79- *Concurrent Pos:* Consult, NMex Bur Mines, 69-83, Ariz Bur Mines, 79, Amax, 80 & Dekalb, 81. *Mem:* hon mem NMex Geol Soc. *Res:* Economic and structural geology; geologic map of Circle Mesa Quadrangle, New Mexico. *Mailing Add:* Dept Natural Sci Western NMex Univ Silver City NM 88061

CUNNINGHAM, JOHN EDWARD, b Chicago, Ill, Mar 18, 20; m 44; c 5. MATERIALS SCIENCE, NUCLEAR MATERIALS TECHNOLOGY. *Educ:* Univ Ill, BS, 44; Univ Tenn, MS, 50. *Prof Exp:* Develop engr aircraft-auto engine components, Thompson Prod, Inc, 44; spec eng detachment construct, Corps Engrs, US Army, 44-46; assoc metallurgist, Clinton Labs, Oak Ridge Nat Lab, 46-48, supvr reactor technol, Fabrication Lab, 48-55, asst dir, 55-68, assoc dir, Metals & Ceramics Div, 68-85; CONSULT, ETE CONSULT ENG, 85- *Concurrent Pos:* Mem & chmn adv bd, Dept Ceramic Eng, Univ Ill, 73-76; assoc ed, Nuclear Eng Mat Handbook, Am Nuclear Soc & Am Soc Metals, 77- *Mem:* Fel Am Nuclear Soc; fel Am Soc Metals; Am Soc Mech Engrs. *Res:* Materials science and engineering related to energy production from nuclear, fossil, solar and geothermal sources; energy transmission, use and conservation; total materials cycle from production of raw materials to use and eventual disposal; metallurgy and physical metallurgical engineering. *Mailing Add:* ETE Consult Eng Oakridge TN 37831

CUNNINGHAM, JOHN JAMES, BODY COMPOSITION, BURN HYPERMETABOLISM. *Educ:* Univ Md, PhD(nutrit sci), 78. *Prof Exp:* ASST PEDIAT ENDOCRINOL & METAB, SHRINER'S BURN INST, BOSTON, 83- *Mailing Add:* 51 Blossom St Boston MA 02114

CUNNINGHAM, JULIE MARGARET, b Gt Brit. ONCOLOGY. *Educ:* Australian Nat Univ, BSc, 76, PhD(immunol), 80. *Prof Exp:* Postdoctoral, Med Univ, SC, 80-81; res fel, Mayo Clin, 81-84; res assoc, Mem Sloan-Kettering Inst, 84-86; SPEC PROJ ASSOC, MAYO CLIN, 90- *Mem:* Soc Neurosci; Am Asn Immunologists. *Res:* Immunohistochemistry of human tumors and correlation with molecular genetics. *Mailing Add:* 200 First St SW Rochester MN 55906

CUNNINGHAM, LEON WILLIAM, b Columbus, Ga, June 9, 27; m 48; c 3. BIOCHEMISTRY, PROTEIN STRUCTURE. *Educ:* Auburn Univ, BS, 47; Univ Ill, MS, 49, PhD(biochem), 51. *Prof Exp:* From asst prof to assoc prof, 53-65, assoc dean, 67-72, PROF BIOCHEM, SCH MED, VANDERBILT UNIV, 65-, CHMN DEPT, 72- *Concurrent Pos:* Fel, Dept Biochem, Univ Wash, 51-53; USPHS spec fel, Netherlands Nat Defense Orgn, 61-62; vis staff, Nat Inst Med Res, Mill Hill, London, Eng, 78; vis prof, Dept Physiol Chem, Univ Utrecht, Netherlands, 80. *Mem:* AAAS; NY Acad Sci; Am Chem Soc; Am Soc Biol Chemists. *Res:* Protein structure and function; glycoproteins; enzyme mechanisms; collagens. *Mailing Add:* Dept Biochem Vanderbilt Univ Sch Med 507 Light Hall Nashville TN 37232

CUNNINGHAM, MADELEINE WHITE, b Greenville, Miss, Feb 24, 46; m 69; c 3. IMMUNOCHEMISTRY. *Educ:* Univ Tenn, MS, 71, PhD(microbiol & immunol), 73. *Prof Exp:* NIH teaching fel immunol, Vet Admin Hosp, Memphis, Tenn, 71-73; adj assoc, dept microbiol, immunol & biochem, 73-76, res assoc, 80-81, asst prof microbiol & immunol, dept microbiol, 81-86, ASSOC PROF MICROBIOL & IMMUNOL, DEPT MICROBIOL, UNIV OKLA HEALTH SCI CTR, 87- *Concurrent Pos:* Okla Med Res Found fel, 73-76; consult microbiol, Clinton Regional Hosp, Okla, 80-84; mem, myocarditis working group, NIH, 85-86; Eloise Gerry fel, Grad Women Sci, 81-82; Res Career Develop Award, NIH, 86-91. *Mem:* Am Soc Microbiol; AAAS; Sigma Xi; Am Heart Asn; Lancefield Soc; Am Asn Immunol. *Res:* Study of antigens of group A streptococci that are immunologically cross-reactive with host tissue antigens, the relationship of these responses by the host to autoimmunity in general, and how they contribute to the development of diseases such as rheumatic carditis or myocarditis. *Mailing Add:* Dept Microbiol & Immunol Univ Okla Health Sci Ctr PO Box 26901 Oklahoma City OK 73190

CUNNINGHAM, MARY ELIZABETH, b Newark, NJ, Apr 21, 31. FLUID DYNAMICS. *Educ:* Mt Holyoke Col, AB, 53; Univ Ill, MS, 55; Univ Ore, PhD(physics), 64; Univ Conn, MD, 82. *Prof Exp:* Sr physicist, Lawrence Livermore Lab, 64-78. *Concurrent Pos:* Reviewer, Am J Physics, Am Phys Soc, 75- *Mem:* Am Phys Soc; NY Acad Sci; AAAS; Am Med Asn; Sigma Xi. *Res:* High temperatures fluid dynamics, 1, 2, and 3 dimensional computer codes for work in this field. *Mailing Add:* 11339 Sutters Ft Way Univ Conn Health Ctr Gold River CA 95670-7211

CUNNINGHAM, MICHAEL PAUL, b St John, NB, Mar 5, 43; m 66; c 1. PHOTOCHEMISTRY, THIN FILMS. *Educ:* St Francis Xavier Univ, BSc, 66; McGill Univ, MSc, 69, PhD(chem), 71. *Prof Exp:* Indust fel, Bristol Meyers Corp, 70-71; CHEMIST, EASTMAN KODAK CO RES LABS, 71- *Res:* Organic photochemistry; chemistry of light-sensitive polymers; photo-imaging processes; photographic sensitizing dyes and dye forming processes; thin films; preparation and characterization of thin films; optical disk. *Mailing Add:* 180 Parklands Dr Rochester NY 14616

CUNNINGHAM, NEWLIN BUCHANAN, b Boone's Path, Va, July 15, 17; m 47; c 4. INDUSTRIAL CHEMISTRY, FIBERS INTERMEDIATES. *Educ:* Emory & Henry Col, BS, 38; WVa Univ, MS, 63. *Prof Exp:* Chemist, E I du Pont de Nemours & Co, Inc, 38-43, asst chief chemist, 43-70, sr chemist, 70-80, staff chemist, Belle Works, 80-82; RETIRED. *Mem:* Sigma Xi; Am Chem Soc. *Res:* Analytical chemistry; waste treatment. *Mailing Add:* 1531 Bedford Rd Charleston WV 25314

CUNNINGHAM, PAUL THOMAS, b Newton, Iowa, Oct 16, 36; m 61; c 3. PHYSICAL CHEMISTRY, SCIENTIFIC & TECHNOLOGICAL MANAGEMENT. *Educ:* Univ Idaho, BS, 58; San Diego State Col, MS, 65; Univ Calif, Berkeley, PhD(phys chem), 68. *Prof Exp:* Mem staff, Argonne Nat Lab, 68-75, group leader, environ chem, 75-76, head anal & environ sect 77-81, mgr anal chem lab, chem eng div, 77-82; GROUP LEADER, ANALYTICAL CHEM, CHEM DIV, LOS ALAMOS NAT LAB, 82. *Honors & Awards:* IR-100 Award. *Mem:* Optical Soc Am; Am Chem Soc; AAAS. *Res:* Atmospheric chemistry; chemical analysis of airborne particulates; molecular spectroscopy; chemical kinetics of reactions related to environmental pollution; analytical chemistry; management of science and technology. *Mailing Add:* Los Alamos Nat Lab CM-NM F628 PO Box 1663 Los Alamos NM 87545

CUNNINGHAM, RICHARD G(REENLAW), b Olney, Ill, Sept 23, 21; m 44; c 3. MECHANICAL ENGINEERING, RESEARCH ADMINISTRATION. *Educ:* Northwestern Univ, BS, 43, MS, 47, PhD, 50. *Prof Exp:* Res engr, Res & Develop Labs, Pure Oil Co, 50-51; proj engr, Dept of Eng Res, Pa State Univ, 51-52, assoc prof eng res, 52-54; res engr, Wood River Res Labs, Shell Oil Co, 54-56, res group leader, 56-60, sr res engr, Mfg Res Dept, 60-61; head dept, Pa State Univ, 62-71, chmn univ fac senate, 67-68, prof mech eng, 61-84, vpres res & grad studies, 71-84, spec asst to pres for comput & info systs, 81-84, EMER VPRES RES & GRAD STUDIES & EMER PROF MECH ENG, PA STATE UNIV, 84- *Concurrent Pos:* Vpres educ, Am Soc Mech Engrs, 73-75. *Honors & Awards:* L P Moody Award, Am Soc Mech Engrs, 74. *Mem:* Fel AAAS; Am Soc Mech Engrs; Am Soc Eng Educ. *Res:* Fluid mechanics; jet pumps; cavitation; two-phase flow. *Mailing Add:* 5140 Grey Mountain Trail N Tucson AZ 85715

CUNNINGHAM, RICHARD PRESTON, b New Bedford, Mass, July 31, 48; m 77. NUCLEIC ACID ENZYMOLOGY, DNA REPAIR. *Educ:* Brown Univ, BA & BSc, 71; Johns Hopkins Univ, PhD(biol), 77. *Prof Exp:* Teaching fel, Sch Med, Yale Univ, 76-80; teaching fel, Sch Med, Johns Hopkins Univ, 80-82; ASST PROF BIOL SCI, STATE UNIV NY, ALBANY, 82- *Mem:* Am Soc Microbiol; Am Soc Biol Chemists. *Res:* The enzymology of base excision repair of DNA in microorganisms; genetic analysis of mutants. *Mailing Add:* Dept Biol Sci State Univ NY 1400 Washington Ave Albany NY 12222

CUNNINGHAM, ROBERT ASHLEY, b Lovington, NMex, May 24, 23; m 48; c 6. MECHANICAL ENGINEERING. *Educ:* Rice Inst, BS, 49, MS, 55. *Prof Exp:* Proj engr design, Hughes Tool Co, 50-52, asst develop engr, 53-58, res engr, Drilling Res, 59-70, dir prod res, 70-78, dir, Advanced Technol Projs & Div Liaison, 78-85. *Concurrent Pos:* Lectr, Dept Mat Sci, Rice Univ, 85- *Mem:* Am Inst Mining, Metall & Petrol Engrs; Soc Petrol Engrs; Am Petrol Inst. *Res:* Lubrication; drilling research; physical properties of solids; theory of fracture of brittle materials; resistance of brittle materials to impact; theory of plasticity; new product development and improvement. *Mailing Add:* Dept Mat Sci Rice Univ 6100 S Main Houston TX 77005

CUNNINGHAM, ROBERT ELWIN, b Parkersburg, WVa, Mar 16, 29; m 54; c 2. POLYMER CHEMISTRY. *Educ:* WVa Univ, BS, 51, PhD(chem), 55. *Prof Exp:* Instr WVa Univ, 54; sr res chemist, Goodyear Tire & Rubber Co, 55-75, res chemist, 75-81, res & develop assoc, 81-88; RETIRED. *Mem:* Am Chem Soc. *Res:* Stereospecific polymerizations of olefins and diolefins; anionic polymerizations initiated by lithium alkyls; preparation and properties of block polymers using anionic techniques. *Mailing Add:* 342 Kenilworth Dr Akron OH 44313

CUNNINGHAM, ROBERT LESTER, b Fullerton, Nebr, July 6, 29; m; c 4. SOIL GENESIS & MORPHOLOGY. *Educ:* Univ Nebr, BS, 59, MS, 61; Wash State Univ, PhD(soil genesis & morphol), 64. *Prof Exp:* Res asst soil genesis, Wash State Univ, 60-64; asst prof soil technol, 64-68; assoc prof, 68-74, PROF SOIL GENESIS & MORPHOL, PA STATE UNIV, 74- *Mem:* Soc Conserv Soc Am; Soil Sci Soc Am. *Res:* Movement of soil materials; arrangement of genetic soil parts; relationships of soils to the ecosystem of which they are a part; characteristics, interpretations and use of Pennsylvania soils; micromorphology examination of certain soil features; application of remote sensing techniques in differentiating soils in the field. *Mailing Add:* 955 Crabapple Dr State College PA 16801-9451

CUNNINGHAM, ROBERT M, b Boston, Mass, July 1, 19; m 45; c 3. METEOROLOGY. *Educ:* Mass Inst Technol, SB, 42, ScD(meteorol), 52. *Prof Exp:* Instr meteorol instruments, Mass Inst Technol, 42-43, asst meteorol, 43-46, res assoc, 46-52; proj scientist, Air Force Geophysics Lab, 74-80, br chief, 61-74, sr scientist meteorol, 74-80; field dir, World Meteorol Orgn, UN, 79-83; RETIRED. *Concurrent Pos:* Mem, Nat Adv Comt Aeronaut Subcomt Icing, 52-57; mem, Weather Modification Panel, World Meteorol Orgn, 66- *Mem:* fel Am Meteorol Soc; Am Geophys Union; Royal Meteorol Soc; Sigma Xi; Can Meteorol & Oceanog Soc. *Res:* Physics of the atmosphere; cloud physics; particle size; water content; cloud dynamics; airborne meteorological instrumentation; weather erosion effects on high speed flight; weather modification; acid fog. *Mailing Add:* 11 Rockwood Lane Lincoln MA 01773

CUNNINGHAM, ROBERT STEPHEN, b Springfield, Mo, June 28, 42; m 64; c 2. COMPUTER SCIENCE, COMPUTER GRAPHICS. *Educ:* Drury Col, BA, 64; Univ Ore, MA, 66, PhD(math), 69, Ore State Univ, MS(comput sci), 82. *Prof Exp:* Asst prof math, Univ Kans, 69-74; assoc prof math & comput sci, Birmingham-Southern Col, 74-82; PROF COMPUT SCI CALIF STATE UNIV, STANISLAUS, 82- *Mem:* Asn Comput Mach; Spec Int Group Comput Graphics; Inst Elec & Electronic Engrs; Math Asn Am. *Res:* Noncommutative ring theory; coding theory; computer graphics. *Mailing Add:* Dept Comput Sci Calif State Univ Stanislaus 801 W Monte Vista Ave Turlock CA 95380

CUNNINGHAM, THOMAS B, b Washington, DC, May 8, 46; m 68; c 2. AEROSPACE ENGINEERING, AUTOMATIC CONTROL SYSTEMS. *Educ:* Univ Nebr, BS, 69; Purdue Univ, MS, 72, PhD(eng), 73. *Prof Exp:* DIR RES ENG AUTOMATIC CONTROL, HONEYWELL INC, 73- *Concurrent Pos:* Adj prof, Univ Minn, 78- *Mem:* Sigma Xi; Am Inst Aeronaut & Astronaut; Inst Elec & Electronics Engrs. *Res:* Applications of modern control and estimation theory to aerospace and industrial problems. *Mailing Add:* 1347 26th Ave NW New Brighton MN 55112

CUNNINGHAM, VIRGIL DWAYNE, b Artesian, SDak, Nov 29, 30; m 56; c 1. ENTOMOLOGY, PLANT PATHOLOGY. *Educ:* NDak State Univ, BS, 58, MS, 60; Iowa State Univ, PhD(entom), 63. *Prof Exp:* Entomologist, Biol Sci Res Ctr, Shell Develop Co, 63-69, supvr entom, 69-70, dept head entom, 70-74, develop mgr animal health, 74-78, dir develop, 78-86; RETIRED. *Res:* Agricultural chemicals; insecticides, herbicides and plant growth regulators. *Mailing Add:* 928 Inverness Dr Paso Robles CA 93446

CUNNINGHAM, W(ALTER) J(ACK), b Comanche, Tex, Aug 21, 17; m 44; c 2. ELECTRICAL ENGINEERING. *Educ:* Univ Tex, AB, 37, AM, 38; Harvard Univ, ScM, 39, PhD(eng sci), 47. *Prof Exp:* Tutor physics, Univ Tex, 35-38; instr physics & commun eng, Harvard Univ, 39-45; from asst prof to prof elec eng, 46-63, assoc chmn, Dept Eng & Appl Sci, 69-72, prof eng & appl sci, 63-81, prof, 81-88, EMER PROF ELEC ENG, YALE UNIV, 88- *Concurrent Pos:* Fel, Trumbull Col, 48-; assoc ed, J Franklin Inst, 62-75; chmn, comt publ, Sigma Xi, 83-87. *Mem:* Inst Elec & Electronics Engrs; Acoust Soc Am; Sigma Xi. *Res:* Systems theory; nonlinear analysis; applied mathematics. *Mailing Add:* Becton Ctr Yale Univ New Haven CT 06520-2157

CUNNINHAM-RUNDLES, CHARLOTTE, b Ann Arbor, Mich, July 12, 43; m; c 1. IMMUNOLOGY, IMMUNOCHEMISTRY. *Educ:* Duke Univ, BS, 65; Columbia Univ, MD, 69; New York Univ, PhD(immunol), 74. *Prof Exp:* Intern med, New York Univ, 69-70, resident, 70-72, asst scientist, 72-74; assoc immunol, Sloan Kettering Inst, 74-86; asst prof immunol & biochem, Cornell Grad Sch Arts & Sci, 76-86; ASSOC PROF MED, MT SINAI MED CTR, NY COL, 86- *Concurrent Pos:* Attending physician, dept med, Med Ctr, New York Univ, 72- & Mem Hosp, 74-86; Senate Exec Comt Mem, Sloan Kettering Cancer Ctr, 77-86; attend physician, Mt Sinai Hosp, 86- *Mem:* Fel Am Col Physicians; Am Asn Immunologists; Am Fedn Clin Res. *Res:* Secretory immune system; immuno deficiency diseases of man; immune complexes; network theory of immunity in man. *Mailing Add:* Dept Med & Pediat Mt Sinai Med Ctr One Gustave L Levy Pl Box 1089 New York NY 10029

CUNNY, ROBERT WILLIAM, b Chicago, Ill, July 22, 24; m 60; c 2. SOIL MECHANICS, EARTH DAM DESIGN. *Educ:* Purdue Univ, BS, 46, MS, 48. *Prof Exp:* Engr, US Army Engr Waterways Exp Sta, 48-50 & Portland Cement Asn, 52-54; engr soils div, US Army Engr Waterways Exp Sta, 54-74, dir, Soil Mech Info Anal Ctr, Geotech Lab, 74-80; CONSULT, 80- *Concurrent Pos:* Mem comt properties soil & rock, Dept Soils, Geol & Found, Hwy Res Bd, Nat Acad Sci-Nat Res Coun, 64- *Mem:* Nat Soc Prof Engrs. *Res:* Performance evaluation of completed earth structures; collection, analysis, evaluation and dissemination of geotechnical information as related to stability of foundations, earth and rockfill dams and navigation and flood control structures; develop computerized data base of geotechnical information. *Mailing Add:* 12 Signal Hill Lane Vicksburg MS 39180

CUNY, ROBERT MICHAEL, b Wettingen, Switz, June 14, 52; Can citizen; m 88; c 2. MOLECULAR BIOLOGY, ZOOLOGY. *Educ:* Swiss Fed Inst Technol, dipl & BEd, 76; Univ Alta, PhD(zool), 82. *Prof Exp:* Res assoc embryol, Ind Univ, Bloomington, 81-84; res assoc ophthal, French Nat Inst Health & Med Res, 85-86; res assoc molecular biol, Univ Sask, Saskatoon, 86-89; INSTR BIOL, LAKELAND COL, VERMILION-LLOYDMINSTER, 89- *Concurrent Pos:* Hon res assoc, Univ Alta, Can, 90-91. *Mem:* Can Soc Cell Biol; Can Entom Soc; Am Soc Develop Biol; Am Soc Zoologists. *Res:* Some newts and salamanders can regulate their lens from the dorsal iris margin-determined the growth factors released by the retina and the pituitary gland that stimulates this process; genes activated during lens regeneration; control of dipteran evolution by changning vegetation patterns. *Mailing Add:* Lakeland Col Bag 6600 2602 59th Ave Lloydminster AB S9V 1Z3 Can

CUPERY, KENNETH N, b Wilmington, Del, May 27, 37; div; c 2. OPTICS. *Educ:* Oberlin Col, BA, 59; Johns Hopkins Univ, MA, 64. *Prof Exp:* Develop engr, Eastman Kodak, 65-66, sr develop engr, 66-69, sr res physicist, 69-72, supvr optics res, Kodak Apparatus Div, Res Lab, 72-80, PROJ MGR, FED SYST, EASTMAN KODAK, 80- *Mem:* Optical Soc Am; Soc Photo Instrument Engrs. *Res:* Modulation transfer function and relation to subjective quality of optical images; MTF based optical test systems; electrooptical imaging systems. *Mailing Add:* 139 Roosevelt Rd Rochester NY 14618

CUPERY, WILLIS ELI, b Wilmington, Del, July 1, 32; m 55; c 3. PESTICIDE CHEMISTRY. *Educ:* Oberlin Col, AB, 54; Univ Ill, PhD(org chem), 58. *Prof Exp:* Res chemist, E I Du Pont de Nemours & Co, Inc, 58-66, sr res chemist, 66-79, res assoc, 79-81, group leader, 81-85; PROPRIETOR, CUPERY WILL CONSULT & SERV, 85- *Concurrent Pos:* Prog dir, Nat Agr Aviation Res & Educ Fedn, 89- *Mem:* Am Soc Testing & Mat. *Res:* Pesticide formulation development, application methods. *Mailing Add:* Cupery Will Consult & Serv 13 Crestfield Rd Wilmington DE 19810-1401

CUPIT, CHARLES R(ICHARD), b New Orleans, La, Sept 8, 30; m 58; c 3. CHEMICAL ENGINEERING. *Educ:* Miss State Univ, BS, 54; Univ Ill, MS, 56, PhD(chem eng), 58; Wichita State Univ, MBA, 86. *Prof Exp:* Res engr, Mobil Oil Co, 54; asst chem eng, Univ Ill, 54-55, instr, 57-58; res engr, Shell Oil Co, 58-65; sr engr, Lockheed Electronics Corp, 65-68; sect head, TRW Systs Group, 68-69; chem eng consult, 69-70; sr scientist chem div, 70-77, ENGR GROUP LEADER, CHEM DIV, RES & DEVELOP, VULCAN CHEM CO, 77- *Mem:* Am Inst Chem Engrs. *Res:* Manage process design and development of chemical plants; chemical plant economic studies; design pilot plant and commercial plant process schemes and equipment; manage pilot plant projects. *Mailing Add:* Chem Div Res & Develop Vulcan Chem Co PO Box 12283 Wichita KS 67277

CUPP, CALVIN R, b Toronto, Ont, June 7, 24; m 49; c 4. METALLURGY, BIOTECHNOLOGY. *Educ:* Univ Toronto, BASc, 50, MASc, 51, PhD(metall), 53. *Prof Exp:* Res metallurgist, Int Nickel Co Can, 53-56; spec assignment, Atomic Energy of Can, Ltd, 57-61; mgr metal physics & metallog, Inco Res & Develop Ctr, Inc, 61-74, res mgr phys/anal group, 74-78, mgr chem dept, 78-82; SCI ADV, SCI COUN OF CANADA, 82-; RETIRED. *Mem:* Fel Am Soc Metals; Soc Indust Microbiol; Am Soc Testing & Mat; Am Phys Soc; Am Chem Soc. *Res:* Physical metallurgy; biotechnology; gases in metals; plastic deformation; radiation effects in metals; surface characterization; electrolytic processes. *Mailing Add:* 1000 King St W Unit 1004 Kingston ON K7M 8H3 Can

CUPP, EDDIE WAYNE, b Highsplint, Ky, Apr 17, 41; m 64; c 1. ENTOMOLOGY, PARASITOLOGY. *Educ:* Murray State Univ, BA, 64; Univ Ill, Urbana, PhD(entom), 69. *Prof Exp:* Res fel, Sch Pub Health & Trop Med, Tulane Univ, 69-70; res biologist, Gulf South Res Inst, La, 70-71; asst prof biol, Univ Southern Miss, 71-74; asst prof, 74-78, ASSOC PROF ENTOM, CORNELL UNIV, 78- *Concurrent Pos:* Adj asst prof, Sch Pub Health & Trop Med, Tulane Univ, 70-71. *Mem:* Entom Soc Am; Am Soc Trop Med & Hyg; Am Mosquito Control Asn; Royal Soc Trop Med & Hyg; Sigma Xi. *Res:* Bionomics of medically important Diptera; ecology of arboviral diseases; dynamics of transmission of filariae. *Mailing Add:* Dept Entom Comstock Hall Cornell Univ Ithaca NY 14850

CUPP, PAUL VERNON, JR, b Corbin, Ky, Oct 18, 42; m 69. PHYSIOLOGICAL ECOLOGY, HERPETOLOGY. *Educ:* Eastern Ky Univ, BS, 65, MS, 70; Clemson Univ, PhD(zool), 74. *Prof Exp:* Asst prof biol, Ga Southern Col, 73-74; asst prof, 74-80, ASSOC PROF BIOL, EASTERN KY UNIV, 80- *Mem:* Sigma Xi; Am Soc Zoologists; Am Inst Biol Sci; Ecol Soc Am; Soc Study Amphibians & Reptiles. *Res:* Thermal tolerance and acclimation and physiological responses to temperature in amphibians and reptiles; aggressive and courtship behavior of amphibians. *Mailing Add:* Dept Biol Sci Eastern Ky Univ Richmond KY 40475

CUPPAGE, FRANCIS EDWARD, b Cleveland, Ohio, Aug 17, 32; m 56; c 3. PATHOLOGY. *Educ:* Case Western Univ, BS, 54; Ohio State Univ, MS, 59, MD, 59. *Prof Exp:* Rotating intern med, Univ Hosps, Cleveland, 59-60, resident path, 60-64; instr, Case Western Reserve Univ, 64-65; asst prof, Ohio State Univ, 65-67; from asst prof to assoc prof, 67-73, PROF PATH, UNIV KANS MED CTR, KANSAS CITY, 73- *Concurrent Pos:* NIH fel path, Case Western Univ, 63-64. *Mem:* Int Soc Nephrol; Am Asn Path & Bact; Int Acad Path; Am Soc Exp Path; Sigma Xi. *Res:* Kidney response of nephron to injury and repair. *Mailing Add:* Dept Path Univ Kans Med Ctr Kansas City KS 66103

CUPPER, ROBERT, b Warren, Pa, Oct 24, 42. CURRICULUM DESIGN, THEORY COMPUT OPERATING SYSTEMS. *Educ:* Univ Pittsburgh, PhD, 74. *Prof Exp:* PROF & CHAIR COMPUTER SCI, ALLEGHENY COL, 77- *Concurrent Pos:* Chmn, Student Chap Comt, Asn Comput Mach, 80-; co-founder, Lib Arts Computer Sci Consortium; vis res prof computer sci, Bowdoin Col, 89-90. *Mem:* Asn Comput Mach. *Mailing Add:* Dept Computer Sci Allegheny Col Meadville PA 16335

CUPPER, ROBERT ALTON, b Tyrone, Pa, Jan 7, 18; m 41; c 4. ORGANIC CHEMISTRY, PETROLEUM CHEMISTRY. *Educ:* Juniata Col, BS, 40. *Prof Exp:* Chemist chem petrol prods, United Ref Co, 41-46; indust fel org chem, Mellon Inst, 46-59; group leader, Chem & Plastic Div, Union Carbide Corp, 60-69, res scientist, Res & Develop Dept, 69-80; CONSULT, FUELS, LUBRICANTS & SYNTHETIC LUBRICANTS, 80- *Mem:* AAAS; Am Chem Soc; Soc Automotive Eng; Am Soc Lubrication Eng; NY Acad Sci. *Res:* Fuels and lubricants; additives and chemical inter-synthetic lubricants; synthesis and formulation; agricultural chemicals; pesticide intermediates and formulants. *Mailing Add:* Rt 1 Box 169A Liberty PA 16930-9506

CUPPLES, BARRETT L(EMOYNE), b Osceola Mills, Pa, Nov 18, 41; m 63; c 3. CHEMICAL ENGINEERING. *Educ:* Pa State Univ, BS, 63, MS, 65, PhD(chem eng), 69. *Prof Exp:* Res engr, Gulf Sci & Technol Co, 68-74, proj engr, 74-78, staff engr, 78-80, develop mgr, Gulf Oil Chem Co, 81-83, mgr, petrochem tech serv, Gulf Oil Prod Co, 83-85, sr engr, olefins & derivatives, Petrochem Technol, Chevron Chem Co, 85-88, DIR, TECH SERV & PAO RES & DEVELOP, CHEVRON CHEMCO, 88- *Mem:* Am Chem Soc; Soc Tribologists & Lubrication Engrs; Soc Auto Engrs. *Res:* Development of chemical processes such as the oligomerization of olefins for use as synthetic functional fluids. *Mailing Add:* 4819 Shore Hills Rd Kingwood TX 77345

CUPPS, PERRY THOMAS, b Granby, Mo, June 18, 16; m 44; c 4. ANIMAL PHYSIOLOGY. *Educ:* Univ Mo, BS, 39; Cornell Univ, PhD(animal physiol), 43. *Prof Exp:* Asst, Cornell Univ, 39-43; from asst prof to assoc prof, Univ Calif, Davis, 46-59, prof animal husb, 59-82; RETIRED. *Concurrent Pos:* NSF fel, 61-62. *Res:* Endocrinology; thyrotrophic hormone assays; normal change of reproductive tract during estrous cycle; joint lesions in horses; estrous cycle in cattle; adrenal physiology; embryo transfer. *Mailing Add:* 828 Miller Dr Davis CA 95616

CURCIO, LAWRENCE NICHOLAS, b Newark, NJ, July 21, 50; m. HAZARD ASSESSMENT, PRODUCT SAFETY. *Educ:* Rutgers Univ, BA, 72, MS, 73, PhD(physiol), 78. *Prof Exp:* Sr toxicologist, Ciba-Geigy Corp, 78-81; staff toxicologist, Exxon Corp, 81-85, STAFF TECH SPECIALIST MKT, EXXON CO, USA, 85- *Mem:* Int Soc Regulatory Toxicol. *Res:* Role of electrolytes in developing kidney disease; role of toxicology in environmental affairs and hazard assessment. *Mailing Add:* Ciba-Geigy Corp 556 Morris Ave Summit NJ 07901

CURD, MILTON RAYBURN, zoology; deceased, see previous edition for last biography

CURETON, GLEN LEE, b Santa Cruz, Calif, Mar 29, 38; m 64, 85; c 3. PHARMACEUTICS. *Educ:* Univ Calif, San Francisco, PharmD, 62; Harvard Univ, MBA, 64. *Prof Exp:* Asst dir life sci res, Stanford Res Inst, 64-65, indust economist, 65-66, health economist, 66-67; dir new prod develop, Chattem Drug & Chem Co, 67-68, dir res & new prod develop, 68-72; dir res & develop, Barnes-Hind Pharmaceuticals, Inc, 72-76; com develop mgr, Cutter Labs, Inc, 76-84; vpres bus develop, Calif Biotechnol Inc, 84-86, Appl Immune Sci, 86-88; SR CONSULT, SRI INT, 88- *Mem:* AAAS; Am Soc Apheresis; Am Pharmaceut Asn; Am Acad Dermat; Asn Res in Vision & Ophthalmol. *Res:* Antacids, anitperspirants and other chemicals, inculding high purity aluminas; development of pharmaceuticals; ophthalmic contact lens solutions; dermatologicals; technoeconomic analyses of life sciences research, pharmaceuticals and cosmetics; licensing and strategic planning; biotechnology, immunology, and medical devices. *Mailing Add:* 714 Raymundo Ave Los Altos CA 94024

CURETON, KIRK J, b Aug 29, 47. PHYSICAL EDUCATION. *Educ:* Carleton Col, BA, 69; Univ Ill, MS, 72, PhD(phys educ), 76. *Prof Exp:* Teaching asst, Univ Ill, 70-71; instr phys educ, Ball State Univ, 71-73; res asst, Phys Fitness Res Lab, Univ Ill, 73-76; dir, Human Performance Lab, 76-87, from asst prof to prof phys educ, 76-91, DIR, EXERCISE PHYSIOL LAB, UNIV, GA, 87-, PROF & HEAD, DEPT EXERCISE SCI, 91- *Concurrent Pos:* Mem, Res Comt, Nat Col Phys Educ Asn Men, 77-78; chairperson, Res Sect, Ga Asn Health, Phys Educ, Recreation & Dance, 79-80; mem, Sports Sci Adv Group, US Olympic Comt, 90; consult. *Honors & Awards:* Mabel Lee Award, Am Alliance Health, Phys Educ, Recreation & Dance, 83. *Mem:* Fel Am Alliance Health, Phys Educ, Recreation & Dance; fel Am Col Sports Med; Am Physiol Soc; Can Asn Appl Sport Sci; fel Human Biol Coun; fel Am Acad Phys Educ. *Res:* Physical fitness. *Mailing Add:* Dept Exercise Sci Univ Ga Athens GA 30602

CURETON, THOMAS (KIRK), JR, b Fernandina, Fla, Aug 4, 01; m; c 3. APPLIED PHYSIOLOGY. *Educ:* Yale Univ, BS, 25; Int YMCA Col, BPE, 29, MPE, 30; Columbia Univ, AM, 36, PhD(educ res), 39. *Hon Degrees:* DSc, Univ Ottawa, 68, Univ Southern Ill, 69. *Prof Exp:* Phys dir & teacher biol & gen sci, Suffield Sch, Conn, 25-29; prof appl physics & animal mech, Springfield Col, 29-41; dir res nat sci & phys educ, 35-42; from assoc prof to prof, 42-74, EMER DIR PHYS FITNESS INST, 70-, EMER PROF PHYS EDUC, UNIV ILL, URBANA-CHAMPAIGN, 74- *Concurrent Pos:* Supvr res, Col Phys Educ, Univ Ill, Urbana-Champaign, 42-74. *Honors & Awards:* Robert-Gulick Award, Phys Educ Soc, 45; Am Col Sports Med Award, 68; Gulick Medal, Am Asn Health, Phys Educ & Recreation, 75; Hetherington Award, Am Acad Phys Educ, 76. *Mem:* Am Physiol Soc; fel Soc Res Child Develop; fel Am Pub Health Asn; fel Am Asn Health, Phys Educ & Recreation; NY Acad Sci. *Res:* Body mechanics; anthropometry; physical fitness; physiology of exercise; aquatics; athletics; gymnastics; sport medicine; physical education. *Mailing Add:* 501 E Washington St Urbana IL 61801

CURIE-COHEN, MARTIN MICHAEL, population genetics, inbreeding; deceased, see previous edition for last biography

CURJEL, CASPAR ROBERT, b Switz, Nov 21, 31; m 55; c 2. MATHEMATICS. *Educ:* Swiss Fed Inst Technol, dipl, 54, DrScMath, 60. *Prof Exp:* From instr to asst prof, Cornell Univ, 60-64; vis assoc prof, 64-65, assoc prof, 65-69, PROF MATH, UNIV WASH, 69- *Concurrent Pos:* Mem, Inst Advan Study, 63-64. *Res:* Algebraic topology; algebra. *Mailing Add:* Dept Math Univ Wash Seattle WA 98195

CURL, ELROY ARVEL, b Marmaduke, Ark, Dec 1, 21. PLANT PATHOLOGY. *Educ:* La Polytech Inst, BS, 49; Univ Ark, MS, 50; Univ Ill, PhD(plant path), 54. *Prof Exp:* Spec res asst plant path, State Natural Hist Surv, Ill, 50-54, asst plant pathologist, 54; asst plant pathologist, 54-57, assoc prof, 57-67, PROF BOT & MICROBIOL, AUBURN UNIV, 67- *Mem:* Am Phytopath Soc; Mycol Soc Am; Sigma Xi. *Res:* Soil microbiology; soil fungus ecology; soil microbe-pesticide interactions; biological control; root diseases of plants. *Mailing Add:* 149 S Ross Auburn AL 36830

CURL, HERBERT (CHARLES), JR, b New York, NY, Feb 26, 28; m 83; c 2. BIOGEOCHEMISTRY. *Educ:* Wagner Col, BS, 50; Ohio State Univ, MS, 51; Fla State Univ, PhD(biol oceanog), 56. *Prof Exp:* Asst, Oceanog Inst, 54-56; res assoc, Woods Hole Oceanog Inst, 56-61; from assoc prof to prof oceanog, Ore State Univ, 61-74; fisheries oceanog coordr, Nat Marine Fisheries Serv, 74-75; ecologist, Environ Res Labs, Boulder, Colo, 75-77, SUPVRY OCEANOGRAPHER, PAC MARINE ENVIRON LAB, NAT OCEANIC & ATMOSPHERIC ADMIN, 77- *Concurrent Pos:* Fulbright grant, Colombia, SAm, 71. *Mem:* Fel AAAS; Am Geophys Union; Am Soc Limnol & Oceanog(secy, 64-66). *Res:* Physiological ecology of marine organisms; biogeochemistry; estuarine and coastal circulation. *Mailing Add:* Pac Marine Environ Lab Bldg 3 7600 Sand Point Way NE Seattle WA 98115

CURL, RANE L(OCKE), b New York, NY, July 5, 29; m 54, 63, 82; c 3. SYSTEMS MODELING. *Educ:* Mass Inst Tech, SB, 51, ScD(chem eng), 55. *Prof Exp:* Asst chem eng, Mass Inst Technol, 54-55; chem engr, Shell Develop Co, 55-61; hon res asst statist & tutor chem eng, Univ Col, London, 61-62; res assoc phys tech, Eindhoven Tech, 62-64; assoc prof chem eng, 64-68, PROF CHEM ENG, UNIV MICH, ANN ARBOR, 68- *Concurrent Pos:* Instr eng exten, Univ Calif, 58-59; adj secy, Int Union Speleol, 81-86; chmn, Eight Int Cong Speleol, 81; consult. *Mem:* AAAS; Am Chem Soc; Am Inst Chem Engrs; hon mem Nat Speleol Soc (pres, 70-74); Int Asn Hydrogeologists; Int Asn Math Geol. *Res:* Automatic process control; fluid dynamics and mass transfer; chemical reactors; stochastic process and computer applications; geomorphologic process; absorption; speleology. *Mailing Add:* Dept Chem Eng Univ Mich Ann Arbor MI 48109-2136

CURL, ROBERT FLOYD, b Alice, Tex, Aug 23, 33; m 55; c 2. PHYSICAL CHEMISTRY. *Educ:* Rice Inst, BA, 54; Univ Calif, PhD, 57. *Prof Exp:* Res fel, Harvard Univ, 57-58; from asst prof to assoc prof, 58-67, PROF CHEM, RICE UNIV, 67- *Concurrent Pos:* NATO fel, Oxford Univ, 64-65; vis res officer, Nat Res Coun Can, 72-73; vis prof, Inst Molecular Sci, Okazaki, Japan, 77 & Univ Bonn, Germany, 85; US Sr Scientist award, Alexander von Humboldt Found, 84. *Honors & Awards:* Clayton Prize, Inst Mech Eng, 58. *Mem:* Am Chem Soc; Optical Soc Am. *Res:* Laser spectroscopy and chemical kinetics. *Mailing Add:* Dept Chem Rice Univ Houston TX 77251

CURL, SAM E, b Tolar, Tex; c 3. ANIMAL SCIENCE. *Educ:* Sam Houston State Univ, BS, 59; Univ Mo, MS, 61; Tex A&M Univ, PhD, 63. *Prof Exp:* Fac mem, dept animal sci, Tex Tech, 61, instr & res, animal physiol & genetics, 61-73, asst, interim & assoc dean & dir res, Col Agr Sci, 68-73; assoc vpres academic affairs, Tex Tech univ, 73-76; pres, Phillips Univ, 76-79; DEAN COL AGR SCI, TEX TECH UNIV, 79- *Concurrent Pos:* Bd trustees, Consortium Int Develop, Mgt Comt chmn, SW Consortium Plant Genetics & Water Resources; mem, Tex Agr Resources Protection Auth, Tex Agr Leadership Coun, Tex Crop & Livestock adv comt. *Mem:* Am Asn Univ Agr Adminr; Am Soc Animal Sci. *Mailing Add:* Col Agr Sci Tex Tech Univ Lubbock TX 79409

CURLEE, NEIL J(AMES), JR, b Schenectady, NY, July 14, 30; m 52; c 6. MECHANICAL & NUCLEAR ENGINEERING. *Educ:* Mass Inst Technol, SB, 52; Carnegie Inst Technol, MS, 55; Univ Pittsburgh, PhD(mech eng), 61. *Prof Exp:* Scientist, Bettis Atomic Power Lab, Westinghouse Elec Corp, 55-62, sr scientist, 62-67, supvr thermal & hydraul anal, 67-72, mgr thermal & hydral lab, 72-76, adv engr, Nuclear Steam Generation Div, 76-90, CONSULT ENGR, WESTINGHOUSE ELEC CORP, 91- *Concurrent Pos:* Instr, Eve Sch, Carnegie Inst Technol, 62. *Mem:* Am Soc Mech Engrs. *Res:* Nuclear reactor dynamics and control; magnetohydrodynamics; transient convective heat transfer. *Mailing Add:* 413 Stratton Lane Pittsburgh PA 15206

CURLESS, WILLIAM TOOLE, b Quincy, Ill, Mar 11, 28; div; c 3. ENVIRONMENTAL SCIENCES, PESTICIDE CHEMISTRY. *Educ:* Univ Ill, BS, 49; Univ Kans, MS, 56. *Prof Exp:* Anal chemist, Victor Chem Works, 49-51, inorg res chemist, 51-53; res chemist, Columbia-Southern Chem Corp, 55-61; res chemist, Spencer Chem Co, 62-66; sr res chemist, Gulf Oil Corp, 66-80; sr res scientist, Farmland Industs, 80-83; RETIRED. *Mem:* Am Chem Soc. *Res:* Nonaqueous solvents; condensed phosphates; phase diagrams; fertilizer technology; pesticide environmental studies in soil; pesticide residue analytical method development. *Mailing Add:* 10241 Woodson Dr Overland Park KS 66207

CURLEY, DENNIS M, IMMUNOLOGY. *Educ:* St John's Univ, Queens, NY, PhD(biol), 71. *Prof Exp:* PROF BIOL, LONG ISLAND UNIV, 62- *Mailing Add:* Dept Biol Long Island Univ University Plaza Brooklyn NY 11201

CURLEY, JAMES EDWARD, b Winchester, Mass, Apr 10, 44; m 66; c 2. ANALYTICAL CHEMISTRY. *Educ:* Univ Mass, Amherst, BS, 66, MS, 68, PhD(chem), 70. *Prof Exp:* Staff chemist, Med Res Lab, Pfizer Ctr Res, 70-73, sr res scientist, 73-75, sect supvr, Pfizer Qual Control, 75-80, proj leader, 80-83, MGR, PFIZER RES CTR, 83- *Mem:* Am Chem Soc; Sigma Xi. *Res:* Analytical methodology for penicillin products; analysis of pharmaceuticals by high pressure liquid chromatography; computer aided analytical chemistry; laboratory robotics. *Mailing Add:* Pfizer Inc RSCH Box 6 Bldg 118 Groton CT 06340-0404

CURLEY, WINIFRED H, b Ft McPhearson, Ga, Dec 15, 51; m 75; c 1. TOXICOLOGY. *Educ:* Queens Col-Charlotte, BS, 74; Univ NC, PhD(biochem), 83. *Prof Exp:* chemist, Nat Inst Environ Health Sci, NIH, Res Triangle Park, NC, 76-78; res fel pharmacol, Georgetown Univ, 78-80; health sci consult, 81-83; tech progs mgr, 83-85, DIR BIO-SCI, V J CICCONE & ASSOC, 85- *Concurrent Pos:* Health effects task group, Nat Sanit Found Standards Develop for Drinking Water Additives. *Mem:* AAAS; NY Acad Sci. *Res:* Federal regulation and evaluation of drinking water additives, pesticides and environmental contaminants; toxicology; biochemistry; pharmacology; health safety; environmental fate and effects of chemicals; water quality and standards, hazardous waste site assessment and cleanup risk and exposure assessment. *Mailing Add:* 1215 Lexham Dr Marietta GA 30068

CURLIN, GEORGE TAMS, b Mayfield, Ky, Sept 13, 39; m 62; c 2. EPIDEMIOLOGY, INFECTIOUS DISEASE. *Educ:* Centre Col Ky, BA, 61; Vanderbilt Univ, MD, 65; Johns Hopkins Univ, MPH, 73. *Prof Exp:* Intern int med, Univ Ky Hosp, 65-66; epidemic intel serv officer, Ctr Dis Control, USPHS, 66-68; fel int med, Johns Hopkins Hosp, 68-69; med epidemiologist cholera, Cholera Res Lab, Dacca, Bangladesh, 69-71; resident int med, Johns Hopkins Hosp, 71-72; head epidemiol div, Cholera Res Lab, 73-77; chief Epidemiol & Biomet Br, Nat Inst Allergy & Infectious Dis, NIH, 77-80; chief, Health Pop & Nutrit Div, Asia Bur, USAID, 80-82, dir, Off Health, 82-84; ASSOC DIR INT HEALTH RES, OFF INT HEALTH, PUB HEALTH SERV, 84- *Concurrent Pos:* Interim dir, Cholera Res Lab, 73-74. *Mem:* AAAS; Am Pub Health Asn; Royal Soc Trop Med & Hyg. *Res:* Epidemiology of enteric disease; field trials of vaccines; assessment of interventions for the control of infectious enteric disease; antitoxin immunity in cholera and other enterotoxigenic diarrheal diseases. *Mailing Add:* Nat Inst Allergy Div Microbiol & Infect Dis 900 Rockville Pike Bldg 31 Rm 7A52 Bethesda MD 20014

CURLIN, LEMUEL CALVERT, b Waxahachie, Tex, Feb 1, 13; m 40; c 3. PHARMACEUTICAL CHEMISTRY. *Educ:* Trinity Univ, Tex, BS, 34; Univ Chicago, MS, 36, PhD(biochem), 40. *Prof Exp:* Chemist, Ideal Lab, Univ Tex, 35-36; lab asst, Univ Chicago, 37-40, Armour & Co fel, 40; asst dir develop lab, Armour & Co, 40-41, res chemist, 41-43; secy, 64-69, chief chemist, L Perrigo Co, 43-78, exec vpres, 69-78; RETIRED. *Concurrent Pos:* Consult drug & cosmetic indust, 78- *Res:* Separation plasma proteins; synthetic thyroid; organic synthesis. *Mailing Add:* 4031 Gulfshore Blvd Naples FL 33940

CURME, HENRY GARRETT, b Niagara Falls, NY, Dec 31, 23; m 48; c 3. PHYSICAL CHEMISTRY. *Educ:* Northwestern Univ, BS, 45, MS, 47; Univ Calif, PhD(chem), 50. *Prof Exp:* Chemist, Manhattan Proj, Tenn Eastman Co, 44-46; asst, Northwestern Univ, 46-47 & Univ Calif, 47-50; res chemist & res assoc, Eastman Kodak Co, 50-65, lab head, 65-70, sr lab head, 70-85, res mgr, 86; RETIRED. *Mem:* Am Chem Soc; Am Asn Clin Chem. *Res:* Interactions between proteins and small molecules; photochemistry; polymer solutions; bulk properties and morphology of hydrophilic polymers; polymer networks; adsorption of polymers; surface properties of polymers; development of analytical techniques for clinical chemistry. *Mailing Add:* 920 Lake Rd Webster NY 14580

CURNEN, EDWARD CHARLES, JR, b Yonkers, NY, Jan 5, 09; m 42, 68; c 7. PEDIATRICS, MICROBIOLOGY. *Educ:* Yale Univ, AB, 31; Harvard Univ, MD, 35. *Prof Exp:* Res bacteriologist, Children's Hosp, Boston, Mass, 35-36, med intern, 36-38, med resident, 38-39; asst, Rockefeller Inst, NY, 39-46; from asst prof prev med to assoc prof prev med & pediat, Sch Med, Yale Univ, 46-52; prof pediat & chmn dept, Univ NC, 52-60; chmn dept, 60-70, Carpentier prof, 60-74, EMER CARPENTIER PROF PEDIAT, COL PHYSICIANS & SURGEONS, COLUMBIA UNIV, 74-, SPEC LECTR, 74- *Concurrent Pos:* Harvard Med Sch fel, Thorndike Mem Lab, Boston City Hosp, 38; asst, Harvard Med Sch, 38-39; asst resident physician, Rockefeller Inst Hosp, 39-46, mem, US Naval Res Unit, 41-46; res worker, Off Sci Res, 41-46; asst physician, New Haven Community Hosp, Conn, 46-48, assoc pediatrician, 48-52; chief pediat serv, NC Mem Hosp, 52-60, chief staff, 59-60; consult pediatrician, US Army Hosp, Ft Bragg, NC & NC State Bd Health, 52-60; dir pediat serv, Babies Hosp, Columbia Presby Med Ctr, 60-70, attend pediatrician, 60-74, consult, 74-83; Schick lectr, Mt Sinai Hosp, New York, 62; consult pediatrician, St Albans Naval Hosp, NY, 63-73 & St Luke's Hosp, New York, 66-80; attend pediatrician, Harlem Hosp, 74-78; assoc, Comn Influenza, US Armed Forces Epidemiol Bd, 47-69; mem, Planning Comt Studies Cardiovasc Effects Oriental Influenza, Nat Heart Inst & Influenza Res Comt, NIH, 57, Bd Sci counr, Div Biol Stand, 59-63, Infectious Dis & Trop Med Training Grant Comt, Nat Int Allergy & Infectious Dis, 60-63; chmn, Allergy & Infectious Dis Panel & mem, Rev Panel, Health Res Coun New York, 62-68, mem, Exec Comt, 68-72. *Mem:* Fel AAAS; Am Acad Pediat; Am Pediat Soc (vpres, 69); fel NY Acad Sci. *Res:* Infectious diseases; bacteriology and virology; hemagglutination phenomena. *Mailing Add:* 445 Sperry Rd Spring Hill Bethany CT 06525

CURNEN, MARY G MCCREA, b Belgium; m 55, 68; c 7. EPIDEMIOLOGY, PEDIATRICS. *Educ:* Cath Univ Louvain, MD, 48; Inst Trop Med, Belg, DTM, 49; Columbia Univ, DrPH(epidemiol), 73. *Prof Exp:* Intern pediat, Cath Univ Louvain Hosp, 47-48; resident, Bellevue Med Ctr, Col Med, NY Univ, 52-53; instr prev med, Sch Med, Yale Univ, 53-54, instr prev med & pediat, 54-56; clin physician child health & pub schs, New Haven City Health Dept, 59-64; res assoc pediat, Yale Univ, 64-66, res assoc med, Yale Univ Virol Lab, Vet Admin Hosp, West Haven & Yale Univ, 67-69; res assoc epidemiol, Sch Pub Health, Columbia Univ, 73-74, from asst prof to assoc prof epidemiol, 74-82; ASSOC PEDIAT, YALE-NEW HAVEN HOSP, 68-; CLIN PROF EPIDEMIOL & PEDIAT, SCH MED, YALE UNIV, 82- *Concurrent Pos:* Clin res fel pediat, Sch Med, Yale Univ, 49-52 & 53-54; exec dir & pediatrician in chg, New Haven Live Poliomyelitis Vaccine Trial, 60. *Mem:* Am Pub Health Asn; Int Epidemiol Asn; Soc Epidemiol Res; hon mem Peruvian Pediat Soc; Sigma Xi; Am Soc Prev Oncol; Int Soc Cancer Registries; Soc Prev Oncol; fel Am Col Epidemiol. *Res:* Cancer epidemiology; virology. *Mailing Add:* 445 Sperry Rd Bethany CT 06525

CURNOW, RANDALL T, EXPERIMENTAL BIOLOGY. *Prof Exp:* VPRES REGULATORY AFFAIRS, GLAXO INC, 90- *Mailing Add:* Glaxo Inc Five Moore Dr Research Triangle Park NC 27709

CURNOW, RICHARD DENNIS, b Omaha, Nebr, Dec 8, 43; m 68; c 2. WILDLIFE BIOLOGY, TERRESTRIAL ECOLOGY. *Educ:* Colo State Univ, BS, 66, MS, 68, PhD(wildlife biol), 70. *Prof Exp:* Res assoc, Dept Fish & Wildlife Biol, Colo State Univ, 70-71; asst leader res, Ohio Coop Wildlife Res Unit, 71-74, staff specialist, US Fish & Wildlife Serv, Div Coop Res, 74-75, actg chief, 75-76, asst dir, Denver Wildlife Res Ctr, 77-78, actg dir res, 78-79, ASST DIR, DENVER WILDLIFE RES CTR, ANIMAL & PLANT

HEALTH INSPECTION SERV, USDA, 79- *Concurrent Pos:* Adj asst prof, Dept Zool, Ohio State Univ, 71-74. *Mem:* Wildlife Soc; Sigma Xi; Audubon Soc. *Res:* Ecological relationships of land use and development and natural resource management to fish and wildlife populations; fate and effects of radionuclides and environmental contaminants in ecological systems; vertebrate pest management research and development. *Mailing Add:* 1981 S Estes St Lakewood CO 80227

CURNUTT, JERRY LEE, b Elwood, Ind, Nov 15, 42; m 63; c 1. PHYSICAL CHEMISTRY. *Educ:* Franklin Col, BA, 64; Univ Nebr, MS, 66, PhD(chem), 68. *Prof Exp:* NASA res traineeship, 66-68; res chemist, Thermal Res Lab, 68-78, res leader, 78-86, TECH INFO SERV, DOW CHEM CO, 86- *Mem:* Am Chem Soc; Am Soc Testing & Mat; Sigma Xi. *Res:* Thermodynamic properties of electrolytes; use of gas adsorption techniques to characterize pore structures in solid materials; catalysis; dimethyl carbonate processes; search hedges. *Mailing Add:* 2800 Mt Vernon Dr Midland MI 48640

CURNUTTE, BASIL, JR, b Portsmouth, Ohio, Mar 1, 23; m 45; c 2. PHYSICS. *Educ:* US Naval Acad, BS, 45; Ohio State Univ, PhD(physics), 53. *Prof Exp:* Res assoc physics, Ohio State Univ, 53-54; from asst prof to prof, 54-88, EMER PROF PHYSICS, KANS STATE UNIV, 88- *Mem:* AAAS; fel Am Phys Soc; Am Asn Physics Teachers; fel Optical Soc Am; Sigma Xi. *Res:* Chemical physics; molecular spectra and structure; atomic spectroscopy; spectroscopy on fast ion beams; study of metastable states; forbidden transitions and energy levels; accelerator-based atomic physics. *Mailing Add:* Dept Physics Cardwell Hall Kans State Univ Manhattan KS 66506

CURNUTTE, JOHN TOLLIVER, III, b Dixon, Ill, Sept 9, 51; m 75; c 3. HEMATOLOGY, CONGENITAL IMMUNODEFICIENCIES. *Educ:* Harvard Univ, AB, 73, MD, 79, PhD(biol chem), 80. *Prof Exp:* Intern pediat, Mass Gen Hosp, 79-80, resident, 80-81; fel med, Children's Hosp, Boston, 81-83; asst prof pediat, Med Sch, Univ Mich, 83-86; asst mem biochem, 86-87, assoc mem, 87-89, ASSOC DIR, GEN CLIN RES CTR, SCRIPPS RES INST, 89-, MEM HEMAT, 89-; ASSOC CLIN PROF, DEPT PEDIAT, UNIV CALIF, SAN DIEGO, 80- *Concurrent Pos:* Res grant, Soc Pediat Res, 84; prin investr, Nat Insts Health res grant, 84-; mem, Study Group Elem Educ, US Dept Educ, 85-86; estab investr, Am Heart Asn, 86-91; guest ed, Hemat/Oncol Clins NAm, 88; med staff, Dept Med, Scripps Clin & Res Found, 88-; mem, Sci Affairs Subcomt Pediat Hemat, Am Soc Heart. *Mem:* Am Soc Clin Investigators; Am Fedn Clin Res; Am Soc Cell Biol; Am Soc Hemat; Soc Pediat Res; Am Soc Biochem & Molecular Biol. *Res:* Molecular and genetic basis of inherited immunodeficiences and their therapies; mechanism by which oxygen radical production by cells in the immune system is regulated; structure-based research to find new anti-inflammatory drugs. *Mailing Add:* 10666 N Torrey Pines Rd SBR 12 La Jolla CA 92037-1093

CUROTT, DAVID RICHARD, b Passaic, NJ, June 3, 37; m 82; c 2. PHYSICS, PHOTOELECTRIC PHOTOMETRY. *Educ:* Stevens Inst Technol, BSc, 59; Princeton Univ, MA, 62, PhD(physics), 65. *Prof Exp:* Res assoc astrophys, Princeton Univ, 65-66, instr physics, 66-67; asst prof, Wesleyan Univ, 67-75; assoc prof, 75-81, PROF PHYSICS & DIR, PLANETARIUM, UNIV NORTHALA, 81- *Mem:* Am Asn Physics Teachers; Am Phys Soc; Sigma Xi; Am Asn Visual Star Observers. *Res:* Photoelectric photometry; experimental and observational cosmology and astrophysics; astronomical electronic instrumentation. *Mailing Add:* Box 5150 Univ North Ala Florence AL 35632

CURPHEY, THOMAS JOHN, b New York, NY, Oct 7, 34; m 59; c 2. ORGANIC CHEMISTRY. *Educ:* Harvard Univ, BA, 56, PhD(chem), 60. *Prof Exp:* Res assoc chem, Univ Wis, 60-62; instr, Yale Univ, 62-64; from asst prof to prof chem, St Louis Univ, 64-75; sr res assoc, 75-80, res assoc prof, 80-85, RES PROF PATH, DARTMOUTH MED SCH, 85- *Concurrent Pos:* Adj prof chem, Dartmouth Col, 75- *Mem:* Am Chem Soc. *Res:* Synthetic organic chemistry; chemical carcinogenesis; chemical and biochemical reaction mechanisms; natural products chemistry. *Mailing Add:* Dept Path Dartmouth Med Sch Hanover NH 03756

CURRAH, JACK ELLWOOD, b Toronto, Ont, Sept 12, 20; m 44; c 3. ANALYTICAL CHEMISTRY. *Educ:* Univ Toronto, BA, 41, MA, 42, PhD(anal & inorg chem), 44. *Prof Exp:* Asst chem, Univ Toronto, 41-43; res chemist, Atomic Energy Proj, Nat Res Coun Can, 44-46; res chemist, Cent Res Lab, Can Industs, Ltd, 46-51, res group leader, 51-56, sect leader, Res Dept, Textile Fibers Div, 56-57 & Process & Prod Res Sect, Tech Dept, 57-62, res sect mgr, Tech Dept, 62-65; tech mgr, Millhaven Fibres Ltd, 65-67; res & tech mgr plastics group, Can Industs Ltd, McMasterville, 67-70, res leader, Cent Res Lab, 70-73, sr res chemist, Chem Res Lab, 74-80; res scientist, Chem Res Lab, Cil Inc, Mississauga, Ont, 80-83; RETIRED. *Mem:* Chem Inst Can; Can Soc Chem Eng; Am Soc Mass Spectrometry. *Res:* Synthetic textile fibers; polyester fibers; mass spectrometry. *Mailing Add:* 1092 Gloucester Dr Burlington ON L7P 2W5 Can

CURRAH, WALTER E, b Tacoma, Wash, Apr 10, 36. ASTROPHYSICS, THEORETICAL PHYSICS. *Educ:* Wash State Univ, BS, 64. *Prof Exp:* Dir, Nat Inst Creativity, 74-79, World Ctr, 80-83; OWNER, CURRAH ENTERPRISES, 83- *Concurrent Pos:* Owner, Currah Res Div, 74- *Res:* Energy independence; chemical reactions; universal equation of molecular structures and astrophysics; writer on subject of energy, atomic and molecular emissions and structures and unified theory of nature. *Mailing Add:* PO Box 9292 Tacoma WA 98409

CURRAN, DAVID JAMES, b Kitchener, Ont, Sept 19, 32; US citizen; m 57; c 6. ANALYTICAL CHEMISTRY. *Educ:* Univ Mass, BS, 53; Boston Col, MS, 58; Univ Ill, PhD(polarography), 61. *Prof Exp:* Asst prof anal chem, Seton Hall Univ, 61-63; from asst prof to assoc prof, 63-78, PROF ANALYTICAL CHEM, UNIV MASS, AMHERST, 78- *Concurrent Pos:* Treas, Div Anal Chem, Am Chem Soc, 91-92. *Mem:* Am Chem Soc; Electrochem Soc; Soc Electroanal Chem; NY Acad Sci; AAAS. *Res:* Electroanalytical chemistry; solid electrodes in potentiometry, coulometry, amperometry and voltammetry; chemical instrumentation; pressure transducers in chemical analysis; electrochemical detectors in flowing streams. *Mailing Add:* 28 Valleyview Circle Amherst MA 01002

CURRAN, DENNIS PATRICK, b Easton, Pa, June 10, 53; m 79; c 2. CHEMISTRY. *Educ:* Boston Col, BS, 75; Univ Rochester, PhD(chem), 79. *Prof Exp:* NIH fel chem, Dept Chem, Univ Wis, 79-81; asst prof, 81-86, assoc prof, 87, PROF CHEM, DEPT CHEM, UNIV PITTSBURGH, 88- *Concurrent Pos:* Dreyfus Award, 81-86; fel Sloan Found, 85-87; Lilly grantee, 85-87; res career develop award, NIH, 87-92. *Mem:* Am Chem Soc; Chem Soc Japan. *Res:* Total synthesis of natural products and development of new synthetic methodology. *Mailing Add:* Dept Chem Univ Pittsburgh Pittsburgh PA 15260

CURRAN, DONALD ROBERT, b Aurora, Ill, Mar 10, 32; m 59. PHYSICS. *Educ:* Iowa State Univ, BS, 53; Wash State Univ, MS, 56, PhD(physics), 60. *Prof Exp:* Physicist, Stanford Res Inst, 56-62; res physicist, Norweg Defense Res Estab, 62-67; physicist, Ernst Mach Inst, WGer, 67-70; PHYSICIST, SRI INT, 70- *Mem:* Am Phys Soc. *Res:* Shock wave and high pressure physics; fluid dynamics; mechanical behavior of solids; fracture mechanics. *Mailing Add:* 723 Canyon Rd Redwood City CA 94062

CURRAN, HAROLD ALLEN, b Washington, DC, Dec 2, 40; m 63; c 2. PALEONTOLOGY, PALEOECOLOGY. *Educ:* Washington & Lee Univ, BS, 62; Univ NC, MS, 65, PhD(geol), 68. *Prof Exp:* Asst dept geol, Univ NC, 63-68; asst prof dept earth, space & graphic sci, US Mil Acad, 68-70; from asst prof to assoc prof, 70-81, PROF GEOL, SMITH COL, 81- *Mem:* Geol Soc Am; Soc Econ Paleont & Mineral; Sigma Xi; Paleont Soc. *Res:* Trace fossils and interaction between trace-making organisms and modern sediments, Atlantic Coastal Plain and the Bahamas; modern and ancient coral reefs; upper cretaceans and cenozoic foraminifera; paleoecology and marine geology. *Mailing Add:* Dept Geol Clark Sci Ctr Smith Col Northampton MA 01063

CURRAN, JAMES W, b Monroe, Mich, Sept 16, 44; m 73; c 2. PUBLIC HEALTH & EPIDEMIOLOGY, MEDICINE. *Educ:* Univ Notre Dame, BS, 66; Univ Mich, MD, 70; Sch Pub Health, Harvard Univ, MPH, 74. *Prof Exp:* Res instr, dept prev & community med, Univ Tenn Med Sch, 71-73; career develop training, Ctr Dis Control, US Pub Health Serv, 73-75; asst comnr health med serv, Columbus City Health Dept, Ohio, 75-78; chief, Oper Res Br, Venereal Dis Control, 78-82, dir AIDS activ, 82-84, chief, AIDS Br, Div Viral Dis, Ctr Infectious Dis, 84-85, DIR, WHO REF CTR AIDS & RETROVIRUSES, 85- *Concurrent Pos:* Clin res investr, Venereal Dis Br, Ctr Dis Control, 71-73; med dir, Influenza Immunication Prog, Franklin County, 76-77; clin res investr & coordr, Oper res Br, Venereal Dis Control Div, Ctr Dis Control, 75-78; clin asst prof, dept prev & community med, Col Med, Ohio State Univ, 76-79; John Forbes fel infectious dis, Fairfield Hosp, Melbourne, Australia, 85; Can Infectius Dis Soc lectureship, Toronto, 86; vis prof, Col Med, Univ Ill, 88. *Honors & Awards:* L Vernon Scott lectr, Univ Okla Health Sci Ctr, 85; William C Watson, Jr Award, 87; Verna & Mars lectr, Baylor Col Med, 88; Oliver Cope lectr, Mass Gen Hosp, 88. *Mem:* Am Epidemiol Soc; Infectious Dis Soc Am; Sigma Xi; Am Venereal Dis Asn; AAAS. *Mailing Add:* Ctr Infectious Dis Aids Prog Ctr Dis Control 1600 Clifton Rd NE Bldg 6-288 Rm 294 Atlanta GA 30333

CURRAN, JOHN FRANKLIN, petroleum & engineering geology, for more information see previous edition

CURRAN, JOHN S, b Camden, NJ, Sept 11, 40; m; c 2. PEDIATRICS. *Educ:* Rutgers Univ, AB, 62; Univ Pa, MD, 66. *Prof Exp:* Med intern, Univ Pa Hosp, 66-67; pediat resident, Children's Hosp of Philadelphia, 67-69; chief newborn serv, US Air Force Hosp, Wiesbaden, Ger, 69-72; ASSOC PROF PEDIAT, CHIEF NEONATAL SECT & DIR REGIONAL NEWBORN INTENSIVE CARE CTR, COL MED, UNIV SFLA, TAMPA, 72- *Mem:* Fel Am Acad Pediat. *Res:* Interactions of shope papilloma virus and rabbit skin cells in vitro; retinopathy of prematurity; evaluation of incidence, natural history and resolution by fundus photography and intravenous flourescein angiography. *Mailing Add:* Dept Pediat Univ SFla Col Med One Davis Blvd Suite 404 Tampa FL 33606

CURRAN, ROBERT KYRAN, b Kingston, Pa, Aug 8, 31; m 55; c 4. PHYSICS. *Educ:* Univ Scranton, BS, 53; Univ Pittsburgh, PhD(physics), 59; New York Univ, MBA, 76. *Prof Exp:* Physicist, Westinghouse Elec Corp, 59-62 & Air Prod & Chem Inc, Pa, 62-64; PHYSICIST, BELL LABS, INC, 64- *Mem:* Sigma Xi. *Res:* Atomic collisions; mass spectrometry; optics; semiconductor technology; digital systems; software systems. *Mailing Add:* 64 Winding Way Stirling NJ 07980

CURRAN, ROBERT M, b Cohoes, NY, May 23, 21. METALLURGICAL ENGINEERING. *Educ:* Rensselaer Polytech Inst, BS, 42, MS, 47. *Prof Exp:* Mgr mat eng, Turbine Dept, Gen Elec Co, 55-84; CONSULT, 84- *Mem:* Fel Am Soc Metals; Am Soc Mech Engrs; Am Soc Testing & Mat. *Mailing Add:* 330 Columbia St Cohoes NY 12047

CURRAN, WILLIAM VINCENT, b Easton, Pa, Nov 30, 29; m 52; c 5. ORGANIC CHEMISTRY. *Educ:* Lafayette Col, BS, 52; Col Holy Cross, MS, 53; NY Univ, PhD, 68. *Prof Exp:* Res worker org chem, Columbia Univ, 54-56; res chemist, Lederle Labs, Am Cyanamid Co, 56-65; asst prof chem, Fairleigh Dickinson Univ, 65-67; res chemist, 68-69, PRIN RES CHEMIST, AM CYANAMID CO, 69- *Concurrent Pos:* Vis prof, Middlebury Col, Vt, 86. *Honors & Awards:* Sci Achievement Award, Am Cyanamid Co, 84. *Mem:* Am Chem Soc; Am Soc Microbiol; Int Soc Heterocyclic Chem. *Res:* Synthesis and chemistry of antibiotics and antibacterial compounds. *Mailing Add:* 27 Harding St Pearl River NY 10965

CURRARINO, GUIDO, b Levanto, Italy, Dec 17, 20. MEDICINE. *Educ:* Univ Genoa, MD, 45. *Prof Exp:* Asst prof radiol & pediat, Med Sch, Univ Cincinnati, 56-60; assoc prof, Med Sch, Cornell Univ, 61-64; PROF RADIOL & PEDIAT, UNIV TEX HEALTH SCI CTR, DALLAS, 65- *Res:* Pediatric radiology. *Mailing Add:* 1935 Motor St Dallas TX 75235

CURRAY, JOSEPH ROSS, b Cedar Rapids, Iowa, Jan 19, 27; m 49; c 3. MARINE GEOLOGY. *Educ:* Calif Inst Technol, BS, 49; Pa State Univ, MS, 51; Univ Calif, PhD(oceanog), 59. *Prof Exp:* Asst & instr geol & mineral, Pa State Univ, 49-51; res geologist, Res Lab, Carter Oil Co, 51-53; asst marine geol & res staff, 53-67, assoc prof oceanog, 67-70, chmn grad dept, 73-75, chmn geol res div, 76-78, PROF OCEANOG & RES GEOLOGIST, SCRIPPS INST OCEANOG, UNIV CALIF, 70- *Concurrent Pos:* Dir, Nertron, Inc, 58-; ed, Marine Geol, 63-66; trustee, Found Ocean Res, 71- *Honors & Awards:* Shepard Medal, Soc Econ Paleontologists & Mineralogists, 70. *Mem:* AAAS; fel Geol Soc Am; Am Asn Petrol Geologists; Soc Econ Paleontologists & Mineralogists; fel Am Geophys Union. *Res:* Sediments; structures; history of continental margin. *Mailing Add:* Scripps Inst Oceanog La Jolla CA 92093-0215

CURRELL, DOUGLAS LEO, b Tulsa, Okla, Feb 5, 27. ORGANIC CHEMISTRY. *Educ:* Univ Colo, BA, 50, MA, 54; Univ Ark, PhD(chem), 56. *Prof Exp:* Res fel chem, Calif Inst Technol, 56-57; from asst prof to assoc prof, 57-65, chmn dept, 65-68, PROF CHEM, CALIF STATE UNIV, LOS ANGELES, 65- *Concurrent Pos:* Vis prof, Univ Rome, 68-69, 70-71 & 78-79. *Mem:* Am Chem Soc; Sigma Xi. *Res:* Mechanisms of organic reactions; hemoglobin chemistry. *Mailing Add:* Dept Chem Calif State Univ Los Angeles CA 90032

CURREN, CALEB, ARCHEOLOGY, ANTHROPOLOGY. *Educ:* Univ Ala, BA, 70, MA, 73. *Prof Exp:* Teaching asst, Univ Ala, 71-73; asst archeol site surv & test excavations for 32 agencies, 73-88; DIR ARCHEOL, ALA TOMBIGGEE COMN, 82-; FAC RES ASSOC, DEPT ANTHROP, SOCIOL & GEOL, OFF CULT & ARCHEOL RES, UNIV W FLA, 87- *Mem:* Soc Prof Archeologists; Soc Am Archeologists. *Res:* Investigations on prehistory and history of the southeast United States; author of 43 articles, reports and books on archeology. *Mailing Add:* Pensacola Arch 1000 College Blvd Pensacola FL 32504

CURRENT, DAVID HARLAN, b Connerville, Ind, July 26, 41; m 66; c 2. EXPERIMENTAL SOLID STATE PHYSICS, NUCLEAR MAGNETIC RESONANCE. *Educ:* Carleton Col, BA, 63; Northwestern Univ, MS, 66; Mich State Univ, PhD(physics), 71. *Prof Exp:* From instr to assoc prof, 66-79, PROF PHYSICS, CENT MICH UNIV, 79- *Mem:* Am Phys Soc; Am Asn Physics Teachers; Am Asn Univ Prof; Int Soc Magnetic Resonance; Sigma Xi. *Res:* Experimental chlorine nuclear quadrupole resonance in insulators, including temperature dependence and phase transitions; optical relaxation in solids and vapors; laser spectroscopy. *Mailing Add:* Dept Physics Cent Mich Univ Mt Pleasant MI 48859

CURRENT, JERRY HALL, b Anderson, Ind, Mar 2, 35; m 61; c 3. PROCESS CONTROL, PROCESS MONITORING. *Educ:* Ind Univ, BS, 57; Univ Wash, PhD(phys chem), 61. *Prof Exp:* NSF fel, Univ Calif, Berkeley, 62-64; asst prof chem, Univ Mich, 64-70; res chemist, Gulf Res & Develop Co, 70-85, SR RES ASSOC & SECT DIR & SR ENG ASSOC, CHEVRON RES CO, 85- *Mem:* Am Chem Soc; Inst Soc Am; Soc Petrol Engrs. *Res:* Gas phase kinetics; x-ray photoelectron spectroscopy; process control; petroleum drilling. *Mailing Add:* Chevron Research & Technol Co 100 Chevron Way Richmond CA 94802-0627

CURRENT, STEVEN P, b Downers Grove, Ill, May 20, 50; m 73; c 2. CATALYSIS, FUEL TECHNOLOGY AND PETROLEUM ENGINEERING. *Educ:* Univ Chicago, BS, 72; Stanford Univ, PhD(org chem), 75. *Prof Exp:* Res assoc, Case Western Reserve Univ, 75-76; res assoc, Mass Inst Technol, 76-77; res chemist, Chevron Res Co, 77-83, res consult, 84-85, sr res chemist, 85-86, SR RES ASSOC, CHEVRON CHEM CO, 87- *Mem:* Am Chem Soc. *Res:* Organic and organometallic chemistry; homogeneous and heterogeneous catalysis, especially of small molecules including carbon monoxide, hydrogen, methanol and olefins; chemicals for enhanced oil recovery, surfactants, foam; polymers; alpha olefins. *Mailing Add:* Chevron Res & Technol Co PO Box 1627 Richmond CA 94802-0627

CURRENT, WILLIAM L, b Durango, Colo, July 23, 49; m 71; c 1. PROTOZOOLOGY, PARASITOLOGY. *Educ:* NMex State Univ, BS, 72; Eatern Wash State Col, MS, 74; Univ Nebr-Lincoln, PhD(life sci), 77. *Prof Exp:* Asst prof 77-80, assoc prof zool, Auburn Univ, 80-84; RES SCIENTIST, LILLY CORP CTR, 84- *Honors & Awards:* Chester A Herrick Award, Ann Midwest Conf Parasitologists, 76. *Mem:* Am Soc Parasitologists; Soc Protozoologists. *Res:* Electron microscopy; host-parasite relationships of myxosporidans, trypanosomatids and sporozoans; cell-to-cell interactions and cellular interactions at the species-to-species interface. *Mailing Add:* Zool Entom Dept Auburn Univ Auburn AL 36830

CURRERI, P WILLIAM, b Milwaukee, Wis, Sept 2, 36; m 75; c 3. SURGERY. *Educ:* Swarthmore Col, AB, 58; Univ Pa, MD, 62. *Prof Exp:* Asst prof, Univ Pa Hosp, 63-66, instr surg, 67-68; chief clin div, US Army Inst Surg Res, Brooke Army Med Ctr, Ft Sam Houston, 68-71; asst prof, Southwest Med Sch, Univ Tex, 71-74; assoc prof, Sch Med, Univ Wash, 74-77; Johnson & Johnson prof surg, Med Ctr, Cornell Univ, New York, 77-81; PROF & CHMN, DEPT SURG, UNIV SALA, MOBILE, 81- *Concurrent Pos:* Fel, Harrison Dept Surg Res, 63-68 & Am Cancer Soc, 67-68; clin instr, Med Sch, Univ Tex, 68-71; NIH res career develop award, 72-77; mem study sect, Surg, Anesthiol & Trauma, NIH, 80-84, chmn, 86-88. *Mem:* Soc Univ Surgeons (secy, 75-79, pres, 80-81); Am Burn Asn (secy, 78-81, vpres, 81-82, pres, 83-84); Asn for Acad Sug; Am Asn for Surg of Trauma; Am Col Surgeons; Sigma X; Int Soc Burn Injuries; Pan-Am Med Asn; Asn Tissue Banks; Int Soc Parenteral Nutrit; Am Trauma Soc; Physicians Sci Soc; Int Soc Surg; Am Surg Asn; Soc Surg Chmn; Surg Infection Soc; Am Soc Parenteral & Enteral Nutrit; Asn Prog Dir Surg. *Res:* Metabolism and nutrition in burns and trauma; hematological alterations after trauma; pulmonary dysfunction following trauma. *Mailing Add:* PO Box 161445 Mobile AL 36616

CURRIE, BRUCE LAMONTE, b Pasadena, Calif, Mar 1, 45; m 65; c 3. MEDICINAL CHEMISTRY, ORGANIC CHEMISTRY. *Educ:* Ariz State Univ, BS, 66; Univ Utah, PhD(org chem), 70. *Prof Exp:* Robert A Welch fel, Inst Biomed Res, Univ Tex, Austin, 69-72, res assoc biochem & endocrinol, 72-74; asst prof, 74-81, ASSOC PROF MEDICINAL CHEM, COL PHARM, UNIV ILL MED CTR, 81- *Concurrent Pos:* Guest scientist, Argonne Nat Lab, 80-83; pharm curric consult, several univs, Thailand & Indonesia. *Mem:* Int Soc Heterocyclic Chem; Am Chem Soc; Endorcrine Soc; Am Asn Col Pharm. *Res:* Design and synthesis of small peptide hormone agonists and antagonists; peptide sequence and structure determination; anticancer nucleoside analogs; conformationally restricted peptides; peptide anti-sickling agents. *Mailing Add:* Dept Medicinal Chem Col Pharm Univ Ill PO Box 6998 M/C 781 Chicago IL 60680

CURRIE, CHARLES H, US citizen. ELECTRICAL ENGINEERING. *Educ:* Ga Inst Technol, BEE, 52, MSEE, 56. *Prof Exp:* Electrician, City Tallahassee, Fla, 41-42; from radar technician to electronics officer, US Air Force, 43-49; electrician, Harrison Wright Construct Co, 49; res asst, Eng Exp Sta, Ga Inst Technol, 50-52, res engr, 52-54, asst proj dir, 54-56; chief electronics engr, Sci-Atlanta, Inc, 56-59, dir & vpres, Mfg Div, 59-63, vpres & mgr prod develop, 64-67, prin engr & mgr, Advan Develop Group, 67-75, prin engr, Microwave Instrumentatation Staff, 75-90; RETIRED. *Mem:* Assoc mem Inst Elec & Electronics Engrs; assoc mem Sigma Xi. *Res:* Radar systems design; analog and digital instrumentation systems; receiving systems; signal sources; electronic circuits; antenna test instrumentation; servo mechanisms; reliability; human engineering; CATV systems; security systems; microwave links; packaging techniques. *Mailing Add:* 3104 Lake Point Circle Acworth GA 30101

CURRIE, IAIN GEORGE, b Glasgow, Scotland, Mar 11, 36; m 58; c 2. MECHANICAL ENGINEERING. *Educ:* Univ Strathclyde, BSc, 60; Univ BC, MASc, 62; Calif Inst Technol, PhD(appl mech), 66. *Prof Exp:* From asst prof to assoc prof, 66-81, PROF MECH ENG, UNIV TORONTO, 81- *Honors & Awards:* Montgomerie-Nielson Medal, 60; Letson Mem Prize, 62. *Mem:* Am Soc Mech Engrs; Sigma Xi; Can Soc Mech Engrs. *Res:* Analytical and experimental fluid mechanics with and without heat transfer. *Mailing Add:* Dept Mech Eng Univ Toronto Toronto ON M5S 1A4 Can

CURRIE, JAMES ORR, JR, b Canton, Ohio, July 28, 43; m 69; c 2. ORGANIC & SYNTHETIC CHEMISTRY. *Educ:* Ohio State Univ, BS, 65; Univ Wash, PhD(org chem), 70. *Prof Exp:* Asst prof org chem, Univ Nev, Reno, 70-71; assoc org synthesis, Utah State Univ, 71-72; from asst prof to assoc prof, 72-85, chmn dept, 76-80, PROF CHEM, PAC UNIV, 85- *Concurrent Pos:* Vis prof, Atlanta Univ, 81-82; comput software ed-auth, Tuloop Sci, 82- *Mem:* Sigma Xi (pres, 76-); Am Chem Soc. *Res:* Synthesis and study of alkyldiazonium salts in virtually aprotic media; synthesis of optically active heterocycles; synthesis of nonbenzenoid aromatic systems. *Mailing Add:* Dept Chem Pac Univ Forest Grove OR 97116

CURRIE, JOHN BICKELL, b Guelph Township, Ont, May 29, 22; m 48; c 2. GEOLOGY. *Educ:* McMaster Univ, BA, 46; Univ Toronto, MA, 47, PhD, 50. *Prof Exp:* Res geologist & sect head, Geol & Geochem Div, Gulf Res & Develop Co, 50-59; from assoc prof to prof geol sci, Univ Toronto, 64-86; RETIRED. *Mem:* Geol Soc Am; Am Asn Petrol Geol; Am Geophys Union. *Res:* Structural geology; mechanical development of geological structures; regional tectonics; petroleum geology. *Mailing Add:* Dept Geol Univ Toronto Toronto ON M5S 1A1 Can

CURRIE, JULIA RUTH, b Freeport, Tex, Dec 13, 44; m 71; c 3. NEUROSCIENCES. *Educ:* Radcliffe Col, AB, 67; Washington Univ, PhD(neurobiol), 74. *Prof Exp:* Res asst neurobiol, Harvard Med Sch, 67-69; res assoc physiol, Med Col, Cornell Univ, 74-76, NIH fel, 75-76; NIH trainee & sr staff assoc anat & ophthal, 76-78, res assoc anat, 78-82, ASST PROF ANAT, COL PHYSICIANS & SURGEONS, COLUMBIA UNIV, 82-; RES SCI IV, DEPT MOLECULAR BIOL, INST BASIC RES. *Mem:* Am Soc Cell Biol; AAAS; Soc Neurosci; Asn Soc Neurochem; Am Asn Anatomists. *Res:* Experimental neuroembryology in general; membrane cell biology, in particular, using cyto chemical, autoradiographic and ultrastructural techniques; vertebrate visual system; nerve regeneration; axonal transport; Alzheimer's disease. *Mailing Add:* Dept Molecular Biol Inst for Basic Res 1050 Forest Hill Rd Staten Island NY 10314

CURRIE, LLOYD ARTHUR, b Portland, Ore, Mar 14, 30; m 59; c 4. PHYSICAL CHEMISTRY, RADIOCHEMISTRY. *Educ:* Mass Inst Technol, BS, 52; Univ Chicago, PhD(phys chem), 55. *Prof Exp:* Asst prof chem, Pa State Univ, 55-62; NUCLEAR CHEMIST, NAT BUR STANDARDS, 62- *Concurrent Pos:* Lectr, Am Univ, 62-66; mem, Int Comn on Radiation Units & Measurements Task Group, 65-; vis prof, Inst Nuclear Sci, Univ Ghent & phys inst, Univ Bern, 70-71. *Honors & Awards:* Silver Medal, Dept Com, 80. *Mem:* Fel Am Inst Chemists; Am Chem Soc; AAAS; Am Geophys Union; Sigma Xi. *Res:* Nuclear reactions; electromagnetic isotope-separation of reaction products; low-level and environmental radioactivity; statistical aspects of nuclear decay and data analysis; application of nuclear methods to trace analysis and physical chemistry. *Mailing Add:* 215 Rolling Rd Gaithersburg MD 20877

CURRIE, MALCOLM R, Spokane, Wash, Mar 13, 27; m 51, 77; c 2. HIGH TECHNOLOGY RESEARCH. *Educ:* Univ Calif, Berkeley, AB, 49, MS, 51, PhD(eng), 54. *Prof Exp:* Res eng, Hughes Aircraft Co, 54-57, head electron dynamics dept, Hughes Res Lab, Culver City, Calif, 57-60, dir physics lab, Hughes Res Lab, Malibu, 60-61, assoc dir, 61-63, vpres & dir, 63- 65, vpres & mgr res & develop div, 65-69; vpres res & develop, Beckman Instruments Inc, 69-73; under secy res & eng dept, Off Secy Defense, Wash, 73-77; pres missile systs group, 77-83, exec vpres, 83-88, CHIEF EXEC OFFICER & CHMN BD, HUGHES AIRCRAFT CO, CANOGA PARK, 88- *Concurrent Pos:* Pres & chief exec officer, Delco Electronics Corp, 86-88; mem bd dirs UNOCAL, Hughes & Delco Electronics; mem bd trustees, Univ SCalif, 89-, Howard Univ Defense Sci Bd, 89- *Honors & Awards:* Intel Medal

Distinguished Serv, NASA. *Mem:* Nat Acad Eng; Am Phys Soc; fel Inst Elec & Electronics Engrs; fel Am Inst Aeronaut & Astronaut; Sigma Xi. *Res:* Microwaves; lasers; plasmas. *Mailing Add:* Hughes Aircraft Co 7200 Hughesterr PO Box 45066 Los Angeles CA 90045-0066

CURRIE, NICHOLAS CHARLES, b Warner Robins, Ga, Apr 7, 45. ELECTROMAGNETIC SCATTERING, RADAR SYSTEMS. *Educ:* Ga Inst Technol, BS, 67, MS, 73. *Prof Exp:* Asst res phys, 67-73, res engr II, 73-77, sr res engr & chief radar exp, 79-84, PRIN RES ENGR, GA TECH RES INST, GA INST TECHNOL, 84- *Concurrent Pos:* Lectr, Continuing Educ Dept, Ga Inst Technol, 76-88; consult, Parks Jaggers, 86 & Computing Devices Co, 87; ed, Artech House, 84 & 87. *Mem:* Sr mem Inst Elec & Electronics Engrs; Sigma Xi. *Res:* Radar scattering, millimeter wave systems, ECCM. *Mailing Add:* 7038 McCurley Rd Acworth GA 30101-1326

CURRIE, PHILIP JOHN, b Toronto, Ont, Mar 13, 49; m 71; c 3. VERTEBRATE PALEONTOLOGY. *Educ:* Univ Toronto, BSc, 72; McGill Univ, MSc, 75, PhD(biol), 81. *Prof Exp:* Mus curator palaeont, Provincial Mus Alberta, 76-81; mus curator, Paleont Mus & Res Inst, 81-82; asst dir res, 82-89, HEAD, DINOSAUR RES, TYRRELL MUS PALAEONT, 89- *Concurrent Pos:* Secy, Alberta Palaeont Adv Comt, 77-89; treas, Palaeont Can, 81-84, prog officer, Soc Vert Paleont, 85-87; conf chmn, Soc Vert Paleont, 88, Mesozoic Terrestrial Ecosystems, 87. *Honors & Awards:* Sir Frederick Hautain Award, 88. *Mem:* Soc Vert Paleont; Paleont Soc; Sigma Xi; Can Soc Petrol Geologists; Am Soc Zoologists. *Res:* Fossil reptiles including Permian Sphenacodonts from Europe and United States; permian eosuchians from Africa and Madagascar; cretaceous dinosaurs from Canada and Asia and their footprints. *Mailing Add:* Tyrrell Mus Palaeont Box 7500 Drumheller AB T0J 0Y0 Can

CURRIE, ROBERT GUINN, geophysics, for more information see previous edition

CURRIE, THOMAS ESWIN, b Bolton, Eng, Feb 16, 26; US citizen; m 58; c 3. CERAMICS ENGINEERING, MATERIALS SCIENCE. *Educ:* Leeds Univ, BSc, 51, PhD(ceramics), 54. *Prof Exp:* Res assoc ceramics, Pa State Univ, 54-57; ceramics engr, US Naval Boiler & Turbine Lab, 57 & Foote Mineral Co, 57-60; res assoc mat sci, Gen Elec Co, 60-87; RETIRED. *Mem:* Fel Am Ceramic Soc; Soc Plastics Engrs; Appliance Engrs Soc. *Res:* Materials characterization; microwave properties of plastic materials, structure of glasses and vitreous enamels. *Mailing Add:* 7725 Nalan Dr Louisville KY 40291

CURRIE, VIOLET EVADNE, b Jamaica, WI; Can citizen. THERAPEUTIC NUTRITION. *Educ:* Howard Univ, BS, 62, MS, 64; Cornell Univ, PhD(nutrit), 74. *Prof Exp:* Lectr biol chem, Univ Toronto, 66-70, assoc prof, 74-75; ASSOC PRIN PHYSIOL & THERAPEUT NUTRIT, CENT COL AGR TECH, MINISTRY AGR & FOOD, 75- *Concurrent Pos:* Dietetic intern, Freedmen's Hosp, Washington, DC, 63; consult, Can Food Serv Supvr Asn, 82-85. *Mem:* Can Dietetic Asn; Am Dietetic Asn; Nutrit Educ Soc; Can Soc Nutrit Sci. *Res:* Developed an in vivo technique using stable magnesium-26 as a biological tracer in measurement of the availability of dietary magnesium; studied the effects of cadmium and of protein deprivation during pregnancy; human anatomy; physiology. *Mailing Add:* Cent Col Agr Technol Foods Tech Huron Park ON N0M 1Y0 Can

CURRIE, WILLIAM DEEMS, b Wallace, NC, Sept 4, 35; m 61; c 2. BIOCHEMISTRY. *Educ:* Davidson Col, BS, 57; Univ NC, MS, 62, PhD(biochem), 64. *Prof Exp:* Res asst biochem, Univ NC, 60-62; appointee, Biol & Med Res Group, Los Alamos Sci Lab, Univ Calif, 64-66; asst prof biol, East Carolina Univ, 66-67; asst prof radiobiol, 67-73, ASSOC PROF RADIOBIOL, MED CTR, DUKE UNIV, 73- *Mem:* AAAS; Am Chem Soc; Biophys Soc; Sigma Xi. *Res:* Control of energy metabolism; nuclear metabolism; effects of hyperbaric oxygen; oxidative phosphorylation; metabolism of cancer cells; seizure activity and energy metabolism. *Mailing Add:* Box 3224 Duke Univ Med Ctr Durham NC 27706

CURRIER, HERBERT BASHFORD, b Richwood, Ohio, Oct 16, 11; m 34; c 2. PLANT PHYSIOLOGY. *Educ:* Ohio State Univ, BS, 32; Utah State Col, MS, 38; Univ Calif, PhD(plant physiol), 43. *Prof Exp:* Asst, Utah State Col, 37-38; asst col agr, Univ Calif, 38-39, assoc bot, 39-42; res chemist, Basic Veg Prod Co, 43-46; asst prof bot & asst botanist, 46-52, assoc prof bot & assoc botanist, 52-58, prof, 58-77, EMER PROF BOT, UNIV CALIF, DAVIS, 77- BOTANIST, EXP STA, 58- *Concurrent Pos:* Guggenheim fel, 54-55 & 61-62; guest prof, Univ Gottingen, 74. *Mem:* Am Soc Plant Physiol; Bot Soc Am; Scand Soc Plant Physiol; Japanese Soc Plant Physiol; Sigma Xi. *Res:* Cellular physiology; phloem translocation; plant water relations; toxicity; structural-functional studies of the phloem tissue of higher plants. *Mailing Add:* 632 Oaks Ave Davis CA 95616

CURRIER, ROBERT DAVID, b Grand Rapids, Mich, Feb 19, 25; m 51; c 2. NEUROLOGY. *Educ:* Univ Mich, AB, 48, MD, 52, MS, 55. *Prof Exp:* From asst prof to assoc prof neurol, Univ Mich, 57-61; assoc prof med, 61-69, chmn Dept Neurol, 77-90, PROF MED, MED CTR, UNIV MISS, 69-, CHIEF DIV NEUROL, 61- *Res:* Clinical, academic and research neurology. *Mailing Add:* Dept Neurol Univ Miss Med Ctr Jackson MS 39216

CURRIER, THOMAS CURTIS, b Sturgis, Mich, Feb 13, 47; m 72. MOLECULAR BIOLOGY, GENETICS. *Educ:* Ind Univ, BS, 70; Univ Wash, MS, 72, PhD(microbiol), 76. *Prof Exp:* Res assoc genetics of cyanobacteria, Dept Energy, Mich State Univ, 76-78; asst prof phytopath bacteria, Kans State Univ, 79-83; group leader molecular genetics, Allied Corp, 83-85; GROUP LEADER BIOCHEM & PHYSIOL, ECOGEN INC, 85-; SR RES SCIENTIST, NAT PROD BIOL, STERLING DRUG INC, 90- *Mem:* Am Soc Microbiologists. *Res:* Genetics and molecular biology of cyanobacteria; the molecular biology of crown gall tumor induction by Agrobacterium tumefaciens and host-parasite interactions of phytopathogenic bacteria; biological disease control. *Mailing Add:* Sterling Drug Inc 25 Great Valley Pkwy Malvern PA 19355

CURRIER, WILLIAM WESLEY, b Seattle, Wash, Sept 18, 47; m 69; c 1. PLANT BIOCHEMISTRY. *Educ:* Univ Wash, BS, 69; Purdue Univ, PhD(biochem), 74. *Prof Exp:* Fel plant path, Mont State Univ, 74-77; from asst prof to assoc prof microbiol & biochem, 77-86, ASSOC PROF AGR BIOCHEM, UNIV VT, 86-, ACTG CHAIR, DEPT AGR BIOCHEM, 84- *Concurrent Pos:* Vis res assoc, Univ Hull, 88. *Mem:* AAAS; Am Soc for Microbiol; Sigma Xi; Am Soc Plant Physiologists; fel Am Inst Chemists. *Res:* Biochemical recognition; plant disease resistance; biological nitrogen fixation; bacterial chemotaxis; plant amino acid transport; pesticide drift. *Mailing Add:* Dept Agr Biochem 111 Hills Bldg Burlington VT 05405

CURRO, FREDERICK A, b Brooklyn, NY, Apr 10, 43; m 71; c 2. PAIN, DENTAL PHARMACOLOGY. *Educ:* St John's Univ, BS, 68; Tufts Univ, DMD, 72; Ohio State Univ, PhD(pharmacol), 76. *Prof Exp:* Asst prof, dept dent mat, dent anat & fixed mouth restoration, Univ Tex Dent Br, Houston, 72-73; asst prof, dept fixed prosthodontics, State Univ Buffalo, 73-84; fel pharmacol, dept anesthesiol, Col Dent & Col Med, Ohio State Univ, 74-77; assoc prof, Sch Dent, Fairleigh Dickinson Univ, 77-83, dir basic sci, Oral Health Res Ctr, 78-82; clin asst prof oral surg, 78-80, PROG COORDR, PAIN & ANXIETY CONTROL PROG, COL DENT, NEW YORK UNIV, 80-, CLIN ASSOC PROF ORAL SURG, 80-; CHMN PHARMACOL, SCH DENT, FAIRLEIGH DICKINSON UNIV, 79-, PROF PHARMACOL, 83- *Concurrent Pos:* Asst attend dentist, dept oral surg, Bellvue Hosp Ctr, 78-; clin coordr, David B Kriser Oral-Facial Pain Ctr, Col Dent, NY Univ, 79-; attend dentist, dept dent & community med, St Joseph's Hosp & Med Ctr, 81-; mem, US Pharmacopeia Adv Panel, 81- *Mem:* Int Asn Dent Res; Am Soc Pharmacol & Exp Therapeut; Am Asn Dent Sch; Am Dent Asn; Am Pain Soc; Int Asn Study Pain; Int Asn Dent Res. *Res:* Evaluation of various drug responses on adrenergic receptors in various blood vessels on animal models; oral facial pain syndrome; evaluation of neurotransmitter serotonin involved in pain pathways and its effect on adrenergic receptors. *Mailing Add:* R&D Reed & Carnrick Pharmaceut Div Block Drug Co Inc Piscataway NJ 08854

CURRO, JOHN GILLETTE, b Detroit, Mich, Oct 5, 42; m 67; c 2. POLYMER PHYSICS. *Educ:* Univ Detroit, BChE, 65; Calif Inst Technol, PhD(mat sci), 69. *Prof Exp:* Res assoc theoret physics, Inst Appl Math & Fluid Dynamics, Univ Md, College Park, 69-70; staff mem, 70-75, DIV SUPVR POLYMER PHYSICS, SANDIA LABS, 75- *Mem:* Fel Am Phys Soc; Am Chem Soc; Soc Rheology. *Res:* Conformation of polymer chains and its relationship to physical properties; viscoelasticity of polymers; statistical mechanics of polymer liquids. *Mailing Add:* Div 1813 Phys Prop Polymer Sandia Nat Labs Albuquerque NM 87185

CURRY, BILL PERRY, b Hopkinsville, KY, Mar 14, 37; wid; c 2. APPLIED MATHEMATICS & APPLIED PHYSICS. *Educ:* Vanderbilt Univ, AB, 59; Univ Tenn, MS, 65; Kennedy-Western Univ, PhD(EE), 90. *Prof Exp:* Res engr gas kinetics, plasmas & magnetohydrodyn phenomena, ARO Inc, Tenn, 60-66; sr physicist high temperature gas physics & nuclear weapons technol, Physics Int Corp, Calif 66-68; sr physicist scattering phenomena, nuclear weapons technol & chem lasers, Lawrence Livermore Lab, Univ Calif, 68-73; res scientist & prog mgr magnetohydrodyn develop, STD Res Corp, Calif, 73-76; physicist electromagnetics & laser diag, ARO Inc, Sverdrup Corp, 76-80; physicist, Arvin/Calspan Field Serv, Inc, 80-87; ARGONNE NAT LAB, ENG PHYSICS DIV, 87- *Concurrent Pos:* consult, NAm Rockwell Corp, 72, lectr chem lasers, 72. *Mem:* Am Phys Soc; AAAS; Sigma Xi; Am Asn Aerosol Res. *Res:* particle beam physics; measurement theory; US patent. *Mailing Add:* Argonne Nat Lab Eng Physics 9700 S Cass Ave Argonne IL 60439

CURRY, BO(STICK) U, b Augusta, Ga, May 10, 51; m 80; c 3. CHEMOMETRICS, COMPUTATIONAL CHEMISTRY. *Educ:* Ga Inst Technol, BS, 73; Univ Calif, Berkeley, PhD(chem), 83. *Prof Exp:* SCIENTIST, HEWLETT-PACKARD CO, 83- *Mem:* Sigma Xi; Am Chem Soc; AAAS; Inst Elec & Electronic Engrs. *Res:* Computational chemistry; artificial intelligence; statistical pattern matching; neural networks applied to chemical analysis; automated chemical analysis; spectroscopic data interpretation. *Mailing Add:* 26417 Mockingbird Lane Hayward CA 94544

CURRY, DONALD LAWRENCE, b Bridgeport, Conn, Mar 26, 36; m 69; c 3. MEDICAL PHYSIOLOGY. *Educ:* Sacramento State Col, AB, 63; Univ Calif, San Francisco, PhD(physiol), 67. *Prof Exp:* Asst res physiologist, Med Ctr, Univ Calif, San Francisco, 67-69, lectr, 68-69; from asst prof to assoc prof, 69-84, PROF PHYSIOL SCI, SCH VET MED, UNIV CALIF, DAVIS 85-, ASSOC DEAN, GRAD STUDIES & RES, 87- *Concurrent Pos:* NSF, NIH-Am Diabetes Asn res grants, 74- *Honors & Awards:* Res Prize, Am Diabetes Asn, 69. *Mem:* Am Physiol Soc; Endocrine Soc. *Res:* Study of factors which control insulin secretion and biosynthesis by the perfused pancreas; including effects of CNS, obesity, brain and gut peptides. *Mailing Add:* Dept Physiol Sci Univ Calif Sch Vet Med Davis CA 95616

CURRY, FRANCIS J, b San Francisco, Calif, July 19, 11; m 48; c 8. INTERNAL MEDICINE. *Educ:* San Francisco Univ, BSc, 36; Stanford Univ, MD, 46; Univ Calif, MPH, 64. *Hon Degrees:* DSc, Univ San Francisco, 84. *Prof Exp:* Asst med bact, Stanford Univ, 41-42, asst path, 42-43; chief tuberc & pulmonary dis sect, Army Specialized Treatment Ctr, Ft Carson, Colo, 53-54, chief tuberc sect, Army & Air Force Specialized Treatment Ctr & coordr clin res, Fitzsimons Army Hosp, Denver, 54-56; chief tuberc control div, Hosps & Ment Health, San Francisco Health Dept, 56-60, asst dir pub health, 60-70, dir pub health, 70-76; RETIRED. *Concurrent Pos:* Dir chest clin, San Francisco Gen Hosp, 56-76; prof med, Univ Calif, 58-, lectr, Sch Pub Health, 60-, prof ambulatory & community med, Sch Med, 66-; assoc prof, Stanford Univ, 62; USPHS clin res grant, 62-76; consult, Nat Jewish Hosp, Denver, Colo, 64-; consult, Tuberc Commun Dis Div & mem, Adv Comt Tuberc Control, USPHS, Ga, 65-; consult, Nat Commun Dis Ctr. *Honors & Awards:* Nat Tuberc Asn Medal, 70. *Mem:* Fel Am Col Chest Physicians; fel Am Col Prev Med; Am Thoracic Soc; fel Am Col Physicians; fel Am Trauma Soc. *Res:* Pulmonary fungus diseases, particularly histoplasmosis and coccidioidomycosis; atypical mycobacteria infections of man; social and behavioral problems of staff and patients and their effect upon each other and upon the maintenance of treatment by patients; tuberculin skin testing on large population groups. *Mailing Add:* 350 Arballo-K4 San Francisco CA 94132

CURRY, GEORGE MONTGOMERY, plant physiology, for more information see previous edition

CURRY, GEORGE RICHARD, b Detroit, Mich, June 4, 32; m 67; c 2. ELECTRICAL ENGINEERING, SYSTEMS ANALYSIS. *Educ:* Univ Mich, BS(elec eng) & BS(math), 55; Mass Inst Technol, SM, 59. *Prof Exp:* Staff engr, Lincoln Lab, Mass Inst Technol, 58-64 & Raytheon Co, 64-66; dir sci, Technol Opers & assoc dir, Santa Barbara Div, Gen Res Corp, Flow Gen, Inc, 66-; VPRES, SCI APPL INT CORP. *Mem:* Inst Elec & Electronics Engrs; Am Inst Aeronaut & Astronaut; Asn Old Crows. *Res:* Design and analysis of military systems; radar systems; ballistic missile defense; satellite and space systems; tactical warfare technology; avionics; communications; strategic defense. *Mailing Add:* Sci Appl Int Corp 5520-C Ekwill St Santa Barbara CA 93111

CURRY, HIRAM BENJAMIN, b Midville, Ga, Sept 19, 27; m 51; c 4. NEUROLOGY. *Educ:* Col Charleston, BS, 47; Med Col SC, MD, 50. *Prof Exp:* Gen practr, 51-57; intern internal med, Vet Admin Hosp, Lake City, Fla, 58; assoc, 63-64, from asst prof to assoc prof, 64-77, PROF NEUROL, MED UNIV SC, 77- PROF FAMILY PRACT & CHMN SECT, 70- *Concurrent Pos:* Nat Inst Neurol Dis & Blindness spec fel, 60-63; prin investr, USPHS grant, 63-70, co-prin investr, 67-69; chief neurol serv, Vet Admin Hosp, Charleston, SC, 66-68; pres prof staff & mem exec comt, Med Univ Hosp, Charleston, SC, 70-; med dir outpatient clins, dir family pract residency prog & mem behav sci comt, Med Univ SC, 70-; coordr, SC Statewide Family Pract Residency Syst, 73-78. *Mem:* Am Acad Neurol; Asn Res Nerv & Ment Dis; Am Acad Gen Pract. *Res:* Cerebral circulation; cerebral vascular disease; epidemiology of vascular disease; health care delivery; medical education. *Mailing Add:* Dept Family Med/Neurol Med Univ SC 171 Ashley Ave Charleston SC 29425

CURRY, HOWARD MILLARD, b Haverhill, Mass, Mar 8, 24; m 80; c 2. ORGANIC CHEMISTRY. *Educ:* Northeastern Univ, BS, 45; Boston Univ, AM, 47, PhD(org chem), 50. *Prof Exp:* Instr chem, Bates Col, 49-50; from assoc prof to prof chem, Wittenberg Univ, 51-89, chmn dept, 74-84; RETIRED. *Concurrent Pos:* Consult, C F Kettering Found, 57-89. *Mem:* Am Chem Soc. *Res:* Nitrogen organics. *Mailing Add:* 125 Crafton Dr South Vienna OH 45369

CURRY, JAMES EUGENE, b Rome, Ga, Oct 15, 26; m 49; c 4. CHEMICAL ENGINEERING, MATERIALS SCIENCE. *Educ:* Ga Inst Technol, BS, 50, MS, 51; Univ Ala, PhD(chem eng), 72. *Prof Exp:* Chem engr, Southern Res Inst, 15-53 & Herty Found Lab, 54-55; area supvr prod, Diamond Alkali Co, 55-57; mat specialist, Army Ballistic Missile Agency, 57-60; CHIEF, NONMETALLIC MAT DIV AEROSPACE, MARSHALL SPACE FLIGHT CTR, NASA, 60- *Mem:* Am Inst Chem Engrs. *Res:* Polymer science. *Mailing Add:* 1201 Nolan Blvd Madison AL 35758

CURRY, JAMES KENNETH, b Amarillo, Tex, Nov 30, 20; m 45; c 3. GEOLOGY. *Educ:* Tex Tech Univ, BS, 42; Univ Calif, Los Angeles, cert meteorol. *Prof Exp:* Geologist, Panhandle Eastern Pipeline Co, 46-48, asst dist geologist, 48-52, dist geologist, 52-54, div geologist, 54-55, mgr geol, 55-64, mgr, Geol Div, 64-69, mgr explor, 69-73, mgr, Rocky Mountain Region, 75-81; CONSULT GEOLOGIST, 81- *Mem:* Am Asn Petrol Geologists; Am Inst Mining, Metall & Petrol Engrs; Am Inst Prof Geologists. *Res:* Exploratory geology in search of oil and gas. *Mailing Add:* 5687 S Geneva St Englewood CO 80111

CURRY, JOHN D, b Xenia, Ohio, Nov 16, 38; m 60; c 2. INORGANIC CHEMISTRY. *Educ:* Wilmington Col, Ohio, BA, 60; Ohio State Univ, PhD(inorg chem), 64. *Prof Exp:* William Ramsay Mem & NATO fels, Oxford Univ, 64-65; RES CHEMIST, MIAMI VALLEY LABS, PROCTER & GAMBLE CO, CINCINNATI, 65- *Mem:* Am Chem Soc; The Chem Soc; Sigma Xi. *Res:* Homogeneous catalysis; reactions of coordinated ligands; phosphorus chemistry; detergent chemistry. *Mailing Add:* 709 Marcia Dr Oxford OH 45056-2531

CURRY, JOHN JOSEPH, III, b Brooklyn, NY, Nov 1, 40; m 66; c 2. NEUROENDOCRINOLOGY. *Educ:* State Univ NY Col Plattsburgh, BS, 62; Adelphi Univ, MS, 64; Univ Calif, Berkeley, PhD(physiol), 69. *Prof Exp:* From instr to asst prof physiol, Sch Med, Boston Univ, 68-73; asst prof, 73-77, ASSOC PROF PHYSIOL & OBSTET & GYNEC, OHIO STATE UNIV, 77-, DIR, MED ACAD PROG, 87- *Mem:* Am Physiol Soc; Endocrine Soc; Soc Neurosci; Am Asn Univ Professors. *Res:* Central nervous system regulation of reproductive behavior and gonadotrophin secretion; sensory and hormonal factors controlling ovulation; mechanisms of hypertensive disease of pregnancy (pre-eclampsia). *Mailing Add:* Dept Physiol 3172 Graves Hall Ohio State Univ 333 W Tenth Ave Columbus OH 43210-1239

CURRY, JUDITH ANN, b Chicago, Ill, Feb 26, 53; c 1. POLAR CLIMATE, ATMOSPHERIC PHYSICS. *Educ:* Northern Ill Univ, BS, 74; Univ Chicago, PhD(geophys sci), 82. *Prof Exp:* Asst scientist, Univ Wis-Madison, 82-86; from asst prof to assoc prof atmospheric sci, Purdue Univ, 86-89; ASSOC PROF METEOROL, PA STATE UNIV, 89- *Concurrent Pos:* NSF presidential young investr award, 88; adj assoc prof, Purdue Univ, 89-; consult, Boeing Com Airplane Group, 90-; mem, Comt Cloud Physics, Am Meteorol Soc, 89-91 & Nat Res Coun Coastal Meteorol, 90-91. *Mem:* Am Meteorol Soc. *Res:* Interactions between clouds, radiation and air chemistry in the high latitude regions; processes that maintain the polar ice caps. *Mailing Add:* 509 Westview Ave State College PA 16803

CURRY, MARY GRACE, b New Orleans, La, June 16, 47. LIMNOLOGY, ECOLOGY. *Educ:* Univ New Orleans, BS, 69, MS, 71; La State Univ, Baton Rouge, PhD(bot), 73. *Prof Exp:* Vis asst prof bot, La State Univ, Baton Rouge, 74; environ scientist life sci, VTN Louisiana, Inc, 74-79; ENVIRON IMPACT OFFICER, JEFFERSON PARISH, LOUISIANA, 79- *Concurrent Pos:* Pvt environ consult; asst prof biol, Loyola Univ, 83-87. *Mem:* Ecol Soc Am; Bot Soc Am; Sigma Xi. *Res:* Taxonomy, ecology and distribution of freshwater leeches; vascular plant taxonomy; environmental science. *Mailing Add:* 3404 Tolmas Dr Metairie LA 70002

CURRY, MICHAEL JOSEPH, b Brooklyn, NY, Aug 15, 20; m 58; c 4. PLASTICS, CHEMISTRY. *Educ:* St John's Col, NY, BS, 41; Univ Wis-Madison, PhD(org chem), 48. *Prof Exp:* Chemist chem warfare, US Naval Res Lab, 42-45; asst prof org chem, Col St Thomas, 48-50; develop chemist, Calco Div, Am Cyanamid Co, 50-51; prod develop engr, Celanese Corp, 51-54, tech serv mgr, 54-59, lab mgr, 60-61, plastics develop dir, 61-67; assoc, Heidrick & Struggles, 67-71; PRES, MICHAEL J CURRY ASSOCS, 71- *Concurrent Pos:* Res dir, Fairleigh Dickinson Univ, 75-, Plastics Inst Am, 83- *Mem:* Am Chem Soc; Soc Plastics Engrs. *Res:* Plastics recycling. *Mailing Add:* 941 St Marks Ave Westfield NJ 07090

CURRY, NORVAL H, b St Francis, Kans, Oct 10, 14. AGRICULTURAL ENGINEERING. *Educ:* Iowa State Univ, MS, 46. *Prof Exp:* Field engr, Struct Play Prod Inst, 40-44; from asst prof to prof agr eng, Iowa State Univ, 46-59; pres, Norval H Curry Consult Engr, 59-79 & Curry Willie & Assoc Consult Engrs, 79-80; RETIRED. *Honors & Awards:* McCormick Medal, 80. *Mem:* Nat Acad Eng; Am Soc Agr Engrs (pres, 69-70). *Mailing Add:* 227 Campus Ave Ames IA 50010

CURRY, R(OBERT) BRUCE, b Iowa City, Iowa, Sept 24, 29; m 56; c 2. AGRICULTURAL ENGINEERING. *Educ:* Kans State Univ, BS, 51; Colo State Univ, MS, 55; Univ Mo, PhD, 60. *Prof Exp:* Design engr bur reclamation, US Dept Interior, 51-52; asst, Colo State Univ, 54-55; instr agr eng, Univ Mo, 55-60; from asst prof to assoc prof, 60-68, PROF AGR ENG, OHIO AGR RES & DEVELOP CTR & OHIO STATE UNIV, 68-, ASST CHMN, 82- *Concurrent Pos:* Vis fel, Clare Hall, Cambridge Univ, 67-68. *Mem:* Fel AAAS; Am Soc Agr Engrs; Am Soc Eng Educ; Soc Comput Simulation; Am Soc Agron. *Res:* Biological modeling and simulation; dynamics of soil-water relationships in plant growth; mathematical modeling and dynamic simulation of plant growth systems; controlled environment for food production; weather data systems. *Mailing Add:* Dept Agr Eng Ohio Agr Res & Develop Ctr Wooster OH 44691

CURRY, STEPHEN H, CLINICAL PHARMACOKINETICS, RESEARCH PHARMACOLOGY. *Educ:* Leicester Col Sci & Technol, Eng, PhD(clin pharmacokinetics), 62. *Prof Exp:* PROF CLIN PHARMACOKINETICS & DIR, DIV CLIN PHARMACOKINETICS, UNIV FLA, 80- *Mailing Add:* Col Pharmacol Univ Fla Box J-494 JHMHC Gainesville FL 32610

CURRY, THOMAS F(ORTSON), b Thomasville, Ga, Nov 22, 26; m 49; c 6. RADAR, COMMUNICATIONS. *Educ:* Ga Inst Technol, BEE, 49; Pa State Univ, MS, 54; Carnegie-Mellon Univ, PhD(elec eng), 59. *Prof Exp:* Teaching asst, Pa State Univ, 49-51; mem tech staff, Bell Tel Labs, NJ, 57-58; dir, Electronics Res Lab, Syracuse Univ Res Corp, NY, 58-64; chmn bd & consult, Microwave/Systs, Inc, 64-65; mgr appl electronics & chief engr, Melpar Div, Am Standard, Inc, 65-70; tech adv to pres, E-Systs, Inc, 70-74; vpres, Microwave Systs, Inc, 74-75; dir, Signal Intelligence Systs, Off Asst Secy Defense, Intel, Dept Defense, 76-77, sr staff specialist, Tactical Reconnaissance, Off Under Secy Defense, Res & Eng, 77-80, assoc dept asst, Secy Navy, 80-83, CHIEF SCIENTIST, E-SYSTS, CTR ADV PLAN & ANALYSIS, 83- *Concurrent Pos:* Consult, Indust & Eng Joint Coun, Dept Defense, 60- *Honors & Awards:* Centennial Medal, Inst Elec & Electronic Engrs, 84. *Mem:* Sigma Xi; fel Inst Elec & Electronic Engrs. *Res:* Sequential decision theory applied to signal detection/estimation; adaptive Bayesian sequential observer for signal recognition, classification, and sorting; computer controlled electronic warfare and SIGINT systems design and analysis. *Mailing Add:* 2403 Beekay Ct Vienna VA 22180

CURRY, THOMAS HARVEY, b Sullivan Co, Ind, Oct 7, 21; m 45; c 3. ORGANIC CHEMISTRY, CHEMICAL ENGINEERING. *Educ:* Purdue Univ, BChE, 42; Ohio State Univ, PhD(org chem), 53. *Prof Exp:* Tech supvr, Holston Ord Works, Tenn Eastman Corp, 43-45; instr chem, Antioch Col, 50-52; from asst prof to assoc prof, Univ Ohio, Athens, 53-61, chmn dept chem eng, 56-61; dean, Col Technol, Univ Maine, 61-67; dir resident res associateships, Nat Acad Sci, 67-69, dir associateships, Comn Human Resources, 69-74, exec secy, Bd Fel & Associateships, 74-77, dir fel, 74-78, RETIRED. *Concurrent Pos:* NSF sci fac fel, 57-60. *Res:* Chemistry of pyrroles and porphyrins. *Mailing Add:* 138 W Leon Ln Cocoa Beach FL 32931

CURRY, WARREN H(ENRY), b Rochester, NY, Jan 13, 24; m 47; c 4. AERONAUTICAL ENGINEERING. *Educ:* Univ Mich, BSE(aeronaut eng) & BSE(math), 45, MS, 46, AeE, 52. *Prof Exp:* Engr, Gen Elec Co, NY, 46-48; res assoc exp aerodyn, Univ Mich, 48-51; staff mem, Sandia Corp, 51-56, supvr eng aerodyn div, 56-65, supvr exp aerodyn div, 65-72, staff mem, 73-86; RETIRED. *Mem:* Assoc fel Am Inst Aeronaut & Astronaut. *Res:* Aeronautical engineering and aerodynamics associated with the development of ballistic vehicles, rocket systems and experimental facilities. *Mailing Add:* 3409 Stardust Ct NE Albuquerque NM 87110

CURRY, WILLIAM HIRST, III, b San Antonio, Tex, Feb 3, 32; m 54; c 4. GEOLOGY. *Educ:* Cornell Univ, BA, 54; Princeton Univ, PhD(geol), 59. *Prof Exp:* Geologist, Marathon Oil Co, 59-65; PRES, CURRY OIL CO, 66- *Concurrent Pos:* Consult geologist, 65- *Mem:* Sigma Xi; Am Asn Petrol Geologists. *Res:* Search for more effective oil and gas exploration tools and combinations of concepts and tools. *Mailing Add:* 3205 Bella Vista Dr Casper WY 82601

CURT, GREGORY, ONCOLOGY. *Prof Exp:* CLIN DIR, NAT CANCER INST, NIH, 89-, ASSOC DIR, CLIN ONCOL PROG, 89- *Mailing Add:* NIH Nat Cancer Inst Clin Oncol Prog Bldg 10 Rm 12N214 Bethesda MD 20892

CURTICE, JAY STEPHEN, b Dallas, Tex, Jan 19, 28; m 57; c 2. PHYSICAL ORGANIC CHEMISTRY. *Educ:* Southern Methodist Univ, BS, 48; Iowa State Univ, PhD(chem), 54. *Prof Exp:* Res chemist, Sinclair Res, Inc, 54-62; from assoc prof to prof, 62-87, EMER PROF CHEM, ROOSEVELT UNIV, 87- *Mem:* Am Chem Soc. *Res:* Mechanisms of free radical reactions; organometallics. *Mailing Add:* 308 W Sixth St Silver City NM 88061-5001

CURTICE, WALTER R, b Rochester, NY, Sept 14, 35; m 58; c 3. ELECTRICAL ENGINEERING. *Educ:* Cornell Univ, BEE, 58, MS, 60, PhD(elec eng), 62. *Prof Exp:* Sr res & develop engr, Raytheon Co, Mass, 62-67; vis asst prof elec eng, Univ Mich, 67-69, assoc prof, 69-72; mem tech staff, RCA Labs, 73-87; MGR, GaAs, CAD, MICROWAVE SEMICONDUCTOR CORP, 87- *Mem:* fel Inst Elec & Electronics Engrs. *Res:* Microwave electronics; solid state devices; linear beam microwave tubes. *Mailing Add:* RCA Labs Princeton NJ 08540

CURTIN, BRIAN THOMAS, b New York, NY, Oct 12, 45. PHYSICAL OCEANOGRAPHY. *Educ:* Boston Col, BS, 67; Ore State Univ, MS, 70; Univ Miami, PhD(phys oceanog), 79. *Prof Exp:* Asst prof phys oceanog, NC State Univ, 76-84; SCI OFFICER, OFF NAVEL RES, 84- *Concurrent Pos:* Prin investr, Nat Aeronaut & Space Admin, 76-78, Dept Interior, 78-84, Nat Oceanic & Atmospheric Admin, 80-83 & NSF, 82-85; adj prof, NC State Univ, 84- *Mem:* Am Meteorol Soc; Am Geophys Union; Am Soc Limnol & Oceanog. *Res:* Physical oceanography of mesoscale oceanic processes including the dynamics of frontal systems, eddies and boundary layers; observational methods and instrumentation. *Mailing Add:* Off Naval Res Code 1125AR 800 N Quincy St Arlington VA 22217-5000

CURTIN, CHARLES BYRON, b Cohoes, NY, Aug 2, 17; m 45; c 2. BIOLOGY. *Educ:* George Washington Univ, BS, 45; Catholic Univ, MS, 47; Univ Pittsburgh, PhD, 56. *Prof Exp:* Asst, Catholic Univ, 46-47; prof biol, Mt St Mary's Col, Md, 47-57; staff ed life sci, McGraw-Hill Encycl Sci & Technol, 57-62; ASSOC PROF BIOL, CREIGHTON UNIV, 62- *Concurrent Pos:* Asst, Univ Pittsburgh, 52-53; consult ed, McGraw-Hill, 71. *Mem:* Am Soc Syst Zool; Ecol Soc Am; Sigma Xi; Am Micros Soc. *Res:* Taxonomy and ecology of Tardigrada; parasitology; microtechnique. *Mailing Add:* 6218 Florence Blvd Omaha NE 68110

CURTIN, DAVID YARROW, b Philadelphia, Pa, Aug 22, 20; m 50; c 3. ORGANIC & STRUCTURAL CHEMISTRY. *Educ:* Swarthmore Col, AB, 43; Univ Ill, PhD(chem), 45. *Prof Exp:* Rockefeller Inst grant, Harvard Univ, 45-46; from instr to asst prof chem, Columbia Univ, 46-51; from asst prof to prof, 51-88, EMER FUSON PROF CHEM, UNIV ILL, URBANA, 88- *Mem:* Nat Acad Sci; fel Royal Soc Chem; Swiss Chem Soc; Am Crystallog Asn; Am Chem Soc. *Res:* Reaction mechanisms and exploratory synthetic organic chemistry; solid state organic chemistry. *Mailing Add:* Three Montclair Rd Urbana IL 61801

CURTIN, FRANK MICHAEL, b Philadelphia, Pa, Sept 16, 51; m 77. COMPUTER LANGUAGES. *Educ:* Pa State Univ, BA, 73 & 74, MPA, 76. *Prof Exp:* Systs engr, Mech Res Inc, Syst Develop Corp, 76-78; sr systs analyst, Auerbach Assocs Inc, 78-80; CHIEF, US FED TRADE COMN, 80- *Concurrent Pos:* Comt mem, Govt Info Technol Assessment, US Cong, 84-85. *Mem:* Am Soc Info Sci; Am Soc Pub Admin. *Res:* Development of software for mini and micro computers; training in database management systems; design and development of specialized information systems for attorneys. *Mailing Add:* Automated Systs Div Rm 603 US Fed Trade Comn 6th & Penn Ave NW Washington DC 20580

CURTIN, RICHARD B, b Yonkers, NY, Mar 29, 40. RESEARCH ADMINISTRATION. *Educ:* Manhattan Col, BEE, 61; Yale Univ, MEng, 62; Trinity Univ, MS, 67. *Prof Exp:* Res engr guid & navig, NAm Aviation, 62; proj engr tempest eng, Security Serv, USAF, 62-65; mgr, EMI prod electronic filters, Sanders Assocs, 67-68; sr res engr electronics res, Southwest Res Inst, 65-67, dir digital electronic systs, Data Systs Dept, 68-87, vpres, Automation & Data Syst Div, 87-91, EXEC VPRES, SOUTHWEST RES INST, 91- *Concurrent Pos:* Chmn, Tech Symp, Asn Old Crows, 79-89. *Mem:* Asn Old Crows; Am Asn Artificial Intel. *Res:* Robotics; machine vision; artificial intelligence; computer integrated manufacturing; automated test systems; software engineering; computer design; training systems. *Mailing Add:* Southwest Res Inst PO Drawer 28510 San Antonio TX 78228-0510

CURTIN, ROBERT H, RESEARCH ADMINISTRATION. *Educ:* US Marine Acad, BS; Harvard Univ, MS. *Prof Exp:* Dir civil eng, USAF, 39-68; dir facil hq, NASA, 68-78; vpres, Ralph M Parsons Co, 78-79; gen mgr, NE Corridor Rail Improvement Proj, Deleuw Cather/Parsons, 79-90; RETIRED. *Mem:* Nat Acad Eng; fel Soc Am Mil Engrs; Am Soc Civil Engrs; Am Inst Plant Engrs; Nat Soc Prof Engrs. *Res:* Design and construction related activities including facilities maintenance, operation and management; airfield and airfield related work for military, space facilities and industry/commercial. *Mailing Add:* 3604 Orlando Pl Alexandria VA 22305-1147

CURTIN, TERRENCE M, b Spencer, SDak, June 9, 26; m 86; c 4. VETERINARY MEDICINE, PHYSIOLOGY. *Educ:* Univ Minn, BS & DVM, 54; Purdue Univ, MS, 63, PhD(vet physiol), 64. *Prof Exp:* Pvt vet practice, SDak, 54-58; exten vet, Purdue Univ, 58-61, NIH fel, 61-64 & grant, 61-65, asst prof vet physiol, 64-66; prof vet physiol & dir continuing educ, Univ Mo-Columbia, 66-68, prof & chmn dept vet physiol & pharmacol, Sch Vet Med, 68-73; head dept, 73-80, dean sch, 79-91, PROF VET SCI, NC STATE UNIV, 73- *Concurrent Pos:* USDA res grant, 65-66. *Honors & Awards:* Martin Litwack Award, 80. *Mem:* Am Vet Med Asn; Am Animal Hosp Asn; Nat Asn State Univs & Land Grant Col; Sigma Xi; Asn Am Vet Med Col. *Res:* Esophagogastric ulcers of swine; mycotoxin induced hepatitis. *Mailing Add:* Col Vet Med NC State Univ Raleigh NC 27606-1428

CURTIN, THOMAS J, b New York, NY; c 4. ENTOMOLOGY, RESEARCH ADMINISTRATION. *Educ:* Manhattan Col, BS, 40; Univ Fla, MS, 51; Univ Md, Col Park, PhD (entom), 59. *Prof Exp:* Consult drugs, Sandoz Pharmaceut, 46-51; consult entom, Strategic Air Command, USAF, 54-59, sect chief entom, USAF Epidemial Lab, Turkey, 59-62, br chief, Tex, 62-68; assoc dir life sci res, Ohio State Univ Res Ctr, 68-73; dir, Off Res & Sponsored Progs Serv, Wayne State Univ, 73-78, adj prof biol, 73-78; dir res, Tex Woman's Univ, 78-83; RETIRED. *Concurrent Pos:* Consult, US Army, 50-51; adv, UN Korean Relief Admin, 52-53 & Turkist Govt, 61-62; del, Int Cong Entom, USAF, 60; Dept Air Force del, US Dept Defense Pesticide Bd, 62-68. *Mem:* AAAS; Am Inst Biol Sci; Entom Soc Am; Am Mosquito Control Asn; Sigma Xi. *Res:* Physiology of insect reproductive systems; insect susceptibility to pesticides; plant hormonal activity. *Mailing Add:* 3127 Mindoro San Antonio TX 78217

CURTIS, ADAM SEBASTIAN G, b London, Eng, Jan 3, 34; m 58; c 2. CELL BIOLOGY. *Educ:* Univ Cambridge, Eng, BA, 55, MA, 58; Univ Edinburgh, PhD(biophys), 57. *Prof Exp:* Hon res asst, Univ Col, London, 57-62, lectr zool, 62-67; PROF CELL BIOL, UNIV GLASGOW, 67- *Concurrent Pos:* Dir, Co Biologists, 61-; ed several jour, 75-; coun mem, Royal Soc Edinburgh, 82-86; chmn, Molecular & Cellular Biol Planning Unit, Glasgow, 91- *Honors & Awards:* Cuvier Medal, Soc Zool France, 75. *Mem:* Am Soc Cell Biol; Am Soc Biophys; Int Soc Develop Biol; Soc Exp Biol (pres, 91-93). *Res:* Cell biology; cell adhesion; cell behaviour; neurobiology; use of microstructures and micromachines. *Mailing Add:* Dept Cell Biol Univ Glasgow Glasgow G12 8QQ Scotland

CURTIS, BRIAN ALBERT, b New York, NY, Nov 25, 36; m 60; c 1. PHYSIOLOGY. *Educ:* Univ Rochester, BA, 58; Rockefeller Inst, PhD(physiol), 63. *Prof Exp:* Guest investr, Physiol Lab, Cambridge Univ, 60-61; from instr to assoc prof physiol, Sch Med & Sch Dent, Tufts Univ, 65-74, asst dean educ planning, 70-73; asst dean undergrad med educ, 74-79, ASSOC PROF PHYSIOL, COL MED, UNIV ILL, PEORIA, 74- *Mem:* Soc Gen Physiol (secy, 67-69); Biophys Soc; Am Physiol Soc. *Res:* Muscle physiology. *Mailing Add:* Dept Basic Sci Col Med Univ Ill Peoria IL 61656

CURTIS, BRUCE FRANKLIN, b Denver, Colo, Dec 16, 18; m 58. GEOLOGY, HYDROLOGY. *Educ:* Oberlin Col, AB, 41; Univ Colo, MA, 42; Harvard Univ, PhD(geol), 49. *Prof Exp:* Field asst, US Geol Surv, 42; from asst geologist to regional geologist, Continental Oil Co, 46-57; from assoc prof to prof geol, Univ Colo, Boulder, 57-83, chmn dept, 61-68; CONSULT GEOL, 57- *Concurrent Pos:* chmn bd, Ampet, Inc, 84-86. *Mem:* Fel Geol Soc Am; Am Asn Petrol Geologists; Int Asn Hydrogeologists; Int Asn Eng Geol; Sigma Xi. *Res:* Subsurface fluids; hydrology; petroleum. *Mailing Add:* 375 Harvard Lane Boulder CO 80303

CURTIS, BYRD COLLINS, b Roosevelt, Okla, Feb 25, 26; m 45; c 6. PLANT BREEDING, PLANT GENETICS. *Educ:* Okla State Univ, BS, 50, PhD(plant breeding, plant genetics), 59; Kans State Univ, MS, 51. *Prof Exp:* Vet instr agr pub sch, Okla, 52-53; from asst prof to assoc prof agron, Okla State Univ, 53-62; prof, Colo State Univ, 63-67; mgr wheat res, Cargill, Inc, 67-81; DIR WHEAT PROG, INT MAIZE & WHEAT IMPROV CTR, INT AGR RES INST, MEXICO, 81- *Mem:* Am Soc Agron; fel Crop Sci Soc. *Res:* Wheat breeding; research management. *Mailing Add:* Cent Int de Mejoramiento de Maiz y Trigo Apdo Posta 1 6-641 Delg Cuauhtemoc Mexico DF 06600 Mexico

CURTIS, CHARLES ELLIOTT, b Bentonville, Ark, Mar 16, 37; m 60; c 3. ENTOMOLOGY. *Educ:* Univ Tex, BA, 59; Univ Ark, MS, 62; Purdue Univ, PhD(entom), 66. *Prof Exp:* Inspector, Food & Drug Admin, Dept Health, Educ & Welfare, 59-60; res asst, Univ Ark, 60-62 & Purdue Univ, 62-66; res entomologist, Sci & Educ Admin, Agr Res, 66-81, RES ENTOMOLOGIST, AGR RES SERV, USDA, 81- *Mem:* Entom Soc Am. *Res:* Pest management of insects in tree nuts, dried fruits, and stone fruits with emphasis on using sex pheromones for mating disruption, and quarantine entomology to help develop export markets. *Mailing Add:* Agr Res Serv USDA 2021 S Peach Ave Fresno CA 93727-5999

CURTIS, CHARLES R, b Ault, Colo, Oct 6, 38; m 66; c 2. PLANT PATHOLOGY, ENVIRONMENTAL SCIENCES. *Educ:* Colo State Univ, BS, 61, MS, 63, PhD(bot sci), 65. *Prof Exp:* Assigned to NASA Ames Res Ctr, 66-67; from asst prof to prof plant path, Univ Md, 67-77; prof & chairperson, Dept Plant Sci, Univ Del, 78-81, spec asst pres & provost, 80-84; PROF & CHAIRPERSON, PLANT PATH, OHIO STATE UNIV, 84- *Concurrent Pos:* Vis prof, Univ Tex, 75; rep, Title XII, Int Strengthening Grant, Univ Del, 80-84. *Honors & Awards:* Cosmos Award, NASA. *Mem:* Am Phytopath Soc. *Res:* Host-parasite relationships; physiology of disease; environmental pollution. *Mailing Add:* Dept Plant Path Ohio State Univ 2021 Coffey Rd Columbus OH 43210-1087

CURTIS, CHARLES WHITTLESEY, b Providence, RI, Oct 13, 26; m 50; c 3. MATHEMATICS. *Educ:* Bowdoin Col, BA, 47; Yale Univ, MA, 48, PhD(math), 51. *Prof Exp:* Asst instr math, Yale Univ, 49-51; from instr to prof, Univ Wis, 51-63; PROF MATH, UNIV ORE, 63- *Concurrent Pos:* Nat Res Coun fel, 54-55; NSF sr fel, 63-64. *Mem:* Am Math Soc; Math Asn Am; London Math Soc; Sigma Xi. *Res:* Representation theory. *Mailing Add:* 3895 Spring Blvd Eugene OR 97405

CURTIS, CYRIL DEAN, b Albion, Ill, Sept 18, 20; m 48; c 2. PHYSICS, PARTICLE ACCELERATOR PHYSICS. *Educ:* McKendree Col, BS, 43; Univ Ill, MS, 47, PhD(physics), 51. *Prof Exp:* Asst, McKendree Col, 41-43 & Univ Ill, 46-50; assoc physicist, Argonne Nat Lab, 51-53; asst prof physics, Vanderbilt Univ, 53-59; scientist, Midwest Univs Res Asn, 59-67; physicist, Fermi Nat Accelerator Lab, 67-86; consult, Loma Linda Univ, 86-90; CONSULT, FERMI NAT ACCELERATOR LAB, 91- *Mem:* Am Phys Soc; Sigma Xi; AAAS. *Res:* Experimental nuclear reactions with charged particles and neutrons; electronic instrumentation; reactor physics; bremsstrahlung production; ion sources and accelerator physics. *Mailing Add:* 230 Woodland Hills Rd Batavia IL 60510

CURTIS, DAVID WILLIAM, b Kalamazoo, Mich, Oct 2, 24; m 55; c 4. APPLIED PHYSICS. *Educ:* Western Mich Univ, BS, 48; Iowa State Univ, PhD(physics), 55. *Prof Exp:* Asst, Ames Lab, AEC, 49-55; sr develop engr, Goodyear Aerospace Corp, 55-57, head theoret group, Aerophys Dept, 57-63, head physics group, Res & Develop Sect, Electronics Eng Div, 63-66; mem tech staff, Advan Develop Oper, Aeronaut Div, Philco-Ford Corp, 66-71; tech consult res & develop, Ariz Eng Div, Goodyear Aerospace Corp, 71-86; RETIRED. *Mem:* Am Phys Soc. *Res:* Engineering analysis; information theory; probability theory in physics and engineering; synthetic array radar; microwave halography; physical optics; data processing systems. *Mailing Add:* 10042 N 26th St Phonix AZ 85028

CURTIS, DORIS MALKIN, petroleum geology; deceased, see previous edition for last biography

CURTIS, DWAYNE H, b Caldwell, Idaho, May 9, 30; m 54; c 4. PHYSIOLOGY. *Educ:* Idaho State Col, BS, 53; Univ Utah, MA, 60, PhD(exp biol), 63. *Prof Exp:* From asst prof to assoc prof, 63-73, PROF BIOL SCI, CALIF STATE UNIV, CHICO, 73- *Concurrent Pos:* Pulmonary function testing technician & consult, N T Enloe Mem Hosp, 69-72. *Res:* Effect of chemical crosslinking agents on the mechanical properties of rat-tail tendon; taxonomy of Myxomycetes in the states of California, Idaho and Oregon; pulmonary physiology function testing. *Mailing Add:* Dept Biol Calif State Univ Chico CA 95929

CURTIS, EARL CLIFTON, JR, b Vt, Oct 15, 32; m 61; c 2. LASERS, SPECTROSCOPY. *Educ:* Univ Vt, BS, 54; Univ Minn, PhD(phys chem), 59. *Prof Exp:* MEM TECH STAFF, ROCKETDYNE DIV, ROCKWELL INT CORP, 60- *Mem:* Am Phys Soc. *Res:* Molecular vibrations; infrared spectroscopy; lasers; diagnostic measurements of and performance analysis of high energy chemical lasers. *Mailing Add:* 2424 Tuna Canyon Topanga CA 90290

CURTIS, FRED ALLEN, b Edmonton, Alta; m 69; c 2. ENVIRONMENTAL DESIGN, INFRASTRUCTURE MANAGEMENT. *Educ:* Univ Man, BSc, 72, MSc, 73. *Prof Exp:* Asst prof environ planning, Queen's Univ, 75-81; assoc prof, 81-84, PROF ENVIRON ENG, UNIV REGINA, 84- *Concurrent Pos:* Prin consult, F A Curtis & Assocs, 75-89 & SOI Group, 89-; vis prof, Shad Valley Prog, Univ BC, 90; instr, Sask Inst Appl Sci & Technol, 91. *Mem:* Sigma Xi. *Res:* Development of software for residential planning, design and management; environmental impact assessment methods; environmental management and mediation procedures. *Mailing Add:* 4037 Hillsdale St Regina SK S4S 3Y8 Can

CURTIS, FREDERICK AUGUSTUS, JR, b Washington, DC, Nov 19, 22; m 46; c 2. AERONAUTICS, MECHANICS. *Educ:* Haverford Col, BS, 44; Worcester Polytech Inst, BS, 48; Calif Inst Technol, MS, 49. *Prof Exp:* Aerodynamicist, Convair, 49, aerodyn engr, Gen Dynamics, 49-50, aerophys engr, 50-52, sr aerophys engr, 52-53, sr & proj aerophys engr, 53-55, aerosysts group engr, Gen Dynamics, Ft Worth, 55-60, chief stability & flight controls, 60-63, mgr stability, guid & control, 63-69, dir eng proj off, Convair Aerospace Div, 69-91; RETIRED. *Concurrent Pos:* Consult subcomt aircraft flight dynamics, Off Advan Res & Technol, NASA, 67-; vpres & F-16 dep prog dir, Plans, Controls, Finances & Contracts, Fort Worth Div, Gen Dynamics, 78- *Mem:* Am Inst Aeronaut & Astronaut; Aerospace Indust Asn Am; Nat Mgt Asn. *Res:* Aerodynamic stability and control; flying qualities; navigation and attack systems functional operation and systems integration; analog computational and servo system design and operations; program management. *Mailing Add:* 6812 Dwight St Ft Worth TX 76116

CURTIS, GARY LYNN, b Belleville, Ill, Jan 21, 44; m 70; c 3. BIOCHEMISTRY, IMMUNOLOGY. *Educ:* Nebr Wesleyan Univ, BA, 66; Univ Nebr, PhD(biochem), 71. *Prof Exp:* Instr biochem, 73-76, asst prof & res asst prof, 76-81, ASSOC PROF BIOCHEM & RES ASSOC PROF OBSTET & GYNEC, MED CTR, UNIV NEBR, 81- *Concurrent Pos:* USPHS trainee, Eppley Cancer Inst, Univ Nebr, 72-75, oncologist, 73-75. *Res:* Chemical carcinogenesis; tumor immunology; reproductive biology. *Mailing Add:* 3216 N 61st St Omaha NE 68104

CURTIS, GEORGE CLIFTON, b St Petersburg, Fla, Dec 10, 26; m 55; c 3. PSYCHIATRY, ANXIETY DISORDERS. *Educ:* Lambuth Col, BA, 50; Vanderbilt Univ, MD, 53; McGill Univ, MSc, 59. *Prof Exp:* Demonstr psychiat, McGill Univ, 57-59; assoc, Univ Pa, 59-60, from asst prof to assoc prof, 60-72; actg chmn, 83-84, chief adult psychiat, 84-87, PROF PSYCHIAT, UNIV MICH, ANN ARBOR, 72-, RES SCIENTIST, 73- *Concurrent Pos:* USPHS res fel psychiat, McGill Univ, 58-59; NIMH career investr, Univ Pa, 61-66; clin asst psychiat, Royal Victoria Hosp, Montreal, Que, 57-59; med res scientist, Eastern Pa Psychiat Inst, 59-72, actg dir clin res, 68-72; consult clin res, Mercy Douglass Hosp, Philadelphia, Pa, 60-62; assoc mem, Albert Einstein Med Ctr, 62-68; dir, Anxiety Disorders Prog, Univ Mich, Ann Arbor, 79- *Mem:* Am Psychiat Asn; Am Psychosom Soc; AAAS; Am Psychopath Asn; Soc Biol Psychiat; Phobia Soc Am. *Res:* Psychiatry; phobias and anxiety disorders; psychobiology; drug treatment, classification, biological markers and brain physiology by position emission tomography in anxiety disorders. *Mailing Add:* Dept Psychiat Univ Mich 1500 E Med Ctr Dr Med Inn Bldg Ann Arbor MI 48109-0840

CURTIS, GEORGE DARWIN, b Galveston, Tex, Apr 30, 28; m 52; c 4. ACOUSTICS, SYSTEMS DESIGN. *Educ:* N Tex Univ, BS, 52. *Prof Exp:* Engr electronics, LTV Corp, 52-60, sr scientist, LTV Res Ctr, 60-67; dir systs comput, Control Data Corp-Pacific, 67-70; res assoc acoust systs, 74-78, RES SCIENTIST, UNIV HAWAII, 78-, INSTR, 88- *Concurrent Pos:* Lectr, Univ Hawaii, 80-; consult acoust systs, 70- *Mem:* Inst Elec & Electronics Engrs; Marine Technol Soc; Acoust Soc Am; Nat Asn Environ Professionals. *Res:* Use of ocean monitoring systems to analyze and improve Tsunami warning system; develop and demonstrate alternative energy systems and methodologies; author of 37 publications. *Mailing Add:* Univ Hawaii - JIMAR 1000 Pope Rd Honolulu HI 96822

CURTIS, GEORGE WILLIAM, b Brussels, Belg, Mar 9, 25; US citizen; m 51; c 4. ASTROPHYSICS, SOLAR PHYSICS. *Educ:* Univ Colo, BS, 49, MS, 52, PhD(astrophys), 63. *Prof Exp:* Physicist, Nat Bur Standards, 52-55; Fulbright fel, Inst Astrophys, Paris, 55-56; observer in charge, High Altitude Observ, 56-59, fel solar physics, Joint Inst Lab Astrophys, 63-64; res scientist, Sacramento Peak Observ, Air Force Cambridge Res Labs, 64-67; res scientist, Nat Ctr Atmospheric Res, 67-73, dep dir high altitude observ, 73-; DIR, VCAR PROJ OFF, 85- *Concurrent Pos:* Sci coordr & exped leader, Solar Eclipse Exped, Nat Ctr Atmospheric Res, 71-73. *Mem:* Am Astron Soc; Royal Astron Soc; Am Inst Physics. *Res:* Low temperature properties of metal and insulating materials; design and development of low temperature instrumentation; acquisition, reduction and analysis of solar observational data on the corona, chromosphere and photosphere, especially spectral data. *Mailing Add:* PO Box 3000 Boulder CO 80307-3000

CURTIS, HERBERT JOHN, b Oak Park, Ill, Aug 18, 18; m 50; c 2. MATHEMATICS. *Educ:* Yale Univ, BA, 39; Ill Inst Technol, MS, 48, PhD(math), 54. *Prof Exp:* Instr math, Ill Inst Technol, 48-54; from asst prof to assoc prof, 54-65, from actg head to head div, 60-64, PROF MATH, UNIV ILL, CHICAGO CIRCLE, 65-, EXEC SECY DEPT, 64- *Mem:* Am Math Soc; Math Asn Am. *Mailing Add:* 1328 Greenwood Ave Wilmette IL 60091

CURTIS, HUNTINGTON W(OODMAN), b US, Jan 31, 21; m 48; c 3. ENGINEERING SCIENCE. *Educ:* Col William & Mary, BS, 42; Univ NH, MS, 48; Univ Iowa, PhD(elec eng), 50. *Prof Exp:* Asst prof elec eng, Univ Va, 50-51, US Mil Acad, 51-53 & Thayer Sch Eng, Dartmouth Col, 53-59; mem syst anal dept, Int Bus Mach Corp, 59-60, mem commun systs ctr, 61-62, mgr tech requirements, Hqs, 62-63, tech adv to vpres res & eng, 63-64, dir, Govt Tech Liaison, Corp Tech & Eng, 64-66, dir sci & technol info, Corp Eng Prog & Technol Staff, 66-71, mem integrated circuit adv technol staff, 71-76, sr tech generalist, 76-80, sr biomed syst architect, 80-84, ENGR CONSULT RES DIV, INT BUS MACH CORP, 85- *Concurrent Pos:* Consult, Mt Wash Meteorol Observ, 55-59, trustee, 57-; mem ionosphere physics adv panel, US Nat Comt Int Geophys Year, 57-59. *Mem:* Sr mem Inst Elec & Electronics Engrs; Sigma Xi. *Res:* Engineering science. *Mailing Add:* IBM Corp 6-112 IBM Research Yorktown Heights NY 10598

CURTIS, JAMES O(WEN), b Lincoln, Nebr, Apr 27, 23; m 46; c 5. AGRICULTURAL ENGINEERING. *Educ:* Univ Ill, BS, 47, MS, 48; Purdue Univ, PhD(agr eng), 62. *Prof Exp:* From instr to prof agr eng, Univ Ill, Urbana-Champaign, 48- 85; RETIRED. *Concurrent Pos:* NSF fel, 59-60; coordr spec projs, Ill Agr Exp Sta, Urbana-Champaign, 86- *Mem:* Am Soc Agr Engrs. *Res:* Analysis and design of timber structures; experimental stress analysis of wooden structural members and frames; strengths of pole anchorage systems. *Mailing Add:* 2005 Shelly Ct Urbana IL 61801

CURTIS, JOHN RUSSELL, b Bessemer City, NC, Nov 7, 34; m 58, 77; c 3. PSYCHIATRY. *Educ:* Univ NC, AB, 56, MD, 60. *Prof Exp:* Clin instr psychiat, Sch Med, Univ Ky, 64-66, consult psychiatrist, Student Health Serv, 64-55, dir psychiat sect, 66-68, asst prof psychiat, 66-68; assoc prof psychiat, Dept Psychol, Univ Ga, 68-74, chief psychiatrist & dir, univ health serv, 68-85; clin prof psychiat, Med Col Ga, 68-88; clin prof psychiat, Emory Sch Med, 80-88; PVT PSYCHIAT PRACT, 85- *Concurrent Pos:* Staff psychiatrist, Clin Res Ctr, NIMH, 64-65, chief ment addiction serv, 65-66; consult psychiatrist, Berea Col, 66-69, Clin Ctr Addiction Res, Ky, 67-69 & Stephens County Pub Health Dept, Ga, 69-; chmn, Gov Adv Coun Mental Health & Mental Retardation, Ga. *Mem:* Fel Am Psychiat Asn; Am Col Health Asn (pres, 78-79). *Res:* College health with emphasis on college mental health; health care delivery systems; community health and mental health. *Mailing Add:* 650 Oglethorpe Ave Suite 6 Athens GA 30606

CURTIS, JOSEPH C, b Manchester, NH, Feb 14, 30; m 59; c 4. CELL BIOLOGY, ENDOCRINOLOGY. *Educ:* Cornell Univ, BA, 51; Brown Univ, PhD(biol), 60. *Prof Exp:* Res assoc biol, Brown Univ, 60-63; from asst prof to prof biol, 63-70, PROF ZOOL, CLARK UNIV, 70- *Concurrent Pos:* USPHS fel, 60-62. *Mem:* Am Soc Zoologists; Am Soc Cell Biol; Electron Microscopy Soc Am. *Res:* Biological effects of ultrasound; cytology; cytochemistry; cell fine structure and biochemistry of steroid-producing tissues; experimental pathology. *Mailing Add:* Dept Biol Clark Univ Worcester MA 01610

CURTIS, LAWRENCE ANDREW, b Hartford, Conn, Apr 14, 42. INVERTEBRATE ZOOLOGY, ESTUARINE ECOLOGY. *Educ:* Nasson Col, BA, 65; Univ NH, MS, 67; Univ Del, PhD(biol sci), 73. *Prof Exp:* Lectr bot & zool, Fairleigh Dickinson Univ, 67-68; instr, 72-73, asst prof, 73-83, ASSOC PROF BIOL SCI, UNIV DEL, 83- *Mem:* AAAS; Am Soc Zool; Ecol Soc Am. *Res:* Influences of parasites on the ecology of intertidal estuarine systems; neogastrapod Ilyanassa obsoleta and its trematode parasites. *Mailing Add:* Sch Life & Health Sci Univ Del Newark DE 19711

CURTIS, LAWRENCE B, b Grand Junction, Colo, 24. PETROLEUM ENGINEERING. *Prof Exp:* Vpres, production eng, Conoco Inc, 49-87. *Honors & Awards:* DeGolyer Medal Offshore Technol Award, Am Soc Petrol Eng, 84; DeGaullier Award, Am Soc Petrol Eng, 84. *Mem:* Nat Acad Eng; hon mem Am Soc Petrol Eng; Am Petrol Inst. *Mailing Add:* 2819 Lakeside Dr Montgomery TX 77356

CURTIS, LORENZO JAN, b St Johns, Mich, Nov 4, 35; m 71. OPTICS, ENGINEERING PHYSICS. *Educ:* Univ Toledo, BS, 58; Univ Mich, MS, 61, PhD(high energy physics), 63. *Prof Exp:* Asst prof physics, 63-68, assoc prof physics & astron, 68-72, assoc prof, 72-80, PROF PHYSICS, UNIV TOLEDO, 80- *Concurrent Pos:* NSF res grant, 64-66, 75-79, fel, Latin Am Sch Physics, Mexico City, 65; vis scientist, Woods Hole Oceanog Inst, 67 & Nobel Inst, Sweden, 70-75; Docent physics, Univ Lund, Sweden, 76-79; Dept Energy res contract, 80- *Honors & Awards:* Outstanding Res Award, Sigma Xi, 80. *Mem:* Am Phys Soc; Optical Soc Am; Europ Phys Soc. *Res:* Atomic and ionic spectra; transition probabilities; collision processes, utilizing beam-foil and pulsed electron beam methods. *Mailing Add:* Dept Physics Univ Toledo Toledo OH 43606

CURTIS, MORTON LANDERS, mathematics; deceased, see previous edition for last biography

CURTIS, MYRON DAVID, b Mt Vernon, Ind, Oct 17, 38; m 66; c 2. INORGANIC CHEMISTRY, ORGANOMETALLIC CHEMISTRY. *Educ:* Wabash Col, AB, 60; Northwestern Univ, PhD, 65. *Prof Exp:* Res chemist, Hooker Chem Corp, 64-66; instr chem, Northwestern Univ, 66-67; from asst prof to assoc prof, 67-77, chmn chem dept, 86-90, PROF CHEM, UNIV MICH, ANN ARBOR, 77- *Concurrent Pos:* Petrol Res Fund grant, 67-69; fac res fel, Univ Mich, 69; Am Metals Climax Found grant, 75-76; NSF grants, 77-90; Off Naval Res grants, 77-80 & 81-84; Petrol Res Fund grants, 77-90. *Mem:* Am Chem Soc; NAm Catalyst Soc. *Res:* Nature of bonding in organometallic compounds; synthesis and structure of organometallic and inorganic systems related to catalysis; structural studies in homogeneous and heterogeneous catalysis, molybdenum compounds and silicones. *Mailing Add:* Dept Chem 2807 Chem Bldg Univ Mich Ann Arbor MI 48109-1055

CURTIS, ORLIE LINDSEY, JR, b Hutchinson, Kans, Feb 27, 34; m 55; c 2. SOLID STATE PHYSICS. *Educ:* Union Col, Nebr, BA, 54; Purdue Univ, MS, 56; Univ Tenn, PhD, 61; Univ Southern Calif, JD, 77. *Prof Exp:* Chief semiconductors, Oak Ridge Nat Lab, 56-63; chief solid state physics, Ventura Div, Northrop Corp, 63-67, asst dir, Solid State Electronics Lab, 67-68, dir, 68-77; PARTNER, KROLOFF, BELCHER, SMART, PERRY & CHRISTOPHERSON, 77- *Concurrent Pos:* Vis lectr, Univ Calif, Berkeley, 70-71. *Mem:* Fel Am Phys Soc; fel Inst Elec & Electronics Engrs. *Res:* Radiation effects on materials, devices, circuits and systems; electronic properties of solids; device physics. *Mailing Add:* PO Box 692050 Stockton CA 95269-2050

CURTIS, OTIS FREEMAN, JR, horticulture, plant physiology; deceased, see previous edition for last biography

CURTIS, PAUL ROBINSON, b Hanumakonda, S India, Aug 14, 31; US citizen; m 61; c 2. MICROBIOLOGY. *Educ:* Col Wooster, BA, 52; Ohio State Univ, MS, 57, PhD(bact), 61. *Prof Exp:* Asst, Ohio State Univ, 55-60; from asst prof to assoc prof biol, Am Univ, 60-70; prof biol, div sci & math, Eisenhower Col, 70-83; PROF BIOL, HUSSON COL, 83- *Mem:* AAAS; Am Soc Microbiol; Sigma Xi; Am Inst Biol Sci. *Res:* Bacteriophages; bacterial and viral genetics; environmental and food microbiology. *Mailing Add:* Husson Col Bangor ME 04401

CURTIS, PHILIP CHADSEY, JR, b Providence, RI, Mar 6, 28; m 50; c 5. MATHEMATICS EDUCATION. *Educ:* Brown Univ, AB, 50; Yale Univ, MA, 52, PhD(math), 55. *Prof Exp:* chmn dept, 71-75, from instr to assoc prof, 55-67, PROF MATH, UNIV CALIF, LOS ANGELES, 67- *Concurrent Pos:* Fulbright travel fel, Aarhus Univ, 69-70; vis prof, Univ Copenhagen, 75-76; res fel, Australian Nat Univ, 86. *Mem:* Am Math Soc; Math Asn Am. *Res:* Functional analysis; Banach algebras; harmonic analysis. *Mailing Add:* Dept Math Univ Calif Los Angeles CA 90024

CURTIS, RALPH WENDELL, b Cuba City, Wis, Oct 20, 36; m 57; c 3. METALLURGICAL CHEMISTRY. *Educ:* Univ Wis-Platteville, BS, 58; Iowa State Univ, PhD(metall), 62. *Prof Exp:* AEC fel metall, Iowa State Univ, 62-63; teacher chem, 63-69, head dept chem, 69-77, ASST V CHANCELLOR, UNIV WIS-PLATTEVILLE, 77- *Mailing Add:* Dept Chem Univ Wis Platteville WI 53818

CURTIS, ROBERT ARTHUR, b Weymouth, Mass, July 15, 54. PHARMACOLOGY. *Educ:* Mass Col Pharm, BS, 77; Univ Mo-Kansas City, PhD(clin pharm), 78. *Prof Exp:* Resident clin pharm, Truman Med Ctr, Mo, 78-79; ASST PROF PHARM PRACT, COL MED, UNIV ILL, 79-, COORDR, RESIDENCY PROG PHARM, UNIV HOSP, 80- *Mem:* Am Soc Hosp Pharmacists; Am Asn Col Pharm; Am Col Clin Pharm. *Res:* Clinical pharmacology trials in the emergency medicine enviroment. *Mailing Add:* Cambridge Neurosci Res One Kendall Sq Bldg 700 Cambridge MA 02139

CURTIS, ROBERT ORIN, b Portland, Maine, Oct 27, 27; m 52; c 3. FOREST MENSURATION. *Educ:* Yale Univ, BS, 50, MF, 51; Univ Wash, PhD(forestry), 65. *Prof Exp:* Forester, Northeastern Forest Exp Sta, 51-54, res forester, 54-62, PRIN MENSURATIONIST, PAC NORTHWEST FOREST & RANGE EXP STA, US FOREST SERV, 65- *Mem:* Soc Am Foresters. *Res:* Mensuration research in forest yield studies, site-growth relationships, related measurement problems, thinning, stocking control and simulation models. *Mailing Add:* Pac Northwest Forest & Range Exp Sta Forestry Sci Lab 3625 93rd Ave SW Olympia WA 98502

CURTIS, ROBIN LIVINGSTONE, b Bellingham, Wash, Jan, 16, 26; m 60; c 2. NEUROANATOMY, PHYSIOLOGICAL PSYCHOLOGY. *Educ:* Wesleyan Univ, BA, 48, MA, 50; Brown Univ, PhD(exp psychol), 54. *Prof Exp:* Chem analyst, E I Du Pont de Nemours & Co, NY, 46-47; psychiat aide, Middletown State Hosp, Conn, 47-48; asst biol, Wesleyan Univ, 49-50; spec biol asst, Brown Univ, 52, asst psychol, 52-53; from instr to assoc prof, NJ Col Med & Dent, 56-67; ASSOC PROF ANAT, MED COL WIS, 67-, ASSOC PROF PHYS MED & REHAB, 79- *Concurrent Pos:* Investr, Jackson Lab, 52-55; USPHS fel, 53-55; Nat Multiple Sclerosis Soc fel anal, Col Med, NY Univ, 55-56; vis prof, Univ Wis, 85. *Mem:* AAAS; Am Psychol Asn; hon mem Am Orthopsychiatric Asn; Am Asn Anat; Am Genetic Asn; Sigma Xi; Soc Neurosci. *Res:* Genetics, behavior and neuroanatomy of hereditary neuromuscular abnormalities in rodents; comparative neurology and animal behavior; plasticity in the rodent central nervous system; motor unit function. *Mailing Add:* Dept Anat Med Col Wis PO Box 26509 Milwaukee WI 53226

CURTIS, RONALD S, b Claremont, NH, Nov 1, 50. SOFTWARE ENGINEERING, DISTRIBUTED SYSTEMS-DEBUGGING. *Educ:* Keene State Col, BA, 72; Univ NH, MS, 74. *Prof Exp:* Grad asst math, Univ NH, 73-74; teacher sci, Windsor, Vt High Sch, 74-76; chmn, Canisius Col, 79-83, asst prof computer sci, 84-89; grad asst computer sci, State Univ NY, Buffalo, 76-77, res asst, 77-79 & 83-84, vis lectr, 89-91, RES SCIENTIST, STATE UNIV NY, BUFFALO, 91- *Concurrent Pos:* Consult, Calspan, 83 & State Educ Dept, 89. *Mem:* Inst Elec & Electronics Engrs; Inst Elec & Electronics Engrs Computer Soc; Asn Comput Mach. *Res:* Software metrics; design of metrics and their application to software design and development; data structures. *Mailing Add:* PO Box 51 Buffalo NY 14216

CURTIS, STANLEY BARTLETT, b Evanston, Ill, Feb 16, 32; m 82; c 4. RADIATION BIOPHYSICS. *Educ:* Carleton Col, BA, 54; Univ Wash, PhD(physics), 62. *Prof Exp:* Res scientist, Biophys Res Div, Lockheed-Calif Co, 62-63; res specialist, Space Physics Group, Boeing Co, 63-65; biophysicist, Donner Lab, 66-78, sr staff biophysicist, Res Med & Radiation Biophys Div, 78-88, dep group leader, Radiation Biophys Group, 85-86, SR BIOPHYSICIST, CELL & MOLECULAR BIOL DIV, LAWRENCE BERKELEY LAB, UNIV CALIF, 88- *Concurrent Pos:* Mem radiation biol panel, NASA, 60; Eleanor Roosevelt fel, 70-71; physics counr, Coun Radiation Res Soc, 72-75; ed, Radiation Res J, 71-74; comt chmn, Int Comn Radiation Units & Measurements, 76-79; sr int Fogarty fel, 81-82; counr, Nat Coun Radiation Protection, 87- *Mem:* Radiation Res Soc; Fed Am Scientists; AAAS. *Res:* Biological effects of ionizing radiation in space; high linear energy transfer effects on biological systems; tumor cell kinetics; risk assessment of high linear energy transfer radiation. *Mailing Add:* Lawrence Berkeley Lab MS 29-100 Berkeley CA 94720

CURTIS, SUSAN JULIA, b Tuskegee, Ala, Jan 30, 45. MEMBRANE BIOCHEMISTRY. *Educ:* Radcliffe Col, Cambridge, Mass, AB, 66; Univ Chicago, MS, 69, PhD(biochem), 73. *Prof Exp:* Fel, Cornell Univ, 73-75; fel, Harvard Univ, 75-78; ASST PROF BIOCHEM, COL MED, HOWARD UNIV, 78- *Concurrent Pos:* Adj asst prof oncol, Howard Univ Cancer Ctr, 81- *Mem:* Sigma Xi; Am Soc Microbiol; Am Chem Soc; AAAS. *Res:* Biochemistry of the structure and function of cell membranes; function of the penicillin-binding proteins of the cell wall of E coli in cell wall synthesis; the role of these proteins in the mechanism of action of penicillin. *Mailing Add:* 3606D Lakefield Dr Greensboro NC 27406-6812

CURTIS, THOMAS EDWIN, b Miami, Okla, Oct 2, 27; m 46; c 4. PSYCHIATRY. *Educ:* Duke Univ, MD, 50; Univ NC-Duke Univ Training Prog, cand, 60. *Prof Exp:* Intern, St John's Hosp, Okla, 50-51; resident psychiat, Fairfield State Hosp, Conn, 52; resident, Dorthea Dix State Hosp, NC, 52-54; from instr to assoc prof, Sch Med, 54-69, prof psychiat, 69-, chmn dept, 73-, PROF PSYCHIAT & CLIN PROF, SCH NURSING, UNIV NC, CHAPEL HILL. *Concurrent Pos:* Resident, NC Mem Hosp, 54-55. *Mem:* AMA; fel Am Psychiat Asn; Am Asn Med Cols. *Res:* Experimental teaching; group analytic psychotherapy; family study and treatment; clinical research in psychotherapy; psychiatric nursing, teaching and administration. *Mailing Add:* Dept Psychiat Sch Med Univ NC Chapel Hill NC 27514

CURTIS, THOMAS HASBROOK, b Detroit, Mich, July 26, 41; m 79; c 2. TELECOMMUNICATION SYSTEMS. *Educ:* Kenyon Col, BA, 63; Yale Univ, MS, 65, PhD(physics), 68. *Prof Exp:* Res physicist, Lawrence Radiation Lab, Univ Calif, 68-70; mem tech staff, 70-80, SUPVR, ECHO CONTROL GROUP, BELL LABS, 80- *Concurrent Pos:* Consult, Lycoming Div, Avco Corp, Conn, 67 & Automation Res Mechanisms, Inc, Calif, 68-69. *Mem:* Inst Elec & Electronics Engrs; AAAS; Am Phys Soc; NY Acad Sci. *Res:* Digital processing for satellite communication; physics of speech and hearing; digital processing of acoustic signals; applications of computers to system control; nuclear physics. *Mailing Add:* AT&T Bell Labs 3G-337 Crawfords Corner Rd Holmdel NJ 07733

CURTIS, VERONICA ANNE, b Reading, Eng, Oct 8, 48; US citizen. NATURAL PRODUCT SYNTHESIS. *Educ:* Univ Ill, Chicago, BA, 71, MS, 73, PhD(chem), 80. *Prof Exp:* Lectr chem, Univ Wis, Whitewater, 77-78; fel, 79-80, ASST PROF CHEM, LOYOLA UNIV, CHICAGO, 80- *Res:* Development of new general methods of deamination and investigations into organic reaction mechansims to improve synthetic methods. *Mailing Add:* Dept Chem Northeastern Univ 5500 N St Louis Ave Chicago IL 60625

CURTIS, WENDELL D, b New York, NY, Jan 18, 45. APPLIED MATHEMATICS, DIFFERENTIAL GEOMETRY. *Educ:* Univ Fla, Gainesville, BS, 66; Univ Mass, Amherst, PhD(math), 70. *Prof Exp:* From asst prof to assoc prof, 70-82, PROF MATH, KANS STATE UNIV, 82- *Concurrent Pos:* Vis prof, Univ Pittsburgh, 76; consult, Lawrence Livermore Lab, 78. *Mem:* Math Asn Am; Soc Indust & Appl Math. *Res:* Non-linear waves. *Mailing Add:* Dept Math Cardwell Hal Kans State Univ Manhattan KS 66506

CURTIS, WESLEY E, b Westfield, NY, Aug 25, 18; m 49; c 2. ENGINEERING, COMPUTER SCIENCE. *Educ:* Alfred Univ, BS, 40. *Prof Exp:* Anal chemist, Atlas Feldspar Corp, 40-42; res assoc, Ansbacher-Siegle Corp, 42, Alfred Univ, 42-44 & Underwater Explosives Res Lab, Woods Hole Oceanog Inst, 44-47; physicist, Ballistic Res Labs, Aberdeen Proving Ground, 47-56; sr opers analyst, Tech Opers, Inc, 56-67; sr res analyst, 67-80, dir spec proj, Hancock/Dikewood Serv, Inc, 80-83; PVT CONSULT, HEALTHCARE MGT, 83- *Mem:* Oper Res Soc Am; AAAS; Sigma Xi. *Res:* Cost Effectiveness analysis; computer system science; nuclear weapon phenomenology. *Mailing Add:* 3329 Santa Clara SE Albuquerque NM 87106

CURTISS, CHARLES FRANCIS, b Chicago, Ill, Apr 4, 21; m 46; c 3. THEORETICAL CHEMISTRY. *Educ:* Univ Wis, BS, 42, PhD(chem), 48. *Prof Exp:* Chemist, Geophys Lab, Carnegie Inst, 42-45; assoc, Allegany Ballistics Lab, George Washington Univ, 45; proj assoc, 48-49, from asst prof to assoc prof, 49-60, PROF CHEM, UNIV WIS-MADISON, 60- *Mem:* AAAS; Am Chem Soc; Am Phys Soc. *Res:* Statistical mechanics; kinetic theory of gases. *Mailing Add:* 1101 Univ Ave Univ Wis Madison WI 53706

CURTISS, HOWARD C(ROSBY), JR, b Chicago, Ill, Mar 17, 30; m 56, 88; c 2. AERONAUTICAL ENGINEERING. *Educ:* Rensselaer Polytech Inst, BAeroE, 52; Princeton Univ, MSE, 57, PhD(aerodyn), 65. *Prof Exp:* Res staff mem aerospace & mech sci, 57-65, lectr, 63-65, from asst prof to assoc prof, 65-71, PROF MECH & AEROSPACE ENG, PRINCETON UNIV, 71- *Concurrent Pos:* Vis lectr, Stevens Inst Technol, 60-61; hon prof, Nanjing Aeronaut Inst, Nanjing, China, 85. *Mem:* Am Inst Aeronaut & Astronaut; Am Helicopter Soc. *Res:* Aerodynamics, stability and control of helicopters and vertical take off and landing aircraft, experimental and theoretical. *Mailing Add:* Mech & Aerospace Eng Princeton Univ Eng Quad D-224 Princeton NJ 08544

CURTISS, LARRY ALAN, b Madison, Wis, Sept 16, 47; m 78; c 2. HYDROGEN BONDING, MOLECULAR INTERACTIONS. *Educ:* Univ Wis, BS, 69; Carnegie-Mellon Univ, MS, 71, PhD(chem), 73. *Prof Exp:* Fel, Batelle Mem Inst, 73-76; SR CHEMIST, ARGONNE NAT LAB, 76- *Mem:* Am Chem Soc; AAAS. *Res:* Quantum mechanical and experimental studies of molecular interactions in the gas phase including hydrogen bonded and ionic complexes; thermodynamics; interfacial materials chemistry. *Mailing Add:* Chem Technol Div Argonne Nat Lab Argonne IL 60439

CURTISS, LINDA K, LIPOPROTEINS. *Educ:* Univ Wash, PhD(immunol), 74. *Prof Exp:* ASSOC MEM, DEPT IMMUNOL, SCRIPP RES INST, 74- *Mailing Add:* Dept Immunol Scripp Res Inst 10666 Torrey Pines Rd La Jolla CA 92037

CURTISS, ROY, III, b New York, NY, May 27, 34; c 6. MICROBIAL GENETICS, MOLECULAR GENETICS. *Educ:* Cornell Univ, BS, 56; Univ Chicago, PhD(microbiol), 62. *Prof Exp:* Jr technical specialist, Biol Div, Brookhaven Nat Lab, 56-58; biologist, Oak Ridge Nat Lab, 63-72, group leader microbial genetics & radiation microbiol group, Biol Div, 69-72; prof microbiol, Univ Tenn-Oak Ridge Grad Sch Biomed Sci, 69-72, assoc dir, 70-71, interim dir, 71-72; prof, Univ Ala, Birmingham, 72-78, sr scientist, Inst Dent Res & Comprehensive Cancer Ctr, 72-83, dir molecular cell biol grad training prog, 73-82, Charles H McCauley prof microbiol, 78-83, dir & sr scientist, Cystic Fibrosis Res Ctr, 81-83, vchmn, dept microbiol, 81-82, actg chmn, microbiol dept, 82-83; DEPT BIOL, WASH UNIV, 83- *Concurrent Pos:* Lectr microbiol, Univ Tenn, 65-72 & Oak Ridge Grad Sch Biomed Sci, 67-69; vis prof, Venezuelan Inst Sci Res, 69, Univ PR, 72 & Cath Univ Chile, 73; lectr, Am Found Microbiol, 69-70, 79-80 & 81-82; ed, J Bact, 70-76; mem, NIH Recombinant DNA Molecule Prog Adv Comt, 74-77; mem, NSF Genetic Biol Study Sect, 75-78; dir, Nat Inst Gen Med Sci Predoctoral & Postdoctoral Training Grants, 75-; mem, NIH Genetic Basis Dis Rev Comt, 80-83, chmn, 81-83. *Honors & Awards:* P R Edwards Award, Am Acad Microbiol, 75. *Mem:* Fel Am Acad Microbiol; hon mem Microbiol Asn Chile; Am Soc Microbiol; Genetics Soc Am; Brit Soc Gen Microbiol; fel AAAS. *Res:* Genetic and biochemcial mechanisms for bacterial pathogenicity; bacterial genetics and molecular biology. *Mailing Add:* Dept Biol Wash Univ Box 1137 St Louis MO 63130

CURTRIGHT, THOMAS LYNN, b Moberly, Mo, Aug 21, 48; m 68; c 3. SUPERSYMMETRY, QUANTUM FIELD THEORY. *Educ:* Univ Mo, BS & MS, 70; Calif Inst Technol, PhD(theoret physics), 77. *Prof Exp:* Res assoc, Univ Calif, Irvine, 76-78; McCormick fel, Enrico Fermi Inst, Univ Chicago, 78-80; from asst prof to assoc prof, dept physics, Univ Fla, 84-88; PROF, DEPT PHYSICS, UNIV MIAMI, 88- *Mem:* Am Phys Soc; Am Math Soc. *Res:* The use of symmetries in physical theories, with emphasis on the interplay between quantum mechanics and symmetries in relativistic field theories; dynamics of extensible systems. *Mailing Add:* Dept Physics Univ Miami Coral Gables FL 33124

CURTSINGER, JAMES WEBB, b South Bend, Ind, Nov, 8, 50; m 83. POPULATION GENETICS, QUANTITATIVE GENETICS. *Educ:* Univ Chicago, BA, 73; Stanford Univ, PhD(biol), 78. *Prof Exp:* Teaching assoc genetics, NC State Univ, 78-80; teaching assoc zool, Univ Tex, Austin, 80-81; ASST PROF GENETICS, UNIV MINN, 81- *Concurrent Pos:* Prin investr, NSF, 83-; res career develop award & prin investr, NIH, 85- *Mem:* Genetics Soc Am; Soc Study Evolution. *Res:* Experimental and theoretical analysis of population genetics and quantitative genetics; the way that natural selection changes populations. *Mailing Add:* Dept Ecol Evol & Behav Univ Minn 318 Church St Minneapolis MN 55455

CURWEN, DAVID, b Ridgewood, NJ, May 25, 35; m 57; c 2. HORTICULTURE, FOOD TECHNOLOGY. *Educ:* Univ Vt, BS, 57; Pa State Univ, MS, 60, PhD(hort), 64. *Prof Exp:* Asst prof & area hort agent veg exten & res, 63-71, assoc prof, 71-80, PROF AGR EXTEN & HORT & AREA EXTEN HORTICULTURIST, UNIV WIS-MADISON, 80- *Mem:* Am Soc Hort Sci; Am Inst Biol Sci. *Mailing Add:* Dept Hort 385 Hort Univ Wis Hancock Exp Sta 1575 Linden Dr Madison WI 53706

CUSACHS, LOUIS CHOPIN, b Orange, NJ, Sept 9, 33; m 80; c 3. COMPUTER SCIENCE, QUANTUM CHEMISTRY. *Educ:* US Naval Acad, BS, 56; Univ Paris, dipl, 61; Northwestern Univ, PhD, 61. *Prof Exp:* Du Pont teaching asst, Northwestern Univ, 58-59; Fulbright exchange prof, Univ Valencia, 61-62; from asst prof to assoc prof phys chem, Tulane Univ, 62-72; prof comput sci, Loyola Univ, La, 72-77; res specialist, Exxon Prod Res Co, 77-85; prof sci & chmn liberal arts & sci, Christian Col Am, Houston, Tex, 85-86, comput consult, 87-88. *Concurrent Pos:* Lectr, Fulbright Prog, Nat Univ Buenos Aires & Univ La Plata, Arg, 74. *Mem:* Asn Christian Therapists; Christian Asn Psychological Studies; Int Soc Quantum Biol (vpres, 73-75, pres, 75-77). *Res:* Scientific computation; electronic structure of molecules; theoretical chemistry and computer modeling; geochemistry. *Mailing Add:* 1107 Jocelyn Houston TX 77023

CUSANO, CRISTINO, b Sepino Italy, Mar 22, 41; m 74. TRIBOLOGY. *Educ:* Rochester Inst Technol, BS, 65; Cornell Univ, MS, 67, PhD(mech eng), 70. *Prof Exp:* From asst prof to assoc prof, 70-83, PROF MECH ENG, UNIV ILL, URBANA-CHAMPAIGN, 83- *Concurrent Pos:* Design engr, Western Elec, 74 & Whirlpool Corp, 77; res engr, Lewis Res Ctr, Cleveland, NASA, 79. *Honors & Awards:* Capt Alfred E Hunt Award, Am Soc Lubrication Engrs, 81 & Al Sonntag Award, 83. *Mem:* Am Soc Mech Engrs; Soc Tribologists & Lubrication Engrs; Am Soc Eng Educ. *Res:* Tribology; thermal-elastohydrodynamic lubrication of line contacts; lubrication behavior of rough surfaces; modeling of rolling element bearing failures; friction and wear behavior of oil-refrigerant mixtures; leakage behavior of labyrinth seals. *Mailing Add:* 1303 Belmeade Dr Champaign IL 61821

CUSANO, DOMINIC A, b Schenectady, NY, May 20, 24; m 45; c 3. SOLID STATE LUMINESCENCE. *Educ:* Union Col, BS, 49, MS, 53; Rensselaer Polytech Inst, PhD(solid state physics), 59. *Prof Exp:* PHYSICIST RES & DEVELOP, GEN ELEC CORP, 49- *Mem:* Am Phys Soc; Electrochem Soc; Sigma Xi. *Res:* Solid state x-ray detectors; solar photovoltaic energy conversion; thin film luminescense; photoemission, incandescent and fluorescent lamp studies; semiconductor hybrid and packaging techniques; chemical vapor deposition processing; wide band gap semiconductors; superconductivity. *Mailing Add:* 2017 Morrow Ave Schenectady NY 12309

CUSANOVICH, MICHAEL A, b Los Angeles, Calif, Mar 2, 42; m 63, 80; c 3. BIOPHYSICS. *Educ:* Univ of the Pac, BS, 63; Univ Calif, San Diego, PhD(chem), 67. *Prof Exp:* NIH fel biochem, Univ Calif, San Diego, 67-68 & Cornell Univ, 68-69; asst prof, 69-74, assoc prof chem, 74-78, assoc prof biochem, 78-79, PROF BIOCHEM & CHEM, UNIV ARIZ, 79-; VPRES RES & DEAN, GRAD COL, UNIV ARIZ, 88- *Concurrent Pos:* NIH Career Develop Award, 75-80; dir, biochem prog, Nat Sci Found, 81-82; consult, Univ Patents Inc, 83-87, vdean, Grad Col, 87-88. *Mem:* Am Chem Soc; Biophys Soc; Am Soc Biol Chem; Protein Soc; Am Soc Photobiol. *Res:* Mechanism of electron transfer as catalyzed by cytochromes and coupled energy conservation; mechanism of action of the visual pigment rhodopsin; myosin ATPase. *Mailing Add:* Dept Biochem Univ Ariz Tucson AZ 85721

CUSCURIDA, MICHAEL, b Waco, Tex, Nov 16, 26. ORGANIC CHEMISTRY. *Educ:* Tex Agr & Mech Col, BS, 48; Baylor Univ, MS, 51, PhD, 55. *Prof Exp:* Assoc res chemist, Midwest Res Inst, 55-57; sr res chemist, Jefferson Chem Co, Inc, 57-65, proj chemist, 65-69, sr proj chemist, 69-82; RES ASSOC, TEXACO CHEM CO, 82- *Mem:* Am Chem Soc. *Res:* Organic chemistry; polyurethane chemistry; polymers of alkylene oxides; inorganic chemistry; chemistry of metal hydrides; high energy fuels based on boron compounds. *Mailing Add:* Texaco Chem Co Box 15730 Austin TX 78761-5730

CUSHEN, WALTER EDWARD, b Hagerstown, Md, Mar 21, 25; m 49; c 2. GOVERNMENT POLICY ANALYSIS. *Educ:* Western Md Col, BA, 48; Univ Edinburgh, Scotland, PhD(metaphys), 51. *Hon Degrees:* ScD, Western Md Col, 66. *Prof Exp:* Mathematician, Ballistic Res Labs, 48 & 51-52; sr programmer, Remington Rand, 52; chmn res group, Opers Res Off, 52-61; assoc prof opers res, Case Inst Technol, 61-63; staff mem, Inst Defense Anal, 63-64; chief, Tech Anal Div, Nat Bur Standards, 64-74; actg asst dir, Off Energy Systs, Fed Power Comn, 74-75; vpres & dir, Mathtech Washington, Mathematica, Inc, 75-81; assoc, J F Coates Assocs, 81-82; res fel, Logistics Mgt Inst, 82-84; OPER RES ANALYST, DEPT DEFENSE, 84- *Concurrent Pos:* Consult, FAA, 63-64; mem, Md Water Sci Adv Bd, 68-76, Gov Sci Adv Coun, 67-73 & Bd Trustees, Western Md Col, 71-75; consult, 81-84; trustee, Wash Oper Res, Mgt Sci Coun, 84-86. *Mem:* Fel AAAS; Opers Res Soc Am (pres, 70-71); Am Acad Polit & Soc Sci. *Res:* Policy analysis on US government problems; agricultural productivity in Egypt; analysis of simulations; constitutional bicentennial analysis. *Mailing Add:* 6910 Maple Ave Chevy Chase MD 20815-5114

CUSHING, BRUCE S, b Bayshore, NY, May 8, 55; m 81. PREDATOR-PREY RELATIONSHIPS, EVOLUTIONARY BIOLOGY. *Educ:* Ariz State Univ, BA, 77; Univ Mont, MA, 80; Mich State Univ, PhD(zool), 84. *Prof Exp:* Instr biol, Biol Sci Prog, Mich State Univ, 84-85; researcher zool, Hawaii Inst Marine Biol, 85-88; RES ASSOC, DEPT BIOL, IND UNIV, 88- *Mem:* Am Soc Mammalogists; Animal Behav Soc; Int Bear Biol Asn. *Res:* Coevolution of perdator-prey interactions with emphasis on the development of a predator's ability to intercept intraspecific reproductive cues emitted by their prey; vulnerability of estrous mice to weasel predation; author of numerous publications. *Mailing Add:* Dept Biol Ind Univ Jordan Hall 142 Bloomington IN 47405

CUSHING, COLBERT ELLIS, b Ft Collins, Colo, Jan 9, 31; m 59; c 3. FRESH WATER ECOLOGY. *Educ:* Colo State Univ, BSc, 52, MSc, 56; Univ Sask, PhD(biol), 61. *Prof Exp:* Fishery biologist, Mont Fish & Game Dept, 56-58; biol scientist, Hanford Labs, Gen Elec Co, 61-65; res scientist, 65-67, SR RES SCIENTIST, PAC NORTHWEST LAB, BATTELLE MEM INST, 67- *Concurrent Pos:* Lectr, Joint Ctr Grad Studies, Wash State Univ, 79-; consult, Nat Acad Sci FREIR Comt, 80; external adv comt, Long Term Ecol Res-NSF; jour ed, 80-83, bull ed, Ecol Soc Am, 83-, J NAm Benthol Soc, 84- *Mem:* Am Soc Limnol & Oceanog; Ecol Soc Am; Am Fisheries Soc; Int Asn Theoret & Appl Limnol; NAm Benthol Soc. *Res:* Stream ecology; radioecological and nutrient cycling studies of aquatic ecosystems; primary productivity. *Mailing Add:* Environ Sci Dept Pac Northwest Lab Battelle Mem Inst Richland WA 99352

CUSHING, DAVID H, b Boston, Mass, Dec 6, 40; m 27; c 4. OPTICS, ELECTRICAL ENGINEERING. *Educ:* Northeastern Univ, BSEE, 69. *Prof Exp:* Prod supvr, Bairdoco, 59-71; TECH DIR, MICROCOATING INC, 71- *Mem:* Soc Photo-Optical Instrumentation Engrs. *Res:* Thin films for spectral control of optical elements; ultra violet and visible filter and reflector areas. *Mailing Add:* 240 Central Way North Reading MA 01864

CUSHING, EDWARD JOHN, b St Louis, Mo, Nov 15, 33; m 55; c 5. ECOLOGY. *Educ:* Wash Univ, AB, 54; Univ Minn, PhD(geol), 63. *Prof Exp:* Res fel geol, Sch Earth Sci, Univ Minn, Minneapolis, 63-64, asst prof, 64-66, from asst prof to prof bot, 66-74; PROF ECOL & BEHAV BIOL, UNIV MINN, MINNEAPOLIS, 75- *Concurrent Pos:* NATO fel, Univ NWales, Bangor, 66-67. *Mem:* AAAS; Geol Soc Am; Ecol Soc Am; Bot Soc Am; Brit Ecol Soc; Sigma Xi. *Res:* Quaternary pollen analysis and paleoecology; plant ecology; glacial geology. *Mailing Add:* Dept Ecol & Behav Biol Univ Minn 318 Church St SE Minneapolis MN 55455

CUSHING, JIM MICHAEL, b N Platte, Nebr, Mar 20, 42. APPLIED MATHEMATICS. *Educ:* Univ Colo, BA, 64; Univ Md, PhD(math), 68. *Prof Exp:* From asst prof to assoc prof, 68-83, PROF MATH, UNIV ARIZ, 84- *Concurrent Pos:* IBM fel, 71-72; Alexander von Humboldt fel, 77-78. *Mem:*

Am Math Soc; Soc Indust & Appl Math; Math Asn Am; Soc Math Biol; Resource Modeling Asn. *Res:* Qualitative theory of integral, differential and integrodifferential equations and applications; population dynamics; mathematical ecology. *Mailing Add:* Dept Math Univ Ariz Bldg 89 Tucson AZ 85721

CUSHING, JOHN (ELDRIDGE), JR, b San Francisco, Calif, Aug 25, 16; m 41; c 2. BIOLOGY. *Educ:* Univ Calif, AB, 38; Calif Inst Technol, PhD(genetics & immunol), 43. *Prof Exp:* Asst immunochem, Calif Inst Technol, 43-45; instr biol, Johns Hopkins Univ, 45-48; from asst prof to prof biol, 48-69, prof, 69-80, EMER PROF IMMUNOL, UNIV CALIF, SANTA BARBARA, 80- *Concurrent Pos:* Guggenheim fel, 58-59. *Mem:* Am Asn Immunologists; Am Soc Naturalists. *Res:* Comparative immunology; evolution studies; blood groups of marine animals; immune reactions of invertebrates. *Mailing Add:* 135 Canyon Acres Dr Santa Barbara CA 93105

CUSHING, KENNETH MAYHEW, b Charlotte, NC, Aug 16, 47; m 85; c 2. FABRIC FILTRATION, AIR POLLUTION CONTROL. *Educ:* Rhodes Col, Memphis, BS, 69; Univ Fla, MS, 72. *Prof Exp:* Assoc, 72-75, res, 75-78, head, Aerosol Physics Sect, 78-81, sr physicist, 81-83, HEAD, FABRIC FILTER RES SECT, SOUTHERN RES INST, 83- *Mem:* Air & Waste Mgt Asn. *Res:* Operating behavior of fabric filters and wet and dry flue gas scrubbing systems; particle sizing measurement techniques including inertial, optical, electrical and diffusional methods. *Mailing Add:* Southern Res Inst PO Box 55305 Birmingham AL 35255-5305

CUSHING, MERCHANT LEROY, b Haverhill, Mass, Jan 31, 10; m 39; c 4. CARBOHYDRATE CHEMISTRY. *Educ:* Univ NH, BS, 31, MS, 33; Columbia Univ, PhD(Org chem), 41. *Prof Exp:* Res chemist, Stein, Hall & Co, NY, 37-42, chief chemist, Starch Explor Lab, 46-48, dir, NY Labs, 48-50, chief chemist, Paper Lab, 52-55; head paper lab, A E Staley Mfg Co, Ill, 55-62; mgr paper res, Fiber Prods Res Ctr, Inc, 62-65; group leader, Paper Appln, CIBA Corp, NJ, 65-69; group leader paper res & develop, Am Maize Prods Co, 69-74; res assoc, Bergstrom Paper Co, 74-78; CONSULT PAPER TECHNOL, 78- *Concurrent Pos:* Captain, Chem Warfare Serv, US Army, 43-46, 50-52. *Mem:* Tech Asn Pulp & Paper Indust. *Res:* Carbohydrate chemistry oriented to new industrial products; modifications of natural raw materials for use in papermaking; enzyme conversion of starches; detackifying stickers during deinking of recycled waste papers; vegetable glue remoistenable adhesives with animal glue equivalent performance. *Mailing Add:* 365 Cleveland St Menasha WI 54952

CUSHING, ROBERT LEAVITT, b Ord, Nebr, Apr 12, 14; m 38; c 3. AGRONOMY. *Educ:* Univ Nebr, BSc, 36, MSc, 38. *Hon Degrees:* DSc, Univ Hawaii, 62. *Prof Exp:* Asst agronomist, Nebr Agr Exp Sta, 38-42 & USDA, 42-43; from asst prof to assoc prof plant breeding, Cornell Univ, 43-47; agronomist, Hawaiian Pineapple Co, 47-49; from assoc prof to prof plant breeding, Cornell Univ, 49-51; asst dir, Pineapple Res Inst Hawaii, 51-52, dir, 52-63, pres, 58-63; dir, Exp Sta, Hawaiian Sugar Planters Asn, 63-79, vpres & secy, 66-79; RETIRED. *Concurrent Pos:* Mem bd regents, Univ Hawaii, 67, chmn, 68; consult, Int Agr Develop Serv & World Bank, 79- *Mem:* AAAS; Am Soc Agron; Am Genetic Asn; Inst Food Technol; Sigma Xi. *Res:* Plant breeding; plant genetics. *Mailing Add:* 2325 Armstrong St Honolulu HI 96822

CUSHLEY, ROBERT JOHN, b Edmonton, Alta, July 12, 36; m 65; c 2. SPECTROSCOPY, PHYSICAL BIOCHEMISTRY. *Educ:* Univ Alta, BSc, 57, MSc, 59, PhD(org chem), 65. *Prof Exp:* Res assoc chem, Sloan-Kettering Inst Cancer Res, 65-68; asst prof, Med Sch, Yale Univ, 68-73, assoc prof, 73-74; assoc prof, 74-79, PROF CHEM, SIMON FRASER UNIV, 79- *Concurrent Pos:* Lectr, Sloan-Kettering Div, Cornell Univ, 67-68. *Mem:* fel Chem Inst Can; Int Soc Magnetic Resonance; Can Biochem Soc; Can Fed Biol Soc; Biophys Soc Can. *Res:* Fourier transform spectroscopy; nuclear magnetic resonance studies of compounds of biological significance; membrane biophysics; Lipoprotein biophysics; dynamic structure of amphipathic peptides. *Mailing Add:* Dept Chem Simon Fraser Univ Burnaby BC V5A 1S6 Can

CUSHMAN, DAVID WAYNE, b Indianapolis, Ind, Nov 15, 39; m 64; c 2. BIOCHEMICAL PHARMACOLOGY. *Educ:* Wabash Col, AB, 61; Univ Ill, Urbana, PhD(biochem), 66. *Prof Exp:* Sr res scientist, 66-70, sr res investr, 70-74, res fel, 74-78, sr res fel, 78-83, asst dir, Dept Pharmacol, 83-87, RES ADV, SQUIBB INST MED RES, 87- *Honors & Awards:* Alfred Burger Award, Am Chem Soc, 82, CIBA Award, 83. *Mem:* AAAS; Am Chem Soc; Am Soc Pharmacol & Exp Therapeut; Am Soc Biol Chemists; Sigma Xi. *Res:* Bacterial hydroxylases and related electron transport systems; non-heme iron proteins; enzymology and pharmacology of vasoactive polypeptides, prostaglandins and enkaphalins. *Mailing Add:* 20 Lake Shore Dr RD No 8 Lawrenceville NJ 08648

CUSHMAN, JOHN HOWARD, b Ames, Iowa, Jan 19, 51; m 74. SOIL PHYSICS, APPLIED MATHEMATICS. *Educ:* Iowa State Univ, BS, 75, MS, 76, PhD(mat & agron), 78. *Prof Exp:* ASST PROF SOIL PHYSICS, DEPT AGRON, PURDUE UNIV, 78- *Mem:* Am Math Soc; Soc Indust & Appl Math; Am Geophys Union; Am Soc Agron; Sigma Xi. *Res:* Numerical and analytical modeling of groundwater and soluble transport; stochastic soil physics; theoretical averaging procedures. *Mailing Add:* Dept Agron Life Sci Bldg Purdue Univ West Lafayette IN 47907

CUSHMAN, MARK, b Fresno, Calif, Aug 20, 45; div; c 1. ORGANIC CHEMISTRY, MEDICINAL CHEMISTRY. *Educ:* Univ Calif, San Francisco, Pharm D, 69, PhD(med chem), 73. *Prof Exp:* NIH fel chem, Mass Inst Technol, 73-75; from asst prof to assoc prof, 75-85, PROF MED CHEM, PURDUE UNIV, 85- *Concurrent Pos:* Sr Fulbright Scholar, Munich Tech Univ, Fed Rep Germany, 83-84. *Mem:* Am Chem Soc; Am Soc Pharmacog; Int Soc Heterocyclic Chem; Int Union Pure & Appl Chem; Am Asn Pharmaceut Scientists; Int Soc Antiviral Res; NY Acad Sci. *Res:* New synthetic methods; organic reaction mechanisms; natural products synthesis; structure-activity relationships in pharmacology and the isolation and structure elucidation of natural products; design and synthesis of novel anticancer, antiviral and antibiotic agents. *Mailing Add:* Dept Med Chem & Pharmacog Purdue Univ West Lafayette IN 47907

CUSHMAN, PAUL, JR, b New York, NY, Feb 4, 30; m 59; c 2. ENDOCRINOLOGY, DRUG ABUSE. *Educ:* Yale Univ, BA, 51; Columbia Univ, MD, 55. *Prof Exp:* Instr med, Columbia Univ, 62-68, assoc, 68-69, asst clin prof, 69-71, NY Heart Asn sr investr, 71-72, asst prof med, 71-77; assoc prof med, psychiat & pharmacol, Med Col Wis, 77-86; assoc prof med, Med Col Va, 82-86, prof med, psychiat & pharmacol, 86-89; PROF PSYCHIAT, STATE UNIV NY, STONY BROOK, 89- *Concurrent Pos:* NIH res fel endocrinol, Univ Rochester, 59-61, St Luke's Hosp, 62-64 & 67-75; consult, Vet Admin Hosp, Batavia, NY, 60-62; from assoc dir to dir endocrinol, St Luke's Hosp, 62-73; assoc attend physician, 65-77, dir clin pharmacol, 73-77; consult, De Paul Rehab Hosp, 77-; lectr, alcohol & drug abuse, 79. *Mem:* AAAS; Endocrine Soc; Am Fedn Clin Res; Am Physiol Soc; fel Am Col Physicians. *Res:* Narcotics, alcohol and marijuana. *Mailing Add:* 1348 Ridge Rd Syosset NY 11791

CUSHMAN, PAUL, JR, b New York, NY, Feb 4, 30; m; c 2. EXPERIMENTAL BIOLOGY. *Educ:* Yale Univ, BA, 51; Columbia Univ, MD, 55; Am Bd Internal Med, cert, 63. *Prof Exp:* Med intern, Barnes Hosp, St Louis, Mo, 55-56; asst res med, Strong Mem Hosp, Rochester, NY, 56-57; USPHS res fel endocrinol, Univ Rochester, 59-61; second asst res, St Luke's Hosp, New York, 61-62; sr res fel, NY Heart Asn, 62-64; from instr to asst prof med, Columbia Univ, 62-71; assoc prof med, pharmacol & psychiat, Med Col Wis, 77-81; from assoc prof to prof, Med Col Va, 81-87; prof psychiat & clin prof med, NY Univ, 87-91; lectr, 89-91, PROF PSYCHIAT, STATE UNIV NY, STONY BROOK, 91- *Concurrent Pos:* Attend physician, St Luke's Hosp, Presby Hosp Vanderbilt Clin, 61-77; staff mem, Milwaukee Gen Hosp, DePaul Rehab Hosp, 77-81; mem sci prog, Am Soc Addiction Med, 80-; dir, Drug Dependency Treatment Prog, McGuire Vet Hosp, Richmond, Va, 81-87; reviewer, Am J Drug & Alcohol Abuse, Advan Substance Abuse, 85-88; internist, Div Alcoholism & Drug Abuse, Bellevue Hosp, New York, 87-; consult, Nat Inst Alcohol Abuse & Alcoholism, 90- *Honors & Awards:* Henry E Sigerist Prize, 67; Caleb Fiske Prize, 73. *Mem:* AAAS; fel Am Col Physicians; Am Fedn Clin Res; Am Physiol Soc; Am Soc Res & Educ Substance Abuse; Am Soc Addiction Med; Am Soc Clin Pharmacol; Endocrine Soc; Sigma Xi. *Res:* Alcoholism and drug abuse. *Mailing Add:* Dept Psychiat Stony Brook Health Serv Ctr Stony Brook NY 11658

CUSHMAN, ROBERT VITTUM, b Middlebury, Vt, Dec 14, 16; m 42; c 3. HYDROGEOLOGY. *Educ:* Middlebury Col, AB, 39; Northwestern Univ, MS, 41. *Prof Exp:* Recorder, US Geol Surv, Alaska, 41; mining geologist, Slide Mines, Inc, Colo, 41-42; mineral economist, US Bur Mines, 42-45; geologist, Water Resources Div, US Geol Surv, NY, 46-47, Conn, 47-61, Ky, 61-71, assoc dist chief, 67-71, dist chief, 71-75; consult ground water, geol & hydrol, 75-86; RETIRED. *Mem:* Geol Soc Am; Am Geophys Union. *Mailing Add:* 20 Court St Middlebury VT 05753

CUSHMAN, SAMUEL WRIGHT, b Bryn Mawr, Pa, Oct 2, 41; m 64; c 3. CELL PHYSIOLOGY, BIOCHEMISTRY. *Educ:* Bowdoin Col, AB, 63; Rockefeller Univ, PhD(physiol chem), 69. *Prof Exp:* Res asst, Inst Clin Biochem, Sch Med, Univ Geneva, 69-71; res asst prof med, Dartmouth Med Sch, 71-73, asst prof med & adj asst prof biochem, 73-80; ASSOC CHIEF, CELLULAR METAB & OBESITY SECT, NAT INST ALLERGY, DIGESTIVE DIS, KIDNEYS, NIH, 76-, CHIEF, EXP DIABETES, METAB & NUTRIT SECT. *Concurrent Pos:* Am Cancer Soc fel, Inst Clin Biochem, Geneva, Switz, 69-71; Am Diabetes Asn career develop award, Dartmouth Med Sch, 72-74. *Mem:* Am Diabetes Asn; Am Soc Cellular Biol; Am Fedn Clin Res; Europ Asn Study Diabetes; Endocrine Soc. *Res:* Adipose tissue and cell structure and function; obesity; mechanism of hormone action, especially insulin and epinephrine; diabetes mellitus. *Mailing Add:* MCNEB-Nat Inst Allergy Digestive Dis Kidney NIH Bldg 10 Rm 5N 102 Bethesda MD 20205

CUSICK, THOMAS WILLIAM, b Joliet, Ill, Sept 18, 43; m 65; c 2. CRYPTOGRAPHY. *Educ:* Univ Ill, Urbana, BS, 64; Cambridge Univ, PhD(math), 67. *Prof Exp:* From asst prof to assoc prof, 68-77, PROF MATH, STATE UNIV NY, BUFFALO, 77- *Mem:* Am Math Soc. *Res:* Diophantine approximation; Cryptography. *Mailing Add:* Dept Math Rm 136 State Univ NY Diefendorf Hall 3435 Buffalo NY 14214

CUSSON, RONALD YVON, theoretical nuclear physics, for more information see previous edition

CUSTARD, HERMAN CECIL, b Cleburne, Tex, Aug 19, 29; m 55; c 3. PHYSICAL CHEMISTRY. *Educ:* Baylor Univ, BS, 57, PhD(chem), 62. *Prof Exp:* Res assoc & activ leader, Field Res Lab, Mobil Res & Develop Corp, 62-83; CHMN, RUBICON EXPLOR CO, 83- *Mem:* Am Chem Soc; Electrochem Soc. *Res:* Dipole moments; electrochemistry of membrane systems; physical chemistry of electrolytic solutions. *Mailing Add:* 4016 S Better Dr Dallas TX 75229

CUSTER, HUBERT MINTER, physics, for more information see previous edition

CUSTER, MICHAEL, b Lowell, Mass, Oct 29, 10. CHEMISTRY. *Educ:* St Anselm's Col, AB, 35; Cath Univ Am, MS, 44. *Prof Exp:* DEPT CHEM, ST ANSELM'S COL, 48- *Mem:* Am Chem Soc. *Mailing Add:* 87 St Anselm's Dr Manchester NH 03102

CUSTER, RICHARD PHILIP, oncology; deceased, see previous edition for last biography

CUSUMANO, JAMES A, b Elizabeth, NJ, Apr 14, 42; m 64, 85; c 2. CATALYSIS. *Educ:* Rutgers Univ, BA, 64, PhD(chem physics), 68. *Prof Exp:* Res chemist heterogeneous catalysis, Exxon Res & Eng Co, 67-74; pres, 74-85, CHMN & DIR, CATALYTICA INC, 85- *Mem:* Am Chem Soc; Catalysis Soc; Am Phys Soc; Sigma Xi; Am Inst Chem Eng. *Res:* Catalysis, chemistry and physics of small metal particle systems; infrared spectroscopic studies of adsorbed species; chemical kinetics of catalytic reactions; theory and measurement of surface thermal transients. *Mailing Add:* 480 Ferguson Dr Bldg 3 Mountain View CA 94043

CUTCHINS, ERNEST CHARLES, b Newsoms, Va, Aug 19, 22. BACTERIOLOGY, VIROLOGY. *Educ:* Va Polytech Inst, BS, 43; Univ Md, MS, 51, PhD(bact), 55; Am Bd Med Microbiol, dipl. *Prof Exp:* Bacteriologist, Walter Reed Army Inst Res, 55-56 & NIH, 56-59; res assoc, Chas Pfizer & Co, Inc, 60-61; ASSOC PROF VIROL, DEPT BIOL, CATH UNIV AM, 61-, CHMN DEPT, 78- *Mem:* Am Asn Immunologists; Am Soc Microbiol. *Res:* Viral inactivation; virus-host interaction; immunology and serology of viruses. *Mailing Add:* Dept Biol Cath Univ Am Washington DC 20017

CUTCHINS, MALCOLM ARMSTRONG, b Franklin, Va, Mar 27, 35; m 54; c 3. AEROSPACE ENGINEERING, ENGINEERING MECHANICS. *Educ:* Va Polytech Inst, BS, 56, MS, 64, PhD(eng mech), 67. *Prof Exp:* Assoc aircraft engr, Lockheed-Ga Co, Ga, 56-61, jr mech engr, 61-62, sr mech engr, 62; instr, Va Polytech Inst, 62-66; assoc prof, 66-79, PROF AEROSPACE ENG, AUBURN UNIV, 79- *Concurrent Pos:* Grants & contracts, NSF, US Forest Serv, USAF, NASA; consult, various companies; chmn, Struct Dynamics Tech Comn, 85-87; gen chmn, 30th Struct, Struct Dynamics & Mat Conf, AIAA, 89. *Honors & Awards:* IR-100 Award, Indust Res, 73. *Mem:* Assoc fel Am Inst Aeronaut & Astronaut; Am Soc Eng Educ; Nat Soc Prof Engrs; Am Acad Mech; Sigma Xi. *Res:* Structural and structural dynamics analysis, damping studies, simulation and aeroservoelasticity; failure analysis, expert witness and simulation; vibration-related studies; one patent. *Mailing Add:* 701 Sanders St Auburn AL 36830

CUTCLIFFE, JACK ALEXANDER, b Quincy, Mass, Aug 10, 29; Can citizen; m 54; c 6. OLERICULTURE. *Educ:* McGill Univ, BSc, 52, MSc, 54. *Prof Exp:* Exten hort crops, Ont Dept Agr, 54-62; res olericult, Agr Can Res Br, Agr Can Res Sta, 62-90, sect head hort, 70-90; RETIRED. *Mem:* Agr Inst Can; Can Soc Hort Sci; Am Soc Hort Sci; Int Soc Hort Sci. *Res:* Conduct research on macro- and micro-nutrient requirements and on management practices to improve efficiency of production of broccoli, cauliflower, brussel sprouts, rutabagas and peas. *Mailing Add:* Three Birchwood St PO Box 1210 Charlottetown PE C1A 5B4 Can

CUTFORTH, HOWARD GLEN, b Topeka, Kans, May 6, 20; m 44; c 3. CHEMISTRY, AEROSPACE SCIENCES. *Educ:* Univ Wichita, BA, 42; Northwestern Univ, PhD(phys chem), 48. *Prof Exp:* Res chemist, Rohm & Haas Co, 47-48 & Phillips Petrol Co, 48-59; sect chief, 59-71, BR MGR, CHEM SYSTS DIV, UNITED TECHNOL CORP, 71- *Mem:* Am Chem Soc. *Res:* Magnetic study of free radicals; chemistry of rare earths; colloid problems in crude oil production; magnetic susceptibility of substituted hexaryl ethanes and related compounds; direct tailoring of standard and development of new solid, composite, rocket fuels; production processing; solid fueled rocket motors. *Mailing Add:* 7011 Via Valverde San Jose CA 95135-1339

CUTHILL, ELIZABETH, b Southington, Conn, Oct 16, 23; m 50; c 1. MATHEMATICS. *Educ:* Univ Buffalo, BA, 44; Brown Univ, MA, 46; Univ Minn, PhD, 51. *Prof Exp:* Chemist, Res & Develop Labs, Socony Vacuum Oil Co, 44-45; asst math, Univ Minn, 46-48 & Conn Col, 48-49; instr, Purdue Univ, 50-52 & Univ Md, 52-53; MATHEMATICIAN, COMPUT, LOGISTICS & MATH DEPT, DAVID W TAYLOR NAVAL SHIP RES & DEVELOP CTR, 53- *Concurrent Pos:* Prof lectr, Am Univ, 67-70. *Mem:* Fel AAAS; Am Math Soc; Soc Indust & Appl Math; Math Asn Am; Asn Comput Mach; Sigma Xi. *Res:* Development of mathematical methods and digital computer programs for solution of problems arising in areas relating to ship research including nuclear reactor design; structural mechanics; hydrodynamics. *Mailing Add:* 12700 River Rd Potomac MD 20854

CUTHILL, JOHN R(OBERT), b Buffalo, NY, Dec 19, 18; m 50; c 1. PHYSICAL METALLURGY. *Educ:* Purdue Univ, BS, 48, PhD(metall), 52. *Prof Exp:* Phys metallurgist, 52-57, RES METALLURGIST, NAT BUR STANDARDS, 57- *Mem:* Am Phys Soc; Am Soc Metals; Am Inst Mining, Metall & Petrol Engrs; Sigma Xi. *Res:* Soft x-ray spectroscopy; electronic structure of metals as related to phase transformations and alloying; electronprobe analysis to determine composition and structure of microconstituents; kinetic data. *Mailing Add:* 12700 River Rd Potomac MD 20854

CUTHRELL, ROBERT EUGENE, b Houston, Tex, Nov 6, 33; m 54; c 4. SURFACE CHEMISTRY, SURFACE PHYSICS. *Educ:* Univ Tex, Austin, BA, 54, BS, 61, PhD(inorg chem), 64. *Prof Exp:* SR STAFF MEM CHEM, SANDIA LABS, 63-68 & 69- *Concurrent Pos:* Res scientist, Air Force Weapons Lab, Kirtland AFB, 70-76, staff develop eng, 76-82; counr & treas, NMex Inst Chemists, 74-78. *Mem:* Fel Chem Soc London; fel Am Inst Chemists; Sigma Xi; Am Chem Soc. *Res:* Ultrahigh vacuum study of the chemistry, physics and mechanics of the surface and the near-surface region of solids; applied research in metal-metal bonding, design of electrical contacts, arc theory, hydrogen embrittlement, friction and wear, atmospheric pollution; coatings for fusion reactors; post magnetron sputter coater design; interferometer construction and film stress measurement. *Mailing Add:* Sandia Labs Orgn 9118- Bldg 880- Rm D43 Albuquerque NM 87115

CUTKOMP, LAURENCE KREMER, b Wapello, Iowa, Jan 24, 16; m 39; c 4. INSECT TOXICOLOGY, ECONOMIC ENTOMOLOGY. *Educ:* Iowa Wesleyan Col, BA, 36; Cornell Univ, PhD(econ entom), 42. *Prof Exp:* Asst entom, Cornell Univ, 42-43; fel zool, Univ Pa, 43-45; res assoc entom, Univ Minn, 45-46; assoc entomologist, TVA, 46-47; from asst prof to prof entom, 47-86, EMER PROF ENTOM, UNIV MINN, ST PAUL, 86- *Concurrent Pos:* Consult, Eli Lilly & Co, 59-64, Onamia Corp, 63-64; entomologist, Int Atomic Energy Agency, Vienna, 65-67 & consult entom & insect biochem, India, 74; consult, Food & Agr Orgn, Tanzania, 82 & Teltech, Inc, Minneapolis, Minn, 86- *Mem:* AAAS; Entom Soc Am; Sigma Xi. *Res:* Insecticidal action; toxicity of insecticides; insect rhythms as related to insecticide effectiveness. *Mailing Add:* Dept Entom Univ Minn 1980 Folwell Ave St Paul MN 55108

CUTKOSKY, RICHARD EDWIN, b Minneapolis, Minn, July 29, 28; m 52; c 3. ELEMENTARY PARTICLE PHYSICS. *Educ:* Carnegie Inst Technol, BS & MS, 50, PhD(physics), 53. *Prof Exp:* Res physicist, Carnegie Inst Technol, 53-54; NSF fel, Inst Theoret Physics, Denmark, 54-55; FROM ASST PROF TO PROF PHYSICS, CARNEGIE MELLON UNIV, 55-, BUHL PROF THEORET PHYSICS, 63- *Concurrent Pos:* Sloan Found res fel, 56-61; NSF res fel, Nordic Inst Theoret Atomic Physics, Denmark, 61-62; overseas fel, Churchill Col, Cambridge, England, 68-69. *Mem:* Fel AAAS; fel Am Phys Soc. *Res:* Quantum field theory; high energy physics. *Mailing Add:* Dept Physics Carnegie-Mellon Univ Pittsburgh PA 15213

CUTLER, CASSIUS CHAPIN, b Springfield, Mass, Dec 16, 14; m 41; c 3. ELECTRICAL ENGINEERING. *Educ:* Worcester Polytech Inst, BA, 37. *Hon Degrees:* Deng, Eorcester Polytech Inst. *Prof Exp:* Dir electronic & comput systs res, Bell Tel Labs, 37-78; EMER PROF APPL PHYSICS, STANFORD UNIV, 79- *Concurrent Pos:* Ed, Spectrum, 66-68. *Honors & Awards:* Edison Medal, Inst Elec & Electronic Engrs, 81; Robert H Goddard Award, Worcester Polytech Inst, 82; Centennial Medal, Inst Elec & Electronic Engrs, 84, Alexander Graham Bell Medal, 91. *Mem:* Nat Acad Sci; Nat Acad Eng; fel Inst Elec & Electronics Engrs; fel AAAS. *Res:* Radio transmitter design; design of radar antennas; microwave electron tube research; satellite communication; microwave acoustics. *Mailing Add:* Hansen Lab Stanford Univ Stanford CA 94305

CUTLER, DOYLE O, b Corpus Christi, Tex, Jan 20, 36. MATHEMATICS. *Educ:* Tex Christian Univ, BA, 59, MA, 61; NMex State Univ, PhD(math), 64. *Prof Exp:* Asst prof math, Tex Christian Univ, 64-65; asst prof, 65-71, assoc prof math, 71-84, PROF MATH, UNIV CALIF, DAVIS, 84- *Mem:* Am Math Soc; Math Asn Am. *Res:* Infinite Abelian groups; algebra. *Mailing Add:* Dept Math 551 Kerr Hall Univ Calif Davis CA 95616

CUTLER, EDWARD BAYLER, b Plymouth, Mich, May 28, 35; div; c 2. MARINE BIOLOGY. *Educ:* Wayne State Univ, BS, 61; Univ Mich, MS, 62; Univ RI, PhD(zool), 67. *Prof Exp:* Asst prof biol, Lynchburg Col, 62-64; oceanog trainee, Coop Oceanog Prog, Marine Lab, Duke Univ, 66-67; from asst prof to assoc prof, 67-77, assoc dean sci & math, 85-88, PROF BIOL, UTICA COL, 77- *Concurrent Pos:* Vis scientist, Univ Tokyo, 79 & Mus Comp Zool, Harvard Univ, 84 & 89-91. *Mem:* AAAS; Am Soc Zoologists; Soc Syst Zool. *Res:* Systematics; zoogeography; ecology; evolution of benthic marine invertebrates, particularly the Sipuncula and Pogonophora. *Mailing Add:* Dept Biol Utica Col Utica NY 13502

CUTLER, FRANK ALLEN, JR, b Pana, Ill, Nov 14, 20; m 46; c 3. CHEMISTRY, RESEARCH ADMINISTRATION. *Educ:* Univ Ill, BS, 42; Univ Minn, PhD(org chem), 48. *Prof Exp:* Jr chemist synthetic rubber res, B F Goodrich Co, 44-46; sr chemist develop res, Merck & Co, Inc, 48-59, tech adv to patent dept, 59-61, coordr compound eval, 61-65, mgr res compounds, 65-77, mgr systs & data handling, 77-87; CONSULT, 87- *Mem:* Am Chem Soc. *Res:* Quinones; cortical steroids; pharmaceuticals; information retrieval; substructure and reaction searching by computer. *Mailing Add:* 41 Faulkner Dr Westfield NJ 07090

CUTLER, GORDON BUTLER, JR, b Cincinnati, Ohio, Oct 16, 47; m 77; c 2. ENDOCRINOLOGY & METABOLISM. *Educ:* Harvard Col, BA, 69; Harvard Med Sch, MD, 73. *Prof Exp:* Sr invest, Develop Endocrinol Br, 78-83, dir pediat endocrinol prog, NIH Clin Ctr Training, 83-89, CHIEF, SECT DEVELOP ENDOCRINOL, NIH, 83-, DIR NAT INST CHILD HEALTH & HUMAN DEVELOP, ADULT ENDOCRINOL PROG, 89- *Honors & Awards:* Commendation Medal, USPHS, 85. *Mem:* Endocrine Soc; Lawson Wilkins Pediat Endocrine Soc; Am Fedn Clin Res; Am Assoc Advan Sci. *Res:* Mechanism and treatment of growth failure and disorders of puberty. *Mailing Add:* Bldg 10 Rm 10-N-262 NIH Bethesda MD 20892

CUTLER, HORACE GARNETT, b London, Eng, Nov 21, 32; US citizen; m 55; c 7. PLANT PHYSIOLOGY, PLANT NEMATOLOGY. *Educ:* Univ Md, BS, 64, MS, 66, PhD(plant path & nematol), 67. *Prof Exp:* Res asst plant physiol, Boyce Thompson Inst Plant Res, 54-59; plant physiologist, Cent Agr Res Sta, WI, 59-62; asst bot, Univ Md, 63-67; plant physiologist, Ga Costal Plain Exp Sta, 67-80, PLANT PHYSIOLOGIST & RES LEADER, MICROBIAL PROD UNIT, RICHARD B RUSSELL RES CTR, AGR RES SERV, USDA, 80-, COORDR NATURAL PROD RES MATRIX. *Honors & Awards:* Carroll E Cox Scholarship Award, 66; Silver Medal, Japanese Soc Chem Regulation of Plants, 87. *Mem:* NY Acad Sci; Japanese Soc Plant Physiol; Sigma Xi; Am Chem Soc; Am Soc Plant Growth Regulators. *Res:* Herbicides; defoliants; growth substances from sugarcane; nematodes and micro-organisms; growth inhibitors from tobacco; mode of action and biochemistry of auxins and other regulators; biologically active natural products from microorganisms, identification and isolation; biologically active natural products; natural products. *Mailing Add:* Richard B Russell Res Ctr USDA PO Box 5677 Athens GA 30613

CUTLER, HUGH CARSON, b Milwaukee, Wis, Sept 8, 12; m 40; c 1. BOTANY. *Educ:* Univ Wis, BA, 35, MA, 36; Univ Wash, PhD(bot), 39. *Prof Exp:* Fel, Wash Univ, 29-40; res assoc, Harvard Univ, 40-43; field technician, Rubber Develop Corp, Brazil, 43-45; res assoc, Harvard Univ, 46-47; cur econ bot, Chicago Natural Hist Mus, 47-52; cur econ bot, Mus Useful Plants, 53-54, from asst dir to exec dir, 54-64, cur useful plants, Mo Bot Garden, 64-77; CONSULT ETHNOBOT & ECON BOT, 77- *Concurrent Pos:* Guggenheim fel, 42-43 & 46-47; assoc prof bot, Wash Univ, 53-72, adj prof anthrop & biol, 72-77. *Res:* Useful cultivated plants of the New World and their wild relatives; evolution of maize and cucurbits; ethnobotany; economic botany. *Mailing Add:* 2300 Tice Creek Dr #2 Walnut Creek CA 94595

CUTLER, JANICE ZEMANEK, b Chicago, Ill, May 3, 42; m 73; c 2. ALGEBRA. *Educ:* Univ Wis-Madison, BS, 64; Univ Ill, Urbana-Champaign, MA, 65, PhD(math), 70. *Prof Exp:* Instr math, Univ Ill, Urbana-Champaign, 70; asst prof math, La State Univ, Baton Rouge, 70-76; CONSULT, 76- *Mem:* Am Math Soc; Math Asn Am. *Res:* Integral representation theory. *Mailing Add:* 405 Balsam Ln Palatine IL 60067

CUTLER, JEFFREY ALAN, b Cambridge, Mass, June 30, 42; m 66; c 2. CLINICAL TRIALS, CARDIOVASCULAR DISEASE. *Educ:* Wesleyan Univ, Conn, BA, 64; Univ Ky, MD, 68; Harvard Univ, MPH, 71. *Prof Exp:* Intern and resident internal med, Med Ctr, Univ Vt, 68-70; resident health serv admin prev med, Harvard Sch Pub Health, 70-71, resident & fel health serv res, Ctr Community Health & Med Care, 71-72; med officer prev med, Sch Health Care Sci, US Air Force, 72-74; asst prof, Med Col Wis, 74-79; med officer epidemiol, Nat Inst Arthritis, Metab & Digestive Dis, 77-80, med officer, Nat Ctr Health Care Tech, 80-81; MED OFFICER, NAT HEART, LUNG & BLOOD INST, 81- *Concurrent Pos:* Fel, Ctr Community Health & Med Care, 71-72; med adv, Health Dept, West Allis, 76-77; clin asst prof, Dept Prev Med, Uniformed Serv Sch Med, 78-; consult subcomt health maintenance systs, Armed Forces Epidemiol Bd, 77-79; mem, coun Epidemiol, Am Heart Asn. *Mem:* Am Col Prev Med; Am Pub Health Asn; Asn Teachers of Prev Med; Soc Epidemiol Res. *Res:* Epidemiology and prevention of cardiovascular diseases; methodology of clinical traits; hypertension. *Mailing Add:* 13520 Collingwood Terr Silver Spring MD 20904

CUTLER, JIMMY EDWARD, b Los Angeles, Calif, Nov 28, 44; m 65; c 2. MICROBIOLOGY. *Educ:* Calif State Univ, Long Beach, BS, 67, MS, 69; Tulane Univ, PhD(microbiol), 72. *Prof Exp:* NIH fel microbiol, Rocky Mountain Lab, Nat Inst Allergy & Infectious Dis, 73-74; asst prof, 74-80, actg head, dept microbiol, 85-86, ASSOC PROF MICROBIOL, MONT STATE UNIV, 80- *Concurrent Pos:* Wellcome vis prof, 81-82. *Mem:* Am Soc Microbiol; Am Asn Immunologists; Med Mycol Soc Am. *Res:* Host-fungal relationships; control and regulation of fungal morphogenesis; fungal virulence factors. *Mailing Add:* Dept Microbiol Mont State Univ Bozeman MT 59715

CUTLER, JOHN CHARLES, b Cleveland, Ohio, June 29, 15; m 42. PUBLIC HEALTH. *Educ:* Western Reserve Univ, BA, 37, MD, 41; Johns Hopkins Univ, MPH, 51; Am Bd Prev Med, dipl, 51. *Prof Exp:* Mem staff, Venereal Dis Res Lab, USPHS, 43-46 & Pan-Am Sanit Bur, 46-48; head, Venereal Dis Demonstration Team, WHO, Southeast Asia, 48-50; Bur State Serv, USPHS, 50-58; asst dir, Nat Inst Allergy & Infectious Dis, 58-59; dist health officer, Allegheny County Health Dept, Pa, 59-60; dep dir, Pan-Am Sanit Bur, 60-67; prof, 67-85, EMER PROF INT HEALTH, GRAD SCH PUB HEALTH, UNIV PITTSBURGH, 85- *Concurrent Pos:* Mem, Int Comt & mem, Bd Dirs, Asn Voluntary Surg Contraception, 71-, pres, 77-78; mem, bd dirs, Family Planning Coun, Western Pa & Med Adv Comt, Planned Parenthood Ctr, Pittsburgh; dir, WHO Collab Ctr for Ref & Res in Prophylactic Methods for Control of Sexually Transmitted Dis, 77-85. *Mem:* Fel Am Pub Health Asn; Am Venereal Dis Asn; Asn Vol Surg Contraception; Asn Planned Parenthood Prof; World Fedn Vol Surg Contraception (pres, 84-85). *Res:* Combined contraceptive venereal disease prophylactic preparation. *Mailing Add:* Grad Sch Pub Health Univ Pittsburgh Pittsburgh PA 15261

CUTLER, LEONARD SAMUEL, b Los Angeles, Calif, Jan 10, 28; m 54; c 4. PHYSICS, MATHEMATICS. *Educ:* Stanford Univ, BS, 58, MS, 60, PhD(physics), 66. *Prof Exp:* Chief engr, Gertsch Prod Inc, 49-55, vpres, 55-57; res engr, Hewlett-Packard Co, 57-62, engr sect leader, 62-64, dir res, 64-67, mgr, Frequency & Time Div, 67-69, dir, Phys Res Lab, 69-81, dir, Phys Sci Lab, 81-84, dir, Instruments & Photo-Metric Lab, 84-87, dir, Superconductivity Lab, 87-90, DISTINGUISHED CONTRIB TECH STAFF, HEWLETT-PACKARD CO, 90- *Concurrent Pos:* Mem adv panel, Nat Bur Standards under Nat Acad Sci contract, 68- *Honors & Awards:* Morris Leed Award, Inst Elec & Electronics Engrs, 84; Achievement Award, Indust Res Inst, 90. *Mem:* Nat Acad Eng; AAAS; Sigma Xi; Inst Elec & Electronics Engrs. *Res:* Frequency measuring devices; quartz and atomic frequency standards; noise theory; frequency stability theory; quantum electronics; atomic beam work; lasers; precision alternating current transformers; magnetic bubble memories; disc memories; superconductivity. *Mailing Add:* Hewlett Packard Co Superconductivity Lab PO Box 10350 Palo Alto CA 94303-0867

CUTLER, LESLIE STUART, b New Brunswick, NJ, Jan 20, 43; m 66; c 2. DEVELOPMENTAL BIOLOGY, PATHOBIOLOGY. *Educ:* Washington Univ, DDS, 68; State Univ NY Buffalo, PhD(path), 73; Hartford Grad Ctr, MS, 84. *Prof Exp:* Fel path, Sch Med, State Univ NY Buffalo, 68-73; from asst prof to assoc prof, 73-81, prof & chmn oral diag, Sch Dent Med, 81-88, ASSOC VPRES ADMIN & RES, HEALTH CTR, UNIV CONN, 88- *Mem:* AAAS; Am Soc Cell Biol; Soc Develop Biol; Histochem Soc; Int Asn Dent Res; Am Asn Med Cols; Asn Acad Health Centers. *Res:* Developmental biology-cytodifferentiation, controls, cell surface-extracellular matrix interations; electron microscopic cytochemistry; animal models of human diseases, sjögren's Syndrome. *Mailing Add:* Assoc Vpres Admin & Res Univ Conn Health Ctr Farmington CT 06030

CUTLER, LOUISE MARIE, b Troy, NY, Oct 1, 21. CHEMISTRY, EDUCATIONAL ADMINISTRATION. *Educ:* Good Counsel Col, BS, 43; Fordham Univ, MS, 52; Columbia Univ, PhD(chem), 62. *Prof Exp:* Instr chem, Good Counsel Acad, 43-44; from instr to assoc prof, 45-62, PROF CHEM, COL WHITE PLAINS, PACE UNIV, 62-, DEAN OF STUDIES, 84- *Concurrent Pos:* Trustee, Good Counsel Col, 63-75; NSF grant, NY Univ, 65. *Mem:* AAAS; Am Chem Soc; Sigma Xi. *Res:* Chelation equilibria and kinetics; solvent effects on RXNS of quaternary ammonium (CPDS); history and philosophy of science. *Mailing Add:* Dept Chem Pace Univ White Plains NY 10603

CUTLER, MELVIN, b Beaumont, Tex, Dec 18, 23; m 50; c 2. SOLID STATE PHYSICS. *Educ:* City Col New York, BS, 43; Columbia Univ, AM, 47, PhD(chem), 51. *Prof Exp:* Mem tech staff, Hughes Aircraft Co, 51-58; mem staff, John Jay Hopkins Lab Pure & Appl Sci, Gen Atomic Div, Gen Dynamics Corp, 58-63; assoc prof, 63-69, PROF PHYSICS, ORE STATE UNIV, 69- *Mem:* Am Phys Soc. *Res:* Electronic properties of disordered systems. *Mailing Add:* Ore State Univ Corvallis OR 97331

CUTLER, PAUL, b Philadelphia, Pa, Mar 27, 20; m 47; c 2. INTERNAL MEDICINE. *Educ:* Univ Pa, BA, 40; Jefferson Med Col, MD, 44. *Prof Exp:* Pvt pract, 50-65; team capt, Care-Medico, Afghanistan, 65-67; prof med & physiol, 68-74, ASSOC DEAN CLIN AFFAIRS, UNIV TEX HEALTH SCI CTR SAN ANTONIO, 71- *Mem:* Fel Am Col Physicians. *Mailing Add:* 21203 Logo Circle Bldg 15A Boca Raton FL 33433

CUTLER, PAUL H, b New York, NY, Nov 17, 26; m 56; c 2. SURFACE PHYSICS, MICROELECTRONICS. *Educ:* Pa State Univ, BS, 48, MS, 55, PhD(physics), 58. *Prof Exp:* Res physicist, Res Lab Gen Tel & Tel, 50-53; from asst prof to assoc prof, 60-66, PROF PHYSICS, PA STATE UNIV, 66- *Concurrent Pos:* Postdoctoral fel, Cavendish Lab, Cambridge Univ, UK, 59-60; Fulbright res fel, Tech Univ, Copenhagen, Denmark, 67-68; consult, Surface Physics Div, Europ Space Agency, 71-, Am Optical Corp, 72-74; vis prof, Inst de Physique, Univ de Liege, 77; vis fel, Wolfson Col, Oxford, 84-85; assoc prof, Univ Paris, Orsay, 86. *Mem:* Fel Am Phys Soc; Am Vacuum Soc. *Res:* Metal-vacuum junction tunneling; application to STM; theory of emission from atomically sharp emitters; theory of laser interaction in STM; electrohydrodynamic theory of stability and cluster formation in EHD sources; theory and measurement of tunneling time. *Mailing Add:* 104 Davey Lab Pa State Univ University Park PA 16802

CUTLER, RICHARD GAIL, b Lovell, Wyo, Aug 6, 35; m 62; c 3. CELL BIOLOGY. *Educ:* Long Beach State Col, BS, 61; Univ Houston, MS & PhD(biophys), 66. *Prof Exp:* Fel geront, Brookhaven Nat Lab, 66-69; asst prof biol, Univ Tex, Dallas, 69-76; RES CHEMIST, GERONTOL RES CTR, NAT INST AGING, BALTIMORE, 76- *Honors & Awards:* Karl August Forster Award, 76. *Mem:* AAAS; Geront Soc; Biophys Soc; Exp Mutagen Soc; Am Aging Asn; Soc Free Radical Res. *Res:* Comparative biochemistry of mammalian aging and life-maintenance processes; physical-chemical accumulation of damage in the genetic apparatus with age; age-dependent alteration of gene expression; level of repair and protective processes as a function of age and innate aging rate in different mammalian species. *Mailing Add:* Gerontol Res Ctr Nat Inst on Aging 4940 Eastern Ave Baltimore MD 21224

CUTLER, ROBERT W P, b New York, NY, Aug 3, 33; m 61; c 3. NEUROLOGY, NEUROCHEMISTRY. *Educ:* Tufts Univ, MD, 57. *Prof Exp:* Assoc neurol, Med Sch, Harvard Univ, 64-68; assoc prof neurol, Univ Chicago, 68-76; PROF NEUROL, DEPT NEUROL, STANFORD UNIV HOSP, CALIF, 76-, ACAD CONSULT, 80- *Concurrent Pos:* Asst, Children's Hosp Med Ctr, Boston, Mass, 64-68; prin investr, USPHS grant, 69. *Mem:* AAAS; Am Soc Neurochem; Am Acad Neurol; Asn Res Nerv & Ment Dis. *Res:* Physiology of cerebrospinal fluid and bloodbrain barrier mechanisms. *Mailing Add:* Dept Neurol Stanford Univ Hosp Stanford CA 94305

CUTLER, ROGER T, b Washington, DC, June 28, 46. GEOPHYSICS. *Educ:* Univ Chicago, BA, 69; Univ Ill, MS, 73, PhD(physics), 75. *Prof Exp:* Res fel high energy physics, Univ Ill, 73-75; fel high energy physics, Argonne Nat Lab, 75-77; res geophysicist, 77-79, SECT DIR GEOPHYSICS, GULF RES & DEVELOP, 79-, SR CHIEF ASST. *Mem:* Soc Explor Geophysicists; Am Phys Soc. *Res:* Applications of inversion theory to geophysics and geology; rock properties research. *Mailing Add:* Chevron Geosci PO Box 42832 Houston TX 77242

CUTLER, ROYAL ANZLY, JR, b Spokane, Wash, March 10, 18; m 43; c 3. PHARMACEUTICAL CHEMISTRY. *Educ:* Whitman Col, AB, 41; Rensselaer Polytech Inst, MS, 42, PhD(org chem), 47. *Prof Exp:* Lab asst chem, Whitman Col, 40-41; asst, Rensselaer Polytech Inst, 41-42, instr chem, 43-44; jr res chemist, Winthrop Chem Co, 44-47, sr res chemist, 47-60, sr res chemist & tech liaison, 60-66, asst dir prod develop, 66-77, qual assurance adminr good mfg pract, 77-81, sr qual assurance adminr good mfg pract, Sterling Winthrop Res Inst, 81-83; RETIRED. *Mem:* Am Chem Soc; Sigma Xi. *Res:* Product development; pharmaceutical chemicals; pharmaceutical dosage forms; antimicrobials and chemical specialties. *Mailing Add:* 36 Euclid Ave Delmar NY 12054

CUTLER, VERNE CLIFTON, b Brookings, SDak, Jan 2, 26; m 48; c 5. ENGINEERING MECHANICS, CIVIL ENGINEERING. *Educ:* Kans State Univ, BS, 50, MS, 51; Univ Wis, PhD(eng mech), 60. *Prof Exp:* Instr civil eng, Kans State Univ, 50-51; from instr to assoc prof, 51-66, chmn dept, 63-74, PROF ENG MECH, UNIV WIS-MILWAUKEE, 66- *Concurrent Pos:* Res consult, A O Smith Corp, 58-60; chmn, mech div, 82-83, N Midwest-Sect, Am Soc Eng Ed, 88-89. *Mem:* Am Soc Eng Educ; Sigma Xi; Am Soc Testing Mat. *Res:* Reinforced plastics; structural analysis properties of structural concrete; experimental stress analysis; fatigue. *Mailing Add:* Civil Eng & Mech Dept Univ Wis Milwaukee WI 53201

CUTLER, WARREN GALE, b Avon, Ill, Jan 31, 22; m 49; c 4. FLUID PHYSICS, TECHNOLOGY TRANSFER. *Educ:* Monmouth Col, BA, 47; Pa State Univ, MS, 53, PhD(physics), 55. *Prof Exp:* Instr physics & math, Monmouth Col, 47-51; asst physics, Pa State Univ, 51-53, res fel, 53-55; assoc prof, Mankato State Col, 55-57, head dept, 55-57; assoc res physicist, Whirlpool Corp, 57-59, mgr mech eng res, 59-60, res dir, 60-68, dir corp res, Res Labs, 68-84, staff vpres, Univ Rels, 84-88; CHMN BD, MICH INFO TECHNOL NETWORK, 88- *Concurrent Pos:* Sr exec course, Sloan Sch Mgt, Mass Inst Technol, 66, coordr IRI vis scientist, Eng Prog, 88- *Mem:* Am Phys Soc; Sigma Xi. *Res:* Physics of high pressure; physical properties of liquid; surface properties of solutions containing surface-active agents; technical management. *Mailing Add:* 218 Crofton Circle St Joseph MI 49085-1839

CUTLER, WINNIFRED BERG, b Philadelphia, Pa, Oct 13, 44; div; c 2. BEHAVIORAL ENDOCRINOLOGY. *Educ:* Ursinus Col, BS, 73; Univ Pa, PhD(biol), 79. *Prof Exp:* Fel, Dept Physiol, Sch Med, Stanford Univ, 79-80; affiliated scientist, Monell Chem Senses Ctr, 81-87; asst prof biol, Beaver Col, 81-82; SCI DIR & CO-FOUNDER, WOMEN'S WELLNESS PROG, HOSP UNIV PA, 86- *Concurrent Pos:* Adj researcher, Dept Obstet & Gynec, Sch Med, Univ Pa, 80- *Mem:* Int Soc Psychoneuroendocrinology; Am Fertil Soc; Sigma Xi; Human Biol Coun; Int Acad Sex Res; Int Menopause Soc. *Res:* Menopause; human pheromones, sexual behavioral and menstrual cycle pattern; reproductive physiology and behavior. *Mailing Add:* 30 Coopertown Rd Haverford PA 19041

CUTLIP, MICHAEL B, b Portsmouth, Ohio, Sept 21, 41; m 66; c 2. CHEMICAL ENGINEERING. *Educ:* Ohio State Univ, BChE & MS, 64; Univ Colo, Boulder, PhD(chem eng), 68. *Prof Exp:* Res assoc chem eng, Univ Colo, 67 68; from asst prof to assoc prof, 68-81, PROF CHEM ENG, UNIV CONN, 81-, HEAD DEPT, 80- *Concurrent Pos:* sr vis fel, Cambridge Univ, Eng, 74, 75 & 83; Extensive consult. *Mem:* Am Inst Chem Engrs; Am Chem Soc; Catalysis Soc; Am Soc Eng Educ. *Res:* Chemical reaction engineering; catalysis; air pollution control; computer-based instruction. *Mailing Add:* Dept Chem Eng U-139 Univ Conn Storrs CT 06268

CUTLIP, RANDALL CURRY, b Hillsboro, WVa, Sept 30, 34; m 57; c 10. VETERINARY PATHOLOGY. *Educ:* Ohio State Univ, DVM, 61; Iowa State Univ, MS, 65, PhD(vet path), 71. *Prof Exp:* RES VET PATH, NAT ANIMAL DIS CTR, USDA, 61- *Mem:* Am Col Vet Pathologists; Am Vet Med Asn; Conf Res Workers in Animal Dis; Am Asn Sheep & Goat Practr. *Res:* Respiratory diseases of cattle, sheep, and goats; identify individual disease entities, establish their etiologies and evaluate their pathogenesis with emphasis on chronic diseases of the sheep lung. *Mailing Add:* 1910 Belair Circle Ames IA 50010

CUTLIP, WILLIAM FREDERICK, b Lincoln, Ill, Oct 20, 36; m 57; c 3. MATHEMATICS, COMPUTER SCIENCE. *Educ:* Eastern Ill Univ, BSEd, 58; Univ Ill, MA, 61; Mich State Univ, PhD(math), 68. *Prof Exp:* Pub sch prin & teacher, Ill, 55-56, teacher, 58-60; from instr to asst prof math, Northern Mich Univ, 61-64; asst prof, 68-70, assoc prof math, 70-79, PROF & DEPT CHMN, CENT WASH UNIV, 79- *Mem:* Asn Comput Mach; Math Asn Am; Nat Coun Teachers of Math. *Res:* Decomposition and synthesis of finite automata. *Mailing Add:* Dept of Math Cent Wash Univ Ellensburg WA 98926

CUTNELL, JOHN DANIEL, b Pittsburgh, Pa, Aug 30, 40; m 68. NUCLEAR MAGNETIC RESONANCE. *Educ:* Lehigh Univ, AB, 62; Univ Wis-Madison, PhD(phys chem), 67. *Prof Exp:* Monsanto Chem Co fel, 67-68; from asst prof to assoc prof physics, 68-83, PROF PHYSICS, SOUTHERN ILL UNIV, CARBONDALE, 83- *Mem:* Am Phys Soc. *Res:* Fourier transform nuclear magnetic resonance; applications to chemical and biophysical problems. *Mailing Add:* Dept of Physics Southern Ill Univ Carbondale IL 62901

CUTRESS, CHARLES ERNEST, b Calgary, Can, Mar 8, 21; nat US; m 46. INVERTEBRATE ZOOLOGY. *Educ:* Ore State Col, BS, 48, MS, 49. *Prof Exp:* USPHS fel, Univ Hawaii, 53-55; assoc cur, Div Marine Invert, US Nat Mus, Smithsonian Inst, 56-65; assoc prof marine biol, Univ PR, Mayaguez, 65-70, prof marine sci 70-90; RETIRED. *Mem:* AAAS; Soc Syst Zool; Am Inst Biol Sci; Am Soc Zoologists. *Res:* Systematics; biology; behavior; marine invertebrates, particularly coelenterates. *Mailing Add:* 838 Providencia Urbanization Bayes Lomas Mayaguez PR 00680

CUTRONA, LOUIS J, b Buffalo, NY, Mar 11, 15; m 38; c 2. ELECTRICAL ENGINEERING. *Educ:* Cornell Univ, BA, 36, Univ Ill, MA, 38, PhD(physics), 40. *Prof Exp:* Instr physics, Duluth Jr Col, 40-42; mem tech staff radar, Bell Tel Labs, 43-45; mem tech staff commun, Fed Tel Commun Lab, 45-47; mem tech staff radar, Sperry Gyroscope Co, 47-49; div head, Willow Run Labs, Mich, 49-56; physicist radar dept, Space Tech Lab, 56-57; prof elec eng, Univ Mich, 57-59, head radar lab, 59-61, prof elec eng, 61-76; vpres appl res div, Conductron Corp, 61-76; RES PHYSICIST, VISIBILITY LAB, UNIV CALIF, SAN DIEGO, 76-; CONSULT, SARCUTRON, INC. *Concurrent Pos:* Physicist, York Safe & Lock Co, 42-43; consult, Wright Air Develop Ctr, US Air Force. *Mem:* Am Phys Soc; fel Inst Elec & Electronics Engrs; fel Optical Soc Am. *Res:* Radar, communication and information theory. *Mailing Add:* Sarcutron Inc 13339 Barbados Way Del Mar CA 92014

CUTSHALL, NORMAN HOLLIS, b Glens Falls, NY, Mar 21, 38; m 61; c 2. CHEMICAL OCEANOGRAPHY. *Educ:* Ore State Univ, BS, 61, MS, 63, PhD(oceanog), 67. *Prof Exp:* Instr oceanog, Ore State Univ, 65-66; chemist, Oak Ridge Nat Lab, 67-68; marine scientist, AEC, 68-69; res assoc oceanog, Ore State Univ, 69-75; mem res staff, Oak Ridge, Tenn, 75-80, prog mgr, low-level waste prog, 80-83, head earth sci sect, 84-88, MGR, OAK RIDGE NAT LAB, WASHINGTON AREA OFF, 88- *Mem:* AAAS; Am Chem Soc; Am Geophys Union; Sigma Xi. *Res:* Environmental radioactivity; geochemistry; economic geology. *Mailing Add:* Oak Ridge Nat Lab Bldg 1505 MS-6038 Oak Ridge TN 37831

CUTSHALL, THEODORE WAYNE, b Lafayette, Ind, Mar 22, 28; m 52; c 2. ORGANIC CHEMISTRY. *Educ:* Purdue Univ, BS, 49; Northwestern Univ, MS, 59, PhD(org chem), 64. *Prof Exp:* res engr, Am Can Co, 49-50 & 53-58, US Army Med; US Army Med Nutrit Lab, 51-52; asst prof, Purdue Univ, 61-66, ASSOC PROF CHEM, IND UNIV- PURDUE UNIV, INDIANAPOLIS, 66- *Concurrent Pos:* Vis lectr, Franklin Col, 62-63, Butler Univ, 65-66 & Northwestern Univ, summers 64-69. *Mem:* Am Chem Soc; Sigma Xi. *Res:* Synthesis of heteroaromatic systems; reaction mechanisms; stereochemistry. *Mailing Add:* Dept Chem Ind Univ-Purdue Univ 1125 E 38th St Indianapolis IN 46205

CUTT, ROGER ALAN, b Rochester, NY, Aug 4, 36; m 59; c 2. PHYSIOLOGICAL PSYCHOLOGY, AUDIOLOGY. *Educ:* Franklin & Marshall Col, BA, 57; Univ Del, MA, 59, PhD(physiol psychol), 62. *Prof Exp:* Res assoc, Otol Res Lab, Presby-Univ Pa Med Ctr, 62-65, dir res, 65-70; med serv specialist, Social & Rehab Serv, Dept Health, Educ & Welfare, Philadelphia, 70-73; comnr med progs, Pa Dept Pub Welfare, Harrisburg, 73-79; CONSULT. *Concurrent Pos:* NIH & John A Hartford Found res grants, 63-70; instr, Dept Otolaryngol, Sch Med, Univ Pa, 62-65, assoc, 65-70, asst prof, 70-73; asst prof community med, Med Col Pa, 73-79. *Mem:* AAAS. *Res:* Basic and medical research in auditory and vestibular physiology, including behavioral hearing tests on animals, temporal bone histology, electrophysiology, electron microscopy and electronystagmography. *Mailing Add:* 5280 Strathmore Dr Mechanicsburg PA 17055

CUTTER, LOIS JOTTER, b Weaverville, Calif, Mar 11, 14; m 42; c 2. SYSTEMATIC BOTANY, PLANT MORPHOLOGY. *Educ:* Univ Mich, BA, 35, MS, 36, PhD, 43. *Prof Exp:* Res asst plant breeding, Parke, Davis & Co, Mich, 42; lectr biol, Univ NC, Greensboro, 63-65, asst prof biol, 65-84; RETIRED. *Mem:* Sigma Xi. *Mailing Add:* 3225 Forsyth Dr Greensboro NC 27407

CUTTS, CHARLES E(UGENE), b Sioux Falls, SDak, May 15, 14; m 46; c 2. STRUCTURAL ENGINEERING. *Educ:* Univ Minn, BCE, 36, MSCE, 39, PhD(civil eng, mech), 49. *Prof Exp:* Instrument man, Chicago, Milwaukee, St Paul & Pac RR Co, 36-38; asst civil eng, Univ Minn, 38-39; from instr to asst prof, Robert Col, Turkey, 39-43; instr math & mech, Univ Minn, 46-47, from instr to asst prof civil eng, 47-50; assoc res engr & assoc prof civil eng, Univ Fla, 50-53; engr res admin, NSF, 53-56; chmn dept eng, 56-69, prof, 56-84, EMER PROF CIVIL ENG, MICH STATE UNIV, 84- *Concurrent Pos:* Mem panel sci equip, NSF, 67-70; univ rep, Hwy Res Bd, 68-79; chmn, Engrs Coun Prof Develop, Region V, 73-74. *Mem:* Am Soc Civil Engrs; Am Concrete Inst; Am Soc Eng Educ (vpres, 70-72); Nat Soc Prof Engrs. *Res:* Prestressed concrete; fatigue of materials; numerical methods of analysis; model analysis. *Mailing Add:* 4599 Ottawa Dr Okemos MI 48864

CUTTS, DAVID, b Providence, RI, Dec 12, 40; c 2. EXPERIMENTAL PARTICLE PHYSICS. *Educ:* Harvard Univ, AB, 62; Univ Calif, Berkeley, PhD(physics), 68. *Prof Exp:* Res assoc physics, State Univ NY Stony Brook, 68-71 & Rutherford High Energy Lab, Eng, 71-73; asst prof, 73-79, assoc prof, 79-85, PROF PHYSICS, BROWN UNIV, 85- *Concurrent Pos:* Vis scientist, Rutherford Lab, Eng, 80-81, Fermi Natural Lab, 88-89. *Mem:* Fel Am Phys Soc. *Res:* Experimental studies of particle interactions and decays using computing and electronic techniques. *Mailing Add:* Dept Physics Brown Univ Providence RI 02912

CUTTS, JAMES HARRY, b Barnsley, Eng, June 12, 26; Can citizen; c 2. HISTOLOGY, HEMATOLOGY. *Educ:* Univ Sask, BA, 48; Dalhousie Univ, MSc, 56; Univ Western Ont, PhD(med res), 58. *Prof Exp:* Chief technician hemat, Gray Nuns Hosp, Regina, Sask, 48-50; technician & head dept, St Paul's Hosp, Vancouver, BC, 50-53; prof hematologist, NS Dept Pub Health, 53-56; from asst prof to assoc prof anat, Univ Western Ont, 60-68; assoc prof, Sch Med, Univ Mo-Columbia, 68-73, chmn dept, 72-73, prof anat, 73-88; RETIRED. *Concurrent Pos:* Nat Cancer Inst Can fel med res, Univ Western Ont, 56-68, Ross Mem fel, 58-60; lectr, Dalhousie Univ, 53-56; adv fac, Belknap Col, 64-72; auth. *Mem:* Am Asn Cancer Res; NY Acad Sci; Am Asn Anatomists; Am Soc Zoologists; Can Asn Anatomists; Royal Micros Soc; Anat Soc Gt Brit & Ireland. *Res:* Oncology; postnatal development of the opossum; comparative hematology. *Mailing Add:* 2260 Oceanforest Dr W Atlantic Beach FL 32233

CUTTS, ROBERT IRVING, b Chicago, Ill, Jan 16, 15; m 48; c 2. PSYCHIATRY. *Educ:* Lewis Inst Technol, BS, 36; Univ Ill, MD, 40. *Prof Exp:* Clin asst internal med, Res & Educ Hosp, Univ Ill, 46-47; staff psychiatrist, Vet Admin Hosp, Ill, 49-53 chief continued treatment serv, 53-57; chief serv, Vet Admin Hosp, Tucson, Ariz, 57-59; pvt practr, 59-66; dir, Southern Ariz Ment Health Ctr, 66-68; chief psychiat serv, Vet Admin Hosp, 68-72; asst prof psychiat, Med Col, Univ Arz, 72-82; RETIRED. *Concurrent Pos:* Clin asst psychiat, Med Sch, Northwestern Univ, 53-55; consult, Southern Ariz Med Health Ctr & Vet Admin Hosp; adj assoc prof, Med Col, Univ Ariz. *Mem:* Fel AAAS; AMA; fel Am Psychiat Asn. *Res:* Outpatient psychiatry. *Mailing Add:* 6332 N Calle De Adelita Tucson AZ 85718

CUTZ, ERNEST, b Nove Zamky, Czech, Dec 16, 42; Can citizen; m 67; c 3. PEDIATRIC PATHOLOGY. *Educ:* Charles Univ, Prague, Czech, MD, 66; FRCPath(C). *Prof Exp:* Res fel, Dept Histol & Embryol, Charles Univ, Prague, Czech, 66-67; resident path, Regular Anticancer Ctr, Toulouse, France, 67-68; res fel path, Dept Path, Res Inst, Hosp Sick Children, Toronto, 68-69; resident, 69-71, res assoc, 71-73, lectr, 73-76, from asst prof to assoc prof, 76-86, PROF PATH, UNIV TORONTO, 87-; SR STAFF PATHOLOGIST, HOSP SICK CHILDREN, TORONTO, 71-, SCIENTIST, RES INST, 71- *Concurrent Pos:* External reviewer, Med Res Coun Can, 75-; consult, Spec Study Sect, NIH, 82-, Nat Heart, Lung & Blood Inst, 85- *Mem:* Am Asn Pathologists; Int Acad Path; Soc Pediat Path; NY Acad Sci; Can Asn Pathologists. *Res:* Pulmonary disorders of infancy and childhood; lung development; neuro-endorine mechanisms of lung function; biogenic amines and regulatory peptides and their role during pulmonary adaptation and in sudden infant death syndrome; pathogenic mechanisms of genetic-metabolic disorders in infancy and childhood. *Mailing Add:* Dept Path Hosp Sick Children 555 Univ Ave Toronto ON M5G 1X8 Can

CVANCARA, ALAN MILTON, b Ross, NDak, Mar 7, 33; m 59; c 2. PALEOBIOLOGY. *Educ:* Univ NDak, BS, 55, MS, 57; Univ Mich, PhD(geol), 65. *Prof Exp:* From asst prof to assoc prof, 63-72, PROF GEOL, UNIV N DAK, 72- *Concurrent Pos:* Fulbright scholar, Australia, 56-57. *Res:* Lower Tertiary mollusks and stratigraphy in north-central United States; Quaternary freshwater and terrestrial mollusks; writing natural history books. *Mailing Add:* Dept Geol & Geol Eng Univ NDak Grand Forks ND 58202

CVANCARA, VICTOR ALAN, b Ross, NDak, Aug 29, 37; m 58; c 4. PHYSIOLOGY. *Educ:* Minot State Col, BS, 59; Univ SDak, MS, 64, PhD(physiol), 68. *Prof Exp:* Sci instr, Seyyida Maatuka Col, Zanzibar, EAfrica, 61-63; teaching asst, Univ SDak, 64-66, res assoc physiol, 67-69; fel, Ore State Univ, 69-71; asst prof, Beloit Col, 71-72; assoc prof physiol, Univ Wis-Eau Claire, 72-80, assoc prof biol, 80- *Mem:* Am Soc Zoologists; Sigma Xi; Am Fisheries Soc. *Res:* Temperature adaptation, protein synthesis and enzyme induction in teleost fishes. *Mailing Add:* Dept Biol Univ Wis Eau-Claire WI 54702

CWALINA, GUSTAV EDWARD, b Baltimore, Md, Feb 6, 09; m 43; c 2. MEDICINAL CHEMISTRY. *Educ:* Univ Md, BS, 31, MS, 33, PhD(pharmaceut chem), 37. *Prof Exp:* Asst chem, Univ Md, 31-37; from asst prof to assoc prof, Creighton Univ, 37-42; from asst prof to prof, 46-75, EMER PROF PHARMACEUT SCI, COL PHARM, PURDUE UNIV, 75-, ASSOC DEAN, 63- *Mem:* Am Chem Soc; Am Pharmaceut Asn. *Res:* Phytochemical investigation of Ipomea pes-caprae; plant analysis. *Mailing Add:* 1634 Northwestern Ave West Lafayette IN 47906

CYBRIWSKY, ALEX, b Pidsosniw, Ukraine, Mar 26, 14; US citizen; m 45; c 4. PHYSICS. *Educ:* Univ Vienna, PhD(physics), 45. *Prof Exp:* Asst physics, Univ Vienna, 43-46; chief chemist, Archer Chem Co, Ky, 50-51; sect head, Reynolds Metals Co, 51-58, proj dir, 58-60; res physicist, Gen Elec Co, 60-62; RES PHYSICIST, ALLIS-CHALMERS MFG CO, 62- *Mem:* Am Phys Soc; Am Soc Metals. *Res:* Nuclear and cosmic rays studies by nuclear photographic emulsions; recovery, finishing and alloying of aluminum; alumina catalyst; thermoelectric and galvano-thermomagnetic phenomena; energy conversion. *Mailing Add:* 4868 S 94 Milwaukee WI 53228

CYGAN, NORBERT EVERETT, b Chicago, Ill, June 5, 30; m 57; c 2. GEOLOGY, PALEONTOLOGY. *Educ:* Univ Ill, PhD(geol), 62. *Prof Exp:* Instr geol, Ohio Wesleyan Univ, 56-59; asst, Univ Ill, 59-62; div stratigrapher, 62-71, MINERALS STAFF GEOLOGIST, CHEVRON OIL CO, 71- *Concurrent Pos:* Lectr, Univ Houston. *Mem:* Am Asn Petrol Geol; fel Geol Soc Am; Sigma Xi. *Res:* Cambrian and Ordovician conodonts; tertiary micro fauna; stratigraphy; uranium geology; paleoecology. *Mailing Add:* 8125 E Geddes A C Arapahoe East CO 80112

CYKLER, JOHN FREULER, b Berkeley, Calif, Oct 28, 16. FIELD IRRIGATION DESIGN. *Educ:* Univ Calif, Berkeley, BS, 40; Univ Wyo, MS, 47. *Prof Exp:* Instr agr eng, Univ Wyo, 40-42 & 46-47; asst agr engr, Agr Exp Sta, Univ Hawaii, 47-50; agr engr, Pineapple Res Inst, Hawaii, 50-56, chief engr, 56-61; dir, res & develop, Lihue Plantation Co, Lihue, Hawaii, 61-66; agr engr, C Brewer & Co, Ltd, Honolulu, 66-70, mgr & engr, Mauna Loa Sugar Co Div, 70-74, sr agr engr, Hawaiian Agron Int Div, 75-80, dir eng, Mauna Loa Macadamia Nut Corp Div, 80-85; RETIRED. *Concurrent Pos:* Eng & cmndg officer, US Navy, 42-46 & 51-53; asst mgr & mgr agr eng, Cainsa Sugar Estate, Uruguay, 63-65. *Mem:* Am Soc Agr Engrs. *Res:* Crop water requirements; machinery design and development for production, harvest and processing of forage crops; computer controls for moisture sensing and varying drying temperatures for batch drying of nuts. *Mailing Add:* 2816 140th Ave NE Bellevue WA 98005

CYMERMAN, ALLEN, b New York, NY, Feb 4, 42; m 66; c 2. PHYSIOLOGY, BIOCHEMISTRY. *Educ:* Rutgers Univ, BA, 63; Jefferson Med Col, MS, 66, PhD(physiol), 68. *Prof Exp:* Res technician clin endocrinol, Temple Univ Hosp & Med Sch, 63-64; sr invest biochem & pharmacol, 68-70, sr invest, Altitude Res Div, 70-82, DIR, ALTITUDE PHYSIOL & MED DIV, US ARMY RES INST ENVIRON MED, 82- *Mem:* Am Physiol Soc; Am Col Sports Med; Sigma Xi. *Res:* Effects of environmental extremes on normal and drug altered physiological and biochemical mechanisms. *Mailing Add:* 78 Delmar Ave Framingham MA 01701

CYNADER, MAX SIGMUND, b Berlin, WGer, Feb 24, 47; Can citizen; m 85; c 3. NEUROBIOLOGY, VISION & COMPUTER VISION. *Educ:* McGill Univ, BSc, 67; Mass Inst Technol, PhD(neurosci), 72. *Prof Exp:* From asst prof to prof psychol, Dalhousie Univ Med Sch, 73-88, from assoc prof to prof physiol, 79-88; PROF & DIR OF RES, DEPT OPHTHAL, UNIV BC, 88- *Concurrent Pos:* E W R Steacie fel, Nat Sci & Eng Res Coun Can, 79-81; vis prof physiol, Univ Calif, San Francisco, 81; Killam res prof, 84-88; fel, Can Inst Advan Res, 86-, BC fel, 89-; Can rep, Int Human Frontier Sci Prog, 90-; Killam res prize, 90. *Mem:* Soc Neurosci; Asn Res Vision & Ophthal; Can Physiol Soc; Int Strabismological Soc; Royal Soc Can (elect, 87); Can Asn Neurosci; Asn Res Otolaryngol. *Res:* Study of neural mechanisms underlying the organization of neocortex and its development, using the visual system as a model system for understanding the function of the brain; artificial intelligence; computervision; strabismus; amblyopia. *Mailing Add:* Dept Ophthal Univ BC Vancouver BC V5Z 3N9 Can

CYNKIN, MORRIS ABRAHAM, biochemistry, microbiology; deceased, see previous edition for last biography

CYPESS, RAYMOND HAROLD, b Brooklyn, NY, June 27, 40; c 2. PARASITOLOGY. *Educ:* Brooklyn Col, BS, 61; Univ Ill, BVS, 65, DVM, 67; Univ NC, Chapel Hill, PhD(parasitol), 71. *Prof Exp:* Res asst parasitol, Univ Ill, Urbana, 63-67; fel parasitol, Univ NC, Chapel Hill, 67-69, NIH spec fel, 69-70; asst prof parasitol & epidemiol, Grad Sch Pub Health, Univ Pittsburgh, 70-74, assoc prof, 74-76; prof, State Univ NY, Upstate Med Ctr, Syracuse, 78-88; prof microbiol & dir diag lab, 77-88, CHMN PREV MED, CORNELL UNIV, 78- *Concurrent Pos:* Adj prof, Grad Sch Pub Health, Univ Pittsburgh, 77-; fel, Fogarty Int, 75 & NIH, Nat Inst Allergy & Infectious Dis, 75-79. *Mem:* Am Soc Parasitol; Am Epidemiol Soc; Am Soc Trop Med & Hyg; Sigma Xi. *Res:* Immunoparasitology and infectious diseases; epidemiology; comparative medicine. *Mailing Add:* 5666 Glade View Memphis TN 38120

CYR, W HOWARD, b New Haven, Vt, Mar 16, 43. RADIATION GENETICS, RISK ANALYSIS. *Educ:* Univ Vt, BA, 65; Pa State Univ, MS, 66, PhD(biophys), 72. *Prof Exp:* Res biophys, Armed Force Inst Path, 70-74; res radiation genetics, Div Biol Effects, Bur Radiol Health, Dept Health, Educ & Welfare, 74-80, Ionizing Radiation & Statist Br & Radiation Biol Br, Div Life Sci, 80-85, asst dir, Dept Health & Human Serv, 85-90, SR RES BIOPHYSICIST, DIV LIFE SCI CTR, DEVICES & RADIOL HEALTH FOOD & DRUG ADMIN, DEPT HEALTH & HUMAN SERV, 90- *Mem:* Sigma Xi; Radiation Res Soc. *Res:* Risk assessment of ionizing radiation at low dose rates; risk assessment of residues of ethylene oxide on medical devices; virus leakage through condoms. *Mailing Add:* FDA-CDRH-OST-DLS HFZ-114 5600 Fisher Lane Rockville MD 20857

CYSYK, RICHARD L, b Baltimore, Md, Mar 16, 42; m; c 3. BIOLOGICAL CHEMISTRY. *Educ:* Univ Md, BS, 65; Yale Univ, PhD(pharmacol), 70. *Prof Exp:* Damon Runyon fel, Dept Cell Biol & Pharmacol, Sch Med, Univ Md, 70-71; sr investr, Drug Eval & Cancer Chemother Div, Microbiol Assocs, Inc, Bethesda, 71-72; staff fel, Lab Chem Pharmacol, Nat Cancer Inst, NIH, 72-77, pharmacologist, 77-78, head, Drug Metab Sect, 78-85, actg chief, Lab Chem Pharmacol, 81-83 & Lab Exp Therapeut & Metab, 85-88, CHIEF, LAB BIOL CHEM, NAT CANCER INST, NIH, 84- *Concurrent Pos:* Adj prof pharmacol, George Washington Univ, 80-; assoc ed, Cancer Treatment Reports & Jour Nat Cancer Inst; mem adv bd, Nat Drug Discovery Prog. *Mem:* Am Asn Cancer Res; Sigma Xi. *Res:* Pharmacology; drug metabolism and design; mechanism of action and biochemistry of antitumor agents. *Mailing Add:* Nat Cancer Inst NIH Lab Biol Chem Bldg 37 Rm 5D02 Bethesda MD 20892

CYWINSKI, NORBERT FRANCIS, b Brown Co, Wis, Aug 24, 29; m 61; c 1. ORGANIC CHEMISTRY. *Educ:* Univ Wis, BS, 54; Northwestern Univ, PhD(org chem), 62. *Prof Exp:* Res chemist, Phillips Petrol Co, 59-65, sr res chemist, 65- 77, staff res chemist, El Paso Prod Co, 77-83; instr, Austin Community Col, 85-87; CHEMIST, SACHEM, 87- *Mem:* Am Chem Soc. *Res:* Bicyclic systems; cyclization and dehydrocyclization reactions; free radical reactions; diolefin synthesis; hydrocarbon oxidation; petrochemicals; nylon intermediates; quaternary ammonium compounds. *Mailing Add:* 1604 Robb Lane Round Rock TX 78663-3032

CZAJKA-NARINS, DORICE M, b Chicago, Ill; m 70; c 2. PHYSIOLOGY. *Educ:* Northwestern Univ, BS; Mass Inst Technol, PhD(nutrit), 66. *Prof Exp:* Postdoctoral nutrit, Rockefeller Univ, 66-68; asst prof nutrit, Mich State Univ, 68-74; assoc prof, Rush Univ, 74-86; prof, Univ Health Sci/Chicago Med Sch, 77-91; prof, Northern Ill, Univ, 84; PROF & CHAIR NUTRIT & FOOD SCI, TEXAS WOMAN'S UNIV, 91- *Concurrent Pos:* Nutrit consult, 80-, Dor-Art Assocs, 86-91. *Mem:* Am Inst Nutrit; Inst Food Technologists; Am Med Writers Asn; Geront Soc Am. *Res:* Growth of infants and nutrition factors affecting growth; trace mineral metabolism; nutrition and the elderly. *Mailing Add:* Dept Nutrit & Food Sci PO Box 24134 Denton TX 76204

CZAMANSKE, GERALD KENT, b Chicago, Ill, Jan 17, 34; div; c 2. PETROLOGY-MINERALOGY. *Educ:* Univ Chicago, BA, 53, BS, 55; Stanford Univ, PhD(geol), 61. *Prof Exp:* NSF fel, Univ Oslo, 60-61; res assoc geochem, Mass Inst Technol, 61-62; asst prof geol, Univ Wash, 62-65; GEOLOGIST, US GEOL SURV, 65- *Concurrent Pos:* Assoc ed, Am Mineral, 80-84 & Can Mineralogist, 88- *Mem:* Mineral Soc Am; Soc Econ Geologists; Mineral Asn Can. *Res:* Petrology of granitoids and mafic intrustions; sulfide mineralogy and phase relationships; use of mafic silicate chemistry to elucidate problems in igneous petrology. *Mailing Add:* US Geol Surv MS 984 345 Middlefield Rd Menlo Park CA 94025

CZANDERNA, ALVIN WARREN, b LaPorte, Ind, May 27, 30; m 53; c 3. SURFACE SCIENCE, SOLAR ENERGY MATERIALS. *Educ:* Purdue Univ, BS, 51, PhD(phys chem), 57. *Prof Exp:* Jr res metallurgist, Aluminum Co Am, Pa, 53; sr res physicist, Parma Res Lab, Union Carbide Corp, 57-63, res scientist, Chem Div, 63-65; assoc prof physics, Clarkson Col Technol, 65-68, mem Inst Colloid & Surface Sci, 65-78, assoc dir, 68-74, coun mem, 66-77, prof physics, 68-78; sr scientist, 78-79, group mgr, 79-82, prin scientist, 82-83, RES FEL, SOLAR ENERGY RES INST, GOLDEN, 83- *Concurrent Pos:* NSF fel, 56-57; consult, Los Alamos Nat Lab, 68-77, Owens-Ill Inc, 69-70, Union Carbide Corp, 69 & 77, Ga Inst Technol, 74-75, Lawrence Livermore Lab, 77, Babcock & Wilcox, 77, Exxon, 81 & W R Grace, 82-84; vis scientist, Fritz-Haber-Inst, Max Planck Soc, Berlin, Ger, 71-72; adj prof chem, Univ Denver, 79-, adj prof physics, 80, adj prof engr, 87- & prof adj, chem eng, Univ Colo, 80-; lectr, Am Vacuum Soc, 73-, vchmn educ comt, 79, chmn, 80-81, mem, bd dir, Surface Sci Div, 68-69, Vacuum Technol Div, 80-81, Appl Surface Sci Div, 87-, vchmn, 89-90, chmn, 90-91; tour speaker, Am Chem Soc, 82-, lectr, 83- *Mem:* Am Chem Soc; Am Vacuum Soc; Catalysis Soc; Sigma Xi; fel NY Acad Sci; fel Am Physics Soc; Mat Res Soc; Am Soc Testing Mat; Electrochem Soc. *Res:* Surface science; organized molecular assemblies; surface analysis; solar materials science; reactions at interfaces; polymer/metal (oxide) interfaces; properties of thin films; desorption; chemisorption; catalysis; oxidation; applications of surface analytical techniques and vacuum ultramicrogravimetry. *Mailing Add:* 2619 S Zephyr Ct PO Box 27209 Lakewood CO 80227

CZAPSKI, ULRICH HANS, b Munich, Ger, July 15, 25; m 54. METEOROLOGY, GEOPHYSICS. *Educ:* Univ Hamburg, PhD(meteorol), 53. *Prof Exp:* Guest staff mem, Int Inst Meteorol, Stockholm, 52; staff mem geophys instruments, Askania-Werke, AG, Berlin, Ger, 53-55; geophysicist, Seismos GmbH, Hannover, 55-61; UN tech expert meteorol & geophys, Mission to Pakistan, World Meteorol Orgn, UN Tech Assistance Admin, 61-62; lectr meteorol, Imp Col, Univ London, 62-64; ASSOC PROF ATMOSPHERIC SCI, STATE UNIV NY ALBANY, 64- *Mem:* Am Meteorol Soc; Am Geophys Union. *Res:* Cooling towers; dispersion of air pollutants; thermal waste use and dispersion; solar energy; transfer and convection; hydrometeorology. *Mailing Add:* Dept Atmospheric Sci State Univ NY 1400 Washington Ave Albany NY 12222

CZARNECKI, CAROLINE MARY ANNE, b Detroit, Mich, Aug 3, 29. VETERINARY ANATOMY. *Educ:* Bemidji State Col, BS, 50; State Col Iowa, MA, 60; Univ Minn, Minneapolis, PhD(anat), 67. *Prof Exp:* Instr high schs, 50-62; teaching asst anat, 62-67, instr, 67, from asst prof to assoc prof, 67-76, PROF VET ANAT, UNIV MINN, ST PAUL, 76- *Concurrent Pos:* Mem, Basic Sci Coun, Am Heart Asn. *Mem:* Sigma Xi (secy-treas, 83-85, pres, 87-88); Am Asn Vet Anat; World Asn Vet Anat; Conf Res Workers Animal Dis; Am Heart Asn. *Res:* Alcoholic cardiomyopathy; drug-induced cardiomyopathy; roundheart disease in turkeys; ultrastructure of avian heart. *Mailing Add:* Dept Vet Biol Univ Minn St Paul MN 55108

CZARNY, MICHAEL RICHARD, b New Britain, Conn, Jan 26, 50; m 78; c 1. ORGANIC CHEMISTRY. *Educ:* Providence Col, BS, 72; Dartmouth Col, PhD(chem), 77. *Prof Exp:* NIH fel, Cornell Univ, 77-78; res chemist org & med chem, Sterling-Winthrop Res Inst, 78-85; SR DEVELOP CHEMIST, PARKE-DAVIS, 85- *Mem:* Am Chem Soc; Sigma Xi. *Res:* Natural products; steroids; synthetic methods. *Mailing Add:* Solvay Animal Health Inc 2000 Rockford Rd Charles City IA 50616-0800

CZECH, MICHAEL PAUL, b Pawtucket, RI, June 24, 45; m 69; c 1. BIOCHEMISTRY. *Educ:* Brown Univ, BA, 67, PhD(biochem), 72; Duke Univ, MA, 69. *Prof Exp:* Fel biochem, Duke Univ, 72-74; from asst prof to prof physiol chem, Div Biol & Med Sci, Brown Univ, 74-81; PROF & CHMN, DEPT BIOCHEM, MED CTR, UNIV MASS, 81- *Honors & Awards:* Res & Develop Award, Am Diabetes Asn, 74, Elliot P Joslin Award, 75, Eli Lilly Award Diabetes, 82; David Rumbough Award, Juvenile Diabetes Found, 85. *Mem:* Endocrine Soc; Am Diabetes Asn; Am Soc Biol Chemists. *Res:* Biochemistry of the modulation of hexose transport activity by insulin and other agents in plasma membranes from isolated fat cells. *Mailing Add:* Dept Biochem Univ Mass Worcester MA 01605

CZEISLER, CHARLES ANDREW, b Chicago, Ill, Nov 7, 52. CIRCADIAN PHYSIOLOGY. *Educ:* Harvard Univ, AB, 74; Stanford Univ, PhD(neuro & bio behav sci), 78; Stanford Med Sch, MD, 81. *Prof Exp:* Proj dir res, Lab Human Chronophysiol, Montefiore Hosp & Med Ctr, Ablert Einstein Col Med, 76-81; JOSIAH T MACY SR FEL HEALTH POLICY, DIV HEALTH POLICY RES & EDUC, HARVARD UNIV, 81-, ASST PROF, DEPT MED, MED SCH. *Concurrent Pos:* Adj instr biol sci, Grad Sch Arts & Sci, Fordham Univ, 78; guest ed, Sleep, 79-80; lectr, Stanford Univ, 79; res assoc, Dept Physiol & Biophysics, Harvard Med Sch, 79-80; appointee, Sleep Schedules, 80-; mem, Task Force Health & Behav, Res Agenda, Inst Med, Nat Acad Sci, 81-82; partic, Consenus Develop Workshop Insomnia & Sleep Disorders, NIH, 81. *Mem:* AAAS; Int Soc Chronobiol; Asn Psychophysiol Study Sleep; Soc Neurosci; Am Physiol Soc. *Res:* The physiology of the human carcadian timing system; neuroendocrine control of hormonal secretory patterns; sleep physiology and pathology; applications of circadian physiology to shift work schedule design and occupational health policy. *Mailing Add:* Brigham & Women's Hosp Dept Med Harvard Med Sch 221 Longwood Ave Fuller Pavilion Boston MA 02115

CZEPIEL, THOMAS P, b Deep River, Conn, May 2, 32; m 55; c 3. PAPER CHEMISTRY. *Educ:* Wesleyan Univ, BA, 54; Inst Paper Chem, Lawrence Univ, MS, 56, PhD(chem), 59. *Prof Exp:* Res group leader, Scott Paper Co, 59-61, mgr tech servs paper mill, Maine, 61-64, tech dir, Detroit, Mich, 64-68 & Pa, 68-72, paper mill supt, Chester, Pa, 72-75, dir venture mfg, Philadelphia, 76-78, plant mgr, Everett, Wash, 79-83, vpres, western region, 83-86, cent region, 87-90, CORP VPRES MFG OPERS, SCOTT PAPER CO, MOBILE, ALA, 90- *Mem:* Am Chem Soc; Tech Asn Pulp & Paper Indust. *Res:* Influence of metal traces on color stability of cellulose; preparation and utilization of synthetic fibers for papermaking; pulp and paper technology. *Mailing Add:* 3294D Dog River Rd Theodore AL 36582

CZEPYHA, CHESTER GEORGE REINHOLD, b New Brunswick, NJ, Sept 14, 27; m 50; c 5. SYSTEMS DEVELOPMENT, SOFTWARE MARKETING. *Educ:* Lehigh Univ, BSEE, 50; Air Force Inst Technol, MSEE, 56. *Prof Exp:* Navigator, USAF, 50-56, proj officer, Missile Develop, Air Force Missile Test Ctr, 56-60, electronics engr, Air Force Cambridge Res Labs, 62-69, master navigator, Southeast Asia, 69-70, wing chief of plans, Repub of Philippines, 71, chief eng systs prog off, Electropmlectronic Systs Div, Hanscom AFB, 71-73; dir metrol, Directorate Metrol, 73; vice commander, Aerospace Guid & Metrol Ctr, 73-76; vice commander, Geophys Lab, 76-79; vice commander software mkt & develop, CNC, 82-85. *Concurrent Pos:* USAF, 50-79. *Mem:* Am Geophys Union; Sigma Xi. *Res:* Sterics as a means of severe weather identification; development of high altitude balloon systems for upper air research. *Mailing Add:* US Peace Corps PO Box 487 Belize City Belize

CZERLINSKI, GEORGE HEINRICH, b Cuxhaven, Ger, Dec 31, 24; US citizen; m; c 2. BIOPHYSICS, BIOCHEMISTRY. *Educ:* Univ Hamburg, BA, 52; Northwestern Univ, MS, 55; Univ Gottingen, PhD(phys chem), 58. *Prof Exp:* Res assoc biophys, Univ Pa, 60-64, asst prof, 64-67; ASSOC PROF BIOCHEM, NORTHWESTERN UNIV, CHICAGO, 67- *Concurrent Pos:* E R Johnson Found fel, Univ Pa, 58-60; vis asst prof biochem, Cornell Univ, 66-67, vis prof chem & biochem, Utah State Univ. *Mem:* AAAS; Biophys Soc; Am Soc Molecular Biol & Biochem; Sigma Xi. *Res:* Instrumentation, theory and application of chemical relaxation for the investigation of the mechanism of enzyme function; rapid kinetics of enzyme reactions; allosteric behavior; temperature-jump method; production and coating of submicron-sized particles; magnetomechanic and magnetocaloric application of microspheres in biology; sorting of live cells. *Mailing Add:* Dept CMS Biol Northwestern Univ Chicago IL 60611-3008

CZERNOBILSKY, BERNARD, b Konigsberg, Ger, Jan 1, 28; US citizen; m 54; c 3. PATHOLOGY. *Educ:* Univ Lausanne, MD, 53. *Prof Exp:* Intern, Jewish Hosp Asn, Cincinnati, Ohio, 53-54, asst resident path, 54-55; asst resident, Israel Beth Hosp, Boston, 57-58, resident, 58-59; chief resident, Mallory Inst, Boston City Hosp, 59-60; assoc pathologist, Women's Med Col Pa, 60-61, from asst prof to assoc prof, 61-65; asst prof surg path, Lab Path Anat, Hosp Univ Pa, 65-69, assoc prof path, 69-70; assoc prof, 71-81, PROF PATH, HEBREW UNIV, JERUSALEM, 81-; CHIEF DEPT PATH, KAPLAN HOSP, ISRAEL, 70- *Concurrent Pos:* Teaching fel path, Harvard Med Sch, 58-59; instr, Sch Med, Tufts Univ, 59-60. *Mem:* NY Acad Sci; Am Asn Pathologists & Bacteriologists; Int Acad Path; Int Soc Gynec Pathologists; Fedn Am Soc Exp Biol. *Res:* Spleen; pancreas; kidneys; breast; cardiovascular system; tissue culture; bilirubin metabolism; gynecologic pathology; ovarian tumors; cytoskeleton. *Mailing Add:* Dept Path Kaplan Hosp Rehovot 76100 Israel

CZEROPSKI, ROBERT S(TEPHEN), b Chicago, Ill, Nov 15, 23; m 47; c 3. CHEMICAL ENGINEERING, ENVIRONMENTAL SCIENCE. *Educ:* Ill Inst Tech, BS, 47. *Prof Exp:* Mgr prod develop, 56-59, dir appl res printing inks, 59-61, asst vpres, Res & Develop, 62-66, asst plant mgr, 66-67, plant mgr, 67-72; ENVIRON & TECH DIR, MARISOL INC, 72- *Mem:* Am Chem Soc. *Res:* Patents; recovery and recycling of industrial wastes. *Mailing Add:* 1167 Tanglewood Lane Scotch Plains NJ 07076

CZERWINSKI, ANTHONY WILLIAM, b St Louis, Mo, Feb 10, 34; m 64; c 3. NEPHROLOGY, PHARMACOLOGY. *Educ:* St Louis Univ, BS, 55, MD, 59. *Prof Exp:* Fel clin nephrology, Vet Admin Hosp, Boston, Mass, 63-64; fel metab, Univ NC, 64-66; from asst prof to assoc prof, 69-77, PROF MED, HEALTH SCI CTR, UNIV OKLA, 77- *Concurrent Pos:* Assoc chief renal sect & chief clin nephrology, Univ Okla Health Sci Ctr & Oklahoma City Vet Admin Hosp, 75- *Mem:* Sigma Xi; Am Soc Nephrology; Am Col Physicians; Int Soc Nephrology; Am Soc Pharmacol & Exp Therapeut. *Mailing Add:* 2508 NW 120 Oklahoma City OK 73120

CZERWINSKI, EDMUND WILLIAM, b Mojave, Calif, Jan 7, 40; m 61; c 2. PROTEIN CRYSTALLOGRAPHY. *Educ:* Univ Calif, Berkeley, AB, 62; Ind Univ, PhD(biochem), 71. *Prof Exp:* Res zoologist, Univ Calif, Berkeley, 62-64; biochemist, Eli Lilly & Co, Ind, 64-68; grad student biochem, Med Sch, Ind Univ, 68-71; res asst prof biochem, Med Sch, Washington Univ, Mo, 72-78; asst prof, 78-89, ASSOC PROF BIOCHEM, UNIV TEX MED BR, GALVESTON, 89- *Mem:* Am Crystallog Asn; Am Chem Soc; AAAS; Sigma Xi; Am Soc Biol Chem; Protein Soc. *Res:* Crystallographic structure determination of proteins and other molecules of biological importance; structure/function relationships of proteins. *Mailing Add:* Dept HBC&G Univ Tex Med Br Galveston TX 77550

CZOP, JOYCE K, b Chestnut Hill, Pa, Apr 26, 45. CELL BIOLOGY, IMMUNOCHEMISTRY. *Educ:* Pa State Univ, BS, 67; Univ Wis, MS, 70, PhD(bacteriol & biochem), 72. *Prof Exp:* Fel immunol, Med Col Ohio, 72-74; trainee, Sch Med, NY Univ, 74-76; res fel, 76-79, asst prof, 79-88, ASSOC PROF MED, HARVARD MED SCH, BRIGHAM & WOMEN'S HOSP, 88- *Mem:* Sigma Xi; AAAS; Am Asn Immunol; Am Asn Path. *Res:* Human beta-glucan receptors; osponic fibronectin. *Mailing Add:* 72 Newcomb St Norton MA 02766

CZUBA, LEONARD J, b East Chicago, Ind, Feb 28, 37; m 60; c 3. FOOD CHEMISTRY AND SAFETY. *Educ:* Ind Univ, AB, 61; Univ Minn, PhD(org chem), 67. *Prof Exp:* Chemist, Sinclair Res, Inc, 61-63; NIH fel org chem, Mass Inst Technol, 67-68; res chemist, 68-72, proj leader, 72-77, mgr, 77-81, asst dir, 81-86, DIR CENT RES, PFIZER INC, 86- *Mem:* Am Chem Soc; NY Acad Sci. *Res:* Synthetic organic chemistry; safety & regulatory affairs for chem products r & d. *Mailing Add:* Cent Res Pfizer Inc Groton CT 06340

CZUBA, MARGARET, b Hohenfels, Ger, Dec 7, 47; Can citizen. ECOTOXICOLOGY, ENVIRONMENTAL PHYSIOLOGY. *Educ:* Univ Toronto, BSc, 69, MSc, 72; Univ Guelph, PhD(environ physiol, 78. *Prof Exp:* Res asst environ physiol, Univ Guelph, 72-77; from res assoc ecotoxicol to asst res officer, 77-85, ASSOC RES OFFICER, NAT RES COUN, 85- *Mem:* Int Soc Soil Sci; Can Soc Plant Physiol; Am Soc Plant Physiol; Am Soc Agron; Can Bot Asn. *Res:* Interaction of persistant metals with ozone, other metals or nutrient elements, as they affect plant function metabolically, toxicologically or mutagenetically in ecosystem; biotransformation of methylated metals and their mechanism of action in aquatic plants, tissue cultures of agronomic species and metal-accumulating wild species; demethylation of Hg and distribution in plants, animals and bacteria in relation to macromolecular production, cell division, toxicity and whole body distribution; mediating effects of plant-incorporated methyl mercury on distribution and energy compound (creatine) status in the rat; production of an aggregation factor in plant culture as indicator of site of toxic sensitivity to mediating; identification of a fluorescent compound (under laser light) which is specific for methyl mercury toxicity, not inorganic species or a few other methylated metals; detoxification molecule interactions with antioxidants in cell systems in relation to tumor marker production and heterogeneity as detected by biological models and cyto chemical methods. *Mailing Add:* Nat Res Coun Can Inst Biol Sci M-54 Ottawa ON K1A 0R6 Can

CZUCHLEWSKI, STEPHEN JOHN, b New York, NY, Apr 17, 44. LASER PHYSICS, LASER-INITIATED PLASMA PHYSICS. *Educ:* Manhattan Col, BS, 65; Yale Univ, MS, 66, PhD(physics), 73. *Prof Exp:* Res asst, Yale Univ, 65-73; res assoc, Kans State Univ, 73-74; MEM STAFF & SECT LEADER, LOS ALAMOS NAT LAB, UNIV CALIF, 75- *Mem:* Am Phys Soc; Inst Elec & Electronic Engrs. *Res:* High-power short-pulse excimer and carbon-dioxide lasers for fusion applications; multiphoton absorption in polyatomic molecules; electron-beam propagation. *Mailing Add:* Los Alamos Nat Lab CLS-5 MS-E543 Los Alamos NM 87545

CZUHA, MICHAEL, JR, b Peninsula, Ohio, July 20, 22; m 51. PHYSICAL CHEMISTRY, ENGINEERING. *Educ:* Kent State Univ, BS, 44. *Prof Exp:* Res chemist polymer chem, Govt Labs, Univ Akron, 46-53, Arnold O Beckman, Inc, 53-56 & Consolidated Electrodynamics Corp, 56-59; sr res chemist, Res Ctr, Bell & Howell, 59-70; sr engr instruments, E I Du Pont de Nemours & Co, Inc, 70-80; CONSULT, MOISTURE ANALYSIS, 80-

Concurrent Pos: Owner & gen mgr, Bubble-O-Meter Co. *Mem:* Sr mem Am Chem Soc; sr mem Instrument Soc Am. *Res:* Chemistry and physics of high polymers; instrumental analysis and physical testing; research and development of analytical laboratory and process instruments; applications of electron spectrometers. *Mailing Add:* 6510 Wheeler Ave LaVerne CA 91750

CZUPRYNSKI, CHARLES JOSEPH, b Chicago, Ill, May 8, 53; m 77; c 2. PHAGOCYTE ANTIBACTERIAL ACTIVITY. *Educ:* Knox Col, BS, 76; Univ Wis-Madison, PhD(med microbiol), 80. *Prof Exp:* Fel immunol, Nat Jewish Hosp & Res Ctr, 80-83; Asst prof, 83-88, ASSOC PROF MICROBIOL & IMMUNOL, DEPT PATHOBIOL SCI, SCH VET MED, UNIV WIS, 88- *Concurrent Pos:* Asst prof, Dept Med Microbiol & Dept Vet Sci, Univ Wis, 83-88, assoc prof, 88- *Honors & Awards:* Presidential Award, Reticuloendothelial Soc, 82. *Mem:* Am Asn Immunologists; Am Asn Vet Immunologists; Am Soc Microbiol; Reticuloendothelial Soc; Asn Gnotobiotics. *Res:* Antibacterial resistance mechanisms; regulation of inflammation; genetically determined resistance to infection; neutrophil and macrophage oxidative response; lymphokines; veterinary infections; inflammatory bowel diseases; respiratory infections; lymphocyte regulation of phagocytes. *Mailing Add:* Dept Pathobiol Sci Sch Vet Med 2015 Linden Dr W Madison WI 53706

CZYZAK, STANLEY JOACHIM, b Cleveland, Ohio, Aug 21, 16; m 42; c 4. ASTROPHYSICS, ATOMIC & MOLECULAR PHYSICS. *Educ:* Fenn Col, BS(chem eng), 35, BS(civil eng), 36; John Carroll Univ, MS, 39; Univ Cincinnati, DSc(physics), 48. *Prof Exp:* Res metallurgist, Aluminum Co of Am, 36; chief metallurgist, Master Metals, Inc, 36-38; res engr, Una Welding, Inc, 38-40; assoc res physicist, Argonne Nat Lab, 48-49; res physicist, Battelle Mem Inst, 49-50; asst prof physics, Univ Detroit, 50-51; chief basic physics sect, Aeronaut Res Lab, Wright Air Develop Ctr, Wright-Patterson AFB, 51-54; from assoc prof to prof physics, Univ Detroit, 54-61; dir gen physics lab, Aerospace Res Labs, 61-66, PROF ASTRON, OHIO STATE UNIV, 66- *Concurrent Pos:* From second lieutenant to brigadier gen, USAF, 40-66. *Mem:* Am Astron Soc; Royal Astron Soc; Int Astron Union. *Res:* Atomic structure calculations; transition probabilities; collision cross-sections; gaseous nebulae and interstellar medium. *Mailing Add:* 800 N Maple Ave Fairborn OH 45324

CZYZEWSKI, HARRY, b Chicago, Ill, Feb 13, 18; m 43; c 3. APPLIED MECHANICS. *Educ:* Univ Ill, BS, 41, MS, 49. *Prof Exp:* Res engr, Caterpillar Tractor Co, Ill, 41-46; asst prof phys metall res, Univ Ill, 47-51; pres & tech dir, Met-Charlton, Inc, 46-83; PRES & SR CONSULT ENGR, ORE TECH SERV CTR, INC, 66- *Concurrent Pos:* Instr, Bradley Polytech, 42-46; mem, Dept Planning & Develop, Sci, Eng & New Technol Comt, 60-67; mem, State Bd Health, Radiation Adv Comt, 60-72; past pres, State Bd Eng Examiners, 61-73; treas, Consult Engrs Coun US, 62-64; gen chmn, Pac Northwest Metals & Mining Conf, 63; rep, Gov's Manpower Coord Comt, 68-71; mem, State Adv Coun Ore, Inst Technol, 71-84; rep from Nat Coun Eng Examiners to bd dirs, Engrs Coun Prof Develop, 72-74; mem adv panel D10-10, Trans Res Bd, Nat Coop Hwy Res Prog, 73-76; mem, Ad Hoc Comt on Eval of Ore Dept Environ Qual Lab Appl Res Div, 76; hon mem pres-elec, Am Coun Independent Labs, Inc, 77-78, pres, 79-81; hon chmn, Pac Northwest Metals & Minerals Conf, 81. *Honors & Awards:* Award, Am Inst Mining, Metall & Petrol Engrs, 42; Eng Excellence Awards, Am Consult Engrs Coun, 70 & 77. *Mem:* Fel Am Inst Chemists; Am Soc Metals; Am Inst Mining, Metall & Petrol Engrs; Sigma Xi. *Res:* Physics, wear and strength of metals; ferrous and non-ferrous physical metallurgy in machinery and industrial equipment field. *Mailing Add:* 2245 SW Canyon Rd Portland OR 97201-2499

D

DAAKE, RICHARD LYNN, b Charles City, Iowa, July 17, 46; m 70; c 4. CHEMISTRY. *Educ:* Houghton Col, NY, BS, 69; Iowa State Univ, Ames, PhD(inorg chem), 76. *Prof Exp:* Teaching asst chem, Iowa State Univ, 71-72, res asst, Ames Lab, 72-76; PROF CHEM, BARTLESVILLE WESLEYAN COL, 76-, CHMN, DIV SCI & MATH, 79- *Concurrent Pos:* Vis chemist, Ames Lab, Iowa State Univ, 79. *Mem:* Am Chem Soc; Am Sci Affil; Sigma Xi. *Res:* Synthesis and structural characterization of air sensitive materials, particularly compounds containing metal-metal bonds; reduced halides and halide hydrides of the group IVB metals, particularly zirconium. *Mailing Add:* Div Sci & Math Bartlesville Wesleyan Col 2201 Silver Lake Rd Bartlesville OK 74006

DAAMS, HERMAN, b Apeldoorn, Netherlands, May 8, 17; Can citizen; m 51; c 1. PHYSICS. *Educ:* Delft Univ Technol, Engr, 49. *Prof Exp:* Asst x-ray diffraction, Delft Univ Technol, 49-52; patent exam gen physics, Patent Off, The Hague, Holland, 52-54; sci officer instrumentation, Delft Univ Technol, 54-57; res officer frequency standards, physics div, Nat Res Coun Can, 57-82; RETIRED. *Mem:* Inst Elec & Electronics Engrs; Can Asn Physicists; Netherlands Phys Soc. *Res:* X-ray diffraction, small angle scattering; instrumentation; atomic frequency standards. *Mailing Add:* 34 Langevin Ave Ottawa ON K1M 1E9 Can

DAANE, ADRIAN HILL, b Stillwater, Okla, June 18, 19; m 44; c 3. INORGANIC CHEMISTRY. *Educ:* Univ Fla, BS, 41; Iowa State Col, PhD(chem), 50. *Prof Exp:* Asst phys chem, Off Sci Res & Develop & Manhattan Proj, Iowa State Col, 42-43, group leader, 43-46, from assoc chemist to sr chemist, AEC, 50-63; from asst prof to prof chem, Iowa State Univ, 50-72; prof chem and head dept, Kans State Univ, 63-72; dean, Col Arts & Sci, 72-79, prof chem, 72-84, dean grad study, 79-84, EMER PROF CHEM & EMER DEAN, UNIV MO-ROLLA, 84- *Concurrent Pos:* Adj prof, LaGrange Col, Ga, 84- *Mem:* Am Chem Soc; Am Inst Mining, Metall & Petrol Engrs; AAAS. *Res:* High temperature chemistry; chemistry of rare earth metals; vapor pressures of metals; chemical education. *Mailing Add:* 2311 Happy Valley Newman GA 30263

DAANE, ROBERT A, b Oostburg, Wis, June 18, 21; m 45; c 1. ENGINEERING MECHANICS, MECHANICAL ENGINEERING. *Educ:* Univ Wis, BSME, 43, PhD(eng mech), 59; Purdue Univ, MS, 49. *Prof Exp:* Engr eng div, Chrysler Corp, 43-46; instr eng mech, Purdue Univ, 46-50; assoc engr, Argonne Nat Lab, 50-54; sr engr, Nuclear Develop Assoc, 54-55; res engr, Beloit Corp, 55-59; assoc prof nuclear eng, Stanford Univ, 59-61; area res dir heat transfer, Beloit Corp, 61-67, dir res, 67-69, vpres res, 69-74; dir res, Tech Systs, Inc, 77-80, res consult, 80-84; CONSULT, WEB DRYING & CHILLING, 85- *Concurrent Pos:* Consult, Atomic Power Develop Assocs, 58-62 & Beloit Corp, 59-61. *Mem:* Am Soc Mech Engrs; Tech Asn Pulp & Paper Indust. *Res:* Stress analysis; nuclear reactor dynamics and safety; heat transfer; numerical analysis and computer simulation of engineering problems; paper drying; industrial research management; development of press drying process; development of web drying computer simulation program. *Mailing Add:* 4447 Indian Trail Green Bay WI 54313

DABBERDT, WALTER F, b New York, NY, Oct 12, 42; m 67; c 2. AIR POLLUTION. *Educ:* State Univ NY, BS, 64; Univ Wis, MS, 66, PhD(meteorol), 69. *Prof Exp:* Res asst meteorol, Univ Wis, Madison, 64-69; Nat Acad Sci-Nat Res Coun vis scientist, US Army Natick Labs, 69-70; assoc dir, atmospheric sci lab, 70-85, SCIENTIST & MGR SURFACE & SOUNDING SYSTS FACIL, NAT CTR ATMOSPHERIC RES, SRI INT, 85- *Concurrent Pos:* Mem comt highways & air qual, Transp Res Bd, Nat Res Coun, 75-81; res fel, Alexander von Humboldt Found, Fed Repub Ger; mem, Col Air Resourced Bd Modeling Adv Comt, 88-, Comt Meteorol Aspects Air Pollution, Am Meteorol Soc, 89-92. *Mem:* Am Geophys Union; Am Meteorol Soc. *Res:* Air pollution; momentum and heat transport in the planetary boundary layer; surface heat budget; urban diffusion modeling; optical propagation in the atmosphere; meteorological instrumentation. *Mailing Add:* Nat Ctr Atmos Res Atmos Technol Div Boulder CO 80307-3000

DABBS, DONALD HENRY, b Forestburg, Alta, Sept 12, 21; m 43; c 3. HORTICULTURE. *Educ:* Univ Alta, BSc, 50, MSc, 52. *Prof Exp:* Horticulturist, Exp Farm, Can Dept Agr, 52-61; from asst prof to prof hort, Univ Sask, 61-86, emer prof, 86-; RETIRED. *Mem:* Am Soc Hort Sci; Can Soc Hort Sci; Int Soc Hort Sci; Agr Inst Can. *Res:* Horticulture. *Mailing Add:* Comp 2061, RR 3 Brinkworthy Pl Ganges BC V0S 1E0

DABBS, JOHN WILSON THOMAS, b Nashville, Tenn, Dec 11, 21; m 45; c 3. NUCLEAR PHYSICS, CRYOGENICS. *Educ:* Univ Tenn, BS, 44, PhD(physics), 55. *Prof Exp:* Asst physics, Metall Lab, Univ Chicago, 44-45; jr physicist, Appl Physics Lab, Johns Hopkins Univ, 45; physicist, Oak Ridge Nat Lab, 46-84; dir, Technol Transfer, Martin Marietta Energy Systs, 84-86; PRES, ALPHA DIGITAL SYSTS INC, 87- *Concurrent Pos:* Fulbright lectr, Argentina, 61; vis scientist, Saclay, France, 67-68. *Mem:* Fel Am Phys Soc; Sigma Xi; Nat Soc Prof Engrs. *Res:* Nuclear polarization; neutron time-of-flight studies; nuclear fission; mechanical engineering. *Mailing Add:* 115 Claymore Lane Oak Ridge TN 37830

DABICH, DANICA, b Detroit, Mich, Aug 6, 30. BIOCHEMISTRY. *Educ:* Univ Mich, BSc, 52; Ohio State Univ, MSc, 55; Univ Ill, PhD(biochem), 60. *Prof Exp:* Anal chemist, Phillips Petrol Co, Okla, 52-53; asst biochem, Edsel B Ford Ford Inst Med Res, Mich, 55-56; res assoc, 61-63, from instr to asst prof, 63-70, ASSOC PROF BIOCHEM MED, WAYNE STATE UNIV, 70- *Concurrent Pos:* NIH fel, Univ Freiburg, 60-61. *Mem:* AAAS; Am Chem Soc; Sigma Xi; Am Soc Biochem & Molecular Biol. *Res:* Biochemistry of development. *Mailing Add:* Dept Biochem Wayne State Univ Col Med Detroit MI 48201

DABORA, ELI K, b Baghdad, Iraq, Sept 24, 28; US citizen; m 57; c 4. GAS DYNAMICS, COMBUSTION. *Educ:* Mass Inst Technol, SB, 51, SM, 52; Univ Mich, PhD(mech eng), 63. *Prof Exp:* Instr mech eng, Northeastern Univ, 52-53; anal engr, Power Generators Inc, 53-55; res engr, Sci Lab, Ford Motor Co, 55-56; res assoc heat transfer, Univ Mich, 57-61, assoc res engr combustion, 61-64, res engr detonations, 64-68; assoc prof, 68-76, PROF MECH & AEROSPACE ENG, UNIV CONN, 76- *Concurrent Pos:* Consult, Feltman Res Lab, Picatinny Arsenal, 70-, Gulf & Western Industs, Electro-Mech Combustion Eng, Avco-Lycoming & Comt Hydrogen Combustion, Nat Res Coun, 85-86; sabbatical, Max Planck Inst fur Stromungsforschung, Goetingen, Ger & Ecoles Nat Superieure Mécanique et d'Aerotechnique, Poitiers, France, 87-88; lectr, Univ Goetingen, Danemark Tech Univ, Univ Heidelberg, Technion, Catholique Univ de Louvain, ENSMA, Univ Orléans, Ben Gurion Univ, Univ Coinbra, 87-88; chmn, Eastern Sect, Combustion Inst, 89-91. *Mem:* Assoc fel Am Inst Aeronaut & Astronaut; Combustion Inst; Am Soc Mech Engrs; Sigma Xi; Soc Automotive Engrs. *Res:* Combustion; heat transfer as related to the design of regenerative heat exchangers; detonation phenomena; standing detonation wave and detonation-boundary interactions; detonations in heterogeneous systems; two-phase phenomena; pollution; ignition. *Mailing Add:* Dept Mech Eng U-139 Univ Conn Storrs CT 06268-3139

DABROWIAK, JAMES CHESTER, b South Bend, Ind, Apr 11, 42; m 66, 80; c 4. CHEMISTRY. *Educ:* Purdue Univ, BS, 65; Western Mich Univ, Kalamazoo, MS, 67, PhD(chem), 70. *Prof Exp:* Fel, Ohio State Univ, 70-72; asst prof, 72-78, ASSOC PROF CHEM, SYRACUSE UNIV, 78- *Mem:* Am Chem Soc; Biophys Soc. *Res:* Metal-drug interactions; drug-DNA interaction, synthesis of metal containing antibiotics. *Mailing Add:* Dept Chem-Bowne Hall Syracuse Univ Syracuse NY 13210

DACEY, GEORGE CLEMENT, b Chicago, Ill, Jan 23, 21; m 54; c 3. RESEARCH ADMINISTRATION, SOLID STATE PHYSICS. *Educ:* Univ Ill, BS, 42; Calif Inst Technol, PhD(physics), 51. *Prof Exp:* Res engr, Westinghouse Res Labs, 42-45; mem tech staff & res physicist, Bell Tel Labs, Inc, 51-58, asst dir solid state electronic res, 58-60, dir, 60-61; vpres res, Sandia Corp, 61-63; exec dir tel & power div, Bel Tel Labs, 63-68, vpres customer equip develop, 68-70, vpres transmission systs, 70-79, vpres opers systs & network planning, 79-81; pres, Sandia Nat Labs, 81-86; CONSULT, 86- *Mem:* Nat Acad Eng; fel Am Phys Soc; fel Inst Elec & Electronics Engrs. *Res:* Semiconductors and transistors; semiconductor devices; microwave electronics. *Mailing Add:* 3256 Montara Dr Bonita Springs FL 33923

DACEY, JOHN ROBERT, b Manchester, Eng, Apr 28, 14; Can citizen; wid; c 3. PHYSICAL CHEMISTRY. *Educ:* Dalhousie Univ, BSc, 36, MSc, 38; McGill Univ, PhD(phys chem), 40. *Hon Degrees:* DSc, Royal Mil Col, 80 & Queens Univ, Kingston Ont, 86. *Prof Exp:* Res asst, Nat Res Coun Can, 40; supt, Defense Res Chem Labs, 47-49, dean sci, 63-67, prof chem, 49-77, prin, 67-80, EMER PROF, ROYAL MIL COL, 80- *Mem:* Am Chem Soc; Chem Inst Can; Sigma Xi. *Res:* Chemical warfare, smoke and flame warfare; surface chemistry, adsorption of gases on charcoal; chemical kinetics, photochemistry of fluorine compounds. *Mailing Add:* Dept of Chem Royal Mil Col Kingston ON K7L 5L0 Can

DACEY, JOHN W H, b Kingston, Ont, Mar 26, 52; m 78; c 3. PHYSIOLOGICAL ECOLOGY. *Educ:* Dalhousie Univ, BSc, 73; Mich State Univ, PhD(zool), 79. *Prof Exp:* Scholar, 79-80, asst scientist, 80-84, assoc scientist, 84-88, TENURED, ASSOC SCIENTIST, WOODS HOLE OCEANOG INST, 88- *Honors & Awards:* Murray F Buell Award, Ecol Soc Am, 79. *Mem:* Ecol Soc Am; Bot Soc Am; AAAS; Am Geophys Union; Am Chem Soc; Am Soc Plant Physiologists. *Res:* Plant-air gas exchange; gas transport in plants; soil gases; atmospheric emissions; biogenic trace gas formation. *Mailing Add:* Woods Hole Oceanog Inst Woods Hole MA 02543

DACHEUX, RAMON F, II, b York, Pa, July 30, 47; m 69; c 2. ELECTROPHYSIOLOGY, NEUROBIOLOGY. *Educ:* Lycoming Col, BA, 69; State Univ NY, Buffalo, MA, 71, PhD(physiol), 78. *Prof Exp:* Res assoc visual, Neurosensory Lab, State Univ NY, Buffalo, 71-72, res assoc retinal, Neurobiol Div, 73-77; res fel, 77-80, instr, 80-81, ASSOC PROF ANAT, HARVARD MED SCH, 81- *Mem:* Asn Res Vis & Ophthalmol; AAAS; Soc Neurosci. *Res:* Intracellular microelectrode recording techniques and intracellular staining methods to investigate the physiology of individually identified neurons of a mammalian (rabbit) retina; examination of the connectivity and synptology of the identified cells to understand cell communication within the retina. *Mailing Add:* Dept Anat Harvard Med Sch 25 Shattuck St Boston MA 02115

DACK, SIMON, b New York, NY, April 19, 08; m 49; c 2. MEDICINE. *Educ:* City Col New York, BS, 28; NY Med Col, MD, 32. *Prof Exp:* Asst, Cardiographic Lab, 34-35, sr clin asst, Cardiac Clin, 35-41; adj physician cardiol, 45-58, chief, Cardiac Clin, 49-55, chief, Div Cardiol, 72-74, CHIEF PRENATAL CARDIAC CLIN, MT SINAI HOSP, 53-; EMER CLIN PROF, MT SINAI SCH MED, 77- *Concurrent Pos:* Res fel, Cardiographic Lab, Mt Sinai Hosp, 34-35; lectr, Columbia Univ, 35-; mem, Congenital Cardiac Group, Mt Sinai Hosp, 45-, assoc physician, 58-70, attend physician, 70-; from asst attend physician to assoc attend physician, Flower Fifth Ave Hosp, 55-66, attend physician, 66-; assoc vis physician, Metrop Hosp, 55-61, vis physician, 61-; assoc attend physician, Bird S Coler Mem Hosp, 55-; asst clin prof, NY Med Col, 55-59, assoc clin prof, 59-; ed in chief, Am J Cardiol, 58-82; assoc clin prof, Mt Sinai Sch Med, 66-67, assoc prof, 67-70, clin prof, 70-77, chief cardiol, Mt Sinai Hosp, 72-74; ed in chief, J Am Col Cardiol, 82- *Mem:* Am Heart Asn; Am Fedn Clin Res; fel Am Col Physicians; fel Am Col Chest Physicians; fel Am Col Cardiol (pres, 56-57). *Res:* Coronary artery disease. *Mailing Add:* 85 East End Ave New York NY 10028

DACRE, JACK CRAVEN, b Auckland, NZ; US citizen; m 50; c 3. TOXICOLOGY, BIOCHEMISTRY. *Educ:* Univ NZ, BSc, 43, MSc, 46; Univ London, PhD(biochem), 50, DSc, 83. *Prof Exp:* Biochemist microbiol, Dairy Res Inst, NZ, 51-54; sr biochemist, Toxicol Res Unit, NZ Med Res Coun, Univ Otago Med Sch, 54-61, toxicologist, Path Res Unit, 61-67, head toxicol, Toxicol Res Unit, 68-70; assoc prof environ med, Sch Med, Tulane Univ, 70-74; PROJ AREA MGR CHEM WEAPONS SYSTS & RES TOXICOLOGIST, HEALTH EFFECTS RES DIV, US ARMY BIOMED RES & DEVELOP LAB, FT DETRICK, MD, 74- *Concurrent Pos:* NZ Med Res Coun proj grant, Med Sch, Univ Otago, NZ, 61-67; vis res biochemist, St Mary's Hosp Med Sch, London, 61-62; vis prof, Sch Med, Univ Calif, San Francisco & Sch Pub Health, Univ Calif, Berkeley, 67-68; vis toxicologist, Nat Commun Dis Ctr, Atlanta, 68; mem, Expert Comt & consult, Expert Adv Panel Food Additives & Pesticide Residues, WHO, Geneva, 67-83; mem, Task Force & consult, US Army Med Res & Develop Command, 73. *Honors & Awards:* I C I Silver Medal & Prize, NZ Inst Chem for Imp Chem Indust, 73; Harvey W Wiley Medal & Cert, Nat Ctr Toxicol Res, 90. *Mem:* Fel NZ Inst Chem; The Chem Soc; Brit Biochem Soc; Soc Toxicol; fel Acad Toxicol Sci. *Res:* Toxicological, biochemical and metabolic fate studies of food additives, food colors, pesticides and environmental chemicals especially munitions products, chemical agents, fungal metabolites and trace metals. *Mailing Add:* 7209 E Sundown Ct Frederick MD 21702-2950

DA CUNHA, ANTONIO BRITO, b Sao Paulo, Brazil, June 17, 25; m 49; c 2. POPULATION GENETICS, CYTOGENETICS. *Educ:* Univ Sao Paulo, PhD, 48. *Prof Exp:* Asst, Univ Sao Paulo, 45-48, privat docent, 48-55, from asst prof to assoc prof, 55-73, actg head dept, 64-66 & 68-70, head dept, 70-73, vpres, 78-82, PROF BIOL, UNIV SAO PAULO, 73- *Concurrent Pos:* Rockefeller Found fels, 49 & 61; vis prof, Univ Tex, Austin, 71; dir, Inst Biol Sci, Univ Sao Paulo, 73-77, vpres, 78-82. *Mem:* Soc Study Evolution (vpres, 62 & 71); Am Soc Naturalists; Genetics Soc Am; Brazilian Soc Genetics; Brazilian Acad Sci; Acad Sci State São Paulo. *Res:* Evolutionary processes; differentiation; chromosomal effects of pathogenic agents; symbiosis; philosophy of science, biology. *Mailing Add:* Rua Itapicuru 777 Apt 21 Sao Paulo SP 05006 Brazil

D'ADAMO, AMEDEO FILIBERTO, JR, b Brooklyn, NY, Apr 15, 29; m 62; c 4. MOLECULAR BIOLOGY, NEUROBIOLOGY. *Educ:* Rutgers Univ, BSc, 50, PhD(chem), 55. *Prof Exp:* Res assoc, Am Cyanamid Co, 53-55, Columbia Univ, 55-56 & Ortho Res Found, 56-58; asst prof biochem, Albert Einstein Col Med, 58-61; asst prof, NJ Col Med, 61-63; assoc prof, Albert Einstein Col Med, 63-68; PROF BIOL, YORK COL, CITY UNIV NEW YORK, 68- *Concurrent Pos:* Adj prof chem, Manhattanville Col, 75. *Mem:* Am Inst Chem; Am Soc Neurochem; AAAS; Am Chem Soc; Sigma Xi. *Res:* Metabolic pathways; control of metabolism; lipogenesis and myelia synthesis; synaptosomal metabolism; effects of hormones on metabolism of the central nervous system. *Mailing Add:* Dept Biol York Col 150-14 Jamaica Ave Jamaica NY 11451

DADE, PHILIP EUGENE, b Hutchinson, Kans, Feb 2, 29; m 56; c 3. AGRONOMY. *Educ:* Kans State Col, BS, 51, MS, 52; State Col Wash, PhD(forage crop prod), 59. *Prof Exp:* Res agronomist, Agr Res Serv, USDA, 57-67; res agronomist, O M Scott & Sons, 67-74, mgr seed res & prod, 74-91; CONSULT, 85- *Mem:* Sigma Xi; Am Soc Agron; Crop Sci Soc Am. *Res:* Pasture and forage crop management; forage and turf grass seed production. *Mailing Add:* 3505 Begonia Dr Salem OR 97304

DADSWELL, MICHAEL JOHN, b Dafoe, Sask, Dec 7, 44; m 68; c 2. FISH ECOLOGY, ZOOGEOGRAPHY. *Educ:* Carleton Univ, Ottawa, BSc, 67, PhD(zoogeog), 73. *Prof Exp:* Res scientist, Huntsman Marine Lab, 73-78; RES SCIENTIST, DEPT FISHERIES & OCEANS, BIOL STA, 78- *Concurrent Pos:* Mem, Shortnose Sturgeon US Nat Marine Fisheries Serv Recovery Team, 76-85; adj prof, Acadia Univ, Wolfville, NS, 81. *Mem:* Am Fisheries Soc; Can Zool Soc; Soc Systs Zool; Crustacean Soc; Am Soc Limnol & Oceanog. *Res:* Biology of scallops (Placopecten); migration and ocean ecology of American shad (Alosa sapadissima); biology of shortnose sturgeon (Acipenser brevirostrum); zoogeography and taxonomy of crustaceans. *Mailing Add:* Dept Biol Sci Acadia Univ Wolfville NS B0P 1X0 Can

DAEHLER, MARK, b Cedar Rapids, Iowa, Mar 21, 34; m 87. OPTICAL SPECTROSCOPY, INOSPHERIC SCIENCE. *Educ:* Coe Col, BA, 55; Univ Wis, MA, 57, PhD(physics), 66. *Prof Exp:* Fel, Los Alamos Sci Lab, Univ Calif, 66-68; staff physicist, Inst Plasmaphysics, Munich, Ger, 68-71; RES PHYSICIST, US NAVAL RES LAB, 71- *Mem:* Optical Soc Am; Am Phys Soc; Inst Elec & Electronic Engrs; Am Geophys Union. *Res:* High resolution interferometric optical spectroscopy; solar and plasma spectroscopy; zodiacal light spectroscopy; infrared rocket photometry; cryogenically-cooled spectroscopic instrumentation; astrophysics of infrared-emitting objects; high frequency radio propagation; ionospheric assessment and forecasting. *Mailing Add:* Code 4181 US Naval Res Lab Washington DC 20375-5000

DAEHLER, MAX, JR, b Cedar Rapids, Iowa, Mar 21, 34; m; c 2. PHYSICS. *Educ:* Coe Col, BA, 55; Univ Wis, MA, 57, PhD, 62. *Prof Exp:* Sr engr res, Autonetics Div, NAm Aviation, Inc, 62-68; sr scientist, Exsar Corp, 68-71; sr res engr, Quantic Industs, 71-79; STAFF ENGR, LOCKHEED MISSILES & SPACE CO, 79- *Mem:* Sigma Xi. *Res:* Electron spin resonance; star-tracking devices; electro-optical sensors; infrared target simulation. *Mailing Add:* 2309 Monserat Ave Belmont CA 94002

DAEHNICK, WILFRIED W, b Berlin, Ger, Dec 30, 28; US citizen; m 60; c 3. EXPERIMENTAL NUCLEAR PHYSICS. *Educ:* Munich Tech Univ, BS, 51; Univ Hamburg, MA, 55, Wash Univ, PhD(physics), 58. *Prof Exp:* Res assoc nuclear physics, Wash Univ, 58-59; res assoc, Princeton Univ, 59-60, instr, 60-62; from asst prof to assoc prof, Univ Pittsburgh, 62-69, chmn, Comput Ctr Exec Comt, 77-80, dir, Sarah Mellon Scaife Nuclear Physics Lab, 78-79, chmn, Univ Senate Budget Policies Comt, 81-84, PROF PHYSICS, UNIV PITTSBURGH, 69-, CHMN EXEC COMT ACAD COMPUT, 84-, ASSOC PROVOST RES & CHAIR, UNIV RES COUN, 89- *Concurrent Pos:* Vis prof, Max Planck Inst, Heidelberg, 73-74, 85; mem, Adv Comt Physics, NSF, 76-79. *Mem:* Fel Am Phys Soc. *Res:* Nuclear structure and nuclear reactions induced by light ions; nuclear instrumentation; weak interactions; pion production. *Mailing Add:* Dept Physics Univ Pittsburgh Pittsburgh PA 15260

DAELLENBACH, CHARLES BYRON, b Minneapolis, Minn, Nov 16, 39; m 62; c 3. EXTRACTIVE & PHYSICAL METALLURGY, MINERALS & METALS CHARACTERIZATION. *Educ:* Univ Wis, BS, 62, MS, 63. *Prof Exp:* Extractive metallurgist, Twin Cities Res Ctr, US Bur Mines, US Dept Interior, 63-64, res extractive metallurgist, 64-66, res metallurgist, 66-67, res metallurgist, Iron Range Demonstration Plant, Keewatin, Minn, 67-69, res metallurgist, Twin Cities Metall Res Ctr, Minn, 69-75, RES SUPVR, CHARACTERIZATION DIV, ALBANY RES CTR, US BUR MINES, ORE, 75- *Concurrent Pos:* Prog chmn, Pac Northwest Met & Minerals Conf, 84, gen chmn, 87- *Mem:* Am Chem Soc; Am Inst Mining, Metall & Petrol Engrs. *Res:* Extractive metallurgy and mineral beneficiation; physical and foundry metallurgy; secondary resources recycling. *Mailing Add:* Albany Res Ctr US Bur Mines 1450 Queen Ave SW Albany OR 97321-2198

DAEMEN, JAAK J K, Belg citizen. ROCK MECHANICS, UNDERGROUND ROCK ENGINEERING. *Educ:* Univ Leuven, Belg, Civ Min Eng, 67; Univ Minn, PhD(geo-eng), 75. *Prof Exp:* Res eng, E I DuPont de Nemours & Co, 75-76; assoc prof eng, Univ Ariz, 82-90; PROF ENG, UNIV NEV, RENO, 90- *Mem:* Int Soc Rock Mech; Am Inst Mining Engrs; Am Soc Civil Engrs; Int Soc Soil Mech; Royal Flemish Eng Soc; Royal Belg Soc Engrs & Industrialists. *Res:* Engineering rock mechanics; geomechanical aspects of underground mining and construction; rock blasting with emphasis on blast vibrations; geological nuclear waste disposal. *Mailing Add:* Dept Mining Eng Univ Nev Reno NV 89557-0139

DAESCHNER, CHARLES WILLIAM, JR, b Houston, Tex, Dec 24, 20; m 48; c 3. PEDIATRICS. *Educ:* Rice Inst, BA, 42; Univ Tex, MD, 45. *Prof Exp:* Intern, Hermann Hosp, Houston, Tex, 45-46; jr resident, St Louis Children's Hosp, 48-49, asst resident, 49-50; sr resident, Boston Children's Hosp, 50-51; from instr to assoc prof, Med Col, Baylor Univ, 51-60; prof pediat & chmn dept, Univ Tex Med Br, 60-90, emer Asbhel Smith prof, 90; RETIRED. *Honors & Awards:* Abraham Jacobi Award, AMA & Am Acad Pediat, 82. *Mem:* Acad Pediat; Soc Pediat Res; Am Pediat Soc. *Res:* Metabolic and renal diseases in children. *Mailing Add:* Dept Pediat Univ Tex Med Br Galveston TX 77551

DAESSLE, CLAUDE, b France, Mar 29, 29; Can citizen; m 64; c 1. ORGANIC CHEMISTRY. *Educ:* Swiss Fed Inst Technol EthETH-Zurich, ChE, 54, DrScPhys, 57. *Prof Exp:* Res chemist, Monsanto Can Ltd, 57-60; OWNER, ORG MICROANAL, 60- *Res:* General organic synthesis, especially natural products. *Mailing Add:* 305 Morrison Ave Montreal PQ H3R 1K8 Can

DAFERMOS, CONSTANTINE M, b Athens, Greece, May 26, 41; m 64; c 2. APPLIED MATHEMATICS. *Educ:* Athens Tech Univ, dipl civil eng, 64; Johns Hopkins Univ, PhD(mech), 67. *Hon Degrees:* Dr, Univ Athens, 87. *Prof Exp:* Fel theoret mech, Johns Hopkins Univ, 67-68; asst prof, Cornell Univ, 68-71; assoc prof, 71-76, PROF APPL MATH, BROWN UNIV, 76- *Mem:* Soc Natural Philos (treas, 75-77, chmn, 77-78; Am Math Soc; Int Soc Interaction Math & Mech (secy, 84-86). *Res:* Stability theory in continuum mechanics; partial differential equations; dynamical systems; hyperbolic conservation laws. *Mailing Add:* Div Appl Math Brown Univ Providence RI 02912

DAFERMOS, STELLA, b Athens, Greece, Apr 14, 40; US citizen; m 64; c 2. TRANSPORTATION SCIENCE. *Educ:* Athens Nat Tech Univ, dipl, 64; Johns Hopkins Univ, PhD(oper res), 68. *Prof Exp:* Instr oper res, Cornell Univ, 68-69, asst prof, 69-71; asst prof appl math, 72-78, assoc prof, 78-82, PROF APPL MATH, BROWN UNIV, 82- *Mem:* Opers Res Soc Am. *Res:* Mathematical programming; equilibrium programming; networks and large scale systems applications to transportation and communication networks. *Mailing Add:* Div Appl Math & Eng Brown Univ Brown Sta Providence RI 02912

DAFFORN, GEOFFREY ALAN, b Kingman, Kans, Feb 4, 44; m 73; c 1. IMMUNOASSAYS. *Educ:* Harvard Univ, BA, 66; Univ Calif, Berkeley, PhD(chem), 70. *Prof Exp:* Fel biochem, Univ Calif, Berkeley, 70-73; asst prof, Univ Tex, Austin, 73-74; from asst prof to assoc prof chem, Bowling Green State Univ, 74-81; sr chemist, 82- 87, RES FEL, SYVA CO, PALO ALTO, CALIF, 87- *Concurrent Pos:* NIH fel, 70-72; adj prof pharmacol, Med Col Ohio, 76-81. *Mem:* Am Chem Soc; Am Asn Advan Sci. *Res:* Transition state analogs and the mechanism of acetylcholinesterase and other esterases; immunoassay systems; diagnostic microbiology. *Mailing Add:* Syva Co 2-218 900 Arastradero Rd Palo Alto CA 94304

DAFNY, NACHUM, b Tel-Aviv, Israel, Mar 5, 34; m 69; c 3. NEUROPHYSIOLOGY, NEUROENDOCRINOLOGY. *Educ:* Hebrew Univ, BSc, 62, MSc, 65, PhD(neurophysiol), 68. *Prof Exp:* Lectr neurosci, Hadassah Med Sch, Hebrew Univ, Israel, 68-69; fel, Calif Inst Technol & Univ Calif, Los Angeles, mem staff, Col Physicians & Surgeons, Columbia Univ, 70-72; asst prof neurosci, 72-74, assoc prof, 74-77, PROF NEUROBIOL & ANAT, UNIV TEX MED SCH, HOUSTON & GRAD SCH BIOMED SCI, UNIV HOUSTON, 77- *Concurrent Pos:* NSF fel, Calif Inst Technol, 69-70; Ford Found fel Brain Res & Anat, Univ Calif, Los Angeles, 70; NIH spec fel neurol, Col Physicians & Surgeons, Columbia Univ, 71-72; sr fac fel, Fogarty Int, 80-81. *Mem:* Am Soc Pharmacol & Exp Ther; Am Asn Anat; Soc Neurosci; Soc Exp Biol & Med; Am Physiol Soc; Int Soc Pain; Int Soc Neuroendocrin. *Res:* Neuroendocrinology; neuropharmacology; neurobiology; neuroimmunology. *Mailing Add:* Dept Neurobiol & Anat Univ Tex Med Ctr Houston TX 77030

DAGA, RAMAN LALL, b Purulia, India, Jan 5, 44; US citizen; m 72; c 2. METALLURGY, MECHANICS. *Educ:* Bihar Inst Technol, BS, 66; NMex Inst Mining & Technol, MS, 68; Case Western Reserve Univ, PhD(metall), 70. *Prof Exp:* Tech leader & res scientist refractory metal res & develop, Gen Elec Co, 70-78; eng specialist, Chem & Metal Div, GTE Prod Corp, Towanda, Pa, 79-82, eng mgr metals res, 82-84; PRES, METADYNE, INC, ELMIRA, NY, 84- *Mem:* Am Soc Metals; Am Soc Mfg Engrs. *Res:* Physical and mechanical metallurgy; structure property relationship; deformation mechanisms; applied metal processing; processing and properties of refractory metals. *Mailing Add:* 27 Oak Hill RD2 Sayre PA 18840

DAGANZO, CARLOS FRANCISCO, b Barcelona, Spain, May 15, 48; m 70; c 3. TRANSPORTATION ENGINEERING, OPERATIONS RESEARCH. *Educ:* Univ Madrid, dipl civil eng, 72; Univ Mich, MSE, 73, PhD(civil eng), 75. *Prof Exp:* Asst prof transp, Mass Inst Technol, 75-76; from asst prof to assoc prof transp eng, 76-88, PROF TRANSP ENG, UNIV CALIF, BERKELEY, 85- *Concurrent Pos:* Assoc ed, Transp Sci, 76-; prin investr, NSF res grants, 77-; consult, Gen Motors, 77-; assoc ed, Transp Res, Series B, Methodological, 78- *Mem:* Transp Res Bd; Inst Transp Engrs; Opers Res Soc Am. *Res:* Theoretical modelling of transportation processes and logistics systems, and development of the necessary analytical tools; mathematical programming, statistical inference, and applied probability and mathematical methods. *Mailing Add:* Dept Transp Eng Univ Calif Berkeley CA 94720

DAGDIGIAN, PAUL J, b Philadelphia, Pa, Oct 18, 45; m 67; c 2. PHYSICAL CHEMISTRY. *Educ:* Haverford Col, BA, 67; Univ Chicago, PhD(chem), 72. *Prof Exp:* Res assoc chem, Columbia Univ, 72-74; from asst prof to assoc prof, 74-81, PROF CHEM, JOHNS HOPKINS UNIV, 81- *Concurrent Pos:* Ed, Laser Chem, 85- *Mem:* Am Chem Soc; fel Am Phys Soc. *Res:* Molecular dynamics of gas-phase collision phenomena; chemical reactions and non-reactive energy transfer processes. *Mailing Add:* Dept Chem Johns Hopkins Univ Baltimore MD 21218

DAGE, RICHARD CYRUS, b Kalamazoo, Mich. BIOCHEMISTRY, ANIMAL PHYSIOLOGY. *Educ:* Ferris State Col, BS, 60; Marquette Univ, PhD(pharmacol), 67. *Prof Exp:* Pharmacologist cardiovasc pharmacol, Lakeside Labs, Div Colgate Palmolive Co, 67-75; sr res pharmacologist cardiovasc pharmacol, Merrell Nat Labs Div Richardson-Merrell Inc, 75-81, sr assoc scientist & head cardiovasc res, Merrell Dow Res Inst, Merrell Dow Pharmaceut Inc, 81-90, HEAD CARDIOVASC DIS RES, MARION MERRELL DOW RES INST, MARION MERRELL DOW INC, 90- *Concurrent Pos:* From clin instr to clin asst prof pharmacol, Med Col Wis, 67-75; adj prof, Dept Pharmacol & Cellular Biophys, Univ Cincinnatti Col Med. *Mem:* AAAS; Am Soc Pharmacol Exp Therapeutics; Am Soc Hypertension. *Res:* Cardiovascular pharmacology; mechanism of hypertension; antihypertensive drugs; cardiotonic drugs; mechanism of action of cardiotonic drugs; autonomic pharmacology; protectants for ischemic insurg, antiarrhythmic drugs. *Mailing Add:* Marion Merrell Dow Res Inst PO Box 156300 Cincinnati OH 45215-6300

DAGG, ANNE INNIS, b Toronto, Ont, Jan 25, 33; m 57; c 3. BEHAVIOR, SCIENCE EDUCATION. *Educ:* Univ Toronto, BA, 55, MA, 56; Univ Waterloo, PhD(mammal), 67. *Prof Exp:* Asst prof zool, Univ Guelph, 68-72; acad dir, 86-89, ACAD ADV, INDEPENDENT STUDIES, UNIV WATERLOO, 89- *Concurrent Pos:* Freelance writer/researcher, 72- *Res:* Behavior ecology of giraffe and camel; analysis of movements and gaits of birds and mammals; relationship of wildlife and people in Canada, especially in cities; sexual differences in behavior of mammals; women and science. *Mailing Add:* 81 Albert St Waterloo ON N2L 3S6 Can

DAGG, CHARLES PATRICK, reproductive biology, for more information see previous edition

DAGG, IAN RALPH, b Winnipeg, Man, Mar 20, 28; m 57; c 3. MICROWAVE PHYSICS. *Educ:* Univ Man, BSc, 49; Pa State Col, MS, 50; Univ Toronto, PhD(physics), 53. *Prof Exp:* Lectr physics, Univ Toronto, 53-54; Nat Res Coun Can fel, Oxford Univ, 54-55; asst res officer appl physics, nat Res Coun Can, 55-59; assoc prof, 59-69, PROF PHYSICS, UNIV WATERLOO, 69-, CHAIR, PHYSICS DEPT, 88- *Concurrent Pos:* Nat Res Coun Can res grants, 60-; Webster fel, Univ Queenslandi, Australia, 82-83. *Mem:* Can Asn Physicists. *Res:* Infrared and Raman spectroscopy; electric field induced and pressure induced absorption in infrared and microwave regions; acoustics, microphone standards and noise control. *Mailing Add:* Dept Physics Univ Waterloo Waterloo ON N2L 3G1 Can

DAGGERHART, JAMES ALVIN, JR, b Shelby, NC, Dec 6, 42; m 63; c 2. GAS DYNAMICS. *Educ:* NC State Univ, BS, 64, PhD(mech eng), 69. *Prof Exp:* Res assoc rarefied gas dynamics, 66-69, ASST PROF MECH & AEROSPACE ENG, NC STATE UNIV, 69- *Concurrent Pos:* NASA res grant, 69-71. *Mem:* Am Inst Aeronaut & Astronaut; Am Vacuum Soc. *Res:* Rarefied gas dynamics; cryoentrainment mechanism for vacuum pump application; high altitude simulation; performance of ionization type vacuum gauges in various environments. *Mailing Add:* 618 Birch Circle Cary NC 27511

DAGGS, RAY GILBERT, b McKees Rocks, Pa, June 26, 04; m 29; c 2. ENVIRONMENTAL PHYSIOLOGY, NUTRITION. *Educ:* Bucknell Univ, BS, 26; Univ Rochester, PhD(physiol), 30. *Hon Degrees:* DSc, Bucknell Univ, 55. *Prof Exp:* Fel physiol, Sch Med & Dent, Univ Rochester, 27-29, from asst to asst prof, 29-36; from asst prof to prof & head dept, Sch Med, Univ Vt, 36-41; maj to colonel, nutrit consult, Med Dept, US Navy Sanit Corps, 41-46; dir res environ med, Army Med Res Lab, Ft Knox, Ky, 46-56; exec secy & treas, Am Physiol Soc, 56-72; RETIRED. *Mem:* Am Physiol Soc. *Res:* Studies on lactation of rats, dogs and humans; effect of sulphydryl compounds on milk production. *Mailing Add:* Monroe Village Apt 5-119 One David Brainerd Dr Jamesburg NJ 08831

DAGGY, TOM, b Mooresville, Ind, July 16, 15; m 39; c 2. BIOLOGY. *Educ:* Earlham Col, BA, 37; Northwestern Univ, MS, 39, PhD(zool), 46. *Prof Exp:* Asst zool, Northwestern Univ, 37-42; instr biol, Maine Twp Jr Col, Ill, 39-42 & Univ Toledo, 42-43; tutor, Olivet Col, 43-47; from assoc prof to prof biol, Davidson Col, 47-80; RETIRED. *Concurrent Pos:* Biol illusr, 38-43. *Mem:* AAAS; Ecol Soc Am; Am Entom Soc; Coleopterists' Soc; Entom Soc Am. *Res:* Ecology and taxonomy of insects; immature stages of Coleoptera. *Mailing Add:* PO Box 628 Davidson NC 28036

DAGIRMANJIAN, ROSE, b Whitinsville, Mass, July 4, 30. PHARMACOLOGY, TOXICOLOGY. *Educ:* Clark Univ, AB, 52; Univ Rochester, MS, 54, PhD(pharmacol), 60. *Prof Exp:* Res assoc pharmacol, Univ Rochester, 52-60; asst prof pharmacol, Col Med, Ohio State Univ, 63-69; assoc prof, 69-75, PROF PHARMACOL, SCH MED, UNIV LOUISVILLE, 75- *Concurrent Pos:* Riker Int Fel, Eng, 60-62; Toxicol Study Sect, USPHS, NIH, 72-76; Miscellaneous External OTC Prep Panel, USPHS, FDA, 75-81; Chairperson, Tech Adv Comt on Poison Prev Packaging, Consumer Prod Safety Comn, 78-79, Toxicol Adv Bd, 79-85; Panel on Qual Criteria for Water Reuse, Nat Res Coun, Nat Acad Sci, 80-82, Comt Toxicol, Comn Life Scis, 84-87, Howard Hughes Doctoral Fel Rev, 89, 90; Behav & Neuroscis Study Sect, Subcomt I, HHS, NIH, 80-84, ad Hoc, 79-80 & 85; Develop Therapeut Contracts Rev Comt, HHS, NCI, 87-90; sci adv bd, Drinking Water Subcomt, Environ Protection Agency, 85-89; toxicol data bank peer rev comt, HHS, NLM, 77-85, sci rev panel, 85- *Mem:* Am Soc Pharmacol & Exp Therapeut; Soc Toxicol; Soc Exp Biol & Med; Soc Neurosci; Sigma Xi. *Res:* Pharmacology of tetrahydrocannabinols; effects of drugs and magnesium on the biogenic amines in the central nervous system; neuropharmacology of heavy metals. *Mailing Add:* Dept Pharmacol/Toxicol Univ Louisville Health Sci Ctr Louisville KY 40292

DAGLEY, STANLEY, biochemistry; deceased, see previous edition for last biography

D'AGOSTINO, MARIE A, MOLECULAR GENETICS, EUKARYOTIC CELLS. *Educ:* Thomas Jefferson Univ, PhD(biochem). *Prof Exp:* Dir genetic eng technol prog, Cedar Crest Col, 84-86; DIR GENETIC BIOTECHNOL PROG, GENETIC VENTURES, INC, 86- *Mem:* Am Soc Cell Biol; AAAS; Sigma Xi. *Mailing Add:* PO Box 637 Conshohocken PA 19428

D'AGOSTINO, PAUL A, b Toronto, Ont, Apr 11, 56; m 79; c 2. MASS SPECTROMETRY. *Educ:* McMaster Univ, BSc, 79, PhD(chem), 83. *Prof Exp:* DEFENSE SCIENTIST, CHEM SECT, DEFENSE RES ESTAB SUFFIELD, 83- *Mem:* Chem Inst Can; Am Soc Mass Spectrometry. *Res:* Analytical chemistry; gas chromatography-mass spectrometry with specialization in the analysis of chemical warfare agents, their degradation products and related compunds. *Mailing Add:* Defense Research Estab Box 4000 Medicine Hat AB T1A 8K6

D'AGOSTINO, RALPH B, b Somerville, Mass, Aug 16, 40; m 65; c 2. MATHEMATICAL STATISTICS. *Educ:* Boston Univ, AB, 62, AM, 64; Harvard Univ, PhD(statist), 68. *Prof Exp:* Lectr, 64-68, from asst prof to assoc prof, 68-77, assoc dean, Grad Sch, 76-78, PROF MATH & STATIST, BOSTON UNIV, 77-, PROF PUB HEALTH, 81-, CHMN, MATH DEPT, 86- *Concurrent Pos:* Statist consult, United Brands, 68-76, Walden Res Corp, 73-, Lakey Clin Found, 73-, US Food & Drug Admin, 75- & Boston City Hosp, 75-; ed, Emergency Health Serv Rev, 81-; lectr law, Boston Univ, 77-; comnr spec citation, Food & Drug Admin, 81. *Mem:* Am Soc Qual Control; Inst Math Statist; Am Statist Asn; Am Pub Health Asn; Sigma Xi; AAAS. *Res:* Experimental statistics. *Mailing Add:* Dept Math Boston Univ 111 Cummington St Boston MA 02215

DAGOSTINO, VINCENT F, b New York, NY, Nov 6, 26; m 35; c 5. BATTERY DEVELOPMENT, ELECTRODIALYSIS. *Educ:* Polytech Inst NY, BS, 51, MS, 57. *Prof Exp:* Chemist, Vaniderstein, 51-52, Airchem, 52-57, Am Cyanamid, 57-61; VPRES, RAI RES CORP, 61- *Mem:* Am Chem Soc; Am Electrochem Soc; N Am Membrane Soc; AAAS. *Res:* Development and research of new ion exchange membranes using radiation grafting techniques used in battery, biological, pollution and industrial manufacturing areas. *Mailing Add:* Four Greenland Dr Huntington Station NY 11746

DAGOTTO, ELBIO RUBEN, b Santa Fe, Arg. LATTICE GAUGE THEORIES, QUANTUM FIELD THEORY. *Educ:* Centro Atomico Bariloche, Arg, lic physics, 82, PhD(physics), 84. *Prof Exp:* Res assoc physics, Univ Ill, Urbana-Champaign, 85-88; RES ASSOC PHYSICS, INST THEORET PHYSICS, SANTA BARBARA, CALIF, 88- *Mem:* Am Phys Soc. *Res:* Particle physics - lattice gauge theories, QCD, QED, Higgs, etc; condensed matter physics - spin models, superconductivity at high temperature, numerical methods, etc. *Mailing Add:* Inst Theoret Physics Univ Calif Santa Barbara CA 93106

DAGUE, RICHARD R(AY), b Little Sioux, Iowa, Feb 20, 31; m 52; c 4. ENVIRONMENTAL ENGINEERING, CIVIL ENGINEERING. *Educ:* Iowa State Univ, BS, 59, MS, 60; Univ Kans, PhD(environ health eng), 67. *Prof Exp:* Surv party chief, Smith Eng Co, Iowa, 53-55; design engr, B H Backlund & Assoc, Nebr, 60-61; instr sanit eng, Iowa State Univ, 61-62; owner engr, Dague Eng Co, Iowa, 62-64; asst prof sanit eng, Kans State Univ, 66-67; from asst prof to assoc prof, Univ Iowa, 67-72; sr environ consult, Henningson, Durham & Richardson, 72-75; prof eng & chmn dept, Univ Iowa, 75-85; PROF & CHAIR, DEPT CIVIL ENG, IOWA STATE UNIV, AMES, 85- *Honors & Awards:* Philip F Morgan Medal, Water Pollution Control Fed, 83. *Mem:* Am Soc Civil Engrs; Water Pollution Control Fedn; Am Water Works Asn; Asn Environ Eng Prof. *Res:* Water quality control, especially biological wastewater treatment. *Mailing Add:* RR 4 Squaw Valley IA 50010

DAHILL, ROBERT T, JR, b Perth Amboy, NJ, Jan 15, 37; m 63; c 1. ORGANIC CHEMISTRY. *Educ:* Tufts Univ, BS, 59; Worcester Polytech Inst, MS, 61; Stevens Inst Technol, PhD(org chem), 64. *Prof Exp:* Group leader, 64-75, asst dir, 75-80, DIR PROD MGT, GIVAUDAN CORP, CLIFTON, 80- *Mem:* Am Chem Soc; Royal Soc Chem; fel Am Inst Chemists; Sigma Xi. *Res:* Synthetic organic chemistry, especially compound of interest to flavor and fragrance industry. *Mailing Add:* PO Box 157 Holmdel NJ 07733

DAHIYA, RAGHUNATH S, b Halalpur, Punjab, India, Oct 4, 31; m 55; c 2. FOOD MICROBIOLOGY. *Educ:* Govt Agr Col, Ludhiana, India, BSc, 51, MSc, 53; NC State Col, PhD(food sci), 62. *Prof Exp:* Agr inspector, Punjab State Agr Dept, India, 53-56; lectr dairy sci & dairy mgr, Govt Agr Col, Ludhiana, 56-58; res assoc food microbiol, NC State Univ, 62-65, asst prof, 65-71; CHIEF MICROBIOLOGIST, NC DEPT AGR, 71- *Mem:* Am Soc Microbiol; Inst Food Technologists; Asn Food & Drug Officials; Int Asn Milk, Food & Environ Sanitarians. *Res:* Microbiological surveillance of foods, drugs and cosmetics for safety; development of faster and more accurate methods of identification of food borne pathogens. *Mailing Add:* 1208 Glen Eden Dr Raleigh NC 27612

DAHIYA, RAJBIR SINGH, b Rattangarh, India, Dec 3, 40; m 66; c 2. MATHEMATICS. *Educ:* Birla Sci Col, Pilani, India, BSc 60, MSc, 62; Birla Inst CSci & Technol, PhD(math), 67. *Prof Exp:* From asst lectr to lectr math, Birla Inst Technol & Sci, Pilani, India, 62-68; from asst prof to assoc prof, 68-78, PROF MATH, IOWA STATE UNIV, 78- *Mem:* Am Math Soc; Soc Indust & Appl Math. *Res:* Integral transforms; special functions and delay differential equations; special functions and use of these various transforms and special funtion for computation techniques in analysis. *Mailing Add:* Dept Math Iowa State Univ Ames IA 50011

DAHL, A(NTHONY) ORVILLE, b Minneapolis, Minn, Apr 18, 10. BOTANY. *Educ:* Univ Minn, BS, 32, MS, 33, PhD(bot), 38, Univ Pa, MA, 71. *Prof Exp:* Asst bot, Univ Minn, 32-38; instr biol, Harvard Univ, 38-41, fac instr, 41-44, tutor, 38-44; from assoc prof to prof bot, Univ Minn, 44-67, chmn dept, 47-57; prof, 67-80, dir, 67-71, EMER PROF, MORRIS ARBORETUM, UNIV PA, 78- *Concurrent Pos:* Res assoc, Karolinska Inst, Sweden, 50-51 & Geol Undersgelse, Denmark, 57-58; res award, NSF, 57-64, NIH, 61-62 & NASA, 63-78; res investr, Univ Pa, 64-66; guest res prof, Bot Inst, Univ Stockholm, 71-72, 73-77, 79-85 & 87, Swedish Natural Sci Res Coun; res grant, Swedish Natural Sci Res Coun, 77. *Honors & Awards:* Res Award, Am Philos Soc, 80. *Mem:* AAAS; NY Acad Sci; Am Bot Soc; Am Soc Naturalists; Int Soc Cell Biol; fel Linnean Soc London. *Res:* Cytology; pollen morphology; atmospheric pollen; cytotaxonomy; electron microscopy. *Mailing Add:* Dept Biol Univ Pa Philadelphia PA 19104-6017

DAHL, ADRIAN HILMAN, b Mott, NDak, Dec 6, 19; div; c 2. BIOPHYSICS. *Educ:* St Olaf Col, BA, 41; Univ Rochester, PhD(biophys), 53. *Prof Exp:* Jr res physicist, Eastman Kodak Co, NY, 41-43, Tenn Eastman Corp, 43, sr physicist, 43-46; physicist AEC, 46-47; asst chief radiation, Instrument Br, Wash Div Prod, 47, chief, 47-49; chief instrumentation sect, Atomic Energy Proj, Sch Med, Univ Rochester, 50-59; prin scientist, Oak Ridge Inst Nuclear Studies, 59-61; prof phys & radiation biol & dir radiol health training prog, Colo State Univ, 61-69; chief astrogeophys, Space Sci Dept, Martin Marietta Corp, 69-71; chief environ sci br, US Energy Res & Develop Admin, 71-77; radiation safety training officer, Los Alamos Nat Labs, 78-82; RETIRED. *Concurrent Pos:* Consult radiation instrumentation, Armed Forces Spec Weapons Proj, 48-52, Fed Civil Defense, 56; Fulbright lectr health physics, Arg, 59-; Int AEC expert radioisotope technol, Indonesia, 61; coordr, Nat Environ Res Park, Idaho Nuclear Eng Lab, 75-77; radiol safety expert, Int Atomic Energy Agency, Bolivia, 82; consult, Yankee Atomic Elec Co, 86 & Benedict Nuclear Pharmaceut Inc, 87-; qualified inspector x-ray mach, State Colo, 89- *Mem:* Assoc Am Phys Soc; assoc Am Asn Physics Teachers; Inst Elec & Electronics Engrs; Health Phys Soc. *Res:* Nuclear radiation effects on photographic emulsions; industrial radiography; photographic dosimetry; circuit design of radiation detection; instruments and components; space physics; planetary geology and atmospheric science; seismology. *Mailing Add:* 2228 Scotch Pine Ct Loveland CO 80538

DAHL, ALAN RICHARD, b Ottawa, Ill, Feb 24, 44; m 82; c 5. TOXICOLOGIST. *Educ:* Princeton Univ, AB, 66; Univ Colo MS, 70; PhD(inorg chem), 71. *Prof Exp:* Res assoc chem, Univ Munich, Ger, 71-72, Tech Univ Berlin, 72-73 & Northwestern Univ, 73-74; SR SCIENTIST TOXICOL, LOVELACE INHALATION TOXICOL RES INST, 77- *Concurrent Pos:* Prin investr, 78-, mem toxicol study sect, NIH, 87-; adj assoc prof, Univ NMex Sch Pharm, 86- *Mem:* Am Chem Soc; Am Soc Pharm & Exp Therapeut; Soc Toxicol; AAAS; Int Soc Study Xenobiotics. *Res:* Enzymes of the respiratory tract involved in metabolism of foreign compounds; toxicokinetics of inhaled particles and vapors; contributions of respiratory tract metabolism to the toxic effects of inhaled compounds and in all other aspects of inhalation toxicology and biochemical toxicology. *Mailing Add:* Lovelace Inhalation Toxicol Res Inst PO Box 5890 Albuquerque NM 87185

DAHL, ALTON, b Clifton, Tex, Jan 8, 37; m 59; c 2. PHYSICAL CHEMISTRY. *Educ:* Tex Lutheran Col, BA, 59; Univ Mich, MS, 61, PhD(chem), 63. *Prof Exp:* Instr chem, Univ Mich, 63-64; chemist, Exp Sta, E I du Pont de Nemours & Co, Inc, 64-68, sr res chemist, 68-70, res mgr, 70-80, res planning consult, 80- 83, tech mgr, 83-85, TECH DIR, EXP STA, E I DU PONT DE NEMOURS & CO, INC, EUROPE, 85- *Mem:* Am Chem Soc. *Res:* Infrared and Raman spectroscopy; polymerization; polymer development. *Mailing Add:* 30 Ave de Miremont CH 1206 Geneva Switzerland

DAHL, ARTHUR LYON, b Palo Alto, Calif, Aug 13, 42; m 75; c 2. CONSERVATION, MARINE ECOLOGY. *Educ:* Stanford Univ, AB, 64; Univ Calif, Santa Barbara, PhD(biol), 69. *Prof Exp:* Vis res assoc bot, Smithsonian Inst, 69-70, assoc cur algae, 70-74; regional ecol adv, South Pac Comn, 74-82, coordr, Regional Environ Prog, 80-82; CONSULT ECOLOGIST, 82- *Concurrent Pos:* Res assoc, Smithsonian Inst, 74-82; chmn, Sci Comt Conserv & Environ, Pac Sci Asn, 75-; mem, Comn Ecol, Int Union Conserv Nature, 79-, Comn Nat Parks & Protected Areas, 81-, chmn, Island Task Force, 86- *Mem:* Int Soc Reef Studies; Int Asn Ecol; Int Phycol Soc; Phycol Soc Am; Pac Sci Asn. *Res:* Development and experimental ecology of benthic marine algae; tropical reef ecosystems and algal communities of coral reefs; conservation and environmental management of islands; environmental education materials. *Mailing Add:* Ocal Pac Unep PO Box 47074 Nairobi Kenya

DAHL, ARTHUR RICHARD, b Des Moines, Iowa, Nov 25, 30; m 52; c 4. GEOLOGY, CIVIL ENGINEERING. *Educ:* Iowa State Col, BS, 54, MS, 58; Iowa State Univ, PhD(geol, eng), 61. *Prof Exp:* Assoc civil eng, Iowa State Univ, 57-61; geologist, Humble Oil & Refining Co, 61-64; geol engr, Esso Prod Res Co, 64-66; sr geologist, 66-67, supv geologist, 67-70, DIST GEOLOGIST, EXXON CO, USA, 70-, MGR METALS EXPLOR, 80- *Concurrent Pos:* Off Naval Res grant, 58-60. *Mem:* Geol Soc Am; Am Asn Petrol Geologists; Am Inst Mining, Metall & Petrol Engrs; Am Geol Inst. *Res:* Geology and engineering of unconsolidated earth materials; exploration and exploitation of base metals and nuclear fuel. *Mailing Add:* 10812 W 28th Pl Lakewood CO 80215

DAHL, BILLIE EUGENE, b Cement, Okla, Oct 24, 29; m 53; c 3. RANGE MANAGEMENT, ANIMAL HUSBANDRY. *Educ:* Okla State Univ, BS, 51; Utah State Univ, MS, 53; Idaho Univ, PhD(range mgt, soils), 66. *Prof Exp:* Range conservationist, Bur Land Mgt, Wyo, 53, range mgr, 54-56; asst range conservationist, Agr Exp Sta, Colo State Univ, 56-66, assoc range conservationist, Eastern Colo Range Sta, 66-67, assoc prof range mgt 67; assoc prof, 67-73, PROF RANGE MGT, TEX TECH UNIV, 73- *Concurrent Pos:* Res asst, Range & Forestry Exp Sta, Univ Idaho, 62-64. *Mem:* Fel Soc Range Mgt; Sigma Xi. *Res:* Grazing management; systems of grazing; legume adaptation studies; range nutrition, improvement, ecology and economics; sand dune stabilization and management. *Mailing Add:* Dept Range & Wildlife Mgt Tex Tech Univ Lubbock TX 79409

DAHL, ELMER VERNON, b Colby, Kans, Apr 17, 21; m 44; c 5. PATHOLOGY, EPIDEMIOLOGY. *Educ:* Univ Southern Calif, BS, 43, MD, 52; Univ Minn, MS, 58. *Prof Exp:* Intern, Walter Reed Army Hosp, USAF, 52-53; chief path br, USAF Sch Aerospace Med, 59-61, comdr epidemiol lab, Lackland AFB, 61-68, 61-68, comdr epidemiol flight, Manila, Philippines, 68-69; res assoc prof, Univ Tex Med Br, 69-75, prof path, 75-76; prof path, Sch Med, Tex Tech Univ, 76-87; RETIRED. *Concurrent Pos:* Fel path, Sch Med, Duke Univ, 53-54, Mayo Found fel, 54-59; consult, USAF Sch Aerospace Med, 61-69, surgeon gen, 61-65, Aeromed Res Lab, Holloman AFB, 64-69. *Mem:* Fel Am Col Path; Int Acad Path; Sigma Xi. *Res:* Infectious diseases; experimental pathology. *Mailing Add:* 5760 Kingsfield Ave El Paso TX 79912

DAHL, HARRY MARTIN, b Yonkers, NY, June 23, 26; m 54; c 3. PETROLEUM GEOLOGY, EXPLORATION GEOLOGY. *Educ:* Hunter Col, AB, 50; Columbia Univ, AM, 51, PhD(geol), 54. *Prof Exp:* Geologist, US AEC, Colo, 54-55, area geologist, 55-56, res geologist, 56-57; res geologist, Texaco Inc, 57-68, proj leader, 69-72, sr res geologist, 73-77, res assoc, 77-81, sr res assoc, 81-84, res consult, 85-86; CONSULT GEOLOGIST, 87- *Concurrent Pos:* Chmn, Coord Comt, Am Petrol Inst Res Proj, 55 & 64-66. *Mem:* Mineral Soc Am; Geol Soc Am; Sigma Xi; Am Asn Petrol Geologists. *Res:* Exploration and recovery techniques in the fields of petroleum and other energy resources. *Mailing Add:* Consult Geologist PO Box 639 Burnet TX 78611

DAHL, HARVEY A, b Waldheim, Sask, Feb 19, 26; US citizen. PHYSICS. *Educ:* Stanford Univ, BS, 51, PhD(physics), 63. *Prof Exp:* From res asst to res assoc, High Energy Physics Lab, Stanford Univ, 56-64; ASST PROF PHYSICS, NAVAL POSTGRAD SCH, 64- *Mem:* Am Phys Soc; Am Asn Physics Teachers; Sigma Xi. *Res:* High energy nuclear physics, especially electromagnetically induced disintegrations of nuclei; electron scattering at high energies. *Mailing Add:* Dept Physics Naval Postgrad Sch 660 Fernwood St Monterey CA 93940

DAHL, JOHN ROBERT, b Jacksonville, Fla, Dec 13, 34; m 78; c 3. NUCLEAR MEDICINE RESEARCH, POSITRON EMISSION TOMOGRAPHY. *Educ:* Marist Col, Poughkeepsie, BA, 63; Polytech Inst, NY, MSc, 83, PhD(bioeng), 90. *Prof Exp:* Technician, Union Carbide Corp, 63-67; res asst, Biophys Lab, Sloan-Kettering Inst Caner Res, 67-71, res radiochemist, 73-86; sr prod chemist, Medi-Physics Inc, 71-73; TECH DIR, CYCLOTRON PET FACIL, NORTH SHORE UNIV HOSP, CORNELL UNIV MED COL, 86- *Concurrent Pos:* Vis investr, Biophys Lab, Sloan-Kettering Inst Cancer Res, 86-91; res assoc, Cornell Univ Med Col, 87-90, asst prof biophys, 90- *Mem:* Inst Elec & Electronic Engrs; Soc Nuclear Med; Int Nuclear Target Develop Soc; NY Acad Sci; Am Chem Soc; AAAS. *Res:* Radionuclide production method development, especially target systems for production of accelerator produced radionuclide for use in biomedical applications; positron emitting radionuclides and very short lived radioactive gases. *Mailing Add:* 24 Pearce Pl Great Neck NY 11021

DAHL, KLAUS JOACHIM, b Berlin, Ger, July 27, 36; Can citizen; m 63; c 2. ORGANIC CHEMISTRY, POLYMER CHEMISTRY. *Educ:* State Sch Eng, Berlin, Ing(HTL), 59; McGill Univ, BSc, 63, PhD(org chem), 66. *Prof Exp:* Fel org chem, Univ Munich, 66-67; CHIEF SCIENTIST, RAYCHEM CORP, 67- *Mem:* Am Chem Soc; Ger Chem Soc; Am Ceramics Soc. *Res:* Synthesis of high temperature resistant polymers; radiation chemistry of polymers; antioxidant and adhesives synthesis; preparation and characterization of ferroelectric ceramics; liquid crystals and polymers for optical displays. *Mailing Add:* Raychem Corp 300 Constitution Dr Menlo Park CA 94025

DAHL, LAWRENCE FREDERICK, b Evanston, Ill, June 2, 29; m 56; c 3. INORGANIC CHEMISTRY & ORGANOMETALLICS. *Educ:* Univ Louisville, BS, 51; Iowa State Univ, PhD(phys chem), 56. *Prof Exp:* AEC fel, Ames Lab, Iowa State Col, 57; from instr to prof, 57-78, R E RUNDEL PROF CHEM, UNIV WIS-MADISON, 78-, HILLDALE PROF CHEM, 91- *Concurrent Pos:* Alfred P Sloan fel, 63-65; Guggenheim fel, 69-70; Brotherton res prof, Univ Leeds, 83; Humboldt Sr US Scientist award, 86-87. *Honors & Awards:* Award in Inorganic Chem, Am Chem Soc-Tex Instruments, 74; R S Nyholm Mem lectr, Royal Soc Chem, 85; P C Reilly Mem Lectr, Univ Notre Dame, 87; H W Davis Lectr, Univ SC, 89; P Chini Mem Lectr, Societa Chimica Italiana, 89; J C Bailer Jr Lectr, Univ Ill, 90. *Mem:* Nat Acad Sci; fel NY Acad Sci; fel AAAS; Am Crystallog Asn; Am Chem Soc; Am Phys Soc; Royal Soc Chem. *Res:* Synthesis, structure and bonding of inorganic compounds and organometallic complexes, especially metal clusters; preparation and stereochemical characterization of new, unusual transition metal cluster systems as models for metal catalysis and chemisorption on transition metal surfaces; experimental and theoretical studies of influence of valence electrons upon molecular geometries of metal clusters. *Mailing Add:* Dept Chem Univ Wis 1101 Univ Ave Madison WI 53706

DAHL, NANCY ANN, b Colby, Kans, July 18, 32; m 52; c 2. NEUROBIOLOGY. *Educ:* Univ Kans, AB, 56, PhD(physiol), 62. *Prof Exp:* Asst instr physiol, Univ Kans, 62-63; res assoc neuropsychiat unit, Med Res Coun Labs, Carshalton, Eng, 63-64; asst prof human develop, 64-66, asst prof, 66-70, ASSOC PROF PHYSIOL, UNIV KANS, 70- *Mem:* Am Physiol Soc; Am Soc Cell Biol; Soc Neurosci; Am Soc Neurochem; Brit Biochem Soc; Sigma Xi. *Res:* Neurobiology; energy flow in the nervous system; mechanisms of ischemic nerve damage; photoreceptor membrane turnover. *Mailing Add:* Dept Physiol & Cell Biol Univ Kans 3038 Haworth Hall Lawrence KS 66045-2106

DAHL, NORMAN C(HRISTIAN), b Seattle, Wash, May 21, 18; m 43; c 2. MECHANICAL ENGINEERING, APPLIED MECHANICS. *Educ:* Univ Wash, BS, 41; Mass Inst Technol, ScD(mech eng), 52. *Hon Degrees:* ScD, Indian Inst Technol, Kanpur, 68. *Prof Exp:* Field serv consult, Off Sci Res & Develop, 44-45; from asst prof to prof mech eng, Mass Inst Technol, 48-68; dep rep India, Ford Found, 68-71, prog adv, 71-73; consult to pres & provost, Mass Inst Technol, 74-77; spec asst to pres, Am Acad Arts & Sci, 77-79; CONSULT, 79- *Concurrent Pos:* Fulbright lectr, Cambridge Univ, 50-51; prog leader, Kanpur Indo-Am Prog, Indian Inst Technol, 62-64. *Mem:* AAAS. *Res:* Stress analysis; educational development; technological development. *Mailing Add:* 302 W Side Rd Block Island RI 02807

DAHL, ORIN I, b Chicago, Ill, July 21, 35. EXPERIMENTAL HIGH ENERGY PHYSICS. *Educ:* Univ Ill, Urbana, BS, 57; Univ Calif, Berkeley, PhD(physics), 62. *Prof Exp:* PHYSICIST, LAWRENCE BERKELEY LAB, UNIV CALIF, 62- *Mem:* Am Phys Soc. *Res:* Study of strong and weak interactions of elementary particles using high speed digital computer techniques. *Mailing Add:* Lawrence Berkeley Lab One Cyclotron Rd Berkeley CA 94720

DAHL, PER FRIDTJOF, b Washington, DC, Aug 1, 32; m 66; c 2. SUPERCONDUCTING MAGNET TECHNOLOGY, HISTORY OF MODERN PHYSICS. *Educ:* Univ Wis, BS, 56, MS, 57, PhD(nuclear physics), 60. *Prof Exp:* Res assoc nuclear physics, Univ Wis, 60; Ford Found res grant, Inst Theoret Physics, Univ Copenhagen, 60-62; proj scientist, Air Force Weapons Lab, 62-63; asst physicist, 63-65, assoc physcist, 65-71, PHYSICIST, BROOKHAVEN NAT LAB, 71- *Mem:* AAAS; Am Phys Soc. *Res:* Applied superconductivity; high energy accelerator design; history of physics. *Mailing Add:* Accelerator Develop Dept Brookhaven Nat Lab Bldg 902A Upton NY 11973

DAHL, PETER STEFFEN, b Port-of-Spain, Trinidad, Nov 17, 48; US citizen; m 73. METAMORPHIC PETROLOGY, GEOCHEMISTRY. *Educ:* Ind Univ, BA, 69, MA & PhD(geol), 77. *Prof Exp:* Anal chemist, O A Labs, Inc, 70-71; electronics technician, US Navy, 71-73; electroplating engr, Delco Electronics, 73-74; asst prof, 77-82, ASSOC PROF GEOCHEM, KENT STATE UNIV, 82- *Concurrent Pos:* Grants, Cottrell Res Corp & Kent State Univ Res Coun, 78-79; NSF grant, 80-82. *Mem:* Mineral Soc Am; Mineral Asn Can; Am Geophys Union; Sigma Xi. *Res:* Mineral-pair geothermometry of metamorphic rocks; oxygen isotope geochemistry of metamorphic iron-formations; Precambrian geology of the western United States; petrology of arbicular granites from South Victoria Land, Antarctica. *Mailing Add:* Dept Geol Kent State Univ 337 McGilvry Hall Kent OH 44242

DAHL, RANDY LYNN, b Devils Lake, NDak, Aug 30, 57. FIBER OPTICS, ANALYSIS. *Educ:* NDak State Univ, BS, 81, MS, 83. *Prof Exp:* Engr, Am Crystal Sugar Res Ctr, 82-83; engr, 83-87, SR ENGR, MOTOROLA, INC, 87- *Concurrent Pos:* Consult, Contental Industs, 84-, Indust Fiber Optics, 87- & Directed Energy, Inc, 90-; writer, Sci Laser, 90. *Mem:* Soc Photo-Optical Instrumentation Engrs. *Res:* Optical polarization; transmitter receiver design for optical radars and fiber optics; natural phenominology; opto-electronics; granted 6 patents; analog. *Mailing Add:* PO Box 3576 Scottsdale AZ 85271

DAHL, ROY DENNIS, b New Britain, Conn, July 5, 39; div; c 2. NEUROBIOLOGY, BIOPHYSICS. *Educ:* Cent Conn State Col, BS, 64; Univ Del, PhD(physiol psychol), 73. *Prof Exp:* Instr biol, Temple Univ, 73-74; res scientist neurobiol, Inst Neurobiol, Nuclear Res Ctr, Julich, WGer, 74-76; asst prof physiol, Sch Med, State Univ NY Buffalo, 76-78; asst prof biol, C W Post Ctr, Greenvale, NY, 78-81; res assoc, Rockefeller Univ, NY, 81-87; RES ASSOC, DEPT PHARMACOL, US UNIV HEALTH SCI, BETHESDA, 87- *Honors & Awards:* Am Philos Soc Award, 73; Nat Acad Sci Award, 74; Individual NRSA Award, Nat Eye Inst, 76-78. *Mem:* Soc Neurosci; Sigma Xi. *Res:* Neurophysiology of learning, especially long-term plastic changes in dentale gyrus. *Mailing Add:* Dept Pharmacol US Univ Health Sci 4301 Jones Bridge Rd Bethesda MD 20814-4799

DAHL, ROY EDWARD, nuclear fuels, nuclear fuel cycle, for more information see previous edition

DAHL, WILLIAM EDWARD, b Pittsburgh, Pa, Oct 7, 36; m 67; c 2. ANALYTICAL CHEMISTRY. *Educ:* Iowa State Univ, BS, 58; Purdue Univ, PhD(anal chem), 65. *Prof Exp:* Sr res chemist, Monsanto Co, 65-71, res specialist, 71-76, sr res specialist, 76-79, sr res group leader, 79-88, mgr, Corp Res Labs, Phys Sci Ctr, 88-90, BIOPROCESS ANALYTICS & QUAL ASSURANCE MGR, MONSANTO CORP RES, MONSANTO CO, 90- *Mem:* Am Chem Soc. *Res:* Method developoment & management in analytical chemistry, physical properties, thermal hazards, electron microscopy, nuclear magnetic resonance, mass spectrometry, computerization & automation. *Mailing Add:* Monsanto Co 700 Chesterfield Village Pkwy St Louis MO 63198

DAHLBERG, ALBERT A, b Chicago, Ill, Nov 20, 08; m 34; c 3. DENTAL ANTHROPOLOGY. *Educ:* Loyola Univ, Ill, BS & DDS, 32. *Prof Exp:* Resident & instr dent surg, Univ Chicago, 32-36; attend dent surgeon, Chicago Mem Hosp, 37-53; assoc prof anthrop, 49-67, actg dir clin, 67-69, res assoc phys anthrop, 49-80, prof & res assoc, Dept Anthrop, 67-80, EMER ASSOC RES PROF DENT & EMER ASSOC RES PROF ANTHROP & EMER PROF COMT EVOLUTIONARY BIOL, UNIV CHICAGO, 80- *Concurrent Pos:* Res assoc, Chicago Natural Hist Mus. *Mem:* Fel AAAS; Am Dent Asn; Am Asn Phys Anthrop; Am Anthrop Asn; fel Am Col Dent. *Res:* Genetics and morphology of the human dentition. *Mailing Add:* Michigan Ave Dent Assoc 122 S Michigan Ave Ste 1212 Chicago IL 60603

DAHLBERG, ALBERT EDWARD, b Chicago, Ill, Sept 19, 38; m 63; c 3. BIOCHEMISTRY. *Educ:* Haverford Col, BS, 60; Univ Chicago, MD, 65, PhD, 68. *Prof Exp:* Res assoc biochem, Nat Cancer Inst, 67-70; from asst prof to assoc prof, 72-82, PROF, DIV BIOL & MED, BROWN UNIV, 82-, chmn, Sect Biochem, 84-85, 86-87. *Concurrent Pos:* Am Cancer Soc fel, Dept Molecular Biol, Aarhus Univ, 70-72; vpres, Mora Pharmaceut, Inc, 84- *Mem:* Am Soc Biochem & Molecular Biol. *Res:* Ribosome structure and function. *Mailing Add:* Div Biol & Med Brown Univ Providence RI 02912

DAHLBERG, DUANE ARLEN, b Parshall, NDak, July 15, 31; m 57; c 3. HEALTH EFFECTS OF ELECTROMAGNETIC ENERGY, ENVIRONMENTAL PROBLEM SOLVING & SUSTAINABLE DEVELOPMENT. *Educ:* Mich Technol Univ, BS, 53, MS, 54; Luther Theol Sem, BD, 60; Mont State Univ, PhD(physics), 67. *Prof Exp:* Res asst neutron physics, Argonne Nat Lab, 53-54; instr, 60-64, asst prof, 67-68, dir, Tri Col, Environ Studies Ctr, 78-81, ASSOC PROF PHYSICS, CONCORDIA COL, MOORHEAD, MINN, 68- *Concurrent Pos:* Dir, environment orgns; chmn, Minn Pollution Control Agency; organizer & dir, Electromagnetics Res Found. *Mem:* Am Asn Physics Teachers; Am Meteorol Soc. *Res:* Neutron cross sections; optical excitation cross sections; solar energy applications; health effects of electromagnetic fields and currents; sustainable systems. *Mailing Add:* Dept Physics Concordia Col Moorhead MN 56562

DAHLBERG, EARL DAN, b Ft Worth, Tex, Aug 17, 48. PHYSICS. *Educ:* Univ Tex, Arlington, BS, 70, MA, 72; Univ Calif, Los Angeles, MS, 74, PhD(physics), 78. *Prof Exp:* Res assoc, Univ Minn, 78-80, asst prof physics, 80-85, assoc prof, 85-90, PROF PHYSICS, UNIV MINN, 90- *Concurrent Pos:* Res fel, Alfred P Sloan Found, 81-85. *Mem:* Am Phys Soc; AAAS; Inst Elec & Electronic Engrs; Am Asn Physics Teachers. *Res:* Magnetic films, magnetic multilayers, force microscopy, and superconductivity. *Mailing Add:* Sch Physics & Astron Univ Minn Minneapolis MN 55455

DAHLBERG, JAMES ERIC, b Chicago, Ill, May 30, 40; m 78; c 2. BIOCHEMISTRY, MOLECULAR BIOLOGY. *Educ:* Haverford Col, BA, 62; Univ Chicago, PhD(biochem), 66. *Prof Exp:* Res assoc, Med Res Coun Lab Molecular Biol, Cambridge, Eng, 66-68; res assoc, Lab Biophys, Univ Geneva, 68-69; from asst prof to assoc prof, 69-74, PROF PHYSIOL CHEM, MED SCH, UNIV WIS-MADISON, 74- *Concurrent Pos:* US Air Force Off Sci Res fel, 66-67; Am Cancer Soc fel, 67-69. *Honors & Awards:* Eli Lilly Award Biol Chem, Am Chem Soc, 74. *Mem:* AAAS; Am Micros Soc; Am Soc Biol Chemists. *Res:* Structure, function and synthesis of small nuclear RNAs; processing of RNA in eukaryotes; formation and function of DNA triplexes. *Mailing Add:* Dept Physiol Chem 589 Med Sci Bldg Univ Wis Madison WI 53706

DAHLBERG, MICHAEL LEE, b Miami, Fla, Jan 19, 39; m 87; c 2. FISHERIES, BIOMETRICS. *Educ:* Ore State Univ, BS, 62, MS, 63; Univ Wash, PhD(fisheries), 68. *Prof Exp:* Res assoc fisheries, Fisheries Res Inst, Univ Wash, 63-68; asst prof, Va Polytech Inst, 68-70; supvry fisheries res biologist, 70-77, SUPVRY MATH STATISTICIAN, NAT MARINE FISHERIES SERV, 77- *Concurrent Pos:* Consult, US Bur Sport Fisheries & Wildlife, 68-69 & Sport Fishing Inst, Washington, DC, 68-70; adv, Int N Pac Fisheries Comn, N Pac Fishery Mgt Coun, US Dept State & Alaska Dept Fish & Game, Univ Alaska. *Mem:* AAAS; Am Inst Fishery Res Biol; Am Fisheries Soc; Am Statist Asn; Sigma Xi. *Res:* Population dynamics of anadromous and marine fish in the North Pacific Ocean, including the solving of fishery research and management problems through information analysis and systems simulation. *Mailing Add:* Auke Bay Fisheries Lab Nat Marine Fisheries Serv Auke Bay AK 99821-0155

DAHLBERG, RICHARD CRAIG, b Astoria, Ore, July 23, 29; m 56; c 3. ENGINEERING MANAGEMENT, NUCLEAR ENGINEERING. *Educ:* Univ Ore, BS, 51; Univ Mich, MS, 54; Rensselaer Polytech Inst, PhD, 64. *Prof Exp:* Prod supvr, Hanford Works, Gen Elec Co, 51-53, mgr reactor physics, Knolls Atomic Power Lab, 54-64, reactor physicist, Gen Atomic Div, Gen Dynamics Co, 64-67, mgr HT&R reactor physics, 67-69; mgr nuclear anal & reactor physics dept, Gulf Energy & Environ Systs, Gulf Oil Corp, 72-; dir, Fuel Eng Div, 75-80, dir, Gen Eng Div, 80-84, DIR, ADV DEF PROJ, GEN ATOMIC CO, 84- *Mem:* Fel Am Nuclear Soc; Sigma Xi; AAAS. *Res:* Reactor physics; nuclear engineering; reactor design. *Mailing Add:* 7733 Esterel Dr La Jolla CA 92037

DAHLE, LELAND KENNETH, b Marietta, Minn, June 19, 26; m 58; c 2. CEREAL CHEMISTRY. *Educ:* St Olaf, BA, 50; Purdue Univ, MS, 53; Univ Minn, PhD(physiol chem), 61. *Prof Exp:* Instr chem, Ausburg Col, 52-56; res cereal chemist, Flour Mills, Peavey Co, 61-69; sr res scientist, 69-72, DIV HEAD CEREAL SCI, CAMPBELL SOUP CO, 72- *Mem:* Am Chem Soc; Am Asn Cereal Chemists; Inst Food Technologists. *Res:* Lipid oxidation; autoxidation phenomena; thiobarbituric acid reaction assay; oxidative stability of pigments; functional effects of flour protein sulfhydryl groups on dough rheology and behavior; functional properties of protein and starch in bread systems; interactions of protein and starch; food texture. *Mailing Add:* 215 Garfield Ave Cherry Hill NJ 08034-1519

DAHLEN, FRANCIS ANTHONY, JR, b American Falls, Idaho, Dec 5, 42; m 67; c 1. GEOPHYSICS. *Educ:* Calif Inst Technol, BS, 64; Univ Calif, San Diego, MS, 67, PhD(geophys), 69. *Prof Exp:* NSF fel geod & geophys, Cambridge Univ, 70-71; from asst prof to assoc prof, 71-80, PROF GEOPHYS, PRINCETON UNIV, 80- *Concurrent Pos:* Alford P Sloan Found fel, 71-73. *Mem:* Am Geophys Union; Seismol Soc Am; Royal Astron Soc; Soc Indust Appl Math. *Res:* Theoretical seismology; free oscillations of the earth; rotation of the earth; lateral heterogeneity of the earth; seismic source mechanisms; mechanics of mountain building. *Mailing Add:* Dept Geol Sci Princeton Univ Princeton NJ 08544

DAHLEN, R(OLF) J(OHN), mechanical engineering; deceased, see previous edition for last biography

DAHLEN, ROGER W, b Iola, Wis, Sept 7, 35; m 86; c 4. MEDICAL PHYSIOLOGY, SCIENCE ADMINISTRATION. *Educ:* Luther Col, Iowa, BA, 56; Univ Iowa, MS, 58, PhD(physiol), 60. *Prof Exp:* Instr physiol, Sch Med, Boston Univ, 60-62; asst prof, NJ Col Med & Dent, 62-68; health scientist adminr, Training Grants & Awards Br, Nat Heart Inst, 68-71, CHIEF, DIV BIOMED INFO SUPPORT, NAT LIBR MED, NIH, 71- *Concurrent Pos:* Mem bd trustees, Upsala Col, 64-68. *Res:* Cardiac electrophysiology; effects of hypothermia. *Mailing Add:* NIH 10958 Trotting Ridge Columbia MD 21044

DAHLER, JOHN S, b Wichita, Kans, May 7, 30; m 54. PHYSICAL CHEMISTRY. *Educ:* Univ Wichita, BS, 51, MS, 52; Univ Wis, PhD(theoret chem), 55. *Prof Exp:* NSF fel, Univ Amsterdam, 55-56; task scientist, Aeronaut Res Lab, Wright Air Develop Ctr, 56-58; from asst prof to assoc prof chem eng, 59-63, PROF CHEM ENG & CHEM, UNIV MINN, MINNEAPOLIS, 63- *Concurrent Pos:* Sloan fel, 64-; NSF sr fel, 65-66; vis prof, Chem Lab III, H C Orsted Inst, Univ Copenhagen, 73. *Mem:* Am Phys Soc; Soc Natural Philos; Am Fedn Scientists. *Res:* Theoretical physical chemistry; equilibrium statistical mechanics; theory of fluid transport phenomena; electronic and atomic collision processes. *Mailing Add:* Depts Chem Eng & Chem Univ Minn Minneapolis MN 55455

DAHLGARD, MURIEL GENEVIEVE, b W Haven, Conn, July 5, 20. ORGANIC CHEMISTRY. *Educ:* Univ Conn, BS, 46, MS, 49; Univ Kans, PhD(org chem), 56. *Prof Exp:* Asst, Univ Conn, 46-48; asst res chemist, Nopco Chem Co, NJ, 48-53; asst, Univ Kans, 53-56; asst res prof, Cancer Res Lab, Univ Fla, 56-61; from asst to assoc prof, Randolph-Macon Woman's Col, 61-82, chmn dept, 75-84, prof chem, 82-87; RETIRED. *Mem:* AAAS; Am Chem Soc. *Res:* Aliphatic diamines; diaryl ethrers; carbohydrates; carcinogenic derivatives of fluorene and biphenyl; antimetabolites; plant lectins. *Mailing Add:* 1216 Krise Circle Lynchburg VA 24503

DAHLGREN, GEORGE, b Chicago, Ill, Apr 12, 29; m 51; c 3. PHYSICAL CHEMISTRY. *Educ:* Ill Wesleyan Univ, BS, 51; Univ Wyo, MS, 56, PhD(chem), 58. *Prof Exp:* Res assoc chem, Cornell Univ, 58-59; from asst prof to assoc prof, Univ Alaska, 59-66, head dept, 64-66; prof, Univ Cincinnati, 66-75, asst head dept, 66-71, head dept, 71-75; dean, Col Arts & Sci, Univ Mo-Kansas City, 75-78; VPRES, FRANKLIN INST, 78-; AT DEPT CHEM, IND UNIV NORTHWEST. *Mem:* Am Chem Soc; fel AAAS. *Res:* Kinetics and thermodynamics. *Mailing Add:* Off Provost Oakland Univ Rochester MI 48063

DAHLGREN, RICHARD MARC, b Riverside, Calif, Sept 20, 52; m 82; c 2. INORGANIC & PHYSICAL CHEMISTRY. *Educ:* Univ Calif, Riverside, BS, 73; Univ Calif, Los Angeles, PhD(chem), 78. *Prof Exp:* staff res chemist, Miami Valley Labs, 78-80, res scientist, Beauty Care Prod Develop Div, 80-84, SECT MGR, BAR SOAP & HOUSEHOLD CLEANING PROD DIV, PROCTER & GAMBLE CO, 84- *Mem:* Am Chem Soc. *Res:* Development of cosmetic and personal cleansing products; studies of the interaction of soaps and cosmetics with the skin; physical chemistry of the excited state, particular emphasis in the areas of photocatalysis and energy-conversion processes. *Mailing Add:* 7613 Windy Knoll Dr Cincinnati OH 45241-1253

DAHLGREN, ROBERT BERNARD, b Walnut Grove, Minn, Jan 27, 29; m 51; c 6. WILDLIFE RESEARCH. *Educ:* SDak State Univ, BS, 50, PhD, 72; Utah State Univ, MS, 55. *Prof Exp:* Small game biologist, SDak Dept of Game, Fish & Parks, 52-61, leader small game & furbearer res proj, 61-65, asst chief, Game Div, 65-67; asst leader, SDak Coop Wildlife Res Unit, US Dept Interior, 67-73, leader, Iowa Coop Wildlife Res Unit, 73-85, leader, Iowa Coop Fish & Wildlife Res Unit, US Dept Interior, 85-87; biologist, 87-89, ASST REGIONAL REFUGE BIOLOGIST, UPPER MISS RIVER REFUGE COMPLEX, 89- *Mem:* Wildlife Soc. *Res:* Population ecology and dynamics of wildlife populations; pesticide effects on game birds. *Mailing Add:* Off Refuge Biol PO Box 2484 La Crosse WI 54602

DAHLGREN, ROBERT R, b Lincoln, Nebr, May 17, 35; m 57; c 2. VETERINARY PATHOLOGY. *Educ:* Okla State Univ, DVM, 63, MS, 66. *Prof Exp:* Instr path, Okla State Univ, 63-66; pathologist, Ralston Purina Co, 66-68; assoc prof path, Univ Nebr, 68-69; PATHOLOGIST, RALSTON PURINA CO, 69- *Concurrent Pos:* Consult comp path, Sch Med, Washington Univ, 67- *Mem:* NY Acad Sci; Am Vet Med Asn. *Res:* Comparative neuropathology and immunology. *Mailing Add:* Wil Res Labs Inc Ashland OH 44805

DAHLIN, DAVID CARL, b Beresford, SDak, Sept 3, 17; m 41; c 3. PATHOLOGY. *Educ:* Univ SDak, BS, 38; Univ Chicago, MD, 40; Univ Minn, MS, 48. *Prof Exp:* Intern & resident path, Ancker Hosp, St Paul, 41 & 42; asst prof, 53-61, PROF PATH, MAYO MED SCH, UNIV MINN, 61-, CHMN DEPT SURG PATH, 70-; CONSULT SURG PATH, MAYO CLIN, 48- *Mem:* Am Soc Clin Path; Am Asn Pathologists & Bacteriologists; AMA; Am Cancer Soc. *Res:* Bone tumors. *Mailing Add:* Mayo Clinic Rochester MN 55905

DAHLKE, WALTER EMIL, b Berlin, Ger, Aug 24, 10; m 40. PHYSICS, ELECTRICAL ENGINEERING. *Educ:* Univ Berlin, Dr phil(physics), 36; Univ Jena, Dr habil(physics), 39. *Prof Exp:* Head microwave lab, Ger Aviation Res Estab, 40-45; head appl res group & adv develop, tube & semiconductor div & inst res, Telefunken AG, Ger, 49-65; NSF vis prof, 64-65, PROF ELEC ENG, LEHIGH UNIV, 65- *Concurrent Pos:* Lectr, Univ Heidelberg, 55-56 & Stuttgart Tech, 56-59; lectr, Karlsruhe Tech, 60-65, hon prof, 61. *Mem:* Fel Inst Elec & Electronics Engrs; Ger Phys Soc; Ger Telecommun Asn. *Res:* Electronic devices; electrical noise. *Mailing Add:* Dept Elec Eng Lehigh Univ Bethlehem PA 18015

DAHLMAN, DOUGLAS LEE, b Bertha, Minn, Sept 29, 40; m 60; c 3. INSECT PHYSIOLOGY. *Educ:* St Cloud State Col, BS, 61; Iowa State Univ, MS, 63, PhD(entom), 65. *Prof Exp:* Res assoc entom, Iowa State Univ, 65; from asst prof to assoc prof entom, 65-82, assoc prof toxicol, Grad Ctr, 79-82, PROF ENTOM, UNIV KY, 82, PROF TOXICOL, GRAD CTR, 82- *Mem:* AAAS; Entom Soc Am; Am Chem Soc; Am Registry Prof Entom. *Res:* Insect parasite-host interactions; physiological and biochemical interactions between insects and naturally occurring substances; insect growth, development and endocrinology. *Mailing Add:* Dept Entom Univ Ky Lexington KY 40546-0091

DAHLQUIST, FREDERICK WILLIS, b Chicago, Ill, June 15, 43; m 63; c 2. BIOPHYSICAL CHEMISTRY. *Educ:* Wabash Col, BA, 64; Calif Inst Technol, PhD(chem), 69. *Prof Exp:* Miller fel biochem, Univ Calif, Berkeley, 69-71; from asst prof to assoc prof, 71-81, PROF CHEM, UNIV ORE, 81- *Concurrent Pos:* Alfred P Sloan Found fel, 75-77. *Honors & Awards:* Fac Res Award, American Cancer Soc, 81. *Res:* Nuclear magnetic resonance studies of biological systems; enzyme mechanisms; mechanism of chemotaxis in bacteria. *Mailing Add:* Dept Chem Univ Ore Eugene OR 97403

DAHLQUIST, WILBUR LYNN, b DeQuincy, La, Sept 2, 42; m 61; c 2. SOLID STATE PHYSICS. *Educ:* McNeese State Col, BS, 63; La State Univ, PhD(physics), 67. *Prof Exp:* From asst prof to assoc prof, 67-75, PROF PHYSICS & HEAD DEPT, MCNEESE STATE UNIV, 75- *Mem:* Am Phys Soc. *Res:* Transport properties of electrons in metals at low temperatures utilizing resonance techniques. *Mailing Add:* Dept Physics McNeese State Univ 4100 Ryan St Lake Charles LA 70609

DAHLSTEN, DONALD L, b Clay Center, Nebr, Dec 8, 33; m 57; c 4. FOREST ENTOMOLOGY. *Educ:* Univ Calif, Davis, BS, 56; Univ Calif, Berkeley, MS, 60, PhD(entom), 63. *Prof Exp:* Lab technician forest entom, Univ Calif, Berkeley, 59-62; asst prof zool, Calif State Col, Los Angeles, 62-63; asst entomologist, 63-65, lectr, 65-68, from asst prof to assoc prof, 68-74, PROF ENTOM, UNIV CALIF, BERKELEY, 74- *Concurrent Pos:* Chair, Div Biol Cont, 80-88, Dept Cons & Res Studies, 89- *Mem:* AAAS; Am Inst Biol Sci; Ecol Soc Am; Soc Am Foresters; Entom Soc Am. *Res:* Biological control; population dynamics of forest insects, especially parasites and predators of bark beetles; neodiprion sawflies and the Douglas fir tussock moth; evaluation of insectivorous birds in the forest and their role in natural control of forest pests. *Mailing Add:* Div Biol Control Univ Calif Berkeley CA 94720

DAHLSTROM, BERTIL PHILIP, JR, b Elizabeth, NJ, Aug 22, 31; m 52; c 3. PHYSICAL CHEMISTRY. *Educ:* Upsala Col, BS, 53; Univ Wis, MS, 55; Temple Univ, MBA, 58. *Prof Exp:* Teach asst chem, Univ Wis, 53-55; res chemist, Rohm & Haas, 55-57, teach asst spec prod, 57-60, staff asst, Prod Control Dept, 60-61, asst mgr spec prod dept, 61-63, mgr, 63-68; gen mgr indust div, vpres & mem, Bd Dirs, Sartomer Resins, Inc, 68-70; owner, Dahlstrom Assocs, Inc, 68-70; develop mgr, Ionac Chem Co, 73-81; PRES, TREAS & DIR, CHEM INSTRUMENTS CORP, 71-; MGR, MGT INFO SYSTS, SYBRON CORP, 81- *Concurrent Pos:* Instr, Temple Univ, 59-61. *Mem:* Am Chem Soc. *Res:* High pressure reactions; stereospecific polymerization; physical properties of high polymers; sales management. *Mailing Add:* 340 Myrtle Ave Woodbury NJ 08096

DAHLSTROM, DONALD A(LBERT), b Minneapolis, Minn, Jan 16, 20; m 42; c 5. CHEMICAL ENGINEERING, METALLURGICAL ENGINEERING. *Educ:* Univ Minn, BS, 42; Northwestern Univ, PhD(chem eng), 49. *Prof Exp:* Petrol & chem engr, Int Petrol Co, Peru, 42-45; from instr to assoc prof chem eng, Northwestern Univ, 46-56; dir res & develop, Eimco Corp, 53-60, vpres & dir, 60-69, vpres & dir, Eimco Envirotech, 69-74; vpres, res & develop, Process Mach Group, Envirotech Corp, 71-79, dir, 74-79 ,sr vpres, Res & Develop, 79-84; adj prof chem eng, 72-84, RES PROF CHEM, METALL & FUELS ENG, UNIV UTAH, 84- *Concurrent Pos:* Adj prof chem, Utah State Univ, 77-84. *Honors & Awards:* Raymond Award, 52, Am Inst Mining, Metall & Petrol Engrs, Richards Award, 76, Taggart Award, 83 & Krumb Lectr, 83, Presidential Citation, 88; Founders Award, Am Inst Chem Engrs, 72, Environ Award, 77. *Mem:* Nat Acad Eng; Am Chem Soc; hon mem Am Inst Mining, Metall & Petrol Engrs (vpres, 76); fel Am Inst Chem Engrs (pres, 64); Am Soc Eng Educ; Am Filtration Soc; Mining & Metall Soc Am. *Res:* Liquid cyclones; filtration; classification; flow of fluids; sedimentation; colloids; centrifugal forces in hydraulic fields; water; waste; sewage treatment; industrial processing; industrial and municipal waste treatment; liquid-solids separation; hydraulics; fine particle technology. *Mailing Add:* Chem Eng Dept Univ Utah Salt Lake City UT 84112

DAHM, ARNOLD J, b Oskaloosa, Iowa, Sept 12, 32; m 68; c 2. PHYSICS. *Educ:* Cent Col Iowa, BA, 58; Univ Minn, Minneapolis, MS, 60, PhD(physics), 65. *Prof Exp:* Fel physics, Univ Pa, 66-68; asst prof, 68-73, assoc prof, 73-80, PROF PHYSICS, CASE WESTERN RESERVE UNIV, 80- *Concurrent Pos:* Fulbright Hays fel, 76-77; Fulbright scholar, 83-84. *Mem:* Fel Am Phys Soc; Am Asn Univ Professors; Sigma Xi. *Res:* Liquid and solid helium, superconductors and two dimensional systems. *Mailing Add:* Dept Physics Case Western Reserve Univ Cleveland OH 44106

DAHM, CORNELIUS GEORGE, b St Louis, Mo, Oct 23, 08; m 35; c 3. SEISMOLOGY. *Educ:* St Louis Univ, AB, 30, MS, 32, PhD(seismol), 34. *Prof Exp:* Instr, St Louis Univ, 34-36; party chief, Root Petrol Co, 36-38; supvr, Magnolia Petrol Co, 38-53; chief geophysicist, Hunt Oil Co, 53-73; CHIEF GEOPHYSICIST, INT FIV, TEX PAC OIL CO, 74- *Mem:* Am Geophys Union; Soc Explor Geophys. *Res:* Seismology and seismic prospecting. *Mailing Add:* Texas Pacific Oil Co Inc 9400 N Central 800 Lockbox 101 Dallas TX 75231

DAHM, DONALD B, b Fargo, NDak, Aug 15, 38; m 60; c 3. PHYSICAL CHEMISTRY. *Educ:* NDak State Univ, BS, 59; SDak State Univ, MS, 61; Ohio State Univ, PhD(phys chem), 66. *Prof Exp:* Asst phys chem, Ohio State Univ, 61-62; sr res engr, Martin Co, 63; asst phys chem, Ohio State Univ, 64-66; res chemist, Res Ctr, Babcock & Wilcox Co, 66-68; res specialist, Dayton Lab, Monsanto Res Corp, 68-71; tech dir, Sinvalco Litho Prod, 71-72; tech dir, Mazer Corp, 72; TECH MGR, DAP, INC, DIV SCHERING-PLOUGH CORP, 73-; VPRES RES & DEVELOP, ILL BRONZE PAINT CO, 78- *Mem:* Am Chem Soc; Nat Painting & Coating Asn; Paint Fedn. *Res:* Chemical kinetics; combustion kinetics; printing inks, chemicals; paints, coatings, caulks, sealants and aerosols. *Mailing Add:* Niles Chem Paint Co PO Box 307 Niles MI 49120-0307

DAHM, DONALD J, b Mahaska Co, Iowa, Oct 26, 41; m 64; c 3. RESIN PRODUCTS. *Educ:* Cent Col, Iowa, BA, 63; Iowa State Univ, PhD(phys chem), 68. *Prof Exp:* Sr res chemist, Cent Res Dept, 68-74, group leader struct chem, 74-76, spec asst to vpres technol, 76-77, mgr, Environ Anal Sci Ctr, 77-81, MGR, RESINS TECHNOL MKT SUPPORT, MONSANTO, CO, 81- *Mem:* Am Chem Soc. *Mailing Add:* 730 Worcester Ave Indian Orchard MA 01151-1089

DAHM, DOUGLAS BARRETT, b Buffalo, NY, Apr 12, 28; m 58; c 3. SYSTEMS MANAGEMENT, TECHNICAL MANAGEMENT. *Educ:* Northwestern Univ, BS, 50; State Univ NY Buffalo, MS, 64. *Prof Exp:* Exec training engr, Caterpillar Tractor Co, 47-51; assoc engr, Cornell Aeronaut Lab, Inc, 56-58, res engr, 58-62, prin engr, 62-67, staff scientist, 67-70, dept head, Environ Systs, 70-76, vpres info sci serv group, 77-80, VPRES PLANS & PROGS, ADVAN TECHNOL CTR, CALSPAN CORP, 80- *Mem:* Nat Security Indust Asn; Armed Forces Commun & Electronics Asn. *Res:* Multi-disciplinary programs in environment and military science; remote sensing; aircraft avionics systems, especially methods to improve the survivability of military aircraft; intelligence information systems; instructional systems design. *Mailing Add:* S 5950 Old Lake Shore Rd Lake View NY 14085

DAHM, KARL HEINZ, b Duisburg, Ger, Jan 20, 35; m 68; c 4. INSECT HORMONES. *Educ:* Univ Gottingen, Diplom-Chemiker, 61, Dr rer nat, 64. *Prof Exp:* Sci asst org chem inst, Univ Gottingen, 62-66; proj assoc zool, Univ Wis-Madison, 66-68; assoc prof biol, 68-76, assoc prof chem, 71-76, PROF CHEM & BIOL, TEX A&M UNIV, 76- *Mem:* Am Chem Soc; Am Entomol Soc; Sigma Xi; AAAS; NY Acad Sci. *Res:* Chemistry of natural products; biochemistry of insect development; insect hormones and pheromones. *Mailing Add:* Dept Biol Tex A&M Univ College Station TX 77843

DAHM, PAUL ADOLPH, b Minneapolis, Minn, Nov 15, 17; m 41; c 4. ENTOMOLOGY, BIOCHEMISTRY. *Educ:* Univ Ill, AB, 40, MA, 41, PhD(entom), 47. *Prof Exp:* From asst prof to prof entom, Kans State Univ, 47-53; prof entom, 53-85, chmn dept zool & entom, 73-74, chmn dept entom, 75-82, CHARLES F CURTISS DISTINGUISHED PROF AGR, IOWA STATE UNIV, 69- *Mem:* Entom Soc Am; Sigma Xi. *Res:* Insecticide toxicology and biochemistry; economic entomology. *Mailing Add:* 1801 20th St Ames IA 51501

DAHMEN, JEROME J, b Johnston, Wash, Nov 22, 19; m 47; c 5. GENETICS, REPRODUCTIVE PHYSIOLOGY. *Educ:* Univ Idaho, BS, 47; Ore State Univ, MS, 52, PhD(genetics), 66. *Prof Exp:* County agent, Exten Serv, 47-55, animal husbandman, 55-58, res prof animal sci, Caldwell Br & supt, Agr Exp Sta, 58-74, PROF ANIMAL SCI & ANIMAL SCIENTIST, DEPT ANIMAL SCI, UNIV IDAHO, 74- *Mem:* Am Soc Animal Sci. *Res:* Management; physiology; breeding, sheep, wool and meat. *Mailing Add:* 1379 Walenta Dr Moscow ID 83843

DAHMS, ARTHUR STEPHEN, b Mankato, Minn, Sept 12, 43; m 66; c 2. BIOCHEMISTRY, MOLECULAR BIOLOGY. *Educ:* Col St Thomas, BS, 65; Mich State Univ, PhD(biochem), 69. *Prof Exp:* NSF & AEC fels, Molecular Biol Inst, Univ Calif, Los Angeles. 69-72, lectr, 71-72; from asst prof to assoc prof, 72-79, PROF CHEM & DIR MOLECULAR BIOL INST, SAN DIEGO STATE UNIV, 79- *Concurrent Pos:* Alexander von Humboldt Found fel, Univ Munich, 79-80; prin investr, NIH & NSF, 72- *Honors & Awards:* Res Award, Am Heart Asn, Am Diabetes Asn. *Mem:* AAAS; Am Chem Soc; Am Soc Biol Chem; Am Soc Microbiol. *Res:* Bioenergetics; structure and function of biological membranes; membrane transport; oxidative phosphorylation; enzymology of carbohydrate biodegradation; hormonal regulation of ion transport phototoxicology. *Mailing Add:* Dept Chem/Molecular Biol Inst San Diego State Univ San Diego CA 92182-0328

DAHMS, THOMAS EDWARD, b Springfield, Mass, May 14, 42; m 64; c 2. PHYSIOLOGY. *Educ:* Col of Wooster, BA, 64; Univ Calif, Santa Barbara, PhD(physiol), 70. *Prof Exp:* Asst res physiologist, Inst Environ Stress, Univ Calif, Santa Barbara, 70-74; asst prof physiol, 74-78, from asst res prof to assoc res prof internal med, 79-88, ASSOC RES PROF ANESTHESIOL, MED SCH, ST LOUIS UNIV, 88- *Mem:* Am Thoracic Soc; Am Physiol Soc. *Res:* Influence of environmental factors on cardiovascular and respiratory systems; control of pulmonary vasoreactivity. *Mailing Add:* 1525 Missouri Ave St Louis MO 63104

DAHMUS, MICHAEL E, b Waterloo, Iowa, Feb 20, 41; m. BIOCHEMISTRY. *Educ:* Iowa State Univ, BS, 63; Calif Inst Technol, PhD(biochem), 68. *Prof Exp:* Fel, Calif Inst Technol, 68-69; asst prof, 69-76, assoc prof, 76-82, PROF BIOCHEM, UNIV CALIF, DAVIS, 76- *Mem:* Am Soc Biol Chemists. *Res:* Biochemical aspects of gene regulation in higher organisms. *Mailing Add:* Dept Biochem & Biophys Univ Calif Davis CA 95616

DAHN, CONARD CURTIS, b Utica, NY, Oct 10, 39; m 62; c 2. ASTRONOMY. *Educ:* Wesleyan Univ, BA, 61; Case Inst Technol, PhD(astron), 67. *Prof Exp:* Astronr, Naval Observ, Washington, DC, 67-70, ASTRONR, NAVAL OBSERV, FLAGSTAFF STA, 70- *Mem:* AAAS; Am Astron Soc; Sigma Xi. *Res:* Astrometry; trigonometric parallax measurements of nearby stars; interstellar matter; studies of the light scattering and/or extinction properties of interstellar dust particles. *Mailing Add:* 780 N Inland Shores Dr Flagstaff AZ 86004

DAHN, JEFFERY RAYMOND, b Bridgeport, Conn, Jan 9, 57; Can citizen; m 87; c 3. SOLID STATE PHYSICS. *Educ:* Dalhousie Univ, BSc, 78; Univ BC, MSc, 80, PhD(physics), 82. *Prof Exp:* Staff mem, Nat Res Coun, Can, 82-85; proj leader, Moli Energy Ltd, Burnaby, BC, 85-88, dir res, 88-90; ASSOC PROF PHYSICS, SIMON FRAZER UNIV, 90- *Honors & Awards:* Innovation Prize, Div Indust & Appl Physics, Can Asn Physicists, 87. *Mem:* Am Phys Soc; Electrochem Soc; Can Inst Synchrotron Radiation. *Res:* Synthesis, characterization and understanding of materials of importance for energy storage. *Mailing Add:* Physics Dept Simon Fraser Univ Burnaby BC V5A 1S6 Can

DAHNEKE, BARTON EUGENE, b San Jose, Calif, Apr 4, 39; m 61; c 7. ENVIRONMENTAL SCIENCES, PARTICLE PHYSICS. *Educ:* Brigham Young Univ, BS, 63; Univ Minn, MS, 65, PhD(mech eng), 67. *Prof Exp:* NIH res fel, Calif Inst Technol, 67-69; res scientist, Inst Aerobiol, 69-72; asst prof biophys, 72-78, assoc prof biophys & chem eng, 78-81, SR SCIENTIST, CHEM ENG DEPT, UNIV ROCHESTER, 81-; SR DESIGN ENGR, EASTMAN KODAK CO, ROCHESTER, 81- *Mem:* Am Inst Chem Engrs. *Res:* Measurement of airborne particulates; diffusion theory and practice; kinetic theory of particles. *Mailing Add:* Dept Chem Eng Univ Rochester Wilson Blvd Rochester NY 14627

DAI, HAI-LUNG, b Taiwan, China, 54. QUANTUM CHEMISTRY, OPTICS. *Educ:* Nat Taiwan Univ, BS, 74; Univ Calif, Berkeley, PhD(chem), 81. *Prof Exp:* Fel, Mass Inst Technol, 81-84; asst prof 84-89, ASSOC PROF CHEM, DEPT CHEM, UNIV PA, 89- *Concurrent Pos:* Prin investr, Lab Res Struct Matter, Univ Pa, 85- *Honors & Awards:* Coblentz Award Spectros, 90. *Mem:* Am Chem Soc; Am Phys Soc. *Res:* Laser spectroscopy and photochemistry of molecules in the gases and on surfaces. *Mailing Add:* Dept Chem Univ Pa Philadelphia PA 19104-6323

DAI, XUE ZHENG (CHARLIE), b Beijing, China, Aug 17, 44; US citizen. CARDIOLOGY, EMERGENCY MEDICINE. *Educ:* Beijing Med Col, MD, 68. *Prof Exp:* Intern med, Beijing Med Col, 68-69, resident orthop surg, 69-70, staff resident med, 70-76, chief resident, 76-80; vis scholar cardiol, 81-83, res fel, 83-86, RES ASSOC CARDIOL, UNIV MINN HOSP & CLINIC, 86- *Concurrent Pos:* Consult lectr, Fujing Res Inst Med Sci, China, 86-; prin investr, Minn Heart & Lung Inst, Univ Minn, 86-, Am Heart Asn, 85-87 & NIH Nat Res Serv Award, 84-86. *Mem:* Am Heart Asn. *Res:* The adrenergic and humoral control of the coronary vessels and microcirculation in normal and hypertrophied myocardium. *Mailing Add:* Box 508 Univ Minn Hosp & Clinic Dept Med Cardiovasc Sect Minneapolis MN 55455

DAIBER, FRANKLIN CARL, b Middletown, NY, Oct 19, 19; m 53; c 2. MARINE BIOLOGY, TIDAL MARSH ECOLOGY. *Educ:* Alfred Univ, AB, 41; Mich State Col, MS, 47; Ohio State Univ, PhD(hydrobiol), 50. *Prof Exp:* Grad asst zool, Mich State Col, 41-42, 46-47; asst prof biol, Alfred Univ, 49-52; from asst prof to assoc prof, 52-68, PROF MARINE STUDIES, UNIV DEL, 69- *Concurrent Pos:* Mem, Biol Comt, Atlantic States Marine Fisheries Comn, 59-71; mem licensing bd panel, Nuclear Regulatory Comn, 72-78; mem, Natural Areas Coun, State of Del, 83- *Mem:* Sigma Xi; Am Soc Ichthyologist & Herpetologists; Ecol Soc Am; Am Fisheries Soc. *Res:* General ecology and utilization of tidal marshes. *Mailing Add:* Col of Marine Studies Univ of Del Newark DE 19711

DAICOFF, GEORGE RONALD, b Granite City, Ill, Nov 10, 30; m 72; c 3. SURGERY. *Educ:* Ind Univ, AB, 53, MD, 56. *Prof Exp:* Extern, Sunnyside Sanitorium, Indianapolis, 54-56; intern, Univ Chicago, 56-57, from jr asst resident to sr asst resident surg, 57-59, instr & sr resident, 61-62, chief resident gen surg & instr, 62, instr thoracic and cardiovasc surg, 63; from asst prof to assoc prof, 63-70, prof thoracic & cardiovasc surg & chief div, Col Med, Univ Fla, 70-77; SURGEON, ALL CHILDREN'S HOSP, ST PETERSBURG, 77- *Concurrent Pos:* Schweppe Found res fel, 63-66; fel cardiovasc surg, Mayo Clin, 66; vis colleague, Royal Postgrad Med Sch, Univ London, 65. *Mem:* Am Col Surg; Am Asn Thoracic Surg; Am Col Cardiol; Int Soc Cardiol; Soc Thoracic Surgeons. *Res:* Cardiovascular surgery. *Mailing Add:* 600 6th St S St Petersburg FL 33701

DAIE, JALEH, b Broojerd, Iran, July 17, 48; m 86. CROP PRODUCTIVITY, CARBON PARTITIONING & METABOLISM IN CROPS. *Educ:* Univ Jundi, BS, 70; Univ Calif-Davis, MS, 75; Utah State Univ, PhD(plant physiol), 81. *Prof Exp:* Soil chemist, Safiabad Agr Res Sta, Iran, 70-73, citrus specialist, 75-76; fel plant physiol, Agr Res Serv, USDA, 80-82; asst prof plant physiol, Utah State Univ, 83-85; assoc prof, 85-89, PROF PLANT PHYSIOL & CHAIR, RUTGERS UNIV, 89- *Concurrent Pos:* Ad hoc reviewer, USDA & NSF, 82-91; prin investr, NSF, Am Soybean Asn & USDA, 82-91; chmn, Working Group Cropping Efficiency & Photosynthesis, Am Soc Hort Sci, 86; chmn, comt status women plant physiol, Am Soc Plant Physiologist, 86; consult, Cytozyme, Inc, 85-88; vis scientist, Div Plant Indust, Commonwealth Sci Indust Res Orgn, 88; proj leader, US Aid, 87-; dir, Grad Prog Plant Biol, Rutgers Univ, 89-91, Henry Rutgers Res fel, 85-87; pres, Rutgers Chap, Sigma Xi, 89-90. *Mem:* Am Soc Plant Physiologists; Am Soc Hort Sci; Sigma Xi. *Res:* Mechanisms and regulation of assimilate partitioning including membrane transport of sugars and induction and incorporation of sugar carriers; drought stress-induced alterations in gene expression and carbon and sugar metabolism; stress physiology. *Mailing Add:* Dept Crops Sci Cook Col Rutgers State Univ NJ PO Box 231 New Brunswick NJ 08903

DAIGH, JOHN D(AVID), b Parsons, Kans, Jan 14, 30; m 51; c 3. ENGINEERING, MATHEMATICS. *Educ:* US Mil Acad, BS, 51; Univ Ill, MS, 57, PhD(civil eng), 57. *Prof Exp:* From instr to assoc prof mech, US Mil Acad, 61-71; instr eng, Eastfield Col, 71-, chmn, Div Math & Eng, 73-; INSTR, MATH SCI DIV, RICHLAND COL. *Mem:* Am Soc Civil Engrs; Nat Soc Prof Engrs; NY Acad Sci; Am Inst Aeronaut & Astronaut; Math Asn Am. *Res:* Structural dynamics; civil and structural engineering; engineering mechanics; fluid mechanics. *Mailing Add:* Math Sci Div Richland Col 12800 Abrams Rd Dallas TX 75243

DAIGLE, DONALD J, b St Louis, Mo, Mar 1, 39; m 63; c 2. CHEMISTRY. *Educ:* Tulane Univ, BS, 61, PhD, 65; Univ Southern Miss, MS, 63. *Prof Exp:* RES CHEMIST, SOUTHERN REGIONAL RES CTR, US DEPT AGR, 65- *Mem:* Am Chem Soc; Am Asn Textile Chemists & Colorists; Sigma Xi. *Res:* Phosphites, phosponates and their respective acids; phosphines, diphosphines and their oxides; polymers containing phosphorus for use in textiles; peanuts; flavonoids; fungi; formulation chemistry. *Mailing Add:* US Dept Agr PO Box 19687 New Orleans LA 70179

DAIGLE, JOSEPHINE SIRAGUSA, b Brooklyn, NY, Nov 16, 26; div; c 1. PHARMACEUTICS. *Educ:* Columbia Univ Col Pharm, NY, BS, 47, MS, 49; Univ Fla, Gainesville, PhD(pharm), 55. *Prof Exp:* Asst & assoc prof pharmaceut chem, Howard Col, Birmingham, Ala, 49-52; instr chem & pharm, Univ Fla, Gainesville, 52-54; asst prof pharm, Loyola Univ, New Orleans, 55-64 & Univ Tenn, Memphis, 64-66; from assoc prof to prof pharmaceut, Xavier Univ La, New Orleans, 77-90; RETIRED. *Mem:* Am Pharmaceut Asn; Am Asn Col Pharm. *Res:* Emulsifiers; suspending agents; equipment testing for making polyphasic preparations; pharmaceuticals; stability of these preparations. *Mailing Add:* 2511 Chelsea Dr New Orleans LA 70131

DAIGNEAULT, AUBERT, b Montreal, Que, Mar 6, 32; m 61; c 1. MATHEMATICAL LOGIC. *Educ:* Univ Montreal, BSc, 54, MSc, 56; Princeton Univ, PhD(math), 59. *Prof Exp:* Lectr math, Royal Mil Col, Can, 58-59; asst prof, Univ Ottawa, 59-60 & Univ Montreal, 60-63; vis asst prof, Univ Calif, Berkeley, 63-64; assoc prof, 64-70, PROF MATH, UNIV MONTREAL, 70- *Concurrent Pos:* Nat Res Coun Can fel, 64-65. *Mem:* Am Math Soc; Math Asn Am; Asn Symbolic Logic; Can Math Cong; Math Soc France. *Res:* Algebriac logic. *Mailing Add:* Dept Math Univ Montreal PO Box 6128 Montreal PQ H3C 3J7 Can

DAIGNEAULT, ERNEST ALBERT, b Holyoke, Mass, Aug 16, 28; m 54; c 4. PHARMACOLOGY. *Educ:* Mass Col Pharm W, BS, 52; Univ Mo-Kansas City, MS, 54; Univ Tenn, PhD(pharmacol), 57. *Prof Exp:* Asst instr pharm, Univ Kansas City, 53-54; from instr to asst prof pharmacol, Univ Tenn, 57-60; from asst prof to prof pharmacol, La State Univ Med Ctr, 62-77; PROF & CHMN DEPT PHARMACOL, E TENN STATE UNIV, 77- *Concurrent Pos:* NIH spec fel, 69-70; consult, Vet Admin Hosp, 78- *Mem:* AAAS; Soc Neurosci; Soc Pharmacol & Exp Therapeut; Am Soc Clin Pharmacol; Acoustical Soc Am. *Res:* Cardiovascular and autonomic nervous system; aspects of tachyphylaxis with sympathominetic amines; clinical pharmacology; auditory pharmacology. *Mailing Add:* Dept Pharmacol Col Med E Tenn State Univ Johnson City TN 37601

DAIGNEAULT, REJEAN, b Montreal, Can, June 28, 41; m 66; c 3. CLINICAL CHEMISTRY, MEDICINAL PLANTS. *Educ:* Univ Montreal, BS, 64, MS, 65, PhD(biochem), 69. *Prof Exp:* Prof biochem, Univ Quebec, Montreal, 69-71; clin biochemist clin chem, Hosp Notre-Dame, 72-87; PROF BIOCHEM, UNIV MONTREAL, 83- *Concurrent Pos:* Consult, Technicon Corp, NY, 74-78, DuPont, Wilmington, 75-80; pres, Asn Quebec Clin Biochemist, 76-78; chmn, Sci Prog-Joint Am Asn Clin Chem Can Soc Clin Chemists Meeting, Boston 1980, 78-80; asst chief ed, Clin Biochem, 79-82; head clin biochemist clin chem, Hosp Notre-Dame, 80-84; chmn, Dept Biochem, Univ Montreal, 83-89. *Mem:* Can Soc Clin Chemists; Am Asn Clin Chem; French Soc Clin Biol; fel Chem Inst Can. *Res:* Two-dimensional electrophoresis of proteins in biological fluids; traditional chinese medicine. *Mailing Add:* Dept Biochem Univ Montreal PO Box 6128 Sta A Montreal PQ H3C 3J7 Can

DAIJAVAD, SHAHVOKH, b July 7, 60. ELECTRICAL & COMPUTER ENGINEERING. *Educ:* BEng, 83, PhD(elec eng), 86. *Prof Exp:* Postdoctorate fel, Univ Calif, Berkeley, 86-87; T J WATSON RES CTR, IBM, 87- *Mem:* Inst Elec & Electronic Engrs. *Res:* Electrical and computer engineering. *Mailing Add:* IBM T J Watson Res Ctr PO Box 218 Yorktown Heights NY 10598

DAIL, CLARENCE WILDING, b Ger, July 7, 07; US citizen; m 31; c 2. PHYSICAL MEDICINE & REHABILITATION. *Educ:* Pac Union Col, AB, 30; Loma Linda Univ, MD, 35. *Prof Exp:* From instr to assoc prof orthop surg & rehab-phys med, 35-59, PROF ORTHOP SURG & REHAB-PHYS MED, SCH MED, LOMA LINDA UNIV, 59- *Concurrent Pos:* Mem staff, Hosps. *Mem:* AMA; Am Cong Rehab Med; Am Acad Phys Med & Rehab; Am Asn Electromyog & Electrodiag. *Res:* Glossopharyngeal breathing; physiology of muscle breathing patterns and swallowing in muscle weakness. *Mailing Add:* 8545 Bluff Rd Banning CA 92220

DAILEY, BENJAMIN PETER, b San Marcos, Tex, Sept 1, 19; m 45; c 3. CHEMISTRY SPECTROSCOPY. *Educ:* Southwest Tex State Teachers Col, BA, 38; Univ Tex, MA, 40, PhD(chem), 42. *Prof Exp:* Res assoc, Explosives Res Lab, Pa, 42-45; fel, Harvard Univ, 46-47; from instr to prof, Columbia Univ, 47-89, chmn dept, 68-70, EMER PROF CHEM, COLUMBIA UNIV, 89- *Concurrent Pos:* NSF sr fel, 63-64; Guggenheim fel, 71-72. *Mem:* Am Chem Soc; fel Am Phys Soc; fel Am Acad Arts & Sci. *Res:* Thermodynamic properties; molecular structure; microwave spectra; nuclear magnetic resonance. *Mailing Add:* Dept Chem Columbia Univ New York NY 10027

DAILEY, C(HARLES) L(EE), b Reedley, Calif, Nov 12, 17; m 49; c 6. AERONAUTICS. *Educ:* Calif Inst Technol, BS, 41, MS, 42, PhD, 54. *Prof Exp:* Mem staff, Douglas Aircraft Co, 42-45; chief, Aeronaut Res Eng Ctr, Univ Southern Calif, 45-47; dir, Aeronaut Dept, Wiancko Eng Co, 57-60, chief plasma propulsion proj, 60; CHIEF PLASMA PROPULSION PROJ, TRW SYSTS, REDONDO BEACH, 60- *Res:* Plasma propulsion; supersonic diffuser instability; airplane aerodynamics. *Mailing Add:* 953 Granvia Altamira Palos Verdes Estates CA 90274

DAILEY, CHARLES E(LMER), III, b Pittsburgh, Pa, Feb 14, 26; m 54; c 3. CHEMICAL ENGINEERING, MANUFACTURING MANAGEMENT. *Educ:* Carnegie-Mellon Univ, BS, 50, MS, 52, PhD(chem eng), 55. *Prof Exp:* Chem engr org chem dept, Plant Tech Sect, Chambers Works, E I Du Pont De Nemours & Co, 54-60, res engr, Jackson Lab, 60-62, sr res engr, 62-64, group supvr org chem dept, Chambers Works, 64-72, chief supvr, 72-77, chief supvr, CD&P dept, Chambers Works, 78-79, tech supt, Chambers Works, 79-81, task force mgr, Du Pont, Wilmington, 81-82, mfg mgr, DuPont Petchem Div, 82-85; RETIRED. *Mem:* Am Inst Chem Engrs; Am Chem Soc. *Res:* Simultaneous heat, mass and momentum transfer; economic evaluation; long-range planning; market analyses; process simulation and hazards; physical properties; organic diisocyanates; organometallic compounds; fluorocarbons; air and water pollution abatement; waste disposal; incident investigation. *Mailing Add:* 1219 Hazeltine Dr Ft Myers FL 33919-7308

DAILEY, FRANK ALLEN, b South Haven, Mich, Dec 8, 46; m 76; c 3. ALLERGY, DIAGNOSTICS. *Educ:* Mich State Univ, BS, 69; Univ Calif, Davis, PhD(biochem), 77. *Prof Exp:* Asst prof biochem, Univ Nev, Reno, 78-79; res assoc, Agron Dept, Wash State Univ, 79-80; res biochemist, 80-82, SR RES BIOCHEMIST, HOLLISTER-STIER, WASH, 82- *Res:* Photosynthesis; lignin degradation; nitrate reduction; immunology; allergy; enzyme kinetics; diagnostics. *Mailing Add:* E14217 23d Veradale WA 99037

DAILEY, GEORGE, b Lexington, Ky, Feb 17, 38; m 70; c 2. APPLIED MECHANICS, MECHANICAL ENGINEERING. *Educ:* Univ Louisville, BME, 61; Kans State Univ, MS, 62, PhD(appl mech), 66. *Prof Exp:* SR ENGR APPL MECH, JOHNS HOPKINS UNIV, 66- *Res:* Dynamic and static finite element analysis of structures and other analytic methods in solid and fluid mechanics. *Mailing Add:* Dept Physics Johns Hopkins Univ Johns Hopkins Rd Laurel MD 20707

DAILEY, HARRY A, b Santa Cruz, Calif, Feb 21, 50; m. HEME BIOSYNTHESIS, MEMBRANE PROTEINS. *Educ:* Univ Calif, Los Angeles, BA, 72, PhD(microbiol), 76. *Prof Exp:* Fel, Dept Biochem, Univ Conn Health Ctr, 76-80; from asst prof to assoc prof microbiol, 80-90, PROF & HEAD MICROBIOL, UNIV GA, 90- *Mem:* Am Soc Microbiol; Am Soc Biochem & Molecular Biol; Biochem Soc. *Res:* Terminal, membrane bound enzymes of the heme biosynthetic pathway; purification, characterization and reconstitution of the enzymes into defined liposomes; regulation of heme biosynthesis during erythropoiesis. *Mailing Add:* Dept Microbiol Univ Ga Athens GA 30602

DAILEY, JOHN WILLIAM, b Keyser, WVa, Feb 10, 43; m 73; c 2. PHARMACOLOGY. *Educ:* Univ Md, BS, 66; Univ Va, PhD(pharmacol), 71. *Prof Exp:* Fel, Karolinska Inst, 70-72; res assoc, Cleveland Clin Found, 72-73; asst prof, Med Sch, George Washington Univ, 73-76; from asst prof to assoc prof pharmacol, Med Ctr, La State Univ, Shreveport, 78-85; ASSOC PROF PHARMACOL, COL MED AT PEORIA, UNIV ILL, PEORIA, ILL, 85- *Concurrent Pos:* Consult pharmaceut, 87- *Mem:* Soc Neurosci; Am Heart Asn; Am Soc Pharmacol & Exp Therapeut; Am Epilepsy Soc. *Res:* Nervous system pharmacology in relation to disease states and mechanisms of drug action. *Mailing Add:* Dept Basic Sci Univ Ill Col Med at Peoria Peoria IL 61656

DAILEY, JOSEPH PATRICK, b Penfield, Ill, Nov 14, 22; m 51; c 5. ORGANIC CHEMISTRY. *Educ:* Univ Ill, BS, 44; Univ Notre Dame, PhD(chem), 50. *Prof Exp:* Chemist, Cent Res Labs, Gen Aniline & Film Corp, 44-46; res chemist, Res Dept, Armour Labs, 49-53, head org chem dept, Armour Pharmaceut Co, 53-59, dir res, 59-67, dir res & develop, 67-71, vpres res & develop, 71-80; sr sci coordr, Revlon Health Care Group, 80-82; PHARMACEUT CONSULT, J P DAILEY ASSOC, 82- *Mem:* AAAS; Am Heart Asn; Am Chem Soc; assoc Royal Soc Med. *Res:* Pharmaceutical organic chemistry; chemotherapy; intravenous fat emulsions, quinolines; ethyleneimines; vinyl ethers; hypothalamic research; lipid metabolism; enzymes; hormones; blood components. *Mailing Add:* Seven Juneberry Lane Ridgefield CT 06877

DAILEY, MORRIS OWEN, FLOW CYTOMETRY, LYMPHOCYTE ACTIVATION. *Educ:* Univ Chicago, PhD(immunol), 76, MD, 77. *Prof Exp:* ASSOC PROF PATH, COL MED, UNIV IOWA, 84- *Mailing Add:* Dept Path Col Med Univ Iowa Iowa City IA 52242

DAILEY, PAUL WILLIAM, JR, b Ft Wayne, Ind, Sept 29, 37; m 59; c 3. METEOROLOGY. *Educ:* Ball State Univ, BS, 59; Purdue Univ, MS, 68. *Prof Exp:* Agr forecaster, Indianapolis, Ind, 66-68, state climatologist, State College, Pa, 68-71, lead forecaster, Des Moines, Iowa, 71-73, meteorologist-in-chg, Cincinnati, Ohio, 73-75, lead forecaster, Des Moines, 75-78, dep meteorologist-in-chg, Nat Oceanic & Atmospheric Admin, Cleveland, 78-81 CHIEF METEOROLOGIST, NAT WEATHER SERV, KANSAS CITY, MO, 81- *Concurrent Pos:* Adj asst prof, Dept Meteorol, Pa State Univ, 69-71. *Mem:* Am Meteorol Soc; Nat Weather Asn. *Res:* Climatology. *Mailing Add:* Nat Weather Ser 10600 W Higgins Rd Rosemont IL 60018

DAILEY, ROBERT ARTHUR, b Charles Town, WVa, May 20, 45; m 71; c 3. REPRODUCTIVE PHYSIOLOGY, ENDOCRINOLOGY. *Educ:* WVa Univ, BS, 67; Univ Wis-Madison, MS, 69, PhD(reproductive physiol), 73. *Prof Exp:* Specialist food inspection, US Army, 69-71; res fel neuroendocrinol, Emory Univ, 73-75; instr, 75-77, from asst prof to assoc prof, 77-84, PROF REPRODUCTIVE PHYSIOL, WVA UNIV, 84- *Concurrent Pos:* Fel, Emory Univ, 73-77. *Mem:* Soc Study Reproductive Biol; Am Soc Animal Scientists; Endocrine Soc; Am Dairy Sci Asn. *Res:* Neuroendocrine regulation of gonadotropic hormone secretion in mammals; factors associated with the control of ovarian follicular growth, maturation and atresia in domestic animals. *Mailing Add:* Div Animal & Vet Sci WVa Univ Morgantown WV 26506-6108

DAILY, FAY KENOYER, b Indianapolis, Ind, Feb 17, 11; m 37. PHYCOLOGY, SYSTEMATICS. *Educ:* Butler Univ, BA, 35, MS, 52. *Prof Exp:* Lab technician, Eli Lilly & Co, Ind, 35-37, Abbott Labs, Ill, 39 & Wm S Merrell Co, Ohio, 40-41; lubrication chemist, Indianapolis Propellor Div, Curtiss-Wright Corp, NY, 45; lectr bot, 47-49, instr immunol & microbiol, 57-58, lectr microbiol, 62-63, mem herbarium staff, 49-87, CURATOR CRYPTOGAMIC HERBARIUM, BUTLER UNIV, 87- *Mem:* Am Inst Biol Sci; Bot Soc Am; Phycol Soc Am; Int Phycol Soc; Sigma Xi; Torrey Bot Club. *Res:* Taxonomy; ecology and distribution of extant and fossil charophytes (algae). *Mailing Add:* 5884 Compton St Indianapolis IN 46220-2653

DAILY, JAMES W(ALLACE), b Columbia, Mo, Mar 19, 13; m 38; c 2. FLUID MECHANICS, HYDRAULIC ENGINEERING. *Educ:* Stanford Univ, AB, 35; Calif Inst Technol, MS, 37, PhD(mech eng), 45. *Prof Exp:* Test engr, Byron Jackson Co, Calif, 35; asst, Calif Inst Technol, 36-37, mgr, Hydraul Mach Lab, 37-40, instr mech eng, 40-46, hydraul engr, Off Sci Res & Develop Proj, 41-46; asst prof hydraul, Mass Inst Technol, 46-49, assoc prof, 49-55, prof, Hydrodynamics Lab, 55-64; prof eng mech & chmn dept, Univ Mich 64-70, prof, 70-80, emer prof fluid mech & hydraul, 80-; RETIRED. *Concurrent Pos:* Vis prof, Delft Technol Univ, 71; vis scientist, E D F Res & Testing Ctr, Paris, 71; mem, US deleg water resources specialists to People's Repub China, 74; vis prof hydraul eng, E China Col, Nanking, 79. *Honors & Awards:* Naval Ord Develop Award, 45. *Mem:* Nat Acad Eng; Am Soc Mech Engrs; Am Soc Civil Engrs; Int Asn Hydraul Res (pres, 67-71); hon mem Japan Soc Civil Engrs. *Res:* General fluid mechanics; hydraulic machinery; pumps and turbines; hydrodynamics of submerged bodies, cavitation; non-Newtonian fluid mechanics. *Mailing Add:* 2968 San Pasqual St Pasadena CA 91107

DAILY, JOHN W, b Pasadena, Calif, Mar 22, 43; m 67, 81; c 1. MECHANICAL ENGINEERING, APPLIED PHYSICS. *Educ:* Univ Mich, BSME, 68, MSME, 69; Stanford Univ, PhD(mech eng), 75. *Prof Exp:* Res asst comput design, Civil Eng Comput Lab, Mass Inst Technol, 60-62; sport car mech, Competition Asn, Cambridge, Mass, 63-65; res asst air pollution, Automotive Lab, Univ Mich, 66-69; heat transfer analyst rocket design, Aerojet Liquid Rocket Co, 70-71; res asst, Dept Mech Eng, Stanford Univ, 71-75; asst prof, 75-81, ASSOC PROF MECH ENG, UNIV CALIF, BERKELEY, 81- *Concurrent Pos:* Investr, Dept of Energy grant, 75-; prin investr, NASA grant, 75-, USAF Off of Sci Res grant, 76- & NSF grant, 77- *Honors & Awards:* Ralph R Teetor Educ Award, Soc Automotive Engrs, 87. *Mem:* Am Inst Aeronaut & Astronaut; Am Soc Mech Engrs; Combustion Inst; Optical Soc Am; Sigma Xi. *Res:* Combustion; fluid mechanisms; optical diagnostics; heat transfer; spectroscopy; air pollution management. *Mailing Add:* Dept Mech Eng Univ Colo Boulder CO 80309

DAILY, WILLIAM ALLEN, b Indianapolis, Ind, Nov 10, 12; m 37. MICROBIOLOGY. *Educ:* Butler Univ, BS, 36; Northwestern Univ, MS, 38. *Prof Exp:* Lab asst, Northwestern Univ, 36-38; lab asst, Univ Cincinnati, 40-41; asst sr microbiologist, Eli Lilly & Co, 41-77; RETIRED. *Concurrent Pos:* Mem staff, Herbarium, Butler Univ, 42- *Mem:* AAAS; Sigma Xi; Soc Phycol Am (secy, 58, secy-treas, 59, pres, 63); Bot Soc Am; Int Asn Plant Taxon. *Res:* Taxonomy of the coccoid Myxophyceae; phytoplankton of Indiana. *Mailing Add:* 5884 Compton St Indianapolis IN 46220

DAIN, JEREMY GEORGE, b New York, NY, July 1, 41; m 74. ORGANIC CHEMISTRY, DRUG METABOLISM. *Educ:* Lowell Technol Inst, BS, 64, MS, 67; Duquesne Univ, PhD(chem), 70. *Prof Exp:* Fel, Univ Pittsburgh, 70-72; SR SCIENTIST, SANDOZ PHARMACEUT, 72-, GROUP LEADER BIOTRANSFORMATION, 80- *Mem:* Am Chem Soc; AAAS; Drug Metab Group. *Res:* Isolation, characterization, identification quantitation and synthesis of drug metabolites; synthesis of labeled compounds; methods of quantitation; in vitro metabolism of xenobiotics. *Mailing Add:* Drug Metab Sandoz Pharmaceut Rte 10 Rm 300 East Hanover NJ 07936-1080

DAIN, JOEL A, b New York, NY, Oct 26, 31; m 56; c 3. BIOCHEMISTRY, NEUROCHEMISTRY. *Educ:* Univ Ill, BS, 53; Cornell Univ, PhD(biochem), 57. *Prof Exp:* Res fel, Col Physicians & Surgeons, Columbia Univ, 56-57, res fel & assoc biochem, 57-59; res assoc biochem glycoproteins, Boston Dispensary & Med Sch, Tufts Univ, 59-62; from asst prof to assoc prof, 62-73, PROF BIOCHEM, UNIV RI, 73- *Mem:* AAAS; Am Chem Soc; Am Soc Microbiol; Am Soc Biol Chemists; Am Soc Neurochem; Int Soc Neurochem. *Res:* Mechanisms of action of nervous tissue glycolipid metabolic enzymes; nonenzymatic protein glycation. *Mailing Add:* Dept Chem Univ RI Kingston RI 02881

DAINIAK, NICHOLAS, tissue culture, hematopoiesis, for more information see previous edition

DAINTY, ANTON MICHAEL, b Cambridge, Eng, Dec 26, 42. SEISMOLOGY. *Educ:* Univ Edinburgh, BSc, 63; Dalhousie Univ, PhD(physics), 67. *Prof Exp:* Fel physics, Univ Toronto, 67-69, lectr, 69-71; res assoc earth & planetary sci, Mass Inst Technol, 71-77; assoc prof, Sch Geophys Sci, Ga Inst Technol, 77-85; VIS SCIENTIST, DEPT EARTH, ATMOSPHERIC & EARTH SCI, MASS INST TECHNOL, 85- *Mem:* Am Geophys Union; Seismol Soc Am. *Res:* Scattering of seismic waves; array processing; expert systems in seismology. *Mailing Add:* Div Geophysical Sci Ga Inst Tech 225 North Ave Atlanta GA 30332

DAINTY, JACK, b Mexborough, Eng, May 7, 19. PLANT PHYSIOLOGY. *Educ:* Cambridge Univ, BA, 40, MA, 43; Univ Edinburgh, DSc(plant physiol), 58. *Prof Exp:* Res scientist, Cambridge Univ, 40-46; res scientist, Atomic Energy Can, Ltd, 46-49; lectr physics, Univ Edinburgh, 49-52, sr lectr biophys, 52-58, reader, 58-63; prof biol, Univ EAnglia, 63-69; prof bot, Univ Calif, Los Angeles, 69-71; prof, 71-77, UNIV PROF BOT, UNIV TORONTO, 77- *Mem:* Can Soc Plant Physiologists; Am Soc Plant Physiologists; Brit Ecol Soc; Brit Soc Exp Biol; Royal Soc Edinburgh. *Res:* Transport of ions and water across the membranes of plant cells. *Mailing Add:* Dept Bot Univ Toronto Toronto ON M5S 1A1 Can

DAIRIKI, NED TSUNEO, b Sacramento, Calif, Dec 16, 36; m 61; c 1. PHYSICS, COMPUTER SCIENCE. *Educ:* Reed Col, BA, 60; Univ Calif, Berkeley, PhD(physics), 68. *Prof Exp:* Res assoc physics, Lawrence Berkeley Lab, 68, STAFF PHYSICIST, LAWRENCE LIVERMORE LAB, UNIV CALIF, 69- *Mem:* Am Phys Soc; AAAS. *Res:* Particle physics; computer language software systems; numerical techniques. *Mailing Add:* 835 Indian Rock Ave Berkeley CA 94707

DAIRIKI, SETSUO, b Penryn, Calif, Feb 11, 22; m 45; c 3. ELECTRICAL ENGINEERING, GENERAL COMPUTER SCIENCES. *Educ:* Stanford Univ, AB, 42; Univ Nebr, MS, 45; Harvard Univ, MA, 47. *Prof Exp:* Instr physics, Univ Nebr, 43-45; engr, Lab for Electronics, Inc, Mass, 48-51; engr, Raytheon Mfg Co, 51-53, sr engr, 54-60; mem staff commun, Lincoln Lab, Mass Inst Technol, 53-54; sr res engr, SRI Int, 60-76; prin engr, Digital Equip Corp, 76-87; CONSULT, 88- *Mem:* Sigma Xi; Inst Elec & Electronics Engrs. *Res:* Electromagnetic theory and applications; communication application. *Mailing Add:* 33 Barry Lane Atherton CA 94027

DAIRMAN, WALLACE M, b Jersey City, NJ. TOXICOLOGY. *Prof Exp:* RES LEADER, HOFFMAN-LA ROCHE INC, 87- *Mailing Add:* Sect Toxicol Hoffman-La Roche Inc Kingsland St Nutley NJ 07110

DAISEY, JOAN M, New York, NY, Oct 4, 41; m 63; c 1. ATMOSPHERIC CHEMISTRY. *Educ:* Georgian Court Col, BA, 62; Seton Hall Univ, PhD(phys chem), 70. *Prof Exp:* Chemist, Carter-Wallace Inc, 63-64 & Wallace & Tiernan Inc, 64-66; asst prof chem, Mt St Mary Col, 70-75; asst res scientist, Med Ctr, NY Univ, 76-77, assoc res scientist, 77-78, res scientist atmospheric chem, 78-79, from asst prof to assoc prof environ med, 79-86; staff scientist 3, 86-89, SR SCIENTIST & PROG LEADER, INDOOR ENVIRON PROG, LAWRENCE BERKELEY LAB, 89- *Concurrent Pos:* Mem, res grants rev panel, US Environ Protection Agency, 83, 84, 85 & 89, chairwoman, Rev Panel Integrated Air Cancer Proj, 84 & 86, mem, rev subcomt, Sci Adv Bd, 87, peer rev comts, 87 & 89, Comt Indoor Air Qual & Total Human Exposure, Sci Adv Bd, 87-; mem, Comt on Advances in Assessing Human Exposures to Airborne Pollutants, Nat Res Coun, 88-90; bd sci counselors, Agency Toxic Substances & Dis Registry, 88-90; bd mem, Int Soc Exposure Anal. *Honors & Awards:* Texaco Res Award, 81. *Mem:* Am Chem Soc; NY Acad Sci; Air & Waste Mgt Asn; Am Conf Gov Indust Hygienist; Sigma Xi; Am Asn Aerosol Res; Int Soc Exposure Anal. *Res:* Human exposures to toxic carcinogenic airborne organic pollutants in indoor and outdoor environments; nature, reactions, and sources of airborne organic pollutants; effects of complex environmental mixtures at the cellular level; nature, reactions, transport, and sources. *Mailing Add:* Indoor Environ Ptog 90-3058 Lawrence Berkery Lab Berkeley CA 94720

DAJANI, ADNAN S, b Jerusalem, Palestine, Apr 7, 36; nat US; m 62; c 3. PEDIATRICS. *Educ:* Am Univ Beirut, BS, 56, MD, 60; Univ NC, MS, 64; Am Bd Pediat, cert, 64. *Prof Exp:* From instr to asst prof, Dept Pediat, Am Univ Beirut, 64-67; from instr to assoc prof, Dept Pediat, Univ Minn, 67-73; PROF, DEPT PEDIAT, WAYNE STATE UNIV, 73- *Concurrent Pos:* Vis prof & lectr, numerous univs, hosps & asn, 72-90; prof dir, Diag Microbiol Lab, 73-83 & 85-88, dir, Div Infectious Dis, Children's Hosp Mich, 73- *Mem:* Am Acad Pediat; AAAS; Am Asn Immunologists; Am Fedn Clin Res; Am Pediat Soc; Am Soc Microbiol; fel Infectious Dis Soc Am; Pediat Infectious Dis Soc; Soc Exp Biol & Med; Soc Pediat Res. *Res:* Piperacillin pharmacokinetics in children; efficacy of Cefotaxime in childhood meningitis; role of glycoproteins of Respiratory Syncytial virus in infection and disease; streptococcal infections in children; treatment of soft tissue, bone and joint infection in children; author of numerous scientific publications. *Mailing Add:* Dept Pediat Wayne State Univ 3901 Beaubien Blvd Detroit MI 48201

DAJANI, ESAM ZAFER, b Jaffa, Palestine, May 30, 40; US citizen; m 64; c 3. MEDICINE, PHARMACY & TOXICOLOGY. *Educ:* Univ Mo-Kansas City, BS, 63; Auburn Univ, MS, 66; Purdue Univ, PhD(pharmacol, toxicol), 68. *Prof Exp:* Sr pharmacologist, Rohm & Haas Res Labs, 68-72; sr res investr, G D Searle Co, 72-74, group leader & chmn gastrointestinal res, 75-79, sect head, 79-80, asst dir, Gastroenterol Clin Res, 80-82, assoc dir, 82-85, dir cytotec sci & med affairs, 85-86, DIR, CLIN RES, G D SEARLE & CO, 86- *Concurrent Pos:* Adj prof pharmacol, Chicago Med Sch, 84-90; prof med, Univ Calif, Los Angeles, 85-; sci reviewer, biomed journels; mem, US Cong Comt Digestive Dis Nat Prob & various other nat & int res comts. *Honors & Awards:* Edgar Queeny Award, 91. *Mem:* Am Soc Pharmacol; NY Acad Sci; Am Pharmaceut Asn; Am Col Gastroenterol; Soc Exp Biol & Med; Am Gastroenterol Asn. *Res:* Conducted and directed preclinical and clinical research in gastroenterology and pharmacology (gastrointestinal, cardiovascular and CNS) and toxicology; extensive expertise in worldwide clinical research and development (phases I-IV). *Mailing Add:* Box V1549 RFD Long Grove IL 60047

DAKIN, JAMES THOMAS, b Pittsburgh, Pa, May 23, 45; m 69; c 2. PHYSICS, ENGINEERING. *Educ:* Harvard Univ, BS, 67; Princeton Univ, MS, 69, & PhD(physics), 71. *Prof Exp:* Instr physics, Princeton Univ, 71; res assoc physics, Stanford Linear Accelerator Ctr, 71-74; asst prof physics, Univ Mass, Amherst, 74-75; res staff physics, 75-78, liaison scientist, 78-82, res staff physics, Corp Res & Develop, 82-88, SR CONSULT PHYSICIST, GEN ELEC LIGHTING, 88- *Mem:* Am Phys Soc; AAAS. *Res:* Fluid mechanics, heat transfer, atomic physics, electrical discharges and lighting. *Mailing Add:* 2867 Weybridge Rd Shaker Heights OH 44120

DAKIN, MATT EITEL, b Skene, Miss, Dec 11, 36; m 57; c 2. ENTOMOLOGY. *Educ:* Delta State Col, BS, 58; Auburn Univ, MS, 60, PhD(entom), 64. *Prof Exp:* Instr biol, Univ Southwestern La, 60-61 & 63-64; asst prof zool, Auburn Univ, 64; from asst prof to assoc prof, 64-71, PROF BIOL, UNIV SOUTHWESTERN LA, 71- *Mem:* Entom Soc Am; Orthopterist Soc; Am Entom Soc; Am Mosquito Control Asn. *Res:* Taxonomy of the Orthoptera of the United States. *Mailing Add:* Box 41435 Univ Southwestern La Lafayette LA 70504

DAKIN, THOMAS WENDELL, physical chemistry; deceased, see previous edition for last biography

DAKSHINAMURTI, KRISHNAMURTI, b Vellore, India, May 20, 28; m 61; c 2. BIOCHEMISTRY & NEUROCHEMISTRY, NUTRITION. *Educ:* Univ Madras, BSc, 46; Univ Rajasthan, India, MSc, 52, PhD(biochem), 57; FRSC. *Prof Exp:* Lectr chem, Christian Med Col, India, 52-56, sr lectr biochem, 56-58; res assoc animal nutrit, Univ Ill, 58-62; res assoc nutrit, Mass Inst Technol, 62-63; assoc dir res, St Joseph Hosp, Lancaster, Pa, 63-65; assoc prof, 65-73, PROF BIOCHEM, UNIV MAN, 73- *Concurrent Pos:* Assoc ed, Can J Biochem, 71-75, Neurosci Lett, 84-; vis prof, Rockefeller Univ, NY, 74-75. *Honors & Awards:* Borden Award, Nutrit Soc Can, 73. *Mem:* Am Inst Nutrit; Brit Biochem Soc; Am Soc Biol Chem; Am Soc Neurochem; Am Soc Cell Biol; Soc Neurosci; fel Am Col Nutrit; Int Brain Res Orgn; Int Soc Neurochem. *Res:* Metabolic control mechanisms; regulation of transcription; metabolic function of Biotin; neurobiology of Pyridoxine and Serotonin. *Mailing Add:* Dept Biochem Univ Man 770 Bannatyne Ave Winnipeg MB R3E 0W3 Can

DAKSS, MARK LUDMER, b New York, NY, Mar 1, 40; m; c 2. OPTICAL FIBER AMPLIFIERS, OPTICAL COMMUNICATIONS. *Educ:* Cooper Union, BEE, 60; Columbia Univ, AM, 62, PhD(physics), 66. *Prof Exp:* Mem res staff, T J Watson Res Ctr, IBM Corp, 66-71; MEM TECH STAFF, GTE LABS, 71- *Concurrent Pos:* Lectr & course coord, Optical Fiber Commun Systs, Northeastern Univ State-of-the-Arts Eng Prog, Dedham, Mass, 74-85; task group single-mode fibers, Electronics Industs Asn, 82-84. *Mem:* Optical Soc Am; Laser & Electro Optics Soc Inst Elec & Electronics Engrs; Inst Elec & Electronic Engrs. *Res:* Fiber-optical communications; nonlinear optics; solid state physics and lasers. *Mailing Add:* GTE Labs Inc 40 Sylvan Rd Waltham MA 02254

DALAL, FRAM RUSTOM, b Madras, India, Jan 24, 35; m 61; c 1. CLINICAL CHEMISTRY, BIOCHEMISTRY. *Educ:* Univ Bombay, BS, 54 & 56, PhD(biochem), 61. *Prof Exp:* Coun Sci & Indust Res fel, Univ Bombay, 60-61; chief technologist, D & P Prod, Ltd, India, 61-64; res assoc microbial biochem, Sch Med, Univ Pa, 64-67; assoc prof microbiol, Dept Chem Technol, Univ Bombay, 67-69; assoc chief biochem, Dept Labs, Albert Einstein Med Ctr, Philadelphia, 69-78; DIR CLIN BIOCHEM, TEMPLE UNIV HOSP, 78- *Mem:* Am Asn Clin Chemists; Am Soc Clin Biochem. *Res:* Diagnostic applications of enzymes in blood and tissues; improvements in automation of enzyme assays; unusual paraproteins in myeloma. *Mailing Add:* Clin Labs Temple Univ Hosp 3401 N Broad St Philadelphia PA 19140

DALAL, HARISH MANEKLAL, b Bombay, India, July 4, 43; m 69; c 2. MATERIALS SCIENCE, QUALITY ASSURANCE. *Educ:* Indian Inst Technol, BTech, 65; Mass Inst Technol, ScD(mat sci), 70. *Prof Exp:* Res asst powder metall, Mass Inst Technol, 65-70, fel, 70-71; mat scientist tribology, SKF Industs, Inc, 72-78; mgr res & develop, KSM Fastening Syst, Moorestown NJ, 78-80; mgr amt process & eval, Advan Energy Prod Div, Gen Elec Co, Valley Forge, Pa, 80-86; pres, Precious Cosmetics, 86-89; PRES, JAYA HIND SCIAKY LTD, INDIA, 89- *Mem:* Am Soc Metals; Am Soc Lubrication Engrs; Am Soc Mech Engrs; Am Ceramic Soc. *Res:* Structure, property relationships in materials, developing processes to generate desired structures in materials and developing/adapting quality assurance techniques to assure consistent product quality; development of thermoelectric materials and generators; design, development and marketing of special purpose machines and robotic systems. *Mailing Add:* 914 Florence Lane Norristown PA 19401

DALAL, NAR S, b Panjab, India, May 11, 41; US citizen; m 66; c 2. PHYSICAL CHEMISTRY, CHEMICAL PHYSICS. *Educ:* Punjab Univ, BSc, 62, MSc, 63; Univ BC, PhD(chem), 71. *Prof Exp:* Killam fel, Univ BC, 71-74; vis scientist & World Trade fel, IBM Res Lab, 74-75; Nat Res Coun res assoc, 76-78; PROF PHYS CHEM, WVA UNIV, 78- *Concurrent Pos:* Anal spectroscopist, Bhabha Atomic Res Ctr, Bombay, 64-67; Benedum Distinguished Scholar Award, 87. *Mem:* Am Chem Soc; Chem Inst Can. *Res:* High resolution EPR and Endor spectroscopy as applied to structural phase transitions in molecular and coal related problems; catalysis, magnetic phase transitions; EPR and Endor spectroscopy and other spectroscopic techniques are being utilized to discover new magnetic materials with high specific heat at cryogenic tempertures, to understand mechanisms of cooperative phase transition in ferroelectric materials, and to elucidate coal to energy conversion process as well as biochemical origin of the black lung disease; high temperature superconductivity. *Mailing Add:* Dept Chem WVa Univ Morgantown WV 26506

DALAL, SIDDHARTHA RAMANLAL, b Ahmedabad, India, Oct 1, 48; m; c 2. STATISTICS. *Educ:* Univ Bombay, BSc, 69; Univ Rochester, MBA, 73, MA & PhD(statist), 75. *Prof Exp:* Asst prof appl & math statist, Rutgers Univ, New Brunswick, NJ, 75-80; mem tech staff, Bellcore, Murray Hill, NJ, 83-89; DIST MGR, STATIST & ECONOMET RES GROUP, BELLCORE, 89- *Honors & Awards:* Outstanding Statist Appln Award, Am Statist Asn, 90. *Mem:* Inst Math Statist; fel Am Statist Asn. *Res:* Development of robust procedures with finite sample optimality, robust estimation of location, of linear regression; topics in Bayes Nonparametric Decision theory, inequality theory, ranking and selection; software reliability; econometrics. *Mailing Add:* Bellcore 445 South St Morristown NJ 07960

DALAL, VIKRAM, b Bombay, India, Feb 4, 44; US citizen. SEMICONDUCTOR PHYSICS, ENGINEERING. *Educ:* Univ Bombay, BS, 64; Princeton Univ, PhD(elec eng), 69, MPA, 75. *Prof Exp:* Res asst physics, Princeton Univ, 66-69; res scientist, RCA Labs, 69-74; mgr physics, Univ Del, 76-81; pres, Res & Develop, Chronar Corp, 81-83; MGR, SEMICONDUCTOR DEVICES, SPIRE CORP, 83- *Mem:* Inst Elec & Electronic Engrs; Am Physical Soc. *Res:* Semiconductor physics; solar energy; energy and environment; economic growth and environmental policy. *Mailing Add:* 1925 Scholl Iowa State Univ Ames IA 50011

D'ALARCAO, HUGO T, b Lisbon, Port, Feb 15, 37; m 56; c 3. ALGEBRA, ARTIFICIAL INTELLIGENCE. *Educ:* Univ Nebr, Lincoln, BA, 59, MA, 61; Pa State Univ, PhD(math), 66. *Prof Exp:* Instr math, Univ Nebr, 61-62; instr, Univ Mass, 62-63; asst prof, State Univ NY Stony Brook, 66-71; assoc prof, 71-78, PROF MATH, BRIDGEWATER STATE COL, 78- *Concurrent Pos:* Vis res prof, Univ Chile, 69. *Mem:* Am Asn Artificial Intel; Asn Comput Mach; Inst Elec & Electronics Engrs Comput Soc; Port Math Soc. *Res:* Algebraic theory of semigroups; robotics. *Mailing Add:* 440 Pleasant Bridgewater MA 02401

DALBEC, PAUL EUCLIDE, b New Bedford, Mass, June 26, 35; m 63; c 1. SOLID STATE PHYSICS. *Educ:* Univ Notre Dame, MS, 59; Georgetown Univ, PhD(physics), 66. *Prof Exp:* Physicist, Phys Sci Lab, Melpar, Inc, Va, 59-60; head, Thin Films Dept, Appl Phys Lab, Gen Instrument Corp, NJ, 60-61; assoc prof, 71-78, PROF PHYSICS, YOUNGSTOWN STATE UNIV, 78- *Concurrent Pos:* Fac improv leave, Inst Recherche sur les Interfaces Solides, Namur, Belg, 78-79. *Mem:* Am Phys Soc; Mats Res Soc; Am Vacuum Soc; Am Soc Testing & Mat; Sigma Xi. *Res:* Photoelectric properties of solids; electronic properties of semiconductor and metallic thin films; vacuum ultraviolet spectroscopy; techniques for preparing thin films including epitaxial growth and sputtering; electron spectroscopy. *Mailing Add:* Dept Physics & Astron Youngstown State Univ Youngstown OH 44555

DALBY, FREDERICK WILLIAM, b Edmonton, Alta, May 5, 28; m 52; c 2. PHYSICS. *Educ:* Univ Alta, BSc, 50; Univ BC, MA, 62; Ohio State Univ, PhD(physics), 55. *Prof Exp:* Res physicist, Radiation Physics Lab, E I du Pont de Nemours & Co, 57-61; assoc prof, 61-66, PROF PHYSICS, UNIV BC, 66- *Res:* Molecular spectroscopy; chemical physics. *Mailing Add:* Dept Physics Univ BC 2075 Westbrook Pl Vancouver BC V6T 1W5 Can

DALBY, PETER LENN, b Flint, Mich, Feb 26, 43; m 67; c 2. VERTEBRATE BIOLOGY. *Educ:* Mich State Univ, BS, 65, MS, 68, PhD(zool), 74. *Prof Exp:* Curatorial technician, Mammal Div, Mich State Univ, 70-73; vis asst prof dept zool, Ohio State Univ, 73; asst prof & cur mammals, Dept Biol, Va Polytech Inst & State Univ, 73-75; vis asst prof dept biol, Univ Va, 75-76; PROF DEPT BIOL, CLARION UNIV, 76- *Concurrent Pos:* Nat Park Serv fel, 75-77; fel Western Pa Conservancy, 78-79. *Mem:* Am Soc Mammalogists; Ecol Soc Am; Wildlife Soc; Sigma Xi; Soc Study Evolution; Soc Syst Zool. *Res:* Ecology of South American grassland rodents; evolution, ecology, and behavior of eastern North American rodents; effects of strip-mining on vertebrate populations; fire ecology. *Mailing Add:* Dept Biol Clarion Univ Clarion PA 16214-1232

DAL CANTO, MAURO CARLO, b Soriano, Italy, Jan 1, 44; US citizen; m 68; c 2. NEUROPATHOLOGY, ELECTRON MICROSCOPY. *Educ:* Sci Lyceum, Livorno, Italy, dipl, 61; Sch Med, Univ Pisa, MD, 67. *Prof Exp:* Asst instr neuropath, Albert Einstein Col Med, 73-74; assoc path, 74-75, asst prof, 75-77, assoc prof, 77-81, dir neuropath, Childrens Mem Hosp, 79-81, DIR NEUROPATH & PROF PATH & NEUROL, SCH MED, NORTHWESTERN UNIV, 81- *Concurrent Pos:* Consult, Evanston Hosp, 74-; reviewer, J Neuropath & Exp Neurol, J Histochem & Cytochem, J Neuroradiol, & Kroc Found, 74-; mem, ad hoc study sects for Alzheimer's Dis Ctr grants, Nat Inst Aging, NIH, 84-85, Neurol C Study Sect, NIH, 85- *Mem:* Am Asn Neuropathologists; AAAS; NY Acad Sci. *Res:* Multiple sclerosis; ultrastructural, immunohistochemical and immunopathological studies of several experimental models of myelin degeneration produced either by autosensitization to myelin proteins or by infection with different viruses. *Mailing Add:* Dept Path Med Sch Northwestern Univ 303 E Chicago Ave Chicago IL 60611

DALE, ALVIN C, b Nashville, Tenn, Aug 12, 19; m 47; c 4. AGRICULTURAL & CIVIL ENGINEERING. *Educ:* Univ Tenn, BS, 41; Iowa State Univ, MS, 42, PhD(agr & civil eng), 50; Univ Chicago, cert, 44. *Prof Exp:* Jr civil engr, US Engrs Off, Ill, 42-43, asst civil engr, 46; from instr to asst prof agr eng, Iowa State Univ, 46-49; from asst prof to assoc prof, 49-56, PROF AGR ENG, PURDUE UNIV, 56- *Honors & Awards:* Award, Metal Bldg Mfrs Asn, 62. *Mem:* Am Soc Agr Engrs; Am Soc Eng Educ; Nat Soc Prof Engrs; Sigma Xi. *Res:* Environmental control and structural areas of farm structures; farm waste handling and disposal. *Mailing Add:* Dept Agr Eng Purdue Univ West Lafayette IN 47907

DALE, BEVERLY A, Detroit, Mich, Oct 19, 42; div; c 2. BIOLOGICAL SCIENCES. *Educ:* Univ Mich, PhD(biochem), 68. *Prof Exp:* from res assoc prof to res prof periodont & oral biol, 86-88, PROF ORAL BIOL, PERIODONT, UNIV WASH, 88- *Concurrent Pos:* NIH fel, 87. *Mem:* Am Soc Cell Biol; Am Asn Dent Res; Soc Investigative Dermat. *Res:* Differentiation and development of epithelia and skin; structural protein markers of epithelial differentiation, especially keratins and filaggrin. *Mailing Add:* Dept Oral Biol SB-22 Univ Wash Seattle WA 98195

DALE, BRIAN, b Goole, Eng, June 19, 30; m 55; c 3. ELECTRONICS, PHYSICS. *Educ:* Univ London, BSc, 51, PhD(physics), 54. *Prof Exp:* Res scientist, Res Labs, Gen Elec Co, Eng, 54-59; mgr device physics, Transitron Elec Corp, Mass, 59-62; chief engr, Sylvania Elec, Mass, 62-69; dir, Electronic Comp Lab, 69-78, dir consumer electronics, 78-81, DIR, ADV COMP LAB, GTE LABS, 81- *Concurrent Pos:* Vis assoc prof, Tufts Univ, 77. *Mem:* Sr mem Inst Elec & Electronics Engrs; Soc Info Display; Sigma Xi. *Res:* Very large scale integrated circuits; semiconductor processing technology and device physics; home computers. *Mailing Add:* 12 Townsend Rd Lynnfield MA 01940

DALE, DOUGLAS KEITH, b Ottawa, Ont, Oct 10, 24; m 51; c 5. MATHEMATICAL STATISTICS. *Educ:* Queen's Univ, Ont, BA, 47, Hons, 49; Univ NC, MS, 58. *Prof Exp:* Lectr math, Queen's Univ, Ont, 46-49; asst to math adv, Dominion Bur Statist, 49-51, chief sampling & anal, 51-60; chief statist div, Nat Energy Bd, 60-62; from asst prof to assoc prof, 62-65, chmn dept, 63-67, PROF MATH, CARLETON UNIV, ONT, 65- *Concurrent Pos:* Lectr, Carleton Univ, Can, 49-52 & 53-60; consult, Can Advert Res Found, 63-64 & Bur Broadcast Measurement, 63- *Mem:* Am Statist Soc; Soc; fel Royal Statist Soc. *Res:* Sampling theory; distribution theory; probability theory. *Mailing Add:* Dept Math Carleton Univ Colonel By Dr Ottawa ON K1S 5B6 Can

DALE, EDWARD EVERETT, JR, b Norman, Okla, Aug 18, 20; m 42; c 2. PLANT ECOLOGY, BOTANY. *Educ:* Univ Okla, BA, 42, MS, 47; Univ Nebr, PhD, 51. *Prof Exp:* Asst bot, Ind Univ, 46-47 & Univ Nebr, 47-50; asst prof biol, Baylor Univ, 51-52 & Tex Christian Univ, 52-57; from asst prof to assoc prof, 57-68, prof bact & chmn dept bot & bact, 75-78, PROF BOT, UNIV ARK, FAYETTEVILLE, 68- *Mem:* Bot Soc Am; Ecol Soc Am; Sigma Xi. *Res:* Grassland ecology; vegetation of Arkansas; bottomland hardwoods; forests of Central US. *Mailing Add:* Dept Bot & Bact Univ Ark Fayetteville AR 72701

DALE, EDWIN, b Louisville, Ky, Oct 17, 33; m 56; c 3. ENDOCRINOLOGY. *Educ:* Eastern Ky State Col, BS, 54; Univ Ky, MS, 56; Univ Iowa, PhD(zool), 60. *Prof Exp:* Res assoc endocrinol, Univ Iowa, 60-61; trainee steroid biochem, Worcester Found Exp Biol, 61-62; from instr anat to asst prof obstet & gynec, Col Med, Univ Ky, 62-66, asst prof zool, 66-68; asst prof, 68-72, ASSOC PROF OBSTET & GYNEC, SCH MED, EMORY UNIV, 72- *Mem:* AAAS. *Res:* Steroid biochemistry; distribution, metabolism and functions of steroids; chemical evaluation of the high-risk obstetric patient; menstrual dysfunction in athletes; correlation of various parameters of evaluation of the fetus; health related aspects of female athletes including nutrition, physical fitness, personality profiles and cardiovascular assessment. *Mailing Add:* Dept Gynec & Obstet Sch Med Emory Univ 69 Butler St SE Atlanta GA 30303

DALE, ERNEST BROCK, b Jackson Co, Okla, Dec 15, 18; m 46; c 3. PHYSICS. *Educ:* Univ Okla, BS, 40, MS, 44; Ohio State Univ, PhD(physics), 53. *Prof Exp:* Engr infrared anal, Phillips Petrol Co, 44-47; res assoc infrared detecting systs, Ohio State Univ, 49-50; proj leader semiconductors, Battelle Mem Inst, 52-57; assoc prof, 57-67, PROF PHYSICS, KANS STATE UNIV, 67- *Res:* Channeling and Rutherford scattering; musical acoustics; solid state physics; thin films; ultra fine particles. *Mailing Add:* E Brock Dale 2120 College Heights Rd Manhattan KS 66502

DALE, GEORGE LESLIE, b Glendale, Calif, July 14, 48; m 68. METABOLISM. *Educ:* Univ Calif, Riverside, BS, 70, Los Angeles, PhD(biochem), 75. *Prof Exp:* Fel biochem genetics, City Hope Med Ctr, 75-77, res scientist, 77-79; assoc mem, Scripps Clin & Res Found, 79-90; ASSOC PROF, OKLA UNIV MED SCH, 90- *Mem:* Am Soc Hematol. *Res:* red cell aging; drug delivery systems; red cell membrane metabolism. *Mailing Add:* Dept Med Univ Okla Health Sci Ctr PO Box 16901 Oklahoma City OK 73190

DALE, GLENN H(ILBURN), b Mountain Park, Okla, Aug 25, 23; m 47; c 3. CHEMICAL ENGINEERING. *Educ:* Univ Okla, BS, 44. *Prof Exp:* Chem engr, Phillips Petrol Co, 44-47, group leader pilot plant, 47-49, process design engr, 50-51, pilot plant supvr, 52-57, sect mgr separations, 58-60, br mgr separations processes, 60-71, staff engr petrol processes, 71-74, staff engr, Refining & Separations Br, 74-80, staff engr, Alt Energy Br, 81-82, sr eng assoc, 82-85; RETIRED. *Mem:* Fel Am Inst Chem Engrs; Am Chem Soc. *Res:* Equipment and process development in the separations field; distillation; adsorption; liquid-liquid extraction; crystallation; filtration; centrifugation; catalytic cracking of heavy oils. *Mailing Add:* 2900 Staats Dr Bartlesville OK 74006-2127

DALE, HOMER ELDON, b Minneapolis, Minn, June 11, 22; m 47; c 4. VETERINARY PHYSIOLOGY. *Educ:* Iowa State Col, DVM, 44, MS, 49; Univ Mo, PhD(physiol), 53. *Prof Exp:* Instr vet physiol, Iowa State Col, 47-49; asst prof, Agr & Mech Col Tex, 49-50; instr vet sci, Univ Wis, 50-51; instr vet physiol, 51-53, assoc prof, 53-56, PROF VET PHYSIOL & PHARMACOL, UNIV MO-COLUMBIA, 56- *Mem:* AAAS; Am Vet Med Asn; Am Physiol Soc. *Res:* Environmental physiology, endocrinology; immunology. *Mailing Add:* 623 Bluff Dale Dr Columbia MO 65201

DALE, J(AMES) D(OUGLAS), b Edmonton, Alta, Dec 5, 39; m 63; c 2. COMBUSTION, ENERGY CONSERVATION & UTILIZATION. *Educ:* Univ Alta, BS, 61, MS, 63; Univ Wash, PhD(mech eng), 69. *Prof Exp:* Sessional lectr mech eng, Univ Alta, 63-64; asst res off, Res Coun Alta, 64-65; from asst prof to assoc prof, 69-80, PROF MECH ENG, UNIV ALTA, 80- *Concurrent Pos:* Vis scholar, Univ Calif, Berkeley, 80-81. *Honors & Awards:* Ralph R Teetor Award, Soc Automotive Engrs, 79. *Mem:* Can Soc Mech Engrs; assoc mem Am Soc Mech Engrs; Soc Automotive Engrs; Combustion Inst (secy, Can sect, 83-85, chmn, 85). *Res:* Natural convection heat transfer to non-Newtonian fluids; air pollution from combustion sources; laser and plasma jet ignition for SI engines; low temp starting diesel engines; envelope energy losses for houses. *Mailing Add:* Dept Mech Eng Univ Alta Edmonton AB T6G 2M7 Can

DALE, JACK KYLE, pharmaceutical chemistry; deceased, see previous edition for last biography

DALE, JAMES LOWELL, b Olney, Ill, Dec 1, 22; m 46; c 1. PLANT PATHOLOGY. *Educ:* Eastern Ill State Col, BS, 52; Univ Ill, MS, 53, PhD(plant path), 56. *Prof Exp:* Asst plant path, Univ Ill, 52-56; exten plant pathologist, Col Agr, 56-57, from asst prof to assoc prof, 57-65, prof, 65-88, EMER PROF PLANT PATH, UNIV ARK, FAYETTEVILLE, 88- *Mem:* Am Phytopath Soc; Am Inst Biol Sci. *Res:* Corn and turf diseases; plant mycoplasma diseases. *Mailing Add:* Dept Plant Path Univ Ark Fayetteville AR 72701

DALE, JOHN IRVIN, III, b Knoxville, Tenn, May 14, 35; m 62, 81; c 1. ORGANIC CHEMISTRY. *Educ:* Carson-Newman Col, BS, 56; Univ NC, MA, 59; Univ Va, PhD(org chem), 63. *Prof Exp:* From chemist to sr chemist, Res Labs, 62-67, sr chemist, Org Chem Div, 67-90, PRIN CHEMIST, ORG CHEM DIV, TENN EASTMAN CO, 90- *Mem:* Am Chem Soc; Am Inst Chem Engrs; Sigma Xi. *Res:* Synthetic organic chemistry; heterocylic and aromatic compounds; textile dyes and intermediates. *Mailing Add:* Rte 6 Jonesboro TN 37659

DALE, VIRGINIA HOUSE, b Rochester, NY, Sept, 9, 51; m; c 2. ECOLOGICAL MODELING, FOREST SUCCESSION. *Educ:* Univ Tenn, Knoxville, BA, 74, MS, 75; Univ Wash, Seattle, PhD(math ecol), 80. *Prof Exp:* Res assoc, dept bot, Univ Wash, 80 & Forest Res Lab, Ore State Univ, 81; asst prof biol & math, Pac Lutheran Univ, 81-82; instr biol, Univ Puget Sound, 82-83; RES ASSOC, ENVIRON SCI DIV, OAK RIDGE NAT LAB, 84- *Concurrent Pos:* Res assoc, Col Forest Resources, Univ Wash, 81-83. *Mem:* Ecol Soc Am; Int Asn Landscape Ecol. *Res:* Impact of disturbances on vegetation including volcanoes (Mt St Helens), wind storm, fire, clearcutting, insect disturbances, air pollution and climate change. *Mailing Add:* Environ Sci Div Oak Ridge Nat Lab Oak Ridge TN 37831-6038

DALE, WESLEY JOHN, b Milwaukee, Wis, Aug 8, 21; m 49; c 1. ORGANIC CHEMISTRY. *Educ:* Univ Ill, BS, 43; Univ Minn, PhD(chem), 49. *Prof Exp:* Asst, Univ Minn, 43, res chemist, Govt Synthetic Rubber Res Prog, 43-46; from asst prof to prof chem, Univ Mo-Columbia, 49-66, asst to dean, Col Arts & Sci, 54-55, chmn dept chem, 61-64; univ res adminr, 69-72, actg provost & dean fac, 71, dean sch grad studies, 66-72, actg chancellor, 76-77, prof chem, Univ Mo-Kansas City, 66-, provost, 72-; DEPT CHEM, PRINCIPIA COL. *Concurrent Pos:* Staff assoc, Sci Facilities Eval Group, Div Instl Prog, NSF, 64, sr staff assoc, Sci Develop Eval Group, 64-66; mem, Bd Dirs, Sci Pioneers, 67 & Inst Community Studies, Kansas City, Mo, 70-73; trustee, Mid-Continent Regional Educational Lab, 72-73; mem, Adv Comt, US Army Command & Gen Staff Col, Ft Leavenworth, Kans, 73. *Mem:* Am Chem Soc; Sigma Xi. *Res:* Chemistry and spectra of vinylaromatic systems; boronic acids; cyclopropanes; carbenes. *Mailing Add:* 310 W 49th St Kansas City MO 64112

DALE, WILLIAM, b Long Branch, NJ, Sept 13, 42; m 78; c 1. POLYMER MECHANICS. *Educ:* Yale Univ, BS, 66; Cas Inst Technol, MS, 70; Case-Western Reserve Univ, PhD(macromolecular sci), 74. *Prof Exp:* Sr res engr, Monsanto Textiles Co, 74-78, TECHNOL SPECIALIST, MONSANTO PLASTICS & RESINS CO, 78- *Concurrent Pos:* Indust res fel, Univ Minn, 90-91. *Mem:* Am Phys Soc; Am Chem Soc; AAAS. *Res:* Mechanical properties of polymers; deformation and fracture morphology adhesives; orientation and anisotropy; processing; composites; structural biological polymers. *Mailing Add:* Monsanto Plastics & Resins Co 730 Worcester St Indian Orchard MA 01151

DALE, WILLIAM ANDREW, b Nashville, Tenn, Mar 13, 20; m 44; c 4. SURGERY. *Educ:* Davidson Col, AB, 41; Vanderbilt Univ, MD, 44; Am Bd Surg, dipl; Am Bd Thoracic Surg, dipl. *Prof Exp:* Intern surg, Strong Mem Hosp, 44-45, asst resident, 45-46, 48-49, chief resident surgeon, 50; instr surg, Univ Rochester, 50-53, instr physiol, 53-54, instr surg anat, 53-55, asst prof physiol, 54-57, surg, 54-58, dir surg exp lab, 53-58; clin asst prof surg, 58; PROF CLIN SURG, VANDERBILT UNIV, 58- *Concurrent Pos:* Vol fel physiol, Univ Rochester, 51-53; asst surgeon, Strong Mem Hosp, 51-58; attend surgeon, Genessee Hosp, 52-58; asst prof surg, Med Col Ala, 52; mem attend surg staff, Vanderbilt Univ Hosp, St Thomas Hosp & Mid-State Baptist Hosp. *Mem:* Fel Am Col Surg; AMA; Soc Univ Surg; Soc Vascular Surg; Int Cardiovasc Soc. *Mailing Add:* 312 Lynwood Blvd Nashville TN 37205-2908

D'ALECY, LOUIS GEORGE, b Staten Island, NY, Nov 21, 41; m 64; c 3. PHYSIOLOGY. *Educ:* NJ Col Med & Dent, DMD, 66; Sch Med, Univ Pa, PhD(physiol), 71. *Prof Exp:* Res asst anat, NJ Col Med & Dent, 62-64; vis scientist physiol, Sch Med, Univ Wash, 69-71, from instr to asst prof physiol, 71-73; asst prof, 73-77, ASSOC PROF PHYSIOL, SCH MED, UNIV MICH, 73- *Concurrent Pos:* NIH fel, 66-71 & spec fel, 71; estab investr, Am Heart Asn, 75-80. *Honors & Awards:* Excellence Award, Am Acad Oral Roentgenol, 66; Irving S Wright Award, Stroke Coun, Am Heart Asn, 72. *Mem:* AAAS; Am Heart Asn; Am Physiol Soc. *Res:* Biological sciences; cardiovascular physiology; control of blood flow to the brain, demonstrating that autonomic nerves, when activated directly or reflexly, produce major brain blood flow changes. *Mailing Add:* Sch Med Dept Physiol Univ Mich 7799 Med Sci II Ann Arbor MI 48109

DALEHITE, THOMAS H, b Memphis, Tenn, Sept 14, 18; m 42; c 3. AERONAUTICAL ENGINEERING, PHYSICS. *Educ:* Miss State Univ, BS, 39. *Prof Exp:* Gen engr, Air Proving Ground Command, 51-58, tech dir, 58-60, dir plans & progs, Off Secy Air Force, Supreme Hq Allied Powers Europe Tech Ctr, Netherlands, 60-62, sci adv foreign technol, Air Force Systs Command, 62-63, dep eng, Off Secy Air Force, 63-66, chief scientist, Air Proving Ground Ctr, 66-68 & Armament Develop & Test Ctr, 68-74; RETIRED. *Concurrent Pos:* Consult to US Defense Rep, NAtlantic Alliance, 60-62, Systs Res Lab, Dayton, 74-77 & Lockheed Missile & Spacecraft, Sunnyvale, 74-78; mem, Coun Air Force Scientist, 62-63, group environ panel, Aeronaut Coord Bd, NASA-Dept Defense, 63-66, tech mgt coun, Air Force Systs Command, 66- & Small Arms Adv Comt, Adv Res Projs Agency, Dept Defense, 68-, dir, Vanguard Banks, Air Force Armament Mus Found. *Honors & Awards:* Exceptional Serv Decorations, Secy Air Force, 62, 65 & 68. *Mem:* Am Ord Asn; Am Inst Aeronaut & Astronaut. *Res:* Non-nuclear munitions, guided missiles systems and high performance aircraft exploratory, advanced and engineering development; global range instrumentation; electromagnetic warfare. *Mailing Add:* 47 Longwood Dr Shalimar FL 32579

DALES, SAMUEL, b Warsaw, Poland, Aug 31, 27; Can citizen; m 52; c 2. VIROLOGY, CELL BIOLOGY. *Educ:* Univ BC, BA, 51, MA, 53; Univ Toronto, PhD(zool, biochem), 57. *Prof Exp:* Nat Cancer Inst Can res fel cell biol, Univ Toronto, 57-60; res assoc cytol & cell biol, Rockefeller Inst, 60-61, asst prof virol & cytol, 61-66; from res assoc prof to res prof microbiol, Post Grad Med Sch, NY Univ & mem, Pub Health Res Inst City of NY, 66-76; prof bacteriol & immunol & chmn, 76-80, PROF MICROBIOL & IMMUNOL, UNIV WESTERN ONT, LONDON, 81- *Concurrent Pos:* Ed, J Cell Biol, 72-; fac fel, Josiah Macy Jr Found, 81-82. *Mem:* Harvey Soc; Am Soc Cell Biol; Electron Micros Soc Am; Am Soc Virol; Can Fedn Biol Soc; Fedn Am Soc Exp Biol; fel Royal Soc Can. *Res:* Cell virus interactions and the early events in the process of infection; cell fine structure and function; virus-induced demyelinating diseases. *Mailing Add:* Dept Microbiol & Immunol Univ Western Ont London ON N6A 5C1 Can

D'ALESANDRO, PHILIP ANTHONY, b Bound Brook, NJ, Apr 2, 1927; m 61. PARASITOLOGY, IMMUNOLOGY. *Educ:* Rutgers Univ, BSc, 52, MSc, 54; Univ Chicago, PhD(microbiol), 58. *Prof Exp:* Res assoc parasitol, Univ Chicago, 58-59; guest investr, Rockefeller Univ, 59-61, from asst prof to assoc prof, 61-75; ASSOC PROF PARASITOL, 75-, ASSOC PROF MICROBIOLOGY, COLUMBIA UNIV, 82- *Concurrent Pos:* USPHS res fel, 59-61 & res grant, 72-; asst ed, J Protozool, 64-65, ed, 80-88; adj prof, Rockefeller Univ, 75-; mem, Trop Med & Parasitol Study Sect, NIH, 76-80, chmn, 77-80; pres, NY Soc Trop Med, 86-87; actg head div trop med, 77- *Mem:* Fel AAAS; Am Soc Parasitol; Soc Protozool; Am Soc Trop Med & Hyg; Sigma Xi. *Res:* Parasitic hemoflagellates; immunological, biochemical and nutritional aspects of the host-parasite relationship. *Mailing Add:* Sch Pub Health Columbia Univ 630 W 168th St New York NY 10032

D'ALESSANDRO-BACIGALUPO, ANTONIO, b Buenos Aires, Arg, Apr 6, 26; m 59; c 2. PARASITOLOGY, TROPICAL MEDICINE. *Educ:* Univ Buenos Aires, MD, 51; Tulane Univ, MPH & TM, 57, PhD, 61. *Hon Degrees:* Prof, Universidad del Valle, Cali, Colombia, 84. *Prof Exp:* Asst parasitol, Med Sch, Univ Buenos Aires, 45-51, chief resident, 52-55, asst prof, 56-61; from asst prof to assoc prof, 61-80, PROF TROP MED, TULANE UNIV, 80- *Concurrent Pos:* Pan Am Health Orgn/WHO fel, 56-57; Dazian Found Med res fel, 57-59; univ fel trop med & pub health, Tulane Univ, 57-61; liaison officer for Tulane Univ, Int Ctr Med Res & Training, Cali, Colombia, 61-63, asst dir, 63-65, assoc dir, 75-76, gen coordr progs, 76-84, chief of mission, 76-84; vis prof, Univ Valle, Colombia, 61-84; consult, NIH, Bogota, 82-; mem sci & tech adv comt & res strengthening group, Spec Prog Trop Dis Res, WHO, 83-87. *Mem:* Am Soc Trop Med & Hyg; Royal Soc Trop Med & Hyg; Am Soc Parasitol; Arg Soc Parasitol; Colombian Soc Parasitol & Trop Med (pres, 80-81); Colombian Acad Med. *Res:* Parasitology, especially American trypanosomiasis, echinococcosis and New World dermal leishmaniasis; clinical, parasitological and epidemiological aspects of helminthic infections. *Mailing Add:* Dept Trop Med Tulane Univ New Orleans LA 70112

DALEY, DANIEL H, b Elmira, NY, Mar 9, 20; c 4. THERMODYNAMICS, AERODYNAMICS. *Educ:* Purdue Univ, BS, 42; Mass Inst Technol, SMAE, 46. *Prof Exp:* USAF, 42-49, 51-, Assoc prof & actg head mech eng, USAF Inst Technol, 52-55, atomic test support group, Eniwetock, 55-56, 483rd Troop Carrier Wing, Japan, 56-58, chief aerodyn sect, B-70 Prof Off, Wright Patterson AFB, 58-61; assoc prof aeronaut, USAF Acad, 61-64, prof & head dept, 64-65; head, Dept Mech Eng, Pakistan Air Force Col Aeronaut Eng, 65-67; PROF AERONAUT & HEAD DEPT, USAF ACAD, 67- *Mem:* Am Inst Aeronaut & Astronaut; Japan Soc Aeronaut & Space Sci. *Res:* Airplane performance; stability and control; thermodynamics; gas dynamics; aerospace propulsion. *Mailing Add:* Dept Aeronaut USAF Acad CO 80840

DALEY, DARRYL LEE, b Detroit, Mich, Sept 28, 50; m; c 3. ANIMAL PHYSIOLOGY, ZOOLOGY. *Educ:* Wayne State Univ, BS, 72, MS, 75; Univ Ill, PhD(biol), 78. *Prof Exp:* POSTDOCTORAL RES ASSOC, SECT NEUROBIOL & BEHAV, CORNELL UNIV, 78-; AT DEPT PHYSIOL, NAT COL CHIROPRACTIC, LOMBARD, ILL. *Mem:* AAAS; Soc Neurosci. *Res:* Neurophysiological mechanisms underlying simple behaviors at the level of single identifiable neurons. *Mailing Add:* Nat Col Chiropractic 200 E Roosevelt Rd Lombard IL 60148

DALEY, HENRY OWEN, JR, b Quincy, Mass, June 18, 36; m 61; c 3. PHYSICAL CHEMISTRY. *Educ:* Mass State Col Bridgewater, BS, 58; Boston Col, PhD(chem), 64. *Prof Exp:* Res chemist, Am Cyanamid Co, 63-64; asst prof phys chem, 64-67, assoc prof, 67-71, PROF CHEM, BRIDGEWATER STATE COL, 71- *Mem:* Am Chem Soc; Sigma Xi. *Res:* Determination of trace metals in an ecosystem. *Mailing Add:* 115 Robinswood Rd South Weymouth MA 02190

DALEY, LAURENCE STEPHEN, b Liverpool, Eng, Sept 21, 36; US citizen; m 70; c 3. PLANT PHYSIOLOGY, BIOCHEMISTRY. *Educ:* Univ Fla, BSA, 64, MSA, 65; Univ Calif, Davis, PhD(plant physiol & biochem), 75. *Prof Exp:* Fel, Univ Ga, 75-77 & Boyce Thompson Inst, 77-78; asst prof, 78-79, ASSOC PROF BIOL SCI, EAST TEX STATE UNIV, 81- *Concurrent Pos:* Vis prof, Mont State Univ, 81. *Mem:* Am Soc Plant Physiologists; AAAS; NY Acad Sci. *Res:* Phosynthesis, carbon fixation; plant cell wall enzymes and their role in plant resistance to fungal attack; horticultural crops, chemistry and physiology of dormancy; cultural practice in CAM plants; value of reiteration in teaching botanical sciences; tropical crop biochemistry. *Mailing Add:* Dept Hort Ore State Univ Corvallis OR 97332

DALGARNO, ALEXANDER, b London, Eng, Jan 5, 28; m 57; c 4. THEORETICAL PHYSICS, ASTROPHYSICS. *Educ:* Univ Col, Univ London, BSc, 47, PhD(physics), 51. *Hon Degrees:* AM, Harvard Univ, 67; DSc, Queen's Univ, Belfast, 80. *Prof Exp:* Prof math, Queen's Univ, Belfast, 51-67; actg dir, Harvard Col Observ, 71-72, prof astron, 67-77, chmn dept, 71-76, PHILLIPS PROF ASTRON, HARVARD UNIV, 67- *Concurrent Pos:* Mem, Smithsonian Astrophys Observ, 67- *Honors & Awards:* Int Acad Quantum Molecular Sci Prize, 67; Hodgkins Medal, Smithsonion Inst, 77; Davisson-Germer Award, Am Phys Soc, 80; Meggars Award, Am Optic Soc, 86; Gold Medal Roy Astron Soc, 86. *Mem:* Fel Am Acad Arts & Sci; fel Am Geophys Union; Am Astron Soc; fel Brit Inst Physics & Phys Soc; fel Royal Soc; fel Am Phys Soc. *Res:* Theoretical atomic and molecular physics; planetary atmospheres; quantum chemistry; astrophysics. *Mailing Add:* Harvard Col Observ 60 Garden St Cambridge MA 02138

DALGLEISH, ARTHUR E, b Wilmington, Calif, Aug 14, 20; m 43; c 2. ANATOMY. *Educ:* La Sierra Col, BA, 45; Loma Linda Univ, MS, 60; Stanford Univ, PhD(anat), 64. *Prof Exp:* Instr, 64-65, asst prof, 65-75, ASSOC PROF ANAT, SCH MED, LOMA LINDA UNIV, 75- *Res:* Development of the limbs of the mouse. *Mailing Add:* Dept Anat & Dent Educ Loma Linda Univ Loma Linda CA 92354

DALGLEISH, ROBERT CAMPBELL, b Paisley, Scotland, Mar 31, 40; US citizen; m 62; c 3. SYSTEMATIC ENTOMOLOGY. *Educ:* San Diego State Col, BS, 62; Cornell Univ, PhD(entom), 67. *Prof Exp:* DIR, EDMUND NILES HUYCK PRESERVE, INC, 66- *Concurrent Pos:* Asst prof biol, Union Col, NY, 69-74; secy-treas, Orgn of Inland Biol Field Stas, 69-73, ed Newslett, 77-79; mem bd gov, Am Inst Biol Sci, 73-79. *Mem:* AAAS; Royal Entom Soc London; Entom Soc Can; Am Entom Soc; Ecol Soc Am. *Res:* Taxonomy and biology of biting lice. *Mailing Add:* Dir Off Grants & Contracts Southeastern Mass Univ North Dartmouth MA 02747

DALIA, FRANK J, b New Orleans, La, Nov 10, 28; m 51; c 4. STRUCTURAL ENGINEERING. *Educ:* Tulane Univ, BS, 49, MS, 52, PhD(econs), 64. *Prof Exp:* Civil engr, F C Gandolfo & Assocs, 49-52; assoc prof, 60-68, PROF CIVIL ENG, TULANE UNIV, 68-; CONSULT ENGR, 55- *Mem:* Am Soc Civil Engrs; Nat Soc Prof Engrs. *Res:* Prestressed concrete; engineering economics. *Mailing Add:* Dept Civil Eng Tulane Univ New Orleans LA 70118

DALINS, ILMARS, b Aizviki, Latvia, Apr 10, 27; US citizen; c 2. SURFACE PHYSICS, SEISMOLOGY. *Educ:* Tex Lutheran Col, BA, 52; Univ Tex, Austin, MA, 53; Univ Cincinnati, PhD(appl sci), 56. *Prof Exp:* Sr electronics scientist dielec mat, Antenna & Radome Group, NAm Aviation Inc, 56; elec propulsion & surface sci, Appl Res, Flight Propulsion Lab, Gen Elec Co, 56-58; sr staff scientist elec propulsion & surface physics, Appl Res, Astronaut Div, Gen Dynamics Corp, 58-60; PHYSICIST SURFACE SCI & SEISMOL, SPACE SCI LAB, MARSHALL SPACE FLIGHT CTR, NASA, 60- *Mem:* Am Inst Physics; Sigma Xi; Inst Elect & Electronics Engrs. *Res:* Surface science of semiconductors (infra-red detector materials) and metals, including molten state (surface science of liquids) for material processing in space applications, solar energy conversion and microelectronics; acoustic-seismic resonance effects produced by rockets (space shuttle) and other atmospheric noise sources. *Mailing Add:* 402 Cumerland Dr SE Huntsville AL 35803

D'ALISA, ROSE M, b Savona, Italy, Mar 14, 48; US citizen. PROTEIN ISOLATION & PURIFICATION, IMMUNOCHEMISTRY. *Educ:* Lehman Col, AB, 69; Columbia Univ, MS, 74, PhD(microbiol), 75. *Prof Exp:* Fac fel virol, Rockefeller Univ, 75-78; res assoc immunol, Columbia Univ, 78-81; sr res scientist, 81-84, ASSOC RES FEL, REVLON HEALTH CARE GROUP, 84- *Concurrent Pos:* Proj mgr, NIH contract. *Mem:* Sigma Xi; NY Acad Sci; Am Asn Immunologists. *Res:* Biochemical characterization of proteins; isolation of proteins from human plasma for therapeutic replacement in deficient individuals; proteins are isolated, purified, characterized, formulated, filtered and lyophilized. *Mailing Add:* 94 Country Club Dr Warrington PA 18976

DALLA, RONALD HAROLD, b Silverton, Colo, Mar 7, 42; m 62; c 3. ALGEBRA. *Educ:* Ft Lewis Col, BA, 64; Univ Wyo, MS, 66, PhD(math), 71; Wash State Univ, MEd, 76. *Prof Exp:* Instr, Colo State Univ, 66-67 & Univ Wyo, 67-70; asst prof, 70-74, assoc prof, 74-78, PROF MATH, EASTERN WASH UNIV, 78-, CHMN, DEPT MATH & COMPUT SCI, 78- *Mem:* Am Math Soc; Math Asn Am; Nat Coun Teachers Math. *Res:* Counting the number of solutions of matric equations over finite fields; finding a canonical form for orthogonal similarity of matrices over finite fields. *Mailing Add:* Dept Math Eastern Wash Univ MS 32 Cheney WA 99004

DALLA BETTA, RALPH A, b Craig, Colo, Mar 7, 45. SURFACE CHEMISTRY. *Educ:* Colo Col, BS, 67; Stanford Univ, PhD(phys chem), 72. *Prof Exp:* Res scientist, Sci Res Staff, Ford Motor Co, 71-75; consult prof, dept chem eng, Stanford Univ, 76-; VPRES, CATALYTICA ASSOCS, 76- *Mem:* Am Chem Soc; NAm Catalysis Soc. *Res:* Catalysis and surface chemistry; catalytic automotive pollution control; adsorption on and infrared spectroscopy of supported metals and zeolites; kinetics of methane synthesis and catalyst deactivation and poisoning; process development and catalyst testing; catalytic combustion. *Mailing Add:* 430 Ferguson Dr Bldg 3 Mountain View CA 94043

DALLA LANA, I(VO) G(IOVANNI), b Trail, BC, July 5, 26; m 56; c 5. CHEMICAL ENGINEERING. *Educ:* Univ BC, BASc, 48; Univ Alta, MSc, 53; Univ Minn, PhD(chem eng), 58. *Prof Exp:* Develop engr, Consol Mining & Smelting Co, 48-51; instr, 52-53, from asst prof to assoc prof, 58-68, PROF CHEM ENG, UNIV ALBERTA, 68- *Mem:* Am Inst Chem Engrs; fel Chem Inst Can. *Res:* Kinetics and heterogeneous catalysis; applications of IR spectroscopy; surface chemistry; chemical economics and design; hydroprocessing of heavy oils. *Mailing Add:* Dept Chem Eng Univ Alta Edmonton AB T6G 2E1 Can

DALLAM, RICHARD DUNCAN, b Kansas City, Mo, Dec 12, 25; m 51; c 2. BIOCHEMISTRY. *Educ:* Univ Mo, AB, 48, MA, 50, PhD(biochem), 52. *Prof Exp:* Res assoc biochem, Univ Mo, 52-53; from instr to assoc prof, 53-68, PROF BIOCHEM, SCH MED, UNIV LOUISVILLE, 68- *Concurrent Pos:* Estab investr, Am Heart Asn, 60-65; vis prof biochem, Univ Leichester, Leichester, England, 66-67. *Mem:* Am Soc Biol Chem; Biophys Soc. *Res:* Oxidation phosphorylation; chemical fraction of mammalian spermatozoa and cellular particulates; cytochemistry of mitochondria; lipoproteins related to enzyme systems; vitamin K; metabolism of dietary sulfates; acetate metabolism. *Mailing Add:* Dept Biochem Health Sci Ctr Sch Med Univ Louisville Louisville KY 40292

DALLDORF, FREDERIC GILBERT, b New York, NY, Mar 12, 32; m 56; c 3. PATHOLOGY. *Educ:* Bowdoin Col, BA, 54; Cornell Univ, MD, 58; Am Bd Path, cert anat & clin path, 65. *Prof Exp:* Intern, path, New York Hosp, 58-59, resident, 59-60; resident, NC Mem Hosp, Chapel Hill, 60-63; projs coordr, Path Div, US Army Biol Labs, Ft Detrick, Md, 63-65; from asst prof to assoc prof, 65-73, PROF PATH, SCH MED, UNIV NC, CHAPEL HILL, 73-, COURSE DIR PATH, 65- *Concurrent Pos:* NIH grant, 60-63; med dir blood bank, NC Mem Hosp, 65-66 & autopsy serv, 66- *Mem:* Am Asn Path. *Res:* Mechanisms of disease; natural history of rheumatic heart disease; endocrinologic aspects of arteriosclerotic vascular disease in man; mechanisms of shock and death in bacterial septicemias; capillary permeability and bacterial toxins; experiment arthritis; cardiac pathology. *Mailing Add:* Dept Path Univ NC CB No 7525 Chapel Hill NC 27514

DALLEY, ARTHUR FREDERICK, II, b Rupert, Idaho, Jan 28, 48; m 69; c 3. GROSS ANATOMY, MORPHOLOGY. *Educ:* Univ Utah, BS, 70, PhD(anat), 75. *Prof Exp:* ASSOC PROF ANAT, SCH MED, CREIGHTON UNIV, 74- *Mem:* Am Soc Zoologists; Am Inst Biol Sci; Am Asn Univ Professors; Am Asn Clin Anatomists. *Res:* Comparative morphology; micro-anatomy; ultrastructure and morphogenesis of vascular and microcirculatory systems, emphasizing venous plexuses sinusoidal structures, arteriovenous anastomoses and lymphaticovenous communications; radiological imaging. *Mailing Add:* Dept Anat Creighton Univ 2500 Cal St Omaha NE 68178

DALLEY, JOSEPH W(INTHROP), b Aberdeen, Idaho, Aug 12, 18; m 43; c 5. ENGINEERING MECHANICS, AERONAUTICAL ENGINEERING. *Educ:* Univ Tex, BS, 47, MS, 51, PhD(eng mech), 59. *Prof Exp:* Stress analyst, McDonnell Aircraft Corp, 47-48; instr mech eng, Univ Tex, Austin, 48-49, instr eng mech, 49-51, asst prof aeronaut eng, 51-59; prof & head, Dept Aeronaut Eng, Univ Wichita, 59-60; prof eng mech & head dept, 60-70, prof aerospace & eng mech, 70-73, assoc dean, col eng, 73-83; CONSULT, DALLEY MCWHERTER & ASSOCS, 85- *Concurrent Pos:* Res engr, Defense Res Lab, Univ Tex, Austin, 48-59; consult, Boeing Aircraft Co, 59-60 & LTV Vought Aeronaut, 60-77; vis prof, USMA West Point, 83-85. *Mem:* Am Inst Aeronaut & Astronaut; fel Soc Exp Stress Anal (pres, 75-76); Am Soc Eng Educ; Sigma Xi; Nat Soc Prof Engrs. *Res:* Experimental mechanics; aircraft structures; structural dynamics. *Mailing Add:* 20814 Highland Lakes Dr Lago Vista TX 78645

DALLMAN, MARY FENNER, CENTRAL NERVOUS SYSTEM, ADRENAL REGULATION. *Educ:* Stanford Univ, PhD(physiol), 67. *Prof Exp:* PROF PHYSIOL, UNIV CALIF, SAN FRANCISCO, 70- *Mailing Add:* Dept Physiol Univ Calif San Francisco CA 94143

DALLMAN, PAUL JERALD, b Washington, DC, July 7, 39. BLOCKING, CLOCKWORKS. *Educ:* Univ Md, BSCE, 63, MMus, 72. *Prof Exp:* Engr II, Wash Suburban Sanit Comn, 63-72; engr, Nat Found & Testing, Md, 73-79; site engr, Greene Acres, 79-86; engr, Joyce Eng Corp, 87-88; proj engr, Greenman-Pederson Inc, 88-89; civil engr, MK Enterprises, 89-90; LECTR & DEMONSTR, SMITHSONIAN INST, 77- *Concurrent Pos:* Reporter & ed, Wash Star Newspaper, 69-76. *Res:* History of home entertainment devices and sound recording technology; mass production of software for these devices; design of buttresses for underground pipe forces. *Mailing Add:* 125 Irving St Laurel MD 20707-4503

DALLMAN, PETER R, b Berlin, Ger, Nov 19, 29; US citizen; m 59; c 3. PEDIATRICS, HEMATOLOGY. *Educ:* Dartmouth Col, BA, 51; Harvard Med Sch, MD, 54. *Prof Exp:* From instr to asst prof pediat, Stanford Univ, 63-68; from asst prof to assoc prof, 68-76, PROF PEDIAT, SCH MED, UNIV CALIF, SAN FRANCISCO, 76- *Concurrent Pos:* NIH award, 66-; res biochem, Wenner-Gren Inst, Stockholm, 67-68. *Honors & Awards:* Nutrit Award, Am Acad Pediat, 82. *Mem:* AAAS; Am Pediat Soc; Soc Pediat Res; Am Acad Pediat. *Res:* Nutritional anemias; iron metabolism. *Mailing Add:* Dept Pediat Sch Med Univ Calif Box 0106 San Francisco CA 94143

DALLMEYER, R DAVID, b St Louis, Mo, Feb 8, 44; m 75. GEOCHRONOLOGY. *Educ:* Long Island Univ, BS, 67; State Univ NY, Stony Brook, MS, 69, PhD(geol), 72. *Prof Exp:* Asst prof geol, 72-76, MEM FAC, UNIV GA, 76- *Mem:* Geol Soc Am; Am Geophys Union. *Res:* Thermal and deformational evolution of metamorphic terranes; emphasis on chronologic development of orogenic events and their interactions. *Mailing Add:* Dept Geol Univ Ga Athens GA 30602

DALLOS, ANDRAS, b Szeged, Hungary, Apr 25, 21; m 47; c 2. ELECTRONIC PHYSICS. *Educ:* Univ Budapest, PhD(electron multiplier), 48; Hungarian Acad Sci, Cand Tech Sci, 62, Dr Tech Sci, 67. *Prof Exp:* Scientist electron tubes, Tungsram Res Lab, Budapest, Hungary, 45-50; dept head electron & vacuum devices, Res Inst Telecommun, Budapest, 51-68; sci adv, Tungsram Res Lab, Budapest, 68-70; sr res & develop physicist, 73-, PRIN ENGR, RAYTHEON CO, WALTHAM, MASS. *Concurrent Pos:* Vis scientist, Univ Ill, Urbana, 64-65; chief tech adv, UNESCO, Univ Islamabad, Pakistan, 70-73. *Honors & Awards:* Kossuth Prize, Hungarian Govt, 55. *Mem:* Sr mem Inst Elec & Electronics Engrs; Soc Info Display. *Res:* Electrochromism in solids as a passive electro-optical transducer for matrix addressed flat panel displays; high current density cathodes for displays; electron beam litography. *Mailing Add:* Raytheon Co Mat & Tech Ctr 190 Willow St Waltham MA 02254

DALLOS, PETER JOHN, b Budapest, Hungary, Nov 26, 34; US citizen; m 77; c 1. BIOPHYSICS & PHYSIOLOGY OF HEARING. *Educ:* Ill Inst Technol, BS, 58; Northwestern Univ, MS, 59, PhD(elec/biomed eng), 62. *Prof Exp:* Asst res engr, Am Mach & Foundry Co, 59, consult engr, Mech Res Div, 59-60; from asst prof to assoc prof audiol, Northwestern Univ, 62-69, chmn, Dept Neurobiol & Physiol, 81-84 & 86-87, assoc dean, Col Arts & Sci, 84-85, PROF AUDIOL & BIOMED ENG & OTOLARYNGOL, NORTHWESTERN UNIV, EVANSTON, 69-, PROF NEUROBIOL & PHYSIOL, 81-, JOHN EVANS PROF NEUROSCI, 86- *Concurrent Pos:* mem commun dis res training comt, Nat Inst Neurol Dis & Stroke, 69-73; vis scientist, Karolinska Inst, Stockholm, Sweden, 77-78; assoc ed, Hearing Res, 77-79, J Neurosci, 85-89 & Trends Neurosci, 86-90; Guggenheim fel, 77; mem, sci rev bd, Am Hearing Res Found, 78-; mem, behav & neurosci rev panel no 5, Nat Inst Neurol & Commun Dis & Stroke, NIH, 81-84, chmn, 82-84, mem, Nat Adv Neurol & Commun Dis & Stroke Coun, 84-87; Jacob Javits neurosci investr award, NIH, 84-90. *Honors & Awards:* Beltone Award, 77; Amplifon Inst Res Prize, 84. *Mem:* Fel AAAS; fel Acoust Soc Am; Int Soc Audiol; fel Inst Elec & Electronics Engrs; Soc Neurosci; Asn Res Otolaryngol (pres-elect, 82-83). *Res:* Biophysics and physiology of hearing; physiological acoustics; biomedical engineering. *Mailing Add:* Auditory Res Lab Northwestern Univ 2299 Sheridan Rd Evanston IL 60201

DALLY, EDGAR B, research instrumentation, cathodes, for more information see previous edition

DALLY, JAMES WILLIAM, b Sardis, Ohio, Aug 2, 29; m 55; c 3. MECHANICAL ENGINEERING, MECHANICS. *Educ:* Carnegie Inst Tech, BS, 51, MS, 53; Ill Inst Tech, PhD(mech), 58. *Prof Exp:* Engr in training, Mesta Mach Co, 51-53; assoc res engr, Armour Res Found, 53-55, res engr, 55-57, sr res engr, 58; asst prof mech, Cornell Univ, 58-61; asst dir res, IIT Res Inst, 61-64; prof mech, Ill Inst Technol, 64-71; prof mech eng, Univ Md, College Park, 71-79; dean eng, Univ RI, 79-82; sr eng, IBM, 82-84; PROF MECH ENG, UNIV MD, 84- *Concurrent Pos:* Adv, NSF, 78-80 & 90-91. *Honors & Awards:* Frocht Award, Soc Exp Stress Anal, 76; William Murray Lectureship Award, Soc Exp Stress Anal, 79. *Mem:* Fel Soc Exp Stress Anal (pres, 71); fel Am Soc Mech Engrs; fel Am Acad Mech (pres); Nat Acad Eng. *Res:* Experimental stress analysis; fracture mechanics; microelectronics packaging. *Mailing Add:* Mech Eng Dept Univ Md College Park MD 20742

DALLY, JESSE LEROY, b Fayette Co, Pa, Sept 3, 23; m 50; c 2. GEOLOGY. *Educ:* WVa Univ, BS, 47; Columbia Univ, MA, 49, PhD(geol), 56. *Prof Exp:* Instr geol, WVa Univ, 49-56; chief paleontologist, Esso Stand, Inc, Turkey, 56-57, supvr cent lab, 57-58; staff geologist, Pan-Am Petrol Corp, Standard Oil Co, Ind, 58-68; vpres, Desanna Corp, 68-78; OWNER, JESSE L DALLY & ASSOC, 78- *Concurrent Pos:* Coop geologist, WVa Geol Surv, 53-56. *Mem:* Geol Soc Am; Soc Econ Paleontologists & Mineralogists; Am Asn Petrol Geologists. *Res:* Stratigraphy; sedimentology; paleontology; basin analysis. *Mailing Add:* 15840 FM 529 Bank One Bldg Suite 225 Houston TX 77095-2504

DALMAN, G(ISLI) CONRAD, b Winnipeg, Man, Apr 7, 17; US citizen; m 41; c 4. MICROWAVE & ENGINEERING ELECTRONICS. *Educ:* City Col New York, BEE, 40; Polytech Inst Brooklyn, MEE, 47, DEE, 49. *Prof Exp:* Lectr elec eng, City Col New York, 52-54; adj prof, Polytech Inst Brooklyn, 53-56; mfg engr electronic tubes, RCA, 40-45; mem tech staff, Bell Tel Labs, 45-47; sect head microwave tubes, Sperry Gyroscope Co, 47-56; dir, Sch Elec Eng, 75-80, prof, 56-87, EMER PROF ELEC ENG, CORNELL UNIV, 87- *Concurrent Pos:* Consult, Electronic Tube Div, Westinghouse Elec Corp, 56-62, Aeronaut Lab, Cornell Univ, 56-57 & TRW, 80-81; consult & founder, Cayuga Assoc Inc; mgr, China Proj, Chiao Tung Univ, Hsinchu, Taiwan, 82-83. *Mem:* Fel AAAS; fel Inst Elec & Electronics Engrs; Sigma Xi. *Res:* Electron devices; solid state microwave and millimeter wave devices; electrical noise problems; microwave subsystems; electrical engineering education; physical electronics. *Mailing Add:* 506 Hanshaw Rd Ithaca NY 14850

DALMAN, GARY, b Grandville, Mich, Oct 1, 36; m 58; c 3. ORGANIC CHEMISTRY. *Educ:* Hope Col, BA, 58; Okla State Univ, PhD(reactions of mercaptans), 63. *Prof Exp:* Res chemist, 62-67, proj leader, Benzene Res Lab, 67-69, group leader, Org Chem Prod Res Dept, 69-70, res mgr, 70-73, res mgr, Western Div Res Labs, 73-79, lab dir, Western Div Res & Develop, 79-83, LAB DIR, AGR PROD RES, DOW CHEM CO, 83- *Mem:* Am Chem Soc; Sigma Xi; AAAS. *Mailing Add:* 1820 Argonne Dr Walnut Creek CA 94598

DALMASSO, AGUSTIN PASCUAL, b Cordoba, Arg, Apr 15, 33; m 60; c 3. MEDICINE, IMMUNOLOGY. *Educ:* Univ Cordoba, MD, 58, DrMedS, 63. *Prof Exp:* Instr physiol, Univ Cordoba, 58-60; head sect immunol, Inst Med Res, Univ Buenos Aires, 66-70; assoc prof, 70-74, PROF LAB MED & PATH, MED SCH, UNIV MINN, MINNEAPOLIS, 74- *Concurrent Pos:* Res fel, Univ Minn, 60-63; res fel immunol, Scripps Clin & Res Found, 63-66; chief, Immunol & Blood Bank, Vet Admin Med Ctr, Minneapolis, 70- *Mem:* Am Asn Immunol; Am Soc Exp Path. *Res:* Role of the thymus in immunology; chemistry and biology of the complement system and of cell membranes. *Mailing Add:* Dept Lab Med & Path Univ Minn Minneapolis MN 55417

DALPE, YOLANDE, b Waterloo, Que, Dec 19, 48. MYCOLOGY, MYCORRHIZAS. *Educ:* Montreal Univ, BSc, 72, MSc, 75; Univ Paul Sabatier, Toulouse, France, DSc, 81. *Prof Exp:* MYCOLOGIST & TAXONOMIST, BIOSYST RES CTR AGR CAN, 81- *Mem:* Can Bot Asn; Asn Can Fr Pour L'avan Des Sci; Mycol Soc Am; Brit Mycol Soc. *Res:* Ericoid mycorrhizae, studies on the genus Oidiodendron; taxonomy of Endogonaceae, their collection, identification and classification, inventory of soil endomycorrhizal fungi. *Mailing Add:* Biosyst Res Ctr Saunders Bldg Exp Farm Ottawa ON K1A 0C6 Can

DALPHIN, JOHN FRANCIS, b Brooklyn, NY, May 7, 40; m 67; c 5. COMPUTER SCIENCE. *Educ:* Clarkson Univ, BME, 62, PhD(math & comput sci), 73; Univ NH, MS, 64. *Prof Exp:* Systs engr, Int Bus Mach Corp, 64-65; instr data processing, US Army Adj Gen Sch, 65-66; assoc prof comput technol & chmn dept, Ind Univ-Purdue Univ, Indianapolis, 67-71; NSF sci fac fel, Clarkson Univ, 71-72; prof comput sci, chmn dept & dir comput serv, State Univ NY Col, Potsdam, 73-77; dean, Sch Eng, Technol & Nursing, Ind Univ-Purdue Univ, Ft Wayne, 77-83; vpres acad affairs, 83-85, prof comp sci, Norwich Univ, 85- 87; prof & chair comp sci dept, Towson State Univ, 87-90; VCHANCELLOR ACAD AFFAIRS, IND UNIV EAST, 90- *Concurrent Pos:* Mem bd dir, Comput Sci Accreditation Bd, 85-87; nat lectr, Asn Comput Mach, 85-90. *Mem:* AAAS; Asn Comput Mach; Am Asn Higher Educ; Math Asn Am; Sr mem, Inst Elec & Electronics Engrs; Sigma Xi. *Res:* Computer science program evaluation and accreditation; curriculum development. *Mailing Add:* 710 Garwood Rd Richmond IN 47374

DALQUEST, WALTER WOELBERG, b Seattle, Wash, Sept 11, 17; m 40; c 1. VERTEBRATE ZOOLOGY. *Educ:* Univ Wash, BS, 40, MS, 41; Univ La, PhD(zool), 51. *Prof Exp:* Res assoc mammals, Mus Natural Hist, Kans, 45-49; fel zool, Univ La, 49-51, asst biochem, 51-52; PROF BIOL, MIDWESTERN UNIV, 52-; AQUATIC BIOLOGIST, STATE GAME & FISH COMN, TEX, 53- *Mem:* Am Soc Mammalogists; Soc Syst Zool. *Res:* Mammals and fishes. *Mailing Add:* 2715 Church St Wichita Falls TX 76308

DALRYMPLE, DAVID LAWRENCE, b Fredericktown, Ohio, Oct 26, 40. SCIENTIFIC COMPUTER APPLICATIONS. *Educ:* Col Wooster, AB, 62; Univ Vt, PhD(org chem), 66. *Prof Exp:* NSF fel, Harvard Univ, 66-67, res fel, 67-68; asst prof chem, Univ Del, 68-74, nuclear magnetic resonance spectroscopist, 75-76; software engr, Nicolet Magnetics, 76-83 & Gen Elec NMR Instruments, 83-85; SR FEL, NICOLET INSTRUMENT CORP, 85- *Mem:* AAAS; Am Chem Soc; Sigma Xi. *Res:* Nuclear magnetic resonance spectroscopy; applications of computers in chemistry. *Mailing Add:* 2934 Cimarron Trail 2 Madison WI 53719-2409

DALRYMPLE, DESMOND GRANT, b May 6, 38; Can citizen; m 65; c 1. MECHANICAL ENGINEERING, NUCLEAR ENGINEERING. *Educ:* Univ Man, BSc, 60; Univ Sask, MSc, 62; Univ Manchester, PhD(appl mech), 67. *Prof Exp:* Mem staff, Appl Res & Develop, Atomic Energy Can Ltd, 62-76, mgr, Mech Equip Develop, 76-79, mgr, Nondestructive Testing, 79-80, dir, Spec Proj Div, 80-85, mgr bus develop, 86-90, TECH CONSULT, 90- *Concurrent Pos:* Develop engr eng res, Atomic Energy Can Res Co, 67-76, br head mech eng develop, 76- *Mem:* Can Nuclear Soc; Can Soc Nondestructive Testing. *Res:* Out-reactor components; recycle fuel manufacture; equipment development. *Mailing Add:* Seven MacDonald Deep River ON K0J 1P0 Can

DALRYMPLE, GARY BRENT, b Alhambra, Calif, May 9, 37; m 59; c 3. ISOTOPE GEOLOGY, GEOCHRONOLOGY. *Educ:* Occidental Col, AB, 59; Univ Calif, Berkeley, PhD(geol), 63. *Prof Exp:* Geologist, Br Theoret Geophys, 63-70, geologist, Br Isotope Geol, 70-81, asst chief geologist, Western Region, 81-84, GEOLOGIST, BR ISOTOPE GEOL, US GEOL SURV, 84- *Concurrent Pos:* Prin investr, Apollo Lunar Samples, 68-71, 90-, vis prof, Stanford Univ, 71-72, 83-85, 90- *Mem:* Fel Am Geophys Union; Am Geophys Union (pres, 90-92); fel Geol Soc Am. *Res:* Isotope geology; potassium-argon dating of young volcanic rocks; geochronology of secular variation and of reversals of the earth's magnetic field; thermoluminescence of geologic materials; origin of Pacific Ocean seamounts; origin of Hawaiian Islands; ages of lunar impact melt rocks. *Mailing Add:* Br Isotope Geol US Geol Surv 345 Middlefield Rd Menlo Park CA 94025

DALRYMPLE, GLENN VOGT, b Little Rock, Ark, Dec 28, 34; m 55; c 2. MEDICINE, RADIOBIOLOGY. *Educ:* Univ Ark, BS, 56, MD, 58. *Prof Exp:* Resident radiol, Med Ctr, Univ Ark, 59; asst, Med Ctr, Univ Colo, 61-62, resident, 62, instr, 62-63; asst prof, Sch Med, Univ Ark, Little Rock, 65-68, assoc prof, Med Ctr, 68-71, head div nuclear med & radiation biol, 69-73, prof radiol, Depts Radiol, Biomet, Physiol & Biophys, Med Ctr, 71-76, chmn dept radiol, 73-76, prof med physics, 74-76; MEM STAFF, RADIOL CONSULT, 76- *Mem:* Am Col Radiol, Soc Nuclear Med; Radiation Res Soc; Radiol Soc NAm; AMA. *Res:* Aspects of radiobiology dealing with the effects of radiation on mammalian cells in culture; effects of radiation on the fetus; mathematical biology and use of computing machinery in biomedical research; clinical radiology and nuclear medicine. *Mailing Add:* Radiol Consult 110 Latvian Village W Apt 110-C Omaha NE 68124

DALRYMPLE, JOEL MCKEITH, b Salt Lake City, Utah, June 18, 39; m 66; c 2. VIROLOGY. *Educ:* Univ Utah, BS, 62, MS, 64, PhD(microbiol), 68. *Prof Exp:* Teaching asst microbiol, Univ Utah, 62-68, res assoc virol, 64-68; res virol, Walter Reed Army Inst Res, 68-73, investr med res virol, 73-80; CHIEF, DEPT VIROL BIOL, VIROL DIV, US ARMY MED RES INST INFECTIOUS DIS, 80- *Concurrent Pos:* Mem, Am Comt Arthropod-borne Viruses, 68-, subcomt interrelationships among catalogued arboviruses, 72- & subcomt appl molecular arbovirol, 75- *Mem:* Am Soc Microbiol; Soc Gen Microbiol; Am Soc Trop Med & Hyg; AAAS; Sigma Xi. *Res:* Biochemical and biophysical investigations of molecular biology of Togavirus antigens and their immunological interactions; ecological and epidemiological investigations of medically important arthropod transmitted viruses. *Mailing Add:* Dept Viral Biol Virol Div US Army Med Res Inst Infectious Dis Ft Detrick Frederick MD 21701-5011

DALRYMPLE, ROBERT ANTHONY, b Camp Rucker, Ala, May 30, 45; m 68; c 1. COASTAL ENGINEERING, MARINE SCIENCES. *Educ:* Dartmouth Col, AB, 67; Univ Hawaii, MS, 68; Univ Fla, PhD(coastal & civil eng), 73. *Prof Exp:* Asst eng, Coastal Eng Lab, Univ Fla, 68-71, res assoc, 71-73; from asst prof to assoc prof civil eng, 73-84, instr appl math, 78-84, asst dean, Col Eng, 79-82, PROF CIVIL ENG, COL MARINE STUDIES, UNIV DEL, 84-, DIR, CTR APPL COASTAL RES, 89- *Concurrent Pos:* Consult, Argonne Nat Lab & Exxon Prod Res, 77-, Amoco Prod Co, 85-; dir, Coastal & Offshore Eng & Res, 77- *Mem:* Fel Am Soc Civil Engrs; Am Geophys Union; Sigma Xi; Int Asn Hydrol Res; Am Shore & Beach Preserv Soc; Soc Indust & Appl Math. *Res:* Nearshore hydrodynamics and sediment transport on beaches; wave forces on structures; wave mechanics. *Mailing Add:* Dept Civil Eng Univ Del Newark DE 19711

DALRYMPLE, RONALD HOWELL, b Elizabeth, NJ, Nov 16, 43; m 74; c 2. MEAT SCIENCE, ANIMAL GROWTH. *Educ:* Del Valley Col, BS, 65; Va Polytech Inst & State Univ, MS, 69; Univ Wis, PhD(animal sci), 72. *Prof Exp:* Vis res scientist meat sci, Fed Inst for Meat Res, Fed Ministry for Nutrit, Agr & Forestry, Ger, 72-74; res scientist, Animal Sci, Agr Div, 74-78, mgr, Animal Prod Develop, Far East Region, 78-81, prin scientist, Nutrit & Physiol, 81-85, SR PROD DEVELOP MGR, GLOBAL ANIMAL INDUST DEVELOP, AM CYANAMID CO, 85- *Mem:* Am Meat Sci Asn; Am Soc Animal Sci; Am Registry Prof Animal Scientists; Am Inst Nutrit. *Res:* Control of muscle growth and lipid deposition in meat producing animals. *Mailing Add:* Agr Div Am Cyanamid Co PO Box 400 Princeton NJ 08543-0400

DALRYMPLE, STEPHEN HARRIS, b Austin, Tex, Dec 2, 32; m 52; c 4. EMULATION, SIMULATION. *Educ:* Univ Tex, BA, 54, MA, 59. *Prof Exp:* Staff mem nuclear, Los Alamos Sci Lab, 54-60; supvr numerical anal, Autonetics Div, NAm Aviation, 60-65; tech mgr, Planning Res Corp, 65-67; dir serv bur, Cent Comput Corp, 69-70; br mgr comput sci, McDonnell Douglas Astronaut Co, 70-81; SR SPECIALIST, SCI APPLNS INT CORP, 81- *Concurrent Pos:* Fel, AEC, 58. *Mem:* Asn Comput Mach; Inst Elec & Electronics Engrs. *Res:* Computer architecture, microprogramming and emulation; software and firmware tools to aid development of emulations; multiple emulations and operating systems; emulation of unusual architectures; systems programming; real-time programming; large scale simulations. *Mailing Add:* 1332 Arloura Way Tustin CA 92680-3544

DALTERIO, SUSAN LINDA, b Worcester, Mass, July 19, 49. REPRODUCTIVE BIOLOGY, ENDOCRINE PHARMACOLOGY. *Educ:* Boston Univ, BA, 71; Assumption Col, Mass, MA, 76; Tufts Univ, PhD(physiol psychol), 78. *Prof Exp:* Res asst, Worcester Found Exp Biol, 72-78; fel, 78-79, res asst prof, 79-84, RES ASSOC PROF, UNIV TEX HEALTH SCI CTR, 84- *Mem:* Soc Study Reproduction; Soc Neurosci; Endocrine Soc; Am Soc Andrology. *Res:* Elucidation of the effects of marihuana on the development and/or function of the male reproductive system in mice; studies also involving the role of gonadal steroids in male sexual behavior and the feedback mechanisms controlling the hypothalmic pituitary testicular axis. *Mailing Add:* Dept Pharmacol Univ Tex Health Sci Ctr San Antonio TX 78284-7764

DALTON, AUGUSTINE IVANHOE, JR, b Richmond, Va, Nov 17, 42; m 65; c 2. PHYSICAL ORGANIC CHEMISTRY. *Educ:* Va Mil Inst, BS, 65; Univ SC, PhD(org chem), 72. *Prof Exp:* Res asst new prod develop, Am Brands, 65-66; mem staff chem, US Army Chem Corps, 66-68; Robert A Welch fel, Rice Univ, 72-73; sr res chemist, 73-81, prin res chemist, 81-82, sr prin res chemist, 82-84, MGR, NEW TECHNOL AIR PROD & CHEM INC, 85- *Mem:* Am Chem Soc; Org Reactions Catalysis Soc; Am Inst Chemists; Sigma Xi; Tech Asn Pulp & Paper Indust. *Res:* Catalytic processes; pulp/paper; oxidations; peroxide reactions; industrial gases; combustion. *Mailing Add:* Air Prod & Chem Allentown PA 18195-1501

DALTON, BARBARA J, b Philadelphia, Pa, 1953. IMMUNOLOGY. *Prof Exp:* SR INVESTR, SMITH KLINE BEECHMAN PHARMACEUT, 83- *Mailing Add:* Dept Cellular Biochem & Immunol Smith Kline Beechman Pharmaceut UW-2101 PO Box 1539 King of Prussia PA 19406-0939

DALTON, COLIN, b Hull, Eng, July 19, 36; m 62; c 2. BIOCHEMICAL PHARMACOLOGY. *Educ:* Univ Hull, BSc, 60; State Univ NY Buffalo, PhD(biochem pharmacol), 65. *Prof Exp:* Sr biochemist, Hoffman-La Roche Inc, 65-70, res group chief biochem pharmacol, 70-76, res planning mgr, 76-; DIR BUS DEVELOP, DEPT PHARMACEUT OPERS, SCHERING-PLOUGH CORP, KENILWORTH, NJ. *Concurrent Pos:* Fel coun on arteriosclerosis, Am Heart Asn, 70. *Mem:* AAAS; Am Soc Pharmacol & Exp Therapeut; Am Heart Asn. *Res:* Mechanism of action of anti-hyperlipidemic and anti-thrombotic drugs; regulation of prostaglandin biosynthesis in different tissues; factors controlling adipose tissue lipolysis; automation of analytical biochemistry procedures. *Mailing Add:* Dir Licensing Sterling Drug Inc Nine Great Valley Pkwy Malvern PA 19355

DALTON, DAVID ROBERT, b Chicago, Ill, Nov 16, 36; m 58; c 3. ORGANIC CHEMISTRY. *Educ:* Northwestern Univ, BA, 57; Univ Calif, Los Angeles, PhD(org chem), 62. *Prof Exp:* Res chemist, Dayton Labs, Monsanto Res Corp, 62-63; instr chem & fel, Ohio State Univ, 64-65; from asst prof to assoc prof, 65-73, PROF CHEM, TEMPLE UNIV, 73- *Concurrent Pos:* NIH fels, 62-; vis prof, Technion, 72, Yale Univ, 78 & Bryn Mawr Col, 89. *Honors & Awards:* Lindback Award, 75; Am Inst Chemists Award, 89. *Mem:* AAAS; Am Chem Soc; Royal Soc Chem. *Res:* Isolation, identification and synthesis of natural products; reaction mechanisms. *Mailing Add:* 143 Gulph Hills Rd Radnor PA 19087

DALTON, FRANCIS NORBERT, b Golden, Colo. SOIL PHYSICS, PLANT PHYSICS. *Educ:* Univ Calif, Riverside, BA, 65; Univ Wis-Madison, MS, 70, PhD(soil physics), 72. *Prof Exp:* Physicist, US Salinity Lab, Agr Res Serv, USDA, Riverside, Calif, 64-66; biometeorol specialist, dept water & soils, Univ Wis-Madison, 67-70, res asst, 70-72, instr soil physics, 72, res assoc, 72-74; consult, R J Dalton & Sons, Pioche, Nev, 74-81; SOIL PHYSICIST, US SALINITY LAB, AGR RES SERV, USDA, RIVERSIDE, CALIF, 82- *Concurrent Pos:* Consult, Univ Wis-Madison, 76; adj assoc prof, Univ Calif, Riverside, 85-; prin investr, US Israel Binat Agr Res & Develop Fund-Time Domain Reflectometry, 85- *Mem:* AAAS; Am Soc Agron; Soil Sci Soc Am; Sigma Xi. *Res:* Physical and mathematical studies of the transport of water and salts in the soil-plant-atmosphere continuum as it relates to the management of plants growing in a saline environment. *Mailing Add:* U S Salinity Lab 4500 Glenwood Dr Riverside CA 92501

DALTON, G(EORGE) RONALD, b Detroit, Mich, Sept 28, 32; m 58; c 3. NUCLEAR ENGINEERING. *Educ:* Univ Mich, BS, 54, PhD(nuclear eng), 60. *Prof Exp:* From asst prof to assoc prof, 60-68, actg chmn, Dept Nuclear Eng Sci, 68-69, PROF NUCLEAR ENG, UNIV FLA, 68- *Concurrent Pos:* Ford Found eng resident, Atomic Power Div, Westinghouse Elec Corp, 65-66; Oak Ridge Fusion Eng Design Ctr, 84-85. *Mem:* Am Soc Eng Educ; Am Nuclear Soc. *Res:* Nuclear systems and numerical analysis; digital computer graphics; neutron transport theory; theory of radiation detection. *Mailing Add:* Dept Nuclear Eng Univ Fla Gainesville FL 32611

DALTON, HARRY P, b Holyoke, Mass, Sept 16, 29; m 53; c 8. BIOCHEMISTRY, MICROBIOLOGY. *Educ:* Am Int Col, AB, 52; Univ Mass, MS, 56, PhD(microbiol, biochem), 62. *Prof Exp:* Asst prof biochem, Hampden Col Pharm, 59-62; chief microbiologist & teaching supvr, Holyoke Hosp, Mass, 56-62 & Montefiore Hosp, 62-66; assoc prof clin path, 66-76, PROF CLIN PATH & MICROBIOL, MED COL VA, 76- *Concurrent Pos:* Res grants, 63-71; asst prof, Med Col Va & Dent Sch, Univ Pittsburgh, 65-66. *Mem:* AAAS; Am Soc Microbiol; Am Soc Med Technol; Am Burn Soc; NY Acad Sci. *Res:* Medical microbiology; mycoplasma; bacterial L-forms; diagnostics systems for rapid identification. *Mailing Add:* 9500 Tuxford Rd Richmond VA 23235

DALTON, HOWARD CLARK, b Brooklyn, NY, Aug 7, 15; m 48. EMBRYOLOGY, GENETICS. *Educ:* Wesleyan Univ, AB, 36, AM, 37; Stanford Univ, PhD(biol), 40. *Prof Exp:* Instr zool, Univ Rochester, 40-41; instr, biol, Brown Univ, 46-47; asst prof, Bates Col, 47-48; fel, Genetics Dept, Carnegie Inst, 48-50; from asst prof to assoc prof, 50-61, prof & chmn dept, NY Univ, 61-67; prof, 67-75, EMER PROF BIOL, PA STATE UNIV, 75- *Mem:* Am Soc Zool; Soc Develop Biol (treas, 59-62); Am Asn Anat; Am Soc Naturalists (Secy, 65-68); Int Soc Develop Biol; Sigma Xi. *Res:* Development of pigment patterns, genetic control of pigment development. *Mailing Add:* 3-3400 Kuhio Hwy Apt A-303 Lihue Kauai HI 96766-1052

DALTON, JACK L, b Hominy, Okla, July 20, 31; m 53; c 3. BIOCHEMISTRY, ORGANIC CHEMISTRY. *Educ:* Chadron State Col, BS, 53; Kans State Univ, MS, 58. *Prof Exp:* Instr chem, Boise Jr Col, 58-64; assoc prof, 64-70, from actg chmn dept to chmn dept, 68-88, PROF CHEM, BOISE STATE UNIV, 71-; EXEC SECY, IDAHO ACAD SCI, 80- *Mem:* AAAS. *Res:* Plant waxes; isolation of natural products. *Mailing Add:* Dept Chem Boise State Univ 1910 University Dr Boise ID 83725

DALTON, JAMES CHRISTOPHER, b Corning, NY, Dec 1, 43; m 71; c 1. ORGANIC CHEMISTRY, PHOTOCHEMISTRY. *Educ:* Calif Inst Technol, BS, 65; Columbia Univ, PhD(chem), 70. *Prof Exp:* Asst prof chem, Univ Rochester, 70-77; assoc prof, 77-84, PROF CHEM, 84-, VPRES PLANNING & BUDGETING, BOWLING GREEN STATE UNIV, 87- *Mem:* Am Chem Soc; Inter-Am Photochem Soc. *Res:* Organic photochemistry; fluorescence and phosphorescence properties of organic compounds; polymer photochemistry. *Mailing Add:* Planning & Budgeting Bowling Green State Univ Bowling Green OH 43403

DALTON, JOHN CHARLES, b Clintwood, Va, Apr 11, 31; m 64; c 2. ZOOLOGY, PHYSIOLOGY. *Educ:* Univ Va, BA, 51; Harvard Univ, AM, 52, PhD(biol), 55. *Prof Exp:* From instr to assoc prof biol, Univ Buffalo, 57-62; exec secy, Metab Study Sect, Div Res Grants, NIH, 62-65, asst chief for rev, Res Grants Rev Br, 65-70, chief prog planning staff, Off Prog Planning & Eval, Bur Health Manpower Educ, 70-72, spec asst to assoc dir, Regional Opers, Bur Health Manpower Educ, 72-73, spec asst to assoc adminr, Opers & Mgt, Health Resources Admin, 74-75, chief planning & eval, Off Prog Develop, Bur Health Manpower, Health Resources Admin, 75-76, dep assoc dir, Prog Activ, Nat Inst Gen Med Sci, 76-78, dir, div extramural activ, Nat Inst Neurol Dis & Stroke, NIH, 78-91. *Res:* Comparative physiology; neurophysiology. *Mailing Add:* Div Extramural Activ Nat Inst Neurol Dis & Stroke NIH Fed Bldg Rm 1016 Bethesda MD 20892

DALTON, LARRY RAYMOND, b Belpre, Ohio, Apr 25, 43; m 66. CHEMICAL PHYSICS. *Educ:* Mich State Univ, BS, 65, MS, 66; Harvard Univ, AM, 71, PhD(chem), 72. *Prof Exp:* From asst prof to assoc prof chem, Vanderbilt Univ, 71-77; from assoc prof to prof chem, State Univ NY Stony Brook, 77-82; prof chem, Carnegie-Mellon Univ, 82-; AT DEPT CHEM, UNIV SOUTHERN CALIF, LOS ANGELES. *Concurrent Pos:* Distinguished consult natural sci, Spring Arbor Col, 71-; consult, Varian Assocs, 71-74, Bruker Physik, 76-77 & IBM, 77-; fel, Alfred P Sloan Found, 74-77; NIH res career develop award, 76-81; res prof biochem, Med Sch, Vanderbilt Univ, 77- *Honors & Awards:* Teacher-Scholar Award, Camille & Henry Dreyfus Found, 75. *Mem:* Am Chem Soc. *Res:* Development of new spectroscopic techniques and application to biomedical research, to the investigation of the dynamics of classical liquids and to the study of solid state organic materials, particularly conducting prolymers. *Mailing Add:* Dept Chem Univ Southern Calif Los Angeles CA 90089

DALTON, LONNIE GENE, b Carter, Okla, July 20, 34; m 60; c 3. GENETICS, AGRONOMY. *Educ:* Okla State Univ, BS, 56, MS, 57; NC State Univ, PhD(genetics), 65. *Prof Exp:* Asst plant breeding, Okla State Univ, 57; asst genetics, NC State Univ, 65; plant breeder, 65-67, DIR RES, PIONEER SORGHUM CO, 67- *Mem:* Am Soc Agron; Am Seed Trade Asn. *Res:* Accumulation of knowledge, perfection of skills and development of material that leads to a major advance in commercial sorghum hybrids. *Mailing Add:* Pioneer Hi-bred Int Inc Sorghum Res PO Box 316 Johnston IA 50131

DALTON, PATRICK DALY, b Salt Lake City, Utah, Oct 11, 22; m 48; c 3. ECOLOGY, BOTANY. *Educ:* Ariz State Univ, BS, 49; Utah State Univ, MS, 51; Univ Ariz, PhD(plant ecol), 61. *Prof Exp:* Range mgr, Soil Conserv Serv, USDA, Utah, 51-52; range mgr, Bur Land Mgt, US Dept Interior, 52-53; instr, Liahona High Sch, Tonga, 53-55; asst prof agr, biol & phys sci, Church Col Hawaii, 55-58; res assoc range res, Univ Ariz, 58-61; asst prof range mgt, Univ Nev, Reno, 61-62; dir range & forest res, UN Korean Upland Proj, 62-63; pres, Latter-day Saint Mission, Tonga, 63-66; PROF BOT, BRIGHAM YOUNG UNIV, HAWAII CAMPUS, LAIE, 66- *Mem:* AAAS; Am Inst Biol Sci; Ecol Soc Am; Bot Soc Am; Am Soc Range Mgt; Sigma Xi. *Res:* Ecology of Southwestern ranges and Pacific Island. *Mailing Add:* 3606 Jonathon Dr Salt Lake City UT 84121

DALTON, PHILIP BENJAMIN, b New York, NY, July 21, 23; m 44; c 2. ORGANIC CHEMISTRY. *Educ:* Univ Ill, BA, 44; Columbia Univ, MS, 47. *Prof Exp:* Org chemist, Sun Chem Corp, 47-50; tech engr, Colgate-Palmolive Co, 50-54; develop engr, Com Develop Dept, GAF Corp, 54-55, sales engr, Acetylene Chem Dept, 55-58, mgr, 58-61, mkt mgr, 61-63, dir com develop, 63-64, vpres develop, 64-67, vpres photo & repro div, 67, exec vpres, 67-77, pres & chief operating officer, 77-79; PRES, DALTON ASSOCS, 79- *Mem:* Am Chem Soc; Soc Chem Indust; Com Develop Asn; Am Inst Chem. *Res:* Reppe chemistry; reactions of acetylene and aldehydes under high pressure. *Mailing Add:* 5636 Country Lakes Dr Sarasota FL 34243-3805

DALTON, ROGER WAYNE, b Sweatman, Miss, Jan 1, 36; m 58; c 3. ANALYTICAL CHEMISTRY, PHYSICAL CHEMISTRY. *Educ:* Delta State Univ, BS, 58; Univ Miss, MS, 60. *Prof Exp:* Qual control engr rockets, Hercules Inc, 60-63, develop chemist pyrotechnics, 63-65, res chemist, propellants, 65-67, sr process engr, 67-76, sr tech engr propellants, 76-83, TECH SPECIALIST, PROPELLANTS & EXPLOSIVES, HERCULES, INC, 83- *Concurrent Pos:* Gas chromatography consult, Hercules, Inc, 67-, liquid chromatography & anal consult, 74- *Mem:* Am Defense Prepavedness Asn. *Res:* Analytical techniques and process applications pertaining to propellants and explosives; chromatographic techniques; x-ray analytical instrumentation; specialist in manufacture of nitrocellulose and nitroglycerin. *Mailing Add:* Hercules Inc PO Box 1 Radford VA 24141-0100

DALVEN, RICHARD, b Brooklyn, NY, Sept 3, 31; m 55. SOLID STATE PHYSICS. *Educ:* Columbia Univ, AB, 53; Mass Inst Technol, PhD(chem physics), 58. *Prof Exp:* Mem res staff, Raytheon Corp, 58-62; mem res staff, RCA Labs, 62-71; LECTR, DEPT PHYS, UNIV CALIF, BERKELEY, 72- *Mem:* Am Phys Soc. *Res:* Electronic structure of solids, particularly semiconductors and superconductors; applications of solid state physics. *Mailing Add:* Dept Physics Univ Calif Berkeley CA 94720

DALVI, RAMESH R, b Bombay, India, Nov 8, 38; US citizen; m 69; c 2. METABOLIC TOXICOLOGY, VETERINARY TOXICOLOGY. *Educ:* Univ Bombay, India, BSc, 62, 64, MSc, 67; Utah State Univ, PhD(toxicol), 72; Am Bd Toxicol, dipl toxicol, 82. *Prof Exp:* Sci res officer food sci & biochem, Bhabha Atomic Res Ctr, 67-69; res fel toxicol, Utah State Univ, 69-72 & Med Sch, Vanderbilt Univ, 72-74; from asst prof to assoc prof, 74-82, PROF TOXICOL, TUSKEGEE UNIV, 82-, DIR, DIAG TOXICOL LAB, 74- *Concurrent Pos:* Consult, Scientists' Inst Pub Info, 81-, Nat Acad Sci, 82 & USAID res projs Senegal, Boston Univ, 85-; mem, Int Adv Bd, Trop Vet, 82; prin investr, minor species proj, Food & Drug Admin, 84-87; external examnr, Univ Mysore, India, 84-; vis prof, Med Res Inst, Fla Inst Technol, 85-; res grants NSF, NIH & USDA; mem, Bd Sci Reviewers, Am Jour Vet Res, 87-89; mem adv bd, J Environ Biol, 90-; consult ed, J Maha Agr Univs,

91- *Mem:* Soc Toxicol; Am Acad Vet & Comp Toxicol; Int Soc Study Xenobiotics; Am Chem Soc; Inst Food Technologists; AAAS. *Res:* Toxicologic implications of the metabolism of drugs and other toxic substances. *Mailing Add:* 1243 Ferndale Dr Auburn AL 36830

DALY, BARTHOLOMEW JOSEPH, b Brooklyn, NY, Jan 3, 29; m 59; c 5. FLUID DYNAMICS. *Educ:* Univ Wyo, BA, 50; Ariz State Univ, MA, 60. *Prof Exp:* Seismologist, Petty Geophys Eng Co, 51-55; geophysicist, Southern Geophys Co, 55-57; party chief, Bible Geophys Co, 57-59; STAFF MEM NUMERICAL FLUID DYNAMICS, LOS ALAMOS NAT LAB, UNIV CALIF, 60- *Res:* Numerical techniques for calculating compressible and incompressible fluid dynamics. *Mailing Add:* 533 Todd Loop Los Alamos NM 87544

DALY, COLIN HENRY, b Glasgow, Scotland, Aug 22, 40. BIOENGINEERING. *Educ:* Univ Glasgow, BSc Hons, 63; Univ Strathclyde, PhD(bioeng), 66. *Prof Exp:* From asst prof to assoc prof, 67-78, PROF MECH ENG, UNIV WASH, 78- *Concurrent Pos:* Res fel, Ctr for Bioeng, Univ Wash, 66-67; affil, Ctr for Res in Oral Biol, Univ Wash, 70-75. *Mem:* Am Soc Mech Engrs; Inst Mech Eng. *Res:* Biomechanics of the skin and other soft tissues; biomechanics of the oral mucosa and the periodontium. *Mailing Add:* Dept Mech Eng Univ Wash Seattle WA 98195

DALY, DANIEL FRANCIS, JR, b Brooklyn, NY, Sept 19, 39; m 73; c 2. ELECTRONICS, SOLID STATE PHYSICS. *Educ:* St Bonaventure Univ, BS, 61; Columbia Univ, MA, 63, PhD(physics), 66. *Prof Exp:* Res asst solid state physics, Columbia Radiation Lab, Columbia Univ, 63-65; res asst, Purdue Univ, 65-66, res assoc, 66-68; mem tech staff, Bell Labs, 68-83; mem tech staff, 84-86, DISTRICT MGR, DESIGN METHODOLOGY, BELLCORE, 86- *Mem:* Inst Elec & Electronics Engrs; Am Asn Physics Teachers. *Res:* Electron paramagnetic resonance; radiation damage in semiconductors; ion implantation; color centers in alkali halides; semiconductor electronics; integrated circuits; very-large-scale-integration computer aided design. *Mailing Add:* Bellcore 445 South St Rm 2E-366 PO Box 1910 Morristown NJ 07960-1910

DALY, DAVID DEROUEN, b St Louis, Mo, Oct 17, 19; m 46; c 4. NEUROLOGY. *Educ:* Stanford Univ, BA, 40; Univ Minn, BS, 42, BM, 44, MD, 45, PhD(neurol), 51; Am Bd Psychiat & Neurol, dipl, 51. *Prof Exp:* Resident neurol, Univ Minn, 45-46, asst, 47-48, instr, 48-49, asst neurol & EEG, Mayo Clin, 49-51, from asst prof to assoc prof neurol, Mayo Found, 56-61; chmn div neurol, Barrow Neurol Inst, Ariz, 61-66; prof, 66-70, SCOTTISH RITE PROF NEUROL, UNIV TEX HEALTH SCI CTR DALLAS, 70-; PROF COMMUN DISORDERS, UNIV TEX, DALLAS, 73- *Mem:* Am Neurol Asn; fel Am Acad Neurol; Soc Clin Neurol (pres, 62); Int League Against Epilepsy (pres, 73-77). *Res:* Electroencephalography; epilepsy; narcolepsy. *Mailing Add:* Dept Neurol Ariz Health Sci Ctr 1501 N Campbell Ave Tucson AZ 85724

DALY, HOWELL VANN, b Dallas, Tex, Oct 30, 33; m 53; c 1. ENTOMOLOGY. *Educ:* Southern Methodist Univ, BS, 53; Univ Kans, MA, 55, PhD(entom), 60. *Prof Exp:* Instr zool, La State Univ, 59-60; from asst prof to assoc prof, 60-71, PROF ENTOM, UNIV CALIF, BERKELEY, 71- *Concurrent Pos:* Res grants, NIH, 62-67 & NSF, 68-75. *Mem:* AAAS; Soc Study Evolution; Soc Syst Zool; fel Royal Entom Soc London; Entomol Soc Am; Sigma Xi. *Res:* Systematic and evolutionary biology; biosystematics of Apoidea; comparative morphology of Hymenoptera. *Mailing Add:* Div Entom Univ Calif Berkeley CA 94720

DALY, JAMES C(AFFREY), b Hartford, Conn, June 10, 38; m 62; c 5. ELECTRICAL ENGINEERING. *Educ:* Univ Conn, BS, 60; Rensselaer Polytech Inst, MEE, 62, PhD(elec eng), 67. *Prof Exp:* Instr elec eng, Rensselaer Polytech Inst, 62-66; mem tech staff, Bell Tel Labs, 66-69; from asst prof to assoc prof, 69-83, PROF ELEC ENG, UNIV RI, 83- *Mem:* Inst Elec & Electronics Engrs. *Res:* Very-large-scale integration design; fiber optics; solar optics; industrial instrumentation. *Mailing Add:* Dept Elec Eng Univ RI Kingston RI 02881

DALY, JAMES EDWARD, b Compton, Calif, Mar 18, 48; m 69; c 1. MATHEMATICAL ANALYSIS. *Educ:* Humboldt State Univ, AB, 70; NMex State Univ, PhD(math), 74. *Prof Exp:* Fel math, NMex State Univ, 74-75; vis asst prof math, Univ Ore, 75-76; INSTR DEPT MATH, COL OF THE REDWOODS, 76- *Mem:* Am Math Soc; Math Asn Am. *Res:* Harmonic analysis; local field singular integrals and multipliers. *Mailing Add:* Dept Math Univ Colo PO Box 7150 Colorado Springs CO 80933

DALY, JAMES JOSEPH, b Detroit, Mich, Sept 6, 35; m 65; c 3. MEDICAL PARASITOLOGY. *Educ:* Wayne State Univ, BS, 61, MS, 64; La State Univ, New Orleans, PhD(parasitol), 68. *Prof Exp:* Asst prof microbiol, Univ Ark Med Ctr, 68-74, ASSOC PROF MICROBIOL, UNIV ARK MED SCI CAMPUS, 75- *Concurrent Pos:* Fel trop med, Int Ctr Med Res & Training, 71-, vis assoc prof, Rockefeller Univ, New York, 77-78. *Mem:* Am Soc Parasitologists; Soc Protozoologists; Wildlife Dis Asn. *Res:* Parasite physiology and biochemistry; protozoology; carbohydrate metabolism of trichomonads; hematozoa of cold-blooded vertebrates; biology of land planaria; sparganosis; C-type viruses of invertebrates and cold-blooded vertebrates. *Mailing Add:* Dept Microbiol & Immunol Univ Ark Med Sci Campus 4301 W Markham St Little Rock AR 72205

DALY, JAMES WILLIAM, b Chicago, Ill, Jan 5, 31; m 53; c 3. OBSTETRICS & GYNECOLOGY, ONCOLOGY. *Educ:* Univ Santa Clara, 48-51; Loyola Univ Chicago, MD, 55; Am Bd Obstet & Gynec, dipl, 66 & 76. *Prof Exp:* Intern, St Mary's Hosp, Gary, Ind, 56; mem staff obstet & gynec, USAF Hosp, Lockbourne AFB, 57-59, chief prof serv, 58-59; resident obstet & gynec, USAF Hosp, San Antonio, Tex, 59-62, chief gynec serv & training officers residency prog, 63-68; assoc prof, Univ Fla, 68-77, chief, Gynec Serv, 70-82, prof obstet & gynec & dir Tumor Div, 77-82; dir, Tumor Clin & Registry, Shand's Teaching Hosp & Clins, 70-82; PROF OBSTET & GYNEC & CHMN DEPT, SCH MED, CREIGHTON UNIV, 82- *Concurrent Pos:* Fel gynec oncol, Univ Tex M D Anderson Hosp & Tumor Inst, 62-63. *Mem:* AMA; Am Radium Soc; fel Am Col Obstet & Gynec; Soc Gynec Oncol. *Res:* Clinical research in cancer of the female genitalia. *Mailing Add:* Dept Obstet/Gynec N-617 Univ Mo Health Sci Ctr Columbia MO 65212

DALY, JOHN, b Philadelphia, Pa, Dec 10, 47. SOCIAL ONCOLOGY. *Educ:* Temple Univ, MD, 73. *Prof Exp:* Assoc prof surg, Sloan-Kettering Cancer Ctr, 80-85; CHIEF, DIV SURG ONCOL, HOSP UNIV PA, 85- *Mailing Add:* Hosp Univ Pa 3400 Spruce St Silverstein Pavilion 4th Floor Philadelphia PA 19104

DALY, JOHN ANTHONY, b New York, NY, Oct 7, 37; m 66; c 1. TECHNOLOGICAL INNOVATION, DEVELOPING COUNTRIES. *Educ:* Univ Calif, Los Angeles, BS, 59, Berkeley, MS, 62 & Irvine, PhD(admin), 75. *Prof Exp:* Sr design engr, Astropower Lab, McDonnell Douglas, 62-65; dep dir, Comput Ctr, Univ Santa Maria, Valparaiso, Chile, 65-67; consult oper res, Ford Found, Cath Univ Valparaiso, 67; sr design engr, Astropower Lab, McDonnell Douglas, 68-70; oper res consult, Health Planning Res Proj, WHO, 70-73; dir, Health Sector Anal Div, Int Health, HEW, 73-76; dir, Sci & Technol Policy Div, Off Sci & Technol, 77-80, SCI PROG DIR, OFF SCI ADV, AID, 81- *Concurrent Pos:* Instr, dept info sci, Univ Calif, Irvine, 68; adj prof, dept social med, Univ Valle, Colo, 71-73; prof, Int Law Inst, 85-88; instr, Dept Tech Mgt, Univ Maryland, 82-89; mem adv bd, J Law & Technol, 87- *Mem:* AAAS; Oper Res Soc Am; Inst Mgt Sci. *Res:* Administration of research in biotechnology; chemistry applied to agriculture; biomass production and conversion; biological control of vectors; global change; biological diversity. *Mailing Add:* 14205 Bauer Dr Rockville MD 20853

DALY, JOHN F, b Jersey City, NJ, June 10, 12; m 41; c 1. OTOLARYNGOLOGY. *Educ:* Fordham Univ, AB, 33; Long Island Col Med, MD, 37. *Prof Exp:* Dir dept otolaryngol, Bellevue Hosp, 47-80, dir otolaryngol, Univ Hosp, 49-80, chmn dept, 49-80, PROF OTOLARYNGOL, SCH MED, NY UNIV, 49- *Concurrent Pos:* Dir Bellevue Hearing & Speech Ctr; consult aural surgeon, NY Eye & Ear Infirmary, 49-80; consult, Hackensack Hosp, NJ, 53-80, Vet Admin Hosp, Manhattan, NY, Bergen Pines County Hosp, NJ & Holy Name Hosp, Teaneck, 54-80 & Phelps Mem Hosp Asn, Tarrytown, 55-80; Brookhaven Mem Hosp Asn, Patchogue, 56-80, St Joseph's Hosp, Stamford, Conn, 58-80, Greenwich Hosp Asn, Conn & Nyack Hosp, NY, 59-80, Elizabeth A Horton Mem Hosp, Middletown, 63-80, Columbus Hosp, NY, Stamford Hosp, Conn, Speech Rehab Inst, NY & Goldwater Mem Hosp, NY; mem bd dirs, Am Bd Otolaryngol, 69-77; chmn tech adv comt hearing & speech, Dept Health, New York; mem proj comt, Nat Inst Neurol Dis & Blindness, 64-70; chmn sci rev comt, Deafness Res Found, 64-79, mem, 64-80; mem, Am Coun Otolaryngol. *Mem:* AAAS; fel Am Col Surg; Am Laryngol, Rhinol & Otol Soc; fel Am Laryngol Asn (pres, 78); fel Am Acad Otolaryngol-Head & Neck Surg (vpres, 68-69, pres, 76); fel Am Otol Soc; fel Am Soc Head & Neck Surg (pres, 66-67); fel Soc Univ Otolaryngolgists (pres, 66-67); Am Asn Univ Prof. *Res:* Audiology; otology; laryngology. *Mailing Add:* 16 N Brae Ct Tenafly NJ 07670

DALY, JOHN FRANCIS, b Kansas City, Mo, Dec 27, 16. ALGEBRA. *Educ:* St Louis Univ, AB, 40, MS, 43. *Prof Exp:* Teacher pvt sch, 43-45; from asst prof to assoc prof, 53-70, PROF MATH, ST LOUIS UNIV, 70- *Mem:* Am Math Soc; Am Soc Eng Educ; Math Asn Am; Sigma Xi. *Res:* Modern abstract algebra; topology; algebraic topology; history of mathematics. *Mailing Add:* Col Arts & Sci St Louis Univ 221 N Grand Ave St Louis MO 63103

DALY, JOHN JOSEPH, JR, b St Paul, Minn, Aug 7, 26; m 50; c 3. ORGANIC CHEMISTRY. *Educ:* Col St Thomas, BS, 50; Univ Md, PhD, 54. *Prof Exp:* Res chemist, 54-61, develop specialist, 62-68, tech assoc, 68-69, mgr tech & mkt develop, 69-75, tech mgr aerosol, 75-78, TECH MGR SOLVENTS & DEVELOP, FREON PROD DIV, E I DU PONT DE NEMOURS CO, INC, 78- *Mem:* Am Chem Soc; Chem Specialties Mfrs Asn. *Res:* Organic intermediates; dyes; textile chemicals; synthetic lubricants and additives; fluorine chemistry and fluorocarbon development; pyrolysis of esters. *Mailing Add:* 4658 Dartmoor Dr Liftwood Estates Wilmington DE 19803

DALY, JOHN M, b Philadelphia, Pa. SURGERY. *Prof Exp:* CHIEF DIV SURG ONCOL & PROF SURG, UNIV PA, 86- *Mailing Add:* Div Surg Oncol Sch Med Univ Pa 3400 Spruce St 4th Floor Silverstein Pavilion Philadelphia PA 19104

DALY, JOHN MATTHEW, b Lexington, Ky, Aug 1, 25; m 53; c 5. INORGANIC CHEMISTRY. *Educ:* Xavier Univ, Ohio, BS, 50, MS, 51; Univ Notre Dame, PhD(chem), 58. *Prof Exp:* Res & develop, Kaiser-Frazer Corp, 51-52; chemist, E I du Pont de Nemours & Co, Inc, 52-53; from asst prof to assoc prof, 53-64, PROF CHEM, BELLARMINE COL, 64-, CHMN DEPT, 59- *Mem:* AAAS; Am Chem Soc. *Res:* Stability of chelates compounds; effect of metal ions on organic reaction mechanisms; pk values or organic acids and bases; benzyne intermediates. *Mailing Add:* 2529 Saratoga Dr Louisville KY 40205-2024

DALY, JOHN T, b Brooklyn, NY, Nov 30, 38. MATHEMATICS MODELING, FUNCTIONAL ANALYSIS. *Educ:* State Univ NY, Binghamton, BA, 60; Syracuse Univ, PhD(math), 69. *Prof Exp:* From asst prof to assoc prof, 68-86, PROF MATH, STATE UNIV NY, OSWEGO, 86- *Concurrent Pos:* Vis scholar math, Yale, 75-76, Univ Calif, San Diego, 83-84. *Mem:* Am Math Soc; Math Soc Am; Soc Indust & Appl Math. *Mailing Add:* Dept Math State Univ NY Oswego NY 13126

DALY, JOHN WILLIAM, b Portland, Ore, June 8, 33. NEUROCHEMISTRY, NATURAL PRODUCTS CHEMISTRY. *Educ:* Ore State Col, BS, 54, MA, 56; Stanford Univ, PhD(org chem), 58. *Prof Exp:* Res biochemist, Nat Inst Arthritis & Metab Dis, 60-69, CHIEF, LAB BIOORG CHEM, NAT INST ARTHRITIS, METAB & DIGESTIVE DIS,

NIH, 69- *Honors & Awards:* Hillebrand Prize, Am Chem Soc, 77. *Mem:* AAAS; Am Chem Soc; Int Soc Neurochemists; Am Soc Pharmacol & Exp Therapeut. *Res:* Natural products and structure elucidation; biochemistry and pharmacology biogenic amines; cyclic nucleotides and other neuroactive compounds; metabolic transformations. *Mailing Add:* Bldg 8 Rm 1A-17 NIH Bethesda MD 20892-0001

DALY, JOSEPH MICHAEL, b Hoboken, NJ, Apr 9, 22; m 51; c 2. PLANT TOXINS, PLANT DISEASES. *Educ:* RI State Col, BS, 44; Univ Minn, MS, 47, PhD(plant physiol), 52. *Prof Exp:* From asst to instr plant path, Univ Minn, 44-50, res fel bot, 50-51, MacMillan res fel, 51-52; asst prof biol, Univ Notre Dame, 52-55; from asst prof to assoc prof bot, Univ Nebr, Lincoln, 55-63, assoc plant pathologist, Exp Sta, 57-63, prof bot & biochem, Univ, 63-66, C Petrus Peterson prof biochem & nutrit, 66-86, prof biochem, 76-86; RETIRED. *Honors & Awards:* E C Stakman Award, 86. *Mem:* Nat Acad Sci; AAAS; fel Am Phytopath Soc; Am Soc Plant Physiol; Am Acad Arts & Sci. *Res:* Biochemistry of disease resistance and microorganisms; biochemistry and physiology of plant diseases. *Mailing Add:* Dept Biochem Univ Nebr Lincoln NE 68588

DALY, KEVIN RICHARD, b San Francisco, Calif, May 31, 31; m 62; c 3. GENETICS. *Educ:* Univ Calif, Davis, BS, 53; Cornell Univ, PhD(genetics), 58. *Prof Exp:* Asst genetics, Cornell Univ, 54-56; USPHS res fel, 58-60; asst prof, 60-64, ASSOC PROF BIOL, CALIF STATE UNIV, NORTHRIDGE, 64- *Concurrent Pos:* NIH spec res fel, 70-71. *Mem:* Soc Study Evolution; Genetics Soc Am; Sigma Xi. *Res:* Quantitative genetics. *Mailing Add:* Dept Biol Calif State Univ Northridge CA 91330

DALY, MARIE MAYNARD, b New York, NY, Apr 16, 21; m 61. BIOCHEMISTRY. *Educ:* Queens Col, NY, BS, 42; NY Univ, MS, 43; Columbia Univ, PhD(chem), 47. *Prof Exp:* Tutor, Queens Col, NY, 43-44; instr, Howard Univ, 47-48; asst, Rockefeller Inst, 51-55; assoc, Columbia Univ Res Serv, Goldwater Mem Hosp, 55-59; asst prof biochem, 60-71, ASSOC PROF BIOCHEM & MED, ALBERT EINSTEIN COL MED, 71- *Concurrent Pos:* Am Cancer Soc fel, Rockefeller Inst, 48-51; estab investr, Am Heart Asn, 58-63; career scientist, Health Res Coun, New York, 62-72, Comn Sci & Technol, 86-89. *Mem:* Fel AAAS; Am Chem Soc; fel NY Acad Sci; Am Soc Biol Chem; fel Am Heart Asn. *Res:* Arterial smooth muscle; creatine transport and metabolism. *Mailing Add:* 12 Copeces Lane PO Box 601 East Hampton NY 11937

DALY, PATRICK JOSEPH, b Mullingar, Ireland, Feb 2, 33; m 62; c 3. NUCLEAR CHEMISTRY. *Educ:* Nat Univ Ireland, BSc, 56, MSc, 59; Oxford Univ, DPhil(nuclear chem), 63. *Prof Exp:* From instr to assoc prof, 63-76, PROF CHEM, PURDUE UNIV, WEST LAFAYETTE UNIV, 70- *Mem:* Am Phys Soc. *Res:* Nuclear reactions and spectroscopy. *Mailing Add:* Dept Chem Purdue Univ West Lafayette IN 47907

DALY, PATRICK WILLIAM, b Toronto, On, May 22, 47; m 85. MAGNETOSPHERIC PHYSICS. *Educ:* Bishop's Univ, BSc, 68; Univ BC, PhD(physics), 73. *Prof Exp:* Fel low temperature, Clarendon Lab, Oxford Univ, Eng, 73-75; res assoc magnetosphere, Nat Res Coun Can, 75-78; MPG Stipendiat Magnetosphere, Max Planck Inst Aeronomy, Ger, 78-83; vis scientist, European Space Agency, 83-84; RES STAFF MEM, MAX PLANCK INST AERONOMY, GER, 84- *Mem:* Am Geophys Union. *Res:* Interaction of solar wind with the earth's magnetosphere; medium energy charged particles. *Mailing Add:* Max Planck Inst Aeronomy D-3411 Katlenburg Lindau 3 Germany

DALY, ROBERT E, b Cambridge, Mass, Mar 1, 37; m 66; c 2. PHARMACY. *Educ:* Mass Col Pharm, BS, 61, MS, 63; Purdue Univ, PhD(med anal chem), 68. *Prof Exp:* From scientist to sr scientist, Parke-Davis Prod Develop, 67-73, sr res assoc, 73-75, from assoc dir to dir, 75-90, SR DIR, ANAL DEVELOP & TECHNOL, PARKE-DAVIS PROD DEVELOP, 90- *Mem:* Am Chem Soc. *Res:* Application of HPTLC and robotics to analysis of specific medicaments in complex pharmaceutical formulations. *Mailing Add:* 283 Washington Valley Rd Randolph NJ 07869

DALY, ROBERT WARD, b Watertown, NY, Oct 1, 32; m 58; c 5. PSYCHIATRY, PSYCHOANALYSIS. *Educ:* St Lawrence Univ, BS, 57; State Univ NY Upstate Med Ctr, MD, 57. *Prof Exp:* Rotating intern clin med, Albert Einstein Med Ctr, 57-58; resident psychiat, State Univ NY Upstate Med Ctr, 58-60, chief resident, 60-61; from instr to assoc prof, 63-75, PROF PSYCHIAT, STATE UNIV NY UPSTATE MED CTR, 75-, PROF MED HUMANITIES, 84- *Concurrent Pos:* Pvt pract psychiat, 63-69 & 70-; supv psychiatrist, Syracuse Psychiat Hosp, 63-66; consult, Maxwell Sch Citizenship, Syracuse Univ, 63-66, Peace Corps, 64-65, Nat Libr Med, 66-68, Psychoanal Rev, 66-86, J Am Med Asn, Arch Gen Psychiat, 77; dir adult psychiat clin, State Univ Hosp, Syracuse, 66-; vis prof med admin, Cornell Univ, 67-82; vis scholar philos, Dept Hist & Philos Sci, Cambridge Univ & NY State fel, Dept Ment Hyg, 69-70; fel, Nat Endowment for Humanities, 74-75; co-dir, Syracuse Consortium Cult Found Med, 78-79, exec dir, 79-; examr, Am Bd Psychiat & Neurol, 80-; dir, Prog Med Humanities, State Univ NY, Health Sci Ctr Syracuse, 84- *Mem:* Fel Am Psychiat Asn; Int Soc Comp Study Civilizations; Soc Health & Human Values. *Res:* Theoretical presupposition of psychiatry, psychoanalysis and medicine; medical ethics; historical, social-cultural and economic determinates of contemporary psychiatric and medical institutions. *Mailing Add:* Dept Psychiat State Univ Hosp 750 E Adams St Syracuse NY 13210

DALY, RUTH AGNES, b Rockville Centre, NY, Jan 21, 58. ASTROPHYSICS, FLUIDS. *Educ:* Boston Col, BA, 79; Boston Univ, MA, 84, PhD(astron & physics), 87. *Prof Exp:* Postdoctoral fel astron, Inst Astron, Cambridge, Eng, 87-88; res assoc astron, 88-89, instr physics, 89-90, ASST PROF PHYSICS, PRINCETON UNIV, 90- *Honors & Awards:* Annie Jump Cannon Spec Commendation Honor, Am Asn Univ Women & Am Astron Soc, 91. *Mem:* Am Astron Soc; Am Phys Soc; Asn Women Sci. *Res:* Theoretical astrophysics and cosmology; formation and evolution of galaxies and larger scale structures; properties of high-redshift galaxies; x-ray background and microwave background; dark matter in galaxies and whether this could be comprised of very low mass stars. *Mailing Add:* Dept Physics Princeton Univ Princeton NJ 08544-0708

DALY, WALTER J, b Michigan City, Ind, Jan 12, 30; m 53; c 2. INTERNAL MEDICINE. *Educ:* Ind Univ, AB, 51, MD, 55; Am Bd Internal Med, dipl pulmonary dis. *Prof Exp:* Instr physiol, 55, from instr to assoc prof med, 62-68; dir, Regenstrief Inst, 76-83, PROF MED, IND UNIV, INDIANAPOLIS, 68-, CHMN DEPT, 70- *Concurrent Pos:* USPHS fel med, Ind Univ, Indianapolis, 60-62. *Mem:* Fel Am Col Physicians; fel Am Col Cardiol; Asn Am Physicians; Am Soc Clin Invest; Am Clin & Climat Asn; Cent Soc Clin Res. *Res:* Pulmonary and cardiopulmonary physiology. *Mailing Add:* Deans Off-Fesler Hall 302 Ind Univ Sch Med 1120 South Dr Indianapolis IN 46223

DALY, WILLIAM HOWARD, b San Francisco, Calif, Jan 22, 39; m 60; c 2. ORGANIC CHEMISTRY, POLYMER CHEMISTRY. *Educ:* Baldwin-Wallace Col, BSc, 60; Polytech Inst Brooklyn, PhD(org chem), 64. *Prof Exp:* Fel polymer synthesis, Univ Mainz, 64-66; asst prof, 66-71, assoc prof, 71-78, PROF ORG POLYMER CHEM, LA STATE UNIV, 78- *Concurrent Pos:* Chmn, La State Univ, 81- *Honors & Awards:* Coates Award, 79. *Mem:* Am Chem Soc; The Chem Soc. *Res:* Modification of condensation polymers; catalysis of cellulose modification; reactivity of polymeric substrates; enzyme immobilization; preparation of polymer reagents. *Mailing Add:* Dept Chem La State Univ Baton Rouge LA 70803

DALZELL, ROBERT CLINTON, b Pittsburgh, Pa, Aug 31, 19; m 44; c 3. MEDICAL MICROBIOLOGY. *Educ:* Univ Pittsburgh, BS, 41; Pa State Univ, MS, 56, PhD(microbiol), 57; Am Bd Microbiol, dipl. *Prof Exp:* Chief clin lab, Army Hosp, Edgewood, Md, 45-48, sci adminr biol prod, Biol Opers, US Army, 57-58; chief biol br, Chem Corps Sch, 58-59, tech adv, Orientation Course, Chem, Biol & Radiol Weapons, 59-62; asst prof microbiol, Morehead State Col, 62-63; asst lab mgr, Melpar, Inc, 63-64; res scientist, Travelers Res Ctr, Inc, 64-65; assoc prof, 65-67, PROF BIOL, KY WESLEYAN COL, 67- *Mem:* Am Acad Microbiol; Am Soc Microbiol. *Res:* Ultrasonic and sonic sound fields; serological studies of plant virus disease; life and medical sciences; immunology; serology. *Mailing Add:* 3844 Bowlds Ct Owensboro KY 42301

DALZELL, WILLIAM HOWARD, b Chatham, NY, Sept 6, 36; m 67; c 2. CHEMICAL ENGINEERING. *Educ:* Mass Inst Technol, BS, 58, SM, 60, ScD, 65. *Prof Exp:* Ford Found res fel, Mass Inst Technol, 65-67, asst prof chem eng, 65-68, 69-70; res group leader, 70-72, SR DEPT MGR, POLAROID CORP, 72- *Concurrent Pos:* Vis lectr, Imp Col, Univ London, 68-69. *Mem:* Am Chem Soc; Am Inst Chem Engrs; Inst Elec & Electronics Engrs. *Res:* Combustion; heat transfer; radiative heat transfer; light scattering; quality control; photographic films. *Mailing Add:* 300 Old Ocean St Marshfield MA 02050-3114

DALZIEL, IAN WILLIAM DRUMMOND, b Glasgow, Scotland, Nov 26, 37; m 60. STRUCTURAL GEOLOGY. *Educ:* Univ Edinburgh, BSc, 59, PhD(geol), 63. *Prof Exp:* Asst lectr geol, Univ Edinburgh, 59-63; vis lectr, Univ Wis, 63-64, asst prof, 64-66; assoc prof, 66-74, PROF GEOL COLUMBIA UNIV, 74-, MEM STAFF, LAMONT-DOHERTY GEOL OBSERV, 68- *Concurrent Pos:* Co-leader exped Somerset Island, Arctic Can, 64-65; consult, Que Cartier Mining Co, 65-; leader expeds, S Shetland Islands, S Omchez Islands, Antarctic Peninsula & Tierra del Fuego, 69-; mem deep sea drilling proj, Joint Oceanog Insts Deep Earth Sampling Antarctic Adv Panel, 71- *Mem:* AAAS; fel Geol Soc Am; fel Geol Soc London; Am Geophys Union. *Res:* Relations of deformation and metamorphism in orogenic belts; development of mountain belts; fault mechanics; stress history of folded rocks; reconstruction of Gondwanaland; development of island arc systems. *Mailing Add:* Inst Geophys Univ Tex 8701 N Pal Austin TX 78759

DAM, CECIL FREDERICK, b Kalamazoo, Mich, June 30, 23; m 50; c 2. PHYSICS. *Educ:* Kalamazoo Col, BA, 48; Cornell Univ, MS, 53; Ohio State Univ, PhD(physics), 56. *Prof Exp:* Asst prof physics & math, Hamline Univ, 57-58; from asst prof to assoc prof, 58-62, PROF PHYSICS, CORNELL COL, 63-, CHMN, DEPT PHYSICS, 80- *Concurrent Pos:* Resident res assoc, Argonne Nat Lab, 65-66. *Mem:* Am Asn Physics Teachers; Optical Soc Am; Am Hist Sci Soc; Am Meteorol Soc; Soc Hist Technol. *Res:* Atmospheric electricity. *Mailing Add:* 8625 D SW 94th St Ocala FL 32676

DAM, RICHARD, b Kalamazoo, Mich, Sept 17, 29; m 59; c 1. BIOCHEMISTRY, NUTRITION. *Educ:* Kalamazoo Col, BA, 51; Cornell Univ, MS, 56, PhD(animal nutrit), 59. *Prof Exp:* Asst biochemist, 58-60, asst prof biochem, 60-76, mem fac life sci, 76-80, ASST PROF LIFE SCI, UNIV NEBR, LINCOLN, 80- *Mem:* AAAS; Am Chem Soc; Am Inst Nutrit; Am Soc Microbiol; Sigma Xi. *Res:* Isolation and characterization proteins; protein complexes and interactions; nutritional utilization proteins. *Mailing Add:* Dept Agr Biochem Univ Nebr Lincoln NE 68583-0718

DAM, RUDY JOHAN, b Semarang, Indonesia, Nov 30, 49; US citizen; m 78; c 2. PHYSICAL CHEMISTRY, ENVIRONMENTAL ENGINEERING. *Educ:* Calif Inst Technol, BS, 72; Univ Ore, PhD(chem), 76. *Prof Exp:* SR RES ASSOC, INSTRUMENT RES & DEVELOP, ENG PHYSICS LAB, E I DU PONT DE NEMOURS & CO, 77- *Res:* Environmental instrument development; industrial toxicology; molecular genetics instrumentation; fluourescence spectroscopy. *Mailing Add:* Eng Physics Lab E357/208 E I du Pont de Nemours & Co Wilmington DE 19898

DAMADIAN, RAYMOND, b New York, NY, Mar 16, 36; m 60; c 3. BIOCHEMISTRY, BIOPHYSICS. *Educ:* Univ Wis, BS, 56; Albert Einstein Col Med, MD, 60. *Prof Exp:* Univ res fel biophys, Harvard Univ, 63-65; sr investr, USAF Sch Aerospace Med, 65-67; asst prof, 67-71, assoc prof biophys, 71-80; PRES, FONAR CORP, MELVILLE, NY, 78- *Concurrent Pos:* res grant, Health Res Coun of City of New York, 67-69, career investr,

67-72. *Honors & Awards:* Lawrence Sperry Award, 84; Nat Medal Technol, Pres US, 88. *Mem:* Am Chem Soc; Biophys Soc; Am Soc Microbiol; AAAS; Am Phys Soc; NY Acad Sci; Sigma Xi. *Res:* Biophysical chemistry of alkali cation accumulation in living cells; physical chemistry of the submolecular structure of cells; nuclear magnetic resonance techniques for cancer detection; applications of nuclear magnetic resonance spectroscopy to the investigation of the basis of a malignant transformation in a living cell; originator of nuclear magnetic resonance scanner; coagulation and thrombus formation; numerous publications and patents. *Mailing Add:* Fonar Corp 110 Marcus Dr Melville NY 11747

DAMAN, ERNEST L, b Hannover, Ger, Mar 14, 23; US citizen; m 45; c 3. MECHANICAL ENGINEERING. *Educ:* Polytech Inst Brooklyn, BME, 43. *Prof Exp:* Develop engr, Foster Wheeler Develop Corp, 46-49, proj engr, 49-53, res engr, 53-55, dept dir res, 55-59, dir res, 59-74, vpres res, 74-81, sr vpres, Foster Wheeler Corp, 81-88, CHMN BD, FOSTER WHEELER DEVELOP CORP, 77- *Concurrent Pos:* Mem exec comt, Pressure Vessel Res Comt, Welding Res Coun, 63-, vchmn, 81, chmn,86-; dir, Metals Properties Coun, 81-; chief exec officer, HDS Fibers Inc, 86-88. *Mem:* Nat Acad Eng; fel AAAS; fel Am Soc Mech Eng; Mat Properties Coun. *Res:* Heat transfer; fluid flow as applied to power generation equipment; metallurgy and structural analysis as applied to power generation; combustion. *Mailing Add:* Foster Wheeler Corp 12 Peach Tree Hill Rd Livingston NJ 07039

DAMAN, HARLAN RICHARD, b New York, NY, Nov 1, 41. MEDICINE, ASTHMATIC DISORDERS. *Educ:* Harvard Col, AB, 63; Albert Einstein Col Med, MD, 67. *Prof Exp:* Intern & resident pediat, Yale Univ-Yale New Haven Hosp, 67; fel allergy-immunol, Univ Colo Med Ctr, 71-73; MEM TEACHING PEDIAT, ALBERT EINSTEIN COL MED, 74- & MT SINAI SCH MED, 75- *Concurrent Pos:* Consult pediat allergy, Lawrence Hosp, Bronxville, 77- *Mem:* Fel Am Acad Allergy; fel Am Acad Pediat; fel Am Col Allergists; fel Am Col Chest Physicians; NY Acad Sci. *Mailing Add:* 769 Kimball Ave Yonkers NY 10704-1534

DAMANN, KENNETH EUGENE, JR, b Chicago, Ill, Apr 27, 44; m 68; c 2. PLANT PATHOLOGY. *Educ:* Eastern Ill Univ, BSE, 66; Univ Ark, MS, 69; Mich State Univ, PhD(bot, plant path), 74. *Prof Exp:* Asst prof, 74-78, ASSOC PROF DEPT PLANT PATH, LA STATE UNIV, BATON ROUGE, 78- *Mem:* Am Phytopath Soc; Am Soc Plant Physiologists. *Res:* Etiology of the ratoon stunting disease of sugarcane; physiology of plant disease. *Mailing Add:* Dept Plant Path La State Univ 472 Life Sci Bldg Baton Rouge LA 70803-1720

DAMASK, ARTHUR CONSTANTINE, b Woodstown, NJ, July 28, 24; m; c 2. BIOPHYSICS. *Educ:* Muhlenberg Col, BS, 49; Iowa State Univ, MS, 54, PhD(physics), 64. *Prof Exp:* Physicist, Frankford Arsenal, 53-65; prof physics, Queens Col, NY, 65-90; RETIRED. *Concurrent Pos:* Guest scientist, Brookhaven Nat Lab; mem, Solid State Sci Panel, Nat Res Coun, 64-74; consult, Lawrence Livermore Lab, Brookhaven Nat Lab & Armed Forces Radiobiol Res Inst; ed, J Semiconductors & Insulators, 73-78, J Phys Chem Solids, 78-; chmn, comt army procurement, Nat Acad Sci, 75-76; mem, Nat Mat Adv Bd, Nat Res Coun, 76-79; vis prof, Am Univ Cairo, 71, NY Univ Ctr Sci & Technol Policy, 79-82; chmn Energy Coun, Nat Mat Adv Bd; pres, Accident & Injury Anal. *Honors & Awards:* Mellon Found Fel, 84. *Mem:* Fel Am Phys Soc; Am Asn Physicists Med; Asn Adv Automotive Med; Soc Automotive Eng. *Res:* Metals physics radiation effects; defects in solids; organic crystals; medical physics, biophysics; accident reconstruction and inqury cunsetion. *Mailing Add:* 29 Brewster Lane Bellport NY 11713

DAMASKUS, CHARLES WILLIAM, b Ill, Oct 28, 24; m 49; c 3. BIOCHEMISTRY, ZOOLOGY. *Educ:* Valparaiso Univ, BA, 49. *Prof Exp:* Res chemist, Baxter Labs, 49-51; sr scientist, Armour Pharmaceut Co, 51-63; clin res scientist, Arnar Stone Labs, Am Hosp Supply Corp, 63-66; PRES, ALPAR LABS, INC, 66- *Mem:* Am Chem Soc; AAAS. *Res:* Pharmaceutical clinical research; pharmaceutical product development; layer lyophilized products; pet product development; layering process for pharmaceuticals; use of Alpha Chymotrypsin in Zonulysis; medicinal uses of Chymotrypsin and trypsin; development of clinical uses of pituitary hormones. *Mailing Add:* 132 S Kensington La Grange IL 60525

DAMASSA, AL JOHN, b Lucca, Italy, Oct 16, 29; US citizen; m 51; c 3. COMPARATIVE PATHOLOGY, ENTOMOLOGY. *Educ:* Calif State Univ, Chico, AB, 56; Univ Calif, Davis, MS, 63, PhD(comp path), 77. *Prof Exp:* RES ASSOC AVIAN MED & ANIMAL MYCOPLASMOSIS, SCH VET MED, UNIV CALIF, DAVIS, 56- *Res:* Avian spirochetosis; vectors and the disease process; avian and animal mycoplasmosis. *Mailing Add:* Dept Epidemiol Univ Calif Sch Vet Med Davis CA 95616

DAMASSA, DAVID ALLEN, b Bayshore, NY, Sept 17, 50; m 74; c 2. ENDOCRINOLOGY, NEUROENDOCRINOLOGY. *Educ:* Stanford Univ, BS, 72, PhD(physiol), 77. *Prof Exp:* Fel, Dept Anat, Brain Res Inst, Univ Calif, Los Angeles, 77-79; asst prof anat, 79-85, ASSOC PROF ANAT & CELLULAR BIOL, MED SCH, TUFTS UNIV, 85- *Concurrent Pos:* Prin investr, NIH, 80- *Mem:* Endocrine Soc; Am Asn Anatomists; Soc Neurosci; Soc Study Reproduction. *Res:* investigation of the neuroendocrine control of reproduction, specifically the effects of steroids on pituitary hormone secretion, reproductive behavior and the ontogeny of brain-pituitary gonadal interactions. *Mailing Add:* Dept Anat & Cellular Biol Med Sch Tufts Univ 136 Harrison Ave Boston MA 02111

D'AMATO, CONSTANCE JOAN, b Shrub Oak, NY, Jan 5, 33. NEUROBIOLOGY, NEUROPATHOLOGY. *Educ:* Tufts Univ, BS, 55. *Prof Exp:* Sr res asst neurobiol, New Eng Deaconness Hosp & Harvard Med Sch, 55-62; res assoc neurobiol & neuropathol, Med Ctr, 62-72, ASST PROF NEUROBIOL, DEPT PATH, UNIV MICH, 72- *Concurrent Pos:* Counr premed students, Univ Mich, 71-; mem, Neurol & Behav Sci Curric, Univ Mich, 80- *Mem:* Soc Neurosci; Am Acad Neurol; Teratology Soc; Asn Am Med Col; Behav Teratology Soc. *Res:* Experimental effects of radiation, trauma, and drugs on developing brain and behavior. *Mailing Add:* Dept Path Box 45 Univ Mich Med Ctr 1301 Catherine Rd Ann Arbor MI 48109-0602

D'AMATO, HENRY EDWARD, b Shrewsbury, Mass, Oct 3, 28; m 53; c 4. PHARMACOLOGY. *Educ:* Col of Holy Cross, AB, 49; Boston Univ, MA, 51, PhD(physiol), 54. *Prof Exp:* From instr to asst prof pharmacol, Tufts Univ, 54-59; pharmacologist, Astra Biol Labs, 59-67, ASSOC SCI DIR, ASTRA PHARMACEUT PROD, INC, 67- *Concurrent Pos:* Lederle med fac award, 56-59. *Res:* Experimental hypothermia; cardiovascular physiology and pharmacology; antispasmodic, antihistaminic and antiarrhythmic drugs; iron metabolism. *Mailing Add:* 12 Forest Dr Holden MA 01520

D'AMATO, RICHARD JOHN, b Springfield, Mass, Sept 24, 40; m 61; c 2. PHYSICAL CHEMISTRY. *Educ:* Am Int Col, BA, 62; Univ Ill, Champaign-Urbana, MS, 65, PhD(phys chem), 71. *Prof Exp:* Res chemist polymer phys chem, Monsanto Co, 65-67; sci specialist, Scott Graphics, Inc, 70-75, sr group leader phys chem, 75-76, mgr prod develop, 76-78; DIR, IMAGING & CUSTOM RES & DEVELOP, JAMES RIVER GRAPHICS, 78- *Mem:* Am Chem Soc; Soc Photog Scientists & Engrs. *Res:* High temperature gas phase reaction kinetics; polymer physical chemistry; physical chemical aspects of adhesion; chemistry and physics of organic photo conductors. *Mailing Add:* 19 Chapel Hill Dr South Hadley MA 01075-1605

DAMBACH, GEORGE ERNEST, b Dayton, Ohio, June 10, 42; m 64; c 2. ELECTROPHYSIOLOGY. *Educ:* Ohio State Univ, BS, 64, PhD(pharmacol), 68. *Prof Exp:* Res fel cardiol, Philadelphia Gen Hosp, 68-69; fel pharmacol, Univ Pa, 69-71, scholar, 71-74; dir grad prog & asst dean curricular affairs, 80-87, ASSOC PROF PHARMACOL, WAYNE STATE UNIV, 74-, ASST DEAN FOR RES & GRAD PROGS, 87- *Mem:* Biophys Soc. *Res:* Cellular electrogenesis; mechanisms of normal and abnormal biological electrical activity; pharmacological modulation of electrogenesis; cardiac arrhythmias; cocaine and arrhythmias. *Mailing Add:* Sch Med Wayne State Univ 540 E Canfield Ave Detroit MI 48202

DAMBERGER, HEINZ H, b Komotau, Czech, Dec 15, 33; m 63; c 4. GEOLOGY. *Educ:* Univ Mainz, Vordiplom, 56; Clausthal Tech Univ, dipl geol, 60, Dr rer nat(geol), 66. *Prof Exp:* Geologist, Saarbergwerke AG, Ger, 60-68; from assoc geologist to geologist, Ill State Geol Surv, 68-76; chief geologist, Explor & Bergbau GMBH, Washington, Pa, 76-77; assoc prof geol, Univ Pittsburgh, 78; GEOLOGIST & HEAD COAL SECT, ILL STATE GEOL SURV, 78- *Concurrent Pos:* Lectr geol & mineral, Sch Mining Eng, Saarbrucken, 61-68; vis scientist geol, Univ Bochum, WGer, 82-83. *Mem:* Geol Soc Am; Int Comt Coal Petrog; Am Asn Petrol Geologists; Soc Mining Engrs. *Res:* Coal geology, including coal quality, coalification, coal mining geology and coal petrography; classification of coal; microstructure of coal; study of geological parameters that control roof stability or roof failure in underground coal mines; coal reserve evaluation. *Mailing Add:* 411 W Indiana Ave Urbana IL 61801

DAMBORG, MARK J(OHANNES), b Ft Dodge, Iowa, Dec 6, 39; m 69. ELECTRICAL ENGINEERING. *Educ:* Iowa State Univ, BS, 62; Univ Mich, MSE, 63, PhD(comput, info & control eng), 69. *Prof Exp:* Asst res engr, Systs Eng Lab, Univ Mich, 65-66, 67-69; from asst prof to assoc prof 69-85, PROF ELEC ENG, UNIV WASH, 85- *Concurrent Pos:* Syst control engr, Elec Energy Systs Div, US Dept Energy, 77-78. *Mem:* Inst Elec & Electronics Engrs; Sigma Xi. *Res:* Stability of nonlinear systems; simulation and control AI applications in electric power systems; database applications in computer aided engineering. *Mailing Add:* Dept Elec Eng Univ Wash Seattle WA 98195

D'AMBROSIO, STEVEN M, b Philadelphia, Pa, May 7, 49; m 81; c 2. TOXICOLOGY, GENETICS. *Educ:* St Joseph's Univ, BS, 71; Tex A&M Univ, PhD(bio-org chem), 75. *Prof Exp:* Res assoc molecular biol, Brookhaven Nat Lab, 75-77; from asst prof to assoc prof, 77-85, PROF PHARMACOL & RADIOL, OHIO STATE UNIV, 85- *Concurrent Pos:* Consult health effects, US Environ Protection Agency, 80, NIH, 84 & Am Fed Aging Res, 84; prin investr, US Environ Protection Agency & NIH, 78, Nat Inst Environ Health Sci, 81. *Mem:* Am Soc Pharmacol & Exp Therpeut; AAAS; Am Soc Photochem; NY Acad Sci; Soc Toxicol; Am Soc Cell Biol. *Res:* Determining the molecular mechanisms of cancer, mutation, teratology and aging, including genetic toxicology, human health effects, means of prevention as well as reversing toxic effects. *Mailing Add:* Dept Radiol Ohio State Univ 400 W 12th Ave Wiseman Hall Rm 103 Columbus OH 43210

D'AMBROSIO, UBIRATAN, b Sao Paulo, Brazil, Dec 8, 32; m 58; c 2. SOCIAL HISTORY OF SCIENCE IN LATIN AMERICA. *Educ:* Univ Sao Paulo, BSc, 54, PhD(math), 63. *Prof Exp:* Regente math, Fac Philos, Univ Sao Paulo, 60-64; res assoc, Brown Univ, 64-65; asst prof, State Univ NY Buffalo, 65-66; assoc prof, Univ RI, 66-68; assoc prof & dir grad studies, State Univ NY Buffalo, 68-74; PROF, INST MATH, UNIV CAMPINAS, BRAZIL, 72- *Concurrent Pos:* Ida Beam vis prof, Univ Iowa, 80-81; head, Unit Improv Educ Systs, Orgn Am States, 80-81; pres, InterAm Comt Math Educ, 79-83; vpres, Int Comn Math Instr, 79-83; Vpres, Brazilian Soc Hist Sci; pro-rector univ develop, Univ Campiras, 86-90. *Mem:* Math Union Italy; Math Soc France; London Math Soc; Am Math Soc; fel AAAS; Brazilian Soc Hist Sci; Latin Am Soc Hist Sci & Technol (pres, 91). *Res:* Historical and epistemological studies of ethnoscience and ethnomathematics and efforts to bring them into the curriculum; social history of mathematics and of its education; culturally based curriculum development. *Mailing Add:* Inst Math Univ Campinas CP 6063 Campinas 13081 SP Brazil

DAME, CHARLES, b Providence, RI, July 18, 29; m 68; c 2. HORTICULTURE, FOOD SCIENCE. *Educ:* Univ RI, BS, 51; Univ Calif, MS, 55. *Prof Exp:* Assoc food sci, Univ Calif, 55-57; res chemist, Standard Fruit & Steamship Co, 57-60; group leader, Gen Foods Corp, 60-68; dir res & qual control, Nat Sugar Refining Co, 68-72; VPRES RES & DEVELOP, STANGE CO, 72-, VP TECH, MCCORMICK STANGE FLAVOR DIV, MCCORMICK & CO INC, 82- *Mem:* Inst Food Technol; Sigma Xi; Am Inst Baking. *Res:* Food science with special interest in spices and their extractives; natural and artificial flavors; artificial colors. *Mailing Add:* Flavor Div 26 McCormick-Stange 204 Wight Ave 230 Schilling Hunt Valley MD 21031

DAME, DAVID ALLAN, b Greenfield, Mass, Oct 4, 31; m 52; c 2. ENTOMOLOGY. *Educ:* Dartmouth Col, AB, 54; Univ Mass, PhD(entom), 61. *Prof Exp:* Res entomologist, Insects Affecting Man & Animals Lab, Arg Res Serv, USDA, 61-88; ENTOM CONSULT, ENTOM SERV, GAINESVILLE, FLA, 88- *Concurrent Pos:* Mem, Expert Comt Parasitic Dis, WHO & Expert Comt Trypanosomsasis, Food Agr Orgn; prof entom, Univ Fla; tech adv, Fla Dept Plant Indust; mem, Fla Coord Comt Mosquito Control. *Honors & Awards:* Recognition Award, Entom Soc Am, 84-85; Meritorious Serv Award, Am Mosquito Control Asn, 91. *Mem:* Soc Vector Ecol; Entom Soc Am; Am Mosquito Control Asn. *Res:* Control of insects of medical importance involving insect behavior, ecology, chemosterilization, radiosterilization, juvenile hormones, biological agents, insecticides, insect propagation and fruit flies. *Mailing Add:* 4729 NW 18th Pl Gainesville FL 32605

DAME, RICHARD EDWARD, b Lewisburg, Pa, May 31, 37; m 61. ENGINEERING, MATHEMATICS. *Educ:* George Washington Univ, BCE, 60, MSE, 63; Cath Univ Am, PhD(eng, math), 72. *Prof Exp:* Civil engr, Fed Aviation Agency, 59-62; sr engr, Dept Defense, DIA, 62-63 & Fairchild Hiller, 64-65; proj engr, Booz Allen Appl Res, 65-66; proj engr, Mega Eng, 66-70, dir eng, 70-; AT DEPT MECH ENG, CATH UNIV AM. *Concurrent Pos:* Dept Energy grant, 77-79; asst prof, Cath Univ Am, 78- *Honors & Awards:* Patent Award, NASA, 74. *Mem:* Am Inst Aeronaut & Astronaut; Am Asn Small Res Co; Consult Engrs Coun. *Res:* Solid mechanics; variational methods; solar energy systems; energy economics. *Mailing Add:* Dept Mech Eng Cath Univ Am 620 Mich Ave NE Washington DC 20064

DAME, RICHARD FRANKLIN, b Charleston, SC, Nov 16, 41; m 67; c 2. MARINE ECOLOGY, ECOSYSTEM ANALYSIS. *Educ:* Col Charleston, SC, BS, 64; Univ NC, Chapel Hill, MA, 67; Univ SC, Columbia, PhD(biol), 71. *Prof Exp:* Teacher biol & physics, St Andrew's High Sch, 66-68; chmn, 81-91,PROF MARINE SCI, COASTAL CAROLINA COL, UNIV SC, 71-, PALMETTO PROF, 90-, PROF MARINE SCI, COLUMBIA, 84- *Concurrent Pos:* Res assoc, Belle W Baruch Inst Marine Biol & Coastal Res, 71- & Inst Ecol, Univ Ga, 78; prin investr, NSF grants, 73-; invited investr, Royal Inst Nature Studies, 86-88, French Inst Sea Res, 91- *Mem:* Am Soc Limnol & Oceanog; Ecol Soc Am; Nat Shellfisheries Asn; Estaurine Res Fedn; Sigma Xi. *Res:* Oyster reefs and mussel beds shown to be major processors of energy and materials in estuaries, speeding up nutrient cycling and coupling the benthos to the water column; esturaries shown to be part of a complex coastal landscape coupling the forest to the sea via material flows and feedback loops. *Mailing Add:* Dept Marine Sci Coastal Carolina Col Conway SC 29526

DAMEN, THEO C, b Oct, 27, 33; m 61; c 4. SUB-PICOSECOND SPECTROSCOPY, SEMICONDUCTOR & SOLID STATE LASER TECHNOLOGY. *Educ:* Univ Eindhoven, Neth, BSEE, 57; NY Univ, BA, 70, MA, 72. *Prof Exp:* MEM RES TECH STAFF, AT&T BELL LABS, 58- *Res:* Experimental verification of theoretical predictions for light-matter interactions, phonons, polaritons, magnons, plasmons, spin flip, surface phonons; hydrodynamic expansion of electron-hole droplets; high density neodymium-pentaphosphate lasers; broad band tunable; semiconductor lasers; ultra fast carbon dioxide-laser modulator; quantum confined stark effect; relaxation dynamics of carriers; excitons and it's spin; ultra high power laser arrays. *Mailing Add:* AT&T Bell Labs Rm 4B-411 Crawfords Corner Rd Holmdel NJ 07733-1988

DAMERAU, FREDERICK JACOB, b Parma, Ohio, Dec 25, 31; m 54; c 2. INFORMATION SCIENCE. *Educ:* Cornell Univ, BA, 53; Yale Univ, MA, 57, PhD(ling), 66. *Prof Exp:* Programmer, Fed Systs Div, 57-61, res staff info sci, 61-68, RES STAFF COMPUTATIONAL LING, RES DIV, IBM CORP, 68- *Concurrent Pos:* Assoc ed, Asn Comput Ling, 74-78; adj fac, Pace Univ, 81- *Mem:* AAAS; Asn Comput Ling; Asn Comput Mach; Ling Soc Am. *Res:* Computer understanding of natural language directed particularly toward data base enquiry. *Mailing Add:* Box 494 North Salem NY 10560-0494

DAMEROW, RICHARD AASEN, b Thief River Falls, Minn, Sept 4, 36; m 57; c 2. NEUTRON SOURCE DEVELOPMENT, RESEARCH MANAGEMENT. *Educ:* Univ Minn, BS, 58, MS, 60, PhD(mass spectros), 63. *Prof Exp:* Mem staff, Sandia Labs, 63-71, div supvr, 71-78, sci adv, US Dept Energy/Mil Appl, 78-79, DIV SUPVR, SANDIA LABS, 79- *Mem:* Am Phys Soc; Am Asn Physics Teachers. *Res:* Mass spectroscopy, precision atomic masses; high magnetic fields; nuclear weapon effects. *Mailing Add:* 2924 Espanola N E Albuquerque NM 87110

DAMEWOOD, GLENN, US citizen. APPLIED PHYSICS. *Educ:* Tex Col Mines, BS, 49. *Prof Exp:* Chief instrumentation, US Army Field Forces Bd No 4, Ft Bliss, Tex, 50-51; asst chief, White Sands Proving Grounds, 51-52; sr engr, El Paso Natural Gas Co, 53-55; proj leader, Dept Physics, Spec Projs Lab, 55-56, mgr indust appln sect, 57-59, dept dir, 59-72, tech vpres & dir, 72, VPRES APPL PHYSICS DIV, SOUTHWEST RES INST, 74- *Res:* Acoustics, analog computer design and nondestructive testing, including radiography; eddy current and magnetic techniques; fluid dynamics, instrumentation and machine design. *Mailing Add:* 10467 Oakland Dr San Antonio TX 78240

DAMIAN, CAROL G, b Salem, Ohio, Oct 30, 39; m 58; c 4. COMMUNICATION AMONG PHYSICS TEACHERS-ORGANIZING ALLIANCES & SELF HELP GROUPS, PROMOTING STUDY OF PHYSICS AMONG STUDENTS OF ALL ETHNIC RACIAL AND SOCIAL GROUPS & FOR BOYS & GIRLS. *Educ:* Ohio State Univ, BS, 76, MA, 84. *Prof Exp:* Teacher chem, physics, anat & physiol, Licking Heights High Sch, 76-80; TEACHER CHEM & PHYSICS, DUBLIN HIGH SCH, 80- *Concurrent Pos:* Consult, Merrill Publ, 90-; NSF/Ohio State Univ writer/consult grant, teaching mat new global sci curric, 90- *Mem:* Nat Sci Teachers Asn; AAAS; Am Asn Physics Teachers; Asn Supervision & Curric Develop; Am Chem Soc. *Res:* How high school students learn science: specifically, chemistry and physics; mathematics readiness among students studying high school chemistry and physics; comparisons of attitudes toward chemistry and physics among high school boys vs high school girls; effectiveness of lab activities in teaching compared to other classroom activities. *Mailing Add:* 4391 Sawmill Rd Columbus OH 43220-2243

DAMIAN, RAYMOND T, b Philadelphia, Pa, Aug 11, 34; m 57; c 3. PARASITOLOGY, IMMUNOLOGY. *Educ:* Univ Akron, BS, 56; Fla State Univ, MS, 58, PhD(parasitol), 62. *Prof Exp:* Res assoc parasitol, Fla State Univ, 62-63; asst prof biol, Emory Univ, 63-67; immunologist, Southwest Found Res & Educ, 67-69, assoc found scientist, Dept Immunol, 69-73; assoc prof, 73-77, PROF ZOOL, UNIV GA, 77- *Concurrent Pos:* Adj assoc prof microbiol, Univ Tex Med Sch, San Antonio, 72-73. *Honors & Awards:* Henry Baldwin Ward Medal, Am Soc Parasitol, 74. *Mem:* Royal Soc Trop Med & Hyg; Am Asn Immunol; NY Acad Sci; Am Soc Parasitol; Am Soc Trop Med & Hyg. *Res:* Immunological parasitology; schistosomiasis; immunology of nonhuman primates. *Mailing Add:* Dept Zool Univ Ga Athens GA 30602

DAMIANOV, VLADIMIR B, b Sofia, Bulgaria, Sept 19, 38; US citizen; m 72; c 1. MACHINE DESIGN, MATERIAL HANDLING EQUIPMENT. *Educ:* Univ Sofia, Bulgaria, BS, 61, MS, 67. *Prof Exp:* Dept mgr, Res & Develop Ctr for Lift Trucks, 61-70; proj engr, Mod Tool & Die Prod, Inc, 72-74; chief engr, Canton Stoker Corp, 74-77; sr engr, McNeil Akron Corp, 77-81; ENGR SPECIALIST, LORAL DEFENSE SYSTS, AKRON, 81- *Mem:* Am Soc Mech Engrs; Am Soc Testing & Mat. *Res:* Lift trucks; uranium enrichment. *Mailing Add:* 3021 Morewood Rd Fairlawn OH 44333

DAMJANOV, IVAN, b Subotica, Yugoslavia, Mar 31, 41; m 64; c 3. PATHOLOGY. *Educ:* Univ Zagreb, MD, 64, MSc, 66, PhD(path), 71. *Prof Exp:* Asst prof path, Sch Med, Univ Zagreb, 71-73; from asst prof to assoc prof path, Sch Med, Univ Conn, Farmington, 73-77; assoc prof, 77-80, prof path, Hahnemann Med Col & Hosp, Philadelphia, 80-86; PROF PATH, JEFFERSON MED COL, PHILADELPHIA, 86- *Mem:* Am Asn Path; Am Asn Cancer Res. *Res:* Developmental aspects of neoplasia. *Mailing Add:* Jefferson Med Col 1040 Walnut St Philadelphia PA 19107

DAMKAER, DAVID MARTIN, b Portland, Ore, Oct 11, 38. ZOOPLANKTON, COPEPODA. *Educ:* Univ Wash, Seattle, BS, 60, MS, 64; George Washington Univ, PhD(zool), 73. *Prof Exp:* Marine biologist zooplankton, Smithsonian Inst, 65-71; fisheries biologist, Nat Marine Fisheries Serv, 71-73, oceanographer biol oceanog, Environ Res Labs, 73-79, FISHERIES BIOLOGIST, NAT OCEANIC & ATMOSPHERIC ADMIN, 79- *Concurrent Pos:* Affil assoc prof oceanog, Sch Oceanog, Univ Wash, 74- *Mem:* Sigma Xi; Crustacean Soc; Challenger Soc; Marine Biol Asn UK; Am Fisheries Soc. *Res:* Systematics and ecology of free-living marine Copepoda; history of the study of Copepoda; effects of environmental pollution, especially of enhanced solar ultraviolet radiation; fisheries enhancement and management (salmonids). *Mailing Add:* Sch Oceanog WB-10 Univ Wash Seattle WA 98195

DAMLE, SURESH B, b Bombay, India, Aug 4, 35; US citizen; m 60; c 1. ORGANIC CHEMISTRY, BIOCHEMISTRY. *Educ:* Univ Bombay, BSc Hons, 58, MSc, 60; Rutgers Univ, New Brunswick, PhD(org chem, biochem), 64. *Prof Exp:* Anal chemist, Shell Refineries, Bombay, India, 54-60; res fel cancer chemother, Med Sch, Univ Pa, 64-65; res chemist, M&T Chem Inc, Rahway, NJ, 65-69 & Am Cyanamid Co, Bound Brook, 69-70; instr biochem, Rutgers Med Sch, 70-73; sr scientist, Indust Chem Div, NL Industs, Hightstown, NJ, 74, sect leader org chem, 75; RES ASSOC ORG CHEM, CHEM PROD DIV, PPG IND, INC, 76- *Mem:* Sigma Xi; The Chem Soc; Am Chem Soc; fel Am Inst Chemists. *Res:* Organic and organometallic chemistry synthesis and mechanisms; cancer chemotherapy; enzyme mechanisms; flame retardants; stabilizers; antioxidants and other plastic additives; Grighard chemistry; organotin, organolead and organosilicon chemistry; bicyclic chemistry. *Mailing Add:* PPG Tech Ctr 440 College Park Dr Monroeville PA 15146-1536

DAMM, CHARLES CONRAD, plasma physics; deceased, see previous edition for last biography

DAMMAN, ANTONI WILLEM HERMANUS, b Utrecht, Netherlands, Apr 21, 32; US citizen; wid; c 2. PLANT ECOLOGY. *Educ:* Univ Wageningen, BSc, 53, MSc, 56; Univ Mich, PhD, 67. *Prof Exp:* Res officer, Res Br, Nfld Dist, Can Dept Forestry, 56-65, res scientist & head tree biol & land classification sect, 65-67; assoc prof, 67-78, head, Ecol Sect, Biol Scis Group, 71-76, PROF PLANT ECOL, UNIV CONN, 78- *Concurrent Pos:* Mem, Nat Adv Comt Forest Land, Can, 64-67 & Subcomt Conserv Terrestial Ecosysts, Can Int Biol Prog, 66-67; adv comt, NSF Peatland Proj, Univ Minn, 80-86; vis prof, Univ Tasmania, Australia, 81-82; vis prof, Univ Lund, Sweden, 86. *Mem:* Am Inst Biol Sci; Int Asn Ecol; Int Soc Vegetation Sci; Bot Soc Am. *Res:* Ecology, vegetation, and biogeochemistry of peatlands, in particular ombrotrophic bogs; phytogeography of eastern North America and in ecological land classification. *Mailing Add:* Dept Ecol & Evolutionary Biol Univ Conn Storrs CT 06268

DAMMANN, JOHN FRANCIS, b Chicago, Ill, Feb 21, 17; m 41; c 4. SURGERY. *Educ:* Harvard Univ, AB, 41; Univ Cincinnati, MD, 43. *Prof Exp:* Intern, Evanston Hosp, Ill, 44; researcher & chief resident, Children's Mem Hosp, Chicago, 44-46; asst prof pediat, Univ Calif, Los Angeles, 50-55; assoc prof surg cardiol, Univ Hosp, 55-60, assoc prof surg cardiol & pediat, Med Ctr, 60-61, prof, 61-68, PROF PEDIAT & BIOMED ENG, MED CTR, UNIV VA, 68-, NIH CAREER RES PROF, 62- *Concurrent Pos:* Fel pediat cardiol, Children's Mem Hosp, Chicago, 48; sr fel, Johns Hopkins Hosp, 49-50; consult, Vet Admin. *Mem:* AAAS; AMA; Am Col Cardiol; Soc Pediat Res; Sigma Xi. *Res:* Critical care medicine; pediatric cardiology; postoperative care and monitoring. *Mailing Add:* Dept Pediat Univ Va Hosp Charlottesville VA 22903

DAMMIN, GUSTAVE JOHN, b New York, NY, Sept 17, 11; m 41; c 3. MEDICINE, PATHOLOGY. *Educ:* Cornell Univ, AB, 34, MD, 38; Univ Havana, cert, 37; Am Bd Path, dipl. *Prof Exp:* Intern med, Johns Hopkins Hosp, 39, asst resident, Peter Bent Brigham Hosp, 40; instr path, Col Physicians & Surgeons, Columbia Univ, 41; from asst prof med & path to prof path & chmn dept, Wash Univ, 46-52; prof, 53-62, Friedman prof, 62-78, EMER PROF PATH, HARVARD MED SCH, 78- *Concurrent Pos:* Niles lectr, Cornell Univ; Held lectr, Beth Israel Hosp, New York; pathologist-in-chief, Barnes Hosp, St Louis, Mo, 51-52 & Peter Bent Brigham Hosp, 52-74; consult path, West Roxbury Vet Admin Hosp, 54-75, actg chief lab serv, 76-77, assoc chief, 78-, actg chief, consult path, 81-; consult, Surgeon Gen, US Dept Army, Far East, 56, 64, 66, Europe, 59, 63, 68, 70 & Surgeon Gen, USPHS; lab consult, Off Civil & Defense Mobilization, 50-60; mem comn enteric infections, Armed Forces Epidemiol Bd, 51- & comn parasitic dis, 54-, dir comn parasitic dis, 59-60, pres bd, 60-73; mem trop med & parasitol study sect, NIH, 59-62, mem cholera adv comt, 65-, mem, Int Ctr Comt, 72; mem panel, Inst Defense Anal, 61-62; mem sci adv bd, Armed Forces Inst Path, 61-71; mem subcomt geog path, Nat Res Coun, 62-65; mem heart spec proj comt, Nat Heart Inst, 63-; chmn Geneva expert comt enteric infections, WHO, 63-; mem sci adv comt, New Eng Regional Primate Res Ctr; mem bd dirs, Gorgas Mem Inst, 67-; mem, Leonard Wood Mem Adv Med Bd, 69-; mem ad hoc comt, Div Med Sci, Nat Acad Sci, 70; mem kidney adv comt, Joint Comn Accreditation Hosps, 72-74; mem US deleg, US-Japan Coop Med Sci Prof, Dept State, 72-74; lectr trop pub health, Harvard Sch Pub Health, 78- *Honors & Awards:* Walter Reed Medallion, 71; Distinguished Pub Serv Medal, Dept Defense, 73. *Mem:* Asn Am Physicians; Am Asn Path; Am Soc Clin Invest; Am Soc Exp Path; Am Soc Trop Med & Hyg. *Res:* Epidemiology of intestinal infections; kidney and other organ transplantation; babesiosis; Lyme disease; parasitic diseases. *Mailing Add:* Dept Path Harvard Med Sch Boston MA 02115

DAMODARAN, KALYANI MUNIRATNAM, b Gudiyattam, India, June 9, 38; m 70; c 1. ORGANIC & MEDICINAL CHEMISTRY. *Educ:* Univ Madras, BS, 59, MS, 61; Indian Inst Sci, Bangalore, PhD(org chem), 70. *Prof Exp:* Teaching asst chem, Loyola Col, Madras, India, 61-62; sr res fel org chem, Indian Inst Sci, Bangalore, 67-69, sr res asst, 69-71; res fel, Temple Univ, 71-72 & Univ Pittsburgh, 72-73; res assoc, Temple Univ, 74-76; res scientist, 76-79, ASSOC SCIENTIST, SOUTHWEST FOUND RES & EDUC, 79- *Mem:* Am Chem Soc. *Res:* Synthetic organic chemistry of natural products and medicinally significant substances. *Mailing Add:* 169 Pennwood Ave Apt 3 Pittsburgh PA 15218

DAMON, DWIGHT HILLS, b Northampton, Mass, Feb 2, 31; m 55; c 2. PHYSICS. *Educ:* Amherst Col, AB, 53; Purdue Univ, MS, 55, PhD(physics), 61. *Prof Exp:* Res physicist, Westinghouse Res Labs, 61-70; assoc prof, 70-74, PROF PHYSICS, UNIV CONN, 74- *Concurrent Pos:* Mem gov bd, Int Thermal Conductivity Conf, 75-; assoc dir, Elec Insulation Res Ctr, 83- *Mem:* Am Phys Soc; AAAS. *Res:* Solid state physics, principally in magnetic and transport properties, electrical and thermal conduction, and galvanomagnetic phenomena. *Mailing Add:* Dept Physics Univ Conn U-46 Storrs CT 06268

DAMON, EDWARD G(EORGE), b Richland, NMex, Feb 21, 27; m 46; c 3. ENVIRONMENTAL BIOLOGY, INHALATION TOXICOLOGY. *Educ:* Eastern NMex Univ, BS, 50; Okla State Univ, MS, 57; Univ NMex, PhD(biol sci), 65. *Prof Exp:* Teacher high sch, NMex, 51-56; instr res cytogenetics, Okla State Univ, 57-58, res assoc, 60-61; sci instr, Jimma Agr Tech Sch Ethiopia, 58-60; head sci dept, High Sch, NMex, 61-62; res biologist, 62-65, ASSOC SCIENTIST, INHALATION TOXICOL RES INST, LOVELACE BIOMED ENVIRON RES INST, 65- *Mem:* AAAS; Bot Soc Am; assoc mem Am Physiol Soc. *Res:* Blast biology; physiological effects of air blast; pulmonary physiology; cultivated sorghums of Ethiopia; radiobiology and inhalation toxicology; computerization of animal inventory and pathology/toxicology data base systems. *Mailing Add:* 9715 Euclid NE Albuquerque NM 87112

DAMON, EDWARD K(ENNAN), b Concord, Mass, Jan 3, 28; m 50; c 2. ELECTRICAL ENGINEERING. *Educ:* Bowdoin Col, BSc, 49; Ohio State Univ, MS, 54. *Prof Exp:* Res assoc elec eng, Antenna Lab, Ohio State Univ, 50-58, from asst supvr to assoc supvr, 58-67, tech dir, Lasers & Optical Propagation, 67-75, assoc prof, 71-83, EMER ASSOC PROF ELEC ENG, OHIO STATE UNIV, 83- *Concurrent Pos:* Dir & treas, Ladar Systs, Inc, 64-70; consult, Data Corp, 64-67, Battelle Mem Inst, 70-73 & 83-85, IIT Res Inst, 81-89, Gen Elec, 83-89, NAm Rockwell, 85-89. *Mem:* Sr mem Inst Elec & Electronics Engrs; Optical Soc Am. *Res:* Electrical physics; microwave circuits and antennas; optical masers; nonlinear optical interactions; atmospheric spectroscopy. *Mailing Add:* Dept Elec Eng Ohio State Univ Elec Engr Bldg 1320 Kinnear Rd Columbus OH 43210

DAMON, JAMES NORMAN, b Baltimore, Md, Aug 31, 45; m 71; c 1. SINGULARITY THEORY, DIFFERENTIAL TOPOLOGY. *Educ:* Dartmouth Col, BA, 67; Oxford Univ, dipl adv math, 69; Harvard Univ, PhD(math), 72. *Prof Exp:* Asst prof math, Tufts Univ, 72-73; Fulbright lectr, Univ Tecnica del Estado, Chile, 73-74; asst prof, Queens Col, 74-76; form asst prof to assoc prof, 76-83, PROF MATH, UNIV NC, CHAPEL HILL, 83- *Concurrent Pos:* Sr Fulbright-Hayes scholar, 73-74; Res Found, City Univ New York grant, 75-76; NSF res grant, 75-; sr vis fel, Univ Liverpool, 79-80; assoc prof, Nat Ctr Sci Res, Univ Nice, France, 84; Fulbright Res Scholar, Univ Warwick, Eng, 88-89; vis prof, Tokyo Inst Technol, 90. *Mem:* Am Math Soc. *Res:* Singularities of smooth mappings and their applications to differential topology, differential equations; smooth and topological stability; topological structure of singularities; stratification techniques; relation to equisingularity problems in algebraic geometry. *Mailing Add:* 360 Phillips Hall 039A Univ NC Chapel Hill NC 27599

DAMON, PAUL EDWARD, b Brooklawn, NJ, Mar 12, 21; m 47; c 2. GEOCHRONOLOGY, GEOCHEMISTRY. *Educ:* Bucknell Univ, BS, 43; Univ Mo, MS, 49; Columbia Univ, PhD, 57. *Hon Degrees:* DSc, Bucknell Univ, 78. *Prof Exp:* Asst prof physics, Univ Ark, 49-54; res assoc geol, Lamont Geol Observ, Columbia Univ, 54-57; from assoc prof to prof geol & geochronology, 57-86, CHIEF SCIENTIST, LAB ISOTOPE GEOCHEM, UNIV ARIZ, 78-; CO-DIR NSF ARIZ ACCELERATOR FACIL RADIOISOTOPE ANAL, 78- *Concurrent Pos:* Consult, Isotopes, Inc, 56-57; distinguished vis res prof, Univ Wash, 74. *Mem:* Fel Geol Soc Am; fel AAAS; Geochem Soc; Am Geophys Union. *Res:* Geochronology, geochemistry and tectonics of ore deposits and volcanic rocks within the North American Cordillera; relationships between radiocarbon, solar activity, geomagnetism and climate; environmental geochemistry. *Mailing Add:* Dept Geosci Gould-Simpson Bldg 77 Univ Ariz Tucson AZ 85721

DAMON, RICHARD WINSLOW, solid state physics; deceased, see previous edition for last biography

DAMON, ROBERT A(RTHUR), b Weston, Ore, July, 21, 32, m 57; c 3. CHEMICAL ENGINEERING. *Educ:* Mont State Col, BS, 55, MS, 59, PhD(chem eng), 61. *Prof Exp:* Eng aide, US Bur Reclamation, 52-53; asst engr, Shell Oil Co, 54; asst instr math, 56-58; asst prof chem eng, Univ Ariz, 59-63; RES ASSOC, CENT RES DIV, CROWN ZELLERBACH CORP, 63- *Concurrent Pos:* Res fel, 55-58. *Mem:* Am Inst Chem Engrs; Am Chem Soc; Sigma Xi; Asn Energy Engrs; Tech Asn Pulp & Paper Indust. *Res:* Exploratory and process research in pulp, paper and wood-chemicals; reaction kinetics; polymer fiber processes; energy processes. *Mailing Add:* 1119 SE 78th Ave Vancouver WA 98664-1707

DAMONTE, JOHN BATISTA, b San Francisco, Calif, May 14, 25; m 51; c 1. ANTENNA DESIGN, RADIO WAVE PROPAGATION. *Educ:* Univ Calif Berkeley, BS, 48 & MS, 62. *Prof Exp:* Microwave engr, Dalmo Victor Co, San Carlos, 50-54, asst dir res, 54-58, mgr microwave eng, 58-66; from asst mgr to mgr antenna systs, 66-76, CONSULT ENGR, LOCKHEED MISSILE & SPACE CO, 76- *Concurrent Pos:* dir, Electronics Conventions Mgt, 82- *Honors & Awards:* Centennial Medal, Inst Elec & Electronics Engrs, 84. *Mem:* Fel Inst Elec & Electronics Engrs (pres, 72). *Res:* Antennas and propagation for satellite communications systems. *Mailing Add:* 1716 Hillman Ave Belmont CA 94002

D'AMORE, MICHAEL BRIAN, b Providence, RI, June 7, 45; m 71; c 4. CATALYSIS, ORGANIC CHEMISTRY. *Educ:* Providence Col, BS, 67; Calif Inst Technol, PhD(org chem), 72. *Prof Exp:* Res chemist, E I du Pont de Nemours & Co, Inc, 73-77, sr res chemist catalysis, 77-84, res assoc, 84-89, SR RES ASSOC & PROJ LEADER, E I DU PONT DE NEMOURS & CO, INC, 89- *Mem:* Am Chem Soc; Catalysis Soc. *Res:* Homogeneous and heterogeneous catalysis; reaction mechanisms of industrial importance. *Mailing Add:* Exp Sta E I du Pont de Nemours & Co Inc Wilmington DE 19880-0262

D'AMORE, PATRICIA ANN, b Everett, Ma, Nov 01, 51; c 1. CELL BIOLOGY, OPHTHALMOLOGIC BASIC SCIENCE. *Educ:* Regis Col, Weston, Ma, BA, 73; Boston Univ, PhD(cell biol), 77; Northwestern Univ, MBA, 87. *Prof Exp:* Teaching fel physiol chem & ophthal, Johns Hopkins Med Sch, 78-79, instr ophthal, 79-80, asst prof, 80-81; RES ASSOC SURG, THE CHILDREN'S HOSP, 81-; asst prof, 82-88, PROF PATH, HARVARD MED SCH, 89-, MEM FAC CELL & DEVELOP BIOL, 83- *Concurrent Pos:* Prin investr, Children's Hosp, NIH grants, 82-; fel, Oliver Wendell Holmes Soc, 85-; estab investr, Am Heart Asn, 86- *Honors & Awards:* Lamport Award, Microcirculatory Soc, 79; Res Award, Meyers Baltimore, 79. *Mem:* Am Soc Cell Biolists; Microcirculatory Soc; Asn Res Vision & Optics; NY Acad Sci; Tissue Cult Asn. *Res:* Mechanisms of vascular endothelial cell growth; determining the etiology of pathologies such as diabetic retinopathy and the retinopathy of prematurity as well as to creating an effective vascular prosthesis. *Mailing Add:* Lab Surg Res Children's Hosp 300 Longwood Ave Boston MA 02115

DAMOS, DIANE LYNN, b Waukegan, Ill, May 29, 49; m 71. HUMAN FACTORS ENGINEERING, AVIATION PSYCHOLOGY. *Educ:* Univ Ill, BS, 70, MA, 73, PhD(psychol), 77. *Prof Exp:* Mem tech staff human factors, Bell Tel Labs, 72-73; asst prof indust eng, State Univ NY Buffalo, 77-81; asst prof psychol, Ariz State Univ, 81-85; ASSOC PROF, DEPT HUMAN FACTORS, UNIV SOUTHERN CALIF, 85- *Concurrent Pos:* Consult, Calspan, 77- & Naval Biodynamics Lab, New Orleans, 77- *Mem:* Human Factors Soc; Asn Aviation Psychologists. *Res:* Multiple task performance, timesharing, human information processing, pilot selection and training. *Mailing Add:* Inst Safety & Systs Mgt Univ Southern Calif Los Angeles CA 90089-0021

DAMOUR, PAUL LAWRENCE, b Concord, NH, Jan 26, 37; m; c 4. PHYSICAL CHEMISTRY. *Educ:* St Anselm's Col, BA, 58; Cath Univ Am, PhD(chem), 63. *Prof Exp:* From instr to asst prof, 62-68, ASSOC PROF CHEM, ST ANSELM'S COL, 68- *Mem:* Am Chem Soc. *Res:* Photochemical reduction of coordinated complexes; electrokinetic properties of membrane electrodes. *Mailing Add:* Dept Chem St Anselms Col Manchester NH 03102

DAMOUTH, DAVID EARL, b Flint, Mich, Jan 27, 37; m 60; c 2. ELECTRONIC PRINTERS, OFFICE AUTOMATION. *Educ:* Univ Mich, BSE(eng physics) & BSE(eng math), 59, MS, 60, PhD(physics), 63. *Prof Exp:* Teaching fel physics, Univ Mich, 59-61; res scientist, NY, 63-67, sr scientist, Advan Eng Div, 67-68, mgr recording processes area, Info Systs Div, 68-71, prin scientist & mgr pattern recognition br, Palo Alto Res Ctr, 71-74, LAB MGR, WEBSTER RES CTR, XEROX CORP, NY, 74- *Res:* Office automation; electronic circuits and systems; electronic imaging systems; pattern recognition. *Mailing Add:* Xerox Webster Res Ctr 800 Phillips Rd, 0128-29E Webster NY 14580-9701

DAMPIER, FREDERICK WALTER, b Winnipeg, Man, Oct 1, 41; US citizen. PHYSICAL CHEMISTRY, ELECTROCHEMISTRY. *Educ:* Univ Man, BSc, 62; Rensselaer Polytech Inst, MS, 64, PhD(phys chem), 68. *Prof Exp:* Fel phys chem, Univ Calif, Berkeley, 68-70; sr scientist electrochem, Technol Ctr, ESB Inc, 70-75 & EIC Corp, 75-78; mem tech staff electrochem,

Power Sources Ctr, GTE Labs, 78-86; prog mgr, GTE Govt Systs Corp, 86-88; PRES, LITHIUM ENERGY ASSOCS, INC, 88- *Mem:* Am Chem Soc; Electrochem Soc; Sigma Xi. *Res:* Non-aqueous electrochemistry, molten salts, preparation and characterization of permselective membranes, primary and secondary lithium batteries. *Mailing Add:* Lithium Energy Assocs PO Box 25 Waverly Sta Belmont MA 02179

DAMRON, BOBBY LEON, b Ocala, Fla, Nov 6, 41; m 68; c 1. POULTRY NUTRITION. *Educ:* Univ Fla, BSA, 63, MSA, 64, PhD(animal sci), 68. *Prof Exp:* Res assoc, 66-68, from asst prof to assoc prof, 68-82, PROF POULTRY SCI, UNIV FLA, 82- *Mem:* Poultry Sci Asn; Sigma Xi; Soc Exp Biol & Med. *Res:* Mineral interrelationships and requirements of chicks, broilers and laying hens; amino acid and energy requirements of laying hens; evaluation of various feed additives and nutrient sources; influence of water quality and water-borne nutrients. *Mailing Add:* Dept Poultry Sci Univ Fla Gainesville FL 32611

DAMSTEEGT, VERNON DALE, b Waupun, Wis, Oct 19, 36; m 62; c 2. PLANT PATHOLOGY. *Educ:* Cent Col, Iowa, BA, 58; Wash State Univ, PhD(plant path), 62. *Prof Exp:* Asst plant path, Wash State Univ, 58-62; res plant pathologist, Crops Div, US Army Biol Labs, 62-71; PLANT PATHOLOGIST, AGR RES SERV, USDA, 71- *Concurrent Pos:* Adj prof biol, Montgomery Col, 74- *Mem:* Am Phytopath Soc. *Res:* Inheritance of resistance and pathogenicity; interaction of host and pathogen; epidemiology of virus diseases; modes of transmission and other general factors affecting disease severity. *Mailing Add:* USDA Agr Res Serv Foreign Dis Weed Sci Res Unit Ft Detrick Bldg 1301 Frederick MD 21702

DAMUSIS, ADOLFAS, b Toscica, Russia, June 16, 08; nat US; m 37; c 4. CHEMISTRY. *Educ:* Univ Vytautas the Great, Lithuania, Chem Eng, 34; Berlin-Charllot Tech Hochschule, Ger, BS & MS, 38, PhD(chem), 40. *Prof Exp:* Asst chem tech, Univ Vytautas the Great, Lithuania, 34-39, assoc prof tech silicates, fac eng, 40-43, dean, 42-43; consult, Lithuanian Dept Bldg Mat, 39-42; mem staff, E I du Pont de Nemours & Co, 47-49 & Sherwin-Williams Co, 49-57; res assoc, Wyandotte Chems Corp, 57-71, BASF Wyandotte Corp, 71-73; res prof, Polymer Inst, Univ Detroit, 73-83; RETIRED. *Concurrent Pos:* Consult, sealant companies, 73- *Mem:* Am Chem Soc; Soc Rheol. *Res:* Silicates; portland cements; adhesives; alumina-iron oxide ratio and shrinkage of portland cements; extender pigments; treatment of carbon blacks; silicate vehicles; amines as curing agents of epoxy-resins; urethane coatings and sealants; polymer research; polymer chemistry in the areas of urethane, epoxies and acrylics. *Mailing Add:* 13255 Oak Ridge Lane Lockport IL 60441-8649

DAMUTH, JOHN ERWIN, b Dayton, Ohio, Nov 22, 42; div. MARINE GEOLOGY & GEOPHYSICS. *Educ:* Ohio State Univ, BS, 65; Columbia Univ, MS, 68, PhD(geol), 73. *Prof Exp:* Res asst, Lamont-Doherty Geol Observ, Columbia Univ, 65-73, res scientist, 73-74, res assoc, 74-82, sr res assoc, 82-83; res geologist, 83-84, SR RES GEOLOGIST, DALLAS RES LAB, MOBIL RES DEVELOP CORP, 84-; ADJ RES SCIENTIST, LAMONT-DOHERTY GEOL OBSERV, COLUMBIA UNIV, 83- *Mem:* Fel Geol Soc Am; Sigma Xi; Am Asn Petrol Geologists; Soc Econ Paleontologists & Mineralogists; Am Geophys Union. *Res:* Structure, sedimentation processes, hydrocarbon potential, and paleosedimentation history of continental margins and the adjacent deep-sea floor; seismic stratigraphy and facies analysis. *Mailing Add:* Dallas Res Lab Mobil Res & Develop Corp PO Box 819047 Dallas TX 75381-9047

DANA, M TERRY, b Oklahoma City, Okla, Nov 17, 43. ATMOSPHERIC PHYSICS, METEOROLOGY. *Prof Exp:* SR RES SCIENTIST ATMOSPHERIC SCI, PAC NORTHWEST DIV, BATTELLE MEM INST, 67- *Mem:* Am Meteorol Soc; Am Geophys Union; Fedn Am Scientists; Union Concerned Scientists. *Res:* Precipitation chemistry; precipitation scavenging. *Mailing Add:* 1848 Wright Ave Richland WA 99352

DANA, MALCOLM NIVEN, b Pomfret, Vt, Dec 8, 22; m 46; c 2. HORTICULTURE. *Educ:* Univ Vt, BS, 48; Iowa State Col, MS, 49, PhD(hort), 52. *Prof Exp:* From asst prof to assoc prof, 52-67, chmn dept, 73-78, PROF HORT, UNIV WIS-MADISON, 67- *Mem:* Am Soc Hort Sci; Weed Sci Soc Am; Am Pomol Soc. *Res:* Weed control in cranberries; cultural studies on strawberries. *Mailing Add:* Dept Hort 393 Hort Bldg Univ Wis Madison WI 53706

DANA, ROBERT CLARK, b Cedar Falls, Iowa, Feb 19, 44; m 67; c 2. ORGANIC CHEMISTRY. *Educ:* Austin Col, BA, 66; Univ Tex, Austin, PhD(org chem), 71; Ohio State Univ, MA, 79. *Prof Exp:* Doc analyst, 71-75, ASST MGR ORG CHEM, CHEMICAL ABSTR SERV, 75- *Mem:* Am Chem Soc. *Mailing Add:* 1755 Churchview Lane Columbus OH 43220-4955

DANA, ROBERT WATSON, b Oklahoma City, Okla; m 67; c 2. REMOTE SENSING, OPTICAL PHYSICS. *Educ:* Univ Wash, BS, 63, MS, 69. *Prof Exp:* Physicist, Pac Southwest Forest & Range Exp Sta, Calif, 69-76, physicist remote sensing, 76-83, TECH INFO SPECIALIST, ROCKY MOUNTAIN FOREST & RANGE EXP STA, FOREST SERV, USDA, 83- *Mem:* Am Soc Photogram; Coun Optical Radiation Measurements. *Res:* Remote sensing of forest and range resources; radiometry photometry; image processing; microdensitometry; atmospheric radiative transfer. *Mailing Add:* 1724 W Stuart Ft Collins CO 80526

DANA, STEPHEN WINCHESTER, geology; deceased, see previous edition for last biography

DANBERG, JAMES E(DWARD), b New Haven, Conn, Oct 13, 27; m 53; c 5. AERONAUTICAL ENGINEERING. *Educ:* Cath Univ, BAE, 49, MAE, 52, DE, 64. *Prof Exp:* Aeronaut engr, Field Eval Div, US Naval Ord Lab, 51-55, aeronaut res engr, Spec Prob Br, Aerophys Div, 55-56, hypersonics group, 56-60, chief, 60-62; prog mgr, Fluid Physics Br, Res Div, Off Adv Res & Technol, Hqs, NASA, 62-67; assoc prof, Univ Del, 67-74, assoc dean, Col Eng, 70-74, prof mech & aerospace Eng, 74-85; AEROSPACE ENGR, BALLISTIC RES LAB, US ARMY, 85- *Concurrent Pos:* Lectr, George Washington Univ, 60; mem heat transfer panel, Comt Aeroballistics, Bur Naval Weapons, 61; exec secy, Res Adv Comt Fluid Mech, NASA, 63-67, US Govt Aberdeen Proving Ground Conducting Res, 88; conducting res, US Gov, Aberdeen Proving Ground, 88. *Mem:* Am Inst Aeronaut & Astronaut; Sigma Xi. *Res:* Hypersonic aerodynamics; viscous flow and aerodynamic heating; heat transfer and skin friction characteristics of turbulent boundary layer including the effects of mass transfer at hypersonic speeds. *Mailing Add:* 801 Baylor Dr Newark DE 19711

DANBURG, JEROME SAMUEL, b Houston, Tex, Dec 21, 40; m 65; c 4. GEOPHYSICAL SOFTWARE SYSTEMS. *Educ:* Mass Inst Technol, BS, 62; Freie Univ, Berlin, dipl, 64; Univ Calif, Berkeley, PhD(physics), 69. *Prof Exp:* Res asst physics, Univ Calif, Berkeley, 66-69; assoc physicist, Brookhaven Nat Lab, 69-72; res geophysicist, Shell Oil Co, 73-75, sr geophysicist, 75-81, res mgr physics & comput sci, 81-86, mgr geophys syst, 86-89, MGR, GEOPHYS DATA PROCESSING, SHELL OIL CO, 89- *Mem:* Am Phys Soc; Soc Explor Geophysicists. *Res:* Application of computing and computer science to the problems of geophysics and geology. *Mailing Add:* 7611 Burning Hills Dr Houston TX 77071

DANBY, GORDON THOMPSON, b Can, Nov 8, 29; m 62; c 2. PHYSICS. *Educ:* Carleton Univ, Can, BSc, 52; McGill Univ, PhD(physics), 56. *Prof Exp:* MEM STAFF, BROOKHAVEN NAT LAB, 57- *Mem:* Am Phys Soc. *Res:* Accelerator development; apparatus for high energy physics, especially magnets and beam transport; experiments with neutrinos; magnetic levitation and applications to high speed vehicles. *Mailing Add:* Brookhaven Nat Lab Bldg 911 B Upton NY 11973

DANBY, JOHN MICHAEL ANTHONY, b London, Eng, Aug 5, 29; m 58; c 2. CELESTIAL MECHANICS. *Educ:* Oxford Univ, BA, 50, MA, 54; Univ Manchester, PhD(astron), 53. *Prof Exp:* Asst prof astron, Univ Minn, 57-61; asst prof, Yale Univ, 61-67; PROF MATH & PHYSICS, NC STATE UNIV, 67- *Concurrent Pos:* Consult, Honeywell, Inc, 58-; consult, Air Force Chart & Info Ctr; pres, Celestial Mech Inst, 75-87, mem ed comt, Celestial Mech, 73-82. *Mem:* Am Astron Soc; Int Astron Union; Am Math Asn. *Res:* Celestial mechanics; stellar dynamics; properties of dynamical systems; use of computers in education. *Mailing Add:* Dept Math NC State Univ Raleigh NC 27695-8205

DANCE, ELDRED LEROY, b Blackfoot, Idaho, Dec 22, 17; m 46; c 3. CHEMISTRY. *Educ:* Univ Calif, BS, 43. *Prof Exp:* Res & develop chemist & chem engr, Dow Chem Co, Pittsburgh, 43-63, sr res chem engr, 63-66, assoc scientist, 66-85; RETIRED. *Mem:* Am Inst Chem Engrs. *Res:* Process evaluation and development including unit operations of reactor design; distillation; refrigeration; process control; heat and mass transfer; fluid mixing; production of plastics and polymers; hollow fiber technology. *Mailing Add:* 37 Hornet Ct Danville CA 94526

DANCHIK, RICHARD S, b Pittsburgh, Pa, Mar 21, 43. ANALYTICAL CHEMISTRY, INDUSTRIAL HYGIENE CHEMISTRY. *Educ:* Duquesne Univ, BS, 65; Wayne State Univ, PhD(anal chem), 68. *Prof Exp:* Res scientist, 68-74, sr scientist, 74-79, MGR ENVIRON HEALTH, ALCOA LABS, ALUMINUM CO, AM, 79- *Concurrent Pos:* Mem, Instrumentation Adv Bd, 85- *Mem:* Am Chem Soc; Sigma Xi; Am Soc Testing & Mat; Am Indust Hyg Asn. *Res:* Indirect spectrophotometric and atomic absorption spectrometric methods of analysis; electroanalytical methods of analysis; investigation and application of selective ion electrodes; ultraviolet and visible absorption spectrometry; development of automated process control systems; environmental chemistry; industrial hygiene chemistry. *Mailing Add:* Alcoa Labs Aluminum Co Am Alcoa Center PA 15069

DANCIK, BRUCE PAUL, b Chicago, Ill, Dec 27, 43; Can citizen; div. POPULATION GENETICS, GENECOLOGY. *Educ:* Univ Mich, BS, 65, MF, 67, PhD(forest genetics), 72. *Prof Exp:* Asst prof bot, Saginaw Valley Col, Mich, 72-73; from asst prof to prof forest genetics, 73-89, PROF & CHAIR, DEPT FOREST SCI, UNIV ALTA, 89- *Concurrent Pos:* Assoc ed, Forestry Chronicle, 76-83; chmn, Forestry Panel, Environ Coun Alta, 77-81; ed, Can J Forest Res, 81-90; ed-in-chief, Nat Res Coun Can Res Journals, 90- *Honors & Awards:* Forestry Achievement Award, Am Inst Forestry, 79. *Mem:* AAAS; Genetics Soc Am; Genetics Soc Can; Soc Study Evolution; Sigma Xi; Am Soc Naturalists. *Res:* Genic (allozyme) variation, molecular genetics, population structure, ecological differentiation, evolution, and mating system dynamics of woody plant species, especially those of western North America. *Mailing Add:* Dept Forest Sci Univ Alta Edmonton AB T6G 2H1 Can

DANCIS, JOSEPH, b New York, NY, Mar 19, 16; m 48; c 2. MEDICINE, BIOCHEMISTRY. *Educ:* Columbia Col, BA, 34; St Louis Univ, MD, 38. *Prof Exp:* Palmer sr investr, 54-56; chmn dept, 74-90, PROF PEDIAT, COL MED, NY UNIV, 63- *Concurrent Pos:* Nat Found res fel, 51-53; Markle scholar, 56-61; res career investr, NIH, 61-74. *Honors & Awards:* Borden Award, 71; Howland Award, 88. *Mem:* Am Acad Pediat; Soc Pediat Res; Am Pediat Soc (pres, 83-84); NY Acad Sci; Harvey Soc. *Res:* Placental physiology; metabolic errors. *Mailing Add:* NY Univ Med Ctr New York NY 10016

DANCY, TERENCE E(RNEST), b Coulsdon, Surrey, Eng, Mar 5, 25; nat US; m 47; c 2. STEELMAKING. *Educ:* Univ London, BSc, 45, PhD(fuel tech) & DIC, 48. *Prof Exp:* Sr sci officer, Iron Making Div, Brit Iron & Steel Res Asn, 47-53; metall liaison officer, Brit Embassy, Washington, DC, 53-54; prin sci officer commercial develop, Brit Iron & Steel Res Asn, 54-56; res supvr, Jones & Laughlin Steel Corp, 56-61, asst dir res, 61-64, asst to vpres eng & plant, 64-66, mgr process eng, 66-71; vpres eng & develop, Sidbec/Dosco, 71-79, sr vpres, 79-85, vpres technol, Sidbec & Sidbec/Dosco, bd dirs, Sidbec/Normines, 76-82; INT STEEL CONSULT, 85- *Concurrent Pos:* UK deleg, Orgn European Econ Co-op, France, 50-52, consult, European Productivity

Agency, 54; UK rep, Comite Int du Bas Fourneau, Belgium, 50-53. *Honors & Awards:* Hunt Award, Am Inst Mining, Metall & Petrol Engrs, 60; Kelly First Award, Asn Iron & Steel Engrs, 64; Medal, Am Iron & Steel Inst, 65 & 81; Howe Mem lectr, Am Inst Mining, Metall & Petrol Engrs, 76; Airey Award, Can Inst Mining & Metall, 83. *Mem:* Fel Am Soc Metals; Am Inst Mining, Metall & Petrol Engrs; Asn Iron & Steel Engrs; Am Iron & Steel Inst; Can Inst Mining & Metall; fel Inst Metals. *Res:* Extractive metallurgy and steel processing; blast furnace; engineering development, iron ore pelletizing, direct reduction, electric furnace operation. *Mailing Add:* PO Box 183 Elkins NH 03233

DANDAPANI, RAMASWAMI, b Nagpur, India, Feb 26, 46; m 77; c 2. COMPUTER SCIENCE. *Educ:* Univ Nagpur, BSc, 64; Indian Inst Sci, BE, 67; Univ Iowa, MS, 69, PhD(comput sci), 74. *Prof Exp:* Teaching & res asst elec eng, Univ Iowa, 68-71, res asst comput sci, 71-74; asst prof, 74-79, ASSOC PROF COMPUT SCI, YOUNGSTOWN STATE UNIV, 79- *Mem:* Inst Elec & Electronics Engrs; Sigma Xi. *Res:* Combinational logic circuits; fault detection in logic circuits; testable design of logic circuits. *Mailing Add:* Elec Engr Dept Univ Colo Colorado Springs CO 80933-0715

DANDEKAR, BALKRISHNA S, b Khed-Ratnagiri, India, Sept 13, 33; m 59; c 2. GEOPHYSICS, AERONOMY. *Educ:* Univ Poona, BSc, 53, MSc, 55; Gujarat Univ, PhD(physics), 62. *Prof Exp:* Lectr physics, Khalsa Col, India, 55; jr res asst, Phys Res Lab, Ahmedabad, 58-61; prof & head dept, Maharashtra Educ Soc Col, Poona, 61-62; res physicist, Wentworth Inst, 62-65; sr res assoc aeronomy, Northeastern Univ, 65-68; RES PHYSICIST, AIR FORCE GEOPHYSICS LAB, 68- *Mem:* Am Geophys Union; Sigma Xi. *Res:* Optical emissions from the upper atmosphere; polar atmospheric processes; ionospheric physics. *Mailing Add:* 92 Oakdale Rd Newton MA 02159

DANDY, JAMES WILLIAM TREVOR, b Preston, Eng, July 18, 29; m 59; c 4. ZOOLOGY, PHYSIOLOGY. *Educ:* Univ Natal, BSc, 53, MSc, 55; Univ Toronto, PhD(zool), 67. *Prof Exp:* Lectr zool, Univ Col Fort Hare, 56-59; asst prof, 65-74, ASSOC PROF ZOOL, UNIV MAN, 74- *Concurrent Pos:* Nat Res Coun Can operating grant, 65-; Fisheries Res Bd Can res grant, 66- *Mem:* AAAS; Marine Biol Asn UK; Brit Soc Exp Biol; Can Soc Zoologists. *Res:* Environmental physiology. *Mailing Add:* Dept Zool Univ Man Winnipeg MB R3T 2N2 Can

DANE, BENJAMIN, b Boston, Mass, Nov 22, 33; m 57; c 2. ANIMAL BEHAVIOR, PHYSIOLOGY. *Educ:* Harvard Univ, AB, 56; Cornell Univ, PhD(zool), 61. *Prof Exp:* Asst biol, Cornell Univ, 59-60; res assoc physiol, Med Ctr, NY Univ, 61-62, instr, 62-63; actg asst prof animal behav, Stanford Univ, 63-66; asst prof, 66-69, ASSOC PROF BIOL, TUFTS UNIV, 69-, CHMN DEPT, 85- *Concurrent Pos:* Consult, Santa Rite Tech Inc, Calif, 64-66. *Mem:* AAAS; Am Soc Zoologists; Animal Behav Soc; Cooper Ornith Soc; Am Ornithologists Union; Sigma Xi. *Mailing Add:* Dept Biol Tufts Univ Medford MA 02155

DANE, CHARLES WARREN, b Washington, DC, Sept 21, 34; m 57; c 2. ECOLOGY, ENDANGERED SPECIES. *Educ:* Cornell Univ, BS, 56, MS, 57; Purdue Univ, PhD(vert ecol), 65. *Prof Exp:* Res biologist avian physiol, Northern Prairie Wildlife Res Ctr, 64-76, migratory bird res specialist, Div Wildlife Res, 76-85, CHIEF, OFF SCI AUTHORITY, US FISH & WILDLIFE SERV, 85- *Mem:* Wildlife Soc; Sigma Xi; AAAS. *Res:* Influence of environmental factors on waterfowl reproductive physiology. *Mailing Add:* 4910 English Dr Annandale VA 22003

DANEHY, JAMES PHILIP, b Ft Wayne, Ind, Apr 27, 12; m 34; c 3. ORGANIC CHEMISTRY. *Educ:* Univ Notre Dame, BS, 33, MS, 34, PhD(org chem), 36. *Prof Exp:* Res chemist, Gen Chem Co, New York, 36-38; res chemist, Harris-Seybold-Potter Co, Cleveland, 38-42; group leader, Corn Prods Refining Co, Argo, Ill, 42-48; keratin group leader, Toni Co, 48-51; asst prof liberal educ, Univ Notre Dame, 51-55, from assoc prof to prof chem, 55-77; sr consult, Bernard Wolnak & Assocs, Chicago, 78-82; CONSULT CHEMIST, 83- *Concurrent Pos:* Fulbright lectr, Univ Col, Cork, 61-62. *Mem:* Am Chem Soc. *Res:* Chemistry of vegetable proteins; zein; keratin; organic sulfur chemistry; mechanisms of reactions of organic divalent sulfur compounds. *Mailing Add:* 1903 Stonehedge Lane South Bend IN 46614-6344

DANEN, WAYNE C, b Green Bay, Wis, Dec 24, 41; m 64; c 4. PHYSICAL ORGANIC CHEMISTRY. *Educ:* St Norbert Col, BA, 64; Iowa State Univ, PhD(org chem), 67. *Prof Exp:* From asst prof to prof chem, Kans State Univ, 67-82; staff mem, 82-83, CLS-4 dep group leader, 83-88, CLS-2 GROUP LEADER, LOS ALAMOS NAT LAB, 89- *Concurrent Pos:* Petrol Res Fund grant, 67-69, 71-75 & 75-78. Res Corp grant, 73-75; NSF grant, 78-81; sabbatical leave, Los Alamos Nat Lab, 77-78 & 81. *Mem:* Am Chem Soc. *Res:* Free radical chemistry; electron spin resonance studies of stable and transient free radicals; laser induced reactions; laser diagnostics; energy research; conventional munitions research. *Mailing Add:* Los Alamos Nat Lab PO Box 1663 MS G738 Los Alamos NM 87545

DANENBERG, PETER V, b Latvia, Europe, 1942. BIOCHEMISTRY. *Prof Exp:* PROF BIOCHEM, UNIV SOUTHERN CALIF, 87- *Mailing Add:* Dept Biochem Cancer Res Lab Univ Southern Calif 1303 N Mission Rd Los Angeles CA 90033

DANEO-MOORE, LOLITA, b Nimes, France, Feb 18, 29; US citizen; m 50; c 4. MICROBIOLOGY. *Educ:* Univ Pa, BA, 60, MS, 63; Rutgers Univ, PhD(microbiol), 65. *Prof Exp:* Assoc ed cancer chemother, Triple I, 60-65; from instr to assoc prof, 66-78, PROF MICROBIOL, SCH MED, TEMPLE UNIV, 78- *Concurrent Pos:* Fels, Arthritis Found, Temple Univ, 65-67, Am Cancer Soc, 67-69 & Res Corp, 72-73; ad hoc mem, oral biol study sect, Nat Inst Dent Res, 78-80; mem, microbial physiol study sect, NIH, 81-85. *Mem:* Am Soc Microbiol; Sigma Xi. *Res:* Regulatory mechanisms; ribosomes; cell membranes; cell growth and division; cell walls ageing; mode of action of antibiotics. *Mailing Add:* Temple Univ Sch Med 3400 N Broad St Philadelphia PA 19140

DANES, ZDENKO FRANKENBERGER, b Prague, Czech, Aug 25, 20; nat US; m 45; c 2. GEOPHYSICS. *Educ:* Charles Univ, Prague, BS, 46, PhD(math), 48, PhD(physics), 49. *Prof Exp:* Asst, State Geophys Inst, Czech, 46-47; asst prof, Physics Inst, Charles Univ, Prague, 48-50; geophys interpreter, Gulf Res & Develop Co, Pa, 53-59; res engr, Boeing Airplane Co, 59-62; prof physics, Univ Puget Sound, 62-84; CONSULT, US GEOL SURV, 84- *Concurrent Pos:* Vis prof, Univ Minn, 62; vis prof, Univ Technol, Vienna, Austria, 74-75; pres, Danes Res Assocs, 78- *Mem:* Am Geophys Union; Soc Explor Geophys. *Res:* Gravity; lunar physics; electromagnetic theory. *Mailing Add:* US Geol Surv Tacoma WA 98416

DANESE, ARTHUR E, b Mt Morris, NY, May 11, 22. MATHEMATICS. *Educ:* Univ Rochester, AB, 47, PhD, 56; Harvard Univ, AM, 49. *Prof Exp:* Instr math, Univ Rochester, 48-54; asst prof, Univ Tenn, 55; mathematician, Eastman Kodak Co, 55-58; asst prof math, Union Univ, NY, 58-64; assoc prof, State Univ NY Buffalo, 64-67; PROF MATH, STATE UNIV NY COL FREDONIA, 67- *Mem:* Am Math Soc; Math Asn Am. *Res:* Operational calculus; mathematical analysis; orthogonal polynomials; special functions; inequalities; classical analysis. *Mailing Add:* Dept Math State Univ NY Col Fredonia NY 14063

DANFORD, DARLA E, b Ponca City, Okla, Nov 8, 45. NUTRITION RESEARCH. *Educ:* Colo State Univ, BSc, 69; Univ Calif, Berkeley, MPH, 71; Harvard Univ, DSc, 80. *Prof Exp:* Res nutritionist, San Francisco Gen Hosp, NIH Clin Study Ctr, Univ Calif, 71-73, asst specialist nutrit, Dept Med, Med Sch, 73-75, assoc specialist, 75-76; nutritionist, Shriver Med Prog, Paul A Denver State Sch & Eunice Kennedy Shriver Ctr Ment Retardation, 77-80; spec consult to dir, USDA Human Nutrit Res Ctr Aging, Tufts Univ, Boston, 80-81; asst prof, Dept Med, dir, Trace Element Lab, NIH Clin Nutrit Res Unit, dir, Masters Prog Clin Nutrit & coordr, Comt Human Nutrit & Nutrit Biol, Med Sch, Univ Chicago, 81-85; proj dir, Comt Inter-Am Conf Food Protection, Food & Nutrit Bd, Nat Acad Sci, Washington, DC, 85-86; spec expert, Nutrit Coord Comt, 86-88 & Div Nutrit Res Coord, 88-89, DIR, DIV NUTRIT RES, OFF DIS PREV, OFF DIR, NIH, 89- *Concurrent Pos:* Lab technician, Vet Admin Hosp, Washington, DC, 79; rep & exec secretariat, Interagency Comt Human Nutrit Res, NIH, 86-, chairperson, Nutrit Coord Comt, 89- *Mem:* Am Dietetic Asn; AAAS; Am Soc Clin Nutrit; Am Inst Nutrit; Am Pub Health Asn; Brit Nutrit Soc; Int Soc Advan Trace Element Res; NY Acad Sci. *Mailing Add:* Div Nutrit Res Bldg 31 Rm 4B63 NIH 9000 Rockville Pike Bethesda MD 20892

DANFORTH, DAVID NEWTON, obstetrics & gynecology, general physiology; deceased, see previous edition for last biography

DANFORTH, RAYMOND HEWES, b Mineola, NY, Nov 24, 44; m 67; c 2. CHEMISTRY. *Educ:* Bates Col, BS, 66; Princeton Univ, PhD(org chem), 71. *Prof Exp:* Res chemist, Nat Patent Develop Corp, 71-73; asst tech dir pulp chem, Berlin-Gorham Div, Brown Co, 73-80; DIR ENVIRON & TECH SERV, JAMES RIVER CORP, 80- *Mem:* Am Chem Soc; Tech Asn Pulp & Paper Indust. *Res:* environmental and technical activities in a kraft pulp mill; purified cellulose and cellulose derivatives. *Mailing Add:* 725 North Rd Gorham NH 03581

DANFORTH, WILLIAM (FRANK), b Washington, DC, Jan 12, 28. FRESH WATER ECOLOGY. *Educ:* Univ Iowa, AB, 48; Columbia Univ, AM, 49; Univ Calif, Los Angeles, PhD(zool), 53. *Prof Exp:* From asst prof to prof biol, Ill Inst Technol, 52-84, emer prof, 84-; RETIRED. *Mem:* AAAS; Soc Protozool; Phycol Soc Am; Sigma Xi. *Res:* Ecology and physiology of planktonic algae. *Mailing Add:* No 5 Cedar Heights Camanche IA 52730

DANFORTH, WILLIAM H, b St Louis, Mo, Apr 10, 26; m 50; c 4. MEDICINE. *Educ:* Princeton Univ, AB, 47; Harvard Med Sch, MD, 51. *Hon Degrees:* Numerous from US Cols & Univs, 71-85. *Prof Exp:* Intern, Barnes Hosp, St Louis, 51-52, from asst resident to resident, 54-57; asst resident, Children's Hosp, St Louis, 55-56; from instr to assoc prof med, 57-67, vchancellor med affairs & pres sch med & assoc hosps, 65-71, CHANCELLOR, WASH UNIV, 71-, PROF INTERNAL MED, SCH MED, 67- *Concurrent Pos:* Univ fel cardiol, Wash Univ, 57-58, NIH fel biochem, 61-63; physician, Vet Admin Hosp, St Louis, 58-61; attend physician, Barnes Hosp, 58-; pres, Independent Cols & Univs Mo, 79-81; mem comt, Asn Am Univs Nat Asn State Univs & Land Grant Cols Joint Health Policy, 78-, chmn, 78-81; mem dirs adv comt, NIH, 80-84; chmn, Asn Am Univs, 82. *Mem:* Inst Med-Nat Acad Sci; Cent Soc Clin Res; Am Soc Clin Invest; fel AAAS; fel Am Acad Arts & Sci. *Mailing Add:* One Brookings Dr Wash Univ St Louis MO 63130

DANG, PETER HUNG-CHEN, b Kwangsi, China, Sept 7, 18; US citizen; m 47; c 4. PHARMACOLOGY, BIOCHEMISTRY. *Educ:* Nat Cent Univ, Nanking, China, BS, 44; Mont State Univ, MS, 59; Univ Chicago, PhD(pharmacol), 64. *Prof Exp:* Instr, Paisha Col, China, 44-45; asst biochemist, biochem & dis, Vet Res Lab, Mont, 56-58; res asst, Univ Chicago, 58-63, res assoc, 63-64; res assoc animal nutrit, drug & dis, Cornell Univ, 64-67; head pharmacol, Food & Drug Res Labs, 67-71, mgr, 72-73; DIR PHARMACOL, DDI PHARMACEUT INC, 73- *Mem:* Am Chem Soc; AAAS; NY Acad Sci. *Res:* Enzymological, immunological, biochemical and physiological effect on chemicals; drug toxicity and pharmokinetic, nutrition and disease; studies on laxative, antispasmotic agents; antiinflammatory agents for arthritis and degenerative diseases. *Mailing Add:* 877 San Ardo Way Mountain View CA 94043

DANG, VI DUONG, US citizen; m 71; c 2. COATING PROCESS, CHEMISTRY. *Educ:* Nat Taiwan Univ, BS, 66; Clarkson Univ, MS, 68, PhD(chem eng), 71. *Prof Exp:* Environ engr environ med, NY Univ Med Ctr, 71-72 & Johns Hopkins Univ, 72-73; chem engr electrochem eng, P R Mallory, Inc, 73-74; chem engr chem & nuclear eng, Brookhaven Nat Lab, 74-80; assoc prof, Catholic Univ Am, 80-84; sr chem engr, Technol Appln Inc, 84-86; US PATENT & TRADEMARK OFF, 86- *Concurrent Pos:* Res assoc, Calif Inst Technol, 72. *Mem:* Am Inst Chem Engrs; Am Chem Soc; Am

Nuclear Soc; AAAS; NY Acad Sci. *Res:* Fundamental research in heat and mass transfer with or without chemical reactions; coal conversion; electrochemical engineering. *Mailing Add:* 6304 Phyllis Lane Bethesda MD 20817

D'ANGELO, GAETANO, b Brooklyn, NY, June 8, 42; m 69; c 1. MEMBRANE TRANSPORT. *Educ:* City Univ New York, BA, 64. *Prof Exp:* Chemist, D H Litter Labs, 64-66; tech specialist, dept chem, 66-70, RES ASST PHYSIOL & BIOPHYS, HEALTH SCI CTR, STATE UNIV NY, STONY BROOK, 70- *Mem:* Am Chem Soc; fel Am Inst Chemists; NY Acad Sci. *Res:* Enzyme regulation of metabolism: protein structure & function. *Mailing Add:* Dept Physiol & Biophys Health Sci Ctr State Univ NY Stony Brook NY 11794-8661

D'ANGELO, HENRY, electrical engineering; deceased, see previous edition for last biography

D'ANGELO, JOHN PHILIP, b Philadelphia, Pa, Mar 5, 51. MATHEMATICS. *Educ:* Univ Pa, BA, 72; Princeton Univ, PhD(math), 76. *Prof Exp:* Moore instr math, Mass Inst Technol, 76-78; from asst prof to assoc prof, 78-86, PROF MATH, UNIV ILL, URBANA, 86- *Concurrent Pos:* Prin investr grants, NSF, 80-88; vis scholar, Princeton Univ, 81, vis assoc prof, 83; mem, Inst for Advan Study, 85-86, Miltag-Leffler Inst, 88; Univ scholar, Univ Ill, 86-88; vis prof, Washington Univ, 90. *Mem:* Am Math Soc. *Res:* Several complex variables, partial differential equations. *Mailing Add:* Dept Math Univ Ill 1409 W Green St Urbana IL 61801

D'ANGELO, NICOLA, b Ripatransone, Italy, Jan 8, 31; US citizen; m 63. PLASMA PHYSICS. *Educ:* Univ Rome, DSc, 53, specialization degree, 55. *Prof Exp:* Res asst physics, Univ Rome, 53-54 & 1st Superior Sanita, 54-56; res assoc, Argonne Nat Lab, 56-59; res physicist, Princeton Univ, 59-66; leader plasma group, Danish Atomic Energy Comn, Riso, 66-68; head diag group, Europ Space Res Inst, 68-70, dir, 70-72; mem staff, Danish Space Res Inst, 72-76; PROF DEPT PHYSICS & ASTRON, UNIV IOWA, 76- *Mem:* Am Geophys Union. *Res:* Cosmic rays; neutron, nuclear plasma and space physics. *Mailing Add:* Dept Physics & Astron Univ Iowa Iowa City IA 52242

D'ANGIO, GIULIO J, b New York, NY, May 2, 22; m 55; c 2. RADIOTHERAPY. *Educ:* Columbia Univ, AB, 43; Harvard Univ, MD, 45; FRCR, 81. *Hon Degrees:* Dr, Univ Bologna, Italy, 83. *Prof Exp:* Intern, Children's Hosp, Boston, Mass, 45-46; resident path, Vet Admin Hosp, West Roxbury, Mass, 48-49; resident radiol, Boston City Hosp, Mass, 49-53; asst, Harvard Med Sch, 53-56, instr, 56-62, clin assoc, 62-64; prof radiol & dir div radiation ther, Univ Minn Hosps, 64-68; div chief, Sloan-Kettering Inst Cancer Res, 68-76; chmn, Dept Radiation Ther, Mem Hosp, 68-76; PROF RADIATION THER & PEDIAT ONCOL, SCH MED, UNIV PA, 76-; VCHMN & CLIN DIR, DEPT RADIATION ONCOL, HOSP-UNIV PA, 89- *Concurrent Pos:* Fel path, Babies Hosp, NY, 48; asst, Boston City Hosp, 53-55, asst physician, 55-64; asst, Children's Hosp, Boston, 56-59, assoc radiologist, 59-62, radiotherapist, 63-64; res assoc Children's Cancer Res Found, 56-59, radiotherapist, 59-64; asst radiologist, Mass Gen Hosp, Boston, 61-62, consult, 62-64; assoc res radiologist, Donner Lab, Univ Calif, Berkeley, 62-63, consult, Donner Lab & Lawrence Radiation Lab, 64-76; consult, Hennepin County Gen Hosp, 66-68; vis radiation therapist, James Ewing Hosp, 68-; attend radiation therapist, Mem Hosp, 68-76, consult, Dept Radiation Ther, 82-86; attend radiologist, NY Hosp, prof radiol, Med Col, Cornell Univ, 68-76; chmn, Nat Wilms' Tumor study; dir, Univ Pa Cancer Ctr, Univ Pa, 76-77; assoc scientist, Sloan-Kettering Inst Cancer Res, 76-81; dir, Children's Cancer Res Ctr, Children's Hosp Pa, 76-89. *Honors & Awards:* Heath Mem Award, Am Cancer Soc, 78; M D Anderson Tumor & Cancer Inst Award, 79; Ann Award, Am Cancer Soc, 78; Tenth Myron Karen Mem Lectr, Children's Hosp Los Angeles, 84; Unicef Health Prize, 87; Distinguished Career Award, Am Soc Pediat Hemat/Oncol, 90. *Mem:* AAAS; Royal Soc Med; Am Asn Cancer Res; Am Col Radiol; Am Radium Soc; Am Soc Clin Oncol; Sigma Xi; Am Roentgen Ray Soc; Am Soc Therapeut Radiologists; Int Soc Pediat Oncol; fel Am Acad Pediat; fel Royal Col Radiologists; Radiol Soc NAm; Soc Pediat Radiol; hon mem Am Soc Pediat Hemat/Oncol. *Res:* Clinical radiotherapy; combined chemotherapy and radiotherapy; mechanisms of cancer growth; late radiation effects. *Mailing Add:* Dept Radiation Oncol Univ Pa Hosp Philadelphia PA 19104

DANGLE, RICHARD L, b New Castle, Pa, July 24, 30; m 56; c 2. NUCLEAR PHYSICS. *Educ:* Westminster Col, Pa, BS, 58; Univ Wis, MS, 60, PhD(physics), 63. *Prof Exp:* Fel physics, Univ Sao Paulo, 63-64; instr, Univ Wis, 64-65; from asst prof to assoc prof, Univ Ga, 65-74, asst dean, Col Arts & Sci, 68-74; PROF PHYSICS & DEAN SCH ARTS & SCI, WGA COL, 74- *Mem:* AAAS; Am Phys Soc; Sigma Xi. *Res:* Low energy nuclear physics, primarily in charged particle spectroscopy. *Mailing Add:* Sch Arts & Sci WGa Col Carrollton GA 30117

DANHEISER, RICK LANE, b New York, NY, Oct 12, 51. ORGANIC CHEMISTRY. *Educ:* Columbia Col, BA, 72; Harvard Univ, MA, 75, PhD(chem), 78. *Prof Exp:* from asst prof to assoc prof, 78-89, PROF CHEM, MASS INST TECHNOL, 89- *Concurrent Pos:* Alfred E Sloan Found fel. *Honors & Awards:* Excellence in Chem Res Award, Stuart Pharmaceut, 85. *Mem:* Am Chem Soc; Chem Soc Brit; Swiss Chem Soc. *Res:* Development of new synthetic methods and their application in the total synthesis of biologically significant natural products. *Mailing Add:* Dept Chem Mass Inst Technol Cambridge MA 02139

DANHOF, IVAN EDWARD, b Grand Haven, Mich, June 24, 28; m 50; c 4. MICROBIOLOGY, ANATOMY. *Educ:* Univ NTex, BA & MA, 49; Univ Ill, PhD(physiol), 53; Univ Tex, MD, 62. *Prof Exp:* Teacher pub sch, 49-50; asst, Univ Ill, 50-53, instr physiol, 53; res assoc, Univ Tex Health Sci Ctr, Dallas, 53-54, from instr to prof physiol, 54-83; RETIRED. *Concurrent Pos:* Dir med educ, Methodist Hosp Dallas, 69; Fulbright lectr, Med Fac, Nangrahar Univ, Afghanistan, 69-70. *Mem:* AAAS; assoc Soc Exp Biol & Med; Am Physiol Soc. *Res:* Motility and hormonal regulation of gastrointestinal tract; nutrition related to gastrointestinal function. *Mailing Add:* 2322 Ingleside Dr Grand Prairie TX 75050

DANIEL, CHARLES DWELLE, JR, b San Antonio, Tex, Oct 30, 25; m 46; c 2. PHYSICS, ELECTRONICS. *Educ:* US Mil Acad, BS, 46; Tulane Univ, MS, 61, PhD(physics), 68; American Univ, BA, 87. *Prof Exp:* Dir missiles & space develop, HQ, Dept of Army, 70-71; comdr gen artillery opers, I Corps Group Artillery, US Army, Korea, 71-72; dir Army res, HQ, Dept of Army, 72-74; dep commandant, Educ Admin, Nat War Col, 74-75; spec asst, Res & Develop Orgn, HQ, US Army Mat Develop & Readiness Command, 74-77; comdg gen, Electronics Res & Develop Acquisition, US Army Electronics Res & Develop Command, 77-79; CONSULT, DEFENSE INDUST, 79- *Res:* Nuclear physics; solid state physics. *Mailing Add:* 4904 Baltan Rd Bethesda MD 20816

DANIEL, CHARLES WALLER, b Annapolis, Md, June 16, 33; m 61; c 3. BIOLOGY. *Educ:* Univ NMex, AB, 54; Univ Hawaii, MS, 60; Univ Calif, Berkeley, PhD(zool), 65. *Prof Exp:* From asst prof to assoc prof, 65-77, PROF BIOL, UNIV CALIF, SANTA CRUZ, 77-, ASSOC ACAD VCHANCELLOR, 87- *Res:* Normal and neoplastic biology of mammary tissues; differentiation in vitro; cellular aging. *Mailing Add:* Dept Biol Univ Calif Santa Cruz CA 95064

DANIEL, DANIEL S, b Basrah, Iraq, July 1, 34; m 65; c 3. PHYSICAL CHEMISTRY, ORGANIC CHEMISTRY. *Educ:* Univ London, BSc, 55, PhD(chem), 59. *Prof Exp:* Res chemist, Carnegie Inst Technol, 60-63; sr res chemist, 64-69, RES ASSOC, RES LAB, EASTMAN KODAK CO, 69- *Concurrent Pos:* Adj fac, Rochester Inst Technol. *Mem:* Am Chem Soc; Royal Soc Chem; Sigma Xi; AAAS; Am Asn Clin Chemists. *Res:* Clinical chemistry; synthetic organic chemistry; biochemistry; electrode phenomena. *Mailing Add:* Res Lab Eastman Kodak Co Rochester NY 14650-1712

DANIEL, DONALD CLIFTON, b Atlanta, Ga, May 21, 42; m 76; c 1. MISSILE AERODYNAMICS, APPLIED MATHEMATICS. *Educ:* Univ Fla, BASE, 64, MSE, 65, PhD(aerospace eng), 73. *Prof Exp:* Res engr astrodyn, Boeing Co, 65-68; res assoc flight mech, Univ Fla, 71-72; AEROSPACE ENGR FLIGHT MECH & AERODYN, USAF ARMAMENT LAB, 72- *Concurrent Pos:* Res asst, Univ Fla, 64-65, NSF trainee, 68-71, res assoc, 71-72. *Honors & Awards:* Tech Achievement Award Basic Res, USAF Systs Command, 78. *Mem:* Assoc fel Am Inst Aeronaut & Astronaut; Inst Elec & Electronics Engrs; fel Royal Aeronaut Soc. *Res:* Missile aerodynamics with emphasis on high angle of attack vortex phenomena and hypersonic flow about lifting bodies. *Mailing Add:* USAF Armament Lab AFATL/AG Eglin AFB FL 32542

DANIEL, EDWIN EMBREY, b Chattanooga, Tenn, Sept 23, 25; m 48, 73; c 3. PHARMACOLOGY. *Educ:* Johns Hopkins Univ, BA, 47, MA, 49; Univ Utah, PhD(pharmacol), 52. *Prof Exp:* Lab instr biol, anat & physiol, Johns Hopkins Univ, 47-49; asst pharmacol, Univ Utah, 49-52; from asst prof to assoc prof, Univ BC, 52-61; prof pharmacol, Univ Alta, 61-74, head dept, 61-72; prof neurosci, 75-86, coordr, Res Prog Control Smooth Muscle Function, 79-86, DIR, HONOR'S BIOL/PHARMACOL & PROF PHYSIOL- PHARMACOL, MCMASTER UNIV, 86- *Honors & Awards:* Upjohn Award, Can Pharmacol Soc, 88; Otto Krayer Award, Am Soc Pharmacol & Exp Therapeut, 89. *Mem:* AAAS; Am Soc Pharmacol & Exp Therapeut; Pharmacol Soc Can; Can Physiol Soc; Biophys Soc; Am Physiol Soc. *Res:* Nature of control of activity of smooth and cardiac muscle, especially membrane phenomena, control of calcium; neural control; cell-to-cell coupling. *Mailing Add:* Dept Neurosci McMaster Univ Hamilton ON Can

DANIEL, HAL J, b Memphis, Tenn, Nov 10, 42; m; c 2. NEUROANATOMY, CRANIO-FACIAL ANATOMY. *Educ:* Univ Tenn, BS, 64, MA, 67; Univ Southern Miss, PhD(speech/hearing sci), 69. *Prof Exp:* PROF SPEECH & HEARING SCI, 69-, ADJ PROF ANTHROP, 81-, ADJ PROF BIOL, EAST CAROLINA UNIV, 87- *Concurrent Pos:* Adj prof anat, East Carolina Univ, 73-78, anthrop, 81-; vis scholar, Univ Wash, 80-81. *Honors & Awards:* Res Award, Sigma Xi, 82. *Mem:* Am Acad Advan Sci; Am Asn Phys Anthropologists; Sigma Xi; Am Speech & Hearing Asn; Human Biol Coun; Europ Soc Biol Soc; Lang Origins Soc. *Res:* Evolution of human communication from an anatomical and physiological point of view; evolution of the auditory system. *Mailing Add:* Sch Allied Health Sci East Carolina Univ Greenville NC 27834

DANIEL, ISAAC M, b Salonica, Greece, Oct 7, 33; US citizen; m 87; c 2. EXPERIMENTAL MECHANICS, MECHANICS OF MATERIALS. *Educ:* Ill Inst Tech, BS, 57, MS, 59, PhD(civil eng), 64. *Prof Exp:* Asst res engr stress anal, IIT Res Inst, 59-60, from assoc res engr to sr res engr, 60-66, mgr, 66-80, sci adv, 75-82; prof & dir mech mat lab, Ill Inst Tech, 82-86; PROF THEORETICAL & APPL MECHS, NORTHWESTERN UNIV, 86- *Concurrent Pos:* Lectr, Soc Exp Stress Anal; mem comt on characterization of org matrix composites, Nat Mat Adv Bd-Nat Res Coun, 78-80. *Honors & Awards:* Award, Am Soc Civil Engrs, 57; Lazan Award, Soc Exp Mech, 84. *Mem:* Soc Exp Mech; Fel Soc Mech; Am Soc Testing & Mat; Am Acad Mech; Am Soc Mech Engrs; Am Soc Nondestructive Testing; Soc Advan Mat Process Engrs. *Res:* Experimental stress analysis; composite materials; fracture mechanics; high-rate testing; nondestructive evaluation; test methods; impact. *Mailing Add:* Civil Eng Dept Northwestern Univ Evanston IL 60208

DANIEL, J(ACK) LELAND, b Spokane, Wash, Dec 15, 24; m 49; c 3. ELECTRON MICROSCOPY, MATERIALS SCIENCE. *Educ:* Univ Wash, Seattle, BS, 45, MS, 48; Ore State Univ, PhD(anal chem), 60. *Prof Exp:* Appl res chemist, Hanford Labs, Gen Elec Co, 48-49, sr chemist, 51-57, specialist, Tech Personnel Placement, 57-59, sr chemist, 59-60, sr scientist, 60-64; res assoc, ceramics, Electron Micros, 65-76, STAFF SCIENTIST, MICROSTRUCTURAL ANAL, ELECTRON MICROS, NUCLEAR WASTE MGT, PAC NORTHWEST LABS, BATTELLE MEM INST, 76- *Concurrent Pos:* Specialist US team first US-Japan info exchange meeting oxide & carbide fuels, AEC, 63; vis chief res scientist, Mitsubishi Atomic Power Indust, Tokyo, 65-66. *Mem:* AAAS; NY Acad Sci; Am Soc Test &

Mat; Electron Micros Soc Am; Int Stereol Soc; Sigma Xi; Mat Res Soc. *Res:* Dynamic and scanning electron microscopy, especially of ceramics, high temperature materials and radioactive specimens; materials research; techniques of instrumental analysis, especially x-ray, quantitative metallography and nuclear ceramic fuels; stereology and quantitative image analysis; nuclear waste reference materials and durability testing. *Mailing Add:* Pacific Northwest Labs Battelle Mem Inst Battelle Blvd Richland WA 99352

DANIEL, JAMES L, PROTEIN PHOSPHORYLATION, CALCIUM MOBILIZATION. *Educ:* Carnegie Mellon Univ, PhD(chem), 73. *Prof Exp:* RES ASSOC PROF CELLULAR & BIOCHEM PHARMACOL, DEPT PHARMACOL, THROMBOSIS RES CTR, TEMPLE UNIV, 81- *Mailing Add:* Thrombosis Res Ctr Temple Univ 3400 N Broad St Philadelphia PA 19140

DANIEL, JAMES WILSON, b Indianapolis, Ind, Sept 16, 40; m 62; c 2. NUMERICAL ANALYSIS, COMPUTER SCIENCES. *Educ:* Wabash Col, BA, 62; Stanford Univ, MA, 63, PhD(math), 65. *Prof Exp:* Res mathematician comput, IBM Corp, 65; res assoc math, Off Naval Res, 66; from asst prof to assoc prof comput sci, Univ Wis, 67-70; assoc prof, 70-74, PROF COMPUT SCI & MATH, UNIV TEX, 74-, CHMN, MATH DEPT, 77- *Concurrent Pos:* Vis mem, Math Res Ctr, Univ Wis, 66-70; vis staff mem, Los Alamos Sci Lab, 70- *Mem:* Am Math Soc; Soc Indust & Appl Math. *Res:* Efficient and accurate numerical methods for approximate solution of differential equations and optimization problems. *Mailing Add:* Dept Math Univ Tex Austin TX 78712

DANIEL, JOHN HARRISON, b Darlington, SC, Sept, 6, 15; m 42; c 2. PHYSICS. *Educ:* The Citadel, BS, 35; Univ Ky, MS, 37; Mass Inst Technol, ScD(physics), 40. *Prof Exp:* Asst, Univ Ky, 35-37; teaching fel, Mass Inst Technol, 37-40; res fel, Nat Cancer Inst, USPHS, 40-41; asst physicist, Div Indust Hygiene, 41-42; res physicist, Firestone Res Lab, Ohio, 46-48; sr physicist, Lab Phys Biol, NIH, 48-53; physicist, Evans Signal Lab, 53-56; analyst, Inst for Defense Anal, 56-80; RETIRED. *Concurrent Pos:* Fel ctr advan eng stud, Mass Inst Technol, 69-70. *Mem:* AAAS; Inst Elec & Electronics Engrs; Am Phys Soc. *Res:* Weapons systems evaluation; missile and air defense; electron theory and surface phenomena of metals; detection of nuclear radiation; electron microscopy; dielectric measurements. *Mailing Add:* 5502 Cromwell Dr Bethesda MD 20816

DANIEL, JOHN SAGAR, b Banstead, Eng, May 31, 42; Can & UK citizen; m 66; c 3. METALLURGY, EDUCATIONAL TECHNOLOGY. *Educ:* Oxford Univ, BA, 64, MA, 69; Univ Paris, France, DSc(metall), 69. *Hon Degrees:* DLitt, Deakin Univ, Australia, 85; DSc, Royal Military Col, Saint-Jean, 88. *Prof Exp:* From asst prof to assoc prof metall, Ecole Polytech, Univ Montreal, 69-73; dir studies, Tele-Universite, Univ Quebec, 73-77; vpres, Athabasca Univ, Alta, 78-80; vrector, Acad Concordia Univ, 80-84; pres, Laurentian Univ, 84-90; VCHANCELLOR, OPEN UNIV, 90- *Concurrent Pos:* Chair Commonwealth Working Group, Instnl Arrangements for coop in distance educ, 88. *Res:* Communications; educational technology; educational administration; open learning; distance learning. *Mailing Add:* Walton Hall Open Univ Milton Keynes MK7 6AA England

DANIEL, JON CAMERON, b Salem, Ore, Oct 17, 42. DEVELOPMENTAL BIOLOGY, CELL BIOLOGY. *Educ:* Calif State Univ, Northridge, BS, 65, MS, 67; State Univ NY Buffalo, PhD(develop biol), 71. *Prof Exp:* Asst prof anat, Northwestern Univ, Chicago, 73-78; ASSOC PROF HISTOL, COL DENT, UNIV ILL, CHICAGO, 78- *Concurrent Pos:* USPHS fel, Univ Pa, 71-73. *Mem:* Soc Develop Biol; Am Soc Cell Biol; Tissue Culture Asn. *Res:* Molecular mechanisms involved in the maintenance of the differentiated state in cell culture of chick embryo chondrocytes. *Mailing Add:* Univ Ill Dent Sch 801 S Paulina St Chicago IL 60612

DANIEL, JOSEPH CAR, JR, b Murphysboro, Ill, Aug 21, 27; m 51; c 6. DEVELOPMENTAL BIOLOGY, REPRODUCTIVE PHYSIOLOGY. *Educ:* St Louis Univ, BS, 49; Univ Mich, MS, 50; Univ Colo, PhD, 56. *Prof Exp:* Lab instr biol, Univ Denver, 50-51; from instr to assoc prof, Adams State Col, 52-60; sr res fel biophys, Med Ctr, Univ Colo, Boulder, 60-62, from asst prof to assoc prof, 62-66, Inst Develop Biol, 66-68 & dept molecular, cellular & develop biol, 68-71; prof & head dept zool, 71-78, PROF ZOOL, UNIV TENN, KNOXVILLE, 79- *Concurrent Pos:* Guest lectr, Univ Colo, 58; assoc dir, Oak Ridge Pop Res Inst, 72-75, consult biol div, Oak Ridge Nat Lab, 73-; Fulbright prof, Univ Nairobi, Kenya, 79-80. *Mem:* Fel AAAS; Am Soc Zool; Soc Study Develop Biol; Am Soc Cell Biol; Soc Study Reproduction. *Res:* Mammalian embryology and reproductive physiology. *Mailing Add:* Dean Scis Old Dominion Univ 5215 Hampton Blvd Norfolk VA 23508

DANIEL, KENNETH WAYNE, b Dallas, Tex, June 12, 44; m 80; c 3. SIGNAL PROCESSING. *Educ:* Ga Inst Technol, BEE, 67, MSEE, 68, George Washington Univ, Prof Degree of Eng, 78. *Prof Exp:* Proj engr, US Naval Security Group Command HQ, 69-70, Nat Security Agency, Ft Meade, MD, 70-72; sr engr, Tetra Tech, Inc, 72-76; engr scientist, Melpar Div, E-Systems, Inc, 76-80; TECH DIR, RADAR & OPTICS DIV, ENVIRON RES INST MICH, 80- *Mem:* Inst Elec & Electronics Engrs. *Res:* Systems analysis and development of reconnaissance and remote sensing systems; synthetic aperture radar; applications of communications theory and digital signal processing; image processing and compression. *Mailing Add:* Environ Res Inst Mich 1501 Wilson Blvd Suite 1105 Arlington VA 22209

DANIEL, LARRY W, b Hartford, Ky, Sept 11, 51; m 75; c 2. LIPID METABOLISM, PROTEIN KINASE C. *Educ:* Western Ky Univ, BS, 73, MS, 75; Univ Tenn, Memphis, PhD(microbiol), 78. *Prof Exp:* ASSOC PROF BIOCHEM, BOWMAN GRAY SCH MED, 78- *Mem:* Am Soc Biol Chemists; Am Asn Cancer Res. *Mailing Add:* Dept Biochem Bowman Gray Sch Med Winston Salem NC 27103

DANIEL, LEONARD RUPERT, b Seattle, Wash, May 13, 26; m 49; c 2. COMPUTER SCIENCES. *Educ:* Ga Inst Technol, BChE, 46, PhD(chem eng), 52. *Prof Exp:* Asst prof math, Ga Inst Technol, 46-52; prof chem, Corpus Christi Univ, 52-55, chmn sci & eng div, 53-55; prof chem, Howard Payne Col, 55-63, chmn sci & math div, 58-63; chmn div sci & math, WGa Col, 63-69; dean, 69-70, PROF CHEM & DIR COMPUT SERV, CLAYTON STATE COL, 70- *Mem:* Asn Comput Mach. *Mailing Add:* Clayton State Col 5900 Lee St PO Box 285 Morrow GA 30260

DANIEL, LOUISE JANE, b Philadelphia, Pa, Oct 28, 12. BIOCHEMISTRY. *Educ:* Univ Pa, BS, 35; Pa State Col, MS, 36; Cornell Univ, PhD(nutrit), 45. *Prof Exp:* Prof chem & physics, Penn Hall Jr Col, 36-42; res assoc, 45-48, from asst prof to prof, 48-73, EMER PROF BIOCHEM, CORNELL UNIV, 73- *Mem:* Am Soc Biol Chemists; Am Inst Nutrit; NY Acad Sci; Soc Exp Biol & Med; Am Chem Soc. *Res:* Functions of folic acid and vitamin B12 in intermediary metabolism; trace mineral interrelationships. *Mailing Add:* 6350 Navarette Ave Atascadero CA 93422-3742

DANIEL, MICHAEL ROGER, b Harrogate, Eng, Feb 26, 37; US citizen; m 60; c 3. CRYOGENICS, MAGNETISM. *Educ:* Univ Leeds, BSc, 59, PhD(physics), 62. *Prof Exp:* Res assoc physics, Northwestern Univ, 62-64; asst prof, Univ Birmingham, 64-65; res fel, Univ Essex, 65-68; sr scientist, 68-80, FEL SCIENTIST PHYSICS, WESTINGHOUSE ELEC CORP, 80- *Mem:* Am Phys Soc; sr mem Inst Elec & Electronics Engrs. *Res:* New superconductors in practical conductor form for magnet application and Josephson junctions as microwave sources; surface acoustic waves on piezoelectoic crystals for signal processing applications; magnetostatic wave devices using magnetic films for microwave applications. *Mailing Add:* Westinghouse Sci & Technol Ctr 1310 Beulah Rd Pittsburgh PA 15235

DANIEL, O'DELL G, b Paris, Ark, Aug 15, 20; m 44; c 3. ANIMAL HUSBANDRY. *Educ:* Univ Md, BS, 49; Okla State Univ, MS, 51, PhD(animal husb), 57. *Prof Exp:* Assoc prof animal husb, Panhandle Agr & Mech Col, 51-56, head div agr, 57-58; prof & head Exten Animal Sci Dept, Univ Ga, 58-83; RETIRED. *Mailing Add:* Rte 1 Box 110D Colbert GA 30628

DANIEL, PAUL MASON, b Philadelphia, Pa, July 12, 24; m 48; c 4. ZOOLOGY. *Educ:* Univ Cincinnati, BS, 49; Miami Univ, Ohio, MS, 54; Ohio State Univ, PhD(zool), 65. *Prof Exp:* Asst biol, Western Col, 49-50; prof sci, Cuttington Col, Liberia, 50-54; teacher pub schs, Ohio, 54-59; from instr to assoc prof, 59-89, EMER PROF ZOOL, MIAMI UNIV, 89- *Concurrent Pos:* NSF faculty fel, 63-64; pres, Ohio Acad Sci, 85. *Mem:* Am Soc Ichthyologists & Herpetologists; Soc Study Amphibians & Reptiles; Sigma Xi. *Res:* Ecology of benthic organisms; herpetology; biological science instruction in public schools; Ohio vertebrate distribution. *Mailing Add:* Dept Zool Miami Univ Oxford OH 45056

DANIEL, PETER FRANCIS, b Caterham, Eng, April 8, 45; m 74; c 3. COMPLEX CARBOHYDRATES. *Educ:* Christ Church, Oxford Univ, BA, 68; MA, 72; DPhil, 72. *Prof Exp:* Res assoc res & teaching, Dept Nutrit & Food Sci, Mass Inst Technol, 72-75; res fel biochem & neurol, Mass Gen Hosp, Boston, 75-76; sr res fel, Eunice Kennedy Shriver Ctr, Mass, 75-76, res assoc, 76-78, from asst to assoc biochemist, 78-87, SR BIOCHEMIST, EUNICE KENNEDY SHRIVER CTR, MASS, 87- *Concurrent Pos:* Assoc biochemist neurol, Mass Gen Hosp, Boston, 88-; instr neurol, Harvard Univ, Cambridge, 89- *Mem:* Biochem Soc UK; Soc Complex Carbohydrates; AAAS; Am Soc Biol Chemists; Am Soc Neurochem. *Res:* The biochemical and clinical basis of mannosidosis; comparison of human, bovine and feline mannosidosis; glycoprotein storage diseases in general; methodology development for separation and analysis of oligosaccharide by high pressure liquid chromatography; the chemical basis for cell recognition during neurogenesis; galactosemia; neuronal cerold-lipofuscinosis (Batten Disease). *Mailing Add:* Eunice Kennedy Shriver Ctr Biochem Div 200 Trapelo Rd Waltham MA 02254

DANIEL, ROBERT EUGENE, b Birmingham, Ala, Jan 22, 38. ANALYTICAL CHEMISTRY. *Educ:* Samford Univ, AB, 60; Univ Ala, MS, 63, PhD(anal chem), 66. *Prof Exp:* Res asst chem, US Army Missile Command, Redstone Arsenal, 64; asst prof, Ala Col, 65-66; asst prof, Samford Univ, 66-71; ANALYTICAL RES CHEMIST, US PIPE & FOUNDRY CO, 71- *Mem:* AAAS; Am Chem Soc. *Res:* Structure and chemical characteristics of phenol complexes of tantalum, niobium, tungsten and molybdenum; kinetic aspects of the complex-forming reaction. *Mailing Add:* 2540 Savoy Rd Birmingham AL 35203

DANIEL, RONALD SCOTT, b Spencer, Iowa, Oct 10, 36; m 74; c 2. ELECTRON MICROSCOPY. *Educ:* San Jose State Col, BS, 60, MA, 63; Univ Minn, Minneapolis, PhD(entom), 68. *Prof Exp:* Teacher, high sch, Calif, 61-64; res asst entom, Univ Minn, 65-66, res fel, 66-68; PROF BIOL SCI, CALIF STATE POLYTECH UNIV, KELLOGG-VOORHIS, 68-, DIR, ELECTRON MICROSCOPE CTR, 71- *Concurrent Pos:* Bd dirs, Sex Info & Educ Coun, US. *Mem:* AAAS; Am Inst Biol Sci; Electron Microscopy Soc Am; Sigma Xi. *Res:* Biological electron microscopy. *Mailing Add:* Dept Biol Sci Calif State Polytech Univ 3801 W Temple Ave Pomona CA 91768

DANIEL, SAM MORDOCHAI, b Veria, Greece, Mar 25, 41; US citizen; m 65; c 2. EXPERT SYSTEM DEVELOPMENT, DIGITAL SYSTEM PROCESSING. *Educ:* Univ Ill, BS, 64, MS, 65, PhD(elec eng), 71. *Prof Exp:* Engr microwaves, Harris Corp, 65-67; sr engr anal, Magnavox Co, 67-73; MEM TECH STAFF SYSTS ANALYSIS & DESIGN, MOTOROLA INC, 73- *Concurrent Pos:* Instr Russian, Fla Inst Technol, 66; Dan Noble fel, Motorola Inc, 81. *Mem:* Inst Elec & Electronics Engrs; Soc Photo-Optical Instrumentation Engrs. *Res:* Mathematical modeling of signals and systems; computer simulation/emulation for signal generation; algorithm performance evaluation & architectural validation; adaptive signal processing, pattern classification and spectral estimation; artificial intelligence systems; architectures; languages and applications. *Mailing Add:* Mem Tech Staff Gov Electronics Group Motorola Inc 2100 E Elliott Rd Tempe AZ 85282

DANIEL, THOMAS BRUCE, b Kansas City, Mo, Sept 7, 26; m 54; c 2. SOLID STATE PHYSICS. *Educ:* Univ Kans, BS, 50, MS, 54, PhD(physics), 59. *Prof Exp:* Asst, Radiation Biophys Proj, AEC, Kans, 51-53, instr physics, 53-56; assoc physicist, Midwest Res Inst, 58-60, sr physicist, 60-62; assoc prof, 62-64, chmn dept, 62-82, PROF PHYSICS, PITTSBURGH STATE UNIV, 64- *Mem:* Am Asn Physics Teachers. *Mailing Add:* Dept Physics Pittsburgh State Univ Pittsburgh KS 66762

DANIEL, THOMAS HENRY, b Rochester, NY, Oct 24, 42; m 79; c 2. OCEAN ENGINEERING. *Educ:* Mass Inst Technol, BS, 64; Univ Hawaii, MS, 73, PhD(phys oceanog), 78. *Prof Exp:* Teacher physics & math, Cameroon Protestant Col, US Peace Corps, 64-66, physics & gen sci, Irondequoit High Sch, Rochester, NY, 66-67, mech drawing & remedial math, Maui High School, Hawaii, 67-68 & chem & math, Seabury Hall, Maui, Hawaii, 68-69; res asst oceanog, Univ Hawaii, 70-75, asst oceanographer, Res Corp, 75-78; res scientist ocean systs, Lockheed Missiles & Space Co, 79-82; lab dir, 82-89, SCI/TECH PROG MGR, NATURAL ENERGY LAB HAWAII AUTHORITY, 90- *Concurrent Pos:* Co-prin investr, Off Naval Res Grant, Univ Hawaii, 74-75, NSF Grants, 75-78; lectr, Univ Hawaii Col Continuing Educ & Community Serv, 78, 82-, Univ Hawaii, Hilo, 83-; adj grad fac, Univ Hawaii, Manoa, 82-; consult, Lockheed Ocean Systs, 85- *Mem:* Am Geophys Union; Am Meteorol Soc; Marine Technol Soc; Sigma Xi. *Res:* Closed and open cycle ocean thermal energy conversion; coldwater tropical mariculture; time series monitoring of oceanographic parameters. *Mailing Add:* Nat Energy Lab Hawaii PO Box 1749 Kailua-Kuna HI 96745

DANIEL, THOMAS MALLON, b Minneapolis, Minn, Oct 27, 28; m 53; c 4. IMMUNOCHEMISTRY, PULMONARY DISEASES. *Educ:* Yale Univ, BS, 51; Harvard Univ, MD, 55. *Prof Exp:* Resident med, Univ Hosps, 55-59; captain, US Army Med Corps, 59-61; fel microbiol, 61-63, instr med, 63-65, sr instr, 65-66, asst prof, 66-69, assoc prof, 69-77, PROF MED, CASE WESTERN RESERVE UNIV, 77- *Concurrent Pos:* Physician, Univ Hosp, 77-; consult, Vet Admin Hosp, 72-, Int Child Care Tuberculosis Control Prog, 74- *Honors & Awards:* Scholar, John & Mary Markle Found, 67. *Mem:* Cent Soc Clin Res; Am Thoracic Soc; Infectious Dis Soc Am; Am Col Chest Physicians; Am Fed Clin Res. *Res:* Immunologic responses to mycobacterial antigens; purification and immunochemistry of mycobacterial antigens; tuberculosis epidemiology and control in developing nations; rapid immunodiagnostic tests for tuberculosis. *Mailing Add:* Dept Med Case Western Reserve Univ 2074 Abington Rd Cleveland OH 44106

DANIEL, THOMAS W, JR, b Wetumpka, Ala, Jan 24, 22. COAL GEOLOGY. *Educ:* Univ Fla, BS, 52. *Prof Exp:* Chief geologist, Energy Resources Div, US Geol Surv, 56-87; RETIRED. *Mem:* Fel Geol Soc Am; Am Inst Prof Geologists. *Mailing Add:* Seven Lavelle Woods Tuscaloosa AL 35404

DANIEL, VICTOR WAYNE, b Baltimore, Md, July 21, 43. MATHEMATICS. *Educ:* Univ NC, Chapel Hill, BS, 65; Univ Va, PhD(math), 70. *Prof Exp:* Asst prof, Univ Wyo, 70-72; ASSOC PROF & CHMN DEPT MATH & COMPUT SCI, SC STATE COL, 73- *Mem:* Am Math Soc; Math Asn Am. *Res:* Study of certain convolution and integral operators on Lebesque spaces of the half-line; determining their invariant subspaces and when they are Volterra, unicellular or similar to the integration operator. *Mailing Add:* 1207 W Platinum Butte MT 59701

DANIEL, WILLIAM A, JR, b Thomaston, Ga, May 13, 14. PEDIATRICS. *Educ:* Northwestern Univ, BS, 36, MD, 40. *Prof Exp:* Intern, Charity Hosp, New Orleans, 39-40; resident pediat, Children's Hosp, Dallas, 40-41 & Children's Mem Hosp, Chicago, 41-42; pvt pract, Ala, 42-66; PROF PEDIAT, SCH MED, UNIV ALA, 66-, DIR ADOLESCENT UNIT, 66- *Concurrent Pos:* Proj dir, Children & Youth Proj 622, 66-; contrib ed, J Pediat; mem staff, Children's Hosp, Birmingham. *Mem:* Am Pediat Soc; Am Acad Pediat; Soc Adolescent Med. *Mailing Add:* 1437 Midlane Ct Montgomery AL 36106

DANIEL, WILLIAM HUGH, b Ark, Dec 31, 19; m 44; c 2. AGRONOMY. *Educ:* Ouachita Col, BA, 41; Univ Ark, BSA, 46; Mich State Univ, MS, 48, PhD(agron, turf mgt), 50. *Prof Exp:* Prof turf mgt, Purdue Univ, 50-85; RETIRED. *Concurrent Pos:* Secy, Midwest Regional Turf Found, Purdue Univ. *Honors & Awards:* Agron Serv Award, Am Soc Agron, 73. *Mem:* Fel Am Soc Agron; Weed Sci Soc Am. *Res:* Turfgrass management; soil science; crop ecology. *Mailing Add:* 2202 Trace Twenty-Two West Lafayette IN 47907

DANIEL, WILLIAM L, b Wyandotte, Mich, Sept 20, 42; m 64; c 3. HUMAN GENETICS, BIOCHEMICAL GENETICS. *Educ:* Mich State Univ, BS, 64, PhD(zool), 67. *Prof Exp:* From asst prof to assoc prof genetics, Ill State Univ, Normal, 67-72; from asst prof to assoc prof genetics, Sch Basic Med Sci & Dept Genetics & Develop, 72-87, ASSOC PROF, DEPTS PEDIAT & CELL, STRUCT BIOL, UNIV ILL, URBANA, 87- *Mem:* AAAS; Am Soc Human Genetics; Genetics Soc Am; Am Genetics Asn; NY Acad Sci; Am Bd Med Genetics. *Res:* Molecular, developmental and evolutionary genetics of acid hydrolase expression. *Mailing Add:* Dept Cell & Struct Biol Univ Ill 506 Morrill Hall 505 S Goodwin Urbana IL 61801

DANIELE, RONALD P, b Philadelphia, Pa, May 13, 42. PULMONARY DISEASES. *Educ:* Hahnemann Univ, MD, 68. *Prof Exp:* Assoc prof, 78-83, PROF, SCH MED & HOSP, UNIV PA, 83- *Mem:* Am Fedn Clin Res; Am Asn Immunologists; Am Asn Pathologists; Am Physiol Soc; Am Thoracic Soc; Am Col Chest Physicians. *Mailing Add:* Dept Med 807 E Gates Bldg Univ Pa Sch Med & Hosp 3400 Spruce St Philadelphia PA 19104

DANIELEY, EARL, b Alamance Co, NC, July 28, 24; m 48; c 3. ORGANIC CHEMISTRY. *Educ:* Elon Col, AB, 46; Univ NC, MA, 49, PhD(org chem), 54. *Hon Degrees:* DSc Catawba Col, 73; LLD, Campbell Col, 74. *Prof Exp:* From instr to assoc prof chem, 46-50, dean col, 53-56, pres, 57-73, PROF CHEM, ELON COL, 52-, THOMAS E POWELL JR PROF CHEM, 82- *Concurrent Pos:* Fel, Johns Hopkins Univ, 56-57; lectr chem, Univ NC, 52, mem bd gov, 83- *Mem:* Am Chem Soc. *Res:* Cyclobutane compounds; porphyrins in petroleum. *Mailing Add:* PO Box 245 Elon College NC 27244-0245

DANIELL, ELLEN, b New Haven, Conn, July 14, 47; m 80. MOLECULAR BIOLOGY. *Educ:* Swarthmore Col, AB, 69; Univ Calif, San Diego, PhD(chem), 73. *Prof Exp:* Fel tumor virol, Cold Spring Harbor Lab, 73-75; asst prof molecular biol, Univ Calif, Berkeley, 76-84; sr dir human resources, 85-88, DIR, PCR BUS DEVELOP, CETUS CORP, 88- *Concurrent Pos:* Fel, Helen Hay Whitney Found, 74-75. *Mem:* Sigma Xi. *Mailing Add:* Cetus Corp 1400 53rd St Emeryville CA 94608

DANIELL, HERMAN BURCH, b Cadwell, Ga, May 25, 29; m 57; c 3. PHARMACOLOGY. *Educ:* Univ Ga, BS, 51, MS, 64; Med Col SC, PhD(pharmacol), 66. *Prof Exp:* Instr pharm, Univ Ga, 62-63; assoc pharmacol, 67-68, asst prof, 68-70, assoc prof, 70-78, PROF PHARMACOL, MED UNIV SC, 78- *Mem:* Am Soc Pharmacol & Exp Therapeut; Sigma Xi. *Res:* Cardiovascular diseases. *Mailing Add:* Dept Pharmacol Med Univ SC 171 Ashley Ave Charleston SC 29425

DANIELL, JEFF WALTER, b Cadwell, Ga, June 29, 24; m 53; c 2. HORTICULTURE. *Educ:* Univ Ga, BS, 48; Clemson Col, MS, 62; Va Polytech Inst, PhD(plant physiol), 66. *Prof Exp:* Teacher agr, Dodge County Bd Educ, Eastman, Ga, 48-54; assoc county agent, Coop Exten Serv, Univ Ga, 54-62; from asst prof to assoc prof, 66-84, assoc horticulturist, 78-84, PROF & HORTICULTURIST FRUIT & NUT CROPS, GA AGR EXP STA, 84- *Mem:* Sigma Xi; Am Soc Plant Physiologists; Am Soc Hort Sci. *Res:* Physiological problems in peaches and pecans, irrigation, chemical thinning of peaches and chemical weed control. *Mailing Add:* Ga Sta Dept Hort Experiment GA 30212

DANIELPOUR, DAVID, b Sept 7, 55. GROWTH FACTORS. *Educ:* State Univ NY, Buffalo, MA, 80; Univ Tex, PhD(biochem), 86. *Prof Exp:* Fel biochem, Sch Med, Univ Tex, 80-87; GUEST RESEARCHER, NIH, BETHESDA, 87- *Mem:* AAAS; Am Soc Cell Biol. *Mailing Add:* NIH Bldg 41 Rm C629 Bethesda MD 20892

DANIELS, ALMA U(RIAH), b Salt Lake City, Utah, Mar 18, 39; m 61; c 3. BIOMATERIALS, ORTHOPEDIC BIOMECHANICS. *Educ:* Univ Utah, BS, 61, PhD(metall), 66. *Prof Exp:* Res asst metall, Univ Utah, 61-65; res chemist indust & biochem dept, Exp Sta, E I du Pont de Nemours & Co, Inc, 65-69, sr res chemist, 69, develop & serv rep, 69-70; res engr, Metcut Res Assoc, Inc, Ohio, 70; sr mat engr, Univ Utah Res Inst, 71-73, head mat sect, 74-77, assoc dir eng dept, 77-78, dir eng dept, UBTL Div, 78-79, DIR ORTHOP BIOENG LAB, DIV ORTHOP SURG, UNIV UTAH SCH MED, 79- *Concurrent Pos:* From adj asst prof to adj assoc prof, dept mat sci & eng, Univ Utah, 72-, adj assoc prof, dept surg, 76-, res assoc prof, dept bioeng, 79- *Mem:* Soc Biomat; Orthopedic Res Soc; Am Soc Testing & Mat; Scoliosis Res Soc. *Res:* Orthopedic implant design and development biomaterials-properties, effects on tissue; surgical and medical devices-test methods, failure analysis; refractory and wear resistant materials and cutting tools-properties, fabrication processes, testing. *Mailing Add:* Orthop Bioeng Lab Univ Utah Health Sci Ctr Salt Lake City UT 84112

DANIELS, EDWARD WILLIAM, b Tracy, Minn, Jan 19, 17; m 43; c 4. CELL BIOLOGY, PHYSIOLOGY ANIMAL. *Educ:* Cornell Col, BA, 41; Univ Ill, MS, 47, PhD(zool, physiol), 50. *Prof Exp:* Teacher high sch, 41-43 & US Navy, 43-46; mem staff, Univ Ill, 46-50; from instr to asst prof physiol, Univ Chicago, 50-54; mem, Toxicol Labs & USAF Radiation Lab, 50-54; biologist, Div Biol & Med Res, Argonne Nat Lab, 54-71, Environ Impact Studies DIv, 71-81, group leader & proj leader, Aquatic Ecol, 78-81; ADJ PROF ANAT, COL MED, UNIV ILL, CHICAGO, 84- *Mem:* Fel AAAS; Soc Protozool; Am Inst Biol Sci. *Res:* Recovery of cells from potentially lethal radiation damage; transplantation tolerance; origin and fate of subcellular components studied in part with electron microscopy and micrurgy; ultrastructure and physiology of large amoebae. *Mailing Add:* 626 S Wright St Naperville IL 60540

DANIELS, FARRINGTON, JR, b Worcester, Mass, Sept 29, 18; m 51; c 3. PHYSIOLOGY. *Educ:* Univ Wis, BA, 40, MA, 42; Harvard Univ, MD, 43, MPH, 52. *Prof Exp:* Intern med, NY Hosp, 44, asst resident, 47-49; head, Physiol Unit, Qm Climatic Res Lab, Mass, 50-53; chief, Stress Physiol Br, Environ Protecti on Div, Qm Res & Develop Ctr, 53-55; asst prof dermat, Med Sch, Univ Ore, 55-61; assoc prof, Univ Ill, 61-62; assoc prof med, Cornell Univ, 62-69, head, Dermat Div, 62-81, prof med, 69-84, prof pub health, 72-84, prof path, grad sch med sci, 77-84, EMER PROF PUB HEALTH & EMER PROF MED, CORNELL UNIV, 84- *Concurrent Pos:* Res fel med, Med Sch, Cornell Univ, 47-49; res fel nutrit, Sch Pub Health, Harvard Univ, 49-50; Army rep physiol panel, Comt Med Sci, Res & Develop Bd, 51-53; mem photobiol comt, Nat Acad Sci, 65-72. *Mem:* AAAS; Am Acad Dermat; Soc Invest Dermat; Human Factors Soc; Sigma Xi. *Res:* Medicine; spinal cord concussion; environmental physiology; human adaptation to heat, cold and ultraviolet radiation; military load-carrying; effects of footwear and clothing on skin; histochemistry of skin, the "sunburn cell"; environmental factors in human skin color and skin cancer; microbiol screening for photosensitive chemicals. *Mailing Add:* 537 Piper Dr Madison WI 53711-1318

DANIELS, JAMES MAURICE, b Leeds, Eng, Aug 26, 24; m 65; c 2. SOLID STATE PHYSICS, MAGNETISM. *Educ:* Jesus Col, Oxford Univ, BA, 45, MA, 49, DPhil(physics), 52. *Prof Exp:* Asst exp officer, Radar Res & Develop Estab, Ministry Supply, 44-46; tech officer physics, Imp Chem Industs, Ltd, 46-47; Imp Chem Industs res fel low temperature physics, Clarendon Lab, Oxford Univ, 52-53; from asst prof to prof physics, Univ BC, 53-61; prof, 61-67, chmn dept, 69-73, actg chmn dept statist, 83-85, EMER PROF PHYSICS, UNIV TORONTO, 67- *Concurrent Pos:* Consult, Pa State Univ, 58; UNESCO expert exp physics, Univ Buenos Aires, 58-59; vis prof, Balseiro Inst Physics, Arg, 60-61; secy-treas & trustee, Inst Particle Physics, 70-73; vis prof, Low Temperature Lab, Helsinki Univ Technol, 76 & physics dept, Columbia Univ, 78; Guggenheim fel; vis sr res physicist, Princeton Univ, 84-85; assoc prof, Ecole Normale Superieure, Paris, 85-86, 88; vis prof, Nat Tsing Hua Univ, Hsinchu, Taiwan, 90. *Mem:* Fel Royal Soc Can; Brit Inst Physics & Phys Soc; Arg Physics Asn; Can Asn Physicists; fel Royal Soc Arts. *Res:* Adiabatic demagnetisation; spatial orientation of atomic nuclei; magnetic resonance and relaxation; Mossbauer spectroscopy. *Mailing Add:* 40 Cranbury Rd Princeton Junction NJ 08550

DANIELS, JEFFREY IRWIN, b San Mateo, Calif, Dec 26, 51; m 82; c 2. RISK ASSESSMENT. *Educ:* Calif State Univ, Northridge, MS, 78; Univ Calif, Los Angeles, BA, 74, DEnv, 81. *Prof Exp:* SR SCIENTIST, ENVIRON SCI DIV, LAWRENCE LIVERMORE NAT LAB, 79- *Concurrent Pos:* Ed, Environ Sci & Eng Soc Newsletter, 79-86. *Mem:* AAAS; Soc Risk Anal (secy, NCalif chap); Sigma Xi; Environ Sci & Eng Soc. *Res:* Environmental, health, and safety analysis and risk assessment of energy technologies, hazardous waste disposal procedures and drinking water constituents. *Mailing Add:* Environ Sci Div L-453 Lawrence Livermore Nat Lab PO Box 5507 Livermore CA 94550-0617

DANIELS, JERRY CLAUDE, IMMUNOLOGY, RHEUMATOLOGY. *Educ:* Univ Tex, MD, 70, PhD(anat & immunol), 73. *Prof Exp:* DIR DIV RHEUMATOLOGY, UNIV TEX MED BR, 75-, PROF INTERNAL MED & MICROBIOL, 85-, VCHMN MED & CLIN AFFAIRS, DEPT MED, 86- *Mailing Add:* Div Rheumatol Dept Internal Med Univ Tex Med Br Galveston TX 77550

DANIELS, JESS DONALD, b Sugar Land, Tex, Oct 20, 34; m 56; c 3. FOREST GENETICS, PLANT BREEDING. *Educ:* Mont State Univ, BS, 57; Univ Idaho, MF, 65, PhD(forest genetics), 69. *Prof Exp:* Forester, US Forest Serv, USDA, 57-63; instr forestry, Col Forestry, Wildlife & Range Sci, Univ Idaho, 65-66; forest geneticist, Western Forestry Res Ctr, Weyerhaeuser Co, 68-82; FOREST GENETICS CONSULT, 82- *Concurrent Pos:* Co-chmn, Working Party on Progeny Testing, Int Union Forestry Res Orgn, 78-82; dir, NW Tree Improv Coop, 86-; tech consult, Nat Provenance Trials & Seed Orchards Prog, Repub Indonesia, 90- *Mem:* Soc Am Foresters. *Res:* Tree improvement, variation inheritance of characteristics related to growth, yield and quality of wood; breeding for yield improvement; technology of seed production and orchard management. *Mailing Add:* 1143 W Roanoke St Centralia WA 98531-2023

DANIELS, JOHN HARTLEY, b Regina, Sask, July 18, 31; m 56; c 3. STRUCTURAL ENGINEERING. *Educ:* Univ Alta, BSc, 55; Univ Ill, MS, 59; Lehigh Univ, PhD(civil eng), 67. *Prof Exp:* Resident engr, Dept Hwys, Alta, Can, 55-56, design engr, 56-58; chief struct engr, Assoc Eng Serv, Ltd, 59-64; res asst, 64-67, from asst prof to assoc prof, 67-76, PROF CIVIL ENG, LEHIGH UNIV, 76- *Concurrent Pos:* Mem, Transp Res Bd. *Mem:* Am Soc Civil Engrs; Int Asn Bridge & Struct Eng; Sigma Xi. *Res:* Plastic design and analysis of multi-story frames; plasticity; composite steel-concrete structures; optimization techniques; fatigue and fracture of welded steel structures. *Mailing Add:* 3139 Patterson Dr Bethlehem PA 18017

DANIELS, JOHN MAYNARD, b Binghamton, NY, Aug 17, 35; m 58; c 3. WATER CHEMISTRY. *Educ:* Trinity Col, BS, 57; Brandeis Univ, PhD(chem), 65. *Prof Exp:* Phys chemist, Nat Bur Standards, 64-66; asst prof chem, Union Col, NY, 66-72; pres, Capitol Dist Water Treat, Inc, 71-80; CONSULT, 80- *Concurrent Pos:* Mkt dir, Encotech Inc, 78- *Mem:* Am Chem Soc; Am Water Works Asn. *Mailing Add:* 722 Rankin Ave Schenectady NY 12308

DANIELS, L B, b Fordyce, Ark, Aug 28, 40; m 63; c 3. FORAGE UTILIZATION. *Educ:* Univ Ark, BS, 62, MS, 63; Univ Mo, PhD (animal sci), 70. *Prof Exp:* From asst prof to assoc prof, 69-79, PROF ANIMAL SCI, UNIV ARK, 79-, ASSOC DIR, AGR EXP STA, 87- *Mem:* Am Dairy Sci Asn; Sigma Xi; Am Soc Animal Sci; Agr Res Inst. *Res:* Fescue toxicity research; carbohydrate metabolism of neonatal calves. *Mailing Add:* Agr Exp Sta 205 Univ Ark Fayetteville AR 72701

DANIELS, MALCOLM, b Wingate, Eng, Jan 27, 30; m 64. PHYSICAL CHEMISTRY. *Educ:* Univ Durham, BSc, 51, PhD(chem), 55. *Prof Exp:* Asst radiation chem & photochem, King's Col, Newcastle, 54-57; resident res assoc, Argonne Nat Lab, 57-60; lectr chem, Univ West Indies, 60-62; assoc prof radiation chem & photochem, Univ PR, 62-65; head radiation prog, PR Nuclear Ctr, 62-65; assoc prof dept chem, 65-73, PROF DEPT CHEM, ORE STATE UNIV, 73- *Mem:* Radiation Res Soc; Royal Soc Chem; Faraday Soc. *Res:* Radiation chemistry, photochemistry and spectroscopy; aqueous solutions of oxyanions; free radicals; transient species; radiation biophysics; transformations of nucleic acids and constituents; photogalvanic effect; artificial photosynthesis. *Mailing Add:* Dept Chem Ore State Univ Corvallis OR 97331

DANIELS, MATHEW PAUL, b New York, NY, Feb 7, 44; m 68; c 2. NEUROBIOLOGY, CELL BIOLOGY. *Educ:* Queens Col, NY, BA, 64; Univ Chicago, PhD(biol), 70. *Prof Exp:* Nat Inst Neurol Dis & Stroke res fel neurobiol, Albert Einstein Col Med, Yeshiva Univ, 70-71; guest worker, 71-73, staff fel, 73-77, RES BIOLOGIST NEUROBIOL, BIOCHEM GENETICS LAB, NAT HEART LUNG & BLOOD INST, NIH, 77- *Concurrent Pos:* Am Cancer Soc res fel, 71-73. *Mem:* Am Soc Cell Biol; Soc Neurosci. *Res:* Development and function of synapses; morphological distribution of neurotransmitter receptors and ion channels; differentiation of excitable membranes in muscle and nerve. *Mailing Add:* Lab Biochem Genetics-Nat Heart Lung & Blood Inst NIH Bethesda MD 20892

DANIELS, PATRICIA D, b New York, NY, Dec 5, 47; m 82. CIRCUITS & SYSTEMS, ELECTRICAL ENGINEERING EDUCATION. *Educ:* Univ Calif, BS, 68, PhD(elec eng computer sci), 74. *Prof Exp:* Assoc engr, Westinghouse Aerospace, 68-69; from asst prof to assoc prof elec eng, Univ Wash, 74-86; PROF ELEC ENG, SEATTLE UNIV, 86-, CHAIRPERSON, 88- *Concurrent Pos:* Specialist engr, Boeing Com Airplane Co, 79-80; mem tech staff, Aerospace Corp, 80-81, consult, 81-83; consult, Boeing Aerospace, 83-86. *Mem:* Am Soc Eng Educ; Inst Elec & Electronics Engrs. *Res:* Simulation, optimal control and estimation; engineering education including curricular innovations, design, university-industry interactions and recruitment and retention programs for under-represented groups. *Mailing Add:* Dept Elec Eng Seattle Univ 12th & E Columbia Seattle WA 98122-4460

DANIELS, PETER JOHN LOVELL, b Stockport, Eng, June 7, 34. ORGANIC BIOCHEMISTRY. *Educ:* Univ Manchester, BSc, 56, MSc, 57; Univ London, PhD(chem), 60. *Prof Exp:* Fel, Royal Inst Tech, Sweden, 60-61; AP Sloane res assoc, Johns Hopkins Univ, 61-62; res med chemist, Schering Corp, 62-73, assoc dir anti-infectives, 73-80, dir chem res, 80-85; CORP LICENSING DIR, UPJOHN CO, 86- *Mem:* Am Chem Soc; Royal Soc Chem; Am Soc Microbiol. *Res:* Antibiotics; synthetic medicinals; microbial biochemistry. *Mailing Add:* Upjohn Co Kalamazoo MI 49001

DANIELS, RALPH, b New York, NY, May 2, 21; m 44; c 3. ORGANIC CHEMISTRY. *Educ:* Brooklyn Col, AB, 44; Harvard Univ, AM, 49, PhD, 50. *Prof Exp:* Res chemist, Givaudan-Delawanna Res Inst, 44-46; fel nonaromatic steroids, Univ Wis, 50-51; instr chem, Purdue Univ, 51-52; from asst prof to prof, Col Pharm, Univ Ill Med Ctr, 52-77, asst dean grad col, 70-72, actg dean grad col, 75-77; ASSOC DEAN, GRAD COL, DIR, OFF RES ADMIN & PROF MED CHEM, UNIV OKLA HEALTH SCI CTR, 77- *Concurrent Pos:* Vis scientist, Univ London, 61-62. *Mem:* Am Chem Soc; Am Asn Col Pharm; AAAS; Nat Coun Univ Res Admin. *Res:* Synthesis of pharmacologically active compounds; natural products; heterocyclic chemistry; radiation protective compounds; reaction mechanisms; spectra of organic compounds. *Mailing Add:* One Gracie Terr Apt 16B New York NY 10028-7955

DANIELS, RAYMOND BRYANT, b Adair, Iowa, Feb 15, 25; m 45; c 2. SOIL GENESIS, GEOMORPHOLOGY. *Educ:* Iowa State Col, BS, 50, MS, 55, PhD(soil genesis), 57. *Prof Exp:* Res soil scientist, Soil Conserv Serv, USDA, NC, 57-68; from assoc prof to prof soil sci, NC State Univ, 68-77; dir soil surv invests, Soil Conserv Serv, USDA, 77-80; PROF, SOIL SCI DEPT, NC STATE UNIV, 81- *Mem:* Mem: Soil Conserv Soc Am; Geol Soc Am; Am Quaternary Asn. *Res:* Soil genesis and stratigraphic geomorphic interrelations. *Mailing Add:* 9112 Leesville Rd Raleigh NC 27612

DANIELS, RAYMOND D(EWITT), b Cleveland, Ohio, Feb 14, 28; m 52; c 1. PHYSICAL METALLURGY, CORROSION. *Educ:* Case Inst Technol, BS, 50, MS, 53, PhD(phys metall), 58. *Prof Exp:* Physicist, Nat Bur Stand, 50-51; asst physics, Case Inst Technol, 51-54; res engr, Linde Co, 54-55; asst & spec lectr metall, Case Inst Technol, 55-58; from asst prof to assoc prof metall eng, 58-64, chmn sch, 62-63, dir sch chem eng & mat sci, 63-65, 69-70 & 86-, assoc dean eng for res & grad study, 65-68, exec dir, res inst, 71-83, dir, off res admin, 73-78, PROF CHEM ENG & MAT SCI, UNIV OKLA, 64- *Concurrent Pos:* Dupont res grant, 58; consult, US Air Force, 58-59, Autoclave Engrs, 59-62 & Avco Corp, 63-66; NSF sci fac fel, Univ Neuchatel, 68-69; mem, Nat Res Coun Comt, 75-78. *Honors & Awards:* Halliburton Distinguished lectr, 83-88; Distinguished Serv Award, Nat Asn Corrosion Engrs, 90. *Mem:* Am Soc Metals Int; Minerals, Metals & Mat Soc; Nat Asn Corrosion Engrs; Am Soc Eng Educ; Am Soc Testing & Mat; Am Inst Chem Engrs; Mat Res Soc; Sigma Xi. *Res:* Physical metallurgy; mechanical properties of metals; embrittlement phenomena; corrosion; administration of research. *Mailing Add:* Sch Chem Eng & Mat Sci Rm T-335 Univ Okla 100 E Boyd Norman OK 73019-0628

DANIELS, ROBERT (SANFORD), b Indianapolis, Ind, Aug 12, 27; m 50; c 4. MEDICINE, PSYCHIATRY. *Educ:* Univ Cincinnati, BS, 48, MD, 51. *Prof Exp:* Asst prof psychiat, Univ Chicago, 57-63, assoc prof & actg chmn dept, 63-68, prof psychiat & social med, assoc dean social & community med & dir ctr health admin studies, 68-71; prof psychiat & dir dept, 71-75, interim dean col med, 72-75, DEAN COL MED, UNIV CINCINNATI, 75- *Concurrent Pos:* Consult, Cook County Hosp, 58-68, Ill State Psychiat Inst, 59-68 & Thresholds, 63-67; chief of staff, Cincinnati Gen Hosp & Holmes Hosp, 72- *Res:* Medical education; group psychotherapy; clinical psychiatric research, medical care organization and financing. *Mailing Add:* La State Univ Sch Med 1542 Tulane Ave New Orleans LA 70112-2822

DANIELS, ROBERT ARTIE, b St Charles, Mo, Feb 14, 49; m 77; c 1. FISHERIES BIOLOGY, AQUATIC ECOLOGY. *Educ:* Univ Calif, Los Angeles, BA, 71; Univ Calif, Davis, MS, 77, PhD(ecology), 79. *Prof Exp:* Researcher, Univ Calif, 80-81; SR SCIENTIST, BIOL SURV, NY STATE MUS, 81- *Mem:* AAAS; Am Fisheries Soc; Am Soc Ichthyologists & Herpetologists; Ecol Soc Am; Am Soc Naturalists. *Res:* Life histories of fishes and aquatic macroinvertebrates and mechanisms of coexistence among sympatric species. *Mailing Add:* 2470 Berne Almont Rd Knox NY 12107

DANIELS, STACY LEROY, b Frankfort, Mich, Aug 29, 37; m 64; c 1. ENVIRONMENTAL SCIENCE & ENGINEERING. *Educ:* Univ Mich, Ann Arbor, BSE, 60, MSE, 61 & 63, PhD(chem eng), 67. *Prof Exp:* Res engr, Chem Dept Res Lab, 66-69, develop engr, Environ Control Syst Tech Serv & Develop, 69-72, sr res engr, 72-74, res specialist, Functional Prod & Systs Dept, 74-76, RES SPECIALIST, ENVIRON SCI RES LAB, DOW CHEM USA, 76- *Concurrent Pos:* Mem res biol lab, US Bur Com Fisheries, 61-65; adj assoc prof, Dept Chem Eng, Univ Mich, 71-81. *Mem:* Am Inst Chem Engrs; Am Water Works Asn; Int Asn Water Pollution Res; Nat Solid Waste Mgt Asn; Sigma Xi. *Res:* Solids-liquid separations (coagulation, flocculation); coordinating information on the assessment of environmental impacts of chemical products; consulting on waste treatment and disposal technologies and other environmental problems; interfacing with Federal and State governments on water, waste-water and solid hazardous waste management regulations; author of 110 publications. *Mailing Add:* Comb Tech Res Dow Chem USA 734 Bldg Midland MI 48667

DANIELS, WILLIAM BURTON, b Buffalo, NY, Dec 21, 30; m 58; c 3. EXPERIMENTAL SOLID STATE PHYSICS, HIGH PRESSURE PHYSICS. *Educ:* Univ Buffalo, BS, 52; Case Western Reserve Univ, MS, 55, PhD, 57. *Prof Exp:* Instr physics, Case Western Reserve Univ, 57-58, asst prof, 58-59; res scientist, Res Labs, Nat Carbon Co, 59-61; from asst prof to prof solid state sci, Princeton Univ, 61-72; chmn dept, 77-80, UNIDEL PROF PHYSICS, UNIV DEL, 72- *Concurrent Pos:* Res collabr, Brookhaven Nat Lab, 66-72; sabbatical vis, Univ Amsterdam, 68, T H Munchen, 69, Res Estab RISØ Denmark, 76, Col de France & Univ Paris VI, 77, IBM Zurich, 77, Max Planch Inst, FKF, Stuttgart, 82 & Univ Groningen, 85; Guggenehim fel, 76-77. *Honors & Awards:* Humboldt Sr Scientist Award, 82. *Mem:* Fel Am Phys Soc. *Res:* Solid state and lattice dynamics at high pressure; solidified rare gases and molecular solids; liquid and plastic crystals; Raman and neutron scattering; electronic structure of solids; non-linear optics. *Mailing Add:* Dept Physics Univ Del Newark DE 19711

DANIELS, WILLIAM FOWLER, SR, b New Bern, NC, Feb 22, 20; m 47; c 3. BIOCHEMISTRY, BIOENGINEERING. *Educ:* La State Univ, BSChE, 42; Univ Fla, MS, 50; Univ Ky, PhD(bact physiol), 57. *Prof Exp:* Consult engr & chemist bldg mat, H C Nutting Co, 47-48; asst indust & eng exp sta, Univ Fla, 49; res bacteriologist water bact, Environ Health Ctr, USPHS, 50-51; asst bact, Univ Ky, 52; res chemist, Hwy Res Lab, State Dept Hwy, Ky, 53-55, head chem sect, 56; prin investr biochem eng, Pilot Plants Div, Ft Detrick, 56-58 & Process Res Div, 58-70; biologist consult, Appl Sci Div, US Adv Mat Concepts Agency, Va, 70-74; phys sci adminr, US Army Mat Develop & Readiness Command, 74-76; gen engr, US Army Med Res & Develop Command, 76-82; RETIRED. *Concurrent Pos:* Consult, 59- *Mem:* Soc Am Microbiologists; Am Chem Soc; Am Inst Chemists; Sigma Xi. *Res:* Paper electrophoresis of carbohydrate compounds; constitution and characterization of asphalt; organic synthesis via microbial oxidation; sterile filtrations; submerged tissue and continuous bacterial fermentation. *Mailing Add:* 7819 Spout Spring Rd Frederick MD 21701

DANIELS, WILLIAM RICHARD, b Dayton, Iowa, Apr 16, 31; m 51; c 4. RADIOCHEMISTRY, NUCLEAR CHEMISTRY. *Educ:* Iowa State Univ, BS, 53; Washington Univ, MA, 55; Univ NMex, PhD(radiochem), 65. *Prof Exp:* Mem staff radiochem, 57-80, assoc group leader, 80-82, dep group leader, 82-84, GROUP LEADER, NUCLEAR & RADIOCHEM, LOS ALAMOS NAT LAB, 84- *Res:* Short-lived radionuclides; nuclear spectroscopy; nuclear waste disposal. *Mailing Add:* 1060 Los Pueblos Los Alamos NM 87544

DANIELS, WILLIAM WARD, b Norfolk, Va, Apr, 13, 26; m 48; c 6. PHYSICAL CHEMISTRY, RESEARCH ADMINISTRATION. *Educ:* La State Univ, BS, 49, MS, 50, PhD(phys chem), 54. *Prof Exp:* Res chem, Dacron Res Lab, 53-58, res supvr, 58-62, sr supvr, Chattanooga Tech Lab, 62-64, res mgr, Carothers Res Lab, 64-67, tech supt, Seaford Tech Lab, 67-69, LAB DIR, DACRON RES LAB, E I DU PONT DE NEMOURS & CO, INC, 69- *Mem:* Am Chem Soc. *Res:* Infrared and raman spectroscopy; fiber structure-property relationships. *Mailing Add:* 1505 St James Pl Kinston NC 28501

DANIELS, WORTH B, JR, b New York, NY, Jan 3, 25; m 62; c 2. INTERNAL MEDICINE. *Educ:* Johns Hopkins Univ, MD, 48. *Prof Exp:* From asst res med to res med, Baltimore City Hosp, 54-57; physician & assoc prof, Johns Hopkins Univ, 58-; RETIRED. *Mem:* Inst Med-Nat Acad Sci; Am Med Asn; Am Col Physicians; Am Soc Internal Med. *Mailing Add:* 210 Ridgewood Rd Baltimore MD 21210-2539

DANIELSON, DAVID MURRAY, b Aurora, Nebr, Feb 24, 29; m 53; c 3. ANIMAL NUTRITION. *Educ:* Univ Nebr, BS, 52, MS, 58; Utah State Univ, PhD(animal prod), 64. *Prof Exp:* PROF ANIMAL NUTRIT, UNIV NEBR, NORTH PLATTE, 58- *Mem:* Am Soc Animal Sci; Am Soc Regist Animal Scientists; Animal Nutrit Res Coun. *Res:* Carbohydrates for baby pig diets; limited feeding of growing-finishing swine diets; source and quality of nitrogen and energy for swine gestation diets; automation in rearing baby pigs; feeding corn, millet, wheat, grain sorghum alone or in combination to finishing swine. *Mailing Add:* W Cent Ctr Rte 4 Box 46A Univ Nebr North Platte NE 69101

DANIELSON, GEORGE EDWARD, JR, b Bozeman, Mont, May 10, 39; m 63; c 4. SPACE INSTRUMENTATION. *Educ:* Principia Col, BS, 61; Univ Ill, Urbana, MS, 63; Univ Rochester, MS, 67. *Prof Exp:* Physicist, United Aircraft Res Lab, 63-65; group supvr, Jet Propulsion Lab, Photo Sci Group, 71-74, asst sect mgr, space photog sect, 74-75, from staff asst to dir, 74-77, dep sect mgr, 77-78; MEM PROF STAFF & SR SCIENTIST, CALIF INST TECHNOL, 78- *Concurrent Pos:* Co-investr, Mariner Venus Mercury Imaging Sci Team, 70-75, Voyager Imaging Sci Team, 74-, Galileo Near Infrared Imaging Spectrometer, 77- & Space Telescope Wide Field/Planetary Camera IDT, 78-; consult, Giotta Halley multicolour camera, 81-; mem staff, Palomar Observ, 81- *Mem:* Am Astron Soc. *Res:* Space photography instrument development; ground based astronomical instrumentation. *Mailing Add:* Div Geol Planetary Sci Calif Inst Technol MS 170-25 Pasadena CA 91125

DANIELSON, GORDON KENNETH, b Burlington, Iowa, Dec 5, 31; m 61; c 7. CARDIOVASCULAR SURGERY. *Educ:* Univ Pa, BA, 53, MD, 56; Am Bd Surg & Am Bd Thoracic Surg, dipl, 63. *Prof Exp:* Asst instr surg, Univ Pa, 57-61, instr, Sch Med, 61-62, assoc, 62-65, asst prof, 65; assoc prof, Col Med, Univ Ky, 65-67; from asst prof to assoc prof, Mayo Grad Sch Med, 67-73, assoc prof, Mayo Med Sch, 73-75, PROF SURG, MAYO MED SCH, 75-, CONSULT CARDIOVASC SURG, MAYO CLIN, 67- *Concurrent Pos:* Univ fel, Harrison Dept Surg Res, Univ Pa, 57-62, Am Cancer Soc fel, 60-61, fel diag & treatment of cancer, 61-62; Markle scholar, 62-67; vis fel, Stockholm, Sweden, 63-64; gen & thoracic surgeon, Hosp Univ Pa, 62-65, asst chief, Surg Div I, 62-65; assoc surg & chief cardiac surg, Univ Hosp, Lexington, Ky, 66-67; consult, St Mary's Hosp, Rochester, Minn, 67- *Honors & Awards:* Spencer Morris Prize in Med; Citations for Int Serv, Am Heart Asn, 75 & 81; Julian Johnson Lectr in Cardiothoracic Surg, Univ Pa, 83; Jennifer B Lalin Mem lectr, Col Physicians & Surgeons, New York, NY, 84; Award for Meritorious Contrib & Outstanding Serv in Cardiol & Open Heart Surg, World Conf Open Heart Surg, 85. *Mem:* Fel Am Col Surg; Am Surg Soc; Am Heart Asn; fel Am Col Cardiol; Soc Thoracic Physicians; Am Asn Thoracic Surg; hon mem Mex Soc Cardiol; hon mem Asn Thoracic & Cardiovasc Surgeons Asia. *Res:* Cardiac physiology; cardiac surgery. *Mailing Add:* Dept Surg W-6 Mayo Grad Sch Med Rochester MN 55905

DANIELSON, IRVIN SIGWALD, VACCINE PRODUCTION. *Educ:* Harvard Univ, PhD(biochem), 33. *Prof Exp:* Dir biol prod, Lederle Lab Div, subsid Am Cyanimid, 42-69. *Mailing Add:* 22 Turner Rd Pearl River NY 10965

DANIELSON, LEE ROBERT, b Summit, NJ, June 19, 46; m 79; c 1. SURFACE PHYSICS, SOLID STATE PHYSICS. *Educ:* Iowa State Univ, BS, 68; Univ Wash, MS, 69; Wash State Univ, PhD(physics), 77. *Prof Exp:* Res assoc, Ore Grad Ctr, 76-78; RES PHYSICIST, THERMO ELECTRON CORP, 79- *Mem:* Am Asn Physics Teachers; Am Vacuum Soc; Mat Res Soc. *Res:* Experimental determinations of surface and solid state properties of materials; ultra-high vacuum physics and direct energy conversion. *Mailing Add:* 1009 Valerie Dr Schenectady NY 12309

DANIELSON, LORAN LEROY, b Havelock, Nebr, July 2, 13; m 33. PLANT PHYSIOLOGY. *Educ:* Univ Iowa, AB, 38, MS, 40, PhD(plant physiol), 41. *Prof Exp:* Chemist & plant physiologist, Delmonte Corp, Ill, 41-44; asst prof bot, Mont State Col, 45; res asst & prof hort, Iowa State Col, 45; plant physiologist, Va Truck Exp Sta, 45-57; ADJ PROF BIOL, UNIV NC, CHARLOTTE, 77- *Concurrent Pos:* Invest leader, Weed Control, Hort crops, 57-72; ed, Weed Sci, 77-80. *Mem:* Am Chem Soc; Am Soc Plant Physiologists; fel Weed Sci Soc Am; Sigma Xi. *Res:* Physiology of vegetable crops; weed control in horticultural crops. *Mailing Add:* 5150 Sharon Rd Charlotte NC 28210-4720

DANIELSON, NEIL DAVID, b Ames, Iowa, July 25, 50; m; c 1. ANALYTICAL CHEMISTRY. *Educ:* Iowa State Univ, BS, 72; Univ Nebr, MS, 74; Univ Ga, PhD(anal chem), 78. *Prof Exp:* Asst prof, 78-83, ASSOC PROF CHEM, MIAMI UNIV, 83- *Concurrent Pos:* Vis scientist, E I du Pont de Nemours Co, 85-86 & consult, Waters Chromatography Div. *Mem:* Am Chem Soc. *Res:* Gas chromatography; high performance liquid chromatography; fluorescence. *Mailing Add:* Hughes Labs Miami Univ Oxford OH 45056

DANIELSON, PAUL STEPHEN, b Passaic, NJ, May 14, 47; m 69; c 2. INORGANIC CHEMISTRY, GLASS CHEMISTRY. *Educ:* Franklin & Marshall Col, AB, 69; Univ Conn, PhD(inorg chem), 74. *Prof Exp:* Res assoc glass mat sci, Vitreous State Labs, Cath Univ Am, 74-76; sr chemist glass chem, Tech Staffs Div, 76-79, sr res chemist glass chem, 79-86, RES ASSOC & SUPVR, GLASS RES GROUP, RES & DEVELOP DIV, CORNING GLASS WORKS, 87- *Mem:* Am Ceramic Soc; Am Chem Soc; Sigma Xi. *Res:* Glass chemistry research; transition metals in glasses; solid state inorganic chemistry. *Mailing Add:* Sullivan Park-FR-5 Corning Inc Corning NY 14831

DANIELSON, ROBERT ELDON, b Brush, Colo, Jan 14, 22; m 49; c 2. SOIL PHYSICS, CROP-WATER RELATIONS. *Educ:* Colo State Univ, BS, 48; Cornell Univ, MS, 49; Univ Ill, PhD(soil physics), 55. *Prof Exp:* Instr soils, 49-52, from asst prof to assoc prof, 55-64, PROF SOIL PHYSICS, COLO STATE UNIV, 64- *Concurrent Pos:* Mem, US Comt on Irrig & Drainage, 76-; vis prof, Mont State Univ, 78; prin invstr res projs. *Mem:* Fel Am Soc Agron; fel Soil Sci Soc Am; Int Soc Soil Sci. *Res:* Soil physical properties; influence on the growth of plants; irrigation; soil aeration and structure; soil-plant-water relations. *Mailing Add:* Dept Agron Colo State Univ Ft Collins CO 80523

DANIELS-SEVERS, ANNE ELIZABETH, b San Francisco, Calif, June 13, 40; m 70; c 5. PHARMACOLOGY. *Educ:* San Francisco Col Women, BA, 57; Univ Calif, San Francisco, MS, 65; Univ Pittsburgh, PhD(pharmacol), 68. *Prof Exp:* Asst prof pharmacol, Sch Med, Hershey Med Ctr, Pa State Univ, Hershey, 70-73; mem coun basic res, Am Heart Asn, 71-75; CONSULT, CHILD GROWTH & DEVELOP, 73-, ENVIRON ANAL, 87- *Concurrent Pos:* Nat Acad Sci resident res fel pharmacol, NASA Ames Res Ctr, Moffett Field, Calif, 68-70; consult, Proj Biosattelite, Univ Calif, Los Angeles, 71. *Mem:* Am Soc Pharmacol & Exp Therapeut; Am Fedn Clin Res. *Res:* Centrally mediated effects of angiotensin including cardiovascular, adrenal and salt-water balance activity; effects of stress on the pituitary-adrenal system; child growth and development; domestic environmental control. *Mailing Add:* 1011 Grubb Rd Palmyra PA 17078

DANIS, PETER GODFREY, b Ottawa, Ont, Apr 12, 09; nat US; m 31; c 10. PEDIATRICS. *Educ:* St Louis Univ, BS, 29, MD, 31, MS, 35. *Prof Exp:* From asst instr to sr instr pediat, 35-47, assoc prof, 48-50, prof, 71-79, EMER PROF PEDIAT & EMER CHMN DEPT, ST LOUIS UNIV HOSP, 79- *Concurrent Pos:* Chmn dept pediat, Univ Hosps, St Louis, 47-56; chief of staff, Cardinal Glennon Mem Hosp for Children, 56-58; mem, Nat Adv Comt United Cerebral Palsy Asn & pediatrician med adv bd spec educ, St Louis County Pub Schs. *Mem:* Am Acad Pediat; AMA; fel Am Acad Cerebral Palsy; assoc Am Acad Neurol; Soc Res Child Develop. *Res:* Neurology; cerebral palsy; convulsive disorders. *Mailing Add:* 19 Conway Lane St Louis MO 63124

DANISHEFSKY, ISIDORE, b Poland, Apr 3, 23; US citizen; m 51; c 2. BIOCHEMISTRY, ORGANIC CHEMISTRY. *Educ:* Yeshiva Univ, BA, 44; NY Univ, MSc, 47, PhD(chem), 51. *Prof Exp:* Res assoc chem, Polytech Inst Brooklyn, 51-52; res assoc biochem, Col Physicians & Surgeons, Columbia Univ, 52-55; from asst prof to assoc prof, 55-66, PROF BIOCHEM, NEW YORK MED COL, 66-, CHMN DEPT, 77- *Concurrent Pos:* Ed, Throbosis Res. *Honors & Awards:* Bernard Revel Mem Award, Yeshiva Univ, 61; Honor Achievement Award, Am Col Angiology, 63. *Mem:* AAAS; Am Soc Biol Chem; Soc Complex Carbohydrates (treas, 72-74); Am Chem Soc; Am

Asn Cancer Res. *Res:* Blood coagulation and anticoagulants; mucopolysaccharide structure and metabolism; nucleotide sugars; carcinogenesis. *Mailing Add:* Dept Biochem New York Med Col Valhalla NY 10595

DANISHEFSKY, SAMUEL, b Bayonne, NJ, Mar 10, 36. ORGANIC SYNTHESIS. *Educ:* Yeshiva Univ, BS, 56; Harvard Univ, Phd(chem), 62. *Prof Exp:* Fel chem, Columbia Univ, 61-63; from asst prof to prof chem, Univ Pittsburgh, 63-79; chmn dept chem, 81-88, PROF CHEM, YALE UNIV, 79-, EUGENE HIGGINS PROF CHEM, 83-, STERLING PROF CHEM, 90- *Concurrent Pos:* Consult, Merck Sharp & Dohme, 73-, Gen Elec Co, 77-; vis prof, Iowa State Univ, 74, Univ Calif, Irvine, 77, Rice Univ, 77, Tex A&M, 86; distinguished vis lectr, Univ Tex, 79. *Honors & Awards:* Guenther Award, 80, Aldrich Award, 86, Am Chem Soc; Reilly Lectr, 79. *Mem:* Nat Acad Sci; Am Chem Soc; Swiss Chem Soc; Japanese Chem Soc; fel AAAS; Am Acad Arts & Sci. *Res:* Synthesis of substances of complex structure with interesting biological activity; origins of diastereofacial and topographic selectivity; pyrrologuinonoid antitumor agents; goal systems with immunochemical implications. *Mailing Add:* Dept Chem Yale Univ New Haven CT 06520

DANK, MILTON, b Philadelphia, Pa, Sept 12, 20; m 54; c 2. SOLID STATE PHYSICS, MATHEMATICS. *Educ:* Univ Pa, BA, 47, PhD(physics), 53. *Prof Exp:* Asst instr physics, Univ Pa, 47-53; res physicist, Owens-Ill Glass Co, 53-56; res physicist, Burroughs Corp, 56; consult physicist, Missile & Space Div, Gen Elec Co, 56-60, mgr superpressure studies, Space Sci Lab, 60-68, consult physicist laser effects, 68-71; CONSULT SOLID STATE PHYSICS, NDA, INC, 71- *Mem:* Am Phys Soc; Am Inst Aeronaut & Astronaut. *Res:* Effect of dynamic high pressures on solids; shock wave phenomena in solids; equations of state of solids at megabar pressures; hypervelocity particle impact effects; laser effects on solids; laser containment for thermonuclear fusion power. *Mailing Add:* NDA Inc 1022 Serpentine Wyncote PA 19095

DANKE, RICHARD JOHN, b New London, Wis, Sept 21, 37; m 62; c 3. ANIMAL NUTRITION. *Educ:* Wis State Col, River Falls, BS, 59; Okla State Univ, Stillwater, MS, 62, PhD(animal nutrit), 65. *Prof Exp:* Proj supvr develop animal nutrit, Monsanto Co, 65-68, res specialist animal nutrit, 68-69, tech specialist comput technol, 69-71; ruminant nutritionist animal nutrit, Farmers Union Grain Terminal Asn, 71-74, res & formulation coordr, 74-75; ruminant res nutritionist, Conagra, Inc, 75-77; vpres & gen mgr, Int Nutrit, Inc, 77-81; MGR, FEED INGREDIENT MKT, ALLIED CHEM CORP, 81- *Mem:* Am Soc Animal Sci; Am Dairy Sci Asn; Poultry Sci Asn; Sigma Xi; AAAS. *Res:* Improving the nutrient availability of feedstuffs for domestic animals; improving the productive efficiency of domestic animals. *Mailing Add:* 7982 Severhill Ct Dublin OH 43017

DANKEL, THADDEUS GEORGE, JR, b Waycross, Ga, Nov 18, 42; m 71. QUANTUM MECHANICS, UNDERWATER SOUND. *Educ:* Duke Univ, BS, 64; Princeton Univ, MA, 66, PhD(math), 69. *Prof Exp:* Asst prof math, Duke Univ, 68-71; assoc prof, 71-77, PROF MATH, UNIV NC, WILMINGTON, 77- *Concurrent Pos:* NSF fel, 64; fel, Woodrow Wilson Found, 64, Southern Regional Adv Panel, 71-; vis assoc prof math, Univ NC, Chapel Hill, 77; vis prof statist, Univ NC, Chapel Hill, 84-85. *Honors & Awards:* Poteat Award, NC Acad Sci, 74. *Mem:* Am Math Soc; Math Asn Am; Soc Indust & Appl Math; Int Assoc Math Physics; Sigma Xi. *Res:* Mathematical problems arising from quantum physics, especially the relations between the theory of stochastic processes and quantum theories; probabilistic methods in underwater sound. *Mailing Add:* 1203 Windsor Dr Wilmington NC 28403

DANKO, JOSEPH CHRISTOPHER, b Homestead, Pa, Jan 12, 27; m 51; c 3. ENERGY DEVICE FABRICATION, STRESS CORROSION. *Educ:* Carnegie-Mellon Inst, BS, 51; Lehigh Univ, MS, 54, PhD(metal eng), 55. *Prof Exp:* Engr chem & metall prog, Gen Elec Co, 51-52; instr metall eng, Lehigh Univ, 52-55; supvr mats develop, Westinghouse Elec Co, 55-64; mgr space power, Gen Elec Co, 64-71, mgr mats processing, 71-78; dir res prog, Elec Power Res Inst, 78-84; vpres & exec dir, Am Welding Inst, 84-86; DIR MATS PROCESSING, UNIV TENN, 86- *Concurrent Pos:* Tech adv comt, Metals Properties Coun, 71-76, mem, 76-; corrosion adv comt, Elec Power Res Inst, 74-80; mem, Welding Res Coun, 82-85; consult, EI Dupont, 85-, EG&G Idaho, 86-, EPRI-NDE Ctr, 87-; chmn, Nat Asn Corrosion Engr, 87-89. *Mem:* Fel Am Soc Metals; Nat Asn Corrosion Engr; Am Welding Soc; Sigma Xi; Am Nuclear Soc. *Res:* Research and development on materials for nuclear reactors on thermoelectric and thermionic direct energy devices, corrosion and stress corrosion cracking of stainless steels in boiling water reactors, welding science and physical and mechanical metallurgy. *Mailing Add:* Ctr Mats Processing Univ Tenn-Knoxville Col Eng 101 Perkins Hall Knoxville TN 37996

DANKS, HUGH VICTOR, b Farnham, Eng, Sept 16, 43; m 71; c 3. ENTOMOLOGY. *Educ:* Univ London, BSc & ARCS, 65, PhD(entom) & DIC, 68. *Prof Exp:* Fel overwintering of Chironomidae, Entom Res Inst, Agr Can, 68-70, res asst arctic Chironomidae, 71; res asst arctic Culicidae, Dept Biol, Univ Waterloo, 71-72; res assoc biol of Tachinidae, Dept Entom, NC State Univ, 72-74; asst prof biol ecol & entom, Dept Biol Sci, Brock Univ, 74-77; entomologist-in-chg, Biol Surv Proj, Entom Soc Can, 77-80; HEAD, BIOL SURV CAN TERRESTRIAL ARTHROPODS, CAN MUS NATURE, 80- *Honors & Awards:* Hewitt Award, Entom Soc Can, 80. *Mem:* Fel Royal Entom Soc London; fel Entom Soc Can. *Res:* Modes of seasonal adaptation in insects; seasonality in aquatic insects, especially Chironomidae; ecology of Chironomidae; arctic insects; arthropod fauna of Canada. *Mailing Add:* Biol Surv Can Mus Nature Ottawa ON K1P 6P4 Can

DANLY, DONALD ERNEST, b Washington, DC, June 14, 29; m 53; c 3. CHEMICAL ENGINEERING. *Educ:* Cornell Univ, BChE, 52; Univ Fla, PhD(chem eng), 58. *Prof Exp:* Chem engr, Textile Intermediates Res & Develop, Monsanto Chem Co, 58-60, sr chem engr, 60-62, proj leader, 62-63, suprv, 63-65, sect head, 65-71, mgr, 71-76, dir, Nylon Intermediates Technol, 76-86, dir, Nylon & Nylon Intermediates Technol, 86-89; ELECTROCHEM ENG CONSULT, 89- *Honors & Awards:* Vittorio de Nora-Diamond & Shamrock Award, Electrochem Soc, 84. *Mem:* Am Inst Chem Engrs; Electrochem Soc. *Res:* Research and development studies on new and improved processes for manufacture of nylon intermediates and nylon fiber products. *Mailing Add:* 3693 Mackey Cove Dr Pensacola FL 32514-8155

DANN, JOHN ROBERT, b Minneapolis, Minn, Sept 6, 21; m 43; c 3. ADHESION THEORY. *Educ:* Univ SDak, BA, 43; Univ Colo, PhD(chem), 52. *Prof Exp:* Chemist, State Chem Lab, SDak, 41-43; chemist, Eastman Kodak Co, 43-44, group leader, 46-48, res chemist, Res Labs, 52-58, res assoc, 58-72, sr res assoc, 72-81; RETIRED. *Mem:* Am Chem Soc; Sigma Xi. *Res:* High polymers; Diels-Alder reaction; chemistry of gelatin; heterocyclic chemistry; theory of adhesion; discovered the first macrocyclic sulfur and oxygen-containing crown ethers in 1957 (See Journal of organic chemistry 26, 1991 (1961)), discovered their silver ion complexing ability in 1958 (See US Patent 3,062,646 filed March 6, 1959. Granted Nov 6, 1962). *Mailing Add:* 371 Olympic View Lane Friday Harbor WA 98250

DANNA, KATHLEEN JANET, b Beaumont, Tex, Aug 21, 45; m 76; c 1. VIROLOGY, MOLECULAR BIOLOGY. *Educ:* NMex Inst Mining & Technol, BS, 67; Johns Hopkins Univ, PhD(microbiol), 72. *Prof Exp:* Res fel, Lab Molecular Biol, Rijksuniversiteit-Gent, Belgium, 72-73; res fel, Dept Biol, Mass Inst Technol, 73-75; asst prof, 75-83, ASSOC PROF MOLECULAR, CELLULAR & DEVELOP BIOL, UNIV COLO, 83- *Concurrent Pos:* NIH res grant, 75-81 & 81-; vis scientist, CSIRO Plant Indust, Canberra, Australia, 84-85. *Honors & Awards:* Wilson S Stone Mem Award Basic Biomed Res, M D Anderson Hosp, Univ Tex, 73. *Mem:* Am Soc Microbiol; Am Women Sci; Am Soc Virol; Intl Plant Molecular Biol. *Res:* Transcriptional control of Simian virus 40 gene expression; Simian virus 40 early proteins with possible roles in transforming cultured cells; plant molecular biology. *Mailing Add:* Dept MCD Biol Univ Colo Campus Box 347 Boulder CO 80309-0347

DANNENBERG, ARTHUR MILTON, JR, b Philadelphia, Pa, Oct 17, 23; m 48; c 3. PATHOLOGY, IMMUNOLOGY. *Educ:* Swarthmore Col, AB, 44; Harvard Univ, MD, 47; Univ Pa, MA, 51, PhD(microbiol), 52; Am Bd Microbiol, dipl. *Prof Exp:* Intern, Albert Einstein Med Ctr, 47-48; res resident, Children's Hosp of Philadelphia, Univ Pa, 48-49, asst prof exp path, Henry Phipps Inst, 56-64, asst prof microbiol, Sch Med & Grad Sch Arts & Sci, Univ Pa, 58-64; assoc prof radiol sci, 64-73, assoc prof path, Sch Med, 64-78, PROF ENVIRON HEALTH SCI, EPIDEMIOL, IMMUNOL, & INFECTIOUS DIS, SCH HYG & PUB HEALTH, JOHNS HOPKINS UNIV, 73-, PROF PATH, SCH MED, 78- *Concurrent Pos:* Nat Tuberc Asn fel exp path, Henry Phipps Inst, Univ Pa, 50-52; Nat Res Coun fel biochem, Univ Utah, 52-54; assoc ed, Am Rev Respiratory Dis, 79-84; lieutenant comdr, Naval Med Res Unit 1, USN; Univ Calif, Berkeley, 54-56. *Mem:* Am Asn Path; Am Asn Immunol; Histochem Soc; Soc Exp Biol & Med; Am Thoracic Soc; Am Soc Microbiol; hon mem Soc Leukocyte Biol. *Res:* Enzymes in pathogenesis of acute and chronic inflammation; tuberculosis; macrophages; allergic, infectious and environmental diseases of skin and lung. *Mailing Add:* Dept Environ Health Sci Johns Hopkins Univ Sch Hyg Baltimore MD 21210

DANNENBERG, E(LI) M, b Bridgeport, Conn, Oct 10, 17; m 44; c 3. CARBON & CARBON BLACK, RUBBER FILLERS. *Educ:* Mass Inst Technol, SB, 39, SM, 40; Univ Louis Pasteur, Strasbourg, France, DSc, 73. *Prof Exp:* Res assoc, Div Indust Coop, Mass Inst Technol, 40-43; res chemist, Labs, Am Cyanamid Co, Conn, 43-44; res chemist, Sprague Elec Co, Mass, 44-45; mgr rubber lab, Godfrey L Cabot, Inc, 45-50, assoc dir res, 50-57, dir res, Cabot Corp, 57-63, dir res & develop, 63-68, vpres & sci dir, Performance Chem Group, 68-74, dir, Chem Res Prog, 74-76, corp res fel, 74-76; sr res scientist, Ga Inst Technol, 77-78; CONSULT, 78- *Honors & Awards:* Rubber Div Award Tech Excellence, Am Chem Soc, 84; Bronze Medal, Societé Chimque de France, 90. *Mem:* Am Chem Soc; fel Plastics & Rubber Inst; fel Am Inst Chemists. *Res:* Surface chemistry of fine pigments; mechanism of rubber reinforcement; carbon black and carbon products; silica; silicates; clays; fillers; rubber plastics; research administration. *Mailing Add:* 2350 Harbour Oaks Dr Longboat Key FL 33548-4165

DANNENBERG, JOSEPH, b New York, NY. PHYSICAL ORGANIC & THEORETICAL CHEMISTRY. *Educ:* Columbia Col, AB, 62; Calif Inst Technol, PhD(chem), 67. *Prof Exp:* USPHS fel theoret chem, Centre du Mecanique Ondulatoire Appliquee, Paris, 66-67; res assoc phys org chem, Columbia Univ, 67-68; from asst prof to assoc prof, 68-77, PROF CHEM, HUNTER COL, 77- *Concurrent Pos:* Vis prof, Univ Paris, 74-75 & Univ Bordeaux, 83; vis distinguished scientist, Nat Res Coun, Ottawa, 85. *Mem:* Am Chem Soc; Am Phys Soc; The Chem Soc; Europ Acad Sci Arts & Letters; Sigma Xi. *Res:* Theoretical energy surfaces for organic reactions; structures of intermediates; mechanisms of free radical reactions; hydrogen bonding interactions; studies of organic solid state. *Mailing Add:* Dept Chem Hunter Col 695 Park Ave New York NY 10021

DANNENBERG, KONRAD K, b Weissenfels, Ger, Aug 5, 12; US citizen; m 44; c 1. ASTRONAUTICS, ENGINEERING. *Educ:* Hanover Tech Univ, MS, 38. *Prof Exp:* Asst combustion eng, Hanover Tech Univ, 38-39; engr, VDO, Univ Frankfurt, 39-40 & Rocket Test Sta, Peenemuende, 41-45; engr, US Army, Ft Bliss, Tex, 45-50, engr develop opers, Redstone Arsenal, Ala, 50-55, dir, Jupiter prog, Army Ballistic Missile Agency, 55-60, Marshall Space Flight Ctr, NASA, 60-70, dep dir missions & payload planning prog develop, 70-73; assoc prof, Univ Tenn Space Inst, Tullohoma, Tenn, 73-78; CONSULT, ALA SPACE & ROCKET CTR, 78- *Honors & Awards:* Durand lectr, Am Inst Aeronaut & Astronaut, 90. *Mem:* Am Inst Aeronaut & Astronaut; Hermann Oberth Soc; World Future Soc; L-5 Soc; Int Asn Educr World Peace. *Res:* Combustion engineering; rocket engine development; space vehicle design, test, check-out and launching; space station design and experiments. *Mailing Add:* 64 Revere Way Huntsville AL 35801-2846

DANNENBURG, WARREN NATHANIEL, b Tulsa, Okla, Jan 9, 26; m 49; c 3. BIOCHEMICAL PHARMACOLOGY. *Educ:* Va Polytech Inst, BS, 48; Agr & Mech Col, Tex, MS, 55, PhD(biochem, nutrit), 57. *Prof Exp:* Res microbiologist, S E Massengill Co, 48-53; res biochemist, R J Reynolds Tobacco Co, 57-60; res asst prof biochem, Bowman Gray Sch Med, 60-65; GROUP MANAGER MOLECULAR BIOCHEM, A H ROBINS CO, 65- *Concurrent Pos:* NIH career develop award, 64-65. *Mem:* Am Chem Soc; Am Inst Nutrit; NY Acad Sci; Soc Exp Biol & Med. *Res:* Therapeutic use and mechanism of action of drugs that affect lipid and carbohydrate metabolism in the intact animal as well as their investigation at the cellular and enzyme levels. *Mailing Add:* 7730 Brentford Dr Richmond VA 23225-2115

DANNER, DEAN JAY, b Milwaukee, Wis, Sept 26, 41; m 68; c 2. BIOCHEMISTRY, MEDICAL GENETICS. *Educ:* Lakeland Col, BS, 63; Univ NDak, MS, 65, PhD(biochem), 68. *Prof Exp:* Instr biochem, Med Units, Univ Tenn, Memphis, 67-70; asst prof, Northwestern State Univ, La, 70-73; PROF PEDIAT, MED SCH, EMORY UNIV, 73-, ASSOC PROF BIOCHEM, 78- *Concurrent Pos:* Am Cancer Soc fel, St Jude Children's Res Hosp, Memphis, Tenn, 67-70. *Honors & Awards:* Young Investr, Postgrad Med Asn, NAm. *Mem:* AAAS; Am Soc Biol Chemists; Am Chem Soc; Am Soc Human Genetics; Soc Inherited Metab Disorders. *Res:* Multienzyme complexes; mitochondria biogenesis; genetic basis for abnormalities of multienzyme complexes in humans. *Mailing Add:* Div Med Genetics Sch Med Emory Univ Atlanta GA 30322

DANNER, RONALD PAUL, b New Holland, Pa, Aug 29, 39; m 60; c 2. ADSORPTION, POLYMER SOLUTION BEHAVIOR. *Educ:* Lehigh Univ, BS, 61, MS, 63, PhD(chem eng), 66. *Prof Exp:* Sr chemist, Eastman Kodak Labs, 65-67; from asst prof to assoc prof, 67-78, PROF CHEM ENG, PA STATE UNIV, UNIV PARK, 78- *Concurrent Pos:* Fac fel, US Gen Accounting Off, 74-75. *Mem:* Am Inst Chem Engrs; Sigma Xi; Am Soc Eng Educ; Am Chem Soc. *Res:* Diffusion and phase equilibria of polymer solutions; correlation and prediction of physical and thermodynamic properties of chemicals; physical adsorption of gas and liquid mixtures. *Mailing Add:* 163 Fenske Lab Dept Chem Eng Pa State Univ University Park PA 16802-4498

DANNER, WILBERT ROOSEVELT, b Seattle, Wash, Feb 28, 24. GEOLOGY. *Educ:* Univ Wash, BSc, 46, MSc, 49, PhD(geol), 57. *Prof Exp:* Asst, Univ Wash, 46-50; instr, Everett Jr Col, 50 & Col of Wooster, 50-54; from instr to assoc prof, 54-67, asst dean fac, 74-79, PROF GEOL, UNIV BC, 67- *Concurrent Pos:* Consult, Permanente Cement Co, 48, Northwest Portland Cement Co, 49-57, Riverside Cement Co, 56 & BC Dept Mines, Geol Surv Can. *Mem:* Geol Soc Am; Soc Econ Paleontologists & Mineralogists; Mineral Soc Am; Am Asn Petrol Geologists; Mineral Asn Can. *Res:* General geology; stratigraphy and regional geology; limestone. *Mailing Add:* Dept Geol Univ BC 2075 Westbrook Pl Vancouver BC V6T 1W5 Can

D'ANNESSA, A(NTHONY) T(HOMAS), b Youngstown, Ohio, July 10, 33; m 57; c 2. METALLURGY. *Educ:* Ohio State Univ, BWE, 58, MSc, 60. *Prof Exp:* Res engr, Columbus Div, NAm Aviation, Inc, 55-59 & Eng Exp Sta, Ohio State Univ, 59-61; sr scientist, Lockheed-Palo Alto Lab Div, Lockheed Aircraft Corp, 61-65, res scientist, Lockheed-Ga Res Lab Div, 65-72; mem tech staff, Bell Tel Labs, 72-84, MEM TECH STAFF, AT&T TECHNOL, 84- *Concurrent Pos:* Consult, 64-; tech adv, consult & mem bd dirs, Weldwire, Inc, Ga, 70-73. *Honors & Awards:* Award, Am Welding Soc, 71. *Mem:* Am Soc Metals; Am Welding Soc. *Res:* Metallurgical engineering; metallurgy, mechanical behavior and solidification mechanics of weldments; development of advanced welding processes and nondestructive testing techniques; optical fiber process development. *Mailing Add:* 3743 Fox Hills Dr SE Marietta GA 30067

DANNHAUSER, WALTER, b Munich, Ger, June 2, 30; nat US; m 53; c 3. PHYSICAL CHEMISTRY. *Educ:* Rutgers Univ, BSc, 51; Brown Univ, PhD(chem), 54. *Prof Exp:* Res chemist, E I du Pont de Nemours & Co, Inc, 54-56; proj assoc chem, Univ Wis, 56-57; asst prof, Univ Buffalo, 57-62; ASSOC PROF CHEM, STATE UNIV NY BUFFALO, 62- *Mem:* AAAS; Am Chem Soc. *Res:* Polymer physical chemistry; dielectrics. *Mailing Add:* Dept Chem Acheson Hall State Univ NY Buffalo NY 14214

DANNIES, PRISCILLA SHAW, b Englewood, NJ, May 3, 45; m 67. ENDOCRINOLOGY, BIOCHEMISTRY. *Educ:* Smith Col, AB, 67; Brandeis Univ, PhD(biochem), 71. *Prof Exp:* Res fel pharmacol, Sch Dent Med & Med Sch, Harvard Univ, 71-74, instr, 74-76; asst prof, 76-80, ASSOC PROF PHARMACOL, SCH MED, YALE UNIV, 81- *Concurrent Pos:* Arthritis Found res fel, 71-74; res career develop award, 80-85; Endorine Study Sect, 85-89. *Mem:* Endocrinol Soc. *Res:* Control of synthesis and secretion of pituitary hormones by hypothalamic releasing factors and steroid hormones. *Mailing Add:* 299 Edwards St New Haven CT 06511

DANNLEY, RALPH LAWRENCE, b Chicago, Ill, June 25, 14; m 50; c 3. ORGANIC CHEMISTRY. *Educ:* Univ Denver, BS, 36; Univ Chicago, PhD(org chem), 43. *Prof Exp:* Res chemist, Nat Defense Res Comt, Univ Chicago, 41-42, instr chem, 42-43; dir oil res, Devoe & Raynolds, Inc, 43-45; from asst prof to prof, 45-80, EMER PROF CHEM, CASE WESTERN RESERVE UNIV, 81- *Mem:* Am Chem Soc. *Res:* Mechanism of polymerization; free radical chemistry; peroxides of elements other than carbon; heterocyclic phosphorus compounds; organic fluorine derivatives; aromatic arylsulfonoxylation. *Mailing Add:* Dept Chem Case Western Reserve Univ Cleveland OH 44106

DANOS, MICHAEL, b Riga, Latvia, Jan 10, 22; nat US; m 49; c 3. FIELD THEORY, NUCLEAR PHYSICS. *Educ:* Tech Univ, Ger, MS, 48; Univ Heidelberg, PhD(physics), 50. *Prof Exp:* Jr asst weak current inst, Tech Univ, Dresden, Ger, 44; asst theoret physics inst, Tech Univ, Hannover, 48-49; asst, Univ Heidelberg, 49-50, asst physics inst, 51; res assoc radiation lab, Columbia Univ, 52-54; PHYSICIST, NAT INST STANDARDS & TECHNOL, 54- *Concurrent Pos:* Guggenheim fel, 59; Sir Thomas Lyle res fel, 70; Alexander V Humboldt sr fel, 74; vis prof var Am & Europ univs; NBS fel, 83. *Honors & Awards:* Silver Medal, Dept of Commerce, 66, Gold Medal, 84. *Mem:* AAAS; Am Phys Soc; Fedn Am Scientists. *Res:* Theoretical nuclear and high energy physics; microwave physics; atomic and molecular physics. *Mailing Add:* Nat Inst Standards & Technol Gaithersburg MD 20899

DANSBY, DORIS, b New Albany, Miss, Apr 22, 18. HEMATOLOGY. *Educ:* Univ Miss, BA, 40; Univ Tenn, Cert Med Technol, 41. *Prof Exp:* Med technologist serol & bact, Baptist Hosp Lab, Tenn, 41-45; res assoc mycol res, Lederle Labs, Am Cyanamid Co, NY, 45-54; pharmacologist hematol, Mead Johnson Res Ctr, 54-63, sr scientist, 63-83; RETIRED. *Mem:* Am Soc Med Technol; NY Acad Sci. *Res:* Microbiology; antibiotics; mycology in relation to microorganisms producing antibiotics; hematologic aspects of the toxicology of potential therapeutic agents. *Mailing Add:* 1100 Erie Evansville IN 47715

DANSE, ILENE H RAISFELD, b New York, NY; m 60, 82; c 2. MEDICINE, ENVIRONMENTAL SCIENCES. *Educ:* Brooklyn Col, BS, 60; NY Univ, MD, 64; Am Bd Internal Med, dipl, 70; Am Bd Toxicol, dipl, 81. *Prof Exp:* Dir, Div Clin Pharmacol & Toxicol, Health Sci Ctr, Sch Med, State Univ NY, Stony Brook, 78-83; sr adv, Environ Health Protection, Chevron Environ Health Ctr, Inc, 82-84; PRIN, ENVIROMED HEALTH SERV, INC, 84-; ASSOC CLIN PROF, OCCUP & ENVIRON MED, UNIV CALIF, SAN FRANCISCO, 86- *Concurrent Pos:* Consult, Dept Energy; consult, Off Technol Assessment, US Congress. *Honors & Awards:* Pharmaceut Mfr Assoc Award. *Mem:* Am Acad Clin Toxicol; Am Col Toxicol; fel Am Col Clin Pharmacol; Am Soc Clin Pharmacol & Therapeut; fel Am Col Physicians; NY Acad Sci. *Res:* Development of programs dealing with actual or potential exposure to toxic substances; community health assessments for environmental pollution; human effects pollutant combinations. *Mailing Add:* 1010 B St San Rafael CA 94901

DANSER, JAMES W, b Long Branch, NJ, Oct 15, 21. GEOLOGY, EXPLORATION. *Educ:* Univ Mo, MA, 50. *Prof Exp:* Raw mat mgr, A P Green, 50-86; RETIRED. *Mem:* Fel Geol Soc Am; Am Inst Prof Geologists. *Mailing Add:* 827 N Calhoun Mexico MO 65265

DANSEREAU, PIERRE, b Montreal, Que, Can, Oct 5, 11; m 35. ECOLOGY, ENVIRONMENTAL SCIENCES. *Educ:* Univ Montreal, BA, 31; Univ Geneva, DSc(plant taxon), 39. *Hon Degrees:* LLD, Univ Sask, 59; DSc, Univ NB, 59, Univ Strasbourg, 70, Univ Sherbrooke & Sir George Williams Univ, 71, Guelph Univ & Univ Western Ont, 73, Mem Univ Nfld, 74, McGill Univ, 76 & Univ Ottawa, 77; DEnvSc, Univ Waterloo, 72, Royal Military Col, 90. *Prof Exp:* Botanist, Montreal Bot Garden, 39-40, asst dir tech serv, 40-42; dir, Prov Biogeog Serv, 43-50; instr bot, Univ Montreal, 40-45, asst prof, 45-50; from asst prof to assoc prof, Univ Mich, 50-55; chmn dept bot & dean fac sci, Univ Montreal, 55-61; asst dir bot, NY Bot Garden, 61-68, head dept ecol, 61-68; prof ecol & urban studies, Univ Montreal, 68-71; prof & dir sci, Ctr Ecol Res Montreal, 71-72; prof ecol, Ctr Environ Sci Res, 72-76, hon prof, 76-80, EMER PROF ECOL, UNIV QUEBEC, MONTREAL, 80- *Concurrent Pos:* Grants, Nat Res Coun, 40, 41, 44, 45-47, 48, 56-57, 59 & 60, Huyck Preserve, 44, Brazilian Res Coun & Nat Geog Coun, 45 & 46, Am Philos Soc, 48, NSF, 54-55 & 62-65; lectr, Macdonald Col, McGill Univ, 42-44; vis prof, Univ Brazil, 46, Univ Paris, 58, Univ NZ, 61, Univ PR, 63 & Univ Lisbon, 77, Univ McGill, 79, 84 & 86, Univ Regina, 80, Univ Calgary, 83 & Univ Concordia, 86; Off Can del, Pac Sci Cong, 49, del, 53 & 61; del, Int Bot Cong, 54, 1st vpres, 59; hon counr, Nat Coun Sci Res, Madrid, 59; hon mem, Acad Sci Toulouse, 60; adj prof biol & geog, Columbia Univ, 61-67; vchmn, Can Comn Int Biol Prog, 69-77; proj dir, Nat Res Coun Ecol Study New Montreal Int Airport Zone, 70-73; vchmn, Can Environ Adv Coun, 72-75; pres, First Int Film Festival Human Environ, 73; observr, UN Env Prog, Nairobi, Kenya, 74; adv environ, Kairouan, Tunisia, Can AID, 74; lectr, Econ Comn Latin Am, Mex, 77; foreign corresp mem, Arg Acad Environ Sci, 84 & Lisbon Acad Sci, 85. *Honors & Awards:* Pariseau Medal, French-Can Asn Advan Sci, 65; Companion Order of Can, Govt Can, 69; Massey Medal, Royal Can Geog Soc, 73; Molson Prize, Can Coun, 75 & Izaak Walton Killam Prize, 85; Nature 83 Prize, Ministere Environ Quebec, 83; Esdras-Minville Prize, Soc St-Jean Baptiste Montreal, 83; Marie-Victorin Prize, Sci, Prov Quebec, 83. *Mem:* Ecol Soc Am; World Soc Ekistics; Asn Am Geogrs; fel Royal Soc Can; Can Geog Soc; emer mem Fr-Can Asn Advan Sci (hon pres, 84). *Res:* Taxonomy and evolution of Cistus, Potentilla, Viola and Acer; comparative structure and dynamics of vegetation in tropical to arctic climates; interdisciplinary research on land use; urban ecology. *Mailing Add:* Univ Quebec Montreal CP 8888 Montreal PQ H3C 3P8 Can

DANTE, MARK F, organic chemistry; deceased, see previous edition for last biography

DANTI, AUGUST GABRIEL, b New Eagle, Pa, Jan 26, 23; m 50; c 1. PHARMACY, CHEMISTRY. *Educ:* Univ Pittsburgh, BS, 50, MS, 52; Ohio State Univ, PhD(pharm), 55. *Prof Exp:* Instr pharm, Univ Pittsburgh, 52-53; asst prof, Wayne State Univ, 56-59; assoc prof, 59-64, head dept allied health sci, 68-71, PROF SCH PHARM, NORTHEAST LA UNIV, 64- *Honors & Awards:* A H Robins Bowl Hygeia, 70; William P O'Brien Clin Pharm Award, 81. *Mem:* AAAS; Am Col Apothecaries; Am Asn Col Pharm; Acad Pharmaceut Sci; Am Pharmaceut Asn; Am Soc Consult Pharmacists; Am Inst Hist Pharm; Sigma Xi. *Res:* The release of medicaments from various pharmaceutical dosage forms. *Mailing Add:* 2211 Ann St Monroe LA 71201

DANTIN, ELVIN J, SR, b Golden Meadow, La, Jan 13, 27; m 47; c 5. CIVIL ENGINEERING, WATER RESOURCES. *Educ:* La State Univ, BS, 49, MS, 52; Stanford Univ, PhD(struct), 60. *Prof Exp:* Instr civil eng, La State Univ, Baton Rouge, 49-52, from asst prof to assoc prof, 52-62, prof & dir div eng res, 62-; PROF CIVIL ENG, FLA STATE UNIV, TALAHASSEE. *Concurrent Pos:* Consult, E E Evans, Consult Engr, La, 54-58, Humble Oil & Refining Co, Tex, 57 & La Dept Pub Works, 61; dir, La Water Resources Res Inst, 70- *Mem:* Am Concrete Inst; Sigma Xi. *Res:* Structural engineering; reinforced concrete; structures. *Mailing Add:* PO Box 2175 Fla State Univ Tallahassee FL 32316-2175

D'ANTONIO, PETER, b Brooklyn, NY, Nov 10, 41; m 66; c 1. STRUCTURAL CHEMISTRY. *Educ:* St John's Univ, BS, 63; Polytech Inst Brooklyn, PhD(phys chem), 67. *Prof Exp:* RES CHEMIST, US NAVAL RES LAB, 67- *Mem:* Am Chem Soc; Sigma Xi; Am Crystallog Asn. *Res:* Molecular structure determinations of substances in the vapor phase and structural analysis of semiconductor thin films using electron diffraction techniques. *Mailing Add:* 12003 Wimbleton 12003 Wimbleton St Upper Marlboro MD 20772

DANTZIG, ANNE H, b Columbus, Ohio, Nov 1, 49; m 76; c 1. BIOCHEMICAL PHARMACOLOGY. *Educ:* Kent State Univ, BS, 71; Johns Hopkins Univ, PhD(biochem), 76. *Prof Exp:* Postdoctoral fel human genetics, Yale Med Sch, 77-82; RES SCIENTIST BIOCHEM PHARMACOL, LILLY RES LABS, 82- *Mem:* Am Soc Biochem & Molecular Biol; Am Soc Cell Biol; Sigma Xi. *Res:* Intestinal absorption of drugs; membrane transport mechanisms; somatic cell genetics. *Mailing Add:* Lilly Res Lab Eli Lilly & Co Indianapolis IN 46285

DANTZIG, GEORGE BERNARD, b Portland, Ore, Nov 8, 14; m 36; c 3. COMPUTER SCIENCE. *Educ:* Univ Md, AB, 36; Univ Mich, MA, 37; Univ Calif, Berkeley, PhD(math), 46. *Hon Degrees:* DSc, Israel Inst Technol, 73; Dr, Univ Linkoping, Sweden, 75, Univ Md, 76, Yale Univ, 78; Cath Univ Louvain, 83, Columbia Univ, 83, Univ Zurich, 83, Carnegie-Mellon Univ, 86. *Prof Exp:* Jr statistician, US Bur Labor Statist, 37-39; statistician, Air Force, 41-45, chief mathematician, Air Force Hq Comptroller, 45-52; res mathematician, Rand Corp, 52-60; prof eng sci & chmn oper res ctr, Univ Calif, Berkeley, 60-66; PROF OPER RES & COMP SCI, STANFORD UNIV, 66- *Honors & Awards:* Nat Medal Sci, 75; Von Neumann Theory Prize, Opers Res Soc & Mgt Sci Soc, 75; Appl Math & Numerical Anal Prize, Nat Acad Sci, 76; Harvey Prize, 85; Coors Am Ingenuity Award, 89; Gibbs lectr, Am Math Soc, 90. *Mem:* Nat Acad Sci; Fel Am Acad Arts & Sci; assoc Am Math Soc; Oper Res Soc; Fel Inst Mgt Sci (pres, 66). *Res:* Existence of similar regions in theory of mathematical statistics; operations research; mathematical theory of optimization in large scale interrelated systems. *Mailing Add:* Dept Opers Res Stanford Univ Stanford CA 94305-4022

DANTZIG, JONATHAN A, b Baltimore, Md, Aug 14, 51; m 76; c 1. MECHANICAL & MATERIALS SCIENCE ENGINEERING. *Educ:* Johns Hopkins Univ, BES, 72, MSE, 75, PhD(mech & mat sci), 77. *Prof Exp:* Res scientist process metall, Olin Metals Res Lab, Olin Corp, 77-79, sr res scientist, 79-82; asst prof, 82-87, ASSOC PROF MECH ENG, UNIV ILL, URBANA- CHAMPAIGN, 87- *Concurrent Pos:* Chmn, Solidification Comt, Metall Soc, 87-88; vis res scientist, Nat Ctr Supercomput Applications, 90-91. *Mem:* Am Soc Metals; Metall Soc; Am Foundrymen's Soc. *Res:* Materials processing; mathematical modeling of solidification processing; inverse problems related to processing; numerical methods for analysis of fluid flow, phase changes and stress analysis. *Mailing Add:* 1206 W Green St Urbana IL 61801

DANTZKER, DAVID ROY, b New York, NY, Mar 19, 43. PULMONARY MEDICINE. *Educ:* State Univ NY, Buffalo, MD, 67. *Prof Exp:* Assoc prof, Univ Mich, Ann Arbor, 75-83; PROF & DIR PULMONARY MED, UNIV TEX HEALTH SCI CTR, HOUSTON, 83- *Mem:* AAAS; Am Fedn Clin Res; Am Physiol Soc; Am Thoracic Soc. *Mailing Add:* Univ Tex Health Sci Ctr 6431 Fannin Suite 1274 Houston TX 77030

DANTZLER, HERMAN LEE, JR, b Walterboro, SC, Nov 19, 45; m 69; c 2. PHYSICAL OCEANOGRAPHY. *Educ:* US Naval Acad, BS, 68; Johns Hopkins Univ, MA, 73, PhD(phys oceanog), 75. *Prof Exp:* Fac oceanog & meteorol, US Naval Acad, 74-78; prog mgr upper ocean variability, US Naval Oceanog Off, 78-80; ocean/atmospheric sci res policy & prog develop & spec asst to chief of naval res on ocean sci, Off Naval Res, 80-82, oceanog & ocean acousts progs officer, Off Oceanogr Navy, Washington, DC, 82-85; staff oceanogr to comdr, Submarine Group Six, Naval Base, Charleston, SC, 85-88; SR PROF STAFF, APPL PHYSICS LAB, JOHNS HOPKINS UNIV, 88- *Concurrent Pos:* Prin investr, Upper Ocean Variability Study, Off Naval Res, 75-78; guest investr, Woods Hole Oceanog Inst, 75; mem, Density Comt, US Polymode Orgn, Mass Inst Technol, 75-77. *Mem:* Am Meteorol Soc; Am Geophys Union. *Res:* Upper ocean variability; ocean acoustics; applied climatology. *Mailing Add:* Appl Physics Lab Johns Hopkins Univ John Hopkins Rd Laurel MD 20723

DANTZLER, WILLIAM HOYT, b Mt Holly, NJ, Aug 25, 35; m 59; c 2. PHYSIOLOGY. *Educ:* Princeton Univ, AB, 57; Columbia Univ, MD, 61; Duke Univ, PhD(zool), 64. *Prof Exp:* Intern med, Univ Wash Hosp, 61-62; asst prof pharmacol, Col Physicians & Surgeons, Columbia Univ, 64-68; assoc prof, 68-74, PROF PHYSIOL, COL MED, UNIV ARIZ, 74- *Concurrent Pos:* Secy, renal sect, Am Physiol Soc, 78-79, chmn, 79-81, chmn, water & electrolyte homeostasis sect, 87-88, treas, comp physiol sect, 87-89; assoc ed, Am J Physiol: Regulatory, Integrative, & Comp Physiol, 81-90, ed, 90- *Honors & Awards:* Alexander von Humboldt Award, 85. *Mem:* Fel AAAS; Am Soc Nephrology; Int Soc Nephrology; Am Soc Zool; Am Physiol Soc. *Res:* Comparative renal physiology; excretion of end products of nitrogen metabolism; control of excretion of ions and water in vertebrates; renal tubular transport of organic anions and cations. *Mailing Add:* Dept Physiol Col Med Univ Ariz Tucson AZ 85724

DANYLUK, STEVEN, b Dec 25, 45; m 64; c 2. MATERIALS SCIENCE. *Educ:* Univ Del, BS, 69; Cornell Univ, PhD(mat sci & eng), 74. *Prof Exp:* Mem staff, Tex Instruments, Inc, 73-74; asst metallurgist, Mat Sci Div, Argonne Nat Lab, 74-79; assoc prof, 79-88, PROF MAT SCI & ENG, UNIV ILL, CHICAGO, 88- *Concurrent Pos:* Sr engr, Jet Propulsion Lab, Calif Inst Technol, 80; resident assoc, Argonne Nat Lab, 80-; consult, Inst Gas Technol, 80-; guest sci, Nat Bur Standards, Ceramics Div, 86-87; Mobil solar Energy Corp, 82; Monsanto Electronic Mat Co, 85-86. *Mem:* Am Soc Metals; Nat Soc Corrosion Engrs; Am Inst Metal Engrs; Am Ceramic Soc; Mat Res Soc. *Res:* Effects of microstructure on mechanical properties of stainless steel; tribology of non-metallic materials. *Mailing Add:* Dept Civil Eng Mech & Metall Univ Ill PO Box 4348 Chicago IL 60680

DANZBERGER, ALEXANDER HARRIS, b New York, NY, Mar 23, 32; m 54, 77; c 7. CHEMICAL ENGINEERING. *Educ:* Mass Inst Technol, BS, 53. *Prof Exp:* Staff engr, Arthur D Little, Inc, 53-59; eng mgr, Union Carbide Corp, 60-70; mgr, Marcon Consults, 71-75 & Dames & Moore, 82-83; vpres, Hydrotechnic Corp, 76-82; prin, Harding Lawson Assocs, 89-90; PRIN, DANZBERGER & ASSOCS, 83- *Concurrent Pos:* Dipl, Am Acad Environ Engrs, 79; chmn, Environ Div, Am Inst Chem Engrs, 85; spec master, Western Dist Okla, US Dist Ct, 86- *Mem:* Fel Am Inst Chem Engrs; Am Soc Mech Engrs; Water Pollution Control Fedn; AAAS; Am Inst Environ Engrs. *Res:* Hazardous waste reduction, hazardous waste treatment and remediation of uncontrolled hazardous waste sites using chemical process engineering and related disciplines. *Mailing Add:* 13245 Willow Lane Golden CO 80401

DANZER, LAURENCE ALFRED, b Chicago, Ill, July 27, 37; m 65; c 1. CLINICAL CHEMISTRY, PHYSICAL CHEMISTRY. *Educ:* Valparaiso Univ, BA, 59; Univ Ky, MS, 63, PhD(chem), 69. *Prof Exp:* Res chemist, Pulmonary Div, Dept Med, Univ Ky, 62-67, chief chemist, clin res ctr, 67-69; asst prof path, Univ Kans Med Ctr, 69-77; clin chemist, 77-85, ADMIN DIR CLIN LABS, MERCY HOSP, DES MOINES, 86- *Concurrent Pos:* Consult, Vet Admin Hosp, Kans City, Mo, 74- *Mem:* Am Chem Soc; Am Asn Clin Chemists; Clin Lab Mgt Asn. *Res:* Protein chemistry; structure in non aqueous solvents; tissue enzyme levels; radioimmunoassay of hormones. *Mailing Add:* Dept Path Mercy Hosp Sixth & Univ Des Moines IA 50314

DANZIG, MORRIS JUDA, b Staten Island, NY, July 31, 25; m 55; c 2. ORGANIC CHEMISTRY, INORGANIC CHEMISTRY. *Educ:* Univ Miami, BS, 49, MS, 51; Tulane Univ, PhD(org chem), 53. *Prof Exp:* Du Pont fel org chem, Univ Minn, 53-54; supvry chemist, North Regional Res Lab, USDA, 54-56; from sr org chemist to head org res, Lord Mfg Co, 56-60; from sr org chemist to sect leader, Am Viscose Corp, 60-64; mgr adv develop chem & plastics, Gen Tire & Rubber Co, 64-68; dir, Corn Prod Corp, 68-74, vpres res & develop, 74-86, corp dir, Sci Develop, 87-88; consult, 88-90; RETIRED, 90- *Concurrent Pos:* Mem Comt Alternatives to Chloroflorocarbons as Solvents, NRC Study Group, 81-; consult, Mich Biotechnol Inst, 88-, Jungbunzlauer, 88-, Phyton Technol, 89-, W C Richards Co, 89- *Mem:* Am Chem Soc; Sigma Xi. *Res:* Industrial product and process research and development. *Mailing Add:* 115 Arrow Wood Dr Northbrook IL 60062

DANZIGER, LAWRENCE, b New York, NY, July 17, 32; m 60; c 2. STATISTICS. *Educ:* City Col New York, BA, 53, MA, 54; NY Univ, PhD(indust eng), 71. *Prof Exp:* Statistician, IBM Corp, 56-57, assoc statistician, 57-58, staff statistician, 58-62, mgr statist anal, 62-68, appl statist res, 68-71, sr statistician, 71-84, mgr statist serv, 84-88, SR STATISTICIAN, IBM CORP, 88- *Concurrent Pos:* Adj prof, Union Col, NY, 67- *Mem:* Am Statist Asn. *Res:* Application of statistical theory to computer components analysis and modelling; industrial reliability and quality control. *Mailing Add:* Three Spur Way Poughkeepsie NY 12603

DANZO, BENJAMIN JOSEPH, b East Liverpool, Ohio, Nov 24, 41. REPRODUCTIVE PHYSIOLOGY. *Educ:* Col Steubenville, BA, 65; Univ Ark, MS, 68; Univ Mich, PhD(zool & reproductive endocrinol), 71. *Prof Exp:* Res assoc, dept obstet & gynec, 71-73, asst prof, 73-79, res asst prof, dept biochem, 79-81, RES ASSOC PROF, SCH MED, VANDERBILT UNIV, 81-, PROF OBSTET & GYNEC, 87- *Concurrent Pos:* Ad Hoc rev, NSF, 78-, NIH, 84, 85, 86, & 87; asst ed, J Andrology, 83- *Mem:* Endocrine Soc; Soc Study of Reproduction; Am Soc Biochem & Molecular Biol; Am Soc Andrology. *Res:* Mechanisms of steroid hormone action; epididymal function. *Mailing Add:* Dept Obstet & Gynec Vanderbilt Univ Sch Med Nashville TN 37232

DAO, FU TAK, b Shanghai, China, Nov 22, 43; US citizen; m 72. PARTICLE PHYSICS. *Educ:* Mass Inst Technol, BA, 66, MS, 67, PhD(particle physics), 70. *Prof Exp:* Fel high energy physics res, Mass Inst Technol, 70-71; res assoc, Fermi Nat Accelerator Lab, 71-74; res assoc & instr, Tufts Univ, 74-76, asst prof physics, 76-79; MEM TECH STAFF, BELL LABS, MURRAY HILL, NJ, 79- *Mem:* Am Phys Soc; Sigma Xi. *Res:* Research in nature of interaction among elementary particles at high energy, and various experimental techniques used in particle physics research. *Mailing Add:* 245 Windmill Ct Bridgewater NJ 08807

DAO, THOMAS LING YUAN, b Soochow, China, Apr 27, 21; nat US; m 54; c 4. MEDICINE. *Educ:* Soochow Univ, BS, 40; St Johns Univ, China, MS, 42, MD, 45. *Prof Exp:* From instr to asst prof surg, Sch Med, Univ Chicago, 51-57; asst prof surg, RES PROF PHYSIOL & ASSOC PROF SURG, SCH MED, STATE UNIV NY BUFFALO, 66-, CHIEF DEPT BREAST SURG, ROSWELL PARK MEM INST, 57-; PROF BIOL, NIAGARA UNIV, 67- *Concurrent Pos:* Sr fel surg, Wash Univ, 49-51. *Mem:* AAAS; Am Col Surg; Endocrine Soc; Am Asn Cancer Res; NY Acad Sci. *Res:* Physiology; endocrine physiology; endocrine aspect of neoplastic disease; chemical carcinogenesis; experimental pathology. *Mailing Add:* 40 Sturbridge Lane Williamsville NY 14221

DAOUD, ASSAAD S, b Zebdani, Syria, Oct 21, 23; US citizen; m 57; c 3. PATHOLOGY. *Educ:* Univ Paris, MD, 51. *Prof Exp:* Intern, St Memmie Hosp, Chalons sur Marne, France, 50-51; intern path, Springfield Hosp, Mass, 52, resident, 52-54; instr path & bact, 54-56, from asst prof to assoc prof path, 56-67, PROF PATH, ALBANY MED COL, 67-; CHIEF LAB SERV, VET ADMIN HOSP, 61- *Concurrent Pos:* Mem coun arteriosclerosis & coun epidemiol, Am Heart Asn. *Mem:* AMA; Int Acad Path. *Res:* Arteriosclerosis. *Mailing Add:* Lab Serv Vet Admin Hosp Albany NY 12208

DAOUD, GEORGES, b Lattakiat, Syria, Mar 27, 27; US citizen; m 54; c 3. CARDIOLOGY. *Educ:* St Joseph Univ, Beirut, MD, 57; Am Bd Internal Med, dipl, 68; Am Bd Cardiovasc Dis, 72. *Prof Exp:* Asst med, 61-67, clin assoc prof med, 67-75, CLIN PROF MED, UNIV CINCINNATI, 75-, CLIN ASSOC PROF PEDIAT, 67-; INSTR & CLINICIAN, DEPT INTERNAL MED & DIR CARDIAC LAB, GOOD SAMARITAN HOSP, 66- *Concurrent Pos:* Fel coun clin cardiol & coun thrombosis, Am Heart Asn.

Mem: Am Fedn Clin Res; fel Am Col Chest Physicians; Am Heart Asn; fel Am Col Cardiol; Int Cardiovasc Soc; Sigma Xi. *Res:* Cardiac hemodynamics; phonocardiography; clinical cardiology. *Mailing Add:* 1030 Grandin Ridge Cincinnati OH 45208

D'AOUST, BRIAN GILBERT, b Powell River, BC, Mar 7, 38; m 59; c 3. CELL & HYPERBARIC PHYSIOLOGY. *Educ:* Queen's Univ, Ont, BSc, 61; Univ Calif, San Diego, PhD(marine biol), 67. *Prof Exp:* Nat Res Coun-Nat Acad Sci res fel physiol, Naval Med Res Inst, Nat Naval Med Ctr, Md, 67-69; asst res physiologist, Univ Calif Naval Biol Lab, Oakland, 70; investr, Virginia Mason Res Ctr, 70-73, sr investr hyperbaric res & diving physiol, 73-78, dir, hyperbaric physiol, 78-83; PRES/OWNER, COMMON SENSING, INC, 80- *Concurrent Pos:* Scientist, Seatransit Exped & Bering Sea Exped, 68; NIH res career develop award, 72-77; asst prof, Univ Wash, 73-81; consult, Corps Engrs, Bur Reclamation, Water Quality. *Mem:* AAAS; Am Soc Limnol & Oceanog; Undersea Med Soc; Am Physiol Soc; Am Fisheries Soc. *Res:* Catabolic processes in oxygen-adapted tissue and physiological studies of gas secretion in teleost swim bladder; natural, evolutionary and dynamic physiological adaptations to high gas and hydrostatic pressure; physiological adjustments to diving; the etiology of decompression sickness; use of isobaric techniques to evaluate critical conditions for decompression sickness. *Mailing Add:* Common Sensing Inc PO Box 373 Clark Fork ID 83811-9998

DAOUST, DONALD ROGER, b Worcester, Mass, Aug 13, 35; m 59; c 3. MICROBIOLOGY. *Educ:* Univ Conn, BA, 57; Univ Mass, MS, 59, PhD(phage), 62. *Prof Exp:* Sr res microbiologist, Merck Sharp & Dohme Res Labs, 62-70, res fel, 70-72, mgr biol qual control, 72-75; dir qual control, Armour Pharmaceut Co, 75-76; vpres, Qual Assurance & Regulatory Compliance, Armour Pharmaceut Co, Phoenix, 76-78; VPRES, QUAL CONTROL, CARTER-WALLACE, INC, 78- *Concurrent Pos:* Vchmn, Qual Control Sect, Pharmaceut Mfg Asn, 88-90, chmn, 90- *Mem:* AAAS; Am Soc Microbiol; Parenteral Drug Asn; Proprietary Asn; Pharmaceut Mfg Asn (Secy, 86-88). *Res:* Microbial production of natural substances. *Mailing Add:* Carter-Wallace Inc Cranbury NJ 08512

DAOUST, HUBERT, b Montreal, Que, Apr 13, 28; m 52; c 4. PHYSICAL CHEMISTRY. *Educ:* Univ Montreal, BSc, 50, MSc, 51, PhD(chem), 54. *Prof Exp:* Res assoc, Baker Lab, Cornell Univ, 53-54; from asst prof to assoc prof, 54-67, PROF CHEM, UNIV MONTREAL, 67- *Mem:* Am Chem Soc; Chem Inst Can. *Res:* Thermodynamics of high polymer solutions including polyelectrolytes and polypeptides; microcalorimetry; osmometry; viscosimetry; ultracentrifugation. *Mailing Add:* Dept Chem Univ Montreal Montreal PQ H3C 3J7 Can

DAOUST, RICHARD ALAN, b Fitchburg, Mass, Jan 26, 48; m 71; c 2. INSECT PATHOLOGY, MICROBIOLOGY. *Educ:* Univ Mass, BS, 70, MS, 74, PhD(environ microbiol), 78. *Prof Exp:* Res entomologist insect path, Ministry Agr, Botswana, Peace Corps Proj, 70-72; postdoctoral fel insect path, Cornell Univ, 78-81, res assoc, Boyce Thompson Inst Plant Res, 81-85; MGR FIELD EVAL, ECOGEN INC, 85- *Concurrent Pos:* Res assoc, Embrapa (CNPAF), Goiania, Goias, Brazil, 81-85. *Mem:* Entomol Soc Am; Soc Invertebrate Path; Sigma Xi; Soc Entom Brazil. *Res:* Basic and applied research in insect pathology and medical entomology; mode of action and efficacy of microbial agents in the control of agricultural and medical pests; control of mosquitoes with entomopathogenic fungi; microbial control of principal insect pests of cowpeas; field evaluation of Bacillus thuringiensis. *Mailing Add:* Ecogen Inc 2005 Cabot Blvd W Langhorne PA 19047-1810

DAOUST, ROGER, b Valleyfield, Que, Oct 13, 24; m 50; c 2. HISTOPATHOLOGY, CANCER. *Educ:* Univ Montreal, BSc, 47, MSc, 50; McGill Univ, PhD(anat), 53. *Prof Exp:* Res assoc, Montreal Cancer Inst, Notre-Dame Hosp, 50-75, res assoc prof med, Univ, 60-67, dir res labs, 67-74, PROF ANAT, UNIV MONTREAL, 67-, GEN DIR, MONTREAL CANCER INST, 74- *Concurrent Pos:* Damon Runyon Mem Fund Cancer Res fel, 52-55; Brit Empire Cancer Campaign exchange fel, London, 55-56; Nat Cancer Inst Can fel, Copenhagen, 56-57. *Mem:* AAAS; Histochem Soc; Am Asn Cancer Res; NY Acad Sci. *Res:* Cytology; histochemistry; cell cycle hepatocarcinogenesis. *Mailing Add:* Off Dir Montreal Cancer Inst Notre Dame Hosp 1560 Sherbrooke E Montreal PQ H2L 4M1 Can

DAPKUS, DAVID CONRAD, b Oct 23, 44; US citizen; m 65. GENETICS. *Educ:* Univ Minn, BS, 68, MS, 69, PhD(zool), 75. *Prof Exp:* Assoc, 75-79, ASST PROF BIOL, WINONA STATE UNIV, 79- *Mailing Add:* Dept Biol Winona State Univ Winona MN 55987

DAPPLES, EDWARD CHARLES, b Chicago, Ill, Dec 13, 06; m 31; c 2. GEOLOGY. *Educ:* Northwestern Univ, BS, 28, MS, 34; Harvard Univ, MA, 35; Univ Wis, PhD(geol), 38. *Prof Exp:* Geologist, Zeigler Coal Co, 28; geologist, Truax-Traer Coal Co, 28-32, mine supt, 31-32; from instr to prof, 36-74, EMER PROF GEOL, NORTHWESTERN UNIV, EVANSTON, 74- *Concurrent Pos:* Asst geologist, State Geol Surv, Ill, 39; consult geologist, Sinclair Oil Co, 45-50 & Pure Oil Co, 51; sr vis scientist, Univ Lausanne, 60-61; prof, Univ Geneva, 70. *Mem:* AAAS; fel Geol Soc Am; fel Soc Econ Geologists; Soc Econ Paleontologists & Mineralogists (pres, 70); hon mem Am Inst Mining, Metall & Petrol Engrs. *Res:* Petrography and deposition of sedimentary rocks. *Mailing Add:* 13035 98th Dr Sun City AZ 85374

D'APPOLONIA, BERT LUIGI, b Capreol, Ont, Nov 6, 39. CEREAL CHEMISTRY. *Educ:* Laurentian Univ, BA, 62; NDak State Univ, MS, 66, PhD(cereal chem), 68. *Prof Exp:* From asst prof to prof cereal chem, 68-85, CHMN, NDAK STATE UNIV, 85- *Mem:* Am Asn Cereal Chemists (secy, 81 & pres, 85); Inst Food Technologists. *Res:* Quality evaluation of hard red spring wheat; investigation of carbohydrates, including starch, pentosans and simple sugars of wheat products; carbohydrates of cereals; starches, pentosans, polysaccharides; bread baking research. *Mailing Add:* 125 Fourth Ave W West Fargo ND 58078

D'APPOLONIA, ELIO, b Provence, Alta, Apr 14, 18. CIVIL ENGINEERING. *Educ:* Univ Alta, BS, 42, MS, 46; Univ Ill, PhD(civil eng), 48. *Hon Degrees:* DrEng, Carnegie-Mellon Univ, 83; DrEng, Univ Genoa, 88. *Prof Exp:* Founder, D'Appolonia Consult Eng Inc, 65-84; consult engr, STS Consults, 84-88, vchmn, 88; CONSULT, 88- *Honors & Awards:* Terzaghi Lectr, 88 & Pickel Award, Am Soc Civil Eng, 91. *Mem:* Nat Acad Eng; Am Soc Civil Engrs; Eng Inst Can; Int Soc Soil Mech; Nat Soc Prof Eng; Am Soc Testing & Mat. *Mailing Add:* 1177 McCulley Dr Pittsburgh PA 15235

DAPSON, RICHARD W, b Flushing, NY, Sept 5, 41; m 62; c 2. HISTOTECHNOLOGY. *Educ:* Cornell Univ, BS, 63, PhD(vert zool), 66. *Prof Exp:* From asst prof to prof biol, Univ Mich-Flint, 66-82, chmn dept, 74-77, David M French distinguished prof, 81-82; mgr res, Richard-Allan Med Industs, 82-84; PRES & DIR, MARKET RES, ANATECH LTD, 84- *Concurrent Pos:* Fac res assoc, Savannah River Ecol Lab, Univ Ga, 72-73; adj prof, Wayne State Univ, 81-82. *Mem:* Nat Soc Histotechnol. *Res:* Age determination; mechanisms of staining with ionic dyes; rate of ageing in wild vertabrate populations; biostastics; histological fixation and tissue processing. *Mailing Add:* 1020 Harts Lake Battle Creek MI 49015

DAR, MOHAMMAD SAEED, b Lahore, Pakistan, Dec 10, 37; US citizen; m 69; c 2. NEUROPHARMACOLOGY, BIOCHEMICAL PHARMACOLOGY. *Educ:* Gordon Col, Rawalpindi, Pakistan, BS, 57, BPharm, 61; Univ Med Sci, Bangkok, Thailand, MS, 66; Va Commonwealth Univ, Richmond, PhD(pharmacol), 70. *Prof Exp:* From asst prof to assoc prof pharmacol, Sch Med, Pahlavi Univ, Shiraz, Iran, 72-79, chmn dept, 77-79; from asst prof to assoc prof, 81-90, PROF PHARMACOL, SCH MED, E CAROLINA UNIV, 90- *Concurrent Pos:* Vis res assoc, Sch Med, Rockefeller Found, Mahidol Univ, Bangkok, 70-72; vis res asst prof, Dept Pharmacol, Sch Med, E Carolina Univ, 79-81. *Mem:* Am Soc Pharmacol & Exp Therapeut; Res Soc Alcoholism; Soc Neurosci; Int Soc Biomed Res Alcoholism; Sigma Xi; Soc Exp Biol & Med. *Res:* Investigation of the neurochemical basis of alcohol-induced motor deficits in rodents; understanding the nature and extent of modulation of brain adenosine system of the alcohol-induced motor deficits; study of the short- and long-term consequences of alcohol consumption on the motor behavioral interactions between alcohol and drugs of abuse such as nicotine and caffeine. *Mailing Add:* 115 Heritage St Greenville NC 27858

DARAVINGAS, GEORGE VASILIOS, b Thessaloniki, Greece, Apr 22, 34; m 68. FOOD SCIENCE, BIOCHEMISTRY. *Educ:* Univ Thessaloniki, 58; Ore State Univ, MS, 63, PhD(food sci), 65. *Prof Exp:* Qual control chemist, Agr Corp Polygyros, Greece, 57-58; chemist, Geigy, Greece, 60-61; res asst, Ore Agr Exp Sta, 61-65; proj leader food res & develop, Hunt Wesson Foods, 65-68; mgr food sci & qual assurance dept, Glidden Durkee, Div of SCM Corp, 68-73; dir res & develop, 73-76, GROUP DIR, GEN MILLS INC, 76- *Concurrent Pos:* Instr, Euclides Univ Prep Inst, Greece, 60-61; lectr, Univ Calif, Riverside & Irvine, 67-68. *Mem:* Am Chem Soc; Inst Food Technol; Am Asn Cereal Chemists. *Res:* Research and development effort including basic food research, exploratory food research, product development, engineering and process development; quality assurance in food processing plants. *Mailing Add:* JFB Tech Ctr, Gen Mills Inc 9000 Plymouth Ave N Minneapolis MN 55427-3879

DARBY, DAVID G, b Oak Park, Ill, Sept 10, 32; m 57; c 2. INVERTEBRATE PALEONTOLOGY. *Educ:* Univ Mich, BS, 59, MS, 61, PhD(paleont), 64. *Prof Exp:* Scientist, US Antarctic Res Prog, 60-61; NSF assoc investr, 62-64; geologist, Mobil Oil Corp, Peru, 64-68; grad sch res grant, 69-71, PROF GEOL, UNIV MINN, DULUTH, 68- *Mem:* Fel AAAS; Geol Soc Am; Soc Econ Paleontologists & Mineralogists, Paleont Soc. *Res:* Precambrian and Paleozoic fossils; Lake Superior sedimentation. *Mailing Add:* Dept Geol Univ Minn Duluth MN 55812

DARBY, DENNIS ARNOLD, b Pittsburgh, Pa, Oct 31, 44; m 72; c 2. SEDIMENTARY PETROLOGY. *Educ:* Univ Pittsburgh, BS, 66, MS, 68; Univ Wis-Madison, PhD(geol), 71. *Prof Exp:* Asst prof geol, Hunter Col, 71-74; asst prof geol sci, adj asst prof oceanog & asst prof geol, Old Dominion Univ, 74-78, chmn dept, 78-81, assoc prof geol sci, oceanog & adj assoc prof, 78-88, PROF GEOL SCI, OLD DOMINION UNIV, 88- *Concurrent Pos:* Lectr geol, NY Univ, 73. *Mem:* Sigma Xi; Soc Econ Paleontologists & Mineralogists; Am Geophys Union; fel Geol Soc Am; Nat Asn Geol Teachers. *Res:* Sand dispersal, provenance; coastal plain sedimentation and stratigraphy; arctic geology and paleoclimatology; environmental impact of dredging and the mobilization of trace metals in estuaries. *Mailing Add:* Dept Geol Sci Old Dominion Univ Norfolk VA 23529-0496

DARBY, ELEANOR MURIEL KAPP, b Easton, Pa, Feb 4, 05; m 28. MEDICAL RESEARCH ADMINISTRATION. *Educ:* Barnard Col, AB, 25; Columbia Univ, PhD(biochem), 42; Univ Pa, MSc, 27. *Prof Exp:* Asst biol, Washington Sq Col, NY Univ, 26-28; asst entomologist, Entom Lab, USDA, 28-31; technician, Dept Med, Col Physicians & Surgeons, Columbia Univ, 32-41, asst dermat, 41-43, instr chem, Dept Nursing & res assoc, Dept Orthop Surg, 46-48; exec secy cardiovasc study sect, Div Res Grants, NIH, 48-58, head confs & publ sect, Grants & Training Br, Nat Heart Inst, 58-62, mem spec projs br, Extramural Progs, 62-69, exec secy therapeut eval comt, Clin Appln Prog, Nat Heart & Lung Inst, 69-73, grants adminr, 69-75; RETIRED. *Concurrent Pos:* Asst, Sch Nursing, NY Hosp, 32-33; fel coun epidemiol, Am Heart Asn. *Mem:* Am Heart Asn; Int Soc Cardiol. *Res:* Salt balance in vertebrate blood; insect physiology, especially fruit flies; blood sedimentation; urinary porphyrins; salicylate metabolism; collagen behavior; spectroscopic analysis; immunochemistry; cardiovascular and cerebrovascular research; cooperative projects; research administration. *Mailing Add:* 309 E Watersville Rd Mt Airy MD 21771

DARBY, JOHN FEASTER, b Chester, SC, Oct 22, 16; m 41; c 2. PLANT PATHOLOGY. *Educ:* Univ Fla, BSA, 48, MSA, 49; Univ Wis, PhD(plant path), 51. *Prof Exp:* Asst plant pathologist, Indian River Field Lab, Univ Fla, 51-54, assoc plant pathologist, 54-64, plant pathologist, 64-66, prof plant path & dir, Agr & Educ Ctr, 66-76, ctr dir, Cent Fla Res & Educ Ctr, Inst Food & Agr Sci, 76-86; RETIRED. *Mem:* Am Phytopath Soc. *Res:* Diseases of vegetable crops. *Mailing Add:* 1324 E 24th St Sanford FL 32771

DARBY, JOSEPH B(RANCH), JR, b Petersburg, Va, Dec 12, 25; m 51; c 4. MATERIALS SCIENCE, NUCLEAR MATERIALS. *Educ:* Col William & Mary, BS, 48; Va Polytech Inst & State Univ, BS, 51; Univ Ill, MS, 55, PhD(metall), 58. *Prof Exp:* Chemist, Allied Dye & Chem Corp, 48-49; metallurgist, Nat Carbon Div, Union Carbide & Carbon Corp, 51-53; phys metallurgist, Argonne Nat Lab, 58-73, group leader, mat sci div, 66-73, asst div dir, 73-74, assoc dir, fusion power prog, 74-86; PROG MGR, DIV MAT SCI, OFF BASIC ENERGY SCI, US DEPT ENERGY, WASHINGTON, DC, 86- *Concurrent Pos:* Union Carbide fel, 53-55; sci res coun sr fel, Dept Phys Metall & Sci Mat, Univ Birmingham, Eng, 70-71; co-ed, J Nuclear Mat, 74-85; adv bd, Mat Lett, 79-, chmn of eds, 85- *Mem:* Fel Am Soc Metals; Am Inst Mining, Metall & Petrol Engrs; AAAS; Sigma Xi. *Res:* Electronic structure of metals and alloys; alloy chemistry; thermodynamic properties of alloys; defects in solids; nuclear materials, glasses. *Mailing Add:* Div Mat Sci US Dept Energy Off Basic Energy Sci Washington DC 20585

DARBY, JOSEPH RAYMOND, b Boonville, Mo, Aug 6, 11; m 46; c 6. ORGANIC CHEMISTRY, POLYMER SCIENCE. *Educ:* St Louis Univ, BS, 33. *Prof Exp:* Control chemist, Monsanto Co, 33-35, res chemist, 35-46, res group leader, 46-63, res mgr, 63-76; PRES, DARBY CONSULT INC, 77- *Mem:* Fel Soc Plastics Engrs; Am Chem Soc; hon mem Soc Testing Mat; Res Soc Am. *Res:* Plasticization, stabilization and modification of all polymers, with particular emphasis on polyvinyl chloride. *Mailing Add:* 335 Papin Webster Grove MO 63119

DARBY, NICHOLAS, b Bristol, Eng, Dec 19, 46; Can citizen; m 74. SYNTHETIC ORGANIC CHEMISTRY. *Educ:* Queen's Univ, Ont, BSc, 68; Univ Alta, PhD(chem), 72. *Prof Exp:* Res chemist prof, Raylo Chem Ltd, 76-79; res leader, Dow Chem Can, Inc, 79-83, res mgr, construct mats & venture develop, mgr, 83-88, SR RES & DEVELOP MGR, SPEC PROJS, DOW CHEM CAN, LTD, 88- *Concurrent Pos:* Can Nat Comt, Int Union Pure & Appl Chem, 87-89, Can nat rep, comt chem & indust, 87-89; bd mem, Can Advan Indust Mat Forum, 88-89; chmn sci adv comt, Inst Chem Sci & Technol, 86-88; bd mem, Can Soc Chem, 88- *Mem:* Chem Inst Can; Am Chem Soc; Soc Chem Indust; Can Pulp & Paper Asn. *Res:* paper chemistry; catalysis; petrochemical production. *Mailing Add:* Vidal St S PO Box 3030 Sarnia ON N7T 7M1 Can

DARBY, RALPH LEWIS, b Youngstown, Ohio, Nov 13, 18; m 42; c 2. INFORMATION SCIENCE, CHEMICAL ENGINEERING. *Educ:* Ohio State Univ, BChE, 42; Univ Chicago, cert meteorol, 43. *Prof Exp:* Res engr chem, Commonwealth Eng Co, Ohio, 46-47; res engr, Graphic Arts Res Div, Battelle Columbus Labs, 47-51, prin chem engr, Info Res Div, 52-63, group dir, Dept Info Res & Econ, 63-65, div chief, 65-70, chief proj mgr, Dept Social & Mgt Sci, Battelle Mem Inst, 70-75, assoc sect mgr, Security Anal & Assessment Sect, 75-81; RETIRED. *Mem:* Am Chem Soc; Am Soc Info Sci; fel Am Inst Chemists. *Res:* Scientific communication; design of information storage and retrieval systems; information center operations and management national security technology assessment. *Mailing Add:* 3112 Leeds Rd Upper Arlington OH 43221

DARBY, ROBERT ALBERT, b Birmingham, Ala, May 6, 30; m 52; c 4. ORGANIC CHEMISTRY. *Educ:* Birmingham-Southern Col, BS, 51; Univ Va, MS, 53, PhD(chem), 57. *Prof Exp:* Res chemist, Exp Sta, E I du Pont de Nemours & Co, Inc, 57-63, sr res chemist, Wash Works, 63-66, sr res supvr, Exp Sta, 66-69, res mgr, Polyolefins Div, Plastics Dept, 69-70, lab dir, Plastics Dept, Res & Develop Div, 70-72, plant mgr, Plastics Dept, Wash Works, Parkersburg, WVa, 72-74, gen mkt mgr, Plastics Dept, Com Resins Div, 74-76, dir res & develop, Feedstocks Div, 76-78, dir, Chem Dyes & Pigments Dept, Res & Develop Div, 78-80, dir, Chem & Pigments Dept, Dept Plans Div, 80-82, dir, Chem & Pigments Dept, Methanol Prods Div, 82-84, dir, Automotive & Fabrication Prods Dept, Res Develop, Wilmington, 84-89, dir, Corp Technol Develop, 89-90; RETIRED. *Mem:* Am Chem Soc; Soc Chem Indust; Sigma Xi. *Res:* Conjugation and unsaturation in 1, 2-diaroylcyclopropanes; fluorocarbon chemistry; fluorocarbon polymers synthesis and properties; nylon and acrylic polymer processes. *Mailing Add:* Five Buckride Dr Wilmington DE 19807-2270

DARBY, RONALD, b La Veta, Colo, Sept 12, 32; m 61; c 2. FLUID DYNAMICS, RHEOLOGY. *Educ:* Rice Univ, BA & BS, 55, PhD(chem eng), 62. *Prof Exp:* NSF fel chem eng, Cambridge Univ, 61-62; sr scientist electrochem, Ling-Temco-Vought Res Ctr, 63-65; from asst prof to assoc prof chem eng, 65-69, Piper prof, 81, PROF CHEM ENG, TEX A&M UNIV, 69- *Concurrent Pos:* Vis prof, Ruhr Univ, Bochum, WGer, 83. *Mem:* Am Inst Chem Engrs; Nat Asn Corrosion Engrs; Soc Rheol. *Res:* Applied electrochemistry and corrosion; heat transfer; viscoelastic fluids; polymers; porous fuel cell electrodes; nucleate boiling; two phase fluid flow; non-Newtonian flow; polymer processing; applied fluid mechanics. *Mailing Add:* Dept Chem Eng Tex A&M Univ College Station TX 77843

DARBY, WILLIAM JEFFERSON, b Little Rock, Ark, Nov 6, 13; m 35; c 3. BIOCHEMISTRY, NUTRITION. *Educ:* Univ Ark, BS, 36, MD, 37; Univ Mich, MS, 41, PhD(biochem), 42. *Hon Degrees:* ScD, Univ Mich, 66, Utah State Univ, 73 & Univ Ark, 84. *Prof Exp:* Asst physiol chem, Sch Med, Univ Ark, 33-37, instr, 37-39; asst res prof pub health nutrit, Sch Pub Health, Univ NC & dir med nutrit, State Bd Health, NC, 43-44; from asst prof biochem & med to assoc prof biochem, 44-48, prof biochem & nutrit & dir, Div Nutrit, 48-79, head, Dept Biochem, 48-71, prof nutrit, 64-79, prof med in nutrit, 65-79, EMER PROF BIOCHEM, SCH MED, VANDERBILT UNIV, 79- *Concurrent Pos:* Rockefeller Found spec fel, Sch Med, Vanderbilt Univ, 42-43; instr, Duke Univ, 43-44; mem food & nutrit bd, Nat Res Coun, 49-71 & steering comt, 55-71; mem sci adv bd, Nat Vitamin Found, 50-54 & tech adv comt, Inst Nutrit Cent Am & Panama, 50-64; mem, Interdept Comt Nutrit Nat Develop, 50-71 & chmn, Comt Food Protection, 54-71; mem comt consults & dir surv, Philippines, Ethiopia, Ecuador, Lebanon, Jordan & Nigeria, 55-66; mem joint exp comt nutrit, Food & Agr Orgn, WHO, 54, 57, 61 & 66, chmn, 57 & 66, chmn joint comt food additives, 56, mem protein adv group, 56-66; mem sci adv comt, Nutrit Found, 58-65, 67-71; mem & chmn adv comt nutrit & metab, Off Surgeon Gen, US Army, 59-60; mem protein adv group, WHO-Food Agr Orgn, UNICEF, 60-62, second citizens' comt, Food & Drug Admin, 61-62 & sci adv comt, United Health Found, 62-70, chmn, 65-70; mem adv comt agr sci, USDA, 63-67 & adv bd, Corn Prod Co, 64-71; pres, Nutrit Found, 72-82; archivist, Am Inst Nutrit, 83- *Honors & Awards:* Mead-Johnson Award, Am Inst Nutrit, 47, Osborne-Mendel Award, 62; Order Rodolfo Robles, Panama Med Asn, 59; Star of Jordan, Hashemite Kingdom Jordan, 63; Goldberger Award, AMA, 64; Underwood-Prescott Mem lectr, Mass Inst Technol, 79; Robert H Hermann, Am Soc Clin Nutrit, 83. *Mem:* Nat Acad Sci; Am Soc Biochem & Molecular Biol; Am Chem Soc; Am Soc Clin Invest; Am Inst Nutrit (pres, 58-59); Asn Am Physicians; Am Pub Health Asn; hon mem Philippine Dietetic Asn; Soc Clin Res (vpres, 48); Nat Med Asn Panama. *Res:* Biochemistry of metabolism and nutrition; clinical nutrition; nutrition surveys and public health nutrition; nutritional anemias; folic acid; zinc metabolism. *Mailing Add:* Dept Biochem Vanderbilt Univ Sch Med Nashville TN 37232-0146

DARCEL, COLIN LE Q, b St Helier, Gt Brit; Feb 5, 25; m 48; c 2. VETERINARY VIROLOGY. *Educ:* Royal Vet Col, London, MRCVS, 47, BSc, 48, PhD(vet sci), 52; Cambridge Univ, BA, 49, MA, 54. *Prof Exp:* Pathologist fowl tumors res, Poultry Res Sta, Eng, 49-52 & Imp Cancer Res Fund, 52-55; virologist, Animal Dis Res Inst, Can Dept Agr, 55-86; VIROLOGIST, PALLISER ANIMAL HEALTH LABS, LTD, 86- *Mem:* Path Soc Gt Brit & Ireland; Brit Soc Immunol; Brit Soc Gen Microbiol; Am Asn Cancer Res; Can Soc Microbiol. *Res:* Immunological and biochemical aspects of pathogenesis of erythroblastosis virus; studies of infectious bovine rhinotracheitis and other cattle viruses. *Mailing Add:* 1819 Seven Ave S Lethbridge AB T1J 1M1 Can

DARCEY, TERRANCE MICHAEL, b Oakland, Calif, Nov 12, 50; m 71; c 3. ELECTRICAL ENGINEERING. *Educ:* Univ Calif, Berkeley, BS, 72; Calif Inst Technol, PhD(eng sci), 79. *Prof Exp:* Res asst, Dept Bioinfo Syst, Calif Inst Technol, 72-79; Swiss NSF fel, Dept Neurol, Univ Zurich, 79-80 , res biomed engr, Electroencephalogram Inst Zurich, 80-81; res assoc, 81-84, ASSOC RES SCIENTIST, DEPT NEUROL, SCH MED, YALE UNIV, 84-; BIOMED ENGR, NEUROL SERV, VET ADMIN MED CTR, WEST HAVEN, CONN, 81- *Res:* Application of engineering and mathematical techniques to analysis and modeling of biological systems; application of electric field theory and time-series analysis to localization of electroencephalographic phenomena. *Mailing Add:* 116B1 Vet Admin Med Ctr West Haven CT 06516

D'ARCY, WILLIAM GERALD, b Calgary, Alta, Aug 29, 31; m 81; c 2. BOTANY. *Educ:* Univ Alta, BA, 54; Univ Fla, MS, 68; Wash Univ, PhD(bot), 72. *Prof Exp:* Asst trade comnr Chicago, Govt Can, 55-57; economist mkt anal, Am Can Co, NY, 58-60; proprietor mfg, Tortola Pure Sparkling Water Co, VI, 60-65; herbarium asst bot, Univ Fla, 65-68; RES BOTANIST, MO BOT GARDEN, 69- *Concurrent Pos:* Asst prof, Dept Biol, Univ Mo, 74-; fac assoc biol, Wash Univ, 74-; NSF grants, 71, 77 & 80; ed, Flora Panama Prog, 72-88. *Mem:* Linnaean Soc London; Am Soc Plant Taxon; Int Asn Plant Taxon. *Res:* Systematics of the Solanaceae. *Mailing Add:* Mo Bot Garden PO Box 299 St Louis MO 63166

DARDEN, COLGATE W, III, b Norfolk, Va, Nov 6, 30; m 52; c 4. ELEMENTARY PARTICLE PHYSICS. *Educ:* Univ Va, BEE, 53, MA, 54; Mass Inst Technol, PhD(nuclear physics), 59. *Prof Exp:* Physicist, Savannah River Lab, 58-62; vpres, Columbia Eng, Inc, 62-64; assoc prof, 64-77, PROF PHYSICS, UNIV SC, 77- *Mem:* Am Phys Soc. *Res:* Positive and negative electron annihilation at high energy. *Mailing Add:* PO Box 2597 West Columbia SC 29171

DARDEN, EDGAR BASCOMB, b Raleigh, NC, Jan 23, 20; m 69. RADIOLOGICAL PHYSICS, HEALTH PHYSICS. *Educ:* Col William & Mary, BS, 41; Univ NC, MS, 50; Univ Tenn, PhD(zool), 57. *Prof Exp:* Instr physics, Col William & Mary, 41-42; from jr biologist to biologist, Oak Ridge Nat Lab, 49-73; mem res staff & radiation safety officer, Univ Tenn-Dept Energy, Comparative Animal Res Lab, 73-81; radiation safety officer, Oak Ridge Assoc Univs, 81-87; RETIRED. *Concurrent Pos:* Consult, Oak Ridge Assoc Univs, 88-, Fed Aviation Admin, 90- *Mem:* Am Phys Soc; Health Physics Soc; Sigma Xi. *Res:* Late somatic effects of radiation; metabolism of internally deposited radioisotopes; radiation dosimetry; radiobiology; physiology; radiological physics. *Mailing Add:* 105 Orchard Lane Oak Ridge TN 37830

DARDEN, LINDLEY, b New Albany, Mass, Dec 12, 45. PHILOSOPHY OF BIOLOGY, HISTORY OF MODERN BIOLOGY. *Educ:* Rhodes Col, BA, 68; Univ Chicago, AM, 69, SM, 72, PhD(conceptual found sci), 74. *Prof Exp:* Instr philos & humanities, Moraine Valley Community Col, Palos Hills, Ill, 69-70; asst prof philos & hist, 74-79, assoc prof, Inst Advan Computer Studies, 85-87, ASSOC PROF PHILOS & HIST, UNIV MD, COLLEGE PARK, 79- *Concurrent Pos:* Nat Endowment Humanities res grant, 76; NSF res grants, 78-79, 85-86 & 90-91; vis scholar, Computer Sci Dept, Stanford Univ, 80 & Dept Hist Sci, Harvard Univ, 82; mem, Rev Panel Grants Humanities, Sci & Technol, Nat Endowment Humanities, 87 & Prog Grants Hist & Philos Sci, 89-; vis assoc prof, Lab Artificial Intel Res, Ohio State Univ, 90-91. *Mem:* Philos Sci Asn; Hist Sci Soc; Am Philos Asn; Am Asn Artificial Intel; Int Soc Hist Philos & Social Studies Biol. *Res:* Reasoning in theory change in science; author of numerous publications and one book. *Mailing Add:* Dept Philos Univ Md College Park MD 20742

DARDEN, SPERRY EUGENE, b Chicago, Ill, Aug 16, 28; m 54; c 4. NUCLEAR REACTIONS & SCATTERING USING POLARIZED PROJECTILES. *Educ:* Iowa State Col, BS, 50; Univ Wis, MS, 51, PhD(physics), 55. *Prof Exp:* Exchange asst, Univ Basel, 55-56; instr physics, Univ Wis, 56-57; from asst prof to assoc prof, 57-65, PROF PHYSICS, UNIV NOTRE DAME, 65- *Concurrent Pos:* Sloan fel, 62-64; guest prof, Univ Basel, 65-66, Mexican Nuclear Ctr, 73 & Univ Birmingham, 85-86. *Mem:* Am Phys Soc. *Res:* Nuclear reactions and scattering processes; polarization effects in nuclear reactions and scattering. *Mailing Add:* Dept Physics Univ Notre Dame Notre Dame IN 46556

DARDEN, WILLIAM H, JR, b Tuscaloosa, Ala, Apr 25, 37; m 59; c 2. ALGOLOGY. *Educ:* Univ Ala, BS, 59, MS, 61; Ind Univ, PhD(bot), 65. *Prof Exp:* From asst prof to assoc prof, 65-73, PROF BIOL, UNIV ALA, 73-, CHMN, 74- *Res:* Cellular differentiation in algae with emphasis on the chemical control of sexual differentiation. *Mailing Add:* Dept Biol Box 870344 Univ Ala Tuscaloosa AL 35487-0344

D'ARDENNE, WALTER H, b Jenkintown, Pa, Oct 31, 32; m 57; c 3. NUCLEAR ENGINEERING, MECHANICAL ENGINEERING. *Educ:* Pa State Univ, BS, 59; Mass Inst Technol, PhD(nuclear eng), 64. *Prof Exp:* From asst prof to assoc prof nuclear eng, Pa State Univ, 64-72; MGR BWR PROD STAND, NUCLEAR ENERGY GROUP, GEN ELEC CO, 72- *Concurrent Pos:* Am Soc Eng Educ-Ford Found residency, Atomic Power Equip Dept, Gen Elec Co, 70-71. *Mem:* Am Nuclear Soc; Sigma Xi. *Res:* Reactor design; nuclear design; thermal-hydraulic design; nuclear safety; system analysis; reactor physics. *Mailing Add:* 1219 Rockhaven Dr San Jose CA 95120

DARDIRI, AHMED HAMED, b Cairo, Egypt, Mar 10, 19; nat US; m 51. VETERINARY VIROLOGY. *Educ:* Cairo Univ, DVM, 40, MVSc, 46; Mich State Univ, MSc, 47, PhD(microbiol), 50. *Prof Exp:* Dir, Poultry Res Exp Sta, Cairo Univ, 40-46; mem, Egyptian Ed Mission to USA, 46-50; sr lectr, Cairo Univ, 50-55; res assoc animal path, Univ RI, 55-56, from asst prof to assoc prof, 56-61; prin res veterinarian, 61-66, leader diag invest, Vet Res Sci Div, 66-84, COLLABR, PLUM ISLAND ANIMAL DIS CTR, AGR RES SERV, USDA, 84- *Concurrent Pos:* Consult, Am Tech Aid to Egypt, Cairo, 51-55; adj prof animal path, Univ RI, 68- *Mem:* Am Vet Med Asn; Am Soc Microbiol; Conf Res Workers Animal Dis; NY Acad Sci; Am Asn Avian Path; Wildlife & Tropical Dis. *Res:* Microbiology; genetics; veterinary science; poultry pathology and health; foreign diseases of animals. *Mailing Add:* PO Box 296 Ozona FL 34660

DARDIS, JOHN G, b Kilkenney, Ireland, May 21, 28; US citizen. ATOMIC PHYSICS, NUCLEAR PHYSICS. *Educ:* Univ Dublin, MA & PhD(physics), 55. *Prof Exp:* Res physicist, Radio Res Sta, Slough, Eng, 55-57; asst prof nuclear physics, Univ Ky, 57-62; res physicist, US Naval Radiol Defense Lab, 62-64, phys sci adminr, Calif, 64-67; physicist, Physics Br, Off Naval Res, 67-76, int rels officer, Off Advan Technol, State Dept, 76-80, physical sci officer, Off Nuclear Export Control, 80-83; SPEC ASST, SCI & TECHNOL, BUR INTEL & RES, 83- *Mem:* Am Phys Soc; Am Asn Physics Teachers; Brit Inst Physics & Phys Soc; Sigma Xi. *Res:* Radio, cosmic ray, radiation transport, plasma, atomic and molecular physics. *Mailing Add:* 1332 Pinetree Rd McLean VA 22101

DARDOUFAS, KIMON C, b Athens, Greece, Apr 25, 16; US citizen; m 59; c 4. ORGANIC CHEMISTRY, CHEMICAL ENGINEERING. *Educ:* Nat Univ Athens, BA, 35; Darmstadt Tech Univ, BSChE, 37; Dresden Tech Univ, MA, 39. *Prof Exp:* Prod supt, Tannerie-Ganferie Dardoufa SA, 42-47, dir & mem, Bd Dirs, 47-50, proj dir new expansion, 50-54; mgr indust exports, Athens Off, Am Merchandising Corp, 54-56; process engr, Gen Aniline & Film Corp, 56-60, proj leader polymer develop, 60-63; MGR RES PILOT PLANT, TECH CTR, ALLIED FIBERS & PLASTICS CO, 63- *Mem:* Am Inst Chem Eng; Am Chem Soc; fel Am Inst Chemists. *Res:* Process development and economic evaluation in chemical industry; development of synthetic fibers; fiber lubrication; fiber technology. *Mailing Add:* 5120 Tyme Rd Richmond VA 23234

DARENSBOURG, DONALD JUDE, b Baton Rouge, La, July 5, 41; m 67. INORGANIC CHEMISTRY, ORGANOMETALLIC CHEMISTRY. *Educ:* Calif State Univ, Los Angeles, BS, 64; Univ Ill, PhD(chem), 68. *Prof Exp:* Res chemist, Texaco Res Ctr, Beacon, NY, 68-69; asst prof chem, State Univ NY Buffalo, 69-73; asst prof chem, Tulane Univ, 73-75, assoc prof, 75-78, prof, 78-82; PROF CHEM, TEX A&M UNIV, 82- *Mem:* Am Chem Soc; AAAS. *Res:* Applications of infrared spectroscopy to inorganic and organometallic systems; mechanisms of photochemical and thermal reactions of organometallic compounds, in particular substituted metal carbonyl derivatives. *Mailing Add:* Dept Chem Tex A&M Univ College Station TX 77843-5000

DARENSBOURG, MARCETTA YORK, b Artemus, Ky, May 4, 42; m 67. INORGANIC CHEMISTRY, ORGANOMETALLIC CHEMISTRY. *Educ:* Union Col, Ky, BA, 63; Univ Ill, Urbana, PhD(inorg chem), 67. *Prof Exp:* Asst prof inorg chem, Vassar Col, 67-69; asst prof, State Univ NY Buffalo, 69-71; from asst prof to assoc prof, 71-79, PROF INOG CHEM TULANE UNIV LA, 79-; PROF CHEM, TEX A&M UNIV. *Concurrent Pos:* Chair-elect & chair, Inorg Div Am Chem Soc, 88-89, 89-90. *Honors & Awards:* Agnes Faye Morgan Res Award, 81. *Mem:* Am Chem Soc; Sigma Xi. *Res:* Ionpairing effects in transition metal-organic chemistry; metal-bound carbon monoxide reduction; transition metal hydrides reaction mechanisms; inorganic and organometallic synthesis. *Mailing Add:* Dept Chem Tex A&M Univ College Station TX 77843

DARITY, WILLIAM ALEXANDER, b Flat Rock, NC, Jan 15, 24; m 50; c 2. PUBLIC HEALTH. *Educ:* Shaw Univ, BS, 48; NC Cent Univ, MSPH, 49; Univ NC, PhD(educ & pub health), 64. *Prof Exp:* Community health educator, City Dept Pub Health, Charlotte, NC, 49-50, Dept Pub Health, Danville, Va, 50-52 & AntiTuberculosis League, Norfolk, 52-53; consult, WHO, 53-56, prof health educ, 56-58, regional adv, 58-64; dir prog develop, NC Fund, 64-65; assoc prof, 65-68, head dept, 69-76, PROF PUB HEALTH, UNIV MASS, AMHERST, 68-, DIR, DIV PUB HEALTH, 76- *Concurrent Pos:* Fac res grant, 66; biomed sci grant, 67; Franklin County biomed sci grant, 69; NIMH grant, 69-70; Nat Inst Child Health & Human Develop grant, 70-72; mem bd dirs, Planned Parenthood Fedn Am, Inc, 67-; mem pub health grant rev comt, NIH, 69-73; mem bd dirs, Drug Abuse Coun, Inc, 72-; assoc, Danforth Found. *Mem:* Am Pub Health Asn; Am Sch Health Asn; Am Nat Coun Health Educ Pub; Int Union Health Educ; Sigma Xi. *Res:* Barriers to utilization of family planning and other health service. *Mailing Add:* 105 Heatherstone Rd Amherst MA 01002

DARKAZALLI, GHAZI, b Damascus, Syria, May 20, 45; m 84; c 2. MECHANICAL ENGINEERING, SOLAR ENERGY. *Educ:* NY Inst Technol, BS, 71; Univ Mass, MS, 72, PhD(mech eng), 77. *Prof Exp:* Mfg engr, Gen Impact Extrusion Mfg, 72-73; res assoc, Energy Alternatives, Univ Mass, 73-76; DIR, SOLAR ENERGY RES FACIL & ASST PROF, DEPT MECH ENG, UNIV TEX, 77- *Concurrent Pos:* Grants, PPG Industs, 76-77 & Lincoln Lab, Mass Inst Technol, 77-; lectr & consult, Energy in Dallas-Ft Worth area, 77- *Mem:* Am Soc Mech Engrs; Int Solar Energy Soc; Am Soc Heating & Air Conditioning Engrs. *Res:* Solar energy related to heating and cooling; wind power for electric generation; thermal properties of materials; energy saving passive solar systems. *Mailing Add:* Nine Reed Lane Bedford MA 01730-1207

DARKOW, GRANT LYLE, b Milwaukee, Wis, Jan 7, 28; m 54; c 4. METEOROLOGY. *Educ:* Univ Wis, BSc, 49, MSc, 58, PhD(meteorol), 64. *Prof Exp:* From asst prof to assoc prof, 61-69, PROF ATMOSPHERIC SCI, UNIV MO-COLUMBIA, 69- *Concurrent Pos:* Trustee, Univ Corp Atmospheric Res, 72-74. *Mem:* Fel Am Meteorol Soc; Sigma Xi. *Res:* Dynamics of severe local storms. *Mailing Add:* Dept Atmospheric Scis Univ Mo Columbia MO 65211

DARLAGE, LARRY JAMES, b Brownstown, Ind, June 1, 45; m 67. ORGANIC CHEMISTRY. *Educ:* Ind Cent Univ, BA, 67; Iowa State Univ, PhD(org chem), 71. *Prof Exp:* Res assoc, Univ Fla, 71-73; instr chem, Pikeville Col, 73-78; div chairperson sci & math, 78-87, VPRES INSTR, BROOKHAVEN COL, 87- *Mem:* Am Chem Soc. *Res:* Comparison of the thermal, photochemical and mass spectral reactions of benzisoxazolin and some related five-membered heterocyclic compounds; characterization of nitrogen in coal liquids. *Mailing Add:* Brookhaven Col 3939 Valley View Lane Farmers Branch TX 75244

DARLAK, ROBERT, b North Tonawanda, NY, Sept 10, 37; m 66; c 3. ORGANIC CHEMISTRY. *Educ:* Univ Miss, BS, 60; WVa Univ, PhD(phys org chem), 64. *Prof Exp:* Res fel org chem under Dr Mel Newman, Ohio State Univ, 64-66; SR RES CHEMIST, EASTMAN KODAK CO, 66- *Mem:* Am Chem Soc. *Res:* Kinetics and synthesis in heterocyclic chemistry; synthesis in polycyclic aromatic compounds; polymer chemistry. *Mailing Add:* Eastman Kodak Co 343 State St Rochester NY 14650

DARLEY, ELLIS FLECK, b Monte Vista, Colo, Nov 2, 15; m 39; c 3. AIR POLLUTION, PLANT PATHOLOGY. *Educ:* Colo State Univ, BS, 38; Univ Minn, PhD(plant path), 45. *Prof Exp:* Asst forestry, Univ Minn, 40-41; instr bot & plant path, Colo State Univ, 41-42; res fel & asst plant path, Univ Minn, 42-45; pathologist, Firestone Plantations Co, Liberia, WAfrica, 45-47; pathologist, Off For Agr Rels, Guatemala, 48-49; from asst plant pathologist to assoc plant pathologist, Citrus Exp Sta, Univ Calif, Riverside, 49-61, lectr plant path, 68-78, plant pathologist, 61-78, EMER PLANT PATHOLOGIST, DEPT PLANT PATH & AIR POLLUTION RES CTR, UNIV CALIF, RIVERSIDE, 78- *Concurrent Pos:* Guggenheim fel, Germany, 63-64. *Mem:* Air Pollution Control Asn; Am Phytopath Soc; Sigma Xi. *Res:* Diseases of Hevea rubber; diseases of palms, especially date; air pollution injury to plants; pollution from forest and agricultural burning. *Mailing Add:* Dept Plant Path Statewide Air Pollution Res Ctr 108 Aldous St Cashmere WA 98815-1101

DARLEY, FREDERIC LOUDON, b Caracas, Venezuela, Nov 25, 18; US citizen; m 45; c 3. SPEECH PATHOLOGY. *Educ:* NMex State Teachers Col, AB, 39; Univ Iowa, MA, 40, PhD(speech path), 50. *Prof Exp:* Instr pub speaking & Eng, Univ Ark, 40-41; instr pub speaking, Univ Calif, Berkeley, 45-47; from instr to assoc prof speech path & audiol, Univ Iowa, 47-61; consult speech path, Mayo Clin, 61-69; prof speech path, Mayo Med Sch, Univ Minn, 69-83. *Mem:* Fel Am Speech Lang & Hearing Asn; Acad Aphasia; Sigma Xi. *Res:* Diagnosis and appraisal of communication disorders; aphasia; motor speech disorders. *Mailing Add:* 1007 NE 17th St Rochester MN 55906

DARLING, BYRON THORWELL, b Napoleon, Ohio, Jan 4, 12; m 46. PHYSICS. *Educ:* Univ Ill, BS, 33, MS, 36; Univ Mich, PhD(physics), 39. *Prof Exp:* Instr math, Mich State Col, 39-41; res physicist, US Rubber Co, Detroit, 41-46; res assoc, Univ Wis & Yale Univ, 46-47; from asst prof to assoc prof physics, Ohio State Univ, 47-53; prof physics, Laval Univ, 55-79; RETIRED. *Concurrent Pos:* Adj prof physics, Univ Fla. *Mem:* AAAS; Am Asn Physics Teachers; Am Phys Soc; NY Acad Sci; Can Asn Physicists. *Res:* Theory of rubber processing; molecular and nuclear theory; elementary particle theory. *Mailing Add:* 8620-64 NW 13th St Gainesville FL 32606

DARLING, CHARLES MILTON, b Mineral Wells, Miss, Feb 7, 34; m 54; c 3. PHARMACEUTICAL CHEMISTRY, MEDICINAL CHEMISTRY. *Educ:* Univ Miss, BS, 55, PhD(pharmaceut chem), 66. *Prof Exp:* Mgr, Woods' Pharm, 55-62; sr res chemist, A H Robins Co, 66-69; assoc prof, 69-75, alumni assoc prof pharmaceut chem, 69-79, assoc prof pharm, 78-79, asst dean, 81-84, PROF PHARM, 75-, ASSOC DEAN, AUBURN UNIV, 84- *Mem:* NY Acad Sci; Am Asn Cols Pharm; Am Chem Soc; Am Pharmaceut Asn; Acad Pharmaceut Sci; Sigma Xi. *Res:* Non-classical histamine antagonists; anticonvulsants. *Mailing Add:* Sch Pharm Auburn Univ Auburn AL 36830

DARLING, DONALD CHRISTOPHER, b Prince Rupert, BC, Dec 8, 51; m 77; c 1. TAXONOMY, PHYLOGENETICS. *Educ:* Queen's Univ, BSc, 74; Univ Utah, MSc, 78, Cornell Univ, PhD(Entom), 83. *Prof Exp:* Asst prof entom, Ore State Univ, 83-85; ASST PROF ZOOL, UNIV TORONTO, 85-; asst cur entom, 85-88, ASST CUR IN CHARGE, ROYAL ONT MUS, 88- *Mem:* Entom Soc Can; Entom Soc Am; Soc Systematic Zool; Willi Hennig Soc. *Res:* Taxonomic, morphological, biological and systematic studies of Hymenoptera, especially Chalcidoidea, with special emphasis on phylogenetic studies as a basis for revised classifications. *Mailing Add:* Dept Entom Royal Ont Mus 100 Queen's Park Toronto ON M5S 2C6 Can

DARLING, EUGENE MERRILL, JR, b Cambridge, Mass, Jan 13, 25. MATHEMATICS, METEOROLOGY. *Educ:* Harvard Univ, AB, 48; Mass Inst Technol, SM, 53. *Prof Exp:* Meteorologist, Pan Am Grace Airways, Inc, 50-51; lectr meteorol, Univ NMex, 52; atmospheric physicist, Air Force Cambridge Res Ctr, 52-62; aerospace technologist, NIMBUS Proj, Goddard Space Flight Ctr, NASA, 62-63, electronics res task group, NASA Hqs, 64 & Electronics Res Ctr, NASA, 64-70; chief data technol br, Dept Transp, 70-76, chief environ technol br, 77-80; RETIRED. *Mem:* Sigma Xi. *Res:* Lima, Peru terminal weather prediction research; analysis of meteorological support for United States Air Force aircraft and missiles; operational aspects of NIMBUS satellite cloud pictures; pattern recognition; imagery processing; management of programs in transportation systems modeling, simulation, environmental analysis, traffic safety and data systems. *Mailing Add:* PO Box 199 Lincoln MA 01773

DARLING, GEORGE BAPST, b Boston, Mass, Dec 30, 05; m 31. PUBLIC HEALTH. *Educ:* Mass Inst Technol, BS, 27; Univ Mich, DrPH, 31. *Hon Degrees:* MA, Yale Univ, 47; LLD, Univ Mich, 75. *Prof Exp:* Res assoc, Dept Health, Detroit, 27-32; assoc exec dir, assoc secy & treas, W K Kellogg Found, 32-34, exec dir & mem finance comt, 34-37, mem corp & bd trustees, assoc dir & comptroller, 37, mem admin comt, 38, pres, 40-43; exec secy comts on mil med, Nat Res Coun, 43-44, vchmn div med sci, 44-45; exec secy, Nat Acad Sci & Nat Res Coun, 46; vchmn div med sci, Nat Res Coun, 47-48; dir med affairs, 46-52, prof human ecol, 52-74, EMER PROF HUMAN ECOL, SCH MED, YALE UNIV, 74- *Concurrent Pos:* Trustee, Grace-New Haven Community Hosp, Conn, 46-59; on leave from Yale Univ as dir, Nat Res Coun Atomic Bomb Casualty Comn, Hiroshima, Japan, 57-72; vis prof, Hiroshima Schs Med & Nursing, Japan; resident scholar, NIH Fogarty Int Ctr Advan Studies, 73-74. *Mem:* AAAS; fel Am Pub Health Asn; Int Acad Polit Sci; NY Acad Sci; Radiation Res Soc Japan. *Res:* Public health administration; epidemiology; statistics; medical administration; professional education. *Mailing Add:* 1171 Whitney Ave Hamden CT 06517-3434

DARLING, GRAHAM DAVIDSON, b Ottawa, Ont, Can, Mar 11, 58. POLYMER-SUPPORTED REAGENTS CATALYSIS SORBENTS & PROTECTING GROUPS, POLYMERS MICROELECTRONICS & PHOTONICS. *Educ:* Univ Ottawa, Can, BSc, 81, BSc(biochem) & BSc(chem), 82, PhD(chem), 87. *Prof Exp:* ASST PROF ORG POLYMER CHEM, MCGILL UNIV, 89- *Concurrent Pos:* Vis scientist org polymer chem, Almaden Res Centre, IBM, San Jose, Calif, 87-89. *Mem:* Sigma Xi; Am Chem Soc; Can Soc Chem. *Res:* Functional polymers for organic synthesis and separations; artificial enzymes; molecular probes; microencapsulation; nonlinear optical materials; microlithographic resists; contractile polymers. *Mailing Add:* Dept Chem McGill Univ 801 Sherbrooke St W Montreal PQ H3A 2K6 Can

DARLING, MARILYN STAGNER, b Boulder, Colo, Apr 20, 35; m 63; c 2. PLANT ECOLOGY. *Educ:* George Washington Univ, BS, 56; Duke Univ, MA, 59, PhD(zool), 66. *Prof Exp:* Instr biol, Hollins Col, Roanoke, 63-64; RES ASSOC PLANT ECOL, DUKE UNIV, 73- *Concurrent Pos:* Instr plant ecol, Duke Univ, 78. *Mem:* Ecol Soc Am; Brit Ecol Soc; Sigma Xi; Bot Soc Am; Am Soc Plant Physiologists. *Mailing Add:* 1306 Kent St Durham NC 27702

DARLING, ROBERT BRUCE, b Johnson City, Tenn, Mar 15, 58. OPTOELECTRONICS, MICROELECTRONICS. *Educ:* Ga Inst Technol, BSEE, 80, MSEE, 82, PhD(elec eng), 85. *Prof Exp:* Res engr, Ga Tech Res Inst, 82-83; asst prof, 85-90, ASSOC PROF ELEC ENG, UNIV WASH, 90- *Mem:* Inst Elec & Electronic Engrs; Am Phys Soc; Am Vacuum Soc; Optical Soc Am; AAAS. *Res:* High-speed optoelectronic materials, devices and subsystems; physics of semiconductors and semiconductor devices, especially photodetectors, field-effect devices and electron transport phenomena; chemistry and physics of device processing, especially vacuum evaporation, sputtering and photolithography; photoconductivity; metal-semiconductor contacts. *Mailing Add:* Dept Elec Eng FT-10 Univ Wash Seattle WA 98195

DARLING, SAMUEL MILLS, b Bradenton, Fla, Jan 13, 17; m 40; c 4. CHEMISTRY. *Educ:* Carleton Col, AB, 39; Western Reserve Univ, MS, 43, PhD(chem), 47. *Prof Exp:* Res supvr, Standard Oil Co, Ohio, 39-67, fuels res supvr, 67-70, supvr lubricant res, 70-75, supvr fuels & lubricant res & develop, 76-77, corp prod safety coordr, 77-83; RETIRED. *Mem:* AAAS; Am Soc Lubrication Eng; Am Chem Soc; Soc Automotive Eng; fel Am Inst Chem. *Res:* Halide catalysis; catalytic cracking and reforming; performance of fuels and lubricants; nitrile chemicals and polymers. *Mailing Add:* 1330 Maydor Lane South Euclid OH 44121

DARLING, STEPHEN DEZIEL, b Appleton, Wis, May 7, 31; m 60; c 2. ORGANIC CHEMISTRY. *Educ:* Univ Wis, BS, 54; Columbia Univ, MA, 57, PhD(org synthesis), 59. *Prof Exp:* Res fel org chem, Columbia Univ, 59-62; asst prof, Univ Southern Calif, 62-68 & Southern Ill Univ, 68-70; assoc prof, 70-80, PROF ORG CHEM, UNIV AKRON, 80- *Mem:* Am Chem Soc; The Chem Soc. *Res:* New synthetic methods; synthesis of terpenes; stereoselective metal reductions; marine natural products; x-ray crystallographic structure determination. *Mailing Add:* Dept Chem Univ Akron Akron OH 44325-0001

DARLINGTON, GRETCHEN ANN JOLLY, b Dayton, Ohio, Jan 24, 42; m 66; c 2. GENETICS. *Educ:* Univ Colo, BA, 64; Univ Mich, PhD(human genetics), 70. *Prof Exp:* Fel biol, Yale Univ, 70-72, res assoc, 72-74; from asst prof to assoc prof human genetics, Cornell Univ Med Col, 74-82, ASSOC PROF PATH, BAYLOR COL MED, 82- *Mem:* Am Soc Human Genetics; AAAS; Tissue Cult Asn; Soc Cell Biol; Sigma Xi. *Res:* Study of the expression of differentiated functions in somatic cell hybrids between cultured mouse hepatoma and a variety of non-hepatic cells; tissue specific gene expression. *Mailing Add:* 7707 Ludington Houston TX 77071

DARLINGTON, SIDNEY, b Pittsburgh, Pa, July 18, 06; m 65; c 2. APPLIED MATHEMATICS. *Educ:* Harvard Univ, BS, 28; Mass Inst Technol, BS, 29; Columbia Univ, PhD(physics), 40. *Hon Degrees:* DSc, Univ NH, 84. *Prof Exp:* Mem tech staff, Bell Tel Labs, Inc, 29-71; ADJ PROF, DEPT ELEC ENG, UNIV NH, 71- *Concurrent Pos:* Expert consult off field serv, Off Sci Res & Develop & tech observer, US Army, 44-45. *Honors & Awards:* Edison Medal & Medal of Honor, Inst Elec & Electronics Engrs. *Mem:* Nat Acad Sci; Nat Acad Eng; fel Inst Elec & Electronics Engrs; assoc fel Am Inst Aeronaut & Astronaut. *Res:* Communication network theory; synthesis of networks which produce prescribed characteristics; smoothing and prediction of stochastic processes; guidance and control of missiles and space vehicles. *Mailing Add:* Dept Elec Eng Eight Fogg Dr Durham NH 03824

DARLINGTON, WILLIAM BRUCE, b Wichita, Kans, July 21, 33; m 57; c 4. PULP & PAPER, DEINKING CHEMISTRY. *Educ:* Baker Univ, AB, 55; Univ Kans, PhD(chem), 61. *Prof Exp:* Chemist, Phillips Petrol Co, 56-57; sr res chemist, Pittsburgh Plate Glass Co, 61-65, res supvr chem div, 65-68, sr supvr, 68-70, head, Electrochem Dept, 70-90, SCIENTIST, PPG INDUSTS, 90- *Concurrent Pos:* Sr res fel, Mead Corp. *Mem:* Am Chem Soc; Electrochem Soc; Tech Asn Pulp & Paper Indust. *Res:* Nonaqueous solution and industrial electrochemistry; fused-salt chemistry; electrometallurgy; alkali metals; halogens; pulp and paper chemicals; deinking chemistry. *Mailing Add:* 77 Timberlane Dr Chillicothe OH 45601

DARNALL, DENNIS W, b Glenwood Springs, Colo, Dec 14, 41; m 63. BIOCHEMISTRY, BIOINORGANIC CHEMISTRY. *Educ:* NMex Inst Mining & Technol, BS, 63; Tex Tech Col, PhD(chem), 66. *Prof Exp:* NIH fel, Northwestern Univ, 66-68; pres & chmn, Bio-Recovery Systs Inc, 86-89; from asst prof to assoc prof, 68-74, assoc dean & dir, Col Arts & Scis, 83-86, PROF CHEM, NMEX STATE UNIV, 74-, DEPT HEAD, 91- *Concurrent Pos:* NIH career develop award, 71-76. *Honors & Awards:* John Dustin Clark Award, Am Chem Soc, 88. *Mem:* AAAS; Soc Res Adminrs; Am Chem Soc; Am Soc Biol Chemists. *Res:* Physical chemistry and chemical modification of proteins; metalloproteins and enzymes; protein subunit interactions; lanthanide ions as probes of protein structure; removal of metal ions from industrial and mining waste waters. *Mailing Add:* Dept Chem NMex State Univ Las Cruces NM 88003

DARNEAL, ROBERT LEE, b Los Gatos, Calif. EARTH SCIENCES, CHEMICAL MICROSCOPY. *Educ:* San Jose State Col, AB, 39; Stanford Univ, MS, 55; Calgary Col Technol, PhD(mineral & geochem), 74. *Prof Exp:* Chemist food chem, Calif Prune & Apricot Growers Asn, 39-40; chemist gas anal, Permanente Metals Corp, 41-42; head dept sci & math, Cloverdale Union High Sch, Calif, 42-43; teaching fel mineral, Stanford Univ, 43-44; instr physics & chem, Western Wash Col, 44-46; Royal Victor fel geol, mineral & petrog, Stanford Univ, 46-49; instr chem & earth sci, San Francisco State Col, 49-51; instr geol, Menlo Jr Col, Calif, 52-54; geophysicist, Div Raw Mat, US AEC, 55-57, physicist, Radiol Physics & Instrumentation Br, Div Biol & Med, 57-73; earth scientist environ progs, Div Biomed & Environ Res, US Energy Res & Develop Admin, 73-79; MGR & PROPRIETOR, ANATECH LABS, 79- *Concurrent Pos:* Lectr earth sci, Frederick Community Col, Md, 72-, instr geol, 81- *Mem:* Am Chem Soc; Mineral Soc Am; Meteoritical Soc; Sigma Xi; Geol Soc Am. *Res:* Determinative mineralogy; forensic chemistry and related forensic sciences. *Mailing Add:* Anatech Labs PO Box 444 Mt Airy MD 21771

DARNELL, ALFRED JEROME, b Denton, Tex, Aug 20, 24; m 47; c 1. PHYSICAL INORGANIC CHEMISTRY. *Educ:* San Diego State Col, BA, 50; Univ Calif, Los Angeles, PhD(chem), 64. *Prof Exp:* Mem tech staff, Rockwell Int, 55-57, proj eng, 57-81, proj scientist, Energy Technol Eng Lab, 81-84; CONSULT,84- *Concurrent Pos:* Consult physics group, Hughes Res Ctr. *Honors & Awards:* Cattrell res award, 49. *Mem:* AAAS; fel Am Inst Chemists; Am Chem Soc; Am Phys Soc; Int Solar Energy Soc; Sigma Xi. *Res:* High temperature physical and inorganic chemistry; ultra high pressure physics and chemistry; low temperature physics; super conductivity; solid state physics and chemistry; electrochemistry; metallurgy; air pollution control; biomass conversion; solar conversion; geothermal. *Mailing Add:* 23030 Burbank Blvd Woodland Hills CA 91367

DARNELL, FREDERICK JEROME, b Washington, DC, May 24, 28; m 52; c 3. SOLID STATE PHYSICS. *Educ:* Yale Univ, BS, 50; Carnegie Inst Technol, MS, 51, PhD(physics), 55. *Prof Exp:* Res physicist, 55-62, res supvr, 62-70, assoc dir res, Cent Res & Develop Dept, 70-79, assoc dir admin, 79-85, DIR ANALYTIC & INFO SCI, CENT RES & DEVELOP DEPT, E I DU PONT DE NEMOURS & CO, 85- *Concurrent Pos:* Mem comt on educ, Am Phys Soc & corp asn adv comn, Am Inst Physics. *Mem:* Am Phys Soc; Sigma Xi; Am Chem Soc. *Res:* Semiconductors; magnetism. *Mailing Add:* Six Bank Swallo Hilton Head Island SC 29926

DARNELL, JAMES EDWIN, JR, b Columbus, Miss, Sept 9, 30. CELL BIOLOGY. *Educ:* Univ Miss, BA, 51; Washington Univ, MD, 55. *Prof Exp:* Sr asst surgeon & virologist, Nat Inst Allergy & Infectious Dis, 56-60; spec fel, Pasteur Inst, Paris, 60-61; from asst prof to assoc prof biol, Mass Inst Technol, 61-64; prof cell biol & biochem, Albert Einstein Col Med, 64-68; prof biol sci, Columbia Univ, 68-74, chmn dept, 71-73; VINCENT ASTOR PROF, ROCKEFELLER UNIV, 74-, VPRES, ACAD AFFAIRS, 90- *Concurrent Pos:* Career res award, Nat Cancer Inst, 62-64; career scientist, Health Res Coun, NY, 65-72; consult, Nat Sci Found, 72-75; sci coun, Sloan

Kettering Inst, 74-83; consult, Rosenthiel Basic Med Sci Res Ctr, 76-78; coun res & clin invest, 76-80; consult, Howard Hughes Med Inst, 81-83; bd sci coun, Nat Inst Arthritis, Diabetes, Digest & Kidney Dis, 83-84. *Honors & Awards:* Harvey Lectr, 73; H T Ricketts Award, Univ Chicago, 79; Abraham White Distinguished Lect & Sci Award, George Washington Univ, 82; Gairdner Int Award, 86. *Mem:* Nat Acad Sci; Am Soc Biol Chem. *Res:* Virology; cellular biology. *Mailing Add:* Rockefeller Univ York Ave & 66th Sts New York NY 10021

DARNELL, REZNEAT MILTON, b Memphis, Tenn, Oct 14, 24; div; c 1. OCEANOGRAPHY, ECOLOGY. *Educ:* Rhodes Col at Memphis, BS, 46; Rice Univ, MA, 48; Univ Minn, PhD(zool), 53. *Prof Exp:* Asst zool, Univ Minn, 48-52; instr, Tulane Univ, 52-55; from asst prof to prof biol, Marquette Univ, 55-69; prof oceanog & biol, 69-72, PROF OCEANOG, TEXAS A&M UNIV, 72- *Concurrent Pos:* Res assoc, Milwaukee Pub Mus, 60-; chmn, Wis State Bd Preserv Sci Areas, 64-68 & Wis Sci Areas Preserv Coun, 68-69; vis prof oceanog, Tex A&M Univ, 68-69; chmn, Conserv Ecosysts Submcomt, US-Int Biol Prog, 72-74; IPA appointee, US Dept Int, 81-84, 85. *Mem:* Fel AAAS; Am Soc Ichthyologists & Herpetologists; Am Soc Limnol & Oceanog; Ecol Soc Am; Soc Study Evolution. *Res:* Ecology of streams, estuaries and oceans; subtropical aquatic ecology; community analysis; organic detritus; nitrogen and energy budgets; ecology and systematics of fishes; ecology and biological resource management of continental shelf. *Mailing Add:* Dept Oceanog Tex A&M Univ College Station TX 77843

DARNELL, W(ILLIAM) H(EADEN), b Roanoke, Va, May 14, 25; m 50; c 1. OCCUPATIONAL HEALTH, SAFETY & NURSING. *Educ:* Va Polytech Inst & State Univ, BS, 50; Univ Wis, MS, 51, PhD(chem eng), 53. *Prof Exp:* Asst, Univ Wis, 51-52; res engr process develop, E I du Pont de Nemours & Co, 53-59, tech supt nylon intermediates, 60-63, prod develop mgr, 63-64, prod mgr, 64-69, res mgr, 69-73, lab adminr, 73-76, environ mgr, 76-85; INDEPENDENT CONSULT, OCCUP HEALTH, 85- *Concurrent Pos:* Regist nurse, 89- *Res:* Atomization; process development; plastics and chemical manufacture; heat transfer, safety and environmental health. *Mailing Add:* Seven Gybe Ho Salem SC 29676-9699

DAROCA, PHILIP JOSEPH, JR, b New Orleans, La, Nov 14, 42; m 66; c 2. ANATOMIC PATHOLOGY. *Educ:* Tulane Univ, BS, 64, MD, 68; Am Bd Path, dipl, 73. *Prof Exp:* Internship, Parkland Mem Hosp, Dallas, Tex, 68-69; residency, Tulane Div, Charity Hosp, 69-73; staff pathologist, Armed Forces Inst Path, 73-75; asst prof, 75-80, ASSOC PROF PATH, TULANE UNIV, 80- *Mem:* Int Acad Pathol; Col Am Pathologists. *Mailing Add:* Dept Path Tulane Med Sch New Orleans LA 70112

DAROFF, ROBERT BARRY, b New York, NY, Aug 3, 36; m 59; c 3. NEUROLOGY, NEURO-OPHTHALMOLOGY. *Educ:* Univ Pa, BA, 57, MD, 61. *Prof Exp:* From asst prof to prof neurol & ophthal, Univ Miami, 68-80; chief, Neurol Clin & asst chief, Neurol Serv, Vet Admin Hosp, Miami, Fla, 68-80, dir, Ocular Motor Neurophysiol Lab, 69-80; GILBERT W HUMPHREY PROF NEUROL & CHMN DEPT, CASE WESTERN RESERVE UNIV, 80-; DIR, DEPT NEUROL, UNIV HOSPS CLEVELAND, 80-; STAFF PHYSICIAN, CLEVELAND VET ADMIN MED CTR, 80- *Concurrent Pos:* Book rev ed, Neuro-ophthal, 81-86; ed-in-chief, Neurol, 87. *Honors & Awards:* Parker Heath Mem Lectr, Am Acad Ophthal, 86. *Mem:* Am Neurol Asn (pres, 90-91); Am Neurotology Soc; Asn Res Vision & Ophthal; Soc Neurosci; hon mem Asn Colombiana Neurol. *Res:* The study of normal human eye movements and their disorders consequent to neurological disease and cerebral dysfunction. *Mailing Add:* Dept Neurol Univ Hosps Cleveland OH 44106

DARON, GARMAN HARLOW, anatomy; deceased, see previous edition for last biography

DARON, HARLOW HOOVER, b Chicago, Ill, Oct 25, 30; m 58; c 4. BIOCHEMISTRY. *Educ:* Univ Okla, BS, 56; Univ Ill, PhD(biochem), 61. *Prof Exp:* NSF res fel biol, Calif Inst Technol, 61-63; asst prof biochem, Tex A&M Univ, 63-67; from asst prof to assoc prof, 67-82, PROF BIOCHEM, AUBURN UNIV, 82- *Mem:* AAAS; Am Soc Biol & Molecular Biol; Am Chem Soc; Sigma Xi. *Res:* Mechanism of enzyme action. *Mailing Add:* Dept Animal Sci Auburn Univ Auburn AL 36849

DARR, J(ACK) E(DWIN), b Shaffersville, Pa, Jan 23, 21; m 47; c 8. ELECTRICAL ENGINEERING. *Educ:* Pa State Univ, BS, 42, MS, 48. *Prof Exp:* Trainee, Bethlehem Steel Co, 42-43, asst foreman power stas, 43-44; asst, Eng Exp Sta, Pa State Univ, 46-48; asst engr, Westinghouse Elec Corp, 48-50, assoc engr, 50-51, eng group leader, 51, sect mgr interceptor armament control systs, Aerospace Div, 51-59, dir eng electronic warfare proj, 59-60, mgr, Astroelectronics Lab, 60-61, mgr airborne weapons control eng, 61-63, proj serv systs opers, 63-65, dep mgr deep submergence systs, 65-67, mgr planning & control, 67-68, mgr opers serv, Underseas Div, 68-71, mem staff ord systs dept, Aerospace & Electronics Systs Div, 71-74, mgr tech servs, Opers Div, Westinghouse Elec Corp, 74-; RETIRED. *Mem:* Inst Elec & Electronics Engrs. *Res:* Project management; military electronic systems. *Mailing Add:* 1405 Margarette Ave Towson MD 21204

DARRAGH, RICHARD T, b Verdun, Que, Apr 30, 31; m 53; c 5. FOOD SCIENCE, PHARMACEUTICAL TECHNOL. *Educ:* Univ Montreal, BS, 52; Cornell Univ, MFS, 55, PhD(biochem), 57. *Prof Exp:* Asst qual controller, Birds Eye Co, Can, 52; food technologist, Continental Can Co, Can, 53; biochemist, 57-63, sect head food prod develop, 63-68, from assoc dir to dir indust food prod develop, 68-77, mgr mfg & prod develop, spec prod, 77-84, mgr pharmaceut res & develop, 84-85, MGR PHARMACEUT RES DEVELOP, MFG, QUAL ASSESSMENT, ENG, LIC & ACQUISITIONS, PROCTER & GAMBLE CO, 85- *Mem:* AAAS; NY Acad Sci; Sigma Xi. *Res:* Industrial research and development in pharmaceutical and food products; development and management of organizations and programs aimed at new business areas; consumer and market research. *Mailing Add:* Proctor & Gamble Plaza PO Box 599 Cincinnati OH 45201

DARRAH, WILLIAM CULP, paleobotany; deceased, see previous edition for last biography

DARRELL, JAMES HARRIS, II, b Riverside, NJ, June 27, 42; m 66; c 2. GEOLOGY, PALYNOLOGY. *Educ:* Ohio Wesleyan Univ, BS, 64; Univ Tenn, Knoxville, MS, 66; La State Univ, Baton Rouge, PhD(geol), 73. *Prof Exp:* Res asst, La Water Resources Res Inst, 67-68; ASST PROF GEOL, GA SOUTHERN COL, 70- *Concurrent Pos:* Consult palynology, SC Geol Surv, 74- & Southeastern Environ Consult Group, 75- *Mem:* Geol Soc Am; Soc Econ Paleontologists & Mineralogists; Am Asn Stratig Palynologists; AAAS; Sigma Xi. *Res:* Palynological biostratigraphy and lithostratigraphy of the coastal plain in Georgia and South Carolina; palynomorph distribution in the Mississippi River Delta; quaternary palynology in Georgia; coastal plain riverswamp sedimentation. *Mailing Add:* Dept Geol Ga Southern Col Landrum Box 8149 Statesboro GA 30460

DARROW, FRANK WILLIAM, b Syracuse, NY, Feb 6, 40; m 61; c 2. CHEMICAL EDUCATION. *Educ:* Williams Col, BA, 61; Univ Pa, PhD(chem), 65. *Prof Exp:* Vis asst prof chem & Great Lakes Cols Asn teaching intern, Earlham Col, 65-66; asst prof, Ithaca Col, 66-71, asst to the Provost, 71-72, actg provost, 72-73, provost, 73-76, chmn dept chem, 82-90, ASSOC PROF CHEM, ITHACA COL, 71- *Concurrent Pos:* Vis mem fac, Evergreen State Col, Olympia, Wash, 76-77, UNIV WIS-MADISON INST CHEM EDUC, 90-91. *Mem:* Am Chem Soc; Nat Sci Teachers Asn. *Res:* Properties of electrolyte solutions and fused salt systems; chemical education. *Mailing Add:* Dept Chem Ithaca Col Ithaca NY 14850

DARROW, ROBERT A, b Syracuse, NY, July 12, 31; m 62; c 3. ENZYMOLOGY, PLANT-MICROBE INTERACTIONS. *Educ:* Amherst Col, AB, 52; Johns Hopkins Univ, PhD(biol), 57. *Prof Exp:* Biochemist, Chem Pharmacol Lab, Nat Cancer Inst, 57-59; Jane Coffin Childs Mem Fund fel, Nat Inst Med Res, London, 59-61; asst biochemist, Mass Gen Hosp, 61-67; assoc biol chem, Harvard Med Sch, 64-67; investr, 67-71, sect head, 71-73, asst mission mgr, Charles F Kettering Res Lab, 73-83, PRIN RES SCIENTIST, BATTELE-KETTERING RES LAB, 83- *Mem:* AAAS; Am Chem Soc; Am Soc Biol Chemists; Am Soc Plant Physiologists; Am Soc Microbiol. *Res:* Mechanism of enzyme action; control mechanisms and enzyme induction; biological nitrogen metabolism; plant growth regulation; plant microbiol interactions. *Mailing Add:* Dept Biol Chem Wright State Univ Dayton OH 45435

D'ARRUDA, JOSE JOAQUIM, b Fall River, Mass, Aug 4, 42; m 65; c 3. THEORETICAL PHYSICS, STATISTICAL MECHANICS. *Educ:* Lowell Technol Inst, BS, 65; Univ Del, MS, 68, PhD(physics), 71. *Prof Exp:* Asst prof physics, Univ Wis Ctr-Richland, 71-74; assoc prof, 74-81, PROF PHYSICS, PEMBROKE STATE UNIV, 81- *Concurrent Pos:* Res assoc, Argonne Nat Lab, 73; vis scientist, Battelle Northwest Lab, 74 & Oak Ridge Nat Lab, 75-81; Energy Workshop Educ leader, US Dept Energy, 76-79; vis assoc prof, Univ Del, 79-80. *Mem:* Am Phys Soc; Am Asn Physics Teachers; Sigma Xi. *Res:* Quantum statistical mechanics; computer solutions to complex molecular problems. *Mailing Add:* Dept Physics Pembroke State Univ Pembroke NC 28372

DARSOW, WILLIAM FRANK, b Mankato, Minn, May 16, 20; m 62; c 2. MATHEMATICS. *Educ:* Univ Minn, BA, 42; Univ Chicago, PhD, 53. *Prof Exp:* Instr math, Ill Inst Technol, 50-51; from instr to asst prof, De Paul Univ, 52-60; ASSOC PROF MATH, ILL INST TECHNOL, 61- *Mem:* Am Math Soc; Math Asn Am. *Res:* Abstract harmonic analysis; topological algebra. *Mailing Add:* 1109 Hohlfelder Rd Glencoe IL 60022

DARST, PHILIP HIGH, b Greensboro, NC, June 8, 43; m 68; c 1. ECONOMIC ENTOMOLOGY. *Educ:* Wake Forest Univ, BS, 66; Clemson Univ, MS, 68; Purdue Univ, PhD(entom), 71. *Prof Exp:* Asst prof biol, Univ Miss, 71-75; field develop rep, Union Carbide Corp, 75-78; res mgr, Cotton Inc, Raleigh, NC, 78-80; pest control adv, Western Farm Serv, Salinas, Calif, 80-82; VINEYARD CONSULT PEST MGT & PLANT NUTRIT, SALINAS, CALIF, 82- *Concurrent Pos:* Forensic Agron. *Mem:* AAAS; Entom Soc Am; Am Inst Biol Sci; Am Reg Prof Entomologists; Sigma Xi. *Res:* Pest management. *Mailing Add:* 23085 Guidotti Pl Salinas CA 93908

DARST, RICHARD B, b Chicago, Ill, Oct 5, 34; m 58; c 5. MATHEMATICS. *Educ:* Ill Inst Technol, BS, 57, MS, 58; La State Univ, PhD(math), 60. *Prof Exp:* Instr math, Mass Inst Technol, 60-62; from asst prof to prof, Purdue Univ, 62-73; PROF MATH, COLO STATE UNIV, 71- *Concurrent Pos:* Vis scholar, Stanford Univ, 81-82; Fulbright Lectr, Egypt, 84. *Mem:* Math Asn Am; Am Math Soc. *Res:* Measure and integration; functional analysis; probability and statistics, approximation, real analysis, operations research. *Mailing Add:* Dept Math Colo State Univ Ft Collins CO 80523

DART, JACK CALHOON, b Concord, Mich, Aug 14, 12; m 40; c 4. CHEMICAL ENGINEERING. *Educ:* Albion Col, AB, 34; Univ Mich, BSE, 35, MSE, 37. *Prof Exp:* Res assoc fac res, Univ Mich, 36; chem engr, Universal Oil Prod Co, 36-37 & Pan-Am Refining Corp, Tex, 37-43; instr chem eng exten, Agr & Mech Col, Tex, 41-42; supvr pilot plant develop group, Magnolia Petrol Co, 43-44 & La Div, Esso Labs, Standard Oil Co, NJ, 44-47; dir develop, Houdry Process Corp, 47-52, mgr res & develop div, 52-55; dir, vpres & gen mgr, Chem Div, 55-57, vpres & gen mgr, Sales & Serv Div, 57-62; OWNER, J C DART & ASSOCS, 62- *Concurrent Pos:* Adj prof chem eng, Cath Univ, Am, 69-85. *Honors & Awards:* Founders Award, Am Inst Chem Engrs, 81. *Mem:* Fel Am Inst Chem; fel Am Inst Chem Engrs; Am Chem Soc. *Res:* Alkylation, isomerization and polymerization of light hydrocarbons; catalytic and thermal cracking of hydrocarbons; hydrocarbon synthesis; azeotropic distillation; hydrogenation; thermal and catalytic reforming; heat transfer, fluid flow, absorption and distillation; ammonia and methanol syntheses; phthalic anhydride. *Mailing Add:* 10101 Gary Rd Potomac MD 20854-4109

DART, SIDNEY LEONARD, b Cape Town, SAfrica, Aug 24, 18; m 42; c 4. PHYSICS. *Educ:* Oberlin Col, AB, 40; Univ Notre Dame, MS, 43, PhD(physics), 46. *Prof Exp:* Asst, Univ Notre Dame, 40-43, instr physics, 43-44, res assoc, 44-46; sr physicist, Am Viscose Corp, 46-53; physicist, Dow

Chem Co, 53-54; prof, 54-84, EMER PROF PHYSICS, CLAREMONT MEN'S COL, 84- *Concurrent Pos:* Consult, Dow Chem Co, 56-59; vis prof & head, Postgrad Physics Dept, Am Col Madurai, SIndia, 67-68 & 77-78; staff physicist, Sci Educ Improv Prog, NSF, New Delhi, 71-72. *Mem:* AAAS; Am Phys Soc; Soc Rheol; Soc Social Responsibility Sci. *Res:* Fundamental physical properties of high polymers including rubber, cork and textile fiber polymers; biophysics of muscle. *Mailing Add:* Dept Physics Joint Sci Dept Claremont CA 91711

DARWENT, BASIL DE BASKERVILLE, b Trinidad, BWI, May 20, 13; m 38; c 1. PHYSICAL CHEMISTRY. *Educ:* McGill Univ, BSc, 41, PhD(phys chem), 43. *Prof Exp:* Asst res chemist, Trinidad Leaseholds, Ltd, 36-40; res assoc, McGill Univ, 43-44; res chemist, Nat Res Coun Can, 44-52; mgr dept phys chem, Olin Industs, Inc, 52-55; res prof, 55-57, prof, 57-74, EMER PROF CHEM, CATH UNIV AM, 74- *Mem:* The Chem Soc; Royal Soc Can. *Res:* Kinetics of elementary gas-phase reactions; photochemistry oxidation; reactions of excited species. *Mailing Add:* PO Box 248 Westover MD 21871

DARWIN, DAVID, b New York, NY, Apr 17, 46; m 68; c 2. STRUCTURAL ENGINEERING, MATERIALS SCIENCE. *Educ:* Cornell Univ, BS, 67, MS, 68; Univ Ill, Urbana-Champaign, PhD(civil eng), 74. *Prof Exp:* Officer eng, US Army Corps Engrs, 67-72; from asst prof to assoc prof, 74-82, PROF CIVIL ENG, UNIV KANS, 82-, DIR, STRUCT ENG & MAT LAB, 82-, DEANE E ACKERS DISTINGUISHED PROF CIVIL ENG, 90- *Concurrent Pos:* Lectr, George Wash Univ, 71-72; NSF grant, Univ Kans Ctr Res Inc, 76-; mem, comt finite elem anal reinforced concrete, Am Soc Civil Engrs, 77-, composite construction, 86-, Property Mat, 80-90, chair, 87-89; mem, comt cracking, Am Concrete Inst, 79-, chair, 79-85, mem bd dir, 88-91; mem, Shear & Torsion, 80-, Bond, 84-, Tech Activ Comt, 85-91, Fracture Mech, 86-; mem, Concrete Mat Res Coun, 85-, chair, 90- *Honors & Awards:* Walter L Huber Res Prize, Am Soc Civil Engrs, 85; Delmar L Bloem Distinguished Serv Award, Am Concrete Inst, 86. *Mem:* Fel Am Soc Civil Engrs; Am Soc Eng Educ; Prestressed Concrete Inst; Post Tensioning Inst; Sigma Xi; Am Inst Steel Construct; fel Am Concrete Inst; Am Soc Eng Educ; AAAS. *Res:* Structural and materials engineering with emphasis on plain, reinforced and prestressed concrete and composite steel-concrete construction. *Mailing Add:* Dept Civil Eng Univ Kans 2006 Learned Hall Lawrence KS 66045-2225

DARWIN, JAMES T, JR, b Decatur, Tex, Apr 13, 33. SIMULATION, HARDWARE-SOFTWARE INTERFACING. *Educ:* Univ Tex, BS, 54, MA, 62, PhD(math), 63. *Prof Exp:* Asst math, Univ Tex, 58-63; asst prof, Auburn Univ, 63-69; assoc prof, Memphis State Univ, 69-75; programmer analyst, Vitro Labs Div, Automation Industs, Inc, 75-78; programmer analyst, 78-82, STAFF ENGR, MCDONNELL-DOUGLAS TECH SERV CO, 82- *Mem:* Math Asn Am; Asn Comput Mach. *Res:* Representation of linear operators on linear spaces; kernels for linear transformations. *Mailing Add:* PO Box 58024 Houston TX 77258

DARWIN, STEVEN PETER, b New Bedford, Mass, Aug 20, 49. BOTANY. *Educ:* Drew Univ, BA, 71; Univ Mass, Amherst, MS, 73, PhD(bot), 76. *Prof Exp:* ASST PROF BIOL, TULANE UNIV, 77- *Concurrent Pos:* Fel, Gray Herbarium, Harvard Univ, 76-77. *Mem:* Int Assoc Plant Taxon; Sigma Xi. *Res:* Taxonomy of flowering plants, especially systematics of Rubiaceae in the Pacific; biogeography; tropical biology; flora of the southeastern United States. *Mailing Add:* Dept Biol Tulane Univ New Orleans LA 70118

DARZYNKIEWICZ, ZBIGNIEW DZIERZYKRAJ, b Dzisna, Poland, May 12, 36; m 66; c 2. CELL BIOLOGY, CYTOCHEMISTRY. *Educ:* Med Acad, Warsaw, MD, 60, PhD(cell biol), 65. *Prof Exp:* Physician, Surg Ward, IVth City Hosp, Warsaw, 60-61; res assoc cytochem, Molecular Enzym Unit, State Univ NY, Buffalo, 65-66; sr res asst histol, Med Acad, Warsaw, 66-68; res assoc cytol, Inst Cell Res, Med Nobel Inst, Stockholm, 68-69; staff scientist, Dept Connective Tissue Res, Boston Biomed Res Inst, 69-74; res assoc cytol, Mem Sloan-Kettering Cancer Ctr, 74-76 & 88-, assoc mem, 78-88; prof, Grad Sch Med Sci, Cornell Univ, 88-90; PROF MED, NY MED COL, VALHALLA, NY, 90-, DIR, CANCER RES INST, 90- *Concurrent Pos:* Am Cancer Soc grant, 70-72; Nat Cancer Inst & Dept Health & Human Serv grants, 78-; mem, Mem Sloan-Kettering Cancer Ctr, 88-90. *Honors & Awards:* Merit Award, Nat Cancer Inst, 86- *Mem:* Cell Kinetics Soc (pres, 86-87); Am Cancer Soc; Soc Anal Cytol. *Res:* Regulation of genome activity in mammalian cells; cell cycle analysis; cell differentiation; flow cytometry. *Mailing Add:* Cancer Res Inst NY Med Col Valhalla NY 10595

DAS, ANADIJIBAN, b Calcutta, India, Mar 1, 34. MATHEMATICAL PHYSICS, GENERAL RELATIVITY. *Educ:* Calcutta Univ, BSc, 53, MS, 55, PhD(math physics), 64; Nat Univ Ireland, PhD(math physics), 61. *Prof Exp:* Asst prof, Carnegie Mellon Univ, 63-66; assoc prof, 66-69, PROF MATH, SIMON FRASER UNIV, 69- *Concurrent Pos:* Vis prof math, Univ Wash, Seattle, 73-74 & Univ BC, 81-82. *Mem:* Am Math Soc; Am Inst Physics; Can Asn Physics; Can Math Soc; Int Soc Relativity & Gravitation. *Res:* Mathematical foundations of quantum mechanics. *Mailing Add:* Dept Math & Statist Simon Fraser Univ Burnaby BC V5A 1S6 Can

DAS, ASHOK KUMAR, b Puri, India, Mar 23, 53; m 83. QUANTUM FIELD THEORIES, SUPERSYMMETRY. *Educ:* Univ Delhi, BSc, 72, MSc, 74; State Univ NY, Stony Brook, PhD(physics), 77. *Prof Exp:* Res assoc physics, City Col New York, 77-79, Univ Md, 79-81 & Rutgers Univ, 81-82; asst prof, 82-86, ASSOC PROF PHYSICS, UNIV ROCHESTER, 86- *Concurrent Pos:* Jr investr, US Dept Energy, 83. *Mem:* Am Phys Soc. *Res:* Quantum field, gauge, supersymmetric and supergravity theories. *Mailing Add:* Dept Phys & Astron Univ Rochester Rochester NY 14627

DAS, BADRI N(ARAYAN), b Calcutta, India, Oct 19, 27; US citizen; m 64; c 2. PHYSICAL METALLURGY. *Educ:* Univ Calcutta, BE, 52; Univ Ill, MS, 58; Ill Inst Technol, PhD(phys metall), 64. *Prof Exp:* Res asst, Univ Ill, 54-60; staff scientist, Tyco Lab, Inc, Mass, 64-67, sr scientist, 67-70; staff metallurgist, Mat Res Ctr, Allied Chem Corp, NJ, 70; METALLURGIST, US NAVAL RES LAB, 70- *Mem:* Am Crystallog Asn; Am Inst Mining, Metall & Petrol Engrs; Sigma Xi; Mat Res Soc. *Res:* Metal physics; x-ray diffraction; material synthesis and characterization of semiconductor, ferroelectric, dielectric, electro-optic and magnetic compounds and alloys; solidification and crystal growth of metal compounds and alloys; superconductors. *Mailing Add:* 12 Thurston Dr Upper Marlboro MD 20772

DAS, DIPAK K, b India, Jan 28, 46; c 2. LIPID BIOCHEMISTRY, FREE RADICAL. *Educ:* Calcutta Univ, PhD(biochem), 77. *Prof Exp:* Prof biochem, 84-86, DIR, CARDIOVASC DIV, UNIV CONN, SCH MED, 84- *Concurrent Pos:* Coun mem, Am Heart Asn. *Mem:* Am Soc Biol Chemists; fel Am Chem Soc; Int Soc Heart Res; Am Heart Asn; Am Phys Soc; Am Soc Biochemists. *Res:* Phospholipids; cellular physiology; cardiovascular research. *Mailing Add:* Cardiovasc Div Sch Med Univ Conn Farmington CT 06032

DAS, GOPAL DWARKA, b Shikarpur, India, Feb 11, 33; m 61; c 2. NEURAL TRANSPLANTATION, NEUROANATOMY. *Educ:* Univ Mysore, BS, 54; Univ Poona, MA, 57; Boston Univ, PhD(exp psychol), 65. *Prof Exp:* Lectr exp psychol, Gujarat Univ, India, 57-61; res asst neuroanat, Mass Inst Technol, 62-63, res assoc, 63-65, lectr, 67-68; res scientist, Max Planck Inst Psychiat, 66-67; asst prof, 68-71, assoc prof, 71-78, PROF NEUROBIOL, PURDUE UNIV, 78- *Mem:* Soc Neuroscience; Int Brain Res Orgn; Int Soc Develop Neurosciences. *Res:* Transplantation of nervous tissue in mammalian brain; neuroembryology in mammals. *Mailing Add:* Dept Biol Sci Purdue Univ West Lafayette IN 47907

DAS, KAMALENDU, b Sylhet, Bangladesh, Feb 2, 44; US citizen; m 69; c 2. OILFIELD CHEMISTRY, FOSSIL FUEL RESEARCH IN GENERAL. *Educ:* Univ Dhaka, Bangladesh, BSc(Honors), 64, MSc, 65; Univ Houston, PhD(chem), 75. *Prof Exp:* Asst prof chem, Women's Col, Sylhet, Bangladesh, 66-68 & Tolaram Col, Dhaka, Bangladesh, 68-71; res group leader, Baker Sand Control, 79-85; RES CHEMIST & ENHANCED OIL RECOVERY PROJ COORDR, MORGANTOWN ENERGY TECHNOL CTR, US DEPT ENERGY, 85- *Mem:* Am Chem Soc; Soc Petrol Engrs; Am Inst Chemists; Int Union Pure & Appl Chem. *Res:* Enhanced oil recovery; oilfield completion and production; shale oil extraction and fossil fuel research in general. *Mailing Add:* 465 Lawnview Circle Morgantown WV 26505

DAS, KIRON MOY, b Bengal, India, Dec 31, 41; m; c 2. MEDICINE, IMMUNOLOGY. *Educ:* Calcutta Univ, MB, BS, 65, DrMed(internal med), 69; Edinburgh Univ, PhD(internal med), 74; Am Bd Internal Med & Gastroenterol, dipl, 75. *Prof Exp:* Clin instr, Health Sci Ctr, State Univ NY Stony Brook, 73-74; fel gastroenterol, Nassau County Med Ctr, 73-74; attend physician & asst prof, 74-79, ASSOC PROF MED, INTERNAL MED & GASTROENTEROL, ALBERT EINSTEIN COL MED, 79- *Concurrent Pos:* Clin res fel, Western Gen Hosp, Univ Edinburgh, 70-73; consult, Food & Drug Admin, 75 & Am Hosp Formulary Serv, 76; clin investr award, USPHS, 76-79; prin investr, Nat Inst Arthritis, Metab & Diag Dis, 78-81 & Irma T Hirschl career scientist award, 80-85. *Mem:* Royal Col Physicians London; Royal Col Physician Edinburgh; Am Gasteroenterol Asn; Am Fed Clin Res; Am Soc Gastrointestinal Endoscopy. *Res:* Inflammatory bowel diseases; immunological disturbances; etiologic agents; pathogenesis of colonic mucosal injury. *Mailing Add:* Dept Med Gastroenterol UMD Robert Wood Johnson Med Sch Ac Health Sci Ctr New Brunswick NJ 08903-0019

DAS, MANJUSRI, b Bengal, India, Dec 27, 46; m 75. EGF-RECEPTOR, TROPHOBLAST-DERIVED GROWTH FACTOR. *Educ:* Univ Calcutta, India, BS, 66, MS, 69, PhD(biochem), 74. *Prof Exp:* Researcher angiotensin-conversion enzyme, Albert Einstein Col Med, NY, 74-76; researcher EGF-receptor, Univ Calif, Los Angeles, 76-78; asst prof, 78-84, assoc prof, 84-88, PROF BIOCHEM & BIOPHYS, SCH MED, UNIV PA, 88- *Concurrent Pos:* NIH res grant, Bethesda, Md, 82-86, Univ Pa, 82- *Mem:* Am Soc Biol Chemists; Am Soc Cell Biol; NY Acad Sci. *Res:* Cell proliferation in eucaryotes; molecular-biological and cytological techniques to study structure and function of peptide growth factors and their receptors. *Mailing Add:* 222 Lantwyn Lane Sch Med Univ Pa Narberth PA 19072

DAS, MIHIR KUMAR, b Purulia, India, Nov 2, 39; m 68; c 1. MACHINE TOOL TECHNOLOGY. *Educ:* Bihar Inst Technol, Ranchi Univ, BS, 61; Birmingham Univ, PhD(mech eng), 65. *Prof Exp:* Gov India res scholar mach tool, Birmingham Univ, 62-66; sr sci officer design & develop, Cent Mech Eng Res Inst, 66-68; chief design engr design & develop, Tech Develop, India, 68-69; Imp Chem Indust res fel mfg, Birmingham Univ, 69-71, sr res fel, 71-81; assoc prof, 81-85, PROF MECH ENG, CALIF STATE UNIV, LONG BEACH, 85-, ASSOC DEAN, SCH ENG. *Concurrent Pos:* Consult, Eng DRD Ltd, 71-81; vis prof, Univ Zulia, Venezuela, 74- & Korea Advan Inst Sci, Seoul, 78-; exchange vis prof, Warsaw Tech Univ, 80- *Mem:* Am Soc Mech Engrs; Am Soc Eng Educ; Inst Mech Engrs England. *Res:* Dynamic metal cutting; machine tool vibration; bar shearing; metal forming; dynamic impact; high energy rate forming; computer-aided design and manufacture; computer in engineering education. *Mailing Add:* Sch Eng Calif State Univ 1250 Bell Flower Blvd Long Beach CA 90840

DAS, MUKUNDA B, b Khulna, Bangladesh, Sept 1, 31; m 56; c 2. SOLID STATE ELECTRONICS, ELECTRICAL ENGINEERING. *Educ:* Univ Dacca, BSc, 53, MSc, 55; Univ London, PhD(transistor electronics), & Imp Col, dipl, 60. *Prof Exp:* Res asst physics, Univ Dhaka, 55-56; lectr elec eng, Imp Col, London, 60-62; sr res officer appl physics, E Regional Lab, Coun Sci & Indust Res, Dhaka, 62-64; from mem sr sci staff to mem prin sci staff semiconductors, Hirst Res Ctr, ASM Ltd, Eng, 64-68; assoc prof, 68-79, PROF ELEC ENG, PA STATE UNIV, 79-, ASSOC DIR, CTR ELECTRONIC MAT & PROCESSING, 87- *Concurrent Pos:* Specialist lectureship, Chelsea Col Sci & Tech, London, 68; vis prof, Wright-Patterson Air Force Base, 83-84, Chalmers Univ Technol, Gothenburg, Sweden, 87; consult, HRB-Singer, Inc, 69 & Device Design, Inst Sci Invest of Venezuela, 76-, Gen Elec Electronics Lab, 84- *Honors & Awards:* Blumlain-Browne-Willan Pemium Award, Inst Elec Engrs, UK, 67; NSF grant, 79, 81, 84, 85, 88, 90. *Mem:* Sr mem Inst Elec & Electronic Engrs; Sigma Xi. *Res:* Design and fabrication of semiconductor devices and integrated circuits; characterization and modeling of metal-oxide-silicon transistors, integrated circuits and GaAs & AlGaAs-GaAs millimeter-wave field effect devices; characterization of LF & HF noise in transistors. *Mailing Add:* Dept Elec Eng Pa State Univ University Park PA 16802

DAS, NABA KISHORE, b Patna, India, Oct 4, 34; m 60; c 1. MICROBIOLOGY, VETERINARY MEDICINE. *Educ:* Bihar Univ, BVS, 59; Univ Mo-Columbia, MS, 63, PhD(microbiol), 67. *Prof Exp:* Vet asst surgeon, Govt Bihar, India, 59-61; asst microbiol, Univ Mo-Columbia, 63-67; sr res microbiologist, Norwich Pharmaceut Co, 67-69; sr res microbiologist, Res Div, W R Grace & Co, 69-79; vet med officer, 79-80, SUPVY RES VET MED OFFICER, DEPT VET MED RES, BUR VET MED, FOOD & DRUG ADMIN, 80- *Mem:* AAAS; NY Acad Sci; Am Soc Microbiol; Sigma Xi. *Res:* Chemotherapy of bacterial infections in experimental animals for drug evaluation; host-parasite relationship and its pathogenesis in model bacterial infections; anaerobic bacteriology; antibiotic resistance development and transfer by bacteria. *Mailing Add:* Ctr Vet Med HFV-133 Food & Drug Admin 5600 Fishers Lane Room 68-12 Rockville MD 20857

DAS, NIRMAL KANTI, b Chittagong, EPakistan, Mar 1, 28; US citizen; m 58; c 2. CELL BIOLOGY. *Educ:* Univ Calcutta, BSc, 48, MSc, 50; Univ Wis, PhD(bot), 57. *Prof Exp:* Lectr biol, Midnapore Col, 51-52; res asst cytochem, Carnegie Inst, 52-53; res assoc bot, Univ Wis, 56-58; from asst res zoologist to assoc res zoologist, Univ Calif, Berkeley, 58-73; assoc prof cell biol, Med Col, Univ Ky, 73-78; health scientist adminr, Nat Inst Aging, 78-81, EXEC SECY, ALLERGY, IMMUNOL & TRANSP RES COMT, NIH, 81- *Mem:* AAAS; Am Soc Cell Biol. *Res:* Quantitative cytochemistry of nucleic acids and proteins; various aspects of normal and abnormal cell growth, proliferation and development. *Mailing Add:* Dept Health Affairs Univ Calif 803 Great Western Bldg 2150 Shattuck Ave Berkeley CA 94720

DAS, PANKAJ K, b Calcutta, WBengal, India, June 15, 37; US citizen; m 67; c 2. SIGNAL PROCESSING DEVICES, ULTRASOUND. *Educ:* Univ Calcutta, BSc, 57, MSc, 60, PHD(elec eng), 64. *Prof Exp:* Instr, Polytech Inst Brooklyn, 64-65, asst prof, 65-68; assoc prof, Univ Rochester, 68-72; assoc prof, 74-77, PROF, RENSSELAER POLYTECH INST, 77- *Concurrent Pos:* Orgn Am States vis prof, Ctr Invest & Advan Studies, Nat Politech Inst, Mexico City, 72-73; vis prof, Elec Eng Dept, Colo State Univ, 80-81, Dept Appl Physics & Info Sci, Univ Calif, San Diego, 81. *Mem:* Am Phys Soc; Inst Elec & Electronics Engrs; Optical Soc Am; Acoustical Soc Am. *Res:* Electron devices; surface acoustic and charge coupled devices; acousto-optic interaction; hot electron microwave devices; application of ultrasound in bioengineering and non-destructive testing; optical signal processing. *Mailing Add:* Elec Comput & Syst Eng Dept Rensselaer Polytech Inst Troy NY 12181

DAS, PARITOSH KUMAR, b Bangladesh, Dec 10, 42; m 73; c 1. FAST KINETICS, ORGANIC PHOTOCHEMISTRY. *Educ:* Dacca Univ, BSc, 63, MSc, 65; Univ Houston, PhD(chem), 77. *Prof Exp:* Lectr chem, Holy Cross Col, 66-69 & Notre Dame Col, 67-69; prof, Ranaghat Col, 69-73; res fel, Univ Houston, 77-78; res assoc, 78-79, ASST PROF SPECIALIST, UNIV NOTRE DAME, 79- *Mem:* Sigma Xi; Am Chem Soc. *Res:* Molecular spectroscopy, photophysics and photochemistry involving organic systems; mechanistic aspects of electron, proton and hydrogen transfer reactions in condensed systems; laser flash photolysis and pulse radiolysis. *Mailing Add:* Radiation Lab Univ Notre Dame PO Box H Notre Dame IN 46556

DAS, PHANINDRAMOHAN, b Bholabo, Bangladesh, Feb 2, 26; m 59; c 3. ATMOSPHERIC PHYSICS, ATMOSPHERIC DYNAMICS. *Educ:* Univ Dacca, BSc, 47, MSc, 48; Univ Chicago, PhD(meteorol), 63. *Prof Exp:* Res asst physics, Banaras Hindu Univ, 49-51; asst meteorologist, India Meteorol Dept, 51-55; asst prof appl physics, Indian Inst Technol, Kharagpur, 55-59; from asst to res assoc meteorol, Univ Chicago, 59-64; asst prof appl physics, Indian Inst Technol, Kharagpur, 64-67; asst prof meteorol, Tex A&M Univ, 67-68; res physicist, Air Force Cambridge Res Lab, 68-70; assoc prof, 70-81, PROF METEOROL, TEX A&M UNIV, 81- *Concurrent Pos:* Collab res, Univ Hawaii, Hilo, 78, Nat Hurricane Res Lab & Nat Oceanic & Atmospheric Admin, Environ Res Lab, Coral Gables, 80; vis prof, Monash Univ, Australia, 84; summer fac res fel, AFOSR/ESD, 87 & AFOSR/Geophys Lab, 89. *Honors & Awards:* Fel, Am Meterol Soc, 87. *Mem:* Am Meteorol Soc; Royal Meteorol Soc. *Res:* Ionospheric physics; radar meteorology; physics and dynamics of clouds; mesoscale and tornado dynamics. *Mailing Add:* 1005 Glade St College Station TX 77840

DAS, PRASANTA, b Calcutta, West Bengal, India, Jan 3, 44; US citizen; m 81. SYSTEMS DESIGN, SYSTEMS & COMPUTER SCIENCE. *Educ:* Burdwan Univ, BS, 66; Univ Vt, MS, 71; Case Western Reserve Univ, PhD(syst eng), 76. *Prof Exp:* Asst prof opers res, Univ Va, 77-80; sr engr systs sci, Hittman Corp, 81-82; systs analyst comput systs, Arbitron Co, subsid Control Data Corp, 82-83; sr systs analyst telecommun eng, MCI Telecommun Corp, 83-84; SYSTS ENGR, DEFENSE COMMAND, CONTROL & COMMUN, ROCKWELL INT CORP, 84- *Mem:* Armed Forces Commun & Electronics Asn. *Res:* Military command, control and communications; electronic warfare; systems analysis and integration. *Mailing Add:* 8606 21st Pl Adelphi MD 20783

DAS, SALIL KUMAR, b Rangoon, Burma, Dec 21, 40; Indian citizen; m 68; c 1. BIOCHEMISTRY, FOOD SCIENCE. *Educ:* Calcutta Univ, BSc, 58 & 59, MSc, 61, DSc(biochem), 74; Mass Inst Technol, ScD(nutrit biochem), 66. *Prof Exp:* Res asst, Mass Inst Technol, 62-66; res assoc physics, Univ Ariz, 66-67; res assoc chem, Grad Inst Tech, Univ Ark, 67-68 & Duke Univ, 68-69; from asst prof to assoc prof, 69-81, PROF BIOCHEM, MEHARRY MED COL, 81- *Concurrent Pos:* NIH grant, 72-; consult UNDP, India, 86. *Honors & Awards:* Cressy Morrison Award, NY Acad Sci, 67; IPA Award, NIH, 85. *Mem:* Fel Am Inst Chemists; Sigma Xi; Am Inst Nutrit; Inst Food Technologists; Int Asn Dent Res; Am Soc Biochem; Biochem Soc. *Res:* Chemistry and metabolism of lipids. *Mailing Add:* Dept Biochem Meharry Med Col 1005 18th Ave N Nashville TN 37208

DAS, SAROJ R, CANCER MARKERS. *Educ:* Univ Bombay, PhD(biochem), 61. *Prof Exp:* tech dir, Metpath, Inc, 80-87; PRES, ANI LYTICS INC, 88- *Mem:* Am Asn Immunologists; Am Asn Clin Chem. *Mailing Add:* Ani Lytics Inc 360 Christopher Lane Gaithersburg MD 20879

DAS, SUBODH KUMAR, b Bhagalpur, India, June 19, 47; m 70; c 2. METALLURGICAL ENGINEERING. *Educ:* Bihar Inst Technol, India, BSc, 69; Indian Inst Technol, M Tech, 72; Univ Mich, Ann Arbor, PhD(metall eng), 74; Univ Pittsburgh, MBA, 82. *Prof Exp:* Jr res asst metall eng, Indian Inst Technol, Kanpur, 70; res asst, Univ Mich, Ann Arbor, 71-74; engr metall eng, Alcoa Labs, Aluminum Co Am, 74-81; RES CONSULT, ANACONDA ALUMINUM CO, ATLANTIC RICHFIELD CO, 81- *Mem:* Sigma Xi; Am Chem Soc; Metall Soc. *Res:* Extractive metallurgy of metals; process metallurgy of aluminum and aluminum alloys; fused salt electrolysis; coal purification; carbon electrode technology; thermodynamics and structure of ordered and disordered carbons; economic analysis of extractive metallurgical industries technology, forecasting and planning. *Mailing Add:* Anaconda Copper Co PO Box 27007 Tucson AZ 85726

DAS, SURYYA KUMAR, b Calcutta, India; US citizen; m 61; c 2. POLYMER CHEMISTRY, PHYSICAL CHEMISTRY. *Educ:* Calcutta Univ, India, BSc, 48, MSc, 50, PhD(chem), 56. *Prof Exp:* Tech officer chem, Allied Resins & Chem, Calcutta, 60-63; sr res chemist, PPG Industs, 63-67, res assoc, 67-73, sr res assoc, 73-82, scientist, 82-85, SR SCIENTIST, PPG INDUSTS, 85- *Concurrent Pos:* Fel, Univ Reading, UK, 56-58 & State Univ NY, Syracuse, 58-60. *Mem:* Am Chem Soc. *Res:* Synthetic polymer chemistry and physical chemistry of polymers. *Mailing Add:* 100 Chapel Knoll Dr Pittsburgh PA 15238-1201

DAS, TARA PRASAD, b India, July 7, 32; US citizen; m 58; c 3. BIOPHYSICS. *Educ:* Patna Univ, BS, 49; Univ Calcutta, MS, 51, PhD(physics), 55. *Prof Exp:* Lectr theoret physics, Saha Inst Nuclear Physics, 53-55; res assoc chem, Cornell Univ, 55-56; res assoc physics, Univ Calif, Berkeley, 56-57; reader, Saha Inst Nuclear Physics, 57-58; res asst prof, Univ Ill, 58-59; res assoc chem, Columbia Univ, 59-60; sr res officer, Atomic Energy Estab, Govt India, 60-61; from assoc prof to prof physics, Univ Calif, Riverside, 61-69; prof, Univ Utah, 69-71; PROF PHYSICS, STATE UNIV NY ALBANY, 71- *Concurrent Pos:* NSF grants, 62-; NIH grants, 72-; NASA grant, 86-87. *Honors & Awards:* Sr US Scientist Award, Alexander von Humbolt Found, WGer, 77-78 & 82; Award for Excellence in Res, State Univ NY, 84; Res Scientist Award, Yamada Sci Found, Japan, 87-89. *Mem:* Fel Am Phys Soc; Biophys Soc. *Res:* Electronic structures of atoms, simple molecules, and solid state, including theory of electron-nuclear hyperfine interactions; nuclear magnetic resonance imaging; electronic structure and properties of biologically important molecules. *Mailing Add:* Dept Physics State Univ NY 1400 Washington Ave Albany NY 12222

DASARI, RAMACHANDRA R, b Krishna, Andhra Prad, India, July 15, 32; m 51; c 2. LASER SPECTROSCOPY, MEDICAL PHYSICS. *Educ:* Andhra Univ, BS, 54; Benares Univ, MS, 56; Aligarh Muslin Univ, PhD(physics), 60. *Prof Exp:* Prof physics, Indian Inst Technol, Kanpur, India, 61-78; sr res officer laser spectros, Nat Res Coun, Ottawa, 78-79; vis scientist laser spectros, Univ Brit Columbia, Vancouver, 79-80; vis prof physics, 80-81, ASST DIR & PRIN RES SCIENTIST LASER SPECTROS, MASS INST TECHNOL, CAMBRIDGE, 81- *Concurrent Pos:* Vis scientist laser spectros, Mass Inst Technol, Cambridge, 66-68; mem, Univ Grants Comn, New Delhi, 72-78, Nat Comt Lasers, India, 75-78. *Res:* Laser spectroscopy of atoms and molecules; atom-perturber collission dynamics; applications in nuclear physics; biomedical echnology programs; laser-optical pumping; Raman and fluorescence studies; picosecond time domain studies. *Mailing Add:* Spectros Lab Mass Inst Technol Cambridge MA 02173

D'ASARO, L ARTHUR, b Buffalo, NY, Jan 20, 27; m 53; c 4. SEMICONDUCTOR DEVICE RESEARCH, PHOTONIC SWITCHING DEVICES. *Educ:* Northwestern Univ, BS, 49, MS, 50; Cornell Univ, PhD(eng physics), 55. *Prof Exp:* DISTINGUISHED MEM TECH STAFF, AT&T BELL LABS, 55- *Mem:* Inst Elec & Electronic Engrs; Am Phys Soc; Sigma Xi. *Res:* Semiconductor device research and development on digital, microwave and photonic devices in silicon and III-V compounds, for example, SEED photonic switch arrays, low noise and power GaAs MESFETS, avalanche photodiodes and semiconductor lasers. *Mailing Add:* AT&T Bell Labs Rm 7B408 Murray Hill NJ 07974-2070

DASCH, CLEMENT EUGENE, b Steubenville, Ohio, Nov 28, 25; m 47; c 4. ENTOMOLOGY, TAXONOMY. *Educ:* Cornell Univ, BS, 49, PhD(entom), 53. *Prof Exp:* From asst prof to assoc prof, 53-61, coordr sci div, 66-72 & 76-79, chmn dept, 53-77, PROF BIOL, MUSKINGUM COL, 61- *Concurrent Pos:* NSF res grant, Univ Mich, 63-64; Mack Found & Am Philos Soc res grants, 64, 72, 80 & 81. *Mem:* AAAS; Soc Syst Zool; Am Inst Biol Sci. *Res:* Systematics and ecology of parasitic wasps of the family Ichneumonidae. *Mailing Add:* 160 Montgomery Blvd New Concord OH 43762

DASCH, ERNEST JULIUS, b Dallas, Tex, July 9, 32; m 60. GEOCHEMISTRY. *Educ:* Sul Ross State Col, BS, 56; Univ Tex, Austin, MA, 59; Yale Univ, MS, 67, PhD(geochem), 69. *Prof Exp:* Geologist, Magnolia Petrol Co, 56 & W F Guyton & Assoc, Tex, 59-61; res asst geochem, Yale Univ, 63-68; Fulbright vis res fel, Australian Nat Univ, 68-70; asst prof, 70-74, ASSOC PROF GEOL, ORE STATE UNIV, 74- *Mem:* AAAS; Geol Soc Am; Am Geophys Union; Geochem Soc; Sigma Xi. *Res:* Strontium and lead geochemistry of selected igneous and sedimentary systems; petrology. *Mailing Add:* NASA Hq Mail Code XEU Washington DC 20546

DASCH, GREGORY ALAN, b Ithaca, NY, Oct 13, 48; m 76; c 2. RICKETTSIOLOGY, INSECT PHYSIOLOGY. *Educ:* Oberlin Col, AB, 70; Yale Univ, PhD(biol), 75. *Prof Exp:* Res asst insect physiol, Yale Univ, 71-75; microbiologist, Dept Microbiol, 75-77, chemist, 77-79, HEAD, RICKETTSIAL DIS PROG, NAVAL MED RES INST, 79- *Concurrent Pos:* Nat Res Coun fel, Naval Med Res Inst, 75-76; adv comt, ATCC virol, 88-

Mem: AAAS; Am Soc Microbiol; NY Acad Sci; Am Soc Rickettsiology & Rickettsial Dis (pres, 87-88); Am Soc Trop Med Hyg; Entom Soc Am. *Res:* Intracellular bacteria, particularly rickettsia, ehrlichia and the symbiotes of insects: evolution, biochemistry, immunology, and host cell interaction; vaccine development; pathogenic mechanisms. *Mailing Add:* Rickettsial Dis Prog Naval Med Res Inst Bethesda MD 20889-5055

DASCH, JAMES R, b Zanesville, Ohio, Feb 17, 56; m 85; c 2. B LYMPHOCYTE ACTIVATION, IMMUNOSUPPRESSION. *Educ:* Kenyon Col, BS, 78; Case Western Reserve Univ, PhD(microbiol), 83. *Prof Exp:* Postdoctoral fel immunol, Dept Biol Sci, Stanford Univ, 83-87; res scientist, 87-89, MGR, DEPT IMMUNOL, CELTRIX LABS, COLLAGEN CORP, 89-, SR SCIENTIST, 91- *Mem:* Am Asn Immunol; AAAS. *Res:* Effects of immuno suppressive cytokines, especially transforming growth factor, on lymphoid system and hematopoietic system; utilize neutralizing monoclonal antibodies to study function of this cytokine; characterization of receptors for this cytokine. *Mailing Add:* 2500 Faber Pl Palo Alto CA 94303

DASCHBACH, JAMES MCCLOSKEY, b Medford, Mass, July 29, 32; m 65; c 8. INDUSTRIAL ENGINEERING. *Educ:* Univ Notre Dame, BS, 54; Southern Methodist Univ, MBA, 61; Okla State Univ, PhD(indust eng), 66. *Prof Exp:* Serv engr, Gen Dynamics, Ft Worth, Tex, 57-60; customer serv engr, Aero Commander Co, 60-64; instr indust eng, Okla State Univ, 64-66; prof indust eng, Col Eng, Notre Dame, 66-84; chmn, 84-89, PROF INDUST ENG, UNIV TOLEDO, OHIO, 84- *Concurrent Pos:* Vis prof, Nat Defense Univ, Ft McNair, 79-80. *Mem:* Sigma Xi; Am Soc Eng Educ; Am Inst Indust Eng; Int Soc Parametric Analysts. *Res:* Modelling state court systems; parametric cost estimating models; reverse engineering techniques. *Mailing Add:* 2025 Orchard Rd Toledo OH 43606-2621

DASCOMB, HARRY EMERSON, b Bath, NY, Aug 12, 16; m 39; c 3. INTERNAL MEDICINE. *Educ:* Colgate Univ, AB, 38; Univ Rochester, MD, 43. *Prof Exp:* Intern med, Univ Rochester, 43; resident physician, Iola Sanatorium, 43-45; intern med, Univ Rochester, 45-46; from instr to prof, med & preventive med, Sch Med, La State Univ, 47-80, asst dean, 75-78, EMER PROF, MED & PREVENTIVE MED, SCH MED, LA STATE UNIV, 80-; PROF MED, SCH MED, UNIV NC, CHAPEL HILL, 80- *Concurrent Pos:* Buswell fel, Sch Med & Dent, Univ Rochester, 46-47; dir, Off Hosp Infections Control, Charity Hosp La, 68-73, med dir, 74-78; named chair, La State Univ, Sch Med; named lecture, infectious dis, La State Univ. *Mem:* AMA; fel Am Col Physicians; Infectious Dis Soc Am. *Res:* Infectious disease. *Mailing Add:* Wake Area Health Ed Ctr 3000 New Bern Ave Raleigh NC 27610

DAS GUPTA, AARON, b Nov 20, 43; m 72; c 2. STRUCTURAL MECHANICS, STRESS ANALYSIS. *Educ:* Indian Inst Technol, BTech (Hons), 63; Nova Scotia Tech Col, M Eng, 68; Va Polytech Inst, PhD(mech eng), 75. *Prof Exp:* Asst engr design, Heavy Eng Corp, India, 63-65; electronics engr, electronic prod, Can Marconi, 67-69; prod engr aerospace struct, Whittaker Corp, 69-70; proj engr design develop, Kingsport Press, 73-75; proj engr stress anal, Sundstrand Aviation, 75; SR MECH ENGR BLAST DYNAMICS, BALLISTIC RES LAB, US ARMY, 75- *Concurrent Pos:* Consult, Iron Ore Co, Can, 66; teaching assoc, Va Polytech Inst, 70-74; lectr, NMex State Univ, 71-72; eng consult, Com Fabrication Co, 72; design consult, AAI Corp, 80 & Black & Veatch Corp, 80-81; consult engr, US Army Meradcom, 80- *Mem:* Am Acad Sci; Sigma Xi; NY Acad Sci; Am Soc Mech Engrs; fel Inst Diag Engrs, UK. *Res:* Penetration and fracture mechanics; blast dynamics; structural dynamics; finite element and finite difference; analysis containment structures; thermal stress analysis; author and coauthor of 50 publications. *Mailing Add:* 104 John St Perryville MD 21903-2630

DASGUPTA, ARIJIT M, b Calcutta, India, Aug 30, 57; US citizen; m 84. POLYMER SYNTHESIS, POLYMER CHARACTERIZATION. *Educ:* Univ Bombay, BSc, 79; Univ Akron, MS, 81, PhD(polymer sci), 85. *Prof Exp:* Sr chemist, Plastics Div, Monsanto Chem Co, 85, sr chemist explor sci & saflex technol, 86-90, res specialist, 90-91, GROUP LEADER SAFLEX TECHNOL, MONSANTO CHEM CO, EUROPE, 91- *Concurrent Pos:* Sen, Tech Community Monsanto, 90- *Mem:* Am Chem Soc; Am Inst Chemists; Am Mgt Asn. *Res:* Synthesis, characterization structure-property relationship evaluation of current and new products; identification and development of second generation products; author of ten publications; awarded six US patents. *Mailing Add:* Monsanto Chem Co 730 Worcester St Springfield MA 01151

DASGUPTA, ASIM, b Calcutta, India, March 19, 51; m 80. VIROLOGY, MOLECULAR BIOLOGY. *Educ:* Calcutta Univ, India, BS, 70, MS, 72; Univ Nebr, Lincoln, PhD(chem), 78. *Prof Exp:* Grad res asst biochem, Univ Nebr, Lincoln, 74-78; fel assoc virol, Mass Inst Technol, 78-80; asst prof, 81-85, ASSOC PROF MICROBIOL & IMMUNOL, UNIV CALIF, LOS ANGELES, 85- *Concurrent Pos:* Prin investr, NIH grant, 81-; fac res award, Am Cancer Soc, 85-89. *Mem:* Am Soc Microbiol; AAAS. *Res:* Mechanism of replication of RNA viruses and the roles of viral and host protein in the process. *Mailing Add:* Dept Microbiol & Immunol Med Sch Univ Calif Los Angeles CA 90024

DASGUPTA, GAUTAM, b Calcutta, WBengal, India, Oct 13, 46; m 70; c 2. COMPUTATIONAL MECHANICS, SINGULAR PHYSICAL PROBLEMS. *Educ:* Bengal Eng Col, India, BEng, 67, MEng, 69; Univ Calif, Berkeley, PhD(civil eng), 74. *Prof Exp:* Teachers trainee appl mech, Bengal Eng Col, India, 67-70; consult math, Berkeley Unified Sch Dist, 71-74; asst res engr civil eng, Univ Calif, Berkeley, 74-77; asst prof, 77-81, ASSOC PROF ENG MECH, COLUMBIA UNIV, 81- *Concurrent Pos:* Lectr appl mech, Govt Polytech, Panaji, Goa, India, 70; consult eng, Bechtel Corp, San Francisco, Calif, 76; prin investr, Grant Columbia Univ, NSF, 77-85, NIH, 85- *Mem:* Am Soc Civil Eng; Am Soc Mech Eng; Am Acad Mech; Soc Indust Appl Math. *Res:* Computational mechanics; dynamic analyses of viscoelastic systems; finite element method for unbounded continua; dynamic impacts; kinematics of cranio-facial growth. *Mailing Add:* Dept Civil Eng Columbia Univ 620 SW Mudd New York NY 10027

DAS GUPTA, KAMALAKSHA, b Calcutta, India, Feb 1, 17; m 47; c 1. PHYSICS. *Educ:* Univ Calcutta, MSc, 40; Univ Liverpool, PhD(physics), 52. *Prof Exp:* Lectr physics, Univ Calcutta, 43-47 & 52-56, reader, 59-61; sr res fel physics, Calif Inst Technol, 61-66; consult, Moon Surveyor Proj, Jet Propulsion Lab, 63-80; PROF PHYSICS, TEX TECH UNIV, 66- *Mem:* Am Phys Soc. *Res:* Electro-magnetic interactions in the region of 1 keV to 2 MeV, coherent interaction of electrons and photons in crystals and search for x-ray laser; electron states of superconductors, Type II, by the method of soft x-ray emission and absorption spectroscopy; total reflection of x-rays from thin films. *Mailing Add:* Dept Physics Tex Tech Univ Lubbock TX 79409

DASGUPTA, RATHINDRA, b Calcutta, West Bengal, India, Nov 12, 48; US citizen; m 77; c 2. SOLIDIFICATION ALLOYS, KINETICS REACTIONS. *Educ:* Regional Eng Col, Durgapur, India, BE, 71; Univ Wis-Milwaukee, MS, 74; Univ Wis-Madison, PhD(metall eng), 78. *Prof Exp:* Lectr metall eng, Univ Wis-Madison, 78-79; from asst prof to prof mat sci, Milwaukee Sch Eng, 79-90, Raymond D Peters Endowed prof mat sci, 87-90; TECH DIR, METAMOLD DIV, AMCAST INDUST CORP, 90- *Concurrent Pos:* Res asst, Motor Casting Co, W Allis, Wis, 80; process metall engr, Wis Centrifugal, Inc, Waukesha, Wis, 84; adj assoc prof, Mat Sci Dept, Univ Wis-Milwaukee, 84-85, adj prof, 86; vis prof, China Steel Corp, Kaohsiung, Taiwan, 85; tech consult, Styberg Eng, Racine, Wis, 89. *Mem:* Am Soc Metals; Am Foundrymen's Soc; Am Inst Mining, Metall & Petrol Engrs. *Res:* Alloy development; metal-matrix composites; evaluation of mold coatings and mold materials; heat treatment. *Mailing Add:* 5450 N Navajo Ave Glendale WI 53217

DAS GUPTA, SOMESH, b Calcutta, India, Sept 3, 35; m 64; c 1. MATHEMATICAL STATISTICS. *Educ:* Univ Calcutta, BSc, 53, MSc, 56; Univ NC, Chapel Hill, PhD(statist), 62. *Prof Exp:* Res scholar statist, Indian Statist Inst, 57-60; res asst, Univ NC, Chapel Hill, 60-62; from instr to asst prof, Columbia Univ, 62-64; from asst prof to assoc prof, Indian Statist Inst, 64-67; assoc prof, Univ Minneapolis, 67-70, chmn statist grad fac, 71-73, prof statist, 70-86; DIV THEORET STATIST & MATH, INDIAN STATIST INST, 82- *Concurrent Pos:* Statist consult, Psychomet Unit, Indian Statist Inst, 57-60; dir, Statist Consult Serv, Columbia Univ, 63-64; NSF grant, Univ Minn, 68-70, US Army res grants, 71-73 & 75-79; vis prof, Stanford Univ, 74; vis prof, Indian Statist Inst, 80-81; mem, adv bd, Ministry Human Resources, Govt India. *Honors & Awards:* Fulbright Award. *Mem:* Fel Inst Math Statist; fel Am Statist Asn; Int Statist Inst; Indian Statist Inst. *Res:* Inference in multivariate analysis; classification and discrimination; nonparametric inference; multiple decision theory; probability inequalities; statistical pattern recognition; author of 2 books. *Mailing Add:* Div Theoret Statist & Math Indian Statist Inst 203 Barrackpore Trunk Rd Calcutta 700035 India

DASGUPTA, SUNIL PRIYA, b Bangladesh; US citizen; m 56; c 1. SOLID STATE PHYSICS, CHEMISTRY. *Educ:* Univ Dacca, BSc, 47, MSc, 49; Univ Delhi, PhD(physics), 58. *Prof Exp:* Sr res officer, Nat Phys Lab, New Delhi, India, 56-62; fel solid state physics, Nat Res Coun Can, Ottawa, 62-65; res assoc relaxation mechanism, Princeton Univ, 65-67; res physicist, 67-81, sr res physicist, 81-88, RES SCIENTIST, HERCULES INC, 88- *Mem:* Tech Asn Pulp & Paper Indust; Am Chem Soc. *Res:* Dielectric properties; relaxation process and molecular structure; transport properties of semiconducting materials; polymer chemistry and polymer physics; surface chemistry and surface physics; several publications in leading journals. *Mailing Add:* 2528 Tigani Dr Wilmington DE 19808

DASH, HARRIMAN HARVEY, b New York, NY, May 26, 10; wid; c 1. BIOCHEMISTRY, QUANTUM CHEMISTRY. *Educ:* City Col New York, BS, 33; Univ Chicago, cert, 44; Polytech Inst Brooklyn, MS, 53. *Prof Exp:* Res chemist, Bellevue Hosp, New York, 34-35; proj chemist, Res Labs, Fordham Univ, 35-38; proj chemist, Consumers Labs, 38-42; biochemist med res, Nat Jewish Hosp, Denver, Colo, 46-47; consult chemist high polymers, 47-58; assoc biochemist, Med Lab, North Shore Hosp, Manhasset, NY, 58-62; chief biochemist, Nassau Hosp, Mineola, 62-66 & Variety Children's Hosp, Miami, Fla, 66-68; chief biochemist, Res Div Miami Heart Inst, 68-76; consult, Am Hosp, Miami, 76-81; RETIRED. *Mem:* AAAS; Am Chem Soc; Am Asn Clin Chemists; Am Inst Physics. *Res:* Clinical biochemistry and application of quantum theory of biochemistry. *Mailing Add:* 6095 Sabal Palm Blvd Tamarac FL 33319

DASH, JAY GREGORY, b Brooklyn, NY, June 28, 23; m 45; c 3. PHYSICS. *Educ:* City Col New York, BS, 44; Columbia Univ, AM, 49, PhD(physics), 51. *Prof Exp:* Mem staff, Los Alamos Sci Lab, 51-60; actg assoc prof physics, 60-61, assoc prof, 61-63, PROF PHYSICS, UNIV WASH, 63- *Concurrent Pos:* Guggenheim fel, 57-58; consult, Los Alamos Sci Lab, 60- & Boeing Co, 61-64; Louis Susman vis prof, Israel Inst Technol, 74-75; exchange prof, Univ d'Aix-Marseille, 77-78, 82 & 85-86; mem adv comt, Nat Sci Found, 76-80; dir, NATO Advanced Study Inst, 79. *Honors & Awards:* Davisson Germer Prize, Am Phys Soc, 85. *Mem:* Fel Am Phys Soc. *Res:* Low temperature, surface physics; physics of two dimensional matter and experimental properties of monolayer films adsorbed on solid surfaces; surface phase transitions; properties of frozen ground, frost heave. *Mailing Add:* Dept Physics Mail Stop FM-15 Univ Wash Seattle WA 98195

DASH, JOHN, b Hazleton, Pa, June 29, 33; m 68, 79; c 3. PHYSICAL METALLURGY. *Educ:* Pa State Univ, BS, 55, PhD(metall), 66; Northwestern Univ, MS, 60. *Prof Exp:* Res metallurgist, Crucible Steel Co, 55-58; res assoc phase transformations, Res Inst Adv Study, Martin Marietta Corp, Md, 60-63; from asst prof to assoc prof, 66-79, PROF PHYSICS, PORTLAND STATE UNIV, 79- *Mem:* Electron Micros Soc Am; Am Soc Metals; Am Electroplaters Soc; Sigma Xi. *Res:* Electron microscopy and electron diffraction; electrodeposition of metals. *Mailing Add:* Dept Physics Portland State Univ Box 751 Portland OR 97207

DASH, SANFORD MARK, b New York, NY, May 26, 43; m 64; c 4. COMPUTATIONAL FLUID DYNAMICS, AEROTHERMOPHYSICS. *Educ:* City Col New York, BSME, 65; NY Univ, MS, 66, PhD(aeronaut & astronaut), 69. *Prof Exp:* Res scientist fluid dynamics, Aerospace Res Labs, NY Univ, 68-70, Advanced Technol Labs, 70-74; sr res scientist, Gen Appl Sci Labs, 74-77; consult, Aeronaut Res Assoc Princeton, 77-80; TECH DIR, PROPULSION GAS DYNAMICS, DIV PROPULSION GAS DYNAMICS, SCI APPLICATIONS INT CORP, 80- *Concurrent Pos:* Adj prof, dept aeronaut, Dowling Col, 70-73; guest lectr, Univ Tenn Space Inst, 78-79; consult, Boeing, Gen Dynamics, Lockheed, Gen Elec & Pratt & Whitney, 83- *Mem:* Am Inst Aeronaut & Astronaut; Int Asn Math & Comput in Simulation. *Res:* Development of computational models for simulation of fluid dynamic problems related to rocket and aircraft plumes, nozzles, wakes and turbulent combustion; over 100 publications in various journals. *Mailing Add:* Sci Applications Int Corp 501 Office Center Dr Suite 420 Ft Washington PA 19034-3211

DASHEK, WILLIAM VINCENT, b Milwaukee, Wis, Aug 28, 39; div; c 2. ENVIRONMENTAL HEALTH & SCIENCE. *Educ:* Marquette Univ, BS, 61, MS, 63, PhD(plant physiol), 66. *Prof Exp:* NIH fel plant physiol & biochem, Mich State Univ, 66-67; asst prof biol, Boston Univ, 67-69; vis asst prof, State Univ NY, Buffalo, 69-70; asst prof, Va Commonwealth Univ, 70-75; vis res biologist, Univ Calif, San Diego, 75-76; sr fel plant biochem, State Univ NY, Syracuse, 76-78; asst prof biol, WVa Univ, 78-84; ASSOC PROF BIOL, DEPT BIOL, CLARK ATLANTA UNIV, 84- *Concurrent Pos:* Reviewer, jour & NSF & USDA grants, 81- *Mem:* Am Soc Plant Physiol; Int Biodeterioration Res Group; Pan Am Biodeterioration Soc; Sigma Xi. *Res:* Cell wall metabolism; pollen physiology and biochemistry; mechanism of action of aflaxtoxins; quantitation, localization and metabolism of plant hormones; air-pollutants and forest productivity; synthesis, secretion and regulation of cellulolytic and ligninolytic enzymes by wood-decaying fungi. *Mailing Add:* Dept Biol Clark Atlanta Univ Atlanta GA 30314

DASHEN, ROGER FREDERICK, b Grand Junction, Colo, May 5, 38; m 64; c 2. THEORETICAL PHYSICS. *Educ:* Harvard Univ, AB, 60; Calif Inst Technol, PhD(physics), 64. *Prof Exp:* Res assoc theoret physics, Calif Inst Technol, 64-65, asst prof, 65-66, prof, 66-69; mem staff, 66-69, prof theoret physics, Inst Advan Study, 69-86, PROF, DEPT PHYSICS, UNIV CALIF, SAN DIEGO, 86-, CHMN, 88- *Concurrent Pos:* Alfred P Sloan Found fel, 66-73, adv, 85-90; vis prof, Princeton Univ, 69-86; mem, Jason, 66-, consult Los Alamos & Fermi Nat Accelerator Lab, 72-75; mem, Defense Sci Bd Panel on Anti-Submarine Warfare & SSBY Security, 80-88; mem, Planning & Steering Adv Comt, Advan Technol Panel, Dept Navy, 84-; mem adv bd, Appl Physics Lab, Univ Washington, Seattle. *Mem:* Nat Acad Sci; Am Acad Arts & Sci. *Res:* Elementary particle theory; quantum field theory; statistical mechanics; waves in random media. *Mailing Add:* Dept Physics 0319 Univ Calif 9500 Gilman Dr San Diego CA 92093-0319

DASHER, GEORGE FRANKLIN, JR, b Russellville, Ky, Aug 5, 22; m 43; c 4. PHYSICAL CHEMISTRY, SURFACE CHEMISTRY. *Educ:* Kalamazoo Col, BA, 43; Univ Mich, MS, 47, PhD(chem), 49. *Prof Exp:* Chemist, Procter & Gamble Co, 49-61; asst dir res, Clairol Inc, Conn, 61-65, dir prod develop, 65-67, vpres res & develop, 67-73; vpres res & develop, Alberto-Culver Co, 73-77; RETIRED. *Concurrent Pos:* Lectr, Xavier Univ, Ohio, 60-61; trustee, William Rainey Harper Col, 77- *Mem:* AAAS; Am Chem Soc; Rheol Soc; Sigma Xi. *Res:* Adsorption at phase boundaries; emulsion stability; surface activity in biological systems; mechanical properties of polymers; flow behavior of liquids; viscoelastic behavior of cosmetic materials. *Mailing Add:* 324 W Br Rd-08 Stratton Mountain VT 05155-0708

DASHER, JOHN, b Ky, June 11, 14; m 38; c 3. METALLURGICAL ENGINEERING. *Educ:* Univ Ala, BS, 35. *Prof Exp:* Asst chem engr, Bur Mines, Md & Ala, 37-42; chief res dept, Tin Processing Corp, 42-46; exec officer, Atomic Energy Comn Proj, Mass Inst Technol, 46-51; asst chief metals group, Chem Construct Co, 51-55; res supvr, Crucible Steel Co Am, Pa, 55-63, proj dir procurement, 63-65; prin engr, Bechtel Group, Bechtel Investments Inc, 65-82; RETIRED. *Concurrent Pos:* Mem adv Bd, Nat Acad Sci, 62; consult, 82- *Mem:* Am Inst Mining, Metall & Petrol Engrs; US Metric Asn. *Res:* Process metallurgy; mineral dressing; coal; nonmetallics; nonferrous and ferrous metals. *Mailing Add:* 1961 Filbert No 301 San Francisco CA 94123

DASHER, PAUL JAMES, b Pleasant Plains, Ill, June 4, 12; m 35; c 2. CHEMISTRY. *Educ:* Univ Ill, BS, 34; Ind Univ, AM, 35, PhD, 37. *Prof Exp:* Res chemist, B F Goodrich Co, 37-39, asst dir rubber res, 39-41, dir develop & eng, Fuel Cell Div, 41-44, tech supt, 44-45, consult, Reclamation Div, 45, dir, New Process Div, 45-46; pres, Merit Chem Co, 46-47 & Summit Indust Prod Co, 46-59; PRES, DASHER RUBBER & CHEM CO, FAIRPORT DEVELOP CO, FAIRPORT TERMINAL CORP, 50-; VPRES, LOWMANS, INC, 50-; CHMN DEPT PHYSICS & PHYS SCI, PALM BEACH JR COL, 69- *Concurrent Pos:* Dir, Lowmans, Inc, 43-50. *Mem:* Am Chem Soc. *Res:* Electrolyte conductance in non-aqueous media; rubber research, emphasizing plant utilization of process; invention of new processes. *Mailing Add:* 1812 SE 14th St Ft Lauderdale FL 33316

DASHIELL, THOMAS RONALD, b Salisbury, Md, Sept 9, 27; m 49; c 2. CHEMICAL & BIOLOGICAL ARMS CONTROL & DISARMAMENT, ENVIRONMENTAL CONTROL & POLLUTION ABATEMENT. *Educ:* Western Md Col, BS, 50; Johns Hopkins Univ, BS, 62. *Prof Exp:* Bacteriologist, Dept Army, Ft Detrick, Frederick, Md, 51-53, lab supvr, 53-59, supvr chem engr, 59-69, asst sci div, 69-70; spec chem tech, Off Secy Defense, Washington, DC, 70-84, dir environ & life sci, 84-88; CONSULT, US ARMS CONTROL & DISARMAMENT AGENCY, 88- *Concurrent Pos:* Consult, var govt agencies & indust, 88- *Mem:* Am Inst Chem Engrs; Am Soc Microbiol; Am Chem Soc; Sigma Xi; NY Acad Sci; AAAS. *Res:* Management of research and development involving policy formulation, program development, fiscal control and program implementation in the areas of chemical and biological process engineering, medical and life sciences and environmental quality research and development; international negotiations; federal government congressional interactions. *Mailing Add:* 504 Thomas Ave Frederick MD 21701

DASHMAN, THEODORE, b Brooklyn, NY, Oct 7, 28; m 57; c 3. BIOCHEMISTRY. *Educ:* Brooklyn Col, BS, 50; Fla State Univ, MS, 64; NY Univ, PhD(biol), 77. *Prof Exp:* SR SCIENTIST, HOFFMANN-LA ROCHE INC, 77- *Concurrent Pos:* Asst dir, Pediat Metab Lab, Dept Pediat, NY Univ Med Ctr. *Mem:* AAAS; Am Chem Soc; Am Soc Pharmacol & Exp Therapeut; Am Inst Nutrit. *Res:* Intermediary metabolism of amino acids; inhibition of enzymes by pharmacologically active compounds. *Mailing Add:* 163 Pinewood Pl Teaneck NJ 07666

DASHNER, PETER ALAN, b Schenectady, NY, April 9, 51; m. CONSTITUTIVE THEORY, POLYMER RHEOLOGY. *Educ:* Southern Univ NY, Buffalo, BS, 73, MS, 75, PhD(eng sci), 76. *Prof Exp:* Asst prof eng sci & mech, Va Polytech Inst & State Univ, 76-77; asst prof theoret appl mech, Cornell Univ, 77-81; LECTR MECH ENG, CALIF STATE POLYTECH UNIV, 81- *Mem:* Soc Rheology; Sigma Xi. *Res:* Theoretical and practical problems relating to the developments of realistic, multiaxial, large deformation constitutive (stress-deformation) models for real materials, both fluid and solid. *Mailing Add:* 6316 Haven Ave Alta Loma CA 91707-3818

DASKAM, EDWARD, JR, b Detroit, Mich, Mar 1, 20; m 48; c 4. ELECTRICAL ENGINEERING, PHYSICS. *Educ:* Univ Tex, BS, 41; NY Univ, MS, 56. *Prof Exp:* Engr radar, Gen Elec Co, 42-44; engr radio commun, Gen Tel Serv Corp, 46-49, dept head, 49-51; engr, Airborne Instruments Lab, Cutler-Hammer, Inc, 51-52, sect head, 52-58, prog mgr, 58-62, dept head, 68-72, prog mgr, 72-75, dept head, 75-82; consult engr, Ail Div, Eaton Corp, Deer Park, 82-85; RETIRED. *Mem:* AAAS; sr mem Inst Elec & Electronics Engrs; Am Phys Soc. *Res:* Electronic systems, including antennas, microwaves, solid state circuits and microminiaturization. *Mailing Add:* 909 Pear Rd Huntington NY 11746

DASKIN, W(ALTER), b Passaic, NJ, July 17, 26; m 51; c 2. MECHANICAL ENGINEERING, FLUID MECHANICS. *Educ:* Cooper Union, BME, 45; Univ Mich, MSE, 48. *Prof Exp:* Engr vibration control, Jayburn Eng Co, NY, 46-47; instr mech eng, Johns Hopkins Univ, 49-54; engr fluid mech, Gen Elec Co, Ohio, 54-56, specialist heat transfer, 56; sr scientist reentry physics, Gen Appl Sci Labs, NY, 56-58, proj scientist, 58-60, sci supvr, 60-64, asst dir res, 64-66; consult appl physics, 66-68, mgr, Aeromech & Mat Lab Sect, 68-70, Technol Eng Sect, Philadelphia, 70-77, Cruise Missile Progs, 78-80 & plans & anal, 80-87, mgr, Advan Technol Progs, 88-90, CONSULT, GEN ELEC CO, LYNN, MASS, 91- *Concurrent Pos:* Res asst, Johns Hopkins Univ, 50-54; consult, Carbide & Carbon Chem Co, Tenn, 52-54 & Gen Elec Co, Ohio, 53-54; adj instr, Polytech Inst Brooklyn, 57. *Mem:* AAAS; Am Soc Mech Engrs; assoc fel Am Inst Aeronaut & Astronaut; Am Phys Soc; Sigma Xi. *Res:* Heat transfer; fluid mechanics; system optimization techniques; management of research and development. *Mailing Add:* 109 Bradlee Ave Swampscott MA 01907

DASLER, ADOLPH RICHARD, b Conklin, Mich, Mar 19, 33; m 58; c 3. PHYSIOLOGY. *Educ:* Western Mich Univ, BA & MA, 60; Mich State Univ, PhD(physiol), 66. *Prof Exp:* Instr med technol & physiol, Clin Labs, Hackley Hosp, 55-60; res asst physiol, Mich State Univ, 61; US Navy, 61- res investr, 62-63, proj officer shipboard toxicol opers protectional systs & head thermal stress sect, 63-64, head heat stress facil, 66-73, dep head environ stress div, 70-73, HEAD THERMAL STRESS BR, BUR MED & SURG, NAVAL MED RES UNIT, US NAVY, 63- MIL OFFICER ENVIRON BIOSCI DEPT, 70-, HEAD HEAT STRESS DIV, 73- *Concurrent Pos:* Officer-in-chg shelter habitability, Nat Naval Med Ctr & Bur Med & Surg, 62-64; US Navy rep med adv subcomt thermal factors in environ, Med Sci Div, Nat Acad Sci-Nat Res Coun, 63-65 & med adv comt environ physiol, 66-; consult, Commandant US Marine Corps, 66-, Naval Ships Eng Ctr, 68-, US Navy Environ Health Ctr, 68- & career & prog planning off, Navy Physiologist, 70-; head thermal stress sect, Navy Med Res & Develop Command, 74- *Mem:* AAAS; Am Soc Heat Refrig & Air-Conditioning Eng; NY Acad Sci; Sigma Xi. *Res:* Environmental physiology; temperature regulation; adaptation and tolerance to heat; etiology, prevention and treatment of heat illnesses. *Mailing Add:* 10005 Wildwood Ct Kensington MD 20895

DASLER, WALDEMAR, b St James, Minn, May 28, 10; m 39; c 3. BIOCHEMISTRY. *Educ:* Univ Wis, BS, 32, MS, 33, PhD(biochem), 38. *Prof Exp:* Chemist, B S Pearsall Co, Ill, 38-41; from res chemist to chief res chemist, Nutrit Res Labs, 41-48; from asst prof to prof, 48-78, EMER PROF, CHICAGO MED SCH, 78- *Mem:* AAAS; Am Chem Soc; Am Inst Nutrit; Soc Exp Biol & Med. *Res:* Sterols; vitamin D; rickets; osteolathyrism; copper determination; experimental cataract. *Mailing Add:* 4047 N Lawler Ave Chicago IL 60641

DASMANN, RAYMOND FREDRIC, b San Francisco, Calif, May 27, 19; m 44; c 3. ECOLOGY, FISH & WILDLIFE SCIENCES. *Educ:* Univ Calif, AB, 48, MA, 51, PhD(zool), 54. *Prof Exp:* Asst zool, Univ Calif, 48-52; instr biol, Duluth Br, Univ Minn, 53-54; from asst prof to prof & chmn div natural resources, Humboldt State Col, 54-58, prof & chmn div, 62-65; Fulbright scholar, S Rhodesia, 59-60; lectr, Univ Calif, 61-62; dir int progs, Conserv Found, 66-70; sr ecologist, Int Union Conserv Nature, 70-77; PROF ENVIRON STUDIES, UNIV CALIF, SANTA CRUZ, 77- *Concurrent Pos:* Consult, UNESCO, 66-70 & Int Union Conserv Nature, SPac, 79. *Honors & Awards:* Browning Award, 74; Order of Golden Ark, 78; Leopold Award, 79; Hon Mem, Int Union Conserv Nature, 88. *Mem:* Wildlife Soc (pres, 70-71); Ecol Soc Am; Am Soc Mammal; Sigma Xi; fel AAAS; fel Calif Acad Sci. *Res:* Conservation of natural resources; human and wildlife ecology. *Mailing Add:* Dept Environ Studies Univ Calif Santa Cruz CA 95064

DAS SARMA, BASUDEB, b India, Jan 1, 23; m 52; c 2. INORGANIC CHEMISTRY. *Educ:* Univ Calcutta, BS, 44, MS, 46, PhD(chem), 51. *Prof Exp:* Lectr chem, Univ Calcutta, 50-53; res assoc inorg chem, Univ Ill, Urbana, 53-55; lectr chem, Univ Calcutta, 55-57; sr chemist, Geol Surv India, 57-65; assoc prof, 66-69, PROF CHEM, WVA STATE COL, 69-, CHMN, 81- *Concurrent Pos:* Res collabr, Oak Ridge Nat Lab, Tenn, 71-72; exchange prof chem, RI Col, 74. *Mem:* Fel Am Inst Chemists; Am Chem Soc; Fedn Am

Scientists; Am Asn Univ Prof; WVa Acad Sci (pres, 83-84). *Res:* Stereo, structural and analytical aspects of coordination chemistry; analytical chemistry and geochemistry of minor and trace metals; solid state reactions in coordination complexes; inorganic pollutants in water; interaction of sulfohydril groups with arsenic compounds; problems and prospects of chemical education in undergraduate colleges; platinum metal complexes with anticancer potential. *Mailing Add:* Dept Chem WVa State Col Institute WV 25112

DASTUR, ARDESHIR RUSTOM, b Bombay, India, Nov 2, 35; Can citizen; m 66. DIGITAL SIMULATION. *Educ:* Univ Bombay, BSc, 55; Univ Toronto, BASc, 58, MASc, 60. *Prof Exp:* Reactor physicist, AEC, Trombay, 60-62 & Can Gen Elec, 62-67; reactor physicist, 67-70, head, Core Physics Sect, 70-86, HEAD, REACTOR PHYSICS, ATOMIC ENERGY CAN LTD, 86- *Honors & Awards:* Atomic Energy Can Ltd. *Mem:* Can Nuclear Soc. *Res:* Reactor physics and numerical methods of reactor analysis; development of the Candu reactor and the methodology presently used for the digital simulation of Candu systems. *Mailing Add:* Atomic Energy Can Ltd Sheridan Park Res Community Mississauga ON L5K 1B2 Can

DATA, JOANN L, b New York, NY, Apr 20, 44; m. CLINICAL PHARMACOLOGY. *Educ:* Washington Univ, MD, 70; Vanderbilt Univ, PhD(pharmacol), 77. *Prof Exp:* Dir clin pharmacol, Hoffman-La Rouch, Inc, 82-85, vpres & dir clin res & develop, 85-90; VPRES & PROJ MGR SPEC ASSIGNMENTS, UPJOHN CO, 90- *Mem:* AMA; Am Fedn Clin Res; Am Soc Pharmacol & Exp Therapeut; Am Soc Clin Pharmacol & Therapeut. *Mailing Add:* Upjohn Co 9115-24-1 Kalamazoo MI 49001

DATARS, WILLIAM ROSS, b Desboro, Ont, June 14, 32; m 59; c 3. PHYSICS. *Educ:* McMaster Univ, BSc, 55, MSc, 56; Univ Wis, PhD(physics), 59. *Prof Exp:* Sci officer physics, Defence Res Bd Can, 59-62; from asst prof to assoc prof, 62-69, PROF PHYSICS, McMASTER UNIV, 69- *Concurrent Pos:* E W R Steacie res fel, 68-70. *Mem:* Am Phys Soc; Can Asn Physicists; fel Royal Soc Can. *Res:* Electronic properties of solids; properties of semi-metals; far infrared properties of solids. *Mailing Add:* Dept Physics McMaster Univ Hamilton ON L8S 4M1 Can

DATERMAN, GARY EDWARD, b Freeport, Ill, June 26, 39; m 60, 85; c 3. FOREST ENTOMOLOGY, INSECT ECOLOGY. *Educ:* Univ Calif, Davis, BA, 62; Ore State Univ, MS, 64, PhD(entom), 69. *Prof Exp:* Res scientist entom, 65-74, asst prof entom, Ore State Univ, 71-79, RES ENTOMOLOGIST & PROJ LEADER, PAC NORTHWEST FOREST & RANGE EXP STA, FOREST SCI LAB, US FOREST SERV, 74-, ASSOC PROF ENTOM, ORE STATE UNIV, 79- *Honors & Awards:* Arthur S Fleming Award, 78. *Mem:* Entom Soc Am; Entom Soc Can. *Res:* Forest insect pheromone identifications and applications with major emphasis on development of operational survey and control applications; tussock moths, Orgyia; spruce budworms, Choristoneura; western pine shoot borer, Eucosma sonomana; tip moths, Rhyacionia; insect behavior; bark beetle semichemicals. *Mailing Add:* USDA Forest Serv Forestry Sci Lab 3200 Jefferson Way Corvallis OR 97331

DATSKO, JOSEPH, b Pa, Feb 4, 21; m 46; c 4. MECHANICAL ENGINEERING. *Educ:* Univ Mich, BSME, 43, MSE, 51. *Prof Exp:* Prod engr, Dynamic Tool Co, 43; chief engr & plant mgr, Stevens Mfg Co, 46-49, vpres & plant mgr, 52-53; instr prod eng, 49-52, from asst prof to assoc prof, 53-63, PROF PROD ENG, UNIV MICH, ANN ARBOR, 63- *Concurrent Pos:* Consult, 52- *Mem:* Am Soc Mech Engrs; Am Soc Metals; Am Soc Eng Educ; Am Foundrymens Soc. *Res:* Correlation of microstructures, mechanical properties and fabricability. *Mailing Add:* Dept Mech Eng Univ Mich Ann Arbor MI 48109

DATTA, DILIP KUMAR, b Jorhat, India, Jan 2, 39; m 68. GEOMETRY. *Educ:* Gauhati Univ, India, BA, 58; Univ Delhi, MA, 60, PhD(math), 63. *Prof Exp:* Lectr math, Assam Eng Col, 63; Brit Govt Commonwealth scholar differential geom, 63-65; asst prof, Univ Calgary, 65-67; from asst prof to assoc prof, 67-80, PROF MATH, UNIV RI, 80- *Mem:* Am Math Soc; Math Soc Can. *Res:* Differential geometry; linear connections; special Riemannian spaces; G-structures. *Mailing Add:* Dept Math Univ RI Kingston RI 02881

DATTA, PADMA RAG, b Jorhat, Assam, India, Feb 27, 27; US citizen; m 52; c 2. BIOCHEMISTRY, ECOLOGY. *Educ:* Univ Mass, MS, 50; WVa Univ, PhD(agr biochem), 56. *Prof Exp:* Res assoc biochem, Fels Res Inst, Sch Med, Temple Univ, 56-57; sr res fel anal & phys chem, Am Spice Trade Asn, Eastern Regional Lab, USDA, 57-61; res chemist, Rohm and Haas Co, 61-63; res biochemist, Food & Drug Admin, US Dept Health, Educ & Welfare, 63-70; BIOCHEMIST, OFF PESTICIDES PROGS, ENVIRON PROTECTION AGENCY, 71- *Concurrent Pos:* Res consult biochem, Sch Med, George Washington Univ, 67- *Mem:* AAAS; Am Chem Soc; NY Acad Sci; Int Soc Technol Assessment; World Future Soc; Sigma Xi. *Res:* Advise on the research, monitoring and regulatory needs for pesticides and other toxic chemicals as they affect environmental ecosystems and human health. *Mailing Add:* 8514 Whittier Blvd Bethesda MD 20817

DATTA, PRASANTA, b Calcutta, India, Oct 10, 29; US citizen; m 57; c 2. BIOCHEMISTRY, MOLECULAR BIOLOGY. *Educ:* Univ Calcutta, BSc, 49, MSc, 51; Univ Wash, PhD(plant biochem), 56. *Prof Exp:* Asst res prof molecular biol, Wash Univ, 61-65; asst prof, to assoc prof, 66-76, PROF BIOCHEM, MED SCH, UNIV MICH, ANN ARBOR, 76- *Concurrent Pos:* Playtex Corp fel, Med Sch, Univ Wash, 57-58; Nat Res Coun Can fel, Nat Res Coun Lab, Ottawa, 58-61; NIH spec fel, Salk Inst, Calif, 72-73; Nat Res Serv Award sr fel, Stanford Univ, 82-83; vis scholar, Stanford Univ, 82-83 & 90; vis prof, Div Biochem, Indian Agr Res Inst, New Delhi, India, 90; mem, Microbial Physiol & Genetics Study Sect, NIH, 90-94. *Mem:* AAAS; Am Soc Biochem & Molecular Biol; Am Soc Microbiol. *Res:* Biochemical and genetic studies of enzyme regulation; gene structure and regulation of expression; control of enzyme function by cellular metabolites; molecular evolution. *Mailing Add:* Dept Biol Chem Univ Mich 4326 Med Sci I Ann Arbor MI 48109-0606

DATTA, RANAJIT K, b Sylhet, Bangladesh, Feb 1, 35. SOLID STATE CHEMISTRY. *Educ:* Univ Calcutta, BSc, 53; Indian Inst Technol, Kharagpur, BS Hons, 56, MTech, 58; Pa State Univ, PhD(geochem), 61. *Prof Exp:* From res asst to res assoc geochem, Pa State Univ, 58-62; res chemist, Large Lamp Eng Sect, Gen Elec Co, 62-64, Lamp Div, 64, SR RES SCIENTIST, ADV TECHNOL DEPT, GE LIGHTING, GEN ELEC CO; DIR, DATTASREE SERV, PELHAM MANOR, NY. *Concurrent Pos:* Res liaison, Gen Elec Co, CNRS, Paris, 70-71. *Mem:* Am Ceramic Soc; Sigma Xi. *Res:* Phase equilibria; thermionic emitters; lamp-cathode chemistry; phosphors; alan/ceramic seals. *Mailing Add:* Adv Technol Dept GE Lighting Nela Park Cleveland OH 44112

DATTA, RANAJIT KUMAR, b Munshiganj, Bangladesh, Apr 1, 33; m 64; c 1. NEUROCHEMISTRY, NEUROPHARMACOLOGY. *Educ:* Univ Calcutta, BSc, 56, MSc, 58, PhD(biochem), 63, DSc(biochem), 75; Am Bd Clin Chem, dipl. *Prof Exp:* Asst res officer, Inst Postgrad Med Educ & Res, Calcutta, India, 60; lectr chem, Jogamaya Devi Col, 61-63; asst prof biochem, Bengal Vet Col, 63-64; sr res officer, USDA Proj, Univ Calcutta, 64-65; sr res scientist, NY State Res Inst Neurochem & Drug Addiction, 65-67; res assoc biochem, Col Physicians & Surgeons, Columbia Univ, 67-68; assoc biophysicist, Beth Israel Med Ctr, 68-85; CONSULT, DATTASREE SERV, PELHAM MANOR, NY, 85- *Concurrent Pos:* Hon lectr, Dept Agr, Calcutta, 64-65; assoc, Dept Path, Mt Sinai Sch Med, 68-71, asst prof, 71-79; ed, Clinician, 70- *Mem:* Nat Acad Clin Biochem; Am Soc Neurochem; Int Soc Neurochem; Brit Biochem Soc; Indian Soc Biol Chem. *Res:* Metabolism of nucleic acids in brain tissue; biochemical properties of brain ribosomes; drug-induced changes in cerebral macromolecules; protein breakdown processes in the brain; methylation of nucleic acids in cancer; clinical and cancer immunology; immunochemistry. *Mailing Add:* 101 Iden Ave Pelham Manor NY 10803

DATTA, RATNA, b Bihar, India, Oct 10, 43; US citizen; m 73. MEDICAL PHYSICS, HEALTH PHYSICS. *Educ:* Ranchi Univ, Bihar, India, BSc, 63; Calcutta Univ, India, MSc, 66, PhD(nuclear physics), 71; Am Bd Radiol, cert, 84. *Prof Exp:* Postdoctoral fel nuclear physics, Saha Inst Nuclear Physics, Calcutta, India, 71-74; NIH sr postdoctoral fel med & health physics, M D Anderson Hosp & Tumor Inst, 74-76; asst prof med physics, Univ Tex Health Sci Ctr, San Antonio, 76-78; asst prof, 78-83, DIR PHYSICS, LA STATE UNIV SCH MED, SHREVEPORT, 78-, ASSOC PROF MED PHYSICS, 83- *Mem:* Am Col Radiol; Am Asn Physicists Med; Am Asn Therapeut Radiol & Oncol; Health Physics Soc. *Res:* Comparative study of different dosimetric devices; estimating carcinogenic risks from low dose of ionizing radiation; comparative study of different treatment techniques in radiation oncology; medical imaging. *Mailing Add:* Dept Radiol Sch Med La State Univ PO Box 33932 Shreveport LA 71130

DATTA, SAMIR KUMAR, b Calcutta, India, June 14, 36; m 66; c 1. ELECTRICAL ENGINEERING. *Educ:* Jadavpur Univ, India, BEE, 58; Manchester Univ, MS, 63, PhD(control systs, electronics), 66. *Prof Exp:* Electronic engr, Askania Werke, Ger, 58-60 & Valvo, GmbH, 60-61; sr develop engr, Speed Variator Dept, Gen Elec Co, 67-68; PROF ELEC ENG, CALIF POLYTECH STATE UNIV, SAN LUIS OBISPO, 68- *Mem:* Inst Elec & Electronics Engrs; Sigma Xi. *Res:* Power electronics; solid state energy conversion; electric motor controls; power systems. *Mailing Add:* 1658 Colina Ct San Luis Obispo CA 93401

DATTA, SUBHENDU KUMAR, b Howrah, India, Jan 15, 36; m 66; c 1. DYNAMICS OF COMPOSITES, MECHANICS OF MATERIALS. *Educ:* Calcutta Univ, BSc, 54, MSc, 56; Jadavpur Univ, PhD(appl math), 62. *Prof Exp:* Vis asst prof aeronaut, Univ Colo, 64-65; asst prof, Indian Inst Technol, 65-67; asst prof math, Univ Man, 67-68; FROM ASST PROF TO PROF MECH, UNIV COLO, 68; FEL, COOP INST RES ENVIRON SCI, 82- *Concurrent Pos:* Prin investr, NSF grant, 70-, ONR grant, 86-; prof appl math, Calcutta Univ, 79-80; consult, Nat Bur Standards, 79- & prin investr, 83-84; chmn, Wave Propagation Comt, Am Soc Mech Engrs, 82-86; fac fel, Univ Colo, 86. *Honors & Awards:* Res Award, Col Eng Appl Sci, 84; Fulbright Res Award, 86. *Mem:* fel Am Acad Mech; fel Am Soc Mech Engrs; Soc Eng Sci; Soc Indust & Appl Math; Earthquake Eng Res Inst; Seismol Soc Am. *Res:* Wave propagation in elastic and inelastic solids; ultrasonic non-destructive evaluation; earthquake engineering; ultrasonic characterization of material properties; dynamics of composites. *Mailing Add:* Dept Mech Eng Univ Colo Boulder CO 80309-0427

DATTA, SURINDER P, b Lahore, India, Apr 29, 33; m 63; c 3. GENETICS, IMMUNOLOGY. *Educ:* Panjab Univ, India, BVetSci, 55; Univ Wis-Madison, MS, 59, PhD(genetics, vet sci), 63. *Prof Exp:* Res asst immunogenetics, Indian Vet Res Inst, 55-57 & Univ Wis-Madison, 58-63; res assoc immunol, Albert Einstein Col Med, 63-64; lectr path, Med Col, Monash Univ, Australia, 64-67; assoc scientist, Oak Ridge Assoc Univs, 67-68; asst prof genetics, 68-72, assoc prof life sci, 72-74, PROF LIFE SCI, UNIV WIS-PARKSIDE, 74- *Concurrent Pos:* AEC res partic fel, Med Div, Oak Ridge Assoc Univs, 69. *Mem:* Am Inst Biol Sci; Genetics Soc Am. *Res:* Immunogenetic analysis of blood groups in cattle, guinea pigs and marmosets; phytohemagglutinins; kidney transplantation in cattle; development of immune competence. *Mailing Add:* Div Sci Col Sci & Soc Univ Wis-Parkside Kenosha WI 53141

DATTA, SYAMAL K, b India Sept 21, 43; nat US citizen; m 76; c 1. SYSTEMIC AUTOIMMUNE DISEASE. *Educ:* Calcutta Univ, India, MBBS & MD, 66; Diplomate Am Bd Int Med, 72. *Prof Exp:* from asst prof to assoc prof, 76-85, FAC, GRAD PROG IMMUOL, TUFTS NEW ENGLAND MED CTR, 75-, PROF MED, 85- *Concurrent Pos:* Prin investr, NIH res proj, 76-; assoc ed, J Immunol, 84-86, 89-; mem, Grants Rev Study Sect, Gen Med A-1, NIH, 87-92; fac res award, Am Cancer Soc, 78-83. *Mem:* Am Assoc Immunol; AAAS; NY Acad Sci. *Res:* Molecular genetics and immunochemistry of pathogenic autoantibodies; novel T helper cells that induce pathogenic autoantibodies; T cell receptor genes, immune response associated genes and retroviral genes-their roles in the development of systemic autoimmune disease. *Mailing Add:* Tufts-New Eng Med Ctr 750 Washington St Boston MA 02111

DATTA, TAPAN K, b India, Sept 3, 39; US citizen; m 64; c 2. TRAFFIC OPERATIONS & CONTROL. *Educ:* Bengal Eng Col, India, BEng, 62, dipl town & regional planning, 65; Wayne State Univ, MS, 68; Mich State Univ, PhD(civil eng), 73. *Prof Exp:* Asst engr structural eng, The Kuljian Corp, 62-65, proj engr, 65-67; asst prof, 73-76, assoc prof, 76-78, prof & chmn civil eng, 79-83, PROF, WAYNE STATE UNIV, 83- *Concurrent Pos:* Chief transp engr, Goodell Grivas, Inc, 68-72, vpres eng, 72-79, pres, 79- *Mem:* Am Soc Civil Engrs; Inst Transp Engrs; Am Pub Works Asn; Transp Res Bd. *Res:* Traffic operations and control; traffic and highway safety; training for professionals in traffic and safety. *Mailing Add:* Dept Civil Eng Wayne State Univ 2152 Eng Bldg Detroit MI 48202

DATTA, TIMIR, b Sept 14, 47; US citizen; m. CONDENSED MATTER PHYSICS. *Educ:* Calcutta Univ, BS, 67; Boston Col, MS, 74; Tulane Univ, PhD(physics), 79. *Prof Exp:* Vis asst prof, Tulane Univ, 81; vis assoc prof, Univ Nebr, Lincoln, 81-86; ASSOC PROF, UNIV SC, COLUMBIA, 86- *Concurrent Pos:* Res assoc, Univ NC, Chapel Hill, 79-81; vis res prof, Univ Ill, Urbana, 90. *Mem:* Am Phys Soc. *Res:* Solid state physics experimental and theory; transport and critical properties of high temperature super conducting systems; general relativity; non-linear response; amorphous media; dynamical chaos. *Mailing Add:* Univ SC Columbia SC 29208

DATTA, VIJAY J, b Jallander, Punjab, India, Sept 30, 44; US citizen; m 71; c 2. COATINGS DEVELOPMENT. *Educ:* India Inst Technol, New Delhi, BS, 67; Newark Col Eng, NJ, MS, 71. *Prof Exp:* TECH MGR, DEVOE MARINE COATING CO, 71- *Mem:* Am Chem Soc; Steel Struct Painting Coun. *Res:* Development of coatings to combat corrosion; water base epoxy; low temperature cure epoxies; epoxy for rusty steel; anti-fouling coatings. *Mailing Add:* 2902 Homewood Pl Louisville KY 40222

D'ATTORRE, LEONARDO, b La Plata, Arg, Feb 2, 20; nat US; m 49. PHYSICS. *Educ:* La Plata Univ, MA, 47, PhD, 52. *Prof Exp:* Instr fluid mech, La Plata Univ, 48-52, assoc prof aerodyn, 52-56; sr res engr, Gen Dynamics/Astronaut, 56-61; prof fluid mech, La Plata Univ, 61-62; staff scientist, Space Sci Lab, Gen Dynamics/Convair, 62-67; staff engr, TRW Systs Group, 67-86; CONSULT, 86- *Concurrent Pos:* Prof, Nat Univ South, Arg, 54-56. *Mem:* Am Math Soc. *Res:* Fluid and classical mechanics; applied mathematics; solution of non-linear systems of partial differential equations applied to steady and unsteady flows; numerical methods. *Mailing Add:* 33655 Scottys Cove Dr Dana Point CA 92629

DATZ, SHELDON, b New York, NY, July 21, 27; m 48; c 2. CHEMICAL DYNAMICS, ATOMIC PHYSICS. *Educ:* Columbia Univ, BS, 50, MA, 51; Univ Tenn, PhD(phys chem), 60. *Prof Exp:* Technician, Substitute Alloy Mat Labs, Div War Res, Columbia Univ, 43-44, Dept Physics, 46-50, asst chem, 50-51; chemist, Chem Div, 51-67, leader molecular beam group, 60-67, assoc dir, 67-74, leader atomic & molecular collisions group, 74-81, DISTINGUISHED RES SCIENTIST, OAK RIDGE NAT LAB, 79-, SECT CHIEF ATOMIC PHYSICS, PHYSICS DIV, 81- *Concurrent Pos:* Consult, Gen Atomic Div, Gen Dynamics Corp, 60-62 & Repub Aviation Corp, Long Island, 61-62; guest scientist, Inst Atomic & Molecular Physics, Netherlands, 62-63; guest prof, Inst Physics, Aarhus Univ, Denmark, 70-71; chmn, Gordon Conf Particle Solid Interactions, 70 & Gordon Conf Dynamics Molecular Collisions, 72; guest prof physics dept, Univ Chicago, 72; chmn, V Int Conf Atomic Collisions Solids, 73; guest scientist, Max Planck Inst Plasma Physics, WGer, 74; mem, Nat Res Coun-Nat Acad Sci Comt Atomic & Molecular Physics, 72-75 & 77-; Union Carbide Corp Res fel, 80-; guest distinguished prof, Tex A&M Univ, 81; guest prof, Univ Pierre & Marie Curie, Paris, France, 85, Fudan Univ, Shanghai, Peoples Repub China, 87; Union Carbide res fel, 79-, Martin Marietta corp fel, 83-88, Martin Marietta sr corp fel, 87-, Fulbright sr res fel, 61-62; Tage Erlander Nat sci prof, Sweden, 90-91. *Mem:* Fel AAAS; Am Chem Soc; fel Am Phys Soc; Sigma Xi. *Res:* Dynamics of atomic and molecular collisions, chemically reactive collisions; high energy atomic collisions; atomic collisions in solids; particle-surface interactions. *Mailing Add:* Oak Ridge Nat Lab PO Box 2008 Oak Ridge TN 37831-6377

DAU, GARY JOHN, b Lewiston, Idaho, Sept 3, 38; m 61; c 2. NUCLEAR & MECHANICAL ENGINEERING. *Educ:* Univ Idaho, BS, 61; Univ Ariz, PhD(nuclear eng), 65; MIT, MS, 87. *Prof Exp:* Sr res scientist, Pac Northwest Labs, Battelle Mem Inst, 64-66, mgr res & develop group, 66-68, mgr nondestructive testing dept, 68-70, mgr appl physics & instrumentation dept, 70-75; prog mgr, Elec Power Res Inst, 75-80, sr prog mgr, 81-89; RETIRED. *Concurrent Pos:* Mem ed adv comt, Nuclear Technol J; mem subgroup on nondestructive examination, Am Soc Mech Engrs. *Mem:* Am Nuclear Soc; Nat Soc Prof Engrs; Am Soc Mech Engrs; Sigma Xi. *Res:* Nondestructive testing; applied physics; instrumentation research; electromagnetics; lasers; holography; ultrasonic and thermal testing techniques. *Mailing Add:* 3958 Duncan Pl Palo Alto CA 94306

DAU, PETER CAINE, b Fresno, Calif, Feb 25, 39; m 65; c 4. IMMUNOLOGY, NEUROLOGY. *Educ:* Stanford Univ, BA, 60, MD, 64. *Prof Exp:* Intern, Los Angeles County Hosp, 64-65; resident neurol, Univ Wis-Madison, 65-66; resident neurol, Univ Chicago, 66-68, fel immunol, 68-69; chief neurol, David Grant US Air Force Hosp, 69-71; sci asst, Ger Cancer Res Ctr, 72-74; assoc neuroimmunologist, Univ Calif, San Francisco, 74-75, instr med, 75-76; res immunologist, Children's Hosp, San Francisco, 75-81; asst clin prof med, Univ Calif, San Francisco, 76-81; prof neurol, Northwestern Univ, 82-; HEAD, DIV ALLERGY & IMMUNOL, EVANSTON HOSP, ILL, 82- *Mem:* AAAS; Am Asn Immunologists; Am Soc Apheresis. *Res:* Myasthenia gravis; multiple sclerosis; polymyositis; scleroderma; atherosclerosis. *Mailing Add:* Dept Med Evanston Hosp 2650 Ridge Ave Evanston IL 60201

DAUB, CLARENCE THEODORE, JR, b Hagerstown, Md, Nov 27, 36; div; c 2. ASTROPHYSICS. *Educ:* Carleton Col, BA, 58; Univ Wis, PhD(astron), 62. *Prof Exp:* Instr astron, Univ Wis, 62; asst prof physics, Iowa State Univ, 62-67; from asst prof to assoc prof, 67-73, PROF ASTRON, SAN DIEGO STATE UNIV, 73- *Concurrent Pos:* Chmn dept, San Diego State Univ, 84-88. *Mem:* Am Astron Soc; Sigma Xi; Astron Soc Pac. *Res:* Physical processes in gaseous media. *Mailing Add:* Dept Astron San Diego State Univ San Diego CA 92182

DAUB, EDWARD E, b Milwaukee, Wis, May 17, 24; m 49; c 5. HISTORY & PHILOSOPHY OF SCIENCE, SCIENCE COMMUNICATIONS. *Educ:* Univ Wis, BS, 45, MS, 47, PhD(hist sci), 66; Union Theol Sem, NY, BD, 50, STM, 60. *Prof Exp:* Assoc prof chem eng, Doshisha, Japan, 57-62; from asst prof to assoc prof hist sci, Univ Kans, 65-71; from assoc prof to prof, 74-90, EMER PROF TECH JAPANESE, UNIV WIS-MADISON, 90- *Concurrent Pos:* NSF res grant, 68-70 & 72-74; Danforth Found Underwood fel, 71; Nat Humanities Inst fel, Univ Chicago, 78-79; mem, United Presbyterian Task Force on Sci, Med & Values, 80-83, Theol & Cosmology, 84-86; Pub Serv Sci Residency, NSF, 80-81. *Mem:* Hist Sci. *Res:* Relations between science and theology; medical ethics; history of chemical engineering; technical Japanese. *Mailing Add:* Eng Prof Develop Univ Wis 1527 University Ave Madison WI 53706

DAUB, GUIDO WILLIAM, b Albuquerque, NMex, Aug 6, 50; m 80; c 4. CLAISEN REARRANGEMENTS IN SYNTHESIS, VINYLIC FLUORIDES. *Educ:* Pomona Col, BA, 72; Stanford Univ, PhD(org chem), 77. *Prof Exp:* Fel, Stanford Univ, 77-78; asst prof, 78-84, ASSOC PROF CHEM, HARVEY MUDD COL, CLAREMONT, CALIF, 84- *Concurrent Pos:* Vis prof, Univ Calif, Irvine, 84-85. *Mem:* Am Chem Soc; AAAS; Sigma Xi. *Res:* Regiochemical and stereochemical aspects of the Claisen rearrangement; applications of the Claisen rearrangement to chemical synthesis; synthesis and reactivity of vinylic fluorides in organic chemistry. *Mailing Add:* 1382 Via Zurita Claremont CA 91711

DAUBEN, DWIGHT LEWIS, b Ft Smith, Ark, Feb 28, 38; m 63; c 1. PETROLEUM ENGINEERING & TECHNOLOGY. *Educ:* Tex Tech Col, BS, 61; Univ Tulsa, MS, 63; Univ Okla, PhD(eng sci), 66. *Prof Exp:* Petrol engr, Chevron Oil Co, Colo, 62-63; res supvr, Amoco Prod Co, Subsid Stand Oil Ind, 66-81; VPRES, K & A ENERGY CONSULTS, INC, 81- *Mem:* Am Inst Mining, Metall & Petrol Engrs. *Res:* Improved methods of recovering oil, particularly miscible processes and polymer flooding. *Mailing Add:* 8916 S 45th Ave Tulsa OK 74132

DAUBEN, JOSEPH W, b Santa Monica, Calif, Dec 29, 44. HISTORY OF MATHEMATICS, SCIENTIFIC REVOLUTION. *Educ:* Claremont McKenna Col, AB, 66; Harvard Univ, AM, 68, PhD(hist sci), 72. *Prof Exp:* Tutor & teaching fel hist sci, Harvard Univ, 67-72; from asst prof to assoc prof, Lehman Col, 72-81, PROF HIST SCI, LEHMAN COL & GRAD CTR, CITY UNIV NEW YORK, 81- *Concurrent Pos:* Vis prof ancient sci, Clark Univ, 71-72, hist sci, Columbia Univ, 79-84 & Oberlin Col, 80-81; Nat Endowment Humanities Younger Humanist fel, 73-74; ed Historia Mathematica, 76-86; mem Inst Advan Study, Princeton, 77-78; Guggenheim fel, 80-81; vis scholar, Harvard Univ, 81; chmn Int Comn Hist Math, 85- *Honors & Awards:* Mead-Swing lectr, Oberlin Col, 77 & 80; Bolzano Medal, Czechoslovak Acad Sci, 78; Harvard lectr, Yale Univ, 82; Lenin Medal, Univ Tashkent, USSR, 86. *Mem:* Fel New York Acad Sci; Sigma Xi; Hist Sci Soc; AAAS; corres mem Int Acad Hist Sci, 85; Int Comn History Math. *Res:* History of mathematics; history of set theory; mathematical logic and model theory; biographies of Georg Cantor and Abraham Robinson; history of science since Scientific Revolution; history and philosophy of science in the 17th and 18th centuries. *Mailing Add:* Dept Hist Lehman Col 250 Bedford Park Blvd W Bronx NY 10468

DAUBEN, WILLIAM GARFIELD, b Columbus, Ohio, 1919; m 47; c 2. PHOTOCHEMISTRY, SYNTHETIC CHEMISTRY. *Educ:* Ohio State Univ, BA, 41; Harvard Univ, AM, 42, PhD(chem), 44. *Hon Degrees:* Dr, Univ Bordeauxi, 80. *Prof Exp:* Res asst, Harvard Univ, 44-45; from instr to assoc prof, 45-57, PROF CHEM, UNIV CALIF, BERKELEY, 57- *Concurrent Pos:* Guggenheim Mem Found fels, 51 & 66; NSF sr postdoc fel, 58; lectr, Am Swiss Found, 62; Miller res prof, Univ Calif, 63; Med Chem Study Sect, NIH; chem panel & chmn, Chem Sect, NSF & Assembly Math & Phys Sci, Nat Res Coun, 77; adv bd, Petrol Res Fund; mem, US Comt Int Union Pure & Appl Chem, 81; Promo sci fel, Japan Soc, 82. *Honors & Awards:* Calif Sect Award, Am Chem Soc, 59; Guenter Award, 73; Alexander von Humboldt Found Award, 80; GN Lewis lectr, Univ Calif, Berkeley, 80; Welch Found lectr, 85; Arthur Cope Scholar Award, Am Chem Soc, 90. *Mem:* Nat Acad Sci; Am Chem Soc; Royal Chem Soc; Swiss Chem Soc; fel Am Acad Arts & Sci; hon mem Pharmaceut Soc Japan. *Res:* Structure and synthesis of alicyclic compounds; stereochemistry, photochemistry and mechanistic chemistry of natural products and related alicyclic compounds. *Mailing Add:* Dept Chem Univ Calif Berkeley CA 94720

DAUBENMIRE, REXFORD, b Coldwater, Ohio, Dec 12, 09; m 38; c 1. PLANT ECOLOGY. *Educ:* Butler Univ, BS, 30: Univ Colo, MS, 32; Univ Minn, PhD(bot), 35. *Prof Exp:* Actg asst prof bot, Univ Tenn, 35-36; asst prof bot, Univ Idaho, 36-46; from asst prof to prof, 46-75, EMER PROF BOT, WASH STATE UNIV, 75- *Concurrent Pos:* Lectr, Biol Sta, Univ Minn, 37; asst prof, Sci Camp, Univ Wyo, 39; consult, US Forest Serv, Nat Park Serv, Bur Reclamation, Brit Col Minis Forestry, Sierra Club & Nature Conserv, KBN Engr & Appl Sci, Lake-Sumter Comm Col, Lake County Water Authority. *Honors & Awards:* Eminent Ecologist, Ecol Soc Am, 79; Barrington Moore Award, Soc Am Forestry, 80. *Mem:* Ecol Soc Am (pres, 67). *Res:* Forest classification for Northern Rockies; factors affecting seasonal growth in temperate and tropical trees; causes of forest distribution in Northern Rockies; ecology of grasslands in Columbia drainage; effect of temperature and drought on seedling survival; water requirement of desert plants; productivity of tropical savanna; potential vegetation of Southeastern USA. *Mailing Add:* 31150 Interlochen Dr Mt Plymouth FL 32776

DAUBENSPECK, JOHN ANDREW, b Denver, Colo, Nov 3, 42; m 86; c 1. RESPIRATORY CONTROL. *Educ:* Swarthmore Col, BSME, 66; Dartmouth Col, PhD(eng sci), 72. *Prof Exp:* NIH res fel & res assoc biomed eng, Univ Southern Calif, 72-74; from asst prof to assoc prof, 74-86, PROF PHYSIOL, DARTMOUTH MED SCH, 86-, ADJ PROF ENG, THAYER SCH ENG, 75- *Concurrent Pos:* Res career develop award, NIH, 76-81. *Mem:* Biomed Eng Soc; Am Physiol Soc. *Res:* Control of breathing, breathing pattern, respiratory mechanics, application of optimal control principles to biological systems and computer simulations of physiological control processes. *Mailing Add:* Dept Physiol HB 7700 Dartmouth Med Sch Hanover NH 03756

DAUBENY, HUGH ALEXANDER, b Nanaimo, BC, Dec 6, 31; m 59; c 3. PLANT BREEDING, HORTICULTURE. *Educ:* Univ BC, BSA, 53, MSA, 55; Cornell Univ, PhD(plant breeding), 58. *Prof Exp:* Teaching asst hort, Univ BC, 53-55; res asst plant breeding, Agassiz, BC, 58-73, RES SCIENTIST SMALL FRUIT BREEDING, RES STA, AGR CAN, 73- *Concurrent Pos:* Chmn, Rubus-Riker Working Group, Int Soc Hort Sci. *Mem:* Int Soc Hort Sci; Am Soc Hort Sci; Am Pomol Soc (pres); Agr Inst Can. *Res:* Strawberry and red rasberry breeding with emphasis on disease and insect resistance. *Mailing Add:* 3558 W 15th Ave Vancouver BC V6R 2Z4 Can

DAUBER, EDWIN GEORGE, b Williamsport, Pa, June 21, 53; m 80; c 1. MICROCONTAMINATION CONTROL, INDUSTRIAL FILTRATION FOR POLLUTION CONTROL & PRODUCT COLLECTION. *Educ:* Lycoming Col, BA, 74; Univ Del, MS & MBA, 81. *Prof Exp:* Teaching asst physics, Univ Del, 74-77 & 78-81, res asst, 77-78; RES & DEVELOP ENGR, W L GORE & ASSOCS, 81- *Mem:* Inst Environ Sci. *Res:* Raman spectroscopy to study molybdenum disulfide; development engineering for product development including patent applications for processes and products used to control the environment inside critically clean enclosures, such as computer disk drives. *Mailing Add:* W L Gore & Assocs Bldg 3 PO Box 1550 100 Airport Rd Elkton MD 21922-1550

DAUBERT, THOMAS EDWARD, b Pottsville, Pa, Oct 14, 37; m 67; c 3. CHEMICAL ENGINEERING. *Educ:* Pa State Univ, BS, 59, MS, 61, PhD(chem eng), 64. *Prof Exp:* Res asst chem eng, 61, from instr to assoc prof, 61-75, PROF CHEM ENG, PA STATE UNIV, 75- *Concurrent Pos:* Year-in-indust prof, E I du Pont de Nemours & Co, 70-71; hon res fel, Univ Col, London, 89. *Mem:* Am Chem Soc; Am Inst Chem Engrs; Am Soc Eng Educ. *Res:* Applied thermodynamics; prediction and determination of physical, thermodynamic and transport properties of pure compounds and mixtures; characterization of petroleum; critical properties; vapor pressure; vapor-liquid equilibrium prediction. *Mailing Add:* 165 Fenske Lab Pa State Univ Pa State Univ Univ Park PA 16802

DAUBIN, SCOTT C, b New London, Conn, Sept 20, 22; m 44; c 7. OCEAN ENGINEERING, ACOUSTICS. *Educ:* US Naval Acad, BS, 44; Princeton Univ, MA, 53, PhD(physics), 54. *Prof Exp:* US naval officer, 44-61; head marine sci, AC Electronics Defense Res Labs, Gen Motors Corp, 61-67; chmn dept ocean eng, Woods Hole Oceanog Inst, 67-71 & Univ Miami, 71-76; PRES, DAUBIN SYSTS CORP, 77-; PRES, MED IMAGING TECHNOL INC, 84- *Concurrent Pos:* Adj prof, Univ Miami, 78- *Mem:* Acoustical Soc Am; Am Asn Physics Teachers; Am Soc Mech Engrs; Marine Tech Soc; Inst Elec & Electronics Engrs. *Res:* Underwater acoustics; instrumentation systems; signal processing; medical ultrasonics; development and application of special purpose computer systems for modelling underwater acoustic fields; non-linear acoustic wind profiling systems; modular MRI medical imaging systems. *Mailing Add:* Daubin Systs Corp 104 Crandon Blvd Suite 400 PO Box 490249 Key Biscayne FL 33149

DAUBLE, DENNIS DEENE, b Walla Walla, Wash, Sept 28, 50; m 71; c 2. ICHTHYOLOGY, ENVIRONMENTAL MANAGER. *Educ:* Ore State Univ, BS, 72; Wash State Univ, MS, 78; Ore State Univ, PhD, 88. *Prof Exp:* Res technician, 73-78, res scientist, 78-87, SR RES SCIENTIST, BATTELLE PAC NORTHWEST LAB, 87- *Mem:* Am Fisheries Soc; Ecol Soc Am; Am Inst Fishery Res Biologists. *Res:* Toxicity and bioaccumulation of complex organic materials to aquatic biota; environmental assessment of power plant activities; ecology and life history of Columbia River fishes. *Mailing Add:* 1911 Hood Ave PO Box 999 Richland WA 99352

DAUCHOT, PAUL J, b Gent Oost-Vlaanderen, Belg, Nov 10, 35; m; c 3. ANESTHESIOLOGY. *Educ:* State Univ Gent, Belg, MD, 60; Am Bd Anesthesiologists, cert, 76. *Prof Exp:* Chief anesthesiologist, Munic Hosp, 65-72; fel anesthesiol, Sch Med, Case Western Reserve Univ, Cleveland, 73-74, from asst prof to prof, 74-83; ANESTHESIOLOGIST, UNIV HOSPS, CLEVELAND, 74-, MEM MED STAFF UNIV SUBURBAN SURG CTR, 85- *Concurrent Pos:* Consult, Vet Admin Med Ctr, 86- *Mem:* Am Heart Asn; Am Med Asn; Am Soc Anesthesiologists; fel Am Col Anesthesiol; Am Soc Pharmacol & Exp Therapeut; Int Anesthesia Res Soc; Soc Cardiovasc Anesthesiologists; Soc Critical Care Med. *Res:* Cardiovascular physiology and pharmacology in anesthesia and geriatrics; autonomic nervous system-cholinergies-anticholinergics; noninvasive cardiovascular monitoring, systolic time intervals, kinetocardiography, echocardiography. *Mailing Add:* Univ Anesthesiol Inc Univ Hosp 2074 Abington Rd Cleveland OH 44106

DAUER, JERALD PAUL, b Toledo, Ohio, Mar 2, 43; m 67; c 3. MATHEMATICS. *Educ:* Bowling Green State Univ, BA, 65; Univ Kans, MA, 67, PhD(math), 70. *Prof Exp:* From asst prof to assoc prof, 70-80, PROF MATH, UNIV NEBR, LINCOLN, 80- *Concurrent Pos:* Vis scholar, Dept Opers Res, Stanford Univ, 75-76. *Mem:* Am Math Soc; Soc Indust & Appl Math; Opers Res Soc Am. *Res:* Optimal control theory; multiobject optimization; operations research; nonlinear optimization. *Mailing Add:* Univ Tenn 615 McCallie Ave Chattanooga TN 37403-2598

DAUERMAN, LEONARD, b New York, NY, May 17, 32; m 75; c 4. ENVIRONMENTAL SYSTEMS & TECHNOLOGY. *Educ:* City Col New York, BS, 53; Purdue Univ, MS, 56; Rutgers Univ, PhD(chem), 62, JD, 73. *Prof Exp:* Instr chem, Rutgers Univ, 58-62; res chemist, Dow Chem, 62-63; res assoc chem eng, NY Univ, 63-66, from asst prof to assoc prof, 66-69; ASSOC PROF CHEM & CHEM ENG, NJ INST TECHNOL, 69-, DIR, LAW & TECHNOL CTR, 79- *Mem:* Combustion Inst; Am Chem Soc; Am Inst Chem Engrs; Am Asn Univ Prof. *Res:* Flue gas desulfurization, magnesia and alkali processes; soot formation in hydrocarbon flames; techniques for detecting low levels of hydrocarbons; development of environnmental law. *Mailing Add:* Dept Chem NJ Inst Technol 323 Martin Luther King Blvd Newark NJ 07102

DAUES, GREGORY W, JR, b St Louis, Mo, Oct 1, 28; m 52; c 4. POLYMER CHEMISTRY, ANALYTICAL CHEMISTRY. *Educ:* Univ St Louis, BS, 50; Univ NMex, MS, 52. *Prof Exp:* Res chemist, 52-59, res group leader, 59-67, mgr res, 67-74, mgr, Polyolefins Res & Mfg, 74-80, GEN SUPT MFG & QUALITY ASSURANCE, MONSANTO CO, 80- *Mem:* Am Chem Soc. *Res:* Plastics applications; spectroscopy. *Mailing Add:* 4417 Leonette Lane Dickinson TX 77539

DAUGHADAY, WILLIAM HAMILTON, b Chicago, Ill, Feb 12, 18; m 45; c 2. MEDICINE. *Educ:* Harvard Univ, AB, 40, MD, 43. *Prof Exp:* Intern med, Boston City Hosp, 44, res fel, Thorndike Mem Lab, 46-47; Instr, 50, asst prof & dir Metab Div, 51-57, from assoc prof to prof, 57-88; res fel med, Wash Univ, 48-49, NIH fel biochem, 49-50; from instr to prof, 50-88, EMER PROF MED & METAB, WASH UNIV SCH MED, 88- *Concurrent Pos:* Dir, Metab Div, Wash Univ Sch Med, 51-86, actg chmn, dept med, 72-73, dir, Diabetes Endocrinol Res Ctr, 75-78 & Diabetes Res & Training Ctr, 78-87; Irene E & Michael M Karl prof metab, 83-88; ed, J Lab & Clin Med, 61-66, J Clin Endocrinol & Metab, 73- 77, assoc ed, J Clin Invest, 77-82; mem, diabetes & metab training grants comt, Nat Inst Arthritis & Metab Dis, 62-66, adv bd, Nat Pituitary Agency, 64-66 & 68-70, Bd Sci Counr, Nat Inst Arthritis & Metab Dis, Endocrine Study Sect, NIH, 67-71, endocrinol & metab, Food & Drug Admin, 76-80 & 82-83, adv comt study influence diabetes control diabetic complications, Nat Inst Arthritis, Metab & Digestive Dis, 79-80, chmn, adv group prev complications mellitus, 76-79, adv comt hormone distrib, Nat Pituitary Agency, NIAMDD, 80-81; chairperson, Task Force Hormone Pharmacol & Therapeut Immunobiol, 77-79, chmn, subspec test comt endocrinol & metab, Am Bd Internal Med, 70-76; assoc physician, Barnes Hosp, consult clin chem, 50-59; comt eval, Nat Pituitary Agency, NAS, 73. *Honors & Awards:* Rollin Turner Woodyatt mem lectr, Chicago Affil Diabetes Asn Inc, 60; Inagural Sir Charles Bickerton mem lectr, Univ Sydney; Fred Conrad Koch Award, Endocrine Soc, 75; Second Dittman lectr, Univ Chicago, 76; Hurst Brown lectr, Univ Toronto, 77; West lectr, Montefiore Hosp, Albert Einstein Col Med, 82; Herman O Mosenthal Mem lectr, Am Diabetes Asn, 82; Joslin Medal, Am Diabetes Asn; Smithkline Biosci Lab Award, Clin Ligand Assay Soc, 87; Solomon A Berson lectr, Israel Endocrine Soc, 87; Lawson Wilkins lectr, Lawson Wilkins Pediat Endocrine Soc, 90. *Mem:* Nat Acad Sci; Am Soc Clin Invest; Am Fedn Clin Res; Asn Am Physicians; Am Diabetes Asn; Endocrine Soc (pres, 71-72); hon mem Japan Endocrine Soc; Am Acad Arts & Sci. *Res:* Pituitary growth hormone; diabetes mellitus; author of 274 publications; insulin-like growth factors. *Mailing Add:* Dept Med Wash Univ Sch Med Box 8127 St Louis MO 63110

DAUGHARTY, HARRY, b Sept 21, 39; m; c 3. MICROBIOLOGY, BIOCHEMISTRY. *Educ:* Med Col Ga, PhD(microbiol), 68. *Prof Exp:* RES MICROBIOLOGIST, CTR DIS CONTROL, 69- *Concurrent Pos:* Supvry microbiologist & chief, Diag Prod Br, Biol Prod Div, Ctr Infectious Dis/Ctr Dis Control, res microbiologist, Div Viral Dis; adj prof, Univ NC; adj instr, DeKalb Col; postdoctoral trainee, Univ Ill, 67-68, Baylor Col Med, 68-69; mem, Nat Comt Clin Lab Standards liaison, 81, Nat Comt Biol Ref for Rubella Antibody Tests, 82; partic, Int Cong Immunol, Toronto; moderator, Protides Biol Fluids, Brugge, Belg. *Mem:* Asn Med Lab Immunologists; Fedn Am Socs Exp Biol; Am Soc Microbiol; Am Asn Immunologists. *Res:* Microbiol antigen localization in tissues; cell receptor function of immune complexes, cytokines and integrity proteins (fibronectin, vimentin); author/coauthor of 67 research and review papers and book chapters; over 25 presentations at scientific meetings. *Mailing Add:* Div Immunol Oncol & Hemat Ctr Dis Control 1600 Clifton RD NE Atlanta GA 30333

DAUGHENBAUGH, RANDALL JAY, b Rapid City, SDak, Feb 10, 48; m 73; c 2. PHYSICAL ORGANIC CHEMISTRY. *Educ:* SDak Sch Mines & Technol, BS, 70; Univ Colo, PhD(chem), 75. *Prof Exp:* res chemist, Air Prod & Chem, Inc, 75-80; mgr process develop, Chem Exchange Indust, 80-83; PRES & RES DIR, HAUSER CHEMICAL RES INC, 83- *Mem:* Am Chem Soc. *Res:* Development of new amine products and manufacturing technology; process research and development of petrochemicals and specialty chemicals; isolation of pharaceuticals from natural products. *Mailing Add:* 11022 N 66th St Longmont CO 80503

DAUGHERTY, DAVID M, b Warren, Ohio, Mar 29, 28; m 48; c 5. ENTOMOLOGY. *Educ:* Ohio State Univ, BSc, 52, MSc, 53; NC State Univ, PhD(entom), 64. *Prof Exp:* Res fel entom, Ohio State Univ, 52-53; agr specialist, Ohio Dept Agr, 53-56; registr inspector econ poisons, Univ Ky, 56-59; res entomologist, USDA, 61-68, leader oilseed insects invests, 68-71, asst dir, Far East Regional Res Off, New Delhi, India, 71-75; dep dir agr, Bur Int Orgn Affairs, US Dept State, 75-77; USDA liaison officer AID, Off Int Coop & Develop, USDA, 77-78, asst chief, Int Progs Staff, Sci & educ Admin, 78-79, dep adminr int res, 80-; WINROCK INT, ARK. *Concurrent Pos:* Res assoc, Univ Mo-Columbia, 61-64, from asst prof to assoc prof, 64-71; attache, Agr Res, US Embassy, New Delhi, 71- *Mem:* Entom Soc Am. *Res:* Biology, ecology and control of insects affecting cereals and forage crops; parasites, predators and diseases of insects and their utilization in insect control; international research in the agricultural sciences. *Mailing Add:* PO Box 58137 Nairobi Kenya

DAUGHERTY, DON G(ENE), b Mendon, Ill, Nov 14, 35; m 59. ELECTRICAL ENGINEERING. *Educ:* Univ Wis, BS, 57, MS, 58, PhD(elec eng), 64. *Prof Exp:* Instr elec eng, Univ Wis, 59-61; from asst prof to assoc prof, 63-76, PROF ELEC ENG, UNIV KANS, 76- *Mem:* Inst Elec & Electronics Engrs; Am Soc Eng Educ. *Res:* Theory and applications of semiconductor devices; methods of engineering education. *Mailing Add:* Dept Elec Eng EECE Univ Kans Lawrence KS 66045

DAUGHERTY, FRANKLIN W, b Alpine, Tex, June 20, 27; m 45; c 3. GEOLOGY, HYDROLOGY. *Educ:* Sul Ross State Col, BS, 50; Univ Tex, MA, 59, PhD(geol), 62. *Prof Exp:* Mgr, Big Bend Mineral Explor Co, 54-58; econ geol consult, 62-63; prof geol, WTex State Univ, 63-77; CONSULT, 77- *Concurrent Pos:* Pres, Pinnacle Resources, Inc, 71; mem, Texas Mining Coun, 82- & adv coun, Big Bend Nat Hist Asn, 84-; chmn, Brewster County Hist

Comm, 83-; mem adv coun, Ctr Big Bend Studies, 89-; pres, D&F Minerals, Inc, 71-82. *Mem:* Am Inst Mining, Metall & Petrol Engrs; Geol Soc Am. *Res:* Economic geology; petrography of alkalic igneous rocks; hydrogeology. *Mailing Add:* PO Box 329 Alpine TX 79831

DAUGHERTY, GUY WILSON, b Richmond, Va, Sept 25, 12; m 40; c 4. CLINICAL MEDICINE. *Educ:* Col William & Mary, BS, 37; Med Col Va, MD, 37; Univ Minn, MD, 43. *Prof Exp:* Practicing physician, WVa, 38-41; first asst, 43-44, mem sect med, 47-62, ASSOC CLIN PROF MED, MAYO CLIN, 62-, PROF MED, MAYO MED SCH, 73- *Concurrent Pos:* Fel coun clin cardiol, Am Heart Asn, 70- *Mem:* AMA; fel Am Col Physicians. *Res:* Cardiovascular renal diseases. *Mailing Add:* Mayo Med Sch 1219 Sixth St SW Rochester MN 55902

DAUGHERTY, KENNETH E, b Pittsburgh, Pa, Dec 27, 38; m 61; c 2. ANALYTICAL CHEMISTRY, HARZARDOUS WASTES. *Educ:* Carnegie Inst Technol, BS, 60; Univ Wash, PhD(anal chem), 64; Claremont Grad Sch, MBE, 71. *Prof Exp:* Res chemist, Rohm & Haas Corp, 64; capt & sect chief of lab, US Army, 64-66; sr tech staff, Amcord Corp, 66-71; assoc prof chem, Univ Pittsburgh, 71-73; dir res & develop, Gen Portland, Inc, 73-77; dir energy & mat sci & adj prof, 77-79, prof chem, 79-86, DISTINGUISHED RES PROF, UNIV NORTH TEX, 86-; DIR, TRAC LABS, 81- *Concurrent Pos:* Adj assoc prof chem, Univ Pittsburgh, 73-77; pres, KEDS Inc & KD Consult, 77- *Honors & Awards:* Achievement Award, Argonne Nat Lab, 87, Res & Develop 100 Award, 89. *Mem:* Am Chem Soc; fel Am Inst Chemists; NY Acad Sci; Sigma Xi; Am Ceramic Soc; Am Soc Test & Mat; Soc Petrol Engrs. *Res:* X-ray fluorescence and diffraction; spectroscopy; cement, forensic and pollution chemistry; instrumental analysis; thermal analysis; materials research; business aspects of research and development; recycling waste materials research; sudden infant death syndrome research; catalyst research. *Mailing Add:* 317 Lakeland Dr Lewisville TX 75067

DAUGHERTY, LEROY ARTHUR, b Scottsbluff, Nebr, Nov 27, 46; m 73; c 1. AGRONOMY, SOIL MORPHOLOGY. *Educ:* Univ Mo-Columbia, BS, 68; Cornell Univ, MS, 72, PhD(soils), 75. *Prof Exp:* Soil scientist, Soil Conserv Serv, USDA, 68-75; ASST PROF SOIL GENESIS, MORPHOL & CLASSIFICATION, NMEX STATE UNIV, 75- *Mem:* Am Soc Agron; Soil Sci Soc Am; Int Soil Sci Soc; Sigma Xi. *Res:* Study relationships between saturated permeability and soil characteristics; effects of soluble calcium, magnesium and carbonates on organic matter accumulation in soils of New Mexico; relationship of soil moisture to native plant communities for use in soil classification. *Mailing Add:* Dept Agron Box 3Q NMex State Univ Las Cruces NM 88003

DAUGHERTY, NED ARTHUR, b Ft Wayne, Ind, Sept 27, 34; m 59; c 2. INORGANIC CHEMISTRY. *Educ:* Purdue Univ, BS, 56; Mich State Univ, PhD(chem), 61. *Prof Exp:* Asst prof chem, Purdue Univ, 61-63 & Colo State Univ, 63-64; mem staff, Los Alamos Sci Lab, Univ Calif, 64-66; asst prof, 66-67, ASSOC PROF INORG CHEM, COLO STATE UNIV, 67- *Mem:* Am Chem Soc. *Res:* Kinetics of oxidation-reduction reactions; coordination compounds; vanadium chemistry. *Mailing Add:* Dept Chem Colo State Univ Ft Collins CO 80523

DAUGHERTY, PATRICIA A, b Mullens, WVa, Oct 24, 22. GENETICS. *Educ:* Seton Hill Col, AB, 44; Columbia Univ, MA, 46; Ohio State Univ, PhD(genetics), 61. *Prof Exp:* Instr biol, Seton Hill Col, 46-48 & WVa Univ, 48-57; assoc prof biol, 61-86, EMER PROF BIOL, E CAROLINA UNIV, 86- *Mem:* Genetics Soc Am. *Res:* Drosophila. *Mailing Add:* 2009 E Fifth St Greenville NC 27858

DAUGHTERS, GEORGE T, II, b Huntington Park, Calif, July 21, 38; m 60; c 3. BIOMEDICAL ENGINEERING, PHYSIOLOGY. *Educ:* Univ Ill, BS, 61; San Jose State Univ, MS, 67. *Prof Exp:* Res asst solid state physics, Fairchild Semiconductor Corp, 61-66; res assoc bioeng, 66-79, SR INVESTR, PALO ALTO MED RES FOUND, 79- *Concurrent Pos:* Instr bioeng, Univ Santa Clara, 72; instr elec eng, Stanford Univ, 80-87. *Mem:* Biomed Eng Soc; Cardiovasc Systs Dynamics Soc; Am Heart Asn. *Res:* Cardiovascular dynamics and physiology. *Mailing Add:* Palo Alto Med Res Found 860 Bryant St Palo Alto CA 94301

DAUL, GEORGE CECIL, b Gretna, La, Oct 27, 16; m 40; c 2. ORGANIC CHEMISTRY. *Educ:* Tulane Univ, AB, 37, AM, 40. *Prof Exp:* Teacher high sch, La, 37-41; mem staff, US Weather Bur, 41-42; chief chemist & inspector trinitrotoluene, Longhorn Ord Works, US War Dept, Tex, 42-45; chemist, Chem Properties Sec, Cotton Fiber Res Div, Southern Regional Res Lab, Bur Agr & Indust Chem, USDA, 45-54; chemist, Res Lab, Courtaulds, Inc, 54-62; dir, Eastern Res Div, ITT Rayonier, Inc, 62-81; RETIRED. *Mem:* Am Chem Soc; Am Asn Textile Chemists & Colorists; Fiber Soc. *Res:* Chemical properties and derivatives of cellulose; cross linking of cellulosic fibers and fabrics; rayon spinning technology; cellulosic films; acetate fibers and plastics. *Mailing Add:* 74 Derbes Dr Gretna LA 70053-4934

DAUM, DONALD RICHARD, b Tionesta, Pa, Dec 25, 33; m 59; c 3. FARM MACHINERY & POWER, MECHANIZATION OF AGRICULTURAL PROCESSES. *Educ:* Pa State Univ, BS, 56, MS, 58. *Prof Exp:* From instr to asst prof agr eng, Univ Ill, 59-66; from asst prof to assoc prof, 66-75, PROF AGR ENG, PA STATE UNIV, 75- *Concurrent Pos:* Comt secy, vchair & chair, Am Soc Agr Engrs, 66-88; consult, Ministry of Agr & Cooperatives, Govt Swaziland, Malkerns, Swaziland, Africa, 82, 84 & 87. *Mem:* Am Soc Agr Engrs; Coun Agr Sci & Technol. *Res:* Pesticide application technology; soil tillage and compaction; mechanization of fruit and vegetable production; machinery management; international agricultural mechanization; author of numerous publications. *Mailing Add:* 202 Agr Eng Bldg Pa State Univ University Park PA 16802

DAUM, SOL JACOB, b New York, NY, Dec 17, 33; m 54; c 2. ORGANIC CHEMISTRY, MEDICINAL CHEMISTRY. *Educ:* NY Univ, AB, 54; Brooklyn Col, AM, 59; Columbia Univ, AM, 60, PhD(chem), 62. *Prof Exp:* Sr technician, Sloan-Kettering Inst Cancer Res, Mem Hosp, New York, 56-59; asst chem, Columbia Univ, 60; assoc res chemist, Sterling Winthrop Res Inst, 62-67, res chemist, 67-74, sr res chemist-group leader, 74-86, sr res chemist, 86-90, RES INVESTR, STERLING RES GROUP, 90- *Mem:* Am Chem Soc; Sigma Xi. *Res:* Natural products; antibiotics; alkaloids; terpenes; heterocycles. *Mailing Add:* Dept Chem Sterling Res Group Rensselaer NY 12144

DAUMIT, GENE PHILIP, b Washington, DC, Sept 4, 43; m 66; c 2. CARBON FIBER PROCESS & PRODUCT TECHNOLOGY, COMPOSITIES TECHNOLOGY. *Educ:* Univ Md, BS, 65; Mass Inst Technol, PhD(org chem), 70. *Prof Exp:* Sr res engr, Fiber Industs, Inc, Div Celanese, 70-74, group leader, 74-78, mgr, Celion Carbon Fibers, 78-85; TECH DIR, BASF STRUCT MAT, INC, 85- *Concurrent Pos:* Sub-comt head, Technol subcomt, Suppliers Advan Composites Mat Asn, 86-; mem, Nat Mat Adv Bd Comt, 88-90; chmn, Panel Carbon Fiber Indust, Am Carbon Soc, 88. *Mem:* Am Chem Soc; Soc Advan Mat & Process Engrs. *Res:* Carbon fibers-composite; development of a unique polyacrylonitrile precursor fiber; development of new carbon fiber process and products; methods to improve carbon fiber handleability; author of numerous publications; awarded ten patents. *Mailing Add:* 527 Medearis Dr Charlotte NC 28211

DAUNORA, LOUIS GEORGE, b Gary, Ind, Feb 22, 32; m 61; c 4. PHARMACEUTICAL CHEMISTRY, PHARMACY. *Educ:* Purdue Univ, BS, 58. *Prof Exp:* Pharmacist asst mgr, Apollo Drugs, 55-58; pharmacist mgr, Perry Pharm, 59-62; res scientist pharm, Ames Co Div, Miles Labs, 62-81; RES & DEVELOP SR FORMULATIONS SCIENTIST, COLMED LABS INC, DENVER, COLO, 85- *Concurrent Pos:* Sponsor, Telecommun Trust Fund Handicapped. *Mem:* Am Pharmaceut Asn; Fedn Sci & Handicapped. *Res:* Pharmaceutical solid systems in research and development applications to product development of human diagnostic devices. *Mailing Add:* 12050 W Carolina Dr Lakewood CO 80228

DAUNT, JOHN GILBERT, physics; deceased, see previous edition for last biography

DAUNT, STEPHEN JOSEPH, b Brooklyn, NY, July 3, 47; div. MOLECULAR SPECTROSCOPY, PLANETARY ATMOSPHERES. *Educ:* Iona Col, NY, BS, 69; Queen's Univ, Ont, PhD(phys chem), 76. *Prof Exp:* Instr math, Dept Math, 75-76, res asst prof, 77-79, ASST PROF ASTRON, DEPT PHYSICS & ASTRON, UNIV TENN, 79-; ASSOC PROF CHEM, DEPT CHEM, CONCORDIA UNIV, MONTREAL, CAN, 80- *Concurrent Pos:* Res assoc physics, Dept Physics & Astron, Univ Tenn, 75-77; res fel, Dept Chem, Concordia Univ, Montreal, Can, 80-; vis scientist, NASA Goddard Space Flight Ctr, 81-; summer fac fel, NASA/Inst Elec & Electronic Engrs, 84 & 85. *Mem:* Am Astron Soc; Am Chem Soc; Am Phys Soc; Optical Soc Am; Soc Appl Spectros. *Res:* Molecular spectroscopy; high resolution infrared and Raman spectra of gases; optical instrumentation design and construction; planetary atmospheres; interstellar molecules; air pollution studies; computer simulation of spectra; molecular dynamical calculations; solid state phase transitions. *Mailing Add:* Molecular Spectros Lab Dept Physics & Astron Univ Tenn Knoxville TN 37996-1200

DAUNTON, NANCY GOTTLIEB, b Washington, DC, Sept 9, 42; m 70. NEUROPSYCHOLOGY, NEUROPHYSIOLOGY. *Educ:* George Washington Univ, AB, 64, MA, 66; Stanford Univ, PhD(neuropsychol), 71. *Prof Exp:* Psychologist neurobiol, 67-71, RES SCIENTIST NEUROSCI, AMES RES CTR-NASA, 71- *Concurrent Pos:* Mgr Ames grant prog vestibular & space sickness res, Ames Res Ctr-NASA, 73- & proj scientist vestibular function res proj spacelab, 77- *Mem:* Soc Neurosci. *Res:* Neurophysiological and neurobehavioral studies of vestibular system function; single unit studies of visual- vestibular interactions; etiology of motion sickness and space sickness; efferent control of sensory input. *Mailing Add:* Biomed Res Div 26-3 Ames Res Ctr-NASA Moffett Field CA 94035

DAUPHINAIS, RAYMOND JOSEPH, b Chicago, Ill, May 30, 25. PHARMACY. *Educ:* Univ Ill, BS, 48; Univ Fla, JD, 54; NY Univ, LLM(trade regulation), 56. *Prof Exp:* Lectr pharm law, Univ Fla, 53-54, asst prof, 54-55; lectr pharmaceut econ, Columbia Univ, 55-56; asst prof pharm, Univ Conn, 57-60; dir legal div, Am Pharmaceut Asn, 60-64; asst to dean col pharm, 64-68, PROF PHARM, WAYNE STATE UNIV, 64-, GRAD OFFICER, 81- *Concurrent Pos:* NSF vis scientist, 64-68; Am Asn Cols Pharm vis lectr, 68- *Honors & Awards:* Food Law Inst Award, 55-56. *Mem:* Am Pharmaceut Asn; Nat Health Lawyers Asn; Am Asn Cols Pharm. *Res:* Pharmaceutical economics; trade regulations; food and drug legislation; health-care administration; professional practice law in healing arts; health care law. *Mailing Add:* 1418 Nicolet Detroit MI 48202

DAUPHINÉ, T(HONET) C(HARLES), b North Battleford, Sask, Nov 8, 13; US citizen; m 38; c 4. CHEMICAL ENGINEERING. *Educ:* Mass Inst Technol, SB, 35, ScD(chem eng), 39. *Prof Exp:* Instr chem eng, Mass Inst Technol, 36-39; process engr res & develop dept, Standard Oil Co, Calif, 39-43, supv eng, Calif Res Corp, 43-46; eastern mgr prod develop dept, Oronite Chem Co, 46-51; mgr sales develop, Hooker Chem Corp, NY, 51-57, mgr prod develop plastics & polymers, 57-59, sr develop engr, Corp Gen Develop, 59-62; vpres prod & eng, Nease Chem Co, 62-64; consult, 64-65; vpres, Chem Process Corp, Conn, 65-68, process design supvr, Badger Am Inc, 68-76, technol mgr, Badger Energy, Inc, 76-78; PRES & TREAS, DESIGN ENTERPRISES, INC, 78- *Concurrent Pos:* Consult, Badger Co, Inc, 48-82, Raytheon Co, 83. *Mem:* Am Chem Soc; Am Inst Chem Engrs; Commercial Develop Asn; fel Am Inst Chemists. *Res:* Inventor, in situ low temperature retorting process for oil shale using radio frequency energy; process design of synthetic fuel plants for methanol from coal, oil and gas from coal; process design various plants for petroleum and petrochemical products, other chemicals; new product development of synthetic detergents, p, m and oxylenes, isophthalic acid, chlorendic acid, fire resistant polyesters and rigid urethane foams; process development for various organic chemicals, including aromatic sulfonation and oxidation; developed universal correlation for calculating pressure drop through bubble cap distillation columns. *Mailing Add:* 57 Alcott St Acton MA 01720-5540

DAUPHINEE, THOMAS MCCAUL, b Vancouver, BC, July 3, 16; m 40; c 3. PHYSICS. *Educ:* Univ BC, BA, 43, MA, 45, PhD, 50. *Prof Exp:* Instr dept physics, Univ BC, 47-48; res officer physics, 45-67, sr res officer, 67-70, prin res officer physics, Nat Res Coun Can, 70-80; CONSULT, 80- *Concurrent Pos:* Measurement consult, 81; pres, Sea-Met Sci Ltd, 81- *Honors & Awards:* Morris E Leeds Award, Inst Elec & Electronics Engrs, 78. *Mem:* Instrument Soc Am; Can Asn Physicists; Inst Elec & Electronics Engrs; Can Meteorol & Oceanog Soc. *Res:* Thermometry; electrical measurements; instrumentation; thermal and electrical properties of matter; oceanographic measurements. *Mailing Add:* Key Creative Inc 36 Ave Rd Ottawa ON K1S 0N9 Can

D'AURIA, JOHN MICHAEL, b New York, NY, Mar 28, 39; m 66; c 2. NUCLEAR CHEMISTRY, TRACE ELEMENT ANALYSIS. *Educ:* Rensselaer Polytech Inst, BS, 61; Yale Univ, MS, 62, PhD(nuclear chem), 66. *Prof Exp:* Res asst chem, Columbia Univ, 66-68; from asst prof to assoc prof, 68-81, PROF CHEM, SIMON FRASER UNIV, 81- *Concurrent Pos:* Assoc, Cent Europ Orgn Nuclear Res, 75-76; res scientist, Triumf, 84-85. *Mem:* Am Phys Soc; Spectros Soc Can; Can Asn Univ Teachers; Sigma Xi; Chem Inst Can; Can Asn Physics. *Res:* Nuclear spectroscopy; on-line isotope separators (ISOL); nuclear reactions; nuclear astrophysics; applications of nuclear techniques; x-ray fluorescence, trace element analysis. *Mailing Add:* Dept Chem Simon Fraser Univ Burnaby BC V5A 1S6 Can

D'AURIA, THOMAS A, b Mt Vernon, NY, Apr 25, 46. COMPUTER SCIENCE, COMPUTER MANAGEMENT & EDUCATION. *Educ:* Polytech Univ, BS, 68. *Prof Exp:* MGT CONSULT, SMA MGT SYST INC, 88- *Concurrent Pos:* Mem bd dirs, Asn Comput Mach, 76- *Mem:* Asn Comput Mach. *Mailing Add:* 50 Main St White Plains NY 10606

DAUSSET, JEAN BAPTISTE GABRIEL, b Toulouse, France, Oct 19, 16. IMMUNOLOGY. *Prof Exp:* Asst fac med, Univ Paris, 46-50; dir lab, Nat Blood Transfusion Ctr, 50-63; prof exp med, Col France, 78-85; DIR & PRES, CENTRE ETUDE POLYMORPHISME HUMAIN, 85- *Honors & Awards:* Nobel Prize in Med, 80. *Mem:* Nat Acad Sci; French Acad Sci & Med. *Mailing Add:* Centre Etude Polymorphisme Humain 27 Rue Juliette Dodu Paris 75010 France

DAUTERMAN, WALTER CARL, b Closter, NJ, June 10, 32; m 63; c 2. INSECT TOXICOLOGY, BIOCHEMISTRY. *Educ:* Rutgers Univ, BS, 54, MS, 57; Univ Wis, PhD(entom), 59. *Prof Exp:* Fulbright res fel, Netherlands, 59-60; res assoc, Cornell Univ, 60-62; from asst prof to assoc prof, 62-72, PROF ENTOM, NC STATE UNIV, 72- *Concurrent Pos:* Sabbatical study leave, Ciba-Geigy, Basel, Switz, 72-73, New S Wales Dept Agr, Rydalmire, Australia, 82-83. *Mem:* Am Chem Soc; Entom Soc Am; Soc Toxicol. *Res:* Mode of action and the selectivity of organophosphorus insecticides. *Mailing Add:* Dept Toxicol NC State Univ Box 7633 Raleigh NC 27695-7633

DAUTLICK, JOSEPH X, b Pittsburgh, Pa, Dec 3, 42; m 66; c 2. CLINICAL BIOCHEMISTRY. *Educ:* Lafayette Col, BS, 64; Bowman Gray Sch Med, MS, 68; Univ Pittsburgh, PhD(biochem), 73. *Prof Exp:* Chief clin chem lab, US Army, Ft Sam Houston, Tex, 68-70; biochemist, 73-75, tech serv supvr, 76-77, tech mgr, Europe, 78-81, worldwide tech serv mgr, Automatic Clin Anal Div, 81-83, planning mgr, 84, PROG MGR, E I DU PONT DE NEMOURS & CO, INC, 85- *Concurrent Pos:* Mem subcomt assay conditions, Nat Comt Clin Lab Standards, 75-77; corresp ed, J Automated Chem, 80- *Mem:* Am Asn Clin Chem; Am Chem Soc. *Res:* Method development and expanded applications for Du Pont automatic clinical analysis relative to clinical chemistry testing of human body fluids; development of automated and semi-automated analyzers and consumables for the determination of medically important analytes and infectious diseases in humans. *Mailing Add:* 25 Alcott Dr Heritage Farms DE 19808

DAUTZENBERG, FRITS MATHIA, b Holland, Feb 13, 40; m 63; c 2. CATALYSIS. *Educ:* Tech Univ, Holland, IR, 64. *Prof Exp:* Res & develop engr, Shell, Holland, 66-72, technologist, France, 72-74, sect head, Holland, 74-78, sr staff, Shell, Can, 78-82, adv, 82-83; dir, 83-87, VPRES, CATALYTICA INC, 87- *Mem:* Am Inst Chem Engr; Am Chem Soc. *Res:* Catalysis. *Mailing Add:* 1962 Laver Ct Los Altos CA 94022-6723

DAUWALDER, MARIANNE, b Long Beach, Calif, Aug 5, 35; m 62; c 5. CELL BIOLOGY. *Educ:* Occidental Col, BA, 56; Univ Tex, PhD(biol, cytol), 64. *Prof Exp:* Res asst zool, Univ Calif, Los Angeles, 59; RES SCIENTIST ASSOC, CELL RES INST, UNIV TEX, AUSTIN, 63- *Mem:* Am Soc Cell Biol. *Res:* Cellular differentiation; cellular mechanisms in plant tropic responses; differential activities of the Golgi apparatus demonstrated with ultrastructural, cytochemical, autoradiographic and immunocytochemical techniques. *Mailing Add:* Dept Bot Univ Tex Austin TX 78713

D'AVANZO, CHARLENE, b Springfield, Mass, Oct 29, 47; m 71. BIOLOGY, ECOLOGY. *Educ:* Skidmore Col, BA, 69; Boston Univ, PhD(marine biol), 75. *Prof Exp:* Fel estuarine ecol, Dalhousie Univ, 75-77; ASSOC PROF ECOL, HAMPSHIRE COL, 77- *Mem:* Am Women Sci; Ecol Soc Am; AAAS. *Res:* Salt marsh ecology, riverine marsh ecology, nitrogen cycling, aquaculture. *Mailing Add:* Sch Natural Sci Hampshire Col Amherst MA 01002

DAVANZO, JOHN PAUL, b Seattle, Wash, Jan 14, 27; m 50; c 6. BIOLOGY. *Educ:* Univ Wash, BSc, 49 & 51; Univ NMex, MSc, 55, PhD, 57. *Prof Exp:* Supvr exp surg lab, Sch Med, Univ Wash, 49-51; res assoc & East NJ Heart Asn fel, Princeton Univ, 57-59; mem res div, Upjohn Co, 59-62; head biochem pharmacol, A H Robins Co, 62-68; dir pharmacol, Ortho Res Found, 68-72, exec dir res basic sci, 72-74; vpres res & develop, Purdue Frederick Co, 74-76; PROF PHARMACOL, SCH MED, EAST CAROLINA UNIV, 76- *Concurrent Pos:* Lectr, Med Col, 62-, res assoc, 64-; vis asst prof obstet & gynec, Univ Kans Med Ctr, 74- *Mem:* Am Soc Pharmacol & Exp Therapeut; Am Physiol Soc; Soc Exp Biol & Med; Soc Toxicol; Endocrine Soc. *Res:* Neurochemistry, especially the effects of drugs on the central nervous system. *Mailing Add:* Pharmacol East Carolina Univ Greenville NC 27858

DAVAR, K(ERSI) S, b Bombay, India, Oct 3, 23; m 50; c 3. CIVIL ENGINEERING, HYDROLOGY. *Educ:* Univ Poona, India, BE, 46; Colo State Univ, MIE, 57, PhD(civil eng), 61. *Prof Exp:* Asst engr, Damodar Valley Corp, India, 50-55; asst prof hydrol & fluid mech, Univ NB, 61-65; assoc prof, Tex Tech Col, 65-66; assoc prof, 66-68, PROF CIVIL ENG, UNIV NB, 68- *Concurrent Pos:* Mem, Can Nat Comt for Int Hydrol Decade, 66-; mem adv comt, Can Ctr Inland Waters, 69- *Mem:* Int Asn Hydraul Res; Am Soc Civil Engrs; Am Water Resources Asn. *Res:* Hydrology, watershed response, snow hydrology, soil moisture; water resources systems, optimization of multipurpose river systems; hydraulics, dispersive processes in streams and channels, resistance to flow in unsymmetrical channels. *Mailing Add:* Dept Civil Eng Univ NB College Hill Box 4400 Fredericton NB E3B 5A3 Can

DAVÉ, BHALCHANDRA A, b Palanpur, India, Dec 10, 31; US citizen; m 65; c 2. MICROBIOLOGY, FOOD SCIENCE. *Educ:* St Xavier's Col, India, BSc, 55, MSc, 58; Univ Calif, Davis, MS, 62, PhD(microbiol), 68. *Prof Exp:* Demonstr, St Xavier's Col, India, 55-58; res asst food sci, Univ Calif, Davis, 59-65, res microbiologist, 66-70; res food scientist, Decco Div, 70-78, supvr, res opers, 78-84, RES DIR, PENNWALT CORP, 85- *Concurrent Pos:* Consult, Food Indust. *Mem:* Inst Food Technologists; Am Soc Microbiol; Am Soc Hort Sci. *Res:* Food microbiology; pectic enzymes; storage of fresh fruits and vegetables; post-harvest pesticide screening, formulation and their Environmental Protection Agency registration; food industry sanitation. *Mailing Add:* Decco Div Pennwalt Corp PO Box 120 1713 S California Ave Monrovia CA 91016-0120

DAVÉ, RAJU S, b Calcutta, India, Aug 29, 58; m 83. ADVANCED PERFORMANCE PLASTICS, FIBER-REINFORCED COMPOSITES. *Educ:* Indian Inst Technol, BSc Hons, 81; Wash State Univ, Pullman, MS, 83; Wash Univ, St Louis, DSc(chem eng), 86. *Prof Exp:* Res asst, Wash State Univ, Pullman, 81-83; teaching & res asst, Unit Oper Lab, Wash Univ, St Louis, 83-86; asst res prof, Mich Molecular Inst, Midland, 86-89; RES SPECIALIST, MONSANTO CHEM CO, 89- *Concurrent Pos:* Adj asst prof, Cent Mich Univ, Mt Pleasant, 87-89 & Mich Technol Inst, Houghton, 87-91. *Mem:* Sigma Xi; Soc Advan Mat & Process Eng. *Res:* Advanced performance plastics which include composites; author of over 20 technical publications. *Mailing Add:* Monsanto Co 730 Worcester St Indian Orchard MA 01151

DAVENPORT, ALAN GARNETT, b Madras, India, Sept 19, 32; Can citizen; m 57; c 4. STRUCTURAL ENGINEERING, AERODYNAMICS. *Educ:* Cambridge Univ, BA, 54, MA, 58; Univ Toronto, MASc, 57; Bristol Univ, PhD(eng), 61. *Hon Degrees:* DSc, Univ Louvain, 79, Tech Univ Denmark, 83, McGill Univ, 84, Univ Waterloo, 86, Univ Toronto, 89. *Prof Exp:* Lectr eng, Univ Toronto, 54-57; asst, Nat Res Coun Can, 57-58; asst, Bristol Univ, 58-61; assoc prof, 61-66, PROF ENG, UNIV WESTERN ONT, LONDON, 66-, DIR, BOUNDARY LAYER WIND TUNNEL LAB, FAC ENG, 65- *Concurrent Pos:* Consult design, World Trade Ctr, New York, 64-65; chmn, Task Force Wind Forces, Am Soc Civil Engrs & mem, Joint Comt Tall Bldgs, Am Soc Civil Engrs-Int Asn Bridge & Struct Engr; mem, Can Aeronaut Res Adv Comt, Can Nat Comt Earthquake Eng & Comt Nat Bldg Code Can; mem, Comt Natural Disasters, Nat Acad Eng & Int Standards Orgn; mem, UN Adv Comt & Can Construct Res Bd; bd dirs, Can Soc Civil Eng & Int Coun Tall Bldgs & Urban Habitat; mem sci comt, Can Meteorol & Oceanog Soc, 88-91. *Honors & Awards:* Duggan Prize, Eng Inst Can, 62, Gzowski Medal, 63 & 78; Alfred Nobel Prize, Am Soc Civil Engrs, 63; Golden Plate Award, Am Acad Achievement, 65; Prize in Appl Meteorol, Can Meteorol Soc, 67; State of the Art Award, Am Soc Civil Engrs, 73 & Can-AM Amity Award, 77; Silver Medal, Asn Prof Engrs Ont, 78; Cancom medal, 83; A B Sanderson Award for Struct Eng, Can Soc Civil Eng, 85; Gold Medal, Inst Struct Engrs UK, 88; Rutherford Lectr, Royal Socs London & Can, 88; Earnest Manning Award Distinction, 90. *Mem:* Foreign assoc Nat Acad Eng; Eng Inst Can; Royal Meteorol Soc; Int Asn Bridge & Struct Engrs; fel Royal Soc Can; Am Soc Civil Engrs; Can Soc Civil Eng; Seismol Soc NAm. *Res:* Meteorology; wind engineering; application of boundary layer wind tunnels to the design of wind sensitive structures, the description of urban wind climates, and other problems involving the action of wind; environmental loads; structural dynamics; earthquake loading. *Mailing Add:* Boundary Layer Wind Tunnel Lab Univ Western Ont Fac Eng Sci London ON N6A 5B9 Can

DAVENPORT, CALVIN ARMSTRONG, b Gloucester, Va, Jan 15, 28; m 63; c 2. BACTERIOLOGY, IMMUNOLOGY. *Educ:* Va State Col, BS, 49; Mich State Univ, MS, 50, PhD(microbiol, pub health), 63. *Prof Exp:* Bacteriologist, Div Labs, State Dept Health, Mich, 52-53, 55-57; assoc prof microbiol, Va State Col, 63-69, head dept, 66-69; assoc prof, 69-72, PROF MICROBIOL, CALIF STATE UNIV, FULLERTON, 72- *Concurrent Pos:* Consult, NIH, Dept Health, Educ & Welfare, 73-; mem bd trustees, Orange County/Long Beach Health Consortium, 72-; mem sr comn, Western Asn Schs & Cols, 76-79 & mem accreditation teams, 72-; consult, Delst Chem Co, 81- & Anaheim Med Labs, 74-79; mem acad planning comt, Health Manpower Proj, Calif Univs & Cols, 74-76; mem coun, Orange County, Calif, Consort, Nursing Educ, 72-73; coordr & mem, Med Technol Prog, Calif State Univ, 69-76, Fullerton Minority Affairs Coun, 81-85, liaison & dir incentive grants prog, Nat Action Coun Minorities Eng, 83-84, chmn, acad affirmative action comt, 82-84, sr mentor, student affirmative action prog, 81-86, AIDS educ comt, 86-; mem, Univ Affirmative Action Resource Person Adv Group, 86-, Univ Health Professions Comt, 87-, Univ Inst Health Educ & Training, 87-, Univ Geront Res Inst, 87-, Univ Minority Biomed Res Support Adv Group, 88-, Univ Substance Abuse Educ Comt, 91-; chair, Dept Curric Comt, 89-, Calif

Asn Pub Health Lab Dirs Acad Assembly, 90- *Mem:* Am Soc Microbiol; Nat Geog Soc; Fedn Am Scientists. *Res:* Antibiotic sensitivities of Neisseria gonorrhoeae; passive ovarian transfer of immunoglobins from mother to its offspring in avian species. *Mailing Add:* Dept Microbiol Calif State Univ Fullerton CA 92634

DAVENPORT, DEREK ALFRED, b Leicester, Eng, Sept 23, 27; m 50. INORGANIC CHEMISTRY, CHEMICAL EDUCATION. *Educ:* Univ London, BSc, 47, PhD, 50. *Prof Exp:* Fel, Ohio State Univ, 51-53; from instr to assoc prof, 53-70, PROF CHEM, PURDUE UNIV, WEST LAFAYETTE, 70- *Concurrent Pos:* Vis prof, Univ Wis-Madison. *Honors & Awards:* Chem Educ Award, Am Chem Soc. *Mem:* Am Chem Soc. *Res:* Borderlands between organic and inorganic chemistry; history of chemistry. *Mailing Add:* Dept Chem Purdue Univ West Lafayette IN 47907-9980

DAVENPORT, FRED M, internal medicine, epidemiology; deceased, see previous edition for last biography

DAVENPORT, HORACE WILLARD, b Philadelphia, Pa, Oct 20, 12; m 45, 69; c 1. PHYSIOLOGY, GASTROENTEROLOGY. *Educ:* Calif Inst Technol, BS, 35, PhD(biochem), 39; Oxford Univ, BA, 37, BSc, 38, DSc(physiol), 61. *Prof Exp:* Instr physiol, Med Sch, Univ Pa, 41-43; instr, Harvard Med Sch, 43-45; prof physiol & head dept, Col Med, Univ Utah, 45-56; prof & chmn dept, 56-78, William Beaumont prof physiol, 78-83, EMER PROF PHYSIOL, MED SCH, UNIV MICH, 83- *Concurrent Pos:* Lilly fel path, Sch Med, Univ Rochester, 39-40; Sterling fel physiol chem, Yale Univ, 40-41; vis prof physiol, Mayo Clin & Mayo Found, 62-63. *Honors & Awards:* Friedenwald Medal, Am Gastroenterol Soc, 80; Daggs Award, Am Physiol Soc, 88. *Mem:* Nat Acad Sci; Am Physiol Soc (pres, 61-62); hon mem Brit Soc Gastroenterol. *Res:* Physiology and pathophysiology of the stomach; history of physiology. *Mailing Add:* Dept Physiol Univ Mich Ann Arbor MI 48109-0622

DAVENPORT, JAMES WHITMAN, b Indianapolis, Ind, July 8, 45; m 68; c 4. SOLID STATE PHYSICS. *Educ:* Brown Univ, ScB, 67; Princeton Univ, ScM, 68; Univ Pa, PhD(physics), 76. *Prof Exp:* Staff, RCA Labs, 68-71; fel surface physics, Univ Pa, 76-77 & Chalmers Univ Technol, 77-78: from asst physicist to assoc physicist, 78-82, PHYSICIST, BROOKHAVEN NAT LAB, 82- *Mem:* Am Phys Soc; Am Vac Soc. *Res:* Surface physics; chemisoption; photoemission from atoms, molecules and solids; interaction of light and matter. *Mailing Add:* Dept Physics Brookhaven Nat Lab Bldg 510A Upton NY 11973

DAVENPORT, JOHN EATON, b Los Vegas, Nev, Aug 26, 44; m 66; c 2. PHYSICAL CHEMISTRY, CHEMICAL PHYSICS. *Educ:* Univ Calif, Santa Barbara, BA, 66, MA, 67; York Univ, PhD(chem), 74. *Prof Exp:* Fel, Molecular Physics Ctr, 73-75, PHYS CHEMIST GAS PHASE KINETICS CHEM LAB, SRI INT, 75- *Mem:* AAAS; Am Chem Soc; Am Phys Soc. *Res:* Gas phase kinetics and dynamics of species and processes important in aeronomy and air-pollution studies. *Mailing Add:* 524 Seventh Ave Menlo Park CA 94025-3446

DAVENPORT, LEE LOSEE, b Schenectady, NY, Dec 31, 15; m 44; c 2. PHYSICS, TELECOMMUNICATIONS. *Educ:* Union Col, NY, BS, 37; Univ Pittsburgh, MS, 40, PhD(physics), 46. *Prof Exp:* Res assoc radar, Mass Inst Technol, 41-46; res fel construct cyclotron, Harvard Univ, 46-50; exec vpres, Perkin-Elmer Corp, 50-57; pres, Sylvania Corning Nuclear Corp, 57-60, vpres planning, Sylvania Elec Prod, Inc, 60-62, pres, GTE Labs, Inc, 62-77, vpres-chief scientist, GTE, 77-80; CONSULT TELECOMMUN, 80- *Concurrent Pos:* Asst dir electronics res lab, Univ Pittsburgh, 46; corp dir, 80- *Mem:* Nat Acad Eng; fel Am Phys Soc; Sci Res Soc Am; fel Inst Elec & Electronics Engrs. *Res:* Research spectroscopy; nuclear physics; microwave radar; particle accelerators; guided missile control system. *Mailing Add:* 61 Winding Lane Greenwich CT 06831

DAVENPORT, LESLEY, b Bristol, UK, Nov 5, 55. MODEL MEMBRANE TECHNIQUES. *Educ:* Univ Salford, UK, BSc, 77, PhD(biochem), 81. *Prof Exp:* Res fel biophys, Johns Hopkins Univ, 81-83, assoc res scientist, 84-85; ASST PROF BIOCHEM, BROOKLYN COL, CITY UNIV NY, 85- *Honors & Awards:* Mem, Royal Soc Chemists. *Mem:* Biochem Soc Gt Brit; Royal Soc Chem; Biophys Soc Gt Brit; Biophys Soc; Am Soc Photobiol; Am Chem Soc; Sigma Xi. *Res:* Lipid-lipid & lipid- protein interactions of biological membranes; fluorescence spectroscopy; membrane heterogeneity; spectroscopic evidence for membrane 'domains'; effects of anesthetics cholesterol and steroid-like molecules on membrane architecture. *Mailing Add:* Chem Dept City Univ NY Brooklyn Col Brooklyn NY 11210

DAVENPORT, LESLIE BRYAN, JR, b Abingdon, Va, July 10, 28; m 52; c 3. VERTEBRATE ZOOLOGY, ECOLOGY. *Educ:* Col Charleston, BS, 47; Va Polytech Inst, MS, 51; Univ Ga, PhD(zool), 60. *Prof Exp:* Teacher, Porter Mil Acad, SC, 47-49; prof biol & head dept, Armstrong State Col, 59-83; CONSULT, 83- *Concurrent Pos:* Wormsloe Found grant ecol res, 64-76. *Mem:* Ecol Soc Am; Am Soc Mammal; Am Ornith Union; Wildlife Soc; Sigma Xi. *Res:* Dynamics and energy relationships of populations of mammals and birds. *Mailing Add:* 726 Windsor Rd Savannah GA 31419

DAVENPORT, PAUL W, b Coalwater, Mich, Feb 14, 51. RESPIRATION. *Educ:* Univ Ky, Lexington, PhD(physiol & biophys), 80. *Prof Exp:* Asst prof, 81-86, ASSOC PROF PHYSIOL SCI, UNIV FLA, 86- *Mailing Add:* Dept Physiol Sci Univ Fla Box J-144 JHMHC Gainesville FL 32610

DAVENPORT, THOMAS LEE, b Bowie, Tex, Oct 16, 47; div; c 2. PLANT PHYSIOLOGY, BIOCHEMISTRY. *Educ:* Trinity Univ, BA, 70; Tex A&M Univ, PhD(plant physiol), 75. *Prof Exp:* Fel, 75-77, ASSOC PROF PLANT PHYSIOL, UNIV FLA, 77- *Concurrent Pos:* Assoc jour ed, Am Soc Hort Sci; foreign & domestic consult; pres, Plant Growth Regulator Soc Am. *Honors & Awards:* Wilson Popanoe Award, Am Soc Hort Sci, 79. *Mem:* Am Soc Plant Physiologists; Am Soc Hort Sci; Int Plant Growth Substances Asn; Plant Growth Regulator Soc Am. *Res:* Growth and developmental aspects of flower induction and young fruit abscission in avocado, 'Tahiti' lime, and other tropical fruit. *Mailing Add:* Trop Res & Educ Ctr 18905 SW 280th St Homestead FL 33031

DAVENPORT, TOM FOREST, JR, b Atlanta, Ga, Apr 25, 30; m 54; c 3. ORGANIC CHEMISTRY, PHARMACEUTICAL CHEMISTRY. *Educ:* Ga Inst Technol, BS, 51; Univ Tex, PhD(chem), 57. *Prof Exp:* Res chemist, Am Oil Co, Tex, 56-60; res chemist, 60-66, prod rep, Com Develop Dept, 66-67, prod mgr, 67-70, mkt develop analyst, Com Develop Dept, 70-74, sr com develop chemist, 74-80, COM DEVELOP ASSOC, ETHYL CORP, 80- *Mem:* Am Chem Soc; Chem Mkt Res Asn. *Res:* Organometallics; hydrogenation; nitrogen compounds; synthesis; market research and development on specialty chemical products, including pharmaceutical chemicals. *Mailing Add:* Ethyl Corp 451 Florida Blvd Baton Rouge LA 70801-1779

DAVENPORT, WILBUR B(AYLEY), JR, b Philadelphia, Pa, July 27, 20; m 45; c 2. ELECTRICAL ENGINEERING. *Educ:* Ala Polytech Inst, BEE, 41; Mass Inst Technol, SM, 43, ScD(elec eng), 50. *Prof Exp:* Asst elec eng, Mass Inst Technol, 41-43, instr, Radar Sch, 43, instr, Elec Eng Dept, 46-49, asst prof, 49-53, group leader, Commun Tech Group, Lincoln Lab, 51-55, assoc div head, Commun & Components Div, 55-57, div head, 57-58, head, Info Processing Div, 58-60, prof elec eng, 60-78, asst dir lab, 63-65, assoc head, Dept Elec Eng, 71-72, dir, Ctr Advan Eng Study, 72-74, head, Dept Elec Eng & Comput Sci, 74-78, prof commun sci & eng, 78-82, EMER PROF COMMUN SCI ENG, DEPT ELEC ENG & COMPUT SCI, MASS INST TECHNOL, 82- *Concurrent Pos:* Vis asst prof, Univ Calif, 50; mem, Carnegie Comn Future of Pub Broadcasting, 77-79; adj res prof elec engr, Naval Postgrad Sch, 82-83; vis prof elec engr, Univ Hawaii, Manoa, 82-; chmn, info technol div, Pac Int Ctr High Technol Res, 84-87. *Honors & Awards:* Pioneer Award, Inst Elec & Electronics Engrs, Aerospace & Electronic Systs Soc, 81. *Mem:* Nat Acad Eng; fel AAAS; fel Inst Elec & Electronics Engrs; fel Am Acad Arts & Sci. *Res:* data networks; communications policy. *Mailing Add:* One Aspen Lane PO Box 3434 Sunriver OR 97707

DAVENPORT, WILLIAM DANIEL, JR, b Corinth, Miss, Apr 19, 47. ANATOMY. *Educ:* Univ Miss, BS, 69, MS, 71; Registry Med Technologists, cert, 68; Med Col Ga, PhD(anat), 76. *Prof Exp:* Med technologist hematol, Magnolia Hosp, 66-69 & Oxford-Lafayette County Hosp, 69-70; instr zool, Ark State Univ, 71-72; from instr to asst prof anat, Med Ctr, Univ Miss, 77-82; asst prof, 82-84, ASSOC PROF ORAL PATH & ANAT, LA STATE UNIV, 84-, DIR RES HISTOL LAB, 82- *Concurrent Pos:* Mem anat sci test construct comt, Coun Nat Bd Exam Am Dent Asn, 78-83. *Mem:* Am Soc Med Technologists; Am Asn Anatomists; Sigma Xi. *Res:* Histology; histochemistry; electron microscopy. *Mailing Add:* Dept Oral Path 1100 Florida Ave New Orleans LA 70119

DAVERMAN, ROBERT JAY, b Grand Rapids, Mich, Sept 28, 41; m 61; c 2. TOPOLOGY. *Educ:* Calvin Col, AB, 63; Univ Wis-Madison, MA, 65, PhD(math), 67. *Prof Exp:* From asst prof to assoc prof, 67-76, PROF MATH, UNIV TENN, KNOXVILLE, 76- *Concurrent Pos:* Vis assoc prof math, Univ Utah, 73-74; vis prof math, Fla State Univ, 81; Fulbright Scholar, 86. *Mem:* Am Math Soc. *Res:* Topology of Euclidean space and of manifolds; flatness and wildness of embeddings; decompositions of manifolds. *Mailing Add:* Dept Math Univ Tenn Knoxville TN 37916

DAVERN, CEDRIC I, b Hobart, Tasmania, Nov 13, 31; m 82; c 6. MOLECULAR GENETICS. *Educ:* Univ Sydney, BSAgr, 53, MSAgr, 56; Calif Inst Technol, PhD(biol), 59. *Prof Exp:* From res officer to sr res officer genetics, Commonwealth Sci & Indust Res Orgn, Australia, 54-59; res assoc molecular biol, Lab Quant Biol, Cold Spring Harbor, NY, 64-65, asst dir, 65-67; from assoc prof to prof biol, Univ Calif, Santa Cruz, 67-76, exec officer, Div Natural Sci, 74-75; dean, Col Med, 76-77, vpres acad affairs, 77-83, PROF GENETICS, UNIV UTAH, 76- *Concurrent Pos:* Eccles vis prof microbiol, Sch Med, Univ Utah, 75-76. *Res:* Molecular mechanisms of DNA replication and its regulation; plant evolutionary biology. *Mailing Add:* 447 S 1200 E Salt Lake City UT 84112

DAVES, GLENN DOYLE, JR, b Clayton, NMex, Feb 12, 36; m 59; c 3. ORGANIC CHEMISTRY. *Educ:* Ariz State Univ, BS, 59; Mass Inst Technol, PhD(org chem), 64. *Hon Degrees:* DPharm, Univ Uppsala, Sweden, 87. *Prof Exp:* Res chemist, Midwest Res Inst, 59-61; res chemist, Stanford Res inst, 64-67; from asst prof to prof chem, Ore Grad Ctr, 67-81, chmn, dept chem, 72-79; prof & chmn chem, Lehigh Univ, 81-88; vis scientist, NIH, 88; PROF CHEM & DEAN SCI, RENSSELAER POLYTECH INST, 89- *Concurrent Pos:* Affil mem, Univ Ore Med Sch, 72-81; chmn manpower & resource comt, Coun Chem Res, 84-87, mem gov bd, 85-86, membership comt, 91. *Mem:* Am Chem Soc; Int Soc Heterocyclic Chem; Coun Chem Res. *Res:* Isolation, structure elucidation and synthesis of compounds with important biological properties: palladium mediated reactions, C-nucleosides, carbohydrate immunochemistry, applications of mass and nmr spectrometries. *Mailing Add:* Sch Sci Rensselaer Polytech Inst Troy NY 12180-3590

DAVES, MARVIN LEWIS, b Lexington, Va, Jan 26, 28; c 3. RADIOLOGY. *Educ:* Wash & Lee Univ, BA, 48; Johns Hopkins Univ, MD, 53. *Prof Exp:* Asst radiologist, Clin Ctr, NIH, 57-59; asst prof radiol & actg head dept, Med Ctr, Univ Ark, 59-60; asst prof, 60-62, actg chmn dept, 61-62, chmn dept, 62-77, PROF RADIOL, MED CTR, UNIV COLO, DENVER, 62- *Mem:* Fel Am Col Radiol; Radiol Soc NAm. *Res:* Cardiology; bone dysplasia. *Mailing Add:* Dept Radiol Univ Colo Med Ctr 4200 E Nineth Ave Denver CO 80262

DAVEY, CHARLES BINGHAM, b Brooklyn, NY, Apr 7, 28; m 52; c 3. SOIL MICROBIOLOGY. *Educ:* NY State Col Forestry, Syracuse Univ, BS, 50; Univ Wis, MS, 52, PhD(soils), 55. *Prof Exp:* Soil scientist, Agr Res Serv, USDA, 57-62; assoc prof, 62-65, prof soil sci, 65-78, head dept forestry, 70-78, CARL ALWIN SCHENCK PROF FORESTRY, NC STATE UNIV, 78-

Concurrent Pos: Vis prof, Oregon State Univ, 69-70. *Honors & Awards:* Barrington Moore Award, Soc Am Foresters. *Mem:* Fel AAAS; fel Am Soc Agron; fel Soil Sci Soc Am (pres, 75-76); Soc Am Foresters; Int Soil Sci Soc. *Res:* Microbial ecology in soil and rhizosphere; Mycorrhizae; tree nutrition; agroforestry; tropical forestry. *Mailing Add:* Dept Forestry NC State Univ Raleigh NC 27695-8008

DAVEY, FREDERICK RICHARD, IMMUNOHEMATOLOGY. *Educ:* State Univ NY, MD, 64. *Prof Exp:* PROF PATH, STATE UNIV NY HEALTH SCI CTR, 71- *Mailing Add:* Dept Path SUNY Health Sci Ctr 750 E Adams St Syracuse NY 13210

DAVEY, GERALD LELAND, b Salt Lake City, Utah, Apr 9, 30; m 54; c 4. APPLIED MATHEMATICS, SYSTEMS ANALYSIS. *Educ:* Stanford Univ, MS, 54, PhD(math), 59. *Prof Exp:* Employee, Hughes Aircraft Co, Calif, 59-61; consult, Systs Anal Consult, 61; dir, Bus Info Systs Div, Hughes Dynamics, Inc, 62-65; vpres, Credit Data Corp, Calif & NY, 65-68, pres, NY, 68-70; sr vpres, TRW Info Serv, Inc, 70; pres, Medlab Comput Serv, Inc, 70-73; PRES, DMB FIN SERV, INC & GERALD L DAVEY & ASSOCS, 74-; PRES, SIERRA PROPERTIES CORP, N SAN JUAN, CALIF, 78- *Concurrent Pos:* Instr math, Exten Div, Univ Calif, Los Angeles, 60-65; consult, TRW Info Serv, Inc, 70-72; mem, State Adv Coun Sci & Technol, Utah, 73-81; pres, Utah Innovation Assoc Inc, 88- *Mem:* NY Acad Sci; AAAS; Soc Indust & Appl Math; Math Asn Am; Soc Mining Engrs; Sigma Xi. *Res:* Systems analysis in placer-mining techniques. *Mailing Add:* PO Box 581039 Salt Lake City UT 84158-1039

DAVEY, JAMES R, b Jackson, Mich, Feb 19, 12. TELEGRAPH & DIGITAL COMMUNICATIONS. *Educ:* Univ Mich, BS, 36. *Prof Exp:* Dept head modern develop, Bell Tel Labs, 36-73; PATENT LITIGATION CONSULT, 73- *Mem:* Inst Elec & Electronics Engrs. *Mailing Add:* Box 612 Rte 518 Princeton NJ 08540

DAVEY, JOHN EDMUND, b Buffalo, NY, July 15, 25; m 50; c 2. PHYSICS. *Educ:* Canisius Col, BS, 49; Univ Notre Dame, MS, 51, PhD(physics), 54. *Prof Exp:* Res physicist, Electron Tubes Br, 54-59, res physicist & sect head, Solid State Electronics Br, 59-69, HEAD SOLID STATE TECHNOL BR, US NAVAL RES LAB, 69- *Mem:* Am Phys Soc; Am Vacuum Soc; sr mem Inst Elec & Electronics Engrs. *Res:* Thermionic, photoelectric and Schottky emission from bulk metals, semiconductors and thin films; structural, electrical and optical properties of thin semiconducting and metal films; x-ray and electron diffraction; electron microscopy; high vacuum techniques. *Mailing Add:* 3212 Wessymton Way Alexandria VA 22309

DAVEY, KENNETH GEORGE, b Chatham, Ont, Apr 20, 32; m 59; c 3. INVERTEBRATE PHYSIOLOGY, PARASITOLOGY. *Educ:* Univ Western Ont, BSc, 54, MS, 55; Cambridge Univ, PhD(zool), 58. *Prof Exp:* Nat Res Coun Can fel, Univ Toronto, 58-59; Drosier fel entom, Gonville & Caius Col, Cambridge Univ, 59-63; assoc prof parasitol, Macdonald Col, McGill Univ, 63-67, dir inst parasitol, 64-74, prof parasitol, 67-74; prof biol & chmn dept, York Univ, 74-81, dean fac sci, 82-85, vpres acad affairs, 86-91, DISTINGUISHED RES PROF BIOL, YORK UNIV, 84- *Concurrent Pos:* Assoc ed, Int J Invert Reproduction, 78-85. *Honors & Awards:* Gold Medal, Entom Soc Can, 84; Fry Medal, Can Soc Zool, 87; Gold Medal, Biol Coun Can, 87. *Mem:* Can Soc Zool (vpres, 79-81, pres, 81-82); fel Entom Soc Can; Sigma Xi; fel Royal Soc Can. *Res:* Reproduction in arthropods; insect physiology and endocrinology; neurosecretion in invertebrates; physiology of helminths. *Mailing Add:* Dept Biol York Univ Downsview ON M3J 1P3 Can

DAVEY, PAUL OLIVER, b West Monroe, La, Apr 5, 31; m 55. PHYSICS. *Educ:* La Polytech Inst, BS, 51; Iowa State Univ, MS, 54; Univ Nebr, PhD(physics), 64. *Prof Exp:* Jr physicist, Ames Lab, AEC, 54-55; asst prof math & physics, Univ Nev, 58-61; assoc prof, 64-74, PROF PHYSICS, STATE UNIV NY COL FREDONIA, 74- *Mem:* Am Phys Soc; Am Inst Physics; Am Asn Physics Teachers; Sigma Xi. *Res:* Theoretical description of light nuclei; theory of the photodisintegration of light nuclei; theoretical physics. *Mailing Add:* 26 Birchwood Dr Fredonia NY 14063

DAVEY, TREVOR B(LAKELY), b Winnipeg, Man, Nov 27, 31; US citizen; m 82; c 4. MECHANICAL & BIOMEDICAL ENGINEERING. *Educ:* Univ Man, BSc, 53; McGill Univ, ME, 57. *Prof Exp:* Engr, Can Gen Elec Co, 53-55; sessional lectr, McGill Univ, 55-57; develop engr, Atomic Energy Can, 57; chmn, Dept Mech Eng, 62-74, PROF, 67-, COORDR, BIOMED ENG PROG, CALIF STATE UNIV, SACRAMENTO, 86- *Concurrent Pos:* Consult, Aerojet Gen Corp, 57-64; co-dir, Bioeng Sect, Cardiopulmonary Dept, Sutter Hosps Med Res Found, 63-72, dir, 72-; NSF fac fel, 74-77. *Mem:* Fel Am Soc Mech Engrs; Am Soc Eng Educ. *Res:* Characteristics of fluid flow through prosthetic heart valves; design of heart assists. *Mailing Add:* Biomed Eng Calif State Univ 6000 J St Sacramento CA 95819

DAVEY, WILLIAM GEORGE, b Wales, Gt Brit, Dec 26, 29; US citizen; m 54; c 2. PHYSICS, NUCLEAR PHYSICS. *Educ:* Univ Birmingham, Eng, BSc, 50, PhD(nuclear physics), 53. *Prof Exp:* Proj physicist fast reactor prog, Atomic Energy Res Estab, Harwell, Eng, 53-57, group leader, Winfrith, Eng, 57-61; sr sci officer reactor technol, Argonne Nat Lab, 61-63, prog mgr, 63-74; asst div leader & reactor prog mgr, 74, assoc div leader, 75, alt div leader, 75-77, DIV LEADER ENERGY DIV, LOS ALAMOS SCI LAB, 77- *Concurrent Pos:* Argonne Univ Asn distinguished vis prof, Univ Tex, 72-73. *Mem:* Fel Am Nuclear Soc. *Res:* Fast reactor physics and cross section evaluations. *Mailing Add:* Alpine Dr Capitan NM 88316

DAVEY, WILLIAM ROBERT, b Evanston, Ill, June 8, 43; c 4. ASTROPHYSICS, COMPUTATIONAL PHYSICS. *Educ:* Iowa State Univ, BS, 66; Univ Colo, PhD(astrophys), 70. *Prof Exp:* Fel astrophys, Univ Waterloo, Can, 70-72; sr tutor physics, Univ Queensland, Australia, 72-74; vis asst prof astrophys, Univ NMex, 74-75; MEM TECH STAFF COMPUT PHYSICS, SANDIA LABS, 75- *Concurrent Pos:* Consult astrophys, Los Alamos Sci Lab, 74-75. *Res:* Stability analysis of models for variable stars; nonlinear hydrodynamics of stellar pulsation; periodic solutions of nonlinear partial differential equations; radiation transport; crater formation and ground shock from nuclear explosions. *Mailing Add:* 809 Hermosa Dr NE Albuquerque NM 87110

DAVEY, WINTHROP NEWBURY, b Jackson, Mich, May 19, 18; m 49; c 3. INTERNAL MEDICINE. *Educ:* Univ Mich, AB, 39, MD, 42. *Prof Exp:* From intern to resident, Hosp, Univ Mich, 44-50, from instr to prof internal med, Sch Med, 44-72, dir med tuberc unit, Hosp, 47-71; assoc dir univ affairs, Training Prog, 71-74, DIR, BUR TRAINING, CTR DIS CONTROL, USPHS, 74- *Concurrent Pos:* Consult tuberc prog, Ctr Dis Control, USPHS, 60-71, mem, Nat Adv Tuberc Control Comt, 61-68, Nat Adv Commun Dis Coun, 68-69 & Nat Adv Heart & Lung Coun, 70-71. *Mem:* Am Col Physicians; Am Pub Health Asn; Asn Am Med Cols; Am Col Chest Physicians; Am Thoracic Soc (pres, 65-66). *Res:* Tuberculosis; pulmonary disease. *Mailing Add:* 660 Weatherly Ln NW Atlanta GA 30328

DAVICH, THEODORE BERT, b McKeenlyville, WVa, Jan 3, 23; m 45; c 2. ECONOMIC ENTOMOLOGY. *Educ:* Ohio State Univ, BSc, 48; Univ Wis, MS, 50, PhD(entom), 53. *Prof Exp:* Asst entom, Univ Wis, 50-53; assoc entomologist, Va Agr Exp Sta, 53-56; entomologist & head lab, Tex Cotton Insects Lab, Agr Res Serv, 56-60, DIR, BOLL WEEVIL RES LAB, SCI & EDUC ADMIN, AGR RES, USDA, 60- *Mem:* Entom Soc Am; Sigma Xi. *Res:* Insects affecting cotton. *Mailing Add:* 110 Forest Hill Dr Starkville MS 39759

DAVID, CARL WOLFGANG, b Hamburg, Ger, June 30, 37; US citizen; m; c 2. PHYSICAL CHEMISTRY. *Educ:* Case Inst Technol, BS, 58; Univ Mich, MS, 60, PhD(chem), 62. *Prof Exp:* Res fel chem, Yale Univ, 62-63; from asst prof to assoc prof chem, 63-74, PROF PHYS CHEM, UNIV CONN, 74- *Mem:* Am Chem Soc; Am Phys Soc. *Res:* Statistical thermodynamics of liquids and liquid mixtures. *Mailing Add:* Dept Chem Univ Conn Storrs CT 06269-3060

DAVID, CHELLADURAI S, b Coonoor, India, June 18, 36; US citizen; m 61; c 2. IMMUNOGENETICS. *Educ:* Berea Col, BS, 61; Univ Ky, MS, 62; Iowa State Univ, PhD(immunogenetics), 66. *Prof Exp:* Res assoc immunogenetics, Iowa State Univ, 66-68; res assoc immunogenetics, 68-73, asst res scientist, Med Sch, Univ Mich, Ann Arbor, 73-75; assoc prof genetics, Med Sch, Wash Univ, St Louis, Mo, 75-77; PROF IMMUNOL, MAYO MED SCH, ROCHESTER, 77-; PROF MICROBIOL, MED SCH, UNIV MINN, MINNEAPOLIS, 78- *Mem:* Transplantation Soc; AAAS; Am Asn Immunol; Genetics Soc Am. *Res:* Immunoglobulin allotypes in fowl; mouse immune response genes; mouse histocompatibility system. *Mailing Add:* Dept Immunol Mayo Med Sch 200 First St SW Rochester MN 55905

DAVID, DONALD J, b St Louis, Mo, June 25, 30; m 52; c 1. POLYMERS. *Educ:* St Marys Univ, BS, 52; Univ Dayton, MS, 77, MS, 79, PhD(mat eng), 81. *Prof Exp:* Res group leader & chief chem, Mobay Chem Co, 59-67; mgr res, Tracor, 67-72; sr res group leader & contract mgr, Monsanto Res Corp, Dayton Lab, 72-79, sr res specialist metall, Mound Lab, 77-79, sr res group leader, Monsanto Plastics & Resins Co, 79-84, MGR RES & DEVELOP, MONSANTO CHEM CO, 84- *Concurrent Pos:* Lectr, Thermal Anal Inst, 65-66, Univ Mo, 69. *Mem:* Am Chem Soc; North Am Thermal Anal Soc; Am Soc Metals; AAAS. *Res:* Polymer fundamentals and characterization; analytical characterization techniques; surface chemistry, instrumentation; mathematical modeling and gas chromstography; polymer analytical chemistry and chromatographic instrumentation. *Mailing Add:* Monsanto Chem Co 730 Worcester St Springfield MA 01151

DAVID, EDWARD E(MIL), JR, b Wilmington, NC, Jan 25, 25; m 50; c 1. ELECTRICAL ENGINEERING. *Educ:* Ga Inst Technol, BEE, 45; Mass Inst Technol, SM & ScD(elec eng), 50. *Hon Degrees:* DEng, Stevens Inst Technol, 71, Polytech Inst Brooklyn, 71, Univ Mich, 71, Carnegie-Mellon Univ, 72, Lehigh Univ, 73, Univ Ill, Chicago Circle, 73, Rose-Hulman Inst Technol, 78, Univ Fla, 81, Rutgers Univ, 83, NJ Inst Technol, 84 & Univ Pa, 85. *Prof Exp:* Asst, Electronics Res Lab, Mass Inst Technol, 46-50; mem, Hartwell Proj, 50; mem tech staff, Bell Tel Labs, 50-70, exec dir res, 65-70; sci adv to the pres & dir off sci & technol, US Govt, 70-73; exec vpres, Gould Inc, Ill, 73-77, consult, 77; pres, Exxon Res & Eng Co, NJ, 77-85; CONSULT, 85- *Concurrent Pos:* Chmn bd trustees, Aerospace Corp, Calif, 75-81; mem, Exec Comt & Corp & Energy Lab Adv Bd, Mass Inst Technol; mem vis comt, Div Phys Sci, Univ Chicago, mem, Adv Coun Humanities Inst; mem, Marshall Scholarships Adv Coun; mem adv bd, Dept Elec Eng, Univ Calif; mem bd trustees, Carnegie Inst Wash; mem bd, Overseers Fac Arts & Sci, Univ Pa; US rep, Sci Comt, NATO; mem adv resource coun, Princeton Univ; mem, White House Sci Coun; trustee, Twentieth Century Fund. *Honors & Awards:* Anak Award, 58; Lancaster Prize, Opers Res Soc; President Award, Am Soc Mech Engrs; Bueche Medal, Nat Acad Eng; Fahrney Medal, Franklin Inst; George W McCarthy Award, Ga Inst Technol, 58; Silver Stein Award, Mass Inst Technol, 91. *Mem:* Nat Acad Sci; Nat Acad Eng; fel AAAS; fel Inst Elec & Electronic Engrs; fel Acoust Soc Am; Am Philos Soc; Nat Acad Pub Admin; Am Acad Arts & Sci. *Res:* Perceptual research in speech and hearing, artificial intelligence, computer simulation, underwater sound, military systems, energy systems, petroleum and synfuels processing. *Mailing Add:* EED Inc PO Box 435 Bedminster NJ 07921

DAVID, FLORENCE N, b Hereford, UK, Aug 23, 09. STATISTICS, MATHEMATICS. *Educ:* Univ London, BSc, 31, PhD(statist), 38, DSc(statist), 50. *Prof Exp:* From lectr to prof statist, Univ Col, Univ London, 35-67; prof statist, Univ Calif, Berkeley, 61-62, 64-65; prof statist, Univ Calif, Riverside, 67-77, chmn dept, 68-77; biostatistician, Univ Calif, Berkeley, 77-91; RETIRED. *Concurrent Pos:* Sr statistician, Ministry of Home Security, UK, 39-45; consult, US Forest Serv, 63-90 & Pac State Hosp, Pomona, Calif, 65-81; ed, Biometrics, 72-77. *Mem:* Fel Am Statist Soc; fel Inst Math Statist; fel Royal Statist Soc; Int Statist Inst; fel AAAS. *Res:* Combinatorial and randomization methods; statistical applications. *Mailing* *Add:* 156 Highland Blvd Kensington CA 94720

DAVID, GARY SAMUEL, b Aurora, Ill, Oct 2, 42; m 69. IMMUNOCHEMISTRY, IMMUNOTHERAPY. *Educ:* Univ Ill, Urbana, BS, 64, PhD(microbiol), 68. *Prof Exp:* US Dept Health, Educ & Welfare fel immunol, City of Hope Med Ctr, 68-70; res assoc, Salk Inst Biol Studies, 70-71; fel, Scripps Clin & Res Found, 71-74, res assoc exp path, 74-77; asst dir, Larson Diagnostics, 78; dir, The Second Antibody, 78-; sr res scientist, 78-81, pres scientist, hybritech, inc, 81-85, RES FEL, HYBRITECH, INC, 85- *Concurrent Pos:* Adv ed, Immunochem, 75-78, Bioconjugate Chem, 89-; adj asst prof, Univ Calif, San Diego, 75-78. *Mem:* Clin Ligand Assay Soc; Am Asn Immunologists; Sigma Xi. *Res:* Immunogenetics; protein chemistry; immunoassay; cellular and developmental immunology; tumor immunology; immunotherapy. *Mailing Add:* Dept Admin PO Box 269006 San Diego CA 92196-9006

DAVID, HENRY P, b Haeen, Ger, May 28, 23; US citizen; m 53; c 2. REPRODUCTIVE BEHAVIOR. *Educ:* Univ Cincinnati, BA, 48, MA, 49; Columbia Univ, PhD(psychol), 51. *Prof Exp:* Sr clin psychol, Topeka State Hosp, 51-52; instr, Western Psychiat Inst, Univ Pittsburgh Med Sch, 52-55; asst prof & head, Dept Psychol, Lafayette Clin, Wayne State Univ Med Sch, 55-56; chief psychol, NJ State Dept Insts & Agencies, 56-63; assoc dir, World Fedn Ment Health, Geneva, 63-65; assoc dir, Int Res Inst, Am Insts Res, 65-71, DIR, TRANSNATIONAL FAMILY RES INST, BETHESDA, MD, 71-; ASSOC CLIN PROF, PSYCHOL, DEPT PSYCHIAT, UNIV MD MED SCH, BALTIMORE, 75. *Concurrent Pos:* Consult, WHO, Geneva, Copenhagen & Washington Offices, 75-, UN Pop Div & Fund, 8; lectr, Univ Bergen, Norway, 86-87. *Honors & Awards:* Distinguished Serv Awards, Am Psychol Asn. *Mem:* Fel Am Psychol Asn; Am Pub Health Asn; AAAS; Sigma Xi; Int Union Sci Study Pop. *Res:* Studies in reproductive behavior; over 300 articles published in various publications, plus books. *Mailing Add:* 8307 Whitman Dr Bethesda MD 20817

DAVID, HERBERT ARON, b Ger, Dec 19, 25; US citizen; m 50; c 1. MATHEMATICAL STATISTICS. *Educ:* Univ Sydney, BSc, 47; Univ London, PhD(statist), 53. *Prof Exp:* Sr lectr statist, Univ Melbourne, 55-57; prof, Va Polytech Inst & State Univ, 57-64; prof, Sch Pub Health, Univ NC, 64-72; dir & head, Statist Lab, 72-84, DISTINGUISHED PROF, COL LIB ARTS & SCCI, IOWA STATE UNIV, 80- *Honors & Awards:* Wilks Award Army Res, 83. *Mem:* Fel Am Statist Asn; fel Inst Math Statist; Int Statist Inst; Biometric Soc (pres, 82-83). *Res:* Order statistics; paired comparisons; inference; design of experiments; competing risks. *Mailing Add:* Dept Statist Iowa State Univ Ames IA 50011-1210

DAVID, ISRAEL A, b Philadelphia, Pa, Oct 25, 25; m 51; c 2. ORGANIC POLYMER CHEMISTRY. *Educ:* Univ Pa, BS, 48; Univ Wis, PhD(org chem), 55. *Prof Exp:* From res chemist to sr res chemist, 54-64, res assoc, 65-77, RES CHEMIST, CENT RES & DEVELOP DEPT, E I DU PONT DE NEMOURS & CO, INC, 77- *Mem:* Am Chem Soc. *Res:* Synthesis and modification of polymers. *Mailing Add:* Cent Res & Develop Dept E I Du Pont de Nemours & Co Inc Wilmington DE 19880-0328

DAVID, JEAN, b Montreal, Que, Dec 19, 21; m 51; c 3. BIOCHEMISTRY, FOOD TECHNOLOGY. *Educ:* Univ Montreal, LSA, 40; Univ Calif, PhD(biochem), 49. *Prof Exp:* Exten specialist hort, Que Dept Agr, 41-44, food technologist, 49-63; from asst prof to assoc prof, 49-74, registr, 69-79, PROF HORT, MACDONALD COL, MCGILL UNIV, 74-, ASSOC DEAN, FAC AGR, 72- *Res:* Food preservation; cold storage; canning and freezing; physiological and biochemical changes in fruits and vegetables after harvest and during storage. *Mailing Add:* Macdonald Col McGill Univ Ste Anne de Bellevue Boie d'urfe' PQ H9X 2Z2 Can

DAVID, JOHN DEWOOD, b Alton, Ill, Dec 1, 42; m 72; c 2. DEVELOPMENTAL GENETICS. *Educ:* Wabash Col, BA, 64; Vanderbilt Univ, PhD(molecular biol), 69. *Prof Exp:* NIH fel, Univ Calif, San Francisco, 69-71, Giannini Found fel, 71-72; asst prof, 72-78, assoc dean res, 82-84, ASSOC PROF BIOL SCI, UNIV MO-COLUMBIA, 78-, CHMN, BIOL SCI, 89- *Concurrent Pos:* Prin investr NIH grants, Univ Mo-Columbia, 78- *Mem:* Soc Develop Biol; Soc Cell Biol; AAAS; Am Tissue Cult Asn. *Res:* Plasma membrane structure and function; role of membrane proteins in skeletal muscle development; intercelluar signalling in the regulation of muscle development. *Mailing Add:* Div Biol Sci 110 Tucker Hall Univ Mo Columbia MO 65211

DAVID, JOHN R, b Eng, Feb 15, 30; US citizen; m 57; c 2. INTERNAL MEDICINE, IMMUNOLOGY. *Educ:* Univ Chicago, BA, 51, BS & MD, 55; Am Bd Internal Med, cert, 68. *Hon Degrees:* Dr, Univ Fed Ceara, 91. *Prof Exp:* From intern to asst resident, Mass Gen Hosp, 55-57; clin assoc, Nat Inst Arthritis & Metab Dis, 57-59; trainee, Rheumatism Res Unit, Eng, 59-60; resident med, Mass Gen Hosp, 60-61; asst prof med, NY Univ, 64-66; from asst prof to assoc prof, 66-73, DIR, PROG TROP MED & INT HEALTH, 81-, JOHN LAPORTE GIVEN PROF & CHMN, DEPT TROP PUB HEALTH, SCH PUB HEALTH, HARVARD MED SCH, 81- *Concurrent Pos:* Univ fel, Sch Med, NY Univ, 61-64; asst physician-in-chief, Robert B Brigham Hosp, 66-82, physician, 66-; sr assoc med, Brigham & Women's Hosp, 80-82, chief, Dept Med, Div Trop Med, 81-, asst chief, Dept Rheumatol-Immunol, 82-, sr physician, Dept Med, 82-; sci adv bd, Int Lab Res Animal Dis, 80-; Burroughs Wellcome vis prof, Johns Hopkins Univ, 83, Royal Soc Med, Eng, 84; mem Sci & Tech Rev Comt, Spec Prog Res & Training Trop Dis, WHO, 81-, consult, sci working group, Dir Commun Dis, 81-, mem, Steering Comt Immunol Tuberculosis, 84-; mem sci adv comt, New Eng Biolabs, 82-; ad hoc consult, Bd Sci Counselors, Rev Lab Parasitic Dis, NIH, 84- *Honors & Awards:* Robert A Cooke Mem Lectr, Am Acad Allergy, 70. *Mem:* Inst Med-Nat Acad Sci; Am Soc Trop Med & Hyg (pres, 89); Am Soc Clin Invest; Am Asn Immunologists; Am Fedn Clin Res; Am Rheumatism Asn; Infectious Dis Soc Am; Soc Exp Biol & Med; Am Asn Physicians; Am Acad Arts & Sci. *Res:* Mechanisms of delayed hypersensitivity and cellular immunity; parasite immunology. *Mailing Add:* Dept Trop Pub Health Harvard Sch Pub Health 665 Huntington Ave Boston MA 02115

DAVID, LARRY GENE, b Searcy, Ark, Apr 11, 38; m 64; c 2. INDUSTRIAL ENGINEERING. *Educ:* Univ Ark, BS, 61, MS, 62; Purdue Univ, PhD(indust eng), 68. *Prof Exp:* Instr, Physics Lab, Univ Ark, 58-60, instr comput prog, 60-61; instr indust eng, Univ Mo-Columbia, 61-64 & Purdue Univ, 67-68; ASSOC PROF INDUST ENG, UNIV MO-COLUMBIA, 68- *Mem:* Am Inst Indust Engrs. *Res:* Manufacturing processes; electric discharge machining; regional blood inventory system; computerized instructional system for classroom supplement. *Mailing Add:* Dept Indust Eng 113 Elec Eng Bldg Univ Missouri Columbia MO 65211

DAVID, MOSES M, b India, Apr 9, 62; m 84; c 2. THIN FILM DEPOSITION, ABRASION RESISTANT & OPTICAL COATINGS. *Educ:* Indian Inst Technol, BTech, 85; McGill Univ, MEng, 87; Clarkson Univ, PhD(chem eng), 90. *Prof Exp:* SR RES ENGR, 3M CO, 90- *Concurrent Pos:* Res award, Sigma Xi, 89; grad fel, IBM, 90. *Mem:* Mat Res Soc. *Res:* Use of laser and plasma methods for depositing thin films and powders and to investigate their applications; RF plasma; excimer laser; carbon oxygen laser and cathodic arc methods. *Mailing Add:* 1671 Century Circle No 119 Woodbury MN 55125

DAVID, OLIVER JOSEPH, b New York, NY, Feb 13, 37; m 62; c 2. SOCIAL PSYCHOLOGY, CHILD DEVELOPMENT. *Educ:* City Col New York, BS, 58; State Univ NY, MD, 62, DSc(med), 72. *Prof Exp:* Resident psychiat, Kings County Hosp, 67-70; from instr to asst prof, 71-81, ASSOC PROF PSYCHIAT, STATE UNIV NY, 81- *Concurrent Pos:* Prin investr, Off Child Develop, 73-76, NY Health Res Coun, 76-78, NIMH, 80-83 & Nat Inst Environ Health Sci res fel, 80-86; dir, Child Behav Res Unit, Kings County Hosp, 73-, Child & Adolescent Psychiat, 86- *Res:* Effects of environmental pollutants on the pediatric brain; systematic effects of interaction on identity formation. *Mailing Add:* 81 Sunnyside Ave Pleasantville NY 10570

DAVID, PETER P, b Szeged, Hungary, June 9, 32; m 61; c 5. QUATERNARY GEOLOGY. *Educ:* Univ Szeged, dipl, 55; McGill Univ, BSc, 59, MSc, 61, PhD(Pleistocene geol), 65. *Prof Exp:* Lectr, 64, from asst prof to assoc prof, 64-81, PROF GEOL, UNIV MONTREAL, 81- *Mem:* Can Quaternary Asn; Geol Soc Am; Geol Asn Can; Int Glaciol Soc; Am Quaternary Asn. *Res:* Quaternary stratigraphy; eolien deposits; development, evolution, morphology, structure and stratigraphy of dunes in Canada; migration rates, absolute chronology and environments; morphology, lithology, stratigraphy and geochemistry of glacial deposits in Gaspesie, Quebec; glaciation and deglaciation models. *Mailing Add:* Dept Geol Univ Montreal PO Box 6128 Montreal PQ H3C 3J7 Can

DAVID, STANISLAUS ANTONY, b Kolar Gold Fields, India, June 3, 43; US citizen; m 71; c 2. METALLURGY ENGINEERING. *Educ:* Mysore Univ, BSc, 63; Indian Inst Sci, BE, 65, ME, 67; Univ Pittsburgh, PhD(metall eng), 71. *Prof Exp:* Postdoctoral fel metall, Univ Pittsburgh, 71-72, res assoc prof metall eng, 72-77; res staff welding, 77-83, GROUP LEADER WELDING, OAK RIDGE NAT LAB, 83- *Concurrent Pos:* Adj prof, Univ Pittsburgh, 80-, Colo Sch Mines, 85- & Ohio State Univ, 86-; Jacquet Lucas Gold Medal & Award, Am Soc Metal Int & Inter Metallog Soc, 87, 88 & 90; Tech Achievement Awards, Martin Marietta Energy Systs, Inc, 87, 88 & 89; vchair, Res & Develop Comt, Am Welding Soc, 89-; chmn, Joining Div Coun, Am Soc Metals, 90- & Tech Divisions Bd, 90- *Mem:* Am Soc Metals Int; AAAS; Metall Soc-Am Inst Mining Metall & Petrol. *Res:* Welding science and solidification; basic and applied research in joining of materials; promote the advancement of welding science through research publications. *Mailing Add:* Oak Ridge Nat Lab Bldg 4508 MS-6095 PO Box 2008 Oak Ridge TN 37831-6095

DAVID, YADIN B, b Haifa, Israel, Nov 25, 45; US citizen; m 68; c 2. CLINICAL ENGINEERING, MEDICAL TECHNOLOGY ASSESSMENT. *Educ:* WVa Univ, BSc, 74, MSc, 75, EdD, 83. *Prof Exp:* Scientist elec eng, Collins Group, Rockwell Int, 75-76; dir biomed eng, WVa Univ Hosp, 76-82; DIR BIOMED ENG, TEX CHILDREN'S HOSP, 82- *Concurrent Pos:* Adj assoc prof, Dept Anesthesiol, Univ Tex Health Sci Ctr, 85-; adj asst prof, Dept Pediat, Baylor Col Med, 87-; chmn, Clin Eng Comt, Eng Med & Biol Soc, 88-; mem, Health Care Eng Policy Comt, Inst Elec & Electronics Engrs, 88- *Mem:* Am Col Clin Eng (pres, 90-91); Eng Med & Biol Soc; Inst Elec & Electronics Engrs. *Res:* Efficient application of medical instrumentation to improve patient care; technology management and medical technology assessment as it relates to standards or developed criterias. *Mailing Add:* 5307 Queensloch Dr Houston TX 77096

DAVIDA, GEORGE I, b Baghdad, Iraq, Aug 2, 44; US citizen; c 3. CRYPTOLOGY. *Educ:* Univ Iowa, BS, 67, MS, 69, PhD(elec eng), 70. *Prof Exp:* Asst prof theoret sci, Univ Wis, 70-75, assoc prof, 75-78; prog dir, NSF, 78-79; assoc prof comput sci, Univ Wis, 79-80; prof, Ga Inst Technol, 80-81; PROF ELEC ENG & COMPUT SCI, UNIV WIS-MILWAUKEE, 81- *Concurrent Pos:* Prin investr, NSF grants, 72-; mem, Pub Cryptography Study Group, Am Coun Educ, 79-81. *Mem:* Asn Computing Machinery; sr mem Inst Elec & Electronics Engrs. *Res:* Data security; methods for protecting data in data bases, operating systems and computer networks. *Mailing Add:* Dept Elec Eng & Comput Sci Univ Wis Milwaukee WI 53201

DAVIDIAN, NANCY MCCONNELL, b Philadelphia, Pa, Feb 16, 41. BIOCHEMISTRY, DRUG METABOLISM. *Educ:* Cornell Univ, BA, 62; Univ NC, PhD(biochem), 69. *Prof Exp:* Res asst endocrinol, Sterling-Winthrop Inst, 62-63; NIH fel, 68-69; res assoc biochem, Univ NC, 69-71, instr, 71-74, res assoc pharmacol, 74-77; mem staff, Off Grants Assocs, 77-79, asst spec progs officer, 79-83, rev activ officer, Off Extramural Res & Training, 83-, PROJ CLEARANCE OFFICER, NIH, 85- *Mem:* Am Chem Soc; Sigma Xi; AAAS. *Res:* Effects of drugs and radiation on intracellular pH; effects of drugs and metabolic disturbances on pH of brain; drug metabolizing enzymes of tumor-bearing and normal animals. *Mailing Add:* Off Extramural Res & Training 5614 Roosevelt St Bldg 31/1B54 NIH Bethesda MD 20817

DAVIDOFF, FRANK F, b New York, NY, Aug 9, 34. INTERNAL MEDICINE. *Educ:* Columbia Univ, MD, 59. *Prof Exp:* PROF MED, SCH MED, UNIV CONN, 74-; CHIEF MED, NEW BRITAIN GEN HOSP, 82- *Mailing Add:* Am Col Physicians Sixth St at Race Philadelphia PA 19106-1572

DAVIDOFF, ROBERT ALAN, b Brooklyn, NY, Oct 5, 34; m 67; c 2. NEUROPHYSIOLOGY, NEUROPHARMACOLOGY. *Educ:* NY Univ, BS, 55, MD, 58. *Prof Exp:* From instr to asst prof psychiat, Sch Med, Ind Univ, 65-69, asst prof neurol, 68-69; from assoc prof to assoc prof pharmacol, 69-76, vchmn neurol, 77-85, PROF NEUROL, SCH MED, UNIV MIAMI, 69-, PROF PHARMACOL, 76-,. *Concurrent Pos:* Res assoc, Vet Admin Hosp, Indianapolis, 65-66, clin investr, 66-69. *Mem:* Am Physiol Soc; Am Soc Neurochem; AAAS; Am Neurol Asn; Soc Neurosci. *Res:* Synaptic transmission; amino acid neurotransmitters. *Mailing Add:* Dept Neurol D4-5 Univ Miami Sch Med PO Box 016960 Miami FL 33101

DAVIDON, WILLIAM COOPER, b Fla, Mar 18, 27; m 47, 63, 87; c 4. MATHEMATICAL PHYSICS, NUMERICAL ANALYSIS. *Educ:* Univ Chicago, BS, 47, MS, 50, PhD(physics), 54. *Prof Exp:* Electronics engr, Mines Equip Co, 43-44; dir res, Nuclear Instrument & Chem Corp, 48-54; res assoc, Univ Chicago, 54-56; assoc physicist, Argonne Nat Lab, 56-61; from assoc prof to prof physics, 69-81, PROF MATH, HAVERFORD COL, 81- *Concurrent Pos:* Fulbright-Hays res grant physics, Aarhus Univ, 66-67 & Inst Theoret Physics, Trondheim, 76-77. *Mem:* Fedn Am Scientists (secy, 60-61); Soc Social Responsibility Sci (pres, 65-67); Am Math Soc; Math Asn Am. *Res:* Foundations of quantum mechanics and special relativity; numerical optimization. *Mailing Add:* Dept Math Haverford Col Haverford PA 19041-1392

DAVIDOVITS, JOSEPH, b Villers-St-Paul, France, Mar 23, 35; m 61; c 3. GEOPOLYMER SCIENCE, APPLIED ARCHAEOLOGY. *Educ:* Mainz Univ, WGer, Dr(polymer sci); Rennes Univ, France, MA & Licence es Sci. *Prof Exp:* PRES, INORG CHEM, CORDI SA-GEOPOLYMER INST, 72-; PROF & DIR APPL ARCHAEOL & ADJ PROF CHEM, INST APPL ARCHAEOL SCI, BARRY UNIV, 83-, FOUNDER & DEVELOPER CHEM, GEOPOLYMERIZATION. *Concurrent Pos:* Sci consult, Lone Star Industs Inc, 83-88; vis prof, Solid State Sci, Pa State Univ, 88- *Honors & Awards:* ACIT Award, Asn Indust Textile Chemists, 64. *Mem:* Am Chem Soc; Am Ceramic Soc; Soc Plastic Engrs; Int Asn Egyptologists. *Res:* Geopolymer science industrial application of mineralogy and geology; applied archaeological science application of archaeological knowledge to solve modern problems in developed and Third World countries; toxic chemical and nuclear waste stabilization; prevention of groundwater contamination using geopolymeric materials. *Mailing Add:* Inst Appl Archaeol Sci Barry Univ Miami Shores FL 33161

DAVIDOVITS, PAUL, b Moldava, Czech, Nov 1, 35; US citizen; m 57; c 2. CHEMICAL PHYSICS. *Educ:* Columbia Univ, BS, 60, MS, 61, PhD(appl physics), 64. *Prof Exp:* Staff engr, Radiation Lab, Columbia Univ, 61-64, res physicist & lectr, 64-65; from asst prof to assoc prof eng & appl sci, Yale Univ, 65-74; PROF CHEM, BOSTON COL, 74- *Concurrent Pos:* Consult, Aerodyne Res Inc, Billerica, Mass. *Mem:* Am Phys Soc; Am Chem Soc. *Res:* Physics of atomic collisions and recombinations; laser physics; atmospheric chemistry. *Mailing Add:* Dept Chem Boston Col Chestnut Hill MA 02167

DAVIDOW, BERNARD, b New York, NY, Aug 4, 19; m 42; c 3. TOXICOLOGY, CLINICAL BIOCHEMISTRY. *Educ:* Fordham Univ, BS, 42; Georgetown Univ, MS, 48, PhD, 50. *Prof Exp:* Chief acute toxicity br, Pharmacol Div, Food & Drug Admin, 46-56; dir labs, New Drug Inst, 56-61; chief food & drug lab, New York City Dept Health, 61-71, dep dir, Bur Labs, 71-72, dir bur labs, 72-86, asst comnr lab serv, 72-86; RETIRED. *Mem:* AAAS; Soc Toxicol; Am Soc Pharmacol & Exp Therapeut; Am Pub Health Asn; NY Acad Sci. *Res:* Toxicity and analysis of chemicals used in foods, drugs and cosmetics; detection of drugs subject to abuse in body fluids; detection of toxic metals in body fluids and food; detection drug abuse, trace toxic metals. *Mailing Add:* 153-24 Booth Mem Ave Flushing NY 11355

DAVIDS, CARY NATHAN, b Edmonton, Alta, Sept 28, 40; m 67; c 2. NUCLEAR PHYSICS, NUCLEAR ASTROPHYSICS. *Educ:* Univ Alta, BSc, 61, MSc, 62; Calif Inst Technol, PhD(nuclear physics), 67. *Prof Exp:* Res fel physics, Calif Inst Technol, 67; res assoc, Mich State Univ, 67-69; asst prof , Ctr Nuclear Studies, Univ Tex, Austin, 69-74; MEM STAFF, ARGONNE NAT LAB 74- *Concurrent Pos:* NSF grant, 70-75; Welch grant, 72-74; Alfred P Sloan fel, 72-76; guest assoc physicist, Brookhaven Nat Lab, 73; vis scholar, Enrico Fermi Inst Univ Chicago, 74-76, sr res assoc, 76-79; sr vis fel, Univ Manchester; vis assoc prof, univ Ill, Chicago, 85; prof, Univ Jyväskylä, Finland, 86. *Mem:* Am Phys Soc. *Res:* Experimental nuclear astrophysics; isotopes far from line of beta-stability, nuclear masses; nuclear spectroscopy and reactions; ion optics. *Mailing Add:* Argonne Nat Lab 203 9700 S Cass Ave Argonne IL 60439

DAVIDSE, GERRIT, b Grijpskerke, Netherlands, Dec 19, 42; US citizen; m 65; c 3. TAXONOMIC BOTANY. *Educ:* Calvin Col, BS, 65; Utah State Univ, MS, 68; Iowa State Univ, PhD(plant taxon), 72. *Prof Exp:* Asst cur, 72-81, ASSOC CUR, MO BOT GARDEN, 81- *Concurrent Pos:* Ed, Ann Mo Bot Garden, 75-82; organizer & ed, Flora Mesoamericana. *Mem:* Am Soc Plant Taxonomists; Bot Soc Am; Int Soc Plant Taxonomists. *Res:* Taxonomy and biosystematics of the Poaceae. *Mailing Add:* 6573 Scanlan St Louis MO 63166

DAVIDSEN, ARTHUR FALNES, b Freeport, NY, May 26, 44; m 66; c 2. ASTROPHYSICS, ULTRAVIOLET ASTRONOMY. *Educ:* Princeton Univ, AB, 66; Univ Calif, Berkeley, MA, 72, PhD(astron), 75. *Prof Exp:* Sci liaison officer x-ray astron, Naval Res Lab, 70-71; from asst prof to prof physics, 75-84, PROF PHYSICS & ASTRON, JOHNS HOPKINS UNIV, 84-, DIR, CTR ASTROPHYS SCI, 85- *Concurrent Pos:* Fel Alfred P Sloan, 76-80; ed, Astrophys Lett, 76-; co-investr, Faint Obj Spectrog, 77- & prin investr, Hopkins Ultraviolet Telescope, 79-; mem sci working group, Adv x-ray Astrophys Fac, 77-83 & chmn rev panel, 84; chmn, comt Space Telescope Inst, John Hopkins Univ, 79-81; dir-at-large, AURA, Inc, 79-82, dir, 82-; mem mgt & opers working group space astron, NASA, 80-83 & Shuttle Sci Working Group, 83-84; mem, Space Telescope Inst Coun & US Nat Comt for Int Astron Union, 82- *Honors & Awards:* Helen B Warner Prize, Am Astron Soc, 79. *Mem:* Am Astron Soc; Int Astron Union; Royal Astron Soc; fel AAAS. *Res:* Far-ultraviolet spectroscopy of quasars and galaxies; ultraviolet astronomy; galactic and extragalactic x-ray astronomy; rocket and satellite x-ray observations; ground-based optical observations; clusters of galaxies; the intergalactic medium. *Mailing Add:* Dept Physics & Astron Johns Hopkins Univ 424 Dunkirk Rd Baltimore MD 21212

DAVIDSON, ALEXANDER GRANT, b Moncton, NB, Sept 23, 27; m 53; c 2. FOREST PATHOLOGY. *Educ:* Univ NB, BSc, 48; Univ Toronto, MA, 51, PhD(forest path), 55. *Prof Exp:* Res officer, Forest Biol Lab, Univ NB, 48-58; head forest path invests, Atlantic Prov Res Br, Can Dept Agr, 58-62; assoc coordr, 62-65, asst prog coordr, 65-72, dis specialist, 72-80, SCI ADV, FOREST INSECT & DIS SURV, CAN FORESTRY SERV, 80- *Mem:* Can Inst Forestry. *Mailing Add:* 1820 Brantwood Drive Ottawa ON K1S 1E8 Can

DAVIDSON, ARNOLD B, b Philadelphia, Pa, June 5, 30; m 52; c 2. NEUROPSYCHOPHARMACOLOGY. *Educ:* Brooklyn Col, BA, 51, MA, 53; Temple Univ, EdD(psychol), 64. *Prof Exp:* High sch teacher, New York City Bd Educ, 51-53; sr investr, Smith Kline & French Labs, 55-73; asst dir pharmacol, 73-81, ASST DIR CLIN INVEST NEUROPSYCHIAT, HOFFMAN-LA ROCHE INC, 81- *Concurrent Pos:* Instr psychol, Peirce Jr Col, Pa, 66-69. *Mem:* AAAS; fel Am Psychol Asn; Am Soc Pharmacol & Exp Therapeut; Behav Pharmacol Soc. *Res:* Interaction of drugs and behavior; psychophysiology; CNS pharmacology. *Mailing Add:* Hoffmann-La Roche Inc 340 Kingsland St Bldg 115-Med Res Neuropsychiat Nutley NJ 07110

DAVIDSON, BETTY, b Brooklyn, NY, Mar 24, 33; m 60; c 2. BIOCHEMISTRY. *Educ:* Brooklyn Col, BS, 53; Univ Chicago, MS, 61, PhD(biochem), 64. *Prof Exp:* Technician biochem, Mt Sinai Hosp, New York, 54-56, State Univ NY Downstate Med Ctr, 56-58 & Med Sch, Northwestern Univ, 58-59; res assoc, Brandeis Univ, 68-70; RESEARCH ASSOC, HARVARD MED SCH, 70- *Concurrent Pos:* Univ fel biochem, Brandeis Univ, 65-68; res fel, Thyroid Res Unit, Mass Gen Hosp, 70-74; asst med, Thyroid Res Unit, Mass Gen Hosp, 74-78. *Mem:* Am Chem Soc; Biophys Soc; Am Soc Biol Chem; Sigma Xi. *Res:* Mechanism of enzyme action; protein structure and function. *Mailing Add:* 56 Beals St Brookline MA 02146

DAVIDSON, BRUCE M, b Ironwood, Mich, Mar 16, 24; m 49; c 3. TRANSPORTATION ENGINEERING. *Educ:* Univ Mich, BSE, 49; Univ Wis, MS, 51, PhD(eng), 56. *Prof Exp:* Instr civil eng, Univ Wis, 49-51 & 53-55; traffic engr, Wis, 55-56; from asst prof to prof civil eng, Univ Wis, Madison, 56-66, asst dean, 57-62, assoc dean, 62-66; prof & chmn dept, Wash State Univ, 66-71; acad dean, US Naval Acad, 71-85; RETIRED. *Concurrent Pos:* Consult, India prog, Agency Int Develop, US Dept State, 69. *Mem:* Am Soc Civil Engrs; Am Soc Eng Educ; Am Rwy Eng Asn. *Res:* Optimization of traffic signal systems; driver response to geometric design of intersections; academic performance of engineering students. *Mailing Add:* 1729 Long Green Dr Annapolis MD 21401

DAVIDSON, CHARLES H(ENRY), b Washington, DC, Dec 10, 20; m 52; c 2. COMPUTER SCIENCE, ELECTRICAL ENGINEERING. *Educ:* Am Univ, AB, 41; Univ Wis, PhM, 43, PhD(physics), 52. *Prof Exp:* Instr physics, Mary Washington Col, 46-47; engr, Continental Elec Co, 47-49; proj assoc elec eng, 52-54, from asst prof to assoc prof, 54-66, dir, Eng Comput Lab, 61-82, prof, 66-87, EMER PROF, ELEC & COMPUT ENG & COMPUT SCI, UNIV WIS, MADISON, 87-,. *Concurrent Pos:* Consult, AC Spark Plug Div, Gen Motors Corp, 59-63; vis prof, Dept Comput Sci, Univ Edinburgh, 68-69. *Mem:* Asn Comput Mach; Am Nat Standards Inst; Inst Elec & Electronics Engrs; Sigma Xi. *Res:* Computer education; computers and society. *Mailing Add:* 2210 Waunona Way Madison WI 53713

DAVIDSON, CHARLES MACKENZIE, b Hamilton, Scotland, June 1, 42; m 65; c 2. FOOD MICROBIOLOGY. *Educ:* Univ Strathclyde, Scotland, BSc, 64; Univ Bath, Eng, PhD(microbiol), 70. *Prof Exp:* Microbiologist qual control, Cerola Fare Ltd, 64-65; microbiologist res & develop, Unilever Res, 65-70; chief microbiologist res & develop, Canada Packers, 70-78; V PRES, SILLIKER LABS CAN, 78- *Mem:* Am Soc Microbiol; Soc Appl Bact. *Res:* Behavior of food borne pathogens in human foods; preservation of food products against microbial spoilage; development of methods for enumeration and identification of bacteria. *Mailing Add:* Silliker Labs Can 2222 S Sheridan Way W Unit 210 Mississauga ON L5J 2M4 Can

DAVIDSON, CHARLES NELSON, b Kankakee, Ill, Oct 19, 37; m 59; c 3. NUCLEAR PHYSICS, NUCLEAR WEAPON EFFECTS. *Educ:* The Citadel, BS, 59; Fla State Univ, PhD(nuclear chem), 62. *Prof Exp:* Chem staff officer, Radiol Div, US Army Combat Develop Command, 62-64, physicist, 64-66, physicist, Effects Div, Inst Nuclear Studies, 66-68, sci adv, 68-79, TECH DIR, HQ, US ARMY NUCLEAR & CHEM AGENCY, 79- *Mem:* Am Chem Soc; Am Nuclear Soc. *Res:* Nuclear weapons effects; nuclear defense; radiac instruments; fallout radiation; low energy nuclear physics; nuclear plans and policy. *Mailing Add:* 10838 Greene Dr Mason Neck VA 22079-3530

DAVIDSON, CHARLES SPRECHER, b Berkeley, Calif, Dec 7, 10. CLINICAL MEDICINE. *Educ:* Univ Calif, AB, 34; McGill Univ, MD & CM, 39. *Hon Degrees:* MA, Harvard Univ, 53. *Prof Exp:* Intern, San Francisco Hosp, 39-40, house officer med, 40-41; asst, Harvard Med Sch, 42-45, instr, 45-47, assoc, 47-49, asst prof, 49-51, asst clin prof, 51-52, from assoc prof to prof, 53-73, William B Castle prof med, 73-76, EMER WILLIAM B CASTLE PROF MED, HARVARD MED SCH, 76- *Concurrent Pos:* Res fel, Harvard Med Sch, 41-42; Fogarty scholar, NIH, 72-73; asst, Boston City

Hosp, 41-43, from jr vis physician to assoc vis physician, 44-63, vis physician, 64-, from res dir to asst dir 2nd & 4th Med Servs, 43-48, assoc dir, 48-73, from asst resident physician to resident physician, Thorndike Mem Lab, 41-43, from asst physician to assoc physician, 43-63, assoc dir, 63-73; vis prof med & prog dir, Clin Res Ctr, Mass Inst Technol, 74-76, sr lectr, 76-; consult, Cushing Vet Admin Hosp, Framingham, Mass, 50-52, Vet Admin Hosp, Boston, 52-72, Lemuel Shattuck Hosp, 55-, nutrit sect, Off Int Res, NIH, Cambridge City Hosp & Mt Auburn Hosp, Cambridge; hon trustee, Med Found & Boylston Med Soc, 61-64. *Honors & Awards:* Mc Collum Award, Am Soc Clin Nutrit; Alexander D Stewart Mem Prize. *Mem:* AMA; Am Soc Clin Investr; Am Fedn Clin Res; master Am Col Physicians; Asn Am Physicians; Am Acad Arts & Sci. *Res:* Nutrition; metabolism; liver diseases in man. *Mailing Add:* 100 Memorial Dr 5-23-A Cambridge MA 02142

DAVIDSON, CHRISTOPHER, b Boise, Idaho, Feb 10, 44. PLANT ANATOMY, PLANT MORPHOLOGY. *Educ:* Whitman Col, AB, 66; Claremont Grad Sch, MA, 68, PhD(bot), 73. *Prof Exp:* cur, Natural Hist Mus, Los Angeles County, 73-81; DIR, IDAHO BOT GARDEN, 81- *Concurrent Pos:* Ed, Madrono. *Mem:* Int Asn Plant Taxonomists; Int Asn Wood Anatomists; Am Bot Soc; Am Asn Plant Taxonomists; Asn Trop Biol; Sigma Xi. *Res:* Anatomy and morphology of flowering plants as it relates to problems of phylogeny and biogeography; wood anatomy of tropical forest trees, especially of roots and buttresses. *Mailing Add:* Idaho Bot Garden PO Box 2140 Boise ID 83701

DAVIDSON, CLIFF IAN, b Passaic, NJ, May 9, 50; m 76; c 2. ENVIRONMENTAL ENGINEERING, AIR POLLUTION. *Educ:* Carnegie-Mellon Univ, BS, 72; Calif Inst Technol, MS, 73, PhD(environ eng), 77. *Prof Exp:* from asst prof to assoc prof, 77-87, PROF CIVIL ENG, ENG & PUB POLICY, CARNEGIE-MELLON UNIV, 87- *Concurrent Pos:* Mem comt Lead Human Environ, Environ Studies Bd, Nat Res Coun, 78-80; mem, Air & Water Qual Tech Adv Comt, Commonwealth Pa. *Honors & Awards:* Lincoln T Work Award, Fine Particle Soc, 76; George T Ladd Award, Carnegie-Mellon Univ, 80; Ralph R Teetor Award, Soc Automotive Engrs, 82. *Mem:* Am Soc Civil Eng; Air & Waste Mgt Asn; Am Asn Aerosol Res. *Res:* Deposition of ambient natural and anthropogenic aerosols onto various surfaces, as functions of particle characteristics, atmospheric state, and surface structure; global atmospheric transport of trace metals; historical trends in air pollutants; indoor air pollution; long-range atmospheric transport of pollutants to remote regions. *Mailing Add:* Dept Civil Eng Schenley Park Pittsburgh PA 15213

DAVIDSON, COLIN HENRY, b Exeter, Gt Brit, Mar 4, 28; Can citizen; m 56; c 2. ARCHITECTURAL RESEARCH, INFORMATION SCIENCE. *Educ:* Brussels City Royal Acad, dipl arch, 51; Mass Inst Technol, MArch, 55. *Prof Exp:* Asst architect, Luccichenti & Monaco, Rome, Italy, 51-54, Architects' Collaborative, Cambridge, 54-55, London County Coun, Eng, 57-60, & Sir Hugh Casson & Partners, London, 60-61; pres, C H Davidson, Res Architects, London, 61-68; assoc prof architect, 68-72, dean, Fac Environ Design, 75-85, PROF ARCHITECT, UNIV MONTREAL, 72-; PRES, INDUSTRIALIZATION FORUM RES CORP, 70- *Concurrent Pos:* Vis prof, Sch Archit, Univ Wash, 64, Wash Univ, 67-69 & Grad Sch Design, Harvard Univ, 73-75; pres, IF Res Corp, 76-; secy, Montreal Bldg Ctr Corp. *Mem:* Int Coun Bldg Res Studies & Doc. *Res:* Technological innovation and industrialized building systems(and accompanying organizational changes); building research and design; science parks; high tech incubators. *Mailing Add:* 146 Springrove Crescent Outremont Montreal PQ H2V 3J2 Can

DAVIDSON, DANIEL LEE, b Columbus, Ohio, Feb 7, 46; m 68, 81; c 2. POLYMER CHEMISTRY. *Educ:* Earlham Col, AB, 68; Univ Akron, PhD(polymer sci), 75. *Prof Exp:* Res chemist, Chem & Plastics Res & Develop, Union Carbide Corp, 74-79; sr res scientist, Arco Chem Co, 79-; PLASTIC & ADDITIVE DIV, CIBA-GEIGY CORP, ARDSLEY, NY. *Mem:* Am Chem Soc; Soc Plastics Engrs. *Res:* Chemical and physical modification of polyolefins for specific novel applications; crosslinking chemistry of polyolefins. *Mailing Add:* Shinkong Syn Fib Crp Eng Plas 7-8 GL 123 Sec 2 Nanking E Rd Taipei Taiwan

DAVIDSON, DARWIN ERVIN, b Rockford, Ill, May 5, 43; m 66; c 2. BOTANY, MYCOLOGY. *Educ:* Ore State Univ, BS, 65; Univ Wyo, MS, 67; Duke Univ, PhD(bot), 71. *Prof Exp:* Res assoc mycol, Univ Wyo, 71-74; supvr malting-brewing res, 74-80, mgr brewing res, 80-85, MGR HOPS BIOTECHNOL, ADOLPH COORS, CO, 85- *Concurrent Pos:* Pres, Hop Res Coun, 82-84 & 90-92. *Mem:* AAAS; Am Soc Brewing Chemists. *Res:* Barley enzyme development during malting and their function during the brewing and fermentation process; product and process improvement; new beverage products; hop chemistry; plant transformation technology. *Mailing Add:* Res & Develop BC600 Adolph Coors Co Golden CO 80401

DAVIDSON, DAVID, radio science, communications engineering; deceased, see previous edition for last biography

DAVIDSON, DAVID EDWARD, JR, b Philadelphia, Pa, Mar 7, 35; m 58; c 2. VETERINARY MEDICINE, RADIOBIOLOGY. *Educ:* Univ Pa, VMD, 59; Univ Rochester, MS, 62; Georgetown Univ, PhD(radiation biol), 70. *Prof Exp:* Vet Corps, US Army, 59-89, vet lab officer, Dept Radiobiol, US Army Med Res Lab, Ft Knox, 59-61, vet lab officer, Div Med Chem, Walter Reed Army Inst Res, 62-72, chief Dept Vet Med, Seato Med Res Lab, Bangkok, Thailand, 72-75, vet lab officer & chief Dept Biol, Div Med Chem, 75-78, chief Dept Parasitol, Div Exp Therapeut, 78-82, dir, Div Exp Therapeut, Walter Reed Army Inst Res, 84-89; DIR, MALARIA UNIV, WHO, 89- *Concurrent Pos:* Sci Working Group Chemother Malaria, WHO, 85-89; US rep Panel VII, group experts chemoprophylaxis, NATO, 80-88; mem, Sci Working Group Chemotherap Chagas Dis, WHO, 84-87, Onchocerciasis Chemother Proj, 87- *Mem:* Am Vet Med Asn; Sigma Xi; hon mem Korean Military Med Asn; hon mem Korean Veterinary Med Asn; Am Soc Trop Vet Med. *Res:* Biological testing of pharmaceuticals in laboratory animals; pharmacology and toxicology of antiradiation, antimalarial, antitrypanosomal, antileishmanial and antischistosomal drugs; parasitology. *Mailing Add:* Malaria Unit WHO Geneva 27 1211 Switzerland

DAVIDSON, DAVID FRANCIS, b New York, NY, Aug 9, 23; m 53; c 3. ECONOMIC GEOLOGY. *Educ:* Lehigh Univ, BA, 48. *Prof Exp:* Geologist, US Geol Surv, 48-54; chief geologist, Spencer Chem Co, Mo, 54-56; staff geologist, US Geol Surv, 56-62, chief off exp geol, Br Geochem Census, 62-67, geologist, Br Foreign Geol, 67-68, dep asst chief geologist resources, 68-71, chief Br Mid East & Asian Geol, 71-84; consult geologist, Interphos Assocs, 84-85; CONSULT, 85- *Mem:* AAAS; Soc Econ Geol; fel Geol Soc Am. *Res:* Geology and geochemistry of selenium and tellurium; geology of ore deposits of sedimentary origin; automatic data processing of geologic data; geochemistry and exploration of phosphate deposits. *Mailing Add:* Interphos Assocs 2605 Mt Laurel Pl Reston VA 22091

DAVIDSON, DAVID LEE, b Houston, Tex, Apr 22, 35; m 65; c 2. MATERIALS SCIENCE, ELECTRON MICROSCOPY. *Educ:* Rice Univ, BS, 58, MS, 63, PhD(mat sci), 68. *Prof Exp:* Metallurgist, Southwest Res Inst, 63-65, sr metallurgist, 68-70; vis prof mat sci, Univ Fed de Rio de Janeiro, Brazil, 70-71; staff scientist mat sci, 71-82, INST SCIENTIST, SOUTHWEST RES INST, 82- *Concurrent Pos:* Instr, Trinity Univ, 69-70, Univ Tex, 72-78, Rice Univ, 78; sr vis fel, Oxford Univ, British Sci Res Coun, 79. *Honors & Awards:* Hetenyi Award, Soc Exp Stress Anal, 81. *Mem:* The Metallurgical Soc; fel Am Soc Metals; Microbeam Anal Soc; Electron Micros Soc Am. *Res:* Fatique and fracture, measuring deformation at very high resolution, for example, at crack tips; developed stereoimaging technique for this purpose; interfacial mech properties in composites. *Mailing Add:* Southwest Res Inst PO Drawer 28510 San Antonio TX 78284

DAVIDSON, DIANE WEST, b New Brighton, Pa, Jan 7, 48. COMMUNITY ECOLOGY, INSECT ECOLOGY. *Educ:* Wilson Col, BA, 69; Univ Denver, MS, 72; Univ Utah, PhD(biol), 76. *Prof Exp:* Inst biol, Dept Zool, Univ Tex, Austin, 76-77; asst prof biol, Dept Biol Sci, Purdue Univ, 77-79; asst prof biol, Dept Biol, 82, ASSOC PROF BIOL, UNIV UTAH, 82- *Concurrent Pos:* NSF grant, 77-88; Guggenheim Fel, 86-87. *Mem:* Ecol Soc Am; Soc Study Evolution; AAAS; Int Union Study Social Insects. *Res:* Evolutionary ecology; geographical ecology, granivory in desert ecosystems; myrmecochory and other ant-plant interactions. *Mailing Add:* Dept Biol Univ Utah Salt Lake City UT 84112

DAVIDSON, DONALD, b Albert Lea, Minn, May 3, 34; m 61; c 5. MECHANICAL ENGINEERING, ELECTRICAL ENGINEERING. *Educ:* Univ Minn, Minneapolis, BS & BME, 57; Washington Univ, St Louis, MSEE, 62, DSc, 77; Univ Mo, Rolla, MS, 70. *Hon Degrees:* Mech Engr, Univ Minn, Minneapolis, 72. *Prof Exp:* Res asst, Heat Transfer Lab, Univ Minn, 56-57; mech engr power plant eng, Iowa Ill Gas & Elec Co, 57; thermodynamics engr design, McDonnell Douglas Corp, 58-74; proj engr design, Pako Corp, 74-78; test engr test mgt, Rosemount Inc, 79-80; develop engr, res & develop, Gen Mills, Inc, 80-82; auto test engr mgt, Stearns Comput Systs, 83-84; auto test engr mgt, Litton Microwave Div, 84-85; THERMAL PHYSICS CONSULT, CONTROL DATA CORP, 85- *Concurrent Pos:* Mgt consult, D Davidson & Assocs, 66- *Mem:* Am Soc Mech Engrs; Inst Elec & Electronic Engrs; Am Soc Heating, Refrigeration & Air Conditioning Engrs; Nat Soc Prof Engrs. *Res:* Electro-mechanics machine design; thermodynamics; heat transfer; research administration and supervision. *Mailing Add:* Control Data Corp 3414 N Flag Ave Minneapolis MN 55427-1733

DAVIDSON, DONALD H(OWARD), b Suffern, NY, Nov 16, 37; m 60; c 1. CHEMICAL ENGINEERING, APPLIED MATHEMATICS. *Educ:* Rensselaer Polytech Inst, BChE, 59; NY Univ, MChE, 62, PhD(chem eng), 67. *Prof Exp:* Chem engr, US Rubber Co, NJ, 60-63, St Regis Paper Co, NY, 63-66 & Petrol & Chem Eng, Shell Develop Co, 67-70; chem engr, Kennecott Copper Corp, 70-78; petrol & chem engr, SAI, 78-79; PETROL & CHEM ENGR, TRW INC, 80- *Mem:* Am Inst Chem Engrs; Am Chem Soc; Soc Petrol Engs. *Res:* Fluid mechanics; heat transfer; applied mechanics; petroleum reservoir engineering. *Mailing Add:* 2687 Linda Marie Dr Oakton LA 22124-1112

DAVIDSON, DONALD MINER, JR, b Minneapolis, Minn, Oct 21, 39; m 66; c 2. ANALYTICAL STRUCTURAL GEOLOGY. *Educ:* Carleton Col, BA, 61; Columbia Univ, MA, 63, PhD(geol), 65. *Prof Exp:* Prof geol, Univ Minn, Duluth, 65-78; prof geol & chmn dept geosci, Univ Tex, El Paso, 78-80; sr res specialist, remote sensing, Exxon Prod Res, Houston, 81-84; prof geol & chmn dept, 84-90, ASST PROVOST, RESOURCE PLANNING, NORTHERN ILL UNIV, 90- *Concurrent Pos:* Consult, Exxon Minerals Co, 71-76; bus ed, Econ Geol Publ Co, 76-81; treas, Soc Econ Geologists, 84- *Mem:* Geol Soc Am; Sigma Xi; Geol Asn Can; Soc Econ Geologists (treas, 86-). *Res:* Structural analysis of Precambrian terraces; environmental remote sensing applications; evolution of geological resources. *Mailing Add:* Provosts Off Northern Ill Univ Dekalb IL 60115

DAVIDSON, DONALD WILLIAM, b Prairie Grove, Ark, June 8, 36; m 69. ECOLOGY. *Educ:* Univ Minn, BA, 59; Rutgers Univ, PhD(plant ecol), 63. *Prof Exp:* Asst bot, Univ Minn, 56-59 & Rutgers Univ, 59-63; asst prof biol, Univ Ala, 63-65; from asst prof to assoc prof, 65-79, PROF BIOL, UNIV WIS-SUPERIOR, 79- *Mem:* Am Inst Biol Sci; Ecol Soc Am. *Res:* Forest ecology; botany; environment. *Mailing Add:* Dept Biol Univ Wis Superior WI 54880

DAVIDSON, DOUGLAS, b North Shields, Eng, Mar 22, 31. CYTOLOGY. *Educ:* Durham Univ, BSc, 53; Oxford Univ, DPhil(cytol), 56. *Prof Exp:* Res fel cytol, Oxford Univ, 56-58; vis biologist, Biol Div, Oak Ridge Nat Lab, 58-61; lectr bot, St Andrew's Univ, 61-64; from asst prof to assoc prof biol, Western Reserve Univ, 64-69; chmn dept, 74-77, PROF BIOL, McMASTER UNIV, 69- *Concurrent Pos:* US AEC grant, 66-68; Nat Res Coun Can grant, 69- *Mem:* Genetics Soc Am; Bot Soc Am; Bot Soc Can; Cell Biol Soc Can. *Res:* Root growth; cell lineage studies in meristems; growth of nuclei; induced mutation in plastid DNA. *Mailing Add:* Dept Biol McMaster Univ Hamilton ON L8S 4K1 Can

DAVIDSON, EDWARD S(TEINBERG), b Boston, Mass, Dec 27, 39; m 64; c 2. ELECTRICAL ENGINEERING, COMPUTER SCIENCE. *Educ:* Harvard Col, BA, 61; Univ Mich, Ann Arbor, MS, 62; Univ Ill, Urbana, PhD(elec eng), 68. *Prof Exp:* Engr, Honeywell, Inc, 62-65; asst prof elec eng, Stanford Univ, 68-73; asst prof, 73-75, assoc prof, 75-, PROF ELEC ENG & COORDR, SCI LAB, UNIV ILL, URBANA. *Concurrent Pos:* NSF grant, NAND Network Design, 69-71; NSF grants, prin investr, 69-76; consult, Hewlett Packard, 75-76, Honeywell, 77 & USA-ECOM, Ft Monmouth, 77- *Mem:* AAAS; Asn Comput Mach; Inst Elec & Electronics Engrs; Sigma Xi; Am Fedn Teachers. *Res:* Computer architecture; pipelined and parallel computers; memory organization and management; microprocessors; logic design; automation. *Mailing Add:* 3316 EE CS Bldg Elec Engr Univ Mich Main Campus Ann Arbor MI 48109-2122

DAVIDSON, ELIZABETH WEST, b Salem, Ohio, Nov 25, 42; m 68; c 1. INVERTEBRATE PATHOLOGY. *Educ:* Mt Union Col, BSc, 64; Ohio State Univ, MSc, 67, PhD(entom), 71. *Prof Exp:* NIH fel acarology, Ohio State Univ, 71-72; instr biol, Univ Rochester, 73-74; res assoc insect path, 74-86, ASSOC RES PROF, ARIZ STATE UNIV, 86- *Concurrent Pos:* Consult, WHO, 72-; ed, Pathogenesis Microbial Dis, 81; ed, Safety Microbial Insecticides, 90. *Mem:* Sigma Xi; Entom Soc Am; Soc Invert Path (pres, 90-92); Am Mosquito Control Asn; Soc Vector Ecologists. *Res:* Pathogenesis of bacterial diseases of invertebrates; effect of invertebrate pathogens on non-target organisms; biological control of vector insects. *Mailing Add:* Dept Zool Ariz State Univ Tempe AZ 85287-1501

DAVIDSON, ERIC HARRIS, b New York, NY, Apr 13, 37. DEVELOPMENTAL BIOLOGY, MOLECULAR BIOLOGY. *Educ:* Univ Pa, BA, 58; Rockefeller Inst, PhD(cell biol), 63. *Prof Exp:* Res assoc cell biol, Rockefeller Univ, 63-65, asst prof cell & develop biol, 65-70; assoc prof, 71-74, PROF BIOL, CALIF INST TECHNOL, 74-, NORMAN CHANDLER PROF CELL BIOL, 80- *Mem:* Nat Acad Sci; fel AAAS. *Res:* Genomic control over cell differentiation; genomic activity underlying early embryological development and oogenesis. *Mailing Add:* Div Biol Calif Inst Technol Pasadena CA 91125

DAVIDSON, ERNEST, b Stuttgart, Ger, June 12, 21; US citizen; m 44; c 1. MECHANICAL ENGINEERING. *Prof Exp:* Design engr, Mitchell Camera Corp, 47-51; design engr, Bausch & Lomb, Inc, 51-56, design engr mgr, 56-61, prod mgr, 61-63, eng div mgr, 63, vpres eng, 63-77, vpres & gen mgr, Appl Res Labs, Inc, 73-79, vpres admin, 79-81; MGT CONSULT, 81- *Concurrent Pos:* Lectr, Los Angeles City Col, 59-, mem adv comt, 64- *Mem:* Instrument Soc Am; Optical Soc Am; Am Soc Testing & Mat. *Res:* Instrumentation for spectrochemical analysis. *Mailing Add:* PO Box 1067 Solano Beach CA 92075

DAVIDSON, ERNEST ROY, b West Terre Haute, Ind, Oct 12, 36; m 56; c 4. THEORETICAL CHEMISTRY. *Educ:* Rose Polytech Inst, BS, 58; Ind Univ, PhD(quantum chem), 61. *Prof Exp:* NSF res fel, Theoret Chem Inst, Univ Wis, 61-62; from asst prof to prof chem, Univ Wash, 62-84; PROF CHEM, IND UNIV, 84- *Concurrent Pos:* Sloan Found fel, 67; Guggenheim Found fel, 73. *Honors & Awards:* Laureate, Int Acad Molecular Quantum Sci, 71. *Mem:* Nat Acad Sci; Am Chem Soc; Am Phys Soc; Int Acad Molecular Quantum Sci. *Res:* Theoretical physical chemistry; quantum mechanics of small molecules. *Mailing Add:* Chem Dept Ind Univ Bloomington IN 47405

DAVIDSON, EUGENE ABRAHAM, b New York, NY, May 27, 30; m 50; c 4. BIOCHEMISTRY. *Educ:* Univ Calif, Los Angeles, BS, 50; Columbia Univ, PhD(biochem), 55. *Prof Exp:* Res assoc, Univ Mich, 55-56, instr biochem, 57-58; from asst prof to prof, Duke Univ, 58-67; Prof Biol Chem & Chmn Dept, Milton S Hershey Med Ctr, PA State Univ, 68-88; PROF & CHMN, DEPT BIOCHEM, GEORGETOWN UNIV, 88- *Concurrent Pos:* Consult Nat Cancer Inst, 60-62 & NIH, 63-; assoc dean educ, Pa State Univ, 75- *Mem:* Am Chem Soc; Am Soc Biol Chem; Am Rheumatism Asn; Brit Biochem Soc. *Res:* Structural chemistry of malignant cells; metabolism of glycoconjugates; hexosamine metabolism and chemistry. *Mailing Add:* Dept Biochem Georgetown Univ Sch Med 3900 Reservoir Rd NW Washington DC 20007

DAVIDSON, FREDERIC M, b Glens Falls, NY, Feb 11, 41; m 68; c 2. ELECTRICAL ENGINEERING. *Educ:* Cornell Univ, BS, 64; Univ Rochester, PhD(physics), 69. *Prof Exp:* Asst prof elec eng, Univ Houston, 68-70; from asst prof to assoc prof, 70-80, PROF ELEC ENG, JOHNS HOPKINS UNIV, 80- *Concurrent Pos:* Vis assoc prof elec eng, Rice Univ, 78-79. *Mem:* Inst Elec & Electronic Engrs; Optical Soc Am; Am Asn Univ Prof. *Mailing Add:* Dept Elec Eng Johns Hopkins Univ Charles & 34th St Baltimore MD 21218

DAVIDSON, GILBERT, b Omaha, Nebr, June 10, 34; m 58; c 3. SCIENCE ADMINISTRATION, OPTICS. *Educ:* Mass Inst Technol, SB, 55, PhD(physics), 59. *Prof Exp:* Res assoc physics, Polytech Sch, Paris, 59-60; sr phys scientist, Am Sci & Eng, Inc, 60-64, sr proj dir geophys, 64-67, vpres educ div, 67-69, geophys div, 69-70 & instrument systs div, 70-73; vpres, Infrared Industs, Inc, 73-77; V PRES, PHOTOMETRICS, INC, 77- *Concurrent Pos:* Fulbright scholar, France, 59-60. *Mem:* Am Vacuum Soc; Am Geophys Union; Instrument Soc Am; Am Phys Soc; Optical Soc Am. *Res:* High-energy physics; atomic physics; geophysics; development of improved opto-electronic devices. *Mailing Add:* Photometrics Inc 4 Arrow Dr Woburn MA 01801

DAVIDSON, GRANT E(DWARD), b Toronto, Ont, Oct 13, 19; m 46; c 3. ELECTRICAL ENGINEERING, ILLUMINATION. *Educ:* Univ Toronto, BASc, 43. *Prof Exp:* Asst testing engr, Illum Lab, Res Div, Ont Hydro, 46-49, asst res engr, 49-52, res engr, 52-57, illum engr, 57-60, sr illum engr, 60-71, supvry elec inspection engr, 71-76, chief elec inspector, 77-83; RETIRED. *Concurrent Pos:* Spec lectr dept univ exten, Univ Toronto; consult, Ont Dept Transp; mem, Int Illum Comn, 57-78 & Int Electrotech Comn, 59-; mem comt, Can Elec Codes, Part I, Part II and Part III, Can Standards Asn Motor Vehicle Lighting Comt, Standards Policy Bd, 85- *Mem:* Fel Illum Eng Soc NAm; fel Illum Eng Soc Gt Brit; fel Chartered Inst Bldg Serv Engrs. *Res:* Radiation optics and illumination; electrical safety. *Mailing Add:* 16 Glen Stewart Ave Toronto ON M4E 1P3 Can

DAVIDSON, H(AROLD), b Brooklyn, NY, May 2, 21. CHEMICAL ENGINEERING. *Educ:* Columbia Univ, BS, 43, ChE, 48. *Prof Exp:* Chem engr, Kolker Chem Works, 46-47; asst, Columbia Univ, 48-49; chem engr, Metal & Thermit Corp, 49-56; eng statistician, Merck & Co, Inc, 57-60; CONTROL SYSTS REP, IBM CORP, 60- *Mem:* Am Chem Soc; Am Inst Chem Engrs; Am Soc Qual Control; Inst Mgt Sci; Am Statist Asn. *Res:* Chemical engineering and applied mathematics; computer applications and computer control systems. *Mailing Add:* 330 W 28th St Apt 20F New York NY 10001-4722

DAVIDSON, HAROLD, b NJ, July 20, 19; m 50; c 3. LANDSCAPE HORTICULTURE. *Educ:* Univ Calif, BS, 49; Mich State Univ, MS, 53, PhD(hort), 57. *Prof Exp:* Coordr nursery & landscape mgt, Mich State Univ, 50-57, from asst prof to prof landscape hort teaching, exp & res, 57-84; RETIRED. *Honors & Awards:* Norman J Coleman Award, Am Asn Nurserymen, 80, L C Chadwick Award, 80. *Mem:* Am Soc Hort Sci; Asn Bot Gardens & Arboretums; Am Hort Soc. *Res:* Physiology of woody plants; photo-period; nutrition; propagation; weed control. *Mailing Add:* 2669 Blue Haven Ct East Lansing MI 48823

DAVIDSON, HAROLD MICHAEL, b Boston, Mass, June 3, 24. BIOCHEMISTRY. *Educ:* Harvard Univ, AB, 44; Univ Ore, MA, 49, PhD(chem), 51. *Prof Exp:* Chemist, Mass Dept Pub Health, 46-47; fel physiol chem, Univ Pa, 50-51; chemist cancer res, Overly Biochem Res Found, 51-52; res assoc, Sch Med, Tufts Univ, 52-59; SCIENTIST & ADMINR RES GRANTS, NIH, 61- *Concurrent Pos:* Am Cancer Soc fel, 53-55; exec secy, Gen Med A Study Sect, Div Res Grants, NIH, 73-; mem Nat Bd, Fed Exec & Prof Asn , 84-87. *Mem:* AAAS; Am Chem Soc; Fed Exec & Prof Asn. *Res:* Carbohydrate metabolism; enzymology; cancer; research grants review. *Mailing Add:* 5301 Westbard Circle No 350 Bethesda MD 20816

DAVIDSON, IVAN WILLIAM FREDERICK, physiology, pharmacology; deceased, see previous edition for last biography

DAVIDSON, J(OHN) P(IRNIE), ASTROPHYSICS. *Educ:* Univ Calif, BA, 48; Wash Univ, AM, 51, PhD, 52. *Prof Exp:* Res assoc physics, Columbia Univ, 52-53; asst prof, Brazilian Ctr Phys Res, Rio de Janerio, 53-55; res scientist, Joint Estab Nuclear Energy Res, Norway, 55-57; from asst prof to prof theoret physics, Rensselaer Polytech Inst, 57-66; chmn dept, 77-88, PROF THEORET PHYSICS, UNIV KANS, 66- *Concurrent Pos:* Vis scientist-in-residence, Naval Radiol Defense Lab, 64-65; visitor, Los Alamos Sci Lab, 74-75. *Mem:* Am Phys Soc; Am Astro Soc. *Res:* Theoretical physics; astrophysics. *Mailing Add:* Dept Physics & Astron Univ Kans Lawrence KS 66045

DAVIDSON, JACK DOUGAN, b Newark, NJ, Jan 31, 18; m 46; c 4. NUCLEAR MEDICINE. *Educ:* Princeton Univ, AB, 40; Columbia Univ, MD, 43; Am Bd Nuclear Med, cert, 72. *Prof Exp:* Intern med, Bellevue Hosp, 44, resident, 47; instr med, Goldwater Mem Hosp, 50-51; asst prof med, Col Physicians & Surg, Columbia Univ, 53-57; head biochem sect, Lab Chem Pharmacol, Nat Cancer Inst, 57-66; chief dept nuclear med, Clin Ctr, NIH, 66-70; assoc prof, Div Nuclear Med, Med Ctr, Duke Univ, 70-76; sr investr, Develop Therapeut Prog, Nat Cancer Inst, 77-82; RETIRED. *Concurrent Pos:* Res fel arteriosclerosis, Goldwater Mem Hosp, 48-50; res fel cancer, Delafield Hosp, 51-57. *Mem:* AAAS; Am Soc Pharmacol; NY Acad Sci; Soc Nuclear Med. *Res:* Nuclear medicine; radioisotopes; anticancer drug research. *Mailing Add:* 4928 Sentinel Dr No 105 Bethesda MD 20816-3507

DAVIDSON, JAMES BLAINE, b Oklahoma City, Okla, Nov 10, 23; m 48; c 3. OCEAN ENGINEERING, ACOUSTICS. *Educ:* US Naval Acad, BS, 46; US Naval Postgrad Sch, BS, 52; Univ Calif, Los Angeles, MS, 53. *Prof Exp:* US Navy, 46-67, anal officer, Key West Testing & Eval Detachment, Oper Testing & Eval Force, 55-57, asst supt, Submarine Construct, Mare Island Naval Shipyard, 57-59, elec officer, Serv Squadron, 7th Fleet, 59-62, proj officer, Off Naval Res, 62-66, dir undersea prog, 66-67; PROF OCEAN ENG, FLA ATLANTIC UNIV, 67- *Concurrent Pos:* Consult, Metallgesellschaft AG, 73-, DWG & Partners; vis prof, Norwegian Tech Inst, Norway, 81-82; People to People Ambassador, Japan, China & Hong Kong on Coastal Zone Mgt, 85. *Mem:* Am Acoust Soc; Am Soc Eng Educ. *Res:* Long range, underwater acoustic propagation and submarine detection; submarine target classification; underwater acoustic imaging; deep ocean mining; manganese nodules; ocean engineering education; beach preservation & restoration; deep ocean habitats using cryogenic sealing; computers in engineering education. *Mailing Add:* Dept Ocean Eng Fla Atlantic Univ Boca Raton FL 33431-0991

DAVIDSON, JAMES MELVIN, b The Dalles, Ore, Apr 16, 34; m 57; c 3. SOIL PHYSICS. *Educ:* Ore State Col, BS, 56, MS, 58; Univ Calif, Davis, PhD(soil physics), 65. *Prof Exp:* Res asst soil physics, Ore State Univ, 56-58; lab technician, Univ Calif, Davis, 58-65; from asst prof to prof, Okla State Univ, 65-74; prof soil sci, 74-86, DEAN RES, INST FOOD & AGR SCI, UNIV FLA, 86- *Mem:* fel Soil Sci Soc Am; fel Am Soc Agron; Am Geophys Union. *Res:* Fluid and solute movement through porous materials; soil management practices for soil and water conservation and good plant root environment. *Mailing Add:* Dept Soil Sci 2169 McCarty Univ Fla Gainesville FL 32611

DAVIDSON, JEFFREY NEAL, b Springfield, Mass, May 7, 50; m 80; c 2. SOMATIC CELL GENETICS. *Educ:* Ind Univ, Bs, 72; Harvard Univ, PhD(biol), 76. *Prof Exp:* Fel somatic cell genetics, Eleanor Roosevelt Inst Cancer Res, 76-79; asst prof, dept med, Health Sci Ctr, Univ Colo, 80-84; asst prof, 84-86, ASSOC PROF, DEPT MICROIMMUNOL, UNIV KY MED CTR, 86- *Concurrent Pos:* Inst fel, Eleanor Roosevelt Inst Cancer Res, 79-83, sr fel, 83-84. *Mem:* Am Soc Microbiol; Am Soc Biochem & Molecular Biol; Sigma Xi. *Res:* Biochemical, genetic, and molecular biological techniques are being used to study multifunctional protein which catalyzes the first three steps of pyrimidine biosynthesis in mammalian cells; genes on human chromosome 21. *Mailing Add:* 2992 Shirlee Dr Lexington KY 40502

DAVIDSON, JOHN ANGUS, b Elizabeth, NJ, July 26, 33; m 57; c 2. ENTOMOLOGY. *Educ:* Columbia Union Col, BA, 55; Univ Md, MS, 57, PhD(entom), 60. *Prof Exp:* From instr to assoc prof biol, Columbia Union Col, 60-66; from asst prof to assoc prof, 66-75, PROF ENTOM, UNIV MD, COLLEGE PARK, 75- *Concurrent Pos:* Prof lectr, Am Univ, 63- *Mem:* Entom Soc Am; Soc Syst Zool. *Res:* Biosystematics of scale insects; biology and control of insect pests of ornamental plants; insect taxonomy and morphology. *Mailing Add:* 7607 Laurel Ridge Ct Laurel MD 20707

DAVIDSON, JOHN EDWIN, b Asheville, NC, Oct 27, 37; m 58; c 2. ANALYTICAL CHEMISTRY, INORGANIC CHEMISTRY. *Educ:* Univ Tenn, BS, 60, MS, 62, PhD(chem), 65. *Prof Exp:* From asst prof to assoc prof, 65-74, PROF CHEM, EASTERN KY UNIV, 74- *Mem:* Am Chem Soc; Sigma Xi; Nat Sci Teachers Asn; Hist Sci Soc. *Res:* Solvent extraction of metal chelates; coordination chemistry; chemical education; experiments in the introductory chemistry laboratory. *Mailing Add:* Dept Chem Moore 337 Eastern Ky Univ Richmond KY 40475

DAVIDSON, JOHN G N, b Murree, Pakistan, Sept 3, 35; Can citizen; m 74; c 2. HORTICULTURE. *Educ:* Univ BC, BSF, 57, MSc, 61; Univ Calif, Berkeley, PhD(plant path), 71. *Prof Exp:* RES SCIENTIST PLANT PATH, AGR CAN, 73- *Concurrent Pos:* Fel mycol, Univ Guelph, Ont, 71-72; fel plant path, Univ Toronto, 73. *Mem:* Can Phytopathol Soc; Am Phytopathol Soc; Can Hort Soc. *Res:* Root rots and damping-off of rapeseed and barley; snow molds of grasses, winter cereals and legumes; didymella stem eyespot of fescue; diseases of Saskatoon berries. *Mailing Add:* Agr Can Res Sta Box 29 Beaverlodge AB T0H 0C0 Can

DAVIDSON, JOHN KEAY, III, b Lithonia, Ga, Mar 30, 22; m 52; c 4. INTERNAL MEDICINE, PHYSIOLOGY. *Educ:* Emory Univ, BS, 43, MD, 45; Univ Toronto, PhD(physiol), 65; Am Bd Internal Med, dipl, 54. *Prof Exp:* Intern surg, Grady Hosp, Atlanta, Ga, 45-46, asst resident internal med, 48-49; chief resident, Emory Univ Hosp, 49-50; resident, New Eng Ctr Hosp, Boston, Mass, 50-51; pvt pract, Columbus, Ga, 51-60; asst prof, Banting & Best Dept Med Res, Univ Toronto, 64-66, from asst prof to assoc prof physiol, 65-68, res assoc internal med, 65-66, clin teacher, 66-68; prof med & dir diabetes unit, 68-87, EMER PROF MED, SCH MED, EMORY UNIV, 90- *Concurrent Pos:* Consult, Vet Admin Hosp, Tuskegee, Ala, 51-60; clin asst, Toronto Gen Hosp, 65-68; dir diabetes sect, Grady Mem Hosp, 68-87. *Honors & Awards:* Starr Medal, Toronto, 63; C H Best Award, Am Diabetes Asn, 87. *Mem:* AMA; Am Diabetes Asn; fel Am Col Physicians; Am Soc Internal Med; Am Physiol Soc; Can Physiol Soc; Endocrine Soc. *Res:* Immunologic insulin resistance; insulin immunity in animals and man; pathophysiology and therapy of obesity-induced diabetes mellitus. *Mailing Add:* 1075 Lullwater Rd NE Atlanta GA 30307

DAVIDSON, JOHN RICHARD, b Derby, Conn, June 17, 29; div; c 3. ENGINEERING MECHANICS, APPLIED MECHANICS. *Educ:* Brown Univ, ScB, 51; Va Polytech Inst & State Univ, MS, 59, PhD(eng mech), 68. *Prof Exp:* Design engr, Nat Adv Comt Aeronaut, 51-53, res engr, 53-57, res engr, Nat Aero & Space Admin, 58-65, sect head, 65-70, asst br head, 70-71, br head, 71-85; CONSULT ENGR, 85- *Concurrent Pos:* Consult, NASA & Light Technol Systs, Van Nuys, Calif; organizer, 10th Int Meeting, Soc Eng Sci, 73; Aerospace Panel chmn, Workshop Potential Appln Aerospace Technol Marine Technol Transp Indust, 78; mem, Struct Panel Res US Maritime Indust, 82 & Struct Integrity Aerospace Vehicles Tech Coop Prog, 82- *Mem:* Am Soc Mech Engrs; Soc Eng Sci; Am Inst Aeronaut & Astronaut; Sigma Xi. *Res:* Fatigue and fracture of aerospace structural materials; probability theory and its application to structural reliability; heat transfer; dynamics; radiation and mechanical impact hazards to space flight; author of numerous publications; inventor of electronically calibratable clock. *Mailing Add:* 114 Glenn Circle Williamsburg VA 23185

DAVIDSON, JON PAUL, b Welwyn Garden City, UK, July 2, 59; m 89. ISOTOPE GEOCHEMISTRY, VOLCANOLOGY. *Educ:* Univ Durham, UK, 81; Univ Leeds, UK, PhD(geol), 84. *Prof Exp:* ASST PROF GEOL & GEOCHEM, UNIV CALIF, LOS ANGELES, 88- *Concurrent Pos:* Vis asst prof geol & geochem, Southern Methodist Univ, Dallas, 84-87 & Univ Mich, Ann Arbor, 87-88; Fullbright travel award, 84. *Mem:* Am Geophys Union; Geochem Soc; Geol Soc Am; Sigma Xi. *Res:* Geochemistry of volcanic rocks particularly from volcanic arcs, with view to evaluating processes of magma genesis and evolution; magma sources and general crust-mantle evolution. *Mailing Add:* Dept Earth & Space Sci Univ Calif Los Angeles CA 90024-1567

DAVIDSON, JOSEPH KILLWORTH, b Columbus, Ohio, Jan 17, 38; m 68; c 1. MECHANICAL ENGINEERING. *Educ:* Ohio State Univ, BME & MSc, 60, PhD(mech eng), 65. *Prof Exp:* From instr to asst prof mech eng, Ohio State Univ, 62-73; assoc prof, 73-78, PROF MECH ENG, ARIZ STATE UNIV, 78- *Concurrent Pos:* Vis res engr, Gleason Works, Rochester, NY, 72; vis prof, Monash Univ, Melbourne, Australia, 80; vis prof, Swiss Fed Inst Technol, Switz, 87. *Mem:* Am Soc Mech Engrs; fel, Am Soc Mech Engrs. *Res:* Kinematics and dynamics of machinery; mechanical vibrations; robotics; mechanical design. *Mailing Add:* Dept Mech & Aero Eng Ariz State Univ Tempe AZ 85287-6106

DAVIDSON, JULIAN M, b Dublin, Eire, Apr 15, 31; m 60; c 3. NEUROPHYSIOLOGY. *Educ:* Hebrew Univ, Israel, MS, 56; Univ Calif, Berkeley, PhD(physiol), 59. *Prof Exp:* USPHS res fel anat, Sch Med, Univ Calif, Los Angeles, 59-60; res fel neurol, Hadassah Univ Hosp, Israel, 61-62; USPHS res fel psychol, Univ Calif, Berkeley, 62-63; from asst prof to assoc prof, 63-80, PROF PHYSIOL, STANFORD UNIV, 80- *Concurrent Pos:* Co-ed, Hormones & Behavior, 69-77; mem endocrinol study sect, NIH, 70-74; Guggenheim vis res fel, Dept Human Anat, Oxford Univ, 70-71; vis fel, Battelle Seattle Res Ctr, 74-75; prof, Nat Univ Greece, Athens, 79-80; vis researcher, NIH, Bethesda, Md, 85-86. *Mem:* Am Physiol Soc; Endocrine Soc; Int Soc Neuroendocrinol; Int Brain Res Orgn. *Res:* Regulation of gonadotropic hormone secretion; reproductive endocrinology, neural and endocrine determinants of reproductive behavior; psychological correlates of consciousness. *Mailing Add:* Dept Physiol Stanford Univ A-167 Physiol Bldg Stanford CA 94305

DAVIDSON, KEITH V(ERNON), metallurgy, materials science, for more information see previous edition

DAVIDSON, KENNETH LAVERN, b Lake Mills, Iowa, May 13, 40; m 60; c 3. METEOROLOGY. *Educ:* Univ Minn, Minneapolis, BS, 62; Univ Mich, MS, 66, PhD(meteorol), 70. *Prof Exp:* Asst res meteorologist, Great Lakes Res Div, Univ Mich, 65-67 & Dept Meteorol & Oceanog, 67-70; ASSOC PROF METEOROL, NAVAL POSTGRAD SCH, 70- *Mem:* Am Meteorol Soc; Am Geophys Union. *Res:* Properties of turbulent flow in the region adjacent to the earth surface responsible for exchange of heat, moisture and momentum between the atmosphere and boundary (ocean or land). *Mailing Add:* Dept Meteorol Naval Postgrad Sch Monterey CA 93940

DAVIDSON, KRIS, b Fargo, NDak, Dec 28, 43. ASTROPHYSICS, ASTRONOMY. *Educ:* Calif Inst Technol, BS, 65; Cornell Univ, PhD(astrophys), 70. *Prof Exp:* Fel astron, Princeton Univ, 70-71, asst prof, 71-74; asst prof astrophys, 74-75, ASSOC PROF ASTROPHYS, SCH PHYSICS & ASTRON, UNIV MINN, MINNEAPOLIS, 75- *Concurrent Pos:* Sloan res fel, 75-79. *Mem:* Am Astron Soc; Int Astron Union; Royal Astron Soc; Astron Soc Pac; AAAS; Sigma Xi. *Res:* Quasars; astronomical x-ray sources; crab nebula and other supernova remnants; very massive stars; infrared dwarf stars; extragalactic nebulae; chemical abundances; accretion disks; other astronomical objects. *Mailing Add:* Sch Physics & Astron Univ Minn 116 Church St SE Minneapolis MN 55455

DAVIDSON, LYNN BLAIR, b Grosse Point Farms, Mich, Sept 22, 40; m 65. DECISION ANALYSIS, PETROLEUM ENGINEERING. *Educ:* Stanford Univ, BS, 62, MS, 64, PhD(petrol eng), 66. *Prof Exp:* Sr reservoir engr, Mobil Oil Libya, Ltd, 66; res engr, Chevron Res Co, Standard Oil Calif, 66-70; opers res analyst, Getty Oil Co, 70-79, prof specialist mgt sci, 79-80, prof specialist strategic planning, 80-81; CONSULT, 81- *Mem:* Soc Petrol Engrs. *Res:* Analysis of decision with uncertain consequences; investment analysis; analysis and modelling of large, fuzzy systems; strategic planning. *Mailing Add:* Direction Focus Suite 1611 3810 Wilshire Blvd Los Angeles CA 90010

DAVIDSON, MARK ROGERS, b Chattanooga, Tenn, July 10, 62; m. SURFACE ANALYSIS, CATALYST DEVELOPMENT. *Educ:* Univ Fla, BS, 84, ME, 86, PhD(chem eng), 90. *Prof Exp:* Lab technician chem, 81-86, POSTDOCTORAL RESEARCHER CHEM ENG, UNIV FLA, 90- *Concurrent Pos:* Consult, var bus, 87- *Mem:* Am Vacuum Soc; Electrochem Soc. *Res:* Development of surface analytical techniques and their application to such problems as spacecraft material degradation, catalyst optimization, semiconductor materials, and wear resistant coatings; design and construction of custom surface analytical and control equipment. *Mailing Add:* Chem Eng Univ Fla Gainesville FL 32611

DAVIDSON, MAYER B, b Baltimore, Md, Apr 11, 35; m 61, 80; c 4. ENDOCRINOLOGY, METABOLISM. *Educ:* Swarthmore Col, AB, 57; Harvard Univ, MD, 61. *Prof Exp:* Res internist, US Army Inst Environ Med, 66-69; from asst prof to prof, Sch Med, Univ Calif, Los Angeles, 69-79; DIR DIABETES PROG, CEDARS-LINAI MED CTR, 79- *Concurrent Pos:* Res fel, Dept Endocrinol & Metab, King County Hosp, Univ Wash, 64-65; USPHS fel, Nat Inst Arthritis & Metab Dis, 65-66. *Mem:* Am Diabetes Asn; Endocrine Soc; Am Soc Clin Invest; Sigma Xi; Am Fedn Clin Res. *Res:* Diabetes; insulin antagonism; mechanism of insulin action; mechanism of sulfonylure agent effect on peripherel tissues. *Mailing Add:* Cedars-Sinai Med Ctr Div Endocrinol 8700 Beverly Blvd Los Angeles CA 90048-0750

DAVIDSON, MELVIN G, b Winnipeg, Man, Apr 7, 38; US citizen; m 62; c 2. NUCLEAR PHYSICS. *Educ:* Whitman Col, AB, 60; Rensselaer Polytech Inst, PhD(theoret physics), 64. *Prof Exp:* Res fel theoret physics, Australian Nat Univ, 64-67; from asst prof to assoc prof, 67-72, PROF PHYSICS & DIR, COMPUT CTR, WESTERN WASH UNIV, 72- *Concurrent Pos:* Consult, Educ Comput Consult Group. *Mem:* Am Phys Soc; Am Inst Physics; Asn Comput Mach. *Res:* Theoretical nuclear physics; nuclear collective model; elementary particle physics. *Mailing Add:* Dept Systs & Computing Western Wash Univ 516 High St Bellingham WA 98225

DAVIDSON, NORMAN RALPH, b Chicago, Ill, Apr 5, 16; m 42; c 4. BRAIN GENES. *Educ:* Univ Chicago, BS, 37, PhD(chem), 41; Oxford Univ, BSc, 39. *Prof Exp:* Res assoc, Nat Defense Res Comt proj, Univ Southern Calif, 41, Div War Res, Columbia Univ, 42 & Univ Chicago, 42; instr chem, Ill Inst Technol, 42; res assoc plutonium proj, Univ Chicago, 43-45; res physicist, Radio Corp Am, 45-46; from instr to prof chem, 46-82, Chandler prof, 82-86, interim chair biol, 89-90, EMER CHANDLER PROF CHEM BIOL, CALIF INST TECHNOL, 86-, EXEC OFFICER BIOL, 90- *Concurrent Pos:* Exec off chem, Calif Inst Technol, 67-73; nat lectr, Biophys soc, 74. *Honors & Awards:* G N Lewis Award, Am Chem Soc, 54, Peter Debye Award, 71; Robert A Welch Award Chem, 89. *Mem:* Nat Acad Sci; Am Chem Soc; Am Soc Biol Chem; Am Soc Microbiol. *Res:* Genes coding for ion channels and neurotransmitter receptors in nerve and muscle. *Mailing Add:* Div Biol 156-29 Calif Inst Technol Pasadena CA 91125

DAVIDSON, PHILIP MICHAEL, b Oakland, Calif, Jan 8, 50; m 72; c 2. FOOD MICROBIOLOGY. *Educ:* Univ Idaho, BS, 72; Univ Minn, MS, 77; Wash State Univ, PhD(food sci), 79. *Prof Exp:* From asst prof to assoc prof, 79-89, PROF FOOD MICROBIOL, UNIV TENN, 89- *Mem:* Inst Food Technologists; Am Soc Microbiol; Am Acad Microbiol. *Res:* Investigation of the activity and mechanism of action of antimicrobial compounds in foods. *Mailing Add:* Dept Food Technol & Sci PO Box 1071 Univ Tenn Knoxville TN 37901

DAVIDSON, RALPH HOWARD, entomology; deceased, see previous edition for last biography

DAVIDSON, RICHARD LAURENCE, b Cleveland, Ohio, Feb 22, 41; m 67. GENETICS. *Educ:* Case Western Reserve Univ, BA, 63, PhD, 67. *Prof Exp:* asst prof, Harvard Med Sch, 70-73, assoc prof microbiol & molecular genetics, 73-81, res assoc human genetics, Children's Hosp Med Ctr, Boston,, 70-81; DIR, CTR GENETICS & BENJAMIN GOLDBERG PROF GENETICS, UNIV ILL MED CTR, 81- *Concurrent Pos:* Air Force Off Sci Res-Nat Res Coun fel, 67-68, Ctr Molecular Genetics, Paris, 67-70; ed-in-chief, Somatic Cell Genetics; co-dir, Cell Cult Ctr, Mass Inst Technol, 75-81; mem, Genetics Study Sect, NIH, 75-81. *Mem:* AAAS; Tissue Cult Asn; Cell Biol Asn. *Res:* Control of differentiation and gene expression in mammalian cells. *Mailing Add:* Ctr for Genetics Univ Ill Med Ctr 808 S Wood St Chicago IL 60612

DAVIDSON, ROBERT BELLAMY, b New York, NY, Jan 24, 47; m 68; c 3. PHYSICAL CHEMISTRY, PHYSICS. *Educ:* City Col New York, BSc, 66; Princeton Univ, MA, 68, PhD(chem), 70. *Prof Exp:* NIH fel chem, Cornell Univ, 69-70; Gibbs instr, Yale Univ, 70-73; asst prof chem, Amherst Col, 73-78; Res & Develop assocs, 78-86; SR SCIENTIST & DIV CHIEF SCIENTIST, SCI APPLNS INT CORP, 86- *Res:* Development of advanced systems, particularly power sources and intelligent control systems; modeling, simulation and analysis of physical processes in natural and artificial environments and in technological artifacts; correlation of molecular electronic structure with chemical and physical properties; foreign technology assessment, particularly Soviet and Japanese science and technology. *Mailing Add:* Sci Applns Int Corp 1710 Goodridge Dr McLean VA 22102-3779

DAVIDSON, ROBERT C, b Ft Payne, Ala, Mar 9, 32; m 56; c 2. MATHEMATICS, PHYSICS. *Educ:* Auburn Univ, BEE, 54; E Tenn State Univ, MA, 64. *Prof Exp:* Design engr, RCA, 54-60; eng mgr, Spec Prod Dept, Sperry Corp, Unisys, 60-87; INSTR MATH, PHYSICS & INSTRUMENTATION, TRI-CITY STATE TECH INST, 87- *Concurrent Pos:* Adj fac math, Tri-City State Tech Inst, 83-87. *Mem:* Sr mem Inst Elec & Electronic Engrs; sr mem Instrument Soc Am. *Mailing Add:* 108 Lovedale Dr Bristol TN 37620

DAVIDSON, ROBERT W, b Buffalo, NY, Nov 19, 21; m 48; c 1. WOOD SCIENCE, WOOD TECHNOLOGY. *Educ:* Mont State Univ, BS, 48; State Univ NY Col Forestry, Syracuse, MS, 56, PhD(wood physics), 60. *Prof Exp:* Salesman wood prod, Yaw-Kinney Co, Inc, 48-53; res asst, State Univ NY, 57-59, from instr to assoc prof, 59-69, asst leader, Org Mat Sci Prog, 69-71, chmn, Dept Wood Prod Eng, 72-79, prog mgr, Trop Timber Info Ctr, 75-79, prof wood physics, Col Environ Sci & Forestry, 69-91; RETIRED. *Concurrent Pos:* NSF res grants, 63-64 & 69-71; assoc prof, Col Forestry, Univ Philippines, 64-65; Soc Wood Sci & Technol & NSF vis scientist, 67; NSF & Govt of India exchange scientist, India, 75. *Mem:* Forest Prod Res Soc; Soc Wood Sci & Technol. *Res:* Physical properties of wood. *Mailing Add:* Col Environ Sci & Forestry State Univ NY Syracuse NY 13210

DAVIDSON, RONALD CROSBY, b Norwich, Ont, July 3, 41; m 63; c 2. PLASMA PHYSICS. *Educ:* McMaster Univ, BSc, 63; Princeton Univ, PhD(plasma physics), 66. *Prof Exp:* Asst res physicist, Univ Calif, Berkeley, 66-68; from asst prof to assoc prof physics, Univ Md, Col Park, 68-73, prof physics & astron, 73-78; asst dir appl plasma physics, Off Fusion Energy, Dept Energy, 76-78; dir, Plasma Fusion Ctr, Mass Inst Technol, Cambridge, 78-88, prof physics, 78-90; PROF ASTROPHYS SCI & DIR, PRINCETON PLASMA PHYSICS LAB, PRINCETON UNIV, 91- *Concurrent Pos:* Consult, Sci Applications, Inc, 80-; dir, Fusion Power Assocs, 80-; Alfred P Sloan fel, 70-71; chmn, Plasma Physics Div, Am Physics Soc, 83-84 & magnetic fusion comt, 82-86. *Honors & Awards:* Distinguished Assoc Award, Dept Energy, 86; Leadership Award, Fusion Power Assocs, 86. *Mem:* Fel Am Phys Soc. *Res:* Plasma turbulence; nonlinear plasma theory; nonneutral plasmas; intense charged particle beams and microwave generation; nonequilibrium statistical mechanics. *Mailing Add:* Plasma Physics Lab Princeton Univ Princeton NJ 08543

DAVIDSON, RONALD G, b Hamilton, Ont, Oct 24, 33; m 57; c 2. PEDIATRICS, MEDICAL GENETICS. *Educ:* Western Ont Univ, MD, 57. *Prof Exp:* Intern, Vancouver Gen Hosp, BC, 57-58, asst resident pediat, 58-59; asst resident path, Children's Hosp Med Ctr, Boston, 59-60 & pediat, Boston City Hosp, 60-61; asst res prof pediat, State Univ NY Buffalo, 64-67, assoc prof, 67-70, prof pediat & assoc chmn dept, 70-74, dir, Div Human Genetics, Children's Hosp, 64-74; PROF PEDIAT & DIR, PROG HUMAN GENETICS, McMASTER UNIV, 75- *Concurrent Pos:* Fel pediat, Johns Hopkins Hosp, 61-63; fel biochem genetics, King's Col, Univ London, 63-64; assoc chief pediat, Roswell Park Mem Inst, 64-67. *Mem:* Am Soc Human Genetics; Soc Pediat Res; Am Pediat Soc. *Res:* Human biochemical genetics; inherited enzyme variants and gene action in the X chromosome. *Mailing Add:* Dept Pediat McMaster Univ Med Ctr Hamilton ON L8N 3Z5 Can

DAVIDSON, ROSS WALLACE, JR, b Columbus, Kans, Aug 12, 02; m 30; c 1. FOREST PATHOLOGY. *Educ:* Univ Ottawa, BS, 27; Univ Iowa, MS, 28. *Hon Degrees:* DSc, Univ Ottawa, 68. *Prof Exp:* Jr mycologist, USDA, 28-31, from asst mycologist to assoc mycologist, Off Forest Path, 31-44, pathologist, 44-51, forest pathologist, Forest Insect & Dis Lab, Forest Exp Sta, 51-57 & Forest Dis Lab, Plant Indust Sta, 57-61; res pathologist, Col Forestry, Colo State Univ, 61-68, res specialist forest path, 68-82; RETIRED. *Mem:* Bot Soc Am; Am Phytopath Soc; Mycol Soc Am. *Res:* Identification of forest fungi from cultural characteristics; fungus damage to wood and wood products. *Mailing Add:* 2201 S Lemay St Ft Collins CO 80521

DAVIDSON, SAMUEL JAMES, b Chicago, Ill, Mar 9, 37; m 60; c 2. BIOCHEMISTRY, CLINICAL CHEMISTRY. *Educ:* Univ Chicago, AB, 56, BS, 59, PhD(biochem), 64. *Prof Exp:* Res assoc biochem, Univ Chicago, 64; res assoc, Sch Med, Tufts Univ, 67-68, from instr to asst prof physiol, 68-74; res assoc, Bio-Res Inst, 74-76; asst dir, 77-81, ACTG DIR, DEPT CLIN BIOCHEM, BOSTON CITY HOSP, 81- *Concurrent Pos:* NIH fel biochem, Brandeis Univ, 65-67. *Res:* Membranes and dynamics of the vacuolar system; biochemistry of aryl hydrocarbon hydroxylase and chemical carcinogenesis. *Mailing Add:* 56 Beals St Brookline MA 02146

DAVIDSON, STEVE EDWIN, b Dumas, Tex, July 2, 30; m 53; c 3. AGRONOMY. *Educ:* WTex State Col, BS, 51; Iowa State Col, MS, 52; Agr & Mech Col Tex, PhD(agron), 56. *Prof Exp:* Asst, Iowa State Col, 51-52 & Agr & Mech Col Tex, 52-56; soil scientist, Soil Conserv Serv, Plant Indust Sta, USDA, Md, 56-58; asst prof, 58-64, PROF BIOL, EVANGEL COL, 64- *Mem:* Am Soc Agron; Soil Sci Soc Am. *Res:* Soil physics; plant physiology. *Mailing Add:* Dept Sci & Sci Tech Evangel Col 111 N Glenstone Springfield MO 65802

DAVIDSON, THEODORE, b Chicago, Ill, Mar 30, 39; m 62; c 1. MATERIALS SCIENCE, PHYSICAL CHEMISTRY. *Educ:* Cornell Univ, BA, 60; Univ Chicago, MS, 62; Rensselaer Polytech Inst, PhD(phys chem), 68. *Prof Exp:* Asst prof mat sci & chem eng, Northwestern Univ, 67-73; SCIENTIST MAT SCI, XEROX CORP, ROCHESTER, NY, 73- *Concurrent Pos:* Consult scientist, 69-73, exec comt mem, Div Org Coatings & Plastics Chem, Am Chem Soc, 78. *Mem:* AAAS; Am Phys Soc; Am Chem Soc; Mats Res Soc. *Res:* Structure of solid polymers in relation to physical properties; physical chemistry of macromolecules; electron and optical microscopy; surface science of polymers; biomaterials; application of physical sciences to archeology. *Mailing Add:* Inst Mat Sci Univ Conn 97 N Eagleville Rd Storrs CT 06268-3136

DAVIDSON, THOMAS RALPH, plant pathology; deceased, see previous edition for last biography

DAVIDSON, WENDY FAY, AUTO-IMMUNITY DISEASE, BIO-CELL DIFFERENTIATION. *Educ:* Australian Nat Univ, PhD(immunol), 77. *Prof Exp:* VIS SCIENTIST, LAB GENETICS, NAT CANCER INST, NIH, 85- *Mailing Add:* Lab Genetics Nat Cancer Inst NIH 9000 Rockville Pike Bldg 37 Rm 2B17 Bethesda MD 20892

DAVIDSON, WILLIAM MARTIN, b Cambridge, Mass, Jan 3, 39; m 64; c 2. ORTHODONTICS. *Educ:* Dartmouth Col, AB, 60; Harvard Univ, DMD, 65; Univ Minn, cert orthod & PhD(anat), 69. *Prof Exp:* Teaching asst anat, Univ Minn, 65-69; asst prof orthod, Univ Conn Health Ctr, Farmington, 69-72, assoc dean dent educ, 72-76, assoc dean grad educ, 73-76; assoc prof orthod & chmn dept, Sch Dent, Univ Miss, 76-78; PROF & CHMN DEPT ORTHOD, UNIV MD, 78- *Concurrent Pos:* Consult, Coun Dent Educ, Am Dent Asn, 74, Am Asn Orthodontists Coun Res. *Mem:* Am Dent Asn; Am Asn Orthod; Int Asn Dent Res. *Res:* Dental education; treatment effects; bone biology; craniofacial defects. *Mailing Add:* Dept Orthod Univ Md Sch Dent 666 W Baltimore St Baltimore MD 21201

DAVIDSON, WILLIAM SCOTT, b Ballater, Scotland, Apr 16, 52; Can citizen; m 74; c 2. RECOMBINANT DNA ENGINEERING, POPULATION GENETICS. *Educ:* Edinburgh Univ, Scotland, BSc, 74; Queen's Univ, Can, PhD(biochem), 78. *Prof Exp:* Fel biochem, Univ Calif, Berkeley, 78-80; centennial fel, Univ Conn Health Ctr, Farmington, 80-81; asst prof, 81-87, ASSOC PROF BIOCHEM, MEM UNIV NFLD, 87- *Mem:* Can Biochem Soc; Sigma Xi; Am Soc Biochem & Molecular Biol; Protein Soc; Genetical Soc; Atlantic Salmon Fedn. *Res:* Molecular, population and evolutionary genetics; organization, structure and regulation of genes and genomes in fish; application of DNA techniques to forensic science. *Mailing Add:* Dept Biochem Mem Univ Nlfd St John's NF A1B 3X9 Can

DAVIDSON-ARNOTT, ROBIN G D, b London, Eng, Mar 2, 47; Can citizen; m 68; c 2. GEOMORPHOLOGY, SEDIMENTOLOGY. *Educ:* Univ Toronto, BA Hons, 70, PhD(geog), 75. *Prof Exp:* Asst prof geog, Scarborough Col, Univ Toronto, 75-76; from asst prof to assoc prof, 76-88, PROF GEOG, UNIV GUELPH, 88- *Mem:* Can Asn Geographers; Soc Econ Paleontologists & Mineralogists; Int Asn Sedimentologists; Geol Asn Can. *Res:* Coastal sedimentology; nearshore processes and sediment transport; sedimentary structures; Great Lakes coastal zone management; formation of coastal sand dunes. *Mailing Add:* Dept Geog Univ Guelph Guelph ON N1G 2W1 Can

DAVIE, EARL W, b Tacoma, Wash, Oct 25, 27; m 52; c 4. BIOCHEMISTRY. *Educ:* Univ Wash, BS, 50, PhD(biochem), 54. *Prof Exp:* From asst prof to assoc prof biochem, Western Reserve Univ, 56-62; assoc prof, 62-66, chmn dept, 75-84, PROF BIOCHEM, SCH MED, UNIV WASH, 66- *Concurrent Pos:* Nat Found Infantile Paralysis fel, Mass Gen Hosp, 54-56; NSF & Commonwealth Fund fel, Inst Molecular Biol, Univ Geneva, 66-67; mem, Nat Bd Med Exam, 71-75; mem, res rev comt, Am Nat Red Cross Blood Prog, 73-77; mem, Nat Hemophilia Found Med & Sci Adv Coun, 74-79; mem, Hemat Study Sect, NIH, 75-79; mem Nat Heart, Lung & Blood Coun, NIH, 85-89; assoc ed, Biochem, 80- *Honors & Awards:* Int Prize, French Asn Hemophiliacs, 83; Waterford Biomed Res Prize, Scripps Clin & Res Found, 85; Robert P Grant Medal, Int Soc Thrombosis & Hemostasis, 89. *Mem:* Nat Acad Sci; Am Acad Arts & Sci; Am Chem Soc; Am Soc Biol Chem (secy, 75-78); Japanese Biochem Soc; foreign mem Royal Danish Acad Sci & Letters; Am Soc Hemat; AAAS. *Res:* Protein chemistry, enzymology and genomic structures of the plasma proteins involved in blood coagulation and fibrinolysis. *Mailing Add:* Dept Biochem Univ Wash Sch Med Seattle WA 98195

DAVIE, JAMES RONALD, b Middlesex, Eng, Mar 13, 51; Can citizen; m 82; c 1. CHROMATIN STRUCTURE, GENE REGULATION. *Educ:* Univ BC, BSc, 75, PhD(biochem), 79. *Prof Exp:* Asst prof enzym & cell biol, dept biochem, 83-88, ASSOC PROF, UNIV MAN, 88- *Concurrent Pos:* Med Res Coun Can Scholar molecular biol. *Mem:* Am Soc Biochem & Molecular Biol; Can Biochem Soc. *Res:* Chronatic structure and function; histone variant and modifications (acetylation, methylation, ubiquitination); chemical carcinogenesis structure and function of sequence; specific nuclear DNA; binding proteins. *Mailing Add:* Dept Biochem Univ Man 770 Bannatyne Ave Winnipeg MB R3E 0W3 Can

DAVIE, JOSEPH MYRTEN, b LaPorte, Ind, Oct 14, 39; m 60; c 3. IMMUNOLOGY. *Educ:* Ind Univ, AB, 62, MA, 64, PhD(bact), 66; Washington Univ, MD, 68. *Prof Exp:* Fel, Sch Med, Washington Univ, 66-67, intern path, 68-69; staff assoc immunol, Nat Inst Allergy & Infectious Dis, NIH, 69-71, resident path, Nat Cancer Inst, 71-72; from assoc prof to prof, dept path, Sch Med, Wash Univ, 72-75, prof & head, microbiol & immunol, 75-87; PRES RES & DEVELOP, G D SEARLE & CO, 87- *Concurrent Pos:* Mem path B study sect, NIH, 74-78; assoc ed, J Immunol, 75-83; sr consult, Marhey Charitable Trust, 85-; mem, Howard Hughes Med Inst Sci Adv Comt, 90- *Mem:* Am Soc Microbiol; Am Asn Immunologists; Am Soc Exp Path; Am Asn Univ Pathologists. *Res:* Cellular basis of immune response and the genetics of immunoglobulin structure. *Mailing Add:* G D Searle & Co 4901 Searle Pkwy Skokie IL 60077

DAVIE, WILLIAM RAYMOND, b Aliquippa, Pa, Mar 16, 24; m 48. ORGANIC CHEMISTRY. *Educ:* Geneva Col, BS, 47; Univ Wis, MS, 49, PhD(org chem), 51. *Prof Exp:* Sampler, J & L Steel Corp, 43-47; asst, Univ Wis, 47-50; asst, 50-51, asst supvr, Agr Res, Pittsburgh Coke & Chem Co, 51-55; supvr, New Prods Res, Chemagro Corp, 55-57; asst supvr, Org Res, Pittsburgh Coke & Chem Co, 57-58, supvr, 59-64; supvr, US Steel Corp, 64-67, sr res chemist, appl res lab, gen org res, chem div, 67-86; sr res chemist, 86-90, STAFF SCIENTIST, ARISTECH CHEM CORP, 90- *Res:* Organic research; agricultural chemical research; resins and protective coatings; phthalocyanine research; dibasic acids and their production; coal carbonization; air and water pollution studies; propagation of trees. *Mailing Add:* 3100 Kane Road Aliquippa PA 15001

DAVIES, CRAIG EDWARD, b Astoria, Ore, Jan 26, 43; m 66; c 3. NUCLEAR ENGINEERING. *Educ:* Ore State Univ, BS, 65; Pa State Univ, MS, 67. *Prof Exp:* Aerodynamics engr, Lockheed Missiles & Space Ctr, 67-69; nuclear plant engr trainee, Westinghouse Elec Corp, Bettis Atomic Power Lab, Naval Reactors Facil, Idaho, 69-70, various mgt position, 70-74, mgr, reactor opers training, 74-75; mgt trainee & student, Westinghouse Elec Corp, Prospective Commanding Officer Sch, Washington, DC, 75; prototype mgr opers & training, Westinghouse Elec Corp, site mgr overall opers, Naval Reactor Facil, Idaho, 78-81, mgr availability assurance, Westinghouse Steam Generator Div, Orlando, Fla, 81-85; PRES, NAM ENERGY SERV, ISSAQUAH, WASH, 85- *Concurrent Pos:* Bd dir, Intermountain Sci Exp Ctr, 78- *Mem:* Am Nuclear Soc. *Res:* Testing and performance evaluation of materials and components associated with power generating nuclear reactor plants. *Mailing Add:* 1340 217th SE Issaquah WA 98027

DAVIES, D K, b Ammanford, Wales, UK, Sept 5, 35; m 62; c 2. PHYSICS. *Educ:* Univ Wales, BSc, 57, PhD(physics), 61. *Prof Exp:* Cent Elec Generating Bd sr res fel physics, Univ Col Swansea, Wales, 60-62; SR PHYSICIST, RES & DEVELOP CTR, WESTINGHOUSE ELEC CORP, 62- *Mem:* Am Phys Soc; fel Brit Inst Physics & Physics Soc. *Res:* Spatial and temporal growth of ionization in gases; atomic collisional processes in gases; vacuum breakdown; analytical spectroscopy; electron transport in gases. *Mailing Add:* Westinghouse Res Labs Beulah Rd Pittsburgh PA 15235

DAVIES, DAVID HUW, b Tredegar, Eng, Oct 29, 42; m 66; c 2. PHYSICAL CHEMISTRY. *Educ:* Univ Col, London, BSc, 64, PhD(phys chem), 67; Univ Pittsburgh, MBA, 72. *Prof Exp:* Sr scientist, Westinghouse Elec Corp, 67-72, mgr thin film mat, Res & Develop Ctr, 72-77; mgr eng & dir, Datascreen Corp, Exxon Corp, 77-79, vpres, Kylex Div, Exxon Enterprises, 79-87; AT 3M MOUNTAIN VIEW LAB, MOUNTAIN VIEW, CALIF, 88- *Concurrent Pos:* Ed, Int Soc Optical Eng Proceedings. *Mem:* Inst Elec & Electronics Engrs; Electrochem Soc; Optical Soc Am; Soc Info Display; Mat Res Soc; Int Soc Optical Eng. *Res:* Solid state display; thin films; electrical packaging; optical memory systems; liquid crystals; optical films; author of 50 publications, several book chapters and 8 patents. *Mailing Add:* 3M Mountain View Lab 420 Bernardo Ave Mountain View CA 94043

DAVIES, DAVID K, b Glamorgan, Wales, Oct 10, 40; m 64; c 2. GEOLOGY. *Educ:* Univ Wales, BSc, 62, PhD(geol), 66, DSc(geol), 91; La State Univ, MS, 64. *Prof Exp:* From asst prof to assoc prof geol, Tex A&M Univ, 66-70, asst to dean, 66-70; from assoc prof to prof geol, Univ Mo-Columbia, 70-77; prof geol, chmn dept & dir Reservoir Studies Inst, Tex Tech Univ, 77-80; PRES, DAVID K DAVIES & ASSOCS INC, 80- *Concurrent Pos:* Distinguished lectr, Soc Petrol Engrs, 84-85. *Honors & Awards:* A I Levorsen Award, Am Asn Petrol Geologists, 78. *Mem:* Am Asn Petrol Geologists; Geol Soc Am; Soc Econ Paleontologists & Mineralogists; Soc Petrol Engrs; Int Asn Sedimentologists. *Res:* Reservoir geology; sandstone diagenesis; volcanic sedimentation; paleogeographic reconstruction; quantitative methods. *Mailing Add:* David K Davies & Assoc 1410 Stonehollow Dr Kingwood TX 77339

DAVIES, DAVID R, b Carmarthen, UK, Feb 22, 27; US citizen; m 51, 85; c 2. STRUCTURAL BIOLOGY, CRYSTALLOGRAPHY. *Educ:* Oxford Univ, BA, 49, DPhil(chem crystallog), 52. *Prof Exp:* Noyes fel chem, Calif Inst Technol, 52-54; res scientist, Albright & Wilson, Eng, 54-55; vis scientist molecular biol, NIMH, 55-62, CHIEF, SECT MOLECULAR STRUCT, NAT INST DIABETES DIGESTIVE & KIDNEY DIS, NIH, 62- *Mem:* Nat Acad Sci; Am Crystallog Asn; Am Soc Biol Chemists; AAAS; Biophys Soc; Am Acad Arts & Sci. *Res:* X-ray crystallography applied to proteins and nucleic acids. *Mailing Add:* NIH Bldg 2 Room 316 Bethesda MD 20892

DAVIES, DONALD HARRY, b Ottawa, Ont, Jan 26, 38; m 78; c 2. PHYSICAL CHEMISTRY. *Educ:* Carleton Univ, BSc, 60; Bristol Univ, PhD(chem), 63. *Prof Exp:* Nat Res Coun Can fel, 63-65; asst prof chem, Dalhousie Univ, 65-69; from asst prof to assoc prof, 69-79, chmn dept, 73-78, PROF CHEM, ST MARY'S UNIV, 79-; PRES, NOVA CHEM, LTD, 78- *Concurrent Pos:* Sessional lectr, Carleton Univ, 65. *Mem:* Fel Chem Inst Can; Royal Soc Chem. *Res:* Surface chemistry and thermodynamics; polymer degradation. *Mailing Add:* Dept Chem St Mary's Univ Halifax NS B3H 3C3 Can

DAVIES, DOUGLAS MACKENZIE, b Toronto, Ont, May 11, 19; m 48; c 2. ENTOMOLOGY. *Educ:* Univ Toronto, BA, 42, PhD(entom), 49. *Prof Exp:* Res fel parasitol, Ont Res Found, 49-51; sessional lectr, McMaster Univ, 50-51, from asst prof to assoc prof zool, 51-63, prof biol, 63-84, EMER PROF, MCMASTER UNIV, 84- *Concurrent Pos:* Meteorol officer, RCAF, 42-45. *Mem:* Entom Soc Am; Am Mosquito Control Asn; fel Royal Entom Soc; Can Soc Zool; N Am Benthol Soc; fel Entom Soc Can; Soc Vector Ecol. *Res:* Ecology, behavior, physiology and systematics of simuliids, other blood sucking flies and muscids. *Mailing Add:* Dept Biol McMaster Univ Hamilton ON L8S 4K1 Can

DAVIES, EMLYN B, b Minnedosa, Man, Feb 23, 27; m 62; c 3. PHYSICS. *Educ:* Univ Man, BSc, 48; Pa State Univ, MS, 51, PhD(physics), 53. *Prof Exp:* Sr res geophysicist, Gulf Res & Develop Co, 53-70, sect supvr, 70-73, mgr geophysics anal dept, 74-85; ASST PRES, CHEVRON GEOSCIENCES CO, HOUSTON, TX, 85- *Mem:* AAAS; Soc Explor Geophys; Acoust Soc Am; Am Phys Soc. *Res:* Elastic wave propagation; information theory. *Mailing Add:* 10914 Burgoyne Rd Houston TX 77042

DAVIES, ERIC, b Liverpool, Eng, Aug 24, 39; m 63; c 2. PLANT PHYSIOLOGY. *Educ:* Wye Col, London Univ, BSc, 62; McGill Univ, PhD(bot), 68. *Prof Exp:* From asst prof to assoc prof bot, 68-82, PROF LIFE SCI, UNIV NEBR-LINCOLN, 82-, MEM FAC LIFE SCI, 76- *Concurrent Pos:* Res prof, Lab Nuclear Med, Univ Brussels, 75-76; vis prof, Purdue Univ, West Lafayette, Ind, 86-87. *Mem:* AAAS; Japanese Soc Plant Physiol; Am Soc Plant Physiologists. *Res:* Molecular biology of plant hormones and plant wound responses. *Mailing Add:* Sch Life Sci Univ Nebr Lincoln NE 68588-0118

DAVIES, FREDERICK T, JR, b Springfield, Mass, Jan 8, 49; m; c 2. HORTICULTURE, PLANT PHYSIOLOGY. *Educ:* Rutgers Univ, BA, 71, MS, 75; Univ Fla, PhD(hort), 78. *Prof Exp:* Grad asst hort, Rutgers Univ, 73-75, Univ Fla, 75-78; asst prof hort, 78-84, assoc prof woody plant physiol, 84-90, PROF WOODY PLANT PHYSIOL, TEX A&M UNIV, 90- *Concurrent Pos:* Vis scientist, USDA, 87; vis assoc prof hort, Ore State Univ, 87; int consult, nursery, greenhouse, tissue cult prod, Mex, Honduras, Peru, Spain; fed grant, NJ State, 74-75. *Honors & Awards:* Distinguished Achievement Award, Am Soc Hort Sci, 89. *Mem:* Am Soc Hort Sci; Int Plant Propagators Soc (pres, 86-87); Plant Growth Regulator Soc; Bot Soc Am; Am Nurseryman Soc; Sigma Xi. *Res:* Woody plant physiology; developmental and stress physiology; plant propagation; mycorrhizal fungi systems; nursery and greenhouse production and management; co-author of one book. *Mailing Add:* Dept Hort Sci Tex A&M Univ College Station TX 77843

DAVIES, GEOFFREY, b Stoke-on-Trent, Eng, Feb 6, 42; m 65; c 3. INORGANIC CHEMISTRY, ANALYTICAL CHEMISTRY. *Educ:* Univ Birmingham, BSc, 63, PhD(chem), 66, DSc(chem), 87. *Prof Exp:* NIH fel chem, Brandeis Univ, 66-68; res assoc, Brookhaven Nat Lab, 68-69; IMP Chem Industs fel, Univ Kent, 69-71; from asst prof to assoc prof, 71-81, prof & dir grad affairs, 81-88, UNIV DISTINGUISHED PROF, NORTHEASTERN UNIV, 88- *Concurrent Pos:* Royal Soc traveling fel, 70; fac fel, Barnett Inst Anal Chem, Northeastern Univ, 72-; ed consult, John Wiley & Sons, Inc, 73- & Macmillan Co, 74- *Mem:* Am Chem Soc. *Res:* Kinetics and mechanisms of solution phase reactions; synthesis and characterization of novel inorganic complexes containing oxygen; synthesis of heteropolymetallic molecules; catalyst synthesis; catalytic mechanisms. *Mailing Add:* Dept Chem Northeastern Univ Boston MA 02115

DAVIES, HAROLD WILLIAM, (JR), b Norton, Kans, Aug 2, 30; m 56; c 3. AQUATIC BIOLOGY, PHYCOLOGY. *Educ:* Emporia State Univ, BS, 56, MS, 58; Mich State Univ, PhD(bot), 66. *Prof Exp:* Instr biol, Kans State Teachers Col, 57-58; instr & ed consult biol sci, State Dept Sec Educ, New South Wales, 61-63; chmn dept, 64-78, 81-83, PROF BIOL SCI, PURDUE UNIV, FT WAYNE, 64- *Mem:* Brit Phycol Soc; Phycol Soc Am; Am Inst Biol Sci; Int Phycol Soc. *Res:* Ecology of the phytopsammon; freshwater algae morphology and ecology; limnology. *Mailing Add:* Dept Biol Sci Purdue Univ 2101 Coliseum Blvd E Ft Wayne IN 46805

DAVIES, HELEN JEAN CONRAD, b New York, NY, Feb 14, 25; m 61; c 2. BIOCHEMISTRY, MICROBIOLOGY. *Educ:* Brooklyn Col, AB, 44; Univ Rochester, MS, 50; Univ Pa, PhD(biochem), 60. *Prof Exp:* Sci worker biochem, Smith Kline & French Labs, Pa, 53-54; res assoc, Grad Sch Med, 54-56, res assoc microbiol, Sch Med, 62-65, from asst prof to assoc prof microbiol, 65-82, asst prof community med, 70-71, assoc prof phys biochem, 71-78, PROF MICROBIOL, UNIV PA, 82- *Concurrent Pos:* Johnson Found res fel, Univ Pa, 60-62; mem bd trustees, Pa State Univ, 73-78. *Mem:* Am Chem Soc; Am Soc Biol Chem; Am Soc Microbiol; Biophys Soc; Asn Women Sci. *Res:* Oxidative enzymes; cytochromes; monoclonal antibodies; respiratory chain systems; enzyme kinetics; steady-state growth of microorganisms; streptococci; recruitment and retention of minority group students and women in biomedical careers; bacterial cytochromes. *Mailing Add:* Dept Microbiol Univ Pa Philadelphia PA 19104-6076

DAVIES, HUW M, b Newtown, UK, Sept 23, 51; m 80. PLANT BIOCHEMISTRY. *Educ:* Oxford Univ, BA, 74, MA, 77; London Univ, PhD(biochem), 77. *Prof Exp:* Res assoc, Mich State Univ-Dept Energy, Plant Res Lab, 77-80; vis res fel, Plant Growth Lab, Univ Calif, Davis, 80-81; PRIN SCIENTIST, CALGENE INC, 81- *Mem:* Am Soc Plant Physiologists. *Res:* Plant biochemistry and physiology especially nitrogen and carbohydrate metabolism, metabolic regulation and properties of plant cell cultures; enzymology and metabolic pathway dynamics. *Mailing Add:* Calgene Inc 1920 Fifth St Suite F Davis CA 95616

DAVIES, JACK, b Eng, Aug 24, 19; m 46; c 2. ANATOMY, EMBRYOLOGY. *Educ:* Univ Iowa, MD, 43; Cambridge Univ, MA, 46; Univ Leeds, MD, 47. *Prof Exp:* Demonstr anat, Univ Leeds, 43-44; demonstr, Cambridge Univ, 44-49, lectr, 49-51; from asst prof to assoc prof, Univ Iowa, 51-55; from assoc prof to prof, Sch Med, Wash Univ, 55-63; prof anat & chmn dept, Sch Med, Vanderbilt Univ, 63-89; RETIRED. *Concurrent Pos:* Fel, Cambridge Univ, 49-51; Markle scholar, 53; consult, USPHS, 58- *Mem:* AAAS; Am Asn Anat; Soc Exp Biol & Med. *Res:* Morphology and physiology of the placenta. *Mailing Add:* 7919 47th Pl W Mukilteo WA 98275

DAVIES, JACK NEVILLE PHILLIPS, b Devizes, Eng, July 2, 15; m 44, 61; c 3. EPIDEMIOLOGY. *Educ:* Bristol Univ, MB & ChB, 39, MD, 48. *Prof Exp:* Lectr physiol & demonstr pharmacol, Bristol Univ, 41-44; pathologist, Uganda Govt, 45-50; prof path, Med Sch, Makerere Univ Col, Uganda, 50-61; reader morbid anat, Postgrad Med Sch, Univ London, 61-63; PROF PATH, ALBANY MED COL, 63- *Concurrent Pos:* Commonwealth Fund fel, Duke Univ, 49-50; consult pathologist, Uganda Govt, 50-61; mem, EAfrican Med Res Coun, 52-61; mem comt, Int Soc Geog Path, 54-60; mem African cancer comt, Int Union Against Cancer, 55-62; Fox lectr, Bristol Univ, 62; Musser lectr, Tulane Univ, 64. *Mem:* Am Asn Path & Bact; Path Soc Gt Brit & Ireland; Brit Med Asn. *Res:* Morbid anatomy; geographic pathology, especially of cancer and cardiovascular disease. *Mailing Add:* Box 110 Rowe Rd Selkirk NY 12158

DAVIES, JAMES FREDERICK, b Winnipeg, Man, Dec 5, 24. ECONOMIC GEOLOGY, PETROLOGY. *Educ:* Univ Man, BSc, 46, MSc, 48; Univ Toronto, PhD(geol), 63. *Prof Exp:* Lectr geol, Univ Man, 47; geologist, Man Dept Mines & Natural Resources, 51-57, chief geologist, 57-67; PROF GEOL, LAURENTIAN UNIV, 67- *Concurrent Pos:* Hon lectr, Univ Man, 57-; mem, Nat Adv Comt Res Geol Sci-Can, 58-62. *Honors & Awards:* Barlow Medal, Can Inst Mining & Metall, 65. *Mem:* Fel Geol Asn Can; fel Mineral Asn Can; Soc Econ Geologists; Can Inst Mining & Metall. *Res:* Regional mapping; compilation, analysis and interpretation of regional geology, tectonic features; geochemistry of Pre-Cambrian rocks, mineral deposits and wall-rocks; studies of Pre-Cambrian volcanic rocks; genesis of base metal sulphide deposits; fabrics of sulphides; deformation and remobilization of sulphide deposits. *Mailing Add:* Laurentian Univ Ramsey Lake Rd Sudbury ON P3E 2E3 Can

DAVIES, JOHN A, b Milwaukee, Wis, May 4, 31; m 63; c 2. PHYSICS. *Educ:* Univ Md, BS, 53, MS, 54, PhD(physics), 60. *Prof Exp:* Res asst physics, Univ Md, 53-60; vis asst prof, Univ Cincinnati, 63; asst prof, 63-69, ASSOC PROF PHYSICS, CLARK UNIV, 69- *Mem:* Am Phys Soc. *Res:* Mathematical physics; lattice dynamics of ionic crystals. *Mailing Add:* Dept Physics Clark Univ Worcester MA 01610

DAVIES, JOHN ARTHUR, b Prestatyn, NWales, Mar 28, 27; Can citizen; m 50; c 6. SOLID STATE SCIENCE, ATOMIC COLLISIONS IN SOLIDS. *Educ:* Univ Toronto, BA, 47, MA, 48, PhD(phys chem), 50. *Hon Degrees:* DSc, Royal Roads Mil Col, Victoria, BC, 84. *Prof Exp:* Asst res officer phys chem, Chalk River Nuclear Labs, 50-54; Can Ramsey fel polymerization kinetics, Leeds, 54-56; assoc & sr res officer nuclear sci, Res Chem Br, Chalk River Nuclear Labs, 56-68, prin res officer, 68-85; PROF PHYSICS & ENG PHYSICS, MCMASTER UNIV, 70- *Concurrent Pos:* Ammanuensis, Physics Inst, Aarhus Univ, 64-65; guest prof, 69-70. *Honors & Awards:* Noranda Award, Chem Inst Can, 65; T D Callinan Award, 68. *Mem:* Am Vacuum Soc; Royal Danish Acad Sci & Letters; Royal Soc Can; Chem Inst Can; Bohmische Phys Soc. *Res:* Ion transport and diffusion processes in solutions; polymerization kinetics; slowing down behaviour of energetic ions in solids; anodic oxidation of metals; channeling of energetic ion beams in crystals; ion implantation. *Mailing Add:* Dept Eng Physics McMaster Univ Hamilton ON L8S 4M1 Can

DAVIES, JOHN CLIFFORD, b Edmonton, Alta, Sept 30, 32; m 61; c 4. GEOLOGY. *Educ:* Univ Manitoba, BSc, 55, MSc, 56, PhD(petrol & geochem), 66. *Prof Exp:* Geologist, Falconbridge Nickel Mines Ltd, 57 & US Steel Corp, 58-59; res geologist, Ont Dept Mines, 60-68; field geologist, Ont Geol Surv, 68-75; head & asst dir, Econ Geol Div, Geol Surv Botswana, 75-77; sr econ geologist, Sask Mining Develop Corp, 77-82, contract, Ontario Geol Surv, 83-85, prof, 85-86; contract, Ontario Geol Surv, 83-85; prof, Univ Sask, 85-86; CONSULT, 86- *Concurrent Pos:* Field geologist, Ethiopian Ministry Mines, 72-74; summer, Polar Continental Shelf Proj, 86 & Borealis Exploration Ltd, Northwest Territories, Can, 87. *Mem:* Geol Asn Can; Soc Econ Geologists. *Res:* Analysis of field data and research programs to identify areas having potential for mineral deposits. *Mailing Add:* 411 Garrison Crescent Saskatoon SK S7H 2Z9 Can

DAVIES, JOHN TUDOR, b Pontypridd, UK, May 9, 37; m 64; c 3. GAS PLASMA TECHNOLOGY, EQUIPMENT DESIGN. *Educ:* Oxford, UK, BA, 59, MA & DPhil, 63. *Prof Exp:* Res assoc, Univ Pittsburgh, 62-65; lectr, Univ Col Swansea, 65-78; process tech mgr, DD Systs, 84-86; sr scientist, 78-84, PROCESS TECH MGR, BRANSON INT PLASMA CORP, 86- *Mem:* Am Vacuum Soc; Int Soc Optical Eng. *Mailing Add:* Branson Int Plasma Corp 31172 Huntwood Hayward CA 94544

DAVIES, JULIAN ANTHONY, b Middlesex, Eng, Aug 10, 55. HOMOGENEOUS CATALYSIS, INDUSTRIAL CHEMISTRY. *Educ:* Univ London, Eng, BSc Hons, 76, PhD(chem), 79; Imperial Col Sci & Technol, ARCS, 76. *Prof Exp:* Res fel, Univ Guelph, 79-81; from asst prof to prof, 81-90, DISTINGUISHED UNIV PROF CHEM, UNIV TOLEDO, 90- *Mem:* Royal Soc Chem; Am Chem Soc; AAAS. *Res:* Homogeneous catalytic processes involving transition metal complexes; chemistry of the platinum metals; coordination chemistry; organometallic chemistry; nuclear magnetic resonance spectroscopy. *Mailing Add:* Dept Chem Univ Toledo 2801 W Bancroft St Toledo OH 43606

DAVIES, KELVIN J A, b London, Eng, Oct 15, 51; Eng & US citizen; m 80; c 2. FREE RADICAL BIOCHEMISTRY, REGULATION OF GENE EXPRESSION. *Educ:* Liverpool Univ, UK, BEd, 74; Univ Wis, BS, 76, MS, 77; Univ Calif, Berkeley, CPhil, 79, PhD(biochem, physiol), 81. *Prof Exp:* Res assoc physiol, Univ Calif, Berkeley, 79-81; res assoc physiol & biophys, Harvard Med Sch, 81-83, instr, 82-83; from asst prof to assoc prof toxicol & biochem, Univ Southern Calif, 83-91, dir grad studies, Inst Toxicol, 85-90; PROF & CHMN, DEPT BIOCHEM & MOLECULAR BIOL, ALBANY MED COL, 91- *Concurrent Pos:* Prin investr, NIH, 83-; ed-in-chief, Free Radical Biol & Med, 84-; secy gen, Oxygen Soc, 87-89. *Honors & Awards:* Harwood S Belding Award, Am Physiol Soc, 82. *Mem:* Am Physiol Soc; Biochem Soc; Soc Free Radical Res; AAAS; NY Acad Sci; Sigma Xi; Am Soc Biochem & Molecular Biol; fel Oxygen Soc (pres-elect, 90-92). *Res:* Regulation of gene expression and protein turnover by oxygen radicals; molecular adaptations to oxidative stress; repair systems for oxidatively modified DNA, proteins & lipids; gene expression during aging. *Mailing Add:* Dept Biochem & Molecular Biol Albany Med Col New Scotland Rd Albany NY 12208

DAVIES, KENNARD MICHAEL, b Los Angeles, Calif, May 26, 41; m 71. PHYSICS. *Educ:* Loyola Univ Los Angeles, BS, 63; Univ Notre Dame, PhD(physics), 70. *Prof Exp:* ASST PROF PHYSICS, CREIGHTON UNIV, 69- *Mem:* Am Phys Soc; Am Asn Physics Teachers. *Res:* Equilibrium and nonequilibrium statistical physics; phase transitions and interphase surfaces; muscle biophysics. *Mailing Add:* Dept Physics & Comput Sci Creighton Univ 2500 Cal St Omaha NE 68178

DAVIES, KENNETH, b Wales, Jan 28, 28; m 58; c 3. IONOSPHERIC PHYSICS. *Educ:* Univ Wales, BSc, 49, PhD, 53. *Prof Exp:* Sect leader, Defence Res Telecommunications Estab, Can, 52-55; asst prof, Brown Univ, 56-58; chief, Res Div, Space Environ Lab, Nat Oceanic & Atmospheric Admin, 58-86, sr scientist, 58-88; GUEST PROF, WVHAN UNIV, 88- *Concurrent Pos:* Adj prof, Univ Colo, 64-; Webster Mem fel, Univ Queensland, 75-76; assoc ed, Radio Sci, 64-67; chmn, US-Int Union Radio Sci Comn G & Working Group V Ionosphere Mapping, Int Union Radio Sci, 84. *Mem:* Fel AAAS; Am Geophys Union; Sigma Xi; fel Inst Elec & Electronics Engrs; Int Sci Radio Union. *Res:* Upper atmosphere physics; radio propagation; satellite to ground radio propagation; plasmaspheric-ionospheric total electron content. *Mailing Add:* 4475 Laguna Pl 207 Boulder CO 80303

DAVIES, KENNETH THOMAS REED, b Pittsburgh, Pa, Nov 26, 34; m 57, 89; c 3. NUCLEAR PHYSICS, THEORETICAL PHYSICS. *Educ:* Carnegie Inst Technol, BS, 57, MS, 59, PhD(physics), 62. *Prof Exp:* Instr physics, Carnegie Inst Technol, 62; THEORET PHYSICIST, OAK RIDGE NAT LAB, 62- *Concurrent Pos:* Sect head & group leader nuclear & atomic theory, Physics Div, Oak Ridge Nat Lab, 82-84. *Mem:* Fel Am Phys Soc. *Res:* Fission and heavy-ion reactions; time-dependent Hartree-Fock theory of nuclear collective dynamics; nonlinear dynamics. *Mailing Add:* Physics Div Oak Ridge Nat Lab PO Box 2008 Oak Ridge TN 37831-6373

DAVIES, MARCIA A, b Omaha, Nebr, Nov 24, 38; m 71. MOLECULAR SPECTROSCOPY. *Educ:* Duchesne Univ, BA, 61; Univ Notre Dame, PhD(inorg chem), 66. *Prof Exp:* Res assoc, Univ Ariz, 66-67; asst prof, 67-74, ASSOC PROF CHEM, CREIGHTON UNIV, 74- *Mem:* Am Chem Soc; Soc Appl Spectros. *Res:* Metal isotope effects in far infrared spectra of metal complexes; magnetic susceptibilities of metal complexes. *Mailing Add:* 1026 Hilcrest Dr Omaha NE 68132-1849

DAVIES, MERTON EDWARD, b St Paul, Minn, Sept 13, 17; m 46; c 3. PLANETARY SCIENCES, PHOTOGRAMMETRY. *Educ:* Stanford Univ, AB, 38. *Prof Exp:* Instr math, Univ Nev, 38-39; math group leader, Douglas Aircraft Co, 40-48; MEM SR STAFF, RAND CORP, 48- *Concurrent Pos:* Mem US deleg, Surprise Attack Conf, Geneva, 58; mem, US Observer Team, Antarctic, 67; imaging sci teams, Mariner 6 & 7, 69, Mariner 9, 71, Mariner 10, 73, Voyager, 77, Galileo, Magellan & Craf. *Honors & Awards:* George W Goddard Award, Soc Photo-Optical Instrument Engineers, 66; Talbert Abrams Award, Am Soc Photogram, 74. *Mem:* AAAS; assoc fel Am Inst Aeronaut & Astronaut; Am Soc Photogram; Am Geophys Union; Am Astron Soc. *Res:* Physics of aerial and space photography; planetary geodesy; inspection for arms control. *Mailing Add:* 1414 San Remo Dr Pacific Palisades CA 90272

DAVIES, MICHAEL SHAPLAND, b Cardiff, Wales, June 6, 39; m 66; c 2. ELECTRICAL ENGINEERING. *Educ:* Cambridge Univ, MA, 61; Univ Ill, MS, 63, PhD(elec eng), 66. *Prof Exp:* Asst prof, 66-76, ASSOC PROF ELEC ENG, UNIV BC, 76- *Mem:* Inst Elec & Electronics Engrs; Can Pulp & Paper. *Res:* Process control systems. *Mailing Add:* Dept Elec Eng Univ BC Vancouver BC V6T 1W5 Can

DAVIES, PETER F, b Warrington, Eng, July 23, 47; m 73; c 1. EXPERIMENTAL PATHOLOGY. *Educ:* Cambridge Univ, PhD(exp path), 75. *Prof Exp:* Sr res fel path, Univ Wash, 77-79; from asst prof to assoc prof path, Harvard Univ, 79-88; PROF PATH, PRITZKER SCH MED, UNIV CHICAGO, 88- *Concurrent Pos:* Vis scientist, Mass Inst Technol, 83; numerous rev comt, NIH; study sect, external adv comt; mem, NIH Res Rev comt B Nat Heart Lung & Blood Inst; prin investr 3 NIH grants. *Honors & Awards:* Sci Res Coun Coop Award, Cambridge Univ. *Mem:* AAAS; fel AMA; AAP. *Res:* Vascular cell biology and pathology; pathophysiology of vascular cells, particularly the endothelium; atheoscleosis research; studies of the effects of blood flow forces upon endothelial cells. *Mailing Add:* Dept Path Univ Chicago Sch Med Box 414 5841 S Maryland Ave Chicago IL 60637

DAVIES, PETER JOHN, b Sudbury, Eng, Mar 7, 40; m 86; c 2. PLANT HORMONES, PLANT SENESCENCE. *Educ:* Univ Reading, BS, 62, PhD(plant physiol), 66; Univ Calif, Davis, MS, 64. *Prof Exp:* Instr biol, Yale Univ, 66-69; from asst prof to assoc prof, 69-83, PROF PLANT PHYSIOL, CORNELL UNIV, 83- *Concurrent Pos:* Vis researcher biochem, Univ Cambridge, 76-77, Univ Col Wales, Aberystwyth, 83-84; vis prof hort, Univ Minn, 84. *Mem:* Am Soc Plant Physiol; Int Plant Growth Substance Asn; Scand Soc Plant Physiol; Plant Growth Regulator Soc Am. *Res:* Use of genotypes in the investigation of physiological and biochemical controls in plant development; mode of action, biochemistry and transport of plant hormones; physiology of senescence in whole plants; control of fruit ripening; author-editor of books on plant hormones, plant development and plant physiology. *Mailing Add:* Sect Plant Biol Plant Sci Bldg Cornell Univ Ithaca NY 14853

DAVIES, PHILIP, IMMUNOLOGY. *Educ:* Univ Wales, BPharm(Hons), 64; Welsh Nat Sch Med, PhD, 67. *Prof Exp:* Res fel, Dept Pharmacol, Welsh Nat Sch Med, Cardiff, 67-68; res assoc, Dept Med, NY Univ Sch Med, 68-70; mem sci staff, Div Cell Path, Med Res Coun Clin Res Ctr, Harrow, Middlesex, Eng, 70-75, mem permanent staff, 75; assoc dir, Dept Inflammation & Arthritis, Merck Sharp Dohme Res Lab, 75-78, dir, Dept Immunol & Inflammation, 78-82, sr dir, 82-85 & 88-90, sr dir, Dept Biochem & Molecular Biol, 85-88, EXEC DIR INFLAMMATION RES, MERCK SHARP DOHME RES LAB, 90- *Concurrent Pos:* Adj prof, Rutgers Univ, 78-; contrib sect ed, J Soc Leukocyte Biol, 81; ed, Biochem J, 84-90; chmn, Immunopharmacol Conf, Fedn Am Socs Exp Biol, 88. *Mem:* Brit Pharmacol Soc; Biochem Soc; Brit Soc Immunol; Soc Leukocyte Biol (treas, 82-86, pres-elect, 87, pres, 88); Am Asn Immunologists; AAAS; Am Fedn Clin Res; Am Rheumatism Asn; Am Thoracic Soc. *Res:* Mononuclear phagocytes; lipoxygenase products in immunity; antiinflammatory drugs. *Mailing Add:* Dept Immunol Merck Inst Therapeut Res Rahway NJ 07065

DAVIES, PHILIP WINNE, BIOPHYSICS OF NERVE & MUSCLE. *Educ:* Univ Pa, PhD(physiol), 43. *Prof Exp:* ASSOC PROF PHYSIOL, SCH MED, JOHNS HOPKINS UNIV, 57- *Mailing Add:* Dept Physiol Sch Med Johns Hopkins Univ 725 N Wolfe St Baltimore MD 21205

DAVIES, REG, b Nottingham, Eng, May 15, 36; c 2. PARTICLE TECHNOLOGY, CHEMICAL ENGINEERING. *Educ:* Univ London, BS, 59; Loughborough Univ Technol, PhD(particle technol), 77. *Prof Exp:* Scientist coal, Nat Coal Bd, Eng, 53-67; scientist particle res, Ill Inst Technol Res Inst, 67-75; CONSULT PARTICLE PROCESSES, E I DU PONT DE NEMOURS & CO, INC, 75- *Concurrent Pos:* Bd mem, Fine Particle Soc, 73-78; sci bd mem for US, 2nd Europ Conf Particle Technol, 78-79; chmn, Tech Comt, Int Fine Particle Res Inst. *Mem:* Fine Particle Soc (pres, 75-76). *Res:* Powder, slurry, dust and mist technology; particle size analysis; particle characterization, specifically the effect of particle characteristics on chemical processes and products. *Mailing Add:* 18 Woodshaw Rd Newark DE 19711

DAVIES, RICHARD EDGAR, b Schenectady, NY, Jan 5, 15; m 43; c 1. ORGANIC CHEMISTRY, PLASTICS. *Educ:* Union Col, BS, 35; Columbia Univ, PhD(chem), 41. *Prof Exp:* Chemist, Catalin Corp of Am, 42-46, Celanese Corp of Am, 46-57 & Air Reduction Inc, 57-65; chemist, 65-66, asst res mgr, 66-78; assoc dir, Eastern Res Div, ITT Rayonier, Inc, 78-81; RETIRED. *Mem:* Am Chem Soc. *Res:* Organic synthesis; cellulose; hydrogenation; synthetic fibers; high polymers. *Mailing Add:* 46 S Murray Ave Ridgewood NJ 07450

DAVIES, RICHARD GLYN, b Congleton, Eng, Nov 1, 34; m 58; c 3. PHYSICAL METALLURGY. *Educ:* Univ Birmingham, BSc, 56, PhD(phys metall), 59; Univ Mich, MBA, 74. *Prof Exp:* Prof metall, Inst Physics, Nat Univ Cuyo, Bariloche, 59-61; STAFF SCIENTIST, FORD MOTOR CO, 62- *Honors & Awards:* Henry Marion Howe Medal, Am Soc Metals, 72. *Mem:* Am Inst Mining, Metall & Petrol Engrs; Am Soc Metals. *Res:* Relationship of metallurgical structure to deformation behavior in ordered systems; nickel base precipitation hardened alloys and ferrous martensites; martensite morphology; high strength low alloy and dual-phase steels. *Mailing Add:* Mat Sci Dept Ford Motor Co PO Box 2053 Dearborn MI 48121

DAVIES, RICHARD O, b Brantford, Ont, July 28, 31; m 54; c 3. MEDICINE, PHARMACOLOGY. *Educ:* Univ Toronto, BSc, 54, MA, 56, PhD(pharmacol), 58, MD, 62. *Prof Exp:* Merck Sharpe & Dome, 76-85; ICI Pharmaceut, 85-89; VPRES, CARDIOVASC CLIN RES & DEVELOP, BRISTOL MYERS SQUIBB, 89- *Concurrent Pos:* Lectr, McGill Univ, 69-76. Concurrent; adj assoc prof med, Thomas Jefferson Univ, Philadelphia, 77-78. *Mem:* Int Soc Hypertension; Can Cardiovasc Soc; Am Fedn Clin Res; Am Soc Clin Pharmacol & Therapeut; Am Col Cardiol. *Res:* Clinical pharmacology in cardiovascular areas, especially angina, beta blockade, hypertension, shock, atherosclerosis and lipids; bioavailability and pharmacokinetics; psychopharmacology. *Mailing Add:* Bristol Myers Squibb Pharmaceut Res Inst Box 4000 Princeton NJ 08543

DAVIES, RICHARD OELBAUM, b New York, NY, Oct 8, 36; m 65; c 1. VETERINARY PHYSIOLOGY. *Educ:* Cornell Univ, DVM, 60; Univ Pa, PhD(physiol), 64. *Prof Exp:* Assoc, 64-66, asst prof, 66-69, assoc prof, 69-79, PROF PHYSIOL, 79-, HEAD, LAB PHARMACOL, SCH VET MED, UNIV PA, 86- *Mem:* AAAS; Am Physiol Soc; Soc Neurosci; Am Soc Vet Physiol & Pharmacol. *Res:* Neurophysiology and respiratory control mechanisms; sleep-disordered breathing; rapid eye movement sleep. *Mailing Add:* Dept Animal Biol 211E Vet Univ Pa Sch Vet Med Philadelphia PA 19104-6046

DAVIES, ROBERT DILLWYN, b Bristol, Eng, Mar 22, 39; m 65; c 2. PHYSICS. *Educ:* Univ Wales, BSc, 62, PhD(physics), 66. *Prof Exp:* Asst lectr physics, Univ Wales, 66-67; res physicist, 67-70, sr res physicist, 70-74, res supvr, 74-82, MGR, EXP STA, E I DUPONT DE NEMOURS & CO, 82- *Res:* Electrical breakdown in gases and vacuum; x-ray physics; surface physics. *Mailing Add:* Exp Sta Lab E I du Pont de Nemours & Co Wilmington DE 19898

DAVIES, ROBERT ERNEST, b Lancashire, Eng, Aug 17, 19; m 61; c 2. BIOCHEMISTRY, ASTRONOMY. *Educ:* Univ Manchester, BS, 41, MS, 42, DS, 52; Univ Sheffield, PhD(biochem), 49; Oxford Univ, MA, 56. *Hon Degrees:* Univ Pa, MA, 71. *Prof Exp:* Demonstr chem, Univ Manchester, 41; asst lectr, Univ Sheffield, 42-45, lectr biochem, 48-54; mem sci staff, Med Res Coun, Univ Sheffield & Oxford Univ, 45-56; prof biochem, Med Sch, 56-62 & Sch Vet Med, 62-70, animal biol, Sch Vet Med & grad group molecular biol, 62-73, BENJAMIN FRANKLIN PROF MOLECULAR BIOL & UNIV PROF, SCH VET MED, UNIV PA, 70- *Concurrent Pos:* Guest prof, Univ Heidelberg, 54; mem fac med, Oxford Univ, 56-59; chair-elect fac senate, Univ Penn, 88, chair, 89-90. *Mem:* Am Soc Biol Chem; Am Biophys Soc; Am Physiol Soc; fel Royal Soc; Brit Biochem Soc; Am Astron Soc. *Res:* Mechanism of secretion of gastric acid; active transports of ions; energy source for contraction of muscle; origin of life. *Mailing Add:* Labs Biochem Dept Animal Biol Univ Pa Sch Vet Med Philadelphia PA 19104-6046

DAVIES, ROBERT MILTON, aquatic ecology, water chemistry, for more information see previous edition

DAVIES, ROGER, b London, Eng, Aug 29, 48; US citizen; m 77; c 2. ATMOSPHERIC RADIATION, CLIMATE THEORY. *Educ:* Victoria Univ, Wellington, NZ, BSc, 70; Univ Wis, PhD(meteorol), 76. *Prof Exp:* Meteorologist, NZ Meteorol Serv, 71-77; scientist, meteorol dept, Univ Wis, 77-79, Space Sci & Eng Ctr, 79-80; from asst prof to assoc prof atmospheric sci, Purdue Univ, 80-87; ASSOC PROF METEOROL, MCGILL UNIV, 87- *Concurrent Pos:* Assoc ed, J Geophys Res, 87- *Mem:* Am Meteorol Soc; Am Geophys Union. *Res:* Atmospheric radiation and climate theory, especially the effect of clouds on three dimensional radiative transfer; applications to remote sensing by satellites; developing computer models of climate change. *Mailing Add:* Dept Meteorol McGill Univ 805 Sherbrooke St W Montreal PQ H3A 2K6 Can

DAVIES, RONALD EDGAR, b Victoria, BC, June 23, 32; Can citizen; m 60; c 6. PHOTOBIOLOGY. *Educ:* Univ BC, BSA, 54, MSA, 56; Agr & Mech Col Tex, PhD(biochem, nutrit), 59. *Prof Exp:* Asst prof poultry sci, Agr & Mech Col Tex, 59-62; asst prof, 62-65, res asst prof biochem, 62-75, ASSOC PROF DERMAT, MED SCH, TEMPLE UNIV, 65-, BIOMED INFO SPECIALIST, 89- *Concurrent Pos:* Consult, Ctr Photobiology, Argus, Inc, 89- *Mem:* Am Soc Photobiol; Am Inst Nutrit; NY Acad Sci; Sigma Xi. *Res:* Cutaneous photobiology and photochemistry; chemical carcinogenesis; photocarcinogenesis; photochemistry of carcinogens. *Mailing Add:* Library Med Sch Temple Univ 3400 N Broad St Philadelphia PA 19140

DAVIES, RONALD WALLACE, b London, Eng, Dec 23, 40; m 63; c 3. ECOLOGY. *Educ:* Univ Col NWales, BSc, 62, Hons, 63, PhD(ecol), 67, DSc, 85; Univ Col SWales, dipl ed, 64. *Prof Exp:* Asst lectr zool, Univ Col NWales, 67-68; Can Int Biol Programme fel ecol, Inst Fisheries, Univ BC, 68-69; from asst prof to assoc prof, 69-80, PROF ZOOL, UNIV CALGARY, 80-, HEAD DEPT, 83- *Mem:* Am Soc Limnol & Oceanog; Can Soc Zool; Ecol Soc Am; Brit Ecol Soc; Sigma Xi; Int Soc Limnol. *Res:* Ecology of freshwater ecosystems, with special emphasis on the invertebrate benthic populations; ecophysiology; ecotoxicology. *Mailing Add:* Dept Biol Sci Univ Calgary Calgary AB T2N 1N4 Can

DAVIES, WARREN LEWIS, b Scranton, Pa, Jan 17, 20; m 46; c 2. MICROBIOLOGY. *Educ:* Pa State Col, BS, 46, MS, 47; Univ Wis, PhD(microbiol), 51. *Prof Exp:* MICROBIOLOGIST, STINE LAB, E I DU PONT DE NEMOURS & CO, 50- *Mem:* Sigma Xi. *Res:* Viruses and bacteria. *Mailing Add:* 1010 Arc Corner Rd Landenberg PA 19350

DAVIES-JONES, ROBERT PETER, b Leicester, Eng, Feb 15, 43; m 66; c 2. GEOPHYSICAL FLUID DYNAMICS. *Educ:* Univ Birmingham, BSc, 64; Univ Colo, PhD(astrogeophys), 69. *Prof Exp:* Fel astrophys, Nat Ctr Atmospheric Res, 69-70; geophysicist, 70-78, METEOROLOGIST, NAT SEVERE STORMS LAB, NAT OCEANIC & ATMOSPHERE ADMIN, 78- *Honors & Awards:* Bronze Medal, US Dept Com, 85; Ed Award, Am Meterol Soc, 86. *Mem:* Am Meteorol Soc. *Res:* Vorticity dynamics of tornadoes and severe thunderstorms and larger-scale weather system. *Mailing Add:* Nat Severe Storms Lab NOAA 1313 Halley Circle Norman OK 73069

DAVIESS, STEVEN NORMAN, b Cedar Rapids, Iowa, Jan 25, 18; m 44; c 2. GEOLOGY. *Educ:* Univ Calif, Los Angeles, BA, 40, MA, 42. *Prof Exp:* From jr geologist to asst geologist, US Geol Surv, 42-46; geologist, Cuban Gulf Oil Co, 46-48, Mozambique Gulf Oil Co, 48-53, staff geologists, Gulf Eastern Co, 56-60, area mgr, Spanish Gulf Oil Co, 60-67; geologic adv, Nuclear Fuels Div, Gulf Oil Corp, 67-70; mgr explor, Gulf Energy & Minerals Co, 70-78; CONSULT GEOLOGIST, 78- *Mem:* Fel Geol Soc London; fel Geol Soc Am; Am Asn Petrol Geologists; Sigma Xi. *Res:* Petroleum geology; economic geology. *Mailing Add:* 3961 S Dexter St Cherry Hills Village CO 80110

DAVIGNON, J PAUL, b New Bedford, Mass, June 21, 37. PHARMACEUTICAL RESEARCH. *Educ:* Univ RI, BS, 60, Pharmacol, 61. *Prof Exp:* Chief, Pharmaceut Resources Br, Nat Cancer Inst, NIH, 64-89; VPRES, PHARMACEUT OPERS, US BIOSCI, 89- *Mailing Add:* US BioSci One Tower Bridge 100 Front St W Conshohocken PA 19428

DAVIGNON, JEAN, b Montreal, Que, July 29, 35; m 61; c 3. MEDICINE. *Educ:* Univ Paris, BA, 53; Univ Montreal, MD, 58; McGill Univ, MSc, 60; FRCPS(C), 63. *Prof Exp:* Resident, Hotel-Dieu Hosp, 60-61; res asst, Dept Physiol, Mayo Clin & Mayo Found, 62-63; res assoc lipid metab, Rockefeller Univ Hosp, 64-67; SR INVESTR & DIR DEPT LIPID METAB & ATHEROSCLEROSIS RES, CLIN RES INST MONTREAL, 67- *Concurrent Pos:* Nat Res Coun Can med res fel, Dept Clin Res, Hotel-Dieu Hosp, Montreal, 58-60; fel internal med, Mayo Clin & Mayo Found, Rochester, Minn, 61-62; Que Ministry Educ scholar, 61-65; Markle scholar acad med, 67; assoc physician, Hotel-Dieu Hosp, 67-70, physician, 70-, head sect vascular med, 72-; asst prof, Dept Med, Univ Montreal, 67-71, assoc prof, 71-77, prof, 77-; asst prof & assoc mem, Dept Exp Med, McGill Univ, 70-; fel coun arteriosclerosis, Am Heart Asn, 68. *Mem:* Am Fedn Clin Res; fel Am Col Physicians; fel Am Col Angiol; Am Heart Asn; NY Acad Sci; fel Am Col Nutrit. *Res:* Peripheral vascular diseases; atherosclerosis; lipid and lipoprotein metabolism in hyperlipidemia. *Mailing Add:* Clin Res Inst Montreal 110 Pine Ave W Montreal PQ H2W 1R7 Can

DAVILA, JULIO C, b Mexico, Dec 1, 21; US citizen; m 48; c 4. CARDIOVASCULAR SURGERY, THORACIC SURGERY. *Educ:* Stanford Univ, BA, 45, MD, 49; Am Bd Surg, dipl, 54; Am Bd Thoracic Surg, dipl, 56. *Prof Exp:* From instr to asst prof surg, Univ Pa, 58-62; clin prof, Sch Med, Temple Univ, 62-65, prof, 65-69; dir cardiopulmonary inst, St Joseph's Hosp, 69-71; chief div thoracic surg, Henry Ford Hosp, 71-76; mem staff, 77-80, CHIEF THORACIC & CARDIOVASC SURGEON, WAUSAU HOSP CTR, WAUSAU WIS, 80- *Concurrent Pos:* Assoc thoracic surgeon, Presby Hosp, St Christopher's Hosp & Fitzgerald Mercy Hosp, 54-61; dir cardiovasc res, Presby Hosp, 57-61, dir res, 61-62; actg chief thoracic surg, Presby Hosp & St Christopher's Hosp, 61-62; fac assoc & mem sem biomat, Columbia Univ, 66-69; clin prof surg, Sch Med, Univ Mich, 72-; consult, Valley Forge Army Hosp, 62-, Philadelphia Vet Hosp, 64-69 & Letterman Gen Hosp, 69-71. *Mem:* Fel Am Col Surg; fel Am Col Cardiol; fel Am Col Chest Physicians; fel Am Col Angiol; Am Asn Thoracic Surg; Soc Thoracic Surgeons. *Res:* Cardiac surgery; circulation physiology research; biomaterials; prostheses for circulation especially artificial heart valves. *Mailing Add:* 534 Alarid St Santa Fe NM 87501

DAVINROY, THOMAS BERNARD, b East St Louis, Ill, Nov 16, 32; m 55; c 3. CIVIL & TRANSPORTATION ENGINEERING. *Educ:* Princeton Univ, BSE, 54, MSE, 60; Univ Calif, Berkeley, DEng(transp eng), 66. *Prof Exp:* Instr mech & graphics, Princeton Univ, 59-61; asst prof, 64-69, ASSOC PROF CIVIL ENG, PA STATE UNIV, 69- *Concurrent Pos:* Soils engr, Moran, Proctor Mueser & Rutledge, NY, 58; eng consult, Tanganyika proj, US Peace Corps, 61; transp engr, James C Buckley Inc, NY, 63; consult, Gannett, Fleming, Corddry & Carpenter, Harrisburg, Pa, 68-69 & Bruinette, Gruger, Stoffberg & Hugo, Pretoria, Repub SAfrica, 70-71. *Mem:* Am Soc Civil Engrs. *Res:* Transportation planning; effects of weather and aircraft characteristics on airport design; air transport planning. *Mailing Add:* Dept Civil Eng Pa State Univ University Park PA 16802

DAVIS, A DOUGLAS, b Mo, Apr 29, 45; m 72; c 2. NUCLEAR PHYSICS, THEORETICAL PHYSICS. *Educ:* Wichita State Univ, BS, 66; Univ Calif, Los Angeles, MS, 68, PhD(physics), 70. *Prof Exp:* Teaching asst physics, Univ Calif, Los Angeles, 66-70, res asst nuclear physics, 68-70; from asst prof to assoc prof, 70-84, PROF PHYSICS, EASTERN ILL UNIV, 84-, CHMN DEPT, 87- *Concurrent Pos:* Govt bureaucrat US Navy, Div Naval Reactors, US Energy Res & Develop Admin, 71-75; vis assoc prof, Univ Ill, 81-82. *Mem:* Am Asn Physics Teachers. *Res:* Computer assisted instruction in physics; scientific writing; nuclear shell structure; nuclear shape; author of textbook on classical mechanics. *Mailing Add:* Dept Physics Eastern Ill Univ Charleston IL 61920

DAVIS, ABRAM, b Cleveland, Ohio, Feb 21, 26; m 52; c 4. SPECTROSCOPY, ANALYTICAL CHEMISTRY. *Educ:* Lake Forest Col, AB, 50; Ill Inst Technol, MS, 52. *Prof Exp:* Res chemist, 51-58, group leader instrumental lab, 58-61, supvr, 61-69, supvr anal, Phys & Instrumental Lab, 69-70, asst sect mgr, 70, SR RES ASSOC, CENT RES, HOOKER CHEM CORP, 70- *Mem:* AAAS; Soc Appl Spectros; Am Chem Soc; Coblentz Soc. *Res:* Instrumental analysis of organic compounds; elucidation of structure and development of analytical methods of product control. *Mailing Add:* 1588 Red Jacket Rd Grand Island NY 14072-1799

DAVIS, ALAN, b Chester, Eng, June 18, 32; m 63; c 2. GEOLOGY, PETROLOGY. *Educ:* Univ London, BSc, Imp Col Sci Technol, ARCS, 56; Pa State Univ, MS, 61; Univ Durham, PhD(geol), 65. *Prof Exp:* Geologist mineral explor, John Taylor & Sons, London, 56; prospecting officer coal explor, Opencast Exec, Nat Coal Bd, 57, scientist coal survey, 58, sr res assoc coal res, Dept of Geol, Univ Newcastle, Eng, 61-65; geologist, res & sr geologist, coal petrol, Geol Surv Queensland, 65-73; assoc prof geol, 73-80, prof geol & asst dir, Coal Res Sect, 80-86, DIR, ENERGY & FUELS RES CTR, PA STATE UNIV, 87- *Mem:* Int Comt Coal Petrol; Geol Soc Am; Am Soc Testing & Mat; Soc Org Petrol. *Res:* Coal and organic petrology; coalification processes; coal utilization; coal characterization. *Mailing Add:* Pa State Univ 513 Deike Bldg Univ Park PA 16802

DAVIS, ALAN, b Gosport, Eng, June 10, 55. CLINICAL RESEARCH. *Educ:* Univ Manchester, Eng, BSc, 76; Southampton Univ, Eng, PhD(neurochem), 79; York Univ, Can, MBA, 90. *Prof Exp:* Res asst pharmacol, Dept Physiol, Southampton Univ, 79; fel, Dept Pharmacol, Toronto Univ, 80-81, asst prof, dept pharmacol & psychiat, 81-83; asst dir clin res, Boehringer Ingelheim Can Ltd, 83-85; DIR, CLIN RES & DEVELOP, SMITH KLINE BEECHAM, 85- *Mem:* Europ Neurosci Asn; Soc Neurosci; Brit Pharmacol Soc; Can Soc Clin Pharmacol; Soc Clin Trials. *Res:* Respirology; hypertension; gastroenterology; CNS. *Mailing Add:* 2030 Bristol Circle Oakville ON L6H 5V2 Can

DAVIS, ALAN LYNN, b Salt Lake City, Utah, Nov 17, 46; m 73. INTEGRATED CIRCUIT DESIGN. *Educ:* Mass Inst Technol, BS, 69; Univ Utah, PhD(comput sci), 72. *Prof Exp:* Asst prof comput sci, Univ Waterloo, 72-73; consult, Burroughs IRC, 73-77; asst prof, Univ Utah, 77-80; consult, Burroughs Corp, 77-82 & Gen Res Corp, 78-80; ASSOC PROF COMPUT SCI, UNIV UTAH, 80- *Concurrent Pos:* Exchange scholar, Acad Sci, USSR, 76-77; vis prof, Novosibirsk State Univ, USSR, 79; vis scholar, Gesellschaft Math & Datenverarb, 80; vis prof, Tech, Haifa, Israel, 80. *Res:* Data-driven computer system which dynamically exploits concurrency in an extensible multiple resource environment. *Mailing Add:* HP Labs 31 PO Box 10490 Palo Alto CA 94303

DAVIS, ALEXANDER COCHRAN, b Ottawa, Ont, Can, Oct 6, 20; nat US; m 43; c 2. ECONOMIC ENTOMOLOGY. *Educ:* Univ Toronto, BSA, 42; Cornell Univ, PhD(entom), 50. *Prof Exp:* Agr scientist, Div Entom, Can Dept Agr, Ottawa, 46; asst entom, Cornell Univ, 47-50; asst prof, 50-54, assoc prof, 54-76, PROF ENTOM, NY STATE AGR EXP STA, GENEVA, 76-, ASSOC DIR EXP STA & ASST DIR RES, 77- *Mem:* Entom Soc Am. *Res:* Vegetable insect control. *Mailing Add:* NY State Exp Sta Geneva NY 14456

DAVIS, ALFRED, JR, b Johnson City, Tex, Feb 10, 19; m 41; c 2. ENGINEERING. *Educ:* Univ Tex, BS, 41. *Prof Exp:* Design engr, Westinghouse Elec Co, Md, 41-45; res engr, Defense Res Lab, 46-62; proj engr, Prod Eng Dept, Tracor, Inc, 62-67, sr engr, 67-82; RETIRED. *Mem:* Sr mem Inst Elec & Electronics Engrs; Nat Soc Prof Engrs. *Res:* Design of specialized microwave and data recording systems; production engineering of frequency standards and countermeasures systems. *Mailing Add:* 6712 Vine Austin TX 78757

DAVIS, ALVIE LEE, b Richardson, Tex, Jan 22, 31; m 61; c 3. BIOCHEMISTRY. *Educ:* Abilene Christian Col, BS, 55; Univ Tex, PhD(biochem), 60. *Prof Exp:* From asst prof to assoc prof, 59-68, PROF CHEM, ABILENE CHRISTIAN UNIV, 68- *Concurrent Pos:* Robert A Welch Found res grant, 62- *Mem:* Am Chem Soc. *Res:* Metabolic inhibitors; correlation of chemical structure and biological activity; nitrogen heterocycles; cyclic benzohydroxamic acids. *Mailing Add:* Dept Chem Abilene Christian Col Abilene TX 79601

DAVIS, ALVIN HERBERT, b Buffalo, NY, Jan 26, 29; m 56; c 2. GENERAL PHYSICS, THEORETICAL PHYSICS. *Educ:* Univ Buffalo, BA, 49, MA, 51; Yale Univ, PhD(physics), 55. *Prof Exp:* Physicist, Livermore Radiation Lab, Univ Calif, 55-57, Naval Res Lab, 57-58, Theoret Div, Goddard Space Flight Ctr, Md, 59-66 & Environ Res Corp, Va, 66-69; physicist & qual assurance eng, Los Alamos Sci Lab, 69-90; CONSULT, 90- *Mem:* AAAS; Am Geophys Union; Am Phys Soc; Sigma Xi; Am Soc Qual Control. *Res:* Development of quality assurance programs for technical activities and technology assessment; geophysics; two-phase flow and fracture in porous media; astrophysics. *Mailing Add:* 52 Osito Los Alamos NM 87544

DAVIS, ANDREW MORGAN, b Port Jefferson, NY, May 18, 50; m 81. COSMOCHEMISTRY. *Educ:* Grinnell Col, BA, 71; Yale Univ, MPh, 73, PhD(geochem), 77. *Prof Exp:* Res assoc, dept geophys sci, 76-78, anal chemist, James Franck Inst, 78-89, SR RES ASSOC, ENRICO FERMI INST, UNIV CHICAGO, 89- *Concurrent Pos:* Res assoc, Field Mus Natural Hist, Chicago, 85- *Mem:* AAAS; Am Geophys Union; Geochem Soc; Meteoritical Soc. *Res:* Early chemical history of the solar system; trace elements in meteorites; isotopic composition of meteorites. *Mailing Add:* Enrico Fermi Inst Univ Chicago 5640 S Ellis Ave Chicago IL 60637

DAVIS, ANTHONY MICHAEL JOHN, b London, Eng, Dec 5, 39; m 65; c 2. VISCOUS FLOWS, WAVES. *Educ:* Univ Cambridge, BA, 64, PhD(appl math), 64; Univ London, DSc, 77. *Prof Exp:* Lectr/reader math, Univ Col, London, 65-84; PROF MATH, UNIV ALA, 85- *Concurrent Pos:* Vis assoc prof, math dept, Univ Denver, 72-73; vis prof, math & oceanog depts, Univ Hawaii, 85; vis res prof, math dept, Univ Toronto, 78, 79, 81, 84-90, Inst Oceanog Univ, BC, 78; vis scientist, chem eng dept, Univ Rochester, 80 & Mass Inst Technol, 82; vis res fel, appl math dept, Univ New S Wales, Australia, 85. *Mem:* Soc Indust & Appl Math; Am Phys Soc; Can Appl Math Soc. *Res:* Creeping flows and biomechanical modeling-hollow boundaries and stenotic tubes; water waves; inertial effects on singularities near solid boundaries; non-Newtonian pipe flows; continental shelf wave scattering by irregular coastlines. *Mailing Add:* Math Dept Univ Ala Tuscaloosa AL 35487-0350

DAVIS, BENJAMIN HAROLD, b Lafayette, Ind, Jan 25, 05; m 33; c 2. PLANT PATHOLOGY. *Educ:* Wabash Col, AB, 28; Cornell Univ, PhD(plant path), 34. *Prof Exp:* Asst plant path, Cornell Univ, 28-29, instr, 29-34; instr sci, Va State Teachers Col, Fredericksburg, 34-35; instr bot, Ohio State Univ, 35-39; from assoc prof to prof, 39-71, chmn dept, 56-71, EMER PROF PLANT PATH, RUTGERS UNIV, 71- *Mem:* Am Phytopath Soc; Mycol Soc Am. *Res:* Diseases of vegetables. *Mailing Add:* 23033 Westchester Blvd No C201 Port Charlotte FL 33980-8460

DAVIS, BERNARD BOOTH, b Parkersburg, WVa, Sept 12, 32; m 56; c 4. GERIATRICS. *Educ:* Wooster Col, BA, 57; Univ Pittsburgh, MD, 61. *Prof Exp:* Asst prof med, Univ Pittsburgh, 67-70, assoc prof, 70-73, prof, 73-76; PROF MED, ST LOUIS UNIV, 76-; CHIEF MED, VET ADMIN MED CTR, ST LOUIS, 76-, DIR GERIAT, 76- *Concurrent Pos:* Chief of staff, Vet Admin, Pittsburgh, 71-76; dir, Renal Div, St Louis Univ, 76-77. *Mem:* Am Col Physics; AAAS; Am Soc Clin Invest; Am Physiol Soc; Am Soc Nephrol. *Res:* Renal metabolism of hormones, drugs and renobiotics. *Mailing Add:* Vet Admin Med Ctr St Louis MO 63125

DAVIS, BERNARD DAVID, b Franklin, Mass, Jan 7, 16; m 55; c 3. MICROBIOLOGY, MEDICINE. *Educ:* Harvard Univ, AB, 36, MD, 40. *Prof Exp:* Officer, USPHS, 42-54; prof pharmacol & chmn dept, Med Sch, New York Univ, 54-57; prof bact & chmn dept, 57-68, Adele Lehman prof bact & immunol, 63-68, dir, bact physiol unit, 68-84, Adele Lehman prof bact physiol, 68-84, EMER ADELE LEHMAN PROF BACT PHYSIOL, HARVARD UNIV, 84- *Concurrent Pos:* Vis prof, Weizmann Inst, Tel Aviv Univ, Nat Taiwan Univ; Fogarty scholar, NIH, 88 & 89. *Honors & Awards:* Waksman Award, Nat Acad Sci, 89; Hoechst-Roussel Award, Am Soc Microbiol, 89. *Mem:* Nat Acad Sci; Am Acad Arts & Sci (vpres, 77-79); Am Soc Microbiol; Am Soc Biol Chemists. *Res:* Antibiotic action; protein synthesis; protein secretion. *Mailing Add:* 23 Clairemont Rd Belmont MA 02178

DAVIS, BERNARD ERIC, b Milwaukee, Wis, Aug 29, 37; m 64; c 2. METALLURGY, CHEMICAL ENGINEERING. *Educ:* Ore State Univ, BS, 60, MS, 65, PhD(metall eng), 70. *Prof Exp:* Process engr, Org Chem Dept, E I du Pont de Nemours & Co, Inc, 60-61; res metallurgist, Teledyne Wah Chang Albany Corp, Teledyne, Inc, 63-67; tech adv qual control, Naval Nuclear Fuel Div, Babcock & Wilcox Co, 70-80; MEM STAFF, LYNCHBURG RES CTR, 80- *Mem:* Am Inst Mining, Metall & Petrol Engrs; Am Soc Metals; Am Inst Chem Engrs. *Res:* Diffusion in metals; liquid extraction of metals. *Mailing Add:* 201 Collington Dr Lynchburg VA 24502

DAVIS, BILL DAVID, b Junction City, Kans, July 22, 37; m 62; c 2. PLANT PHYSIOLOGY. *Educ:* Kans State Teachers Col, BS, 59, MS, 61; Purdue Univ, PhD(biol), 65. *Prof Exp:* Teacher, Manhattan Jr High Sch, Kans, 60-62; asst prof, 65-69, dept chmn, 78-81, ASSOC PROF BIOL SCI, RUTGERS UNIV, 69- *Mem:* AAAS; Am Soc Plant Physiol; Bot Soc Am. *Res:* Regulation of hydrolase activity in plants, especially alpha- and beta-amylase and acid phosphatase; physiological interactions during seed germinations; application of learning theory to design of undergraduate courses. *Mailing Add:* Dept Biol Sci Rutgers Univ New Brunswick NJ 08903

DAVIS, BILLY J, b Hobart, Okla, Oct 27, 32; m 53; c 4. ICHTHYOLOGY, WATER POLLUTION. *Educ:* Southwestern State Col, Okla, BS, 54, MS, 57; Okla State Univ, PhD(ichthyol), 66. *Prof Exp:* Pub sch teacher, Okla, 54-63; from asst prof to assoc prof zool, 66-73, PROF ZOOL, LA TECH UNIV, 73- *Mem:* Soc Study Amphibians & Reptiles; Am Soc Ichthyologists & Herpetologists; Herpetologists League; Asn Southwestern Naturalists; Asn Southeastern Biologists. *Res:* Ecological and taxonomic studies of North American freshwater fishes and herptiles, particularly the gross morphology of the cyprinid fish brain in relation to behavior; water pollution biology. *Mailing Add:* Dept Zool La Tech Univ Ruston LA 71272

DAVIS, BOB, b Rockdale, Tex, Oct 26, 38; m 60; c 2. PRICE THEORY, MARKETING. *Educ:* Tex A&M Univ, BS, 61, MS, 62; NC State Univ, PhD(econ), 70. *Prof Exp:* Teaching asst agr bus, Dept Agr Econ, Tex A&M Univ, 61-62; agr economist, Econ Res Serv, US Dept Agr, 62-71; assoc prof bus admin, Col Bus, E Tex State Univ, 71-73, Univ Houston, Victoria, 73-76 & Dept Agr Econ, Tex Tech Univ, 77-88; PROF AGR BUS & CHMN, DEPT AGR, SOUTHWEST TEX STATE UNIV, 88- *Concurrent Pos:* Consult, USAID, Guatemala, 84 & 86. *Mem:* Am Agr Econ Asn; Nat Asn Cols & Teachers Agr; Southern Econ Asn. *Res:* Vegetable marketing, spatial equilibrium, interregional competition and demand studies; economics of wind erosion and soil conservation; international economic development and training; agricultural finance. *Mailing Add:* Dept Agr Southwest Tex State Univ San Marcos TX 78666

DAVIS, BRIAN CLIFTON, b St Louis, Mo, Jan 5, 42; m 66; c 4. PHYSICAL ORGANIC CHEMISTRY. *Educ:* Calif State Univ, Northridge, BS, 67; Univ Wash, PhD(org chem), 72. *Prof Exp:* Res assoc biochem, Chem Dept, Univ Wash, 72-74; res chemist org chem, 74-81, sr res chemist, 81-86, PROD MGR FUEL PLANNING, SUN OIL CO, 86- *Concurrent Pos:* Fel, Nat Sci Found, 68. *Mem:* Am Chem Soc; Soc Automotive Engrs. *Res:* Synthesis of commercially useful products from raw materials available from petroleum refinery streams; development and evaluation of alternative fuels. *Mailing Add:* Sun Refining & Marketing Ten Penn Ctr 1801 Market St Philadelphia PA 19103-1699

DAVIS, BRIAN K, physiology of reproduction, for more information see previous edition

DAVIS, BRIAN KENT, b Laramie, Wyo, Dec 2, 39; m 62; c 3. MUTAGENESIS, RISK ASSESSMENT. *Educ:* Univ Wis-Madison, BA, 62, MA, 63; Univ Wash, PhD(genetics), 70. *Prof Exp:* Sec sch teacher sci, US Peace Corps, 64-65; res fel genetics, Univ Calif, San Diego, 70-72, asst prof, 72; asst prof genetics, Va Polytech Inst & State Univ, 73-80; asst prof, Col Med, King Faisal Univ, Saudi Arabia, 80-81; res assoc, Allied Corp, 81-86; toxicologist, Calif Dept Food & Agr, 86-88, Calif Dept Health Serv, 88-91; TOXICOLOGIST, DEPT TOXIC SUBSTANCES CONTROL, CALIF ENVIRON PROTECTION AGENCY, 91- *Concurrent Pos:* Fel Am Cancer Soc, 70-72. *Mem:* Environmental Mutagen Soc. *Res:* Regulatory toxicology and risk assessment. *Mailing Add:* 4945 Puma Way Carmichael CA 95608

DAVIS, BRIANT LEROY, b Brigham City, Utah, Nov 18, 36; m 57; c 4. ATMOSPHERIC CHEMISTRY & PHYSICS. *Educ:* Brigham Young Univ, BS, 58, MS, 59; Univ Calif, Los Angeles, PhD(geol), 64. *Prof Exp:* Asst res geophysicist, Inst Geophys & Planetary Physics, Univ Calif, Los Angeles, 61-62; from asst prof to assoc prof geol & geol eng, 63-70, asst prof geophys, 70-74, RES PROF GEOPHYS, SDAK SCH MINES & TECHNOL, 75-, RES GEOPHYSICIST, INST ATMOSPHERIC SCI, 66-, SR SCIENTIST & HEAD CLOUD PHYSICS GROUP, INST, 77- *Concurrent Pos:* Dean, Grad Div, SDak Sch Mines & Technol, 87- *Mem:* Crystallog Asn; Am Meteorol Soc; Sigma Xi; Am Geophys Union. *Res:* Air pollution chemistry & physics; nucleation processes; asbestos analysis; x-ray diffraction quantitive analysis methods. *Mailing Add:* Grad Div SDak Sch of Mines & Technol Rapid City SD 57701

DAVIS, BRUCE ALLAN, b Saskatoon, Sask, Aug 17, 41. ANALYTICAL BIOCHEMISTRY. *Educ:* Univ Sask, BAS, 62; PhD(chem), 66. *Prof Exp:* Res fel org chem, Univ Munich, 66-68; fel mass spectrometry, Univ Cambridge, 68-70; lectr & res fel org chem, Univ Sask, 70-72; lab scientist biol psychiat, Psychiat Res Div, Dept Health, Sask, 72-87; SR RES SCIENTIST, NEUROPSYCHIAT UNIT, DEPT PSYCHIAT, UNIV SASK, 87- *Concurrent Pos:* Mem staff, Dept Org Chem, Univ Sask Hosp. *Mem:* Can Biochem Soc. *Res:* Development and application of analytical methods for amines and their acid metabolites in biological materials, particularly in the brain and biological fluids, with reference to psychiatric and neurological diseases. *Mailing Add:* Cancer & Med Res Bldg A138 Univ Sask Neuropsychiat Unit Saskatoon SK F7N 0W0 Can

DAVIS, BRUCE HEWAT, b May 4, 51; m 86; c 2. HEMATOPATHOLOGY, IMMUNOLOGY. *Educ:* Univ Conn, MD, 77. *Prof Exp:* ASST PROF PATH & DIR HEMETOPATH, DARTMOUTH HITCHCOCK MED CTR, 86- *Honors & Awards:* Upjohn Med Res Award. *Mem:* AAP; Am Soc Cell Bio; Am Soc Clin Pathologists; CAP; Soc Anal Cytology; Int Acad Path. *Res:* Neutrophil activation; development of flow cytometric assays to diagnostic pathology, neutrophil activation and cell proliferation. *Mailing Add:* Dartmouth Hitchcock Med Ctr Hanover NH 03756

DAVIS, BRUCE W, b Glendale, Calif, July 19, 37; m 64; c 4. CHEMICAL ENGINEERING, PETROLEUM ENGINEERING. *Educ:* Univ Southern Calif, BS, 60, MS, 62; Univ Calif, Riverside, PhD(phys chem), 64. *Prof Exp:* Teaching asst chem, Univ Southern Calif, 60; Army Res Off Durham assoc critical phenomena, Univ NC, 64-66; asst prof chem, Ga Inst Technol, 66-72; vis asst prof, Cornell Univ, 72-73; sr res chemist, 73-80, SR RES ASSOC, CHEVRON OIL FIELD RES CO, 80- *Concurrent Pos:* Petrol Res Fund grants, 68-71; Army Res Off Durham grants, 68-71; lectr, Dept Petrol Eng, Univ Southern Calif, 88-90; adj assoc prof, Univ Southern Calif, 90- *Mem:* Am Chem Soc; Sigma Xi; Soc Petrol Engrs. *Res:* Thermodynamics of the gas, liquid and adsorbed states; properties of fluids in porous media; oil recovery research; properties of emulsions and microemulsions; thermal recovery of heavy oils. *Mailing Add:* 1920 Las Lanas Lane Fullerton CA 92633

DAVIS, BRUCE WILSON, b Owen Co, Ind, Dec 22, 21; m 48; c 3. APPLIED MATHEMATICS. *Educ:* Ind Univ, AB, 48; Purdue Univ, MS, 55. *Prof Exp:* Teacher high sch, Ind, 48-50; mathematician & math consult, Ord Plant, US Navy, 50-55; sr res engr, Allison Div, Gen Motors Corp, Ind, 55-56; asst head math div, Avionics Facil, US Navy, 56-58; engr, head math & comput facil & chief adv studies, Defense Systs & Allison Div, Gen Motors Corp, 58-63; systs anal supvr & proj dir, 63-71, mgr space systs prog off, 72-79, SECT MGR, SPACE SYSTS & APPLN SECT, 79-, DEPT MGR, DEFENSE & SPACE SYSTS DEPT, COLUMBUS LABS, BATTELLE MEM INST, 79- *Mem:* Am Inst Aeronaut & Astronaut; Sigma Xi. *Res:* Control systems; guidance; synthesis of electrical circuitry; operations research; systems analysis; space launch system analyses. *Mailing Add:* 683 Sternberger Pl Columbus OH 43214

DAVIS, BRYAN TERENCE, b Hawkhurst, Eng, Dec 19, 35; m 61; c 3. LUBRICANT & FUEL ADDITIVES. *Educ:* Oxford Univ, BA, 60, BSc, 61, MA, 77. *Prof Exp:* Res chemist, Castrol Ltd, 60-67, sr res chemist, 67-69; sr res chemist, Edwin Cooper Ltd, 69-74; chief chemist res, Ethyl Corp, 75-77, res supvr, Petroleum Chem Res, 77-83, supvr info & library serv/recruiting, 84-89, MGR RECRUITING, ETHYL CORP, 89- *Concurrent Pos:* Mem, Col Placement Coun. *Mem:* Am Chem Soc. *Res:* Synthesis evaluation and commercialization of petroleum additives including improved antioxidants; rust inhibitors, antiwear and dispersant additives for lubricants; additives for both lubricants and diesel fuels to prevent malfunction of equipment due to gelation and crystallization at low temperatures. *Mailing Add:* PO Box 14799 Baton Rouge LA 70898

DAVIS, BURL EDWARD, b Doniphan, Mo, Aug 30, 35; m 27; c 4. IN-SITU PROCESSES, SYNTHETIC FUEL PROCESS CHEMISTRY. *Educ:* Southern Ill Univ, BA, 64; Univ Wyo, MS, 66. *Prof Exp:* Instr anal chem, Natural Resources Inst, Univ Wyo, 63-66; res chemist, Koppers Co, 66-68; sr proj chemist, Gulf Oil Co, 68-85; EXEC VPRES, ENERGY INT, 85- *Mem:* Am Chem Soc; Sigma Xi. *Res:* Synthetic fuel process chemistry with an emphasis on underground coal gasification; development of the multi-disciplinary technology related to in-situ processes; commercialization of alternate energy technologies. *Mailing Add:* Energy Int Corp 135 William Pitway Pittsburgh PA 15235

DAVIS, BURNS, b Fulton, Ky, Mar 15, 31; m 54; c 2. POLYMER CHEMISTRY, ORGANIC CHEMISTRY. *Educ:* Murray State Col, BS, 53, MA, 59; Univ Louisville, PhD(polymer chem), 62. *Prof Exp:* Teacher high sch, Ky, 56-58; from res chemist to sr res chemist, 62-71, RES ASSOC, TENN EASTMAN CO, 71- *Mem:* Am Chem Soc. *Res:* Condensation and vinyl polymers. *Mailing Add:* 2019 Bruce St Kingsport TN 37664

DAVIS, BURTRON H, b Points, WVa, Dec 21, 34; m 66; c 3. PHYSICAL CHEMISTRY. *Educ:* WVa Univ, BS, 58; St Joseph's Col, Pa, MS, 62; Univ Fla, PhD(chem), 65. *Prof Exp:* Analyst, Atlantic Refining Co, Pa, 59-62; res assoc, Johns Hopkins Univ, 65-66; sr res chemist, Res & Develop Corp, Mobil Oil Corp, NJ, 66-70; assoc prof chem, Potomac State Col, WVa Univ, 70-77; sr chemist, Inst Mining & Minerals Res, 77-84, ADJ PROF, METALL ENG & MAT SCI DEPT, UNIV KY, 78-; ASSOC DIR, CTR APPL ENERGY RES, 84- *Mem:* Am Chem Soc; Catalysis Soc; Mat Res Soc; Am Ceramic Soc. *Res:* Heterogeneous catalysis; coal liquefaction. *Mailing Add:* 113 Lakeview Dr Georgetown KY 40324

DAVIS, CARL F, b Milo, Maine, May 12, 19; m 45; c 1. ELECTRICAL ENGINEERING, MATHEMATICS. *Educ:* Univ Maine, BA, 42; US Air Force Inst Technol, BS, 55; Univ Ill, MS, 56, PhD(elec eng), 60. *Prof Exp:* Commun & electronics staff officer, USAF, 48-60, assoc prof elec eng, USAF Acad, 60-68, chief electronics div, Air Force Weapons Lab, 68-76; sr assoc prin engr, Govt Aerospace Systs Div, Harris Corp, 76-85; RETIRED. *Mem:* Sigma Xi. *Res:* Nuclear radiation effects on weapon systems; transient radiation effects on electronics; electromagnetic pulse effects on systems; research and exploratory development of arming and fuzing systems for nuclear re-entry vehicles; physical security of high value facilities. *Mailing Add:* PO Box 3553 Indialantic FL 32903

DAVIS, CARL GEORGE, b St Louis, Mo, Nov 30, 37; m 67; c 3. SCIENCE EDUCATION, COMPUTER SCIENCE. *Educ:* Ga Inst Technol, BS, 61; Univ Ala, Tuscaloosa, MS, 66, MS & PhD(eng mech), 72. *Prof Exp:* Instr, Ord Guided Missile Sch, US Army, 61-64, aerospace engr, Missile Command, 64-69, sr proj engr, Ballistic Missile Defense Advan Technol Ctr, 72-81, dir data processing, 81-85, prog element dir, Strategic Defense Command, 85-86, dir, Battle Mgt Div, 86; CHMN, COMPUTER SCI DEPT, UNIV ALA, HUNTSVILLE, 86- *Honors & Awards:* Outstanding Contrib Award, Inst Elec & Electronic Engrs, 90. *Mem:* Fel Inst Elec & Electronic Engrs; Inst Elec & Electronic Engrs Computer Soc. *Res:* Techniques and tools for the development of large, complex software systems; requirements and design approaches for real-time software. *Mailing Add:* Computer Sci Dept Univ Ala Huntsville AL 35899

DAVIS, CARL LEE, b Dunnville, Ky, Sept 20, 24; m 46; c 5. NUTRITION, BIOCHEMISTRY. *Educ:* Western Ky State Col, BS, 52; Univ Ky, MS, 54; Univ Ill, PhD(dairy sci), 59. *Prof Exp:* Instr dairy sci, Western Ky State Col, 52-53; res assoc, Univ Ky, 53-54; res assoc, 55-59, asst prof, 60-69, PROF NUTRIT, UNIV ILL, URBANA-CHAMPAIGN, 69- *Concurrent Pos:* Am Feed Mfrs Award for Dairy Cattle Res, Am Dairy Sci Asn, 71- *Mem:* Am Dairy Sci Asn; Am Inst Nutrit. *Res:* Nutrition of ruminant animals, especially as related to milk production. *Mailing Add:* Dept Sci Canville Area Community Col 2000 E Main St Danville IL 61832

DAVIS, CECIL GILBERT, b Los Angeles, Calif, Feb 20, 25; m 57; c 2. THEORETICAL PHYSICS, ASTROPHYSICS. *Educ:* Occidental Col, BA, 50; Univ Calif, Los Angeles, MA, 51, PhD(physics), 57. *Prof Exp:* Staff mem physics, Los Alamos Sci Lab, 57-60, Physics Sect, Convar, 60-61 & Theoret Physics Sect, Gen Atomic, 61-65; STAFF MEM PHYSICS, LOS ALAMOS SCI LAB, 65- *Mem:* Am Astron Soc; Int Astron Union. *Res:* Variable star research; atmospheric sciences. *Mailing Add:* Los Alamos Sci Lab MS D406 PO Box 1663 Los Alamos NM 87545

DAVIS, CHANDLER, b Ithaca, NY, Aug 12, 26; m 48; c 3. OPERATORS & MATRICES. *Educ:* Harvard Univ, BS, 45, MA, 47, PhD(math), 50. *Prof Exp:* Teaching fel math, Harvard Univ, 47-48; instr, Univ Mich, 50-54; lectr, Columbia Univ, 55-57; mem, Inst for Advan Study, 57-58; assoc ed, Math Reviews, 58-62; assoc prof, 62-65, PROF MATH, UNIV TORONTO, 65- *Concurrent Pos:* Ed-in-chief, Math Intelligencer, 91- *Mem:* Soc Indust & Appl Math; Am Math Soc; Can Math Soc (vpres, 91-); Math Asn Am. *Res:* Estimates of spectral resolutions; inequalities for operators. *Mailing Add:* Math Dept Univ Toronto Toronto ON M5S 1A1

DAVIS, CHARLES (CARROLL), b Azusa, Calif, Nov 24, 11; m 36; c 2. LIMNOLOGY, INVERTEBRATE ZOOLOGY. *Educ:* Oberlin Col, AB, 33; Univ Wash, MS, 35, PhD(zool), 40. *Prof Exp:* Asst, Scripps Inst, Univ Calif, 38-40; instr sci, Nat Training Sch, Mo, 40-42; biologist, State Dept Res & Educ, Md, 42-43; chemist, US Navy Powder Factory, 43-44; instr biol chem, Jacksonville Jr Col, 44-47; asst prof zool, Univ Miami, 47-48; from asst prof to prof biol, Case Western Reserve Univ, 48-68; prof, 68-77, EMER PROF BIOL, MEM UNIV NFLD, 84- *Concurrent Pos:* Instr, San Diego Eve Jr Col, 39; Gjeste prof, Univ Tromso, 75-76; vis prof, Addis Ababa Univ, 86-87. *Mem:* AAAS; Am Soc Limnol & Oceanog; Ecol Soc Am; Int Asn Limnol; Marine Biol Asn UK; Can Soc Zoologists; Plankton Soc Japan. *Res:* Crab fishery, marine and fresh-water plankton; pelagic copepoda of the Northeast Pacific Ocean; pollution; Lake Erie ecology; hatching mechanisms; invertebrates; Newfoundland freshwater and marine zooplankton. *Mailing Add:* Dept Biol Mem Univ Nfld St John's NF A1B 3X9 Can

DAVIS, CHARLES A, b Beaumont, Tex, Dec 15, 33; m 56; c 2. WILDLIFE ECOLOGY. *Educ:* Tex A&M Univ, BS, 56, MS, 58; Okla State Univ, PhD(zool), 64. *Prof Exp:* Asst prof biol, Ark State Col, 63-66; assoc prof, Southwestern State Col, 66-67; from asst prof to assoc prof, 67-78, PROF WILDLIFE SCI, DEPT FISH & WILDLIFE SCI, NMEX STATE UNIV, 78-, HEAD DEPT, 81- *Mem:* Wildlife Soc; Cooper Ornith Soc. *Res:* Ecology birds. *Mailing Add:* Dept of Fish & Wildlife Sci NMex State Univ Main Campus Las Cruces NM 88003-0003

DAVIS, CHARLES ALAN, b Holland, Mich, March 31, 50. NUCLEAR STRUCTURE. *Educ:* Transylvania Univ, BA, 72; Univ Wis, PhD(physics), 81. *Prof Exp:* Res assoc, Triumf Group, Univ Alta, 81-84; RES ASSOC, TRIUMF GROUP, UNIV MAN, 84- *Mem:* Am Phys Soc. *Res:* Nuclear spectroscopy; intermediate energy proton induced reactions; scattering and fundamental symmetries in N plus N scattering. *Mailing Add:* Triumf 4004 Wesbrook Mall Vancouver BC V6T 2A3 Can

DAVIS, CHARLES ALFRED, b Marion, NC, Mar 6, 39; m 62. APPLIED MATHEMATICS. *Educ:* NC State Univ, BS, 61, MApMath, 63, PhD(appl math), 68. *Prof Exp:* Instr math, NC State Univ, 65-67; from asst prof to assoc prof, 67-79, chmn dept math, 67-80, asst dean & registr, 80-89, PROF MATH, MEREDITH COL, 79- *Mem:* Math Asn Am. *Res:* Mathematics as applied to heat transfer with particular application of integral transforms. *Mailing Add:* 1422 Ridge Rd Raleigh NC 27607

DAVIS, CHARLES EDWARD, DIAGNOSTIC MICROBIOLOGY. *Educ:* Univ Tex, MD, 58. *Prof Exp:* PROF PATH, SCH MED, UNIV CALIF, SAN DIEGO, 79-, DIR MICROBIOL, MED CTR, 84- *Res:* Cell biology of African trypanosomes. *Mailing Add:* Dept Path H-811F Med Ctr Univ Calif 225 Dickinson St San Diego CA 92103

DAVIS, CHARLES FREEMAN, JR, b Chicago, Ill, Aug 1, 25; m 56; c 4. PHYSICS. *Educ:* Northwestern Univ, BS, 48, MS, 49; Mass Inst Technol, PhD(physics), 54. *Prof Exp:* Opers analyst, Opers Res Off, Johns Hopkins Univ, 54-56; staff mem res lab electronics, Mass Inst Technol, 56-58; head microwave devices sect, Device Res Dept, Tex Instruments, Inc, 58-60, br mgr microwave & photo sensors, Diode Dept, 60-63; eng mgr microwave diode dept, Sylvania Elec Prod, Inc, Gen Tel & Electronics Corp, Mass, 63-67; prin res scientist, Res Div, Raytheon Corp, 67-69, prin engr, Missile Systs Div, 69-70, independent consult, 70-74; mem staff, Physics Dept, Mass Inst Technol, 74-81; PRIN ENG, TEXTRON DEFENSE SYSTS, WILMINGTON, MASS, 81- *Mem:* Am Chem Soc; Am Phys Soc; Inst Elec & Electronics Engrs; AAAS; Sigma Xi. *Res:* Nuclear hardening of semiconductors & microwave physics; masers; low temperature techniques; microwave and photosensitive semiconductor devices; lasers. *Mailing Add:* 195 Deacon Haynes Rd Concord MA 01742-4711

DAVIS, CHARLES HOMER, b Glendale, Ariz, Feb 13, 12; m 37; c 2. AGRONOMY. *Educ:* Univ Ariz, BS, 35; Iowa State Col, MS, 36, PhD(plant physiol, crops), 39. *Prof Exp:* Asst agronomist, Exp Sta, Univ Ariz, 37-43; assoc agronomist, Guayule Res Proj, Bur Plant Indust, USDA, 43-45; agronomist, Bur Reclamation, US Dept Interior, 45-52; fieldman, Ariz Fertilizer Inc, 52-62, agronomist, 62-69; chief agronomist, Farm Builders Div, Am Biocult Inc, 69-71; CONSULT AGRONOMIST, 71- *Mem:* Am Soc Agron; Soil Sci Soc Am. *Res:* Water and growth relations of plants; use of water by plants and by irrigation projects in relation to yields; chemical and cultural weed control. *Mailing Add:* 4415 N 31st Dr Phoenix AZ 85017

DAVIS, CHARLES MITCHELL, JR, b Washington, DC, July 2, 25; m; c 4. UNDERWATER ACOUSTICS. *Educ:* Cath Univ Am, BA, 51, MS, 54, PhD, 62. *Prof Exp:* Gen physicist, US Naval Ord Labs, 51-62; from asst prof to prof physics, Am Univ, 62-70; supv res physicist phys acoust, Naval Res Lab, 70-78; chief scientist, Dynamic Systs Inc, 78-82; VPRES, OPTICAL TECHNOLOGIES, INC, 82- *Concurrent Pos:* Adj prof physics, Am Univ, 71-78. *Mem:* Sigma Xi; Acoust Soc Am; Am Phys Soc; Optical Soc Am; Instrument Soc Am; Soc Photo-Optical Instrumentation Engrs. *Res:* Fiber optic sensors; detection of underwater sound by acoustic-optic techniques; measurement of electrical current and voltage with optical interferometers; radiation and scattering of underwater sound. *Mailing Add:* Optical Technologies Inc 360 Herndon Pkwy Suite 1200 Herndon VA 22070

DAVIS, CHARLES PACKARD, b Concord, Mass, May 24, 22; m 46; c 2. MECHANICAL ENGINEERING. *Educ:* Rensselaer Polytech Inst, BSME, 48. *Prof Exp:* Instr, Rensselaer Polytech Inst, 48-53; develop engr, Gen Elec Co, 51-58; prof mech eng, Calif State Polytech Col, 58-61, head, Aeronaut Eng Dept, 61-74, prof aeronaut eng, 61-82, prof civil eng, 76-82; RETIRED. *Mem:* Am Soc Eng Educ. *Res:* Shock vibration; mechanical design. *Mailing Add:* 677 Buchon St San Luis Obispo CA 93407

DAVIS, CHARLES PATRICK, b Dayton, Ohio, Aug 29, 45; m 68; c 1. MICROBIOLOGY, INTERNAL MEDICINE. *Educ:* St Edward's Univ, BS, 67; Univ Tex, Austin, MA, 69, PhD(microbiol), 75; Univ Tex Med Br, MD, 86. *Prof Exp:* Res asst microbiol, US Army Inst Dent Res, 69-71; USPHS fel microbiol, Univ Wis-Madison, 75-76; asst prof, Dept Microbiol, 76-80, ASSOC PROF MICROBIOL, UNIV TEX MED BR GALVESTON, 80- *Honors & Awards:* Lubben Award for Excellence in Urol, 86. *Mem:* Am Soc Microbiol; Asn Gnotobiotics; Sigma Xi; Am Acad Microbiol. *Res:* Host-parasite relationships in mammalian gastrointestinal tracts; light microscopy, transmission and scanning electron microscopy along with anaerobic culture techniques to discern intimate host-bacterial cell attachments; pathogenic mechanisms of urinary tract pathogens; cancer induction by bacteria. *Mailing Add:* Dept Microbiol Univ Tex Med Br Galveston TX 77550

DAVIS, CHARLES STEWERT, b Akron, Ohio, Apr 22, 35; m 57; c 4. CLINICAL PHARMACOLOGY. *Educ:* WVa Univ, BS, 57; Purdue Univ, MS, 59, PhD(med chem), 61; State Univ NY Upstate Med Ctr, MD, 70. *Prof Exp:* Asst prof med chem, Purdue Univ, 61-63; unit leader, Eaton Labs, Norwich Pharmacol Co, 63-71, asst dir clin pharmacol, 71-73, assoc med dir, 72-75; DIR MED CTR, DRAKES BRANCH, VA, 75- *Concurrent Pos:* Fel, Univ Va, 60-61; NIH res grant, 62. *Mem:* Am Chem Soc. *Res:* Pathophysiology of endocrine diseases and the evaluation of drugs altering these conditions. *Mailing Add:* Found Christian Med Care PO Box 368 Drakes Branch VA 23937

DAVIS, CHESTER L, b Charleston, WVa, July 2, 23; m 51; c 4. COMPUTER SCIENCE, APPLIED MATHEMATICS. *Educ:* Western Mich Univ, AB, 47; Univ Mich, AM, 53; Mich State Univ, PhD, 65. *Prof Exp:* Instr math & eng mech, Gen Motors Inst, 47-55; mathematician, Curtiss-Wright Res Labs, 55-56; sr math programmer, Data Processing Dept, Gen Motors Res Labs, 56-58; prof math & eng mech & chmn dept, Tri-State Col, 58-61; assoc prof mech eng, Univ Toledo, 63-66, dir comput ctr, 64-66; dir, Ogden Col Comput Lab, 70-82, PROF MATH & COMPUT SCI, WESTERN KY UNIV, 67- *Mem:* Am Asn Comput Mach; Comput Soc; Int Elec & Electronics Engrs; AAUP. *Mailing Add:* Dept Comput Sci Western Ky Univ Bowling Green KY 42101

DAVIS, CLARENCE DANIEL, obstetrics & gynecology; deceased, see previous edition for last biography

DAVIS, CLAUDE GEOFFREY, b Morristown, NJ, May 9, 51; m 79. MEMBRANE BIOCHEMISTRY. *Educ:* Swarthmore Col, BA, 73; Univ Calif, San Francisco, PhD(immunol), 82. *Prof Exp:* NIH FEL, HEALTH SCI CTR, UNIV TEX, DALLAS, 83- *Mem:* Am Soc Cell Biol; NY Acad Sci. *Res:* Structure-function relationships of the low density lipo-protein receptor. *Mailing Add:* Box 896 Euless TX 76039

DAVIS, CLYDE EDWARD, b Glenns Ferry, Idaho, June 24, 37; m 59; c 2. INORGANIC CHEMISTRY. *Educ:* Col Idaho, BS, 59; Ore State Univ, MS, 62; Colo State Univ, PhD(inorg chem), 68. *Prof Exp:* Teacher chem, Casper Col, 61-64; asst prof, Calif State Polytech Col, 67-68; NIH fel, Res Sch Chem, Australian Nat Univ, 68-69; from asst prof to assoc prof, 69-82, PROF CHEM, HUMBOLDT STATE UNIV, 82-, DEPT CHAIR, 89- *Mem:* Am Chem Soc. *Res:* Kinetics and mechanisms of base hydrolysis of coordination compounds; reactions of coordinated ligands of biological interest; sequential analysis of polypeptides by cobalt complexes; photovoltaic and photogalvanic cells; borate wood preservatives. *Mailing Add:* Dept Chem Humboldt State Univ Arcata CA 95521

DAVIS, COURTLAND HARWELL, JR, b Alexandria, Va, Feb 14, 21; m 42, 85; c 6. NEUROLOGICAL SURGERY. *Educ:* George Washington Univ, BA, 41; Univ Va, MD, 44; Am Bd Neurol Surg, dipl, 54. *Prof Exp:* Rotating intern, US Marine Hosp, New Orleans, La, 44-45; asst resident neurosurg, Univ Va, 45-46; NIH fel neuropath, 48-49, instr, Duke Hosp, 49-50, from asst resident to resident neurosurg, Duke Univ, 50-52; from instr to prof surg, Bowman Gray Sch Med, 52-87; mem staff, NC Baptist Hosp, 52-87, assoc chief prof, 75-82, chief prof, 82-86; EMER PROF NEUROSURG, BOWMAN GRAY SCH MD, 87- *Concurrent Pos:* NIH fel neurol, Med Ctr Duke Univ, 49-50; instr, Duke Hosp, 49-50; co-chmn res comt, Nat Asn Retarded Children, 58; vis prof, Kuala Lumpur, Malaysia & Vellore, Madras, SIndia, 66; vis neurosurgeon, HOPE, Cartagena, Colombia, SAm, 67 & Jamaica, WI, 71; pres, Bowman Gray Med Found, 61 & Children's Ctr Physically Handicapped, 67-69; mem bd trustees, Found Int Educ Neurol Surg, Inc, 69-, vchmn, 79-89, chmn, 89-; chmn med care eval comt & deleg comt, Piedmont Med Found, 75-; mem bd dirs, Forsyth County Rehab House, Inc & Forsyth County Sheltered Workshop, 76-79. *Mem:* Fel Am Col Surg; Am Acad Neurol Surg; Asn Res Nervous & Ment Dis; Soc Brit Neurol Surgeons; Soc Neurol Surgeons. *Res:* Applied research in neurology and neurosurgery. *Mailing Add:* Sect of Neurosurg Bowman Gray Sch Med Winston-Salem NC 27103

DAVIS, CRAIG BRIAN, b Washington, DC, Nov 29, 38; m 70; c 5. ECOLOGY. *Educ:* Colo State Univ, BS, 64, MS, 67; Univ Calif, Davis, PhD(bot), 72. *Prof Exp:* Asst prof biol, Alma Col, 70-73; from asst prof to prof bot & coordr environ studies, Iowa State Univ, 73-84; PROF & DIR, SCH NATURAL RESOURCES, OHIO STATE UNIV, 84- *Concurrent Pos:* Consult, ecol, Inst Ecol, 74; assoc dean agr, Ohio State Univ, 84- *Mem:* Ecol Soc Am; Sigma Xi; Int Asn Aquatic Vascular Plant Biologists. *Res:* Vegetative structure and nutrient cycling in aquatic ecosystems; higher vascular plant and substrate components of aquatic ecosystems; environmental impact of pollutants on aquatic ecosystems. *Mailing Add:* Bot 108 Bot/Zool Bldg Ohio State Univ Columbus OH 43210

DAVIS, CRAIG H, b Pittsburgh, Pa, Mar 31, 35; m 57; c 2. GENETICS. *Educ:* Ore State Univ, BS, 57; Univ Wash, MS, 62, PhD(biochem), 65. *Prof Exp:* Fel, Calif Inst Tech, 65-67; ASST PROF BIOL, SAN DIEGO STATE UNIV, 67- *Mem:* AAAS. *Res:* Genetics of alcohol preference in mice. *Mailing Add:* Dept Biol San Diego State Univ San Diego CA 92182

DAVIS, CRAIG WILSON, b Ypsilanti, Mich, Mar 19, 49; m 73; c 3. MOLECULAR PHARMACOLOGY, NEUROPHARMACOLOGY. *Educ:* Univ Mich, Ann Arbor, BS, 72; Emory Univ, PhD(pharmacol), 77. *Prof Exp:* Staff fel, NIH, 76-79; ASSOC PROF PHARMACOL, SCH MED, UNIV SC, 80- *Mem:* Am Soc Pharmacol & Exp Therapeut. *Res:* Neurochemical mechanisms involved in the functioning of the nervous system; potential role of cyclic nucleotides. *Mailing Add:* Dept Pharmacol Med Sci Bldg Sch Med Univ SC Garners Ferry Rd Columbia SC 29208

DAVIS, CURRY BEACH, b Hornell, NY, Oct 6, 39; m 64; c 4. ORGANIC CHEMISTRY. *Educ:* Alfred Univ, BS, 61; Col Forestry, State Univ NY, PhD(org chem), 69. *Prof Exp:* CHEMIST DEVELOP LAB, ARIZ CHEM CO, 69- *Mem:* Am Chem Soc. *Res:* Tall oil rosin; fatty acids and terpene chemistry. *Mailing Add:* 2517 High St Bayview Heights Panama City FL 32405-1224

DAVIS, CURTISS OWEN, b Los Angeles, Calif, Jan 1, 45. BIOLOGICAL OCEANOGRAPHY. *Educ:* Univ Calif, Berkeley, BA, 66; Univ Wash, MS, 69, PhD(oceanog), 73. *Prof Exp:* Sr oceanogr, Univ Wash, 73, res assoc oceanog, 73-74; res investr, Great Lakes Res Div, Univ Mich, Ann Arbor, 74-76, asst res oceanographer, 76-79; RES OCEANOGRAPHER & ACTG DIR, TIBURON CTR ENVIRON STUDIES, SAN FRANCISCO STATE UNIV, 80- *Mem:* AAAS; Am Soc Limnol & Oceanog; Estuarine Res Fedn; Phycol Soc Am. *Res:* Phytoplankton nutrient uptake kinetics; continuous culture of phytoplankton; dynamics of coastal marine and estuarine systems. *Mailing Add:* 4800 Oak Grove Dr Pasadena CA 91109

DAVIS, D WAYNE, b Ponce de Leon, Mo, Nov 7, 35; m 65; c 3. ENVIRONMENTAL BIOLOGY. *Educ:* Univ Mo, BS, 61; Univ Sask, MA, 63; Univ Ark, PhD(ecol), 69. *Prof Exp:* From asst prof to assoc prof, 68-74, chmn dept, 78-80 & 86-89, chmn col fac, 80-81, PROF BIOL, SCH OF OZARKS, 74- *Concurrent Pos:* Cur, Table Rock Vis Ctr, 76; vis prof, Univ Sask, 78. *Mem:* Nat Audubon Soc; Nature Conservancy. *Res:* Rodent and avian ecology; black walnut and shortleaf pine culture; establishing of native grasses on Ozark glades and prairies; wildlife habitat improvement; growth of black spruce. *Mailing Add:* Dept Math & Sci Sch Ozarks Point Lookout MO 65726

DAVIS, DANIEL LAYTEN, b Waynesville, NC, Apr 25, 38; m 60; c 3. PLANT PHYSIOLOGY, PLANT GENETICS. *Educ:* Berea Col, BS, 60; Mich State Univ, MS, 62; NC State Univ, PhD(plant physiol), 67. *Prof Exp:* Asst prof physiol, Miss State Univ, 67; from asst prof to prof agron, Univ Ky, 67-87, actg dir, Tobacco & Health Res Inst, 87; MGR AGR SCI DIV, BOWMAN GRAY TECH CTR, RJ REYNOLDS TOBACCO CO, 88- *Concurrent Pos:* Univ Ky Res Found tobacco & health res & serv contract, Agr Res Serv, USDA, 68-70. *Mem:* Am Soc Plant Physiol; Am Soc Agron; Crop Sci Soc Am; Phytochem Soc NAm. *Res:* Physiological and biochemical aspects of plant terpenoids, and other natural products as related to tobacco and health research. *Mailing Add:* Agr Sci Ctr Bowman Gray Tech Ctr RJ Reynolds Tobacco Co Winston-Salem NC 27102

DAVIS, DARRELL LAWRENCE, b Corral, Idaho, Feb 17, 27; m 55; c 4. PHYSIOLOGY. *Educ:* Ore State Col, BS, 49, MA, 52; St Louis Univ, PhD(physiol), 56. *Prof Exp:* Asst, Ore State Col, 49-51; asst, St Louis Univ, 51-56; asst res prof, Med Col Ga, 57-60, from asst prof to assoc prof physiol, 60-71; assoc prof, 71-74, PROF PHYSIOL, UNIV SFLA, 74- *Concurrent Pos:* Cardiovasc trainee, Med Col Ga, 56-57. *Mem:* Am Physiol Soc. *Res:* Cardiovascular and temperature regulation. *Mailing Add:* Dept Physiol Col Med Univ SFla Tampa FL 33612

DAVIS, DAVID, b Liverpool, Eng. Oct 5, 27; US citizen; m 52; c 3. PSYCHIATRY. *Educ:* Glasgow Univ, MB, ChB, 49, MD, 74; Royal Col Physicians & Surgeons, DPM, 54. *Prof Exp:* House officer, Stobhill Hosp, Glasgow, Scotland, 49-50; registrar psychiat, St Crispin & South Ockendon Hosps, Eng, 52-55; registrar, Bethlem Royal & Maudsley Hosps, London, Eng, 57-59; from asst prof to prof psychiat, Sch Med, Univ Mo-Columbia, 60-86, chief, Sect Gen Psychiat, 61-68, chmn, Dept Psychiat, 68-69 & 75-76, assoc chmn dept, 71-75 & 77-86, chief psychiat, Univ Med Ctr, 77-86, dir psychiat educ, 81-86; vchmn, 86-89, PROF PSYCHIAT, MED BR, UNIV TEX-GALVESTON, 89-; EMER PROF PSYCHIAT, SCH MED, UNIV MO-COLUMBIA, 86- *Concurrent Pos:* Fulbright travel scholar psychiat, Wash Univ, 55-57; Am Fund Psychiat teaching fel, 61-62; vis fac, Lab Community Psychiat, Harvard Med Sch, 65-67; co-chmn, Inter Forum Educators Community Psychiat, 67-69; Nat Inst Ment Health vis scientist alcohol res, Nat Ctr Prev & Control Alcoholism, 69-70; consult, Vet Admin, Columbia Univ, Mo, 70; vis prof psychiat, Univ Edinburgh, Scotland, 76-77. *Mem:* Fel AAAS; fel Royal Col Psychiat; fel Am Col Psychiat; fel Am Psychiat Asn; fel Royal Soc Health; Sigma Xi. *Res:* General psychiatry; social and community psychiatry; alcoholism; medical education. *Mailing Add:* Dept Psychiat Univ Tex Graves Bldg D29 Galveston TX 77550

DAVIS, DAVID, b Poland, Dec 20, 20; nat US; m 46; c 2. PLANT PATHOLOGY. *Educ:* Cornell Univ, BA, 46; Univ Ill, PhD, 50. *Prof Exp:* Plant pathologist, Conn Agr Exp Sta, 50-55; plant pathologist res lab, Merck Sharp & Dohme Div, 55-60; res assoc, NY Bot Garden, 60-70; DIR RES, PHYTA LABS, 70- *Mem:* Am Phytopath Soc; Am Inst Biol Sci. *Res:* Evaluation of pesticides; biological control of plant disease. *Mailing Add:* Phyta Labs Ghent NY 12075

DAVIS, DAVID A, b Springfield, Tenn, Mar 24, 18; m 41; c 2. MEDICINE. *Educ:* Vanderbilt Univ, BA, 38, MD, 41. *Prof Exp:* Instr surg, Tulane Univ, 47-49; assoc prof anesthesiol, Med Col Ga, 49-52; from prof surg in chg anesthesiol to clin prof anesthesiol, Sch Med, Univ NC, Chapel Hill, 52-71; PROF ANESTHESIOL, MED CTR, DUKE UNIV, 71- *Concurrent Pos:* Pres, Monitor Instruments, 69- *Mem:* AMA; Am Soc Anesthesiol; Asn Univ Anesthetists. *Res:* Muscle relaxants; circulatory effects of anesthetic agents; local anesthetic drugs. *Mailing Add:* PO Box 250 Med Ctr Duke Univ Kings Mill Rd Springfield TN 37172

DAVIS, DAVID G, b Dickinson, NDak, July 21, 35; m 72; c 2. PLANT PHYSIOLOGY, TISSUE CULTURE. *Educ:* NDak State Univ, BS, 60, MS, 62; Wash State Univ, PhD(bot), 65. *Prof Exp:* Asst bot, NDak State Univ, 60-62 & Wash State Univ, 62-63 & 63-65; AEC fel, Univ Minn, 65-67; RES PHYSIOLOGIST PLANT PHYSIOL, BIOSCI RES LAB, AGR RES SERV, 67- *Concurrent Pos:* Summer fel, NSF Grad Res Fel, 63; adj prof bot, NDak State Univ, 69-; sabbatical leave, Dept Plant Genetics, Weizmann Inst Sci, Israel, 79-80. *Mem:* Am Asn Plant Physiol; Int Asn Plant Tissue Cult; Weed Sci Soc Am; Scand Soc Plant Physiol; Sigma Xi. *Res:* Metabolism of pesticides; pesticide additives in plant tissue cultures; growth and development of cells, tissues and organs in vitro as affected by pesticides, organic solvents, growth regulators and natural products and environment; anatomical and morphological aspects of herbicide application. *Mailing Add:* Biosci Res Lab Agr Res Serv State Univ Sta Fargo ND 58105

DAVIS, DAVID GALE, b Leicester, NC, July 21, 35; m 57; c 3. GENETICS, TOXICOLOGY. *Educ:* Western Carolina Col, BS, 57; Univ NC, MA, 60; Univ Ga, PhD(zool), 63. *Prof Exp:* Asst zool, Univ NC, 57-60 & Univ Ga, 60-61; res fel, Univ Ga & Oak Ridge Nat Lab, 61-63; res assoc, Oak Ridge Nat Lab, 63-64; sr lectr zool, Monash Univ, Australia, 64-67; asst prof, Univ Western Ont, 67-68; NIH spec res fel, Lab Genetics, Univ Wis-Madison, 68-69; ASSOC PROF BIOL, UNIV ALA, TUSCALOOSA, 69- *Mem:* AAAS; Genetics Soc Am; Am Inst Biol Sci; Am Genetic Soc. *Res:* Chromosome mechanics; recombination; developmental genetics; teratogenesis. *Mailing Add:* Dept Biol Univ Ala PO Box 870344 Tuscaloosa AL 35487-0344

DAVIS, DAVID WARREN, b Mankato, Minn, July 6, 30; m 72. PLANT BREEDING, VEGETABLE CROPS. *Educ:* Univ Hawaii, BS, 51; Univ Ill, MS, 56; Ore State Univ, PhD(hort), 63. *Prof Exp:* Res asst genetics, Univ Ill, 55-56; asst-in-training, Exp Sta, Hawaiian Sugar Planter's Asn, 56-57, asst geneticist, 58-61; agriculturist, Lihue Plantation Co, Hawaii, 57-58; res horticulturist, Veg & Ornamental Crops Res Br, Agr Res Serv, USDA, 63-65; assoc prof veg crop breeding, 65-68, PROF HORT, INST AGR, UNIV MINN, ST PAUL, 69- *Concurrent Pos:* Mem coun Agr Sci & Technol. *Mem:* Coun Agr Sci & Technol; fel AAAS; Am Soc Hort Sci. *Res:* Genetic improvement of plant populations for adaptability to commercial and consumer needs in vegetable agriculture; pest and environmental stress resistance; raw product quality. *Mailing Add:* Dept Hort Sci Univ Minn 1970 Folwell Ave St Paul MN 55108

DAVIS, DEAN FREDERICK, b Johnson City, Tenn, May 4, 22; m 47; c 2. RESEARCH ADMINISTRATION. *Educ:* Univ Tenn, AB, 48, MS, 49. *Prof Exp:* Project leader agr res, USDA, Stored-Prod Insects Lab, 51-58, dir, 58-61, asst br chief, 61-66, asst dir mkt qual res div, 66-70, dir, 70-72, area dir, Sci & Educ Admin-Agr Res, USDA, 72-84; AGR CONSULT, 84-; PROG MGR, CARIBBEAN BASIN ADV GROUP, UNIV OF FLA, 85- *Honors & Awards:* Super Serv, USDA. *Mem:* Entom Soc Am; Caribbean Food Crop Soc. *Res:* Insect resistant food packaging; pest management; tropical agriculture. *Mailing Add:* 2825 NW 21st Ave Gainesville FL 32605

DAVIS, DENNIS DUVAL, b Cleveland, Ohio, Nov 9, 41; m 63; c 3. ORGANIC CHEMISTRY, ORGANOMETALLIC CHEMISTRY. *Educ:* Case Inst Technol, BS, 63; Univ Calif, Berkeley, PhD(chem), 66. *Prof Exp:* From asst prof to assoc prof, 66-75, PROF CHEM, NMEX STATE UNIV, 76- *Mem:* Am Chem Soc. *Res:* Physical-organic chemistry; electrophilic aliphatic substitution reactions; photochemistry; free radical chemistry; chemistry of organotransition metal compounds. *Mailing Add:* Dept Chem NMex State Univ Box 3C Las Cruces NM 88003-4992

DAVIS, DENNY CECIL, b Toppenish, Wash, Dec 21, 44; m 72; c 1. AGRICULTURAL ENGINEERING. *Educ:* Wash State Univ, BS, 67; Cornell Univ, MS, 69, PhD(agr eng), 73. *Prof Exp:* Asst prof agr eng, Univ Ga, 73-76; asst prof, 76-83, assoc prof, 83-, PROF AGR ENG, WASH STATE UNIV. *Concurrent Pos:* Consult, Cornell Univ, 74-75, legal consult, 80-81; res grant, Sci & Educ Admin, USDA, 79-82, Wash Cranberry Comn, 83- & Wash Tree Fruit Comn, 85- *Mem:* Am Soc Agr Engrs; Inst Food Technologists; Am Soc Eng Educ; Am Soc Heating, Refrigerating & Air Conditioning Engrs. *Res:* Food quality and properties measurement; finite element applications; food texture; storage of agricultural products; dynamics of machinery; food processing; energy conservation; solar energy. *Mailing Add:* Dept Agr Eng Wash State Univ Pullman WA 99164-6120

DAVIS, DICK D, b Hobart, Okla, Aug 7, 33; m 52; c 5. AGRONOMY. *Educ:* Univ Okla, BS, 60; Okla State Univ, MS, 63, PhD(bot), 65. *Prof Exp:* Asst prof, 64-69, assoc prof, 69-77, PROF AGRON, NMEX STATE UNIV, 77- *Mem:* Am Soc Agron; Crop Sci Soc Am. *Res:* Cotton breeding; pest resistant strains and hybrids. *Mailing Add:* Dept Agron NMex State Univ University Park NM 88001

DAVIS, DONALD ECHARD, b Charleston, Ill, Jan 12, 16; m 40; c 2. PLANT PHYSIOLOGY. *Educ:* Eastern Ill State Univ, BEd, 36; Ohio State Univ, MS, 40, PhD(bot), 47. *Hon Degrees:* PedD, Eastern Ill State Col, 56. *Prof Exp:* Inspector, Ravenna Ord Plant, US War Dept, 42-43; instr bot, Ohio State Univ, 46-47; from asst prof to assoc prof bot, Auburn Univ, 48-49, assoc botanist, Exp Sta, 49-50; AEC agr res fel, Univ Tenn, 51-52; from assoc prof to prof bot, Auburn Univ, 52-68, assoc botanist, 52-55, botanist, exp sta, 55-82, alumni prof bot & microbiol, 68-82, EMER PROF BOT, AUBURN

UNIV, 82- *Concurrent Pos:* Vis prof, Univ Ill, 65; mem comt on persistent pesticide residues, Nat Acad Sci-Nat Res Coun, 68-69, mem consult comt to rev the use of 2, 4, 5-T, Nat Acad Sci, 70; mem, Nat Adv Panel Weed Res; ed, Weed Sci, Weed Sci Soc Am, 73-76, 83-86; coordr spec projs, Off Vpres Res, Auburn Univ, 86-91. *Honors & Awards:* Res Award, Weed Sci Soc Am, 74; Charles A Black Award, Coun Agr Sci & Technol, 90- *Mem:* Fel AAAS; Am Soc Plant Physiol; fel Weed Sci Soc Am; Am Inst Biol Sci. *Res:* Ecological plant physiology; aquatic pollution; physiology of herbicidal action; pesticide problems. *Mailing Add:* Dept Bot & Microbiol Auburn Univ Auburn AL 36849

DAVIS, DONALD MILLER, b Ft Knox, Ky, May 7, 45; m 68; c 1. TOPOLOGY. *Educ:* Mass Inst Technol, BS, 67; Stanford Univ, PhD(math), 72. *Prof Exp:* Actg asst prof math, Univ Calif, San Diego, 71-72; asst prof, Northwestern Univ, 72-74; from asst prof to assoc prof, 74-84, PROF MATH, LEHIGH UNIV, 84- *Concurrent Pos:* NSF res grant, 75- ; vis assoc prof math, Northwestern Univ, 78-79, Univ Warwick, 82, Inst Adv Study, 83, Math Sci Res Inst, Berkeley, 89, Oxford Univ, 90. *Mem:* Am Math Soc; Math Asn Am. *Res:* Topics in homotopy theory, including homotopy groups of spheres, various questions about projective spaces, and cobordism theory. *Mailing Add:* Dept Math Lehigh Univ Bethlehem PA 18015

DAVIS, DONALD RAY, b Concord, NC, Aug 21, 39; m 63, 90; c 2. PLANETARY SCIENCE, ASTRODYNAMICS. *Educ:* Clemson Univ, BS, 62; Univ Ariz, PhD(physics), 67. *Prof Exp:* Staff scientist & mgr astrodyn, TRW Syst Group, 67-70; res scientist planetary sci, IIT Res Inst, 71-72; SR SCIENTIST & DIV MGR, PLANETARY SCI INST & SCI APPLN INC, 72- *Concurrent Pos:* NSF fel, 64-65; NASA fel, 65-67; lectr, Dept Planetary Sci, Univ Ariz, 72-73. *Honors & Awards:* Medal of Freedom, Apollo XIII Team Award. *Mem:* Am Astron Soc; Am Inst Aeronaut & Astronaut; Sigma Xi; Am Asn Physics Teachers. *Res:* Origin, evolution and present state of the solar system; dynamical astronomy; hypervelocity impact studies; trajectory analysis; numerical modeling. *Mailing Add:* Planetary Sci Inst 2421 E Sixth St Tucson AZ 85719-5234

DAVIS, DONALD RAY, b Oklahoma City, Okla, Mar 28, 34; m; c 2. ENTOMOLOGY. *Educ:* Univ Kans, BA, 56; Cornell Univ, PhD(entom), 62. *Prof Exp:* Assoc cur, 61-64, chmn, Dept Entom, 76-81, CUR LEPIDOPTERA, SMITHSONIAN INST, 64- *Concurrent Pos:* Smithsonian Res Found grant, 66-67, 73-74, 78-84; Nat Geog grant, 81; Smithsonian Res Award, 73, 78-81 & 86-88. *Honors & Awards:* Jordan Medal Award, Lepidop Soc, 77. *Mem:* Entom Soc Am; Nat Speleol Soc; Soc Syst Zool; Lepidop Soc (pres, 85). *Res:* Systematics, phylogeny and biology of the Microlepidoptera, particularly the superfamily Tineoidea and all members of the Monotrysia; biology of leaf mining Lepidoptera; biology of cave dwelling Lepidoptera. *Mailing Add:* Dept Entom Smithsonian Inst Washington DC 20560

DAVIS, DONALD ROBERT, b La Jara, Colo, Mar 19, 41; div. NUTRITION, BIOCHEMISTRY. *Educ:* Univ Calif, Los Angeles, PhD(chem), 65. *Prof Exp:* NSF fel chem, Calif Inst Technol, 65-67; asst prof, Univ Calif, Irvine, 67-74; RES ASSOC NUTRIT, CLAYTON FOUND BIOCHEM INST, UNIV TEX, AUSTIN, 74-, DIR, ROGER J WILLIAMS INST, 87- *Concurrent Pos:* Instr part-time, Calif Inst Technol, 65-67; ed-in-chief, J Appl Nutrit, 86-90; consult, Ctr Improv Human Functioning Int, Wichita, Kans, 89- *Mem:* AAAS; Acad Orthomolecular Med; Am Col Nutrit; Soc Nutrit Educ; Int Acad Nutrit Preventive Med. *Res:* Variability of nutrient requirements; statistics of nutrient allowances; nutritional qualities of foods and diets; unrecognized functions of nutrients. *Mailing Add:* Clayton Found Biochem Inst Univ Tex Austin TX 78712-1096

DAVIS, DONALD WALTER, b San Francisco, Calif, May 30, 20; m 43; c 6. ENTOMOLOGY. *Educ:* Univ Calif, BS, 41, PhD, 50. *Prof Exp:* Asst entom, Univ Calif, 46-50; res entomologist, Calif Spray-Chem Corp, 50-54; assoc prof, Teaching & Exp Sta, 54-67, PROF ENTOM, DEPT BIOL, UTAH STATE UNIV, 67- *Concurrent Pos:* Pres, Pac Br, Entom Soc Am, 83-84. *Mem:* Entom Soc Am; Sigma Xi. *Res:* Biology and control of agricultural pests; spider mites of the genus Tetranychus; agricultural entomology. *Mailing Add:* Dept Biol Utah State Univ Logan UT 84322-5305

DAVIS, DUANE M, b Indianapolis, Ind, July 10, 33; m 57, 75; c 3. MISSILE SYSTEMS ENGINEERING, CONTROL SYSTEMS. *Educ:* Purdue Univ, BS, 55, PhD(aeronaut), 67; Univ Pittsburgh, MS, 57. *Prof Exp:* Engr, Bettis Atomic Power Div, Westinghouse Elec Co, 55-57; from instr to assoc prof aeronaut, USAF Acad, 66-73; chief aerospace dynamics br, Air Force Flight Dynamics Lab, 73-75, dep chief structures div, Air Force Flight Dynamics Lab, 75-77, chief eng, Laser Guided Bomb, Systs Prog Off, Armament Div, 77-78, dir laser guided bombs, Systs Prog Off, 78-79, dep dir precision guided munitions, Systs Prog Off, 79, dep surface attack, Armament Div, 79-82, armament consult, 82-86, ASSOC PRIN ENGR, SVERDERP TECH INC TEAS GROUP, EGLIN AFB, FLA, 86- *Concurrent Pos:* Mem, Missile Systs Technol Comt, Am Inst Aeronaut & Astronaut. *Mem:* Am Inst Aeronaut & Astronaut. *Res:* Optimization; weapons system selection and specification; optimal tactics for weapons system employment. *Mailing Add:* Sverderp Tech Inc TEAS Group PO Box 1935 Eglin AFB FL 32542

DAVIS, E(ARL) JAMES, b St Paul, Minn, July 22, 34; div; c 2. CHEMICAL ENGINEERING. *Educ:* Gonzaga Univ, BS, 56; Univ Wash, PhD(chem eng), 60. *Prof Exp:* Design engr, Union Carbide Chem Co, WVa, 56; res engr, Boeing Airplane Co, Wash, 57; from asst prof to assoc prof chem eng, Gonzaga Univ, 60-68; from assoc prof to prof chem eng, Clarkson Col Technol, 68-78, chmn dept, 73-74; prof & chmn, Dept Chem & Nuclear Eng, Univ NMex, 78-80; prof & dir eng div, Inst Paper Chem, 80-; PROF CHEM ENG, UNIV WASH. *Concurrent Pos:* Petrol Res Fund grants, 61-65, 85-88; res grants, NSF, 63-64, 66-68, 69-80, 86-89, Sigma Xi, 69, Bur Mines, 89-91 & Dept Energy, 89-92; res fel, Imp Col, London, 64-65; mem, Inst Colloid & Surface Sci, Clarkson Col Technol, assoc dir, 74-78; consult Unilever Res, Port Sunlight, Eng, 74-79, Sandia Nat Lab, 78-80, 86-88, Los Alamos Sci Lab, 78-80, DuPont, 87 & Philip Morris, 85-91; Nat Acad Sci-Chinese Acad Sci vis scholar award, 89. *Honors & Awards:* Burlington Northern Award for Distinguished Res, 88. *Mem:* Am Chem Soc; Am Inst Chem Engrs; Am Soc Eng Educ; Sigma Xi; Am Asn Aerosol Res (treas, 90-92); AAAS. *Res:* Fluid mechanics and heat transfer; surface and colloid science; aerosol physics and chemistry. *Mailing Add:* Dept Chem Eng BF-10 Univ Wash Seattle WA 98195

DAVIS, EARLE ANDREW, JR, b Aliquippa, Pa, June 1, 19; m 43. NEUROPHYSIOLOGY. *Educ:* Grove City Col, BS, 41; Univ Pittsburgh, MS, 43; Univ Ill, PhD(physiol ecol), 53. *Prof Exp:* Asst prof biol, Elmhurst Col, 46-49; prof, Buena Vista Col, 52-53; prof, West Liberty State Col, 53-55; LECTR ANAT, UNIV CALIF, IRVINE-CALIF COL MED, 55- *Mem:* AAAS; Am Asn Anat; Int Col Surg; Soc Neurosci; Sigma Xi. *Res:* Human gross and microanatomy; effects of metabolic rate on distribution of animals; sites of action of hallucinogenic drugs in cat brain. *Mailing Add:* 10881 Coronel Rd Santa Anna CA 92705

DAVIS, EDGAR GLENN, b Indianapolis, Ind, May 12, 31; m 53; c 3. PHARMACEUTICALS. *Educ:* Kenyon Col, BA, 53; Harvard Univ, MBA, 55. *Prof Exp:* Assoc acct, Eli Lilly & Co, 58-62, chief financial anal, 62, chief budgeting & profit planning, 62-63, mgr, 63-66, mgr econ studies, 66-67, mgr Atlanta sales dist, 67-68, dir pharmaceut mkt res & sales manpower planning div, 68-69, dir pharmaceut mkt plans div, 69-74, exec dir, worldwide pharmaceut mkt planning, 74-75, exec dir, corp affairs, 75-76, VPRES CORP AFFAIRS, ELI LILLY & CO, 76- *Concurrent Pos:* Sr adj fel, Hudson Inst; chmn, Pan Am Econ Leadership Conf; trustee, Kenyon Col. *Mem:* Inst Med-Nat Acad Sci. *Mailing Add:* Lilly Corp Ctr Indianapolis IN 46285

DAVIS, EDWARD ALEX, b Houston, Tex, Jan 2, 31; m 52; c 4. OPERATIONS RESEARCH, ENVIRONMENTAL ASSESSMENT. *Educ:* Rice Univ, BA, 55, MA, 56, PhD(nuclear physics), 61. *Prof Exp:* Proj officer, Nuclear Power Off, US Army Engr Res & Develop Labs, 57-58; res physicist, Texaco, Inc, 61-62; asst prof physics, Okla City Univ, 62-64; mem fac, Eve Sch, 66-71, PRIN STAFF PHYSICIST OPERS RES, JOHNS HOPKINS UNIV, 64- *Mem:* AAAS; Opers Res Soc Am; Am Asn Physics Teachers; Sigma Xi; Inst Elec & Electronics Engrs. *Res:* cooling towers; nuclear well-logging; tactical military operations analysis; command and control systems; transit systems analysis; applied mathematics; simulation; probabilistic models; air quality; environmental systems; design and analysis of instruction systems; applications of personal computers. *Mailing Add:* Appl Physics Lab Univ Leicester Univ Rd Leicester LEI 7RH England

DAVIS, EDWARD ALLAN, b San Francisco, Calif, Oct 26, 17; m 56; c 3. MATHEMATICS. *Educ:* Univ Calif, Berkeley, BA, 40, MA, 44, PhD(math), 51. *Prof Exp:* Teaching asst, assoc & jr instr for Army Air Force in math, Univ Calif, Berkeley, 40-47; from instr to asst prof, Univ Nev, 47-55; from asst prof to assoc prof, 55-69, actg dean col sci, 75-76, PROF MATH, UNIV UTAH, 69- *Concurrent Pos:* Fund for Adv Ed res fel, Stanford Univ & Univ Chicago, 53-54; assoc prog dir spec proj in sci educ, NSF, DC, 61-62, consult, 62-67, 71-73, prog dir, Student & Coop Prog, 67-70. *Mem:* Am Math Soc; Math Soc Am; Nat Coun Teachers Math. *Res:* Mathematical economics; teacher training programs in mathematics. *Mailing Add:* Dept Math Univ Utah Salt Lake City UT 84112

DAVIS, EDWARD DEWEY, b Philadelphia, Pa, Sept 24, 33. MATHEMATICS. *Educ:* Univ Pa, BA, 55, MA, 57; Univ Chicago, PhD(math), 61. *Prof Exp:* Instr math, Northwestern Univ, 61-62; lectr & res assoc, Yale Univ, 62-64; asst prof, Purdue Univ, 64-67; asst prof, 67-68, assoc prof math, 68-80, PROF MATH, STATE UNIV NY, ALBANY, 80- *Mem:* Am Math Soc. *Res:* Commutative algebra, especially Northerian rings; algebraic geometry. *Mailing Add:* Dept Math & Statist State Univ NY 1400 Washington Ave Albany NY 12222

DAVIS, EDWARD E, neurophysiology, endocrinology, for more information see previous edition

DAVIS, EDWARD LYON, b Fall River, Mass, July 15, 29. BOTANY. *Educ:* Harvard Univ, BA, 51; Univ Mass, MS, 53; Wash Univ, PhD(bot), 56. *Prof Exp:* Instr bot, Univ Mass, 53; asst, Mo Bot Garden, 54-56; from asst prof to assoc prof, 57-72, PROF BOT, UNIV MASS, AMHERST, 72-, HEAD DEPT, 75- *Concurrent Pos:* Vis lectr, Smith Col, 58-59. *Mem:* Bot Soc Am. *Res:* Anatomy, systematics and evolution of higher plants; morphogenesis. *Mailing Add:* Dept Bot Univ Mass Amherst MA 01002

DAVIS, EDWARD MELVIN, b Springfield, Mass, Mar 22, 13; m 40; c 3. ORGANIC CHEMISTRY. *Educ:* Rensselaer Polytech Inst, ChE, 34, MS, 35, PhD(org chem), 38. *Prof Exp:* Res chemist, Joseph E Seagram & Sons, Inc, 39-40; develop chemist, Biofen Labs, 40-41 & Kroger Grocery & Baking Co, Inc, 41-44; vpres mfg, Solo Marx Rubber Co, 44-56; from asst dir to assoc dir chem develop, Hilton-Davis Chem Co Div, Sterling Drug Inc, 56-75, dir chem develop, 75-78; RETIRED. *Concurrent Pos:* Environ consult. *Mem:* Sigma Xi. *Mailing Add:* 2197 Section Rd Cincinnati OH 45237

DAVIS, EDWARD NATHAN, b Danville, Ill, Oct 9, 11; m 37; c 2. CHEMICAL ENGINEERING. *Educ:* Univ Ill, BS, 33; DePaul Univ, MS, 41. *Prof Exp:* Control chemist, US Rubber Reclaiming Co, 35-36; rubber res chemist, Underwriters Labs, Inc, 37-38, exec engr, 38-54, managing engr, 54-64, chief engr, fire protection, 64-65, vpres, 65-77; RETIRED. *Mem:* Soc Fire Protection Engrs; Nat Soc Prof Engrs; fel Am Inst Chemists; Am Chem Soc; Am Soc Testing & Mats. *Res:* Standards preparation and reserach relating to fire hazards and protection of materials and devices. *Mailing Add:* Two Exeter on Oxford Rolling Meadows IL 60008-1910

DAVIS, EDWIN ALDEN, b New Haven, Conn, Dec 28, 23; m 51; c 2. WATERSHED MANAGEMENT. *Educ:* Univ Conn, BS, 45; Yale Univ, PhD(plant physiol), 49. *Prof Exp:* Res fel plant physiol, Carnegie Inst, Dept Plant Biol, Stanford, Calif, 49-52; plant physiologist, Agr Res, Dow Chem Co, 53-57; plant physiologist, Agr Res Serv, USDA, 58-63, prin plant physiologist, Forest Serv, USDA, 63-84, CONSULT ROCKY MOUNTAIN FOREST & RANGE EXP STA, FOREST SERV, USDA, 84- *Mem:* Bot Soc Am; Am Soc Plant Physiol; Weed Sci Soc; Soil Conserv Soc Am; Agron Soc. *Res:* Water augmentation research to increase stream flow from chaparral watersheds by converting chaparral to plants that use less water; brush control methods and the effects of vegetation conversions on water yield and quality and the environment. *Mailing Add:* Forestry Sci Lab Rocky Mountain Forest & Range Exp Sta Ariz State Univ Tempe AZ 85287

DAVIS, EDWIN NATHAN, microbiology; deceased, see previous edition for last biography

DAVIS, ELDON VERNON, b Burwell, Nebr, Aug 4, 23; m 45; c 1. VIROLOGY. *Educ:* Univ Nebr, BS, 49, MS, 51; Univ Pa, PhD, 57. *Prof Exp:* Jr res scientist, Upjohn Pharmaceut Co, 50-52, scientist & head virus labs, 52-54; asst, Univ Pa, 54-56, res assoc, 56-57, asst prof, 57-58; virologist & head labs, USPHS, 58-63; sr virologist, Midwest Res Inst, 63-64; from virologist to res scientist, 64-69, SR RES SCIENTIST, NORDEN LABS, INC, 69- *Mem:* AAAS; NY Acad Sci; Tissue Cult Asn; Sigma Xi. *Res:* Cell physiology, tissue culture; growth of mammalian cells in submerged culture; kinetics of viral growth; viral isolation and characterization; vaccines. *Mailing Add:* 7221 South St Unit 18 Lincoln NE 68520

DAVIS, ELDRED JACK, b Kelleyville, Okla, Oct 14, 30; m 64; c 5. INTERMEDIARY METABOLISM, BIOENERGETICS. *Educ:* Abilene Christian Col, BSc, 56; Fla State Univ, MSc, 58; McGill Univ, PhD(biochem), 63. *Prof Exp:* Res biochemist, Lederle Labs, Am Cyanamid Co, 58-61; from asst prof to assoc prof, 65-72, PROF BIOCHEM, SCH MED, IND UNIV, INDIANAPOLIS, 72- *Concurrent Pos:* USPHS fel, Lab Biochem, Univ Amsterdam, 63-65; vis asst prof, Univ Amsterdam, 68; vis prof, Univ Oslo, 71-72, Univ Helsinki, 78-79; consult, NIH; sci reviewer, Am Heart Asn, Ind, 73-78; travel fel, NATO, 75-79; assessor, Australian Res Grants Scheme, 84-; distinguished vis prof, Univ Adelaide, Australia, 86. *Mem:* AAAS; Am Soc Biol Chemists; Brit Biochem Soc. *Res:* Intermediary metabolism and its regulation: Perfused organs, mitochondria, reconstructed cell-free systems, identification, characterization of new malic enzymes; mitochondria and metabolic control; gluconeogenesis; anaplerotic pathways in mammalian systems. *Mailing Add:* Dept Biochem Ind Univ Sch Med Indianapolis IN 46223

DAVIS, ELIZABETH ALLAWAY, b New York, NY, Jan 5, 41; m 68; c 2. DEVELOPMENTAL BIOLOGY, PLANT TISSUE CULTURE. *Educ:* Mt Holyoke Col, BA, 62; Brandeis Univ, PhD(develop biol), 69. *Prof Exp:* Postdoc student, biol labs, Harvard Univ, 68-69; ASST PROF BIOL, UNIV MASS, BOSTON, 69- *Mem:* AAAS; Am Soc Plant Physiol; Soc Develop Biol; Tissue Cult Asn; Bot Soc Am; Int Asn Plant Tissue Cult. *Res:* Physiology, genetics, multiplication, differentiation and gene expression in chloroplasts and plant cells; plant morphogenesis; plant tissue culture. *Mailing Add:* Dept Biol Univ Mass Boston Harbor Campus Boston MA 02125-3393

DAVIS, ELIZABETH YOUNG, b Ft Collins, Colo, Apr 23, 20; m 41; c 3. NUTRITION, PHYSIOLOGY. *Educ:* Colo State Univ, BS, 41; Auburn Univ, MS, 57, PhD(nutrit biochem), 64. *Prof Exp:* instr foods & nutrit, Auburn Univ, 57-60; assoc prof nutrit res, Tuskegee Inst, 64-66; prof nutrit, Auburn Univ, 66-73, res coordr home econ, 69-73; group leader human nutrit & social sci, USDA, 77-78, asst dep adminr, Human Nutrit, Food & Social Sci Coop State Res Serv, 79-83; RETIRED. *Honors & Awards:* Distinguished Serv Award, Am Home Econ Asn Found, 84. *Mem:* Am Chem Soc; Am Dietetic Asn; Am Inst Nutrit; Inst Food Technol; Am Home Econ Asn. *Res:* Lipid metabolism; biosynthesis of carnitine; incorporation of ethionine ethyl into phospholipids; ethylated ethanolamines in phospholipids; metabolism of N-alkyl amines. *Mailing Add:* 1365 Gatewood Dr Apt 501 Auburn AL 36830-2989

DAVIS, ELMO WARREN, b Idaho, Sept 9, 20; m 51; c 4. GENETICS, HORTICULTURE. *Educ:* Univ Idaho, BS, 48; Univ Calif, Davis, MS, 49, PhD(genetics), 52. *Prof Exp:* Assoc prof hort & assoc olericulturist, Kans State Univ, 52-53; geneticist, USDA, 53-56, res horticulturist, 57-66; mem jr bd exec, 67-69, dir agr res & develop, Gilroy Foods, Inc, Philippines, Seychelles, Cent Am, US, W Europe, Morocco, Egypt, Brazil, Venezuela, 66-76, mem bd dirs, 70-76,; corp agr scientist, McCormick & Co Inc, 76-80, dir agr sci & technol, 80-85; agr scientist, Gilroy Foods, Inc, 85-88; CONSULT HERBS & SPICES, 88- *Concurrent Pos:* Agr res & prod herbs & spices, Australia, Tonga. *Mem:* Fel Am Soc Hort Sci; Am Genetics Asn; Int Soc Hort Sci. *Res:* Breeding, culture, seed production and dehydration of onions and garlic; horticultural character of allium species; hybrid onions; culinary herbs and spices, sesame, vanilla, onions, garlic, capsicum, basil, etc, in US and foreign. *Mailing Add:* 1650 Vallejo Dr Hollister CA 95023

DAVIS, ELWYN H, b Leon, Iowa, Jan 10, 42; m 63; c 2. COMBINATORICS ALGEBRA. *Educ:* Univ Mo-Columbia, BSE, 64, MA, 66, PhD(math), 69. *Prof Exp:* Assoc prof, 69-77, PROF MATH, PITTSBURG STATE UNIV, 77-, CHAIR, MATH DEPT, 89- *Concurrent Pos:* NSF grant convexity conf, 71; vis instr math, Pa State Univ, 75-76; math consult, Phillips Petrol, 84- *Mem:* Am Math Soc; Math Asn Am; Soc Petrol Engrs. *Res:* Projective planes; nearfields and generalizations; construction of finite projective planes via coordinating systems; generalizations of projective planes and the coordinating systems; computer assisted instruction; solution of large sparse linear systems. *Mailing Add:* Dept Math Pittsburg State Univ Pittsburg KS 66762

DAVIS, ERNST MICHAEL, b Victoria, Tex, Oct 12, 33; m 73; c 2. AQUATIC POLLUTION CONTROL, INDUSTRIAL WASTE MANAGEMENT. *Educ:* NTex State Univ, BA, 56, MA, 62; Univ Okla, PhD(sanit eng), 66. *Prof Exp:* Res assoc, Univ Tex, Austin, 66-68, asst prof sanit eng, 69-70; from asst prof to assoc prof, 70-85, PROF ENVIRON SCI, UNIV TEX SCH PUB HEALTH, HOUSTON, 85- *Concurrent Pos:* NSF grant, Univ Tex, Austin, 66-67, Fed grant sanit eng, 66-68, Off Water Resources Res grant limnol invests, 67-68, Fed Water Pollution Control Admin grant, 67-69, Tex Water Qual Bd grant, 68, Tex Water Develop Bd Grant, Univ Tex Sch Pub Health, Houston, 70-72 & NASA grant, 71-72; consult, Environ Protection Agency, 70-74; res fels, NTex State Univ; var indust res contracts, 70- *Mem:* Am Acad Environ Engrs; Water Pollution Control Fedn; Int Asn Water Pollution Res; Soc Environ Toxicol & Chem; Am Water Resources Asn; Biodeterioration Soc, UK. *Res:* Water pollution abatement methodology; water resources management; industrial wastewater treatment; fate of organics in waste treatment units; microbiol indicators in wastewater. *Mailing Add:* Univ Tex Sch Pub Health PO Box 20186 Houston TX 77225

DAVIS, EUGENIA ASIMAKOPOULOS, b Chicago, Ill, Oct 15, 38; m 60; c 2. FOOD SCIENCE, ORGANIC CHEMISTRY. *Educ:* Univ Chicago, BS, 59, MS, 60; Free Univ Brussels, PhD(org chem), 67. *Prof Exp:* Asst scientist chem, Res Inst, Ill Inst Technol, 60-62; res asst, Dept Internal Med, 67-70, res assoc, Dept Foods, 70-72, from asst prof to assoc prof, 72-82, PROF, DEPT FOOD SCI & NUTRIT, UNIV MINN, 82- *Concurrent Pos:* Agr Exp Sta Proj grant, 75-88; NSF grant, 76-78; co-ed, J Food Microstruct, 81-; Continental Baking grant, 84-; mem, Wis Milk Mkt Bd, 85-88; Kraft Co Inc grant, 87- *Mem:* Am Chem Soc; AAAS; Inst Food Technologists; Am Asn Cereal Chemists; Sigma Xi. *Res:* Heat and mass transfer studies; scanning and transmission; electron microscope ultra structural studies; spectroscopic and x-ray microanalysis of foods heated in controlled environment ovens equipped with conventional and microwave heating elements. *Mailing Add:* Dept Food Sci & Nutrit Univ Minn 1334 Eckles Ave St Paul MN 55108

DAVIS, FLOYD ASHER, b Chester, Pa, June 10, 34; m 62; c 2. NEUROLOGY, NEUROSCIENCES. *Educ:* Franklin & Marshall Col, BS, 56; Univ Pa, MD, 60. *Prof Exp:* Asst prof neurol, Col Med, Univ Ill, 68-71; ASSOC PROF NEUROL & BIOMED ENG, RUSH MED COL, 71- *Concurrent Pos:* From asst attend neurologist to assoc attend neurologist, Presby-St Luke's Hosp, Chicago, 68-74, sr attend neurologist, 74- *Mem:* Am Acad Neurol; Am Neurol Asn; Soc Neurosci; AAAS; NY Acad Sci; Sigma Xi. *Res:* Pathophysiology of impulse conduction in demyelinated nerve as it relates to multiple sclerosis. *Mailing Add:* Rush-Presby St Lukes Med Ctr 1725 W Harrison St Chicago IL 60612

DAVIS, FRANCES MARIA, CELL KINETICS, MITOSIS. *Educ:* Yale Univ, PhD(microbiol), 77. *Prof Exp:* ASST PROF CELL BIOL CHEMOTHER RES, M D ANDERSON HOSP & TUMOR INST, 81- *Mailing Add:* Tanox Biosysts Inc 1031 Stella Link Houston TX 77025

DAVIS, FRANCIS CLARKE, b Sumter, SC, 41; m; c 1. TRANSLATIONAL REGULATION, HISTONE REGULATION. *Educ:* Univ Tenn, PhD(biochem), 68. *Prof Exp:* ASSOC PROF CELL BIOL, UNIV FLA, 78- *Mem:* AAAS; Sigma Xi; Am Soc Cell Biol. *Res:* Molecular biology of development. *Mailing Add:* Dept Microbiol & Cell Sci Univ Fla Gainesville FL 32611-0116

DAVIS, FRANCIS KAYE, b Scranton, Pa, May 4, 18; m 41; c 2. PHYSICAL METEOROLOGY. *Educ:* Westchester Univ, BS, 39; Mass Inst Technol, MS, 44; NY Univ, PhD, 57. *Prof Exp:* Asst, Mass Inst Technol, 43-44; PROF PHYSICS, DREXEL UNIV, 46-, DEAN COL SCI, 70- *Concurrent Pos:* Staff meterologist, WFIL-WFIL-TV, 47-71; meteorol consult, 50-; assoc, Johns Hopkins Univ, 54-56; consult, C W Thornthwaite Assocs, 54-70, Day & Zimmerman, 55-, City of Philadelphia, 56-, Am Mach & Foundry Co, 57- & Radio Corp Am, 58- *Mem:* Fel Am Meteorol Soc; Am Phys Soc; Am Asn Physics Teachers; Am Acad Forensic Scientists. *Res:* Storm damage; atmospheric pollution; physics of fog formation; upper atmosphere physics. *Mailing Add:* 31 Fairway Winds Pl Hilton Head Island SC 29928

DAVIS, FRANK, biochemistry, for more information see previous edition

DAVIS, FRANK FRENCH, b Pendleton, Ore, July 19, 20; m 48; c 2. BIOCHEMISTRY. *Educ:* Univ Hawaii, BS, 50; Univ Calif, PhD(biochem), 55. *Prof Exp:* Jr res biochemist, Univ Calif, 54-56, asst, 56-57; from asst prof to assoc prof biochem, 57-64, PROF BIOCHEM, RUTGERS UNIV, 64- *Mem:* AAAS; Am Chem Soc; Am Soc Biol Chemists; NY Acad Sci; Sigma Xi. *Res:* Enzyme modification for therapeutic purposes. *Mailing Add:* 1407 Rifle Range Rd El Cerrito CA 94530-2506

DAVIS, FRANK W, ENGINEERING. *Prof Exp:* RETIRED. *Mem:* Nat Acad Eng. *Mailing Add:* 939 Coast Blvd Apt 9-C La Jolla CA 92037

DAVIS, FRANKLIN A, b Des Moines, Iowa, Apr 1, 39; m 67; c 2. ORGANIC CHEMISTRY. *Educ:* Univ Wis, BS, 62; Syracuse Univ, PhD(org chem), 66. *Prof Exp:* Res asst org chem, Syracuse Univ, 63-66; Welch fel, Univ Tex, 66-68; from asst prof to assoc prof, 68-79, PROF CHEM, 79-, GEORGE S SASIN PROF ORG CHEM, DREXEL UNIV, 86- *Concurrent Pos:* Nat Prog Chmn, Org Div, ACS, 89-92. *Mem:* Am Chem Soc. *Res:* Mechanistic and synthetic organic chemistry; selective oxidations and asymmetric oxidations using N-sulfonyloxaziridines; synthetic methods: the application of unusual and unusually reactive organosulfur and organosulfur-nitrogen in organic synthesis; the chemistry of sulfenic acids (RSOH) and sulfur-nitrogen derivatives. *Mailing Add:* Dept Chem Drexel Univ Philadelphia PA 19104

DAVIS, FREDERIC I, b New York, NY, June 24, 37; m 58; c 2. MATHEMATICAL PHYSICS, NUMERICAL ANALYSIS. *Educ:* Mass Inst Technol, SB, 58; Brandeis Univ, MA, 66, PhD(theoret physics), 72. *Prof Exp:* From asst prof to assoc prof, 68-83, chair math dept, 82-88, PROF MATH, US NAVAL ACAD, 83- *Mem:* Am Math Soc; Math Asn Am; Soc Indust & Appl Math; Am Phys Soc. *Mailing Add:* Dept Math US Naval Academy Annapolis MD 21402

DAVIS, GARY EVERETT, b San Diego, Calif, Oct, 13, 44; m 83; c 1. ECOSYSTEM DYNAMICS, POPULATION DYNAMICS. *Educ:* San Diego State Col, BS, 67, MS, 71. *Prof Exp:* Park ranger, Lassen Volcanic Nat Park, Nat Park Serv, 64-67; teaching asst biol, San Diego State Col, 67-68; park ranger, Virgin Islands Nat Park, Nat Park Serv, 68-69; biol lab asst, Sixth US Army Med Lab, 69-71; res marine biologist, Everglades Nat park, 71-80, RES MARINE BIOLOGIST, CHANNEL ISLANDS NAT PARK, NAT PARK SERV, 80- *Concurrent Pos:* Biologist, Tektite I Prog, Dept Interior, 68-69; consult, Bahamas Nat Trust, 80, World Wildlife Fund, 83 & Int Union Conserv Nature, 83. *Mem:* Am Acad Underwater Sci (pres, 86-87); Am Inst Fishery Res Biol; AAAS; Am Fisheries Soc; Nat Areas Asn (bd dir, 88-91). *Res:* Population dynamics of system components; evaluate and develop strategies and tactics for natural area management, especially involving coastal marine fisheries. *Mailing Add:* 1901 Spinnaker Dr Ventura CA 93001

DAVIS, GEORGE DIAMENT, b Ithaca, NY, May 7, 26; m 50; c 1. PHYSIOLOGY. *Educ:* Princeton Univ, AB, 46; Yale Univ, PhD(physiol), 51. *Prof Exp:* From instr to assoc prof, 51-63, prof, 63-85, EMER PROF PHYSIOL, LA STATE UNIV MED CTR, 85- *Concurrent Pos:* Vis prof, La State Univ-Agency Int Develop Contract, Sch Med, Univ Costa Rica, 61-63. *Mem:* AAAS; Am Acad Neurol; Am Physiol Soc; NY Acad Sci; Sigma Xi. *Res:* Neurophysiology; biophysical instrumentation. *Mailing Add:* 5182 St Ferdinand Dr New Orleans LA 70126-3412

DAVIS, GEORGE H, b Detroit, Mich, July 25, 21; m 49, 61; c 1. HYDROLOGY, GEOLOGY APPLIED. *Educ:* Univ Ill, BS, 42. *Prof Exp:* Asst geol, State Geol Surv, Ill, 42-46; geologist, US Geol Surv, 46-68, res hydrologist, 68-79, asst dir, 79-81; CONSULT GEOLOGIST/HYDROLOGIST, 81- *Concurrent Pos:* First officer, Int Atomic Energy Agency, Vienna, 66-68; ed, Water Resources Res, 69-76; chmn int hydrol decade work group on ground water, UNESCO, 70-74; vchmn int hydrol decade work group on nuclear techniques in hydrology, Int Atomic Energy Agency, 69-74; chmn work group on water & energy, Int Hydrol Prog, 76-; ed, J Hydrol, 77-; lectr, Univ Md Univ Col, 81. *Honors & Awards:* Meinzer Award, Geol Soc Am, 72. *Mem:* Geol Soc Am; Am Geophys Union; Int Water Resources Asn (treas); Int Asn Hydrogeologists. *Res:* Geologic occurrence of ground water. *Mailing Add:* 10408 Insley St Silver Spring MD 20902

DAVIS, GEORGE HERBERT, b Pittsburgh, Pa, Aug 30, 42; m 65; c 3. STRUCTURAL GEOLOGY. *Educ:* Col Wooster, BA, 64; Univ Tex, Austin, MA, 66; Univ Mich, Ann Arbor, PhD(struct & econ geol), 71. *Prof Exp:* From asst prof to assoc prof, 70-82, PROF GEOSCI, UNIV ARIZ, 82-VPROVOST, 86- *Concurrent Pos:* Dir summer field camp, Univ Ariz, 70-74, chmn dept, 82-86; fel, NSF. *Mem:* Geol Soc Am; Sigma Xi; AAAS. *Res:* Metamosphic core complexes, detachment faulting; thin-skinned deformation at Bryce National Park. *Mailing Add:* Bus & Finance Univ Ariz Tucson AZ 85721

DAVIS, GEORGE KELSO, b Pittsburgh, Pa, July 2, 10; m 36; c 6. ANIMAL NUTRITION. *Educ:* Pa State Univ, BS, 32; Cornell Univ, PhD(nutrit), 37. *Prof Exp:* Asst animal nutrit, Cornell Univ, 34-37; asst chem, Mich State Univ, 37-42; prof nutrit, Inst Food & Agr Sci, 42-79, dir div sponsored res, 70-75, res prof & dir nuclear sci, 60-65, dir biol sci, 65-70, DISTINGUISHED EMER PROF, UNIV FLA, 79- *Concurrent Pos:* Eli Lilly lectr; hon prof, Univ Chile; mem, Frasch Found Awards Comt; mem, Animal Nutrit Comt; mem geochem environ comt, Nat Res Coun; sect convenor, Spec Comt Int Biol Prog; pres, Nat Nutrit Consortium, 77-78; mem, Int Coun Sci Unions; prof nutrit, Univ Hawaii, 85. *Honors & Awards:* Borden Award, Am Inst Nutrit, 64; Spencer Award, Am Chem Soc, 80; Conrad A Elvehjem Award, 85. *Mem:* Nat Acad Sci; Am Chem Soc; Am Inst Nutrit (pres, 75-76); Am Soc Biol Chem; Am Soc Animal Sci; Soc Environ Geochem & Health (pres, 76). *Res:* Relation of nutrition to development of disease, trace substances, vitamins, mineral elements; use of radioactive tracers; metabolism of copper and cardiovascular disorders; trace element metabolism in relation to cardiovascular calcification and urolithiases. *Mailing Add:* 2903 SW Second Ct Gainesville FL 32601

DAVIS, GEORGE MORGAN, b Bridgeport, Conn, May 21, 38; m 61, 75; c 2. MALACOLOGY. *Educ:* Marietta Col, BA, 60; Univ Mich, MS, 62, PhD(zool), 65. *Hon Degrees:* DSc, Marietta Col, 89. *Prof Exp:* Res assoc malacol, Univ Mich, 61-65; chief malacol sect, US Army Med Command, Japan, 65-70; assoc cur mollusca, 70-78, CUR MOLLUSCA, ACAD NATURAL SCI PHILADELPHIA, 78-, CHMN DEPT MALACOL, 72- *Concurrent Pos:* Adv, US Educ Comn, Japan, 66-67; consult, Agency Int Develop, Mekong River Proj, 71; partic, US-Japan Coop Med Sci Prog, 71; co-ed-in-chief, Malacologia, Int J Malacol, 72-88; exec secy-treas, Int Malacol, 74-; mem steering comt, Am Coun Syst Malacologists, 73-79; adj prof, Univ Pa, 77-; assoc prof, Jefferson Med Univ, 70-81; adj prof, Dept Biol & Pathobiol, Univ Penn; treas, Asn Systs Collections, 77-81, vpres, 81-82, pres, 82-84; reviewer, Inst Mus Serv, 81-; consult, MAP prog, Am Asn Mus, 82-; ed-in-chief, Malacologia, 88-; mem, bd dir, Am Type Culture Collections, 84-, chmn, 87-90; coun mem, Am Inst Biol Sci, 86-; secy, Invetebrate Tech Comt, PA Biol Survey, 89- *Mem:* Fel AAAS; Am Inst Biol Sci; Am Soc Zool; Am Malacol Union (pres, 76-77); Japanese Soc Parasitol. *Res:* Systematic studies of freshwater, amphibious, brackish-water snails; medical malacology; morphological, molecular genetic studies of snails and clams. *Mailing Add:* Dept Malacol 19th & Pkwy Acad Natural Sci Philadelphia PA 19103

DAVIS, GEORGE THOMAS, organic chemistry; deceased, see previous edition for last biography

DAVIS, GEORGE THOMAS, b Montour Falls, NY, Oct 20, 33; m 59; c 6. POLYMER CHEMISTRY. *Educ:* Cornell Univ, BChE, 56; Princeton Univ, PhD(chem), 63. *Prof Exp:* Chem engr, Esso Res & Eng Co, 56-60; actg asst prof chem, Univ Va, 63-64; res assoc polymer physics, 64-66, CHEMIST, NAT INST STANDARDS & TECHNOL, 66- *Mem:* Am Chem Soc; Am Phys Soc. *Res:* Secondary oil recovery; diffusion in solids; structure of polymers; polymer crystallization; piezoelectric polymers. *Mailing Add:* Nat Inst Standards & Technol Gaithersburg MD 20899

DAVIS, GERALD GORDON, b Owen Sound, Ont, July 7, 37; m 59; c 6. POLYMER CHEMISTRY. *Educ:* Queen's Univ, Ont, BSc, 60, MSc, 61; Oxford Univ, DPhil(chem), 63. *Prof Exp:* Res assoc, Cornell Univ, 63-64; asst prof chem, Queen's Univ, Ont, 64-66; scientist, 66-68, tech dir polymer chem, Glidden Co Div, 70- 85, GEN MGR, WALKER BROTHERS DIV, GLIDDEN CO (CAN) LTD, 85- *Mem:* Am Chem Soc; fel Chem Inst Can; Can Res Mgt Asn. *Res:* Physical chemistry of polymers and coatings. *Mailing Add:* Walker Brothers Div Glidden Co (Can) Ltd Burnaby BC V5J 1J2 Can

DAVIS, GERALD TITUS, b Kingsport, Tenn, Sept 2, 32; m 52; c 2. PAPER CHEMISTRY. *Educ:* King Col, BA, 54; Univ Tenn, PhD(chem), 58. *Prof Exp:* Res scientist, 58-70, assoc dir res, 70-77, mgr, prod develop, 77-80, mgr, process develop, 80-82, LAB DIR, MEAD CORP, 82- *Mem:* Tech Asn Pulp & Paper Ind. *Res:* Printing and speciality coated papers. *Mailing Add:* Central Res Mead Corp Eighth & Hickory Sts Chillicothe OH 45601

DAVIS, GORDON WAYNE, b Bremerton, Wash, June 7, 45; m 74; c 2. ANIMAL SCIENCE, MEAT SCIENCE. *Educ:* Wash State Univ, BS, 68, 69; Tex A&M Univ, MS, 74, PhD(animal sci), 77. *Prof Exp:* Instr animal sci, Tex A&M Univ, 72-76; asst prof food technol & sci, Univ Tenn, 77-80; asst prof animal sci, 80-86, ASSOC PROF, TEX TECH UNIV, 86- *Concurrent Pos:* Res fel, Tex A&M Univ, 76-77. *Mem:* Am Soc Animal Sci; Inst Food Technol; Am Meat Sci Asn; Nat Asn Col Teachers Agr. *Res:* Pork and beef quality; effect of slaughter and processing methods on meat quality; forage finished beef and beef and pork meat blends; cholesteral; lean lite beef; beef tenderness. *Mailing Add:* 4214 88th Pl Lubbock TX 79423

DAVIS, GRAHAM JOHNSON, b Trenton, NC, Oct 5, 25; m 49; c 1. AQUATIC ECOLOGY. *Educ:* ECarolina Col, BS, 49; George Peabody Col, MA, 50; Univ NC, PhD(bot), 56. *Prof Exp:* Asst prof biol, Brenau Col, 50-53 & Univ Tenn, 56-57; plant physiologist, Crops Res Div, USDA, 57-59; from asst prof to prof, 59-87, EMER PROF BIOL, ECAROLINA UNIV, 87- *Mem:* Am Soc Limnol & Oceanog; Sigma Xi. *Res:* Physiological ecology of aquatic macrophytes; estuarine pollution; upgrading wastewaters with floating vascular plant systems. *Mailing Add:* Biol Dept ECarolina Univ Greenville NC 27834

DAVIS, GRAYSON STEVEN, Washington, DC, May 29, 47; m 80; c 1. MORPHOGENESIS, CELL ADHESION. *Educ:* George Washington Univ, BS, 70; Univ Va, PhD(biol), 81. *Prof Exp:* Asst prof biol, 81-87, ASSOC PROF BIOL, VALPARAISO UNIV, 87- *Mem:* Am Soc Zoologists. *Res:* Physical analysis of amphibian gastrulation; role of cell-cell adhesion in morphogenesis. *Mailing Add:* Dept Biol Valparaiso Univ Valparaiso IN 46383

DAVIS, GREGORY ARLEN, b Portland, Ore, Jan 29, 35; div; c 3. STRUCTURAL GEOLOGY, TECTONICS. *Educ:* Stanford Univ, BS, 56, MS, 57; Univ Calif, Berkeley, PhD(geol), 61. *Prof Exp:* From asst prof to assoc prof, 61-71, chmn, 77-81, PROF GEOL, UNIV SOUTHERN CALIF, 71- *Concurrent Pos:* Vis assoc prof, Univ Wash, 71; dist lectr, Am Asn Petrol Geologists, 74-75; consult, Wash Pub Power Supply Syst, 77-89; geologist, US Geol Surv, 83-87; chair, Geol Soc Am, Struct Geol & Tectonics Div, 87-88. *Mem:* Geol Soc Am; Am Geophys Union; Geol Asn Can. *Res:* Thrust belt and extensional tectonics; regional tectonics, NAM Cordillera, Northeastern China. *Mailing Add:* Dept Geol Sci Univ Southern Calif Los Angeles CA 90089-0740

DAVIS, GUY DONALD, b Newport News, Va, June 15, 52; m 90; c 1. SURFACE SCIENCE, ADHESIVE BONDING. *Educ:* Rensselaer Polytech Inst, BS, 74; Univ Wis-Madison, MS, 75 & 79, PhD(mat sci), 82. *Prof Exp:* scientist, 80-85, sr scientist, 85-88, GROUP LEADER, MARTIN MARIETTA LABS, 88-, STAFF SCIENTIST, 88- *Concurrent Pos:* Mem adv bd, Surface and Interface Anal, 87-; vchmn, Local Arrangement Comt, Am Vacuum Soc, 86, mem bd dir, Appl Surface Sci Div, 89-; second vchmn, E42 Comt, Am Soc Testing & Mat, 87-; mem, Sci Coun, Md Acad Sci, 87- *Mem:* Adhesion Soc (treas, 88-); Am Vacuum Soc; Sigma Xi; Am Soc Testing & Mat. *Res:* Surface analysis; adhesive bonding; industrial failure analysis; corrosion and corrosion inhibition of aluminum and painted steel; in-process monitoring/control. *Mailing Add:* Martin Marietta Labs 1450 S Rolling Rd Baltimore MD 21227

DAVIS, HALLOWELL, b New York, NY, Aug 31, 96; m 23, 44, 83; c 3. NEUROPHYSIOLOGY, AUDIOLOGY. *Educ:* Harvard Univ, AB & MD, 22. *Hon Degrees:* ScD, Colby Col, 54, Northwestern Univ, 62, Wash Univ, 73, Upstate Med Ctr, Syracuse, 79, Univ Mich, 83. *Prof Exp:* From instr to assoc prof physiol, Harvard Med Sch, 23-46, actg head dept, 42-43; res prof otolaryngol, 46-65, from assoc prof to prof, 46-65, EMER PROF PHYSIOL, SCH MED, WASH UNIV, 65-; EMER DIR RES & RES ASSOC, CENT INST FOR DEAF, 65- *Concurrent Pos:* Dir res, Cent Inst for Deaf, 46-65; mem div med sci, Nat Res Coun, 47-53, exec secy comt hearing & bioacoust, Armed Forces, Nat Res Coun, 53-59. *Honors & Awards:* Shambaugh Prize, 53; Beltone Award, 66; Nat Medal Sci, 75. *Mem:* Nat Acad Sci; Am Acad Arts & Sci; Am Physiol Soc (treas, 41-48, pres, 58-59); Acoust Soc Am (pres, 53); Am Electroencephalog Soc (pres, 49); Am Philos Soc. *Res:* Central nervous and auditory physiology; audiology; psychoacoustics; electroencephalography; electric response audiometry. *Mailing Add:* 7526 Cornell Ave University City MO 63130

DAVIS, HAMILTON SEYMOUR, medicine; deceased, see previous edition for last biography

DAVIS, HARMER E, b Rochester, NY, July 11, 05; c 3. CIVIL & TRANSPORTATION ENGINEERING. *Educ:* Univ Calif, BS, 28, MS, 30. *Prof Exp:* Asst civil eng, 28-30, from instr to assoc prof, 30-48, dir, Inst Transp & Traffic Eng, 48-77, chmn dept, 54-59, prof, 48-73, EMER PROF CIVIL ENG, UNIV CALIF, BERKELEY, 73-, EMER DIR, INST TRANSP & TRAFFIC ENG, 73- *Concurrent Pos:* Res & develop consult, 30-48; vchmn, Hwy Res Bd, Nat Acad Sci-Nat Res Coun, 58, chmn, 59; mem, Div Eng &

Indust Res, Nat Res Coun, 60-, mem exec comt, 62-65; consult transp, 73- *Honors & Awards:* Crum Award, Nat Acad Sci-Nat Res Coun, 59 & George S Bartlett Award, 70; James Laurie Prize, Am Soc Civil Engrs, 67; T M Matson Award, Inst Traffic Engrs, 74. *Mem:* Nat Acad Eng; Am Soc Civil Engrs; Am Soc Testing & Mat; Am Concrete Inst. *Res:* Soil mechanics; portland cement concrete; bituminous materials; transportation planning and functioning. *Mailing Add:* 1645 Skycrest Dr 12 Walnut Creek CA 94595

DAVIS, HAROLD, electrical engineering, for more information see previous edition

DAVIS, HAROLD LARUE, b Philadelphia, Pa, Nov 18, 25; m 57; c 2. NUCLEAR PHYSICS. *Educ:* Carnegie Inst Technol, BS, 49; Cornell Univ, PhD(physics), 54. *Prof Exp:* Proj engr, Pratt & Whitney Aircraft, 54-57; managing ed, Nucleonics, 57-69; ed Physics Today, Am Inst Physics, 69-85; RETIRED. *Concurrent Pos:* Adj asst prof atomic physics, Hartford Grad Ctr, Rensselaer Polytech Inst, 55-57. *Mem:* Am Nuclear Soc; Am Phys Soc. *Res:* Reactor design and development; neutron cross sections; high energy physics and accelerators; particle detectors; scientific data processing. *Mailing Add:* Three Rutherford Pl New York NY 10003

DAVIS, HAROLD LLOYD, b Portage, Ohio, Sept 18, 30; m 59; c 2. THEORETICAL SOLID STATE PHYSICS, SURFACE PHYSICS. *Educ:* Bowling Green State Univ, BS, 54; Ohio State Univ, PhD(physics), 59. *Prof Exp:* Mem staff solid state, Sandia Corp, 59-63; mem staff syst anal, Bellcomm Corp, 63-64; SR STAFF MEM & GROUP LEADER THEORY SECT, SOLID STATE DIV, OAK RIDGE NAT LAB, 64- *Concurrent Pos:* Lectr physics, Univ Tenn, 68-72. *Mem:* Fel Am Phys Soc; AAAS. *Res:* Low energy electron diffraction; electronic band structure of metals; magnetism of rare-earth and transition metals; crystal field theory of rare-earth compounds; actinide metals and their compounds. *Mailing Add:* Solid State Div Bldg 3025 MS 6032 Oak Ridge Nat Lab PO Box 2008 Oak Ridge TN 37831

DAVIS, HARRY FLOYD, b Colby, Kans, Oct 2, 25; m 61; c 2. MATHEMATICS. *Educ:* Mass Inst Technol, PhD(math), 54. *Prof Exp:* Instr math, Mass Inst Technol, 49-54; asst prof, Miami Univ, 54-55 & Univ BC, 55-58; assoc prof, Royal Mil Col, Can, 58-60; assoc prof, 60-66, PROF MATH, UNIV WATERLOO, 66- *Mem:* Math Asn Am; Soc Indust & Appl Math. *Res:* Applied algebra. *Mailing Add:* Dept Appl Math Univ Waterloo Waterloo ON N2L 3G1 Can

DAVIS, HARRY I, electronics engineering; deceased, see previous edition for last biography

DAVIS, HARRY L, internal medicine; deceased, see previous edition for last biography

DAVIS, HARRY R, JR, PHARMACOLOGY. *Educ:* Univ Maine, BS, 77; George Washington Univ, MS, 79; Univ Chicago, PhD(path), 82. *Prof Exp:* Res assoc, Dept Path, Univ Chicago, 82-84, res assoc & asst prof, 84-86; SR PRIN SCIENTIST, ATHEROSCLEROSIS RES, SCHERING-PLOUGH RES, 87- *Concurrent Pos:* Sr res scientist, Atherosclerosis Res, Pharm Div, Ciba-Geigy Corp, 87. *Mem:* Am Heart Asn; Sigma Xi. *Res:* Pharmacology; numerous publications. *Mailing Add:* Pharm Res Dept Schering-Plough Res 60 Orange St Bloomfield NJ 07003

DAVIS, HARVEY SAMUEL, b Columbus, Ohio, July 6, 36; m 64; c 2. TOPOLOGY. *Educ:* Eastern Ill Univ, BS, 59; Univ Miami, MS, 61; Univ Ill, PhD(math), 65. *Prof Exp:* Asst math, Univ Miami, 59-61; asst, Univ Ill, 61-65; asst prof, 65-69, ASSOC PROF MATH, MICH STATE UNIV, 69- *Mem:* Am Math Soc; Sigma Xi. *Res:* Point set topology; differential manifolds; continuum classification by set functions. *Mailing Add:* Dept Math Mich State Univ A333 Wells Hall East Lansing MI 48824

DAVIS, HARVEY VIRGIL, b Paces, Va, Dec 19, 32; m 59; c 4. PHARMACOLOGY, OCCUPATIONAL HEALTH. *Educ:* Cent State Col, Wilberforce, Ohio, BS, 55; Wayne State Univ, MS, 66, PhD(pharmacol & physiol), 68. *Prof Exp:* Indust toxicologist, SysteMed Corp, Dayton, Ohio, 68-71; sr toxicologist, Standard Oil Co, Chicago, Ill, 71-76; dir environ health & hyg, Velsicol Chem Corp, Chicago, 76-87; PRES, DAVIS ENVIRON SERV, 87- *Mem:* Soc Toxicol; NY Acad Sci; AAAS; Am Col Toxicol. *Mailing Add:* 29 W 122 Morris Ct Warrenville IL 60555

DAVIS, HENRY MCRAY, b Whitakers, NC, Jan 21, 28; m 49; c 2. ANALYTICAL CHEMISTRY, PHYSICAL CHEMISTRY. *Educ:* NC Col Durham, BS, 52; Howard Univ, MS, 61. *Prof Exp:* Anal chemist, Div Colors & Cosmetics Technol, Food & Drug Admin, 60-65, res chemist, Cosmetics Br, 65-68, supvry chemist, Cosmetics Compos Sect, 68-75, chief, Prod Comp Br, Div Cosmetic Technol, 75-86; RETIRED. *Concurrent Pos:* Chief sect, Food & Drug Admin, 70-75. *Mem:* Asn Off Anal Chem; Am Chem Soc; Soc Cosmetic Chem. *Res:* Methodology in cosmetic analysis, adsorption, ion-exchange and partition chromatographic techniques; infrared, ultraviolet and x-ray fluoresence spectroscopy; chrono-potentiometry of the iodide-iodine-triiodide system. *Mailing Add:* 14013 Cricket Lane Silver Spring MD 20904

DAVIS, HENRY MAUZEE, b Sherman, Tex, Oct 25, 02. PHYSICAL CHEMISTRY. *Educ:* Univ Okla, BS, 29, MS, 30; Univ Minn, PhD(phys chem), 34. *Prof Exp:* Asst chem, Univ Okla, 29-30 & Univ Minn, 30-34; in charge dept, Itasca Jr Col, 34-36; asst ceramics, 36-37, res assoc, 37-38, from asst prof to prof metall, 38-62, EMER PROF METALL, PA STATE UNIV, 62- *Concurrent Pos:* Dir metall & mat sci div, US Army Res Off, 62-74; adj prof mat eng, NC State Univ, Raleigh, 63-81. *Mem:* Fel Am Ceramic Soc; Am Chem Soc; fel Am Soc Metals; Am Inst Mining, Metall & Petrol Engrs. *Res:* Phase equilibria in metallic and refractory systems; physical chemistry of metallurgical systems; equilibria and kinetics of gas-metal reactions. *Mailing Add:* 2219 New Hope Dr Chapel Hill NC 27514

DAVIS, HENRY WERNER, b Cambridge, Mass, Aug 31, 36; m 62; c 2. COMPUTER SCIENCE. *Educ:* Rice Univ, BA, 59; Univ Colo, MA, 61, PhD(math), 65. *Prof Exp:* Engr, Martin Co, 62-63; res assoc math, Spec Res Numerical Anal, Duke Univ, 65-67; asst prof, Univ NMex, 67-69; assoc mathematician, Brookhaven Nat Lab, 69-74; asst prof comput sci, Hofstra Univ, NY, 74-75; assoc prof, 75-82, PROF COMPUT SCI, WRIGHT STATE UNIV, DAYTON, OHIO, 82- *Mem:* Asn Comput Mach; Am Asn Artificial Intelligence. *Res:* Artificial intelligence. *Mailing Add:* Dept Comput Sci Wright State Univ Dayton OH 45435

DAVIS, HERBERT JOHN, b Windsor, Eng, Dec 20, 38; Can citizen; m 60; c 2. PHYSICAL CHEMISTRY, ELECTROCHEMISTRY. *Educ:* London Univ, BSc Hons, 60; ARCS, 60. *Prof Exp:* Sr chemist mat sci, Mullard's Mat Res Labs, 60-62; res scientist fuel cells, Energy Conversion Ltd, 62-63; group leader, Ferranti-Packard Ltd, 63-71; res mgr, Unican Security Systs, Ltd, 71-74; PRIN SCIENTIST, NORANDA TECHNOL CTR, 74- *Mem:* Electrochem Soc; Chem Inst Can; Inst Elec & Electronic Engrs. *Res:* Dielectrics; electrostatic precipitation; pollution control; batteries; non-aqueous solvents; photovoltaic devices; photoconductive materials; energy transmission and storage; water electrolysis. *Mailing Add:* Noranda Technol Ctr 240 Hymus Blvd Point Claire PQ H9R 1G5 Can

DAVIS, HERBERT L, JR, b Hendersonville, NC, Aug 18, 35; m 57; c 2. CYTOGENETICS, RADIATION BIOLOGY. *Educ:* Berry Col, BS, 57; Emory Univ, MS, 61. PhD(radiation biol), 65. *Prof Exp:* Instr biol, Western Md Col, 59-61; instr biol & chem, Berry Col, 61-62; from asst prof to assoc prof, 65-80, PROF BIOL, EMORY UNIV, 80-; DEAN, SCH SCI & ALLIED HEALTH , KENNESAW COL, 82- *Concurrent Pos:* McCandless Fund res grant, 66-67; chmn, Div Natural Sci & Math, Kennesaw Col, 80- *Mem:* AAAS; Am Soc Biologists. *Res:* Role of metabolic energy compounds on recovery from radiation damage; effects of inhibitors of DNA synthesis on recovery from radiation damage. *Mailing Add:* Dean Sci-Allied Health Kennesaw State Col PO Box 444 Marietta GA 30061

DAVIS, HORACE RAYMOND, b Clayton, Mo, Oct 8, 22; m 46; c 4. ORGANIC CHEMISTRY. *Educ:* Univ Ill, BA, 43; Univ Minn, PhD(chem), 49. *Prof Exp:* Res chemist, M W Kellogg Co, 49-51, res supvr, 51-57; group supvr, 57-72, MGR CHEM PROCESS DEVELOP, MINN MINING & MFG CO, 72-, GROUP SUPVR, 3-M CTR, 77- *Mem:* AAAS; Am Chem Soc; fel Am Inst Chem; Royal Soc Chem. *Res:* Fluorocarbon chemistry; nucleophilic substitution; chemistry of high polymers; chemical process development. *Mailing Add:* 16 Braintree Crescent Penfield NY 14526-9518

DAVIS, HOWARD TED, b Hendersonville, NC, Aug 2, 37; m 60; c 2. PHYSICAL CHEMISTRY, CHEMICAL PHYSICS. *Educ:* Furman Univ, BS, 59; Univ Chicago, PhD(phys chem), 62. *Prof Exp:* NSF fel, 59-62, post doc fel, Free Univ Brussels, 62-63; from asst prof to assoc prof, 63-69, dir grad studies, Dept Chem Eng & Mat Sci, 75-78, assoc head, 78-80, PROF CHEM ENG, UNIV MINN, 69-, HEAD, DEPT CHEM ENG & MAT SCI, 80- *Concurrent Pos:* Gen Motors scholar, 58-59; mem fac study team, Inst Defense Anal, 62; vis comt, chem eng dept, Purdue Univ, 78; consult, Argonne Nat Labs, 70-74, Pillsbury, 77-82; adv comt, dept food sci & nutrit, Univ Minn, 78-80, adv bd, 3M External Technol Eval, 81-85, adv coun chem eng, Cornell Univ, 86-; chmn, Gordon Conf Chem Interfaces, 84, comt eval grad sch, Dean Holts Performance, 87, eval comt, chem eng dept, Tex A&M, 88; vis prof, Mex Inst Petrol, 79; Am Inst Chem Engr Award comt, 85-; Nat Res Coun Panel, Microgravity Prog, NASA, 85; chem eng adv coun, Cornell Univ & vis comt, Univ Mich, 86-; fel, Minn Supercomputer Ctr, Univ Minn, 87- & prog dir, Ctr Interfacial Eng, 88-; mem, Indust Tech Adv Comt & Res Coord Coun, Gas Res Inst, 88-, rev comt, Chem Eng, Mat Sci & Petrol Eng Depts, Univ Southern Calif, 89. *Honors & Awards:* Lacey Lectr, Calif Inst Technol & Reilly Lectr, Notre Dame Univ, 82; Kurt Wohl Distinguished Lectr, Univ Del, 87; Donald Katz Distinguished Lectr, Univ Mich, 88; George Taylor Distinguished Serv Award, 89; Louis T Pirkey Centennial Lectr, Univ Tex, Austin, 90; Kelly Distinguished Lectr, Purdue Univ, 90; Juan O Gonzales Lectr Award, 91. *Mem:* Nat Acad Eng; Am Chem Soc; Am Inst Chem Engrs; AAAS; Soc Petrol Engrs; Sigma Xi; Am Phys Soc. *Res:* Thermodynamics and transport phenomena; statistical mechanics; molecular dynamics; liquid electronics; heat and mass transfer in processed food; large scale scientific computation; porous media; microsstructural fluids; wetting hydrodynamics; enhanced oil recovery; surfactant colloids; cold stage electron microscopy; nuclear magnetic resonance; x-ray and light scattering studies of fluid microstructures. *Mailing Add:* Dept Chem Eng & Mat Sci Univ Minn Minneapolis MN 55455

DAVIS, HUBERT GREENIDGE, physical chemistry, chemical engineering; deceased, see previous edition for last biography

DAVIS, HUGH MICHAEL, b Buffalo, NY, Aug 31, 58; m 80; c 1. IMMUNOLOGY. *Educ:* Gannon Univ, BS, 79; Villanova Univ, MS, 83, PhD(chem), 85. *Prof Exp:* Jr chemist, Congoleum Corp, 79-80; postdoctoral fel, Centocor, Inc, 85-86, res scientist, 86-88; sr res scientist, Rorer Cent Res, 88-89; SECT MGR IMMUNOCHEM, RHONE-POULENC RORER CENT RES, 90- *Concurrent Pos:* Lectr, Villanova Univ, 85- *Mem:* Am Chem Soc; AAAS; Am Inst Chemists; Am Soc Biochem & Molecular Biol; Am Soc Bone & Mineral Res. *Res:* Immunoradiological and in vitro assay development for human anti-mouse antibody, von Villebrand factor fragment, calcitonin, estrogen, serotonin and bisphosphonates. *Mailing Add:* 1106 Aurora Dr West Chester PA 19380

DAVIS, JACK, b New York, NY, May 21, 37; m 59; c 2. PHYSICS. *Educ:* Northeastern Univ, BS, 58, MS, 60; Univ London, DIC & PhD(physics), 67. *Prof Exp:* Res assoc physics, Univ Md, 60-61 & Northeastern Univ, 61-63; staff scientist, Avco Corp, 63-67, sr staff scientist, 67-68; sci specialist theoret physics, EG&G, Inc, Mass, 68-70, dept mgr phys sci, 70; SUPVRY RES PHYSICIST & HEAD RADIATION HYDRODYN BR, US NAVAL RES LAB, 70- *Concurrent Pos:* Res assoc, Imp Col, Univ London, 65-67, fel, Spectros Lab, 67; adj prof, George Mason Univ, 74-76. *Mem:* AAAS; Am

Phys Soc; fel, Am Phys Soc. *Res:* Plasma spectroscopy; atomic collision theory; radiation transport; non-local thermodynamic equilibrium radiation hydrodynamic models applicable to laser and particle beam interactions with matter, exploding wires; imploding plasmas; x-ray lasers and atomic processes in ultra-dense plasmas and strong fields. *Mailing Add:* Code 4720 Plasma Physics Div Naval Res Lab Washington DC 20375

DAVIS, JACK H, b Clinton, SC, Nov 22, 39; m 62; c 2. SOLID STATE PHYSICS. *Educ:* Clemson Univ, BS, 62, PhD(physics), 66. *Prof Exp:* From instr to asst prof, 66-69, ASSOC PROF PHYSICS, UNIV ALA, HUNTSVILLE, 69- *Mem:* Am Phys Soc; Sigma Xi. *Res:* Metal whiskers growth, strength and electrical properties. *Mailing Add:* 307 Shady Brook Huntsville AL 35801

DAVIS, JAMES ALLEN, b Glasgow, Ky, Oct 17, 40; m 67; c 2. BIOLOGY, ORGANIC CHEMISTRY. *Educ:* WKy Univ, BS, 62; Univ Akron, MS, 67. *Prof Exp:* Res scientist electrophysiol, Goodyear Aerospace Corp, 64-67; asst res dir, Dept Biochem & Microbiol, Carrtone (Century) Labs, 67-68; group leader microbiol fermentations, Baxter Labs, 68-69; SR RES SCIENTIST, MAT SCI, FIRESTONE TIRE & RUBBER CO, 69- *Mem:* Am Chem Soc. *Res:* Research programs designed to improve product quality and performance; increased technical knowledge and development of new products and processes. *Mailing Add:* Tech Ctr Bldg 21-6 Owens-Corning Fiberglas Granville OH 43023

DAVIS, JAMES AVERY, b New Orleans, La, Aug 13, 39; div. MATHEMATICAL STATISTICS. *Educ:* Occidental Col, AB, 61; Calif State Col Long Beach, MA, 63; Univ NMex, PhD(math), 67. *Prof Exp:* Res specialist, NAm Aviation, Inc, 62-67; res assoc probability & statist, Univ Montreal, 67-68; STAFF MEM APPL MATH, SANDIA CORP, 68- *Concurrent Pos:* Fel, Univ Montreal, 67-68. *Mem:* Am Math Soc. *Res:* Probability theory; stochastic processes; convergence rates for sums or random variables; probability; statistics; combinatorial analysis; cryptology. *Mailing Add:* Appl Math Div 1423 Sandia Labs Albuquerque NM 87115

DAVIS, JAMES IVEY, QUANTUM OPTICS, LASER APPLICATIONS. *Educ:* Calif Inst Technol, BS, 62; Univ Calif, Los Angeles, MS, 65, PhD(physics), 69. *Prof Exp:* Appl physicist, Hughes Aircraft Co, 62-66; instr advan optics, Univ Ghana, 66-67; grad student physics, Univ Calif, Los Angeles, 67-69, NSF fel, 69-70; dept mgr, Hughes Aircraft Co, 70-74; prog dir, 74-87, LAB EXEC OFFICER, LAWRENCE LIVERMORE NAT LAB, 87- *Concurrent Pos:* Master fel, Hughes Aircraft Co, 63-65. *Mem:* Am Phys Soc; AAAS. *Res:* High technology application; laser isotope separation, fusion, photochemistry and radar; high energy laser weapons and low energy laser systems for countermeasures and intelligence gathering; formal scientific training in many-body theory and collective effects; laser-matter interactions and beam propagation; laser physics; quantum electronics; turbulence theory. *Mailing Add:* 4114 Sugar Maple Dr Danville CA 94506-4639

DAVIS, JAMES N(ORMAN), b Dallas, Tex, Oct 24, 39; m 65; c 3. NEUROBIOLOGY, CLINICAL NEUROLOGY. *Educ:* Cornell Univ, BA, 61, MD, 65. *Prof Exp:* PROF NEUROL, PHARMACOL & NEUROBIOL, DUKE UNIV, 72-, DIR STROKE CTR, 80-; CHIEF NEUROL, DURHAM VET ADMIN CTR, 74- *Concurrent Pos:* Vis researcher, Univ Goteborg, Sweden, 72; Fulbright fel, 72; consult, NIH, NSF, Can Heart Asn, & Vet Admin, 76- *Mem:* Am Neurol Asn; Soc Neurosci; Am Soc Pharmacol & Exp Therapeuts; Am Acad Neurol; Am Soc Neurochem. *Res:* Recovery of function after stroke and related brain injury; modulation of ischemic brain damage. *Mailing Add:* Duke Univ PO Box 2900 Durham NC 27710

DAVIS, JAMES OTHELLO, b Tahlequah, Okla, July 12, 16; m 41; c 2. PHYSIOLOGY, MEDICINE. *Educ:* Northeastern Okla State Col, BS, 37; Univ Mo, MA, 39, PhD(zool), 42, BS(med), 43; Wash Univ, MD, 45. *Prof Exp:* Asst zool, Univ Mo, 37-42, asst anat, 42-43; intern, Barnes Hosp, St Louis, 45-46; investr, NIH, USPHS & Baltimore City Hosps, 47-49, Lab Kidney & Electrolyte Metab, Nat Heart Inst, 49-57, chief sect exp cardiovasc dis, 57-66; prof & chmn dept, 66-83, EMER PROF PHYSIOL, SCH MED, UNIV MO, COLUMBIA, 83-, CONSULT, DEPT, 83- *Concurrent Pos:* Fel, Sch Med, Wash Univ, 46; assoc prof, Sch Med, Temple Univ, 55-56; vis assoc prof, Sch Med, Johns Hopkins Univ, 61-64; vis prof, Sch Med, Univ Va, 64; chmn, Nat Coun High Blood Pressure Res, Am Heart Asn, 72-74; pres, Cardiovasc Sect Am, Physiol Soc, 80-81; teaching med, grad students & postdoctoral fel, Univ Mo & NIH. *Honors & Awards:* Sigma Xi Res Award, Univ Mo, 71; Franz Volhard Award, Int Soc Hypertension, 74; Ciba Award, Hypertension Res, Am Hearth Asn, 76; Carl J Wiggers Award, 76; A Festchrift Award. *Mem:* Nat Acad Sci; Am Physiol Soc; Am Soc Nephrology; Int Soc Nephrology; Int Soc Hypertension (vpres, 78-80, pres, 80-82); Endocrine Soc; Int Am Soc Hypertension. *Res:* Cardiovascular, renal and endocrine physiology; physiology of congestive heart failure and hypertension; research administration. *Mailing Add:* Dept Physiol MA415 Med Sci Bldg Univ Mo Sch Med Columbia MO 65212

DAVIS, JAMES ROBERT, b York, Nebr, Dec 28, 29; m 56; c 3. PLANT PATHOLOGY. *Educ:* Univ Calif, Riverside, AB, 56; Univ Calif, Davis, MS, 61, PhD(plant path), 67. *Prof Exp:* Lab technician I, Univ Calif, Riverside, 56, lab technician II, Univ Calif, Davis, 56-68; from asst prof to assoc prof, 68-77, PROF PLANT PATH, UNIV IDAHO, 77- *Mem:* Am Phytopath Soc. *Res:* Studies in stone fruit pathology relating to canker diseases induced by low temperature, fungi and bacteria; soil-borne diseases of potato. *Mailing Add:* Br Exp Sta Univ Idaho PO Box AA Aberdeen ID 83210

DAVIS, JAMES ROYCE, b Rison, Ark, Apr 6, 38; m 59; c 3. MICROBIAL PHYSIOLOGY, CLINICAL MICROBIOLOGY. *Educ:* Tex Col Arts & Indust, BS, 60; Univ Houston, MS, 62, PhD(microbiol), 65. *Prof Exp:* Lectr microbiol, Univ Houston, 64-65; asst prof, Dent Br, Univ Tex, 65-68; ASSOC PROF MICROBIOL, BAYLOR COL MED, 68- *Concurrent Pos:* Dir microbiol sect, Clin Lab, Methodist Hosp, 68-; Chmn combined prog med technol, Tex Med Ctr, 70-, assoc dir labs. *Mem:* Am Soc Microbiol. *Res:* Steroid metabolism by microorganisms; streptomycete taxonomy; chemical composition and microbial flora of saliva; rapid and automated techniques for diagnostic microbiology. *Mailing Add:* Dept Path Methodist Hosp 6565 Fannin Houston TX 77030

DAVIS, JAMES WENDELL, b Tulsa, Okla, Aug 22, 27; m 52; c 4. BIOCHEMISTRY, STATISTICS. *Educ:* Univ Tulsa, BS, 49; Ore State Univ, MS, 52, PhD(biochem), 54. *Prof Exp:* Instr chem, Univ Tulsa, 49-50; assoc biochemist, Oak Ridge Nat Lab, 54-57; from sr instr to assoc prof, 57-86, PROF BIOCHEM, SCH MED, ST LOUIS UNIV, 86-, DIR, CANCER EDUC PROG, 81- *Mem:* Am Chem Soc; Am Asn Cancer Educ; Am Statist Asn; Am Soc Biol Chemists; Sigma Xi. *Res:* Catecholamines. *Mailing Add:* 1402 S Grand Blvd St Louis MO 63104

DAVIS, JAY C, b Haskell, Tex, July 12, 42; m 63; c 2. NUCLEAR ANALYTICAL TECHNIQUES, ACCELERATOR TECHNOLOGY. *Educ:* Univ Tex, Austin, BA, 63, MA, 64; Univ Wis-Madison, PhD(physics), 69. *Prof Exp:* Teaching asst physics, Univ Tex, Austin, 63-64; teaching asst, Univ Wis-Madison, 64-65, res asst, 65-69, AEC fel nuclear physics, 69-70, res assoc, 70-71; physicist, Lawrence Livermore Nat Lab, 71-74, proj mgr, 74-78, opers mgr, 79-82, dep assoc dir, 84-88, div leader, 85-88, DIR, CTR ACCLERATOR MASS SPECTROMETRY, LAWRENCE LIVERMORE NAT LAB, 88- *Concurrent Pos:* Postdoctoral fel, AEC. *Mem:* Am Phys Soc; AAAS; fel NSF. *Res:* Development and application of nuclear analytical technologies and radiation sources for material, science, biomedical dosimetry and clinical applications, geosciences, radiocarbon dating and arms control; research management; program development. *Mailing Add:* Lawrence Livermore Lab L397 Livermore CA 94550

DAVIS, JEFFERSON C, b Sulphur Springs, Ark, Dec 8, 32. AEROSPACE MEDICINE. *Educ:* Univ Mo, BS, 55, MD, 57; Univ Calif, MPH, 63. *Prof Exp:* Chief aerospace med, Ellsworth AFB, SDak, 59-61; instr, USAF Sch Aerospace Med, Brooks AFB, Tex, 61-62; resident, USAF Sch Aerospace Med, Hq Strategic Air Command, Omaha, Nebr & Univ Calif, 62-65; instr altitude & hyperbaric med, USAF Sch Aerospace Med, Brooks AFB, Tex, 65-68; comdr, USAF Hosp, Phu Cat Air Base, Vietnam, 68-69; dep chief biodynamics & bionics, Aerospace Med Res Lab, Wright-Patterson AFB, Ohio, 69-70; chief aerospace med br, Brooks AFB, Tex, 70-74, chief hyperbaric med br, USAF Sch Aerospace Med, 74-77, vcomdr, Aerospace Med Div, 77-79; DIR, HYPERBARIC MED, METHODIST HOSP, SAN ANTONIO, TEX, 79- *Concurrent Pos:* Co-ed, Hyperbaric Oxygen Ther, Undersea Med Soc, Nat Libr Med, 78-; ed, Undersea & Hyperbaric Med, 79-80. *Honors & Awards:* Scientist of the Year, Air Force Asn, 72; Gary Wratten Award, Asn Mil Surgeons, 75; Charles W Shilling Award, Undersea Med Soc, 85. *Mem:* Aerospace Med Asn (pres, 81-82); Undersea Med Soc (vpres, 75-76, pres, 79-80); Soc USAF Flight Surgeons (pres, 76-77); Am Col Prevention Med (pres, 81-82). *Res:* Hyperbaric medicine-advanced techniques of treatment of decompression sickness, gas embolism, gas gangrene, carbon monoxide poisoning and wound healing enhancement. *Mailing Add:* 4499 Medical Dr San Antonio TX 78229

DAVIS, JEFFERSON CLARK, JR, b Jacksonville, Fla, Mar 20, 31; m 54; c 3. MAGNETIC RESONANCE, SPECTROSCOPY. *Educ:* Univ Ariz, BS, 53, MS, 54; Univ Calif, Berkeley, PhD(chem), 59. *Prof Exp:* Asst, Lawrence Radiation Lab, Univ Calif, 56-59; from instr to asst prof chem, Univ Tex, 59-65; assoc prof, 65-71, chmn dept, 78-82, PROF CHEM, UNIV SFLA, 72- *Concurrent Pos:* Mem adv coun col chem, Int Conf Educ in Chem; ed, Exam Comt, Am Chem Soc, 68-88, chmn, Bd of Trustees, 86-88; consult, Multi-Media Proj, 74-; vis prof, Calif Inst Technol, 75; college chem consult serv, Am Chem Soc, 83-; mem Nat Chem Olympian Comt, Am Chem Soc, 84-90, tech adv comt, Grad Rec Exam Bd, 88- *Mem:* AAAS; Am Chem Soc; Am Inst Chem; Sigma Xi. *Res:* Application of nuclear magnetic resonance spectroscopy to chemical problems; molecular association in solutions; chemical education. *Mailing Add:* Dept Chem Univ SFla Tampa FL 33620

DAVIS, JEFFREY ARTHUR, b Meriden, Conn, July 26, 43; m 76. OPTICS. *Educ:* Rensselaer Polytech Inst, BS, 65; Cornell Univ, PhD(physics), 70. *Prof Exp:* Actg asst prof physics, Univ Calif, Los Angeles, 69-71; asst prof, Ill Inst Technol, 71-77; assoc prof, 77-80, PROF PHYSICS, SAN DIEGO STATE UNIV, 80- *Mem:* Am Phys Soc; Soc Photo-Optical Instrumentation Engrs; Optical Soc Am. *Res:* optical pattern recognition; spatial light modulators; development of a highly technical optics education program. *Mailing Add:* Dept Physics San Diego State Univ San Diego CA 92182

DAVIS, JEFFREY ROBERT, b Boise, Idaho, June 22, 35; m 56; c 3. MATHEMATICS. *Educ:* Rensselaer Polytech Inst, BEE, 57, MS, 59; Washington Univ, PhD(math), 63. *Prof Exp:* Instr, Univ Calif, Berkeley, 63-65; asst prof, 65-74, ASSOC PROF MATH, UNIV NMEX, 74- *Mem:* Am Math Soc. *Res:* Toeplitz operators; Wiener-Hopf equations. *Mailing Add:* 4507 Tenth St Northwest Albuquerque NM 87107

DAVIS, JERRY COLLINS, b Richland, Ga, Nov 29, 43; m 66; c 2. ENTOMOLOGY, ZOOLOGY. *Educ:* Mars Hill Col, NC, BS, 65; Univ Tenn, Knoxville, MS, 67; Ohio State Univ, PhD(entomol), 70. *Prof Exp:* Prof biol & vpres develop, Cumberland Col, Ky, 74-77; vpres & prof biol, N Greenville Col, Tigerville, SC, 76-77; PRES, ALICE LLOYD COL, 77- *Concurrent Pos:* Fel, Ohio State Univ, 70; dir, Ky Acad Sci & Appalachian Leadership & Community Outreach, Inc, 77-78. *Mem:* Entomol Soc Am. *Res:* Systematics and biology of a group of tenebrionid beetles. *Mailing Add:* Col Ozarks Point Lookout MO 65726

DAVIS, JIMMY HENRY, b Lexington, Tenn, Mar 22, 48; m 74; c 1. ORGANOMETALLIC & PHYSICAL CHEMISTRY. *Educ:* Union Univ, Tenn, BS, 70; Univ Ill, Urbana, PhD(chem), 76. *Prof Exp:* Fel struct chem, Univ Fla, 76-78; asst prof phys chem, Jackson, Tenn, 78-87, DEAN, BMH CAMPUS, UNION UNIV, MEMPHIS, TENN, 87- *Mem:* Sigma Xi; Am Chem Soc. *Res:* Investigation of the electron spin resonance spectra of seven-coordinated complexes. *Mailing Add:* BMH Campus Union Univ 999 Monroe Memphis TN 38104

DAVIS, JOEL L, b Newton, Mass. PSYCHOBIOLOGY, NEUROSCIENCE. *Educ:* Reed Col, BA, 63; Univ Ore, MS, 67; Univ Calif, Irvine, PhD(psychobiol), 69. *Prof Exp:* Res fel, Brain Res Inst, NY Med Col, 69-70; asst prof psychol, Northern Ill Univ, 70-76; res psychobiologist, Vet Admin Med Ctr, Sepulveda, Calif, 76-80, chief aging & behav biol, 81-84; sci officer, 84-85, prog mgr, neurobiol sect, 85-86, PROG MGR, BIOL INTELLIGENCE, OFF NAVAL RES, 86- *Concurrent Pos:* Exchange fel, Czech Acad Sci-Nat Acad Sci, 71, 82 & 84; adj asst prof, dept psychiat, Med Sch, Univ Calif, Los Angeles, 76-80, adj assoc prof, 81- *Mem:* Soc Neurosci. *Res:* Biological basis of learning and memory; neuropeptides and model systems. *Mailing Add:* Off Naval Res Code 1142 BI 800 N Quincy St Arlington VA 22217

DAVIS, JOHN ALBERT, JR, b Pocatello, Idaho, Apr 22, 36; m 60; c 2. CHEMICAL & PETROLEUM ENGINEERING. *Educ:* Univ Kans, BS, 58, PhD(chem eng), 63; Univ Mich, MSE, 59. *Prof Exp:* Res engr, Marathon Oil Co, Denver, 63-64, adv res engr, 64-70, sr res engr, 70, mgr prod sci dept, 70-72, mgr eng dept, 72-74, res assoc res & chem, 74-77, sr staff engr, Nat Gas Div, 77-79, mgr, liquified natural gas, 79-84, mgr natural gas div, 85-86, DIR PETROL TECHNOL, MARATHON OIL CO, 86- *Mem:* Am Inst Chem Engrs; Am Chem Soc; Soc Petrol Engrs. *Res:* Cryogenics; low temperature phase behavior and thermodynamics; secondary oil recovery processes. *Mailing Add:* Marathon Oil Co PO Box 269 Littleton CO 80169

DAVIS, JOHN ARMSTRONG, b Tupelo, Miss, May 4, 50; m 69; c 2. AQUATIC ECOLOGY, WATER CHEMISTRY. *Educ:* Univ Miss, BA, 72, MS, 75; Auburn Univ, PhD(aquatic ecol), 78. *Prof Exp:* Teaching asst zool, Univ Miss, 72-75; res asst water chem, Auburn Univ, 75-77; vpres & sci dir, Breedlove Assoc, 77-83; PRES ENVIRON SERV, G PERMITTING INC, GAINESVILLE, FLA, 83- *Mem:* AAAS; Am Fisheries Soc; Am Soc Limnol & Oceanog; Sigma Xi; Am Soc Testing & Mat. *Res:* Effects of pollution on aquatic organisms; indices of water quality. *Mailing Add:* Environ Serv & Permitting Inc PO Box 5489 Gainesville FL 32602

DAVIS, JOHN CLEMENTS, b Neodesha, Kans, Oct 21, 38; m 78; c 2. GEOLOGY. *Educ:* Univ Kans, BS, 61; Univ Wyo, MS, 63, PhD(geol), 67. *Prof Exp:* Instr stratig, Idaho State Univ, 64-66; res assoc sedimentary petrol, 66-77, SR SCIENTIST, KANS GEOL SURV, 77- CHIEF ADVAN PROJS, 70-; PROF CHEM & PETROL ENG, UNIV KANS, 74- *Concurrent Pos:* Vis prof, Wichita State Univ, 69; assoc prof chem & petrol eng, Univ Kans, 70-74, prof, 74-, adj prof geog, 74-; vis sr fel geog, Univ Nottingham, Eng, 72-73; dir, Terradata Inc, 78- *Mem:* Am Asn Petrol Geol; Pattern Recognition Soc; Int Asn Math Geol (pres, 84-); Soc Econ Paleont & Mineral. *Res:* Quantitative analysis of sedimentary rock parameters, especially by statistical means and by optical data processing; probabilistic exploration for petroleum. *Mailing Add:* Dept Chem Eng Univ Kans Lawrence KS 66045

DAVIS, JOHN DUNNING, b Freeport, Maine, June 13, 29; m 55; c 2. ENVIRONMENTAL SCIENCES. *Educ:* Bowdoin Col, AB, 52; Univ Boston, MEd, 54; Univ NH, MS, 61, PhD(zool), 63. *Prof Exp:* Sci teacher high sch, NH, 54-56, 57-60; from instr to asst prof zool, Smith Col, 63-70; pres, Normandeau Assocs, Inc, 70-79; RETIRED. *Concurrent Pos:* NSF grant, 64. *Mem:* AAAS; Am Soc Zool; Nat Shellfisheries Asn; Ecol Soc Am; Sigma Xi. *Res:* Systematics and biology of marine Pelecypod mollusks. *Mailing Add:* Neils Point Rd South Harpswell ME 04079

DAVIS, JOHN EDWARD, JR, b Welch, WVa, Nov 18, 22; m 49; c 1. ZOOLOGY. *Educ:* Univ Va, BA, 48, MA, 50, PhD, 55. *Prof Exp:* Instr biol, Washington & Lee Univ, 49-51; asst, Univ Va, 51-54; instr, Washington & Lee Univ, 54-56; from asst prof to prof, Wake Forest Col, 56-68; prof biol, 68-71, chmn, Dept Biol, 68-71, actg provost, 71-72, dean, Sch Arts & Sci, 72-73, prof, 73-86, coordr pre-med studies, 78-86, EMER PROF BIOL, JAMES MADISON UNIV, 86- *Mem:* AAAS; Sigma Xi; NY Acad Sci. *Res:* Embryology; histology. *Mailing Add:* 1106 Ridgewood Rd Harrisonburg VA 22801

DAVIS, JOHN F, b Montreal, Que, Can, Oct 2, 17; m 45; c 2. BIOENGINEERING & BIOMEDICAL ENGINEERING. *Educ:* McGill Univ, BE, 42, ME, 49, MD, 50. *Prof Exp:* Asst prof psychiat, McGill Univ, 60-83; PVT PRACT PSYCHIAT, 74- *Concurrent Pos:* Consult psychiat, Royal Victoria Hosp, 60-83. *Mem:* Inst Elec & Electronics Engrs; AAAS; Biomed Eng Soc; Am Asn Med Instrumentation; Can Psychiat Asn. *Res:* Computer applications in psychiatry. *Mailing Add:* 4773 Sherbrooke Montreal PQ H3Z 1G5 Can

DAVIS, JOHN GRADY, JR, b Durham, NC, Nov 23, 38; m 60; c 2. ENGINEERING MECHANICS, MECHANICAL ENGINEERING. *Educ:* NC State Univ, BS, 62; Va Polytech Inst & State Univ, MS, 65, PhD(eng), 73. *Prof Exp:* Mat engr thermal protection systs, 62-66, mat engr, 66-81, HEAD, MAT PROCESSING & APPLNS BR, LANGLEY RES CTR, NASA, 76- *Concurrent Pos:* NASA coordr composites prog grant & adj prof, Va Polytech Inst & State Univ, 74-; NASA monitor composites res minority & women, Old Dominion Univ, 75-, mgr casts proj, 76-81. *Mem:* Am Soc Testing & Mat. *Res:* Graphite-polyimide materials; composite structures; materials manufacturing; nondestructive evaluation; stress analysis; environmental effects. *Mailing Add:* NASA Langley Res Ctr MS 241 Three Ames Rd Hampton VA 23665

DAVIS, JOHN LITCHFIELD, b Weymouth, Mass, Mar 5, 32; m 61; c 2. SOLID STATE PHYSICS. *Educ:* Bowdoin Col, AB, 53; Univ Md, PhD(physics), 65; George Wash Univ, MEA, 74. *Prof Exp:* Physicist, US Naval Ord Lab, 58-67; physicist, Anal Serv, Inc, 67-81; physicist, Naval Surface Weapons Ctr, 81-85; PHYSICIST, NAVAL RES LAB, WASHINGTON, DC, 85- *Mem:* Am Phys Soc. *Res:* Physics of solid surfaces; infrared detectors; lead salts; epitaxial growth; molecular beam epitaxy. *Mailing Add:* Naval Res Lab Code 6812 4555 Overlook Ave SW Washington DC 20375

DAVIS, JOHN MARCELL, b Kansas City, Mo, Oct 11, 33; m 60, 86; c 2. PSYCHIATRY. *Educ:* Princeton Univ, AB, 56; Yale Univ, MD, 60. *Prof Exp:* Intern, Mass Gen Hosp, 60-61; resident psychiat, Yale Univ, 61-64; clin assoc, NIMH, 64-66, specialist psychopharmacol, 66-69; chief unit clin pharmacol, Lab Clin Sci, NIH, 69-70; assoc prof pharmacol & prof psychiat, Vanderbilt Univ, 70-73; prof psychiat, Sch Med, Univ Chicago, 73-80; GILMAN PROF PSYCHIATRY, SCH MED, UNIV ILL, 80-; DIR RES, ILL STATE PSYCHIAT INST, 73- *Concurrent Pos:* Dir, Clin Div, Tenn Neuropsychiat Inst, Cent State Hosp, 70-73. *Mem:* Am Psychiat Asn; AMA; Am Col Neuropsychopharmacol; Psychiat Res Soc; Am Col Psychiat; Sigma Xi. *Res:* Biology of depression; psychopharmacology; biochemical factors in mania, depression and schizophrenia; catecholamines, serotonin in mental diseases; lithium treatment of mania; treatment of schizophenia. *Mailing Add:* Ill State Psychiat Inst 1601 W Taylor St Rm 309 W Chicago IL 60612

DAVIS, JOHN MOULTON, b Nottingham, UK, Aug 28, 38; US citizen; m 76; c 1. SOLAR PHYSICS, X-RAY ASTRONOMY. *Educ:* Univ Leeds, BSc, 60, PhD(physics), 64. *Prof Exp:* Demonstr physics, Univ Leeds, 63-64; res assoc space physics, Lab Nuclear Sci, Mass Inst Technol, 64-67 & Ctr Space Res, 67-70; from sr scientist to sr staff scientist, Am Sci & Eng, Inc, 70-84; dir space res, 84-85, CHIEF, SOLAR PHYSICS BR, MARSHALL SPACE FLIGHT CTR, 86- *Honors & Awards:* Skylab Acheivement Award, NASA, 74. *Mem:* Am Geophys Union; Am Astron Soc. *Res:* Study of the physics of the solar corona through direct observation of its structure and evolution using soft x-ray images and vector magnetic field measurements; solar activity; solar-terrestrial relations; satellite, rocket and balloon instrumentation with emphasis on grazing incidence optics and hard x-ray imaging. *Mailing Add:* NASA Marshall Space Flight Ctr Code ES 52 Huntsville AL 35812

DAVIS, JOHN R(OWLAND), b Minneapolis, Minn, Dec 19, 27; m 47; c 4. AGRICULTURAL ENGINEERING. *Educ:* Univ Minn, BS, 49, MS, 51; Mich State Univ, PhD(agr eng), 59. *Prof Exp:* Hydraul engr, US Geol Surv, 50-51; instr agr eng, Mich State Univ, 51-55; asst prof, Purdue Univ, 55-57; lectr & specialist, Univ Calif, Davis, 57-62; hydraul engr, Stanford Res Inst, 62-64; prof agr eng, Univ Nebr, 64-71, dean, Col Eng & Archit, 65-71; head dept, 71-75, assoc dean, Sch Agr & dir, Agr Exp Sta, 75-85, PROF AGR ENG, ORE STATE UNIV, 71-, DIR SPEC PROGS, 85- *Concurrent Pos:* Consult, USDA, 60 & Stanford Res Inst, 64; vpres, Nat Col Athletic Asn, 79-83, secy-treas, 83-85 & pres, 85- *Mem:* Am Soc Agr Engrs. *Res:* Hydrology; hydraulics, irrigation systems; methods of irrigation water application; irrigation feasibility; engineering education; teaching methods. *Mailing Add:* Gill 103 Ore State Univ Corvallis OR 97331

DAVIS, JOHN ROBERT, b Mattoon, Ill, July 10, 29; m 52; c 3. PATHOLOGY. *Educ:* Univ Iowa, BA, 52, MD, 59. *Prof Exp:* Instr path & resident anat path, Col Med, Univ Iowa, 60-63, assoc path, 63-64, asst prof, 64-67; assoc prof, 67-76, PROF PATH, COL MED, UNIV ARIZ, 76- *Concurrent Pos:* Consult pathologist, Vet Admin Hosp, Tucson, 68- *Mem:* AAAS; AMA; Col Am Path; Int Acad Path; Am Soc Cytol; Am Asn Path & Bact. *Res:* Experimental pathology and immunopathology of schistosomiasis, gynecopathology and miscellaneous infectious diseases; surgical pathology; cytopathology computer pattern analysis. *Mailing Add:* Dept Path Univ Ariz Med Ctr Tucson AZ 85721

DAVIS, JOHN SHELDON, b Fall River, Mass, Aug 24, 46; m 69; c 3. ELECTRICAL ENGINEERING. *Educ:* Southeastern Mass Univ, BS, 68, MS, 75. *Prof Exp:* Staff engr, 68-70, proj engr, 70-76, res prog mgr estimation theory, 76-79, supvy engr, systs anal & synthesis for combat control systs, 79-, HEAD, SYSTS MGT & ASSESSMENT DIV, NAVAL UNDERWATER SYSTS CTR, NEWPORT, RI. *Mem:* Inst Elec & Electronics Engrs. *Res:* Application of advanced statistical estimation techniques and modern control theory methods to critical Navy problems; advanced systems synthesis, test and evaluation. *Mailing Add:* Head Systs Mgt & Assessment Div Naval Underwater Systs Ctr Newport RI 02948

DAVIS, JOHN STAIGE, IV, b New York, NY, Oct 28, 31; m 56; c 5. INTERNAL MEDICINE. *Educ:* Yale Univ, BA, 53; Univ Pa, MD, 57; Am Bd Internal Med, recert, 78. *Prof Exp:* From instr to asst prof internal & prev med, 61-67, assoc prof internal med, 67-72, PROF INTERNAL MED, SCH MED, UNIV VA, 72-, HEAD DIV RHEUMATOL, 67-, TROLINGER PROF RHEUMATOL, 83- *Concurrent Pos:* Sr investr, Arthritis Found, 64-69; Markle scholar acad med, 64-69; vis prof, chair immunol, Univ Milan Italy, 66-77; chmn, Coun Subspecialty Socs, ACP; prof invite, Univ Geneva, Switz, 78-79. *Mem:* Am Clin & Climat Asn; Am Rheumatism Asn; Am Fedn Clin Res; AMA; Am Asn Immunol; Am Bd Int Med; fel Am Col Phys. *Res:* Immune complexes in connective tissue diseases, especially the roles of rheumatoid factors, complement and cryoglobulins. *Mailing Add:* Dept Internal Med Univ Va Sch Med Box 412 Charlottesville VA 22908

DAVIS, JOHN STEWART, b Fargo, NDak, Nov 11, 52; m 74; c 2. ENDOCRINOLOGY, REPRODUCTION. *Educ:* Minot State Col, BA, 75; Univ NDak, MS, 77, PhD(physiol), 79. *Prof Exp:* Teaching asst, dept physiol, Sch Med, Univ NDak, 75-79; fel reproductive biol & biochem, Endocrine Lab, Sch Med, Univ Miami, 79-81, res assoc, dept obstet & gynec, 81-82, asst prof, 82-83; ASST PROF, DEPT INTERNAL MED, DIV ENDOCRINOL & METAB, COL MED, UNIV SFLA, 83-; RES PHYSIOLOGIST, RES SERV, 83-, ASSOC PROF, DEPT INT MED, DIV ENDOCRINOLOGY & METALS, 88- *Concurrent Pos:* NIH grants, 82-86, & 83-89; mem admis comt, Col Med, Univ SFla, 83-84; mem res & develop comt, James A Haley Vet Hosp, 83- & res animals studies subcomt, 84-; prin investr, Vet Admin Merit Rev, 85-88, 88-92. *Mem:* Soc Study Reproduction; Endocrine Soc; Sigma Xi; AAAS. *Res:* Reproductive endocrinology; mechanisms of hormone action; evaluation of local reproductive endocrine regulatory mechanisms and ovarian function; role of membrane phospholipids in ligand-receptor-mediated cell activation; protein phosphorylation. *Mailing Add:* Res Serv Vet Admin Med Ctr 901 George Washington Blvd Wichita KS 67211

DAVIS, JOHNNY HENRY, b Crossville, Ala, May 3, 20; m 40; c 2. AGRONOMY. *Educ:* Ala Polytech Inst, BS, 43; Purdue Univ, MS, 52, PhD(plant breeding, genetics), 53. *Prof Exp:* Asst county agent, Ala Agr Exten Serv, 43-50; jr agronomist, Purdue Univ, 50-53; geneticist, Miss Agr Exp Sta, 53-54; agronomist, Tex Res Found, 54-56, assoc head high plains sta, 57-58; asst dir, High Plains Res Found, 59-60; assoc agronomist, LA State Univ-Iberia Liverstock Exp Sta, 60-68, supt, 74-84; RETIRED. *Concurrent Pos:* Assigned to La State Univ-US AID Mission, Managua, Nicaragua, 71-72. *Mem:* Am Soc Agron; Genetics Soc Am. *Res:* Plant breeding; crop production and management. *Mailing Add:* 641 Chippenham Dr Baton Rouge LA 70808

DAVIS, JON PRESTON, b Eau Claire, Wis, Nov 26, 45; m 69; c 3. SCATTERING THEORY. *Educ:* Marquette Univ, BS, 68; Columbia Univ, PhD(chem), 75. *Prof Exp:* Fel chem, Univ Alberta, 75-77; lectr, Univ Troms, Norway, 78-79; vis prof, Univ Mo-Columbia, 79-80; ASST PROF CHEM, PA STATE UNIV, 80- *Mem:* Am Chem Soc; Sigma Xi. *Res:* Reaction dynamics of small systems forming collision complexes; radiative association of polyatomic molecules. *Mailing Add:* 429 Krewson Terr Willow Grove PA 19090

DAVIS, JOSEPH B, b Forsyth Co, NC, Dec 2, 33; m 78; c 1. SPECTRACOPY, CHROMATOGRAPHY. *Educ:* Western Carolina Col, BS, 60; Clemson Univ, MS, 63, PhD(anal chem), 65. *Prof Exp:* Lab instr chem, Fla Presby Col, 60-61; asst, Clemson Univ, 62-65; from asst prof to assoc prof, 65-73, prof chem, Winthrop Col, 73-, chmn dept, 72-79; PRES, DAVIS & STALLINGS INC, 87- *Concurrent Pos:* Editor, C-P News, Am Chem Soc, 80-88. *Mem:* Am Chem Soc. *Res:* Analytical instrument design and development; spectroscopy; electro-chemistry; chromatography; chemical research development; chemical software development; analysis via chemical equilibrium. *Mailing Add:* Dept Chem Winthrop Col Rock Hill SC 29730

DAVIS, JOSEPH HARRISON, b Flushing, NY, Apr 16, 24; m 52; c 7. FORENSIC MEDICINE. *Educ:* Long Island Col Med, MD, 49. *Prof Exp:* Intern surg, Univ Calif Hosp, San Francisco, 49-50, asst resident path, 50-51; asst surgeon, USPHS, 51-52, sr asst surgeon, USPHS Hosp, Seattle, Wash, 52-54, New Orleans, La, 54-55; instr path, Sch Med, La State Univ, 55-56; asst med examr, 56-57, MED EXAMR, OFF MED EXAMR, MIAMI, 57-; PROF PATH, SCH MED, UNIV MIAMI, 60- *Concurrent Pos:* Consult pathologist, Fed Aviation Agency. 60-80. *Mem:* AMA; Am Acad Forensic Sci; Am Soc Clin Path. *Res:* Accident causation and prevention with emphasis on human factors involved. *Mailing Add:* Dade County Off Med Examr Sch Med Univ Miami One Bob Hope Rd Miami FL 33136

DAVIS, JOSEPH RICHARD, b Chicago, Ill, May 13, 36; m 58; c 3. TOXICOLOGY, ONCOLOGY. *Educ:* Univ Ill, BS, 56, MD, 58, MS, 59; Baylor Univ, PhD(pharmacol), 61. *Prof Exp:* Instr pharmacol, Col Med, Baylor Univ, 60-61; from asst prof to assoc prof, 61-69, PROF PHARMACOL, STRITCH SCH MED, LOYOLA UNIV CHICAGO, 69- *Concurrent Pos:* Fel, Am Cancer Soc, 58-61; Lederle Med Fac Award, 62-65. *Honors & Awards:* Bordon Res Award, 58; Wiiliam B Peck Sci Res Award, 70. *Mem:* Am Asn Cancer Res; Am Soc Pharmacol & Exp Therapeut; Am Physiol Soc; Endocrine Soc; Soc Study Reproduction; Sigma Xi. *Res:* Biochemical approaches to cancer chemotherapy; protein synthesis of the normal and cryptorchid testis; pharmacology of the testicular capsule; urinary delta-aminolevulinic acid in lead poisoning; biology of azopurines; cholesterol and atherosclerosis; pharmacology. *Mailing Add:* Dept Pharmacol Stritch Sch Med Loyola Univ Chicago 2160 S First Ave Maywood IL 60153

DAVIS, JOYCE S, medicine, pathology, for more information see previous edition

DAVIS, KAREN PADGETT, b Blackwell, Okla, Nov 14, 42; m 63, 69; c 1. HEALTH ECONOMICS, HEALTH POLICY. *Educ:* Rice Univ, BA, 65, PhD(econ), 69. *Prof Exp:* Asst prof econ, Rice Univ, 68-70; econ policy fel social security, Brookings Inst, 70-71, res assoc, 71-74, sr fel, 74-77; asst secy, Dept Planning & Eval Health, HEW, 77-80, adminr, Health Resources Admin, Pub Health Serv, 80-81; PROF & CHMN, DEPT HEALTH POLICY & MGT, JOHNS HOPKINS SCH HYG & PUB HEALTH, 83- *Concurrent Pos:* Mem, Health Tech Study Sect, HEW, 72-76; assoc ed, Milbank Mem Fund Quarterly, 72-77; consult, NSF, 73-74 & Nat Acad Sci, 74-75; vis lectr econ, Harvard Univ, 74-75; mem, Health Adv Panel, US Cong, Off Technol Assessment, 75-77; consult, WHO, 81-86; co-chair, Joint Policy Comt, Am Pub Health Asn, 82-83, chair, Action Bd, 83-84; US regional ed, Health Policy, 85-90; adminr, Agency Health Care Policy & Res, US Dept Health & Human Serv, 90-91. *Mem:* Inst Med-Nat Acad Sci; Am Econ Asn; Am Pub Health Asn; Asn Health Serv Res. *Res:* Health economics; health care financing; health care delivery; author of six books. *Mailing Add:* Sch Hyg & Pub Health Johns Hopkins Univ 624 N Broadway Baltimore MS 21205

DAVIS, KENNETH BRUCE, JR, b Texarkana, Ark, Mar 13, 40; m 69, 87; c 1. VERTEBRATE PHYSIOLOGY. *Educ:* Univ Ark, BA, 63, MS, 65; La State Univ, PhD(physiol, vert zool), 70. *Prof Exp:* Instr gen zool, La State Univ, 68-69; from asst prof to assoc prof, 69-80, PROF BIOL, MEMPHIS STATE UNIV, 80- *Concurrent Pos:* Res award, Assoc Southeastern Biologist, 76; res physiologist, Southeastern Fish Cult Lab, US Fish & Wildlife Serv, Marion, Ala, 78-79. *Mem:* AAAS; Am Inst Biol Sci; Am Soc Zool; Am Ornith Union. *Res:* Role of annual and diurnal rhythms of hormones in the regulation of daily and seasonal behavior of sub-mammalian vertebrates, particularly fish; intra-extracellular electrolyte distribution and osmoregulation; stress second reproductive physiology in fish. *Mailing Add:* Dept Biol Memphis State Univ Memphis TN 38152

DAVIS, KENNETH JOSEPH, b Spartanburg, SC, Aug 10, 37; m 59; c 4. MATHEMATICS. *Educ:* Wofford Col, BS, 59; Univ Tenn, MA, 62, PhD(math), 63. *Prof Exp:* Res scientist, Marshall Space Flight Ctr, NASA, 63-65; assoc prof math, Old Dom Col, 65-66; assoc prof, 66-77, PROF MATH, ECAROLINA UNIV, 77- *Mem:* Am Math Soc; Math Asn Am. *Res:* Analytic number theory; control theory. *Mailing Add:* Dept Math ECarolina Univ 312 Rutledge Greenville NC 27834

DAVIS, KENNETH LEON, b New York, NY, Sept 10, 47; m 72; c 2. PSYCHIATRY. *Educ:* Yale Univ, BA, 69; Mt Sinai Sch Med, MD, 73; Am Bd Psychiat & Neurol, dipl, 80. *Prof Exp:* Res asst, Dept Pharmacol, Mt Sinai Sch Med, 71-73; intern, Stanford Univ, 73-74, fel clin pharmacol, 74-76; clin psychiat consult, Santa Clara Valley Med Ctr, 76-79; ASSOC PROF PSYCHIAT & PHARMACOL, MT SINAI SCH MED, 79-; CHIEF, DEPT PSYCHIAT, VET ADMIN MED CTR, 79- *Concurrent Pos:* Consult, Nassau County Mental Health Bd, 69-73; resident, Sch Med, Stanford Univ, 73-76, life sci res assoc, 75-76; dir & founder, Stanford Comprehensive Care Clin, Stanford Hosp, 74-79; asst dir, Stanford Psychiat Clin Res Ctr, Vet Admin Med Ctr, 75-79, res assoc, 76-79; A E Bennett Clin Sci Res Award, 77; dir, Schizophrenia Biol Res Ctr, Vet Admin Med Ctr, 81- *Honors & Awards:* Saul Horowitz Jr Mem Award, 77-78; Solomon Silner Award, 81. *Mem:* Am Psychiat Asn; Soc Biol Psychiat; Acad Psychosomatic Med; Am Col Neuropsychopharmacol; NY Acad Med. *Res:* Biological basis of senile dementia of the Alzheimer's type; depression and schizophrenia. *Mailing Add:* 130 W Kingsbridge Rd Bronx NY 10468

DAVIS, L(LOYD) WAYNE, b Medicine Lodge, Kans, July 16, 29; m 63; c 3. CORPORATE MANAGEMENT, NUCLEAR WEAPONS EFFECTS. *Educ:* Univ Kans, BS, 52; Univ NMex, MS, 59. *Prof Exp:* Staff mem, Systs Anal Dept, Sandia Corp, 52-56, consult, 56-57; res physicist, Dikewood Corp, 57-60, sr res physicist, 60-64, head Weapons Effects Div, 64-67, dep tech dir, 67-69, asst vpres, 69-72, vpres, 72-77, secy, 70-80, dir, 71-82, sr vpres, 77-80, pres & chmn bd, 80-82, vpres Kaman Sci Corp & gen mgr, Dikewood Div, Albuquerque, 82-83; SCI CONSULT, 83- *Mem:* Sr mem Inst Elec & Electronic Engrs; Sigma Xi. *Res:* Nuclear weapons effects and phenomenology effects on personnel and complex military systems; developed urban nuclear-casualty prediction model for high-yield nuclear bursts from Japanese data base over many years. *Mailing Add:* Sci Consult 4411 Altura Ave NE Albuquerque NM 87110

DAVIS, LANCE A(LAN), b Ridley Park, Pa, Nov 19, 39; m 62; c 3. MATERIALS SCIENCE, PHYSICS. *Educ:* Lafayette Col, BS, 61; Yale Univ, ME, 63, PhD(eng & appl sci), 66. *Prof Exp:* Res staff scientist, Yale Univ, 66-68; res physicist, Mat Res Ctr, Allied Signal Corp, 68-74, mgr, Strength Physics Dept, 74-78, mgr, Metglas Develop Sect, Corp Develop Ctr, 78-80, dir, Mat Lab, 81-84, VPRES RES & DEVELOP, ALLIED SIGNAL CORP, 84- *Concurrent Pos:* Mem, Comt Technol Potential Amorphous Mat, Nat Mat Adv Bd, 78. *Mem:* Am Phys Soc; fel Am Soc Metals; Am Inst Mining, Metall & Petrol Engrs; Mat Res Soc; Sigma Xi. *Res:* Physical properties of materials at high pressure, particularly elastic, anelastic and plastic properties; elasticity, plasticity and fracture of metallic glasses; technology of metallic glasses. *Mailing Add:* Corp Technol Allied-Signal Inc PO Box 1021R Morristown NJ 07962

DAVIS, LARRY ALAN, b Delano, Calif, June 1, 40; m 60; c 3. CROP PHYSIOLOGY. *Educ:* La Verne Col, BA, 62; Univ Calif, Davis, MS, 64, PhD(plant physiol), 68. *Prof Exp:* Agronomist & asst mgr, Alina Farms Corp, 64-65; agronomist & mgr, Calcot Pty, Ltd, Australia, 68-70; agronomist in charge & pres, Alina Farms Corp, 70-80; MANAGING PARTNER, FRIENDSHIP FARMS, 80- *Concurrent Pos:* Adj lectr biol, Calif State Col, Bakersfield, 72. *Mem:* AAAS; Am Soc Plant Physiol; Am Inst Biol Sci; Am Soc Agron; Crop Sci Soc Am; Sigma Xi. *Res:* Control of fruit growth and development in the cotton plant by means of plant hormones and the identification and measurement of these hormones by gas-liquid chromatography and other methods. *Mailing Add:* Rte One Box 290 McFarland CA 93250

DAVIS, LARRY DEAN, b Marathon, Iowa, June 23, 35; m 58; c 2. HUMAN PHYSIOLOGY. *Educ:* Univ Dubuque, BS, 57; Univ Wis, PhD(physiol), 61. *Prof Exp:* From instr to assoc prof, 63-75, PROF PHYSIOL, SCH MED, UNIV WIS-MADISON, 75- *Concurrent Pos:* USPHS fel, 61-63, career develop award, 75-80. *Mem:* Am Physiol Soc; Soc Exp Biol & Med. *Res:* Cardiac physiology, specifically cellular transmembrane potential changes involved in origin of arrhythmias and antiarrhythmic actions. *Mailing Add:* Dept Physiol Univ Wis Sch Med 1300 University Ave Madison WI 53706

DAVIS, LARRY ERNEST, b New York, NY, Aug 16, 40; m 68; c 2. NEUROLOGY, VIROLOGY. *Educ:* Stanford Univ, BS, 63, MD, 66. *Prof Exp:* Officer virol, Ctr Dis Control, 68-70; instr neurol, Med Sch, Johns Hopkins Univ, 73-75; from asst prof to assoc prof neurol & microbiol, 75-83, PROF NEUROL, UNIV NMEX, 83- *Honors & Awards:* Moore Award, Am Asn Neuropathologists, 79, Weil Award, 84. *Mem:* Am Neurol Asn; Am Acad Neurol; Am Soc Microbiol; Am Asn Neuropathologists; fel Am Col Physicians. *Res:* Infections of brain and inner ear; Reye's syndrome. *Mailing Add:* Dept Neurol Univ NMex Albuquerque NM 87131

DAVIS, LAWRENCE CLARK, b London, Eng, Aug 16, 45; US citizen; m 67; c 3. BIOCHEMISTRY, MOLECULAR BIOLOGY. *Educ:* Haverford Col, BS, 66; Yeshiva Col, PhD(molecular biol), 70. *Prof Exp:* Res assoc neurobiol, A Ribicoff Res Ctr, Norwich Hosp, Conn, 71-73; res assoc biochem, Univ Wis-Madison, 73-75; from asst prof to assoc prof, 75-84, PROF BIOCHEM, KANS STATE UNIV, 84- *Concurrent Pos:* NIH fel, Univ Wis-Madison, 70-71, NSF fel, 75. *Mem:* AAAS; Am Soc Microbiol; Am Soc Biol Chemists; Biophys Soc; Am Genetic Asn; Am Asn Biol Chemists. *Res:* Interacting protein systems; biological nitrogen fixation; diffusion processes; protein structure-function relations. *Mailing Add:* Dept Biochem Willard Hall Kans State Univ Manhattan KS 66506

DAVIS, LAWRENCE H, b Hot Springs, Ark, Aug 18, 29. MATHEMATICS EDUCATION, ALGEBRA. *Educ:* Northwestern State Univ, BS, 53, MS & MSE, 58; La State Univ, PhD(math educ), 76. *Prof Exp:* From instr to assoc prof, 58-76, PROF MATH, SOUTHEASTERN LA UNIV, 76-, HEAD DEPT, 87- *Concurrent Pos:* NSF fel, Fla State Univ, 66; mem, Nat Coun Teachers Math. *Mem:* Math Asn Am. *Mailing Add:* Dept Math Southeastern La Univ Box 326 Univ Sta Hammond LA 70402

DAVIS, LAWRENCE S, b Missoula, Mont, July 27, 34; m 59; c 3. FOREST ECONOMICS. *Educ:* Univ Mich, BS, 56, MF, 60; Univ Calif, Berkeley, PhD(agr econ), 64. *Prof Exp:* Res forester, US Forest Serv, 56-57; res specialist, Univ Calif, Berkeley, 60-64; from asst prof to assoc prof forest econ, Va Polytech Inst, 64-70; assoc prof forest sci, Utath State Univ, 70-72, prof, 72-, head, dept forestry & outdoor redreation, 70-; AT DEPT FORESTRY MGT, UNIV CALIF, BERKELEY. *Mem:* Soc Am Foresters. *Res:* Resource policy; recreation economics. *Mailing Add:* Dept Forestry Mgt Utah Calif Berkeley CA 94720

DAVIS, LAWRENCE WILLIAM, JR, physics; deceased, see previous edition for last biography

DAVIS, LEODIS, b Stamps, Ark, Sept 25, 33; m 62; c 2. BIOCHEMISTRY. *Educ:* Univ Kansas City, BS, 56; Iowa State Univ, MS, 58, PhD(biochem), 60. *Prof Exp:* Res asst biochem, Iowa State Univ, 60-61; asst prof, Tenn State Univ, 61-63; asst prof, Col Med, Howard Univ, 63-68; vis prof chem, 68-69, assoc prof, 69-76, PROF CHEM, UNIV IOWA, 76-, CHMN CHEM, 79- *Mem:* Am Chem Soc; Am Soc Biol Chem; Sigma Xi. *Res:* Mechanism of pyridoxal phosphate requiring enzymes. *Mailing Add:* Dept Chem Univ Iowa Iowa City IA 52242-1219

DAVIS, LEONARD GEORGE, b Chicago, Ill, Nov 23, 46; m 69; c 3. NEUROSCIENCE, NEUROCHEMISTRY. *Educ:* Univ Ill-Urbana, BS, 69; Northwestern Univ, MS, 73; Univ Ill, Chicago, PhD(biochem), 77. *Prof Exp:* Clin res, Beaumont Army Med Ctr, 75-76; res assoc neurochem, Mo Inst Psychiat, Univ Mo-Columbia, 76-80; GROUP LEADER, MOLECULAR NEUROBIOL GROUP, DUPONT MERCK PHARMACOL CO, 80- *Concurrent Pos:* NIMH traineeship, Northwestern Univ Adminr, 71-73; fel, Ill State Psychiat Inst, 74; USPHS neurosci trainee, Univ Ill Med Ctr Admin, 74-75. *Mem:* Soc Neurosci; Am Soc Neurochem; AAAS. *Res:* Gene regulation in neural systems; implications of hormones in mental health. *Mailing Add:* Neurobiol Group DuPont Merck Pharmacol Co Exp Sta E328/146A Wilmington DE 19880-0328

DAVIS, LEROY, b Orangeburg, SC; m 73; c 2. MICROBIAL PHYSIOLOGY. *Educ:* SC State Col, BS, 71; Purdue Univ, MS, 72, PhD(molecular biol), 79. *Prof Exp:* Instr, 73-76, from asst prof to assoc prof, 79-85, PROF BIOL, SC STATE COL, 85- *Concurrent Pos:* Extramural assoc, NIH, 81; vis res assoc, Brookhaven Nat Lab, 82. *Mem:* Am Soc Microbiol; AAAS; Am Asn Univ Prof; NY Acad Sci. *Res:* Regulation of amino acid and RNA biosynthesis in histidine regulatory mutants of salmonella typhimurium. *Mailing Add:* 450 Meadowlark NE Orangeburg SC 29115

DAVIS, LEROY THOMAS, b New Castle, Pa, Feb 12, 25; m 48; c 3. FAMILY MEDICINE. *Educ:* Westminster Col, BS, 48; Syracuse Univ, MS, 51, PhD(zool), 54; NY Med Col, MD, 61; Am Bd Family Pract, dipl, 73. *Prof Exp:* Instr physiol, NY Med Col, 54-57; intern, Easton Hosp, Pa, 61-62; pvt pract, Md, 62-72; asst prof med, Med Sch, Univ Md, 71-72, assoc prof family pract, 72-78; PVT PRACT, 78- *Mem:* Fel Am Acad Family Physicians; Soc Teachers Family Pract. *Res:* Classification and computerization of the health problems in family practice. *Mailing Add:* Toll House Med Ctr 801 Toll House Ave Frederick MD 21701

DAVIS, LEVERETT, JR, b Elgin, Ill, March 3, 14; m 43; c 1. SPACE PHYSICS. *Educ:* Ore State Col, BS, 36; Calif Inst Technol, MS, 38, PhD(physics), 41. *Prof Exp:* Instr & res staff mem, 41-46, from asst prof to prof, 46-81, group supvr, 44-46, EMER PROF PHYSICS, CALIF INST TECHNOL, 81- *Concurrent Pos:* NSF sr fel, Max Planck Inst Physics, Univ Gottingen, 57-58. *Honors & Awards:* Exceptional Sci Achievement Medal, NASA, 70. *Mem:* Am Phys Soc; Am Astron Soc; Am Geophys Union; Int Astron Union. *Res:* Planetary magnetic fields, interplanetary medium; cosmic rays; astrophysics; hydromagnetics; polarization of starlight. *Mailing Add:* Dept Physics Calif Inst Technol 412 Downs Pasadena CA 91125

DAVIS, LLOYD CRAIG, b Council Bluffs, Iowa, May 29, 41; m 63; c 3. PHYSICS. *Educ:* Iowa State Univ, BS, 63, PhD(physics), 66; Calif Inst Technol, MS, 66. *Prof Exp:* Res assoc physics, Iowa State Univ, 66-67; res assoc, Univ Ill, Urbana, 67-69; mem sci res staff, 69-87, MGR, PHYSICS DEPT, FORD MOTOR CO, 87- *Concurrent Pos:* NSF fel, Univ Ill, Urbana, 67-68; guest scientist, Deutsches Elektronen-Synchrotron, Hamburg, Ger, 81. *Mem:* Fel Am Phys Soc; Sigma Xi. *Res:* Magnetic levitation of high-speed ground vehicles; electron tunneling in solids; electron energy levels of solids in a magnetic field; sound velocity and lattice vibrations in semiconductors; photoemission, electron energy loss, absorption and Auger spectra of 3d metals; electronic structure of alloy semiconductors; applications of superconductors; scanning tunneling microscopy; mechanical properties of composite materials. *Mailing Add:* S 3028 Sci Lab Ford Motor Co Dearborn MI 48121-2053

DAVIS, LLOYD EDWARD, b Akron, Ohio, Aug 23, 29; m 72; c 2. PHARMACOLOGY. *Educ:* Ohio State Univ, DVM, 59; Univ Mo-Columbia, PhD(pharmacol), 63. *Prof Exp:* Instr vet pharmacol, Univ Mo-Columbia, 59-63, assoc prof pharmacol, Sch Med, 63-69; prof, Ohio State Univ, 69-72; vis prof, Univ Nairobi, Kenya, 72-74; prof clin pharmacol, Colo State Univ, 74-78; PROF CLIN PHARMACOL, UNIV ILL, URBANA, 78- *Concurrent Pos:* Chmn & mem, Adv Panel Vet Med, Gen Rev Comt & Standards Comt, US Pharmacopeia; chmn, Subcomt Vet Pharmacol; ed, Topics Drug Ther, J Am Vet Med Asn. *Mem:* Am Soc Pharmacol & Exp Therapeut; Soc Exp Biol & Med; Am Col Vet Pharmacol Therapeut (pres, 77-79); Am Asn Vet Clinicians; Am Soc Clin Pharmacol Therapeut; Am Vet Med Asn. *Res:* Comparative pharmacology; drug metabolism and pharmacokinetics in domestic animals; clinical pharmacology; Species differences in disposition and fate of drugs; influence of disease on drug disposition; evaluation of drug therapy in animal patients; biomedical ethics. *Mailing Add:* Col Vet Med Univ Ill 1102 W Hazelwood Dr Urbana IL 61801

DAVIS, LOUIS E(LKIN), b Brooklyn, NY, Sept 10, 18; m 42; c 2. MANAGEMENT, ORGANIZATION. *Educ:* Ga Inst Technol, BSME, 40; Univ Iowa, MS, 42. *Prof Exp:* Asst prof indust eng, Ga Inst Technol, 42-43; supvr, Bell Aircraft Corp, Ga, 43-44; sr indust engr, Western Elec Co, NY, 44-46; asst chief indust eng, Conmar Prods Corp, NJ, 46-47; from asst prof to prof indust eng, Univ Calif, Berkeley, 47-66; prof indust eng, Univ Calif, Los Angeles, 66-71, chmn, Grad Sch Mgt, 67-71, dir, Ctr Orgn Studies, 71, prof orgn sci, 66-89, EMER PROF ORGN SCI, UNIV CALIF, LOS ANGELES, 89- *Concurrent Pos:* Consult, Europ Productivity Agency, France, 57-58, Maritime Transport Res, Nat Acad Sci-Nat Res Coun, 58-62, US Forest Serv, 62-64; adv, Exec Off Pres, 62-64; Lucas prof, Univ Birmingham, 62-63; comnr, Calif State Comn Manpower, Automation & Technol & Manpower Adv Comn, 63-67; res fel, Tavistock Inst Human Rels, London & Inst Indust Social Res, Norway, 66; res socio-tech scientist, Univ Calif, Los Angeles, 76-; chmn, Ctr Qual Working Life, Inst Indust Rels, 75-82; pres, Davis Group Inc, Orgn Consults, 79-88; vis distinguished prof, Univ Queensland, 81. *Honors & Awards:* Hayhow Award, Am Col Hosp Admin. *Mem:* Am Inst Indust Engrs; Human Factors Soc; Brit Ergonomics Res Soc; Brit Inst Indust Psychol; Inst Mgt Sci. *Res:* Human performance skills and measurement; job and organization design; socio-technical systems. *Mailing Add:* Grad Sch Mgt Univ Calif Los Angeles CA 90024

DAVIS, LUCKETT VANDERFORD, b Smyrna, Tenn, Oct 16, 32. ENTOMOLOGY, ENVIRONMENTAL BIOLOGY. *Educ:* Mid Tenn State Col, BS, 55; Duke Univ, MA, 58, PhD(zool), 62. *Prof Exp:* Instr biol, Vanderbilt Univ, 60-61; asst prof, Univ Southwestern La, 61-63; asst prof biol sci, Univ of the Pac, 63-64; assoc prof, 64-69, PROF BIOL, WINTHROP COL, 69-, CHMN DEPT, 77- *Mem:* Entom Soc Am; Ecol Soc Am; Am Soc Zool; Am Soc Limnol & Oceanog. *Res:* Ecology of insects of marine-influenced ecosystems. *Mailing Add:* Dept Biol Winthrop Col Rock Hill SC 29733

DAVIS, LUTHER, JR, b Mineola, NY, July 12, 22; m 51; c 3. SOLID STATE ELECTRONICS. *Educ:* Mass Inst Technol, BS, 42, PhD(physics), 49. *Prof Exp:* Mem staff, Radiation Lab, Mass Inst Technol, 42-45, res assoc physics, 45-49; mem staff, Raytheon Co, 49-64, asst mgr res div, 64-69, gen mgr res div, 69-87; RETIRED. *Mem:* Am Phys Soc; fel Inst Elec & Electronics Engrs. *Res:* Semiconductors; dielectrics; ferromagnetics. *Mailing Add:* Two Winthrop Terr Wayland MA 01778

DAVIS, LYNNE, b Swansea, Wales, July 4, 33. RESPIRATORY PHYSIOLOGY. *Educ:* McGill Univ, PhD(physiol), 69. *Prof Exp:* PROF INTERNAL MED, WAYNE STATE UNIV, 80- *Mem:* Am Physiol Soc; Am Fedn Clin Res; Am Thoracic Soc; Royal Col Physicians Can. *Mailing Add:* ACOS Res & Develop Vet Admin Med Ctr Allen Park MI 48101

DAVIS, MARC, b Canton, Ohio, Sept 8, 47; m 69. ASTROPHYSICS. *Educ:* Mass Inst Technol, SB, 69; Princeton Univ, MA, 71, PhD(physics), 73. *Prof Exp:* Instr physics, Princeton Univ, 73-75; PROF ASTRON & PHYSICS & CHAIR, DEPT ASTRON, UNIV CALIF, BERKELEY, 75- *Concurrent Pos:* Alfred P Sloan res fel, Sloan Found, 75-77. *Mem:* Nat Acad Sci; Am Astron Soc. *Res:* Statistical properties of galaxy clustering; techniques for detection of galaxy haloes. *Mailing Add:* Dept Astron Univ Calif Berkeley CA 94720

DAVIS, MARGARET BRYAN, b Boston, Mass, Oct 23, 31. PALEOECOLOGY, PALYNOLOGY. *Educ:* Radcliffe Col, AB, 53; Harvard Univ, PhD(biol), 57. *Hon Degrees:* MS, Yale Univ, 74. *Prof Exp:* Fel, dept biol, Harvard Univ, 57-58, dept geosci, Calif Tech Inst, 59-60, res, Dept Zool, Yale Univ, 60-61; res assoc, dept bot, Univ Mich, 61-64, assoc res biologist, Great Lakes Res Div, 64-70, assoc prof, dept zool, 66-70, res biologist, Great Lakes Res Div & prof zool, 70-73; prof biol, Yale Univ, 73-76; prof ecol & head, Dept Ecol & Behav Biol, 76-81, prof, 81- 83, REGENTS PROF ECOL, UNIV MINN, 83- *Concurrent Pos:* mem, US nat comt, Int Union Quaternary Res, Nat Acad Sci & Nat Res Coun, 66-74, vchmn, 72-74, ad hoc mem, 76-80, comt int exchange persons, 72-75, comt water qual policy, 74-76, working group, Holocene & Pleistocene Hist & US-USSR Bilateral Agreement Environ Protection, 82, comt plaentary biol, Space Sci Bd, 82-83, res comt sci, math & tech educ, 84-; Nat Acad Sci deleg, Int Union Quaternary Res Cong, 69, 73, 77 & 82; mem, bd dirs, Int Asn Great Lakes Res, 70-73; vis prof, Quaternary Res Ctr, Univ Wash, 73; mem, adv panel ecol sci, NSF, 76-79; mem, eval comt ecol grad prog, Rutgers Univ, 81; vis investr, environ studies prog, Univ Calif, Santa Barbara, 81-82; mem, sci adv comt, Biol, Behav & Soc Sci Directorate, NSF; mem adv panel, Geol Records Global Changes, NSF, 81-, comt Global Change, 87-90. *Mem:* Nat Acad Sci; fel Geol Soc Am; fel AAAS; Am Quaternary Asn (pres, 78-80); Am Soc Limnol & Oceanog; Ecol Soc Am (pres, 87-88); Int Soc Veg Sci; Sigma Xi. *Res:* Paleoecology and biogeography of American forest communities, especially Late-Quaternary history as recorded by fossil pollen; watershed-lake interactions and ecosystem development over long time intervals; sedimentary processes in lakes; biotic responses to climatic changes; author and co-author of numerous publications. *Mailing Add:* Dept Ecol Evolution & Behav Univ Minn 318 Church St SE Minneapolis MN 55455-0302

DAVIS, MARJORIE, developmental biology, invertebrate zoology; deceased, see previous edition for last biography

DAVIS, MARSHALL EARL, b Richmond, Calif, Aug 14, 31; m 56; c 2. FUEL TECHNOLOGY. *Educ:* WTex State Col, BS, 57; Purdue Univ, PhD(phys & nuclear chem), 63. *Prof Exp:* Res staff mem, Fission Prod Characterization, Oak Ridge Nat Lab, 62-65; chemist, Texaco Inc, 65-70, res chemist, 70-76, sr res chemist, Fuels Res Sect, 76-89, technologist, 89-90, SR

TECHNOLOGIST, TEXACO INC, 90- *Mem:* Sigma Xi; Am Chem Soc. *Res:* Sorption of high temperature vapors by metal surfaces; fuel and fuel additive development; combustion; air pollution; corrosion. *Mailing Add:* Texaco Inc PO Box 509 Beacon NY 12508

DAVIS, MARTIN (DAVID), b New York, NY, Mar 8, 28; m 51; c 2. RECURSIVE FUNCTION, COMPUTER SCIENCE. *Educ:* City Col New York, BS, 48; Princeton Univ, MA, 49, PhD(math), 50. *Prof Exp:* Res instr math, Univ Ill, 50-51, res assoc control syst lab, 51-52; mem sch math, Inst Advan Study, 52-54; asst prof math, Univ Calif, Davis, 54-55, Ohio State Univ, 55-56 & Rensselaer Polytech Inst, 56-57; assoc prof, Grad Div, Hartford Univ, 57-59; res scientist, Inst Math Sci, NY Univ, 59-60; from assoc prof to prof math, Belfer Grad Sch Sci, Yeshiva Univ, 60-65; PROF MATH, COURANT INST, NY UNIV, 65- *Concurrent Pos:* Vis prof, Belfer Grad Sch Sci, Yeshiva Univ, 70-71, Univ Calif, Berkeley, 76-77 & Univ Calif, Santa Barbara, 78-79; Guggenheim Found fel, 83. *Honors & Awards:* Chauvenet Prize & Ford Prize, Math Asn Am, 75; Steele Prize, Am Math Soc, 75. *Mem:* Am Math Soc; Asn Symbolic Logic; Asn Comput Mach. *Res:* Recursive functions; Hilbert's tenth problem; theorem proving by computing machine; theory of software testing; history of computing. *Mailing Add:* Courant Inst NY Univ 251 Mercer St New York NY 10012

DAVIS, MARTIN ARNOLD, organic chemistry, pharmaceutical research; deceased, see previous edition for last biography

DAVIS, MARVIN LESTER, b Brockton, Mass, Apr 26, 16; m 62. SYNTHETIC ORGANIC CHEMISTRY. *Educ:* Univ Rochester, BA, 37, MS, 39. *Prof Exp:* Chemist, Eastman Kodak Co, 39-54, supvr synthetic org chem, 54-61, patent liaison synthetic org chem, 63-78; RETIRED. *Mem:* Am Chem Soc. *Res:* Organic synthesis relating to photography. *Mailing Add:* 225 Greystone Lane Rochester NY 14618

DAVIS, MERRITT MCGREGOR, b Ont, Feb 16, 23; m 47; c 2. CIVIL ENGINEERING. *Educ:* Queen's Univ, Ont, BSc, 45; Purdue Univ, MSc, 49. *Prof Exp:* Div maintenance engr, Hwy Dept, Ont, 49-51, res engr, 51-55, design engr, 56; from asst prof to assoc prof hwy eng, Univ Toronto, 56-88; RETIRED. *Concurrent Pos:* Assoc mem, Hwy Res Bd, Nat Acad Sci-Nat Res Coun. *Mem:* Eng Inst Can; Am Asn Automotive Med; Sigma Xi. *Res:* Frost action in soils; bituminous pavement mixes; traffic accident causes and crash injury performance. *Mailing Add:* 112 Three Valleys Dr Don Mills ON M3A 3B9 Can

DAVIS, MICHAEL, b Bronxville, NY, Nov 14, 42; m 67; c 2. LEARNING & MEMORY, BEHAVIOR ANALYSIS. *Educ:* Northwestern Univ, BA, 65; Yale Univ, PhD(psychol), 69. *Prof Exp:* Res assoc, 69-70, from asst prof to assoc prof, 70-84, PROF, PSYCHIAT & PSYCHOL, YALE UNIV SCH MED, 84- *Concurrent Pos:* Res career develop award, NIMH, 75 & 80. *Honors & Awards:* Res scientist award, NIMH, 85. *Mem:* Soc Neurosci; Am Psychol Asn; Am Col Neuropsychopharmacol; Soc Psychophysiol. *Res:* Pharmacological, anatomical, physiological and behavioral analysis of learning and memory. *Mailing Add:* Yale Univ 34 Park St New Haven CT 06508

DAVIS, MICHAEL ALLAN, b Boston, Mass, May 19, 41; m 63; c 2. NUCLEAR MEDICINE, DIAGNOSTIC RADIOLOGY RESEARCH. *Educ:* Worcester Polytech Inst, BS, 62, MS, 64; Harvard Sch Pub Health, SM, 65, ScD(radiobiol), 69; Northeastern Univ, MBA, 79; Univ Mass Med Sch, MD, 85. *Prof Exp:* Res assoc radiol, Harvard Med Sch, 68-70, prin assoc radiol, 70-75, asst prof, 75-78, assoc prof radiol, 78-80; PROF RADIOL & NUCLEAR MED, MED SCH, UNIV MASS, 80- *Concurrent Pos:* Assoc radiol, Peter Bent Brigham Hosp, 68-80; dir radiopharm, Children's Hosp Med Ctr, 71-80; clin assoc prof, Col Pharm & Allied Health Professions, Northeastern Univ, 73-77, adj prof radiopharmaceut sci, 77-80; consult radiopharm, Brigham & Women's Hosp, 81-83; lectr radiol, Harvard Med Sch, 81-; affil prof biomed eng, Worcester Polytech, 86-; adj prof radiol, Tuft Univ Sch of Vet Med, 84- *Honors & Awards:* Stauffer Award, Asn Univ Radiol. *Mem:* Soc Nuclear Med; Am Chem Soc; Sigma Xi; Radiol Soc NAm; Am Roentgen Ray Soc; Soc of Magnetic Resonance in Med; Soc of Magnetic Resonance Imaging; Assoc of Univ Radiol. *Res:* Design, synthesis and biologic evaluation of short-lived radiodiagnostic agents with particular interest in cardiovascular disease; use of radionuclides in all areas of biomedical research; contrast agents for computed tomography and nuclear magnetic resonance. *Mailing Add:* Dept Radiol Mass Med Sch 55 Lake Ave N Worcester MA 01655

DAVIS, MICHAEL E, b Bellefontaine, Ohio, Sept 30, 52; m 81; c 2. ANIMAL BREEDING, STATISTICS. *Educ:* Ohio State Univ, BS, 74; Colo State Univ, MS, 77, PhD(animal husbandry), 80. *Prof Exp:* Fel, Univ Wis-Madison, 80-82; asst prof, 82-87, ASSOC PROF ANIMAL SCI, DEPT ANIMAL SCI, OHIO STATE UNIV, 87- *Mem:* Am Soc Animal Sci. *Res:* Beef cattle genetics; estimation of breeding values in beef bulls; examination of selection and inbreeding effects in beef cattle; study of factors affecting lifetime efficiency of beef cows and postweaning efficiency of beef calves; twinning in beef cattle; selection for IGFI in beef cattle. *Mailing Add:* Dept Animal Sci Ohio State Univ 2029 Fyffe Rd Columbus OH 43210

DAVIS, MICHAEL EDWARD, b Jacksonville, Ill, May 17, 22; m 47; c 6. GEOLOGY. *Educ:* Kans State Univ, BS, 50, MS, 51. *Prof Exp:* Instr geol, Kans State Univ, 51-52; asst prof, St Joseph's Col, Ind, 52-54; geologist, Knox-Bergman-Shearer, Colo, 54-57; PROF GEOL, ST JOSEPH'S COL, IND, 57-, CHMN DEPT, 61- *Concurrent Pos:* Glacial geologist, Ind Geol Surv, 62-64. *Mem:* Geol Soc Am; Nat Asn Geol Teachers; Am Asn Petrol Geologists; Asn Prof Geol Scientists. *Res:* Use of photogeologic methods in interpretation of glacial features; photogeologic procedures as applied to geomorphic and structural phenomena; science education; remote sensing. *Mailing Add:* Dept Geol St Joseph's Col Rensselaer IN 47978

DAVIS, MICHAEL I, b London, Eng, July 17, 36; m 70; c 1. STRUCTURAL CHEMISTRY. *Educ:* Univ London, BSc, 58, PhD(chem), 62. *Prof Exp:* From instr to asst prof chem, Univ Tex, Austin, 61-68; assoc prof, 68-71, PROF CHEM, UNIV TEX, EL PASO, 71- *Mem:* Am Crystallog Asn; The Chem Soc. *Res:* Molecular structure studies by gas phase lectron diffraction. *Mailing Add:* 1717 Elm St El Paso TX 79930

DAVIS, MICHAEL JAY, b Denver, Colo, Mar 9, 47; m 80; c 1. PHYTOBACTERIOLOGY, MYCOPLASMOLOGY. *Educ:* Colo State Univ, BS, 73, MS, 75; Univ Calif, Berkeley, PhD(plant path), 78. *Prof Exp:* Res assoc plant path, Colo State Univ, 73-75; fel, Univ Calif, Berkeley, 78-79; asst prof path, Rutgers Univ, 79-81; asst prof, 81-85, ASSOC PROF PLANT PATH, UNIV FLA, 85- *Mem:* Am Phytopath Soc; Am Soc Microbiol; Sigma Xi; Am Soc Sugar Cane Technologists; Int Soc Plant Path. *Res:* Plant pathogenic, fastidious prokaryotes; bacterial diseases of plants. *Mailing Add:* 18905 SW 280th St Homestead FL 33031

DAVIS, MICHAEL MOORE, b Geneva, NY, Dec 6, 38; m 61; c 4. RADIO ASTRONOMY. *Educ:* Yale Univ, BS, 60; State Univ Leiden, PhD(astron), 67. *Prof Exp:* Asst scientist, Nat Radio Astron Observ, WVa, 67-72; sci coordr, NSF, 72-73; sr res assoc, Astron & Ionosphere Ctr, 74-77, head, Radio Astron Group, 77-85; asst dir, 85-88, DIR, ARECIBO OBSERV, 88- *Concurrent Pos:* mem, NSF Subcomt Radio Astron, Nat Acad Sci Radio Astron Frequency Allocations Subcomt, NASA Sci Working Group on Search for Extraterrestrial Intel, Dept Energy/NASA Satellite Power Syst Assessment Subcomt & Nat Acad Sci Astron & Astrophys Survey Comt, Radio Astron Panel. *Mem:* Int Astron Union; Am Astron Soc; Sigma Xi. *Res:* Extragalactic radio sources; quasar absorption lines, pulsars; search extraterrestrial intelligence. *Mailing Add:* Arecibo Observ Box 995 Arecibo PR 00613

DAVIS, MICHAEL WALTER, b Norristown, Pa, Apr 26, 49; m 81. TRANSFORMATION GROUPS. *Educ:* Princeton Univ, AB, 71, PhD(math), 75. *Prof Exp:* Moore instr math, Mass Inst Technol, 74-76; mem, Inst Advan Study, 76-77 & 82-83; asst prof math, Columbia Univ, 77-82; ASSOC PROF MATH, OHIO STATE UNIV, 82- *Res:* Transformation groups on homotopy spheres; groups generated by reflections. *Mailing Add:* Math Dept Ohio State Univ 231 W 18th Ave Columbus OH 43210

DAVIS, MILFORD HALL, b Chicago, Ill, Aug 20, 25; div; c 2. SPACE PHYSICS. *Educ:* Yale Univ, BS, 49; Calif Inst Technol, MS, 50, PhD(physics), 55. *Prof Exp:* Physicist, Rand Corp, 55-67 & Nat Ctr Atmospheric Res, 67-73; PROG DIR, UNIV SPACE RES ASN, 76- *Mem:* Am Phys Soc. *Res:* Cloud microphysics. *Mailing Add:* Univ Space Res Asn PO Box 391 Boulder CO 80306

DAVIS, MILTON W(ICKERS), JR, b Frederick, Md, Apr 5, 23; m 48; c 2. CHEMICAL ENGINEERING. *Educ:* Johns Hopkins Univ, BE, 43; Univ Calif, Berkeley, MS, 49, PhD(chem eng), 51. *Prof Exp:* Asst radiation lab, Univ Calif, 47-50; res engr atomic energy div, E I Du Pont de Nemours & Co, 51-53, res supvr, Savannah River Plant, 54-62; prof, 62-88, dir, Environ Res Inst, 66-70, EMER PROF CHEM ENG, UNIV SC, 88- *Concurrent Pos:* Chair prof, Am Inst Chem Eng, 62-88. *Mem:* Fel Am Inst Chem Eng; Am Chem Soc; Sigma Xi. *Res:* Liquid extraction; ion exchange; nuclear chemistry and physics; chemical reaction kinetics; heterogeneous catalysis; 6 United States patents. *Mailing Add:* PO Box 5867 Hilton Head Island SC 29938

DAVIS, MONTE V, b Cove, Ore, Apr 29, 23; m 73; c 2. NUCLEAR PHYSICS. *Educ:* Linfield Col, BA, 49; Ore State Univ, MA, 51, PhD(physics), 56. *Prof Exp:* Sr reactor physicist, Gen Elec Co, 51-57; group leader & proj engr, Atomics Int Div, NAm Aviation, Inc, 57-61; prof nuclear eng & dir nuclear reactor lab, Univ Ariz, 61-73; dir, Neely Nuclear Res Ctr, 73-80, PROF NUCLEAR ENG, GA INST TECHNOL, 73- *Concurrent Pos:* Consult, indust & US Govt, 61-; pres, MND Inc, 74- *Mem:* Fel Am Nuclear Soc; Am Phys Soc; Int Solar Energy Soc; Am Soc Nondestructive Testing. *Res:* Direct energy conversion; neutron interactions; nondestructive testing; materials science. *Mailing Add:* 1207 Reeder Cir NE Atlanta GA 30306

DAVIS, MONTIE GRANT, b Nashville, Tenn, Aug 15, 36; m 62; c 2. PHYSICAL CHEMISTRY, OPTICAL PHYSICS. *Educ:* Vanderbilt Univ, BS, 58; Univ Tenn, MA, 66, PhD(physics), 68. *Prof Exp:* Radiol physicist, Tenn Dept Pub Health, 60-63; res physicist spectroscopy, ARO Inc, 63-69; pvt consult, ARO Inc & USAF, 69-71; physics instr, Univ Tenn, 71-75, assoc dean arts & sci, 75-79; prof physics, Utah Space Inst, 79-81; PROF PHYSICS, VOL STATE COMMUNITY COL, 81- *Concurrent Pos:* Consult, ARO Inc & USAF, 71-, DuPont, 81- *Mem:* Am Phys Soc; Sigma Xi. *Res:* Spectral characteristics of molecular diatomic spectral, in particular those molecules associated with atmospheric pollution such as nitric oxide. *Mailing Add:* 107 Hillcrest Dr Hendersonville TN 37075

DAVIS, MORRIS SCHUYLER, b Brooklyn, NY, Dec 14, 19; m 45; c 6. CELESTIAL MECHANICS. *Educ:* Brooklyn Col, BA, 46; Univ Mo, MA, 47; Yale Univ, PhD(astron), 50. *Prof Exp:* Asst instr math, Univ Mo, 46-47; asst astron, Yale Univ, 47-49; from instr to asst prof math & astron, Univ Ky, 50-52, in charge observ, 50-52; from asst prof to assoc prof astron, Univ NC, 52-56; res assoc astron & dir comput ctr, Yale Univ, 56-66; pres & dir, Triangle Univs Comput Ctr, Research Triangle Park, NC, 66-70; Morehead prof astron, 70-85, EMER MOREHEAD PROF ASTRON, UNIV NC, CHAPEL HILL, 85- *Concurrent Pos:* Tech adv & writer, Morehead Planetarium, 52-56; dir, Triangle Univs Comput Ctr & adj prof, Univ NC, Chapel Hill, NC State Univ & Duke Univ, 66-70; mem bd trustees, Nat Accelerator Lab, 77-83. *Mem:* Fel AAAS; Am Astron Soc; Int Astron Union. *Res:* Computer science; celestial mechanics; numerical analysis; astrometry. *Mailing Add:* CB No 3255 Dept Physics & Astron Univ NC Chapel Hill NC 27514

DAVIS, MORTON DAVID, b Bronx, NY, May 31, 30; m 63; c 2. MATHEMATICS. *Educ:* Univ Colo, AB, 52; Univ Calif, Berkeley, MA, 56, PhD(math), 61. *Prof Exp:* Mathematician, Int Bus Corp, 59-61; asst prof math, Rutgers Univ, 61-65; mem fac, 65-70, assoc prof, 70-78, PROF MATH, CITY COL NEW YORK, 78- *Concurrent Pos:* Res assoc economet, Princeton Univ, 61-63; consult, Mathematica, NJ, 62-65. *Res:* Game theory; infinite games of perfect information; n-person games; game theory models in disarmament. *Mailing Add:* 25 Brinkerhoff Ave Teaneck NJ 07666

DAVIS, MYRTIS, b Bessemer, Ala, Oct 16, 18. MATHEMATICS. *Educ:* Birmingham-Southern Col, AB, 39; La State Univ, MA, 48. *Prof Exp:* Teacher pub schs, Ala, 39-47; asst prof math, Southeastern La Col, 48-49; instr, Miss Southern Col, 49-50; from instr to asst prof, Nicholls State Col, 50-54; assoc prof, Wesleyan Col, Ga, 54-61; from assoc prof to prof math, Greensboro Col, 61-81, chmn dept math & sci, 71-81; RETIRED. *Mem:* Am Math Asn; Nat Coun Teachers Math. *Mailing Add:* 1309 Cardinal Pl Greensboro NC 27408

DAVIS, NANCY TAGGART, b Harrisburg, Pa, Mar 30, 44; m 70. PATHOLOGY, ANIMAL BEHAVIOR. *Educ:* Rollins Col, BS, 66; Univ Pa, PhD(path), 73. *Prof Exp:* Res trainee path, Penrose Lab, Philadelphia Zool Garden, 67-73; asst prof, Stockton State Col, 73-78; ADJ RES ASSOC COMP PATH, PENROSE LAB, PHILADELPHIA ZOOL GARDEN, 77-; ASSOC PROF PATH & EPIDEMIOL, STOCKTON STATE COL, 78- *Mem:* Am Heart Asn. *Res:* Comparative pathology; reproductive physiology; relationship between psychosocial stimuli and disease; diet and disease in zoo animals. *Mailing Add:* Dept Professional Studies Stockton State Col Pomona NJ 08240

DAVIS, NEIL MONAS, b Philadelphia, Pa, Apr 17, 31; m 59; c 2. PHARMACY. *Educ:* Philadelphia Col Pharm, BS, 53, MS, 55, PharmD, 70. *Prof Exp:* Res hosp pharm, Jefferson Med Col Hosp, 53-55, asst dir pharm, 58-61, dir pharm serv, 61-65; instr pharm, Philadelphia Col Pharm & Jefferson Med Col, 61-65; dir pharm serv, Univ Hosp, 65-76, from asst prof to assoc prof, 65-77, PROF PHARM, SCH PHARM, TEMPLE UNIV, 77- *Concurrent Pos:* Ed, Hosp Pharm. *Honors & Awards:* Fel, Am Soc Hosp Pharmacists, 88. *Mem:* Am Pharmaceut Asn; Am Soc Hosp Pharmacist. *Res:* Hospital pharmacy administration and editing; hospital and clinical pharmacy education; medication errors in hospitals. *Mailing Add:* 1143 Wright Dr Huntingdon Valley PA 19006

DAVIS, NORMAN DUANE, b San Diego, Calif, May 7, 28; m 52; c 2. MICROBIOLOGY. *Educ:* Univ Ga, BSc, 53; Ohio State Univ, MSc, 55, PhD, 57. *Prof Exp:* Instr bot, Univ Ga, 57-58; from asst prof to assoc prof bot & microbiol, Auburn Univ, 58-67, prof, 67-90, DIR CELL SCI CTR, AUBURN UNIV, 87-, EMER PROF BOT & MICROBIOL, 90- *Concurrent Pos:* Prof alumni, Auburn Univ, 82- *Mem:* AAAS; fel Am Acad Microbiol; Am Soc Microbiol; Am Chem Soc. *Res:* Industrial and applied microbiology; mycotoxicology. *Mailing Add:* Dept Bot & Microbiol Auburn Univ Auburn AL 36830

DAVIS, NORMAN RODGER, b Toronto, Ont, June 30, 43. BIOCHEMISTRY. *Educ:* Univ Toronto, BSc, 66, PhD(biochem), 70. *Prof Exp:* Wellcome fel, Meat Res Inst, Langford, Eng, 70-71; asst prof biochem, 71-74, ASSOC PROF DENT, UNIV ALTA, 74- *Res:* Chemistry and structure of collagen crosslinks; mechanism of collagen mineralization; mechanism of desmosine and isodesmosine crosslink formation in elastin. *Mailing Add:* 7404 106th St Edmonton AB T6E 4V9 Can

DAVIS, NORMAN THOMAS, b DosCabezas, Ariz, Mar 23, 27; m 50; c 4. ENTOMOLOGY. *Educ:* Ariz State Univ, BS, 49; Iowa State Col, MS, 51; Univ Wis, PhD, 54. *Prof Exp:* From instr to assoc prof, 54-68, PROF BIOL, UNIV CONN, 80- *Mem:* Am Entom Soc. *Res:* Insect morphology; physiology; neurobiology. *Mailing Add:* Biol Sci Group Univ Conn Storrs CT 06268

DAVIS, OSCAR F, b Oak Park, Ill, June 19, 28; m 51; c 4. PSYCHIATRY, PHARMACOLOGY. *Educ:* Roosevelt Univ, BS, 49; Loyola Univ, Ill, MS, 52, PhD(pharmacol), 54, MD, 58; Am Bd Psychiat, dipl, 66. *Prof Exp:* Assoc pharmacol, Stritch Sch Med, Loyola Univ, Ill, 53-58; intern, Michael Reese Hosp, 58-59; clin instr psychiat, Col Med, Univ Ill, 60-62; clin asst prof, 62-67, CLIN ASSOC PROF PSYCHIAT & PHARMACOL, CHICAGO MED SCH, 67-; DIR CHILD & ADOLESCENT PSYCHIAT, MT SINAI HOSP, 62-; ASSOC PROF PSYCHIAT, MED SCH, NORTHWESTERN UNIV, 81- *Concurrent Pos:* Asst prof psychiat, 70-81; fel, Inst Juvenile Res, Ill, 61-63; resident psychiat, Univ Ill Res & Educ Hosp, 59-62, Passavant Mem Hosp, Children's Mem Hosp & Evanston Hosp; consult pharmacologist, 52-63; sr attend physician, Northwestern Memorial & Evanston Hosps, 76- *Mem:* Am Acad Child Psychiat; fel Am Psychiat Asn; Am Fedn Clin Res; Am Soc Pharmacol & Exp Therapeut; Sigma Xi. *Res:* Child, adolescent and adult psychiatry; infantile development and psychosomatic diseases in childhood; psychopharmacology; child psychiatry and adolescent. *Mailing Add:* 993 Forest Ave Glencoe IL 60022

DAVIS, OWEN KENT, b Nampa, Idaho, Mar 13, 49; m 71; c 2. PALEOECOLOGY, PALEOCLIMATOLOGY. *Educ:* Col Idaho, Caldwell, DS, 71; Wash State Univ, Pullman, MS, 74; Univ Minn, Minneapolis, PhD(ecol), 81. *Prof Exp:* asst prof, 82-88, ASSOC PROF GEOSCI, UNIV ARIZ, 88- *Mem:* AAAS; Am Asn Stratig Palynologists; Am Quaternary Asn; Ecol Soc Am. *Res:* Holocene climate and the astronomical theory of climate change; late Quaternary climate of western North America; paleoecology of extinct Pleistocene megafauna; palynology; biogeography; ecology. *Mailing Add:* Dept Geosci Univ Ariz Tucson AZ 85721

DAVIS, P(HILIP) C, b Cornish, Ark, Apr 18, 21; m 44; c 2. CHEMICAL ENGINEERING. *Educ:* Univ Chicago, SB, 43; Univ Kans, BS, 48, MS, 49, PhD(chem eng), 52. *Prof Exp:* Process design engr, 52, engr process develop, 53-55, supvr, 55-59, sr assoc, 59-64, head eng & math sci, 64-67, SR ENG ADV RES & DEVELOP, ETHYL CORP, 74-, ASSOC DIR PROCESS DEVELOP, 81- *Mem:* Am Inst Chem Engrs; Sigma Xi. *Res:* Vapor-liquid phase equilibria; chemical processes; computer techniques. *Mailing Add:* 4199 Downing Dr Baton Rouge LA 70809

DAVIS, PAMELA BOWES, b Jamaica, NY, July 20, 49; c 2. PULMONARY MEDICINE, CYSTIC FIBROSIS. *Educ:* Smith Col, AB, 68; Duke Univ, PhD(physiol), 73, MD, 74. *Prof Exp:* Resident internal med, Duke Hosp, 74-75; pulmonary med fel, NIH, 75-79; asst prof med, Univ Tenn Sch Med, 79-81; asst & assoc & prof pediat & med, 81-89, CHIEF PEDIAT PULMONARY DIV, SCH MED, CASE WESTERN RESERVE UNIV, 85-, PROF PEDIAT & MED, 89- *Concurrent Pos:* Dir, Cystic Fibrosis Res Ctr, Case Western Reserve Univ, 85-, prin investr, 86-; Nat Inst Diabetes & Digestive & Kidney Dis, B Sandy Sect, 88-, chmn, 89- *Mem:* Am Fed Clin Res (pres, 89-90); Am Thoracic Soc; Soc Pediat Res; AAAS. *Res:* Autonomic physiology of the lung; airway epithelial cells; cystic fibrosis. *Mailing Add:* Pediat Pulmonary Div Rainbow Babies & Children Hosp 2101 Adelbert Rd Cleveland OH 44106

DAVIS, PAUL COOPER, b Glenville, WVa, Mar 14, 37; m 60; c 4. ELECTRICAL ENGINEERING. *Educ:* Univ WVa, BSEE, 59; Mass Inst Technol, MS, 61; Lehigh Univ, PhD(elec eng), 68. *Prof Exp:* MEM TECH STAFF, BELL TEL LABS, READING BR, 62- *Mem:* Inst Elec & Electronics Engrs. *Res:* Design of custom linear and digital integrated circuits. *Mailing Add:* AT&T Bell Labs 2525 N 12th St Reading PA 19604

DAVIS, PAUL JOSEPH, b Chicago, Ill, Oct 28, 37; m 62; c 3. ENDOCRINOLOGY. *Educ:* Westminster Col, Mo, BA, 59; Harvard Univ, MD, 63. *Prof Exp:* Clin assoc endocrinol, Nat Inst Child Health & Human Develop, NIH, 67-69, sr staff assoc, 69-70; head, Endocrinol Div, Baltimore City Hosps, 70-75; head, Endocrinol Div & prof med, Med Sch, State Univ NY, Buffalo, 75-90; PROF & CHMN, DEPT MED, ALBANY MED COL, NY, 90- *Concurrent Pos:* Assoc prof med, Johns Hopkins Univ, 74-75; chief med serv, Vet Admin Med Ctr, Buffalo, 80-90; bd sci counr, Nat Inst on Aging, NIH, 77-81; mem, Merit Rev Bd Endocrinol, VA Cent Off, 91-94. *Mem:* Endocrine Soc; Am Thyroid Asn; Am Diabetes Asn; Am Fedn Clin Res; Am Soc Biochem Molecular Biol. *Res:* Mechanisms of action of thyroid hormone. *Mailing Add:* Dept Med Albany Med Col Albany NY 12208

DAVIS, PAUL WILLIAM, b Albany, NY, Oct 14, 44; m 66; c 2. APPLIED MATHEMATICS. *Educ:* Rensselaer Polytech Inst, BS, 66, MS, 67, PhD(math), 70. *Prof Exp:* Res mathematician appl math, Tex Instruments Corp, 66; actg chief appl math prog, US Army Res Off, 74-75; from asst prof to assoc prof, math sci, Worcester Polytech Inst, 70-80, chmn dept, 78-83, secy fac, 84-88, PROF MATH SCI & HEAD DEPT, WORCESTER POLYTECH INST, 80- *Concurrent Pos:* Consult, Dept Civil Eng, Worcester Polytech Inst, 71, Digital Equip Corp, 85, 88-89, Package Indust, Inc, 81-89, CMX Systs, Inc, 89-; US Army Res Off, 73 & 74, Dept Gynec, Mass Gen Hosp, 77; ed, Soc Indust Appl Math News, 75-86. *Mem:* Soc Indust Appl Math; Am Math Soc; Inst Elec & Electronic Engrs; Sigma Xi; Math Asn Am; AAAS. *Res:* Measurement system design and error detection for electric power systems; flying height measurement for magnetic storage systems. *Mailing Add:* Dept Math Worcester Polytech Inst 100 Institute Rd Worcester MA 01609

DAVIS, PAULS, b Cesis, Latvia, Mar 18, 21; nat US; m 50; c 2. ORGANIC CHEMISTRY. *Educ:* Univ Tübingen, dipl chem, 47, DSc(org chem), 49. *Prof Exp:* Res assoc org chem, Mass Inst Technol, 50-53; res chemist, Burke Res Co, 53-59; from res chemist to sr res chemist, 59-64, res assoc, 64-74, sr res assoc, 74-78, RES FEL, BASF-WYANDOTTE CORP, 78- *Mem:* Am Chem Soc; Sigma Xi; Royal Soc Chem. *Res:* Metalorganic compounds; synthetic fat-soluble vitamines; organic peroxides and ozonides; metalorganic polymerization catalysts; fire retardant polymers; organic halogenous compounds. *Mailing Add:* 30027 White Gibraltar MI 48173-9425

DAVIS, PEYTON NELSON, b Lodi, Calif, Apr 4, 25; m 47; c 2. NUTRITION, BIOCHEMISTRY. *Educ:* Colo State Univ, BSc, 48; Univ Calif, Davis, MSc, 63, PhD(nutrit), 66. *Prof Exp:* Res asst, Exp Sta, Colo State Univ, 48-52; lab technician, Univ Calif, Davis, 52-66; sr scientist, Vivonex Corp, 66-70; sr nutritionist, Dept Food Sci, Stanford Res Inst, 70-74; chief nutritionist, Shaklee Corp, Emeryville, 74-77; mem staff, Dept Nutrit, Nat Live Stock & Meat Bd, 77-81; ASSOC PROF, NAT COL CHIROPRACTICE, 81- *Mem:* Fel Am Col Nutrit; Animal Nutrit Res Coun; Am Inst Nutrit; NY Acad Sci; Sigma Xi. *Res:* Feeding values of proteins; amino acid requirements; unidentified growth factors; vitamin requirements; pigmentation of shanks; mineral utilization and requirements; effects of chelating agents; chemically defined diets; carbohydrate effect upon atherosclerosis; purine utilization; fabricated foods. *Mailing Add:* 615 Maiden Lane Glen Ellyn IL 60137

DAVIS, PHILIP J, b Lawrence, Mass, Jan 2, 23; m 44; c 4. MATHEMATICS. *Educ:* Harvard Univ, PhD(math), 50. *Prof Exp:* Mathematician, Nat Bur Standards, 51-58, chief numerical anal, 58-63; PROF APPL MATH, BROWN UNIV, 63- *Concurrent Pos:* Guggenheim fel, 56-57. *Honors & Awards:* Chauvenet Prize, Math Asn Am, 63; Lester R Ford Award, Math Asn Am, 83; George Polya Award, Math Asn Am, 87. *Mem:* Math Asn Am. *Res:* Numerical analysis; interpolation and approximation theory; philosophy of science. *Mailing Add:* Dept Appl Math Brown Univ Providence RI 02912

DAVIS, PHILIP K, b Effingham, Ill, Aug 29, 31; m 55; c 2. ENGINEERING MECHANICS, CIVIL ENGINEERING. *Educ:* Univ Tex, BS, 58, MS, 59; Univ Mich, MSE & PhD(eng mech), 63. *Prof Exp:* Res engr, Struct Mech Res Lab, Balcones Res Ctr, Univ Tex, 59; stress analyst, Boeing Airplane Co, Kans, 59-60; instr eng mech, Univ Mich, 63-64; from asst prof to prof eng mech & mat, 64-71, chmn dept, 71-78, actg dean, Col Eng & Technol, 78-79, chmn dept civil eng & mech, 79-87, PROF CIVIL ENG & MECH, SOUTHERN ILL UNIV, 79- *Concurrent Pos:* Lectr, Univ Wichita, 59-60; NSF grant, 66-67; NASA grant, 68-70; grants, US Dept Energy, 81-90, Nat Mine Land Reclamation Ctr, 90-92. *Mem:* Am Soc Eng Educ; Am Soc Mech Engrs; Soc Mining Engrs; Am Acad Mech; Soc Eng Sci; Sigma Xi; Soc Mining Engrs; Asn Chmn Dept Mech. *Res:* Motion of solid bodies in rotating viscous fluids; viscous and inviscid flows; liquid squeeze film motion;

cushioning of large structures against seismic inputs, hydrocyclones; coal water mixtures, slurry flows and biomechanics, fine particle coal cleaning and dewatering and heavy-liquid cyclones. *Mailing Add:* Civil Engr & Mech Southern Ill Univ Carbondale IL 62901

DAVIS, PHILLIP BURTON, b Owosso, Mich, Oct 11, 51; m 73; c 2. INTEGRATED NATURAL RESOURCE MANAGEMENT. *Educ:* Univ Mich-Flint, BA, 74; Mich State Univ, MS, 75, PhD(resource develop), 79. *Prof Exp:* Teaching asst ecol, Univ Mich-Flint, 71-73; res asst, Wildlife Dept, Mich State Univ, 74-75, res proj coordr, Resource Develop Dept, 76-79, grad teaching asst, Watershed Mgt, 78-79; ASST PROF NATURAL RESOURCES MGT, DEPT RANGE, WILDLIFE & FORESTRY, UNIV NEV, RENO, 79- *Concurrent Pos:* Lab asst, Dept Biol, Univ Mich-Flint, 71; mem staff, Inst Sci & Technol, Willow Run Labs, 72; terrestrial ecologist, Mich State Univ, 74. *Mem:* Soc Range Mgt; Wildlife Soc; Soc Am Foresters; Sigma Xi; Nat Wildlife Fedn. *Res:* Coordinated development, use and management of renewable natural resources in accordance with ecological principles for the benefit of man. *Mailing Add:* 3015 Heights Dr Reno NV 89503

DAVIS, PHILLIP HOWARD, b Ft Collins, Colo, Mar 14, 46; m 68; c 6. CHEMICAL EDUCATION, ELECTRON PARAMAGNETIC RESONANCE. *Educ:* Colo State Univ, BS, 68; Univ Southampton, UK, MPhil, 70; Univ Ill, Urbana, PhD(chem), 72. *Prof Exp:* Asst prof chem, Univ Ill, Urbana, 72-74 & Univ Sask, 74-77; res assoc, State Univ NY, Albany, 77-78; from asst prof to assoc prof, 78-87, PROF CHEM, UNIV TENN, MARTIN, 87-, CHAIR DEPT, 90- *Concurrent Pos:* Lectr, Univ Ill, Urbana, 81; res assoc, Ill ESR Res Ctr, 88. *Mem:* Am Chem Soc; AAAS; Sigma Xi. *Res:* Chemical education; electron paramagnetic resonance studies of the solid state. *Mailing Add:* Dept Chem Univ Tenn Martin TN 38238

DAVIS, R(OBERT) E(LLIOT), b Chicago, Ill, Dec 1, 22; m 47; c 4. SOLID STATE PHYSICS, ENGINEERING. *Educ:* Purdue Univ, BSEE, 44, MS, 49. *Prof Exp:* Res physicist, Westinghouse Elec Corp, 49-55, supvr engr solid state device develop & appln, 55, sect mgr, 55-58, dept mgr, 58-60, mgr semiconductor dept, Res Labs, 60-62, mgr adv develop, Semiconductor Div, 62-65; vpres, engr & mem bd dirs, Pa Electronics Technol, Inc, 65-87; RETIRED. *Concurrent Pos:* Private consultant; chmn bd, Al-Anon Family Groups, Inc, 86-89, exec comt, 89- *Mem:* Inst Elec & Electronic Engrs. *Res:* Semiconductor physics; material preparation; electrical and optical measurements on materials and devices; device design and application. *Mailing Add:* Box 13 Murrysville PA 15668-0013

DAVIS, RALPH ANDERSON, b Huntington, Ind, Aug 14, 17; m 40; c 2. CHEMISTRY. *Educ:* Huntington Col, AB, 39; Ind Univ, MA, 42. *Prof Exp:* Teacher high sch, Ind, 39-41; asst chem, Ind Univ, 41-42; from res chemist to sr res chemist, Dow Chem Co, 42-74, sr res specialist, 74-80, res assoc, 80-82; RETIRED. *Concurrent Pos:* Consult, 82- *Mem:* Am Chem Soc; AAAS; Sigma Xi; NY Acad Sci. *Res:* Organic and inorganic fluorides and other halides; catalysis; low temperature distillation; fluorine and high energy oxidizers. *Mailing Add:* 1160 Poseyville Rd RR 7 Midland MI 48640-8922

DAVIS, RALPH LANIER, b Ala, Sept 10, 21; m 43; c 3. GENETICS, PLANT BREEDING. *Educ:* Ala Polytech Inst, BS, 43; Purdue Univ, MS, 48, PhD(genetics, breeding), 50. *Prof Exp:* Asst dean grad sch, Sch Agr, Purdue Univ, 65-71, prof agron, 50-86, assoc dir, Div Sponsored Progs, 66-86; RETIRED. *Concurrent Pos:* Vis prof, Ore State Univ, 59-60 & NC State Col, 63; ed, Crop Sci, Crop Sci Soc Am, 64-67; patent mgr, Off Patent Mgt, 74-80. *Mem:* Fel AAAS; fel Am Soc Agron; Crop Sci Soc Am (pres-elect & actg pres, 62, pres, 63). *Mailing Add:* 800 Cove Keay Dr Unit C1 Clearwater FL 34620

DAVIS, RAYMOND, JR, b Washington, DC, Oct 14, 14; m 48; c 5. CHEMISTRY. *Educ:* Univ Md, BS, 37, MS, 40; Yale Univ, PhD(phys chem), 42. *Hon Degrees:* DSc, Univ Pa, 90. *Prof Exp:* Res chemist, Dow Chem Co, Mich, 37-38; res chemist, Monsanto Chem Co, AEC, Ohio, 46-48; scientist, Brookhaven Nat Lab, 48-84; RES PROF, ASTRON DEPT, UNIV PA, 84- *Honors & Awards:* Boris Prejel Prize, NY Acad Sci, 54; Comstoc Prize, Nat Acad Sci, 78; Nuclear Applns Award, Am Chem Soc, 79; Tom Bonner Prize, Am Phys Soc, 88. *Mem:* NAS; fel Am Phys Soc; Am Geophys Union; fel AAAS; Am Acad Arts & Sci; Meteoritical Soc. *Res:* Nuclear chemistry; meteorites and cosmic rays; neutrino detection; lunar sample studies. *Mailing Add:* 28 Bergen Lane Blue Point NY 11715

DAVIS, RAYMOND E, b Hobbs, NMex, Nov 7, 38; m 60; c 3. PHYSICAL CHEMISTRY. *Educ:* Univ Kans, BS, 60; Yale Univ, PhD(phys chem), 65. *Prof Exp:* From cancer res scientist to sr cancer res scientist Center Crystallog Res, Roswell Park Mem Inst, 64-66; from asst prof to assoc prof, 66-76, PROF CHEM, UNIV TEX, AUSTIN, 76- *Mem:* Am Crystallog Asn; Am Chem Soc. *Res:* X-ray diffraction; molecular structure studies of organometallic and small ring organic compounds; molecular structure studies of transition metal phosphine complexes and macrocyclic compounds and their complexes; packing in molecular crystals. *Mailing Add:* Dept Chem Univ Tex Austin TX 78712-0667

DAVIS, RAYMOND F, b St Petersburg, Fla, Nov 18, 28; m 49, 82; c 4. EXOBIOLOGY. *Educ:* Univ Fla, BS, 56; Brantridge Univ, Eng, PhD(chem), 76. *Prof Exp:* Anal chemist, US Naval Air Sta, Pub Works Utilities, Jacksonville, Fla, 53-54; bacteriologist, & serologist, vet sci dept, Agr Exp Sta, Gainesville, Fla, 55-56; chief microbiologist, Permachem Corp, W Palm Beach, Fla, 56- 61; dir res, Bard-Parker Co, Inc, Danbury, Conn, 63-64; consult serv, Danbury, Conn, 64-65; CONSULT, RAYMOND F DAVIS INC ENVIRON UNLTD, 67- *Concurrent Pos:* Prin investr advan develop, Grumman Aerospace Corp, Bethpage, NY, 65-70; chemist, Dept Pub Works, Suffolk County, NY, 71-78; sr chemist, Dept Environ Control, Suffolk County, NY, 78-82. *Mem:* Nat Asn Environ Prof; Water Pollution Control Fedn. *Res:* Chemical and microbiological analysis and testing of environmental constituants and contaminants; design of facilities and systems for contamination control. *Mailing Add:* Eniron Unltd Inc Star Rte 5 Box 103E Beaufort SC 29902

DAVIS, RICHARD A, b Chicago, Ill, June 15, 25; m 60; c 2. NEUROSURGERY, MEDICAL SCIENCES. *Educ:* Princeton Univ, AB, 47; Northwestern Univ, MD, 51, MS, 56. *Prof Exp:* Asst neurol, Nat Hosp, London, 56-57; asst prof, 61-66, ASSOC PROF NEUROSURG, SCH MED, UNIV PA, 66-, RES ASSOC PHYSIOL, 59- *Concurrent Pos:* USPHS res grant, 57-; Am Cancer Soc fel, 62-65; staff surgeon, Hosp Univ Pa, 59- *Mem:* Am Col Surg; Am Asn Neurol Surg. *Res:* Central nervous system control of gastric secretion; clinical research in depth electrode recordings from brain; decerebrate rigidity in animals and man. *Mailing Add:* Dept Neurosurg Univ Pa Philadelphia PA 19104

DAVIS, RICHARD ALBERT, JR, b Joliet, Ill, Sept 11, 37; m 62; c 2. GEOLOGY. *Educ:* Beloit Col, BS, 59; Univ Tex, MA, 61; Univ Ill, Urbana, PhD(geol), 64. *Prof Exp:* Alumni Res Found fel geol, Univ Wis, 64-65; from asst prof to assoc prof, Western Mich Univ, 65-73; chairperson geol, 73-82, assoc dean res & grad affairs, 82-84, prof geol, 73-88, GRAD RES PROF, UNIV SFLA, 88- *Concurrent Pos:* Sr Fulbright fel, Univ Melbourne, 76; vis prof, Duke Univ & Univ NC, 84-85, Univ Copenhagen, 90. *Mem:* Soc Econ Paleontologists & Mineralogists; Geol Soc Am; Am Asn Petrol Geologists; Int Asn Sedimentologists. *Res:* Beach and nearshore processes; morphodynamics of barrier island systems with emphasis on inlets and beaches; clostic deposition of systems. *Mailing Add:* Dept Geol Univ SFla Tampa FL 33620

DAVIS, RICHARD ARNOLD, b Cedar Rapids, Iowa, Apr 19, 42; m; c 1. GEOLOGY, PALEONTOLOGY. *Educ:* Cornell Col, BA, 63; Univ Iowa, MS, 65, PhD(geol), 68. *Prof Exp:* NSF fel paleont, Univ Col Swansea, Wales, 68-69; asst prof geol & cur geol, Univ Cincinnati, 69-75; dir sci educ, 75-78, PALEONTOLOGIST, CINCINNATI MUS NATURAL HISTORY, 75-, CUR COLLECTIONS, 78- *Concurrent Pos:* Adj assoc prof biol sci, Univ Cincinnati, 84-86. *Mem:* Geol Soc Am; Paleont Soc; Am Asn Zool Nomenclature; Palaeont Asn; Soc Econ Paleont & Mineral; Paleont Res Inst; Int Paleont Asn; Systs Asn. *Res:* Paleobiology of ammonoid cephalopods; biology and paleobiology of nautiloid cephalopods; problematic fossils. *Mailing Add:* Cincinnati Mus Natural Hist 1720 Gilbert Ave Cincinnati OH 45202

DAVIS, RICHARD BRADLEY, b Iowa City, Iowa, Nov 6, 26; m 57; c 3. INTERNAL MEDICINE, HEMATOLOGY. *Educ:* Yale Univ, BS, 49; Univ Iowa, MD, 53; Univ Minn, Minneapolis, PhD(internal med), 64. *Prof Exp:* From instr to asst prof med, Univ Minn, Minneapolis, 59-69; assoc prof, 69-73, actg dir, 73-76, dir, Hemat Div, 76-79, head, Hemat Sect, Dept Int Med, 79-82, PROF MED, MED SCH, UNIV NEBR, 73- *Concurrent Pos:* USPHS res career develop award, 61-69; vis investr, Sir William Dunn Sch Path, Oxford Univ, 64-65; dir, Nebr Regional Hemophilia Ctr, 83-85. *Mem:* Am Soc Hemat; Am Fedn Clin Res; Soc Exp Biol & Med; Am Soc Exp Path; Sigma Xi; Am Asn Hist Med; Cent Soc Clin Res. *Res:* Electron microscopy of blood platelets, platelet aggregating agents, freeze drying of blood platelets. *Mailing Add:* Univ Nebr Hosps Omaha NE 68105

DAVIS, RICHARD LAVERNE, b Minneapolis, Minn, May 20, 32; m 64; c 3. PATHOLOGY, NEUROPATHOLOGY. *Educ:* Univ Minn, Minneapolis, BA, 53, BS & MD, 56. *Prof Exp:* Intern med, Bellevue Hosp, New York, 56-57; resident path, Univ Minn, Minneapolis, 57-60; Nat Inst Neurol Dis & Blindness training fel neuropath, Armed Forces Inst Path, 60-61; assoc pathologist, Armed Forces Inst Path, 61-65; pathologist, Lab Serv, US Naval Hosp, 65-69; from assoc prof to prof path, Univ Southern Calif, 69-80, chief, Cajal Lab Neuropath, 69-80; PROF PATH, NEUROL SURG & NEUROL, UNIV CALIF, SAN FRANCISCO, 80- *Concurrent Pos:* Nat Cancer Inst training fel, Univ Minn, Minneapolis, 58-60; consult, Washington Hosp Ctr, DC, 61-65, Nat Naval Med Ctr, Md, 62-65, Long Beach Vet Admin Hosp, 70-; clin instr, Sch Med, George Washington Univ, 63-65. *Mem:* Am Asn Neuropath; Am Asn Path; Am Acad Neurol; Soc Exp Biol & Med. *Res:* Histochemistry; Schwartzman phenomenon; experimental tumors; radiation effects on central nervous system; brain tumors. *Mailing Add:* Dept Path HSW 501 Univ Calif San Francisco CA 94143-0506

DAVIS, RICHARD RICHARDSON, b Ala, Dec 7, 23; m 45; c 3. AGRONOMY. *Educ:* Auburn Univ, BS, 47; Purdue Univ, MS, 49, PhD, 50. *Prof Exp:* Res asst, Purdue Univ, 47-50; from asst prof to assoc prof, Ohio Agr Exp Sta, 50-59, prof agron, Ohio Agr Res & Develop Ctr, 59-78, prof, Ohio State Univ, 62-78, assoc chmn, 62-69, asst dir ctr, 69-78; ASST TO VPRES & PROF AGRON, MISS STATE UNIV, 78- *Mem:* Fel AAAS; fel Am Soc Agron; fel Crop Sci Soc Am (pres, 74); Int Turfgrass Soc (pres, 73). *Res:* Turfgrass management; weed control; pasture management. *Mailing Add:* Div Agr-Forestry/Vet Med Miss State Univ PO Box 5386 Mississippi State MS 39762

DAVIS, ROBERT, b Delhi, NY, Feb 22, 31; m 56; c 3. INSECT ECOLOGY. *Educ:* Univ Ga, BSA, 56, MS, 61, PhD(zool), 63, Ga Southern Univ, MBA, 91. *Prof Exp:* Entomologist, Ga State Dept Entom, 56; med entomologist, Third US Army Med Lab, 56-59; forest entomologist, Southeastern Forest Exp Sta, USDA, 59-60, res entomologist, Southern Grain Insects Res Lab, 63-65; asst prof, Univ Ga, 65-69; dir Stored-Prod Insects Res & Develop Lab, Agr Res Serv, 69-90; VPRES & RES DIR, PEST MGT CONSULT & ASSOC INC, 90- *Concurrent Pos:* Adj assoc prof entom, Univ Ga, 69-90. *Mem:* Am Entom Soc; Sigma Xi. *Res:* Acarology; taxonomy and ecology of Eriophyidae; fumigation and ecology of insects attacking stored products. *Mailing Add:* Pest Mgt Consult & Assoc Inc PO Box 30217 Savannah GA 31410

DAVIS, ROBERT A(RTHUR), b Weehawken, NJ, June 15, 26; m 47, 62; c 3. SPACE SCIENCES, AERODYNAMICS. *Educ:* NY Univ, BAeE, 46, MAeE, 48, DEngSci, 55. *Prof Exp:* Asst aeronaut, Col Eng, NY Univ, 46-47, instr, 47-50; sr engr & group leader aerodyn, Sparrow I Proj, Sperry Gyroscope Co, NY, 50-55; leader satellite opers, Systs Opers Dept, Rand Corp, Calif, 55-62; group dir advan systs, Advan Progs Div, Aerospace Corp, 62-88; prin engr, Archit Planning & Technol Div, 80-88; RETIRED.

Concurrent Pos: Consult long range planning & mil space policy to Asst Secy Defense & Defense Sci Bd, 78-83; leader, USAF mission anal on future mil space activ, 66-78; co-leader, Pres Space Task Group, 69; lectr, Am Inst Aeronaut & Astronaut & Aerospace Corp on space syst eng, 81-87; adv to NASA/Univ Space Res Assoc, 84; Am Inst Aeronaut & Astronaut liaison to Int Astronaut Fedn, 84-87; consult to Nat Comn on Space & Nat Res Coun, 85-86; consult to Smithsonian Inst on hist of space, 88. *Mem:* Fel Am Inst Aeronaut & Astronaut; Int Acad Astronaut (pres, 87). *Res:* Space and missiles research and development; flight test data analysis; operations research; space systems engineering; international space activities. *Mailing Add:* 4510 Santa Lucia Dr Woodland Hills CA 91364

DAVIS, ROBERT BENJAMIN, b Fall River, Mass, June 23, 26; div; c 2. MATHEMATICS. *Educ:* Mass Inst Technol, SB, 46, SM, 47, PhD(math), 51. *Prof Exp:* From asst to instr math, Mass Inst Technol, 46-51; asst prof, Univ NH, 51-56; assoc prof math & educ, Syracuse Univ, 56-63, prof math educ, 63-72; CURRIC LAB & ASSOC DIR COMPUT-BASED EDUC RES LAB, UNIV ILL, URBANA, 72-; DIR, MADISON PROJ, 57- *Concurrent Pos:* Ed, J Math Behavior. *Mem:* Am Math Soc; Math Asn Am; Am Psychol Asn; AAAS. *Res:* Third order partial differential equations; mathematical physics; mathematics education; computer-assisted instruction. *Mailing Add:* 501 S First Ave Highland Park NJ 08904-2123

DAVIS, ROBERT BERNARD, b Miami, Fla, Dec 4, 35; m 60; c 4. FIBER TECHNOLOGY, POLYMER CHEMISTRY. *Educ:* Univ Miami, BEd, 57, BS, 59, MS, 61; Mass Inst Technol, PhD(org chem), 65. *Prof Exp:* Res chemist, Textile Fibers Dept, Pioneering Res Div, E I du Pont de Nemours & Co, Del, 65-68; asst prof org chem, Northeastern Univ, 68-72; asst dir, 72-77, assoc dir FRL, Albany Int Co, 77-80, DIR, ALBANY INT RES CO, 80- *Mem:* Am Chem Soc. *Res:* Materials science; membrane technology; fiber technology; biomaterials; polymer science. *Mailing Add:* Three Lavelle Ln Framingham MA 01701

DAVIS, ROBERT CLAY, b Dallas, Tex, June 8, 41; m 67; c 1. PURE MATHEMATICS. *Educ:* Southern Methodist Univ, BA, 63; Tulane Univ, PhD(math), 67. *Prof Exp:* Asst prof, 67-73, ASSOC PROF MATH, SOUTHERN METHODIST UNIV, 73- *Mem:* Math Asn Am; Sigma Xi. *Res:* Category theory; universal algebra. *Mailing Add:* Dept Math Southern Methodist Univ Dallas TX 75275

DAVIS, ROBERT DABNEY, b Kershaw, SC, Apr 13, 39. MATHEMATICS. *Educ:* NC State Univ, BS, 61, MS, 63; Fla State Univ, PhD(math), 69. *Prof Exp:* Instr math, Univ Richmond, 63-65; asst prof, 69-74, ASSOC PROF MATH, UNIV NEV RENO, 74- *Mem:* Am Math Soc; Math Asn Am. *Res:* Abstract Algebra; ramification series of ramified v-rings. *Mailing Add:* Dept Math Univ Nev Reno NV 89557

DAVIS, ROBERT E, GEOLOGICAL ENGINEERING. *Educ:* Univ Ariz, MA, 55. *Prof Exp:* Admin, US Geol Surv, 70-76, chief, Off Sci Publ, 78-81; gen partner, 76-80, PRES, HOSKINSON & DAVIS, INC, 80- *Mem:* Fel Geol Soc Am; Asn Eng Geologists; Asn Earth Sci Ed. *Mailing Add:* Hoskinson & Davis Inc 1725 K St Suite 614 Washington DC 20006

DAVIS, ROBERT EDWARD, b Brooklyn, NY, Jan 27, 39; m 62; c 2. PLANT PATHOLOGY, MICROBIOLOGY. *Educ:* Univ RI, BSc, 61; Cornell Univ, PhD(plant path), 67. *Prof Exp:* Lectr plant virol, Univ Minn, 65; resident res assoc, 66-67, res plant pathologist, plant virol lab, 67-85, RES LEADER, MICROBIOL & PLANT PATH LAB, AGR RES SERV, USDA, 85- *Concurrent Pos:* Exchange scientist plant physiol & biochem, Nat Inst Agron Res, Bordeaux, France, 73-74. *Mem:* Fel Am Phytopath Soc; Am Soc Microbiol; AAAS. *Res:* Role of spiroplasmas, mycoplasmas, other unusual procaryotes and viruses in the etiology of plant diseases; molecular detection/identification of pathogens. *Mailing Add:* Microbiol & Plant Path Barc W 252 Bldg 011A Beltsville MD 20705

DAVIS, ROBERT ELLIOTT, b Salt Lake City, Utah, Mar 21, 30; m 55; c 3. INORGANIC CHEMISTRY. *Educ:* Univ Utah, AB, 51, PhD(chem), 54. *Prof Exp:* Res anal chemist, M W Kellogg Co, NJ, 54-57; res proj engr, Am Potash & Chem Corp, Calif, 57-60, head new prod sect, 60-63 & process chem sect, Trona Res Lab, 63-67, mgr heavy chem res, 67-69; adminr & res consult, 69-70, mgr chem extraction sect, 70-76; mgr process eng sect, Tech Ctr, Kerr McGee Corp, 76-79, mgr process technol, 79-83, mgr tech ctr, 83-87, tech adv, 87-90, MGR, TECHNOL EVAL, KERR MCGEE CORP, 90- *Mem:* Am Chem Soc; Am Inst Mining, Metall & Petrol Engrs; Am Inst Chem Engrs. *Res:* Coordination chemistry; solvent extraction; alkali metals, particularly cesium, rubidium and lithium; chemical process development; synthetic fuel from coal; solution mining; super-critical fluid extraction. *Mailing Add:* 4840 NW 62nd Ter Oklahoma City OK 73122

DAVIS, ROBERT F(OSTER), b Greensboro, NC, Apr 12, 42; m 69; c 1. CERAMIC SCIENCE, THIN FILM PROCESSING AND CHARACTERIZATION. *Educ:* NC State Univ, BS, 64; Pa State Univ, MS, 66; Univ Calif, Berkeley, PhD(ceramic eng), 70. *Prof Exp:* Res asst mat res, Lawrence Berkeley Lab, 67-70; glass scientist, Res Ctr, Corning Glass Works, 70-72; PROF MAT ENG, NC STATE UNIV, 72- *Honors & Awards:* Alcoa Distinguished Res Award. *Mem:* Fel Am Ceramic Soc; Sigma Xi; Mat Res Soc. *Res:* Diffusion and high temperature deformation of ceramic materials; growth and characterization of wide band gap electronic materials; ceramic/metal coating deposition and interface adherence. *Mailing Add:* Mat Eng 203 Page NC State Univ Raleigh NC 27650

DAVIS, ROBERT FOSTER, JR, b Crowell, Tex, Apr 24, 37. PLANT PHYSIOLOGY. *Educ:* NTex State Univ, BA, 62, MA, 64; Wash State Univ, PhD(bot), 68. *Prof Exp:* Asst bot, zool & physiol, NTex State Univ, 60-64; asst, Wash State Univ, 64-66, res asst plant physiol, 66-68; NIH trainee, Neurophysiol, Col Physicians & Surgeons, Columbia Univ, 68-69; asst prof, 69-74, chmn dept, 78-85, ASSOC PROF BOT, RUTGERS UNIV, 74- *Mem:* AAAS; Am Soc Plant Physiol. *Res:* Ion transport; electrophysiology; water relations; biophysics of flux processes; salinity tolerance. *Mailing Add:* Dept Bot Rutgers Univ Newark NJ 07102

DAVIS, ROBERT GENE, b Doddsville, Miss, Mar 2, 32; m 54; c 4. PLANT PATHOLOGY, FOOD SOIL & WATER. *Educ:* Miss State Univ, BS, 53, MS, 68; La State Univ, Baton Rouge, PhD(plant path), 70. *Prof Exp:* Farm mgr, 57-65; fel plant path, Tex A&M Univ, 70-71; plant pathologist, Miss Agr & Forestry Exp Sta, 71-82; MFR BIOL PROD PLANT RES & LAB ANALYSIS FOOD, WATER & WASTEWATER, DAVIS RES INC, 82- *Mem:* Am Soc Microbiol; Soc Indust Microbiol. *Res:* Biological remediation of polluted sites; biological products; plant growth and protection products; food and water microbiology. *Mailing Add:* Davis Res Inc PO Box 40 Avon MS 38723-0040

DAVIS, ROBERT HARRY, b Wilkes Barre, Pa, July 16, 27; m 54; c 2. ENDOCRINOLOGY. *Educ:* Kings Col, BS, 50; Rutgers Univ, PhD(endocrinol), 58. *Prof Exp:* Med technician, Wilkes Barre Gen Hosp, 50-51; biologist, Warner-Chilcott Labs, 51-55; asst zool, Rutgers Univ, 55-56; sect head, Wm S Merrell Co, 58-60; sr physiologist, Neuroendocrine Res Unit, Willow Brook State Hosp, Staten Island, NY, 60-63; assoc prof endocrinol, Villanova Univ, 63-66; chief reprod endocrinol, Thomas M Fitzgerald Mercy Hosp, Darby, Pa, 66-69; assoc prof obstet & gynec, physiol & biophys, Hahnemann Med Col, 69-75; PROF PHYSIOL SCI, PA COL PODIATRIC MED, 75- *Concurrent Pos:* Sr reprod teratologist, Merck Inst Therapeut Res, Pa, 66. *Mem:* Am Soc Cytol; Soc Exp Biol & Med; Endocrine Soc; Soc Study Reprod; Am Physiol Soc. *Res:* Nutrition and endocrines; anti-inflammation; physiology of reproduction; endocrinology of mental retardation and pineal gland; peritoneal fluid cytology; teratology. *Mailing Add:* Pa Col Podiatric Med 8th & Race Sts Philadelphia PA 19107

DAVIS, ROBERT HOUSER, b Long Island, NY, Mar 20, 26; m 56. NUCLEAR PHYSICS. *Educ:* Univ Nebr, BS, 49; Univ Wis, MS, 50, PhD(physics), 55. *Prof Exp:* Res assoc nuclear physics, Rice Inst, 55-57; from asst prof to assoc prof physics, 57-64, PROF PHYSICS, FLA STATE UNIV, 64- *Mem:* Am Phys Soc. *Res:* Experimental nuclear physics; high vacuum technology; thin film physics. *Mailing Add:* Dept Physics Fla State Univ Tallahassee FL 32306

DAVIS, ROBERT JAMES, b Omaha, Nebr, Oct 26, 29; m 53; c 4. ASTRONOMY. *Educ:* Harvard Univ, AB, 51, AM, 56, PhD(astron), 60. *Prof Exp:* ASTROPHYSICIST, SMITHSONIAN ASTROPHYS OBSERV, 56- *Mem:* Am Astron Soc; Int Astron Union. *Res:* Space, television, extragalactic and stellar astronomy; machine-readable astronomical data files. *Mailing Add:* Ctr for Astrophys 60 Garden St Cambridge MA 02138

DAVIS, ROBERT LANE, b Henderson, Ky, Oct 23, 36; m 57; c 3. ENGINEERING MECHANICS. *Educ:* Univ Evansville, BS, 58; Univ Md, MS, 62, PhD(mech eng), 65. *Prof Exp:* Staff engr, Naval Ord Lab, 58-62; instr eng, Univ Md, 62-65; asst dean, 68-, PROF ENG MECH, & DEAN SCH ENG, UNIV MO, ROLLA, 78- *Concurrent Pos:* Ford Found fel, 63-64; consult, Pressure Sci Inc, 64-66, Nooter Corp, 67-74, D K Eng Assoc, 69-76. *Mem:* Sigma Xi; Am Soc Mech Engrs. *Res:* Materials processing; plasticity; finite element techniques; high pressure mechanics. *Mailing Add:* Hobson Star Rte Box 60-D Rolla MO 65401

DAVIS, ROBERT LLOYD, b New York, NY, May 23, 19; m 49; c 1. MATHEMATICS. *Educ:* Univ Chicago, BS, 49, MS, 51; Univ Mich, PhD(math), 57. *Prof Exp:* Res asst math for social sci, Univ Mich, 51-54, instr math, 54-56; asst prof, Univ Va, 56-59; NSF sci fac fel, Stanford Univ, 59-60; from asst prof to assoc prof, 60-70, PROF MATH, UNIV NC, CHAPEL HILL, 70- *Concurrent Pos:* Assoc prof, Univ Nice, France. *Mem:* Am Math Soc; Math Asn Am; Sigma Xi. *Res:* Algebra and combinatorial theory; incidence and matrix algebras; theory of relations or graphs; representations. *Mailing Add:* Dept Math Univ NC Chapel Hill NC 27514

DAVIS, ROBERT PAUL, b Malden, Mass, July 3, 26; m 53; c 3. BIOCHEMISTRY, MEDICINE. *Educ:* Harvard Univ, AB, 47, MD, 51, AM, 55; Am Bd Internal Med, dipl, 58; Am Bd Internal Med, dipl, nephrology, 76. *Hon Degrees:* AM, Brown Univ, 55. *Prof Exp:* Asst protein chem, Univ Lab Phys Chem, Harvard Med Sch, 48-51; med house officer, Peter Bent Brigham Hosp, 51-52, asst med, 52-55, sr asst res physician, 55-56, chief res physician, 56-57; asst prof med, Sch Med, Univ NC, 57-59; from asst prof to assoc prof, Albert Einstein Col Med, 59-67; prof, 67-84, EMER PROF MED SCI, BROWN UNIV, 85- *Concurrent Pos:* Soc of Fels jr fel phys chem & kinetics, Harvard Univ, 52-55, asst, 55-56, fel med, 56-57; Willard O Thompson Mem traveling scholar, Am Col Physicians, 65; asst vis physician, Bronx Munic Hosp Ctr, 59-65, assoc vis physician, 66-67; career scientist, Health Res Coun City New York, 62-67; physician in chief, Miriam Hosp, 67-74, dir renal & metab dis, 74-78; trustee, New England Organ Bank, 69-, treas, 70- *Mem:* Fel AAAS; Am Col Physicians; Am Soc Nephrology; Am Soc Pediat Nephrology; Am Soc Artificial Internal Organs; Am Fed Clin Res; Am Philos Asn; Biophys Soc; Soc Gen Physiol; Am Physiol Soc; Soc Cell Biol; NY Acad Sci; Am Soc Transplant Physicians; Harvey Soc; NY Acad Sci. *Res:* Enzyme kinetics; intermediary metabolism; renal physiology. *Mailing Add:* One Randall Sq Providence RI 02904

DAVIS, ROBERT WILSON, b Grinnell, Iowa, Oct 20, 10; m 38; c 4. ANATOMY, PATHOLOGY. *Educ:* Colo State Univ, DVM, 35, MS, 52. *Prof Exp:* Jr vet, US Bur Animal Indust, 35-36; asst to dep state vet, Mont, 35-37; asst prof anat & med, 37-40, assoc prof anat, 41-42, prof, 46-70, head dept, 48-70, CENTENNIAL PROF ANAT, COLO STATE UNIV, 70- *Concurrent Pos:* Prof oral biol, Sch Dent, Univ Colo, 74-75; vis prof anat, Univ Ill, Urbana-Champaign, 75-78; vis prof anat, Purdue Univ, WLafayette, 78; prof anat, Sch Vet Med, Tufts Univ, Boston, 79-81. *Mem:* Am Asn Anatomists; Am Vet Med Asn; Am Asn Vet Anatomists (pres, 54); World Asn Vet Anatomists; Nat Asn Dent Res. *Res:* Anatomy, physiology and pathology dealing with domestic and big game animals; antler and bone growth in deer, emphasis connective and mineralized tissues. *Mailing Add:* 1728 W Vine Dr Ft Collins CO 80521

DAVIS, ROBIN EDEN PIERRE, b Twickenham, Eng, Feb 19, 34; m 61; c 3. HIGH ENERGY PHYSICS. *Educ:* Oxford Univ, BA, 55, MA, 59, DPhil(nuclear physics), 62. *Prof Exp:* Res assoc, Enrico Fermi Inst, Univ Chicago, 62-64; res assoc, Northwestern Univ, 64-66, asst prof, 66-70; assoc prof, 70-76, PROF PHYSICS, UNIV KANS, 76-, PROF ASTRON, 80- *Mem:* Am Phys Soc. *Res:* Experimental high energy physics. *Mailing Add:* Dept Physics & Astron Univ Kans Lawrence KS 66045

DAVIS, RODERICK LEIGH, b Victoria, BC; m 78; c 2. AGROMETEOROLOGY. *Educ:* Univ BC, BSc, 73; McMaster Univ, MSc, 78. *Prof Exp:* Agrometeorologist, 73-80, head climat sect, 80-84, mgr, air resources sect, 84-85, MGR, ENVIRON SERV SECT, PROV BC, 85- *Concurrent Pos:* Mem, Can Expert Comt Agrometeorol, 80; vpres, Victoria & Islands Br, BC Inst Agrology, 83-84, pres, 84-85. *Mem:* Agr Inst Can; Can Meteorol & Oceanog Soc; Am Meteorol Soc. *Res:* Plant-climate relationships; growth-yield modelling, frost protection, water relationships; environmental quality impacts of waste discharges. *Mailing Add:* Ministry Environ 780 Blanchard St Victoria BC V8V 1X5 Can

DAVIS, ROGER (EDWARD), b Milwaukee, Wis, Aug 7, 29; m 58; c 3. BEHAVIORAL BIOLOGY. *Educ:* Univ Mich, BSc & MSc, 54; Univ Wis, PhD(zool), 61. *Prof Exp:* Res assoc, Dept Fisheries, 61-63, from asst res zoologist to assoc res zoologist, 63-69, assoc prof psychol, 69-73, PROF PSYCHOL, UNIV MICH, ANN ARBOR, 74-, RES PSYCHOBIOLOGIST, MENT HEALTH RES INST, 70- *Concurrent Pos:* USPHS training fel, 64-, NIMH res develop award, 68. *Mem:* AAAS; Animal Behav Soc; Am Soc Zoologists. *Res:* Animal behavior; learning, memory physiology; fish brain behavior. *Mailing Add:* 1402 Washington Hts Ann Arbor MI 48104

DAVIS, ROGER ALAN, LIPOPROTEIN ASSEMBLY, BLUE ACID METABOLISM. *Educ:* Wash State Univ, PhD(chem), 71. *Prof Exp:* Prof physiol, Med Sch, La State Univ, 84-86; PROF PHYSIOL, MED SCH, UNIV COLO, 86- *Res:* Cell biology; biochemistry. *Mailing Add:* Dept Med & Physiol Univ Colo Health Sci Ctr 4200 E 9th Ave Denver CO 80262

DAVIS, ROMAN, b Swidnica, Poland, Aug 15, 48; US citizen; m 85; c 1. ORGANO-SULFUR CHEMISTRY, PROCESS RESEARCH-SCALE UP. *Prof Exp:* Chemist, Dynapol, 75-76; from chemist to sr chemist, Syntex Res, 77-89; SR SCIENTIST, GLAXO RES INST, 89- *Mem:* Am Chem Soc. *Res:* Process research and scale-up of medicinal agents; enantioselective synthesis; enzymic resolution and polymerimmobilized reagents in organic synthesis; organosulfur chemistry for the elaboration of organic molecules. *Mailing Add:* Glaxo Res Five Moore Dr Research Triangle Park NC 27709

DAVIS, RONALD STUART, b Lethbridge, Alta, Aug 3, 41. REACTOR PHYSICS, NUCLEAR PHYSICS. *Educ:* Univ Alta, BS, 63; Univ BC, MS, 65, PhD(physics), 68. *Prof Exp:* Nat Res Coun Can fel theoret physics, Oxford Univ, 68-70; APPL MATHEMATICIAN NUCLEAR REACTORS, CHALK RIVER NUCLEAR LAB, ATOMIC ENERGY CAN LTD, 70- *Res:* Complex energy and angular momenta in nuclear reactions; nuclear three-body problem, especially formalism; applications in particle physics; economics and control of nuclear power reactors; risk of nuclear weapons proliferation; artificial intelligence, mostly machine vision. *Mailing Add:* Sta No 68 Chalk River Lab Seven Hillcrest Ave Chalk River ON K0J 1J0 Can

DAVIS, RONALD WAYNE, b Maroa, Ill, July 17, 41; m 69; c 2. MOLECULAR GENETICS, ELECTRON MICROSCOPY. *Educ:* Eastern Ill Univ, BS, 64; Calif Inst Technol, PhD(chem), 70. *Prof Exp:* NIH traineeship chem, Calif Inst Technol, 64-70; NIH fel, Harvard Univ, 70-72; from asst prof to assoc prof, 72-80, PROF BIOCHEM, STANFORD UNIV, 80-, PROF DEPT GENETICS, 90- *Concurrent Pos:* Dreyfus Teacher-Scholar Grant, 76. *Honors & Awards:* Young Fac Res Award, Sigma Xi, 76; Microbiol, & Immunol Award, Eli Lilly & Co, 76; Nat Acad Sci Award, Mol Biol, 81. *Mem:* Nat Acad Sci; Am Soc Biol Chemists; Genetic Soc Am; Am Chem Soc. *Res:* Higher cell gene isolation and cloning in heterologous cells; transcription and the regulation of gene expression; evolution of DNA sequences; development of electron microscopic and heteroduplex mapping techniques; human genome mapping. *Mailing Add:* Dept Biochem Beckman Ctr B400 Stanford Univ Sch Med Stanford CA 94305-5307

DAVIS, ROWLAND HALLOWELL, b Boston, Mass, Dec 8, 33;; div. BIOCHEMISTRY, MICROBIOLOGY. *Educ:* Harvard Univ, AB, 54, PhD(biol), 58. *Prof Exp:* Resident tutor, Dunster House, Harvard Univ, 55; NSF res fel biol, Calif Inst Technol, 58-60; from asst prof to prof bot, Univ Mich, Ann Arbor, 60-75; chmn dept, 77-79, PROF MOLECULAR BIOL & BIOCHEM, UNIV CALIF, IRVINE, 75- *Concurrent Pos:* Res grants, 61-; mem genetic biol panel, NSF, 67-70; assoc ed, Genetics, 71-75, Microbiol Rev, 78- & Molecular Cellular Biol, 81-83; mem genetics study sect, NIH, 73-77, chmn, 81-85; co-chair, Gordon Conference Polyamines, 89. *Mem:* Am Soc Biol Chemists; AAAS; Genetics Soc Am; Am Soc Microbiol. *Res:* Biochemical genetics; genetics and compartmentation of metabolic pathways of Neurospora crassa, with emphasis on arginine and polyamine synthesis; heterokaryosis in Neurospora crassa. *Mailing Add:* Dept Molecular Biol & Biochem Univ Calif Irvine CA 92717

DAVIS, RUSS E, b San Francisco, Calif, Mar 8, 41; c 1. PHYSICAL OCEANOGRAPHY. *Educ:* Univ Calif, Berkeley, BS, 63; Stanford Univ, PhD(chem eng), 67. *Prof Exp:* Asst res geophysicist, Inst Geophys & Planetary Physics, 67- 68, asst prof, 68-74, assoc prof, 74-77, PROF OCEANOG, SCRIPPS INST OCEANOG, UNIV CALIF, SAN DIEGO, 77- *Concurrent Pos:* Assoc ed, J Atmospheric & Ocean Technol; US sci steering comt, World Ocean Circulation Exp, 87-90. *Mem:* Nat Acad Sci; fel Am Geophys Union. *Res:* Fluid dynamics, including surface and internal waves; motion in rotating and stratified fluids; statistical analysis of dynamical systems; author of more than 50 scientific publications; oceanographic instrumentation. *Mailing Add:* Phys Ocean Res Div 0230 Scripps Inst Oceanog Univ Calif San Diego CA 92093-0230

DAVIS, RUSSELL PRICE, b Hanford, Calif, Aug 24, 28; m 51; c 2. ZOOLOGY. *Educ:* Univ Redlands, BA, 50; Long Beach State Col, MA, 56; Univ Ariz, PhD(zool), 63. *Prof Exp:* Teacher biol, Santa Ana Col, 56-59; asst prof, Southern Ore Col, 63-64; ASSOC PROF BIOL, UNIV ARIZ, 64- *Mem:* Am Soc Mammal. *Res:* Mammalogy, especially the natural history of bats. *Mailing Add:* Dept Ecol Univ Ariz Tucson AZ 85721

DAVIS, RUTH MARGARET, b Sharpsville, Pa, Oct 19, 28; m 55. COMPUTER SCIENCES. *Educ:* Am Univ, BA, 50; Univ Md, MA, 52, PhD(math), 55. *Hon Degrees:* DEng, Carnegie-Mellon Univ, 79. *Prof Exp:* Mathematician, US Bur Standards, 50; res assoc inst fluid dynamics & appl math, Univ Md, 52-55; mathematician, David Taylor Model Basin, 55-58, head opers res div, 57-61; staff asst off of spec asst intel & reconnaissance, Off Dir Defense Res & Eng, US Dept Defense, 61-67; assoc dir res & develop, Nat Libr Med, 67-68, dir, Lister Hill Nat Ctr Biomed Commun, 68-70; dir ctr comput sci & technol, Nat Bur Standards, Dept Com, 70-72, dir Inst Comput Sci & Technol, 72-77; dep under secy defense for res & eng, US Dept Defense, 77-79; asst secy of energy, Dept of Energy, 79-81; PRES, CHIEF EXEC OFFICER & FOUNDER, PYMATUNING GROUP, INC, 81- *Concurrent Pos:* Lectr, Univ Md, 55-56 & Am Univ, 57-58; consult, Off Naval Res, 57-58; mem, Nat Acad Pub Admin, 74-; mem, Md Govt Sci Adv Comt, 74-77, vchairperson, 75-76; mem nat adv coun, Elec Power Res Inst, 74-78, vchmn, 75-76; adj prof eng, Univ Pittsburgh, 81-; mem, Bd Dirs, Air Prod & Chem, Inc; Control Data Corp; Premark Internat, Inc; Prin Financial Group, Inc; United Telecommun, Inc; Varian Assoc, Inc & Brit Technol Group; mem, Bd Trustees, Aerospace Corp, Consol Edison Co NY; Inst Defense Anal; mem bd overseers, Univ Pa, Sch Eng & Sci, 81-87, Sch Eng Ad Bd, Univ Calif, Berkeley; vchmn & bd mem, Aerospace Corp; regent's lectr, Sch Eng & Appl Sci, Univ Calif, Berkeley, 87. *Honors & Awards:* Gold Medal, Dept Com, 72; Rockefeller Pub Serv Award Prof Accomplishment & Leadership, 73; Nat Civil Serv League Award, 76; Ada Augusta Lovelace Award Computer Sci, 84; Director's Choice Award, Nat Women's Econ Alliance, 89. *Mem:* Nat Acad Eng; fel AAAS; Coun Libr Resources; Am Math Soc; Math Asn Am; Nat Acad Pub Admin; fel Am Inst Aeronaut & Astronaut; fel Soc Info Display. *Res:* Automation; electronics; computers; energy. *Mailing Add:* Pymatuning Group Inc 2000 15th St Suite 707 Arlington VA 22201

DAVIS, SAMUEL HENRY, JR, b Houston, Tex, Dec 19, 30; m 67; c 3. CHEMICAL ENGINEERING, APPLIED MATHEMATICS. *Educ:* Rice Inst, BA, 52, BS, 53; Mass Inst Technol, ScD(chem eng), 57. *Prof Exp:* Engr chem eng, Gen Elec Co, 56-57; PROF CHEM ENG & MATH SCI, RICE UNIV, 57-, CHMN CHEM ENG, 77- *Mem:* Am Inst Chem Engrs. *Res:* Analysis of chemical systems used in atmospheric control in spacecraft. *Mailing Add:* Dept Appl Math Rice Univ Box 1892 Houston TX 77251

DAVIS, SELBY BRINKER, b Washington, DC, Dec 26, 14; m 48; c 3. RESEARCH ADMINISTRATION, PHARMACEUTICAL CHEMISTRY. *Educ:* George Wash Univ, BS, 37; Harvard Univ, AM, 41, PhD(org chem), 42. *Prof Exp:* Res chemist, Socony-Vacuum Oil Co, 37-39; res fel, Harvard Univ, 42-43; group leader, Chemother Div, Am Cyanamid Co, 43-56, head, Med Chem Dept, Lederle Labs, 56-71, head, Chem Res Dept, Cent Nervous Syst Res Sect, 71-74, dir, Cardiovascular-Renal Res Sect, 74-76, dir Clin Res Info Serv, 76-80; RETIRED. *Mem:* AAAS; Am Chem Soc; NY Acad Sci; Brit Chem Soc. *Res:* Stereochemistry; structure of natural products; chemotherapeutic agents. *Mailing Add:* 1508 Willow Crest Dr Richardson TX 75081-3041

DAVIS, SHELDON W, b Kenton, Ohio, Aug 21, 51. SET THEORETICAL TOPOLOGY. *Educ:* Ohio Univ, BS & MS, 73, PhD(math), 76. *Prof Exp:* Asst prof math, Auburn Univ, 77-78; from asst prof to assoc prof, 78-86, PROF MATH, MIAMI UNIV, 86- *Mem:* Am Math Soc; Math Asn Am. *Mailing Add:* Dept Math & Statist Miami Univ Oxford OH 45056

DAVIS, SPENCER HARWOOD, JR, b Philadelphia, Pa, Apr 2, 16; m 45. PLANT PATHOLOGY. *Educ:* Westminster Col, Pa, BS, 37; Univ Pa, PhD(plant path), 44. *Prof Exp:* Asst instr, Univ Pa, 42-43; res assoc, Off Sci Res & Develop, 43-45; asst prof, Univ Del, 46-48; assoc prof, 48-54, extension specialist, 54-80, EMER PROF, RUTGERS UNIV, 80- *Mem:* Am Phytopath Soc; Int Soc Arboriculture (vpres, 63, pres, 64); Am Soc Consult Arborists (exec dir, 70-). *Res:* Diseases of ornamentals. *Mailing Add:* 315 Franklin Rd N New Brunswick NJ 08902

DAVIS, STANLEY GANNAWAY, b Hancock Co, Ind, Aug 18, 22. PHYSICAL CHEMISTRY. *Educ:* Purdue Univ, BS, 42; Univ Chicago, PhD(chem), 55. *Prof Exp:* Jr chemist & asst radiation chem, Metall Lab, Univ Chicago, 43-44; jr chemist anal chem, Clinton Eng Works, Tenn Eastman Corp, 45-46; asst res chemist high temperature chem, Inst Eng Res, Univ Calif, 55-57; asst prof chem, 57-63, ASSOC PROF CHEM, RUTGERS UNIV, 63- *Res:* Chemical physics; high-temperature thermodynamics. *Mailing Add:* Dept Chem Rutgers Univ 311 N Fifth St Camden NJ 08102

DAVIS, STANLEY NELSON, b Rio de Janeiro, Brazil, Aug 6, 24; US citizen; m 82; c 6. HYDROGEOLOGY. *Educ:* Univ Nev, BS, 49; Univ Kans, MS, 51; Yale Univ, PhD(geol), 55. *Prof Exp:* Instr geol, Univ Rochester, 53-54; from asst prof to prof, Stanford Univ, 54-67; prof, Univ Mo-Columbia, 67-73, chmn dept, 70-72; prof, Ind Univ, 73-75; head dept, 75-79, PROF HYDROL & WATER RESOURCES, UNIV ARIZ, 75- *Concurrent Pos:* Vis prof, Univ Chile, 60-61 & Univ Hawaii, 66. *Honors & Awards:* Sci Award, Asn Ground Water Sci & Eng, 80; De Meinzer Award, Geol Soc Am, 89. *Mem:* AAAS; Geol Soc Am; Am Water Resources Asn; Am Geophys Union; assoc Soc Econ Paleont & Mineral; Asn Ground Water Sci & Eng. *Res:* Ground-water geology; isolation of hazardous waste, dating ground water, minor strains induced by ground-water movement, chemistry of ground water and trace radionuclides in ground water. *Mailing Add:* Dept Hydrol & Water Resources Univ Ariz Tucson AZ 85721

DAVIS, STARKEY D, b Atlanta, Tex, Jan 29, 31. PEDIATRICS, INFECTIOUS DISEASES. *Educ:* Baylor Univ, BA, 53, MD, 57. *Prof Exp:* Rotating intern, Confederate Mem Med Ctr, Shreveport, La, 57-58; instr prev med, Emory Univ, 58-60; resident pediat, Baylor Univ, 60-62; fel, Sch Med, Univ Wash, 62-64, from instr to prof, 64-75; PROF & CHMN, DEPT PEDIAT, MED COL WIS, 83-, PEDIATRICIAN-IN-CHIEF, CHILDREN'S HOSP WIS, 83- *Concurrent Pos:* Res fel pediat, Univ Wash, 62-63, spec res fel, 64-65. *Mem:* Am Asn Immunol; Am Soc Microbiol; Infectious Dis Soc Am; Am Pediat Soc. *Mailing Add:* MACC Fund Res Ctr Dept Pediat Med Col Wis 8701 Watertown Plank Rd Milwaukee WI 53226

DAVIS, STEPHEN H(OWARD), b New York, NY, Sept 7, 39; m 66. MATHEMATICS, FLUID DYNAMICS. *Educ:* Rensselaer Polytech Inst, BEE, 60, MS, 62, PhD(math), 64. *Prof Exp:* Teaching asst math, Rensselaer Polytech Inst, 60-61; mathematician, Rand Corp, 64-66; lectr math, Imp Col London, 66-68; from asst prof to prof mech, Johns Hopkins Univ, 68-79; PROF, DEPT ENG SCI & APPL MATH, NORTHWESTERN UNIV, 79-, WALTER P MURPHY PROF, 87- *Concurrent Pos:* Lectr exten , Univ Calif, Los Angeles, 65 & Univ Southern Calif, 66; asst ed, J Fluid Mech, 69-75, assoc ed, 75-89; vis prof, Dept Math, Monash Univ, Melbourne, Australia, 73; vis prof, Dept Chem Eng, Univ Ariz, 77, Dept Aerodyn & Mech Eng, 82, Ecole Polytechnique Federale de Lausanne, 87 & 88; mem, US Nat Comt Theoret & Appl Math, 78-87; chmn, Div Fluid Dynamics, Am Phys Soc, 78-79 & 87-88; mem coun, Am Phys Soc, 80-81, Soc Indust & Appl Math, 83-87. *Honors & Awards:* Fel, Am Phys Soc. *Mem:* Am Phys Soc; Soc Indust & Appl Math. *Res:* Theory and applications of hydrodynamic stability; interfacial phenomena; solidification phenomena; bifurcation theory. *Mailing Add:* Dept Eng Sci & Appl Math Northwestern Univ Evanston IL 60208

DAVIS, STEVEN LEWIS, b Pocatello, Idaho, Sept 26, 41; m 63; c 3. ENDOCRINOLOGY. *Educ:* Univ Idaho, BS, 64, MS, 66; Univ Ill, PhD(animal sci), 69. *Prof Exp:* NIH fel, Univ Mich, 69-70; asst prof path, M S Hershey Med Ctr, Pa State Univ, 70-73; ASSOC PROF ANIMAL SCI, UNIV IDAHO, 73- *Concurrent Pos:* NIH res grant, 72-77; consult, Merck, Sharp & Dohme Inst, 75. *Mem:* Am Soc Animal Sci; Endocrine Soc. *Res:* The regulation of prolactin, growth hormone and thyrotropin secretion. *Mailing Add:* Dept Animal Sci Ore State Univ Corvallis OR 97331

DAVIS, STUART GEORGE, b Lethbridge, Alta, June 15, 17; m 40; c 1. PHYSICAL CHEMISTRY. *Educ:* Univ Alta, BSc, 39, MSc, 40; McGill Univ, PhD(chem), 42. *Prof Exp:* Lectr, 42-43, from asst prof to assoc prof, 51-82, EMER PROF CHEM, UNIV ALTA, 82- *Honors & Awards:* Prize, Asn Prof Eng, 39; Stiernotte Prize, 39. *Mem:* Fel, Chem Inst Can. *Res:* Adsorption. *Mailing Add:* Dept Chem Univ Alta Edmonton AB T6G 2G2 Can

DAVIS, SUMNER P, b Burbank, Calif. PHYSICS. *Educ:* Univ Calif, Los Angeles, AB, 47; Univ Ill, AM, 48; Univ Calif, Berkeley, PhD(physics), 52. *Prof Exp:* Instr physics, Mass Inst Technol, 52-55, res staff mem, 55-59; lectr, 59-60, from asst prof to assoc prof, 60-67, PROF PHYSICS, UNIV CALIF, BERKELEY, 67- *Concurrent Pos:* NATO sr fel sci, 67-68. *Mem:* Am Phys Soc; Optical Soc Am; Am Astron Soc. *Res:* Optical spectroscopy; molecular spectra; atomic energy level analysis; hyperfine structure in atomic spectra; astrophysical spectroscopy. *Mailing Add:* Dept Physics Univ Calif Berkeley CA 94720

DAVIS, TERRY CHAFFIN, b Pearisburg, Va, Apr 12, 32; m 59. FOREST PATHOLOGY. *Educ:* Va Polytech Inst, BS, 59, MS, 61; WVa Univ, PhD(plant path), 65. *Prof Exp:* Asst prof forestry & forest path, 65-70, asst prof bot & plant path, 70-75, ASST PROF BOT & MICROBIOL, AUBURN UNIV, 75- *Mem:* Am Phytopath Soc; Mycol Soc Am. *Res:* Microorganisms which cause tree diseases. *Mailing Add:* Dept Forestry Auburn Univ Auburn AL 36849

DAVIS, THOMAS ARTHUR, b Columbia, SC, Aug 12, 39; m 62; c 3. CHEMICAL ENGINEERING, BIOMEDICAL ENGINEERING. *Educ:* Univ SC, BS, 61, PhD(eng sci), 67. *Prof Exp:* Jr chem engr, Textile Chem Pilot Plant, Deering Milliken Res Corp, 63-64; asst chem eng, Univ SC, 64-67; res chem engr, Southern Res Inst, 67-68, head, Biomed Eng Sect, 77-78; sr staff engr, Corp Res Lab, Exxon Res & Eng Co, 78-84; res dir, Graver Water Co, 84-89; PVT CONSULT, 89- *Concurrent Pos:* Ann guest lectr, Ctr Prof Advan, 72- *Mem:* Am Inst Chem Engrs; NAm Membrane Soc. *Res:* Membrane processes research, particularly electrodialysis for desalination of brackish water, chemical processing and water pollution control; electro-regeneration of ion-exchange resins; development of artificial kidney devices, activated carbon, liquid membrane systems. *Mailing Add:* 74AA Sand Hill Rd Annandale NJ 08801

DAVIS, THOMAS AUSTIN, b Belgian Congo, May 31, 34; US citizen; m 59; c 2. MATHEMATICS. *Educ:* Denison Univ, AB, 56; Univ Mich, MS, 57; Cambridge Univ, PhD(math), 63. *Prof Exp:* Asst prof math, DePauw Univ, 63-69, assoc prof, asst dean univ & dir grad studies, 69-73; PROF MATH & DEAN UNIV, UNIV PUGET SOUND, 73- *Concurrent Pos:* Am Coun Educ acad admin intern, Princeton Univ, 71-72. *Mem:* Math Asn Am. *Res:* Banach algebras and Fourier analysis. *Mailing Add:* Off Dean Fac Univ Puget Sound 1500 N Warner Tacoma WA 98416

DAVIS, THOMAS HAYDN, b Philadelphia, Pa, Sept 4, 39; m 71; c 3. PHYSICS. *Educ:* Lehigh Univ, BS, 61; Carnegie-Mellon Univ, MS, 66, PhD(physics), 71. *Prof Exp:* Assoc engr nuclear reactor physics, Westinghouse Bettis Atomic Power Lab, 61-65; PROG MGR ELECTRONICS, MICROWAVES & OPTICS, MAJOR APPLIANCE LAB, GEN ELEC CO, 71-, MGR ENG PHYSICS LAB, 79- *Concurrent Pos:* Adj asst prof, Dept Physics, Univ Louisville, 73-78. *Mem:* Am Phys Soc. *Mailing Add:* 4824 Nottinghamshire Dr Louisville KY 40229

DAVIS, THOMAS MOONEY, b Claysville, Pa, Mar 12, 34; m 56; c 2. GEOPHYSICS, GEODESY. *Educ:* Pa State Univ, BS, 55, PhD(geophys), 74. *Prof Exp:* Field party chief, Airborne Geomagnetic Surv, US Naval Oceanog Off, 55-66, head, Earth Physics Br, 68-70, head, math modeling Proj, 70-87, chief scientist, 87-90; CONSULT, 90- *Concurrent Pos:* Vis prof, Univ SC, 70; adj prof, Univ NDak, 80-81. *Mem:* Am Geophys Union; Soc Explor Geophysicists. *Res:* Potential fields; mathematical methods of physical geodesy; survey design; Fourier analysis; numerical models of oceanographic data. *Mailing Add:* PO Box 279 McNeill MS 39457

DAVIS, THOMAS NEIL, b Greeley, Colo, Feb 1, 32; m 51; c 3. GEOPHYSICS. *Educ:* Univ Alaska, BS, 55, PhD(geophys), 61; Calif Inst Technol, MS, 57. *Prof Exp:* Assoc prof geophys, Geophys Inst, Univ Alaska, 61-64; aerospace technologist, Goddard Space Flight Ctr, NASA, 64-65; asst dir, Geophys Inst, 65-70, actg dir, Geophys Inst & div geosci, 76-77, prof, 65-81, EMER PROF GEOPHYS, GEOPHYS INST, UNIV ALASKA, 82- *Concurrent Pos:* Nat Acad Sci-Nat Res Coun resident res assoc, Goddard Space Flight Ctr, 62-64; dir, Univ Alaska Press, Rairbanks, 84-; mem, bd dir, Alaska Power Authority, 87- *Honors & Awards:* Int Geophys Medal, Soviet Geophys Comt, USSR Acad Sci, 85. *Mem:* AAAS; Int Asn Geomagnetism & Aeronomy. *Res:* Aurora and geomagnetism; Alaska energy resources; history of Alaska science and education; seismology; rocket and satellite instrumentation. *Mailing Add:* Neil Davis Alaska Assocs 375 Miller Hill Rd Fairbanks AK 99709

DAVIS, THOMAS PEARSE, b Kansas City, Mo, May 21, 26; m 74; c 6. RADIATION PHYSICS, OPTICAL PHYSICS. *Educ:* Purdue Univ, BS, 50; Univ Rochester, MS, 55, PhD(biophys), 59. *Prof Exp:* Res asst optics & biophys, Atomic Energy Proj, Univ Rochester, 50-59; from asst prof to assoc prof radiol physics & biophys, Dept Radiation Biol & Biophys, Univ Rochester, 59-67; sr scientist biophys & optics, E H Plesset Assocs, 67-70, sr sci specialist physics & optics, Santa Barbara Div, 70-83, SCI SPEC III, EG&G/ENERGY MEASURMENTS, INC, SANTA BARBARA OPERS, 83- *Concurrent Pos:* Vis lectr, Univ Calif, Santa Barbara, 85. *Mem:* Sigma Xi. *Res:* Studies of radiation sensors and data analysis; electro-optics research and laser effects studies. *Mailing Add:* 6589 Camino Ventcroso Goleta CA 93117

DAVIS, THOMAS WILDERS, b Nyack, NY, Aug 1, 05; m 42; c 3. PHYSICAL CHEMISTRY. *Educ:* NY Univ, BS, 25, MS, 26, PhD(chem), 28. *Prof Exp:* Res chemist, Combustion Utilities Corp, NJ, 28-29; from instr to prof chem, 29-73, chmn dept, 55-64, EMER PROF CHEM, NY UNIV, 73- *Concurrent Pos:* Instr, Rand Sch Social Sci, 33; res assoc, Metall Lab, Univ Chicago, 42; sr chemist, Clinton Lab, Tenn, 46-47; res chemist, Brookhaven Nat Lab, 48, 50, US Naval Ord Test Sta, 52 & Oak Ridge Nat Lab, 54; res chemist, Argonne Nat Lab, 55, 56, res assoc, 56; vis prof & mem senate, Univ Leeds, 64-65. *Mem:* Am Chem Soc; fel Am Inst Chemists; Radiation Res Soc; Fedn Am Sci. *Res:* Reaction kinetics and mechanisms of reaction; photochemistry and radiation chemistry; thermodynamics; uses of radioactive tracers. *Mailing Add:* 253 Pineola St Burnsville NC 28714-3213

DAVIS, THOMAS WILLIAM, b Belvidere, Ill, Mar 14, 46; m 71; c 1. COMPUTERS, ROBOTICS. *Educ:* Milwaukee Sch Eng, BS, 68; Univ Wis, MS, 71. *Prof Exp:* Dir comput eng technol, Milwaukee Sch Eng, 75-77, chair, Dept Elec Eng & Comput Sci, 77-84, dir, Acad Comput Ctr, 78-83, dean & res, 84-87, PROF ELEC ENG, MILWAUKEE SCH ENG, 73-, VPRES ACAD, 85- & DEAN OF FAC, 90- *Concurrent Pos:* Dir, Milwaukee Air, Inc, 72-73; lectr comput sci, Univ Wis-Milwaukee, 73-76; consult to over 150 cos; dir, Matarah Industs, Inc, 86-, Michaels Machine Co, Inc, 88-, Jordon Controls, Inc; pres, Milwaukee Coun Eng Sci Socs. *Mem:* Sr mem Inst Elec & Electronic Engrs; Am Soc Eng Educ; sr mem Soc Mfg Engrs; Asn Comput Mach. *Res:* Computer aided circuit analysis and topology; robotics; engineering education; programming; microprocessors. *Mailing Add:* 5590 Gray Log Ct Grafton WI 53024

DAVIS, V TERRELL, b Long Branch, NJ, July 14, 11; m 36, 76; c 3. PSYCHIATRY. *Educ:* Wash Univ, MD, 36; Am Bd Psychiat & Neurol, cert, 46. *Prof Exp:* Asst surgeon drug addiction, USPHS Hosp, Ft Worth, Tex, 39-41, sr asst surgeon clin psychiat, 41-43, exec officer, 43-44; clin dir gen hosp admin, USPHS, Staten Island, NY, 44-45; chief psychiat & neurol, USPHS Hosp, Ellis Island, 45-49; med dir psychiat & Neurol, USPHS, Staten Island, 49-54; clin assoc prof neuropsychiat & asst dir, Psychiat Inst, Univ Wis, 54-56; prof psychiat, Col Med & Dent NJ, Newark, 59-72; CONSULT, 83- *Concurrent Pos:* Chief med officer, Fed Penitentiary, Lewisburg, Pa, 44; consult, US Immigration Serv, 45-54; dir, Staten Island Ment Health Clin, 53-54; asst dir div ment health, State Dept Pub Welfare, Wis, 54-56; clin dir, Wis Diag Ctr, Madison, 54-56; dir ment health & hosps, NJ, 56-69; WHO fel, 62; clin prof, Med Sch, Rutgers Univ, 68; ment health prog consult, Region II & med dir, NIMH, 69-72; clin prof psychiat & human behav, Jefferson Med Col, 72-76, hon clin prof, 76-; chief, Dept Psychiat, Wilmington Med Ctr, Wilmington, Del, 72-83; consult, NIMH. *Mem:* Life fel Am Psychiat Asn; Asn Mil Surg US; Am Pub Health Asn; Nat Asn State Ment Health Prog Dirs (pres, 63-65); fel Am Col Psychiat. *Res:* Psychotherapy; psychiatric training; relation of injury to psychiatric illness; relation of psychiatry to employment; psychoanalysis; mental health program administration; forensic psychiatry; psychiatric emergency treatment. *Mailing Add:* 1004 Park Plaza 1100 Lovering Ave Wilmington DE 19806

DAVIS, VIRGINIA EISCHEN, b Okarche, Okla, Nov 9, 25; m 47; c 2. PHARMACOLOGY, BIOCHEMISTRY. *Educ:* Okla State Univ, BS, 47, MS, 49; Rice Univ, PhD(biochem), 60. *Prof Exp:* Biochemist, Res Serv, Vet Admin Hosp, Houston, 49-57; res assoc biochem, Rice Univ, 60-61; res biochemist, Res Serv, 61-63, assoc dir metab res lab, 63-71, DIR NEUROCHEM & ADDICTION RES LAB, VET ADMIN HOSP, 71-; RES ASSOC PROF BIOCHEM & MED, BAYLOR COL MED, 69- *Concurrent Pos:* Res asst prof, Baylor Col Med, 65-69; lectr, Sch Pharm, Univ Houston, 67-69. *Mem:* AAAS; Am Soc Pharmacol & Exp Therapeut; Res Soc Alcoholism; Soc Neurosci; NY Acad Sci. *Res:* Biochemical mechanisms of drug action; neuroamine metabolism; alcoholism. *Mailing Add:* Neurochem & Addiction Res Lab Vet Admin Med Ctr Bldg 211 Rm 108 2002 Holcombe Blvd Houston TX 77211

DAVIS, W KENNETH, b Seattle, Wash, July 26, 18. NUCLEAR ENERGY. *Educ:* Mass Inst Technol, BA, 40, MS, 42. *Prof Exp:* Mgrdevelop & eng, Standard Oil, 54-58; vpres, Bechtel Power Corp, 58-81; dep secy, US Dept Energy, Washington, DC, 81-83; CONSULT MGT & ENG, BECHTEL POWER CORP & CHMN, BOLD, INC, 83- *Concurrent Pos:* Dep dir, Atomic Energy Comn, 54-55, dir reactor develop, 55-58; chmn & pres, Atomic Indust Forum, 64-67; chmn, Bold Inc, 83-90. *Honors & Awards:* Arthur S Flemming Award, 56; Robert E Wilson Award, Am Inst Chem Engrs, 69; Walter H Zinn Award, Am Nuclear Soc, 83. *Mem:* Nat Acad Eng; fel Am Nuclear Soc (vpres, 78-81); fel Am Inst Chem Engrs (pres, 81); fel AAAS; Brit Nuclear Soc. *Mailing Add:* Bechtel Power Corp PO Box 3965 San Francisco CA 94119

DAVIS, WALLACE, JR, b Pawtucket, RI, Dec 17, 18; m 42; c 3. CHEMISTRY. *Educ:* Brown Univ, ScB, 41; Univ Rochester, PhD(phys chem), 47. *Prof Exp:* Res chemist, Kellex Corp, Tenn, 47; res chemist, Oak Ridge Gaseous Diffusion Plant, 47-58, group leader, 58-72, SR STAFF MEM, OAK RIDGE NAT LAB, UNION CARBIDE CORP, 79- *Concurrent Pos:* Guest scientist, Atomic Energy Res Estab, Harwell, Eng, 65-66. *Mem:* Fel AAAS; fel Am Inst Chem; Am Chem Soc; Am Inst Chem Eng; Sigma Xi. *Res:* Photochemistry of gases; thermodynamic and phase properties of uranium-fluorine-fluorocarbon systems; isotope separation; gas-solid reaction kinetics; radiation chemistry; radiochemical reprocessing of nuclear fuels; thermodynamics of solvent extraction; foam separation; analysis of environmental impacts of nuclear energy. *Mailing Add:* 601 Florida Ave Oak Ridge TN 37830

DAVIS, WALTER LEWIS, b Philadelphia, Pa, Sept 30, 42; m 67; c 2. CELL BIOLOGY. *Educ:* Abilene Christian Col, BS, 65; Baylor Univ, MS, 69, PhD(anat, physiol), 71. *Prof Exp:* Instr, dept biol, Eastfield Col, Dallas, Tex, 70-71; fel, dept biochem & biophys, Univ Pa Med Sch, 71-74; from asst prof to assoc prof, 75-82, PROF ANAT, BAYLOR COL DENT, DALLAS, TEX, 83- *Concurrent Pos:* Consult path, Baylor Univ Med Ctr, 73-; adj prof biol, Southern Methodist Univ, 74-; vis prof, dept biol, NTex State Univ, 84- *Mem:* Sigma Xi; AAAS; Am Asn Anatomists; Am Soc Cell Biol; Electron Micros Soc Am; Histochem Soc. *Res:* The roles of cells and their organelles in the regulation of cellular and extracellular calcium homeostasis; events controlling mineralization phenomena in bones and teeth; epithelial transport and the mechanism of action of vasopressin; electron microscopic histochemistry. *Mailing Add:* Dept Microscopic Anat Baylor Col Dent 800 Hall St Dallas TX 75226

DAVIS, WARD BENJAMIN, b Alhambra, Calif, Apr 13, 33; m 57; c 1. TECHNICAL WRITING, TECHNICAL MANAGEMENT. *Educ:* Calif Inst Technol, BS, 54; Univ Southern Calif, MS, 65, PhD(biochem), 73. *Prof Exp:* Instr chem, Whittier Col, 56-62; res chemist, Don Baxter, Inc, 62-64; instr chem & math, Glendale Col, 64-66; consult, Aseptic-Thermo Indicator Co, 64-72, res dir, 72-74; assoc dir corp relations, Calif Inst Technol, 74-83, coordr biol labs, 83-85. *Concurrent Pos:* Consult, 85- *Mem:* Am Chem Soc. *Res:* Physical and chemical characterization of high-density lipoproteins; chemical and physical technology of specialty (indicator) inks; development of operating and quality control procedures from industrial research and engineering programs; analysis of industrial research management. *Mailing Add:* PO Box 834 Rosemead CA 91770-0834

DAVIS, WAYNE ALTON, b Ft Macleod, Alta, Nov 16, 31; m 59, 90; c 3. COMPUTER SCIENCE. *Educ:* George Washington Univ, BSE, 60; Univ Ottawa, MSc, 63, PhD(elec eng), 67. *Prof Exp:* Sci officer, Defence Res Bd, Defence Res Telecommun Estab, 60-69; sci officer, Dept Commun, Commun Res Ctr, 69; from assoc prof to prof, 69-71, EMER PROF COMPUT SCI, UNIV ALTA, 91- *Concurrent Pos:* Lectr, Univ Ottawa, 65-69; nat leader, Working Panel S-3, Subgroup S, Tech Coop Prog, 66-69; lectr, Carleton Univ, 67; mem reactor safety adv comt, Atomic Energy Control Bd, 67-69; vis scientist, Graphics Sect, Nat Res Coun Can, 75-76; vis prof & lectr, Harbin Shipbuilding Eng Inst, Dept Comput Sci, Harbin, Heilongjiang, China, 81-90. *Mem:* Inst Elec & Electronics Engrs; Asn Comput Mach; Can Info Processing Soc (vpres, 77-78, pres, 78-79). *Res:* Digital image processing; computer vision; data base techniques for spatial data; display of 3D data; texture edges; digital pattern recognition in image processing. *Mailing Add:* Box 1098 Summerland BC V0H 1Z0 Can

DAVIS, WAYNE HARRY, b Morgantown, WVa, Dec 31, 30; m 58; c 3. MAMMALOGY, ORNITHOLOGY. *Educ:* WVa Univ, AB, 53; Univ Ill, MS, 55, PhD(zool), 57. *Prof Exp:* Asst zool, Univ Ill, 53-54 & biol, 54-57; res fel, Univ Minn, 57-59; instr biol, Middlebury Col, 59-62; from asst prof to assoc prof, 62-70, prof zool, 70-80, PROF BIOL SCI, UNIV KY, 80- *Mem:* Fel AAAS; Am Soc Mammal. *Res:* Life history of bats; human ecology; competition in cavity nesting birds. *Mailing Add:* Sch Biol Sci Univ Ky Lexington KY 40506-2251

DAVIS, WILBUR MARVIN, b Calumet City, Ill, Apr 13, 31; m 56; c 2. PHARMACOLOGY. *Educ:* Purdue Univ, BS, 52, MS, 53, PhD(pharmacol), 55. *Prof Exp:* From asst prof to prof pharmacol, Sch Pharm, Univ Okla, 55-63; assoc prof, 64-65, chmn dept, 64-83, PROF PHARMACOL, SCH PHARM, UNIV MISS, 65- *Concurrent Pos:* Prin or co-prin investr, 14 fed grants or contracts. *Honors & Awards:* Lalor Found Award, 58. *Mem:* Soc Toxicol; Am Soc Pharmacol & Exp Therapeut; Soc Neurosci; Am Ornith Union. *Res:* Neuropsychopharmacology; behavioral and CNS pharmacology; pharmacology and toxicology of dependence-producing drugs. *Mailing Add:* Dept Pharmacol Sch Pharm Univ Miss University MS 38677

DAVIS, WILFORD LAVERN, b Fairport, Mo, May 6, 30; m 54; c 2. RESEARCH ADMINISTRATION, APPLIED STATISTICS. *Educ:* Univ Mo, BS, 56; NC State Univ, MS, 58. *Prof Exp:* Asst statistician, NC State Col, 56-58; from assoc scientist to scientist, Westinghouse Elec Corp, 58-61; res statistician, Atlantic Richfield, 61-65; chief engr, 65-68, chief comput sci, 68-71, assoc dir systs, 71-73, dir corp systs & data processing, Owens-Ill, Inc, 73-87; RETIRED. *Concurrent Pos:* Lectr, Villanova Univ, 62-65; instr, Univ Toledo, 66-68, adj assoc prof, 71-72. *Mem:* Am Statist Asn; Sigma Xi. *Res:* Mathematical applications; applied and theoretical mathematical statistics; computer application in research and business. *Mailing Add:* 29570 Gleneagle Perrysburg OH 43551

DAVIS, WILLIAM ARTHUR, b East St Louis, Ill, July 29, 47; m 68; c 2. ELECTROMAGNETICS, RADIO ENGINEERING. *Educ:* Univ Ill, BS, 69, MS, 70, PhD(elec eng), 74. *Prof Exp:* Instr electromagnetic radio, Univ Ill, 69-74; asst prof, Air Force Inst Technol, 74-78; asst prof, 78-84, ASSOC PROF VA POLYTECH INST & STATE UNIV, 84- *Concurrent Pos:* Engr, Air Force Weapons Lab, 79, Gen Elec Co, 81 & Nat Res Lab, 85; prin investr, Va Polytech Inst & State Univ, 80- *Mem:* Inst Elec & Electronics Engrs. *Res:* Electromagnetic scattering with an emphasis on numerical and transient techniques; radio engineering; nonlinear circuits. *Mailing Add:* Dept Elec Eng Vpl & Su Va Polytech Inst & State Univ Blacksburg VA 24061

DAVIS, WILLIAM C, b Red Bluff, Calif, Feb 12, 33; m 56; c 2. MICROBIOLOGY, IMMUNOLOGY. *Educ:* Chico State Col, BA, 55; Stanford Univ, MA, 59, PhD(med microbiol), 67. *Prof Exp:* Fel, Med Ctr, Univ Calif, San Francisco, 66-68; from asst prof to assoc prof, 68-78, PROF VET MICROBIOL & PATH, COL VET MED, WASH STATE UNIV, 78- *Mem:* AAAS; Transplantation Soc; Am Asn Path & Bact. *Res:* Transplantation immunology; autoimmunity; host defense failure syndromes; flow cytometry; monoclonal antibodies; leukucyte differentiation antigens. *Mailing Add:* Dept Vet Microbiol & Path Wash State Univ Col Vet Med Pullman WA 99164-7040

DAVIS, WILLIAM CHESTER, b Manchester, NH, Dec 22, 25; div; c 4. PHYSICS, EXPLOSIVES. *Educ:* Tufts Col, BS, 49; Johns Hopkins Univ, PhD(physics), 54. *Prof Exp:* Mem staff, 54-90, PRES, ENERGETIC DYNAMICS, LOS ALAMOS NAT LAB, 90- *Concurrent Pos:* Vis prof physics, Univ Calif, 57; fel Los Alamos Lab. *Mem:* Sigma Xi; Am Phys Soc; Optical Soc Am; Soc Photo-Optical Instrumentation Engrs; Am Asn Physics Teachers. *Res:* High explosives; shock and detonation waves; ultra-high-speed photography. *Mailing Add:* 693 46th St Los Alamos NM 87544

DAVIS, WILLIAM DONALD, b Miami, Fla, Aug 6, 21; m 90; c 5. PHYSICAL CHEMISTRY. *Educ:* Univ Miami, BS, 42; Univ Pittsburgh, PhD(phys chem), 49. *Prof Exp:* Asst phys chem, Off Naval Res Proj, Univ Pittsburgh, 46-48; res assoc phys sect, Knolls Atomic Power Lab, 49-59, physicist, Electronics Sci & Eng, Res & Develop Ctr, Gen Elec Co, 59-84; RETIRED. *Concurrent Pos:* Adj prof, Rensselaer Polytech Inst, 67-84. *Mem:* Sigma Xi; Am Phys Soc; Am Vacuum Soc; Am Soc Mass Spectrometry. *Res:* Combustion and reaction in solution calorimetry; gas discharges; vacuum and arc physics; diffusion of gases in metals; mass spectroscopy. *Mailing Add:* 6021 County Farm Rd Ballston Spa NY 12020

DAVIS, WILLIAM DUNCAN, JR, b Brookhaven, Miss, Apr 4, 18; m 49; c 3. MEDICINE, GASTROENTEROLOGY. *Educ:* Tulane Univ, BS, 39, MD, 43; Am Bd Internal Med, dipl, 50; Am Bd Gastroent, dipl, 54. *Prof Exp:* Intern, City Hosp, Cleveland, 43, resident, 44-45; instr med, 45-54, from asst prof to assoc prof, 54-65, PROF CLIN MED, TULANE UNIV, 65- *Concurrent Pos:* Mem staff dept internal med, Ochsner Clin & Found Hosp, 45-, head sect gastroenterol, 53-78, head dept internal med, 68-78, trustee, Alton Ochsner Med Found; consult med, Charity Hosp & Vet Admin Hosp, New Orleans & surgeon gen, Army Subcomt Gastroenterol, 63-69. *Mem:* AMA; Fel Am Col Physicians; Am Fedn Clin Res; Am Gastroenterol Asn; Am Asn Study Liver Dis. *Res:* Hemochromatosis; liver disease; gastric secretion. *Mailing Add:* 14376 Thompson Rd Folsom LA 70437

DAVIS, WILLIAM EDWIN, JR, b Toledo, Ohio, Nov 17, 36; m 68; c 2. ORNITHOLOGY. *Educ:* Amherst Col, BA, 59; Univ Tex, MA, 62; Boston Univ, PhD(paleont), 66. *Prof Exp:* From instr to asst prof, 65-80, 71-80, PROF SCI, BOSTON UNIV, 80- *Concurrent Pos:* Chmn, Div Sci & Math, Col Basic Studies. *Mem:* Am Ornithologist's Union; Asn Field Ornithologists (pres, 87-89); Nuttall Ornith Club (pres, 88-). *Res:* Research on tropical birds, Belize, Peru, Trinidad and Australia; heron vocalizations and foraging behavior; winter bird foraging ecology. *Mailing Add:* 127 E St Foxboro MA 02035

DAVIS, WILLIAM ELLSMORE, JR, b Denver, Colo, June 22, 27; m 54; c 3. CARCINOGENESIS, REGULATORY AFFAIRS. *Educ:* Stanford Univ, BS, 51, MS, 53. *Prof Exp:* Biochemist clin res, Vet Admin Hosp, Oakland, Calif, 53-58; radiation biologist, US Naval Radiol Defense Lab, 58-69; res biologist, Vet Admin Hosp, San Francisco, 69-71; cancer biologist, Life Sci Div, SRI Int, 81-86, sr cancer biologist, Toxicol Lab, 81-86; clin studies coordr, Bio-Response Inc, 87; TECH ASST, ENVIRON PROTECTION AGENCY, SAN FRANCISCO, 88- *Honors & Awards:* US Naval Radiol Defense Lab Silver Medal Award for Sci Achievement. *Mem:* AAAS. *Res:* Radiation biology; tissue transplantation; carcinogenesis; toxicology; author of 45 publications. *Mailing Add:* Environ Protection Agency Air & Toxics Div A-2-3 75 Hawthorne St San Francisco CA 94105

DAVIS, WILLIAM F, b New York, NY, Sept 14, 22; m 87. HIGH SPEED PRODUCTION EQUIPMENT, SYSTEMS CREATION & DEVELOPMENT. *Educ:* NY Univ, ME, 45. *Prof Exp:* Chief engr, R Hoe Co, 48-50, Wood Newspaper Mach, 50-51; exec engr, Mach & Tool Co, 51-53, dir eng, 53-60; CHIEF EXEC OFFICER & DIR DEVELOP, DAVIS CONSULTS & SYNERGETICS, 60- *Concurrent Pos:* Corp Consult, Norton-Simon Corp, 65-72, Am Can, Printing Div, 71-72. *Mem:* Am Soc Mech Engrs; Soc Automotive Engrs. *Res:* Research and development on many high speed publications and newspaper printing presses biodemequipment and other special purpose machinery; developing an experimental aircraft and an advanced web printing press; development of chemical processes and equipment. *Mailing Add:* Voelbel Rd Hightstown NJ 08520

DAVIS, WILLIAM JACKSON, b Portsmouth, Va, Feb 7, 42; m 64; c 4. PSYCHOBIOLOGY, ENVIRONMENTAL STUDIES. *Educ:* Univ Calif, Berkeley, BA, 64; Univ Ore, PhD(biol), 67. *Prof Exp:* USPHS fel neurobiol, Univ Ore, 67-68; fel biol, Stanford Univ, 68-70; PROF BIOL, UNIV CALIF, SANTA CRUZ, 69- *Concurrent Pos:* Prin investr, NIH & NSF, 69-; dir, Environ Studies Inst, 80-; advisor, Repub Nauru, 81- *Honors & Awards:* Humboldt Prize, 82. *Mem:* AAAS; Soc Neurosci; Soc Exp Biol; Am Soc Zool. *Res:* Brain; neural mechanisms of learning; environment; ocean. *Mailing Add:* Dept Biol Univ Calif Santa Cruz CA 95064

DAVIS, WILLIAM JACKSON, b Warrenton, Va, Sept 29, 30; m 59. AQUATIC BIOLOGY. *Educ:* Va Polytech Inst & State Univ, BS, 53; Univ Kans, PhD(zool), 59. *Prof Exp:* Asst, Kans Biol Surv, 53-57; asst, Univ Kans, 53-59; asst prof biol, St Cloud State Col, 59 & Western Mich Univ, 59-63; assoc marine scientist, VA Inst Marine Sci, 63-68, sr scientist & head dept ichthyol, 68-72, asst dir & head Fisheries & Serv, 72-77; chief scientist, South Atlantic Fishery Mgt Coun, Charleston, SC, 77-83; FISHERIES BIOLOGIST, SULTAN QABOOS UNIV, SULTANATE OMAN, 86- *Mem:* Am Fisheries Soc. *Res:* Fisheries science and management; limnology. *Mailing Add:* PO Box 32484 Al Khod Oman

DAVIS, WILLIAM JAMES, b Wilmington, Del, Sept 8, 40; m 63; c 3. PHYSIOLOGY, ENDOCRINOLOGY. *Educ:* Univ Del, AB, 62; Northwestern Univ, MS, 65, PhD(biol), 68. *Prof Exp:* Tech asst rocket propellants, Thiokol Chem Corp, 63; asst biol, Northwestern Univ, 63-67; asst prof zool, Univ Tenn, Knoxville, 68-73; from asst prof to assoc prof, 73-84, PROF BIOL, LAMBUTH COL, 84- *Concurrent Pos:* NIH biomed support grant, 69. *Mem:* Am Soc Zoologists; Am Inst Biol Sci; Sigma Xi. *Res:* General, cellular and comparative physiology; cellular and comparative endocrinology; the biology of pigment cell effectors. *Mailing Add:* Dept Biol Lambuth Col Jackson TN 38301

DAVIS, WILLIAM POTTER, JR, b Cleveland, Ohio, Aug 27, 24; m 47; c 5. PHYSICS, ACADEMIC ADMINISTRATION. *Educ:* Oberlin Col, AB, 48; Univ Mich, MS, 49, PhD(physics), 54. *Hon Degrees:* MA, Dartmouth Col, 67. *Prof Exp:* Instr physics, Univ Mich, 54-55; from instr to prof, Dartmouth Col 55-85, assoc provost, 67-70, actg dean, Thayer Sch Eng, 69-70, budget officer, 70-74, treas, 74-75, EMER PROF PHYSICS & EMER TREAS, DARTHMOUTH COL, 85- *Concurrent Pos:* Assoc prog dir studies & curriculum improv sect, Pre-Col Educ in Sci Div, NSF, 65-66. *Mem:* AAAS; Sigma Xi. *Res:* Cosmic rays; gas discharges; plasma physics. *Mailing Add:* 21 Col Hill RR 1 Box 246 Lebanon NH 03766

DAVIS, WILLIAM R, b Los Angeles, Calif, Nov 30, 23; m 45; c 6. CONTROL SYSTEMS, ELECTRONICS. *Educ:* Calif Inst Technol, BS, 44, MS, 47; Stanford Univ, PhD(elec eng), 66. *Prof Exp:* Res engr, Hughes Aircraft Co, 47-51 & Santa Barbara Pac Mercury Res Ctr, 51-52, group supvr, Detroit Controls Res Div, 52-56; sr staff scientist, 56-78, prog mgr precision attitude systs, 66-70, mgr spec syst anal, 71-74, proj mgr, Navstar GPS/ Satellite Users, 75-79, SYST ENG, REMOTELY PILOTED VEHICLE PROG, LOCKHEED MISSILES & SPACE CO, LOCKHEED AIRCRAFT CORP, 80- *Res:* Development of systems for precise determination and control of space vehicle attitude; optical sensors; gyros; airborne computers, particularly development of space precision attitude reference and pointing system; satellite system synthesis and analysis; satellite guidance and attitude determination. *Mailing Add:* 1359 Belleville Sunnyvale CA 94087

DAVIS, WILLIAM ROBERT, b Oklahoma City, Okla, Aug 22, 29; m 70. PHYSICS. *Educ:* Univ Okla, BS, 53, MS, 54; Hannover Tech Univ, Dr rer nat, 56. *Prof Exp:* Physicist, Trisophia Enterprises, Okla, 56-57; from asst prof to assoc prof, 57-66, PROF PHYSICS, NC STATE UNIV, 66- *Concurrent Pos:* Consult, Trisophia Enterprises, 57-59, Regulus Corp, Okla, 58-59, Lab for Electronics, Inc, Calif, 62-63 & Res Triangle Inst, NC, 64-; Guggenheim fel theoret physics, 70-71. *Mem:* AAAS; fel Am Phys Soc; Am Asn Physics Teachers. *Res:* Theoretical mechanics; electrodynamics; field theory; symmetry properties; the general theory of relativity. *Mailing Add:* Dept Physics NC State Univ PO Box 5383 State Col Sta Raleigh NC 27650

DAVIS, WILLIAM S, b Los Angeles, Calif, Sept 16, 30; m 60; c 2. PLANT TAXONOMY, PLANT CYTOGENETICS. *Educ:* Whittier Col, AB, 51, MS, 59; Univ Calif, Los Angeles, PhD(bot), 64. *Prof Exp:* Asst prof, 63-71, PROF BIOL, UNIV LOUISVILLE, 71- *Mem:* AAAS; Am Soc Plant Taxon; Int Asn Plant Taxon; Bot Soc Am. *Res:* Experimental taxonomy and cytotaxonomy of plants. *Mailing Add:* Dept Biol Univ Louisville Louisville KY 40292

DAVIS, WILLIAM SPENCER, b Harrisonburg, Va, Sept 23, 25; m 60; c 2. FISH BIOLOGY, ECOLOGY. *Educ:* Va Polytech Inst, BS, 50, MS, 53. *Prof Exp:* Fishery res biologist, US Bur Commercial Fisheries, NC, 54-59, Wash, 60-67; fishery biologist, Br Resources Mgt, Bur Commercial Fisheries, Dept Interior, 67-70; biol sci administr, Environ Protection Agency, 70-76, aquatic biologist, 76-88; RETIRED. *Mem:* Am Fisheries Soc; Am Inst Fishery Res Biol; Am Soc Ichthyol & Herpet; Ecol Soc Am; Am Chem Soc. *Res:* Effect of human use of freshwater estuarine, oceanic and great lakes environments on communities of aquatic organisms; develop and evaluate means of abating and preventing pollution there; control of the discharge of dredged or fill material in water. *Mailing Add:* 8523 Durham Ct Springfield VA 22151

DAVIS, WILLIAM THOMPSON, b Champaign, Ill, Nov 3, 31; m 52; c 2. TECHNICAL MANAGEMENT. *Educ:* Univ Ill, BS, 56, DVM, 58; Univ Chicago, MBA, 67. *Prof Exp:* Pvt pract, Wyo, 58-62; resident vet, Newhall Land & Farm Co, 62-63 & Abbott Labs, 63-67; asst dir animal health res, Ciba Corp, 67-69; dir animal health clin res, E R Squibb & Sons, Inc, 69-73; dir licensing, Pfizer Inc, 73-90; ASST DIR, LIC PATENT TRADEMARK & COPYRIGHT OFF, UNIV CALIF, BERKELEY, 90- *Concurrent Pos:* Mem, Nat Mastitis Coun. *Mem:* Am Vet Med Asn; Indust Vet Asn; Licensing Exec Soc; Am Mgt Asn. *Res:* Animal health research including pharmaceuticals and biologicals, especially final developmental stages, product promotion, liaison with marketing, research organization and management; acquisition of new animal health, pharmaceutical, chemical and over the counter health care products, processes and technology from world wide sources for development and marketing. *Mailing Add:* 1320 Harbor Bay Pkwy Suite 150 Alameda CA 94501

DAVISON, ALLAN JOHN, b Cape Town, SAfrica, July 8, 36; Can citizen; m 58; c 2. NEURO-TOXICOLOGY, ROLES OF METALS AND PHENOLIC COMPOUNDS. *Educ:* Univ Cape Town, BSc, 57; Rutgers State Univ, MS, 62, PhD(biochem), 64. *Prof Exp:* Analyst mat testing, City Engr's Dept, Cape Town, SAfrica, 55-56; jr lectr physiol chem, Med Sch, Univ Cape Town, 57-60, sr lectr, 64-70; teaching asst exp physiol, Rutgers State Univ, 60-64; PROF BIOCHEM, SCH KINESIOLOGY, SIMON FRASER UNIV, BC, 71- *Concurrent Pos:* Prin investr, Bioenergetics Res Group, Med Res Coun, 64-71, Res Group on Cellular Mechanisms of Oxygen Toxicity, Can Natural Sci & Eng Res Coun, 74-; consult, Epidemiol Environ Carcinogenesis Sect, BC Cancer Res Ctr, 85-; mem consumer adv panel, Can Standards Asn, 85-; mem exec coun, Oxygen Soc Am, 88- *Mem:* Fel Oxygen Soc Am; Am Soc Biochem & Molecular Biol; Soc Free Radical Res Int; Can Biochem Soc; Can Standards Asn. *Res:* Roles of metals and free radicals in oxygen mediated cytotoxicity and genotoxicity; toxicity of catechols and iron, manganese, vanadium and copper quinones in the etiology of parkinsonism, aging and cancer; tocopherol, carotene and ascorbate as protective agents against oxygen derived active species; carcinogenesis and cancer prevention; computer communication and remote data bases; environmental carcinogenesis, dietary carcinogens and anticarcinogens. *Mailing Add:* BC Cancer Res Ctr 601 W Tenth St Vancouver BC V5Z 1L3 Can

DAVISON, BEAUMONT, b Atlanta, Ga, May 30, 29; m 52; c 2. MANUFACTURING ENGINEERING EDUCATION. *Educ:* Vanderbilt Univ, BE, 50; Syracuse Univ, MEE, 52, PhD(elec eng), 56. *Prof Exp:* Instr & res assoc, Syracuse Univ, 51-56; asst prof elec eng, Case Inst Technol, 56-59; exec vpres, Indust Electronic Rubber Co, 59-67; chmn dept elec eng, Ohio Univ, 67-69, dean col eng & technol, 69-71, vpres regional higher educ, 71-74; dean eng, Calif State Polytech Univ, Pomona, 74-83; pres, Tri-State Univ, Angola, IN, 83-89; CONSULT, ENG EDUC, 89- *Mem:* Am Soc Eng Educ; Soc Mfg Engrs. *Res:* Microwave devices; electronic circuit component development; rubber technology. *Mailing Add:* 4553 Oak Trail Dr Sarasota FL 34241

DAVISON, BRIAN HENRY, b Lancaster, Pa, Aug 20, 57. BIOCHEMICAL ENGINEERING, FERMENTATION. *Educ:* Univ Rochester, BS, 79; Calif Inst Technol, PhD(chem eng), 85. *Prof Exp:* BIOCHEM ENGR RES STAFF, CHEM TECHNOL DIV, OAK RIDGE NAT LAB, 85- *Concurrent Pos:* Adj prof, Dept Chem Eng, Univ Tenn, Knoxville, 87-91. *Mem:* Am Inst Chem Engrs; Am Chem Soc. *Res:* Experiments and modeling of bioreaction for fermentation, particularly in immobilized cell fluidized beds; fermentation products include ethanol, lactic, and butanol; biosorption and mixed culture dynamics. *Mailing Add:* Oak Ridge Nat Lab PO Box 2008 Oak Ridge TN 38731-6226

DAVISON, CLARKE, b Washington, DC, Nov 8, 27; m 53, 64; c 2. FORENSIC CHEMISTRY. *Educ:* George Washington Univ, BSc, 48, MS, 49; Harvard Univ, PhD(biochem), 54. *Prof Exp:* Asst pharmacol, George Washington Univ, 47-49, from asst prof to prof, 53-67; sect head metab chem, Sterling Winthrop Res Inst, 67-76; forensic chemist, Lee County Sheriff's Dept, Ft Myers, 76-89; RETIRED. *Mailing Add:* 1214 SE 14th Terr Ft Myers FL 33990

DAVISON, E(DWARD) J(OSEPH), b Toronto, Ont, Sept 12, 38; m 66; c 4. CONTROL SYSTEM THEORY. *Educ:* Univ Toronto, BASc, 60, MA, 61, Cambridge Univ, PhD(control eng), 64. *Hon Degrees:* ScD, Cambridge Univ, 77. *Prof Exp:* Asst prof elec eng, Univ Toronto, 64-66; asst prof, Univ Calif, Berkeley, 66-68; assoc prof, 68-74, PROF ELEC ENG, UNIV TORONTO, 74- *Concurrent Pos:* Athlone fel, Eng, 61-63, E W R Steacie Mem fel, Nat Res Coun Can, 74-77; Killam res fel, 79-80 & 81-82; dir, Elec Eng Consociates Ltd, Toronto, 77-; pres, Control Systs Soc, Inst Elec & Electronics Engrs, 83, consult ed, Trans on Automatic Control, 85; assoc ed, Automatica, 74-90, Large Scale Syst: Theory Appl, 79-90, Optimal Control Appl & Methods, 83-87; hon prof, Beijing Inst Aeronaut & Astronaut, 86; vchmn theory comt, Int Fedn Automatic Control, chmn, 87-90, vchmn tech bd, 90-93 & coun mem, 90-93. *Honors & Awards:* Centennial Medal, Inst Elec & Electronic Engrs, 84. *Mem:* Fel Inst Elec & Electronic Engrs; fel Royal Soc Can; distinguished mem Inst Elec & Electronic Engrs Control Systs Soc. *Res:* Large scale system theory; multivariable systems; computational methods; optimization theory; applications of system theory to biology. *Mailing Add:* Dept Elec Eng Univ Toronto Toronto ON M5S 1A4 Can

DAVISON, FREDERICK CORBET, b Atlanta, Ga, Sept 3, 29; m 52; c 2. VETERINARY PATHOLOGY, PHYSIOLOGY. *Educ:* Univ Ga, DVM, 52; Iowa State Univ, PhD(path, biochem, physiol), 63. *Prof Exp:* Pvt pract vet med, Ga, 52-58; res assoc, Iowa State Univ, 58-59, asst prof physiol, 59-63; asst dir, Dept Sci Activ, Am Vet Med Asn, 63-64; dean, Sch Vet Med, Univ Ga, 64-66; vchancellor, Univ Syst Ga, 66-67; pres, Univ Ga, 67-87; PRES & CHIEF EXEC OFFICER, NSF, 87- *Concurrent Pos:* Assoc, Inst Atomic Res, Iowa State Univ, 58-63; proj leader, US AEC, 59-63; mem prof ed comn, Inst Lab Animal Resources, Nat Acad Sci-Nat Res Coun, 63-; mem Coun Biol & Therapeut Agents, Am Vet Med Asn, 64- *Mem:* Am Soc Vet Physiol & Pharmacol; Sigma Xi. *Res:* Comparative toxicity of the lanthanide series of rare earths. *Mailing Add:* 721 Chipande Dr Augusta GA 30909

DAVISON, J LESLIE, b Glasgow, Scotland, May 26, 44; Can citizen; m 65; c 4. COMBINATORICS & FINITE MATHEMATICS. *Educ:* Univ St Andrews, Scotland, BS, 66; Univ Dundee, Scotland, PhD(math), 69. *Prof Exp:* Lectr math, Univ St Andrews, Scotland, 68-69; from asst prof to assoc prof, 69-82, PROF MATH, LAURENTIAN UNIV, 82- *Concurrent Pos:* Vis prof, Univ Warwick, Coventry, Eng, 77-78. *Mem:* Am Math Soc; Can Math Soc. *Res:* Number theory; discrete mathematics; combinatorics; author of 12 publications. *Mailing Add:* Dept Math & Computer Sci Laurentian Univ Ramsey Lake Rd Sudbury ON P3E 2C6 Can

DAVISON, JOHN (AMERPOHL), b Janesville, Wis, June 25, 28; m 50, 65; c 3. PHYSIOLOGY. *Educ:* Univ Wis, BS, 50; Univ Minn, PhD, 55. *Prof Exp:* Instr zool, Wash Univ, 54-57; asst prof, Fla State Univ, 57-60 & La State Univ, 60-64; from asst prof to assoc prof, Rensselaer Polytech Inst, 64-67; ASSOC PROF ZOOL, UNIV VT, 67- *Mem:* Am Soc Zoologists; Soc Gen Physiol. *Res:* Body size and metabolism; cell form; developmental physiology of amphibia and fresh-water invertebrates. *Mailing Add:* Dept Zool Univ Vt Burlington VT 05405

DAVISON, JOHN BLAKE, b Waynesboro, Pa, Jan 7, 46; m 74; c 2. ELECTROCHEMISTRY, CATALYSIS. *Educ:* Va Polytech Inst, BS, 67; Univ Md, PhD(inorg chem), 74. *Prof Exp:* Res chemist catalysis, Hooker Chem Corp, 76-78; res chemist Catalysis, Occidental Res Corp, 78-82; semiconductor mat, J C Schumacher Co, 82-85; metal finishing, Dynachem, 85-86; MGR, CHEM TECHNOL, ATHENS CORP, 86- *Mem:* Am Chem Soc. *Res:* Reprocessing, recycling and repurifying ultrapure chemicals; on-line analytical instrumentation for ultra trace analysis; atomic absorption spectroscopy; inorganic electrochemistry; printed circuit chemicals and semiconductor chemicals; electrocatalysis and homogeneous catalysis; organometallic polymers. *Mailing Add:* Athens Corp 1922 Avenida Deloro Oceanside CA 92056

DAVISON, JOHN PHILIP, biochemistry, for more information see previous edition

DAVISON, KENNETH LEWIS, b Hopkins, Mo, Dec 27, 35; m 57; c 3. NUTRITION, PHYSIOLOGY. *Educ:* Univ Mo, BS, 57; Iowa State Univ, MS, 59, PhD(nutrit), 61. *Prof Exp:* Res assoc nutrit, Cornell Univ, 61-65; res physiologist, Metab & Radiation Res Lab, 65-88; RES PHYSIOLOGIST, BIOSCI RES LAB, US DEPT AGR, AGR RES SERV, 88- *Concurrent Pos:* Adj prof, NDak State Univ, 68- *Mem:* Am Soc Animal Sci; Am Dairy Sci Asn; Am Inst Nutrit; AAAS. *Res:* Accumulation of nitrate in plants and toxicity to animals; pesticide metabolism by animals. *Mailing Add:* USDA/ARS Biosci Res Lab 1605 W Col St Fargo ND 58105

DAVISON, LEE WALKER, b Moscow, Idaho, Aug 10, 37. APPLIED MECHANICS. *Educ:* Univ Idaho, BS, 59; NY Univ, MS, 61; Calif Inst Technol, PhD(appl mech), 65. *Prof Exp:* Mem tech staff, Bell Tel Labs, 59-62; res fel, Johns Hopkins Univ, 65-66; staff mem, Sandia Nat Labs, 66-68, supvr shock wave physics, 68-72, supvr, Explosives Physics Res Div, 80-82, supvr, Appl Mech Div, 80-82, mgr, Solid Dynamics Dept, 82-88, mgr, Eng Anal Dept, 88-91, RES SCIENTIST & PROG COORDR, SANDIA NAT LABS, 91- *Mem:* Am Soc Mech Eng; Am Phys Soc; Soc Natural Philos; Am Acad Mech. *Res:* Modern continuum mechanics; propagation of strong shock waves in solids; dynamic fracture; explosives. *Mailing Add:* 7900 Harwood Ave NE Albuquerque NM 87110

DAVISON, PETER FITZGERALD, b London, Eng, Nov 12, 27; m 54; c 2. PHYSICAL CHEMISTRY. *Educ:* Univ London, BSc, 49, PhD(chem), 54. *Prof Exp:* Chemist, Chester Beatty Res Inst, Inst Cancer Res, Eng, 51-56; Indust Cellulose Res Ltd, Can, 56-57 & Dept Biol, Mass Inst Technol, 57-69; DIR DEPT FINE STRUCT RES, BOSTON BIOMED RES INST, 69- *Mem:* Am Chem Soc; Am Soc Biol Chem; AAAS; Asn Res Vision & Ophthal. *Res:* Physical chemistry of nucleoproteins; nucleic acids; proteins and synthetic polymers; collagen organization. *Mailing Add:* 193 Cedar St Lexington MA 02173

DAVISON, RICHARD READ, b Marlin, Tex, Apr 3, 26; m 51; c 3. CHEMICAL ENGINEERING. *Educ:* Tex Tech Col, BS, 49; Tex A&M Univ, MS, 58, PhD(chem eng), 62. *Prof Exp:* Engr gas indust, Lion Oil Co, 49-55; res scientist, 55-61, from instr to assoc prof, 58-68, PROF CHEM ENG, TEX A&M UNIV, 68- *Mem:* Am Chem Soc; Am Inst Chem Engrs; Int Solar Energy Soc; Sigma Xi. *Res:* Solvent extraction process for conversion of saline water and sewage effluent; aquifer energy storage thermodynamics of solutions and solar energy utilization; methanol as motor fuel. *Mailing Add:* Dept Chem Eng Tex A&M Univ College Station TX 77843

DAVISON, ROBERT WILDER, b Albany, NY, Dec 30, 20; m 49; c 2. PHYSICAL CHEMISTRY. *Educ:* Union Col, BS, 42; Mass Inst Technol, PhD(phys chem), 50. *Prof Exp:* Res chemist, Res Ctr, Hercules Powder Co, 50-66; from res chemist to sr res chemist, 66-71, RES SCIENTIST, HERCULES RES CTR, HERCULES, INC, 71- *Mem:* Am Chem Soc; Sigma Xi. *Res:* Surface chemistry; colloidal systems; chemical additives for paper; physical properties of paper. *Mailing Add:* 616 Northside Dr North Hills Wilmington DE 19809

DAVISON, SOL, b Los Angeles, Calif, Apr 3, 22; m 45; c 2. POLYMER SCIENCE. *Educ:* Pomona Col, BA, 47; Univ Notre Dame, PhD(chem), 51. *Prof Exp:* AEC fel, Univ Notre Dame, 51-52; res assoc radiobiol & phys chem studies on foods, Mass Inst Technol, 52-56; chemist, Shell Develop Co, 56-87; RETIRED. *Mem:* Am Chem Soc; Sigma Xi. *Res:* Reaction kinetics and photochemistry; radiation chemistry and radiobiology; testing of polymers; physics and technology of elastomers and thermoplastics. *Mailing Add:* PO Box 820609 Houston TX 77282-0609

DAVISON, SYDNEY GEORGE, b Stockport, Eng, Sept 6, 34; m 59; c 2. QUANTUM SURFACE CHEMISTRY. *Educ:* Univ Manchester, BSc, 58, MSc, 62, PhD(math), 64, DSc(math), 82. *Prof Exp:* Teacher, Salford Grammer Sch, 58-59; nuclear physicist, Nuclear Power Div, G E C/Simon-Carvers, Eng, 59-60; demonstr math, Univ Manchester, 61-64; fel physics, Univ Waterloo, 64-65, from asst prof to assoc prof, 65-70, dir quantum theory group, 68-70; prof physics, Clarkson Univ, 70-72; prof physics, Bartol Res Found, Franklin Inst, 72-74; vis prof, 74-78, PROF APPL MATH, UNIV WATERLOO, 78- *Concurrent Pos:* Grants, Natural Sci & Eng Res Coun Can, 65-70, 75-; ed, Progress Surface Sci, Pergamon, Oxford, 68-; vis prof physics, Univ Tex, Dallas, 80-81; dir, Guelph-Waterloo Surface Sci & Technol Group, 82-84. *Mem:* Fel Am Phys Soc; fel Brit Inst Physics; fel NY Acad Sci. *Res:* Quantum theory of crystal surfaces. *Mailing Add:* 313 Hiawatha Dr Waterloo ON N2L 2V9 Can

DAVISON, THOMAS MATTHEW KERR, b Scotland, Jan 18, 39; Can citizen; m 64; c 2. MATHEMATICS. *Educ:* Sir George Williams Univ, BSc, 60; Univ Toronto, MA, 62, PhD(math), 65. *Prof Exp:* Asst prof math, Univ Waterloo, 64-66; Nat Res Coun Can res fel, Kings Col, London, 66-68; from asst prof to assoc prof, 68-81, PROF MATH SCI, MCMASTER UNIV, 81- *Concurrent Pos:* Ed, Ont Math Gazette, 70-73. *Mem:* Am Math Soc; Can Math Cong; Math Asn Am. *Res:* Algebra, number theory, and functional equations. *Mailing Add:* Dept Math McMaster Univ Hamilton ON L8S 4L8 Can

DAVISON, WALTER FRANCIS, b Chicago, Ill, Apr 28, 26; div; c 1. MATHEMATICS. *Educ:* Calif Inst Technol, BS, 51; Univ Va, PhD(math), 56. *Prof Exp:* Instr math, Univ Mich, Ann Arbor, 56-60; sr staff scientist, Collins Radio Co, Tex, 60-61; mem tech staff electronics, Space Gen Corp, Calif, 61-62; branch head electrooptics, Tex Instruments, 62; sr scientist, Int Tel & Tel Fed Labs, Calif, 63-66; ASSOC PROF MATH, CALIF STATE UNIV, NORTHRIDGE, 66- *Concurrent Pos:* Rackham fac fel, Univ Mich, Ann Arbor, 58. *Mem:* Am Math Soc; NY Acad Sci. *Res:* Closure operators; mosaic spaces; frechet equivalence; optical receivers; electrooptical demodulation; noise theory; quasigroups. *Mailing Add:* Dept Math Calif State Univ Northridge CA 91324

DAVISSON, CHARLOTTE MEAKER, b Phillipston, Mass, Oct 7, 14; m 47; c 1. ION IMPLANTATION. *Educ:* Wellesley Col, BA, 36; Smith Col, MA, 39; Mass Inst Technol, PhD(nuclear physics), 48. *Prof Exp:* Physicist, Eclipse-Pioneer Div, Bendix Aviation Corp, 42-44; res asst, Manhattan Proj, Columbia Univ, 44-46; res physicist, Condensed Matter & Radiation Sci Div, US Naval Res Lab, DC, 54-86; RETIRED. *Mem:* Fel AAAS; Am Phys Soc; NY Acad Sci. *Res:* Interaction of gamma rays with matter; gamma-ray albedos; computer programming for cyclotron studies, neutron detector efficiency, x-ray analysis and other cyclotron research projects, including ion implantation. *Mailing Add:* 400 Cedar Ridge Dr Oxon Hill MD 20745-1411

DAVISSON, JAMES W, b Pittsburgh, Pa, Oct 23, 14. CRYSTAL PHYSICS. *Educ:* Princeton Univ, BA, 38; Mass Inst Technol, PhD, 42. *Prof Exp:* Physicist, Naval Res Lab, 47-75; RETIRED. *Mem:* Am Phys Soc; NY Acad Sci. *Mailing Add:* 400 Cedar Ridge Dr Oxon Hill MD 20745

DAVISSON, LEE DAVID, b Evanston, Ill, June 16, 36; m 57, 75; c 3. ELECTRICAL ENGINEERING. *Educ:* Princeton Univ, BSE, 58; Univ Calif, Los Angeles, MSE, 61, PhD(eng), 64. *Prof Exp:* Res engr, Philco Corp, 60-62; mem tech staff elec eng, Hughes Aircraft Corp, 62-64; from asst prof to assoc prof, Princeton Univ, 64-70; from assoc prof to prof, Univ Southern Calif, 69-76; chmn dept, 80-85, PROF, UNIV MD, 76-; PRES, TECHNO-SCIENCES, INC, 75- *Concurrent Pos:* Consult, NASA, 65-70; Bell Tel Labs, 68-69; sr staff engr, Hughes Aircraft Co, 69-74; second vpres, Info Theory Group, 78, first vpres, 79, pres, 80; co-chmn, Int Info Theory Symp, 79. *Honors & Awards:* Princeton Eng Soc award, 65; outstanding young elec engr award, Eta Kappa Nu, 68; prize paper award, Inst Elec & Electronics Engrs, 76. *Mem:* Fel Inst Elec & Electronics Engrs. *Res:* Communication theory. *Mailing Add:* 7833 Walker Dr Suite 620 Greenbelt MD 20770

DAVISSON, M T, b Grafton, WVa, Dec 23, 31; m 55; c 4. CIVIL ENGINEERING. *Educ:* Univ Akron, BCE, 54; Univ Ill, MS, 55, PhD(civil eng), 60. *Prof Exp:* Designer struct eng, Clark, Dietz, Painter & Assocs, 55-56; from instr to prof civil eng, Univ Ill, Urbana, 59-81; CONSULT, 81- *Concurrent Pos:* Indust consult, 58-81; mem Deep Found Inst. *Honors & Awards:* Alfred A Raymond Award, 59; Collingwood Prize, Am Soc Civil Engrs, 64. *Mem:* Am Soc Civil Engrs; Am Concrete Inst; Am Rwy Eng Asn; Am Soc Testing & Mat; Nat Soc Prof Engrs; Int Soc Soil Mech & Found Eng; Sigma Xi; Deep Found Inst. *Res:* Deep foundations; soil dynamics; settlement of structures; waterfront structures; heavy construction. *Mailing Add:* PO Box 125 Savoy IL 61874

DAVISSON, MURIEL TRASK, b Tremont, ME, Apr 19,41; m 67; c 1. MAMMALIAN GENETICS, CYTOGENETICS. *Educ:* Mount Holyoke Col, AB, 63; Pa State Univ, PhD(genetics), 69. *Prof Exp:* Res assoc, 71-80, assoc staff scientist, 80-85, STAFF SCIENTIST, JACKSON LAB, 85- *Concurrent Pos:* Cytogenetic consult, Ctr Human Genetics, Bar Harbor, Me, 71- *Mem:* AAAS; Genetics Soc Am; Am Genetic Asn; Am Soc Human Genetics. *Res:* Mouse models of human disease; genetic characterization of new mouse mutant; gene mapping; mouse cytogenetics. *Mailing Add:* The Jackson Lab 600 Main St Bar Harbor ME 04609-0800

DAVITIAN, HARRY EDWARD, b Flushing, NY, June 19, 45; m 73; c 2. PHYSICS. *Educ:* Mass Inst Technol, BS, 66; Cornell Univ, MS, 69, PhD(appl physics), 73. *Prof Exp:* Res assoc, Cornell Univ, 73; scientist, Inst Energy Anal, Oak Ridge Assoc Univs, 74-75; group leader, Nat Ctr Anal Energy Systs, Brookhaven Nat Lab, NY, 75-80; PRES, ENTEK RES, INC, 80- *Concurrent Pos:* Mem res & develop subgroup, Modeling Resources Group, Comn Nuclear & Alt Energy Resources, Nat Res Coun, 76-77; secy, Independent Power Producers NY, 88, vpres, 90-91; chmn, Energy Comt, Long Island Asn, 88- *Res:* Energy systems analysis; electric utility economics. *Mailing Add:* 11 Satterly Rd East Setauket NY 11733

DAVITT, HARRY JAMES, JR, b Philadelphia, Pa, Oct 10, 39; m 64; c 3. FUEL TECHNOLOGY & PETROLEUM ENGINEERING. *Educ:* Pa State Univ, BS, 61; Purdue Univ, MS, 66, PhD(chem eng), 69. *Prof Exp:* Res engr, Sun Oil Co, 68-73; supt, Oil Sands Technol, Great Can Oil Sands, 73-75; sr prof engr, Sunoco Energy Develop Co, 75-85; GROUP LEADER, RECOVERY PROCESSES, ORYX ENERGY CO, 85- *Mem:* Sigma Xi. *Res:* Fuel cells; extraction of bitumen from tar sands; enhanced oil recovery. *Mailing Add:* PO Box 830936 Richardson TX 75083-0936

DAVITT, RICHARD MICHAEL, b Wilmington, Del, Mar 13, 39; m 66; c 2. MATHEMATICS. *Educ:* Niagara Univ, BS, 63; Lehigh Univ, MS, 66, PhD(math), 69. *Prof Exp:* Instr, St Francis de Sales High Sch, Toledo, Ohio, 58-60; asst math, Lehigh Univ, 64-65; from instr to asst prof, Lafayette Col, 67-70; asst prof, 70-74, ASSOC PROF MATH, UNIV LOUISVILLE, 74- *Mem:* Am Math Soc. *Res:* Lattice-ordered groups; finite p-groups. *Mailing Add:* Dept Math Univ Louisville Louisville KY 40292

DAW, GLEN HAROLD, b Baltimore, Md, June 20, 54; m 82; c 4. IMAGE RECOVERY & RECONSTRUCTION, BACKSCATTER-SURFACE ROUGHNESS ANALYSIS. *Educ:* NMex State Univ, BS, 79, MS, 82; Univ NMex, PhD(optical sci), 87. *Prof Exp:* Res asst, Physics Dept, NMex State Univ, 79-81; teaching asst mech physics & astron, Univ NMex, 81-82; physicist, Chem Laser Br, USAF Weapons Lab, 82-87; OPTICAL SCIENTIST, KAMAN SCI CORP, 87- *Concurrent Pos:* Assoc zone coun rep, Soc Physics Students, 80-82. *Mem:* Am Phys Soc; Optical Soc Am; Inst Elec & Electronic Engrs. *Res:* Wavefront correction via phase retrieval and image sharpening; chemical and reactor pumped laser systems; cryogenic liquid laser and flashlamps via electron beam and optical pumping; backscatter and surface micro-roughness analysis. *Mailing Add:* 1286 Amsterdam Dr Colorado Springs CO 80907

DAW, HAROLD ALBERT, b Granger, Utah, Oct 25, 25; m 53; c 7. PHYSICS. *Educ:* Univ Utah, BS, 48, MA, 52, PhD, 56. *Prof Exp:* Chief lab asst, Elementary Physics Lab, Univ Utah, 47-48, 50-52; res physicist from Utah dept astrophys & phys meteorol, Johns Hopkins Univ, 53-54; from asst prof to assoc prof physics, NMex State Univ, 54- 61, assoc physicist, Phys Sci Lab, 54-61, head dept physics, 61-70, from actg dean to assoc dean arts & sci, 70-74, from asst acad vpres to assoc acad vpres, 74-89, prof physics, 61-90, EMER PROF PHYSICS, NMEX STATE UNIV, 90- *Honors & Awards:* Distinguished Serv Citation, Am Asn Physics Teachers, 71, Robert A Millikan Award, 75. *Mem:* AAAS; Am Asn Physics Teachers. *Res:* Optics; maser optics; physics instructional equipment. *Mailing Add:* Dept Physics NMex State Univ Las Cruces NM 88003

DAW, JOHN CHARLES, b Tulsa, Okla, July 18, 31; c 7. MEDICAL EDUCATION. *Educ:* Case Western Reserve Univ, BA, 59, PhD(physiol), 66. *Prof Exp:* Res assoc physiol, Univ Va, 66-67; USPHS fel, 67-68; instr, Univ Va, 68; asst prof lab educ, Mt Sinai Sch Med, 68-69, assoc dir lab, 68-71, asst prof med educ & res asst prof physiol, 69-71; asst prof physiol, 71-77, adj asst prof physiol, 77-87, ASSOC DIR MED SCI TEACHING LABS, SCH MED, UNIV NC, CHAPEL HILL, 71-, ADJ ASSOC PROF PHYSIOL, 87- *Mem:* Asn Multidiscipline Educ in Health Sci. *Res:* Glycogen metabolism; carbohydrate metabolism; cardiac metabolism; experimental shock; experimental burn; surgical research. *Mailing Add:* Univ NC Med Sch Chapel Hill NC 27599-7520

DAW, NIGEL WARWICK, b London, Eng, Dec 12, 33; US citizen; m 63; c 2. NEUROPHYSIOLOGY, PSYCHOPHYSICS. *Educ:* Cambridge Univ, BA, 56, MA, 61; Johns Hopkins Univ, PhD(biophys), 67. *Prof Exp:* Assoc scientist, Polaroid Corp, 58-63; fel neurobiol, Harvard Med Sch, 67-69, NIH fel, 67-68, spec fel, 68-69; from asst prof to assoc prof, 69-77, PROF NEUROBIOL, CELL BIOL & OPHTHAL, MED SCH, WASHINGTON UNIV, 77- *Concurrent Pos:* Consult visual sci study sect, Div Res Grants, NIH, 70-73, Behav & Neural Sci fel, study sect, 81; ed, J Neurophysiol, 78-83 & 90-, Investigative Ophthal, 83- 88, J Physiol, 85- & J Neurosci, 90- *Mem:* AAAS; Am Physiol Soc; Asn Res Vision & Ophthal; Soc Neurosci. *Res:* Effects of synaptic transmitter drugs on the visual system; development of vision; mechanisms of directional sensitivity in the visual system; neurophysiology and psychophysics of color vision. *Mailing Add:* 78 Aberdeen Pl St Louis MO 63105

DAWBER, THOMAS ROYLE, b Duncan, BC, Can, Jan 18, 13; nat US; m 37; c 2. INTERNAL MEDICINE, CARDIOLOGY. *Educ:* Haverford Col, AB, 33; Harvard Univ, MD, 37, MPH, 58. *Prof Exp:* Intern, US Marine Hosps, Va, 37-38, med resident, 38-40, chief med, Mass, 41-42, asst, NY, 42, chief, Mass, 44-49, mem staff diabetes res, Boston City Hosp, 49-50; chief heart dis epidemiol study, US Dept Health, Educ & Welfare, 50-66; prog planning officer, Med Ctr, 66-68, assoc prof, 66-76, PROF MED, SCH MED, BOSTON UNIV, 76- *Mem:* Am Col Chest Physicians; Am Col Physicians; Am Heart Asn. *Res:* Epidemiology of coronary heart disease and hypertension; public health. *Mailing Add:* Dept Med Sch Med Boston Univ 80 E Concord St Boston MA 02118

DAWE, ALBERT ROLKE, b Milwaukee, Wis, June 1, 16; m 42, 73; c 3. COMPARATIVE PHYSIOLOGY. *Educ:* Yale Univ, BA, 38; Harvard Univ, MA, 51; Univ Wis, PhD(zool, physiol), 53. *Prof Exp:* Instr zool, Exten Div, Univ Wis, 46-50, asst prof physiol, Med Sch, 53-56; biologist, Off Naval Res, Chicago, 56-58, chief scientist, 58-67, dep dir & chief scientist, 67-77; RETIRED. *Concurrent Pos:* From adj assoc prof to adj prof physiol, Stritch Sch Med, Loyola Univ, 64-78; chief biol consult, World Book Encycl; mem bd assocs, John Crerar Libr. *Mem:* Am Physiol Soc; Soc Cryobiol; Int Hibernation Soc. *Mailing Add:* 2900 Maple No 18E Downers Grove IL 60515

DAWE, HAROLD JOSEPH, b Deckerville, Mich, June 15, 12; m 33; c 4. CHEMISTRY. *Educ:* Cent Mich Col, BS, 32; Univ Mich, MS, 34, PhD(chem), 40. *Prof Exp:* Prod develop engr, Acheson Colloids Co, 40-50, tech dir, 50-56; res dir, Acheson Industs Inc, 56-64, vpres res, 64-76; RETIRED. *Mem:* AAAS; Am Chem Soc; fel Am Inst Chem; Am Soc Lubrication Engrs; fel Chem Inst Can. *Res:* Colloid chemistry of dispersed solids; relation between the stability of suspensions and their interfacial free surface energies. *Mailing Add:* 26 Lark Spur Hendersonville NC 28792

DAWES, CLINTON JOHN, b Minneapolis, Minn, Sept 23, 35; m 62; c 3. PHYCOLOGY, ALGALCYTOLOGY. *Educ:* Univ Minn, Minneapolis, BS, 57; Univ Calif, Los Angeles, MA, 58, PhD(sci), 61. *Prof Exp:* Teaching asst, Bot Labs, Univ Calif, Los Angeles, 57-59; NSF fel, 63-64; asst prof bot, 64-70, assoc prof biol, 70-72, PROF BIOL, UNIV SFLA, 72-, CHMN DEPT, 85- *Concurrent Pos:* NSF grants, 65-66, 69-71, 70-72 & 78; Sea grant, 71-73; Fla Sea grant, 74, 75 & 76, Fla Dept Natural Resources, 78-82, Gas Res Inst, 81-; grants, Fla Dept Natural Res, 78-82, Gas Res Inst, 82-86, US-Israel Binat, 88-91, AID, 89-92; distinguished scholar, Univ SFla, 85; Fulbright scholar, 89-90. *Mem:* Bot Soc Am; Phycol Soc Am (pres, 79); Int Phycol Soc; Sigma Xi; Electron Micros Soc Am. *Res:* Algal ultrastructure, especially cell development; cell wall fine structure and cytology of coenocytic algae; physiological ecology of marine algae and seagrasses; floristics of Florida algae. *Mailing Add:* Dept Biol Univ SFla Tampa FL 33620

DAWES, DAVID HADDON, b Cornwall, Ont, Aug 14, 38; m 62; c 5. PHYSICAL CHEMISTRY, POLYMER SCIENCE. *Educ:* McGill Univ, BSc, 59, PhD(phys chem), 62. *Prof Exp:* Nat Res Coun Can fel chem, 62-64; res chemist, E I Du Pont de Nemours & Co, 65-67, sr res chemist, 67-74, res mgr, 74-81 & 84-86, mgr, Environ Sci, Du Pont Can, 82-84; CONSULTM, 84- *Mem:* Fel Chem Inst Can. *Res:* Radiolysis of gases; product and process research and development. *Mailing Add:* RR One Westbrook ON K0H 2X0 Can

DAWES, JOHN LESLIE, b Olney, Eng, Dec 31, 42; m 69. ORGANOMETALLIC CHEMISTRY, ORGANIC CHEMISTRY. *Educ:* Univ Leicester, BS, 64, PhD(chem), 67. *Prof Exp:* NSF Ctr of Excellence vis asst prof, La State Univ, 67-69; chemist, 69-77, SR RES CHEMIST, TEX EASTMAN CO, 77- *Mem:* Am Chem Soc; Catalysis Soc. *Res:* Inorganic chemistry; homogeneous and heterogeneous catalysis related to the preparation of organic compounds. *Mailing Add:* 115 Fredrick Ct Longview TX 75601

DAWES, WILLIAM REDIN, JR, b Charlotte, NC, Oct 10, 40; m 64; c 2. HIGH ENERGY PHYSICS, SOLID STATE PHYSICS. *Educ:* Univ NC, BS, 62; Univ Ariz, MS, 64, PhD(physics), 68. *Prof Exp:* Mem tech staff, Bell Tel Labs, 68-70, supvr, Silicon Device Technol Group, 70-75; supvr integrated circuit technol, 75-83, dept mgr 83-88, SUPERVISOR, POWER SYSTEM ELECTRONICS, SANDIA NAT LABS, 88- *Mem:* Sigma Xi. *Res:* Silicon device technology, especially radiation hardened. *Mailing Add:* 112 White Oaks Dr NE Albuquerque NM 87122

DAWID, IGOR BERT, b Czernowitz, Romania, Feb 26, 35. BIOCHEMISTRY, DEVELOPMENTAL BIOLOGY. *Educ:* Univ Vienna, PhD(chem), 60. *Prof Exp:* Lab asst chem, Univ Vienna, 59-60; vis lectr biochem, Mass Inst Technol, 60-62; fel, Carnegie Inst, 62-67, mem staff, Dept Embryol, Carnegie Inst Washington, 67-78; head develop biochem sect, Lab Biochem, Nat Cancer Inst, 78-82; CHIEF LAB MOLECULAR GENETICS, NAT INST CHILD HEALTH & HUMAN DEVELOP, NIH, 82- *Concurrent Pos:* Vis mem, Max Planck Inst Biol, Univ Tubingen, 65-67; from asst prof to assoc prof biol, Johns Hopkins Univ, 67-73, prof, 73-; ed-in-chief, Develop Biol, 75-80, assoc ed, Cell, ed bd, Develop Growth & Differentiation, 82-, Nucleic Acids Res, 83-, Genes & Develop, 86-, assoc ed, Annual Revs Biochem, 84-; vis prof, Technion, Haifa, Israel; Mider Lect Ser, 87. *Mem:* Nat Acad Sci; Am Soc Cell Biol; Am Soc Biol Chemists; Int Soc Develop Biologists; AAAS. *Res:* Biochemistry of development; function of nucleic acids in development; gene isolation, ribosomal DNA. *Mailing Add:* Bldg 6 Room 408 NIH Bethesda MD 20892

DAWKINS, WILLIAM PAUL, b Houston, Tex, July 25, 34; m 57; c 2. CIVIL ENGINEERING. *Educ:* Rice Univ, BA & BS, 57, MS, 62; Univ Ill, Urbana, PhD(civil eng), 66. *Prof Exp:* Sr engr, Proj Mohole, Brown & Root, Inc, Tex, 62-63; res asst civil eng, Univ Ill, Urbana, 63-65, res assoc, 65-66; asst prof, Univ Tex, 66-69; assoc prof, 69-74, PROF CIVIL ENG, OKLA STATE UNIV, 74- *Mem:* Am Soc Civil Engrs. *Res:* Application of computers to solutions of civil engineering problems, particularly to problems of structural analysis. *Mailing Add:* Sch Civil Eng Okla State Univ Main Campus Stillwater OK 74078

DAWSON, ARTHUR DONOVAN, b Wilmington, Del, May 5, 43; m 66; c 1. ORGANIC CHEMISTRY, POLYMER CHEMISTRY. *Educ:* Temple Univ, AB, 67, 68, PhD(org chem), 76. *Prof Exp:* Control analyst, Smith Kline & French Lab, Inc, 68-69; teaching asst, Dept Chem, Temple Univ, 69-72, res fel, 72-76; develop scientist, Household Prod Div, Lever Bros, 75-78; sr prod develop chemist, Johnson & Johnson Personal Prod Co, 78-80; sr res scientist, Becton Dickinson Consumer Prods, 80-82, mgr, Chem Prod Develop, 83-87, mgr, Tech Serv, 87-90, PATENT AGENT, CORP PATENT DEPT, BECTON DICKINSON, 91- *Concurrent Pos:* Fel, NIH, 74-75; Chmn, North Jersey Am Chem Soc, 81. *Mem:* Am Chem Soc; Am Oil Chem Soc; Sigma Xi; Royal Soc Chem; Am Diabetes Asn; Soc Cosmetics Chemists. *Res:* Product development of feminine hygiene products; study of adhesives; non-woven fabrics; surface chemistry; paper & pulp chemistry; technical service of syringe products. *Mailing Add:* 575 Ridgewood Ave Glen Ridge NJ 07028

DAWSON, CHANDLER R, b Denver, Colo, Aug 24, 30; m 54; c 3. OPHTHALMOLOGY. *Educ:* Princeton Univ, AB, 52; Yale Univ, MD, 56. *Prof Exp:* USPHS epidemiologist, Commun Dis Ctr, 57-60; resident ophthal, 60-63, asst clin prof, 63-66, asst res prof, 66-69, assoc prof in residence, 69-75, PROF OPHTHALMOL, SCH MED, UNIV CALIF, SAN FRANCISCO, 75- *Concurrent Pos:* Fel, Middlesex Hosp Med Sch, London, 63-64; dir, WHO Collab Ctr Prevention Blindness & Trachoma; assoc dir, Francis I Proctor Found, 70- *Honors & Awards:* Knapp Award, AMA, 67 & 69; Medaille Trachome, 78. *Mem:* Am Soc Microbiol; Am Acad Ophthal & Otolaryngol; Asn Res Vision & Ophthalmol. *Res:* Epidemiology of infectious eye diseases and cataracts; prevention of blindness; pathogenesis of virus diseases of the eyes; electron microscropy of eye diseases. *Mailing Add:* Dept Ophthal Univ Calif Sch Med San Francisco CA 94143

DAWSON, CHARLES ERIC, ichthyology, for more information see previous edition

DAWSON, CHARLES H, b Uniontown, Pa, Dec 11, 16; m 38; c 2. ELECTRICAL ENGINEERING, MATHEMATICS. *Educ:* Cornell Univ, EE, 38; Univ Rochester, MS, 41; Iowa State Col, PhD(elec eng), 52. *Prof Exp:* Instr eng, Univ Rochester, 38-42, 45, from asst prof to prof, 46-58; prof & head dept, Univ RI, 58-59; sr eng specialist, Philco Western Develop Labs, 59-62; sr res engr, Radio Systs Lab, Stanford Res Inst, 62-67, staff scientist, 68-70 & 72-73, STAFF SCIENTIST, RADIO PHYSICS LAB, SRI INT, 74- *Concurrent Pos:* UNESCO assignment, specialist in telecommun, Univ Brasilia, 67-68; prof, Univ PEI, 70-72. *Mem:* Inst Elec & Electronics Engrs; Asn Comput Mach; Am Asn Eng Educ. *Res:* Statistical detection theory; digital computer systems; real time programming; simulation. *Mailing Add:* 208 Ravenswood Ave Menlo Park CA 94025

DAWSON, CHRISTOPHER A, PHYSIOLOGY. *Prof Exp:* PROF PHYSIOL, VET ADMIN MED CTR, MED COL WIS, 86- *Mailing Add:* Dept Physiol Vet Admin Med Ctr Res Serv 151A Med Col Wis 5000 W National Ave Milwaukee WI 53295-1000

DAWSON, D(ONALD) E(MERSON), b Detroit, Mich, Oct 3, 25; m 52; c 1. ENGINEERING SCIENCE. *Educ:* Wayne State Univ, BS, 48; Pa State Univ, MS, 56, PhD(eng mech), 58. *Prof Exp:* Engr commun, Mich Bell Tel Co, 48-50; physicist aerosol physics, Battelle Mem Inst, 51-52; res assoc, Pa State Univ, 52-53; res engr eng sci, E I du Pont de Nemours & Co, 58-62; res mgr, Armour Res Found, 62-63; assoc prof eng sci, 63-66, prof, 66-84, EMER PROF MATH & MECH, MICH TECHNOL UNIV, 84- *Mem:* AAAS; Am Phys Soc; Soc Indust & Appl Math. *Res:* Applied and engineering mathematics; dynamics; vibrations. *Mailing Add:* RR 1 Box 49 Houghton MI 49931

DAWSON, DANIEL JOSEPH, b Philadelphia, Pa, July 17, 46. ORGANIC POLYMER CHEMISTRY. *Educ:* Univ NC, Chapel Hill, BS, 67; Calif Inst Technol, PhD(org chem), 71. *Prof Exp:* NIH fel org chem, Harvard Univ, 71-72; sr res chemist & mgr process chem, Dynapol, 72-78, assoc dir process develop, 78-80, dir chem res & develop, 80-82; RSM, 82-86, MGR SYNTHETIC DEVELOP, IBM RES, SAN JOSE, CALIF, 86- *Mem:* Am Chem Soc. *Res:* Synthesis of high technology functional polymers including polymers for electronic applications. *Mailing Add:* IBM Res K92-801 650 Harry Rd San Jose CA 95120

DAWSON, DAVID CHARLES, b Pittsburgh, Pa, Feb 5, 44; m 67; c 2. PHYSIOLOGY. *Educ:* Univ Pittsburgh, BS, 66, PhD(biol), 71. *Prof Exp:* NIH fel membrane physiol & biophys, Sch Med, Yale Univ, 71-73; asst prof physiol & biophys, Sch Med, Univ Iowa, 73-79, assoc prof, 79-80; assoc prof, 81-87, PROF PHYSIOL, SCH MED, UNIV MICH, 87- *Mem:* Am Physiol Soc; Biophys Soc; Am Soc Zoologists; AAAS; Soc Gen Physiologists; Mt Desert Island Biol Lab. *Res:* Molecular basis for ion transport by epithelia and hormonal control; control of amphibian pituitary secretion by light. *Mailing Add:* 6811 Med Sci II Univ Mich Med Sch Ann Arbor MI 48109

DAWSON, DAVID FLEMING, b Denton, Tex, Sept 16, 26; m 48; c 6. MATHEMATICAL ANALYSIS. *Educ:* Univ Tex, PhD(math), 57. *Prof Exp:* Asst prof math, Univ Mo, 57-59; from asst prof to assoc prof, 59-64, PROF MATH, N TEX STATE UNIV, 64- *Mem:* Am Math Soc. *Res:* Analysis; continued fractions; summability theory. *Mailing Add:* 1015 Ector Denton TX 76201

DAWSON, DAVID LYNN, b Denver, Colo, Sept 16, 42; m 63; c 3. GROSS ANATOMY, SURGICAL ANATOMY. *Educ:* Adams State Col, BA, 64; Southern Ill Univ, MA, 67, PhD(zool), 75. *Prof Exp:* Instr zool, Southern Ill Univ, 71; asst prof & chmn biol, Univ SDak, 71-74; asst prof, 75-80, assoc prof anat, Sch Med, Marshall Univ, 80-86, assoc prof surg, 84-86; CLIN ANATOMIST, DES MOINES UAMC, 86- & IOWA METHODIST MED CTR, 86- *Concurrent Pos:* Adj prof anat, Simpson Col, 88-; historian, Am Asn Clin Anatomists, 86- *Mem:* Am Asn Anatomists; Sigma Xi; Am Asn Clin Anatomists. *Res:* Surgical anatomy. *Mailing Add:* VA Med Ctr 30th & Euclid Des Moines IA 50310

DAWSON, DONALD ANDREW, b Montreal, Que, June 4, 37; m 64; c 2. MATHEMATICS, STATISTICS. *Educ:* McGill Univ, BSc, 58, MSc, 59; Mass Inst Technol, PhD(math), 63. *Prof Exp:* Sr engr commun theory, Raytheon Corp, Mass, 62-63; asst prof math, McGill Univ, 63-66; vis asst prof, Univ Ill, 66-67; PROF MATH, CARLETON UNIV, 67- *Concurrent Pos:* Assoc ed, Ann Probability, 88-90; co-ed-chief, Can J Math, 89-; mem, adv bd Stochastic Processes & Their Applications. *Mem:* Am Math Soc; Can Math Soc; Inst Math Statist; Can Statist Soc; Bernoulli Soc; Fel Royal Soc Can. *Res:* Probability theory; statistical mechanics; stochastic processes; communications theory; functional analysis; probability theory. *Mailing Add:* Dept Math Carleton Univ Ottawa ON K1S 5B6 Can

DAWSON, EARL B, b Perry, Fla, Feb 1, 30; m 51; c 4. BIOCHEMISTRY, NUTRITION. *Educ:* Univ Kans, BA, 55; Univ Mo, MA, 60, Tex A&M Univ, PhD(biochem & nutrit), 64. *Prof Exp:* Teacher, Salem Acad, NC, 57-58; technician biochem, R J Reynolds Tobacco Co, 58; lab instr physiol, Univ Mo, 58-60; instr biochem, Tex A&M Univ, 60-63; from instr to asst prof, 63-65, ASSOC PROF BIOCHEM, UNIV TEX MED BR GALVESTON, 68- *Concurrent Pos:* Biochemist consult, Interdept Comt Nutrit Nat Defense, NIH, 65. *Mem:* AAAS; Am Inst Nutrit; Am Soc Clin Nutrit; Am Soc Exp Biol Med; Am Inst Physics; Am Col Nutrit. *Res:* Mammalian renal physiology; changes in oxidative enzyme activity associated with cataract formation; placental changes in enzyme activity associated with toxemia of pregnancy; clinical nutrition; trace elements. *Mailing Add:* Dept Obstet & Gynec Univ Tex Med Br Galveston TX 77550

DAWSON, FRANK G(ATES), JR, b Alliance, NC, July 6, 25; m 47; c 3. NUCLEAR ENGINEERING. *Educ:* NC State Col, BEE, 50; Univ Wash, PhD(nuclear eng), 73. *Prof Exp:* Test engr, Gen Elec Co, 50-51, sr tech engr, Aircraft Nuclear Propulsion Dept, 51-55, prin engr, 56-60, tech specialist, Hanford Labs, 60, mgr appl physics, 60-63, mgr reactor physics, 63-65; mgr reactor physics, Pac Northwest Labs, 65-68, mgr physics & eng div, 68-71, asst lab dir, 71-73, dir, Battelle Energy Prog, 73-76, DIR, CORP TECH DEVELOP, BATTELLE MEM INST, 76- *Concurrent Pos:* Actg asst prof, Univ Cincinnati, 58-60; mem adv comt reactor physics, Atomic Energy Comn, 64-70; lectr, Univ Calif, Los Angeles, 67-68; mem, Europ-Am Comt Reactor Physics, 70-; US rep, panel on plutonium utilization, Int Atomic Energy Agency, 68, chmn panel on plutonium recycling in thermal power reactors, 71; US deleg, UN Int Conf Peaceful Uses of Atomic Energy, 71. *Res:* Reactor physics and design; energy. *Mailing Add:* Battelle Europe Seven Route de drize Carouge-Geneva 1227 Switzerland

DAWSON, GLADYS QUINTY, b Trenton, NJ, Sept 3, 24; m 52. INORGANIC CHEMISTRY. *Educ:* Univ Ill, BS, 46, PhD(chem), 51. *Prof Exp:* Prin chemist chem res, Battelle Mem Inst, 51-52; instr chem, Pa State Univ, 52-58; patent liaison, E I du Pont de Nemours & Co, 58-63; from teaching assoc to assoc prof, 63-84, EMER ASSOC PROF CHEM, MICH TECHNOL UNIV, 85- *Mem:* Am Chem Soc. *Res:* Chemical education. *Mailing Add:* Royalewood Addition RR 1 Box 49 Houghton MI 49931

DAWSON, GLYN, b New Mills, Eng, Mar 24, 43; m 66; c 2. BIOCHEMISTRY. *Educ:* Univ Bristol, BSc, 64, PhD(biochem), 67. *Prof Exp:* Fel, Univ Pittsburgh, 67-68; res assoc, Mich State Univ, 68-69; asst prof, 69-75, ASSOC PROF PEDIAT & BIOCHEM, UNIV CHICAGO, 75- *Mem:* AAAS; Am Soc Biol Chem; Soc Complex Carbohydrates; Am Soc Neurochem. *Res:* Regulation of glycosphingolipid glycoprotein and phospholipid metabolism in human cells especially cultured cells-fibroblasts, glial, neuroblastomas by neurotransmitters, hormones etc; inborn errors of metabolism; carbohydrate structure; mass spectrometry. *Mailing Add:* Dept Pediat & Biochem Box 82 Univ Chicago 950 E 59th St Chicago IL 60637

DAWSON, HORACE RAY, b Wills Point, Tex, Mar 29, 35; m 56; c 2. PHYSICS. *Educ:* NTex State Univ, BA, 57, MA, 61; Univ Ark, PhD(physics), 68. *Prof Exp:* Instr physics, ETex State Univ, 61-63; assoc prof, 66-76, PROF PHYSICS & HEAD DEPT, ANGELO STATE UNIV, 76- *Mem:* Am Asn Physics Teachers. *Res:* Atomic collisions; radiative lifetimes of atoms; gamma ray spectroscopy. *Mailing Add:* Dept Physics Angelo State Univ San Angelo TX 76909

DAWSON, J W, b Toronto, Ont, Dec 30, 28; m 52; c 4. INTERNAL MEDICINE, ENDOCRINOLOGY. *Educ:* Univ Toronto, MD, 53; FRCP(C). *Prof Exp:* ASSOC DEAN, UNIV CALGARY, 67-, PROF MED, FAC MED, 68-, DIR DIV CONTINUING MED EDUC, 71- *Concurrent Pos:* Consult internist, Calgary Assoc Clin. *Mem:* Fel Am Col Physicians; Can Soc Clin Invest. *Res:* Pituitary cytology; thyroidology; medical education. *Mailing Add:* 1228 Kensington Rd NW Calgary AB T2N 4P9 Can

DAWSON, JAMES CLIFFORD, b Toronto, Ont, Apr 19, 41; US citizen; m 71. SEDIMENTOLOGY, COASTAL PROCESSES. *Educ:* Univ Calif, Los Angeles, BS, 65, MS, 67; Univ Wis, Madison, PhD(geol), 70. *Prof Exp:* Asst prof, 70-74, assoc prof geol, 74-80, dir, Inst for Man & Environ, 76-82, PROF ENVIRON SCI, STATE UNIV NY COL, PLATTSBURGH, 80- *Concurrent Pos:* Prin investr, State Univ NY Res Found fel & grant-in-aid, 71 & 73; Forest Preserve Specialist, NY Gov Comn on Adriondacks in the Twenty-First Century, 89-90. *Mem:* AAAS; Nat Asn Environ Educ; Nat Asn Geol Teachers; Sigma Xi; Soc Econ Paleontologists & Mineralogists. *Res:* Carboniferous sedimentology of the Falkland Islands; upper Jurassic and lower Cretaceous sedimentology of the Colorado Plateau; shoreline processes and water quality management of Lake Champlain; chert petrography; Adirondack wilderness management; Adriondack land acquisition and open space protection; Adriondak ecological health trends. *Mailing Add:* Ctr Earth & Environ Sci State Univ NY Plattsburgh NY 12901

DAWSON, JAMES THOMAS, b Dover, Ohio, Nov 10, 47; m 76; c 2. PHYCOLOGY, MEDICAL MYCOLOGY. *Educ:* Kent State Univ, BS, 70; Univ Ky, PhD(biol), 77. *Prof Exp:* Technician chem, Dover Chem Corp, 70-71; lab supvr biol, Univ Ky, 77-78; from asst prof to assoc prof, 78-89, PROF BIOL, PITTSBURG STATE UNIV, 89- *Mem:* Phycological Soc Am; Sigma Xi; Int Phycol Soc; Brit Phycol Soc. *Res:* Taxonomy and physiology of freshwater algae, particularly the green algae; taxonomy and distribution of higher aquatic plants; physiology and identification of medically important fungi. *Mailing Add:* Dept Biol Pittsburg State Univ Pittsburg KS 66762

DAWSON, JEAN HOWARD, b Stacy, Minn, Apr 14, 33; m 54; c 5. WEED SCIENCE. *Educ:* Univ Minn, BS, 55; Univ Calif, MS, 57; Ore State Univ, PhD(weed control), 61. *Prof Exp:* Res asst agron, Univ Calif, 55-57; res agronomist & res leader, 57-90, COLLABR WEED SCI, USDA, 90- *Concurrent Pos:* Consult weed sci, Food & Agr Orgn UN, Alfalfa Improv Proj, Arg, 74-75; coop weed res in Costa Rica & El Salvador, Int Plant Protection Ctr, Ore State Univ, 78; instr & co-dir weed sci course, Food & Agr Orgn, UN, Argentina, 79; instr, Course on Herbicide Action, Purdue Univ, 81-; consult, Interam Inst Agr Coop, Arg, 88; mem, Sci Rev Panel, USAID Proj, Moracco, 90. *Mem:* Weed Sci Soc Am (pres); fel, Western Soc Weed Sci; fel, Weed Sci Soc Am. *Res:* Principles and practices of weed control in irrigated crops; dodder control in alfalfa; herbicide mode of action and weed biology studies. *Mailing Add:* Irrig Agr Res & Exten Ctr USDA Prosser WA 99350

DAWSON, JEFFREY ROBERT, b Lakewood, Ohio, Oct 5, 41; m 64; c 3. IMMUNOLOGY, BIOCHEMISTRY. *Educ:* Rensselaer Polytech Inst, BS, 64; Case Western Reserve Univ, PhD(biochem), 69. *Prof Exp:* NIH fel biochem, 69-71, instr, 71-72, assoc, 72-74, asst prof, 74-77, ASSOC PROF IMMUNOL, MED CTR, DUKE UNIV, 77- *Mem:* Am Asn Immunol; Sigma Xi; Brit Soc Immunol; Am Asn Cancer Res. *Res:* Human immunity to ovarian cancer. *Mailing Add:* Div Immunol Duke Univ Med Ctr Box 3010 Durham NC 27710

DAWSON, JOHN E, b Hamilton, Ohio, Oct 19, 24; m 69; c 3. REPRODUCTIVE PHYSIOLOGY. *Educ:* Univ Cincinnati, BA, 47, MS, 56, PhD(endocrinol), 62. *Prof Exp:* Teacher high sch, Ky, 47-53 & Ohio, 53-55; asst prof gen zool & sci educ, San Jose State Col, 58-64; asst prof gen biol & physiol, 64-67, prof biol, 64-75, assoc prof physiol, 67-76, prof, 76-82, EMER PROF BIOL, OHIO NORTHERN UNIV, 82- *Concurrent Pos:* Res fel, Univ Wash, 68-69. *Mem:* AAAS; Sigma Xi; Nat Educ Asn; Am Asn Univ Prof. *Res:* Effect of chronic administration of sodium tolbutamide on pregnant albino rats and their offspring; effect of luteinizing hormone on membrane permeability of frog oocytes. *Mailing Add:* 7951 Deer Creek Dr Sacramento CA 95823

DAWSON, JOHN FREDERICK, b Springfield, Ohio, Jan 4, 36; m 58; c 1. PHYSICS. *Educ:* Antioch Col, BS, 58; Stanford Univ, PhD(physics), 63. *Prof Exp:* Res assoc physics, McGill Univ, 62-64; asst prof, Antioch Col, 64-65; vis asst prof, Oberlin Col, 65-66; res assoc, Lowell Technol Inst Res Found, 66-68; from asst prof to assoc prof, 74-80, PROF PHYSICS, UNIV NH, 80- *Mem:* Am Phys Soc. *Res:* Theoretical nuclear physics. *Mailing Add:* Dept Physics Univ NH Durham NH 03824

DAWSON, JOHN HAROLD, b Englewood, NJ, Sept 19, 50; m 73; c 1. BIO-INORGANIC CHEMISTRY, BIO-ORGANIC CHEMISTRY. *Educ:* Columbia Univ, AB, 72; Stanford Univ, PhD(chem), 76. *Prof Exp:* Res & teaching asst chem, Stanford Univ, 72-76; fel chem, Calif Inst Technol, 76-78; from asst prof to assoc prof, 78-86, PROF CHEM & BIOCHEM, UNIV SC, 87-, CAROLINA RES PROF, 88- *Concurrent Pos:* Camille & Henry Dreyfus teacher/scholar, 82-87; Alfred P Sloan Res Fel, 83-87; res career develop award, NIH, 83-88; ad hoc mem, Molecular & Cellular Biophys Study Sect, NIH, 85 & Metallobiochem Study Sect, NIH, 86-87; vis prof, Dept Chem, Mass Inst Technol, 89; mem, NSF Biophys, Prog Rev Panel, 88-89, NSF Instrumentation & Instrumentation Develop Prog Rev Panel, 90-, Howard Hughes Found Predoctoral Fel Rev Panel Struct Biol & Biochem, 90- & chair, 91. *Mem:* Am Chem Soc; Sigma Xi; Am Soc Biochem & Molecular Biol; fel AAAS. *Res:* Biochemistry; spectroscopy and mechanism of action of heme iron oxygen-utilizing metallo-enzymes and model systems; electron transfer mechanisms; magnetic circular dichroism spectroscopy; metalloporphyrins. *Mailing Add:* Dept Chem Univ SC Columbia SC 29208

DAWSON, JOHN MYRICK, b Champaign, Ill, Sept 30, 30; m 57; c 2. PLASMA PHYSICS. *Educ:* Univ Md, BS, 52, MS, 54, PhD(physics), 57. *Prof Exp:* Res physicist, Proj Matterhorn, Princeton Univ, 56-62, plasma physics lab, 62-64, assoc head theoret group, 64-66, head, 66-73, lectr, 60-73; PROF PHYSICS, UNIV CALIF, LOS ANGELES, 73-, DIR, CTR PLASMA PHYSICS & FUSION ENG, 76- *Concurrent Pos:* Consult, RCA Corp, 62-63, Boeing Co, 64, Northwestern Univ, 72 TRW Systs, 73-86; Fulbright fel, Inst Plasma Physics, Univ Nagoya, 64-65; sci adv, Dir Naval Res Lab, 69-71; chmn, Plasma Physics Div, Am Phys Soc, 70-71; chmn, Gordon Conf Plasma, 70 & Gordon Conf Lasers, 77; guest, Soviet Acad Sci, 71 & Japanese Atomic Energy Agency, 78. *Honors & Awards:* James Clerk Maxwell Prize Plasma Physics, 77; Jessie & John Danz Lectr, Univ Wash,74. *Mem:* Nat Acad Sci; fel Am Phys Soc; Sigma Xi; fel AAAS; fel NY Acad Sci. *Res:* Plasma, atomic and molecular physics; plasma simulation. *Mailing Add:* 359 Arno Way Pacific Palisades CA 90272

DAWSON, JOHN WILLIAM, JR, b Wichita, Kans, Feb 4, 44; m 70. MATHEMATICAL LOGIC. *Educ:* Mass Inst Technol, BS, 66; Univ Mich, PhD(math), 72. *Prof Exp:* Instr math, Pa State Univ, 72-75; from asst prof to assoc prof, 75-85, PROF MATH, PA STATE UNIV, YORK CAMPUS, 85- *Concurrent Pos:* Mem, Inst Adv Study, Princeton, 82-84; vis scholar, Stanford Univ, 86. *Mem:* Am Math Soc; Asn Symbolic Logic; Hist Sci Soc. *Res:* History of modern logic; axiomatic set theory. *Mailing Add:* 393 Waters Rd York PA 17403

DAWSON, KENNETH ADRIAN, b Belfast, NIreland, Apr 23, 58. STATISTICAL MECHANICS OF COMPLEX FLUIDS, APPLICATION OF STATISTICAL MECHANICS TO BIOPHYSICAL PROBLEMS. *Educ:* Queen's Univ, Belfast, NIreland, BSc, 80, MSc, 81; Univ Oxford, UK, PhD(theoret chem), 84. *Prof Exp:* Postdoctoral fel, Cornell Univ, 85-88; ASST PROF CHEM, UNIV CALIF, BERKELEY, 88-, ADJ PROF BIOPHYS, 89- *Concurrent Pos:* Res visitor, Inst Sci Study, 83-84; vis lectr, Univ Ulm, WGer, 84-85; new blood lectr, Univ Leeds, Eng, 86-87. *Honors & Awards:* Harrison Mem Prize, Royal Soc Chem, 87; Supercomput Award, Phys Sci & Math Div, IBM, 91, Computer Sci-Distrib & Coop Processing, 91. *Mem:* Am Chem Soc; Am Phys Soc. *Res:* Methods of statistical mechanics to study complex fluids; mechanics of polymers and proteins. *Mailing Add:* Dept Chem Univ Calif Berkeley CA 94720

DAWSON, KERRY J, b Buckhannon, WVa, Sept 4, 46; c 1. LANDSCAPE ARCHITECTURE, ARBORETUM MANAGEMENT. *Educ:* Univ Fla, BLA, 69; Univ Calif, MLA, 73. *Prof Exp:* Prof land archit, La State Univ, 79-80; PROF LAND ARCHIT, UNIV CALIF, 80-, DIR ARBORETUM, 85- *Concurrent Pos:* Researcher, Agr Exp Sta, 80- *Honors & Awards:* Am Soc Landscape Archit Award, 88. *Mem:* Am Soc Landscape Archit. *Res:* The planning, design and management of national parks, natural areas, historic sites and wildlife reserves. *Mailing Add:* Dept Environ Design Univ Calif Davis CA 95616

DAWSON, LAWRENCE E, b Mich, July 23, 16; m 45; c 1. FOOD SCIENCE. *Educ:* Mich State Univ, BS, 42, MS, 46; Purdue Univ, PhD(agr mkt), 49. *Prof Exp:* Asst, Purdue Univ, 46-49; prof, 49-83, actg chmn, food & human nutrit dept, 79-82, EMER PROF FOOD SCI, MICH STATE UNIV, 83- *Honors & Awards:* Res Achievement Award, Poultry & Egg Nat Bd, 61. *Mem:* Fel Poultry Sci Asn; fel Inst Food Technologists; World Poultry Sci Asn; Am Poultry Hist Soc (pres). *Res:* Poultry, egg and fish products technology, flavor, composition and preservation. *Mailing Add:* Dept Food Sci & Human Nutrit Mich State Univ East Lansing MI 48824

DAWSON, M JOAN, b Greenville, SC, Feb 9, 44; c 1. MUSCLE PHYSIOLOGY, MAGNETIC RESONANCE TECHNIQUE. *Educ:* Columbia Univ, BSc, 69; Univ Pa, PhD(pharmacol), 72. *Prof Exp:* Resident coun tutor, biol, chem & physics, Off Dean Res Life, Univ Pa, 70-72; res fel neurobiol, Lab Neurobiol, Columbia Univ, 72-73; teaching fel physiol, Univ Col London, 74-77, Sharpey scholar, 78-80, lectr, 80-84, hon lectr, dept med, Univ Col Hosp, 82-84; ASSOC PROF PHYSIOL, UNIV ILL, 85- *Mem:* Physiol Soc; Soc Magnetic Resonance Med; AAAS; Biophys Soc. *Res:* Studies of the relation between metabolism and function in living tissues, anesthetized animals and human subjects using nuclear magnetic resonance spectroscopy techniques. *Mailing Add:* Dept Physiol Univ Ill 524 Burrill Hall Urbana IL 61801

DAWSON, MARCIA ILTON, b Detroit, Mich, May 4, 42; div. BIO-ORGANIC CHEMISTRY. *Educ:* Univ Mich, Ann Arbor, BS, 64; Stanford Univ, PhD(org chem), 68. *Prof Exp:* Fel org chem, Calif Inst Technol, 68; NIH fel biochem, Harvard Univ, 69; fel org chem, Calif Inst Technol, 70-71; NIH spec fel, Harvard Univ, 71-72; org chemist, 72-76, sr biochemist, 77-80, PROG DIR, LIFE SCI DIV, SRI INT, 80- *Honors & Awards:* Frank R Blood Award. *Mem:* Am Chem Soc; Sigma Xi. *Res:* Application of organic synthesis to biochemical problems, particularly in the areas of retinoids, targeted delivery with drug-monoclonal antibody conjugates, oligonucleotides and biochemistry. *Mailing Add:* SRI Int Life Sci Div Menlo Park CA 94025

DAWSON, MARGARET ANN, POLLUTANT PHYSIOLOGY, COMPARATIVE PHYSIOLOGY. *Educ:* Univ Mich, MS, 70. *Prof Exp:* RES PHYSIOLOGIST, NAT MARINE FISHERIES SERV, 74- *Mailing Add:* Milford Lab Nat Fisheries Serv Milford CT 06460

DAWSON, MARY (RUTH), b Highland Park, Mich, Feb 27, 31. VERTEBRATE PALEONTOLOGY. *Educ:* Mich State Univ, BS, 52; Univ Kans, PhD(zool), 57. *Prof Exp:* Instr zool, Univ Kans, 56-57; from instr to asst prof, Smith Col, 58-61; asst prog dir, NSF, 61-62; res assoc, 62-63, from asst cur to cur & chmn, earth sci, 63-71,CUR, CARNEGIE MUS, 71-; ADJ PROF, EARTH & PLANETARY SCI, UNIV PITTSBURGH, 70- *Concurrent Pos:* Fulbright scholar, 52-53; Am Asn Univ Women fel, 57-58; actg dir, Carnegie Mus, 82-83. *Honors & Awards:* Arnold Guyot Award, Nat Geog Soc, 81. *Mem:* Paleont Soc; fel Geol Soc Am; Soc Vert Paleont (pres, 73-74); Paläontologische Ges; Sigma Xi; fel Arctic Inst NAm. *Res:* Paleontology of lagomorphs and rodents; early tertiary holarctic faunas; arctic tertiary mammals. *Mailing Add:* Sect Vert Fossils Carnegie Mus 4400 Forbes Ave Pittsburgh PA 15213

DAWSON, MURRAY DRAYTON, b Christchurch, NZ, Dec 16, 25; US citizen; m 52; c 6. SOILS, PLANT PHYSIOLOGY. *Educ:* Univ NZ, BAgSc, 49, MAgSc, 51; Cornell Univ, MS, 52, PhD(soils), 54. *Prof Exp:* Res scientist field husb, Can Exp Sta, Swift Current, Sask, 50-51; prof soils, Ore State Univ, 54-; AT IADS, ARLINGTON, VA. *Concurrent Pos:* Consult prof, Univ Agr, Bangkok, 59-60; consult, Ford Found, Chiang Mai Univ, 74- *Mem:* Am Soc Agron; Sigma Xi. *Res:* Legume establishment growth and development on infertile acid soils, with special reference to sulfur nutrition and molybdenum; multiple cropping research. *Mailing Add:* 1941 S Arlington Ridge Rd Arlington VA 22202

DAWSON, PETER HENRY, b Derby, Eng, May 28, 37; m 63; c 2. CHEMICAL PHYSICS. *Educ:* Univ London, BSc, 58, PhD(phys chem), 61. *Prof Exp:* Fel, Nat Res Coun Can, 61-63; phys chemist, Res Lab, Gen Elec Co, NY, 63-69; res assoc, Ctr Res Atoms & Molecules, Laval Univ, 69-74; MEM STAFF PHYSICS DIV, NAT RES COUN, CAN, 74- *Mem:* Am Phys Soc; Am Vacuum Soc; Am Soc Mass Spectrometry. *Res:* Mass spectrometry and partial pressure analysis; development of quadrupole field instruments; vacuum techniques; surface physics; secondary ion mass spectrometry of surfaces. *Mailing Add:* Inst Micro-Struct Sci Nat Res Coun Bldg M50 Ottawa ON K1A 0R6 Can

DAWSON, PETER J, b Wolverhampton, Eng, Feb 17, 28. PATHOLOGY. *Educ:* Cambridge Univ, BA, 49, MB, BCh, 52, MA, 53, MD, 60; Univ London, dipl clin path, 55; Am Bd Path, dipl, 68; FRCPath, 75. *Prof Exp:* House physician, Royal Berkshire Hosp, Reading, Eng, 52-53; house surgeon, Victoria Hosp for Children, London, 53; demonstr path, St George's Hosp Med Sch, London, 53-54, registr morbid anat & asst lectr, Postgrad Med Sch, 58-60; vis asst prof path, Sch Med, Univ Calif, San Francisco, 60-62; lectr, Med Sch, Univ Newcastle, 62-64; from assoc prof to prof path, Med Sch, Univ Ore, 64-76, head div surg path & cytol, 74-76; prof path & dir lab surg path, Univ Chicago, 77-89; PROF PATH & LAB MED, UNIV SFLA, 89- *Mem:* Am Asn Cancer Res; Path Soc Gt Brit & Ireland; Royal Col Path; Am Asn Pathologists; Int Acad Path. *Res:* Breast cancer. *Mailing Add:* Dept Path 12901 Bruce B Downs Blvd Tampa FL 33612

DAWSON, PETER SANFORD, b Philadelphia, Pa, Apr 16, 39; c 2. POPULATION BIOLOGY, GENETICS. *Educ:* Wash State Univ, BS, 60; Univ Calif, Berkeley, PhD(genetics), 64. *Prof Exp:* Proj asst, Univ Wis, Madison, 65-66; asst prof zool, Univ Ill, Urbana, 66-69; assoc prof, 69-75, chmn genetics prog, 77-84, PROF ZOOL, ORE STATE UNIV, 75- *Mem:* Genetics Soc Am; Soc Study Evolution. *Res:* Genetic structure of populations; evolution of interspecific competitive ability and cannibalistic behavior; adaptation of populations to heterogeneous and to suboptimal environments; behavioral genetics of flour beetles. *Mailing Add:* Dept Genetics Ore State Univ Corvallis OR 97331

DAWSON, PETER STEPHEN SHEVYN, b Birmingham, Eng, Apr 10, 23; m 57; c 2. MICROBIAL BIOCHEMISTRY, PHYSIOLOGY. *Educ:* Birmingham Univ, Eng, BSc, 44 & 47, MSc, 48, Dipl, 49; Univ Sask, PhD(biochem), 64. *Prof Exp:* Res asst chem, Tame & Rae Drainage Bd, Birmingham, Eng, 44-48; sr sci officer microbiol, Chem Res Lab, Dept Sci & Indust Res, Teddington, Eng, 50-57; sr res officer biotechnol, Prairie Regional Lab, Nat Res Coun Can, 57-88; RETIRED. *Concurrent Pos:* Ed, Microbial Growth, Dowden, Hutchinson & Ross, 74. *Mem:* Assoc Royal Inst Chem; assoc Soc Gen Microbiol. *Res:* Metabolism and physiology of microbes and cells in continuous, asynchronous/synchronous growth; development of rationale, procedures, processes and equipment for fundamental and applied microbiology based upon the cell. *Mailing Add:* 2322 Wiggins Saskatoon SK S7J 1W9 Can

DAWSON, PETER THOMAS, b Billingham, Eng, May 16, 38; m 65; c 2. SURFACE CHEMISTRY. *Educ:* Univ Birmingham, Eng, BSc, 59; Cambridge Univ, PhD(surface chem), 63. *Prof Exp:* Res assoc chem, Ames Lab, Iowa State Univ, AEC, 63-67; from asst prof to assoc prof, 67-83, PROF CHEM, McMASTER UNIV, 83- *Concurrent Pos:* Assoc mem, Inst Mat Res & Dept Metall & Mat Sci, McMaster Univ, 67- & Dept Physics, 69- *Mem:* Royal Soc Chem; Am Vacuum Soc. *Res:* Fundamental studies on the interactions between gases and well-characterized surfaces emphasizing systems of catalytic interest, alloy surfaces and nitride films; ultra-high vacuum techniques used include field emission, mass spectrometry, low energy electron diffraction and electron spectroscopy. *Mailing Add:* Dept Chem McMaster Univ Hamilton ON L8S 4L8 Can

DAWSON, ROBERT LOUIS, b Rochester, NY, Oct 18, 36; m 65. POLYMER CHEMISTRY. *Educ:* Univ Rochester, BS, 58; Harvard Univ, PhD(org chem), 62. *Prof Exp:* RES CHEMIST, E I DU PONT DE NEMOURS & CO, INC, 62- *Mem:* Am Chem Soc. *Res:* Development of new polymers and applications. *Mailing Add:* 1220 Elderon Dr Wilmington DE 19808

DAWSON, STEVEN MICHAEL, b London, Ont, July 12, 62. METALLURGY & PHYSICAL METALLURGICAL ENGINEERING, PHYSICAL CHEMISTRY. *Educ:* McGill Univ, BEng, 85; Univ Toronto, MASc, 87, PhD(metall eng), 90. *Prof Exp:* Teaching asst chem, metall & mat, Univ Toronto, 86-88, res assoc metall res, 86-90; consult metall design, Mountford Assocs, 86-90; sr res engr, 90-91, STAFF ENGR METALL RES, LTV STEEL CO, 91- *Concurrent Pos:* Vchmn, Instrumentation Subcomt Process Technol Div, Iron & Steel Soc, Am Inst Mining, Metall & Petrol Engrs, 91. *Honors & Awards:* Frank B McKune Award, Iron & Steel Soc, Am Inst Mining, Metall & Petrol Engrs, 90. *Mem:* Iron & Steel Soc; Am Inst Mining, Metall & Petrol Engrs. *Res:* Optimization of the steelmaking and continuous casting operations; physical chemistry; process metallurgy; instrumentation; process control. *Mailing Add:* 6801 Brecksville Rd Independence OH 44131

DAWSON, THOMAS LARRY, b Logan, WVa, Nov 7, 34; div; c 2. PHYSICAL & POLYMER CHEMISTRY, ANALYTICAL & ENVIRONMENTAL ANALYTICAL CHEMISTRY. *Educ:* Berea Col, AB, 56; Univ Ky, MS, 58, PhD(phys chem), 60. *Prof Exp:* Chemist, Am Viscose Corp, 56; chemist, 60-70, res scientist, 70-75, GROUP LEADER & TECHNOL MGR, ENG, MFG & TECHNOL SER DIVS, UNION CARBIDE CORP, 75- *Mem:* Am Chem Soc. *Res:* Kinetics and mechanisms of reactions in solutions; reactions on polymers; polymer synthesis; kinetics of free radical polymerization; analytical chemistry. *Mailing Add:* 103 Monroe St South Charleston WV 25303-1604

DAWSON, WALLACE DOUGLAS, JR, b Louisville, Ky, Mar 15, 31; m; c 3. GENETICS, EVOLUTION. *Educ:* Western Ky State Col, BS, 54; Univ Ky, MS, 59; Ohio State Univ, PhD(genetics), 62. *Prof Exp:* From asst prof to prof, Univ SC, 62-77, dept head, 74-77, George Bunch chair prof biol, 77-80; CONSULT, 80- *Mem:* AAAS; Am Genetic Asn; Am Soc Mammal; Soc Study Evolution; Genetics Soc Am; Sigma Xi. *Res:* Developmental genetics and evolution of rodents; speciation and endocrinology of Peromyscus. *Mailing Add:* Dept Biol Univ SC Columbia SC 29208

DAWSON, WILFRED KENNETH, b Quebec, Que, Oct 6, 27; m 54; c 2. NUCLEAR PHYSICS. *Educ:* Laval Univ, BScA, 51; Queen's Univ, Can, PhD(physics), 55. *Prof Exp:* Defense serv sci officer nuclear physics, Defense Res Bd, Can, 55-59; from asst prof to assoc prof physics, 59-70, prof nuclear physics, Univ Alta, 70-; HEAD, TECHNOL DIV, TRIUMF, 82- *Mem:* Am Phys Soc; Can Asn Physicists. *Res:* Fast neutron time-of-flight spectroscopy; nuclear stripping reactions; photonuclear reactions. *Mailing Add:* Dept Physics Univ Alta Edmonton AB T6G 2M7 Can

DAWSON, WILLIAM RYAN, b Los Angeles, Calif, Aug 24, 27; m 50; c 3. COMPARATIVE PHYSIOLOGY. *Educ:* Univ Calif, Los Angeles, PhD(zool), 53; Univ Western Australia, DSc, 71. *Prof Exp:* Asst zool, Univ Calif, Los Angeles, 51-52; USPHS fel, 53; from instr to prof zool, 53-81, D E S BROWN PROF BIOL SCI, UNIV MICH, ANN ARBOR, 81- *Concurrent Pos:* Guggenheim fel, 62-63; res fel, Australian-Am Educ Found, 69-70; mem environ biol panel, NSF, 67-69, adv comt res, 73-77, regulatory biol panel, 79-82. *Honors & Awards:* Henry Russell Award, Univ Mich, 59; Painton Award, Cooper Ornith Soc, 63; Brewster Medal, Am Ornith Soc, 79. *Mem:* Ecol Soc Am; Am Ornith Union; Cooper Ornith Soc; Am Soc Zool (pres, 86); Am Physiol Soc. *Res:* Temperature regulation and water balance of birds and mammals; reptile physiology; avian paleontology. *Mailing Add:* Dept Biol Univ Mich Ann Arbor MI 48109

DAWSON, WILLIAM WOODSON, b Nashville, Tenn, May 21, 33; m 55; c 3. PHYSIOLOGY, BIOPHYSICS. *Educ:* Vanderbilt Univ, BA, 55; Fla State Univ, MS, 57, PhD(psychol), 61. *Prof Exp:* Asst res prof psychol, Auburn Univ, 63-64; Joseph P Kennedy prof, George Peabody Col, 64-65; assoc prof ophthal, 65-69, PROF OPHTHAL & PHYSIOL, COL MED, UNIV FLA, 69-, MEM, CTR NEUROBIOL SCI, 66- *Concurrent Pos:* Assoc investr, Nat Inst Neurol Dis & Blindness res grants, 60-61, prin investr, 66-; Army Med Res & Develop Command, 63-64; Consult, Donner Lab Biophys, Univ Calif, Berkeley, 61 & Stanford Res Inst, 62; mem vision comt, Nat Acad Sci-Nat Res Coun, 68-; dir, Ctr Res on Human Prostheses, 71-; mem adv panel, NSF Neurobiol Prog, 72-75; prin investr, NSF res grants, 72-83, NIH res grants, 74-; res fel Dept Navy, Naval Ocean Systs Ctr, Kailua, Hawaii, 85. *Mem:* fel AAAS; Radiation Res Soc; Am Physiol Soc; Sigma Xi; Int Asn Aquatic Animal Med; Int Soc Clin Electrophysiol Vision; NY Acad Sci; Soc Neurosci. *Res:* Sense receptor physiology, especially cutaneous quality, vision and receptor biology; clinical eye electrophysiology; physiology of vision; control of application of laser speckle; light and retinal pathology; eye disease models. *Mailing Add:* Dept Ophthal Univ Fla Box J 284 JHMHC Gainesville FL 32611

DAX, FRANK ROBERT, b Pittsburgh, Pa, Apr 26,50; m 77; c 2. METALWORKING PROCESSES. *Educ:* Carnegie-Mellon Univ, BS, 72, ME, 77. *Prof Exp:* Engr, Value Eng, 72-75; res metallurgist, Crucible Res Ctr, Colt Indust, 77-82; res metallurgist, Cytemp Specialty Steel, Cyclops Indust, 82-84, mgr tech serv, 84-97, DIR ENG TECHNOL, DYNAMET POWDER PRODS, 87- *Concurrent Pos:* Lectr, Univ WVa, 90. *Honors & Awards:* IR-100 Award, 81. *Mem:* Am Soc Metals Int; Int Soc Productivity Enhancement (vpres, 91-92); Metall Soc; Instrument Soc Am; Soc Mfg Engrs; Am Powder Metall Inst. *Res:* Models of metalworking processes, primarily forging and shape rolling; three dimensional simulations. *Mailing Add:* 600 Mayer St Dynamet Powder Prods Bridgeville PA 15017

DAY, ARDEN DEXTER, b West Rutland, Vt, Mar 16, 22; m 45, 82; c 3. GENERAL AGRICULTURE & AGRONOMY. *Educ:* Cornell Univ, BS, 50; Mich State Univ, PhD(plant breeding), 54. *Prof Exp:* Teaching asst agr, Mich State Univ, 51-53; asst prof agron, Univ Ariz, 54-56, from assoc prof to prof plant sci, 60-88, assoc head, 77-88; RETIRED. *Mem:* Fel Am Soc Agron; fel Crop Sci Soc Am; Am Asn Univ Prof. *Res:* Effective utilization of municipal wastes in commercial agriculture, and of plants in disturbed land reclamation; development of improved plant genotypes for saline environments. *Mailing Add:* Dept Plant Sci Univ Ariz Tucson AZ 85721

DAY, BENJAMIN DOWNING, b Nassawadox, Va, July 19, 36; m 58; c 2. ENGINEERING. *Educ:* Wesleyan Univ, BA, 58; Cornell Univ, PhD(theoret physics), 64. *Prof Exp:* Asst res physicist, Univ Calif, Los Angeles, 63-64, asst prof physics, 64-65; resident res assoc, Argonne Nat Lab, 65-67, asst physicist, 67-69, assoc physicist, 70-79, sr scientist, 80-84; MEM TECH STAFF, AT&T BELL LABS, 84- *Concurrent Pos:* Guggenheim fel, 71. *Mem:* Am Phys Soc. *Res:* Theoretical nuclear physics, especially the nuclear many-body problem. *Mailing Add:* 18 S Park St Hinsdale IL 60521

DAY, BILLY NEIL, b Arthur, WVa, Oct 23, 30; m 53; c 5. ANIMAL SCIENCE. *Educ:* WVa Univ, BS, 52, MS, 54; Iowa State Col, PhD(animal husb), 58. *Prof Exp:* Instr animal husb, Iowa State Col, 57-58; from asst prof to assoc prof, 58-68, PROF ANIMAL SCI, UNIV MO-COLUMBIA, 68- *Honors & Awards:* Animal Physiol & Endocrinol Award, Am Soc Animal Sci, 82. *Mem:* Am Soc Animal Sci; Brit Soc Study Fertil; Soc Study Reproduction. *Res:* Physiology of reproduction in domestic animals. *Mailing Add:* 159 Animal Sci Res Ctr Univ Mo Columbia MO 65211

DAY, CECIL LEROY, b Dexter, Mo, Oct 4, 22; m 48; c 2. AGRICULTURAL ENGINEERING. *Educ:* Univ Mo, BS, 45, MS, 48; Iowa State Univ, PhD(eng), 57. *Prof Exp:* From instr to prof, 45-85, chmn dept, 69-82, EMER PROF AGR ENG, UNIV MO-COLUMBIA, 85- *Mem:* Am Soc Agr Engrs. *Mailing Add:* Dept Agr Eng Univ Mo Columbia MO 65211

DAY, D(ELBERT) E(DWIN), b Avon, Ill, Aug 16, 36; m 56; c 2. CERAMICS, MATERIALS SCIENCE. *Educ:* Mo Sch Mines, BS, 58; Pa State Univ, MS, 60, PhD(ceramic tech), 61. *Prof Exp:* Asst prof ceramic eng, Mo Sch Mines, 61-62; from asst prof to assoc prof ceramic eng, 64-67, dir, Indust Res Ctr, 67-72, PROF CERAMIC ENG, UNIV MO-ROLLA, 67-, SR INVESTR, MAT RES CTR, 74-, CURATOR'S PROF, 81- *Concurrent Pos:* Dir, Mats Res Ctr, Univ Mo-Rolla, 83- *Honors & Awards:* Outstanding Young Ceramic Engr, Nat Inst Ceramic Engrs, 71. *Mem:* Fel Am Ceramic Soc; Am Soc Testing & Mat; Am Soc Eng Educ; Nat Inst Ceramic Engrs; Brit Soc Glass Technol; Sigma Xi; Mats Res Soc. *Res:* Mass transport, anelasticity, structure and electrical properties of vitreous solids; crystal structure and properties of refractory cements; biomaterials. *Mailing Add:* PO Box 357 Rolla MO 65401

DAY, DAVID ALLEN, b Ann Arbor, Mich, Nov 22, 24; m 45; c 5. ENGINEERING. *Educ:* Cornell Univ, BCE, 45; Univ Ill, MS, 51. *Prof Exp:* Field supt, Raymond Concrete Pile Co, 46-47; tech engr, Gen Paving Co, Ill, 47-48; from instr to assoc prof civil eng & in-charge construct option, Univ Ill, 48-58; prof & chmn dept, Univ Denver, 58-60, dean col eng, 60-68, prof civil eng, 68-74; sr staff civil engr, Stearns-Roger, Inc, 74-77, proj civil engr, 77-81, proj engr, 81-85; DIR, URBAN DRAINAGE & FLOOD CONTROL DIST BD, 77- *Concurrent Pos:* Consult, State of Ill, 51-58 & Bridge Off, Ill Div Hwys, 52- 56, construct eng, 85-; mem proj adv comt, Ill Div Hwys & US Bur Pub Rds, 54-58; assoc, Gen Contractors Am, 59-70. *Mem:* Am Soc Civil Engrs; Am Soc Eng Educ; Nat Soc Prof Engrs. *Res:* Construction engineering, structural design problems; economics and planning of construction operations; structural and construction materials. *Mailing Add:* 2758 E Geddes Ave Littleton CO 80122

DAY, DONAL FOREST, b New Brunswick, NJ, Feb 24, 43; m 73; c 2. BIOENGINEERING. *Educ:* Univ NH, BSc, 65; McGill Univ, PhD(microbiol), 73. *Prof Exp:* Fel microbiol, Dept Microbiol & Immunol, McGill Univ, 73-75; asst prof microbiol, Dept Microbiol, Univ Guelph, 75-79; from asst prof, to assoc prof, 79-88, PROF MICROBIOL AUDBON SUGAR INST, LA STATE UNIV, 88- *Mem:* Am Soc Microbiol; Am Chem Soc; Can Soc Microbiologists; NY Acad Sci; Sigma Xi; Am Soc Sugar Cane Technologists. *Res:* Ethanol production by immobilized cells, cellulose degradation; use of enzymes in and the development of new industrial processes based on enzymes or microorganisms; industrial polysaccharides. *Mailing Add:* Audubon Sugar Inst La State Univ Baton Rouge LA 70803

DAY, DONALD LEE, b Leedey, Okla, Aug 14, 31; m 54; c 3. AGRICULTURAL ENGINEERING. *Educ:* Okla State Univ, BS, 54, PhD(agr eng), 62; Univ Mo, MS, 58. *Prof Exp:* Instr agr eng, Tex Tech Col, 57-58; asst, Univ Mo, 58-59 & Okla State Univ, 59-62; from asst prof to assoc prof, 62-71, PROF AGR ENG, UNIV ILL, URBANA, 71- *Concurrent Pos:* Nat Ctr Urban & Indust Health res grant, 66-69; consult exec, UN WHO, Romania, 73-75 & Int Exec Serv Corps, Mex, 78. *Honors & Awards:* Cert of Serv, Am Soc Agr Engrs, 75. *Mem:* Am Soc Agr Engrs; Am Soc Eng Educ. *Res:* Livestock housing and environmental factors; livestock waste disposal; recycling agricultural wastes; air pollution in and around livestock buildings. *Mailing Add:* Dept Agr Eng Univ Ill 1304 W Pennsylvania Ave Urbana IL 61801

DAY, EDGAR WILLIAM, JR, b New Albany, Ind, Sept 7, 36; m 59; c 4. PESTICIDE RESIDUE, ENVIRONMENTAL CHEMISTRY. *Educ:* Univ Notre Dame, BS, 58; Iowa State Univ, PhD(anal chem), 63. *Prof Exp:* Sr anal chemist, Eli Lilly & Co, 63-81, res assoc, 81-85, res adv, 85-90, RES FEL, DOW ELANCO, 91- *Mem:* Am Chem Soc; Soc Environ Toxicol & Chem.

Res: gas chromatography; general organic chemical analysis; pesticide formulation and residue analysis; radiochemistry; organic mass spectrometry; high-performance liquid chromatography; environmental fate and risk analysis associated with pesticides. *Mailing Add:* Dow Elanco PO Box 708 Greenfield IN 46140-0708

DAY, ELBERT JACKSON, b Cullman, Ala, Mar 11, 25; m 48; c 3. POULTRY NUTRITION. *Educ:* Auburn Univ, BS, 52, MS, 53, PhD(poultry nutrit), 56. *Prof Exp:* Asst animal nutritionist, Auburn Univ, 55-56; from asst prof to assoc prof, 56-59, nutrionalist, 56-59, PROF POULTRY SCI, MISS STATE UNIV, 60- *Mem:* Poultry Sci Asn; Am Inst Nutrit. *Res:* Effects of dietary modifications on pigmentation of broilers; proper dietary balance of protein and energy for poultry rations; improvement in performance of poultry rations with feed additives. *Mailing Add:* Dept Poultry Sci Miss State Univ Box 5188 Mississippi State MS 39762

DAY, EMERSON, b Hanover, NH, May 2, 13; m 37; c 5. INTERNAL MEDICINE. *Educ:* Dartmouth Col, AB, 34; Harvard Univ, MD, 38. *Prof Exp:* Intern, Med Serv, Presby Hosp, New York, 38-40; Libman fel med, Johns Hopkins Hosp, 40-42; asst resident med, NY Hosp, 42; med dir int div, TransWorld Airline, New York, 45-47; from asst prof to assoc prof pub health & prev med, Med Col, Cornell Univ, 47-54, prof prev med & chief div, Sloan-Kettering Div, 54-64, head dept prev med, Mem Ctr, 54-63; dir, Strang Clin, 63-69, pres, Prev Med Inst, 66-69; vpres & med dir, Medequip Inc, 69-76 & sr med consult, 76-82; prof med, 76-81, assoc dir, Cancer Ctr, 76-81, EMER PROF MED, MED SCH NORTHWESTERN UNIV, 81- *Concurrent Pos:* Dir, Kips Bay-Yorkville Cancer Detection Ctr, 47-50 & Strang Cancer Prev Clin, 50-63; consult, Off Sci Res & Develop, US Navy, 47-50 & med, 81-; attend physician, Mem Hosp, 50-63; assoc, Sloan-Kettering Inst, 52-54, mem, 54-64; mem advan technol comt, Ill Regional Med Prog; mem bd trustees, Prev Med Inst, 66-; dir, Int Health Eval Asn, 73-; attend physician, Northwestern Mem Hosp, 76-81, sr physician, 81-; physician affil staff, Evanston & Glenbrook Hosp, Evanston, Ill, 76-; med dir, Portes Cancer Prev Ctr, 78-79; med consult serv, 81- *Honors & Awards:* Bronze Medal, Am Cancer Soc, 56; Papanicolaou Award, 78. *Mem:* Fel Am Col Physicians; fel Am Asn Med Systs & Informatics; fel NY Acad Med; fel NY Acad Sci (pres, 65); hon fel Am Soc Cytol (pres, 58). *Res:* Preventive medicine; automated multiphase health testing; medical care systems; cancer detection and prevention; cardiology; aviation physiology; neurology; pain mechanisms. *Mailing Add:* The Portes Ctr 222 N La Salle St No 240 Chicago IL 60601-1005

DAY, EMMETT E(LBERT), b Paris, Tex, July 21, 15; m 37; c 2. MECHANICAL ENGINEERING. *Educ:* ETex State Teachers Col, BA, 36; Mass Inst Technol, BS, 45, MS, 47. *Prof Exp:* Instr, Pub Sch, Tex, 36-40; indust specialist, War Prod Bd, 40-42; instrument specialist, San Antonio Arsenal, 42-43; asst instr mech eng, Mass Inst Technol, 44-46, instr, 46-47; form asst prof to prof mech eng, Univ Wash, 47-54, prof 54-85, assoc chmn Mech Eng Dept, 69-84; RETIRED. *Concurrent Pos:* Ed, Mach Design Bull, Am Soc Eng Educ, 54-57. *Mem:* Soc Exp Stress Anal (pres, 75); fel Am Soc Mech Engrs (vpres, 62-66); Am Soc Eng Educ; fel Soc Exp Mech. *Res:* Experimental stress analysis; materials of engineering. *Mailing Add:* Dept Mech Eng Univ Wash Seattle WA 98195

DAY, EUGENE DAVIS, b Cobleskill, NY, June 24, 25; m 46; c 1. IMMUNOLOGY. *Educ:* Union Univ, NY, BS, 49; Univ Del, MS, 50, PhD(biochem), 52. *Prof Exp:* Res assoc, Roscoe B Jackson Mem Lab, Maine, 52-54; sr cancer res biochemist, Roswell Park Mem Inst, 54-58, assoc cancer res biochemist, 58-62, asst res prof chem, Roswell Park Div, Sch Grad Studies, Buffalo, 57-62; assoc prof, 62-65, prof, 65-88, prof exp surg, 77-88, EMER PROF IMMUNOL, DUKE UNIV, 88-; IMMUNOCHEM CONSULT, 88- *Concurrent Pos:* Javits neurosci investr, NIH, 84- *Mem:* Am Soc Neurochem; Am Asn Immunol; Sigma Xi; AAAS. *Res:* Cancer immunology; immunochemistry; radio immunoassays; subcellular fractionations; neuroimmunology. *Mailing Add:* 60 Huntleigh Dr Loudonville NY 12211-1174

DAY, FRANK PATTERSON, JR, b Bristol, Va, July 12, 47; m 69; c 3. ECOLOGY. *Educ:* Univ Tenn, BS, 69; Univ Ga, MS, 71, PhD(ecol), 74. *Prof Exp:* Res asst, Int Biol Prog, Univ Ga, 70-74; from asst prof to assoc prof, 74-86, PROF BIOL SCI, OLD DOMINION UNIV, 86- *Mem:* Am Inst Biol Sci; Ecol Soc Am; Bot Soc Am; Torrey Bot Club. *Res:* Ecosystem dynamics on Virginia barrier islands; root production, decomposition, and nutrient cycling. *Mailing Add:* Dept Biol Sci Old Dominion Univ Norfolk VA 23529

DAY, HAROLD J, b Milwaukee, Wis, May 22, 29; m 53; c 2. CIVIL ENGINEERING. *Educ:* Univ Wis, BS, 52, MS, 53, PhD(civil eng), 63. *Prof Exp:* Res asst, Hydraul & Sanit Lab, Univ Wis, 52-53, 62-63, instr civil eng, 59-60; proj engr, Scott Paper Co, 53-59; from asst prof to assoc prof civil eng, Carnegie-Mellon Univ, 63-70; chmn dept, 70-76, PROF ENVIRON CONTROL, UNIV WIS-GREEN BAY, 70- *Mem:* Am Geophys Union; Am Soc Civil Engrs; Sigma Xi. *Res:* Fluid mechanics; hydrology; water resources; hydraulics. *Mailing Add:* 2377 S Webster Green Bay WI 54301

DAY, HARRY GILBERT, b Monroe Co, Iowa, Oct 8, 06; m 33, 69; c 3. NUTRITIONAL BIOCHEMISTRY. *Educ:* Cornell Col, AB, 30; Johns Hopkins Univ, ScD(biochem), 33. *Hon Degrees:* ScD, Cornell Col, 67. *Prof Exp:* Nat Res fel, Johns Hopkins Univ, 33-34; Gen Educ Bd fel, Yale Univ, 34-36; assoc biochem, Sch Hyg & Pub Health, Johns Hopkins Univ, 36-40; from asst prof to prof, 40-50, chmn dept chem, 52-62, assoc dean res & advan studies, 67-72, prof chem, 50-76, EMER PROF CHEM, IND UNIV, BLOOMINGTON, 76- *Concurrent Pos:* Mem select comt on generally regarded as safe substances, Fedn Am Soc Exp Biol, Bethesda, Md, 73-83. *Mem:* Fel AAAS; Am Chem Soc; fel Am Inst Nutrit (pres, 71). *Res:* Evaluation of health aspects of food ingredients; history of nutrition. *Mailing Add:* Dept Chem Ind Univ Bloomington IN 47405

DAY, HARVEY JAMES, b Souderton, Pa, Mar 2, 29; m 51; c 3. INTERNAL MEDICINE, HEMATOLOGY. *Educ:* Villanova Univ, BS, 49; Hahnemann Med Col, MD, 53; Am Bd Internal Med, dipl. *Prof Exp:* Intern, Abington Mem Hosp, Pa, 53-54, resident med, 54-56; resident, Ohio State Univ Hosp, 56-57, instr, 57-58; instr hemat, 58-60, assoc med, 61-63, assoc prof, 63-64, PROF MED, SCH MED, TEMPLE UNIV, 66-, DIR HEMAT, 61- *Concurrent Pos:* Consult, Abington Mem Hosp, 59-, chief hemat, 65; res assoc, Inst Thrombosis Res, Riks Hosp, Norway, 66-68; Fulbright res scholar, Univ Oslo, 66-68. *Mem:* Assoc Am Col Path; fel Am Col Physicians; Am Soc Hemat; Am Fedn Clin Res; Int Soc Hemat. *Res:* Platelets in thrombosis; blood platelets. *Mailing Add:* 3401 N Broad St Philadelphia PA 19140

DAY, HERMAN O'NEAL, JR, b Dallas, Tex, Dec 4, 25; m 52; c 4. PHYSICAL CHEMISTRY. *Educ:* ETex State Univ, BS, 45; Univ Tex, AM, 48, PhD(phys chem), 51. *Prof Exp:* Asst defense res lab, Univ Tex, 48-50; chemist, Oak Ridge Nat Lab, 51-56; chemist, Plant Tech Sect, Chambers Works, Org Chem Dept, E I Du Pont De Nemours & Co, Inc, 56-60, Jackson Lab, 60-63, Process Dept, Chambers Works, Org Chem Dept, 63-69, sr process chemist, 69-70, sr chemist, Petrol Chem Div Tech Sect, Chambers Works, Chem & Pigments Dept, 70-84; RETIRED. *Mem:* Am Chem Soc; Sigma Xi. *Res:* The pressure-volume-temperature relationships of gases and liquids; solution chemistry of the heavy elements; radiation corrosion in homogeneous nuclear reactors; manufacture of tetraalkyl lead. *Mailing Add:* 108 Dakota Ave Shipley Heights Wilmington DE 19803

DAY, IVANA PODVALOVA, b Prague, Czech, Oct 29, 32; m 73. PHARMACOLOGY, BIOLOGY. *Educ:* Charles Univ, Prague, MSc, 64, PhD(pharmacol), 70. *Prof Exp:* Sr pharmacologist, Res Inst Pharm & Biochem, Prague, Czech, 64-70; fel, Inst Mario Negri, Milan, Italy, 71; fel, Univ Minn, 71-73; SR PHARMACOLOGIST, CNS DIS THER SECT, MED RES DIV, AM CYANAMID CO, 73- *Mem:* Am Chem Soc; Am Soc Pharmacol & Exp Therapeut; Col Int Neuro-Psychopharmacol. *Res:* Pharmacology of the central nervous system. *Mailing Add:* CNS Dis Ther Sect Med Res Div Am Cyanamid Co Pearl River NY 10965

DAY, JACK CALVIN, b Stamford, Tex, June 17, 36; m 68. ORGANIC CHEMISTRY. *Educ:* Univ Calif, Berkeley, AB, 61; Univ Calif, Los Angeles, PhD(org chem), 67. *Prof Exp:* Res assoc chem, Columbia Univ, 67-68; ASST PROF CHEM, HUNTER COL, 68- *Mem:* Chem Soc London; Am Chem Soc; Sigma Xi. *Res:* Synthesis and study of new severely hindered organic bases; synthesis of a noncoordinating buffer system; control of orientation in elimination reactions. *Mailing Add:* 30 Fifth Ave New York NY 10011

DAY, JAMES MEIKLE, b Wickham, NB, Oct 20, 24; US citizen; m 51. ENVIRONMENTAL CHEMISTRY. *Educ:* Univ NH, BS, 45; Univ Ill, MS, 52; Univ Ark, PhD(inorg & nuclear chem), 54. *Prof Exp:* Chem engr, E I du Pont de Nemours & Co, NY, 45-46; chemist, Burgess Battery Co, Ill, 46-48; chem engr, Anderson Phys Lab, 48-51; res assoc, Univ Wis, 53-54; radiation chemist, Phillips Petrol Co, Okla, 54-56, nuclear chemist, Idaho, 56-58, reactor core physicist, 58-59; sr res chemist, Indust Reactor Lab, NJ for Am Tobacco Co, Va, 59-62; mgr chem & physics res, Whirlpool Corp, Mich, 62-70, staff chemist, 70-74; sr engr, Northeast Utilities Serv Co, 75-90; RETIRED. *Mem:* Am Chem Soc; Sigma Xi. *Res:* Water chemistry; water treatment, chlorination and other disinfection methods; pollution control. *Mailing Add:* PO Box 57 Bentonville AR 72712

DAY, JANE MAXWELL, b Avon Park, Fla, Mar 12, 37; m 58; c 2. TOPOLOGY. *Educ:* Univ Fla, BA, 58, MS, 61, PhD(math), 64. *Prof Exp:* Asst prof math, Univ Fla, 64-66; Am Asn Univ Women fel, Inst Advan Study, 66-67; from asst prof to assoc prof math, Col Notre Dame, 67-76, chmn dept, 69-71 & 74-75, prof, 76-82; ASSOC PROF MATH, SAN JOSE STATE UNIV, 82- *Concurrent Pos:* Lectr, Women & Math Lectureship Prog; sect officer, Math Asn Am, 75-78. *Mem:* Am Math Soc; Math Asn Am; Asn Comput Mach; Nat Coun Teachers Math; Am Asn Univ Professors. *Res:* Topological algebra, including semigroups and acts with periodic properties, and nonassociative structures. *Mailing Add:* Dept Math & Comput Sci San Jose State Univ San Jose CA 95192

DAY, JESSE HAROLD, b Bend, Ore, Oct 27, 16; m 38. PHYSICAL CHEMISTRY. *Educ:* Reed Col, BA, 42; Case Sch Appl Sci, MS, 45, Case Inst Technol, PhD(phys chem), 48. *Prof Exp:* Instr chem, Case Inst Technol, 46-48; from asst prof to assoc prof, Ohio Univ, 48-58, chmn dept, 58-63, actg dean, 68-69 & 71-72, prof chem, 63-87, assoc dean Col Arts & Sci, 67-68 & 69-87; RETIRED. *Concurrent Pos:* Res chemist, Rubber Reserve Bd, 44-45 & Glenn L Martin Co, 46-47; vis prof, Univ Idaho, 64; ed, J Soc Plastic Eng. *Mem:* Fel Am Inst Chem; Am Chem Soc; Soc Plastics Eng; fel NY Acad Sci. *Res:* Thermochromism; vinyl polymerization; chemistry of fulvenes. *Mailing Add:* 61 Briarwood Dr Athens OH 45701

DAY, LAWRENCE EUGENE, b Findlay, Ohio, Feb 12, 33; m 57; c 3. MICROBIOLOGY. *Educ:* Miami Univ, Ohio, AB, 55; Mich State Univ, MS, 60, PhD(microbiol), 63. *Prof Exp:* Res microbiologist, Med Res Lab, Chas Pfizer & Co, Inc, Conn, 63-66; res microbiologist antibiotic develop, 66-70, new fermentation prod team leader, 70-73, mgr antibiotic cult develop, 73-82, HEAD, FERMENTATION PROD RES, ELI LILLY & CO, 82- *Concurrent Pos:* Prin investr, Eli Lilly & Co. *Mem:* Am Soc Microbiol. *Res:* Genetics of antibiotic biosynthesis; scale-up and manufacture of human insulin by recombinant DNA technology. *Mailing Add:* 7823 Traders Cove Lane Indianapolis IN 46254

DAY, LEROY E(DWARD), b Doswell, Va, Jan 2, 25; m 47; c 3. AERONAUTICAL ENGINEERING. *Educ:* Ga Inst Technol, BAeroEng, 46; Univ Calif, Los Angeles, MS, 55; Mass Inst Technol, MS, 60. *Prof Exp:* Test engr guided missile, US Naval Missile Ctr, 48-51, head controls br develop & testing guid systs, 51-56, head guid div, 56-59, head inertial guid div, develop res testing guid systs missiles, 59-60, dep head, Missile Progs Dept, 60-62; chief, Proj Gemini, NASA, 62-63, dir, Gemini Test Prog, 63-66, dir, Apollo Test, 66-69, mgr space shuttle task group, 69-71, dep dir space shuttle prog, 71-80, dir systs eng & integration, Manned Space Flight, 80-81; CONSULT, 59-, AEROSPACE MGT, US & FOREIGN FIRMS, 81- *Concurrent Pos:* Lectr, Univ Calif, Los Angeles, 59; Sloan fel, Mass Inst Technol, 59-60. *Honors & Awards:* Distinguished Serv Medal, NASA, 81. *Mem:* Nat Space Club; Am Astronaut Soc; Int Acad Astronaut,. *Res:* Automatic control and guidance of missiles and space vehicles; manned space flight. *Mailing Add:* 11709 Magruder Lane Rockville MD 20852

DAY, LEWIS RODMAN, fisheries management; deceased, see previous edition for last biography

DAY, LOREN A, b Evanston, Ill, Oct 9, 36; m 65; c 2. PHYSICAL CHEMISTRY. *Educ:* Oberlin Col, BA, 58; Yale Univ, PhD(chem), 63. *Prof Exp:* Fel molecular biol div, Max Planck Inst Virus Res, 64-68; assoc, Dept Biochem, 69-75, assoc mem, 76-80, MEM & HEAD DEVELOP & STRUCT BIOL, PUB HEALTH RES INST, NY, 81- *Concurrent Pos:* NIH fel, 64-67; Res Career Develop Award, NIH, 72-77; res prof biochem, Sch Med, NY Univ, 76-; mem, Biophysics & Biophys Chem Study Sect, NIH, 78-82. *Mem:* Biophys Soc; Am Soc Biol Chemists; Am Soc Microbiol; Am Chem Soc. *Res:* Physical chemistry of macromolecules, the structures of viruses and viral nucleic acids, and protein-nucleic acid interactions. *Mailing Add:* Pub Health Res Inst 455 First Ave New York NY 10016

DAY, MAHLON MARSH, b Rockford, Ill, Nov 24, 13; m 39, 52; c 5. MATHEMATICS. *Educ:* Oregon State Col, BS, 35; Brown Univ, ScM, 38, PhD(math), 39. *Prof Exp:* Keene fel, Inst Adv Study, 39-40; from instr to prof math, 40-83, head dept, 58-65, EMER PROF MATH, UNIV ILL, URBANA, 83- *Concurrent Pos:* Res assoc, Brown Univ, 44-46; NSF sr res fel, 56-57. *Mem:* Fel AAAS; Am Math Soc; Math Asn Am. *Res:* Linear spaces; ordered systems; geometry of normed spaces; amenable semigroups. *Mailing Add:* Dept Math Univ Ill 1409 W Green St Urbana IL 61801

DAY, MARION CLYDE, JR, b Malvern, Ark, Aug 7, 27; m 50; c 3. INORGANIC CHEMISTRY, PHYSICAL CHEMISTRY. *Educ:* San Jose State Col, AB, 50; Iowa State Univ, PhD(chem), 55. *Prof Exp:* From asst prof to assoc prof, 55-69 prof, 69-, EMER PROF INORG CHEM, LA STATE UNIV. *Mem:* Am Chem Soc. *Res:* Spectroscopic and conductance studies of ion-solvent and ion-ion interactions in solvents of low dielectric constant; alkali metal aluminum alkyls in non-polar and mixed solvent systems. *Mailing Add:* 724 Druid Circle Baton Rouge LA 70808

DAY, MICHAEL HARDY, b St Louis, Mo, April 4, 50; m 72; c 1. ATOMIC PHYSICS, MOLECULAR PHYSICS. *Educ:* Univ Mo, Columbia, BS, 72; Univ Wis, Madison, PhD(physics), 77. *Prof Exp:* Instr physics, Univ Wis, 77-78; res fel, Univ Sussex, UK Atomic Energy Authority, 78-80; asst prof physics, Kans State Univ, 80; AMERITECH SERVS. *Mem:* Am Phys Soc. *Res:* Theoretical studies of ionization and charge exchange in ion-atom collisions; interaction of fast ions with solids. *Mailing Add:* Ameritech Servs Bldg 40 Gould Ctr 2850 Golf Rd Rolling Meadows IL 60008

DAY, NOORBIBI KASSAM, US citizen; div; c 2. IMMUNOBIOLOGY. *Educ:* Trinity Col, Dublin, BA, 56; McGill Univ, PhD, 67. *Prof Exp:* Fel pediat, Univ Minn, 67-71, asst prof path, 71-73; assoc prof biol, Cornell Univ, 73-77; lab head, Mem Sloan-Kettering Cancer Ctr, 73-77, assoc mem, 73-79; prof biol, Cornell Univ Grad Sch Med Sci, 79-82; mem, Okla Med Res Found, 82-; PROF, DEPT MICROBIOL, ALL CHILDREN'S HOSP, UNIV SFLA, ST PETERSBURG. *Concurrent Pos:* NIH spec fel, 70-71; vis scientist molecular path, Scripps Clin, Calif, 71; rheumatoid arthritis fel, 71-73; estab investr, Am Heart Asn, 72-77. *Mem:* Can Soc Immunol; Can Soc Microbiol; Am Soc Microbiol; Am Asn Immunol; Am Soc Exp Path. *Res:* Perturbations of the Complement system in human diseases; isolated deficiencies of the Complement system in man and experimental animals; development of C-phylogenetic and ontogenetic perspectives; effector biology; mouse mammory tumor virus and human breast cancer; feline leukemia and treatment of. *Mailing Add:* Dept Pediat Univ SFla 801 Sixth St S St Petersburg FL 33701

DAY, PAUL PALMER, computer sciences, physics, for more information see previous edition

DAY, PETER RODNEY, b Chingford, Essex, Eng, Dec 27, 28; nat US citizen; m 51; c 3. PLANT BREEDING, FUNGAL GENETICS. *Educ:* Univ London, BS, 50, PhD(botany-genetics), 50. *Prof Exp:* sr scientific officer, John Innes Inst, Eng, 46-63; assoc prof genetics, Ohio State Univ, 63-64; dept head, Conn Agr Experiment Station, 64-79; dir, Plant Breeding Inst, Eng, 79-87; DIR, CTR AGR MOLECULAR BIOL, RUTGERS UNIV, 87- *Concurrent Pos:* Mem, Comt Genetic Vulnerability Major Crops, Nat Res Coun, 70-72, chmn, Comt Global Genetic Resources, 86-; mem, Recombinant DNA Adv Comt, NIH, 76-79. *Mem:* Brit Soc Plant Pathol (pres, 86); Am Phytopathol Soc; Inst Biol UK; Genetic Soc; Int Genetics Fedn (secy, 83-); Nat Inst Agr Botany (coun mem, 80-87). *Res:* Interested in molecular biology and genetics and their application to crop plant and livestock improvement and to solving environmental problems. *Mailing Add:* Ctr Agr Molecular Biol Cook Col Rutgers Univ Box 231 New Brunswick NJ 08903

DAY, REUBEN ALEXANDER, JR, b Atlanta, Ga, Feb 3, 15; m 43; c 4. ANALYTICAL CHEMISTRY. *Educ:* Emory Univ, AB, 36, MS, 37; Princeton Univ, PhD(chem), 40. *Prof Exp:* From instr to prof, 40-81, chmn dept, 57-68, EMER PROF CHEM, EMORY UNIV, 81- *Concurrent Pos:* Res assoc, Metall Lab, Univ Chicago, 43; chemist, Clinton Labs, Oak Ridge, Tenn, 43-44; sr chemist, Oak Ridge Nat Lab, Tenn, 48. *Mem:* Emer Am Chem Soc. *Mailing Add:* 1189 Houston Mill Rd NE Atlanta GA 30329

DAY, RICHARD ALLEN, b Kellogg, Iowa, Apr 4, 31; m 56; c 2. BIOCHEMISTRY. *Educ:* Iowa State Univ, BS, 53; Mass Inst Technol, PhD(organic chem), 58. *Prof Exp:* Damon Runyon Mem Fund grant biol, Mass Inst Technol, 57-59; from asst prof to assoc prof chem, 59-68, from asst prof to assoc prof biol chem, Col Med, 60-72, PROF CHEM, UNIV CINCINNATI, 68-, PROF BIOL CHEM, COL MED, 72- *Concurrent Pos:* NIH career develop award, 69-74. *Mem:* Am Chem Soc; Am Soc Biol Chem; AAAS; Sigma Xi. *Res:* Protein structure by gas chromatography and mass spectrometry; affinity labels of active sites; interactions of ionens with DNA and chromatin. *Mailing Add:* 7415 Silver Creek Rd Cleves OH 45002

DAY, ROBERT J(AMES), b Newark, NJ, Feb 2, 10; wid; c 1. PHYSICAL CHEMISTRY, FUEL TECHNOLOGY. *Educ:* Union Univ, NY, BS, 33; Pa State Univ, PhD(fuel tech), 49. *Prof Exp:* Instr petrol prod, Petrol & Nat Gas Eng Dept, Pa State Univ, 38-47, consol fel, 47-49; res engr, Consolidation Coal Co, 49-52; res dir chem process & fuel utilization, Philadelphia & Reading Corp, 52-56; chem process develop analyst, Planning Dept, Tex Power & Light Co, 56-58; tech dir water & oil well conditioning, United Chem Corp NMex, 58-59; chief chemist, Chem Eng Co, Inc, Tex, 60; staff engr, Armament Div, Universal Match Corp, 60-61; sr engr-scientist, Missile & Space Systs Div & Aircraft Div, McDonnell Douglas Corp, 61-70; CONSULT, 70- *Concurrent Pos:* Mem res adv comt, Anthracite Inst & Anthracite Res Adv Comt, Pa State Univ, 52-56; mem, Gov Fuel Res Adv Comt, 55. *Mem:* Am Chem Soc; Am Inst Mining, Metall & Petrol Engrs; Am Inst Aeronaut & Astronaut. *Res:* Chemical process development; propellants development; propulsion systems design; reaction kinetics; hydrogenation; gasification; fuels processing and utilization; carbon and graphite properties and production; petroleum production; water conditioning; surfactants; sintering; calcining. *Mailing Add:* Apt 26 3030 Merrill Dr Torrance CA 90503

DAY, ROBERT JAMES, b Los Angeles, Calif, Feb 7, 41. ANALYTICAL CHEMISTRY, ELECTROCHEMISTRY. *Educ:* Univ Of Calif, Riverside, BA, 62; Univ NC, PhD(anal chem), 66. *Prof Exp:* Res chemist, Org Chem Dept, E I du Pont de Nemours & Co, 66-67; mem staff, TRW Systs Group, 67-80, PROJ SCIENTIST, TRW SPACE & TECHNOL GROUP, 80- *Mem:* AAAS; Am Chem Soc. *Res:* Magnetic resonance spectroscopy; electroanalytical chemistry; gas chromatography; electrobiochemistry; exobiology. *Mailing Add:* 22 Pony Lane Rolling Hills CA 90274

DAY, ROBERT WILLIAM, b Worcester, Mass, Feb 7, 24; m 45; c 2. THERMODYNAMICS. *Educ:* Univ Mass, BS, 48; Rensselaer Polytech Inst, MME, 54. *Prof Exp:* Instr mech eng, Rensselaer Polytech Inst, 48-54; from asst prof to assoc prof, 54-69, prof mech eng, 69-81, PROF & ASSOC DEPT HEAD, MECH ENG, UNIV MASS, AMHERST, 81- *Concurrent Pos:* Engr, Gen Elec Co, NY, 50-52; engr, Boeing Airplane Co, 54; consult, Kollmorgan Optical Co, 57; Western Elec Fund Award, 70. *Mem:* Am Soc Eng Educ; Am Soc Mech Engrs; Sigma Xi; Aerospace Indust Asn Am. *Res:* Application of thermodynamics and heat transfer to aircraft and space vehicles. *Mailing Add:* 265 E Plaeasant St Amherst MA 01002

DAY, ROBERT WINSOR, b Framingham, Mass, Oct 22, 30; c 2. EPIDEMIOLOGY. *Educ:* Univ Chicago, MD, 56; Univ Calif, Berkeley, MPH, 58, PhD(epidemiol), 62. *Prof Exp:* Trainee epidemiol, Univ Calif, Berkeley, 57-60; res specialist ment retardation, Sonoma State Hosp, Calif, 60-62; asst prof prev med, Univ Calif, Los Angeles, 62-64; chief hereditary defects unit, Calif State Dept Pub Health, 64-65, chief bur maternal & child health, 65-66, chief dept dir, 66-67; assoc prof prev med & dir div health serv, 68-70, chmn dept health serv, 70-72, PROF HEALTH SERV, UNIV WASH, 70-, DEAN, SCH PUB HEALTH & COMMUNITY MED, 72-; DIR, FRED HUTCHINSON CANCER RES CTR, SEATTLE, 81- *Concurrent Pos:* Res staff, Pac State Hosp, Pomona, Calif, 62-64; consult, Porterville State Hosp, 62-63 & State Dept Pub Health, Berkeley, 63-64; lectr, Sch Pub Health, Univ Calif, Berkeley, 64-67; assoc clin prof, Univ Calif, San Francisco, 66-67; vis assoc prof, Univ Mich, 68. *Mem:* Am Soc Human Genetics; Am Pub Health Asn; Soc Pediat Res; Am Epidemiol Soc. *Res:* Genetics and epidemiology; population; health services and medical care. *Mailing Add:* 1124 Columbia St Seattle WA 98104

DAY, STACEY BISWAS, b London, Eng, Dec 31, 27; US citizen; m 52, 73; c 2. INTERNATIONAL HEALTH. *Educ:* Royal Col Surgeons, Ireland, MD, 55; McGill Univ, PhD(exp surg), 64; Univ Cincinnati, DSc(surg), 70. *Prof Exp:* Res fel surg & physiol, Univ Minn, 56-60; clin asst med, St George's Hosp, London, Eng, 60-61; demonstr & prosector anat, McGill Univ, 61-62, lectr surg, 64-66; clin investr, Hoechst Pharmaceut, Inc, Ohio, 66-68; regional med dir for New Eng, Hoffmann-La Roche Inc, NJ, 68-69; asst prof res surg, Col Med, Univ Cincinnati, 69-71; from asst prof to assoc prof path & conservator, Bell Mus Path, Univ Minn, Minneapolis, 71-73; prof biol sci, Grad Sch Med Sci, Cornell Univ, 73-80; mem & dir dept biomed commun & med educ, Sloan-Kettering Inst Cancer Res, 73-80; prof biopsychosocial med, prof & chmn, Dept Community Med, Col Med Sci, Calabar Univ, Nigeria, 82-85; adj prof, Family Med & Community Health, Col Med Sci, Ariz Univ, Tucson, 85-88; prof & dir, Int Ctr Health Sci, 85-89, dir & founding prof, 87-89, EMER DIR, WHO COLLABORATING CTR COMMUNITY BASED & MULTIPROF EDUC HEALTH PERSONNEL, MEHARRY MED COL, NASHVILLE, 90-; CLIN PROF, DIV BEHAV MED, NY MED COL, 80- *Concurrent Pos:* Asst prof, NJ Col Med, 68-69; assoc dir basic med res, Shriners Hosp, Burns Inst, Cincinnati, 69-71; mem, Inst, Admin Coun & Coun Field Coordr, Sloan-Kettering Inst Cancer Res; mem, World Priorities Pop Comt; consult, Pan-Am Health Orgn; exchange scientist to Soviet Union, US-USSR Agreement for Health, 76; vpres res & sci affairs, Mario Negri Found, NY, 74-80; vpres int health affairs, Am Rural Health Asn, 77-80; pres, Int Found Biosocial Develop & Human Health, 77-; founding officer & vpres, Am Inst Stress, 79; Consult, dict sci biog, Health Commun, Chile, Inst Creative Health, Lyford Cay; adv dean med col, consult & planning group curric develop, fac med & health sci, ABHA, Prov of Asir, Saudi Arabia, 81; consult adv, Health Serv Mgt Bd & dir, SAMA Found, Nigeria, 82-84; consult, Community Health Care, Sage Mem Hosp, Navajo Nation Reservation, Ganado, AZ, 84; hon chmn, Am Friends Lambo Found, 84-90; consult, Expert Comt, Div of Strengthening Health Resources, WHO, 85-90; consult & vis prof int health, Meharry Med Col, Nashville, 85; ed-in-chief, Biosci Commun & J Health Commun & Biopsychosocial Health; liaison officer, Nat Asn Equal Opportunity Higher Educ/Agency Int Develop, 85-90; mem bd, African Health Consult Serv, Nigeria, 86-; mem, Expert Comt Health Manpower Develop, WHO, Geneva, 86-90; Fulbright prof, Postgrad Med Col, Charles Univ, Prague, 89-90; vis prof, Kyushu Univ, Japan, 90-91. *Honors & Awards:* Reuben Harvey Triennial Prize & Medal, Royal Col Physicians, Ireland, 57; Moynihan Prize Medal, Royal Col Surgeons, Eng, 60; Arris & Gale Award, Royal Col Surgeons, 72; WHO Medal, 87; Citation Cong Record Int Health, 87. *Mem:* Fel Zool Soc London; fel Royal Micros Soc; Asn Surgeons Gt Brit & Ireland; Harvey Soc; Int hon fel Japanese Found Biopsychosocial Health; fel World Acad Arts & Sci; fel Lambo Found. *Res:* Biopsychosocial medicine with focus on primary health care; biosocial development and integration of interdisciplinary modalities centered upon human behavior; international health; theoretical foundations of information based knowledge. *Mailing Add:* Dept Med NY Med Col Valhalla NY 10595

DAY, STEPHEN MARTIN, b New York, NY, Dec 17, 31; m; c 6. SOLID STATE PHYSICS. *Educ:* La State Univ, BS, 57; Rice Univ, MA, 59, PhD(physics), 61. *Prof Exp:* From asst prof to assoc prof, Univ Ark, Fayetteville, 61-71, chmn dept, 69-75, prof, 71-83, assoc dean, Col Arts & Sci, 79-83; DEAN, ARTS & SCI, MIAMI UNIV, OXFORD, OHIO, 83- *Mem:* Am Phys Soc; Sigma Xi. *Res:* Low temperature properties of solids; magnetic resonance phenomena of solids, and superionic properties of solids. *Mailing Add:* Dept Physics Miami Univ 121 Culler Hall Oxford OH 45056

DAY, THOMAS BRENNOCK, b New York, NY, Mar 7, 32; m 53; c 9. HIGH ENERGY PHYSICS. *Educ:* Univ Notre Dame, BS, 52; Cornell Univ, PhD(physics), 57. *Prof Exp:* Res assoc, Univ Md, College Park, 57-58, from asst prof to assoc prof, 58-64, prof physics, 64, vchancellor acad planning & policy, 70; PRES, SAN DIEGO STATE UNIV, 78-, PROF PHYSICS, 78- *Concurrent Pos:* Engr, Bendix Aviation Corp, 52-53; consult, US Govt, 58- *Mem:* Am Phys Soc. *Res:* Theoretical investigations in elementary particle physics and quantum mechanics; experimental work in high energy physics. *Mailing Add:* San Diego State Univ San Diego CA 92182

DAY, WALTER R, JR, b Fairfield, Ala, Aug 12, 31; m 58; c 2. ELECTRICAL ENGINEERING. *Educ:* Auburn Univ, BSEE, 53; Ga Inst Technol, MSEE, 57. *Prof Exp:* Sr engr, Sperry Electronic Tube Div, 56-66; mem tech staff, Plasma Physics Lab, Princeton Univ, 66-67; sr scientist, Electron Tube Div, Litton Industs, Inc, 67-78; eng mgr, Solid State Western Div, Varian Assocs, 78-90; DIR MICROWAVE COMPONENTS, FARINON DIV, HARRIS CORP, 90- *Mem:* Inst Elec & Electronics Engrs. *Res:* Microwave electron devices including linear beam oscillators and amplifiers; electron beams and focusing systems; solid-state microwave devices and materials. *Mailing Add:* 1288 Belair Way Menlo Park CA 94025

DAY, WILLIAM H, b Wilmington, Del, Sept 29, 34; m 59; c 1. ECONOMIC ENTOMOLOGY. *Educ:* Univ Del, BS, 55; Cornell Univ, PhD(entom), 65. *Prof Exp:* Res asst entom, Cornell Univ, 55-61; res entomologist, 65-71, res leader, 71-78, RES ENTOMOLOGIST, PARASITE RES LAB, AGR RES SERV, USDA, 78- *Mem:* Entom Soc Am; Am Entom Soc (pres, 71-72); Int Orgn Biol Control. *Res:* Applied biological control of insect pests; population dynamics of insect parasites, predators and pest species; insect pests of forage and vegetable crops; reproductive modes of insect parasites; insect sampling and plot design. *Mailing Add:* Parasite Res Lab Agr Res Serv USDA 501 S Chapel St Newark DE 19713-3814

DAYAL, RAMESH, b New Delhi, India, July 15, 42, Can & US citizen; m 72; c 2. NUCLEAR WASTE MANAGEMENT. *Educ:* Univ Stuttgart, Vordiplom, 66, Diplom, 68; Dalhousie Univ, PhD(geochem), 74. *Prof Exp:* Asst prof marine geochem, Marine Sci Res Ctr, State Univ NY Stony Brook, 74-79; group leader, Low Level Waste Res, Dept Nuclear Energy, Brookhaven Nat Lab, 79-85; CONSULT, 85- *Concurrent Pos:* Adj assoc prof, Marine Sci Res Ctr, State Univ NY, Stony Brook, 79-83; proj coordr, Nuclear Regulatory Comn, 80-85. *Mem:* Int Asn Cosmochem & Geochem; Am Geophys Union; Geochem & Meteorol Soc; Mat Res Soc. *Res:* Geochemical aspects of nuclear waste disposal; silica geochemistry; geochemical C-14 isolation research; cement-based materials in radioactive waste management; natural analogue studies. *Mailing Add:* Civil Res Dept Ont Hydro 800 Kipling Ave Toronto ON M8Z 5S4 Can

DAYAL, YOGESHWAR, b New Delhi, India, Aug 30, 39; m 64; c 3. GASTROINTESTINAL PATHOLOGY. *Educ:* India Inst Med Sci, MS, 61, MD, 67; Am Bd Path, dipl & cert anat path, 73. *Prof Exp:* From asst prof to clin assoc prof, 72-87, CLIN PROF PATH, TUFTS UNIV SCH MED, 87- *Concurrent Pos:* From asst pathologist to sr pathologist, New England Med Ctr Hosp, 71- *Mem:* Int Acad Path; Am Gastroenterology Asn. *Res:* Endocrine tumors and hyperplasias. *Mailing Add:* Dept Path New England Med Ctr 750 Washington St Boston MA 02111

DAYAN, JASON EDWARD, b Newburgh, NY, Aug 6, 23; m 52; c 3. ORGANIC CHEMISTRY. *Educ:* Yale Univ, BS, 43, PhD(org chem), 49. *Prof Exp:* Lab instr, Yale Univ, 46-47; process develop chemist, Gen Aniline & Film Corp, NY, 48-54, supvr intermediate area, 54-56, azoic area, 56-59, chief chemist, Intermediate Area, NJ, 59-62; asst to plant mgr, Geigy Chem Corp, 62-71, asst to vpres, Prod, Plastics & Additives Div, Ciba-Geigy Corp, 71-74; asst to gen mgr, Colors & Auxiliaries Div, BASF, 74-77, dir technol, Colors & Intermediates Group, 77-78, mfg mgr, Colors & Auxiliaries Div, 78-82, dir group technol, 82-85, dir group technol, Dyestuffs & Pigments Group, Coatings & Colorants Div, 85-91; RETIRED. *Concurrent Pos:* Mem adv comt prof educ, NY State Educ Dept, 58-59; chmn, US oper comt, Environ & Toxicol Asn Dye Mfg Indust, 84-86; consult, 91- *Mem:* Am Asn Textile Chemists & Colorists; Am Chem Soc; Sigma Xi. *Res:* Dyes, drugs; drug and dye intermediates; specialty chemicals; textile, leather and paper auxiliaries; ultra violet absorbers; pigments and brightening agents. *Mailing Add:* 41 Fieldstone Dr Morristown NJ 07960

DAYANANDA, MYSORE ANANTHAMURTHY, b Mysore City, India, July 1, 34; m 72; c 1. MATERIALS SCIENCE, METALLURGICAL ENGINEERING. *Educ:* Univ Mysore, BSc hons, 55; Indian Inst Sci, Bangalore, dipl, 57; Purdue Univ, MS, 61, PhD(metall eng), 65. *Prof Exp:* Sr res asst metall, Indian Inst Sci, Bangalore, 57-58; res assoc metall eng, 65-66, from asst prof to assoc prof mat sci & metall eng, 66-75, PROF MAT ENG, PURDUE UNIV, 75- *Concurrent Pos:* Vis prof, Inst Metall, Univ Münster, WGer, 80. *Mem:* Am Inst Mining, Metall & Petrol Engrs; Am Soc Metals; Micro-Beam Anal Soc; Am Soc Eng Educ; AAAS; Sigma Xi. *Res:* Diffusion in multicomponent alloys and metallurgical systems; application of electron microprobe and scanning electron microscope in science and engineering. *Mailing Add:* Sch Mat Eng Purdue Univ West Lafayette IN 47907

DAYBELL, MELVIN DREW, b Berkeley, Calif, May 8, 35; m 59; c 2. LOW TEMPERATURE PHYSICS. *Educ:* NMex State Univ, BS, 56; Calif Inst Technol, PhD(physics), 62. *Prof Exp:* from asst prof physics to assoc prof, NMex State Univ, 61-68; ASSOC PROF PHYSICS & ENG, UNIV SOUTHERN CALIF, 68- *Concurrent Pos:* Vis staff mem, Los Alamos Sci Lab, 66-68. *Mem:* Am Phys Soc; Inst Elec & Electronics Engrs. *Res:* High energy physics; ultra low temperature physics; Kondo effect. *Mailing Add:* Dept Physics Univ Southern Calif MC 0484 Los Angeles CA 90089-0484

DAYHOFF, EDWARD SAMUEL, b New York, June 26, 25; wid; c 2. PHYSICS, MILITARY SYSTEMS. *Educ:* Columbia Univ, AB, 46, MA, 47, PhD(physics), 52. *Prof Exp:* Lab asst, Radiation Lab, Columbia Univ, 48-52; physicist, Bur Standards, 52-55, Naval Ord Lab, 55-66, chief, Div Electronics & Electromagnetics, Naval Surface Weapons Ctr, White Oak, 66-71, res physicist, 71-79; dir, Pressure Sci Inc, 80-85. *Concurrent Pos:* Consult, US Bur Standards, 57-63 & Med Sch, Georgetown Univ, 76-83; sr res assoc, Univ Md, 82-83. *Mem:* Am Phys Soc; Optical Soc Am. *Res:* Free polarized electrons; spectrum of hydrogen atom; stellar image detection; medical instrumentation; computers; lasers; microwave propagation; magnetic resonances; military surveillance and countermeasures. *Mailing Add:* 1618 Tilton Dr Silver Spring MD 20902

DAYKIN, PHILIP NORMAN, computer science, mathematics, for more information see previous edition

DAYKIN, ROBERT P, b Cleveland, Ohio, Oct 11, 20. METALLURGICAL ENGINEERING. *Educ:* Oberlin Col, AB, 42. *Prof Exp:* Vpres res metall, Ladish Co, 74-82, vpres qual & technol, 82-85; RETIRED. *Mem:* Fel Am Soc Metals. *Res:* Forging superalloys, refractory and reactive metals. *Mailing Add:* 3174 Menomonee R Pkwy Wauwatosa WI 53222

DAYLOR, FRANCIS LAWRENCE, JR, b Fall River, Mass, Dec 17, 32; m 56; c 3. NATURAL PRODUCTS CHEMISTRY, FLAVOR CHEMISTRY. *Educ:* Fordham Univ, BS, 54. *Prof Exp:* Chemist, Beech Nut Life Savers, Inc, 56-59; assoc chemist, 60-63, res chemist, 63-67, sr chemist, 67-69, mgr gen prod, 68-72, mgr flavor develop, 72-85, PRIN SCIENTIST PHILIP MORRIS RES CTR, 85- *Mem:* Inst Food Technologists; Am Soc Heating, Refrig & Air-Conditioning Engrs; Am Chem Soc. *Res:* Flavor research and development; composition of tobacco and smoke to determine flavor; flavor formulation. *Mailing Add:* Philip Morris Inc Res Ctr Po Box 26583 Richmond VA 23261

DAYNES, RAYMOND AUSTIN, IMMUNOLOGY. *Educ:* Purdue Univ, PhD(med microbiol & cellular immunol), 72. *Prof Exp:* PROF IMMUNOL & HEAD EXP PATH, SCH MED, UNIV UTAH, 75- *Res:* Immunology of ultraviolet radiation carcinogenesis; lymphocyte recirculation dynamics; interleukin-1. *Mailing Add:* 1747 Orchard Dr Salt Lake City UT 84106

DAY-SALVATORE, DEBRA-LYNN, b Hoboken, NJ, Oct 23, 53. PEDIATRICS, PERINATAL GENETICS. *Educ:* Harvard Univ, BA, 75; NY Univ GSAS, MS, 79, PhD(pharmacol), 82; Case Western Reserve Univ, Sch Med, MD, 86. *Prof Exp:* Res asst genetics, NY Univ Med Ctr, 77-78, res fel pharmacol, 78-79; sr res asst med, Case Western Reserve Univ, 79-82, res assoc molecular biol/microbiol, 82-84; exec vpres consult, Lysantian Mgt, Inc, 84-86; mem staff, Dept Pediat, Cleveland Clinic Found, 86-89; ASST PROF PEDIAT, UNIV MED & DENT NJ-ROBERT WOOD JOHNSON MED SCH, 90- *Concurrent Pos:* Consult, Hudson Regional Health Comn, 77-78. *Mem:* AAAS; AMA; Am Soc Cell Biol; NY Acad Sci; Am Acad Pediat; Am Soc Human Genetics. *Res:* Inborn errors of metabolism; regulation of gene expression in neoplasia. *Mailing Add:* Dept Pediat Div Med Genetics Robert Wood Johnson Med Sch One Wood Johnson Pl New Brunswick NJ 08903-0019

DAYTON, BENJAMIN BONNEY, b Rochester, NY, Feb 25, 14; m 43; c 2. THEORETICAL ATOMIC PHYSICS, NUCLEON STRUCTURE. *Educ:* Mass Inst Technol, BS, 37; Univ Rochester, MS, 48. *Prof Exp:* Tech dir, Vacuum Div, Bendix, 68-69, chief scientist, Sci Instr Div, 69-71, consult, 71-72; consult, Xerox Corp, Webster, NY, H J Ross Assoc, Inc, Miami, Fla, 74-78 & Babcock & Wilcox Co, Alliance, Ohio, 82-84; teacher physics & math, Blue Ridge Tech Col, 80-82; CONSULT, PICKERING FIRM, MEMPHIS, 84- *Concurrent Pos:* US ed, Vacuum, 59-72; chmn, Adv Comt, Am Nat Standards Inst, 64-71; mem, Adv Panels, Nat Bur Standards, 65-71; staff lectr, George Washington Univ, 66-68 & Rochester Inst Technol, 67-69; tech dir, Int Union Vacuum Sci Technol & Appln, 66-72. *Mem:* Hon mem Am Vacuum Soc (pres, 61); Am Phys Soc; Am Sci Affil. *Res:* High vacuum equipment and systems; theory of molecular flow of gases; theory of photon transmission between emitter and absorber; author of 29 publications; holder of six patents. *Mailing Add:* 209 S Hillandale Dr East Flat Rock NC 28726-2609

DAYTON, BRUCE R, b Glen Cove, NY, Oct 11, 37; m 84; c 2. FORESTRY, MARINE SCIENCES. *Educ:* State Univ NY Col Environ Sci & Forestry, Syracuse Univ, BSF, 59; Univ NC, Chapel Hill, MA, 65, PhD(bot), 68. *Prof Exp:* from asst prof to assoc prof, 70-85, PROF BIOL, STATE UNIV NY COL ONEONTA, 85- *Mem:* Am Museum Natural Hist; Am Inst Biol Sci;

Ecol Soc Am; Am Forestry Asn. *Res:* Forest ecology; the influence of soils and topography on plant distribution; primary productivity of terrestrial ecosystems; nutrient cycles in terrestrial ecosystems; coastal plant communities. *Mailing Add:* RD 2 Box 482 CB Oneonta NY 13820

DAYTON, JAMES ANTHONY, JR, b Chicago, Ill, Dec 22, 37; m 65; c 3. ELECTRON OPTICS. *Educ:* Ill Inst Technol, BS, 59; Univ Iowa, MS, 60; Univ Ill, PhD(elec eng), 65. *Prof Exp:* Res electronics engr, Aeronaut Lab, Cornell Univ, 65-67; aerospace technologist, Lewis Res Ctr, 67-83, commun technol prog mgr, NASA HQ, Washington, DC, 83-84, head, Microwave Amplifier Sect, 84-85, CHIEF, ELECTRON BEAM TECHNOL BR, LEWIS RES CTR, NASA, 85- *Concurrent Pos:* Adj prof, Cleveland State Univ, 69- *Mem:* Inst Elec & Electronics Engrs. *Res:* Acoustic wave propagation in plasmas; diagnostic techniques in hypersonic flow machines; gas discharge tubes; space communications; multi stage depressed collectors; submm wave oscillators; electron beam focusing; microwave tubes. *Mailing Add:* 10423 Lake Ave Cleveland OH 44102

DAYTON, PAUL K, b Tucson, Ariz, Aug 4, 41; m 69; c 2. COASTAL ECOLOGY. *Educ:* Univ Ariz, BS, 63; Univ Wash, PhD(zool), 70. *Prof Exp:* PROF OCEANOG, SCRIPPS INST OCEANOG, 70- *Honors & Awards:* George Mercer Award, Ecol Soc Am, 74. *Mem:* Fel AAAS; Ecol Soc Am; Am Soc Naturalists. *Res:* Coastal ecology. *Mailing Add:* Scripps Inst Oceanog A-001 La Jolla CA 92093-0201

DAYTON, PETER GUSTAV, b Szigishoava, Rumania, Mar 9, 26; m 53; c 1. PHARMACOLOGY. *Educ:* Mass Inst Technol, BS, 50; Univ Paris, DS, 53. *Prof Exp:* Fel med, Col Med NY Univ, 54-64, res scientist, 65-67; asst prof pharmacol, Emory Univ, 67-71, assoc prof med & chem, 67-72, prof med, 72-78, adj prof chem, 72-78; guest scientist, NY Psychiat Inst, 78-79; VIS SCIENTIST, DEPT PHARMACOL & TOXICOL, DARTMOUTH MED SCH, 79- *Concurrent Pos:* Chemist, Pharmacol Lab, Nat Heart Inst, Md; vis prof, Dept Psychiat, Univ Pa, 79- *Honors & Awards:* Martini Prize, German Pharmacol Soc, 70. *Mem:* AAAS; Harvey Soc; Am Soc Pharmacol & Therapeut; Am Chem Soc; Soc Exp Biol & Med. *Res:* Chemical pharmacology; carbo-hydrate metabolism; clinical pharmacology; radioimmunoassay. *Mailing Add:* RR 3 Box 480 Lyme NH 03768-9777

DAYWITT, JAMES EDWARD, b Cedar Rapids, Iowa, Jan 3, 47; m 85; c 2. COMPUTATIONAL FLUID DYNAMICS. *Educ:* Iowa State Univ, BS, 70, MS, 74, PhD(aerospace eng), 77. *Prof Exp:* Aerospace engr, McDonnell-Douglas Aircraft Corp, 70-71; res asst, Iowa State Univ, 71-75; res assoc, Ames Res Ctr, NASA, 75-77; vis scientist, Inst Comput Appl Sci & Eng, 77-78; PROG MGR, RE-ENTRY SYSTS OPERS, GEN ELEC, CO, 78- *Concurrent Pos:* Adj prof fluid mech, Drexel Univ, 79-80 & numerical methods, Univ Pa, 81-84; gen chmn, 18th Fluid Dynamics, Plasmadynamics & Lasers Conf & 8th Appl Aerodynamics Conf, Am Inst Aeronaut & Astron, mem tech comt, Fluid Dynamics, 81-84, Appl Aerodynamics, 87-90, chmn, Greater Philadelphia Sect, 90-; assoc ed, J Spacecraft & Rockets, 90- *Honors & Awards:* Special Recognition Award, Am Inst Aeronaut & Astron, 90. *Mem:* Am Inst Aeronaut & Astronaut. *Res:* Computational techniques with emphasis on finite-difference methods for the analysis of the flow field about supersonic and hypersonic missiles and earth and planetary re-entry vehicles. *Mailing Add:* Gen Elec Re-Entry Systs Div Rm 6418 3198 Chestnut St Philadelphia PA 19101

DAZA, CARLOS HERNAN, b Cali, Colombia, Apr 9, 31; m 60; c 3. NUTRITION, PUBLIC HEALTH. *Educ:* Nat Univ Colombia, MD, 54; Columbia Univ, MS, 62, MPH, 63. *Prof Exp:* Med nutritionist, Nat Inst Nutrit, Colombia, 55; dir, Secy Pub Health, Valle del Cauca, Colombia, 56-57; adv nutrit, Nutrit & Health Educ Prog, Interam Coop Pub Health Serv, Colombia, 58-60; chief med care, Secy Pub Health, Valle del Cauca, Colombia, 63-65; MED OFFICER NUTRIT ADV, PAN-AM HEALTH ORGN-WHO, 66- *Mem:* Am Pub Health Asn; Latin Am Nutrit Soc; Am Inst Nutrit. *Res:* Operational research, development of nutrition manpower and public health nutrition services, formulation and implementation of national food and nutrition policies. *Mailing Add:* Pan-Am Health Orgn WHO 525 23rd St NW Washington DC 20037

DAZZO, FRANK BRYAN, b Miami, Fla, Apr 8, 48. MICROBIOLOGY. *Educ:* Univ Fla, BS, 70, MSc, 72, PhD(microbiol), 75. *Prof Exp:* Res assoc, Univ Wis, 76-77; from asst prof to assoc prof, 78-88, PROF, MICH STATE UNIV, 88- *Honors & Awards:* Enil Truog Award, Soil Sci Soc Am, 76. *Mem:* Am Soc Microbiol; AAAS. *Res:* Molecular basis for successful infection of legumenoius roots by nitrogen. *Mailing Add:* Dept Microbiol Mich State Univ 178 Giltner Hall East Lansing MI 48824

D'AZZO, JOHN JOACHIM, b New York, NY, Nov 30, 19; m 53; c 1. ELECTRICAL ENGINEERING, CONTROL ENGINEERING. *Educ:* City Col New York, BEE, 41; Ohio State Univ, MS, 50; Univ Salford, PhD(eng), 78. *Prof Exp:* Jr engr qual control, Western Elec Co, NJ, 41-42; proj engr res & develop, Wright Air Develop Ctr, 42-45, PROF ELEC ENG & DEPT HEAD, AIR FORCE INST TECHNOL, WRIGHT-PATTERSON AFB, 47- *Concurrent Pos:* Vis assoc prof, Dayton Campus, Ohio State Univ, 64; vis prof, Univ Salford, 76-79. *Mem:* Am Soc Eng Educ; fel Inst Elec & Electronics Engrs; assoc fel Am Inst Aeronaut & Astronaut; Sigma Xi. *Res:* Feedback control systems; servomechanisms; digital control systems; aircraft flight control systems; advanced control systems. *Mailing Add:* Dept Elec & Computer Eng Air Force Inst Technol Wright-Patterson AFB OH 45433-6583

DE, GOPA SARKAR, Indian citizen. ACOUSTICS, SOLID STATE PHYSICS. *Educ:* Calcutta Univ, BSc, 67, MSc, 69; Univ Calif, San Diego, MS, 72, PhD(physics), 77. *Prof Exp:* Res asst condensed matter, Univ Calif, San Diego, 71-77; postdoctoral condensed matter physics, Ga Tech, 77-79; postdoctoral condensed matter, Emory Univ, Atlanta, 79-80; res physicist seismic data processing, Western Geophys, Houston, 80-81; res physicist acoust boreholes, 82-87, SR RES PHYSICIST ACOUST BOREHOLES, CHEVRON OIL FIELD RES CO, 87- *Mem:* Am Phys Soc; Soc Explor Geophysicists. *Res:* Theoretical, numerical and field data analyses of full waveform acoustic logging in boreholes. *Mailing Add:* Chevron Oil Field Res Co 1300 Beach Blvd La Habra CA 90631

DEA, PHOEBE KIN-KIN, b Canton, China, June 17, 46; US citizen; m 67; c 2. PHYSICAL CHEMISTRY. *Educ:* Univ Calif, Los Angeles, BS, 67; Calif Inst Technol, PhD(chem), 72. *Prof Exp:* Fel, Nucleic Acid Res Inst, ICN Pharmaceut, Inc, 72-74, head anal & biophys chem, ICN Pharmaceut, Inc, 74-76; from asst prof to assoc prof, 76-79, PROF CHEM, CALIF STATE UNIV, LOS ANGELES, 84- *Concurrent Pos:* Vis assoc, Calif Inst Technol, 82- *Mem:* Sigma Xi; Am Chem Soc. *Res:* Nuclear magnetic resonance spectroscopy; conformational studies on compounds of biological interest; membrane structure and dynamics; applications of analytical instrumentation. *Mailing Add:* Dept Chem Calif State Univ Los Angeles CA 90032-8202

DEACETIS, WILLIAM, b Joliet, Ill, June 1, 28; m 63. ORGANIC CHEMISTRY. *Educ:* Univ Ill, BS, 50; Univ Wis, MS, 52, PhD(chem), 54. *Prof Exp:* Asst alumni res found, Univ Wis, 50-54; fel, Univ Calif, Berkeley, 54-56; res chemist high energy fuels, Olin Mathieson Chem Co, 56-57; res chemist, org chem, Shell Develop Co, Calif, 57-66, prod develop chemist, Shell Chem Co, NY, 66-70; mem staff chem econ handbook, Union Camp Corp, 70-71, mgr surface-active agents sect, 71-73, mkt specialist chem div, 74-80; SR CONSULT, CHEM SYSTS, INC, 80- *Mem:* Am Chem Soc. *Res:* Product and market development; market research; fuel additives; surfactants; resins; commercial development; strategic planning; specialty chemicals. *Mailing Add:* 807 High St Modesto CA 95354-0249

DEACON, DAVID A G, b Ont, Can, June 1, 53; m 79; c 1. FREE ELECTRON LASERS. *Educ:* Mass Inst Technol, BSc, 75, Stanford Univ, PhD(physics), 79. *Prof Exp:* Res assoc, High Energy Physics Lab, 79-81; RES PHYSICIST, DEACON RES, 81-, PRIN INVESTR, 83- *Concurrent Pos:* Vis scientist, Ctr Nat Studies, Soclay, France, 79-81. *Res:* Free electron laser physics; developing oscillator harmonics as coherent XUV source for scientific applications; characterizing UV photon induced absorption in multilayered dielectric mirrors and UV optics using undulator radiation. *Mailing Add:* Deacon Res 2440 Embarcadero Way Palo Alto CA 94303

DEACON, JAMES EVERETT, b White SDak, May 18, 34; m 84; c 2. VERTEBRATE ZOOLOGY, AQUATIC ECOLOGY. *Educ:* Midwestern Univ, BS, 56; Univ Kans, PhD(vert zool, bot), 60. *Prof Exp:* Asst, Univ Kans, 56-59, 60; asst prof zool, Univ Nev, Las Vegas, 60-64, asst res prof, Desert Res Inst, 64-65, assoc prof zool, 65-68, prof biol, 68-87, chmn dept biol sci, 75-82, DISTINGUISHED PROF, UNIV NEV, LAS VEGAS, 88- *Concurrent Pos:* NSF res grant, 64-65, US Fish & Wildlife Serv, US Bur Reclamation, Environ Protection Agency, Nat Park Service & Nev Dept Wildlife res grants, 65- *Honors & Awards:* Wildlife Conserv Award, Nat Wildlife Fedn, 70; Award of Excellence, Am Fisheries Soc, 87. *Mem:* Am Soc Ichthyologists & Herpetologists; Am Fisheries Soc; Desert Fishes Coun; fel AAAS; Am Inst Fish Res Biol; Soc Conserv Biol. *Res:* Fishes of Nevada; ecology of desert fish; biology of endangered and exotic fishes. *Mailing Add:* Dept Biology Univ Nevada 4505 Maryland Pkwy Las Vegas NV 89154-4004

DE ACOSTA, ALEJANDRO, b Buenos Aires, Arg, Feb 1, 41. PROBABILITY THEORY. *Educ:* Univ Buenos Aires, Lic en Ciencias, 65; Univ Calif, Berkeley, PhD(statist), 69. *Prof Exp:* Actg asst prof, dept statist, Univ Calif, Berkeley, 70; instr math, Mass Inst Technol, 70-72; asst prof, fac sci, Univ La Plata, Arg, 72-74; assoc prof, fac sci, Univ Buenos Aires, 75-83; PROF MATH & STATIST, CASTE WESTERN RESERVE UNIV, 83- *Concurrent Pos:* Vis prof math, Univ Wis, Madison, 81-83; researcher, Venezolano Inst Invest, 76-81; NSF grant, 84-88. *Mem:* Am Math Soc; Inst Math Statist. *Mailing Add:* Dept Math & Statist Case Western Reserve Univ Cleveland OH 44106

DEADY, MATTHEW WILLIAM, b Chicago, Ill, Oct 22, 53. ELECTRON SCATTERING, RADIATIVE CORRECTIONS. *Educ:* Univ Ill, BS (math), BS (physics), 75, MS, 77; Mass Inst Technol, PhD(physics), 81. *Prof Exp:* Chief operator, Ill Accelerator Lab, 74-77; res asst, Bates Accelerator Lab, Mass Inst Technol, 77-81; asst prof physics, mount holyoke col, 81-87; AT DEPT PHYSICS, BARD COL, 88- *Res:* Investigations of electron scattering from nuclei in the region of large energy loss are being done in order to provide insights into the behavior of nucleons in the environment of the nucleus. *Mailing Add:* Dept Physics Bard Col Annandale NY 12504

DEAHL, KENNETH LUVERE, b Pikeville, Ky, Aug 24, 43; c 3. PHYTOPATHOLOGY. *Educ:* Fairmont State Col, BA, 65; WVa Univ, MS, 67, PhD(plant path), 71. *Prof Exp:* RES PLANT PATHOLOGIST, AGR RES SERV, USDA, 71- *Mem:* Am Phytopath Soc; Mycol Soc Am; Am Soc Microbiol. *Res:* Host-parasite interaction between Phytophthora infestans and Solanum tuberosum; nature of disease resistance; biochemical response to infection; diseases of mushrooms. *Mailing Add:* Agr Res Serv USDA Veg Lab-Beltsville Agr Res Ctr Lab 215 Bldg 004 Beltsville MD 20705

DEAL, ALBERT LEONARD, III, b Hickory, NC, Aug 31, 37; m 63. MATHEMATICAL ANALYSIS. *Educ:* Univ NC, BS, 59, MA, 62, PhD(differential equations), 65. *Prof Exp:* From instr to assoc prof, 62-71, PROF MATH, VA MIL INST, 71- *Mem:* AAAS; Soc Indust & Appl Math; Math Asn Am; Am Math Soc; Sigma Xi. *Res:* Linear differential and difference boundary problems. *Mailing Add:* Dept of Math Va Mil Inst Lexington VA 24450

DEAL, BRUCE ELMER, b Lincoln, Nebr, Sept 20, 27; m 50; c 3. PHYSICAL CHEMISTRY. *Educ:* Nebr Wesleyan Univ, AB, 50; Iowa State Col, MS, 53, PhD(chem), 55. *Prof Exp:* Asst, Ames Lab, AEC, 50-55; res chemist, Kaiser Aluminum & Chem Corp, 55-59 & Rheem Semiconductor Corp, Calif, 59-63; res chemist, Fairchild Camera & Instrument Corp, 63-87; prin technologist, Nat Semiconductor Corp, 88-89; VPRES DEVELOP, ADVAN PROD

TECHNOL, 89- *Concurrent Pos:* Consult prof, elec eng dept, Stanford Univ, 77-; lectr, Continuing Educ Inst, 82-89; adj prof, elect eng dept, Santa Clara Univ, 88- *Honors & Awards:* Electronics Div Tech Award, Electrochem Soc, 74; Dielectrics & Insulation Div Award, Electrochem Soc, 82. *Mem:* AAAS; Int Electrochem Soc (vpres, 85-88, pres, 88-89); fel Inst Elec & Electronics Engrs. *Res:* Surface physics and chemistry of solids; electrochemistry; semiconductor materials and processing. *Mailing Add:* 638 Towle Pl Palo Alto CA 94306

DEAL, CARL HOSEA, JR, physical chemistry; deceased, see previous edition for last biography

DEAL, DON ROBERT, b Dayton, Ohio, Sept 26, 37; m 62; c 2. BOTANY, POMOLOGY. *Educ:* Capital Univ, BA, 60; Miami Univ, MA, 65; Cornell Univ, PhD(plant path), 69. *Prof Exp:* Teacher high schs, Ohio, 60-63; assoc prof, 69-76, PROF BIOL, GLENVILLE STATE COL, 76- *Mem:* Am Inst Biol Sci; Nat Asn Biol Teachers; Sigma Xi. *Res:* New grape, tree-fruit cultivars and rootstocks for productivity, quality and disease resistance in West Virginia test conditions. *Mailing Add:* 7517 Sunhill Dr Sciotoville OH 45662

DEAL, DWIGHT EDWARD, b Staten Island, NY, Apr 18, 38; m 65. ENVIRONMENTAL GEOLOGY. *Educ:* Rensselaer Polytech Inst, BS, 59; Univ Wyo, MS, 63; Univ NDak, PhD(geol), 70. *Prof Exp:* Asst prof geol, Sul Ross State Univ, 67-74; DIR, CHIHUAHUAN DESERT RES INST, 73- *Concurrent Pos:* Chief geologist, Geofactors, Tex, 71-; res scientist assoc, Bur Econ Geol, Univ Tex, Austin, 72-73; asst prof geol, Earlham Col, 81-82. *Mem:* AAAS; Geol Soc Am; Am Asn Petrol Geol; Am Quaternary Asn; Nat Speleol Soc. *Res:* Groundwater, quaternary, environmental and general exploration geology of Mexico and southwestern United States; applied glacial geology in North Dakota; glacial, fluvial and groundwater processes; erosional history of the Rio Conchos-Rio Grande drainages of the Chihuahuan Desert, United State of America and Mexico. *Mailing Add:* PO Box 3154 Carlsbad NM 88221

DEAL, ELWYN ERNEST, b Appling Co, Ga, Oct 10, 36; m 60; c 4. AGRONOMY. *Educ:* Univ Ga, BSA, 58, MS, 60; Rutgers Univ, PhD(turf mgt), 63. *Prof Exp:* Asst turf mgt, Ga Coastal Plain Exp Sta, USDA, 58-60; Rutgers Univ, 60-63 & NC State Univ, 63-64; from asst prof to assoc prof turf mgt, Univ Md, College Park, 64-86, asst dir, Coop Ext Serv, 69-86, adminr, Ext Serv, Col Agr, 80-86; ASST DIR AGR NATURAL RESOURCES, COOP EXTEN SERV, CLEMSON UNIV, 86- *Mem:* Am Soc Agron; Coun Agr Sci & Technol; Sigma Xi. *Res:* Turf grass management including species selection and adaptation; physiology, ecology, weed control, mowing, fertilization, irrigation and growth control of lawn, golf course and highway roadside turf. *Mailing Add:* 108 Barre Hall Clemson Univ Clemson SC 29634

DEAL, ERVIN R, US citizen. MATHEMATICS. *Educ:* Nebr Wesleyan Univ, AB, 51; Kans State Univ, MS, 53, Univ Mich, PhD(math), 62. *Prof Exp:* Asst prof, 59-62, ASSOC PROF MATH, COLO STATE UNIV, 62- *Mem:* AAAS; Math Asn Am; Am Math Soc; Sigma Xi. *Res:* Functional analysis. *Mailing Add:* 1112 Lynnwood Dr Ft Collins CO 80521

DEAL, GEORGE EDGAR, b Marion, Ind, July 31, 20; m 45; c 6. HEALTH CARE ADMINISTRATION. *Educ:* Ind Univ, BS, 41, MS, 42; George Washington Univ, DBA(health care admin), 70. *Prof Exp:* From asst to pres, Bionetics Res Labs, 63-68; dir, AMA Grad Mgt Sch, 68-70; officer, US HEW, 70-79 & US Dept Health & Human Serv, 79-89; RETIRED. *Concurrent Pos:* Living res medium, US Gemini Space Prog, 65-66; assoc ed, Opers Res Soc Am, 70-75; alt US rep, NATO Conf Opers Res, 71-80; mem, Int Consults Found, 71-85; consult, World Bank, 81- *Mem:* Cath Acad Sci (vpres, 87-). *Res:* Costs of medical care in the US; scientific approach to ethics. *Mailing Add:* 6245 Park Rd McLean VA 22101

DEAL, GLENN W, JR, b Kannapolis, NC, Apr 3, 22; m 41; c 1. CHEMISTRY. *Educ:* Catawba Col, AB, 48; Appalachian State Teachers Col, MS, 56. *Prof Exp:* Student chem analyst, Tidewater Assoc Oil Refinery, 46; chmn dept sci, Ro County High Sch, 46-60; ASSOC PROF CHEM, CATAWBA COL, 60- *Mem:* Am Chem Soc. *Res:* Education; analytical chemistry. *Mailing Add:* Box 264 China Grove NC 28023-1900

DEAL, RALPH MACGILL, b Charlotte, NC, May 29, 31; m 53, 76; c 3. PHYSICAL CHEMISTRY & COMPUTER SCIENCE. *Educ:* Oberlin Col, BA, 53; Johns Hopkins Univ, PhD(chem), 58. *Prof Exp:* Res assoc, Monadnock Res Inst, 58-59; Imp Chem Industs fel, Univ Keele, 59-61; NIH trainee biophys chem, Univ Ill, 61-62; asst prof, 62-68, assoc prof, 68-79, PROF CHEM & COMPUT SCI, KALAMAZOO COL, 79- *Concurrent Pos:* Consult, Dept Energy, 79-81. *Mem:* Am Chem Soc; Soc Comput Simulation. *Res:* Modeling physical processes; computer-assisted-learning. *Mailing Add:* Dept Chem Kalamazoo Col Kalamazoo MI 49007

DEAL, WILLIAM CECIL, JR, b Lake Providence, La, Mar 21, 36; m 57; c 2. PHYSICAL BIOCHEMISTRY. *Educ:* La Col, BS, 58; Univ Ill, Urbana, PhD(phys chem), 62. *Prof Exp:* From asst prof to assoc prof, 62-71, PROF BIOCHEM, MICH STATE UNIV, 71- *Concurrent Pos:* Lectr sch med, Univ Wash, 65; spec res fel, Univ Munich, 69-70. *Mem:* Am Soc Biol Chem. *Res:* Roles of phosphorylation-dephosphorylation of proteins in regulation of proliferation of lymphocytes; structure and function of phosphorylated proteins in lymphocytes; roles of protein kinase C and protein tyrosine kinases in regulation of lymphocyte functions. *Mailing Add:* Dept Biochem Rm 209 Biochem Bldg Mich State Univ East Lansing MI 48824

DEAL, WILLIAM E, JR, b Ft Sam Houston, Tex, Sept 24, 25; m 49; c 4. HYDRODYNAMICS, HIGH PRESSURE PHYSICS. *Educ:* Univ Tex, BS, 47, MA, 48, PhD(physics), 51; Univ NMex, MM, 77. *Prof Exp:* Mem staff physics, 50-60, group leader, 60-65, asst div leader, 65-70, alt div leader, 70-72, div leader, 72-79, dep assoc dir, 79-80, CONSULT, LOS ALAMOS NAT LAB, UNIV CALIF, 80- *Mem:* Fel Am Phys Soc. *Res:* Near ultraviolet absorption spectra of polynuclear aromatics; shock hydrodynamics; explosives. *Mailing Add:* PO Box 567 Los Alamos NM 87544

DE ALBA MARTINEZ, JORGE, b Aguascalientes, Mex, Mar 28, 20; m 43; c 4. REPRODUCTIVE PHYSIOLOGY, ANIMAL HUSBANDRY. *Educ:* Univ Md, BS, 41; Cornell Univ, MS, 42, PhD(animal physiol), 44. *Prof Exp:* Mgr, Hacienda Sierra Hermosa, Mex, 44-49; physiologist, Inter-Am Inst Agr Sci, Costa Rica, 50-51, head dept animal indust, 52-63; dean & founder col agr, Univ Sonora, Mex, 51-52; livestock adv, Bank of Mex, 63-67; Ford Found fel animal sci, Cornell Univ, 67-73; dir Animal Prod Training Ctr, Mex Asn Animal Prod, Tampico, 73-83; chmn, Animal Prod Dept, Trop Agr Res & Training Ctr, Turrialba, Costa Rica, 83-86; LIVESTOCK TRAINING OFFICER, BANK OF MEX, FIRA, 88- *Concurrent Pos:* Guggenheim fel, 56-57; Kellogg authorship grant, 60-61. *Mem:* Mex Asn Animal Prod (first pres, 64-66); Latin Am Asn Animal Prod (first pres, 66-); Am Soc Animal Sci; Costa Rican Soc Criollo Breeder (first pres, 87-88). *Res:* Tropical pasture research for beef and dairy production. *Mailing Add:* Juarez 86 Coyoacan DF 21 Mexico

DEALY, JAMES BOND, JR, b Medford, Mass, Sept 7, 20; wid; c 6. RADIOLOGY. *Educ:* Yale Univ, AB, 42; Columbia Univ, MD, 45; Am Bd Radiol, dipl, 52. *Prof Exp:* Fel med, Harvard Univ, 50-52, fel radiol, 52, asst, 53-54, instr, 55, from asst clin prof to assoc clin prof, 56-66, actg head dept, 56-63, chmn exec comt, 60-63; PROF RADIOL, TUFTS UNIV, 67-; SR RADIOLOGIST, NEW ENG MED CTR HOSP, 75- *Concurrent Pos:* Assoc radiologist, Peter Bent Brigham Hosp, 53-55, actg radiologist in chief, 56, radiologist in chief, 57-66; consult, US Vet Amin, 56-, Brookline Health Dept, 56-66, Mass Ment Health Ctr, 56-66 & Pondville Hosp, 60-; mem comt radiol, Nat Acad Sci-Nat Res Coun, 63-69; chief diag radiol serv, Lemuel Shattuc Hosp, 67-75; mem consult staff, New Eng Med Ctr Hosps, 68-75. *Mem:* Radiol Soc NAm; Asn Univ Radiol; Am Col Radiol; Am Roentgen Ray Soc; Am Radium Soc. *Res:* Diagnostic and therapeutic radiology; biological effects of ionizing irradiation. *Mailing Add:* New Eng Med Ctr Hosp 750 Washington St Boston MA 02111

DEALY, JOHN EDWARD, b Cardiff, Calif, Oct 25, 30; m 55; c 3. WILDLIFE HABITAT ECOLOGY, FOREST ECOLOGY. *Educ:* Ore State Univ, BS, 55, MS, 59, PhD(forestry), 75. *Prof Exp:* Range scientist, USDA, Juneau, Ark, 58-63, assoc plant ecologist, 64-72, plant ecologist, 72-79, prin plant ecologist, Pac Northwest Forest & Range Exp Sta, 79-82, prin forest ecologist, Forestry Sci Lab, Forest Serv, 82-88; CONSULT, 81- *Concurrent Pos:* Assoc prof, Eastern Ore State Col, 69-82 & Wash State Univ, 75-82; referee ed, Wildlife Soc, 77- & Soc Range Mgt, 78- *Mem:* Wildlife Soc; Ecol Soc Am; AAAS; Soc Am Foresters. *Res:* Analysis of forest ecosystems; relating wild ungulate behavior to habitat characteristics in order to develop timber harvest options suitable for maintaining optimum ungulate habitat. *Mailing Add:* PO Box 1860 Corvallis OR 97339

DEALY, JOHN MICHAEL, b Waterloo, Iowa, Mar 23, 37; m 64; c 1. RHEOLOGY, POLYMER PHYSICS. *Educ:* Univ Kans, BSChE, 58; Univ Mich, MSE, 59, PhD(chem eng), 63. *Prof Exp:* Fel chem eng, Univ Mich, 64; from asst prof to assoc prof, 64-73, PROF CHEM ENG, MCGILL UNIV, 73- *Concurrent Pos:* Res vis, Univ Cambridge, 70; vis prof, Univ Del, 78-79 & Univ Wis, 85-87. *Mem:* Can Soc Chem Eng; Am Inst Chem Engrs; Soc Rheology (pres, 87-89); Soc Plastics Engrs. *Res:* Rheological properties of polymer melts and solutions and other fluids; development of new techniques for measuring rheological properties of molten polymers and of relationships between these properties and processing behavior. *Mailing Add:* Chem Eng Dept McGill Univ 3480 Univ St Montreal PQ H3A 2A7 Can

DEAM, JAMES RICHARD, b Springfield, Ohio, Jan 9, 42; m 66; c 1. COMPUTER AIDED ENGINEERING, PROCESS SIMULATION. *Educ:* Univ Cincinnati, BS, 64; Okla State Univ, MS, 66, PhD(chem eng), 69. *Prof Exp:* Coop engr, Int Harvester Co, 60-63; sr res engr, Mgt Info & Systs Dept, Monsanto Co, 68-76; mgr, Mgt Info Systs, Monsanto Can LTD, 76-79; mgr eng, 79-85, MGR ENG TECH, MONSANTO CHEM CO, 86- *Mem:* Am Inst Chem Engrs; Am Chem Soc; Sigma Xi. *Res:* Computer aided engineering; chemical process simulation; physical property estimation; process control; expert systems. *Mailing Add:* Monsanto Co 800 N Lindbergh St Louis MO 63167

DEAMER, DAVID WILSON, JR, b Santa Monica, Calif, Apr 21, 39; m 62; c 3. BIOPHYSICS. *Educ:* Duke Univ, BS, 61; Ohio State Univ, PhD(biochem), 65. *Prof Exp:* Trainee biophys, Univ Calif, Berkeley, 65-66, USPHS fel, 66-67, asst prof zool, 67-70, assoc prof, 70-75, PROF ZOOL, UNIV CALIF, DAVIS, 75- *Concurrent Pos:* Guggenheim fel, 86-87. *Mem:* Am Soc Biol Chem; Biophys Soc; Am Soc Cell Biol; AAAS. *Res:* Membrane structure; ion flux mechanisms; prebiotic evolution. *Mailing Add:* Dept Zool Univ Calif Davis CA 95616

DEAN, ANDREW GRISWOLD, b Rochester, NY, Apr 4, 38; m 63; c 1. EPIDEMIOLOGY. *Educ:* Oberlin Col, AB, 60; Harvard Med Sch, MD, 64; Harvard Sch Pub Health, MPH, 72. *Prof Exp:* Rotating internship, King County Hosp, Seattle, 64-65; Peace Corps physician, USPHS, 65-67; staff mem, Pac Res Sect, Nat Inst Allergy & Infectious Dis, Honolulu, 67-71; epidemiologist, Burkitt's Lymphoma Proj, WHO, Arua, Uganda, 72-73; actg state epidemiologist & actg dir commun dis, Ark Dept Health, 74-76; STATE EPIDEMIOLOGIST & DIR DIS PREV & CONTROL, MINN DEPT HEALTH, 76- *Concurrent Pos:* Epidemic Intel Serv officer, Ctr Dis Control, Atlanta, Ga, 74-76; mem immunol-epidemiol segment, Working Group Comt, Virus Cancer Prog, Nat Cancer Inst, 75-77; dir, Pac Ctr Geog Dis Res, Honolulu, 76-78. *Mem:* Am Soc Microbiol; Am Soc Trop Med & Hyg. *Res:* A new statewide disease detection system; epidemiology of diseases of public health interest. *Mailing Add:* 2877 McClave Dr Doraville GA 30340

DEAN, ANN, b Brooklyn, NY, Dec 18, 44; m 76; c 3. MOLECULAR BIOLOGY. *Educ:* Bucknell Univ, AB, 66; George Washington Univ, PhD(biochem), 81. *Prof Exp:* Res chemist, Lab Chem Biol, 66-84, RES CHEMIST, LAB CELLULAR & DEVELOP BIOL, NAT INST DIABETES DIGESTIVE-KIDNEY DIS, NIH, 84- *Mem:* Am Soc Biochem & Molecular Biol; AAAS. *Res:* Regulation of gene expression during development;

molecular biology of transcriptional activation of genes; chromatin structure in differential gene expression. *Mailing Add:* Lab Cellular & Develop Biol Bldg 6 Rm B1-08 Nat Inst Diabetes Digestive & Kidney Dis NIH Bethesda MD 20892

DEAN, ANTHONY MARION, b Savannah, Ga, Aug 26, 44; m 66; c 4. PHYSICAL CHEMISTRY. *Educ:* Spring Hill Col, BS, 66; Harvard Univ, AM, 67, PhD(phys chem), 70. *Prof Exp:* Asst prof phys chem, Univ Mo-Columbia, 70-75, assoc prof, 75-79; sr staff chemist, 79-82, res assoc, 82-90, SR RES ASSOC, EXXON RES & ENG CO, 90- *Mem:* Am Chem Soc; Combustion Inst; Am Phys Soc. *Res:* Gas phase kinetics; laser diagnostics of flames; kinetics and mechanism of combustion reactions; detailed kinetic modeling of high temperature reactions. *Mailing Add:* Corp Res Lab Exxon Res & Eng Co Rte 22 E Annandale NJ 08801

DEAN, BILL BRYAN, b Omaha, Nebr, Dec 1, 48; m 71; c 3. HORTICULTURE, PLANT PHYSIOLOGY. *Educ:* Wash State Univ, BS, 71, MS, 74, PhD(hort), 76. *Prof Exp:* Asst prof hort, Mich State Univ, 76-78; asst prof, 78-80, ASSOC PROF HORT, WASH STATE UNIV, 86- *Concurrent Pos:* Dir agr, UI Group, Kennewick, Wa. *Mem:* Potato Asn Am; Am Soc Hort Sci. *Res:* International marketing of vegetable crops; asparagus variety evaluation; carrot seed quality; oxidation of potato tissue following damage; biosynthesis of suberin; wound healing in plants. *Mailing Add:* Dept Hort Wash State Univ IAREC MA Rte 2 Box 2953A Prosser WA 99350

DEAN, BURTON VICTOR, b Chicago, Ill, June 3, 24; m 58; c 4. OPERATIONS RESEARCH. *Educ:* Northwestern Univ, BS, 47; Columbia Univ, MS, 48; Univ Ill, PhD(math), 52. *Prof Exp:* Instr math, Columbia Univ, 47-49 & Hunter Col, 49-50; mathematician, Nat Security Agency, 52-55; res mathematician, Opers Res, Inc, 55-57; assoc prof opers res, Case Western Reserve Univ, 57-65, prof orgn sci & chmn opers res group, 65-67, chmn dept, 67-76 & 79-85, prof opers res, Weatherhead Sch Mgt, 57-85; PROF & CHMN DEPT ORGN & MGT, SCH OF BUS, SAN JOSE STATE UNIV, CALIF, 85- *Concurrent Pos:* Indust & govt consult, 57-; vis prof, dept indust & mgt eng, Technion, Israel Inst Technol, 62-63, Univ Louvain, Belg, Tel-Aviv Univ & Ben-Gurion Univ, Israel, 78 & dept indust eng & eng mgt, Stanford Univ, 85; ed, Mgt Sci, Dept Res & Develop & Innovation, 70-86, assoc ed, 90-, dept ed, Trans Eng Mgt, Inst Elec & Electronics Engrs, 85-; mem coun, Inst Mgt Sci, 66-67, Col Res & Develop, 73-85, Col Eng Mgt, 78-85, Opers Res Soc Am Educ Sci Sect Coun, 73-76 & Col Innovation Mgt & Entrepreneurship, 89-; assoc ed, J Opers Res Soc India, 68-74, Opers Res Letters, 81-84; assoc, Inst Pub Admin, New York, NY & Washington, DC, 72-79; pres & mem coun, Omega Rho, 77-86; ed, J Business Venturing, 85-, Prod & Opers Mgt J, 90- *Honors & Awards:* Centennial Medal, Case Inst Technol, 81, Eng Mgt Soc, 84. *Mem:* Fel AAAS (vpres, sect indust sci, 71); Inst Mgt Sci; Opers Res Soc Am; Am Soc Qual Control; Inst Elec & Electronics Engrs; Sigma Xi; Prod & Opers Mgt Soc. *Res:* Applications of operations research to industrial management; innovation management; research planning and corporate strategy; manufacturing and operations management; corporate planning; entrepreneurship and venturing; management of research and development; author or co-author of numerous books, chapters and papers. *Mailing Add:* 161 Gabarda Way Portola Valley CA 94028-7444

DEAN, CHARLES E(ARLE), b Pickens Co, SC, May 23, 1898; m 27; c 2. COMMUNICATION ENGINEERING. *Educ:* Harvard Univ, AB, 21; Columbia Univ, AM, 23; Johns Hopkins Univ, PhD(physics), 27. *Prof Exp:* Tech asst, Bell Tel Labs, 21-24; asst physicist, Johns Hopkins Univ, 24-26; tech writer, Am Tel & Tel Co, 27-29; tech writer & ed, Hazeltine Corp, 29-63; ed, Electronics & Commun in Japan, Scripta Publ Corp, Washington, DC, 63-78; RETIRED. *Mem:* Fel Inst Elec & Electronics Engrs. *Res:* Electronics engineering. *Mailing Add:* 5400 Vantage Pt Rd No 1101 Columbia MD 21044

DEAN, CHARLES EDGAR, b Monticello, Fla, Apr 24, 29; m 57; c 1. PLANT BREEDING, PLANT GENETICS. *Educ:* Univ Fla, BS, 53, MS, 57; NC State Univ, PhD(field crops), 59. *Prof Exp:* From asst agronomist to agronomist, NFla Exp Sta, 59-69, chmn dept, 79, PROF AGRON & AGRONOMIST, UNIV FLA, 70- *Concurrent Pos:* USDA res grant, 64-68. *Mem:* Genetic Asn Am; Am Soc Agron; Sigma Xi. *Res:* Breeding and genetics of tobacco. *Mailing Add:* Dept Agron Univ Fla 304 Newell Hall Gainesville FL 32601

DEAN, DAVID, b Paterson, NJ, Nov 12, 26; m 49; c 3. BIOLOGICAL OCEANOGRAPHY. *Educ:* Lehigh Univ, AB, 49; Rutgers Univ, PhD(zool), 57. *Prof Exp:* Instr biol, Norwich Univ, 49; from instr to asst prof zool, Univ Conn, 57-66; dir, Darling Ctr, 66-79, prof oceanog, 70-81, prof zool, 66-88, EMER PROF ZOOL, UNIV MAINE, 88- *Concurrent Pos:* Co-investr, Maine Yankee Atomic Power Co grants, 69-79; prin investr, Sea Grant Prog, 71-83 & NIH, 84- *Mem:* Marine Biol Asn UK. *Res:* Definition of marine benthic communities and the interrelationships of their components; marine worms; formation and larvae of benthic communities. *Mailing Add:* Ira C Darling Ctr Univ Maine Walpole ME 04573

DEAN, DAVID A, b Shreveport, La, Jan 14, 41. IMMUNITY TO PARASITES. *Educ:* Howard Univ, PhD(biol), 71. *Prof Exp:* Head schistosomiasis br, Naval Med Res Inst, 81-88; HEAD IMMUNOL DEPT, NAVAL MED RES UNIT-3, 88- *Mem:* Am Soc Parasitologists; Am Asn Immunologists; Am Soc Trop Med & Hyg. *Mailing Add:* Naval Med Res Unit No 3 FPO New York New York NY 09527-1600

DEAN, DAVID CAMPBELL, b Buffalo, NY, Apr 19, 31; m 56; c 3. INTERNAL MEDICINE, CARDIOLOGY. *Educ:* Bowdoin Col, BA, 52; Johns Hopkins Univ, MD, 56; Am Bd Internal Med, dipl, 66; Am Bd Cardiovasc Dis, dipl, 67. *Prof Exp:* Res fel med, Harvard Univ, 59-61; clin instr, State Univ NY, Buffalo, 61-65, clin assoc, 65-66, asst prof, 66-72, clin assoc prof, 72-78; chief cardiopulmonary lab, Vet Admin Hosp, 62-86; PROF MED, STATE UNIV NY, BUFFALO, 78-; DIR CARDIAC REHAB, BUFFALO VET ADMIN HOSP, 86-, DIR, LIPID CLIN, 88- *Concurrent Pos:* Res fel cardiol, Mass Gen Hosp, Boston, 60-61; clin asst, Buffalo Gen Hosp, 61-68, clin assoc, 68-74, asst physician, 74-88, co-dir cardiac rehabilitation, 86-, physician, 88-; staff mem & chief echocardiography, St Joseph Hosp, 76-; impartial specialist in cardiol; mem, Workmans Compensation Bd, NY. *Mem:* Fel Am Col Physicians; fel Am Col Cardiol; Am Fedn Clin Res; Am Heart Asn; fel Am Col Chest Physicians. *Res:* Electrophysiology of the heart and the electrical control of the heart; lipids; cardiac rehabilitation. *Mailing Add:* 4955 N Bailey Ave Buffalo NY 14226

DEAN, DAVID DEVEREAUX, b Cincinnati, Ohio, Oct 24, 52; m 77; c 1. METALLOPROTEASES, CONNECTIVE TISSUE BIOCHEMISTRY. *Educ:* Randolph-Macon Col, BS, 75; Univ NC, PhD(bot-biochem), 81. *Prof Exp:* Res assoc biochem, 81-83, assoc, 83-86, RES ASST PROF MED-BIOCHEM, SCH MED, UNIV MIAMI, 86- *Mem:* Sigma Xi; Am Chem Soc; Am Rheumatism Asn; AAAS. *Res:* Mechanisms of connective tissue degradation in arthritis; crystal deposition diseases; ovulation and cervical dilation as related to metalloprotease production and their regulation by endogeneous inhibitors. *Mailing Add:* 12962 SW 88th Lane Miami FL 33186-1730

DEAN, DAVID LEE, b Grand Rapids, Mich, Oct 12, 46; m 84; c 5. TECHNICAL MANAGEMENT. *Educ:* Mich State Univ, BS, 67, PhD(chem), 72. *Prof Exp:* Res assoc chem, State Univ NY Buffalo, 72-73; res chemist, E I Du Pont de Nemours & Co, Inc, 73-77; asst prof chem, Univ NC, Wilmington, 77-80; asst prof chem, Eastern Wash Univ, 80-82; scientist, Morton Thiokol, Inc, 83-86; PRIN SCIENTIST, ATLANTIC RES CORP, 86- *Mem:* Am Chem Soc; Soc Advan Mat & Process. *Res:* Development of high performance composite cases and bondlines for solid fuel rocket motors; synthesis of thermally stable high performance polymers, and development of processes to make them into high performance composite aerospace hardware. *Mailing Add:* McDonald Douglas 689 Discovery Dr MS 12D1 Huntsville AL 35806

DEAN, DONALD E, b Flushing, NY, June 3, 27; m 51; c 4. COSMETIC CHEMISTRY, PHARMACEUTICAL CHEMISTRY. *Educ:* Oberlin Col, BA, 50; Stevens Inst Technol, MS, 57. *Prof Exp:* Chemist, Am Cyanamid Co, 50-52, anal method develop, 52-54; mgr, Control & Anal Lab, Shulton, Inc, 54-65; mgr anal & qual control, 65-69 & cosmetic & toiletries res, 69-73, dir prod develop, Leeming /Pacquin Div, 73-75, DIR PROD DEVELOP, CONSUMER PROD OPERS, PFIZER, INC, 75- *Mem:* Am Chem Soc; NY Acad Sci; Soc Cosmetic Chemists. *Res:* Organic analysis; instrumental analysis by infrared and ultra-violet absorption; vapor phase chromatography; development of product forms for use as cosmetics, pharmaceuticals and household products; documentation of product claims through controlled testing. *Mailing Add:* Nine Yorke Rd Mountain Lakes NJ 07046

DEAN, DONALD HARRY, b San Diego, Calif, Nov 20, 42; m 64; c 1. MOLECULAR GENETICS. *Educ:* Tex Christian Univ, BS, 65, MS, 68; Univ Mich, PhD(cell biol, molecular biol), 72. *Prof Exp:* Fel molecular genetics, Sch Med, Wash Univ, 73 & Rosenstiel Res Ctr, Brandeis Univ, 73-75; asst prof microbiol, 75-79, ASSOC PROF MICROBIOL, OHIO STATE UNIV, 79-, DIR, BACILLUS GENETIC STOCK CTR, 78- *Concurrent Pos:* Ed, Microbial Genetics Bull, 78-80. *Mem:* Am Soc Microbiol; Soc Insect Path; Soc Indust Microbiol. *Res:* Molecular genetics of Bacillus subtilis and Bacillus thuringiensis and their temperate bacteriophage; restriction enzymes; recombinant molecules; genetic engineering. *Mailing Add:* Dept Genetics Ohio State Univ Main Campus 963 Biol Sci Bldg Columbus OH 43210

DEAN, DONALD L(EE), b Litchfield, Ill, Nov 25, 26; div; c 2. ENGINEERING. *Educ:* Univ Mo, BS, 49, MS, 51; Univ Mich, PhD(civil eng), 55. *Prof Exp:* Instr civil eng, Mo Sch Mines, 49-51, admin asst to dean, 52-53, asst prof civil eng, 54-55; assoc prof, Univ Kans, 55-60; prof & chmn civil eng & eng mech, Univ Del, 60-65, . H Fletcher Brown prof civil eng, 62-65; prof & head dept, NC State Univ, 65-78; dean eng, Ill Inst Technol, 78-80; OWNER, PYRAMID ELECTRONICS, 82- *Concurrent Pos:* Mem, Gov Air Pollution Comn, Del, 61-65. *Honors & Awards:* Walter L Huber Prize, Am Soc Civil Engrs, 67. *Mem:* Fel Am Soc Civil Engrs; Int Asn Shell Struct. *Res:* Structural design and analysis; latticed and ribbed shells; applied mathematics; mechanics in relation to the behavior of complex structural systems. *Mailing Add:* 1645 Redwood Ave Sarasota FL 34237

DEAN, DONALD STEWART, botany, for more information see previous edition

DEAN, DONNA JOYCE, b Danville, Ky, Apr 22, 47. ENDOCRINOLOGY, TOXICOLOGY. *Educ:* Berea Col, BA, 69; Duke Univ, PhD(biochem), 74. *Prof Exp:* Vis res fel biochem, Dept Biol, Princeton Univ, 74-77; res biochemist biochem endocrinol, NIH, 77-79; scientist/adminr, Food & Drug Admin, 80-82; SUPVRY HEALTH SCIENTIST ADMINR, NIH, 82- *Concurrent Pos:* Nat Cancer Inst fel, Princeton Univ, 74-77; NIH fel, 77-79; lectr, NIH, 83- *Mem:* AAAS; Grad Women in Sci; Am Chem Soc; Am Asn Pathologists; Am Inst Nutrit. *Res:* Glycoprotein biochemistry, endocrinology, cell biology, especially role of glycoproteins and glycolipids in disease processes; toxicology and pharmacology of drugs and food additives; physiology and pathophysiology of gastrointestinal and hepatic diseases; nutrition and disease prevention; cell and molecular biology. *Mailing Add:* Div Res Grants Westwood 235 NIH Bethesda MD 20892

DEAN, EUGENE ALAN, b Freeport, Tex, Nov 30, 31; m 56; c 3. PHYSICS. *Educ:* Univ Tex, El Paso, 58; NMex State Univ, MS, 64; Tex A&M Univ, PhD(physics), 69. *Prof Exp:* Physicist, Schellenger Res Lab, 58-61, dir spec proj, 63-65; instr physics, Univ Tex, El Paso, 61-63; asst prof, Tex Southern Univ, 67-69; ASSOC PROF PHYSICS, UNIV TEX, EL PASO, 69- *Concurrent Pos:* Consult, Globe Exploration Co, 65-68; asst prof, Univ Houston, 69. *Mem:* Am Asn Physics Teachers. *Res:* Atmospheric acoustics; plasma physics. *Mailing Add:* Dept Physics Univ Tex El Paso TX 79968-0515

DEAN, FREDERICK CHAMBERLAIN, b Boston, Mass, May 22, 27; m 50; c 3. WILDLIFE MANAGEMENT. *Educ:* Univ Maine, BS, 50, MS, 52; State Univ NY, PhD, 57. *Prof Exp:* From asst prof to prof wildlife mgt, Univ Alaska, 54-90, leader, Alaska Coop Park Studies Unit, 72-82; RETIRED. *Concurrent Pos:* Asst leader, Coop Wildlife Res Unit, Univ Alaska, 54-72, head dept wildlife & fisheries, 54-73, ed, Biol Papers, 55-72, assoc dean, Col Biol Sci & Renewable Resources, 72-73; fel systs anal in ecol, Dept Bot, Univ Tenn, 68-69; fel, Oak Ridge Nat Lab, 68-69. *Mem:* Wildlife Soc; Am Soc Mammal; Bear Biol Asn. *Res:* Wildlife management, population ecology; grizzly bears, general ecology; subarctic ecology; wildland park management. *Mailing Add:* 810 Ballaine Rd Fairbanks AK 99709

DEAN, HERBERT A, b Damon, Tex, Sept 1, 18; c 2. ENTOMOLOGY. *Educ:* Tex A&M Univ, BS, 40, MS, 49. *Prof Exp:* Instr, Tex Agr Exp Sta, Col Sta, Tex A&M Univ, 46-49, from asst entomologist to assoc entomologist, 49-67, assoc prof entom, Agr Res & Ext Ctr, 67-81; RETIRED. *Mem:* Entom Soc Am. *Res:* Citrus mite control with selective miticides; biological control of citrus insects and mites; oil tolerance studies with citrus; citrus pest management through integrated control methods; author of 95 publications. *Mailing Add:* 704 S Georgia Ave Weslaco TX 78596

DEAN, JACK HUGH, b Joplin, Mo, Dec 6, 41; m 62; c 3. IMMUNOLOGY, MICROBIOLOGY. *Educ:* Calif State Univ, Long Beach, BS, 65, MS, 68; Univ Ariz, PhD(molecular biol), 72. *Prof Exp:* Microbiologist, Mem Hosp, Long Beach, 65-67; lab supvr, Kaiser Found Hosp, San Diego, 67-68; immunologist, Litton Bionetics, 72-74, assoc dir, 74-76, dir dept immunol, 76-; AT CHEM INDUST INST TOXICOL. *Concurrent Pos:* Lectr immunol, George Mason Univ, 75-76; mem, NIH Grants Rev Study Sect, 78- *Mem:* Am Soc Microbiol; Am Asn Cancer Res; Am Asn Immunologists; AAAS. *Res:* Tumor immunobiology; study of anti-tumor immunity and general immunocompetence in tumor bearing mice and humans; modulation of immunocompetence by environmental chemicals; regulation and control of cell-mediated immune responses. *Mailing Add:* Dept Toxicol Sterling-Winthrop Res Inst 81 Columbia Turnpike Rensselaer NY 12144

DEAN, JACK LEMUEL, b Keota, Okla, Mar 15, 25; m 49; c 2. PHYTOPATHOLOGY. *Educ:* Okla State Univ, BS, 49, MS, 52; La State Univ, PhD(bot, phytopath), 65. *Prof Exp:* Plant pathologist, Sugar Crops Field Sta, USDA, 51-66, plant pathologist, Sugarcane Field Sta, Canal Pt Fla, 66-87; PROF PLANT PATH, UNIV FLA, 83- *Mem:* Am Phytopath Soc; Int Soc Sugarcane Technol; Am Inst Biol Sci. *Res:* Pathology of sugarcane and sugar sorghum. *Mailing Add:* 92 Dayton Rd Lake Worth FL 33467

DEAN, JEFFREY STEWART, b Lewiston, Idaho, Feb 10, 39; m 59, 77; c 2. ARCHAEOLOGY, DENDROCHRONOLOGY. *Educ:* Univ Ariz, BA, 61, PhD(anthrop), 67. *Prof Exp:* Res assoc dendrochronology, 64-66, from instr to asst prof, 66-72, assoc prof, 72-77, PROF DENDROCHRONOLOGY, UNIV ARIZ, 77- *Concurrent Pos:* Vis scientist, adj prof anthrop, Southern Ill Univ, 85-86. *Mem:* Fel AAAS; fel Am Anthrop Asn; Soc Am Archaeol. *Res:* Archaeological theory and method; archaeology of the Southwestern United States; archaeological tree-ring dating; dendroclimatology of the Southwestern United States. *Mailing Add:* Lab Tree-Ring Res Univ Ariz Tucson AZ 85721

DEAN, JOHN AURIE, b Sault Ste Marie, Mich, May 9, 21; m 43, 81; c 5. ANALYTICAL CHEMISTRY. *Educ:* Univ Mich, BS, 42, MS, 44, PhD(chem), 49. *Prof Exp:* Chemist, Chrysler Corp, 44-45; lectr, Univ Mich, 46-48; assoc prof, Univ Ala, 48-50; from asst prof to assoc prof, 50-58, prof 58-81, EMER PROF CHEM, UNIV TENN, KNOXVILLE, 81- *Concurrent Pos:* Consult, Nuclear Div, Union Carbide Corp, 53-74 & Stewart Labs, Inc, 68-81; Ed, McGraw-HillBook Co, 68-; lectr, Peoples Repub China, 85. *Honors & Awards:* Charles H Stone Award, Carolina-Piedmont Sect Am Chem Soc, 74; Distinguished Serv Award, Soc Appl Spectros, 91. *Mem:* Am Chem Soc; Soc Appl Spectros; Archaeol Inst Am; US Naval Inst. *Res:* Flame emission and atomic absorption spectrometry; solvent extraction; instrumental methods of analysis; voltammetry. *Mailing Add:* 201 Mayflower Dr Knoxville TN 37920-5871

DEAN, JOHN GILBERT, b Pawtuxet, RI, Feb 16, 11; m 42; c 3. CHEMISTRY. *Educ:* Brown Univ, PhB, 31, MS, 32; Columbia Univ, PhD(chem), 36. *Prof Exp:* Dir res div, Permutit Co, 36-40; teacher sci, Sarah Lawrence Col, 40-42; Int Nickel Co Sr fel, Mellon Inst, 42-45, in-chg develop & res div, Indust Chem Sect, 46-51; CHEM CONSULT & MGR, DEAN ASSOCS, 52- *Concurrent Pos:* Dir div co-op res, Sch Eng, Columbia Univ, 52-54; dir res & develop div, Nickel Processing Corp Div, Nat Lead Co, 53-60. *Mem:* Am Chem Soc; Am Inst Mining, Metall & Petrol Engrs. *Res:* Management of chemical and metallurgical research; inorganic syntheses; ion exchange; catalysis; air and water pollution control; chemical economics; chemistry and chemical metallurgy of the metals, especially nickel, cobalt, copper, zinc, molybdenum, cadmium, mercury and the precious metals. *Mailing Add:* 2005 Gridley Ave Reno NV 89503

DEAN, JOHN MARK, b Cedar Rapids, Iowa, Oct 2, 36; m 60; c 3. FISH & WILDLIFE SCIENCES, FISHERIES ECOLOGY. *Educ:* Cornell Col, BA, 58; Purdue Univ, MS, 60, PhD(biol sci), 62. *Prof Exp:* Res assoc, Marine Lab, Duke Univ, 62-63; biol scientist, Hanford Labs, Gen Elec Co, 63-64 & Pac Northwest Labs, Battelle Mem Inst, 65-70; assoc prof marine sci, Belle W Baruch Inst Marine Biol & Coastal Res, Univ SC, 73-77, assoc prof biol, 70-77, dir marine sci prog, 76-82, PROF MARINE SCI & BIOL, BELLE W BARUCH INST MARINE BIOL & COASTAL RES, UNIV SC, 77- *Concurrent Pos:* Mem fac fisheries, Hokkaido Univ, Hakadate, Japan, 77, Nagasaki Univ, Japan, 85, SC Coastal Coun, 79-83, S Atlantic Fishery Mgt Coun, 87-90. *Mem:* Fel AAAS; Ecol Soc Am; Am Soc Zoologists; Am Fish Soc; Southeastern Estuarine Res Soc; Est Res Fedn; Am Inst Fishery Res Biologists. *Res:* Physiological ecology of estuarine fish; age and growth of fishes; fisheries management. *Mailing Add:* 127 S Edisto Ave Columbia SC 29205

DEAN, JURRIEN, b New York, NY, Mar 22, 47; m 76; c 4. INTERNAL MEDICINE, MEDICAL GENETICS. *Educ:* Columbia Univ, BA, 69, MD, 73. *Prof Exp:* Med resident, Presby Hosp, New York City, 73-75; res assoc biochem, Lab Chem Biol, Nat Inst Diabetes & Digestive & Kidney Dis, NIH, 75-79, sr investr develop biol, 79-83 & Lab Cellular & Develop Biol, 83-91, SECT CHIEF, MAMMALIAN DEVELOP BIOL SECT, LAB CELLULAR & DEVELOP BIOL, NAT INST DIABETES & DIGESTIVE & KIDNEY DIS, NIH, 91- *Concurrent Pos:* Comn officer, USPHS, 75-; fel med genetics, Johns Hopkins Univ, 77-78. *Mem:* Am Soc Biochem & Molecular Biol; Am Soc Clin Invest; Am Soc Cell Biol; Am Soc Develop Biol; Am Soc Human Genetics. *Res:* Mechanisms of gene expression during mammalian oogenesis and early development; genes that encode the mouse and human zonae pellucidae. *Mailing Add:* Lab Cellular & Develop Biol Nat Inst Diabetes & Digestive & Kidney Bldg 6 Rm B1-26 NIH Bethesda MD 20892

DEAN, NATHAN WESLEY, b Johnson City, Tenn, Dec 10, 41; m 63; c 1. SCIENCE AND SOCIETY. *Educ:* Univ NC, BS, 63, Cambridge Univ, PhD(physics), 68. *Prof Exp:* Vis scientist, Europ Orgn Nuclear Res, Geneva, Switz, 67-68; from instr to asst prof physics, Vanderbilt Univ, 68-70; asst prof, Iowa State Univ, 70-74, assoc prof, 74-78, prof physics, 78-80, asst dean sci & humanities & asst dir, Sci & Human Res Inst, 74-80; prof physics & asst vpres res, 80-84, ACTG VPRES RES, UNIV GA, 84- *Concurrent Pos:* Assoc physicist, Ames Lab, US Dept Energy, 70-74, physicist, 74-78, sr physicist, 78-80; mem admin staff, US Atomic Energy Comn, Germantown, Md, 73. *Mem:* Am Phys Soc. *Res:* Theories of elementary particles and strong interactions; interactions of science with religion and society. *Mailing Add:* 412 Clarkson Dr Vestal NY 13850

DEAN, PHILIP ARTHUR WOODWORTH, b Oldham, Lancashire, Eng, June 10, 43; m 65; c 2. INORGANIC CHEMISTRY. *Educ:* Univ London, BSc, 64; Royal Col Sci, London, ARCS, 64; Univ London, PhD(chem), 67; Imperial Col, London, DIC, 67. *Prof Exp:* Asst prof chem, Wash Univ, 70-71; from asst prof to assoc prof, 71-84, PROF CHEM, UNIV WESTERN ONT, 84- *Mem:* Chem Inst Can; Royal Soc Chem; Am Chem Soc. *Res:* Chemistry of metal ions in weakly co-ordinating solvents; inorganic aspects of biological activity of metal ions; applications of NMR in inorganic and organometallic chemistry. *Mailing Add:* Dept Chem Univ Western Ont London ON N6A 5B7 Can

DEAN, RICHARD A, b Brooklyn, NY, Dec 22, 35; m 57; c 3. TECHNICAL MANAGEMENT. *Educ:* Ga Inst Technol, BS, 57; Univ Pittsburgh, MS, 63, PhD, 70. *Prof Exp:* Engr, Westinghouse Nuclear Energy Systs, Pittsburgh, 60-66, mgr thermal-hydraul eng, 66-70; tech dir LWR Fuel Div, Gulf Gen Atomic, 70-71, vpres, 71-74; div dir, 74-76, vpres, Ga Technol Inc, 76-86; SR VPRES, GENERAL ATOMICS. *Mem:* Am Soc Mech Engrs; Am Nuclear Soc. *Res:* Boiling heat transfer and two phase flow; flow instability; nuclear reactor and fuel design and development; nuclear fuel reprocessing; management of uranium litigation; high temperature gas cooled reactor technology; nuclear waste management. *Mailing Add:* General Atomics 3550 General Atomics Ct San Diego CA 92121-1194

DEAN, RICHARD ALBERT, b Columbus, Ohio, Oct 9, 24; m 48, 79; c 1. MATHEMATICS. *Educ:* Calif Inst Technol, BS, 45; Denison Univ, BA, 47; Ohio State Univ, MA, 48, PhD(math), 53. *Hon Degrees:* DSc, Denison Univ, 73. *Prof Exp:* Instr physics, Middlebury Col, 47; Bateman fel math, 54-55, from asst prof to prof, 55-87, EMER PROF MATH, CALIF INST TECHNOL, 87- *Concurrent Pos:* Mem Math Adv Comts, Calif State Dept Educ, 60- *Mem:* Am Math Soc; Math Asn Am. *Res:* Abstract algebra; partially ordered sets and lattices; combinatorics and finite mathematics. *Mailing Add:* 6349 Stone Bridge Rd Santa Rosa CA 95409

DEAN, RICHARD RAYMOND, b Pittsburgh, Pa, June 23, 40; m 64; c 2. PHARMACOLOGY. *Educ:* Duquesne Univ, BS, 62; Univ Mich, PhD(pharmacol), 66. *Prof Exp:* Res investr, G D Searle & Co, 66-71, sr res investr, 71-72, group leader cardiovasc pharmacol, 72-76, asst dir, Clin Cardiovasc-Renal Sect, 76-78, assoc dir, 78-80; MGR, CARDIOVASC PHARMACOL DEPT, ABBOTT LABS, 80- *Concurrent Pos:* Mem coun basic sci, Am Heart Asn. *Mem:* Am Soc Clin Pharmacol & Therapeut; Am Chem Soc. *Res:* Laboratory and clinical evaluation of drugs on the cardiovascular system. *Mailing Add:* Abbott Labs 6 RF North Chicago IL 60064

DEAN, ROBERT CHARLES, JR, b Atlanta, Ga, Apr 13, 28; m 51; c 5. MECHANICAL ENGINEERING, BIOTECHNOLOGY. *Educ:* Mass Inst Technol, SB & SM, 49, ScD(mech eng), 54. *Prof Exp:* Engr, Ultrasonic Corp, Mass, 49-51; asst prof mech eng, Mass Inst Technol, 51-56; head adv eng dept, Ingersoll-Rand Co, NJ, 56-60; dir res, Thermal Dynamics Corp, NH, 60-61; founder & pres, Creare Inc, 61-75; founder, chmn & prin engr, Creare Innovations, 76-79; founder & chmn bd, Verax Corp, 79, pres, 79-83, dir sci & technol, 79-83; founder & chmn, Synosys Inc, 87-89; FOUNDER & PRES, DEAN TECHNOL INC, 89- *Concurrent Pos:* Mem turbine & compressor subcomt, NACA, 54-55; venture capital panel, Com Tech Adv Bd, Dept Com; adj prof eng, Thayer Sch Eng, Dartmouth Col, 61-, Northeastern Univ. *Honors & Awards:* Prod Eng Master Designer Award, 67. *Mem:* Nat Acad Eng; Am Chem Soc; fel Am Soc Mech Engrs; Am Inst Chem Engrs. *Res:* Research and development of advanced bioprocessing for manufacture of biochemicals; impact; vibration; rock drilling; heat transfer; acoustics; materials testing; total replacement artificial hearts; medical instruments; office, agricultural and forest machinery; numerous articles and publications. *Mailing Add:* PO Box 318 Norwich VT 05055

DEAN, ROBERT GEORGE, b Laramie, Wyo, Nov 1, 30; m 54; c 2. COASTAL ENGINEERING, OCEAN ENGINEERING. *Educ:* Univ Calif, Berkeley, BS, 54; Agr & Mech Col Tex, MS, 56; Mass Inst Technol, ScD(civil eng), 59. *Prof Exp:* Asst prof civil eng, Mass Inst Technol, 59-60; res engr coastal res, Calif Res Corp, Standard Oil Co, Calif, 60-63, sr res engr, 63-65; actg assoc prof oceanog, Univ Wash, 65-66; chmn, Coastal & Oceanog Eng Dept, Univ Fla, Gainesville, 66-72, prof, 66-75; Unidel prof, Dept Civil

Eng & Col Marine Studies, Univ Del, Newark, 75-82; GRAD RES PROF, COASTAL & OCEANOG ENG DEPT, UNIV FLA, GAINESVILLE, 82- *Concurrent Pos:* Assoc ed, Marine Sci Commun; United Gas fel eng oceanog, 54-55; John R Freeman fel hydraul, 58-59; coastal & ocean eng consult, various firms & govt agencies; mem, Coastal Eng Res Bd, US Army CEngr, 68-80, Marine Bd, Nat Res Coun, 81 & Comt Natural Disasters, 82; chmn working group, Nat Acad Eng, 81; dir, Div Beaches & Shores, Fla Dept Natural Resources, Tallahassee, 85-87. *Honors & Awards:* John G Moffatt-Frank E Nichol Harbor & Coastal Eng Award, Am Soc Civil Eng, 87. *Mem:* Nat Acad Eng; Am Soc Civil Engrs; AAAS; Am Geophys Union; Sigma Xi. *Res:* Physical oceanography; nonlinear water wave mechanics; interaction of waves with structures; general coastal engineering problems; potential flow applications; problems associated with nearshore and offshore siting of power plants; author of many publications in the areas of wave theories, beach erosion problems, tidal inlets and coastal structures. *Mailing Add:* Dept Coastal & Oceanog Eng Univ Fla 336 Weil Hall Gainesville FL 32611

DEAN, ROBERT REED, b Bedford Ind, Apr 18, 14; m 49; c 2. CHEMISTRY. *Educ:* Ind Univ, 34-37; Northwestern Univ, 41. *Prof Exp:* Chemist, Johns-Manville Prod Corp, 38-40 & US Gypsum Res Lab, 40-41; inspector, US Army Ord, 41-42; chemist, Am Cyanamid, 48-49, Diamond Alkali Co, 49-50 & Va Carolina Chem Corp, 50-53; mgr mkt res, Westvaco Chlor-alkali Div, FMC Corp, NY, 53-56, dir mkt res & develop, Res Labs, WVa, 56-58, mgr com develop, 58-59, mgr inorg chem div mkt res develop, 59-71; CONSULT COM DEVELOP & INDUST MKT RES, 71- *Mem:* Am Chem Soc; Am Inst Chem; Chem Mkt Res Asn; Com Chem Develop Asn. *Res:* Introduction and commercial development of new chemicals; application of existing chemicals to new uses; evaluation of markets for all commercial chemicals. *Mailing Add:* 3137 NW 57th Pl Gainesville FL 32606

DEAN, ROBERT WATERS, b West Chester, Pa, June 6, 29; m 62; c 5. FOOD SCIENCE, NUTRITION. *Educ:* Bates Col, BS, 51; Rutgers Univ, MS, 57, PhD(food chem, technol), 59. *Prof Exp:* Chemist gelatin desserts, Standard Brands, Inc, 51-53; technologist, Dessert Prod & Beverages, Gen Foods Corp, 53-55; res chemist, Wilson & Co, 59-61; prod develop chemist, Glidden Co, 61-62; res chemist, Peter Eckrich & Sons, 62-64; asst tech dir, F & F Labs, 64-67; chief chemist, Paradise Fruit Co, 67-70; chemist, Growers Processing Serv, Inc, 71-72; COORDR, NUTRIT & TECH SERVS, MIDWEST REGION, FOOD & NUTRIT SERV, USDA, 72- *Mem:* Am Chem Soc; Soc Nutrit Educ; Inst Food Technologists; AAAS. *Res:* Determination of pigments responsible for color of fresh meat; child nutrition; food stamps; food distribution. *Mailing Add:* Food & Nutrit Serv USDA 50 E Washington Chicago IL 60602-2102

DEAN, ROBERT YOST, b Portland, Ore, Jan 13, 21; m 54; c 2. MATHEMATICS. *Educ:* Willamette Col, BA, 42; Calif Inst Technol, MS, 42, PhD(math), 52. *Prof Exp:* Mem staff, Inst Math, Case Inst Technol, 46-48; sr mathematician, Hanford Labs Oper, Gen Elec Co, 52-63, mgr math subsect, 64; math sect, Appl Math Dept, Pac Northwest Labs, Battelle Mem Inst, 64-68; prof math & chmn dept, Cent Wash Univ, 68-86; RETIRED. *Concurrent Pos:* Lectr, Ctr Grad Study, Univ Wash, 52- *Mem:* Math Asn Am; Am Math Soc; Soc Indust & Appl Math. *Res:* Applied mathematics. *Mailing Add:* 6888 Starboard Lane Gig Harbor WA 98335

DEAN, SHELDON WILLIAMS, JR, b Flushing, NY, July 3, 35; m 60; c 3. CHEMICAL ENGINEERING, SURFACE CHEMISTRY. *Educ:* Middlebury Col, AB, 58; Mass Inst Technol, SB, 58, ScD(chem eng), 62. *Prof Exp:* Res asst chem eng, Mass Inst Technol, 60-61; mem sr staff res lab, Int Nickel Co, 61-64; eng specialist metal finishing, Metals Res Lab, 64-66, group supvr, metal finishing & bonding, 66-74, group supvr, Corrosion Group, Metals Res Lab, Olin Corp, 74-75; sr corrosion engr, 75-77, mgr mat eng, 77-84, CHIEF MAT ENG, CORPORATE ENG, AIR PROD & CHEM INC, 84- *Honors & Awards:* Sea Horse Inst Award, 66; Hamden Jaycee DSA Award, 70; Award of Merit, Am Soc Testing & Mat, 86; Honor Award, Am Inst Chemists, 90; F N Speller Award, Nat Asn Corrosion Eng, 91. *Mem:* Electrochem Soc; fel Am Inst Chem Eng; Am Chem Soc; Nat Asn Corrosion Eng; Am Soc Testing & Mat. *Res:* Corrosion inhibition; electrochemical corrosion testing; stress corrosion cracking; corrosion kinetics. *Mailing Add:* Air Prods & Chem Inc Allentown PA 18195

DEAN, STEPHEN ODELL, b Niagara Falls, NY, May 12, 36; div; c 3. FUSION ENERGY DEVELOPMENT. *Educ:* Boston Col, BS, 60; Mass Inst Technol, SM, 62; Univ Md, PhD(physics), 71. *Prof Exp:* Physicist, Atomic Energy Comn, 62-68; physicist lasers, Naval Res Lab, 68-72; div dir fusion, Atomic Energy Comn, Energy Res & Develop Admin, Dept Energy, 72-79; PRES, FUSION POWER ASSOCS, 79- *Concurrent Pos:* Dir fusion energy, Sci Appl Inc, 79- *Honors & Awards:* Res Publ Award, US Naval Res Lab, 71; Spec Achievement Award, US Energy Rest Develop Admin, 76, 77; Distinguished Assoc Award, US Dept Energy, 88. *Mem:* Am Physical Soc; fel Am Nuclear Soc; Am Soc Asn Exec. *Res:* Fusion plasmas; laser interactions with plasmas; magnetic fusion programs. *Mailing Add:* Fusion Power Assocs Two Professional Dr Suite 248 Gaithersburg MD 20879

DEAN, SUSAN THORPE, b Nashville, Tenn. COMPUTER SCIENCE EDUCATION. *Educ:* Vanderbilt Univ, BA, 70; Univ Ala, Birmingham, MS, 74, PhD(comput sci), 83. *Prof Exp:* Programmer analyst, Univ Ala, Birmingham, 70-77, from instr to asst prof comput sci, 75-85; ASSOC PROF COMPUT SCI, SAMFORD UNIV, 85- *Mem:* Asn Comput Mach; Inst Elec & Electronics Engrs. *Res:* Graph theory; computational complexity; computer science education. *Mailing Add:* Dept Math Eng & Comput Sci Samford Univ 800 Lakeshore Dr Birmingham AL 35229

DEAN, THOMAS SCOTT, b Sherman, Tex, July 6, 24; m 45; c 3. SOLAR TECHNOLOGY, ENGINEERING. *Educ:* N Tex State Univ, BS, 45, MS, 47; Univ Tex, PhD(eng mech), 63. *Prof Exp:* Owner, Thomas Scott Dean, Architect & Engr Dynamics Consult, Tex, 50-60; lectr archit eng, Univ Tex, 60-64; chmn archit sci, Okla State Univ, 64-75, prof archit, 70-75; PROF ARCHIT & ENG, UNIV KANS, 76- *Concurrent Pos:* Lectr, Southern Methodist Univ, 55-59; consult, Tex Indust, Inc, 55-59; fac fel, Latin Am Studies Inst, 63; mem, Hot Weather Res Inst, 63-64; archit eng consult, Brackenridge Field Lab, 63-64; vis prin lectr, NE London Polytech Inst, 73-74; mem prof adv panel, Gen Servs Admin, 77-78; mem, Nat Voluntary Lab Accred Prog Comn, Dept Com, 78-82. *Mem:* Am Inst Archit; Am Soc Heat, Refrig & Air-Conditioning Engrs. *Res:* Design criteria for research structures; innovative teaching methods; societal aspects of technology; solar energy systems; energy conservation. *Mailing Add:* Sch Archit Eng Univ Kans Lawrence KS 66045

DEAN, WALTER ALBERT, physical metallurgy; deceased, see previous edition for last biography

DEAN, WALTER E, JR, b Wilkes-Barre, Pa, July 12, 39; m 61; c 2. GEOCHEMISTRY, SEDIMENTOLOGY. *Educ:* Syracuse Univ, AB, 61; Univ NMex, MS, 64, PhD(geol), 67. *Prof Exp:* Res asst geol, Univ NMex, 63-64; res assoc bot, Univ Minn, 67-68; from asst prof to assoc prof geol, Syracuse Univ, 68-75; GEOLOGIST, US GEOL SURV, COLO, 76- *Mem:* Geol Soc Am; Soc Econ Paleont & Mineral; Am Soc Limnol & Oceanog; Int Asn Sedimentology; AAAS; Sigma Xi; Am Geophys Union. *Res:* Geochemistry of evaporite and carbonate deposits; geochemistry of lakes and lake sediments; geochemistry of marine sediments. *Mailing Add:* US Geol Surv Fed Ctr Mail Stop 939 Denver CO 80225

DEAN, WALTER KEITH, b Big Timber, Mont, Nov 2, 17; m 41; c 4. INORGANIC CHEMISTRY. *Educ:* Univ Wis, BS, 39; Univ Mo- Rolla, MS, 41. *Prof Exp:* Asst chem, Sch Mines & Metall, Univ Mo-Rolla, 39-41; Chemist, 41-69, res supvr, 69-82, ASSOC SCIENTIST, SCI PROD DIV, MALLINCKRODT CHEM WORKS, 82- *Concurrent Pos:* lectr, Univ Col, Wash Univ, 53-62. *Mem:* Am Chem Soc; Am Soc Testing & Mat. *Res:* Chromatography chemicals; high-purity silicon; rhenium; analytical reagents; ultrapure organic solvents. *Mailing Add:* 457 N Elizabeth Ave St Louis MO 63135-3509

DEAN, WALTER LEE, b Lenoir City, Tenn, Dec 13, 28; m 53; c 3. ORGANIC CHEMISTRY, INORGANIC CHEMISTRY. *Educ:* Maryville Col, BS, 50; Univ Tenn, MS, 53, PhD(org chem), 56. *Prof Exp:* RES CHEMIST, BUCKEYE CELLULOSE CORP, 56- *Mem:* Am Chem Soc; Soc Petrol Engrs. *Res:* Organic chemistry and synthesis; natural products; cellulose derivatives; graft polymer synthesis; nonwoven binders; petroleum recovery; absorbent structures; superabsorbent materials. *Mailing Add:* 5226 Quince Rd Memphis TN 38117-6840

DEAN, WARREN EDGELL, b Richwood, WVa, Aug 1, 32; m 59; c 3. PHYSICAL CHEMISTRY, INORGANIC CHEMISTRY. *Educ:* WVa Inst Technol, BS, 54; WVa Univ, MS, 57, PhD, 59. *Prof Exp:* Instr chem, WVa Univ, 58-59; with Columbia-Southern Chem Corp, 59; supvr process res, 59-74, supt org area, 74-78, tech supt, 79-80, plant tech mgr, 80-86, LAB MGR, CHEM DIV, NATRIUM PLANT, PPG INDUSTS, INC, 86- *Mem:* Am Chem Soc. *Res:* Chelate compounds of alkanol-substituted ethylendiamines; process research; inorganic heavy chemicals. *Mailing Add:* 1252 Wren Dr New Martinsville WV 26155-2832

DEAN, WILLIAM C(ORNER), b Pittsburgh, Pa, Nov 21, 26; m 50; c 2. ELECTRICAL ENGINEERING. *Educ:* Carnegie Inst Technol, BS, 49, MS, 50, PhD, 52. *Prof Exp:* Res geophysicist, Gulf Res & Develop Co, 52-60; sr eng specialist, United Electrodynamics & Teledyne, Inc, 60-62, proj eng, 62, chief res sect, 62-66, mgr seismic data lab, Teledyne, Inc, 66-70, MGR SEISMIC ARRAY ANAL CTR, TELEDYNE-GEOTECH, 70- *Mem:* Inst Elec & Electronics Engrs; Soc Explor Geophys; Sigma Xi. *Res:* Circuit theory; information theory; data processing; analog and digital computers; potential fields; wave propagation. *Mailing Add:* 1301 Namassin Rd Alexandria VA 22308

DEAN, WILLIAM L, MEMBRANE PROTEIN, CALCIUM TRANSPORT. *Educ:* Univ Mich, PhD(biochem), 76. *Prof Exp:* ASSOC PROF BIOL CHEM, UNIV LOUISVILLE, 79- *Mailing Add:* Dept Biochem Univ Louisville Health Sci Ctr Louisville KY 40292

DEANE, DARRELL DWIGHT, b Anacortes, Wash, Nov 9, 15; m 43; c 3. DAIRY BACTERIOLOGY. *Educ:* Univ Idaho, BS, 38; Univ Nebr, MSc, 39; Pa State Univ, PhD(bact), 42. *Prof Exp:* Asst prof dairy husb, Univ Nebr, 46-51; asst prof dairy indust, Iowa State Univ, 51-61; from assoc prof to prof, 61-79, EMER PROF DAIRY MFG, UNIV WYO, 79- *Mem:* Am Dairy Sci Asn. *Res:* Ripening of cheddar cheese; lactic streptococcus bacteriophages; microbial spoilage of cottage cheese; cheese packaging; direct acid manufacture of sour cream and cottage cheese; manufacture of low fat cheese; keeping quality of fluid milk. *Mailing Add:* 3124 Silverwood Dr Ft Collins CO 80525-2848

DEANE, NORMAN, b Newark, NJ, Aug 20, 21; m 56; c 2. INTERNAL MEDICINE, NEPHROLOGY. *Educ:* Temple Univ, BA, 43, MD, 46; Am Bd Internal Med, dipl. *Prof Exp:* Instr physiol, Col Med, NY Univ, 49-51, from instr to asst prof med, Post Grad Med Sch, 53-61, asst prof clin med, 54-56, lectr physiol, Col Med, 55-61, dir clins, Univ Hosp & Med Ctr, 56-60; ASSOC PROF MED & ASSOC ATTEND PHYSICIAN, NEW YORK MED COL, FLOWER & FIFTH AVE HOSPS, 61- *Concurrent Pos:* Vis investr, Med Dept, Brookhaven Nat Lab, 50-51; asst vis physician, Fourth Med Div, NY Univ, 53-61, assoc attend physician, Univ Hosp & Med Ctr, 58-61; Polachek fel med res, 56-61; assoc vis physician, Metrop Hosp, 61-69, vis physician, 70-; attend physician, Lenox Hill Hosp, 67-79; mem adv coun, NY State Kidney Dis Inst, 67-70; mem renal dialysis comt, Health & Hosp Planning Coun Southern NY, 67-; mem renal dis subcomt, New York Metrop Regional Med Prog, 68-70; mem hypertension & renal dis comt, NJ Regional Med Prog, 68-74; mem nephrology adv comt, New York Medicaid Prog, 68-72; consult physician, Englewood Hosp, 68-78, Hackensack Hosp, 69-76 & St Vincent's Hosp & Med Ctr, 71-; mem med bd, Metrop Hosp, 70-; med dir, Manhattan Kidney Ctr, New York Nephrol Found, 71-73; med dir, Nat

Nephrology Found, 73-, pres, 81-; pres, Renal Physicians Asn, 79-81, counr, 81-; consult physician nephrology, Bronx Lebanon Hosp Ctr, 81- *Mem:* Am Physiol Soc; fel Am Col Physicians; Am Soc Nephrology; Am Soc Artificial Internal Organs; Soc Exp Biol & Med. *Res:* Nephrology; integration of normal cardio-pulmonary, renal and liver function in man and alterations produced by disease; chemical compositon and mechanisms of control of the internal environment of the body; computerized medical information systems. *Mailing Add:* Manhattan Kidney Ctr 40 E 30th St New York NY 10016

DEANGELIS, DONALD LEE, b Baltimore, Md, June 7, 44; div; c 1. ECOLOGY. *Educ:* Mass Inst Technol, BS, 66; Yale Univ, PhD(plasma physics), 72. *Prof Exp:* Presidential intern ecol, Oak Ridge Nat Lab, Union Carbide, 72-73; res assoc ecol, Inst Soil Sci & Forest Fertil, Goettingen Univ, WGer, 73-74; res staff mem, 76-83, RES ASSOC ECOL, OAK RIDGE NAT LAB, UNION CARBIDE, 74-, SR SCIENTIST, 83- *Concurrent Pos:* Vis scientist, Cornell Univ, 82; guest scholar, Kyoto Univ, 88. *Mem:* Ecol Soc Am; Fel AAAS; Am Phys Soc; Sigma Xi. *Res:* Population and ecosystem modeling and the analysis of the mathematical properties of ecological models. *Mailing Add:* Environ Sci Div Oak Ridge Nat Lab PO Box X Oak Ridge TN 37831

DEANHARDT, MARSHALL LYNN, b Anderson, SC, Sept 12, 48; m 70; c 2. ANALYTICAL CHEMISTRY. *Educ:* Clemson Univ, BS, 70; NC State Univ, PhD(anal chem), 75. *Prof Exp:* From asst prof to assoc prof chem, George Mason Univ, 75-85; assoc prof, 85-90, PROF CHEM, LANDER COL, 90- *Concurrent Pos:* Consult, Naval Res Lab, Washington, DC. *Mem:* Am Chem Soc. *Res:* Molten salt and aqueous electrochemistry; chemical instrumentation; computer interfacing of laboratory instrumentation. *Mailing Add:* Dept Sci & Math Lander Col Greenwood SC 29649

DEANIN, RUDOLPH D, b Newark, NJ, June 7, 21; m 46; c 2. POLYMER CHEMISTRY. *Educ:* Cornell Univ, AB, 41; Univ Ill, MS, 42, PhD(org chem), 44. *Prof Exp:* Lab asst org chem, Cornell Univ, 40-41; jr sci aide, Regional Soybean Indust Prod Res Lab, USDA, 41-42; asst chemist, Nat Defense Res Comt Proj, Univ Ill, 42, asst instr, 42-43, from res asst to spec res asst, Off Rubber Res, 43-47; res chemist, proj leader & group leader, Allied Chem Corp, 47-60; dir chem res & develop, DeBell & Richardson, Inc, Conn, 60-67; PROF PLASTICS, UNIV LOWELL, 67- *Concurrent Pos:* Vis lectr, Univ Mass, 63-64, Brown Univ, 65-66 & Northeastern Univ, 67-68. *Honors & Awards:* Res Award, Plastics Inst Am, 89; Memorial Lectureship, Am Leather Chemists Asn, 89. *Mem:* Am Chem Soc; Soc Plastics Eng; Adhesion Soc; New England Soc Coating Technol. *Res:* Polymerization, compounding, properties, applications and economics of polymers and plastics. *Mailing Add:* Plastics Eng Dept Univ Lowell Lowell MA 01854

DEANS, HARRY A, b Dallas, Tex, June 17, 32; m 56; c 3. CHEMICAL ENGINEERING. *Educ:* Rice Inst, BA, 53, BS, 54, MS, 56; Princeton Univ, PhD(chem eng), 60. *Prof Exp:* Assoc prof, 59-70, prof chem eng, Rice Univ, 70-, assoc, Brown Col, 77-; AT DEPT CHEM ENG, UNIV HOUSTON, TEX. *Concurrent Pos:* Fulbright lectr, Israel, 64; Consult, Esso Prod Res Co. *Mem:* Am Inst Chem Engrs. *Res:* Multiphase, multicomponent fluid flow in porous media; gas-liquid and gas-solid chromatography of chemically reactive systems; chemisorption; bubble-column chromatography; tracer methods in petroleum reservoir evaluation; continuous chromatographic separation technology. *Mailing Add:* Dept Chem Eng Univ Houston University Park 4800 Calhoun Rd Houston TX 77204

DEANS, ROBERT JACK, b Ft Wayne, Ind, Dec 4, 27; m 50; c 5. ANIMAL HUSBANDRY. *Educ:* Ohio State Univ, BSc, 49; MSc, 50; Mich State Univ, PhD(animal husb), 56. *Prof Exp:* Instr animal husb, Ohio State Univ, 50-52; from instr to assoc prof, 52-74, PROF ANIMAL HUSB, MICH STATE UNIV, 74- *Concurrent Pos:* Adv, Univ Nigeria, 65. *Mem:* Am Soc Animal Sci; Am Meat Sci Asn. *Res:* Effects of endocrine-like substances in meat-animal production; beef and lamb carcass investigations. *Mailing Add:* Dept Animal Sci Mich State Univ 102 Anthony Hall East Lansing MI 48824

DEANS, SIDNEY ALFRED VINDIN, b Montreal, Que, Dec 31, 18; m 59; c 1. ORGANIC CHEMISTRY. *Educ:* McGill Univ, BSc, 39, PhD, 42. *Prof Exp:* Res chemist, Can Industs, Ltd, 42-46; develop chemist, Ayerst, McKenna & Harrison Ltd, 47-56; assoc dir chem develop, Union Carbide Can, Ltd, 56-62; tech dir, Pfizer Can Inc, 62-85; RETIRED. *Res:* Synthesis of ethers; polymerization; penicillin; estrogenic sulfates; tetraethylthiuram disulfide; synthetic organic chemicals; petrochemicals; pharmaceuticals. *Mailing Add:* 361 Lethbridge Ave Montreal PQ H3P 1E6 Can

DEAR, ROBERT E A, b Bristol, UK, June 5, 33; US citizen; m 56; c 3. LUBRICANT ADDITIVES, FLUOROCHEMICAL SPECIALTIES. *Educ:* Univ Western Ont, PhD(org chem), 63. *Prof Exp:* Sr res chemist, Allied Chem Corp, 64-70; group leader, CIBA-GEIGY Corp, 70-79; res dir, Elco Corp, Subsid of Detrex Corp, 79-83; VPRES PRODS TECHNOL, TECH DIV, PENNZOIL PROD CO, WOODLANDS, TX, 83- *Mem:* Am Chem Soc; Royal Soc Chem. *Res:* Synthesis and development of lubricant additives in the industrial and automotive fields. *Mailing Add:* Pennzoil Prod Co Tech Div PO Box 7569 Woodlands TX 77387

DEARBORN, DELWYN D, b Miller, SDak, Feb 5, 33; m 56; c 3. ANIMAL BREEDING. *Educ:* SDak State Univ, BS, 54, MS, 59; Univ Nebr, PhD(animal sci), 71. *Prof Exp:* Assoc county exten agent, SDak State Univ, 56-59, exten livestock specialist, 59-66; exten livestock specialist, Univ Nebr, 66-70; prof animal sci & head dept, 71-74, DEAN, COL AGR & BIOL SCI, SDAK STATE UNIV, 74- *Mem:* Am Soc Animal Sci; Coun Agr Sci & Technol. *Res:* Beef production. *Mailing Add:* 514 W Third St North Platte NE 69101

DEARBORN, JOHN HOLMES, b Bangor, Maine, Feb 26, 33; m 60; c 2. MARINE ZOOLOGY, MARINE ECOLOGY. *Educ:* Univ NH, BA, 55; Mich State Univ, MS, 57; Stanford Univ, PhD(biol), 65. *Prof Exp:* Asst, Stanford Univ, 58-64; from asst prof to assoc prof, 66-76, PROF ZOOL, UNIV MAINE, 76- *Concurrent Pos:* NSF fel, 65-66; mem higher inverts adv comt, Smithsonian Oceanog Ctr, 68- *Mem:* AAAS; Ecol Soc Am; Am Soc Zool; Soc Syst Zool. *Res:* Marine invertebrate zoology and ecology, especially Antarctic benthos; polar and deep sea echinoderms. *Mailing Add:* Dept Zool Univ Maine Orono ME 04469-0146

DEARDEN, BOYD L, b Coalville, Utah, Sept 1, 43; m 71; c 1. WILDLIFE ECOLOGY, SYSTEMS ECOLOGY. *Educ:* Univ Utah, BS, 65, MS, 67; Colo State Univ, PhD(syst ecol), 74. *Prof Exp:* Asst prof, 74-78, ASSOC PROF WILDLIFE ECOL, DEPT FORESTRY, WILDLIFE & FISHERIES & GRAD PROG ECOL, UNIV TENN, 78- *Mem:* Wildlife Soc; Am Soc Mammal; Am Inst Biol Sci; Soc for Comput Simulation Ecol. *Res:* Development of systems and analysis and simulation techniques for resource management applications. *Mailing Add:* Dept Forestry Univ Tenn Knoxville TN 37996

DEARDEN, DOUGLAS MOREY, b Echo, Utah, Aug 25, 23; m 48; c 3. ZOOLOGY, GENETICS. *Educ:* Univ Utah, BA, 47, MA, 49; Univ Calif, Berkeley, 49; Univ Minn, PhD(sci educ, zool), 59. *Prof Exp:* Asst biol & genetics, Univ Utah, 47-49; from instr to assoc prof biol, 51-69, PROF BIOL, UNIV MINN, MINNEAPOLIS, 69- *Mem:* AAAS; Nat Asn Res Sci Teaching. *Res:* Drosophila and human genetics. *Mailing Add:* Dept Sci-Bus-Math 106 Nicholson Univ Minn 216 Pillsbury Dr SE Minneapolis MN 55455-0117

DEARDEN, LYLE CONWAY, b Salt Lake City, Utah, Apr 27, 22; m 43; c 3. ANATOMY. *Educ:* Univ Utah, BA, 47, MA, 49, PhD(vert zool), 55. *Prof Exp:* Asst comp anat, Univ Utah, 47-49; asst instr comp anat & embryol, Univ Kans, 49-50; instr zool, Univ Mass, 50-53; instr biol, St Mary of Wasatch Col, 53-54; instr zool, Univ Mass, 54-55; from instr to asst prof anat, Med Sch, Univ Southern Calif, 55-59; assoc prof, Sch Med, George Washington Univ, 59-63; prof, Calif Col Med, 63-66; PROF ANAT, UNIV CALIF, IRVINE-CALIF COL MED, 66-, PROF RADIOL SCI, 77- *Concurrent Pos:* Instr biol, Univ Utah, 53-54; assoc, NSF, 59-63, consult, 63-; vis prof, Univ Rome, 73-74. *Mem:* Pan-Am Soc Anat; Am Asn Anat. *Res:* Histology and EM GI system cartilage. *Mailing Add:* Dept Anat Col Med Univ Calif Irvine CA 92664

DEARLOVE, THOMAS JOHN, b Syracuse, NY, Jan 28, 41; m 64. POLYMER CHEMISTRY. *Educ:* Norwich Univ, BS, 63; Rensselaer Polytechnic Inst, PhD(org chem), 67. *Prof Exp:* Adhesives chemist, Harry Diamond Labs, US Govt, Washington, DC, 67-69; SECT MGR, POLYMER DEPT, GEN MOTORS RES LABS, 69- *Mem:* Am Chem Soc; Adhesion Soc; Sigma Xi; Soc Automotive Engrs. *Res:* Fundamental aspects of adhesion; formulation and testing of adhesives; chemistry of silane primers; formulation of polyurethane sealants, the mechanism of tin catalysts in polyurethanes, rapid injection molding of epoxy resins, the chemistry of chopped fiber composites and properties of structural composits. *Mailing Add:* Polymers Dept Gen Motors Res Lab Warren MI 48090

DEARMAN, HENRY HURSELL, b Statesville, NC, Aug 28, 34; m 61. PHYSICAL CHEMISTRY. *Educ:* Univ NC, BS, 56; Calif Inst Technol, PhD(chem), 60. *Prof Exp:* Res assoc chem, Enrico Fermi Inst Nuclear Studies, Univ Chicago, 60-61; res chemist, Chemstrand Res Ctr, Inc, Monsanto Co, 61-62; from asst prof to assoc prof chem, 62-74, PROF CHEM, UNIV NC, CHAPEL HILL, 74- *Mem:* Am Chem Soc; Am Phys Soc. *Res:* Paramagnetic resonance spectra; electronic spectra of organic molecules and inorganic transition metal complexes. *Mailing Add:* Dept Chem Univ NC Chapel Hill NC 27514-2820

DEARMON, IRA ALEXANDER, JR, b Charlotte, NC, Sept 18, 20; m 50; c 4. STATISTICS. *Educ:* Va Polytech Inst, BS, 43, MS, 48. *Prof Exp:* Sanitarian, State Health Dept, Va, 47-50; statistician, Biomath Div, Ft Detrick, 54-60, res analyst, Chem Corps Opers Res Group, Army Chem Ctr, 60-64, opers res analyst, Opers Res Group, US Army Munitions Command, Edgewood Arsenal, 64-73; opers res analyst, US Army Armament Command, Rock Island, Ill, 73-76; biostatistician, Frederick Cancer Res Ctr, Litton Bionetics, Inc, 76-81; RETIRED. *Concurrent Pos:* Asst prof math, Harford Jr Col, 66-68; exec dir, Pine Bluff Arsenal, Ark, 68-72. *Mem:* Opers Res Soc Am; Biomet Soc; Am Defense Preparedness Asn. *Res:* Planning and coordination of experimental designs for scientific personnel in all areas of cancer research including clinical trials; statistical analyses on data collected in studies to identify candidate carcinogens, on chemotherapy fermentation, on bioassay, on animal health, on animal breeding and in life survival testing. *Mailing Add:* 1612 Jennings Ct Frederick MD 21702

DEARMOND, M KEITH, b Ft Wayne, Ind, Dec 10, 35; m 85; c 2. PHYSICAL CHEMISTRY, INORGANIC CHEMISTRY. *Educ:* DePauw Univ, BA, 58; Univ Ariz, PhD(phys chem), 63. *Prof Exp:* Res assoc magnetic resonance spectros, Univ Ill, 63-64; from asst prof to assoc prof, NC State Univ, 64-75, asst head, 79-81, prof chem, 75-87; PROF & HEAD CHEM DEPT, NMEX STATE UNIV, 87- *Concurrent Pos:* NIH trainee, 63-64; vis prof, Univ Bologna, Italy, 79; officer, NSF Prog, 82-83; vis prof, Univ Paris. *Mem:* Am Chem Soc; Sigma Xi; NAm Photochem Soc. *Res:* Luminescence of transition metal complexes to elucidate nonradiative processes; electrochemical and photoelectrochemical studies of metal complexes; electron spin resonance of transition metal complexes. *Mailing Add:* Dept Chem NMex State Univ Las Cruces NM 88003

DEARTH, JAMES DEAN, b Marietta, Ohio, Dec 22, 46. SYNFUELS RESEARCH, COMPUTER MODELING. *Educ:* Univ Calif, Berkeley, BS, 68; Mass Inst Technol, SM, 70, ScD, 79. *Prof Exp:* Sr prin res engr, Atlantic Richfield Co, 74-81; SR STAFF ENGR, EXXON RES & ENG CO, 81- *Mem:* Am Inst Chem Engrs. *Res:* Predevelopment activities on a proprietary process for shale oil generation by thermal pyrolysis of oil shale rock; operations center on a small continuous bench-scale reactor. *Mailing Add:* 4500 Sherwood Commons Apt No 1008 Baton Rouge LA 70816

DEAS, JANE ELLEN, b New Orleans, La, Mar 9, 33. MEDICAL MICROBIOLOGY, PARASITOLOGY. *Educ:* Loyola Univ, La, BS, 55; La State Univ, MS, 72, PhD(microbiol), 75. *Prof Exp:* Technologist pharmacol, Med Ctr, La State Univ, 55-56; technologist anat, Med Sch, Tulane Univ, 56-59; med res specialist, 59-74, ASST PROF TROP MED, MED SCH, LA STATE UNIV, NEW ORLEANS, 74- *Mem:* Am Soc Trop Med & Hyg; Electron Micros Soc Am; Am Soc Microbiol; Am Soc Parasitologists; Sigma Xi. *Res:* Erythrocyte membrane proteins and their relationships to parasitic mechanisms. *Mailing Add:* Dept Trop Med & Med Parasitol La State Univ Sch Med 1901 Perdido St New Orleans LA 70112

DEAS, THOMAS C, b Augusta, Ga, Aug 5, 21; m 43; c 2. ANESTHESIOLOGY. *Educ:* Univ Ga, BS, 42; Ga Sch Med, Md, 45; Am Bd Anesthesiol, dipl, 56. *Prof Exp:* Intern, US Naval Hosp, Parris Island, SC, 45-46, med officer, Post Med Detachment, 46-48; resident anesthesiol, Univ Hosp, Augusta, Ga, 48-49; anesthesiologist, Navy Hosp Ship Consolation, 49-51 & Naval Hosp, Jacksonville, Fla, 51, resident, Philadelphia, 51-52, chief anesthesiol serv, 56-63, anesthesiologist, Bainbridge, Md, 52-55; PROF ANESTHESIOL, TEMPLE UNIV, 63- *Concurrent Pos:* Consult, US Naval Hosp, Philadelphia, 63. *Mem:* Fel Am Col Anesthesiol; Am Soc Anesthesiol; Royal Soc Med; Am Soc Testing & Mat. *Mailing Add:* 421 Wister Rd Wynnewood PA 19096-1808

DEASON, TEMD R, b York, Ala, Oct 13, 31; m 65; c 2. ALGOLOGY. *Educ:* Univ Ala, BS, 54, MS, 58; Univ Tex, PhD(bot), 60. *Prof Exp:* Asst prof biol sci, Univ Del, 60-61; from asst prof to assoc prof biol, 61-69, PROF BIOL, UNIV ALA, 69- *Honors & Awards:* US Sr Scientist Teaching & Res Award, Alexander von Humboldt Found, W Ger, 74. *Mem:* Am Phycol Soc; Int Phycol Soc. *Res:* Morphology and taxonomy of soil algae; electron microscopy. *Mailing Add:* Dept Biol Univ Ala Box 870344 Tuscaloosa AL 35487-0344

DEATH, FRANK STUART, b Winnipeg, Man, Apr 15, 32; m 54; c 5. METALLURGY. *Educ:* Univ BC, BASc, 55, MASc, 56. *Prof Exp:* Engr, Tonawanda Labs, Linde Div, Union Carbide Corp, 56-60, group leader melting & refining process, 60-65, lab div head phys chem steelmaking, Newark Labs, 65-68, asst mgr mkt develop, 68-70, mgr AOD Steelmaking, Tarrytown Labs, 70-73, asst mgr, Gas Prod Div, 73-77, mgr gas prod div, 77-86, DIR INDUST GASES DEVELOP, LINDE DIV, UNION CARBIDE CORP, 83-, DIR TECHNOL, 86- *Mem:* Am Inst Mining, Metall & Petrol Engrs. *Res:* Metallurgical science; physical chemistry of steelmaking; thermochemical processes involving new techniques; metal-producing industry. *Mailing Add:* 55 Bucyruss Ave Carmel NY 10512

DEATON, BOBBY CHARLES, b Pittsburg, Tex, Jan 20, 36; m 60; c 2. SOLID STATE PHYSICS. *Educ:* Baylor Univ, BA, 57, MS, 59; Univ Tex, PhD(physics), 62, MS, 82. *Prof Exp:* Sr res scientist, 61-66, staff scientist, Gen Dynamics, 66-72; PROF, TEX WESLEYAN COL, 67-, HEAD DEPT PHYSICS, 68- *Concurrent Pos:* Res scientist, Univ Tex, 63-64, adj instr, Univ Tex, Arlington, 66-67; res grant plate tectonics, Res Corp, 81. *Honors & Awards:* J Clark Streett Am Asn Univ Prof Award, 88. *Mem:* Am Phys Soc; Sigma Xi; Am Asn Univ Prof. *Res:* Solid state physics involving Fermi surfaces of metals, superconductivity and study of materials at high pressures and temperatures; plate tectonics of Caribbean Region, especially motion of Polochic Fault; quantification of color in geology. *Mailing Add:* Dept Physics Tex Wesleyan Col Ft Worth TX 76105

DEATON, EDMUND IKE, b Sulphur Springs, Tex, Aug 18, 30; m 76; c 3. DATA BASE DESIGN, NUMERICAL ANALYSIS. *Educ:* Hardin-Simmons Univ, BA, 50; Univ Tex, MA, 56, PhD(math), 60. *Prof Exp:* comput sci, Okla State Univ, 80-82; PROF COMPUT SCI, SAN DIEGO STATE UNIV, 60- *Mem:* Math Asn Am; Asn Comput Mach. *Res:* Data base design. *Mailing Add:* Dept Math San Diego State Univ San Diego CA 92182

DEATON, JAMES WASHINGTON, b Manning, Ark, June 29, 34; m 56; c 1. POULTRY SCIENCE. *Educ:* Univ Ark, BS, 56; Tex A&M Univ, MS, 59, PhD(poultry sci), 64. *Prof Exp:* Poultry serviceman, Paymaster Feed Mills, Tex, 56-58; res asst poultry nutrit, Tex A&M Univ, 58-59, poultry nutrit & genetics, 61-63, res assoc poultry nutrit & statist eval data, 63-64; poultry serviceman, DeKalb Agr Asn, Inc, 59-61; Tex Turkey Fedn grant poultry dis, Tex Agr Exp Sta, 64; res poultry husbandmen, S Cent Res Lab, 64-68, DIR S CENT POULTRY RES LAB, AGR RES SERV, USDA, 68- *Mem:* AAAS; Poultry Sci Asn; World Poultry Sci Asn; Int Soc Biometeorol. *Res:* Poultry management and environmental research, including poultry nutrition, genetics and disease aspects and physiological relationships. *Mailing Add:* USDA Agr Res Serv PO Box 5367 Mississippi State MS 39762

DEATON, LEWIS EDWARD, b Washington, DC, July 5, 49; m 81; c 2. ANIMAL PHYSIOLOGY. *Educ:* Col William & Mary, BS, 70, MA, 74; Fla State Univ, PhD(physiol), 79. *Prof Exp:* Res asst physiol, Univ Tenn, Knoxville, 79-81 & State Univ NY, Stony Brook, 81-83; res asst physiol, Whitney Lab, Univ Fla, 83-85; adj asst prof, Iowa State Univ, 85-86; ASST PROF, UNIV SOUTHWESTERN LA, 86- *Mem:* AAAS; Am Soc Zoologists; Am Soc Cell Biol. *Res:* Comparative physiology of salt and water balance in invertebrates; invertebrate neurohormones. *Mailing Add:* Dept Biol Univ Southwestern La Lafayette LA 70504

DEATS, EDITH POTTER, wid. FETAL-MATERNAL INTERRELATIONS. *Educ:* Univ Minn, MD, 26, DMSc, PhD, 34. *Prof Exp:* Prof path, Dept Obstet & Gynec, Univ Chicago, 33-67; RETIRED. *Mailing Add:* 1920 Virginia Ave Apt 901 Ft Myers FL 33901

DEAVEN, DENNIS GEORGE, b Hershey, Pa, Nov 24, 37; m 65; c 2. METEOROLOGY, COMPUTER SCIENCE. *Educ:* Penn State Univ, BS, 68, MS, 70, PhD(meteorol), 74. *Prof Exp:* Res asst meteorol, Penn State Univ, 65-72; scientist, Nat Ctr Atmospheric Res, 72-77; RES METEOROLOGIST, US WEATHER SERV, 77- *Mem:* Am Meteorol Soc; Sigma Xi. *Res:* Numerical weather prediction; dynamic meteorology; atmospheric air quality. *Mailing Add:* 7021 Groveton Dr Clinton MD 20735

DEAVEN, LARRY LEE, b Hershey, Pa, Oct 28, 40. CELL BIOLOGY, CYTOGENETICS. *Educ:* Pa State Univ, BS, 62, MS, 64; Univ Tex, PhD(biomed sci), 69. *Prof Exp:* Instr biol, Pa State Univ, 64; fel, M D Anderson Hosp, Univ Tex, 65-69, Nat Cancer Inst fel, 69-71; fel, 71-73, DEPT DIR CELLULAR & MOLECULAR RADIOBIOL, LOS ALAMOS SCI LAB, 73-; GENETICIST, US DEPT ENERGY, 80- *Concurrent Pos:* Assoc ed, Radiation Res, 81- *Mem:* AAAS; Am Chem Soc; Am Soc Cell Biol; Genetics Soc Am; Soc Anal Cytol. *Res:* Chromosome structure and physiology; DNA content and structure in mammalian metaphase chromosomes; automated cytogenetics; chromosome damage by environmental pollutants. *Mailing Add:* Los Alamos Nat Lab CHGS MS M885 Los Alamos NM 87545

DEAVER, BASCOM SINE, JR, b Macon, Ga, Aug 16, 30; m 51; c 3. EXPERIMENTAL SOLID STATE PHYSICS. *Educ:* Ga Inst Technol, BS, 52; Wash Univ, MA, 54; Stanford Univ, PhD(physics), 62. *Prof Exp:* Physicist, Air Force Spec Weapons Command, 54-57 & Stanford Res Inst, 57-62; NSF fel, Stanford Univ, 62-63, res assoc low temperature physics, 63-65; assoc prof, 65-73, PROF PHYSICS, UNIV VA, 73- *Concurrent Pos:* Physicist, Stanford Res Inst, 63-65; Alfred P Sloan fel, 66-68. *Mem:* Am Phys Soc. *Res:* Superconductivity; superconducting electronics; low temperature physics. *Mailing Add:* Dept Physics Univ Va McCormick Rd Charlottesville VA 22901

DEAVER, FRANKLIN KENNEDY, b Springdale, Ark, Jan 10, 18; m 45; c 2. MECHANICAL ENGINEERING, THERMODYNAMICS. *Educ:* Univ Ark, BSChE, 39, MSME, 60; Univ Minn, PhD(mech eng), 69. *Prof Exp:* Asst mgr construct, Pioneer Co, Ark, 39-49, mgr construct, 49-55; from asst prof to prof, 55-84, head dept, 69-80, EMER PROF MECH ENG, UNIV ARK, FAYETTEVILLE, 84- *Concurrent Pos:* Consult. *Mem:* Am Soc Mech Engrs; Am Soc Eng Educ; Nat Soc Prof Engrs; Sigma Xi. *Res:* Convective heat transfer; solar energy; heat transfer from an oscillating horizontal cylinder to a liquid; heat transfer and temperature fields in a horizontal fluid cylinder; energy transfer to a solar photovoltaic collector. *Mailing Add:* Rte 9 Box 276 Fayetteville AR 72701

DEAVERS, DANIEL RONALD, b Normal, Ill, Oct 23, 43; m 66; c 1. PHYSIOLOGY. *Educ:* San Diego State Univ, BS, 67, MS, 69; Cornell Univ, PhD(environ physiol), 75. *Prof Exp:* Fel physiol, Dalton Res Ctr, Univ Mo, 75-76; cardiovasc trainee, USPHS, 76-78; instr physiol, Univ Louisville, 78-80; ASSOC PROF & CHMN DEPT PHYSIOL & PHARMACOL, UNIV OSTEOP MED & HEALTH SCI, DES MOINES, 80- *Mem:* Am Physiol Soc; Am Heart Asn; Sigma Xi. *Res:* Cardiovascular physiology. *Mailing Add:* Univ Osteop Med Surg 3200 Grand Ave Des Moines IA 50312

DEB, ARUN K, b India, May, 1936; m 62; c 1. ENGINEERING. *Educ:* Univ Calcutta, BS, 57, PhD(civil eng), 68; Univ Wis, MS, 61. *Prof Exp:* USAID fel, Univ Wis, 60-61; asst prof environ eng, Bengal Eng Col, Univ Calcutta, 61-71; sr res fel, Univ Col London, 71-73; vis prof, Univ Notre Dame, 74; mgr environ systs, 74-82, VPRES, WESTON, 82- *Concurrent Pos:* USAID fel, 61; NSF res grant, 77, 80. *Honors & Awards:* Gold Medal, Inst Engr, Pub Health Engr, 73. *Mem:* Am Soc Civil Engrs; Water Pollution Control Fedn; Am Water Works Asn; Am Acad Environ Engrs. *Res:* Water and wastewater resources management; mathematical modeling and econometric modeling; systems analysis and optimization of environmental engineering systems; water-distribution systems analysis; water supply, water reuse and wastewater systems planning, evaluation and design. *Mailing Add:* Roy F Weston Inc Weston Way West Chester PA 19382

DEBACKER, HILDA SPODHEIM, b Bucharest, Rumania, July 17, 24; US citizen; m 48. NEUROANATOMY, MICROSCOPIC ANATOMY. *Educ:* Cornell Univ, AB, 45; Polytech Inst Brooklyn, MS, 49; Med Col SC, MS, 57, PhD(anat), 67. *Prof Exp:* Res asst chem, Calco Chem Div, Am Cyanamid Co, NJ, 45-46; lab asst polymer chem, Polytech Inst Brooklyn, 48; lit searcher pharmaceut chem, Warner Inst Therapeut Res, NY, 49-50; res librn, Wallace & Tiernan Co, Inc, NJ, 50-52; teaching asst, 53-55, instr, 62-67, assoc, 67-68, asst prof, 68-72, ASSOC PROF ANAT, MED UNIV SC, 72- *Mem:* AAAS; Am Asn Anatomists; Am Chem Soc. *Res:* Microscopic structure and function of living liver; microscopic observations of conjunctival circulation in health and disease in man; conjunctival circulation during cerebral ischemia and during electric shock therapy. *Mailing Add:* Dept Cell Biol Med Univ SC 171 Ashley Ave Charleston SC 29425

DEBAKEY, LOIS, b Lake Charles, La. MEDICAL EDUCATION, WRITING & LECTURING. *Educ:* Tulane Univ, BA, 49, MA, 59, PhD(lit), 63. *Prof Exp:* Asst prof english, Tulane Univ, 60-63, from asst prof to prof sci commun, 63-68, LECTR SCI COMMUN, SCH MED TULANE UNIV, 68-; PROF SCI COMMUN, BAYLOR COL MED, 68- *Concurrent Pos:* Adj prof sci commun, Sch Med, Tulane Univ, 81-; dir ann course sci commun, Am Col Surgeons, 69-76; mem, comt teaching tech & sci writing, Nat Coun Teachers of English, rev comt, Biomed Libr, 73-77 & comt on col & exec coun, Southern Asn Cols & Schs, 75-80; consult, Tex State Tech Inst, 74, Nat Asn Standard Med Vocab & Legal Writing Comt, Am Bar Asn, 83-; usage panel, The Am Heritage Dict, 80-; nat adv coun, USC Ctr for Continuing Med Educ, 81-; bd regent, Nat Libr Med, 81-86, consult, 86-, Literature Select Tech Rev Comt, 88-, co-chmn, Permanent Paper Task Force, 88, Outreach Planning Panel, 88-89, founding bd, Friends of Nat Libr Med, 85-, nat adv bd, John Muir Med Film Festival, 90- *Honors & Awards:* Bausch & Lomb Sci Award; John P McGovern Award, Med Libr Asn, 83. *Mem:* Soc Tech Commun; Nat Asn Sci Writers; Int Soc Gen Semantics; Dictionary Soc N Am; hon mem Med Libr Asn; Soc Preserv English Lang & Lit; fel Am Col Med Informatics. *Res:* Biomedical communication-teaching, writing, editing, publishing, ethics; communication; linguistics; influence of science on literature and of literature on science; literacy; author of numerous articles in medical journals and the public press and two books. *Mailing Add:* Baylor Col Med One Baylor Plaza Houston TX 77030

DEBAKEY, MICHAEL ELLIS, b Lake Charles, La, Sept 7, 08; m 36, 75; c 5. SURGERY. *Educ:* Tulane Univ, BS, 30, MD, 32, MS, 35. *Hon Degrees:* Numerous from US & foreign univs, 61-76. *Prof Exp:* From instr to assoc prof surg, Tulane Univ, 37-48; vpres med affairs, 68-69, chief exec officer, 68-69, pres, 69-79, PROF SURG & CHMN DEPT, BAYLOR COL MED, 48-, CHANCELLOR, 79- *Concurrent Pos:* Mem nat adv health coun, NIH 61-65, mem nat adv gen med sci coun, 65, mem nat adv coun on regional med progs, 65-, mem nat adv coun, Nat Heart Lung & Blood Inst, 82-, mem Texas Sci & Technol Coun, 84-86; mem comt on epidemiol & vet follow-up studies, Nat Res Coun; chmn, President's Comn Heart Dis, Cancer & Stroke, 64; chmn sci adv bd, Delta Regional Primate Res Ctr, Tulane Univ, 65; mem consult staff, Dept Surg, Tex Children's Hosp, Houston, 70-; mem patron's comt, Damon Runyon Mem Fund for Cancer Res, 71; dir, Nat Heart & Blood Vessel Res & Demonstration Ctr, & DeBakey Heart Ctr, Houston; bd dir, Found Biomed Res. *Honors & Awards:* Vishnevsky Medal, Inst Surg, Acad Med Sci, USSR, 62; Lasker Award, 63; Accademia Internazionale di Pontzen di Lettere Scienze ed Arti Gran Collare Accademico d'Oro, 69; Nat Medal Sci, 89. *Mem:* Inst Med-Nat Acad Sci; Am Col Surg; Am Surg Asn; Soc Vascular Surg (pres, 54); Int Cardiovasc Soc (pres, 59); Asn Int Vascular Surg (pres, 83); hon mem Acad Med Sci, USSR. *Res:* Cardiovascular and thoracic surgery; cardiovascular diseases, including aortic diseases; replacement of excised segments of arteries by homografts and plastic prostheses; venous thrombosis; aneurysms and occlusive diseases of arteries; peripheral vascular diseases, heart transplantation, and development of the artificial heart. *Mailing Add:* One Baylor PLaza Houston TX 77030

DEBAKEY, SELMA, b Lake Charles, La. SCIENTIFIC COMMUNICATION, MEDICAL EDUCATION. *Educ:* Tulane Univ, BA. *Prof Exp:* Dir dept med commun, Alton Ochsner Med Found, 42-68; PROF SCI COMMUN, BAYLOR COL MED, 68- *Concurrent Pos:* Ed, Selected writings from Ochsner Clin, 42-54; ed, Ochsner Clin Reports, 55-58; mem comt judges, Mod Med Monogr Award, 58-60; co-ed, Bull Am Med Writers Asn, 61-64; ed, Cardiovasc Res Ctr Bull, 72-84; consult, Nat Asn Standard Med Vocab. *Mem:* AAAS; Asn Teachers Tech Writing; Soc Tech Commun; Soc Health & Human Values; Am Med Writers Asn; Coun Biol Ed. *Res:* Biomedical communications-teaching writing. *Mailing Add:* One Baylor Plaza Houston TX 77030

DEBANY, WARREN HARDING, JR, b New Rochelle, NY, July 12, 55; m; c 1. DIGITAL LOGIC TEST, FAULT-TOLERANT COMPUTING. *Educ:* State Univ NY, Buffalo, BS, 77; Syracuse Univ, NY, MS, 83, PhD(computer & info sci), 85. *Prof Exp:* Res asst, Dept Elec Eng, State Univ NY, Buffalo, 77; ELECTRONICS ENGR, USAF ROME LAB, 77- *Concurrent Pos:* Mem, Test Technol Comt Prog Comt, Inst Elec & Electronics Engrs. *Mem:* Sr mem Inst Elec & Electronics Engrs. *Res:* Digital logic design; integrated diagnostics; design-for-testability; built-in-test; testability measurement techniques; hardware description languages; logic and fault simulation; automatic test generation; fault-tolerant computer architecture development and performance assessment; applications of statistical techniques to digital logic testing. *Mailing Add:* USAF Rome Lab RL/ERDA Griffiss AFB NY 13441-5700

DE BARBADILLO, JOHN JOSEPH, b York, Pa, Jan 27, 42; m 65; c 2. MANUFACTURE & USE OF REACTIVE METAL ADDITIVES FOR TREATMENT OF MOLTEN STEEL, RECOVERY OF METALS FROM INDUSTRIAL WASTE. *Educ:* Lehigh Univ, BS, 63, MS, 65, PhD(metall & mat sci), 67. *Prof Exp:* Engr, Inco Inc, P D Merica Res Lab, 67-70, sect & dept mgr, 70-84; res & develop mgr, 84-89, DIR RES & DEVELOP PLANNING, INCO ALLOYS INT, 89- *Concurrent Pos:* Key reader, Metall Transactions, 80-84; mem, Nat Mat Adv Bd-Comt Cobalt & Bur Mines Res Ctrs, 84; mem, US Bur Mines Generic Ctrs Rev Bd, 85-, Tech Div Bd, Am Soc Metals, 86-90 & Joint Comt Metall Transactions, Metall Soc, 91- *Honors & Awards:* Hunt Award, Am Inst Mech Engrs, 75. *Mem:* Fel Am Soc Metals Int; Metall Soc; Sigma Xi. *Res:* Commercialization of nickel containing alloys, products and processes; manufacture of nickel alloys. *Mailing Add:* Inco Alloys Int PO Box 1958 Huntington WV 25720

DEBARDELEBEN, JOHN F, organic chemistry, organophosphorus compounds; deceased, see previous edition for last biography

DEBARI, VINCENT A, b Jersey City, NJ, Feb 1, 46; m 70; c 3. IMMUNOCHEMISTRY, CELL BIOPHYSICS. *Educ:* Fordham Univ, BS, 67; Newark Col Eng, MS, 70; Rutgers Univ, PhD, 81. *Prof Exp:* Res chemist, Sonneborn Div, Witco Chem Corp, 67-70 & Activated Carbon Div, 70-73; res chemist, 73-81, dir, Renal Lab, 81-89, DIR, RES & EDUC, LUPUS TREAT & RES CTR, ST JOSEPH'S HOSP & MED CTR, PATTERSON, NJ, 88-, DIR, RHEUMATOLOGY LAB & RES DEPT MED, 89- *Concurrent Pos:* Consult, Rutgers Univ, 81, Biomed Clin Labs, Inc, 85, Micromembranes Inc, 86-89, Cancer Lab, Gen-Care Biomed Res Corp, 89-; adj asst prof, Dept Biol, Seton Hall Univ, 81-89, adj assoc prof, 89-, assoc prof, Dept Med, 88-; vis specialist, Montclair State Col, 82; assoc grad fac, Rutgers Univ, 84-89. *Honors & Awards:* Bernard F Gerulat Award, 90. *Mem:* Fel Am Inst Chemists; Sigma Xi; Am Asn Clin Chem; fel Nat Acad Clin Biochemists; Biophys Soc; Clin Ligand Assay Soc; Am Col Rheumatology; Am Fedn Clinic Res. *Res:* Biochemistry of neutrophil; characterization of auto-antibodies; anti-DNA antibodies; oxygen free radicals; immunochemistry; molecular immunology. *Mailing Add:* Rheumatology Lab St Joseph's Hosp & Med Ctr 703 Main St Paterson NJ 07503

DEBAS, HAILE T, SURGERY. *Educ:* Univ Col, Addis Ababa, BS, 58; McGill Univ, MD & CM, 63; FRCS(C); Am Bd Surg, dipl, 82. *Prof Exp:* Rotating intern surg, Ottawa Civic Hosp, 63-64; asst resident, Vancouver Gen Hosp, 64-65, assoc resident, 66-67, chief resident, 68-69; res fel, Univ BC, 65-66 & 69-70, Univ Glasgow, 67-68; from asst prof to assoc prof, Univ BC, 71-80; prof residence surg, Univ Calif, Los Angeles, 81-85; prof & chief, Sect Gastrointestinal Surg, Univ Wash, Seattle, 85-87; PROF & CHMN, DEPT SURG, UNIV CALIF, SAN FRANCISCO, 87- *Concurrent Pos:* Pvt pract, Whitehorse, Yukon Territories, Burn Lake, BC, Can, 70-71; Med Res Coun fel gastroenterol, Univ Calif, Los Angeles/Vet Admin Wadsworth Med Ctr, 71-74; dep chief, Shaughnessy Vet Hosp, Vancouver, BC, 71-72; dir, Gastroenterol Clin, Vancouver Gen Hosp, 74-80, chief C-5 surg serv, 78-80; dir res, Univ BC, 76-78; key investr, Ctr Ulcer Res & Educ, Los Angeles, 80-; attend surgeon, Harbor-Univ Calif, Los Angeles Med Ctr, Torrance, 82-85, Ctr Health Sci Hosp, 82-85; counr, Asn Acad Minority Physicians, 87-; mem, External Adv Bd, Mich Peptide Res Ctr, 88-; prin investr training grant, NIH, 88- *Mem:* Inst Med-Nat Acad Sci; fel Am Col Surgeons; Am Gastroenterol Asn; Am Surg Asn; Int Col Surgeons; Can Gastroenterol Asn; Int Hepato-Biliary-Pancreatic Asn (pres, 91-); AMA; Am Asn Endocrine Surgeons. *Res:* Regulation of gastric function by calcitonin gene-related peptide; mechanisms of inhibition of the exocrine pancreas. *Mailing Add:* Dept Surg Univ Calif 513 Parnassus Ave S-320 San Francisco CA 94143-0104

DE BAULT, LAWRENCE EDWARD, b Yoarum Tex, Feb 15, 41; m 65; c 1. CYTODIFFERENTIATION, RECEPTOR EXPRESSION. *Educ:* Victoria Col, AA, 61; Univ Tex, Austin, BSSEd, 64; Univ Stockholm, Sweden, PhD(cell physiol), 69. *Prof Exp:* Res asst, Inst cell Res & Med Genetics, Med Nobel Inst, Karolinska Inst, 64-67 & Children's Cancer Res Found, Harvard Med, 67-69; asst prof, dept psychiat, Univ Iowa Col Med, 69-75 & dept anat, 70-75, sr res scientist, 75-80; assoc prof, 80-83, PROF DEPT PATH, UNIV OKLA HEALTH SCI CTR, 83-, DIR RES PROGR, 84- *Concurrent Pos:* NIH trainee cytochem, Inst Cell Res & Med Genetics, Med Nobel Inst, Karolinska Inst, 64-66, personal fel, 66-67; NSF Int Travel Award for NATO Advan Study Inst Microbeam Irradiation & Cell Biol, Stresa, Italy, 70; adj assoc prof, dept pediat, Univ Okla Health Sci Ctr, 80-; sci dir diagnostic electron, dept path, Okla Children's Mem Hosp, Univ Okla Health Sci, 80-, Flow Cytometry Lab, 81- & Histol Lab Immunoperoxidase, 84-86. *Res:* Endothial cell biology. *Mailing Add:* Dept Path Health Sci Ctr Univ Okla PO Box 26901 Oklahoma City OK 73190

DEBAUN, ROBERT MATTHEW, b New York, NY, June 23, 24; m 50; c 4. INFORMATION SCIENCE, STATISTICS. *Educ:* Fordham Univ, BS, 47, MS, 49, PhD(chem), 51. *Prof Exp:* Instr org chem, Col Pharm, St John's Univ, NY, 50-51; assoc scientist, Nat Dairy Res Labs, Inc, 51-54; proj leader, Cent Res Labs, Gen Foods Corp, 54-55; statistician, 55, group leader, Math Anal Group, Stamford Labs, 56-58, ref catalysts group, 58-61, mgr tech comput corp hq, 61-65, adv plan, Corp Data Processing, 65, dir, 65-70, dir opers anal, 70-71, mgr info syst plan, 71-80, DIR INFO SYST, AM/FAR EAST DIV, AM CYANAMID CO, 80- *Mem:* Am Statist Asn; Inst Mgt Sci; Sigma Xi. *Res:* Applied statistics; management sciences. *Mailing Add:* 15 Lois Ct Wayne NJ 07470

DEBE, MARK K, b Duluth, Minn, June 28, 47; m 69; c 2. SURFACE SCIENCE, MICROGAVITY MATERIALS PROCESSING. *Educ:* Univ Minn, BA, 69; Univ Wis, PhD(physics), 74. *Prof Exp:* Lectr, Univ Liverpool, Eng, 76-77; asst prof physics, Univ Tex, San Antonio, 77-78; sr physicist, 78-82, res specialist, 82-86, SR RES SPECIALIST, 3 M CO, 86- *Concurrent Pos:* Postdoctoral surface sci, Univ Liverpool, Eng, 75-76; prin investr, Space Shuttle exp, NASA, 85-88. *Mem:* Am Phys Soc; Am Vacuum Soc. *Res:* Surface science; organic thin films; vapor depositions; microgravity materials processing; physical vapor transport. *Mailing Add:* Corp Res 201 2N 19 3M Center 3M Co St Paul MN 55144

DEBELL, ARTHUR GERALD, optics; deceased, see previous edition for last biography

DEBELL, DEAN SHAFFER, b Woodbury, NJ, July 6, 42; m 64; c 3. FORESTRY. *Educ:* Juniata Col, BS, 63; Duke Univ, MF, 64, PhD(forest soils, plant physiol), 70. *Prof Exp:* Forester, Gifford Pinchot Nat Forest, US Forest Serv, 64-65, res forester, Southeastern Forest Exp Sta, 65-70; res forester, Cent Res Lab, Crown Zellerbach Corp, 70-75; RES FORESTER & PROJ LEADER SILVICULT, PAC NORTHWEST FOREST & RANGE EXP STA, US FOREST SERV, 75- *Concurrent Pos:* Vis prof forest biol, Belle W Baruch Forest Sci Inst, Clemson Univ, 80-81; vis forest scientist, Bioenergy Develop Corp, C Brewer & Co Ltd, Hilo, Hawaii, 85. *Mem:* Soc Am Foresters; Soil Sci Soc Am; Northwest Sci Asn. *Res:* Silviculture of conifers and hardwoods on the west-side of Cascade Range in Oregon and Washington. *Mailing Add:* Forestry Sci Lab 3625 93rd Ave SW Olympia WA 98502

DEBENEDETTI, PABLO GASTON, b Buenos Aires, Arg, Mar 30, 53; m 87; c 1. SOLIDS FORMATION FROM SUPERCRITICAL FLUIDS, METASTABLE LIQUIDS. *Educ:* Buenos Aires Univ, BS, 78; Mass Inst Technol, MSc, 81, PhD(chem eng), 84. *Prof Exp:* Asst prof, 85-90, ASSOC PROF CHEM ENG, PRINCETON UNIV, 90- *Concurrent Pos:* Res engr, O de Nora, Impianti Elettrochimici, Milan, Italy, 78-80. *Honors & Awards:* Presidential Young Investr Award, Nat Sci Found, 87. *Mem:* Sigma Xi; Am Inst Chem Engrs; Am Chem Soc; Am Phys Soc; AAAS. *Res:* Thermodynamics and statistical mechanics of supercritical and near critical mixtures; metastable liquids; nucleation in supercritical fluids; fluctuation theory. *Mailing Add:* Dept Chem Eng Princeton Univ Princeton NJ 08544-5263

DEBENEDETTI, SERGIO, b Florence, Italy, Aug 17, 12; nat US; m 44; c 3. EXPERIMENTAL PHYSICS. *Educ:* Univ Florence, PhD(physics), 33. *Prof Exp:* Asst prof, Univ Padua, 34-38; fel, Curie Lab, Univ Paris, 38-40; res assoc, Bartol Res Found, 40-43; assoc prof physics, Kenyon Col, 43-44; sr physicist, Monsanto Co, Ohio, 44-45; prin physicist, Clinton Labs, Oak Ridge Nat Lab, 46-48; assoc prof, Wash Univ, 48-49; prof, 49-83, EMER PROF PHYSICS, CARNEGIE-MELLON UNIV, 83- *Concurrent Pos:* Mem exped cosmic rays, Eritrea, EAfrica, 33; fel, Curie Lab, Univ Paris, 34-35; vis prof, Univ Rio de Janeiro, 52; Fulbright scholar, Univ Turin, 56-57; guest lectr, Brazilian Ctr Phys Sci, 61; lectr, Univ Calif, Berkeley, 63. *Mem:* Fel Am Phys Soc. *Res:* Nuclear and high energy physics; cosmic rays; radioactivity; short-lived isomers; positrons; mesons; Mossbaur spectroscopy; history of science. *Mailing Add:* 122 Hastings St Pittsburgh PA 15206

DEBER, CHARLES MICHAEL, b Brooklyn, NY, Apr 20, 42; m 71; c 1. BIOLOGICAL CHEMISTRY. *Educ:* Polytech Inst Brooklyn, BS, 62; Mass Inst Technol, PhD(org chem), 67. *Prof Exp:* NIH res fel, Harvard Med Sch, 67-69, assoc biol chem, 70-74; vis scientist, Inst Enzyme Res, Univ Wis, 75; from asst prof to assoc prof, 76-84, PROF BIOCHEM DIV, RES INST, HOSP FOR SICK CHILDREN, TORONTO & BIOCHEM DEPT, UNIV TORONTO, 85- *Mem:* Am Peptide Soc (pres, 91-93). *Res:* Conformations of biological macromolecules studied by nuclear magnetic resonance spectroscopy; peptide-membrane interactions; molecular aspects of transmembrane signalling and transport by membrane proteins; site-directed mutagenesis of transmembrane segments. *Mailing Add:* Biochem Dept Res Inst Hosp for Sick Children Toronto ON M5G 1X8 Can

DEBERRY, DAVID WAYNE, b Gonzales, Tex, Feb 5, 46; m 67; c 2. ELECTROCHEMISTRY, CHEMICAL KINETICS. *Educ:* Univ Tex, Austin, BS, 69; Rice Univ, PhD(chem), 74. *Prof Exp:* Res asst, Tracor, Inc, Austin, Tex, 66-69; res scientist, 69-70, sr res scientist, 73-78, PRIN SCIENTIST, RADIAN CORP, AUSTIN, TEX, 78- *Mem:* Electrochem Soc; Am Chem Soc; Nat Asn Corrosion Engrs; Sigma Xi. *Res:* Electrochemistry, particularly electrochemical kinetics; chemically modified electrodes; corrosion and corrosion inhibition; electron transfer kinetics. *Mailing Add:* 4001 Knollwood Austin TX 78731

DE BETHUNE, ANDRE JACQUES, b Schaerbeek-Brussels, Belgium, Aug 20, 19; m 49; c 10. CHEMISTRY. *Educ:* St Peter's Col, BS, 39; Columbia Univ, PhD(phys chem), 45. *Prof Exp:* Res chemist, Columbia Univ, 42-45; Nat Res fel, Mass Inst Technol, 45-47; from asst prof to assoc prof, 47-55, chmn, Dept Chem, 65-67 & 74-76, prof, 55-88, EMER PROF CHEM, BOSTON COL, 88- *Concurrent Pos:* Theoret ed jour, Electrochem Soc, 57-80; Guggenheim fel, Yale Univ, 60-61. *Mem:* Am Chem Soc; Electrochem Soc. *Res:* Population kinetics, bioethics of population control; electrochemistry; isotope separation; kinetic theory of gases; hydrogen energy. *Mailing Add:* 160 Bristol Ferry Rd Portsmouth Chestnut Hill RI 02871

DEBEVEC, PAUL TIMOTHY, b Cleveland, Ohio, May 30, 46. NUCLEAR PHYSICS. *Educ:* Mass Inst Technol, BS, 68; Princeton Univ, MA, 70, PhD(physics), 72. *Prof Exp:* Res assoc physics, Argonne Nat Lab, 72-74; asst prof, Ind Univ, Bloomington, 74-77; assoc prof, 77-83, PROF PHYSICS, UNIV ILL, URBANA, 83- *Concurrent Pos:* Vis assoc prof, Inst Physics, Louvain-La-Neuve, 79-80; mem subcomt electromagnetic interactions, Nuclear Sci Adv Comt, 81; mem panel nuclear physics, Nat Acad Sci, 84; mem prog adv comt, Los Alamos Meson Physics Facil, 86-89. *Mem:* Am Phys Soc; Am Asn Physics Teachers. *Res:* Low and medium energy nuclear reactions and structure. *Mailing Add:* Dept Physics Univ Ill 1110 W Green St Urbana IL 61801

DEBIAK, TED WALTER, b McKeesport, Pa, Sept 21, 44; m 69; c 1. ACCELERATOR PHYSICS. *Educ:* Pa State Univ, BS, 66; Univ Pittsburgh, PhD(chem), 73. *Prof Exp:* Instr phys chem, 75-76, res assoc nuclear chem, State Univ NY Stony Brook, 74-76; SR RES SCIENTIST, GRUMMAN, 76- *Mem:* Am Phys Soc; Inst Elec & Electronics Engrs; Am Nuclear Soc. *Res:* Negative ion sources and transportation of charged particles in accelerators. *Mailing Add:* 57 S Windhorst Ave Bethpage NY 11714

DEBIAS, DOMENIC ANTHONY, b Tresckow, Pa, Aug 31, 25; m 51; c 5. PHYSIOLOGY. *Educ:* Temple Univ, AB, 49, AM, 50; Jefferson Med Col, PhD(physiol), 56. *Prof Exp:* Instr & asst biol, Temple Univ, 49-50, instr & asst physiol, Sch Dent, 50-51, res assoc bact, 51; instr physiol, Sch Med, Univ Pa, 55; instr, Sch Dent, Temple Univ, 55; res assoc, Div Endocrine & Cancer Res, Jefferson Med Col, 56-57, from instr to prof physiol, 57-75; asst dean, 78-86, PROF PHYSIOL & PHARMACOL & CHMN DEPT, PHILADELPHIA COL OSTEOP MED, 75-, ASST DEAN, 89- *Mem:* Sigma Xi; Am Physiol Soc; Endocrine Soc; Am Heart Asn; Int Asn Heart Res; NY Acad Sci. *Res:* Physiology of the endocrine glands; endocrine regulation of cardiopulmonary functions; interrelationship of endocrine systems and nervous system; environmental physiology; somatic component to MI. *Mailing Add:* 725 Pecan Dr Philadelphia PA 19115

DE BLAS, ANGEL LUIS, b Madrid, Spain, Jan 31, 50. NEUROBIOLOGY. *Educ:* Univ Madrid, BS & MS, 72; Ind Univ PhD(biochem), 78. *Prof Exp:* Res fel, NIH, 78-81; asst prof, 81-86, ASSOC PROF, STATE UNIV NY STONY BROOK, 86- *Concurrent Pos:* Comn Cult Exchange between Spain & US fel, 74-75; John E Fogarty fel, 78-81; NIH res grant, 81-88, regular mem neurol sci study sect, 85-89; prin investr, State Univ NY, Stony Brook, 81-; March of Dimes Birth Defects Found award, 82-87; NSF res grant, 84-87; Epilepsy Found award, 85-86; Esther A & Joseph Klingenstein neuroscience award, 85-88; grant reviewer, Molecular & Cellular Neurobiology Prog, NSF; vis prof, CNRS, Gif-sur-Yvette, France, 87-88. *Honors & Awards:* Span Ministry Educ & Sci Prize, 73. *Mem:* Soc Neurosci; Am Soc Neurochem; Span Biochem Soc; Am Soc Biol Chemists; NY Acad Sci; Int Soc Neurochem. *Res:* Cellular and molecular neurobiology; monoclonal antibodies to synaptic molecules; gamma-aminobutyric acid and benzodiazepine receptors; endogenous benzodiazepines; role of glial cells in brain development; nerve cells in culture. *Mailing Add:* Div Molecular Biol & Biochem Univ Mo Sch Basic Life Sci Kansas City MO 64110-2499

DEBLOIS, RALPH WALTER, b Benton Harbor, Mich, Jan 11, 22; m 59; c 3. PHYSICS. *Educ:* Univ Mich, BS, 43, MS, 47; Rensselaer Polytech Inst, PhD(physics), 63. *Prof Exp:* Asst, Manhattan Proj, Columbia Univ, 43-46; fel, Univ Mich, 46-49; instr physics, Am Univ, Beirut, 50-53; physicist, Gen Elec Res & Develop Ctr, 54-87; adj prof, Rensselaer Polytech Inst, 87-89; RETIRED. *Honors & Awards:* I-R 100 Award, Indust Res Mag, 70. *Mem:* AAAS; Am Vacuum Soc; Am Phys Soc. *Res:* Ferromagnetic domain structure; submicron particle analysis; molecular-beam epitaxy. *Mailing Add:* 2509 Antonia Dr Schenectady NY 12309

DEBNATH, LOKENATH, b Hamsadi, India, Sept 30, 35; m 69; c 1. APPLIED MATHEMATICS, PURE MATHEMATICS. *Educ:* Univ Calcutta, BS, 54, MS, 56, PhD(pure math), 64; Imp Col, Univ London, dipl & PhD(appl math), 67. *Prof Exp:* Lectr math, Calcutta & Burdwan Univs, 57-65; res fel appl math, Imp Col, Univ London, 65-67; sr fel, Cambridge Univ, 67-68; assoc prof, ECarolina Univ, 68-69, prof math, 69-82; actg chair statist, 89-90, CHAIR & PROF, DEPT MATH, UNIV CENT FLA, ORLANDO, 83- *Concurrent Pos:* Ed, Calcutta Math Soc Bulletin, 73-; vis prof, Univ Md, 74; vis prof, Univ Calcutta, 69, 72, 75, 78 & 81, consult, Ctr Advan Study Appl Math, 72-, vis prof Univ Oxford, 80; vis scientist, US-India Exchange of Scientists Prog, NSF, 75; ed, Bulletin of Calcutta Math Soc, 71-; managing ed, Int J Math & Math Sci, 77-; reviewer NSF res proposals, 77-; prof physics, ECarolina Univ, 72-, res prof, Univ Grants Comn, 78-; sr Fulbright fel oceanic turbulence, 78; dir res conf on spec functions, NSF-Conf Bd of Math Sci, 79; dir grad studies, ECarolina Univ, 79-80, Res Conf Nonlinear waves, 82; pres, Calcutta Math Soc, 86-89; dir, NSF Conf Nonlinear Dispersive Wave Systs, Univ Cent Fla, 91; author or co-author of 7 graduate books. *Mem:* Am Math Soc; Am Phys Soc; Soc Indust & Appl Math; fel British Inst Math; Sigma Xi. *Res:* Nonlinear waves; dynamics of oceans; complex analysis; generalized functions with applications; fluid dynamics; elasticity-wave motions; vibrations; elements of theory of elliptic and associated functions with applications; elements of general topology; magnetohydrodynamics; rotating and stratified flows; oceanography. *Mailing Add:* Dept Math Univ Cent Fla Orlando FL 32816-6990

DEBNATH, SADHANA, b Jagatpur, India, June 1, 38; m 69; c 1. ANALYTICAL CHEMISTRY, PHYSICAL CHEMISTRY. *Educ:* Gujarat Univ, India, BS, 58, MS, 60; Univ Calcutta, PhD(chem), 64. *Prof Exp:* Fel chem, Univ Minn, 64-67; scientist, Nat Chem Lab, Poona, India, 67-69; res assoc chem, Sch Med, ECarolina Univ, 72-76, instr biochem, 76-83; ADJ PROF CHEM, UNIV CENT FLA, 84- *Concurrent Pos:* Fulbright scholar, 64. *Honors & Awards:* Sigma Xi Bislinghoff Res Award, ECarolina State Univ, 74. *Mem:* Indian Asn Cultivation Sci; Sigma Xi. *Res:* Acid base chemistry; dyes; kinetic study of organic peroxides, and organic synthesis; fatty acids; metabolism; bacterial pigment; pteridine-peptides; polorographic studies in non-acqueous media; acid-base reaction in non-aqueous media; characterization of pigment peptides. *Mailing Add:* Dept Chem Univ Cent Fla Orlando FL 32816-6990

DEBNEY, GEORGE CHARLES, JR, b Beaumont, Tex, Feb 19, 39; m 62, 66; c 3. APPLIED MATHEMATICS, THEORETICAL PHYSICS. *Educ:* Rice Univ, BA, 61; Univ Tex, Austin, PhD(math), 67. *Prof Exp:* Teaching asst math, Univ Tex, Austin, 62-63, res asst relativity theory, Relativity Ctr, 63-64, 65-66; mem tech staff, TRW Systs Group, 66-68; from asst prof to assoc prof math, Va Polytech Inst & State Univ, 68-89; SR SCIENTIST, SCI APPLN INT CORP, 89- *Concurrent Pos:* Vis prof, Math Inst, Oxford Univ, 76-77. *Mem:* Am Phys Soc; Math Asn Am. *Res:* Differential geometry; exact solutions of the field equations in general relativity theory; mathematical modelling and simulation. *Mailing Add:* SACI 1710 Goodridge Dr McLean VA 22102

DEBOER, BENJAMIN, b Grand Rapids, Mich, Sept 13, 11; m 39; c 5. PHARMACOLOGY. *Educ:* Calvin Col, AB, 33; Univ Mo, MA, 39, PhD(physiol), 42. *Prof Exp:* Asst zool, Univ Mo, 34-38, from instr to asst prof pharmacol, 41-46; from asst prof to assoc prof, St Louis Univ, 46-51; prof, 51-76, EMER PROF PHYSIOL & PHARMACOL, SCH MED, UNIV NDAK, 76- *Mem:* Am Physiol Soc; Am Soc Pharmacol; Soc Exp Biol & Med. *Res:* Barbituate hypnosis; narcotic analgesics; hyperbaric pharmacology. *Mailing Add:* Univ NDak Sch Med Grand Forks ND 58202

DEBOER, EDWARD DALE, b Hammond, Ind, May 24, 55; m 77; c 3. INDUSTRIAL STARCH PRODUCTS, STATISTICAL EXPERIMENTAL DESIGN. *Educ:* Calvin Col, BS, 77. *Prof Exp:* Jr chemist, 77, entry level scientist, Corn Processing Div, 77-81, scientist, Res Dept, 81-84, PROJ LEADER, AM MAIZE-PROD CO, 84- *Mem:* Am Chem Soc. *Res:* Chemistry of corn starch and related products; oxidations, etherification, and esterification; design starch products for the textile, paper, adhesive and oil well drilling industries; modifications of cyclodextrins. *Mailing Add:* 21911 Merrill Ave Sauk Village IL 60411

DEBOER, GERRIT, b Ureterp, Neth, May 14, 42; Can citizen; m 65; c 3. MEDICAL BIOPHYSICS, BIOSTATISTICS. *Educ:* Univ Western Ont, BSc Hons, 65; Univ Toronto, MSc, 67, PhD(med biophys), 70. *Prof Exp:* Asst physics, Univ Toronto, 65-69; instr phys chem, Johns Hopkins Univ, 70-73; ASST PROF MED BIOPHYS, UNIV TORONTO, 73-; PHYSICIST, ONT CANCER INST, 73-, HEAD DEPT BIOSTATIST, 75-; DIV HEAD CLIN TRIALS & EPIDEMIOL, TORONTO BAYVIEW REGIONAL CANCER CTR. *Concurrent Pos:* Fel, Med Res Coun Can, 70-73; sr investr demonstration model grant, Ont Ministry Health, 75-78, co-investr, 78-81; co-investr, Ont Cancer Treat & Res Found grants, 80-83. *Mem:* Biophys Soc; Can Orgn Advan Comput Health; Soc Clin Trails. *Res:* Cancer research; biophysics; ultraviolet photochemistry of nucleic acid derivatives; physical chemistry of nucleic acids; computerized handling of cancer records; computerized medical audits; cancer epidemiology; cost-effectiveness analysis. *Mailing Add:* Toronto Bayview Regional Cancer Ctr 2075 Bayview Ave Toronto ON M4N 3M5 Can

DEBOER, JELLE, b Zeist, Holland, Aug 23, 34; m 63; c 3. GEOPHYSICS. *Educ:* Univ Utrecht, PhD(geol), 63. *Hon Degrees:* MA, Wesleyan Univ, 74. *Prof Exp:* Assoc prof geol, 68-74, Seney prof geol & prof earth & environ sci, 74-80, STEARNS PROF GEOL, WESLEYAN UNIV, 80- *Concurrent Pos:* Consult, nuclear & hydropower projs. *Mem:* Geol Soc Am; Am Geophys Union. *Res:* Geotectonics; use of paleomagnetism and rock magnetism for structural analyses. *Mailing Add:* Parmelee Rd Haddam CT 06438

DEBOER, KENNETH F, b Verdi, Minn, Nov 6, 38; m 59; c 5. BIOLOGY. *Educ:* Univ Minn, Minneapolis, BA, 61; Mankato State Col, MA, 68; Iowa State Univ, PhD(reproductive physiol), 70. *Prof Exp:* From asst prof to assoc prof biol, Western State Col, Colo, 70-78; PROF ANAT, PALMER COL CHIROPRACTIC, 78- *Concurrent Pos:* Am Philos Soc res grant, 71-73, Found Chiropractic Educ & Res grant, 81-82, 85-87; vis scientist, Cancer Ctr, Univ Calif, San Diego. *Mem:* Am Asn Univ Profs; Nat Asn Animal Breeders. *Res:* Clinical efficacy of chiropractic; endocrinology; neurophysiology; biochemistry; biotechnology; monoclonal and polyclonol antibodies in studies of neurohormones and immune system in response to neutral stimuli. *Mailing Add:* Dept Res Palmer Col Chiropractic Davenport IA 52803

DE BOER, P(IETER) C(ORNELIS) TOBIAS, b Leiden, Netherlands, May 21, 30; nat US; m 56; c 3. MECHANICAL & AEROSPACE ENGINEERING. *Educ:* Technol Univ Delft, Ing, 55; Univ Md, PhD(physics), 62. *Prof Exp:* Res asst aerodyn & hydrodyn, Delft Univ Technol, 54-55; res assoc, Inst Fluid Dynamics & Appl Math, Univ Md, 57-62, res asst prof, 62-64; from asst prof to assoc prof, 64-73, PROF MECH & AEROSPACE ENG, CORNELL UNIV, 73-, ASSOC DIR, 82- *Concurrent Pos:* Consult, Conelec, Inc, 65-71 & Allied Chem Automotive Div, 72-; NATO fel, 68; vis prof, von Karman Inst Fluid Dynamics, Belgium, 68 & Cornell Aeronaut Lab, 68-69; engr & consult, Ford Motor Co, 71-73; combustion engr, Gen Elec Co, 78-79; vis prof, Technol Univ Delft, 85-86; corresp, Royal Neth Acad Sci, 88- *Mem:* Am Soc Mech Engrs; Int Asn Hydrogen Energy; Am Phys Soc; Am Inst Aeronaut & Astronaut; Netherlands Royal Inst Eng; Combustion Inst. *Res:* Physics of fluids; ionized gases; relaxation phenomena; combustion processes. *Mailing Add:* Upson Hall Cornell Univ Ithaca NY 14853

DE BOER, SOLKE HARMEN, b Haren, Neth, Jan 5, 48; Can citizen; m 78; c 2. PLANT PATHOLOGY, PHYTOBACTERIOLOGY. *Educ:* Univ BC, Bsc, 70, MSc, 72; Univ Wis-Madison, PhD(plant path), 76. *Prof Exp:* Int Agr Ctr fel, Inst Phytopath Res, Neth, 76; Nat Res Coun Can fel, Univ BC, 76-77; RES SCIENTIST PLANT PATH, RES BR AGR CAN, 77- *Concurrent Pos:* Sabbatical leave, Inst Phytopath Res, Neth, 84-85. *Mem:* Am Phytopathol Soc; Can Phytopathol Soc; Am Soc Microbiol; Potato Asn Am. *Res:* Serology and ecology of plant pathogenic bacteria, especially Erwinia carotovora and Corynebacterium sepedonicum. *Mailing Add:* Agr Can Res Sta 6660 NW Marine Dr Vancouver BC V6T 1X2 Can

DE BOLD, ADOLFO J, b Parana, Arg, Feb 14, 42; Can citizen; m 68; c 5. RESEARCH ADMINISTRATION, EDUCATION ADMINISTRATION. *Educ:* Nat Univ Cordoba, Arg, BSc, 68; Queen's Univ, Can, MSc, 72, PhD(path), 73; FRCP(C), 88. *Prof Exp:* Resident clin labs, Clin Hosp, Cordoba, Arg, 66-68, chief resident, 67-68; from asst prof to prof path, Queen's Univ, Kingston, Can, 74-86; DIR RES, OTTAWA HEART INST, CAN, 86-; PROF PATH, UNIV OTTAWA, CAN, 86- *Concurrent Pos:* Dir electron micros & lab scientist, Hotel Dieu Hosp, Kingston, Can, 74-86; adj prof, Queen's Univ, 86-; hon prof, Nat Univ Tucuman, Arg, 88; vis prof, Cleveland Clinic Asn, 90. *Honors & Awards:* Gairdner Found Int Award, 86; McLaughlin Med, Royal Soc Can, 88. *Mem:* Int Acad Path; Am Soc Cell Biol; Can Soc Cell Biol; Soc Exp Biol & Med; NY Acad Sci; Can Soc Anatomists. *Res:* Combined morphological and biochemical approach to biomedical research; Discovery of atrial natriuretic factor, which are hormones produced by the heart with major effect on blood pressure and blood volume regulation. *Mailing Add:* Dir Res Heart Inst Ottawa Civic Hosp 1053 Carling Ave Ottawa ON K1Y 4E9 Can

DEBOLD, JOSEPH FRANCIS, b Boston, Mass, Nov 3, 47; m 69. PSYCHOBIOLOGY, NEUROENDOCRINOLOGY. *Educ:* Univ Calif, Los Angeles, BA, 69, Univ Calif, Irvine, PhD(biol sci), 76. *Prof Exp:* Res asst psychol, Calif State Univ, Northridge, 69-71; Nat Inst Child Health & Human Develop fel, Univ Calif, Irvine, 71-75; instr & res assoc zool, Mich State Univ, 75-77; asst prof psychol, Carnegie Mellon Univ, 77-79; ASSOC PROF PSYCHOL, TUFTS UNIV, 79- *Concurrent Pos:* Vis res assoc neurosci, Children's Hosp, 81-85. *Mem:* Soc Neurosci; Sigma Xi; AAAS. *Res:* Neural and hormonal bases of sexual differentiation and behavior; mechanisms of steroid action within the brain; hormone-alcohol interactions on behavior. *Mailing Add:* Dept Psychol Tufts Univ Medford MA 02155

DEBONA, BRUCE TODD, b Medford, Mass, Dec 26, 45; m 73; c 2. POLYMER CHEMISTRY, ORGANIC CHEMISTRY. *Educ:* Rensselaer Polytech Inst, BS, 67; Univ Pa, PhD(org chem), 71. *Prof Exp:* Sr res chemist polymer chem, Advan Res Group, Coatings & Resins Div, PPG Industs, 70-74; res chemist, Allied Chem Corp, 74-80, res assoc polymer chem, Corp Res Lab, 80-87; SR RES ASSOC POLYMER CHEM, R&T LAB, ALLIED SIGNAL INC, 87- *Concurrent Pos:* Res fel, Univ Pa, 67-70, asst instr, 67-69. *Mem:* Am Chem Soc; Am Inst Chemists. *Res:* Synthesis of high polymers; structure property relationships; mechanism and stereochemistry of polymer forming reactions; organic synthesis; development of engineering thermoplastics, synthetic fibers and polymer composites. *Mailing Add:* 46 Musiker Ave Randolph NJ 07869

DEBONO, MANUEL, b Melliha, Malta, Sept 3, 36; US citizen; m 60; c 5. ORGANIC CHEMISTRY. *Educ:* Univ San Francisco, BS, 59; Univ Ore, PhD(chem), 63. *Prof Exp:* SR RES SCIENTIST, ELI LILLY & CO, 63- *Mem:* Am Chem Soc; Sigma Xi. *Res:* Natural products; steroids; fertility control agents; photochemical syntheses; structure and modification of antibiotics especially glycopeptides, lipopeptides and macrolides. *Mailing Add:* Lilly Res Labs Natural Prod Res Div Indianapolis IN 46285

DEBONS, ALBERT FRANK, b Brooklyn, NY, Nov 4, 29; m 55; c 3. PHYSIOLOGY. *Educ:* Syracuse Univ, BS, 53; George Washington Univ, MS, 55, PhD(physiol), 58. *Prof Exp:* Asst med res, George Washington Univ, 53-58; res assoc, Brookhaven Nat Labs, 58-61; prin scientist, Vet Admin Hosp, Birmingham, Ala, 61-64, prin scientist, Vet Admin Med Ctr, Brooklyn, NY, 64-90; RETIRED. *Mem:* Endocrine Soc; Am Fed Clin Res; Nuclear Med Soc. *Res:* Nuclear medicine; endocrines as related to intermediary metabolism; mechanism of hormone action; obesity studies; atherosclerosis. *Mailing Add:* 22899 Alfalfa Market Rd Bend OR 97701-9346

DEBOO, PHILI B, b Bombay, India, Dec 5, 34; US citizen; m 60; c 1. GEOLOGY, PALEONTOLOGY. *Educ:* Univ Bombay, BSc, 53; La State Univ, MS, 55, PhD(geol), 63. *Prof Exp:* Asst prof geol, Eastern NMex Univ, 63-65; from asst prof to assoc prof, 65-77, PROF GEOL, MEMPHIS STATE UNIV, 77- *Mem:* Geol Soc Am. *Res:* Mid-Tertiary biostratigraphy of the Central Gulf Coastal Plain. *Mailing Add:* Dept of Geol Memphis State Univ Memphis TN 38152

DE BOOR, CARL (WILHELM REINHOLD), b Stolp, Ger, Dec 3, 37; c 4. MATHEMATICS. *Educ:* Univ Mich, PhD(numerical anal), 66. *Prof Exp:* Assoc sr res mathematician, Gen Motors Res Labs, 60-64; from asst prof to assoc prof math sci, Purdue Univ, 66-72; PROF MATH & COMPUT SCI, MATH RES CTR, UNIV WIS-MADISON, 72- *Concurrent Pos:* Vis staff mem, Los Alamos Sci Labs, 70- *Mem:* Fel Am Acad Arts Sci; Soc Indus Appl Math. *Res:* Numerical analysis; approximation theory; approximation by splines and other piecewise polynomial functions. *Mailing Add:* Ctr Math Sci Univ Wis 610 Walnut St Madison WI 53705

DEBOUCK, CHRISTINE MARIE, b Brussels, Belg, Jan 16, 56; m 84. MOLECULAR BIOLOGY. *Educ:* Free Univ, Brussels, Belgium, BS & MS, 78, PhD(molecular biol), 82. *Prof Exp:* Vis fel biochem, Nat Cancer Inst, NIH, Md, 82-83; postdoctoral molecular genetics, SmithKline & French Labs, Pa, 83-85, assoc sr investr, 85-88, sr investr, 88-89; ASST DIR MOLECULAR GENETICS, SMITHKLINE BEECHAM PHARMACEUTICALS, PA, 89- *Concurrent Pos:* Co-prin investr grant, NIH, 86-, prin investr grant, 87- & 88-91; co-dir, SB AIDS prog, SmithKline Beecham Pharmaceuticals, 87-; mem, AIDS & Related Res Study Sect 3, NIH-DRG, 90- *Mem:* Am Soc Microbiol; Int Soc Antiviral Res; Int AIDS Soc. *Res:* Application of recombinant DNA technology to the isolation, expression and functional characterization of proteins of therapeutic interest, in particular essential proteins of the AIDS virus (protease, reverse transcriptase, tat transactivator). *Mailing Add:* SmithKline Beecham Pharmaceuticals 709 Swedeland Rd King of Prussia PA 19406

DEBOW, W BRAD, b Butler, Mo, Mar 15, 48; m 74; c 1. DISTRIBUTED CONTROL SYSTEMS, NUCLEAR REACTOR CONTROL SYSTEMS. *Educ:* Northrop Univ, Inglewood, BSEE, 74, MSEE, 76. *Prof Exp:* Mem tech staff, Hughes Aircraft, Inglewood, Calif, 74-76; SR ENG SPECIALIST, EG&G IDAHO, INC, IDAHO FALLS, 76- *Concurrent Pos:* Control systs consult, Proj Mgt Off, Spec Isotope Separation Proj, Lawrence Livermore Nat Lab, 86-90; nat deleg, Instrument Soc Am, 89- *Mem:* Instrument Soc Am. *Res:* Application of distributed process control systems to high reliability nuclear reactors and nuclear materials processing systems. *Mailing Add:* EG&G Idaho Inc PO Box 1625 Mail Stop 3423 Idaho Falls ID 83415-3423

DEBRA, DANIEL BROWN, b New York, NY, June 1, 30; m 54; c 6. ENGINEERING MECHANICS. *Educ:* Yale Univ, BE, 52; Mass Inst Technol, SM, 53, PhD(eng mech), 62. *Prof Exp:* Proj engr boiler auxiliary equip, Thermix Corp, 53-54; supvr dynamics & control anal, Lockheed Missiles & Space Co, 56-64; from res engr to assoc prof guid control, 64-73, prof Aeronaut & Astronaut, 73-90, DIR GUID & CONTROL LAB, STANFORD UNIV, 64-, EDWARD C WELLS PROF AERONAUT & ASTRONAUT, 90- *Honors & Awards:* Indust Res 100 Award, Indust Res Mag; Thurlow Award, Inst Navig. *Mem:* Nat Acad Eng; fel Am Inst Aeronaut & Astronaut; fel Am Astronaut Soc; corresp mem, Int Inst Prod Eng Res, France; Soc Mfg Engrs; Soc Automotive Engrs; Inst Elec & Electronics Engrs; Am Soc Mech Engrs. *Res:* Guidance and attitude control of aerospace vehicles; precision manufacturing. *Mailing Add:* Dept Aeronaut & Astronaut Stanford Univ Stanford CA 94305-4035

DE BREMAECKER, JEAN-CLAUDE, b Antwerp, Belg, Sept 2, 23; m 52; c 2. GEODYNAMICS, FRACTURE PROPAGATION. *Educ:* Univ Louvain, Belg, Mining Engr, 48; La State Univ, MSc, 50; Univ Calif, PhD(geophys), 52. *Prof Exp:* Res scientist geophysics, Inst Sci Res Cent Africa, 52-56, sr scientist, 56-58; from asst prof to assoc prof, 59-65, PROF GEOPHYSICS, RICE UNIV, 65- *Concurrent Pos:* Boese fel, Columbia Univ, 55-56 & Harvard Univ, 58-59; mem, Nat Belgium Comt of Geodesy & Geophys, 55-58; vis mem, Tex Inst Comput Mech, Univ Tex, Austin, 77; vis prof, Univ Paris, 80-81. *Mem:* Am Geophys Union; Int Asn Seismol & Phys Earth's Interior (secy gen, 71-79). *Res:* Seismology; gravimetry; magnetism; tectonophysics. *Mailing Add:* 2128 Addison Houston TX 77030

DEBREU, GERARD, b Calais, France, July 4, 21; US citizen; m 45; c 2. MATHEMATICAL ECONOMICS. *Educ:* Univ Paris, DSc, 56. *Hon Degrees:* Dr rer pol, Univ Bonn, 77; DEcon, Univ Lausanne, 80; Dr, Northwestern Univ, 81, Univ Soc Sci Toulouse, 83 & Yale, 87. *Prof Exp:* Res assoc econ, Ctr Nat Res Sci, Paris, 46-48; Rockefeller fel, US, Sweden & Norway, 48-50; res assoc, Cowles Comn Res Econ, Univ Chicago, 50-55; assoc prof econ, Cowles Found Res Econ, Yale Univ, 55-61; prof, Miller Inst Basic Res Sci, 73-74, fac res lectr, 84-85, PROF ECON, UNIV CALIF, BERKELEY, 62-, PROF MATH, 75-, UNIV PROF, 85- *Concurrent Pos:* Assoc ed, Int Econ Rev, 59-69; fel, Ctr Adv Study Behav Sci, Stanford, 60-61; vis prof, Yale Univ, 61 & 76, Univ Louvain, 68-69 & 71-72, & Univ Canterbury, Christchurch, NZ, 73, Univ Sydney, 87; Guggenheim fel, Univ Louvain, 68-69; Erskine fel, Univ Canterbury, Christchurch, NZ, 69 & 87; overseas fel, Churchill Col, Cambridge, Eng, 72; sr US scientist award, Alexander von Humboldt Found, Univ Bonn, 77; Aisenstadt chair, Univ Montreal, 85. *Honors & Awards:* Nobel Prize in Econ Sci, 83; Fisher-Schultz Lectr, 69; Snyder Lectr, Univ Calif, Santa Barbara, 84; Frisch Mem Lectr, Fifth World Congress, Econometric Soc. *Mem:* Nat Acad Sci; Economet Soc (vpres & pres, 69-71); fel AAAS; fel Am Acad Arts & Sci; fel Econ Asn (pres 90); Am Philos Soc; foreign assoc Fr Acad Sci. *Res:* Theory of general economic equilibrium - preference, utility, and demand theory. *Mailing Add:* Dept Econ Univ Calif Berkeley CA 94720

DEBROY, CHITRITA, b New Delhi, India. MOLECULAR BIOLOGY. *Educ:* Delhi Univ, BS, 67, MS, 69; Jawaharlal Nehru Univ, MPhil, 75, PhD(life sci), 78. *Prof Exp:* Staff scientist molecular biol, Boston Biomed Res Inst, 78-80; res assoc, 81-88, SR RES ASSOC MOLECULAR BIOL, PA STATE UNIV, 88- *Mem:* Am Microbiol Soc; AAAS. *Res:* Recombinant DNA technology in molecular biology; cloning and sequencing genes; developing probes for identification of pathogens; expressing genes in bacteria and eukaryotic; cell lines. *Mailing Add:* 624 Stoneledge Rd State College PA 16803

DEBROY, TARASANKAR, b Calcutta, India, Oct 29, 46; m 80; c 2. TRANSPORT PHENOMENA, MATHEMATICAL MODELING. *Educ:* Regional Eng Col, Durgapur, India, BS, 69; Indian Inst Sci, Bangalore, India, PhD(metallurgy), 74. *Prof Exp:* Fel metallurgy, Imp Col Sci & Technol, London, 74-77; res assoc metallurgy, Univ Ky, 77-78, Mass Inst Technol, 78-79; from asst prof to assoc prof metall, 80-89, PROF MAT SCI & ENG, PA STATE UNIV, 89- *Concurrent Pos.* Prin invest, NSF, 82-87, US Dept Energy, 83-, Am Iron & Steel Inst, 83-86, Eng Found, 81-82, Martin Marietta Energy Syst Inc, 87-89; co-prin invest, Am Foundrymens Soc, 81-82, Off Naval Res 88- *Mem:* Metallurgical Soc Am Inst Mining, Metallurgical & Petrol Engrs; Am Soc Metals; Am Welding Soc. *Res:* Application of rate phenomena in process metallurgy, particularly in the areas of high temperature thermochemical processing, material joining, material synthesis and chemical vapor deposition. *Mailing Add:* Dept Mat Sci & Engineering Penn State Univ University Park PA 16802

DEBRUIN, KENNETH EDWARD, b Oskaloosa, Iowa, May 29, 42; m 69; c 1. ORGANIC CHEMISTRY. *Educ:* Iowa State Univ, BS, 64; Univ Calif, Berkeley, PhD(org chem), 67. *Prof Exp:* NIH fel, Princeton Univ, 67-69; asst prof, 69-74, ASSOC PROF CHEM, COLO STATE UNIV, 74- *Concurrent Pos:* Vis prof, Louis Pasteur Inst, 77-78. *Mem:* Am Chem Soc; Royal Soc Chem. *Res:* Mechanisms of organic reactions; phosphorus stereochemistry; enzyme mechanisms. *Mailing Add:* Dept Chem Colo State Univ Ft Collins CO 80523

DEBRUNNER, LOUIS EARL, b Cincinnati, Ohio, Dec 9, 35; m 58; c 2. FOREST ECOLOGY. *Educ:* Univ Cincinnati, BS, 57; Yale Univ, MF, 59; Duke Univ, DFor, 67. *Prof Exp:* ASST PROF FORESTRY, AUBURN UNIV, 61- *Mem:* Soc Am Foresters. *Res:* Regeneration of forest stands; forest recreation. *Mailing Add:* Sch Agr Dept Forestry Auburn Univ Main Campus Auburn AL 36849

DEBRUNNER, MARJORIE R, b Auburn, Nebr, Feb 21, 27. ORGANIC CHEMISTRY. *Educ:* Nebr State Teachers Col, Kearney, BScEd, 48; Univ Nebr, MS, 51, PhD(chem), 53. *Prof Exp:* Res chemist, 53-72, res div head, 72-76, mgr tech develop, 76-81, TECH MGR, E I DU PONT DE NEMOURS & CO, 81- *Mem:* AAAS; Sigma Xi; Am Chem Soc; The Chem Soc. *Res:* Synthetic hydrocarbon and fluorocarbon elastomers. *Mailing Add:* 11 Binford Lane Devonshire Wilmington DE 19810

DEBRUNNER, PETER GEORG, b Sitterdorf, Switz, Mar 11, 31; m 55; c 3. PHYSICS. *Educ:* Swiss Fed Inst Technol, PhD(physics), 60. *Prof Exp:* Res assoc, 60-63, res asst prof, 63-68, assoc prof physics, 68-73, PROF PHYSICS, UNIV ILL, URBANA, 73- *Mem:* Am Phys Soc; Biophys Soc. *Res:* Angular correlation; Mossbauer and electron paramagnetic resonance spectroscopy; biological physics. *Mailing Add:* Univ Ill Dept Physics 1110 W Green St Urbana IL 61801

DEBRUNNER, RALPH EDWARD, b Cincinnati, Ohio, Oct 11, 32; m 56; c 2. ORGANIC CHEMISTRY. *Educ:* Univ Cincinnati, BS, 54, PhD(org chem), 60. *Prof Exp:* Res chemist, Chemstrand Corp, Ala, 54-55; res chemist spec proj dept, Monsanto Chem Co, Mass, 60-61, sr res chemist, Dayton Lab, Monsanto Res Corp, Monsanto Co, 61-66, group leader, 66-68, group leader, Chemstrand Res Ctr, NC, 68-72, sr tech serv rep, 72-77, sr group leader appln res, nonwoven bus group, Monsanto Textile Co, 77-85; sr tech specialist prod develop, James River Corp, 85-87, mkt tech mgr indust prod, Nonwoven Div, 87-90; RES FEL, FIBERWEB NAM, 90- *Mem:* Am Chem Soc; Am Soc Testing & Mat; Int Nonwovens Disposables Asn. *Res:* Synthesis of thermally and oxidatively stable polymers, and preparation of high performance composites; applications research and product development for nonwoven fabrics. *Mailing Add:* Fiberweb NAm 545 N Pleasantburg Dr Greenville SC 29607

DE BRUYN, PETER PAUL HENRY, b Amsterdam, Holland, July 28, 10; nat US; m 31; c 2. HISTOLOGY. *Educ:* Univ Amsterdam, PhD, 38, MD, 41. *Prof Exp:* Asst, Histol Lab, Univ Amsterdam, 36-39; first asst, Inst Prev Med, Leyden, Holland, 39-40, chief bact dept, 40-41; from instr to assoc prof, 41-52, chmn dept, 46-61, PROF ANAT, UNIV CHICAGO, 52- *Concurrent Pos:* Stokvis-fonds fel, Univ Chicago, 38; mem staff, AEC, 46. *Mem:* Am Asn Anat; Am Soc Cell Biol. *Res:* Lipids of leucocytes; locomotion of leucocytes; histopathology of radiation effects; lymphatic tissue; vital staining of nuclei; fine structural studies on transmural migration of blood cells and malignant cells; leukopoietic mechanisms, sinusoidal endocytom of protein. *Mailing Add:* Dept Anat Univ Chicago 1025 E 57th St Chicago IL 60637

DE BRUYNE, PETER, b Middelburg, Netherlands, Mar 7, 28; US citizen; m 60; c 3. ELECTRICAL ENGINEERING. *Educ:* London Univ, BSc, 47; Inst Electronics Technol, London, MIET, 56. *Prof Exp:* Group leader radio lab, Van Der Heem, Holland, 56-58; chief engr nuclear physics, Tracerlab Inc, Europe & US, 58-63; corp app res, Physics Dept, Harvard Univ, 63-68; sr engr, LLTV, RCA Div Burlington, 68-71; mgr prod planning, Philips, Comput Peripherals, Holland, 71-72; RES ASSOC, SWISS FED INST TECHNOL, 72- *Concurrent Pos:* Consult, physics dept, Harvard Univ, 68-71; mem comt, Carnahan conf security technol, Univ Ky, 78- *Mem:* Sigma Xi; NY Acad Sci. *Res:* High energy physics research with wire spark chambers for inelastic e-P scattering experiments; low light level television target identification and signature analysis; new detection and measurement methods in physics and in communication; author of over 40 publications; electro-optic reconnaissance; radar vehicle detection; ultrasonic graphic tablets for computers; signature verification. *Mailing Add:* Inst fur Komm Technik ETH-Zentrum-KT Zurich CH 8092 Switzerland

DEBS, ROBERT JOSEPH, b Chicago, Ill, Mar 31, 19; m 46; c 2. MICROWAVE PHYSICS. *Educ:* Mass Inst Technol, PhD(physics), 52. *Prof Exp:* Jr engr, Westinghouse Elec Corp, 42-44; develop engr, Raytheon Mfg Co, 44-46; res assoc electronics & nuclear sci, Mass Inst Technol, 46-52; microwave lab, Stanford Univ, 52-58; res assoc, West Coast Labs, Gen Tel & Electronics Corp, 58-63; res scientist, Ames Res Ctr, NASA, 63-76, consults, 77-82; RETIRED. *Mem:* Am Phys Soc. *Res:* Low-intensity, low-temperature behavior of solar cells under charged-particle bombardment; superconducting magnetometry; low-level radioactivity counting. *Mailing Add:* 3145 Flowers Lane Palo Alto CA 94306

DEBUCHANANNE, GEORGE D, waste disposal; deceased, see previous edition for last biography

DEBUDA, RUDOLF G, communications theory; deceased, see previous edition for last biography

DE BUHR, LARRY EUGENE, b Rock Rapids, Iowa, Nov 21, 48; m 77; c 2. PLANT SYSTEMATICS, PLANT ANATOMY. *Educ:* Iowa State Univ, BS, 71; Claremont Grad Sch, MA, 73, PhD(bot), 76. *Prof Exp:* Fac mem biol, Cottey Col, 77-80; asst prof, & asst dir ctr acad develop, 80-88, ASST PROF BIOL & EDUC, UNIV MO, KANSAS CITY, 88- *Mem:* Bot Soc Am; Am Soc Plant Taxonomists; Int Asn Plant Taxon; Sigma Xi. *Res:* Plant systematics, anatomy and morphology of Sarraceniaceae, Droseraceae and Crossosomataceae. *Mailing Add:* Mo Bot Garden PO Box 299 St Louis MO 63166

DEBUS, ALLEN GEORGE, b Chicago, Ill, Aug 16, 26; m 51; c 3. HISTORY OF MEDICAL CHEMISTRY. *Educ:* Northwestern Univ, BS, 47; Ind Univ, AM, 49; Harvard Univ, PhD(hist sci), 61. *Hon Degrees:* DSc, Cath Univ Louvain, Belg, 85. *Prof Exp:* Res & develop chemist, Abbott Labs, 51-56; from asst prof to prof hist sci, 61-78, dir, Morris Fishbein Ctr Study Hist Sci & Med, 71-77, MORRIS FISHBEIN PROF HIST SCI & MED, UNIV CHICAGO, 78- *Concurrent Pos:* Overseas fel, Churchill Col, Cambridge, Eng, 66-67 & 69; Guggenheim fel, 66-67; mem, Inst Advan Studies, Princeton, 72-73; fel, Inst Res Humanities, Univ Wis, Madison, 81-82; lectr, Univ Coimbra, Port, 83; vis distingushed prof, Ariz State Univ, 84; consult, Lit & Sci Prog, Ga Inst Technol, 85-86. *Honors & Awards:* Bowdoin Award, Harvard Univ, 57 & 58; Edward Kremers Award, Am Inst Hist Pharm, 78; Pfizer Award, Hist Sci Soc, 78; Dexter Award, Am Chem Soc, 87. *Mem:* Int Acad Hist Med; Int Acad Hist Sci; Hist Sci Soc; Am Asn Hist Med; fel AAAS; Soc Hist Alchemy & Chem; foreign mem Acad Ciencias Lisboa, 87. *Res:* Renaissance and Early Modern science and medicine; the Scientific Revolution; early chemistry and medical chemistry (pre-Lavoisier, history of pharmacy, alchemy). *Mailing Add:* Morris Fishbein Ctr Study Hist Sci & Med Univ Chicago 1126 E 59th St Chicago IL 60637

DEBUSK, A GIB, b Lubbock, Tex Jan 15, 27; m 47; c 6. GENETICS, BIOTECHNOLOGY. *Educ:* Univ Wash, BS, 50; Univ Tex, MA, 52, PhD(genetics), 54. *Prof Exp:* Res scientist, Biochem Inst, Univ Tex, 51-52, Genetics Found, 52-55; instr, Dept Biol Sci, Northwestern Univ, 55-57; from asst prof to assoc prof, 57-69, assoc dir, Inst Molecular Biophys, 62-63, assoc chmn, Dept Biol Sci, 66-67, assoc chmn grad studies, Dept Biol Sci, 72-76, actg chmn dept, 75-76, dir genetics training prog, 62-80, chmn dept, 76-83, PROF BIOL SCI, FLA STATE UNIV, 69-, UNIV SERV PROF, 85-; PRES, FLA BIOTECHNOL INC, 83- *Concurrent Pos:* Vis prof, Southwestern Univ, 54-55; mem staff, Brookhaven Nat Lab, 56; vis fel, Inst Advan Studies, Australian Nat Univ, 68; chmn, bd dirs, Receptor Molecules Inc, 84-86; consult, Southern Technol & Develop, 87- *Mem:* Am Chem Soc; Biophys Soc; Am Soc Microbiol; Genetics Soc Am; Am Soc Biol Chemists. *Res:* Molecular and biochemical genetics; cellular transport; receptor molecules; biotechnology. *Mailing Add:* Fla Biotechnol Inc 1380 Blountstown Hwy 20 Tallahassee FL 32304

DEBYE, NORDULF WIKING GERUD, b Czech, Aug 2, 43; US citizen; m 67; c 1. PHYSICAL INORGANIC CHEMISTRY, PHYSICAL CHEMISTRY. *Educ:* Rice Univ, BA, 65; Cornell Univ, PhD(phys chem), 70. *Prof Exp:* Nat Res Coun-Nat Bur Standards res assoc struct inorg chem, Nat Bur Standards, 70-72; asst prof chem, Colo Women's Col, 72-76; asst prof, 76-83, ASSOC PROF CHEM, TOWSON STATE UNIV, 83-, DEPT CHMN, 86- *Concurrent Pos:* Vis instr chem, Towson State Col, 75-76. *Mem:* Am Chem Soc; NAm Therm Anal Soc; Sigma Xi; NY Acad Sci; Coblentz Soc; Int Confederation Thermal Anal. *Res:* Structural spectroscopy of organometallic compounds; scanning calorimetry; thermodynamic properties of organometallics. *Mailing Add:* Dept Chem Towson State Univ Baltimore MD 21204

DEBYLE, NORBERT V, b Green Bay, Wis, May 1, 31; m 54; c 2. FORESTRY, WATERSHED MANAGEMENT. *Educ:* Univ Wis, BS, 53, MS, 57; Univ Mich, PhD(forestry), 62. *Prof Exp:* Conserv aide wildlife mgt, Wis State Dept Conserv, 53-54; res asst forest & wildlife mgt, Univ Wis, 56-57; lectr forestry, Univ Mich, 58-59; res forester, USDA Forest Serv, Nev, 61-64 & Utah, 64-85, ecologist, Intermountain Res Sta, Utah, 85-87; RETIRED. *Concurrent Pos:* Adj assoc prof, Col Natural Resources, Utah State Univ, 70-; Temp assoc prof, Utah State Univ, 89-90, 91-92. *Mem:* Fel AAAS; Wildlife Soc; Soc Am Foresters; Sigma Xi. *Res:* Aspen ecology and management; water yield and quality improvement from mountain watersheds; effects of forest harvest and wildland fire on soils, water and flora; wildlife ecology. *Mailing Add:* Dept Forest Resources Col Nat Resources Utah State Univ Logan UT 84322-5215

DECAMP, MARK RUTLEDGE, b Orange, NJ, Aug 21, 46; m 70; c 2. ORGANIC CHEMISTRY. *Educ:* Williams Col, BA, 68; Princeton Univ, MA, 70, PhD(chem), 72. *Prof Exp:* Fel, Cornell Univ, 72-73; NIH fel, Univ Rochester, 73-75; asst prof, 75-80, ASSOC PROF, DEPT NATURAL SCI, UNIV MICH-DEARBORN, 80- *Mem:* Am Chem Soc; Nat Asn Adv Health Prof. *Res:* Organic synthesis; mechanisms of thermal reactions; reactive intermediates. *Mailing Add:* Dept Natural Sci Univ Mich 4901 Evergreen Rd Dearborn MI 48128

DECAMP, PAUL TRUMBULL, b Seoul, Korea, Feb 26, 15; US citizen; m 45; c 6. SURGERY. *Educ:* Wheaton Col, BS, 35; Univ Pa, MD, 41. *Prof Exp:* Asst path, Sch Med, Baylor Univ, 42-44; asst surg, Ochsner Clin, Sch Med, Tulane Univ, 44-48, from instr to asst prof, 48-59, cancer coordr, 48-50, staff surgeon, 50-90, assoc prof clin surg, 59-90; RETIRED. *Mem:* Am Asn Thoracis Surg; Am Col Surg; Soc Vascular Surg; Int Soc Surg; Int Cardiovasc Soc. *Res:* Venous thrombosis and embolism; venous pressure in post-phlebitic and related conditions; cancer of the lung; hypertension due to renal arterial stenosis; cerebrovascular insufficiency. *Mailing Add:* 7811 St Charles Ave New Orleans LA 70118

DE CAMP, WILSON HAMILTON, b Evanston, Ill, Sept 22, 36; div; c 3. PHYSICAL ORGANIC CHEMISTRY. *Educ:* Ind Univ, BS & MS, 60; Univ Md, PhD(chem), 70. *Prof Exp:* Res assoc chem, Clarkson Col, 70; res fel biol, Nat Res Coun Can, 70-72; res assoc chem, Univ Ill, 72-73; res assoc chem, Inst Natural Prod Res, Univ Ga, 73-78; asst prof chem, Cumberland Col, 78-79; chemist, Div Drug Chem, 79-85, CHEMIST, DIV ANTI-INFECTIVE DRUG PROD, FOOD & DRUG ADMIN, 85- *Mem:* Am Crystallog Asn; Am Chem Soc; AAAS; Sigma Xi; Am Pharm Asn; Am Asn Pharmaceut Scientists. *Res:* Determination of molecular structure by x-ray crystallography; absolute configuration of natural products; molecular mechanisms of drug-receptor interactions; conformational analysis of heterocyclic systems; X-ray powder diffraction. *Mailing Add:* 7640 Provincial Dr Apt 110 McLean VA 22102

DE CANI, JOHN STAPLEY, b Canton, Ohio, May 8, 24; wid. APPLIED STATISTICS. *Educ:* Univ Wis, BS, 48; Univ Pa, MBA, 51, PhD(statist), 58. *Prof Exp:* Instr statist, Univ Pa, 48-58; from asst prof to assoc prof, 58-72, chmn dept statist & opers res, 71-78, PROF STATIST, UNIV PA, 72- *Concurrent Pos:* Consult, US Naval Air Develop Ctr, 57-; Fulbright grant, Norweg Sch Econ & Bus Admin, 59-60; consult, McNeil Labs, Inc, 67-, Nat Asn Advan Colored People Legal Defense Fund, Inc, 67- & Merck, Sharp & Dohme Res Labs, 78-; lectr, Brazilian Inst, 71- *Mem:* Fel Am Statist Asn; Inst Math Statist; Economet Soc; Biomet Soc; Soc Indust & Appl Math. *Res:* Paired comparison experiments; analysis of categorical data; mathematical programming. *Mailing Add:* Dept of Statist Univ Pa 3000 Steinberg-Dietrich Hall Philadelphia PA 19104

DE CARLO, CHARLES R, b Pittsburgh, Pa, May 7, 21; m 46; c 4. MATHEMATICS, ACADEMIC ADMINISTRATION. *Educ:* Univ Pittsburgh, BE, 43, PhD(math), 51. *Prof Exp:* Lectr math, Univ Pittsburgh, 47-51; from asst dir to dir, Appl Sci Div, Int Bus Mach Corp, 51-57, dir sales serv, 57-58, dir mkt prog, 58, mgr mkt & serv, 58-59, asst gen mgr, Data Systs Div, 59-61, corp dir & ed, 61-63, corp dir & ed, Systs Res & Develop, 63-65, dir automation res, 65-69; pres, Sarah Lawrence Col, 69-80; RETIRED. *Concurrent Pos:* Fac mem, Am Studies Inst, 68; consult, US Off Educ; mem bd directors, Nat Comn of Resources for Youth, 69- & Inst for Archit & Urban Studies & Comn on Independent Cols & Univs, 74- *Mem:* AAAS; Am Acad Arts & Sci; Soc Indust & Appl Math; Instrument Soc Am; Economet Soc. *Res:* Application of computers and automata to science and business. *Mailing Add:* Off of the Pres Sarah Lawrence Col Bronxville NY 10708

DE CARLO, JOHN, JR, b Philadelphia, Pa, July 9, 18; m 47; c 3. RADIOLOGY. *Educ:* Temple Univ, AB, 40; Jefferson Med Col, MD, 44; Am Bd Radiol, dipl, 50. *Prof Exp:* Dir dept radiol, Baltimore City Hosps, 50-61; dir radiol, St Joseph's Hosp, 62-87; RETIRED. *Concurrent Pos:* Asst prof, Univ Md, 50-61; instr, Johns Hopkins Univ, 58-61. *Mem:* AMA; fel Am Col Radiol; Radiol Soc NAm. *Res:* Diagnostic roentgenology. *Mailing Add:* 701 Seabrook Ct Baltimore MD 21204-2912

DECARO, THOMAS F, b Brooklyn, NY, Mar 10, 19; m 50; c 2. CELLULAR & VERTEBRATE PHYSIOLOGY. *Educ:* Rutgers Univ, BS, 48; Univ NH, MS, 50; Univ Pa, PhD, 63. *Prof Exp:* Researcher, Smith, Kline & French Labs, 49-51; instr biol, St Michael's Col, 51-54; asst prof, Villanova Univ, 54-65; assoc prof, Widener Univ, 66-76; assoc prof biol, Lincoln Univ, 77-89; RETIRED. *Concurrent Pos:* NIH fel, Sch Med, Univ Pa, 65-66. *Mem:* AAAS; Am Soc Zoologists; Am Soc Cell Biol. *Res:* Muscle physiology. *Mailing Add:* 335 N Franklin St Westchester PA 19380

DECASTRO, ARTHUR, b Newark, NJ, July 8, 11; m 51; c 1. ANALYTICAL CHEMISTRY. *Educ:* Newark Col Eng, BS, 31. *Prof Exp:* Chemist, Calco Chem Co, 31-33; control chemist, Nopco Chem Co, Harrison, 33-36, anal res chemist, 36-38, chief chemist, Anal Res Lab, 38-58, lab dir anal res & phys testing labs, 58-73. *Concurrent Pos:* Consult, 73- *Mem:* Fel Am Inst Chemists; Am Chem Soc; Am Soc Testing & Mat. *Res:* Surfactants and processed chemicals used in detergent, textile, tanning, paper, paint and metal working. *Mailing Add:* 14 Midvale Dr New Providence NJ 07974-2532

DECEDUE, CHARLES JOSEPH, b New Orleans, La, July 5, 44; m 66; c 2. PROTEIN CHEMISTRY, ANALYTICAL BIOCHEMISTRY. *Educ:* Univ New Orleans, BS, 66; La State Univ, PhD(microbiol), 71. *Prof Exp:* Fel microbiol, Duke Univ, 71-74; res assoc biochem, Univ NC, 74-76; from asst scientist to assoc scientist, Univ Kans, 76-88, dir, biochem res serv labs, 76-89, sr scientist, 88-89, EXEC DIR, HIGUCHI BIO SCI CTR, UNIV KANS, 89- *Concurrent Pos:* Vis scientist, dept med, Univ Wash, 83; Courtesy prof, 88-; sci adv, off food & drug admin, Kansas City, 85-87. *Mem:* Am Chem Soc; Sigma Xi; AAAS; Protein Soc. *Res:* Analytical biochemistry, including assay techniques, structure analysis, chemical composition of proteins. *Mailing Add:* Higuchi Biosci Ctr Univ Kans Lawrence KS 66047

DE CHAMPLAIN, JACQUES, b Quebec, Que, Mar 13, 38; m 61; c 3. PHYSIOLOGY, PHARMACOLOGY. *Educ:* Univ Montreal, BA, 57, MD, 62; McGill Univ, PhD(invest med), 65. *Prof Exp:* Vis res assoc, NIMH, 65-67; Med Res Coun Que fel, 67-68; from asst prof to prof, 68-76, EDWARDS FOUND PROF PHYSIOL, UNIV MONTREAL, 75- *Concurrent Pos:* Markle scholar, 68; mem, Coun High Blood Pressure. *Mem:* AAAS; Am Fedn Clin Res; Can Physiol Soc; Int Brain Res Orgn; Int Soc Hypertension; Can Soc Hypertension. *Res:* Studies on the role of peripheral and central sympathetic nervous system in the regulation of normal blood pressure and in the pathogenesis of hypertensive diseases. *Mailing Add:* Dept Physiol PO Box 6128 Univ Montreal Fac Med Sta A Montreal PQ H3C 3J7 Can

DE CHAZAL, L(OUIS) E(DMOND) MARC, b St Denis, Reunion, Nov 23, 21; m 51; c 2. CHEMICAL ENGINEERING, NUCLEAR ENGINEERING. *Educ:* La State Univ, BS, 49, MS, 51, PhD(chem eng), 53. *Prof Exp:* From asst prof to assoc prof, 53-66, PROF CHEM ENG, UNIV MO, COLUMBIA, 66- *Concurrent Pos:* Consult, Mo Farmers Asn, 53-54; prin sci officer, Brit Atomic Res Estab, 59-60; res assoc, Brit AEC, 66-67. *Mem:* AAAS; Am Chem Soc; Am Inst Chem Engrs; Am Soc Testing & Mat. *Res:* Solvent extraction; solids mixing; non-Newtonian fluid mechanics; bubble and drop phenomena. *Mailing Add:* 1031 Eng Bldg Univ Mo Columbia Columbia MO 65211

DECHENE, LUCY IRENE, b Petaluma, Calif, Dec 25, 50. COMMUTATIVE RING THEORY, GRAPH THEORY. *Educ:* Univ San Francisco, BS, 73; Univ Calif, Riverside, MS, 75, PhD(math), 78; Guild Carillonneurs NAm, cert, 77. *Prof Exp:* Teaching asst math, Univ Calif, Riverside, 76-77; asst prof, 78-84, ASSOC PROF MATH, FITCHBURG STATE COL, 84- *Concurrent Pos:* Carillon instr, Smith Col, 80-81; consult, Vanguard-Lion Assocs, 83. *Mem:* NY Acad Sci; Sigma Xi; Am Math Soc; Math Asn Am; Asn Women Math; Soc Indust & Appl Math. *Res:* Adjacent extensions of rings; role of histamine in the postviral fatigue syndrome. *Mailing Add:* Fitchburg State Col Box 6167 Fitchburg MA 01420

DECHER, RUDOLF, b Wuerzburg, WGer, Aug 22, 27; US citizen; m 56; c 2. SPACE SCIENCE, GRAVITATIONAL PHYSICS. *Educ:* Univ Wuerzburg, WGer, MA, 50, PhD(physics), 54. *Prof Exp:* Res physicist indust res & develop, Dynamit Nobel AG, WGer, 55-60; dep br chief, R F Systs, Astrionics Lab, NASA, 60-63, staff mem syst eng, 63-66, staff lab dir res & develop planning, 66-69, asst to lab dir space sci, 69-70, chief astrophysics div, Space Sci Lab, 70-86, asst dir, Space Sci Lab, 86-89, CHIEF ASTROPHYSICS DIV, NASA-MARSHALL SPACE FLIGHT CTR, 89- *Mem:* Am Phys Soc; Am Inst Aeronaut & Astronaut. *Res:* Experimental tests of gravitational theories involving space flight; infrared astronomy and detectors; superconducting devices. *Mailing Add:* Marshall Space Flight Ctr ES-01 Huntsville AL 35812

DECICCO, BENEDICT THOMAS, b Rahway, NJ, Feb 7, 38; m 60; c 6. MICROBIAL PHYSIOLOGY, ENVIRONMENTAL MICROBIOLOGY. *Educ:* Rutgers Univ, AB, 60, MS, 62, PhD(bact), 64. *Prof Exp:* Asst bact, Rutgers Univ, 60-62, sr lab technician, 62-63, res asst, 63-64; from asst prof to assoc prof, 64-77, chmn dept, 73-77, PROF BACT, CATH UNIV AM, 78-, CHMN DEPT, 86- *Concurrent Pos:* Consult, var pharmaceut co; mem, Consumer Prod & Qual Assurance Comt, Nat Registry Microbiologists. *Mem:* AAAS; Sigma Xi; Am Soc Microbiol (pres DC chap, 83-84). *Res:* Microbial contamination of water systems and pharmaceuticals; environmental microbiology; single cell protein; antimicrobial agents; microbial resistance. *Mailing Add:* Dept Biol Cath Univ Am Washington DC 20064

DECIUS, JOHN COURTNEY, b San Francisco, Calif, Feb 13, 20; m 48; c 3. PHYSICAL CHEMISTRY. *Educ:* Stanford Univ, AB, 41; Harvard Univ, PhD(chem physics), 47. *Prof Exp:* Res assoc & supvr underwater explosives res lab, Oceanog Inst, Woods Hole, 44-47; res assoc chem, Brown Univ, 47-49; from asst prof to assoc prof, 49-56, PROF CHEM, ORE STATE UNIV, 56- *Concurrent Pos:* Guggenheim fel, Fulbright res scholar, Oxford Univ, 55-56, Sloan Found fel, 56-60; NSF fel, King's Col, London, 63-64. *Mem:* Fel Am Phys Soc; Am Chem Soc. *Res:* Molecular structure; vibrational spectra of solid state; energy transfer in gases. *Mailing Add:* SR 60 Idleyld Park OR 97447

DECK, CHARLES FRANCIS, b Norfolk, Va, June 5, 30; m 56; c 3. INORGANIC CHEMISTRY, RADIOCHEMISTRY. *Educ:* St Louis Univ, BS, 52; Wash Univ, PhD (chem), 56. *Prof Exp:* Asst, Wash Univ, 52-56; from res chemist to sr res chemist, 57-79, res assoc, 79-85, SR RES ASSOC, BASF WYANDOTTE CORP, 85- *Mem:* Am Chem Soc. *Res:* Industrial process study; chemical reaction kinetics. *Mailing Add:* 2805 Trenton Dr Trenton MI 48183

DECK, HOWARD JOSEPH, b Cincinnati, Ohio, Sept 25, 38; m 60; c 2. ELECTRICAL ENGINEERING. *Educ:* Univ Cincinnati, EE, 61, MS, 63; Mich State Univ, PhD(elec eng), 68. *Prof Exp:* Instr elec eng, Mich State Univ, 63-65 & 67-68; asst prof, Ohio Univ, 68-74; ASSOC PROF ELEC ENG, UNIV SALA, 74- *Mem:* Inst Elec & Electronics Engrs; Am Soc Eng Educ. *Res:* Electromagnetic field theory; antenna and circuit theory; reduction of backscattered energy from receiving antennas; active receiving antennas. *Mailing Add:* Dept Elec Eng Univ SAla 307 University Dr Mobile AL 36688

DECK, JAMES DAVID, b Atlanta, Ga, Nov 6, 30; m 55; c 4. HISTOLOGY. *Educ:* Davidson Col, BS, 51; Princeton Univ, MA, 53, PhD(biol), 54. *Prof Exp:* From instr to asst prof, 54-62, ASSOC PROF ANAT, SCH MED, UNIV VA, 62- *Concurrent Pos:* USPHS fel anat, Harvard Univ, 65-66. *Mem:* Am Asn Anat; Am Asn Univ Prof; Am Soc Cell Biol; AAAS. *Res:* Influence of nerves in amphibian regeneration; experimental production of regenerates in an non-regenerating system by implantations of microinfusions; histamine and antihistamines in regneration; skin structure and response to injury; structure and function of mammalian aortic valves; nerves in mammalian regeneration; scanning and conventional electron microscopy of connective tissues; stress and atherogenesis. *Mailing Add:* Dept Anat & Cell Biol Box 439 Univ Va Med Ctr Charlottesville VA 22908

DECK, JOSEPH CHARLES, b Canton, Ohio, July 16, 36; m 61; c 4. COMPUTER AIDED INSTRUCTION. *Educ:* Duquesne Univ, BS, 60; Univ Ill, MS, 64, PhD(phys chem), 66. *Prof Exp:* Chemist, Gulf Res & Develop Co, 60-62; chmn dept, Univ Louisville, 75-78, assoc dean, 83-85, actg dean, 85-86, from asst prof to prof chem, 66-; DEAN, COL ARTS & SCI, SOUTHEASTERN MASS UNIV. *Concurrent Pos:* Vis prof, Univ Ill, 78-79. *Mem:* Am Chem Soc. *Res:* Magnetic resonance; molecular structure; intermolecular and intramolecular interactions. *Mailing Add:* Southeastern Mass Univ North Dartmouth MA 02747

DECK, JOSEPH FRANCIS, b St Louis, Mo, Mar 19, 07; m 36; c 3. CHEMISTRY. *Educ:* St Louis Univ, AB, 28, MS, 30; Univ Kans, PhD(chem), 32. *Hon Degrees:* DSC, Santa Clara Univ, 86. *Prof Exp:* Res chemist, Stewart Inso Bd Corp, 32-35; qual supvr, US Gypsum Co, 35; prof, 36-77, EMER PROF CHEM, UNIV SANTA CLARA, 77- *Concurrent Pos:* Anal chemist, Richmond-Chase Co, 40- *Mem:* Am Chem Soc. *Res:* Food analyses; synthetic organic chemistry; synthesis of heterocyclic ring compounds; synthesis of stable free radicals. *Mailing Add:* Dept Chem Santa Clara Univ Santa Clara CA 95053

DECK, ROBERT THOMAS, b Philadelphia, Pa, Aug 6, 35. QUANTUM OPTICS & NONLINEAR OPTICS. *Educ:* La Salle Col, BA, 56; Univ Notre Dame, PhD(physics), 61. *Prof Exp:* Res assoc theoret physics, Bartol Res Found, 61-63; res assoc & instr, Univ Mich, 63-65; from asst prof to assoc prof, 65-74, PROF PHYSICS, UNIV TOLEDO, 74-, PROF ENG PHYSICS, 80- *Mem:* Am Phys Soc; Am Asn Physics Teachers. *Res:* High and low energy theoretical nuclear physics; quantum electrodynamics; elementary particle theory; quantum and nonlinear optics. *Mailing Add:* Dept Physics Univ Toledo Toledo OH 43606

DECK, RONALD JOSEPH, b New Orleans, La, May 16, 34. SOLID STATE PHYSICS. *Educ:* Loyola Univ, La, 56; La State Univ, MS, 58, PhD(physics), 61. *Prof Exp:* Asst prof physics, La State Univ, 61-62; assoc physicist, Oak Ridge Nat Lab, 62-63; asst prof, 63-77, ASSOC PROF PHYSICS, TULANE UNIV LA, 77- *Mem:* Am Inst Physics. *Res:* Galvanomagnetic and thermomagnetic effects in metals at low temperatures; behavior of superconducting alloys. *Mailing Add:* Dept Physics & Astron Tulane Univ La New Orleans LA 70118

DECKARD, EDWARD LEE, b Cynthiana, Ind, Aug 7, 43; m 67; c 3. CROP PHYSIOLOGY, MOLECULAR GENETICS. *Educ:* Purdue Univ, BS, 65; Univ Ill, PhD(agron), 70. *Prof Exp:* From asst prof to assoc prof, 70-82, PROF AGRON, NDAK STATE UNIV, 82- *Mem:* Am Soc Agron; Crop Sci Soc Am; Am Soc Plant Physiologists. *Res:* Development of biochemical and physiological selection criteria that could be used by plant breeders in crop improvement; present emphasis on nitrogen metabolism and tissue culture. *Mailing Add:* Dept Agron NDak State Univ State Univ Sta Fargo ND 58105

DECKER, ALVIN MORRIS, JR, b Manocs, Colo, Oct 12, 18; m 43; c 2. AGRONOMY. *Educ:* Colo Agr & Mech Col, BS, 49; Utah State Agr Col, MS, 51; Univ Md, PhD(agron), 53. *Prof Exp:* From instr to assoc prof, 52-67, PROF AGRON, UNIV MD, COLLEGE PARK, 67- *Honors & Awards:* Res Award, Northeast Br, Am Soc Agron, 73; Sigma Xi Res Achievement Award, 77. *Mem:* Fel Am Soc Agron; Sigma Xi. *Res:* Forage crop management and breeding. *Mailing Add:* Dept Agron Univ Md College Park MD 20742

DECKER, ARTHUR JOHN, b Butte, Mont, Oct 16, 41; m 70; c 1. OPTICAL PHYSICS. *Educ:* Univ Wash, BS, 63; Univ Rochester, AM, 66; Case Western Reserve Univ, PhD(elec eng & appl physics), 77. *Prof Exp:* OPTICAL PHYSICIST, NASA LEWIS RES CTR, 66- *Concurrent Pos:* adj prof, OAI. *Mem:* Optical Soc Am. *Res:* Use of coherent optics for measurements related to aerospace propulsion; holography; speckle interferometry and optical information processing; neural networks; visualization. *Mailing Add:* 25041 Mitchell Dr N Olmsted OH 44070

DECKER, C(HARLES) DAVID, b Oxnard, Calif, Feb 12, 45; m 68; c 3. LASERS, QUANTUM ELECTRONICS. *Educ:* Wabash Col, AB, 67; Harvard Univ, AM, 69; Rice Univ, PhD(elec eng), 74. *Prof Exp:* Physicist, Arthur D Little, Inc, 68-69; advan res engr, GTE Sylvania Electrooptics Orgn, 73-75, mgr, Res & Develop Dept, 75-79, res mgr, GTE Labs, Inc, 79-82; dir, RCA Advan Technol Labs, 82-83; dir, Fundamental Res Lab, 83-86, VPRES & DIR RES, GTE LABS, INC, WALTHAM, MASS, 86- *Mem:* Am Phys Soc; Optical Soc Am; Inst Elec & Electronic Engrs; Sigma Xi. *Res:* Dye lasers; nonlinear optics; photochemical interactions. *Mailing Add:* GTE Labs Inc 40 Sylvan Rd Waltham MA 02254

DECKER, CLARENCE FERDINAND, b Taintor, Iowa, Nov 9, 25; m 48; c 2. BIOETHICS. *Educ:* Western Mich Univ, BS, 50; Mich State Univ, MS, 52; PhD(physiol & pharmacol), 54; MDiv, Seabury Western Theol Sem, 65. *Prof Exp:* Teaching asst animal physiol & pharmacol, Mich State Univ, 50-52, US Pub Health Serv res fel biochem, 54-56; asst sci, div biol & med res, Argonne Nat Lab, 56-59, assoc sci, 59-61; chief biochemist, endocrinol sect, Presby-St Luke's Hosp, Chicago, 61-62; chief, biochem sect, Radioisotope Serv, Hines Vet Admin Hosp, Ill, 62-67; from assoc prof & chmn to prof & chmn, biol dept, Point Park Col, Pittsburgh, Pa, 67-75; CONSULT, HUMAN SUBJECTS REV POLICY COORD COMT & BIOHAZARDS AND GENETIC ENG LAB INSPECTION COMT, OHIO STATE UNIV, 76- *Concurrent Pos:* Consult biochemist, dept pathol, Presby-St Luke's Hosp, Chicago, 62-67; assoc ed, dept biochem, Chem Abstr Serv, 75-82. *Mem:* Sigma Xi; Am Chem Soc; NY Acad Sci. *Res:* Cardiovascular physiology; toxicity and metabolism of heavy metals and trace elements; skeletal retention of radioactive isotopes of alkaline earth series; bioethics and human subject research; public policy on scientific matters; ceruloplasmin and copper metabolism. *Mailing Add:* 545 Woodsfield Dr Columbus OH 43214

DECKER, CLIFFORD EARL, JR, b Valdese, NC, July 20, 41; m 64; c 2. INDUSTRIAL HYGIENE, ANALYTICAL CHEMISTRY. *Educ:* Univ NC, BS, 63, MSPH, 65. *Prof Exp:* Scientist air pollution, Res Triangle Inst, 65-66; health serv officer, Nat Ctr Air Pollution Control, 66-68; sr scientist, 68-75, dept mgr Air Pollution, 75-83, DIR, CTR ENVIRON MEASUREMENTS & QUAL ASSURANCE, RES TRIANGLE INST, 83- *Mem:* Air Pollution Control Asn; Am Chem Soc. *Res:* Areas of atmospheric chemistry, air pollution studies in urban and nonurban areas, field measurements, quality assurance, testing and evaluation of instrumentation, hazardous waste, industrial hygiene and measurement method methodology. *Mailing Add:* 1219 Huntsman Drt Durham NC 27713

DECKER, DANIEL LORENZO, b Provo, Utah, Sept 22, 29; m 54; c 7. SOLID STATE PHYSICS, HIGH PRESSURE PHYSICS. *Educ:* Brigham Young Univ, BS, 53, MS, 55; Univ Ill, PhD(physics), 58. *Prof Exp:* Asst math, Brigham Young Univ, 52-53, asst physics, 53-55 & asst appl math, 55; asst physics, Univ Ill, 55-56; from asst prof to assoc prof, 58-67, PROF PHYSICS, BRIGHAM YOUNG UNIV, 67-, CHAIR, DEPT PHYSICS, 88- *Concurrent Pos:* Vis staff mem, Los Alamos Sci Lab, 64-70; vis scientist, Argonne Nat Lab, 71-72; res scientist, Centre d'Etudes Nucleaires de Saclay, France, 77-79; vis scientist, Argonne Nat Lab, 86. *Mem:* Fel Am Phys Soc; Am Asn Physics Teachers. *Res:* High pressure physics; Mossbauer measurements; diffusion; neutron diffraction; superconductivity. *Mailing Add:* 1321 E 820 N Provo UT 84606

DECKER, DAVID GARRISON, obstetrics & gynecology; deceased, see previous edition for last biography

DECKER, FRED WILLIAM, b Portland, Ore, July 5, 17; m 42; c 3. PHYSICS, FORENSIC METEOROLOGY. *Educ:* Ore State Col, BS, 40, PhD(physics), 52; NY Univ, MS, 43. *Prof Exp:* Jr meteorologist, US Weather Bur, Calif, 40-41; instr meteorol, NY Univ, 41-44; instr physics, Multnomah Col, 46; from instr to asst prof, 46-81, EMER PROF, PHYSICS & ATMOSPHERIC SCI, ORE STATE UNIV, 81-; CONSULT, FORENSIC METEOROLOGIST, 85- *Concurrent Pos:* consult to litigants, US Bur Reclamation, 52-, consult, 64-65; consult, Adv Comt Weather Control, 54-57; TV meteorologist, KOIN, Portland, Ore, 68 & KOAC-KOAP, Portland & Corvallis, Ore, 68-70; consult, Fed Water Pollution Control Agency, 68-69 & AEC, 69-70; ed, Universitas, Univ Prof for Acad Order, 75-; atty-gen, Ore, 79-81; prin investr, US Army Res, NSF & NASA, 52-81; dep asst secy educ, off Educ Res & Improvement, Wash, DC, 81-85. *Mem:* Fel AAAS; Nat Weather Asn; Univ Prof for Acad Order (pres, 74); fel Royal Meteorol Soc; Am Inst Physics; Am Meteorol Soc; Am Geophys Union; Am Phys Soc; Am Assoc Physics Teacher; Assoc Am Weather Observers; Mont Pelerin Soc. *Res:* Atmospheric ozone measurements; meteorological optics; short-period weather forecasting; weather modification evaluation; mesometeorology; weather radar; writer on science museums; meteorological instruction. *Mailing Add:* Forensic Expert Consult 827 NW 31st St Corvallis OR 97330-5163

DECKER, JAMES FEDERICK, b Albany, NY, Aug 5, 40; m 64; c 2. RESEARCH ADMINISTRATION, PLASMA PHYSICS. *Educ:* Union Col, BS, 62; Yale Univ, MS, 63, PhD(physics), 67. *Prof Exp:* Mem tech staff, Bell Telephone Lab, 67-73; chief exp res, 73-78, dir appl plasma physics, 78-82, spec asst dir, 82-84, dir sci comput staff, 84-87, DEPT DIR, OFF ENERGY RES, DEPT ENERGY, 87- *Concurrent Pos:* Consult, Int Atomic Energy Agency 75-78; chmn, Comte Res & Develop, Int Energy Agency, 85-87, Fed coord coun sci, Eng & Technol Comt Supercomputers, 83-88; mem, Indust res roundtable, Nat Acad Sci, 85-88; Inst Elect & Electronics Engrs Comt, 83-88, Econ Policy Comte Res & Develop; head deleg, Int Thermonuclear Exp Reactor, Int Atomic Energy Agency, 88. *Mem:* Am Phys Soc; Inst Elec & Electronics Engrs; AAAS. *Res:* Energy basic research program: High energy & nuclear physics; basic energy sci; health & environmental research; magnetic fusion energy. *Mailing Add:* 9612 Trailridge Terrace Potomac MD 20854

DECKER, JANE M, plant cytotaxonomy; deceased, see previous edition for last biography

DECKER, JOHN D, b Middletown, NY, July 10, 22; m 45; c 1. ANATOMY, EMBRYOLOGY. *Educ:* Univ Fla, BS, 50, MS, 52; State Univ NY Upstate Med Ctr, PhD(anat), 65. *Prof Exp:* From instr to assoc prof biol, Hartwick Col, 52-65; res assoc, Washington Univ, 65-67; asst prof, 67-69, ASSOC PROF ANAT, MED CTR, UNIV MO-COLUMBIA, 69- *Res:* Neuroembryology and behavior. *Mailing Add:* Dept Anat Univ Mo Med Ctr M228 Med Sci Bldg Columbia MO 65201

DECKER, JOHN LAWS, b Brooklyn, NY, June 27, 21; m 54; c 4. RHEUMATOLOGY, INTERNAL MEDICINE. *Educ:* Univ Richmond, BA, 42; Columbia Univ, MD, 51. *Prof Exp:* Instr med, Columbia Univ, 54-55; clin & res fel, Mass Gen Hosp, 55-58; from instr to assoc prof, Univ Wash, 58-65; clin dir, Metab & Digestive Dis, 76-80, chief, Arthritis & Rheumatism Br, Nat Inst Arthritis, Diabetes, Digestive & Kidney Dis, 65-83, dir, clin ctr & assoc dir clin care, 83-90, EMER SCIENTIST, NIH, 90- *Concurrent Pos:* Gov, Am Col Physicians, 82-86. *Honors & Awards:* Philip Hench Award, Asn Military Surg US; Gold Medal, Am Col Rheumatology; Dirs Award, NIH. *Mem:* Am Col Rheumatology (pres, 72-73); Am Col Physicians. *Res:* Clinical studies of rheumatoid arthritis and systemic lupus erythematosus. *Mailing Add:* 4703 Broad Brook Dr Bethesda MD 20814

DECKER, JOHN P, b Chicago, Ill, Aug 16, 25; m 51; c 2. MOLECULAR PHYSICS, SPECTROSCOPY. *Educ:* Univ Ark, BSEE, 49, MS, 53; Tex A&M Univ, PhD(physics), 64. *Prof Exp:* Engr, Ark Power & Light, 49-51; off engr, F E Woodruff, 52-53; instr physics, Ark Agr & Mech Col, 53-56; from instr to asst prof, Tex A&M Univ, 56-63; from assoc prof to prof, Sam Houston State Col, 63-65; head dept, Stephen F Austin State Univ, 65-79, prof physics, 65-86; RETIRED. *Mem:* Am Asn Physics Teachers; Inst Elec & Electronics Engrs. *Res:* Investigations of the ultraviolet absorption spectra of sulfer dioxide with isotopic substitution and of selenium dioxide; development of gas lasers. *Mailing Add:* 2027 Sheffield Rd Nacogdoches TX 75961

DECKER, JOHN PETER, applied synecology; deceased, see previous edition for last biography

DECKER, L(OUIS) H, b Monticello, NY, Nov 23, 13; m 38; c 2. METALLURGY. *Educ:* Rensselaer Polytech Inst, ChE, 35. *Prof Exp:* Mem methods dept, Rome Div, Revere Copper & Brass, Inc, 35-39, metall chemist, Res Dept, 39-43, chief chemist, Refining Div, 43-45, supvr chem & metall sect, Res Dept, 45-64, asst mgr metall dept, Res & Develop Ctr, 64-67, res

& develop mgr, New Bedford Div, 67-77, consult, Res & Develop Ctr, 78-80; RETIRED. *Concurrent Pos:* Consult, Revere Copper Prods, 85. *Mem:* Am Soc Metals. *Res:* Metallurgy of copper alloys. *Mailing Add:* 7980 Inwood Dr Rome NY 13440

DECKER, LUCILE ELLEN, b Grand Rapids, Mich, Jan 4, 27; m 48; c 2. ENZYME METABOLISM, ABSTRACTING-BIOCHEMISTRY. *Educ:* Western Mich Univ, Kalamazoo, BS, 48; Mich State Univ, East Lansing, MS, 52, PhD(chem), 56. *Prof Exp:* Chemist, Am Cyanamid Co, Kalamazoo, 46-50; Pub Health Serv teaching fel, dept chem, Mich State Univ, 50-52, res instr, dept food & nutrit, 52-54, Pub Health Serv res fel, 54-56; res assoc, dept biochem, Med Sch, Northwestern Univ, 58-61; supv biochemist, Pulmonary Dis Res Lab, Hines Vet Admin Hosp, Ill, 61-67; asst prof biochem & biophys, Stritch Sch Med, Loyola Univ, Chicago, 63-67; assoc prof, dept biol sci, Point Park Col, Pittsburgh, 67-71, prof, dept nat sci, 71-75; assoc ed, 75-78, SR ASSOC ED, CHEM ABSTR SERV, COLUMBUS, OHIO, 78- *Concurrent Pos:* Adj prof, Foods & Nutrit Col, Pa State Univ, Col Park, 67-75. *Mem:* Sigma Xi; Am Chem Soc; Am Soc Info Sci; Royal Soc Chem. *Res:* Biochemical methods, immunology, enzymology, pharmacology; multiple forms of transaminases and their coenzymes, trace metal toxicity and metabolism; pyridine ring metabolism, nutrition, metabolic modeling. *Mailing Add:* Chem Abstr Serv PO Box 3012 2540 Olentangy River Rd Columbus OH 43210-0012

DECKER, QUINTIN WILLIAM, b Rochester, NY, Aug 22, 30; m 59; c 3. SYNTHETIC ORGANIC CHEMISTRY. *Educ:* Univ Buffalo, BA, 53, MA, 56, PhD(org chem), 58, Univ Charleston, BS, 78. *Prof Exp:* Proj scientist, Eastman Kodak Co, 54-55; proj scientist, Chem & Plastics Div, 57-81, patent mgr, 81-83, GROUP MGR TECHNOL INFO SERV, UNION CARBIDE CORP, SOUTH CHARLESTON, 83- *Mem:* Am Chem Soc. *Res:* Aliphatic silanes compounds; aliphatic olefin, hydroxyl, carbonyl and amine reactions; hydrogenation reactions and catalysts; dialdehyde tissue fixation; textile chemicals; surfactants. *Mailing Add:* 1006 Sand Hill Dr St Albans WV 25177

DECKER, R(AYMOND) F(RANK), b Afton, NY, July 20, 30; m 51; c 4. METALLURGICAL ENGINEERING. *Educ:* Univ Mich, BS, 52, MS, 55, PhD(metall eng), 58. *Prof Exp:* Asst high temperature metall, Eng Res Inst, Univ Mich, 54-58; res metallurgist, Alloy Studies & Develop, Inc, Ltd, 58-59, sr metallurgist, 59-60, sect head nickel & stainless steels, 60-62, group leader nonferrous metals, 62-67, asst to mgr, Paul D Merica Res Lab, 67-69, asst mgr, 69-76, vpres US res & develop, 76-78, vpres corp tech & diversification ventures, 78-82; vpres res & corp rels, Mich Technol Univ, 82-86; PRES & CEO UNIV SCI PARTNERS, 86-; CHMN, THIXOMAT, INC, 88- *Concurrent Pos:* Adj prof, Polytech Inst Brooklyn, 62-66 & NY Univ, 68; mem res & technol adv panel mat aircraft, NASA, 70-71, NSF, 78-; rep sci bd, Biogen, SAm, 80-82; mem Nat Mat Adv Bd, 80-88, chmn Mich Energy Resources & Res Assoc, 85-86, mem exec comm, Strategic Hwy Res Prog, Nat Res Coun, 88- *Honors & Awards:* IR-100 Award, 64; Campbell lectr, Am Soc Metals, 85. *Mem:* Nat Acad Eng; fel Am Soc Metals (vpres, 85, pres, 87); Am Inst Mining, Metall & Petrol Engrs; Sigma Xi; hon mem Am Soc Metals Int. *Res:* High temperature alloys and transformations; alloy, maraging and stainless steels; cast irons; nickel, copper, aluminum and magnetic alloys; extractive, powder and process metallurgy; welding; electrochemistry, corrosion, ceramics, paints, plastics. *Mailing Add:* 721 E Huron St Ann Arbor MI 48104-1526

DECKER, RICHARD H, b Grand Rapids, Mich, Aug 12, 34; m 60; c 3. BIOCHEMISTRY. *Educ:* Hope Col, AB, 56; Univ Ill, Urbana, MS, 58; Okla State Univ, PhD(biochem), 60. *Prof Exp:* Res assoc & lectr biochem, Mayo Grad Sch Med, Univ Minn, 62-71; head, Sect Infectious Dis, Abbott Labs, 71-74, head, Hepatitis Res Labs, 81-82, dir, exp biol res, 82-90, RES FEL IMMUNOCHEM, ABBOTT LABS, 74-, SR RES FEL, 85- *Concurrent Pos:* NIH fel tryptophan metab, Sch Med, Univ Wis, 60-62, res career develop award, 66-71; sr res fel, Volwiler Soc; mem, Worldwide Hepatitis Support, 90-92. *Mem:* AAAS; Am Chem Soc; Soc Invest Dermat; Am Soc Exp Path; Fedn Am Socs Exp Biol. *Res:* Mechanism of cellular adhesion, acantholysis; phosphoproteins; control of epidermal metabolism; determination of steroids, viral antigens via radioimmunoassay, physical and immunopurification of antibodies, antigens; hepatitis A tissue culture; acquired immune deficiency syndrome research-human immunodeficiency virus virology and immunology; hepatitis B and C research. *Mailing Add:* Abbott Labs North Chicago IL 60064

DECKER, ROBERT DEAN, b Uniondale, Ind, July 7, 33; m 57; c 2. BOTANY, ENVIRONMENTAL BIOLOGY. *Educ:* Purdue Univ, BS, 59, MS, 61; NC State Univ, PhD(bot), 66. *Prof Exp:* Lectr bot, Butler Univ, 61-62; asst prof, 66-69, ASSOC PROF BIOL & BOT, UNIV RICHMOND, 69- *Mem:* Bot Soc Am; Nat Asn Biol Teachers; Am Jr Acad Sci. *Res:* Plant morphogenesis. *Mailing Add:* Dept Biol Univ Richmond Richmond VA 23173

DECKER, ROBERT SCOTT, b Orange, Calif, Oct 9, 42; m 65; c 2. DEVELOPMENTAL BIOLOGY, CELL BIOLOGY. *Educ:* Calif State Col, Long Beach, BA, 65; Univ Iowa, MS, 67, PhD(zool), 70. *Prof Exp:* Vis asst prof biol, Claremont Cols, 70-71; USPHS fel, Univ Calif, San Francisco, 71-73; asst prof cell biol, Southwestern Med Sch, Univ Tex Health Sci Ctr, Dallas, 73-78, assoc prof cell biol, 78-; PROF MED, CELL BIOL & ANAT, DEPT MED, NORTHWESTERN UNIV. *Concurrent Pos:* Mem, Basic Sci Coun, Am Heart Asn; estab investr, Am Heart Asn; spec res fel, Cambridge Univ. *Mem:* AAAS; Am Soc Zool; Soc Develop Biol; Am Soc Cell Biol. *Res:* Role of cell junctions in development; mechanisms of junctional assembly during differentiation; lysosomal alterations during myocardial ischemia. *Mailing Add:* Dept Med Northwestern Univ 303 E Chicago Ave Chicago IL 60611

DECKER, ROBERT WAYNE, b Williamsport, Pa, Mar 11, 27; m 50; c 4. ACTIVE VOLCANISM. *Educ:* Mass Inst Technol, BS, 49, MS, 51; Colo Sch Mines, DSc(geol), 53. *Prof Exp:* Asst geologist, Bethlehem Steel Co, 49-50; geologist, New World Explor Co, 52-54; asst prof geol, Univ Ill, 54; from asst prof to assoc prof, Dartmouth Col, 54-67, chmn dept geol, 63-65, prof geophys, 67-79; scientist-in-charge, Hawaiian Volcano Observ, US Geol Survey, 79-84; CONSULT, 84- *Concurrent Pos:* Geophysicist, US Geol Surv, 57-, vis scientist, Nat Ctr Earthquake Res, 69-70; assoc prof, Inst Technol Bandung, Indonesia, 59-60; res affil, Hawaii Inst Geophys, Univ Hawaii, 64-; Am Philos Soc-NSF grants, Iceland, 66-77; mem earth sci grants rev panel, NSF, 71-73; pres, Int Asn Volcanology & Chem of the Earth's Interior, 75-79; chmn, Geosci Adv Panel, Los Alamos Nat Lab, 77-79; chmn, Panel Thermal Regimes Continental Sci Drilling Comt, Nat Acad Sci, 80-81; mem bd dir, Hawaii Natural Hist Asn, Hawaii Nat Park, 80-84; actg scientist-in-charge, Mount St Helens' Proj, US Geol Surv, 80; distinguished vis prof geophysics, Univ Hawaii, Hilo, 89-90. *Mem:* Fel Geol Soc Am; fel Am Geophys Union. *Res:* Physics of volcanoes; structural geology; applied geophysics. *Mailing Add:* 4087 Silver Bar Rd Mariposa CA 95330

DECKER, ROLAN VAN, b Bartlesville, Okla, Nov 4, 36; m 61; c 3. PHYSICAL BIOCHEMISTRY. *Educ:* Okla State Univ, BS, 58; Purdue Univ, PhD(biochem), 65. *Prof Exp:* From asst prof to assoc prof, 65-74, PROF CHEM, SOUTHWESTERN OKLA STATE UNIV, 74- *Mem:* Am Chem Soc. *Res:* Protein-ion and protein-small molecule interactions and their effects on protein conformation; topology of subcellular particles. *Mailing Add:* Dept Chem Southwestern State Univ Weatherford OK 73096

DECKER, WALTER JOHNS, b Tannersville, NY, June 13, 33; m 61; c 3. TOXICOLOGY. *Educ:* State Univ NY Albany, BA, 54, MA, 55; George Washington Univ, PhD(biochem), 66. *Prof Exp:* US Army, 55-75, res asst biochem, Walter Reed Army Inst Res, 55-56, res biochemist, 57-60 & 62-65, chief indust hyg sect, 406th Med Gen Lab, Japan, 56-57, chief lab serv, Dept Med Res & Develop, William Beaumont Gen Hosp, 65-71, chief chem, Fifth US Army Area Med Lab, Ft Sam Houston, Tex, 71-75; assoc prof, Dept Pharmacol & Toxicol, med br, Univ Tex, 76- 83 & Dept Pediat, 76-86; CONSULT TOXICOL, TOXICOL CONSULT SERV, 83- *Concurrent Pos:* Consult to surgeon, White Sands Missile Range, 66-71; lectr, Univ Tex, El Paso, 67-71; adj asst prof, Univ Tex Health Sci Ctr, 72-74, adj assoc prof, 74-75. *Mem:* Am Chem Soc; fel Am Acad Clin Toxicol (secy-treas, 78-84); Sigma Xi; Soc Toxicol. *Res:* Analytical, forensic, and clinical toxicology; industrial hygiene. *Mailing Add:* Toxicol Consult Serv 9220 McCombs Suite A El Paso TX 79924

DECKER, WAYNE LEROY, b Madison County, Iowa, Jan 24, 22; m 43; c 1. METEOROLOGY. *Educ:* Cent Col, Iowa, BS, 43; Iowa State Univ, MS, 47, PhD(agr climatol), 55. *Prof Exp:* Meteorologist climatol, US Weather Bur, 47-49; agr climatologist, 49-67, PROF ATMOSPHERIC SCI, UNIV MO-COLUMBIA & CHMN DEPT, 67- *Concurrent Pos:* Adv agr meteorol coop res, Agr Res, Sci & Educ Admin, USDA, 77-79; chmn, comt on climate & weather fluctuations & agr product, Nat Acad Sci, 75-76; mem exec comt, Coun Agr Sci & Technol, 80-87; dir, Coop Inst Appl Meteorol, 86- *Mem:* Fel Am Meteorol Soc; Am Geophys Union; Agron Soc; Int Soc Biomet (treas, 91); AAAS. *Res:* Impact of climate and meteorological variations on agricultural production; development of methods for assessing the impacts of climate change and variabilities on agricultural productions; risk analysis of climatic variabilities to economic activities. *Mailing Add:* Dept Atmospheric Sci 701 Hitt Rm 106 Univ Mo Columbia MO 65211

DECKER, WINSTON M, b Deckerville, Mich, Jan 10, 23; m 45; c 2. VETERINARY MEDICINE. *Educ:* Mich State Univ, DVM, 46. *Prof Exp:* Pvt pract, 46-47; chief vet, Kalamazoo City County Health Dept, 47-49; pub health vet, Mich Dept Health, 50-55, asst to state health comnr, 55-60; chief spec projs sect, Milk & Food Br, Div Environ Eng & Food Protection, Pub Health Serv, 60-65, prog planning officer, 65-66, asst prog officer, Bur State Serv, 66-67, dir, Off Res & Develop, Bur Dis Prev & Environ Control, 67-69; dir sci activ & asst exec vpres, Am Vet Med Asn, 69-76; Wash rep, Am Vet Med Asn & Asn Am Vet Med Cols, 77-87; RETIRED. *Mem:* Am Vet Med Asn; Am Pub Health Asn. *Res:* Relationships of environmental factors to cause, control or prevention of chronic and communicable disease and veterinary medicine's capability and productivity in meeting society's requirements upon it. *Mailing Add:* 5536 37th St E Bradenton FL 34203

DECKERS, JACQUES (MARIE), b Antwerp, Belgium, Aug 25, 27; Can citizen; m 58; c 7. PHYSICAL CHEMISTRY. *Educ:* Univ Louvain, Candidat, Sci Chim, 49, Lic Sci Chim, 41, DSc(phys chem), 56. *Prof Exp:* Asst, Univ Louvain, 56-58; res assoc chem eng, Princeton Univ, 58-61; assoc prof, 61-66, prof chem, Univ Toronto, 66-; AT DEPT CHEM, ERINDALE COL, MISSISAUGA, ONT. *Honors & Awards:* Belgium Govt Travel Award, 57. *Mem:* Am Chem Soc; Am Phys Soc; Chem Inst Can; Sigma Xi. *Res:* Combustion; ions in in flames; high energy particles in chemical reactions; molecular beams; electric discharges. *Mailing Add:* RR 2 Grand Valley ON L0N 1G0 Can

DECKERT, CHERYL A, b Philadelphia, Pa, Feb 22, 48; div. MATERIAL SCIENCE ENGINEERING, INORGANIC CHEMISTRY. *Educ:* Drexel Univ, BS, 69; Univ Ill, MS, 71, PhD(chem), 73. *Prof Exp:* Res assoc molecular biol, Univ Ill, 73-74; staff mem chem, David Sarnoff Res Ctr, RCA Corp, 74-80; mgr printed circuit res & develop, Shipley Co, 80-85; Europ tech mgr, Shipley Europe Ltd, Coventry, Eng, 85-87; Tech mkt mgr, Shipley Co, Inc, 87-88, res & develop mgr, 88-91; CONSULT, 91- *Concurrent Pos:* NSF fel, 69, Univ Ill fel, 70-71. *Mem:* Am Chem Soc; Sigma Xi; Electrochem Soc; Am Electroplaters Soc; Inst Interconnecting & Pkg Electronical Circuits. *Res:* Photoresist adhesion to semiconductor devices electroless and electro-plating; chemical etching; process and product development for printed circuit board fabrication; quality improvement in PCB fabrication; electrodeposited photoresists for electronic device imaging; molded interconnect fabrication. *Mailing Add:* Shipley Co Inc 2300 Washington St Newton MA 02162

DECKERT, CURTIS KENNETH, b Whittier, Calif, Jan 3, 39; m 64; c 2. MECHANICAL ENGINEERING, TECHNICAL MANAGEMENT. *Educ:* Univ Ariz, BSME, 60; Univ Southern Calif, MSME, 62, MBA, 68; Calif Coast Univ, PhD, 89. *Prof Exp:* Assoc engr optical-mech eng, Systs Support Div, Nortronics Co, 60-66; sr engr, Gilfillan Div, Int Tel & Tel Corp, 66 & Calif Comput Prods, 66-70; develop engr, Aeronutronic Div, Ford Motor Co, 72-75; sr mech engr med eng, Diagnostics Div, Abbott Labs, 75-76; PRES, TECH & MGT CONSULT, CURT DECKERT ASSOCS, INC, 76- *Mem:* Soc Photo-optical Instrumentation Engrs; Am Sci Asn; Am Mgt Asn; Asn Mgt Consult; Inst Mgt Consult. *Res:* Research and development management strategy for high technology business; medical systems; optical systems for instrumentation and entertainment; mechanical systems; photographic and other optical and automation systems. *Mailing Add:* 18061 Darmel Pl Santa Ana CA 92705

DECKERT, FRED W, b Berlin, Ger, Jan 16, 43; US citizen; m 65; c 2. SAFETY & ENVIRONMENTAL SERVICES, MANAGEMENT. *Educ:* Gonzaga Univ, BS, 66; Univ Wis-Madison, MS, 68, PhD(biochem), 70; Univ Tüingen, PhD(toxicol), 72. *Prof Exp:* Jr chemist med chem, Cent Res Labs, 3M Co, St Paul, 66; res asst biochem, Univ Wis-Madison, 66-70; res asst biochem pharmacol, Inst Toxicol, Univ Tüingen, Ger, 70-72; res assoc develop pharmacol, Med Ctr, Univ Ill, Chicago, 72-73; group leader biochem pharmacol & drug metab, Res Labs, Rohm & Haas Co, 73-77, sr group leader biochem toxicol, 77-79, head subchronic, reproductive & biochem toxicol, 79-81, mgr, anal & biochem toxicol res sect, 81-82, sr group leader, residue, metab & environ fate, 83-85, mgr, agr-chem qual assurance, 86-87; MGR, RES & DEVELOP, ENVIRON & TOXICOL SERV, AIR PRODS & CHEMS, INC, 88- *Concurrent Pos:* Asst to assoc prof pharmacol & toxicol, Dept Pharmacol, Jefferson Med Col, Philadelphia, 77-; assoc prof, Toxicol Ctr, Rutgers Univ, 81- *Mem:* Soc Toxicol; Int Soc Study Xenobiotics; Am Chem Soc; Am Indust Health Coun. *Res:* Biochemical-toxicology and metabolic fate of rodenticides; biotransformation and biochemical-toxicology of agricultural and industrial chemicals; biochemistry of 4-hydroxycoumarin anticoagulants; warfarin dicumarol; nutrition and detoxication mechanisms; drug interactions; xenobiotic metabolism; enzyme induction; environmental fate/property studies on agricultural chemicals. *Mailing Add:* Air Prod & Chem Inc 7201 Hamilton Blvd Allentown PA 18195

DECKERT, GORDON HARMON, b Freeman, SDak, May 18, 30; m 51; c 2. PSYCHIATRY, ACADEMIC ADMINISTRATION. *Educ:* Northwestern Univ, BS, 52, MD, 55. *Hon Degrees:* DSc, Albany Med Col, 80, Georgetown Univ, 83. *Prof Exp:* Fel med, Mayo Clin & Found, 56-57; clin investr psychiat, Vet Admin Hosp, Oklahoma City, 63-66; career teacher, NIMH, 66-68; prof psychiat & behav sci & chmn dept, Health Sci Ctr, 69-86, DAVID ROSS BOYD PROF, UNIV OKLA, 80-, PROG DIR, OMME-21, DEANS OFF, 90- *Concurrent Pos:* Chief of staff, Univ Hosp, Univ Okla, 72-74; chmn psychiat test comt & mem exec comt, Nat Bd Med Examr, 74-78. *Honors & Awards:* Aescalapian Award, 70. *Mem:* Am Psychiat Asn; Asn Am Med Cols; AMA; AAAS; Sigma Xi. *Res:* Psychophysiology of imagery; nonverbal behavior; effectiveness of medical education. *Mailing Add:* Dept Psychiat & Behav Sci PO Box 26901 Oklahoma City OK 73190

DECKKER, B(ASIL) E(ARDLEY) L(EON), thermodynamics, fluid mechanics; deceased, see previous edition for last biography

DECLARIS, N(ICHOLAS), b Drama, Greece, Jan 1, 31; m 56. ELECTRICAL ENGINEERING, MEDICAL SCIENCE & INFORMATICS. *Educ:* Agr & Mech Col, Tex, BS; Mass Inst Technol, ScD. *Prof Exp:* Res engr, Calif Res Corp, 52; asst, Electronic Res Lab, Mass Inst Technol, 52-56; from asst prof to prof elec eng & appl math, Cornell Univ, 56-67; head dept, 67-76, PROF ELEC ENG & RES PROF, INST FLUID DYNAMICS & APPL MATH, UNIV MD, COLLEGE PARK, 67-, PROF PATH, MED SCH, BALTIMORE, 83- *Concurrent Pos:* Consult, Melpar, Inc, 54, Spencer-Kennedy Labs, 55-56, Gen Elec Co, 56-59 & Int Bus Mach Corp, 58-70; assoc ed, Transactions, Inst Elec & Electronics Engrs, 58-68; res collabr, NIH, 82-; assoc dir, Md Inst Emergency Med Serv, 85-87; adv NSF, 87-88; ed, Electrical Sci Acad Press Inc, 65-; prof epidemiol & prevent med, Med Sch, Baltimore, 87- *Mem:* Am Soc Eng Educ; fel Inst Elec & Electronics Engrs; Sigma Xi; Am Math Asn. *Res:* System theory and engineering; medical science and informatics; computer-based systems; biomedical engineering and neuroscience; applied mathematics. *Mailing Add:* Beaufort Dr Fulton MD 20759

DECLUE, JIM W, CARDIOVASCULAR, ENDOCRINE. *Educ:* Univ Mo, PhD(physiol), 73. *Prof Exp:* ASSOC PROF PHYSIOL & SECT HEAD, PHYSIOL IN BIOMED SCI, SOUTHERN ILL UNIV, EDWARDSVILLE, 77- *Mailing Add:* Southern Ill Univ Sch Dent Med 2800 College Ave Alton IL 62002

DECOOK, KENNETH JAMES, b Hebron, Ind, June 7, 25; c 4. GEOLOGY, HYDROLOGY. *Educ:* Univ Ariz, BS, 51, PhD(water resouces admin), 70; Univ Tex, MA, 57. *Prof Exp:* Hydrologic asst, Surface Water Br, US Geol Surv, Ariz Dist, 50-51, geologist, Ground Water Br, Ariz & Tex Dists, 51-58; res assoc, Inst Water Utilization, Univ Ariz, 58-59; consult ground water hydrologist, Water Develop Corp, 58-61; asst dist engr, San Carlos Irrig & Drainage Dist, Ariz, 61-63; geologist & hydrologist, W S Gookin & Assocs, 63-65; res assoc, 65-70, ASSOC HYDROLOGIST, WATER RESOURCES RES CTR, UNIV ARIZ, 70-, ASSOC PROF GEOSCI, 74-, EMER HYDROLOGIST. *Mem:* Asn Prof Geol Scientists; Am Water Resources Asn; Soc Econ Paleontologists & Mineralogists. *Res:* Arid zone hydrology; cretaceous stratigraphy; ground water geology and hydrology; water resources management; economic water allocation and transfer; legal institutions related to water resources. *Mailing Add:* Box 1144 Tucson AZ 85702

DECORA, ANDREW WAYNE, b Rock Springs, Wyo, July 25, 28; m 53; c 2. PHYSICAL ORGANIC CHEMISTRY, SPECTROSCOPY. *Educ:* Univ Wyo, BS, 50, MS, 57, PhD(kinetics), 62. *Prof Exp:* Chemist, Laramie Petrol Res Ctr, US Bur Mines, 50-51, supvry chemist, 53-59; asst kinetics, Univ Wyo, 59-61; proj leader, Off Res & Develop, US Dept Interior, Washington, DC, 61-74, asst to asst dir explor & extraction, 74; dir, Laramie Energy Technol Ctr, Energy Exec Serv, US Dept Energy, 75-81; PRES, SNOWY RANGE MGT SERV, INC, 81- *Concurrent Pos:* Adj prof chem & chem eng, Univ Wyo, 75-82; consult energy, environ, educ & mgt, 81-; pres, Citizens Bank, Laramie Wyo, 84. *Mem:* AAAS; Sigma Xi; Am Chem Soc; Am Bar Asn. *Res:* Gas chromatography; kinetics and mechanisms of organic reactions; thermal and photochemical reactions of sulfur and nitrogen compounds; mass, infrared, nuclear magnetic resonance, spectroscopy; energy development research, particularly in situ processes for oil shale, tar sands, and underground gasification of coal. *Mailing Add:* 1408 Baker St Laramie WY 82070-2927

DECORPO, JAMES JOSEPH, b Lawrence, Mass, Aug 8, 42; m 68; c 1. PHYSICAL CHEMISTRY, ANALYTICAL CHEMISTRY. *Educ:* Northeastern Univ, BS, 64; Pa State Univ, PhD(phys chem), 69. *Prof Exp:* Fel, Rice Univ, 69-70; Nat Acad Sci-Nat Res Coun fel, 70-72; res chemist, 72-75, SUPVRY CHEMIST, NAVAL RES LAB, 75- *Mem:* AAAS; Am Chem Soc; Am Soc Mass Spectrometry. *Res:* Mass spectrometry; development of analytical instrumentation; surface reactions and contaminant characterization; gas phase ion chemistry atmosphere control of enclosed areas and the development of life support system. *Mailing Add:* 4888 Old Well Rd Annandale VA 22003

DECOSSAS, KENNETH MILES, b New Orleans, La, Aug 14, 25. COST ENGINEERING, EQUIPMENT & PLANT DESIGN. *Educ:* Tulane Univ, La, BE, 44. *Prof Exp:* Chem engr, Southern Regional Res Ctr, Agr Res Serv, USDA, 44-48, proj leader & chem engr, 48-54, unit supvr, indust anal, 54-57, supvry chem engr, 57-61, invest head, eng costs & design, 61-74, chem engr, technol & econ anal res, 74-82, food & feed eng, 82-85, chem engr, food & feed processing, 85-88; RETIRED. *Mem:* Emer mem Am Inst Chem Engrs; Nat Soc Prof Engrs; emer mem Am Asn Cost Engrs; emer mem Am Oil Chemists Soc; Sigma Xi. *Res:* Chemical engineering cost analysis, design and processing related to invention; development and commercialization of new processes and products of agricultural utilization; author of 60 research papers. *Mailing Add:* 6529 Gen Diaz St New Orleans LA 70124-3201

DECOSSE, JEROME J, b Valley City, NDak, Apr 19, 28; m 57; c 5. MEDICINE. *Educ:* Col St Thomas, BS, 48; Univ Minn, MD, 52; State Univ NY Upstate Med Ctr, PhD(anat), 69. *Prof Exp:* Intern surg, Roosevelt Hosp, New York, 52-53, asst resident, 53-55; fel exp med, Sloan-Kettering Inst, 55-56; asst & chief resident, Roosevelt Hosp, 58-60; sr resident Mem Ctr, 60-62; asst, Med Col, Cornell Univ, 62-63; asst prof & cancer coordr, State Univ NY Upstate Med Ctr, 63-66; from assoc prof to prof surg, Sch Med, Case Western Reserve Univ, 66-71; prof surg & chmn dept, Med Col Wis, 71-78; prof surg & chmn dept, Mem Sloan-Kettering Cancer Ctr, 78-; CHIEF SURG, MEM SLOAN-KETTERING CANCER CTR. *Concurrent Pos:* Markle scholar, 64; consult, NIH, 68- *Honors & Awards:* Borden Award, 52. *Mem:* Am Col Surg; Am Gastroenterol Asn; Am Soc Cell Biol; Soc Surg Alimentary Tract; Soc Head & Neck Surgeons; Sigma Xi. *Res:* Cancer, cell biology and immunology; gastrointestinal physiology. *Mailing Add:* Prof VChmn Surg Dept NY Hospital Cornell Med Ctr 525 E 68th St New York NY 10021

DECOSTA, EDWIN J, b Chicago, Ill, Mar 25, 06; m 35; c 4. OBSTETRICS & GYNECOLOGY. *Educ:* Univ Chicago, BS, 26, MD, 29; Am Bd Obstet & Gynec, dipl. *Prof Exp:* Mem fac, 46-52, EMER PROF OBSTET & GYNEC, MED SCH, NORTHWESTERN UNIV, CHICAGO, 52- *Concurrent Pos:* Emer Attend obstetrician & gynecologist, Prentice Maternity Hosp, 52-; emer attend gynecologist, Cook County Hosp, 58. *Mem:* Am Gynec & Obstet Soc; fel Am Col Obstet & Gynec. *Res:* Endocrinology. *Mailing Add:* 1540 N Lakeshore Dr Ave Chicago IL 60610

DECOSTA, PETER F(RANCIS), b New Bedford, Mass. TECHNOLOGY PLANNING & MANAGEMENT. *Educ:* Univ Mass, Dartmouth, BS, 62; Univ Conn, Storrs, MA, 66; Mast Inst Technol, MS, 70; Univ Wis-Madison, PE, 81. *Prof Exp:* Chemist, Portsmouth Naval Shipyard, 62-64; teaching fel nuclear sci, Univ Conn, Storrs, 64-66; qual assurance engr, US Food & Drug Admin, Boston, 66-69; advan eng fel, Mass Inst Technol, 69-70; systs analyst, US Environ Protection Agency, Washington, DC, 70-75; opers res analyst, 76-83, phys scientist, 83-86, GEN ENGR, US ARMY NATICK RES, DEVELOP & ENG CTR, 86- *Concurrent Pos:* Planner, City Planning Bd, New Bedford, Mass, 75-76; instr, Mass Bay Community Col, 83-, Worcester Polytech Inst, 84- & Northeastern Univ, 87-; NSF-AAAS fel, Mass Inst Technol Laser Lab, 84. *Mem:* Am Inst Chem Engrs; Am Chem Soc; fel Am Inst Chemists; Sigma Xi. *Res:* Physical science; chemical engineering; systems engineering; science education; author of 22 publications. *Mailing Add:* 39 Deane St New Bedford MA 02746-2306

DE COURCY, SAMUEL JOSEPH, JR, b Newport, RI, June 13, 18; m 44; c 1. MEDICAL MICROBIOLOGY. *Educ:* Yale Univ, BMus, 43; Univ RI, BS, 44; Univ Del, MS, 51; Fla State Christian Univ, ScD, 72. *Prof Exp:* Control chemist, US Naval Torpedo Sta, 44-46; fisheries biologist & chief collab, US Fish & Wildlife Serv, 49; microbiologist, Biochem Res Found, Franklin Inst, 51-60; instr, Rutgers Univ, 61-62; staff res microbiologist, Vet Admin Hosp, Philadelphia, 60-72; assoc surg res, Grad Hosp, Univ Pa, 73-74; assoc in biosci, Albert Einstein Med Ctr, 74-78; MICROBIOL CONSULT, 84- *Concurrent Pos:* Mem staff, Philadelphia Gen Hosp, 64-73; dir, Delmont Labs, Inc, 70-84; microbiol consult, 84- *Mem:* Am Soc Microbiol; Am Chem Soc; NY Acad Sci; Reticuloendothelial Soc; Int Soc Quantum Biol; Sigma Xi. *Res:* Microbial drug resistance; ribosomal vaccines; biological response modifiers in immunotherapy of cancer; cell-mediated immunity. *Mailing Add:* 2522 Traynor Ave Northridge Claymont DE 19703

DECOURSEY, DONN G(ENE), b Auburn, Ind, Oct 21, 34; m 57; c 2. HYDROLOGIC MODELING, AGRICULTURAL RESEARCH. *Educ:* Purdue Univ, BS, 57, MS, 58; Ga Inst Technol, PhD(civil eng), 70. *Prof Exp:* Engr, Ind Flood Control & Water Resources Comn, 58-61; res hydraul engr, Southern Plains Watershed Res Ctr, Agr Res Serv, Chickasha, Okla, 61-71, dir ctr, 71-74, res hydraul engr & dir, Sedimentation Lab, Sci & Educ Admin, Oxford, Miss, 74-81, RES LEADER & NAT TECH ADV, HYDROL RES GROUP, AGR RES SERV, USDA, FT COLLINS, COLO, 81- *Concurrent Pos:* Fel, Colo State Univ, Ft Collins, 78-79. *Mem:* Am Soc Civil Engrs; Am Soc Agr Engrs; Am Geophys Union; Soil Conserv Soc Am; Sigma Xi; Am Water Resources Asn. *Res:* Hydrologic, hydraulic, erosion and sedimentation engineering; statistical and process oriented hydrologic model development; agricultural hydrologic systems. *Mailing Add:* PO Box E Ft Collins CO 80522

DECOURSEY, RUSSELL MYLES, b Indianapolis, Ind, Jan 17, 00; m 30; c 2. ENTOMOLOGY, ZOOLOGY. *Educ:* DePauw Univ, AB, 23; Univ Ill, AM, 25, PhD(entom), 27. *Prof Exp:* Asst prof entom, La Univ, 27-29; from asst prof to prof, 29-70, EMER PROF ENTOM & ZOOL, UNIV CONN, 70- *Mem:* Fel AAAS; emer mem Am Soc Zoologists; Entom Soc Am. *Res:* Bionomics of nymphs of Hemiptera; Pentatomidae. *Mailing Add:* 24 Storrs Heights Rd Storrs CT 06268

DECOURSEY, W(ILLIAM) J(AMES), b Rimbey, Alta, Sept 14, 30; m 57; c 4. MASS TRANSFER. *Educ:* Univ Alta, BSc, 51; Univ London, DIC & PhD, 55. *Prof Exp:* Process engr, Sherritt Gordon Mines, Ltd, Alta, 55-57, res & develop engr, 57-60; from asst prof to assoc prof, 61-71, PROF CHEM ENG, UNIV SASK, 71- *Mem:* Can Soc Chem Eng; Am Inst Chem Engrs; Chem Inst Can. *Res:* Research in mass transfer, particularly mass transfer with chemical reaction; theoretical development of enhancement factors; mathematical modeling and comparison with experimental data; computer programs for design calculations; stage efficiencies. *Mailing Add:* Dept Chem Eng Univ Sask Saskatoon SK S7N 0W0 Can

DECOWSKI, PIOTR, b Lwow, Poland, Jan 17, 40; Dutch & Polish citizen; m 72; c 2. NUCLEAR PHYSICS. *Educ:* Univ Warsaw, Poland, MSc, 61, PhD(physics), 67. *Prof Exp:* Asst physics, Univ Warsaw, Poland, 61-67, from asst prof to prof, 67-90; PROF PHYSICS, SMITH COL, 90- *Concurrent Pos:* Researcher, Joint Inst Nuclear Res, USSR, 63-64 & Mich State Univ, 75-76; adj asst prof physics, Univ Warsaw, Poland, 67-73; vis prof, Nuclear Physics Ctr, WGer, 81-84 & Univ Utrecht, Neth, 86-90. *Mem:* Am Phys Soc; Am Asn Physics Teachers; Sigma Xi; Europ Phys Soc; Polish Phys Soc. *Res:* Experimental nuclear physics; mechanism of nuclear reactions induced by heavy ions and nucleons; nuclear structure. *Mailing Add:* 229 Riverside Dr Northampton MA 01060

DECRAENE, DENIS FREDRICK, b Geneseo, Ill, Aug 28, 42. ELECTROCHEMISTRY. *Educ:* St Ambrose Col, BS, 64; Univ Ill, PhD(chem), 69. *Prof Exp:* Sr res chemist, Res Dept, Diamond Shamrock, 69-75, asst mgr res & develop, Chemetals Div, 75-78; mgr res & develop, 78-84, DIR BUS DEVELOP, CHEMETALS CORP, 84- *Mem:* Sigma Xi; Am Inst Mining Engrs; Am Soc Metals. *Res:* Electrochemical processes; solid-state inorganic reactions; hydrometallurgy; fused salt electrochemistry; chemistry of manganese; copper and cobalt. *Mailing Add:* C/Chemetals 7310 Ritchie Hwy Glen Burnie MD 21060

DECROSTA, EDWARD FRANCIS, JR, b Hudson, NY, Sept 20, 26; m 53; c 2. HEAT TRANSFER, FLUID FLOW. *Educ:* Rensselaer Polytech Inst, BChemEng, 50, MBA, 78, 79; Siena Col, MS, 60. *Prof Exp:* Plant chemist, Universal Watch Corp, 51-64; chemist, Albany Felt Co, 65-69, mgr technol develop, 69-71, dir res & develop, 73-79; dir res technol & develop, Papermaking Prod Group, Albany Inst Corp, 80-82, sr res scientist, 82-83; CONSULT, PROCESS INDUSTS, 83-, TECH CONSULT, COLUMBIA ASSOCS. *Concurrent Pos:* Lectr, Pulp & Paper Asn, 72-81; chmn eng div, Tech Asn Pulp & Paper Indust. *Honors & Awards:* Eng Div Award, Tech Asn Pulp & Paper Indust, 87, TAPPI FELLOW, 89. *Mem:* Am Chem Soc; Can Pulp & Paper Asn; fel Tech Asn Pulp & Paper Indust; NY Acad Sci. *Res:* Papermachine dryer section ventilation caused by moving fabrics; fluid flow through porous media as related to water removal from paper webs and fabrics on the paper machine; thermally activated electrochemical cells of the concentration type; physical characteristics of papermachine clothing; mechanical and chemical conditioning of papermachine clothing. *Mailing Add:* 28 James St Hudson NY 12534

DEDDENS, J(AMES) C(ARROLL), b Louisville, Ky, Mar 25, 28; m 53; c 4. MECHANICAL & NUCLEAR ENGINEERING. *Educ:* Univ Louisville, BME, 52 & MME, 53. *Prof Exp:* Nuclear engr, Atomic Energy Div, Babcock & Wilcox Co, 53-54, nuclear engr & group supvr, 54-58, prof engr, Indian Point Reactor Proj, 58-62, proj mgr, 62-64, asst coordr utility mkt, 64-66, mgr nuclear serv, Nuclear Power Generation Dept, Power Generation Div, 66-74, mgr, Eng Dept, 74-78, mgr proj mgt nuclear, power generation div, 78-80, mgr proj mgt, Math Bus Integration Dept, 80-83; vpres, River Bend Nuclear Group, 83-87, SR VPRES, GULF STATES UTILITIES CO, 87- *Concurrent Pos:* US observer, Int Conf Peaceful Uses of Atomic Energy, Geneva, Switz, 64. *Mem:* Am Nuclear Soc; Am Soc Mech Engrs. *Res:* Nuclear power plant design, construction and operations; nuclear products marketing. *Mailing Add:* Gulf States Utilities Co PO Box 220 St Francisville LA 70775

DEDDENS, JAMES ALBERT, b Cincinnati, Ohio, Sept 7, 43; m 65; c 2. MATHEMATICS, MATHEMATICAL STATISTICS. *Educ:* Univ Cincinnati, BS, 65; Ind Univ, Bloomington, MA, 67, PhD(math), 69. *Prof Exp:* Asst prof math, Univ Mich, Ann Arbor, 69-70; asst prof, 70-75, assoc prof math, Univ Kans, 75-77; assoc prof, 77-78, PROF MATH, UNIV CINCINNATI, 79- *Concurrent Pos:* NSF grants, 69-71 & 72-79; asst prof math, State Univ NY Buffalo, 74-75; assoc prof biostatist, Univ Cincinnati, 82-; vis assoc prof, Univ NC, 84-85, statist consult, 85-; sr res associateship, NRC/NIOSH, 90. *Mem:* Am Statist Soc. *Res:* bio statistics; survival analysis; time series. *Mailing Add:* Dept Math Univ Cincinnati Cincinnati OH 45221

DEDECKER, HENDRIK KAMIEL JOHANNES, b Vorst, Belgium, Sept 4, 15; nat US; m 40; c 1. PHYSICAL CHEMISTRY, RESEARCH MANAGEMENT. *Educ:* Univ Amsterdam, BS, 36; Univ Utrecht, PhD(chem), 41. *Prof Exp:* Res chemist, Nat Sci Orgn, Neth, 41-45; head corrosion res, Shell Petrol Co, 45-47, sect head crude distilling, 47-49, dir res, Rubber Stichting, 49-55; mgr polymer res & develop, Tex-US Chem Co, 56-67; planning coordr, Uniroyal Int, 67-71; managing dir, Uniroyal Plastics Europe, 71-74; corp develop mgr, 74-79, corp strategic planning dir, Uniroyal, Inc, 79-85; PROF BUSINESS STRATEGY, NEW HAVEN UNIV, 85- *Mem:* Am Chem Soc; Strategic Planning Inst. *Res:* Polymer research and development; testirng of materials; rubber technology; market development. *Mailing Add:* 710A Heritage Village Southbury CT 06488

DEDHAR, SHOUKAT, b Kenya, Mar 20, 52; Can citizen; m 82; c 1. CELL BIOLOGY, PHARMACOLOGY. *Educ:* Univ Aberdeen, Scotland, BSc Hons, 74; Univ BC, MSc, 82, PhD(path), 84. *Prof Exp:* Postdoctoral fel, La Jolla Cancer Res Found, 85-87; RES SCIENTIST, BC CANCER AGENCY, 87-; ASST PROF PATH, UNIV BC, 87- *Concurrent Pos:* Sr scientist, Reichmann Res Ctr, Toronto, 91-; assoc prof med biophys, Univ Toronto, 91-; panel mem, Nat Cancer Inst, Can, 91. *Mem:* Am Asn Cancer Res; Am Soc Cell Biol. *Res:* Cell-cell and cell-xtracellular matrix interaction in oncogenesis and differentiation; integrin biochemistry and cell biology; signal transduction mechanisms; growth factors and angiogenesis; tumor invasion and metastasis; integrin-cytoskeleton interactions. *Mailing Add:* Cancer Prog Sunny Brook Health Sci Ctr Reichmann Res Bldg 2075 Bayview Ave North York ON M4N 3M5 Can

DEDINAS, JONAS, b Sakiai, Lithuania, Aug 22, 29; US citizen; m 58; c 1. DRUG METABOLISM, PHOTOCHEMISTRY. *Educ:* Johns Hopkins Univ, BE, 54; Univ Del, MChE, 58; Carnegie Inst Technol, PhD(chem), 65. *Prof Exp:* Petrol engr, Gulf Res & Develop Co, 56-62; sr res chemist, 65-70, res assoc photochem, 71-86, ASSOC BIOANAL CHEMIST, DRUG METAB, EASTMAN PHARMACEUT, EASTMAN KODAK CO, 87- *Honors & Awards:* Frank R Blood Award, Soc Toxicol, 77. *Mem:* Am Chem Soc. *Res:* Petroleum refining and thermal dealkylation; radiation chemistry of tetramethylsilane; photochemistry of aromatic ketones, photosynthesis of decafluorobenzopinacole, thermal reactions of benzopinacoles with azo dyes; mass spectroscopy; plasma chemistry, metabolism of aliphatic ketones, n-hescane, drugs; drug metabolism. *Mailing Add:* 1007 Lakeview Dr Westchester PA 19382-8044

DEDMAN, JOHN RAYMOND, b Washington, DC, Aug 24, 47; m 69; c 3. CELL BIOLOGY, CELL PHYSIOLOGY. *Educ:* Concord Col, BS, 69; NTex State Univ, MS, 72, PhD (zool/biochem), 74. *Prof Exp:* Fel, cell biol, Baylor Col Med, 74-76 & asst prof, 76-80; ASSOC PROF PHYSIOL & CELL BIOL, MED SCH, UNIV TEX, HOUSTON, 80- *Concurrent Pos:* Fel NIH grant, 76; Res Career Develop Award, NIH, 79. *Honors & Awards:* Gail Patrick Verde Award, Am Diabetes Asn, 79. *Mem:* Am Soc Cell Biologists; Am Soc Biol Chemists; Am Chem Soc; NY Acad Sci. *Res:* Understanding the mechanism of action of intracellular calcium, calcium-binding proteins and calcium regulated processes; relating to many diseased states of calcium dysfunction in muscle development and cancer. *Mailing Add:* Dept Physiol & Cell Biol Univ Tex Med Sch 6431 Fannin Houston TX 77025

DEDOMINICIS, ALEX JOHN, b New York, NY, May 13, 35; m 70; c 2. POLYMER CHEMISTRY. *Educ:* NY Univ, BS, 56, MS, 61, PhD(org chem), 62. *Prof Exp:* Res chemist, 62-68, res supvr, 68-78, mgr technol sales, 78- 86, competitive intel, 86-90, MGR, CREATIVITY, E I DU PONT DE NEMOURS & CO, INC, 86-, RES FEL, 90. *Res:* Thiophene chemistry; fiber forming polymers. *Mailing Add:* 2638 Majestic Dr Brandywood Wilmington DE 19810

DEDRICK, KENT GENTRY, b Watsonville, Calif, Aug 9, 23; div; c 1. THEORETICAL PHYSICS. *Educ:* San Jose State Col, BA, 46; Stanford Univ, MS, 49, PhD(physics), 55. *Prof Exp:* Res asst nuclear reactor technol, Univ Mich, 54-55; res assoc theoret physics, W W Hansen Labs, Stanford Univ, 56-59; staff mem, Stanford Linear Accelerator Ctr, 60-62; math physicist, Stanford Res Inst, 62-75; consult, 75-80; RES SPECIALIST, CALIF STATE LANDS COMN, 80- *Mem:* Am Phys Soc; AAAS; Sigma Xi. *Res:* Electromagnetic theory; nuclear theory; aerosol scattering; paramagnetic resonance; environmental sciences; high energy physics; applied mathematics; tidal phenomena; estuarine hydraulics, tidal wetlands. *Mailing Add:* 1807 13th St Sacramento CA 95814

DEDRICK, ROBERT L(YLE), b Madison, Wis, Jan 12, 33; m 55; c 3. CHEMICAL ENGINEERING. *Educ:* Yale Univ, BE, 56; Univ Mich, MSE, 57; Univ Md, PhD(chem eng), 65. *Prof Exp:* Asst prof mech eng, George Washington Univ, 59-62, asst prof eng & appl sci, 62-63, assoc prof, 65-66; actg chief, 66-67, CHIEF CHEM ENG SECT, BIOMED ENG & INSTRUMENTATION BR, NIH, 67- *Mem:* AAAS; Am Soc Eng Educ; Am Chem Soc; Am Inst Chem Engrs; Am Soc Artificial Internal Organs. *Res:* Pharmacokinetics; cancer chemotherapy; transport, thermodynamics and kinetics in living systems; biomaterials; artificial internal organs; risk estimation. *Mailing Add:* 1633 Warmer Ave McLean VA 22101

DE DUVE, CHRISTIAN RENE, b Thames-Ditton, Eng, Oct 2, 17; Belg citizen; m 43; c 4. BIOCHEMISTRY, CYTOLOGY. *Educ:* Univ Louvain, MD, 41, PhD, 45, MSc, 46. *Hon Degrees:* MD, Univ Turin, 69, State Univ Leiden, 70, Univ Sherbrooke, 70, Univ Lille, 73, Cath Univ Chile, 74, Univ Paris V, 74, State Univ Ghent, 75, State Univ Liege, 75, Gustavus Adolphus Col, 75, Univ Rosario, 76, Univ Aix-Marseille II, 79, Univ Keele, 81, Cath Univ Leuven, 84. *Prof Exp:* Therese & Johan Anderson Stiftelse fel, Stockholm, 46-47; Rockefeller Found fel, Washington Univ, 47-48; lectr, 47-51, prof & dept head, 51-85, emer prof physiol chem, Fac Med, Cath Univ Leuvain, 85-; Andrew W Mellon prof, 74-88, EMER ANDREW W MELLON PROF BIOCHEM CYTOL, ROCKEFELLER UNIV, 88- *Concurrent Pos:* Vis prof, Albert Einstein Col Med, 61-62, State Univ Ghent, 62-63, Free Univ Brussels, 63-64, Univ Queensland, 72 & State Univ Liege,

72-73; mem adult develop & aging res & training rev comt, Nat Inst Child Health & Human Develop, 70-72; mem adv comt med res, WHO, 74-78; mem sci adv comt, Max Planck Inst Immunobiol, 75-79; prof biochem cytol, Rockefeller Univ, 62-74; spec adv, Comn Europ Communities, 81-; mem, sci comt, Ludwig Inst Cancer Res, 85-, adv comt, Clin Res Inst Montreal, 85, sci adv comt, Mary Imogene Bassett Res Inst, 86-90, biol adv comt, NY Hall Sci, 86-, int adv bd, Basel Inst Immunol, 89- *Honors & Awards:* Nobel Prize in Physiol or Med, 74; Prix Pfizer, Royal Acad Med Belg, 57; Prix Francqui, Belg, 60; Prix Quinquennal des Science Medicales, Belg Govt, 67; Dr H P Heineken Prize, Royal Neth Acad Sci, 73; Harden Medal, Gt Brit, 78; Theobald Smith Award, 81; Jimenez Diaz Award, Spain, 85; E B Wilson Award, Am Soc Cell Biol, 89. *Mem:* Foreign assoc Nat Acad Sci; Am Acad Arts & Sci; Royal Acad Med Belg; Royal Acad Belg; Leopold Carol Ger Acad Researchers Natural Sci; Pontifical Acad Sci; Paris Acad Sci; Europ Acad Arts, Sci & Humanities; Athens Acad Sci; fel AAAS; Am Chem Soc; Am Soc Biol Chemists; Am Soc Cell Biol; Int Soc Cell Biol; NY Acad Sci. *Res:* Carbohydrate metabolism; action of insulin and glucagon; tissue fractionation; intracellular distribution of enzymes; lysosomes; peroxisomes. *Mailing Add:* Rockefeller Univ 1230 York Ave New York NY 10021

DEE, DIANA, b Los Angeles, Calif, Sept 7, 44. PHYSICAL CHEMISTRY. *Educ:* Reed Col, BA, 65; Univ Calif, Los Angeles, PhD(chem), 70. *Prof Exp:* Asst prof chem, Reed Col, 71-72; physicist, R&D Assocs, 72-76; AREA LEADER, TRW S&TG, 76- *Mem:* Am Phys Soc. *Res:* Chemical laser design; effects of high-energy particles on molecules and ions in the upper atmosphere; nuclear weapons effects; high-explosive and incendiary materials; electronic states of diatomic molecules. *Mailing Add:* TRW Systs One Space Park O1-1070 Redondo Beach CA 90278

DEEDS, WILLIAM EDWARD, b Lorain, Ohio, Feb 23, 20; wid; c 4. PHYSICS. *Educ:* Denison Univ, AB, 41; Calif Inst Technol, MS, 43; Ohio State Univ, PhD(physics), 51. *Prof Exp:* Asst, Calif Inst Technol, 41-43, jr physicist, 43-46; asst prof physics, Denison Univ, 46-48; asst, Ohio State Univ, 49, Texas Co fels, 51-52; from asst prof to prof physics, 52-88, EMER PROF PHYSICS, UNIV TENN, KNOXVILLE, 89- *Concurrent Pos:* Consult, Redstone Arsenal, 55-56, Chemstrand Corp, 56-62, Oak Ridge Nat Lab, 62-, Adaptronics, 79, White's Electronics, 83-84 & Exxon, 89-91. *Honors & Awards:* IR-100 Award, 83. *Mem:* Am Phys Soc; Am Asn Physics Teachers. *Res:* Theoretical physics; electromagnetism; molecular spectroscopy; optics; acoustics. *Mailing Add:* 3601 Montlake Dr Knoxville TN 37920-2846

DEEDWANIA, PRAKASH CHANDRA, b India, Aug 28, 48; US citizen; m 77; c 1. CARDIOLOGY, PULMONARY MEDICINE. *Educ:* Univ Rajasthan, Jaipur, India, MB & BS, 69; Am Bd Internal Med, cert internal med, 75, cert pulmonary dis, 76, cert cardiol, 77. *Prof Exp:* Chief cardiol, Vet Admin Med Ctr-Univ Va, Salem, 77-78; dir, Non-Invasive Lab & co-dir, Cardiac Care Unit & Cath Lab, Vet Admin Med Ctr-Univ Ill, Chicago, 78-80; SECT CHIEF CARDIOL, DIR, MED INTENSIVE CARE UNIT-CORONARY CARE UNIT & CHMN CRITICAL CARE, VET ADMIN MED CTR-UNIV CALIF, SAN FRANCISCO, FRESNO, 80- *Concurrent Pos:* Prin investr var clin studies, 80- & Silent Myocardial Ischenia Prog, 85-; lectr, int med educ progs, 81- & health educ technol, 83-; mem, clin coun, Am Heart Asn. *Mem:* Fel Am Col Cardiol; fel Am Heart Asn; fel Am Col Physicians; fel Am Col Chest Physicians; Am Fedn Clin Res; NY Acad Sci. *Res:* Coronary artery disease; silent myocardial ischemia; congestive heart failure; cardiac arrhythmia. *Mailing Add:* 2615 W Clinton Ave Fresno CA 93703

DEEG, EMIL W(OLFGANG), b Selb, Ger, Sept 20, 26; m 53; c 4. CERAMICS, OPTICS. *Educ:* Univ W rzburg, Dipl, 54, Dr rer nat (phy sci), 56. *Prof Exp:* Res asst glass & ceramics, Max Planck Inst Silicates, 54-59; mem tech staff ceramic eng, Bell Tel Labs, Inc, Pa, 59-60; res assoc glass & ceramics, Jenaer Glaswerk Schott & Gen, Mainz, WGer, 60, dir res, 61-65; assoc prof physics & solid state sci, Am Univ Cairo, 65-67; mgr ceramic res, Am Optical Corp, Southbridge, 67-73, dir mat & process res, 73-77, tech adv glass technol, 77-78; mgr mat res & develop, Anchor Hocking Corp, 78-80; mgr glass technol, Bausch & Lomb, 80-82; mgr, glass & fiber develop, Mead Off Syst, Plano Tex, 82-84; PROJ MGR, AMP INC, 84- *Concurrent Pos:* Mem Ger Standards Comt, 60-65; mem int comn on glass, 63-81, optics, 64-67; mem adv comt glass mfg in space, NASA, 70-75. *Honors & Awards:* Hall of Fame, Eng, Sci & Tech, 88. *Mem:* Fel Am Ceramic Soc; Optical Soc Am; Nat Inst Ceramic Engrs. *Res:* Theory of ceramic manufacturing processes; physics of highly disordered solids; optical, laser and chalcogenide glasses; mechanical properties of glass and ceramics; diffusion and heat transfer; glass manufacturing in space; strength of ophthalmic lenses; container and tableware glasses; fiber optics; liquid crystal polymers, simulation and computer characterization of surface defects. *Mailing Add:* 501 Ohio Ave Lemoyne PA 17043

DEEGAN, LINDA ANN, b Montgomery, Ala, Nov 20, 54. FISHERIES, ECOSYSTEMS. *Educ:* Northeastern Univ, BS, 76; Univ NH, MS, 79; La State Univ, PhD(marine sci), 85. *Prof Exp:* Teaching asst zool, Univ NH, 76-79; res asst marine sci, La State Univ, 79-81 & 83-84, res assoc, 81-82, fel, 84-85; ASST PROF MARINE FISHERIES, UNIV MASS, 85- *Concurrent Pos:* Prin investr, Sigma Xi, 82; consult, La State Univ, 84, prin investr, 84-85. *Mem:* Ecol Soc NAm; Estuarine Res Fedn; Am Fisheries Soc; Sigma Xi. *Res:* Marine and estuarine fish population dynamics in the context of overall ecosystem dynamics; management proposals for the species and their habitats. *Mailing Add:* Dept Forestry & Wildlife Mgt Univ Mass Amherst MA 01003

DEEGAN, MICHAEL J, b Camden, NJ, Sept 23, 42; c 2. PATHOLOGY. *Educ:* St Joseph's Col, Pa, BS, 64; Univ Md, MD, 68. *Prof Exp:* Chief clin path serv, Walter Reed Army Med Ctr, Washington, DC, 73-74; asst prof path & dir clin immunol lab, Univ Mich, 74-77, protein chem sect, William Pepper Lab, Univ Hosp, Univ Pa, 77-78; assoc prof path & lab med, Sch Med & dir, Diag Immunol Lab, Univ Hosp, Emory Univ, 78-79; head, Immunopath Div, Henry Ford Hosp, 80-90, vchmn path, 85-90; DIR LABS, RIVERSIDE METHODIST HOSP, 90- *Concurrent Pos:* Consult path, US Vet Admin Hosp, Ann Arbor, Mich, 74-77; mem inteflex partic fac comt, Med Sch, Univ Mich, 74-77. *Mem:* AAAS; Am Soc Clin Pathologists; Am Fedn Clin Res; Int Acad Path; Am Asn Pathologists. *Res:* Membrane phenotype analysis of malignant lymphomas; cell-mediated immune responses in malignancy; B lymphocyte ontogeny; analysis of autoantibody patterns in disease. *Mailing Add:* Riverside Methodist Hosp 3535 Olentangy River Rd Columbus OH 43214

DEEGAN, ROSS ALFRED, b Montreal, Que, Aug 19, 41; m 66; c 1. SOLID STATE PHYSICS. *Educ:* Loyola Col, Montreal, BSc, 61; McGill Univ, BEng, 63; McMaster Univ, PhD(physics), 67; Duquesne Univ, MBA, 74. *Prof Exp:* Nat Res Coun Can overseas fel, Cavendish Lab, Cambridge Univ, 66-68; res asst prof physics, Univ Ill, Urbana, 68-71; res geophysicist, Gulf Res & Develop Co, 71-73, dir geophys anal, 73-77, mgr geophys sci, 77-88; MGR FRONTIER EXPLOR, CHEVRON CAN RESOURCES, 88- *Mem:* Am Phys Soc; Soc Exploration Geophysicists; Can Asn Physicists. *Res:* Solid state theory, especially band theory of transition metals; seismology for petroleum exploration. *Mailing Add:* Chevron Can Resources 500 Fifth Ave SW Calgary AB T2P 0L7 Can

DEEHR, CHARLES STERLING, b Kalamazoo, Mich, Apr 12, 36; m 60; c 4. SPACE PHYSICS, AERONOMY. *Educ:* Reed Col, AB, 58; Univ Alaska, MS, 61, PhD(geophys), 68. *Prof Exp:* Asst geophysicist, 64-68, from asst prof to assoc prof, 68-81, prof geophysics, 81-88, EMER PROF PHYSICS, GEOPHYSICS INST, UNIV ALASKA, 89-; SCI DIR, POKER FLAT RES RANGE, 89- *Concurrent Pos:* Fel, Royal Norweg Indust & Sci Res Coun, Norweg Inst Cosmic Physics, 69-71 & 79. *Honors & Awards:* Nansen Medal, Norweg Acad Sci, 85. *Mem:* Am Geophys Union; Arctic Inst NAm. *Res:* Aeronomy with emphasis on spectrophotometric studies of atmospheric emissions and scattered light; space physics with emphasis on the ionospheric signatures of magnetospheric events. *Mailing Add:* Geophys Inst Univ Alaska Fairbanks AK 99775-0800

DEELEY, ROGER GRAHAM, b Birmingham, Eng, Dec 9, 46; Can citizen; m; c 2. MOLECULAR ENDOCRINOLOGY OF STEROIDS, MOLECULAR DEVELOPMENTAL BIOLOGY. *Educ:* Sheffield Univ, Eng, BSc, 68, PhD(biochem), 71. *Prof Exp:* Vis fel, Lab Chem Biol, NIH, 71-73, vis assoc, Cellular Reg Sect, Lab Biochem, Nat Cancer Inst, 73-76, vis scientist, 76-77, expert, 77-80; from assoc prof to prof molecular biol, Dept Biochem, 80-87, STAUFFER RES PROF MOLECULAR BIOL, FAC MED, QUEEN'S UNIV, 87-, DIR BASIC ONCOL, CANCER RES LABS, 87- *Concurrent Pos:* Assoc ed, Can J Biochem & Cell Biol, 89-; chmn genetics grant rev comt, Med Res Coun, 89-91. *Honors & Awards:* Development Award, Med Res Coun Can, 80; Terry Fox Team Develop Award, Nat Cancer Inst Can, 89. *Mem:* Am Soc Biol Chemists; Am Soc Microbiol. *Res:* Pretranslational and translational mechanisms involved in developmental and hormonal regulation of gene expression in the liver, particularly those mechanism responsible for the acquisition of competence to respond to steroid hormones and in the regulation of genes involved in lipoprotein synthesis and catabolism. *Mailing Add:* Cancer Res Lab Queens Univ Botterell Hall Rm 314 Kingston ON K7L 3N6 Can

DEELY, JOHN JOSEPH, b Cleveland, Ohio. MATHEMATICAL STATISTICS. *Educ:* Ga Inst Technol, BEE, 55; Purdue Univ, MS, 58, PhD(math statist), 65. *Prof Exp:* Aeronaut res scientist, Nat Adv Comt Aeronaut, 55-56; asst math, Purdue Univ, 56-58, instr, Ft Wayne exten, 58-60, res asst, 60-61, instr, 61-65; mem tech res staff, Western Elec-Bell Labs, Sandia Corp, 65-68; sr lectr, Dept Math, 68-73, PROF MATH, UNIV CANTERBURY, 73- *Concurrent Pos:* Consult, Holloman AFB, 61, Naval Avionics Facility, Indianapolis, 63-64 & others in New Zealand. *Mem:* Royal Statist Soc; Inst Math Statist; Am Statist Asn; Int Statist Inst. *Res:* Applications of statistical decision theory to applied problems, especially multiple decision problems, empirical Bayes and Bayes procedures; sample surveys. *Mailing Add:* Dept Math Univ Canterbury Christchurch New Zealand

DEEM, MARY LEASE, b Sycamore, Ill, June 10, 37; m 63; c 1. ORGANIC & ORGANOMETALLIC SYNTHESIS & REACTIONS. *Educ:* Univ Mich-Ann Arbor, AB, 58; Univ Wis-Madison, MS, 60, PhD(chem), 64. *Prof Exp:* Mem staff, Union Carbide Corp, 56, 60-61 & 73-78; fel, Univ Wis-Madison, 63-65; instr, Douglass Col, Rutgers Univ, 65-67; mem staff, Air Reduction Co, Inc, 67-68; VIS RES SCIENTIST, LEHIGH UNIV, 79- *Concurrent Pos:* Mem staff, Merck, 70; vis fel, Princeton Univ, 72-74 & 79. *Mem:* Am Chem Soc; fel Royal Soc Chem; Sigma Xi; fel Am Inst Chemists. *Res:* Organic and organometallic-synthesis and reactions; Sp3 carbon-hydrogen bond activation; hydrogen transfer; chemistry of medium, small and strained-ring compounds. *Mailing Add:* Sycamore Hill Rd Bernardsville NJ 07924

DEEM, WILLIAM BRADY, b Parkersburg, WVa, Mar 23, 39; m 63; c 1. GENERAL COMPUTER SCIENCES. *Educ:* Lehigh Univ, BS, 61; Univ Wis, MS, 62, PhD(chem eng), 65. *Prof Exp:* Res asst chem eng, Univ Wis, 61-65; engr, Esso Res & Eng Co, NJ, 65; chem engr, Feltman Res Labs, Picatinny Arsenal, Dover, 65-66, tech opers officer, Ammunition Eng Directorate, 66-67; sr eng assoc, 67-81, SECT HEAD, PLANT COMPUT DIV, EXXON RES & ENG CO, 81- *Mem:* Am Inst Chem Engrs; Instrument Soc Am; Sigma Xi. *Res:* Control of petroleum and petrochemical processes; mathematical modeling; computer control incentives, applications and project implementation; plant monitoring. *Mailing Add:* Exxon Res & Engr Co 180 Park Ave Florham Park NJ 07932

DEEMER, MARIE NIEFT, m; c 2. THIONUCLEOTIDES. *Educ:* Univ Southern Calif, PhD(biochem), 48. *Prof Exp:* Res biochemist, Nat Arthritis Inst, 58-82; RETIRED. *Res:* Nucleic acid chemistry. *Mailing Add:* 2539 Ridgeview Rd Arlington VA 22207

DEEMING, TERENCE JAMES, b Birmingham, Eng, Apr 25, 37. PHYSICS, APPLIED MATHEMATICS. *Educ:* Univ Birmingham, BSc, 58; Cambridge Univ, PhD(astron), 61. *Prof Exp:* Asst observer, Cambridge Univ, 61-62; vis lectr, 62-64, from asst prof to assoc prof astron, Univ Tex, Austin, 67-77; sr scientist, 77-85, dir res, 85-87, VPRES, RES & DEVELOP, DIGICON GEOPHYS CORP, HOUSTON, 87- *Mem:* AAAS; Int Astron Union; Soc Exp Geophysists; Inst Elec & Electronics Engrs. *Res:* Geophysics; statistics; data analysis; astrophysics. *Mailing Add:* Icarus Res PO Box 540205 Houston TX 77098

DEEMS, ROBERT EUGENE, b Zanesville, Ohio, May 23, 27; m 53; c 2. PLANT PATHOLOGY. *Educ:* Marietta Col, AB, 49; Ohio State Univ, MS, 51, PhD(plant path), 56. *Prof Exp:* Plant pathologist, Velsicol Corp, 51-52; PLANT PATHOLOGIST, AM CYANAMID CO, 56- *Mem:* Am Phytopath Soc. *Res:* Plant disease control chemicals; agriculture attendant upon specialty; pesticide development. *Mailing Add:* 13 Lawnside Dr Trenton NJ 08638

DEEN, DENNIS FRANK, b North Platte, Nebr, Apr 22, 44; m 65; c 2. RADIATION BIOLOGY. *Educ:* Chadron State Col, BS, 66; Univ Kans, PhD(radiation biophys), 75. *Prof Exp:* Sci teacher, Syracuse-Dunbar Sch Syst, 66-69 & Wellsville Sch Syst, 69-70; researcher radiation biophysics, 75-77, asst res radiation biophysicist, 77, from asst to assoc prof, 77-85, PROF NEUROL SURG & RADIATION ONCOL, UNIV CALIF, SAN FRANCISCO, 85-, ASSOC DIR, BRAIN TUMOR RES CTR, 88- *Concurrent Pos:* Vis fel health physics, Brookhaven Nat Lab, Upton, NY, 71; AEC fel radiation sci & protection, 70-73; USPHS trainee, 73-75. *Mem:* AAAS; Radiation Res Soc; Am Asn Cancer Res. *Res:* Radiobiology, particularly drug-radiation interactions in mammalian cell cultures and in animals. *Mailing Add:* Brain Tumor Res Ctr Univ Calif San Francisco 783 HSW Box 0520 San Francisco CA 94143

DEEN, HAROLD E(UGENE), b Detroit, Mich, Aug 7, 26; m 48; c 4. PETROLEUM CHEMISTRY, CHEMICAL ENGINEERING. *Educ:* Wayne Univ, BS, 48; Purdue Univ, MS, 51. *Prof Exp:* Asst foreman, Rinshed Mason Co, 48-50; engr, Process Res Div, Esso Res & Eng Co, 51-54, engr, Enjay Labs Div, 54-58, group head, 58-61, sr engr, 61-63, sect head, 63-67, mkt coordr, 67-69, sr eng assoc, Esso Res & Eng Co, 69-77, chief scientist, Exxon Chem Co, 77-86; CONSULT, 87- *Mem:* Am Chem Soc; Soc Automotive Engrs. *Res:* Additive research for crank case motor oils; diesel lubricants; automatic transmission fluids. *Mailing Add:* 216 Oak Lane Cranford NJ 07016

DEEN, JAMES ROBERT, b Dallas, Tex, Mar 1, 44; m 71; c 7. REACTOR PHYSICS, THERMAL REACTOR PHYSICS. *Educ:* Univ Tex, Austin, BEngSci, 66, MSME, 70, PhD(nuclear eng), 73. *Prof Exp:* Sr engr boiling water reactor physics, Gen Elec Co, 72-74, sr engr boiling water develop, 74-76; asst nuclear engr reactor physics, 76-80, NUCLEAR ENGR, ARGONNE NAT LAB, 80- *Mem:* Am Nuclear Soc; Sigma Xi. *Res:* Reactor physics, especially in research and test reactor reduced enrichment and new production reactor core design. *Mailing Add:* Reactor Analysis Div 9700 S Cass Ave Argonne IL 60439

DEEN, MOHAMED JAMAL, b Georgetown, Guyana, June 1, 55; m 79; c 2. MICROELECTRONICS, DEVICE PHYSICS. *Educ:* Univ Guyana, BSc, 78; Case Western Reserve Univ, MS, 82 & PhD(elec eng & appl physics), 86. *Prof Exp:* Instr physics, Univ Guyana, 78-80; res engr elec eng, Lehigh Univ, 83-85, asst prof, 85-86; asst prof, 86-89, ASSOC PROF ENG SCI, SIMON FRASER UNIV, 89- *Concurrent Pos:* Fulbright scholar, Laspau Found, 80; scholar, Am Vacuum Soc, 83; vis scientist astron, Nat Res Coun, 86; adv & consult radio astron, Nat Res Coun, 86-; assoc mem, Cancer Imaging Sect, BC Cancer Res Ctr, 87- *Mem:* Am Phys Soc; Inst Elec & Electronic Engrs; Advan Systs Inst; Soc Photo-Optical Instrumentation Engrs. *Res:* Low temperature electronics; microelectronics; device physics and reliability; superconducting electronics; novel high frequency devices. *Mailing Add:* Sch Eng Sci Simon Fraser Univ Burnaby BC V5A 1S6 Can

DEEN, ROBERT C(URBA), geotechnical engineering, traffic engineering; deceased, see previous edition for last biography

DEEN, THOMAS B, b Lexington, Ky, Apr 4, 28; m 54; c 4. PUBLIC TRANSIT, HIGHWAYS. *Educ:* Univ Ky, BSCE, 51. *Prof Exp:* Asst city traffic engr, Nashville, Tenn, 56-57, dir metrop area transp study, 57-59; appl sci rep, IBM Corp, 59-60; dir planning, Nat Capital Transp Agency, 60-64; sr vpres, Alan M Voorhees & Assocs, 64-78, chmn & pres, PRC Voorhees, 78-80; EXEC DIR, TRANSP RES BD, 80- *Concurrent Pos:* Consult, ENO Found Transp, Inc, Saugatuck, Conn, 81-83; mem, bd dir, API Inc, Fairfax, Va, 82-87; mem, adv comt, Transp Ctr, Mass Inst Technol, 83-; vchmn board dir, Padco, Inc, Washington, DC, 75-; mem Bd Trustees, Am Pub Works Res Found. *Honors & Awards:* Pike Johnson Award, 75. *Mem:* Inst Transp Engrs; Am Pub Works Asn. *Res:* Transportation planning; transportation systems management; transportation systems analysis. *Mailing Add:* Transp Res Bd 2101 Constitution Ave NW Washington DC 20418

DEEN, WILLIAM MURRAY, b Seattle, Wash, Jan 30, 47. BIOMEDICAL ENGINEERING, MEMBRANE TRANSPORT. *Educ:* Columbia Univ, BS, 69; Stanford Univ, MS, 71, PhD(chem eng), 73. *Prof Exp:* Nat Kidney Found res fel, Dept Med, Univ Calif & Vet Admin Hosp, 73-75, adj asst prof renal physiol, Depts Physiol & Med, Univ Calif, San Francisco, 75-76; from asst prof to assoc prof, 80-85, PROF CHEM ENG, MASS INST TECHNOL, 85- *Concurrent Pos:* Western Elec Fund Award, Am Soc Eng Educ, 81-82. *Mem:* Am Physiol Soc; Am Inst Chem Engrs; Am Soc Nephrology. *Res:* Membrane transport processes; renal and microcirculatory physiology; pharmacokinetics and toxicokinetics. *Mailing Add:* Dept Chem Eng Rm 66-509 Mass Inst Technol Cambridge MA 02139

DEENEY, ANNE O'CONNELL, microbiology, biochemistry, for more information see previous edition

DEEP, IRA WASHINGTON, b Dover, Tenn, July 26, 27; m 52; c 5. PLANT PATHOLOGY. *Educ:* Miami Univ, BA, 50; Univ Tenn, MS, 52; Ore State Univ, PhD(plant path), 56. *Prof Exp:* Instr bot, Ore State Univ, 53-57, from asst prof to assoc prof bot & plant path, 57-68, asst dean grad sch, 65-68; chmn dept, 68-84, PROF PLANT PATH, OHIO STATE UNIV, 68- *Concurrent Pos:* Staff biologist, Comn Undergrad Educ Biol Sci, Washington, DC, 66-67. *Mem:* AAAS; Am Inst Biol Sci; Am Phytopath Soc. *Res:* Field crops diseases. *Mailing Add:* Ohio State Univ Dept Plant Path 2021 Coffey Rd Columbus OH 43210

DEEPAK, ADARSH, b Sialkot, India, Nov 13, 36. ATMOSPHERIC OPTICS, METEOROLOGY. *Educ:* Delhi Univ, BS, 56, MS, 59; Univ Fla, PhD(aerospace eng), 69. *Prof Exp:* Lectr physics, D B & K M Cols, Delhi Univ, 59-63; instr phys sci, Univ Fla, 65-68, res assoc physics, 70-71; Nat Res Coun fel, Marshall Space Flight Ctr, NASA, 72-74; res assoc prof physics & geophys sci, Old Dominion Univ, 74-77; pres, Inst Atmospheric Optics & Remote Sensing, Hampton, Va, 77-84; PRES, SCI & TECHNOL CORP, HAMPTON, VA, 79- *Concurrent Pos:* Consult eng sci, Wayne State Univ, 70-72; mem panel remote sensing & data acquisition, NASA/OAST Technol Workshop, 75; NSF travel grant to visit Indian insts, 76; adj prof physics, Col William & Mary, 79-80; leader, US Deleg, Int Workshop Appln Remote Sensing Rice Prod, India, 81. *Mem:* Optical Soc Am; Am Meteorol Soc; Am Chem Soc; Am Geophys Union; AAAS. *Res:* Remote sensing of atmospheric particulate and gaseous pollutants and motions, using laser doppler, optical scattering and photographic techniques from space, airborne and ground platforms; theory of radiative transfer in scattering atmospheres, fogs and clouds; inversion methods for remotely sensed data. *Mailing Add:* Sci & Technol Corp PO Box 7390 Hampton VA 23666-7390

DEERE, DON U(EL), b Corning, Iowa, Mar 17, 22; m 44; c 2. ROCK MECHANICS, CIVIL ENGINEERING. *Educ:* Iowa State Col, BS, 43; Univ Colo, MS, 49; Univ Ill, PhD(civil eng), 55. *Prof Exp:* Jr mine engr, Phelps Dodge Corp, Ariz, 43-44; mine engr explor dept, Potash Co of Am, NMex, 44-47; from asst prof to assoc prof civil eng, Col A&M, Univ PR, 46-50, head dept, 50-51; partner, Found Eng Co, PR, 51-55; from assoc prof to prof civil eng & geol, Univ Ill, Urbana-Champaign, 55-76; AFFIL PROF, DEPT GEOL & CIVIL ENG, UNIV FLA, 72- *Concurrent Pos:* Consult found eng & eng geol, Nat Acad Sci; pres, Comn Standardization Lab & Field Tests, Int Soc Rock Mech, 68-73; int consult, hydroelec eng, 72-; chmn, US Nat Comt Tunneling Technol, 72; counr, Geol Soc Am, 76-79. *Mem:* Nat Acad Sci; Nat Acad Eng; AAAS; Geol Soc Am; Am Soc Civil Engrs; Asn Prof Geol Scientists; Soc Mining Engrs. *Res:* Stability of natural slopes; author of approximately 50 papers on engineering geology, applied mechanics, dam foundations & tunnels. *Mailing Add:* 6834 SW 35th Way Gainesville FL 32608

DEERING, CARL F, b Royal Oaks, Mich, July 27, 59. ASYMMETRIC SYNTHESIS, REACTION SCALEUP. *Educ:* Mich Technol Univ, BS, 81; Univ Mich, MS, 84, PhD(org & med chem), 87. *Prof Exp:* Scientist, 87-90, SR SCIENTIST CHEM DEVELOP, PARKE-DAVIS, 90- *Mem:* Am Chem Soc. *Res:* Asymmetric synthesis of optically active gamma lactones; vinyl stannane chemistry as pertaining to organic synthesis; scale up of organic synthetic reactions from the laboratory scale to manufacturing. *Mailing Add:* 188 W 19th St Holland MI 49423

DEERING, REGINALD, b Brooks, Maine, Sept 21, 32; m 56; c 4. MOLECULAR BIOPHYSICS, DNA ENZYMOLOGY. *Educ:* Univ Maine, BS, 54; Yale Univ, PhD(biophys), 58. *Prof Exp:* Asst prof physics, Southern Ill Univ, 57-58; Fulbright res grant biophys, Univ Oslo, 58-59; res assoc, Yale Univ, 59-61; from asst prof to assoc prof, NMex Highlands Univ, 61-64; assoc prof, 64-69, prof biophysics, 69-81, PROF MOLECULAR & CELL BIOL, PA STATE UNIV, 81- *Concurrent Pos:* Vis prof, Stanford Univ, 74-75 & Univ Calif, Berkeley, 83-84. *Mem:* Am Soc Biol Chemists; Am Soc Microbiol; Am Soc Photobiol. *Res:* Genetics, enzymology, and regulation of DNA repair; recombinant DNA; effects of radiation and chemical mutagens on DNA of cells; development of dictyostelium discoideum. *Mailing Add:* Molecular & Cell Biol Pa State Univ 201 Althouse Lab University Park PA 16802

DEERING, WILLIAM DOUGLESS, b Burleson, Tex, Dec 18, 33; m 54; c 3. PHYSICS. *Educ:* Tex Christian Univ, BA, 56; NMex State Univ, MS, 60, PhD(physics), 63. *Prof Exp:* Res assoc atmospheric physics, Grad Res Ctr Southwest, 63-65; asst prof physics, 65-67, ASSOC PROF PHYSICS, NTEX STATE UNIV, 67- *Mem:* Am Phys Soc. *Res:* Statistical mechanics; plasma physics. *Mailing Add:* Dept Physics NTex State Univ Denton TX 76203

DEES, J W, b Feb 20, 33; m; c 3. RESEARCH ADMINISTRATION, ELECTRICAL ENGINEERING. *Educ:* Tri-state Col, BS, 53 & 54; Univ Cincinnati, MS, 55. *Prof Exp:* Assoc engr, Int Telephone & Telegraph Labs, 55-60; group engr, Martin Marietta, 60-67, sr staff engr & proj mgr, 67-71; dir, Electromagnetics Lab, 71-81, PRIN RES ENGR, GA TECH RES INST, GA INST TECHNOL, 71-, ASSOC VPRES RES & DIR OFF CONTRACT ADMIN, 81- *Concurrent Pos:* Lectr metal-oxide-metal detector & millimeter wave technol, China Asn Sci & Technol, 84; mem, Nat Coun Univ Res Adminr, Res Adv Coun, Ga Inst Technol, prog comt, Int Microwave Symp, 74 & Estes Park Millimeter Wave Conf, 65; chmn, Millimeter Wave & Submillimeter Wave Conf, 63, prof prog comt, SConf, 83, Panel Antennas, Solar Power Satellite Symp, Toulouse, France, 86; co-leader, Goodwill People to People Prog, United States Microwave Electronics Leaders Deleg, 80; exec secy, Int Conf Submillimeter Waves, 74 & 76; invited partic, workshop space shuttle, NASA & Goddard Space Flight Ctr, 74; served on satellite commun, Dept Civil Aviation, 71; mem tech prog comt & session chmn, Int Microwave Symp, Microwave Theory & Technique, Inst Elec & Electronics Engrs, 82. *Mem:* Fel Inst Elec & Electronic Engrs; Soc Res Admin; Nat Coun Univ Res Admin; Nat Soc Prof Engrs. *Res:* Author of 35 scientific journals. *Mailing Add:* Off Contract Admin Ga Inst Technol Atlanta GA 30332-0420

DEES, SUSAN COONS, b Hancock, Mich, May 26, 09; m 35; c 4. PEDIATRICS. *Educ:* Goucher Col, AB, 30; Johns Hopkins Univ, MD, 34; Univ Minn, MS, 38. *Prof Exp:* House officer med serv, Johns Hopkins Hosp, 34-35, asst dispensary physician, 38-39; asst pediatrician, Duke Hosp, 39-48; from asst prof pediat to prof pediat & allergy, 48-79, EMER PROF PEDIAT & ALLERGY, SCH MED, DUKE UNIV, 79-; CONSULT PEDIAT ALLERGY, 79- *Concurrent Pos:* Asst res, Strong Mem Hosp, 35-36; intern, Baltimore City Hosp, 36-37. *Honors & Awards:* B Ratner Award, Am Acad Pediat; Distinguished Clinician, Am Acad Allergy Immunol, 90. *Mem:* Fel Am Med Asn; fel Am Col Allergists; Am Pediat Soc; Am Acad Pediat; Am Acad Allergy & Immunol. *Res:* Allergy. *Mailing Add:* Dept Pediat Duke Univ Med Ctr PO Box 2193 Durham NC 27710

DEESE, DAWSON CHARLES, b Raleigh, NC, Dec 7, 29. BIOCHEMISTRY, FOOD SCIENCE AND TECHNOLOGY. *Educ:* Agr & Tech Col NC, BS, 52; Tuskegee Inst, MS, 54; Univ Wis, PhD(chem, biochem), 61. *Prof Exp:* Asst gen chem, Agr & Tech Col NC, 52 & Tuskegee Inst, 53; asst biochem, Univ Wis-Madison, 55-60, res assoc biochem & vet sci, 60-64, instr, 64-68, assoc prof nutrit sci, 69-80, chmn dept, 70-77, ASSOC PROF HUMAN BIOL, CHEM & NUTRIT, UNIV WIS-GREEN BAY, 80- *Mem:* Fel AAAS; Am Chem Soc; fel Am Inst Chemist; NY Acad Sci; Sigma Xi; Inst Food Technol. *Res:* Digestive enzymes in the ruminant animal and ruminal metabolism of plant macromolecules; pectolytic enzymes in fungal disease and disease resistance of plants; auxin metabolism and plant growth substances; nutritional sciences; chemistry. *Mailing Add:* 221 Huth St Suite 323 Green Bay WI 54302-3454

DEETER, CHARLES RAYMOND, b Norcatur, Kans, Dec 30, 30; m 57; c 2. NUMERICAL MATHEMATICS. *Educ:* Ft Hays Kans State Col, BS, 52, MS, 56; Univ Kans, PhD(discrete harmonic kernels), 63. *Prof Exp:* From asst prof to assoc prof, 60-70, PROF MATH, TEX CHRISTIAN UNIV, 70- *Mem:* Math Asn Am; Am Math Soc; Soc Indust & Appl Math; Asn Comput Mach. *Res:* Discrete analytic functions. *Mailing Add:* Dept Math Tex Christian Univ Ft Worth TX 76129

DEETS, GARY LEE, b Sunbury, Pa, Feb 28, 43; m 61; c 3. POLYMER CHEMISTRY. *Educ:* Bloomsburgh State Col, BS, 65; Univ Pittsburgh, PhD(org chem), 69. *Prof Exp:* Sr res chemist, 69-77, res specialist, GROUP LEADER, SPRINGFIELD LAB, MONSANTO CO, 81- *Mem:* Am Chem Soc. *Res:* Mechanistic studies in pyridine nitrogen-oxide chemistry; fundamental and exploratory studies of the flame retardancy of styrene based polymeric systems; paper resins; toner resins. *Mailing Add:* Monsanto Co 730 Worcester St Springfield MA 01151

DEETZ, LAWRENCE EDWARD, III, b Elyria, Ohio, May 7, 51; m 78; c 2. ENDOCRINOLOGY, STATISTICAL DESIGN. *Educ:* Ohio State Univ, BS, 73; Univ NH, MS, 75; Pa State Univ, PhD(animal nurtit), 79. *Prof Exp:* Res assoc animal nutrit, Univ Ky, 79-81; res assoc, animal nutrit, health & nutrit res div, Eastman Chem Div, Eastman Kodak, 81-88; RES NUTRITIONIST, HOECHEST-ROUSSEL AGR-VET CO, 88- *Mem:* Am Soc Animal Sci; Am Inst Nutrit; Sigma Xi. *Res:* Applied and basic research in animal nutrition; feeding trials designed to measure performance of ruminant animals, specifically beef and dairy cattle; determining mechanisms of action by measuring biological parameters. *Mailing Add:* Hoechst-Roussel Agr-Vet Co Animal Health Prods Rte 202-206 North Somerville NJ 08876

DEEVER, DAVID LIVINGSTONE, b Dayton, Ohio, Aug 31, 39; m 61; c 2. PURE MATHEMATICS. *Educ:* Otterbein Col, BA & BS, 61; Ohio State Univ, PhD(math), 66. *Prof Exp:* Asst prof math, Westmar Col, 66-71; ASSOC PROF MATH & CHMN DEPT, OTTERBEIN COL, 71- *Mem:* Math Asn Am; Am Math Soc. *Res:* Set theory and transfinite arithmetic. *Mailing Add:* 77 N West St Westerville OH 43081

DEEVEY, EDWARD SMITH, JR, lymnology, paleolimnology; deceased, see previous edition for last biography

DEFACIO, W BRIAN, b Palestine, Tex, Dec 14, 36; m 64; c 2. INVERSE SCATTERING THEORY, CONDENSED MATTER THEORY. *Educ:* Tex A&M Univ, BS, 63, MS, 65, PhD(theoret physics), 67. *Prof Exp:* From asst prof to assoc prof, 67-83, PROF PHYSICS, UNIV MO, COLUMBIA, 83- *Concurrent Pos:* Vis asst prof physics, Tex A&M Univ, 67; vis assoc prof physics, Iowa State Univ, 74-75, vis assoc prof math, 79-80; assoc ed, J Math Physics, 79-81; prin investr, Off Naval Res Grant, Iowa State Univ, 81-85; prof theoret physics, Chalmers Univ Technol, Sweden, 83-84; vis prof, Univ Sci & Technol at Langvedoc, Montpellier, France, 84 & 87. *Mem:* Sigma Xi; Am Phys Soc; Am Math Soc; Int Asn Math Physicists; Am Asn Physics Teachers; Soc Indust & Appl Math. *Res:* Mathematical physics. *Mailing Add:* Dept Physics Univ Mo Columbia MO 65211

DEFANTI, DAVID R, b Wakefield, RI, Nov 12, 32; m 58; c 2. PHARMACOLOGY. *Educ:* Colgate Univ, AB, 55; Univ RI, MS, 57, PhD(pharmacol), 62. *Prof Exp:* Asst prof, 61-73, PROF PHARMACOL, UNIV RI, 73-, DIR, CRIME LAB, 80- *Concurrent Pos:* Co-prin investr, NIH Grant, 62-64; consult, RI State Labs, Sci Criminal Invest, 63-70, asst dir, 70-73, dir, 73-; prin investr, RI Heart Asn Grant, 64-; prin investr, Law Enforcement Assistance Admin Grant, 70-; dir breath tests alcohol training, State of RI, 70- *Mem:* AAAS; Am Acad Forensic Sci; Int Asn Identification; Int Narcotic Enforcement Officers Asn; Sigma Xi. *Res:* Cardiovascular pharmacology; medico-legal toxicology. *Mailing Add:* Dockray Rd Wakefield RI 02879

DEFANTI, THOMAS A, b New York, NY, Sept 18, 48; m; c 1. COMPUTER GRAPHICS & ANIMATION. *Educ:* City Univ New York, BA, 69; Ohio State Univ, MS, 70, PhD(comput & info sci), 73. *Prof Exp:* PROF COMPUT SCI, UNIV ILL, CHICAGO, 73- *Concurrent Pos:* Assoc ed, Comput Graphics & Appl, Inst Elec & Electronics Engrs, 85-; Univ Scholar, Univ Ill, 89-92. *Honors & Awards:* Outstanding Contrib Award, Asn Comput Mach, 88. *Mem:* Asn Comput Mach (secy, 77-81). *Res:* Real-time interactive computer graphics, computer animation; script and programming languages; scientific visualization. *Mailing Add:* Dept Elec Eng MC/154 Univ Ill Box 4348 Chicago IL 60880

DE FAZIO, THOMAS LUCA, b Philadelphia, Pa, May 21, 40; m 69. MECHANICAL PRODUCT & SYSTEM DESIGN, ASSEMBLY SYSTEM DESIGN. *Educ:* Mass Inst Technol, SB, 61, SM, 63, ScD(mech eng), 71. *Prof Exp:* Engr, Northern Res & Eng Corp, 63; defense res staff engr, Instrumentation Lab, Mass Inst Technol, 63-67, proj engr, Dept Earth & Planetary Sci, 67-71; from asst prof to assoc prof mech eng, Fla Inst Technol, 71-78; TECH STAFF MEM, CHARLES STARK DRAPER LAB, INC, 78- *Concurrent Pos:* Assoc prin engr, Harris Corp, 77. *Mem:* Am Soc Mech Engrs; Am Geophys Union; Am Acad Mech. *Res:* Design of assembly systems; rational determination of assembly sequences; task planning; work planning; automation; robotics; mechanical design; author of numerous publications; awarded 11 patents. *Mailing Add:* 103 Chapman St Watertown MA 02172

DEFELICE, EUGENE ANTHONY, b Beacon, NY, Dec 24, 27; m 57; c 2. INTERNAL MEDICINE, CARDIOLOGY. *Educ:* Columbia Univ, BS, 51; Boston Univ Sch Med, MD, 56; Nat Bd Med Examiners, dipl. *Prof Exp:* Lectr & fel pharmacol, Sch Med, Boston Univ, 54-57; from assoc prof to prof, New Eng Col Pharm, 56-58; pvt pract internal med, North Miami, Fla, 58-61; asst dir clin res, Warner-Lambert Res Inst, NJ, 61-64; dir clin pharmacol, Bristol Labs, NY, 64-66; dir clin res, Norwich Hosp, Conn, 66-67; dir clin pharmacol, Sandoz Pharmaceut, 67-68, exec dir clin res, 69-70, dir sci affairs & commercial develop, Sandoz Inc, 70-73, dir corp-sci affairs, 76, vpres corp-sci develop, 76-77, vpres int med liaison, 77-81, vpres & med adv, 82-83; clin prof anesthesiol, Sch Med, Univ Calif, Los Angeles, 78-83; clin assoc prof med, 77-84, CLIN PROF MED, ROBERT WOOD JOHNSON MED SCH, RUTGERS, 85- *Concurrent Pos:* Res assoc anesthesiol, Boston City Hosp, 57; intern, Newton-Wellesley Hosp, 57-58; consult internal med & med res, Morristown, NJ, 61-87 & E Schodack, NY, 88-; consult, Strasenburgh Labs, NY, 66-67; mem bd dirs, Delmark Co, Minneapolis, 73-77 & Vital Assists, Inc, Salt Lake City, 74-77. *Mem:* Am Soc Clin Pharmacol & Therapeut; AMA; fel Acad Psychosom Med; fel Am Geriat Soc; fel NY Acad Sci. *Res:* Clinical pharmacology, internal medicine/cardiology; geriatrics; psychosomatic medicine. *Mailing Add:* Stablegate PO Box 38 2373 Payne Rd East Schodack NY 12063

DE FELICE, LOUIS JOHN, b Morristown, NJ, Aug 17, 40; m 62; c 2. BIOPHYSICS, ELECTROPHYSIOLOGY. *Educ:* Fla State Univ, BS, 62, MS, 64; Univ Calgary, PhD(physics), 67. *Prof Exp:* Med Res Coun Can fels, McGill Univ, 67-69; Med Res Coun Can fel, State Univ Leiden, 69-70, asst prof physiol, 70-71; from asst prof to assoc prof, 71-78, PROF ANAT, EMORY UNIV, 78- *Concurrent Pos:* Asst prof elec eng, Ga Inst Technol, 71-74, assoc prof, 74-; Fulbright-Hayes fel, 78. *Mem:* Biophys Soc; Hist of Sci Soc; Am Physiol Soc. *Res:* Biophysics of biological and synthetic membranes; heart cell membranes and intercellular communication; noise, impendance and single channnel studies in excitable membranes. *Mailing Add:* Dept Anat Emory Univ Sch Med 1648 Pierce Dr Rm 100 Atlanta GA 30322

DEFENBAUGH, RICHARD EUGENE, b Kansas City, Mo, Sept 6, 46; m 78; c 2. MARINE ECOLOGY, INVERTEBRATE ZOOLOGY. *Educ:* Kans State Teachers Col, BA, 67; Tex A&M Univ, MS, 70, PhD(zool), 76. *Prof Exp:* Teaching asst, Tex A&M Univ, 67-74, res asst, 74, lectr biol, 74-75; staff biol oceanogr, 75-81, chief environ studies, Bur Land Mgt, 81-82, CHIEF ENVIRON STUDIES SECT, MINERALS MGT SERV, US DEPT INTERIOR, 82- *Mem:* AAAS; Am Soc Zoologists. *Res:* Marine and coastal ecology; impacts of offshore oil and gas activities on marine and coastal environments; Gulf of Mexico benthic invertebrates; taxonomy of hydroids. *Mailing Add:* 1201 Elmwood Park Blvd Mailstop 5430 New Orleans LA 70123-2394

DEFENDI, VITTORIO, b Treviglio, Italy, Nov 16, 28; m 55; c 3. PATHOLOGY. *Educ:* Univ Pavia, MD, 51. *Prof Exp:* Instr path dept, Univ Pavia, 51-52; pathologist virus sect, Lederle Labs, NY, 56-58; assoc path, Med Sch, Univ Pa, 58-64, assoc prof, 64-68, Wistar prof, 68-74; PROF PATH & CHMN DEPT, SCH MED, NY UNIV, 74- *Concurrent Pos:* Brit coun scholar, Post-grad Med Sch, Univ London, 52-53; Fulbright fel, Med Sch, Univ Vt, 53-54; res fel, Detroit Inst Cancer Res, 54-56; Leukemia Soc scholar, 62-66; fac res award, Am Cancer Soc, 68-72; assoc mem, Wistar Inst, 58-64, mem staff, 64-74; Am Cancer Soc res prof, 73- *Mem:* Am Soc Cell Biol; Am Soc Exp Path; Histochem Soc; Am Asn Immunol; Am Asn Cancer Res. *Res:* Viral oncology; tumor biology; mechanism of immunological defense. *Mailing Add:* Dept Path NY Univ Sch Med 550 First Ave New York NY 10016

DEFEO, JOHN JOSEPH, b Southington, Conn, Mar 14, 22; m 55; c 8. PHARMACOLOGY. *Educ:* Univ Conn, BS, 51; Purdue Univ, MS, 53, PhD(pharmacol), 54. *Prof Exp:* From instr to asst prof pharmacol, Univ Pittsburgh, 54-57; assoc prof, 57-64, PROF PHARMACOL, COL PHARM, UNIV RI, 64-, CHMN, DEPT PHARMACOL & TOXICOL, 80- *Mem:* AAAS; Am Soc Pharmacol; Fedn Am Soc Exp Biol; NY Acad Sci. *Res:* Cardio-vascular area. *Mailing Add:* 607 Curtis Corner Rd Peace Dale RI 02879

DE FEO, VINCENT JOSEPH, b New York, NY, Oct 1, 25; div; c 2. ENDOCRINOLOGY, REPRODUCTIVE BIOLOGY. *Educ:* Juniata Col, BS, 49; Rutgers Univ, MS, 51; Ohio State Univ, PhD(physiol), 54. *Prof Exp:* Asst zool, Rutgers Univ, 50; asst physiol, Ohio State Univ, 50-52, asst prof, 54-55; NIH fel embryol, Carnegie Inst, Washington, DC, 55-57; asst prof anat, Col Med, Univ Ill, 57-63; assoc prof anat & obstet & gynec, Sch Med, Vanderbilt Univ, 63-66; assoc prof, 66-68, chmn dept, 69-73, PROF ANAT & REPRODUCTIVE BIOL, UNIV HAWAII, MANOA, 83- *Mem:* Am Asn Anat; Soc Scientific Study of Sex; Soc Study Reproduction. *Res:* Anatomy and physiology of reproduction; ovum-uterine relationship; neuroendocrinology; human sexuality. *Mailing Add:* Dept Anat Univ Hawaii Burns Sch Med 1960 E West Rd Honolulu HI 96822

DEFERRARI, HARRY AUSTIN, b Stoneham, Mass, May 31, 37; m 62; c 3. UNDERWATER ACOUSTICS. *Educ:* Cath Univ Am, BME, 59, PhD(acoust), 66; Mass Inst Technol, MS, 62. *Prof Exp:* Res scientist vibrations anal, David Taylor Model Basin, 61-62; res scientist acoust, Chesapeake Instrument Corp, 65-67; PROF OCEAN ENG, ROSENSTIEL SCH MARINE & ATMOSPHERIC SCI, UNIV MIAMI, 67- *Mem:* Acoust Soc Am. *Res:* Ocean acoustics, the study of acoustical transmission through the ocean; the relationships between acoustic and oceanographic fluctuations. *Mailing Add:* Dept Marine Sci Univ Miami Coral Gables FL 33124

DEFFEYES, KENNETH STOVER, b Oklahoma City, Okla, Dec 26, 31; m 62; c 1. OCEANOGRAPHY. *Educ:* Colo Sch Mines, GeolE, 53; Princeton, MS, 56, PhD(geol), 59. *Prof Exp:* Geologist, Shell Develop Co, Tex, 58-63; asst prof, Univ Minn, 63-64; assoc prof oceanog, Ore State Univ, 65-67; assoc prof geol & geophys sci, 67-74, PROF GEOL & GEOPHYS SCI, PRINCETON UNIV, 74- *Mem:* Mineral Soc Am; Am Asn Petrol Geologists; Sigma Xi. *Res:* Chemical oceanography; sedimentation. *Mailing Add:* Dept Geol Princeton Univ Princeton NJ 08540

DE FIEBRE, CONRAD WILLIAM, b Brooklyn, NY, Jan 19, 24; m 46; c 6. MICROBIOLOGY, BIOCHEMISTRY. *Educ:* Rensselaer Polytech Inst, BS, 49; Univ Wis, MS, 50, PhD(bact), 52. *Prof Exp:* Asst bact, Univ Wis, 49-52; res microbiologist, E I Du Pont de Nemours & Co, Inc, 52-61; res dir, Wilson Labs Div, Wilson Pharmaceutical & Chem Corp, 61-67, vpres res, 67-69; dir, Ross Labs Div, Abbott Labs, 69-71, vpres res & develop, 71-85; RETIRED. *Concurrent Pos:* Chmn bd, Infant Formula Coun, 76-77 & 81-85; actg dir food indust ctr, Ohio State Univ, 85-86, mem adv comt, Nisonger Ctr Ment Retardation & Develop Disabilities, 87- *Mem:* Am Soc Microbiol; Am Chem Soc; fel Am Inst Chemists. *Res:* Bacteriology; fermentation; enzymes; natural products; pharmaceutical development. *Mailing Add:* 6486 Strathaven Ct N Worthington OH 43085

DEFIGIO, DANIEL A, b Republic, Pa, June 4, 39; m 66; c 2. MYCOLOGY. *Educ:* California State Col, Pa, 61; WVa Univ, MA, 64; Ill State Univ, PhD(mycol), 70. *Prof Exp:* High sch teacher, Pa, 61-64; instr biol, Washington & Jefferson Col, 64-66; teaching asst, Ill State Univ, 66-70; ASSOC PROF BIOL, EDINBORO STATE COL, 70- *Mem:* AAAS; Am Inst Biol Sci; Bot Soc Am. *Res:* Taxonomic analysis of the genus Hymenochaete. *Mailing Add:* Dept Biol & Health Sci Edinboro Univ Edinboro PA 16444

DE FIGUEIREDO, RUI J P, b Pangim, Goa, India, Apr 19, 29; m 61; c 5. INTELLIGENT SYSTEMS, INFORMATION SCIENCE & SYSTEMS. *Educ:* Mass Inst Technol, SB, 50 & SM, 52; Harvard Univ, PhD(appl math), 59. *Prof Exp:* Consult atomic energy, Govt Portugal, 55-58; head, Appl Math & Physics Div, Portuguese Nuclear Energy Res Ctr, SacavéM, Portugal, 59-62; assoc prof elec eng, Purdue Univ, 62-64; vis assoc prof elec eng, Univ Ill, 64-65; assoc prof elec eng, Rice Univ, 65-67, prof elec & computer eng & math sci, 68-90; PROF ELEC & COMPUTER ENG & MATH, UNIV CALIF, IRVINE, 90- *Concurrent Pos:* Vis prof & res fel, Math Res Ctr, Univ Wis-Madison, 72-73; vis prof elec eng, Univ Calif, Berkeley, 73; vis res prof, Swiss Fed Inst Tech ETH, Zurich, 81; chair, Nonlinear Circuits & Systs Tech Comt, Inst Elec & Electronic Engrs Circuits & Systs Soc, chair Houston Chap, 76-90; mem adv comt, Inst Elec & Electronic Engrs Ocean Eng Soc; mem NSF panels; consult to indust. *Honors & Awards:* Citation of Merit for Outstanding Serv, Portuguese Govt, 60; Cert Recognition for Creative Develop & Tech Innovation, NASA, 87, Cert for Inventive Contrib, 88. *Mem:* Fel Inst Elec & Electronic Engrs; NY Acad Sci; Soc Indust & Appl Math; Sigma Xi. *Res:* Intelligent signal and image processing; artificial neural networks; nonlinear systems analysis; machine vision. *Mailing Add:* Dept Elec & Computer Eng Univ Calif Irvine CA 92717

DEFILIPPI, LOUIS J, b New York, NY, July 15, 49. BIOTECHNOLOGY, BIOCHEMISTRY. *Educ:* Queens Col, BA, 71; Univ Mich, PhD(biochem), 76. *Prof Exp:* Postdoctoral res biochem, Cornell Univ, 76-78; RES SCIENTIST, DEPT BIOTECHNOL I, ALLIED SIGNAL RES & TECHNOL, 78- *Mem:* Am Soc Biochem & Molecular Biol; Am Chem Soc. *Mailing Add:* Dept Biotechol I Allied-Signal Res & Technol 50 E Algonquin Rd Box 5016 Des Plaines IL 60017-5016

DE FILIPPI, R(ICHARD) P(AUL), b New York, NY, Mar 26, 36; m 58; c 2. SEPARATION PROCESSES. *Educ:* Amherst Col, AB, 57; Mass Inst Technol, SM, 59, ScD(chem eng), 62. *Prof Exp:* Org chemist, Am Cyanamid Co, 56, chem engr, 57-58; chem engr, Arthur D Little Co, Inc, 59; res engr hydrogen processing petrol refining, Calif Res Corp, 62-65; supvr membrane separations sect, Abcor, Inc, 65-66, prog mgr, 66-68, prog dir, 68-70, vpres & gen mgr res & develop div, 70-74; mem spec staff, Arthur D Little, Inc, 74-77, unit mg, 77-80; PRES, CF SYSTS, INC, 80- *Concurrent Pos:* Mem bd dirs, Abcor, Inc, 72-73; chmn bd dirs, Walden Res Corp, 72-73; vis fel, Imp Col Sci & Technol, London, 73-74; chmn, Cambridge Health Policy Bd, 77-78, mem, 78- *Mem:* Am Inst Chem Engrs; Am Chem Soc. *Res:* Membrane technology; mass transfer; supercritical fluids. *Mailing Add:* CF Systs Corp 500 W Cummings Park Suite 6600 Woburn MA 01801

DEFILIPPS, ROBERT ANTHONY, b Chicago, Ill, Mar 4, 39. PLANT TAXONOMY. *Educ:* Univ Ill, BS, 60; Southern Ill Univ, MS, 62, PhD(bot), 68. *Prof Exp:* Collabr, Flora Dominica Proj, Smithsonian Inst, 68-69; res assoc, Flora Europaea Proj, Univ Reading, Eng, 69-74; PLANT CONSERV UNIT STAFF, DEPT BOT, SMITHSONIAN INST, 74- *Mem:* Heliconia Soc Int. *Res:* Endangered plant species of the world; ornamental and medicinal flora of the Guianas, Dominica and the Philippines. *Mailing Add:* Dept Bot Nat Mus Natural Hist Smithsonian Inst Washington DC 20560

DEFOGGI, ERNEST, b Butler, Pa. APPLICATION OF COMPUTER SYSTEM ELECTRICAL CHARACTERISTICS TO THE BRAIN FUNCTIONS FOR ELIMINATION OF STRESS. *Educ:* Univ Mo, BSEE, 60. *Prof Exp:* Field engr, Sperry Corp, Long Island, NY, 60; radar engr, Fed Aviation Agency, Washington, DC, 61-67; elec qual control engr, US Dept Defense, Binghamton, NY, 67-80; RESEARCHER, BRAIN ELEC ACTIVITY, 55- *Res:* Electrical activity of the brain using advanced electronic techniques towards developing a simplified instant approach for anyone to experience the "timeless/stress free/mystical/non-knowing" state of mind irregardless of the environmental stress sensory inputs; perform analysis-correlation of various energy systems characteristics to the human mind electrical parameters. *Mailing Add:* Rte 1 Box 514-A Newport NC 28570

DEFOLIART, GENE RAY, b Stillwater, Okla, June 23, 25; m 50; c 3. ENTOMOLOGY. *Educ:* Okla State Univ, BS, 48; Cornell Univ, PhD(entom), 51. *Prof Exp:* From asst prof to assoc prof entom, Univ Wyo, 51-59; assoc prof, 59-66, chmn dept, 68-76 & 82-83, PROF ENTOM, RUSSELL LABS, UNIV WIS-MADISON, 66- *Mem:* Entom Soc Am; Am Mosquito Control Asn; Am Soc Trop Med & Hyg; AAAS. *Res:* Medical and veterinary entomology; with emphasis on the epidemiology; transmission and vector biology of mosquito-borne arboviruses; insects as a high-protein source of food and animal feed. *Mailing Add:* Dept Entom 237 Russell Labs Univ Wis 1630 Linden Dr Madison WI 53706

DEFORD, DONALD DALE, b Alton, Kans, Dec 28, 18; m 42; c 2. ANALYTICAL CHEMISTRY. *Educ:* Univ Kans, AB, 40, PhD(chem), 48. *Prof Exp:* From instr to assoc prof chem, 48-60, chmn dept, 62-69, asst to provost, 69-70, assoc dean faculties, 70-71, PROF CHEM, NORTHWESTERN UNIV, 60-, ASST VPRES RES, 71- *Mem:* AAAS; Instrument Soc Am; Am Chem Soc; Sigma Xi. *Res:* Analytical instrumentation; gas chromatography. *Mailing Add:* Dept of Chem Northwestern Univ Evanston IL 60201

DEFORD, JOHN W, b Lincoln, Nebr, Mar 12, 36; m 58; c 3. SOLID STATE PHYSICS. *Educ:* Carleton Col, BA, 57; Univ Ill, MS, 59, PhD(physics), 62. *Prof Exp:* Asst prof physics, 62-67, ASSOC PROF PHYSICS, UNIV UTAH, 67- *Mem:* Am Phys Soc. *Res:* Radiation damage in metals. *Mailing Add:* Dept Physics Univ Utah 201b J Fletcher Bldg Salt Lake City UT 84112

DE FORD, RONALD K, b San Diego, Calif, Jan 22, 02. STRUCTURAL GEOLOGY. *Educ:* Colo Sch Mines, EM, 21, MS, 22. *Prof Exp:* Prof geol, Univ Tex, Austin, 48-72; RETIRED. *Mem:* Fel Geol Soc Am. *Mailing Add:* 1604 Rabb Rd Austin TX 78704

DEFOREST, ADAMADIA, b Harrisburg, Pa, July 16, 33; m 57. VIROLOGY. *Educ:* Hood Col, AB, 55; Wash State Univ, MS, 57; Temple Univ, PhD, 68; Am Bd Med Microbiol, dipl. *Prof Exp:* Grad teaching asst, dept zool, Wash State Univ, 55-57, res assoc, dept vet microbiol, 57-60; res assoc dept microbiol & immunol, Temple Univ Sch Med, 61-62, instr, 62-64, grad teaching asst, 64-67, from instr to asst prof pediat virol, 67-78, ASSOC PROF PEDIAT VIROL, TEMPLE UNIV SCH MED, 78-, ASSOC PROF MICROBIOL & IMMUNOL, 82- *Concurrent Pos:* Pediat virol, St Christopher's Hosp for Children. *Honors & Awards:* Difco Award, Am Pub Health Asn, 89. *Mem:* Am Soc Microbiol; Am Fedn Clin Res; Am Pub Health Asn; Pan Am Group for Rapid Viral Diag; fel Am Acad Microbiol. *Res:* Protective effects of breast milk against virus infections; bacterial-viral interrelationships in cystic fibrosis; role of cytomegalovirus infections in chronic interstitial pneumonitis of infancy; antibody responses of infants and children to live attenuated virus vaccines; immune responses in virus infections. *Mailing Add:* St Christopher's Hosp Children Erie Ave at Front St Philadelphia PA 19134-1095

DEFOREST, ELBERT M, b Natoma, Kans, July 17, 17; m 42; c 2. PETROLEUM ENGINEERING. *Educ:* Univ Tulsa, BS, 40. *Prof Exp:* Petrol engr, Gulf Oil Corp, 40-41; chemist, E I du Pont de Nemours & Co, 41-42, develop process engr, 42-46; sr process engr, Spencer Chem Co, 46-47, mgr process eng, 47-49; sr proj engr, Pan Am Petrol Corp, Standard Oil Co, Ind, 49-50, supt chem mfg, 50-52; mgr new proj, Frontier Chem Co Div, 52-59, mgr res & develop, 59-67, vpres chem div res & develop, Vulcan Mat Co, Kans, 67-82, vpres metallics div, Ohio, 67-82; RETIRED. *Mem:* AAAS; Am Inst Chem Engrs; Am Chem Soc; Am Inst Mining, Metall & Petrol Engrs; Electrochem Soc. *Res:* Distillation technology and fluid mechanics; hydrocarbon chlorination technology; chemical plant design. *Mailing Add:* 3047 Benjamin Ct Wichita KS 67204

DEFOREST, PETER RUPERT, b Los Angeles, Calif, Aug 24, 41; m 66; c 2. FORENSIC SCIENCE & CRIMINALISTICS. *Educ:* Univ Calif, Berkeley, BS, 64, DCrim, 69. *Prof Exp:* From asst prof to assoc prof, 69-80, PROF CRIMINALISTICS, JOHN JAY COL, CITY UNIV NEW YORK, 80- *Concurrent Pos:* Consult sci interpretation of phys evidence. *Mem:* Am Acad Forensic Sci; Am Chem Soc; NY Acad Sci; AAAS. *Res:* Instrumental methods for the chemical individualization of physical evidence; tagged antibody methods for simultaneous detection of several antigens in dried bloodstains; pyrolysis supercritical fluid chromatography of polymer characterization. *Mailing Add:* Dept Sci 445 W 59th St New York NY 10019

DE FOREST, RALPH EDWIN, b Detroit, Mich, Mar 19, 19; m 50; c 3. MEDICINE. *Educ:* Wayne State Univ, BS, 41, MD, 43; Univ Minn, MS, 51. *Prof Exp:* Fel orthop surg, Mayo Found, 44-46, fel phys med & rehab, 48-51; dir dep med physics & AMA, 50-66, dir postgrad progs,66-70, dir dept med instrumentation, 70-72, asst dir dept grad med educ, 72-77, dir dept continuing med eval, AMA, 77-82; RETIRED. *Concurrent Pos:* Chmn, President's Task Force on Physically Handicapped, 69-70; secy, Liaison Comt Continuing Med Educ, 77-81. *Mem:* Am Acad Phys Med & Rehab; Sigma Xi; Am Cong Rehab Med. *Res:* Internal derangements of the knee; effects of physical agents on lymph flow; effects of ultrasound on bone. *Mailing Add:* 211 Ninth St Wilmette IL 60091

DEFORREST, JACK M, b Huntingdon, Pa, Apr 2, 51. RENAL PHYSIOLOGY. *Educ:* Pa State Univ, PhD(physiol), 76. *Prof Exp:* Sr res investr, 83-86, RES GROUP LEADER PHARMACOL, SQUIBB INST MED, 86- *Mailing Add:* Dept Pharmacol Squibb Inst Med PO Box 4000 Princeton NJ 08540

DEFOTIS, GARY CONSTANTINE, b Chicago, Ill, June 30, 47. MAGNETISM, MAGNETIC PHASE TRANSITIONS. *Educ:* Univ Ill, BSc, 68; Univ Chicago, PhD(phys chem), 77. *Prof Exp:* Res asst chem, Argonne Nat Lab, 68 & Univ Chicago, 69-76; res assoc, Univ Ill, 77-78; asst prof chem, Mich State Univ, 78-80; from asst prof to assoc prof, 80-89, PROF CHEM, COL WILLIAM & MARY, 89- *Concurrent Pos:* Prin investr, NSF, PRF & others; consult. *Mem:* Am Chem Soc; Am Phys Soc; Inst Elec & Electronics Engrs Magnetics Soc. *Res:* Solid state chemistry; paramagnetism, ferromagnetism and antiferromagnetism; phase transitions and critical behavior in mixed and dilute magnetic systems and in lower dimensional systems. *Mailing Add:* Dept Chem Col William & Mary Williamsburg VA 23185

DEFOUW, DAVID O, b Grand Rapids, Mich, June 3, 45; m 67; c 1. ANATOMY, PHYSIOLOGY. *Educ:* Hope Col, BA, 67; Mich State Univ, MS, 70, PhD(anat), 72. *Prof Exp:* From instr to asst prof, 72-78, ASSOC PROF ANAT, NJ MED SCH, NEWARK, 78- *Mem:* Microcirculation Soc; Am Asn Anatomists. *Res:* Pulmonary ultrastructure and function; neurogenic control of the microcirculation, especially the pulmonary circulation. *Mailing Add:* Dept Anat NJ Med Sch Newark NJ 07103

DEFRANCE, JON FREDRIC, b Battle Creek, Mich, Mar 17, 43; m 73; c 2. NEUROPHYSIOLOGY, NEUROPHARMACOLOGY. *Educ:* Mich State Univ, BS, 67; Wayne State Univ, PhD(neurophysiol), 73. *Prof Exp:* From instr to asst prof anat, Wayne State Univ, 73-76; ASSOC PROF NEUROBIOL, MED SCH, UNIV TEX, 76-, ASSOC PROF ANAT, 80- *Mem:* Soc Neurosci. *Res:* Electrophysiology of limbic system. *Mailing Add:* Dept neurobiol & Anat Univ Tex Med Sch Houston PO Box 20708 Houston TX 77225

DE FRANCE, JOSEPH J, b New York, NY, Aug 22, 09; m 35; c 2. AUDIO, COMMUNICATIONS. *Educ:* City Col New York, BS, 30, EE, 31. *Prof Exp:* Lab engr, City Col New York, 31-39; instr high sch, NY, 39-42; prof radio, electronics, Signal Corps Training Sch, 42-43; chief engr, Int Div, Trans World Airline, 46-47; dept head electronic technol, 47-64, prof electronics, 64-74, EMER PROF ELECTRONICS, NEW YORK CITY COMMUNITY COL, 74- *Concurrent Pos:* Lectr, RCA Insts, 39-42; electronics officer, US Coast Guard, 43-46; consult, Cleveland Inst Electronics, Ohio, 60- *Mem:* Am Soc Eng Educ; Am Tech Educ Asn; Inst Elec & Electronics Engrs. *Res:* Electronics communications; technical textbooks and pamphlets. *Mailing Add:* 143 Jackson St Garden City NY 11530

DE FRANCESCO, LAURA, b Arlington, Mass, Sept 22, 48. CYTOPLASMIC INHERITANCE, CELLULAR BIODYNAMICS. *Educ:* Tufts Univ, BS, 70; Univ Calif, San Diego, PhD(biol), 75. *Prof Exp:* NIH fel, Calif Inst Technol, 76-77, Am Cancer Soc fel, 77-79; ASST RES SCIENTIST, BIOL DIV, CITY HOPE RES INST, 80- *Mem:* Am Soc Cell Biol. *Res:* Mechanisms governing the expression and inheritance of the mitochondrial genome of animal cells, with particular emphasis on recombination, and the role it may play in maintaining homogenity of this highly repeated genome. *Mailing Add:* Dept Biol Beckman Res Inst City of Hope 1450 E Duarte Rd Duarte CA 91010

DEFRANCO, RONALD JAMES, b Baltimore, Md, Feb 18, 43; m 74. APPLIED MATHEMATICS. *Educ:* Univ Dubuque, BS, 65; Univ Ariz, MS, 68, PhD(math), 73. *Prof Exp:* Instr math, Univ Ariz, 73-76; asst prof math, Univ Portland, 76-78; programmer & analyst, 78-80, TECH SPECIALIST, CUBIC CORP, 80- *Mem:* Am Math Soc; Math Asn Am. *Res:* Numerical solutions for multiple Volterra integral equations. *Mailing Add:* 8024 Linda Vista Rd No 1G San Diego CA 92111

DEFRANK, JOSEPH J, b Rochester, New York, 46; m 81. BIOTECHNOLOGY, MOLECULAR BIOLOGY. *Educ:* St John Fisher Col, BS, 68; Univ Miami, PhD(biochem), 75. *Prof Exp:* Dir microbiol, Bio-Tech Resources, Inc, 77-81; sr res chemist, indust chem div, biotechnol group, Barberton Tech Ctr, PPG Indust, Inc, 81-82, Mellon Inst, 82-83; RES CHEMIST, BIOTECHNOL DIV, US ARMY CHEM RES, DEVELOP & ENG CTR, ABERDEEN PROVING GROUND, MD, 84- *Concurrent Pos:* Res assoc, dept biochem, Univ Miami, 76-77. *Mem:* AAAS; NY Acad Sci; Am Chem Soc; Sigma Xi; Am Soc Microbiol; Soc Indust Microbiol. *Res:* Detection and decontamination of chemical and biological warfare agents; use of biosensors and enzymes. *Mailing Add:* SMCCR-RSB US Army Chem Res, Develop & Eng Ctr Aberdeen Proving Ground MD 21010-5423

DEFRIES, JOHN CLARENCE, b Delrey, Ill, Nov 26, 34; m 56; c 2. BEHAVIORAL GENETICS. *Educ:* Univ Ill, BS, 56, MS, 58, PhD(genetics), 61. *Prof Exp:* Asst prof genetics in dairy sci, Univ Ill, 61-66; assoc prof genetics in dairy sci, Univ Ill, Urbana, 66-67; assoc prof, 67-70, PROF BEHAV GENETICS, UNIV COLO, BOULDER, 70-, PROF PSYCHOL, 76-, DIR, INST BEHAV GENETICS, 80- *Concurrent Pos:* USPHS res fel genetics, Univ Calif, Berkeley, 63-64; co-founder & ed, Behavior Genetics, 70-78; consult ed, J Learning Disabilities, 87-; sci & corresp mem, Rodin Remediation Acad. *Mem:* AAAS; Soc Study Social Biol; fel Int Acad Res Learning Disabilities; Behav Genetics Asn (secy, 74-77, pres, 82-83); Soc Res Child Develop; Int Soc Study Individual Differences; Am Psychol Soc. *Res:* Genetics of specific cognitive abilities and reading disability in humans; genetics of ethanol-induced narcosis in laboratory mice. *Mailing Add:* Inst Behav Genetics Univ Colo Campus Box 447 Boulder CO 80309-0447

DE GASTON, ALEXIS NEAL, b Seattle, Wash. HIGH EXCELLERATION PROTECTION FROM HIGH EXCELLERATION FORCES, CHEMICAL BIOLOGICAL RADIATION WARFARE PROTECTION. *Educ:* Calif Inst Technol, BSS, 59; Brigham Young Univ, PhD(physics), 65; Univ Autonoma Ciudad Juarez, MD, 78. *Prof Exp:* Psychiatrist, Utah State Ment Hosp, 81; instr med, Family Pract, Loma Linda Univ, 81-82; pvt pract, A Neal de Gaston, MD, 81-84; med dir rehab, Cent Utah Rehab & Med Ctr, 84-86 & med clin, A Neal de Gaston PC, 84-87; prin staff scientist aerospace med, McDonnell Douglas Corp, 87-90; PVT PRACT MED, 90- *Concurrent Pos:* Radiation specialist, Irvine Med Ctr, Univ Calif, 74-76; consult med physics/radiol physics, 75-77. *Mem:* Am Acad Family Pract; Am Asn Physicists Med; AMA. *Res:* Investigation into the etiology of the differential rotation of the sun; radar signal processing; upper atmospheric physics; protecting man subject to high accelerations. *Mailing Add:* 116 1/2 E Hillview Ave Winslow AZ 86047

DEGEN, VLADIMIR, b Prague, Czech, Apr 24, 31; Can citizen; m 61; c 4. AERONOMY. *Educ:* Univ Toronto, BA, 58, MA, 60; Univ Western Ont, PhD(physics), 66. *Prof Exp:* NASA fel, Lab Atmospheric & Space Physics, Univ Colo, Boulder, 67-70; asst prof, Univ Alaska, Fairbanks, 70-74, assoc prof physics, 74-87; RETIRED. *Mem:* Am Geophys Union. *Res:* Spectroscopy of earth's auroral and airglow emissions. *Mailing Add:* 713 Depauw Dr Fairbanks AK 99709

DEGENFORD, JAMES EDWARD, b Bloomington, Ill, June 11, 38; m 59; c 6. ELECTRICAL ENGINEERING. *Educ:* Univ Ill, BS, 60, MS, 61, PhD(elec eng), 64. *Prof Exp:* Res asst elec eng, Univ Ill, Urbana-Champaign, 60-64, res assoc, 64-65; sr engr, appl physics sect, 65-69, fel engr, microwave physics group, 69-78, adv engr, 78-84, MGR, G&As ENG, WESTINGHOUSE ELEC CORP, 84- *Mem:* Fel Inst Elec & Electronics Engrs; Microwave Theory & Techniques Soc (secy-treas, 75). *Res:* Investigation of transmission systems and detection techniques suitable for use at submillimeter wavelengths; microwave integrated circuits and monolithic microwave circuits. *Mailing Add:* Westinghouse Adv Tech Lab PO Box 1521 MS K13 Baltimore MD 21203

DEGENHARDT, KEITH JACOB, b Goodsoil, Sask, Jan 18, 50; m 73; c 1. PLANT PATHOLOGY. *Educ:* Univ Alta, BS, 71, MS 73; Univ Sask, PhD(plant path), 78. *Prof Exp:* Lectr biol sci, Kelsey Inst Appl Arts & Sci, 77-78; res scientist cereal path, Res Br, Agr Can, 78-84; AGR CONSULT, CEREAL & OILSEED DIS, 84- *Mem:* Can Phytopath Soc; Am Phytopath Soc. *Res:* The biology, epidemiology, and control of the Tilletia species (bunt) and Urocystis species (stem smut) on wheat and rye; biology and control of other cereal and oil seed deseases. *Mailing Add:* Gen Delivery Hughenden AB T0B 2E0 Can

DEGENHARDT, WILLIAM GEORGE, b Queens Co, NY, Apr 16, 26; m 58. VERTEBRATE ZOOLOGY. *Educ:* Syracuse Univ, AB, 50; Northeastern Univ, MS, 53; Tex A&M Univ, PhD(zool), 60. *Prof Exp:* Teaching fel, Northeastern Univ, 51-53; grad asst, Tex A&M Univ, 53-54, instr, 55-60; cur reptiles & amphibians, Mus Southwestern Biol, 61-85; from asst prof to assoc prof, 60-75, prof, 75-85, EMER PROF BIOL, UNIV NMEX, 85-; EMER CUR REPTILES & AMPHIBIANS, MUS SOUTHWESTERN BIOL, 85- *Concurrent Pos:* Collabr, Nat Park Serv, 55-59, 64-, ranger-naturalist, 60. *Mem:* Am Soc Naturalists; Am Soc Ichthyologists & Herpetologists; Ecol Soc Am; Soc Study Amphibians & Reptiles; Herpetologists League. *Res:* Herpetology; vertebrate ecology. *Mailing Add:* Dept Biol Univ NMex Albuquerque NM 87131

DEGENKOLB, HENRY JOHN, structural engineering; deceased, see previous edition for last biography

DE GENNARO, LOUIS D, b New York, June 1, 24; m 49; c 5. ZOOLOGY. *Educ:* Fordham Univ, BS, 48; Boston Col, MS, 50; Syracuse Univ, PhD, 59. *Prof Exp:* Asst, Boston Col, 48-49; from instr to assoc prof, 49-62, PROF BIOL, LE MOYNE COL, 62- *Concurrent Pos:* Res prof, State Univ NY Upstate Med Ctr, 74-; NSF & NIH grants, 62, 64, 88 & 90. *Mem:* Aerospace Med Asn; Soc Develop Biol; Am Asn Anatomists. *Res:* Experimental embryology; tissue and organ culture; electron microscopy. *Mailing Add:* Dept Biol Le Moyne Col Syracuse NY 13214

DEGGINGER, EDWARD R, b Chicago, Ill, Oct 12, 26; m 49; c 2. ORGANIC CHEMISTRY. *Educ:* Univ Ill, BS, 49; Univ Wis, MS, 51, PhD(chem), 52. *Prof Exp:* Res supvr, Solvay Process Div, 52-67, tech assoc, Corp, 67-72, RES GROUP LEADER, ALLIED CHEM CORP, 72- *Mem:* Am Chem Soc. *Res:* Urethanes, pesticides and flame retardants; nylon composites. *Mailing Add:* 40 Laura Lane Morristown NJ 07960

DEGHETT, VICTOR JOHN, b New York, NY, May 26, 42. ETHOLOGY & STATISTICS, BIOMETRICS & BIOSTATISTICS. *Educ:* Univ Dayton, BA, 64; Bowling Green State Univ, PhD(animal behav), 72. *Prof Exp:* Instr psychol, Univ Dayton, 66-67; from instr to asst prof, 71-77, assoc prof, 77-86, PROF PSYCHOL, STATE UNIV NY POTSDAM, 86- *Mem:* Animal Behav Soc; Am Soc Mammal; Sigma Xi. *Res:* Vertebrate behavioral and morphological development; quantitative methods in ethology; rodent ultrasound production; evolution of behavior; acoustic communication. *Mailing Add:* Dept Psychol State Univ NY Potsdam NY 13676

DEGIOVANNI-DONNELLY, ROSALIE F, b Brooklyn, NY, Nov 22, 26; m 61; c 2. MICROBIAL GENETICS, BIOCHEMISTRY. *Educ:* Brooklyn Col, BA, 47, MA, 53; Columbia Univ, PhD(zool), 61. *Prof Exp:* Mem staff, Allergy Lab, Univ Hosp, Bellevue Med Ctr, New York, 47-51; technician, Sch Pub Health, Columbia Univ, 52-54, asst microbial genetics & biochem of nucleic acids, Col Physicians & Surgeons, 54-62; sr scientist, Bionetics Res Labs, Inc, Va, 62-67; asst prof lectr, 68-71, asst res prof, 71-77, ASSOC PROF LECTR MICROBIOL, MED CTR, GEORGE WASHINGTON UNIV, 77-; RES BIOLOGIST, GENETIC TOXICITY BR, BUR SCI, FOOD & DRUG ADMIN, 68- *Mem:* AAAS; Am Soc Microbiol; NY Acad Sci; Environ Mutagen Soc; Sigma Xi. *Res:* Genotypic and phenotypic effects of specific chemicals on microbes; mutation mechanisms. *Mailing Add:* Microbiol Lab George Washington Univ Washington DC 20057

DEGIROLAMI, UMBERTO, NEUROPATHOLOGY. *Educ:* Univ Miami, MD, 68. *Prof Exp:* PROF PATH & NEUROL, UNIV MASS, 82- *Mailing Add:* Dept Path Univ Mass Med Sch 55 Lake Ave N Worcester MA 01605

DE GIUSTI, DOMINIC LAWRENCE, b Treviso, Italy, March 31, 11; nat US; m 38, 74; c 3. PARASITOLOGY. *Educ:* St Thomas Col, Minn, BS, 36; Univ Mich, MS, 38; Univ Wis, PhD(zool), 42. *Prof Exp:* Instr biol, Col St Thomas, 36-38, asst prof biol, 42-43 & 46-47; teaching asst, Biol Sta, Univ Mich, 40-41; asst prof prev med, Col Med, NY Univ, 43-45; res assoc pharmacol, Univ Minn, 47; asst prof biol, Catholic Univ, 47-49; assoc prof, 49-58, chmn dept biol, 67-72, chmn dept comp med, Sch Med, 78-79, prof parasitol, Lib Arts Col & prof parasitol microbiol, 59-79, prof comp med, Sch Med, 74-79, prof immunol-microbiol, 79-81, EMER PROF, WAYNE STATE UNIV, 81- *Concurrent Pos:* Markle Found fel, Cent Am; lectr, Med Sch, Georgetown Univ, 47-48; consult, USPHS, 48, Off Surgeon-Gen, 48-49, Off Sci Res & Develop & Vet Admin Hosp, 73-; Fulbright fel, Zool Sta, Univ Naples, 52-53; univ assoc, Dept Path, Detroit Gen Hosp & Dept Med & Path, Hutzel Hosp & Vet Admin Hosp, 72- *Mem:* Am Soc Parasitologists; Am Zool Soc; Am Soc Trop Med & Hyg; fel NY Acad Sci; fel AAAS. *Res:* Life cycles of acanthocephala; histology and cytology of trematodes; Blood protozoa of cold blooded vertebrates; life cycles of gregarines. *Mailing Add:* Dept Immunol-Microbiol Sch Med Wayne State Univ Detroit MI 48201

DEGNAN, JOHN JAMES, III, b Philadelphia, Pa, Dec 10, 45; m 69; c 2. ENGINEERING PHYSICS. *Educ:* Drexel Inst Technol, BS, 68; Univ Md, MS, 70, PhD(physics), 79. *Prof Exp:* Physicist carbon laser commun, NASA-Goddard Space Flight Ctr, 68-72, sr physicist, 72-75, sr physicist laser ranging & lidar, 75-79, head, Advan Electro-Optical Instrument Sect, 79-89, DEP MGR, CRUSTAL DYNAMICS SECT, NASA-GODDARD SPACE FLIGHT CTR, 89- *Concurrent Pos:* Lab instr, Drexel Inst Technol, 67-68; dep mem, Adv Group Electron Devices, NASA; adj prof physics, Am Univ, Washington, DC, 88- *Honors & Awards:* Moe I Schneebaum Mem Award, 87. *Mem:* Optical Soc Am; Am Inst Physics; Am Geophys Union; Int Laser Commun Soc. *Res:* Lasers, optics, lidar, spectroscopy, CO2 laser communications; short pulse laser ranging; geophysics; geodesy. *Mailing Add:* Crystal Dynamics Proj Code 901 NASA Goddard Space Flight Ctr Greenbelt MD 20771

DEGNAN, KEVIN JOHN, EPITHELIAL TRANSPORT, REGULATION. *Educ:* New York Univ, PhD(physiol & biophys), 75. *Prof Exp:* ASST PROF MED SCI, BROWN UNIV, 81- *Mailing Add:* Dept Physiol Biophysics Brown Univ Prog Med 97 Waterman St Providence RI 02912

DE GOES, LOUIS, b Maui, Hawaii, June 23, 14; m 44; c 3. ENGINEERING GEOLOGY. *Educ:* Colo Sch Mines, GeolE, 41; Stanford Univ, MS, 50. *Prof Exp:* Master navigator, USAF, 41-44, chief navigator, Pac Div Mil Air Transp Serv, 45-48, assoc prof, USAF Inst Technol, 50-53 & Res Studies Inst, Air Univ, 54-56, dir terrestrial sci lab, Air Force Cambridge Res Labs, 57-61, dep technol & subsysts, Foreign Technol Div, Air Force Systs Command, 62-66; exec secy, Polar Res Bd, Nat Acad Sci, 67-81; CONSULT, 81- *Concurrent Pos:* Mem, Southeast Asia Geog Exped, 55-56; panels earth physics, oceanog & polar res, US Dept Defense, 57-61, arctic adv comt, 70; proj scientist, Fletcher's Ice Island Res Prog, Int Geophys Yr, 57-58. *Mem:* Fel Geol Soc Am; fel Arctic Inst NAm; Int Glaciol Soc; AAAS; Am Geophys Union. *Res:* Site selection and construction in permafrost; polar geomorphology; applied glaciology; polar logistics; technical writing; climate dynamics; polar mineral and living resource development. *Mailing Add:* 17215 NE Eighth St Bellevue WA 98008

DEGOWIN, RICHARD LOUIS, b Iowa City, Iowa, May 14, 34; m 57; c 2. INTERNAL MEDICINE, HEMATOLOGY. *Educ:* Univ Chicago, MD, 59. *Prof Exp:* Nat Heart Inst fel, 62-63; from res asst to res assoc internal med, Univ Chicago, 63-65, asst prof med, 65-68; assoc prof med & radiation res, 68-73, MEM STAFF, RADIATION RES LAB, UNIV IOWA, 68-, PROF MED & RADIOBIOLOGY, COL MED, 73-, DIR, CANCER CTR, 78- *Concurrent Pos:* Proj supvr hemat, Argonne Cancer Res Hosp, 65-68; USPHS career develop, 68-69; consult physician, Vet Admin Hosp, 69- *Mem:* Am Soc Clin Invest; fel Am Col Physicians; Am Soc Hemat; Radiation Res Soc; Int Soc Exp Hemat; Am Assoc Cancer Res. *Res:* Hematology and radiobiology; hemopoietic stem cell kinetics; erythropoietin and erythropoiesis; postirradiation recovery of hemopoiesis; bone marrow stromal cells; drug-induced hemolysis; malaria. *Mailing Add:* Dept Med Univ Iowa Hosp Iowa City IA 52242

DE GRAAF, ADRIAAN M, b Naaldwijk, Neth, Aug 4, 35; m 64; c 2. SOLID STATE PHYSICS. *Educ:* Swiss Fed Inst Technol, PhD(physics), 62. *Prof Exp:* Vis prof, Univ Sao Paulo, 62-64; vis prof, Centro Brasileiro de Pesquisas Fisicas, Rio de Janeiro, 64-65; sr res scientist, Sci Lab, Ford Motor Co, 65-68; assoc prof physics, Wayne State Univ, 68-74, prof physics, 74-, prof astron, 80-, mem Res Inst Eng Sci, 80-; AT DIV MATERIAL RES, NSF, WASHINGTON, DC. *Concurrent Pos:* Ford Found grant, 64-65; NSF grant, 72- *Mem:* Fel Am Phys Soc. *Res:* Theoretical solid state physics; properties of noncrystalline materials. *Mailing Add:* Div Mat Res NSF Washington DC 20550

DEGRAAF, DONALD EARL, b Grand Rapids, Mich, June 17, 26; m 48; c 3. PHYSICS, PHYSICS EDUCATION. *Educ:* Univ Mich, PhD(physics), 57. *Prof Exp:* from instr to prof, 56-90, EMER PROF PHYSICS, UNIV MICH, FLINT, 90- *Concurrent Pos:* Pres, Crystal Press, 90. *Mem:* Am Sci Affiliation; Am Asn Physics Teachers. *Res:* Physics education and curriculum design; helical physics. *Mailing Add:* Dept Physics Univ Mich-Flint Flint MI 48502-2186

DEGRADO, WILLIAM F, b Exeter, Calif, Sept 12, 55. BIORGANIC CHEMISTRY, PEPTIDE CHEMISTRY. *Educ:* Kalamazoo Col, BA, 78; Univ Chicago, PhD(org chem), 81. *Prof Exp:* RESEARCH LEADER, CENT RES & DEVELOP DEPT, E I DU PONT DE NEMOURS & CO, 81- *Honors & Awards:* Protein Soc Young Investr Award; Du Vigneaud Award. *Mem:* Am Chem Soc; Biophys Soc; Protein Soc. *Res:* Peptide and protein structure and design; peptide synthesis; biorganic chemistry. *Mailing Add:* Exp Sta PO Box 80328 E I du Pont de Nemours & Co Wilmington DE 19880-0328

DEGRAFF, ARTHUR C, JR, b New York, NY, June 18, 29; m 56; c 2. MEDICAL. *Educ:* Wesleyan Univ, BA, 51; NY Univ, MD, 55. *Prof Exp:* Intern, Bellevue IV Med Serv, NY, 55-56, residency, 56-57; residency, Bronx Vet Admin Hosp, NY, 57-58; fel, Dept Med, Tulane Univ, New Orleans, 58-59 & Dept Med, Cardio-Pulmonary Lab, Southwestern Med Sch, Univ Tex, 61-62; instr med, Southwestern Med Sch, Univ Tex, Dallas, 61-63, asst prof, 63-65; active staff, Hartford Hosp, Conn, 65-74; from asst clin prof to assoc clin prof, Sch Med, Univ Conn Health Ctr, Farmington, 76-89; PULMONARY RES SPECIALIST, DEPT MED, HARTFORD HOSP, SCH MED, UNIV CONN, 69-; CLIN PROF MED, SCH MED, UNIV CONN HEALTH CTR, FARMINGTON, 90- *Concurrent Pos:* Consult, Newington Vet Admin Hosp, Conn, 61-65, Newington Children's Hosp, Conn, 65-80, Windham Mem Hosp, Wilimantic, Conn, 71-, Rockville Gen Hosp, Rockville, Conn, 76-84 & Charlotte Hungerford Hosp, Torrington, Conn, 83-; clin asst prof med, Yale Univ, New Haven, Conn, 65-; med dir, Pulmonary Lab, Hartford Hosp, Conn, 65-; pvt consult pract pulmonary dis, Hartford, Conn, 67- *Mem:* Sigma Xi; Am Phsyiol Soc; fel Am Col Physicians; Am Thoracic Soc; Am Heart Asn; Am Fedn Clin Res; fel Am Col Chest Physicians. *Res:* Asthma; several publications and abstracts. *Mailing Add:* Dept Med Hartford Hosp Sch Med Univ Conn 85 Seymour St Hartford CT 06106

DEGRAFF, BENJAMIN ANTHONY, b Columbus, Ohio, Dec 23, 38; m 60; c 2. CHEMICAL KINETICS. *Educ:* Ohio Wesleyan Univ, BA, 60; Ohio State Univ, MS, 63, PhD(phys chem), 65. *Prof Exp:* Fel, Harvard Univ, 65-67; asst prof phys chem, Univ Va, 67-72; assoc prof, 72-76, head dept, 76-80, PROF CHEM, JAMES MADISON UNIV, 76- *Mem:* Am Chem Soc; Am Phys Soc. *Res:* Kinetics of fast reactions; energy transfer processes; photochemistry. *Mailing Add:* Dept Chem James Madison Univ Harrisonburg VA 22807

DEGRAND, THOMAS ALAN, b Knoxville, Tenn, June 16, 50. LATTICE GAUGE THEORY, COMPUTATIONAL PHYSICS. *Educ:* Univ Tenn, BS, 72; Mass Inst Technol, PhD(physics), 76. *Prof Exp:* Res assoc physics, Stanford Linear Accelerator Ctr, 76-78; res assoc, Univ Calif, Santa Barbara, 78-79, asst prof-in-residence, 79-81; asst prof, 81-87, ASSOC PROF PHYSICS, UNIV COLO, 87- *Concurrent Pos:* Sci collabr, Los Alamos Nat Lab, 81- *Mem:* Am Phys Soc. *Res:* Elementary particle physics; theory and phenomenology of quark confinement. *Mailing Add:* Dept Physics Univ Colo Boulder CO 80309

DE GRANDPRE, JEAN LOUIS, b Montreal, Que, May 25, 29; c 3. SYSTEMS ANALYSIS, SIMULATION. *Educ:* Univ Montreal, BA, 48, BSc, 52, MSc, 54. *Prof Exp:* Engr, Spec Weapons Dept, Canadair Ltd, 54-55, coordr systs, 55-57, engr, Sparrow II 6-D Digital Simulation, 57-59, engr, Flight Simulation, 59-60; systs coordr & chief programmer, Sperry Gyroscope Can Ltd, 60-61; comp coordr, Adv Prog Develop, Reentry Syst Anal, Missiles & Space Systs Div, Douglas Aircraft, 61-65, sect chief, 63-64, br mgr methodology, 64-65; MEM TECH STAFF, GEN RES CORP, 65-, PRIN SCIENTIST, 88- *Concurrent Pos:* Quebec Prov Govt Grant, 50-54; Societe St Jean Baptiste grant, 52-54. *Mem:* Inst Elec & Electronics Engrs; sr mem Am Inst Aeronaut & Astronaut; Asn Comput Mach; Can Asn Physicists; assoc mem Opers Res Soc Am; Soc Computer Simulation. *Res:* US and Soviet defense systems evaluation; radars and interceptor evaluations; tracking and filtering; theoretical discrimination; counter-insurgency analysis; simulation of radars, interceptors, tank breakup, defense systems, maneuvering reentry vehicles; non-nuclear kill defense systems; real-time missile simulation design and implementation; nuclear directed energy weapons; comparison of nuclear codes for survivability analysis; development of SDI level communications model embedded in large BM-C3 simulation. *Mailing Add:* Gen Res Corp 635 Discovery Dr Huntsville AL 35806

DEGRASSE, ROBERT W(OODMAN), b Yakima, Wash, July 4, 29; m 52; c 4. ELECTRONICS, DATA PROCESSING. *Educ:* Calif Inst Technol, BS, 51; Stanford Univ, MS, 54, PhD(elec eng), 58. *Prof Exp:* Res engr guided missile develop, Jet Propulsion Lab, Calif Inst Technol, 51-53; asst electronics res lab, Stanford Univ, 53-55, res assoc microwave tube res, 55-57; mem tech staff, Bell Tel Labs, 57-60; dir res & develop, Microwave Electronics Corp, 60-64; vpres, Quantum Sci Corp, 64-69; vpres & tech dir, Quantor Corp, 69-78; dir res & develop, NCR Micrographics Div, 78-82; sr vpres, 82-87, DE GRASSE ASSOC, QUANTUM GROUP INT, 88- *Mem:* Inst Elec & Electronics Engrs; Sigma Xi. *Res:* Microfilm and electronic data processing information systems; input-output theory data base for technology forecasting; satellite communications microwave lasers, memories, solid state devices and tubes. *Mailing Add:* 26755 Robleda Ct Los Altos Hills CA 94022

DEGRAW, JOSEPH IRVING, JR, b Washington, DC, May 26, 33; m 57; c 2. MEDICINAL CHEMISTRY. *Educ:* Univ Calif, Berkeley, BS, 56; Stanford Univ, PhD(org chem), 61. *Prof Exp:* Org chemist, Merck & Co, 56-57; org chemist, Cancer Chemother & Med Chem, 57-74, PROG MGR, MED CHEM, STANFORD RES INST, 74- *Mem:* Am Chem Soc; NY Acad Sci; Int Soc Heterocyclic Chem. *Res:* Folic acid antagonists; components of white snakeroot, indole alkylating agents; indole compounds; tryptamines, pyrimidines, pteridines, analgesics, histamine releasers and piperidines; synthesis of alkaloids; antileprotic drugs. *Mailing Add:* 880 Hanover Ave Sunnyvale CA 94087

DEGRAW, WILLIAM ALLEN, b Washington, DC, Apr 26, 39; m 61; c 3. ZOOPHYSIOLOGY, ENDOCRINOLOGY. *Educ:* Allegheny Col, BS, 61; Colo State Univ, MS, 65; Wash State Univ, PhD(zoophysiol), 72. *Prof Exp:* Jr biologist pharmacol, Wallace & Tiernan, Inc, 61-62; from asst prof to assoc prof, 69-80, PROF BIOL, UNIV NEBR, OMAHA, 80- *Mem:* Am Ornithologist's Union; Cooper Ornith Soc; Inland Bird Banders Asn. *Res:* Avian physiology and endocrinology: seasonal adaptations in lipid metabolism, plasma volume changes during molting, seasonal fluctuations in responsiveness to glucagon. *Mailing Add:* Dept Biol Univ Nebr Omaha NE 68182-0040

DE GRAY, RONALD WILLOUGHBY, b Hartford, Conn, Feb 10, 38. MATHEMATICS. *Educ:* Univ Conn, BA, 60, MA, 62; Syracuse Univ, PhD(math), 69. *Prof Exp:* Asst prof math, Utica Col, 69-75; asst prof math, Denison Univ, 75-79; ASSOC PROF MATH & DIR COMPUT CTR, ST JOSEPH COL, 82- *Concurrent Pos:* Vis asst prof math, Saint Joseph Col, 81. *Mem:* AAAS; Am Math Soc; Math Asn Am; Sigma Xi; Asn Comput Mach. *Res:* Probability. *Mailing Add:* St Joseph Col 1678 Asylum Ave West Hartford CT 06117-2791

DE GROAT, WILLIAM C, b Trenton, NJ, May 18, 38; m 59; c 3. AUTONOMIC PHYSIOLOGY, UROLOGY. *Educ:* Philadelphia Col Pharm & Sci, BSc, 60, MSc, 62; Univ Pa, PhD, 65. *Prof Exp:* NSF Fel, Australian Nat Univ, Canberra, 66-67; vis res fel, John Curtin Sch Med Res, Canberra, 67-68; from asst prof to assoc prof pharmacol, Med Sch, Univ Pittsburgh, 68-77, actg chmn, Dept Pharm, 78-80, adj prof pharm, 78-88, prof psychol, 82-86, PROF PHARMACOL, MED SCH, UNIV PITTSBURGH, 77-, PROF, DEPT BEHAV NEUROSCI, 86- *Concurrent Pos:* Pharmacol fel, Riker Pharm Co, 66-67; res career develop award, NIH, 72-77; mem neurobiol study sect, NIH, 83-88; mem, Ctr Neurosci, Univ Pittsburgh, 84-; assoc ed, J Autonomic Nervous Syst, 85-90; mem exec coun, Soc Neurosci, 88-92; vis scientist, Nat Inst Alchol Abuse & Alcoholism, NIH, 89-90. *Mem:* AAAS; NY Acad Sci; Am Soc Pharmacol & Exp Therapeut; Soc Neurosci; Int Brain Res Orgn; Am Gastroenterol Asn; Urodynamics Soc; Int Med Soc Paraplegia; Sigma Xi. *Res:* Analysis of visceral reflex pathways and visceral pain mechanisms; plasticity in reflex pathways during postnatal development and following neural injury. *Mailing Add:* Dept Pharmacol Univ Pittsburgh Med Sch W 1352 Biomedical Science Tower Pittsburgh PA 15261

DEGROAT, WILLIAM CHESNEY, JR, b Trenton, NJ, May 18, 38; m 59; c 3. PHARMACOLOGY. *Educ:* Philadelphia Col Pharm, BSc, 60, MSc, 62; Univ Pa, PhD(pharmacol), 65. *Prof Exp:* Lab instr, Philadelphia Col Pharm, 60-62; USPHS fel pharmacol, Sch Med, Univ Pa, 65-66; hon fel physiol, John Curtin Sch Med Res, Australian Nat Univ, 66-68; from asst prof to assoc prof, 68-77, actg chmn dept, 78-80. PROF PHARMACOL & BEHAV NEUROSCIENCE, SCH MED, UNIV PITTSBURGH, 77- *Concurrent Pos:* Chemist, Vet Admin Hosp, Philadelphia, 61-62; NSF fel, 66; Riker Int fel pharmacol, 66-67; mem, pharmacol test comt, Nat Bd Med Examrs, 78-83 & NIH Neurobiol Study Sect, 83-88; assoc ed, J Autonomic Nerv Syst, 85-; vis scientist, NIH, 88-89; exec coun, Soc Neuroscience; distinguished res award, Univ Pittsburgh, 90. *Mem:* AAAS; Am Soc Pharmacol & Exp Therapeut; NY Acad Sci; Soc Neurosci; Urodynamics Soc; Am Gastroenterol Asn; Int Brain Res Orgn; Sigma Xi; Int Med Soc Paraplegia. *Res:* Central autonomic mechanisms; nervous control of the urinary bladder and pharmacology of transmission in peripheral autonomic ganglia. *Mailing Add:* Dept Pharmacol Sch Med Biomed Sci Tower Univ Pittsburgh W1352 Pittsburgh PA 15261

DEGROOT, DOUG, b Austin, Tex, Mar 13, 51; c 1. ARTIFICIAL INTELLIGENCE, PARALLEL PROCESSING. *Educ:* Univ Tex, BS, 76, PhD(comput sci), 81. *Prof Exp:* Mem tech staff, Comput Automation, 78-81; mgr, Parallel Processors Group, IBM, 81-85; vpres, res & develop, Quintus Comput Systs, Palo Alto, Calif, 85-87; BR MGR, TEX INSTRUMENTS, 87- *Concurrent Pos:* Dir, comput sci prog, NY Univ, 83-85. *Res:* Artificial intelligence; parallel processing; logic programming; distributed processing; distributed knowledge; computer architecture and processor design. *Mailing Add:* Texas Instruments Comp Sci Ctr PO Box 655474, Ms 238 Dallas TX 75265

DEGROOT, LESLIE JACOB, b Ft Edward, NY, Sept 20, 28; c 5. ENDOCRINOLOGY. *Educ:* Union Col, NY, BS, 48; Columbia Univ, MD, 52. *Prof Exp:* Intern & asst res med, Presby Hosp, New York, 52-54; pub health physician, US Oper Mission, Afghanistan, Nat Cancer Inst, 55-56; resident, Mass Gen Hosp, 57-58; asst, Harvard Med Sch, 58-59, instr, 59-62, assoc, 62-66; assoc prof exp med, Mass Inst Technol & assoc dir dept nutrit & food sci, Clin Res Ctr, 66-68; PROF ENDOCRINOL & CHIEF SECT, THYROID STUDY UNIT, PRITZKER SCH MED, UNIV CHICAGO, 68- *Concurrent Pos:* Clin fel, Nat Cancer Inst, 54-55; clin & res fel med, Mass Gen Hosp, 56 & 58-60, asst, 60-64, asst physician, 64-66. *Mem:* Asn Am Physicians; Am Thyroid Asn; Endocrine Soc; Am Soc Clin Invest; Am Fedn Clin Res. *Res:* Biochemistry and physiology of thyroid, thyroid hormone receptors; auto-immune thyroid disease. *Mailing Add:* Thyroid Study Unit Dept Med & Radiol Univ Chicago Box 138 5841 S Maryland Ave Chicago IL 60637

DEGROOT, MORRIS H, b Scranton, Pa, June 8, 31; m 79; c 2. STATISTICS. *Educ:* Roosevelt Univ, BS, 52; Univ Chicago, MS, 54, PhD(statist), 58. *Prof Exp:* From asst prof to assoc prof math & indust admin, 63-66; head dept statist, 66-72, prof statist, 66-84, UNIV PROF STATIST & INDUST ADMIN, CARNEGIE-MELLON UNIV, 84- *Concurrent Pos:* Prof, Inst Advan Studies in Mgt, Belg, 71; assoc ed, J Am Statist Asn, 70-74, book rev ed, 71-75, theory & methods ed, 76-78, assoc ed, Ann Statist, 74-75, exec ed, Statist Sci, 85-; Nat Res Coun, 73-78 & 80-86, mem, Comm Behav, Soc Sci & Educ, Comt Nat statist, 75-79, & assoc chmn, 78-79 & Comt Environ Monitoring, 75; mem, Coun, Inst Math Statist, 75-78 & 81-85 & Exec Comt, 85-; NSF, mem, Adv Comt, Div Math Sci,; mem, Nat Res Coun, Numerical Data adv bd, 84-, comt Appl & Theoret Statist, 87-, bd math Sci, 88-, expert task group, Strategic Hwy Res prog, 88. *Mem:* Fel Am Inst Math Statist; fel AAAS; Int Statist Inst; fel Am Statist Asn; fel Royal Statist Soc; fel Economet Soc. *Res:* Statistics and probability. *Mailing Add:* Carnegie-Mellon Univ 5000 Forbes Ave Carnegie-Mellon Univ Pittsburgh PA 15213

DEGROOT, RODNEY CHARLES, b Racine, Wis, Dec 24, 34; m 54; c 3. PLANT PATHOLOGY, MYCOLOGY. *Educ:* Univ Wis, BS, 58, MS, 60; State Univ NY Col Forestry, PhD(forest path), 63. *Prof Exp:* Instr bot, Syracuse, 62-63; sr scientist bot, NY State Mus & Sci Serv, 63-68; prin pathologist, Forest & Wood Prod Dis Lab, 68-77, plant pathologist, Prod Lab, 77-78, proj leader, 79-88, PATHOLOGIST, WOOD PROTECTION RES, US FOREST SERV, 88- *Mem:* Forest Prod Res Soc; Am Wood Prosesses Assoc. *Res:* Forest pathology; wood decay; wood protection. *Mailing Add:* 1408 Arrowood Dr PO Box 5130 Madison WI 53704

DE GROOT, SYBIL GRAMLICH, b Evanston, Ill, Feb 28, 28; div; c 3. HUMAN FACTORS ENGINEERING, INDUSTRIAL ENGINEERING. *Educ:* Ohio State Univ, BA, 47, MA, 49, PhD(eng psychol), 68. *Prof Exp:* Res asst human performance, Inst Coop Res, Johns Hopkins Univ, 50-51; psychophysicist of vision submariners' vision, US Naval Med Res Lab, New London, Conn, 51-52; staff psychologist exp psychol & syst design, Dunlap & Assocs, Inc, Darien, Conn, 58-62; res assoc children's vision, Bur Educ Res & Serv, Ohio State Univ, 62-66; res assoc human performance, Div Info & Comput Sci, Ohio State Univ, 67-68; asst prof human factors psychol, Mont State Univ, 68-72; assoc prof psychol, 72-74, from assoc prof to prof indust systs,74-87, PROF & DIR RES SCI, OFF INSTNL RES, FLA INT UNIV, 87- *Concurrent Pos:* Proj dir, HEW grant, 64-66; NSF grant, 67-68; consult, Mont State Off Water Resources, 71-72; res dir, Univ grant to Sch Hotel, SFla Hotel Owner's Asn, 73-74; prin investr, Air Force Off Sci Res grant, 77-78; consult, Pratt & Whitney, GPD Div, West Palm Beach, Fla, 78- 80; consult, Fla State Atty Off, 80-83; NASA-Am Soc Eng Educ summer design fel, Langley Air Force Base, Va, 75, Wright-Patterson Air Force Base, Ohio, 76, & Navy-Am Soc Eng Educ sr summer res fel, Naval Training Sch, Fla, 86. *Mem:* Human Factors Soc; Int Ergonomics Asn; Am Soc Eng Educr; Am Psychol Asn; Soc Women Engrs. *Res:* Performance evaluation; training and system design; professional and managerial women in industry; human perception; sensor displays and controls; human factors engineering; strength-endurance standards; flight simulators. *Mailing Add:* Naval Submarine Med Res Lab PO Box 900 Groton CT 06349-5900

DE HAAN, FRANK P, b Paterson, NJ, Nov 1, 34; m 58; c 4. PHYSICAL CHEMISTRY, INORGANIC CHEMISTRY. *Educ:* Calvin Col, AB, 57; Purdue Univ, PhD(chem), 61. *Prof Exp:* From asst prof to assoc prof, 61-74, chmn dept, 70-74, PROF CHEM, OCCIDENTAL COL, 74- *Concurrent Pos:* NSF sci fac fel, 68-69. *Mem:* Am Chem Soc; AAAS; Sigma Xi. *Res:* Mechanisms of Friedel-Crafts and related reactions. *Mailing Add:* 10937 Goss St Sun Valley CA 91352

DE HAAN, HENRY J(OHN), b St Clair Co, Ill, Nov 23, 20; m 43. SPEECH INTELLIGIBILITY, BEHAVIORAL NEUROSCIENCE. *Educ:* Wash Univ, AB, 42, AM, 49; Univ Pittsburgh, PhD(psychol), 60. *Prof Exp:* Postdoctoral, Vet Admin Hosp, Coatesville, Pa, 60-62; res scientist, HUMRRO, George Washington Univ, Washington, DC, 62-64; res psychologist, Armed Forces Radiobiol Res Inst, Bethesda, Md, 65-69 & US Army Res Inst, Alexandria, VA, 69-86; RETIRED. *Concurrent Pos:* Fac mem, Grad Sch, USDA, Washington, DC, 67-77. *Mem:* Sigma Xi; Soc Neurosci; Am Psychol Am; Psychonomic Soc; Int Neuropsychol Soc; Am Psychol Soc; Int Primatological Soc. *Res:* Behavioral neuroscience; biological factors in perception, cognition and behavior; intelligibility and comprehension of speech, both human and machine speech; speech science and technology. *Mailing Add:* 5403 Yorkshire St Springfield VA 22151

DEHAAN, ROBERT LAWRENCE, b Chicago, Ill, Nov 18, 30; m 57; c 2. BIOLOGY, EMBRYOLOGY. *Educ:* Univ Calif, Los Angeles, BA, 52, MA, 54, PhD(zool), 56. *Prof Exp:* Jr res physiologist, Univ Calif, Los Angeles, 54-56; res embryologist, Carnegie Inst Wash, 56-73; prof anat & physiol, 73-76, WILLIAM P TIMMIE PROF EMBRYOL, DEPT ANAT & CELL BIOL, EMORY UNIV, 76-, DIR GRAD DIV BIOL & BIOMED SCI, 89-, DIR CTR ETHICS IN PUB POLICY & PROF, 90- *Concurrent Pos:* Instr, Woods Hole Marine Biol Lab, 62-64 & 65; from assoc prof to prof biol, Johns Hopkins Univ, 64-73; mem biol study sect, NIH, 70-74; extramural res assoc, Carnegie Inst Wash, 73-; southern regional res rev & adv subcomt, Am Heart Asn, 75-79; foreign investr, Deleg Gen Res Sci & Tech, Univ Paris-Sud, Orsay, France, 80-81; consult, US Senate Comt Human Resources, 79; Josiah Macy Found fac scholar, 80-81; dir & prin investr, Prog Cardiac Cell Function, 81- *Mem:* Am Soc Cell Biol; Biophys Soc; Soc Develop Biol; Tissue Cult Asn; Int Soc Develop Biol (int secy, 72-81); fel AAAS. *Res:* Physiological development of the heart; cell migration and morphogenetic movements; biophysics of excitable cells; differentiation and physiology of cardiac pacemakers; heart cell culture. *Mailing Add:* Dept Anat & Cell Biol Emory Univ Atlanta GA 30322

DE HAAS, HERMAN, b Northbridge, Mass, Jan 6, 24; m 51; c 4. BIOLOGICAL CHEMISTRY, NUTRITION. *Educ:* Westminster Col, Pa, BS, 47; Univ Mich, MS, 50, PhD(biol chem), 55. *Prof Exp:* Asst, Univ Mich, 47-52, res assoc biochem, 54-55; from asst prof to assoc prof chem, Westminster Col, 55-59; from asst prof to prof, 59-90, EMER PROF BIOCHEM, UNIV MAINE, 90- *Mem:* Am Chem Soc; Am Sci Affiliation; Sigma Xi. *Res:* Protein nutrition; amino acids of marine gastropods. *Mailing Add:* 17 Spencer St Orono ME 04473

DE HAËN, CHRISTOPH, b St Gallen, Switz, Sept 6, 40; m 67; c 3. ENDOCRINOLOGY, RADIOLOGY. *Educ:* Swiss Fed Inst Technol, Zurich, MS, 66, PhD(molecular biol), 69. *Prof Exp:* Sr fel, Univ Wash, 70-73, res assoc, 73-74, res asst prof biochem & med, 74-79, res assoc prof, 79-87, res prof, 87-88; head, Biochem Dept, 88-89, DIR PRECLIN RES, BRACCO, MILAN, 89- *Concurrent Pos:* Res career develop award, NIH, 79. *Mem:* Am Soc Biol Chemists; AAAS; Am Fed Clin Res; Am Chem Soc; Sigma Xi; Soc Chim Italy. *Res:* Mechanisms of hormone action, with particular emphasis on insulin action and its relationship to diabetes mellitus; protein hydrodynamics; radiological contrast agents for x-ray, magnetic resonance imaging and echography. *Mailing Add:* Preclinical Res & Dev Bracco Industria Chimica SpA Via Egidio Folli 50 Milan I-20134 Italy

DEHART, ARNOLD O'DELL, b Davy, WVa, May 8, 26; m 51; c 3. BEARING TECHNOLOGY, TRIBOLOGY. *Educ:* Univ WVa, BSME, 50. *Prof Exp:* Sr res engr, Dept Mech Develop Res Labs, Gen Motors Co, 51-64, supvry res engr, 64-76, sr engr, 76-77, sr staff res engr, 77-86, prin res engr, 86-88; CONSULT, 88- *Honors & Awards:* Cliffard Steadman Award, Inst Mech Engrs, 84; Arch T Colwell Award, Soc Automotive Engrs, 84; Sea Ralph Teetor Award, 85. *Mem:* Soc Automotive Engrs. *Res:* General bearing, fluid film and rolling contact hydrodynamics; hydrostatics; tribology; system kinematics; mechanisms; sliding surface bearings; lubrication, high temperature tribology. *Mailing Add:* 7000 Trail Blvd Naples FL 33963

DEHART, ROBERT C(HARLES), b Laramie, Wyo, Aug 16, 17; m 41, 70; c 2. STRUCTURAL ENGINEERING. *Educ:* Univ Wyo, BS, 38; Ill Inst Technol, MS, 40, PhD(civil eng), 55. *Prof Exp:* Design engr, Standard Oil Co, Ind, 40-46; assoc prof civil eng, Mont State Univ, 46-53; struct analyst, Armed Forces Spec Weapons Proj, 53-58; mgr, struct mech, 58-59, dir struct res dept, 59-70, VPRES, SOUTHWEST RES INST, 70- *Concurrent Pos:* Lectr, George Washington Univ, 55-58. *Mem:* NY Acad Sci; Am Soc Mech Engrs; Am Soc Civil Engrs; Sigma Xi. *Res:* Structural dynamics; theoretical and applied mechanics; air, underwater and underground shock; underwater vehicles. *Mailing Add:* 403 La Jara Blvd San Antonio TX 78209

DE HARVEN, ETIENNE, b Brussels, Belg, Mar 5, 28; m 53; c 4. IMMUNO-ELECTRON MICROSCOPY. *Educ:* Univ Brussels, MD, 53. *Prof Exp:* Asst, Free Univ Brussels, 55-62 & in chg cytol, Electron Micros Lab, Anat Inst, 58-62; asst prof biol, 62-69, PROF BIOL, SLOAN-KETTERING DIV, CORNELL UNIV, 69-; PROF PATHOL, UNIV TORONTO, 81- *Concurrent Pos:* Fel, Inst Cancer, France, 55-56; Belg Am Educ Found & Damon Runyon res fels, 56-57; vis res fel, Sloan-Kettering Inst Cancer Res, 56-57, vis assoc, 59 & 61, assoc, 62-64, assoc mem, 64-68; assoc, Nat Fund Sci Res, Belg, 59-61; guest investr, Rockefeller Inst, 62; mem, Div Cytol, Sloan-Kettering Inst Cancer Res, 68-81. *Mem:* Electron Micros Soc Am; NY Acad Sci; Int Soc Cell Biol. *Res:* Electron microscope pathology; immuno-electron microscopy. *Mailing Add:* Dept Path Electro Micros Lab Univ Toronto Banting Inst 100 College St Toronto ON M5G 1L5

DEHAVEN-HUDKINS, DIANE LOUISE, b East Stroudsburg, Pa, Aug 3, 54; m. NEUROCHEMISTRY, NEUROPHARMACOLOGY. *Educ:* Millersville State Col, BA, 76; Syracuse Univ, PhD(biophyschol), 81. *Prof Exp:* Fel, Biol Sci Res Ctr, Univ NC, 81-85; sr res assoc, 85-86, staff scientist, Nova Pharmaceut Corp, 86-88; SR RES SCIENTIST, EASTMAN PHARMACEUT, 88- *Mem:* Soc Neurosci; Soc Toxicol; Am Soc Pharmacol & Exp Therepeut. *Res:* Effects of drugs on central nervous system chemistry and function; neurochemical and behavioral sequelae of toxicant exposure; neuropharmacology. *Mailing Add:* Dept Enzymol & Receptor Biochem Sterling Res Group 25 Great Valley Pkwy Great Valley PA 19355

DE HEER, JOSEPH, b Eindhoven, Neth, Jan 24, 22; nat US. PHYSICAL CHEMISTRY. *Educ:* Delft Inst Technol, Chem Eng, 47; Univ Amsterdam, PhD(physics, math), 50. *Prof Exp:* Res assoc physics & astron, Ohio State Univ, 50-52; from asst prof to assoc prof, 52-63, PROF CHEM, UNIV COLO, BOULDER, 63- *Concurrent Pos:* Guggenheim fel, Quantum Chem Group, Univ Uppsala, 59-60; Fulbright res scholar, Univ Copenhagen, 63-64. *Mem:* Fel Am Phys Soc. *Res:* Quantum mechanics of molecules; valence theory; thermodynamics. *Mailing Add:* Dept Chem Campus Box 215 Univ Colo Boulder CO 80309

DEHER, KEVIN L, b Springfield, NJ, Aug 10, 52. BIOCHEMISTRY. *Educ:* Villanova Univ, BS, 74; Pa State Univ, MA, 77, PhD(biochem), 80. *Prof Exp:* Postdoctoral immunol & genetics, NIH, 79-83; asst prof, Dept Biochem, Sch Med, WVa Univ, 83-87; staff scientist, Weis Ctr Res, Danville, Pa, 87-91; RES CHEMIST, PULMONARY TOXICOL BR, BIOCHEM SECT, ENVIRON PROTECTION AGENCY, 91- *Mem:* Am Soc Biochem & Molecular Biol; Am Asn Immunol; Am Soc Cell Biol. *Mailing Add:* Pulmonary Toxicol Br Environ Toxicol Div Environ Protection Agency Biochem Sect MD-82-HERL Research Triangle Park NC 27711

DE HERTOGH, AUGUST ALBERT, b Chicago, Ill, Aug 24, 35; m 57, 86; c 3. PLANT PHYSIOLOGY, HORTICULTURE. *Educ:* NC State Univ, BS, 57, MS, 61; Ore State Univ, PhD(plant physiol), 64. *Prof Exp:* Asst plant physiologist, Boyce Thompson Inst, 64-65; from asst prof to prof hort, Mich State Univ, 65-78; head dept, hort sci, 78-88, PROF HORT SCI, NC STATE UNIV, 78- *Concurrent Pos:* Fel Am Soc Hort Sci, 81; assoc ed, Hort Sci, 73-76. *Honors & Awards:* Alex Laurie Award, Am Soc Hort Sci, 76; Futura Award, Prof Plant Growers Asn, 89. *Mem:* Int Soc Hort Sci; fel Am Soc Hort Sci; Sigma Xi. *Res:* Influence of environmental and growth regulator factors on growth and development of bulbous and tuberous plants; geophytes; author of various publications. *Mailing Add:* Dept Hort Sci NC State Univ Col Agr Raleigh NC 27695-7609

DEHLINGER, PETER, b Berlin, Ger, Oct 3, 17; nat US; m 41; c 2. GEOPHYSICS, OCEANOGRAPHY. *Educ:* Univ Mich, BS, 40; Calif Inst Technol, MS, 43, PhD(geophysics), 50. *Prof Exp:* Seismologist, Shell Oil Co, Inc, 43-48; geophysicist, Battelle Mem Inst, 50-53; from assoc prof to prof geophys, Tex A&M Univ, 54-62; prof geophys oceanog, Ore State Univ, 62-68; dir Marine Sci Inst, 68-75, mem, 76-86, prof geophys, 68-87, EMER PROF GEOPHYS, UNIV CONN, 87- *Concurrent Pos:* Indust consult, 57-66; with US Geol Surv, 57-68; univ assoc, Univ Hawaii, 64-69; head geophys prog, Ocean Sci & Technol Div, Off Naval Res, Washington, DC, 66-68; adj prof, Am Univ, 66-67,. *Mem:* AAAS; fel Geol Soc Am; Am Geophys Union; Seismol Soc Am; Soc Explor Geophys; Planetary Soc. *Res:* Earthquake seismology; gravity and seismic measurement at sea; tectonics of oceanic crustal and mantle structures. *Mailing Add:* 4121 221st Ct SE Issaquah WA 98027

DEHM, HENRY CHRISTOPHER, b Newark, NJ, Apr 1, 21; m 48, 67; c 2. POLYMER CHEMISTRY. *Educ:* Univ Denver, BS, 48, MS, 49; Univ Wis, PhD(irg chem), 54. *Prof Exp:* Asst chemist, Univ Denver, 46-49; asst chemist, Univ Wis, 49-52, G D Searle res fel, 52-54; res chemist, Res Ctr, Hercules Powder Co, Hercules Inc, 54-61, tech specialist, 61-66, sr tech specialist, 66-85, sr scientist, Chem Propulsion Div, 85-86; PRES, H C DEHM ASSOCS, 86- *Concurrent Pos:* Lectr, Brigham Young Univ, 63-66; adj prof chem & chem eng, Univ Utah, 67- *Mem:* Am Chem Soc; Sigma Xi. *Res:* Synthesis and stereochemistry of steroids; organic synthesis; halogenation; catalysis; metal complexes; polymers; ultrahigh energy compounds; fundamental chemistry of solid propellants mechanical and ballistic properties; alternate energy systems; nitrene insertion; fundamentals of adhesion; composite structures. *Mailing Add:* 3003 E 4505 S Salt Lake City UT 84117

DEHM, RICHARD LAVERN, b Pontiac, Ill, Sept 11, 27; m 52; c 3. ELECTROPHOTOGRAPHY, QUALITY ASSURANCE. *Educ:* Ill Wesleyan Univ, BS, 50; Univ Ill, MS, 52, PhD(chem), 54; Univ Rochester, MBA, 69. *Prof Exp:* Asst, Univ Ill, 51-54; anal chemist, Eastman Kodak Co, 54-60, tech assoc, 60-65, asst dir indust lab, 65-70, asst supt photochem div, 70-72, dir proj develop div, 72-77, asst supt, Rollcoating Div, 79-83, dir qual assurance, Copy Prods Div, 83-89; RETIRED. *Mem:* Am Chem Soc. *Res:* Emission, x-ray and absorption spectroscopy; x-ray diffraction; concentration techniques; trace element analysis; microprobe; spark source mass spectroscopy; electrophotography. *Mailing Add:* 115 Drumlin View Dr Mendon NY 14506

DEHMELT, HANS GEORG, b Goerlitz, Ger, Sept 9, 22. ATOMIC PHYSICS. *Educ:* DrRerNat, Gottingen Univ, 50. *Hon Degrees:* Dr, Univ Ruprecht Karl, 86 & Univ Chicago, 87. *Prof Exp:* Res fel, Inst H Kopfermann, 50-52; res assoc, Duke Univ, 52-55; from vis asst prof to assoc prof, 55-61, PROF & EXP PHYSICIST, DEPT PHYSICS, UNIV WASH, 61- *Concurrent Pos:* Res fel, Univ Goettingen, 50-52; consult, Varian Assocs, 56-80. *Honors & Awards:* Nobel Prize in Physics, 89; Davisson-Germer Prize, Am Phys Soc, 70; Humboldt Prize, Alexander von Humboldt Stiftung, 74; Morris Loeb Lectr, Harvard Univ, 77; Int Soc Magnetic Resonance Award, 80; Count Rumford Prize, AAAS, 85. *Mem:* Nat Acad Sci; Am Acad Arts & Sci; fel AAAS; Am Phys Soc; Am Optical Soc; Sigma Xi. *Res:* Nuclear quadrupole resonance; nuclear and electron paramagnetic resonance; optical detection of free atom orientation; spin exchange resonance; spectroscopy of stored ions; free electron/positron magnetic moment from geonium spectroscopy. *Mailing Add:* Dept Physics FM 15 Univ Wash Seattle WA 98105

DEHMER, JOSEPH LEONARD, b St Charles, Mo, Jan 21, 45; m 70. CHEMICAL PHYSICS, ATOMIC & MOLECULAR PHYSICS. *Educ:* Wash Univ, BS, 67; Univ Chicago, PhD(chem physics), 71. *Prof Exp:* Res assoc, 71-72, from asst physicist to physicist, 73-80, SR PHYSICIST, ARGONNE NAT LAB, 80- *Concurrent Pos:* Consult, Nat Bur Standards, 80-86; mem, Comt Atomic & Molecular Sci, Nat Res Coun, 83-86 & Comt Line Spectra Elements & Atomic Spectros, 87-90; mem nominating comt, Div Atomic, Molecular & Optical Physics, Am Phys Soc, 84 & 87, publ comt, 89-91; chmn, Gordon Res Conf Electron Spectros, 86; mem, Int Adv Bd & Prog Comt, Fourth Int Conf Electron Spectros, 88-89. *Mem:* fel Am Phys Soc; Am Chem Soc; Radiation Res Soc. *Res:* Multiphoton processes in atoms and molecules; molecular photoionization dynamics; laser photophysics and photochemistry; effects of radiation interaction with matter; measurement science. *Mailing Add:* Bldg 203 Argonne Nat Lab Argonne IL 60439

DEHMER, PATRICIA MOORE, b Chicago, Ill, Sept 18, 45; m 70. ATOMIC & MOLECULAR PHYSICS, CHEMICAL PHYSICS. *Educ:* Univ Ill, BS, 67; Univ Chicago, PhD(chem physics), 72. *Prof Exp:* Res assoc, 72-75, asst chemist, 75-78, chemist, 78-85, SR CHEMIST, ARGONNE NAT LAB, 85- *Concurrent Pos:* Publ comt, Am Phys Soc, 80-82, Comt Opportunities in Physics, 84-86, Comt Const & Bylaws, 86-88, 90, Counr at large, 87-90, Audit Comt, 87-88, Comt Status of Women in Physics, 87-89, Exec Comt, 89-90, Budget Comt, 89-90, Comt on Comts, 89-90, Arthur L Schawlow Prize Comt, 91, Maria Goeppert-Mayer Award Comt, 91-; chmn, Gordon Res Conf Multiphoton Processes, 86, Coun Gordon Res Conf, 87-89, Nat Sci Found Adv Comt Physics, 88-91; secy-treas, Div Atomic, Molecular & Optical Physics, Amn Phy Soc, 81-84, exec comt, 79, 81-84, 86-89, vchmn, 90, chmn, 91. *Mem:* Am Chem Soc; Am Phys Soc; Opt Soc Am. *Res:* Experimental atomic and molecular physics; electronic structure of molecules, free radicals and clusters; photoionization dynamics; multiphoton processes; Rydberg state and excited state reactions; clustering and nucleation. *Mailing Add:* Argonne Nat Lab Bldg 203 Argonne IL 60439

DEHN, JAMES THEODORE, b New York, NY, Oct 24, 30; m 68; c 3. CHEMICAL PHYSICS, PHYSICAL MATHEMATICS. *Educ:* Fordham Univ, AB, 55; Georgetown Univ, MS, 59, PhD(physics), 61. *Prof Exp:* NASA physics grant, 62-64; instr physics, St Peter's Col, 65-66; res assoc, Pa State Univ, 66-68; PHYSICIST, BALLISTICS RES LAB, 68- *Mem:* Am Phys Soc; Sigma Xi; Combustion Inst. *Res:* Mathematical and chemical physics involved in terminal ballistics; vulnerability and survivability. *Mailing Add:* 1224 Grafton Shop Rd Bel Air MD 21014

DEHN, JOSEPH WILLIAM, JR, b Brooklyn, NY, Feb 18, 28; m 53; c 2. ORGANIC CHEMISTRY. *Educ:* Columbia Col, BA, 49; Stevens Inst Technol, MS, 53; Polytech Inst Brooklyn, PhD(org chem), 64. *Prof Exp:* Sr org chemist, Cent Res Labs, Interchem Corp, 49-64; group leader org chem, Wallace & Tiernan, Inc, 65-67; sr scientist, Cent Res Labs, Shulton Inc, 68-70; sr org chemist, Process Chem Div, Diamond-Shamrock Chem Co, Morristown, NJ, 71-87; sr chemist, Atlantic Industs Div, Jepson Corp, Nutley, NJ, 88-90; SR CHEMIST, PALL CORP, GLEN COVE, NY, 90- *Mem:* AAAS; Am Chem Soc; Am Inst Chemists; NY Acad Sci; Am Asn Textile chemists & Colorists; Sigma Xi. *Res:* Synthesis of dyes, pigments, organic intermediates and heterocyclic and organometallic compounds; polymers; surfactants; leather tanning compounds; textile chemicals; filters. *Mailing Add:* 52 Berkshire Rd Great Neck NY 11023-1416

DEHN, RUDOLPH A(LBERT), b East Rutherford, NJ, Aug 12, 19; m 45; c 3. ELECTRICAL ENGINEERING. *Educ:* Newark Col Eng, BSEE, 41. *Prof Exp:* Develop engr, Gen Elec Co, 41-45, res assoc, 45-60, consult engr, 60-66, mgr tube res, microwave tube bus sect, 66-72, mem eng staff, res & develop, 72-82; RETIRED. *Mem:* Inst Elec & Electronics Engrs. *Res:* New methods of generating microwaves; design of attendant transmission networks; application of microwave energy for materials processing. *Mailing Add:* 1054 Hickory Rd Schenectady NY 12309

DEHNÉ, EDWARD JAMES, b Ft Clark, NDak, June 28, 11; m 32; c 3. OCCUPATIONAL MEDICINE, PUBLIC HEALTH. *Educ:* Univ NDak, BS, 35; Univ Ore, MD, 37; Johns Hopkins Univ, MPH, 41, DPH, 55; Am Bd Prev Med, dipl & cert pub health, 49, dipl & cert occup med, 55; Am Bd Indust Hyg, dipl & cert, 62. *Prof Exp:* Indust med pract, Idaho, 38-39; dir,

Coos County Health Dept, Ore, 39-41; asst post surgeon, US Army, Ft McDowell, Calif, 41-43, health officer, Mil Govt Pub Health, Eng, 43-44 & Civil Affairs Ctr, Europ Civil Affairs Div, Eng, 44, chief prev med, Europ Theater Opers, 44, chief surg, Forward Echelon, France, 44, exec officer, Western Europe, 44-45, chief prev med, Off Mil Govt, Ger, 45-47, dir off occup health, Edgewood Arsenal, Md, 48-50, dir off occup health, Hq, Third US Army, Ga, 50-51, dir off occup health, Caribbean, CZ, 51-55, comdr, US Army Environ Hyg Agency, 55-59, chief prev med, Second US Army, 59-60 & Brooke Army Med Ctr, 60-63, dir health & welfare, US Civil Admin, Ryukyu Islands, 63-65, dir health, Fifth US Army, 65-66; health officer, State of Nev, 66-68; med dir WVa regional med prog, Med Ctr, Univ WVa, 68-69; dir NJ tuberc serv, State of NJ Dept Health, 69-71; CHIEF MED CONSULT OFFICER, STATE OF NEV, 71- *Concurrent Pos:* Consult, Off Surgeon Gen, 57; mem fac Vietnam surv, Sch Pub Health, Univ Calif, Los Angeles, 68-69. *Mem:* Fel Am Col Physicians; fel Am Col Prev Med; fel Am Pub Health Asn; fel Am Occup Med Asn; Int Health Soc; fel World Asn Social Psychiat; fel Royal Soc Health. *Res:* Identifying, evaluating and minimizing environmental health hazards; industrial hygiene; rehabilitation; effects of environmental temperature upon susceptibility to toxic agents; international health and behavioral sciences. *Mailing Add:* Dept Human Resources 1050 E William St Carson City NV 89710

DEHNE, GEORGE CLARK, b Pittsburgh, Pa, Oct 2, 37; m 58; c 2. SPECTROSCOPY, ELECTROCHEMISTRY. *Educ:* Allegheny Col, BS, 59; Purdue Univ, MS, 61, PhD(anal chem), 63. *Prof Exp:* From asst prof to assoc prof, 63-74, chmn dept, 69-74, dir Computer Ctr, 80-86, PROF CHEM, CAPITAL UNIV, 74- *Concurrent Pos:* Col Pharm, Ohio State Univ, 72-77. *Mem:* AAAS; Am Chem Soc; Am Soc Testing & Mat; Sigma Xi. *Res:* Absorption spectroscopy; analytical instrumentation and spectrophotometry; heteropoly chemistry; computer applications in chemistry. *Mailing Add:* Dept Chem Capital Univ Columbus OH 43209

DEHNEL, PAUL AUGUSTUS, b Pomona, Calif, Dec 31, 22; m 43, 71; c 2. COMPARATIVE PHYSIOLOGY, INVERTEBRATE ZOOLOGY. *Educ:* San Diego State Col, BA, 43; Univ Calif, Berkeley, MA, 48; Univ Calif, Los Angeles, PhD, 54. *Prof Exp:* Instr zool, San Diego State Col, 48-50; assoc, Univ Calif, Los Angeles, 53-55; from asst prof to prof, 55-86, EMER PROF MARINE INVERT, DEPT ZOOL, UNIV BC, 86- *Res:* Electrolyte balance; water regulation; respiration marine invertebrates. *Mailing Add:* 830 Burley Dr West Vancouver BC V7T 1Z6 Can

DEHNER, EUGENE WILLIAM, b Burlington, Iowa, May 26, 14. ZOOLOGY. *Educ:* St Benedict's Col, BSc, 37; Cornell Univ, MSc, 43, PhD(zool), 46. *Prof Exp:* From instr to prof biol, St Benedict's Col, 46-71, chmn dept, 53-83, prof, 71-83, EMER PROF BIOL, BENEDICTINE COL, 84- *Concurrent Pos:* Res assoc, Clayton Found Biochem Inst, Univ Tex, 66-67. *Mem:* AAAS; Wilson Ornith Soc; Am Soc Zoologists; Sigma Xi. *Res:* Functional anatomy of birds, especially the adaptation of bodily dimensions to the habits of these animals; wing dimensions of two doves of different flight habits; respiratory volume and specific gravity of diving and surface feeding ducks. *Mailing Add:* Dept Biol Benedictine Col Atchison KS 66002-1499

DEHNHARD, DIETRICH, b Rengshausen, Ger, Apr 29, 34; m 58; c 3. NUCLEAR PHYSICS. *Educ:* Univ Marburg, dipl, 61, DrPhil(nuclear physics), 64. *Prof Exp:* Resident res assoc nuclear res, Argonne Nat Lab, 64-66; from asst prof to assoc prof, 66-77, PROF, SCH PHYSICS & ASTRON, UNIV MINN, MINNEAPOLIS, 77- *Concurrent Pos:* Visitor, Max Planck Inst Nuclear Physics, Heidelberg, Ger, 71-72, Swiss Inst Nuclear Res, Villigen, Switz, 81 & Tokyo Inst Technol, 85. *Mem:* Am Phys Soc. *Res:* Nuclear structure studies by use of intermediate energy pions and protons. *Mailing Add:* Sch Physics & Astron Univ Minn 116 Church St SE Minneapolis MN 55455

DE HOFF, GEORGE R(OLAND), b Baltimore, Md, Oct 16, 23; m 48; c 3. CHEMICAL & PLASTICS ENGINEERING. *Educ:* Johns Hopkins Univ, BE, 48, MS, 50. *Prof Exp:* Chem engr, Nat Plastics Prod Co, 50-53; res engr, Plastics Dept, E I du Pont de Nemours & Co, Wilmington, Del, 53-62, tech rep, 62-67, consult, 67-72; sr engr, Plastics & Resins Dept, Mfg Div, Wash Works, 72-79, staff engr, 79-85; CONSULT, GRD CONSULT, 85- *Concurrent Pos:* Plastics processing consult, GRD Consulting. *Mem:* Soc Plastics Engrs. *Res:* Behavior of various thermoplastics in processing equipment, including relating the rheological properties of a resin to its processing characteristics; thermoplastic polymer processing. *Mailing Add:* 107 Tanglewood Place Parkersburg WV 26101

DEHOFF, PAUL HENRY, JR, b York Twp, Pa, Mar 12, 34; m 56; c 1. SOLID MECHANICS. *Educ:* Pa State Univ, BS, 56, MS, 58; Purdue Univ, PhD(solid mech), 65. *Prof Exp:* Mech engr, York Div, Bendix Corp, 58-61; instr aeronaut & astronaut eng sci, Purdue Univ, 61-65; res engr, exp sta, E I du Pont de Nemours & Co, Del, 65-66; asst prof mech eng, Bucknell Univ, 66-71, assoc prof & coordr grad studies, 71-78; PROF & CHAIR, MECH ENG & ENG SCI, UNIV NC, CHARLOTTE, 78- *Concurrent Pos:* Consult, Scott Paper Co, Philadelphia, 73-78, Med Col Ga, 77- & Univ Fla. *Mem:* Am Soc Mech Engrs; Am Soc Eng Educ; Sigma Xi; Am Acad Mech. *Res:* Solid deformations of polymeric material; creep and relaxation behavior under large strains; anisotropic behavior of polymers; biomechanical response of soft biological tissues; stress analysis of metal-ceramic systems. *Mailing Add:* Dept Mech Eng & Eng Sci Univ NC Charlotte NC 28223

DEHOFF, ROBERT THOMAS, b Sharon, Pa, Jan 15, 34; m 57; c 2. METALLURGY. *Educ:* Youngstown State Univ, BE, 55; Carnegie-Mellon Univ, MS, 58, PhD(metall eng), 59. *Prof Exp:* From asst prof to assoc prof, 59-70, PROF METALL ENG, UNIV FLA, 70- *Concurrent Pos:* Consult, Hanford Labs, Gen Elec Co, Wash, 63-64 & Bausch & Lomb, 76-78; ed, Bull Int Soc Stereology, 64-67. *Honors & Awards:* Res Award, Sigma Xi, 64. *Mem:* Am Inst Mining, Metall & Petrol Engrs; Int Soc Stereol (vpres, 75-83); fel Am Soc Metals; Int Metallog Soc; Sigma Xi; Am Ceramic Soc; Int Soc Stereology. *Res:* Quantitative microscopy, sintering, multicomponent diffusion; thermodynamics; microstructural processes. *Mailing Add:* Dept Mat Sci & Eng Univ Fla Gainesville FL 32611

DE HOFFMANN, FREDERIC, physics; deceased, see previous edition for last biography

DEHOLLANDER, WILLIAM ROGER, b Grand Rapids, Mich, Nov 15, 18; m 47; c 5. PHYSICAL CHEMISTRY. *Educ:* Univ Wash, PhD(phys chem), 51. *Prof Exp:* Sr scientist, Hanford Atomic Prods Oper, 51-59, specialist, Atomic Prod Equip Dept, 59-67, mgr process develop, Nuclear Fuels Dept, 67-75, consult engr, Boiling Water Reactor Systs Dept, 75-77, mgr chem eng, 77-81, consult engr, Gen Elec Co, 81-82; RETIRED. *Mem:* Am Chem Soc. *Res:* Gas chromatography; pyrophoricity; gas-solid reactions; kinetics; thermodynamics; reactor fuels and materials computer process control; hydraulics; radiation buildup and control. *Mailing Add:* 1626 Fairlawn Ave San Jose CA 95125-4930

DEHORATIUS, RAPHAEL JOSEPH, b Philadelphia, Pa, Sept 16, 42; m; c 2. RHEUMATOLOGY, CLINICAL IMMUNOLOGY. *Educ:* Jefferson Med Col, MD, 68. *Prof Exp:* Asst prof med, Univ NMex, 74-76; from asst prof to prof med, Jefferson Med Col, 76-82; interim chmn med, 83-84, PROF MED & PATH, HAHNEMANN UNIV, 82- *Mem:* Am Asn Immunologists; Am Fedn Clin Res; Am Rheumatism Asn; Am Col Physicians. *Res:* Immunologic processes as they relate to rheumatic diseases. *Mailing Add:* Hahnemann Univ M5117 Broad & Vine St Philadelphia PA 19102

DEHORITY, BURK ALLYN, b Peoria, Ill, Sept 3, 30; m 53, 87; c 4. RUMINANT NUTRITION, MICROBIOLOGY. *Educ:* Blackburn Col, BA, 52; Univ Maine, MS, 54; Ohio State Univ, PhD(biochem), 57. *Prof Exp:* Asst prof animal nutrit, Univ Conn, 57-58; from asst prof to assoc prof, 59-70, PROF ANIMAL SCI, OHIO AGR RES & DEVELOP CTR, 70-, ASSOC CHMN, 81- *Mem:* Am Soc Microbiol; Am Soc Animal Sci; Am Dairy Sci Asn; Soc Protozoologists. *Res:* Rumen microbiology. *Mailing Add:* Dept Animal Sci Ohio Agr Res & Develop Ctr Wooster OH 44691-4096

DEIBEL, ROBERT HOWARD, b Chicago, Ill, Dec 20, 24; m 49; c 5. BACTERIOLOGY. *Educ:* Univ Chicago, MS, 52, PhD, 62. *Prof Exp:* Bacteriologist, Am Meat Inst Found, 52-64; assoc prof bact, Cornell Univ, 64-66; from assoc prof to prof, 66-85, ADJUNCT PROF BACT, UNIV WIS-MAD, 85- *Mem:* Inst Food Technologists; Brit Soc Gen Microbiol. *Res:* Bacterial metabolism and taxonomy; food microbiology. *Mailing Add:* 1219 Rutledge St Madison WI 53703

DEIBEL, RUDOLF, b Berlin, Ger, Apr 27, 24; US citizen; m 57; c 3. PEDIATRICS, VIROLOGY. *Educ:* Univ Berlin, Cand Med, 50; Univ Freiburg, Dr Med, 53. *Prof Exp:* Intern med, Wenckebach Hosp, Berlin, 53-55; asst path, Inst Path, Univ Freiburg, 55-56; vis scientist virol, NY State Dept Health, 58-59; res asst pediat & virol, Children's Hosp, 56-61; assoc med virologist, NY State Dept health, 62-67, dir, Virus Labs, 67-89; prof, Albany Med Col, 78-89; RETIRED. *Concurrent Pos:* Asst prof pediat & microbiol, Albany Med Col, 65-69, assoc prof, 69-78; prof, Grad Sch Pub Health, State Univ NY, Albany, 85- *Mem:* Am Asn Immunol; Am Soc Trop Med & Hyg. *Res:* Pathogenesis, immunology and epidemiology of virus infections in man. children. *Mailing Add:* 10 S Helderberg Pkwy Slingerlands NY 12159

DEIBERT, MAX CURTIS, b Lansing, Mich, May 19, 37; m 61; c 2. CHEMICAL ENGINEERING, ENVIRONMENTAL ENGINEERING. *Educ:* Cornell Univ, BChE, 60; Mass Inst Technol, ScD(chem eng), 64. *Prof Exp:* Asst prof chem eng, Mass Inst Technol, 64-70; tech dir, Anacon, Inc, 70-71; chief scientist, Environ Res & Technol Inc, 71-76, mgr, Billings Mont Off, 76-81; DIR QUAL ASSURANCE, NORTHERN TIER PIPELINE CO, 81- *Concurrent Pos:* Ford fel, 64-66; sr res chem engr, Monsanto Res Corp, Mass, 64-69; res group leader, Ionics, Inc, Mass, 69-70. *Mem:* Am Chem Soc; Am Inst Chem Engrs; Electrochem Soc; Am Pollution Control Asn. *Res:* Process instrumentation; electrochemistry; surface chemistry; process control; materials properties; infrared technology; instrumentation science and technology; pollution control systems development; drilling fluids development; energy development consulting. *Mailing Add:* 1220 S Tracy Ave Bozeman MT 59715-5656

DEICHMANN, WILLIAM BERNARD, toxicology; deceased, see previous edition for last biography

DEIFT, PERCY ALEC, b Durban, SAfrica, Sept 10, 45. MATHEMATICAL PHYSICS. *Educ:* Univ Natal, MSc, 71; Princeton Univ, PhD(math physics), 76. *Prof Exp:* From instr to assoc prof, 76-87, PROF MATH, COURANT INST MATH SCI, NY UNIV, 87- *Mem:* Am Math Soc; Sigma Xi. *Res:* Application of the methods of functional analysis to spectral theoretic problems in mathematical physics; inverse spectral problems. *Mailing Add:* NY Univ Courant Inst Math Sci 251 Mercer St New York NY 10012

DEILY, FREDRIC H(ARRY), b Evanston, Ill, June 9, 26; m 47; c 2. MECHANICAL ENGINEERING. *Educ:* Northwestern Univ, BS, 47, PhD(mech eng), 51. *Prof Exp:* Asst, Nat Defense Res Coun-Off Sci Res & Develop Proj, Northwestern Univ, 46-47, dept mech eng, 47-51; res engr, Carter Oil Co, Standard Oil Co, NJ, 51-58, group head, Jersey Prod Res Co, 58-59, res assoc, 60-63; eng adv, Imp Oil Ltd Can, 59-60, drilling adv, Int Petrol Co, Peru, 63-64, RES ASSOC, EXXON PROD RES CO, TEX, 64- *Concurrent Pos:* Mem, Drilling Domain Adv Comt, Am Petrol Inst, 60-63. *Mem:* Am Inst Mining, Metall & Petrol Engrs; Soc Petrol Engrs. *Res:* Oil well drilling research and engineering; rock behavior and failure; underground stress distribution; drilling fluid rheology and mechanics; drilling optimization. *Mailing Add:* 13410 Perthshire Houston TX 77070

DEINES, PETER, b Münden, Ger, Apr 2, 36. GEOCHEMISTRY. *Educ:* Pa State Univ, MSc, 64, PhD(geochem), 67. *Prof Exp:* Res asst geochem, 66-67, from asst prof to assoc prof geochem, 67-80, PROF GEOCHEM & MINERALOGY, PA STATE UNIV, UNIVERSITY PARK, 80- *Res:* Isotope geochemistry; variations in stable isotopes of carbon, oxygen, and hydrogen. *Mailing Add:* Dept Geol Pa State Univ Main Campus University Park PA 16802

DEINET, ADOLPH JOSEPH, b Elberfeld, Germany, Oct 29, 20; nat US; m 42; c 2. ORGANIC CHEMISTRY. *Educ:* Univ Va, BS, 43, PhD(chem), 46. *Prof Exp:* Res chemist antimalarials, Off Sci Res Develop, 44-46; res chemist, Tenneco Chem, Inc, 46-70, supvr synthetic org res & develop, Heyden Div, 70-75, group leader, 75-78, lab mgr, Process Develop Lab, Org & Polymers Div, 78-83; RETIRED. *Mem:* Am Chem Soc. *Res:* General synthetic organic research and development. *Mailing Add:* Rte Four Box 205 Luray VA 22835

DEININGER, ROBERT W, b Monroe, Wis, Aug 15, 27; m 60; c 2. GEOLOGY, PETROLOGY. *Educ:* Univ Wis, BS, 50, MS, 57; Rice Univ, PhD(geol), 64. *Prof Exp:* Geologist, Tidewater Oil Co, 57-58; instr geol, Univ Conn, 60-62; from instr to asst prof, Univ Ala, 62-66; from asst prof to prof, 66-89, EMER PROF GEOL, MEMPHIS STATE UNIV, 89- *Mem:* Geol Soc Am. *Res:* Relationships among tectonics, metamorphism and igneous activity. *Mailing Add:* 4707 Sequoia Rd Memphis TN 38117

DEININGER, ROLF A, b Ulm, Ger, Feb 13, 34; nat US; m 61; c 2. CIVIL ENGINEERING, ENVIRONMENTAL HEALTH. *Educ:* Stuttgart Tech Univ, Dipl Ing, 58; Northwestern Univ, MS, 61, PhD(civil eng), 65. *Prof Exp:* From asst prof to assoc prof, 63-72, PROF ENVIRON HEALTH, SCH PUB HEALTH, UNIV MICH, ANN ARBOR, 73- *Concurrent Pos:* Consult, UN, WHO, UNESCO, Environ Protection Agency, Pan Am Health Orgn. *Mem:* Am Soc Civil Eng; Am Water Works Asn; Opers Res Soc Am; Asn Comput Mach; Water Pollution Control Fedn. *Res:* Computer aided design of waste water collection and treatment systems; design of water distribution systems; optimal lake level control. *Mailing Add:* Dept Pub Health Univ Mich Main Campus Ann Arbor MI 48109-2029

DEINZER, MAX LUDWIG, b Weehauken, NJ, June 19, 37; m 68; c 3. ORGANIC CHEMISTRY. *Educ:* Rutgers Univ, BS, 60; Univ Ariz, MS, 63; Univ Ore, PhD(org chem), 69. *Prof Exp:* Res chemist, E I du Pont de Nemours & Co, 69-71 & Environ Protection Agency, 71-73; PROF AGR CHEM, ORE STATE UNIV, 73- *Mem:* Am Chem Soc; AAAS; Am Soc Mass Spectrometry. *Res:* Environmental and agricultural chemistry; mass spectrometry. *Mailing Add:* Dept Agr Chem Ore State Univ Corvallis OR 97331

DEIS, DANIEL WAYNE, b Martinez, Calif, May 9, 43; m; c 4. SOLID STATE PHYSICS. *Educ:* Stanford Univ, BS, 64; Duke Univ, PhD(physics), 68; Univ Pittsburgh, MBA, 85. *Prof Exp:* Instr physics, Duke Univ, 67-68; sr engr cyrogenics, Westinghouse Res & Develop Ctr, 68-73, fel eng cryogenics, 73-75; mech engr, Lawrence Livermore Lab, 75-79; fel eng electrotechnol, Westinghouse Res & Develop Ctr, 79-81, mgr, high current systs, 81-83, prog mgr, electromagnetiC launchers, 84-87, MGR, ENG SCI DEPT, WESTINGHOUSE RES & DEVELOP CTR, 87- *Res:* Cryogenics; superconductivity; fusion magnets; machine design; pulsed power. *Mailing Add:* 204 Royal Oak Ave Pittsburgh PA 15235

DEISCHER, CLAUDE KNAUSS, b Emmaus, Pa, Oct 14, 03; m 29; c 1. CHEMISTRY. *Educ:* Muhlenberg Col, BS, 25; Univ Pa, MS, 28, PhD(chem), 33. *Prof Exp:* Teacher pub schs, 21-24 & 25-27; from instr to assoc prof, 28-71, asst chmn dept, 52-65, EMER PROF CHEM, UNIV PA, 71- *Concurrent Pos:* Actg cur, E F Smith Mem Libr, 55-71. *Mem:* Am Inst Chemists; Am Chem Soc; Hist Sci Soc. *Res:* Quantitative inorganic analysis; history and literature of chemistry; chemistry of rare elements. *Mailing Add:* 158 Idris Rd Merion Station PA 19066-1611

DEISENHOFER, JOHANN, b Zusamaltheim, Bavaria, Sept 30, 43. BIOCHEMISTRY. *Educ:* Tech Univ Munich, dipl, 71, PhD, 74, Habil, 87; Max Planck Inst Biochem, PhD, 74. *Prof Exp:* Res assoc, Max Planck Inst Biochem, 74-88; INVESTR, HOWARD HUGHES MED INST, UNIV TEX, 88-, REGENTAL PROF & PROF BIOCHEM, SOUTHWESTERN MED CTR, DALLAS, 89- *Honors & Awards:* Nobel Prize in Chem, 88; Biol Physics Prize, Am Phys Soc, 86; Otto Bayer Preis, 88. *Mem:* Nat Acad Sci; AAAS; Am Crystallog Asn; Biophys Soc. *Res:* Protein crystallography; structure and function of biological macromolecules; crystallographic software. *Mailing Add:* Howard Hughes Med Inst & Dept Biochem Univ Tex SW Med Ctr 5323 Harry Hines Blvd Dallas TX 75235-9050

DEISHER, ROBERT WILLIAM, b Bradford, Ill, Aug 20, 20; m 48; c 3. MEDICINE, PEDIATRICS. *Educ:* Knox Col, AB, 41; Univ Wash, MD, 44. *Prof Exp:* From instr to assoc prof, 49-62, PROF PEDIAT, SCH MED, UNIV WASH, 62- *Mem:* Soc Pediat Res; Am Acad Pediat; Am Pediat Soc. *Res:* Child growth and development; adolescence; school problems; delinquency; mental retardation. *Mailing Add:* Dept Pediat Univ Wash Seattle WA 98195

DEISS, WILLIAM PAUL, JR, b Shelbyville, Ky, Feb 1, 23; m 48; c 3. INTERNAL MEDICINE. *Educ:* Univ Notre Dame, BS, 42; Univ Ill, MD, 45. *Prof Exp:* Intern, Univ Wis, 45-46, resident internal med, 48-51; Arthritis & Rheumatism Found res fel, 51-54; from asst prof to assoc prof med & biochem, Duke Univ, 54-58; from assoc prof to prof, Ind Univ, Indianapolis, 56-68; chmn dept, 68-84, prof, 68-90, CLIN PROF MED, UNIV TEX MED, BR, 90- *Concurrent Pos:* In chg med serv, Durham Vet Admin Hosp, 56-58; consult, NIH, 60-64 & 68-72. *Mem:* Fel Am Col Physicians; Am Soc Clin Invest; Am Fedn Clin Res; Asn Am Physicians; Endocrine Soc. *Res:* Endocrinology and metabolism; bone and thyroid chemistry and physiology. *Mailing Add:* 2878 Dominque Dr Galveston TX 77551

DEISSLER, ROBERT G(EORGE), b Greenville, Pa, Aug 1, 21; m 50; c 4. FLUID DYNAMICS, HEAT TRANSFER. *Educ:* Carnegie Inst Technol, BS, 43; Case Inst Technol, MS, 48, PhD(fluid & thermal sci), 89. *Prof Exp:* Engr mat develop, Goodyear Aircraft Corp, Ohio, 43-44; aerospace res spec fluid dynamics & chief fundamental heat transfer br, 47-72, STAFF SCIENTIST & SCI CONSULT FLUID PHYSICS, LEWIS RES CTR, NASA, 72- *Concurrent Pos:* Fel, Lewis Res Acad, Lewis Res Ctr, NASA, 83- *Honors & Awards:* Heat Transfer Mem Award, Am Soc Mech Engrs, 64; Max Jacob Award, Am Soc Mech Engrs & Am Inst Chem Engrs, 75; Exceptional Serv Award, Nat Adv Comt Aeronaut - NASA, 57. *Mem:* Am Phys Soc; fel Am Inst Aeronaut & Astronaut; fel Am Soc Mech Engrs; Soc Natural Philos; Sigma Xi. *Res:* Fluid turbulence; turbulent heat transfer; thermal radiation; vortex flows; heat transfer in powders; meteorological and astrophysical flows; nonlinear dynamics and chaos. *Mailing Add:* NASA Lewis Res Ctr 21000 Brookpark Rd Cleveland OH 44135

DEIST, ROBERT PAUL, b Reading, Pa, Nov 28, 28; m 55; c 3. PHARMACEUTICAL CHEMISTRY. *Educ:* Albright Col, BS, 50. *Prof Exp:* Chemist, Allied Chem Co, 50-51; CHEMIST & SR RES SCIENTIST, WYETH LABS INC, AM HOME PROD CORP, 53- *Mem:* Am Chem Soc. *Mailing Add:* 110 W Valley Hill Rd Malvern PA 19355-2214

DEITCH, ARLINE D, b New York, NY, Mar 12, 22; m 42. CELL BIOLOGY, CYTOPATHOLOGY & FLOW CYTOMETRY. *Educ:* Brooklyn Col, BA, 44; Columbia Univ, MA, 46, PhD, 54. *Prof Exp:* Asst zool, Columbia Univ, 44-48, lectr, 49-50, asst surg, 53-55, vis fel, 55-56, from res assoc surg to res aasoc microbiol, 58-62, from asst prof microbiol to asst prof path, 62-73, assoc prof path, Columbia Univ, 73-84; ADJ PROF UROL, UNIV CALIF, DAVIS, 84- *Concurrent Pos:* USPHS fel, Nat Inst Neurol Dis & Blindness, 55-56. *Mem:* Histochem Soc; Tissue Cult Asn; Sigma Xi; Soc Anal Cytol. *Res:* Flow cytometry, as an adjunct to pathological stage and grade as two method of monitoring patients with urological and other cancers; method for assessing the infertile male patient. *Mailing Add:* 911 Commons Dr Sacramento CA 95825-6651

DEITEL, MERVYN, b Toronto, Can, Aug 18, 36; m 62; c 2. OBESITY SURGERY. *Educ:* Univ Toronto, MD, 61; FRCS(C), 67. *Prof Exp:* Staff, Dept Surg, 69-87, PROF SURG, UNIV TORONTO, 87-; SURGEON, ST JOSEPH'S HEALTH CTR, TORONTO, 68- *Concurrent Pos:* Prof, nutrit sci, Univ Toronto, 87- *Mem:* Cent Surg Asn; Soc Surg Alimentary Tract; Soc Surg Oncol; Soc Head & Neck Surgeons; Am Asn Endocrine Surgeons. *Res:* Developed concepts for obesity surgery; 150 articles published in various medical journals, and also published three textbooks. *Mailing Add:* 2238 Dundas W Suite 201 Toronto ON M6R 3A9 Can

DEITERS, JOAN A, b Cincinnati, Ohio, Apr 28, 34. INORGANIC CHEMISTRY. *Educ:* Col Mt St Joseph, BA, 63; Univ Cincinnati, PhD(chem), 67; Creighton Univ, MChr Sp, 85. *Prof Exp:* Fel chem, Univ Mass, Amherst, 67-68; from instr to prof chem, Col Mt St Joseph, 68-80; assoc prof, 80-86, PROF, VASSAR COL, 86- *Concurrent Pos:* Petrol Res Found-Am Chem Soc res grant, 69-71; vis lectr, Vassar Col, 78-80; Texaco res award, Mid-Hudson Sect, Am Chem Soc, 91. *Mem:* Am Chem Soc; Sigma Xi. *Res:* Ion-molecule hydrogen bonds; spectroscopy and structure of pentacoordinated compounds; interaction of halide ions with carbon tetrahalides; computer simulation of structures of phosphorus compounds; molecular orbital calculations of pentacoordinate silicon anions. *Mailing Add:* Dept Chem Vassar Col Box 143 Poughkeepsie NY 12601

DEITRICH, L(AWRENCE) WALTER, b Pittsburgh, Pa, Oct 17, 38; m 64; c 1. NUCLEAR REACTOR DESIGN & SAFETY. *Educ:* Cornell Univ, BME, 61; Rensselaer Polytech Inst, MS, 63; Stanford Univ, PhD(mech eng), 69. *Prof Exp:* Engr, Knolls Atomic Power Lab, Gen Elec Co, 61-64; asst mech engr, 69-72, Argonne Nat Lab, 69-72, mech engr, 72-81, mgr, fuel behav sect, reactor anal & safety div, 74-79, spec asst off dir, 79-80, assoc dir, 80-90, SR MECH ENGR, ARGONNE NAT LAB, 81-, ASSOC DIR, REACTOR ENG DIV, 90- *Concurrent Pos:* Chair, Nuclear Reactor Safety Div, Am Nuclear Soc, 87-88. *Mem:* Am Nuclear Soc; Am Soc Mech Engrs. *Res:* Nuclear reactor safety and design; heat transfer; fluid mechanics; engineering mechanics; modeling and analysis of reactor fuel elements; analysis of hypothetical reactor accidents. *Mailing Add:* Reactor Eng Div 9700 S Cass Ave Argonne IL 60439

DEITRICH, RICHARD ADAM, b Monte Vista, Colo, Apr 22, 31; m 54; c 3. PHARMACOLOGY, BIOCHEMISTRY. *Educ:* Univ Colo, BS, 53, MS, 54, PhD(pharmacol), 59. *Prof Exp:* Nat Heart Inst fel physiol chem, Sch Med, Johns Hopkins Univ, 59-61, instr, 61-63; from asst prof to assoc prof, 63-76, PROF PHARMACOL, UNIV COLO MED CTR, DENVER, 76- *Concurrent Pos:* Lederle fac award med, 63-65; Nat Inst Neurol Dis & Blindness res grant, 63-66; Nat Inst Gen Med Sci res career develop award, 65-75; Nat Inst Alcohol Abuse & Alcoholism res grants, 66-; Nat Coun Alcoholism res grant, 73-74; vis prof, Inst Med Chem, Univ Berne, 73-74; dir, Univ Colo Alcohol Res Ctr, 77-, Nat Inst AIC Abuse & Alcoholism Career Sci Award, 86- *Honors & Awards:* Res Excellence Award, Res Soc Alcoholism, 84. *Mem:* Am Chem Soc; Res Soc Alcoholism (pres, 81-83); Am Soc Pharmacol & Exp Therapeut; Am Soc Biol Chem; Am Soc Neurochem; Int Soc Biomed Res Alcoholism (treas, 86-). *Res:* Mechanism of enzyme action; biochemical basis of drug action; metabolic pathways for aldehydes; mechanism of sedative addiction; behavioral pharmacogenetics. *Mailing Add:* Dept Pharmacol Univ Colo Sch Med Denver CO 80262

DEITRICK, JOHN E, b New York, NY, Apr 3, 40; c 4. RESEARCH ADMINISTRATION, EDUCATION ADMINISTRATION. *Educ:* Princeton Univ, AB, 62; Cornell Univ, MD. *Prof Exp:* Lewis Ledyard fel surg res, Med Col, Cornell Univ, 71-72; instr surg, NY Hosp, 73-74; asst prof surg, Med Col, Cornell Univ, 74-75; assoc, 75-79, DIR GEN SURG, LEISINGER MED CTR, 79- , CHMN RES, LEISINGER CLINIC, 84-; ASSOC PROF SURG, PA STATE UNIV, 80- *Concurrent Pos:* Mem bd gov, Leisinger Med Ctr, 84- *Res:* Biliary kinetics in primates; studies of bile and cholesteal metabolism during pregnancy in baboons; effects of feeding chenodeoxy choleic acid on bile composition. *Mailing Add:* RD 2 Box 443 Danville PA 17821

DEITZ, LEWIS LEVERING, b Harford Co, Md, June 22, 44. SYSTEMATICS, HOMOPTERA. *Educ:* Univ Md, BS, 67, MS, 68; NC State Univ, Raleigh, PhD(entom), 73. *Prof Exp:* Res asst entom, NC State Univ, 71-73, res assoc, 73-75; hemipterist, Dept Sci & Indust Res, Auckland, New Zealand, 75-79; res assoc, 79-80, asst prof, 80-86, ASSOC PROF, ENTOM

DEPT & DIR INSECT COLLECTION, NC STATE UNIV, 86- *Mem:* Entom Soc Am; Entom Soc NZ; Sigma Xi; Systs Asn NZ; Soc Syst Zool. *Res:* Systematic research on homopterous insects, treehoppers, scale insects; major publications on identification and biology of soybean arthropods, systematics of New World treehoppers and scale insects, W M Maskell's Homoptera and bibliography of Membracoidea. *Mailing Add:* Dept Entom Box 7613 NC State Univ Raleigh NC 27695-7613

DEITZ, VICTOR REUEL, b Downingtown, Pa, Apr 13, 09; m 40; c 4. SURFACE CHEMISTRY. *Educ:* Johns Hopkins Univ, PhD(chem), 32. *Prof Exp:* Asst, Johns Hopkins Univ, 32-33; Nat Res Coun fel, Univ Ill, 33-35, spec res assoc, 36-37; with res lab, Gen Elec Co, 37-38; res assoc, US Bur Standards, US Cane Sugar Res Proj, 39-46, chemist, 46-63; chemist, 63-73, RES CONSULT, US NAVAL RES LAB, 73- *Concurrent Pos:* Guggenheim fel, Imp Col, Univ London, 57-58. *Mem:* Am Chem Soc; Sugar Indust Technologists. *Res:* Adsorption and adsorbents; carbon fibers; sugar refining with bone char; carbon filters and charcoal adsorbents. *Mailing Add:* 3310 Winnett Rd Chevy Chase MD 20815-3202

DEITZ, WILLIAM HARRIS, b Amsterdam, NY, June 14, 25; m 46; c 2. MICROBIOLOGY, BACTERIOLOGY. *Educ:* Hartwick Col, BS, 49; Univ Mass, MS, 51. *Prof Exp:* Res asst bact, Sterling Winthrop Res Inst, 52-62, res assoc, 62-68, res biologist, 68-70; dir alumni rels, Hartwick Col, 70-; dir develop, Southern Calif Col Optom; asst dir, Inst Advan, Western New Eng Col; RETIRED. *Mem:* Am Soc Microbiol; assoc mem Sigma Xi. *Res:* Antibacterial agents; methods of bioassay; antibiotics; mode of action of nalidixic acid. *Mailing Add:* PO Box 815 South Orleans MA 02662-0815

DEITZER, GERALD FRANCIS, b Buffalo, NY, June 5, 42; m 66; c 1. PHOTOBIOLOGY, CHRONOBIOLOGY. *Educ:* State Univ NY, Buffalo, BA, 66; Univ Ga, PhD(bot), 71. *Prof Exp:* Teaching asst plant anat, State Univ NY, Buffalo, 66-67; teaching asst biol, Univ Ga, 67-70; res asst photobiol, Univ Freiburg, WGer, 71-75; plant physiologist photobiol, Smithsonian-Radiation Biol Lab, 75-87; ASSOC PROF HORT, UNIV MD, 87-, FAC, MOLECULAR CELL BIOL PROG, 89- *Concurrent Pos:* Daad fel, 71-73; officer, Washington Sect, Am Soc Plant Physiologists, exec comt, 89-; officer, NCR-101 comt, Grth Chmbr Use & Technol, 89-; BARD grant panel chmn, cell & molecular biol, 90. *Mem:* Am Soc Plant Physiologist; Am Soc Photobiol; Am Asn Advan Sci; Am Soc Hort Sci; Am Bamboo Soc; Sigma Xi. *Res:* Molecular mechanisms regulating the control of plant reproduction by light; interaction of light with the biological clock that determines photoperiodic sensitivity. *Mailing Add:* Dept Hort Univ Md Col Park MD 20742

DEIVANAYAGAN, SUBRAMANIAN, b Tamilnadu, India, Nov 15, 41; US citizen; m 71; c 2. ERGONOMICS, WORK PLACE DESIGN. *Educ:* Annamalai Univ, India, BE, 63, MS, 66; Tex Tech Univ, PhD(indust eng), 73. *Prof Exp:* Grad engr, Enfield India Ltd, 63-64; lectr mech eng & indust eng, Thiagarajar Col Eng, India, 66-69; asst prof indust eng, Tex Tech Univ, 73-76; from asst prof to prof indust prof, Univ Tex Arlington, 76-86; PROF INDUST ENG, TENN TECH UNIV, 86- *Concurrent Pos:* Consult, several orgn, 76-; prin investr, several res projs, 76- *Honors & Awards:* Jerome H Ely Award, Human Factors Soc, 85. *Mem:* Am Inst Indust Engrs; Human Factors Soc; Soc Mfg Engrs; Soc Automotive Engrs; Am Soc Eng Educ; Aerospace Med Asn; Int Found Indust Ergonomics & Safety Res; Sigma Xi. *Res:* Biomechanics; ergonomics; work place design; occupational safety and health; rehabilitation; human factors; robotic work applications; production management; manufacturing systems design and integration; author of various publications. *Mailing Add:* Dept Indust Eng Box 5011 Tenn Technol Univ Cookeville TN 38505

DEJAIFFE, ERNEST, b Fernwood, Pa, Mar 28, 12; m 35; c 1. CIVIL ENGINEERING. *Educ:* Pa State Univ, BS, 33, MS, 47. *Prof Exp:* Teacher, Altoona Sch Dist, Pa, 35-46 & 47-52; from instr to prof eng, Pa State Univ, 46-76; RETIRED. *Concurrent Pos:* Design engr, Gwin Engrs, 64. *Honors & Awards:* English Speaking Union Award, 61. *Mem:* Am Soc Eng Educ. *Res:* Engineering education and technical institutes. *Mailing Add:* 214 21st Ave Altoona PA 16601

DEJARNETTE, FRED ROARK, b Rustburg, Va, Oct 21, 33; m 51; c 2. AEROSPACE ENGINEERING. *Educ:* Ga Inst Technol, BS, 57, MS, 58; Va Polytech Inst, PhD(aerospace eng), 65. *Prof Exp:* Aerodyn engr, Douglas Aircraft Co, Inc, 58-61; asst prof aerospace eng, Va Polytech Inst, 61-63; aerospace engr, Langley Res Ctr, NASA, 63-65; assoc prof aerospace eng, Va Polytech Inst, 65-70; assoc prof, 70-71, grad adminr, 73-83, assoc head, 80-83, PROF MECH & AEROSPACE ENG, NC STATE UNIV, 71- *Concurrent Pos:* Consult, McDonnell Douglas Corp, 77- & Res Triangle Inst, 79- *Mem:* Am Inst Aeronaut & Astronaut; Soc Automotive Eng. *Res:* Aerodynamics and computational fluid dynamics. *Mailing Add:* Dept Mech Eng NC State Univ Raleigh Main Campus Box 7910 Raleigh NC 27695-7910

DEJMAL, ROGER KENT, b Hubbell, Nebr, Dec 26, 40; m 70; c 2. PHYSIOLOGY. *Educ:* Westmont Col, BA, 63; Ore State Univ, MS, 67, PhD(insect physiol), 69. *Prof Exp:* Asst prof physiol, Sioux Falls Col, 69-70; vol food biochemist, Inst Fisheries Develop, Peace Corps, Santiago, Chile, 70-72; MEM FAC CHEM & LIFE SCI FOR NURSING STUDENTS, UMPQUA COMMUNITY COL, 72- *Res:* Protein synthesis and deposition during egg formation in insects; protein-lipid interaction in fish meat. *Mailing Add:* Dept Sci Umpqua Community Col Box 967 Roseburg OR 97470

DE JONG, DIEDERIK CORNELIS DIGNUS, b Haarlem, Neth, Apr 23, 31; US citizen; m 66; c 1. PLANT TAXONOMY. *Educ:* Univ Guelph, BSA, 58, MSA, 60; Mich State Univ, PhD(bot), 64. *Prof Exp:* From instr to asst prof bot, Miami Univ, 63-66; asst prof, McMicken Col Arts & Sci, Univ Cincinnati, 66-71, from asst prof to assoc prof, 71-81, PROF BIOL & BOT, RAYMOND WALTERS COL, UNIV CINCINNATI, 81- *Honors & Awards:* Swed Knighthood of North Star, 81. *Res:* Taxonomic studies in the family Asteraceae. *Mailing Add:* 5052 Collinwood Pl Cincinnati OH 45227-1412

DEJONG, DONALD WARREN, b Doon, Iowa, Oct 14, 30; m 52; c 5. PLANT CYTOCHEMISTRY, AGRICULTURAL BIOTECHNOLOGY. *Educ:* Calvin Col, AB, 51; Univ Ga, PhD(bot), 65. *Prof Exp:* Jr high sch teacher, 51-54, sr high sch teacher, Mich, 54-55; high sch teacher, Guam, 55-57; instr biol, Pac Island Cent, Truk, 57-59; high sch instr, Mont, 59-61; instr, NSF Inst Sci Teachers, 61-62; res asst histochem, Univ Ga, 63-64; res biologist, Plant Enzyme Pioneering Res Lab, Agr Res Serv, USDA, 65-68, res chemist, Mid-Atlantic Area, Southern Region, 68-79; assoc prof bot, NC State Univ, 68-79; assoc staff scientist plant biochem, Int Harvester Co, Chicago, 79-82; plant physiologist, Beltsville Agr Res Ctr, 82-83; res leader, 84-86, RES LEADER PLANT PHYSIOL, TOBACCO RES LAB, USDA-AGR RES SERV, OXFORD, NC, 86- *Concurrent Pos:* Nat Acad Sci-Nat Res Coun resident associateship, 65-67. *Mem:* Soc Green Veg Res; Am Inst Biol Sci; Int Soc Plant Molecular Biol; Phytochem Soc NAm; Am Chem Soc. *Res:* Enzyme levels and isoenzyme patterns associated with physiological stages of growth in plants; biochemical properties of subcellular organelles during senescence of leaves; leaf protein fractionation and utilization; pigments and processes in photosynthesis; genetic engineering of leaf protein traits; molecular biology of gene expression in tobacco plants. *Mailing Add:* Crops Res Lab USDA Agr Res Serv Oxford NC 27565-1168

DE JONG, GARY JOEL, b Bellingham, Wash, June 21, 47; m 67; c 2. ANALYTICAL CHEMISTRY. *Educ:* Univ Calif, Riverside, BS, 69; Ore State Univ, PhD(anal chem), 73. *Prof Exp:* SCIENTIST CHEM, BRISTOL RES LABS, ROHM AND HAAS CO, 73- *Mem:* Am Chem Soc. *Res:* chromatographic methods for particule size analysis; instrumentation automation; instrumental-computer interfaces; thermal analysis of polymers; polymer analysis by spectroscopic and chromatographic methods; gel permeation chromatography; evaluation of hazardous materials by thermal methods. *Mailing Add:* 119 N State St Newtown PA 18940-2028

DE JONG, PETER J, b Zeeland, Mich, July 22, 37; m 59; c 2. MICROBIOLOGY. *Educ:* Hope Col, BA, 59; Univ Wis, PhD(bacteriol), 64. *Prof Exp:* Res microbiologist, Procter & Gamble Co, 64-68; prof biol, Saginaw Valley State Col, 68-76; PROF BIOL, COE COL, 76- *Mem:* Am Soc Microbiol; Am Inst Biol Sci; AAAS; Sigma Xi. *Res:* Physiology of the streptomycetes, more specifically sporulation and germination of the spores of the streptomycetes. *Mailing Add:* Dept Biol Coe Col Cedar Rapids IA 52402

DE JONG, RUDOLPH H, b Amsterdam, Netherlands, Aug 10, 28; US citizen; m 56, 76. MEDICINE, ANESTHESIOLOGY. *Educ:* Stanford Univ, BS, 51, MD, 54; Am Bd Anesthesiol, dipl, 61. *Prof Exp:* Asst prof anesthesia, Med Ctr, Univ Calif, San Francisco, 61-65; assoc prof anesthesiol & pharmacol, Sch Med, Univ Wash, 65-70, prof, 70-76; sr ed, JAMA, 76-78; Richard Saltonstall prof res anesthesiol, Sch Med, Tufts Univ, 78-82; prof anesthesiol & pharmacol, Sch Med, Univ Cincinnati, 82-; DEP COMDR, COL OFF, CHEM DEFENSE, US ARMY RES INST, ABERDEEN PROVING GROUND. *Concurrent Pos:* Consult, Med Serv, Vet Admin & US Navy, 61-76; res career develop award, Univ Wash, 66-76. *Mem:* AAAS; AMA; Am Soc Anesthesiol. *Res:* Physiology and pharmacology of central and peripheral nervous system, especially problems related to pain. *Mailing Add:* Dept Anesthesiol Med Col GA Augusta GA 30912-2700

DEJONG, RUSSELL NELSON, neurology; deceased, see previous edition for last biography

DEJONGH, DON C, b Burnips, Mich, May 10, 37; m 60; c 3. ORGANIC CHEMISTRY. *Educ:* Hope Col, BA, 59; Univ Mich, MS, 61, PhD, 63. *Prof Exp:* Res assoc, Mass Inst Technol, 62-63; from asst prof to prof chem, Wayne State Univ, 63-72; prof chem, Univ Montreal, 72-78; PRES, FINNIGAN INST, 78- *Concurrent Pos:* Vis prof, Univ Montreal, 71. *Mem:* Am Chem Soc; Am Soc Mass Spectrometry; AAAS; Can Inst Chem. *Res:* Application of mass spectrometry to structure problems in chemistry, biological sciences, and environmental sciences. *Mailing Add:* 2296 Bryant St Palo Alto CA 94301-3909

DEJONGHE, LUTGARD C, b Antwerp, Belg, Jan 3, 41; m 73; c 2. MATERIAL ENGINEERING. *Educ:* Higher Tech Inst, Antwerp, Belgium, BA, 63; Univ Del, Newark, MA, 68; Univ Calif, Berkeley, PhD(mat sci), 70. *Prof Exp:* Res fel mat sci, Harvard Univ, 70-72; asst prof mat sci, Cornell Univ, 72-78; staff sr scientist mat sci, Lawrence Berk Lab, Berkeley, Calif, 78-88; PROF MAT SCI, UNIV CALIF, BERKELEY, 88- *Concurrent Pos:* Vis prof, Fed Univ San Carlos, 80; resident prof mat sci, Univ Calif, Berkeley, 78-88. *Honors & Awards:* R M Fulrath Award, Am Ceramic Soc, 85. *Mem:* Fel Am Ceramic Soc. *Res:* Science of ceramic processing; advanced structural ceramics; properties and processing; novel electrochemical systems; ceramic superconductors. *Mailing Add:* Dept Metall & Mat Univ Calif Berkeley CA 94720

DEKAZOS, ELIAS DEMETRIOS, b Merkovouni, Greece, Sept 14, 20; US citizen; m 55; c 3. PLANT PHYSIOLOGY, POMOLOGY. *Educ:* Univ Thessaloniki, dipl, 44; Univ Calif, MS, 53, PhD(plant physiol), 57. *Prof Exp:* Prof agr, Pedagogical Acad Tripolis, Greece, 49-51; tech adv, Agr Bank, Greece, 45-54; res assoc & instr bot, Univ Chicago, 58-60; lectr biochem, Loyola Univ, Ill, 61-62; plant physiologist, Mkt Qual Res Div, USDA, Beltsville, 63-67, plant physiologist, Human Nutrit Res Div, Agr Res Ctr, 67-70, plant physiologist, Russell Agr Res Ctr, 70-81; dept head, 81-89, prof, 81-89, EMER PROF POMOL, AGR UNIV ATHENS, GREECE, 90- *Concurrent Pos:* Pres, Hellenic Acad Soc Ill, 62; chmn & presiding officer, NATO-Advan Studies Inst, Sounion, Greece, 81; visitor, People's Rep China, Int Sci Exchange Prog, 85. *Mem:* Am Soc Hort Sci; Am Soc Plant Physiol; Inst Food Technologists; NY Acad Sci; Sigma Xi. *Res:* Fruit production and postharvest horticulture; control of flowering; anthocyanins; objective methods of quality evaluation; color and texture; quality sorting; growth regulators in fruit production; fruit maturity; storage; postharvest ripening of immature fruit; dehydration-explosion puffing. *Mailing Add:* 408 Sandstone Dr Athens GA 30605

DE KIMPE, CHRISTIAN ROBERT, b Brussels, Belgium, Aug 1, 37; Can citizen; m 64; c 3. PHYSICAL CHEMISTRY, MINERALOGY. *Educ:* Univ Louvain, Ing Chim & Ind Agr, 59, Dr Sc Agr(phys chem & clay mineral), 61. *Prof Exp:* Res assoc clay mineral, Fonds Nat De Recherche Sci, 63-65; res asst clay genesis, Ministere Educ Nat, 65-67; res scientist soil genesis, 67-89, RES COORDR, AGR CAN, 90- *Concurrent Pos:* Ed, Can J Soil Sci, 80-83; lectr, McGill Univ, 80-85; treas, Int Asn Study of Clays; Can Soc Soil Sci travel award, 86. *Honors & Awards:* Travel Awards, Can Soc Soil Sci, 86. *Mem:* Can Soc Soil Sci: Clay Minerals Soc; Int Soc Soil Sci; Int Asn Study of Clays. *Res:* Soil genesis, clay formations; relations between soil; physical condition, water movement and natural fertility; effect of amorphous substances and organic matter on soil aggregation. *Mailing Add:* Agr Can SJC Bldg Cent Experimental Farm Ottawa ON K1A 0C5 Can

DEKKER, ANDREW, b Amsterdam, Neth, Sept 26, 27; US citizen; m 56; c 3. CYTOLOGY, HEPATOLOGY & ONCOLOGY. *Educ:* Cornell Univ, AB, 50; State Univ Groningen, Neth, MD, 58. *Prof Exp:* From asst prof to asoc prof, 63-83, PROF PATH, UNIV PITTSBURGH, 84- *Mem:* Int Acad Path; Am Soc Clin Pathologists; Am Soc Cytol; Col Am Path. *Res:* Diagnostic cytopathology. *Mailing Add:* Dept Path Univ Pittsburgh Sch Med DeSoto at O'Hara Sts Pittsburgh PA 15213

DEKKER, CHARLES ABRAM, b Chicago, Ill, Apr 9, 20; m 47; c 5. BIOCHEMISTRY. *Educ:* Calvin Col, AB, 41; Univ Ill, PhD(biochem), 47. *Prof Exp:* Asst biochem, Univ Ill, 41-43, teaching asst chem, 43-44, biochem, 46-47; res asst, Yale Univ, 47-49; fel Am Cancer Soc, Cambridge Univ, 49-51; from asst prof to assoc prof & from lab asst to assoc res biochemist, 51-62, PROF BIOCHEM & RES BIOCHEMIST, VIRUS LAB, UNIV CALIF, BERKELEY, 62-, MAJOR ADV, DEPT BIOCHEM, 80- *Mem:* Am Chem Soc; Am Soc Biol Chem; Sigma Xi. *Res:* Chemistry and enzymology of nucleic acids and their derivatives. *Mailing Add:* Dept Biochem Univ Calif Barker Hall Berkeley CA 94720

DEKKER, DAVID BLISS, b Evanston, Ill, May 28, 19; m 42; c 2. MATHEMATICS. *Educ:* Univ Calif, AB, 41, PhD(math), 48; Ill Inst Technol, MS, 43. *Prof Exp:* Asst math, Ill Inst Technol, 41-43; mathematician, Lockheed Aircraft Corp, Calif, 43-44; asst math, Univ Calif, 46-48; from instr to asst prof, 48-59, assoc, 59-81, EMER PROF MATH & COMPUT SCI, UNIV WASH, SEATTLE, 81- *Concurrent Pos:* Dir, Comput Ctr, Univ Wash, 56-66. *Mem:* Am Math Soc. *Res:* Foundations of geometry; metric differential geometry; hypergeodesic curvature and torsion; generalizations of hypergeodesics; numerical analysis. *Mailing Add:* 2524 E Miller Seattle WA 98112

DEKKER, EUGENE EARL, b Highland, Ind, July 23, 27; m 58; c 3. BIOCHEMISTRY. *Educ:* Calvin Col, AB, 49; Univ Ill, MS, 51, PhD(biochem), 54. *Prof Exp:* Res assoc biochem, Univ Ill, 54; instr, Sch Med, Univ Louisville, 54-56; from instr to assoc prof, Univ Mich, 56-70, asst chmn dept, 75-78, assoc chmn dept, 78-88, PROF BIOCHEM, SCH MED, UNIV MICH, ANN ARBOR, 70- *Concurrent Pos:* Life ins med res fel, Med Sch, Univ Mich, Ann Arbor, 56-58; Lederle med fac award, 58-61; NIH res fel, Univ Calif, Berkeley, 65-66, Scripps Clin & Res Found, La Jolla, Calif, 73-74, sr res fel, Univ Wash, Seattle, 81-82. *Mem:* Am Chem Soc; Am Soc Biol Chem; Am Soc Plant Physiol; AAAS; Oxygen Soc; Protein Soc. *Res:* Mechanism of action and comparative biochemistry of enzymes; metabolism of nitrogen compounds in animals, plants and bacteria; enzyme structure-function interrelationships; biochemistry of aging. *Mailing Add:* Dept Biochem/0606 Univ Mich Med Sch Ann Arbor MI 48109-0606

DEKKER, JACOB CHRISTOPH EDMOND, b Hilversum, Neth, Sept 6, 21; nat US; m 50. MATHEMATICS. *Educ:* Syracuse Univ, PhD(math), 50. *Prof Exp:* Instr math, Univ Chicago, 51-52; from instr to asst prof, Northwestern Univ, 52-55; vis asst prof, Univ Chicago, 55-56; mem, Inst Advan Study, 56-58; assoc prof, Univ Kans, 58-59; from asst prof to prof, 59-86, EMER PROF MATH, RUTGERS UNIV, 86- *Concurrent Pos:* Lectr, Nat Univ Mex, 61; consult, Int Bus Mach Corp, 58 & 60; mem, Inst Advan Study, 67-68. *Mem:* NY Acad Sci; Math Asn Am; Am Math Soc; Asn Symbolic Logic. *Res:* Recursive functions; combinatorial theory. *Mailing Add:* 56 Jefferson Rd Princeton NJ 08540

DEKLOET, SIWO R, b Maarssen, Neth, Feb 22, 33; m 64; c 4. BIOCHEMISTRY, MOLECULAR GENETICS. *Educ:* Univ Utrecht, BS, 53, MS, 56, PhD(biophys, chem), 61. *Prof Exp:* Res assoc cell biol, Rockefeller Univ, 61-62; res scientist, Philips Res Labs, Eindhoven, Neth, 63-67; assoc prof, 67-80, PROF BIOL SCI, INST MOLECULAR BIOPHYS, FLA STATE UNIV, 80- *Res:* Metabolism and properties of nucleic acids; genetics of fungi; metabolism of nucleotide analogues; mechanism of action of antibiotics; molecular evolution of birds. *Mailing Add:* Dept Biol Sci Fla State Univ Tallahassee FL 32306

DE KOCK, CARROLL WAYNE, b Oskaloosa, Iowa, Apr 14, 38; m 59; c 2. INORGANIC CHEMISTRY. *Educ:* Calvin Col, BS, 60; Iowa State Univ, PhD(chem), 65. *Prof Exp:* Fel chem, Argonne Nat Lab, 65-67; assoc prof, 67-80, PROF CHEM, ORE STATE UNIV, 80-, DEPT CHMN, 84- *Res:* Organometallic chemistry of molybdenam and early transition elements. *Mailing Add:* Dept Chem Ore State Univ Corvallis OR 97331

DEKOCK, ROGER LEE, b Oskaloosa, Iowa, Oct 4, 43; m 65; c 2. STRUCTURE & BONDING. *Educ:* Calvin Col, BA, 65; Univ Wis-Madison, PhD(inorg chem), 70. *Prof Exp:* Res assoc, Univ Fla, 69-70, Univ Birmingham, Eng, 70-72; asst prof inorg chem, Am Univ, Beirut, 72-76; assoc prof, 76-80, PROF CHEM, CALVIN COL, 80- *Concurrent Pos:* Fulbright res scholar, Free Univ Amsterdam, 82-83. *Mem:* Am Chem Soc. *Res:* Hartree-Fock-Slater calculations on transition metal complexes containing heavy metals; studies on the three-center two-electron chemical bond. *Mailing Add:* Dept Chem Calvin Col Grand Rapids MI 49546

DEKORNFELD, THOMAS JOHN, b Iregszemcse, Hungary, June 19, 24; US citizen; m 52; c 4. ANESTHESIOLOGY. *Educ:* George Washington Univ, BS, 48, MS, 49; Harvard Med Sch, MD, 53. *Prof Exp:* Instr anesthesiol, Sch Med, Univ Wis, 56-57; from asst chief to chief, Baltimore City Hosp, Md, 57-63; dir clin therapeut, Parke, Davis & Co, Mich, 63-64; assoc prof anesthesiol, 64-68, PROF ANESTHESIOL, MED SCH, UNIV MICH, ANN ARBOR, 68- *Mem:* AMA; Am Soc Anesthesiol; Am Soc Pharmacol; Sigma Xi. *Res:* Clinical pharmacology; clinical and laboratory investigation of new narcotic and non-narcotic analgesics. *Mailing Add:* 2581 Hawthorne Ann Arbor MI 48104

DE KORTE, AART, b Rotterdam, Neth, Sept 4, 34; US citizen. PHYSICAL CHEMISTRY. *Educ:* NY Univ, BA, 58; Yale Univ, MS, 60, PhD(chem), 65. *Prof Exp:* Instr chem, Queens Col, NY, 64-66; asst prof, 66-71, assoc prof, 71-77, chmn dept, 78-81, PROF CHEM, FAIRLEIGH DICKINSON UNIV, 77- *Mem:* Am Chem Soc; NY Acad Sci. *Res:* Fluorescence; science, technology and society. *Mailing Add:* Dept Chem Fairleigh Dickinson Univ Teaneck NJ 07666-1996

DEKORTE, JOHN MARTIN, b Grand Rapids, Mich, Sept 20, 40; m 63; c 2. INORGANIC CHEMISTRY. *Educ:* Hope Col, BA, 62; Purdue Univ, PhD(inorg chem), 69. *Prof Exp:* From asst prof to assoc prof, 66-81, PROF CHEM, NORTHERN ARIZ UNIV, 81- *Concurrent Pos:* Res grant, Petrol Res Fund-Am Chem Soc, 68; res grant, Res Corp, 70; consult, John Wiley & Sons Inc, WB Saunders Co; mem prog comt, Am Chem Soc Educ Div, 80-85; consult chem textbooks, 87 & 90; mem, Standardization Inorganic Chem Exam Comt, Am Chem Soc, 78-, comt chair, 88-; mem, Nat Chem Olympiad Exam Comt, Am Chem Soc, 90- *Mem:* Am Chem Soc; Sigma Xi. *Res:* Kinetics and mechanisms of inorganic oxidation-reduction reactions. *Mailing Add:* Dept Chem Box 5698 Northern Ariz Univ Flagstaff AZ 86011-5698

DE KORVIN, ANDRE, b Berlin, Ger, Dec 13, 35; US citizen; m 67. MATHEMATICS. *Educ:* Univ Calif, Los Angeles, BA, 62, MA, 63, PhD(math), 67. *Prof Exp:* Mathematician, IBM Corp, San Jose, Calif, 63-64, res mathematician, Sci Ctr, Los Angeles, 64-67; asst prof math, Carnegie-Mellon Univ, 67-69; vis asst prof, Ind State Univ, Terre Haute, 68-69, from asst prof to assoc prof, 69-75, prof math & comput sci, 75-; PROF APPL MATH SCI, UNIV HOUSTON. *Concurrent Pos:* Vis prof, Ind Univ-Purdue Univ, 80-81. *Mem:* AAAS; Am Math Soc; Math Asn Am; NY Acad Sci. *Res:* Information science and systems; intelligent systems; vector measures; information sciences; computer sciences; probability theory. *Mailing Add:* Dept Applied Math Sci Univ Houston Downtown One Main St S629 Houston TX 77002

DE KRASINSKI, JOSEPH S, b Mszana, Poland, June 15, 14; m 47; c 4. FLUID MECHANICS. *Educ:* Univ London, BSc, 44, PhD(aeronaut), 64. *Prof Exp:* Res scientist, Royal Aircraft Estab, Farnborough, Eng, 44-46; head aerodyn res, Argentine Aeronaut Res Inst, 47-67; assoc prof, 66-76, prof 76-85, EMER PROF FLUID MECH, UNIV CALGARY, 85- *Concurrent Pos:* Prof, Univ Cordoba, 55-67. *Mem:* Am Inst Aeronaut & Astronaut; Can Soc Mech Engrs. *Res:* Aerodynamics; gas dynamics; boundary layer theory; separated flows; atmospheric aerodynamics; wind tunnelling; experimental techniques in wind tunnels; wind tunnel design. *Mailing Add:* 3319 24th St NW Calgary AB T2M 3Z8 Can

DEKSTER, BORIS VENIAMIN, b Leningrad, USSR, Oct 8, 38; Can citizen; m; c 1. BODIES OF CONSTANT WIDTH. *Educ:* Leningrad Univ, MS, 62; Steklov Inst Math, PhD (math), 71. *Prof Exp:* Res fel math, Univ Toronto, 74-76, asst prof, 76-79 & 81-86; asst prof, Univ Notre Dame, 79-81; from asst prof to assoc prof, 86-90, FULL PROF MATH, MT ALLISON UNIV, CAN, 90. *Concurrent Pos:* Can Natural Sci & Eng Res Coun Can grant, 81- *Mem:* Am Math Soc; Can Math Soc. *Res:* Bodies of constant width; Jung's theorem; Borsuk's problem; simplexes in spheres and hyperbolic spaces. *Mailing Add:* Dept Math & Comput Sci Mt Allison Univ Sackville NB E0A 3C0 Can

DELABARRE, EVERETT MERRILL, JR, b Boston, Mass, Mar 3, 18; m 53; c 6. MEDICINE. *Educ:* Columbia Univ, BA, 40, MD, 43. *Prof Exp:* Asst path, Columbia Univ, 44; clin instr med, Yale Univ, 54-57; assoc chief med, Mem Med Ctr, Williamson, WVa, 57-64; asst chief, Vet Admin Hosp, 64-65; asst prof, 65-70, ASSOC PROF PHYS MED & REHAB, TUFTS UNIV, 70-; ASST CHIEF REHAB MED SERV, BOSTON VET ADMIN MED CTR, 81- *Mem:* Asn Mil Surg US; Am Col Physicians. *Res:* Internal medicine rehabilitation. *Mailing Add:* Vet Admin Med Ctr 150 S Huntington Ave Boston MA 02130

DELACERDA, FRED G, b Natchitoches, La, Sept 19, 37; m 72. HUMAN FACTORS, BIOMECHANICS. *Educ:* La Tech Univ, BS, 60; La State Univ, MS, 69, PhD(exercise physiol), 71; Univ Tenn, BS, 73; Okla State Univ, BS, 79. *Prof Exp:* Syst analyst indust, Tex Eastern Gas Transmission, Inc, 61-62; instr math & physics, Sabine Parish Sch Syst, 64-67; asst prof health, Tenn Technol Univ, 71-72 & Univ Southwestern La, 73-75; asst prof, 75-80, ASSOC PROF HEALTH, OKLA STATE UNIV, 80- *Concurrent Pos:* Gooch Allied Health Scholarship, Med Ctr, Univ Tenn, 72-73; consult physiotherapy, Teachers' Conf Phys Disabled Children, La State Univ, Eunice, 74-75, Wagoner Community Hosp, 76-77, Logan County Mem Hosp, 76-77, Williams Clin, Pawnee Manor Nursing Home, 77-, Hominy City Hosp, 76- & Okla Dept Corrections. *Mem:* Am Soc Biomech; Am Soc Mech Engrs; Human Factor Soc; Am Inst Aeronaut & Astronaut. *Res:* Biomechanics and human factor engineering. *Mailing Add:* Sch Hpels Okla State Univ Stillwater OK 74074

DE LA CRUZ, VIDAL F, b Jersey City, NJ, Jan 31, 59. IMMUNOLOGY. *Prof Exp:* SR RES SCIENTIST, MED IMMUNE INC, 88- *Mailing Add:* Med Immune Inc 19 Firstfield Rd Gaithersburg MD 20878

DE LA FUENTE, ROLLO K, b Mt Province, Philippines, Oct 6, 33; m 61; c 2. PLANT PHYSIOLOGY. *Educ:* Univ Philippines, BS, 55; Univ Hawaii, MS, 59, PhD(bot) 64. *Prof Exp:* From asst instr to instr bot, Univ Philippines, 55-61; res asst, Univ Hawaii, 63-64; res fel plant physiol, Purdue Univ, 64-69; from asst prof to assoc prof, 69-87, PROF PLANT PHYSIOL, KENT STATE UNIV, 87- *Concurrent Pos:* Mem adv panel, Physiol Cell Metab Biol Sect, NSF, 81-84. *Mem:* Sigma Xi; Am Soc Plant Physiologists; AAAS; Am Soc Hort Sci; Scand Soc Plant Physiol. *Res:* Mechanism and significance of the polar transport of the plant hormone indoleacetic acid; the possible role that calcium plays in regulating this pheonomenon. *Mailing Add:* Dept Biol Sci Kent State Univ Kent OH 44242

DELAGI, EDWARD F, b New York, NY, Nov 4, 11; m 41; c 2. MEDICINE. *Educ:* Fordham Univ, BS, 34; Hahnemann Med Col, MD, 38. *Prof Exp:* Asst chief paraplegic sect, Vet Admin Hosp, Bronx, 51-53, chief ward sect phys med & rehab, 51-56; adj, Jewish Hosp Chronic Dis, Brooklyn, 56-58; from assoc prof to prof, 59-85, EMER PROF REHAB MED, ALBERT EINSTEIN COL MED, 85- *Concurrent Pos:* Attend physician, Vet Admin Hosp, Bronx, 56-; dir, Frances Schervier Hosp & Home, Riverdale, 56-; vis physician, Bronx Munic Hosp Ctr, 56-; chief, St Josph's Hosp, Yonkers, 57- & Misericordia Hosp, Bronx, 58- *Mem:* AAAS; fel AMA; fel Am Acad Phys Med & Rehab; fel Am Cong Rehab Med; fel Am Col Physicians. *Res:* Physical medicine and rehabilitation; kinesiology; electromyography; electrodiagnostic methods. *Mailing Add:* 100 Gerard Dr East Hampton NY 11937

DELAGI, RICHARD GREGORY, b New York, NY; m 52; c 4. MECHANISMS OF SOLID-PHASE BONDING OF METALLIC MATERIALS, PRODUCT & PROCESS DEVELOPMENT. *Educ:* Columbia Univ, ME, 53. *Prof Exp:* Engr metal develop, US Atomic Energy Comn, 51-54; res engr metall, Metals & Controls Corp, 54-56; mgr eng metall, M&C Nuclear Inc, 56-59; mem tech staff metall, Tex Instruments Inc, 60-70, mgr prod develop mat sci, 70-80, sr mem tech staff, 80-85, FEL MAT SCI, TEX INSTRUMENTS INC, 86- *Concurrent Pos:* Mem, Joint Fuel Element Develop Comn, USAEL/AECL, 53-56, NSF Electronic Mat, 82-; consult, Extrusions Corp, 76-81, Avitar Corp, 78-85, Ilford Ltd, Div, Ciba-Geigy, 82-86; consult, Flex Watt Corp, Wareham, Mass, 85-, dir, 88- *Mem:* Am Soc Mat; Am Inst Aeronaut & Astronaut; Am Defense Preparedness Asn. *Res:* Rolling and roll bonding of laminate metal composites; development of engineered material systems including metal matrix and metal-ceramic composites; thermomechanical processing of refractory metals and ordered intermetallic compounds. *Mailing Add:* Tex Instruments 34 Forest St M/S 10-10 Attleboro MA 02703

DE LA HABA, GABRIEL LUIS, b PR, June 29, 26; m 61. BIOLOGICAL CHEMISTRY. *Educ:* Johns Hopkins Univ, AB, 46, PhD(biol), 50. *Prof Exp:* USPHS fel, NY Univ, 51-52; USPHS fel, Yale Univ, 52-53, instr biochem, 53-54, asst prof, 54-55; sr asst scientist, USPHS, 55-58; res assoc, Johns Hopkins Univ, 58-59; res assoc anat, Univ, 59-68, ASSOC PROF ANAT, SCH MED, UNIV PA, 68- *Mem:* Am Soc Biol Chemists. *Res:* Biochemistry of development. *Mailing Add:* Dept Anat G-3 Univ Pa Sch Med Philadelphia PA 19174

DELAHAY, PAUL, b Sas Van Gent, Neth, Apr 6, 21; nat US; m 62. CHEMICAL PHYSICS. *Educ:* Univ Brussels, BS, 41, MS, 45, Liege, 44; Univ Ore, PhD(chem), 48. *Prof Exp:* Instr chem, Univ Brussels, 45-46; res assoc, Univ Ore, 48-49; from asst prof to Boyd prof, La State Univ, 49-65; prof chem, NY Univ, 65-87, Frank G Gould prof sci, 74-87; RETIRED. *Concurrent Pos:* Guggenheim fels, Cambridge Univ, 55 & NY Univ, 71-72; Fulbright prof, Univ Paris, 62-63. *Honors & Awards:* Heyrovsky Medal, Czech Acad Sci, 65; Turner Prize, Electrochem Soc, 51, Palladium Medal Award, 67; Award in Pure Chem, Am Chem Soc, 55, Southwest Award, 59. *Res:* Photoelectron emission spectroscopy of solutions; electron scattering in gases; solvated electron. *Mailing Add:* Six Rue Benjamin Godard Paris 75116 France

DELAHAYES, JEAN, b Rouen, France, June 9, 36; m 57; c 1. PHYSIOLOGY. *Educ:* Univ Paris, Lic es Sci, 64, Dipl, 66, Dr es Sci(physiol), 68. *Prof Exp:* Teaching asst physiol, Univ Paris, 65-68; from instr to asst prof, Ohio State Univ, 69-74; ASSOC PROF PHYSIOL, MED COL GA, 74- *Concurrent Pos:* Head pharmacol, Clin-Midy Res Ctr, Montpellier, France, 75-78. *Res:* Movements of ions across the cell membrane in heart during activity; problems of excitation cantiaction coupling; movements of Ca-45 associated with heart activity. *Mailing Add:* 21 Rue Ampere Pantoise France

DE LA HUERGA, JESUS, BIOMEDICAL RESEARCH, LIVER DISEASE. *Educ:* Univ Santo Domingo, MD, 40; Northwestern Univ, PhD(path). *Prof Exp:* Instr, Northwestern Univ, 48-81; RETIRED. *Mailing Add:* 1414 Lincoln St Evanston IL 60201

DELAHUNTA, ALEXANDER, b Concord, NH, Dec 3, 32; m 55; c 4. VETERINARY ANATOMY, NEUROLOGY. *Educ:* State Univ NY Vet Col, DVM, 58; Cornell Univ, PhD(vet anat), 64; Am Col Vet Internal Med, dipl neurol. *Prof Exp:* Vet practr, Cilley's Animal Hosp, NH, 58-60; from instr to assoc prof, 60-73, PROF VET ANAT, STATE UNIV NY VET COL, CORNELL UNIV, 73-, CHMN DEPT CLIN SCI, 77- *Mem:* Am Vet Med Asn; Am Asn Vet Anatomists; Am Asn Anatomists; Am Asn Vet Neurologists. *Res:* Clinical neurology; neuropathology; neuroanatomy. *Mailing Add:* 126 Judd Falls Rd Ithaca NY 14850

DELAHUNTY, GEORGE, b Upper Darby, Pa, May 5, 52; m 78. ICHTHYOLOGY. *Educ:* Duquesne Univ, BS, 74; Marquette Univ, PhD(biol), 79. *Prof Exp:* AT DEPT BIOL, GOUCHER COL. *Mem:* Am Soc Zoologists; AAAS; Am Soc Ichthyologists & Herpetologists. *Res:* Environmental control of metabolism and reproduction in teleosts; role of the pineal organ as possible tranducer of photo period information; the role of the pineal in the circadian organization of the animal. *Mailing Add:* Dept Biol Sci Goucher Col Towson MD 21204

DE LA IGLESIA, FELIX ALBERTO, b Cordoba, Arg, Nov 27, 39; Can citizen; m 64; c 3. EXPERIMENTAL PATHOLOGY, TOXICOLOGY. *Educ:* Nat Univ Cordoba, BSc, 56, MD, 64; Acad Toxicol Sci, dipl, 83. *Prof Exp:* Instr path, Nat Univ Cordoba, 59-62, res asst electron micros, 63-64; res fel, Hosp Sick Children, Toronto, 64-66; scientist, Warner-Lambert Res Inst Can, 66-68, dir toxicol, 68-72, dir, 72-76; dir, Dept Toxicol, 77-81, dir, Dept Path & Exp Toxicol, 81-83, VPRES PATH & EXP TOXICOL, PHARMACEUT RES DIV, WARNER-LAMBERT/PARKE DAVIS, 83- *Concurrent Pos:* Consult, Ontario Ministry Health, 76 & Ministry of Health Fed Repub Germany, 78, sci adv on benefit/risk of hipolipidemic agents, Berlin, 79-80, expert consult, path & toxicol, exp path, comp path, cell biol & carcinogenesis, NIH-Nat Cancer Inst, 79-; mem, Pharmacol-Morphol Sci Adv & Fel Prog comt, Pharmaceut Mfrs Asn Found, 80-, adv comt, Can Ctr Toxicol, 82-, Environ Health Sci rev comt, Nat Inst Environ Health Sci, 83-86, Int Study Group Res in Cardiac Metab; res fel, Med Res Coun Can, external consult, 82-; adj prof path, Univ Toronto Fac Med, 81-, adj prof toxicol, Univ Mich Sch Pub Health, 82-, adj prof path, Med Col Ohio, Toledo, 85-, adj res sci path dept, Univ Mich Sch Med; ed, Current Topics Toxicol, Drug Metab Rev, 81-, Toxicol Path, 83-88, assoc ed, 81-82, 88-91; mem, Prof Standards Eval Bd, Acad Toxicol Sci, 87-90; mem res comt, Mich Eye Bank & Transplant Ctr. *Honors & Awards:* J Carveth Sci Award. *Mem:* Int Acad Path; Am Col Toxicol; Soc Toxicol Pathologists (pres, 87-88); Am Asn Pathologists; Soc Toxicol; Micros Soc Can (secy, 72-76). *Res:* Liver diseases; nutrition and pathology; adverse hepatic effects of therapeutic agents; chemical carcinogenesis; drug safety evaluation; subcellular pathology and toxicology; published 116 articles, 18 chapters and books. *Mailing Add:* 2307 Hill St Ann Arbor MI 48104

DE LA MARE, HAROLD ELISON, b Burley, Idaho, Aug 5, 22; m 52; c 6. ORGANIC CHEMISTRY, POLYMER CHEMISTRY. *Educ:* Utah State Univ, BS, 44; Purdue Univ, PhD(org chem), 51. *Prof Exp:* Chemist, Eastman Kodak Co, NY, 44-46; asst org chem, Purdue Univ, 46-48 & 49-50, instr chem, 50-51; res chemist, 51-72, sr staff res chemist, 72-80, RES ASSOC, SHELL DEVELOP CO, 80- *Mem:* Am Chem Soc; NY Acad Sci. *Res:* Organic peroxide chemistry and oxidation reactions; free radical-metal ion interactions; coordination complex catalysis in elastomer synthesis; anionic polymerization; epoxy and photocure resins, epoxy reaction injection molding, thermostat composites. *Mailing Add:* 5426 Meadowcreek Dr No 1012 Dallas TX 75248

DELAMATER, EDWARD DOANE, b Plainfield, NJ, Jan 24, 12; m 43; c 5. MICROBIOLOGY, ELECTRON MICROSCOPY. *Educ:* Johns Hopkins Univ, MA, 37; Columbia Univ, PhD(bact, dermat), 41, MD, 42. *Prof Exp:* Asst bot, Johns Hopkins Univ, 33-36; asst dermat, Columbia Univ, 36-42, mycologist, Vanderbilt Clin, Med Ctr, 36-42; fel dermat, Mayo Found, Univ Minn, 46-47, asst prof bact & mycol, 47-48, mycologist, Mayo Clin, 46-48; assoc res prof dermat, Sch Med, Univ Pa, 48-51, from assoc res prof to res prof microbiol, 50-63, res prof dermat, 51-63, dir sect cytol & cytochem, 53-63, consult, Pepper Lab Hosp, 48-63; prof & chmn dept microbiol, New York Med Col, 63-66; hosp epidemiologist, New York City Dept Health, 65-66; prof microbiol & dean col sci, Fla Atlantic Univ, 66-68, distinguished prof sci, 68-78; RETIRED. *Concurrent Pos:* Consult, Smith, Kline & French Lab, 48-51, Off Surgeon Gen, 48-63, Children's Hosp, Philadelphia, 49-63 & Skin & Cancer Hosp, 58-63; Guggenheim fel, 53; mem med staff, Eastern Shore Hosp Ctr, Cambridge, 77-84. *Mem:* Mycol Soc Am; Am Col Physicians; Bot Soc Am; Genetics Soc Am; Am Soc Trop Med & Hyg. *Res:* Cytology and cytochemistry of micro-organisms; medical mycology; life cycles of spirochetes; structure and ultrastructure of fish scales. *Mailing Add:* Rte 1 Box 97 Oxford MD 21654

DELAMATER, GEORGE (BEARSE), industrial chemistry; deceased, see previous edition for last biography

DE LAMIRANDE, GASTON, b Montreal, Que, Dec 22, 23; m 50; c 3. BIOLOGICAL CHEMISTRY. *Educ:* Univ Montreal, BA, 43, BSc, 46, MSc, 47, PhD(chem), 49. *Prof Exp:* Lectr org chem, Univ Montreal, 46-49; res assoc biochem, Montreal Cancer Inst, Nat Cancer Inst Can, Notre Dame Hosp, 49-60; res assoc prof biol chem, 60-67, RES PROF BIOL CHEM, FAC MED, UNIV MONTREAL, 67-, VDEAN, FAC MED, 75- *Concurrent Pos:* Damon Runyon Found Cancer Res fel, 52-55; res scientist, Montreal Cancer Inst, Nat Cancer Inst Can, Notre Dame Hosp, 60-67, assoc dir, 67-75. *Mem:* AAAS; Am Asn Cancer Res; NY Acad Sci; Can Physiol Soc; Can Biochem Soc. *Res:* Chemistry and metabolism of nucleic acids and proteins; enzymology; cancer. *Mailing Add:* 9925 Verville Montreal PQ H3L 3E4 Can

DE LA MONEDA, FRANCISCO HOMERO, b Dec 20, 39. SOLID STATE DEVICE MODELING. *Educ:* Rensselaer Polytech Inst, BS, 61, Syracuse Univ, MS, 65; Univ Fla, PhD(elec eng), 70. *Prof Exp:* Staff engr, Syst Commun Div, 70-74, adv engr, Data Syst Div, 74-78, Gen Technol Div, 78-81, ADV ENGR, GEN PROD DIV, IBM, 81- *Mem:* Inst Elec & Electronics Engrs. *Res:* Analytical and numerical models for bipolar and metal-oxide-semiconductor transitors and development of measurement techniques for constants and parameter of such models. *Mailing Add:* IBM Corp Gen Prod Div 900 S Rita Rd Tucson AZ 85744

DE LANCEY, GEORGE BYERS, b Cresson, Pa, Oct 19, 40; c 3. CHEMICAL ENGINEERING. *Educ:* Univ Pittsburgh, BS, 62, MS, 65, PhD(chem eng), 67. *Hon Degrees:* MEng, Stevens Inst Technol, 78. *Prof Exp:* Res engr, Jones & Laughlin Steel Corp, 63-65, consult, 65-66; res assoc math, Nat Bur Stand, 67-68; from asst prof to assoc prof, 68-75, PROF CHEM ENG, STEVENS INST TECHNOL, 75-, SR ACAD ADV, INT PROG, 80- *Mem:* Am Inst Chem Engrs. *Res:* Interfacial mass transfer and heterogeneous catalysis. *Mailing Add:* Dept Chem & Chem Eng Steven Inst Technol Hoboken NJ 07030-5991

DELAND, EDWARD CHARLES, b Lusk, Wyo, May 16, 22; m 52; c 1. MATHEMATICS, PHYSIOLOGY. *Educ:* SDak Sch Mines & Technol, BS, 43; Univ Calif, Los Angeles, PhD(math), 56. *Prof Exp:* Engr, Corning Glass Works, 43-46; asst prof physics, San Diego State Col, 46-48; sr scientist, Rand Corp, 56-72; ADJ PROF THORACIC SURG, ANESTHESIOL & BIOMATH, UNIV CALIF, LOS ANGELES, 72- *Concurrent Pos:* Dir analog comput facil, Univ Calif, Los Angeles, 53-56, instr eng, 56-62; res colloid sci, Univ Cambridge, 65; NIH spec res resources bd fel, 67-68; consult, Nat Ctr Health Serv Res & Develop, 68-70, Nat Lob Med, Nat Bur Standards & NASA. *Mem:* Math Asn Am; Asn Comput Mach; Int Asn Analog Comput; Soc Advan Med Systs; Bioeng Soc. *Res:* Computer design and applications; applied mathematics; pedagogy; biochemistry; physiology; computer systems; mathematical models; biochemistry of heart cells; hemoglobin. *Mailing Add:* Dept Anesthesiol Univ Calif Los Angeles Med Ctr BH533A Los Angeles CA 90024

DELAND, FRANK H, b Jackson, Mich, July 2, 21; m 49; c 4. NUCLEAR MEDICINE. *Educ:* Univ Mich, Ann Arbor, BS, 47; Univ Louisville, MD, 52; Univ Minn, St Paul, MS, 56. *Prof Exp:* Fel path, Mayo Found, 53-56; dir lab med, Lakeland, Fla, 57-67; assoc prof nuclear med, Johns Hopkins Univ, 67-70; prof, Univ Fla, 70-74; prof radiation med, Med Ctr, Univ Ky, 74-85; CHIEF STAFF, VET ADMIN MED CTR, SYRACUSE, NY, 85- *Concurrent Pos:* Instr path, Ohio State Univ, 65-67; NIH fel, Johns Hopkins Univ, 66-67; res grants, US Dept Health, Educ & Welfare, 67-71 & Vet Admin, 71-; ed, J Nuclear Med, 75-85. *Honors & Awards:* Distinguished Serv Award, Soc Nuclear Med. *Mem:* Am Soc Clin Path; Col Am Path; Int Acad Path; Soc Nuclear Med; Sigma Xi. *Res:* Application of radionuclides for in-vivo study and diagnosis of neoplastic, inflammatory and congenital disease; automation of microbiology by means of in-vitro radionuclide methodology. *Mailing Add:* Vet Admin Med Ctr 800 Irving Ave Syracuse NY 13210

DELANEY, BERNARD T, b Cornwall, NY, April 2, 44; m 69; c 2. ENVIRONMENTAL HEALTH ENGINEERING. *Educ:* Univ NMex, BS, 68, MS, 72; Univ Tex, PhD(environ health eng), 76. *Prof Exp:* Proj mgr, Exxon Res & Eng, 76-78; vpres, Fred C Hart & Assoc, 78-82; exec vpres, Ground Water Technol, 82-84; discipline coordr, CH2M Hill, Inc, 84-86; PRES, FIRST ENVIRON INC, 86- *Concurrent Pos:* Proj engr, Air Correction Div, Universal Oil Prods,68-69; specialist - US Army, 69-71; eng technician, Air Force Weapons Lab, 71-72; asst res engr, Civil Eng Res Facil, 72-73. *Mem:* Air Pollution Control Asn; Am Inst Chem Engrs; Am Soc Civil Engrs. *Res:* Environmental health engineering. *Mailing Add:* Moretrench Am Corp 80 Jackson St Keyport NJ 07735

DELANEY, EDWARD JOSEPH, b Evergreen Park, Ill, Apr 28, 53; m 77. CHEMISTRY. *Educ:* Loyola Univ, Bs, 74, Univ Ill Med Ctr, PhD(med chem), 80. *Prof Exp:* Teaching asst gen chem, Univ Ill Med Ctr, 74-80; FEL, NORTHWESTERN UNIV, 80- *Mem:* Am Chem Soc; Sigma Xi. *Mailing Add:* 137 Cherry Brook Dr RD 5 Princeton NJ 08540

DELANEY, JOHN P, b St Paul, Minn, Oct 1, 30; m 60; c 2. SURGERY, PHYSIOLOGY. *Educ:* Univ Minn, BS, 53, MD, 55, PhD(physiol), 63. *Prof Exp:* Univ fel, 59-67, from asst prof to assoc prof, 67-80, PROF SURG, UNIV MINN, MINNEAPOLIS, 80- *Concurrent Pos:* USPHS fel, 63- *Res:* Splanchnic blood flow; cause of peptic ulcer. *Mailing Add:* Dept Surg Box 89 Univ Minn Hosps Minneapolis MN 55455

DELANEY, MARGARET LOIS, b Buffalo, NY, July 7, 55. MARINE GEOCHEMISTRY, PALEOCEANOGRAPHY. *Educ:* Yale Univ, BS, 77; Mass Inst Technol-Woods Hole Oceanog Inst, PhD(oceanog), 83. *Prof Exp:* Asst res marine scientist, Scripps Inst Oceanog, 83-84; asst prof, 83-90, ASSOC PROF MARINE SCI, UNIV CALIF, SANTA CRUZ, 90- *Mem:* Am Geophys Union; Geochem Soc; Sigma Xi. *Res:* Marine geochemistry, paleoceanography, geochemical cycles, biogenic calcite and aragonite composition; effects of diagenesis on marine sediments. *Mailing Add:* Inst Marine Sci Univ Calif Santa Cruz CA 95064

DELANEY, PATRICK FRANCIS, JR, b Fall River, Mass, Mar 11, 33; m 54; c 4. METABOLISM, ACADEMIC ADMINISTRATION. *Educ:* Providence Col, AB, 54; Brown Univ, AMT, 61, PhD(biol), 64. *Prof Exp:* Teacher, Mass High Schs, 55-59, chmn, dept sci & math, 60-61; asst biol, Brown Univ, 61-63; asst prof biochem & physiol, Col Holy Cross, 64-67, assoc prof biol, 67-69; prof biol & chmn dept, Lindenwood Cols, 69-79, dean, Lindenwood Col For Men, 71-79; vpres Acad Affairs, Fitchburg State Col, 79-90; VPRES ACAD AFFAIRS, ELMS COL, 90- *Concurrent Pos:* Nat Inst Arthritis & Metab Dis res grant, 64-70; consult, Dept Internal Med, Med Sch, St Louis Univ, 70-71; mem, Am Conf Acad Deans & Coun Cols Arts & Sci; NSF grants; Fund Improv Postsecondary Educ grants. *Mem:* AAAS; NY Acad Sci; Am Asn Higher Educ; Sigma Xi; Am Mgt Asn; Am Asn State Cols & Univs; Am Asn Col Teacher Educ. *Res:* Renal metabolism; mitochondrial protein synthesis; hormonal control of metabolism; effect of acidosis on kidney metabolism. *Mailing Add:* Elms Col 291 Springfield St Chicopee MA 01013

DELANEY, ROBERT, b Pittsfield, Mass, May 20, 28; m 53; c 8. BIOCHEMISTRY. *Educ:* Boston Col, BS, 52; Albany Med Col, PhD(biochem), 63. *Prof Exp:* Res scientist, Div Labs & Res, NY State Dept Health, 60-63; Nat Inst Arthritis & Metab Dis fel, Med Ctr, Duke Univ, 63-66, res assoc biochem, 65-66; from asst prof to assoc prof, 66-74, PROF BIOCHEM, SCH MED, UNIV OKLA, 74- *Mem:* AAAS; Am Soc Biochem & Molecular Biol; Sigma Xi. *Res:* Protein structure of acid proteases carboxylated proteins and bacterial toxins. *Mailing Add:* Dept Biochem & Molecular Biol PO Box 26901 Univ Okla Health Sci Ctr Oklahoma City OK 73190

DELANEY, ROBERT MICHAEL, b Wood River, Ill, Nov 13, 31. THEORETICAL PHYSICS. *Educ:* St Louis Univ, BS, 53, PhD(physics), 58. *Prof Exp:* From asst prof to assoc prof, 56-68, PROF PHYSICS, ST LOUIS UNIV, 68- *Mem:* Am Phys Soc; Sigma Xi. *Res:* Thermal neutron scattering; relativistic wave equations; high energy physics; scattering of polarized particles; resonant states in quantum mechanics. *Mailing Add:* 248D Glandore Dr Manchester MO 63021

DELANGE, ROBERT J, b Richfield, Utah, Mar 30, 37; m 60; c 6. BIOCHEMISTRY, PROTEIN CHEMISTRY. *Educ:* Brigham Young Univ, BS, 61; Univ Wash, PhD(biochem), 65. *Prof Exp:* Fel, 65-67,from asst prof to assoc prof, 67-77, PROF BIOL CHEM, SCH MED, UNIV CALIF, LOS ANGELES, 77- *Concurrent Pos:* Guggenheim fel, 73-74. *Mem:* Am Soc Biochem & Molecular Biol; AAAS. *Res:* Structure-function relationships of chromosomal and other proteins; amino acid sequences; role of naturally occurring derivatives of amino acids; antioxidation; antimicrobiols. *Mailing Add:* Dept Biol Chem Univ Calif Sch Med Los Angeles CA 90024-1737

DELANGLADE, RONALD ALLAN, b Indianapolis, Ind, May 20, 36; m 62; c 2. PLANT MORPHOLOGY. *Educ:* Wabash Col, BA, 58; Purdue Univ, MS, 61, PhD(morphol), 64. *Prof Exp:* Instr biol, Wabash Col, 61-64; from asst prof to assoc prof, Eastern Ky Univ, 64-67; from asst prof to assoc prof, 67-77, chmn dept, 74-91, PROF BIOL, WITTENBERG UNIV, 77- *Mem:* AAAS; Bot Soc Am. *Res:* Plant succession in old-fields; leaf morphogenesis. *Mailing Add:* Dept Biol Box 720 Springfield OH 45501

DELANO, ERWIN, b New York, NY, June 27, 26; m 55. OPTICS. *Educ:* Yale Univ, BS, 50; Univ Rochester, MS, 56, PhD(optics), 66. *Prof Exp:* Engr, Bausch & Lomb Optical Co, 51-53, sect head optical design, 53-58, dept head optical design & comput, Bausch & Lomb Inc, 58-63; from asst prof to assoc prof, 63-69, chmn dept, 66-72, 74-81 & 83-84, PROF PHYSICS, ST JOHN FISHER COL, 69- *Concurrent Pos:* Optical consult. *Mem:* Fel Optical Soc Am; NY Acad Sci. *Res:* Geometrical optics, theory and practice of lens design; methods of synthesis for dielectric multilayer interference filters; primary aberrations of fresnel lenses. *Mailing Add:* 271 Village Lane Rochester NY 14610

DE LA NOUE, JOEL JEAN-LOUIS, b Ferryville, Tunisia, Mar 18, 38; m 59, 80; c 3. CELL PHYSIOLOGY, NUTRITION. *Educ:* Univ Laval, Quebec, BSc, 60, DSc (biol),68. *Prof Exp:* Asst prof biol, Laval Univ, 64-68; fel cell physiol, Univ Sheffield, 68-69; res assoc zool, Univ Wash, 69-70; adj prof biol, Laval Univ, 70-72, from assoc prof to prof biol, 80-88, dir nutrit, 78-88, HEAD, FOOD SCI TECH DEPT, LAVAL UNIV, 89- *Concurrent Pos:* NATO fel, 68-70. *Mem:* Can Soc Zoologists; AAAS; French Can Asn Advan Sci; Nutrit To-Day Soc; Aquacult Asn Can. *Res:* Wastewater treatment through selective biological utilization of nutrients; invertebrate and fish nutrition. *Mailing Add:* Dept Food Sci & Technol Laval Univ Ste Foy Quebec PQ G1K 7P4 Can

DELANSKY, JAMES F, b Philipsburg, Pa, July 27, 34; m 59. ELECTRICAL ENGINEERING. *Educ:* Pa State Univ, BS, 62; Cornell Univ, MS, 64, PhD(elec eng, math), 68. *Prof Exp:* Technician, Electronic Tube Div, Westinghouse Elec Corp, 57 & Remington Rand Corp, 58; asst prof elec eng, 68-74, ASSOC PROF ELEC ENG, PA STATE UNIV, 74- *Mem:* Inst Elec & Electronics Engrs. *Res:* Circuit theory; synthesis of passive and active networks, including networks of several variables. *Mailing Add:* Dept Elec Eng 304 EEW Pa State Univ University Park PA 16802

DELAP, JAMES HARVE, b Carbondale, Ill, Feb 6, 30; m 59; c 4. KINETICS, SPECTROSCOPY. *Educ:* Southern Ill Univ, BA, 52; Duke Univ, MA, 59, PhD(chem), 60. *Prof Exp:* Chemist, Universal Match Corp, 55-56; res chemist, Chemstrand Res Ctr, 60-62; from asst prof to assoc prof, 62-72, PROF CHEM, STETSON UNIV, 72- *Concurrent Pos:* Mem rev panel, NSF, 68-69; mem vis comt, Southern Asn Cols & Schs, 68-75; reader adv placement, Educ Testing Serv, 72-81 & 85, contribr, grad rec exam chem, 75-81. *Honors & Awards:* Fulbright lectr, Tribhuvan Univ, Katmandu, Nepal, 70-71 & Univ Liberia, Monrovia, Liberia, 78-79. *Mem:* Am Chem Soc; Sigma Xi; Soc Physics Students; Union Concerned Scientists. *Res:* Mechanisms of photochemical and chemiluminescent reactions; demonstrations for science classroooms. *Mailing Add:* Box 8277 Stetson Univ Deland FL 32720

DE LA PENA, RAMON SERRANO, b San Jacinto, Pangasinan, Philippines, Oct 2, 36; US citizen; m 63; c 4. AGRONOMY, PLANT PHYSIOLOGY. *Educ:* Univ Philippines, BSc, 58; Univ Hawaii, MSc, 64, PhD(agron), 67. *Prof Exp:* Asst agronomist, 67-76, assoc agronomist, 76-83, AGRONOMIST & PROF, UNIV HAWAII, 83- *Concurrent Pos:* Rice Training Off, Rice Training Ctr, Univ Hawaii, 68-72; dir econ develop, County of Kauai, 73-74; vis sci, Philippine Root Crop Res Ctr, 76; prin investr, Coop State Res Serv USDA, 76-79; consult, AID, Thailand, 74, World Bank, Thailand, 78, Universe Tankships, 80 & Nat Acad Sci, Indonesia, 81. *Mem:* Am Soc Agron; Soil Sci Soc Am; Int Soil Sci Soc; Int Soc Trop Root Crops; Int Soc Hort Sci. *Res:* Management, breeding, production and physiology of tropical root and tuber crops, legumes and cereals; weed control in tropical root crops with emphasis on taro, Colocasia esculenta. *Mailing Add:* 6155 Kala Kea Place Kapaa HI 96746-8643

DELAPP, TINA DAVIS, b Los Angeles, Calif, Dec 18, 46; m 69; c 2. HIGHER EDUCATION. *Educ:* Ariz State Univ, BSN, 69; Univ Colo, MS, 72; Univ Southern Calif, EdD, 86. *Prof Exp:* Staff nurse, Tempe Community Hosp, 68-69, Alaska Native Hosp, 69-70; health aid instr, Dept Health Care, Yukon-Kuskokwim Health Corp, 70-71; asst prof nursing, Bacone Col, 72-74; instr nursing, Alaska Methodist, Univ, 75-76; PROF NURSING & ASSOC DEAN, UNIV ALASKA, ANCHORAGE, 76- *Concurrent Pos:* Site vis, Nat League Nursing, 88-; mem, Nat Coun State Bd Nursing-Comt Nursing Pract & Educ, 89-, Alaska Bd Nursing, 89- *Res:* Higher education; stress experienced by ill adults. *Mailing Add:* 13101 S Bragaw Anchorage AK 99516

DELAPPE, IRVING PIERCE, b Boston, Mass, Oct 28, 15; m 42; c 3. SCIENCE ADMINISTRATION. *Educ:* Harvard Univ, BS, 42, AM, 46, PhD, 53. *Prof Exp:* Teaching fels, Harvard Med Sch, 47 & Harvard Biol Labs, 47-48; asst epidemiol, Sch Pub Health, 48; asst prof bact & pub health, Mich State Col, 48-54; adminstr, Am Cyanamid Co, 54-60; exec secy microbiol panel, 60-62, asst chief fel & career awards, 62-65, chief biochem & physiol br, 65-76, chief parasitol & med entomol, 74-76, chief molecular microbiol & parasitol, 77-86, CHIEF, MOLECULAR MICROBIOL, PHYSIOL BR,

NAT INST ALLERGY & INFECTIOUS DIS, NAT INSTS HEALTH, 86- *Res:* Physiology of Histomonas meleagridis, Trichomonas gallinarum, Entamoeba histolytica studies; alliance-prudent use of antibiotics. *Mailing Add:* 8907 Ridge Pl Bethesda MD 20817-3363

DE LA SIERRA, ANGELL O, b Santurce, PR; m 60; c 5. NEUROSCIENCES, CELL BIOPHYSICS. *Educ:* Univ PR, Rio Piedras, BS, 54; City Univ New York, MS, 58; St John's Univ, NY, PhD(cell biophys), 63. *Prof Exp:* Res analyst biophys, Smithsonian Inst, 63-64; res chemist, Armed Forces Radiobiol Res Inst, Dept Defense, 64-65; NIH fel biophys, Col Med, Georgetown Univ, 65-67; vis prof, Col Med, Univ PR, San Juan, 67, mem fac, 67-68; dir fac natural sci, Univ PR, Cayey, 68-72; CONSULT, 72- *Concurrent Pos:* Numerous books published, 63-87; med malpract lawyer. *Mem:* NY Acad Sci; Sigma Xi; Am Bar Asn. *Res:* Electromyogram recordings of facial musculature and "demeanor"; neuropharmacology of behavior. *Mailing Add:* Dept Biol Univ Col Cayey PR 00633

DE LA TORRE, JACK CARLOS, b Paris, France, Dec 2, 37; US citizen; m 87. NEUROSURGERY, NEUROANATOMY. *Educ:* Am Univ, BS, 61; Univ Geneva, PhD(neuroanat), 68; Univ Juarez, MD, 79. *Prof Exp:* From instr to assoc prof neurol surg & psychiat, Pritzker Sch Med, Univ Chicago, 69-78; assoc prof neurol surg, Sch Med, Univ Miami, 79-82 & Med Sch, Northwestern Univ, 82-83; PROF NEUROL SURG, FAC MED, UNIV OTTAWA, 83- *Concurrent Pos:* Conf chmn, Bio Actions & Med Applns of Dimethyl Sulfoxide, NY Acad Sci, 82; fel, Stroke Coun, Am Heart Asn, 91. *Honors & Awards:* Paddington Award, 83. *Mem:* NY Acad Sci; Inter Am Col Physicians & Surgeons; Soc Neurosci; Am Acad Neurol; Int Brain Res Orgn; Royal Soc Med Gt Brit; Cajal Club. *Res:* Strokes; dementia; nuclear magnetic resonance spectroscopy; central nervous system regeneration; memory. *Mailing Add:* Div Neurosurg 451 Smyth Rd Ottawa ON K1H 8M5 Can

DE LATOUR, CHRISTOPHER, b Nov 24, 47; US citizen; m 70. ENVIRONMENTAL PHYSICS. *Educ:* Georgetown Univ, BS, 69; Mass Inst Technol, SM, 71, PhD(physics), 74. *Prof Exp:* Staff res scientist environ physics, Francis Bitter Nat Magnet Lab, Mass Inst Technol, 75-77; LECTR, PHYSICS DEPT, CALIF POLYTECH STATE UNIV, 77- *Mem:* Am Phys Soc; Sigma Xi. *Res:* Improvement and maintenance of environmental quality; development of advanced methods of water purification. *Mailing Add:* PO Box 14438 San Francisco CA 94114-0438

DE LAUBENFELS, DAVID JOHN, b Pasadena, Calif, Dec 5, 25; div; c 4. GEOGRAPHY, BOTANY. *Educ:* Colgate Univ, AB, 49; Univ Ill, AM, 50, PhD(geog), 53. *Prof Exp:* From asst prof to assoc prof geog & geol, Univ Ga, 53-59; assoc prof, 59-71, PROF GEOG, SYRACUSE UNIV, 71- *Concurrent Pos:* Bowman fel, Johns Hopkins Univ, 55-56. *Mem:* Asn Am Geogrs; Am Geog Soc; Ecol Soc Am; Int Soc Plant Morphologists. *Res:* Vegetation geography; morphology and taxonomy of conifers; climatology. *Mailing Add:* Dept Geog Syracuse Univ Syracuse NY 13244-1160

DELAUER, R(ICHARD) D(ANIEL), aeronautics; deceased, see previous edition for last biography

DELAUNE, RONALD D, b Baton Rouge, La, June 16, 43. BIOGEOCHEMISTRY. *Educ:* La State Univ, BS, 65, MS, 68; Wageningen Univ, Neth, PhD(soil sci & geol), 88. *Prof Exp:* PROF SOIL SCI, LA STATE UNIV, 68- *Mem:* Sigma Xi; Soc Wetland Scientists. *Res:* Sedimentation rates in salt marshes; sea level rise and land lost in Louisiana Gulf Coast marshes; environmental aspects of coastal wetlands. *Mailing Add:* La State Univ Wetlands Soils & Sediments Ctr Wetland Resources Baton Rouge LA 70803

DELAWARE, DANA LEWIS, b Gardiner, Maine, Mar 16, 51; m 79; c 1. CARBOHYDRATE CHEMISTRY, ANTIBIOTICS. *Educ:* Marist Col, BA, 73; Purdue Univ, PhD(med chem & pharmacog), 78. *Prof Exp:* Teaching asst chem, Purdue Univ, 73-75, res asst, 76; res asst, Univ Ariz, 76-78; res asst, Univ Ill, 78-80, vis asst prof, 78-79; ASST PROF ORG CHEM, NORTHEAST MO STATE UNIV, 80- *Mem:* Am Chem Soc; AAAS; Sigma Xi. *Res:* Investigation of structural activity-relationships of carbohydrate derived antibiotics and synthesis of inhibitors of Aldolase enzymes. *Mailing Add:* Div Sci Nohrteast Mo St Univ Kirksville MO 63501

DE LAY, ROGER LEE, b Waukee, Iowa, Aug 3, 45; m 67; c 2. RUMINANT NUTRITION. *Educ:* Iowa State Univ, BS, 67; Colo State Univ, MS, 69, PhD(ruminant nutrit), 72. *Prof Exp:* Res nutritionist, 72-76, develop mgr ruminant prods, 76-80, develop mgr animal prods, int, 80-84, group leader Animal Nutrit & Physiol Discovery, 84-86, sr prod develop mgr, 86-87, MGR LATIN AM ANIMAL PROD DEVELOP, AM CYANAMID CO, 87- *Mem:* Am Soc Animal Sci. *Res:* Growth promotion and feed utilization in beef cattle and sheep; animal feed and health products that will result in improved rate, efficiency and/or quality of animal production. *Mailing Add:* Am Cyanamid Co PO Box 400 Princeton NJ 08540

DELBECQ, CHARLES JARCHOW, b Toledo, Ohio, Aug 19, 21; m 47; c 3. SOLID STATE CHEMISTRY. *Educ:* Univ Toledo, BS, 43; Univ Ill, PhD(chem), 49. *Prof Exp:* Assoc chemist, 49-62, SR CHEMIST, ARGONNE NAT LAB, 62- *Mem:* Am Phys Soc; Sigma Xi. *Res:* Study of imperfections and impurities, and the trapping of electrons and holes in solids, especially the alkali halides. *Mailing Add:* 1591 Placito Embate 107-7 Green Valley AZ 85614

DEL BEL, ELSIO, b Worthington, Ont, Can, Jan 28, 20; US citizen. ORGANIC CHEMISTRY. *Educ:* Clarkson Tech Col, BS, 41; Columbia Univ, MA, 48; Pa State Univ, PhD(fuel technol), 51. *Prof Exp:* Metallurgist, Rochester Ord Dist, War Dept, 41-42; chemist pharmaceut prod, Ayerst Labs, 45-47; res org chemist, Res Div, Consol Coal Co, 51-71; supvry chem engr, Pittsburgh Energy Res Ctr, Bur Mines, 72-74, Dept Energy, 74-82; RETIRED. *Res:* Chemistry of phenolic compound, aryl mercaptans and coal chemicals; conversion of coal and organic waste to liquid fuels. *Mailing Add:* 5697 Library Rd Bethel Park PA 15102

DEL BENE, JANET ELAINE, b Girard, Ohio, June 3, 39. THEORETICAL CHEMISTRY. *Educ:* Youngstown State Univ, BS, 63, BA, 65; Univ Cincinnati, PhD(chem), 68. *Prof Exp:* Fel chem, Theoret Chem Inst, Univ Wis, 68-69 & Mellon Inst, 69-70; from asst prof to assoc prof, 70-76, PROF CHEM, YOUNGSTOWN STATE UNIV, 76- *Concurrent Pos:* Grants, Am Chem Soc, 72, Nat Inst Gen Med Sci, 74, Camille & Henry Dreyfus Found, 74 & Nat Inst Gen Med Sci, 80, 85; consult basic med sci, Col Med, Northeastern Ohio Univ, 76, res prof biochem & molecular biol, 77-; mem grad fac, molecular path & biol area comt, Kent State Univ, 78-; mem selection comt, Langmuir Award, Am Chem Soc, 78-83 & comput chem, 84-85; vis prof, Carnegie Mellon Univ, 80-81 & Ohio State Univ, 88-89; consult study group, NIH, 88; mem rev panel, NSF, 90. *Mem:* fel AAAS; Am Chem Soc; NY Acad Sci; Sigma Xi. *Res:* Molecular orbital theory of chemical bonding, structure, and spectroscopy; computational chemistry; molecular orbital theory of hydrogen bonding and ion-molecule interactions; basis set and electron correlation effects on interaction energies. *Mailing Add:* Dept Chem Youngstown State Univ Youngstown OH 44555

DEL BIANCO, WALTER, b Firenze, Italy, Feb 28, 33; m 67. NUCLEAR PHYSICS. *Educ:* Univ Rome, Dr, 58; Univ Pa, PhD, 61. *Prof Exp:* Res assoc, Univ Pa, 61-62; res physicist, Nuclear Physics Div, Max Planck Inst Chem, 62-65; from asst prof to assoc prof, 65-77, PROF PHYSICS, UNIV MONTREAL, 77-, AT NUCLEAR PHYSICS LAB. *Mem:* Am Phys Soc; Can Asn Physicists. *Res:* Photonuclear reactions. *Mailing Add:* Nuclear Physics Lab Univ Montreal PO Box 6128 Montreal PQ H3C 3J7 Can

DELCAMP, ROBERT MITCHELL, b Lexington, Ky, Apr 18, 19; m 40; c 3. ORGANIC CHEMISTRY. *Educ:* Transylvania Col, AB, 39; Univ Cincinnati, MA, 41, PhD(chem), 54. *Prof Exp:* Shift supvr, E I du Pont de Nemours & Co, 41-44; shift supvr, Tenn Eastman Corp, 44-45; instr, 45-48, from asst prof to prof, 48-58, asst dean, 62-67, assoc dean, 67-78, actg dean, 76-78, prof org chem, 58-65, PROF CHEM ENG, UNIV CINCINNATI, 65- *Concurrent Pos:* Consult, Darling & Co, 47-48, Charles Straus & Assocs, 49-51 & Tanner's Coun Res Lab, Cincinnati, 54-68. *Mem:* Am Chem Soc; Am Soc Eng Educ; Am Inst Chem Engrs; Sigma Xi. *Res:* Base-catalyzed condensation of aromatic aldehydes with compounds containing active hydrogen; enzymatic assay using epidermis as substrate; microbiological conversion of raw materials into useful organic chemicals; resin modification of leather; fermentation processes. *Mailing Add:* 160 Wentworth Ave Cincinnati OH 45220

DEL CASTILLO, JOSE, b Salamanca, Spain, Dec 25, 20; US citizen; m 55; c 2. NEUROPHYSIOLOGY. *Educ:* Lit Univ Salamanca, MB, 45; Univ Madrid, MD, 47. *Prof Exp:* Asst prof physiol, Fac Med, Lit Univ Salamanca, 45-46; Sandoz res fel, Physiol Inst, Bern, 48-50; asst lectr, Univ Col, Univ London, 50-52, lectr biophys, 52-53, lectr, 53-56; vis prof physiol, State Univ NY Downstate Med Ctr, 56-57; chief sect clin neurophysiol, Nat Inst Neurol Dis & Blindness, Md, 57-59; assoc prof pharmacol, 59-60, PROF PHARMACOL & HEAD DEPT, SCH MED, UNIV PR, 60-, DIR LAB NEUROBIOL, 67- *Concurrent Pos:* Nat Inst Neurol Dis & Blindness career res prof grant, 62. *Mem:* Am Physiol Soc; Int Brain Res Orgn; Brit Physiol Soc. *Res:* Synaptic physiology and pharmacology; comparative neurophysiology. *Mailing Add:* Dept Pharmacol Univ PR Sch Med G PO Box 5067 San Juan PR 00936

DEL CERRO, MANUEL, b Buenos Aires, Arg, Aug 20, 31; m 57; c 2. NEUROSCIENCES, DEVELOPMENTAL BIOLOGY. *Educ:* Univ Buenos Aires, MD, 58. *Prof Exp:* Sr instr histol, Sch Med, Univ Buenos Aires, 58-61, assoc prof, 61-64; res assoc neuroanat, Univ Rochester, 65-69, sr res assoc, 69-71, assoc prof, 71-80, PROF NEUROBIOL & ANAT, CTR VISUAL SCI, STRONG MEM HOSP, UNIV ROCHESTER, 80-, ASSOC PROF OPHTHALMOL, 80- *Concurrent Pos:* Prof natural sci, Univ Buffalo, NY, 90- *Mem:* Int Brain Res Asn; Soc Neurosci; fel Royal Microscopical Soc; Int Soc Eye Res; Asn Res Vision Ophthalmol. *Res:* Eye research; experimental study of neurogenesis; growth and regeneration of retina and eye; action of ultrasounds and microwaves on neural development; study of retinal transplants into blind retinas, as a mean to treat retinal impairment using clinical, physiological, quantitative light and electron microscopy approaches; study of the effects of the AIDS virus on fetal human brain and retina. *Mailing Add:* Dept Neurobiol & Anat Box 603 Univ Rochester Med Sch Rochester NY 14642

DELCO, EXALTON ALFONSO, JR, b Houston, Tex, Sept 4, 29; m 52; c 4. VERTEBRATE ECOLOGY. *Educ:* Fisk Univ, AB, 49; Univ Mich, MS, 50; Univ Tex, PhD(zool), 62. *Prof Exp:* From instr to assoc instr biol, Tex Southern Univ, 50-57; res asst zool, Univ Tex, 58-60; from asst prof to assoc prof, 59-63, vpres acad affairs 67-85, PROF BIOL, HUSTON-TILLOTSON COL, 63-, DEAN, 67- *Concurrent Pos:* Co-investr, NSF Grant, 58-62; vpres acad affairs, Austin Community Col, 85- *Honors & Awards:* Stoye Prize, Am Soc Ichthyologists & Herpetologists, 60. *Mem:* Fel AAAS; Am Fisheries Soc; Am Soc Ichthyologists & Herpetologists; Ecol Soc Am. *Res:* Vertebrate speciation and ethology, especially isolating mechanisms in the vertebrate group. *Mailing Add:* Dept Acad Affairs Austin Community Col PO Box 140526 Austin TX 78714

DELCOMYN, FRED, b Copenhagen, Denmark, June 4, 39; US citizen; m 69; c 3. NEUROSCIENCE. *Educ:* Wayne State Univ, BS, 62; Northwestern Univ, MS, 64; Univ Ore, PhD(physiol), 69. *Prof Exp:* Res assoc zool, Univ Glasgow, 69-71; lectr physiol, Inst Physiol, Univ Glasgow, 71-72; asst prof entom, 72-77, ASSOC PROF ENTOM, UNIV ILL, 77-, ASSOC PROF DEPT PHYSIOL, 77- *Concurrent Pos:* Assoc prof, Dept Physiol, Univ Ill, 77-; vis prof, Univ Alta, Dept Physiol, 80-81; sr Fulbright Scholar, Dept Biol, Univ Kaiserslautern, WGer, 87-88. *Mem:* Soc Exp Biol; fel AAAS; Am Soc Zoologists; Soc Neurosci. *Res:* Physiological basis of behavior in invertebrates; neural basis of coordination; locomotion. *Mailing Add:* Dept Entom Univ Ill 505 S Goodwin Urbana IL 61801

DELCOURT, PAUL ALLEN, b Clifford, Mich, Feb 10, 49; m 71; c 1. QUATERNARY PALEOECOLOGY, GEOLOGY. *Educ:* Albion Col, BA, 71, La State Univ, MS, 74; Univ Minn, PhD(geol), 78. *Prof Exp:* ASSOC PROF QUATERNARY PALEOECOLOGY, DEPT GEOL SCI & GRAD PROG ECOL, UNIV TENN, KNOXVILLE, 78- *Honors & Awards:* Res Award, Asn Southeastern Biologists, 74; Murray Buell Award, Ecol Soc Am, 78. *Mem:* Ecol Soc Am; Am Quaternary Asn. *Res:* Quantitative landscape development; quaternary paleovegetation and geomorphology; modern vegetation-pollen calibrations; pollen and plant-macrofossil identification. *Mailing Add:* Prog Quaternary Studies Dept Geol Sci Univ Tenn Knoxville TN 37996

DELEANU, ARISTIDE ALEXANDRU-ION, b Pitesti, Romania, Apr 12, 32; US citizen; m 56. ALGEBRAIC TOPOLOGY, CATEGORY THEORY. *Educ:* Bucarest Univ, MS, 55, PhD(math), 61; Inst Civil Eng, MS, 55. *Prof Exp:* Asst prof math, Univ Bucarest, 55-58; res fel, Inst Math, Romanian Acad, 60-66, div head, 66-68; vis assoc prof, 68-71, PROF MATH, SYRACUSE UNIV, 71- *Concurrent Pos:* NSF res grant, 73-75. *Honors & Awards:* Simion Stoilow Prize, Romanian Acad Sci, 66. *Mem:* Am Math Soc. *Res:* Generalized completions in the sense of J Frank Adams with applications to localizations and completions in homotopy theory; categorical aspects of the theory of shape in the sense of Karol Borsuk. *Mailing Add:* Dept Math Syracuse Univ 15 Carnegie Hall Syracuse NY 13244-1150

DELEEUW, J H, b Amsterdam, Netherlands, Jan 4, 29; Can citizen; m 59; c 3. AEROSPACE ENGINEERING. *Educ:* Delft Univ Technol, Dipl eng, 53; Ga Inst Technol, MS, 52; Univ Toronto, PhD(aerophys), 58. *Prof Exp:* Res engr aerodyn, Nat Aeronaut Res Inst, Netherlands, 52-53; from asst prof to prof aerospace eng, 58-76, asst dir res, 70-76, PROF & DIR INST AEROSPACE STUDIES, UNIV TORONTO, 76- *Mem:* Am Inst Aeronaut & Astronaut; Can Aeronaut & Space Inst. *Res:* Rarefied plasma and gasdynamics; rocket sounding of upper atmosphere. *Mailing Add:* Inst for Aerospace Studies Univ Toronto Toronto ON M5S 1A1 Can

DELEEUW, SAMUEL LEONARD, b Grand Rapids, Mich, Aug 2, 34; m 56; c 3. ENGINEERING MECHANICS, CIVIL ENGINEERING. *Educ:* Mich State Univ, BS, 56, MS, 58, PhD(appl mech), 61. *Prof Exp:* Asst prof civil eng, Yale Univ, 60-65; prof civil eng & chmn dept, 65-86, PROF CIVIL ENG, UNIV MISS, 86- *Concurrent Pos:* Int Bus Mach Corp res assoc, Mass Inst Technol, 62-63. *Mem:* Am Soc Civil Engrs; Am Soc Eng Educ; Am Soc Mech Engrs; Nat Soc Prof Engrs; Miss Eng Soc; Miss Acad Sci. *Res:* Bending and buckling of viscoelastic columns and plates with large deflections; transportation; water resources. *Mailing Add:* Dept Civil Eng Univ Miss University MS 38677

DELELLIS, RONALD ALBERT, ENDOCRINOLOGY, IMMUNOHISTOCHEMISTRY. *Educ:* Tufts Univ, MD, 66. *Prof Exp:* PROF PATH, SCH MED, TUFTS UNIV, 84- *Mailing Add:* Dept Path Tufts Univ Sch Med 136 Harrison Ave Boston MA 02178

DE LEMOS, CARMEN LORETTA, b Santo Domingo, Dominican Repub, May 6, 37. CELL BIOLOGY. *Educ:* Hunter Col, BA, 60; NY Univ, MS, 66, PhD(biol, basic med sci), 72. *Prof Exp:* Asst res scientist, 60-72, instr cell biol, 72-77, RES ASST PROF, NY UNIV, 77- *Mem:* Am Soc Cell Biol; Am Asn Anat; Am Soc Zool; Soc Develop Biol. *Res:* Development of human gastric mucosa; topography of membrane glyco proteins of subcellular organelles by fractionation, electrophoresis, enzymatic analysis and electron microscopic lectin binding techniques. *Mailing Add:* Dept Cell Biol NY Univ 550 First Ave New York NY 10016

DELEON, ILDEFONSO R, b Marathon, Tex, Jan 23, 47; m 70; c 3. MASS SPECTROMETRY, GAS CHROMATOGRAPHY. *Educ:* Sul Ross State Univ, Alpine, Tex, BS, 72. *Prof Exp:* Prog mgr & res chemist, Ctr Bio-Org Studies, Univ New Orleans, 76-88; ENVIRON AFFAIRS SUPVR, NOSWB, NEW ORLEANS, 88- *Concurrent Pos:* Co-investr, 80-84, Ctr Bio-Org Studies, Univ New Orleans, prin investr, 84- *Res:* Environmental affairs, bio-enviroanalytical research. *Mailing Add:* SPL Inc 7939 Devine Ave New Orleans LA 70127-1121

DELEON, MORRIS JACK, b Seattle, Wash, June 21, 41; div; c 1. NUMBER THEORY. *Educ:* Univ Calif, Los Angeles, BA, 63; Univ Ill, MS, 64; Pa State Univ, PhD(math), 68. *Prof Exp:* PROF MATH, FLA ATLANTIC UNIV, 68- *Mem:* Am Math Soc; Math Asn Am. *Res:* Integer sequences; diophantine and congruence equations. *Mailing Add:* Dept Math Fla Atlantic Univ Boca Raton FL 33431-0991

DE LEON, VICTOR M, b Trujillo Alto, PR, Nov 13, 39. CELLUAR MOLECULAR BIOLOGY. *Educ:* Cornell Univ, PhD(celluar molecular biol), Cornell Univ, 74. *Prof Exp:* RES ASSOC BIOL, CORNELL MED COL, 74-; PROF BIOL, HOSTOS COMMUNITY COL, CITY UNIV NY, 82- *Concurrent Pos:* Site vis, Minority Biomed Res Support Prog, NIH, 83-87; chmn, Dept Biol, Hostos Community Col, 85-87, dean Academic Affairs, 87-89. *Mem:* Am Soc Cell Biol; AAAS. *Res:* Biology of RNA's. *Mailing Add:* Hostos Community Col 475 Grand Concourse Bronx NY 10451

DE LEONIBUS, PASQUALE S, oceanography, meteorology, for more information see previous edition

DELERAY, ARTHUR LOYD, b Sonora, Calif, June 27, 36; m 61; c 2. CHEMICAL & NUCLEAR ENGINEERING. *Educ:* Univ Calif, Berkeley, BSE, 59; Princeton Univ, MSE & MA, 62, PhD(chem eng), 66. *Prof Exp:* Staff chem engr, MB Assocs, 64-70; prof, Chabot Col, 70-76, Valley Campus, 76-88; PROF, LAS POSITAS COL, 88- *Concurrent Pos:* Consult solar energy; pres, Acad Senate. *Honors & Awards:* NSF grant. *Mem:* Sigma Xi. *Res:* Fast neutron moderation; pyrotechnics; unique radar reflectors; miniature rocketry; solar heating applications. *Mailing Add:* Dept Sci & Math Las Positas 3033 Collier Canyon Rd Livermore CA 94550

DELESPESSE, GUY JOSEPH, b Bois-de-Lessines, Belg, June 29, 41; m 64; c 2. IMMUNOLOGY. *Educ:* Brussels Free Univ, MD, 66, PhD(med), 76. *Prof Exp:* Asst prof immunol, Brussels Free Univ, 76-81; prof, Univ Man, 81-86; PROF IMMUNOL, UNIV MONTREAL, 86- *Mem:* Am Soc Immunol; Brit Soc Immunol; Can Soc Immunol; Am Soc Allergy & Clin Immunol. *Res:* Regulation of human synthesis. *Mailing Add:* Dept Med Univ Montreal Notre Dame Hosp Res Ctr 1560 E Sherbrooke St Montreal PQ H2L 4M1 Can

DELEVIE, ROBERT, b Amsterdam, Holland, July 21, 33; m 60; c 2. ELECTROCHEMISTRY, BIOPHYSICS. *Educ:* Univ Amsterdam, Drs, 60, PhD, 63. *Prof Exp:* Vis asst prof, La State Univ, 63-65; from asst prof to assoc prof, 65-73, PROF CHEM, GEORGETOWN UNIV, 73- *Concurrent Pos:* US ed, J Electroanal Chem, 70-80. *Mem:* Am Chem Soc; Biophys Soc; Electrochem Soc; Royal Neth Chem Soc; Royal Dutch Acad Sci. *Res:* Mechanisms of electrode kinetics; phenomena of nucleation and growth; double layer effects; ion transport through membranes; electrochemical and biophysical instrumentation. *Mailing Add:* Dept Chem Georgetown Univ 37th & O St NW Washington DC 20057

DELEVORYAS, THEODORE, b Chicopee Falls, Mass, July 22, 29; m 81; c 5. PALEOBOTANY. *Educ:* Univ Mass, BS, 50; Univ Ill, MS, 51, PhD, 54. *Prof Exp:* Rockefeller fel, Nat Res Coun, Univ Mich, 54-55; asst prof bot, Mich State Univ, 55-56; from instr to asst prof, Yale Univ, 56-60; assoc prof, Univ Ill, 60-62; from assoc prof to prof biol, Yale Univ & from assoc cur to cur paleobot, Peabody Mus Natural Hist, 62-72; chmn dept bot, 74-80, chmn, Div Biol Sci, 82-89, PROF BOT, UNIV TEX, AUSTIN, 72- *Concurrent Pos:* Guggenheim Found Fel, 64-65; Pres, Int Orgn Palaeobotany, 78-81. *Honors & Awards:* Fel, Linnean Soc. *Mem:* AAAS; Int Soc Plant Morphologists; Brit Palaeont Asn; Bot Soc Am (treas, 67-72, vpres, 73, pres, 74); fel Linnean Soc. *Res:* Morphology and evolution of fossil and living vascular plants. *Mailing Add:* Dept Bot Univ Tex Austin TX 78713-7640

DELFIN, ELISEO DAIS, b San Andreas, Manila, Phillipines, Sept 28, 25; m 64. BIOLOGY, ENTOMOLOGY. *Educ:* Cent Philippines Univ, BS, 49, BSE, 51; Univ Colo, Boulder, MA, 54; Ohio State Univ, PhD(entom), 60. *Prof Exp:* Instr biol, Cent Philippines Univ, 49-52, asst prof, 60-64, chmn dept life sci, 62-64; fel acarology, Univ Calif, Berkeley, 64-65; asst prof zool, San Jose State Col, 65-67; assoc prof, 67-71, chmn dept, 69-77, prof biol, Ind Cent Col, 71-80, PROF BIOL, IND CENT UNIV, 80- *Concurrent Pos:* Consult, Philippines Bur Plant Indust, Iloilo City Br, 60-64. *Mem:* AAAS; Am Inst Biol Sci. *Res:* Biology of mites and ticks and of central Indiana; leafhoppers, especially Cicadellidae. *Mailing Add:* Dept Bio Scis 1N Central Col Indianapolis IN 46227

DELFLACHE, ANDRE P, b Brussels, Belg, 1923; nat US; m 50; c 4. CIVIL ENGINEERING, GEOPHYSICS. *Educ:* Univ Brussels, CEngMines, 47, ScD(soil mech), 64. *Prof Exp:* Asst prof eng geol, Univ Brussels, 47-48; vis prof geol, La State Univ, 48-49; engr, Explor Div, Petrofina, Brussels, 49-54; seismologist, United Geophys Co, Calif, 54; Europ mgr subsurface explor in Europe & NAfrica, Independent Explor Co, Tex, 54-58; PROF CIVIL ENG, LAMAR UNIV, 58- *Concurrent Pos:* Consult soils eng & found, 58; NSF res grant, 73-78. *Mem:* Am Soc Civil Engrs; Sigma Xi. *Res:* Geotechnique; marine geophysics; determination of geotechnical properties of soils and sediments by seismic methods; land subsidence in the Houston-Galveston region of Texas. *Mailing Add:* 730 Thicket Houston TX 77079

DELGADO, JAIME NABOR, b El Paso, Tex, July 28, 32; m 54; c 1. PHARMACEUTICAL CHEMISTRY, MEDICINAL CHEMISTRY. *Educ:* Univ Tex, BS, 54, MS, 55; Univ Minn, PhD(pharmaceut chem), 60. *Prof Exp:* Asst pharmaceut chem, Univ Minn, 55-57; from asst prof to assoc prof, 59-72, PROF PHARMACEUT CHEM, COL PHARM, UNIV TEX, AUSTIN, 72- *Concurrent Pos:* Mem res rev comt, Nat Inst Drug Abuse, HEW, 77- *Mem:* AAAS; Am Pharmaceut Asn; Am Chem Soc; Acad Pharmaceut Sci; Sigma Xi. *Res:* Synthesis of organic medicinals and structure-activity studies; natural products; mechanisms of drug-receptor interactions; anticholinergics. *Mailing Add:* Dept Pharmaceut Chem Bldg 202c Univ Tex Austin TX 78712

DELGASS, W NICHOLAS, b Jackson Heights, NY, Oct 14, 42; m 67; c 2. CATALYSIS, KINETICS. *Educ:* Univ Mich, BS, 64; Stanford Univ, MS, 66, PhD(chem eng), 69. *Prof Exp:* Asst prof eng & appl sci, Yale Univ, 69-74; assoc prof, 74-78, PROF CHEM ENG, PURDUE UNIV, 78- *Concurrent Pos:* Consult, Am Cyanamid Co, 70-76 & Amoco Oil, Inc, 77- *Honors & Awards:* Giuseppe Paravano Mem Award, 85. *Mem:* Am Inst Chem Engrs; Am Chem Soc. *Res:* Heterogeneous catalysis, especially application of Mossbauer and photoelectron spectroscopy, secondary ion mass spectrometry and transient kineteics to the study of catalysts and interactions of gases with catalytically active surfaces. *Mailing Add:* Sch Chem Eng Purdue Univ West Lafayette IN 47907-9980

DEL GRECO, FRANCESCO, b Italy, Aug 23, 23; nat US; m 56; c 1. MEDICINE. *Educ:* Univ Rome, Italy, MD, 46. *Prof Exp:* Res fel exp med, Cleveland Clin, 51-52, res assoc, 52-54, asst staff, 55-57; from intern to chief med resident, Passavant Mem Hosp, Chicago, 58-60, dir metab unit, 58-61; dir, Clin Res Ctr, 61-81, chief, sect nephrology-hypertension, 73-85, PROF MED, MED SCH, NORTHWESTERN UNIV, 67-, DIR, DIALYSIS CTR, 61-, ATTEND PHYSICIAN, 64- *Concurrent Pos:* Danish govt scholar, 51; Nat Heart Inst fel, NIH, 52-53; vol asst, Postgrad Med Sch, London, 54; hon res fel, St Thomas' Hosp, 55; vis attend physician, Vet Admin Lakeside Med Ctr, Chicago, 60-; asst to assoc prof, Northwestern Univ, 60-67; chmn med adv bd, Kidney Found Ill, 75-79; mem cardiovasc & renal study sect, Nat Heart & Lung Inst, 75-79; mem coun circulation, coun arteriosclerosis, coun kidney in cardiovasc dis & coun high blood pressure res, Am Heart Asn. *Mem:* Am Physiol Soc; Soc Exp Biol & Med; Am Soc Nephrology; Am Col Physicians; Int Soc Hypertension; Am Soc Clin Pharmacol & Therapeut. *Res:* Renal and cardiovascular physiology; hypertension. *Mailing Add:* Dept Med & Nephrol-Hypertension Northwestern Univ Med Sch Superior St Rm 446 Chicago IL 60611

DEL GROSSO, VINCENT ALFRED, b Newark, NJ, Aug 9, 25; m 51; c 5. ACOUSTICAL & OPTICAL PHYSICAL OCEANOGRAPHY. *Educ:* Northeastern Univ, BS, 47; Cath Univ Am, PhD, 68. *Prof Exp:* Phys chemist, Naval Res Lab, 48-52, physicist & head ultrasonics sect, 52-72, physicist & head, Ocean Instrumentation Sect, 72-74, consult, Ocean Technol Div, 74-81; sr engr, 81-87, PRIN SCIENTIST, E G & G-WASC, 87- *Concurrent Pos:* Res assoc, Univ Calif, Los Angeles, 70-71. *Mem:* AAAS; sr mem Inst Elec & Electronics Engrs; sr mem Am Chem Soc; fel Acoust Soc Am; Soc Photo-Optical Instrumentation Engrs. *Res:* Ultrasonics; underwater sound; physical chemistry; underwater optics and imaging. *Mailing Add:* Rte One Box 226 Edelen Dr Bryantown MD 20617

DEL GUERCIO, LOUIS RICHARD M, b New York, NY, Jan 15, 29; m 57; c 8. SURGERY, PHYSIOLOGY. *Educ:* Fordham Univ, BS, 49; Yale Univ, MD, 53. *Prof Exp:* From instr to prof surg, Albert Einstein Col Med, 60-71, assoc dir, Clin Res Ctr, 63-67, dir, 67-71; prof surg, Col Med & Dent NJ & dir surg, St Barnabas Med Ctr, 71-76; PROF & CHMN DEPT SURG, NY MED COL, 76- *Concurrent Pos:* Am Thoracic Soc teaching & res fel, 59-60; Health Res Coun NY res grants, 64-66 & career scientist award, 66; res grants, Am Heart Asn, 65-68 & NIH, 65-70; from assoc attend to attend, Bronx Munic Hosp Ctr, 60-65, pres med bd, 69; mem tech rev comt artificial heart test & eval facil, NIH, 68, mem surg study sect, Div Res Grants, 70; mem comt on shock, Nat Acad Sci-Nat Res Coun, 69-; mem merit rev bd surg, US Vet Admin, 72-; consult surg devices, Food & Drug Admin, 74- & health care technol study sect, Nat Ctr Health Serv Res, Dept Health & Human Serv, 80-83. *Honors & Awards:* Morgagni Medal, NY Acad Med. *Mem:* Am Surg Asn; Am Thoracic Soc; AMA; fel Am Col Surgeons; Soc Univ Surgeons. *Res:* Cardiorespiratory physiology in clinical practice; thoracic surgery; surgery of biliary pancreatic system. *Mailing Add:* Dept Surg NY Med Col Munger Pavilion Valhalla NY 10595

D'ELIA, CHRISTOPHER FRANCIS, b Bridgeport, Conn, Aug 7, 46; m 73; c 1. MARINE ECOLOGY, BIOLOGY. *Educ:* Middlebury Col, AB, 68; Univ Ga, PhD(zool), 74. *Prof Exp:* Scholar biol, Univ Calif, Los Angeles, 73-75; vis asst prof, Univ Southern Calif, 75; fel biol, Woods Hole Oceanog Inst, 75-77; from asst prof to assoc prof, 77-88, PROF BIOL, CHESAPEAKE BIOL LAB, UNIV MD, 88-, DIR, SEA GRANT COL, 89- *Concurrent Pos:* Consult, Century Eng, Inc Towson, Md, 80-81, Proyectos Marinos, SC, Mex, DF, 80-38 & L & Design Assocs, Huntington, NY, 83; distinguished Ruth Patrick Scholar, Acad Natural Sci, Philadelphia, 82-83; mem, Md Sea Grant Prog Sci Adv Comt, 80-84, Biol Oceanog Adv Panel Nat Sci Found Ocean Sci Div, 82-84, Point Source working Group, Chesapeake Bay Comn, 85 & Natural Resources & Environ Affairs Comt, Tri-county Coun Southern Md, 85-; mem bd dir, Md Chap, Nat Asn Environ Prof, 84-85; mem, Patuxent River Adv Group, 80-, actg chmn 80, chmn, 81-82; distinguished Patrick Scholar, Acad Nat Sci, Philadelphia, 83; prog dir, Biol Oceanog Prog, NSF, 87-89; mem bd marine div, Nat Asn State Univ Land Grant Col, 90-; chair, Pub Affairs Comn, Ecol Soc Am, 90- *Mem:* AAAS; Am Soc Limnol & Oceanog; Estuarine Res Fedn (vpres, 89-91); Sigma Xi; Int Soc Coral Reef Studies; Ecol Soc Am. *Res:* Marine and estuarine nutrient dynamics; coral reef ecology; algal endosymbiosis. *Mailing Add:* Sea Grant Col 1224 H J Patterson Hall Univ Md College Park MD 20742

DELIA, THOMAS J, b Brooklyn, NY, Nov 19, 35; m 64; c 4. ORGANIC CHEMISTRY, MEDICINAL CHEMISTRY. *Educ:* Col Holy Cross, BS, 57; Va Polytech Inst, MS, 59, PhD(org chem), 62. *Prof Exp:* Res assoc bio-org chem, Sloan-Kettering Inst Cancer Res, 62-66; from asst prof to assoc prof, 66-70, PROF CHEM, CENT MICH UNIV, 70- *Concurrent Pos:* Nat Acad Sci fel, Czech, 71-72; Fulbright Sr Scholar, Australia, 82-83. *Mem:* Sigma Xi; Am Chem Soc; Int Soc Heterocyclic Chem. *Res:* Synthesis of heterocyclic compounds as antitumor, antiviral and antimalarial agents. *Mailing Add:* Dept Chem Cent Mich Univ Mt Pleasant MI 48859

DELIHAS, NICHOLAS, b New York, NY, Sept 22, 32; m 61; c 3. MOLECULAR BIOLOGY. *Educ:* Queens Col, NY, BS, 54; Yale Univ, PhD(biophys), 61. *Prof Exp:* Jr tech specialist, Biol Dept, Brookhaven Nat Lab, 54-57; res asst biophys, Yale Univ, 57-58; res assoc, Sloan-Kettering Inst Cancer Res, 60-62; from assoc scientist to scientist, Med Dept, Brookhaven Nat Lab, 62-71; asst prof, State Univ NY, Stony Brook, 71-74, assoc prof & dir, Multidisciplinary Labs, Sch Basic Health Sci, 74-80, assoc prof, 80-83, assoc dean, 80-89, PROF SCH MED, STATE UNIV NY, STONY BROOK, 83- *Concurrent Pos:* Res assoc, Sloan-Kettering Div, Cornell Univ, 61-62; NIH spec fel, Ciba Res Labs, Basel, Switz, 64-65; adj prof biol, Southampton Col, 70-80; univ grant-in-aid, State Univ NY, 73-; Nat Inst Gen Med Sci res grant, 73-83; grant, NSF, 83- *Mem:* Am Soc Biol Chemists; Am Soc Microbiol; Biophys Soc; Am Soc Cell Biol; AAAS; Sigma Xi. *Res:* Small RNA structure and function; small RNAs as regulators of gene expression. *Mailing Add:* Sch Med State Univ NY Stony Brook NY 11794

DE LILLO, NICHOLAS JOSEPH, b New York, NY, July 14, 39; m 66; c 4. MATHEMATICAL LOGIC, COMPUTER SCIENCE. *Educ:* Manhattan Col, BS, 60; Fordham Univ, MA, 62; NY Univ, PhD(math), 71. *Prof Exp:* From instr to assoc prof, 63-84, PROF MATH MANHATTAN COL, 84- *Concurrent Pos:* From adj asst prof to adj assoc prof, Lehman Col, 72-83. *Mem:* Math Asn Am; Am Math Soc; Asn Symbolic Logic; Asn Comput Mach; Sigma Xi. *Res:* Model theory; theory of abstract computability. *Mailing Add:* Dept Math & Comput Sci Manhattan Col New York NY 10471

DELINGER, WILLIAM GALEN, b Hyannis, Nebr, Apr 29, 39. SOLID STATE PHYSICS. *Educ:* Chadron State Col, BS, 61; SDak Sch Mines & Technol, MS, 65; Univ Iowa, PhD(physics), 72. *Prof Exp:* Teacher physics, Western Ill Univ, 63-66; NASA trainee physics, Univ Iowa, 68-72; asst prof, 72-80, ASSOC PROF PHYSICS, NORTHERN ARIZ UNIV, 80- *Mem:* Am Asn Physics Teachers; Int Solar Energy Soc; Asn Comput Mach; Am Inst Aeronaut & Astronaut. *Res:* Solar energy; mathematical modeling to describe the performance of solar collectors and storage systems as well as solar instrumentation. *Mailing Add:* Dept Physics Northern Ariz Univ Box 6010 Flagstaff AZ 86011

DE LISI, CHARLES, b New York, NY, Dec 9, 41; m 68; c 2. BIOPHYSICS. *Educ:* City Col New York, BA, 63; NY Univ, PhD(physics), 69. *Prof Exp:* Engr, Sperry Rand Corp, 64-65; fel biophysics, Yale Univ, 69-72, sr lectr eng & appl sci, 71-72; staff scientist biophys, Los Alamos Sci Lab, Univ Calif, 72-77; vis scientist, 75-77, spec asst to off dir, 79-80, chief, Lab Math Biol, 82-84, chief sect theoret immunol, 81-85, SR INVESTR BIOPHYS, NAT CANCER INST, NIH, 77-; PROF & CHMN, DEPT BIOMATH SCI, MT SINAI SCH MED, 87- *Concurrent Pos:* Dir, Off Health & Environ Res, Dept Energy, 85-87. *Honors & Awards:* Gordon Res Conf Award, Soc Math Biol. *Res:* Biophysics; immunology; physical chemical properties of biopolymers; biomathematics. *Mailing Add:* Dept Bio Statist Mt Sinai Sch Med One Gustave Levy Plaza New York NY 10029

DELISI, DONALD PAUL, b Pittsburgh, Pa, Nov 15, 44; m 71; c 1. ATMOSPHERIC & OCEANIC WAVES, STRATIFIED SHEAR FLOWS. *Educ:* Princeton Univ, BS, 66; Univ Calif, Berkeley, MS, 67, PhD(mech eng & fluid mech), 72. *Prof Exp:* Nat res coun resident res assoc, Geophys Fluid Dynamics Lab, Nat Oceanic & Atmospheric Admin, 72-74; res scientist, Flow Res Co, 74-77; staff scientist, Phys Dynamics, Inc, 77-86; VPRES & TREAS & SR RES SCIENTIST, NORTHWEST RES ASSOC, INC, 86- *Mem:* Am Geophys Union; Am Meteorol Soc; Am Phys Soc; Am Inst Aeronaut & Astronaut. *Res:* Observation and dynamics of the stratosphere and mesosphere; physical oceanography of the upper ocean (from the surface to approximately 300 meters depth) including the study of internal waves, shear, turbulence and fine and micro-structure; laboratory studies of stratified shear flows. *Mailing Add:* PO Box 3027 Northwest Res Assoc Inc Bellevue WA 98009-3027

DELISLE, CLAUDE, b Quebec, Que, Nov 15, 29; m 58; c 5. OPTICS. *Educ:* Univ Montreal, BA, 51; Laval Univ, BScA, 58, PhD(optics), 63. *Prof Exp:* Lectr physics, Laval Univ, 62-63; res assoc optics, Inst Optics, NY, 63-65; from asst prof to assoc prof, 65-73, PROF PHYSICS, LAVAL UNIV, 73- *Concurrent Pos:* Nat Res Coun Can grants, 66-; consult, Ctr Psychol Res Inc, Montreal, 69-71; prof consult, Ministry of Educ, Que, 73-74; assoc ed, Can J Physics, 78-88. *Mem:* Fel Optical Soc Am; Can Asn Physicists; Fr-Can Asn Advan Sci; Int Soc Optical Eng. *Res:* Coherence, interferometry; hybrid bistability with acousto-optic modulator; metrology; optical testing; optical design. *Mailing Add:* LROL Physics Dept Laval Univ Quebec PQ G1K 7P4 Can

DELIVORIA-PAPADOPOULOS, MARIA, b Athens, Greece, Feb 23, 31; nat US; m; c 2. MECHANISMS OF BRAIN CELL INJURY IN THE FETUS AND NEWBORN, MECHANISMS OF HYPOXIC-ISCHEMIC INJURY. *Educ:* Sorbonne Univ, Athens Br, BA, 52; Nat Univ Sch Med, Athens, MD, 57; Am Bd Pediat, dipl, 71; Univ Pa, MA, 73. *Prof Exp:* Res pediat, Children's Hosp, Univ Athens Med Ctr, 58-59; intern pediat, St Francis & County Hosp, Wichita, Kans, 59-60; res pediat, Jewish Hosp Brooklyn & Kings County Hosp, Downstate Univ, 60-61; res psychiat, Colo State Hosp, Univ Colo Med Ctr, 61-62; fel neonatal pathophysiol, Hosp Sick Children, Res Inst, Univ Toronto, 62-64, res pediat, 64; fel neonatal & fetal physiol, Univ Colo Med Ctr, 65-66; fel physiol, Sch Med, Univ Pa, 67-70, instr, dept pediat, 67-70, asst prof physiol & pediat, 70-73, assoc prof pediat & obstet/gynec, 73-77, PROF PEDIAT & OBSTET/GYNEC, SCH MED, UNIV PA, 76-, PROF PHYSIOL, 77- *Concurrent Pos:* Instr, dept pediat, Univ Colo Med Ctr, 66-67; fac mem, Grad Sch Arts & Sci, Univ Pa, 74-, staff neonatologist, Hosp Univ Pa, 68-, dir newborn servs, 74-; sr physician, Children's Hosp Philadelphia, 74-; consult pediat, Pa Hosp, 76-; NIH spec res fel, 66-68, NIH young investr award, 68-70, NIH career develop award, 70-75; consult, Bur Community Environ Mgt, Dept HEW, 72, Ministry Educ, Govt Greece, 77; vis prof, J Pediat Educ Prog, Downstate Univ Med Ctr, 80; sci reviewer, J Pediat, Pediat Res, J Appl Physiol, Europ J Pediat, NEng J Med, Develop Pharmacol & Therapeut; sci reviewer grants, NIH, March Dimes, Am Heart Found; NIH study sect mem, Human Embryol & Develop, Maternal & Child Health Res Comt, Study Sect fel comt, NIH Nat Heart, Lung & Blood Inst Res Comt. *Mem:* Am Pediat Soc; Am Physiol Soc; Soc Pediat Res; Soc Gynec Invest; Can Pediat Res Soc; fel Am Acad Pediat; Am Fedn Clin Res; Hellenic Med Soc; NY Acad Sci. *Res:* Neonatal oxygen transport; cellular and molecular aspects of tissue oxygenation; the interrelation between energy metabolism and the functional state of the newborn brain to understand cerebrovascular disturbances in newborn; cerebral energy metabolism of the fetus and newborn. *Mailing Add:* Sch Med Univ Pa A-201 Richards Bldg 37th & Hamilton Walk Philadelphia PA 19104

DELIYANNIS, PLATON CONSTANTINE, b Athens, Greece, Aug 21, 31. MATHEMATICS. *Educ:* Nat Tech Univ Athens, dipl eng, 54; Univ Chicago, MS, 55, PhD(math), 63. *Prof Exp:* Res asst, Univ Chicago, 55-56; from instr to asst prof math, Ill Inst Technol, 61-65; dir, Ctr Advan Studies, Greek Atomic Energy Comn, 66-69; asst prof, 69-74, chmn dept, 75-80, ASSOC PROF MATH, ILL INST TECHNOL, 74- *Mem:* Am Math Soc; Math Asn Am; NY Acad Sci; Soc Indust & Appl Math. *Res:* Functional analysis and operator algebras; topological groups and representations; abstract quantum theory. *Mailing Add:* Dept Math Ill Inst Technol 3300 Federal St Chicago IL 60616

DELK, ANN STEVENS, b Lynchburg, Va, May 25, 42. ENZYMOLOGY, NUCLEIC ACID CHEMISTRY. *Educ:* Mary Baldwin Col, Va, 63; Univ Calif, Berkeley, PhD(biochem), 70. *Prof Exp:* Chemist, Nat Cancer Inst, 63-64; res asst, Syntex Inst Molecular Biol, 64-65; asst genetics, Univ Calif, Berkeley, 71-72, assoc biochem, 72-80; staff res assoc internal med, Univ Calif, Davis, 80-81; SR SCIENTIST & DIR DIAG RES & DEVELOP, HANA BIOLOGICS, BERKELEY, 82- *Concurrent Pos:* Staff res assoc clin enzymol, Vet Admin, Martinez, 80-81; consult, Hanna Biologics & Am Bionetics, 81- *Mem:* Am Soc Biol Chem. *Res:* Development of immuno and other assays for diagnosis of conditions of medical significance in animals and humans; fetal development; transplant function; bacterial, urinal and mycoplasma infections; diabetes; drug monitoring. *Mailing Add:* Four Willow Lane Suite 101 Kensington CA 94707

DELKER, GERALD LEE, b Portland, Ore, Dec 9, 47; m 72; c 3. INORGANIC CHEMISTRY. *Educ:* Univ Calif, Riverside, BS, 70, Univ Ill, Champaign-Urbana, PhD(inorganic chem), 76. *Prof Exp:* Fel, Univ Calif, Davis, 76-77, Occidental Col, Los Angeles, 77-79; SR CHEMIST & LAB SUPVR, ANALYSIS RES LABS, INC, 79-; SR SCIENTIST, IOLAB CORP, CLAREMONT, 90- *Mem:* Am Chem Soc. *Res:* Methods development for the analysis of various substances or environments; general chemical analyses. *Mailing Add:* 3046 Treefern Dr Duarte CA 91010

DELL, CURTIS G(EORGE), b Buffalo, NY, Mar 18, 24; m 49; c 4. ELECTRICAL ENGINEERING. *Educ:* Rutgers Univ, BS, 48; Cornell Univ, MEE, 51. *Prof Exp:* CONSULT, CONDUX INC, 85- *Concurrent Pos:* Lectr, Univ Del, 52-54. *Mem:* Sigma Xi; Instrument Soc Am; Inst Elec & Electronics Engrs. *Res:* Industrial research, development and marketing of laboratory and process analytical instruments. *Mailing Add:* 211 Comet Lane Newark DE 19711-2919

DELL, GEORGE F, JR, b Columbus, Ohio, Dec 28, 31; m 65; c 2. PARTICLE ACCELERATOR PHYSICS. *Educ:* Ohio State Univ, BSc, 53, MSc, 55, PhD(physics), 62. *Prof Exp:* Res assoc nuclear physics, Purdue Univ, 63-65; res fel high energy physics, Harvard Univ, 65-70; from asst physicist to physicist, 70-78, physics assoc solid state physics, 78-80, physics assoc, accelerator dept, 80-85, physicist, accelerator dept, 85-86, PHYSICIST, ACCELERATOR DEVELOP, BROOKHAVEN NAT LAB, 86- *Mem:* Am Phys Soc. *Res:* Design and simulation for high energy particle accelerators; neutron induced damage in insulators; detector development for high energy physics. *Mailing Add:* Accelerator Develop Dept Bldg 1005S Upton NY 11973

DELL, M(ANUEL) BENJAMIN, b Chelsea, Mass, Apr 12, 19; m 62; c 2. CARBON ELECTRODES, ALUMINUM SMELTING. *Educ:* Tufts Col, BS, 40; Pa State Col, MS, 50, PhD(fuel technol), 51. *Prof Exp:* Res chemist bituminous prod, Barrett Div, Allied Chem & Dye Corp, 47-48; res investr emulsions, Flintkote Co, 51-53; res chemist, Metal Prod Labs, Aluminium Co Am, 53-60, sci assoc, 60-81, tech consult, 81-83; RETIRED. *Concurrent Pos:* Consult, 84- *Mem:* AAAS; Am Inst Mining, Metall & Petrol Engrs; Am Chem Soc. *Res:* Fuels; chemistry of coal tar, asphalt, coal; carbon electrodes; aluminum smelting; clay-stabilized emulsions. *Mailing Add:* 144 Woodshire Dr Pittsburgh PA 15215

DELL, ROGER MARCUS, b Los Angeles, Calif, Oct 20, 36; m 62; c 3. MATHEMATICAL ANALYSIS. *Educ:* Calif State Univ Long Beach, BS, 59, MA, 62; Univ Calif, Los Angeles, PhD(math), 71. *Prof Exp:* Engr, Douglas Aircraft Co, 56-58; comput scientist, Hughes Aircraft Co, 59-60; instr math & physics, Calif State Univ, Long Beach, 60-62; comput scientist, Gen Elec Co, 62-63; instr math & physics, Deep Springs Col, 63-66; teaching asst math, Univ Calif, Los Angeles, 67-69; instr math & physics, Deep Springs Col, 70-74; ASSOC PROF MATH, LINFIELD COL, 74- *Mem:* Math Asn Am. *Mailing Add:* 677 Tanglewood Circle McMinnville OR 97128

DELL, TOMMY RAY, b New Orleans, La, May 21, 37; m 57; c 2. FOREST BIOMETRY. *Educ:* La State Univ, BS, 59; Univ Ga, MS, 64, PhD, 69. *Prof Exp:* Res forester, Southlands Exp Forest, Int Paper Co, Ga, 60-61; chief biomet br, 63-77, PROJ LEADER, INST QUANTITATIVE STUDY, SOUTHERN FOREST EXP STA, US FOREST SERV, NEW ORLEANS, 77- *Concurrent Pos:* Mem adv bd, Forest Sci, 73-77. *Mem:* Soc Am Foresters; Biomet Soc; Am Statist Asn. *Res:* Design and analysis of studies on biological topics, particularly forest resources; research on biometrical procedures, timber growth and yield forecasting. *Mailing Add:* 353 Homestead Ave Metairie LA 70005

DELLA-FERA, MARY ANNE, b Wilmington, Del, Mar 29, 54. NEUROPHYSIOLOGY. *Educ:* Univ Del, BA, 75; Univ Pa, VMD, 79, PhD(neurosci), 80. *Prof Exp:* Fel neurosci, NIH, 80-81; res assoc, Univ Pa, 81, res asst prof, 82; res specialist, 82-84, consult, 82, RES GROUP LEADER, NUTRIT CHEM DIV, MONSANTO CO, 84- *Concurrent Pos:* Alfred P Sloan Found Fel, 81; adj res asst prof, dept prev med, Washington Univ Sch Med, 82- *Honors & Awards:* Donald Lindsley Prize, Soc Neurosci, 81. *Mem:* AAAS; Am Vet Med Asn; Soc Neurosci; Am Physiol Soc. *Res:* Role of the central nervous system in the control of food intake and regulation of energy balance; role of particular brain peptides in feeding behavior; central nervous system control of gastrointestinal function. *Mailing Add:* Dept Clin Studies Univ Pa Sch Vet Med 314 Levy Bldg 16002 4010 Locust St Philadelphia PA 19104

DELLARIA, JOSEPH FRED, JR, b Midland, Mich, March 23, 56; m 80; c 4. CHEMISTRY. *Educ:* Hope Col, BA, 78; Univ Minn, PhD(chem), 82. *Prof Exp:* Fel, Calif Inst Technol, 82-83; fel, Harvard Univ, 83-84; chemist I, 84-86, chemist II, 86-89, RES INVESTR, ABBOTT LABS, 89- *Mem:* Am Chem Soc. *Res:* Designing & synthesizing inhibitors of enzymes to develop orally active pharmaceuticals; inhibitors of renin & five lipoxygenase; development of synthetic methodology necessary to synthesize the enzyme inhibitors. *Mailing Add:* Abbott Labs Dept 47K Abbott Park IL 60064

DELLA TORRE, EDWARD, b Milan, Italy, Mar 31, 34; US citizen; m 56; c 3. ELECTRICAL ENGINEERING, PHYSICS. *Educ:* Polytech Inst Brooklyn, BEE, 54; Princeton Univ, MS, 56; Rutgers Univ, MS, 61; Columbia Univ, DEngSc(elec eng), 64. *Prof Exp:* Assoc prof elec eng, Rutgers Univ, 56-67; mem staff, Bell Tel Labs, 67-68; assoc prof, McMaster Univ, 68-70, assoc chmn dept, 70-72, chmn dept, 72-78, prof elec eng, 70-79; prof & chmn, Elec & Comput Eng, Wayne State Univ, 79-82; PROF ELEC ENG & COMPUT SCI, GEORGE WASHINGTON UNIV, 82- *Mem:* Inst Elec & Electronics Engrs; Sigma Xi; Am Physics Soc; fel Inst Elec & Electronics Engrs; Asn Comput Mach. *Res:* Theoretical and experimental study of magnetic materials; electromagnetic theory; application of information and computer sciences; numerical solutions of partial differential equations. *Mailing Add:* George Washington Univ Washington DC 20052

DELLENBACK, ROBERT JOSEPH, b Los Angeles, Calif, June 3, 28; m 58; c 4. MEDICAL PHYSIOLOGY, ACADEMIC ADMINISTRATION. *Educ:* Univ Calif, Los Angeles, BA, 50, MA, 53, PhD(zool, physiol), 55. *Prof Exp:* Asst zool & physiol, Univ Calif, Los Angeles, 50-55; from instr to asst prof physiol, Col Physicians & Surgeons, Columbia Univ, 56-71; assoc prof biochem, Dent Sch, Fairleigh Dickinson Univ, 72-74; admin cordr, Elizabeth Morrow Sch, 74-78; PRES INVESTMENTS, R G DELL CORP, 78- *Concurrent Pos:* Eli Lilly fel & Nat Res Coun fel, Zool Inst, Univ Wo6rzburg 55-56. *Mem:* Am Physiol Soc; Soc Exp Biol & Med; Harvey Soc; Am Soc Zoologists. *Res:* Hemorrhage; cardiac output; blood volume. *Mailing Add:* PO Box 1748 Lakeville CT 06039a

DELLER, JOHN JOSEPH, b Pittsburgh, Pa, Nov 3, 31; m 53; c 3. INTERNAL MEDICINE, ENDOCRINOLOGY. *Educ:* Univ Pittsburgh, BSc, 53, MD, 57; Am Bd Internal Med, dipl, 67. *Prof Exp:* US Army, 56-76, resident internal med, Walter Reed Gen Hosp, DC, 58-61, chief gen med serv, Letterman Gen Hosp, 64-68, asst chief dept med, 68-70, chief dept med, Letterman Army Med Ctr, 70-76; dir primary care internal med res prog, Univ Calif, San Francisco, 76-77; dir educ & res, 78-81, PRACTR INTERNAL MED & ENDOCRINOL, EISENHOWER MED CTR, RANCHO MIRAGE, CALIF, 81- *Concurrent Pos:* Fel endocrinol & metab, Univ Calif, San Francisco, 63-64, asst clin prof, 67-72, assoc clin prof, 72-77. *Mem:* Fel Am Col Physicians; Endocrine Soc; Am Geriat Soc. *Mailing Add:* 34 Sierra Madre Way Rancho Mirage CA 92270

DELLEUR, JACQUES W(ILLIAM), b Paris, France, Dec 30, 24; nat US; m 57; c 2. HYDRAULIC ENGINEERING, HYDROLOGY. *Educ:* Nat Univ Colombia, SAm, CE & MinE, 49; Rensselaer Polytech Inst, MSCe, 50; Columbia Univ, DEngSc, 55. *Prof Exp:* Engr, R J Tipton & Assocs Inc, 50-51; asst civil eng & eng mech dept, Columbia Univ, 52-53, instr civil eng, 53-55; from asst prof to assoc prof hydraul eng, 55-63, head hydromech & water resources area, 65-76, assoc dir, Water Resources Res Ctr, 71-90, head, Hydraulics & Systs Eng Area, 81-90, PROF HYDRAUL ENG, SCH CIVIL ENG, PURDUE UNIV, 63- *Concurrent Pos:* Consult, Res Staff, Gen Motors Corp, 57-62; vis fac, Univ Grenoble, France, 61-62 & 83-84; Freeman fel, Am Soc Civil Engrs, 61; deleg, Univs Coun Water Resources, 64-83; researcher, French Nat Hydraulics Lab, France, 68-69, 76-77; tech adv, Nationwide Urban Runoff Prog, Environ Protection Agency/US Geol Survey, 79-82; sr scientist exchange to France, NSF, Nat Ctr Sci Res, 83 & 84. *Mem:* Am Soc Civil Engrs; AAAS; Am Water Resources Asn; Am Geophys Union; Int Asn Hydraul Res; Am Asn Eng Educ; Int Water Resources Asn. *Res:* Hydrology of water resources, stochastic and urban hydrology, hydraulic engineering; open channel hydraulics. *Mailing Add:* Sch Civil Eng Purdue Univ Lafayette IN 47907

DELLICOLLI, HUMBERT THOMAS, b Utica, NY, July 8, 44; m 67; c 1. PHYSICAL CHEMISTRY, AGRICULTURAL CHEMISTRY. *Educ:* Clarkson Col Technol, BS, 66, PhD(chem), 71. *Prof Exp:* Res chemist phys chem, Edgewood Arsenal, US Army, 71-73; res chemist lignin chem, Charleston Res Ctr, 73-75, PROD DEVELOP MGR AGR CHEM, POLYCHEM DEPT, WESTVACO CORP, 75- *Concurrent Pos:* Consult, US Army Chem Systs Lab, 71- *Mem:* Am Chem Soc; AAAS. *Res:* Colloid and surface science; macromolecular chemistry; biocolloids; pesticide formulation chemistry; controlled release of bioactive materials; agricultural chemistry. *Mailing Add:* Westvaco PO Box 2941105 North Charleston SC 29411

DELLINGER, THOMAS BAYNES, b Crawfordsville, Ind, Jan 31, 26; m 52; c 4. PETROLEUM ENGINEERING. *Educ:* Purdue Univ, BS, 48; Univ Tulsa, MS, 62, PhD(petrol eng), 70. *Prof Exp:* Engr, Creole Petrol Corp, 48-59; res engr, Jersey Prod Res Co, 60-62; engr, Fenix & Scisson, Inc, 63-67; res engr, Mobil Res & Develop Corp, 70-88; RETIRED. *Mem:* Soc Petrol Engrs. *Mailing Add:* PO Box 380163 Duncanville TX 75138

DELLMANN, H DIETER, b Berlin, Ger, June 6, 31; m 55; c 2. HISTOLOGY, NEUROENDOCRINOLOGY. *Educ:* Nat Vet Sch, Alfort, France, DVM, 54, MS, 55; Univ Munich, PhD(anat, histol & embryol), 61. *Prof Exp:* Mem staff vet med, Chemie Gruenenthal, Stolberg, Ger, 55-57; res asst anat, histol & embryol vet fac, Univ Munich, 57-61, pvt docent, 62-63, univ docent, 64; vis prof & actg chmn dept fac vet med, Cairo Univ, 63-64; from assoc prof to prof anat, histol & embryol, Sch Vet Med, Univ Mo-Columbia, 64-75; PROF VET ANAT, COL VET MED, IOWA STATE UNIV, 75- *Concurrent Pos:* Mem, Int Comn Vet Anat Nomenclature, 59-; Fulbright travel grant, 64-65; assoc prof, Louis Pasteur Univ, Strasbourg, France, 70-71, Univ Marseille, France, 81-82 & Univ Montpellier, France, 88-89. *Mem:* Am Asn Anatomists; World Asn Vet Anat; Europ Asn Vet Anat; Ger Anat Soc; Europ Soc Comp Endocrinol; Int Soc Neuroendocrinol; Am Asn Vet Anatomists. *Res:* Hypothalamus; neurosecretion; hypothalamo-hypophyseal systems; CNS regeneration; circumventricular organs; peripheral nerve grafts. *Mailing Add:* Dept Vet Anat Iowa State Univ Col Vet Med Ames IA 50011

DELL'ORCO, ROBERT T, b St Louis, Mo, Jan 12, 42; m 67; c 3. CELL BIOLOGY. *Educ:* Rockhurst Col, BA, 63; Univ Kans, PhD(microbiol), 69. *Prof Exp:* Nat Cancer Inst Can fel, 69-70; res biologist, 70-73, from asst scientist, to assoc scientist, 73-81, SCIENTIST, SAMUEL ROBERTS NOBLE FOUND, INC, 81-, HEAD, CELL BIOL SECT, 73- *Concurrent Pos:* Cellular physiol rev group, NIH, 82-83. *Mem:* Soc Exp Biol & Med; Tissue Cult Asn (secy, 80-84); Am Soc Cell Biol; Gerontol Soc Am. *Res:* Metabolism and physiology of human diploid cells maintained in culture as applicable to their use as a model system for the study of senescence on a cellular level. *Mailing Add:* Samuel Roberts Noble Found Inc PO Box 2180 Ardmore OK 73402

DELL'OSSO, LOUIS FRANK, b Brooklyn, NY, Mar 16, 41. OCULAR MOTOR NEUROPHYSIOLOGY. *Educ:* Brooklyn Polytech Inst, BEE, 61; Univ Wyo, PhD(biomed eng), 68. *Prof Exp:* Electronics engr, Repub Aviation, Inc, 61-63; res engr, Westinghouse Res Lab, 63, sr bioengr, 66-70; electronics engr, Naval Med Res Inst, 64; asst prof biomed eng surg, Univ

Miami, 70-72, from asst prof to prof neurol, 72-80; biomed engr, Vet Admin Med Ctr, Miami, 71-72, health scientist, co-dir ocular motor neurophysiol lab, 72-80; PROF NEUROL & BIOMED ENG, CASE WESTERN RESERVE UNIV, 80-; BIOMED ENGR & DIR OCULAR MOTOR NEUROPHYSIOL LAB, VET ADMIN MED CTR CLEVELAND, 80- *Concurrent Pos:* Consult bioeng, Westinghouse Res Lab, 66-67 & 70-71; prin investr, NSF & NIH internal grants, Univ Miami, 72-78; co-investr & co-prin investr, merit rev grants, Vet Admin, 73-; vis prof, var US & foreign univs, 77-85; site visitor, Nat Inst Neurol & Commun Dis & Stroke, NIH, 84. *Mem:* AAAS; Soc Neurosci; Asn Res Vision & Ophthal; sr mem Inst Elec & Electronics Engrs. *Res:* Neural control of eyes and pathological control in neurological and ophthalmological patients; oscillations in ocular motor control systems; ocular motor oscillations and control. *Mailing Add:* Ocular Motility Lab 127A Vet Admin Med Ctr Cleveland OH 44106

DELLUVA, ADELAIDE MARIE, b Bethlehem, Pa, Sept 2, 17. BIOCHEMISTRY. *Educ:* Bucknell Univ, BS, 39, MS, 40; Univ Pa, PhD(physiol chem), 46. *Prof Exp:* Asst instr physiol chem, 43-46, instr, Sch Med, 46-54, asst prof biochem, 54-69, from asst prof to assoc prof, Sch Vet Med, 69-78, assoc chmn, Dept Animal Biol, 71-73, actg chmn, 73-75, PROF BIOCHEM, SCH VET MED, UNIV PA, 78- *Mem:* Am Soc Biol Chemists; AAAS. *Res:* Organic synthesis of compounds with C-13 and C-14 for use in researches in physiological chemistry; various phases of metabolism using isotopes; carbon dioxide assimilation; adrenaline biosynthesis; lactate metabolism; energy source for contraction of muscle; alanine metabolism; gastric urease; anion transport in mitochondria; energetics of muscle contraction. *Mailing Add:* Dept Animal Biol Univ Pa Sch Vet Med Philadelphia PA 19104

DELLWIG, LOUIS FIELD, b Washington, DC, Feb 13, 22; m 48; c 3. REGIONAL GEOLOGY, REMOTE SENSING. *Educ:* Lehigh Univ, BA, 43, MS, 48; Univ Mich, PhD(geol), 54. *Prof Exp:* Instr, Univ Mich, 51-53; from asst prof to assoc prof, 53-63, dir, Remote Sensing Lab, 75-76, PROF GEOL, UNIV KANS, 63- *Concurrent Pos:* Fulbright grant, Univ Heidelberg, 71. *Mem:* Am Asn Petrol Geol; Soc Econ Paleont & Mineral; Am Soc Photogram; Sigma Xi. *Res:* Evaporites; use of radar in geologic mapping. *Mailing Add:* Dept Geol Univ Kans 120 Lindley Hall Lawrence KS 66045-2124

DELMASTRO, ANN MARY, b Brooklyn, NY, Nov 29, 45; m 75; c 1. ANALYTICAL CHEMISTRY. *Educ:* Fordham Univ, BS, 67; Cornell Univ, MS, 69, PhD(anal chem), 71. *Prof Exp:* SR CHEMIST ANALYTICAL CHEM, ALLIED CHEM CORP, 72- *Mem:* Am Chem Soc; Soc Appl Spectros; Am Soc Mass Spectrometry; Sigma Xi. *Res:* Development of analytical methods in x-ray spectrometry, emission spectrometry, and mass spectrometry for the analysis of samples related to the chemical reprocessing of nuclear fuels. *Mailing Add:* 2670 Homestead Lane Idaho Falls ID 83404

DELMASTRO, JOSEPH RAYMOND, b Joliet, Ill, Sept 10, 40; m 75; c 1. ANALYTICAL CHEMISTRY. *Educ:* Northern Ill Univ, BS, 62; Northwestern Univ, PhD(anal chem), 67. *Prof Exp:* Res chemist electrochem, Idaho Nuclear Corp, 67-71; sr chemist, Allied Chem Corp, 71-73, assoc scientist, 73-76, group leader, 76-79; GROUP LEADER, EXXON NUCLEAR IDAHO CO, 79- *Res:* Theory of electroanalytical techniques; coulometry; electroanalysis; general analytical chemistry methods development; thermal analysis; chromatography. *Mailing Add:* 2670 Homestead Lane Idaho Falls ID 83404

DELMER, DEBORAH P, b Indianapolis, Ind, Dec 7, 41; m 65; c 1. PLANT PHYSIOLOGY, BIOCHEMISTRY. *Educ:* Ind Univ, AB, 63; Univ Calif, San Diego, PhD(biol), 68. *Prof Exp:* NIH fel chem, Univ Colo, 68-69; NIH fel biol, Univ Calif, San Diego, 69-70, asst res biologist, 70-73; from asst prof to prof biochem, Dept Biochem, Mich State Univ, 74-82; sr prin scientist, Arco Plant Cell Res Inst, 83-86; PROF, DEPT BOT, HEBREW UNIV, 87- *Mem:* Am Soc Plant Physiol. *Res:* Biochemistry of sucrose synthesis and metabolism in higher plants; circadian rhythms in fungi; biosynthesis of cellulose and other plant polysaccharides; glycoprotein synthesis in higher plants; plant cell wall structure. *Mailing Add:* Dept Bot Hebrew Univ Jerusalem Israel

DELMONTE, DAVID WILLIAM, b Auburn, NY, May 29, 30; m 58; c 6. ORGANIC CHEMISTRY, POLYMER CHEMISTRY. *Educ:* Villanova Univ, BS, 52; Univ Notre Dame, PhD(org chem), 57. *Prof Exp:* Chemist, Pioneering Res Lab, Textile Fibers Dept, E I du Pont de Nemours & Co, Inc, 52-53; asst, Univ Notre Dame, 53-54; res chemist, 56-69, mkt develop rep, 69-73, mkt develop supvr, 73-75, develop mgr, 75-81, MGR, FIBERS TECH CTR, HERCULES INC, 81- *Mem:* Am Chem Soc. *Res:* Condensation polymerization of esters and phosphorus compounds; isocyanate chemistry; polyurethane foam preparation and characterization; adhesives; polymer syntheses; terpenes and rosin acids chemistry; printing inks; polyolefin fibers. *Mailing Add:* Hercules Inc Res Ctr Bldg 8136 Wilmington DE 19894-0002

DELMONTE, LILIAN, b Hamburg, Ger, Mar 19, 28; US citizen. HEMATOLOGY, IMMUNOLOGY & ONCOLOGY. *Educ:* Mt Holyoke Col, AB, 50; NY Univ, MSc, 52; Sorbonne Univ, Dr es Sci(hemat), 57. *Prof Exp:* Med writer, Cyanamid Int, Am Cyanamid Co, 59-62; from instr to asst prof anat, Baylor Col Med, 64-74; vis investr, 74-78, res assoc, Sloan-Kettering Inst, 78-82. *Concurrent Pos:* USPHS spec fel anat, Baylor Col Med, 62-64; Int Cancer Soc-Eleanor Roosevelt-Am Cancer Soc fel; Nat Cancer Inst spec fel, 72-73; sr vis scientist, Cancer Res Inst, Walter & Eliza Hall Inst Med Res, Melbourne, Australia, 71-73; free-lance med writer, 82- *Mem:* Am Med Writers Asn; Am Soc Hemat; Int Soc Exp Hemat. *Res:* Hemopoietic self-regulatory mechanisms; leukemic cell regulation; transplantation immunobiology. *Mailing Add:* 440 E 62nd St New York NY 10021

DEL MORAL, ROGER, b Detroit, Mich, Sept 13, 43; c 2. ECOLOGY, BOTANY-PHYTOPATHOLOGY. *Educ:* Univ Calif, Santa Barbara, BA, 65, MA, 66, PhD(biol), 68. *Prof Exp:* From asst prof to assoc prof, 68-82, PROF BOT, UNIV WASH, 83- *Concurrent Pos:* Deleg, Int Bot Cong, 69, 75 & 87; NSF grants, 70-72, 79-81, 81-84, 84-87, 89-94 & 90; consult, AEC, 72-75 & 85, Seattle Parks Dept, 73-75 & 79-80, US Forest Serv, 75, 80 & 81, Nat Park Serv, 78, 80 & 85, King County Parks Dept, 79-80 & King County Human Resources Dept, 85; sr res fel, CSIRO & Univ Melbourne, 76-77. *Mem:* Brit Ecol Soc; Bot Soc Am; Ecol Soc Am. *Res:* Competition; statistical ecology; phytosociology; vegetation structure; vegetation control; volcano ecology; succession. *Mailing Add:* Dept Bot KB-1S Univ Wash Seattle WA 98195

DELNORE, VICTOR ELI, b Yuba City, Calif, July 20, 43; m 70; c 2. REMOTE SENSING. *Educ:* Rensselaer Polytech Inst, BEE, 65; Univ Miami, MS, 67; Old Dominion Univ, PhD(geophys), 76. *Prof Exp:* Res oceanographer, Nat Oceanic & Atmospheric Admin, 70-74; asst prof phys oceanog, Rutgers Univ, 76-79; SR CONSULT ENGR, NASA LANGLEY RES CTR, PRC KENTRON, 79- *Concurrent Pos:* NSF fel, Univ Miami, 66-67; prin investr, Storms Response Exp, Univ Wash, NASA & Nat Oceanic & Atmospheric Admin, 80-81; artic sea ice & iceberg radar exps, NASA, 79, 82 & 85. *Mem:* Am Geophys Union; Am Inst Aeronaut & Astronaut. *Res:* Microwave remote sensing of the earths oceans and atmosphere; wind shear detection. *Mailing Add:* NASA-Langley Res Ctr Mailstop 490 Hampton VA 23665-5225

DELOACH, BERNARD COLLINS, JR, b Birmingham, Ala, Feb 19, 30; m 51; c 2. PHYSICS. *Educ:* Ala Polytech Inst, BS, 51, MS, 52; Ohio State Univ, PhD(physics), 56. *Prof Exp:* Mem tech staff, Bell Labs, 56-63, head, Gallium Arsenide Laser Dept, 63-80, head, Solid State Mat & Devices Dept, 80-89. *Honors & Awards:* David Sarnoff Award, Inst Elec & Electronics Engrs, 75; Stuart Ballantine Award, Franklin Inst, 75. *Mem:* Fel Inst Elec & Electronics Engrs. *Res:* Solid state millimeter wave sources; semiconductor junction electroluminescence; solid state physics; microwave applications; gallium arsenide solid state lasers. *Mailing Add:* 4273 Steed Terr Winter Park FL 32792

DELOACH, CULVER JACKSON, JR, b Greensboro, NC, July 25, 32; m 57; c 3. ENTOMOLOGY, BIOLOGICAL CONTROL OF WEEDS. *Educ:* Auburn Univ, BS, 54, MS, 60; NC State Univ, PhD(entom), 64. *Prof Exp:* Asst prof entom, Univ Hawaii in Japan, 64-65; res entomologist, Biol Control Insects Lab, Mo, 65-71 & Biol Control Aquatic Weeds Lab, Hurlingham, Arg, 71-74, RES ENTOMOLOGIST, GRASSLAND SOIL & WATER RES LAB, AGR RES SERV, USDA, 74- *Concurrent Pos:* Consult, Inst for Plant Protection, Yugoslavia, 85- & Nat Inst Technol, Argentina, 88. *Mem:* Entom Soc Am; Weed Sci Soc Am; Int Orgn Biol Control; Hyacinth Control Soc. *Res:* Biological control of insect pests of Brassica crops; biological control of aquatic weeds and weeds of southern US rangelands and pastures by introducing foreign control insects. *Mailing Add:* Grassland Soil & Water Res Lab PO Box 748 808 E Blackland Rd Temple TX 76502

DELOACH, JOHN ROOKER, Humboldt, Tenn, July 5, 46; m 68; c 2. DRUG CARRIERS. *Educ:* Union Univ, BS, 68; Memphis State Univ, PhD(biochem), 75. *Prof Exp:* Res assoc med biochem, Sch Med, Univ Pittsburgh, 75-77; vis asst prof, Tex A&M Col Med, 77-78; res chemist, Vet Toxicol Entom Res Lab, 78-, RES CHEMIST, FOOD ANIMAL PROTECTION RES LAB, USDA. *Concurrent Pos:* Consult, Int Atomic Energy Agency, 80-85. *Mem:* AAAS; NY Acad Sci; Entom Soc Am; Sigma Xi. *Res:* Drug, pesticide, enzyme encapsulation in erythrocytes; interaction of mycotoxins with immune system; physiology of blood digestion by arthropods. *Mailing Add:* Food Animal Protection Res Lab USDA Rte 5 Box 810 College Station TX 77845

DELOATCH, EUGENE M, b Piermont, NY, Feb 3, 36; m 67; c 1. CONTROLS. *Educ:* Tougaloo Col, BS, 59; Lafayette Col, BS, 59; Polytech Inst Brooklyn, MS, 67, PhD(bio eng), 71. *Prof Exp:* Systs Eng, NY State Elec & Gas Co, 59-60; fac, 60, CHMN, ELEC & BIO ENGR, HOWARD UNIV, 75- *Concurrent Pos:* Chmn math, Urban Ctr Manhattan, State Univ NY, 68-69; consult, City Univ NY, 70; sci rev bd, Lister Hill-Ctr Biomed Comn, NIH, 80-83; mem acad adv bd, Indust Res Inst, 81-84; mem tech adv bd, Whirlpool Corp, 82-83. *Mem:* AAAS; Am Soc Eng Educ; Sigma Xi. *Res:* Enginering applications in the areas of biology and medicine. *Mailing Add:* 2400 Fairhill Dr Suitland MD 20746

DELONG, ALLYN F, DRUG METABOLISM, PHARMACOLOGY & TOXICOLOGY. *Educ:* Ursinus Col, BS, 64; Loyola Univ, MS, 66, PhD(biochem), 68. *Prof Exp:* Postdoctoral Univ Pa, 68-70; biochem res, Rorer-Rhone-Poulonc, 70-82; PHARMACEUT RES, ELI LILLY & CO, 82- *Concurrent Pos:* Adj prof pharmacol, Sch Med, Ind Univ, 85- *Mem:* Am Soc Clin Pharmacol & Exp Therapeut; Am Chem Soc; Am Asn Pharmaceut Scientist; Sigma Xi. *Res:* Biotransformation, disposition, and pharmacokinetics of drugs in humans and animal species. *Mailing Add:* Drug Metab Res Lilly Lab Clin Res Eli Lilly & Co 1001 W Tenth St Indianapolis IN 46202

DE LONG, CHESTER WALLACE, b Seattle, Wash, Feb 27, 25; m 56; c 3. BIOCHEMISTRY, PHYSIOLOGY. *Educ:* Univ Wash, BS, 48; State Col Wash, PhD(chem), 56. *Prof Exp:* Biochemist isotope metab, Hanford Atomic Power Labs, Gen Elec Co, 48-52; asst chem, State Col Washington, 54-55; biochemist, US Govt, 56-68; chief training grants & career develop prog, Educ Serv, 68-71, chief prog & career develop rev div, Med Res Serv, 71-74, chief, Rev Div, Manpower Grants Serv, 74-79, COORDR, REV & LIAISON, OFF ACAD AFFAIRS, DEPT MED SURG, US VET ADMIN, 79- *Concurrent Pos:* Vis res assoc, Phys Biol Lab, Nat Inst Arthritis & Metab Dis, 62-63; ex officio Vet Admin rep, Nat Adv Res Resources Coun, 73-80. *Mem:* Sigma Xi; Asn Mil Surgeons of US; fel AAAS; Am Chem Soc; Sigma Xi. *Res:* Intermediary metabolism; radioisotope toxicology and metabolism; chemistry of penicillin formation; electron paramagnetic resonance of biological materials; biochemistry of amino acids; research training; health manpower education; program management. *Mailing Add:* 213 W Columbia St Falls Church VA 22046

DELONG, KARL THOMAS, b Philadelphia, Pa, July 18, 38; m 63; c 5. ECOLOGY. *Educ:* Oberlin Col, AB, 60; Univ Calif, Berkeley, PhD(ecol), 65. *Prof Exp:* Asst prof biol, Ripon Col, 65-66; from asst prof to assoc prof, 66-70, PROF BIOL, GRINNELL COL, 70- *Mem:* Ecol Soc Am. *Res:* Prairie and oak woodland restoration. *Mailing Add:* Dept Biol Grinnell Univ PO Box 805 Grinnell IA 50112

DELONG, LANCE ERIC, b Denver, Colo, Nov 12, 46; m 83; c 3. LOW TEMPERATURE SOLID STATE PHYSICS. *Educ:* Univ Colo, Boulder, BA, 68; Univ Calif, San Diego, MS, 69, PhD(physics), 77. *Prof Exp:* Asst prof physics, Univ Va, 77-79; asst prof, 79-83, ASSOC PROF PHYSICS, UNIV KY, 83- *Concurrent Pos:* Prin investr, Cottrell grant, Res Corp, 82-83, 88-; co-prin investr res contract, US Dept Energy, 82-85, lab assoc, Ames Lab, 85-; mem user group, High Flux Isotope Reactor, Oak Ridge Nat Lab, 83-; co-ed conf proc, Rare Earth Res Conf, 85-88; scientist-in-residence, Argonne Nat Lab, 85-86, nonpaid staff mem, 86-; consult & prog dir, Div Mat Res, NSF, 88 & 89; chmn bd dir & treas, Rare Earth Res Conf, Inc, 88-91, gen conf chmn, 91; mem, Int Steering Comt Conf on f-Element. *Honors & Awards:* Res Opportunity Award, Res Corp, 88. *Mem:* Am Phys Soc; Am Asn Physics Teachers. *Res:* Low temperature solid state physics; electrical and magnetic properties of rare earth, actinide and transition metal materials at low temperatures, high pressures and high magnetic fields; properties of materials at high pressure and high magnetic fields. *Mailing Add:* Dept Physics & Astron Univ Ky Lexington KY 40506-0055

DELONG, STEPHEN EDWIN, b Evansville, Ind, Sept 22, 43; m 65; c 3. PETROLOGY. *Educ:* Oberlin Col, AB, 65; Univ Tex, Austin, MA, 69, PhD(geol), 71. *Prof Exp:* Res fel geochem, Calif Inst Technol, 71-72; from asst prof to assoc prof geol, 73-90, assoc vpres acad affairs, 79-81, actg vpres, 81-82, PROF GEOL, ACTG VPRES RES & DEAN GRAD STUDIES, STATE UNIV NY, ALBANY, 90- *Mem:* Am Geophys Union; Am Soc Mass Spectrometry. *Res:* Petrology, geochemistry and tectonics of oceanic crust and subduction zones. *Mailing Add:* Dept Geol Sci State Univ NY Albany NY 12222

DELONG, WILLIAM T, b Bethlehem, Pa, Dec 9, 21. METALLURGICAL ENGINEERING. *Educ:* Lehigh Univ, BS, 43. *Prof Exp:* Dir res, Teledyne McKay, 75, vpres, 75-84; CONSULT, 84- *Honors & Awards:* Comfort A Adams Lectr, Am Soc Welding, 74. *Mem:* Fel Am Welding Soc (pres, 81-82); fel Am Soc Metals. *Res:* Welding; awarded 15 US patents. *Mailing Add:* 820 Upland Rd York PA 17403

DELORENZO, ROBERT JOHN, b Clifton, NJ, June 26, 47; c 3. NEUROSCIENCE, NEUROLOGY. *Educ:* Yale Univ, BS, 69, PhD(neuropharmacol), 73, MD, 74, MPH, 74. *Prof Exp:* Resident, Neurol Dept, Yale Univ, 74-76, chief resident, 76-77, asst prof, 77-80, ASSOC PROF NEUROL, MED SCH, YALE UNIV, 80- *Concurrent Pos:* Mem staff res career develop grant, NIH, 76-81; mem, Epilepsy Res Found, 77; Consult, West Haven Veterans Hosp, 77-, Conn Mental Health Ctr, 79- & Parent Educ Res; prin investr, NIH res grant, 77-; attend physician, Yale New Haven Hosp, 77-; vis prof, Sch Med, Wash Univ, 81. *Honors & Awards:* Fisk Award. *Mem:* Am Soc Neurochem; Soc Neuroscience; Int Soc Neurochemistry; Am Acad Neurol; NY Acad Sci. *Res:* Mechanisms involved in mediating the effects of calcium on synaptic function; antionvulsant and neuroleptic drugs; role of calmodulin in synaptic function. *Mailing Add:* Dept Neurol Med Col Va Richmond VA 23298-0599

DELORENZO, RONALD ANTHONY, b Schenectady, NY, Sept 28, 41; m 67; c 2. PHYSICAL INORGANIC CHEMISTRY. *Educ:* St John's Univ, BS, 63; Lowell Technol Inst, PhD(chem), 70. *Prof Exp:* From asst prof to assoc prof, 70-84, comput coordr, 73-84, PROF CHEM, MID GA COL, 84- *Concurrent Pos:* Feature ed, J Chem Educ. *Mem:* Am Chem Soc; AAAS; Am Asn Univ Prof. *Res:* Thermodynamic studies of coordination compounds in mixed solvents; electron paramagnetic resonance studies of photochemically generated radicals; random number generation; chemical education; computer assisted instruction software development. *Mailing Add:* Dept Chem Mid Ga Col Cochran GA 31014-1599

DE LORGE, JOHN OLDHAM, b Jacksonville, Fla, Dec 14, 35; m 60; c 3. RESEARCH ADMINISTRATION. *Educ:* Jacksonville Univ, BA, 60; Hollins Col, MA, 62; Univ NC, PhD(exp psychol), 64. *Prof Exp:* Teaching asst psychol, Univ NC, 61-64, instr, 64-65; res psychologist, Evansville St Hosp, Ind, 65-66; asst prof psychol, Univ S Ala, Mobile, 66-69; RES PSYCHOLOGIST, NAVAL AEROSPACE MED RES LAB, 69- *Concurrent Pos:* Adj prof, Evansville Col, 65, Pensacola Jr Col, 70-72, Univ W Fla, 72-; vis assoc prof, Univ Rochester, 77-82; consult, Elec Power Res Inst & Off Health & Environ Res, Dept of Energy, 78-84; mem bd dirs, Bioelectromagnetic Soc, 81-84 & 90-93. *Mem:* Am Psychol Soc; Bioelectromagnetic Soc; Psychonomic Soc; AAAS; Appl Behav Anal. *Res:* Biological effects of physical agents such as electromagnetic radiation (microwaves), chemical warfare antidotes and antimotion sickness drugs; technical director of research. *Mailing Add:* Naval Aerospace Med Res Lab code 02 NAS Pensacola FL 32508-5700

DELORIT, RICHARD JOHN, b Door Co, Wis, May 22, 21; m 42; c 2. WEED SCIENCE, SEED TAXONOMY. *Educ:* Univ Wis-River Falls, BS, 42; Univ Wis-Madison, MS, 48, PhD(agron), 59. *Prof Exp:* Teacher, Wis High Schs, 42-53; critic teacher voc agr, Univ Wis-River Falls & River Falls High Sch, 53-55; from asst prof to prof agron, Univ Wis-River Falls, 56-85, dean col agr, 57-64, vchancellor, 64-90; RETIRED. *Concurrent Pos:* Mem, Coun Agr Sci & Technol. *Mem:* Weed Soc Am; Am Soc Agron. *Res:* Crop production; seed taxonomy. *Mailing Add:* 1010 County Trunk M River Falls WI 54022

DELOS, JOHN BERNARD, b Ann Arbor, Mich, Mar 24, 44; m 65; c 3. MOLECULAR PHYSICS, THEORETICAL CHEMISTRY. *Educ:* Univ Mich, BSChem, 65; Mass Inst Technol, PhD(chem), 70. *Prof Exp:* Fel chem, Univ Alta, 70 & Univ BC, 70-71; from asst prof to assoc prof, 71-83, PROF PHYSICS, COL WILLIAM & MARY, 83- *Concurrent Pos:* Vis scientist, FOM Inst Atomic & Molecular Physics, 79-80; vis fel, JILA, 86-87. *Mem:* Am Phys Soc; Fedn Am Scientists. *Res:* Atomic and molecular collision theory; structure and properties of small molecules; order and chaos in class and quantum systems. *Mailing Add:* Dept Physics Col William & Mary Williamsburg VA 23185

DELOUCHE, JAMES CURTIS, b La, Oct 30, 30; m 54; c 3. BOTANY. *Educ:* Southwestern La Inst, BS, 51; Iowa State Univ, MS, 52, PhD(econ bot), 55. *Prof Exp:* From asst agronomist to assoc agronomist, 57-65, agronomist seed technol, 65-77, PROF AGRON, MISS STATE UNIV, 65- *Concurrent Pos:* State seed analyst, Miss Dept Agr, 58-77. *Mem:* Am Soc Agron; Asn Off Seed Anal. *Res:* Economic botany; quality evaluation of seed; seed physiology; dissemination and germination of weed seeds. *Mailing Add:* Box 5267 Mississippi State MS 39762

DELOVITCH, TERRY L, B-CELL ACTIVATION, DIABETES. *Prof Exp:* PROF IMMUNOL, UNIV TORONTO, 80- *Res:* Molecular biology. *Mailing Add:* Dept Med Res Univ Toronto 112 College St Toronto ON M5G 1L6 Can

DELP, CHARLES JOSEPH, b St Louis, Mo, May 9, 27; div; c 4. PLANT PATHOLOGY, FUNGICIDE DEVELOPMENT MANAGEMENT. *Educ:* Colo State Univ, BS, 50; Univ Calif, PhD(plant path), 53. *Prof Exp:* Res asst, Univ Calif, Davis, 50-53; from sr res investr to sr res assoc, E I du Pont de Nemours & Co, Inc, 53-85; CONSULT, 86- *Concurrent Pos:* Consult fungicide resistance problems. *Mem:* fel Am Phytopath Soc; Am Inst Biol Sci; Am Assoc Adv Sci. *Res:* Environmental influence on grape powdery mildew; discovery and development of chemicals to control plant diseases; Third World agricultural research. *Mailing Add:* 4894 W Brigantine Ct Wilmington DE 19808-1824

DELPHIA, JOHN MAURICE, b Kans, Mar 29, 25; m 49; c 5. VERTEBRATE MORPHOLOGY. *Educ:* St Benedicts Col, BS, 49; Kans State Col, MS, 50; Univ Nebr, PhD(zool, anat), 59. *Prof Exp:* Instr zool, NDak Agr Col, 50-51, comp anat, embryol & histol, 52-54, asst prof, 55; asst prof, 64-70, ASSOC PROF ANAT, COL MED, OHIO STATE UNIV, 70- *Concurrent Pos:* NIH grants, 57, 58 & 59. *Mem:* Am Asn Anatomists. *Res:* Developmental and gross anatomy of avian lungs and airsacs; abnormalities in bovine endocrine glands. *Mailing Add:* Dept Anat Ohio State Univ Col Med 1645 Neil Ave Columbus OH 43210

DEL REGATO, JUAN A, b Camaguey, Cuba, Mar 1, 09; US citizen; m 39; c 3. THERAPEUTIC RADIOLOGY, CANCER. *Educ:* Univ Paris, MD, 37. *Hon Degrees:* DSc, Colo Col, 67, Hahnemann Med Col, 79, Wis Med Col, 81. *Prof Exp:* Asst roentgentherapist, Radium Inst Paris, 36-37; asst radiotherapist, Chicago Tumor Inst, 37-38; radiotherapist, Warwick Cancer Clin, Washington, DC, 39-40, Nat Cancer Inst, Baltimore, Md, 41-42 & Ellis Fischel Cancer Hosp, 43-49; prof clin radiol, Univ Colo, & dir, Penrose Cancer Hosp, Colorado Springs, 49-74; prof, 74-83, EMER PROF RADIOL, COL MED, UNIV SFLA, 83- *Concurrent Pos:* Mem adv comt, PR Nuclear Ctr & Nat Adv Cancer Coun, 67-71, Milheim Found Cancer Res, 61-85; pres, Int Club Radiotherapists, 62-66; trustee, Am Bd Radiol, 75-85; pres & bd dirs, Fedn Am Oncol Soc, 76-77; historian, Am Inst Radiol, 79-90; distinguished physician, Vet Admin, emer distinguished physician, 82- *Honors & Awards:* Laureat, French Acad Med, 48 & 79; Gold Medals, Inter-Am Col Radiol, Am Col Radiol, Radiol Soc NAm & Am Soc Therapeutic Radiologists; Prix Bruninghaus, French Acad Med, 79; Beclere Medal, Ctr Antoine Beclere, Paris, 80; Distinguished Serv Award, Am Cancer Soc, 83. *Mem:* Inter-Am Col Radiol (pres, 67-71); fel Am Col Radiol; Am Roentgen Ray Soc; Radiol Soc NAm (vpres, 59-60); Am Radium Soc (vpres, 63-64, treas, 65, pres, 68-69); hon mem, Am Asn Physicists Med; hon mem, Arthur Purdy Stout Soc Pathologists. *Res:* Therapeutic radiology; radiotherapy of cancer of the maxillary sinuses; transvaginal roentgentherapy for cancer of the cervix; total body irradiation for chronic leukemia; pathways of spread of malignant tumors; radiotherapy of soft tissue sarcomas; radiotherapy of inoperable cancer of the prostate; treatment of cancer of the breast; author. *Mailing Add:* Dept Radiol Univ SFla Col Med Tampa FL 33612

DEL RIO, CARLOS EDUARDO, b Caguas, PR, Oct 13, 28; m 52; c 4. ENDODONTICS INSTRUMENT EVALUATION, RADIOGRAPHIC IMAGERY. *Educ:* Creighton Univ, BS, 52, DDS, 56; Am Bd Oral Med, cert & dipl oral med, 69; Am Bd Endodontics, cert & dipl endodontics, 76. *Prof Exp:* Dent officer, US Army, 56-77, chief, div clin res, Inst Dent Res, Walter Reed Med Ctr, 71-77; assoc prof, 77-82, head, grad endodontics, 77-80, PROF ENDODONTICS, UNIV TEX HEALTH SCI CTR, SAN ANTONIO, 82-, CHMN DEPT, 80- *Concurrent Pos:* Resident oral med, Univ Pa, 66 & endodontics, Martin Army Hosp, 71; instr oral med, Sch Dent Med, Univ Pa, 64-66 & Sch Dent, Univ NC, 69-70; prof lectr oral biol, George Washington Univ, 71-77; vis lectr endodontics, Sch Dent Med, Univ Pa, 76-83; consult endodontics, US Army, 77 & Audie Murphy Vet Admin Hosp, San Antonio, 77-; extraordinary prof endodontics, Autonomous Univ Nuevo Leon, Monterrey, Mex, 81-; dir, Am Bd Endodontics, Am Asn Endodontists, 86- *Mem:* Int Asn Dent Res; Am Dent Asn; Am Asn Endodontists; Am Acad Oral Med; Am Asn Dent Schs; fel Am Col Dentists. *Res:* Efficiency of endodontic instruments in cleaning and shaping the root canal; design of instrument delivery system in endodontics; inadvertent contamination of the root canal by endodontic instruments; condensation of gutta-percha with ultrasonics; development of a model to study periapical pathology; development of sterile abscesses in rats; imaging in endodontics; steroid action on exposed pulp; action of condiments on the pouch mucosa of golden Syrian hamsters; evaluation of methyl cynoacrylate and acrylate amide as an oval bandage; cell culture of scrapings of human oral epithelium. *Mailing Add:* Univ Tex Health Sci Ctr 7703 Floyd Curl Dr San Antonio TX 78284-7892

DELSANTO, PIER PAOLO, b Torino, Italy, Mar 14, 41; m 70; c 2. NUCLEAR & THEORETICAL PHYSICS. *Educ:* Univ Torino, Italy, Dr(physics), 63, specialization nuclear physics, 65. *Prof Exp:* Fel physics & instr math, Univ Torino, Italy, 64-66, Alexander von Humboldt Stiftung fel physics, Univ Frankfurt, 66-68, res assoc, 69; from asst prof to assoc prof physics, Univ PR, Mayaguez, 69-77, prof, 77-; NAVAL RES LAB, WASHINGTON, DC. *Concurrent Pos:* Scientist I, PR Nuclear Ctr, 70-75; vis prof, Univ Melbourne, 73 & Univ Cagliari, Sardinia, 75-76. *Mem:* Am Phys Soc; Europ Phys Soc. *Res:* Nuclear reaction theory; intermediate energy nuclear physics; calculations of photonuclear reaction cross sections. *Mailing Add:* Dip Fisica Del Polirecnico C Duca Degli Abruzzi 24 Torino 10129 Italy

DELSEMME, ARMAND HUBERT, b Verviers, Belg, Feb 1, 18; m; c 3. ASTROPHYSICS. *Educ:* Univ Liege, BA, 38, MA & MEd, 40, PhD(physics), 51. *Prof Exp:* Res worker, Res Labs, Belge de l' Azote Corp, Belg, 46-51, head phys res lab, 51-53, mgr & head res, 53-56; dir, Belgian Congo Astron Observ & sci adv, Belgian Inst Sci Res Cent Africa, 56-61; head basic studies div, Directorate Sci Affairs, Orgn Econ Coop & Develop, France, 61-66; prof, 66-86, DISTINGUISHED PROF ASTROPHYS, UNIV TOLEDO, 86- *Concurrent Pos:* Corresp mem, Belg Nat Comt Astron, 58-64, assoc mem, 64-68; prof, Belg State Univ, Elizabethville, Congo, 60; guest investr, CRB adv fel, Mt Wilson & Palomar Observ, 60-61; chmn int govt comn, Int Ctr Geothermal Res, 64-66; vpres comn 15 for phys study of comets, Int Astron Union, 70-73, pres comn 15 for phys study of comets, minor planets and meteorites, 73-76, chmn comt world inventory cometary archives, 75-79; consult, Var NASA & Nat Acad Sci Panels; distinguished vis scientist, Jet Propulsion Lab, Calif Inst Technol, 79-80; chmn, steering group, Int Comet Halley Watch, 84-85; interdisciplinary scientist, Comet Rendezvous & Asteroid Flyby Mission, Jet Propulsion Lab, NASA, 87- *Mem:* Am Astron Soc; Int Astron Union; Astron Soc France; Belg Phys Soc; Royal Belg Acad Overseas Sci; Belg Royal Acad. *Res:* Molecular spectroscopy; thermodynamics; geophysics; volcanology; stellar astrophysics; cometary phenomena; air glow. *Mailing Add:* Dept Physics & Astron Univ Toledo Toledo OH 43606

DELSON, ERIC, b New York, NY, Jan 18, 45; m 67; c 1. PALEOANTHROPOLOGY, PRIMATOLOGY. *Educ:* Harvard Univ, AB, 66; Columbia Univ, PhD(geol), 73. *Prof Exp:* Asst prof anthrop, Univ Pittsburgh, 72-73; asst prof, 73-75, assoc prof, 76-79, PROF ANTHROP, LEHMAN COL, 80- *Concurrent Pos:* Res grants, Wenner-Gren Found Anthrop Res, 71, 78, 80 & 83, Nat Geog Soc, 71, NSF, 74 & 79-88, City Univ New York Res Award Prog, 75-88, L S B Leakey Found, 78, 80 & 83, Found Res Origin Man, 80; asst prof, assoc prof & prof anthrop, City Univ NY Grad Ctr, 74-, co-ordinator phys anthrop, 78-79, 83- & prof biol, 88-; res assoc, Dept Vert Paleont, Am Mus Natural Hist, 75-, co-cur, ancestors exhib, 80-84; paleoanthrop deleg to People's Repub China, Nat Acad Sci, 75; vis prof, Yale Univ, 80; assoc ed, Contributions Primatol, 80-; mem, Anthrop Adv Panel, NSF, 80-82; J S Guggenheim Mem fel, 81-82; consult, AMS Press, 80-, Stonehenge Press, 81, exhib, NY Zool Soc, 89-90, exhib, Am Mus Natural Hist, 89-; mem, Nat Deleg Int Quaternary Asn Cong, Moscow, USSR, 82; sci assoc, Inst Human Origins, 82-87; assoc ed, J Human Evolution, 83-85, 90-, ed, 86-89; exec adv bd, Encycl of Human Biol, Dict of Sci & Technol, 89-; co-ed Encycl of Human Evolution & Prehist, 86-88, Paleonthropology Ann, 90-; mem permanent coun, Int Asn Human Paleont, 87-92; nat prog vis scientist, comm scholarly commun with People's Rep China, Nat Acad Sci, 87-88. *Mem:* Soc Vert Paleont; Am Asn Phys Anthropologists; Int & Am Primatol Soc; Soc Study Evolution; Am Anthrop; Paleont Soc; AAAS; Sigma Xi; Int Asn Human Paleont. *Res:* Primate paleontology and primatology, especially Cercopithecidae; biological and cultural evolution of humans; later Cenozoic stratigraphy and biochronology; evolutionary biology and systematic methodology; primate classification and systematics; vertebrate paleontology. *Mailing Add:* Dept Anthrop Lehman Col Bronx NY 10468

DELTON, MARY HELEN, b Troy, NY, Mar 6, 46; m 76. ORGANIC CHEMISTRY. *Educ:* Ariz State Univ, BS, 67; Univ Calif, Los Angeles, MS & PhD(org chem), 70. *Prof Exp:* NSF fel, Northwestern Univ, 70-71; asst prof chem, Mt Holyoke Col, 71-76; adj asst prof, Wayne State Univ, 76-78; asst prof chem, Oakland Univ, 78-80; SR RES CHEMIST, EASTMAN KODAK, 80- *Mem:* Am Chem Soc; Sigma Xi. *Res:* Silver halide photoscience. *Mailing Add:* 50 Buggywhip Trail Honeoye Falls NY 14472

DEL TORO, VINCENT, b New York, NY, Sept 17, 23; m 65. ELECTRICAL ENGINEERING, CONTROL SYSTEMS. *Educ:* City Col New York, BEE, 46; Polytech Inst Brooklyn, MEE, 50. *Prof Exp:* Tutor, 46-50, from instr to assoc prof, 50-62, asst dean, 62-65, PROF ELEC ENG, CITY COL NEW YORK, 63-, ASSOC DEAN SCH ENG, 65- *Concurrent Pos:* Consult control systs, Gen Dynamics/Convair, 57-59; consult, Bendix Corp, 77-78. *Mem:* Sr mem Inst Elec & Electronics Engrs; Am Soc Eng Educ. *Res:* Self-adaptive control systems; feedback and nonlinear control systems; synchronous induction motors; analog computers. *Mailing Add:* Dept Elec Eng City Col New York Convent Ave at 138th St New York NY 10031

DELUCA, CARLO J, b Bagnoli del Trigno, Italy, Oct 12, 43; nat US; m; c 1. BIOMEDICAL ENGINEERING. *Educ:* Univ BC, BASc, 66, MSc, 68, Queens Univ, PhD(biomed eng), 72. *Prof Exp:* Lab instr elec eng, Univ NB, 67-68, lectr computer sci, 68-69; lectr biomed eng, Queens Univ, 69-70, lab instr anat, 70-71, lectr anat, 71-72, asst prof anat, 72-73; res assoc orthop surg, Harvard Med Sch, 73-79. prin res assoc, 79-84; lectr mech eng, Mass Inst Technol, 73-85; res assoc orthop surg, Children's Hosp, Boston, 73-84, Neuro Muscular Res Lab, 80-84; PROF BIOMED ENG & DIR NEUROMUSCULAR RES CTR, BOSTON UNIV, 84-, RES PROF NEUROL, SCH MED, 85- *Concurrent Pos:* Proj dir, Liberty Mutual Res Ctr, 73-; grants, var insts, corp & orgn, 74-92; adj assoc prof biomed eng, Boston Univ, 77-84, interim dean, Col Eng, 86-89, interim chmn biomed eng, 86-; affil scientist, New Eng Reg Primate Ctr, 77-81; mem, Rev Comt Rehab Eng Res & Develop, Vet Admin, 77-80, Comt New Appln, Methodologies & Quant Anal EMG, Int Fed Soc Electroencephalography & Clin Neurophysiol, 81-85, Comt electrophysiol Tech Eng in Med & Biol Soc, 84-; res mem, Div Health Sci & Technol, Harvard-Mass Inst Technol, 78-84; health sci specialist, Vet Admin, 84-; sr lectr mech eng, Mass Inst Technol, 85- *Mem:* Sigma Xi; fel Inst Elec & Electronics Engrs; Int Soc Electrophysiol Kinesiol (secy, 80-84, vpres, 85-88, pres, 88-90); AAAS; Soc Neurosci; Orthop Res Soc; Biomed Res Soc; Rehab Eng Soc; Am Soc Eng Educ; Acad Int Med Studies. *Res:* Motor control of normal and abnormal muscles; motor unit properties of muscles; objective evaluation of muscle fatigue in humans; objective evaluation of low back pain; eight patents; numerous technical publications. *Mailing Add:* NeuroMuscular Res Ctr Boston Univ 44 Cummington St Boston MA 02215

DE LUCA, CHESTER, b Bristol, Pa, Sept 7, 27; m 54; c 5. BIOCHEMISTRY, CELL BIOLOGY. *Educ:* Georgetown Univ, BS, 52; Johns Hopkins Univ, PhD(biochem), 56. *Prof Exp:* Jr instr biol, Johns Hopkins Univ, 52-54, Nat Cancer Inst fel, 54-56; Am Cancer Soc fel, Rockefeller Inst, 56-59; instr pediat, Johns Hopkins Univ, 59-62; sr cancer res scientist, Roswell Park Mem Inst, 62-65; asst prof, 65-71, ASSOC PROF ORAL BIOL, SCH DENT MED, STATE UNIV NY BUFFALO, 71-, ADJ PROF NUTRITION, SCH HEALTH RELATED PROFESSIONS, 89- *Concurrent Pos:* Res assoc pediat, Sinai Hosp, Baltimore, Md, 59-62. *Mem:* Am Soc Cell Biol; Nutrit Today Soc; Sigma Xi. *Res:* Glycosaminoglycans as markers of connective tissue disease; applied and clinical nutrition relative to oral health. *Mailing Add:* Dept Oral Biol Sch Dent Med State Univ NY 3435 Main St Buffalo NY 14214

DELUCA, DOMINICK, CELLULAR IMMUNOLOGY. *Educ:* Univ Calif, Los Angeles, PhD(microbiol), 74. *Prof Exp:* ASST PROF IMMUNOL, MED UNIV SC, 80- *Mailing Add:* Dept Microbiol & Immunol Univ Ariz Pharm-Microbiol Bldg 90 Tucson AZ 85721

DE LUCA, DONALD CARL, b Amsterdam, NY, May 24, 36; m 66; c 2. ORGANIC CHEMISTRY, BIOCHEMISTRY. *Educ:* Rensselaer Polytech Inst, BS, 57; Univ Minn, PhD(org chem), 63. *Prof Exp:* Res fel biochem, Sch Med, Johns Hopkins Univ, 64-66; asst prof biochem, Sch Med, 66-75, ASSOC PROF BIOCHEM, COL MED, UNIV ARK MED SCI, LITTLE ROCK, 75- *Mem:* AAAS; Am Chem Soc; Royal Soc Chem. *Res:* Photochemical oxidations; chemical and biological oxidative processes of nitrogen containing compounds; nitrosoureas. *Mailing Add:* Dept Biochem Univ Ark for Med Sci Little Rock AR 72205

DELUCA, HECTOR FLOYD, b Pueblo, Colo, Apr 5, 30; m 54; c 4. VITAMINS & HORMONES. *Educ:* Univ Colo, BA, 51; Univ Wis, MS, 53, PhD, 55. *Hon Degrees:* DSc, Univ Colo, Boulder, 74; DSc, Med Col Wis, 80. *Prof Exp:* Res asst biochem, Univ Wis-Madison, 51-55, proj assoc, 55, fel, 56-57, from instr to assoc prof, 57-65, chmn dept, 70-86, PROF BIOCHEM & HARRY STEENBOCK RES PROF, UNIV WIS-MADISON, 65-, CHMN DEPT, 91- *Concurrent Pos:* Vis scientist, Strangeways Res Labs, Eng, 60; lectr, Royal Col Physicians & Surgeons Can, 77; Lenore Richards vis prof, Northwestern Hosp, Minneapolis, Minn, 77; distinguished lectr med sci, Mayo Clin, Rochester, 78; plenary lectr, 5th Int cong Hormonal Steroids, New Delhi, 78. *Honors & Awards:* Andre Lichtwitz Prize, 68; Mead Johnson Award, 68; Nicolas Andry Award, 71; Osborne & Mendel Award, Am Inst Nutrit, 73; Roussel Prize of France, 74; Gairdner Found Award, Can, 74; Dixon Medal, Irish Med Coun, 75; Atwater Award, 79; William Rose lectr, 80; Harvey lectr, 80; R A Morton Lectr, Biochem Soc, 81; Roger L Corson Medal, Franklin Inst, 85; William F Neuman Award, Am Soc Bone & Mineral Res. *Mem:* Nat Acad Sci; Am Chem Soc; Am Soc Biol Chemists; Am Inst Nutrit; Endocrine Soc; Am Acad Arts & Sci; Am Physiol Soc; Am Soc Bone & Mineral Res; AAAS; Soc Exp Biol & Med (pres). *Res:* Mechanism of action and metabolism of vitamins and hormones, especially vitamin D; issued 100 patents; over 900 publications in the fields of vitamin A, vitamin D, parathyroid hormone and calcitonin. *Mailing Add:* Dept Biochem Univ Wis 420 Henry Mall Madison WI 53706

DE LUCA, LUIGI MARIA, b Maglie, Italy, Feb 25, 41; US citizen; m 65; c 2. CARCINOGENESIS, METABOLISM OF NUTRIENTS. *Educ:* Univ Pavia, PhD(org chem), 64. *Prof Exp:* Res assoc, Maggiore Hosp, Milan, 64-65; res assoc nutrit biochem & food sci, Mass Inst Technol, 71-73; res chemist, 71-73, SECTION CHIEF CARCINOGENESIS, NAT CANCER INST, BETHESDA, 73- *Concurrent Pos:* Vis prof, Mass Inst Technol, 71-73, Univ Tokyo, Japan, 76 & Univ Pavia, Italy, 88; Rockefeller Found scholar, Bellagio, 88. *Honors & Awards:* Mead-Johnson Award, Am Inst Nutrit, 78. *Mem:* Am Soc Cell Biol; Am Inst Nutrit; Am Soc Biol Chemists; Am Soc Clin Nutrit; Am Asn Cancer Res; NY Acad Sci. *Res:* The mechanism of vitamin A in controlling differentiation, in inhibiting or modulating the response of tissues to cancer causing agents and in enhancing cell adhesiveness. *Mailing Add:* Nat Cancer Inst NIH Bldg 37 Rm 3A-17 9000 Rockville Pike Rm 3A-17 Bethesda MD 20892

DELUCA, MARLENE, biochemistry, for more information see previous edition

DELUCA, PATRICK JOHN, b Springfield, Mass, July 27, 44; m 66; c 2. MICROBIOLOGY, CYTOLOGY. *Educ:* St Michael's Col, Vt, BA, 66; Fordham Univ, MS, 68, PhD(biol), 75. *Prof Exp:* Lab asst biol, Fordham Univ, 66-69, teaching fel, 69-70; asst prof, 70-77, assoc prof biol, 77-80, PROF BIOL, MT ST MARY COL, NY, 80-, PRE-MED ADV, 77- *Concurrent Pos:* Instr, Dutchess Community Col, 75. *Mem:* Am Inst Biol Sci; Sigma Xi; Torrey Bot Club. *Res:* Ultrastructural characteristics of certain green soil algae and how these characteristics might be used for taxonomic purposes. *Mailing Add:* 20 Blake St Newburgh NY 12550

DELUCA, PATRICK PHILLIP, b Scranton, Pa, Sept 7, 35; m 56; c 6. PHARMACEUTICS. *Educ:* Temple Univ, BS, 57, MS, 60, PhD(pharm), 63. *Prof Exp:* Jr anal chemist, Smith Kline & French Labs, 57-59; sr res pharmacist, Ciba Pharmaceut Co, 63-66, plant mgr, Somerville Opers, 66-69, plant mgr, Cormedics Corp, 69-70, dir develop & control, 70; assoc prof, Univ Ky, 70-75, asst dean col pharm, 71-75, assoc dean, 75-87, PROF PHARM, UNIV KY, 75- *Concurrent Pos:* Mem, US Pharm-FDA Nat Coord Comt on Large Vol Parenterals. *Honors & Awards:* Lunsford-Richardson Pharm Award, 60 & 62. *Mem:* Am Pharmaceut Asn; fel Acad Pharmaceut Sci; NY Acad Sci; fel Am Asn Pharmaceut Scientists. *Res:* Pharmaceutical technology; kinetics and stabilization; lyophilization; sterile products; intravenous administration; formulation of proteins; drug delivery system; over 100 publications. *Mailing Add:* Col Pharm Univ Ky Lexington KY 40536-0082

DELUCA, PAUL MICHAEL, JR, b Albany, NY, Apr 22, 44; m 66; c 2. MEDICAL PHYSICS, NUCLEAR PHYSICS. *Educ:* LeMoyne Col, BS, 66; Univ Notre Dame, PhD(physics), 71. *Prof Exp:* Fel radiol, 71-73, adj asst prof, 73-75, asst prof radiol, 75-81, assoc prof med physics, 81-85, PROF MED PHYSICS, UNIV WIS-MADISON, 85- *Mem:* Am Inst Physics; Health Physics Soc; Am Asn Physicists Med. *Res:* Fast neutron physics; high linear energy transfer therapy beams; dosimetry and charged particle spectroscopy. *Mailing Add:* Dept Radiol Univ Wis Clin Sci Ctr 600 Highland Ave Madison WI 53792

DELUCA, ROBERT D(AVID), b Passaic, NJ, Jan 11, 41; m 63; c 4. METALLURGY, SOLID STATE PHYSICS. *Educ:* Stevens Inst Technol, BE, 62, MS, 64, PhD(metall), 66. *Prof Exp:* Sr metallurgist, Corning Glass Works, 66-76; MGR MAT SCI, JOHNSON & JOHNSON DENTAL PRODS CO, 76- *Honors & Awards:* Philip B Hoffman Award. *Mem:* Am Phys Soc; Int Asn Dent Res; Am Inst Mining, Metall & Petrol Engrs. *Res:* Strengthening mechanisms in metals; physical metallurgy; dental materials; amalgam alloys; metal-ceramic systems; manufacturing processes for high strength, lowloss, optical waveguide fibers; electronic materials; semiconducting glass-ceramics; metal-glass systems. *Mailing Add:* Johnson & Johnson Dent Care 501 George St New Brunswick NJ 08903

DE LUCCIA, JOHN JERRY, b Philadelphia, Pa, Mar 15, 35; m 57; c 3. MATERIALS SCIENCE, CORROSION. *Educ:* Drexel Univ, BSc, 61; Univ Pa, MSc, 67, PhD(metall & mat sci), 76. *Prof Exp:* Instr metall eng, Drexel Univ, 61-63; metallurgist, 63-72, supvry mat engr, 72-90, SR MAT ENGR, AERO MAT LAB, NAVAL AIR DEVELOP CTR, 90- *Concurrent Pos:* Adj prof mat & metall eng, Univ Col, Drexel Univ, 66-; N Am Coordr Corrosion Fatigue Coop Study, Adv Group for Aerospace Res & Develop, NATO, 77-; vis prof mat eng, Drexel Univ, 78-79. *Honors & Awards:* Templin Award, Am Soc Testing & Mat, 70; Sci Achievement Award, Naval Air Develop Ctr, 78; Sauveur Award, Am Soc Metals, 82. *Mem:* Fel Am Soc Metals; Nat Asn Corrosion Engrs; Naval Civilian Adminr Asn; Sigma Xi. *Res:* Corrosion and degradation of aerospace materials; stress corrosion cracking and hydrogen embrittlement of high strength alloys; corrosion and embrittlement control of aerospace alloys; crack arrestment inhibiting compounds; hydrogen embrittlement measurements and electrochemical assessments. *Mailing Add:* Drexel Univ Mat & Metall Eng Dept 32 & Chestnut St Philadelphia PA 19104

DE LUCIA, FRANK CHARLES, b St Paul, Minn, June 21, 43; m 66; c 2. MOLECULAR PHYSICS, QUANTUM ELECTRONICS. *Educ:* Iowa Wesleyan Col, BS, 64; Duke Univ, PhD(physics), 69. *Prof Exp:* From instr & res assoc to prof physics, Duke Univ, 69-90, chmn, physics dept, 87-88; PROF PHYSICS, OHIO STATE UNIV, 90-, CHMN, PHYSICS DEPT, 90- *Mem:* Inst Elec & Electronics Engrs; Am Phys Soc; Optical Soc Am. *Res:* Microwave spectroscopy; millimeter and submillimeter waves. *Mailing Add:* Dept Physics Ohio State Univ 174 W 18th Ave Columbus OH 43210

DELURY, DANIEL BERTRAND, b Walker, Minn, Sept 19, 07; Can citizen; m 41; c 2. MATHEMATICS, STATISTICS. *Educ:* Univ Toronto, BA, 29, MA, 30, PhD(math), 36. *Prof Exp:* Instr math, Univ Sask, 31-34; lectr, Univ Toronto, 36-43, asst prof, 43-45; from assoc prof to prof statist, Va Polytech Inst, 45-47, dir dept math statist, Ont Res Found, 47-58; prof math, 58-77, EMER PROF MATH, UNIV TORONTO, 77- *Concurrent Pos:* Chmn dept math, Univ Toronto, 58-68. *Mem:* Can Statist Soc; fel Am Statist Asn; Am Math Soc; Inst Math Statist; Biomet Soc. *Res:* Population dynamics; estimation of biological populations. *Mailing Add:* 131 Beecroft Rd Suite 201 Willowdale ON M2P 1G2 Can

DELUSTRO, FRANK ANTHONY, b Brooklyn, NY, May 8, 48; m 74. CELLULAR IMMUNOLOGY. *Educ:* Fordham Univ, BS, 70; State Univ NY, PhD(immunol), 76. *Prof Exp:* Res assoc immunol, Dept Basic & Clin Immunol & Microbiol, Med Univ SC, 76-78, instr immunol, Div Rheumatol & Immunol, 78-80, asst prof, Med Div, 80-83; mgr, Immunol Connective Tissue Res Lab, 83-85, mgr clin sci, 85, DIR MED AFFAIRS, COLLAGEN CORP, 85- *Concurrent Pos:* Am Cancer Soc fel, 76-78. *Mem:* Sigma Xi; Am Asn Immunol; Reticuloendothelial Soc; Soc Exp Biol Med. *Res:* Elucidation of connective tissue biology, especially as immunity and monocyte function relate to collagen, autoimmunity and fibrosis. *Mailing Add:* 2517 DeKovan Ave Belmont CA 94002

DELVAILLE, JOHN PAUL, b Riverside, Calif, Oct 5, 31; div; c 1. PHYSICS. *Educ:* Univ Calif, Berkeley, BA, 54; Cornell Univ, PhD(exp physics), 62. *Prof Exp:* Instr & res assoc physics, Cornell Univ, 62-63, from actg asst prof to asst prof, 63-70; res mem, Ctr Space Res, Mass Inst Technol, 70-74, res affil, 74-80; physicist, Ctr Astrophys, Smithsonian Astrophys Observ, 74-80; LINCOLN LABS, MASS INST TECHNOL, LEXINGTON, 81- *Concurrent Pos:* Consult computer mfg processes. *Mem:* AAAS; Am Asn Physics Teachers; Am Phys Soc. *Res:* Cosmic rays; extensive air showers; gamma-ray and x-ray astronomy; atomic collisions and radar analysis. *Mailing Add:* Lincoln Lab Mass Inst Technol PO Box 73 Lexington MA 02173

DEL VALLE, FRANCISCO RAFAEL, b Laredo, Tex, Oct 19, 33; Mex citizen; m 61; c 2. PHYSICAL CHEMISTRY OF FOODS. *Educ:* Mass Inst Technol, BS, 54, MS, 56 & 57, PhD(food sci), 65. *Prof Exp:* Prof chem eng, Inst Technol Monterrey, 61-69; res mgr, Molinos Azteca, SA, Monterrey, Mex, 69-71; tech mgr, Mexicana de Jugos y Sabores, SA, Monterrey, Mex, 71-73; prof food sci, Inst Technol, Monterrey, 73-77 & Independent Univ, Chihuahua, 77-84; PRES, FOUND INVEST CIENC ALIM & NUTRIT, 80-; PROF, CHEM ENG DEPT, NMEX STATE UNIV, LAS CRUCES, 86- *Concurrent Pos:* Vis scientist, Swedish Inst Food Preserv Res, Gothemburg, Sweden, 66 & Nat Polytech, Quito, Ecuador, 85; vis prof, Inst Food Technol, Campinas, Brazil, 76, dept nutrit & food sci, Mass Inst Technol, 78, Dominican Inst Indust Technol, Dominican Repub, 81, dept food sci & human nutrit, Univ Del, 81-82 & Col Agr, Home Econ, NMex State Univ, 82 & Inst CIEPE, San Felipe, Venezuela, 86; vis lectr, dept nutrit and food sci, Mass Inst Technol, 79-84; sci consult, var Mex co, 79- & Res & Develop Labs, McCormick & Co, Hunt Valley, Md, 81; adj prof chem, Univ Tex, El Paso, 84- *Honors & Awards:* Nat Tech Prize, President Mex, 78; Tomas Valles Vivar Prize, Sci & Tech, 83. *Mem:* Mex Acad Sci Invest; Mex Nat Acad Eng; Inst Food Technologists; Mex Asn Food Technol; Latin Am Soc Nutrit. *Res:* High nutrition, low cost foods from vegetable, animal and marine sources. *Mailing Add:* Dept Chem Eng NMex State Univ Box 3805 Las Cruces NM 88003-3805

DEL VALLE, LUIS A, water pollution control, hydraulics, for more information see previous edition

DEL VECCHIO, VITO GERARD, b Dunmore, Pa, May 13, 39. BIOCHEMICAL GENETICS. *Educ:* Univ Scranton, BS, 61; St John's Univ, NY, MS, 63; Hahnemann Med Col, PhD(biochem genetics), 67. *Prof Exp:* USPHS res fel microbiol, Univ Geneva, 66-68; assoc prof biol, Stonehill Col, 68-69; from asst prof to assoc prof, 73-77, PROF BIOL, UNIV SCRANTON, 77- *Mem:* AAAS; Genetics Soc Am; Am Soc Microbiol. *Res:* Tyrosinase genetics; biochemistry and immunochemistry of Neurospora crassa morphogenesis; biochemical control of morphogenesis of Neurospora isozymes of Neurospora crassa. *Mailing Add:* Dept Biol Univ Scranton Scranton PA 18510

DELVIGS, PETER, b Riga, Latvia, June 28, 33; US citizen; m 65; c 1. ORGANIC POLYMER CHEMISTRY. *Educ:* Western Reserve Univ, BA, 59; Univ Minn, Minneapolis, PhD(org chem), 63. *Prof Exp:* Res assoc biochem, Cleveland Clin Found, 63-67; AEROSPACE CHEMIST, NASA LEWIS RES CTR, CLEVELAND, 67- *Mem:* AAAS; Am Chem Soc; NY Acad Sci. *Res:* Synthesis, characterization and properties of new thermally stable polymers; development of fiber-reinforced resin composites. *Mailing Add:* 21290 Parkwood Ave Cleveland OH 44126-2735

DEL VILLANO, BERT CHARLES, b Yeadon, Pa, Apr 9, 43; m 63; c 1. IMMUNOLOGY, CANCER. *Educ:* Lehigh Univ, BA, 65; Univ Pa, PhD(microbiol), 71. *Prof Exp:* Fel immunol, Scripps Clin Res Found, 71-73, asst, 73-75; mem staff immunol, Cleveland Clin, 75-80; dir sales, Centocor, 80-84, vpres mkt sales, 84-89. *Concurrent Pos:* Basil OConnor grant, March of Dimes Nat Found, 74. *Mem:* Am Soc Microbiol; Int Asn Comp Res Leukemia & Related Dis; Am Asn Immunologists. *Res:* Expression of viral proteins in normal and malignant tissues and the immune response against these viral proteins; tumor markers; CA 19-9 tumor associated antigen. *Mailing Add:* 183 Bay View Dr San Carlos CA 94070

DELWICHE, CONSTANT COLLIN, b Wis, Nov 26, 17; m 43; c 4. COMPARATIVE BIOCHEMISTRY. *Educ:* Univ Wis, BS, 40; Univ Calif, PhD, 49. *Prof Exp:* Chmn dept soils & plant nutrit, 65-72, PROF GEOBIOL, DEPT LAND, AIR & WATER RESOURCES, UNIV CALIF, DAVIS, 60- *Mem:* AAAS; Am Soc Plant Physiologists; Am Soc Microbiologists; Sigma Xi. *Res:* Inorganic nitrogen transformation; nitrogen fixation; inorganic energy metabolism; isotope distribution; mass spectrometry. *Mailing Add:* LAWR-Hoagland Univ Calif Davis CA 95616

DELWICHE, EUGENE ALBERT, b Green Bay, Wis, Nov 26, 17; m 49; c 4. BACTERIOLOGY. *Educ:* Univ Wis, BS, 41; Cornell Univ, PhD, 48. *Prof Exp:* From asst prof to assoc prof, 48-55, PROF BACT, CORNELL UNIV, 55- *Concurrent Pos:* Consult, Oak Ridge Nat Labs, 51-59; Guggenheim fel, Karolinska Inst, Sweden, 64. *Mem:* Am Soc Biol Chemists; Am Acad Microbiol; Am Soc Microbiol; Can Soc Microbiol. *Res:* Physiology, biochemistry, nutrition and intermediary metabolism of bacteria and other microorganisms. *Mailing Add:* Stocking Hall Cornell Univ Ithaca NY 14853

DEMAEYER, BRUCE R, ELECTRICAL ENGINEERING. *Prof Exp:* MEM STAFF, AMERITECH MOBILE COMMUN. *Mailing Add:* Ameritech Mobile Commun 1515 Woodfield Suite 1400 Schaumburg IL 60173

DEMAGGIO, AUGUSTUS EDWARD, b Malden, Mass, Apr 22, 32; m 54; c 3. BOTANY, PHYSIOLOGY. *Educ:* Mass Col Pharm, BS, 54, MS, 56; Harvard Univ, AM, 58, PhD, 60. *Hon Degrees:* AM, Dartmouth Col, 70. *Prof Exp:* From asst prof to assoc prof pharmacog, Col Pharm, Rutgers Univ, 59-64; from asst prof to assoc prof 64-66, PROF BIOL, DARTMOUTH COL, 66- *Concurrent Pos:* Waksman Found fel, France; univ fac fel, Rutgers Univ & mem staff, Nat Ctr Agr Res, France, 62-63; res fel, Harvard Univ, 69-70; vis prof, Yale Univ, 73-74. *Mem:* Fel AAAS; Bot Soc Am; Am Soc Plant Physiol; Torrey Bot Club; Am Soc Pharmacog. *Res:* Experimental botany; morphogenesis; chloroplast biochemistry; plant chemistry. *Mailing Add:* Dept Biol Sci Dartmouth Col Hanover NH 03755

DEMAIN, ARNOLD LESTER, b Brooklyn, NY, Apr 26, 27; m 52; c 2. INDUSTRIAL MICROBIOLOGY. *Educ:* Mich State Univ, BS, 49; MS, 50; Univ Calif, PhD(microbiol), 54. *Prof Exp:* Asst yeast physiol, Univ Calif, 52-54; res microbiologist, Merck, Sharp & Dohme Res Labs, 54-64, head fermentation res, 64-69; PROF INDUST MICROBIOL, MASS INST TECHNOL, 69- *Concurrent Pos:* Labatt lectr, Univ Western Ont, 77. *Honors & Awards:* Waksman Award, Am Soc Microbiol, 75, Cetus Biotechnol Award, 90; Charles Thom Award, Soc Indust Microbiol, 78; Hotpack Award, Can Soc Microbiol, 78; Rubbo Award, Australian Soc Microbiol, 79. *Mem:* Soc Indust Microbiol; Am Soc Microbiol; Am Chem Soc. *Res:* Industrial microbiology; penicillin and cephalosporin biosynthesis; biotechnology; enzyme synthesis; antibiotic biosynthesis; regulation of fermentation processes. *Mailing Add:* Dept Biol Mass Inst Technol Rm 56-123 Cambridge MA 02139

DE MAINE, PAUL ALEXANDER DESMOND, b Koster, SAfrica, Oct 11, 24; US citizen; m 55. COMPUTER SCIENCE. *Educ:* Univ Witwatersrand, BA, 48; Univ BC, PhD(chem), 56. *Prof Exp:* Res spectros, Univ Chicago, 54-55; res math & chem, Univ Cambridge, 55-56; res spectros, King's Col, London, 56; res electrochem, Univ Ottawa, Ont, 57; from assoc prof to prof chem, State Univ Col Educ, Albany, 57-60; prof, Univ Miss, 60-63; vis scientist, Univ Ill, 63-64; assoc specialist, Univ Calif, Santa Barbara, 64-65; assoc specialist, Ctr Comput Sci & Technol, Nat Bur Standards, 65-67; assoc prof comput sci, Pa State Univ, University Park, 68-70, prof, 70-81; prof comput sci, Univ Ala, Huntsville, 81-82; PROF COMPUT SCI & ENG, AUBURN UNIV, 82- *Concurrent Pos:* A von Humboldt Found sr US scientist award, 74-75 & 76-77. *Mem:* Am Chem Soc; Asn Comput Mach; Inst Elec & Electronics Engrs. *Res:* Computerized information storage and retrieval; data processing; transportable software; deductive systems for chemistry; artificial intelligence. *Mailing Add:* Comput Sci Dept Auburn Univ Auburn Univ AL 36849

DEMAIO, DONALD ANTHONY, SCIENCE INFORMATION, BIOLOGICAL CHEMISTRY. *Educ:* Rutgers Univ, MS, 74. *Prof Exp:* SR INFO RES SCIENTIST, BRISTOL MYERS-SQUIBB INST MED RES, 82- *Mailing Add:* Bristol Myers-Squibb Inst Med Res PO Box 4000 Princeton NJ 08543

DEMAN, JOHN MARIA, b Rotterdam, Neth, Apr 13, 25; Can citizen; m 54; c 3. FOOD CHEMISTRY. *Educ:* Univ Neth, Chem Eng, 51; Univ Alta, PhD(dairy chem), 59. *Prof Exp:* Res chemist, Unilever Res Labs, Holland, 49-54; res asst, Univ Alta, 54-59, asst prof, 59-64, assoc prof dairy & food chem, 64-69; chmn dept, 69-80, PROF FOOD SCI, UNIV GUELPH, 69- *Honors & Awards:* Dairy Res Inc Award, Am Dairy Sci Asn, 74; W J Eva Award, Can Inst Food Sci & Technol, 75. *Mem:* Can Inst Food Technol (past pres); Inst Food Technologists; Am Oil Chemists' Soc; Am Dairy Sci Asn; fel Inst Food Sci Technol (UK). *Res:* Food texture and rheology; instrumentation for food texture measurement; fat crystallization and polymorphism; triglyceride composition and structure; food contaminants; cereal and oilseed technology. *Mailing Add:* Dept Food Sci Univ Guelph Guelph ON N1G 2W1 Can

DEMANCHE, EDNA LOUISE, b Marionville, Mo, Aug 1, 15. SCIENCE EDUCATION, ENVIRONMENTAL & EARTH SCIENCES. *Educ:* Col Mt St Vincent, BS, 40; Univ Notre Dame, MS, 64, PhD(plant physiol), 69. *Prof Exp:* Parochial sch teacher, Hawaii, 40-59; sci consult, 59-67; admin tech & prof asst sci educ res & develop, Foundational Approaches in Sci Prog, Univ Hawaii, 67-73, dir, Hawaii Nature Study Proj, 73-80; ASSOC SUPT CUR RES & DEVELOP, CATH SCH DEPT, STATE OF HAWAII, 80- *Concurrent Pos:* Consult sci fair projs, pub & pvt schs, Hawaii, 59-68; grad fel, NSF, 62-63; traveling team lectr, Hawaii, 65-66; exec secy, Hawaiian Acad Sci, 80-83; mem adv bds, Hawaii Sci Teachers Asn & Hawaii Acad Sci, 80- *Mem:* Nat Cath Educ Asn; Nat Sci Teachers Asn; Nat Asn Surv & Curric Develop. *Res:* Factors associated with changes in phyllotaxy; development of new laboratory-oriented procedures in ecology centered on local environmental phenomena and accompanying procedures for classroom implementation; marine sciences; right/left brain and childhood development patterns relative to learning styles; application to lesson planning with practicism. *Mailing Add:* 3351 Kalihi St Honolulu HI 96819

DEMAR, ROBERT E, b Keene, NH, Nov 7, 31; m 59. VERTEBRATE PALEONTOLOGY. *Educ:* Harvard Univ, AB, 53; Univ Chicago, MS, 60, PhD(vert paleont), 61. *Prof Exp:* Geologist, US Geol Surv, 53-54; asst, 56-58, from instr to assoc prof, 58-74, PROF GEOL, UNIV ILL, CHICAGO CIRCLE, 74-, HEAD DEPT, 79- *Concurrent Pos:* Assoc ed, Paleobiol, 75-80. *Mem:* AAAS; Soc Study Evolution; Geol Soc Am; Soc Vert Paleont; Paleont Soc. *Res:* Late Paleozoic vertebrates; jaw mechanics of synapsid reptiles; functional and evolutionary models in paleobiology; jaw mechanics and the organization of dentitions. *Mailing Add:* Dept Geol Univ Ill Chicago Chicago IL 60680

DE MARCO, F(RANK) A(NTHONY), b Italy, Feb 14, 21; nat US; m 48; c 11. CHEMICAL ENGINEERING. *Educ:* Univ Toronto, BASc, 43, MASc, 43, PhD(chem eng), 51. *Prof Exp:* Instr, Univ Toronto, 43-46; from asst prof to prof chem & head dept, Assumption Univ, 46-57, actg head eng dept, 57-59, assoc dean arts & sci, 58; dean fac appl sci, Univ Windsor, 59-64, vpres univ, 63-76, prof chem & chem eng, 70-88, sr vpres univ, 76-88; RETIRED. *Concurrent Pos:* Chmn staff comt, Essex Col, Assumption, 56-59, prin, 59-63; chmn bd gov, St Clair Col Appl Arts & Technol. *Mem:* Am Soc Eng Educ; fel Chem Inst Can; Eng Inst Can. *Res:* Coordination compounds; electrorefining of copper; glueline studies; cyanine dye synthesis; solubilization of hydrocarbons. *Mailing Add:* 7750 Matchette Rd Windsor ON N9J 2J4 Can

DEMARCO, JOHN GREGORY, b New York, NY, May 13, 39; m 62; c 6. ORGANIC CHEMISTRY. *Educ:* Iona Col, BS, 62; Fordham Univ, PhD(org chem), 67. *Prof Exp:* Res chemist textile, Bound Brook, NJ, 66-71, develop chemist, Charlotte, NC, 71-73, sales rep intermediate chem, 73-77, chief chemist speciality chem, 77-83, PROJ SUPT URETHANE CHEM & SUPV CONTINUOUS PROCESS PILOT PLANT, AM CYANAMID, 83- *Res:* Reaction mechanism and kinetics of reductive triazolation. *Mailing Add:* Am Cyanamid Co 12600 Eckel Rd Perrysburg OH 43551

DEMARCO, RALPH RICHARD, physics, mathematical statistics, for more information see previous edition

DE MARCO, RONALD ANTHONY, b Newark, NJ, Apr 10, 44; m 68; c 2. FLUORINE CHEMISTRY. *Educ:* Montclair State Col, BA, 66; Univ Idaho, MS, 69, PhD(chem), 72. *Prof Exp:* RES CHEMIST FLUORINE CHEM, NAVAL RES LAB, 72-, HEAD, ADVAN INORG MAT, 76- *Mem:* Am Chem Soc; Sigma Xi. *Res:* Synthesis, characterization and chemistry of fluorine-containing compounds of non-metallic elements; synthesis and characterization of conducting polymers, chemistry of non-implantation into covalent materials. *Mailing Add:* 5597 Blake House Ct Burke VA 22015-2015

DEMAREE, RICHARD SPOTTSWOOD, JR, b Akron, Ohio, July 1, 42; m 65, 81. CYTOLOGY, PARASITOLOGY. *Educ:* Purdue Univ, BS, 64; Ind State Univ, MA, 66; Colo State Univ, PhD(zool), 69. *Prof Exp:* Head, Electron Micros Br, Path Div, US Army Med Res & Nutrit Lab, Fitzsimons Gen Hosp, Denver, Colo, 69-72; from asst prof to assoc prof biol sci, 72-80, PROF BIOL SCI, CALIF STATE UNIV, 80- *Mem:* Am Soc Trop Med & Hyg; Electron Micros Soc Am; Am Soc Parasitol; Soc Protozool. *Res:* Parasite ultrastructure and morphogenesis; ultrastructural cytology and pathology; high altitude research. *Mailing Add:* Dept Biol Sci Calif State Univ First & Normal St Chico CA 95929

DEMAREE, THOMAS L, b Cherokee, Iowa, Sept 28, 53; m 90. TECHNICAL MANAGEMENT. *Educ:* Univ Mont, BS, 76. *Prof Exp:* Mgr, Yellowstone-Tongue Areawide Planning Orgn, 77-79; lab mgr, 80-88, TECH DIR, PAC ADHESIVES CO, 88- *Concurrent Pos:* Guest lectr, Forestry Sch, Ore State Univ, 85- *Res:* Development of foamed adhesives for wood products; development of foamed adhesives equipment. *Mailing Add:* Pac Adhesives Co Box 17518 Portland OR 97217

DEMAREST, HAROLD HUNT, JR, b New York, NY, Dec 20, 46; m 68; c 1. ROCK PHYSICS, EQUATIONS OF STATE. *Educ:* Reed Col, BA, 69; Columbia Univ, MA, 71; Univ Calif, Los Angeles, PhD(planetary physics), 74. *Prof Exp:* Postdoctoral scholar geophys, Univ Calif, Los Angeles, 74-75; res assoc geophys, Univ Chicago, 75-79; asst prof geol, Ore State Univ, 79-83; PRES, ASTRO RES INC, 83- *Concurrent Pos:* Vis staff mem, Los Alamos Sci Lab, 72-75. *Mem:* AAAS; Am Geophys Union; Sigma Xi. *Res:* Physical properties of rocks and minerals; geophysics; application of statistics to geological and geophysical problems; elasticity; equations of state. *Mailing Add:* 3621 NW Sylvan Dr Corvallis OR 97330

DEMAREST, JEFFREY R, b Nyack, NY, May 10, 46; m 81. EPITHELIAL TRANSPORT, ECOLOGICAL PHYSIOLOGY. *Educ:* Monmouth Col, BS, 73; Univ Calif, Berkeley, PhD(physiol), 80. *Prof Exp:* Res biophysicist Bodega Marine Lab, 79-81; res assoc, Univ NC, Chapel Hill, 81-87; asst res physiologist, Univ Calif, Berkeley, 84-88; ASST PROF PHYSIOL & CELL BIOL, UNIV ARK, FAYETTEVILLE, 88- *Concurrent Pos:* Vis scientist, Marine Biol Lab, Woods Hole, 85 & 86; grants, NIH, 85, 87, 90; consult, Ctr Ulcer Res & Educ, 87; lectr, Univ Calif, Davis, 88. *Mem:* Am Physiol Soc; Am Soc Zoologists; Biophys Soc; AAAS; Sigma Xi. *Res:* Salt and water transport across biological membranes; cellular and molecular mechanisms of gastric secretion; cellular and organismal ion and water balance; environmental physiology of the evolution of osmoregulation. *Mailing Add:* Dept Biol Sci 632 SE Univ Ark Fayetteville AR 72701

DEMAREST, KEITH THOMAS, NEUROENDOCRINOLOGY, NEUROPHARMACOLOGY. *Educ:* WVa Univ, PhD(pharmacol), 78. *Prof Exp:* ASST PROF MED PHARMACOL, MICH STATE UNIV, 80- *Res:* Neurochemistry. *Mailing Add:* Rte 202 Raritan NJ 08869

DE MARGERIE, JEAN-MARIE, b Prud'homme, Sask, Dec 11, 27; m 55; c 5. OPHTHALMOLOGY. *Educ:* Univ Ottawa, BA, 47; Laval Univ, BMed, 49, MD, 52; Oxford Univ, DPhil(ophthal), 59; FRCPS(C), 61. *Prof Exp:* Asst prof, Queen's Univ, 60-66; head, Dept Ophthal, Univ Sherbrooke, 66-77, vdean res fac med, 73-79, dean fac med, 79-83, PROF OPHTHAL, UNIV SHERBROOKE, 66- *Concurrent Pos:* Consult, Ont Hosp, 60-66 & Armed Forces Hosp, Can, 64-66; head Dept Ophthal, Hotel-Dieu Hosp, Kingston, Ont, 64-66; mem, Dept Ophthal, Univ Hosp, Sherbrooke, 66-; consult, Sherbrooke Hô-Dieu Hosp, St Vincent-de-Paul & Sherbrooke Hosp, 66-; exec mem, Med Res Coun Can, 73-79, vpres, 76-79, actg pres, 77-78; exec mem, Health Res Coun Que, 74-78. *Mem:* Can Med Asn; Can Ophthal Soc; fel Am Col Surgeons; fel Acad Ophthal & Otolaryngol; Asn Res Vision & Ophthal; Can Ophthal Soc (pres, 75-76). *Res:* Anatomy and pathology of ocular fundus; arterial hypertension; toxic retinopathies; ocular photography; fluorescein photography of the eye; diabetic retinopathies. *Mailing Add:* 1427 Bd Portland Sherbrooke PQ J1J 1S4 Can

DEMARIA, ANTHONY JOHN, b Italy, Oct 30, 31; US citizen; m 53; c 1. APPLIED & LASER PHYSICS. *Educ:* Univ Conn, BSEE, PhD, 65; Rensselaer Polytech Inst, MS, 60. *Prof Exp:* Res acoustic engr, Anderson Labs, Conn, 56-57; staff physicist phys electronics, Hamilton Standard Div, 57-58, prin scientist & group leader, res labs, 58-70, chief scientist, Electromagnetics Labs, 70-73, mgr electromagnetic & physics labs, 73-81, ASST DIR RES ELECTRONICS & ELECTRO-OPTICS TECHNOLS, UNITED TECHNOLS RES CTR, 81- *Concurrent Pos:* Adj prof, Rensselaer Polytech Inst, 68-77; mem bd dirs, Laser Indust Asn; consult lasers, Nat Acad Sci; consult, Nat Bur Standards, 76-79; ed, J Quantum Electronics, 77-82; chmn bd adv group electronic devices, Dept Defense, 79-85; Distinguished Fairchild Scholar, Calif Inst Technol, 82-83. *Honors & Awards:* R P I Davies Medal & Award, 80; Lieberman Award, Inst Elec & Electronics Engrs, 79; Ives Medal & Award, Optical Soc Am, 88. *Mem:* Nat Acad Eng; Am Phys Soc; fel Optical Soc Am (pres, 79-82); fel Inst Elec & Electronics Engrs. *Res:* Utilization of laser devices; interaction of elastic waves with coherent light radiation; generation, measurement and application of picosecond light pulses; gas laser research and applications; acoustic-optics; laser physics and devices; optics. *Mailing Add:* United Technol Res Ctr East Hartford CT 06108

DE MARIA, F JOHN, b Sliema, Malta, Apr 30, 28; Can citizen; m 58; c 4. OBSTETRICS & GYNECOLOGY. *Educ:* Royal Univ Malta, MD, 52; FRCS(C), 63. *Prof Exp:* House officer obstet & gynec, Postgrad Med Sch, Univ London, 54-56; registr, Durham & Newcastle, 57-59; Ramsay res fel physiol, Univ St Andrew's, 59-61; asst prof obstet & gynec, Univ BC, 61-66; ASSOC PROF OBSTET & GYNEC, MED SCH, UNIV ORE, 66- *Concurrent Pos:* Los Angeles County res fel, 60-61; obstetrician & gynecologist, Vancouver Gen Hosp, 61-66. *Mem:* Fel Am Col Obstet & Gynec; Royal Col Obstet & Gynec. *Res:* Neuroendocrinology; placental enzymes and pre-eclampsia; temporal correlation between the hypothalamus and uterus; surgery of infertility. *Mailing Add:* 621 Ninth St Lake Oswego OR 97034-2220

DEMARINIS, BERNARD DANIEL, b New York, NY, Aug 7, 46; m 71; c 2. COMMUNICATIONS ENGINEERING, FIBER OPTIC COMMUNICATIONS. *Educ:* City Col New York, BEE, 68; Polytech Inst NY, MSEE, 71; Fairleigh Dickinson Univ, MBA, 81. *Prof Exp:* Microwave engr, Wheeler Labs Div, Hayeltine Corp, 68-70; engr, RCA Corp, 70-72; sr engr, ITT-DCD, 72-76; sr managing assoc, Booz Allen & Hamilton, 76-90; DEPT HEAD, MITRE CORP, 90- *Concurrent Pos:* chapter pres, Princeton Sect, Inst Elec & Electronics Engrs, 80; regional vpres, Armed Forces Electronics Asn, 85-; bd dir, Am Defense Preparedness Asn, 88-90, Nat Security Indust Asn, 88-91; instr continuing educ, Monmouth Col, 88-90. *Honors & Awards:* Distinguished Young AFCEAN, Armed Forces Electronics Asn, 81. *Mem:* Armed Forces Electronics Asn; Inst Elec & Electronics Engrs; Am Defense Preparedness Asn; Nat Security Indust Asn. *Res:* Manage/direct research and development in fiberoptic/radio and satellite communications systems; responsible for advanced development of radar and electronic warfare (EW) systems for government sponsors. *Mailing Add:* 87 Monmouth Rd Spotswood NJ 08884

DEMARQUE, PIERRE, b Fez, Morocco, July 18, 32; Can citizen; m 58; c 2. ASTROPHYSICS. *Educ:* McGill Univ, BSc, 55; Univ Toronto, MA, 57, PhD(astron), 60. *Hon Degrees:* MA, Yale Univ, 68. *Prof Exp:* Mem staff appl math, Canadair Ltd, Mont, 55-56; asst prof astron, La State Univ, 59-60 & Univ Ill, 60-62; from asst prof to assoc prof, Univ Toronto, 62-66; from assoc prof to prof, Univ Chicago, 66-68; chmn dept astron, 68-74, MUNSON PROF NATURAL PHILOS & ASTRON, YALE UNIV, 68- *Honors & Awards:* Warner Prize, Am Astron Soc, 67. *Mem:* Am Astron Soc; Royal Astron Soc; Int Astron Union. *Res:* Stellar structure and evolution; stellar atmospheres; star clusters and galaxies; cosmology. *Mailing Add:* Dept Astron Yale Univ PO Box 6666 New Haven CT 06511

DEMARR, RALPH ELGIN, b Detroit, Mich, Jan 17, 30. MATHEMATICS. *Educ:* Univ Idaho, BS, 52; Wash State Univ, MA, 54; Univ Ill, PhD(math), 61. *Prof Exp:* Mem tech staff math, Bell Tel Labs, Inc, 54-56; Ford Found study grant, Moscow State Univ, 61-62; asst prof, Univ Wash, 62-68; assoc prof, 68-73, PROF MATH, UNIV NMEX, ALBUQUERQUE, 73- *Concurrent Pos:* Vis lectr, Leningrad State Univ, 69, Tashkent State Univ, 70 & Acad Sci, Moscow, 78. *Mem:* Am Math Soc. *Res:* Functional analysis; probability; statistics. *Mailing Add:* Dept Math Univ NMex Albuquerque NM 87131

DEMARS, ROBERT IVAN, b New York, NY, Apr 10, 28. MICROBIOLOGY. *Educ:* City Col New York, BS, 49; Univ Ill, PhD(bact), 53. *Prof Exp:* Res fel biol, Calif Inst Technol, 53-54; instr microbiol, Med Sch, Washington Univ, 54-56; microbiologist, NIH, 56-59; from asst prof to assoc prof, 59-69, PROF MED GENETICS, UNIV WIS-MADISON, 69- *Res:* Intermediate stages in the multiplication of bacterial viruses; genetics of cultivated animal cells; differentiation in early embryos. *Mailing Add:* Dept Human Oncol Univ Wis Genetics Bldg Madison WI 53706

DEMARTINI, EDWARD EMILE, b San Francisco, Calif, Aug 12, 46; m 73. BEHAVIORAL ECOLOGY, ICHTHYOLOGY. *Educ:* Univ San Francisco, BSc, 68, MSc, 70; Univ Wash, Seattle, PhD(zool), 76. *Prof Exp:* Sr scientist & prin investr marine biol, Lockheed Marine Biol Lab, 76-77; independent consult marine biol, Army Corp Engrs, 77-78; asst res biologist, 78-81, ASSOC RES BIOLOGIST, MARINE SCI INST, UNIV CALIF, SANTA BARBARA, 81- *Concurrent Pos:* NSF fel, 70-73. *Mem:* Am Soc Naturalists; Ecol Soc Am; Soc Study Evolution. *Res:* Behavioral ecology and evolution of the lower vertebrates, especially studies of the spacing patterns, mating systems and interspecific associations among marine reef fishes. *Mailing Add:* Nat Marine Fisheries Serv SW Fisheries Ctr Honolulu Lab 2570 Dole St Honolulu HI 96822

DE MARTINI, JAMES CHARLES, b Los Angeles, Calif, Apr 29, 42; m 69; c 4. IMMUNOPATHOLOGY, VIRAL PATHOLOGY. *Educ:* Univ Calif, Davis, BS, 64, DVM, 66, PhD(comp path), 72; Am Col Vet Path, dipl, 73. *Prof Exp:* Vet path, Intermountain Labs, Inc, Salt Lake City, Utah, 72-74; clin asst prof path, Dept Path, Col Med, Univ Utah, Salt Lake City, 73-74; from asst prof to assoc prof, 74-85, PROF PATH, COL VET MED & BIOMED SCI, COLO STATE UNIV, FT COLLINS, 85- *Concurrent Pos:* Intern, dept clin sci, Sch Vet Med, Univ Calif, Davis, 66-68; vis prof, dept microbiol & path, Wash State Univ, Pullman, Wa, 80-81; prin investr, Small Ruminants Collab Res Support Prog, 80-; vis scientist, ILRAD, Nairobi, Kenya, 87-88. *Mem:* AAAS; Am Asn Pathologists; Am Col Vet Pathologists; Am Asn Vet Immunologists; Sigma Xi; Am Soc Virol. *Res:* Mechanisms of pathogenesis and host immune responses in retrovirus infections of domestic ruminants. *Mailing Add:* Dept Path Colo State Univ Ft Collins CO 80523

DEMARTINI, JOHN, b San Francisco, Calif, Oct 11, 33; m 55; c 7. INVERTEBRATE ZOOLOGY. *Educ:* Humboldt State Col, BA, 55, MA, 60; Ore State Univ, PhD(zool), 64. *Prof Exp:* Instr high sch, Calif, 56-59; instr zool, bot & plant taxon, Humboldt State Col, 59-61, from asst prof to assoc prof zool, 63-72, PROF BIOL, HUMBOLDT STATE UNIV, 72- *Concurrent Pos:* Mem adv bd underwater parks & reserves, Calif Dept Parks & Recreation, 68-, chmn, 72-73, spec consult, 75; mem ad hoc comt abalone res, Calif Dept Fish & Game, 71-73, mem sea otter sci adv comt, 78- *Mem:* Wildlife Dis Asn; Am Soc Parasitologists. *Res:* Comparative and functional invertebrate morphology; marine ecology and invertebrate reproductive cycles; parasitology. *Mailing Add:* Dept Biol Sci Humboldt State Univ Arcata CA 95521

DEMARTINIS, FREDERICK DANIEL, b Philadelphia, Pa, Dec 10, 24; m 66; c 3. PHYSIOLOGY. *Educ:* Temple Univ, BA, 48, MA, 50; Jefferson Med Col, PhD(physiol), 59. *Prof Exp:* Asst biol, Temple Univ, 48-50, instr physiol, Sch Dent, 50-54; asst, Jefferson Med Col, 57-58; instr, 58-61, assoc, 61-62, from asst prof to assoc prof, 62-77, PROF PHYSIOL, MED COL PA, 77- *Mem:* AAAS; Am Physiol Soc; NY Acad Sci; Sigma Xi. *Res:* Control of adipose tissue development; regulation of adipocyte growth in tissue culture; regulation of thyroid gland function. *Mailing Add:* 76 Powderhorn Dr Phoenixville PA 19460

DEMARTINO, RONALD NICHOLAS, b Bayonne, NJ, Jan 14, 43; m 68; c 2. CHEMISTRY. *Educ:* Fairleigh Dickinson Univ, BS, 64; Fordham Univ, PhD(org chem), 69. *Prof Exp:* Res chemist, Nat Starch & Chem Corp, 68-73; sr res chemist, 73-83, STAFF SCIENTIST, CELANESE CORP, 83- *Concurrent Pos:* Mem chem technol adv, Comt Union County Tech Inst, 75- *Mem:* Am Chem Soc; Nat Geog Soc. *Res:* Synthesis of modified polysaccharides and wholly aromatic liquid crystalline polymers; surface modification of polyethylene terephtalate to improve comfort and stain release properties; synthesis of modified polyesters exhibiting improved pilling performance. *Mailing Add:* 11 Mandeville Dr Wayne NJ 07470-6535

DEMAS, JAMES NICHOLAS, b Washington, DC, Dec 28, 42; m 65; c 2. PHOTOCHEMISTRY. *Educ:* Univ NMex, BS, 64, PhD(chem), 70. *Prof Exp:* Res assoc chem, Univ Southern Calif, 70-71; from asst prof to assoc prof, 71-85, PROF CHEM, UNIV VA, 85- *Concurrent Pos:* NSF fel, 70; vis staff scientist, Los Alamos Nat Lab, 83-84. *Mem:* Am Chem Soc; Sigma Xi; Optical Soc Am. *Res:* Photochemistry and luminescence of transition metal complexes and organic dyes; chemical, instrumental and mathematical methods for measuring or evaluating photochemical and optical properties of materials; microcomputers. *Mailing Add:* Dept Chem Univ Va Charlottesville VA 22901

DEMASON, DARLEEN AUDREY, b Battle Creek, Mich, June 4, 51; m 85; c 2. PLANT MORPHOLOGY, PLANT ANATOMY. *Educ:* Univ Mich, BS, 73; Univ Calif, Berkeley, MS, 76, PhD(bot), 78. *Prof Exp:* PROF BOT, UNIV CALIF, RIVERSIDE, 78- *Mem:* Bot Soc Am; Am Soc Plant Physiol; Sigma Xi. *Res:* Development and control of development in plant systems with special emphasis on unique structural features of monocotyledons. *Mailing Add:* Dept Plant Sci Univ Calif Riverside CA 92521

DEMASSA, THOMAS A, b Detroit, Mich, Nov 6, 37; m 59; c 3. ELECTRICAL ENGINEERING. *Educ:* Univ Mich, BS, 60, MS, 61 & 63, PhD(elec eng), 66. *Prof Exp:* Res assoc, Univ Mich, 64-66; from asst prof to assoc prof elec eng, 66-73, PROF ELEC & COMPUT ENG, ARIZ STATE UNIV, 73- *Concurrent Pos:* Consult; lectr. *Mem:* Am Soc Eng Educ; sr mem Inst Elec & Electronics Engrs; Sigma Xi. *Res:* Solid state devices; electronics. *Mailing Add:* 1746 Fairfield Mesa AZ 85203

DEMASTER, DOUGLAS PAUL, b Sheboygan, Wis, Mar 27, 51; m 75. POPULATION DYNAMICS, MARINE MAMMALOGY. *Educ:* Univ Wis-Madison, BA, 73; Univ Minn, Minneapolis, 78. *Prof Exp:* Res asst, Antarctic Seal Prog, Univ Minn, 73-79; res biologist, US Fish & Wildlife Serv, 79-80; RES BIOLOGIST, NAT MARINE FISHERIES SERV, 80- *Concurrent Pos:* US Rep, Antarctic Seal Conf, 77, prin investr, Off Polar Prog, NSF, 78-81. *Mem:* Wildlife Soc; Am Soc Mammologists. *Res:* Population dynamics of marine mammals that occur off the coast of California; population assessment of California sea lions, harbor seal, and pilot whales. *Mailing Add:* Southwest Fisheries Ctr 8604 La Jolla Shores Dr La Jolla CA 92038

DE MASTER, EUGENE GLENN, b Sheboygan, Wis, Oct 20, 43; m 68; c 2. BIOCHEMISTRY. *Educ:* Dordt Col, Iowa, BA, 65; Wayne State Univ, PhD(biochem), 72. *Prof Exp:* CHEMIST ALCOHOLISM, MINNEAPOLIS VET MED CTR, 73- *Mem:* Am Chem Soc; Res Soc Alcoholism. *Res:* Alterations in biochemical processes induced by chronic alcohol consumption; role of vanadium oxyanions and tungstic acids in biological systems; drug alcohol interactions. *Mailing Add:* Vet Admin Hosp 151 One Veterans Dr Minneapolis MN 55417-2300

DE MATTE, MICHAEL L, b Bridgeport, Ohio, Nov 3, 37; m 62; c 1. ORGANIC CHEMISTRY. *Educ:* Wheeling Col, BS, 59; WVa Univ, MS, 61, PhD(org chem), 66. *Prof Exp:* Res chemist, 66-73, SR RES CHEMIST, WESTVACO, INC, 73- *Mem:* Am Chem Soc. *Res:* Polyaromatic synthesis; reaction mechanisms in heterocyclic N-oxides; chemistry of paper coatings. *Mailing Add:* 5056 W Running Brook Columbia MD 21044

DE MAYO, BENJAMIN, b Atlanta, Ga, Aug 4, 40; m 71; c 2. EXPERIMENTAL SOLID STATE PHYSICS. *Educ:* Emory Univ, BS, 62; Yale Univ, MS, 64; Ga Inst Technol, PhD(physics), 69. *Prof Exp:* Res assoc metal physics, Univ Ill, Urbana-Champaign, 69-71; from asst prof to assoc prof physics, 71-81 PROF PHYSICS, WEST GA COL, 81- *Concurrent Pos:* Vis prof, Ga Tech, 81-82. *Mem:* Am Asn Physicists Med; Am Phys Soc; Am Soc Metals. *Res:* Metals and alloys; magnetism; low temperature physics; Mossbauer spectroscopy; hydrogen in metals. *Mailing Add:* Dept Physics W Ga Col Carrollton GA 30118

DEMBER, ALEXIS BERTHOLD, b Dresden, Ger, May 30, 12; nat US; m 43; c 2. PHYSICS. *Educ:* German Univ, Prague, PhD(physics), 35. *Prof Exp:* Instr & asst physics, Istanbul Univ, 35-36; res fel, Calif Inst Technol, 37-44; chief res sect, Friez Instrument Div, Bendix Aviation Corp, Md, 44-47; chief radiation br, Eng Res & Develop Labs, Va, 47-48; head photog develop br, US Naval Ord Test Sta, 48-49, head instrument develop div, 49-56, asst head test dept, 56-59; tech consult to dir test & eval, US Naval Missile Ctr, Pac Missile Range, 59, dep head, Astronaut Dept, 59-60, head, 60-69, head, Electro-Optics Div, Lab Dept, 69-76; RETIRED. *Concurrent Pos:* Mem working group optical instrumentation, Panel Test Range Instrumentation, Res & Develop Bd, 48-51, chmn, 51-52; mem inter-range instrumentation group, 56-58, vchmn, 58-59; consult, Spec Comt Adequacy Range Facil, Dept Defense, 57-58; mem starlight study group naval appln space technol, 62; sea bed study group adv, Sea-based Deterrent Systs, 64. *Mem:* Am Phys Soc; Optical Soc Am; Inst Elec & Electronics Engrs; Sigma Xi. *Res:* Visible and infrared optics; ballistic and meteorological instrumentation; low temperature crystal physics; semiconductors; military space systems. *Mailing Add:* 4275 Varsity St Ventura CA 93003

DEMBICKI, HARRY, JR, b Poughkeepsie, NY, Oct 18, 51; m 70. ORGANIC GEOCHEMISTRY. *Educ:* State Univ NY Col, New Paltz, BS, 73; Ind Univ, PhD(org geochem), 77. *Prof Exp:* Res asst geochem, Ind Univ, 73-74, assoc instr, 74-76; geochem consult, Ind Geol Surv, 76-77; res scientist org geochem, Explor Res Div, Continental Oil Co, 77-84. *Mem:* Geochem Soc. *Res:* Application of biological marker hydrocarbons to petroleum exploration; thermodynamics and kinetics of the pyrolysis of kerogen; computer systems for data management; design of specialized analytical systems. *Mailing Add:* 5977 S Gallop Denver CO 80201

DEMBITZER, HERBERT, b New York, NY, June 18, 34; wid; c 3. CELL BIOLOGY, EXPERIMENTAL PATHOLOGY. *Educ:* NY Univ, AB, 55, MS, 58, PhD(biol), 67. *Prof Exp:* Res asst biol, NY Univ, 58-62, asst res scientist, 62-63; electron microscopist, Montefiore Hosp, 63-67; asst prof, 73-83, ASSOC PROF PATH, ALBERT EINSTEIN COL MED, 83-; RES ASSOC, MONTEFIORE HOSP, 67- *Mem:* Am Soc Cell Biol; Electron Micros Soc Am. *Res:* Fine structure and histochemistry of cellular development; structure and development of cell junctions. *Mailing Add:* 94 Rivergate Dr Wilton CT 06897

DEMBOWSKI, PETER VINCENT, b Brooklyn, NY, Sept 24, 46; m 69; c 2. NEW PRODUCT DEVELOPMENT, PRODUCTIVITY. *Educ:* Polytech Inst Brooklyn, BS, 67, MS, 69. *Prof Exp:* Gen mgr chem & metall prod, Gen Elec Co, 76-87; vpres & gen mgr, Advan Mat Div, Mat Res Corp, 87-88; vpres opers, Wyman-Gordon Co, 88-89; DIR, PROG OPERS, J I CASE, 89- *Concurrent Pos:* Adj prof, Union Col, Schenectady, NY, 75-78; pres, Sprinkler Mining Co, 85-87. *Mem:* Sigma Xi; Am Soc Mat; Am Inst Metall Engrs. *Res:* Application of technology to business and manufacturing; physical and mechanical metallurgy applications; thermal processing; manufacturing and enginering systems and methods; product development and introduction. *Mailing Add:* Seven S 600 County Line Rd Richmond Heights OH 44143

DEMBURE, PHILIP PITO, b Chilimanzi, Zimbabwe, Dec 15, 41; US citizen; m 70; c 1. BIOCHEMISTRY. *Educ:* Xavier Univ, La, BS, 68; State Univ NY, Buffalo, MA, 74, PhD(biochem), 78. *Prof Exp:* Res asst biochem genetics, State Univ NY, Buffalo, 70-78; instr, 78-80, ASST PROF PEDIAT, SCH MED, EMORY UNIV, 81- *Mem:* AAAS; Sigma Xi. *Res:* Genetics; analytical chemistry. *Mailing Add:* Div Med Genetics 2040 Ridgewood Dr Atlanta GA 30322

DEMEDICIS, E M J A, b Etterbeek, Belg, Dec 17, 37; m 62; c 3. BIOCHEMISTRY, GENETICS. *Educ:* Univ Louvain, Lic in Sci, 59. PhD(org chem), 62. *Prof Exp:* Asst org chem, Lab Gen & Org Chem, Univ Louvain, 63-67; asst lectr, Dept Chem, Fac Sci, fel biochem, 69-71, PROF BIOCHEM, FAC MED, UNIV SHERBROOKE, 71- *Concurrent Pos:* Clin biochem. *Mem:* Chem Soc Belg; Can Biochem Soc; Chem Inst Can. *Res:* Manganese in archaebacteria; enzymes; human pyruvate kinase gene. *Mailing Add:* Univ Hosp Ctr Univ Sherbrooke Sherbrooke PQ J1H 4N5 Can

DEMEIO, JOSEPH LOUIS, b Hurley, Wis, Sept 9, 17; m 41; c 1. MEDICAL MICROBIOLOGY. *Educ:* Marquette Univ, BS, 50, MS, 53; Univ Wis, PhD(med microbiol), 58. *Prof Exp:* Virologist, Naval Med Res Unit 4, 51-54, immunologist, 57-58; asst, Univ Wis, 54-57; asst chief diag reagents, Commun Dis Ctr, USPHS, 58-59; res assoc, Merrell-Nat Labs, 59-77; MEM STAFF, SALK INST, 78- *Concurrent Pos:* Assoc, Moravian Col, 73-75. *Mem:* AAAS; NY Acad Sci; Am Pub Health Asn; Soc Exp Biol & Med; Am Asn Immunol. *Res:* Antigenic relationships among arboviruses. *Mailing Add:* The Salk Inst Govt Serv Div PO Box 250 Swiftwater PA 18370

DE MELLO, F PAUL, b Goa, India, July 20, 27. DYNAMIC ANALYSIS OF POWER SYSTEMS. *Educ:* Mass Inst Technol, BS, 47, MS, 48. *Prof Exp:* Sr engr, Gen Elec Co, 55-69; prin engr dynamics & control & secy-treas, 69-73, vpres-secy, 73-87, PRIN CONSULT, POWER TECHNOLOGIES, INC, 87- *Concurrent Pos:* Mem, Syst Controls Subcomt & Joint Working Group Plant Response, Inst Elec & Electronics Engrs. *Honors & Awards:* Ralph Cordiner Award, 69. *Mem:* Nat Acad Eng; fel Instrument Soc Am; fel Inst Elec & Electronics Engrs; Comput Inst Appln Sci & Eng. *Res:* Dynamic analysis of electric power systems; planning, design and the operation of power plants and electric transmission systems; author of numerous technical publications. *Mailing Add:* Power Technologies Inc PO Box 1058 Schenectady NY 12301

DE MELLO, W CARLOS, b Florianopolis, Brazil, Sept 11, 31; m 56; c 4. PHYSIOLOGY. *Educ:* Univ Rio de Janeiro, MD, 55. *Prof Exp:* Asst prof physiol, Sch Med, Univ Rio de Janeiro, 57-58, assoc researcher physiol & biophys, 58-66; vis assoc prof, 63-64, assoc prof, 66-70, PROF PHARMACOL, SCH MED, UNIV PR, 70-, CHMN DEPT, 72- *Concurrent Pos:* Guest fel physiol, State Univ NY, 58-59; Rockefeller Found fel physiol, Nat Inst Med Res, Eng, 65-66; Nat Heart Inst res grants, 65-89. *Mem:* Am Physiol Soc; NY Acad Sci. *Res:* Electrophysiology of the heart; ionic mechanisms of cardiac electrogenesis; excitatory and inhibitory processes in Ascaris; membrane biophysics and physiology; cell communication. *Mailing Add:* Univ PR Sch Med Med Sci Campus Rio Piedras PR 00936

DEMEMBER, JOHN RAYMOND, b Elmira, NY, Oct 30, 42; m 64; c 4. ORGANIC CHEMISTRY. *Educ:* Niagara Univ, BS, 64; George Washington Univ, PhD(chem), 68. *Prof Exp:* Instr, Mt Vernon Col, 64-68; instr, Case Western Reserve Univ, 68-69, fel, 68-70; scientist chem, 70-76, SR SCIENTIST CHEM, POLAROID CORP, 76- *Mem:* Sigma Xi; Am Chem Soc. *Res:* Electronic energy transfer; carbonium ions; Raman and nuclear magnetic resonance spectroscopy; chemistry of photographic systems. *Mailing Add:* Ten Liberty Tree Lane Shrewsburg MA 01545-1558

DEMENT, JACK (ANDREW), b Portland, Ore, Feb 6, 20. DISEASE MODELING, LUMINESCENCE. *Educ:* Reed Col, 38-41; Am Bd Bio-Analysts, dipl nuclear physics, 55; Phila Col Pharm Sci, CE, 83, St John's Univ, 84. *Hon Degrees:* DSc, Western States Col, 55. *Prof Exp:* RES CHEMIST & HEAD, DE MENT LABS, PORTLAND, ORE, 41-; PRES, POLYPHOTON CORP, 55- *Concurrent Pos:* Assoc ed, Minerologist, 40-51; consult secy war, Bikini Atomic Tests, UN Group, 46; res asst biophys & pharmacol, Ore Health Sci Univ, 48-66; reviewer, Nat Sci Found, 79-80. *Honors & Awards:* Gravity Res Found Award. *Mem:* Sigma Xi. *Res:* New weapons; radiology; directed energy; new methods for radioactive decontamination; new laser systems; radioactive photos; optoexplosives; explotron. *Mailing Add:* 4847 SE Division St Portland OR 97206

DEMENT, JOHN M, b Henderson, NC, Aug 5, 49. ENVIRONMENTAL SCIENCES. *Educ:* NC State Univ, BS, 71; Harvard Univ, MS, 74; Univ NC, Chapel Hill, DPhil(indust hyg-epidemiol), 80; Am Bd Indust Hyg, cert, 76. *Prof Exp:* Indust hygienist, Environ Invest Br, Div Field Studies & Clin Invest, Nat Inst Occup Safety & Health, Ohio, 71-75, asst chief, Indust Hyg Sect, Industrywide Studies Br, Div Surveillance, Hazard Eval & Field Studies, 75-77, doctoral res, 77-79, dep dir, Div Respiratory Dis Studies, WVa, 79-81; health & safety mgr, Off Dir, 81-88, DIR, OFF OCCUP HEALTH & TECH SERV, OFF DIR, NAT INST ENVIRON HEALTH SCI, NIH, 88- *Concurrent Pos:* Mem, Epidemiol Comt, Am Indust Hyg Asn, 83-87, vchmn, 85-86, chmn, 86-87; mem, Fed Asbestos Task Force, 86-; vchair elect, Am Conf Govt Indust Hygienist, 87-88, vchair, 88-89, chair, 89-90. *Mem:* Am Indust Hyg Asn; Am Acad Indust Hyg. *Mailing Add:* Off Occup Health & Tech Serv NIH Nat Inst Environ Health Sci Bldg 101 Rm B350 PO Box 12233 Research Triangle Park NC 27709

DEMENT, WILLIAM CHARLES, b Wenatchee, Wash, July 29, 28; m 56; c 3. NEUROPHYSIOLOGY, POLYSOMOGRAPHY. *Educ:* Univ Wash, BS, 51; Univ Chicago, MD, 55, PhD(physiol), 57. *Prof Exp:* Intern, Mt Sinai Hosp, 57-58, res fel psychiat, 58-63; assoc prof, 63-67, PROF PSYCHIAT, SCH MED, STANFORD UNIV, 67-, DIR SLEEP RES LABS, DEPT PSYCHIAT, 63-, DIR SLEEP DISORDERS CLIN & LAB, 70- *Concurrent Pos:* Med adv, Am Narcolepsy Asn, 76-; chmn, US Surgeon General's Coun, Proj Sleep, 79-; co-ed-in-chief, J Sleep, 77-; mem, Int Sci Adv Bd, Ctr Design Indust Schedules, 81-; merit award, Nat Inst Aging, 86-; career scientist award, NIH, 90- *Honors & Awards:* Hofheimer Prize, Am Psychiat Asn, 64, Rush Silver Medal, 72, 78 & 79, Rush Bronze Medal, 81, Distinguished Psychiatrist Award, 84; Thomas W Salmon Medal, NY Acad Med, 69; Distinguished Scientist Award, Soc Exp Biol & Med, 71; Cert of Merit, AMA, 72, 75; Nathaniel Kleitman Prize, Asn Sleep Dis Ctrs, 82 & 87. *Mem:* Inst Med-Nat Acad Sci; Am Psychiat Asn; Asn Sleep Disorders Ctr (first pres, 76-); Asn Psychophysiol Study Sleep; Psychiat Res Soc; Am Col Neuropsychopharmacol; Am Sleep Dis Asn (pres, 75); Sleep Res Soc; Am Physiol Soc; Geront Soc Am. *Res:* Physiology of dreaming and sleep; electroencephalography; author or co-author of over 400 scientific publications. *Mailing Add:* Dept Psychiat Stanford Univ Sch Med Stanford CA 94305

DEMENTI, BRIAN ARMSTEAD, b Richmond, Va, Mar 3, 38; c 2. BIOCHEMISTRY, TOXICOLOGY. *Educ:* Hampden-Sydney Col, BS, 61; Univ Richmond, MS, 64; Med Col Va, PhD(biochem), 77. *Prof Exp:* Teacher sci, Henrico County Pub Sch Syst, Va, 67-68; res polymer chemist, Fibers Div, Allied Chem Corp, 68-73; TOXICOLOGIST, STATE VA HEALTH DEPT, 77- *Mem:* Am Chem Soc; Soc Occup & Environ Health; Sigma Xi. *Res:* Enzyme regulation; regulation of key hepatic glycolytic and gluconeogenic enzymes; synthesis of novel pyri-midines as potential antiviral and antileukemic agents; polyester polymer chemistry. *Mailing Add:* 7519 Oakmont Dr Richmond VA 23228

DE MEO, EDGAR ANTHONY, b Yonkers, NY, Jan 14, 42; m 68; c 2. ELECTRICAL ENGINEERING, SOLID STATE PHYSICS. *Educ:* Rensselaer Polytech Inst, BEE, 63; Brown Univ, ScM, 65, PhD(elec eng), 68. *Prof Exp:* Mem tech staff, Bell Tel Labs, 63; res asst elec eng, Brown Univ, 64-66; instr, US Naval Acad, 67-69; from asst prof to assoc prof, Brown Univ, 69-76; proj mgr, Solar Energy Prog, 76-79, MGR, SOLAR POWER SYSTS PROG, ELEC POWER RES INST, 80- *Mem:* Inst Elec & Electronics Engrs; AAAS; Am Solar Energy Soc; Sigma Xi. *Res:* Millimeter wave devices; anisotropic magnetic materials; magnetic resonance investigations; far infrared spectroscopy; photovoltaic devices; solar and wind power. *Mailing Add:* 2791 Emerson Palo Alto CA 94304

DEMERDASH, NABEEL A O, b Apr 26, 43; US citizen; m 69; c 3. ELECTRIC MACHINERY & POWER ELECTRONICS, COMPUTATIONAL ELECTROMAGNETICS. *Educ:* Cairo Univ, BScEE, 64; Univ Pittsburgh, MSEE, 67, PhD(elec eng), 71. *Prof Exp:* Teaching asst elec eng, Cairo Univ, 64-66 & Univ Pittsburgh, 66-68; engr, Westinghouse Elec Corp, 68-72; from asst prof to prof elec eng, Va Polytech Inst & State Univ, 72-83; PROF ELEC ENG, CLARKSON UNIV, 83- *Concurrent Pos:* Consult, Gen Motors Res Labs, Warren, Mich, 78, Honeywell Avionics Div, St Petersburg, Fla, 79-80 & Sundstrand Corp, Rockford, Ill, 80 & 86-; chmn, DC & Permanent Magnet Mach Subcomt, Inst Elec & Electronics Engrs Power Eng Soc, 87- *Honors & Awards:* Cert of Recognition, NASA, 79. *Mem:* Fel Inst Elec & Electronics Engrs; Inst Elec & Electronics Engrs Power Eng Soc; Inst Elec & Electronics Engrs Magnetics Soc; Inst Elec & Electronics Engrs Aerospace & Electronic Systs Soc; Inst Elec & Electronics Engrs Power Electronics Soc; Am Soc Eng Educ; Sigma Xi; Inst Elec & Electronics Engrs Indust Applications Soc; Inst Elec & Electronics Engrs Educ Soc. *Res:* Computational electromagnetics; three dimensional finite element computation of electromagnetic fields in electric machinery and devices including coupled electromagnetic-electromechanical systems problems; power electronics applied to the operation and control of electric machinery and electromechanical actuators; computer-aided simulation and design optimization of solid state controlled electric motor drives and machines; coupled problems of dynamic interaction between the power electronic and electromechanical components, in particular for aerospace applications and in electric machinery-power system dynamics. *Mailing Add:* Dept Elec & Computer Eng Clarkson Univ Potsdam NY 13699-5720

DEMERJIAN, KENNETH LEO, b Cambridge, Mass, Sept 10, 45; m 68; c 2. PHYSICAL CHEMISTRY, ATMOSPHERIC SCIENCE. *Educ:* Northeastern Univ, BS, 68; Ohio State Univ, MS, 70, PhD(phys chem), 73. *Prof Exp:* Res assoc, Calspan Corp, 73-74; br chief phys chem & atmospheric sci, Nat Oceanic & Atmospheric Admin, US Environ Protection Agency, 74-81, dir Meteorol Lab, 81-84; DIR, ATMOSPHERIC SCI RES CTR, STATE UNIV NY, ALBANY, 84-, PROF ATMOSPHERIC SCI, 88- *Concurrent Pos:* Mem, Nat Res Counc comts, Nat Acad Sci, 87-, Nat Sci Adv Comt, Desert Res Inst, 88-, adv panel, US Cong, 87-89, adv bd, Int Joint Comn, 87-, Elec Power Res Inst, 80-87; assoc ed, Atmospheric Environ, 82-, Environ Sci & Technol, 82-84; mem, Working Group 2 Atmospheric Modeling, US Air Qual Agreement, 80-82, vchmn, 81-82; vchmn working group, NATO, 82-84; co-chmn, Gordon Res Conf, Environ Sci-Air, 81; mem, Task Forces Air Qual Criteria Doc, Environ Protection Agency, adv, lab reorgn planning, Off Res & Develop. *Mem:* AAAS; Am Chem Soc; Am Meteorol Soc; Am Geophys Union. *Res:* Chemistry and mechanistic processes of clean and polluted tropospheres; Kinetic and decomposition pathway studies of atmospheric species; computer models for simulating air quality and atmospheric processes; instrumentation development for the measurement of trace atmospheric constituents. *Mailing Add:* Atmospheric Sci Res Ctr State Univ NY Albany 100 Fuller Rd Albany NY 12205

DEMERS, LAURENCE MAURICE, b Lawrence, Mass, May 9, 38; m 62; c 5. CLINICAL PATHOLOGY, ENDOCRINOLOGY. *Educ:* Merrimack Col, AB, 60; State Univ NY Upstate Med Ctr, PhD(biochem), 70. *Prof Exp:* Lalor Found fel, Lab Human Reprod & Reprod Biol, Harvard Med Sch, 70-73; PROF PATH, ASSOC DIR DIV CLIN PATH & DIR CLIN CHEM & CORE ENDOCRINE LAB, HERSEY MED CTR, PA STATE UNIV, 74- *Concurrent Pos:* Fogarty Sr Int Fel, 82. *Honors & Awards:* Ames Award, Am Asn Clin Chemists. *Mem:* AAAS; Am Soc Clin Pathologists; NY Acad Sci; Am Asn Clin Chem; Endocrine Soc; Nat Acad Clin Biochem; Sigma Xi. *Res:* Biochemical endocrinology with particular emphasis on hormonal regulation of carbohydrate metabolism in reproductive tissue; mechanism of action of the prostaglandins; reproductive endocrinology with emphasis on arachidonic acid metabolism and prostaglandin production in disorders of human reproduction; immunoassay tests development and automation; cancer research focusing on hormonally dependent forms of cancer including breast, prostate, and endometrium from an etiologic and therapeutic point of view. *Mailing Add:* Dept Path M S Hershey Med Ctr Hershey PA 17033

DEMERS, SERGE, b Verdun, Que. ASTRONOMY, ASTROPHYSICS. *Educ:* Univ Montreal, BSc, 61; Univ Toronto, MA, 63, PhD(astron), 66. *Prof Exp:* Astronr, US Naval Observ, 67-68 & Cerro Tololo Inter Am Observ, 68-69; prof astron, Laurentian Univ, Sudbury, Ont, 69-76; PROF PHYSICS & ASTRON, UNIV MONTREAL, 76- *Mem:* Am Astron Soc; Can Astron Soc. *Res:* Astronomy of nearby galaxies. *Mailing Add:* Dept Physics Univ Montreal PO Box 6128 Sta A Montreal PQ H3C 3J7 Can

DEMERSON, CHRISTOPHER, b St John, NB, May 16, 42; m 72; c 3. MEDICINAL CHEMISTRY. *Educ:* Univ NB, BSc, 64, PhD(chem), 68. *Prof Exp:* Fel, Univ NB, 68-69; res chemist, 69-72, res assoc, 72-80, SECT HEAD, AYERST LABS, 80- *Concurrent Pos:* Mem adv bd chem technol, Dawson Col, 75-77. *Mem:* Am Chem Soc; Chem Inst Can. *Res:* Synthesis of the alkaloid delphinine; development of synthetic methods; medicinal chemistry and the structure-activity relationship of drugs. *Mailing Add:* 1025 Blvd Laurentien St Laurent Montreal PQ H4R 1J6 Can

DEMET, EDWARD MICHAEL, b Elmhurst, Ill, July 27, 49; m 83. BIOCHEMISTRY, PSYCHOPHARMACOLOGY. *Educ:* Univ Ill, BS, 71; Ill Inst Technol, PhD(biochem), 76. *Prof Exp:* Res technician pharmacol, Univ Chicago, 72-73, res technician toxicol, 73-75, sr res technician, 75-76, res assoc, 76-77, instr, 77-78, res asst prof psychiat, 78-80; res chemist, VA Med Ctr, Brentwood, 80-; asst prof, Neuropsychiatric Inst, Univ Calif, Los Angeles, 80-; PROF PSYCHIAT, UNIV CALIF, IRVINE. *Concurrent Pos:* USPHS fel, Univ Chicago, 76-77, 78-79; NSF-KKI sci exchange fel, Budapest, Hungary, 77. *Mem:* AAAS; Soc Neurosci. *Res:* Neurochemistry of psychiatric disorders and subcellular biochemistry. *Mailing Add:* Dept Psychiat Univ Calif Irvine CA 92717

DEMETER, STEVEN, b Budapest, Hungary, Jan 12, 47; US citizen; m 84; c 2. NEUROLOGY. *Educ:* Brooklyn Col, BS, 69; NY Med Col, MD, 73. *Prof Exp:* Med intern, Beth Israel Med Ctr, New York, 73-74; neurol resident, Albert Einstein Col Med, Bronx, NY, 74-77; instr neurol, NY Med Col, 77-79; res fel neurol, Col Med, Univ Iowa, Iowa City, 79-81; res fel, Univ Rochester, 81-84, asst prof, Ctr Brain Res, 84-87, instr neurol & psychiat, 81-87, asst prof, Neurol & Psychiat, 87-90, CLIN ASST PROF NEUROL, SCH MED, UNIV ROCHESTER, NY, 90- *Concurrent Pos:* Asst attend physician, Metrop Hosp Ctr, New York, 77-79; asst attend physician neurol & assoc attend physician psychiat, Strong Mem Hosp, Rochester, 81-87, attend physician neurol & assoc attend physician psychiat, 87-90; clin physician II, Rochester Psychiat Ctr, 85-; mem, med comt, Tourette Syndrome Asn, 85-, mem, bd dirs, 87- *Mem:* Am Acad Neurol; AAAS; Soc Neurosci; Am Asn Anatomists; AMA. *Res:* Connectional organization of the cerebral cortex in non-human primates, with emphasis on interhemispheric relations and memory; anatomic correlates of normal and abnormal higher brain function in humans. *Mailing Add:* Univ Rochester Med Ctr Box 605 Rochester NY 14642

DEMETRESCU, M, b Bucharest, Romania, May 23, 29; m 69; c 1. NEUROPHYSIOLOGY, BIOMEDICAL ENGINEERING. *Educ:* Bucharest Polytech Inst, MEE, 54; Romanian Acad Sci, PhD(electrophysiol), 57. *Prof Exp:* Prin investr neurophysiol & EEG, Inst Endocrinol, Romanian Acad Sci, 58-66; fel neurophysiol, Brain Res Inst, Univ Calif, Los Angeles, 66-67; asst res physiologist, 67-71, assoc res physiologist, 78-79, ASST PROF PHYSIOL, MED SCH, UNIV CALIF, IRVINE, 72-, ASSOC CLIN PROF, 79- *Honors & Awards:* Victor Babes Prize, Romanian Acad Sci, 62. *Mem:* AAAS; sr mem Inst Elec & Electronics Engrs; Am Physiol Soc; Soc Neurosci. *Res:* Neurophysiology of active inhibition at thalamo-cortical level; control of cortical excitability by diffuse subcortical mechanisms; clinical electroencephalography; electrophysiology of bladder stimulation in paraplegics; new electronic-electrophysiologic research methods. *Mailing Add:* Dept Physiol & Res Lab Univ Calif Res Lab 17871 Fitch St Suite B Irvine CA 92714

DEMETRIADES, STERGE THEODORE, b Athens, Greece, June 30, 28; US citizen; m 56; c 3. SCIENCE ADMINISTRATION, PLASMA PHYSICS. *Educ:* Bowdoin Col, AB, 50; Mass Inst Technol, MS, 51; Calif Inst Technol, ME, 58. *Prof Exp:* Res engr, Mass Inst Technol, 51-53; ord engr, Ballistic Res Labs, US Army Ord Corps, 53-54; res engr, Lear, Inc, 54-55; res engr, Astronaut Dept, Aerojet-Gen Corp, 58-59; res engr, Northrop Corp, 59-60, head space propulsion & power lab, 60-62, head plasma labs, 62-63; chief scientist, Res Labs, Rocket Power Inc, 63-64; PRES & DIR, STD RES CORP, 64-, PRES STD INT RES & DEVELOP CORP, 69- *Concurrent Pos:* Consult, Aerojet-Gen Corp, 55-62, Air Logistics Corp, 58-59, Hughes Tool Co, 59-60, Marquardt Corp, 60, McGraw-Hill, Inc, 61-, Jet Propulsion Lab, Calif Inst Technol, 62-63 & Space Sci Lab, Litton Industs, Inc, 63-64. *Mem:* Assoc fel Am Inst Aeronaut & Astronaut; Sigma Xi. *Res:* Electrostreaming birefringence; powered space flight; plasma accelerators and propulsive fluid accumulator engines; experimental magnetogasdynamics; measurement of plasma properties and energy conversion; magnetohydrodynamic power generation; scientific technology development; research and development program assessment and national science and technology planning. *Mailing Add:* 2065 Vista Ave Sierra Madre CA 91024

DEMETRIOU, CHARLES ARTHUR, b Memphis, Tenn, Nov 13, 41; m 67; c 2. BIOMATHEMATICS, APPLIED MATH. *Educ:* Christan Brothers Col, BS, 65; Univ Miss, MS, 67. *Prof Exp:* Instr math, Christian Brothers Col, 67-70; instr math, Univ Tenn, Martin, 70-71; PROF & CHMN MATH & SCI, STATE TECH INST, MEMPHIS, 71- *Concurrent Pos:* Res assoc, Univ Tenn, Memphis, 67-69, res consult, 71-81. *Res:* Application of cluster analysis to electrocardiographic lead section; application of quadratic programming to the inverse problem in electrocardiography. *Mailing Add:* 7783 Deer Run Cv Cordova TN 38018-2905

DEMEYER, FRANK R, b San Francisco, Calif, Nov 7, 39; m 67. MATHEMATICS. *Educ:* Univ Seattle, BS, 61; Univ Ore, MA, 63, PhD(math), 65. *Prof Exp:* Asst prof math, Purdue Univ, 65-68; from asst prof to assoc prof, 68-75, PROF MATH, COLO STATE UNIV, 75- *Mem:* Math Asn Am. *Res:* Mathematical economics; abstract algebra; decision problems in welfare economics. *Mailing Add:* Dept Math Colo State Univ Ft Collins CO 80523

DE MICHELI, GIOVANNI, b Milano, Italy, Nov 26, 55; m. VERY LARGE SCALE INTEGRATION CIRCUITS, COMPUTER-AIDED DESIGN. *Educ:* Politecnico di Milano, Italy, Dr Eng, 79; Univ Calif, Berkeley, MS, 80, PhD(elec eng & computer sci), 83. *Prof Exp:* Asst prof, Dept Electronics, Politecnico di Milano, Italy, 83-84; res staff mem, Int Bus Mach T J Watson Res Ctr, 84-85, proj leader, 85-86; asst prof, 87-89, ASSOC PROF ELEC ENG, STANFORD UNIV, 89- *Concurrent Pos:* Tech chmn, Int Conf Computer Design, Inst Elec & Electronics Engrs, 88, gen chmn, 89; NSF initiation award, 87, presidential young investr award, 88; co-dir & lectr, Advan Study Inst, 87. *Honors & Awards:* Distinguished Serv Award, Inst Elec & Electronics Engrs, Computer Soc, 90. *Mem:* Sr mem Inst Elec & Electronics Engrs; Inst Elec & Electronics Engrs, Computer Soc; Inst Elec & Electronics Engrs, Circuit & Systs Soc. *Res:* Algorithms and computer-aided design tools for synthesis, verification and simulation of electronic very large scale integration (VLSI) circuits; author of numerous publications and one book; one patent issued. *Mailing Add:* Ctr Integrated Systs Stanford Univ Stanford CA 94305

DEMILLO, RICHARD A, b Hibbing, Minn, Jan 26, 47; m; c 4. MATHEMATICS, COMPUTER SCIENCE. *Educ:* Col St Thomas, BA, 69; Ga Inst Technol, PhD(comput sci), 72. *Prof Exp:* Res asst comput sci, Los Alamos Sci Lab, Univ Calif, 69-70; res assoc, Ga Inst Technol, 69-72; asst prof, Univ Wis-Milwaukee, 72-76; assoc prof, 76-80, prof comput sci, Ga Inst Technol, 80-87, dir, Software Eng Res Ctr, 85-87; PROF COMPUT SCI & DIR NSF SOFTWARE ENG RES CTR, PURDUE UNIV, 87-, DIR CCR DIV, 89- *Concurrent Pos:* Consult, US Army Electronics Command, Commun/ADP Lab, 74-, US Army Comput Systs Command, Army Inst Res Mgt, Info & Comput Sci, Mgt Sci Am, Math Res Ctr, Univ Wis & UNESCO, Off Naval Res. *Mem:* Am Math Soc; Math Asn Am; Soc Indust & Appl Math; AAAS; Asn Symbolic Logic; Asn Comput Mach. *Res:* Theoretical computer science; software engineering; privacy, security, and cryptography. *Mailing Add:* Dept Comput Sci Purdue Univ West Lafayette IN 47907-1398

DEMING, JODY W, b Houston, Tex, July 2, 52. DEEP-SEA MICROBIOLOGY, THERMOPHILY. *Educ:* Smith Col, BA, 74; Univ Md, PhD (microbiol), 81. *Prof Exp:* Res assoc, NASA/Goddard Space Flight Ctr, 74-77; grad res asst, Dept Microbiol, Univ Md, 77-81; assoc res scientist, Chesapeake Bay Inst, Johns Hopkins Univ, 81-86, res scientist, 86-88; ASSOC PROF OCEANOG, SCH OCEANOG, UNIV WASH, 88- *Concurrent Pos:* Postdoctoral fel, Scripps Inst Oceanog, 81-82 & Marine Pollution Control Off, Nat Oceanic Atmospheric Admin, 82-83; from asst prof to assoc prof biol, Johns Hopkins Univ, 83-88; mem, Alvin Rev Comt, 84-87, Polar Progs Rev Panel, NSF, 87, US Task Group/US-France Bilateral Agreement Coop Oceanog, 87-89, Arctic Systs Sci Ocean/Atmosphere/Ice Interaction Workshop, 90 & Biotechnol Sci Adv Comt, US Environ Protection Agency, 90-93; staff scientist, Ctr Marine Biotechnol, Univ Md, 86-88; NSF presidential young investr award, 89-94; organizer, NATO Conf Deep-Sea Food Chains: Their Relation Global Carbon Cycle, 91. *Mem:* Am Soc Microbiol; Am Soc Limnol & Oceanog; Am Geophys Union; AAAS; Sigma Xi; Oceanog Soc. *Res:* Investigation into the existence, modes of behavior, and effects on global geochemical cycles of bacteria in marine environments characterized by extremes in temperature and hydrostatic pressure. *Mailing Add:* Sch Oceanog WB-10 Univ Wash Seattle WA 98195

DEMING, JOHN MILEY, b Prescott, Ariz, May 28, 25; m 51; c 5. SOILS, MICROENCAPSULATION. *Educ:* Univ Ariz, BS, 48, MS, 49; Purdue Univ, PhD(agron), 51. *Prof Exp:* Chemist, proj leader, group leader, sr group leader & fel, Monsanto Agr Co, 51-88; RETIRED. *Mem:* Sigma Xi. *Res:* Early development of commercial pre-emergence herbicides; extensive formulation development of commercial pesticides; 96 patents foreign and domestic including commercial microencapsulation methodology. *Mailing Add:* 550 Innsbrook Estates Wright City MO 63390

DEMING, QUENTIN BURRITT, b New York, NY, July 24, 19; m 49; c 2. MEDICINE. *Educ:* Dartmouth Col, AB, 41; Columbia Univ, MD, 43. *Prof Exp:* From intern to asst resident med, Presby Hosp, 44-48, from instr to asst prof, Stanford Univ, 50-53; from asst prof to assoc prof, Columbia Univ, 53-59; assoc prof, 59-64, PROF MED, ALBERT EINSTEIN COL MED, 64- *Concurrent Pos:* Res fel, Sch Med, Stanford Univ, 48-50; Markle Found scholar, 50-55; asst vis physician, Goldwater Mem Hosp, Columbia, 53-; vis physician, Bronx Munic Hosp Ctr, 59- *Mem:* Am Soc Clin Invest; Harvey Soc; Am Fedn Clin Res; Am Heart Asn. *Res:* Hormonal aspects of edema formation; hypertension; atherosclerosis. *Mailing Add:* Dept Med Albert Einstein Col Med 1300 Morris Park Ave New York NY 10461

DEMING, ROBERT W, b Wabasha, Minn, Apr 5, 28; m 60; c 2. MATHEMATICS. *Educ:* Univ Minn, BS, 52, MA, 60; NMex State Univ, PhD(math), 65. *Prof Exp:* Instr math, Univ Minn, Duluth, 57-62; asst prof, Idaho State Univ, 65-67; assoc prof math, 67-77, PROF MATH & COORDR APPL MATH ECON, STATE UNIV NY COL OSWEGO, 77- *Mem:* Math Asn Am; Am Math Soc; Sigma Xi. *Res:* Applications of algebraic topology to the theory of uniform and uniform-like spaces. *Mailing Add:* 157 Ellen St Oswego NY 13126

DEMING, STANLEY NORRIS, b Corpus Christi, Tex, May 7, 44; m 67; c 2. ANALYTICAL CHEMISTRY. *Educ:* Carleton Col, BA, 66; Purdue Univ, MS, 70, PhD(chem), 71. *Prof Exp:* Asst prof chem, Emory Univ, 70-74; asst prof, 74-76, ASSOC PROF CHEM, UNIV HOUSTON, 76- *Mem:* AAAS; Am Chem Soc; Sigma Xi. *Res:* Automated development of analytical chemical methods; optimization in chemistry; process optimization; laboratory scale continuous flow processes; high performance liquid chromatography. *Mailing Add:* Dept Chem Univ Houston University Park Houston TX 77004

DE MIRANDA, PAULO, b Goa, India; m 66; c 4. BIOCHEMICAL PHARMACOLOGY, DRUG DEVELOPMENT. *Educ:* Univ Oporto, Lic, 60; Univ Wis, MS, 63; Marquette Univ, PhD(pharmacol), 65. *Prof Exp:* Instr pharmacol, Sch Med, Marquette Univ, 66; asst vis prof, Fac Med, Univ Valle, Colombia, 66-69; sr res scientist, 69-78, group leader, 79-86, SECT HEAD, DIV EXP THER, WELLCOME RES LABS, 87- *Concurrent Pos:* Ford Int Fel, 61-62. *Mem:* AAAS; Am Soc Pharmacol & Exp Therapeut; AAAS; Int Soc Study Xenobiotics; Int Soc Antiviral Res. *Res:* Biochemistry of nucleic acid antagonists; drug metabolism and disposition; pharmacokinetics; antiviral chemotherapy publications and patents. *Mailing Add:* Wellcome Res Labs 3030 Cornwallis Rd Research Triangle Park NC 27709

DEMIREL, T(URGUT), b Bursa, Turkey, Mar 2, 24; m 50; c 2. GEOTECHNICAL ENGINEERING, MATERIAL SCIENCE. *Educ:* Univ Ankara, ChemEng, 49; Iowa State Univ, MSc, 59, PhD(soil eng), 62. *Prof Exp:* Hwy mat engr, Gen Directorate Turkish Hwys, 49-51, sr res engr, 51-56; asst soil eng, 57-58, res assoc, 58-63, from asst prof to assoc prof, 63-70, prof soil eng, 70-88, PROF EMER SOIL ENG, IOWA STATE UNIV, 88- *Concurrent Pos:* Assoc mem, Hwy Res Bd, Nat Acad Sci-Nat Res Coun, 51; vis prof, Mid East Tech Univ, Ankara, 72-73; consult, UN, UNESCO, 77-78. *Res:* Load bearing capacity and physiochemical properties of soils; effects of chemical treatments on load bearing capacity of soils; soil-water interaction; concrete materials. *Mailing Add:* 482 A Town Eng Bldg Iowa State Univ Ames IA 50011

DEMIS, DERMOT JOSEPH, b New York, NY, Apr 19, 29. DERMATOLOGY, PHARMACOLOGY. *Educ:* Union Univ, NY, BS, 50; Univ Rochester, PhD(pharmacol), 53; Yale Univ, MD, 57. *Prof Exp:* Chief dept dermat, Walter Reed Army Inst Res, Washington, DC, 60-64; assoc prof med & dir div dermat, Sch Med, Wash Univ, 64-67; PROF DERMAT, ALBANY MED COL, 67-, CHMN, DEPT DERMAT. *Concurrent Pos:* Mem pharmacol & exp therapeut study sect, NIH, 60-68, mem med study sect, 68-72; mem subcomt dermat, US Army Surgeon Gen Adv Comt Med, 63-64; consult, Barnes & Jewish Hosps, St Louis, Mo & Scott AFB, Ill, 64-67, Albany Med Ctr Hosp, 67-, Albany Mem Hosp, 67- & Child Hosp, 67-; USPHS res career develop award, 65. *Mem:* Am Soc Clin Invest; Am Dermat Asn; NY Acad Sci; Am Soc Pharmacol & Exp Therapeut; Am Col Physicians; Sigma Xi. *Res:* Investigative dermatology; role of vasoactive amines in pathophysiologic processes; physiologic control and pathologic alterations of microcirculation; mucopolysaccharide metabolism. *Mailing Add:* Dermat Res Assoc 195 Delaware Ave Delmar NY 12054

DEMKOVICH, PAUL ANDREW, b Zborova, Czech, Nov 19, 22; nat US; m 47; c 5. PETROLEUM CHEMISTRY, ANALYTICAL CHEMISTRY. *Educ:* Univ Chicago, BS, 47, MS, 48. *Prof Exp:* Group leader, Res & Develop Dept, Amoco Oil Co, Standard Oil Co Ind, 47-65, supvr, Anal Lab, 65-74, supt lab inspection, 74-81, mgr, Lab Serv Div, 81-85; RETIRED. *Mem:* Am Chem Soc. *Mailing Add:* 7520 Magoun Ave Hammond IN 46324

DEMLING, ROBERT HUGH, BURN TRAUMA SURGERY. *Educ:* Marquette Med Sch, MD. *Prof Exp:* PROF SURG, HARVARD MED SCH, 82- *Concurrent Pos:* Dir, Longwood Area Trauma Ctr. *Mailing Add:* Longwood Area Trauma Ctr 75 Francis St Boston MA 02115

DEMMERLE, ALAN MICHAEL, b Port Jefferson, NY, Nov 4, 33; m 61; c 2. ELECTRONICS ENGINEERING. *Educ:* Carnegie Inst Technol, BS, 55; Columbia Univ, MS, 58. *Prof Exp:* Engr circuit design, Westinghouse Elec, 55-56; engr, US Naval Res Lab, 57-60; engr telemetry processing, Goddard Space Flight Ctr, NASA, 60-66; CHIEF COMPUT SYSTS LAB, NIH, 66-; PRES, ROLF INST, 88- *Concurrent Pos:* Dir, Aspin Res Inst, 81- *Res:* New applications of the computer technology to facilitate laboratory & clinical biomedical research; on-line laboratory automation and real time clinical evaluation and care. *Mailing Add:* 7104 Edgevale St Bethesda MD 20815

DEMMING, W(ILLIAM) EDWARDS, b Sioux City, Iowa, Oct 14, 00; m 32; c 3. MATHEMATICAL STATISTICS. *Educ:* Univ Wyo, BS, 21; Univ Colo, MS, 24; Yale Univ, PhD(physics), 28. *Hon Degrees:* LLD, Univ Wyo, 58; ScD, Rivier Col, 81, Ohio State Univ, 82, Md Univ, 83, Clarkson Univ, 83; Dir Eng, Univ Miami, 85, George Wash Univ, 87, Fordham Univ, 88, Univ Colo, 88, Univ Ala, 88, Ore State Univ, 89. *Prof Exp:* Instr elec eng, Univ Wyo, 21-22; from instr to asst prof physics, Colo Sch Mines, 22-24; asst prof, Univ Colo, 24-25; instr, Yale Univ, 25-27; physicist, Bur Chem & Soils, USDA, 27-39; sampling adv, US Bur Census, 39-45; PROF STATIST, NY UNIV, 46-; DISTINGUISHED PROF, COLUMBIA UNIV, 85- *Concurrent Pos:* Spec lectr, Bur Standards, 30-41; head dept math & statist, USDA Grad Sch, 33-53; consult to various countries, 40-; adv sampling, US Bur Budget, 45-53; statist consult, NY Univ, 46-; adv sampling tech, Supreme Command of Allied Powers, Tokyo, 47 & 50, High Comn Germany, 52 & 53; mem, UN Sub-Comn Statist Sampling, 47-52; teacher & consult, Union Japanese Scientists & Engrs, Japan Indust, 50; exchange scholar, Germany, 52 & 53; lectr, Univ Kiel, Inst Social Res, Frankfurt, Tech Acad, Wuppertal-Elberfeld, Gutenberg Tech Univ & Australian Inst Econ Res, Vienna, 53. *Honors & Awards:* Shewhart Medal, Am Soc Qual Control, 56. *Mem:* AAAS; fel Am Statist Asn (vpres, 41); fel Inst Math Statist (pres, 45); Opers Res Soc Am; hon mem Int Statist Inst. *Res:* Fundamentals of statistical inference; statistical logic in administration; applications in demography, sociology, medicine and industry; statistical theory of failure; application to depreciation; application to complex apparatus; schizophrenia. *Mailing Add:* 4924 Butterworth Pl Washington DC 20016

DE MONASTERIO, FRANCISCO M, b Buenos Aires, Arg, Jan 26, 44; US citizen; m 75. NEUROPHYSIOLOGY, NEUROANATOMY. *Educ:* Fed Univ Rio de Janeiro, Brazil, MD, 69, DSc, 72. *Prof Exp:* Vis scientist, 73-80, MED OFFICER & CHIEF, SECT VISUAL PROCESSING, NAT EYE INST, NIH, 81- *Concurrent Pos:* Pres, assembly scientist, Nat Eye Inst, NIH, 77-78; ad hoc mem, Visual Disorders Study Sect, NIH, 79-80. *Mem:* Brazilian Soc Physiol; Asn Res in Vision & Ophthalmol; AAAS; Found Advan Educ Sci. *Res:* Anatomical and physiological properties of neurons of the visual system of primates and other mammals; clinical electrophysiology and psychophysics of the visual system. *Mailing Add:* Nat Eye Inst Bldg 9 Rm 1E108 Bethesda MD 20892

DEMOND, JOAN, b Los Angeles, Calif. MARINE BIOLOGY, INVERTEBRATE ECOLOGY. *Educ:* Univ Calif, Los Angeles, BA; Mills Col, MA. *Prof Exp:* Fishery aide, Nat Oceanic & Atmospheric Admin, Nat Fisheries Serv, Honolulu; sci asst, Inter-Am Trop Tuna Comn; res biologist, Scripps Inst Oceanog, Univ Calif; marine zoologist, Div Mollusks, US Nat Mus; res assoc, Univ Calif, Los Angeles, mus scientist, Dept Earth & Space Sci & instr phys sci; marine biologist ecol; CONSULT. *Concurrent Pos:* NSF grant; consult marine zool, Univ Hawaii & Pac Sci Bd Projs. *Mem:* AAAS; Underwater Photog Soc; Western Soc Naturalists; Western Soc Malacologists; Am Malacol Union. *Res:* Ecology, systematics of Indo-Pacific reef-dwelling mollusks; zoogeographical distribution; phylogenetic and areal abundance of zooplankton of central Pacific; ecology & taxonomy of mollusks of Gulf of Mexico, west coast of North America and Hawaii; ocean ecology and environmental impacts upon marine ecosystems; ecology, systematics and zoogeography of the South Pacific. *Mailing Add:* PO Box 5064 Santa Monica CA 90405-0064

DEMONEY, FRED WILLIAM, b Oak Park, Ill, Nov 25, 19; m 44; c 5. PHYSICAL METALLURGY, CRYOGENICS. *Educ:* Ill Inst Technol, BS, 41; Univ Minn, MS, 51, PhD(phys metall), 54. *Prof Exp:* Design engr, Kimberly-Clark Corp, 41-44; asst, Ill Inst Technol, 44-45; engr, Parten Mach Co, 45-47; prod eng supt, Maico Co, Inc, 47; instr phys metall, Univ Minn, 47-51, res assoc mech & metals, 51-54; res metallurgist, Magnesium Dept, Dow Chem Co, 54-55; res engr, DMR, Kaiser Aluminum & Chem Corp, 55-57, br head mech metall & tech supvr mech metall, Fabrication & Appln Res Dept, 66-69, tech supvr, 69-71, prog mgr, 71-72; pres, Mont Col Mineral Sci & Technol, 72-85; RETIRED. *Concurrent Pos:* Consult, Twin City Testing & Eng Lab, 51-54; vchmn, Cryogenic Eng Conf, 66-69; pres & chmn bd, Mont Energy & MHD Res & Develop Inst, 77-78. *Mem:* Am Soc Metals; Am Inst Mining, Metall & Petrol Engrs; Am Soc Mech Engrs; Sigma Xi; NY Acad Sci; Mining & Metall Soc Am. *Res:* Engineering properties of aluminum for cryogenic service; pressure vessel materials and applications; rolling and extrusion technology; formability; terminal ballistics; aluminum armor; dynamic response of materials; properties of materials at cryogenic temperature; effect of microstructure on properties of metals; MHD. *Mailing Add:* PO Box 1083 Freeland WA 98249

DEMONSABERT, WINSTON RUSSEL, b New Orleans, La, June 12, 15; m 55; c 1. CHEMISTRY, RESEARCH ADMINISTRATION. *Educ:* Loyola Univ, BS, 37; Tulane Univ, AM, 45, PhD(chem), 52. *Prof Exp:* Prof, Warren Easton High Sch, 40-44 & Behrman High Sch, 44- 48; from assoc prof to prof chem, Loyola Univ, La, 48-66; phys scientist adminr, Proj Off & chief chemist, Community Studies Div, Pesticides Prog, Dept Health & Human Serv, Nat Commun Dis Ctr, Ga, 66-69, health scientist adminr & dir contract liaison br, Nat Ctr Health Serv Res & Develop, 69-73, head extramural prog, bur drugs, 73-79, sci coord interagency affairs, Off Comnr, Food & Drug Admin, 79-83; CONSULT SCIENTIST, 83- *Concurrent Pos:* Assoc prof, Tulane Univ, 57-58; prof lectr anal chem, Cath Univ Am, 75-76. *Mem:* Fel AAAS; fel Am Inst Chemists; Soc Res Adminr; Am Chem Soc; Sigma Xi. *Res:* Health services research; chemical instrumentation, gas chromatography, spectrophotometry, and polarography; zirconium chemistry; complex-ion formation; chemical toxicology; drug research. *Mailing Add:* 4317 Lake Trail Dr Kenner LA 70065

DEMOOY, CORNELIS JACOBUS, b Rotterdam, Neth, July 1, 26; m 53; c 2. SOIL FERTILITY, PLANT NUTRITION. *Educ:* Univ Wageningen, BS, 51, MS, 53; Iowa State Univ, PhD(soil fertil), 65. *Prof Exp:* Assoc soil surv classification & genesis, Univ Wageningen, 52-53; res scientist, Commonwealth Sci & Indust Res Orgn, 53-60; res assoc soil fertil, Iowa State Univ, 60-65; asst prof soil sci, Univ Utrecht, 65-67; from asst prof to assoc prof soil fertil, Iowa State Univ, 67-72; PROF AGRON, COLO STATE UNIV, 72- *Concurrent Pos:* Consult, Sir Alexander Gibb & Partners, Cent Africa, 66; agr res adv, US AID Mission to Pakistan, 72-76; consult, Castlewood Corp, Littleton, Colo, 79, Mich State Univ, WAfrica, 79, Colo State Univ, Indonesia, 80, Botswana, 87, Mich State Univ, Senegal, 80 & Nigeria, Botswana, 81; interdisciplinary environ res award, Colo State Univ, 78, team leader, Bean/ Cowpea CRSP Botswana, 82-86. *Mem:* Am Soc Agron; Netherlands Royal Soc Agr Sci; Int Soil Sci Soc; Sigma Xi. *Res:* Soil-plant-water relationships; nutritional requirements of crops; optimum fertilization; soil classification and genesis; quantitative evaluation of soil properties and soil fertility factors in terms of land use potentialities. *Mailing Add:* 309 Mae St Ft Collins CO 80525

DEMOPOULOS, HARRY BYRON, b New York, NY, Feb 14, 32; m 55; c 4. EXPERIMENTAL PATHOLOGY. *Educ:* State Univ NY, MD, 56. *Prof Exp:* Intern, Kings County Hosp, Brooklyn, NY, 56-57; resident path, Bellevue Hosp, New York, 57-60, asst pathologist, 60-61; from asst prof to assoc prof path, Univ Southern Calif, 63-67; dir cancer ctr planning, 73-76, ASSOC PROF PATH, SCH MED, NY UNIV, 67- *Concurrent Pos:* NIH res training grant, Sch Med, NY Univ, 57-61; Nat Cancer Inst res career develop award, Sch Med, Univ Southern Calif, 63-66. *Mem:* Am Chem Soc; Am Soc Exp Path. *Res:* Immune suppression; molecular pathology of membranes; role of free radical reactions in altering lipids, proteins and associated nucleic acids; free radical pathology and antioxidants in aging, central nervous system disorders, cancer and arteriosclerosis. *Mailing Add:* Dept Path 550 First Ave New York NY 10016

DEMOPOULOS, JAMES THOMAS, b New York, NY, Dec 26, 28; m 69; c 3. PHYSICAL MEDICINE & REHABILITATION. *Educ:* NY Univ, BS, 50; State Univ NY, MD, 56. *Prof Exp:* From intern to resident, Kings County Hosp, Brooklyn, NY, 56-58; resident rehab med, Inst Rehab Med, NY Univ Med Ctr, 59-61, assoc dir outpatient serv, 61-65; assoc dir, Dept Rehab Med, City Hosp, Elmhurst, 65-67; dir dept rehab med, Hosp Joint Dis & Med Ctr, 67-; prof rehab med, Mt Sinai Sch Med, 72-; dir dept rehab med, Beth Israel Med Ctr, 76-; CHMN DEPT REHAB MED, SCH MED, TEMPLE UNIV HEALTH SCI CTR. *Concurrent Pos:* Fel stroke coun, Am Heart Asn, 69-; chmn adv comt amputee, orthotic, neuromuscular & pediat-orthop progs, NY Dept Health, 70-; consult, NY Bur Handicapped Children, 70-; fel, Arthritis Found, 72- *Mem:* Fel Am Acad Phys Med & Rehab; fel Am Cong Rehab Med; fel Am Rheumatism Asn. *Res:* Prosthetic and orthotic devices; development of predictors of outcome in rehabilitation as guides to management. *Mailing Add:* Temple Univ Health Sci Ctr Broad & Ontario Sts Philadelphia PA 19140

DE MORI, RENATO, b Italy, Aug 5, 41; c 1. AUTOMATIC SPEECH RECOGNITION, SIMULATION. *Educ:* Politecnico di Torino, PhD, 67. *Prof Exp:* Assoc prof, Politecnico di Torino, 69-76; prof, Univ Turin, 76-86, chmn, 77-79; assoc prof, Concordia Univ, 82-83, prof, 83-85, chmn, 84-85; PROF & DIR, MCGILL UNIV, 86- *Concurrent Pos:* Mem, Sci Coun, Ctr Nat Etudes Telecommun, 81-; mem, Sci Coun, Ctr Res Info Montreal, 84-, vpres res, 87-; mem & chair, AI Asn Comt, Nat Res Coun Can, 87-; assoc ed, Inst Elec & Electronics Engrs Trans on Pattern Anal & Mach Intel, 88-; mem, Sci Adv Bd, Info Technol Res Ctr, 89; proj leader, Inst Robotics & Intel Systs, 90-93. *Mem:* Inst Elec & Electronics Engrs; Asn Comput Mach; Am Asn Artificial Intel. *Res:* Computer arithmetic and architecture; automatic speech recognition; industrial applications of artificial intelligence; author of various publications. *Mailing Add:* Sch Computer Sci McGill Univ 3480 University St Montreal PQ H3A 2A7 Can

DEMORT, CAROLE LYLE, b Independence, Mo, Apr 1, 42. PHYCOLOGY, ECOLOGY. *Educ:* Park Col, BA, 64; Univ Mo-Kansas City, MS, 66; Ore State Univ, PhD(bot), 69. *Prof Exp:* Asst prof biol, St Mary's Col, Ind, 69-74; ASSOC PROF NAT SCI, UNIV N FLA, 74- *Mem:* Am Inst Biol Sci; Phycol Soc Am; Int Phycol Soc; Bot Soc Am; Am Soc Limnol & Oceanog. *Res:* Relative nutritional value of phytoplankton species as food for shellfish and shrimp larvae; developmental morphology and biochemical analysis of estuarine phytoplankton species; toxic effects of heavy metals on marine phytoplankton species. *Mailing Add:* Univ Natural Sci Univ NFla PO Box 17074 Jacksonville FL 32216

DEMOS, PETER THEODORE, b Toronto, Ont; m 41; c 3. PHYSICS. *Educ:* Queen's Univ, Ont, BSc, 41; Mass Inst Technol, PhD(physics), 51. *Hon Degrees:* LLD, Trent Univ, Ont, 81. *Prof Exp:* Instr math & physics, Queen's Univ, Ont, 41-42; mem, Ballistics Res Staff, Nat Res Lab, Ont, 42-44; mem staff, Can Army Res Estab, Que, 44-46; asst physics, Mass Inst Technol, 46-51, mem staff, Lab Nuclear Sci, 51-52, lectr & assoc dir lab, 52-61, dir, 61-75, emer prof physics, 61-89, dir, Bates Linear Accelerator, 75-83. *Concurrent Pos:* Mem bd trustees, Assoc Univ Inc, 59-69; mem, NRC Comt, Rad Sources US Army Lab, Natick, Mass, 60-65; mem, Nat Res Coun Eval Panels, Nat Bur Standards Radiation Res Ctr, 65-68, 77-83, chmn, 67-; dir, Bates Linear Accelerator, Mass Inst Technol, 75-83. *Mem:* AAAS; Am Phys Soc; Am Acad Arts & Sci; Fedn Am Sci; Union Concerned Scientists. *Res:* Linear accelerator development; electro and photonuclear studies. *Mailing Add:* 49 Orchard St Belmont MA 02178

DEMOSS, JOHN A, BIOCHEMISTRY. *Prof Exp:* PROF & CHMN, UNIV TEX HEALTH SCI CTR, HOUSTON. *Mailing Add:* Dept Biochem & Molecular Biol Univ Tex Health Sci Ctr PO Box 20708 Houston TX 77225

DEMOTT, BOBBY JOE, b Kans, Nov 6, 24; m 47; c 5. DAIRY CHEMISTRY. *Educ:* Kans State Univ, BS, 49; Univ Idaho, MS, 51; Mich State Univ, PhD(dairy), 54. *Prof Exp:* Instr, Milk Factory Tests, Univ Idaho, 49-51; asst market milk, buttermaking & cheese, Mich State Univ, 51-54; asst prof dairy indust, Colo State Univ, 54-57; ASSOC PROF DAIRY INDUST, UNIV TENN, KNOXVILLE, 57- *Mem:* Am Dairy Sci Asn; Inst Food Technologists; Int Asn Milk Food & Environ Sanitarians. *Res:* Homogenized milk; adding iron to milk; calcium in foods; whey utilization. *Mailing Add:* PO Box 1071 Knoxville TN 37901-1071

DEMOTT, HOWARD EPHRAIM, b Bloomsburg, Pa, Oct 24, 13; m 40; c 1. BOTANY. *Educ:* Bloomsburg State Col, BS, 35; Bucknell Univ, MS, 40; Univ Va, PhD, 66. *Prof Exp:* Teacher high schs, NY & Pa, 35-48; from instr to prof, 48-81, EMER PROF BIOL, SUSQUEHANNA UNIV, 81- *Mem:* Bot Soc Am. *Res:* Plant morphology and morphogenesis. *Mailing Add:* 902 N Ninth St Selinsgrove PA 17870-1711

DEMOTT, LAWRENCE LYNCH, geology, paleontology; deceased, see previous edition for last biography

DEMPESY, COLBY WILSON, b Chicago, Ill, Mar 12, 31; m 52; c 5. PHYSICS, NEUROBIOLOGY. *Educ:* Oberlin Col, BA, 52; Rice Inst, MA, 55, PhD(physics), 57. *Hon Degrees:* MA, Amherst Col, 67. *Prof Exp:* From instr to assoc prof, 57-67, chmn dept, 72-74, prof physics, Amherst Col, 67-, mem fac, neurosci prog, 74-; AT DEPT NEUROSURG, TULANE UNIV, NEW ORLEANS. *Concurrent Pos:* Fulbright res fel to Japan, 65-66; vis prof, Tulane Univ, 74-; res scientist, Charity Hosp, New Orleans, 74. *Mem:* Am Phys Soc; Am Asn Physics Teachers; AAAS; Sigma Xi. *Res:* Low temperature physics; neurophysiology. *Mailing Add:* Tulane Med Sch 1430 Tulane Ave New Orleans LA 70112

DEMPSEY, BARRY J, b Galesburg, Ill, Mar 17, 38; m 63; c 2. CIVIL ENGINEERING. *Educ:* Univ Ill, Urbana, BS, 60, MS, 66, PhD(civil eng), 69. *Prof Exp:* Asst resident engr hwy construct, State Ill Hwy Dept, 60 & 63-64; res asst, 64-69, asst prof, 69-74, assoc prof civil eng, 74-79, PROF CIVIL ENG, UNIV ILL, URBANA, 79- *Concurrent Pos:* Prin investr, Ill Div Hwy, 68-; consult, NY State Dept Transp, 70-71; mem, Transp Res Bd, Nat Res Coun-Nat Acad Sci, 70-; US Army CEngrs, Monsanto, 73- & Fed Hwy Admin, 76- *Honors & Awards:* A W Johnson Mem Award, 70; K B Woods Award, 78. *Mem:* Am Soc Civil Engrs; Soc Am Mil Engrs; Soil Sci Soc Am; Am Soc Testing & Mat. *Res:* Transportation materials with major emphasis on the investigation of the influence climatic factors have on construction, design, behavior and performance of pavement systems. *Mailing Add:* Two Rolling Hills RR 1 Box 156 White Heath IL 61884

DEMPSEY, DANIEL FRANCIS, b Buffalo, NY, July 23, 29; m 60; c 3. PHYSICS. *Educ:* Canisius Col, BS, 51; Univ Notre Dame, PhD(physics), 57. *Prof Exp:* From asst prof to assoc prof physics, 56-77, dept head, 71-77, PROF PHYSICS, CANISIUS COL, 77- *Mem:* Am Asn Physics Teachers. *Res:* Focusing trajectories for charged and neutral particles; magnetic fields. *Mailing Add:* 6641 Powers Rd Orchard Park NY 14127

DEMPSEY, JOHN NICHOLAS, b St Paul, Minn, June 16, 23; m 48; c 3. PHYSICAL CHEMISTRY. *Educ:* Col St Thomas, BS, 48; Univ Iowa, PhD(phys chem), 51. *Prof Exp:* Asst, Univ Iowa, 48-50; res chemist, Ethyl Corp, 51-52; res physicist, Minneapolis Honeywell Regulator Co, 52-56, res sect head, 56-60, from asst dir res to dir res, Honeywell Inc, 60-65, vpres, Corp Res Ctr, 65-67, vpres sci & eng, 67-72; vpres tech serv, Bemis Co, Inc, 72-75, vpres sci & technol, 75-88; RETIRED. *Concurrent Pos:* Mem Patent & Trademark Adv Comt, US Dept Com, 76; dir indust res, AAAS. *Mem:* Am Inst Chemists; Am Chem Soc; Indust Res Inst; AAAS; NY Acad Sci. *Res:* X-ray crystal structures of coordination; complex compounds; intermetallic compounds and bond orders. *Mailing Add:* 4926 Westgate Rd Minnetonka MN 55343

DEMPSEY, JOHN PATRICK, b Te Kuiti, NZ, Oct 13, 53; m 80; c 2. ENGINEERING MECHANICS, CIVIL ENGINEERING. *Educ:* Auckland Univ, NZ, BE(1st hons), 75, PhD(appl mech), 78. *Prof Exp:* Postdoc civil eng, Northwestern Univ, 78-80; from asst prof to assoc prof, 80-90, PROF, DEPT CIVIL ENG, CLARKSON UNIV, 90- *Concurrent Pos:* Sabbatical, US Army Cold Regions Res & Eng Lab, Hanover, NH, 86-87; John W Graham Jr fac res award, Clarkson Univ, 86; vis prof, Depts Eng Sci & Mech Eng, Univ Auckland, NZ, 90. *Honors & Awards:* OMAE Achievement Award, Am Soc Mech Engrs, 90. *Mem:* Am Soc Civil Engrs; Am Soc Mech Engrs; Am Soc Testing & Mat; Int Glaciol Soc; Soc Indust & Appl Math; Soc Exp Mech. *Res:* Engineering mechanics; fracture mechanics; contact mechanics; ice mechanics. *Mailing Add:* Dept Civil & Environ Eng Clarkson Univ Potsdam NY 13699-5710

DEMPSEY, MARTIN E(WALD), b Chicago, Ill, Mar 28, 21; m 45; c 3. SURFACE ACOUSTIC WAVE TECHNOLOGY, ULTRASONICS. *Educ:* Purdue Univ, BS, 48, MS, 49, PhD(elec eng), 55. *Prof Exp:* Engr voice commun lab, Purdue Univ, 48-49, chief engr, 50-55; mem tech staff, Bell Tel Labs, 49-50; dir aid to hearing res, Zenith Radio Corp, 55-59; dir psychoacoust res, Beltone Res Labs, 59-61; mgr advan phys develop, GTE Automatic Elec Labs, 61-71; mgr, GTE Labs, Waltham Res Ctr, 71-78, prin investr, 78-82; RETIRED. *Concurrent Pos:* Consult dept commun disorders, Northwestern Univ, 58-62. *Mem:* Acoust Soc Am; Am Vacuum Soc; Electrochem Soc; NY Acad Sci; Optical Soc Am; Sigma Xi. *Res:* Communication engineering; psychoacoustics; materials science. *Mailing Add:* 1381 Newport Ave Arroyo Grande CA 93420

DEMPSEY, MARY ELIZABETH, b St Paul, Minn, Sept 23, 28. BIOCHEMISTRY. *Educ:* St Catherine Col, BS, 50; Wayne State Univ, MS, 52; Univ Minn, PhD(enzym), 61. *Prof Exp:* Res biochemist, Minneapolis Vet Admin Hosp, 52-56; instr clin biochem, 56-58, from instr to assoc prof biochem, 61-75, PROF BIOCHEM, MED COL, UNIV MINN, MINNEAPOLIS, 75- *Concurrent Pos:* Am Heart Asn res fel, 61-63; res grants, Minn Heart Asn, 61-64 & 77-78, Nat Heart Inst, 64-80, Muscular Dystrophy Asn Am, Inc, 65-70, NSF, 70-74 & Am Heart Asn, 71-74. *Mem:* Am Chem Soc; Am Soc Biol Chem; Soc Exp Biol & Med. *Res:* Enzymology; steroid and sterol; arteriosclerosis; bioenergetics muscle contraction; oxygen-18 methodology; cholesterol biosynthesis; regulation of lipid biosynthesis. *Mailing Add:* Dept Biochem Univ Minn Box 68 Mayo Bldg Minneapolis MN 55455-0100

DEMPSEY, WALTER B, b San Francisco, Calif, Nov 21, 34; m 81; c 4. BIOCHEMISTRY. *Educ:* Univ San Francisco, BS, 56; Univ Mich, MS, 58, PhD(biol chem), 60. *Prof Exp:* Instr biochem, Univ Florida Gainesville, 62-63, asst prof, 63-67; from asst prof to assoc prof, 67-72, PROF BIOCHEM & INTERNAL MED, UNIV TEX SOUTHWESTERN MED SCH, 72-; RES CAREER SCIENTIST, VET ADMIN HOPS, DALLAS, 74- *Concurrent Pos:* NSF fel, Univ Calif Berkeley, 60-62; prin investr, NIH grants, 63-, US Vet Admin res grants, 69-; sr postdoctoral fel, Univ Edinburgh, Scotland, 73-74. *Mem:* AAAS; Fedn Am Socs Exp Biol; Am Soc Microbiol; Am Chem Soc; Am Soc Biol Chemists; Am Assoc Adv Slavic Studies; AAAS. *Res:* Mechanisms controlling promiscuity of R factors and other bacterial plasmids; bacterial genetics; microbiology of pyridoxine. *Mailing Add:* Dept of Biochem Univ Tex Health Sci Ctr Dallas TX 75235

DEMPSEY, WESLEY HUGH, b Waltham, Mass, Dec 2, 26; m 51; c 4. GENETICS, PLANT BREEDING. *Educ:* Cornell Univ, BS, 49; Univ Calif, MS, 50, PhD(genetics), 54. *Prof Exp:* Res asst genetics & plant breeding, Univ Calif, 51-54; PROF BIOL, CALIF STATE UNIV, CHICO, 54- *Concurrent Pos:* NSF sci fac fel, Univ Wis, 63-64; adj prof plant breeding, Pa State Univ, 70-71; vis prof bot, Bot Dept, Univ Canterbury, Christchurch, NZ, 80-81. *Mem:* Am Soc Hort Sci; Bot Soc Am; AAAS; Am Inst Biol Sci. *Res:* Tomato genetics, inheritance on consistency; red cotyledon in lettuce; pectic substances in tomatoes; electron microscopy of tomato mutants; population cytogenetics of Trimerotropis. *Mailing Add:* Dept Biol Chico State Univ Chico CA 95929-0515

DEMPSKI, ROBERT E, b Centermoreland, Pa, July 3, 34; m 66; c 3. PHARMACEUTICS. *Educ:* Philadelphia Col Pharm, BS, 56; Univ Wis, PhD(pharm), 60. *Prof Exp:* Instr pharm, Univ Wis, 56-58; res assoc pharm res, Merck & Co, Inc, 60-66, unit head, 66-72, sr res fel pharm res, 72-82, DIR, MERCK SHARP & DOHME, WEST POINT, 82- *Mem:* Am Pharmaceut Asn; Acad Pharmaceut Sci; Am Asn Pharmaceut Sci. *Res:* Design of new dosage forms for medicinals; study of new methods of dermatologic therapy; solid dosage form research and development. *Mailing Add:* 1629 Arran Way Dresher PA 19025

DEMPSTER, GEORGE, b Edinburgh, Scotland, June 28, 17; m 42; c 2. MEDICAL MICROBIOLOGY. *Educ:* Univ Edinburgh, MB, ChB, 40, BSc, 41, MD, 52. *Prof Exp:* Intern, Peel Hosp, Scotland, 40; asst, Pub Health Lab, Edinburgh, 41-42, lectr bact, 42-46 & 48-50; res assoc, Connaught Med Res Lab, Toronto, 50-55; prof virol, Univ Sask, 55-77, head dept, 56-72, prof microbiol, 77-80; RETIRED. *Concurrent Pos:* Dir bact, Univ Hosp, Saskatoon, 55-72, virologist, 55-80. *Mem:* Can Soc Microbiol; Can Med Asn; Can Pub Health Asn. *Res:* Virology; epidemic respiratory disease; influenza; atypical pneumonia and adenoviruses; neuro-tropic viruses, particularly Coxsackie viruses. *Mailing Add:* 564 St Patrick St Victoria BC V8S 4X3 Can

DEMPSTER, LAURAMAY TINSLEY, b El Paso, Tex, May 11, 05; m 27; c 2. TAXONOMIC BOTANY. *Educ:* Univ Calif, MA, 27. *Prof Exp:* Asst bot, 33-35, herbarium botanist, 51-61, res geneticist bot, NSF grants, 59-69, RES ASSOC, DEPT BOT, UNIV CALIF, BERKELEY, 69- *Mem:* Am Soc Plant Taxon. *Res:* Taxonomy of flowering plants; currently the genus Galium in western North America, Mexico and South America. *Mailing Add:* Jepson Herbarium Univ Calif Berkeley CA 94720

DEMPSTER, WILLIAM FRED, b Berkeley, Calif, Dec 10, 40; div; c 2. SYSTEMS DESIGN & SYSTEMS SCIENCE. *Educ:* Univ Calif, Berkeley, BA, 63. *Prof Exp:* Pres, Feedback Systs, Inc, 77-79 & Inst Ecotechnics, 83-89; exped chief, Inst Ecotechnics, Amazon River Exped, 80-82; DIR SYSTS ENG, SPACE BIOSPHERES VENTURE, 85- *Mem:* NY Acad Sci; Am Mus Natural Hist. *Res:* Systems for the creation of artificial biospheres, especially including the means of sealing large airtight glazed structures and processes not requiring external material supply; surveying; computer programming; energy systems. *Mailing Add:* PO Box 689 Oracle AZ 85623

DEMSKEY, SIDNEY, statistical analysis, for more information see previous edition

DEMSKI, LEO STANLEY, b Pittsburgh, Pa, Mar 29, 43; m 65; c 2. NEUROANATOMY, ANIMAL BEHAVIOR. *Educ:* Miami Univ, AB, 65; Univ Rochester, PhD(anat, neurobiol), 69. *Prof Exp:* Asst prof anat, Sch Med, Univ NMex, 71-74; assoc prof, Sch Med, La State Univ, 74-77; assoc prof, 77-84, PROF, SCH BIOL SCI, UNIV KY, 84- *Concurrent Pos:* NIMH fel, Am Mus Natural Hist, NY, 69-71. *Mem:* Animal Behav Soc; Am Soc Ichthyologists & Herpetologists; Am Soc Zool. *Res:* Identification of neuroendocrine systems controlling sexual behavior in fishes and other vertebrates. *Mailing Add:* Dept Biol Sci Univ Ky Lexington KY 40506

DEMUTH, GEORGE RICHARD, b Sherwood, Ohio, Nov 16, 25; m 51; c 4. PEDIATRICS. *Educ:* Univ Cincinnati, MD, 50. *Prof Exp:* From instr to assoc prof, 56-65, assoc dean, 68-81, dir integrated premed prog, 73-75, PROF PEDIAT, UNIV MICH, ANN ARBOR, 65-, MED SCIENTIST TRAINING PROG, 79- *Mem:* AAAS; Soc Pediat Res; Asn Am Med Cols. *Mailing Add:* Dept Pediat Univ Mich Med Sch Ann Arbor MI 48109

DEMUTH, HOWARD B, b Junction City, Kans, June 22, 28; m 51; c 4. ELECTRICAL ENGINEERING, COMPUTER SCIENCE. *Educ:* Univ Colo, BS, 49; Stanford Univ, MS, 54, PhD(elec eng), 57. *Prof Exp:* Staff mem electronic comput, Los Alamos Sci Lab, 49-53; res asst network synthesis, Stanford Univ, 53-54; res engr, Stanford Res Inst, 54-56; proj engr, Int Bus Mach Res Lab, 56-58; staff mem digital comput & control systs, Los Alamos Nat Lab, 58-79; prof elec eng, Univ Tulsa, 79-85; PROF ELEC ENG, UNIV IDAHO, 85- *Concurrent Pos:* Prof, Univ NMex, 58-79; vis lectr, Univ Colo, 62-63; vis prof, Univ Hawaii, 68-69. *Mem:* Inst Elec & Electronics Engrs. *Res:* Digital data communication; digital control systems; information and communication theory. *Mailing Add:* Dept Elec Eng Univ Idaho Moscow ID 83843

DEMUTH, JOHN ROBERT, b St Louis, Mo, Nov 15, 24. ORGANIC CHEMISTRY. *Educ:* Washington Univ, BA, 49; Univ Ill, MS, 52, PhD(chem), 55. *Prof Exp:* Asst, Washington Univ, 49-50; asst, Univ Ill, 50-55; from assoc prof to prof chem, 55-89, chief adv, Dept Chem, 75-80, EMER PROF CHEM, UNIV NEBR, LINCOLN, 89- *Mem:* Am Chem Soc. *Mailing Add:* Dept Chem Univ Nebr 731 Hamilton Hall Lincoln NE 68588

DEMYER, MARIAN KENDALL, b Greensburg, Ind; c 3. CHILD PSYCHIATRY, PSYCHIATRY. *Educ:* Ind Univ, BS, 49, MD, 52; Am Bd Psychiat & Neurol, cert psychiat, 59, cert child psychiat, 60. *Prof Exp:* Intern med & surg, Detroit Receiving Hosp, Mich, 52-53; resident psychiat, Univ Med Ctr, 53-56, from instr to assoc prof psychiat, 57-69, dir clin res ctr early childhood schizophrenia, 61-74, actg coordr child psychiat, 74, chief clin res sect, 74-80, PROF PSYCHIAT, INST PSYCHIAT RES, IND UNIV-PURDUE UNIV, INDIANAPOLIS, 69-, PRIN INVESTR, 80- *Concurrent Pos:* Fel child psychiat, Eastern Pa Psychiat Inst, 56-57; NIMH fel, Ind Univ, 61-74; Ind Dept Ment Health fel, Ind Univ-Purdue Univ, Indianapolis, 74-77; dir children's serv, Carter Hosp, Indianapolis, 57-61; assoc ed, J Child Schizophrenia & Autism, 71-76. *Mem:* Am Psychiat Asn; Sigma Xi. *Res:* Neurological correlates of mental illness; magnetic resonance images in schizophrenia. *Mailing Add:* Inst Psychiat Res-PR 103 Ind Univ Sch Med 791 Union Dr Indianapolis IN 46223

DE MYER, WILLIAM ERL, b South Charleston, WVa, Aug 7, 24; m 52; c 2. NEUROLOGY. *Educ:* Ind Univ, BS, 49, MD, 52; Am Bd Psychiat & Neurol, dipl, 58, cert child neurol, 68. *Prof Exp:* Intern, Univ Mich Hosp, 52-53; resident, Med Ctr, 53-56, from instr to assoc prof, 54-68, PROF NEUROL, SCH MED & DIR NEUROANAT LAB, IND UNIV, 68- *Concurrent Pos:* NIH spec fel, 62; mem med fac training prog, Univ Pa, 56-57; consult, LaRue Carter Hosp, Wishard Mem Hosp & Vet Admin Hosp, Indianapolis, 65- *Honors & Awards:* Frederich Bachman Lieber Award, 81. *Mem:* AAAS; Int Soc Cranio-Facial Biol; Am Asn Neuropath; World Fedn Neurol; fel Am Acad Neurol; Child Neurol Soc. *Res:* Quantitative neuroanatomy; relation of congenital malformations of face and brain; developmental neuroanatomy and teratology. *Mailing Add:* Pediat Neurol Ind Univ Sch Med Riley Hosp Indianapolis IN 46223

DE NAULT, KENNETH J, b Los Angeles, Calif, Apr 3, 43; div; c 2. GEOLOGY, MINERALOGY. *Educ:* Stanford Univ, BS, 65, PhD(geol), 74; Univ Wyo, MS, 67. *Prof Exp:* Geologist uranium explor, Getty Oil Co, 68-69 & Union Oil Co, 70; asst prof, 73-79, ASSOC PROF GEOL, UNIV NORTHERN IOWA, 80- *Concurrent Pos:* NSF res grant, 85-86. *Mem:* Am Asn Petrol Geologists; Geol Soc Am; Sigma Xi; Mineral Soc Am; Am Fern Soc. *Res:* Mineralogy, petrology, and crystallography; investigating the mineralogy and petrology of the Rattlesnake Hills eruptive center, Wyoming; computer aided instruction in geology. *Mailing Add:* 3750 Ranchero Rd RR 2 Cedar Falls IA 50613

DENAVIT, JACQUES, b Paris, France, Oct 1, 30; US citizen; m 54; c 3. ENGINEERING. *Educ:* Univ Paris, Baccalaureat, 49; Northwestern Univ, MS, 53, PhD, 56. *Prof Exp:* Asst prof, dept mech eng, Northwestern Univ, 58-60, assoc prof mech eng & astronaut sci, 60-65, prof, 65-; AT LAWRENCE LIVERMORE LAB. *Mem:* Am Soc Mech Engrs; fel Am Phys Soc. *Res:* Plasma physics and kinetic theory. *Mailing Add:* Lawrence Livermore Lab Box 808 Livermore CA 94550

DENBER, HERMAN C B, b New York, NY, Oct 9, 17; m 44; c 1. PSYCHIATRY, BIOCHEMICAL PHARMACOLOGY. *Educ:* NY Univ, BA, 38, MS, 63, PhD(biol), 67; Univ Geneva, BMS, 41, MD, 43; Am Bd Psychiat & Neurol, dipl, 55. *Prof Exp:* Assoc res scientist, Manhattan State Hosp, 54-55, dir psychiat res, 55-72, prof psychiat & pharmacol, Col Med, Univ Fla, 72-74; prof psychiat, Sch Med, Univ Louisville, 74-80; prof, 80-82, adj prof psychiat, Sch Med, Univ Ottawa, Can, 82-87; PROF PSYCHIAT, SCH MED, SOUTH ILL UNIV, SPRINGFIELD, 87- *Concurrent Pos:* Instr psychiat, Col Physicians & Surgeons, Columbia Univ, 56-60; assoc clin prof, New York Med Col, 60-66, prof, 66-72; vis prof psychiat, Univ Lausanne, Switz, 72; adj prof med, Mt Sinai Med Sch, NY, 90- *Mem:* Am Col Psychiatrists; fel AAAS; NY Acad Sci; Sigma Xi; hon mem Royal Belg Soc Ment Med. *Res:* Psychopharmacology; clinical psychiatry; molecular biology; psychoanalysis; social psychiatry; electroencephalography; neuropathology. *Mailing Add:* 113 Tidy Island Blvd Bradenton FL 34210

DEN BESTEN, IVAN EUGENE, b Corsica, SDak, Jan 11, 33; m 60; c 2. PHYSICAL CHEMISTRY. *Educ:* Calvin Col, AB, 57; Northwestern Univ, PhD(chem), 61. *Prof Exp:* Instr chem, 61-62, from asst prof to assoc prof, 62-70, PROF CHEM, BOWLING GREEN STATE UNIV, 70- *Mem:* AAAS; Am Chem Soc; NY Acad Sci. *Res:* Catalysis by metals and metal oxides; photocatalysis and photochemistry; reactions in discharges; surface reactions. *Mailing Add:* Dept Chem Bowling Green State Univ Main Campus Bowling Green OH 43403

DENBO, JOHN RUSSELL, b Evansville, Ind, June 12, 47; m 71. PHYSIOLOGY, NEUROANATOMY. *Educ:* Eastern Ill Univ, BS, 69; Southern Ill Univ, MA, 72, PhD(physiol), 74. *Prof Exp:* From asst prof to assoc prof physiol, Dept Biol, Grinnell Col, 74-85; adminr, Harrisburg Med Ctr, Ill, 85-88; asst mgr, Shelps County; REP HOSP ROLLA, MISSOURI, 88- *Mem:* Sigma Xi; AAAS. *Res:* Elucidation of the controlling mechanisms of movement in the acellular slime mold Physarum polycephalum. *Mailing Add:* 1224 S Roosevelt St Harrisburg IL 62946

DENBOW, CARL (HERBERT), b Zanesville, Ohio, Dec 13, 11; m 39; c 3. MATHEMATICS. *Educ:* Univ Chicago, BS, 32, MS, 34, PhD(math), 37. *Prof Exp:* From instr to assoc prof math, Ohio Univ, 36-46; assoc prof math & mech, US Naval Postgrad Sch, 46-50; prof, 50-82, chmn dept, 54-55 & 66-67, EMER PROF MATH, OHIO UNIV, 82- *Concurrent Pos:* Ford fac fel, Harvard Univ, 55-56; coordr, Int Teacher Develop Prog, 61, 62; chmn AID surv team, Cambodia, 62; dir training prog, Peace Corps, Cameroon, 63, 64. *Mem:* Am Math Soc; Math Asn Am. *Res:* Applied mathematics; foundations of mathematics; philosophy of science and mathematics. *Mailing Add:* 61 Columbia Dr Athens OH 45701

DENBOW, DONALD MICHAEL, b Washington, DC, Apr 25, 53; m 79; c 2. ENVIRONMENTAL PHYSIOLOGY, POULTRY MANAGEMENT. *Educ:* Univ Md, BS, 75, MS, 77; NC State Univ, PhD(physiol), 80. *Prof Exp:* Fel, 79-80, asst prof, 80-86, ASSOC PROF POULTRY SCI, VA POLYTECHN INST STATE UNIV, 86- *Mem:* Poultry Sci Asn; AAAS; Sigma Xi. *Res:* Central nervous system; peripheral regulation of food intake; investigation of the physiology of heat stress. *Mailing Add:* Dept Poultry Sci Va Polytech Inst & State Univ Blacksburg VA 24061

DENBURG, JEFFREY LEWIS, b Brooklyn, NY, Oct 5, 44; m 81; c 3. NEUROBIOLOGY, BIOCHEMISTRY. *Educ:* Amherst Col, BA, 65; Johns Hopkins Univ, PhD(biol), 70. *Prof Exp:* Fel neurobiol, Cornell Univ, 70-72; res fel, Australian Nat Univ, 72-77; asst prof zool, 77-82, ASSOC PROF BIOL, UNIV IOWA, 82- *Concurrent Pos:* NIH res develop award, 79-84. *Mem:* AAAS; Soc Neurosci. *Res:* Identification of the genetic mechanisms responsible for the formation of neuronal connections and the isolation of macromolecules mediating these mechanisms. *Mailing Add:* Dept Biol Univ Iowa Iowa City IA 52242

DENBY, LORRAINE, US citizen. STATISTICS. *Educ:* Aquinas Col, BS, 69; Univ Mich, Ann Arbor, MS, 72, PhD(statist), 84. *Prof Exp:* MEM TECH STAFF, AT&T BELL LABS, 72- *Concurrent Pos:* Chair, Sect Statist Graphics, Am Statist Asn, 90. *Mem:* Fel Am Statist Asn. *Res:* Robust estimation, regression modelling, censored data, graphical methods, data analysis applications and methodology and statistical education. *Mailing Add:* AT&T Bell Labs Murray Hill NJ 07922

DENBY, LYALL GORDON, b Regina, Sask, Oct 4, 23. POMOLOGY. *Educ:* Univ BC, BSA, 45, MSA, 50. *Prof Exp:* Nursery sales & mgr, Hyland Barnes Nursery, 47-50; head veg crops, 50-59, head veg & ornamentals, 59-70, RES SCIENTIST POMOL & GRAPES, SUMMERLAND RES STA, CAN DEPT AGR, 70- *Honors & Awards:* Sadler Mem Gold Medal, 45. *Mem:* Am Soc Hort Sci; Am Pomol Soc. *Res:* Tree fruit varieties; rootstocks and tree training; grape culture and breeding. *Mailing Add:* 6204 Happy Valley Rd Summerland BC V0H 1Z0 Can

DENCE, MICHAEL ROBERT, b Sydney, Australia, June 17, 31; m 67; c 2. GEODYNAMICS, METEORITICS. *Educ:* Univ Sydney, BSc, 53. *Prof Exp:* Geologist, Falconbridge Nickel Mines, Ltd, 53-54; res asst tectonics, Geophys Lab, Univ Toronto, 56-58; tech officer, Dept Mines & Tech Surv, 59-61, res scientist, Dom Observ, 62-70, res scientist, 70-81, res dir, Dept Energy, Mines & Resources, Govt Can, 81-86; EXEC DIR, ROYAL SOC CAN, 87- *Concurrent Pos:* Vis prof, Univ Teubingen, 72-73. *Honors & Awards:* Barringer Medal, 88. *Mem:* Am Geophys Union; fel Meteoritical Soc; fel Geol Asn Can; fel Royal Soc Can; fel AAAS. *Res:* Meteor crater studies; meteoritics; rock deformation; metamorphism; rock physics; tectonophysics; planetary sciences; lunar sample studies; nuclear waste management; global dynamics. *Mailing Add:* Royal Soc Can PO Box 9734 207 Queen St Ottawa ON K1G 5J4 Can

DENDINGER, JAMES ELMER, b Long Beach, Calif, Nov 13, 43; m 85; c 1. INVERTEBRATE PHYSIOLOGY. *Educ:* Calif State Univ, Long Beach, BA, 69, MS, 71; Univ Mass, PhD(zool), 75. *Prof Exp:* Mem res & develop staff biol, Hyland Labs, Costa Mesa, Calif, 68-69; biomed res technician, Vet Admin Hosp, Long Beach, 69-70; asst physiol, Dept Biol, Calif State Univ, Long Beach, 70-71; res asst parasitol, Dept Zool, Univ Mass, 71-75; asst prof, James Madison Univ, 75-81, assoc prof, 81-88, dir, Off Sponsored Progs, 90-91, PROF BIOL, JAMES MADISON UNIV, 88- *Concurrent Pos:* Technician immunohemat, Walter Reed Army Med Ctr, Washington, DC, 65-67 & Georgetown Univ Med Ctr, 66-67. *Mem:* Am Soc Zoologists; AAAS; Sigma Xi. *Res:* Regulation of intermediary metabolism in invertebrate animals; crustacean physiology. *Mailing Add:* Dept Biol James Madison Univ Harrisonburg VA 22807

DENDINGER, RICHARD DONALD, b Minot, NDak, Feb 21, 36; m 57; c 3. INORGANIC CHEMISTRY, NUCLEAR CHEMISTRY. *Educ:* Minot State Col, BS, 58; NDak State Univ, MS, 66; SDak State Univ, PhD(inorg chem & nuclear chem), 74. *Prof Exp:* Chem, physics, math & biol teacher, Bottineau Spec Sch Dist, NDak, 58-62; assoc prof inorg, nuclear & gen chem, 65-80, PROF CHEM, ST CLOUD STATE UNIV, 80- *Concurrent Pos:* Chem traineeship, NSF, 62. *Mem:* Am Chem Soc. *Res:* Photochemical effects on polar solutions of tungsten, vanadium and molybdenum polydentate chelate compounds, a method of preparation of lower oxidation state compounds by photoredox. *Mailing Add:* Dept Chem St Cloud State Univ St Cloud MN 56301-4498

DENDROU, STERGIOS, b Djibouti, Africa, March 25, 50; US citizen. COASTAL FLOODING METEOROLOGY. *Educ:* Polytech Inst Technol, Switz, dipl, 73; Purdue Univ, MS, 74, PhD(hydraul & hydrol), 77. *Prof Exp:* Asst prof, water resources, dept civil eng, Purdue Univ, 77-78; sr engr, Camp Dresser & McKee Inc, 79-86; PRIN & DIR SCI SOFTWARE, ZEI ENG INC, 86- *Concurrent Pos:* Consult, Fed Emergency Mgt Agency, 85-87. *Honors & Awards:* Walter L Huber Res Prize, Am Soc Civil Engrs, 85. *Mem:* Am Geophys Union; Am Soc Civil Engrs; AAAS; NY Acad Sci; Sigma Xi. *Res:* Advanced scientific software development for personal computers, workstations and supercomputers; advanced numerical algorithm architectures; groundwater movement and contamination, saturated and unsaturated; simulation of coastal surges from hurricanes and northeasters; non-linear dynamics plasticity theory; chaotic phenomena; fractals. *Mailing Add:* ZEI Eng Inc PO Box 1344 Annandale VA 22003

DENDY, JOEL EUGENE, JR, b Anderson, SC, Mar 24, 45; m 67; c 2. NUMERICAL ANALYSIS. *Educ:* Rice Univ, BA, 67, PhD(math), 71. *Prof Exp:* Asst prof math, Univ Denver, 71-73; STAFF MEM MATH, LOS ALAMOS NAT LAB, 73- *Mem:* Soc Indust & Appl Math. *Res:* Finite difference and finite element methods for the numerical solution of partial differential equations; multigrid methods. *Mailing Add:* Los Alamos Nat Lab MS B 284 Los Alamos NM 87545

DENEAU, GERALD ANTOINE, b Oxford, Mich, May 9, 28; m 52; c 3. PHARMACOLOGY. *Educ:* Univ Western Ontario, BA, 50, MS, 52; Univ Mich, PhD(pharmacol), 57. *Prof Exp:* Instr & assoc prof pharmacol, Med Ctr, Univ Mich, 57-65; sr pharmacologist, Southern Res Inst, 65-72; assoc prof pharmacol, Sch Med, Univ Calif, Davis, 72-75; res scientist, NY State Div Subst Abuse Serv, Brooklyn, 76-; COUNR, CHEM DEPENDENCY CTR, MISS BAPTIST MED CTR, JACKSON. *Mem:* Am Soc Pharmacol. *Res:* Physical dependence and tolerance development to addicting drugs; general pharmacology of all centrally-acting drugs. *Mailing Add:* Chem Dependency Ctr Miss Baptist Med Ctr 1225 N State St Jackson MS 39202

DENEKAS, MILTON OLIVER, b Dempster, SDak, Apr 20, 18; m 49; c 3. CHEMISTRY. *Educ:* Hope Col, BA, 40; Univ Mich, PhD(chem), 47. *Prof Exp:* Jr res chemist, Upjohn Co, Mich, 45-47; from instr to asst prof chem, Univ Tulsa, 47-57; res chemist, Esso Prod Res Co, 57-74, sr res chemist, 74-79, res specialist, 79-85; RETIRED. *Mem:* Am Chem Soc. *Res:* Isoaromatization studies; drugs for amebiasis; amino acids and protein hydrolysates; synthesis of omega-omega diaminoalkanes; petroleum reservoir wettability studies; geochemistry; tertiary petroleum recovery; secondary petroleum recovery; oil well drilling fluids. *Mailing Add:* 12431 Kimberley Lane Houston TX 77024

DENELL, ROBIN ERNEST, b Peoria, Ill, Aug 6, 42; m 66; c 2. GENETICS. *Educ:* Univ Calif, Riverside, BA, 65; Univ Tex, Austin, MA, 68, PhD(genetics), 69. *Prof Exp:* NIH trainee, Univ Tex, Austin, 69-70; NIH fel genetics, Univ Calif, San Diego, 70-72; Ford Found fel genetics, Univ Edinburgh, 72-73; asst prof biol, 73-76, assoc prof, 77-83, PROF BIOL, KANS STATE UNIV, 84- *Mem:* Fel AAAS; Soc Develop Biol; Genetics Soc Am. *Res:* Insect developmental genetics. *Mailing Add:* Div Biol Ackert Hall Kans State Univ Manhattan KS 66506

DENELSKY, GARLAND, b Des Moines, Iowa, Nov 13, 38; m 62; c 3. SMOKING CESSATION, COPING SKILLS. *Educ:* Grinnell Col, BA, 60; Purdue Univ, MS, 62, Purdue Univ, PhD(psychol), 66. *Prof Exp:* Teaching assoc, psychol, Purdue Univ, 62-66; assessment psychologist, Cent Intel Agency, 66-67, res psychologist, 67-71; DIR PSYCHOL TRAINING PROG & SMOKING CESSATION PROG & STAFF PSYCHOLOGIST, CLEVELAND CLIN FOUND, 71- *Concurrent Pos:* Intern psychol, Univ Ore Med Sch, 63; psychol consult, Wade Park Vet Admin Hosp, Cleveland, 82- *Mem:* Am Psychol Asn. *Res:* Psychology of smoking and process of smoking cessation; performance anxiety problems especially with performing artists, an approach to psychological diagnosis and treatment based upon a classification of coping skills. *Mailing Add:* Cleveland Clin Found Sect Psychol P57 One Clin Ctr 9500 Euclid Ave Cleveland OH 44195-5001

DENENBERG, HERBERT S, b Nov 20, 29; m 58. PUBLIC HEALTH. *Educ:* Creighton Univ, JD, 55; Johns Hopkins, BS, 58; Harvard Univ, LLM, 58; Univ Pa, PhD, 62. *Hon Degrees:* LLD, Allentown Col, 89. *Prof Exp:* From asst prof to assoc prof, Univ Pa, Harry J Loman prof ins, 62-74; comnr ins, Pa, 71-74; asst prof, Univ Iowa, 89; REPORTER, WCAU-TV, 75- *Concurrent Pos:* Consult, US Dept Labor, Satellite Broadcasters Asn, Dept Transp. *Mem:* Inst Med-Nat Acad Sci. *Res:* Various topics relating to consumer rights and consumer protection; received nineteen Emmy awards. *Mailing Add:* WCAU-TV City Ave & Monument Rd Bala Cynwyd PA 19131

DENENBERG, VICTOR HUGO, b Apr 3, 25; m 50, 75; c 4. BEHAVIOR-PSYCHOBIOLOGY. *Educ:* Bucknell Univ, BA, 49; Purdue Univ, MS, 51, PhD(psychol), 53. *Prof Exp:* Res assoc human res off, George Washington Univ, 52-54; from asst prof to prof psychol, Purdue Univ, 54-69; PROF PYSCHOBIOL, UNIV CONN, 69-, COORDR BIOBEHAV SCI GRAD DEGREE PROG, 85- *Concurrent Pos:* Carnegie fel, Roscoe B Jackson Mem Lab, 55, vis investr, 56-; NIH spec fel, Cambridge Univ, 63-64. *Mem:* AAAS; Am Psychol Asn; Int Soc Develop Psychobiol (pres, 70). *Res:* Early experience and brain laterality; ontogeny of behavior; animal behavior and early experience. *Mailing Add:* Biobehav Sci Grad Degree Prog U-154 Univ Conn Storrs CT 06269-4154

DENES, PETER B, b Budapest, Hungary, Nov 9, 20. COMMUNICATION SCIENCES, ELECTRICAL ENGINEERING. *Educ:* Manchester Univ, England, BSc, 41, MSc, 43; Univ London, PhD(eng), 60. *Prof Exp:* Demonstr elec eng, Manchester Col, 41-44; res engr, Welwyn Elec Labs, 44-46; lectr phonetics, Univ Col, Univ London, 46-61; mem tech staff, AT&T Bell Labs, 61-67, head, Speech & Commun Res Dept, 67-81, head, Comput Enhanced Commun Res Dept, 67-81, AT&T CO, MORRIS PLAINS, NJ. *Concurrent Pos:* Mem, med & sci comt, Nat Inst for Deaf, London, 47-61; mem subcomt hearing aids & audiometers, Brit Standards Inst, 52-61; physicist, audiol unit, Royal Nat Throat, Nose & Ear Hosp, London, Eng, 52-61; vis fel, Columbia Univ, 53; assoc ed, Acoust Soc Am J, 69-75. *Mem:* Fel Acoust Soc Am. *Res:* Speech communication; automatic speech recognition; speech synthesis; hearing and deafness; graphics; digital type settings; man-machine interface. *Mailing Add:* 481 Long Hill Rd Gillette NJ 07933

DE NEUFVILLE, RICHARD LAWRENCE, b New York, NY, May 6, 39; m 84; c 2. AIRPORT SYSTEMS DESIGN, TRANSPORTATION. *Educ:* Mass Inst Technol, SB & SM, 61, PhD(civil eng systs), 65. *Prof Exp:* White House fel, Washington, DC, 65-66; from asst prof to assoc prof, Civil Eng Dept, Mass Inst Technol, 66-74, dir, Civil Eng Systs Lab, 70-74, dir urban data processing, 70-78, geog data technol, 82-90, PROF & CHMN, TECHNOL & POLICY PROG, MASS INST TECHNOL, 75- *Concurrent Pos:* Vis prof, London Grad Sch Bus, 73-74, Univ Calif, Berkeley, 74-77 & Ecole Centrale de Paris,

81-82; Guggenheim fel, 81-82; adj prof, Ecole Nationale des Ponts et Chaussees, Paris, 86-; US-Japan leadership fel, Nat Inst Sci & Technol, Japan, 90-91. *Honors & Awards:* Systs Sci Prize, NATO, 73; Risk & Ins Prize, Risk & Ins Soc, 78; Sizer Award for Educ Contrib, 88. *Mem:* Am Soc Civil Engrs; Opers Res Soc Am; AAAS. *Res:* Systems design for major large-scale constructed facilities, incorporating both the technical aspects and the policy issues such as economics; airport design and management. *Mailing Add:* Technol & Policy Rm E40-251 Mass Inst Technol Cambridge MA 02139

DE NEVERS, NOEL HOWARD, b San Francisco, Calif, May 21, 32; m 55; c 3. CHEMICAL ENGINEERING. *Educ:* Stanford Univ, BS, 54; Univ Mich, MS, 56, PhD(chem eng), 59. *Prof Exp:* Res engr, Calif Res Corp, Standard Oil Co Calif, 58-63; PROF CHEM ENG, UNIV UTAH, 63- *Concurrent Pos:* Officer air progs, Environ Protection Agency, 71-72. *Mem:* Am Inst Chem Engrs; Air Pollution Control Asn. *Res:* General and multifluid flow; thermodynamics and thermodynamic properties; interaction of technology and society; teaching of technology to nontechnologists, air pollution control technology; explosions and fires; accident investigation. *Mailing Add:* Dept Chem Eng Univ Utah Salt Lake City UT 84112

DENFORD, KEITH EUGENE, b London, Eng, Feb 10, 46; Can citizen; m 68; c 2. PHYTOCHEMISTRY, CHEMOSYSTEMATICS. *Educ:* Univ London, BSc, 67, PhD(chemosyst), 70. *Prof Exp:* Fel phytochem, 70-71, from asst prof to prof 71-86, CHMN DEPT BOT, UNIV ALTA, 84-, PROF PHYTOCHEM, 86- *Concurrent Pos:* Forensic consult, Royal Can Mounted Police, 71- *Mem:* Can Bot Asn (pres, 87-88); Phytochem Soc; Linnean Soc London. *Res:* Phytochemical studies of vascular plant populations with respect to their evolutionary relationships with special reference to glaciation and glacial refugia in western North America. *Mailing Add:* Dept Bot Univ Alta Edmonton AB T6G 2M7 Can

DENGO, GABRIEL, b Heredia, Casta Rica, Mar 9, 22; m 50; c 3. PETROLOGY, STRUCTURAL GEOLOGY. *Educ:* Univ Costa Rica, BS, 44; Univ Wyo, BA, 45, MA, 46; Princeton Univ, MA, 48, PhD(geol), 49. *Prof Exp:* Res assoc, Princeton Univ, 50; geologist, Direccion de Geol, Ministerio de Minas e Hidrocarburos, Venezuela, 50-52 & Union Oil Co Calif, 53-62; consult, Orgn Am States, 62-64; dep secy gen, Permanent Secretariat Cent Am Econ Integration Treaty, 64; mem res staff, Cent Am Res Inst Indist Technol, 65-70, assoc dir, 70-75, dir, 75-79; RETIRED. *Concurrent Pos:* Consult, Ministry Energy & Mines, Venezuala, 79-80, Guatemalan Electrification Inst, 81-89. *Mem:* Fel Geol Soc Am; Mineral Soc Am; Asn Petrol Geologists; Ger Geol Asn; Soc Econ Geologists. *Res:* Economic geology; regional geology of Central America and Northern South America; tectonics. *Mailing Add:* Apartado 468 Guatemala City Guatemala

DENHAM, JOSEPH MILTON, b Port Jervis, NY, Jan 21, 30; m 51; c 2. ORGANIC CHEMISTRY. *Educ:* Pa State Univ, BS, 51; Ohio Univ, MS, 56, PhD(chem), 59. *Prof Exp:* Instr math & chem, Orange County Community Col, 51-52, 54; from asst prof to assoc prof chem, 58-69, PROF CHEM, HIRAM COL, 69- *Concurrent Pos:* Vis asst prof, Ohio Univ, 60; NSF res partic, Univ Fla, 63, vis asst prof, 64; res fel, Col Pharm, Ohio State Univ, 71-72; vis prof chem, Case Western Reserve Univ, 78-81. *Mem:* Am Chem Soc; Smithsonian Assocs. *Res:* Synthesis, organophosphorus compounds; heterocyclic; small ring compounds. *Mailing Add:* Dept Chem Hiram Col Hiram OH 44234

DENHARDT, DAVID TILTON, b Sacramento, Calif, Feb 25, 39; m 61; c 3. MOLECULAR BIOLOGY. *Educ:* Swarthmore Col, BA, 60; Calif Inst Technol, PhD(biophys, physics), 65. *Prof Exp:* From instr to asst prof biol, Harvard Univ, 64-70; from assoc prof to prof biochem, McGill Univ, 70-80; dir Cancer Res Lab, Univ Western Ont, 80-88; PROF & CHMN BIOL SCI, RUTGERS UNIV, 88- *Mem:* Am Microbiol Soc; Am Soc Biochem & Molecular Biol; Am Asn Cancer Res; Am Soc Cell Biol; fel Royal Soc Can. *Res:* Mechanisms of invasion and metastasis of mammalian cells; structure and function of tissue inhibition of metalloproteinases (TIMP) and osteopontin (OPN); expression of cell cycle dependent genes; control of eukaryote DNA replication; systems physiology of TIMP and OPN. *Mailing Add:* Dept Biol Sci Nelson Lab Rutgers Univ PO Box 1059 Piscataway NJ 08854

DEN HARTOG, J(ACOB) P(IETER), engineering mechanics; deceased, see previous edition for last biography

DENHOLM, ALEC STUART, b Glasgow, Scotland, Aug 24, 29; US citizen; c 1. ELECTRICAL ENGINEERING. *Educ:* Univ Glasgow, BSc, 50, PhD(elec eng), 56. *Prof Exp:* Assoc res officer elec eng, Nat Res Coun Can, 54-59; sr engr, Ion Physics Corp, 59-63, dir eng & develop, 63-68, vpres, 68-69; vpres, Energy Sci Inc, 70-81, chief res exec, 76-81; MGR RES & DEVELOP, EATON SED, 94- *Mem:* Inst Elec & Electronics Engrs. *Res:* Electrical discharges in vacuum and gases; development of high voltage equipment. *Mailing Add:* Eaton Corp 108 Cherryhill Dr Beverly MA 01915

DENHOLM, ELIZABETH MARIA, CHEMOTAXIS MONOCYTES & NEUTROPHILS, MACROPHAGE ACTIVATION. *Educ:* Univ Akron, MS, 77; Wake Forest Univ, PhD(exp path), 85. *Prof Exp:* Fel, Bowan Gray Sch Med, Wake Forest Univ, 85-86; fel, Dept Path, Univ Mich, 87-89; ASST PROF, DEPT MED, ALBANY MED COL, 89- *Mem:* Am Asn Pathologists; Am Thoracic Soc; AAAS. *Res:* Chemotactic factors in chronic inflammation; monocyte-macrophage activation; pulmonary fibrosis. *Mailing Add:* Div Hemat Albany Med Col 47 New Scotland Ave Box A52 Albany NY 12208

DENIG, WILLIAM FRANCIS, b New York, NY, Apr 17, 53; m 82; c 2. IONOSPHERIC PHYSICS, AURORAL PHYSICS. *Educ:* Siena Col, BS, 75; Utah State Univ, Logan, MS, 79, PhD(physics), 82. *Prof Exp:* Teaching asst, Utah State Univ, 76-79, res asst, 77-82; res scholar, Southeast Ctr Elec Eng Educ, 82-84; RES PHYSICIST, USAF, 84- *Concurrent Pos:* Prin investr auroral photog exp, 83-85. *Mem:* Am Geophys Union; Am Phys Soc; Sigma Xi. *Res:* Physics of upper atmosphere and the interaction between large space structures, for example the space shuttle and the natural space environment. *Mailing Add:* 104 Ash St Townsend MA 01469

DENINE, ELLIOT PAUL, b Marlboro, Mass, Sept 9, 35; m 65; c 3. TOXICOLOGY, PHARMACOLOGY. *Educ:* Memphis State Univ, BS, 59; Boston Univ, PhD(path), 71. *Prof Exp:* Lab technician toxicol, Worcester Found Exp Biol, 61-62; assoc pathologist, Worcester Mem Hosp Res Lab, 62-65; mem staff toxicol lab, Arthur D Little, Inc, 65-71; chief pathologist, Foster D Snell, Inc, 71-73; sr pharmacologist toxicol, Southern Res Inst, 73-81; CHIEF PATH-TOXICOL SECT, NORWICH-EATON PHARMACEUTS, 81- *Mem:* Soc Toxicol; Soc Pharmacol & Environ Pathologists; Am Asn Cancer Res; NY Acad Sci. *Res:* Toxicological-pathological action of chemotherapeutics, industrial chemicals, nutrients and protheses, especially potential hazard of interactions resulting from combined exposure; in vitro mechanism studies. *Mailing Add:* 9442 Bainswood Dr Cincinnati OH 45249

DENIS, KATHLEEN A, b Buffalo, NY, Sept 10, 52; m 75; c 2. B LYMPHOCTE DEVELOPMENT, HUMAN HYBRIDOMA FORMATION. *Educ:* Cornell Univ, BS, 74; Univ Tex Med Br, MA, 77; Univ Pa, PhD(immunol), 81. *Prof Exp:* Fel, Dept Microbiol & Immunol, Univ Calif, Los Angeles, 81-83, sr fel, 83-86, sr assoc, Howard Hughes Med Inst, 86-89; SR RES SCIENTIST, SPECIALTY LABS, INC, 89- *Concurrent Pos:* Asst res immunologist, Dept Microbiol, Univ Calif, Los Angeles, 86-89. *Mem:* AAAS; Am Asn Immunologists; Clin Immunol Soc. *Res:* Use tissue culture and animal models to study B lymphocyte development; Ig gene expression; immunization of human lymphocytes for hybridoma formation; human immunodeficiency; autoimmunity and infectious diseases. *Mailing Add:* Specialty Lab Inc 2211 Michigan Ave Santa Monica CA 90404-3900

DE NISCO, STANLEY GABRIEL, b New York, NY, Sept 24, 18; m 41; c 4. NUTRITION. *Educ:* Fordham Univ, BA, 40; NY Univ, MA, 46, PhD, 61. *Prof Exp:* Chemist charge qual control, Sheffield Farms, Inc, 40-42; chemist res veg oils, Best Foods, Inc, 42-44; asst dir appl res, Stand Brands, Inc, 44-51; admin asst tech serv, Chas Pfizer & Co, Inc, 51-53; VPRES & DIR SCI DEPT, TED BATES & CO, INC, 53- *Mem:* AAAS; Am Chem Soc; Inst Food Technol; NY Acad Sci; NY Acad Med. *Res:* Biochemistry; food chemistry; food technology. *Mailing Add:* PO Box 173 East Sandwich MA 02537

DENISEN, ERVIN LOREN, b Austin, Minn, Nov 10, 19; m 43; c 3. HORTICULTURE. *Educ:* Univ Minn, BS, 41; Iowa State Univ, MS, 47, PhD(hort, plant physiol), 49. *Prof Exp:* Voc agr instr high sch, Minn, 41-42; from instr to assoc prof hort, 46-65, chmn dept, 67-73, PROF HORT, IOWA STATE UNIV, 65- *Concurrent Pos:* Consult, US AID, Uruguay, 63 & Proj Unicorn, Inc; chmn small fruits sect, Nat Clonal Repository, USDA, 75-86. *Mem:* Fel AAAS; Am Pomol Soc; fel Am Soc Hort Sci; Am Soc Plant Physiol; Int Soc Hort Sci. *Res:* Small fruits breeding; physiology of horticulture crops; chemical weed control; air pollution; mechanical harvesting of strawberries. *Mailing Add:* 2137 Friley Rd Ames IA 50011

DENISON, ARTHUR B, b Oakland, Calif, June 17, 36; m 60; c 3. EXPERIMENTAL PHYSICS, MOLECULAR PHYSICS. *Educ:* Univ Calif, Berkeley, AB, 59; Univ Colo, PhD(physics), 63. *Prof Exp:* From asst prof to assoc prof physics, 63-73, PROF PHYSICS, UNIV WYO, 73- *Concurrent Pos:* Vis prof, Max Planck Inst Med Res, 67, 68, & 69. *Mem:* Am Inst Physics. *Res:* Nuclear magnetic resonance, electron paramagnetic resonance and ultrasonic nuclear magnetic resonance in solids; nuclear magnetic resonance of paraffin hydrocarbons in liquid and solid state; excited state molecular physics optically modified mass spectra; muon physics. *Mailing Add:* Dept Physics & Astron Univ Wyo Laramie WY 82071

DENISON, FRANK WILLIS, JR, b Temple, Tex, Jan 5, 21; m 48; c 4. MICROBIOLOGY. *Educ:* Univ Tex, PhD(bact), 52. *Prof Exp:* Sr res microbiologist, Abbott Labs, 52-55, group leader microbiol, 55-63, asst mgr microbiol physiol, 63-64, mgr microbial chem, 64-74, mgr admin & sci servs, 75-76, assoc dir licensing, 76-79, dir licensing, 79; technol develop liaison mgr, Upjohn Co, 80-84, far east licensing dir, 85-90; CONSULT, 90- *Concurrent Pos:* Lectr, Lake Forest Col, 58-61. *Mem:* Am Chem Soc. *Res:* Antibiotics and submerged fermentations with special reference to pilot plant equipment; chemistry of microbial products and intermediary metabolism. *Mailing Add:* 4260 Sunnybrook Dr Kalamazoo MI 49008

DENISON, JACK THOMAS, b Gainesville, Fla, Mar 8, 26; m 78; c 1. POLYMER CHEMISTRY. *Educ:* Univ Calif, Los Angeles, BS, 48, PhD(chem), 51. *Prof Exp:* Res chemist, E I Du Pont de Nemours & Co, Inc, 51-57, from res supvr to sr res supvr, 57-70, sr res assoc, 70-90, RES FEL, TEXTILE FIBERS DEPT, E I DU PONT DE NEMOURS & CO, INC, ORANGE, TEX, 90- *Mem:* Am Chem Soc; Sigma Xi. *Res:* Thermodynamics of electrolytes in nonaqueous media; high polymers; high temperature chemistry; digital computation; process research in low and high pressure polyethylene synthesis; process research in elastomers. *Mailing Add:* 2300 Sunset Orange TX 77630-3210

DENISON, JOHN SCOTT, b Waco, Tex, June 18, 18; m 41; c 2. ELECTRICAL ENGINEERING. *Educ:* NMex State Univ, BSEE, 48; Agr & Mech Col, Tex, MSEE, 49. *Prof Exp:* From instr to assoc prof 49-67, acting head dept elec eng, 66-67, dir elec power inst, 67-76, PROF ELEC ENG, TEX A&M UNIV, 67-; PRES, ELEC POWER ENGRS, INC. *Concurrent Pos:* Consult, 51- *Mem:* Inst Elec & Electronics Engrs. *Res:* Electrical transmission and distribution. *Mailing Add:* 800 Chisholm Trail Mill Creek Apt 8 Salado TX 76571

DENISON, ROBERT HOWLAND, vertebrate paleontology; deceased, see previous edition for last biography

DENISON, RODGER ESPY, b Ft Worth, Tex, Nov 11, 32; m 57; c 2. GEOLOGY. *Educ:* Univ Okla, BS, 54, MS, 58; Univ Tex, PhD, 66. *Prof Exp:* Geologist, Okla Geol Surv, 58-61 & Crustal Studies Lab, Tex, 62-64; geologist, Mobil Oil Corp, 64-68, sr res geologist, Mobil Res & Develop Corp, Field Res Lab, 68-74; CONSULT GEOLOGIST, 74- *Mem:* Am Asn Petrol Geologists; Geol Soc Am. *Res:* Petrology and geochronology of basement rocks in southern midcontinent and their influence on later geologic history. *Mailing Add:* 15141 Kingstree Dallas TX 75248

DENISON, WILLIAM CLARK, b Rochester, NY, June 1, 28; m 48; c 4. BOTANY. *Educ:* Oberlin Col, AB, 50, AM, 52; Cornell Univ, PhD(mycol), 56. *Prof Exp:* Asst prof bot, Swarthmore Col, 55-58; vis asst prof, Univ NC, 58-59; asst prof biol, Swarthmore Col, 59-66; ASSOC PROF BOT & CUR MYCOL HERBARIUM, ORE STATE UNIV, 66- *Mem:* AAAS; Int Lichenological Asn; Mycol Soc Am; Air Pollution Control Asn. *Res:* Mycology, especially Pezizales, lichens, edible fungi; forest ecology. *Mailing Add:* Dept Bot Ore State Univ Corvallis OR 97331

DENK, RONALD H, b Buffalo, NY, Sept 17, 37. ORGANIC CHEMISTRY, LASERS. *Educ:* Canisius Col, BS, 59; Villanova Univ, MS, 61; Va Polytech Inst, PhD(chem), 67. *Prof Exp:* Chemist, Texaco, Inc, 66-67, sr chemist, 67-70; asst prof chem, Genesee Community Col, 70-79; chem safety officer, Roswell Park Mem Inst, 79-81; RES MGR, NUCLEAR SCI & TECHNOL FACIL, STATE UNIV NY, BUFFALO, 82- *Concurrent Pos:* Coordr lasers & holographics, Buffalo Mus Natural Sci, 75-; founder & current pres, Western NY Holographers Asn. *Mem:* Am Chem Soc; Sigma Xi; Am Inst Chemists. *Res:* Synthesis of potential anticarcinogenic agents; mechanism of the thermal decomposition of 1-bromo-2-(1-naphthyl) naphthyl carbinol and related diarylcarbinols; Ziegler-Natta copolymerization of alpha-olefins. *Mailing Add:* 100 Colden Ct Cheektowaga NY 14225

DENKER, MARTIN WILLIAM, b Paterson, NJ, June 20, 43; m 69; c 2. PSYCHIATRY. *Educ:* Johns Hopkins Univ, BA, 65, MD, 68. *Prof Exp:* Asst prof psychiat, 72-75, actg chmn, 73-75, ASSOC PROF PSYCHIAT, COL MED, UNIV SFLA, 75- *Concurrent Pos:* Dir, Lab for Appl Math in Behav Systs, 75-; prog dir, Eval Men Health Care Delivery Systs Proj, 76-82, Col Med, Univ SFla, chief, Div Family & Social Psychiat, Dept Psychiat, 82. *Mem:* AAAS; Soc Gen Systs Res; Soc Psychother Res; Soc Comput Simulation; Int Neural Network Soc. *Res:* Family sociology research-mathematical models and computer simulation; clinical family therapy; neural networks. *Mailing Add:* Dept Psychiat Col Med 12901 Bruce B Downs Blvd Tampa FL 33612

DENKO, CHARLES W, b Cleveland, Ohio, Aug 12, 16; m 50; c 3. BIOCHEMISTRY. *Educ:* Geneva Col, BS, 38; Pa State Col, MS, 39, PhD(physiol chem), 43; Johns Hopkins Univ, MD, 51; Am Bd Nutrit, dipl, 52. *Prof Exp:* Asst physiol chem, Pa State Univ, 40-43; instr biochem, Sch Med, Univ WVa, 43; res chemist, Res Lab, SMA Corp, Ohio, 44-45; prof chem, Geneva Col, 48; intern, Univ Ill, 51-52; resident, Dept Med, Univ Chicago, 52-53, res assoc, 55-56; from instr to asst prof internal med, Med Sch, Univ Mich, 56-59; asst prof med, Ohio State Univ, 59-66, with div rheumatic dis, Univ Hosp, 59-66, res assoc, Res Ctr, 66-68; asst prof clin med, Sch Med, Case Western Reserve Univ, 70-71; dir res, Fairview Gen Hosp, 68-86; ASSOC PROF CLIN MED, CASE WESTERN RESERVE UNIV, 77- *Concurrent Pos:* Fel, Univ Chicago, 53-55; sr investr, Arthritis & Rheumatism Found; vis res fel, Australian Nat Univ, 74-75; mem comt orphan drugs, NIH, 83-90; mem res comt, Can Arthritis & Rheumatism Soc, 85-87, mem med adv comt, Lupus Found, NEO. *Mem:* AAAS; Am Chem Soc; Am Col Rheumatology; Am Soc Clin Pharmacol & Therapeut; NY Acad Sci. *Res:* Synthesis and biochemical effects of organic gold compounds; nutritional biochemistry; metabolism of connective tissue; clinical rheumatology; experimental arthritis pharmacology; clinical pharmacology; prostaglandin pharmacology; endorphine physiology, growth factors and neurohormones in rheumatic disorders. *Mailing Add:* Rheumatol Univ Hosp CWRU Cleveland OH 44106

DENKO, JOHN V, b Ellwood City, Pa, July 22, 23; m 47; c 3. PATHOLOGY. *Educ:* Univ Chicago, BS, 46, MD, 47; Am Bd Path, dipl, 52. *Prof Exp:* From intern to resident path, St Luke's Hosp, Chicago, 47-49; instr, Col Med, Univ Ill, 50-52; clin instr path, Univ Wash, 52-54; PATHOLOGIST & DIR LABS, NORTHWEST TEX HOSP, 54-; ASSOC PROF PATH, SCH MED, TEX TECH UNIV, 73-, ASSOC CLIN PROF, 74- *Concurrent Pos:* Dep chief path, USPHS Hosp, Seattle, 52-54; consult, Vet Admin Hosp, Amarillo, 54-; ed med bull, Panhandle Dist Med Soc, 58-; pres, Amarillo Found Health & Sci Educ, Inc, 67-; counr, Tex Med Asn, 72-; mem bd dirs, Health Systems Agency, Tex Dist One, 76- *Mem:* Fel Am Soc Clin Path; fel Col Am Path; AMA; Soc Nuclear Med. *Res:* Clinical pathology and surgical diagnostic pathology. *Mailing Add:* 2613 S Hughes Amarillo TX 79109

DENLEY, DAVID R, b Ill, Dec 28, 50. SURFACE SCIENCE, X-RAY SPECTROSCOPY. *Educ:* Univ Chicago, BA, 73, Univ Calif, Berkeley, PhD(physics), 79. *Prof Exp:* SR RES PHYSICISTS, SHELL DEVELOP CO, 79- *Mem:* Am Phys Soc; Sigma Xi. *Res:* Interest include x-ray absorbtion microscopy and scanning tunneling microscopy. *Mailing Add:* 5226 Pine Cliff Dr Houston TX 77084

DENLINGER, DAVID LANDIS, b Lancaster, Pa, Nov 20, 45; m 67; c 2. INSECT PHYSIOLOGY. *Educ:* Pa State Univ, BS, 67; Univ Ill, PhD(entom), 71. *Prof Exp:* Neth Min Agr & Fisheries res fel entom, Agr Univ, Wageningen, 71-72; NIH res scientist entom, Int Centre Insect Physiol & Ecol, Nairobi, Kenya, 72-74; res fel biol, Harvard Univ, 74-76; from asst prof to assoc prof, 76-84, PROF ENTOM, OHIO STATE UNIV, 84- *Mem:* Entom Soc Am; Fel AAAS; Asn Trop Biol; Am Soc Zool. *Res:* Environmental and physiological mechanisms regulating insect dormancy and reproduction; tropical biology. *Mailing Add:* Dept Entom Ohio State Univ 1735 Neil Ave Columbus OH 43210-1220

DENLINGER, EDGAR J, b June 17, 39; m; c 2. ELECTRICAL ENGINEERING. *Educ:* Pa State Univ, BS, 61; Univ Pa, MS, 64, PhD(elec eng), 69. *Prof Exp:* Mem staff, Appl Res Dept, RCA, 61-65; res asst, Univ Pa, 65-67; staff mem, Lincoln Labs, Mass Inst Technol, 67-73; mem tech staff, 73-81, sr mem tech staff, 82-83, group head, Signal Conversion Syst Res, TV Res Lab, 83-86, HEAD, MICROWAVE SYSTS RES GROUP, MICROWAVE RES LAB, DAVID SARNOFF RES CTR 87- *Mem:* Fel Inst Elec & Electronics Engrs; Sigma Xi. *Mailing Add:* David Sarnoff Res Ctr Cn 5300 Rm 3-231 Princeton NJ 08543-5300

DENMAN, EUGENE D(ALE), b Farmington, Mo, Mar 15, 28; m 52; c 2. ELECTRICAL ENGINEERING, SYSTEMS ENGINEERING. *Educ:* Washington Univ, St Louis, BS, 51; Vanderbilt Univ, MS, 55; Univ Va, DSc, 63. *Prof Exp:* Engr electronic equip, Magnavox Co, 51-52; engr, Sperry Gyroscope Co, 54-56; sr engr, Midwest Res Inst, 56-60; sr physicist, res labs eng sci, Univ Va, 60-63; asst prof elec eng, Vanderbilt Univ, 63-69; prof elec & systs eng, Univ Houston, 69-88; RETIRED. *Concurrent Pos:* Consult, Lockheed Electronics Co, 74-86. *Mem:* Inst Elec & Electronics Engrs. *Res:* Electromagnetic propagation, microwave tubes; radiological physics; system engineering; numerical methods in engineering; mathematical modeling; control systems; algebra; mathematical analysis. *Mailing Add:* PO Box 622 Farmington MO 63640

DENMAN, HARRY HARROUN, b Riverside, NJ, Jan 7, 25; m 50; c 3. THEORETICAL MECHANICS, NONLINEAR DYNAMICS. *Educ:* Drexel Inst Technol, BSEE, 48; Univ Cincinnati, MS, 50, PhD(theoret physics), 52. *Prof Exp:* Mem res staff, Digital Comput Lab, Mass Inst Technol, 52-54; from asst prof to assoc prof 54-70, PROF PHYSICS, WAYNE STATE UNIV, 71- *Concurrent Pos:* Consult, Avco Corp, 56, Gen Elec Co, 59 & Ford Motor Co, 66-; ed, J Indust Math Soc, 66-69. *Mem:* Am Phys Soc; Sigma Xi. *Res:* Applied mathematics; group theory; nonlinear problems. *Mailing Add:* Dept Physics Wayne State Univ Detroit MI 48202

DENMARK, HAROLD ANDERSON, b Lamont, Fla, July 3, 21; m 47; c 2. ACAROLOGY. *Educ:* Univ Fla, BSA, 52, MS, 53. *Prof Exp:* Interim instr, Univ Fla, 53; entomologist, 54-55, actg chief entomologist, 56-58, CHIEF ENTOM, FLA DEPT AGR & CONSUMER SERV, 58- *Mem:* Entom Soc Am. *Res:* Phytoseiidae; monographs and description of new species of phytoseiid mites in relation to ecological and biological control projects. *Mailing Add:* 10930 NW 12th Pl Gainesville FL 32606

DENN, MORTON M(ACE), b Passaic, NJ, July 7, 39; div; c 3. CHEMICAL ENGINEERING. *Educ:* Princeton Univ, BSE, 61; Univ Minn, PhD(chem eng), 64. *Prof Exp:* Fel chem eng, Univ Del, 64-65, asst prof chem eng & comput sci, 65-68, from assoc prof to prof chem eng, 68-77, Allan P Colburn prof chem eng, 77-81; PROF CHEM ENG, UNIV CALIF, BERKELEY, 81- *Concurrent Pos:* Guggenheim fel, 71-72; Fulbright Lectr & Harry Pierce prof, Technion, Israel, 79-80; vis Chevron energy prof, Calif Inst Technol, 80; prog leader polymers & composites, Ctr Advan Mat, Lawrence Berkeley Lab, 83-; ed, Am Inst Chem Engrs jour, 85-91; vis prof, Univ Melbourne, Australia, 85. *Honors & Awards:* Prof Progress Award, Am Inst Chem Engrs, 77; William N Lacey Lectr, Calif Inst Technol, 79; P C Reilly Lectr, Univ Notre Dame, 80; William H Walker award, Am Inst Chem Engrs, 84; Bicentennial Commemoration Lectr, La State Univ, 84; Bingham Medal, Soc Rheology, 86; M Van Winkle Lectr, Univ Tex, Austin, 87; Stanley Katamem Lectr, City Col NY, 90. *Mem:* Nat Acad Eng; fel Am Inst Chem Engrs; Brit Soc Rheol; Polymer Processing Soc; Soc Rheol. *Res:* Polymer processing; rheology; non-Newtonian fluid mechanics; process simulation. *Mailing Add:* Dept Chem Eng Univ Calif Berkeley CA 94720

DENNARD, ROBERT H, b Terrell, Tex, Sept 5, 32. DIGITAL SEMICONDUCTOR DEVICES. *Educ:* Southern Methodist Univ, BS, 54, MS, 56; Carnegie Inst Technol, PhD(elec eng), 58. *Hon Degrees:* DSc, State Univ NY, 90. *Prof Exp:* Staff engr, Res Div, 58-63, res staff mem, 63-71, mgr field effect transistor devices & circuits, 71-79, FEL, THOMAS J WATSON CTR, IBM CORP, 79- *Honors & Awards:* Cledo Brunetti Award, Inst Elec & Electronics Engrs, 82; Achievement Award, Indust Res Inst, 89; Harvey Prize, Technion, Israel Inst Technol, 90. *Mem:* Nat Acad Eng; Inst Elec & Electronics Engrs. *Res:* Scaling of digital devices and integrated circuits to very small dimensions; invention of the dynamic Random Access Memory cell. *Mailing Add:* Thomas J Watson Ctr IBM Corp PO Box 218 Yorktown Heights NY 10598

DENNEN, DAVID W, b Clarks Summit, Pa, Mar 20, 32; m 54; c 3. BIOCHEMISTRY, MICROBIOLOGY. *Educ:* Mass Inst Technol, BS, 54; Ind Univ, MS, 64, PhD(microbiol), 66. *Prof Exp:* Assoc phys chemist, Res Labs, Eli Lilly & Co, 54-56, phys chemist, 59-64, sr microbiologist, Analyt Develop, Greenfield Labs, 66-67 & Antibiotic Develop Div, 67-69, mgr, 69-71, dir 71-73; mgr dir, Lilly Pharmachemie GmbH, 73-75; dir, Antibiotic Tech Serv, Antibiotic Mfg & Develop Div, 75-79, managing dir, 80-82, exec dir, 82-83, VPRES BIOCHEM DEVELOP & BIOSYNTHETIC OPERS, LILLY RES CTR, ELI LILLY & CO, ENG, 84- *Mem:* Am Chem Soc; Am Soc Microbiol. *Res:* Reaction kinetics; control mechanisms in cellular systems; enzyme regulation during morphogenesis; biosynthesis of antibiotics; industrial scale-up of biotechnology products; the scale-up and manufacture of human insulin derived from recombinant DNA. *Mailing Add:* Biogen Inc 14 Cambridge Ctr Cambridge MA 02142-1401

DENNEN, WILLIAM HENRY, b Gloucester, Mass, Apr 8, 20; m 44; c 3. GEOLOGY, GEOCHEMISTRY. *Educ:* Mass Inst Technol, SB, 42, PhD(geol), 49. *Prof Exp:* From instr to assoc prof geol, Mass Inst Technol, 49-67; chmn dept, Univ Ky, 67-74, actg dean, Grad Sch & coordr res, 70-72, chmn dept, 67-74 & 77-81, prof geol, 67-85; RETIRED. *Concurrent Pos:* Cert Prof Geologist, 84. *Mem:* Geol Soc Am; Soc Appl Spectros; Asn Eng Geologists. *Res:* Applications of spectrography to petrological problems; mineral exploration; engineering geology. *Mailing Add:* 21 High St Nahant MA 01908

DENNER, MELVIN WALTER, b North Washington, Iowa, Aug 27, 33; m 65; c 2. CELL BIOLOGY, DEVELOPMENTAL BIOLOGY. *Educ:* Upper Iowa Univ, BS, 61; Univ Ky, MS, 63; Iowa State Univ, PhD(parasitol), 68. *Prof Exp:* NSF res fel, Univ Ky, 61-63; asst zool, Iowa State Univ, 63-64, res fel, 63-65, NIH fel, 64-66, instr, 66-68; from asst prof to assoc prof biol, 68-74, assoc div chmn sci & math, 77-78, chmn, 78-85, PROF BIOL, IND STATE UNIV, EVANSVILLE, 74-; CHMN DIV SCI & MATH, UNIV SOUTHERN IND, 85- *Concurrent Pos:* Asst, Vanderbilt Univ, 77 & Clemson Univ, 79; chmn sci comt, Evansville Mus, 84-85. *Mem:* Am Micros Soc (treas, 82-); Int Soc Invert Path; Sigma Xi; Am Soc Parasitol; fel AAAS. *Res:* Biological control of grasshopper populations through the use of the parasite Mermis migrescens to reduce reproductive potential. *Mailing Add:* Sch Sci & Eng Univ Southern Ind Evansville IN 47712

DENNERT, GUNTHER, b Hettstedt, Ger, Oct 7, 39; m 68; c 2. IMMUNOLOGY, MOLECULAR BIOLOGY. *Educ:* Univ Koln, Ger, PhD(genetics, biochem & physiol), 67. *Prof Exp:* Asst genetics, Max Planck Inst Biol, Tubingen, Ger, 66-68; fel biochem, Univ Uppsala, Sweden, 68-69; fel immunol, Univ Koln, Ger, 69-70; res assoc immunol, Salk Inst, 70-73, assoc res prof immunol, 78-84; PROF MICROBIOL, UNIV SOUTHERN CALIF MED SCH, 84- *Concurrent Pos:* Fel, Deutsche Forschungsgemeinschaft, 68-69, Jane Coffins Child Mem Fund, 70-72 & Edna McConnel Clark Found, 72-73; Nat Cancer Inst grant, 73- *Mem:* Am Asn Immunologists; Ger Asn Biochemists. *Res:* Cellular immunology; tumor immunology; lymphocytes: functional heterogeneity, interactions, specificity, antigen receptors; immunostimulatory drugs. *Mailing Add:* Dept Microbiol Univ Southern Calif Med Sch 2025 Zonal Ave Los Angeles CA 90007

DENNEY, DONALD BEREND, b Seattle, Wash, Apr 3, 27; m 56. ORGANIC CHEMISTRY. *Educ:* Univ Wash, BS, 49; Univ Calif, PhD(chem), 52. *Prof Exp:* Asst, Univ Calif, 49-51; res chemist, E I du Pont de Nemours & Co, 52-53; fel Hickrill Chem Res Found, 53-54; instr chem, Yale Univ, 54-55; from asst prof to assoc prof, 55-62, PROF CHEM, RUTGERS UNIV, 62- *Concurrent Pos:* Fel, A P Sloan Found, 55-59; mem NIH med panel B, 65-67, 75. *Mem:* Am Chem Soc. *Res:* Organic reaction mechanisms; polymeric materials. *Mailing Add:* Dept Chem Rutgers Univ PO Box 939 Piscataway NJ 08855-0939

DENNEY, DONALD DUANE, b Boone, Iowa, Nov 30, 30; c 3. PSYCHIATRY. *Educ:* Willamette Univ, BA, 53; Univ Ore, MS & MD, 57. *Prof Exp:* Resident psychiat, 59-62, from instr to assoc prof, 63-74, resident internal med, 68-70, PROF PSYCHIAT, SCH MED, UNIV ORE, 74- *Concurrent Pos:* USPHS career teachers' award, 62-64, career develop award, 64-68. *Mem:* Am Psychiat Asn. *Res:* Neurophysiology; psychiatric consultation in internal medicine. *Mailing Add:* 3181 SW Sam Jackson Park Rd Portland OR 97201

DENNEY, JOSEPH M(YERS), b Auburn, Wash, May 25, 27; m 50; c 3. PHYSICS, METALLURGY. *Educ:* Calif Inst Technol, BS, 51, MS, 52, PhD(metall, physics), 54. *Prof Exp:* Res engr solid state physics, Atomics Int, 52-54; res assoc phys metall, Gen Elec Res Lab, 55-57; res fel, Calif Inst Technol, 57-59; dir solid state physics lab, TRW Space Technol Labs, 59-69; PRES, DIGITAL DEVELOP CORP, SAN DIEGO, 69- *Concurrent Pos:* Head radiation effects, Hughes Res Labs, 58-59; consult, Nat Acad Sci, Washington, DC, 70. *Mem:* Assoc fel Am Inst Aeronaut & Astronaut; sr mem Inst Elec & Electronics Engrs; Sigma Xi. *Res:* Radiation effects in solids; alloy theory; physics of semiconductors; space environment; satellite and spacecraft design and development; electronics. *Mailing Add:* 408 Via Almar Palos Verdes Estates CA 90274

DENNEY, RICHARD MAX, b Portland, Ore, Jan 28, 46; c 3. HUMAN GENETICS, IMMUNOCHEMISTRY. *Educ:* Reed Col, Ore, BA, 68; Stanford Univ, Calif, PhD, 73. *Prof Exp:* Fel, Oxford Univ, Eng, 73-75; fel, Yale Univ, 75-77; ASSOC PROF HUMAN GENETICS, DEPT HUMAN BIOL CHEM & GENETICS, UNIV TEX MED BR, GALVESTON, 77- *Mem:* AAAS; Am Soc Cell Biol; Am Soc Neurosci. *Res:* Immunochemistry & molecular biology of monoamine oxidases. *Mailing Add:* Dept Human Biol Chem & Genetics Univ Tex Med Br 408 Gail Borden Galveston TX 77550

DENNING, DOROTHY ELIZABETH ROBLING, b Grand Rapids, Mich, Aug 12, 45; m 74. DATA SECURITY, USER INTERFACES. *Educ:* Univ Mich, AB, 67, AM, 69; Purdue Univ, PhD(comput sci), 75. *Prof Exp:* From asst prof to assoc prof comput sci, Purdue Univ, West Lafayette, 75-83; SR COMPUT SCIENTIST, SRI INT, 83-87; MEM RES STAFF, DIGITAL EQUIP CORP, 87- *Mem:* Asn Comput Mach; Inst Elec & Electronics Engrs; Sigma Xi; Int Asn Cryptologic Res (pres, 83-86). *Res:* Cryptography and related protection mechanisms that control access to and the dissemination of information in computer systems. *Mailing Add:* 30 Bear Gulch Dr Portola Valley CA 94025

DENNING, GEORGE SMITH, JR, b Chicago, Ill, Dec 4, 31; m 55; c 5. BIO-ORGANIC CHEMISTRY. *Educ:* Washington & Lee Univ, BS, 54; Cornell Univ, PhD(org chem), 60. *Prof Exp:* From res assoc to instr biochem, Med Col, Cornell Univ, 60-62; res assoc peptide chem, Norwich Eaton Pharmaceut, 62-71, chief, amino acid chem sect, 71-80, chief, org chem sect, 80-83, assoc dir, anti-infective prod develop, 83-85, assoc dir , skeletomuscular prod develop, 85-86, ASSOC MGR, LICENSING & ACQUISITIONS, NORWICH EATON PHARMACEUT, 86- *Concurrent Pos:* Scholar, Dept Biol Chem, Sch Med, Univ Calif, Los Angeles, 69-70. *Mem:* AAAS; Am Chem Soc; Royal Soc Chem; NY Acad Sci. *Res:* Chemistry of amino acids and peptides; organic synthesis; method development for drug analysis. *Mailing Add:* Norwich Eaton Pharm Inc 17 Eaton Ave Eaton NY 13815

DENNING, PETER JAMES, b New York, NY, Jan 6, 42; m 64; c 2. ELECTRICAL ENGINEERING, COMPUTER SCIENCE. *Educ:* Manhattan Col, BEE, 64; Mass Inst Technol, MS, 65, PhD(elec eng), 68. *Hon Degrees:* LLD, Concordia Univ, 84; ScD, Manhattan Col, 85. *Prof Exp:* Asst prof elec eng, Princeton Univ, 68-72; assoc prof comput sci, Purdue Univ, 72-75, prof comput sci, 75-, dept head, 79-; dir, 83-90, RES FEL, RES INST ADVAN COMPUTER SCI, 90- *Concurrent Pos:* Ed, Elsevier Ser on Operating & Prog Systs, 73-80; assoc ed, ACTA Informatica, 73-; lectr, Technol Transfer, Inc, 76-85; ed-in-chief, Comput Survs, 77-79. *Honors & Awards:* Distinguished Serv Award, Asn Computer Mach, 89 & Computer Res Asn, 89. *Mem:* NY Acad Sci; fel Inst Elec & Electronic Engrs; Asn Computer Mach (vpres, 78-80, pres, 80-82); fel AAAS. *Res:* Computer system organization; analysis of queueing phenomena; parallel computation; data security and reliabiltiy. *Mailing Add:* MS 230-5 NASA Ames Res Ctr Moffett Field CA 94035

DENNING, RICHARD SMITH, b Plainfield, NJ, Apr 20, 40; m 63; c 2. NUCLEAR ENGINEERING. *Educ:* Cornell Univ, BS, 63; Univ Fla, MSE, 65, PhD(nuclear eng), 67. *Prof Exp:* Researcher nuclear eng, 67-75, assoc sect mgr, 75-77, res leader, 77-81, SR RES LEADER, NUCLEAR ENG, COLUMBUS LABS, BATTELLE MEM INST, 81- *Concurrent Pos:* Comt on the Safety of DOE Reactors, NAS, 86-88; adv comt, Nuclear Fac Safety, DOE, 88- *Mem:* AAAS; Am Nuclear Soc. *Res:* Reactor safety and risk research including core meltdown behavior, reactor kinetics, transient thermal-hydraulics; criticality and shielding analysis. *Mailing Add:* Battelle Mem Inst 505 King Ave Columbus OH 43201-2693

DENNIS, CLARENCE, b St Paul, Minn, June 16, 09; m 39, 77; c 4. SURGERY, PHYSIOLOGY. *Educ:* Harvard Univ, BS, 31; Johns Hopkins Univ, MD, 35; Univ Minn, MS, 38, PhD(surg, physiol), 40. *Prof Exp:* Instr physiol, Univ Minn, 38-39, from instr to prof surg, 40-47; prof surg, State Univ NY Downstate Med Ctr, 51-72, chmn dept, 52-72, spec asst technol, Off Dir, Nat Heart & Lung Inst, 72-74; PROF SURG, STATE UNIV NY, STONY BROOK, 74- *Concurrent Pos:* Dir surg, Univ Div, Kings County Hosp, 51-59, surgeon-in-chief, 59-72, Univ Hosp, 67-72; mem, Surg Study Sect, NIH, 62-66; mem bd dirs, Nat Soc Med Res, 67-, pres, 77-; pres US chap, Int Soc Surg, 74-77; clin prof surg, Georgetown Univ Sch Med, 72-75, prof lectr, 75-; assoc chief staff res & develop, Vet Hosp, Northport, NY, 75- *Mem:* Soc Univ Surgeons (secy, 50-52); Am Surg Asn (vpres, 71-72); Soc Clin Surg; Soc Vasc Surg (pres, 65-66); Am Soc Artificial Internal Organs (past pres). *Res:* Mechanical support in acute heart failure; artificial heart; gastrointestinal physiology. *Mailing Add:* Dept Surg State Univ NY Health Sci Ctr T-19 Stony Brook NY 11794

DENNIS, DAVID THOMAS, b Preston, Eng, Nov 2, 36; m 60; c 2. PLANT BIOCHEMISTRY, PLANT MOLECULAR BIOLOGY. *Educ:* Univ Leeds, BSc, 59, PhD(biophys), 62. *Prof Exp:* Fel, Biosci Div, Nat Res Coun Can, 62-63; fel chem, Univ Calif, Los Angeles, 63-65; scientist, Unilever Res Inst, Colworth House, Eng, 65-68; assoc prof, 68-76, PROF PLANT PHYSIOL, QUEEN'S UNIV, ONT, 76-, PROF & HEAD BIOL, 84- *Mem:* Can Soc Plant Molecular Biol; Can Biochem Soc; Can Soc Plant Physiol; Am Soc Plant Physiologists; Am Soc Biol Chemists. *Res:* Regulation of metabolism in plants; extraction, purification and kinetics of plant enzymes; isoenzymes and cell compartments in plants; plant molecular biology. *Mailing Add:* Dept Biol Queen's Univ Kingston ON K7L 3N6 Can

DENNIS, DON, b Baltimore, Md, Feb 22, 30; m 49; c 6. BIOCHEMISTRY. *Educ:* Univ Md, BS, 52; Brandeis Univ, PhD(biochem), 59. *Prof Exp:* Biologist pharmacol, Nat Cancer Inst, 52-55; NIH fel & biochemist org chem, Harvard Univ, 59-61; asst prof, 61-70, ASSOC PROF CHEM, UNIV DEL, 70-, PROF CHEM & BIOCHEM, 84- *Mem:* Fedn Am Socs Exp Biol. *Res:* Enzymology; mechanism of action of enzymes. *Mailing Add:* Dept Chem & Biochem Univ Del Newark DE 19716

DENNIS, EDWARD A, b Chicago, Ill, Aug 10, 41; m 69; c 3. BIOCHEMISTRY, PHYSICAL ORGANIC CHEMISTRY. *Educ:* Yale Univ, BA, 63; Harvard Univ, MA, 65, PhD(chem), 68. *Prof Exp:* NIH res fel, Harvard Med Sch, 67-69; from asst prof to prof chem, Univ Calif, San Diego, 70-84, vchmn, 84-87. *Concurrent Pos:* Mem NSF adv panel, 81-85; Guggenheim fel, 83-84. *Mem:* Am Soc Biol Chemists; Biophys Soc; NY Acad Sci; Am Chem Soc; Sigma Xi; fel AAAS. *Res:* Mechanism of phospholipases; nuclear magnetic resonance studies of enzymes, phospholipids and membrane structure; enzymes of phospholipid biosynthesis; mechanism of organophosphorus reactions and phosphate ester hydrolysis, pseudo-rotation, prostaglandins, and macrophages. *Mailing Add:* Dept Chem (0601) Univ Calif at San Diego La Jolla CA 92093-0601

DENNIS, EMERY WESTERVELT, bacteriology, parasitology; deceased, see previous edition for last biography

DENNIS, EMMET ADOLPHUS, b Bassa, Liberia, June 3, 39; m 63; c 2. PARASITOLOGY. *Educ:* Cuttington Col, Liberia, BS, 61; Ind Univ, Bloomington, MA, 65; Univ Conn, PhD(parasitol), 67. *Prof Exp:* Chmn div sci & math, Cuttington Col, Liberia, 67-69; asst prof, 69-74, ASSOC PROF ZOOL, RUTGERS UNIV, NEW BRUNSWICK, 74- *Concurrent Pos:* Assoc investr, NIH Training Grant, 70-74; dir, Liberian Inst Biomed Res, Liberia, 75- *Mem:* Am Soc Parasitologists; Sigma Xi. *Res:* Pathogenesis in host-trematode interactions with particular reference to nutritional balance or imbalance and cellular resistance. *Mailing Add:* Dept Biol Rutgers Univ New Brunswick NJ 08903

DENNIS, FRANK GEORGE, JR, b Lyons, NY, Apr 12, 32; m 54. POMOLOGY, PLANT PHYSIOLOGY. *Educ:* Cornell Univ, BS, 55, PhD(pomol), 61. *Prof Exp:* NSF fel, 61-62; from asst prof to assoc prof pomol, NY State Agr Exp Sta, Cornell Univ, 62-68; assoc prof, Dept Hort, 68-72, PROF, DEPT HORT, MICH STATE UNIV, 72- *Concurrent Pos:* Vis prof, Univ Bristol, 74-75, Nat Ctr Sci Res, Meudon, France, 82 & Jamaica Col Agr, 83; Fulbright fel, Inst Agron et Veterinaire, Hassan II, 90; vis prof, Univ Minn, 90. *Mem:* Fel Am Soc Hort Sci; Sigma Xi; Am Soc Plant Physiol; Int Soc Hort Sci. *Res:* Fruit set and development; dormancy; flowering; plant growth substances. *Mailing Add:* Dept Hort Mich State Univ East Lansing MI 48824-1325

DENNIS, JACK BONNELL, b Elizabeth, NJ, Oct 13, 31; m 56; c 3. COMPUTER SCIENCE. *Educ:* Mass Inst Technol, SB & SM, 54, ScD(elec eng), 58. *Prof Exp:* Asst, 54-58, from instr to prof, 58-87, EMER PROF ELEC ENG, MASS INST TECHNOL, 87-; PRES, DATAFLOW COMPUTER CORP, 88- *Honors & Awards:* Eckert-Manokly Award, Asn Comput Mach/Inst Elec & Electronics Engrs, 84. *Mem:* Fel Inst Elec & Electronics Engrs; Asn Comput Mach. *Res:* Hardware, software, performance, programmability and reliability issues for general purpose computer systems. *Mailing Add:* Lab Comput Sci Mass Inst Technol Cambridge MA 02139

DENNIS, JOE, b Sherman, Tex, Dec 5, 11; m 35; c 3. BIOCHEMISTRY. *Educ:* Austin Col, AB, 33; Univ Tex, AM, 37, PhD(biol chem), 42. *Hon Degrees:* DSc, Austin Col, 65. *Prof Exp:* Tutor biol chem, Sch Med, Univ Tex, 34-36, instr, 36-38; from instr to assoc prof chem, 38-47, head dept, 50-69, prof, 47-76, EMER PROF CHEM, TEX TECH UNIV, 76- *Mem:* AAAS; Am Chem Soc. *Res:* Protein denaturation; blood potassium and calcium. *Mailing Add:* 2718 29th St Lubbock TX 79410

DENNIS, JOHN EMORY, JR, b Coral Gables, Fla, Sept 24, 39; m 60; c 1. MATHEMATICS. *Educ:* Univ Miami, BS, 62, MS, 64; Univ Utah, PhD(math), 66. *Prof Exp:* Asst prof math, Univ Utah, 66-68; from asst prof to prof comput sci, Col Eng, Cornell Univ, 68-79; NOAH HARDING PROF MATH SCI, RICE UNIV, 79- *Mem:* Am Math Soc; Soc Indust & Appl Math; Asn Comput Mach; Math Prog Soc. *Res:* Numerical analysis; optimization. *Mailing Add:* Dept Math/Sci Rice Univ PO Box 1892 Houston TX 77251

DENNIS, JOHN GORDON, structural geology, tectonics; deceased, see previous edition for last biography

DENNIS, JOHN MURRAY, b Willards, Md, Jan 31, 23; m 47; c 4. DIAGNOSTIC RADIOLOGY. *Educ:* Univ Md, BS, 43, MD, 45. *Prof Exp:* Fel radiol, Univ Pa, 50-51; assoc radiol, Univ Md, 51-53, prof & chmn, 53-73, dean, Sch Med, 73-90, EMER DEAN, SCH MED, UNIV MD, 90- *Concurrent Pos:* Chmn, Comn Med Discipline, Md, 69-73, Bd Chancellors, Am Col Radiol, 75-76 & Accreditation Coun Continuing Med Educ, 85-86; trustee, Am Bd Radiol, 69-83 & Nat Bd Med Examiners, 76-79. *Honors & Awards:* Gold Medal, Am Col Radiol & Am Roentgen Ray Soc. *Mem:* Am Col Radiol (pres, 76-77); Am Roentgen Ray Soc; Physiol Soc NAm. *Res:* Clinical research in diagnostic radiology. *Mailing Add:* 803 Huntsman Rd Towson MD 21204

DENNIS, KENT SEDDENS, b New Eagle, Pa, June 25, 28. PHYSICAL CHEMISTRY, POLYMER CHEMISTRY. *Educ:* Grove City Col, BS, 50; Western Reserve Univ, MS, 53, PhD(phys chem), 54. *Prof Exp:* Res assoc, Western Reserve Univ, 51-54; phys chemist, Dow Chem Co, 54-62, proj leader, 62-65, sr res chemist, 65-73, res specialist, 73, sr res specialist, 73-81, res assoc, 81-86; RETIRED. *Mem:* Am Chem Soc; Sigma Xi; NY Acad Sci. *Res:* Polymer chemistry; anionic polymerization; polymer characterization; new polymer development; polymer flammability; block polymers. *Mailing Add:* 5800 Highland Dr Midland MI 48640

DENNIS, MARTHA GREENBERG, b New Haven, Conn, Dec 17, 42; m 69; c 3. COMPUTER SCIENCE. *Educ:* Smith Col, BA, 64; Harvard Univ, MS, 68, PhD(appl math), 71. *Prof Exp:* Sr mem tech staff, Comput Sci Corp, San Diego, Calif, 70-72; sr systs analyst, Systs, Sci & Software, La Jolla, 72-74; sr systs analyst, Computervision Corp, San Diego, 74-76; dir, software eng, Linkabit Corp, San Diego, 76-85, asst vpres, eng, 86-87; VPRES SYSTS & SOFTWARE, PACIFIC COMMUNICATIONS SCIENCES INC, SAN DIEGO, 87- *Mem:* Sigma Xi; Asn Comput Mach; Inst Elec & Electronics Engrs. *Res:* Real-time microprocessor applications for communication; communications systems. *Mailing Add:* 2731 Glenwick Pl La Jolla CA 92037

DENNIS, MARY, b Toledo, Ohio, June 11, 26. INORGANIC CHEMISTRY. *Educ:* Madonna Col, BA, 55; Creighton Univ, MS, 59; Univ Notre Dame, PhD(chem), 62. *Prof Exp:* From instr to assoc prof, 62-76, PROF CHEM, MADONNA COL, 76- *Mem:* Am Chem Soc. *Res:* Coordination compounds. *Mailing Add:* 7327 Joy Detroit MI 48204

DENNIS, MELVIN B, JR, b Bellingham, Wash, Dec 4, 37; m 65; c 3. LABORATORY ANIMAL MEDICINE, EXPERIMENTAL SURGERY. *Educ:* Wash State Univ, BS & DVM, 61, Am Col Lab Animal Med, dipl. *Prof Exp:* Res assoc dept med, Univ Wash, 71-77, res asst prof, 77-80, asst prof lab animal med, div animal med, 80-87; vet med officer, Seattle Vet Admin Hosp, 83-89; ASSOC PROF LAB ANIMAL MED, UNIV WASH, 87- *Concurrent Pos:* Resident, Lab Animal Med, Div Animal Med, Univ Wash, 77-79; prin investr, Med Serv Vet Admin. *Mem:* Acad Surg Res (pres, 85-86); Am Asn Lab Animal Sci; Am Vet Med Asn; Soc Biomat; Am Soc Artificial Internal Organs. *Res:* Blood access devices; implantation of indwelling vascular catheters; infections involving implanted vascular catheters; artificial internal organs; hemodialysis; animal models of human disease. *Mailing Add:* Dept Comp Med SB-42 Univ Wash Seattle WA 98195

DENNIS, PATRICIA ANN, CELL BIOLOGY, TRANSPORT. *Educ:* Brigham & Women's Hosp, PhD(biochem), 85. *Prof Exp:* RES FEL, SCH MED, HARVARD UNIV, 85- *Mailing Add:* Organogenesis Inc 83 Rodgers St Cambridge MA 02142

DENNIS, PATRICK P, b Minneapolis, Minn, Nov 19, 42; m 69; c 2. MOLECULAR BIOLOGY, BIOCHEMISTRY. *Educ:* Univ Wis-Eau Claire, BSc, 65; Univ Minn, PhD(genetics), 69. *Prof Exp:* Res assoc, Roswell Park Mem Inst, Buffalo, NY, 70-71, Univ Tex, Dallas, 71-73 & Enzyme Inst, Univ Wis-Madison, 73-75; asst prof microbiol, 75-77, ASST PROF BIOCHEM, UNIV BC, 77- *Concurrent Pos:* Vis prof, Microbiol Inst, Univ Copenhagen, 76 & 78. *Res:* Regulation of ribosome synthesis in Escherichia Coli; ribosome genetics; molecular mechanisms of antibiotic action. *Mailing Add:* Dept Biochem Univ BC Fac Med 2194 Health Sci Mall Vancouver BC V6T 1W5 Can

DENNIS, ROBERT E, b Adrian, Mich, Oct 15, 20; m 48; c 3. AGRONOMY. *Educ:* Mich State Univ, BS, 42, MS, 53, PhD(plant sci), 58. *Prof Exp:* Teacher pub schs, Mich, 46-51; instr agron, Mich State Univ, 51-59; prof agron & exten agronomist, Univ Ariz, 59-83; RETIRED. *Concurrent Pos:* Consult, Saudi Arabia, Mex, Pakistan, Sudan & Niger; consult agronomist, 83- *Mem:* Fel Am Soc Agron; fel Crop Sci Soc. *Res:* Production of agronomic plants in an irrigated desert environment. *Mailing Add:* 836 N Corinth Tucson AZ 85721

DENNIS, SUSANA R K DE, CLINICAL PHARMACOLOGY. *Educ:* Univ Buenos Aires, Arg, MD, 61. *Prof Exp:* CORP MED DIR, HAZLETON LAB AM, INC, 67-; MEM STAFF HOFFMANN-LA ROCHE. *Mailing Add:* Hoffmann-La Roche 201 Lyons Ave Newark NJ 07112

DENNIS, TOM ROSS, b Macon, Ga, Jan 17, 42; m 69. ASTRONOMY. *Educ:* Univ Mich, BA, 63; Princeton Univ, PhD(astrophys), 70. *Prof Exp:* From asst prof to assoc prof, 78-90, PROF ASTRON, MT HOLYOKE COL, 90- *Mem:* Am Astron Soc; AAAS; Sigma Xi. *Res:* Observational optical astronomy; photometry of galaxy clusters. *Mailing Add:* 11 Buttonfield Lane South Hadley MA 01075

DENNIS, WARREN HOWARD, b Louisville, Ky, Aug 15, 25; m 51; c 5. PHYSIOLOGY, BIOPHYSICS. *Educ:* Univ Louisville, BChE, 47, MS, 55, PhD(phys chem, biophys), 59. *Prof Exp:* Lectr med math, Univ Louisville, 53-59, instr community health, 59-61, from asst prof to assoc prof physiol, 61-64; from assoc prof to prof physiol, 64-77, ASSOC PROF PHYSIOL & PREV MED, UNIV WIS-MADISON, 77- *Concurrent Pos:* Estab investr, Am Heart Asn, 61-66. *Mem:* AAAS; Am Physiol Soc; Biophys Soc; Biomet Soc. *Res:* Electro-physiology of ion transporting systems with emphasis on gastric secretion; chemical biomedical engineering applications to mass transfer systems. *Mailing Add:* Dept Physiol Serv Mem Inst Univ Wis-Madison 1300 Univ Ave Madison WI 53706

DENNIS, WILLIAM E, ELECTROCHEMISTRY. *Educ:* London Univ, PhD(Phys chem). *Prof Exp:* Chief metallurgist, GEC Atomic Energy Div; dep gen mgr, Redbourn Steel Works RTB, Scunthorpe, head develop oper & dep chief tech officer, 63-66, prod coordr, 64-66; staff mem, Jones & Laughlin Steel Corp, 66-69, dir res, 68- 72, gen mgr, Qual Control & Tech Servs, 72-80; VPRES MFG & TECHNOL, AM IRON & STEEL INST, 80- *Honors & Awards:* Frontiers of Process Metall Award, Am Inst Mining Engrs, 73; Howe Mem Lectr, 91. *Mem:* Iron & Steel Soc Am Inst Mining, Metall & Petrol Engrs; fel Am Soc Metals. *Res:* High temperature electrochemistry; thermodynamics of liquid iron-carbon-chromium alloys. *Mailing Add:* Am Iron & Steel Inst 1133 15th St NW Suite 300 Washington DC 20005

DENNISON, BRIAN KENNETH, b Louisville, Ky, Aug 14, 49. RADIO ASTRONOMY. *Educ:* Univ Louisville, BS, 70; Cornell Univ, MS, 74, PhD(astron), 76. *Prof Exp:* Res asst, Nat Astron & Ionosphere Ctr, 71-74, Cornell Univ, 74-76; res assoc, 76-77, asst prof, 77-82, ASSOC PROF PHYSICS, VA POLYTECH INST, 82- *Concurrent Pos:* Comt mem, User's Comt, Nat Radio Astron Observ, 81-83; sr Fulbright scholar res, Sweden, 84-85. *Mem:* Am Astron Soc; Int Astron Union. *Res:* Observational study of low-frequency variability in extragalactic radio sources; theoretical and observational studies of the intracluster medium of clusters of galaxies; observational studies of interstellar scattering of radiowaves. *Mailing Add:* Physics Dept Va Polytech Inst & State Univ Blacksburg VA 24061

DENNISON, BYRON LEE, b Clarksburg, WVa, Dec 8, 30; div; c 2. ELECTRICAL ENGINEERING EDUCATION, DIGITAL LOGIC HARDWARE. *Educ:* Univ WVa, BSEE, 53; Va Polytech Inst, MSEE, 62; Worcester Polytech Inst, PhD(elec eng), 67. *Prof Exp:* Sr elec engr, Govt & Indust Div, Philco Corp, 53-58; assoc prof elec eng, Va Polytech Inst, 58-66; prof elec eng, Lowell Technol Inst, 66-74, actg head dept, 67-68, head dept, 68-72; vis prof, Va Polytech Inst & State Univ, 77-79, Southeastern Mass Univ, 79-81; adj prof, 81-85, VIS PROF, UNIV LOWELL, 85- *Concurrent Pos:* Consult, Polysci Corp, 62-63 & Worcester Found Exp Biol, 66-68. *Mem:* Inst Elec & Electronics Engrs; Am Soc Eng Educ; Sigma Xi. *Res:* Application of control system theory and simulation techniques to the study of biological systems, particularly to the study of the pupillary control system. *Mailing Add:* 70 Patten Rd Rte 4 Westford MA 01886

DENNISON, CLIFFORD C, b Riffle, WVa, Mar 26, 22; m 42; c 5. ZOOLOGY. *Educ:* Marshall Col, BA, 52; Marshall Univ, MA, 61; Univ Fla, EdD(biol curriculum & develop), 69. *Prof Exp:* Teacher biol, Lee Col, Tenn, 55-61; assoc prof life & phys sci, Monroe Community Col, 64-65; ASSOC PROF LIFE & PHYS SCI, LEE COL, TENN, 65- *Concurrent Pos:* Head res & develop, Clean Water Soc, Ltd, 74-75; pres, Dennison Technol, Inc. *Mem:* AAAS; Am Inst Biol Sci. *Res:* Comparative evaluation of two approaches to teaching physical science materials to junior high school students; water purification equipment and processes. *Mailing Add:* Dept Natural Sci Lee Col Cleveland TN 37311

DENNISON, DAVID KEE, b Monett, Mo, July 15, 52; m 74. HUMAN ANATOMY. *Educ:* William Jewell Col, BA, 74; Univ Kans, PhD(anat), 78. *Prof Exp:* Assoc investr immunol, Howard Hughes Med Inst, 78-80; DIR, FLOW CYTOMETRY LAB CELL BIOL, COL MED, BAYLOR UNIV, 80- *Mem:* Am Asn Immunologists. *Res:* Regulation of cell mediated immune responses; flow cytometry. *Mailing Add:* Dept Periodont Univ Tex Dent Br 6516 John Freeman Ave Houston TX 77225

DENNISON, DAVID SEVERIN, b Ann Arbor, Mich, Mar 19, 32; m 54; c 2. BIOPHYSICS. *Educ:* Swarthmore Col, BA, 54; Calif Inst Technol, PhD(biophys), 58. *Prof Exp:* Asst, Calif Inst Technol, 54-58; from instr to assoc prof biol, 58-70, PROF BIOL, DARTMOUTH COL, 70- *Concurrent Pos:* NSF sci fac fel, 64; NIH spec res fel, 69. *Mem:* Biophys Soc; Am Soc Photobiology; AAAS. *Res:* Sensory physiology; stimulus-response relationships in Phycomyces sporangiophores. *Mailing Add:* Dept Biol Sci Dartmouth Col Hanover NH 03755

DENNISON, JOHN MANLEY, b Keyser, WVa, Apr 13, 34; m 57, 89; c 1. GEOLOGY. *Educ:* WVa Univ, BS, 54, MS, 55; Univ Wis, PhD(geol), 60. *Prof Exp:* From asst prof to assoc prof geol, Univ Ill, 60-65; assoc prof, Univ Tenn, 65-67; chmn dept, 69-74, PROF GEOL, UNIV NC, CHAPEL HILL, 67- *Concurrent Pos:* Coop geologist, WVa Geol Surv, 60-; distinguished lectr, Am Asn Petrol Geologists, 74. *Mem:* Geol Soc Am; Am Asn Petrol Geologists; Paleont Soc; Am Inst Prof Geologists; Soc Econ Paleontologists & Mineralogists. *Res:* Appalachian structural geology and stratigraphy; palinspastic maps; energy resources. *Mailing Add:* Dept Geol Univ NC Chapel Hill NC 27514

DENNISTON, JOSEPH CHARLES, b Mineola, NY, June 3, 40; m 63; c 1. MEDICAL PHYSIOLOGY. *Educ:* Univ Pa, VMD, 67; Baylor Col Med, PhD(cardiovasc physiol), 74. *Prof Exp:* Asst chief lab animal med, Pine Bluff Arsenal, US Army, 67-68, adv large animal dis, AID, South Vietnam, 68-69, res microbiologist infectious dis, Med Res Inst Infectious Dis, 69-71, res physiologist high altitude res, Res Inst Environ Med, 74-76, res physiologist, Aviation Med, US Army Aeromed Ctr, 72nd Med Detachment, 76-78 & Med Res & Develop Command, 78-81, chem defense staff officer, 81-, STAFF OFFICER, US ARMY MED RES & DEVELOP COMMAND, FT DETRICK, MD. *Concurrent Pos:* Vis assoc prof physiol, Baylor Col Med, 75-; consult to Surgeon Gen, Dept of Army, 75- *Mem:* Am Vet Med Asn; AAAS; Am Asn Vet Anesthesiol; Sigma Xi. *Res:* Incidence and evidence of ischemic heart disease in the military; effects of high altitude on cardiac mechanics and performance; pathophysiology of high altitude pulmonary edema; interaction of carbon monoxide and altitude in aviator performance; medical aspects of chemical defense. *Mailing Add:* US Army Res Inst Environ Med Natick MA 01760

DENNISTON, ROLLIN H, II, b Chicago, Ill, Dec 16, 14; m 41; c 3. ANIMAL BEHAVIOR. *Educ:* Univ Wis, BA, 36, MA, 37; Univ Chicago, PhD(zoophysiol), 41. *Prof Exp:* Instr zool & physiol, Univ Ariz, 41-43; from instr zool, physiol & physics to asst prof zoophysiol, Univ Wyo, 43-48, from assoc prof to prof physiol, 48-70, dir res & develop, 65-70, chief of party, AID-Univ Wyo, Kabul, Afghanistan, 70-73, prof zool & physiol, 73-82, EMER PROF ZOOL & PHYSIOL, UNIV WYO, 82- *Concurrent Pos:* New York Zool Soc grants, 47-52; Ford Found grant & fel psychol, Yale Univ, 52-53; NIMH grant, 53-60, sr res fel neurophysiol, 60-61; NSF-Nat Res Coun conf grant, 60-61; dir dept, NDEA, 60-; res consult, Jackson Lab, 63; BLM res contract, 80-83; adj prof, Col Human Med, Univ Wyo, 83- *Mem:* Am Psychol Asn; Am Soc Zoologists; Am Physiol Soc; Animal Behav Soc; Psychonomic Soc. *Res:* Endocrinology; behavior; reproduction neurophysiology; vertebrate ecological ethology. *Mailing Add:* 701 S 17th St Laramie WY 82070

DENNO, KHALIL I, b Baghdad, Iraq, Dec 25, 33; US citizen; m 61; c 3. ALTERNATIVE ENERGY SOURCES, MAGNETOHYDRODYNAMICS. *Educ:* Univ Baghdad, BSc, 55; Rensselaer Polytech Inst, MEE, 59; Iowa State Univ, PhD(elec eng), 67. *Prof Exp:* Fac mem elec eng, Univ Baghdad, 59-64; asst prof, Univ La, 67-68; assoc prof, 69-74, PROF, 74-, DISTINGUISHED PROF ELEC ENG, NJ INST TECHNOL, 87- *Concurrent Pos:* Researcher, NJ Inst Technol, 69- *Honors & Awards:* Res Excellence Award, Inst Elec & Electronics Engrs, 82. *Mem:* Fel Inst Elec Engrs; sr mem Inst Elec & Electronics Engrs; Am Soc Eng Educ; Sigma Xi; Plasma & Nuclear Sci Soc; Am Nuclear Soc. *Res:* Modes of advanced alternative energy sorces including: magnetohydrodynamics using exhaust plasma of fusion reactors and characterization of ferromagnetic fluids for redox batteries as insulation barriers; author of over one hundred published papers. *Mailing Add:* NJ Inst Technol 323 High St Newark NJ 07012

DENNY, CHARLES STORROW, b Brookline, Mass, Sept 17, 11; c 3. GEOLOGY. *Educ:* Harvard Univ, AB, 34, PhD(geol), 38. *Prof Exp:* Instr geol, Dartmouth Col, 38-42; asst prof, Wesleyan Univ, 42-44; assoc geologist, US Geol Surv, 44-47, geologist, 47-81. *Mem:* Fel Geol Soc Am. *Res:* Glacial geology; geomorphology. *Mailing Add:* 104 Hilltop Pl New London NH 03257

DENNY, CLEVE B, b Dallas, Tex, Oct 12, 25; m 51; c 2. FOOD MICROBIOLOGY. *Educ:* Univ Tex, BA, 49. *Prof Exp:* From jr bacteriologist to asst chief bacteriologist, Nat Food Processors Asn, 49-65, head bact sect, 65-73, asst to dir res serv, 73-75, mgr res servs, 75-78, dir res servs, 78-87, asst vpres res serv, Wash Lab, 88-90; CONSULT FOOD, 91- *Concurrent Pos:* Assoc referee for bact testing of canned foods, Asn Off Anal Chemists, 68-, for microbiol methods sugar & sugar prod, 74-; mem, Am Pub Health Asn Intersoc-Agency Comt Microbiol Exam Foods, 72-84; chmn, Food Processors Inst Curric Comt, 78-90 & Comt Microbiol Food, Nat Res Coun Adv Bd, Mil Personnel Supplies, 82-84; Nat Adv Comt, Microbiol Criteria for Foods, 88-90. *Honors & Awards:* Bill Williams Award, 86. *Mem:* Am Soc Microbiol; Inst Food Technologists; Am Pub Health Asn; Am Soc Mech Eng. *Res:* Antibiotic preservation of food; radiation sterilization of food; food poisoning; canned food spoilage organisms; heat inactivation of bacterial toxins; hydrogen peroxide sterilization of food containers; flexible package integrity. *Mailing Add:* 6230 Valley Rd Bethesda MD 20817

DENNY, FLOYD WOLFE, JR, b Hartsville, SC, Oct 22, 23; m 46; c 3. PEDIATRICS. *Educ:* Wofford Col, BS, 44; Vanderbilt Univ, MD, 46; Am Bd Pediat, dipl. *Hon Degrees:* DSc, Wofford Col, 85. *Prof Exp:* Asst prof pediat, Univ Minn, 52-53; asst prof, Sch Med, Vanderbilt Univ, 53-55; from asst prof to assoc prof prev med & pediat, Western Reserve Univ, 55-60; prof pediat & chmn dept, 60-81, ALUMNI DISTINGUISHED PROF PEDIAT, SCH MED, UNIV NC, CHAPEL HILL, 73- *Concurrent Pos:* Asst to pres, Armed Forces Epidemiol Bd, 55-57, mem comn streptococcal dis, 54-70; dep dir, 59-63; mem comn acute respiratory dis, 60-73, dep dir, 63-67, dir, 67-73. *Honors & Awards:* Lasker Award, 54. *Mem:* Inst Med-Nat Acad Sci; Am Pediat Soc (pres, 80-81); Am Soc Clin Invest; Am Soc Microbiol; Infect Dis Soc Am (pres, 79-80); Soc Pediat Res (pres, 68-69). *Res:* Streptococcal infections; rheumatic fever; viral infections; mycoplasma infections. *Mailing Add:* Wing D CB 7240 Box 3 Univ NC Sch of Med Chapel Hill NC 27599

DENNY, GEORGE HUTCHESON, b Westfield, NJ, May 30, 28; m 67; c 2. ORGANIC CHEMISTRY. *Educ:* Washington & Lee Univ, BS, 50; Johns Hopkins Univ, MA, 51, PhD(chem), 54. *Prof Exp:* Jr instr chem, Johns Hopkins Univ, 51-53; res chemist, Nylon Res Div, E I Du Pont de Nemours & Co, 56-58; USPHS fel chem, Wayne State Univ, 58-59; asst prof, Arlington State Col, 59-60; asst prof, Va Polytech Inst & State Univ, 60-63; assoc prof, Peabody Col, 63-64; res assoc, 64-65, sr chemist, 65-70, RES FEL, MERCK SHARP & DOHME RES LABS, 70- *Mem:* Am Chem Soc. *Res:* Medicinal chemistry. *Mailing Add:* 3621 Cottage Club Lane Naples FL 33942-2705

DENNY, J(OHN) P(ALMER), b Pittsburgh, Pa, Mar 7, 21; m 45; c 3. METALLURGY. *Educ:* Colo Sch Mines, EMet, 42; Univ Utah, MS, 48, PhD, 50. *Prof Exp:* Res engr, Battelle Mem Inst, 45-47; phys metallurgist, Gen Elec Co, NY, 50-59; sect chief, Beryllium Corp, 59-69; mgr beryllium develop, Berylco Div, Kawecki Berylco Industs, Inc, 69-80, mgr prod develop, Cabot Corp, 80-85; RETIRED. *Honors & Awards:* Bradley Stoughton Award Metall, Am Soc Metals, 84. *Mem:* Am Soc Metals; Am Inst Mining, Metall & Petrol Engrs. *Res:* Physical metallurgy; alloy development; precipitation hardening; low melting alloys; high temperature materials. *Mailing Add:* 412 Marshall Dr Reading PA 19607

DENNY, JOHN LEIGHTON, b Birmingham, Ala, Oct 11, 31; m 56. IMAGE SCIENCE, GRAPHICS. *Educ:* Stanford Univ, BA, 53, Univ Calif, Berkeley, PhD(statist), 62. *Prof Exp:* Asst prof, Ind Univ, 62-64; asst prof, Univ Calif, Riverside, 65-66; assoc prof, 67-69, PROF MATH, UNIV ARIZ, 70- *Concurrent Pos:* Actg head statist, Univ Ariz, 86-90; invited lectr, Inst Math Statist, 79-82. *Mem:* Optical Soc Am; Inst Math Statist; Int Statist Inst; Am Statist Asn. *Res:* Statistical cytogenetics; stochastic differential equations; image reconstruction. *Mailing Add:* Dept Statist Univ Ariz Tucson AZ 85721

DENNY, WAYNE BELDING, b Oberlin, Ohio, Feb 4, 14; m 39; c 2. PHYSICS. *Educ:* Oberlin Col, AB, 35; Yale Univ, PhD(ed), 41. *Prof Exp:* Instr high sch, NY, 35-38; ed, Univ Conn, 40-41; instr physics, Emory Univ, 41-43; vis lectr, Oberlin Col, 43-46, asst prof, 46-48; assoc prof physics, Grinnell Col, 48-55, prof, 55-79; RETIRED. *Concurrent Pos:* Fulbright prof, Robert Col, Istanbul, 58-59; Ahmednagar Col, India, 65-66; vis prof physics, Silliman Univ, Philippines, 71-72. *Mem:* Audio Eng Soc. *Res:* Electronics; acoustics. *Mailing Add:* Dept Physics Grinnell Col Grinnell IA 50112

DENNY, WILLIAM F, b Tryon, Okla, Aug 15, 27; m 49; c 2. HEMATOLOGY, INTERNAL MEDICINE. *Educ:* Cent State Col, Okla, BS, 49; Univ Okla, MD, 53. *Prof Exp:* Intern, George Washington Univ Hosp, 53-54; resident med, Sch Med, Univ Okla, 54-56, instr, 56, chief resident, 56-57; asst prof, Sch Med, Univ Ark, 61-67; assoc prof med, 67-80, PROF INTERNAL MED, COL MED, UNIV ARIZ, 80-; CHIEF MED SERVS, VET ADMIN HOSP, 67- *Concurrent Pos:* Clin res investr, Consol Vet Admin Hosp, Ark, 61-64, chief hemat sect, Med Serv, 64-67. *Mem:* AAAS; AMA; Am Fedn Clin Res; Asn Prog Dir Internal Med; fel Am Col Physicians. *Res:* Erythropoietin measurements in anemic and non-anemic individuals; non-immune hemolytic mechanisms; graduate medical education; stress mechanics. *Mailing Add:* 5640 N Placita Arizpe Tucson AZ 85718

DENNY, WILLIAM F, Shreveport, La, Feb 14, 46; m 83; c 1. SCIENTIFIC SOFTWARE, DIFFERNTIAL EQUATIONS. *Educ:* La Tech Univ, BS, 68; Univ Okla, MA, 70, PhD(math), 74, Univ SW La, PhD(computer sci), 88. *Prof Exp:* from asst prof to assoc prof, 75-89, PROF MATH & COMPUTER SCI, MCNEESE STATE UNIV, 89- *Concurrent Pos:* Consult, Certainteed, 78-80, Cities Serv, 80-82. *Mem:* Asn Comput Mach; Soc Indust & Appl Math; Am Math Soc; Math Asn Am; Inst Elec & Electronics Engrs Computer Soc. *Mailing Add:* 602 N Lake Court Dr Lake Charles LA 70605

DENO, DON W, b Wilmette, Ill, Sept 3, 24. POWER FREQUENCY, ELECTRIC & MAGNETIC FIELDS. *Educ:* Cornell Univ, BS, 49; Univ Pa, MS, 68, PhD(elec eng), 74. *Prof Exp:* SR RES ENGR, GEN ELEC CO, 49- *Mem:* Fel Inst Elec & Electronics Engrs. *Mailing Add:* Elec Field Measurments Co Box 326 West Stockbridge MA 01266

DENO, NORMAN C, b Chicago, Ill, Feb 15, 21; m 44; c 3. PHYSICAL ORGANIC CHEMISTRY. *Educ:* Univ Ill, BS, 42; Univ Mich, MS, 46, PhD(chem), 48. *Prof Exp:* Asst chem, Univ Mich, 42-45; res assoc, Ohio State Univ, 48-50; prof chem, 50-81, EMER PROF, PA STATE UNIV, 81- *Res:* Reaction mechanisms. *Mailing Add:* 139 Lenor Dr University Park PA 16802

DENONCOURT, ROBERT FRANCIS, b Manchester, NH, Sept 13, 32; m 55; c 6. AQUATIC ECOLOGY, ICHTHYOLOGY. *Educ:* Springfield Col, BS, 55, MEd, 57; Union Col, NY, MS, 61; Cornell Univ, PhD(vert zool, ichthyol), 69. *Prof Exp:* Teacher sci, Gilboa-Conesville Cent Sch, NY, 57-59; asst prof zool, Ithaca Col, 59-65; teaching & res asst ichthyol, Cornell Univ, 66-68; res ichthyologist aquatic ecol, Ichthyol Assocs, 68-69; PROF BIOL & ZOOL, YORK COL PA, 69- *Concurrent Pos:* Biol consult fishes, Pa Power & Light Co, 71-88; biol consult aquatic fauna, P H Glatfelter Co, 75-88 & USDA, Environ Impact Studies, Soil Conserv Serv, US CEngrs, Dept Interior, 74-80. *Mem:* Am Fisheries Soc; Am Soc Ichthyol & Herpetol; Benthological Soc Am; Nat Geog Soc. *Res:* Taxonomy, distribution and life history of freshwater fishes, particularly West Virginia and Pennsylvania; studies of fishes and macroinvertebrates as affected by natural and man-caused disturbances, flood, warm water, toxic chemicals and sewage; environmental impact studies. *Mailing Add:* York Col Pa Country Club Rd York PA 17405

DENOON, CLARENCE ENGLAND, JR, b Richmond, Va, Feb 25, 15; m 42; c 2. ORGANIC CHEMISTRY. *Educ:* Univ Richmond, BS, 34, MS, 35; Univ Ill, PhD(org chem), 38. *Hon Degrees:* DSc, Univ Richmond, 86. *Prof Exp:* Asst, Univ Ill, 35-37; res chemist, Exp Sta, E I du Pont de Nemours & Co, 38-42; res dir, Landers Corp, 42-45; mgr spec prod dept, Rohm & Haas Co, 45-58, asst mgr chem & plastics div, 58-62, mgr indust chem div, 62-65, mkt mgr chem div, 65-66, vpres, 66-76, dir, 72-76; dir, Sartomer Indust, 76-84; vpres, Tri-ex Corp, 78-83; MANAGING PARTNER, ENGLEHOLD GROUP, 82- *Mem:* Sigma Xi. *Res:* High polymers; corporate management. *Mailing Add:* 18980-0143 Wycombe PA 18980-0143

DE NOTO, THOMAS GERALD, b Rochester, NY, Apr 2, 43. CHEMICAL ENGINEERING, TECHNICAL MANAGEMENT. *Educ:* Univ Rochester, BS, 64, MS, 67, PhD(chem eng), 72. *Prof Exp:* Engr, Res Div, Polaroid Corp, 71-72, chem engr, 72-74, sr engr, 74-80, prin engr, Films Div, 80-87, sr prin eng & prod mgr, 88-89, SR PRIN ENG & TECH MGR, INDUST COATING DIV, POLAROID CORP, 90- *Concurrent Pos:* Vis scholar, Univ Calif, Berkeley, 68-70 & Argonne Nat Lab, 70; NATO fel, 73 & 76; reviewer, Am Inst Chemists Jour, 78- *Mem:* Am Inst Chem Engrs. *Res:* Filtration of polymeric fluids, agitation of fluids and scale-up criteria; thin film coating. *Mailing Add:* 288 Concord Ave Lexington MA 02173

DENOYER, JOHN M, b Kalaw, Burma, May 19, 26; US citizen; m 51; c 4. GEOPHYSICS. *Educ:* Chico State Col, AB, 53; Univ Calif, MA, 55, PhD(geophys), 58. *Prof Exp:* Asst seismol, Univ Calif, 54-57; from instr to assoc prof geol, Univ Mich, 57-65, actg head acoust & seismics lab, 63-65; dep dir nuclear test detection off, Advan Res Projs Agency, Dept Defense, Washington, DC, 65-67; asst dir res, US Geol Surv, Dept Interior, 67-69; dir earth observations prog, NASA, 69-72; dir Earth Observations Syst Prog, US Geol Surv, Dept Interior, 72-79, res geophysicist, 79-91; RETIRED. *Concurrent Pos:* Mem staff, Inst Defense Anal, Washington, DC, 62-63. *Honors & Awards:* Henry Russel Award, 64; William T Pecora Award, NASA & US Dept Interior, 79. *Mem:* AAAS; Geol Soc Am; Seismol Soc Am; Am Geophys Union. *Res:* Wave propagation; signal processing; energy in seismic waves; strain energy in crustal deformation; crustal structure; remote sensing; gravity and magnetic fields. *Mailing Add:* 600 Austin Lane Herndon VA 22070

DENOYER, LINDA KAY, b Hollywood, Calif. ASTROPHYSICS, COHERENCE. *Educ:* Univ Wis, BA; Cornell Univ, PhD(astrophys), 72. *Prof Exp:* Instr astron, Univ Toronto, 69-71 & Cornell Univ, 71-72; res assoc, Univ Ill, 72-75; asst prof physics, Colgate Univ, 75-76; Am Asn Univ Women fel, Cavendish Lab, Cambridge, 76-77; lectr & res assoc, 77-80, SR RES ASSOC & FEL, CORNELL UNIV, 80- *Concurrent Pos:* NSF vis prof women, 82-83. *Mem:* Am Astron Soc. *Res:* Molecular abundances and excitation conditions in shocked clouds; maximum entropy and maximum likelihood restoration techniques; coherence techniques. *Mailing Add:* Spectrum Sq Asn Cornell Univ 755 Snyder Hill Ithaca NY 14850

DENSEN, PAUL M, b New York, NY, Aug 1, 13; m 39; c 2. HEALTH RESEARCH & ADMINISTRATION. *Educ:* Brooklyn Col, BA, 34; Johns Hopkins Univ, DSc(hyg), 39. *Prof Exp:* Instr in charge biostatist, Dept Prev Med & Pub Health, Sch Med, Vanderbilt Univ, 39-41, from asst prof to assoc prof, 41-46; chief med res statist div, US Vet Admin, 47-49; assoc prof biostatist, Grad Sch Pub Health, Univ Pittsburgh, 49-52, prof biomet, 52-54; dir res & statist, Health Ins Plan Greater New York, 54-59; dep comnr health, New York City Health Dept, 59-66, dep health serv adminr, 66-68; dir, Ctr Community Health & Med Care, 68-84, EMER PROF COMMUNITY HEALTH & MED CARE, SCH MED, HARVARD UNIV, 84- *Honors & Awards:* Lasker Group Award, Am Pub Health Asn, 47; Cutter Lectr, Harvard Sch Pub Health, 63; Centennial Scholar, Johns Hopkins Univ, 76; Statist Sect Award, Am Pub Health Asn, 79; Outstanding Civilian Serv Medal, Dept Army, 81 & 88; Lowell J Reed Lectr, Statist Sect, Am Pub Health Asn, 85; Distinguished Career Award, Asn Health Serv Res, 86; Gov Citiation, 87. *Mem:* Nat Acad Sci; fel Am Statist Asn; fel Am Pub Health Asn; Am Epidemiol Soc; AAAS. *Res:* Medical, hospital, public health statistics; health services. *Mailing Add:* PO Box 405 165 Fremont Rd Sandown NH 03873

DENSHAW, JOSEPH MOREAU, b Trenton, NJ, Jan 26, 28; m 53; c 6. ELECTRICAL ENGINEERING. *Educ:* Univ Pa, BS, 52, MS, 63. *Prof Exp:* Engr, Minneapolis Honeywell Regular Co, 52-56; engr, Missile & Space Div, Gen Elec Co, 56-62, mgr, 62-64, develop engr, 64-65, mgr electro-optics, 65-70, prod develop engr, Reentry & Environ Systs Div, 70-74; instrument engr, United Engrs & Constructors Inc, 74-76; prof engr, Honeywell, Inc, 76-80, mgr prod assurance, 77-84; mgr control equip eng, Gen Elec Co, 84-90; RETIRED. *Mem:* Inst Elec & Electronics Engrs (secy, 62-63, treas, 63-64, vchmn, 64-65); Optical Soc Am; Am Inst Physics; Instrument Soc Am. *Res:* Synthesize and analyze information systems, particularly development of new techniques in signal processing and control theory applications; instrumentation engineering; development of microprocessor based control systems for radiation monitoring; coal gasification; methane reforming; naphtha hydrotreating. *Mailing Add:* 907 Jode Rd Audubon PA 19403-1933

DENSLEY, JOHN R, b Swansea, S Wales, UK, June 6, 43; m. ELECTRICAL INSULATION, HIGH VOLTAGE ENGINEERING. *Educ:* Queen Mary Col & Univ London, BS, 64, PhD(elec eng), 67. *Prof Exp:* SR RES OFFICER, NAT RES COUN, 68- *Honors & Awards:* Thomas W Dakin Award, Inst Elec & Electronicsr Engrs, 87. *Mem:* Fel Inst Elec & Electronics Engrs; assoc mem Inst Elec Engrs UK. *Res:* Aging of electrical insulation and long term deterioration of polymeric insulating materials under the action of high voltage. *Mailing Add:* Nat Res Coun Montreal Rd Ottawa ON K1A 0R8 Can

DENSON, DONALD D, b Beverly, Mass, July 11, 45. PHARMACOLOGY, NEUROSCIENCE. *Educ:* Univ Ga, BS, 67, PhD(org chem), 70. *Prof Exp:* Res chemist, Air Force Rocket Propulsion Lab, 70-71 & Air Force Mat Lab, 71-72; res chemist, Stanford Res Inst, 72-78; from asst prof to assoc prof anesthesia, Univ Cincinnati Med Sch, 78-90, dir, 78-90, assoc prof pharmacokinetics, 82-90; ASST PROF ANESTHESIOL & DIR ANESTHESIOL LABS, EMORY UNIV, 90- *Concurrent Pos:* Consult, Div Res Pediat & Neurophysiol, Univ Cincinnati Med Sch, 71-90. *Mem:* Am Soc Anesthesiologists; Am Heart Asn; NY Acad Sci; Am Soc Regional Anesthesia; Asn Univ Anesthetists; fel Am Col Clin Pharmacol. *Res:* Neuropharmacology of anesthetic drugs; anesthetic metabolism; pharmacokinetics of local anesthetics; drug interactions; clinical pharmacokinetics of analgesics. *Mailing Add:* Dept Anesthesiol Emory Clin 1365 Clifton Rd NE Atlanta GA 30322

DENT, ANTHONY L, b Indian Head, Md, Apr 19, 43; c 3. HETEROGENEOUS CATALYSIS, NEW PRODUCT DEVELOPMENT. *Educ:* Morgan State Univ, BS, 66; Johns Hopkins Univ, PhD(phys chem), 70. *Prof Exp:* From asst prof to assoc prof chem eng, 76-78, dir catalyst res, Carnegie Mellon Univ, 78-79; sr chemist catalyst res, PQ Corp, 79-80, res & develop supvr catalyst res, 81-85, mgr, 85-87, PRIN SCIENTIST CATALYST RES, PQ CORP, 87- *Concurrent Pos:* Vis res eng, E I du Pont de Nemours & Co, 72; NES scholar adv comt, Nat Res Coun, 71-78; vchmn, Catalysis Soc, 75-77; tech prog comt, Am Inst Chem Eng, 77-78. *Mem:* Catalysis Soc; Am Chem Soc; Am Inst Chemists. *Res:* Development of catalysts for chemical process including olefin polymerization and other petrochemical applications. *Mailing Add:* PQ Corp 280 Cedar Grove Rd Conshohocken PA 19426-2240

DENT, BRIAN EDWARD, b Binghamton, NY, June 29, 43; m 69. GEOPHYSICS. *Educ:* Rensselaer Polytech Inst, BS, 65; Calif State Univ, Northridge, MS, 67; Stanford Univ, PhD(geophys), 74. *Prof Exp:* Instr physics, State Tech Univ, Chile, 67-69; res geophysicist seismol, Newmont Mining Ltd, 74-75; res geophysicist seismol, Cities Serv Co, 75-84; RES GEOPHYSICIST, DIGICON GEOPHYS CORP, 84- *Mem:* Soc Explor Geophysicists; Europ Asn Explor Geophysicists; Am Geophys Union. *Res:* Seismology applied to petroleum exploration, especially statics and migration; borehole seismology; gravity modeling; tectonophysics, especially impact cratering. *Mailing Add:* Digicon Geophys Corp 3701 Kirby Dr Suite 600 Houston TX 77098

DENT, JAMES (NORMAN), b Martin, Tenn, May 10, 16; div; c 2. ZOOLOGY. *Educ:* Univ Tenn, AB, 38; Johns Hopkins Univ, PhD(zool), 41. *Prof Exp:* Asst anat, Johns Hopkins Univ, 38-41, res assoc, 41-42; asst prof biol, Marquette Univ, 45-46 & Univ Pittsburgh, 46-49; from assoc prof to prof, 49-86, EMER PROF ZOOL, UNIV VA, 86- *Concurrent Pos:* Mem Johns Hopkins Univ exped, Jamaica, BWI, 41; consult, Biol Div, Oak Ridge Nat Lab, 55-70; Guggenheim fel, Univ St Andrews, 59-60; USAID, Philippines, 63; USPHS spec res fel, Harvard Univ, 69-70; Fulbright lectr, Banaras Hindu Univ, 76 & Univ Calcutta, 76; vis fel, Univ Calif, Berkeley, 77. *Mem:* AAAS; Am Soc Zoologists. *Res:* Developmental physiology; comparative endocrinology. *Mailing Add:* 1940 Thomson Rd Charlottesville VA 22903

DENT, PETER BORIS, b Prague, Czech, May 16, 36; Can citizen; m 62; c 3. IMMUNOLOGY, PEDIATRICS. *Educ:* Univ Toronto, MD, 60; FRCP(C), 65. *Prof Exp:* From instr to asst prof pediat, Univ Minn, 64-68; from asst prof to assoc prof, 68-76, PROF PEDIAT, MCMASTER UNIV, 76- *Concurrent Pos:* Consult, Ont Cancer Treat & Res Found, 75- *Mem:* Can Soc Immunol; Am Asn Immunol; Am Asn Cancer Res; Can Soc Clin Oncol; Am Soc Clin Oncol. *Res:* Rheumatology; clinical immunology. *Mailing Add:* Dept Pediat McMaster Univ Med Ctr 1200 Main St W Hamilton ON L8N 3Z5 Can

DENT, SARA JAMISON, b Lockhart, SC, Feb 5, 22. ANESTHESIOLOGY. *Educ:* Univ SC, MD, 45. *Prof Exp:* From asst prof to assoc prof, 55-65, PROF ANESTHESIOL, DUKE UNIV, 65-, STAFF ANESTHESIOLOGIST, UNIV HOSP, 55- *Concurrent Pos:* Attend, Vet Admin Hosp, 55- *Mem:* Am Soc Anesthesiol; AMA. *Res:* General anesthesiology. *Mailing Add:* Anesthesiology Duke Hosp Box 3094 Durham NC 27710

DENT, THOMAS CURTIS, b Canton, Ohio, June 6, 28; m 48; c 2. BOTANY, PLANT TAXONOMY. *Educ:* Univ Akron, BAEd, 62; Univ Okla, MNS, 64, PhD(plant taxon), 69. *Prof Exp:* Teacher, Hoover High Sch, Ohio, 61-68; instr biol, Kent State Univ, 68-69; PROF BIOL, GORDON COL, 69-, CHMN DEPT, 72- *Mem:* Am Inst Biol Sci; Am Soc Plant Taxon; Nat Asn Biol Teachers. *Res:* Relationships of sugar maple in North America. *Mailing Add:* Dept Biol Gordon Col Wenham MA 01984

DENT, WILLIAM HUNTER, JR, b Philadelphia, Pa, Sept 30, 36; m 59; c 3. MATHEMATICS. *Educ:* Maryville Col, BA, 57; Univ Ky, MS, 62; Univ Tenn, PhD(math), 72. *Prof Exp:* Instr math, Univ Ky, 60-64; from instr to asst prof, 64-72, ASSOC PROF MATH, MARYVILLE COL, 72-, CHMN DEPT MATH & PHYSICS, 77- *Mem:* Math Asn Am; Nat Coun Teachers Math. *Res:* Topology. *Mailing Add:* Dept Math Physics & Comput Sci Maryville Col Maryville TN 37801

DENTAI, ANDREW G, b Budapest, Hungary, June 8, 42. EPILAXIAL CRYSTAL GROWTH, PHOTONICS. *Educ:* Univ Veszprem, Hungary, dipl, 66; Rutgers Univ, MS, 73, PhD(ceramic sci), 74. *Prof Exp:* DISTINGUISHED MEM TECH STAFF, AT&T BELL LABS, 69- *Mem:* Sr mem Inst Elec & Electronics Engrs. *Res:* Epilaxial crystal growth of three and four compounds for photonic and electronic devices. *Mailing Add:* Bell Labs 1363 Holmdel-Keyport Rd Holmdel NJ 07733

DENTEL, STEVEN KEITH, b Washington, DC, Nov 4, 51; m 84; c 2. SEPARATION PROCESSES, COLLOID SCIENCE. *Educ:* Brown Univ, BS, 74; Cornell Univ, MS, 80, PhD(environ eng), 83. *Prof Exp:* Student asst, Max Planck Inst fuer Stroemungsforgchung, Goettingen, Ger, 74; mech engr, Petro-Tex Chem, Houston, 75 & Versar, Inc, Springfield, Va, 76; grad res asst civil & environ eng, Cornell Univ, 76-83; asst prof, 83-89, ASSOC PROF ENVIRON ENG, CIVIL ENG DEPT, UNIV DEL, 89- *Concurrent Pos:* Prin investr, NSF & Am Water Works Asn Res Found, 83-; mem, Task Group Drinking Water Additives Prog, NSF, 87-89; chair, Task Group Standard Methods Exam Water & Wastewater, 88-; vis researcher, Nat Ctr Sci Res, Polytech Inst Lorraine, Nancy, France, 90. *Mem:* Am Water Works Asn; Int Asn Water Pollution Res & Control; Water Pollution Control Fedn; Asn Environ Eng Professors; Am Chem Soc; Am Soc Civil Engrs. *Res:* Modeling and optimization of contaminant removal processes in water and wastewater treatment; coagulation, destabilization and dewatering phenomena in particulate suspensions. *Mailing Add:* Dept Civil Eng Univ Del Newark DE 19716

DENTLER, WILLIAM LEE, JR, CELL MOTILITY, MICROTUBULE ASSEMBLY. *Educ:* Univ Minn, PhD(cell biol), 73. *Prof Exp:* ASST PROF CELL BIOL, UNIV KANS, 76-, PROF, 88- *Mailing Add:* Dept Physiol & Cell Biol Univ Kans Lawrence KS 66045

DENTON, ARNOLD EUGENE, b Remington, Ind, Mar 18, 25; m 50; c 3. BIOCHEMISTRY. *Educ:* Purdue Univ, BS, 49; Univ Wis, MS, 50, PhD, 53. *Prof Exp:* Head pet food res div, Res Lab, Swift & Co, 53-55, head biochem res div, 55-58; dir basic res, 58-66, vpres basic res, 66-70, vpres, Tech Admin, 70-78, VPRES RES & TECHNOL, CAMPBELL SOUP CO, 70-, PRES, CAMPBELL INST RES & TECHNOL, 78- *Mem:* Am Chem Soc; Inst Food Technol; Am Inst Nutrit; NY Acad Sci. *Res:* Amino acid and vitamin assays; vitamin stability in foods and feeds; protein digestibility; by-product utilization; commercial applications of enzymes; meat tenderness; chemistry of flavors; nutritional value of foods. *Mailing Add:* Six Walnut Ct Moorestown NJ 08057

DENTON, DENICE DEE, b El Campo, Tex, Aug 27, 59. SOLID STATE SENSORS, MICROELECTRONICS. *Educ:* Mass Inst Technol, BS, 82, MS, 82, PhD(elec eng), 87. *Prof Exp:* Instr physics, Univ Mass-Boston, 83-84; Hertz fel, Mass Inst Technol, 84-87; ASST PROF ELEC ENG, UNIV WIS-MADISON, 87- *Concurrent Pos:* Mem, NSF Adv Comt, Div Elec & Commun Systs, 88-, Educ Comt Inst Elec & Electronics Engrs Components, Hybrid & Mfg Tech Soc, 90-; chair, NSF Comt Visitors Solid State & Microstructures Prog, 90-; consult, AAAS Proj 2061: Sci All Am, 90-; Presidential Young Investr, NSF, 87-92. *Mem:* Inst Elec & Electronics Engrs; Am Phys Soc; Sigma Xi; Mat Res Soc. *Res:* Investigation of the long term reliability implications of the use of polymers in microelectronics, the design of an integrated smart moisture sensor, and the use of micromachining in solid state actuator design. *Mailing Add:* Dept Elec & Computer Eng Univ Wis 1415 Johnson Dr Madison WI 53706

DENTON, JOAN E, EARTH SCIENCES. *Educ:* Univ San Francisco, BS, 76; Univ Nev, MS, 73; Univ Calif, PhD(biol), 79. *Prof Exp:* Grad fel, US Environ Protection Agency, 77-79; res assoc, Lab Exp Oncol, Sch Med, Ind Univ, 79-82; MGR, SUBSTANCE EVAL SECT, STATIONARY SOURCE DIV, AIR RESOURCES BD, 82- *Mem:* Am Physiol Soc; Sigma Xi; Air Pollution Control Asn. *Res:* Earth sciences; numerous publications. *Mailing Add:* Stationary Source Div Air Resources Bd 1102 Q St PO Box 2815 Sacramento CA 95812

DENTON, JOHN JOSEPH, b Newkirk, Okla, Nov 24, 15; m 49; c 3. MEDICINAL CHEMISTRY. *Educ:* Okla Agr & Mech Col, BS, 37; Univ Ill, PhD(org chem), 41. *Prof Exp:* Asst chem, Univ Ill, 37-40, spec asst, Off Sci Res & Develop, 40-41; res chemist, Calco Chem Div, Am Cyanamid Co, 41-45, group leader pharmaceut res, 45-50, sect dir, 50-52, dir, Bound Brook, 52-54, tech dir fine chem div, NY, 54-56, dir org chem res, Lederle Labs, 56-71, dir cardiovasc-renal res, 71-73, dir new prod acquisitions, 73-81; RETIRED. *Mem:* Am Chem Soc; NY Acad Sci; Royal Soc Chem. *Res:* Medicinal chemistry. *Mailing Add:* 565 Upper Blvd Ridgewood NJ 07450

DENTON, M BONNER, b Beaumont, Tex, June 15, 44. ARRAY DETECTORS, SPECTROSCOPY. *Educ:* Lamar State Col Technol, BS & BA, 67; Univ Ill, PhD(chem), 72. *Prof Exp:* From asst prof to assoc prof, 71-80, PROF, DEPT CHEM, UNIV ARIZ, 80- *Concurrent Pos:* Corresp ed, J Automatic Chem, 78-; Alfred P Sloan res fel, 76-80. *Mem:* Am Chem Soc; Soc Appl Spectros. *Res:* New techniques for spectrochemical analysis including trace level spectroscopic techniques; advanced detectors in spectroscopic analysis; instrumentation design; automation and computer control of experimental systems; new design principles for quadruple mass spectrometers and photoionization. *Mailing Add:* 13790 E Lookout Lane Tucson AZ 85749-8812

DENTON, MELINDA FAY, b Horton, Kans, Mar 27, 44; m; c 1. PLANT SYSTEMATICS. *Educ:* Kans State Teachers Col, BS, 65; Univ Mich, AM, 67, PhD(bot), 71. *Prof Exp:* Vis asst prof & actg cur bot, Mich State Univ, 71-72; asst prof, 72-78, assoc prof, 78-83, PROF, UNIV WASH, 83-, CHMN BOT, 87-, CUR, 72- *Concurrent Pos:* Assoc ed, Am Midland Naturalist, 78-81; ed, Syst Bot, 83-85. *Honors & Awards:* George R Cooley Award, Am Soc Plant Taxonomists, 78. *Mem:* Am Soc Plant Taxonomists (secy, 76-79); Int Asn Plant Taxon; Bot Soc Am; Soc Study Evolution; Asn Trop Biol; Sigma Xi. *Res:* Systematic studies of vascular plants; phytogeography; evolutionary biology; floristics of the Pacific Northwest. *Mailing Add:* Dept Bot Univ Wash Seattle WA 98195

DENTON, RICHARD T, b York, Pa, July 13, 32; m 53; c 11. ELECTRICAL ENGINEERING, SOLID STATE PHYSICS. *Educ:* Pa State Univ, BS, 53, MS, 54; Univ Mich, PhD(elec eng), 60. *Prof Exp:* Res asst comput circuits, Pa State Univ, 53-54; mem tech staff, Bell Tel Labs, Inc, 54-56; res assoc solid state physics res inst, Univ Mich, 56-59; mem tech staff, 59-68, head, Shore Technol Dept, 68-76, HEAD, UNDERSEA TECHNOL DESIGN DEPT, BELL TEL LABS, 76- *Mem:* Am Phys Soc; fel Inst Elec & Electronics Engrs. *Res:* Digital signal processing; microwave properties of magnetic materials; microwave ultrasonic devices and study of ultrasonic properties of solids; optical processing; undersea cable transmission system. *Mailing Add:* Denton Vacuum Pin Oak Ave Cherry Hill NJ 08003

DENTON, TOM EUGENE, b Montgomery, Ala, May 29, 37; m 59; c 2. GENETICS. *Educ:* Huntington Col, BA, 59; Univ Ala, MS, 63, PhD(biol), 66. *Prof Exp:* From asst prof to assoc prof, Samford Univ, 65-74, prof biol, 74-88; DEPT HEAD BIOL, AUBURN UNIV, 88- *Mem:* Sigma Xi; AAAS. *Res:* Human genetics and cytogenetics. *Mailing Add:* 6249 Oliver Dr Montgomery AL 36117-3536

DENTON, WILLIAM IRWIN, b Paterson, NJ, July 5, 17; m 41; c 5. CHEMICAL ENGINEERING. *Educ:* Case Western Reserve Univ, BS, 38, MS, 39. *Prof Exp:* Res engr, Am Gas Asn, 39-40; sr chemist, Socony-Mobil Oil Co, 40-53; sect chief org chem res dept, 53-59, dir process develop, 59-70, dir process technol, 71-75, SAFETY & HAZARD MGR, OLIN CORP, 76- *Concurrent Pos:* Vpres, Sprayed Reinforced Plastics Co, 59-60. *Mem:* Am Chem Soc; Am Inst Chem Engrs; Soc Plastics Engrs; Instrument Soc Am. *Res:* New chemical processes; catalytic processes; petrochemicals; urethanes; alkoxylations; isocyanates; hazard identification. *Mailing Add:* 83 Chipman Dr Cheshire CT 06410-3102

DE NUCCIO, DAVID JOSEPH, b Lawrence, Mass, May 29, 35; m 60; c 2. MEDICAL PHYSIOLOGY. *Educ:* Merrimack Col, BA, 59; Boston Col, MS, 61; Univ Tenn, Memphis, PhD(med sci), 69. *Prof Exp:* From instr to assoc asst prof human anat & physiol, 64-77, PROF HUMAN ANAT & PHYSIOL, CENT CONN STATE UNIV, 77-, COORDR GRAD ANESTHESIA PROG & HEALTH SCI PROG, 83- *Concurrent Pos:* Consult health educ; chief health professions adv & coordr, Med Technol Prog, Cent Conn State Univ, 75-87, coordr, Health Prof Progs, 81-87; mem, Nat Comn Nurse Anesthesia Educ. *Mem:* AAAS; Am Soc Zoologists; Asn Adv Health Professions; Sigma Xi. *Res:* Effects of sex steroids upon the incidence and severity of epileptic seizures in rats; assessment of pre-, inter-, and post-operative techniques for anesthetic management for surgical patients. *Mailing Add:* Dept Biol Sci Cent Conn State Univ New Britain CT 06050

DENYES, HELEN ARLISS, environmental biology & evolution; deceased, see previous edition for last biography

DENZEL, GEORGE EUGENE, b Seattle, Wash, Nov 1, 39; m 58; c 3. MATHEMATICS. *Educ:* Univ Wash, BS, 60, MS, 63, PhD(math), 65. *Prof Exp:* Res instr math, Dartmouth Col, 65-67; asst prof statist, Univ Mo-Columbia, 67-70; dir grad prog, 73-77, chair, 79-82, ASSOC PROF MATH, YORK UNIV, 70- *Mem:* Am Math Soc; Inst Math Statist. *Res:* Markov processes and potential theory; Martingale theory. *Mailing Add:* York Univ Dept Math 4700 Keele St Toronto ON M3J 1P5 Can

DEODHAR, SHARAD DINKAR, b Poona, India, Nov 17, 29; US citizen; m 55; c 3. PATHOLOGY, BIOCHEMISTRY. *Educ:* Pa State Univ, MS, 52; Western Reserve Univ, PhD, 56, MD, 60. *Prof Exp:* Res biochemist, Mt Sinai Hosp, Cleveland, Ohio, 56-60; pathologist, 64-69, DIR IMMUNOL, CLEVELAND CLIN FOUND, 69- *Concurrent Pos:* Res fel path, Univ Hosps Cleveland, 60-64; Am Col Cardiol young investr award, 63- *Res:* Immunopathology; experimental hypertension. *Mailing Add:* Dept Immunopath Cleveland Clin Found Cleveland OH 44106

DEONIER, D L, b La Russell, Mo, June 27, 36; m 65; c 2. ENTOMOLOGY, ECOLOGY. *Educ:* Kans State Col Pittsburg, BS, 59; Iowa State Univ, MS, 61, PhD(entom), 66. *Prof Exp:* Med entomologist, US Army, SEATO Med Res Lab, Thailand, 65-66, instr med entom, Brooke Army Med Ctr, Ft Sam Houston, Tex, 66; asst prof zool, 66-71, ASSOC PROF ZOOL, MIAMI UNIV, 71- *Mem:* Entom Soc Am; Soc Syst Zool; Royal Entom Soc London; NAm Benthological Soc (treas, 75-80, pres-elect, 81). *Res:* Systematics of Diptera; taxonomy and ecology of Ephydridae; insect ecology. *Mailing Add:* Dept Zool Miami Univ Oxford OH 45056

DEONIER, RICHARD CHARLES, b Lakeport, Calif, Apr 9, 42; m 74; c 2. MOLECULAR GENETICS, PLASMID MOLECULAR BIOLOGY. *Educ:* Okla State Univ, BS, 64; Univ Wis, PhD(chem), 70. *Prof Exp:* Fel, Calif Inst Technol, 71-73; asst prof chem, 73-77, asst prof, 77-79, ASSOC PROF MOLECULAR BIOL, UNIV SOUTHERN CALIF, 79- *Mem:* Am Chem Soc; AAAS; Am Soc Microbiol. *Res:* Molecular genetics of transposons in prokaryotes; molecular genetics of plasmids and episomes; specialized recombination. *Mailing Add:* 3256 Peppermint St Newbury Park CA 91320

DE PACE, DENNIS MICHAEL, b Monticello, NY, Jan 19, 47. ANATOMY. *Educ:* State Univ NY Buffalo, BA, 68, PhD(anat), 74. *Prof Exp:* ASST PROF ANAT, HAHNEMANN MED COL & HOSP, PHILADELPHIA, 74- *Res:* Autonomic nervous system including morphology, histochemistry and electron microscopy of autonomic neurons; autonomic innervation of the bone marrow. *Mailing Add:* Dept Anat Hahnemann Med Col & Hosp 230 N Broad St Philadelphia PA 19102

DE PAIVA, HENRY ALBERT RAWDON, b Edmonton, Alta, Feb 29, 32; m 64; c 3. PROJECT MANAGEMENT. *Educ:* Univ Alta, BS, 55; Univ Ill, MSc, 60, PhD(civil eng), 61. *Prof Exp:* Res bridge engr, Bridge Dept, Dept Hwys, Alta, 55-57; from asst prof to assoc prof, 61-68, asst dean, 65-67, actg dean, 67-68, head dept, 69-72, actg vpres capital resources, 71-72, vpres serv, 72-85, PROF CIVIL ENG, UNIV CALGARY, 68- *Concurrent Pos:* Nat Res Coun sr res fel, 68-69. *Honors & Awards:* Gznoski Medal, Eng Inst Can, 71; State of Art of Civil Eng Award, Am Soc Civil Engrs, 74. *Mem:* Proj Mgt Inst; Am Soc Civil Engrs; Eng Inst Can; Can Soc Civil Engrs; Asn Prof Engrs Geologists & Geophysicist. *Res:* Reinforced and prestressed concrete structures; project management. *Mailing Add:* Dept Civil Eng Univ Calgary Calgary AB T2N 1N4 Can

DEPALMA, ANTHONY MICHAEL, b Jersey City, NJ, June 12, 48; m 71; c 2. MECHANICAL ENGINEERING, COOLING TOWER PERFORMANCE. *Educ:* Polytech Inst Brooklyn, BS, 70; Stevens Inst Technol, MS, 76. *Prof Exp:* Engr, Curtiss-Wright Corp, 70-72, engr qual control, 72-73, proj engr, 73-77; thermal engr, Res-Cottell Inc, 77-80, mgr thermal eng, 80-86, prod mgr, 86-87; DIR RES & DEVELOP, CUSTODIS-ECODYNE INC, 87- *Mem:* Am Soc Mech Engrs; Cooling Tower Inst. *Res:* The accurate prediction of cooling tower performance. *Mailing Add:* 155 Norway Lane South Plainfield NJ 07080

DEPALMA, JAMES JOHN, b Rochester, NY, Oct 30, 27; m 52; c 4. OPTICS, MATHEMATICS. *Educ:* Univ Rochester, BS, 55, MS, 57. *Prof Exp:* Sr res physicist, Eastman Kodak Co, 52-66, lab head, 66-77, sr lab head, Physics Div, Res Labs, 77-, mem sr staff, 67-; DEPT PHYSICS, COMMUNITY COL FINGER LAKES, CANANDAIGUA, NY. *Concurrent Pos:* Instr, Rochester Inst Technol, 57-67. *Honors & Awards:* Optical Soc Am Award, 62. *Mem:* Optical Soc Am; Soc Photog Sci & Eng. *Res:* Optical and photographic physics, image science, and photometry; psychophysics; optical filters; emulsion optics. *Mailing Add:* Dept Sci & Sci Technol Community Col Finger Lakes Lincoln Hill Rd Canandaigua NY 14424

DEPALMA, PHILIP ANTHONY, b Boston, Mass, Mar 2, 30; m 63; c 1. MEDICAL MICROBIOLOGY. *Educ:* Boston Univ, AB, 60, AM, 62, PhD(microbiol), 66. *Prof Exp:* From instr to asst prof biol, Boston Univ, 65-75; from asst prof to prof, 75-83, ASSOC PROF BIOL, SALEM STATE COL, 83- *Mem:* AAAS; Am Soc Microbiol. *Res:* Biochemistry of morphogenesis in fungi; biochemistry and electron microscopy of the cell wall of Candida albicans; relationship between virulence and morphology in the fungal pathogen, Candida albicans. *Mailing Add:* Dept Biol Salem State Col Salem MA 01970

DEPALMA, RALPH G, b New York, NY, Oct 29, 31; m 55; c 4. SURGERY. *Educ:* Columbia Univ, AB, 53; NY Univ, MD, 56. *Prof Exp:* Intern, Columbia-Presby Hosp, New York, 56-57; asst resident, St Lukes Hosp, 57-58 & 61-62; resident, Univ Hosps Cleveland, 62-64; from instr to assoc prof, Case Western Reserve Univ, 64-71, prof surg, 71-; assoc surgeon, Univ Hosps Cleveland, 70-; CHMN DEPT SURG, GEORGE WASHINGTON UNIV.

Concurrent Pos: Fel Coun Atherosclerosis, Am Heart Asn, 70- *Mem:* Am Col Surg; Soc Univ Surg; Am Heart Asn; Europ Soc Exp Surg. *Res:* Vascular surgery, atherogenesis; lipid metabolism; cellular and subcellular changes in shock; electron microscopy. *Mailing Add:* George Washington Univ Washington DC 20052

DEPAMPHILIS, MELVIN LOUIS, b Pittsburgh, Pa, Apr 15, 43; m; c 2. MOLECULAR BIOLOGY. *Educ:* Univ Pittsburgh, BS, 64; Univ Wis, PhD(biochem), 70. *Prof Exp:* Fel biochem, Univ Wis, 70-71 & Stanford Univ Med, 71-73; from asst prof to prof biol chem, Harvard Med Sch, 73-86; mem, Dept Cell & Develop Biol, 87, HEAD LAB VIRAL & CELLULAR REPLICATION, ROCHE INST, 87- *Concurrent Pos:* Predoctoral fel, NIH, 67-69, postdoctoral fel, 70-71 & 72-73; postdoctoral fel, NSF, 71-72; estab investr, Am Heart Asn, 74-79, res study comt, physiol chem, 84-88; virol study sect, NIH, 81-85; distinguised lectr, Roswell Park Cancer Inst, 90. *Mem:* Am Soc Biol Chemists; AAAS; Sigma Xi; Am Soc Microbiol. *Res:* To understand at the molecular level how mammalian cells replicate their chromosomes and how chromosome replication is related to the control of cell proliferation. *Mailing Add:* Dept Cell & Develop Biol Roche Inst Molecular Biol Roche Res Ctr Nutley NJ 07110

DE PANGHER, JOHN, b Oakland, Calif, Apr 6, 18; m 45; c 3. PHYSICS. *Educ:* Univ Calif, Berkeley, AB, 41, PhD(physics), 53. *Prof Exp:* Physicist, Radiation Lab, Univ Calif, Berkeley, 42-45; physicist, Naval Ord Test Sta, China Lake, Calif, 45-46; physicist, Radiation Lab, Univ Calif, Berkeley, 50-53; sr physicist, Gen Elec Co, 53-62; staff scientist, Lockheed Missiles & Space Co, 62-70; sr health physicist, Stanford Univ, 70-84; RETIRED. *Mem:* Am Phys Soc; Health Physics Soc. *Res:* Neutron dosimetry; radiological and radiation damage physics. *Mailing Add:* 809 Newell Rd Palo Alto CA 94303

DEPAOLA, DOMINICK PHILIP, b Brooklyn, NY, Dec 29, 42; m 69; c 1. DENTAL RESEARCH, NUTRITION. *Educ:* St Francis Col, NY, BS, 64; NY Univ, DDS, 69; Mass Inst Technol, PhD(nutrit biochem), 74. *Prof Exp:* Fel, Mass Inst Technol, 70-74; clin instr oral diag, Sch Dent Med, Tufts Univ, 73-74; assoc prof nutrit & prev dent, Sch Dent, Med Col Va, Va Commonwealth Univ, 74-78; prof pharmacol & oral biol, Sch Dent, Fairleigh Dickinson Univ, 78-81, dir, Div Oral Biol, dir res & grants admin & dir nutrit, asst dean postgrad affairs & res, Oral Health Res Ctr, 78-81; actg chmn, Dept Community Dent, Univ Tex, 82, coordr curric, 83, prof, Health Sci Ctr, Dent Sch, 81-88, dean, 83-88; DEAN, NJ DENT SCH, UNIV MED & DENT, NJ, 88- *Concurrent Pos:* Chmn, Comt Nutrit Dent & Med Educ, Va Coun Health & Med Care, 75-; Dept Health & Human Serv res grants, Nat Inst Dent Res, 75- & 78- & 83-; comm nutrit, dent, med, prog, Am Soc Clin Nutrit. *Mem:* AAAS; Int Asn Dent Res; Teratolog Soc; Nutrit Today Soc; Am Dent Asn; Sigma Xi; fel Am Col Dentists; hon fel Am Acad Oral Med; Am Asn Dent Schs; Am Soc Clin Nutrit; Am Inst Nutrit. *Res:* Biochemical development of the cranio-facial complex; the effects of nutrition on development; the effects of nutrition on the etiology, progression and therapy of oral disease; integration of basic science research and clinical dentistry in dental education; integration of nutrition education with community health programs; ethiological aspects of clefts of the lip and palate; biologic aspects of congenital malformations; interactions of nutrition with host defense mechanisms; clinical and laboratory evaluation of preventive-oriented products of the food and pharmaceutical industry; applied nutrition in the practice of dentistry; nutrition in health and disease. *Mailing Add:* 3302 Gaston Ave Dallas TX 75246

DEPAOLI, ALEXANDER, b Italy, Mar 20, 36; US citizen; m 61; c 2. VETERINARY PATHOLOGY, TOXICOLOGIC PATHOLOGY. *Educ:* Mich State Univ, BS, 59, DVM, 61; Univ Calif, MS, 67; George Washington Univ, PhD(comp path), 74. *Prof Exp:* Resident vet path, US Army Med Unit, 63-65; staff pathologist, Div Vet Path, Armed Forces Inst Path, 67-69; asst chief exp path, Dept Path, SEATO Med Lab, 70-72; asst chief vet path, Armed Forces Inst Path, 72-74, chief, Div Vet Path, 74-76; chief, Path Div, US Army Inst Infectious Dis, 76-82; dir path, Squibb Inst Med Res, 82-89, EXEC DIR, DRUG SAFETY ASSESSMENT, BRISTOL-MYERS SQUIBB PHARMACEUT RES INST, 89- *Concurrent Pos:* Consult comp path, WHO, 75. *Mem:* Am Vet Med Asn; Am Asn Pathologists; Am Col Vet Pathologists; Int Acad Path; Soc Toxicol Pathologists. *Res:* Toxicologic pathology; drug development. *Mailing Add:* Squibb Inst Med Res PO Box 191 New Brunswick NJ 08903

DE PASQUALI, GIOVANNI, b Theresienstadt, Czech, Jan 20, 17; US citizen; m 46. INORGANIC CHEMISTRY, RADIOCHEMISTRY. *Educ:* Univ Vienna, Bachelor Law, 38; Vienna Tech Univ, MS, 43, dipl eng, 47. *Prof Exp:* Asst prof mineral, Tech Inst Appl Mineral & Petrog, Austria, 49-50; res asst, 53-57, res asst prof, 61-73, RES ASSOC PROF PHYSICS, UNIV ILL, URBANA, 73- *Res:* Mossbauer spectroscopy; biomolecules; preparation of inorganic and organic compounds for Mossbauer spectroscopy and biochemistry; one and two dimensional conductors. *Mailing Add:* RR 1 Sidney IL 61877

DEPASS, LINVAL R, b Kingston, Jamaica, Jan 14, 48; US citizen. CARCINOGENESIS, TERATOGENESIS. *Educ:* Georgetown Univ, BS, 68; Univ Miami, MS, 73; Univ Ark, PhD(toxicol), 78; Am Bd Toxicol, cert. *Prof Exp:* Toxicologist, group leader & mgr, Bushy Run Res Ctr, Union Carbide Corp, 78-84; TOXICOLOGIST, SECT LEADER & DEPT HEAD, INST TOXICOL SCI, SYNTEX RES, PALO ALTO, CALIF, 84- *Concurrent Pos:* Adj asst prof, Univ Pittsburgh, 80-84. *Mem:* Teratology Soc; Soc Toxicol; Am Col Toxicol; AAAS. *Res:* Oral and dermal toxicology; acute, subchronic and chronic toxicity tests, reproduction studies and carcinogenesis bioassays. *Mailing Add:* Inst Toxicol Sci Syntex Res 3401 Hillview Ave Palo Alto CA 94304

DEPATIE, DAVID A, b St Albans, Vt, Mar 24, 34; m 58; c 2. LASER PHYSICS, NONLINEAR OPTICS. *Educ:* Univ Vt, BA, 56, MS, 58; Yale Univ, PhD(physics), 64. *Prof Exp:* Instr physics, Univ Vt, 57-58; res asst, Yale Univ, 62-64; asst prof, Amherst Col, 64-66; vis staff mem, Los Alamos Sci Lab, 66-67; asst prof physics, Univ Vt, 67-72; RES PHYSICIST, USAF WEAPONS LAB, 73- *Mem:* Am Phys Soc; Optical Soc Am. *Res:* Experimental studies of nonlinear opitcal properties of matter and applications to phase conjugation; laser spectroscopy; atmospheric propagation; opto-acoustic spectroscopy. *Mailing Add:* 1012 Maria Court NE Albuquerque NM 87123

DE PENA, JOAN FINKLE, b Lincoln, Nebr, Dec 3, 23; div; c 2. BIOLOGICAL ANTHROPOLOGY, PHYSICAL ANTHROPOLOGY. *Educ:* Univ Nebr, BA, 45; Ind Univ, PhD(anthrop), 58. *Prof Exp:* From instr to asst prof anthrop, St Louis Univ, 58-63, asst prof anat & orthod, Med & Dent Schs, 63-66; assoc prof anthrop, Univ Man, 66-88, head, Dept Anthrop, 81- 87; RETIRED. *Concurrent Pos:* Consult & lectr health orgn res prog, St Louis Univ, 58-66 & Peace Corps training progs Latin Am, 60-63; consult, Nat Asn Home Builders, St Louis, Mo & Washington, DC, 63-64; Can Comn Int Biol Prog grant human adaptability, Igloolik Proj, Univ Man, 68-75, coordr comt human develop, 72- *Mem:* AAAS; fel Am Anthrop Asn; NY Acad Sci; Am Asn Phys Anthropologists; fel Royal Anthrop Inst Gt Brit & Ireland. *Res:* Physical growth and skeletal maturation; human adaptability; human evolution; medical anthropology; circumpolar population biology. *Mailing Add:* 787 Waverley St Univ Man Winnipeg MB R3B 2E9 Can

DE PENA, ROSA G, b Bairamcea, Rumania, Sept 1, 21; Arg citizen; m 46; c 2. ATMOSPHERIC CHEMISTRY, CLOUD PHYSICS. *Educ:* Univ Buenos Aires, PhD(chem), 50. *Prof Exp:* Instr phys chem, Univ Buenos Aires, 55-60, from instr to assoc prof meteorol, 60-67; res assoc, 69-73, assoc prof, 73-78, PROF METEOROL, PA STATE UNIV, UNIVERSITY PARK, 78- *Concurrent Pos:* Sci investr, Arg Coun Sci & Technol Res, 61-67; Arg Nat Coun Sci & Technol Res fel, Univ Clermont-Ferrand, 63-64; sabbatical leave, Univ Frankfurt, 78-79. *Mem:* Am Geophys Union; Sigma Xi; AAAS; Am Meteorol Soc; Air Pollution Control Asn. *Res:* Nucleation and growth of particles from gas phase reactions; scavenging of gases and particles by clouds and raindrops; chemistry of precipitation. *Mailing Add:* Dept Meteorol 514 Walker Bldg Pa State Univ University Park PA 16802

DEPEW, CREIGHTON A, b Minneapolis, Minn, Mar 30, 31; m 52; c 5. MECHANICAL ENGINEERING. *Educ:* Univ Calif, Berkeley, BS, 56, MS, 57, PhD(mech eng), 60. *Prof Exp:* From asst prof to assoc prof, 60-72, PROF MECH ENG, UNIV WASH, 72- *Res:* Heat transfer and fluid mechanics. *Mailing Add:* Dept Mech Eng Univ of Wash Seattle WA 98195

DEPEYSTER, FREDERICK A, b Chicago, Ill, Nov 8, 14; m 48; c 2. MEDICINE. *Educ:* Williams Col, BA, 36; Univ Chicago, MD, 40. *Prof Exp:* From asst instr to prof, Col Med, Univ Ill, 46-71; prof surg, 71-85, EMER PROF SURG, RUSH MED COL, 85- *Concurrent Pos:* Asst attend surgeon, Presby Hosp, 48-58, assoc attend surgeon, Presby-St Luke's Hosp, 58-62, attend surgeon, 62-; from assoc attend surgeon to attend surgeon, Cook County Hosp, 48-, clin prof, Cook County Grad Sch Med, 59-69. *Mem:* AMA; Am Col Surgeons; Am Asn Cancer Res; Int Soc Surg; Western Surg Asn. *Res:* Surgery of the gastrointestinal tract; cancer of colon and rectum; behavior of experimental cancer in animals. *Mailing Add:* 696 Prospect Ave Winnetka IL 60093-2320

DEPHILLIPS, HENRY ALFRED, JR, b New York, NY, Apr 16, 37; m 59; c 3. PHYSICAL CHEMISTRY. *Educ:* Fordham Univ, BS, 59; Northwestern Univ, PhD(chem), 65. *Prof Exp:* From asst prof to assoc prof, 63-73, chmn dept, 71-76 & 80-85, PROF CHEM, TRINITY COL, CONN, 73- *Concurrent Pos:* Mem corp, Marine Biol Lab, Woods Hole. *Mem:* AAAS; Am Chem Soc; Soc Appl Spectros. *Res:* Physical biochemistry; scientific examination of works of art; structure-function relationships in respiratory proteins. *Mailing Add:* Dept Chem Trinity Col Hartford CT 06106

DE PHILLIS, JOHN, b New York, NY, Dec 21, 36; m 60; c 3. MATHEMATICS. *Educ:* Stevens Inst Technol, ME, 58; Univ Calif, Berkeley, MA, 62, PhD(math), 65. *Prof Exp:* Asst prof math, San Francisco State Col, 62-65; from asst prof to assoc prof, 65-73, chmn dept, 70-76, PROF MATH, UNIV CALIF, RIVERSIDE, 73- *Concurrent Pos:* Vis assoc mathematician, Brookhaven Nat Lab, 68-69; vis res fel, Math Inst, Florence, Italy, 72-73; vis res mathematician, Univ Karlsruhe, Karlsruhe, WGer, 81-82; vis researcher, IBM Sci Ctr, Bergen, Norway, 86-87. *Mem:* Am Math Soc; Soc Indust & Appl Math; Math Asn Am. *Res:* Functional analysis; operator algebras; convexity; iterative analysis; complexity of computation; acceleration of convergence schemes; multilinear algebra; computer science. *Mailing Add:* Dept Math Univ Calif 900 University Ave Riverside CA 92521

DEPINTO, JOHN A, b Youngstown, Ohio, Jan 4, 37; m 62; c 3. MICROBIOLOGY, BIOCHEMISTRY. *Educ:* Youngstown Univ, BA, 58; Univ Ill, PhD(microbiol), 65. *Prof Exp:* Nat Acad Sci-Agr Res Serv fel microbiol chem, Northern Regional Res Lab, USDA, Ill, 65-66; from asst prof to assoc prof, 66-75, PROF BIOL, BRADLEY UNIV, 75-, ASSOC DEAN, LIB ARTS & SCI, 78- *Mem:* AAAS; Am Soc Microbiol. *Res:* Mechanism of action of microbial amylases. *Mailing Add:* Dept Biol Bradley Univ Peoria IL 61625

DE PLANQUE, GAIL, b Orange, NJ, Jan 15, 45. RADIATION PHYSICS, RADIATION DOSIMETRY. *Educ:* Immaculata Col, AB, 67; Newark Col Eng, MS, 73; NY Univ, PhD, 83. *Prof Exp:* physicist, 67-82, DEP DIR, ENVIRON MEASUREMENTS LAB, US DEPT ENERGY, 82- *Concurrent Pos:* Chmn health physics soc stand working group, Am Nat Stand Inst, 73-75 & 80-, co-chmn, Comt for Int Intercomparison of Environ Dosimeters, 74-, US expert deleg, Int Orgn Stand Comt for Develop of an Int Stand on Thermoluminescence Dosimetry; bd dirs, Am Nuclear Soc, 77-80 & 84-87. *Mem:* Am Phys Soc; Asn Women Sci; fel Am Nuclear Soc; Health Physics Soc; AAAS. *Res:* Physics of radiation, radiation shielding and transport, radiation protection, solid state dosimetry, reactor and personnel monitoring; design and testing of radiation instrumentation and calibration facilities. *Mailing Add:* 13 Bowdoin St Maplewood NJ 07040

DEPOCAS, FLORENT, b Montreal, Que, Jan 1, 23; m 52; c 2. PHYSIOLOGY. *Educ:* Univ Montreal, BSc, 46, PhD(biochem), 51. *Prof Exp:* Biochemist, Sacred Heart Hosp, Montreal, 51-52; asst res officer, Div Biosci, Nat Res Coun Can, 52-58, assoc res officer, 58-61, sr res officer, 61-69, asst dir, Div Biol Sci, 69-79, assoc dir, 79-84; RETIRED. *Concurrent Pos:* Assoc dir assessments, Can Coun Animal Care, 86-90. *Mem:* Am Physiol Soc; Can Physiol Soc; Can Biochem Soc. *Res:* Biochemistry and physiology of acclimation to cold in small mammals; noradrenaline-induced calorigenesis. *Mailing Add:* 606-35 Holland Ave Ottawa ON K1Y 4S2 Can

DEPOE, CHARLES EDWARD, b Southampton, NY, Sept 18, 27; m 52. BOTANY, FRESH WATER ECOLOGY. *Educ:* NC State Col, BS(ornamental hort) & BS(zool), 56, MS, 58, PhD(bot), 61. *Prof Exp:* Tech asst, Long Island Agr & Tech Inst, 53-54; asst prof biol, 61-66, ASSOC PROF BIOL, NORTHEAST LA UNIV, 66- *Concurrent Pos:* Chmn, La Jr Acad Sci, 66- *Mem:* Fel AAAS; Bot Soc Am; Soc Study Evolution; Ecol Soc Am; Am Soc Plant Taxonomists; Sigma Xi. *Res:* Distribution, ecology and productivity of aquatic macrophytes. *Mailing Add:* 100 Curve Dr Monroe LA 71203-4235

DEPOMMIER, PIERRE HENRI MAURICE, b Montcy-St-Pierre, France, Dec 15, 25; m 56; c 3. ELEMENTARY PARTICLE PHYSICS. *Educ:* Univ Lille, Lic es Sci, 46; Univ Paris, Lic es Sci, 48; Univ Grenoble, PhD(physics), 61. *Prof Exp:* Lectr, Fac Sci, Univ Grenoble, 61-65, prof, 65-69; dir, Lab Nuclear Physics, 69-82, PROF FAC SCI, UNIV MONTREAL, 69- *Mem:* Am Phys Soc; Brit Inst Physics & Phys Soc; French Phys Soc; Europ Phys Soc; fel Royal Soc Can. *Res:* Experiments in weak and strong interactions. *Mailing Add:* Lab Nuclear Physics Univ Montreal CP 6128 Sta A Montreal PQ H3C 3J7 Can

DEPOORTER, GERALD LEROY, b Everett, Wash, Mar 4, 40. CERAMICS. *Educ:* Univ Wash, BS, 61; Univ Calif, Berkeley, MS, 63, PhD(eng), 65. *Prof Exp:* Staff mem, Los Alamos Nat Lab, 65-87; asst prof, 78-79, ASSOC PROF METALL & MAT ENG, COLO SCH MINES, 87- *Mem:* Am Chem Soc; Am Ceramic Soc; Sigma Xi. *Res:* Advance ceramics research; computer modeling. *Mailing Add:* 12509 W Idaho Dr Lakewood CO 80228

DEPOY, PHIL EUGENE, b Frankfort, Ind, Sept 30, 35; m 60; c 2. NAVAL ANALYSES. *Educ:* Purdue Univ, BS, 57; Mass Inst Technol, MS, 58; Stanford Univ, PhD(chem Eng), 74. *Prof Exp:* Dir, SE Asia Combat Anal Div, Oper Eval Group, Ctr for Naval Anal, 68-69, dir systs Eval Group, 69-74, vpres & dir, Oper Eval Group, 74-84, sr vpres & dir res, 84-85, PRES, CTR NAVAL ANALYSIS, 85- *Concurrent Pos:* Chmn, US rep, NATO Systs Sci Panel, 76-79; vchmn, US Army Sci Bd, 78-81; mem Air Force Sci Adv Bd, 83-87. *Res:* Support of tactical development and evaluation to assessments of advanced systems and alternative manpower policies. *Mailing Add:* 4220 Elizabeth Ln Alexandria VA 22003

DEPP, JOSEPH GEORGE, b Pittsburgh, Pa, Dec 13, 43; m 68; c 2. RESEARCH & DEVELOPMENT MANAGEMENT. *Educ:* Carnegie Inst Technol, BS, 65, MS, 66; Carnegie-Mellon Univ, PhD(nuclear theory), 70. *Prof Exp:* Physicist radio physics, 71-76, mgr electro optics prog, 76-79, DIR, SPEC SYSTS OFF, SRI INT, 79- *Mem:* Am Phys Soc. *Res:* Apply state-of-the-art technology across the electromagnetic spectrum to the design and development of sensor systems. *Mailing Add:* 4815 Blue Ridge Dr San Jose CA 95129

DEPREE, JOHN DERYCK, b Zeeland, Mich, Dec 5, 33; m 54; c 3. MATHEMATICS. *Educ:* Hope Col, BA, 55; Univ Colo, MS, 58, PhD(math), 62. *Prof Exp:* Inst appl math, Univ Colo, 57-62; asst prof math, Ore State Univ, 62-65; assoc prof, Va Polytech Inst & State Univ, 65-68; PROF MATH, NMEX STATE UNIV, 68-, HEAD DEPT, 75- *Mem:* Am Math Soc; Math Asn Am. *Res:* Theory of analytic functions; entire functions; integral equations. *Mailing Add:* 303 Hermosa Dr Albuquerque NM 87108

DEPRIMA, CHARLES RAYMOND, b Paterson, NJ, July 10, 18; m 43, 51, 87. APPLIED MATHEMATICS. *Educ:* NY Univ, AB, 40, PhD(math), 43. *Prof Exp:* Instr math, Washington Sq Col, NY Univ, 41-43, lectr, Grad Sch, 43-46, res scientist, Appl Math Panel, 42-46; from asst prof to prof appl mech, 46-64, prof math, 64-86, EMER PROF, CALIF INST TECHNOL, 86- *Concurrent Pos:* Vis prof, Univ Calif, Los Angeles, 48-; head math br, Off Naval Res, 51-52; vis prof, NY Univ, 63-64; ed, Pac J Math, 73-86; consult ed, Springer Verlag. *Mem:* Am Math Soc; Math Asn Am. *Res:* development of theory of hyperbolic partial differential equations; mathematical theory of compressible gases and supersonic nozzle flows; water waves; contributions to non-linear functional analysis; investigations of hypernormal operators of hilbert spaces. *Mailing Add:* PO Box 657 Gualala CA 95455-0657

DEPRISTO, ANDREW ELLIOTT, b Newburgh, NY, Nov 3, 51; m 73; c 2. HETEROGENEOUS CATALYSIS. *Educ:* State Univ NY, Oneonta, BS, 72; Univ Pittsburgh, MS, 73; Univ Md, PhD(chem physics), 76. *Prof Exp:* Teaching asst physics, Univ Pittsburgh, 72-73; teaching & res asst physics & chem, Univ Md, 73-76; NSF fel chem, Princeton Univ, 76-77, res assoc, 77-79; asst prof, Univ NC, Chapel Hill, 79-82; from asst prof to assoc prof, 82-87, PROF CHEM, IOWA STATE UNIV, 87- *Concurrent Pos:* Prin investr, NSF, 81-; Alfred P Sloan fel, 84; John S Guggenheim Fel, 87. *Mem:* Am Phys Soc; Am Chem Soc; Am Vacuum Soc. *Res:* Dynamics and interactions for molecular reactions at transition metal surfaces; density functional theory. *Mailing Add:* Dept Chem Iowa State Univ A 111 Gilman Ames IA 50011

DEPRIT, ANDRE A(LBERT) M(AURICE), b St Servais, Belg, Apr 10, 26; m 59; c 1. ASTRONOMY. *Educ:* Univ Louvain, MA, 48, MSc, 53, PhD(math), 57. *Prof Exp:* Lectr celestial mech, Lovanium Univ, Congo, 57-58; lectr, Univ Louvain, 58-62, prof, 62-64; mem staff, Boeing Sci Res Labs, 64-71; prof astron, Univ SFla, 71-72; prof math sci, Univ Cincinnati, 78-79; mathematician, Nat Inst Standards & Technol, 79-82, sr res mathematician, 82-83, SR FEL, CTR COMPUTATIONAL & APPL MATH, 83- *Concurrent Pos:* NATO advan res fel, 63; vis lectr, Univ Wash, 65-67; vis prof, Univ Liege, 70; Nat Acad Sci sr res fel, 71; C P Taft prof, Univ Cincinnati, 79; vis prof, Univ Namur, 81, Univ Zaragoza, 86- *Honors & Awards:* Agathon De Potter Prize, Royal Acad Sci, Belg, 57; Adolphe Wattrems Prize, 71; James Craig Watson Golden Medal, Nat Acad Sci; Silver Medal, Dept Com. *Mem:* AAAS; fel Am Inst Aeronaut & Astronaut; Royal Astron Soc; Am Astron Soc; Int Astron Union; Sigma Xi. *Res:* Celestial and analytical mechanics; periodic orbits in the three body problem; axiomatic foundations of Hamiltonian formalisms; computer science. *Mailing Add:* Nat Inst Standards & Technol Admin Bldg A 302 Gaithersburg MD 20899

DEPROSPO, NICHOLAS DOMINICK, b New York, NY, July 16, 23; m 60; c 1. ANATOMY, ACADEMIC ADMINISTRATION. *Educ:* NY Univ, BA, 46, MA, 47, PhD(biol ed), 57. *Prof Exp:* From instr to assoc prof biol, Seton Hall Univ, 47-57, actg dean, Col Arts & Sci, 71-74, actg asst vpres acad affairs, 72-73, dean, Col Arts & Sci, 74-79, vpres planning, 79-84, assoc chancellor policy & planning 84-86, interim provost, 86, PROF BIOL, SETON HALL UNIV, 57-, EXEC ASST TO CHANCELLOR, 86-, DEAN SCH GRAD MED EDUC, 87- *Concurrent Pos:* mem, State Panel Sci Adv, NJ, 81-87, Adv Grad Med Educ Coun, NJ, 90- *Mem:* Asn Instnl Res. *Res:* Interrelationships between the pineal gland and other endocrine glands; comparative vertebrate anatomy; mammalian endocrinology. *Mailing Add:* Off of the Dean Seton Hall Univ Sch Grad Med Educ South Orange NJ 07079-2698

DEPUE, ROBERT HEMPHILL, b Pittsburgh, Pa, Aug 15, 31; wid; c 2. CANCER, OCCUPATIONAL HEALTH. *Educ:* Carnegie Inst Technol, BS, 53; Hahnemann Med Col, PhD(microbiol), 63. *Prof Exp:* Res asst biochem, Univ Ill, 53-54; fel biophys, Mellon Inst, 62-65; USPHS officer, Nat Cancer Inst, 65-85; CONSULT EPIDEMIOLOGIST, 86- *Concurrent Pos:* Assoc prof epidemiol, Univ Southern Calif, 78-80. *Mem:* Soc Epidemiol Res. *Res:* relationship of sex hormones to cancer and birth defects; occupational and environmental cancer; biophysics of muscle proteins; electron microscopy; viral oncology; research administration; cancer epidemiology; enzymology; molecular biology; statistics; legal epidemiology. *Mailing Add:* 8612 Bunnell Dr Potomac MD 20854

DEPUIT, EDWARD J, b July 19, 48; m; c 2. DISTURBED LAND REHABILITATION, RANGE PLANT ECOLOGY. *Educ:* Mich Technol Univ, BS, 70; Utah State Univ, MS, 73, PhD(range sci), 74. *Prof Exp:* Res assoc reclamation, Mont State Univ, 74-80; ASSOC PROF RANGE MGT, UNIV WYO, 80- *Concurrent Pos:* Comt mem, Steering Comt Soil & Water Resources, Res Priorities for the Nation, 81- *Mem:* Soc Range Mgt; Soil Conserv Soc Am; Am Soc Surface Mining & Reclamation. *Res:* Rehabilitation of drastically disturbed lands, with emphasis on western mined land reclamation; range plant ecology, ecophysiology and improvements. *Mailing Add:* Range Mgt Div Univ Wyo PO Box 3354 Laramie WY 82071

DE PUY, CHARLES HERBERT, b Detroit, Mich, Sept 10, 27; m 49; c 4. ORGANIC CHEMISTRY. *Educ:* Univ Calif, Berkeley, BS, 48; Columbia Univ, AM, 52; Yale Univ, PhD(chem), 53. *Prof Exp:* Res fel, Univ Calif, Los Angeles, 53-54; from asst prof to prof chem, Iowa State Univ, 54-64; chmn dept, 66-68, PROF CHEM, UNIV COLO, BOULDER, 64- *Concurrent Pos:* NIH fel, Univ Basel, 69-70; vis prof, Univ Ill, 54 & Univ Calif, Berkeley, Guggenheim fel, 60 & 77-78; consult, Marathon Oil Co; Alexander von Humdoldt fel, Berlin, 89- *Mem:* Am Chem Soc. *Res:* Organic reaction mechanisms and stereochemistry; organic gas-phase ion molecule chemistry. *Mailing Add:* Dept Chem Univ Colo Campus Box 215 Boulder CO 80309

DERANLEAU, DAVID A, b Seattle, Wash, Apr 9, 34. BIOPHYSICS. *Educ:* San Francisco State Col, BA, 56; Stanford Univ, MS, 58; Univ Wash, PhD(biochem), 63. *Prof Exp:* NIH fel biochem, Univ Wash, 63-65; NIH fel, Swiss Fed Inst Technol, 65-67, Ciba res fel, 67-68; asst prof biochem, Univ Wash, 69-75; dir res, Hayes Prod, 76; RES ASSOC, UNIV BERN, 77- *Mem:* Am Chem Soc; Am Soc Biol Chemists; Sigma Xi. *Res:* Blood cell shape changes, signal transduction, locomotion and product release. *Mailing Add:* Theodor Kocher Inst Univ Bern CH3012 Bern Switzerland

DERBY, ALBERT, b Antwerp, Belg, Nov 12, 39; US citizen; m 62; c 3. DEVELOPMENTAL BIOLOGY, ENDOCRINOLOGY. *Educ:* City Col New York, BS, 61; NY Univ, MS, 64; City Univ New York, PhD(biol), 69. *Prof Exp:* Teacher gen sci, New York Bd Educ, 62-64; NIH training grant develop biol, Yale Univ, 68-70; ASSOC PROF BIOL, UNIV MO, ST LOUIS, 70- *Mem:* AAAS; Am Soc Zool; Soc Develop Biol. *Res:* Developmental study, both in vivo and in vitro of the biochemistry and endocrinology of amphibian metamorphosis; wound healing. *Mailing Add:* 8001 Natural Brige Rd St Louis MO 63121

DERBY, BENNETT MARSH, b Brooklyn, NY, May 5, 29; div. NEUROPATHOLOGY, NEUROLOGY. *Educ:* Hamilton Col, AB, 52; Univ Va, MD, 56. *Prof Exp:* From intern to asst resident, Univ Va Hosp, 56-58; asst resident, NY Hosp, 58-59; res fel infectious dis, Vanderbilt Univ Hosp, 59-60, resident neurol, 60-61; vis fel, Neurol Inst NY, 61-63; res & clin fel neuropath, Mass Gen Hosp, 63-65; resident path & neuropath, Bellevue Med Ctr, 68-69, PROF CLIN NEUROL, SCH MED, NY UNIV, 71- *Concurrent Pos:* Dir neurol, Vet Admin Hosp, NY, 65- & neuropathologist, 66-; attend neurologist, Univ Hosp, 65-; vis neurologist, Bellevue Hosp, 65- *Mem:* Asn Res Nervous & Ment Dis; AAAS. *Res:* Long-term delineation of interrelation of systemic and neurological disease, using sophisticated techniques of internal medicine and clinical-pathological correlation with adjunct electron microscopy. *Mailing Add:* Dept Neurol & Path NY Univ Sch Med Bellevue Hosp 120 E 34th St New York NY 10016

DERBY, JAMES VICTOR, b Keene, NH, Sept 11, 44. PROJECT MANAGEMENT, ANALYTICAL CHEMISTRY. *Educ:* Oberlin Col, BA, 66; Univ Hawaii, PhD(anal chem), 70. *Prof Exp:* Anal develop chemist, St Joe Zinc Co, 70-74, chief anal develop chemist, 74-77, supt metall control, Smelting Div, 77-79, mgr res, St Joe Minerals Corp, 79-87, VPRES TECHNOL, ZINC CORP AM, 87- *Mem:* Am Chem Soc; Am Soc Testing

& Mat; Am Inst Mining Engrs; Can Inst Mining. *Res:* Project management of new non-ferrous processes plus management of professional services including analytical labs, engineering, technical services and environmental affairs; specific experience in analytical chemistry, wet and instrumental, quality assurance, and project trouble-shooting, technology sales and process development. *Mailing Add:* Zinc Corp Am 300 Frankfort Rd Monaca PA 15061

DERBY, STANLEY KINGDON, b Bangor, Mich, Sept 12, 20; m 43; c 3. ATOMIC SPECTROSCOPY. *Educ:* Univ Chicago, BS, 44; Univ Mich, MS, 48, PhD, 57. *Prof Exp:* From asst prof to prof, 55-64, PROF PHYSICS, WESTERN MICH UNIV, 64- *Mem:* Am Asn Physics Teachers. *Res:* Faraday effects; analysis of biological material by ultraviolet emission spectroscopy; holography. *Mailing Add:* 1308 Brelton Dr Kalamazoo MI 49001

DERBYSHIRE, JOHN BRIAN, b Manchester, Eng, Apr 15, 33; m 55; c 2. VETERINARY VIROLOGY. *Educ:* Univ London, BSc, 55, PhD(microbiol), 60; Royal Col Vet Surgeons, MRCVS, 55. *Prof Exp:* Asst lectr vet path, Univ London, 55-56; sci officer path, Agr Res Coun, Inst Res Animal Dis, 56-64; assoc prof vet sci, Univ Wis, 64-65; prin sci officer microbiol, Agr Res Coun, 65-71; PROF VIROL, UNIV GUELPH, 71- *Concurrent Pos:* Bd mem, WHO/Food & Agr Orgn Prog Comp Virol, 75-79; ed, Can J Vet Res, 86- *Mem:* Soc Gen Microbiol; Can Soc Microbiologists; Can Vet Med Asn; Conf Res Workers Animal Dis (pres, 80); Am Soc Microbiol; Am Soc Virol. *Res:* Immunological responses of swine to enteric viruses; recombinant fowlpox virus vaccines; biology of porcine interferons. *Mailing Add:* Dept Vet Microbiol & Immunol Univ Guelph Guelph ON N1G 2W1 Can

DERBYSHIRE, WILLIAM DAVIS, b Paterson, NJ, June 26, 24; m 47; c 2. PHYSICS. *Educ:* Stevens Inst Technol, ME, 45; Purdue Univ, MS, 51, PhD(physics), 58. *Prof Exp:* Engr, Gen Elec Co, 45-47; instr physics, Stevens Inst Technol, 47-48; asst, Purdue Univ, 48-56; asst prof, 56-61, ASSOC PROF PHYSICS, COLO STATE UNIV, 61- *Mem:* Am Phys Soc. *Res:* Ferromagnetism; statistical physics. *Mailing Add:* 709 Garfield Ft Collins CO 80524

DERDERIAN, EDMOND JOSEPH, b Sofia, Bulgaria, June 23, 42; m 73; c 2. PHYSICAL CHEMISTRY, SURFACE SCIENCE. *Educ:* Colby Col, AB, 66; Pa State Univ, PhD(chem), 74. *Prof Exp:* RES SCIENTIST & TECHNOL MGR, UNION CARBIDE CORP, 76- *Concurrent Pos:* Assoc, ERDA, Ames Labs, Iowa State Univ, 74-76. *Mem:* Am Chem Soc; Am Soc Testing & Mat. *Res:* Microemulsions, surface and colloid science; heterogeneous catalysis; solution properties of polymers; fuel and lubricant additives. *Mailing Add:* 1864 Louden Heights Rd Charleston WV 25314-1565

DEREMER, RUSSELL JAY, b Bell, Calif, May 2, 40; div; c 2. PHYSICS. *Educ:* Occidental Col, AB, 61; Ind Univ, MS, 63, PhD(high energy physics), 66. *Prof Exp:* From asst prof to assoc prof physics, Calif State Col, San Bernardino, 66-78, assoc dean activ & housing, 68-78; ASSOC PROF PHYSICS & DEAN OF STUDENTS, WHITMAN COL, 78- *Mem:* Am Phys Soc; Am Asn Physics Teachers; Sigma Xi. *Res:* High energy physics. *Mailing Add:* Phys Dept Whitman Col Walla Walla WA 99362

DERENIAK, EUSTACE LEONARD, b Standish, Mich, Dec 29, 41; m 68; c 2. INFRARED ASTRONOMY, OPTICAL DESIGN. *Educ:* Mich Technol Univ, BS, 63; Univ Mich, MS, 65; Univ Ariz, PhD(optics), 76. *Prof Exp:* Engr, Rockwell Int, 65-72, Ball Brother Res, 72-73; tech asst fourier optics, Univ Ariz, 73-76; res assoc radiol imaging, Ariz Health Ctr, 76-78; PROF CORONA CURRENT DETECTORS, UNIV ARIZ, 78-, PROF, OPTICAL DETECTION LAB. *Concurrent Pos:* Vis prof, Hanscom AFB, Lexington, Mass, 81. *Mem:* Optical Soc; Soc Photo-Optical Instrumentation Engrs; Radiol Soc; Optical Soc Am. *Res:* Infrared detectors with emphasis on understanding physics; schottky barrier diodes; charge transfer devices, used in the infrared spectrum. *Mailing Add:* Optical Detection Lab Univ Ariz Optical Sci Bldg Tucson AZ 85721

DE RENOBALES, MERTXE, b Bilbao, Spain, July, 18, 48. INSECT BIOCHEMISTRY, INSECT CUTICULAR LIPIDS. *Educ:* Univ Basque Country, lic, 75; Univ San Francisco, MS, 76; Univ Nev, Reno, PhD(biochem), 79. *Prof Exp:* Res assoc biochem, Wash State Univ, 79-80; res assoc, 80-84, ASST PROF BIOCHEM, UNIV NEV, RENO, 84- *Concurrent Pos:* Vis prof, Uppsala Univ, Sweden, 86; prin investr, Univ Nev, Reno. *Mem:* Am Soc Biol Chemists; Entom Soc Am; AAAS; Am Chem Soc. *Res:* Metabolism of lipids in insects; cuticular hydrocarbons; polyunsaturated, medium chain and methyl branched fatty acid biosynthesis; characterization of enzymes from these pathways; developmental changes; pesticide penetration across the cuticle; hemolymph transport. *Mailing Add:* Inasmet Dept Chem Technol Camino De Partuetxe 12 E-20009 San Sebastian Spain

DE RENZO, EDWARD CLARENCE, biochemistry; deceased, see previous edition for last biography

DERENZO, STEPHEN EDWARD, b Chicago, Ill, Dec 31, 41; m 66; c 2. NUCLEAR MEDICINE. *Educ:* Univ Chicago, BS, 63, MS, 65, PhD(physics), 69. *Prof Exp:* Res asst, Enrico Fermi Inst, 64-68; PHYSICIST, LAWRENCE BERKELEY LAB, UNIV CALIF, 68- *Concurrent Pos:* Prof in residence, Elec Eng & Comput Sci Dept, Univ Calif, Berkeley, 88- *Mem:* Am Phys Soc; Inst Elec & Electronics Engrs. *Res:* Instrumentation for nuclear particle detection and radionuclide medical imaging; solid state photodectectors and scintillators for positron emission tomography; search for new heavy-atom scintillators. *Mailing Add:* Bldg 55 Rm 146 Lawrence Berkeley Lab One Cyclotron Rd Berkeley CA 94720

DERESIEWICZ, HERBERT, b Czech, Nov 5, 25; nat US; m 55; c 3. THEORETICAL MECHANICS, APPLIED MECHANICS. *Educ:* City Col New York, BME, 46; Columbia Univ, MS, 48, PhD(mech), 52. *Prof Exp:* Res engr, Sr Staff, Appl Physics Lab, Johns Hopkins Univ, 50-51; res assoc civil eng, 51-53, asst prof, 53-55, from asst prof to assoc prof mech eng, 55-62, PROF MECH ENG, COLUMBIA UNIV, 62-, CHEM DEPT, 81-87, 90- *Concurrent Pos:* Fulbright sr res scholar, Italy, 60-61; Fulbright lectr, Israel, 66-67, vis prof, 73-74. *Mem:* AAAS. *Res:* Theory of elasticity; vibrations of crystals; thermoelasticity; elastic contact theory; mechanics of granular media; wave propagation in porous media; soil consolidation. *Mailing Add:* Dept Mech Eng Columbia Univ New York NY 10027

DERFER, JOHN MENTZER, b Navarre, Ohio, Aug 9, 20; m 44. INDUSTRIAL ORGANIC CHEMISTRY. *Educ:* Col Wooster, AB, 42; Ohio State Univ, PhD(org chem), 46. *Prof Exp:* Asst, Ohio State Univ, 42-45, res assoc & ed res proj, Am Petrol Inst, 45 & Air Res & Develop Command Proj 572, 45-47, res assoc, Univ Res Found, 47-55, assoc dir petrol inst res proj, 55-59; mgr res labs, Glidden Co, 59-61, dir res, Org Chem Div, 61-66, mgr explor res, Org Chem Group, Glidden-Durkee Div, 66-71, asst tech, Dir, Org Chem Group, 71-75, asst dir res & develop, Org Chem Div, SCM Corp, 75-82; RETIRED. *Honors & Awards:* Dwight P Joyce Award, 67. *Mem:* Am Chem Soc; Tech Asn Pulp & Paper Indust; Am Oil Chemists Soc; NY Acad Sci. *Res:* Synthesis of low molecular weight hydrocarbons; synthesis and infrared spectra of cyclic hydrocarbons; knocking characteristics of hydrocarbons; pre-flame reactions of fuels; terpene chemistry; rosin; fatty acids; naval stores; essential oils; flavor and perfume chemicals. *Mailing Add:* 9136 August Rd Star Rte Box 107 Jacksonville FL 32226-2906

DERGARABEDIAN, PAUL, b Racine, Wis, Jan 19, 22; m 47; c 4. MECHANICAL ENGINEERING, PHYSICS. *Educ:* Univ Wis, BS, 48, MS, 49; Calif Inst Technol, PhD(mech eng, physics), 52. *Prof Exp:* Actg head hydrodyn br, US Naval Ord Test Sta, 52-55; mgr, Syst Design and Anal Dept, TRW Systs, Redondo Beach. 55-65, starr mgr, 65-80; SR ENGR, AEROSPACE CORP, EL SEGUNDO, CALIF. *Concurrent Pos:* Vis prof, Calif Inst Technol, 71-72. *Mem:* Am Astron Soc (pres, 69-71); Sigma Xi. *Res:* Mechanism of cavitation; water-entry impact; rotational non-viscous flow; missile systems design and analysis; powered-flight mechanics; space technology; meteorology and theoretical analysis of tornadoes and hurricanes. *Mailing Add:* Aerospace Corp M4-919 PO Box 92957 Los Angeles CA 90009

DERGE, G(ERHARD) (JULIUS), b Lincoln, Nebr, Feb 11, 09; m 37; c 2. EXTRACTIVE METALLURGY. *Educ:* Amherst Col, AB, 30; Princeton Univ, PhD(phys chem), 34. *Prof Exp:* Metallurgist, metals res lab, 34-39, from asst prof to assoc prof metall, 39-49, prof, 49-51, Jones & Laughlin prof, 51-64, prof, 64-77, EMER PROF METALL, CARNEGIE-MELLON UNIV, 77- *Concurrent Pos:* Ed, Metall Trans, 58- *Mem:* Am Chem Soc; Am Inst Mining, Metall & Petrol Engrs; fel Am Soc Metals; fel Metall Soc; Am Ceramic Soc. *Res:* Kinetics and mechanisms of slag-metal reactions, especially ferrous systems, constitution and properties of non-aqueous melts, including slag, matte, fused salts and related high temperature systems; especially by electrochemical measurements, diffusion in such melts. *Mailing Add:* Eight Longfellow Rd Pittsburgh PA 15215

DERI, ROBERT JOSEPH, b White Plains, NY, Dec 16, 57; m 79; c 1. INTEGRATED-OPTICS, SEMICONDUCTOR OPTICAL DEVICE PHYSICS. *Educ:* Cornell Univ, BS, 79; Univ Rochester, MA, 83 , PhD(physics), 86. *Prof Exp:* Res scientist, Eastman Kodak Res Labs, 79-82; MEM TECH STAFF, BELL COMMUN RES, 86- *Concurrent Pos:* Co-op student, AVCO-Everett Res Labs, 77 & 78; vis researcher, Fujitsu Labs, Japan, 89. *Mem:* Am Phys Soc; Inst Elec & Electronics Engrs. *Res:* Integrated optoelectric devices for fiber optic communications; electronic measurements at high frequency; integrated optic devices. *Mailing Add:* Bell Commun Res 331 Newman Springs Rd Red Bank NJ 07701-7020

DERIEG, MICHAEL E, b Jan 24, 35; US citizen; m 62; c 3. ORGANIC CHEMISTRY. *Educ:* Univ Nebr, BS, 56, MS, 58, PhD(org chem), 60. *Prof Exp:* Asst org chem, Univ Nebr, 55-59; res chemist, Celanese Corp, 60-61; appointee, Mass Inst Technol, 61-62; sr chemist, 62-71, tech coordr, Div Animal Health, 71-73, asst dir animal health res, 73-75, ASST DIR CHEM RES, HOFFMANN-LA ROCHE INC, 75-; HOFFMAN-LA ROCHE, INC, CALDWELL, NJ. *Mem:* Am Chem Soc; The Chem Soc; fel Am Inst Chemists; NY Acad Sci; Sigma Xi. *Res:* Naphthenic acids; exocyclic olefins; macrocyclic and pyrimidine nucleoside antibiotics; reaction mechanisms; heterocyclic chemistry, especially benzodiazepines, benzodiazocines, quinazolines; polyether antibiotics, anthelmintics and coccidiostats. *Mailing Add:* Hoffmann-La Roche Inc 3 Cleveland Rd Caldwell NJ 07006

DE RIJK, WALDEMAR G, b Venlo, Neth, Mar 5, 45; US citizen; m 74; c 2. DENTISTRY, ATOMIC PHYSICS. *Educ:* Univ Amsterdam, BA, 68; Univ Nebr, PhD(physics), 74, DDS, 77. *Prof Exp:* Res asst, Lab Atomic Physics, Found Fundamental Res on Matter, Amsterdam, 62-68; res asst atomic physics, Univ Nebr, 68-74; asst prof dent, Creighton Univ, 77-79; sr staff fel, NIH, 79-81; chief res scientist, Am Dental Asn, 81-85; assoc prof dent & dental mat, Sch Dent, Univ Md, 85-87; RES SCIENTIST, NAT BUR STANDARDS, DENTAL & MED MAT GROUP, 87-; ASSOC PROF RESTORATIVE DENTISTRY, UNIV TEX HEALTH SCI CTR, SAN ANTONIO, 89- *Concurrent Pos:* Staff dentist, Plaza Dent Group, Lincoln, 77-79. *Honors & Awards:* E Hatton Award, Int Asn Dent Res, 76. *Mem:* AAAS; Int Asn Dent Res; Am Phys Soc; Soc Biomed Mat Res; Acad Dent Mat. *Res:* Restorative dentistry; diagnostic procedures and dental materials, particularly myography of masticatory muscles and adaptation of dental direct restorative materials. *Mailing Add:* 2506 Rim Rock Trail San Antonio TX 78251-2507

DERINGER, MARGARET K, cancer, biology; deceased, see previous edition for last biography

DERISO, RICHARD BRUCE, b Macon, Ga, Oct 2, 51; m 75; c 2. POPULATION DYNAMICS. *Educ:* Auburn Univ, BSIE, 72; Univ Fla, MS, 75; Univ Wash, PhD(biomath), 78. *Prof Exp:* Vis res asst prof, Univ NC, Chapel Hill, 78-80; pop dynamicist, Int Pac Halibut Comn, 80-88; CHIEF SCIENTIST, INTER-AM TROP TUNA COMN, SCRIPPS INST OCEANOG, 88- *Concurrent Pos:* Adj assoc prof pop dynamics, Univ Wash, 84-; adj assoc prof ecol, Scripps Inst Oceanog, 90- *Honors & Awards:* W F Thompson Award, Am Inst Fisheries Res Biologists, 81. *Mem:* Am Fisheries Soc; Am Inst Fisheries Res Biologists. *Res:* Application of mathematics and statistics to problems concerning natural resources and interpretation and analysis of the dynamics of animal populations. *Mailing Add:* Inter-Am Trop Tuna Comn Scripps Inst Oceanog La Jolla CA 92093

DERKITS, GUSTAV, b Philadelphia, Pa, Apr 19, 50; m 75; c 1. SEMICONDUCTOR PHYSICS, SEMICONDUCTOR DEVICE PHYSICS. *Educ:* St Joseph's Col, Philadelphia, BS, 72; Univ Pittsburgh, MS, 76, PhD(physics), 80. *Prof Exp:* Res assoc, Dept Physics, Univ Pittsburgh, 80-82; MEM TECH STAFF, BELL LABS, MURRAY HILL, 82- *Mem:* Am Phys Soc; Inst Elec & Electronics Engrs; Sigma Xi. *Res:* Semiconductor physics, especially low temperature uses of tunnelling to examine impurity states; device-related physics. *Mailing Add:* 55 Holmes Oval New Providence NJ 07974

DERMAN, CYRUS, b Philadelphia, Pa, July 16, 25; m 61; c 2. STATISTICS, OPERATIONS RESEARCH. *Educ:* Univ Pa, AB, 48, AM, 49; Columbia Univ, PhD(math statist), 54. *Prof Exp:* Instr math, Syracuse Univ, 54; prof indust eng, 55-68, PROF OPERS RES, COLUMBIA UNIV, 68- *Concurrent Pos:* Vis prof, Israel Inst Technol, 61-62, Stanford Univ, 65-66, Univ Calif, Davis, 75-76 & Univ Calif, Berkeley, 79. *Mem:* Fel Inst Math Statist; fel Am Statist Asn. *Res:* Applied probability theory. *Mailing Add:* 302A SW Mudd Bldg Columbia Univ New York NY 10027

DERMAN, SAMUEL, b New York, NY, Mar 11, 31. ELECTRONICS ENGINEERING. *Educ:* City Col NY, BEE, 53; Columbia Univ, MSEE, 59; NY Univ, PhD(physics), 73. *Prof Exp:* ASSOC PROF ELECTRONICS TECHNOL, CITY COL NEW YORK, 79- *Mem:* Am Soc Eng Educ; Am Asn Physics Teachers. *Res:* Electronics technology. *Mailing Add:* Dept Eng Technol City Col New York New York NY 10031

DER MATEOSIAN, EDWARD, b New York, NY, Aug 6, 14; m 47; c 2. NUCLEAR PHYSICS. *Educ:* Columbia Univ, BA, 35, MA, 41. *Prof Exp:* Res chemist, Barrett Co, 38-41; asst physics, Ind Univ, 41-42; physicist, US Naval Res Lab, 42-46 & Argonne Nat Lab, 47-49; PHYSICIST, BROOKHAVEN NAT LAB, 49- *Mem:* Fel Am Phys Soc. *Res:* Radioactive decay; nuclear energy levels; high spin states and deexcitation of compound nucleus following heavy ion reactions. *Mailing Add:* Three Browns Lane Bellport NY 11713

DERMER, OTIS CLIFFORD, b Hoytville, Ohio, Nov 11, 09; m 35; c 3. INDUSTRIAL ORGANIC CHEMISTRY. *Educ:* Bowling Green State Univ, BS, 30; Ohio State Univ, PhD(chm), 34. *Hon Degrees:* DSc, Bowling Green State Univ, 60. *Prof Exp:* From instr to prof, 34-72, head dept, 49-71, Regents Serv prof, 72-75, EMER PROF CHEM, OKLA STATE UNIV, 75- *Mem:* Am Chem Soc; AAAS; Royal Soc Chem. *Res:* Chemicals from petroleum and natural gas; organic nomenclature; chemical literature. *Mailing Add:* Dept Chem Okla State Univ Stillwater OK 74078

DERMIT, GEORGE, b Istanbul, Turkey, Feb 9, 25; US citizen; m 50; c 2. SEMICONDUCTOR DEVICES PHYSICS, MATERIALS SCIENCE. *Educ:* Robert Col, Istanbul, BS, 47; Cornell Univ, MS, 49; Polytech Inst Brooklyn, PhD(physics), 61. *Prof Exp:* Engr, Sylvania Elec Prod, Inc, 52-54; sr engr, Link Aviation, Inc, 54-56; chief scientist, Gen Transistor Corp, 56-59; sect head, Gen Tel & Electronics Lab, Inc, 59-63; owner, G Dermit Electronics, 63-83; CONSULT, 83- *Concurrent Pos:* Assoc ed, Solid State Technol, 84-86. *Mem:* Am Inst Physics. *Res:* Semiconductor devices, integrated circuits, design, manufacturing; metals and alloys for electronics. *Mailing Add:* 198-31 27th Ave Flushing NY 11358

DERMODY, WILLIAM CHRISTIAN, b Lompoc, Calif, Sept 22, 41; m 64; c 1. ONCOLOGY, ENDOCRINOLOGY. *Educ:* Calif State Polytech Univ, BS, 64; Utah State Univ, MS, 68, PhD(animal physiol), 69. *Prof Exp:* NIH fel reproductive physiol, Cornell Univ, 69-70; sr res physiologist, Park-Davis & Co, 70-76; head, Hormone Markers Sect, Biol Markers Lab, Frederick Cancer Res Ctr, 76-81; dir biotechnol res & develop, AM Dade, Miami, Fla, 81-84; biomed mkt mgr, Miles Sci/ICN, 84-86; tech resources mgr, Disco Labs, 86-88; assoc dir sci, Am Type Cult Collection, 88-90; PRIN OWNER, BIOWORLD ASSOCS, 90- *Concurrent Pos:* Consult, Dept Anat, Med Sch & Dept Pediat & Commun Dis, Med Ctr, Univ Mich, 71-76. *Mem:* NY Acad Sci. *Res:* Molecular endocrinology; oncology. *Mailing Add:* BioWorld Assocs 11611 Pleasant Meadow Dr North Potomac MD 20878

DERMOTT, STANLEY FREDERICK, b Ormskirk, Eng, Aug 14, 42; m 65; c 2. PLANETARY SCIENCE, ASTRONOMY. *Educ:* Univ Col London, BSc, 64; Univ London, PhD(physics), 75. *Prof Exp:* Demonstr physics, Royal Mil Col Sci, 67-71; sr demonstr, Univ Newcastle upon Tyne, 72-74, sr res assoc, 75-76; res assoc & lectr, 77-80, SR RES ASSOC, CTR RADIOPHYS & SPACE RES, CORNELL UNIV, 80- *Mem:* Fel Royal Astron Soc. *Res:* Dynamical evolution of the solar system; dynamics of planetary rings; tidal and resonant interactions between planets and satellites; structure of the asteroid belt. *Mailing Add:* Dept Astron Univ Fla Gainsville FL 32611

DE ROCCO, ANDREW GABRIEL, b Westerly, RI, July 31, 29; div; c 1. CHEMICAL PHYSICS, BIOPHYSICS. *Educ:* Purdue Univ, BS, 51; Univ Mich, MS, 53, PhD(chem physics), 56. *Prof Exp:* Instr chem, Univ Mich, 54-56, Nat Acad Sci fel physics, 56-57, instr chem, 57-60, asst prof, 60-62; from asst prof to prof, Inst Molecular Physics, Univ Md, College Park, 63-76, prof molecular physics, Inst Phys Sci & Technol, 76-79; prof natural sci & dean fac, Trinity Col, 79-84; PRES, DENISON UNIV, 84- *Concurrent Pos:* Vis prof, Univ Colo, 62-63 & Tufts Univ, 68-69; distinguished vis prof, USAF Acad, 75-76, mem, Defense Educ Study Group, 76-77; mem staff, Phys Sci Lab, Div Comput Res & Technol, NIH, 69-79. *Mem:* AAAS; Sigma Xi; Biophys Soc; Am Phys Soc. *Res:* Statistical mechanics, especially liquid crystals; membrane phase transitions; circadian clocks; mathematical biology. *Mailing Add:* 505 S Parkview Ave 304 Granville OH 43209

DEROME, JACQUES FLORIAN, b Montreal, Que, Apr 20, 41; m 67. DYNAMIC METEOROLOGY. *Educ:* McGill Univ, BS, 63, MS, 64; Univ Mich, PhD(meteorol), 68. *Prof Exp:* Res fel meteorol, Mass Inst Technol, 68-69; res scientist, Dynamic Prediction Res Unit, Can Atmospheric Environ Serv, 69-72; assoc prof, 72-85, PROF METEOROL, MCGILL UNIV, 85-, CHMN DEPT, 87- *Honors & Awards:* President's Prize, Can Meterol & Oceanog Soc, 83. *Mem:* Can Meteorol Soc; prof mem Am Meteorol Soc. *Res:* Numerical weather predictions; structure and stability of large-scale waves in the atmosphere; climate dynamics. *Mailing Add:* Dept Meteorol 805 Sherbrooke W Montreal PQ H3A 2K6 Can

DEROOS, FRED LYNN, b Minneapolis, Minn, Oct 23, 47; m 76. ANALYTICAL CHEMISTRY, MASS SPECTROMETRY. *Educ:* Univ SDak, Vermillion, BA, 69; Univ Nebr-Lincoln, PhD(anal chem), 76. *Prof Exp:* Asst prof & dir mass spectrometry, Univ Pa, 76-77; RES CHEMIST ANALYTICAL MASS SPECTROMETRY, BATTELLE MEM INST, 77- *Mem:* Am Chem Soc; Am Soc Mass Spectrometry. *Res:* Analytical mass spectrometry; chemical instrumentation design and modification; kinetic energy release accompanying metastable transitions in the mass spectrometer; chemical ionization mass spectrometry. *Mailing Add:* 3387 Oxford Bay Woodbury MN 55125-2725

DEROOS, ROGER MCLEAN, b Fresno, Calif, Aug 11, 30; m 55; c 3. ZOOLOGY. *Educ:* Univ Calif, Berkeley, BA, 55, PhD(zool), 61; Utah State Univ, MS, 58. *Prof Exp:* From asst prof to assoc prof zool, 61-70, PROF BIOL SCI, UNIV MO, COLUMBIA, 70- *Concurrent Pos:* Assoc dir, Div Biol Sci, Univ Mo, 71. *Mem:* AAAS; Soc Exp Biol & Med. *Res:* Comparative endocrinology; adrenal cortex functions and control; reproductive physiology. *Mailing Add:* 14 Eubanks Ct Columbia MO 65203

DEROSE, ANTHONY FRANCIS, b Chicago, Ill, June 7, 20; m 52; c 6. MEDICINAL CHEMISTRY, PHARMACEUTICAL CHEMISTRY. *Educ:* Univ Ill, BS, 41, MS, 43. *Prof Exp:* Asst pharmacog & pharmacol, Col Pharm, Univ Ill, 41-43; res biochemist, Res Div, Abbott Labs, 46-60, supvr res serv, 60-64, mgr dept sci bldg serv, Sci Div, 64-66, res pharmaceut chemist, New Prod Div, 66-70, chief pharmacist & mgr res & develop pilot plant opers, 70-80, sr res pharmacist, Hosp Prod Div, 80-86; RETIRED. *Concurrent Pos:* Consult. *Mem:* AAAS; Am Chem Soc; Sigma Xi. *Res:* Chemical constitution; pharmacognosy and pharmacology of medical plants; isolation and chemistry of substances of biochemical origin; antibiotic and vitamin research; pharmaceuticals. *Mailing Add:* 253 Stanley Ave Waukegan IL 60085-2115

DEROSIER, DAVID J, b Milwaukee, Wis, Feb 22, 39; m 62; c 2. BIOPHYSICS, MOLECULAR BIOLOGY. *Educ:* Univ Chicago, BS, 61, PhD(biophys), 65. *Prof Exp:* Visitor, Lab Molecular Biol, Cambridge Univ, 65-69; from asst prof to assoc prof, Univ Tex, Austin, 69-73; from assoc prof to prof physics, 73-79, chmn dept biol, 84-87, PROF BIOL, ROSENSTIEL BASIC MED SCI RES CTR, BRANDEIS UNIV, 79- *Concurrent Pos:* Air Force Off Sci Res fel, 65-66; Am Cancer Soc fel, 66-67; NSF fel, 67-68; vis prof, Yale Univ, 87; visitor, molecular biol res lab, Cambridge, Eng, 88; Guggenheim fel, 88-89. *Honors & Awards:* Merit Award, NIH, 90. *Mem:* AAAS; Am Soc Cell Biol; Biophys Soc. *Res:* Determination and interpretation of the three-dimensional structure of complexes of biological macromolecules, in particular bacterial flagella and actin containing structures. *Mailing Add:* Rosenstiel Basic Med Sci Res Ctr Brandeis Univ Waltham MA 02254

DEROSSET, ARMAND JOHN, b New York, NY, Jan 10, 15; m 39; c 5. CHEMISTRY. *Educ:* Lafayette Col, BS, 36; Univ Wis, PhD(chem), 39. *Prof Exp:* Jr chemist, State Hwy Comn Wis, 36-39; res chemist, Universal Oil Prod Co, 36-64, asst dir res, 64-74, assoc dir res, 74-76, dir separation res, Corp Res Ctr, 76-80; consult, 80-82; chem counr, 82-89, ADJ SCIENTIST, MOTE MARINE LAB, 89- *Mem:* Am Chem Soc; Newcomen Soc. *Res:* Process and catalyst research in the petroleum refining and petrochemical field of hydrotreating and separation via adsorbents; sulfur removal from stack gases; coal liquefaction; adsorptive sampling of toxins (red tide) and pollutants (coprostanol) from natural waters. *Mailing Add:* 3223 Village Green Dr Sarasota FL 33579

DE ROSSET, WILLIAM STEINLE, b Chicago, Ill, Apr 1, 42; m 71; c 3. PHYSICS, BALLISTICS. *Educ:* Johns Hopkins Univ, BA, 64; Univ Ill, MS, 66, PhD(physics), 70. *Prof Exp:* res physicist ballistics, 71-85, SUPVRY PHYS SCIENTIST, BALLISTICS RES LAB, LABCOM, 85- *Mem:* Am Phys Soc. *Res:* Penetration mechanics; shock physics; material properties at high strain rates; modeling of dynamic material failure mechanisms; ballistic testing. *Mailing Add:* Ballistics Res Lab Attn: SLCBR-TB-P Aberdeen Proving Ground MD 21005-5066

DEROTH, LASZLO, b Budapest, Hungary, Oct 26, 41; m 70; c 3. PHYSIOLOGY, CARDIOLOGY. *Educ:* Univ Montreal, DVM, 72; Univ Guelph, MSc, 75, PhD(biomed), 77. *Prof Exp:* Lectr physiol, 72-75, asst prof biomed, 75-80, chmn, Dept Med, 81-82, assoc dean res, 81-85, PROF BIOMED, FAC VET MED, UNIV MONTREAL, 84- *Honors & Awards:* Scherring Award, 72. *Mem:* Am Asn Vet Physiologists, Pharmacologists & Biochemists; Am Physiol Soc; Acad Vet Cardiol; World Asn Vet Physiologists, Pharmacologists & Biochemists. *Res:* Veterinary clinical physiology; veterinary cardiology; fetal and neonatal physiology; autonomic pharmacology; stress pathophysiology. *Mailing Add:* Fac Vet Med Univ Montreal St-Hyacinthe PQ J2S 7C6 Can

DEROUSSEAU, C(AROL) JEAN, b Rice Lake, Wis, Aug 29, 47. OSTEOLOGY, COMPARATIVE ONTOGENY. *Educ:* Univ Chicago, BA, 70; Northwestern Univ, MA, 74, PhD(anthrop), 78. *Prof Exp:* Instr anthrop, Northwestern Univ, 74-78; staff assoc ophthal, Columbia Univ, 78-80; adj prof anthrop, Hunter Col, City Univ New York, 80-89; asst prof anthrop, NY Univ, 80-89. *Concurrent Pos:* Co-prin investr NIH grant. *Mem:* Am Asn Phys Anthropologists; Human Biol Coun; Paleopath Asn; NY Acad Sci; Geront Soc Am. *Res:* Documentation of age-changes in the adult rhesus monkey including joint degeneration, loss of bone, presbyopia and other age-related disorders; patterns of growth and aging are species-specific. *Mailing Add:* 5030 Pinetree Dr Miami Beach FL 33140

DEROW, MATTHEW ARNOLD, b New York, NY, Apr 29, 09; m 41; c 2. MICROBIOLOGY. *Educ:* City Col New York, BS, 29; Columbia Univ, AM, 30; Boston Univ, MD, 34, PhD(med sci, biochem), 41; Army Med Sch, dipl, 43. *Prof Exp:* Teaching fel biochem, 35-37, instr bact & immunol, 36-56, from asst prof to assoc prof, 57-75, PROF MICROBIOL, SCH MED, BOSTON UNIV, 75- *Concurrent Pos:* Chemist & bacteriologist, Natick, Mass, 35-53; vis physician, Allergy Clin, Mass Mem Hosp, 36-43; path consult, Norfolk County Hosp, 42-; China Med Bd fel trop med, 56. *Mem:* NY Acad Sci; AAAS. *Res:* Vitamin assays; blood groups; immunochemistry; allergy; chemotherapeutic agents and antibiotics; bacterial toxins and enzymes; medical parasitology. *Mailing Add:* 200 Ledgewood Dr Apt 504 Stoneham MA 02180-3622

DERR, JOHN SEBRING, b Boston, Mass, Nov 12, 41; c 3. GEOPHYSICS. *Educ:* Amherst Col, BA, 63; Univ Calif, Berkeley, MA, 65, PhD(geophys), 68. *Prof Exp:* Field asst geol mapping, US Geol Surv, 62-63; geophysicist, Pan Am Petrol Corp, 64; res assoc lunar seismol, Mass Inst Technol, 68-70; res scientist geophys, Martin Marietta Aerospace, 70-74; chief opers seismol, US Geol Surv, Denver, Colo, 74-79; chief tech reports unit, Menlo Park, Calif, 80-82; mem staff, Earth Quake Info Serv, Denver, Co, 83-89; CONSULT, 89- *Concurrent Pos:* Co-investr seismol, Pioneer Venus '78, Comp Atmospheric Struct Exp, 76-78. *Mem:* Sigma Xi; Am Geophys Union; Seismol Soc Am; AAAS; Soc Sci Explor. *Res:* Seismology, free oscillations; seismicity; earthquake prediction; planetary seismology; luminous phenomena; international data services. *Mailing Add:* Albuquerque Seismol Lab Albuquerque NM 87115

DERR, ROBERT FREDERICK, b Philadelphia, Pa, Feb 15, 34; m 68; c 3. BIOCHEMISTRY. *Educ:* Pa State Univ, BS, 55; Univ Minn, MS, 58, PhD(biochem), 61. *Prof Exp:* Fel, Univ Minn, 61-62; biochemist, Northern Grain Insects Res Lab, 62-64 & Minneapolis Vet Hosp, 64-69; Nat Inst Gen Med Sci spec fel, 69-71; RES BIOCHEMIST, MINNEAPOLIS VET HOSP, 71- *Concurrent Pos:* Ed, Biochem Archives, 85- *Mem:* Am Soc Biol Chemists; Am Inst Nutrit. *Res:* Alcoholism; control analysis. *Mailing Add:* Vet Admin Med Ctr One Veterans Dr Minneapolis MN 55417

DERR, RONALD LOUIS, b Chicago, Ill, July 13, 38; m 76; c 3. COMBUSTION ENGINEERING. *Educ:* Purdue Univ, BS, 60, MS, 62, PhD(mech eng), 67. *Prof Exp:* Res asst, Purdue Univ, 61-62; develop engr solid propellant combustion res, Aerojet Corp, 62-63; res asst, Purdue Univ, 63-67; tech specialist, Lockheed Aircraft Corp, 67-72, chief combustion sect, 72-73; assoc head aerothermochem div, 73-75, head, 75-83, head eng sci div, 83-86, HEAD RES DEPT, NAVAL WEAPONS CTR, 86- *Concurrent Pos:* Mem steering comt, Joint Army-Navy-NASA-Air Force Combustion Working Group, 74-, chmn steering comt, 75-77 & 83-85, chmn, Propulsion Syst Hazards Subcomt, 79-83; prin US rep, NATO AC/310, 85-88; chmn, Pilot NATO Insensitive Munitions Info Ctr, 88-90, NATO Insensitive Munitions Info Ctr, 90-; US coordr & mem Adv Group Aerospace Res & Develop, Propulsion & Energetics Panel, 91- *Honors & Awards:* L Thompson Award, Naval Weapons Ctr, 84. *Mem:* Assoc fel Am Inst Aeronaut & Astronaut. *Res:* Basic and applied research in the area of solid propellant combustion including ignition, steady state combustion, combustion instability, and deflagration to detonation phenomena. *Mailing Add:* Res Dept Naval Weapons Ctr Code 38 China Lake CA 93555

DERR, VERNON ELLSWORTH, b Baltimore, Md, Nov 22, 21; m 43; c 4. ATMOSPHERIC PHYSICS. *Educ:* St Johns Col, AB, 48; Johns Hopkins Univ, PhD, 59. *Prof Exp:* Asst appl physics lab, Johns Hopkins Univ, 50, res assoc, Radiation Lab, 51-59; prin res scientist, Martin Co, 59-67; res scientist, Nat Oceanic & Atmospheric Admin, 67-77, chief atmospheric spectro, Wave Propagation Lab, 67-81, dep dir, 81-83, dir, Environ Res Labs, 83-89, SR SCIENTIST, ENVIRON RES LABS, NAT OCEANIC & ATMOSPHERIC ADMIN, 89- *Concurrent Pos:* Adj prof, Rollins Col, 59-67 & Univ Colo, 71- *Mem:* Inst Elec & Electronics Engrs; Am Geophys Union; Optical Soc Am; Am Meteorol Soc. *Res:* Microwave, infrared and optical spectroscopy; atmosperic radiation and climate; atmospheric optics; cloud physics; statistical decision theory; statistical mechanics; lidar development; environmental forecasting by artificial intelligence. *Mailing Add:* R453 Wave Propagation Lab Nat Oceanic & Atmospheric Admin Boulder CO 80302

DERR, WILLIAM FREDERICK, b Reading, Pa, June 27, 39; m 61; c 2. PLANT ANATOMY. *Educ:* Lebanon Valley Col, BS, 60; Univ Wis, MS, 62, PhD(bot), 64. *Prof Exp:* Asst bot, Univ Wis, 60-64; from asst prof to assoc prof, 64-72, PROF BIOL, CALIF STATE UNIV, CHICO, 72- *Concurrent Pos:* NSF res grant, 65-67. *Mem:* AAAS; Bot Soc Am. *Res:* Seasonal development of cambium, ontogeny, and structure of phloem; histochemical studies of differentiating cells. *Mailing Add:* Dept Biol Calif State Univ Chico CA 95929

DERRICK, FINNIS RAY, b Ballentine, SC, May 1, 11; m 37; c 2. ZOOLOGY. *Educ:* Univ SC, BS, 34, MS, 37, PhD(zool), 55. *Prof Exp:* Teacher pub sch, SC, 38-41; instr biol, Augusta Jr Col, 41-46; head dept biol, Appalacian State Univ, 46-80, prof biol, 46-80; RETIRED. *Res:* Aquatic biology; conservation. *Mailing Add:* Rte 7 Box 207 Boone NC 28607

DERRICK, JOHN RAFTER, b Clayton, Ga, Jan 17, 22; m 51; c 6. SURGERY. *Educ:* Clemson Col, BS, 43; Tulane Univ, MD, 46. *Prof Exp:* Instr chest & cardiovasc surg, Sch Med, Emory Univ, 56-57; assoc prof thoracic surg & actg chief div thoracic & cardiovasc surg, 57-67, PROF THORACIC SURG & CHIEF DIV THORACIC & CARDIOVASC SURG, UNIV TEX MED BR, GALVESTON, 67- *Mem:* Am Asn Thoracic Surg; Am Col Angiol; Am Col Cardiol; Am Col Surgeons; Am Fedn Clin Res; Sigma Xi. *Res:* Cardiovascular and thoracic surgery. *Mailing Add:* Dept Surg Univ Tex Med Br Galveston TX 77550

DERRICK, MALCOLM, b Hull, Eng, Feb 15, 33; m 57, 65; c 1. EXPERIMENTAL HIGH ENERGY PHYSICS. *Educ:* Univ Birmingham, BSc, 54, PhD(physics), 59; Oxford Univ, MA, 61. *Prof Exp:* Instr physics, Carnegie Inst Technol, 57-60; sr res officer nuclear physics, Oxford Univ, 60-63; from asst physicist to sr physicist, Argonne Nat Lab, 63-74, dir, High Energy Physics Div, 74-81. *Mem:* Fel Am Phys Soc. *Res:* Elementary particle interactions and decays; neutrino physics; electron-positron annihilation. *Mailing Add:* Argonne Nat Lab 9700 S Cass Ave Argonne IL 60439

DERRICK, MILDRED ELIZABETH, b Augusta, Ga, Oct 11, 41. PHYSICAL CHEMISTRY. *Educ:* ECarolina Univ, AB, 63; Emory Univ, MS, 65, PhD(chem), 70. *Prof Exp:* Instr gen & phys chem, Salem Col, NC, 65-67; teacher gen & org chem & chmn sci dept, Davidson County Community Col, 70-76; assoc prof, 76-78, PROF CHEM, VALDOSTA STATE COL, 78- *Concurrent Pos:* Teaching fel, Univ Ga, 79 & 80; vis lectr, Univ Ga, 80; res asst, UPAC Solubility Proj, Emory Univ, 82-85; Jessie Ball Dupont Fac Enrichment Grant, Fla State Univ, 87 & 89. *Mem:* Am Chem Soc; Sigma Xi; Am Asn Univ Prof. *Res:* Thermodynamics of solutions; physical properties of microemulsions; history of women in chemistry. *Mailing Add:* Dept Chem Valdosta State Col Valdosta GA 31698

DERRICK, ROBERT P, b Anderson, SC, July 31, 31; div; c 3. DRIVES & CONTROL SYSTEMS, PULP & PAPER INDUSTRY. *Educ:* Vanderbilt Univ, BE, 54. *Prof Exp:* Dir, Prototype Develop Processing Indust, Westinghouse Elec Co, 54-64; design group mgr, process control & elec eng dept, Simons Eastern Co Consult, 64-84; SR STAFF ENGR, CORP ENG DIV, UNION CAMP CORP, 84- *Honors & Awards:* Pulp & Paper Indust Award, Inst Elec & Electronics Engrs, 75; Tech Asn Pulp & Paper Indust Eng Award, 78. *Mem:* Fel Inst Elec & Electronics Engrs; fel Tech Asn Pulp & Paper Indust. *Mailing Add:* Union Camp Corp Eng Div Po Box 2310 Savannah GA 31402

DERRICK, WILLIAM RICHARD, b Oklahoma City, Okla, May 18, 38; m 60; c 3. MATHEMATICS. *Educ:* Okla State Univ, BS, 58, MS, 60; Ind Univ, PhD(math), 66. *Prof Exp:* Programmer, Int Bus Mach Corp, 60-61; asst prof math, Univ Utah, 66-71; vis assoc prof, Ariz State Univ, 71-72; assoc prof, 72-75, dept chmn, 80-82, PROF MATH, UNIV MONT, 75-, DEPT CHMN, 87- *Concurrent Pos:* Fulbright fel, 75; NSF res fel, 77-78. *Mem:* Am Math Soc; Resource Modeling Asn. *Res:* Complex analysis, particularly quasiconformal mappings in space and differential equations; epidemic theory. *Mailing Add:* Dept Math & Sci Univ Mont Missoula MT 59812

DERRICK, WILLIAM SHELDON, anesthesiology; deceased, see previous edition for last biography

DERRICKSON, CHARLES M, b Simpson, Ky, Apr 26, 27; m 49; c 3. ANIMAL SCIENCE. *Educ:* Univ Ky, BS, 51, MS, 56; Mich State Univ, PhD(animal sci), 65. *Prof Exp:* Asst county agent exten, Univ Ky, 52-57, supt, Robinson Agr Exp Substa, 57-65; assoc prof, 65-70, PROF ANIMAL SCI & DEAN, SCH APPL SCI & TECHNOL, MOREHEAD STATE UNIV, 75-, HEAD DEPT AGR, 68- *Concurrent Pos:* Mem, Ky State Bd Agr, 70- *Mem:* Am Soc Animal Sci. *Res:* Basic and applied research in the field of animal nutrition. *Mailing Add:* Technol Upo 721 Morehead State Univ University Blvd Morehead KY 40351

DERRYBERRY, OSCAR MERTON, occupational medicine, public health, for more information see previous edition

DERSE, PHILLIP H, b Milwaukee, Wis, Apr 6, 20. ORGANIC BIOCHEMISTRY. *Educ:* Univ Wis-Madison, PhD(biochem), 47. *Prof Exp:* PRES, DERSE & SCHROEDER ASSOC, LTD, 81- *Mem:* AAAS; Am Biol Soc. *Mailing Add:* Derse & Schroeder Assoc 1202 Ann St Madison WI 53713-2410

DERSHEM, HERBERT L, b Troy, Ohio, Mar 26, 43; m 68; c 3. COMPUTER SCIENCE. *Educ:* Univ Dayton, BS, 65; Purdue Univ, MS, 67, PhD(comput sci), 69. *Prof Exp:* From asst prof to assoc prof, 69-81, PROF MATH & COMPUT SCI, HOPE COL, 81-, CHMN, DEPT COMPUT SCI, 77- *Mem:* Asn Comput Mach; Math Asn Am; Inst Elec & Electronics Engrs. *Res:* Computer science education; programming languages; programming techniques; mathematical software. *Mailing Add:* Dept Comput Sci Hope Col Holland MI 49423

DERSHWITZ, MARK, b Dearborn, MI, Jan 27, 55; m. ANESTHESIOLOGY, DRUG METABOLISM. *Educ:* Northwestern Univ, PhD(pharmacol),82, MD, 83. *Prof Exp:* ANESTHESIOLOGIST, DEPT ANESTHESIA, MASS GEN HOSP, 83- *Mem:* Biophysical Soc; Am Soc Anesthesiologists; Int Anesthesiol Res Soc; Am Soc Pharmacol & Exp Therapeut; Am Soc Clin Pharmacol & Therapeut. *Res:* Clinical pharmacology of intravenous anesthetic, sedative and antiemetic agents; diagnosis and treatment of malignant hyperthermia. *Mailing Add:* Dept Anesthesia Mass Gen Hosp Fruit St Boston MA 02114

DERTOUZOS, MICHAEL L, b Athens, Greece. COMPUTER SCIENCE, ELECTRICAL ENGINEERING. *Educ:* Mass Inst Technol, PhD, 64. *Prof Exp:* Chmn bd, Computek, Inc, 68-74; DIR, LAB COMPUTER SCI & PROF COMPUTER SCI & ELEC ENG, MASS INST TECHNOL, 74- *Concurrent Pos:* Mem, Computer Sci & Technol Bd, Nat Res Coun. *Mem:* Nat Acad Eng;

fel Inst Elec & Electronics Engrs; corresp mem Athens Nat Acad Arts & Sci. *Res:* Computer science and technology; author of various publications. *Mailing Add:* Mass Inst Technol 1548 Technol Sq Rm 105 Cambridge MA 02139

DERUBERTIS, FREDERICK R, b Pittsburgh, Pa, Oct 8, 39; c 3. ENDOCRINOLOGY, METABOLISM. *Educ:* Univ Pittsburgh, BS, 62, MD, 65. *Prof Exp:* From asst prof to assoc prof med, Univ Pittsburgh Sch Med, 73-82; clin investr endocrinol, 73-76, chief endocrinol, 76-79, CHIEF MED, VET ADMIN MED CTR PITTSBURG, 79-; VCHMN DEPT MED, UNIV PITTSBURGH SCH MED, 79-, PROF MED, 80- *Mem:* Am Col Physicians; Am Fed Clin Res; AAAS; Am Soc Clin Invest. *Res:* Study of the mechanisms by which diabetes mellitus biochemically alters the renal glomerulus and ultimately leads to renal failure. *Mailing Add:* Vet Admin Med Ctr Univ Drive C Pittsburgh PA 15240

DERUCHER, KENNETH NOEL, b Massena, NY, Jan 24, 49; m 78; c 1. CIVIL ENGINEERING MATERIALS, ENGINEERING EDUCATION. *Educ:* Tri-State Univ, Ind, BSCE, 71; Univ NDak, Grand Forks, MSCE, 73; Va Polytech Inst, PhD(civil eng), 77. *Prof Exp:* Asst prof civil eng, Univ Md, 76-79; res consult, Civil Design Corp, 79-80; assoc prof, 80-85, PROF CIVIL ENG, STEVENS INST TECHNOL, HOBOKEN, NJ, 85-, HEAD DEPT, 82- *Concurrent Pos:* Prin investr, 77-; ed-in-chief, Civil Eng for Pract & Design Engrs, 82-86; mem, Marine Sci Adv Panel, Nat Acad Sci, 82-84; consult civil eng in litigation, 84-; mem, Transp & Mat Handling Comt, Gov Comn Sci & Technol, NJ, 85-; dean grad sch, Stevens Inst Technol, Hoboken, NJ. *Honors & Awards:* Charles B Dudley Award, Am Soc Testing & Mat, 82. *Mem:* Sigma Xi; Am Soc Eng Educators. *Res:* Micro and macro cracking of concrete and concrete making materials under various compressive stress fields and bridge and pier protective systems and devices and artificial soils. *Mailing Add:* Stevens Inst PO Box 1355 Hoboken NJ 07030

DERUSSO, PAUL M(ADDEN), b Albany, NY, Sept 9, 31; m 53; c 3. ELECTRICAL & SYSTEMS ENGINEERING. *Educ:* Rensselaer Polytech Inst, BEE, 53, MEE, 55; Mass Inst Technol, EE, 58, ScD(elec eng), 59. *Prof Exp:* From asst prof to assoc prof, 59-64, chmn systs eng div, 67-74, PROF ELEC ENG, RENSSELAER POLYTECH INST, 64-, ASSOC DEAN ENG, 74- *Concurrent Pos:* Consult, Gen Elec Co, 60-66; Du Pont Year-In-Indust prof, 66-67; field reader, off educ, US Dept HEW, 70-80; ad hoc vis, Eng Coun Prof Develop, 77-83. *Mem:* Fel Inst Elec & Electronics Engrs; Am Soc Eng Educ. *Res:* Systems engineering, especially automatic control systems. *Mailing Add:* Jonsson Eng Ctr Rensselaer Polytech Inst Troy NY 12181

DERVAN, PETER BRENDAN, b Boston, Mass, June 28, 45; m 75; c 2. ORGANIC CHEMISTRY, BIOCHEMISTRY. *Educ:* Boston Col, BS, 67; Yale Univ, PhD(chem), 72. *Prof Exp:* NIH fel chem, Stanford Univ, 72-73; from asst prof to assoc prof, 73-88, BREN PROF CHEM, CALIF INST TECHNOL, 88- *Concurrent Pos:* Alfred P Sloan Found fel, 77; Camille & Henry Dreyfus teacher scholar, Camille & Henry Dreyfus Found, 78; John Simon Guggenheim mem fel, 83; vis prof, Eidgenössische Tech Sch, Zurich, 83; mem, Bio-Organic & Natural Prod Chem Study Sect, NIH, 84; mem, Adv Panel Chem Life Sci, NSF, 85, Adv Comt Chem, 86- 89, chmn, 88-89; T Y Shen vis prof med chem, Mass Inst Technol, 87; Morris S Kharasch vis prof, Univ Chicago, 88; mem sci adv bd, Robert A Welch Found, 88-; mem, Bd Chem Sci & Technol, Nat Res Coun, 88-, co-chmn, 91-; Alexander Todd vis prof, Univ Cambridge, Eng, 89; mem, Am Chem Soc Joint Bd-Coun Comt Sci, 89; Walker-Ames vis prof, Univ Wash, 90. *Honors & Awards:* Hanes Willis Lectr, Univ NC, 85; McElvain Lectr, Univ Wis, 85, Perlman Lectr, 88; Abbott Lectr, Yale Univ, 86, Univ NDak, 89; R C Fuson Lectr, Univ Nev, 87; Camille & Henry Dreyfus Lectr, Dartmouth Col, 87; Harrison Howe Award, 88; H Smith Broadbent Lectr, Brigham Young Univ, 88; Rosetta Briegel Barton Lectr, Univ Okla, 88; DeWitt Stetton Jr Lectr, NIH, 88; Clifford B Purvis Lectr, McGill Univ, 89; Walker-Ames Lectr, Univ Wash, 90; Max Hoffer Mem Lectr, Hoffmann La Roche, 91; Alexander von Humboldt Sr US Scientist Award, 90. *Mem:* Nat Acad Sci; fel Am Acad Arts & Sci. *Res:* Physical organic, especially reaction mechanisms; biophysical organic, especially DNA recognition. *Mailing Add:* Div Chem 164-30 Calif Inst Technol 391 S Holliston Ave Pasadena CA 91125

DERVARTANIAN, DANIEL VARTAN, b Boston, Mass, July 16, 33; m 64; c 2. BIOCHEMISTRY. *Educ:* Boston Univ, AB, 56; Northeastern Univ, MSc, 59; Univ Amsterdam, ScD, 65. *Hon Degrees:* ScD, Univ Amsterdam, 65. *Prof Exp:* Res fel biochem, Univ Amsterdam, 61-65; res assoc, Univ Wis-Madison, 65-68; from asst prof to assoc prof, 68-78, PROF BIOCHEM, UNIV GA, 78-, ASSOC DIR BIOCHEM, SCH CHEM SCI, 85- *Concurrent Pos:* Res career develop award, Nat Inst Gen Med Sci, 71-76; mem, Int Union Pure & Appl Chem & Int Union Biochem subcomt Cytochrome Nomenclature, 73-82. *Mem:* Am Soc Microbiol; Am Soc Biol Chem. *Res:* Role of respiratory chain in energy conservation; function and structure of metal (redox-active) containing proteins by electron spin resonance spectroscopy; study of bacterial mutants in energy coupling. *Mailing Add:* Dept Biochem Univ Ga Boyd Grad Studies Res Athens GA 30602

DERZKO, NICHOLAS ANTHONY, b Kapuskasing, Ont, Jan 19, 40; m 66; c 4. APPLIED MATHEMATICS. *Educ:* Univ Toronto, BS, 62; Calif Inst Technol, PhD(math, physics), 65; York Univ, MBA, LLB, 88. *Prof Exp:* Asst prof, 65-70, ASSOC PROF MATH, UNIV TORONTO, 70- *Concurrent Pos:* Software sci appln, Clarke, Derzko & Assocs, Ltd; law pract, corp & intellectual property. *Mem:* Soc Indust & Appl Math; Am Math Soc; Can Math Soc. *Res:* Matrix theory; mathematical scattering theory; Monte Carlo methods; mathematical economics. *Mailing Add:* Dept Math Univ Toronto Toronto ON M5S 1A1 Can

DE SA, RICHARD JOHN, b New York, NY, Aug 4, 38; m 59; c 3. ENZYMOLOGY. *Educ:* St Bonaventure Univ, BS, 59; Univ Ill, PhD(biochem), 64. *Prof Exp:* Trainee biochem, Johnson Res Found, Univ Pa, 64-65 & Cornell Univ, 65-68; from asst prof to assoc prof biochem, Univ Ga, 68-74; PRES, ON-LINE INSTRUMENT SYSTS, 80- *Mem:* Am Soc Biol Chemists; Am Inst Chemists; AAAS. *Res:* Bioluminescence of marine organisms; enzyme kinetics, particularly flavin enzymes; instrumental design and construction; computerization and electro-mechanical enhancement and/or simplification of kinetic and spectrophotomeric instruments; specializing in high speed optical techniques. *Mailing Add:* Rte 2 Box 111 Jefferson GA 30549

DESAI, BIPIN C, b Rangoon, Burma, Oct 28, 39; Can citizen; m 74; c 3. DATABASE SYSTEMS, OFFICE INFORMATION SYSTEMS. *Educ:* Jadaupur Univ, BEE, 62; Purdue Univ, MSEE, 65; McGill Univ, PhD(elec eng), 77. *Prof Exp:* Program engr, Gen Electric, 63-64; systs engr, Xerox Corp, 65-66 & CAE Electronics Ltd, 66-70; ASSOC PROF COMPUT SCI, CONCORDIA UNIV, 70- *Concurrent Pos:* Sessional lectr, Loyola Col, 67-70. *Mem:* Inst Elec & Electronics Engrs; Asn Comput Mach; Brit Comput Soc. *Res:* Architectural improvement for database, operating system and office automation applications. *Mailing Add:* Concordia Univ 7141 Sherbrooke St W Montreal PQ H4B 1R6 Can

DESAI, BIPIN RATILAL, b Hansot, India, Oct 6, 35; m 61. HIGH ENERGY PHYSICS. *Educ:* Univ Bombay, BSc, 54; Univ Ill, MS, 57; Univ Calif, Berkeley, PhD(physics), 61. *Prof Exp:* Physicist, Lawrence Radiation Lab, Univ Calif, Berkeley, 61; res assoc physics, Ind Univ, 61-63 & Univ Wis, 63-64; asst res physicist, Univ Calif, Los Angeles, 64-65; from asst prof to assoc prof, 65-71, PROF PHYSICS, UNIV CALIF, RIVERSIDE, 71- *Concurrent Pos:* Prin investr, Dept Energy contract; vis scientist, CERN Europ Orgn Nuclear Res, Geneva, Switz, 70 & 83; fel, Cavendish Lab, Univ Cambridge, Eng, 70; vis scientist, CEN Saclay, France, 73, Rutherford High Energy Lab, Eng; vis prof, Univ Paris VI, France, 87; prin investr, NSF grant. *Mem:* Fel Am Phys Soc. *Res:* Quark confinement; quantum chromo dynamics; supersymmetry models; elementary particle masses. *Mailing Add:* Dept Physics Univ Calif Riverside CA 92521

DESAI, INDRAJIT DAYALJI, b Nairobi, Kenya, Jan 7, 32; Can citizen. NUTRITION, BIOCHEMISTRY. *Educ:* Nat Dairy Res Inst, India, IDD, 50; Gujarat Univ, India, BSc, 54, MSc, 58; Univ Calif, Davis, PhD(nutrit), 63. *Prof Exp:* Instr dairy sci, Gujarat Univ, India, 54-55; res assoc nutrit, Cornell Univ, 63-64; res assoc biochem, 65-67, asst prof human nutrit, 67-70, assoc prof, 70-74, coordr div human nutrit, 71-72, 76-77, 81-82, dir continuing educ nutrit & dietetics, 71-84, PROF HUMAN NUTRIT, HEALTH SCI CTR, UNIV BC, 74- *Concurrent Pos:* US Educ Found, 58-61; vis prof, Med Sch, Univ Sao Paulo, Ribeirao Preto, Brazil, 77-; chmn, Comn IV, Comt 4, Int Union Nutrit Sci, 82-90; contrib ed, Nutrit Reviews, 86-90; assoc ed, Nutrit Notes, 87- *Honors & Awards:* Gilmore Award, 63; Pediat Soc Res Award, 82. *Mem:* Am Inst Nutrit; Nutrit Today Soc; Can Soc Nutrit Sci; Indian Dairy Sci Asn; Sigma Xi. *Res:* Nutritional aspects of vitamins; nutritional studies of vitamins E and A; lipid peroxidation; ceroids and biology of aging; food habits and assessment of nutritional status; clinical studies on vitamin E in infants and nutritional aspects of human milk; international studies on malnutrition in developing countries, specifically Brazil. *Mailing Add:* Sch Family & Nutrit Sci Univ BC Vancouver BC V6T 1W5 Can

DESAI, KANTILAL PANACHAND, b Mota-Samadhiala, India, Feb 7, 29; US citizen; m 60; c 3. EXPLORATION GEOPHYSICS. *Educ:* Univ Bombay, BS, 52; Colo Sch Mines, GpE, 56, MS, 57; Univ Tulsa, PhD(petrol eng), 68. *Prof Exp:* Trainee well logging, Seismog Serv Corp, 58, log analyst, Birdwell Div, 58-61, area engr, 61-62; from geophysicist to sr res geophysicist, Res Ctr, Sinclair Oil Co, 62-69, sr res engr, Atlantic Richfield Co, 69; sr res engr, Field Res Lab, Mobil Oil Corp, 69-84; PETROL ENG SPECIALIST, ARAMCO, 85- *Mem:* Soc Prof Well Log Analysts (pres, 74-75); Soc Petrol Engrs. *Res:* Design and development of laboratory measuring systems which precisely and sequentially measure both the longitudinal and shear velocities of a rock sample under triaxial pressure. *Mailing Add:* 6006 Hunters View Dallas TX 75232

DESAI, PARIMAL R, b India, Sept 16, 36; US citizen. BIOLOGY. *Educ:* Univ Bombay, India, BSc hons, 56; Drexel Univ, Philadelphia, MS, 62; Northwestern Univ, PhD(microbiol-immunol), 70. *Prof Exp:* Res asst immunochem res, Evanston Hosp, 63-69; res assoc, 70-74, sr res assoc, 75-77, assoc med staff, sr res immunochemist & assoc dir immunochem res, 77-89, ASSOC DIR, H M BLIGH CANCER BIOL RES LAB, UHS/CHICAGO MED SCH, 89-, ASSOC PROF MICROBIOL/IMMUNOL, 90- *Mem:* Am Asn Immunol; Am Asn Cancer Res; AAAS; NY Acad Sci; Indian Immunol Soc. *Res:* Isolation and immunochemical characterization of human carcinoma-associated T and Tn antigens and antibodies, their role in patients' humoral and cellular immune reactivities, and disease status; author of over 130 publications. *Mailing Add:* H M Bligh Cancer Biol Res Lab Chicago Med Sch 3333 Green Bay Rd North Chicago IL 60064

DESAI, PRAMOD D, b Sangamner, India, Sept 11, 39; US citizen; m 67; c 3. THERMOPHYSICS. *Educ:* Tex A&M Univ, PhD(chem), 68. *Prof Exp:* Res assoc thermodynamics, Univ Calif, Berkeley, 68-72; ASSOC SR RES THERMOPHYS, PURDUE UNIV, 73- *Mem:* Am Chem Soc; Sigma Xi. *Res:* Collection, correlation analysis and synthesis of thermodynamic and thermophysical properties of metals, alloys, steels, polymers, minerals, rocks and other technologically important materials; generation of reference data values. *Mailing Add:* Dept CINDAS 2595 Yeager Rd West Lafayette IN 47906

DESAI, PRATEEN V, b Baroda, India, Aug 14, 36. MECHANICAL ENGINEERING. *Educ:* Univ Baroda, BEng, 59; Va Polytech Inst, MS, 63; Tulane Univ, PhD(mech eng), 67. *Prof Exp:* Asst engr, Nat Mach Mfrs, India, 59-61; asst prof, 66-71, ASSOC PROF MECH ENG, GA INST TECHNOL, 71- *Concurrent Pos:* Consult, Lockheed-Ga Co, 68- *Mem:* Am Soc Eng Educ. *Res:* Thermal sciences; turbulent boundary layers; fluid vibrations; fluidics; biomechanics; whiplash studies. *Mailing Add:* Sch Mech Eng Ga Inst Technol Atlanta GA 30332

DESAI, RAJENDRA C, b Junagadh, India, Nov 7, 23; US citizen; m 55; c 4. HEMATOLOGY. *Educ:* Univ Bombay, MB, BS, 49; Boston Univ, PhD(physiol), 55. *Prof Exp:* Indian Coun Med Res fel hemat, 49-52; res fel, New Eng Ctr Hosp, Boston, 52-55; Fulbright scholar, 52-55; Damon Runyon fel, 53-55; hematologist, Nat Med Col, India, 56-57; res assoc hemat, Sch Med, Stanford Univ, 57-62; asst prof med, Sch Med, Boston Univ & Univ Hosp, 62-65; chief hemat, Orange County Gen Hosp, 65-69; DIR ONCOL, MERCY GEN HOSP, 78-; ASST CLIN PROF, SCH MED, UNIV CALIF, IRVINE, 69- *Concurrent Pos:* Anna Fuller Fund travel award, Far East, 60. *Mem:* Am Fedn Clin Res; fel Am Col Angiol; Microcirculatory Soc; fel Int Soc Hemat. *Res:* Microcirculation; transplantation immunity; kinetics of cell transfer across placenta; clinical and therapeutic aspects of various blood disorders. *Mailing Add:* 1001 Thornton Rd Suite 306 PO Box 512 Lithia Springs GA 30057

DESAI, RASHMI C, b Amod, India, Nov 21, 38; m 63; c 2. THEORETICAL PHYSICS, THEORETICAL CHEMISTRY. *Educ:* Univ Bombay, BSc, 57; Cornell Univ, PhD(appl physics), 66. *Prof Exp:* Trainee physics, Atomic energy Estab, Trombay, Bombay, India, 57-58, sci officer theoret physics, 58-62; res assoc statist physics, Mass Inst Technol, 66-68; coordr grad studies, 83-87, asst prof, 68-78, PROF PHYSICS, UNIV TORONTO, 78- *Concurrent Pos:* Vis assoc, Calif Inst Technol, 75; vis scientist, IBM Res Lab, San Jose, Calif, 81-82. *Mem:* Am Phys Soc; Am Asn Physics Teachers; Can Asn Physicists; Mat Res Soc. *Res:* Nonequilibrium and equilibrium statistical mechanics; molecular transport phenomena in liquids and gases; surface physics; dynamics of phase transitions; inelastic neutron and light scattering; kinetic theory of molecular fluids. *Mailing Add:* Dept Physics Univ Toronto Toronto ON M5S 1A7 Can

DESAI, VINODRAI RANCHHODJI, organic polymer chemistry, for more information see previous edition

DE SALVA, SALVATORE JOSEPH, b New York, NY, Jan 14, 24; m 48; c 8. NEUROLOGY, PHARMACOLOGY. *Educ:* Marquette Univ, BS, 47, MS, 49; Loyola Univ, PhD, 57. *Prof Exp:* Asst anat, Marquette Univ, 47-49; asst neuroanat, Sch Med, Univ Ill, 50-51, instr, 51-52; asst prof anat & physiol, Chicago Col Optom, 51-53; pharmacologist, Armour Lab, 53-59; head pharmacol sect, Biol Res Lab, Colgate Palmolive Co, 59-66, sr res assoc, 66-72, mgr pharmacol & toxicol, 72-76, assoc dir res, 76-80, dir, 80-88, worldwide dir, 88-90, CORP DIR, HUMAN-ENVIRON SAFETY, COLGATE PALMOLIVE CO, 90- *Concurrent Pos:* Biochemist, Milwaukee County Gen Hosp, 50-51; lectr pharmacol, Stritch Sch Med, Loyola Univ, 57- *Mem:* AAAS; Soc Exp Biol & Med; NY Acad Sci; Soc Pharmacol & Exp Therapeut; Inst Elec & Electronic Engrs; Sigma Xi; Int Regulatory Soc Pharmacol & Toxicol; Soc Toxicol. *Res:* Forebrain of primate; cytoarchitectonic of cerebral cortex in man, primate and squirrel; anti-convulsion and brain excitability; brain excitability and endocrine; interdependencies; pulmonary pharmacology; analgesimetry; dental pharmacology; experimental toxicology of surfactants; experimental dermatology; pharmacokinetics; risk assessments. *Mailing Add:* 83 De Mott Lane Somerset NJ 08873

DESANCTIS, ROMAN WILLIAM, b Cambridge Springs, Pa, Oct 30, 30; m; c 4. CARDIOLOGY. *Educ:* Univ Ariz, BS, 51; Harvard Univ, MD, 55; Am Bd Internal Med, cert, 62. *Hon Degrees:* DSc, Wilkes Col. *Prof Exp:* Dir, Coronary Care Unit, Mass Gen Hosp, 67-81; clin prof, 73-74, PROF MED, HARVARD MED SCH, 74-; DIR CLIN CARDIOL, MASS GEN HOSP, 81-, ACTG CHIEF, CARDIAC UNIT, 89- *Honors & Awards:* Somers Mem Lectr, Portland, Ore, 72; Hamilton Southworth Lectr, Columbia Presby Med Sch, 79; Robert Flinn Mem Lectr, St Joseph's Hosp, Phoenix, Ariz, 84. *Mem:* Inst Med-Nat Acad Sci; Am Heart Asn; fel Am Col Physicians; fel Am Col Cardiol; Asn Am Physicians; Asn Univ Cardiologists. *Res:* Ischemic heart disease, acute and chronic; diseases of the aorta; cardiac arrhythmias; the medical-surgical cardiovascular interface. *Mailing Add:* Ambulatory Care Ctr Mass Gen Hosp 15 Parkman St Suite 467 Boston MA 02114

DESANTIS, JOHN LOUIS, b Bradford, Pa, Sept 7, 42; m 67; c 3. PRODUCT DEVELOPMENT, TECHNICAL SERVICES. *Educ:* St Bonaventure Univ, BS, 65, MS, 67. *Prof Exp:* Develop engr, Air Reduction Co, 68-71; res & develop chemist, Graphite Prod Corp, 71-72; sr res & develop chemist, Man-Gill Chem Co, 72-77; TECH DIR, MID-STATE CHEM & SUPPLY CORP, 77- *Mem:* Am Chem Soc; Soc Tribologists & Lubrication Engrs; Am Soc Metals; fel Am Inst Chemists. *Res:* Metal working and metal finishing chemicals; iron phosphates, lubricants, cleaners, paint strippers and anti-rust products. *Mailing Add:* 668 Hawthorne Dr Carmel IN 46032-9411

DESANTIS, MARK EDWARD, b Vineland, NJ, May 9, 42; m 69; c 1. NEUROBIOLOGY. *Educ:* Villanova Univ, BS, 63; Creighton Univ, MS, 66; Univ Calif, Los Angeles, PhD(anat), 70. *Prof Exp:* Res assoc neurophysiol, Naval Aerospace Med Res Inst, 70-71; from instr to assoc prof anat, Georgetown Univ, 71-78; assoc prof, 78-82, PROF BIOL SCI, WASH, ALASKA, MONT & IDAHO MED EDUC PROG, UNIV IDAHO, 82- *Concurrent Pos:* Res assoc, Nat Res Coun, 70-71; mem, Biol & Neurosci Subcomt, NIMH, 80; vis prof, Yale Univ, 88, Friday Harbor Labs, Univ Wash, 83, 84, 86, Shannon Point Marine Ctr, Western Wash Univ, 88. *Mem:* AAAS; Soc Neurosci; Am Asn Anatomists; Sigma Xi; Am Soc Zoologists; Electron Micros Soc Am. *Res:* Development, degeneration and regeneration of nervous tissue; structure and function of muscle receptors and their central nervous system projections; echinoderm and mustelid neuroendocrine systems. *Mailing Add:* Dept Biol Sci & WAMI Med Prog Univ of Idaho Moscow ID 83843

DESANTO, DANIEL FRANK, b New Rochelle, NY, June, 21, 30; m 65. FLUID-STRUCTURE INTERACTION, VIBRATION. *Educ:* NY Univ, BAeroE, 52, MAeroE, 53, DEngSc, 61. *Prof Exp:* Res asst, Col Eng, NY Univ, 52-56, assoc res scientist, 59-63; flight test res engr, Grumman Aircraft Eng Corp, 56-59; prin aerodynamicist, Cornell Aeronaut Lab, Inc, 63-70; sr engr, Res & Develop Ctr, 70-85, FEL ENGR, WESTINGHOUSE MACH TECHNOL DIV, 85- *Concurrent Pos:* Instr, Manhattan Col, 56; adj asst prof, Col Eng, NY Univ, 62-63. *Mem:* Am Soc Mech Engrs; Am Inst Aeronaut & Astronaut; Acoust Soc Am. *Res:* Experimental and theoretical studies of flow-induced vibrations and dynamic response of structures to unsteady forcing functions including fluid coupling; acoustics. *Mailing Add:* 15 Morris St Export PA 15632

DESANTO, JOHN ANTHONY, b Wilkes-Barre, Pa, May 25, 41; m 64; c 3. THEORETICAL PHYSICS, ACOUSTICS. *Educ:* Villanova Univ, BS & MA, 62; Univ Mich, MS, 63, PhD(physics), 67. *Prof Exp:* Res physicist, Acoust Div, Naval Res Lab, 67-78, res physicist, Ocean Sci Div, 78-81; sr scientist & prog mgr theoret physics, Electro Magnetic Applications, Inc, 81-; prof math, Univ Denver, 82-83; PROF MATH, COLO SCH MINES, 83- *Concurrent Pos:* Exchange scientist, Admiralty Res Lab, Teddington, UK, 74-75. *Mem:* Fel Acoust Soc Am; Am Phys Soc; Soc Indust & Appl Math; Sigma Xi; Inst Elec & Electronics Engrs. *Res:* Sound scattering from periodic surfaces and from random rough surfaces; propagation of sound in both deterministic and stochastic waveguides; theoretical foundations of acoustics and oceanography; electromagnetic scattering and propagation and theory. *Mailing Add:* De Santo Res 7692 S Saulsbury Ct Littleton CO 80123

DESANTO, ROBERT SPILKA, b New Rochelle, NY, Sept 21, 40; m 64; c 2. ENVIRONMENTAL MANAGEMENT, ENVIRONMENTAL AUDITING. *Educ:* Tufts Univ, BS, 62; Columbia Univ, PhD, 68. *Prof Exp:* Lectr zool, Columbia Univ, 62-64; asst prof, Conn Col, 68-72, dir summer marine sci prog, 69-72; chief ecol scientist, C E Maguire, Inc, 72-74, staff assoc, 74; co-founding dir & chief ecol scientist, Environ Serv, Comsis Corp, 74-77; chief ecologist, 77-83, CHIEF SCIENTIST, DELEUW, CATHER CO, 83- *Concurrent Pos:* Pres, Thames Sci Ctr, 72-73; ed-in-chief, Environ Mgt, 74-78, ed, 79-, Environ Auditor, 88-; Triaminobenzene environ comt; chmn environ mediation & commun subcomt. *Mem:* Sci Res Soc; Syst Safety Soc; Am Soc Zool; Soc Wetland Sci; Soc Ecol Restoration Mgt. *Res:* Environmental monitoring, analysis, and management relative to environmental impacts and applied ecology; forensic ecology; environmental auditing; mediation. *Mailing Add:* Eight Sylvan Glen East Lyme CT 06333

DE SAPIO, RODOLFO VITTORIO, b New York, NY, Aug 16, 36. MATHEMATICS. *Educ:* Univ Chicago, MS, 61, PhD(math), 64. *Prof Exp:* Instr math, Stanford Univ, 64-66; asst prof, Univ Calif, Los Angeles, 66-69; assoc prof, Belfer Grad Sch Sci, Yeshiva Univ, 69-71; ASSOC PROF MATH, UNIV CALIF, LOS ANGELES, 71- *Concurrent Pos:* Mem, Inst Advan Study, 68-69; NSF grants. *Mem:* Am Math Soc; Math Asn Am; NY Acad Sci. *Res:* Topology and geometry of manifolds; algebraic topology, including homotopy theory and homology theory as applied to classification questions in differential topology. *Mailing Add:* Dept Math Univ Calif Los Angeles CA 90024

DE SAUSSURE, GERARD, b Geneva, Switz, Nov 22, 24; US citizen; m 55; c 2. NUCLEAR PHYSICS. *Educ:* Swiss Fed Inst Technol, dipl, 49; Mass Inst Technol, PhD(physics), 54. *Prof Exp:* Res asst physics, Mass Inst Technol, 52-54; PHYSICIST, NEUTRON PHYSICS DIV, OAK RIDGE NAT LAB, 55- *Mem:* Am Phys Soc; Am Nuclear Soc; Ital Phys Soc. *Res:* Measurement of neutron transport parameters by the pulsed neutron source technique; measurement of neutron cross sections, especially of fissionable isotopes, by the time of flight technique. *Mailing Add:* 100 Windham Rd Oak Ridge TN 37830

DESAUSSURE, RICHARD LAURENS, JR, b Macon, Ga, Dec 29,17; m 52. NEUROSURGERY. *Educ:* Univ Va, AB, 39, MD, 42; Am Bd Neurol Surg, 46-49. *Prof Exp:* Intern, Univ Va Hosp, 42-43, resident neurol surg, 46-49; resident neuropath & neurophysiol, Cincinnatti Gen Hosp, 47-48; asst chief neurosurg, Kennedy Vet Admin Hosp, 49-50; asst prof, Univ Tenn, Memphis, 50-65, clin assoc prof, 65-70, prof neurosurg, Med Sch, 71-87, asst dean grad med educ, 80-87; RETIRED. *Concurrent Pos:* Pvt pract, 50-; mem exec comt, Baptist Mem Hosp, 62 & 65-67, pres staff, 66- chief of staff, 69; mem, adv coun neurol surg, Am Col Surgeons, 64-68, chmn, 68; mem, Am Bd Neurol Surg, 66-72, secy, 70-73; mem, Am Bd Med Specialties, 70-73. *Mem:* Am Med Asn; Cong Neurol Surg (pres, 61-62); Am Col Surgeons; Am Acad Neurol Surg; Am Asn Neurol Surgeons (vpres 67-68, pres-elect, 74, pres, 75-76); Soc Neurol Surgeons; Soc Computerized Tomography. *Res:* Concussion experiments. *Mailing Add:* 4290 Heatherwood Lane Memphis TN 38117

DESAUTELS, EDOUARD JOSEPH, b Winnipeg, Man, Jan 18, 38; m 61; c 3. COMPUTER SCIENCES. *Educ:* Univ Man, BSc, 60; Univ Ottawa, MSc, 64; Purdue Univ, PhD(comput sci), 69. *Prof Exp:* Sci programmer comput sci, IBM Sci Ctr, Ottawa, 60-62; systs programmer comput sci, Coop Comput Lab, Mass Inst Technol, 62-63; fel comput sci, IBM Systs Res Inst, New York, 63-64; instr comput sci, Purdue Univ, 66-69; from asst prof to assoc prof, 69-82, PROF COMPUT SCI, UNIV WIS-MADISON, 83- *Concurrent Pos:* Consult, Bur Purchases & Serv, Printing Bur, State of Wis, 75. *Mem:* Asn Comput Mach; Inst Elec & Electronics Engrs; Can Info Processing Soc. *Res:* Programming systems; minicomputer and microcomputer systems; computer networks; intelligent terminals; personal computers; managing computing. *Mailing Add:* Comput Sci Dept Univ Wis 1210 W Dayton Madison WI 53706

DESBOROUGH, GEORGE A, b Panama, Ill, Jan 15, 37; m 66. GEOLOGY, MINERALOGY. *Educ:* Southern Ill Univ, BA, 59, MA, 60; Univ Wis, PhD(geol), 66. *Prof Exp:* Res assoc geol, Univ Wis, Madison, 64-66, fel, 66; geologist, Br Astrogeol, 66-67 SR RES GEOLOGIST, BR CENT MINERAL RESOURCES, US GEOL SURV, 67- *Concurrent Pos:* Consult, Ray-O-Vac Res & Develop, 62 & US Forest Serv Regional Off, 64; assoc prof, Dept Geol, Colo State Univ, 70; Mineral Soc Am rep, Int Comn Ore Micros, 71-73, secy, Comn, 74-78, chmn, Comn, 78-82; assoc ed, Am Mineral, 79-80. *Mem:* Fel Mineral Soc Am; fel Geol Soc Am; Mining Soc Gt Brit; Soc Econ Geologists; Am Inst Mining, Metall & Petrol Eng. *Res:* Mineralogy and origin of ore deposits. *Mailing Add:* Br Cent Mineral Resources Bldg 25 Fed Ctr Denver CO 80225

DESBOROUGH, SHARON LEE, b Carbondale, Ill, Dec 22, 35. PLANT GENETICS. *Educ:* Southern Ill Univ, BA, 58, MA, 59; Univ Wis, PhD(genetics), 67. *Prof Exp:* Res assoc genetics, Univ Wis, 60-62, fel hort, 68-69; res fel, 69-72, from asst prof to assoc prof, 72-83, PROF HORT, UNIV MINN, 83- *Mem:* Am Genetics Asn; Genetics Soc Am; Am Potato Asn; Europ Potato Asn. *Res:* Potato-biochemical genetics, improvement of potato protein, isoenzymes and proteins from tubers; evolutionary species relationships. *Mailing Add:* 111 Hort Univ Minn St Paul MN 55108

DESCARRIES, LAURENT, b Montreal, Que, Jan 27, 39; m 62; c 3. NEUROBIOLOGY, NEUROANATOMY. *Educ:* Univ Paris, BA, 56; Univ Montreal, MD, 61; FRCP(C), 66. *Prof Exp:* Intern med, Maisonneuve Hosp, Montreal, 60-61; resident, Notre Dame Hosp, 61-62 & Maisonneuve Hosp, 62-63; from asst prof to assoc prof, 69-77, PROF PHYSIOL, FAC MED, UNIV MONTREAL, 77-, MEM CTR RES SCI NEUROL, 70- *Concurrent Pos:* Fel neurol & neuropath, Mass Gen Hosp, Boston, 63-67; clin & res fel neurol, Harvard Med Sch, 63-64, res fel neuropath, 64-65 & exp neuropath, 65-67; Med Res Coun Can fel, 63-67, centennial fel, 67-69, scholar, 69-74; fel neurobiol, Commissariat Atomic Energy, Ctr Nuclear Studies, France, 67-69. *Mem:* Can Asn Neurosci; Can Asn Anat; Am Asn Anat; Soc Neurosci. *Res:* Neurocytology; transmitters and modulators; monoaminergic systems; neuronal plasticity. *Mailing Add:* Ctr for Res in Neurol Sci Univ Montreal Montreal PQ H3C 3J7 Can

DESCHAMPS, GEORGES ARMAND, b Vendome, France, Oct 18, 11; US citizen; m 43; c 3. ELECTRICAL ENGINEERING, APPLIED MATHEMATICS. *Educ:* Univ Paris, Licence, 33, Agrege(math), 34. *Prof Exp:* Sr scientist fed telecom lab, ITT Corp, 58; prof elec eng & dir, electromagnetics lab, 58-83, EMER PROF ELEC ENG, UNIV ILL, 83- *Concurrent Pos:* Chmn, US Comn B Fields & Waves, Int Union Radio Sci, 79- *Mem:* Nat Acad Eng; Am Phys Soc; fel Inst Elec & Electronics Engrs; fel AAAS. *Res:* Physics; electromagnetics. *Mailing Add:* 920 W Charles St Champaign IL 61821

DESCHNER, ELEANOR ELIZABETH, b Jersey City, NJ, Oct 18, 28; c 2. CELL KINETICS, CARCINOGENESIS. *Educ:* Notre Dame Col, Staten Island, BA, 49; Fordham Univ, MS, 51, PhD(biol), 54. *Prof Exp:* Asst, Fordham Univ, 51-52; jr tech specialist, Brookhaven Nat Lab, 52, res collabr, 54; USPHS fel, Nat Cancer Inst & Brit Empire Cancer Campaign, Res Unit Radiobiol, Mt Vernon Hosp, Eng, 54-57; res assoc, Columbia Univ, 58-59; res assoc, 60-63, asst prof radiol, 63-76, asst prof radiobiol med, 68-76, ASSOC PROF RADIOBIOL IN MED, CORNELL UNIV, 76-; ASSOC MEM, SLOAN-KETTERING INST CANCER RES, 75- *Concurrent Pos:* AEC fel, 52-54; assoc, Sloan-Kettering Inst Cancer Res, 71-75, assoc mem & head, Lab Digestive Tract Carcinogenesis, 75-; asst radiobiologist, Mem Hosp Cancer & Allied Dis, 72-78, assoc radiobiologist, 78-; mem path subcomt, Nat Large Bowel Cancer Proj; sci comt cancer res unit, Clin St Michel, Brussels, Belg, 77-80; mem, Int Study Group Gastric Cancer, 78- *Honors & Awards:* Auxiliary Lectr, Am Col Gastroenterol, 81. *Mem:* AAAS; Radiation Res Soc; Sigma Xi; Am Asn Cancer Res; Am Gastroenterol Asn. *Res:* Carcinogenesis; autoradiographic methods for early detection of colon cancer; cell kinetic studies of the gastrointestinal tract during carcinogenesis; nutritional modification of cell proliferation and tumorigenesis. *Mailing Add:* Mem Sloan-Kettering Cancer Ctr 1275 York Ave New York NY 10021

DESELM, HENRY RAWIE, b Columbus, Ohio, Nov 1, 24; m 48; c 2. PLANT ECOLOGY. *Educ:* Ohio State Univ, PhD(bot), 53. *Prof Exp:* Asst bot, Ohio State Univ, 50-53, asst instr, 54; instr biol, Mid Tenn State Col, 54-56; instr & res assot bot, 56-60, from asst prof to assoc prof, 60-73, PROF BOT, UNIV TENN, KNOXVILLE, 73- *Mem:* Fel AAAS; Ecol Soc Am; Bot Soc Am. *Res:* Natural vegetation distribution and control; primary production; mineral and fission product cycling; calciphiles; ecological races; remote sensing of environment. *Mailing Add:* Dept Bot & Grad Prog Ecol Univ Tenn Knoxville TN 37996-1100

DE SELMS, ROY CHARLES, b San Pedro, Calif, Dec 17, 32; m 59; c 2. ORGANIC CHEMISTRY. *Educ:* Univ Wash, BS, 54; Stanford Univ, PhD(chem), 59. *Prof Exp:* Res chemist, Res Labs, Eastman Kodak Co, NY, 59-62; fel alkaloid biosynthesis, Univ Calif, Berkeley, 62-63; res chemist, Ortho Div, Chevron Chem Co, 63-66; sr res chemist, 66-70, RES ASSOC, EASTMAN KODAK CO, 70- *Concurrent Pos:* Instr, Univ Calif, Berkeley, 64-66; assoc instr, Eve Exten, Univ Rochester, 67-76. *Mem:* Am Chem Soc. *Res:* Organic chemistry of heterocyclics, carbenes, alicyclics, pesticides and photoreproduction. *Mailing Add:* Eastman Kodak Res Labs B82 Kodak Park Rochester NY 14650

DESER, STANLEY, b Poland, Mar 19, 31. THEORETICAL PHYSICS. *Educ:* Brooklyn Col, BS, 49; Harvard Univ, MA, 50, PhD(physics), 53. *Hon Degrees:* DPhil, Stockholm Univ, 78. *Prof Exp:* Mem & Jewett fel, Inst Advan Study, Princeton, 53-55; NSF fel, Inst Theoret Physics, Denmark, 55-57; lectr physics, Harvard Univ, 57-58; from assoc prof to prof, 58-80, ANCELL PROF PHYSICS, BRANDEIS UNIV, 80- *Concurrent Pos:* Res assoc, Radiation Lab, Univ Calif, 54; Fulbright & Guggenheim fels, 66-67; vis prof, Univ Sorbonne, 66-67, 71-72; Fulbright prof, Univ of Repub, Uruguay, 70; Loeb lectr, Harvard Univ, 75-76; NATO sr res fel, 76; vis sr scientist, Europ Orgn Nuclear Res, Geneva, 61-62, 76 & 80-81; vis fel, All Souls, Oxford, 77; vis prof, Col France, 76, 84, Boston Univ & Inst Adv Study, 87-88. *Mem:* Fel Am Phys Soc; fel Am Acad Arts & Sci. *Res:* Elementary particle physics; field theory; relativity. *Mailing Add:* Dept Physics Brandeis Univ Waltham MA 02254

DE SERRES, FREDERICK JOSEPH, b Dobbs Ferry, NY, Sept 24, 29; m 54; c 5. GENETIC TOXICOLOGY, ENVIRONMENTAL MUTAGENESIS & CARCINOGENESIS. *Educ:* Tufts Univ, BS, 51; Yale Univ, MS, 53, PhD(bot), 55. *Hon Degrees:* Dr, Catholic Univ Louvain, Belg, 87. *Prof Exp:* Res assoc, Biol Div, Oak Ridge Nat Lab, 55-57, sr staff biologist, 57-72, coordr environ mutagenesis prog, 69-72, lectr, Univ Tenn-Oak Ridge Grad Sch Biomed Sci, 71-72; chief, Environ Mutagenesis Br, Nat Inst Environ Health Sci, 72-76, assoc dir genetics, 76-86; DIR, CTR LIFE SCI & TOXICOL, RES TRIANGLE INST, 86- *Concurrent Pos:* Mem comt for RBE of neutrons, Int Comn Radiol Protection Task Group, 69-72; adv comt, Environ Mutagen Info Ctr, 69-72; comt assessment nitrate accumulation in environ, Agr Bd, Div Biol & Agr, Nat Res Coun, 70-72, Genetics Soc Am rep, 70-73, mem comt radiation preserv of food, Nat Acad Sci-Nat Res Coun, 72-75; mem, Mammalian Genetics Panel, Sci Adv Bd, Nat Ctr Toxicol Res, 72-74; chmn panel mutagenesis & carcinogenesis, US-Japan Coop Med Sci Prog, 72-87; US coordr biol & genetic consequences proj, US-USSR Environ Protection Agreement, 72-77; chmn subcomt environ mutagenesis, Dept HEW Comt to Coord Toxicol & Related Progs, 72-87; assoc ed, Mutation Res, 73-; adj prof path, Univ NC, Chapel Hill, 73-; mem ad hoc adv group on pre-mkt predictive testing, Off Technol Assessment, 74; consult, Genetics Study Sect, Div Res Grants, NIH, 67, NASA Biosci Exp Surv, 68, Joint Food & Agr Orgn-Int Atomic Energy Agency-WHO Expert Comt Irradiated Food, 69, DDT Adv Comt, Environ Protection Agency, 71 & Panel Vapor Phase Org Air Pollutants, Nat Acad Sci-Nat Res Coun, 71-75; chmn, Coord Comt, Int Prog Evaluation Short-Term Tests Carcinogenicity, 78-; mem, Steering Comt & Assessment Panel, Gene-Tox Prog, Off Toxic Substances, Environ Protection Agency, 79-; sect ed, J Environ Path & Toxicol, 79-82 & Toxicol & Indust Health, 84-, consult ed, Environ Res, 81-; chmn, mutagenesis task group, Int Prog Chem Safety, WHO-Int Labor Orgn-UN Environ Prog, Geneva, Switz, 81- & hazard eval & risk assessment group, Nat Inst Environ Health Sci, 81-; vis prof, Univ Zagazig, Egypt & Ain-Shams Univ, Cairo, 82 & Case Western Reserve Univ Sch Med, 83. *Honors & Awards:* Ann Award, Environ Mutagen Soc, 79. *Mem:* AAAS; Radiation Res Soc; Environ Mutagen Soc (vpres, 72-73, pres, 73-76); Genetics Soc Am; Am Asn Cancer Res; Europ Environ Mutagen Soc; hon mem Japanese Environ Mutagen Soc; Int Asn Environ Mutagen Socs (vpres, 85-). *Res:* Environmental mutagenesis; mutagenicity of carcinogens, radiation and chemical mutagenesis; space biology; genetic toxicology; microbial genetics; author of 270 scientific publications. *Mailing Add:* Ctr Life Sci & Toxicol Res Triangle Inst PO Box 12194 Research Triangle Park NC 27709

DE SESA, MICHAEL ANTHONY, b Boston, Mass, Feb 21, 27; m 53; c 2. RESEARCH ADMINISTRATION. *Educ:* Boston Col, BS, 49; Mass Inst Technol, PhD(anal chem), 53. *Prof Exp:* Res chemist, Raw Mat Develop Lab, AEC, 53-54, anal group leader, 54-58, chem group leader, 58-59, dept head inorg process develop, Feed Mat Prod Ctr, 59-62, asst tech dir uranium chem & metall, 62-63, assoc tech dir, 64; asst sect chief mech, chem & thermal properties, Res & Adv Develop Div, Avco Corp, 64-67; head, Anal & Phys Res Dept, NL Industs, Inc, 67-73, dir, Cent Res Lab, 70-75, dir res & develop, Indust Chem Div, 76-78, dir technol, NL Chem, 79-89; VPRES RES, RHEOX INC, 90- *Mem:* Am Chem Soc; Asn Res Dirs (secy, 82-84, vpres, 84-85, pres, 85-86); Indust Res Inst. *Res:* Instrumental methods of analysis; uranium hydrometallurgy; uranium compounds and metal; ablative plastics; development of chemical products; process development. *Mailing Add:* 33 Haddon Park Fair Haven NJ 07701

DESESSO, JOHN MICHAEL, b Phillipsburg, NJ, Mar 8, 47; m 73; c 5. TERATOLOGY, TOXICOLOGY. *Educ:* Hamilton Col, AB, 68; Va Commonwealth Univ, PhD(anat), 75. *Prof Exp:* Asst prof anat, Sch Med, Univ Cincinnati, 75-81; systs scientist, 81-86, LEAD SCIENTIST, MITRE CORP, 86- *Concurrent Pos:* Adj asst prof anat, Sch Med, Univ Cincinnati, 81-82, adj assoc prof anat & cell biol, 82-84; ad hoc mem, Safety & Occup Health Study Sect, Nat Inst Occup Safety & Health, 81-84; adj assoc prof anat, Sch Med & Dent, Georgetown Univ, 82-; adj assoc prof pediat, Children's Hosp, Univ Cincinnati, 89- *Mem:* Am Asn Anatomists; Am Soc Cellular Biol; Teratology Soc; Soc Toxicol; Soc Study Reproduction; Soc Risk Anal; Am Fertility Soc; Int Soc Study Xenobiotics; Electron Micros Soc Am; Sigma Xi. *Res:* Experimental and human teratology, especially mechanisms of chemical teratogenicity; embryology and experimental production of malformations; regulatory mechanisms in development; toxicological mechanisms; risk assessment. *Mailing Add:* The MITRE Corp 7525 Colshire Dr McLean VA 22102-3481

DESFORGES, JANE FAY, b Melrose, Mass, Dec 18, 21; m 48; c 2. MEDICINE. *Educ:* Wellesley Col, BA, 42; Tufts Univ, MD, 45; Am Bd Hemat, dipl; Am Bd Internal Med, dipl. *Prof Exp:* Intern path, Mt Auburn Hosp, 45-46; from intern to resident med, Boston City Hosp, 46-50, res fel hematol, 50-52; from asst prof to assoc prof, 52-72, PROF MED, SCH MED, TUFTS UNIV, 72-; SR PHYSICIAN & HEMATOLOGIST, NEW ENG MED CTR HOSP, 73- *Concurrent Pos:* NIH res fel hemat, Salt Lake City Hosp, 47-48; physician in-chg immunohemat lab, Hosp, Tufts Univ, 52-68, asst dir I & III Med Serv, 52-68, assoc dir, 68-72, actg dir clin labs, 67-69; assoc dir, Tufts Univ Hemat Labs, Boston City Hosp, 56-72; assoc ed, New Eng J Med, 60-; mem adv comt, Oak Ridge Assoc Univs, 72-75; mem drug experience comt, Food & Drug Admin, Bur Drugs, 72-73; mem automation med lab sci rev comt, Nat Inst Gen Med Sci, 71-76, chmn, 74-76; mem bd gov, Am Bd Internal Med, 80-86; Austin S Weisburger prof, Case Western Reserve Univ, 82; Paul M Aggeler vis prof, Univ Calif, San Francisco, 83; P K Bondy vis prof, Sch Med, Yale Univ, 85; consult hemat, Choate Mem Hosp, Lawrence Mem Hosp, Malden Hosp, Melrose-Wakefield Hosp, and others. *Honors & Awards:* K J R Wightman Lectr, Univ Toronto, 82; William B Castle Lectr, Boston City Hosp, 83; Hiken Lectr, Wash Univ, 86. *Mem:* Inst Med-Nat Acad Sci; Am Asn Physicians; master Am Col Physicians; Am Fedn Clin Res; Am Soc Clin Path; Am Soc Hemat (pres, 84-85); Int Soc Hemat; NY Acad Sci. *Res:* Hematology. *Mailing Add:* Tufts-New Eng Med Ctr Hosp 171 Harrison Ave Boston MA 02111

DESHARNAIS, ROBERT ANTHONY, b Cambridge, Mass, Mar 29, 55; m 78. POPULATION GENETICS, THEORETICAL ECOLOGY. *Educ:* Univ Mass, Boston, BA, 76; Univ RI, MS, 79, PhD(zool), 82. *Prof Exp:* Killam fel biol, Dalhousie Univ, 82-83; res assoc pop biol, Rockefeller Univ, 83-88; ASST PROF BIOL, CALIF STATE UNIV, LOS ANGELES, 88- *Mem:* AAAS; Genetics Soc Am; Am Math Soc; Sigma Xi; Am Mus Nat Hist. *Res:* Mathematical and empirical research in the areas of population genetics, ecology and demography; use of laboratory populations of flour beetles to test predictions from theoretical models. *Mailing Add:* Dept Biol Calif State Univ 5151 State University Dr Los Angeles CA 90032

DESHAW, JAMES RICHARD, b Monticello, Iowa, June 19, 42; m 65; c 1. BIOLOGY, ENVIRONMENTAL BIOLOGY. *Educ:* Loras Col, BS, 65; Tex A&M Univ, MS, 67, PhD(biol), 70. *Prof Exp:* Asst prof, Sam Houston State Univ, 70-73, ASSOC PROF BIOL, 73-, DIR FAC RES & GRANTS, 81-, PROF, SAM HOUSTON STATE UNIV, 85- *Mem:* AAAS; Am Soc Indust Microbiol; Am Soc Microbiol; Sigma Xi. *Res:* Primary productivity, physical-chemical and biological aspects of streams and reservoirs. *Mailing Add:* Dept Life Sci Sam Houston State Univ Huntsville TX 77341

DE SHAZER, LARRY GRANT, b Washington, DC, Nov 3, 34; m 60, 73; c 5. PHYSICS. *Educ:* Univ Md, BS, 56; Johns Hopkins Univ, PhD(physics), 63. *Prof Exp:* Physicist, Aerospace Group, Hughes Aircraft Co, 63-66; from asst prof to assoc prof physics, elec eng & mat sci, Univ Southern Calif, 66-78; sr staff physicist, Hughes Res Labs, 78-80, head, laser optical materials, 80-85; DIR, SOLID STATE LASERS, SPECTRA TECHNOL INC, 85- *Concurrent Pos:* Vis prof, Univ Rennes, France, 84; assoc ed, Inst Elec & Electronics Engrs J Quantum Electronics, 85-; chmn, Tunable Solid State Laser Conf, 86. *Mem:* Royal Inst Gt Brit; Am Phys Soc; Soc Photo-Optical Instrumentation Engrs; Sigma Xi; fel Optical Soc Am. *Res:* Solid state lasers; spectroscopy of rare-earth ions and organic dyes; nonlinear absorption spectroscopy; propagation of high-power optical beams; physics of dielectric thin films; fiber optics. *Mailing Add:* 13621 171st Ave NE Redmond WA 98052

DESHAZO, MARY LYNN DAVISON, b Carthage, Tex, Aug 14, 29; m 64. BIOCHEMISTRY. *Educ:* ETex Baptist Col, BS, 49; Univ Houston, MEd, 57; Tex A&M Univ, PhD(biochem), 68. *Prof Exp:* Teacher pub schs, Tex, 51-57; asst prof, 57-67, assoc prof, 68-72, PROF CHEM, SAM HOUSTON STATE UNIV, 72-, ASST DEAN COL ARTS & SCI, 84- *Mem:* Am Chem Soc; Sigma Xi; Am Inst Chemists. *Res:* Proteolytic enzymes of Aeromonas proteolytica. *Mailing Add:* Dept Chem Sam Houston State Univ Huntsville TX 77341

DESHMUKH, DIWAKAR SHANKAR, b Nagpur, India, Aug 16, 36; m 67; c 2. BIOCHEMISTRY. *Educ:* Univ Nagpur, BSc, 59, MSc, 61; Indian Inst Sci, Bangalore, PhD(biochem), 68. *Prof Exp:* Sr res fel biochem, Indian Inst Sci, Bangalore, 66-68; fel nutrit, Nat Res Coun Can, 68-69; fel biochem, Med Sch, Temple Univ, 69-73, instr, 73-74; ASSOC RES SCIENTIST NEUROCHEM, NY STATE BASIC RES INST MENT RETARDATION, 74- *Mem:* AAAS; NY Acad Sci; Sigma Xi; Am Soc Neurochem; Int Soc Neurochem; Am Soc Biol Chem; Am Asn Ment Deficiency. *Res:* Biogenesis of myelin membrane of central nervous system under normal and diseased conditions by studying metabolism of the myelin-specific compounds during different stages of myelin development. *Mailing Add:* Dept Neurochem State Basic Res Inst Ment Retard Staten Island NY 10314

DESHOTELS, WARREN JULIUS, b New Orleans, La, Jan 3, 26; m 51; c 11. PHYSICS, SOLAR ENERGY. *Educ:* Tulane Univ, BS, 45, MS, 47; St Louis Univ, PhD(physics), 53. *Prof Exp:* Dir physics, Xavier Univ, La, 48-53; chief instrumentation engr, Jackson & Church Co, Mich, 53-55; sr physicist, Clevite Res Ctr, Ohio, 55-64; chmn dept, 65-77, ASSOC PROF PHYSICS, MARQUETTE UNIV, 64- *Mem:* AAAS; Am Phys Soc; Am Asn Physics Teachers. *Res:* Solid state and electron physics. *Mailing Add:* 8916 N Tennyson Dr Bayside WI 53217-1963

DESHPANDE, ACHYUT BHALCHANDRA, Indian citizen; m 68; c 3. POLYMER CHEMISTRY, POLYMER TECHNOLOGY. *Educ:* Univ Poona, BS, 55, BS Hons, 56, MS, 57, PhD(polymer chem), 64. *Prof Exp:* Sr sci asst polymer, Nat Chem Lab, India, 62-68, scientist A, 68-73; sr res chemist, Gaylord Res Inst, Whippany, NJ, 73-82; sr res scientist, Synfax Mfr Co, Fairfield, NJ, 82-89; SR POLYMER CHEMIST, SPECIALTY TOHER CORP, FAIRFIELD, NJ, 89- *Concurrent Pos:* Jr res fel, Coun Sci & Indust Res, New Delhi, 59-62; trainee Colombo plan, Overseas Tech Coop Agency, Japan, 66-68; res assoc, Polytech Inst New York, 74-75. *Mem:* Am Chem Soc; NY Acad Sci. *Res:* Linear and stereo regular polymerization with Ziegler-Natta polymerization catalysts and charge transfer complex catalysts and other initiating systems; synthesis of flame resistant fiber polymers, coating composition, contact lens composition, emulsion polymerization and telomerization; specialty polymers by emulsion and dispersion processes for copying machine toner composition. *Mailing Add:* Am Chem Soc 113 Fairview Ave South Orange NJ 07079

DESHPANDE, KRISHNANATH BHASKAR, b India, Nov 1, 21; m 50; c 2. PHYSICAL CHEMISTRY, INORGANIC CHEMISTRY. *Educ:* Bombay Univ, BS, 43, MS, 46, PhD(phys chem), 51. *Prof Exp:* Instr chem, Bombay Univ, 43-46, lectr inorg & phys chem, 47-57; curators grant, Univ Mo, 57, Am Petrol Inst fel, 57-60; res assoc phys chem, Univ NC, 60-65; dir NSF acad year inst, 69-71, PROF CHEM, FISK UNIV, 61-, COORDR ADVAN INSTNL DEVELOP PROG, 74- *Concurrent Pos:* Chadraseniya Kayastha Prabhu scholar, 57-60; Fulbright travel grant, US Dept State, 57-60; Am Petrol Inst fel, Oak Ridge Inst Nuclear Studies, 60; res chemist, US AEC, 60-65; dir US Off Educ-Inst Advan Studies in Phys Sci, 70-72; consult, Univ Tenn-AEC Agr Res Lab, Tenn. *Mem:* Am Chem Soc; Clay Minerals Soc. *Res:* Colloid chemistry of systems containing soaps and organic solvents; electrochemistry of clay-electrolyte systems; surface chemistry; ion exchange in inorganic exchangers using radioisotopes; ion exchange thermodynamics. *Mailing Add:* Dept Chem Fisk Univ 17th Ave North Nashville TN 37205

DESHPANDE, MOHAN DHONDORAO, fluid mechanics, biomechanics, for more information see previous edition

DESHPANDE, NARAYAN V(AMAN), b Bhadwan, India, May 4, 38; m 66; c 3. MECHANICAL ENGINEERING, APPLIED MECHANICS. *Educ:* Univ Poona, BEng, 61; Univ Rochester, MS, 64, PhD(fluid mech), 66. *Prof Exp:* Eng asst, Tata Thermal Power Co, India, 61-62; res assoc fluid instabilities, Culham Lab, UK Atomic Energy Authority, 66-67; asst prof mech & aerospace sci, Univ Rochester, 67-73; MEM RES STAFF, XEROX CORP, ROCHESTER, 73- *Concurrent Pos:* Ctr Naval Anal res grant, 68-70; consult, Xerox Corp, 72-73. *Mem:* Am Soc Mech Engrs; Am Acad Mech; Sigma Xi. *Res:* Fluid mechanics and magnetohydrodynamics; the stability of magnetohydrodynamic boundary layer type flows and the two-stream instability; study of a boundary layer over a moving surface; air bearings; heat transfer; contact problems in elasticity; magnetic recording technology; three dimensional fluid flow calculations with multivalued free surfaces. *Mailing Add:* 101 Highledge Dr Penfield NY 14526

DESHPANDE, NILENDRA GANESH, b Karachi, Pakistan, Apr 18, 38; m 60; c 2. ELEMENTARY PARTICLE PHYSICS. *Educ:* Madras Univ, BSc, 59, MA, 60, MSc, 61; Univ Pa, PhD(physics), 65. *Prof Exp:* Res fel theoret physics, Imp Col, Univ London, 65-66; mem, Inst Math Sci, Madras, India, 66; res assoc physics, Northwestern Univ, Evanston, 66-67, asst prof, 67-73; assoc prof, Univ Tex, Austin, 73-75; assoc prof, 75-84, PROF PHYSICS, UNIV ORE, 84-, DIR, INST THEORET SCI, 87- *Concurrent Pos:* Vis prof, Univ Warsaw, Poland, 74-75 & Chalmers Inst, Gothenburg, Sweden, 75. *Honors & Awards:* Outstanding Jr Investr, Dept Energy, 81-86. *Mem:* Am Phys Soc; fel Am Physiol Soc, 87; Sigma Xi. *Res:* CP violation; neutrino physics; weak interactions and gauge theories; grand unified theories; physics beyond the standard model. *Mailing Add:* Dept Physics Univ Ore Eugene OR 97403

DESHPANDE, SHIVAJIRAO M, b Maharashtra, India, Dec 20, 36. SOLID STATE PHYSICS, SEMICONDUCTOR PHYSICS. *Educ:* Osmania Univ, India, MS; Indian Inst Technol, India, MTech, 60; Univ Vt, PhD(physics), 71. *Prof Exp:* Sr engr, Motorola, Bell Systs & other Insts, 69-78; STAFF ENGR, IBM CORP, 78- *Concurrent Pos:* Staff mem, Bhaba Atomic Res, 60-66. *Mem:* Sigma Xi; Int Sci Cong; Am Vacuum Soc. *Res:* Discontinous thin films; room-temperature oxidation rate constants for molybdenum films; semiconductor processes; plasma CVD; plasma etching-metalization (sputtering/evaporation); multilayer metal and dielectric technology; HDA technology; electromigration. *Mailing Add:* IBM Corp 5600 Cottle Rd San Juan CA 95193

DESIDERATO, ROBERT, JR, b New York, NY, Aug 21, 39; m 68; c 2. PHYSICAL CHEMISTRY. *Educ:* Columbia Univ, AB, 61; Rice Univ, PhD(chem), 66. *Prof Exp:* Res assoc crystallog, Univ Pittsburgh, 65-66; asst prof, 66-74, ASSOC PROF CHEM, N TEX STATE UNIV, 75- *Mem:* Am Chem Soc; Am Crystallog Asn. *Res:* X-ray structure and analysis of biologically important compounds and transition metal complexes; computer programming. *Mailing Add:* Dept Chem NTex State Univ Denton TX 76203

DESIDERIO, ANTHONY MICHAEL, b Philadelphia, Pa, Sept 26, 43; m 66; c 2. SYSTEMS ANALYSIS, APPLIED RESEARCH. *Educ:* La Salle Col, BA, 65; Univ Notre Dame, MA, 67, PhD(physics), 70. *Prof Exp:* Mem prof staff, Ctr Naval Anal, 71-78; MEM RES STAFF, SYST PLANNING CORP, 78- *Mem:* Am Phys Soc; Sigma Xi; Opers Res Soc Am. *Res:* Analysis of real world problems related to optimization of systems; analysis of factors and cost-benefit analysis. *Mailing Add:* 1215 Ingleside McLean VA 22101

DESIDERIO, DOMINIC MORSE, (JR), b McKees Rocks, Pa, Jan 11, 41; m 65; c 2. BIOCHEMISTRY, MASS SPECTROMETRY. *Educ:* Univ Pittsburgh, BA, 61; Mass Inst Technol, SM, 64, PhD(anal chem), 65. *Prof Exp:* Organic control chemist, Pittsburgh Coke & Chem Co, 58-60; res chemist, Univ Pittsburgh, 60-61; teaching asst, Mass Inst Technol, 61-62, res asst, 62-65; res chemist, Am Cyanamid Co, 66-67; from asst prof to assoc prof chem, Inst Lipid Res & Dept Biochem, Baylor Col Med, 67-78; PROF NEUROL & BIOCHEM & DIR CHARLES B STOUT NEUROSCI MASS SPECTROMETRY LAB, UNIV TENN, MEMPHIS, 78- *Concurrent Pos:* Intra-Sci Res Found fel, 71-75. *Mem:* Am Chem Soc; Am Soc Mass Spectrometry; AAAS; Am Soc Biol Chemists; Soc Neurosci. *Res:* Analytical neurochemistry; mass spectrometry; neurosciences; peptides. *Mailing Add:* Univ Tenn Memphis 800 Madison Ave Memphis TN 38163

DESIENO, ROBERT P, b Scranton, Pa, Sept 1, 33; m 62; c 2. PHYSICAL CHEMISTRY. *Educ:* Union Col, NY, BS, 55, MS, 62; Univ Calif, Davis, PhD(chem), 66. *Prof Exp:* Chemist, Gen Elec Co, 57-62; res asst, Univ Calif, 62-65; sr chemist, Rohm and Haas Corp Res Labs, 65-68; asst prof, 68-72, assoc prof chem, Westminster Col, Pa, 72-80; prof chem & dir, Comput Ctr, Davidson Col, NC, 80-83; AT DEPT COMPUT SCI, SKIDMORE COL, SARATOGA SPRINGS, NY, 83- *Mem:* AAAS; Am Chem Soc; Sigma Xi. *Res:* Science and literature; spectroscopic properties of electrically exploded wires; computers and national defense; history of science. *Mailing Add:* Dept Comput Sci Skidmore Col Saratoga Springs NY 12866

DE SIERVO, AUGUST JOSEPH, b Passaic, NJ, Feb 4, 40; m 73; c 3. BIOCHEMISTRY. *Educ:* Rutgers Univ, BA, 63, MS, 66, PhD(microbiol), 68. *Prof Exp:* Asst scientist microbiol, Warner Lambert Res Inst, 63-64; instr, Med Sch, NY Univ, 68-69; asst prof, 70-76, ASSOC PROF MICROBIOL, UNIV MAINE, ORONO, 76- *Mem:* Am Soc Microbiol; Sigma Xi; AAAS. *Res:* Membrane biosynthesis and function; role of phospholipids and phospholipid-associated enzymes in membrane structure and function; plant lipids. *Mailing Add:* Dept Biochem Microbiol & Molecular Biol Univ Maine Orono ME 04469

DESILETS, BRIAN H, b Leominster, Mass, Oct 7, 27; m 70; c 1. PHYSICS. *Educ:* Marist Col, BA, 50; St John's Univ, NY, MS, 54; NY Univ, MS, 58; Cath Univ Am, PhD(physics), 64. *Prof Exp:* Teacher, Bishop Dubois High Sch, 50-54; instr math & physics, Marist Col, 54-56, asst prof physics, 56-60; lectr elec eng, Cath Univ Am, 63-64; from asst prof to prof physics, Marist Col, 65-74; adv engr, 74-83, develop eng mgr, 83-88, SR ENGR MGR, IBM CORP, 88- *Concurrent Pos:* Res assoc, Cath Univ Am, 62-64; consult, X-Ray Labs, IBM Corp, NY, 57-60. *Mem:* Am Asn Physics Teachers. *Res:* Solid state physics; x-ray studies; microwave attenuation; photoconductivity; reactive ion etching; semiconductor device process development. *Mailing Add:* Six Lake Oniad Dr Wappingers Falls NY 12590

DESILVA, ALAN W, b Los Angeles, Calif, Feb 8, 32; m 59; c 3. PLASMA PHYSICS. *Educ:* Univ Calif, Los Angeles, BS, 54; Univ Calif, Berkeley, PhD(physics), 61. *Prof Exp:* Res physicist, Lawrence Radiation Lab, Univ Calif, Berkeley, 61-62; NSF fel plasma spectros, Culham Lab, UK Atomic Energy Authority, 62-63, res assoc, 63-64; from asst prof to assoc prof, 64-74, PROF PHYSICS, UNIV MD, COLLEGE PARK, 74- *Concurrent Pos:* US-Japan Coop Sci Prog fel, Inst Plasma Physics, Nagaya Univ, Japan, 70-71; mem, Exec Comt, Div Plasma Physics, Am Phys Soc. 82-83. *Honors & Awards:* Alexander von Humboldt US Sr Scientist Award, 84. *Mem:* Fel Am Phys Soc; Inst Elec & Electronics Engrs. *Res:* Plasma physics, including hydromagnetic wave phenomena and radiations from plasmas; interactions of electromagnetic radiation with plasmas; collision free shockwaves and turbulence in plasmas; diagnostics; fluctuations in collisional plasmas. *Mailing Add:* Lab Plasma Res Univ Md College Park MD 20742

DESILVA, CARL NEVIN, b British Guiana, Aug 6, 23; US citizen; m 54; c 6. ENGINEERING MECHANICS. *Educ:* Columbia Univ, BS, 49, MS, 50; Univ Mich, PhD(mech), 55. *Prof Exp:* Res asst, Univ Mich, 52-55, res assoc, 55-57; unit chief mech res, Boeing Airplane Co, 57-60; from assoc prof to prof aeronaut & eng mech, Univ Minn, 60-66; chmn dept mech eng sci, 66-76, PROF MECH ENG, WAYNE STATE UNIV, 66- *Concurrent Pos:* Consult, Boeing Airplane Co, 60-62, Honeywell Corp, 62-63, Gen Mills, Inc, 63-64 & Geophys Corp Am, 65-66. *Mem:* Am Math Soc; Am Soc Mech Engrs; Soc Natural Philos. *Res:* Solutions of problems in the classical theory of elastic shells; development of nonlinear theories of elastic shells and of non-Newtonian fluids; constitutive equations of viscoelastic materials with memory; analysis of fluid suspensions as applied to blood flow. *Mailing Add:* Dept Mech Eng Wayne State Univ 2100 W Eng Detroit MI 48202

DE SILVA, JOHN ARTHUR F, b Colombo, Sri Lanka (Ceylon), Sept 23, 33; US citizen; m 59; c 1. ANALYTICAL CHEMISTRY, PHARMACEUTICAL CHEMISTRY. *Educ:* Univ Ceylon, BSc, 56; Rutgers Univ, MS, 58, PhD(soil chem), 61. *Prof Exp:* Res chemist, Agr Div, Am Cyanamid Co, NJ, 61-63; sr res chemist, Dept Clin Pharmacol, Hoffmann-La Roche Inc, 63-70, res group chief, Bioanal Method Develop, 70-74, res sect head bioanal methods develop, 74-78, asst dir, Dept Biochem & Drug Metab, 78-79, asst dir dept pharmacokinetics & biopharmaceutics, 79-85; dir, Pharmaceut Anal & Drug Stability, Zenith Labs Inc, 85-87; ed jour pharmaceut sci, Am Pharmaceut Asn, 87-91; MGR, DRUG METAB OPERS, DEPT DRUG METAB & PHAMACEUT, SCHERING-PLOUGH RES, BLOOMFIELD, NJ, 91- *Mem:* Am Chem Soc; fel Am Inst Chem; Sigma Xi; fel Am Pharmaceut Asn; fel Am Asn Pharmaceut Scientists. *Res:* Drug analysis in biological fluids; drug metabolism; biopharmaceutics; pharmacokinetics; pharmaceutical analysis; sr author in 48 and co-author in 38 peer reviewed scientific publications. *Mailing Add:* 419 Harding Dr 340 Kingsland St Bldg 86 Rm 802-c South Orange NJ 07079

DE SIMONE, DANIEL V, b Chicago, Ill, May 4, 30; m 55; c 3. SCIENCE POLICY. *Educ:* Univ Ill, BS, 56; NY Univ, JD, 60. *Prof Exp:* Staff mem commun technol, Bell Tel Labs, 56-62; consult technol innovation, Asst Secy Sci & Technol, US Dept Com, 62-64; dir, Off Invention & Innovation, Nat Bur Standards, 64-69; dir, US Metric Study for Cong, 69-71; sci policy asst, Off Sci & Technol, Exec Off of the President, 71-73, exec secy, Fed Coun Sci & Technol, 72-73; dep dir, Cong Off Technol Assessment, 73-80; pres, Innovation Group, Arlington, Va, 80-90; CONSULT, 90- *Concurrent Pos:* Mem & rapporteur, Interagency Comt E-W Trade, 63; exec dir, Nat Inventors Coun, 63-69; exec secy, Panel on Invention & Innovation, US Dept Com, 65-67; consult, Nat Comn Technol, Automation & Econ Progress, 66 & Arms Control & Disarmament Agency, 66-69; mem, Panel on Venture Capital, US Dept Com, 68-70; adv, Dept Sci Affairs, Orgn Am State, 68-70. *Honors & Awards:* Gold Medal Distinguished Achievement Fed Serv, US Dept Com, 69. *Mem:* Inst Elec & Electronics Engrs; AAAS. *Res:* Technology assessments concerning energy, natural resources, health care, transportation, food, agriculture, nutrition, telecommunications, electronic funds transfer, international trade and development, national security, and federal research and development programs and priorities. *Mailing Add:* PO Box 7064 Arlington VA 22207

DESIO, PETER JOHN, b Boston, Mass, June 29, 38. ORGANOCADMIUM CHEMISTRY. *Educ:* Boston Col, BS, 60; Univ NH, PhD(org chem), 64. *Prof Exp:* Asst gen & org chem, Univ NH, 60-63; res assoc, Mat Div, Mass Inst Technol, 64-66; assoc prof org chem, 66-80, chmn, Chem Dept, 76-80, actg dir fire sci, 77-78, PROF ORG CHEM, UNIV NEW HAVEN, 80- *Mem:* Am Chem Soc; Am Asn Univ Prof; Soc Col Sci Teachers; Royal Soc Chem; Sigma Xi; Nat Sci Teachers Asn. *Res:* Structure, particularly ring-chain tautomerism in acids and alcohols; organo-cadmium reactions as well as other organometallics. *Mailing Add:* Dept Chem & Chem Eng Univ New Haven 300 Orange Ave West Haven CT 06516

DESJARDINS, CLAUDE, b Fall River, Mass, June 13, 38; m 62; c 3. PHYSIOLOGY. *Educ:* Univ RI, BS, 60; Mich State Univ, MS, 64, PhD(animal physiol), 67. *Prof Exp:* Instr reproductive physiol, Mich State Univ, 60-67; staff fel, Jackson Lab, Maine, 67-68; from assoc prof to prof physiol, Okla State Univ, 68-71; from assoc prof to prof, Univ Tex, Austin, 75-86; PROF PHYSIOL, UNIV VA, 87-, PROF UROL, 91- *Concurrent Pos:* Consult reproductive endocrinol, NIH, NASA, NSF; assoc ed, Am J Physiol: Endocrinol & Metab, 82-88, ed, 91-96; ed-in-chief J Androl, 89-92; dept chair, Ctr Res in Reproduction, Univ Va Med Ctr, 91. *Honors & Awards:* Abbott Microcirculation Award, 88. *Mem:* Soc Neurosci; Soc Study Reproduction (secy, 77-80, pres, 81-82); Brit Soc Study Fertil; Am Physiol Soc; Endocrine Soc; Am Soc Cell Biol; Am Soc Andrology; Microcirculatory Soc. *Res:* Endocrinology; meuroendocrine control of reproduction and metabolism; testicular function and physiology of the male reproductive system; microcirculation and blood flow in endocrine glands. *Mailing Add:* Dept Physiol Box 449 Jordan Bldg Med Sch Univ Va Charlottesville VA 22903

DESJARDINS, CLAUDE W, b Montreal, Que, Sept, 26, 50; m 72; c 1. TECHNICAL MANAGEMENT. *Educ:* Univ Montreal, BSc, 72. *Prof Exp:* Plant inspector, Dept Health & Welfare, Can, 74-78, chemist, 79; REGIONAL CHEMIST, DEPT FISHERIES & OCEANS, CAN, 79- *Mem:* Asn Off Anal Chem; Can Inst Chem; Can Inst Food Sci & Technol. *Res:* Polychlorinated biphenyl, mirex and other organic chemistry contaminants in fish and fish products; mercury, lead, cadmium and other heavy metals; histamine, sulfite, nitrates; quantification of polychlorinated biphenyl isomers by gas-liquid chromatography-exchange chromatography capillary columns. *Mailing Add:* Dept Fisheries & Ocean Gov Can 789 Roland Therriem Longueuil PQ J4H 4A6 Can

DESJARDINS, PAUL ROY, b Cheyenne, Wyo, Aug 7, 19; m 47; c 3. PLANT PATHOLOGY. *Educ:* Colo State Univ, BS, 42; Univ Calif, Berkeley, PhD(plant path), 52. *Prof Exp:* Res asst, Univ Calif, Berkeley, 47-50, jr plant pathologist, Univ Calif, Riverside, 52-53, asst plant pathologist, 53-59, assoc prof, 61-70, PROF PLANT PATH, UNIV CALIF, RIVERSIDE, 70- *Concurrent Pos:* Res assoc, Nat Acad Sci-Nat Res Coun, 64-65. *Mem:* AAAS; Am Phys Soc; Electron Micros Soc Am; Torrey Bot Club; Sigma Xi. *Res:* Seed transmission; electron microscopy, purification and serological studies of plant viruses; cytological studies of virus infected tissues; virus structure and morphology; virus nucleic acid and disease of citrus, avocado and other plants. *Mailing Add:* Div Plant Path Univ Calif Riverside CA 92507

DESJARDINS, RAOUL, b Montreal, Que, Oct 8, 33; m 61; c 2. PHARMACEUTICAL RESEARCH, CLINICAL PHARMACOLOGY. *Educ:* Univ Montreal, BA, 53, MD, 58; Baylor Col Med, MS, 64, PhD(pharmacol), 66; Rutgers Univ, MBA, 90. *Prof Exp:* Asst prof pharmacol, Baylor Col Med, 66; dir clin res, Ortho Res Found, 66-72; pres, Raoul Desjardins Assocs, Inc, 72-73; PRES, RES CONSULTS, INC, 73- *Concurrent Pos:* Int med dir, Int Med Exchange & Develop, 75-, chmn, Bd Gov; vis prof, Chung Ang Univ, Seoul, Korea, 75-; bd dirs, Fed Inst Health, 88- *Mem:* Fel Am Col Clin Pharmacol; fel NY Acad Med; Drug Info Asn; Am Col Clin Pharmacol & Clin Chemother; Am Acad Clin Toxicol; Am Fedn Clin Res. *Res:* Consultation in planning and execution of clinical research programs concerning safety and effectiveness of new drugs or other human health care products. *Mailing Add:* Res Consults 135 Talmage Rd Mendham NJ 07945

DESJARDINS, STEVEN G, b Fall River, Mass, Apr 14, 58; m 81; c 1. THEORETICAL CHEMISTRY. *Educ:* Clark Univ, BA, 80; Brown Univ, PhD(chem), 85. *Prof Exp:* Postdoctoral & lectr chem, Univ Tex, Austin, 85-86; ASST PROF CHEM, WASH & LEE UNIV, 86- *Concurrent Pos:* Vis asst prof, Brown Univ, 87 & 88. *Mem:* Am Chem Soc; Am Phys Soc; Sigma Xi. *Res:* Computer simulation of chemical systems, especially biochemical molecules; lipid bilayers & protein folding dynamics. *Mailing Add:* Dept Chem Wash & Lee Univ Lexington VA 24450

DESKIN, WILLIAM ARNA, b Mo, Aug 16, 24; m 49; c 3. INORGANIC CHEMISTRY. *Educ:* Northeast Mo State Teacher Col, BS & AB, 48; Univ Mo, MA, 50; Univ Iowa, PhD(chem), 57. *Prof Exp:* Phys chemist, Chem Labs, US Army Chem Ctr, Md, 50-51; prof chem & physics, Upper Iowa Univ, 52-54; from asst prof to assoc prof, 56-63, chmn dept, 61-80, PROF CHEM, CORNELL COL, 63- *Concurrent Pos:* Resident res assoc, Argonne Nat Lab, 63-64. *Mem:* Am Chem Soc; AAAS. *Res:* Coordination compounds of transition metals with sulfur containing ligands. *Mailing Add:* 506 College Blvd Mt Vernon IA 52314

DESKINS, WILBUR EUGENE, b Morgantown, WVa, Feb 20, 27; m 53; c 2. MATHEMATICS. *Educ:* Univ Ky, BS, 49; Univ Wis, MS, 50, PhD(math), 53. *Prof Exp:* Instr math, Univ Wis, 53; from instr to asst prof, Ohio State Univ, 53-56; from asst prof to prof, Mich State Univ, 56-71; chmn dept, 71-87, PROF MATH, UNIV PITTSBURGH, 71-, ASSOC DEAN, CAS, 88- *Concurrent Pos:* Staff assoc, Comprehensive Sci Math Prog, Cent Midwestern Regional Educ Lab, 69-80; sr res fel, Univ London, 69; mem, Steering Comt, Pittsburgh Math Collab, 86-, Adv Comt, NSF Proj, Children's Mus, Washington DC, 85-89 & Adv Comt, Pittsburgh Sci Inst, 89- *Mem:* Am Math Soc; Math Asn Am. *Res:* Algebra; group theory. *Mailing Add:* Dept Math 722 Thackeray Hall Univ Pittsburgh Pittsburgh PA 15260

DESLATTES, RICHARD D, JR, b New Orleans, La, Sept 21, 31; m 56; c 5. RADIOLOGY, PHYSICS. *Educ:* Loyola Univ, La, BS, 52; Johns Hopkins Univ, PhD(physics), 59. *Prof Exp:* Instr, Loyola Univ, La, 54-55; res assoc physics, Fla State Univ, 56-58 & Cornell Univ, 58-62; physicist, Nat Bur Standards, 62-68; CHIEF, QUANTUM METROL DIV & SR RES FEL, PHYSICS LAB, NAT INST STANDARDS & TECHNOL, 68- *Concurrent Pos:* Dir, Div Physics, NSF, 80-81; Alexander von Humboldt Found Sr US Scientist Award, Soc Heavy Ion Res, Phys Inst, Univ Heidelberg, Ger & Laue-Langevin Inst, Grenoble, France, 83-84. *Honors & Awards:* Sun-Amco Medal, 90. *Mem:* AAAS; Am Phys Soc. *Res:* X-ray spectroscopy of solids; atomic spectroscopy of high excited systems; x-ray diffraction microscopy. *Mailing Add:* Physics Lab Nat Inst Standards & Technol Gaithersburg MD 20899

DESLAURIERS, ROXANNE MARIE LORRAINE, b Montreal, Que, Oct 2, 47; m 70; c 2. PHYSICAL BIOCHEMISTRY. *Educ:* Univ Laval, BSc, 68; Univ Ottawa, PhD(biochem), 72. *Prof Exp:* Asst res officer, Nat Res Coun Can, 72-80, assoc res officer biochem, 80-85, sr res officer, 86-90, SECT HEAD, NAT RES COUN CAN, 90- *Concurrent Pos:* Asst prof, Dept Physiol & Biophys, Univ Ill Med Ctr, 75-; adj prof, Dept Pathol, Fac Med, Univ Ottawa, 87- *Mem:* Biophys Soc; Can Biophys Soc; Soc Magnetic Resonance Med; Int Soc Magnetic Resonance. *Res:* Structural and metabolic studies on live organisms using nuclear magnetic resonance spectroscopy. *Mailing Add:* Inst Biol Sci Bldg 40 Nat Res Coun Can Ottawa ON K1A 0R6 Can

DESLOGE, EDWARD AUGUSTINE, b St Louis, Mo, Aug 31, 26; m 58; c 5. PHYSICS. *Educ:* Univ Notre Dame, BS, 47; St Louis Univ, MS, 55, PhD(physics), 57. *Prof Exp:* Instr physics, Yale Univ, 58-59; from asst prof to assoc prof, 59-69, PROF PHYSICS, FLA STATE UNIV, 69- *Mem:* Am Phys Soc. *Res:* Classical and relaturatic mechanics; thermal and statistical physics. *Mailing Add:* Dept Physics Fla State Univ Tallahassee FL 32306

DESLONGCHAMPS, PIERRE, b St Lin, Que, May 8, 38; m 87; c 2. ORGANIC CHEMISTRY. *Educ:* Univ Montreal, BSc, 59; Univ NB, PhD(org chem), 64. *Hon Degrees:* Dr, Univ Pierre et Marie Curie, Paris, 83, Bishops Univ, Univ de Montreal & Univ Laval, Quebec, 84, Univ of New Brunswick, NB, 85. *Prof Exp:* Res fel chem, Harvard Univ, 64-65; asst prof, Univ Montreal, 66-67; adj prof, 67-68, PROF CHEM, UNIV SHERBROOKE, 72- *Concurrent Pos:* A P Sloan fel, 70-72; Sci Prize of Que, 71-72; E W R Steacie fel, 71-74; Izaak Walton Killam mem scholarships in sci, 76-77; Guggenheim Found fel, 79-; fel, Royal Soc London, 83; officer, Order Can, 89. *Honors & Awards:* Steacie Prize, 74; Medaille Vincent, Fr-Can Asn Advan Sci, 79; Merck, Sharp & Dohme Lects Award, Chem Inst Can, 76; Medaille Pariseau, Fr-Can Asn Advan Sci, 79; Marie-Victorin Prize, Province Quebec, 87. *Mem:* Am Chem Soc; fel Royal Soc Can; fel Chem Inst Can; AAAS; Fr-Can Asn Advan Sci; Swiss Chem Soc; Chem Inst Can; Ordre des Chimistes du Que. *Res:* Organic synthesis; author on, stereo electronic effects in organic chemistry. *Mailing Add:* Dept Chem Univ Sherbrooke Sherbrooke PQ J1K 2R1 Can

DES MARAIS, DAVID JOHN, b Richmond, Va, Jan 12, 48; m 70; c 2. GEOCHEMISTRY, BIOCHEMISTRY. *Educ:* Purdue Univ, BS, 69; Ind Univ, Bloomington, MS, 72, PhD(geochem), 74. *Prof Exp:* Assoc geochem, Ind Univ, 74-75; res fel, Inst Geophys & Planetary Physics, Univ Calif, Los Angeles, 75-76; RES CHEMIST GEOCHEM, NASA-AMES RES CTR, 76- *Concurrent Pos:* Mem bd dirs, Cave Res Found, Yellow Springs, Ohio, 73-75 & 78-81; lunar sample prin investr, NASA, 77-79; assoc fel, Ames Res Ctr, 83; mem, Int Comt, Int Symposia Environ Biogeochem, ICS/IUGS Working Group on Terminal Proterozoic Syst, 89- *Honors & Awards:* NASA Except Sci Achievement Medal, 88. *Mem:* Geochem Soc; Am Geophys Union. *Res:* Geochemistry of carbon, nitrogen and hydrogen in lunar materials; light isotope analytical chemistry; light isotope biogeochemistry; geochemical cycle of carbon. *Mailing Add:* NASA-Ames Res Ctr Mail Stop 239-4 Moffett Field CA 94035

DESMARTEAU, DARRYL D, b Garden City, Kans, May 25, 40; m 62; c 3. INORGANIC CHEMISTRY, FLUORINE CHEMISTRY. *Educ:* Wash State Univ, BS, 63; Univ Wash, PhD(chem), 66. *Prof Exp:* Asst prof chem, Univ Wash, 66-67 & Northeastern Univ, 67-71; from asst prof to prof chem, Kans State Univ, 71-82; head dept chem & geol, 82-89, TOBEY-BEUDROT PROF CHEM, CLEMSON UNIV, 89- *Concurrent Pos:* Alfred P Sloan fel, 75-77; Alexander von Humboldt res fel, 79-80. *Honors & Awards:* Award for Creative Work in Fluorine Chem, Am Chem Soc, 83; Alexander von Humboldt Sr Distinguished US Scientist Award, 88. *Mem:* Am Chem Soc; Sigma Xi. *Res:* Synthesis and properties of nonmetal fluorine compounds; fluorocarbon derivatives of non-metals; swallowing heterocycles; compounds of phosphorus, sulfur and strong oxidizers such as peroxides, fluoroxides, hypohalites, nitrogen halides and xenon compounds; fluorinated superacids, polymer electrolytes and selective fluorination. *Mailing Add:* Chem Dept Clemson Univ 359 Hunter Lab Clemson SC 29634-1905

DESMOND, MARY ELIZABETH, b La Junta, Colo, May 29, 40. DEVELOPMENTAL BIOLOGY, NEUROBIOLOGY. *Educ:* Marquette Univ, BA, 63; Univ Colo, Boulder, MA, 71, PhD(biol), 73. *Prof Exp:* Res technician clin med, Med Sch, Univ Colo, 64-68, res technician cell biol, Dept Anat, 68-69, teaching asst biol dept, 68-72, instr biol & continuing educ, 72-73; chmn dept sci, St Scholastica Acad, 73-75; fel develop biol, Dept Zool, Austin Tex, 75-77; asst prof embryol & anat, 77-84, ASSOC PROF COMPARATIVE VERTEBRATE ANAT & DEVELOP BIOL, VILLANOVA UNIV, 84- *Concurrent Pos:* Co-investr grant, Carnegie Embryol Lab, Davis, Calif, 77, Dept Anat, Univ Utah, 83-86. *Mem:* Soc Develop Biologists; Int Soc Develop Biologists; AAAS; Am Asn Anatomists; Soc Neurosci. *Res:* Mechanisms involved in the morphogenesis of the vertebrate brain; growth parameters of the human brain during embryogenesis; occlusion of the spinal cord during embryogenesis; the role of glycoaminoglycans during skull morphogenesis. *Mailing Add:* Dept Biol Villanova Univ Villanova PA 19085

DESMOND, MURDINA MACFARQUHAR, b Isle of Lewis, Scotland, Nov 14, 16; US citizen; m 48; c 2. PEDIATRICS. *Educ:* Smith Col, BA, 38; Temple Univ, MD, 42. *Prof Exp:* From instr to prof, 48-87, EMER PROF PEDIAT, BAYLOR COL MED, 87- *Concurrent Pos:* Fel pediat, Sch Med, George Washington Univ, 47-58. *Honors & Awards:* Wyeth Award, 58; Myrtle Wreath Award, 85; Award Tex Perinatal Soc, 87; Apgar Award, 87. *Mem:* Soc Pediat Res; Am Pediat Soc; Am Acad Ment Deficiency; Am Acad Pediat; Am Med Workers Asn. *Res:* Neonatology; transition of infant from intrauterine to extrauterine life; relation of newborn area to later development. *Mailing Add:* Dept Pediat Baylor Col Med Houston TX 77030

DESNICK, ROBERT JOHN, b Minneapolis, MN, July 12, 43. HUMAN & MOLECULAR GENETICS, SOMATIC CELL GENETICS. *Educ:* Univ Minn, Minneapolis, BA, 65, PhD(genetics), 70, MD, 71. *Prof Exp:* From asst prof to prof pediat, Univ Minn, 71-77, asst prof, Dight Inst Human Genetics, 71, assoc prof genetics & cell biol, 75-77; ARTHUR J & NELLIE Z COHEN PROF PEDIAT & GENETICS, MT SINAI SCH MED, NEW YORK, 77-, CHIEF, DIV MED & MOLECULAR GENETICS, 77- *Concurrent Pos:* Mem med adv bd, Nat Tay-Sachs & Allied Dis Asn, 75-, Nat Found Jewish Genetic Dis, 81-, Am Porphyria Found, 84-, Mucolipidoses Found, 84-, Nat Mucopolysaccharidosis Soc & Familial Dysautonomia Found, 90-; NIH Career Develop Award, 75-80; hon mem, Japanese Soc Inherited Metab Dis, 85. *Honors & Awards:* Ross Award Pediat Res, 72; E Mead Johnson Award, Res Pediat, Am Acad Pediat, 81. *Mem:* Am Soc Clin Invest; Am Asn Physicians; Am Soc Human Genetics; Am Soc Biol & Molecular Biol; Harvey Soc (secy, 84-89); Soc Inherited Metab Dis (pres, 89-90); Fifth Int Cong Inborn Errors Metab (pres, 90-); Am Soc Microbiol. *Res:* Molecular genetics of human lysosomal enzymes, lysosomal storage diseases, human heme biosynthetic enzymes and porphyrias. *Mailing Add:* Mt Sinai Sch Med Fifth Ave at 100th St New York NY 10029

DESNOYERS, JACQUES EDOUARD, b Ottawa, Ont, Jan 28, 35; m 64; c 3. PHYSICAL CHEMISTRY. *Educ:* Univ Ottawa, Can, BSc, 58, PhD(phys chem), 61. *Prof Exp:* NATO & Ramsay fels, Battersea Col Tech & Manchester Univ, Eng, 61-62; lectr, Univ Sherbrooke, 62-63, from asst prof to prof phys chem, 63-82; sci dir, Nat Inst Sci Res, 83-91; PROF, INST NAT RES SCI ENERGY, 91- *Honors & Awards:* Lash Miller Award, Electrochem Soc, 70; Huffman Mem Award, Calorimetry Conf, 78; Brit Petrol Prize, 80; Archambault Medal, Asn Can Advan Sci, 84. *Mem:* Am Chem Soc; Electrochem Soc; Chem Inst Can. *Res:* Theoretical and experimental studies of thermodynamic properties of aqueous solutions and colloidal solutions and in particular surfactants and microemulsions; extractions of bitumen from oil sands. *Mailing Add:* Inst Nat Res Sci Energy 1650 Montee St Julie Case Postale 1020 Varennes PQ J3X 1S2 Can

DESOER, CHARLES A(UGUSTE), b Brussels, Belg, Jan 11, 26; nat US; m 51, 66; c 3. ENGINEERING, SYSTEMS THEORY. *Educ:* Univ Liege, Belg, Dipl, 49; Mass Inst Technol, ScD, 53. *Hon Degrees:* DSc, Univ Liege, 77. *Prof Exp:* Mem tech staff, Bell Tel Labs, Inc, 53-58; assoc prof elec eng, 58-62, PROF ELEC ENG, UNIV CALIF, BERKELEY, 62- *Concurrent Pos:* Miller res prof, 67-68; Guggenheim fel, 70-71. *Honors & Awards:* Educ Award, Am Autom Control Coun, 83; Control Syst Sci & Eng Award, Inst Elec & Electronics Engrs, 86. *Mem:* Nat Acad Eng; fel AAAS; Am Math Soc; Soc Indust & Appl Math; fel Inst Elec & Electronics Engrs. *Res:* System theory; control and circuits. *Mailing Add:* Dept Elec Eng & Comput Sci Univ Calif Berkeley CA 94720

DESOMBRE, EUGENE ROBERT, b Sheboygan, Wis, May 6, 38; m 60; c 2. BIOCHEMISTRY, ENDOCRINOLOGY. *Educ:* Univ Chicago, BS, 60, MS, 61, PhD(org chem), 63. *Prof Exp:* Res assoc, Ben may Inst, 63-65, from instr to assoc prof, 66-80, dir, Biomed Comput Facil, 80-86, PROF, BEN MAY INST, UNIV CHICAGO, 80- *Concurrent Pos:* Mem Breast Cancer Task Force, Nat Cancer Inst, 74-80, chmn, 78-80; mem adv comts, Am Cancer Soc, 78-82, 85-90; assoc ed, Cancer Res, 75-80, 82-; bd govs, Int Assoc Breast Cancer Res, 80-85, 89- *Mem:* AAAS; Endocrine Soc; Am Asn Cancer Res; Am Chem Soc. *Res:* Organophosphorous chemistry; steroid mechanism of action; estrogen endocrinology; hormone dependent cancer. *Mailing Add:* Ben May Inst Univ Chicago 5841 S Maryland Ave Chicago IL 60637

DESOR, JEANNETTE ANN, b Baltimore, Md, July 11, 42. PSYCHOPHYSICS, EXPERIMENTAL PSYCHOLOGY. *Educ:* Cornell Univ, AB, 64; PhD(exp psychol), 69. *Prof Exp:* Res asst psychol, Dept Psychol, Yale Univ, 65; teaching asst psychol, Cornell Univ, 65-69; staff scientist exp psychol, Monell Chem Senses Ctr, Univ Pa, 70-75; mgr sensory eval, res & develop, Warner-Lambert Co, 75-78; mgr behav sci, Res & Develop, Gen Foods Corp, 78-82, prin scientist, 82-88; PVT CONSULT, 88- *Concurrent Pos:* Fel chemoreception, Univ Pa, 70-72; NIH fel, 71 & 72; res investr, Vet Admin Hosp, Philadelphia, 73-75; asst prof, Dept Otorhinol & Human Commun, Sch Med, 73-75, staff therapist, Behav Weight Control Prog, Dept Psychiat, Hosp Univ Pa, 74-75; bd dir, Mace-Fremont, Inc, 80- *Mem:* Soc Ingestive Behav; AAAS; Asn Chemoreception Sci; Human Factor Soc. *Res:* Human sensory systems; development of human taste preferences; clinical assessment of taste; functions of taste and olfaction in nutrition; measurement techniques for sensory evaluation; crowding; product development; one patent. *Mailing Add:* 10630 Breezewood Circle Woodstock MD 21163

DESOWITZ, ROBERT, b New York, NY, Jan 2, 26; m 54; c 2. PARASITOLOGY, IMMUNOLOGY. *Educ:* Univ Buffalo, BA, 48; Univ London, PhD(parasitol), 51, DSc(parasitol), 60. *Prof Exp:* Prin sci officer, Colonial Med Res Serv, WAfrican Inst Trypanosomiasis Res, 51-60; prof parasitol & head dept, Univ Singapore, 60-65; chief dept parasitol, SEATO Med Res Lab, Bangkok, Thailand, 65-68; PROF TROP MED & PUB HEALTH, LEAHI HOSP, SCH MED, UNIV HAWAII, MANOA, 68- *Concurrent Pos:* Mem expert comt parasitic dis, WHO, 64- *Mem:* Am Soc Trop Med; fel Royal Soc Trop Med & Hyg; hon fel Malaysian Soc Trop Med. *Res:* Host-parasite relationships, especially malaria; immunologic response to malaria, trypanosomiasis and filariasis. *Mailing Add:* Leahi Hosp Univ Hawaii Sch Med Honolulu HI 96816

DESPAIN, ALVIN M(ARDEN), b Salt Lake City, Utah, July 2, 38; m 57; c 2. COMPUTER SCIENCE, ELECTRICAL ENGINEERING. *Educ:* Univ Utah, BS, 60, MS, 64, PhD(elec eng), 66. *Prof Exp:* Engr trainee elec eng, Southern Calif, Edison Co, 57; res asst elec eng & physics, Univ Utah, 57-60, res engr, 60-66, asst res prof elec eng, 66-67; asst prof, Utah State Univ, 66-69, asst dir electrodyn labs, 67-73, assoc prof elec eng, 69-76; assoc prof, 76-80, PROF COMPUT SCI, DIV ELEC ENG & COMPUT SCI, UNIV CALIF, BERKELEY, 80- *Concurrent Pos:* Vis assoc prof, Stanford Univ, 72-73. *Mem:* Inst Elec & Electronics Engrs; Am Soc Eng Educ; Asn Comput Mach; AAAS; Sigma Xi. *Res:* Communication-information theory; data communications, computer design, computer architecture, microprocessors and computer hardware. *Mailing Add:* 503 Evans Hall Berkeley CA 94720

DESPAIN, LEWIS GAIL, b Salt Lake City, Utah, Feb 7, 28; m 56; c 4. SPACE PHYSICS. *Educ:* Univ Utah, BS, 55, MS, 57, PhD(physics), 62. *Prof Exp:* Sr scientist, Space Sci Div, Jet Propulsion Lab, Calif Inst Technol, 61-64, group supvr, 64-66, assoc proj scientist, 66-67; supvr, Space Physics Group, Boeing Co, Seattle, 67-73, dir, Radiation Effects Lab, 73-76, staff scientist, inertial upper stage syst integration, 76-78 & solar elec power syst, 79, adv defense missile early warning syst, 80, staff scientist short range attack missile advan concepts, 81-83 & charge of nuclear surviveability/vulnerability of AWACS aircraft, Boeing Aerospace Co, 83-90; RETIRED. *Mem:* Am Asn Physics Teachers. *Res:* Direction of specialized group which conducts theoretical and experimental research in space physics. *Mailing Add:* 33304 Tenth Ct SW Federal Way WA 98023

DESPER, CLYDE RICHARD, b Greenwood, Ark, Dec 14, 37; m; c 5. POLYMER SCIENCE. *Educ:* Mass Inst Technol, BS, 59, MS, 60; Univ Mass, Amherst, PhD(chem), 67. *Prof Exp:* Res assoc, Fabric Res Labs, Inc, Mass, 60-62; res chemist, US Army Natick Labs, Mass, 66-68; RES CHEMIST POLYMERS, US ARMY MAT TECHNOL LAB, 68- *Mem:* Am Phys Soc; Am Chem Soc; Am Crystallog Asn; Sigma Xi. *Res:* Investigation of solid state morphology and properties in polymers, primarily by x-ray diffraction; crystallography; preferred orientation effects; microphase segregation. *Mailing Add:* Army Mat Technol Lab Watertown MA 02172-0001

DESPLAN, CLAUDE, French citizen. DEVELOPMENTAL MOLECULAR GENETICS, DROSOPHILA. *Educ:* Teachers Training Col, St Cloud, France, Master, 74; Univ Paris VII, Dr, 79, PhD(molecular biol), 83. *Prof Exp:* Postdoctoral fel biochem, Univ Calif, San Francisco, Calif, 84-87; ASST PROF GENETICS, ROCKEFELLER UNIV, NEW YORK, NY, 87-; ASST INVESTR GENETICS & HEAD LAB MOLECULAR GENETICS, HOWARD HUGHES MED INST, ROCKEFELLER UNIV, 87- *Concurrent Pos:* Postdoctoral fel, Fogarty Int Ctr, 84-85 & Europ Molecular Biol Orgn, 86-87; Andre Meyer fel, Rockefeller Univ, 87- *Mem:* AAAS; Genetics Soc Am; Am Soc Microbiol; Harvey Soc. *Res:* Investigation of questions of transcriptional control in the context of early Drosophila development. *Mailing Add:* Rockefeller Univ 1230 York Ave New York NY 10021

DESPOMMIER, DICKSON, b New Orleans, La, June 5, 40; div; c 2. PARASITOLOGY, MOLECULAR PARASITOLOGY. *Educ:* Fairleigh Dickinson, BS, 62; Columbia Univ, MS, 64; Univ Notre Dame, PhD(microbiol), 67. *Prof Exp:* Asst parasitol, Sch Pub Health & Admin Med, Columbia Univ, 62-63; asst biol, Univ Notre Dame, 64-65; USPHS guest investr parasitol, Rockefeller Univ, 67-70; from asst prof to assoc prof, 70-77, PROF PARASITOL, SCH PUB HEALTH & DEPT MICROBIOL, COLUMBIA UNIV, 77- *Concurrent Pos:* NIH career develop award, 71-76. *Mem:* AAAS; Am Soc Parasitol; Am Soc Cell Biol; Am Soc Trop Med & Hyg; Am Soc Immun; Int Comn Trichinellosis. *Res:* Effects of the immune state on the biology of nematode parasites in mammalian hosts; molecular biology of the muscle phase of trichinosis; mechanisms of host immunity to trichinosis. *Mailing Add:* Dept Trop Med Columbia Univ Sch Pub Health New York NY 10032

DESPRES, THOMAS A, b Grand Rapids, Mich, Nov 13, 32; m 59; c 2. MECHANICAL & METALLURGICAL ENGINEERING. *Educ:* Univ Mich, BSIE, 56, MSME, 59, PhD(mech eng), 64. *Prof Exp:* Test & eval missile components, Redstone Arsenal, 57-58; teaching fel mech & metall eng, 58-63, from asst prof to assoc prof, 63-76, PROF MECH ENG, UNIV MICH, DEARBORN, 76- *Mem:* Sigma Xi. *Res:* Metallurgical and solid state research in fatigue of metals; mechanical properties of metals and application of electron microscopy to metallurgical and mechanical properties study. *Mailing Add:* 135 Ft Dearborn Dearborn MI 48124

DES PREZ, ROGER MOISTER, b Chicago, Ill, Mar 14, 27; m 65; c 7. INTERNAL MEDICINE, PULMONARY DISEASES. *Educ:* Dartmouth Col, AB, 51; Columbia Univ, MD, 54; Am Bd Internal Med, dipl, 62. *Prof Exp:* Intern, asst med resident & chest med resident, NY Hosp, 54-57; physician, Ft Defiance Tuberc Sanatorium, 57-59; from instr to asst prof med, Med Col, Cornell Univ, 57-63; assoc prof, 63-68, PROF MED, SCH MED, VANDERBILT UNIV, 68- *Concurrent Pos:* Teaching fel, Am Trudeau Soc, 56-57, Edward L Trudeau fel, 59-63; physician outpatients, NY Hosp, 59-63; chief med serv, Vet Admin Hosp, Nashville, Tenn, 63-; ed chest sect, Yearbk Med, Asn Am Physicians. *Mem:* Asn Vet Admin Chiefs Med; Am Soc Clin Invest; Am Orthop Asn; Asn Am Physicians. *Res:* Patterns of tissue injury; immunologic injury to rabbit platelets. *Mailing Add:* Dept Med Vet Admin Med Ctr 1310 24th Ave Nashville TN 37203

DESROCHERS, ALAN ALFRED, b Northampton, Mass, June 1, 50. CONTROL SYSTEMS ENGINEERING. *Educ:* Univ Lowell, BSEE, 72; Purdue Univ, MSEE, 73, PhD(elec eng), 77. *Prof Exp:* Assoc engr, Lockheed Missiles & Space Co, 73-75; asst prof comput & syst eng, Boston Univ, 77-80; from asst prof to assoc prof, 80-90, PROF ELEC, COMPUT & SYST ENG, RENSSELAER POLYTECH INST, 90- *Concurrent Pos:* Fac res assoc, Auburn Univ, 78. *Mem:* Inst Elec & Electronics Engrs; Am Soc Eng Educ. *Res:* Modeling and control of dynamic systems with application to manufacturing, automation and robotics. *Mailing Add:* Elec Comput & Syst Eng Dept Rensselaer Polytech Inst Troy NY 12180-3590

DESROSIERS, RONALD CHARLES, b Manchester, NH, June 16, 48; m 70; c 2. VIROLOGY. *Educ:* Boston Univ, BA, 70; Mich State Univ, PhD, 75. *Prof Exp:* Fel, Yale Univ, 75-77, res assoc, 77; instr, 77-79, asst prof, 79-85, ASSOC PROF MICROBIOL & MOLECULAR GENETICS, SCH MED, HARVARD UNIV, 85-, CHMN, DIV MICROBIOL, NEW ENG PRIMATE RES CTR, 84- *Concurrent Pos:* Actg chmn, div microbiol, New Eng Regional Primate Res Ctr, Harvard Univ, 83-84; res fel, NIH, 75-77, Med Found Boston, Inc, 79-81, Leukemia Soc Am, 82-84. *Mem:* Am Soc Virol. *Res:* Molecular basis for the pathogenicity of viruses of monkeys; molecular basis for oncogenicity of herpes viruses in New World primates; investigation of Old World primate viruses related to virus that causes acquired immune deficiency syndrome in humans. *Mailing Add:* New Eng Regional Primate Res Ctr Harvard Med Sch 1 Pine Hill Dr Southborough MA 01772

DESS, HOWARD MELVIN, b Chicago, Ill, Dec 23, 29; m 51; c 3. INORGANIC CHEMISTRY. *Educ:* Ind Univ, BS, 51; Univ Mich, MS, 53, PhD(chem), 55. *Prof Exp:* Res chemist, Electrometall Co, 55-56 & Pennsalt Chem Corp, 56-58; res chemist, Union Carbide Corp, 58-63, group leader crystal prod res, Speedway Lab, Linde Div, 63, res & develop supvr, Crystal Prod Dept, Electronics Div, 63-68, group mgr, 68; asst tech dir, Nat Lead Co, 68, dir res, Hightstown Cent Res Lab, 68-70, MGR COM CRYSTALS DEPT, TITANIUM PIGMENT DIV, NL INDUSTS, INC, 70- *Mem:* Am Chem Soc; Am Asn Crystal Growers; Am Ceramic Soc; Electrochem Soc. *Res:* Crystal growth. *Mailing Add:* 316 Goldfinch Dr Bridgewater NJ 08807

DESSAUER, HERBERT CLAY, b New Orleans, La, Dec 30, 21; m 49; c 3. MEDICAL SCIENCES, ZOOLOGY. *Educ:* La State Univ, PhD(biochem), 52. *Prof Exp:* Teaching fel biochem, Sch Med, 49-50, asst, 50-51, from instr to assoc prof, 51-63, actg head dept, 77-78, PROF BIOCHEM, SCH MED, LA STATE UNIV MED CTR, NEW ORLEANS, 63- *Concurrent Pos:* Consult, Vet Admin Hosp, New Orleans, La, 62-79; mem panel for advan sci educ, NSF, 65-67, mem panel for systematics, 72-75; mem alpha helix exped to New Guinea, 69; mem task force fluoridation, New Orleans Health Planning Coun, 71-72; mem nat comt resources herpet, 73-75; mem, NSF panel, Res Initiation & Support, 76; res assoc, Dept Herpet, Am Mus Nat Hist, 78-; consult, La Nature Ctr, 78-; deleg, US State Dept, Strategy Conf Conserv Biol Diversity, 81; dir, Soc Study Amphibians Reptiles, 81; mem, Australian Workshop Frozen Tissue Collections Mus, 82; co-chmn, NSF Workshop Frozen Tissue Collections, 82-83; mem panel, Res & Develop, La Bd Regents, 82; mem, res adv bd, Audubon Zool Park, 86-, Coun Collections, Asn Systematics Collections 86-, Workshop on Living & Culture Collections, 88. *Mem:* Fel AAAS; Am Physiol Soc; Am Soc Ichthyologists & Herpetologists (pres, 86); Soc Study Amphibians Reptiles; fel Herpetologists League; Am Genetic Asn; AMA. *Res:* Biochemistry of the lizard Anolis carolinensis; comparative biochemistry; blood proteins; protein taxonomy; structural gene expression; reptilian genetics and zoogeography; parthenogenetic lizards; frozen tissue collection; protein and nucleic acid evidence is used to examine problems concerning the origin and genetics of unisexual vertebrates and upon the genetics of endangered species such as whooping cranes. *Mailing Add:* 7100 Dorian St New Orleans LA 70126

DESSAUER, JOHN HANS, b Aschaffenburg, Ger, May 13, 05; nat US; m 36; c 3. CHEMISTRY, CHEMICAL ENGINEERING. *Educ:* Munich Tech Inst, BS, 26; Aachen Tech Inst, Master, 27; DIngSc, 29. *Hon Degrees:* LHD, Le Moyne Col, 63; DSc, Clarkson Col, 75; LLD, Fordham Univ, 76. *Prof Exp:* Res chemist, Ansco, NY, 29-35; res chemist & dir, Haloid Co, 35-51, vpres in chg res & prod develop & dir, 46-58, exec vpres res & eng, Xerox Corp, 59-66, vchmn bd & exec vpres res & advan eng, 66-70, dir, 59-73, dir, Rank Xerox Ltd, 59-70; trustee, NY State Sci & Technol Found, 72-78; RETIRED. *Concurrent Pos:* Emer trustee, Fordham Univ. *Honors & Awards:* Phillipp Medal, Inst Elec & Electronics Engrs, 74; Indust Res Inst Medal, 68. *Mem:* Nat Acad Eng; fel NY Acad Sci; fel Am Inst Chemists; Am Chem Soc; Am Phys Soc. *Res:* Photo research; organic chemistry; xerography. *Mailing Add:* 37 Parker Dr Pittsford NY 14534

DESSAUER, ROLF, b Nurnberg, Ger, Nov 3, 26; nat US; wid; c 2. ORGANIC CHEMISTRY. *Educ:* Univ Chicago, BS, 48, MS, 49; Univ Wis, PhD(chem), 52. *Prof Exp:* From res chemist to sr res chemist, Org Chem Dept, 52-68, res assoc, 69-78, res assoc, 78-85, sr res assoc, Chem, Dyes & Pigments Dept, Photo Prod Dept, E I Du Pont De Nemours & Co, Inc, 86-87; SR RES ASSOC, DX IMAGING CO, LIONSVILLE, PA, 87- *Mem:* Am Chem Soc; Soc Photog Sci & Eng. *Res:* Steroids; dyes; photochemistry; photochromism; imaging systems; liquid crystals; chemical marketing; development and marketing of novel non-silver imaging systems; electrophotography. *Mailing Add:* DX Imaging 101 Gordon Dr Lionville PA 19353

DESSEL, NORMAN F, b Ida Grove, Iowa, July 9, 32; m 55; c 3. PHYSICS, APPLIED STATISTICS. *Educ:* Univ Iowa, BA, 57, MA, 58, PhD(physics, sci educ), 61. *Prof Exp:* From asst prof to assoc prof physics, San Diego State Univ, 61-68, chmn, Dept Phys Sci, 69-73, prof physics & phys sci, 68-76, PROF NATURAL SCI, SAN DIEGO STATE UNIV, 76- *Concurrent Pos:* Consult, US Naval Electronics Lab, 62-69; mgr grad traineeships & fels, NSF, 76-77, consult, anal fel, 77- *Mem:* AAAS; Optical Soc Am; Am Asn Physics Teachers. *Res:* Coherent optical information processing and holography, laser communications, physics education and energy research and development; merit evaluations on national grant programs. *Mailing Add:* Dept Natural Sci San Diego State Univ San Diego CA 92182

DESSER, SHERWIN S, b Winnipeg, Man, Sept 2, 37; m 63; c 3. ZOOLOGY, PARASITOLOGY. *Educ:* Univ Man, BSc, 59, MSc, 63; Univ Toronto, PhD(parasitol), 67. *Prof Exp:* Res fel parasitol, Univ Toronto, 67 & Hebrew Univ, Israel, 67-68; from asst prof to assoc prof, 68-81, assoc chmn dept zool, 84-90, PROF PARASITOL, UNIV TORONTO, 81-, CHMN DEPT ZOOL, 90- *Concurrent Pos:* Ed, J Parasitol, 84-88. *Mem:* Am Soc Parasitol; Soc Protozool; Royal Soc Trop Med & Hyg; Can Soc Zoologists; Wildlife Dis Asn. *Res:* Protozoology; Haemosporidia, Myxosporea, Coccidia; parasitology. *Mailing Add:* Dept Zool Univ Toronto Toronto ON M5S 1A1 Can

DESSLER, ALEXANDER JACK, b San Francisco, Calif, Oct 21, 28; m 52; c 4. SPACE PHYSICS. *Educ:* Calif Inst Technol, BS, 52; Duke Univ, PhD(physics), 56. *Prof Exp:* Res assoc, Duke Univ, 55-56; mem staff, Lockheed Missile & Space Co, 56-62; prof, Southwest Ctr Advan Studies, 62-63; chmn dept space sci, Rice Univ, 63-69, prof space physics & astron, 63-82, mgr campus bus affairs, 74-76, chmn, Space Physics & Astron Dept, 79-82; dir, 82-84, prof space physics & astron, 82-86, DIR, SPACE SCI LAB & DEPT CHMN, SPACE PHYSICS & ASTRON DEPT, NASA MARSHALL SPACE FLIGHT CTR, 87- *Concurrent Pos:* Mem, US Nat Comt, Int Sci Radio Union, 60-63 & 67-70; ed, J Geophys Res, 65-69; sci adv, Nat Aeronaut & Space Coun, 69-70; ed, Rev Geophys & Space Physics, 69-74 & Geophysics Res Lett, 86-89; consult, NSF, 76; assoc ed, Rev Geophys, 63-68, Space-Solar Power Rev, 80-85. *Honors & Awards:* James B Macelwane Award, Am Geophys Union, 63; Soviet Geophys Comt Medal, 84. *Mem:* Fel AAAS; fel Am Geophys Union; Int Sci Radio Union; Univs Space Res Asn (pres, 75-81); Int Asn Geomagnetism & Aeronomy (vpres, 79-83); Am Astron Soc; Sigma Xi. *Res:* Geomagnetism; theory of planetary magnetospheres; plasma physics; hydromagnetism; interplanetary physics. *Mailing Add:* Dept Space Physics & Astron Rice Univ PO Box 1892 Houston TX 77251-1892

DESSOUKY, DESSOUKY AHMAD, b Mitgamr, UAR, Jan 18, 32; m 69; c 1. ANATOMY, ELECTRON MICROSCOPY. *Educ:* Ain Shams Univ, Cairo, MD, 56, MS, 60; Tulane Univ, PhD(anat), 64. *Prof Exp:* Instr anat, Sch Med, Ain Shams Univ, 57-61; from instr to asst prof, 64-67, resident obstet & gynec, 67-70, asst prof, 70-74, ASSOC PROF OBSTET & GYNEC, GEORGETOWN UNIV, 74- *Mem:* AAAS; Am Asn Anat; Am Col Obstet & Gynec; Am Fertil Soc. *Res:* Fine structure of corpus luteum, uterine wall and uterine blood vessels in normal and abnormal pregnancy; electron microscopy of female reproductive system. *Mailing Add:* 1715 N George Mason Dr Arlington VA 22205

DESSOUKY, MOHAMED IBRAHIM, b Cairo, Egypt, Dec 20, 26; US citizen; m 58; c 4. PRODUCTION MANAGEMENT. *Educ:* Cairo Univ, BSc, 48; Purdue Univ, MS, 54; Ohio State Univ, PhD(indust eng), 56. *Prof Exp:* Res fel, oper res, Case Inst Technol, 56-57; asst prof indust eng, Cairo Univ, 57-63; assoc prof oper res, Thayer Sch Eng, Dartmouth Col, 64-65; prof mgt sci, Nat Inst Mgt Develop, 65-68; assoc prof indust eng, Univ Ill, Urbana, 68-90; PROF & CHAIR, DEPT INDUST ENG, NOTHERN ILL UNIV, DEKALB, 90- *Concurrent Pos:* Consult, UN Indust Develop Orgn, 65-, Construction Eng Res Lab, 71-84, Westinghouse Elec Corp, 83-85 & Kalkan Foods, 84-; vis prof, Univ Khartoum, 67-68; sr scientist, Kuwait Inst Sci Res, 78-82. *Mem:* Sr mem Am Inst Indust Eng; Oper Res Soc Am; Inst Mgt Sci; Egyptian Engrs Soc. *Res:* Production planning and control, project network analysis, scheduling, simulation, multiple criteria decision making, quality control and applications of operations research in developing countries, especial. *Mailing Add:* 116 Eng Bldg Northern Ill Univ DeKalb IL 60115

DESSY, RAYMOND EDWIN, b Reynoldsville, Pa, Sept 3, 31; m 59. CHEMICAL INSTRUMENTATION, MICROELECTRONIC SENSORS. *Educ:* Univ Pittsburgh, BS, 53, PhD, 56. *Hon Degrees:* DSc, Hampden-Sydney Col, 86. *Prof Exp:* Fel & instr, Ohio State Univ, 56-57; from asst prof to assoc prof chem, Univ Cincinnati, 57-66; PROF CHEM, VA POLYTECH INST & STATE UNIV, 66- *Concurrent Pos:* Sloan fel, 62-64; NSF sr fel, 63-64. *Honors & Awards:* Computers in Chem Award, Am Chem Soc, 86. *Mem:* Am Chem Soc. *Res:* Design and automation of new chemical instruments for spectroscopy and chromotography; solid state detector development; development of courses to teach small computer automation to chemists; biosensors and bioanalytical separations. *Mailing Add:* Dept Chem 0212 Va Polytech Inst & State Univ Blacksburg VA 24061-0212

DESTEFANO, ANTHONY JOSEPH, b Middletown, Conn, June 6, 49; m. MASS SPECTROMETRY. *Educ:* Villanova Univ, BS, 71; Cornell Univ, MS, 73, PhD(chem), 76. *Prof Exp:* Staff chemist mass spectrometry, Miami Valley Labs, Procter & Gamble, 76-83; staff chemist, 83-85, SECT HEAD, PHYS CHEM SECT, NORWICH-EATON PHARMACEUT, INC, 85- *Mem:* Am Chem Soc; Am Soc Mass Spectrometry. *Res:* Organic mass spectrometry including GC/MS, field desorption, fastatom bombardment and chemical ionization. *Mailing Add:* Norwich Eaton Pharmaceut Inc PO Box 191 Norwich NY 13815

DESTEVENS, GEORGE, b Tarrytown, NY, Aug 21, 24; m 50. ORGANIC CHEMISTRY. *Educ:* Fordham Univ, BS, 49, MS, 50, PhD, 53. *Hon Degrees:* DSc, Drew Univ, 88. *Prof Exp:* Instr chem, Fordham Univ, 52-53; lectr org chem, Marymount Col, NY, 52; res chemist, Remington Rand Corp, 53-55; sr res scientist, Ciba Pharmaceut Co, 55-61, Ciba-Geigy Pharmaceut Res Labs, 55-61, dir med chem res, 62-66, dir chem res, Chem Res Dept, Ciba Pharmaceut Prod Inc, 66-67, vpres res, Ciba Pharmaceut Co, 67-70, exec vpres & dir res, Ciba-Geigy Pharmaceut Res Lab, 70-79; RES PROF CHEM, DREW UNIV, 79- *Honors & Awards:* Walter J Hartung Memorial Award Outstanding Contrib Med Res, 79. *Mem:* fel NY Acad Sci; Sigma Xi. *Res:* Cyanine dyes as therapeutics; diuretics; analgesics; tranquilizers; chemistry of heterocyclics; spectral properties or organic compounds. *Mailing Add:* Two Warwick Rd Summit NJ 07901

DESU, MANAVALA MAHAMUNULU, b Ongole, India, Nov 12, 31; m 58; c 2. STATISTICS. *Educ:* Andhra Univ, India, BA, 50; Madras Univ, MA, 53; Univ Minn, PhD(statist), 66. *Prof Exp:* Lectr math, SSN Col, Narasaraopet, India, 53-55, math & statist, SV Univ Tirupati, India, 59-62; from lectr to assoc prof, 65-82, PROF STATIST, STATE UNIV NY, BUFFALO, 82- *Mem:* Inst Math Statist; Am Statist Asn; Royal Statist Soc; Biomet Soc; Int Statist Inst. *Res:* Nonparametric methods; survival data analysis; biostatistics; ranking and selection procedures. *Mailing Add:* Dept Statist State Univ NY 249 Farber Hall Buffalo NY 14214

DESU, SESHU BABU, b Tenali, Andhra, India, June 22, 55; m 84; c 2. ELECTRONIC CERAMICS, FERROELECTRICS. *Educ:* Andhra Univ, India, BS, 74, MS, 76; Indian Inst Technol, Kanpur, MTech, 79; Univ Ill, Urbana, PhD(ceramic eng), 82. *Prof Exp:* Mem tech staff, AT&T Labs, 82-86; group leader, Gen Elec, 86-88; ASSOC PROF MAT, VA POLYTECHNIC INST & STATE UNIV, 88- *Concurrent Pos:* Consult, PPG Indust, Inc, 89-90, Arcova, Va, 89-, Solar Prod, Inc, 90- *Mem:* Am Ceramic Soc; Mat Res Soc; Electrochem Soc; Am Vacuum Soc. *Res:* Processing of electronic ceramics including thin films and fibers; interrelations between processing, structure and properties of materials and the development of new processes for namo-materials. *Mailing Add:* Mat Eng Dept Va Polytechnic Inst Blacksburg VA 24061

DESUA, FRANK CRISPIN, b Monessen, Pa, Oct 26, 21; m 45; c 1. MATHEMATICS. *Educ:* Univ Pittsburgh, BS, 44, PhD(math), 56. *Prof Exp:* Instr math, Univ Pittsburgh, 44-54; asst prof, Ohio Univ, 54-56 & Univ Pittsburgh, 56-57; mem tech staff, Bell Tel Labs, 57-58; assoc prof math, Col William & Mary, 58-60; prof & chmn dept, Simmons Col, 60-70 & Sweet Briar Col, 70-75; prof math, Fla Inst Technol, 75-87; RETIRED. *Mem:* Am Math Soc; Math Asn Am. *Res:* Foundations of mathematics; symbolic logic; meta-mathematics; point-set topology. *Mailing Add:* 6687 Flamingo Rd Melbourne Village FL 32904

DE SYLVA, DONALD PERRIN, b Rochester, NY, July 20, 28; m 50; c 2. BIOLOGICAL OCEANOGRAPHY, ICHTHYOLOGY. *Educ:* Cornell Univ, BS, 52, PhD(vert zool), 58; Univ Miami, MS, 53. *Prof Exp:* Asst fish collection, Cornell Univ, 51-52, asst oceanog, 52; asst zool, Univ Miami, 52-53; asst biomet & statist, Univ Calif, 53-54; res instr oceanog, Univ Miami, 54, res instr fisheries, 54-56; asst vert zool, Cornell Univ, 56-57; asst prof biol sci, Univ Del, 58-61; from asst prof to assoc prof marine sci, 61-75, PROF MARINE SCI, UNIV MIAMI, 75- *Concurrent Pos:* Res aide, Univ Miami, 53; scientist chg, Pac Billfish Exped, Lou Marron & Univ Miami, Chile, 56; consult, Dr Edward C Raney, Cornell Univ, 57 & Comt on the Effects of Herbicides in Vietnam, Nat Acad Sci, 71-73; mem Atomic Safety & Licensing Bd, USAEC/US Nuclear Regulatory Comn, 71-85. *Mem:* AAAS; Am Fisheries Soc; Coun Biol Educ; Am Soc Ichthyol & Herpet; Am Inst Fisheries Res Biol. *Res:* Life history, systematics and ecology of marine and estuarine fishes; fisheries and estuarine ecology; ecology of mangrove fishes; early life history of fishes; ecology of poisonous fishes and pelagic fishes. *Mailing Add:* Marine Div Sch Marine & Atmospheric Sci Univ Miami 4600 Rickenbacker Causeway Miami FL 33149

DE TAKACSY, NICHOLAS BENEDICT, b Budapest, Hungary, Feb 24, 39; Can citizen; m 62; c 3. NUCLEAR PHYSICS. *Educ:* Loyola Col, Can, BSc, 59; Univ Montreal, MSc, 63; McGill Univ, PhD(nuclear theory), 66. *Prof Exp:* Lectr physics, Loyola Col, Can, 61-63; res fel nuclear theory, Calif Inst Technol, 66-67; asst prof, Loyola Col, Que, 67-68; from asst prof to assoc prof nuclear theory, 68-77, chmn dept, 79-82, PROF NUCLEAR THEORY, MCGILL UNIV, 77- *Concurrent Pos:* Res fel, McGill Univ, 67-68. *Mem:* Am Phys Soc; Can Asn Physicists. *Res:* Nuclear structure of transitional nuclei; low energy pion-nucleon interaction. *Mailing Add:* Dept Physics 3600 University St Montreal PQ H3A 2T8 Can

DETAR, DELOS FLETCHER, b Kansas City, Mo, Jan 18, 20; m 43; c 4. ORGANIC CHEMISTRY, PHYSICAL CHEMISTRY. *Educ:* Univ Ill, BS, 41; Univ Pa, MS, 43, PhD(org chem), 44. *Prof Exp:* Res chemist, Pioneering Res Sect, Rayon Dept, Tech Div, E I du Pont de Nemours & Co, 44-46; fel, Univ Ill, 46; from instr to asst prof chem, Cornell Univ, 46-53; from assoc prof to prof, Univ SC, 53-60; PROF CHEM, FLA STATE UNIV, 60- *Concurrent Pos:* NSF sr fel, Harvard Univ, 56-57; vis prof, Univ Calif, Berkeley, 60. *Mem:* Am Chem Soc; fel Royal Soc Chem; Sigma Xi; AAAS. *Res:* Mechanisms of organic reactions; peptides and enzyme models; steric effects; molecular mechanics. *Mailing Add:* Dept Chem Fla State Univ Tallahassee FL 32306-3006

DETAR, REED L, b Oil City, Pa, May 14, 32; m 59; c 4. PHYSIOLOGY. *Educ:* Susquehanna Univ, AB, 54; Hahnemann Med Col, MS, 59; Univ Mich, PhD(physiol), 68. *Prof Exp:* Assoc scientist, Warner-Lambert Res Inst, 59-63; asst prof physiol, 69-75, assoc prof physiol, 75-80, RES ASSOC PROF PHARMACOL, DARTMOUTH MED SCH, 80- *Mem:* Assoc mem Am Physiol Soc. *Res:* Local control of vascular smooth muscle. *Mailing Add:* Dept Physiol Dartmouth Med Sch Hanover NH 03756

DETELS, ROGER, b Brooklyn, NY, Oct 14, 36; m 63; c 2. EPIDEMIOLOGY. *Educ:* Harvard Univ, BA, 58; NY Univ, MD, 62; Univ Wash, MS, 66. *Prof Exp:* Med officer, US Naval Med Res, Taiwan, 66-69, Epidemiol Br, Nat Inst Neurolog Dis & Stroke, NIH, 69-71; assoc prof epidemiol, 71-73, actg head, Div Epidemiol, 71-72, div head, 72-80, dean, 80-85, PROF EPIDEMIOL, SCH PUB HEALTH, UNIV CALIF, LOS ANGELES, 73- *Concurrent Pos:* Prin investr, Epidemiol Study Chronic Obstructive Respiratory Dis, Nat Inst Pub Health Serv, Calif, 72-, Studies Immune Factors in Multisclerosis, 75-82; prog dir, Cancer Epidemiol Training Prog, Environ Epidemiol Training grant, Nat Cancer Inst, 75-, Nat Inst Environ Health Sci, 78-; co-prin investr, Health Status Am Men, Nat Inst Child Health & Human Develop, 79-83, Epidemiol Study Phenoxy Herbicide incl Agent Orange, 81-83; mem, Adv Panel Clin Sci, Nat Res Coun, 81-84, Sci Rev Panel Health Res Environ Protection Agency, 80- 85; prin investr, Multicenter Aids Cohort Study Los Angeles Nat Inst Allergy & Infectious Dis, 83- & AIDS- Barrier Contraceptive Study, Nat Inst Child Health & Human Develop, 86- *Mem:* Am Pub Health Asn; Soc Epidemiol Res (pres, 77-78); Am Epidemiol Soc; Int Epidemiol Asn (treas, 84-90, pres, 90-93); Am Thoracic Soc; Sigma Xi; Asn Sch Pub Health (secy-treas, 84-85). *Res:* Epidemiologic research into factors causing and which determining the clinical course of multiplesclerosis; relationship between chronic exposure to air pollution and lung function; health outcomes of vasectomy; control of hypertension; natural history of AIDS and HIV infection, barrier contraceptives in prevention of AIDS. *Mailing Add:* Sch Pub Health Univ Calif Los Angeles CA 90024-1772

DETEMPLE, DUANE WILLIAM, b Portland, Ore, July 22, 42; m 66; c 2. MATHEMATICS EDUCATION. *Educ:* Portland State Col, BS, 64; Stanford Univ, MS, 66, PhD(math), 70. *Prof Exp:* Assoc engr sr, Lockheed Missiles & Space Co, 66-67; from asst prof to assoc prof, 70-85, PROF MATH, WASH STATE UNIV, 85- *Concurrent Pos:* Vis assoc prof, Claremont Grad Sch, 81-82. *Mem:* Math Asn Am; Nat Coun Teachers Math. *Res:* Univalent function theory; geometric probability; combinatorics; graph theory; elementary geometry; mathematics education. *Mailing Add:* Wash State Univ Pullman WA 99164-3113

DETEMPLE, THOMAS ALBERT, b New York, NY, Sept 2, 41; m 66; c 2. ELECTRICAL ENGINEERING, QUANTUM ELECTRONICS. *Educ:* San Diego State Univ, BS, 65, MS, 69; Univ Calif, Berkeley, PhD(elec eng), 71. *Prof Exp:* Res physicist, Navy Electronics Lab Ctr, San Diego, 66-69; res asst elec eng, Univ Calif, Berkeley, 69-71; asst prof optical sci, Optical Sci Ctr, Univ Ariz, 71-72; asst prof, 72-74, assoc prof, 74-79, PROF ELEC ENG, UNIV ILL, 79- *Mem:* Inst Elec & Electronics Engrs; Am Inst Physics. *Res:* Quantum optics; gaseous electronics; superradiance, far infrared sources and techniques; gas lasers. *Mailing Add:* Dept Elec Eng Univ Ill 200 Eerl Urbana IL 61801

DETENBECK, ROBERT WARREN, b Buffalo, NY, Feb 11, 33; m 54; c 2. PHYSICS, OPTICS. *Educ:* Univ Rochester, BS, 54; Princeton Univ, PhD(nuclear physics), 62. *Prof Exp:* Instr physics, Princeton Univ, 58-59; asst prof, Univ Md, 59-67; assoc prof, 67-70, PROF PHYSICS, UNIV VT, 70- *Mem:* AAAS; Am Phys Soc; Am Asn Physics Teachers; Optical Soc Am; Inst Elec & Electronics Engrs. *Res:* Quantum optics; instrumentation; light scattering from aerosols. *Mailing Add:* Dept Physics Cook Bldg Univ Vt Burlington VT 05405

DETERLING, RALPH ALDEN, JR, b Williamsport, Pa, Apr 29, 17; wid; c 4. HOSPITAL ADMINISTRATION. *Educ:* Stanford Univ, BA, 38, MD, 42; Univ Minn, MS, 46, PhD(surg), 47. *Prof Exp:* From asst prof to assoc prof surg, Col Physicians & Surgeons, Columbia Univ, 48-53, assoc prof clin surg & dir surg res labs, 53-59; chmn dept surg, 59-75, PROF SURG, SCH MED, TUFTS UNIV, 59-; MED DIR RISK MGT, NEW ENG MED CTR, 85- *Concurrent Pos:* From asst attend surgeon to assoc attend surgeon, Presby Hosp, 48-59; consult surg, Manhattan Vet Admin Hosp & Paterson Gen Hosp, US Naval Hosp, 52-59, Boston Vet Admin Hosp, 60-, St Elizabeth's Hosp, Brighton, 61-, Mt Auburn Hosp, Cambridge, 62-, Nashoba Community Hosp, Ayer, 75-, Framingham Union Hosp & Emerson Hosp, Concord, 76-; vis surgeon, Francis Delafield Hosp, 53-59; dir first surg serv, Boston City Hosp, 59-70; surgeon-in-chief, Tufts-New Eng Med Ctr Hosps, 59-75; surg consult, Lemuel Shattuck Hosp, Boston, 59- & Choate Hosp & Waltham Weston Hosp, 78. *Honors & Awards:* Malmo Surg Found Award, 63; Outstanding Achievement Award, Univ Minn, 64. *Mem:* Am Surg Soc; Int Cardiovasc Soc (secy-gen, 63, pres, 71); Int Soc Surg; AMA; Am Acad Arts & Sci. *Res:* Blood vessel replacement; cardiovascular surgical technics; application of hypothermia to cardiac surgery; extracorporeal circulation for cardiac surgery; surgical applications of laser; organ homotransplantation. *Mailing Add:* 750 Washington St Boston MA 02111

DE TERRA, NOËL, b New York, NY, Dec 31, 33; m 62. CELL BIOLOGY. *Educ:* Barnard Col, Columbia Univ, BA, 55; Univ Calif, Berkeley, PhD(zool), 59. *Prof Exp:* Fel, Rockefeller Inst, 59-61, res assoc biochem genetics, 61-62; asst res zoologist, Univ Calif, Los Angeles, 62-63, asst res biophysicist, Lab Nuclear Med & Radiation Biol, Sch Med, 63-67; asst mem, Inst Cancer Res, 67-75; res assoc prof anat, Hahnemann Med Col, 75-81; independent investr, Marine Biol Lab, 81-85; CONSULT, 85- *Mem:* Am Soc Cell Biol; Soc Protozool. *Res:* The role of the cell surface in control of cell division and morphogenesis in the eiliate Stentor; the role of the cell surface in determining the position of the macronucleus in Stentor. *Mailing Add:* The Penington 215 E 15th St New York NY 10003

DETERS, DONALD W, b Quincy, Ill, Dec 20, 44. BIOENERGETICS, MITOCHONDRIAL BIOGENESIS. *Educ:* St Louis Univ, BS, 66; Univ Calif, Irvine PhD(chem), 72. *Prof Exp:* Fel biochem, Cornell Univ, 72-74; asst scientist biochem, Bioctr, Univ Basel, Switz, 74-78; asst prof microbiol, Univ Tex, Austin, 79-85; ASST PROF BIOL, BOWLING GREEN STATE UNIV, 85- *Mem:* Am Soc Microbiol; AAAS. *Res:* Enzymes, particularly those in mitochrondria that participate in oxidation and adenosine triphosphate synthesis; mitochondrial gene expression and genetic structure. *Mailing Add:* Dept Biol Sci Bowling Green State Univ Bowling Green OH 43403

DETERT, FRANCIS LAWRENCE, b San Diego, Calif, Apr 13, 23. ORGANIC CHEMISTRY. *Educ:* Univ Santa Clara, BS, 44; Stanford Univ, MS, 48, PhD(org chem), 50. *Prof Exp:* Researcher, Hickrill Chem Res Found, 50-52; res chemist, Calif Res Corp, 52-62; mem staff petrochem prod develop, Calif Chem Co, 62-65; petrochem prod tech specialist, Chevron Chem Co, 65-71, chem mkt res, 71-80; RETIRED. *Mem:* Am Chem Soc. *Res:* Diazonium coupling of furans; chemistry of cycloheptatrienone; synthetic detergents; chemical process development; gas odorants, xylenes and polybutenes; chemical market research and economics. *Mailing Add:* 14036 Reed Ave San Leandro CA 94578-2726

DETHIER, BERNARD EMILE, b Boston, Mass, June 5, 26; m 52; c 6. METEOROLOGY. *Educ:* Calif Inst Tech, BS, 46, MS, 47; Johns Hopkins Univ, PhD(geog), 58. *Prof Exp:* Dir, Climat Div, Patterson Weather Serv, 47-48; from asst prof to assoc prof math, Nazareth Col, 48-52; from asst prof to assoc prof agr climatol, 58-69, prof, 69-88, EMER PROF METEOROL, CORNELL UNIV, 88- *Concurrent Pos:* Climatologist & dir, NY Climate Prog, 82-; dir, Northeast Regional Climate Prog, 81-; asst dir, Nat Climate Progs Pff, 82-84; mem Am Meterol Comt Appl Climat, 85- *Mem:* Fel Am Meteorol Soc; Sigma Xi; Am Asn State Climatologists. *Res:* Impact of climate variability on agriculture and climatology of the Northeast. *Mailing Add:* PO Box 745 Blue Hill ME 04614

DETHIER, VINCENT GASTON, b Boston, Mass, Feb 20, 15; m 60; c 2. INSECT PHYSIOLOGY. *Educ:* Harvard Univ, AB, 36, AM, 37, PhD(biol), 39. *Hon Degrees:* ScD, Providence Col, 64, Ohio State Univ, 70, Univ Mass, 84, Dr, Univ Pau, France, 86. *Prof Exp:* Entomolgist, G W Pierce Lab, NH, 37-38; asst, Cruft Physics Lab, Harvard Univ, 39; from instr to asst prof biol, John Carroll Univ, 39-41; res physiologist, Army Chem Corps, 46; prof zool & entom, Ohio State Univ, 47; from assoc prof to prof biol, Johns Hopkins Univ, 48-58; prof zool & psychol, Univ Pa, 58-67, assoc, Inst Neurol Sci, Sch Med, 58-67; Class 1877 prof biol, Princeton Univ, 67-75; GILBERT WOODSIDE PROF ZOOL, UNIV MASS, AMHERST, 75- *Concurrent Pos:* Res fel, Atkins Inst, Cuba, 39-40; Belg-Am Educ Found fel, Belg Congo, 52; Fulbright sr scholar, London Sch Hyg & Trop Med, 54-55; Guggenheim fels, State Agr Univ, Wageningen, 64-65 & Univ Sussex, Eng, 72-73; mem, Nat Res Coun Comt, Sanit Eng, 51-59, NSF Adv Panel, Molecular Biol, 56-60, NIH Trop Med & Parasitol Study Sect, 56- 61, Nat Res Coun Adv Bd Quartermaster Res & Develop, 57-65, NSF Adv Panel Phycol, 61-63, NIH Exp Psychol Study Sect, 63-65, nat adv coun, Monell Chem Senses Ctr, 82-; consult, Off Surg Gen, Med Res & Develop Command, 57-65, Educ Testing Serv, Princeton, NJ, 68-74; steering comt, Am Inst Biol Sci, Biol Sci Curric study, 61-63, exec comt, 62-63; ed, J Insect Physiol, 58-66; assoc mem, Armed Forces Epidemiol Bd, Comn Parasitic Dis, 61-63; dir res, Insect Physiol & Behav Group, Int Ctr Insect Physiol & Ecol, Nairobi, Kenya, 70-72; pres bd trustees, Chapin Sch, Princeton, NJ, 71-74; first vpres bd trustees, Kneisel Hall Sch Music, 76- *Honors & Awards:* Frank Allison Linville/RH Wright Award in Olfactory Res, 89. *Mem:* Nat Acad Sci; Am Philos Soc; Royal Entom Soc London; Am Soc Zool (pres, 68); Am Acad Arts & Sci; Royal Acad Arts London. *Res:* Chemoreception in insects; chemistry of food plant choice by larvae; life histories of Lepidoptera; neural basis of feeding behavior. *Mailing Add:* Dept Zool Univ Mass Amherst MA 01003

DETHLEFSEN, LYLE A, b Oakes, NDak, Feb 27, 34; m 57; c 2. RADIOBIOLOGY, ONCOLOGY. *Educ:* Colo State Univ, BS, 60, DVM, 62; Univ Pa, PhD(molecular biol), 66. *Prof Exp:* Lectr radiation biol, St Joseph's Col (Pa), 65; asst prof radiol, Sch Vet Med, Univ Pa, 66-72, asst prof radiol sci, Sch Med, 68-72; assoc prof, 72-76, PROF RADIOL, COL MED, UNIV UTAH, 76- *Concurrent Pos:* USPHS res career develop award, 69-72; consult, Breast Cancer Task Force, Nat Cancer Inst, NIH, 74-77, Exp Therapeut Study Sect, 77-81 & Cancer Res Manpower Rev Comt, 84-88; res prof anat, Univ Utah, 79; assoc ed, Radiation Res J, 77-80 & Cancer Res, 81-84; spec subj ed, Int Encycl Pharmacol & Therapeut, 80-85. *Mem:* AAAS; Radiation Res Soc; Am Soc Cell Biol; Am Asn Cancer Res; Soc Anal Cytol. *Res:* Cell and molecular biology and radiobiology; tumor growth; tumor cell kinetics. *Mailing Add:* Dept Radiol Univ Utah Med Ctr Salt Lake City UT 84132

DETHLEFSEN, ROLF, b Niebuell, Ger, Aug 30, 34; US citizen; m; c 4. PLASMA PHYSICS, ELECTRICAL POWER ENGINEERING. *Educ:* Braunschweig Tech Univ, Dipl Ing, 61; Mass Inst Technol, MS, 62, ScD(physics of fluids), 65. *Prof Exp:* Staff scientist space sci lab, Convair Div, Gen Dynamics Corp, Calif, 65-68; assoc dir, Advan Tech Ctr, Allis Chalmers Mfg Co, 68-72; consult engr, res & develop, Gould Brown Boveri, 72-79, sr proj mgr, Brown Boveri Elec Corp, 72-79; PRIN SCIENTIST, MAXWELL LABS, CALIF, 83- *Concurrent Pos:* NATO fel, 61-62. *Mem:* Am Vac Soc; Inst Elec & Electronics Engrs; Asn Ger Engrs. *Res:* Electric arcs; circuit breakers; electrical power equipment; pulse power systems; rail guns; space power systems. *Mailing Add:* 13476 Samantha Ave San Diego CA 92129

DETIG, ROBERT HENRY, b Pittsburgh, Pa, Oct 6, 35. ELECTRONICS, ELECTRICAL ENGINEERING. *Educ:* Carnegie Inst Technol, BS, 57, MS, 58, PhD(elec eng), 62. *Prof Exp:* Staff engr, satellite commun agency, US Army, 62-63; asst prof elec eng, Carnegie Inst Technol, 63-65; scientist, res div, Xerox Corp, 65-68; mgr electronics res br, Olivetti Corp Am, 68-72; treas, Med Graph Inc, 72-73; self employed, 74-76; PROJ ENGR, ELECTRONICS FOR MED, 76- *Mem:* Inst Elec & Electronics Engrs. *Res:* Reprographics; non-impact printing; information display; chemical and surface physics; medical electronics; electrostatics. *Mailing Add:* Olin Corp Div Spec Prod 201 Roosevelt Pl Palisades Park NJ 07650

DETITTA, GEORGE THOMAS, b Jersey City, NJ, Nov 29, 47; m 69. CRYSTALLOGRAPHY, BIOPHYSICS. *Educ:* Villanova Univ, BS, 69; Univ Pittsburgh, PhD(crystallog, biochem), 73. *Prof Exp:* SR RES SCIENTIST MOLECULAR BIOPHYS, MED FOUND BUFFALO RES LABS INC, 73- *Concurrent Pos:* Prin investr, NIH grants, 77- *Honors & Awards:* Sidhu Award, Pittsburgh Diffraction Conf, 78. *Mem:* Am Crystallog Asn. *Res:* Elucidation of the relationships between molecular structure and molecular function in the biological milieu. *Mailing Add:* 337 Argonne Dr Buffalo NY 14217

DETMERS, PATRICIA ANNE, b Riverside, Calif, Nov 13, 53; m 81; c 1. CELL BIOLOGY, LEUKOCYTES. *Educ:* Pomona Col, BA, 74; Univ Pa, PhD(biol), 79. *Prof Exp:* Fel, biol dept, Washington Univ, 79-81; fel, anat dept, Albert Einstein Col Med, 81-84; ASST PROF BIOL, ROCKEFELLER UNIV, 85- *Concurrent Pos:* Lectr, Univ Pa Col Gen Studies, 79, Albert Einstein Col Med, 82. *Mem:* Am Soc Cell Biol; Am Assoc Immunologists. *Res:* Regulation of the ligand binding activity of adhesion promoting receptors on human leukocytes; factors regulating receptor-mediated phagocytosis by human leukocytes. *Mailing Add:* Lab Cell Physiol & Immunol Rockefeller Univ 1230 York Ave New York NY 10021

DETOMA, ROBERT PAUL, b Milton, Mass, Sept 5, 44; m 73. PHYSICAL CHEMISTRY, MOLECULAR SPECTROSCOPY. *Educ:* St Anselm's Col, BA, 66; Johns Hopkins Univ, PhD(chem), 73. *Prof Exp:* Assoc res scientist, Dept Biol, Johns Hopkins Univ, 73-77; asst prof chem, Univ Richmond, 77-78; ASSOC PROF CHEM, LOYOLA COL, 78- *Concurrent Pos:* NIH fel, 75-77. *Mem:* Sigma Xi; Am Chem Soc; Am Phys Soc; Am Soc Photobiol. *Res:* Time-dependent spectroscopy, excited state solvation, radiationless transitions, adiabatic photophysical reactions, biological application of fluorescence, photophysics, diffusion limited processes, chemometrics. *Mailing Add:* Dept Chem Loyola Col Baltimore MD 21210

DE TOMMASO, GABRIEL LOUIS, b Providence, RI, Aug 19, 34; m 56; c 4. ORGANIC CHEMISTRY, POLYMER CHEMISTRY. *Educ:* Univ RI, BS, 56; Univ Ill, PhD(org chem), 60. *Prof Exp:* Org polymer chemist, Air Reduction Co, Inc. 59-62; sr chemist, 62-76, proj leader, 76-80, SECT MGR, ROHM AND HAAS CO, INC, SPRING HOUSE, 80- *Res:* Emulsion polymerization of vinyl monomers; films; polymer structure; application of aqueous coatings for coil, general product finishing and building products. *Mailing Add:* 249 Laurel Lane Lansdale PA 19446

DE TORNYAY, RHEBA, b Petaluna, CA, Apr 17, 26. MEDICAL EDUCATION, NURSING. *Educ:* San Francisco State Univ, AB, 51, MA, 54; Standford Univ, EdD, 67. *Prof Exp:* Prof & chmn, Dept Nursing, San Francisco State Univ, 59-67; assoc prof, Sch Nursing, Univ Calif, San Francisco, 68-71; dean & prof, Sch Nursing, Univ Calif, Los Angeles, 71-75; EMER DEAN & PROF, UNIV WASH, SEATTLE, 75- *Concurrent Pos:* Dir, Western Coun Higher Educ Nursing, 76-79; trustee, Robert Wood Johnson Found, 91- *Mem:* Inst Med-Nat Acad Sci; Am Nurses Asn; Am Acad Nursing (pres, 73-75). *Mailing Add:* Sch Nursing SM-24 Univ Wash Seattle WA 98195

DETRA, RALPH W(ILLIAM), b Thompsontown, Pa, Mar 23, 25; m 47; c 2. AERONAUTICAL ENGINEERING. *Educ:* Cornell Univ, BS, 46, MAeroE, 51; Swiss Fed Inst Technol, DrScTech, 53. *Prof Exp:* Flight test proj engr, US Naval Air Test Ctr, Md, 46-49; instr mech eng, Cornell Univ, 53-55; prin res scientist, Avco-Everett Res Lab, 55-59, vpres missile prog, Avco Res & Adv Develop Div, 59-65, vpres & asst gen mgr missile systs, 65-66, vpres & gen mgr, Avco Systs Div, 66-71, pres, Tyco Labs, Inc, 71-72, vpres energy technol, Avco-Everett Res Lab, Inc, 72-81, PRES, AVCO-EVERETT RES LAB, INC, AVCO CORP, 81- *Mem:* Am Inst Aeronaut & Astronaut. *Res:* Hypervelocity flight; high temperature gasdynamics. *Mailing Add:* 15 Frost St Arlington MA 02174

DETRAY, DONALD ERVIN, b Napoleon, Ohio, Nov 9, 17; m 40; c 2. VETERINARY MEDICINE. *Educ:* Ohio State Univ, DVM, 40. *Prof Exp:* Pvt pract, 40-47; vet, Mex-US Comn Eradication Foot & Mouth Dis, 47-48; res vet, USDA, 49-51, E Africa, 51-61, asst to dir, Animal Dis & Parasite Res Div, Agr Res Serv, Md, 61-63, from asst dir to assoc dir, 63-66; regional livestock adv, USAID, Lagos, Nigeria, 66-68, Nairobi, Kenya, 68-70, Ethiopia, 70-73; CONSULT, USDA, 73-; CONSULT, USAID, 73- *Mem:* Am Vet Med Asn; US Animal Health Asn; Conf Res Workers Animal Dis; Asn Advan Agr Sci Africa; Am Asn Trop Vet Med. *Res:* Bovine and porcine brucellosis; rinderpest of cattle and African swine fever; research administration in animal diseases and parasites. *Mailing Add:* 84 Pukihae St Apt 305 Hilo HI 96720-2402

DETRAZ, ORVILLE R, b Chicago, Ill, Aug 23, 30; m 55; c 3. THREE PHASE POWER SYSTEMS. *Educ:* Purdue Univ, BS, 58, MS, 60. *Prof Exp:* Engr, ITT Indust Labs, Ft Wayne, 65-68; sr engr, Ind & Mich Elec Co, 68-80; ASSOC PROF CIRCUITS ELECTRONICS POWER & MACH, EET DEPT, PURDUE UNIV, 68- *Concurrent Pos:* Vis prof, Tri-State Univ-Angola, Ind, 76-77; ad hoc visitor, Accreditation Bd Eng & Technol, 82-87. *Mem:* Inst Elec & Electronics Engrs; Am Soc Elec Eng. *Res:* Author of electrical machines and transformers for engineering technology; electronic devices; circuit analysis. *Mailing Add:* 2536 Eastbrook Dr Ft Wayne IN 46805

DETRE, THOMAS PAUL, b Budapest, Hungary, May 17, 24; m 56; c 2. PSYCHIATRY. *Educ:* Gym Piarist Fathers, BA, 42; Univ Rome, MD, 52; Am Bd Psychiat & Neurol, dipl, 59. *Prof Exp:* Consult psychologist, Salvator Mundi Int Hosp, Rome, Italy, 51-53; chief resident psychiat, 57-58, from instr to prof, Sch Med, Yale Univ, 57-72, dir psychiat inpatient serv, Yale-New Haven Hosp, 60-68, psychiatrist-in-chief, 68-72; prof psychiat & chmn dept, Univ Pittsburgh, 72-82, assoc sr vchancellor health sci, 82-85, pres, Med & Health Care Div, 86- 90, DISTINGUISHED SERV PROF HEALTH SCI, UNIV PITTSBURGH, 82-, SR VPRES HEALTH SCI, 85- & PRES, MED CTR, 90- *Concurrent Pos:* Consult, Fairfield Hosp, Newton, Conn, 58-61, Vet Admin Hosp, W Haven, 61-72 & Norwich Hosp, 62-72; dir, Western Psychiat Inst & Clin, 72-; assoc examr, Am Bd Psychiat & Neurol; chmn, res scientist develop review comt, NIMH, 81-84; chmn, psychopharmacologic drugs adv comt, Food & Drug Admin, 84-86; counr, Am Col Neuropsychopharmacol, 90-93; secy adv comt Health Res Policy, US Dept Vet Affairs, 90. *Mem:* AAAS; fel Am Psychiat Asn; fel Am Col Psychiatrists; fel Am Col Neuropsychopharmacol; Pan-Am Med Asn; fel Acad Behav Med Res. *Res:* Clinical research in psychopharmacology; hospital psychiatry. *Mailing Add:* 3811 O'Hara St Pittsburgh PA 15213

DETRICK, JOHN K(ENT), b Denver, Colo, Mar 10, 20; m 47; c 1. CHEMICAL ENGINEERING. *Educ:* Univ Denver, BS, 41; Univ Cincinnati, MS, 43. *Prof Exp:* Res engr, East Lab, Res & Develop Div, Explosives Dept, E I du Pont de Nemours & Co, Inc, 42-47 & Burnside Lab, 47-49, supvr, East Lab, 49-53, tech asst res & develop staff, Del, 53-57, supt, Repauno Develop Lab, 57-67 & East Lab, 67-68, spec asst dept eng sect, 68-70, res assoc, Exp Sta, Petrochem Dept, 70-78, sr financial analyst, 78-79; financial consult, 79-82; RETIRED. *Mem:* Am Chem Soc. *Res:* Process development in polymer intermediates. *Mailing Add:* 731 Park Ave Woodbury Heights NJ 08097

DETRICK, ROBERT SHERMAN, b Denver, Colo, Feb 3, 18; m 41, 74; c 6. CHEMISTRY. *Educ:* Univ Denver, BS, 39; Rensselaer Polytech Inst, MChE, 40. *Prof Exp:* From jr fel to sr fel, Mellon Inst, 40-51; mgr tar prod res, 51-54, from asst mgr to mgr lab res, 54-57, asst mgr tech planning, 57-58, mgr applns eval res, 58-66, mgr develop, Assoc Opers, 66-67, mgr systs develop, 67-71, mgr environ health & safety, 71-79, mgr external res, Res Dept, Koppers Co, Inc, 79-83; RETIRED. *Mem:* Am Chem Soc. *Res:* Reinforced plastics and environmental control. *Mailing Add:* 177 Pineneedle Dr Bradenton FL 34210

DETRIO, JOHN A, b Miami, Fla, June 13, 37; m 63; c 3. SOLID STATE PHYSICS. *Educ:* Spring Hill Col, BS, 59; Univ Ala, MS, 61. *Prof Exp:* Res asst, Univ Ala, 59-61; physicist, Army Rocket & Guided Missile Agency, Ala, 61; physicist, Army Munition Command, Picatinny Arsenal, 61-66; res physicist, Univ Dayton, 66-75, RES GROUP LEADER OPTICAL MAT, RES INST, UNIV DAYTON, 75- *Mem:* Optical Soc Am; Sigma Xi; Am Phys Soc; Am Soc Testing & Mat. *Res:* Electroluminescence; solid state lasers; optical properties of solids; photoconductivity; atomic spectra; electrical properties of solids; infrared, ultraviolet and visible spectroscopy and photometry. *Mailing Add:* Res Inst KL 542 Univ Dayton Dayton OH 45469-0170

DE TROYER, ANDRE JULES, b Souvret, Hainaut, Belg, Nov 28, 48; m 74; c 2. RESPIRATORY MUSCLES, CHEST WALL MECHANICS. *Educ:* Athenee Royal Gosselies, BS, 66; Univ Brussels, MD, 73, PhD, 80. *Prof Exp:* Resident internal med, St Pierre Univ Hosp, Brussels, 73-78; fel respiratory physiol, McGill Univ, Montreal, Can, 80-82; prof & chmn physiol, Univ Brussels, Belg, 84-87; prof physiol & med, Mayo Med Sch, Rochester, Minn, 87-89; from asst prof to assoc prof, 78-84, PROF MED, CHEST SERV, ERASME UNIV HOSP, BRUSSELS, BELG, 84-87 & 89- *Concurrent Pos:* Dir, Lab Cardio-Respiratory Physiol, Med Sch, Univ Brussels, Belg, 89- *Honors & Awards:* Brompton Lectr, London Sch Med, UK, 83; Cournand lectr, Europ Soc Respiratory Physiopath, 83. *Mem:* Am Thoracic Soc; Am Physiol Soc; Am Soc Clin Invest. *Res:* Assessment of the actions of the respiratory muscles, of the coordination of these muscles and of the pathways of neural activation. *Mailing Add:* Chest Serv Erasme Univ Hosp 808 Rte De Lennik Brussels 1070 Belgium

DETTBARN, WOLF DIETRICH, b Berlin, Ger, Jan 30, 28; US citizen; m 60; c 2. NEUROCHEMISTRY, NEUROPHARMACOLOGY. *Educ:* Univ Gottingen, Dr med, 53. *Prof Exp:* Intern, Med Sch, Univ Gottingen, 54; res assoc, Ciba, Switz, 54-55; res assoc, Physiol Inst Med Sch, Univ Saarland, 55-58; res assoc neurol, Col Physicians & Surgeons, Columbia Univ, 58-61, from asst prof to assoc prof, 61-68; PROF PHARMACOL, SCH MED, VANDERBILT UNIV, 68- *Concurrent Pos:* Mem Marine Biol Lab Corp, Woods Hole, 63; consult, US Army Res & Develop Command, 81, 82, Nat Res Coun, Comt Toxicol Anticholinesterase Chem. *Honors & Awards:* Career Develop Award; Nat Res Coun Awards, Interacad Exchange to Yugoslavia. *Mem:* Am Soc Pharmacol & Exp Therapeut; Am Physiol Soc; Soc Gen Physiol; Am Soc Neurochem; Soc Neurosci. *Res:* Neurophysiology; trophic function of the neuron; peripheral nerve; ion flux; membrane permeability. *Mailing Add:* Dept Pharmacol Med Ctr S Vanderbilt Univ Sch Med 2100 Pierce Ave Nashville TN 37212

DETTINGER, DAVID, b Little Falls, NY, June 1, 19; m 52; c 2. RADIO ENGINEERING. *Educ:* St Lawrence Univ, BS, 41. *Prof Exp:* Engr, Hazeltine Electronics Corp, 42-45 & Teleregister Corp, 45-47; vpres & chief engr, Wheeler Labs, 47-61; head, Commun Dept, 61-75, FAC MEM, MITRE INST, MITRE CORP, 75-, STAFF MEM, DIV CONTINUING EDUC, MITRE CORP, BEDFORD, MA. *Mem:* Life mem Inst Elec & Electronics Engrs; Sigma Xi. *Res:* Communications engineering; microwaves; antennas; communication systems. *Mailing Add:* Mitre Corp Burlington Rd Box 208 Bedford MA 01730

DETTMAN, JOHN WARREN, b Oswego, NY, July 14, 26; m 50; c 3. MATHEMATICAL ANALYSIS. *Educ:* Oberlin Col, AB, 50; Carnegie Inst Technol, MS, 52, PhD(math), 54. *Prof Exp:* Instr math, Carnegie Inst Technol, 53-54; mem tech staff, Bell Tel Labs, 54-56; from asst prof to assoc prof math, Case Inst Technol, 56-64; prof, 64-88, EMER PROF MATH, OAKLAND UNIV, 88- *Concurrent Pos:* NSF fac fel, 62-63; sr res fel, Univ Glasgow, 70-71; NSF grants, 68, 70, 72, 74 & 77; vis prof, Univ Dundee, Scotland, 85, Univ Auckland, NZ, 86. *Mem:* Soc Indust & Appl Math; Am Math Soc; Math Asn Am; Sigma Xi. *Res:* Differential equations; functional analysis. *Mailing Add:* Rte 4 Box 565B Boone NC 28607

DETTRE, ROBERT HAROLD, b Philadelphia, Pa, Aug 20, 28; m 57; c 2. PHYSICAL CHEMISTRY. *Educ:* Lafayette Col, BS, 52; Johns Hopkins Univ, MA, 54, PhD(phys chem), 57. *Prof Exp:* From res chemist to sr res chemist, 57-70, from res assoc to sr res assoc, 70-85, RES FEL, E I DU PONT DE NEMOURS & CO, INC, 85- *Mem:* AAAS; Am Chem Soc. *Res:* Thermodynamics of solutions; calorimetry; surface chemistry. *Mailing Add:* RD 6 Box 175B Hockessin DE 19707

DETTY, WENDELL EUGENE, synthetic organic & natural products chemistry; deceased, see previous edition for last biography

DETWEILER, DAVID KENNETH, b Philadelphia, Pa, Oct 23, 19; m 65; c 6. COMPARATIVE CARDIOLOGY, CARDIOVASCULAR TOXICOLOGY. *Educ:* Univ Pa, VMD, 42, MS, 49. *Hon Degrees:* DSc, Ohio State Univ, 66; DVM, Univ Vienna, 68 & Univ Torino, 69. *Prof Exp:* Asst instr physiol & pharmacol, Univ Pa, 42, instr, 43-45, assoc, 45-47, asst prof, 47-51, assoc prof physiol, 51-62, prof physiol & pharmacol & head lab, 62-68, chmn, Dept Vet Med Sci, Grad Sch Med, 56-70, assoc prof, Grad Sch Arts & Sci & Grad Sch Med, 56-62, dir cardiovasc studies unit, 60-90, prof physiol & head lab, Sch Vet Med, 68-90, chmn, Grad Group Comp Med Sci, 71-90, emer prof physiol, 90-; RETIRED. *Concurrent Pos:* Guggenheim fel, Univ Zurich, 55-56; consult, WHO, 58, Food & Drug Admin, 70-90; guest prof, Univ Munich, 63, Univ Berlin, 68, & Hannover Vet Sch, Ger, 73; mem, Expert Panel Vet Educ, Food & Agr Orgn, 63; mem, Physiol Training Grant Comt, Nat Inst Gen Med Sci, 67-70; mem, Coun Basic Sci, Am Heart Asn. *Honors & Awards:* Gaines Award Medal, Am Vet Med Asn, 60; D K Detweiler Prize, World Small Animal Vet Asn, 82. *Mem:* Inst Med-Nat Acad Sci; fel AAAS; Am Physiol Soc; Am Vet Med Asn; Am Heart Asn. *Res:* Comparative cardiology; electrocardiography; cardiovascular physiology; cardiovascular toxicology; over 160 publications. *Mailing Add:* 4636 Larchwood Ave Philadelphia PA 19143

DETWEILER, STEVEN LAWRENCE, b Yonkers, NY, Sept 4, 47; m 71; c 2. GENERAL RELATIVITY, GRAVITATIONAL WAVES. *Educ:* Princeton Univ, AB, 69; Univ Chicago, PhD(physics), 75. *Prof Exp:* Fel physics, Univ Md, 74-76; Tolman fel, Calif Inst Technol, 76-77; from asst prof to assoc prof, Yale Univ, 77-83; assoc prof physics, 83-88, PROF PHYSICS, UNIV FLA, 88- *Mem:* Am Phys Soc; Am Astron Soc. *Res:* Gravitational wave sources; astrophysics of black holes. *Mailing Add:* 215 Williamson Univ Fla Gainesville FL 32611

DETWEILER, W KENNETH, b Quakertown, Pa, Jan 27, 23; m 44; c 2. ORGANIC CHEMISTRY. *Educ:* Ursinus Col, BS, 47; Lehigh Univ, MS, 49, PhD(chem), 51. *Prof Exp:* Group leader org chem res, Union Carbide & Carbon Chem Corp, 51-55; sect head new uses & new areas res, Exxon Corp, 55-66, asst dir lubricants & petrol specialties, 66-67, coordr hq mkt, 67-68, coordr prod res, Exxon Res & Eng Co, 68-83; RETIRED. *Mem:* Am Chem Soc; Soc Automotive Eng; Am Inst Chem Eng. *Res:* Synthesis and characterization of heterocyclic nitrogen derivatives, acyl aldehydes, polyglycol ethers, alcohols and phosphate esters; study of the effect of structure upon plasticizing action; synthetic lubricants; diesel fuels and diesel lubricants; recruiting; industrial lubricants and greases; new uses and new areas research; worldwide petroleum research management. *Mailing Add:* PO Box 149 Center Harbor NH 03226-0149

DETWILER, DANIEL PAUL, b Woodbury, Pa, Feb 16, 27; wid; c 3. SOLID STATE PHYSICS. *Educ:* Swarthmore Col, AB, 49; Yale Univ, MS, 50, PhD(physics), 52. *Prof Exp:* Res physicist, Lab, Franklin Inst, 52-54; from asst prof to assoc prof physics, State Univ NY Col Ceramics, Alfred, 54-60, chmn dept, 59-60; prof, Wilkes Col, 60-66, chmn dept, 61-66, chmn div natural sci & math & dir res & grad ctr, 63-66; physics specialist, NSF-India Liaison Off, New Delhi, 66-69; staff physicist, Comn Col Physics, 69-70; chmn, Dept Physics, 70-79, PROF PHYSICS, CALIF STATE UNIV, BAKERSFIELD, 79- *Mem:* AAAS; Am Asn Physics Teachers; Am Phys Soc; Sigma Xi. *Res:* Low temperature physics; semiconductivity; dielectrics; internal friction. *Mailing Add:* Dept Physics & Geol Calif State Col Bakersfield CA 93311-1099

DETWILER, JOHN STEPHEN, b Pittsburgh, Pa, Sept 23, 46; m 67; c 2. BIOMEDICAL ENGINEERING. *Educ:* Carnegie Inst Technol, BSEE, 67; Carnegie-Mellon Univ, MS, 68, PhD(elec eng), 71. *Prof Exp:* Biomed engr, Vet Admin Hosp, Pittsburgh, 71-76; asst prof eng med, Carnegie-Mellon Univ, 71-79; res asst prof obstet, Res Instr Med, Univ Pittsburgh, 76-79; MGR SYST ENG, BROWN BOVERI CONTROL SYSTS, 80- *Mem:* Inst Elec & Electronics Engrs. *Res:* Neural control of fetal heart rate and information extraction from clinical measurement of heart rate variability. *Mailing Add:* 5723 Solway St Pittsburgh PA 15217

DETWILER, PETER BENTON, b Stamford, Conn, Apr 17, 44; m; c 1. PHYSIOLOGY, BIOPHYSICS. *Educ:* St Lawrence Univ, BS, 66; Georgetown Univ, PhD(pharmacol), 70. *Prof Exp:* Staff fel physiol, NIH, 71-75; res fel biophys, Univ Cambridge, Eng, 76-78; from asst prof to assoc prof, 78-86, PROF PHYSIOL & BIOPHYS, UNIV WASH, 86- *Mem:* Biophys Soc; Soc Gen Physiol. *Res:* Molecular mechanisms underlying signal transduction in vertebrate photoreceptors. *Mailing Add:* Dept Physiol & Biophys Univ Wash Seattle WA 98195

DETWILER, THOMAS C, b Hannibal, Mo, Dec 28, 33; m 62; c 2. BIOCHEMISTRY. *Educ:* Univ Ill, BS, 57, PhD(nutrit), 60. *Prof Exp:* res assoc, Oak Ridge Nat Lab, 60-61; res assoc, Philadelphia Gen Hosp, Pa, 61-63; res assoc, McCollum-Pratt Inst, Johns Hopkins Univ, 63-65; from instr to assoc prof, 65-75, PROF BIOCHEM, STATE UNIV NY HEALTH SCI CTR BROOKLYN, 75- *Concurrent Pos:* Vis assoc prof biochem, Univ Wash, 72-73; mem, Hemat Study Sect, NIH, 78-82; vis scientist, Mayo Clin, 83; mem path res study comt, Am Heart Asn, 85- *Honors & Awards:* Brinkhouse Award, Int Congress of Thrombosis & Hemostasis. *Mem:* Am Soc Biochem; Am Asn Univ Prof; AAAS; NY Acad Sci; Sigma Xi; Am Heart Asn. *Res:* Biochemistry of platelets; stimulus-response coupling mechanisms, including receptors and second messengers; structure-function relationships of secreted platelet proteins. *Mailing Add:* Dept Biochem State Univ Ny Health Sci Ctr 450 Clarkson Ave Brooklyn NY 11203

DETWYLER, ROBERT, b Middletown, NY, Apr 16, 29; m 51; c 2. ZOOLOGY. *Educ:* State Univ NY Col New Paltz, BS, 54; Univ NH, MS, 59, PhD(zool), 63. *Prof Exp:* Instr campus elem sch, NY State Teachers Col, New Paltz, 54-55; teacher pub sch, 55-57; instr, Univ NH, 62-63; asst prof biol, Nasson Col, 63-65; from asst prof to prof, 65-75, chmn dept, 68-78, CHARLES A DANA PROF BIOL, NORWICH UNIV, 75- *Mem:* Am Inst Biol Sci; AAAS; Am Fisheries Soc; Oceanic Soc; Sigma Xi. *Res:* Tropical marine ecology; fish embryology; biology of the sea snail, Liparis atlanticus. *Mailing Add:* Dept Biol Norwich Univ Northfield VT 05663

DETZ, CLIFFORD M, b New York, NY, Dec 16, 42; m 66; c 2. RESEARCH MANAGEMENT, PETROLEUM PROCESSING. *Educ:* Brown Univ, AB, 64; Univ Chicago, MS, 66, PhD(chem physics), 70. *Prof Exp:* Sr res scientist, Res Dept, Linde Div, Union Carbide Corp, 70-75, group leader, 75-78; group leader, 78-82, MGR PROCESS DEVELOP, CHEVRON RES CO, STANDARD OIL CALIF, 82- *Mem:* Am Chem Soc; Am Inst Chem Engrs; Soc Mining Engrs. *Res:* Petroleum refining process and catalyst development; environmental control process development; methane conversion; extractive metallurgy; separation and purification process research. *Mailing Add:* 842 Butternut Dr San Rafael CA 94903-3126

DEUBEN, ROGER R, b Detroit, Mich, Sept 9, 38; m 66; c 3. DENTISTRY, ENDOCRINOLOGY. *Educ:* Mich State Univ, BS, 60, MS, 64; Univ Pittsburgh, PhD(pharmacol), 69. *Prof Exp:* Physiologist, Mich State Univ, 61-63; instr physiol, Sch Dent Med, Univ Pittsburgh, 65-66, from asst prof to assoc prof pharmacol, 69-77; assoc prof physiol & pharmacol & chmn dept, Sch Dent, 77-89, ASSOC DEAN ACAD AFFAIRS, UNIV DETROIT, 89- *Mem:* AAAS; Int Asn Dent Res; Soc Exp Biol & Med; Am Asn Dent Schs. *Res:* Studies on the central effects of angiotensin II and its possible role in the etiology of hypertension; regulation of anterior pituitary hormone secretion; role of prostaglandins in the above fields. *Mailing Add:* Off Dean 2985 E Jefferson Ave Detroit MI 48207

DEUBERT, KARL HEINZ, b Weissensee, Ger, Feb 1, 29; nat US; m 50. ENVIRONMENTAL CHEMISTRY. *Educ:* Univ Halle, MS, 53, PhD(agr zool), 55. *Prof Exp:* Asst agr zool, Univ Halle, 53-60; sr asst phytopath, Cent Biol Inst, Berlin, Ger, 61-65; prof biol, Nat Univ Honduras, 65-67; res assoc phytopath, 67-74, from asst prof to prof residue anal, 74-91, EMER PROF RESIDUE ANALYSIS, CRANBERRY EXP STA, UNIV MASS, 91- *Mem:* Am Inst Biol Sci. *Res:* Pesticide residues in soil, water and tissues; hydrology; soil chemistry as related to movement of chemicals. *Mailing Add:* Cranberry Exp Sta PO Box 569 Univ Mass Buzzard Bay MA 02532

DEUBLE, JOHN LEWIS, JR, b New York, NY, Oct 2, 32; m 55; c 2. AIR TOXICS MONITORING, HAZARDOUS WASTE MANAGEMENT. *Educ:* Univ Pac, BS, 59. *Prof Exp:* Sr chemist, Aerojet-Gen Corp, 59-66; asst dir res, Lockheed Propulsion Co, 66-72, mgr bus planning 73; mgr, appl sc prog, Systs, Sci & Software, 74-79; gen mgr, Wright Energy Nev Corp, 80-81; asst vpres western opers, Energy Resources Co, 81-82; dir hazardous waste mgt proj, Aerovironment, Inc, 84-86; CONSULT HAZARDOUS WASTE, SYST ENG, 87-; AIR TOXICS MGR, ERC ENVIRON & ENERGY SERV CO, 89- *Honors & Awards:* Technol Award, Am Ord Asn, 70. *Mem:* Am Chem Soc; fel Am Inst Chemists; Am Inst Chem Engrs; Am Meteorol Soc; Am Waste Mgt Asn; Am Soc Test Mat; NY Acad Sci; Hazardous Mat Control Res Inst. *Res:* Developed and pioneered the use of chemical(non-radioactive) tracers-gaseous, aqueous and particulate in environmental and energy applications; developed air toxics sampling techniques. *Mailing Add:* 369 Cerro Encinitas CA 92024

DEUBLER, EARL EDWARD, JR, b Sayre, Pa, May 19, 27; m 51; c 4. ICHTHYOLOGY. *Educ:* Moravian Col, BS, 50; Cornell Univ, PhD(vert zool), 55. *Prof Exp:* Asst, Cornell Univ, 50-55; asst fisheries res biol, US Fish & Wildlife Serv, 55-56; from asst prof to assoc prof zool, Inst Fisheries Res, Univ NC, 56-61; assoc prof biol, Hartwick Col, 61-62, fisheries biol, Univ Mass, 62-63 & zool, Inst Fisheries Res, Univ NC, 63-67; from actg chmn dept to chmn dept, Hartwick Col, 70-75, chmn fac, 75-76, actg pres, 76-77, dean fac, 80-84, from assoc prof to distinguished prof biol, 67-89; RETIRED. *Mem:* NY Acad Sci. *Res:* Systematic ichthyology; ecology; life histories of marine and brackish water fishes. *Mailing Add:* 355 Chestnut St Oneonta NY 13820

DEUBLER, MARY JOSEPHINE, b Philadelphia, Pa, May 4, 17. VETERINARY MEDICINE. *Educ:* Univ Pa, VMD, 38, MS, 41, PhD(path), 44. *Prof Exp:* Asst instr vet path, Sch Vet Med, Univ Pa, 40-44; res assoc parasitol, Jefferson Med Col, 44-46; instr bact, Sch Vet Med, Univ Pa, 46-51, asst prof path in med, 51-88; RETIRED. *Mem:* Am Vet Med Asn; Sigma Xi. *Res:* Virus, feline enteritis; anatomy and histology of the horse's eye; veterinary clinical bacteriology and mycology. *Mailing Add:* 2811 Hopkinson House Washington Square S Philadelphia PA 19106

DEUBNER, RUSSELL L(EIGH), b Jamestown, Ohio, Aug 10, 19; m 44; c 2. ELECTRONICS ENGINEERING. *Educ:* Ohio State Univ, BMetE, 42. *Prof Exp:* Asst metall, Ohio State Univ, 40-42; res engr & asst to supvr Res Div, graphic arts, Battelle Mem Inst, 45-50, mgr, Battelle Develop Corp, 50-58, admin dept econ, Battelle Mem Inst, 57-58; gen mgr, Chrome Plating Div, Gen Develop Corp, 58-61 & Ohio Semiconductors, Inc, 61-62; chmn & treas, Sci Columbus Inc, Div Esterline Corp, 62-69, pres, 69-81; RETIRED. *Mem:* Am Soc Metals. *Res:* Metallurgy of case hardening chromium alloys; improvements in photo-engraving processes; development of xerography; printing; electroplating; patent and invention management; diversification studies and new product development; semiconductor product sales and production; Hall effect and thermoelectric devices and systems; corporate financing; electric power measurement (transducers and meters). *Mailing Add:* 8480 Gullane Ct Dublin OH 43017

DEUPREE, JEAN DURLEY, b Washington, DC, Nov 9, 42. PHARMACOLOGY. *Educ:* Ferris State Col, BS, 65; Mich State Univ, PhD(biochem), 70. *Prof Exp:* Res fel pharmacol, Med Ctr, Univ Wis, 70-72; asst prof, 72-79, ASSOC PROF PHARMACOL, UNIV NEBR MED CTR, 79- *Concurrent Pos:* Guest res, NIH, 84. *Mem:* AAAS; Am Soc Pharmacol & Exp Therapeut; Am Soc Neurochem; Sigma Xi; Soc Neurosci. *Res:* Epilepsy; mechanism of action of anticonvulsant drugs; neurotransmitter receptors; uptake, storage and release of neurotransmitters from storage granules. *Mailing Add:* Dept Pharmacol Univ Nebr Med Ctr 600 S 42nd St Omaha NE 68188-6260

DEUPREE, ROBERT G, b Washington, DC, Aug 5, 46; m 68; c 2. COMPUTER HYDRODYNAMICAL MODELLING. *Educ:* Univ Wis, BA, 68; Univ Colo, MS, 70; Univ Toronto, PhD(astron), 74. *Prof Exp:* Fel astron res, Dept Astrophys Sci, Princeton Univ, 74-75; staff mem, Los Alamos Sci Lab, 75-78; asst prof astron, Boston Univ, 78-80; STAFF MEM RES, LOS ALAMOS NAT LAB, 80- *Concurrent Pos:* Consult, Los Alamos Nat Lab, 78-80. *Mem:* Am Astron Soc; Int Astron Union. *Res:* Numerical modelling of geophysical and astrophysical events; studies related to Nuclear Test Treaties. *Mailing Add:* ESS-5 MS-F665 Los Alamos Nat Lab PO Box 1663 Los Alamos NM 87545

DEURBROUCK, ALBERT WILLIAM, b Kansas City, Mo, Jan 2, 32; m 61; c 2. MINING ENGINEERING, MINERAL DRESSING. *Educ:* Univ Idaho, BS, 57. *Prof Exp:* Mining engr, US Bur Mines, 57-61, supvry mining methods res engr, 61-64, supvry mining engr, 64-77; dir, coal prep div, Dept Energy, 77-89; COAL CONSULT, 90- *Concurrent Pos:* US del, Int Coal Prep Cong, 77-; US ed, Int J Coal Prep. *Honors & Awards:* Percy W Nicholls Award, Am Inst Mining Metall & Petrol Engrs, 90. *Mem:* Am Inst Mining Metall & Petrol Engrs; Am Mining Cong; Am Filtration Soc. *Res:* Research on flotation characteristics of American coals; pyritic sulfur reduction potential of conventional and non-conventional coal washing devices; liquid-solid separation; mineral benefication. *Mailing Add:* 6915 Hilldale Dr Pittsburgh PA 15236

DEUSCHLE, KURT W, b Kongen, Ger, Mar 14, 23; nat US; m 75; c 3. MEDICINE. *Educ:* Kent State Univ, BS, 44; Univ Mich, MD, 48; Am Bd Internal Med, dipl. *Prof Exp:* Intern med, Colo Med Ctr, 48-49; asst resident internal med, Syracuse Med Ctr, 49-50; resident, Col Med, State Univ NY Upstate Med Ctr, 50-52, instr, 51-52 & 54-55, dir tumor clin & cancer coordr, 54-55; chief tuberc, Ft Defiance Indian Hosp, USPHS, 52-54; asst prof pub health & prev med & dir, Navajo-Cornell Field Health Proj, Med Col, Cornell Univ, 55-60; prof community med & chmn dept, Med Col, Univ Ky, 60-68; Ethel H Wise prof & chmn dept, 68-90, PROF COMMUNITY MED, MT SINAI SCH MED, 90- *Concurrent Pos:* Consult, Off Technol Assessment, Cong US, 79-; Sr Health Fel Award, Commonwealth Fund, 66-67. *Honors & Awards:* Meyer Bodansky lectr, Univ Tex, 70; Merriman Lectr, Univ NC, Chapel Hill, 75; Award Excellence Domestic Health, Am Public Health Asn, 75; Distinguished Serv Award, Am Col Prev Med, 75; Duncan Clark Award, Asn Teachers Prev Med, 90. *Mem:* Inst Med Nat Acad Sci; Am Pub Health Asn; AAAS; Am Bd Prev Med; Am Thoracic Soc; Asn Am Med Cols; fel Am Col Prev Med (pres, 74); NY Acad Sci; Pan Am Med Asn. *Res:* Clinical investigations of antituberculous chemotherapeutic agents; cross-cultural medical research in areas of low economic development. *Mailing Add:* Dept Community Med Mt Sinai Sch Med New York NY 10029

DEUSER, WERNER GEORG, b Duesseldorf, Ger, Oct 31, 35; m 59; c 2. MARINE GEOCHEMISTRY. *Educ:* Univ Bonn, Vordiplom, 57; Pa State Univ, MS, 61, PhD(geochem), 63. *Prof Exp:* Consult geochemist, Nuclide Corp, 61-63; res scientist, Geol Surv WGer, 63-64; res geochemist, Nuclide Corp, 64-66; res assoc, 66-67, assoc scientist, 67-82, SR SCIENTIST, WOODS HOLE OCEANOG INST, 82- *Concurrent Pos:* Vis prof, Univ Catolica de Chile, 85 & Hebrew Univ, Jerusalem, 86; trustee, Bermuda Biol Sta, 82- *Mem:* AAAS; Geochem Soc; Am Geophys Union. *Res:* Mass spectrometry applied to geological problems; isotope geology; radioactive age determinations; stable isotope geochemistry, chemical and geological oceanography, deep-sea sedimentation. *Mailing Add:* Dept Chem Woods Hole Oceanog Inst Woods Hole MA 02543

D'EUSTACHIO, DOMINIC, b Pittsburgh, Pa, Feb 19, 04; m 45; c 2. PHYSICS. *Educ:* Columbia Univ, BS, 26; NY Univ, PhD(physics), 36. *Prof Exp:* Res assoc, Columbia Univ, 27-28; asst & instr physics, NY Univ, 29-35; instr, Polytech Inst Brooklyn, 36-42; chief crystal res sect, Ft Monmouth, NJ, 42-44; res dir, Bliley Mfg Co, 44-47; head physics res, Pittsburgh-Corning Corp, 47-60, res dir res & eng ctr, 60-65, vpres res, 65-69; prof mat sci, Univ PR, Mayaguez, 69-75; CONSULT, 75- *Mem:* AAAS; Acoust Soc Am; Am Phys Soc. *Res:* Hyperfine structure; surface properties of crystalline solids; high vacuum techniques; physical properties of glass; thermal conductivity at low temperatures; acoustics; portland cement and related materials; energy conservation, particularly as it applies to housing; use of solar energy. *Mailing Add:* 1431 Arboretum Dr Chapel Hill NC 27514-9117

D'EUSTACHIO, PETER, b Olean, NY, Sept 14, 49; m 76; c 1. SOMATIC CELL GENETICS, IMMUNOGENETICS. *Educ:* Oberlin Col, BA, 70; Rockefeller Univ, PhD(immunol), 76. *Prof Exp:* Res assoc cell biol, Rockefeller Univ, 76-77; fel somatic cell genetics, Dept Biol, Yale Univ, 77-81; asst prof, 81-88, ASSOC PROF BIOCHEM, NY UNIV MED CTR, 88- *Mem:* AAAS; Genetics Soc Am; Am Soc Human Genetics. *Res:* Somatic cell genetic analysis of immune function; high-resolution mapping of mammalian chromosomes and chromosome segments. *Mailing Add:* Dept Biochem NY Univ Med Ctr New York NY 10016

DEUSTER, RALPH W(ILLIAM), b Paterson, NJ, June 28, 20; m 46; c 3. MECHANICAL ENGINEERING, NUCLEAR PHYSICS. *Educ:* Purdue Univ, BSME, 42; Princeton Univ, MS, 50; Univ Pittsburgh, Cert Bus Mgt, 67. *Prof Exp:* Proj mgr, Armour Res Found, Ill, 54-55; with Babcock & Wilcox Co, 55-70; vpres & gen mgr, Reactor Fuels Div, Nuclear Fuel Serv, Inc, 70-73, pres & dir, 73-83; sr vpres, Getty Oil Develop Co, Sydney, Australia, 83-85; CONSULT, RADIATION ATTENUATING DEVICES CO, 88- *Mem:* Am Nuclear Soc; Am Soc Mech Engrs; US Coun Energy Awareness; Sigma Xi. *Res:* Economics of nuclear reactor fuels; beta spectroscopy. *Mailing Add:* 101 Tayloe Circle Williamsburg VA 23185

DEUTCH, BERNHARD IRWIN, b New York, NY, Sept 29, 29; m 63; c 2. PHYSICS. *Educ:* Cornell Univ, BA, 51, MS, 53; Univ Pa, PhD(physics), 59. *Prof Exp:* Asst biophys, Cornell Univ, 51-53; asst nuclear physics, Univ Pa, 54-58; physicist, Bartol Res Found, 58-62; amanuensis physics, Univ Aarhus, 62-64; physicist, Nobel Inst Physics, 64-65; lectr, 65-70, GROUP LEADER PHYSICS, UNIV AARHUS, 70- *Concurrent Pos:* Vis, Orsay, Brookhaven Nat Lab, 67, Niels Bohr Inst, 69, Fu Dan Univ, 74, Yale Univ, 76, Fu Dan Univ, Shanghai, 78 & Osaka Univ, 83; chmn, Int Hyperfine Conf, Aarhus, 71; assoc ed, J Nuclear Physics, 73-; ed, J Hyperfine Interactions, 74-; hon prof, Fudan Univ, Shanghai, 83. *Mem:* AAAS; Sigma Xi. *Res:* Atomic and nuclear polarization; nuclear structure; hyperfine interactions; biophysics; elementary particle physics; antimatter as atomic antihydrogen; 1 United States patent. *Mailing Add:* Inst Physics Univ Aarhus Aarhus C DK 8000 Denmark

DEUTCH, JOHN MARK, b Brussels, Belg, July 27, 38; US citizen; c 3. PHYSICAL CHEMISTRY, STATISTICAL MECHANICS. *Educ:* Amherst Col, BA, 61; Mass Inst Technol, BEng, 61, PhD(phys chem), 66. *Hon Degrees:* DSc, Amherst Col, 78, Univ Lowell, 86. *Prof Exp:* Systs analyst, Off Secy Defense, 61-65; Nat Acad Sci-Nat Res Coun fel, Nat Bur Stand, 65-66; asst prof, Princeton Univ, 66-70; assoc prof, Mass Inst Technol, 70-72, chmn dept, 77-78, dir energy res, 78-79, dean sci, 82-85, provost, 85-90, PROF CHEM, MASS INST TECHNOL, 72-, INST PROF, 90- *Concurrent Pos:* Consult, Bur Budget, Exec Off President, 66-68; mem, Defense Sci Bd, 75- & Army Sci Adv Panel, 76-78; Undersecy Energy, 79-80; trustee, Urban Inst, Wellesley Col, Mus Fine Arts, Boston, President's Nuclear Safety Oversight Comt, 80-81 & President's Comn on Strategic Forces, 82-83; mem, President's Foreign Intel Adv Bd, 90- *Mem:* Am Phys Soc; Am Chem Soc. *Res:* Liquids; transport processes; light scattering; polymer theory. *Mailing Add:* Dept Chem Rm 6-208 Mass Inst Technol Cambridge MA 02139

DEUTSCH, CAROL JOAN, CELLULAR PHYSIOLOGY. *Educ:* Yale Univ, PhD(chem), 72. *Prof Exp:* ASSOC PROF PHYSIOL, SCH MED, UNIV PA, 85- *Res:* Biophysics; cellular physiology of lymphocytes and mechanisms of activation; role and regulation of ion gradiants in cell function. *Mailing Add:* Dept Physiol Univ Pa Hamilton Walk Philadelphia PA 19174

DEUTSCH, DANIEL HAROLD, b New York, NY, Aug 29, 22; m 46; c 2. ORGANIC CHEMISTRY, NUCLEAR CHEMISTRY. *Educ:* Calif Inst Technol, BS, 48, PhD(chem, math), 51. *Prof Exp:* Fel cyclobutane chem, Calif Inst Technol, 51; pres & dir, Calif Found Biochem Res, 51-73; CONSULT, CALBIOCHEM, 70-; PRES & DIR, DANIEL H DEUTSCH LABS, INC, 77- *Concurrent Pos:* Vpres, Calbiochem, 58-70, dir, 58-72. *Mem:* Am Chem Soc; NY Acad Sci; AAAS; Am Nuclear Soc. *Res:* Commercial production of research biochemicals; medicinal chemistry; physical chemistry; ion-exchange resin kinetics; limitations on the second law of thermodynamics; isotope separation methodology; variant model for the stellar red shift; the liquid state; turbulence in liquids. *Mailing Add:* Daniel H Deutsch Labs Inc 141 Kenworthy Dr Pasadena CA 91105-1012

DEUTSCH, EDWARD ALLEN, b New York, NY, July 13, 42; div; c 2. INORGANIC CHEMISTRY. *Educ:* Univ Rochester, BS, 63; Stanford Univ, PhD(inorg chem), 67. *Prof Exp:* USPHS fel, Univ Calif, San Diego, 67-68; asst prof inorg chem, Univ Chicago, 68-73; from assoc prof to prof inorg chem, Univ Cincinnati, 78-89, prof radiol, 80-89; DIR, NUCLEAR MED RES & DEVELOP, MALLINCKRODT MED INC, 89- *Concurrent Pos:* Vis scientist, Argonne Nat Lab, 69- & Brookhaven Nat Lab, 80- & Consiglio Nazionale Richerche, Italy, 86- *Mem:* AAAS; Am Chem Soc; Soc Nuclear Med; Ital Soc Nuclear Biol & Med. *Res:* Chemistry of technetium as applied to nuclear medicine; catalytic processes; electron transfer reactions; chemistry of transuranium elements; kinetics and mechanisms of homogeneous reactions; inorganic models for biochemical systems; chemistry of technetium as applied to nuclear medicine. *Mailing Add:* Mallinckrodt Med Inc 675 McDonnell Blvd PO Box 5840 St Louis MO 63134

DEUTSCH, ERNST ROBERT, b Frankfurt, Ger, May 13, 24; Can citizen; m 49; c 3. GEOPHYSICS. *Educ:* Univ Toronto, BA, 46, MA, 49; Univ London, PhD(geophys) & DIC, 54. *Prof Exp:* Seismic explor, Texaco Explor Co, Alta, 49-51; res asst physics, Imp Col, Univ London, 54-57; res engr, Imp Oil Ltd, Alta, 57-63; from assoc prof to prof, 63-90, EMER PROF GEOPHYS, MEM UNIV NFLD, 90- *Concurrent Pos:* Vis lectr, Univ Western Ont, 57; mem assoc comt geod & geophys, Nat Res Coun Can, 64-70; assoc ed, J Can Soc Exp Geophys, 78-86; vis prof, Univ Bergen, Norway, 86. *Mem:* Soc Explor Geophys; Geol Asn Can; Can Geophys Union; Am Geophys Union; Soc Geomagnetism & Geoelec Japan; Can Soc Explor Geophys. *Res:* Magnetic properties and domain structure of rocks; paleomagnetism and application to hypotheses of polar wandering and plate tectonics; paleogeography of the North Atlantic; gravity studies and crustal structure. *Mailing Add:* Dept Earth Sci Mem Univ Nfld St John's NF A1B 3X5 Can

DEUTSCH, FRANK RICHARD, b New Britain, Conn, Dec 18, 36; m 65; c 2. GENERAL MATHEMATICS. *Educ:* Univ Hartford, BS, 60; Northwestern Univ, MS, 61; Brown Univ, PhD(appl math), 66. *Prof Exp:* Mathematician, United Aircraft Res Lab, 60-61; res asst, Brown Univ, 61-65; PROF, PA STATE UNIV, 65- *Concurrent Pos:* Ed, J Approximation Theory, 84-; vis prof, Univ Gottingen, 71-72, Univ Frankfurt, 78-79 & Tex A&M Univ, 87; prin investr, NSF, 66-71, 77-78 & 81-83. *Mem:* Am Math Soc; Soc Indust & Appl Math. *Res:* Theory of best approximation; analysis and functional analysis. *Mailing Add:* Dept Math Pa State Univ Univ Park PA 16802

DEUTSCH, HAROLD FRANCIS, b Sturgeon Bay, Wis, Sept 2, 18; m 42; c 2. CHEMISTRY. *Educ:* Univ Wis, PhB, 40, PhD(physiol chem), 44. *Prof Exp:* Asst cancer res, McArdle Lab, 40-41, teaching asst physiol chem, 42-44, res assoc, Sch Med, 44-45, asst prof, 45-46, assoc prof, 47-56, PROF PHYSIOL CHEM, SCH MED, UNIV WIS-MADISON, 56- *Concurrent Pos:* Rockefeller Found fel natural sci, Univ Stockholm, 50-51; vis prof, Univ Brazil, 50, Univ Sao Paulo, 54, Univ Hokkaido, 71 & Tech Univ Munich, 74-75; vis scientist, Rockefeller Found, 60, Max-Planck Inst Biochem, Munich, 60 & 64, Max-Planck Inst Cell Biol, 87, Univ Hokkaido, 81, Univ Gerais, Brazil, 88 & Univ Osaha, 89; Alexander von Humboldt prof, 87. *Mem:* Am Chem Soc; Am Soc Biol Chem. *Res:* Separation and characterization of plasma, erythrocyte and tissue protein; immunochemistry. *Mailing Add:* Dept Physiol Chem Sch Med Univ Wis Madison WI 53706

DEUTSCH, JOHN LUDWIG, b New York, NY, May 5, 38; m 61; c 2. PHYSICAL CHEMISTRY, SPECTROSCOPY. *Educ:* Tulane Univ, BS, 59; Oxford Univ, DPhil(ultraviolet spectros), 63. *Prof Exp:* NSF fel & tutor chem, Oxford Univ, 63-64; actg chmn dept, 74-75, assoc prof, 66-77, PROF CHEM, STATE UNIV NY COL GENESEO, 77- *Concurrent Pos:* Vis researcher, Phys Chem Lab, Oxford Univ, 80, 81 & 83; vis scientist, Physics Inst, Univ Stockholm, 73, Herzberg Inst Astrophysics, Ottawa, 86 & 87; vis prof, Univ Rochester, NY, 87-88; vis asst prof, Pomona Col, 64-66; Rhodes scholar; postdoctoral fel, NSF. *Mem:* AAAS; Am Chem Soc; Royal Soc Chem; Sigma Xi; NY Acad Sci; Am Inst Chem. *Res:* Electronic spectra of simple molecules and nuclear magnetic resonance spectroscopy. *Mailing Add:* Dept Chem State Univ NY Col Geneseo NY 14454

DEUTSCH, LAURENCE PETER, b Boston, Mass, Aug 7, 46. INFORMATION SCIENCE. *Educ:* Univ Calif, Berkeley, BA, 69, PhD(comput sci), 73. *Prof Exp:* Programmer comput sci, Univ Calif, Berkeley, 65-68; syst architect, Berkeley Comput Corp, 68-71; mem res staff, Comput Sci, Xerox Palo Alto Res Ctr, 71-79, prin scientist, 79-83, res fel, 83-86; CHIEF SCIENTIST, PARCPLACE SYSTS, 86- *Mem:* Asn Comput Mach; Inst Elec & Electronics Engrs. *Res:* Programming languages and systems; user interfaces. *Mailing Add:* PO Box 60264 Palo Alto CA 94306

DEUTSCH, MARSHALL EMANUEL, b New York, NY, Aug 17, 21; m 47; c 3. CLINICAL CHEMISTRY, NUTRITION. *Educ:* City Col New York, BS, 41; NY Univ, PhD(physiol sci), 51. *Prof Exp:* Asst & assoc biochem, NY Univ-Bellevue Med Ctr, 47-51; jr assoc, Henry Ford Hosp, 51-53; head chem microbiol, Warner-Chilcott Res Labs, 53-55, sr scientist biochem, 55-58; dir prod develop, G W Carnrick Co, 58-59, dir res & develop, 59-60; dir life sci res, Becton, Dickinson & Co, 60-66; tech dir, NEN-Picker Radiopharmaceut, 66-68; consult, Farbwerke Hoechst AG, 68, tech dir, Picker-Hoechst, Inc, 69-70; vpres & tech dir, Mead Diag, Inc, 70-72; vpres & tech dir, Thyroid Diag, 72-86; chmn & tech dir, Marshall Diag, 86-90; VPRES & TECH DIR, J&S MED ASSOCS, 90- *Concurrent Pos:* Vpres & tech dir, CIS Radiopharmaceut, 72-75; consult, UN Capital Develop Fund, 77; consult, Agency Int Develop, 79. *Mem:* Am Chem Soc; fel AAAS; Clin Ligand Assay Soc; Am Asn Clin Chem; Sigma Xi; Mycol Soc Am; NY Acad Sci. *Res:* Thyroid and anti-thyroid drugs; effects of viricides; chemical kinetics; diagnostic reagents; pharmaceutical products; nutrition. *Mailing Add:* 41 Concord Rd Sudbury MA 01776-2328

DEUTSCH, MARTIN, b Vienna, Austria, Jan 29, 17; US citizen; m 39; c 2. NUCLEAR PHYSICS. *Educ:* Mass Inst Technol, BS, 37, PhD(physics), 41. *Hon Degrees:* Univ Algiers, DHC, 59. *Prof Exp:* Instr physics, Mass Inst Technol, 41-45; scientist, Los Alamos Sci Lab, Univ Calif, 44-46; from asst prof to assoc prof, 45-53, dir lab nuclear sci, 75-80, prof, 53-87, EMER PROF PHYSICS, MASS INST TECHNOL, 87- *Honors & Awards:* Rumford Medal, AAAS, 85. *Mem:* Nat Acad Sci; fel Am Acad Arts & Sci; fel Am Phys Soc. *Res:* Study of radioactive radiations; study of the fission process; nuclear spectroscopy; elementary particle physics. *Mailing Add:* Dept Physics Mass Inst Technol Cambridge MA 02138

DEUTSCH, MIKE JOHN, b Denver, Colo, Apr 4, 20; m 42; c 5. BIOCHEMISTRY. *Educ:* Univ Denver, BS, 42. *Prof Exp:* Biochemist, 57-61, SUPVRY BIOCHEMIST, FOOD & DRUG ADMIN, 61- *Mem:* Asn Off Agr Chemists. *Res:* Vitamin methodology. *Mailing Add:* Food & Drug Admin Div Nutrit 200 C St SW Washington DC 20204

DEUTSCH, ROBERT W(ILLIAM), b Far Rockaway, NY, Mar 21, 24; m 49; c 2. NUCLEAR ENGINEERING. *Educ:* Mass Inst Technol, BS, 48; Univ Calif, PhD(physics), 53. *Prof Exp:* Physicist high energy nuclear physics, Radiation Lab, Univ Calif, 50-53; res assoc theoret nuclear physics, Knolls Atomic Power Lab, 53-57; chief physicist, Gen Nuclear Eng Corp, 57-61; consult physics, Martin Co, 61-63; prof & chmn dept nuclear sci & eng, Cath Univ, 63-71; pres, Gen Physics Corp, 66-87; PRES, RWD TECHNOLOGIES, 88- *Concurrent Pos:* Bd adv, Univ Md Technol Advan Prog, 88-; bd trustees, Goucher Col, 90-, Am Nuclear Soc Prof Develop Accreditation Conn, 89- *Mem:* AAAS; fel Am Nuclear Soc; Am Soc Eng Educ. *Mailing Add:* RWD Technologies 10480 Little Patuxent Pkwy Suite 1200 Columbia MD 21044

DEUTSCH, S(ID), b New York, NY, Sept 19, 18; m 41; c 3. BIOENGINEERING, ELECTRICAL ENGINEERING. *Educ:* Cooper Union, BEE, 41; Polytech Inst Brooklyn, MEE, 47, DEE, 55. *Prof Exp:* Technician elec motor, Rite-Way Fur Machine Co, 35-40; designer electromech equip, Otis Elevator Co, 40-41 & Allied Process Engrs, 41-43; instr physics, Hunter Col, 43-44; instr TV, Madison Inst Technol, 46-50; engr electronics, Polytech Res & Develop Co, 50-54; instr commun, Polytech Inst Brooklyn, 51-56, proj engr electronics, Microwave Res Inst, 54-60, prof elec eng, 56-72; prof, Rutgers Med Sch, 72-79; adj prof, Rutgers Univ, 72-79; prof biomed eng, Tel Aviv Univ, 79-84. *Concurrent Pos:* Designer, Fairchild Camera & Instrument Co, 43-44; instr, City Col New York, 55-57; consult, Polytech Res & Develop Co, 57, Budd Electronics Co, 58-60 & Rockefeller Inst, 61-64; vis prof elec eng, Univ SFla, 83- *Mem:* Fel Inst Elec & Electronics Engrs; fel Soc Info Display. *Res:* Communications and electronics; information and network theory; biomedical electronics; numerous books published. *Mailing Add:* 3967 Oakhurst Blvd Sarasota FL 34233-1434

DEUTSCH, STANLEY, b Brooklyn, NY, Apr 4, 30; m 54; c 2. ANESTHESIOLOGY, PHARMACOLOGY. *Educ:* NY Univ, BA, 50; Boston Univ, MA, 51, PhD(physiol), 55, MD, 57; Am Bd Anesthesiol, dipl. *Prof Exp:* Rotating intern, Grad Hosp, Univ Pa, 57-58, resident anesthesiol, Hosp Univ Pa, 58-61, assoc, 63-64, asst prof, Univ Pa, 63-65; assoc in anesthesia, Harvard Med Sch, 65-68, asst prof, 68-69; assoc surg, Peter Bent Brigham Hosp, 65-69; prof anesthesiol, Univ Chicago & chmn dept, Michael Reese Hosp, 69-71; prof anesthesiol & chmn dept, Sch Med, Univ Okla Health Sci Ctr, 71-82; prof, dept anesthesiol, Univ Tex Med Sch, Houston, 82-; PROF ANESTHESIOL, GEORGE WASHINGTON UNIV, WASHINGTON, DC, 82- *Concurrent Pos:* Consult, Vet Admin Hosp, Philadelphia, Pa, 60-61 & 63-65; mem sect anesthesia, Nat Acad Sci-Nat Res Coun, 70-72. *Mem:* AAAS; NY Acad Sci; Am Soc Anesthesiol; Asn Univ Anesthetists. *Res:* Cardiovascular and renal effects of anesthesia and surgery; pharmacology of anesthetics and drugs used in association with anesthesia; cardiovascular and renal physiology. *Mailing Add:* Dept Anesthesiol George Washington Univ 901 23rd St NW Washington DC 20037

DEUTSCH, THOMAS F, b Vienna, Austria, Apr 24, 32; nat US; m 90. LASERS, OPTICS. *Educ:* Cornell Univ, BEngPhys, 55; Harvard Univ, AM, 56, PhD(appl physics), 61. *Prof Exp:* From sr res scientist to prin res scientist, Res Div, Raytheon Co, Mass, 60-74; staff mem, Lincoln Lab, Mass Inst Technol, 74-84; STAFF MEM, WELLMAN LAB, MASS GEN HOSP, 84-; ASSOC PROF, HARVARD MED SCH, 87- *Honors & Awards:* Lamb Prize, Optical Soc, 91. *Mem:* Fel Am Phys Soc; Inst Elec & Electronics Engrs. *Res:* Medical applications of lasers; optical diagnostic techniques for medicine; laser-tissue interactions; laser induced shock waves. *Mailing Add:* Wellman Lab-2 Mass Gen Hosp Boston MA 02114

DEUTSCH, WALTER A, b Paris, Ark, Aug 29, 45. DNA REPAIR. *Educ:* Tex A&M Univ, PhD, 74. *Prof Exp:* ASSOC PROF BIOCHEM, LA STATE UNIV, 79- *Mem:* Am Soc Biol Chemists; AAAS. *Mailing Add:* Dept Biochem La State Univ Baton Rouge LA 70803-1806

DEUTSCHER, MURRAY PAUL, b New York, NY, Sept 1, 41; m 66; c 2. BIOCHEMISTRY, MOLECULAR BIOLOGY. *Educ:* City Col New York, BS, 62; Albert Einstein Col Med, PhD(biochem), 66. *Prof Exp:* Am Cancer Soc fel, Stanford Univ, 66-68; from asst prof to assoc prof, 68-75, PROF BIOCHEM, UNIV CONN HEALTH CTR, 75- *Concurrent Pos:* Vis scientist, Weizmann Inst Sci, 69, Nat Inst Med Res, London, 75, Stanford Univ, 86; exec ed, Nucleic Acids Res, 79-85, sr exec ed, 86-90. *Mem:* AAAS; Am Soc Biol Chemists; Am Chem Soc; Am Soc Microbiol. *Res:* Protein biosynthesis; enzymology of RNA metabolism. *Mailing Add:* Dept Biochem Univ Conn Health Ctr Farmington CT 06030

DEUTSCHMAN, ARCHIE JOHN, JR, b Chicago, Ill, Nov 21, 17; m 46; c 5. ORGANIC CHEMISTRY. *Educ:* Univ Ill, BS, 39; Lawrence Col, MS, 41, PhD(chem), 43. *Prof Exp:* Spec asst, Univ Ill, 44-45; chief chemist, Graham, Crawley & Assocs, Chicago, 45-46; sr res chemist, Phillips Petrol Co, Okla, 46-47; dir chem res, Spencer Chem Co, 47-57; prof agr biochem, 57-75, PROF NUTRIT & FOOD, UNIV ARIZ, 75- *Mem:* Am Chem Soc; Sigma Xi. *Res:* High pressure reactions; vapor phase and solution; furfural polymer control; butadiene chemistry; polymerizations; electrical conductivity of paper; hydrothermal crystal growth; copper recovery. *Mailing Add:* 4860 N Camino Real Tucson AZ 85718-5922

DEV, PARVATI, b New Delhi, India, Dec 6, 46; m 75; c 2. BIOMECHANICS, COMPUTER SIMULATION. *Educ:* Indian Inst Technol, Kharagpur, BTech, 68; Stanford Univ, MS, 69, PhD(elec eng), 75. *Prof Exp:* Staff scientist, Neurosci Res Prog, Mass Inst Technol, 72-76; asst prof biomed eng, Col Eng, Boston Univ, 76-78; res scientist, Develop Ctr, Vet Admin Med Ctr, Palo Alto, Calif, 79-82; vpres res & develop, Contour Med Systs, Mountain View, Calif, 82-88, VPRES ADVAN TECHNOL, CEMAX, FREMONT, CA, 88-; DIR, SUMMIT, STANFORD UNIV SCH MED, STANFORD, CALIF, 90- *Concurrent Pos:* Res assoc, Dept Psychol, Mass Inst Technol, 73-76; assoc ed, Transactions Biomed Eng, Inst Elec & Electronics Engrs, 80-83. *Mem:* Inst Elec & Electronics Engrs; Soc Neurosci; Asn Comput Mach; Am Asn Artificial Intel. *Res:* Mathematical analysis and quantitative modelling of neuromuscular systems; biomechanics of the hand; computer graphics for surgical planning; multimedia in computer-aided medical education. *Mailing Add:* 573 Suzanne Ct Palo Alto CA 94306

DEV, VAITHILINGAM GANGATHARA, b Kalayarkurichi, India, Mar 8, 37; m 66; c 2. GENETICS, CYTOGENETICS. *Educ:* Univ Madras, BVSc, 59; Univ Mo, MS, 61, PhD(animal husbandry), 65. *Prof Exp:* Mem scientists pool, Coun Sci & Indust Res, India, 66-69; res assoc, Columbia Univ, 72-73, asst prof human genetics & develop, 74-77; asst prof path, Univ Tex Health Sci Ctr, Dallas, 78-81; assoc prof med genetics, Univ SAla, Mobile, 81-84; assoc res prof, Vanderbilt Univ, 84-89; DIR, GENETICS ASSOCS, 90- *Concurrent Pos:* Trainee, Columbia Univ, 70-72. *Res:* Chromosome structure; polymorphisms; linkage. *Mailing Add:* Genetics Assocs 121 21st Ave W Suite 300 Nashville TN 37203

DEV, VASU, b Lahore, Punjab, Mar 18, 33; m 63; c 2. ORGANIC CHEMISTRY, MEDICINAL CHEMISTRY. *Educ:* Punjab Univ, BSc, 51, Hons, 53, MSc, 54; Univ Calif, Davis, PhD(chem), 62. *Prof Exp:* Chemist, Drug Res Lab, India, 55-56 & Govt Med Col, India, 56-59; res assoc org chem, Univ Chicago, 63-64; asst prof med chem, Univ Tenn, 64-65; from asst prof to assoc prof, 65-72, chmn dept, 70-78, PROF CHEM, CALIF STATE POLYTECH UNIV, POMONA, 73- *Mem:* Am Chem Soc; Royal Soc Chem; Sigma Xi; Am Pharmacog Soc. *Res:* Reaction mechanisms, natural products and organic synthesis; synthesis and study of products possessed with pharmacodynamic properties. *Mailing Add:* Dept Chem Calif State Polytech Univ Pomona CA 91768

DEVAN, JACKSON H, b Chicago, Ill; m 50; c 4. CORROSION SCIENCE. *Educ:* Stanford Univ, BS, 52; Univ Tenn, MS, 60. *Prof Exp:* Proj officer, Air Force Mat Lab, 52-54; metallurgist, 54-55, group leader, 55-86, SR STAFF MEM, OAK RIDGE NAT LAB, 87- *Concurrent Pos:* Vis scientist, Harwell Lab, UK, 86-87. *Mem:* Fel Am Soc Metals Int; Sigma Xi; Electrochem Soc. *Res:* Corrosion resistant high temperature materials; effects of liquid metals, molten salts, and sulfur-containing gases on ceramics, graphics, and intermetallic compounds. *Mailing Add:* Metals & Ceramics Div Oak Ridge Nat Lab PO Box 2008 Oak Ridge TN 37831

DEVANEY, JOSEPH JAMES, b Boston, Mass, Apr 29, 24; m 54; c 1. LASERS, MATHEMATICAL PHYSICS. *Educ:* Mass Inst Technol, SB, 47, PhD(theoret physics), 50. *Prof Exp:* Mem staff, Theoret Div, 50-72, Laser Div, 72-80, MEM STAFF, PARTICLE TRANSP GROUP, LOS ALAMOS NAT LAB, 80- *Concurrent Pos:* Adj prof math, Univ NMex, 56-59, adj prof physics, 59-; mem, Gov Policy Bd Air & Water Pollution, 70. *Mem:* Am Phys Soc; Am Nuclear Soc; Sigma Xi. *Res:* Theoretical nuclear physics; laser systems analysis; particle transport. *Mailing Add:* X6-MS-B226 PO Box 1663 Los Alamos Nat Lab Los Alamos NM 87545

DE VAUCOULEURS, GERARD HENRI, b Paris, France, Apr 25, 18. ASTRONOMY. *Educ:* Univ Paris, BSc, 36, Lic es sci, 39, D Univ, 49; Australian Nat Univ, DSc, 57. *Prof Exp:* Res fel, Physics Res Lab, Sorbonne & Inst Astrophys, Nat Ctr Sci Res, France, 43-50 & Australian Nat Univ, 51-54; observer, Yale-Columbia Southern Sta, Australia, 54-57; astronr, Lowell Observ, Ariz, 57-58; res assoc, Harvard Col Observ, 58-60; from assoc prof to prof, 60-81, Ashbel Smith prof astron, 81-83, Blumberg Centennial prof astron, 83-88, EMER PROF, UNIV TEX, AUSTIN, 88- *Concurrent Pos:* Sci prog, British Broadcasting Corp, French Sect, London, 50-51. *Honors & Awards:* Herschel Medal, Royal Astron Soc, 80; Russel Lectr Prize, Am Astron Soc, 88; Janssen Medal Prize, French Astron Soc, 88. *Mem:* Nat Acad Sci; Royal Astron Soc; Soc Astron France; French Phys Soc; Int Astron Union; Am Astron Soc; corresp mem Nat Acad Sci Arg. *Res:* Extragalactic research; cosmology. *Mailing Add:* Dept Astron RLM Hall 16-320 Univ Tex Austin TX 78712

DEVAULT, DON CHARLES, b Battle Creek, Mich, Dec 10, 15; m 48; c 2. PHOTOSYNTHESIS, ELECTRON TRANSPORT. *Educ:* Calif Inst Technol, BS, 37; Univ Calif, Berkeley, PhD(chem), 40. *Prof Exp:* Jr res assoc chem, Stanford Univ, 40-42, assoc investr, Off Sci Res & Develop Cyclotron Proj, Dept Physics, 42; instr chem, Inst Nuclear Studies, Univ Chicago, 46-48; assoc prof chem & physics, Col of the Pac, 49-58; consult, Biophys Electronics, Inc & Bionic Instruments, Inc, 59-63; fel biophys, Johnson Res Found, Med Sch, Univ Pa, 63-64, assoc, 64-67, asst prof, 67-77; vis assoc prof, 77-84, VIS PROF BIOPHYS, UNIV ILL, URBANA-CHAMPAIGN, 84- *Concurrent Pos:* NSF grants, 67-69 & 71-83. *Mem:* Biophys Soc; Am Soc Biol Chemists; Am Soc Photobiol. *Res:* Electron transport phenomena in biology; energy transduction from electron transport as in respiration and photosynthesis. *Mailing Add:* Dept Physiol & Biophys Univ Ill 524 Burrill Hall 407 S Goodwin Ave Urbana IL 61801

DEVAY, JAMES EDSON, b Minneapolis, Minn, Nov 23, 21; m 47; c 6. PLANT PATHOLOGY. *Educ:* Univ Minn, BS, 49, PhD(plant path), 53. *Prof Exp:* Asst plant path, Univ Minn, 49-52, from instr to assoc prof, 53-57; from asst prof to assoc prof, 57-65, assoc dean, Div Biol Sci, 76-79, PROF PLANT PATH, UNIV CALIF, DAVIS, 65-, CHMN DEPT, 80- *Mem:* Fel AAAS; fel Am Phytopath Soc; Mycol Soc Am; Scand Soc Plant Physiol. *Res:* Physiology and biochemistry of host-parasite relationships; diseases of cotton. *Mailing Add:* Dept Plant Path Univ Calif Davis CA 95616

DE VEBER, LEVERETT L, b Toronto, Ont, Jan 27, 29; m 54; c 6. HEMATOLOGY, IMMUNOLOGY. *Educ:* Univ Toronto, MD, 53; FRCP(C), 61. *Prof Exp:* Jr intern, St Michael's Hosp, Toronto, 53-54; sr intern med, Shaughnessey Hosp, Univ BC, 54-55; sr house officer pediat, Univ Manchester, 57-58; registr, Univ Liverpool, 58; jr asst resident, NY Univ, 59-60; asst resident pediat & path, Univ Toronto, 60-61; lectr, 62-65, from asst prof to assoc prof pediat & path chem, 65-74, PROF PEDIAT, UNIV WESTERN ONT, 74-, DIR Rh RES LAB, 64- *Concurrent Pos:* Mead-Johnson fel immunohemat, Univ Man, 61-62; Can Life Ins Co fel, 62-64; dir immunohemat div, Rh Serv & Blood Bank, Victoria Hosp, London, 62-74; Western Can Rh Prev Prog investr, 64-; Nat Health & Welfare res investr, 64-68, chief investr, 68- *Mem:* Can Soc Immunol (treas, 68-); Can Fedn Biol Sci. *Res:* Prevention of Rh sensitization with Rh immune globulin; detection of early Rh sensitization with serological and lymphocyte culture techniques; detection of Rh antigen on amniotic cells with fluorescent antibody technique; cellular immunity in children with acute lymphoblastic leukemia; families of children with cancer, specifically death at home; bereavement counselling; pain studies; psycho social studies; long term cancer survivors. *Mailing Add:* Dept Pediat Univ Western Ont Fac Med London ON N6A 5C1 Can

DE VEDIA, LUIS ALBERTO, b Buenos Aires, Arg, Apr 12, 41; m; c 2. WELDING TECHNOLOGY, FRACTURE MECHANICS. *Educ:* Nat Univ La Plata, Arg, elec engr, 68; Nat Atomic Energy Comn, Arg, cert, 69; Cranfield Inst Technol, UK, MSc, 74. *Prof Exp:* Head welding & NDT Div, Nat Atomic Energy Comn, Arg, 68-77; dir, Simet Eng, Arg, 77-80; vdir, Res Inst Sci & Technol Mat, Arg, 80-86; EXEC DIR, LATIN AM WELDING FOUND, 86- *Concurrent Pos:* Consult & lectr, Orgn Am States & IEAE, 74-; invited lectr, Univs & Acad Orgn, Latin Am, 74-, Arg Steel Inst, 82-; prof, Nat Tech Univ, Arg, 75-78, Nat Univ Mar del Plata, Arg, 80-86; vis prof, Nat Univ Mar del Plata, Arg, 86-88, Nat Univ La Plata, 87-90, Nat Univ Buenos Aires, Arg, 90. *Mem:* Welding Inst UK; Am Welding Soc; Am Soc Mech Engrs; Inst Physics UK. *Res:* Fracture mechanics and its applications to assess defects significance in welded joints under static or dynamic loading; relationship between weld metal microstructure and fracture toughness in order to develop a rational basis for welding consumables design; mathematical modelling of welded joints to predict temperature distribution and thermal and residual stresses. *Mailing Add:* Fundacion Latinoamericana de Soldadura Calle 18 No 4113 Villa Lynch Buenos Aires 1672 Argentina

DE VEGVAR, PAUL GEZA NEUMAN, b White Plains, NY, Oct 27, 57; m 88. LOW TEMPERATURE PHYSICS, MESOSCOPIC PHYSICS. *Educ:* Harvard Univ, BA, 79; Cornell Univ, MS, 82, PhD(physics), 86. *Prof Exp:* Postdoctoral, 86-88, MEM TECH STAFF, AT&T BELL LABS, 88- *Mem:* Am Phys Soc. *Res:* Quantum transport properties of electrons in small devices at low temperatures. *Mailing Add:* AT&T Bell Labs 600 Mountain Ave 1D222 Murray Hill NJ 07974

DEVENEY, JAMES KEVIN, b Boston, Mass, Feb 28, 45; m 70. PURE MATHEMATICS. *Educ:* Boston Col, BS, 70; Fla State Univ, PhD(math), 74. *Prof Exp:* Vis prof math, Kans State Univ, 74; asst prof, 74-80, ASSOC PROF MATH, VA COMMONWEALTH UNIV, 81- *Concurrent Pos:* Vis prof math, La State Univ, 80. *Mem:* Am Math Soc; Math Asn Am. *Res:* Structure of field extensions and related p-adic fields, with an emphasis on Galois theory. *Mailing Add:* Dept Math Sci Va Commonwealth Univ Richmond VA 23284

DEVENS, W(ILLIAM) GEORGE, b Ft Eustis, Va, Mar 2, 26; m 48; c 8. ENGINEERING & EDUCATION. *Educ:* US Mil Acad, BS, 46; Univ Ill, Urbana-Champaign, MS, 53. *Prof Exp:* US Army CEngrs, 44-66; Dep commandant cadets, Norwich Univ, 66-67; asst prof eng & assoc dir, 67-69, PROF ENG & DIR DIV ENG FUNDAMENTALS, VA POLYTECH INST & STATE UNIV, 69- *Mem:* Soc Am Mil Engrs; Am Soc Eng Educ; Nat Soc Prof Engrs. *Res:* Educational technology. *Mailing Add:* Dept Eng Va Polytech Inst & State Univ Blacksburg VA 24061

DEVENUTO, FRANK, b Giovinazzo, Italy, July 28, 28; nat US; m 57; c 4. ORGANIC CHEMISTRY, BIOLOGICAL CHEMISTRY. *Educ:* Univ Rome, PhD(org chem, biol chem), 51. *Prof Exp:* Asst org res, Univ Rome, 48-51; sr chemist antibiotic res, Leo-Penicillin Co, 51-52; consult, chemist abrasive res, Ace Abrasive Labs, NY, 52-53; chief sect, Steroid Hormones Res, US Army Med Res Lab, 55-74; supv res chemist, 74-84, MED RES CONSULT, LETTERMAN ARMY INST RES, 84- *Mem:* AAAS; Soc Exp Biol & Med; NY Acad Sci; Am Soc Biol Chemists. *Res:* Metabolism and mechanism of action of steroid hormones; carbohydrate metabolism; blood research; hemoglobin and blood substitutes. *Mailing Add:* 1320 Ponderosa Dr Petaluma CA 94952

DEVER, DAVID FRANCIS, b Quebec, Que, Oct 9, 31; m 57; c 4. PHYSICAL CHEMISTRY. *Educ:* Spring Hill Col, BS, 53; Fla State Univ, MS, 55; Ohio State Univ, PhD(phys chem), 59. *Prof Exp:* Off Naval Res fel thermodyn, Ohio State Univ, 60-61, univ fel molecular spectros, 62-63; asst prof chem, Univ Miami, 61-62; res photochemist, US Bur Mines, 63-64; proj leader air pollution res, 64-66; chmn, Dept Chem, Col Petrol & Minerals, Dharhan, Saudi Arabia, 66-69; chmn, Div Natural Sci & Math, 69-72, PROF CHEM, MACON JR COL, 72- *Concurrent Pos:* Hazardous mats consult, Fire Dept. *Honors & Awards:* M D Marshall Award, 53; Nat Award, Mfg Chemists Asn, 79. *Mem:* AAAS; Am Chem Soc; Soc Technol & Humanities. *Res:* Thermodynamics; photochemistry; kinetics; IR laser chemistry; high temperature spectroscopy. *Mailing Add:* Dept Chem Macon Col Macon GA 31206

DEVER, DONALD ANDREW, b Sudbury, Ont, Sept 18, 26; m 47; c 2. PLANT SCIENCE. *Educ:* Ont Agr Col, BSA, 49; Univ Wis, MSc, 50, PhD(entom plant path), 53. *Prof Exp:* Tech asst, Dom Parasite Lab, Can Sci Serv, Ont, 45-48, proj leader, 49; asst, Univ Wis, 49-53, asst prof, 53-56; dist res entomologist, Res & Develop, Calif Spray-Chem Corp, 56-62; tech dir, Niagara Brand Chem Div, FMC Mach & Chem Ltd, 62-68, mgr int develop, Niagara Chem Div, FMC Corp, NY, 68-69; secy-gen, 69-85, PRES, CAN GRAINS COUN, 85- *Concurrent Pos:* Dir, Man Res Coun; Western Transportation Adv Coun, Vancouver; Dir, Biomass Energy Inst, Winnipeg. *Mem:* Entom Soc Am; Can Entom Soc; Can Agr Chem Asn (pres, 67-68). *Res:* Administration; maximize export and domestic sales of Canadian grain through market analysis and development. *Mailing Add:* Can Grains Coun 360-760 Main St Winnipeg MB R3C 3Z3 Can

DEVER, G E ALAN, b Kingston, Ont, June 9, 41; m 63; c 2. PUBLIC HEALTH ADMINISTRATION. *Educ:* State Univ NY, Buffalo, BA, 67, MA, 68; Univ Mich, PhD(geog), 70. *Prof Exp:* Asst prof, Univ Md, Baltimore, 70-71; asst prof, Ga State Univ, 71-74, res grants, 72-77; chief epidemiologist & dir, Health Serv Res & Statist Sect, 72-77, dir, Health Serv Anal, Div Phys Health, Ga Dept Human Resources, 77-81; ASSOC PROF EPIDEMIOL, MED SCH, MERCER UNIV, 81- *Concurrent Pos:* Consult, Appl Statist Training Inst, Nat Ctr Health Statist, US Dept HEW, 74; pres & dir, Health Serv Anal, Inc, 81- *Mem:* Am Pub Health Asn; Can Pub Health Asn; Pop Ref Bur. *Res:* Health program planning and evaluation; epidemiology; nutrition. *Mailing Add:* Dept Family & Community Mercer Univ Sch Med 1400 Coleman Ave Macon GA 31207

DEVER, JOHN E, JR, b Camden, NJ, Sept 24, 32; m 52; c 2. PLANT PHYSIOLOGY, BIOCHEMISTRY. *Educ:* Rutgers Univ, BA, 60; Ore State Univ, MS, 62; Mich State Univ, PhD(plant physiol, biochem), 67. *Prof Exp:* Lectr biol, Flint Col, Mich, 67; asst prof bot, 67-74, ASSOC PROF BIOL, FT LEWIS COL, 74- *Mem:* Am Soc Plant Physiol; Am Inst Biol Sci; Scand Soc Plant Physiol. *Res:* Cell wall structure and metabolism; exocellular enzymes in root cell walls. *Mailing Add:* Dept Biol & Agr Ft Lewis Col Durango CO 81301

DEVEREUX, OWEN FRANCIS, b Lexington, Mass, Aug 23, 37; m 57, 69; c 4. CORROSION, METALLURGY. *Educ:* Mass Inst Technol, SB, 59, SM, 60, PhD(metall), 62. *Prof Exp:* Res chemist, Chevron Res Corp, 62-64, Corning Glass Works, 64-66 & Chevron Res Corp, 66-68; assoc prof metall, 68-76, PROF METALL, INST MAT SCI, UNIV CONN, 76-, DEPT HEAD, 83- *Mem:* Metall Soc; Electrochem Soc; Corrosion Soc. *Res:* Thermodynamics; transport phenomena; oxidation corrosion; application of modelling techniques and of AC and DC polarization methods to corrosion electrochemistry. *Mailing Add:* Dept Metall Univ Conn 97 N Eagleville Rd Box U-136 Rm 111 Storrs CT 06269-3136

DEVEREUX, RICHARD BLYTON, b Philadelphia, Pa, Oct 23, 45; m 70; c 2. CARDIOLOGY, CLINICAL RESEARCH. *Educ:* Yale Univ, BA, 67; Univ Pa, MD, 71. *Prof Exp:* Intern & resident internal med, NY Hosp, 71-74; fel cardiol, Univ Pa, 74-76; maj US Army Med Corps, US Army Hosp, Heidelberg, Ger, 76-78; asst prof, 78-83, ASSOC PROF MED, CORNELL UNIV MED COL, 83- *Concurrent Pos:* Dir, Echocardiography Lab, NY Hosp, 78-; consult, Rockefeller Univ Hosp, 81-; Coun Circulation and High Blood Pressure Res, Am Heart Asn; dir, Am Soc Hypertension, 85-87. *Honors & Awards:* Res Award, Nat Marfan Found, 87. *Mem:* Am Heart Asn; Am Soc Hypertension; Am Soc Echocardiography; fel Am Col Cardiol; Am Fed Clin Res; fel Am Col Physicians. *Res:* Usage of echocardiography; restriction fragment length polymorphism and ambulatory blood pressure monitoring in determining etiology, pathogenesis and prognosis of structural heart diseases including hypertension, chronic regurgitant valvular disease, mitral valve prolapse and Marfan syndrome. *Mailing Add:* Div Cardiol NY Hosp Cornell Med Ctr 525 E 68th St Box 222 New York NY 10021

DEVEREUX, THEODORA REYLING, b Glen Cove, NY, Sept 11, 44; m 70; c 1. PULMONARY PHARMACOLOGY. *Educ:* Duke Univ, BS, 66, MA, 71. *Prof Exp:* RES BIOLOGIST, NAT INST ENVIRON HEALTH SCI, 71- *Mem:* Am Soc Pharmacol Exp Therapeut; Am Asn Cancer Res. *Res:* Relationships between oncogene activation and susceptibility to carcinogenesis in inbred mice; relationships between DNA alkylation by tobacco specific nitrosamines in certain lung cell types and pulmonary tumorigenesis in mice. *Mailing Add:* Nat Inst Environ Health Sci PO Box 12233 Md-D4-04 Research Triangle Park NC 27709

DEVGAN, SATINDERPAUL SINGH, Rahawan, Punjab, India, Apr 4, 39; US citizen; m 73; c 2. POWER TRANSMISSION & DISTRIBUTION, COMPUTER APPLICATIONS TO POWER SYSTEM ANALYSIS. *Educ:* G N Eng Col, BSc, 61; Ill Inst Technol, MS 65, PhD(elec eng), 70. *Prof Exp:* Elec engr, J C Textile Mills, Pvt Ltd, Phagwara, India, 62-64; proj engr, Scam Instruments Corp, Skokie, Ill, 65-67; from asst prof to assoc prof, 70-78, PROF & HEAD, DEPT ELEC ENG, TENN STATE UNIV, NASHVILLE, 79- *Concurrent Pos:* Dir, Div Student Serv, Sch Eng & Technol, Tenn State Univ, 75-79. *Mem:* Sr mem Inst Elec & Electronics Engrs; Am Soc Eng Educ. *Res:* Electric power system analysis; power distribution system reliability analysis; static VAR compensator applications and engineering design. *Mailing Add:* 846 Percy Warner Blvd Nashville TN 37205

DEVGUN, JAS S, WASTE MANAGEMENT. *Educ:* Panjab Univ, BSc, 72, MSc, 73; Univ NB, PhD(phys chem), 79. *Prof Exp:* Fel chem eng, Univ NB, 79-80; safety eng analyst, Atomic Energy Can Ltd, CANDU Opers, Mississauga, 81-83; scientist & officer, Chalk River Environ Authority, 83-85, sr res scientist, Waste Mgt Technol Div, Chalk River Nuclear Labs, 86-88; proj mgr & environ systs engr, EAIS Div, 88-90, ASSOC DIR, RES & DEVELOP PROG COORD OFF, CHEM TECHNOL DIV, ARGONNE NAT LABS, 91- *Concurrent Pos:* Coordr, Low & Intermediate Level Radioactive Wastes Mgt Working Party, Can, 83-88. *Mem:* Am Health Physics Soc; Am Nuclear Soc; Sigma Xi. *Res:* Environmental impact analysis; radioactive waste disposal; contaminant migration; geochemistry; environmental monitoring; waste technologies. *Mailing Add:* Res & Develop Prog Cord Off Chem Technol Div Argonne Nat Lab 9700 S Cass Ave Argonne IL 60439

DEVILLEZ, EDWARD JOSEPH, b Covington, Ky, Apr 12, 39; m 59; c 6. BIOLOGICAL SCIENCE EDUCATION. *Educ:* Xavier Univ, Ohio, BS, 59; Univ Miami, MS, 61; Univ Ill, Urbana, PhD(physiol), 64. *Prof Exp:* Res assoc NSF fel, Friday Harbor Labs, Univ Wash, 64-65; from asst prof to assoc prof, 65-68, PROF ZOOL, MIAMI UNIV, 72- *Mem:* Am Soc Zoologists; Nat Sci Teachers Asn; Nat Asn Biol Teachers. *Res:* Science education. *Mailing Add:* Dept Zool Miami Univ 288 Biol Sci Bldg Oxford OH 45056

DEVILLIS, JEAN, b Tunis, Tunisia, June 24, 35. DEVELOPMENTAL NEUROBIOLOGY. *Educ:* Univ Calif, Los Angeles, PhD(plant biochem), 62. *Prof Exp:* PROF CELL BIOL, UNIV CALIF, LOS ANGELES, 78-, HEAD DIV NEUROCHEM, 82- *Mem:* Soc Neurosci; Soc Develop Biol; Am Soc Cell Biol; Am Soc Pharmacol & Exp Therapeut; Int Soc Neurochem; Am Soc Neurochem. *Mailing Add:* Mental Retardation Res Ctr Univ Calif 760 Westwood Plaza Los Angeles CA 90024-1759

DEVIN, CHARLES, JR, acoustics, hydrodynamics, for more information see previous edition

DEVINATZ, ALLEN, b Chicago, Ill, July 22, 22; m 52, 56; c 2. MATHEMATICS. *Educ:* Ill Inst Technol, BS, 44; Harvard Univ, AM, 47, PhD(math), 50. *Prof Exp:* Instr math, Ill Inst Technol, 50-52; NSF fel, Inst Advan Study, 52-53, mem, 53-54; asst prof math, Univ Conn, 54-55; from asst prof to assoc prof, Wash Univ, 55-60; NSF sr fel, 60-61; prof, Wash Univ, 61-67; PROF MATH, NORTHWESTERN UNIV, EVANSTON, 67- *Concurrent Pos:* Vis mem, Weizmann Inst, 80, Inst Hautes Etudes Sci, 83 & Inst Appln, Rome, 88; vis scholar, Univ Calif, Berkeley, 85. *Mem:* Am Math Soc. *Res:* Analysis. *Mailing Add:* Dept Math Northwestern Univ Evanston IL 60208

DEVINE, CHARLES JOSEPH, JR, b Norfolk, Va, Feb 23, 23; c 5. SURGERY. *Educ:* Washington & Lee Univ, BA, 43; George Washington Univ Sch Med, MD, 47. *Prof Exp:* Intern, Brady Inst, Johns Hopkins Hosp, 47-48; urol fel, Cleveland Clin, 48-50; urol residency, USN Hosp, Pa, 51-52; chmn dept surg, DePaul Hosp, 65-66; pres med staff, Med Ctr Hosp, Inc, 69-70; PROF & CHMN UROL, EASTERN VA MED SCH, 70-; CHIEF UROL, CHILDREN'S HOSP OF THE KING'S DAUGHTERS, 77- *Concurrent Pos:* Chief urol, Med Ctr Hosp, Inc, 79-80; lectr urol, Navy Med Ctr, 70-, Aukland, Wellington, New Zealand, Melborne & Adelaide, Australia, 82; asst ed, J Urol, 78-; found lectr & sect guest, Royal Australasian Col Surgeons, 82; consult, Proj Hope, Univ Alexandria, Egypt, 77. *Mem:* AAAS; AMA; Am Urol Asn; Am Asn Genitourinary Surgeons; Soc Int D'Urologie. *Res:* Reconstructive surgery of the genitourinary tract. *Mailing Add:* 844 Kempsville Rd Norfolk VA 23510

DEVINE, JOHN EDWARD, b Du Bois, Pa, Feb 14, 23; m 53; c 5. MECHANICAL ENGINEERING. *Educ:* Carnegie Inst Technol, BS, 48. *Prof Exp:* Res engr, Alcoa Labs, Aluminum Co Am, 48-60, construct proj engr, Alcoa Bldg, 60-65, head, Bldg Serv Dept, Alcoa Tech Ctr, 65-67, eng adv, 67-69, mgr, Facil Eng Dept, Alcoa Labs, 70-79, eng mgr, 79-83; RETIRED. *Concurrent Pos:* Consult engr, 83-90. *Mem:* Am Soc Mech Engrs; Sigma Xi; Am Soc Heating, Refrig & Air Conditioning Engrs. *Res:* Quenching of metals; design of laboratory buildings, systems and equipment. *Mailing Add:* 519 Chester Dr Lower Burrell PA 15068

DEVINE, MARJORIE M, b East Machias, Maine, May 19, 34. NUTRITION. *Educ:* Univ Maine, BS, 56, MS, 62; Cornell Univ, PhD(nutrit), 67. *Prof Exp:* Teacher high sch, Conn, 56-58 & high sch, Maine, 58-62; instr food sci, Univ Maine, 62-64; asst prof, 67-73, assoc prof human nutrit & food, 73-78, PROF NUTRIT SCI, CORNELL UNIV, 78-, COORDR UNDERGRAD PROG, 74-, ASSOC DIR ACAD AFFAIRS, DIV NUTRIT SCI, 76- *Res:* Ascorbic acid metabolism; nutrition education. *Mailing Add:* Cornell Univ 334 Martha Van Rensselaer Hall Ithaca NY 14850

DEVINE, THOMAS EDWARD, PLANT GENETICS, PLANT BREEDING. *Educ:* Fordham Univ, BS; Pa State Univ, MS; Iowa State Univ, PhD(plant breeding), 67. *Prof Exp:* Asst prof plant breeding, Cornell Univ, 67-69; RES GENETICIST, AGR RES SERV, USDA, 67- *Honors & Awards:* Res Award, Am Soc Agron, 81. *Mem:* Coun Am Genetics Asn; Crop Sci Soc Am; fel Am Soc Agron; Am Genetic Asn; Sigma Xi. *Res:* Development of resistant disease and insect resistant alfalfa; genetics and breeding of nitrogen fixation by soybeans; interaction of soybean cultivars with nitrogen fixing Rhizobium; extension of the range of adaptation of soybeans to marginal edaphic environments; soybean genetic mapping; coevolution of legume host and rhizobial microsymbiont. *Mailing Add:* 14925 Belle Ami Dr Laurel MD 20707

DEVINEY, MARVIN LEE, JR, b Kingsville, Tex, Dec 5, 29; m 75; c 3. INSTRUMENTAL ANALYSIS. *Educ:* Southwest Tex State Univ, BS, 49; Univ Tex, MA, 52, PhD(phys chem), 56. *Prof Exp:* Develop chemist, Plant Lab, Celanese Corp Am, Tex, 56-58; res chemist, Indust Chem Res Lab, Shell Chem Co, 58-66; sr scientist, United Carbon Co Div, 66-68, mgr phys & anal res, Ashland Chem Co Div, 68-71, mgr phys chem res sect, 71-78, res assoc & supvr appl surface chem, 78-84, RES ASSOC & SUPVR ELECTRON MICROS, ADV COMPOSITES & SPEC CONTRACTS, RES & DEVELOP DIV, ASHLAND OIL, INC, 84- *Concurrent Pos:* Mem sci adv comt, Am Petrol Inst Res Proj, 60; adj prof, Univ Tex, San Antonio, 73-75; mem, chem educ adv comts, Columbus Tech Inst, Ohio, 74-84 & Cent Ohio Tech Col, 75-82; cert prof, chemist, Am Inst Chem, 78; bus mgr, Am Chem Soc, Petrol Chem Div Preprints, 86. *Honors & Awards:* Two Best Paper Award, Rubber Chem Div, Am Chem Soc, 67-70. *Mem:* Am Chem Soc; fel Am Inst Chem; Sigma Xi; NAm Catalysis Soc; Soc Advan Mat Process Eng; Mat Res Soc; Electron Micros Soc Am; Am Soc Composites. *Res:* Surface chemistry; homogeneous and heterogeneous catalysis; electron microscopy; electron spectroscopy for chemical analysis; x-ray diffraction analysis; x-ray energy dispersive spectroscopy; advanced aerospace composites; fiber-polymer interfacial research; rubber and carbon technology; industrial petrochemicals; physical separation methods; physico-chemical measurements; electrochemistry; radiochemical methods; quality improvement. *Mailing Add:* Res & Develop Div Ashland Chem Co PO Box 2219 Columbus OH 43216

DEVINS, DELBERT WAYNE, b Warwick, Okla, Sept 6, 34; m 66; c 1. NUCLEAR PHYSICS. *Educ:* Fresno State Col, BS, 58; Univ Southern Calif, MS, 62, PhD(physics), 64. *Prof Exp:* Vis asst prof physics, Univ Southern Calif, 64-65; from asst prof to assoc prof, 65-76, prof physics, Ind Univ, Bloomington, 76-82; VPRES, COALSCAN, INC, 82- *Concurrent Pos:* Sr sci officer, Rutherford High Energy Lab, Eng, 65-66; vis prof physics, Melbourne Univ, Australia, 78-; vis prof, Alexandria Univ, Egypt, 85. *Mem:* AAAS; Am Phys Soc; Am Asn Physics Teachers. *Res:* Intermediate energy reactions; knock-out, transfer reactions; neutron polarization; sequential decay; reaction mechanisms. *Mailing Add:* Koppers Process Tech PO Box 1824 Baltimore MD 21203

DEVITA, VINCENT T, JR, b Bronx, NY, Mar 7, 35; m 57; c 1. INTERNAL MEDICINE, PHARMACOLOGY. *Educ:* Col William & Mary, BS, 57; George Washington Univ, MD, 61; Am Bd Internal Med, dipl, 68 & 74, cert hemat, 72, cert oncol, 73 & 74. *Hon Degrees:* DSc, Col William & Mary, 82, Ohio State Univ, 83, George Washington Univ, 84, NY Med Col, 87, Georgetown Univ, 89. *Prof Exp:* Intern med, Med Ctr, Univ Mich, 61-62; resident, Gen Hosp, Med Serv, George Washington Univ, 62-63; clin assoc, Lab Chem Pharmacol, Nat Cancer Inst, 63-65; resident med, Yale New Haven Med Ctr, 65-66; sr investr, Med Br, 66-68, chief solid tumor serv, 68-71, chief med br, 71-74, dir, Div Cancer Treatment, 74-80, clin dir, Nat Cancer Prog, Nat Cancer Inst, 75-88; physician-in-chief, 88-91, ATTEND PHYSICIAN, DEPT MED, MEM SLOAN-KETTERING CANCER CTR, 88-, MEM, SLOAN-KETTERING INST, 88- *Concurrent Pos:* Prof, Sch Med, George Washington Univ, 75-89; assoc ed, Jour Nat Cancer Inst, 68-74; sci ed, Cancer Chemother Reports, 70-74; adv ed, J Radiation Oncol, Biol & Physics, 75-; assoc ed, Cancer Clin Trials, 77-; mem bd med/sci consult, Cancer Nursing, 77-; mem bd sci coun, Cancer Chemother & Pharmacol, 77-; assoc ed, Am J Med, 78-81; adv panel mem, WHO, 76-; mem panel consults clin oncol, Int Union Against Cancer, 79-82; mem, Awards Assembly, Gen Motors Cancer Res Found, 81-85; prof med, Med Col, Cornell Univ, New York, 89-; vis physician, Rockefeller Univ Hosp, New York, 89-; pres, US Bd Coord Coun Cancer Res, 89-; mem bd dirs, Found Biomed Res, 89- *Honors & Awards:* Albert & Mary Lasker Med Res Award, 72; Esther Langer Award, 76; Jeffrey Gottlieb Award, 76; Karnofsky Prize, 79; Griffuel Prize, Asn Develop Res Cancer, Paris, 80; James Ewing Award, Soc Surg Oncol, 82; Medal of Honor, Am Cancer Soc, 85; Stratton Lectr, Am Soc Hemat, 85; Surgeon General's Exemplary Serv Medal, 88. *Mem:* Inst Med-Nat Acad Sci; Am Soc Hemat; fel Am Col Physicians; Am Soc Clin Invest; Soc Surg Oncol; Am Asn Cancer Res; Am Soc Clin Oncol; Am Fedn Clin Res; AMA; fel NY Acad Med. *Res:* Pharmacology of anti tumor agents in relation to tumor cell kinetics; chemotherapy of the lymphomas. *Mailing Add:* Mem Sloan-Kettering Cancer Ctr 1275 York Ave New York NY 10021

DEVITO, CARL LOUIS, b New York, NY, Oct 21, 37; m 65; c 1. MATHEMATICAL ANALYSIS. *Educ:* City Col New York, BS, 59; Northwestern Univ, PhD(math), 67. *Prof Exp:* Instr math, DePaul Univ, 65-67; asst prof, 67-71, ASSOC PROF MATH, UNIV ARIZ, 71- *Concurrent Pos:* Invited speaker, Liege Colloquium, Belg, 70; adj prof, Naval Postgrad Sch, 86-88. *Mem:* Am Math Soc; Sigma Xi. *Res:* Theory of locally convex, topological vector spaces, particularly the completions of these spaces for various topologies and the relations among these completions, and study of the weakly compact subsets of these spaces; applications of mathematics to certain aspects of space science. *Mailing Add:* Dept Math Univ Ariz Tucson AZ 85721

DEVITO, JUNE LOGAN, b Alta, Jan 12, 28; m 53; c 4. NEUROANATOMY, NEUROPHYSIOLOGY. *Educ:* Univ BC, BA, 47, MA, 49; Univ Wash, PhD(physiol), 54. *Prof Exp:* Asst embryol & physiol, Univ BC, 47-49; res assoc physiol & biophys, 54-55, actg instr, 55-58, res instr, 58-60, instr, 60-63, res instr neurosurg, 63-65, res staff mem, Regional Primate Res Ctr, 71-78, res asst prof, 65-78, RES ASSOC PROF NEUROL SURG, SCH MED, UNIV WASH, 78- *Concurrent Pos:* USPHS fel, 61-62. *Mem:* AAAS; Am Asn Anat; Soc Neurosci. *Res:* Central pathways subserving weight discrimination; effects of sensory stimulation on activity; supplementary motor area projections; corticothalamic connections of sensory cortices; septo-hippocampal pathways; autonomic nervous system; basal ganglia connections. *Mailing Add:* Dept Biol Structure Univ of Wash Seattle WA 98195

DEVLETIAN, JACK H, b Boston, Mass, May 23, 41; m 69; c 3. WELDING RESEARCH. *Educ:* Univ Mass, BS, 63; Univ Wis, MS, 66, PhD(metall eng), 72. *Prof Exp:* Equip develop engr, Union Carbide Corp, Linde Div, 63-64; mat & standards engr, Raytheon Missile Systems Div, 66-67; welding engr, Rocketdyne Div Rockwell Int, 67-68; assoc prof mat sci, Youngstown State Univ, 72-79; ASSOC PROF MAT SCI, ORE GRAD CTR, 79- *Honors & Awards:* Charles H Jennings Mem Award, Am Welding Soc, 76. *Mem:* Am Welding Soc; Am Foundrymen's Soc. *Res:* Physical metallurgy of weldments and solidification mechanics. *Mailing Add:* Dept Mat Sci & Eng Ore Grad Ctr 19600 NW Von Newman Dr Beaverton OR 97006

DEVLIN, JOHN F, b Rockville Centre, NY, Nov 24, 44. METAL PHYSICS. *Educ:* Harpur Col, State Univ NY, Binghamton, BA, 66; Mich State Univ, MS, 68, PhD(physics), 70. *Prof Exp:* Res assoc, Univ Groningen, Neth, 70-72; staff physicist, Battelle Mem Inst, 72-76; asst prof, 76-82, ASSOC PROF PHYSICS, UNIV MICH, DEARBORN, 82- *Mem:* Am Phys Soc; Am Asn Physics Teachers; Sigma Xi. *Res:* Theoretical solid state physics; structural properties of metals and alloys. *Mailing Add:* Dept Natural Sci Univ Mich Dearborn MI 48128-1491

DEVLIN, JOSEPH PAUL, b Hale, Colo, Jan 11, 35; m 57; c 4. PHYSICAL CHEMISTRY. *Educ:* Regis Col, Colo, BS, 56; Kans State Univ, PhD(phys chem), 60. *Prof Exp:* Res assoc spectros, Univ Minn, 60-61; from asst prof to assoc prof, 61-70, chmn dept, 76-81, PROF PHYS CHEM, OKLA STATE UNIV, 70- *Concurrent Pos:* AEC grant, 66-71; NSF grant, 62-89. *Honors & Awards:* Sigma Xi Lectr, 84. *Mem:* Am Chem Soc. *Res:* Vibrational spectra and structures of disordered solids; matrix idolated high temperature species, solid state charge transfer complexes and proton transfer in protic solids. *Mailing Add:* Dept Chem Okla State Univ Stillwater OK 74074

DEVLIN, RICHARD GERALD, JR, b Philadelphia, Pa, Aug 4, 42; m 65; c 2. IMMUNOLOGY, CLINICAL PHARMACOLOGY. *Educ:* LaSalle Col, BA, 64; Villanova Univ, MS, 66; Univ Md, PhD(immunol), 69. *Prof Exp:* Asst biol, Villanova Univ, 64-66; asst zool, Univ Md, 66-69; NIH fel biol, Univ Pa, 69-71; sr scientist, Mead Johnson & Co, 71-74, sr investr biochem, 74-75; mem staff, Merck Inst Therapeut Res, 75-78; asst dir clin pharmacol, Squibb Inst Med Res, 78-80, assoc dir clin pharmacol, 81-87, dir clin field opers, 87-89; dir maj prod support group, 89, DIR ANTIVIRAL CLIN RES, BRISTOL-MEYERS SQUIBB CO, 90- *Concurrent Pos:* Instr, Pa State Univ, Ogontz, 70; adj asst prof, Ind State Univ, Evansville & assoc fac, Ind Univ Med Sch, 73-75; vis asst prof, dept pharmacol, Rutgers Col Pharm, 80-81. *Mem:* Am Asn Immunologists; AAAS; Can Soc Immunol; Am Col Clin Pharm; Am Asn Clin Pharm & Ther; Am Asn Pharmaceut Scientist; Am Soc Microbiol; Int Soc Immunopharmacol. *Res:* Developmental, transplantation and radiation biology; cell physiology; tumor immunology; immunopharmacology; cellular immunology; clinical pharmacology, cardiovascular, dermatological, anti-infective, anti-arthritic, anti-viral (AIDS). *Mailing Add:* Bristol-Meyers Squibb Co PO Box 4000 Princeton NJ 08543-4000

DEVLIN, ROBERT B, b Denver, Colo, Mar 31, 47; m 75; c 1. BIOLOGY, GENETICS. *Educ:* Univ Tex, El Paso, BS, 69; Univ Va, PhD(biol), 76. *Prof Exp:* Asst prof biol, Emory Univ, 79-86; RES BIOLOGIST, US ENVIRON PROTECTION AGENCY, 86- *Mem:* Soc Develop Biol; Am Soc Cell Biologists; AAAS. *Res:* Factors responsible for control of gene expression during development and differentiation; muscle differentiation and drosophila development as model systems. *Mailing Add:* Health Effect Res Lab US Environ Protection Agency Research Triangle Park NC 27711

DEVLIN, ROBERT MARTIN, b US citizen; Oct 13, 31; m; c 3. PLANT PHYSIOLOGY. *Educ:* State Univ NY Albany, BS, 59; Dartmouth Col, MA, 61; Univ Md, PhD(biol), 63. *Hon Degrees:* Dr, Szczecin Univ, Poland. *Prof Exp:* Asst prof plant physiol, NDak State Univ, 63-65; assoc prof, 65-74, PROF PLANT PHYSIOL, AGR EXP STA, LAB EXP BIOL, UNIV MASS, 74- *Concurrent Pos:* Consult, plant growth regulator & herbicide use & physiol, var utilities & corp; mem bd sci adv, Am Coun Sci & Health; mem bd dir, Coun Agr Sci & Technol. *Mem:* Am Soc Plant Physiologists; Am Soc Hort Sci; Int Weed Sci Soc; Plant Growth Regulator Soc Am (pres, 74-75); fel Weed Sci Soc Am; Am Coun Sci & Health; Coun Agr Sci & Technol. *Res:* Plant hormone effects, herbicide metabolism and mode of action. *Mailing Add:* Dept Plant Physiol Lab Exp Biol Univ Mass Agr Exp Sta East Wareham MA 02538

DEVLIN, THOMAS J, b Pa, Aug 23, 35; m 62; c 3. PHYSICS. *Educ:* La Salle Col, BA, 57; Univ Calif, Berkeley, MA, 59, PhD(physics), 61. *Prof Exp:* Fel, Lawrence Berkeley Lab, 61-62; from instr to asst prof physics, Princeton Univ, 62-67; assoc prof, 67-78, PROF PHYSICS, RUTGERS UNIV, 78- *Concurrent Pos:* Guest scientist, Fermilab, 80-81 & 88-90 & Europ Orgn Nuclear Res, 70-71. *Mem:* Am Phys Soc. *Res:* Experimental high energy physics; hyperon magnetic moments and polarization; CP-violation; colliding beams experiments; detectors; data acquisition systems; online computing with parallel processors. *Mailing Add:* Dept Physics Rutgers Univ PO Box 849 Piscataway NJ 08855-0849

DEVLIN, THOMAS MCKEOWN, b Philadelphia, Pa, June 29, 29; m 53; c 2. BIOCHEMISTRY. *Educ:* Univ Pa, BA, 53; Johns Hopkins Univ, PhD(physiol chem), 57. *Prof Exp:* Asst org chem, Sharples Corp, Pa, 47-49; asst biophys, Johnson Found, Univ Pa, 49-53; res assoc enzyme chem, Merck Inst, 57-61, sect head, Biol Cancer Res, 61-66, dir enzymol, 66-67; PROF & CHMN DEPT BIOL CHEM, HAHNEMANN UNIV SCH MED 67- *Concurrent Pos:* Vis res scientist, Brussels, 64-65; actg dean, Col Allied Health Prof, Hahnemann Univ, 72-74 & 80-81. *Mem:* Soc Exp Biol & Med; Am Soc Biochem & Molecular Biol; Am Asn Cancer Res; Biochem Soc; Am Soc Cell Biol; Biophys Soc. *Res:* Oxidative phosphorylation and electron transport; mitochondrial physiology and biogenesis, calcium transport mechanisms and prostaglandins; biochemistry of membranes; biochemical control mechanisms; tissue anoxia. *Mailing Add:* Dept Biol Chem Hahnemann Univ Sch Med Philadelphia PA 19102

DEVOE, HOWARD JOSSELYN, b White Plains, NY, Dec 10, 32; m 80; c 3. PHYSICAL CHEMISTRY. *Educ:* Oberlin Col, AB, 55; Harvard Univ, PhD(chem), 60. *Prof Exp:* Univ fel, Univ Calif, Berkeley, 60-61; res chemist, Phys Chem Sect, NIMH, 61-68; ASSOC PROF CHEM, UNIV MD, COLLEGE PARK, 68-, DIR LOWER DIV PROG, 88- *Mem:* Am Chem Soc. *Res:* Molecular interactions in nucleic acids; dye aggregation; hydrophobic interactions; theory of optical properties of molecular aggregates and biopolymers; solubility and partitioning of nonpolar solutes. *Mailing Add:* Dept Chem & Biochem Univ Md College Park MD 20742

DEVOE, IRVING WOODROW, b Brewer, Maine, Oct 4, 36; m 60; c 5. MEDICAL MICROBIOLOGY, ELECTRON MICROSCOPY. *Educ:* Aurora Col, BS, 64; Univ Ore, PhD(microbiol), 68. *Prof Exp:* Fel, MacDonald Col, McGill Univ, 68-69; asst prof microbiol, Aurora Col, 69-70; from asst prof to assoc prof microbiol, MacDonald Col, McGill Univ, 70-78; assoc prof & chmn dept microbiol & immunol, fac med, McGill Univ, 78-84; pres, DeVoe-Holbein Can Inc, 82-83, MANAGING DIR, DE VOE-HOLBEIN INT INC, 83-, SR SCIENTIST, DEVOE-HOLBEIN TECHNOL BV, 83- *Concurrent Pos:* Res assoc, Argonne Nat Lab, 69-70; consult shoreline surv, Fisheries Res Bd Can, 71; med scientist, Dept Med & Microbiol, Royal Victoria Hosp, Montreal, 79- *Mem:* AAAS; Am Soc Microbiol; Can Soc Microbiol. *Res:* Bacterial physiology; role of inorganic ions in gram-negative bacterial cell walls; carrier mediated transport; phagocytosis of bacterial pathogens; pathogenesis of meningococcus; role of iron in microbiol infections; removal of toxic, radioactive and precious metals from water. *Mailing Add:* Dept Microbiol & Immunol McGill Univ Fac Med 3775 University St Montreal PQ H3A 2B5 Can

DEVOE, JAMES ROLLO, b Sterling, Ill, Sept 27, 28; m 56; c 3. SPECTROSCOPY. *Educ:* Univ Ill, BS, 50; Univ Minn, MS, 52; Univ Mich, PhD(chem), 59. *Prof Exp:* Consult, Subcomt Radiochem, Nat Acad Sci-Nat Res Coun, 60-61; phys chemist, Anal & Inorg Chem Div, 61-63, chief radiochem sect, Anal Chem Div, 63-66, chief activation anal sect & radiochem anal sect, 66-69, chief tech support group & radiochem anal sect, 69-73, leader, Instrument Develop Group, Ctr Anal Chem, 80-84, CHIEF SPEC ANALYTICAL INSTRUMENTATION SECT, NAT INST STANDARDS & TECHNOL, 73-, CHIEF, INORG ANALYSIS RES, DIV, NAT BUR STANDARDS, 84- *Concurrent Pos:* Mem subcomt low background counting, Nat Acad Sci-Nat Res Coun, 63-69; working comt lunar probe, NASA, 64-65. *Honors & Awards:* Distinguished Serv Award, Nat Bur Standards, 64, Silver Medal, 70. *Mem:* Am Chem Soc. *Res:* Radioisotope techniques in analysis; Mossbauer effect; activation analysis; low level radiation detection; analytical radiochemistry; laboratory automation with digital computers; spectroscopic techniques; laser spectroscopy. *Mailing Add:* 17708 Parkridge Dr Gaithersburg MD 20878-1113

DEVOE, RALPH GODWIN, b New York, NY, May 19, 45; m 76; c 2. PHYSICS. *Educ:* Harvard Col, BA, 67; Univ Chicago, MA, 71, PhD(physics), 77. *Prof Exp:* Res assoc physics, Lawrence Berkeley Labs, Univ Calif, Berkeley, 76-77; res assoc, 77-78, RES STAFF MEM PHYSICS, IBM RES LABS, 78- *Mem:* Am Phys Soc; fel Am Optical Soc. *Res:* High resolution laser spectroscopy of atoms, molecules and solids; trapping and optical cooling of atoms. *Mailing Add:* IBM Almaden Res Ctr K01-801 650 Harry Rd San Jose CA 95120

DEVOE, ROBERT DONALD, b White Plains, NY, Oct 7, 34; m 89; c 2. PHYSIOLOGY, BIOPHYSICS. *Educ:* Oberlin Col, AB, 56; Rockefeller Inst, PhD(biophys), 61. *Prof Exp:* From instr to asst prof physiol, Sch Med, Johns Hopkins Univ, 61-69, dir year I prog, 70-73 & 75-76, assoc prof physiol, 69-83, assoc prof neurosci, 80-83; PROF OPTOM, DEPT VISUAL SCI, SCH OPTOM, IND UNIV, BLOOMINGTON, 83-, CHMN, 85- *Concurrent Pos:* Alexander von Humboldt Found SR US Scientist Award, 73-74; guest prof, Inst Zool, Technische Hochschule, Darmstadt, 77 & 80. *Mem:* AAAS; Soc Neurosci; Asn Res Vision & Ophthal; Sigma Xi. *Res:* Electrophysiology of vision; arthropod vision; visual motion detection. *Mailing Add:* Sch Optom Ind Univ 800 E Atwater Bloomington IN 47405

DE VOLPI, ALEXANDER, b New York, NY, Feb 28, 31; m 55, 78; c 4. REACTOR PHYSICS, NUCLEAR ENGINEERING. *Educ:* Washington & Lee Univ, BA, 53; Va Polytech Inst, MS, 58, PhD(physics), 67. *Prof Exp:* PHYSICIST, ARGONNE NAT LAB, 60- *Mem:* AAAS; Am Phys Soc; Am Nuclear Soc. *Res:* Arms control and treaty verification; nuclear parameters of use in nuclear reactor physics design, especially fission parameters; development of a fast neutron hodoscope used in nuclear reactor safety research; environmental aspects of nuclear power; relationships between science and society. *Mailing Add:* Eng Physics Div Argonne Nat Lab D208 Argonne IL 60439

DEVONS, SAMUEL, b Bangor, NWales, UK, Sept 30, 14; m 38; c 4. SCIENCE EDUCATION. *Educ:* Cambridge Univ, BA, 35, MA, 39, PhD(physics), 39. *Hon Degrees:* MSc, Univ Manchester, 59. *Prof Exp:* Sci officer, Air Ministry, Ministry Supply, 39-45; fel & dir studies, Trinity Col, Cambridge, 46-49; prof physics, Imp Col, Univ London, 50-55; Langworthy prof & dir phys labs, Univ Manchester, 55-60; chmn dept, 63-67, prof, 60-84, EMER PROF PHYSICS, COLUMBIA UNIV, 85- *Concurrent Pos:* Lectr, Cambridge Univ, 46-49, UNESCO tech aide, Mission to Arg, 57; vis prof, Columbia Univ, 59-60; Royal Soc vis prof, Andhra Univ, India, 67-68; dir, Barnard Col-Columbia Univ Hist Physics Lab, vis prof, Barnard Col, 69-; Racah vis prof, Hebrew Univ, Jerusalem, 73-74; bd gov, Weizman Inst Israel, 71- *Honors & Awards:* Rutherford Medal & Prize, Brit Inst Physics & Phys Soc, 70; Rutherford Mem Lectr, Royal Soc, Australia, 89. *Mem:* Fel Am Phys Soc; fel Royal Soc; Brit Inst Physics & Phys Soc (vpres, 53-55); Am Asn Physics Teachers; History Sci Soc. *Res:* Radar; nuclear and elementary particle physics; history of physics. *Mailing Add:* Nevis Lab Columbia Univ PO Box 137 Irvington NY 10533

DEVOR, ARTHUR WILLIAM, b El Paso Co, Colo, Apr 13, 11; m 38; c 2. BIOCHEMISTRY. *Educ:* McPherson Col, BS, 35; Kans State Col, MS, 37; Univ Southern Calif, PhD(biochem), 47. *Prof Exp:* Asst, Kans State Col, 36-41; instr chem, Erie Ctr, Univ Pittsburgh, 41-43; instr, Adelphi Col, 43-45; asst biochem, Univ Southern Calif, 45-46; assoc, Univ SDak, 47; prof chem & head dept, NDak State Teachers Col, Minot, 47-48; from asst prof to assoc prof biochem, SDak State Col, 48-51; from asst prof to prof physiol chem, Col Med, Ohio State Univ, 52-75, emer prof physiol chem, 75-79; prof chem, SDak State Univ, 80-81, dir, Water Resource Lab, 81; RETIRED. *Mem:* AAAS; Am Chem Soc. *Res:* Oxidation of monosaccharides; dehydration of bile acids; blood proteins; chemical education; sulfonated alpha-naphthol as carbohydrate test; dialysis; lyophilization; cerebrosides; sulfonated resorcinol as carbohydrate test; studies on nondialyzable materials in human urine. *Mailing Add:* 47 King Arthur Blvd Westerville OH 43081

DEVOR, KENNETH ARTHUR, b Erie, Pa, Feb 15, 43. BIOCHEMISTRY. *Educ:* Ohio State Univ, BS, 65; Univ Calif, Riverside, MS, 67, PhD(biochem), 70. *Prof Exp:* Fel biochem, Johns Hopkins Univ, 70-72, Inst Genetics, Cologne, WGer, 72-73 & Max Planck Inst Biol, 73-74; asst prof biochem, Calif State Univ, Los Angeles, 74-; AT DEPT CHEM, LE MOYNE-OWEN COL, MEMPHIS. *Mem:* Am Chem Soc; AAAS. *Res:* Structure, synthesis and metabolism of biological membranes; phospholipid metabolism. *Mailing Add:* 980 Blvd Laval No 204 Laval PQ H7S 2K2 Can

DEVORE, GEORGE WARREN, b Laramie, Wyo, Apr 29, 24; m 52; c 2. GEOLOGY. *Educ:* Univ Wyo, BA, 46; Univ Chicago, PhD(geol), 52. *Prof Exp:* From instr to asst prof geol, Univ Chicago, 50-60; assoc prof, Fla State Univ, 60-64, prof geol & chmn dept, 64-79; CONSULT, 88. *Concurrent Pos:* Geologist, US Geol Surv, 48-60. *Mem:* Mineral Soc Am; Am Clay Minerals Soc. *Res:* Geochemistry and petrology; the distribution of elements in minerals and minerals in rocks. *Mailing Add:* Dept Geol Fla State Univ Tallahassee FL 32306

DEVORE, THOMAS CARROLL, b Muscatine, Iowa, Mar 25, 47; m 69; c 2. PHYSICAL CHEMISTRY. *Educ:* Univ Iowa, BS, 69; Iowa State Univ, PhD(phys chem), 75. *Prof Exp:* Asst prof chem, Univ Iowa, 75-76; res chem, Univ Fla, 76-77; from asst prof to assoc prof, 77-90, PROF CHEM, JAMES MADISON UNIV, 90- *Mem:* Am Chem Soc. *Res:* Vibrational infrared and electronic spectrum of high temperature molecules (molecule stable at high temperatures); structure and bonding of high temperature molecules. *Mailing Add:* Dept Chem James Madison Univ Harrisonburg VA 22807

DEVORKIN, DAVID HYAM, b Los Angeles, Calif, Jan 6, 44; m 70; c 1. MUSEUM & PLANETARIUM ADMINISTRATION. *Educ:* Univ Calif, Los Angeles, AB, 66; San Diego State Univ, MS, 68; Yale Univ, MPhil, 70; Univ Leicester, PhD(hist sci), 78. *Prof Exp:* From asst prof to assoc prof astron, Cent Conn State Col, 70-80; from assoc cur to cur, hist astron, 81-84 chmn dept space sci & explor, 84-86, CUR, NAT AIR & SPACE MUS, SMITHSONIAN INST, 86- *Concurrent Pos:* Planetarium dir, Cent Conn State Col, 74-80; res assoc, Ctr Hist Physics, Am Inst Physics, 76-78; chmn, Adv Comt Smithsonian Videohist Prog, 86-; Otto E Neugebauer vis fel, Inst Advan Study, Princeton Univ, 91. *Mem:* Am Astron Soc; Int Astron Union; Hist Sci Soc; Royal Astron Soc; Astron Soc Pac; Soc Hist Technol. *Res:* History of twentieth century astronomy; origins of space science; oral history. *Mailing Add:* Dept Space Hist Nat Air & Space Mus Smithsonian Inst Washington DC 20560

DEVOTO, RALPH STEPHEN, b San Francisco, Calif, Oct 23, 34; m 87; c 3. PLASMA PHYSICS. *Educ:* Mass Inst Technol, SB, 58, SM, 60; Stanford Univ, PhD(aeronaut eng), 65. *Prof Exp:* Asst prof aeronaut eng, Stanford Univ, 66-67 & 69-72; assoc prof mech eng, Ga Inst Technol, 72-74; PHYSICIST, LAWRENCE LIVERMORE NAT LAB, 74- *Concurrent Pos:* NSF fel, Institut fur Plasmaphysik, Munich, 67-69. *Mem:* Am Phys Soc; Am Nuclear Soc; Sigma Xi. *Res:* Confinement and production of plasmas for production power by nuclear fusion. *Mailing Add:* L-644 PO Box 808 Lawrence Livermore Nat Lab Livermore CA 94550

DEVREOTES, PETER NICHOLAS, b Long Branch, NJ, April 22, 48; m 80. CELL-CELL INTERACTIONS, CHEMOTAXIS. *Educ:* Univ Wis, BS, 71; Johns Hopkins Univ, PhD(biophys), 77. *Prof Exp:* Res fel, Univ Chicago, 77-79; from asst prof to assoc prof, 79-87, PROF BIOCHEM, DEPT BIOL CHEM, JOHNS HOPKINS UNIV, 87- *Mem:* Sigma Xi. *Res:* Dictyostelium; discoideum as a model system for investigation of chemotaxis, cell-cell interactions and transmembranesignal transduction. *Mailing Add:* Dept Biol Chem Johns Hopkins Univ 725 N Wolfe St Baltimore MD 21205-2185

DE VRIES, ADRIAAN, b Harderwijk, Neth, June 21, 31; nat US; m 57; c 3. RESEARCH INVOLVING HUMAN SUBJECTS, RADIOACTIVE MATERIAL USAGE. *Educ:* State Univ Utrecht, BS, 52, Drs(chem), 57, PhD(chem), 63. *Prof Exp:* Asst phys chem, State Univ Utrecht, 54-58, scientist x-ray crystallog, 58-59; res assoc, Roswell Park Mem Inst, 59-61; scientist x-ray crystallog & phys chem, State Univ Utrecht, 61-62, scientist first class, 62-65; res assoc, 65-71, sr res fel x-ray studies liquid crystals, liquids & crystals, Liquid Crystal Inst, 71-86, ASST DEAN, DIV RES & SPONSORED PROGS, KENT STATE UNIV, 86- *Concurrent Pos:* Prin investr, Air Force Off Sci Res, 68-74 & NSF, 72-86; adj assoc prof physics dept, Kent State Univ, 75-; chmn, 6th Int Liquid Crystal Conf, 76; vis prof, Univ Utrecht, 77; Ger Acad Exchange serv res fel, 82, radiation safety officer, 87-88, chmn, Human Subj Rev Bd, Kent State Univ, 86- & Radiation Safety Comt, 88- *Honors & Awards:* Unilever Chem Prize, Univ Utrecht, 57. *Mem:* Sigma Xi. *Res:* Study of the structure of liquid crystals, liquids and crystals, mainly through x-ray diffraction techniques; classification of liquid crystals on the basis of structure; characterization of liquid crystal phases. *Mailing Add:* Res Off Kent State Univ Kent OH 44242-0001

DEVRIES, ARTHUR LELAND, b Conrad, Mont, Dec 12, 38; m 68; c 1. COMPARATIVE PHYSIOLOGY, BIOCHEMISTRY. *Educ:* Univ Mont, BA, 60; Stanford Univ, PhD(biol), 68. *Prof Exp:* Assoc res physiologist, Scripps Inst Oceanog, 70-76; asst prof, 76-80, ASSOC PROF PHYSIOL & BIOPHYS, UNIV ILL, URBANA, 80- *Mem:* AAAS; Soc Cryobiol; Am Soc Zoologists; Sigma Xi. *Res:* Cryobiology, especially the role of peptide and glycopeptide antifreezes in the survival of polar fishes; cold adaptation in the lower vertebrates; research known for the discovery of the glycopeptide and peptide antifreeze in cold-water fishes. *Mailing Add:* Dept Physiol & Biophysics Univ Ill 17 Burrill Hall Urbana IL 61801

DEVRIES, DAVID J, b Grand Rapids, Mich, Sept 22, 42; m 65. MATHEMATICS. *Educ:* Calvin Col, AB, 65; Pa State Univ, MA, 66, PhD(math), 69. *Prof Exp:* Asst prof math, Hobart & William Smith Cols, 69-71; asst prof math, Mars Hill Col, 71-74, chmn dept math & physics, 71-76, assoc prof, 74-, coordr info systs, 76-; AT DEPT MATH, GA COL, MILLEDGEVILLE. *Mem:* Am Math Soc; Math Asn Am. *Res:* General prime number theory; algebraic number theory. *Mailing Add:* Dept Math & Comp Sci Ga Col Milledgeville GA 31061

DEVRIES, FREDERICK WILLIAM, b New York, NY, Feb 5, 30; m 59; c 3. CHEMICAL ENGINEERING, ENVIRONMENTAL SCIENCES. *Educ:* Columbia Univ, AB, 49, BS, 50, MS, 51. *Prof Exp:* From jr engr to prod asst, E I du Pont de Nemours & Co, Inc, 51-53, process supvr, 53-55, sr supvr & engr, 56-61, supvr, 61- 64, semiworks supvr, 64-66, eng supvr, 66-71, tech rep, 71-78, tech serv consult, 78-84, sr tech serv consult, 84-90; PVT CONSULT, 90- *Mem:* Am Chem Soc; Am Inst Chem Engrs; Soc Mining Engrs; Sigma Xi; fel SAfrican Inst Mining & Metall; Mining & Metall Soc Am. *Res:* Hydrogen peroxide in mineral processing; recovery of uranium mineral leaching processes; sodium cyanide in mineral processing; process design and economic evaluation; applications research. *Mailing Add:* Chem-Mining Consult Ltd 25 Hillandale Rd Chadds Ford PA 19317

DEVRIES, GEORGE H, b Paterson, NJ, Dec 22, 42; m 65; c 2. NEUROBIOLOGY, NEUROCHEMISTRY. *Educ:* Wheaton Col, BS, 64; Univ Ill, PhD(biochem), 69. *Prof Exp:* Teaching fel neurochem, Albert Einstein Sch Med, 69-71; from asst prof to assoc prof, 72-83, PROF BIOCHEM, MED COL VA, 83- *Concurrent Pos:* Mem study sect, Ment Retardation Res Comt, Nat Inst Child Health & Human Develop, NIH, 85-88, Ment Rev Bd Neurobiol, 90- & Nat Neuroribonucleosis Study Sect, 90- *Honors & Awards:* Javits Neurosci Award, Nat Inst Health. *Mem:* Am Soc Neurochem; Soc Neurosci; Int Soc Neurochem; Am Soc Cell Biol; Am Soc Biol Chemists. *Res:* Molecular basis of neuron-glial interaction; membrane isolation and characterization; membrane mitogens; neural influences in demyelinating disease; insulin receptors in Schulmann cells in diabetic neuropathy. *Mailing Add:* 2329 Tuscora Rd Med Col Va 1101 E Marshall Richmond VA 23235

DE VRIES, JOHN EDWARD, b Fenton, Ill, Oct 4, 19; m 46; c 2. CHEMISTRY. *Educ:* Hope Col, AB, 41; Univ Ill, PhD(chem), 44. *Prof Exp:* Asst chem, Univ Ill, 41-44; res chemist, Manhattan Proj, Stand Oil Co, Ind, 44-46; from asst prof to assoc prof chem, Kans State Univ, 46-51, head gen anal res sect, Naval Ord Test Sta, 51-55; sr chemist & mgr anal serv, Stanford Res Inst, 55-64; prof, 64-84, EMER PROF CHEM, CALIF STATE UNIV, HAYWARD, 84. *Mem:* AAAS; Am Chem Soc. *Res:* Organic analytical reagents; spectroscopy in chemical analysis; spectrophotometry; analytical applications of phenanthrolinium and related compounds; analysis of rocket propellants; nitrogen compounds; polarography; environmental analysis; pesticide residue analysis; gas chromatography. *Mailing Add:* 886 Garland Dr Palo Alto CA 94303

DEVRIES, K LAWRENCE, b Ogden, Utah, Oct 27, 33; m 58; c 2. MECHANICAL ENGINEERING, MATERIALS SCIENCE. *Educ:* Univ Utah, BS, 59, PhD(physics), 62. *Prof Exp:* Design engr hydraulics, Convair Aircraft, Tex, 57; from asst prof to assoc prof, 60-68, chmn dept, 70-81, PROF MECH ENG, UNIV UTAH, 68-, ASSOC DEAN, 84- *Concurrent Pos:* Consult, 60-; dir polymer prog, NSF, 75-76. *Mem:* AAAS; Am Soc Mech Engrs; Am Phys Soc; Soc Exp Stress Anal; Am Chem Soc. *Res:* Mechanical behavior of materials; molecular phenomena associated with deformation and failure; biomedical and dental materials. *Mailing Add:* MEB 2220 Univ Utah Salt Lake City UT 84112

DEVRIES, MARVIN FRANK, b Grand Rapids, Mich, Oct 31, 37; m 59; c 3. COMPUTER AIDED MANUFACTURING. *Educ:* Calvin Col, Mich, BS, 60; Univ Mich, BSME, 60, MSME, 62; Univ Wis, Madison, PhD(mech eng), 66. *Prof Exp:* From instr to assoc prof, 62-77, PROF MECH ENG, UNIV WIS, MADISON, 77- *Concurrent Pos:* Assoc dir, Indust Res Prog, Univ Wis, 75-77; vis prof, Cranfield Inst Technol, Eng, 79-80; prog dir, NSF, 87-90. *Mem:* Soc Mfg Engrs (pres, 85-86); Am Soc Mech Engrs; Int Inst Prod Res. *Res:* Manufacturing processes; material removal processes and computer-aided manufacturing; microscopic aspects of manufacturing; macroscopic approach to manufacturing systems. *Mailing Add:* Dept Mech Eng Univ Wis 1513 University Ave Madison WI 53706

DEVRIES, RALPH MILTON, b Los Angeles, Calif, Apr 16, 44. NUCLEAR PHYSICS. *Educ:* Univ Calif, Los Angeles, BS, 65, PhD(physics), 71. *Prof Exp:* Fel nuclear physics, Ctr Nuclear Studies, Saclay, France, 71-72 & Univ Wash, 73; asst prof nuclear physics, Univ Rochester, 74-76, assoc prof, 76-78; mem staff, Los Alamos Nat Lab, 78-; AT OFF SCI & TECHNOL POLICY, WASHINGTON, DC; VPRES RES & INTERNAT PROGS, OFF RES, UNIV WYO, 85- *Mem:* Am Phys Soc. *Res:* Heavy-ion interactions; instrument development; complex computer calculations of theoretical reaction models; antiproton-nucleus interactions. *Mailing Add:* Univ Wyo PO Box 3355 Laramie WY 82071-3355

DE VRIES, RICHARD N, b Cortland, Nebr, May 24, 32; m 58; c 2. ENVIRONMENTAL ENGINEERING, WATER RESOURCES. *Educ:* Univ Nebr, BS, 58, MS, 63; Utah State Univ, PhD(water resources eng), 69. *Prof Exp:* Draftsman, Lincoln Tel & Tel Co, 54; storage engr, Northern Natural Gas Co, Nebr, 58-60; dist engr & mgr, Sanit Dist 1, Lancaster County, 60-62; asst prof civil eng, Univ Nebr, Lincoln, 63-66, 68-69; assoc prof, 69-77; RETIRED. *Concurrent Pos:* Co-investr, Res Grants, 64-66; prin investr res grants, Univ Nebr, 68-69, Okla State Univ, 69-70 & Okla Water Resources Res Inst, 70-71. *Mem:* Am Soc Civil Engrs; Nat Soc Prof Engrs; Am Geophys Union; Am Water Resources Asn; Am Soc Eng Educ; Sigma Xi. *Res:* Bioenvironmental and water resources engineering; hydrology; hydraulics; urban planning. *Mailing Add:* 406 Hardy Dr Edmond OK 73013

DEVRIES, ROBERT CHARLES, b Evansport, Ohio, Oct 10, 22; m 43; c 5. MINERALOGY. *Educ:* DePauw Univ, BA, 48; Pa State Univ, PhD(mineral), 53. *Prof Exp:* From asst to res assoc mineral, Pa State Univ, 50-54; mem tech staff, Metall & Ceramic Div, Res Lab, Gen Elec Co, 54-61; assoc prof ceramics, Rensselaer Polytech Inst, 61-65; mem tech staff, Inorg Mat Lab, Res & Develop Ctr, Gen Elec Co, 65-88; CONSULT, 88- *Honors & Awards:* Coolidge fel, Res & Develop Ctr, Gen Elec Co, 81. *Mem:* Fel Am Ceramic Soc; fel Am Mineral Soc; Am Chem Soc; AAAS; Am Asn Crystal Growth. *Res:* Phase equilibria studies at high temperature and pressures and crystal-chemical relationships in systems of geologic and ceramic interest; crystal growth; microstructure; property relationships in ceramics; diamond synthesis and characterization. *Mailing Add:* 17 Van Vorst Dr Burnt Hills NY 12027

DEVRIES, RONALD CLIFFORD, b Chicago, Ill, Dec 4, 36; m 63; c 3. ELECTRICAL ENGINEERING, COMPUTER ENGINEERING. *Educ:* Northwestern Univ, BS, 59; Univ Ariz, MS, 62, PhD(elec eng), 68. *Prof Exp:* Coop student, Wells-Gardner & Co, 56-58; jr engr, Data-Stor Div, Cook Elec Co, 59-60; asst, Univ Ariz, 61-64; asst prof elec eng, San Diego State Col, 64-66; from asst prof to assoc prof elec eng, 67-80, PROF ELEC & COMPUT ENG, UNIV NMEX, 80- *Concurrent Pos:* Consult, Sandia Nat Lab, 77- *Mem:* Inst Elec & Electronics Engrs, Computer Soc; Sigma Xi. *Res:* Logic design; computer organization and arithmetic; fault detection and tolerance; nuclear safety and security; automated fault-tree generation. *Mailing Add:* Dept Elec Eng No 124 Univ NMex Albuquerque NM 87131

DEVRIES, YUAN LIN, LYMPHOKINES, GENE EXPRESSION. *Educ:* Univ Calif, Davis, PhD(biochem), 71. *Prof Exp:* PRIN SCIENTIST, CENT RES & DEVELOP, E I DU PONT DE NEMOURS, 82- *Mailing Add:* Dept Biochem & Biophys FDA 8800 Rockville Pike Bldg 29 Rm 111 Bethesda MD 20892

DEW, JOHN N(ORMAN), b Okemah, Okla, Feb 27, 22; m 53; c 5. CHEMICAL ENGINEERING. *Educ:* Univ Okla, BS, 43; Univ Mich, MSE, 49, PhD(chem eng), 53. *Prof Exp:* Res engr, Dept Physics, Univ NMex, 43-46, Res & Develop Div, NMex Sch Mines, 46-47; prod res, Conoco Inc, 53-54, res engr, 54-55, res group leader, 55-57, supvr res engr, 57-61, supvr res scientist, 61-66, asst dir, Prod Res Div, 66-67, asst mgr, 67-68, mgr proj develop, 68-74, dir fuels technol, Res & Eng, 74-78, dir spec proj, Res & Develop Dept, 78-85; CONSULT, 85- *Concurrent Pos:* Asst, Univ Okla, 47-48; res fel, Univ Mich, 49-53; distinguished lectr, Soc Petrol Engrs, 74-75. *Mem:* Am Soc Petrol Engrs; Am Inst Chem Engrs. *Res:* Reaction kinetics; technological forecasting, research planning and coordination, project evaluations; oil shale technology and economics. *Mailing Add:* 800 Dalewood Ave Ponca City OK 74601

DEW, WILLIAM CALLAND, b Belle Valley, Ohio, Dec 30, 16; m 42; c 2. DENTISTRY. *Educ:* Ohio State Univ, DDS, 41. *Prof Exp:* Intern prosthodontics, Ohio State Univ, 41-42, from instr to assoc prof, 42-60, asst dean, Col Dent, 64-70, assoc dean, 70-76, prof dent, Col Dent, 60-, secy, 59-; RETIRED. *Concurrent Pos:* Ed, Dent Alumni Asn Quart, 77-; chmn, Callahan Mem Comn; mem, Boucher Prosthodontic Conf. *Mem:* Pierre Fauchard Acad; Am Dent Asn; Am Col Dent. *Res:* Fixed and removable prosthodontics; operative dentistry; dental materials. *Mailing Add:* 254 Croswell Columbus OH 43214

DEWALD, GORDON WAYNE, b Jamestown, NDak, July 22, 43; m 65; c 3. CYTOGENETICS. *Educ:* Jamestown Col, BS, 65; Univ NDak, MS, 68, PhD(cytogenetics), 72. *Prof Exp:* From res asst to res assoc, Mayo Clin, 72-77, assoc consult, 77-78, consult, 79-, assoc prof, Mayo Med Sch, 80-; AT DEPT NURSING, ROCHESTER COMMUNITY COL, MINN. *Concurrent Pos:* Nat Defense fel, 68-71; March of Dimes, Basil O'Connor Starter Res Grant, 74-77; NIH, Res Career Develop Grant. *Mem:* Sigma Xi; Am Soc Human Genetics. *Res:* Chromosomes in malignant pleural effusions and in hematologic disorders: leukemics and lymphomas; development of improved methods in laboratory cytogenetics. *Mailing Add:* Cytogenetic Lab Hilton 354 Rochester MN 55901

DEWALD, HORACE ALBERT, b Emlenton, Pa, Oct 25, 22; m 55; c 6. ORGANIC CHEMISTRY. *Educ:* Allegheny Col, BS, 44; Univ Ill, PhD(chem), 50. *Prof Exp:* Chemist, Eastman Kodak Co, 44-46; res chemist, Gen Aniline & Film Corp, 50-52; Parke Davis & Co fel med chem, Mellon Inst, 52-57; from res chemist to sr res chemist, 57-72, SR RES ASSOC WARNER-LAMBERT/PARKE-DAVIS, 72- *Honors & Awards:* Excellence in Indust Chem Res Award, Am Chem Soc, Univ Mich Sect, 73. *Mem:* Am Chem Soc. *Res:* Addition of Grignard reagents to olefinic hydrocarbons; synthesis of new detergents; preparation of potential chemotherapeutic agents; anti-cancer, anti-fertility; central nervous system. *Mailing Add:* 241 Woodhave Dr Pittsburgh PA 15228-1550

DEWALD, ROBERT REINHOLD, b Twining, Mich, Aug 31, 35; m 63; c 1. PHYSICAL CHEMISTRY. *Educ:* Cent Mich Univ, BS, 58; Mich State Univ, PhD(chem), 63. *Prof Exp:* From asst prof to assoc prof, 65-77, PROF CHEM, TUFTS UNIV, 77- *Mem:* Am Chem Soc. *Res:* Kinetics of fast reactions in solution; properties of metal nonaqueous solutions. *Mailing Add:* 149 Horn Pond Brook Rd Winchester MA 01890

DE WALL, GORDON, b Muskegon, Mich, Feb 6, 41; m 63; c 3. ORGANIC CHEMISTRY. *Educ:* Calvin Col, BS, 62; Univ Mich, MS, 64, PhD(chem), 67. *Prof Exp:* Res chemist, 67-70, res mgr, 70-72, dir prod, 72-77, vpres prod, 77-86, DIR MFR, BAXTER BURDICK & JACKSON, DIV, 86- *Concurrent Pos:* Mem res & develop exec comt, Nat Safety Coun, 73-80. *Mem:* Am Chem Soc. *Res:* Synthesis of organic compounds; small chemical plant operations. *Mailing Add:* PO Box 256 Fruitport MI 49415-0256

DEWALL, RICHARD A, b Appleton, Minn, Dec 16, 26; c 3. THORACIC SURGERY, CARDIOVASCULAR SURGERY. *Educ:* Univ Minn, BA, 49, BS, 50, MB, 52, MD, 53, MS, 61. *Hon Degrees:* DSc, Wright State Univ, Dayton, 84. *Prof Exp:* From instr to asst prof surg, Univ Minn, 59-62; prof & chmn dept, Chicago Med Sch & Mt Sinai Hosp, 62-66; chief surg, Cox Heart Inst, 66-75; coordr surg residency, Kettering Hosp, 66-75; CLIN PROF SURG, WRIGHT STATE UNIV, 75- *Concurrent Pos:* Estab investr, Am Heart Asn, 60-65. *Mem:* AMA; Am Col Surgeons; Soc Univ Surg; Am Asn Thoracic Surgeons. *Res:* Open heart surgery. *Mailing Add:* 421 Thornhill Rd Dayton OH 45419

DEWALLE, DAVID RUSSELL, b St Louis, Mo, June 18, 42; m 65; c 3. FOREST MICROCLIMATOLOGY, WATERSHED MANAGEMENT. *Educ:* Univ Mo-Columbia, BS, 64, MS, 66; Colo State Univ, PhD(watershed mgt), 69. *Prof Exp:* From asst prof to assoc prof, 69-80, chmn, Forest Sci Prog, Sch Forest Resources, 80-83, PROF FOREST HYDROL, PA STATE UNIV, 80- *Concurrent Pos:* Vis prof, Pingree Park Campus, Colo State Univ, 78 & 81; consult, US Dept Interior, 78, US Forest Serv, 84 & US Environ Protection Agency, 85; vis fel, Sch Forestry, Univ Canterbury, NZ, 79; assoc ed, Water Resources Bull, 85- *Mem:* Am Geophys Union; Soc Am Foresters; Am Meteorol Soc; Am Water Resources Asn. *Res:* Tracing stream flow sources with natural isotopes, tree ring chemistry and environmental pollution, stream water quality and trout. *Mailing Add:* Sch Forest Resources Pa State Univ University Park PA 16802

DEWAMES, ROGER, b Menin, Belg, Dec 9, 31; US citizen; m 56; c 6. SOLID STATE PHYSICS. *Educ:* St Mary's Univ, Tex, BS & BA, 56; Tex Christian Univ, MA, 58; Tex A&M Univ, PhD(physics), 61; Pepperdine Univ, MBA, 81. *Prof Exp:* Mem tech staff theoret physics, Sci Ctr, NAm Aviation, Inc, 61-74, prog develop mgr, Sci Ctr, 74-75, dir physics & chem, Rockwell Int Corp, 75-, dir res & technol, 81-; PRIN SCIENTIST, SCI CTR, ROCKWELL INT, THOUSAND OAKS, CA, 85- *Mem:* Fel Am Phys Soc; Sigma Xi. *Res:* Particle scattering; lattice dynamics; magnetism; molecular spectroscopy; collective excitation in solids; signal processing; semiconductors; superconductors. *Mailing Add:* Sci Ctr Rockwell Int 1049 Camino Dos Rios PO Box 1085 Thousand Oaks CA 91360

DEWAN, EDMOND M, b Forest Hills, NY, Feb 17, 31; m 59; c 2. THEORETICAL PHYSICS, APPLIED MATHEMATICS. *Educ:* Duke Univ, BS, 53; Yale Univ, PhD(theoret physics), 57. *Prof Exp:* Physicist, Microwave Physics Lab, Air Force Cambridge Res Labs, 59-63; adj asst prof biol, Brandeis Univ, 63; theoret physicist, Data Sci Lab, USAF Cambridge Res Labs, 64-72, theoret physicist, Aeronomy Div, 72-86, 0PTICAL PHYSICIST, AIR FORCE GEOPHYS LAB, HANSCOM AFB, 86- *Concurrent Pos:* Res asst, Yale Univ, 53-57; res assoc, Brandeis Univ, 61-63; consult psychiat res, Mass Gen Hosp, Boston, 63-64, res assoc, 64; res assoc, Harvard Med Sch, 64. *Honors & Awards:* Aerospace Educ Found 1st Prize, Nat Air Force Symp Sci, 63. *Mem:* AAAS; Inst Elec & Electronics Engrs; Am Geophys Union; Am Soc Cybernet (vpres, 68-); Sigma Xi. *Res:* Plasma physics; special relativity; mathematics of nonlinear oscillations and applications to physical and biological oscillations; theory of sleep and rapid eye movement state; electroencephalographic analysis; control of human ovulation cycles by photic stimulation; atmospheric physics; vertical transport in the stratosphere; turbulence and waves in stratified media; atmospheric gravity waves; mesospheric airglow structure. *Mailing Add:* GL OPS Hanscom Field Bedford MA 01731

DEWAN, SHASHI B, b Punjab, India, Apr 16, 41. POWER ELECTRONICS. *Educ:* Univ Roorkee, MEE, 62; Univ Toronto, MSc, 64, PhD(elec eng), 66. *Prof Exp:* PROF ELEC ENG, UNIV TORONTO, 66-; PRES, INVERPOWER CONTROLS, 80- *Concurrent Pos:* Consult power electronics, 65- *Honors & Awards:* Bill Newell Power Electronics Award, Inst Elec & Electronics Engrs, 79. *Mem:* Fel Inst Elec & Electronics Engrs. *Mailing Add:* Dept Elec Eng Univ Toronto 35th St George St Toronto ON M5S 1A1 Can

DEWANJEE, MRINAL K, b Chittagong, Bangladesh, Mar 19, 41; US citizen. NUCLEAR MEDICINE. *Educ:* Comilla Victoria Col, Bangladesh, BSc, 60; Dacca Univ, Bangladesh, MSc, 63; McGill Univ, Can, PhD(nuclear & radiochem), 67. *Prof Exp:* Res assoc radiol, Harvard Med Sch, 71-73; assoc prof, Tufts Med Sch, New Eng Med Ctr, 73-76; dir, radiopharmaceut lab & nuclear med res, Mayo Clin, 76-83; prof lab med, Mayo Med Sch, 83-87; PROF RADIOL & DIR, RADIOPHARMACEUT LAB, UNIV MIAMI-JACKSON MEM HOSP, 87- *Concurrent Pos:* Mem, Radiopharmaceut Sci Coun, Soc Nuclear Med; consult, Vet Admin Med Ctr, Miami. *Mem:* Am Chem Soc; Am Phys Soc; Soc Nuclear Med; AAAS; Sigma Xi; Am Heart Asn; Am Soc Art Int Orgns. *Res:* Diagnosis of cardiovascular diseases and development of thromboresistant biomaterials for development of cadiovascular prostheses by immobilization of platelet-active drugs; use of radiolabeled monoclonal antibodies for noninvasive imaging of tumor in experimental models and patients. *Mailing Add:* Dept Radiol D-57 Univ Miami PO Box 016960 Miami FL 33101

DEWAR, GRAEME ALEXANDER, b Hawkesbury, Ont, Jan 28, 51; m 81; c 2. MAGNETISM, MAGNETO-ELASTIC EFFECTS. *Educ:* Bishop's Univ, BSc, 72; Simon Fraser Univ, PhD(physics), 80. *Prof Exp:* Instr physics, Princton Univ, NJ, 77-81; vis asst prof physics, Univ Miami, 81-82, from asst prof to assoc prof, 82-89; ASSOC PROF PHYSICS, UNIV NDAK, 89- *Concurrent Pos:* Consult, Airtron Div Litton Industs, 81-82. *Honors & Awards:* Gov Generals Medal, Bishops Univ-Govt Can, 72. *Mem:* Am Phys Soc. *Res:* Development of a new method for measuring the magnetization of thick magnetic films; elucidation of the nature of certain interactions between magnons and phonons in ferromagnetic materials. *Mailing Add:* Physics Dept Box 8008 Univ NDak Grand Forks ND 58202

DEWAR, MICHAEL JAMES STEUART, b Ahmednagar, India, 18; US citizen; m 44; c 2. CHEMISTRY. *Educ:* Oxford Univ, BA, 40, PhD(chem), 42, MA, 43. *Prof Exp:* Imp Chem Industs res grant, Oxford Univ, 42-45 & fel, 45; phys chemist, Courtaulds Ltd, 45-51; prof chem & head dept, Queen Mary Col, London, 51-59; prof, Univ Chicago, 59-63; Robert A Welch prof chem, Univ Tex, Austin, 63-90; GRAD RES PROF CHEM, UNIV FLA, GAINESVILLE, 90- *Concurrent Pos:* William Pyle Philips vis, Haverford Univ, 64, 70; vis prof, Yale Univ, 57; Arthur D Little vis prof, Mass Inst Technol & Marchon vis lectr, Newcastle-upon-Tyne, Eng, 66; Kharasch vis prof, Univ Chicago, 71; Firth vis prof, Univ Sheffield, Eng, 72; Five-Col lectr, Mass, 73; spec lectr, Univ London, 74; consult, Monsanto Chem Ltd, Eng, 54-59 & Monsanto Co, US, 59-; Distinguished Bicentennial prof, Univ Utah, 76; res scholar lectr, Drew Univ, 84. *Honors & Awards:* Reilly lectr, Univ Notre Dame, 51; Tilden lectr, Brit Chem Soc, 54; Howe Award, Am Chem Soc, 62, SW Regional Award, 78, James Flach Norris Award, 84, William H Nichols Award, 86 & Auburn-Kosolapoff Award, 88; Falk Plaut lectr, Columbia & Daines Mem lectr, Univ Kans, 63; Glidden Co lectr, Western Reserve Univ 64 & Kent State Univ, 67; Gnehm lectr, Eidg Technische Hochschule, Zurich, 68; Barton lectr, Univ Okla, 69; Kahlbaum lectr, Univ Basel, 70; Benjamin Rush lectr, Univ Pa & Venable lectr, Univ NC, Chapel Hill, 71; Foster lectr, State Univ NY, Buffalo, 73; Sprague lectr, Univ Wis, 74; Bircher lectr, Vanderbilt Univ, 76; Pahlavi lectr, Iran, 77; Michael Faraday lectr, Northern Ill Univ, 77; Robert Robinson Medal & lectr, The Chem Soc, 74; Priestly lectr, Pa State Univ, 81; Davy Medal, Royal Soc, 82; J Clarence Karcher lectr, Univ Okla, 84; Coulson lectr, Univ Ga, 88; Tetrshedron Prize, 89. *Mem:* Nat Acad Sci; Am Chem Soc; fel Am Acad Arts & Sci; Brit Chem Soc (hon secy, 57-59); fel Royal Soc. *Res:* Interpretation of structure and chemical reactivity in terms of fundamental physical theory; organic, inorganic, physical and theoretical chemistry. *Mailing Add:* Dept Chem Univ Fla Gainesville FL 32611

DEWAR, NORMAN ELLISON, b Rochester, NY, Nov 14, 30; m 55; c 3. MICROBIOLOGY, BIOCHEMISTRY. *Educ:* Syracuse Univ, BS, 52; Purdue Univ, MS, 55, PhD(microbiol, biochem), 59. *Prof Exp:* Head div microbiol, Vestal Labs, Inc, 59-62, dir res, Vestal Div, 62-69, vpres res, Vestal Div, 69-76, VPRES & TECH DIR, VESTAL DIV, CHEMED CORP, 76- *Concurrent Pos:* Chmn disinfectant & sanitizers div, Chem Specialties Mfrs Asn, 71-72; mem antimicrobial prog adv comt, Environ Protection Agency, 73-74. *Mem:* Fel Royal Soc Health; Am Soc Testing & Mat; Am Soc Microbiol; Soc Indust Microbiol; Am Chem Soc. *Res:* Physiology and biochemistry of microorganisms; chemoautotrophic carbon dioxide assimilation; development of environmental biocidal agents; environmental microbiology. *Mailing Add:* 9522 Benchmark Lane Blue Ash OH 45242

DEWAR, ROBERT LEITH, b Melbourne, Australia, Mar 1, 44; m 69; c 1. THEORETICAL PLASMA PHYSICS. *Educ:* Univ Melbourne, BSc, 65, MSc, 67, PhD(astrophys sci, plasma physics), 70. *Prof Exp:* Fel, Ctr Theoret Physics, Dept Physics & Astron, Univ Md, 70-71; res assoc, Plasma Physics Lab, Princeton Univ, 71-73; res fel, Dept Theoret Physics, Res Sch Phys Sci, Australian Nat Univ, 74-76, sr res fel, 76-77; mem prof res staff, Plasma Physics Lab, Princeton Univ, 77-79, res physicist, 79-81, prin res physicist, 81-82; SR FEL, DEPT THEORET PHYSICS, RES SCH PHYS SCI ENG, AUSTRALIAN NAT UNIV, 82- *Mem:* Fel Am Phys Soc; fel Australian Inst Physics. *Res:* Theoretical plasma physics; hydromagnetic stability; canonical transformations; nonlinear phenomena and turbulence. *Mailing Add:* Dept Theoret Physics Res Sch Phys Sci Eng Australian Nat Univ GPO Box 4 ACT 2601 Canberra Australia

DEWART, GILBERT, b New York, NY, Jan 14, 32. SEISMOLOGY, GLACIOLOGY. *Educ:* Mass Inst Technol, BS, 53, MS, 54; Ohio State Univ, PhD(geol), 68. *Prof Exp:* Seismologist, US-Int Geophys Yr Antarctic Exped, Arctic Inst NAm, Calif Inst Technol, 56-58, res engr, Seismol Lab, 58-59, exchange scientist with Soviet Antarctic Exped, 59-61, res engr, Seismol Lab, 61-63; prin investr seismol gravimetry, Inst Polar Studies, Ohio State Univ, 64-70, res assoc, 68-71, adj asst prof geol, 69-71; geophys consult & mining geologist, 72-79; PRES & FOUNDER, ESD GEOPHYSICS, 80- *Concurrent Pos:* Disaster consult, Am Red Cross, 85-; geol lectr, Calif State Univ, Northridge, 87-88. *Honors & Awards:* Antarctic Serv Cert, US Nat Comt, Int Geophys Yr, 56-58, 59-61. *Mem:* AAAS; Soc Explor Geophys; Seismol Soc Am; Am Geophys Union; Am Inst Prof Geologists; Int Glaciological Soc. *Res:* Seismic study of earth's crust; seismic effects of underground nuclear explosions; seismic properties of glacier ice and other anisotropic media; gravity and magnetic surveys in polar regions; geotechnical and glaciological applications of acoustic emission; microseismic studies of Yellowstone caldera. *Mailing Add:* PO Box 331 Pasadena CA 91102

DE WEER, PAUL JOSEPH, b Avelgem, Belg, July 15, 38; m 65; c 3. BIOPHYSICS, PHYSIOLOGY. *Educ:* Cath Univ Louvain, BS, 59, MD, 63, MS, 64; Univ Md, Baltimore, PhD(biophys), 69. *Prof Exp:* Res assoc endocrinol, Med Sch, Cath Univ Louvain, 63-65; from instr to asst prof biophys, Sch Med, Univ Md, Baltimore, 70-73; assoc prof, 73-78, PROF CELL BIOL & PHYSIOL, SCH MED, WASHINGTON UNIV, 78- *Concurrent Pos:* NIH res career develop award, 70-73; assoc ed, Am J Physiol, 76-81; Sen Jacob Javits Award, 84-; physiol study sect mem, NIH, 84-88. *Mem:* Biophys Soc; Am Physiol Soc; Soc Gen Physiol (pres, 84-85); fel AAAS. *Res:* Active transport of ions through cell membranes. *Mailing Add:* Dept Physiol Univ Pa Sch Med 4-D Richards Bldg Philadelphia PA 19104-6085

DEWEES, ANDRE AARON, b Herrin, Ill, Feb 17, 39; m 63; c 3. POPULATION GENETICS. *Educ:* Southern Ill Univ, BA, 61, MA, 63; Purdue Univ, PhD(genetics), 68. *Prof Exp:* From asst prof to assoc prof, 61-75, PROF BIOL, SAM HOUSTON STATE UNIV, 75- *Mem:* AAAS; Genetics Soc Am; Biomet Soc; Sigma Xi. *Res:* Insect population genetics; role of crossing over between linked genes in evolution. *Mailing Add:* Life Sci Dept Sam Houston State Col Huntsville TX 77341

DEWEESE, DAVID D, b Columbus, Ohio, Mar 16, 13; m 38, 75; c 2. OTOLARYNGOLOGY, SURGERY. *Educ:* Univ Mich, AB, 34, MD, 38; Am Bd Otolaryngol, dipl, 43. *Prof Exp:* Resident & instr otolaryngol, Med Sch, Univ Mich, 42-44; chmn dept, Portland Clin, 44-62; prof & chmn dept, 62-79, EMER PROF OTOLARYNGOL, MED SCH, UNIV ORE, 79- *Concurrent Pos:* Asst examr, Am Bd Otolaryngol, 57-59, mem bd dirs, 60-79, pres, 72-76; mem res training comt communicative dis, NIH, 63-67; mem adv coun, Nat Inst Neurol Dis & Stroke, 68-69; mem bd regents, Maryhurst Col. *Honors & Awards:* Am Acad Ophthal & Otolaryngol Award, 59. *Mem:* Am Acad Ophthal & Otolaryngol; Am Otol Soc; Am Laryngol, Rhinol & Otol Soc (pres, 74-75); Am Laryngol Asn; Am Broncho-Esophagol Asn. *Res:* Deafness and hearing loss; vertigo and dizziness; diseases of ear, throat and nose. *Mailing Add:* 1200 SW 61st Dr Portland OR 97221

DEWEESE, JAMES A, b Kent, Ohio, Apr 5, 25; m 50, 62; c 6. SURGERY. *Educ:* Univ Rochester, MD, 49. *Prof Exp:* From instr to assoc prof surg, Med Ctr, Univ Rochester, 55-69, dir surg res, 58-62, chmn div cardiothoracic surg, 75-91, chmn sect, vasc surg, 83-90, PROF SURG, MED CTR, UNIV ROCHESTER, 69-, SURGEON, 68- *Concurrent Pos:* Asst surgeon, Strong Mem Hosp, 56-58, assoc surgeon, 58-75, surgeon, 75-; consult, Rochester Gen Hosp, 59- & Batavia & Bath Vet Hosps, 63- *Mem:* fel Am Col Surgeons; Soc Vascular Surg (pres, 77-78); Int Cardiovasc Soc (pres, 83-84); Sigma Xi. *Res:* Cardiovascular diseases; hypothermia in cardiac surgery; venous thrombosis, phlebography, arterial reconstructions and venous reconstructions. *Mailing Add:* 601 Elmwood Ave Rochester NY 14642

DEWEESE, MARION SPENCER, b Corydon, Ind, Aug 17, 15; m 41; c 3. SURGERY. *Educ:* Kent State Univ, AB, 35; Univ Mich, MD, 39, MS, 48. *Prof Exp:* From instr to asst prof surg, Univ Mich, 48-51; pvt pract, Calif, 51-53; assoc prof surg, Univ Mich, 53-64; prof surg & chmn dept, Sch Med, Univ Mo-Columbia, 64-74; CLIN PROF SURG, UNIV MICH, ANN ARBOR, 74-; AT ST JOSEPH MERCY HOSP, 74- *Concurrent Pos:* Chief surg serv, Ann Arbor Vet Admin Hosp, 53-56; consult, 56-64; mem, Am Bd Surg; pvt pract surg, Ann Arbor, 74-; chief staff, St Joseph Mercy Hosp, Ann Arbor, 79-81. *Mem:* AMA; Am Surg Asn; fel Am Col Surgeons. *Res:* Vascular disease; diseases of aorta and peripheral arteries; thromboembolism; thermal injury and wound healing. *Mailing Add:* 2229 Glendaloch Rd Ann Arbor MI 48104

DEWERD, LARRY ALBERT, b Milwaukee, Wis, July 18, 41; m 63; c 3. MEDICAL PHYSICS, SOLID STATE PHYSICS. *Educ:* Univ Wis-Milwaukee, BS, 63; Univ Wis-Madison, MS, 65, PhD(physics, radiol), 70. *Prof Exp:* Res assoc mat sci, Univ Wash, 70-72, res asst prof dosimetry, 72-75; vis asst prof, Ctr Radiol Physics, Univ Wis-Madison, 75-76, clin asst prof med, physics, 76-79, clin assoc prof med physics & dir, Midwest Ctr Radiol Physics & Accredited & Dosimetry Calibration Lab, 79-86, mgr prod develop, RMI, 87-90, CLIN PROF, UNIV WIS-MADISON, 90- *Concurrent Pos:* Mem screening & diag, Wis Coun Cancer Control, 77-; mem, Radiotherapy, Mammography, Chamber Calibration Tech Group, 78-; mem mammography QA, Am Cancer Soc, 83-89. *Mem:* Am Asn Physicists Med; Health Physics Soc; Am Phys Soc; Soc Photo-Optical Instrumentation Engrs; Am Asn Physics Teachers. *Res:* Radiological systems in diagnostic radiology-medical physics; electrophotographic imaging; mammography; solid state dosimetry; luminescence and physics of the solid state. *Mailing Add:* 13 Pilgrim Circle Madison WI 53706

DEWET, JAN M J, genetics, for more information see previous edition

DE WETTE, FREDERIK WILLEM, b Bussum, Neth, June 29, 24; m 52; c 3. SOLID STATE PHYSICS. *Educ:* State Univ Utrecht, Drs, 50, Dr(theoret physics), 59. *Prof Exp:* Res assoc, State Univ Utrecht, 50-52; vis lectr physics, Brown Univ, 52-53; res assoc chem, Univ Md, 53-55; asst prof, State Univ Utrecht, 55-60; res asst prof physics, Univ Ill, 60-62; res physicist, Neth Reactor Ctr, 62-63; resident res assoc, Solid State Sci Div, Argonne Nat Lab, 63-65; chmn, Dept Physics, 69-74, assoc dean, Col Nat Sci, 76-80, PROF PHYSICS, UNIV TEX, AUSTIN, 65- *Concurrent Pos:* Consult, Argonne Nat Lab, 65-72; trustee, Argonne Univs Asn, 72-74; Kramers prof, State Univ Utrecht, 74-75; vis scientist, Max-Planck Inst Solid State Res, Stuttgart, W Ger, 83-84. *Honors & Awards:* Humboldt Award, 83-84. *Mem:* Fel Am Phys Soc; European Phys Soc. *Res:* Surface physics, dynamical, thermodynamical and scattering properties of crystals and crystal surfaces. *Mailing Add:* Dept Physics Univ Tex Austin TX 78712

DEWEY, BRADLEY, JR, b Pittsburgh, Pa, Apr 10, 16; m 40; c 5. CHEMICAL ENGINEERING. *Educ:* Harvard Univ, BS, 37; Mass Inst Technol, ScD(chem eng), 41. *Prof Exp:* Indust res, Dewey & Almy Chem Co, 40-43, dir develop dept, 45-50, vpres, 50-56; pres, Cryovac Div, W R Grace & Co, 56-64, sr vpres chem group, 64-70; pres & treas, Thermal Dynamics Co, 70-80; RETIRED. *Mem:* Am Chem Soc; Am Inst Chem Engrs. *Res:* Creaming of latex; colloid chemistry. *Mailing Add:* 43 Occom Ridge Hanover NH 03755

DEWEY, C(LARENCE) FORBES, JR, b Pueblo, Colo, Mar 27, 35; m 63. FLUID MECHANICS, LASER PHYSICS. *Educ:* Yale Univ, BE, 56; Stanford Univ, MS, 57; Calif Inst Technol, PhD(aeronaut), 63. *Prof Exp:* Mem tech staff aerodyn, Aeronutronic Div, Philco Corp, 57-59; NSF fel aeronaut, Calif Inst Technol, 59-63; asst prof aerospace sci, Univ Colo, 63-68; assoc prof mech eng, 68-76, PROF MECH ENG, MASS INST TECHNOL, 76- *Concurrent Pos:* Consult, Rand Corp, 60-76, Swissident, Inc, 64-65, Adv Group Aerospace Res & Develop, NATO, 66-67, Eng & Develop Co Colo, 66-68, Xenon Corp, 69-, Sausum Clin, 70 & others; lectr, Univ Calif, Los Angeles, 64; vis scientist, Inst Plasmaphysik, Ger, 66-67; chmn bd, Sensoresearch Corp, 72-79; trustee, Cardiovasc Trust of Boston, 74-80; consult med, Mass Gen Hosp, 76-80; vis prof path, Harvard Med Sch, 78-79; mem bd, Concurrent Comput Corp, 81- *Mem:* Am Phys Soc; Am Inst Aeronaut & Astronaut. *Res:* Gas physics; ionization and collision phenomena; applied laser technology using wavelength-tunable lasers; fluid mechanics in biomedical engineering, including noninvasive diagnostic techniques; intelligent systems; endothelial cell biology; image analysis applied to cell biology. *Mailing Add:* Dept Mech Eng Rm 3-250 Mass Inst Technol Cambridge MA 02139

DEWEY, DONALD HENRY, b Geneva, NY, Apr 25, 18; m 47; c 2. HORTICULTURE, POMOLOGY. *Educ:* Cornell Univ, BS, 39, PhD(veg crops), 50. *Prof Exp:* Jr olericulturist, Cheyenne Hort Field Sta, USDA, 39-45, horticulturist, Fresno Lab, 45-51; actg chmn dept, 77-78, prof, 51-83, EMER PROF HORT, MICH STATE UNIV, 83- *Concurrent Pos:* Orgn Europ Econ Coop sr vis fel, Ditton Lab, Eng, 60; vis prof, Technion Univ, Israel, 80, Dookie Col, Victoria, Australia. *Mem:* Fel Am Soc Hort Sci; Int Soc Hort Sci. *Res:* Harvesting; handling and storage of fruits and vegetables; controlled atmosphere storage of fruit; physiology of maturation, ripening and senescence. *Mailing Add:* 1423 Lakeside East Lansing MI 48823-2454

DEWEY, DONALD O, b Portland, Ore, July 9, 30; m 52; c 3. CONSTITUTIONAL HISTORY. *Educ:* Univ Ore, BA, 52; Univ Utah, MS, 54; Univ Chicago, PhD(hist), 60. *Prof Exp:* Managing ed, Condon Globe Times, Ore, 52-53; city ed, Ashland Daily Tidings, Ore, 53-54; assoc ed, Papers James Madison, 57-62; dean lett & sci, 70-84, PROF HIST, CALIF STATE UNIV, LOS ANGELES, 62-, DEAN, NATURAL & SOCIAL SCI, 84- *Res:* Constitutional history. *Mailing Add:* Calif State Univ 5151 State University Dr Los Angeles CA 90032

DEWEY, DOUGLAS R, b Brigham City, Utah, Oct 23, 29; m 49; c 3. CYTOGENETICS, PLANT BREEDING. *Educ:* Utah State Univ, BS, 51, MS, 54; Univ Minn, PhD(plant genetics), 56. *Prof Exp:* Res agronomist, 56-62, res geneticist, US Dept Agr, 62-; res prof plant sci, Utah State Univ, Logan, 90; RETIRED. *Mem:* Am Soc Agron; Bot Soc Am; Crop Sci Soc Am. *Res:* Genetics, cytology, interspecific and intergeneric hybridization of species of the Triticeae tribe of grasses. *Mailing Add:* 809 N 400 E Logan UT 84321

DEWEY, FRED MCALPIN, b Akron, Ohio, Sept 9, 39; m 59; c 10. ORGANIC CHEMISTRY. *Educ:* Colo State Univ, BS, 61; Univ Colo, PhD(chem), 65. *Prof Exp:* Res chemist, Harry Diamond Labs, DC, 65-67 & Air Force Rocket Propulsion Lab, Calif, 67-68; from asst prof to assoc prof, 68-78, PROF CHEM, METROP STATE COL, 78- *Concurrent Pos:* Vis Fulbright prof, DeLaSalle Univ, Manila, 84-85. *Mem:* Am Indust Hyg Asn. *Res:* Synthesis; stereochemistry. *Mailing Add:* Metrop State Col Denver Box 52 PO Box 173362 Denver CO 80217-3362

DEWEY, J(OHN) L(YONS), chemical engineering, for more information see previous edition

DEWEY, JAMES EDWIN, b Geneva, NY, Jan 15, 17; m 43; c 1. ENTOMOLOGY. *Educ:* Cornell Univ, BS, 40, PhD(entom), 44; Univ Tenn, MS, 41. *Prof Exp:* Instr exten entom, NY State Col Agr & Life Sci, Cornell Univ, 44-45, from asst prof to assoc prof, 45-54, prof insect toxicol, 54-84, leader Chem-Pesticides Prog, 64-84; RETIRED. *Mem:* Entom Soc Am; Am Chem Soc. *Res:* Insect toxicology; insect resistance to insecticides; synergism; bioassay; insecticide formulation; fruit insect control. *Mailing Add:* 1556 Ellis Hollow Rd Ithaca NY 14850

DEWEY, JAMES W, b Glen Ridge, NJ, Jan 21, 43; m 67; c 2. EARTHQUAKE SEISMOLOGY, REGIONAL TECTONICS. *Educ:* Univ Calif, Berkeley, AB, 65, MA, 67, PhD(geophys), 71. *Prof Exp:* Res geophysicist, Nat Oceanic & Atmospheric Admin, 71-73; RES GEOPHYSICIST, US GEOL SURV, 73- *Concurrent Pos:* Am ed, Geophys J, Royal Astron Soc, 77-81; assoc ed, J Geophys Res, 84-86. *Mem:* Seismol Soc Am; Am Geophys Union; Royal Astron Soc; AAAS. *Res:* Earthquake location algorithms; seismotectonics of US, Central and South America; fault zones in Iran, Turkey, Kurile Islands, and Australia; midplate earthquakes; seismic hazard mitigation. *Mailing Add:* US Geol Surv MS 967 Fed Ctr Box 25046 Denver CO 80225

DEWEY, JOHN FREDERICK, b London, Eng, May 22, 37; m 61; c 2. STRUCTURAL GEOLOGY. *Educ:* Univ London, BSc, 58, DIC & PhD(struct geol), 60; Cambridge Univ, MA, 64; ScD, Oxford Univ, 88, MA, 85, DSc, 89. *Prof Exp:* From asst lectr to lectr geol, Univ Manchester, 60-64; lectr, Cambridge Univ, 64-70; prof geol, State Univ NY Albany, 70-; AT SCI LAB, DURHAM, ENG. *Concurrent Pos:* Sr res assoc, Lamont-Doherty Geol Observ, 67-; commonwealth fel, Mem Univ, 70; comt mem, Int Geodynamics Proj, 71-; assoc ed bull, Geol Soc Am. *Honors & Awards:* Daniel Pidgeon Fund Award, Geol Soc London, 64 & Murchison Fund Award, 71. *Mem:* Fel Geol Soc London; fel Geol Asn Can; Am Geophys Union. *Res:* Stratigraphic-structural evolution of the Appalachian-Caledonian orogen and the Alpine orogen of Europe; significance of plate tectonics for the evolution of continental margins and orogenic belts. *Mailing Add:* Dept Earth Sci Univ Oxford Parks Rd Oxford 0X1 3PR England

DEWEY, JOHN MARKS, b Portsmouth, Eng, Mar 23, 30; Can citizen; m 51; c 2. SHOCK & BLAST WAVES, FLOW VISUALIZATION. *Educ:* Univ London, BSc, 50, PhD(physics), 64. *Prof Exp:* Sci master physics, De La Salle Col, Channel Islands, 50-53, head sci dept, 53-56; scientist, Planning & Reporting Sect, Suffield Exp Sta, Can, Univ Victoria BC, 56-58, scientist, Physics Sect, 58-63, leader aerophys group, 63-64, head aerophys & shock tube sect, 64-65; FROM ASST PROF TO PROF PHYSICS, UNIV VICTORIA, BC, 65-, dean acad affairs, 73-77, dean grad studies & res, 77-85, assoc vpres res, 82-85; dep minister, Univs Sci & Commun BC, 85-86. *Concurrent Pos:* Pres, Can Asn Grad Schs, 84, Can Asn Univ Res Admin, 85. *Mem:* Am Phys Soc; fel Inst Physics; Can Asn Physicists. *Res:* Physics of blast waves; effect of blast waves; shock tube flows; shock wave reflections; high speed photography. *Mailing Add:* Dept Physics Univ Victoria Victoria BC V8W 3P6 Can

DEWEY, KATHRYN G, b New York, NY, Mar 9, 52; c 2. HUMAN LACTATION, COMMUNITY NUTRITION. *Educ:* State Univ NY, Albany, BS, 73; Univ Mich, MS, 76, PhD(biol sci), 80. *Prof Exp:* ASSOC PROF NUTRIT, UNIV CALIF, DAVIS, 80- *Mem:* Am Inst Nutrit; Am Soc Clin Nutrit; Comt Nutrit Anthrop. *Res:* Maternal and child nutrition, especially human lactation and infant growth; international nutrition, particularly the influence of agricultural change on diet and nutritional status. *Mailing Add:* Dept Nutrit Univ Calif Davis CA 95616

DEWEY, MAYNARD MERLE, anatomy; deceased, see previous edition for last biography

DEWEY, THOMAS GREGORY, b Pittsburgh, Pa, June 2, 52; m; c 3. BIOCHEMISTRY. *Educ:* Carnegie-Mellon Univ, BSc, 74; Univ Rochester, MS, 77, PhD(chem), 79. *Prof Exp:* NIH fel, Cornell Univ, 79-81; asst prof, 81-87, ASSOC PROF CHEM, UNIV DENVER, 87- *Concurrent Pos:* Sr fel, NIH, Duke Univ, 89. *Mem:* Am Chem Soc; Biophys Soc. *Res:* Thermodynamic and kinetic investigations of membrane bound enzymes; fractal dynamics of biochemical systems. *Mailing Add:* Dept Chem Univ Denver Denver CO 80208

DEWEY, VIRGINIA CAROLINE, biochemistry, microbiology, for more information see previous edition

DEWEY, WADE G, b Los Angeles, Calif, Aug 10, 27; m 51; c 4. PLANT BREEDING, GENETICS. *Educ:* Utah State Univ, BS, 53; Cornell Univ, PhD(plant breeding), 56. *Prof Exp:* Asst prof agron, Utah State Univ, 56-58; geneticist, USDA, 58-59; from asst prof to assoc prof, 59-66, PROF AGRON, UTAH STATE UNIV, 66- *Mem:* Am Soc Agron. *Res:* Plant breeding and genetics of the cereal crops, particularly winter wheat. *Mailing Add:* Dept Plant Sci Utah State Univ Logan UT 84322-4820

DEWEY, WILLIAM, b Nov 4, 29. RADIATION ONCOLOGY. *Educ:* Univ Wash, BS, 51; Univ Rochester, PhD(radiation biol), 58. *Prof Exp:* Assoc prof biophys, Postgrad Sch Med & Grad Sch Biomed Sci, Univ Tex, Houston, 61-65; prof radiation biol, Dept Radiol & Radiation Biol, Colo State Univ, Ft Collins, 65-81; PROF RADIATION ONCOL & DIR RADIATION ONCOL RES LAB, DEPT RADIATION ONCOL, UNIV CALIF, SAN FRANCISCO, 81- *Concurrent Pos:* Sabbatical leave, Swiss Inst Exp Cancer Res, Lausanne, Switz, 72-73; mem, NIH Radiation Res Study Sect, 75-79 & bd dirs, Far West Lab, 88-90; sr assoc ed hyperthermia, Int J Radiation Oncol Biol & Phys, 82-90; assoc ed hyperthermia, Radiation Res & Int J Hyperthermia, 84- *Honors & Awards:* Failla Lectr, Radiation Res Soc, 89. *Mem:* Radiation Res Soc (pres, 79). *Res:* Cellular responses to combinations of hyperthermia and radiation. *Mailing Add:* Dept Radiation Oncol Radiation Oncol Res Lab Univ Calif CED-200 Box 0806 San Francisco CA 94143-0806

DEWEY, WILLIAM LEO, b Albany, NY, Oct 21, 34; m 60; c 6. PHARMACOLOGY, BIOCHEMISTRY. *Educ:* St Bernardine of Siena Col, BS, 57; Col St Rose, MS, 64; Univ Conn, PhD(pharmacol), 66. *Prof Exp:* Asst res biologist, Sterling-Winthrop Res Inst, 59-64; asst pharmacol, Sch Pharm, Univ Conn, 64-66; from instr to asst prof, Schs Med & Pharm, Univ NC, Chapel Hill, 68-72; assoc prof, 72-76, PROF PHARMACOL, MED COL VA, VA COMMONWEALTH UNIV, 76-, HEAD CNS DIV, 78-, ASST DEAN, SCH GRAD STUDIES & ASSOC DEAN, SCH BASIC HEALTH SCI. *Concurrent Pos:* Fel pharmacol, Sch Med & Pharm, Univ NC, Chapel Hill, 66-68. *Mem:* AAAS; Am Soc Pharmacol & Exp Therapeut; Am Pharmaceut Asn; Am Chem Soc. *Res:* Agents affecting central nervous system, especially hallucinogens, analgesics, stimulants and tranquilizers. *Mailing Add:* Med Col Va MCV Box 568 Richmond VA 23298-0568

DEWHIRST, LEONARD WESLEY, b Marquette, Kans, Sept 28, 24; m 46; c 3. VETERINARY PARASITOLOGY. *Educ:* Kans State Col, BS, 49, MS, 50, PhD(parasitol), 57. *Prof Exp:* Res asst parasitol, Kans State Col, 49-52, instr zool, 52-57; asst prof & asst animal pathologist, Univ Ariz, 57-60, assoc prof, 60-63, prof vet sci & animal pathologist, Agr Exp Sta, 63-74; prof vet microbiol & asst dean students, Col Vet Med, Univ Mo-Columbia, 74-76; PROF VET SCI, ASSOC DEAN, COL AGR & DIR AGR EXP STA, UNIV ARIZ, 76- *Mem:* AAAS; Am Soc Parasitol; Am Micros Soc; Am Soc Trop Med & Hyg. *Res:* Agricultural parasitology; biology, treatment and control of parasites of domestic animals. *Mailing Add:* 740 E Agave Pl Tucson AZ 85718

DEWHURST, HAROLD AINSLIE, b Ottawa, Ont, June 18, 24; m 48; c 4. PHYSICAL CHEMISTRY. *Educ:* McGill Univ, BS, 46, PhD(phys chem), 50. *Prof Exp:* Res chemist, Can Atomic Energy Proj, 47-50; res fel, Univ Edinburgh, 50-52; res assoc phys chem, Univ Notre Dame, 52-54; res lab, Gen Elec Co, 54-60, liaison scientist chem, 60-61, personnel & admin, metall & ceramics res, 62-64, mgr struct & reactions studies, 64-65, mgr gen chem lab, Res & Develop Ctr, 65-68, mgr mat sci & eng, 68; mgr res group, 68-72, dir corp res & develop, 72-75, dir res & develop explor res, 75-78, dir strategic tech planning, 78-80, DIR SCI AFFAIRS, OWENS-CORNING FIBERGLAS CORP, 80- *Mem:* Am Chem Soc; Faraday Soc; Sigma Xi. *Res:* Kinetics of reactions; photochemistry of aqueous solutions; electric discharge chemistry; physical radiation chemistry; polymers; metallurgy; inorganic metals; glass fibers; textiles. *Mailing Add:* 6206 Brooksong Way Blacklick OH 43004

DEWHURST, PETER, b Great Harwood, Eng, Mar 7, 44. METAL MANUFACTURING, ROBOTICS. *Educ:* Univ Manchester, BSc, 70, MSc, 71, PhD(mech eng), 73. *Prof Exp:* Fac mem mech eng, Univ Salford, 73-80; vis prof, Mfg Group, Univ Mass, 80-81, prof, 81-85; PROF INDUST & MFG ENG, UNIV RI, 85-, DIR, GRAD STUDIES MFG ENG, 85- *Concurrent Pos:* Referee, Int J Appl Mech, Int J Prod Res & Am Soc Mech Engrs; assoc ed, Am Soc Mech Engrs Mfg Rev, J Design & Mfg, Int J Systs Automation & Applns. *Honors & Awards:* Nat Medal of Technol, 91; F W Taylor Medal, Col Int Pour Res Prod, 80. *Mem:* Sr mem Soc Mfg Engrs; Col Int Pour Res Prod. *Res:* Mechanics of metal forming; metal cutting; computer aided numerical control of machine tools; design for manufacture and the application of robots in assembly procedures. *Mailing Add:* Dept Indust & Mfg Eng Univ RI 103 Gilbreth Hall Kingston RI 02881

DEWHURST, WILLIAM GEORGE, b Co Durham, Eng, Nov 21, 26; m 60; c 2. NEUROCHEMISTRY, AFFECTIVE DISORDERS. *Educ:* Oxford Univ, BA, 47, BM & BCH, 50, MA, 62; London Univ, DPM, 61; FRCP(C), 78; FRCPsychiatrists, 82. *Prof Exp:* Jr & sr house surgeon, London Hosp, London Univ, 50-51, med, 51-52; captain internal med, Royal Army Med Corps, 52-54; jr registr, London Hosp, 54-56, registr, 57-58; registr psychiat, Maudsley Hosp, 65-69, sr registr, 58-61; lectr, Inst Psychiat, 62-64; sr lectr & consult physician, Inst Psychiat & Bethlem & Maudsley Hosps, 65-69; assoc prof, Univ Alta, 69-72, chmn, 75-90, co-dir, Neurochem Res Unit, 79-90, PROF PSYCHIAT, UNIV ALTA, 72-, EMER DIR, NEUROCHEM RES UNIT, 90- *Concurrent Pos:* Chmn, Royal Col Can Test Comt Psychiat, 71-79, exec secy, 71-80; examr, RCPS(Can), 73-; consult, psychiatrist, Royal Alexander Hosp, 76- & Edmonton Gen Hosp, 77-; chmn, Can Asn Prof Psychiat & test comt psychiat, Med Coun Can, 77-79; hon prof, pharm & pharmaceut sci, Univ Alta, 79- & div oncol, 83-; mem, bd dirs, Ctr Geront, Univ Alta, 85-; mem, adv comt, Can Psychiat Res Found, 85-90. *Mem:* Can Col Neuropharmacol (pres, 82-84); Can Psychiat Asn (pres, 83-84); fel Am Col Psychiatrists; fel Am Psychiat Asn; Can Asn Prof Psychiat (pres, 77-79); fel Am Psychopath Asn. *Res:* Investigations of physiology and neurochemistry of normal and abnormal mood and their pharmacological manipulation; development of new drugs; receptor studies. *Mailing Add:* Dept Psychiat Univ Alta 1E701 Mackenzie Ctr Edmonton AB T6G 2B7 Can

DEWIRE, JOHN WILLIAM, elementary particle physics; deceased, see previous edition for last biography

DE WIT, MICHIEL, b Amsterdam, Neth, June 6, 33; US citizen; m 57; c 5. LASERS. *Educ:* Ohio Univ, BS, 54; Yale Univ, PhD(physics), 60. *Prof Exp:* MEM TECH STAFF, TEX INSTRUMENTS, INC, 59- *Concurrent Pos:* Mem res associateships eval panel, Nat Res Coun, 74 & 76. *Mem:* Am Phys Soc; Optical Soc Am; Inst Elec & Electronics Engrs. *Res:* Studies of defects in solids via electron paramagnetic resonance, luminescence, Zeeman and Raman spectroscopy; laser and nonlinear optical device development; design and development of transversal filters with charge coupled devices. *Mailing Add:* 8418 Birchcroft Dr Dallas TX 75243

DEWIT, ROLAND, b Amsterdam, Neth, Feb 28, 30; US citizen; m 54; c 2. DISLOCATION THEORY, FRACTURE MECHANICS. *Educ:* Ohio Univ, BS, 53; Univ Ill, MS, 55, PhD(physics), 59. *Prof Exp:* Asst, Univ Ill, 53-59, res assoc, 59; asst res physicist, Univ Calif, Berkeley, 59-60; PHYSICIST, NAT INST STANDARDS & TECHNOL, 60- *Mem:* Am Phys Soc; Metall Soc; Am Soc Metals Int; Am Soc Mech Engr; Am Soc Testing & Mat; Soc Exp Mech; Sigma Xi. *Res:* Theory of dislocations and fracture mechanics. *Mailing Add:* Metall Div Nat Inst Standards & Technol Gaithersburg MD 20899

DEWITT, BERNARD JAMES, b Oak Harbor, Wash, Jan 29, 17; m 42; c 2. CHEMISTRY. *Educ:* Hope Col, AB, 37; Carnegie Inst Technol, DSc(phys chem), 41. *Prof Exp:* Res chemist, Pittsburgh Plate Glass Co, 40-51; sr supvr anal & phys chem, Columbia-South Chem Corp, 51-59, mgr, 59-64; asst dir res, PPG Indust, 64-78, dir analytical & electrochem res, 78-83; RETIRED. *Mem:* Am Chem Soc; Am Indust Hyg Asn; Am Soc Testing & Mat. *Res:* Chlorination aliphatic organics; silica pigments; physical testing of rubber products; analysis of chlorinated hydrocarbons and particle size analysis; environmental chemistry. *Mailing Add:* 337 Mineola Ave Akron OH 44313-7860

DEWITT, BRYCE SELIGMAN, b Dinuba, Calif, Jan 8, 23; m 51; c 4. THEORETICAL PHYSICS. *Educ:* Harvard Univ, BS, 43, MA, 47, PhD(physics), 50. *Prof Exp:* Mem, Inst Advan Study, NJ, 49-50; Fulbright fel, Tata Inst Fundamental Res, India, 51-52; sr physicist, Radiation Lab, Univ Calif, 52-55; res prof physics, Univ NC, Chapel Hill, 56-59, prof, 60-64, Agnew Hunter Bahnson, Jr prof, 64-72, dir res, Inst Field Physics, 56-72; prof physics, Univ Tex, Austin & dir Ctr Relativity, 73-87; JANE & ROLAND BLUMBERG PROF PHYSICS, 86- *Concurrent Pos:* Mem, Inst Advan Study, 54, 64, 66; Fulbright lectr, France, 56; consult, Gen Atomic Div, Gen Dynamics Corp, 59; mem, Int Comt Relativity & Gravitation, 59-72; NSF sr fel, 64; Fulbright lectr, Japan, 64-65; leader, Tex-Mauritanian Eclipse Exped, 73; mem fels panel, Nat Res Coun, 74-; Guggenheim vis res fel, All Souls Col, Oxford, 75-76; group leader, Inst Theoret Physics, Univ Calif, Santa Barbara, 80-81, Joint Inst Nuclear Res, Dubna, USSR, 88. *Honors & Awards:* Dirac Medal, UNESCO Int Atomic Energy Agency, 87. *Mem:* Nat Acad Sci; Fel Am Phys Soc; Sigma Xi. *Res:* Non-Abelian gauge field theory; quantum theory of gravity; astrophysics. *Mailing Add:* Dept Physics Univ Tex Austin TX 78712

DEWITT, CHARLES WAYNE, JR, b Akron, Ohio, Oct 16, 21; m 46; c 3. IMMUNOLOGY. *Educ:* Morris Harvey Col, BS, 49; Ohio State Univ, MS, 51, PhD(bact), 52. *Prof Exp:* Res scientist bact, Upjohn Co, 52-55, res assoc infect dis, 55-61; asst prof, Depts Microbiol & Surg, Sch Med, Tulane Univ, 61-62, from assoc prof to prof surg, 62-68; PROF PATH & SURG, MED CTR, UNIV UTAH, 68- *Concurrent Pos:* Mem ed bd, Transplantation. *Mem:* AAAS; Am Soc Microbiol; fel Am Acad Microbiol; Am Asn Immunol; Transplantation Soc. *Mailing Add:* Dept Path Univ Utah Med Ctr 50 N Medical Dr Salt Lake City UT 84132

DEWITT, DAVID P, b Bethlehem, Pa, Mar 2, 34; c 3. HEAT TRANSFER. *Educ:* Duke Univ, BS, 55; Mass Inst Technol, SM, 57; Purdue Univ, PhD(mech eng), 63. *Prof Exp:* Instr thermodyn, Col Eng, Duke Univ, 57-59; physicist, Nat Bur Standards, 63-64; dep dir, Thermophys Properties Res Ctr, 65-72, assoc prof, 72-78, PROF MECH ENG, PURDUE UNIV, 78- *Concurrent Pos:* Guest worker, Fed Physics-Technol Inst, Braunschweig, WGer, 70-71. *Mem:* Am Soc Mech Eng; Am Inst Aeronaut & Astronaut; Sigma Xi. *Res:* Heat transfer and thermophysical properties of matter, especially thermal radiation properties of materials, experimental procedures and techniques. *Mailing Add:* Sch Mech Eng Purdue Univ West Lafayette IN 47907

DE WITT, HOBSON DEWEY, b New Bern, NC, July 12, 23; m 48; c 3. ORGANIC CHEMISTRY. *Educ:* Erskine Col, BA, 44; Vanderbilt Univ, PhD(chem), 51. *Prof Exp:* Chemist, E I Du Pont de Nemours & Co, 44; res chemist, Southland Paper Mills, 50-52, Chemstrand Corp, 52-56; chmn dept, 56-83, PROF CHEM, WESTMINSTER COL, 56- *Mem:* Am Chem Soc; Sigma Xi; fel Am Inst Chem; Hist Sci Soc; Soc Hist Technol. *Res:* Synthesis of amino acid derivatives; monomers; polymers of textile interest; resolution of optically active compounds; history of science. *Mailing Add:* Dept Chem Westminster Col New Wilmington PA 16172

DEWITT, HUGH EDGAR, b Memphis, Tenn, May 27, 30; m 56; c 3. PLASMA PHYSICS, ASTROPHYSICS. *Educ:* Stanford Univ, BS, 51; Cornell Univ, PhD(physics), 57. *Prof Exp:* PHYSICIST, LAWRENCE LIVERMORE LAB, 57- *Concurrent Pos:* Fulbright prof, Inst Math Sci, India, 63-64; lectr, Univ Calif, Berkeley, 66-68; vis prof, Nat Univ Mex, 69, Univ Iowa, 70-71; lectr, Univ Calif, Davis-Livermore, 76-77. *Honors & Awards:* Pub Serv Award, Fedn Am Scientists, 88. *Mem:* Fel Am Phys Soc; fel AAAS. *Res:* Theoretical physics; statistical mechanics of liquids and strongly coupled plasmas; numerical simulation of planetary and stellar interiors; thermonuclear reaction; application of integral equations. *Mailing Add:* Lawrence Livermore Lab Box 808 Livermore CA 94550

DE WITT, HUGH HAMILTON, b San Jose, Calif, Dec 28, 33; m 56, 81; c 3. ICHTHYOLOGY, MARINE BIOLOGY. *Educ:* Stanford Univ, BA, 55, MA, 60, PhD(biol), 66. *Prof Exp:* Res assoc biol sci, Hancock Found, Univ Southern Calif, 62-67; asst prof marine biol, Univ SFla, 67-69; asst prof zool, 69-70, from asst prof oceanog to prof oceanog, 70-81, chmn, Dept Oceanog, 76-79, PROF ZOOL, UNIV MAINE, ORONO, 81- *Concurrent Pos:* Co-prin investr & field leader, Southern Ocean Res Cruises, 62-67, 75-76; mem, comt biol & ecol of Antarctic fishes, 79-82, Biomass Res Group. *Honors & Awards:* Mountain named in honor. *Mem:* Am Soc Ichthyologists & Herpetologists (pub secy, 74-79); Sigma Xi; Ichthyol Soc Japan; Soc Systematic Zool; Soc Francaise Ichtyologie; Indian Soc Icthyologists. *Res:* Freshwater fishes of southeastern Asia; polar marine fishes; taxonomy, biology and zoogeography of Antarctic fishes and of deep-water fishes of western north Atlantic. *Mailing Add:* Dept Zool Univ Maine Orono ME 04469-0146

DE WITT, JOHN WILLIAM, JR, b Pawhuska, Okla, Dec 16, 22; m 47; c 1. FISHERIES, LIMNOLOGY. *Educ:* Ore State Univ, PhD, 63. *Prof Exp:* Fishery biologist, US Fish & Wildlife Serv, 48-49; prof fisheries, Humboldt State Univ, 49-78, chmn dept, 75-78; RETIRED. *Concurrent Pos:* Dir fisheries training & overseas support progs, Peace Corps, Chile, 66-69; fishery biologist, Lake Nasser Develop Ctr, Food & Agr Orgn, UN, UAR, 69-71; mem, Calif State Regional Water Qual Control Bd, 73-76. *Mem:* Am Fisheries Soc; Water Pollution Control Fedn; Am Soc Ichthyol & Herpet; Am Soc Limnol & Oceanog; Sigma Xi. *Mailing Add:* 300 Luman Rd 196 Phoenix OR 97535

DE WITT, ROBERT MERKLE, b Wolcott, NY, May 31, 15. ZOOLOGY. *Educ:* Univ Mich, BS, 40, MS, 41, PhD, 53. *Prof Exp:* Teacher high schs, NY, 41-46; asst prof biol, Sampson Col, 46-49; teaching asst, Univ Mich, 49-50, teaching fel, 50-52, instr, 52-54, Lloyd fel, 53-54; asst prof, 54-59, actg chem dept, 69-71, assoc prof biol, 59-80, EMER ASSOC PROF ZOOL, UNIV FLA, 80-, VCHMN DEPT, 80- *Mem:* AAAS; Am Malacol Union; Am Micros Soc; Ecol Soc Am; Am Soc Zool. *Res:* Morphological and physiological factors of adaptation in amphibious snails; reproduction, ecology, population biology and systematics of fresh-water mollusks. *Mailing Add:* 1602 NW 19th Circle Gainesville FL 32605

DEWITT, WILLIAM, b Washington, DC, Nov 28, 39; m; c 2. MOLECULAR BIOLOGY, BIOCHEMISTRY. *Educ:* Williams Col, BA, 61; Princeton Univ, MA, 63, PhD(biol), 66. *Prof Exp:* Res assoc biochem, Mass Inst Technol, 66-67; asst prof biol, 67-74, assoc prof, 74-77, PROF BIOL & CHMN DEPT, WILLIAMS COL, 77- *Mem:* Sigma Xi. *Res:* Synthesis of hemoglobin and other specific proteins during amphibian development. *Mailing Add:* Dept Biol Williams Col Williamstown MA 01267

DEWITT-MORETTE, CECILE, b Paris, France, Dec 21, 22; m 51; c 4. THEORETICAL PHYSICS. *Educ:* Univ Caen, Lic es sci, 43; Univ Paris, dipl, 44, PhD(theoret physics), 47. *Prof Exp:* Mem, Inst Advan Studies, Ireland, 46-47, Univ Inst Theoret Physics, Denmark, 47-48 & Inst Advan Study, Princeton, 48-50; teacher res, Inst Henri Poincare, France, 50-51; res assoc & lectr, Univ Calif, 52-55; vis res prof, Univ NC, Chapel Hill, 56-67, dir inst field physics, 58-66, lectr physics, 67-71; prof astron, 72-83, PROF PHYSICS, UNIV TEX, AUSTIN, 83- *Concurrent Pos:* Dir & founder, Summer Sch Theoret Physics, Les Houches, France, 51-72. *Honors & Awards:* Chevalier Ordre National Du Merite, France, 81. *Res:* Theory of field; elementary particles; mathematical physics; gravitation. *Mailing Add:* Dept Physics Univ Tex Austin TX 78712

DE WOLF, DAVID ALTER, b Dordrecht, Neth, July 23, 34; c 3. THEORETICAL PHYSICS, ELECTRICAL ENGINEERING. *Educ:* Univ Amsterdam, BSc, 55, MSc, 59; Univ Eindhoven, Dr Tech, 68. *Prof Exp:* Res physicist, Nuclear Defense Lab, Edgewood Arsenal, Md, 62; mem tech staff, David Sarnoff Res Ctr, RCA Corp, 62-82; assoc ed, J Optical Soc Am, 69-81. *Mem:* AAAS; fel Inst Elec & Electronics Engrs; Optical Soc Am; Neth Phys Soc. *Res:* Wave propagation and multiple scattering (in random media); electron optics. *Mailing Add:* Dept Elec Eng Va Polytech Inst & State Univ Blacksburg VA 24061

DEWOLF, GORDON PARKER, JR, b Lowell, Mass, Aug 17, 27; m 55; c 3. TAXONOMIC BOTANY. *Educ:* Univ Mass, BSc, 50; Tulane Univ, MSc, 52; Univ Malaya, MSc, 54; Cambridge Univ, PhD, 59. *Prof Exp:* Asst bot, Tulane Univ, 50-52; asst, Bailey Hortorium, Cornell Univ, 53-54, res assoc, 54-56; sr sci officer, Royal Bot Gardens, Eng, 59-61; from assoc prof to prof bot, Ga Southern Col, 61-67, actg head dept biol, 66-67; hort taxonomist, Arnold Arboretum, Harvard Univ, 67-70, horticulturist, 70-75; COORDR, HORT PROG, MASS BAY COMMUNITY COL, 76- *Concurrent Pos:* Hort consult, Horticulture Magazine. *Mem:* Int Asn Plant Taxonomists. *Res:* Ficus of Africa and America; taxonomy of cultivated plants. *Mailing Add:* Dept Math & Sci Mass Bay Comm Cl 50 Oakland St Wellesley MA 02181

DEWOLF, JOHN T, b Oakland, Calif, Jan 12, 43; m 68; c 2. STRUCTURAL ANALYSIS, STRUCTURAL DESIGN. *Educ:* Univ Hawaii, BSCE, 66; Cornell Univ, ME, 67, PhD(struct eng), 73. *Prof Exp:* Struct engr, Albert Kahn Assoc, Detroit, 67-69; PROF CIVIL ENG, UNIV CONN, 73- *Concurrent Pos:* Acad vis, Imp Col, London, 79. *Mem:* Fel Am Soc Civil Engrs; Am Concrete Inst. *Res:* Structural engineering and applied mechanics, with major emphasis on application to structural design. *Mailing Add:* Dept Civil Eng Univ Conn Main Campus Box U-37 261 Glenbrook Rd Storrs CT 06268

DEWOLFE, BARBARA BLANCHARD OAKESON, b San Francisco, Calif, May 14, 12; m 50, 60. VERTEBRATE ZOOLOGY. *Educ:* Univ Calif, AB, 33, PhD(zool), 38. *Prof Exp:* Asst zool, Univ Calif, 33-35, 36-37; Palmer fel, Wellesley Col, 38-39; instr, Placer Jr Col, 39-42; instr zool, Col Agr & jr zoologist, Exp Sta, Univ Calif, 42-43; instr zool, Smith Col, 43-45; lectr, 46, from asst prof to prof, 46-77, EMER PROF ZOOL, UNIV CALIF, SANTA BARBARA, 77- *Mem:* Cooper Ornith Soc; fel Am Ornith Union; Sigma Xi. *Res:* Environment, annual cycle and migration in white-crowned sparrows; vertebrate cycles and microclimates; song dialects. *Mailing Add:* Dept Biol Sci Univ Calif Santa Barbara CA 93106

DEWS, EDMUND, b Medford, Ore, Oct 1, 21; m 53, 59; c 4. ACQUISITION POLICY, EDUCATIONAL ADMINISTRATION. *Educ:* Stanford Univ, BA, 43; Univ Wash, dipl, 44; Univ Calif, Los Angeles, MA, 47; Oxford Univ, BA, 51, MA, 54; Air War Col, dipl, 70. *Prof Exp:* Chief air weather serv upper air forecast sect, Oper Crossroads, Kwajalein-Bikini, 46; instr math & physics, Southern Ore State Col, 46; res meteorologist, Air Force Cambridge Res Ctr, 55-56, br chief & aeronaut res adminstr geophys, 57-58; vpres & managing ed, Pergamon Press, Inc, 59-60; mem, 60-86, CONSULT, ECON & STATIST DEPT, RAND CORP, 86- *Concurrent Pos:* Mem, Secy of Air Force's Spec Study Group on Air Power in Southeast Asia, 67-68; mem, Task Group on Role Tech Report, Fed Coun Sci & Technol, 67-68. *Mem:* Inst Elec & Electronics Engrs. *Res:* Political economy of research and development; computer integrated manufacturing; military systems analysis; atmospheric sciences, general; history of science and technology; educational administration. *Mailing Add:* 470 Siskiyou Blvd Ashland OR 97520-2136

DEWS, PETER BOOTH, b Ossett, Eng, Sept 11, 22; nat US; m 49; c 4. PHYSIOLOGY, PSYCHOBIOLOGY. *Educ:* Univ Leeds, MB, ChB, 44; Univ Minn, PhD(physiol), 52. *Hon Degrees:* MA, Harvard Univ, 59. *Prof Exp:* Lectr pharmacol, Univ Leeds, 46-47; Wellcome res fel, Wellcome Res Lab, 48-49; Mayo Found fel, Univ Minn, 50-51, res assoc biomet, Mayo Found, 52; from instr to assoc prof pharmacol, 53-62, STANLEY COBB PROF PSYCHIAT & PSYCHOBIOL, HARVARD MED SCH, 62- *Mem:* Inst Med-Nat Acad Sci; Am Soc Pharmacol; Am Physiol Soc; Brit Pharmacol Soc; Physiol Soc Gt Brit; Am Asn Arts & Sci. *Res:* Pharmacology of the central nervous system; psychobiology. *Mailing Add:* 220 Longwood Ave Boston MA 02115-6078

DEWSBURY, DONALD ALLEN, b Brooklyn, NY, Aug 11, 39; m 63; c 2. ANIMAL BEHAVIOR, HISTORY & PHILOSOPHY OF SCIENCE. *Educ:* Bucknell Univ, AB, 61; Univ Mich, PhD(psychol) 65. *Prof Exp:* Fel psychol, Univ Calif, Berkeley, 65-66; from asst prof to assoc prof, 66-73, PROF PSYCHOL, UNIV FLA, 73- *Mem:* Fel Animal Behav Soc (treas, 73-78, pres, 76-80); fel Am Psychol Asn; Psychonomic Soc; Sigma Xi; AAAS; Cheiron Soc; Hist Sci Soc. *Res:* Evolution and adaptive significance of animal behavior; evolution of reproduction behavior in muroid rodents and the role of behavior in reproductive success; history of study of animal behavior. *Mailing Add:* Dept Psychol Univ Fla Gainesville FL 32611

DEWYS, WILLIAM DALE, b Zeeland, Mich, Sept 14, 39; m 61; c 3. MEDICAL ONCOLOGY, INTERNAL MEDICINE. *Educ:* Calvin Col, BS, 60; Univ Mich, MD, 64; Am Bd Internal Med, cert, 72; Am Bd Med Oncol, cert, 73. *Prof Exp:* Intern, Presby-St Luke's Hosp, 64-65; resident, Univ Mich, 65-66; res assoc, Nat Cancer Inst, 66-68; resident & fel, Univ Rochester, 68-71, asst prof, 71-73; from assoc prof to prof med & chief, med oncol sect, Sch Med, Northwestern Univ, 73-79; head, nutrit sect & chief, Clin Invest Br, Nat Cancer Inst, 80-82, assoc dir prev prog, 82-85; SR PHYSICIAN, CAPITAL AREA PERMANENT MED GROUP, 85- *Concurrent Pos:* Clin fel, Am Cancer Soc, 70-71, jr fac clin fel, 71-73; assoc physician, Strong Mem Hosp, 71-73; asst attend physician, Rochester Gen Hosp, 71-73; attend staff physician, Northwestern Mem Hosp, 73-79 & Vet Admin Res Hosp, 73-79; affil staff physician, Evanston Hosp, 73-79; chmn clin progs task force, Ill Cancer Coun, 75-79; mem cancer clin invest rev comt, Nat Cancer Inst, 76-80; attend staff physician, Fairfax Hosp, Falls Church, Va, 86, Arlington Hosp, Arlington, Va, 86-; attend physician, Hospice Northern Va, Arlington, Va, 86, Holy Cross Hosp, Silver Spring, Md, 87-, Northern Va Doctors Hosp, Arlington, Va, 90- *Mem:* Am Asn Cancer Res; Am Fedn Clin Res; Am Soc Clin Oncol; fel Am Col Physicians; Soc Clin Trials; Am Soc Hemat. *Res:* Adverse systemic effects of cancer; experimental chemotherapy of cancer; prevention of cancer; hematology. *Mailing Add:* 6830 Hillmead Rd Bethesda MD 20817

DEXTER, DEBORAH MARY, b Oakland, Calif, Sept 28, 38. MARINE BIOLOGY, BENTHIC ECOLOGY. *Educ:* Stanford Univ, BA, 60, MA, 62; Univ NC, Chapel Hill, PhD(zool), 67. *Prof Exp:* From asst prof to assoc prof, 67-73, PROF BIOL, SAN DIEGO STATE UNIV, 73- *Concurrent Pos:* Fulbright scholar, 85. *Mem:* Ecol Soc Am. *Res:* Marine invertebrate zoology and ecology; population ecology of benthic invertebrates; sandy beach ecology. *Mailing Add:* Dept Biol San Diego State Univ San Diego CA 92182

DEXTER, LEWIS, CARDIOLOGY, INTERNAL MEDICINE. *Educ:* Harvard Univ, MD, 36. *Prof Exp:* EMER PROF INTERNAL MED, UNIV MASS, 81- *Mailing Add:* 108 Upland Rd Brookline MA 02146

DEXTER, RALPH WARREN, b Gloucester, Mass, Apr 7, 12; m 38; c 2. ECOLOGY, HISTORY OF BIOLOGY. *Educ:* Mass State Col, BS, 34; Univ Ill, PhD(ecol), 38. *Prof Exp:* Asst zool, Univ Ill, 34-37; from instr to assoc prof, 37-48, PROF BIOL, KENT STATE UNIV, 48- *Concurrent Pos:* Res contract, US AEC, 56-62; vis prof, Stone Lab, Ohio State Univ, 53-55, 67, 69 & 71. *Mem:* Marine Biol Asn India; Ecol Soc Am; Am Malacol Union (vpres, 64-65 & pres, 65-66); Am Ornith Union; Am Soc Zool. *Res:* Ecology of marine communities, molluscs and crustaceans (anostracan phyllopods); life history of chimney swift; history of nineteenth century American naturalists; studies of bird-banding; history of biology. *Mailing Add:* 1228 Fairview Dr Kent OH 44240

DEXTER, RICHARD NEWMAN, b Port Huron, Mich, Sept 2, 33; m 61; c 2. INTERNAL MEDICINE. *Educ:* Harvard Univ, AB, 55; MD, Cornell Univ, 59; Am Bd Internal Med, dipl, 67, 74 & 80, cert endocrinol & metab, 73. *Prof Exp:* From intern to resident internal med, Univ Minn Hosps, 59-62; res assoc, NIH, Bethesda, 62-64; resident, Peter Bent Brigham Hosp, Boston, 64-65; fel endocrinol & instr med, Sch Med, Vanderbilt Univ, 65-67; from asst prof to assoc prof, 67-73, PROF MED, SCH MED, IND UNIV, INDIANAPOLIS, 73- *Mem:* Endocrine Soc; fel Am Col Physicians. *Res:* Endocrinology; metabolism. *Mailing Add:* Dept Med Ind Univ Med Ctr Indianapolis IN 46202

DEXTER, RICHARD NORMAN, b Ashland, Wis, Nov 22, 27; m 58; c 4. SOLID STATE PHYSICS. *Educ:* Mich State Univ, BS, 49; Univ Wis, MS, 51, PhD(physics), 55. *Prof Exp:* Staff microwave res, Lincoln Lab, Mass Inst Technol, 52-55, solid state physics, 53-55; asst from asst prof to assoc prof, 55-61, PROF PHYSICS, UNIV WIS-MADISON, 61- *Concurrent Pos:* Alfred P Sloan Found fel, 55-59. *Mem:* AAAS; fel Am Phys Soc. *Res:* Experimental solid state physics; energy band structure determinations; optical properties in vacuum ultraviolet. *Mailing Add:* Dept Physics Sterling Hall Univ Wis Madison WI 53706

DEXTER, STEPHEN C, b East Orange, NJ, Sept 18, 42; m 71; c 2. BIOLOGICAL CORROSION, CORROSION ENGINEERING. *Educ:* Univ Del, BS, 65, MAS, 68, PhD(metall), 71. *Prof Exp:* Postdoctoral, Woods Hole Oceanog Inst, 71-72, asst scientist, ocean eng, 72-75; asst prof, 76-80, ASSOC PROF APPL OCEAN SCI & MAT SCI, UNIV DEL, 81- *Concurrent Pos:* Res grants, Off Sea Grant, Nat Oceanic & Atmospheric Admin, 71-76 & 80-92, Nat Data Buoy Off, 72-76, Alcoa Found, 73-74, Dept Energy, 76-78, Off Naval Res, 77-81 & 86-89, NSF, 84-86, Alcoa Lab, 84-86, Elec Power Res Inst, 88-90; consult, various indust, 80-; chmn, Int Symp Biologically Induced Corrosion, 85, tech comt biol corrosion, 86-90; res comt, Nat Asn Corrosion Engrs, 82-, chmn, 88-91, bd dir, 88-91. *Mem:* Am Soc Metals; Nat Asn Corrosion Engrs; Sigma Xi. *Res:* Microbiologically influenced corrosion; electrochemical corrosion; surface chemistry of microbiological fouling; interactions of structural materials with seawater. *Mailing Add:* Col Marine Studies Univ Del Lewes DE 19958

DEXTER, THEODORE HENRY, b Preston, Cuba, June 1, 23; US citizen; m 52; c 3. INDUSTRIAL INORGANIC & ELECTROCHEMICALS. *Educ:* Tulane Univ, BS, 44, MS, 47; Univ Ill, PhD(chem), 50. *Prof Exp:* Asst chem, Tulane Univ, 43-44; chemist, E I du Pont de Nemours & Co, 44-45; asst chem, Tulane Univ, 46-47; Gen Aniline asst, Illinois, 47-49; chief inorg res sect, Chem Div, Olin Mathieson, 49-60; res supvr, Hooker Chem Div of Occidental Chem Corp, 60-75, prog leader res & develop mineral prod, 75-76, sr res chemist, res & develop electrochem, 76-85; PRES, DEXTER CONSULT SERV, 86. *Concurrent Pos:* Chmn, Chem Environ Res; comt, Soap Detergent Asn, 78-79. *Mem:* Am Chem Soc; Electrochem Soc; Sigma Xi. *Res:* Inorganic specialty and heavy chemicals; applications to detergents, pulp, paper textiles, leather; fused salts; coordination compounds; aqueous inorganic process stream purifications. *Mailing Add:* 850 Hillside Dr Lewiston NY 14092

DEY, ABHIJIT, b Patna, India, Oct 2, 45. GEOPHYSICS, MATHEMATICS. *Educ:* Indian Inst Technol, Kharagpur, BSc Hons, 65, MSc, 67; Univ Calif, Berkeley, PhD(eng geosci), 72. *Prof Exp:* Assoc res geophysicist, Univ Calif, Berkeley, 72-77; sr geophysicist, Geothermal Div, 77-80, staff geophysicist, 80-87, SR STAFF GEOPHYSICIST, MINERALS EXPLOR, CHEVRON RESOURCES, 87- *Concurrent Pos:* Consult, var mining co & geotech eng co, 72-77. *Mem:* Soc Explor Geophysicists; Europ Asn Explor Geophysicists; AAAS; Australian Soc Explor Geophysicists. *Res:* Interpretation of electrical, electromagnetic, seismic and pot field data; mineral, geothermal and geochemical exploration; quantitative numerical modelling of two and three-dimensional geological structures. *Mailing Add:* 463 Jackson St Albany CA 94706

DEY, SUDHANSU KUMAR, b Calcutta, India, Nov 8, 44; m 70; c 1. REPRODUCTIVE BIOLOGY, ENDOCRINOLOGY. *Educ:* Univ Calcutta, BSc, 65, MSc, 67, PhD(physiol), 72. *Prof Exp:* Lectr physiol, City Col Calcutta, 70-72; Ford Found fel reproductive biol, 73-75, res assoc, 75-77, from asst prof to assoc prof gynec & obstet, 77-84 asst prof, 79-86, PROF PHYSIOL, KANS UNIV MED CTR, 86-, PROF GYNEC & OBSTET, 84- *Concurrent Pos:* Sr res fel reproductive biol, Dept Physiol, Univ Calcutta, 72. *Mem:* Brit Soc Study Fertil; Soc Study Reproduction; Am Physiol Soc; Sigma Xi. *Res:* Physiology and metabolism of preimplantation and early postimplantation mammalian embryos; physiology and mechanism of ovum implantation; molecular biology; toxicology. *Mailing Add:* Dept Gynec & Obstet MRRC 314 Univ Kans Med Ctr 39th St & Rainbow Blvd Kansas City KS 66103

DEYO, RICHARD ALDEN, b New Orleans, La, May 26, 49; m 78; c 2. INTERNAL MEDICINE, AMBULATORY CARE. *Educ:* Grinnell Col, BA, 71; Pa State Univ, MD, 75; Univ Wash, MPH, 81. *Prof Exp:* Resident, int med, Univ Tex Teaching Hosps, San Antonio, 75-79; Robert W Johnson Clin Scholar, Univ Wash, 79-81; asst prof, med, Univ Tex Health Sci Ctr, San Antonio, 82-86; ASSOC PROF, MED, UNIV WASH, SEATTLE, 86- *Concurrent Pos:* Dir, Health Servs Res & Develop Field Prog, Vet Admin Med Ctr, Seattle, 86-; assoc dir, Robert W Johnson Clin Scholars Prog, Univ Wash; prin investr, Back Pain Outcome Assessment Team; chmn res comt, Soc Gen Int Med, 88, Nat Coun, 89- *Honors & Awards:* Nellie Westerman Prize Res Med Ethics, Am Fedn Clin Res, 82- *Mem:* Soc Gen Int Med; Am Pub Health Asn; fel Am Col Physicians; Physicians Social Responsibility. *Res:* Research concerning methodology and applications of questionnaire instruments for measuring health-related quality-of-life; res in the epidemiology and management of low back pain and chronic pain. *Mailing Add:* HSR&D Seattle VA Med Ctr (152) 1660 S Columbian Way Seattle WA 98108

DEYOE, CHARLES W, b Two Buttes, Colo, Mar 12, 33; m 56; c 5. BIOCHEMISTRY, NUTRITION. *Educ:* Kans State Univ, BS, 55; Tex A&M Univ, MS, 57, PhD(biochem), 59. *Prof Exp:* Asst prof biochem & poultry nutrit, Tex A&M Univ, 60-62; from asst prof to prof feed technol, 62-68, PROF GRAIN SCI, KANS STATE UNIV, 68-, HEAD DEPT GRAIN SCI & INDUST & DIR, FOOD & FEED GRAIN INST, 77- *Concurrent Pos:* Feed technol res scientist, Agr Exp Sta, Kans State Univ, 75-; dir, Int Grain Prog, 80. *Honors & Awards:* Agr Excellence Sci Award, Nat Agri-Bus Asn, 87. *Mem:* Am Chem Soc; Am Soc Animal Sci; Poultry Sci Asn; Am Asn Cereal Chemists; Inst Food Technologists; Sigma Xi. *Res:* Nutrition of farm animals and biochemistry related to animal processes; feed technology and chemical relationships between foodstuffs and nutritive values; milling quality of wheats. *Mailing Add:* Dept Grain Sci & Indust Kans State Univ Manhattan KS 66506

DE YOUNG, DAVID SPENCER, b Colorado Springs, Colo, Nov 29, 40. THEORETICAL ASTROPHYSICS. *Educ:* Univ Colo, BA, 62; Cornell Univ, PhD(physics), 67. *Prof Exp:* Res scientist, Los Alamos Sci Lab, 67-69; scientist, Nat Radio Astron Observ, 69-80; astronomer, 80-82, assoc dir, 82-87, DIR, KITT PEAK NAT OBSERV, 88- *Concurrent Pos:* Consult, Los Alamos Sci Lab, 70-74; mem exec comt, Bd Trustees, Aspen Ctr Physics, 74-80, mem adv bd, 80-; chmn, Radio Astron Exp Selection Panel, 79-91; mem, San Diego Supercomput Ctr, 85-, chmn stg comt, 88-91. *Mem:* Int Astron Union; Am Astron Soc; Am Phys Soc; Int Union Radio Sci. *Res:* Origin and evolution of extended extragalactic radio sources; physics of galactic nuclei and quasi-stellar objects; evolution of galaxies and galaxy clusters. *Mailing Add:* Kitt Peak Nat Observ 950 N Cherry Ave Tucson AZ 85719

DE YOUNG, DONALD BOUWMAN, b Grand Rapids, Mich, July 29, 44; m 66; c 3. ASTRONOMY. *Educ:* Mich Technol Univ, BS, 66, MS, 68; Iowa State Univ, PhD(physics), 72; Grace Theol Sem, MDiv, 81. *Prof Exp:* Teaching asst physics, Mich Technol Univ, Houghton, 66-68; res worker, Inst Atomic Res, AEC, Ames, Iowa, 68-72; PROF PHYSICS, GRACE COL, WINONA LAKE, IND, 72- *Mem:* Sigma Xi; Am Asn Physics Teachers; Am Phys Soc; Creation Res Soc. *Res:* Mossbauer effect studies of transition metal borides. *Mailing Add:* Physics Dept Grace Col Winona Lake IN 46590

DEYOUNG, EDWIN LAWSON, b Milwaukee, Wis, Jan 14, 29; m 56. ORGANOMETALLIC CHEMISTRY. *Educ:* Univ Louisville, BS, 52, MS, 53; Univ Ill, PhD(org chem), 56. *Prof Exp:* Res chemist, Reynolds Metals Co, Ky, 52-53, Minn Mining & Mfg Co, 54, Shell Develop Co, Colo, 55, Whiting Res Labs, Standard Oil Co Ind, 56-60, R B & P Chem Co, Wis, 60-62 & Universal Oil Prod Co, Des Plaines, 62-68; chmn, Dept Phys Sci, 68-80, PROF CHEM, CHICAGO CITY COL, LOOP BR, 68- *Mem:* Am Inst Chem; Sigma Xi. *Res:* Metal organic compounds; synthesis and reactions; pi-complexes and metal aromatic compounds. *Mailing Add:* Harold Washington Col 30 E Lake St Chicago IL 60601

DEYOUNG, JACOB J, b Grand Rapids, Mich, May 14, 26; m 57; c 4. ORGANIC CHEMISTRY. *Educ:* Hope Col, AB, 50; Wayne State Univ, MS, 52, PhD, 58. *Prof Exp:* Chemist, Merck Chem Co, 52-53; from asst prof to assoc prof, 57-77, PROF CHEM, ALMA COL, 77- *Mem:* Am Chem Soc; Sigma Xi. *Res:* Organic synthesis; isolation and synthesis of natural products from plants. *Mailing Add:* 1104 Michigan Ave St Louis MI 48880

DEYRUP, JAMES ALDEN, b Englewood, NJ, Oct 13, 36; m 61; c 1. ORGANIC CHEMISTRY. *Educ:* Swarthmore Col, BA, 57; Univ Ill, PhD(org chem), 61. *Prof Exp:* NSF fel, Univ Zurich, 61-62; instr chem, Harvard Univ, 62-65; from asst prof to assoc prof chem, 65-74, asst dean preprof educ, 71-74, asst dean, Cols Med & Dent, 74-80, PROF CHEM, UNIV FLA, 74- *Res:* Chemistry of heterocyclic compounds including macromolecular heterocyclic catalysis. *Mailing Add:* Dept Chem Univ Fla Gainesville FL 32611

DEYRUP-OLSEN, INGRITH JOHNSON, b Nyack, NY, Dec 22, 19; wid. ZOOLOGY. *Educ:* Columbia Univ, AB, 40, PhD(physiol), 44. *Prof Exp:* Instr physiol, Col Physicians & Surgeons, Columbia Univ, 42-47, lectr zool, Barnard Col, 47, from asst prof to prof, 47-64; from res prof to prof, 64-90, EMER PROF ZOOL, UNIV WASH, 90- *Concurrent Pos:* Guggenheim fel, 53-54; Fulbright fel Denmark, 53-54; Arctic Res Lab, 56-57. *Mem:* AAAS; Soc Gen Physiol; Am Soc Zool; Am Physiol Soc; NY Acad Sci; Nat Asn Biol Teachers. *Res:* Physiology; circulation; kidney; water and electrolyte exchange of tissue slices; fluid exchange in molluscs; mechanism of mucus secretion in molluscs. *Mailing Add:* Dept Zool Univ Wash Seattle WA 98195

DEYSACH, LAWRENCE GEORGE, b Milwaukee, Wis, July 9, 36; m 66. MATHEMATICAL BIOLOGY. *Educ:* Marquette Univ, BS, 57; Harvard Univ, AM, 63. *Prof Exp:* Lectr math biol, Univ Chicago, 67-69; med statistician, G D Searle & Co, 69-80, sr biostatistician, 80-84, supvr, Med Statist, 84-87, head clin trial statist, 87-89, SR CONSULT BIOSTATISTICIAN, G D SEARLE & CO, 89- *Mem:* Am Statist Asn; Biomet Soc; Sigma Xi. *Res:* Topological dynamics; mathematical ecology; experimental design of clinical trials; pharmacokinetics; stochastic processes in medicine. *Mailing Add:* 944 Wesley Evanston IL 60202

DE ZAFRA, ROBERT LEE, b White Plains, NY, Feb 15, 32; m; m. ATOMIC PHYSICS, ATMOSPHERIC CHEMISTRY. *Educ:* Princeton Univ, AB, 54; Univ Md, PhD(physics), 58. *Prof Exp:* Res asst physics, Princeton Univ, 55; res asst, Univ Md, 57-58; instr, Univ Pa, 58-61; from asst prof to assoc prof, 61-84, actg dir, Inst Atmospheric Res, 87-89, PROF PHYSICS, STATE UNIV NY, STONY BROOK, 84- *Concurrent Pos:* Adj prof, Radiation Lab, Columbia Univ, 62-64; prin investr sponsored res, NSF, 63-67, 85-88, NASA, 76-78, 82-93 & Chem Mfrs Asn, 76-88. *Mem:* Am Phys Soc; Am Geophys Union; Sigma Xi. *Res:* Experimental research on physical and chemical effects of natural and anthropogenic trace gases on earth's atmosphere; atomic and molecular physics; development of remote sensing techniques. *Mailing Add:* Dept Physics State Univ NY Stony Brook NY 11794

DE ZEEUW, CARL HENRI, b East Lansing, Mich, Dec 6, 12; m 39; c 5. WOOD TECHNOLOGY, MECHANICS. *Educ:* Mich State Col, BA, 34, BS, 37; State Univ NY Col Forestry, Syracuse Univ, MS, 39, PhD(wood anat), 49. *Prof Exp:* From instr to emer prof wood technol, State Univ NY Col Environ Sci & Forestry, 47-88; RETIRED. *Concurrent Pos:* Consult, Col Forestry, Univ Philippines, State Univ NY & US Int Coop Admin, 59-61, Food , Agr Orgn, Philippines, 66, Venezuela, 68, Arg, 71, Brazil, 77 & Burma, 80. *Mem:* Forest Prod Res Soc; Am Soc Testing & Mat; Int Asn Wood Anat; Soc Wood Sci & Technol; fel Int Acad Wood Sci. *Res:* Interrelationship of gross wood anatomy, ultra structure and composition of the woody cell wall and the physical properties of wood; descriptive anatomy of tropical woods. *Mailing Add:* 309 E Genesee St Fayetteville NY 13066

DE ZEEUW, JOHN ROBERT, biochemistry, for more information see previous edition

DEZELSKY, THOMAS LEROY, b Saginaw, Mich, Mar 3, 34; m 71, 85; c 5. HEALTH SCIENCE. *Educ:* Cent Mich Univ, BS, 56; Univ Mich, MA, 59; Ind Univ, HSD, 66. *Prof Exp:* Teacher, Copemish High Sch, Mich, 56-57 & Mt Pleasant Jr High, Mich, 57-61; asst prof health & phys educ, Wis State Univ-Oshkosh, 61-66; assoc prof, Brigham Young Univ, 66-68; ASSOC PROF EXERCISE & WELLNESS, ARIZ STATE UNIV, TEMPE, 68- *Mem:* Am Sch Health Asn; Am Pub Health Asn. *Res:* Suicide behavior; drug and substance abuse; mental health problems of aged populations. *Mailing Add:* Dept Exercise Sci Ariz State Univ Tempe AZ 85287

DEZENBERG, GEORGE JOHN, b Tientsin, China, Jan 12, 35; US citizen; m 60; c 4. ELECTRICAL ENGINEERING. *Educ:* Auburn Univ, BEE, 60; Univ Ark, MS, 62; Ga Inst Technol, PhD(elec eng), 66. *Prof Exp:* RES ELECTRONIC ENGR, PHYS SCI LAB, US ARMY MISSILE COMMAND, 65- *Mem:* Inst Elec & Electronics Engrs; Optical Soc Am. *Res:* Carbon dioxide laser research, including Q-switching, mode locking, multipath, and mode control techniques. *Mailing Add:* 910 San Ramon Ave Huntsville AL 35802

DE ZOETEN, GUSTAAF A, b Tjepoe, Indonesia, July 5, 34; m 61; c 3. PLANT PATHOLOGY, PLANT VIROLOGY. *Educ:* Univ Wageningen, MSc, 60; Univ Calif, Davis, PhD(plant path), 65. *Prof Exp:* Tech asst plant physiol, Western Prov Res Sta, SAfrica, 57-58; lab technician plant path, Univ Calif, Davis, 62 & 64-65, asst res plant pathologist, Univ Calif, Berkeley, 65-67; asst prof, 67-70, assoc prof, 70-74, PROF PLANT PATH, UNIV WIS-MADISON, 74- *Concurrent Pos:* Vis prof, Dept Plant Pathol, Univ Calif, Davis, 78-79. *Mem:* Am Phytopath Soc; Royal Soc Agr Sci Neth. *Res:* Physiology and cytology of virus host relationships; electron microscopy; plant virus replication. *Mailing Add:* Dept Plant Path Univ Wis 793A Russell Lab Madison WI 53706

DHALIWAL, AMRIK S, b Punjab, India, Nov 17, 34; m 62; c 2. HORTICULTURE, VIROLOGY. *Educ:* Punjab Univ, India, BSc, 55; Utah State Univ, MSc, 59, PhD(biol), 62. *Prof Exp:* Res assoc plant virol, Utah State Univ, 62-65, asst prof bot, 65-66; from asst prof to assoc prof, 66-75, PROF BIOL, LOYOLA UNIV CHICAGO, 75- *Mem:* AAAS; Am Inst Biol Sci; Am Physiol Soc. *Res:* Cytogenetics and plant breeding of field and horticultural crops; post harvest physiology, pathology and biochemistry and in vitro synthesis of plant viruses; chemotherapy of viruses. *Mailing Add:* Dept Biol Loyola Univ 6525 N Sheridan Rd Chicago IL 60626

DHALIWAL, RANJIT S, b Bilaspur, India, June 21, 30; m 58; c 1. APPLIED MATHEMATICS. *Educ:* Punjab Univ, India, MA, 55; Indian Inst Technol, Kharagpur, PhD(appl math), 60. *Prof Exp:* Lectr math, Guru Nanak Col Dabwali, Punjab, India, 55-56, lectr math, Guru Nanak Eng Col, Ludhiana, 56-61; lectr math, Indian Inst Technol, New Delhi, 61-63, asst prof, 63-66; assoc prof, 66-71, PROF MATH, UNIV CALGARY, 71- *Concurrent Pos:* Visitor, Imp Col, Univ London, 64-65; vis prof, City Univ London & sr res assoc, Glasgow Univ, 71-72; mem, Alta Heritage Coun, 87-90. *Mem:* Am Acad Mech; Am Math Soc; Soc Indust & Appl Math; Can Math Cong; Indian Math Soc. *Res:* Theory of plates; elastodynamics; micropolar elasticity; thermoelasticity; fracture mechanics. *Mailing Add:* Dept Math Univ Calgary Calgary AB T2N 1N4 Can

DHALLA, NARANJAN SINGH, b Punjab, India, Oct 10, 36. PHYSIOLOGY, PHARMACOLOGY. *Educ:* Univ Pa, MS, 63; Univ Pittsburgh, PhD(pharmacol), 65; Inst Chem, India, FIC. *Prof Exp:* Res assoc biochem, Sch Med, St Louis Univ, 66, asst prof pharmacol, 66-68; from asst prof to assoc prof, 68-74, PROF PHYSIOL, FAC MED, UNIV MAN, 74- *Concurrent Pos:* Secy gen, Int Soc Heart Res. *Mem:* Am Soc Pharmacol & Exp Therapeut; Am Physiol Soc; Can Physiol Soc; Pharmacol Soc Can; fel Inst Chem India. *Res:* Pathophysiology of heart, muscle contraction, membrane transport and autonomic nervous system. *Mailing Add:* Dept Physiol Univ Man Fac Med St Boniface Hosp Res Ctr 351 Tache Ave Winnipeg MB R2H 2A6 Can

DHAMI, KEWAL SINGH, b Punjab, India, Jan 10, 33; US citizen; m 59; c 1. ORGANIC CHEMISTRY, POLYMER CHEMISTRY. *Educ:* Panjab Univ, India, BSc, 54, MSc, 55; Univ Western Ont, PhD(org chem), 64. *Prof Exp:* US Air Force assoc chem, Ohio State Univ, 64-65; sr res chemist, Res & Develop Div, Polymer Corp Ltd, Ont, 65-69; sr res chemist, ITT Wire & Cable Div, Int Tel & Tel Corp, 69-85; SR POLYMER CHEMIST, SURPRENANT WIRE & CABLE, 85-, SR SCIENTIST & MGR, IRRADIATION DEPT, 87- *Mem:* Am Chem Soc; Soc Plastic Engrs. *Res:* C13 nuclear magnetic resonance studies of organic compounds; preparation and properties of new polymers; mechanical and physical properties of polymers; degradation of polymeric compounds; synthesis of new sulfur compounds; irradiation crosslinking of polymers; preparation of new high temperature radiation sensitive crosslinking agents. *Mailing Add:* Teledyne Thermatics PO Box 909 Elm City NC 27822

DHANAK, AMRITLAL M(AGANLAL), b Bhavnagar, India, July 13, 25; nat US; m 54; c 3. MECHANICAL ENGINEERING, SOLAR ENERGY. *Educ:* Univ Calif, MS, 51, PhD(mech eng), 56. *Prof Exp:* Res engr, Univ Calif, 52-56; proj engr, Gen Elec Co, 56-58; assoc prof, Rensselaer Polytech Inst, 58-61; PROF MECH ENG, MICH STATE UNIV, 61- *Concurrent Pos:* Sr staff scientist & consult, Avco Mfg Corp, 59-62. *Mem:* Sigma Xi; Am Soc Mech Engrs (chmn, Technol & Soc Div, 78-79); Int Solar Energy Soc. *Res:* Heat transfer; thermodynamics and fluid dynamics with specializations in solar energy. *Mailing Add:* 333 University Dr East Lansing MI 48823

DHAR, SACHIDULAL, b Muktagachha, Bangladesh, July 16, 43. HIGH ENERGY PHYSICS, SOFTWARE SYSTEMS DESIGN. *Educ:* Univ Dacca, BSc, 64, MSc, 65; Duke Univ, PhD(physics), 74. *Prof Exp:* Sci officer physics, Pakistan Atomic Energy Comn, 65-67; lectr & res assoc, State Univ NY. Albany, 74-75; res assoc physics, Univ Mass, Amherst, 75-79; software engr, Raytheon Co, Wayland, 80-81; MEM TECH STAFF, MITRE CORP, BEDFORD, 81- *Res:* Experimental and phenomenological study of elementary particle interactions at high energies. *Mailing Add:* Mitre Corp Mail Stop E025 Burlington Rd Bedford MA 01730

DHARAMSI, AMIN N, b Dar-Es-Salaam, Tanzania, Jan 2, 51; US citizen; m 79; c 2. OPTOACOUSTICS, SPECTROSCOPY. *Educ:* Univ Nairobi, BSEE, 73; Univ Alta, MSEE, 77, PhD(elec eng), 81. *Prof Exp:* From instr to asst prof, 80-86, ASSOC PROF ELEC ENG, OLD DOMINION UNIV, 86- *Concurrent Pos:* Prin investr, NSF Grant, 83-85, NASA grant, 90-91. *Mem:* Inst Elec & Electronics Engrs; Am Phys Soc. *Res:* Quantum electronics (i.e. interaction of light with matter); have published numerous articles in refereed journals on subjects such as novel laser sources, transient electronic spectroscopy, applied optics, optoacoustics, thermoluminescense, shock physics, and theoretical computations of excited electronic states. *Mailing Add:* Dept Elec & Computer Eng Old Dominion Univ Norfolk VA 23529-0246

DHARANI, LOKESWARAPPA R, b JN Kote, Kannataka, India, Oct 25, 47; m 76; c 2. COMPOSITE MATERIALS, FRACTURE & FAILURE ANALYSIS. *Educ:* Univ Mysore, India, BE, 70; Indian Inst Technol, India, MTech, 72; Cranfield Inst Technol, UK, MSc, 75; Clemson Univ, PhD(eng mech), 82. *Prof Exp:* Res asst struct, Indian Inst Technol, 70-71; aeronaut engr fatigue & fracture, Hindustan Aeronaut, Ltd, 72-76, dep design engr, 76-78; grad res asst mech, Clemson Univ, 79-82; asst prof, 82-88, ASSOC PROF MECH, UNIV MO-ROLLA, 88-, SR INVESTR MAT ENG, MAT RES CTR, 88- *Concurrent Pos:* Mem, Damage Tolerance Sub-comt, Aeronaut Res & Develop Bd, Govt India, 78-79; vis scientist, Air Force Mat Lab, Ohio, 86-88; consult, Silver Dollar City, Inc, 87-88, RETEK, 88-89 & Universal Technologies, Corp, 88-90; co-dir, Friction Lab, Univ Mo-Rolla, 88- *Honors & Awards:* Keramos, Am Ceramic Soc, 88. *Mem:* Am Soc Mech Eng; Soc Advan Mat & Process Eng; Am Ceramic Soc; Am Soc Eng Educ; Am Acad Mech; Soc Automotive Engrs. *Res:* Micromechanics analysis of fracture and failure of polymer, metal and ceramic matrix composites; development of non-asbestos brake linings for cars and light trucks. *Mailing Add:* 129 Mech Eng Bldg Univ Mo Rolla MO 65401-0249

DHESI, NAZAR SINGH, b Dhesian Kahna, India, Sept 2, 23; Can citizen; m 55; c 2. SEED TESTING. *Educ:* Punjab Univ-Lahore, BSc, 45; Utah State Univ, MS, 50; Kans State Univ, PhD(agron), 52. *Prof Exp:* Asst veg botanist plant breeding, Punjab Agr Dept, 55-58, seed testing officer, 58-60, veg botanist plant breeding, 60-64, dep dir agr, exten, 64-65; biologist seed testing, 65-71, BIOLOGIST METHOD DEVELOP SEED TESTING, AGR CAN, 72- *Concurrent Pos:* Vis fel, Ohio State Univ, 59; assoc grad fac dept crop sci, Guelph Univ, Ont, 78-79 & 85-86. *Mem:* Am Soc Agron; Can Seed Grower's Asn. *Res:* Development of laboratory test methods for the identification of crop cultivars required for monitoring the genetic purity of seed lots offered for sale in Canada or produced for export. *Mailing Add:* Lab Serv Div Bldg 22 Carling Ave CEF Ottawa ON K1A 0C6 Can

D'HEURLE, FRANCOIS MAX, b France, Nov 23, 25; m 50; c 3. THIN FILMS, ELECTRONIC MATERIALS. *Educ:* Ecole des Arts et Metiers, Paris, BSc, 46; Mich Technol Univ, MSc, 48; Ill Inst Technol, PhD(Metall), 58. *Prof Exp:* SCIENTIST, INT BUS MACH, 58-; PROF, ROYAL INST TECHNOL, STOCKHOLM, 85- *Honors & Awards:* Cledo Brunetti Award, Inst Elec & Electronics Engrs, 88; Gaede-Langmuir Award, Am Vacuum Soc, 90. *Mem:* Fel Inst Elec & Electronics Engrs; Am Inst Mining Metall & Petrol Engrs; Am Vacuum Soc; Mat Res Soc; Electrochem Soc. *Res:* Kinetics of chemical reactions in the solid state, diffusion and nucleation processes; formation of intermetallic compounds, oxides, carbides; behavior of solids in thin layers, interfaces, crystal boundaries, capillarity. *Mailing Add:* IBM Corp, Watson Res Ctr Div PO Box 218 Yorktown Heights NY 10562

DHILLON, BALBIR SINGH, b Jhaj, India, Aug 5, 47; m 79; c 2. RELIABILITY ENGINEERING, INDUSTRIAL ENGINEERING. *Educ:* Univ Wales, BSc, 72, MSc, 73; Univ Windsor, PhD(indust eng), 75. *Prof Exp:* Fel reliability eng, Univ Windsor, 75-76; reliability engr, C A E Electronics Ltd, 76-77; tech supvr reliability eng, Ont Hydro, 77-79; from asst prof to assoc prof, 80-85, PROF, UNIV OTTAWA, 85-, ACTG DIR & DIR ENG MGT PROG, 88- *Concurrent Pos:* Adv ed, Microelectronic & Reliability, 77-; referee, Inst Elec & Electronic Engrs on Reliability, 76- & Can Univs res grants comt, Nat Res Coun Can, 78-; assoc ed, Int J Energy Syst, 85- *Honors & Awards:* Austin J Bonis Reliability Award, Am Soc Qual Control. *Res:* Reliability, especially common cause failures; reliability of multi state devices and reliability modeling; human reliability; author of 12 texbooks on reliability eng & related areas and 210 articles. *Mailing Add:* Dept Mech Eng Univ Ottawa Ottawa ON K1N 6N5 Can

DHINDSA, DHARAM SINGH, b Nov 25, 34; m 61; c 3. REPRODUCTIVE ENDOCRINOLOGY, CARDIOLOGY. *Educ:* Punjab Univ, DVM, 56; Mont State Univ, MS, 63; Univ Ill, Urbana, PhD(physiol), 67. *Prof Exp:* Vet asst surgeon, Punjab State Govt, India, 56-57; lectr physiol, Punjab Vet Col, 57-58, in-chg rabies vaccine prep, 58-59, lectr parasitol, 59-61; res asst animal sci, Mont State Univ, 61-63; res asst, Univ Ill, 63-66, res assoc, Col Vet Med, 67; res fel, Heart Res Lab, Sch Med, Univ Ore, 67-72, res assoc, Dept Med, 71-74, instr, Health Sci Ctr, 74-75; EXEC SECY, REPRODUCTIVE BIOL, NIH, 75-, REFERRAL OFFICER, DIV RES GRANTS, 84- *Concurrent Pos:* Chmn, Dept Animal Sci, Ore Zool Res Ctr, 71-75. *Honors & Awards:* NIH Award Merit, 88. *Mem:* AAAS; Soc Study Reproduction; Am Physiol Soc; Int Primatological Soc; Am Soc Animal Sci. *Res:* Physiology, especially reproductive endocrinology, cardiovascular pre- and postnatal; reproductive biology. *Mailing Add:* Div Res Grant NIH Westwood Bldg Bethesda MD 20892

DHINDSA, K S, b India, Jan 24, 32; Can citizen; m 74. CELL BIOLOGY, EXPERIMENTAL BIOLOGY. *Educ:* Panjab Univ, India, BSc, 51, Hons, 53, MSc, 54; Univ Helsinki, PhD(zool), 70. *Prof Exp:* Demonstr zool, Panjab Univ, 53-56, lectr, Col Faridkot, 55-61, sr lectr biol, Dairy Sci Col, 61-62; asst zool, Univ Calif, Los Angeles, 62-63; asst biol, Inst Biol Res, Culver City, Calif, 63-65; asst prof zool, Panjab Agr Univ India, 65-66; asst prof, 67-74, ASSOC PROF BIOL, CONCORDIA UNIV, 73- *Concurrent Pos:* Res grant, Loyola Col Montreal, 69; Nat Res Coun Can res grants, 69 & 71-72. *Mem:* AAAS; fel Royal Micros Soc; Am Soc Zoologists; Can Soc Cell Biol; NY Acad Sci; Toxic Soc Can. *Res:* Effect of psychotherapeutic drugs and food additives on brain and endocrine metabolism; RNA synthesis in brain and other tissues in mammals using histochemical and radiographic methods. *Mailing Add:* Dept Biol Concordia Univ 1455 De Maisonneuve Blvd W Montreal PQ H3G 1M8 Can

DHIR, SURENDRA KUMAR, b Sonekatch, India, Aug 9, 37; m 65; c 2. ENGINEERING MECHANICS, PHYSICS. *Educ:* Birla Inst Technol & Sci, Ranchi, India, BS, 60; Munich Tech Univ, Dr Ing, 64. *Prof Exp:* Design engr, Allgemeine Elektricit-tsgesellschaft, Frankfurt, 64-65; sr proj engr struct dept, 65-71, head numerical struct mech br, David Taylor Naval Ship Res & Develop Ctr, 71-86; DIR CAPITAL GOODS TECHNOL CTR, EXPORT ADMIN, DEPT COM, 86- *Honors & Awards:* Gold Medal, Dept Com, 90. *Mem:* Sigma Xi. *Res:* Theoretical and experimental mechanics; coherent optics; computer application in numerical mechanics. *Mailing Add:* 5808 Plainview Rd Bethesda MD 20817

DHIR, VIJAY KUMAR, b Giddarbaha, India, Apr 14, 43; m 73; c 2. ENGINEERING. *Educ:* Punjab Eng Col, BSc, 65; Indian Inst Technol, Kanpur, MTech, 68; Univ Ky, PhD(mech engr), 72. *Prof Exp:* Asst develop engr pump design, Jyoti Pumps, Ltd, 68-69; engr auto part design, TATA Eng & Locomotive Co, 69; res assoc & lectr, Univ Ky, 72-74; from asst prof to assoc prof, 74-82, PROF ENG, UNIV CALIF, LOS ANGELES, 82- *Concurrent Pos:* Res grants, US Nuclear Regulatory Comn, 76- & Elec Power Res Inst, 78-, NSF, Dept Energy, Gas Res Inst. *Mem:* Fel Am Soc Mech Engrs; Am Nuclear Soc; Am Inst Astronaut & Aeronaut; Sigma Xi. *Res:* Nuclear engineering; thermal and hydraulics of nuclear reactors and reactor safety; heat transfer during phase change; boiling and condensation, thermal and hydrodynamic instability. *Mailing Add:* Dept Mech Eng 5732 Boelter Univ Calif 405 Hilgard Los Angeles CA 90024

DHRUV, ROHINI ARVIND, b Bombay, Maharashtra, Mar 26, 50; m 74; c 1. BIOANALYTICAL RESEARCH. *Educ:* St Xavier's Col, Univ Bombay, BSc, 71; E Stroudsburg State Col, Pa, MEd, 73; Rutgers Univ, PhD(microbiol), 80. *Prof Exp:* Asst chem, E Stroudsburg State Col, 71-72; res intern immunol, Rutgers Univ, 72-73, teaching asst biol, 73-77; instr biol, Montclair State Col, 78; res investr, 78-84, SR RES INVESTR, BIOANALYTIC RES & DEVELOP, E R SQUIBB & SONS, 84- *Mem:* Am Soc Microbiol. *Res:* Chemical immunological or microbiological assays for the detection of different antibiotics and other drugs in tissues, biological fluids, pharmaceutical formulations, water and feeds. *Mailing Add:* 25 Dutch Rd East Brunswick NJ 08816

DHUDSHIA, VALLABH H, b Shahpur, India, July 1, 39; US citizen; m 65; c 2. RELIABILITY, MAINTAINABILITY. *Educ:* Gujarat Univ, BS, 63; Ill Inst Technol, MS, 65; New York Univ, PhD(indust eng & opers res), 73. *Prof Exp:* Res engr solid mech, Foster Wheeler Corp, 65-72; eng specialist reliability, 72-79, MEM RES STAFF, XEROX CORP, 79- *Mem:* Am Soc Mech Engrs; Am Soc Qual Control. *Res:* Reliability of advanced marking technology; reliability and maintainability assurance, management, assessment, and evaluation. *Mailing Add:* 2605 Graphic Pl Plano TX 75075

DHURANDHAR, NINA, b Bombay, India, Jan 6, 37; m 63; c 2. PATHOLOGY, CYTOLOGY. *Educ:* G S Med Col, Bombay, MB, 62; Univ London, DCP, 65; Am Bd Path, dipl anat path, 73. *Prof Exp:* From instr to asst prof, 66-79, ASSOC PROF PATH, SCH MED, TULANE UNIV, 79- *Concurrent Pos:* Vis pathologist, Charity Hosp New Orleans, 73-, dir autopsy path, 75-78; staff pathologist, Tulane Med Ctr Hosp, 76-; training fine needle aspiration, Salgrenska Hosp, Goteborg, Sweden, 84 & Karolinska-Radiumhemmut, Stockholm, Sweden, 85; fac, Colposcopy Update & Gynec Laser Surg Course, New Orleans, 86; chairperson, Local Comt 35th Ann Sci Meeting, Am Soc Cytol, 87; fac path update, New Orleans, 90. *Mem:* Int Acad Path; Am Soc Cytol; Col Am Pathologists. *Res:* Human papillomer virus and cervical squamous neoplasia; fine needle aspiration cytology. *Mailing Add:* Dept Path Tulane Univ Sch Med New Orleans LA 70112

DHURJATI, PRASAD S, b Kurnool, India, Jan 15, 56; US citizen. BIOCHEMICAL ENGINEERING, EXPERT SYSTEMS APPLICATIONS. *Educ:* Indian Inst Technol, Kanpur, BS, 77; Purdue Univ, PhD(chem eng), 82. *Prof Exp:* From asst prof to assoc prof, 82-91, PROF CHEM ENG, UNIV DEL, 91- *Concurrent Pos:* NSF presidential young investr award, 86; vis scientist, Pasteur Inst, Paris, 88-89. *Mem:* Am Inst Chem Engrs; Am Chem Soc; Am Soc Eng Educ. *Res:* Understanding of intracellular phenomena that influence formation of biologically active product in genetically engineered microorganisms; application of expert systems to fault diagnosis and supervision in the chemical process industries. *Mailing Add:* Dept Chem Eng Univ Del Newark DE 19716

DHYSE, FREDERICK GEORGE, b Hinkley, Utah, June 23, 18; m 50; c 2. BIOCHEMISTRY, INFORMATION SCIENCE. *Educ:* Univ Calif, BA, 41. *Prof Exp:* Chemist control & prod res, Cutter Labs, Calif, 41-46; chemist vitamins in canned foods, Gerber Prods Co, 46-47; biochemist endocrinol sect, Nat Cancer Inst, 48-50, res analyst document sect, 50-63, biochemist cancer studies, 53-66, SCI INFO OFFICER, NAT INST CHILD HEALTH & DEVELOP, NIH, 66- *Mem:* AAAS; Am Chem Soc. *Res:* Vitamin function; biotin; hormonal activity; antivitamins, antimetabolites in cancer; machine and punch-card applications to problems of scientific literature. *Mailing Add:* 8603 Bunnell Dr Potomac MD 20854

DIACHUN, STEPHEN, b Phenix, RI, Aug 20, 12; m 38; c 2. PLANT PATHOLOGY. *Educ:* RI State Col, BS, 34; Univ Ill, MS, 35, PhD(plant path), 38. *Prof Exp:* Asst bot, Univ Ill, 34-37; asst plant path, 37-46, from asst prof to assoc prof, 47-53, PROF PLANT PATH, EXP STA, UNIV KY, 53- *Concurrent Pos:* Assoc plant pathologist, Exp Sta, Univ Ky, 46-50. *Mem:* AAAS; Am Phytopath Soc; Bot Soc Am. *Res:* Relation of environment to bacterial and fungus leaf spots of tobacco; growth of bacteria on roots; distribution of tobacco mosaic virus in plants; inoculation methods with streak virus of tobacco; virus diseases of forage legumes. *Mailing Add:* 220 Tahoma Rd Lexington KY 40503

DIACONIS, PERSI, b New York, NY, Jan 31, 45. MATHEMATICAL STATISTICS. *Educ:* City Col New York, BS, 71; Harvard Univ, MA, 73, PhD(statist), 74. *Prof Exp:* Asst prof, 74-80, ASSOC PROF STATIST, STANFORD UNIV, 80- *Res:* Probabilistic number theory; data analysis; foundations of inference. *Mailing Add:* Math Sci Ctr Harvard Univ Cambridge MA 02138

DIAKOW, CAROL, b New York, NY. ANIMAL BEHAVIOR, NEUROENDOCRINOLOGY. *Educ:* City Col New York, BS, 63; NY Univ, MS, 67, PhD(biopsychol), 69. *Prof Exp:* Lectr biol, City Col New York, 63; fel biopsychol, Dept Animal Behav, Am Mus Natural Hist, 63-69; fel neurophysiol & neuroendocrinol, Inst Animal Behav, Rutgers Univ, 69-71; fel neurophysiol & behav, Rockefeller Univ, 71, res assoc neuroanat & behav, 71-73; from asst prof to assoc prof, 73-83, PROF BIOL, ADELPHI UNIV, 84-, CHAIR BIOL, 91- *Concurrent Pos:* Staff assoc, Psychobiol Prog, NSF, 81-82. *Honors & Awards:* Sergei Zlinkoff Award Med Res. *Mem:* Animal Behav Soc; NY Acad Sci; Int Soc Psychoneuroendocrinol; Soc Neurosci; AAAS; Sigma Xi; Soc Study Amphibians & Reptiles. *Res:* Neuroendocrine bases of mating behavior in female vertebrates. *Mailing Add:* Dept Biol Adelphi Univ Garden City NY 11530

DIAL, NORMAN ARNOLD, b Kell, Ill, June 17, 26; m 58; c 4. ZOOLOGY. *Educ:* Ill Col, AB, 49; Univ Ill, MS, 52, PhD(zool), 60. *Prof Exp:* prof, Life Sci Dept, Ind State Univ, 60-88; RETIRED. *Concurrent Pos:* Adj prof anat, Sch Med, Terre Haute Ctr Med Educ, 60-88. *Res:* Effects of heavy metals and toxic chemicals on embryonic development. *Mailing Add:* 904 N Fourth St Terre Haute IN 47807

DIAL, WILLIAM RICHARD, b Washington Court House, Ohio, Aug 30, 14; m 36; c 1. ORGANIC CHEMISTRY. *Educ:* Ohio Wesleyan Univ, AB, 36; Univ Ill, PhD(org chem), 39. *Prof Exp:* Org chemist, Columbia Chem Div, Pittsburgh Plate Glass Co, 39-52; org chemist, Columbia-Southern Chem Corp, 52-59, mgr appl org res, Chem Div, Pittsburgh Plate Glass Co, 59-64, asst dir res, 64-69, asst dir res, Chem Div, PPG Industs, Inc, 69-77; RETIRED. *Mem:* Am Chem Soc; Soc Plastic Engrs; Soc Plastic Indust. *Res:* Synthetic resins and plastics; plasticizers; casting synthetic resins; structure of gossypol; chlorinated hydrocarbons; polymerization catalysts; reinforcing rubber and paper fillers; phosgene derivatives. *Mailing Add:* 438 Roslyn Ave Akron OH 44320

DIAMANDOPOLUS, GEORGE TH, b Crete, Greece, Nov 21, 29. PATHOGENESIS OF NEUROPLASIA. *Educ:* Univ Vt, MD, 55. *Prof Exp:* PROF PATH, HARVARD MED SCH, 81- *Mem:* Am Asn Pathologists; Am Asn Cancer Res. *Mailing Add:* Dept Path Harvard Med Sch 25 Shattuck St Boston MA 02115

DIAMANTE, JOHN MATTHEW, b New York, NY, Jan 14, 40; m 81; c 1. CLIMATE SYSTEM DYNAMICS, TIDAL DYNAMICS & SEA LEVEL. *Educ:* New York Univ, BE, 61, MS, 63, PhD(aeronaut & astronaut), 69. *Prof Exp:* Asst res scientist, Dept Meteorol, NY Univ, 67-69; head, Anal Sect Command Control & Commun Dept, Systs Group TRW Inc, 69-71; sr scientist, Wolf Res & Develop Group, Wash Anal Sci Ctr, EG&G, Inc, 71-74; sr staff scientist, Bus & Technol Systs Inc, 74-77; tech adv, Tides & Water Levels Div, Nat Ocean & Atmospheric Admin/Off Ocean & Atmospheric Res, 77-81, spec asst to chief scientist, 82-84, spec asst to assoc adminr, 84-85, climate coordr, 85-87, mgr, Prog Develop Climate & Global Change, 87-90; CHIEF, SCI & EMERGING ISSUES TEAM, OFF INT ACTIV, ENVIRON PROTECTION AGENCY, 90- *Honors & Awards:* Alexander Klemin Award Aeronaut Eng, 61. *Mem:* Am Geophys Union; Am Meterol Soc; AAAS; Sigma Xi. *Res:* Measurement and modeling of ocean tides and mean sea surface using satellite radar altimeter data; development of techniques using in situ sea level systems, VLBI and GPS geodetic systems to determine absolute sea level changes. *Mailing Add:* 1614 Sherwood Rd Silver Spring MD 20902

DIAMANTIS, WILLIAM, b New York, NY, May 27, 23; m 50; c 2. PHARMACOLOGY. *Educ:* Fordham Univ, BS, 44, MS, 55, PhD(biol), 59. *Prof Exp:* Asst instr physiol & pharmacol, Col Pharm, Fordham Univ, 44-45, instr pharmacog, 45-58, asst prof pharmacog, microbiol & physiol, 58-60; sr pharmacologist, 60-78, sect leader pharmacol, 78-80, DIR PHARMACOL, WALLACE LABS DIV, CARTER-WALLACE, INC, 80- *Mem:* AAAS; NY Acad Sci; Am Soc Pharmacol & Exp Therapeut; Soc Exp Biol Med; Inflammation Res Asn. *Res:* Autonomic pharmacology; gastrointestinal pharmacology; inflammation; antipyresis; chemical mediators in anaphylaxis; pulmonary pharmacology; pharmacology of platelet function. *Mailing Add:* Wallace Labs Div Carter-Wallace Inc PO Box 1001 Cranbury NJ 08512

DIAMENT, PAUL, b Paris, France, Nov 14, 38; US citizen; m 63; c 3. ELECTRICAL ENGINEERING, PHYSICS. *Educ:* Columbia Univ, BS, 60, MS, 61, PhD(elec eng), 63. *Prof Exp:* PROF ELEC ENG, COLUMBIA UNIV, 63- *Concurrent Pos:* Res assoc, Stanford Univ, 66-67; vis assoc prof, Tel Aviv Univ, 70-71; consult, Appl Sci Labs, 70-71, Air Force Cambridge Res Labs, 73-74, State Univ NY Downstate Med Ctr, 74-75, Riverside Res Inst, 77-81, 86-87, KMS Fusion, 83-85 & IBM Res Ctr, 74, 82, 85. *Mem:* Am Phys Soc; Inst Elec & Electronics Engrs; Optical Soc Am. *Res:* Relativistic electron beams; optics; laser radiation statistics; nonlinear wave interactions; microwaves; antennas; signal processing; electromagnetic theory; plasma physics; free-electron lasers; stochastic processes; author of one textbook. *Mailing Add:* Dept Elec Eng Columbia Univ New York NY 10027

DIAMOND, BRUCE I, b Far Rockaway, NY, June 7, 45; m 76; c 1. PHARMACOLOGY. *Educ:* Bradley Univ, BA, 67; Long Island Univ, MS, 69; Chicago Med Sch, PhD(pharmacol), 75. *Prof Exp:* Res asst psychopharmacol, Cent Islip State Hosp, 67-70; res scientist, Neuropsychiat Hosp, Borda, Arg, 70-71; res assoc anesthesia, Mt Sinai Hosp, Chicago, 72-81; asst prof pharmacol, Rush-St Lukes-Presby Med Ctr, 78-81; assoc prof, 81-88, PROF PSYCHIAT, MED COL GA, 88- *Concurrent Pos:* Asst prof, Chicago Med Sch, 77-81; res pharmacol, Augusta, Va, 81-; Ill Dept Ment Health grants, 78-81. *Mem:* Soc Neurosci; Soc Biol Psychiat; Am Acad Neurol; fel Am Col Clin Pharmacol; Am Epilepsy Soc. *Res:* Neuropharmacology and mechanisms of action of central nervous system agents and putative neurotransmitters; animal models of neuropsychiatric diseases. *Mailing Add:* Dept Psychiat Med Col Ga Augusta GA 30912

DIAMOND, DAVID J(OSEPH), b New York, NY, Dec 31, 40; m 62; c 3. REACTOR SAFETY, REACTOR PHYSICS. *Educ:* Cornell Univ, BEP, 62; Univ Ariz, MS, 64; Mass Inst Technol, PhD(nuclear eng), 68. *Prof Exp:* Nuclear engr, Westinghouse Astronuclear Lab, 63-64; NUCLEAR ENGR, BROOKHAVEN NAT LAB, 68- *Concurrent Pos:* Adj prof, Polytech Inst NY, 77-78; consult, Singer Link, 82-90, Elec Power Res Inst, 84-88, Ont Nuclear Safety Rev, 87, Gen Physics, 90- *Mem:* Am Nuclear Soc; Sigma Xi. *Res:* Reactor safety of light water and heavy water power and research reactors; responsible for evaluation and application of computer codes used for core performance and plant transients. *Mailing Add:* Brookhaven Nat Lab Upton NY 11973

DIAMOND, EARL LOUIS, b Tiffin, Ohio, Nov 8, 28; m 60; c 1. BIOSTATISTICS EPIDEMIOLOGY. *Educ:* Univ Miami, AB, 50; Univ NC, MA, 52, PhD(statist), 58. *Prof Exp:* Asst math statist, Univ NC, 52-54, biostatist, 56-57, asst prof, 57-59; sr asst sanitarian, Commun Dis Ctr, USPHS, 54-56; asst prof, 59-64, assoc prof, 64-69, PROF EPIDEMIOL & BIOSTATIST, SCH HYG & PUB HEALTH, JOHNS HOPKINS UNIV, 69- *Mem:* Am Statist Asn; Biomet Soc; Inst Math Statist; Soc Epidemiol Res. *Res:* Biostatistics; epidemiology. *Mailing Add:* Sch Hyg & Pub Health 615 N Wolfe St Johns Hopkins Univ Baltimore MD 21205

DIAMOND, FRED IRWIN, b Brooklyn, NY, Dec 13, 25; m 56; c 3. SIGNAL PROCESSING, ARTIFICIAL INTELLIGENCE. *Educ:* Mass Inst Technol, BEE, 50; Syracuse Univ, MEE, 53, PhD(elec eng), 66. *Prof Exp:* Electronic engr, 50-61, sr scientist, 61-73, tech dir, 73-81, CHIEF SCIENTIST, ROME AIR DEVELOP CTR, AIR FORCE SYSTS COMMAND, US AIR FORCE, 81- *Concurrent Pos:* Instr, Utica Col, 56-58; lectr, Syracuse Univ, 59-61; exec chmn, Commun Subgroup, Australian, Can, NZ, UK & US, Tech Coop Prog, 83-; NATO adv group for aerospace res & dev, 75-86. *Mem:* Fel Inst Elec & Electronics Engrs; AAAS; NY Acad Sci. *Res:* Exploratory development in electronics, electromagnetics, materials, communications, sensors, and information processing as applied to surveillance, communications, command and control systems. *Mailing Add:* 1715 Lincoln Lane Rome NY 13440

DIAMOND, HAROLD GEORGE, b Wurtsboro, NY, Feb 15, 40; m 63; c 2. NUMBER THEORY, ANALYSIS. *Educ:* Cornell Univ, AB, 61; Stanford Univ, PhD(math), 65. *Prof Exp:* NSF fel, Swiss Fed Inst Technol, 65-66 & Inst Adv Study, 66-67; from asst prof to assoc prof, 67-71, PROF MATH, UNIV ILL, URBANA, 71- *Concurrent Pos:* Sabbatical vistor, Univ Nottingham, 73-74 & Univ Tex, 81-82, 88-89; ed, Am Math Monthly, 87-91. *Mem:* Am Math Soc; Math Asn Am. *Res:* Analytic number theory; mathematical analysis; asymptotic distribution of multiplicative arithmetical functions; application of Banach algebras and harmonic analysis in number theory; sieve theory. *Mailing Add:* Dept Math 374 Altgeld Hall Univ Ill 1409 W Green St Urbana IL 61801

DIAMOND, HERBERT, b Chicago, Ill, July 28, 25; m 48; c 1. NUCLEAR CHEMISTRY. *Educ:* Univ Chicago, PhB, 47, BS, 48. *Prof Exp:* From asst chemist to chemist, 49-72, CHEMIST NUCLEAR CHEM, ARGONNE NAT LAB, 72- *Mem:* Am Chem Soc; Am Phys Soc. *Res:* Production and characterization of new nuclides; nuclear half-lives; cross sections; decay schemes; nuclear aspects of cosmic and geochemical problems; separations chemistry; nuclear waste management; plutonium polymer characterization. *Mailing Add:* Argonne Nat Lab Bldg 200 M119 Argonne IL 60439

DIAMOND, HOWARD, b Detroit, Mich, Aug 11, 28; m 51; c 4. ELECTROCHEMISTRY, PHYSICAL CHEMISTRY. *Educ:* Univ Mich, BS, 52, MS, 54, PhD(physics & elec eng), 60. *Prof Exp:* Res assoc, Eng Res Inst, Univ Mich, 57-59, from instr to assoc prof, 58-70; pres, Transidyne-Gen Corp, 67-86, chmn, 80-86; PRES, DIAMOND GEN CORP, 86- *Concurrent Pos:* NSF grant, 62-63; consult, Lear Siegler Inc, Mich, 60-62 & Electrovoice Corp, 63- *Mem:* Nat Acad Sci; AAAS; Am Phys Soc; Sigma Xi. *Res:* Semiconductor and dielectric theory; theory of ferroelectricity and piezoelectricity. *Mailing Add:* Diamond Gen PO Box 8014 Ann Arbor MI 48107

DIAMOND, IVAN, b Brooklyn, NY, May 7, 35; m 62; c 3. BIOCHEMISTRY. *Educ:* Univ Chicago, AB, 56, BS, 57, MD, 61, PhD(neuroanat), 67. *Prof Exp:* Intern med, New Eng Ctr Hosp, Boston, Mass, 61-62; resident neurol, Univ Chicago, 62-65, instr, Sch Med, 65-67; asst, Beth Israel Hosp, Boston, 67-69; from asst prof to assoc prof, 68-80, PROF NEUROL, PEDIAT & PHARMACOL, MED CTR, UNIV CALIF, SAN FRANCISCO, 80-, VCHMN, DEPT NEUROL & DIR, ERNEST GALLO CLIN & RES CTR, 85- *Concurrent Pos:* USPHS res fel neurol, Univ Chicago, 64-65, spec res fel biol chem, Harvard Med Sch, 67-69 & career develop award, 69. *Honors & Awards:* Joseph A Capps Prize, Inst Med Chicago, 66; Royer Award, 82. *Mem:* AAAS; Am Acad Neurol; Am Fedn Clin Res; Am Soc Clin Invest; Am Soc Biol Chem; Am Neurol Asn. *Res:* Bilirubin metabolism and neurotoxicity; biochemistry of synaptic function; regulation of receptor function; interaction of alcohol with the nervous system; alcoholic neurologic disorders. *Mailing Add:* Dept Neurol Univ Calif San Francisco CA 94143

DIAMOND, JACOB J(OSEPH), b New York, NY, July 25, 17; m 44; c 2. PHYSICAL CHEMISTRY, SCIENCE ADMINISTRATION. *Educ:* Brooklyn Col, BA, 37. *Prof Exp:* Chemist, Nat Bur Standards, 40-70, mgr protective equip prog, Law Enforcement Standards Lab, 70-72, chief, Law Enforcement Standards Lab, 72-79; RETIRED. *Concurrent Pos:* Assoc, George Washington Univ, 47-48; ed, Bibliog High Temperature Chem & Physics of Mat in Condensed State, 57-70; assoc mem, Comn High Temperatures & Refractories, Int Union Pure & Appl Chem, 59-67; mem, Nat

Adv Comt Law Enforcement Equip & Technol, 76-79. *Honors & Awards:* Student Medal, Am Inst Chem, 37; Silver Medal, Dept of Com, 76. *Mem:* Fel AAAS; emer mem Am Chem Soc; Am Ceramic Soc; Int Asn Chiefs of Police. *Res:* Performance standards and guidelines for law enforcement equipment; protective, communications, security and investigative equipment; instrumental methods of chemical analysis; analysis of silicates; optical glass; flame photometry; image-furnace research; vaporization of refractory materials; high temperature chemistry. *Mailing Add:* 109 Ridgewood Dr Longwood FL 32779-3312

DIAMOND, JARED MASON, b Boston, Mass, Sept 10, 37; c 2. PHYSIOLOGY, ECOLOGY. *Educ:* Harvard Univ, BA, 58; Cambridge Univ, Eng, PhD(physiol), 61. *Prof Exp:* Fel, NSF, 58-61 & 61-62; jr fel, Soc Fel, Harvard Univ, 62-65, assoc biophys, Med Sch, 65-66; assoc prof physiol, Med Ctr, 66-68, PROF PHYSIOL, SCH MED, UNIV CALIF, LOS ANGELES, 68- *Concurrent Pos:* Fel physiol, Trinity Col, Cambridge, Eng, 61-65; res assoc, dept ornith, Am Museum Natural Hist, 73-, Los Angeles County Mus Natural Hist, 85-; fel, MacArthur Found, 85- *Honors & Awards:* Bowditch Prize, Am Physiol Soc, 76; Kroc Found Lectr, Western Asn Physicians, 78; Burr Award, Nat Geog Soc, 79. *Mem:* Nat Acad Sci; fel Am Acad Arts & Sci; fel Am Ornithologists Union; Am Physiol Soc; Biophys Soc; Am Soc Naturalists; Am Philos Soc. *Res:* Biological membranes; ecology & evolutionary biology, mainly bird faunas of New Guinea and other southwest Pacific islands; author of over 300 articles, and books. *Mailing Add:* Dept Physiol Sch Med Univ Calif Los Angeles CA 90024

DIAMOND, JOSHUA BENAMY, b Chicago, Ill. MAGNETISM PHYSICS. *Educ:* Univ Calif, Berkeley, BA, 65; Univ Pa, PhD(physics), 71. *Prof Exp:* Res assoc, Kamerlirgh Onnes Lab, Univ Leiden, 71-72; fel, Mass Inst Technol, 72-73; asst prof physics, New England Aeronaut Int, 73-75 & Univ Pittsburgh, Bradford, 75-80; ASST PROF PHYSICS, SIENA COL, 80- *Mem:* Am Phys Soc; Am Asn Physics Teachers; Sigma Xi. *Res:* Theory of magnetism in alloys; electronic bend structure. *Mailing Add:* Physics Dept Siena Col Londonville NY 12211

DIAMOND, JULIUS, b Philadelphia, Pa, Apr 12, 25; m 53, 74; c 3. MEDICINAL CHEMISTRY. *Educ:* Univ Pa, BS, 45; Temple Univ, MA, 53, PhD(chem), 55. *Prof Exp:* Org res chemist, Wyeth Labs, Pa, 50-56; dir labs, G F Harvey Co, NY, 56-57; proj leader, Wallace Labs, NJ, 57-59; tech dir, Lincoln Labs, Ill, 59-62; group leader, William H Rorer Inc, Pa, 62-67; dir basic res, Cooper Labs, Inc, NJ, 73-79; dir chem res, Berlex Labs, NJ, 79-81; dir new drug acquisitions, Key Pharm, Fla, 82-86; consult pharmaceut prod acquisitions & develop, Schering-Plough Corp, NJ, 86-87; RETIRED. *Concurrent Pos:* Lectr, Pa State Univ, 64-65. *Mem:* Am Chem Soc; fel Am Inst Chemists; Acad Pharmaceut Sci; Sigma Xi; NY Acad Sci. *Res:* Pharmaceuticals; medicinal chemistry; drug metabolism; gastrointestinal drugs; antiinflammatory drugs; cardiovascular drugs; synthetic analgesics; bronchopulmonary drugs; dental antiplaque agents. *Mailing Add:* 179 Lake Dr Mountain Lakes NJ 07046

DIAMOND, LEILA, b Newark, NJ, July 19, 25. CHEMICAL CARCINOGENESIS, CELL BIOLOGY. *Educ:* Univ Wis-Madison, BA, 45; Cornell Univ, PhD(biol), 61. *Prof Exp:* From asst prof to assoc prof, 64-77, PROF, WISTAR INST, 78- *Concurrent Pos:* USPHS fel, Inst Virol, Univ Glasgow, 61-63; mem, Chem Path Study Sect, NIH, 86-90; assoc ed, carcinogenesis, 85-, J Nat Cancer Inst, 87; prof path & lab med & assoc fac, Sch Med Univ Pa, 82- *Mem:* Am Asn Cancer Res; Am Soc Cell Biol; Tissue Cult Asn; AAAS. *Res:* Tissue culture; cell transformation; mutagenesis; tumor virology. *Mailing Add:* Wistar Inst 3601 Spruce St Philadelphia PA 19104-4268

DIAMOND, LOUIS, b Baltimore, Md, July 13, 40; m 67; c 2. PHARMACOLOGY. *Educ:* Univ Md, BSc, 61, MSc, 64, PhD(pharmacol), 67. *Prof Exp:* From asst prof to assoc prof, Univ Ky, 67-80, prof pharmacol, 80-86; PROF & DEAN, SCH PHARM, UNIV COLO, 86- *Mem:* AAAS; Am Soc Pharmacol & Exp Therapeut; Am Therapeut Soc. *Res:* Pulmonary pharmacology. *Mailing Add:* Sch Pharm Univ Colo Boulder CO 80309-0297

DIAMOND, LOUIS KLEIN, b New York, NY, May 11, 02; m 29; c 2. MEDICINE, PEDIATRICS. *Educ:* Harvard Univ, AB, 23, MD, 27. *Prof Exp:* Asst pediat & path, Harvard Med Sch, 30-32, asst pediat, 32-33, instr, 33-35, assoc, 35-41, from asst prof to prof, 41-68; assoc prof, Med Sch, San Francisco Med Ctr, Univ Calif, 68-70, resident prof, 70-76, adj prof pediat, 76-84; RETIRED. *Concurrent Pos:* Res fel pediat, Harvard Med Sch, 27-28; mem res staff, Children's Hosp, 28-30, asst physician, 31, assoc physician, 32-33, assoc vis physician, 34-38, sr physician, 38-45, assoc physician-in-chief, 46-; med dir, Nat Blood Prog, Am Nat Red Cross, 48-50; mem, Hemat Study Sect, NIH, 49-53 & Human Embryo & Develop Study Sect, 54-59, 64-68; mem, Subcomt Blood & Related Probs, Nat Res Coun. *Honors & Awards:* Mead-Johnson Award, Am Acad Pediat, 46; Carlos J Findlay Gold Medal, Cuba, 51; Merit Award, Neth Red Cross, 59; Karl Landsteiner Award, 62; Theodore Roosevelt Award, 64; George R Minot Award, 65. *Mem:* Am Acad Pediat; Soc Pediat Res; Am Pediat Soc; Am Soc Clin Invest; AMA. *Res:* Children's diseases, especially diseases of the blood. *Mailing Add:* 12320 Montana Ave No 201 Los Angeles CA 90049

DIAMOND, LOUIS STANLEY, b Philadelphia, Pa, Feb 6, 20; m 39; c 3. ZOOLOGY, ENTOMOLOGY. *Educ:* Univ Pa, AB, 40; Univ Mich, MS, 41; Univ Minn, PhD, 58. *Prof Exp:* Wildlife res biologist, Patuxent Res Refuge, US Fish & Wildlife Serv, 51-53; vet parasitologist, Animal Dis & Parasite Res Br, Agr Res Serv, USDA, 53-58; MED PARASITOLOGIST SEC CHIEF, LAB PARASITIC DIS, NAT INST ALLERGY & INFECTIOUS DIS, NIH, US DEPT HEALTH & HUMAN SERV, 59- *Concurrent Pos:* Guest scientist & founding mem, Ctr for Studies on Amoebiasis, Mex City, 77; sr vis scientist & consult, Sch Med, Keio Univ, Tokyo, 78; consult, Expert Adv Panel Parasitic Dis, WHO, 68-87. *Mem:* Am Soc Parasitol; Am Micros Soc; Am Soc Trop Med & Hyg; Soc Protozoologists; Sigma Xi. *Res:* Axenic cultivation of entamoeba and trichomonads of human and veterinary importance; pathobiology of human pathogen entamoeba histolytica; host parasite relationships of indigenous viruses of E histolytica; cryobiology. *Mailing Add:* NIH Bldg 4 Rm 126 Bethesda MD 20892

DIAMOND, MARIAN C, b Glendale, Calif, Nov 11, 26; m 50, 82; c 4. NEUROBIOLOGY. *Educ:* Univ Calif, Berkeley, AB, 48, MA, 49, PhD(anat), 53. *Prof Exp:* Res asst neurobiol, Harvard Col, 52-53; res asst & instr, Cornell Univ, 54-58; lectr anat, Sch Med, San Francisco, 59-61, res assoc psychol, Univ Calif, Berkeley, 61-65, from asst prof to assoc prof anat, 65-74, from asst dean to assoc dean, Col Lett & Sci, 68-72, dir, Laurence Hall Sci, 90, PROF ANAT, UNIV CALIF, BERKELEY, 74- *Concurrent Pos:* Res grants, NSF, 61-62, 66-71 & 71-74, NIH, 66-69, WHO, 78 & biomed, Univ Calif, Berkeley, 78-90; vis fel, Nat Univ Australia, Canberra, 77; vis lectr, Fudan Univ, Shanghai, People's Rep China, 85, Nairobi Univ, Kenya, EAfr, 88; fel, WHO, Australia. *Honors & Awards:* Hall of Fame, Am Asn Univ Women. *Mem:* Fel AAAS; Soc Neurosci; Int Soc Psychoneuroendocrinol; Am Asn Anat; Int Soc Develop Neurosci; hon mem Sigma Xi. *Res:* Environmentally induced anatomical and chemical brain changes; pituitary hormones and sex steroid hormones and brain development. *Mailing Add:* Dept Integrative Biol Univ Calif 345 Malford Hall Berkeley CA 94720

DIAMOND, MILTON, b New York, NY, Mar 6, 34; wid; c 4. SEXOLOGY & REPRODUCTIVE BIOLOGY, AIDS. *Educ:* City Col New York, BS, 55; Univ Kans, PhD(anat & psychol), 62. *Prof Exp:* From instr to asst prof anat, Sch Med, Univ Louisville, 62-67; assoc prof anat, 67-71, PROF ANAT & REPROD BIOL, SCH MED, UNIV HAWAII, 71- *Concurrent Pos:* NIH grant, Univ Louisville & Univ Hawaii, 62-77; consult, State of Hawaii Dept Educ, 71-; Pop Coun grant, Univ Hawaii, 72-73, NIH grants, 73-76; res prof psychiat & behav sci, State Univ NY Stony Brook, 75-78; Consult, Dutch Univ Health, Swedish (Noah's Arc) AIDS Soc; grants, foreign & domestic, 80-; exec dir, Hawaii AIDS task group; pres, Western region, Soc Sci Study Sex, 87. *Honors & Awards:* Lederle Med Fac Award, 68-71; Creative Prog Award, Nat Univ Exten Asn "Human Sexuality" TV Ser, 73. *Mem:* fel Soc Sci Study Sex; fel Int Acad Sex Res; Am Asn Sex Educ, Counr & Therapists; Am Soc Sexology. *Res:* Sexology; sexual development; reproduction; contraception and abortion; sex education; pornography; genetics of homosexuality and bisexuality; AIDS. *Mailing Add:* Dept Anat & Reprod Biol Univ Hawaii Sch Med 1951 East-West Rd Honolulu HI 96822

DIAMOND, RAY BYFORD, b Louisa, Ky, Jan 10, 33; m 57; c 3. SOIL FERTILITY. *Educ:* Ohio State Univ, BS, 54; Univ Fla, MSA, 61, PhD(soils), 63. *Prof Exp:* Agriculturist, Tenn Valley Authority, 63-64, agronomist fertilizer usage, 64-76, fertilizer use specialist, USAID contract, 74-76; regional coordr, Africa, 76-79, CHIEF COORDR MKT DEVELOP & FERTILIZER EVALUATIONS, INT FERTILIZER DEVELOP CTR, 79- *Concurrent Pos:* Vis scientist, Int Rice Res Inst, 82-83; res & training consult, Govt Bangladesh, 84-85. *Mem:* Am Soc Agron; Soil Sci Soc Am; Int Soil Sci Cong. *Res:* Evaluation of agronomic performance of fertilizer products and fertilization practices under a variety of soil, climatic and cropping conditions; data collection and analysis and technical assistance on fertilizer marketing; distribution and usage relevant to developing countries. *Mailing Add:* Rte 8 Box 107 Florence AL 35630

DIAMOND, RICHARD MARTIN, b Los Angeles, Calif, Jan 7, 24; div; c 4. NUCLEAR STRUCTURE & CHEMISTRY. *Educ:* Univ Calif, Los Angeles, BS, 47; Univ Calif, Berkeley, PhD(nuclear chem), 51. *Prof Exp:* Instr chem, Harvard Univ, 51-54; asst prof, Cornell Univ, 54-58; MEM SR STAFF, LAWRENCE BERKELEY LAB, UNIV CALIF, 58- *Concurrent Pos:* Guggenheim fel, Denmark, 66-67; mem, US Physics Deleg USSR, 66; chmn, Gordon Conf on Nuclear Chem, 65 & Conf on Ion Exchange, 69; mem rev comt, Physics Div, Oak Ridge Lab, 72-74, Neutron-Gamma Facil & Isotope Separator, Brookhaven Nat Lab, 83 & 88, Pi-Gamma Spectrometer, 83, NSRC, Can, Chalk River Nuclear Lab, Nuclear Physics, 88-90; chmn rev comt, Univ Isotope Separator at Oak Ridge, Oak Ridge Nat Lab, 74-75; Fulbright travel grant, Australia, 77; mem prog adv comt, Ind Cyclotron Facil, 80-83, Tandem-Linac Facil, Argonne Nat Lab, 83-86, Holifield Henryalon Res Facil, 88-90; vis fel, Japan Soc Prom Sci, 81; chmn, subcomt high spin & nuclei far from stability, Dept Energy-NSF, 83, co-organizer int conf nuclear physics, 80, co-organizer, workshop nat gamma-ray facil, 87. *Honors & Awards:* Tom W Bonner Award, Am Physical Soc, 80. *Mem:* Am Chem Soc; fel Am Phys Soc; AAAS. *Res:* Nuclear spectroscopy; coulomb excitation; high-spin nuclear structure. *Mailing Add:* Nuclear Sci Div Lawrence Berkeley Lab Berkeley CA 94720

DIAMOND, SEYMOUR, b Chicago, Ill, Apr 15, 25; m 48; c 3. NEUROLOGY, CLINICAL PHARMACOLOGY. *Educ:* Chicago Med Sch, BM, 48, MD, 49. *Prof Exp:* DIR, DIAMOND HEADACHE CLIN, LTD, 73- *Concurrent Pos:* Adj assoc prof neurol, Chicago Med Sch, 70-82, adj prof pharmacol, 83-; exec dir, Nat Migraine Found, 80; exec officer res group on headache & head pain, World Fedn Neurol, 80-; pres, Interstate Postgrad Med Asn, 81. *Honors & Awards:* Physicians Recognition Award, AMA, 70-73 & 74-77; #1 Regent Award, Am Asn Study Headache, 84. *Mem:* AMA; Am Soc Clin Pharmacol & Therapeut; Int Asn Study Pain; World Fedn Neurol; Am Asn Study Headache (past pres). *Res:* Research using pharmaceutical preparations for the prophylactic and abortive treatment of headache; behavioral modification and pain. *Mailing Add:* 5252 N Western Ave Chicago IL 60625

DIAMOND, SIDNEY, b New York, NY, Nov 10, 29; m 53; c 2. ENGINEERING MATERIALS, CHEMISTRY. *Educ:* Syracuse Univ, BS, 50; Duke Univ, MF, 51; Purdue Univ, PhD(soil chem), 63. *Prof Exp:* Hwy res engr, US Bur Pub Roads, 53-62; asst, Purdue Univ, 62-63; res chemist, US Bur Pub Roads, 64-65; assoc prof, 65-70, PROF ENG MAT, PURDUE UNIV, 70- *Concurrent Pos:* Ed, Cement & Concrete Res, 76-; pres, Sidney Diamond & Assocs, Inc, 77-; mem, Panel on Status of Res & Develop US Cement & Concrete Industs, Nat Mat Adv Bd, 77-80; chmn, Fourth Int Conf Effect Alkalies, 78, Seventh Int Cong Chem of Cement, Paris, 80 & Symp Effects Flyash in Cement & Concrete, Boston, 81; mem, adv bd, Ctr Cement Composite Mat, Univ Ill, 87-90, Nat Acad Sci-Nat Res Coun Panel Concrete Durability, 85-87 & adv comt, Concrete & Struct, Strategic Highway Res Prog, 87-88; chmn, Int Symp Durability Glass Fiber Reinforced Concrete, Chicago, 85. *Honors & Awards:* Copeland Award, Am Ceramic Soc. *Mem:*

Fel Am Ceramic Soc (trustee, 78-81); Am Concrete Inst; Mat Res Soc. *Res:* Physics and chemistry of cement and concrete; microstructural characterization of inorganic systems by scanning electron microscopy; cement hydration; concrete durability. *Mailing Add:* G219 Civil Eng Bldg Purdue Univ West Lafayette IN 47907-1284

DIAMOND, STEVEN ELLIOT, b Brooklyn, NY, Sept 15, 49; m 79. CLINICAL DIAGNOSTICS, CLINICAL CHEMISTRY. *Educ:* Rensselaer Polytech Inst, BS, 71; Stanford Univ, PhD(chem), 75. *Prof Exp:* Res chem, Allied Signal, Inc, 75-78, sr res chem, 78-81, res assoc, 81-84, sr res assoc, 84-88, mgr biochem, 84-88; DIR, CHEM RES & DEVELOP, BAXTER DIAGNOSTICS, INC, 88- *Mem:* Am Chem Soc; NY Acad Sci; AAAS; Am Asn Clin Chem. *Res:* Bioassay biosensor analytical methods and synthetic chemical research and development; areas of study include immunoassays, nucleic acid probe assays and enzyme; metabolite; ion reagent development. *Mailing Add:* 2139 Oakland Hills Way Coral Springs FL 33071-6599

DIANA, JOHN N, b Lake Placid, NY, Dec 19, 30; m 54, 66; c 3. PHYSIOLOGY. *Educ:* Norwich Univ, BA, 52; Univ Louisville, PhD(physiol), 65. *Prof Exp:* Biochemist, Inst Med Res, Louisville, Ky, 54-56; physiologist, US Army Med Res Lab, Ft Knox, Ky, 56-58; from res asst to res assoc cardiovasc physiol, Sch Med, Univ Louisville, 58-65; asst prof, Col Human Med, Mich State Univ, 66-68; from assoc prof to prof physiol & biophys, Col Med, Univ Iowa, 68-78; prof & chmn physiol & biophys, Sch Med, La State Univ, 78-85; DIR, CARDIOVASC RES PROG, UNIV KY, COL MED, 85-, ASSOC DEAN, RES & BASIC SCI, 87- *Concurrent Pos:* Fel, Sch Med, Univ Okla, 65-66; Am Heart Inst fel, 65-67; fel coun circulation, Am Heart Asn, 71-; mem, Cardiovasc Res Study Comt A, AMA, 80-85, Cardiovasc-Renal Study Sect, NHLBI, 80-84, Clin Sci Study Sect, 86- *Mem:* AAAS; Am Fedn Clin Res; Am Physiol Soc; Microcirc Soc (pres, 77-78). *Res:* Cardiovascular research, especially venomotor activity, transcapillary fluid movement and capillary permeability in skeletal muscle, intestine and myocardium during infusion of vasoactive agents; effects of hypertension, hypotension, shock and heart failure on capillary permeability; testing the hypothesis that microvascular permeability is a regulated variable; permeability to fluid and solutes is modulated in accordance with cellular and organism homeostasis. *Mailing Add:* c/o T&H Res Inst Cooper & Alumni Dr Univ Ky Col Med Lexington KY 40546

DIANA, LEONARD M, b Columbia, Pa, Jan 26, 23; m 50, 80; c 3. PHYSICS. *Educ:* Ga Inst Technol, BS, 48; Univ Pittsburgh, PhD(physics), 53. *Prof Exp:* Asst, Univ Pittsburgh, 48-50, instr, 49-52, res assoc, 51-53; proj physicist res & develop, Standard Oil Co, 53-59; physicist, Am Tobacco Co, 59-62; assoc prof physics, Univ Richmond, 62-65; assoc prof, 65-71, assoc dean, Col Sci, 75-81, PROF PHYSICS, UNIV TEX, ARLINGTON, 71- *Concurrent Pos:* Va state chmn, Vis Scientists Prog Physics, High Schs, 63-65; specialist, US Agency Int Develop, India, 66; adv coun, Am Phys Soc, 82-87. *Mem:* Fel AAAS; fel Am Phys Soc; Am Asn Physics Teachers. *Res:* Experimental nuclear physics; instrumentation; positron annihilation; liquid structure; position scattering from atoms and molecules. *Mailing Add:* Dept Physics Univ Tex Box 19059 Arlington TX 76019-0059

D'IANNI, JAMES DONATO, b Akron, Ohio, Mar 11, 14; m 40; c 1. POLYMER CHEMISTRY. *Educ:* Univ Akron, BS, 34; Univ Wis, PhD(org chem), 38. *Hon Degrees:* DSc, Univ Akron, 79. *Prof Exp:* Asst chem, Univ Wis, 34-37; Procter & Gamble fel, Univ Wis, 37-38; res chemist, Goodyear Tire & Rubber Co, 38-51, asst to vpres res & develop, 52-61, chem prod liaison, 61-63, assoc dir res, 63-65, asst dir, 65-68, dir res, 65-77, asst to vpres res, 78; CONSULT, 78- *Concurrent Pos:* Instr, Univ Akron, 41-46; chief polymer res br, Off Rubber Reserve, Reconstruct Finance Corp, 46-47; chmn, Gordon Res Conf Elastomers, 52. *Honors & Awards:* Charles Goodyear Medal Award, Am Chem Soc, 77. *Mem:* Am Chem Soc (pres, 80); Am Inst Chem; Am Inst Chem Eng; Royal Inst Chem. *Res:* Catalytic hydrogenation of hydroxyamides and lignin; synthesis of polymerizable monomers; synthetic rubber; vinyl resin; chemical derivatives of rubbers. *Mailing Add:* 860 Sovereign Rd Akron OH 44303

DIANZANI, FERDINANDO, b Grosseto, Italy, Sept 12, 32; m 61; c 2. BIOLOGY OF INTERFERON CLINICAL TRIALS, HUMAN RETROVIROLOGY. *Educ:* Univ Siena, MD, 59, PhD(microbiol), 65, PhD(virol), 68. *Prof Exp:* Asst prof microbiol, Med Sch, Univ Siena, 59-70; prof microbiol, Med Sch, Univ Turin, 70-76; prof, 76-81, ADJ PROF MICROBIOL, MED SCH, UNIV TEX, 81; PROF & CHMN VIROL, MED SCH, UNIV ROME, 81- *Concurrent Pos:* Co-ed, J Biol Regulators & Homeostatic Agents, 87-; vis prof award, Wellcome Found Am Acad Microbiol, 89; mem, Ital Nat AIDS Comt & Nat Health Coun, 91; mem, Int Coun, Int Soc Antiviral Res; memm, Int Coun, Int Soc Antiviral Res. *Honors & Awards:* Prem Int G Lenghi, Nat Acad Lincei, 73. *Mem:* Am Soc Microbiol; Int Soc Interferon Res (pres, 90-92); Int Soc Antiviral Res; Soc Exp Biol & Med; Am Asn Immunologists. *Res:* Biology and epidemiology of measles virus and coxsackie viruses; pathogenesis of viral infection; interferon; defense mechanisms of viral and neoplasias; mechanisms of cellular immunity; etiopathogenesis of AIDS; virology. *Mailing Add:* Inst Virol Univ Rome Rome 00185 Italy

DIAS, JERRY RAY, b Oakland, Calif, Oct 26, 40; m 58; c 3. PHYSICAL ORGANIC CHEMISTRY, CHEMICAL ENGINEERING. *Educ:* San Jose State Univ, BS, 65; Ariz State Univ, PhD(phys & org chem), 70. *Prof Exp:* Test technician, Fuel Filtration Filters, Inc, 59-61; engr asst semiconductors, Amelco, Inc, 61-62; supvr electroplating, Huggins Microwave Labs, 62-64; res asst chem, San Jose State Univ, 64-66; engr electrochem, Fairchild Corp, Mountain View, 66-67; NIH fel chem, Ariz State Univ, 68-70; fel, Stanford Univ, 70-72; from asst prof to assoc prof, 72-83, PROF CHEM, UNIV MO, KANSAS CITY, 83- *Concurrent Pos:* Consult, Mobay Chem Corp, 78 & Region VII, Environ Protection Agency, 79 & 80; Fulbright-Hays sr scholar, Yugoslavia, 81. *Mem:* Am Chem Soc; Am Electroplaters Soc; Am Soc Testing & Mat; Electrochem Soc; Nat Soc Prof Engrs; fel Am Inst Chemists. *Res:* Organic chemical mechanisms and synthesis; material science; mass spectrometry, structural elucidation, natural products and chemical graph theory. *Mailing Add:* Dept Chem Univ Mo Kansas City MO 64110

DIASIO, ROBERT BART, b New York, NY, Jan 20, 46; m 70; c 3. ONCOLOGY, PHARMACOLOGY. *Educ:* Univ Rochester, BA, 67; Yale Univ, MD, 71. *Prof Exp:* Intern internal med, Strong Mem Hosp, Univ Rochester, 71-72, resident, 72-73; clin assoc med oncol, Nat Cancer Inst, 73-75, res assoc clin pharmacol, 75-76; from asst prof to assoc prof med & pharmacol, Med Col Va, 76-84; PROF MED & PHARMACOL & DIR CLIN PHARMACOL, UNIV ALA, BIRMINGHAM, 84-, CHMN DEPT PHARMACOL, 89- *Mem:* Am Fedn Clin Res; Am Soc Clin Oncol; Am Asn Cancer Res; Am Soc Clin Investr; Am Soc Pharmacol & Exp Therapeut; Am Soc Clin Pharmacol Ther; Am Col Clin Pharmacol. *Res:* Biochemical and clinical pharmacology of antineoplastic drugs. *Mailing Add:* Dept Pharmacol Univ Ala Box 600-VH Birmingham AL 35294

DIASSI, PATRICK ANDREW, b Morristown, NJ, July 1, 26; m 52. ORGANIC CHEMISTRY. *Educ:* St Peter's Col, NJ, BSc, 46; Rutgers Univ, MSc, 50, PhD(chem), 51. *Prof Exp:* Asst, Rutgers Univ, 47-50; res assoc, 51-63, res supvr, 63-66, sr res assoc, 66-68, asst dir dept org chem, 68-71, dir dept chem process develop, 71-72, dir chem & microbiol, 72-77, ASSOC DIRS SQUIBB INST MED RES, 72-, VPRES CHEM, RES & DEVELOP, 77- *Mem:* AAAS; Am Chem Soc; Royal Soc Chem. *Res:* Chemistry of natural products. *Mailing Add:* 744 Norgate Westfield NJ 07090-3427

DIAUGUSTINE, RICHARD PATRICK, b Hackensack, NJ, Jan 15, 42; m 65; c 2. ONCOLOGY, MOLECULAR PHARMACOLOGY. *Educ:* Northeastern Univ, BS, 64; Tulane Univ, PhD(pharmacol & biochem), 68. *Prof Exp:* NIH fel pharmacol, Sch Med, Univ Iowa, 68-70; staff fel biochem, 70-74, res chemist molecular pharmacol & biochem, 74-77, HEAD ENDOCRINOL GROUP LAB PULMONARY FUNCTION & TOXICOL, NAT INST ENVIRON HEALTH SCI, NIH, 77- *Concurrent Pos:* Adj asst prof med, Med Ctr, Duke Univ, 75- *Mem:* Am Soc Pharmacol & Exp Therapeut. *Res:* Formation and secretion of polypeptide hormones by chemically-induced lung carcinomas; early modulation of gene expression in chemical oncogenesis; secretions and secretory cells of the terminal airways of the lung. *Mailing Add:* NIH Nat Inst Environ Health Sci PO Box 12233 Research Triangle Park NC 27709

DIAZ, ARTHUR FRED, b Calexico, Calif, Dec 25, 38; m 62; c 5. ORGANIC CHEMISTRY, ELECTROCHEMISTRY. *Educ:* San Diego State Univ, BS, 60; Univ Calif, Los Angeles, PhD(chem), 65. *Prof Exp:* Res assoc, Univ Calif, Los Angeles, 65-69; mem prof res staff, TRW Inc, 69-70; asst prof, Univ Calif, San Diego. 70-74; prog mgr, NSF, 74-75; MEM RES STAFF, IBM CORP, 75- *Mem:* Am Chem Soc; Electrochem Soc. *Res:* Thin polymer films on electrode surfaces. *Mailing Add:* 3864 Wellington Sq San Jose CA 95136

DIAZ, CARLOS MANUEL, b Chile, Apr 16, 32; m 56; c 5. EXTRACTIVE METALLURGY, PYROMETALLURGY. *Educ:* Univ Chile, BSc, 54; Columbia Univ, MSc, 58; Univ London, PhD(extractive metall), 66. *Prof Exp:* Asst prof extractive metall, Univ Chile, 55-58, assoc prof, 58-66, prof, 67-75, head, Dept Mines, 67-73, dir, Sch Eng, 69-73; process engineer 75-78, SECT HEAD PYROMETAL, J ROY GORDON RES LAB, INCO METALS CO, 78- *Concurrent Pos:* Dir mining & extractive metall, Latin Am Prog Adv Sci & Technol, 70-72; vis prof, Nat Univ Eng, Peru, 70, Nat Univ, Colombia, 70 & 78, Univ San Luis Potosi, Mex, 78; consult, UNESCO, 79. *Honors & Awards:* Ismael Valdes Prize, Chilean Inst Engrs, 57; Gold Medal, Chilean Inst Mining Engrs, 85. *Mem:* Chilean Inst Mining Engrs; Can Inst Mining & Metall Engrs. *Res:* Pyrometallurgical operations for treating sulfide and oxide nickel and copper ores with the objective of improving existing processes. *Mailing Add:* 210 Radley Rd Mississauga ON L5G 2R7 Can

DIAZ, FERNANDO G, b Mexico City, Mex, Dec 29, 46; c 2. MICROSURGERY, LASER SURGERY. *Educ:* Univ Mexico, BSc, 63, MD, 70; Univ Kans, MA, 73; Univ Minn, PhD, 80; Cent Mich Univ, MSA, 88. *Prof Exp:* Instr anat, Univ Mex, 65-68, neuroanat, Univ Kans, 71-73; resident surg, Univ Kans, 71-73, neurosurg, Univ Minn, 73-77; fel microsurg, Univ Minn, 77-78; attending neurosurgeon, Henry Ford Hosp, 78-87; assoc prof neurosurg, Univ Mich, 86-88; CHMN, NEURO INST, SANTA FE HEALTH CARE, 88-; PROF & CHMN NEUROL SURG, WAYNE STATE UNIV, DETROIT, 90- *Concurrent Pos:* Vis prof, Microneurosurg Loyola Univ & Cerebrovasc Review, Univ Kans, 80, Univ Osaka & Univ Kyoto, Japan, 82, Univ Madrid, Spain & Univ Newcastle, Eng, 83, Univ Paris, France & Univ Verona, Italy, 85; site reviewer, Cleveland Clin Grant Rev, 81; fel Stroke Coun, Am Heart Asn; consult, Henry Ford Hosp, 88-; chmn credential comt, Cong Neurol Surgeons; Organon Lab grant, 79; mem, Cong Neurol Surgeons. *Mem:* Am Heart Asn; fel Am Col Surgeons; Int Col Surgeons (vpres, 68-); Am Asn Neurol Surgeons; Neurosurg Soc Am. *Res:* Cerebrovascular research in cerebral blood flow hemodynamics and cerebral reperfusion; methods to extend cerebral tissue viability during or after ischemia; applications of laser technology to surgical reconstructive and abortive procedures. *Mailing Add:* Dept Neurol Surg UHC-6E Wayne State Univ 4201 St Antoine Blvd Detroit MI 48201

DIAZ, HENRY FRANK, b Cuba, July 15, 48; US citizen; m 69; c 2. ATMOSPHERIC SCIENCES, CLIMATOLOGY. *Educ:* Fla State Univ, BS, 71; Univ Miami, MS, 74; Univ Colo, Boulder, PhD(geog & climat), 85. *Prof Exp:* Meteorologist, Ctr Exp Design & Data Anal, Environ Data Serv, 74-75, Nat Climat Ctr, 75-83 & Climate Res Prog, Environ Res Lab, 83-85, actg dir, Climate Res Prog, ESG, 85-87, CLIMATE RES DIV, CLIMATE MONITORIAL & DIAG LAB, NAT OCEANIC & ATMOSPHERIC ADMIN, 88- *Concurrent Pos:* Vis scholar, Scripps Inst Oceanog, 82; vis scholar, Univ Mass, 88-89. *Mem:* Am Meteorol Soc; Am Geophys Union; Am Quarternary Asn. *Res:* Climatology and climatic variation; developing long-term, historical climatic data bases; spatial and temporal characteristics of climatic variation in North America. *Mailing Add:* 3520 Cloverleaf Dr Boulder CO 80304

DIAZ, LUIS A, b Cascas, Cajamarca, Peru, Sept 16, 42; US citizen; m 69; c 3. DERMATOLOGY. *Educ:* Universidad Nacional de Trujillo, Trujillo, Peru, MD, 68, Am Bd Dermat, dipl, 74. *Prof Exp:* Clin asst instr, Dept Dermat, State Univ NY, 71-74; fel, Dept Immunol & Dermat, Mayo Grad Sch Med, Rochester, 74-76; from asst prof to assoc prof, Dept Dermat, Univ Mich, 76-82; from assoc prof to prof, Dept Dermat, Johns Hopkins Univ, 82-88; PROF & CHMN, DEPT DERMAT, MED COL WIS, 88- *Concurrent Pos:* Consult physician dermat, Univ Hosp, Univ Mich, 76-82, Wayne County Gen Hosp, 76-82; chief, Dermat Serv, Vet Admin Med Ctr, Ann Arbor, 79-82; attend physician dermat, Johns Hopkins Univ, 82-88, vis prof, 88, Froedtert Mem Lutheran Hosp, 89-, Milwaukee County Med Ctr, 89-, Vet Admin Med Ctr, 89- *Mem:* Soc Investigative Dermat; fel Am Acad Dermat; Am Med Asn; AAAS; Am Fedn Clin Res; Dermat Found; Sigma Xi; Soc Exp Biol & Med; NY Acad Sci; Am Asn Immunologists; Am Dermat Asn. *Res:* Dermatology; immunochemistry; immunocytochemistry; monoclonal antibody production; molecular biology; cell culture. *Mailing Add:* Dept Dermat Med Col Wis 8701 Watertown Plank Rd Milwaukee WI 53226

DIAZ, LUIS FLORENTINO, b Lima, Peru, Apr 20, 46; US citizen; m 68; c 2. ENVIRONMENTAL ENGINEERING, HEAT TRANSFER. *Educ:* San Jose State Univ, BS, 72; Univ Calif, Berkeley, MS, 73, PhD(environ eng), 76. *Prof Exp:* Engr, Pac Gas & Elec Co, 68-72; res asst, Univ Calif, Berkeley, 72-75; PRES, CAL RECOVERY SYSTS INC, 75- *Concurrent Pos:* Consult, WHO, 78, Calif Solid Waste Mgt Bd, 76, City of Berkeley, 75, World Bank, Asian Develop Bank, US Aid & UN Indust Develop Orgn. *Mem:* Am Soc Mech Engrs; Sigma Xi; Soil Conserv Soc; Am Soc Agr Engrs. *Res:* Material and energy recovery from wastes; composting; biogasification and waste heat utilization. *Mailing Add:* Cal Recovery Systs Inc 160 Broadway Richmond CA 94804

DIAZ, NILS JUAN, b Moron, Cuba, Apr 7, 38; US citizen; m 60; c 3. NUCLEAR SPACE POWER & PROPULSION, ADVANCED ENERGY CONVERSION. *Educ:* Univ Villanova, Cuba, BS, 60; Univ Fla, MS, 64, PhD(nuclear eng), 69. *Prof Exp:* Res assoc, Nuclear Eng Sci Dept, Univ Fla, 65-69, asst prof & reactor sup, 69-74, assoc prof & dir, Nuclear Facil, 74-79, prof & dir, 79-84, DIR RES, INNOVATIVE NUCLEAR SPACE POWER INST, UNIV FLA, 85-, PROF & DIR, COL ENG, 86- *Concurrent Pos:* Prin investr grants & contracts, NSF, Dept Energy, Dept Defense, Strategic Defense Initiative Orgn & NASA, 71-91; sr consult, Fla Power Corp, 72-81, Fla Power & Light Co, 75-81, Exxon Nuclear Corp, 76-80, Startup Nuclear Serv, Inc, 83-85; pres & chief engr, Fla Nuclear Assocs, Inc, 76-; dir, Qual Assurance Task Force, Wash Pub Power Supply Syst, 78-79; prin adv, Nuclear Safety Coun, Spain, 81-83; assoc dean res, Sch Eng, Calif State Univ, Long Beach, 84-86; chmn, Eng Res Coun, Am So Eng Educ, 86. *Mem:* Am Nuclear Soc; Am Soc Mech Engrs; Am Soc Eng Educ; AAAS. *Res:* Advanced nuclear space power and propulsion, including innovative ultra high temperature reactor design and energy conversion systems; non-destructive examination and novel imaging techniques. *Mailing Add:* 202 Nuclear Sci Ctr Univ Fla Gainesville FL 32611

DIAZ, PEDRO MIGUEL, anesthesiology, for more information see previous edition

DIAZ, ROBERT JAMES, b Chester, Pa, Oct 16, 46; m 71; c 2. MARINE ECOLOGY. *Educ:* La Salle Col, BA, 68; Univ Va, MS, 71, PhD(marine biol), 76. *Prof Exp:* Res asst ecol & pollution, Va Inst Marine Sci, 71-74, asst marine sci, 74-76, res marine biol, CEngrs, Waterways Exp Sta, 77-78,; asst prof, 78-80, ASSOC PROF, COL WILLIAM & MARY, 80-; ASSOC MARINE SCI, VA INST MARINE SCE, 78- *Res:* Tidal freshwater and estuarine ecology; application of multivariate methods in ecology; taxonomy of oligochaetes; habitat evaluation; secondary productivity; sediment profile photography and image analysis. *Mailing Add:* Sch Marine Sci Col William & Mary Gloucester Pt VA 23062

DI BARTOLO, BALDASSARE, b Trapani, Italy, Jan 5, 26; m 68; c 2. SOLID STATE PHYSICS. *Educ:* Univ Palermo, DSc(indust eng), 50; Inst Telecommun, Rome, dipl, 51; Mass Inst Technol, PhD(physics), 64. *Prof Exp:* Design engr, Microlambda, Italy, 53-56; microwave engr, Studio Tecnico di Consulenza Elettronica, 56-57; vis fel physics, Mass Inst Technol, 57-58, res staff mem, Lab Insulation Res, 58-63; sr scientist & dir Spectros Lab, Mithras Div, Sanders Assocs, Inc, 64-68; assoc prof, 68-75, PROF PHYSICS, BOSTON COL, 75- *Concurrent Pos:* Lectr, Cybernetics Study Ctr, Naval Inst, Naples, 54-56; dir, Int Sch Atomic & Molecular Spectros, 73- *Mem:* AAAS; Am Phys Soc; Ital Phys Soc; Int Photochem Soc; Am Chem Soc. *Res:* Solid state and molecular spectroscopy; maser and laser theory; theory of atomic and crystal spectra; information theory; microwave technique; flash photolysis; photoacoustic spectroscopy. *Mailing Add:* Dept Physics Boston Col Chestnut Hill MA 02167

DIBB, DAVID WALTER, b Draper, Utah, July 4, 43; m 66; c 4. SOIL FERTILITY. *Educ:* Brigham Young Univ, BSc, 70; Univ Ill, PhD(soil fertil & plant nutrit), 74. *Prof Exp:* Res assoc soils, NC State Univ, 74-75; southern midwest dir mkt develop & res, Potash Inst, 75-77, southcentral dir res, educ & mkt develop, 77-82, latin am coordr & southeast dir, 82-85, vpres domestic prods, 85-87, sr vpres, 87-, PRES, POTASH & PHOSPHATE INST. *Concurrent Pos:* Mem, Agron Sci Found bd trustees, Am Soc Agron, 82-85, pres, 83-85, chmn, Budget & Finance Comt, 87-88; adj prof, Purdue Univ, 85-88; mem, Fertil Indust Adv Comt, Food & Agr Orgn, 89- *Honors & Awards:* Al Lang Award. *Mem:* Fel Am Soc Agron; Soil Sci Soc Am; Int Soil Sci Soc; AAAS; Soil & Water Conserv Soc; Coun Agr Sci Technol. *Res:* Corn growth as affected by form of nitrogen; subsoil management and its effect on soybean growth; crop management for maximum economic yields. *Mailing Add:* Potash & Phosphate Inst 2801 Buford Hwy NE Suite 401 Atlanta GA 30329

DIBBEN, MARTYN JAMES, b Gosport, Eng, Jan 26, 43; m 68; c 2. SYSTEMATIC BOTANY, MYCOLOGY. *Educ:* Univ London, BSc, 65, MA, 66; Duke Univ, PhD(bot), 74. *Prof Exp:* Instr bot, Duke Univ, 72-74; res fel lichenology, Harvard Univ, 74-75; CHMN, DEPT BOT, MILWAUKEE PUB MUS, 75- *Concurrent Pos:* Mem, Int Standing Comt Mycological Nomenclature, 75-78 & Wis State Sci Areas Preserv Coun, 75-84; adj prof, Univ Wis-Milwaukee, 78-; dep dir, Org Flora Neotropica, UNESCO Comn, 80-; ed newsletter, Int Asn Lichenol, 81- *Honors & Awards:* Am Trauma Soc Award, 77; New York Bot Garden Award, 79; Joan Marr Pick Award, 82; Lois Almond Award, 85. *Mem:* Am Bryological & Lichenological Soc; Am Inst Biol Sci; Int Asn Plant Taxon; Mycol Soc Am; Sigma Xi. *Res:* Taxonomy of bryophytes, fleshy fungi and lichens; biochemical systematics of lichens; edible and poisonous mushrooms; lichen flora of the neotropics; world distribution and systematics of pertusariaceae (lichens). *Mailing Add:* Bot Div Milwaukee Pub Mus 800 W Wells St Milwaukee WI 53233

DIBBLE, JOHN THOMAS, b Kenosha, Wis, Apr 28, 50. POLLUTANTS, RESOURCE RECOVERY. *Educ:* Univ Wis, Stevens Pt, BS, 72; Western Ky Univ, MS, 74; Rutgers Univ, PhD(microbiol), 78. *Prof Exp:* Lab tech, Western Ky Univ, 74; res intern appl environ microbiol, Cook Col, Rutgers Univ, 74-78; proj microbiologist, 78-87, MGR ENVIRON RES, TEXACO RES, 87- *Concurrent Pos:* Consult, 77-78. *Mem:* Am Soc Microbiol; Soc Indust Microbiol; Soc Petrol Indust Biologists. *Res:* Treatment and disposal of waste water and solid waste; permit negotiations, public comment and proposed legislation; environmental assessment, oil spill response and the restoration of perturbed ecosystems. *Mailing Add:* 17210 Windypine Dr Spring TX 77379

DIBBLE, MARJORIE VEIT, b Brooklyn, NY, Jan 11, 28; m 54; c 2. NUTRITION, FOOD SCIENCE. *Educ:* Hunter Col, BS, 49; Univ Tenn, Knoxville, MS, 50. *Prof Exp:* Instr foods & nutrit, Syracuse Univ, 51-57; instr foods, Teachers Col, Columbia Univ, 57-58; from asst prof to assoc prof, 58-73, chmn dept, 63-86, dean, Col Human Develop, 73-74, PROF NUTRIT & FOOD SCI, SYRACUSE UNIV, 73-, ASSOC DEAN, 86- *Mem:* Am Dietetic Asn; fel Am Pub Health Asn. *Res:* Nutritional status surveys of adolescent and older age adults; relationship of nutrition to growth and development. *Mailing Add:* Dept Nutrit Syracuse Univ Syracuse NY 13244-1250

DIBBLE, WILLIAM E, b Schenectady, NY, Dec 25, 30. PHYSICS. *Educ:* Calif Inst Technol, BS, 54, PhD(physics), 60. *Prof Exp:* From asst prof to assoc prof, 61-71, PROF PHYSICS, BRIGHAM YOUNG UNIV, 71- *Mem:* AAAS; Am Phys Soc; Sigma Xi. *Res:* Small-angle x-ray scattering. *Mailing Add:* Dept Physics 296 ESC Brigham Young Univ Provo UT 84602

DIBELER, VERNON HAMILTON, b Elizabeth, NJ, July 20, 18; m 43; c 3. CHEMICAL PHYSICS. *Educ:* Duke Univ, BS, 39, MA, 40; Columbia Univ, PhD(chem), 50. *Prof Exp:* Asst chemist, Duke Univ, 39-41; from asst phys chemist to phys chemist, Nat Bur Stand, 42-75; RETIRED. *Honors & Awards:* Meritorious Serv Award, 52; Gold Medal Award, Dept Com, 69. *Mem:* AAAS; Am Chem Soc; Am Phys Soc; Am Inst Chemists. *Res:* Separation and use of isotopes in chemical research; mass spectrometry, theory and analytical application; bond dissociation energies and dissociation of isotopically-substituted molecules by electron and photon impact. *Mailing Add:* 272 Second St S Naples FL 33940-8600

DIBELLA, EUGENE PETER, b New York, NY, June 19, 28; m 50; c 4. ORGANIC CHEMISTRY. *Educ:* Fordham Univ, BS, 48, MS, 50, PhD(chem), 53. *Prof Exp:* Asst chem, Fordham Univ, 49-53; res chemist, Heyden Chem Corp, Tenneco Chem Inc, 53-66, supvr, Heyden Chem Div, 66-70, group leader org synthesis, Intermediates Div, 70-75, supvr specialty chem, 75-77, mgr org systhesis, 77-82; mgr org synthesis, Nuodex, Inc, 82-88; MGR ORG SYNTHESIS, HULS AM INC, 88- *Concurrent Pos:* Instr, Fairleigh Dickinson Univ, 53- *Mem:* Am Chem Soc; AAAS; Sigma Xi. *Res:* Synthetic organic chemistry; flame retardants; chlorination technology; organophosphates; polyesters; plastics additives. *Mailing Add:* 19 Ralston Ave Piscataway NJ 08854

DIBENEDETTO, ANTHONY T, b New York, NY, Oct 27, 33; m 55; c 5. CHEMICAL ENGINEERING, MATERIALS SCIENCE. *Educ:* City Col New York, BChE, 55; Univ Wis, MS, 56, PhD(chem eng), 60. *Prof Exp:* Chem engr, Bakelite Co, Union Carbide Corp, 54-55; from instr to assoc prof chem eng, Univ Wis, 56-66; from assoc prof to prof, Wash Univ, 66-71, dir, Mat Res Lab, 68-71; prof & head chem eng, 71-76, vpres grad educ & res, 79-81, vpres acad affairs, 81-86, UNIV PROF CHEM ENG, UNIV CONN, 86- *Honors & Awards:* Educ Serv Award, Plastic Inst Am, 72. *Mem:* Soc Plastics Engrs; Am Inst Chem Engrs; Am Chem Soc. *Res:* Physical properties of organic high polymers; polymer composite materials. *Mailing Add:* Dept Chem Eng Univ Conn Storrs CT 06268

DIBENNARDO, ROBERT, b New York, NY, Oct 19, 41; m 69; c 2. MORPHOMETRICS, FORENSIC ANTHROPOLOGY. *Educ:* Hunter Col, BA, 63, MA, 66; City Univ New York, PhD(anthrop), 73. *Prof Exp:* Lectr, Hunter Col, 65-68; lectr, 68-73, asst prof, 73-81, assoc prof, 81-88, PROF ANTHROP, LEHMAN COL, CITY UNIV NEW YORK, 88- *Concurrent Pos:* Vis Scientist, Am Mus Natural Hist, 77-; coordr, Anthrop-Biol Prog, Lehman Col, 75-; asst dir, Metro Forensic Anthrop Team, 80- *Mem:* Am Asn Phys Anthropologists. *Res:* Multivariate morphometrics; forensic anthropology; human genetics; dental anthropology; quantitative methods. *Mailing Add:* Dept Anthrop Lehman Col Bedford Park Blvd W Bronx NY 10468

DIBERARDINO, MARIE A, b Philadelphia, Pa, May 2, 26. DEVELOPMENTAL BIOLOGY, DEVELOPMENTAL GENETICS. *Educ:* Chestnut Hill Col, BS, 48; Univ Pa, PhD(zool), 62. *Hon Degrees:* LLD, Chestnut Hill Col, 90. *Prof Exp:* Res assoc embryol, Inst Cancer Res, 60-64, asst mem, 64-67; from assoc prof to prof anat, 67-81, PROF PHYSIOL, MED

COL PA, 81- *Concurrent Pos:* Prin investr, Nat Sci Found, 68-77 & NIH, 78-93; adv bd, Int Review Cytol, 76-, Series Develop Biol, Plenum Press, 82-; assoc ed, J Exp Zool, 84-86; mem, Fogarty Int Fel Study Group, NIH, 84; mem bd dirs, Int Soc Differentiation, 85-; mem bd adv, Robert Chambers Lab Cellular Microsurg, NY, 84-; mem exec comt, Int Soc Differentiation, 78-85 & 87-90. *Honors & Awards:* Lindback Award, Med Col Pa, 78. *Mem:* Am Soc Zoologists; Soc Develop Biol (treas, 75-78); fel AAAS; Am Soc Cell Biol; Int Soc Develop Biologists; Int Soc Differentiation. *Res:* Genetic potential of nuclei during normal differentiation and cancer being studied by means of nuclear transplantation into eggs and oocytes; genomic activation. *Mailing Add:* Dept Physiol & Biochem Med Col Pa Philadelphia PA 19129

DI BIASE, STEPHEN AUGUSTINE, b Rochester, NY, Apr 6, 52; m 75; c 1. ORGANIC CHEMISTRY. *Educ:* St John Fisher Col, BS, 74; Pa State Univ, PhD(org chem), 78. *Prof Exp:* Anal technician photog chem, Eastman Kodak Co, 72-74; res chemist, 78-81, RES SUPVR LUBRICANT ADDITIVES, LUBRIZOL CORP, 81- *Mem:* Am Chem Soc. *Res:* Development of new synthetic methods with subsequent application to the preparation of lubricant additives and other synthetic intermediates. *Mailing Add:* 504 E 266th St Euclid OH 44132

DIBLE, WILLIAM TROTTER, JR, b Oakmont, Pa, Sept 7, 25; m 48; c 4. CHEMISTRY. *Educ:* Pa State Col, BS, 49; Univ Wis, PhD, 52. *Prof Exp:* Asst, Univ Wis, 49-52; prod planning mgr, Int Minerals & Chem Corp, 52-64, dir mkt, 64; PRES, TERRA CHEMS, INTL, 64- *Mem:* Sigma Xi. *Res:* Boron determination in plants and soils; response of alfalfa to and its distribution in the plant. *Mailing Add:* 4327 Country Club Blvd Sioux City IA 51104

DIBNER, MARK DOUGLAS, b New York, NY, Nov 7, 51; m 83; c 1. NEUROBIOLOGY, NEUROPHARMACOLOGY. *Educ:* Univ Pa, BA, 73; Cornell Univ, PhD(neurobiol), 77; Widener Univ, MBA, 85. *Prof Exp:* Fel pharmacol, Med Sch Univ Colo, 77-79, res fel & lectr, Univ Calif, San Diego, 79-80; prin scientist neurobiol, E I Du Pont de Nemours & Co, 80-86; ADJ ASSOC PROF TECHNOL MGT, FUQUA SCH BUS, DUKE UNIV, 86- *Concurrent Pos:* Instr Phychol, Metrop State Col, Denver, 79 & Nat Univ, San Diego, 80. *Mem:* Soc Neurosci; AAAS; Sigma Xi; Am Soc Pharmacol & Exp Therapeut. *Res:* Strategies used for developing biotechnology in the pharmaceutical industry; receptors for neurotransmitters in brain and on cultured cells and ways drugs and other factors regulate these receptors; US competitiveness in biotechnology. *Mailing Add:* N Car Biotechnol Ctr PO Box 13547 Research Triangle Park NC 27709-3547

DIBOLL, ALFRED, b San Diego, Calif, Aug 30, 30; m 67; c 3. DEVELOPMENTAL ANATOMY, PLANT ANATOMY. *Educ:* San Diego State Col, BS, 56; Claremont Grad Sch, MA, 59; Univ Tex, PhD(bot), 64. *Prof Exp:* Sr lab technician, Univ Calif, Riverside, 51-53, asst prof bot, Univ Calif, Los Angeles, 63-68; head fac biol, 68-72, PROF BIOL & CHMN DIV NATURAL SCI & MATH, MACON COL, 72- *Mem:* Am Inst Biol Sci. *Mailing Add:* Div Natural Sci & Math Macon Col Col Sta Dr Macon GA 31297

DI BONA, GERALD F, INTERNAL MEDICINE. *Prof Exp:* PROF & CHMN, UNIV IOWA, 85- *Mailing Add:* Dept Internal Med Col Med Univ Iowa Iowa City IA 52242

DIBONA, PETER JAMES, b Philadelphia, Pa, Dec 31, 41; m 71; c 4. SOLID STATE PHYSICS, SURFACE PHYSICS. *Educ:* Villanova Univ, BS, 64; Univ Del, MS, 67, PhD(physics), 74. *Prof Exp:* Physicist explosives, Naval Surface Weapons Ctr, 76-80; prin develop engr, SR PRIN DEVELOP ENGR, HONEYWELL, INC, 85- *Concurrent Pos:* Resident res assoc, Nat Res Coun, Picatinny Arsenal, 73-75. *Honors & Awards:* Civilian Commendation, Picatinny Arsenal, 74. *Mem:* Am Phys Soc; Am Defense Preparedness Asn. *Res:* X-ray photoelectron spectroscopy; detonation physics; radiation damage; explosives applications; flash radiography; weapons design; weapons development. *Mailing Add:* 15809 Park Terrace Dr Eden Prairie MN 55346

DICARLO, ERNEST NICHOLAS, b Philadelphia, Pa, Jan 27, 36; m 75; c 4. CHEMICAL PHYSICS. *Educ:* St Joseph's Univ, Pa, BS, 58; Princeton Univ, PhD(chem), 62. *Prof Exp:* Asst chem, Princeton Univ, 58-59, res asst, 60-62; res chemist, Gulf Res & Develop Co, 62-63; from asst prof to assoc prof, 63-69, PROF CHEM, ST JOSEPH'S COL, PA, 69. *Mem:* Am Chem Soc. *Res:* Dipole moments; microwave absorption of liquids; nuclear magnetic resonance; electron spin resonance. *Mailing Add:* Dept Chem St Joseph's Univ 5600 City Ave Philadelphia PA 19131

DI CARLO, FREDERICK JOSEPH, b New York, NY, Nov 24, 18; m 43; c 3. CHEMISTRY, TOXICOLOGY. *Educ:* Fordham Univ, BS, 39, MS, 41; NY Univ, PhD(org chem), 44. *Prof Exp:* Asst chem, NY Univ, 41-44; res assoc, Squibb Inst Med Res, 44-45; res chemist, Fleischmann Labs, Standard Brands, Inc, 45-46, dept head, 47-53, head biochem div, 53-60; sr res assoc, Warner-Lambert Res Inst, 60-70, dir, Dept Drug Metab, 70-77; consult, Off Toxic Substances, 77-84, SR SCI ADV, US ENVIRON PROTECTION AGENCY, 84- *Concurrent Pos:* Adj prof, Col St Elizabeth, 68-77; ed-in-chief, Drug Metab Rev; co-ed, Drug & Chem Toxicol Series; dir, toxicol short-courses, Am Chem Soc; assoc ed, Xenobiotica; mem coun, Int Soc Study Xenobiotics, 86-89, finance comt, 90-91. *Mem:* Am Soc Biol Chem; Am Soc Pharmacol & Exp Therapeut; Reticuloendothelial Soc (secy-treas, 65-67 & pres, 69); Am Chem Soc; Int Soc Biochem Pharmacol; Int Soc Study Xenobiotics (secy, 84-85). *Res:* Drug metabolism; mechanisms of toxicity; penicillin; nucleic acids; amylases; fermentation; yeast derivatives; natural products; host defense mechanisms. *Mailing Add:* Xenobiotics Inc Denville NJ 07834

DICARLO, JAMES ANTHONY, b Buffalo, NY, Jan 15, 38; m 66; c 5. MATERIALS SCIENCE, CERAMICS SCIENCE & ENGINEERING. *Educ:* Canisius Col, BS, 59; Univ Pittsburgh, PhD(physics), 64. *Prof Exp:* Res assoc solid state physics, Brookhaven Nat Lab, 65-67; physicist, 67-85, DEP CHIEF CERAMICS, LEWIS RES CTR, NASA, 85- *Concurrent Pos:* Nat Mat adv bd. *Mem:* Am Ceramics Soc; Am Phys Soc. *Res:* Crystal imperfections and their interactions in metals; radiation effects in solids; deformation and fracture of fiber-reinforced metal; ceramic matrix composites; ceramic fibers. *Mailing Add:* Lewis Res Ctr MS106-1 Nat Aeronaut & Space Admin Cleveland OH 44135

DICCIANI, NANCY KAY, b Philadelphia, Pa, Oct 18, 47. KINETICS & CATALYSIS, TRANSPORT PROCESSES. *Educ:* Villanova Univ, BS, 69; Univ Va, MS, 70; Univ Pa, PhD(chem eng), 77, MBA, 86. *Prof Exp:* Supt water treatment, City Philadelphia, 72-74; res engr, Air Prods & Chems, Inc, 77-78, res mgr, 78-81, dir res, 81-84, dir res & develop, 84-86, gen mgr, 86-88, dir com develop, 88-91; GLOBAL BUS DIR PETROL CHEM, ROHM & HAAS CO, 91- *Mem:* Am Inst Chem Engrs; Soc Women Engrs. *Res:* Mass transfer; three phase fluid dynamics; heterogeneous kinetics and catalysis; separations science. *Mailing Add:* Rohm & Haas Co Independence Mall W Philadelphia PA 19105

DICE, J FRED, b Fowler, Calif, Aug 16, 47; m; c 1. CELL BIOLOGY, BIOCHEMISTRY. *Educ:* Univ Calif, Santa Cruz, BA, 69; Stanford Univ, PhD(biol), 73. *Prof Exp:* Res assoc physiol, Med Sch, Harvard Univ, 73-75; asst prof biol, Univ Calif, Santa Cruz, 74-78; from asst prof to assoc prof physiol & biophys, Med Sch, Harvard Univ, 78-85; ASSOC PROF-FULL PROF PHYSIOL, MED SCH ,TUFTS UNIV, 85- *Concurrent Pos:* Res Career Develop Award, 82-87; ed, biochem J, 84-91, Exp Geront, 90-, J Biol Chem, 90- *Honors & Awards:* Merit Award, NIH, 90. *Mem:* NY Acad Sci; AAAS; Am Soc Cell Biol; Biochem Soc; Am Soc Biol Chem; Geront Soc Am. *Res:* Intracellular protein degradation; regulation by hormones, mechanisms of breakdown, influence of protein conformation on degradative rates; altered protein degradation in aging. *Mailing Add:* Dept Physiol Tufts Univ Sch Med 136 Harrison Ave Boston MA 02111

DICE, JOHN RAYMOND, b Ann Arbor, Mich, Jan 11, 21; m 44; c 4. MEDICINAL CHEMISTRY. *Educ:* Univ Mich, BS, 41, MS, 42, PhD(chem), 46. *Prof Exp:* Shift foreman, Manhattan Proj, Tenn Eastman Co, 44-45; asst prof chem, Univ Tex, 46-51; from res chemist to sr res chemist, Warner Lambert/Parke Davis, 51-61, lab dir, 61-62, group dir, 63-69, asst dir chem res, 70-76, dir chem develop, 76-85; RETIRED. *Mem:* Am Chem Soc; NY Acad Sci. *Res:* Chemistry of phenanthrenes; chemotherapy of virus diseases and cancer; chemistry of peptides and protein; drugs affecting the central nervous system. *Mailing Add:* Three Breckan Rd Brunswick ME 04011

DICE, STANLEY FROST, b Pittsburgh, Pa, July 26, 21. MATHEMATICS. *Educ:* Oberlin Col, AB, 42; Univ Pittsburgh, MLitt, 51, PhD(math), 58. *Prof Exp:* Instr math & physics, WLiberty State Col, 51-52; instr math, Univ Detroit, 52-55; from instr to asst prof, Bucknell Univ, 55-62; asst prof, Carleton Col, 62-66; assoc prof, 66-85, EMER ASSOC PROF MATH, WITTENBERG UNIV, 85- *Mem:* Math Asn Am. *Res:* Summability of divergent series. *Mailing Add:* 9896 Bustleton Ave Apt B416 Philadelphia PA 19115

DICELLO, JOHN FRANCIS, JR, b Bradford, Pa, Dec 18, 38; m 62; c 2. CANCER RESEARCH, RADIATION EFFECTS. *Educ:* St Bonaventure Univ, BS, 60; Univ Pittsburgh, MS, 62; Tex A&M Univ, PhD(physics), 68. *Prof Exp:* Instr physics, St Bonaventure Univ, 62-63; res scientist, 67-69, res assoc radiol physics, Columbia Univ, 69-73; mem staff, Los Alamos Nat Lab, 73-; instr, Univ NMex, Los Alamos, 81-82; DEPT PHYSICS, CLARKSON UNIV, POSTDAM, NY. *Concurrent Pos:* Univ fel, Texas A&M Univ, 63-65, Assoc Western Univ & Atomic Energy, Comm, 65-67, Northwest Col, Univ Asn Sci & Dept Energy, 89; honorific title, NATO, 75. *Mem:* Am Asn Physicists Med; Radiation Res Soc; Am Phys Soc; Am Cancer Soc; Sigma Xi; Inst Electronics & Elec Engrs. *Res:* Medical physics; microdosimetry; nuclear structure; biophysics. *Mailing Add:* Dept Physics Clarkson Univ Potsdam NY 13699

DI CENZO, COLIN D, b Hamilton, Ont, July 26, 23; m 50; c 6. ELECTRICAL & COMPUTER ENGINEERING. *Educ:* Univ NB, BSc, 52, MSc, 57; Imp Col, London, DIC, 53. *Hon Degrees:* DSc, McMaster Univ, 91. *Prof Exp:* Lectr elec eng, Royal Mil Col, Ont, 54-57; proj engr & dep head, Sonar Group, Royal Can Naval Hq, Ottawa, 57-60, head, Underwater Fire Control, 60-62, systs engr, Hydrofoil Ship HMCS Bras d'Or, 64-65; from assoc prof to prof elec eng, 65-79, dir undergrad studies & vchmn, Fac Eng, 68-75, assoc dean eng, 75-79, prof elec & comput eng, 79-80, EMER PROF ELEC & COMPUT ENG, MCMASTER UNIV, 80-; PRIN, C DICENZO ASSOCS, 87- *Concurrent Pos:* Assoc ed, Trans: Indust Electronics & Control Instrumentation, Inst Elec & Electronics Engrs, 75-78; vis mem, Inst Engrs Australia, 79; distinguished visitor, Inst Engrs India, 80; adv, Jadavpur Univ India, Petrol Directorate Govt Nfld & Labrador, 81-84 & Nfld Inst Cold Ocean Sci; sr adv, James F Hickling Mgt Consults Ltd, 83-85; chmn, Colpat Enterprises Inc, 83-89; dir, Continuing Eng Educ, George Washington Univ, 84-87 & Can Inst Advan Eng Studies, 85-; foreign vis expert, People's Repub China, 88-90; res adv, Chengdu Univ, People's Repub China, 91. *Honors & Awards:* Can Decoration Centennial Medal, 67; Decorated Order Can, 72; Eng medal, Asn Prof Engrs Ont, 76; Julian C Smith Medal, Eng Inst Can, 77; Queen's Jubilee Medal, 77. *Mem:* Fel Eng Acad Can; fel Inst Elec & Electronics Engrs; fel Eng Inst Can (sr vpres, 78, pres, 79-80); Can Soc Elec Eng (pres, 76-78). *Res:* Engineering and management systems; author of over 50 scientific publications. *Mailing Add:* C diCenzo Assocs 28 Millen Ave Hamilton ON L9A 2T4 Can

DICHTER, MICHAEL, petroleum chemistry, polymer chemistry; deceased, see previous edition for last biography

DICHTL, RUDOLPH JOHN, b Chicago, Ill, May 16, 39; m 60; c 4. TECHNICAL MANAGEMENT. *Educ:* Ill Inst Technol, BS, 61; Air Force Inst Technol, MS, 66. *Prof Exp:* Staff scientist, Syst Div, USAF, 61-81; assoc prof aerospace studies, Univ Evansville, 77-78; assoc prof aerospace studies

& physics, Univ Colo, 78-81; SR PROG MGR, BALL AEROSPACE SYSTS GROUP, BALL CORP, 81- *Concurrent Pos:* Exchange scientist, High Frequency Physics Res Inst, Fed Repub Ger, 71-72; investr & mem, Shroud of Turin Res Proj, Inc, 77-80. *Res:* Aerospace; physics. *Mailing Add:* 41 Pineview Lane Boulder CO 80302-9414

DICIOCCIO, RICHARD A, b Buffalo, NY, Jan 25, 46. EXPERIMENTAL BIOLOGY. *Prof Exp:* CANCER RES SCIENTIST IV, ROSWELL PARK MEM INST, 84- *Mailing Add:* Dept Gynec Oncol Roswell Park Mem Inst 666 Elm St Buffalo NY 14263

DICK, BERTRAM GALE, b Portland, Ore, June 12, 26; m 56; c 3. PHYSICS. *Educ:* Reed Col, BA, 50; Oxford Univ, MA, 57; Cornell Univ, PhD(physics), 58. *Prof Exp:* Res assoc physics, Univ Ill, 57-59; from asst prof to assoc prof, 59-65, from actg chmn dept to chmn, 64-67, PROF PHYSICS, UNIV UTAH, 65-, DEAN GRAD SCH, 87- *Concurrent Pos:* Consult, Minn Mining & Mfg Co, 60-67; vis prof, Munich Tech Univ, 67-68; vis scientist, Max-Planck-Inst Solid State Res, Stuttgart, 76-77. *Mem:* Fel Am Phys Soc; AAAS; Fedn Am Scientists. *Res:* Solid state physics; ionic crystals; phonons and phonon-defect interactions; paraelectric defects; color centers. *Mailing Add:* Dept Physics Univ Utah Salt Lake City UT 84112

DICK, CHARLES EDWARD, b Fort Wayne, Ind, Apr 24, 37; m 58; c 2. ATOMIC AND MOLECULAR PHYSICS. *Educ:* Ill Benedictine Col, BS, 58; Univ Notre Dame, PhD(physics), 63. *Prof Exp:* PHYSICIST, NAT BUR STAND, 62- *Concurrent Pos:* Fel, US Rubber, Univ Notre Dame, 62. *Mem:* AAAS; Sigma Xi; Am Phys Soc; Soc Photo-Optical Instrumentation Engrs. *Res:* Low energy electron scattering; bremstrahlung production; low energy electromagnetic interactions; x-ray analysis; applications of x-rays; medical and industrial radiographic systems. *Mailing Add:* Physics Dept Nat Inst Standards & Technol Gaithersburg MD 20899

DICK, DONALD EDWARD, b Little Rock, Ark, Apr 23, 42; m 63. INDUSTRIAL INSTRUMENTATION, COMPUTERS. *Educ:* Calif Inst Technol, BS, 64; Univ Wis-Madison, MS, 65, PhD(elec eng), 68. *Prof Exp:* Res asst elec eng, Univ Wis-Madison, 64-66; asst prof elec eng, Univ Colo, Boulder, 68-73; asst prof phys med & rehab, Med Ctr, Univ Colo, Denver, 68-73; sect head res & develop, Unirad Corp, 73-75; res assoc anat, Bioeng Sect, Med Ctr, Univ Colo, Denver, 76-78; sr engr, Life Instruments Corp, 78-; mgr eng, Armco Autometrics, 81-88; DIR APPL ENG, SORICON, 88- *Concurrent Pos:* NIH grant biomed sci, 69-70. *Mem:* Instr Soc Am. *Res:* Instrumentation and process control for on-line measurement and monitoring of industrial processes, using ultrasound, x-ray fluorescence, and laser technologies; integrated with microprocessors and PC computers, and applied in mining and steel industries. *Mailing Add:* Soricon 4725 Walnut St Boulder CO 80301

DICK, ELLIOT C, b Miami, Fla, June 30, 26; m 50, 67; c 4. MICROBIOLOGY, EPIDEMIOLOGY. *Educ:* Univ Minn, BA, 50, MS, 53, PhD(bact), 55; Univ Wis, cert epidemiol, 65; Am Bd Med Microbiol, dipl, 68. *Prof Exp:* Asst prof bact, Univ Kans, 55-59; asst prof med, Sch Med, Tulane Univ, 59-61; from asst prof to assoc prof, 61-72, PROF PREV MED, UNIV WIS-MADISON, 72- *Concurrent Pos:* Grant, Univ Kans, 56-59, Kans State Bd Health grant, 57-58; USPHS grants, Univ, Wis, 61-70 & 71-, S C Johnson, Inc grants, 66-67 & 70-, Smith Kline & French grants, 66-71 & 74, NASA grants, 67-80 & NSF grants, 76-; USPHS career develop award, 64-68; vis scientist, Delta Regional Primate Ctr, Tulane Univ, 67; mem, WHO Collab Comts Rhinoviruses; consult, S C Johnson & Son, Inc, 62-, Smith Kline & French Labs, 65-, NIH, 65-, NASA, 65-, Abbott Labs, 65-, Sterling Drug Co, Albany, NY, 73-, NSF, 80- & Kimberly-Clark Corp, 81- *Honors & Awards:* Elizabeth M Watkins Res Award, Univ Kans, 58, USPHS Career Develop Award, 64-68; NIH Career Res Develop Award, 46-70. *Mem:* Soc Exp Biol & Med; fel Infectious Dis Soc Am; NY Acad Sci; Am Soc Microbiol; fel Explorers Club. *Res:* Etiology, pathogenesis and epidemiology of respiratory infections; role of viruses in asthma; transmission of respiratory viruses in isolated Antarctic populations and among human volunteers; virology. *Mailing Add:* Dept Prev Med Univ Wis-Madison Madison WI 53706

DICK, GEORGE W, b Toronto, Ont, June 12, 31; m 53; c 3. ELECTRONICS ENGINEERING. *Educ:* Univ Toronto, BASc, 53, MASc, 57, PhD(eng), 60. *Prof Exp:* Mem staff elec res & develop, Apparatus Engr Lab, Can Gen Elec Co, 53-55; mem staff commun res, 59-65, MEM TECH STAFF, BELL TEL LABS, 68-, SUPVR. *Concurrent Pos:* Assoc prof elec eng, Univ Toronto, 65-68. *Mem:* Assoc Inst Elec & Electronics Engrs. *Res:* Communication principles; display communications techniques including gas plasma display panels, light emitting diodes and associated data transmission electronics. *Mailing Add:* Supvr Bell Tel Lab 2525 N 12th St Reading PA 19604

DICK, HENRY JONATHAN BIDDLE, b Portland, Ore, Aug 30, 46; m 90. PETROLOGY. *Educ:* Univ Pa, BA, 69; Yale Univ, MPhil, 71, PhD(geol), 76. *Prof Exp:* Investr petrol, 75-76, scientist, 76-80, assoc scientist, 80-90, SR SCIENTIST, WOODS HOLE OCEANOG INST, 90- *Mem:* Geol Soc Am; Am Geophys Union. *Res:* Petrology of ultramafic and related rocks, petrology and structure of the ocean lithosphere; fracture zone tectonics. *Mailing Add:* Dept Geol & Geophys Woods Hole Oceanog Inst Woods Hole MA 02543

DICK, HENRY MARVIN, b Duchess, Alta, June 1, 31; m 57; c 3. ORAL PATHOLOGY. *Educ:* Univ Alta, DDS, 57; Univ Man, MSc, 68. *Prof Exp:* Gen pract dent, 57-66; assoc prof, 68-77, PROF ORAL PATH & ACTG CHMN DEPT, UNIV ALTA, 77- *Res:* Immunopathology of periodontal disease. *Mailing Add:* Dept Oral Biol Univ Alta Edmonton AB T6G 2N8 Can

DICK, JAMES GARDINER, b Perth, Scotland, Oct 22, 20; Can citizen; m 56; c 2. ANALYTICAL CHEMISTRY, ELECTROCHEMISTRY. *Prof Exp:* Chief chemist & metallurgist, Can Bronze Co, Ltd, 47-54, dist mgr prod, res & sales, Montreal Bronze, Ltd, 54-58, mgr mfg prod, control & res, Can Bronze Co, Ltd, 58-63; from asst prof to assoc prof chem & mat sci, 63-71, vchmn dept, 71-74, chmn dept, 74-76, PROF CHEM, CONCORDIA UNIV, 71-, DIR SCI INDUST RES, 75- *Concurrent Pos:* Mgr, Roast Labs Registered, 41-47; lectr, Sir George Williams Univ, 47-63; pres, Technitrol Ltd, 63-68, consult, 68-; consult, Can Metal Co Ltd, Metals & Alloys Cp Ltd & Ingot Medal Co Ltd, 68-; pres, Methodologies Consult Ltd, 81- *Mem:* NY Acad Sci; Sigma Xi; Chem Inst Can; Spectros Soc; Am Chem Soc. *Res:* X-ray spectrochemical analysis; environmental chemistry; electrochemical kinetics; polarography. *Mailing Add:* 2760 Carousel Crescent Suite 708 Gloucester ON K1T 2N4 Can

DICK, JERRY JOEL, b Langdon, NDak, May 12, 42; m 65; c 5. PHYSICS, MATERIAL SCIENCE. *Educ:* Ore State Univ, BA, 63; Southern Ore Col, MA, 70; Wash State Univ, PhD(physics), 74. *Prof Exp:* Exp physicist shockwaves in solids, Stanford Res Inst, 66-69 & Physics Int Co, 69; asst physicist, Wash State Univ, 74-77; STAFF MEM DETONATION PHYSICS, LOS ALAMOS SCI LAB, 77- *Res:* Shock initiation of detonation in reactive solids; dynamic failure and elastic-plastic wave propagation in solids; phase transition and transport phenomena under dynamic high pressure. *Mailing Add:* 3504 Arizona Ave Los Alamos NM 87544

DICK, KENNETH ANDERSON, b Vancouver, BC, July 30, 37; US citizen; m 59; c 2. THERMAL PHYSICS. *Educ:* Univ BC, BSc, 60, MSc, 63, PhD(physics), 66. *Prof Exp:* Res assoc physics, Johns Hopkins Univ, 66-69, asst physicist, Kitt Peak Nat Observ, 69-74; vis res scientist, Johns Hopkins Univ, 74-75; vis assoc prof, Inst Phys Sci & Technol, Univ Md, 74-78; staff scientist, Bendix Corp, 78-83; physicist, Chem Res & Develop Ctr, Aberdeen Proving Ground, Md, 83-85; chief, Human Factors & Infrared Br, 85-90, CHIEF, SENSOR TECHNOL BR, US ARMY COMBAT SYSTS TEST ACTIVITY, 90- *Mem:* Optical Soc Am. *Res:* Atomic and molecular spectroscopy; terrestrial aeronomy and planetary astronomy; atmospheric optics. *Mailing Add:* Att: STECS-EN-PS-DICK US Army Combat Syst Tests Activity Aberdeen Proving Ground MD 21005-5059

DICK, MACDONALD, II, b Del, July 24, 41; m 75; c 2. PEDIATRIC CARDIOLOGY. *Educ:* Williams Col, BA, 63; Univ Va, MD, 67. *Prof Exp:* Instr pediat, Children's Hosp, 74-77, asst prof, Harvard Med Sch, 77; from asst prof to assoc prof, 77-87, PROF PEDIAT, UNIV MICH, 87- *Mem:* Soc Pediat Res; Am Pediat Soc; Am Col Cardiol; Am Acad Pediat; Int Soc Heart Res. *Res:* Cardiac electrophysiology. *Mailing Add:* Dept Pediat Div Pediat Cardiol F1310 Univ Mich C S Mott Childrens Hosp Box 0204 Ann Arbor MI 48109-0204

DICK, RICHARD DEAN, b Angola, Ind, Sept 27, 30; m 56; c 3. SHOCK WAVE PHYSICS, EQUATION OF STATE. *Educ:* Ariz State Univ, BS, 57, MS, 60, PhD(physics), 68. *Prof Exp:* Staff mem condensed matter physics, Los Alamos Sci Lab, 59-75, dep group leader shock & weapon physics, 75-81, STAFF MEM ROCK FRAGMENTATION, LOS ALAMOS NAT LAB, 81- *Mem:* Am Phys Soc; Am Geophys Union. *Res:* Dynamic high pressure equation of state properties of solids and liquids; radiography of high speed events using a 30 mega electron volts flash x-ray machine; fragmentation of oil shale using explosives. *Mailing Add:* Los Alamos Nat Lab Box 1663 MS C-335 Los Alamos NM 87545

DICK, RICHARD IRWIN, b Sanborn, Iowa, July 18, 35; m 58; c 4. ENVIRONMENTAL ENGINEERING, CIVIL ENGINEERING. *Educ:* Iowa State Univ, BS, 57; Univ Iowa, MS, 58; Univ Ill, Urbana, PhD(sanit eng), 65. *Prof Exp:* Sanit engr, USPHS, 58-60, Clark, Daily, Dietz & Assocs, Consult Engrs, 60-62; from instr to prof sanit eng, Univ Ill, Urbana-Champaign, 62-70; vis engr, Water Pollution Res Lab, Eng, 70-71; prof civil eng, Univ Del, 72-77; JOSEPH P RIPLEY PROF ENG, CORNELL UNIV, 77- *Concurrent Pos:* Gov bd & exec comt, Int Asn Water Pollution Res, 74-78; actg ed, J Environ Eng, Am Soc Civil Engrs, 79-85. *Honors & Awards:* Harrison Prescott Eddy Medal, Water Pollution Control Fedn, 68; Thomas R Camp lectr, Boston Soc Civ Engrs, 81. *Mem:* Asn Environ Eng Prof (pres, 73); AAAS; Water Pollution Control Fedn; Am Pub Health Asn; Am Water Works Asn; Inst Water Pollution Control; Am Acad Environ Engrs; Am Soc Civil Engr; Int Asn Water Polluton Res. *Res:* Unit operations and processes used in water and waste water treatment; treatment and disposal of sludges. *Mailing Add:* Dept Civil Eng Cornell Univ Main Campus 218 Hollister Hall Ithaca NY 14853

DICK, RONALD STEWART, b Queens, NY, Jan 14, 34; m 58; c 2. MATHEMATICAL STATISTICS. *Educ:* Queens Col, BS, 55; Columbia Univ, MA, 57, PhD(math statist), 68. *Prof Exp:* Math statistician, US Census Bur, Washington, DC, 56-57; lectr math, Queens Col, NY, 57-62; sr mem tech staff, ITT Defense Commun, NJ, 60-68; asst prof math, C W Post Col, Long Island Univ, 63-68; assoc prof mgt, George Washington Univ, 68-75; supvr opers anal, Social Security Admin, 75-80; STATISTICIAN, CHI ASSOC, 80- *Concurrent Pos:* Asst, Columbia Univ, 56-57; engr, Sperry Gyroscope Co, NY, 57-59; reliability engr, Am Bosch Arma, 59-60; vis lectr, Stevens Inst Technol, 63. *Mem:* Opers Res Soc Am; Inst Math Statist; Am Statist Asn. *Res:* Queuing theory with balking; reliability of repairable complex systems. *Mailing Add:* CHI Assoc Inc 2000 N 14th St Suite 740 Arlington VA 22201

DICK, STANLEY, b Brooklyn, NY, Aug 13, 36. FUNGAL GENETICS & MORPHOGENESIS. *Educ:* Brooklyn Col, AB, 56; Harvard Univ, MA & PhD(biol), 60. *Prof Exp:* Northern Atlantic Treaty Orgn fel, Dept Bot, Univ Col, Univ London, 60-61; USPHS fel, Bot Inst, Univ Cologne, 61-62; res assoc biol, Harvard Univ, 62-64; asst prof bot, Ind Univ, Bloomington, 64-71; from asst prof to assoc prof, 71-78, PROF BIOL, FITCHBURG STATE COL, 78-, CHMN DEPT, 73- *Concurrent Pos:* NSF res fel, Orgn Trop Studies, San Jose, Costa Rica, 70; vis prof biol, Harvard Univ, 72, 73 & 74; vis res assoc bot, Univ Wis-Madison, 76 & 77, vis prof & NSF fel, 78 & 84-88. *Mem:* AAAS; Genetics Soc Am; Mycol Soc Am; Am Soc Microbiol. *Res:* Genetics, physiology and morphogenesis in higher basidiomycetes; hormonal, chemical and mechanical induction of fruiting bodies and tumors; fungal phenoloxidases; mutation expression and somatic recombination in heterokaryons; biochemical genetics of incompatibility; biotechnology of edible mushrooms. *Mailing Add:* Dept Biol Fitchburg State Col 160 Pearl St Fitchburg MA 01420

DICK, WILLIAM EDWIN, JR, b Waynesboro, Va, Oct 31, 36; m 58, 83; c 3. CHEMISTRY. *Educ:* NC State Univ, BS, 58; Purdue Univ, MS, 62, PhD, 66. *Prof Exp:* Chemist, Northern Utilization Res Lab, USDA, 65-84; CHEMIST, PRAXIS BIOLOGICS, INC, 86- *Mem:* Am Chem Soc; Sigma Xi. *Res:* Modification and conjugation of bacterial carbohydrate antigens for use in vaccines. *Mailing Add:* Praxis Biologics Inc 300 E River Rd Rochester NY 14623

DICKAS, ALBERT BINKLEY, b Sidney, Ohio, Sept 4, 33; m 62; c 4. RESEARCH ADMINISTRATION, GEOLOGY. *Educ:* Miami Univ, BA, 55, MS, 56; Mich State Univ, PhD(geol), 62. *Prof Exp:* Develop geologist, Magnolia Petrol Co, 56-59; instr geol, Mich State Univ, 59-62; explor geologist, Standard Oil Co Calif, 63-66; from asst prof to assoc prof, 66-76, PROF GEOL, UNIV WIS-SUPERIOR, 76- DIR, OFF EXTRAMURAL PLANNING, 69- *Mem:* Am Asn Petrol Geologists; Soc Econ Paleontologists & Mineralogists; Int Asn Great Lakes Res; Geol Soc Am. *Res:* Geologic, economic, physical and geophysical aspects of present and future development of the Lake Superior Basin. *Mailing Add:* Dept Geol Univ Wis Superior WI 54880

DICKASON, ELVIS ARNIE, b Ore, Oct 9, 19; m 46; c 2. ENTOMOLOGY. *Educ:* Ore State Univ, BS, 47, MS, 49; Mich State Univ, PhD(entom), 59. *Prof Exp:* From instr to assoc prof entom, Ore State Univ, 49-70; interim dir, Int Progs, Univ Nebr, Lincoln, 73-75, prof entom & head dept, 70-84; RETIRED. *Concurrent Pos:* Grant-in-aid entom, IRI Res Inst, Salvador, Brazil, 66-67. *Mem:* Entom Soc Am; Sigma Xi. *Res:* Applied entomology; insect ecology. *Mailing Add:* 2800 NW 29th St No 3 Corvallis OR 97330-3543

DICKE, FERDINAND FREDERICK, b New Bremen, Ohio, Aug 25, 99; m 29; c 3. ENTOMOLOGY. *Educ:* Ohio State Univ, BS, 27. *Prof Exp:* Field asst, Bur Entom & Plant Quarantine, USDA, Mich, 27-28, jr entomologist, 28-29, asst entomologist, Va, 29-33, Arlington Farm, 33-42 & Beltsville Res Ctr, 42, assoc entomologist, Ohio, 42-50, entomologist, Entom Res Div, Agr Res Serv, 50-63; entomologist, Pioneer Hi-Bred Int Inc, 63-83; assoc prof entom, 57-61, PROF ENTOM, IOWA STATE UNIV, 61- *Concurrent Pos:* Consult & collabr, Agr Res Serv, USDA & Iowa State Univ, 84- *Mem:* AAAS; Entom Soc Am; Sigma Xi. *Res:* Development of crop varieties resistant to insects; corn insect vectors of diseases. *Mailing Add:* 1430 Harding Ave Ames IA 50010

DICKE, ROBERT HENRY, b St Louis, Mo, May 6, 16; m 42; c 3. ASTROPHYSICS. *Educ:* Princeton Univ, AB, 39; Univ Rochester, PhD(physics), 41. *Hon Degrees:* DSc, Univ Edinburgh, Scotland, 72, Univ Rochester & Ohio Northern Univ, 81, Princeton Univ, 89. *Prof Exp:* Staff mem radiation lab, Mass Inst Technol, 41-46; from asst prof to prof physics, 46-57, Cyrus Fogg Brackett prof, 57-75, chmn dept physics, 67-70, Albert Einstein prof sci, 75-84, EMER ALBERT EINSTEIN PROF SCI, PRINCETON UNIV, 84- *Concurrent Pos:* Vis prof, Harvard Univ, 54-55; mem adv panel physics, NSF, 59-61, chmn adv comt radio astron, 67-69; chmn adv comt atomic physics, Nat Bur Standards, 61-63; mem comt physics, NASA, 63-70, chmn, 63-66, mem lunar ranging exp team, 66-; chmn physics panel, Adv to Comt Int Exchange of Persons, Fulbright Fels, 64-66; mem, Nat Sci Bd, 70-76, Inst Advan Study, 70-71; mem vis comt, Nat Bur Stand, 75-80, chmn, 78. *Honors & Awards:* Rumford Premium Award, Am Acad Arts & Sci, 67; Nat Medal Sci, 71; Comstock Prize, Nat Acad Sci & Medal Exceptional Sci Achievement, NASA, 73; Cresson Medal, Franklin Inst, 74. *Mem:* Nat Acad Sci; Am Philos Soc; Am Acad Arts & Sci; fel Am Phys Soc; fel Am Geophys Union; Royal Astron Soc. *Res:* Gravitation; relativity; astrophysics; solar physics; cosmology. *Mailing Add:* Joseph Henry Labs Dept Physics Jadwin Hall Princeton Univ Princeton NJ 08544

DICKEL, HÉLÈNE RAMSEYER, b Cambridge, Mass, Mar 19, 38; m 61; c 2. ASTRONOMY. *Educ:* Mt Holyoke Col, AB, 59; Univ Mich, MA, 61, PhD(astron), 64. *Prof Exp:* Res assoc astron, Univ Ill Observ, Urbana, 65-70; vis scientist astron, Div Radio Physics, Commonwealth Sci & Indust Res Orgn, Australia, 70-71; res assoc astron, 71-77, RES ASSOC PROF ASTRON, UNIV ILL OBSERV, URBANA, 77- *Concurrent Pos:* grants, NSF, 72-80, N Atlantic Treaty Orgn, 80-83, Am Astron Soc, 82-85, Int Sci Found, 85-86; Sponsor, Career Explor Astron, Mount Holyoke Col, Univ Mass, Wheaton Col & Oberlin Col, 74-; vis astronomer, Huygens Lab, Leiden, Neth, 77-78, Harlow Shapley vis lectr, 81-; collabr, Los Alamos Nat Lab, NMex, 85-86. *Mem:* Int Astron Union; Int Sci Radio Union; Am Astron Soc; Astron Soc Pac; Sigma Xi; Asn Women Scientists. *Res:* Radio studies of physical conditions and star formation in molecular clouds which are composed of dust and molecules and embedded massive stars surrounded by ionized hydrogen. *Mailing Add:* 1005 S Busey Urbana IL 61801

DICKENS, BRIAN, b Manchester, Eng, Mar 15, 37; US citizen. PHYSICAL CHEMISTRY. *Educ:* ARIC, London, 58; Univ Minn, MS, 60, PhD(phys chem), 62. *Prof Exp:* Res assoc chem, Harvard Univ, 60-62; chemist, Chloride Tech Serv, Manchester, Eng, 62-64; phys chemist, Naval Propellant Plant, Md, 64-66; PHYS CHEMIST, NAT BUR STANDARDS, 66- *Concurrent Pos:* Res chemist, Am Dent Asn, 73-74; NATO sr scientist, Res Ctr Macromolecules, Strasbourg, France, 78-79. *Mem:* Am Chem Soc; Royal Soc Chem. *Res:* Inks, composites; polymer degradation; polymer oxidation. *Mailing Add:* 19223 Dunbridge Way Gaithersburg MD 20879-2909

DICKENS, CHARLES HENDERSON, b Thomasville, NC, Nov 22, 34; m 65; c 2. SCIENCE POLICY, SCIENCE EDUCATION. *Educ:* Duke Univ, BS, 57, MEd, 64, DEduc(math educ), 66. *Prof Exp:* Res technician, Nat Security Agency, 57-58 & 60-62; teacher high sch, NC, 62-63; instr educ, Wake Forest Col, 65-66, asst prof math & educ, 66-67; planning specialist, Planning & Eval Unit Off Assoc Dir Educ, NSF, 67-69, assoc prog dir, Student-Originated Studies Prog, Div Undergrad Educ Sci, 69-73, proj mgr, Exp Projs & Prob Assessment Group, Off Exp Projs & Progs, 73, study dir, 73-80, sr study dir, Supply & Educ Anal Group, 80-83, sect head, Sci & Tech Personnel Studies Sect, 83-86, sect head, Surveys & Analysis Sect, Div Sci Resources Studies, 86-89; SR STAFF ASSOC, FED COORD COUN SCI, ENG & TECHNOL, OFF SCI & TECHNOL POL, 89- *Concurrent Pos:* Consult, Am Political Sci Asn & US Civil Serv Comn Cong fel, 71-72; NSF rep, Higher Educ Adv Panel, Nat Ctr Educ Statist, 81-89. *Mem:* AAAS; NY Acad Sci; Am Statist Asn. *Res:* Statistics of science and engineering funding and personnel; statistics of higher education; characteristics of science and engineering faculty; supply and demand of scientists and engineers; science and engineering expenditures, research facilities and equipment; organization of federal interagency programs. *Mailing Add:* Office of Science & Technology Policy 730 Jackson Pl NW Washington DC 20506

DICKENS, ELMER DOUGLAS, JR, b Charleston, WVa, Dec 26, 42; c 2. POLYMER SCIENCE. *Educ:* Morris Harvey Col, BS, 65; WVa Univ, MS, 67, PhD(physics), 70. *Prof Exp:* From res physicist to sr res physicist, 70-74, group leader new prod, 74-76, sect leader new ventures, 76-78, sect mgr, 78-81, mgr corp res, 81-83, RES FEL, B F GOODRICH RES CTR, 83- *Concurrent Pos:* NASA fel, 67-70. *Mem:* Combustion Inst; Am Asn Artificial Intel; Am Inst Physics. *Res:* Mathematical models of physical systems; polymer combustion; computer modeling of fires; flammability testing; physics of combustion; magnetic materials. *Mailing Add:* 4160 Maple Dr Richfield OH 44286

DICKENS, JUSTIN KIRK, b Syracuse, NY, Nov 2, 31; m 57; c 4. NUCLEAR PHYSICS, SCIENCE COMMUNICATION. *Educ:* Univ Southern Calif, AB, 55, PhD(physics), 62; Univ Chicago, MS, 56. *Prof Exp:* Physicist, 62-78, Tech Asst to Assoc Dir Phys Scis, 78-79, PHYSICIST, LOW ENERGY NUCLEAR PHYSICS, OAK RIDGE NAT LAB, 80- *Mem:* AAAS; Am Phys Soc; Am Nuclear Soc; Sigma Xi. *Res:* Nuclear reaction mechanisms, experimental and theoretical; experimental reactor safety research. *Mailing Add:* Oak Ridge Nat Lab Oak Ridge TN 37831

DICKENS, LAWRENCE EDWARD, b North Kingstown, RI, Dec 8, 32; m 52; c 7. ELECTRICAL & ELECTRONICS ENGINEERING. *Educ:* Johns Hopkins Univ, BSEE, 60, MSEE, 62, DEng, 64. *Prof Exp:* Field Engr Radio Div, Bendix Corp, 53-56, asst proj engr, 56-58, proj engr, 58-60; res staff asst electronic design, Radiation Lab, Johns Hopkins Univ, 60-62 res assoc microwave semiconductors, 62-64, res scientist, 64-65; res scientist, Advan Technol Corp, 65-69; chief engr, Ford Microelectronics Inc, 83-86; vpres, Herley Microwave Systs Inc, 86-89; adv engr, 69-89, SR ADV ENGR, WESTINGHOUSE ELEC CORP, 89- *Concurrent Pos:* Consult, Radio Div, Bendix Corp, 60-63, Appl Microwave Elec Corp, 61, Am Electronics Labs, Inc, 63, Res Div, Electronic Commun, Inc, 63-64 & Pinkerton Electro-Security Co, 65-66. *Mem:* Inst Elec & Electronics Engrs. *Res:* Development of low noise communications systems and components; microwave and millimeter wave semiconductor components and their applications. *Mailing Add:* 8805 Blairwood Rd Apt A2 Baltimore MD 21236

DICKENSON, DONALD DWIGHT, b Paris, Ill, Jan 15, 25; m 46; c 4. PLANT PHYSIOLOGY, PLANT GENETICS. *Educ:* Univ Ill, BS, 49, MS, 50; Univ Minn, PhD(grass breeding), 57. *Prof Exp:* Plant breeder, 53-62, asst dir agr res, 62-66, DIR AGR RES, HOLLY SUGAR CORP, 66- *Mem:* Am Soc Agron; Am Soc Sugar Beet Technol; Am Phytopath Soc; Crop Sci Soc Am; Soil Sci Soc Am. *Res:* Assessment of agronomic problems with appropriate action-all areas; coordinate state & federal research for agronomic problems. *Mailing Add:* 2203 Northglen Colorado Springs CO 80909

DICKER, DANIEL, b Brooklyn, NY, Dec 30, 29; m 51; c 2. AEROELASTICITY. *Educ:* City Col New York, BCE, 51; NY Univ, MCE, 55; Columbia Univ, EngScD(appl mech), 61. *Prof Exp:* Engr, Bogert-Childs, 51-52 & 54-55; proj engr, Praeger-Kavanagh, 55-58; res asst eng, Columbia Univ, 59-60, from instr to asst prof, 60-62; from asst prof to assoc prof, 62-68, asst dean, Grad Sch, 65-69, exec officer, Col Eng, 70-71, dir postgrad exten prog, Col Eng, 74-78, PROF APPL MATH & ENG, STATE UNIV NY STONY BROOK, 68- *Concurrent Pos:* NSF grant, 63-70; NATO fel, Imp Col, Univ London, 69-70; hon res fel, Harvard Univ, 78-79; vis prof, Univ WI, 86. *Honors & Awards:* Norman Medal, Am Soc Civil Engrs, 67, Arthur M Wellington Prize, 72. *Mem:* Soc Indust & Appl Math; fel NY Acad Sci; Sigma Xi; fel Am Soc Civil Engrs. *Res:* Hydrodynamics; approximate solutions of boundary value problems; heat conduction; structural dynamics; transient flow in porous media; extension of the method of separation of variables in partial differential equations to nonhomogenous and discontinuous boundary-value problems; engineering theory for small arch dams. *Mailing Add:* Dept Appl Math & Statistics State Univ of NY Stony Brook NY 11794-3600

DICKER, PAUL EDWARD, b Philadelphia, Pa, June 17, 25; m 53; c 4. ELECTRICAL ENGINEERING. *Educ:* Swarthmore Col, BSc, 45; Ohio State Univ, MSc, 48. *Prof Exp:* Asst math, Ohio State Univ, 46-47, instr elec eng, 47-48; instr, Princeton Univ, 48-51; res assoc, Eng Res Inst, Univ Mich, 51-54; from asst prof to assoc prof elec eng, Vanderbilt Univ, 54-59; instr elec eng, Tenn State Univ, 59-64; chief engr, Aladdin Electronics, 59-65, asst gen mgr, 65-67, mgr opers & dir prod develop, 67-70, vpres, Telecommun Prod, 70-79, vpres mkt & prod develop, Vernitron Div, AIE, 79-88; PRES, QAD ELECTRONICS, INC, 69-; SR ENGR, SCHOTT CORP, 88- *Concurrent Pos:* plant engr, Kaiser-Frazer Corp, 52-; consult, Avco Mfg Co, Miniature Electronic Components & Essex Electronics. *Mem:* Sr mem Inst Elec & Electronics Engrs. *Res:* Electronics; magnetic components. *Mailing Add:* 6110 Elizabethan Dr Nashville TN 37205

DICKERHOOF, DEAN W, b Akron, Ohio, Nov 9, 35; m 58; c 2. INORGANIC CHEMISTRY. *Educ:* Univ Akron, BS, 57; Univ Ill, MS & PhD(inorg chem), 61. *Prof Exp:* From asst prof to assoc prof, 61-73, PROF CHEM, COLO SCH MINES, 74- *Mem:* Am Chem Soc; The Chem Soc. *Res:* Synthesis and characterization of inorganic and organometallic compounds; analysis of coal and coal-derived liquids. *Mailing Add:* Dept Chem Colo Sch Mines Golden CO 80401-1888

DICKERMAN, CHARLES EDWARD, b Carbondale, Ill, Mar 9, 32; m 54; c 4. NUCLEAR SCIENCE. *Educ:* Southern Ill Univ, BA, 51, MA, 52, Univ London, DIC, 53; Univ Iowa, PhD(physics), 57. *Prof Exp:* PHYSICIST, ARGONNE NAT LAB, 57- *Mem:* Am Phys Soc; Am Nuclear Soc. *Res:* Nuclear safety of fast reactors. *Mailing Add:* 4802 Highland Ave Downers Grove IL 60515

DICKERMAN, HERBERT W, b New York, NY, Aug 3, 28; m 63; c 4. BIOCHEMISTRY, MOLECULAR ENDOCRINOLOGY. *Educ:* State Univ NY, MD, 52; Johns Hopkins Univ, PhD(biol), 60. *Prof Exp:* Instr med, Sch Med, Johns Hopkins Univ, 60-63; investr clin biochem, Nat Heart Inst, 63-66; assoc prof med, Sch Med, Johns Hopkins Univ, 66-75; DIR, WADSWORTH CTR LABS & RES, NY STATE DEPT HEALTH, 75-; PROF BIOMED SCI, STATE UNIV NY ALBANY, 85- *Concurrent Pos:* Estab investr, Am Heart Asn, 60-66; NIH res career develop award, 66-71; Soc Scholars, Johns Hopkins Univ, 90. *Mem:* Am Soc Biol Chemists; Endocrine Soc. *Res:* Folate and vitamin B-12 metabolism, biochemical control mechanisms; ribonucleic acid metabolism; steroid hormone mechanism of action; mechanisms of induction by estrogen of rat creatine kinase B gene; operation of estrogen dependent functions in human breast cancer cell lines by altering estrogen metabolism. *Mailing Add:* Wadsworth Ctr Lab & Res NY State Dept Health Albany NY 12201-0509

DICKERMAN, RICHARD CURTIS, b Chicago, Ill, Jan 31, 34; m 60; c 2. GENETICS, ZOOLOGY. *Educ:* Col Wooster, BA, 56; Univ Tex, PhD(genetics), 62. *Prof Exp:* Asst, Genetics Found, Univ Tex, 61-62; asst prof biol, Kans State Teachers Col, 62-65; resident res assoc, Div Biol & Med Res, Argonne Nat Lab, 65-67; asst prof, 67-69, ASSOC PROF BIOL, CLEVELAND STATE UNIV, 69-, DIR ADMIS & REC, 79- *Mem:* AAAS; Genetics Soc Am. *Res:* Genetic studies of x-irradiated Drosophilia oocytes; genetic studies of irradiated Drosophilia populations; radiation and chemical induced mutations in Drosophilia. *Mailing Add:* Admis & Rec Cleveland State Univ Cleveland OH 44115

DICKERSON, CHARLESWORTH LEE, b Fredericksburg, Va, Dec 14, 27; m 59; c 2. ORGANIC CHEMISTRY. *Educ:* Col William & Mary, BS, 49; Univ Va, MS, 51, PhD(org chem), 54. *Prof Exp:* Res chemist, Am Enka Corp, 54-55; sr chemist appl res, S C Johnson & Son, Inc, 57-65, prod res, 65-67, tech librn-info specialist, 67, sr res chemist & info specialist, 67-74, toxicol coordr, 74-78, sr prod safety coordr, 78-83, group leader prod safety, 83-87; RETIRED. *Mem:* AAAS; Am Chem Soc; Spec Libr Asn (treas, 68-73, vpres, 73-74 & pres, 74-75); Nat Microfilm Asn; Am Col Toxicol; Soc Toxicol; Int Soc Regulatory Toxicol & Pharmacol. *Res:* Action of Grignard reagents on 2, 3-unsaturated-1, 4-diketones and related furans; high polymer compositions and properties; synthesis of insecticides; ocular and respiratory toxicology. *Mailing Add:* 1345 Marie St Racine WI 53404

DICKERSON, CHESTER T, JR, b Lewes, Del, Apr 26, 39; m 60; c 2. WEED SCIENCE, OLERICULTURE. *Educ:* Univ Del, BS, 62, MS, 64; Cornell Univ, PhD(weed sci), 68. *Prof Exp:* Supvr weed sci, 68-71, tech mgr, 71-73, reg mgr prod develop, 73-75, res mgr, Monsanto Japan Ltd, 75-81, DIR AGR AFFAIRS, MONSANTO CO, WASH, DC, 81- *Mem:* Weed Sci Soc Am. *Res:* Herbicide research & development. *Mailing Add:* 11313 Willowbrook Dr Oldfield Potomac MD 20851

DICKERSON, DONALD ROBERT, b Champaign, Ill, Jan 21, 25; m 51; c 2. ORGANIC GEOCHEMISTRY. *Educ:* Univ Ill, BS, 50, MS, 59, PhD(food sci), 62. *Prof Exp:* Chemist, Clark Microanal Lab, Urbana, Ill, 50-52, chief chemist, 52-53; asst chemist, Ill State Geol Surv, 53-62, assoc chemist, 62-69, chemist, 69-70, organic chemist, 70-86, prin staff org chemist, 86-90; RETIRED. *Mem:* Am Chem Soc. *Res:* Organic geochemistry of oil and black shale; organic matter as environmental pollutants. *Mailing Add:* 703 S Pine St Champaign IL 61820

DICKERSON, GORDON EDWIN, b La Grande, Ore, Jan 30, 12; m 33; c 4. ANIMAL BREEDING. *Educ:* Mich State Univ, BS, 33; Univ Wis, MS, 34, PhD(animal genetics), 37. *Prof Exp:* Asst genetics, Univ Wis, 34-35, instr genetics & dairy records, 35-39, instr genetics & dairy husb, 39-41; geneticist, Regional Swine Breeding Lab, USDA, 41-47; prof animal husb, Univ Mo, 47-52; geneticist, Kimber Farms, 52-65; geneticist, Res Br, Can Dept Agr, 65-67; prof, Univ Nebr, Lincoln & geneticist, US Meat Animal Res Ctr, Agr Res Serv, USDA, 67-87, EMER PROF ANIMAL SCI, UNIV NEBR & COLLABR, AGR RES SERV, USDA, 87- *Honors & Awards:* Breeding-Genetics Award, Am Soc Animal Sci, 70, Morrison Award, 78. *Mem:* Fel AAAS; fel Am Soc Animal Sci; Genetics Soc Am; Am Genetic Asn; Biomet Soc; Am Dairy Sci Asn; Poultry Sci Asn. *Res:* Animal genetics; experimental design; effectiveness of selection and breeding systems in swine, dairy and beef cattle, poultry, mice. *Mailing Add:* A218 ANS Univ Nebr Lincoln NE 68583-0908

DICKERSON, JAMES PERRY, b Bunn, NC, May 8, 39; m 70. ORGANIC CHEMISTRY. *Educ:* Univ NC, Chapel Hill, BS, 62; Wayne State Univ, PhD(org chem), 66. *Prof Exp:* Sr res chemist, Ash Stevens Inc, 66-68; sr res chemist, R J Reynolds Industs, Inc, 68-78, group leader, 78-80, mgr res & develop, R J Reynolds Tobacco Co, 80-89; SCI INSTR, SUNNY COUNTY COMMUNITY COL, 89- *Mem:* Am Chem Soc; Sigma Xi. *Res:* Carbohydrates; organic synthesis; natural products; amino sugars; tobacco science. *Mailing Add:* Res & Develop 2015 Greenbriar Rd Winston-Salem NC 27104-2325

DICKERSON, L(OREN) L(ESTER), JR, b Fitzgerald, Ga, May 11, 18; m 53; c 3. ELECTRONICS ENGINEERING, PHYSICS. *Educ:* Emory Univ, BS, 39; Mass Inst Technol, ScD(chem eng), 42. *Prof Exp:* Res chem engr, Arthur D Little, Inc, 42-44; res assoc metall, Mass Inst Technol, 44-45; res supvr process metall, Reynolds Metals Co, 45-46, proj engr process develop, 46-48; pvt consult & mfg printing & embroidering methods, 48-53; dir reduction res, Reynolds Metals Co, 53-57, assoc dir & leader fundamental res dept, 57-60; consult & prin sci investr, Army Missile Command, 60-70; aerospace engr & coordr nuclear effects studies, Ballistic Missile Defense, Dept Defense, Dover, 70-76, physicist, Army Metrol & Calibration Ctr, 76-77, Army Standards Lab, 77-78, electronics engr, Develop & Eval Comput & Automation, Long-range Res & Develop Plan, Army Munitions & Chem Command, 78-90; RETIRED. *Concurrent Pos:* Comput consult. *Mem:* Fel AAAS; Am Inst Mining, Metall & Petrol Engrs; Am Chem Soc; Electrochem Soc; Int Soc Gen Semantics; NY Acad Sci. *Res:* Electrical and control engineering; electronics; automation by computers; instrumentation; chemical, metallurgical and nuclear engineering. *Mailing Add:* 167 Diamond Spring Rd Denville NJ 07834

DICKERSON, OTTIE J, b Mulberry, Ark, Sept 4, 33; m 55; c 2. ADMINISTRATION, PLANT PATHOLOGY. *Educ:* Univ Ark, BS, 55, MS, 56; Univ Wis, PhD(plant path), 61. *Prof Exp:* From asst prof to prof nematol, Kans State Univ, 61-78, chmn crop protection curric comt, 72-78; HEAD DEPT PLANT PATH & PHYSIOL, CLEMSON UNIV, 78- *Mem:* Am Phytopath Soc; Soc Nematol. *Res:* Biology and control of plant parasitic nematodes. *Mailing Add:* Dept Plant Path & Physiol Clemson Univ Clemson SC 29634-0377

DICKERSON, R(ONALD) F(RANK), b Middletown, NY, Jan 27, 22; m 44; c 2. METALLURGICAL ENGINEERING. *Educ:* Va Polytech Inst, BS, 47, MS, 49. *Prof Exp:* Prin metallurgist reactor mat, Battelle Mem Inst, Battelle Develop Corp, 48-53, asst div chief, 53-56, div chief, 56-62, asst mgr metall & physics dept, 62-64, staff mgr, Pac Northwest Lab, 64-68, asst dir BMI Int, 68-70, mgr spec projs, 70-72, vpres & gen mgr, 72-87; RETIRED. *Mem:* Am Soc Metals; Sigma Xi. *Res:* Fuel alloy development for reactor cores, melting and casting or uranium, zirconium, thorium, niobium, molybdenum and other rare metals; metallography of these materials and irradiation damage. *Mailing Add:* 377 Springs Dr Columbus OH 43214

DICKERSON, RICHARD EARL, b Casey, Ill, Oct 8, 31; m 56; c 5. BIOCHEMISTRY, MOLECULAR EVOLUTION. *Educ:* Carnegie Inst Technol, BS, 53; Univ Minn, PhD(phys chem), 57. *Prof Exp:* NSF fel, Univ Leeds, 57-58 & Cavendish Lab, Cambridge Univ, 58-59; asst prof phys chem, Univ Ill, 59-63; from assoc prof to prof chem, Calif Inst Technol, 63-81; PROF BIOCHEM & GEOPHYS, UNIV CALIF, LOS ANGELES, 81-, DIR, MOLECULAR BIOL INST, 83- *Mem:* Nat Acad Sci; AAAS; Am Inst Physics; Am Soc Biol Chem; Am Crystallog Asn; Sigma Xi; Am Asn Arts & Sci. *Res:* X-ray crystallography and molecular structure of DNA and complexes with drugs and proteins; evolution at the molecular level; molecular biology. *Mailing Add:* Molecular Biol Inst Univ Calif Los Angeles CA 90024

DICKERSON, ROGER WILLIAM, JR, food engineering & technology, heat transfer, for more information see previous edition

DICKERSON, STEPHEN L(ANG), b Rockford, Ill, Jan 6, 40; m 62; c 2. MECHANICAL ENGINEERING. *Educ:* Ill Inst Technol, BS, 62; Univ Calif, Berkeley, MS, 63; Mass Inst Technol, ScD(eng), 65. *Prof Exp:* From asst prof to assoc prof, 65-73, PROF MECH ENG, GA INST TECHNOL, 74- *Concurrent Pos:* Pres, Urban Transp Co, 75- *Mem:* Am Soc Mech Engrs; Soc Automotive Engrs; Sigma Xi. *Res:* Automatic control; system design; transportation systems; robotics. *Mailing Add:* Sch Mech Eng Ga Inst Technol Atlanta GA 30332

DICKERSON, WILLARD ADDISON, b Granville Co, NC, Aug 25, 42; m 64; c 2. ECOLOGY, BIOLOGICAL CONTROL. *Educ:* Univ Mo, BS, 67, MS, 71. *Prof Exp:* Entomologist, Shade Tobacco Insects, US Dept Agr, Agr Res Serv, 68-69, res entomologist, Biol Control Insects Lab, 69-77, res entomologist, Boll Weevil Eradication Res, Serv, 77-87, ADMINR, PLANT PROTECTION SECT, NC DEPT AGR, RALEIGH, 88- *Mem:* Entomol Soc Am; Int Orgn Biol Control Noxious Animals & Plants; Can Entomol Soc. *Res:* Biological control of insects; development and evaluation of insect traps and pheromones; wide area insect population, evaluation and control. *Mailing Add:* 4116 Reedy Creek Rd Raleigh NC 27607

DICKERSON, WILLIAM H, infectious diseases; deceased, see previous edition for last biography

DICKERT, HERMAN A(LONZO), b Newberry, SC, Jan 16, 03; m 27; c 2. ENGINEERING, MATERIALS SCIENCE. *Educ:* Newberry Col, AB, 23, ScD, 55; Univ NC, MA, 25. *Prof Exp:* Res chemist, Food & Nutrit Lab, State Dept Agr, NC, 24-25; res chemist, Cellulose & Rayon, E I du Pont de Nemours & Co, 25-31, supvry work, Rayon Div, 34-45; dir, A French Textile Sch, 45-58, prof textile eng, 58-70, EMER PROF TEXTILE ENG, GA INST TECHNOL, 70- *Concurrent Pos:* Pres, Nat Coun Textile Sch Deans, 52; Fulbright lectr, Tech Sch Indust Eng, Barcelona, 63-64; consult & lectr, Israel Inst Technol, 64; consult, Int Exec Serv Corp, Fortaleza, Brazil, 73. *Res:* Food and nutrition studies; motion picture film base; synthetic and natural fibers including manufacturing processes for synthetics, fabricating, dyeing, finishing and application to plastic laminates; industrial chemistry. *Mailing Add:* 374 E Paces Ferry Rd NE Atlanta GA 30305

DICKEY, DAVID S, b Chicago, Ill, May 2, 46; m 68; c 2. MIXING-EQUIPMENT DESIGN, MIXING-SCALE-UP & TESTING. *Educ:* Univ Ill, Urbana, BS, 68; Purdue Univ, MS, 71, PhD(chem eng), 72. *Prof Exp:* Assoc sr res eng, Gen Motors Res Labs, 72-75; tech dir, Chemineer, Inc, 75-85; dir technol, Div Harsco Corp, Patterson-Kelley Co, 85-90; TECH DIR, AM REACTOR CORP, 90- *Concurrent Pos:* Chmn, Equip Testing Procedures Comt, Am Inst Chem Engrs, 75-85; mem, Nat Res Coun, Eval Panel, Nat Inst Standards & Technol, Ctr Chem Eng, 83-85. *Mem:* Am Inst Chem Engrs; Am Chem Soc; Am Soc Mech Engrs. *Res:* Design of mixing and agitation equipment for chemical processing; scale-up of laboratory and pilot-plant results to large scale equipment; testing and evaluation of mixing, agitation, heat transfer, impeller power and other aspects of stirred tank reactors. *Mailing Add:* Am Reactor Corp 1700 Dalton Dr New Carlisle OH 45344-2307

DICKEY, E(DWARD) T(HOMPSON), b Oxford, Pa, Nov 16, 96; m 44; c 2. ELECTRONICS. *Educ:* City Col New York, BS, 18. *Prof Exp:* Engr, Radio Res, Radio Corp Am, 18-24, sample testing & equip develop, T & T Dept, 24-29, field res radio competitive anal, 29-40, admin asst radio ed work on res projs, 41-61; pub acquisition, Libr, Plasma Physics Lab, 62-66, film scanner, Jadwin Hall, Princeton Univ, 68-72; RETIRED. *Mem:* Sr mem Inst Elec & Electronics Engrs; fel Radio Club Am. *Res:* Development of radio test equipment; hi-speed transoceanic reception equipment; radio receiver circuit development; Poulsen arc research. *Mailing Add:* 511 Scenic Dr Trenton NJ 08628-2205

DICKEY, FRANK R(AMSEY), JR, b San Antonio, Tex, Apr 10, 18; m 44; c 3. ELECTRICAL ENGINEERING. *Educ:* Univ Tex, BS, 39; Harvard Univ, MS, 46, PhD(appl physics), 51. *Prof Exp:* Engr, Gen Elec Co, 50-53; mem res staff, Melpar, Inc, 53-54; consult engr, 54-76, SR CONSULT ENGR, HEAVY MIL ELECTRONICS DEPT, GEN ELEC CO, 76- *Mem:* Fel Inst Elec & Electronics Engrs. *Res:* Radar and communication systems. *Mailing Add:* 112 Cornwall Dr Dewitt NY 13214

DICKEY, FREDERICK PIUS, b Zanesville, Ohio, Sept 1, 16; m 42; c 3. PHYSICS. *Educ:* Muskingum Col, BS, 38; Ohio State Univ, MS, 39, PhD(physics), 46. *Prof Exp:* From instr to prof physics, Ohio State Univ, 46-86; RETIRED. *Mem:* Am Phys Soc. *Res:* Infrared; microwaves; lasers. *Mailing Add:* 29435 Blosser Rd Logan OH 43138-9506

DICKEY, HOWARD CHESTER, b Durand, Mich, June 26, 13; m 39; c 4. DAIRY HUSBANDRY. *Educ:* Mich State Univ, BS, 34; WVa Univ, MS, 36; Univ Iowa, PhD(dairy husb), 39. *Prof Exp:* Asst dairy husb, WVa Univ, 34-36; asst, Iowa State Univ, 36-39; from asst prof to assoc prof animal husb, Colo State Univ, 39-45; assoc prof dairy husb, Univ Vt, 45-47; prof & head, dept animal indust, 47-57, prof & head dept, 57-76, EMER PROF ANIMAL SCI, UNIV MAINE, 76- *Concurrent Pos:* Animal nutrit consult, Jackson Lab. *Mem:* Am Dairy Sci Asn; Am Genetics Soc; Am Soc Animal Sci; Sigma Xi. *Res:* Dairy cattle nutrition and inheritance; animal nutrition; relationship between the curd tension in milk and its rate and percentage of digestibility. *Mailing Add:* 548 College Ave Orono ME 04473

DICKEY, JOHN SLOAN, JR, b Washington, DC, Jan 24, 41; m 63, 78; c 1. PETROLOGY. *Educ:* Dartmouth Col, AB, 63; Univ Otago, NZ, MSc, 66; Princeton Univ, PhD(geol), 69. *Prof Exp:* Field geologist, Brit Newf Explor Co Ltd, 63-64; res assoc lunar geol, Smithsonian Astrophys Observ, 69-70; res fel petrol, Geophys Lab, Carnegie Inst Wash, 70-72; asst prof geol, Mass Inst Technol, 72-76, assoc prof, 76-79; prog dir, NSF, 79-81; chmn dept & Jessie Page Heroy prof geol, Syracuse Univ, 81-88; DEAN, DIV SCI, MATH & ENG, TRINITY UNIV, 88- *Mem:* Geochem Soc; Mineral Soc Am; Am Geophys Union; AAAS. *Res:* Petrology of mafic and ultramafic igneous rocks; economic petrology; experimental petrology.. *Mailing Add:* Dean Off Trinity Univ 715 Stadium Dr San Antonio TX 78284

DICKEY, JOSEPH FREEMAN, b Orange County, NC, Apr 1, 34; m 59; c 2. REPRODUCTIVE PHYSIOLOGY, ELECTRON MICROSCOPY. *Educ:* NC State Col, BS, 56, MS, 62; Pa State Univ, PhD(dairy sci), 65. *Prof Exp:* From asst prof to prof, 65-84, DISTINGUISHED PROF DAIRY SCI, CLEMSON UNIV, 85- *Concurrent Pos:* Sabbatical leave NC State Univ, 87. *Mem:* Am Dairy Sci Asn; Am Soc Animal Sci; Soc Study Reproduction; Am Asn Univ Professors; Sigma Xi; Int Embryo Transfer Soc. *Res:* Physiology and endocrinology of reproduction in mammals; histochemistry and electron microscopy of reproductive tissues; causes of embryonic mortality; embryo manipulation and culture. *Mailing Add:* Dept Animal Dairy & Vet Sci Clemson Univ Clemson SC 29634-0363

DICKEY, JOSEPH W, b Quantico, Va, Feb 26, 39; c 2. DYNAMIC SYSTEMS. *Educ:* Cath Univ Am, PhD(physics), 76. *Prof Exp:* Cong sci fel, US House Representatives, 83-84; res scientist, 65-83, SR RES SCIENTIST, DAVID TAYLOR RES LAB, 84- *Concurrent Pos:* Vis prof, US Naval Acad, 81-82. *Mem:* Acoust Soc Am. *Res:* Structural dynamics and complex dynamic systems; development of mathematical models to describe, and techniques to modify, the vibration response and radiation of structures; physical acoustics and acoustic emission from fluid flow. *Mailing Add:* 282 Cape St John Rd Annapolis MD 21401-5067

DICKEY, PARKE ATHERTON, b Chicago, Ill, Mar 3, 09; m 35; c 4. PETROLEUM GEOLOGY. *Educ:* Johns Hopkins Univ, PhD(geol), 32. *Prof Exp:* Petrographer, Lago Petrol Corp, Venezuela, 30-31; geologist, Trop Oil Co, Colombia, 32-38; geologist chg oil & gas, Pa State Geol Surv, 38-42; geologist, Forest Oil Corp, 42-44; geologist, Quaker State Oil Ref Co, 44-46; head geol res, Carter Oil Co, 46-56, asst chief res, 56-58; geologist, Creole Petrol Corp, Venezuela, 58-60; mgr geol div, Jersey Prod Res Co, 60-61; prof geol & head dept earth sci, 61-72, EMER PROF GEOL, UNIV TULSA, 74- *Concurrent Pos:* Lectr, Oil & Gas Consults Int, 68-90. *Mem:* Soc Petrol Engrs; hon mem Am Asn Petrol Geologists; fel Geol Soc Am; Am Geophys Union. *Res:* Petroleum geology and geochemistry. *Mailing Add:* 8050 N Yale Owasso OK 74055

DICKEY, RICHARD PALMER, b Napoleon, Ohio, Feb 10, 35; m 57; c 4. OBSTETRICS & GYNECOLOGY. *Educ:* Ohio State Univ, BA, 56, MMSc, 65, PhD, 70; Western Reserve Univ, MD, 60; Am Bd Obstet & Gynec, dipl, 67; Am Bd Reprod Med, cert, 76. *Prof Exp:* Res assoc obstet & gynec, Ohio State Univ, 65-70, asst prof, 70-72, instr pharmacol, dir div reprod endocrinol & dir family planning, 71-72; assoc prof obstet & gynec & chief sect reprod endocrinol, La State Univ Med Ctr, New Orleans, 72-76; CONSULT, 86- *Concurrent Pos:* NIH trainee gynec endocrinol, 63-66; mem adv comt obstet & gynec devices, Food & Drug Admin; chmn subcomt conception control devices, 74-77. *Mem:* AAAS; AMA; Am Col Obstet & Gynec. *Res:* Reproductive endocrinology; infertility; population control and family planning; mechanisms of actions of steroids in contraception; neuropharmacology of gonadotropic secretion. *Mailing Add:* 6020 Bullard New Orleans LA 70128-2813

DICKEY, ROBERT SHAFT, plant pathology; deceased, see previous edition for last biography

DICKEY, RONALD WAYNE, b Compton, Calif, Mar 12, 38; m 65. APPLIED MATHEMATICS. *Educ:* Univ Calif, Los Angeles, AB, 59; NY Univ, MS, 62, PhD(math), 65. *Prof Exp:* Vis mem math, Courant Inst, NY Univ, 65-67; from asst prof to assoc prof, 67-74, PROF MATH, UNIV WISMADISON, 74- *Concurrent Pos:* Sci Res Coun sr vis fel, Univ Newcastle, 71-72. *Mem:* Am Math Soc; Soc Indust & Appl Math. *Res:* Nonlinear differential and integral equations and their relations to problems in elasticity. *Mailing Add:* Dept Math Univ Wis 713 Van Vleck Madison WI 53706

DICKHOFF, WALTON WILLIAM, b Watertown, Wis, Apr 8, 47; c 6. ENDOCRINOLOGY, PHYSIOLOGY. *Educ:* Univ Calif, Berkeley, AB, 70, PhD(physiol), 76. *Prof Exp:* Actg asst prof, 77-78, 80 & 81, RES ASSOC, SCH FISHERIES, UNIV WASH, 75-, PROF SCH FISHERIES, 87-; RES PROF PHYSIOLOGIST, NAT MARINE FISHERIES SERV, 87- *Concurrent Pos:* NIH fel, Univ Wash, 76-77. *Mem:* Am Soc Zoologists; AAAS; Endocrine Soc; Am Fisheries Soc; Zool Soc Japan. *Res:* Evolution of endocrine and neuroendocrine control systems; hormonal control of development reproduction and metabolism in lower vertebrates; comparative physiology; finfish aquaculture. *Mailing Add:* Sch Fisheries HF-15 Univ Wash Seattle WA 98195

DICKIE, HELEN AIRD, medicine; deceased, see previous edition for last biography

DICKIE, JOHN PETER, b Waseca, Minn, Apr 4, 34; div; c 4. BIOCHEMISTRY, ORGANIC CHEMISTRY. *Educ:* Univ Minn, BA, 56; Univ Wis, MS, 58, PhD(biochem), 60. *Prof Exp:* Wis Alumni Res Found fel chem, Univ Wis, 60-61; res fel, Mellon Inst, 61-68; group mgr, Res Dept, Koppers Co, Inc, 68-74; dir basic res, Carnation Co, 75-78; PRES, JOHN P DICKIE ASSOCS, 81- *Concurrent Pos:* Res scientist instr, Univ Minn, 86-88. *Mem:* Am Chem Soc. *Res:* Antimycins; chemistry and origin of natural bitumens; antibiotic structure; mechanism of antibiotic action. *Mailing Add:* 1261 Ingerson Rd St Paul MN 55112-3714

DICKIE, LLOYD MERLIN, b Canning, NS, Mar 6, 26; m 52; c 3. ECOLOGY. *Educ:* Acadia Univ, BSc, 46; Yale Univ, MSc, 48; Univ Toronto, PhD(zool), 53. *Prof Exp:* Asst biol, Acadia Univ, 43-46; asst, Yale Univ, 46-48; demonstr, Univ Toronto, 48-51; asst scientist biol, Biol Sta, St Andrews, NB, 51-53, in chg clam & scallop invest, 53-57, assoc scientist in chg population dynamics studies, Groundfish Invest, 57-65, dir, Marine Ecol Lab, Fisheries Res Bd Can, 65-74, RES SCIENTIST MARINE ECOL LAB, BEDFORD INST OCEANOG, 74- *Concurrent Pos:* Assoc prof biol, Dalhousie Univ, 65-74, prof oceanog & chmn dept, 74-77, adj prof, 78-84. *Mem:* Ecol Soc Am; Brit Ecol Soc; fel Royal Soc Can. *Res:* Marine ecology; population dynamics of molluscan shellfish and marine demersal species; fisheries biology. *Mailing Add:* Biol Sci Br Bedford Inst Oceanog Dartmouth NS B2Y 4A2 Can

DICKIE, RAY ALEXANDER, b Minot, NDak, Jan 19, 40; m 65; c 2. POLYMER CHEMISTRY, ORGANIC COATINGS. *Educ:* Univ NDak, BS, 61; Univ Wis-Madison, PhD(phys chem), 65. *Prof Exp:* Res fel, Glasgow Univ, 66; chemist, Stanford Res Inst, 67-68; PRIN RES SCIENTIST ENG & RES STAFF, FORD MOTOR CO, 68- *Mem:* Am Chem Soc; Soc Rheol. *Res:* Structure and mechanical properties of polymer blends and elastomers; chemistry and physics of organic coatings, adhesives and corrosion protection; polymer and surface characterization. *Mailing Add:* PO Box 710 Northville MI 48167-0710

DICKIESON, ALTON C, b New York, NY, Aug 16, 05. ELECTRICAL DESIGNS. *Prof Exp:* Mem tech staff, dir, exec dir & vpres, AT&T Bell Labs, 23-70; RETIRED. *Mem:* Nat Acad Eng; fel Inst Elec & Electronics Engrs; assoc fel Am Inst Aeronaut & Astronaut. *Mailing Add:* 10330 W Thunderbird Blvd C103 Sun City AZ 85351

DICKINSON, ALAN CHARLES, b New Britain, Conn, Jan 10, 40; m 63; c 3. PHYSICAL CHEMISTRY. *Educ:* Tufts Univ, BS, 61; Ind Univ, PhD(phys chem), 65. *Prof Exp:* Teacher chem pub schs, Springfield, Mass, 65-68; from asst prof to assoc prof chem, 68-90, PROF CHEM, AM INT COL, 90- *Concurrent Pos:* Fel, Univ Ala, 78-79. *Mem:* Am Chem Soc. *Res:* Photochemistry and electron spin resonance spectroscopy. *Mailing Add:* Dept Chem Am Int Col Springfield MA 01109

DICKINSON, BRADLEY WILLIAM, b St Marys, Pa, Apr 28, 48; m 83; c 1. SIGNAL PROCESSING, SYSTEM THEORY. *Educ:* Case Western Reserve Univ, BS, 70; Stanford Univ, MSEE, 71, PhD(elec eng), 74. *Prof Exp:* From asst prof to assoc prof, 74-85, PROF ELEC ENG, DEPT ELEC ENG, PRINCETON UNIV, 85-, ASSOC DEAN, SCH ENG & APPL SCI, 91- *Concurrent Pos:* Assoc ed, Trans on Automatic Control, Inst Elec & Electronics Engrs, 80-81, Transactions on Information Theory, 84-86; fel, Inst Elec Electronics Engr, 87; ed, Math of Control, Signals & Systems, 88- *Mem:* Inst Elec & Electronics Engrs; Soc Indust & Appl Math. *Res:* Applications of system theory and statistical analysis to signal processing; neural networks, signal analysis and applied linear algebra. *Mailing Add:* Dept Elec Eng Princeton Univ Princeton NJ 08544-5263

DICKINSON, DALE FLINT, b Galveston, Tex, Oct 11, 33; m 63; c 4. PHYSICS. *Educ:* Univ Tex, BS, 55; Univ Calif, PhD(physics), 65. *Prof Exp:* Res assoc radio astron lab, Univ Calif, 66-67; res assoc, Smithsonian-Harvard Ctr Astrophys, 67-78, lectr astron, Harvard Univ, 75-78; vis assoc prof physics, Williams Col, 78-80; AT JET PROPULSION LAB, 80-; AT LOCKHEED PALO ALTO RES LAB, PALO ALTO, CA. *Mem:* Am Astron Soc. *Res:* Spectral line radio astronomy; interstellar molecules, maser emission in long-period variable stars; synthetic aperture radar. *Mailing Add:* OR6 92-20 254 Lockheed Palo Alto Res Lab 3251 Hanover St Palo Alto CA 94304

DICKINSON, DAVID (JAMES), b Denver, Colo, Sept 16, 20; m 44; c 2. MATHEMATICS. *Educ:* Univ Denver, BA, 42; Columbia Univ, AM, 47; Univ Mich, PhD(math), 54. *Prof Exp:* Mem staff radiation lab, Mass Inst Technol, 42-45; lectr, Columbia Univ, 45-46; instr math, Pa State Univ, 50-58; prof, 58-81, EMER PROF MATH, UNIV MASS, AMHERST, 81- *Mem:* Am Math Soc. *Res:* Classical analysis; special functions; orthogonal polynomials; theoretical linguistics. *Mailing Add:* PO Box 235 Ashfield MA 01330

DICKINSON, DAVID BUDD, b New York, NY, Jan 10, 36; m 61; c 3. PLANT PHYSIOLOGY. *Educ:* Univ NH, BS, 57; Univ Ill, PhD(hort), 62. *Prof Exp:* From instr to assoc prof plant physiol, 61-74, head dept hort, 86-88, PROF PLANT PHYSIOL, UNIV ILL, URBANA, 74- *Mem:* AAAS; Am Chem Soc; Am Soc Plant Physiol; Am Soc Hort Sci. *Res:* Plant carbohydrate metabolism; plant cell wall biogenesis; metabolic transformations in germinating pollen, seeds and mechanisms regulating germination. *Mailing Add:* Dept Hort Univ Ill 1201 S Dorner Dr Urbana IL 61801

DICKINSON, DEAN RICHARD, b Highland Park, Ill, May 21, 28; m; c 1. CHEMICAL ENGINEERING. *Educ:* Cornell Univ, BChE, 51; Univ Wis, MS, 54, PhD(chem eng), 58. *Prof Exp:* Sr chem engr, Hanford Labs, Gen Elec Co, 58-64 & Pac Northwest Lab, Battelle Mem Inst, 65-71; PRIN ENGR, WESTINGHOUSE HANFORD CO, 71- *Mem:* Am Inst Chem Engrs; Am Nuclear Soc; Sigma Xi. *Res:* Coolant systems of nuclear reactors; corrosion, fluid flow and heat transfer; gas entrainment, liquid mixing and thermal striping in nuclear reactors; aerosol behavior. *Mailing Add:* 1314 Hains Washington Way Richland WA 99352

DICKINSON, EDWIN JOHN, b Carlisle, Eng, Oct 15, 33; m 59; c 5. FLUID MECHANICS. *Educ:* Cambridge Univ, BA, 55, MA, 58; Laval Univ, PhD(fluid mech), 65. *Prof Exp:* Engr servomechanisms, A V Roe, Manchester, 56-58; res asst aeronaut, Laval Univ, 59-61, teaching asst fluid mech, 61-63; res asst boundary layers, Univ Poitiers, 63-64; asst prof appl math, 64-67, ASSOC PROF APPL MATH, LAVAL UNIV, 67- *Concurrent Pos:* Defense Res Bd Can res grant, 65. *Mem:* Assoc fel Royal Aeronaut Soc; Can Aeronaut & Space Inst. *Res:* Experimental determination of turbulent skin friction, particularly the development of floating element skin friction balances for this purpose. *Mailing Add:* Dept Mech Eng Fac Sci Laval Univ Quebec PQ G1K 7P4 Can

DICKINSON, ERNEST MILTON, veterinary medicine, for more information see previous edition

DICKINSON, FRANK N, b Boston, Mass, July 25, 30; m 57; c 3. ANIMAL HUSBANDRY, GENETICS. *Educ:* Univ Mass, BS, 53; Univ Ill, MS, 58, PhD(dairy sci), 62. *Prof Exp:* Dairy husbandman, Agr Res Serv, USDA, Univ Ill, 57-61; anal statistician, Biomet Serv, Md, 61-65; asst prof animal sci, Univ Mass, 65-67; invest leader, Dairy Herd Improv Invests, Agr Res Ctr, 67-72, chief, animal improv, 72-88; CHIEF EXEC OFFICER, NAT DAIRY HERD IMPROV ASN, 88- *Mem:* Nat Asn Animal Breeders; Am Dairy Sci Asn; Am Jersey Cattle Club; Am Dairy Goat Asn; Am Jersey Cattle Club. *Res:* Biological statistics; improvement of production in dairy cattle; dairy cattle genetics; dairy record keeping procedure; national coop dairy herd improvement prog. *Mailing Add:* Nat Diary Herd Improv Asn Inc 3021 Dublin Granville Rd Columbus OH 43231

DICKINSON, HELEN ROSE, b Skowhegan, Maine, Feb 6, 45. BIOPHYSICS, PHYSICAL CHEMISTRY. *Educ:* Univ Maine, BS, 67; Ore State Univ, PhD(biophys), 72. *Prof Exp:* Res assoc chem, Ill Inst Technol, 73-77; res assoc biophys, Case Western Reserve Univ, 77-81; Sr Chemist, Marine Colloids Div, FMC Corp, 81-82; Monsanto Co, 84-88; CHEMIST, CHICAGO COL OSTEOP MED, 90- *Concurrent Pos:* Nat Inst Gen Med Sci, NIH fel, 75-77. *Mem:* Am Chem Soc; Biophys Soc; AAAS. *Res:* Conformations of polysaccharides and proteins; interactions of biological molecules; physical chemistry. *Mailing Add:* 429 S Wesley No 304 555 31st St Oak Park IL 60302

DICKINSON, JAMES M(ILLARD), b Waterloo, Iowa, July 31, 23; m 47; c 3. PHYSICAL METALLURGY. *Educ:* Iowa State Univ, BS, 49, PhD(phys chem & metall), 53. *Prof Exp:* Jr chemist, Phys Metall Res, Atomic Energy Comn, Ames Lab, Iowa State Col, 46-53; metallurgist & mem staff, Los Alamos Nat Lab, Univ Calif, 53-74, asst group leader, 74-81, group leader mat technol, 81-87; RETIRED. *Mem:* Am Soc Metals; Am Inst Mining & Metall Engrs; Soc Advan Mat Process Eng; Am Defense Preparedness Asn. *Res:* General physical metallurgy of the less common metals; very high temperature metallurgy; phase diagrams, metal reduction, casting and fabrication; graphite fabrication, properties and structure; powder metallurgy; ceramics. *Mailing Add:* 8700 E University No 629 Mesa NM 85207

DICKINSON, JOHN OTIS, b Champaign, Ill, Aug 29, 24; m 47; c 3. VETERINARY PHARMACOLOGY. *Educ:* Univ Ill, BS, 48 & 61, DVM, 63, MS, 65, PhD(vet med sci), 68. *Prof Exp:* Instr vet physiol & pharmacol, Univ Ill, 63-64, Nat Inst Arthritis & Metab Dis fel, 64-66; from asst prof to prof vet physiol & pharmacol, Wash State Univ, 66-87; RETIRED. *Mem:* Am Soc Vet Physiol & Pharmacol; Soc Study Reproduction; Am Col Vet Toxicologists; Am Vet Med Asn. *Res:* Oral histamine metabolism in ruminants; effect of chlorinated hydrocarbon pretreatment on toxicity of carbamate insecticides; mechanism of allium poisoning; cacodylic acid and monososium acid methane arsonate toxicity in cattle; pyrrolizidine alkaloids and milk transfer. *Mailing Add:* 2135 Wooded Knolls Dr Philomath OR 97370

DICKINSON, JOSHUA CLIFTON, JR, b Tampa, Fla, Apr 28, 16; m 36; c 3. ORNITHOLOGY. *Educ:* Univ Fla, BS, 40, MS, 45, PhD, 50. *Prof Exp:* Asst biol, Univ Fla, 40-42, instr biol & geol, 46-50, from asst prof to prof zool, 50-79, cur biol sci, Fla State Mus, 53-59, actg dir, 59-61, dir & cur ornith, Fla State Mus, 61-79, PROF ZOOL, UNIV FLA, 73-, EMER DIR, FLA STATE MUS, 79- *Concurrent Pos:* Mem, Univ Fla Exped, Honduras, 46; Mus comp zool res fel & Gen Ed Bd fel, Harvard Univ, 51-52; vis investr, Woods Hole Oceanog Inst, 52; mem, Fla State Mus Exped, Baffin Island, 55, Bahamas Islands, 58, 59, 60, 61, 66, & 67, Sombrero Island, 64 & Navassa Island, 67; mem mus adv panel, Nat Endowment Arts, 70-72, co-chmn, 72-74; mem, Nat Coun Arts, 76-82. *Honors & Awards:* Pres Medallion, 79; James L Short Award, 86. *Mem:* Am Soc Naturalists; Am Soc Zoologists; Am Ornithologists Union; Am Asn Mus (secy, 70-73); Asn Systs Collections, Inc (pres, 72-75); Sigma Xi. *Res:* Taxonomy and zoogeography; general ecology. *Mailing Add:* Fla State Mus 238A Dickinson Hall Univ Fla Gainesville FL 32603

DICKINSON, LEONARD CHARLES, b Glasgow, Ky, Dec 12, 41; m 66; c 2. PHYSICAL & BIOPHYSICAL CHEMISTRY, POLYMER CHEMISTRY. *Educ:* Bellarmine Col, Ky, AB, 63; Univ Wis-Madison, PhD(phys chem), 69. *Prof Exp:* Sci res fel, Univ Leicester, 69-70; asst prof chem, 70-73 & 76-89, sr res fel, 73-75, PROJ DIR, UNIV MASS, AMHERST, 75-, NMR DIR, 89- *Mem:* Sigma Xi; Am Chem Soc; Soc Lit & Sci. *Res:* Solids nuclear magnetic resonance of network polymers; solids nuclear magnetic resonance of blends; magnetic resonance line shapes; active site structures of hematin enzymes; metal replacement in hematin enzymes; artificial polymeric enzymes; photosynthetic reaction center tetrapyrrole function; water soluble polymers; single crystal electron paramagnetic resonance of metalloenzymes. *Mailing Add:* Dept Polymer Sci & Eng Univ Mass Amherst MA 01003

DICKINSON, PETER CHARLES, b London, Eng, Sept 30, 39; m 65; c 1. STATISTICS. *Educ:* Univ London, BS, 65; Rutgers Univ, MS, 67, PhD(statist), 69. *Prof Exp:* Engr, Res & Develop Labs, E M I Ltd, Eng, 58-64; teaching asst statist, Rutgers Univ, 65-69; asst prof, 69-76, ASSOC PROF STATIST, UNIV SOUTHWESTERN LA, 76- *Mem:* Biomet Soc; Inst Math Statist; Am Statist Asn; Am Soc Qual Control. *Res:* Non-parametric statistics; experimental design; order statistics. *Mailing Add:* Dept Statist Univ Southwestern La Lafayette LA 70504

DICKINSON, ROBERT EARL, b Millersburg, Ohio, March 26, 40; m 74; c 1. LAND SURFACE PROCESSES, CLIMATE CHANGE. *Educ:* Harvard Univ, AB, 61; Mass Inst Technol, MS, 62, PhD(meteorol), 66. *Prof Exp:* Res assoc, dept meteorol, Mass Inst Technol, 66-68; scientist, 68-73, sr scientist, 73-90, head climate sect, 75-81, dep dir AAP div, Nat Ctr Atmospheric Res, 81-86; PROF ATMOSPHERIC PHYSICS, UNIV ARIZ, 90- *Concurrent Pos:* Mem steering Comts, UN Univ Proj Climatic, Biotic & Human Interactions in Humid Tropics & Int Satellite Land Surface Climatology Proj, 83-86; mem climate res comt, global change comt & comt earth sci, Nat Res Coun, 85-90. *Honors & Awards:* Meisinger Award, Am Meteorol Soc, 73, Jule Churner Award, 88. *Mem:* Nat Acad Sci; fel AAAS; Am Geophys Union; fel Am Meteorol Soc. *Res:* Atmospheric processes; upper atmosphere, radiation, large scale dynamics, climate and climate change; land processes; man's impact on climate; climate effects of land use change. *Mailing Add:* Inst Atmospheric Physics Univ Ariz Tucson AZ 85721

DICKINSON, STANLEY KEY, JR, b Clarksburg, WVa, Feb 16, 31; c 1. GEOCHEMISTRY, SOLID STATE SCIENCE. *Educ:* WVa Univ, BS, 53, MS, 55; Harvard Univ, PhD(geol), 68. *Prof Exp:* Res & develop officer, Air Force Cambridge Res Ctr, 54-56; teaching fel & res asst, Harvard Univ, 57-62; phys sci, Air Force Cambridge Res Labs, Rome Air Develop Ctr, 62-77; Phys sci admin, 77-85, dir sci, 85-86, DEP RES & GEOPHYSICS, HQ AIR FORCE SYSTS COMMAND, 87- *Concurrent Pos:* Vis scientist, Diamond Res Lab, Johannesburg, 72. *Mem:* Sigma Xi. *Res:* Inorganic equilibria and phase relationships; solid state science; thermodynamics; physical inorganic chemistry; mineral synthesis and stability relations; general inorganic geochemistry; crystallography; x-ray diffraction analysis; stratigraphy; paleontology. *Mailing Add:* Rte Box 242 Ashburn VA 22011

DICKINSON, WADE, b Hickory Twp, Pa, Oct 29, 26; m 52; c 3. BIOPHYSICS, NUCLEAR PHYSICS. *Educ:* US Mil Acad, BS, 49. *Prof Exp:* Proj officer aircraft nuclear propulsion, Wright Air Develop Ctr, 51-53; nuclear physicist, Rand Corp, 53-54; pres & chief exec, W W Dickinson Corp, 62-71; PRIN ENGR & CONSULT, BETCHEL CORP, 54-; PRES & CHIEF EXEC, AGROPHYSICS, INC, 68-, PETROLPHYSICS PARTNERS, 76- *Concurrent Pos:* Consult, Rand Corp, 54-56; tech consult, 85th Cong Joint Comt Atomic Energy, 57-58; vis lectr, Univ Calif, Berkeley, 86-; cardiol consult, Mt Zion Med Ctr, Univ Calif, San Francisco. *Mem:* Am Phys Soc; Soc Petrol Engrs. *Res:* Weight gain stimulation and behavior control in meat animals by internal devices; human contraceptive devices; ultrasonic measurement of heart performance; high speed, precision seed planters; nondestructive tests of seed vigor and genotype; nuclear engineering and physics; petroleum drilling technology. *Mailing Add:* Petrolphysics Inc 2101 Third St San Francisco CA 94107

DICKINSON, WILLIAM BORDEN, b Norfolk, Va, Jan 20, 26. MEDICINAL CHEMISTRY. *Educ:* Emory Univ, AB, 46, MS, 47; Univ Wis, PhD, 50. *Prof Exp:* Res assoc, 50-57, assoc mem-res chemist, 57-65, group leader, 58, sr res chemist, Sterling-Winthrop Res Inst, 65-88; CONSULT, 88- *Mem:* Am Chem Soc. *Res:* Ester condensations; hindered ketones; steroids; heterocyclic compounds; synthetic therapeutic agents; organic nomenclature. *Mailing Add:* Dutch Village Apt 1-DR Menands Albany NY 12204

DICKINSON, WILLIAM CLARENCE, b St Joseph, Mo, Mar 15, 22; m 47; c 2. ENERGY CONVERSION. *Educ:* Univ Calif, AB, 45; Mass Inst Technol, PhD(physics), 50. *Prof Exp:* Asst physics, Mass Inst Technol, 48-50; mem staff, Physics Div, Los Alamos Sci Lab, 50-54; prof physics & head dept, Univ Indonesia, 54-57; mem staff, Neutronics Div, Lawrence Radiation Lab, Univ Calif, Livermore, 57-61, mem staff, Test Div, Lawrence Radiation Lab, 61-73, solar projs leader, Lawrence Livermore Lab, 73-82; RETIRED. *Concurrent Pos:* Lectr, Univ NMex, 53-54; consult, 82- *Mem:* Am Phys Soc; Am Asn Physics Teachers; Int Solar Energy Soc. *Res:* Nuclear magnetic resonance; nuclear reactions; neutron physics; nuclear radiation detectors; solar thermal collectors; solar thermal conversion. *Mailing Add:* 54 Panoramic Way Berkeley CA 94704

DICKINSON, WILLIAM JOSEPH, b Pasadena, Calif, June 17, 40; m 63; c 2. DEVELOPMENTAL GENETICS. *Educ:* Univ Calif, Berkeley, BA, 63; Johns Hopkins Univ, PhD(biol), 69. *Prof Exp:* Sci investr, Cellular Radiobiol Br, US Navy Radiol Defense Lab, Calif, 63-65; asst prof biol, Reed Col, 69-72; from asst prof to assoc prof, 72-81, PROF BIOL, UNIV UTAH, 81- *Concurrent Pos:* Vis scientist, Dept Zool, Univ BC, 71 & Dept Genetics, Univ Hawaii, 78-79; vis fel, Molecular Embryology Group, Cambridge Univ, 91-92. *Mem:* AAAS; Genetics Soc Am; Soc Develop Biol; Am Soc Naturalists; Am Soc Zoologists. *Res:* Genetic and molecular analysis of developmentally specific gene regulation; evolution of patterns of gene regulation; the evolutionary significance of regulatory genes. *Mailing Add:* Dept Biol Univ Utah Salt Lake City UT 84112

DICKINSON, WILLIAM RICHARD, b Tenn, Oct 26, 31; m 53, 70; c 4. GEOLOGY. *Educ:* Stanford Univ, BS, 52, MS, 56, PhD(geol), 58. *Prof Exp:* From asst prof to prof geol, Stanford Univ, 58-79; PROF GEOSCI, UNIV ARIZ, 79- *Concurrent Pos:* Guggenheim fel, 65. *Mem:* AAAS; Geol Soc Am; Soc Econ Paleontologists & Mineralogists; Am Geophys Union; Am Asn Petrol Geologists. *Res:* Petrology; structural geology; sedimentology; plate tectonics. *Mailing Add:* Dept Geosc Univ Ariz Tucson AZ 85721

DICKINSON, WILLIAM TREVOR, b Toronto, Ont, Aug 30, 39; m 63; c 2. AGRICULTURAL ENGINEERING, SOIL SCIENCE. *Educ:* Univ Toronto, BSA, 61, BASc, 62, MSA, 64; Colo State Univ, PhD(hydroll hydraul), 67. *Prof Exp:* Lectr eng, 63-64, from asst prof to assoc prof, 67-78, PROF HYDROL, UNIV GUELPH, 78- *Concurrent Pos:* Res assoc, Inst Hydrol, Wallingford, Berkshire, Gt Brit, 72; Coord instrnl develop, Off Educ Pract, Univ Guelph, 79-82; vis prof, Dept Geog & Watershed Ecosysts Grad Prog, Trent Univ, Peterbourgh, Ont, 88-89; Int Hydrol Decade Distinguished Lect, 74. *Mem:* Can Asn Univ Teachers; Soil Conserv Soc Am. *Res:* Characterization of hydrological variables; development of watershed response models; soil erosion and fluvial sedimentation models. *Mailing Add:* Sch Eng Univ Guelph Guelph ON N1G 2W1 Can

DICKISON, HARRY LEO, b Ashland, Ky, Sept 3, 12; m 35; c 2. PHARMACOLOGY. *Educ:* Vanderbilt Univ, AB, 35, MS, 36, PhD(org chem), 39. *Prof Exp:* Asst pharmacol, Med Sch, Vanderbilt Univ, 39-41, res assoc, 41-45, asst prof, 45-47; dir pharmacol res, Bristol Labs, 47-58, asst dir res, 51-58, dir labs, 58-66, dir res US, 66-68, gen mgr vet prod, 68-76; RETIRED. *Concurrent Pos:* Lectr, State Univ NY Syracuse, 49-76. *Mem:* Am Soc Pharmacol & Exp Therapeut; Am Fedn Clin Res; Am Chem Soc; Soc Toxicol. *Res:* Hypnotics; anesthetics; analgetics; metabolic fate of drugs; antibiotics; antihistaminics; terpenoid amines; isomeric thujyl amines. *Mailing Add:* 4639 Prince Edward Dr Jacksonville FL 32210

DICKISON, WILLIAM CAMPBELL, b Jamaica, NY, Mar 12, 41; m 63; c 1. PLANT MORPHOLOGY, PLANT ANATOMY. *Educ:* Western Ill Univ, BSEd, 62; Ind Univ, AM, 64, PhD(bot), 66. *Prof Exp:* Asst prof, Va Polytech Inst & State Univ, 66-69; from asst prof to assoc prof, 69-84, PROF BOT, UNIV NC, CHAPEL HILL, 84- *Mem:* Bot Soc Am; Int Soc Plant Morphol; Int Soc Plant Taxon; Sigma Xi. *Res:* Morphology and phylogeny of vascular plants; comparative morphology and relationships of the Dilleniaceae and allies. *Mailing Add:* Dept Biol Univ NC Chapel Hill NC 27514

DICKLER, HOWARD B, b Chicago, Ill, Jan 2, 42. FC RECEPTORS. *Educ:* George Washington Univ, MD, 68. *Prof Exp:* SR INVESTR, IMMUNOL BR, NIH, 74- *Mem:* Am Asn Immunologists; Am Fedn Clin Res; Am Soc Clin Invest. *Mailing Add:* Div Allergy Immunol & Transplant NIAID NIH Westwood Bldg Rm 755 Bethesda MD 20892

DICKMAN, ALBERT, b Lake Placid, NY, Nov 3, 03; m 28; c 1. LABORATORY MEDICINE. *Educ:* Johns Hopkins Univ, AB, 24; Univ Pa, MS, 31, PhD(bact), 33; Am Bd Microbiol, dipl. *Prof Exp:* Teacher pub schs, Md, 24-25; asst to cur, Com Mus, Pa, 25-26; teacher biol, Pub Schs, 28-41; DIR, DICKMAN LABS, 33- *Concurrent Pos:* Mem ad comt lab procedures, Pa Dept Health & Nat Adv Serol Coun, USPHS; mem, Conf State & Prov Pub Health Dirs. *Mem:* Fel Am Pub Health Asn; Am Asn Clin Chem; Am Chem Soc; Am Soc Microbiol. *Res:* Medical laboratory science; serology; general clinical laboratory field. *Mailing Add:* Foxcroft Square Apt 802 Jenkintown PA 19046

DICKMAN, JOHN THEODORE, b Hamilton, Ohio, Oct 27, 27; m 56; c 2. BIOCHEMISTRY. *Educ:* Ohio State Univ, BS, 50, MS, 57, PhD(physiol chem), 60. *Prof Exp:* Instr high sch, Ohio, 50-53; asst ed, Biochem Sect, 60-63, asst dept head, Biochem Edit Dept, 63-64, dept head, 64-65, asst managing ed, Abstract Issues, 65-69, mgr ed, 69-71, asst managing ed publ, 71-79, SR ASST ED OPERS, CHEM ABSTRACTS SERV, 79- *Concurrent Pos:* Teaching assoc, Ohio State Univ, 65-71; bd dirs, Doc Abstr Inc, 83-88. *Mem:* Am Chem Soc; Coun Biol Educ; Asn Earth Sci Educ (pres, 84). *Res:* Science education; toxicology; lipid chemistry; chemical documentation. *Mailing Add:* 2661 Montcalm Rd Columbus OH 43221-3452

DICKMAN, MICHAEL DAVID, b Pittsburgh, Pa, June 30, 40; m 62; c 2. LIMNOLOGY, AQUATIC BIOLOGY. *Educ:* Univ Calif, Santa Barbara, BA, 62; Univ Ore, MSc, 65; Univ BC, PhD(limnology), 68. *Prof Exp:* Asst prof, Univ Ottawa, 69-74; assoc prof biol, 74-85, PROF BIOL, BROCK UNIV, 85- *Concurrent Pos:* Thord Gray fel, Univ Uppsala, 69; dir, Lower Rideau River Pollution Abatement prog-Carleton Regional Munic, 73-75; bd dir, Can Environ Law Asn, 80-82; vpres, Brock Univ Fac bd, 80-81, pres, 81-82, Senate, 81-84; hon prof, geol dept, McMaster Univ; vchmn, Environ Ecol Adv Comt Regional Niagara, 84-90; educ exchange res prog, Nanjing, China; mem, Ont Environ Appeal bd, 90- *Mem:* Soc Canadian Limnologists (vpres, 81-83, pres, 83-85). *Res:* Paleolimnology of Canadian shield lakes and the Black Sea; acid rain, lake Meromixis, aquatic ecology and wetland pollution abatement; benthic invertebrate toxicology. *Mailing Add:* Dept Biol Brock Univ St Catharines ON L2S 3A1 Can

DICKMAN, RAYMOND F, JR, b Cincinnati, Ohio, July 8, 37; m 63. TOPOLOGY. *Educ:* Univ Miami, BS, 61, MS, 63; Univ Va, PhD(math), 66. *Prof Exp:* From asst prof to assoc prof math, Univ Miami, 66-71; assoc prof 71-74, asst vpres acad affairs, 74-75, PROF MATH, VA POLYTECH INST & STATE UNIV, 74- *Mem:* Am Math Soc; Math Asn Am. *Res:* Mappings; extensions of spaces and mappings; compactifications; unicoherence. *Mailing Add:* 2102 Linden Ct No R4 Blacksburg VA 24060

DICKMAN, ROBERT LAURENCE, b New York, NY, May 16, 47; m 75; c 2. ASTROPHYSICS, MICROWAVE ENGINEERING. *Educ:* Columbia Col, AB, 69; Columbia Univ, MA, 72, PhD(physics), 76. *Prof Exp:* Res assoc physics, Rensselaer Polytech Inst, 75-78; mem tech staff physics, Aerospace Corp, Los Angeles, 78-80; fac res assoc radio astron, 80-85, ASSOC PROF ASTRON, UNIV MASS, AMHERST, 85- *Honors & Awards:* Ernest F Fullam Award, Dudley Observ, 86. *Mem:* Am Phys Soc; Am Astron Soc; Int Astron Union. *Res:* Molecular radio astronomy; star formation; interstellar cloud dynamics; millimeter-wave receiver design and engineering. *Mailing Add:* Five Col Radio Observ Grad Res Ctr Univ MA Amherst MA 01003

DICKMAN, SHERMAN RUSSELL, b Buffalo, NY, Jan 15, 15; m 41; c 3. BIOCHEMISTRY, SCIENCE EDUCATION. *Educ:* Pa State Col, BS, 36; Univ Ill, MS, 37, PhD(soil chem), 40. *Prof Exp:* First asst soil serv anal, Univ Ill, 40-41, spec asst chem, 41-45; asst chem, Argonne Nat Lab, Chicago, 45-46; res assoc med biochem, Col Physicians & Surgeons, Columbia Univ, 46-47; from asst prof to prof, 47-87, EMER PROF BIOCHEM, SCH MED, UNIV UTAH, 87- *Mem:* AAAS; Am Chem Soc; Am Soc Biol Chemists; Sigma Xi. *Res:* Writer. *Mailing Add:* 1560 Indian Hills Dr Salt Lake City UT 84108

DICKMAN, STEVEN RICHARD, b Brooklyn, NY, June 24, 50; m 81; c 1. TIDAL DYNAMICS. *Educ:* Columbia Univ, BA, 72; Univ Calif, Berkeley, MA, 74, PhD(geophys), 77. *Prof Exp:* Asst prof, 77-84, ASSOC PROF GEOPHYS, STATE UNIV NY, 77- *Mem:* Am Geophys Union; Int Astron Union. *Res:* Theoretical prediction of ocean tidal effects on wobble and length of day. *Mailing Add:* Dept Geol Sci State Univ NY Binghamton NY 13902-6000

DICKS, JOHN BARBER, JR, physics; deceased, see previous edition for last biography

DICKSON, ARTHUR DONALD, b Lowellville, Ohio, Apr 26, 27; m 52; c 7. PSYCHIATRY. *Educ:* Carnegie Inst Technol, BSc, 50; Univ Minn, PhD(phys chem), 55. *Prof Exp:* Naval Ord res fel, Univ Minn, 54-55; res chemist, Electrochem Dept, E I du Pont de Nemours & Co, Inc, 55-57; res chemist, 57-66, Univac, 66-72, AT MINN MINING & MFG CO, 72- *Concurrent Pos:* Dir, Aquamotion, Inc, Minneapolis. *Mem:* Am Vacuum Soc; Sigma Xi; Am Chem Soc. *Res:* Infrared spectroscopy; high pressure gas absorption spectra; colorimetry of fluorescent pigments; thermographic reaction kinetics; vacuum deposition of dielectric films; durable pavement markings; retroreflective sheeting; organizational effectiveness. *Mailing Add:* 2593 Western Ave St Paul MN 55113

DICKSON, DAVID ROSS, b Chicago, Ill, Oct 22, 31; m 71; c 4. SPEECH PATHOLOGY. *Educ:* Grinnell Col, BA, 53; Northwestern Univ, MA, 54, PhD(speech path), 61. *Prof Exp:* Assoc prof speech path, Sch Speech, Northwestern Univ, 58-68, lectr cleft palate, Dent Sch, 65-68; prof speech & anat, Univ Pittsburgh, 68-75; PROF PEDIAT, UNIV OF MIAMI & COORDR TRAINING, MAILMAN CTR FOR CHILD DEVELOP, 75- *Mem:* Fel Am Speech & Hearing Asn; Am Cleft Palate Asn; Sigma Xi. *Res:* Speech science; cleft palate; anatomy and physiology. *Mailing Add:* 414 Papaga St No 50 Goodland FL 33933

DICKSON, DON ROBERT, b Devil's Slide, Utah, May 19, 25; m 53; c 7. METEOROLOGY. *Educ:* Univ Utah, BS, 50, MS, 53. *Prof Exp:* Lectr meteorol, Univ NMex, 51-52; physicist, US Geol Surv, 53-55; asst prof meteorol, Okla State Univ, 55-57; asst prof, 57-65, actg head dept, 63-67, chmn dept, 67-72, assoc prof, 65-86, EMER PROF METEOROL, UNIV UTAH, 86- *Concurrent Pos:* Consult, Kennecott Copper Corp, 67-73 & Hales & Co, 74- *Mem:* Am Meteorol Soc; Am Geophys Union; Int Soc Biometeorol; Royal Meteorol Soc; Sigma Xi; Asn Biomed Appln Electromagnetism. *Res:* Physical meteorology; neurometeorology studies; microclimatology; meteorological instruments; atmospheric pollution; severe storm damage; physical environmental correlates of idiopathic pain and discomfort in various clinical subgroups, especially arthritics and victims of orthopaedic pain. *Mailing Add:* Dept Meteorol Univ Utah Salt Lake City UT 84112

DICKSON, DONALD WARD, b Dickson, Tenn, Dec 9, 38; m 62; c 2. NEMATOLOGY, PLANT PATHOLOGY. *Educ:* Austin Peay State Univ, BS, 63; Okla State Univ, MS, 65; NC State Univ, PhD(plant path), 68. *Prof Exp:* Plant pathologist, 68-70, NEMATOLOGIST, UNIV FLA, 70- *Mem:* Soc Nematol (secy, 77-80, vpres, 80-81, pres, 81-82); Am Phytopath Soc; Europ Soc Nematol; Orgn Trop Am Nematol. *Res:* Biological control of nematodes, nematode management; nematode population ecology, biology of root-knot nematodes and biochemical systematics of nematodes. *Mailing Add:* Dept Entom & Nematol Bldg 970 Univ Fla Hull Rd Gainesville FL 32611-0740

DICKSON, DOUGLAS GRASSEL, b Montclair, NJ, Nov 11, 24; m 52; c 2. MATHEMATICAL ANALYSIS. *Educ:* Wesleyan Univ, BA, 47; Harvard Univ, AM, 49; Columbia Univ, PhD(math), 58. *Prof Exp:* Instr math, Dartmouth Col, 48-49; lectr, Hunter Col, 49-51; lectr, Columbia Univ, 51-52, instr, 52-57; from asst prof to assoc prof, 58-74, PROF MATH, UNIV MICH, 74- *Concurrent Pos:* Ed, Mich Math J, 88- *Mem:* Am Math Soc. *Res:* Complex analysis. *Mailing Add:* Dept Math Univ Mich Ann Arbor MI 48109

DICKSON, DOUGLAS HOWARD, b St Thomas, Ont, Feb 16, 42; m 69; c 2. MICROSCOPIC ANATOMY. *Educ:* Univ Western Ont, BA, 65, MSc, 68, PhD(anat), 71. *Prof Exp:* Med Res Coun Can fel, Dept Ophthal, Univ Melbourne, 71-72; assoc prof, 72-81, prof anat, 81-, ASSOC DEAN MED RES, DALHOUSIE MED SCH, 87- *Mem:* Can Fedn Biol Soc; Can Asn Anat; Am Asn Anat; Micros Soc Can; Asn Res Vision & Ophthal; Asn Can Med Col. *Res:* Fine structural, by transmission and scanning electron microscopy, studies of normal and pathological ocular tissues. *Mailing Add:* Dept Anat Dalhousie Univ Halifax NS B3H 4H7 Can

DICKSON, FRANK WILSON, b Oplin, Tex, Nov 29, 22; m 45; c 4. GEOLOGY, GEOCHEMISTRY. *Educ:* Univ Calif, Los Angeles, BA, 50, BS, 53, PhD, 55. *Prof Exp:* Res geologist, Univ Calif, Los Angeles, 52-55 & Shell Develop Co, Tex, 55-56; from asst prof to assoc prof geol, Univ Calif, Riverside, 56-69; prof geochem, Stanford Univ, 69-79, chmn dept, 76-79; staff scientist, Oak Ridge Nat Lab, 79-84; ADJ PROF GEOCHEM, UNIV NEV, RENO, 85- *Concurrent Pos:* Shell Oil fel, 53-54; Fulbright res scholar, 62-63;

Guggenheim fel, 62-63. *Mem:* Geol Soc Am; Geochem Soc; Nat Asn Geol Teachers; Soc Econ Geol. *Res:* Field and laboratory genesis of ore deposits; equilibria in systems of metal sulfides and water, as functions of temperature and pressure; rock-aqueous solution reactions and geochemical applications; exploration and environmental geochemistry. *Mailing Add:* 110 E Sky Ranch Blvd Sparks NV 89431

DICKSON, JAMES FRANCIS, III, b Boston, Mass, May 4, 24; m 77. MEDICINE, ELECTRICAL ENGINEERING. *Educ:* Dartmouth Col, AB, 44; Harvard Med Sch, MD, 47; Am Bd Surg, dipl, 56. *Prof Exp:* Res assoc, Mass Inst Technol, 61-65; dir eng in biol & med, Dept Health & Human Serv, NIH, 65-70, dep asst secy health, Dept HEW, 75-77, actg asst secy health, 77, asst surgeon gen, 78-90; ASSOC DIR, HEALTH POLICY INST, 90- *Concurrent Pos:* NIH spec fel eng, Mass Inst Technol, 61-65; consult, President's Comn Technol, Automation & Econ Progress, 65 & President's Adv Coun Mgt Improv, 71. *Mem:* Inst Med-Nat Acad Sci; Am Col Surg. *Res:* Surgery; control systems engineering; science and technology policy. *Mailing Add:* 53 Bay State Rd Boston MA 02215

DICKSON, JAMES GARY, b Chattanooga, Tenn, Apr 26, 43; m 71; c 1. WILDLIFE MANAGEMENT, FORESTRY. *Educ:* Univ of the South, BS, 65; Univ GA, MS, 67; La State Univ, PhD(forestry), 74. *Prof Exp:* Res asst, La State Univ, 72-74; asst prof, La Tech Univ, 74-76; RES WILDLIFE BIOLOGIST, SOUTHERN FOREST EXP STA, USDA FOREST SERV, 76- *Concurrent Pos:* Mem, Nongame Subcomt Southeast Sect Wildlife Soc, 76-; grad fac, Stephen F Austin State Univ, 77- *Mem:* Wildlife Soc; Am Ornithologists Union; Sigma Xi; Am Soc Mammalogists. *Res:* Seasonal populations and vertical distribution of bird communities; bird-habitat relationships, specifically effects of snags, edge and herbicides, on bird habitat and populations. *Mailing Add:* USDA Forest Serv PO Box 7600 SFA Nacogdoches TX 75962

DICKSON, KENNETH LYNN, b Jacksboro, Tex, Nov 20, 43; m 66, 89; c 2. AQUATIC BIOLOGY, ENVIRONMENTAL SCIENCES. *Educ:* NTex State Univ, BS, 66, MS, 68; Va Polytech Inst & State Univ, PhD(zool), 71. *Prof Exp:* From asst prof to assoc prof zool, Va Polytech Inst & State Univ, 70-78, asst dir, Ctr Environ Studies, 70-78; res scientist, 78-79, PROF, DEPT BIOL SCI, NTEX STATE UNIV, 81-, DIR, INST APPL SCI, 79- *Concurrent Pos:* Consult, Nat Acad Sci, 71- *Mem:* Am Fisheries Soc; Soc Environ Toxicol & Chem; Sigma Xi. *Res:* Limnology of reservoirs and rivers; microbiotic cycles in reservoirs; development of biological pollution monitoring systems; effects of pollution on aquatic organisms; biological diversity indices of community structure; effects of carbon, nitrogen and phosphorus on aquatic communities; fate and effects of chemicals in aquatic life. *Mailing Add:* Inst Appl Sci Box 13078 NTex State Univ FDR 1 Denton TX 76203

DICKSON, LAWRENCE JOHN, b Seattle, Wash, Oct 25, 47. MATHEMATICAL ANALYSIS, APPLIED MATHEMATICS. *Educ:* Seattle Univ, BS, 68; Princeton Univ, PhD(math), 71. *Prof Exp:* Mathematician, Naval Torpedo Sta, Keyport, Wash, 68; instr math, Purdue Univ, West Lafayette, 71; teaching fel, Univ New South Wales, 73-74; sr engr, Boeing Com Airplane Co, Subsid Boeing Co, 75-78; consult, 78-81; PROGRAMMER, SUPERSET, INC, 81- *Concurrent Pos:* NSF fel, 68. *Mem:* Am Math Soc; Math Asn Am. *Res:* Aerodynamic flow analysis; curve and surface fitting; generalized Poisson kernel on unbounded domains in complex vector spaces; convex set theory; hyperbolic geometry models. *Mailing Add:* 7404 Eads Ave La Jolla CA 92037

DICKSON, LAWRENCE WILLIAM, b Saskatoon, Sask, Dec 1, 56; m 83; c 2. CHEMICAL KINETICS, RADIATION CHEMISTRY. *Educ:* Univ Sask, BSc, 77; Univ Toronto, PhD(phys chem), 82. *Prof Exp:* RES SCIENTIST, RES CO, ATOMIC ENERGY CAN, LTD, 81- *Res:* Nuclear fuel damage and fission product transport in severe reactor accident scenarios. *Mailing Add:* Atomic Energy Can Ltd Chalk River ON K0J 1J0 Can

DICKSON, LEROY DAVID, b New Brighton, Pa, June 26, 34; m 54; c 4. ELECTRICAL ENGINEERING. *Educ:* Johns Hopkins Univ, BES, 60, MSE, 62, PhD(elec eng), 68. *Prof Exp:* Res assoc optical data processing, Barton Lab, Johns Hopkins Univ, 61-68; mem tech staff laser atmospheric transmission, Bell Tel Labs, NJ, 68; staff engr, 68-74, adv engr, 74-81, SR ENGR, COMPUT APPLN LASER TECHNOL, IBM CORP, 81- *Mem:* Optical Soc Am; Soc Photo-Optical Instrumentation Engrs; Laser Inst Am. *Res:* Optical data processing; holography; laser technology; applications of laser and laser systems to information processing; holographic scanning. *Mailing Add:* 4805 Connell Dr Raleigh NC 27612

DICKSON, MICHAEL HUGH, b Eng, Apr 2, 32; nat US; m 58; c 3. PLANT BREEDING. *Educ:* McGill Univ, BS, 55; Mich State Univ, MS, 56, PhD(veg breeding), 58. *Prof Exp:* Asst prof hort, Univ Guelph, 58-64; from asst prof to assoc prof, 64-76, PROF HORT, NY STATE AGR EXP STA, GENEVA-CORNELL, 76- *Honors & Awards:* Campbell Award, 77. *Mem:* Fel Am Soc Hort Sci; Am Soc Agron. *Res:* Breeding and genetics of snap beans, with special emphasis on disease resistance and plant efficiency; breeding cabbage, especially insect and blackrot resistance and male sterility. *Mailing Add:* Hort Sci NY State Agr Exp Sta Geneva NY 14456

DICKSON, RICHARD EUGENE, b Carbondale, Ill, Sept 13, 32; m 56; c 3. TREE PHYSIOLOGY. *Educ:* Southern Ill Univ, BS, 60, MS, 62; Univ Calif, Berkeley, PhD(plant physiol), 68. *Prof Exp:* Res asst, Univ Calif, Berkeley, 62-68; res plant physiologist, NCent Forest Exp Sta, 68-70, res plant physiologist, 70-81, PRIN PLANT PHYSIOLOGIST, FORESTRY SCI LAB, US FOREST SERV, 81- *Mem:* AAAS; Am Soc Agron; Ecol Soc Am; Crop Sci Soc Am; Am Soc Plant Physiol; Soc Am Foresters. *Res:* Water relations studies on ecology of bottomland hardwoods; water and oxygen relationships of swamp trees and herbaceous plants; water relations studies on walnut; translocation of organic compounds; physiology of wood formation; carbon metabolism; carbon and nitrogen allocation in trees; carbon and nitrogen interactions. *Mailing Add:* Forestry Sci Lab PO Box 898 Rhinelander WI 54501

DICKSON, ROBERT BRENT, CANCER RESEARCH, PHARMACOLOGY. *Educ:* Yale Univ, PhD(pharmacol), 80. *Prof Exp:* STAFF FEL, NAT CANCER INST, NIH, 80- *Res:* Cell biology. *Mailing Add:* Dept Anat & Cell Biol Georgetown Univ 3900 Reservoir Rd Washington DC 20007

DICKSON, ROBERT CARL, b Coevr d'Alene, Idaho, Apr 22, 43. MOLECULAR BIOLOGY. *Educ:* Univ Redlands, BS, 65; Univ Calif, Los Angeles, PhD(molecular biol), 70. *Prof Exp:* Fel phage assembly, Calif Inst Technol, 70-72, med microbiol, Univ Calif, Los Angeles, 72-73 & genetic regulation, Univ Calif, San Diego, 73-75; from ast prof to assoc prof, 79-88, PROF BIOCHEM, COL MED, UNIV KY, 88- *Mem:* Sigma Xi; Am Soc Microbiol; AAAS; Am Soc Biol Chemists. *Res:* Eucaryatic gene regulation; sphingolipid function. *Mailing Add:* Dept Biochem Col Med Univ KY Lexington KY 40536

DICKSON, SPENCER E, b Topeka, Kans, Dec 17, 38; div; c 3. MATHEMATICS. *Educ:* Univ Kans, BA, 60; NMex State Univ, MS, 61, PhD(math), 63. *Prof Exp:* From asst prof to assoc prof math, Univ Nebr, 63-67; leave from Univ Nebr, Off Naval Res res assoc, Univ Ore, 66-67; assoc prof, Univ Southern Calif, 67-68; from assoc prof to prof math, Iowa State Univ, 68-79; RETIRED. *Mem:* Am Math Soc. *Res:* Homological algebra with applications to ring theory and theory of modules. *Mailing Add:* 111 Lynn Ave Apt 505 Ames IA 50010

DICKSON, STANLEY, b New York, NY, Sept 3, 27; m 50; c 3. SPEECH PATHOLOGY, AUDIOLOGY. *Educ:* Brooklyn Col, BA, 50, MA, 54; Univ Buffalo, EdD, 61. *Prof Exp:* Speech clinician, New York City Bd Educ, 50-51; speech clinician, Rochester Bd Educ, 51-52; exec dir, Rochester Hearing & Speech Ctr, 52-56; assoc prof, 56-60, PROF SPEECH PATH & AUDIOL, STATE UNIV NY COL BUFFALO, 60- *Concurrent Pos:* Speech clinician, Cerebral Palsy Ctr Rochester, 51-52; coordr workshop audiol, State Univ NY Col Geneseo, 54; lectr, Univ Rochester & Nazareth Col, 54-56; consult, Edith Hartwell Clin, LeRoy, NY, 54-56; consult, Children's Rehab Ctr, Buffalo, NY, 63-64, audiol consult, 68-81; proprietor, Williamsville Audiol, Tonawanda Audiol. *Mem:* Am Speech & Hearing Asn; fel Am Speech-Lang-Hearing Asn; NY Acad Sci; Am Acad Sci; fel Am Acad Audiol; Am Audiol Soc. *Res:* Developmental speech and hearing disorders and factors related to its inception and remediation. *Mailing Add:* Dept Speech-Lang Path & Audiol SUNY Coll at Buffalo 1300 Elmwood Ave Buffalo NY 14222

DICKSON, WILLIAM MORRIS, b Denver, Colo, Oct 22, 24; m 46, 79; c 2. VETERINARY PHYSIOLOGY. *Educ:* Colo Agr & Mech Col, DVM, 49; State Col Wash, MS, 53; Univ Minn, PhD, 61. *Prof Exp:* Instr vet physiol & pharmacol, State Col Wash, 49-53, asst prof, 53-55; instr, Univ Minn, 55-57; from asst prof to assoc prof, 57-66, chmn dept, 72-76, PROF VET PHYSIOL & PHARMACOL, WASH STATE UNIV, 66- *Concurrent Pos:* Endocrinol consult, Hanford Biol Lab, 62-63. *Mem:* Am Physiol Soc; Soc Study Reproduction. *Res:* Veterinary endocrinology; reproductive physiology. *Mailing Add:* Dept Vet Physiol Wash State Univ Pullman WA 99163

DICKSTEIN, JACK, b Philadelphia, Pa, Dec 14, 25; m 50; c 3. ORGANIC & POLYMER CHEMISTRY. *Educ:* Pa State Univ, BS, 46; Temple Univ, MA, 51; Rutgers Univ, PhD(polymer chem), 58. *Prof Exp:* Res chemist, Lederle Labs, Am Cyanamid Co, 46-48; asst, Temple Univ, 49-51; chemist, Parenteral Formulations, E R Squibb & Sons Div, Olin Mathieson Chem Corp, 51-56; prof org chem, Alma White Col, 57-58; group leader adhesives, Borden Chem Co, 58-60, develop mgr, Monomer Polymer Labs, 60-61, develop mgr, Thermoplastics Div, 61-67, dir res, 67-74; group mgr & dir res & develop, Haven Chem Co, 74-77; vpres & dir res & develop, Seal Inc, 77-79; PRES, MONOMER-POLYMER, DAJAC LABS, 79- *Concurrent Pos:* Mem, Smithsonian Inst. *Mem:* AAAS; Am Chem Soc; Franklin Inst; Am Inst Chem; NY Acad Sci; Acad Natural Sci. *Res:* High polymers; emulsions; medicinals; adhesives; physical chemistry; research and development in specialty monomers and polymers and medicinal and diagnostic agents. *Mailing Add:* 318 Keats Rd Huntingdon Valley PA 19006

DICORLETO, PAUL EUGENE, b Hartford, Conn, May 19, 51; m 75; c 1. VASCULAR CELL BIOLOGY, MEMBRANE BIOCHEMISTRY. *Educ:* Rensselaer Polytech Inst, BS, 73; Cornell Univ, PhD(biochem), 78. *Prof Exp:* Res fel, Dept Path, Sch Med, Univ Wash, 78-81; PROJ SCIENTIST, DEPT CARDIOVASC RES, CLEVELAND CLIN FOUND, 81- *Mem:* Am Soc Cell Biologists; NY Acad Sci. *Res:* Interaction of growth factors with the cells of the artery wall and the degree to which cell-cell interactions play a role in the development of the atherosclerotic plaque. *Mailing Add:* Res Div Cleveland Clin Found 9500 Euclid Ave Cleveland OH 44195

DI CUOLLO, C JOHN, b Scotch Plains, NJ, Jan 1, 35; m; c 3. BIOCHEMISTRY, MICROBIOLOGY. *Educ:* Va Polytech Inst, BS, 56; Rutgers Univ, PhD(biochem), 60. *Prof Exp:* Sr res biochemist, Colgate Palmolive Co, 60-63; sr res biologist, Smith-Kline Corp, 63-67, asst dir microbiol, 67-71, mgr bioanal sect, 71-73, mgr develop oper, Animal Health Prods, 73-84; vpres sci & technol, Pittman-Moore, 84-86; dir, worldwide regulatory affairs, QA-QC, 86-90, VPRES, REGULATORY AFFAIRS, COMPLIANCE & SCI SUPPORT SERV, SMITH-KLINE BEECHAM ANIMAL HEALTH PROD, 90- *Mem:* AAAS; Am Soc Microbiol; Am Chem Soc; Asn Qual Participation; assoc Am Col Vet Toxicologists. *Res:* Applications of radioisotopes to biological problems; metabolic fate of drugs; development of automation in the field of microbiology; development of tissue residue methods; perform safety studies on potential animal health products; development of biologics and diagnostic products for veterinary and animal health use; responsible for quality assurance and quality control; worldwide regulatory affairs. *Mailing Add:* 353 Colonial Ave Collegeville PA 19426

DICUS, DUANE A, b Okanogan, Wash, Nov 23, 38; m 57; c 3. ELEMENTARY PARTICLE PHYSICS. *Educ:* Univ Wash, BS, 61, MS, 63, Univ Calif, Los Angeles, PhD(physics), 68. *Prof Exp:* Res scientist, Boeing Co, 63-64; asst prof, Univ Calif, Los Angeles, 68-69; res assoc, Mass Inst Technol, 69-71, Univ Rochester, 71-73; from asst prof to assoc prof, 73-84, PROF PHYSICS, UNIV TEX, AUSTIN, 84- *Mem:* Fel Am Phys Soc. *Res:* Fundamental and phenomenological studies of the field theories of strong, electromagnetic, and weak interactions and the application of particle physics to astrophysics and cosmology. *Mailing Add:* Physics Dept Univ Tex Austin TX 78712

DIDCHENKO, ROSTISLAV, b Strzalkowo, Poland, Dec 24, 21; US citizen; m 53. CARBON FIBERS. *Educ:* Technische Hochschule, Ger, dipl chem, 49; Harvard Univ, PhD(inorg chem), 54. *Prof Exp:* Res chemist, Nat Carbon Co, 54-60, group leader, 60-80, sr res mgr, Carbon Prod Div, Union Carbide Corp, 80-84; RETIRED. *Honors & Awards:* S B Meyer Award, Am Ceramic Soc, 56. *Mem:* Am Chem Soc. *Res:* Chemistry of carbon and carbonaceous materials carbon fibers; high temp chemistry; fused salt electrochemistry; semiconductor materials. *Mailing Add:* 7700 Webster Rd Cleveland OH 44130

DIDDLE, ALBERT W, b Hamilton, Mo, July 1, 09; m 42; c 2. GYNECOLOGY. *Educ:* Univ Mo, AB, 30, MA, 33; Yale Univ, MD, 36. *Prof Exp:* Asst anat, Univ Mo, 31-33; asst, Yale Univ, 33-36; instr obstet & gynec, Univ Iowa, 39-40, assoc, 40-42; assoc prof, Southwestern Med Sch, Univ Tex, 45-48; regional consult, Vet Admin, Ga, 48-55; clin prof, Univ Tenn, Knoxville, 56-68, chmn dept, 56-72, prof obstet & gynec, 68-72, emer clin prof & chmn obstet & gynec, 74; RETIRED. *Concurrent Pos:* Consult, Vet Admin, Tex, 46-48 & Oak Ridge Hosp, 49-60. *Honors & Awards:* Humanitarian Award, Res Ctr & Hosp, Univ Tenn, 81. *Mem:* Am Gynec & Obstet Soc; fel Am Col Obstet & Gynec; Continental Gynec Soc; Sigma Xi. *Res:* Anatomy; endocrinology; oncology. *Mailing Add:* 7209 Sheffield Dr Knoxville TX 37909

DIDIO, LIBERATO JOHN ALPHONSE, b Sao Paulo, Brazil, May 7, 20; m 60; c 4. ANATOMY. *Educ:* Univ Sao Paulo, BS, 39, MD, 45, DSc(anat), 49, PhD(anat), 52. *Hon Degrees:* Prof Hon Causa, Univ Mogi das Cruzes, Univ Catalica, Chile, Univ Trujillo; Dr Hon Causa, Univ Nova Lisboa, Univ Rio de Janeiro. *Prof Exp:* Instr physiol, Fac Med, Univ Sao Paulo, 42-43, from instr to assoc prof anat, 43-53; prof topog anat & head dept, Cath Univ Minas Gerais, 53-54, prof anat & chmn dept, Med Sch, Fed Univ Minas Gerais, 54-63; prof, Med, Dent & Grad Sch, Northwestern Univ, Chicago, 63-67; prof anat & chmn dept, Med Col Ohio, 67-88, dean, Grad Sch, 72-86; RETIRED. *Concurrent Pos:* Rockefeller Found fel, Sch Med, Univ Wash; intern, Hosp Med Sch, Univ Sao Paulo, 44-45; vis prof, Univ Messina, 55, Sch Med, Univ Brazil, 57 & Univ Parma, 58; mem Nat Coun Sci Res, Brazil, 56-60; guest investr, Rockefeller Inst Med Res, NY & Harvard Med Sch, 60-61; Brazilian del, Int Cong Electron Micros, Philadelphia, 62; treas, Brazilian Asn Schs Med, 62-64, secy, 64-66; mem, Order Med Merit Presidency Brazilian Repub, Int Anat Nomenclature Comt, 60-86; secy gen, Fed Comt Anat Terminology, 90- *Honors & Awards:* William H Rorer Award, 70; Ipiranga Medal, Govt State, Sao Paulo, Brazil, 79 & Fortaleza, 88. *Mem:* Am Asn Anat; Electron Micros Soc; NY Acad Sci; Pan Am Asn Anat (pres, 69-72, hon pres, 72-); Int Fedn Asn Anatomists (secy, 80-85, secy gen, 85-, pres, 86-89). *Res:* Gross anatomy; surgical anatomy; coronary circulation; electron microscopy of the prostate; pineal body and heart musculature. *Mailing Add:* Off Pres Med Col Ohio PO Box 10008 Toledo OH 43699-0008

DIDISHEIM, PAUL, b Paris, France, June 3, 27; US citizen; m 52; c 3. HEMATOLOGY, CARDIOVASCULAR DISEASES. *Educ:* Princeton Univ, BA, 50; Johns Hopkins Univ, MD, 54. *Prof Exp:* Life Ins Med Res Fund fel, Sch Med, Univ Pittsburgh, 55-57; Nat Heart Inst spec res fel, Nat Blood Transfusion Ctr, Paris, France, 57-58; from instr to asst prof med, Col Med, Univ Utah, 58-65; mem staff, Sect Clin Path, Mayo Med Sch, 65-70, asst prof exp path, 66-70, assoc prof clin path, Mayo Grad Sch Med, 70-73, from assoc prof to prof lab med, 73-86, mem staff, Dept Lab Med, Mayo Clin, 71-86, dir Thrombosis Res Lab, 65-86; HEAD BIOMAT PROG, DEVICES & TECHNOL BR, DIV HEART & VASCULAR DIS, NAT HEART, LUNG & BLOOD INST, 86- *Concurrent Pos:* Chief coagulation lab & assoc physician, Salt Lake County Gen Hosp, 58-65; chief coagulation subdiv, Div Hemat, Dept Med, Col Med, Univ Utah, 58-65; attend, Salt Lake City Vet Admin Hosp, Utah, 59-65; mem med adv coun, Nat Hemophilia Found, 65-77, mem res rev panel, 70-74; mem ad hoc comt standardization human antihemophilic factor assay, Nat Acad Sci-Nat Res Coun, 63; mem comp plasma fractionation & coagulation res, Am Nat Red Cross, 67-72; mem adv bd, Coun on Circulation, Am Heart Asn, 68-, mem orgn comt coun thrombosis, 70-71, mem exec comt, 71-, mem path res study comt, 73-76; mem contract rev panels, Nat Heart & Lung Inst, 74-; mem NIH site visit teams, Prog Proj & Clin Res Ctrs, 75-, Surg & Bioeng Study Sect, 82-86. *Honors & Awards:* Borden Award, Johns Hopkins Univ, 54. *Mem:* Am Fedn Clin Res; Am Asn Pathologists; Am Physiol Soc; Am Soc Hemat; Int Soc Hemat; Int Soc Thrombosis Haemost; Soc Exp Biol & Med. *Res:* Blood coagulation; hemostasis; thrombosis; hemorrhagic diseases; artificial surfaces in contact with blood. *Mailing Add:* Rm 312 Federal Bldg Nat Heart Lung & Blood Inst 7550 Wisconsin Ave Bethesda MD 20814

DIDOMENICO, MAURO, JR, b New York, NY, Jan 12, 37; m 64; c 2. SOLID STATE PHYSICS. *Educ:* Stanford Univ, BS, 58, MS, 59, PhD(elec eng), 63. *Prof Exp:* Mem tech staff, Bell Labs, 62-66, supvr, 66-70, dept head, 70-82; div mgr, AT&T, 82-84; EXEC DIR, BELLCORE, 84- *Mem:* Fel Am Phys Soc; fel Inst Elec & Electronics Engrs; Sigma Xi. *Res:* Quantum electronics; lasers, photodetectors; nonlinear optical phenomena; ferroelectricity; luminescence in semiconductors; optical communications. *Mailing Add:* 157 Mine Mt Rd Bernardsville NJ 07924

DIDWANIA, HANUMAN PRASAD, b Bhagalpur, India, Mar 13, 35; US citizen; m 69; c 2. PHYSICAL CHEMISTRY, POLYMER CHEMISTRY. *Educ:* Univ Bihar, Bsc, Hons, 55; Inst Paper Chem, MS, 65, PhD(chem), 68; Rider Col, MBA, 71. *Prof Exp:* Shift-in-charge pulp & paper mfg, Orient Paper Mills Ltd, India, 55-61; tech supt, Sirpur Paper Mills Ltd, 61-63; Inst Int Educ develop fel, 65-66; res scientist, Union Camp Corp, 68-73; res assoc, Am Can Co, 73-77, mgr new prod develop, 77-80; PRIN SCIENTIST, PULP & PAPER, SMURFIT RES CTR, CONTAINER CORP AM/JEFFERSON SMURFIT, 82- *Concurrent Pos:* consult, UN Tokten Proj, 81. *Mem:* AAAS; Am Chem Soc; Tech Asn Pulp & Paper Indust; assoc Am Inst Chem Engrs; Indian Inst Eng. *Res:* Pulping, bleaching, soda recovery; physics and chemistry of papermaking; chemical modification of cellulose; effect of hydroxyethylation of fibers on strength properties of paper; paper product development; tissue and towel manufacturing; fiber technology systems; paperboard manufacturing. *Mailing Add:* 2660 Normandy Pl Lisle IL 60532

DIEBEL, ROBERT NORMAN, b Chico, Calif, May 15, 27; m 51; c 3. ANALYTICAL CHEMISTRY, RADIOCHEMISTRY. *Educ:* Univ Ore, BS, 50, MS, 54. *Prof Exp:* Chemist, Hanford Labs, Gen Elec Co, 60-65; sr scientist, Pac Northwest Labs, Battelle Mem Inst, 65-71; lead chemist, Atlantic Richfield Hanford Co, 71-77, SR CHEMIST, HANFORD OPERS, ROCKWELL INT, 77- *Mem:* AAAS; Am Chem Soc; Sigma Xi. *Res:* Nuclear chemical analytical techniques; quality assurance and chemical standards; chemical assay and impurity analysis on plutonium and uranium. *Mailing Add:* 1925 Everest Richland WA 99352

DIEBOLD, GERALD JOSEPH, b Louisville, Ky, May 20, 43; m 77; c 2. PHYSICAL & ANALYTICAL CHEMISTRY. *Educ:* Univ Notre Dame, BS, 65; Boston Col, PhD(physics), 74. *Prof Exp:* Fels, Boston Col, 74-75, Columbia Univ, 76-77 & Stanford Univ, 77-78; PROF CHEM, BROWN UNIV, 78- *Mem:* Am Chem Soc; Am Phys Soc. *Res:* Molecular photodissociation; optoacoustic effect; trace detection techniques. *Mailing Add:* Dept Chem Brown Univ Providence RI 02912

DIEBOLD, JOHN BROCK, b New York, NY, Mar 22, 44. EXPLORATION SEISMOLOGY, MARINE GEOPHYSICS. *Educ:* Univ Colo, BA, 74; Columbia Univ, PhD(marine geophys), 80. *Prof Exp:* Consult, Gulf Res & Develop Corp, 81-83 & 84, sr geophysicist, 83-84; ASSOC RES SCIENTIST & RES ASSOC, LAMONT-DOHERTY GEOL OBSERV, COLUMBIA UNIV, 84- *Concurrent Pos:* Lectr, Columbia Univ, 83; vis prof, Chevron Oil Field Res Co, 85. *Mem:* Am Geophysicists Union; Soc Explor Geophysicists. *Res:* Innovative methods in exploration seismology; wide angle, continuous coverage and large offset velocity surveys. *Mailing Add:* 85 N Midland Ave Nyack NY 10960

DIEBOLD, ROBERT ERNEST, b Rhinelander, Wis, Aug 31, 37; m 57, 82; c 4. EXPERIMENTAL HIGH ENERGY PHYSICS. *Educ:* Univ NMex, BS, 58; Calif Inst Technol, MS, 60, PhD(physics), 63. *Prof Exp:* NSF fel, Europ Orgn Nuclear Res, Geneva, Switz, 62-64; res assoc, Stanford Linear Accelerator Ctr, 64-69; physicist, Argonne Nat Lab, 69-75, assoc dir, High Energy Physics Div, 77-79, assoc lab dir high energy physics, 79-80, sr physicist, 75-87, dir, High Energy Physics Div, 81-84; dir, SSC Div, 87-89, DIR, SSC SCI & TECHNOL DIV, US DEPT ENERGY, 89- *Concurrent Pos:* Mem, high energy adv comt, Brookhaven Nat Lab, 72-74, prog adv comt, Fermi Nat Accelerator Lab, 73-76 & Stanford Linear Accelerator Ctr, 74-76; high energy physics rev comt, Lawrence Berkeley Lab, 74-78 & High Energy Physics Adv Panel, Energy Res & Develop Admin, 75-78, subpanel on accelerator res & develop & review & planning, US High Energy Physics Prog, 80. *Mem:* Am Phys Soc; AAAS. *Res:* Experimental high energy physics using spectrometer systems to study hadronic interactions; design of proton storage rings. *Mailing Add:* 7908 Lewinsville Rd McLean VA 22102-2407

DIECK, RONALD LEE, US citizen. INORGANIC CHEMISTRY, POLYMER CHEMISTRY. *Educ:* Ripon Col, AB, 68; Ariz State Univ, PhD(chem), 72. *Prof Exp:* Res chemist, Res & Develop Ctr, Armstrong Cork Co, 73-77; specialist prod develop, Valox Bus Sect, Gen Elec Co, 77-79, mgr qual control & anal, 79-80; tech dir, Thermofit Bus Sect, Raychem Corp, 80-81, bus develop mgr, Med Prod Group, 81-89; VPRES, RND DOW CORNING WRIGHT, 89- *Mem:* AAAS; Soc Plastics Engrs; Am Chem Soc; Sigma Xi. *Res:* Synthesis and characterization of compounds of the lanthanides and of inorganic polymers and oligomers; processing and physical properties of engineering thermoplastics; radiation cured polymer systems; polymer formulation; medical devices. *Mailing Add:* 1221 Oak River Rd Memphis TN 38119-3318

DIECKE, FRIEDRICH PAUL JULIUS, b June 27, 27; m 55; c 2. PHYSIOLOGY. *Educ:* Univ Würzburg, Dr rer nat, 53. *Prof Exp:* Jr res zoologist, Univ Calif, Los Angeles, 53-55; asst comp physiol, Univ Würzburg, 55-56; from instr to asst prof, Col Med, Univ Tenn, Memphis, 56-59; guest investr, Rockefeller Inst, 59; assoc prof physiol, Sch Med, George Washington Univ, 59-63; prof physiol & biophys, Col Med, Univ Iowa, 63-75; PROF PHYSIOL & CHMN DEPT, COL MED & DENT NJ-NJ MED SCH, 75- *Concurrent Pos:* Actg head, Col Med, Univ Iowa, 73. *Mem:* AAAS; Biophys Soc; Fedn Am Socs Exp Biol; Soc Neurosci; Am Physiol Soc; Sigma Xi. *Res:* Ion transport in nerve and smooth muscle; excitation-contraction coupling and muscle. *Mailing Add:* 16 Sheridan Dr Short Hills NJ 07078

DIECKERT, JULIUS WALTER, b Houston, Tex, June 15, 25; m 50; c 5. BIOCHEMISTRY. *Educ:* Agr & Mech Col, Tex, BS, 49, MS, 51, PhD(biochem), 55. *Prof Exp:* Biochemist, Res Found, Agr & Mech Col, Tex, 53-55; res chemist, Southern Utilization Res Br, USDA, 55-60; assoc prof, 60-70, PROF BIOCHEM & NUTRIT, TEX A&M UNIV, 70- *Concurrent Pos:* Asst prof, Med Sch, Tulane Univ, 59. *Honors & Awards:* USDA Award, 58. *Mem:* Am Chem Soc; Sigma Xi; Am Soc Plant Physiol. *Res:* Chemistry of natural products, isolation and identification of lipids, saponins and proteins; glass paper chromatography, theory and application of natural products; plant cytochemistry, histochemical and cytochemical organization of seeds; protein chemistry; macromolecules of Avian eggshell. *Mailing Add:* 1605 Lenert Circle College Station TX 77840

DIEDERICH, DENNIS A, b Greenleaf, Kans, July 11, 36; m. MEDICAL. *Educ:* St Benedicts Col, Atchison, Kans, BS, 57; St Louis Univ, MD, 61. *Prof Exp:* From asst prof to assoc prof, 69-78, PROF, DEPT MED, DIV NEPHROLOGY, MED CTR, UNIV KANS, 78- *Concurrent Pos:* USPHS fel, Dept Biochem, Med Ctr, Univ Kans, 64-65, spec res fel, 67-69. *Mem:* Fel Am Col Physicians. *Res:* Medicine; numerous publications. *Mailing Add:* Dept Med Univ Kans Med Ctr 39th & Rainbow Blvd Kansas City KS 66103

DIEDERICH, FRANCOIS NICO, b Ettelbruck, Luxemburg, July 9, 52; m 75; c 2. AROMATIC & HETEROAROMATIC COMPOUNDS. *Educ:* Univ Heidelberg, dipl chem, 77, Dr rer Nat(chem), 79, Dr hab(org chem), 85. *Prof Exp:* Res fel excimer interactions, Max Planck Inst Med Res, Heidelberg, 79; res fel reactive intermediates, Dept Chem, Univ Calif, Los Angeles, 79-81; res assoc host-guest chem, Dept Org Chem, Max Planck Inst Med Res, Heidelberg, 81-85; assoc prof, 85-89, PROF ORG CHEM, DEPT CHEM & BIOCHEM, UNIV CALIF, LOS ANGELES, 89- *Honors & Awards:* Otto Hahn Medal, Max Planck Soc, 79. *Mem:* Soc Ger Chemists; Soc Ger Naturalists & Physicians; Am Chem Soc; AAAS; Sigma Xi. *Res:* Synthetic host-guest chemistry; complexation of organic compounds in aqueous and organic solutions and in the solid state; hosts that catalyze and mimic the action of enzymes; hosts that complex enantioselectively; novel aromatic and heteroaromatic systems with material properties; molecular and polymeric carbon allotropes. *Mailing Add:* Dept Chem & Biochem Univ Calif 405 Hilgard Ave Los Angeles CA 90024

DIEDRICH, DONALD FRANK, b Passaic, NJ, May 17, 32; c 1. BIOCHEMICAL PHARMACOLOGY. *Educ:* Univ Ill, BS, 54; Univ Wis, MS, 56, PhD(biochem), 59. *Prof Exp:* Res fel physiol, Col Med, Univ Cincinnati, 59; res instr, Col Med & Dent, Univ Rochester, 59-63; from asst prof to assoc prof pharmacol, 63-74, acad ombudsman, 73-74, PROF PHARMACOL & TOXICOL, COL MED & DENT, UNIV KY, 74- *Concurrent Pos:* Vis scientist, Max Planck Inst Biophys, 71; Fulbright award, 71; vis prof, Eidgenossische Technische Hochschule, Zurich, 78-79, Univ Hamburg, WGer, 84 & 85; sr distinguished scientist, A von Humboldt Found, 88-89. *Mem:* AAAS; Am Chem Soc; Am Soc Pharmacol & Exp Therapeut; NY Acad Sci; Am Asn Univ Professors. *Res:* General cellular biology; transport of metabolites across cell membranes; carbohydrate chemistry; drug-receptor interrelationships. *Mailing Add:* Dept Pharmacol Univ Ky Col Med Lexington KY 40506

DIEDRICH, JAMES LOREN, b St Paul, Minn, Jan 13, 25; m 50, 82; c 6. POLYMER CHEMISTRY. *Educ:* Univ Minn, BChem, 45; Ind Univ, AM, 47, PhD(chem), 49. *Prof Exp:* Res assoc, Northwestern Univ, 49-50; instr chem, Loyola Univ, Ill, 50-51; res chemist, A B Dick Co, 51-52 & Minn Mining & Mfg Co, 52-58; serv engr, Northwest Orient Airlines, Inc, 58-60; proj leader, Borden, Inc, Mass, 60-67; SR RES ASSOC POLYMERS, H B FULLER CO, 67- *Mem:* Am Chem Soc. *Res:* Emulsion polymerization. *Mailing Add:* 1251 Stryker Ave St Paul MN 55118

DIEFENDORF, RUSSELL JUDD, b Mt Vernon, NY, Aug 28, 31; m 52; c 4. MATERIALS SCIENCE. *Educ:* Univ Rochester, BS, 53; Univ Toronto, PhD(phys chem), 58. *Prof Exp:* Scientist graphite, Missile & Ord Dept, Gen Elec Co, 58-59, scientist vapor deposition, Res Lab, 60-65; from assoc prof to prof mat sci, Rensselaer Polytech Inst, 65-90; MCALLISTER TRUSTEE PROF, CLEMSON UNIV, 90- *Honors & Awards:* Humboldt-Preis, Alexander von Humboldt-Stiftung, 74. *Mem:* Am Chem Soc; Am Ceramic Soc; Am Soc Metals; Am Helicopter Soc; Soc Advan Mat & Process Engrs. *Res:* Mechanical properties; structure to properties; graphite; pyrolytic materials; boron and carbon fibers; composite structures and materials; high temperature composites; gas phase kinetics. *Mailing Add:* Dept Ceramic Eng Clemson SC 29634-0907

DIEGELMANN, ROBERT FREDERICK, b July 10, 43; c 5. COLLAGEN METABOLISM, WOUND HEALING. *Educ:* Georgetown Univ, PhD(microbiol), 69. *Prof Exp:* ASSOC PROF BIOCHEM, MED COL VA, 79- *Res:* Regulation of collagen biosynthesis during tissue repair; biochemical studies invoved with all phases of in flammation, connectice tissue deposition and remodeling; regulation of collagen expression by nonfibroblast cell types; regulatory mechanisms involved in the pathologic deposition of connective tissue by other cell types; collagen synthesis by hepatocytes as related to liver fibrasis; smooth muscle cells as related to intestinal stricture formation in Crohn's disease; exploration of the mechanisms responsible for the optimal healing responses seen in the fetus, human subjects, animal models, reconstituted tissue culture and cell culture systems; in vitro protein synthesis and c-DNA hybridization techniques; slab gel electrophoresis, high performance liquid chromatogrphy, gel filtration chromatography, absorption chromatography and enzyme analysis. *Mailing Add:* Med Col Va Box 629 Richmond VA 23298-0629

DIEGLE, RONALD BRUCE, b Marion, Ohio, Jan 17, 47; m 69; c 2. CORROSION SCIENCE, METALLURGY & BATTERY ENGINEERING. *Educ:* Case Inst Technol, BS, 69; Rensselaer Polytech Inst, MS, 72, PhD(mat eng), 74. *Prof Exp:* Assoc staff, Gen Elec Corp, Res & Develop Ctr, 69-74; sr res metallurgist, Battelle Mem Inst, 74-81; DIV SUPV EXPLOR BATTERIES, SANDIA NAT LAB, 81- *Honors & Awards:* A B Campbell Award, Nat Asn Corrosion Engrs, 76. *Mem:* Electrochem Soc; Nat Asn Corrosion Engrs. *Res:* Development of lithium batteries; characterization and analysis of the corrosion of alloys used in energy and weapons systems; corrosion of amorphous alloys. *Mailing Add:* 1043 Red Oaks Loop NE Albuquerque NM 87122

DIEGNAN, GLENN ALAN, b Paterson, NJ, Aug 12, 47. PHYSICAL ORGANIC CHEMISTRY, ANALYTICAL CHEMISTRY. *Educ:* Bucknell Univ, BS, 69; Duke Univ, PhD(phys org chem), 73. *Prof Exp:* Mgr anal chem, Biometric Testing Inc, 74-75; sr anal chemist pharmaceut anal, Hoechst-Roussel Pharmaceut Inc, 75-76; sr scientist pharmaceut anal, Hoffmann-La Roche Inc, 76-79; sr res scientist, biopharmaceutics, CIBA-Geigy Corp, 79-87; prin res scientist, Cytogen Corp, 87-90; GROUP LEADER, ANALYTICAL CHEM, ANAQUEST DIV, BOC GROUP, 90- *Concurrent Pos:* Chmn, Chromatography group, Am Chem Soc, NJ, 85-86; asst to USP rev prog, 76-79. *Mem:* Am Chem Soc; AAAS. *Res:* Biopharmaceutics; organic structure determination by spectroscopic methods; development of chromatographic methods for identification and quantification of drugs; laboratory automation; preformulation and analytical characterization of monoclonal antibody conjugates. *Mailing Add:* 162 Hillside Ave Bridgewater NJ 08807

DIEHL, ANTONI MILLS, b Minneapolis, Minn, Nov 5, 24; m 48; c 4. PEDIATRICS, CARDIOLOGY. *Educ:* Univ Minn, BS, 46, MB, 47; MD, 48; Am Bd Pediat, dipl & cert cardiol, 53. *Prof Exp:* Intern, Univ Mich, 47-48; resident pediat, Univ Minn Hosp, 48-50, fel pediat cardiol, 50-51; instr pediat cardiol, 53-55, from asst prof to prof pediat, 55-78, CLIN PROF PEDIAT, MED SCH, UNIV MO, KANSAS CITY, 78- *Concurrent Pos:* Med Dir, Children's Cardiac Ctr, 53-64; pvt pract cardiol, 78-; teacher, Children's Mercy Hosp, Kansas City, Mo, 79- *Mem:* Am Heart Asn; fel Am Acad Pediat; fel Am Col Cardiol; fel Am Col Chest Physicians; AMA. *Res:* Pediatric cardiology; rheumatic fever. *Mailing Add:* 348 Med Plaza 4320 Wornall Rd Kansas City MO 64111

DIEHL, FRED A, b Staunton, Va, Aug 15, 36; m 58; c 4. DEVELOPMENTAL BIOLOGY. *Educ:* Bridgewater Col, BA, 60; Western Reserve Univ, PhD(biol), 65. *Prof Exp:* NIH fel biol, Univ Brussels, 64-65; res fel, Univ Va, 65-66; instr, Western Reserve Univ, 66-67; asst prof, 69-73, ASSOC PROF BIOL, UNIV VA, 73- *Concurrent Pos:* Chief reader, Advan Placement Biol Prog, Univ Va, 73- *Mem:* Am Soc Zoologists; Soc Develop Biol. *Res:* Studies of form genesis and regulation in cnidaria; specific pathways of cellular differentiation, cellular migration and controlling mechanisms of morphogenesis examined in Hydra, Cordylophora and various other hydroids. *Mailing Add:* Dept Biol Univ Va 270 Gilmer Hall Charlottesville VA 22903

DIEHL, HARVEY, b Detroit, Mich, Nov 2, 10; m 36; c 6. ANALYTICAL CHEMISTRY. *Educ:* Univ Mich, BS, 32, PhD(chem), 36. *Prof Exp:* Instr anal chem, Cornell Univ, 36-37; instr, Purdue Univ, 37-39; from asst prof to assoc prof, 39-47, distinguished prof, 65-81, PROF ANALYTICAL CHEM, IOWA STATE UNIV, 47-, EMER PROF SCI & HUMANITIES, 81- *Honors & Awards:* Anachem Award, Asn Anal Chemists, 66; Fisher Award, Am Chem Soc, 56, Gold Medal, 61. *Mem:* AAAS; Am Chem Soc. *Res:* Coordination and chelate ring compounds; electro analysis; tridentate compounds of cobalt; chemical structure of vitamin B12; organic reagents for iron, copper, calcium, magnesium, cobalt and beryllium; perchlorate chemistry; high-precision coulometric titrations; the value of the Faraday; properties and structure of fluorescein and calcein. *Mailing Add:* Dept Chem Iowa State Univ Ames IA 50011

DIEHL, JOHN EDWIN, b Sunbury, Pa, Feb 7, 29; m 53; c 4. BIOCHEMISTRY. *Educ:* Susquehanna Univ, AB, 52; Pa State Univ, MS, 54, PhD(biochem), 60. *Prof Exp:* Sr res biochemist, Va Inst Sci Res, 59-64; asst prof chem, Dickinson Col, 64-65; from asst prof to assoc prof, 65-73, PROF CHEM & HEAD DEPT, SHEPHERD COL, 73- *Concurrent Pos:* Gen chmn, Tobacco Chemists' Res Conf, Va, 62. *Mem:* AAAS; Am Chem Soc; NY Acad Sci. *Res:* Enzymes; sheep erythrocyte sphingolipides and tobacco leaf proteins; histone chemistry and neurochemistry. *Mailing Add:* PO Box 544 Shepherdstown WV 25443-0544

DIEHL, RENEE DENISE, b Chambersburg, Pa, May 12, 55; m 89. SURFACE PHYSICS, ADSORPTION. *Educ:* Juniata Col, BS, 76; Univ Wash, MS, 79, PhD(physics), 82. *Prof Exp:* Postdoctoral res asst, Univ Liverpool, 82-83, lectr physics, 83-89; ASSOC PROF PHYSICS, PA STATE UNIV, 90- *Mem:* Am Phys Soc; Inst Physics; Europ Phys Soc. *Res:* Structural studies of surfaces, primarily adsorption systems, using low-energy electron diffraction, helium scattering, and other techniques; phase transitions, surface potentials, two dimensional solids and fluids. *Mailing Add:* 104 Davey Lab Pa State Univ University Park PA 16827

DIEHN, BODO, b Hamburg, Ger, June 22, 34. HAZARDOUS WASTE MANAGEMENT & REMEDIATION, BIOPHYSICAL CHEMISTRY. *Educ:* Univ Hamburg, BS, 60; Univ Kans, PhD(phys chem), 64. *Prof Exp:* Res assoc biochem, Univ Ariz, 64-66; from asst prof to prof chem, Univ Toledo, 66-79; dir, Legis Sci Off, State Mich, 79-83; prof zool, Mich State Univ, 80-83; PRES, STB ASSOC ENG SERVS, 84- *Concurrent Pos:* Adj assoc prof biochem, Med Col Ohio Toledo, 70-75, adj prof, 75-79. *Honors & Awards:* Outstanding Res Award, Sigma Xi, 78. *Mem:* Am Chem Soc; Biophys Soc; Am Soc Photobiol; Fedn Am Scientists. *Res:* Radio and hot-atom chemistry; photosynthesis and phototaxis; origin of life; membrane phenomena. *Mailing Add:* 7419 E Palm Ln Scottsdale AZ 85257-1567

DIEHR, PAULA HAGEDORN, b Philadelphia, Pa, Sept 26, 41; m; c 1. BIOSTATISTICS. *Educ:* Harvey Mudd Col, BS, 63; Univ Calif, Los Angeles, MS, 67, PhD(biostatist), 70. *Prof Exp:* Sci programmer, ITT Fed Labs, 63-66; from instr to assoc prof, 70-84, PROF BIOSTATIST, UNIV WASH, 84-, AFFIL PROF HEALTH SERV, 87- *Concurrent Pos:* Researcher health serv qual res, Nat Ctr Health Serv Res, HEW, 75-76. *Mem:* Am Statist Asn; Biomet Soc; Am Pub Health Asn; Asn Health Serv Res. *Res:* Statistical methods in health services research; evaluation of health care technology; utilization of health and mental health services; statistical re-evaluation of common health problems. *Mailing Add:* SC-32 Dept Biostatist Univ Wash Seattle WA 98195

DIEKHANS, HERBERT HENRY, b West Palm Beach, Fla, Mar 4, 25; m 55; c 2. MATHEMATICS. *Educ:* Univ Ala, BS, 50, MA, 51; Univ Ill, Urbana, PhD(math), 64. *Prof Exp:* Instr math, Ohio Univ, 55-60; instr, Univ Ill, Urbana, 64; assoc prof, 64-70, PROF MATH, IND STATE UNIV, TERRE HAUTE, 70- *Mem:* Am Math Soc; Math Asn Am. *Res:* Mathematical analysis. *Mailing Add:* 110 Terre Vista Dr Terre Haute IN 47803

DIEKMAN, JOHN DAVID, b Bridgeport, Conn, Jan 1, 43; m 68. ORGANIC CHEMISTRY. *Educ:* Princeton Univ, AB, 65; Stanford Univ, PhD(chem), 69. *Prof Exp:* Res fel, Australia Nat Univ, 69-70; sr chemist, 70-71, proj mgr prod develop, 71-72, mgr toxicol & regist, 72-73, dir prod develop, 73-75, vpres, Crop Protection Div, 84-86, VPRES RES & DEVELOP, ZOECON CORP, 75-, VPRES & GEN MGR, INT DIV, 81-, GEN MGR, CROP PROTECTION DIV, 84-, PRES, CROP PROTECTION DIV, 86- *Mem:* Am Chem Soc; Entom Soc; AAAS. *Res:* Innovative approaches to pest control. *Mailing Add:* 140 Catalpa Atherton CA 94027

DIEL, JOSEPH HENRY, b Tulsa, Okla, Aug 3, 37; m 68; c 2. COMPUTER APPLICATIONS, BIOMEDICAL RESEARCH. *Educ:* Univ Tulsa, BS, 59, MS, 60; NMex State Univ, PhD(math), 74. *Prof Exp:* Physicist, Phys Sci Lab, NMex State Univ, 60-65; physicist, 75-85, SUPVR SCI COMPUT APPLN, LOVELACE BIOMED & ENVIRON RES INST, INC, 85- *Concurrent Pos:* Adj prof, Univ NMex, 80-81. *Mem:* Inst Elec & Electronics Engrs Comput Soc; Soc Indust & Appl Math; AAAS; Radiation Res Soc. *Res:* Computer applications to biomedical research; metabolism, dosimetry and toxicity of internally deposited radioactive materials; use of computerized data collection and analysis; computer simulation of biological systems. *Mailing Add:* Lovelace Found PO Box 5890 Albuquerque NM 87185

DIEM, HUGH E(GBERT), b Arendtsville, Pa, Mar 31, 22; m 48; c 2. SPECTROSCOPY. *Educ:* Ohio Wesleyan Univ, AB, 47; Ohio State Univ, MA, 49. *Prof Exp:* Asst chem, Ohio State Univ, 47-49; res chemist, Colgate-Palmolive-Peet Co, 50-51; develop engr, B F Goodrich Chem Co, 51-54, res chemist, 57-58, proj leader, 59-61, sect leader olefin rubbers, 61, res assoc, 68-83, sr res assoc, 83-87, CONSULT, B F GOODRICH CO, 87- *Concurrent Pos:* Gov bd, Coblentz Soc, 79-83. *Mem:* AAAS; Am Chem Soc; Coblentz Soc. *Res:* Polymer chemistry; relation of polymer properties to physical structure; polymer reactions; vibrational spectroscopy; analytical chemistry. *Mailing Add:* 142 Hall Dr Wadsworth OH 44281

DIEM, JOHN EDWIN, b Bridgeport, Conn, Dec 7, 37; m 63; c 2. STATISTICS. *Educ:* Pa State Univ, BA, 61, MA, 62; Purdue Univ, PhD(math), 65. *Prof Exp:* Asst prof math, 66-72, assoc prof biostat, 73-80, PROF STATIST, TULANE UNIV, 81- *Mem:* Am Statist Asn; Biometric Soc; Am Thoracic Soc; Soc Epidemiol Res; AAAS. *Res:* Biostatistics; pulmonary disease; epidemiology. *Mailing Add:* 2020 Short St New Orleans LA 70118

DIEM, KENNETH LEE, b Milwaukee, Wis, Apr 17, 24; m 50; c 2. ZOOLOGY. *Educ:* Lawrence Col, BS, 48; Utah State Univ, MS, 52, PhD(wildlife mgt), 58. *Prof Exp:* Game technician, N Kaibab Deer Herd, Ariz Game & Fish Comn, 51-54; from asst prof to prof zool & game mgt, 57-77, DIR NAT PARK RES CTR, UNIV WYO, 77- *Concurrent Pos:* Nat Res Coun grant, Univ Wyo, 58; NY Zool Soc grant, 61-63; US Nat Park Serv grant, 63; consult environ impact assessment, Thorne Ecol Inst, 70-; vchmn Conserv & Land Use Study Comn, Wyo, 73-75; actg admin archivist, Am Heritage Ctr, Univ Wyo, 82-83; pres, Eisenhower Consortium, 84-86. *Mem:* Fel AAAS; Am Ornith Union; Wildlife Soc; Ecol Soc Am; Sigma Xi. *Res:* Animal ecology; dynamics of wildlife populations; big game and avian populations. *Mailing Add:* Dept Zool Univ Wyo PO Box 3166 Univ Sta Laramie WY 82071

DIEM, MAX, b Karlsruhe, WGer, Nov 6, 47; m 74; c 2. CHEMICAL INSTRUMENTATION, COMPUTER INTERFACING. *Educ:* Univ Karlsruhe, BS, 70; Univ Toledo, MS, 75, PhD(chem), 76. *Prof Exp:* Res assoc & fel phys chem, Syracuse Univ, 76-78; asst prof chem, Southeast Mass Univ, 78-79; ASST PROF PHYS CHEM & BIOPHYS CHEM, CITY UNIV NEW YORK, HUNTER COL, 79- *Concurrent Pos:* Consult, Instruments SA, 81-87. *Mem:* Sigma Xi; Am Chem Soc. *Res:* Elucidation of solution structures of biogically interesting molecules via newly developed spectroscopic techniques. *Mailing Add:* Dept Chem Hunter Col NY City Univ New York Box 315 695 Park Ave New York NY 10021

DIEMER, EDWARD DEVLIN, b Pittsburgh, Pa, Nov 4, 33; m 56; c 6. METEOROLOGY. *Educ:* St Louis Univ, BS, 55, M Pr Gph, 60, PhD(meteorol, math), 65. *Prof Exp:* Forecaster, US Weather Bur, Mo, 59-64, asst regional meteorologist, Utah, 65-66, chief sci serv div, 66-71, METEOROLOGIST-IN-CHARGE, WEATHER SERV FORECAST OFF, ALASKAN REGION, NAT WEATHER SERV, NAT OCEANIC & ATMOSPHERIC ADMIN, 71- *Concurrent Pos:* Lectr math, Univ Alaska, 66- *Mem:* Nat Weather Asn. *Res:* Weather forecasting; synoptic meteorology. *Mailing Add:* 5326 Wandering Dr Anchorage AK 99502

DIEMER, F(ERDINAND) P(ETER), b New York, NY, Oct 16, 20; m 52; c 9. ELECTRICAL & ELECTRONICS ENGINEERING, SYSTEMS DESIGN & SCIENCE. *Educ:* Cooper Union, BSEE, 48; NY Univ, MSEE, 50, PhD(studies). *Prof Exp:* Commun engr, Telephonics Corp, 41-44; group leader instruments & electronics, Celanese Corp, NJ, 46-49; asst mgr res lab & staff adv to vpres, Fleischman Labs, Standard Brands, Inc, 49-51; sr develop engr control inst div, Burroughs Corp, 51-52; proj coordr & sr res engr, Am Bosch Arma Corp, 52-54; mgr appl physics, G M Giannini & Co, Inc, 54-57, tech dir, Giannini Res Lab, Calif, 56-57; dir eng & tech consult, Cal-Tronics Corp, 57-58; proj mgr & sr staff engr, Hughes Aircraft Co, 58-60; vpres & dir eng, Daystrom, Inc, 60-61; asst dir advan prog & dir eng, Martin-Marietta Corp, 61-66; exec engr, TRW Systs, Inc, Washington, DC, 66-67, mgr command & control, 67-71; phys sci adminr ocean sci, Off Naval Res, Arlington, Va, 71-80; DIR, INST APPL SCI, NTEX STATE UNIV, 79-, PROF, DEPT BIOL SCI, 81- *Concurrent Pos:* Consult, Indust Eng Dept, Columbia Univ, 50-51, Electrodata Div, Burroughs Corp, 57 & Macro Econ Anal; lectr, Univ Southern Calif, 56-, Univ Calif, Los Angeles, 57- & Am Univ; dep dir, Int Tech Conf, NATO, 78; dir, Primars Int Sci Conf, Univ Manchester, Eng, 79. *Mem:* Inst Elec & Electronics Engrs; NY Acad Sci; AAAS. *Res:* Information and control systems; data processing computer technology; program management; technical consultation; military weapons systems; ocean science. *Mailing Add:* US Dept Energy 1000 Independence Ave Ei-43 Washington DC 20585

DIEN, CHI-KANG, organic chemistry, for more information see previous edition

DIENEL, GERALD ARTHUR, b Boston, Mass, Sept 19, 45. NEUROCHEMISTRY. *Educ:* Pa State Univ, Bs, 67; Harvard Univ, MA, 69, PhD(biochem), 78. *Prof Exp:* Fel, Cornell Univ Med, 78-81, instr biochem, dept neurol, 81-84; LAB CEREBRAL METAB, NIMH, BETHESDA, MD, 84- *Mem:* Soc Neurosci; Am Soc Neurochem. *Res:* Brain development and metabolism in disease states; effects of brain injury or disease of protein synthesis and degradation. *Mailing Add:* Lab Cerebral Metab NIMH Bldg 36 Rm 1A05 Bethesda MD 20892

DIENER, ROBERT G, b Brookville, Pa, Apr 12, 38; m 61. AGRICULTURAL ENGINEERING. *Educ:* Pa State Univ, BS, 60, MS, 63; Mich State Univ, PhD(agr eng), 66. *Prof Exp:* Technol & develop engr, Int Harvester Co, Ill, 60-61; asst, Pa State Univ, 61-62 & Mich State Univ, 63-65; asst prof agr eng, Mich State Univ, 65-68; from asst prof to assoc prof, 68-77, PROF AGR ENG, W VA UNIV, 77- *Concurrent Pos:* Agr engr, USDA, 65-68. *Mem:* Am Soc Agr Engrs; Soc Rheol. *Res:* Mechanical viscoelastic behavior of engineering; agricultural materials; mechanization of harvest of fruits and vegetables. *Mailing Add:* 1287 Woodruff Pl Morgantown WV 26506

DIENER, ROBERT MAX, b Zurich, Switz, Jan 15, 31; US citizen; m 54; c 4. TOXICOLOGY, PATHOLOGY. *Educ:* Cornell Univ, BS, 53; Mich State Univ, DVM, 60, MS, 61; Am Col Vet Path, dipl. *Prof Exp:* Instr clin vet med, Mich State Univ, 60-61; sr vet, 61-62, asst dir toxicol, 63-69, dir, 69-77, exec dir toxicol & path, 77-84, SR ADV SAFETY EVALUATION, CIBA PHARMACEUT CO, 84- *Concurrent Pos:* Assoc ed, J Am Col Toxicol; mem ed bd, J Toxicol Path & J Toxicol Environ Health. *Mem:* Am Vet Med Asn; Am Asn Lab Animal Sci; fel Am Col Vet Toxicol; Soc Toxicol; Int Acad Path; Am Col Toxicol (vpres, 87, pres, 88). *Res:* Risk assessment and toxicological evaluation of pharmaceutical products; evaluation of procedures for safety evaluation. *Mailing Add:* Ciba-Geigy Pharmaceut Co 556 Morris Ave Summit NJ 07901

DIENER, THEODOR OTTO, b Zurich, Switz, Feb 28, 21; nat US; m 50, 68; c 3. VIROLOGY, PHYTOPATHOLOGY. *Educ:* Swiss Fed Inst Technol, Dr sc nat(plant path), 48. *Prof Exp:* Asst, Swiss Fed Inst Technol, 46-48; plant pathologist, Swiss Fed Exp Sta, Wine, Fruit & Hort, Waedenswil, 49; asst plant pathologist, RI State Col, 50; from asst plant pathologist to assoc plant pathologist, Wash State Univ, 50-59; res plant pathologist, Plant Virol Lab, USDA, 59-88, COLLABR, USDA, 88-; DISTINGUISHED PROF BOT, UNIV MD, COLLEGE PARK, 88-, SR STAFF SCIENTIST, CTR AGR BIOTECHNOL, 88- *Concurrent Pos:* Assoc ed, Virol, 64-67 & 74-76, ed, 68-71; regents' lectr, Univ Calif, 70; Andrew D White prof-at-large, Cornell Univ, 79-81, distinguished lectr, Boyce Thompson Inst Plant Res, 87; ed, Viroids, Plenum Publ Corp, 87. *Honors & Awards:* Campbell Award, Am Inst Biol Sci, 68; Super Serv Award, USDA, 69 & Distinguished Serv Award, 77; Alexander Von Humbolt Award, Alexander Von Humboldt Found, 75; Ruth Allen Award, Am Phytopath Soc, 76; James Law Distinguished Lectr, NY State Col Vet; Wolf Agr Prize, 87; Nat Medal Sci, 87; EC Stakman Award, Univ Minn, St Paul, 88. *Mem:* Nat Acad Sci; Am Acad Arts & Sci; AAAS; fel Am Phytopath Soc; fel NY Acad Sci; Leopoldina, Ger Acad Nat Scientists. *Res:* Plant viruses and virus diseases; physiology of virus diseases; nature and properties of viroids, a novel class of pathogens; viroids and viroid diseases. *Mailing Add:* Microbiol & Plant Plant Lab US Dept Agr Beltsville MD 20705

DIENER, URBAN LOWELL, b Lima, Ohio, May 26, 21; m 56. PHYTOPATHOLOGY, MYCOTOXICOLOGY. *Educ:* Miami Univ, AB, 43; Harvard Univ, AM, 45; NC State Univ, PhD(plant path), 53. *Prof Exp:* Indust mycologist, Sindar Corp, 45-47; asst plant pathologist, SC Agr Exp Sta, 47-48; asst plant pathologist, 52-57, from assoc prof to prof, 57-87, EMER PROF PLANT PATH, AUBURN UNIV, 87- *Concurrent Pos:* Chmn, Task Force on mycotoxins, Coun Agr Sci & Technol, 79; consult sci & technol, Thailand, 84, 85 & 87; Fulbright lectr & sr res scholar, Brazil, 85; consult & expert witness mycotoxicology; sr ed, Res Bull, ed, Mycotoxicol Newsletter, 90, 91. *Honors & Awards:* Golden Peanut Res Award, 72. *Mem:* Fel AAAS; fel Phytopath Soc; Am Soc Microbiol; Int Soc Plant Path. *Res:* Mycotoxicology; fungus ecology; aflatoxin in peanuts and corn and other mycotoxins of food crops; author or co author of 19 book chapters, 64 referred in press papers, 70 abstracts, 3 AES bulletin, 39 extension and popular articles; fungus toxins. *Mailing Add:* Dept Plant Path Auburn Univ AL 36849

DIENES, GEORGE JULIAN, b Budapest, Hungary, Apr 28, 18; nat US; m 40; c 1. SOLID STATE PHYSICS. *Educ:* Carnegie Inst Technol, BS, 40, MS, 42, DSc(phys chem), 47; Columbia Univ, MA, 46. *Prof Exp:* Instr chem, Wash & Jefferson Col, 40-41; asst, Carnegie Inst Technol, 41-43; res chemist, Ridbo Labs, NJ, 43-44; group leader, Physics Div, Bakelite Corp, 44-49; res specialist, NAm Aviation, 49-51; sr physicist, 51-87, EMER SR PHYSICIST, BROOKHAVEN NAT LAB, UPTON, 87- *Concurrent Pos:* Mem solid state sci panel, Nat Acad Sci-Nat Res Coun, 53-; assoc ed, J Phys Chem Solids, 74- *Mem:* AAAS; Soc Rheol (secy-treas, 49-53); fel Am Phys Soc; Radiation Res Soc. *Res:* Theory of diffusion in crystals; flow and mechanical properties of high polymers; molecular weight distributions; solid state physics; imperfections in crystals; radiation effects in solids; phase transitions; shock waves; equation of state. *Mailing Add:* Brookhaven Nat Lab 510-B Upton NY 11973

DIERCKS, FREDERICK O(TTO), b Rainy River, Ont, Sept 8, 12; US citizen; m 37; c 2. PHOTOGRAMMETRY, CARTOGRAPHY. *Educ:* US Mil Acad, BS, 37; Mass Inst Technol, MSCE, 39; Syracuse Univ, MS, 50. *Prof Exp:* CEngrs, US Army, 37-67, engr photogram res, Wright Field, Ohio, 37-38, co comdr photomapping, Ft Belvoir, Va, 39-41, topog engr, Ft Jackson, SC, 41-42, officer-in-chg hydrographic surv, Nicaragua Canal Surv, 42, battalion comdr, Eng Aviation Battalion, Geiger Field, Wash, 42-44, topog engr battalion, France & Ger, 44-45, officer-in-chg topog eng res, Ft Knox, Ky, 45-47, battalion comdr geod surv, Philippines, 47-48, officer-in-chg eng intel & mapping, Gen Hq, Far East Command, Tokyo, 48-49, topog engr ed, Ft Belvoir, 50-52, engr intel mapping, Hq, US Army Europe, Heidelberg, 53-

56, commanding officer, Army Map Serv, DC, 57-61, asst dir mapping, charting & geod, Defense Intel Agency, 61-63, dep engr, Eng Sect, Eighth US Army, Korea, 63-64, dir, Coastal Eng Res Ctr, DC, 64-67; assoc dir aeronaut charting & cartog, US coast & geod surv, 67-74, consult, Nat Ocean Surv, Nat Oceanic & Atmospheric Admin, 75-87; RETIRED. *Concurrent Pos:* Lectr, Cath Univ, 50-51; mem nat atlas comt & adv comt on cartog, Nat Acad Sci, 57-61; US mem comn cartog, Pan-Am Inst Geog & Hist, Orgn Am States, 61-67, alt US mem, dir coun & vchmn, US Nat Sect, 70-74, exec secy, 75-87. *Honors & Awards:* Grand Cross, Order of King George II, Greece, 59; Comdr, Most Exalted Order of the White Elephant, Thailand, 65; Legion of Merit, USA, 67; Luis Struck Award, Am Soc Photogram, 69; Colbert Medal, Soc Am Mil Engrs, 72. *Mem:* Fel Am Soc Civil Engrs; fel Soc Am Mil Engrs; hon mem Am Soc Photogram (pres, 70-71); Am Cong Surv & Mapping. *Res:* Geodetic and topographic surveying and mapping instruments and methods; photogrammetric plotting equipment and techniques; wave theory, shore processes, tides, inlet and estuary dynamics; coastal works design and construction techniques. *Mailing Add:* 9313 Christopher St Fairfax VA 22031

DIERENFELDT, KARL EMIL, b Eureka, SDak, Mar 17, 40. PHYSICAL CHEMISTRY, NUCLEAR CHEMISTRY. *Educ:* SDak State Univ, BS, 62; Univ Calif, Davis, PhD(phys chem), 66. *Prof Exp:* Asst prof, 66-73, ASSOC PROF CHEM, CONCORDIA COL, MOORHEAD, MINN, 73- *Mem:* Am Chem Soc. *Res:* Production of light fragments in high energy nuclear reactions; free radicals in solid organic glasses. *Mailing Add:* Dept Chem Concordia Col Moorhead MN 56560

DIERKS, RICHARD ERNEST, b Flandreau, SDak, Mar 11, 34; m 56; c 3. VIROLOGY, VETERINARY MICROBIOLOGY. *Educ:* Univ Minn, BS, 57, DVM, 59, MPH & PhD(microbiol), 64; Am Col Vet Prev Med, dipl; Am Col Vet Microbiol, dipl. *Prof Exp:* Field vet, Minn State Livestock Sanit Bd, 59; NIH fel, Univ Minn, 59-64; vet officer in chg rickettsial dis lab, Spec Projs Unit, Lab Br Commun Dis Ctr, USPHS, Ga, 64-66, chief rabies invests lab, Vet Pub Health Sect, Epidemiol Br, 66-68; from assoc prof to prof vet med, Vet Med Res Inst, Col Vet Med, Iowa State Univ, 68-74; prof vet med & head vet res lab, Col Agr, Mont State Univ, 74-76; DEAN COL VET MED, UNIV ILL, URBANA, 76- *Concurrent Pos:* Nat Inst Allergy & Infectious Dis grants, 68-71 & 73-75; Air Force Off Sci res grant, 69-71; NIH res career develop award, 69-74; Agr Res Serv, USDA grant, 70-72; vis prof, Fed Res Inst Animal Virus Dis, Tübingen, WGer, 71-72; Jensen Salbery grant, 73-75. *Mem:* Am Vet Med Asn; Am Asn Avian Path; Am Soc Microbiol; Soc Exp Biol & Med; Am Asn Immunol; Sigma Xi. *Res:* Rabies; Rhabdoviruses; bovine respiratory viruses; rickettsial diseases; viral immunology; equine viruses; purification and concentration of viruses and subviral proteins. *Mailing Add:* Univ Fla Box J-125 JHMHC Gainesville FL 32610-0125

DIERS, DONNA KAYE, b Sheridan, Wyo, May 11, 38. NURSING. *Educ:* Univ Denver, Colo, BSN, 60; Yale Univ, Conn, MSN, 64. *Prof Exp:* Staff nurse, Yale Psychiat Inst, New Haven, Conn, 60-62; instr psychiat nursing, Sch Nursing, Yale Univ, 64-67, from asst prof to assoc prof, 67-79, dean, 72-85, prof, 79-90, ANNIE W GOODRICH PROF NURSING, SCH NURSING, YALE UNIV, 90- *Concurrent Pos:* Chmn, Prog Nursing Res, Sch Nursing, Yale Univ, 69-72, prof, Inst Social & Policy Studies, 76, chair, Adult Health Div, 90- *Honors & Awards:* Virginia Henderson Award, for contrib to Nursing Res, Conn Nurses Asn, 81; Jesse M Scott Award, Am Nurses Asn, 86. *Mem:* Inst Med-Nat Acad Sci; fel Am Acad Nursing; Am Nurses Asn; Am Asn Col Nursing. *Mailing Add:* Yale Univ Sch Nursing 855 Howard Ave PO Box 9740 New Haven CT 06536-0740

DIERSCHKE, DONALD JOE, b Rowena, Tex, Nov 30, 34; m 56; c 2. REPRODUCTIVE BIOLOGY, ENDOCRINOLOGY. *Educ:* Tex A&M Univ, BS, 56; Mont State Univ, MS, 57; Univ Calif, Davis, PhD(reprod endocrinol), 65. *Prof Exp:* Assoc prof biol, Okla City Univ, 65-69; USPHS spec res fel physiol, Sch Med, Univ Pittsburgh, 69-71, res asst prof, 71-73; PROF, WIS REGIONAL PRIMATE RES CTR & PROF MEAT & ANIMAL SCI, UNIV WIS, 73- *Concurrent Pos:* NIH res grants, 67-69, 75-78, 77-80, 79-83 & 83-86; Eli Lilly res grant, 68-69; Fogarty sr int fel, 80-81; ed-in-chief, Biol of Reproduction, 85-89. *Mem:* Am Soc Animal Sci; Soc Study Reproduction; Am Physiol Soc; Endocrine Soc; Int Soc Neuroendocrinol. *Res:* Endocrine mechanisms regulating ovarian function; neuroendocrine regulation of gonadotropin secretion; hormonal/neural control of puberty and postpartum infertility; hormonal changes associated with aging. *Mailing Add:* Wis Regional Primate Res Ctr Madison WI 53715

DIERSSEN, GUNTHER HANS, b Hamburg, Ger, Jan 10, 26; US citizen; m 59; c 4. CRYSTAL CHEMISTRY, PHYSICS. *Educ:* Tech Univ Denmark, MS, 54. *Prof Exp:* Develop engr biochem, Danish Fermentation Indust, Ltd, 54-58; advan prod engr, Gen Elec Co, 56-60 & Harshaw Chem Co, 60-63; sr res physicist, 63-68, res specialist, 68-74, sr res specialist mat sci, 74-81, STAFF SCIENTIST, CENT RES LAB, 3M CO, 81- *Mem:* Am Asn Crystal Growth; Am Inst Chem Engrs. *Res:* Materials science and processing. *Mailing Add:* Process Technol Labs 3M Ctr 208-1 St Paul MN 55144

DIESCH, STANLEY L, b Blooming Prairie, Minn, May 16, 25; m 56; c 2. VETERINARY PUBLIC HEALTH & MICROBIOLOGY. *Educ:* Univ Minn, St Paul, BS, 51, DVM, 56; Univ Minn, Minneapolis, MPH, 63. *Prof Exp:* Teacher veterans agr training, Belle Plaine Schs, Minn, 51-52; pvt pract, 57-62; USPHS traineeship, 62-63; asst prof prev med & environ health, Col Med, Univ Iowa, 63-66; asst prof vet microbiol & pub health, Col Vet Med, Univ Minn, 66-70, assoc prof vet microbiol & pub health, Col Vet Med & Epidemiol, Sch Pub Health, 70-73, actg dir int vet med progs, 85-90, PROF LARGE ANIMAL CLIN SCI, COL VET MED, UNIV MINN, 73-, PROF EPIDEMIOL, SCH PUB HEALTH, 73-, DIR INT VET MED PROGS, 90- *Concurrent Pos:* Chmn epidemiol sect, Leptospirosis Res Conf, 66-67; consult, Meat Hyg Training Ctr, Chicago, 67-68; adv vet pub health to Venezuela, Pan-Am Health Orgn, 68; USPHS res grant, 68-76; mem, Coun Pub Health & Regulatory Vet Med, Am Vet Med Asn, 77-, chmn, 77; chmn vet med comt, Minn-Uruguay, Partners of the Americas, 81; adv, Animal Dis Surveillance, USDA; mem, Conf Pub Health Vets. *Mem:* Am Vet Med Asn; Am Col Vet Prev Med; US Animal Health Asn; Asn Teachers Vet Pub Health & Prev Med. *Res:* Epidemiology of leptospirosis in animals and man; zoonotic diseases; animal disease surveillance. *Mailing Add:* Dept Clin & Pop Sci Col Vet Med Univ Minn 1365 Gortner Ave St Paul MN 55108

DIESEM, CHARLES D, b Galion, Ohio, July 5, 21; m 45; c 3. VETERINARY ANATOMY. *Educ:* Ohio State Univ, DVM, 43, MSc, 49, PhD(anat), 56. *Prof Exp:* From instr to assoc prof, 47-61, PROF VET ANAT, OHIO STATE UNIV, 61- *Concurrent Pos:* Health comnr, Ohio, 54-76. *Mem:* Am Asn Anat; Am Vet Med Asn; Conf Res Workers Animal Dis; Sigma Xi. *Res:* Gross anatomy and histology; ophthalmology and hematology. *Mailing Add:* 1872 Berkshire Rd Columbus OH 43221

DIESEN, CARL EDWIN, b Cloquet, Minn, Aug 21, 21; m 49; c 2. NUMERICAL ANALYSIS. *Educ:* Univ Minn, Minneapolis, BA, 42, MA, 49. *Prof Exp:* Mgr electronic data processing, Bell Aircraft Corp, 51-60, eng systs, Hughes Aircraft Co, 60-61, electronic data processing, Gen Dynamics Astronaut, 61-64, digital comput, Telecomput Serv, Inc, 64-66 & data processing & comput, Ling-Temco-Vought, Inc, 66-67; CHIEF COMPUT CTR DIV, US GEOL SURV, 67- *Mem:* Asn Comput Mach; Math Asn Am; Data Processing Mgt Asn. *Res:* Numerical analysis for digital computation and for business information systems design and data processing. *Mailing Add:* 7066 Catalpa Rd Frederick MD 21701-7134

DIESEN, RONALD W, b Highland, Ill, Oct 16, 31; m 51; c 2. PHYSICAL CHEMISTRY. *Educ:* Southern Ill Univ, BA, 53; Univ Wash, PhD(phys chem), 58. *Prof Exp:* From res chemist to sr res chemist, 58-72, SR RES SPECIALIST, DOW CHEM CO, 72- *Mem:* Am Chem Soc; Sigma Xi. *Res:* High temperature kinetics; shock tube applications; free radical and combustion reactions; mass spectrometry; laser photochemistry; photochemical recoil spectroscopy; atmospheric chemistry; oxidative catalysis. *Mailing Add:* 5802 Flaxmoor Midland MI 48640-2214

DIESTEL, JOSEPH, b Westbury, NY, Jan 27, 43; m 64; c 2. MATHEMATICS. *Educ:* Univ Dayton, BS, 64; Cath Univ Am, PhD(math), 68. *Prof Exp:* Res scientist, Tech Opers Res, Inc, Washington, DC, 67-68; sr scientist, Consultec, Inc, 68; asst prof math WGa Col, 68-70; fel, Univ Fla, 70-71; assoc prof, 71-75, PROF MATH, KENT STATE UNIV, 75- *Concurrent Pos:* Eng consult, Southwire Int, Inc, Ga, 69-70; Fulbright fel, Univ Col, Dublin, 77-78. *Mem:* Am Math Soc; Irish Math Soc. *Res:* Functional analysis; measure and integration. *Mailing Add:* Dept Math Kent State Univ Kent OH 44242

DIESTLER, DENNIS JON, b Ames, Iowa, Oct 23, 41; m 76. THEORETICAL CHEMISTRY. *Educ:* Harvey Mudd Col, BS, 64; Calif Inst Technol, PhD(chem), 68. *Prof Exp:* Asst prof chem, Univ Mo, St Louis, 67-69; asst prof, 69-72, assoc prof, 72-79, PROF CHEM, PURDUE UNIV, 79- *Honors & Awards:* Sr US Scientist Award, Alexander Von Humboldt Found, 75. *Mem:* Am Chem Soc; Am Phys Soc. *Res:* Theoretical and computational studies of fluids in porous media. *Mailing Add:* 820 Hillcrest Rd Lafayette IN 47906

DIETER, GEORGE E(LLWOOD), JR, b Philadelphia, Pa, Dec 5, 28; m 52; c 2. ENGINEERING METALLURGY. *Educ:* Drexel Inst Technol, BS, 50; Carnegie Inst Technol, DSc(metall), 53. *Prof Exp:* Res coordr, Ballistics Res Lab, Aberdeen Proving Ground, 53-55; res engr, Eng Res Lab, E I du Pont de Nemours & Co, 55-58, res supvr, 58-62; prof metall eng & head dept, Drexel Univ, 62-69, dean col eng, 69-73; prof eng, Carnegie-Mellon Univ, 73-77; DEAN COL ENG, UNIV MD, 77- *Concurrent Pos:* Mem, Nat Metals Adv Bd, 79-81 & ad panel, Nat Eng Lab, Nat Bur Standards, 83-86; chmn, Eng Dean's Coun, 87-89. *Honors & Awards:* A E White Award, Am Soc Metals & Sauveur Award; Educ Award, Soc Mfg Engrs. *Mem:* Fel Am Soc Metals; Am Inst Mining, Metall & Petrol Engrs; fel Am Soc Eng Educ; Nat Soc Prof Engrs; fel AAAS; Soc Mfg Engrs; Fedn Mat Soc (pres, 90-92). *Res:* Mechanical metallurgy; materials processing; engineering design. *Mailing Add:* Col Eng Univ Md College Park MD 20742

DIETER, MICHAEL PHILLIP, b Joplin, Mo, Jan 1, 38; m 62; c 3. LEUKEMOGENESIS, METALS TOXICOLOGY. *Educ:* Notre Dame Univ, BS, 60; Univ Mo, MA, 65, PhD(zool), 68. *Prof Exp:* Asst zool, Univ Mo, 62-67; staff fel physiol, NIH, 67-69, sr staff fel, 69-71; physiologist, US Dept Interior, 71-77; health scientist adminr, Nat Inst Aging, 78-79, Nat Cancer Inst, 79-80, PHYSIOLOGIST, NAT INST ENVIRON HEALTH SCI, NIH, 80- *Concurrent Pos:* Ed, Environ Toxicol Chem. *Mem:* AAAS; Am Physiol Soc; Am Soc Exp Biol & Med. *Res:* Effects of industrial, pharmaceutical and environmental chemicals on mammalian physiology; petroleum hydrocarbons; heavy metals; industrial pollutants; comparative toxicology; leukemogenesis; blood and tissue enzymes; molecular biology of proto-oncogenes; carcinogenesis. *Mailing Add:* Nat Inst Environ Health Sci NIH Box 12233 Research Triangle Park NC 27709

DIETER, RICHARD KARL, b Philadelphia, Pa, Apr 21, 51; m 79; c 2. PHOTOCHEMISTRY. *Educ:* Lehigh Univ, BA, 73; Univ Pa, PhD(chem), 78. *Prof Exp:* Fel chem, Cornell Univ, 78-79; asst prof chem, Boston Univ, 79-85; ASSOC PROF CHEM, CLEMSON UNIV, 85- *Mem:* Am Chem Soc. *Res:* Organic synthesis; new synthetic methodology; organic photochemistry; asymmetric synthesis. *Mailing Add:* Dept Chem Clemson Univ Clemson SC 29634-1905

DIETERICH, DAVID ALLAN, b Cleveland, Ohio, Sept 9, 46; m 68; c 2. PHOTOGRAPHIC SCIENCE. *Educ:* Col Wooster, BA, 68; Univ Ill, PhD(org & phys chem), 73; Simon Sch, MBA, 86. *Prof Exp:* Sr res chemist, 73-80, res lab head, res labs, 80-82, mgr, res lab, Kodak Pvt Ltd, 82-84, mgr prod develop, 84-86, dir prod planning & consumer prods, 86, DIR WORLDWIDE BUS PLANNING, NEW PHOTOG SYSTS, EASTMAN KODAK CO, 86- *Mem:* Am Chem Soc; AAAS. *Mailing Add:* 228 Overbrook Rd Rochester NY 14618-3648

DIETERICH, ROBERT ARTHUR, b Salinas, Calif, Mar 22, 39; m 67; c 2. WILDLIFE DISEASES, EQUINE MEDICINE & SURGERY. *Educ:* Univ Calif, Davis, BS, 61, DVM, 63. *Prof Exp:* Private practice, Calif, 63-67; vet & zoophysiologist, Inst Arctic Biol, Univ Alaska, 67-76, prof vet sci, 77-90; RETIRED. *Concurrent Pos:* Proj mgr, Wildlife Dis Proj, UN, Kabete, Kenya, 74-75; vis prof, Sch Vet Med, Univ Calif, Davis, 84-85. *Mem:* Am Vet Med Asn; Wildlife Dis Asn; Am Asn Equine Practr. *Res:* Wild animal diseases; control of brucellosis in Alaskan reindeer; various pathological lesions found in arctic mammals; equine medicine and surgery. *Mailing Add:* 1483 Green Valley Rd Watsonville CA 95076

DIETERT, MARGARET FLOWERS, b Pittsfield, Mass, July 16, 51; m 75; c 2. PLANT TISSUE CULTURE, BRYOLOGY. *Educ:* Mt Holyoke Col, AB, 73; Univ Tex, PhD(bot), 77. *Prof Exp:* assoc fel plant path, Cornell Univ, 77-80, res assoc plant path, 80-82; lectr Biol, 82, asst prof 82-88, ASSOC PROF BIOL, WELLS COL, 88- *Concurrent Pos:* Consult, Corning Glass, 80-81; Little Brown Consult & Co, 87- *Mem:* Bot Soc Am; Sigma Xi. *Res:* Bryophyte ecology; plant-parasite interactions; plant tissue culture. *Mailing Add:* Dept Life Sci Wells Col Aurora NY 13026

DIETERT, RODNEY REYNOLDS, b Ft Lee, Va, Dec 6, 51; m 75; c 2. IMMUNOGENETICS, IMMUNOTOXICOLOGY. *Educ:* Duke Univ, BS, 74; Univ Tex, PhD(zool), 77. *Prof Exp:* Asst prof, 77-83, ASSOC PROF IMMUNOGENETICS, CORNELL UNIV, 83- *Concurrent Pos:* Ed, CRC Critical Reviews in Poultry Biol. *Mem:* Poultry Asn Am; Soc Exp Biol & Med; Am Asn Immunologists; Int Soc Develop Comp Immunol. *Res:* Avian immunogenetics; cell surface antigens; genetic regulation of chicken fetal antigen expression; chicken macrophage regulation; avian immunotoxicology. *Mailing Add:* Dept Poultry & Avian Sci Cornell Univ 216 Rice Hall Tower Rd Ithaca NY 14853-5601

DIETHORN, WARD SAMUEL, b Waukegan, Ill, Sept 8, 27; m 53; c 2. NUCLEAR ENGINEERING, CHEMISTRY. *Educ:* Lake Forest Col, BS, 50; Carnegie Inst Technol, MS, 53, PhD(chem), 56. *Prof Exp:* Asst div chief, Radio-isotope & Radiation Div, Battelle Mem Inst, 56-60; from asst prof to assoc prof, 60-64, PROF NUCLEAR ENG, PA STATE UNIV, 64- *Mem:* AAAS; Am Chem Soc; Am Nuclear Soc. *Res:* Radiation dosimetry; nuclear reactor materials; radioisotope technology. *Mailing Add:* Dept Nuclear Eng Pa State Univ University Park PA 16802

DIETLEIN, LAWRENCE FREDERICK, b New Iberia, La, Feb 9, 28; m 58; c 3. RESEARCH ADMINISTRATION, INTERNAL MEDICINE. *Educ:* La State Univ, BS, 48; Harvard Univ, MA, 49, MD, 55. *Prof Exp:* From intern to asst resident, Harvard Med Serv, Boston City Hosp, 55-57; instr med, Sch Med, Tulane Univ, 57-59; from resident to chief med res, USPHS Hosp, New York, 59-61, chief outpatient serv, New Orleans, La, 61-62; chief space med br, 62-65, asst div chief, Crew Systs Div, 65-66, chief Biomed Res Off, 66-68, asst dir res, Med Res & Opers Directorate, Manned Space Craft Ctr, 68-72, dep dir, 72-76, actg dir, 76-77, ASST DIR LIFE SCI, SPACE & LIFE SCI DIRECTORATE, JOHNSON SPACE CTR, NASA, 78- *Concurrent Pos:* Res fel, Sch Med, Tulane Univ, 57-59; mem comt hearing & bioacoust, Nat Res Coun, 62-; mem, US/USSR Joint Working Group on Space Biol & Med, 72-; Space Med Adv Panel, Am Inst Biolog Sci, NASA, 83- *Honors & Awards:* Hubertus Strughold Award, Aerospace Med Asn, 75; Melbourne W Boynton Award, Am Astronaut Soc, 75; John Jeffries Award, Am Inst Aeronaut & Astronaut, 75; Louis H Bauer Founders Award, Aerospace Med Asn, 76; Meritorious Serv Medal, USPHS, 76. *Mem:* AMA; NY Acad Sci; fel, Aerospace Med Asn; Int Acad Astronaut. *Res:* Human physiology, particularly cardiovascular, musculoskeletal and vestibular as affected by the space environment. *Mailing Add:* 7702 Glenheath Houston TX 77061-2127

DIETMEYER, DONALD L, b Wausau, Wis, Nov 20, 32; m 57; c 4. COMPUTER ENGINEERING. *Educ:* Univ Wis, BS, 54, MS, 55, PhD(elec eng), 59. *Prof Exp:* From asst prof to assoc prof, 58-67, PROF ELEC ENG, UNIV WIS-MADISON, 67-, ASSOC DEAN, 82- *Concurrent Pos:* Sr assoc engr, Int Bus Mach Corp, 63-64, consult, 64- *Honors & Awards:* Am Soc Elec Eng Western Elec Fund Award, 72. *Mem:* Inst Elec & Electronics Engrs; Asn Comput Mach; Sigma Xi; Am Soc Eng Educ. *Res:* Digital computer use and design; design automation computer hardware description languages. *Mailing Add:* 1513 University Ave Madison WI 53706-1512

DIETRICH, JOSEPH JACOB, b Bismarck, NDak, Oct 31, 32; m 59; c 4. INORGANIC CHEMISTRY, MATHEMATICS. *Educ:* Iowa State Univ, PhD, 57. *Prof Exp:* Sr res chemist, Columbia-South Corp, 57-59; res chemist, Spencer Chem Co, 60, sr staff mem, 61-63; sr res engr, Diamond Alkali Co, 64-66, group leader, T R Evans Res Ctr, 66-69, mgr org prod & processes, 69-71, assoc dir org polymers, 71-73, dir res, 73-77, dir tech dev, electrolytic systs div, Diamond Shamrock Corp, 77-85, dir technol, commun develop, Europe, 85-90; PRES, ELTECH INT, 90- *Mem:* Am Chem Soc; Soc Plastics Engrs; Electrochem Soc. *Res:* Heterocyclic compounds; organometallic compounds; condensation polymers; high pressure polymerization; polymer development. *Mailing Add:* 6958 Pennywhistle Circle Painesville OH 44077

DIETRICH, MARTIN WALTER, b Chicago, Ill, Feb 2, 35; m 60; c 3. ENVIRONMENTAL CHEMISTRY, ANALYTICAL CHEMISTRY. *Educ:* Northwestern Univ, BA, 57; Washington Univ, St Louis, PhD(chem), 62. *Prof Exp:* Res chemist, 61-67, GROUP LEADER, MONSANTO CO, 67- *Mem:* Am Chem Soc; Sci Res Soc Am. *Res:* Toxicology; mass spectrometry; environmental science; radiochemistry. *Mailing Add:* Monsanto Co 800 N Lindbergh EHL St Louis MO 63166

DIETRICH, RICHARD VINCENT, b La Fargeville, NY, Feb 7, 24; m 46; c 3. PETROLOGY. *Educ:* Colgate Univ, AB, 47; Yale Univ, MS, 50, PhD(geol), 51. *Prof Exp:* Geologist, State Geol Surv, Iowa, 47; from asst prof to prof geol, Va Polytech Inst, 51-69, assoc mineral technologist, 52-56, assoc dean col arts & sci, 66-68, actg dean, 68-69; dean arts & sci, 69-75, PROF GEOL, CENT MICH UNIV, 69- *Concurrent Pos:* Fulbright res scholar, Mineral-Geol Mus, Univ Oslo, Norway, 58-59; ed, Mineral Ind J, 53-61, Managing ed, Econ Geol, 66-73, consult ed, Mineral Rec, 69-74, exec ed, Rocks & Minerals, 80-88, petrol adv ed, 88- *Honors & Awards:* Book Award, NY Acad Sci, 80. *Mem:* Fel Geol Soc Am; Soc Econ Geol; Geol Soc Finland; Norweg Geol Soc; Asn Earth Sci Ed (pres, 71); Sigma Xi. *Res:* Petrology of northwestern Adirondacks; geology of Blue Ridge; petrology of Migmatites and banded gneisses; dolomite-chert petrogenesis; feldspar geothermometry; Zr under high T-P conditions; tourmaline mineralogy; author/coauthor of 18 books on geology, especially petrology and mineralogy. *Mailing Add:* Brooks 311 Cent Mich Univ Mt Pleasant MI 48859

DIETRICH, SHELBY LEE, b Lexington, Ky, Apr 27, 24; m 51; c 3. PEDIATRICS. *Educ:* Univ Mich, BA, 45; Univ Mich, MD, 49. *Prof Exp:* Dir pediat, Calif Pediat Ctr, 57-60; dir, Hemophilia Projs, 64-68, asst med dir pediat, 70-75, dir dept res & spec prog, 70-78, actg asst med dir, 74-75, ASST MED DIR OUTPATIENT SERV, ORTHOP HOSP, 78- *Concurrent Pos:* Consult pediatrician & sch physician, Pasadena Unified Sch Dist, 51-70, mem adv comt, Genetically Handicapped Persons Prog, Calif, 75-; assoc clin prof pediat, Univ Southern Calif, 76-; consult ed, Phys Ther, Am Phys Ther Asn, 76- *Honors & Awards:* Physicians Recognition Award, AMA, 74. *Mem:* Fel Am Acad Pediat; fel Am Acad Cerebral Palsy; Ambulatory Pediat Soc; Orthop Res Soc. *Res:* Effectiveness of various treatment methods in hemophila; multidisciplinary treatment of spina bifida; bone architecture and biosynthesis of collagen in osteogenesis imperfecta. *Mailing Add:* Ten Congress St No 340 Pasadena CA 91105-3023

DIETRICH, WILLIAM EDWARD, b Easton, Pa, Nov 20, 42; m 66; c 3. PLANT PHYSIOLOGY. *Educ:* LaSalle Col, BA, 64; Univ Pa, PhD(biol), 70. *Prof Exp:* Res assoc photosynthesis, Brookhaven Nat Lab, 69-71; asst prof, 71-75, ASSOC PROF BIOL, INDIANA UNIV PA, 75- *Mem:* Am Soc Plant Physiologists; Sigma Xi; AAAS. *Res:* Tree photosynthesis and adaptations of the leaf to light; photosynthetic endosymbiosis. *Mailing Add:* Dept Biol Ind Univ Indiana PA 15705

DIETRICK, HARRY JOSEPH, b Cleveland, Ohio, Aug 15, 22; m 43; c 2. PHYSICAL CHEMISTRY. *Educ:* Western Reserve Univ, BS, 48, MS, 50, PhD(chem), 51. *Prof Exp:* Anal chemist, Cosma Labs Co, 40-43 & 46-49; asst phys chem, Western Reserve Univ, 49-51, res assoc, 51-53; tech mgr 53-55, sr tech mgr, 55-59, sect leader, 59-63, mgr aerospace & indust prod res, 63-64, mgr tire res, 65-74, dir tech admin & spec asst to vpres technol, 74-85, dir, Tire Res & Develop Admin, B F Goodrich Tire Co, 80-86; dir, Uniroyal Goodrich Tire Co, 86-87; RETIRED. *Mem:* Fel AAAS; Am Chem Soc; Acoust Soc Am; fel Am Inst Chem. *Res:* Electrode potentials; primary batteries; ultrasonics; physical chemistry of high polymers; space materials; tires. *Mailing Add:* 7354 Brookside Pkwy Cleveland OH 44130

DIETSCHY, JOHN MAURICE, b Alton, Ill, Sept 23, 32; m 59; c 4. INTERNAL MEDICINE, GASTROENTEROLOGY. *Educ:* Wash Univ, AB, 54, MD, 58. *Prof Exp:* Asst med, Sch Med, Boston Univ, 62-63; fel gastroent, Univ Tex Health Sci Ctr, 61-63, metab, 63-65, from asst prof to assoc prof internal med, 65-69, PROF MED, UNIV TEX HEALTH SCI CTR, DALLAS, 71- *Concurrent Pos:* USPHS trainee gastroenterol, Sch Med, Boston Univ, 61-63; Markle scholar acad med, 66-71; consult metab study sect, NIH, 71, 74-78; consult, Monsanto Co, 75-82; mem NIH task force eval, Nat Inst Arthritis, Metab, Diabetes & Digestive Dis, 78-79; ed, Clin Gastroenterol Monograph Series, ed-in-chief, Sci & Pract Clin Med; pres, S Soc Clin Invest, 82. *Honors & Awards:* McKenna Medal, Can Asn Gastroenterol; Distinguished Achievement Award, Am Asn Gastroenterol, 78; Heinrich Weiland Prize, 83. *Mem:* AAAS; Am Fedn Clin Res (pres, 75); Am Soc Clin Invest; Am Gastroenterol Asn (pres, 86); Am Soc Biol Chem. *Res:* Mechanisms of control of cholesterol synthesis in liver and other tissues; mechanisms of lipoprotein transport; mechanisms of intestinal absorption of bile acids; intestinal transport of sugars. *Mailing Add:* 5411 Stonegate Dallas TX 75209

DIETZ, ALBERT (GEORGE HENRY), b Lorain, Ohio, Mar 7, 08; m 36; c 2. STRUCTURAL ENGINEERING, MATERIALS SCIENCE. *Educ:* Miami Univ, AB, 30; Mass Inst Technol, SB, 32, ScD(mat), 41. *Prof Exp:* Designer & job foreman, Peter Dietz, Lorain, Ohio, 32-33; mill foreman, Nat Tube Co, 33-34; asst, 34-36, instr, 36-41, asst prof struct eng, Dept Bldg Eng & Construct, 41-45, assoc prof struct design, 45-51, prof struct eng, 51-53, prof bldg eng & construct, Dept Civil Eng & Dept Archit, 53-73, EMER PROF BLDG ENG & CONSTRUCT & SR LECTR, DEPT ARCHIT, MASS INST TECHNOL, 73- *Concurrent Pos:* Sr consult engr, Forest Prod Lab, USDA, 42; mem, Eng Ed Mission, Japan, 51; bldg res adv bd, Nat Acad Sci-Nat Res Coun; sr res fel, East-West Ctr, Honolulu, Hawaii, 73-75; vis prof, Univ Hawaii, Honolulu, 74-75 & Univ Mich, Ann Arbor, 76. *Honors & Awards:* Templin Award, Am Soc Testing & Mat, Award of Merit, 57, Voss Award, 74; Derham Int Award, Plastics Inst Australia, 62; New Eng Award, Eng Socs New Eng, 69; Int Award Plastics Sci & Eng, Soc Plastics Engrs, 71; Construct Man of the Quarter Century, Bldg Res Adv Bd, Nat Acad Sci-Nat Acad Eng, 77. *Mem:* Fel AAAS; fel & hon mem Am Soc Civil Engrs; fel & hon mem Am Soc Testing & Mat; Soc Plastics Engrs; fel Am Acad Arts & Sci; fel Royal Soc Arts. *Res:* Construction materials, especially wood, plastics, and composites; materials for developing countries; industrialized buildings; solar energy for buildings. *Mailing Add:* 19 Cambridge St Winchester MA 01890

DIETZ, DAVID, b Brooklyn, NY, Mar 10, 46; m 66; c 3. MAGNETOFLUIDDYNAMICS, RADIATION TRANSPORT. *Educ:* Univ Calif, Los Angeles, BS, 66; Ind Univ, Bloomington, MS, 69, AM, 70, PhD(math physics), 75. *Prof Exp:* Head, Nuclear Radiation Div, US Naval Eval Facil, 76-80; tech adv, Test & Eval Ctr, 80-82, RES PHYSICIST, USAF WEAPONS LAB, PHILLIPS LAB, 82- *Concurrent Pos:* Lectr, Dept Math & Physics, Univ NMex, 80-; partic guest, Lawrence Livermore Nat Lab, 88-90; guest scientist, Los Alamos Nat Lab, 90- *Mem:* Am Phys Soc; Am Math Soc; Soc Indust & Appl Math; Sigma Xi. *Res:* Plasma physics; magnetofluiddynamics; electrodynamics; computational techniques and supercomputers. *Mailing Add:* High Energy Plasma Br WSEP USAF Phillips Lab Kirtland AFB NM 87117-6008

DIETZ, EDWARD ALBERT, JR, b Pa, Oct 2, 45; m 67; c 2. ANALYTICAL CHEMISTRY. *Educ:* Geneva Col, Pa, BS, 67; Univ Mich, Ann Arbor, PhD(chem), 70. *Prof Exp:* Res chemist stable isotope geochem, Gulf Oil Corp, 70-71; Welch fel, Univ Tex, Arlington, 71-72; sr anal chemist, Ansul Co, 72-76; res chemist, 76-80, sr res chemist, 80-86, ASSOC SCIENTIST, OCCIDENTAL CHEM CORP, 86- *Mem:* Am Chem Soc. *Res:* Separation science, environmental analysis and development of analytical techniques for use in corporate research and development effort; trace analysis for organic materials in air, water, and soil. *Mailing Add:* 3325 Baseline Rd Grand Island NY 14072-1010

DIETZ, FRANK TOBIAS, b Bridgeport, Conn, Aug 13, 20; m 45; c 2. GEOPHYSICS, ACADEMIC ADMINISTRATION. *Educ:* Bates Col, BS, 42; Wesleyan Univ, MA, 46; Pa State Univ, PhD(physics), 51. *Prof Exp:* Asst physics, Wesleyan Univ, 42-44; mem staff, Radiation Lab, Mass Inst Technol, 45; asst, Wesleyan Univ, 45-47; instr physics, Pa State Univ, 47-49, asst, 49-51; res assoc, Woods Hole Oceanog Inst, 51-54; asst prof marine physics & res assoc phys oceanog, Narragansett Marine Lab, Univ RI, 54-56, from asst prof to assoc prof physics, 56-64, assoc dean, Col Arts & Sci, 74-76, prof oceanog, 68-76, prof, 64-84, EMER PROF PHYSICS, UNIV RI, 85- *Concurrent Pos:* Vis assoc prof, Inst Marine Sci, Univ Miami, Fla, 63-64. *Mem:* Am Asn Physics Teachers; fel Acoust Soc Am; Am Geophys Union; Sigma Xi. *Res:* Underwater acoustics. *Mailing Add:* 60 Spring Hill Rd Kingston RI 02881

DIETZ, GEORGE ROBERT, b Schofield Barracks, Hawaii, Jan 15, 31; m 52; c 4. NUCLEAR SCIENCE. *Educ:* US Mil Acad, BS, 52; Ga Inst Technol, MSNS, 62. *Prof Exp:* Proj officer, AEC food irradiation prog, US Army Radiation Lab, Mass, 60-64; chief facilities eng sect, Div Isotopes Develop, AEC, 64-69; asst vpres opers, Radiation Mach Corp, 69-70, mgr radiation serv, Nuclear Div, Radiation Int, Inc, 70-73; VPRES, ISOMEDIX, INC, 73- *Mem:* Am Nuclear Soc. *Res:* Preservation of foods by ionizing energy and related processing facilities; all aspects of radioisotope applications to commercial radiation facilities and processing. *Mailing Add:* Isomedix Inc 11 Apollo Dr Whippany NJ 07981

DIETZ, GEORGE WILLIAM, JR, b New York, NY, Apr 14, 38. BIOCHEMISTRY, MOLECULAR BIOLOGY. *Educ:* Williams Col, BA, 59; Yale Univ, PhD(biochem), 65. *Prof Exp:* NSF fel, Inst Biophys & Biochem, Paris, 65-67; mem staff res, Dept Biochem & Molecular Biol, Med Col, Cornell Univ, 67-69, asst prof biochem, 69-77; ASST RES SCIENTIST, DIV NEUROSCI, CITY OF HOPE RES INST, 78- *Mem:* AAAS; Am Chem Soc; Am Soc Microbiol; Harvey Soc; Am Soc Biol Chemists. *Res:* Neurochemistry, metabolism of neurotransmitters; control mechanisms in the expression of transport processes; biochemistry of transport in microorganisms. *Mailing Add:* Dept Med Genetics City of Hope Med Ctr 1500 E Duarte Rd Duarte CA 91010

DIETZ, JESS CLAY, b Fod du Lac, Wis, Oct 10, 14; m 42; c 3. SANITARY & ENVIRONMENTAL ENGINEERING. *Educ:* Univ Wis, BS, 39, MS, 41, PhD(sanit eng), 47. *Prof Exp:* Instr eng, Univ Wis, 41-42; prof sanit eng, Univ Ill, Urbana, Ill, 47-57; sr vpres, Clark, Dietz & Assoc, 57-79; CONSULT, 45-47 & 79- *Concurrent Pos:* Environ consult, 47-57. *Mem:* Am Soc Civil Engrs; Am Pub Health Asn; Am Water Works Asn; Water Pollution Control Asn. *Res:* Anaerobic treatment of industrial and domestic wastes; DRP control of activated sludge process. *Mailing Add:* 469 Village Pl Longwood FL 32779-6028

DIETZ, JOHN R, b Omaha, Nebr, July 24, 51; m 77; c 2. SALT & WATER BALANCE, HYPERTENSION. *Educ:* Univ Nebr, Lincoln, BS, 73, Univ Nebr, Omaha, PhD(physiol), 79. *Prof Exp:* Res assoc physiol, dept physiol, Col Med, Univ Mo, 79-82; ASSOC PROF PHYSIOL, COL MED, UNIV SFLA, 82- *Mem:* Am Physiol Soc; Soc Exp Biol & Med. *Res:* Salt and water balance; renin secretion; natriuretic hormone; hypertension. *Mailing Add:* Dept Physiol Col Med Univ SFla Tampa FL 33612

DIETZ, JOHN W, b Rainelle, WVa, Dec 17, 34; m 62; c 3. CHEMICAL ENGINEERING, METALLURGICAL ENGINEERING. *Educ:* WVa Univ, BS, 56; Cornell Univ, PhD(chem eng), 60. *Prof Exp:* Res chem engr, Exp Sta, 60-68, res supvr, 68-73, tech supt, 73-78, planning assoc, 78-83, bus anal mgr, 83-85, MGR, TECHNOL ASSESSMENT, E I DU PONT DE NEMOURS & CO, INC, 85- *Mem:* Am Inst Chem Engrs. *Res:* High temperature inorganic chemistry; process development. *Mailing Add:* 602 Black Gates Rd Wilmington DE 19803

DIETZ, MARK LOUIS, b Clarksville, Tenn, Oct 11, 57; m 84. CHEMICAL SEPARATIONS, TRACE METAL ANALYSIS. *Educ:* Ind Univ, BS, 79; Univ Ariz, PhD(chem), 89. *Prof Exp:* Instr chem, Ind Univ, Pa, 82-84; res asst, Strategic Metals Recovery Res Facil, Univ Ariz, 85-88, res asst, Ctr Separation Sci, 88; POSTDOCTORAL ASSOC CHEM, CHEM DIV, ARGONNE NAT LAB, 89- *Concurrent Pos:* Consult, Eichrom Industs, Inc, 90- *Mem:* Am Chem Soc; Sigma Xi. *Res:* Chemical separations, particularly to the development of improved separations methodology for nuclear waste processing and to the development of new chromatographic methods for the isolation of various radionuclides from biological and environmental samples. *Mailing Add:* Chem Div Argonne Nat Lab 9700 S Cass Ave Argonne IL 60439

DIETZ, RICHARD DARBY, b Rahway, NJ, Sept 28, 37. ASTRONOMY. *Educ:* Calif Inst Technol, BS, 59; Univ Colo, PhD(astrogeophys), 65. *Prof Exp:* Asst astronr, Univ Hawaii, 65-69; asst prof astron, 69-73, assoc prof earth sci, 73-77, PROF ASTRON, UNIV NORTHERN COLO, 77- *Mem:* AAAS; Astron Soc Pac; Am Astron Soc; Am Asn Physics Teachers; Meteoritical Soc; Am Asn Variable Star Observers. *Res:* Ground-based observational infrared astronomy. *Mailing Add:* Dept Earth Sci Univ Northern Colo Greeley CO 80639

DIETZ, ROBERT AUSTIN, b New York, NY, Feb 14, 22; m 60; c 4. CYTOTAXONOMY, ECOLOGY. *Educ:* Washington Univ, PhD(bot, zool), 52. *Prof Exp:* Instr bot, Univ Tenn, 52-54; assoc prof biol, 54-70, PROF BOT, TROY STATE UNIV, 70- *Concurrent Pos:* Bache fel, Nat Acad Sci, 57. *Mem:* AAAS; NY Acad Sci; Sigma Xi. *Res:* Variation in southeastern liliaceous genera; orchid genetics; historical ecology; coastal plain historical ecology; Central American orchid speciation. *Mailing Add:* Dept Biol Troy State Univ Troy AL 36081

DIETZ, ROBERT SINCLAIR, b Westfield, NJ, Sept 14, 14; m 55; c 2. GEOLOGY, OCEANOGRAPHY. *Educ:* Univ Ill, BS, 37, MS, 39, PhD(geol), 41. *Hon Degrees:* DSc, Ariz State Univ, 88. *Prof Exp:* Asst, Ill State Geol Surv, 35-37 & Scripps Inst Oceanog, Univ Calif, San Diego, 37-39; oceanogr, Navy Electronics Lab, 46-52, 54-58; oceanogr, US Coast & Geog Surv, 58-65; oceanogr, Inst Oceanog, Environ Sci Serv Admin, Nat Oceanic & Atmospheric Admin, Md, 65-67, Fla, 67-70, oceanogr, Atlantic Oceanog & Meteorol Labs, 70-77; prof geol, 77-85, EMER PROF GEOL, ARIZ STATE UNIV, 86- *Concurrent Pos:* Fulbright scholar, Univ Tokyo, 52-53; Alexander von Humboldt scholar, WGer, 78; mem, London Br, Off Naval Res, 54-58; res assoc, Scripps Inst Oceanog, Univ Calif, San Diego. *Honors & Awards:* Bucher Medal, Am Geophys Union, 72; Shepard Medal for Marine Geol, 79; Penrose Medal, Geol Soc Am, 88; Gold Medal, US Dept Com, 70. *Mem:* Geol Soc Am; Am Geophys Union; Meteoritical Soc; hon mem Geol Soc London; Mineral Soc Am. *Res:* Marine geology plate tectonics, meterorities and oceanography; underwater sound; sediments and structure of sea floor; submarine processes; nature of continental shelves and slopes; bathyscaph and deep submersibles; selenography and meteoritics; astroblemes; plate tectonics; sea floor spreading; continental drift. *Mailing Add:* Dept Geol Ariz State Univ Tempe AZ 85287

DIETZ, RUSSELL NOEL, b New York, Dec 11, 38; m 61; c 3. GASEOUS TRACER TECHNOLOGY, GAS CHROMATOGRAPHY. *Educ:* Polytech Inst Brooklyn, BChE, 60, PhD(chem eng), 64. *Prof Exp:* Jr res assoc, polymer studies, Battelle Mem Inst, 60, comput data correlation, M W Kellogg Co, 61-62 & radiation chem, Brookhaven Nat Lab, 62-64; chem eng, radiation chem, process chem & tracer studies, 64-85, HEAD, TRACER TECH CTR, BROOKHAVEN NAT LAB, 85 - *Concurrent Pos:* Consult, Radiation Dynamics, Inc, 73-76; vpres, AIM, Inc, 84 - *Honors & Awards:* R L Templin Award, Am Soc Testing & Mat, 77; Fed Lab Consortium Spec Award Excellence Technol , 86; Engrs Week Spec Recognition Award Advan Tracer Technol, 87. *Mem:* Am Chem Soc; Air Pollution Control Asn. *Res:* Correlation of unsaturated hydrocarbon from pyrolysis; gamma irradiation fixation of nitrogen, synthesis of ozone and treatment of waste water; correlation of sulphur emissions from oil fired boilers of power plants; application of perfluorocarbon tracers to atmospheric transport, air infiltration into homes and building, detection of clandestine bombs, and geophysical tracing for enhanced oil recovery. *Mailing Add:* Brookhaven Nat Lab Upton NY 11973

DIETZ, SHERL M, b Ames, Iowa, Nov 29, 27; m 51; c 2. PLANT PATHOLOGY. *Educ:* Ore State Univ, BS, 50; Wash State Univ, PhD(plant path), 63. *Prof Exp:* Agr res aide plant path, ARS/USDA, 54-57, res plant pathologist, 57-66, regional coordr plant introd, 66-74, RES LEADER & TECH ADV PLANT GERMPLASM, AGR RES SERV, USDA, 74- *Concurrent Pos:* Consult, US Agency Int Develop Germplasm, Pakistan, 81, germplasm syst, China, 87; mem, Plant Explor Comt, US Drug Admin, 15 yr Plant Germplasm Task Force, bd agr study team, US Plant Germplasm. *Mem:* Am Phytopath Soc; Soc Econ Bot; Sigma Xi. *Res:* Stem smut of grasses; screening plants for disease resistance; plant germ plasm, introduction, increase, evaluation, documentation, maintenance, distribution and administration. *Mailing Add:* 7015 NE Lake St Pullman WA 99163

DIETZ, THOMAS HOWARD, b Tacoma, Wash, Jan 22, 40; m 62; c 2. ANIMAL PHYSIOLOGY. *Educ:* Wash State Univ, BS, 63, MS, 65; Ore State Univ, PhD(physiol), 69. *Prof Exp:* Instr zool, Ore State Univ, 68-69; NIH trainee biophys, Cardiovasc Res Inst, Univ Calif, San Francisco, 69-71; from asst prof to assoc prof, 71-82, chmn dept, 84-87, PROF ZOOL & PHYSIOL, LA STATE UNIV, BATON ROUGE, 82- *Concurrent Pos:* NSF, Physiol Proc Panel, 90- *Mem:* Am Physiol Soc; Sigma Xi; fel AAAS; Am Soc Zool. *Res:* Osmotic and ionic regulation in animals; mechanism of ion transport; control of ion and water balance. *Mailing Add:* Dept Zool & Physiol La State Univ Baton Rouge LA 70803

DIETZ, THOMAS JOHN, b Tarrytown, NY, May 30, 63; m 91. BIOCHEMICAL ADAPTATION, HEAT SHOCK OR STRESS RESPONSE. *Educ:* State Univ NY Albany, BS, 85; Wash Univ, PhD(molecular biol & biochem), 90. *Prof Exp:* Postdoctoral res assoc, Wash Univ, 90 & Scripps Inst Oceanog, 90-91; NSF POSTDOCTORAL FEL, SCRIPPS INST OCEANOG & ORE STATE UNIV, 91- *Mem:* Am Soc Cell Biol; Am Soc Gravitational & Space Biol; Am Acad Underwater Sci. *Res:* Biochemical adaptations of organisms to stressful and unique environments; ubiquitous stress response system in a number of organisms in order to evaluate when environmental change effects adaptation on an organism. *Mailing Add:* Dept Zool Ore State Univ Cordley Hall Corvallis OR 97331-2914

DIETZ, WILLIAM C, b Chicago, Ill, Apr 17, 19. ENGINEERING ADMINISTRATION. *Educ:* Aeronaut Inst, BSc, 40. *Prof Exp:* Vpres F16 eng, Gen Dynamics Corp, 72-78, vpres, Tomahawk Missile, 79-82, vpres spec proj, 82-90, DIV VPRES & SR TECH STAFF MEM, GEN DYNAMICS CORP, 90- *Honors & Awards:* Sylvanius Reed Award, Am Inst Aeronaut & Astronaut. *Mem:* Nat Acad Eng; Am Inst Aeronaut & Astronaut. *Mailing Add:* 6317 Klamath Rd Ft Worth TX 76116

DIETZ, WILLIAM H, b Philadelphia, Pa, Oct 6, 44; m 66; c 2. PEDIATRICS. *Educ:* Wesleyan Univ, BA, 66; Univ Pa, MD, 70; Mass Inst Technol, PhD(nutrit), 81. *Prof Exp:* ASSOC PROF, TUFTS UNIV SCH MED. *Concurrent Pos:* Asst med, Boston Children's Hosp, 79-; asst dir, Clin Res Ctr,

Mass Inst Technol, 78-; dir clin nutrit, Boston Floating Hosp, New Eng Med Ctr; res assoc, Shriners Burns Inst. *Mem:* Am Acad Pediat; Am Soc Trop Med Hyg; Am Soc Clin Nutrit. *Res:* Childhood obesity, identification, morbidity, therapy; body composition, protein and glucose metabolism among obese adolescents; assessment of nutritional status. *Mailing Add:* Box 213 New Eng Med Ctr 750 Wash St Boston MA 02111

DIEWERT, VIRGINIA M, b Grassland, Alta, Feb 12, 43; m; c 2. EMBRYOLOGY. *Educ:* Univ Alta, DDS, 66; Northwestern Univ, MSc, 70. *Prof Exp:* PROF ORTHOD, UNIV BC, 86- *Concurrent Pos:* Orthodonist in clinical practice. *Res:* Craniofacial biology; cleft lip and palate. *Mailing Add:* Univ BC 2199 Wesbrook Mall Vancouver BC V6T 1Z7 Can

DIFATE, VICTOR GEORGE, b Mt Vernon, NY, Aug 5, 43; m 66. ORGANIC CHEMISTRY, ANTIMICROBIOL CHEMISTRY. *Educ:* Iona Col, BS, 65; NY Univ, MS, 67, PhD(chem), 71. *Prof Exp:* Instr chem, Grad Sch Arts & Sci, NY Univ, 71-74; sr res chemist, 74-79, proj leader, 79, COMMERCIAL DEVELOP MGR, MONSANTO CO, 79- *Concurrent Pos:* Assoc res scientist, NY Univ, 71-74. *Mem:* Am Chem Soc. *Res:* Development of new applications for sorbates; widely used food and feed preservatives; food microbiology; development of environmentally acceptable antimicrobial compounds that are biodegradable and of lower toxicity to higher organisms; organic reaction mechanisms; organic synthesis; animal nutrition; food and feed preservatives. *Mailing Add:* 131 N Bemiston Ave Clayton MO 63105-3810

DIFAZIO, LOUIS T, b Brooklyn, NY, Jan 22, 38; m 61; c 2. DRUG METABOLISM, PHARMACEUTICAL CHEMISTRY. *Educ:* Rutgers Univ, BS, 59; Univ RI, PhD(pharmaceut chem), 64. *Prof Exp:* Res scientist, Colgate-Palmolive Co, 64-65; res scientist, E R Squibb & Sons, Inc, 66-67, sect head gen pharm, 67-71 & preformulation studies, 71-73, asst dir clin pharmacol, Squibb Inst Med Res, 73-75, dir drug metab, 75-77, DIR QUAL CONTROL, SQUIBB INST MED RES, E R SQUIBB & SONS, 77- *Mem:* Am Chem Soc; Am Pharmaceut Asn; Am Col Clin Pharmacol. *Res:* Synthesis of potential psychotherapeutic agents; factors effecting bioavailability of drugs; physical-chemical studies of pharmaceutical agents and products; factors effecting absorption, blood levels and distribution of drugs; product quality control. *Mailing Add:* Squibb Tech Oper PO Box 191 One Squibb Dr New Brunswick NJ 08903-0191

DIFEO, DANIEL RICHARD, JR, b Baltimore Md, July 10, 48; m 78; c 3. ANALYTICAL CHEMISTRY, PLANT CHEMISTRY. *Educ:* Bloomsburg State Col, BA, 70; Univ Tex, Austin, PhD(bot), 77. *Prof Exp:* Fel phytochem, La State Univ, 77; res assoc biomolecular anal, Univ Tex Med Sch, 77-80; mkt engr, Finnigan Mat Corp, 80-82; MASS SPECTRONOMY MGR, SOUTHERN PETROL LABS, 82- *Mem:* Am Soc Mass Spectrometry; Phytochem Soc NAm; Am Soc Pharmacog; Water & Waste Water Anal Asn. *Res:* Analysis of biomolecules and drugs by means of gas chromatography/ mass spectronometry; structural elucidation of plant natural products and their pharmacology; analysis of environmental and petroleum samples by gas chromatography and mass spectronomy. *Mailing Add:* 2101 Ashley Pl Ponca City OK 74604

DI FERRANTE, DANIELA TAVELLA, biology, biochemistry, for more information see previous edition

DI FERRANTE, NICOLA MARIO, medicine, biochemistry, for more information see previous edition

DIFFENDAL, ROBERT FRANCIS, JR, b Hagerstown, Md, June 20, 40; m 67. INVERTEBRATE PALEONTOLOGY, GEOLOGY. *Educ:* Franklin & Marshall Col, Pa, AB, 62; Univ Nebr, MS, 64, PhD(geol), 71. *Prof Exp:* From instr to asst prof geol, St Dominic Col, 66-70; from asst prof to assoc prof geol, Doane Col, Nebr, 70-80, chmn sci div, 78-79; PROF & RES GEOLOGIST, CONSERV & SURV DIV, UNIV NEBR, LINCOLN, NEBR, 80- *Concurrent Pos:* Res geologist, Nebr Geol Surv, Univ Nebr, 75-80. *Mem:* Geol Soc Am; Nat Asn Geol Teachers; Paleont Soc; Am Asn Univ Prof; Sigma Xi. *Res:* Stratigraphy, sedimentology, and geomorphology of Cenozoic deposits of Nebraska; classification and stratigraphic distribution of microfossils from Pennsylvanian rocks in Nebraska; geologic mapping of tertiary outcrops in Western Nebraska. *Mailing Add:* Conserv & Surv Div Univ Nebr Lincoln NE 68588-0517

DIFFLEY, PETER, b St Paul, Minn, Mar 15, 46; m 74; c 1. IMMUNOPARASITOLOGY. *Educ:* Tulane Univ, BS, 68; Univ Mont, MA, 74; Univ Mass, Amherst, PhD(zool), 78. *Prof Exp:* Fel, Yale Sch Med, 78-81; asst prof, Tex Tech Univ, 81-; ASST PROF, DEPT BIOL SCI, UNIV NOTRE DAME, IND. *Mem:* Am Soc Parasitologists; Am Soc Microbiol; AAAS; Sigma Xi. *Res:* Effector immune responses to infectious disease agents and tumor and parasite/tumor evasion strategies, specifically, the models being used to study these relationships include rodent responses to African trypanosomissis, leishmaniasis and murine melanoma. *Mailing Add:* Dept Biol Sci Univ Notre Dame Notre Dame IN 46556

DIFFORD, WINTHROP CECIL, b East Liverpool, Ohio, Nov 12, 21; m 44; c 3. GEOLOGICAL OCEANOGRAPHY. *Educ:* Mt Union Col, BS, 43; WVa Univ, MS, 47; Syracuse Univ, PhD(geol), 54. *Prof Exp:* Asst geol, Ohio State Univ, 42-43; instr, WVa Univ, 47-48; asst area geologist, US Bur Reclamation, Nebr, 48-49, area geologist, Colo, 50-51; from asst prof to prof geol, Dickinson Col, 54-66; assoc prof geol, asst dean & dir grad studies, Col Arts & Sci, Univ Bridgeport, 66-68; dean grad col & dir summer session, Univ Wis-Stevens Point, 68-78, prof geol, 68-86; RETIRED. *Concurrent Pos:* Asst state geologist, Pa, 55-56; vis scientist, Am Geol Inst, 63-65; intern acad admin, Ellis L Phillips Found, 65-66. *Mem:* Fel Geol Soc Am; Marine Tech Soc; fel Royal Soc Arts; fel Explorers Club. *Res:* Oceanography; engineering geology. *Mailing Add:* 7252 Sixth St Custur WI 54423

DIFOGGIO, ROCCO, b Chicago, Ill, Aug 12, 52. NEAR-INFRARED ANALYSIS, IMAGE ANALYSIS. *Educ:* Ill Inst Technol, BS, 74; Univ Chicago, MS, 77, PhD(physics), 80. *Prof Exp:* Res assoc, Univ Chicago, 80-81; res physicist, Shell Oil, 81-88; DIR, NEAR-INFRARED PROJS, WESTERN ATLAS INT, 88- *Honors & Awards:* Nottingham Prize, Conf Phys Electronics, 80. *Mem:* Am Phys Soc; Soc Appl Spectroscopists; Soc Petrol Engrs; Soc Prof Well Log Analysts; Soc Core Analysts. *Res:* Near-infrared analysis of fuels, crude oil, and other hydrocarbons; digital color image analysis of geological samples; methods for removing or analyzing pore fluids of reservoir rocks. *Mailing Add:* PO Box 1407 Houston TX 77251-1407

DIFRANCESCO, LORETTA, NUTRITION. *Prof Exp:* SR SCIENTIST, KRAFT GEN FOODS CORPS, 91- *Mailing Add:* Dept Nutrit & Physiol Gen Foods Tech Ctr Kraft Gen Foods Corp 250 North St T22-1 White Plains NY 10625

DIFRANCO, JULIUS V, b New York City, NY, June 16, 25; m; c 4. RADAR ANALYSIS, SIGNAL PROCESSING. *Educ:* Columbia Univ, BS, 50, MS, 56. *Prof Exp:* Design engr, Liquidometer, 50-58; res dept mgr, Unisys, 58-87; INDEPENDENT RADAR CONSULT, 87- *Mem:* Fel Inst Elec & Electronics Engrs. *Res:* Detection theory; estimation theory; radar systems; electronic scanning antennas; signal processing and signal processing distortion; co-author of one book. *Mailing Add:* 32 Candlewood Path Dix Hills NY 11746

DI FRANCO, ROLAND B, b New York, NY, July 26, 36; m 65; c 2. COMPUTER GRAPHICS. *Educ:* Fordham Univ, BS, 58; Rutgers Univ, MS, 60; Ind Univ, PhD(math), 65. *Prof Exp:* Asst prof math, Fordham Univ, 65-66 & Swarthmore Col, 66-72; assoc prof, 72-78, chairperson dept, 75-78, PROF MATH, UNIV OF THE PAC, 78- *Concurrent Pos:* NSF sci fac fel, Univ Calif, Berkeley, 69-70; Danforth assoc, 70-86; vis scholar, Univ Calif, Berkeley, 77; vis prof, Harvey Mudd Col, 82, 84-85; consult, Honeywell, Inc, 85-88. *Mem:* Am Math Soc; Math Asn Am; Sigma Xi. *Res:* Differential equations; mathematics of vidio simulation; computer graphics. *Mailing Add:* Dept Math Univ of the Pac 3601 Pacific Ave Stockton CA 95211

DI GANGI, FRANK EDWARD, b West Rutland, Vt, Sept 29, 17; m 46; c 2. CHEMISTRY. *Educ:* Rutgers Univ, BSc, 40; Western Reserve Univ, MSc, 42; Univ Minn, PhD(pharmaceut chem), 48. *Prof Exp:* From asst prof to assoc prof pharmaceut chem, 48-57, asst dean, 69-77, assoc dean student affairs, 76-78, PROF MED CHEM, COL PHARM, UNIV MINN, MINNEAPOLIS, 57- *Concurrent Pos:* Reviewer, J Pharmaceut Sci, Am Pharmaceut Asn, 60-; chmn pharm continuing educ prog, Col Pharm, Univ Minn, 60-61, mem planning comt, 68-70, drug abuse prog, 69-70, comput mgt info serv comt, 76-79 & admin adv coun, 78-; mem bd dirs, Sigma Xi, 72-76; mem planning comt, Univ Health Sci, Univ Minn, 76-, bldg adv comt, 77-, public relations adv comt, 79- & dir res, Biomed Res Support Grants, 80- *Mem:* Fel Am Pharmaceut Asn; Am Asn Univ Prof; fel Am Chem Soc; Am Asn Cols Pharm; Sigma Xi (secy-treas, 52-55). *Res:* Synthetic medicinal chemistry; chemistry of medicinal plants and drugs acting on central nervous system; health careers educational programs in hospitals, high schools and community colleges in metropolitan and outstate areas. *Mailing Add:* Univ Minn Col Pharm Minneapolis MN 55455

DIGAUDIO, MARY ROSE, b Brooklyn, NY. PROTOZOOLOGY, ALGOLOGY. *Educ:* St John's Univ, BS, 61; Fordham Univ, MS, 63; NY Univ, PhD(biol), 75. *Prof Exp:* ASST PROF, ST FRANCIS COL, NY, 77- *Mem:* Sigma Xi; Am Soc Cell Biol; NY Acad Sci; AAAS; Soc Protozoologists. *Res:* Pollutants and their effects on aquatic biota. *Mailing Add:* 1038 85th St Brooklyn NY 11228

DIGBY, JAMES F(OSTER), b Aug 11, 21. SYSTEMS ANALYSIS, INTERNATIONAL AFFAIRS. *Educ:* La Polytech Inst, BS, 41; Stanford Univ, MA, 42. *Prof Exp:* Res engr, Watson Labs, USAF, 45-49; res engr, Rand Corp, 49, head opers dept, 55-59, spec asst to head eng div, 59-60, assoc mem res coun, 61-62, prog mgr, Int Security Affairs, 63-65; asst to pres, 65-66, prog mgr & proj leader, 66-86, CONSULT, NATO FORCE PLANNING, 86- *Concurrent Pos:* Consult & comt mem, President's Sci Adv Comt, Fed Aviation Agency & Dept Defense; mem President's task force on air traffic control, 61; exec dir, Calif Sem on Int Security & Foreign Policy, 76-88; dir, Europ-Am Inst for Security Res, 75- *Res:* Evaluation of defense weapon systems; military strategy. *Mailing Add:* 20773 Big Rock Dr Malibu CA 90265-5311

DIGBY, PETER SAKI BASSETT, b London, Eng, Jan 15, 21; m 47; c 4. MARINE BIOLOGY, PHYSIOLOGY. *Educ:* Cambridge Univ, BA, 43, MA, 47; Univ London, DSc, 67. *Prof Exp:* Res asst agr entom, Agr Adv Serv, Cambridge Univ, 42-46; res marine biol, Marine Lab, Plymouth & Exped to Spitsbergen, 46-48; entom, Oxford Univ, 48-50; in field res grant, Marine Plankton Exped to Greenland, 50-51; entom, Oxford Univ, 51-52; from lectr to sr lectr biol, St Thomas's Hosp, Med Sch, 52-67; prof zool, 67-86, RES, MCGILL UNIV, 86- *Concurrent Pos:* Res grants, Develop Comn, 46-48, Agr Res Coun, 48-50, Browne Fund, 50-51, Percy Sladen Mem Fund, 56 & Nat Res Coun, Can, 68-82. *Mem:* Marine Biol Asn UK; Soc Exp Biol; Am Physiol Soc; Linnean Soc; Am Soc Zoologists; Zool Soc London. *Res:* Marine biology, ecology and physiology; plankton; geographical aspects of biology, especially in arctic; physiological mechanisms, especially pressure sensitivity; physiology of calcification in marine plants and animals and in vertebrate bone and teeth and its apparent electrochemical basis; organic semiconductors; 9eneral physiology and ecology of insects and marine organisms. *Mailing Add:* Dept Biol McGill Univ 1205 Ave Dr Penfield Montreal PQ H3A 1B1 Can

DIGENIS, GEORGE A, b Athens, Greece, Sept 28, 35; US citizen; m 61; c 3. MEDICINAL CHEMISTRY, NUCLEAR MEDICINE. *Educ:* Am Univ Beirut, BSc, 59; Univ Wis-Madison, MSc, 62, PhD(med chem), 64. *Prof Exp:* Asst prof pharmaceut chem, Am Univ Beirut, 64-67; from asst prof to assoc prof, 67-74, PROF MEDICINAL CHEM, COL PHARM & PROF NUCLEAR MED, COL MED, UNIV KY, 74- *Mem:* Am Asn Pharmaceut Scientists. *Res:* Natural products chemistry; organic and medicinal chemistry; biochemistry; bio-organic chemistry; radiopharmaceuticals. *Mailing Add:* Col Pharm Univ Ky Med Sch Lexington KY 40536-0082

DI GEORGE, ANGELO MARIO, b Philadelphia, Pa, Apr 15, 21; m 51; c 3. ENDOCRINOLOGY, PEDIATRICS. *Educ:* Temple Univ, AB, 43, MD, 46, MS, 52. *Prof Exp:* Intern, Temple Univ Hosp, 47; pediat resident, Temple Univ Hosp & St Christopher's Hosp for Children, 49-52; instr pediat, Sch Med, Temple Univ, 52-57, assoc, 57-58, from asst prof to assoc prof, 58-67; from asst attend pediatrician & endocrinologist to assoc attend pediatrician & endocrinologist, 52-61, dir, Clin Res Ctr, 63-82, CHIEF ENDOCRINE & METAB SERV, ST CHRISTOPHER'S HOSP FOR CHILDREN, 61-; PROF PEDIAT, SCH MED, TEMPLE UNIV, 67- *Concurrent Pos:* Nat Inst Arthritis & Metab Dis fel, Jefferson Med Col, 52-54; asst chief pediat, Philadelphia Gen Hosp, 56-66; consult & lectr, US Naval Hosp, Philadelphia, 67-80. *Honors & Awards:* Graeme Mitchell lectr; Brenneman lectr; Frederick Packard lectr; William Beaumont Lectr. *Mem:* Am Pediat Soc; Am Soc Human Genetics; Am Acad Pediat; Endocrine Soc; Am Diabetes Asn; Soc Pediat Res; Sigma Xi; AAAS; NY Acad Sci; Lawson Wilkins Pediat & Endocrinol Soc (pres, 83-84); Nat Comn on Orphan Dis. *Res:* Endocrinologic, metabolic and genetic disorders of growth and development. *Mailing Add:* St Christopher's Hosp for Children Erie Ave at Front St Philadelphia PA 19134-1095

DIGERNESS, STANLEY B, b 1941; m; c 2. ENERGY METABOLISM, CALCIUM HOMEOSTASIS. *Educ:* Univ Ala, Birmingham, PhD(biochem), 75. *Prof Exp:* ASST PROF, DEPT SURG, UNIV ALA, BIRMINGHAM, 81- *Res:* Biochemistry of myocardium. *Mailing Add:* Univ Ala 708 Zeigler Bldg Univ Sta Birmingham AL 35294

DIGGINS, MAUREEN RITA, b Omaha, Nebr, July 17, 42. FRESH WATER ECOLOGY, COMPARATIVE PHYSIOLOGY. *Educ:* Mt Marty Col, BA, 66; Northwestern Univ, MS, 68, PhD(biol sci), 71. *Prof Exp:* Instr, St Agnes Sch, 62-65; from instr to asst prof biol, Mt Marty Col, 67-74, assoc prof, 74-, chairperson, Div Natural Sci, 78-; AT DEPT BIOL, AUGUSTA COL. *Mem:* AAAS; Ecol Soc Am; NAm Benthological Soc; Sigma Xi. *Res:* Eutrophication of prairie lakes; benthic life of South Dakota rivers; species diversity of benthos; quality of rural drinking water in South Dakota. *Mailing Add:* Dept Biol Augustana Col 29th St & Summitt Sioux Falls SD 57197

DIGGS, CARTER LEE, b Deltaville, Va, Dec 31, 34; m 56; c 3. TROPICAL MEDICINE, VACCINE DEVELOPMENT. *Educ:* Randolph Macon Col, BS, 56; Med Col Va, MD, 60; Johns Hopkins Univ, PhD(immunol), 68, Am Bd Prev Med & Pub Health, cert. *Prof Exp:* Intern & resident path, Med Col Va, 60-62; researcher parasitic dis, Dept Med Zool, Walter Reed Army Inst Res, 62-64; fel microbiol, Johns Hopkins Univ, 64-68; chief, Dept Parasitic Dis, SEATO Med Res Lab, Bangkok, 68-70; dep dir, Div Commun Dis & Immunol, Walter Reed Army Inst Res, 70-73, chief immunol, 73-80, dir, Div Commun Dis & Immunol, 80-84, assoc dir plans, 84-87, assoc dir plans & overseas opers, 87-88, proj officer, Agency Int Develop Malaria Vaccine Develop Prog, 88-90; ADJ ASSOC PROF, GEORGE WASHINGTON UNIV, 90- *Concurrent Pos:* Lectr, Johns Hopkins Univ, 73-85; mem, Study Group Trop Med & Parasitol, NIH, 78-80. *Mem:* Am Asn Immunologists; Am Soc Trop Med & Hyg; Soc Exp Biol & Med; AAAS; Am Soc Microbiol. *Res:* Effector mechanisms of immunity against malaria and trypanosomiasis; experimental immunization; malaria vaccine development. *Mailing Add:* Walter Reed Army Inst Res Washington DC 20012

DIGGS, GEORGE MINOR, JR, b Charlottesville, Va, Feb 4, 52; div. PLANT TAXONOMY & SYSTEMATICS. *Educ:* Col William & Mary, BS, 74, MA, 76; Univ Wis-Madison, PhD(bot), 81. *Prof Exp:* ASSOC PROF BIOL & BOT, AUSTIN COL, 81- *Mem:* Am Soc Plant Taxonomists; Int Soc Plant Taxonomy; AAAS. *Res:* Systematics of Arbuteae (Ericaceae) and Comarostaphylis (Ericaceae) with emphasis on field research (Mexico and Central America) and numerical techniques. *Mailing Add:* Dept Biol Austin Col Sherman TX 75090

DIGGS, JOHN W, b Mar 23, 36; m; c 3. MEDICAL RESEARCH. *Educ:* Lane Col, BS, 56; Howard Univ, MS, 69, PhD, 72. *Prof Exp:* Res asst, Dept Physiol, Southern Ill Univ, Carbondale, 56-58; biol sci asst, Dept Surg Metab, Walter Reed Army Inst Res, Washington, DC, 58-60, res physiologist, 60-62, physiologist, Div Basic Surg Res, 62-65, res physiologist, Div Surg, 65-69 & Div Biochem, 69-73, sr res physiologist, Div Aug Biochem, 73-74; health sci adminr, Nat Inst Neurol & Commun Dis & Stroke, NIH, Bethesda, 74-75, health sci adminr/exec secy, 75-78, chief, Sci Rev Br, Extramural Activ Prog, 78-80, dep dir, Extramural Activ Prog, 80-82, dir, Div Extramural Activ, Nat Inst Allergy & Infectious Dis, NIH, 82-90, DEP DIR EXTRAMURAL RES, NIH, BETHESDA, MD, 90- *Concurrent Pos:* Mem & chmn numerous comts, 73- *Honors & Awards:* Blacks in Govt Spec Recognition Award, 85; Distinguished Sr Prof Award, Int Personnel Mgt Asn, 86; Spec Recognition Award, Asn Black Hosp Pharmacists, 87; Presidential Meritorious Exec Rank Award, 87. *Mem:* Int AIDS Soc; Am Soc Microbiol; Soc Neurosci; AAAS; Am Zool Soc; Nat Inst Sci; Sigma Xi. *Mailing Add:* Extramural Res NIH Bldg 1 Rm 144 Bethesda MD 20892

DIGHE, SHRIKANT VISHWANATH, b Murud, India, Nov 29, 33; m 64; c 2. ORGANIC CHEMISTRY, BIOPHARMACEUTICS. *Educ:* Univ Bombay, BSc, 55, MSc, 57; Univ Cincinnati, PhD(org chem), 65; Johns Hopkins Univ, MAS, 77. *Prof Exp:* Res asst med drugs, Haffkine Inst, India, 57-58; res chemist, W R Grace & Co, 65-71; res assoc, Sch Med, Johns Hopkins Univ, 71-73; chemist, 73-79, chief biopharmaceut rev br, Div Biopharmaceut, 80-84, DIR, DIV BIOEQUIVALENCE, FOOD & DRUG ADMIN, 85- *Mem:* Am Chem Soc; NY Acad Sci; fel Am Inst Chem; Am Asn Pharmaceut Scientists; Controlled Release Soc; Sigma Xi. *Res:* Organometallic chemistry; chemistry of metal carbonyls, sulfur compounds and polymer synthesis; coordination chemistry; organic and organometallic synthesis; biopharmaceutics; pharmacokinetics. *Mailing Add:* 9811 Wildwood Rd Bethesda MD 20814

DI GIACOMO, ARMAND, b NY, Jan 26, 29; m 54; c 4. POLYMER CHEMISTRY. *Educ:* Long Island Univ, BS, 50; Princeton Univ, MA, 52, PhD(chem), 53. *Prof Exp:* Res assoc, E I du Pont de Nemours & Co Inc, 53-90; RETIRED. *Mem:* Sigma Xi. *Res:* Polymer physical chemistry; polymerization theories, mechanisms and kinetics; glass and crystallization transitions; solution properties; structure-property relations; kinetics over solid catalysts. *Mailing Add:* 1140 Webster Dr Wilmington DE 19803

DIGIORGIO, JOSEPH BRUN, b San Francisco, Calif, Aug 4, 32; m 57, 85; c 6. PHYSICAL ORGANIC CHEMISTRY. *Educ:* Johns Hopkins Univ, BE, 54, MA, 57, PhD, 60. *Prof Exp:* Res assoc org chem, Johns Hopkins Univ, 59, NIH fel, 60-61; NIH fel, vis scientist & Nat Res Coun Can fel, Nat Res Coun Can, Ottawa, 61-63; res assoc, Johns Hopkins Univ, 63-64; from asst prof to assoc prof, 64-70, PROF CHEM, CALIF STATE UNIV, SACRAMENTO, 70- *Mem:* AAAS; Am Chem Soc; Royal Soc Chem; Sigma Xi; Am Soc Testing & Mat; Am Asn Univ Professors. *Res:* Steroids and natural products; relationship between reactivity and geometry; conformational analysis; infrared and Raman spectroscopy of complex organic compounds. *Mailing Add:* Dept Chem Calif State Univ 6000 J St Sacramento CA 95819

DI GIROLAMO, RUDOLPH GERARD, b Brooklyn, NY, Jan 26, 34; c 1. MARINE BIOLOGY, MICROBIOLOGY. *Educ:* Mt St Mary's Col, Md, BS, 55; Univ Wash, PhD(marine biol, microbiol), 69. *Prof Exp:* Res assoc microbiol, St John's Univ, NY, 55-56; res assoc microbiol & virol, Univ Wash, 64-69, res assoc prof sanit eng, 69; from asst prof to prof biol & environ sci, Col Notre Dame, Calif, 70-80, chmn dept, 73-80; INSTR BIOL SCI, LOS RIOS COMMUNITY COL, 80-; CONSULT MARINE RESOURCES, 80-; PROF BIOL SCI, NAT UNIV SACRAMENTO, CALIF, 84- *Concurrent Pos:* Chmn comt environ sci, Col Notre Dame, Calif, 70-80, dir marine resources ctr, 73-80; consult fish dis remedies, Halox-Am Corp, 73-; mem task force shellfish in San Francisco Bay, Asn Bay Area Govts, 77; mem comn mussel quarantine in San Francisco Bay, Calif Dept Pub Health, 77; mem environ concerns comt, Am Fish Soc; Pfizer Res fel. *Mem:* Am Soc Malacologists; Am Soc Microbiol; Am Fisheries Soc; NY Acad Sci; Int Oceanog Found; AAAS. *Res:* Uptake and survival of enteroviruses and bacteria in shell fish and other marine food products; diseases of marine tropical fish; aquatic microbial pollution; lab manual microbiology; field guide intertidal animals of California. *Mailing Add:* One Shoal Ct No 112 Sacramento CA 95831-1446

DI GIULIO, RICHARD THOMAS, b Edmonton, Alta, Oct 13, 50; US citizen; div; c 1. AQUATIC TOXICOLOGY, ECOTOXICOLOGY. *Educ:* Univ Tex, BA, 72; La State Univ, MS, 78; Va Polytech Inst & State Univ, PhD(environ toxicol), 82. *Prof Exp:* Res assoc ecotoxicol, Sch Forestry & Environ Studies, 82-85, asst prof, 85-91, ASSOC PROF ECOTOXICOL, SCH ENVIRON, DUKE UNIV, 91- *Concurrent Pos:* Consult, Am Cyanamid Corp, 85-, Ciba-Giegy Corp, 86-, FMC Corp, 86-, Hydrosyts Inc, 90-, Rhone Poulene Inc, 90-, Gradient Inc, 91-, Sci Adv Bd, US Environ Protection Agency, 91- *Mem:* Soc Environ Toxicol & Chem; Soc Toxicol; AAAS. *Res:* Aquatic biochemical toxicology; metabolism and mechanisms of action of contaminants in aquatic animals; free radical biology. *Mailing Add:* Sch Environ Duke Univ Durham NC 27706

DIGLIO, CLEMENT ANTHONY, b Syracuse, NY, Dec 21, 43; m 65; c 5. MICROVASCULAR ENDOTHELIUM, SMOOTH MUSCLE CELLS. *Educ:* Niagara Univ, BS, 65; State Univ NY, PhD(virol), 73. *Prof Exp:* fel, Vet Sch Med, Univ Pa; asst prof, ASSOC PROF PATH, SCH MED, WAYNE STATE UNIV, 78- *Concurrent Pos:* Consult, Lab Med, VA Hosp, Squibb Inst Med Res & Henri Beaufour Inst; res asst prof, NY Med Col. *Mem:* AAAS; Am Soc Microbiol; Am Asn Pathologists; Am Soc Cell Biol; Am Heart Asn; Tissue Cult Asn. *Res:* Cardiovascular cell biology; studies on the pathogenesis of hypertension vascular disease using tissue culture techniques; development of cell models from the microvasculative to study response to vasctive agents. *Mailing Add:* Sch Med Wayne State Univ Canfield Ave Scott Hall Detroit MI 48201

DIGMAN, ROBERT V, b Wendel, WVa, Jan 9, 30; m 53; c 4. ORGANIC CHEMISTRY. *Educ:* Alderson-Broaddus Col, BS, 51; Univ Maine, MS, 53; Pa State Univ, PhD(chem), 63. *Prof Exp:* Instr chem, Alderson-Broaddus Col, 54-56; res asst petrol chem, Pa State Univ, 56-59; from asst prof to assoc prof chem, Marshall Univ, 59-65; chmn, Dept Natural & Appl Sci, Alderson-Broaddus Col, 65-78, prof Charles McClung Switzer Chair chem, 67-75, dean instr, 75-82, ACAD DEAN, ALDERSON-BROADDUS COL, 82- *Concurrent Pos:* consult-evaluator, 83-; accreditation rev coun mem, CIHE NCent Asn Cols & Schs, 90- *Mem:* Am Chem Soc; AAAS; fel Am Inst Chemists; Sigma Xi. *Res:* Vapor-phase oxidation of hydrocarbons; reactions of epoxides. *Mailing Add:* 11 Greystone Dr Philippi WV 26416

DIGNAM, MICHAEL JOHN, b Toronto, Ont, May 25, 31; m 53; c 5. SURFACE CHEMISTRY, ELECTROCHEMISTRY. *Educ:* Univ Toronto, BA, 53, PhD(phys chem), 56. *Prof Exp:* Demonstr chem, Univ Toronto, 53-54; res chemist, Aluminum Labs, Ltd, 56-58; lectr chem, 58-59, from asst prof to assoc prof, 59-66, PROF CHEM, UNIV TORONTO, 66-, CHMN CHEM, 88- *Concurrent Pos:* Assoc ed, J Electrochem Soc, 76-81. *Mem:* Electrochem Soc; Am Chem Soc; Chem Inst Can; Can Asn Physicists. *Res:* Surface science; polarimetric and conventional spectroscopies of adsorbed species, dispersed systems and ordered molecular layers; electrochemistry. *Mailing Add:* Dept Chem Univ Toronto Toronto ON M5S 1A1 Can

DIGNAM, WILLIAM JOSEPH, b Manchester, NH, Aug 11, 20; m 47; c 4. OBSTETRICS & GYNECOLOGY. *Educ:* Dartmouth Col, AB, 41; Harvard Univ, MD, 43. *Prof Exp:* Intern, Boston City Hosp, 44; resident obstet & gynec, Univ Kans, 47-50; resident endocrinol, Duke Univ, 48; instr obstet & gynec, Univ Calif, San Francisco, 51-53; from asst prof to assoc prof, 53-66, PROF OBSTET & GYNEC, CTR HEALTH SCI, SCH MED, UNIV CALIF, LOS ANGELES, 66- *Mem:* AMA; Am Fedn Clin Res; Endocrine Soc; Am Gynec Soc; Soc Gynec Invest. *Res:* Gynecologic endocrinology. *Mailing Add:* Sch Med Ob-Gyn 22-154 Univ Calif Ctr Health Sci 405 Hilgard Ave Los Angeles CA 90024

DI GREGORIO, GUERINO JOHN, b Philadelphia, Pa, May 11, 40; m 63. PHARMACOLOGY. *Educ:* Pa State Univ, BS, 62; Hahnemann Med Col, PhD(pharmacol), 66, MD, 78. *Prof Exp:* From instr to assoc prof, 66-80, PROF PHARMACOL, HAHNEMANN MED COL, 80- *Mem:* Am Soc Exp Path; Am Soc Clin Toxicol; Am Chem Soc. *Res:* Isolation of biologically active compounds from plants; structure activity relationships of autonomic nervous system; synthesis of quinoline compounds; pharmacokinetics of drugs of abuse; salivary secretion of drugs; drug analysis and toxicology. *Mailing Add:* Dept Pharmacol MS 431 Hahnemann Med Col 230 N Broad St Philadelphia PA 19102

DIILIO, CHARLES CARMEN, b Philadelphia, Pa, May 10, 12; m 37; c 3. MECHANICAL ENGINEERING. *Educ:* Pa State Univ, BS, 34, MS, 35. *Prof Exp:* Machine designer, Yale & Towne Mfg Co, 35-37; instr mech eng, Rensselaer Polytech, 37-41; from asst prof to prof, 41-76, mem grad sch fac 49-76, EMER PROF MECH ENG, PA STATE UNIV, 76- *Concurrent Pos:* Examr mech eng, Pa State Registr Bd Prof Engrs, 58-62; consult, Consumers Res Inc, 50-56, Curtis Wright Corp & Fairchild Stratos Corp. *Mem:* Am Soc Mech Engrs. *Res:* Heat power; thermodynamics; refrigeration; air conditioning; internal combustion engines; rocket motors; fluid mechanics; heat transfer. *Mailing Add:* 430 Philmont Dr Lancaster PA 17601

DIJKERS, MARCELLINUS P, b Utrecht, Neth, May 3, 47. REHABILITATION. *Educ:* Katholieke Univ Nijmegen, BS, 68, MA, 71; Wayne State Univ, PhD(sociol), 78. *Prof Exp:* Assoc dir data mgt, Detroit Receiving Hosp, Univ Health Ctr, 78-81; INSTR SOCIOL, WAYNE STATE UNIV, 82-, ASST PROF, PHYS MED & REHAB, 83- *Concurrent Pos:* Dir res, Rehab Inst, Wayne State Univ, 81- *Mem:* Am Sociol Asn; Am Cong Rehab Med. *Res:* Attitudes toward the disabled; rehabilitation professional staff; rehabilitation of spinal cord injuries. *Mailing Add:* Rehab Inst Mich 261 Mack Blvd Detroit MI 48201

DIKE, PAUL ALEXANDER, b Mt Vernon, Iowa, Nov 26, 12; m 45; c 3. GEOLOGY. *Educ:* Johns Hopkins Univ, BA, 37, Bryn Mawr Col, MA, 50. *Prof Exp:* Instr geol, Univ Pa, 46-53; master parochial sch, Pa, 53-54; foreman phys testing lab, E L Conwell & Co, Pa, 54-55; asst prof geol & dir dept, Temple Univ, 55-62; asst prof sci, 62-70, chmn dept phys sci, 71-77, ASSOC PROF GEOL, GLASSBORO STATE COL, 70- *Concurrent Pos:* Lectr, Drexel Inst Technol, 48-62; tutor, Bryn Mawr Col, 50-55; consult, E L Conwell & Co, 55-, Ambric Testing & Eng Assocs, Inc, 56-, Rowle & Henderson, 57- & Asphalt Technol, 62-; lectr, Gloucester County Col, 68-70. *Res:* Field studies with magnetometer, earth resistivity and seismology locating mineral deposits; geology of Central Delaware County, Pennsylvania; Coastal Plains materials. *Mailing Add:* 118 Wildwood Ave Pitman NJ 08071

DIKEMAN, ROXANE NORRIS, b Beemer, Nebr, Nov 14, 42; div; c 2. ENZYMOLOGY, BIOCHEMISTRY. *Educ:* Univ Nebr, BSc, 64, MSc, 67; La State Univ, PhD(biochem), 78. *Prof Exp:* Res technologist atherosclerosis, Cleveland Clin Res Found, 69-72; res biochemist microbiol group, med prod div, Mallinckrodt, Inc, 78-82; res scientist, NutraSweet Res & Develop, G D Searle & Co, 82-85; res scientist enzymol, res & develop, 86-87, group leader, biochem prod develop, 87-89, MGR, BIOCHEM DEVELOP, NUTRASWEET CO, 89- *Mem:* Am Chem Soc. *Res:* Enzymology, proteases, enzymatic peptide synthesis, lipid biochemistry, atherosclerosis, microbial biochemistry, protein isolation and characterization; polysaccharide chemistry; food product development. *Mailing Add:* NutraSweet Co 601 E Kensington Mt Prospect IL 60056

DIKSIC, MIRKO, b Cvetkovic, Croatia, Yugoslavia, May 22, 42; Can citizen; m 68; c 2. RADIOPHARMACEUTICALS, BIOLOGICAL MODELLING. *Educ:* Univ Zagreb, Crotia, Yugoslavia, BS, 66, MS, 68, PhD(radio-nuclear chem), 70. *Prof Exp:* Asst prof, 79-86, ASSOC PROF RADIOPHARMACEUT, DEPT NEUROL & NEUROSURG, MCGILL UNIV, 86-, DIR, RADIOCHEM-CYCLOTRON UNIT, MONTREAL NEUROL INST, 86- *Concurrent Pos:* Assoc fac mem, Dept Chem, McGill Univ, 79-, Med Physics Unit, 83-87; prin investr, MRC grants & NIH grant, 81-; consult, Med Res Coun Can, Nat Sci & Eng Res Coun Can, NIH, Nat Cancer Inst, 82-; vis lectr, Heritage Found Alta, Univ Alta, 85; bd dirs Radiopharmaceut Sci Coun, Soc Nuclear Med, 88-90; vis prof, Nat Tsing-Hua Univ, Hsinchu, Taiwan, 90. *Mem:* Am Chem Soc; Int Soc Cerebral Blood Flow & Metabolism; Int Isotope Soc; Soc Neurosci. *Res:* Development of a radiochemical synthesis for radiopharmaceuticals used in the evaluation of brain receptor density; enzyme inhibition; choline high affinity uptake system; tumor chemotherapeutic drug uptake; autoradiographic and PET evaluation of same radiopharmaceuticals; peripheral benzodiazepine receptors in rat brain tumor models and human brain tumors; nitrosourea uptake in human brain tumors; brain tumor protein synthesis rate measurements; serotonin synthesis rate measurements in rat and human brain; evaluate the influence of lesions induced by some neurotoxins on the rate of serotonin axonal transport and on the rate of synthesis by an in vivo autoradiographic method. *Mailing Add:* Med Cyclotron McGill Univ 3801 University St Montreal PQ H3A 2B4 Can

DIKSTEIN, SHABTAY, b Budapest, Hungary, Apr 10, 31; Israeli citizen; m 51, 90; c 2. BIOCHEMISTRY. *Educ:* MSc, 55, PhD, 59. *Prof Exp:* Asst, dept pharmacol, Med Sch, Jerusalem, 55-57; lectr, Israel Inst Biol Res, 57-60; instr, dept pharmacol, Med Sch, 60-61; res fel, 61-65, sr lectr appl pharmacol, 65-70, assoc prof, 70-80, PROF, DEPT APPL PHARMACOL, SCH PHARM, HEBREW UNIV, 80-, HEAD CELL PHARMACOL UNIT, 78- *Concurrent Pos:* Travel grants, Europ Molecular Biol Orgn, 67 & 69; vis sr lectr & reader, Inst Ophthalmol, Univ London, 67 & 68; vis assoc prof, Stanford Univ, Calif, 69 & 70; fel res prev blindness, Stanford Univ & Columbia Univ, 73; vpres & chmn med sci comt, Health Resorts Authority, 73-83; vis reader, Kennedy Inst Res, London, 79; secy, Int Comt Standardization Medicocosmetic Measurements & Prods, 79-81; expert pharmacol & toxicol, Ministry Health, France, 81- *Honors & Awards:* Foreign Fel Award, NIH, 60. *Mem:* Int Soc Eye Res (vpres, 74-76); hon mem Arg Pharmacol Asn. *Res:* Survival of corneal endothelium pump, artificial tear fluid; medicocosmetic measurement on consumer level including individual treatment and aging of the skin; author of over 110 research papers and 16 invited papers. *Mailing Add:* Unit Cell Pharmacol Hebrew Univ Jerusalem Sch Pharm PO Box 12065 Jerusalem 91120 Israel

DI LAVORE, PHILIP, III, b Lawrence, Mass, Apr 24, 31; m 53; c 4. PHYSICS. *Educ:* Dakota Wesleyan Univ, AB, 54; Univ Mich, MS, 61, PhD(physics), 67. *Prof Exp:* Teacher high schs, SDak, 53-54 & Mich, 56-58; lectr physics, Univ Mich, 62 & 64; asst prof, Univ Md, 65-71, assoc chmn dept physics & astron, 69-71; assoc prof, 71-75, assoc dean, Sch Grad Studies, 76-77, asst vpres acad affairs, 77-81, dir, comput resource develop, 83-86, PROF PHYSICS, IND STATE UNIV, 75- *Concurrent Pos:* Staff physicist, Comn Col Physics, 67-69; coordr, Nat Tech Physics Proj, 72-76; ed, Physics Technol Modules, 74-75; prof physics, Ind Univ, 86-88. *Mem:* Am Asn Physics Teachers; Am Phys Soc; Sigma Xi. *Res:* Level-crossing spectroscopy; optical pumping. *Mailing Add:* Dept Physics Ind State Univ Terre Haute IN 47809

DILCHER, DAVID L, b Cedar Falls, Iowa, July 10, 36; m 61; c 2. PALEOBOTANY. *Educ:* Univ Minn, BS, 58, MS, 60; Yale Univ, PhD(biol), 64. *Prof Exp:* Instr biol, Yale Univ, 65-67; from asst prof to prof paleobot, Ind Univ, Bloomington, 67-90; GRAD RES PROF PALEOBOT, UNIV FLA, GAINESVILLE, 90- *Concurrent Pos:* Sigma Xi grants-in-aid, 61, 62 & 66; NSF fel, 64-65, res grants, 66-; Guggenheim fel, 72-73; distinguished vis res scholar, Adelaide Univ, 80; Guggenheim fel, 87-88. *Honors & Awards:* Sonneborn Outstanding Teaching & Res Award, 88. *Mem:* Int Orgn Paleobot; Orgn Trop Biol; AAAS; Bot Soc Am; Am Inst Biol Sci; Geol Soc Am; fel Linnean Soc. *Res:* Cretaceous and Tertiary plant fossils of North America; angiosperm evolution with particular reference to foliar and reproductive biology, anatomy and morphology of fossil and modern flowering plants. *Mailing Add:* Fla Mus Natural Hist Paleobot Lab Dept Natural Sci Univ Fla Gainesville FL 32611-2035

DILEONE, GILBERT ROBERT, b Providence, RI, Oct 30, 35; m 65; c 2. T-CELLS, NATURAL SUPPRESSOR CELLS. *Educ:* Boston Univ, AB, 58; Univ RI, MS, 60, PhD(biol sci), 64; Cath Univ, cert, 91. *Prof Exp:* Res assoc immunol, Brown Univ, 64-66; consult, New Eng Monoclonal Resources, 83-85; assoc biol, Dept Cancer Res, 66-83, RES ASST, CLIN HEMAT DEPT, RI HOSP, 85- *Mem:* AAAS; Am Soc Microbiol; Tissue Cult Asn; Reticuloendothelial Soc; Sigma Xi. *Res:* Natural suppressor (NS) cell function in the peripheral immune system; origin of NS cells; mechanism of NS activity; creation of a recombinant cDNA library of NS cells; origin and function of human chronic lymphocytic leukemia cells. *Mailing Add:* 24 Old Lyme Dr Warwick RI 02886

DILIDDO, BART A(NTHONY), b Cleveland, Ohio, Mar 5, 31; m 55. CHEMICAL ENGINEERING. *Educ:* Fenn Col, BS, 54, Ill Inst Technol, MS, 56, Case Inst Technol, PhD(chem eng), 60. *Prof Exp:* Mem tech staff, Res & Develop Ctr, BF Goodrich Co, 56-58, res engr, 60-62, sr res engr, Avon Lake Tech Ctr, 62-64, res sect leader, 64-67, proj mgr, 67-70, process mgr, 70-72; dir, Latex & Specialty Chemicals, Cleveland, BF Goodrich Chem Co, 72-73, dir res & develop, 73-75, div vpres, BF Goodrich Tire Div, 75-78, Special Projs, BF Goodrich Chem Div, 78, Plastics, 78-79, sr vpres & gen mgr, Plastics, 79-80; pres chem group, 80-85, EXEC VPRES, BF GOODRICH CO, 85- *Concurrent Pos:* Dir, Chlorine Inst; chmn, Vinyl Inst. *Mem:* Am Inst Chem Engrs; Soc Plastics Industs. *Res:* New chemical processes. *Mailing Add:* PO Box 577 Bath OH 44210

DILIDDO, REBECCA MCBRIDE, b Canton, Ohio, June 25, 51; m 78. ROOT PHYSIOLOGY, CELL MOTILITY. *Educ:* Milligan Col, BS, 77; Ohio State Univ, PhD(bot), 77. *Prof Exp:* Asst prof bot & cell biol, Eastern Ore State Col, 78-79; ASSOC PROF BOT & CELL BIOL, SUFFOLK UNIV, 81- *Concurrent Pos:* Asst prof, Hiram Col, 78-80; assoc, Mass Inst Technol, 80-81. *Mem:* Am Soc Plant Physiologists. *Res:* Hormonal, environmental, physiological plantgrowth; root growth and the roles of auxin and hydrogen plus ions in the control of elongation and geotropism. *Mailing Add:* Dept Biol Suffolk Univ Beacon Hill Boston MA 02108

DILIELLO, LEO RALPH, b Baltimore, Md, Jan 17, 32; m 54; c 3. MICROBIOLOGY. *Educ:* Univ Md, BS, 54, MS, 56, PhD, 58. *Prof Exp:* Res asst, Univ Md, 54-58; from assoc prof to prof microbiol, State Univ NY Agr & Tech Col Farmingdale, 68-88; FOOD INDUST & DAIRY ANALYSIS CONSULT. *Mem:* Am Soc Microbiol; Sigma Xi; Am Inst Food Technol; Am Soc Microbiol. *Res:* Veterinary medicine; oral bacteria; dairy and food microbiology; medical microbiology; medical-veterinary aspects; human-animal diseases involving Vibrios; synergistic etiologies of human disease; microbial interactions. *Mailing Add:* Dept Biol Sci Col Technol State Univ NY Farmingdale NY 11735

DILL, ALOYS JOHN, b Jersey City, NJ, Jan 8, 40; m 62; c 2. ELECTROCHEMISTRY. *Educ:* Hunter Col, AB, 62; City Col New York, MA, 64; City Univ New York, PhD(chem), 68. *Prof Exp:* res chemist, Paul D Merica Res Lab, 67-80, SR SCIENTIST, INCO RES & DEVELOP CTR, INT NICKEL CO, INC, 80- *Mem:* Am Chem Soc; Electrochem Soc; Am Electroplaters Soc; Sigma Xi. *Res:* Electrodeposition of metals. *Mailing Add:* 32 Drewry Lane Tappan NY 10983-1106

DILL, CHARLES WILLIAM, b Greenville, SC, June 1, 32; m 54; c 3. FOOD SCIENCE, ORGANIC CHEMISTRY. *Educ:* Berea Col, BS, 54; NC State Col, MS, 57, PhD(food sci), 62. *Prof Exp:* Instr food chem, NC State Col, 59-62; asst prof dairy sci, Univ Nebr, 62-66; assoc prof, 66-75, PROF ANIMAL SCI, TEX A&M UNIV, 75- *Mem:* AAAS; Inst Food Technol; Am Dairy Sci Asn; Sigma Xi. *Res:* Thermal denaturation and degradation of proteins; instrumentation and control in systems for heating fluid products; flavor stability of fats and oils. *Mailing Add:* Dept Animal Sci Tex A&M Univ College Station TX 77843

DILL, DALE ROBERT, b Towanda, Kans, Jan 10, 34; m 55; c 3. ORGANIC CHEMISTRY, PHARMACEUTICAL CHEMISTRY. *Educ:* Univ Kans, BS, 55, PhD(pharmaceut chem), 58. *Prof Exp:* Res assoc pharmaceut chem, Univ Kans, 58-59; sr res chemist, 59-63, res specialist, 63-67, group leader, 67-74, SR GROUP LEADER, MONSANTO CO, 74- *Concurrent Pos:* NIH grant, 58-59. *Mem:* Am Chem Soc; NY Acad Sci; Tech Asn Pulp & Paper Indust. *Res:* Anti-hypertensive drugs; plasticizers; paper chemicals; cancer drugs. *Mailing Add:* Six Wrenwood Ct Webster Groves MO 63119-2795

DILL, DAVID BRUCE, physiology; deceased, see previous edition for last biography

DILL, EDWARD D, b Salt Lake City, Utah, July 6, 41; m 63; c 4. INORGANIC CHEMISTRY, PHYSICAL CHEMISTRY. *Educ:* Southwestern State Col, BS, 63; Univ Ark, PhD(chem), 69. *Prof Exp:* ASSOC PROF CHEM, SOUTHWESTERN OKLA STATE UNIV, 68-, CHMN, DEPT CHEM, 79- *Mem:* Am Chem Soc. *Res:* Inorganic synthesis; x-ray structure analysis. *Mailing Add:* Dept Chem Southwestern Okla State Univ Weatherford OK 73096

DILL, ELLIS HAROLD, b Pittsburgh Co, Okla, Dec 31, 32; m 53; c 2. CIVIL ENGINEERING. *Educ:* Univ Calif, PhD(civil eng), 57. *Prof Exp:* Prof aeronaut & astronaut, Univ Wash, 56-77; DEAN ENG, RUTGERS UNIV, 77- *Mem:* Soc Natural Philos. *Res:* Applied mechanics. *Mailing Add:* Dean Eng Rutgers Univ New Brunswick NJ 08903

DILL, FREDERICK H(AYES), JR, b Sewickley, Pa, Mar 1, 32; m; c 3. ELECTRICAL ENGINEERING, PHYSICS. *Educ:* Carnegie Inst Technol, BS, 54, MS, 56, PhD(elec eng), 58. *Prof Exp:* MEM RES STAFF, THOMAS J WATSON RES CTR, IBM CORP, 58- *Concurrent Pos:* Mackay vis lectr, Univ Calif, Berkeley, 68-69; mem bd dirs, Inst Electric & Electronic Engrs, 90-91. *Mem:* Nat Acad Eng; Electron Devices Soc; fel Inst Elec & Electronics Engrs. *Res:* Semiconductor junction phenomena and devices. *Mailing Add:* IBM Res Ctr PO Box 218 Yorktown Heights NY 10598

DILL, JAMES DAVID, theoretical chemistry, for more information see previous edition

DILL, JOHN C, b Vancouver, BC, Nov 20, 39; m 85; c 3. COMPUTER SCIENCE, COMPUTER GRAPHICS. *Educ:* Univ BC, BASc, 62; NC State Univ, MS, 64; Calif Inst Technol, PhD(info sci), 70. *Prof Exp:* Comput scientist, Res Labs, Gen Motors Tech Ctr, 70-80; sr res assoc, Col Eng, Cornell Univ, 80-84; systs mgr, Microtel Pacific Res, 84-87; PROF ENG SCI, SIMON FRASER UNIV, 87- *Mem:* Asn Comput Mach; Inst Elec & Electronics Engrs. *Res:* Computer graphics; computer-aided design. *Mailing Add:* Sch Eng Sci Simon Fraser Univ Burnaby BC V5A 1S6 Can

DILL, KENNETH AUSTIN, b Oklahoma City, Okla, Dec 11, 47. POLYMER CHEMISTRY. *Educ:* Mass Inst Technol, SB & SM, 71; Univ Calif, San Diego, PhD(biol), 78. *Prof Exp:* Res asst biomed eng, Mass Inst Technol, 70-71; res asst biol, Univ Calif, San Diego, 71-78; Damon Runyon-Walter Winchell fel, 79-81, fel chem, Stanford Univ, 78-81; asst prof chem, Univ Fla, 81-83; PROF, PHARMACEUT CHEM, UNIV CALIF, SAN FRANCISCO, 83- *Mem:* Am Chem Soc; Am Phys Soc; Biophys Soc. *Res:* Physical properties of membranes, micelles, macromolecules; protein stability and folding. *Mailing Add:* 1125 Harte St Montara CA 94037

DILL, LAWRENCE MICHAEL, b Vancouver, BC, Apr 4, 45; m 67; c 2. BEHAVIORAL ECOLOGY. *Educ:* Univ BC, BSc, 66, MSc, 68, PhD(ecol), 72. *Prof Exp:* Biologist, Dept Fisheries, Can, 67-69; asst prof ecol, York Univ, Toronto, 72-74; PROF BEHAV ECOL, SIMON FRASER UNIV, BURNABY, 74- *Concurrent Pos:* Hon assoc prof animal resource ecol, Univ BC, 80-; mem, Pop Biol Grant Selection Comt, Natural Sci & Eng Res Coun, Can, 81-83; chmn, Can Comt Int Ethological Congress, 81-83; assoc ed, Can J Zool, 85-88, Copeia, 88-, Ethology, 90- *Mem:* Can Soc Zoologists; Asn Study Animal Behav; Ecol Soc Am; Int Soc Behav Ecol (pres, 92-); Am Soc Ichthyologists & Herpetologists. *Res:* Behavioral strategies and tactics used by animals, particulary in predator-prey interactions and territoriality utilizing primarily fish and invertebrates as experimental subjects. *Mailing Add:* Dept Biol Sci Simon Fraser Univ Burnaby BC V5A 1S6 Can

DILL, NORMAN HUDSON, b Wilmington, Del, Apr 6, 38. BOTANY, PLANT ECOLOGY. *Educ:* Univ Del, BA, 60; Rutgers Univ, MS, 62, PhD(plant ecol), 64. *Prof Exp:* Prof biol, 64-66, PROF BIOL & NATURAL RESOURCES, DEL STATE COL, 66- *Concurrent Pos:* Consult, USAID-NSF Sci Educ Improv Prog for India, 68 & 69; mem bd experts, Rachel Carson Trust for Living Environ. *Mem:* AAAS; Ecol Soc Am; Am Inst Biol Sci; Am Nature Study Soc; Nat Asn Biol Teachers. *Res:* Forest productivity; vegetation management; plant geography; use of wild shrubs in songbird management; effects of periodic cicada on forest ecosystems; plants for noise and air pollution abatement; agricultural drainage effects. *Mailing Add:* Dept Biol Sci Del State Col Dover DE 19901

DILL, ROBERT FLOYD, b Denver, Colo, May 25, 27; m 45; c 4. MARINE GEOLOGY. *Educ:* Univ Southern Calif, BS, 50, MS, 52; Univ Calif, San Diego, PhD, 64. *Prof Exp:* Oceanogr, US Navy Electronics Lab, 51-68, marine geologist, Naval Undersea Res & Develop Ctr, 68-71; marine geologist, Nat Oceanic & Atmospheric Admin, 71-74 & Dept Interior, US Geol Surv, 74-75; dir, West Indies Lab, Fairleigh Dickenson Univ, 75-81; mem staff, Nat Oceanic & Atmospheric Admin, 81-83; serv scientist, Dept Interior's Geol Surv & Minerals Mgt Serv & Nat Oceanic & Atmospheric Admin, 84-86; PRES, DILL GEOMARINE CONSULTS, 86- *Concurrent Pos:* Consult & owner, Gen Oceanog Inc, 53-71; mem fac, San Diego State Col, 68-71; investr, Inst Oceanog Res, Univ Baja Calif; dipl, Mex Comt Eng for Ocean Resources, 70; adj prof, George Washington Univ, 71-74; mem, Gov Comn Water Resources Virgin Islands; adj prof geol, Univ SC, 85- & Humboldt State Univ, 89-; dir geol prog, Caribbean Marine Res Ctr. *Honors & Awards:* Leverson Award, Am Asn Petrol Geologists, 69. *Mem:* Fel Geol Soc Am; Marine Technol Soc; Am Asn Petrol Geologists; Soc Econ Paleontologists & Mineralogists; Sigma Xi; Am Acad Underwater Scientists; Am Geophys Union; Am Inst Prof Geologists. *Res:* Marine sedimentation, erosion, and sediment distribution patterns in present and ancient seas; mass physical properties of sediments, Pleistocene sea level fluctuations, comparative studies of ancient and recent carbonate reefs, submarine canyons, scientific application of scuba and submersibles; deep sea mineral development. *Mailing Add:* 610 Tarento Dr San Diego CA 92106

DILL, RUSSELL EUGENE, b Rising Star, Tex, July 27, 32; m 54; c 3. PHYSIOLOGY, NEUROANATOMY. *Educ:* NTex State Col, BA, 53, MS, 57; Univ Ill, PhD(physiol), 60. *Prof Exp:* Instr anat, Univ Tex Med Br Galveston, 61-65; from asst prof to assoc prof, 65-74, chmn dept, 74-85, PROF ANAT, COL DENT, BAYLOR UNIV, 74- *Concurrent Pos:* Fel anat, Univ Tex Med Br Galveston, 60-61. *Mem:* Int Asn Dent Res; Am Asn Anat; Soc Neurosci; Sigma Xi. *Res:* Experimental neurology; extrapyramidal motor systems; central nervous system pharmacology. *Mailing Add:* Dept Anat Col Dent Baylor Univ Dallas TX 75246

DILLARD, CLYDE RUFFIN, b Norfolk, Va, Apr 17, 20. INORGANIC CHEMISTRY. *Educ:* Va Union Univ, BS, 40; Univ Chicago, MS, 48, PhD, 49. *Prof Exp:* Res chemist, Manhattan Proj, Chicago, 43-45, asst chem, 45-48; prof, Tenn Agr & Indust State Univ, 48-53 & Morgan State Col, 53-59; from assoc prof to prof, 59-86, assoc dean faculties, 70-72, EMER PROF CHEM, BROOKLYN COL, 86- *Concurrent Pos:* Asst ed, Appl Spectros, Soc Appl Spectros, 63-; Fulbright res scholar, Osaka Univ, 65-66. *Mem:* AAAS; Am Chem Soc; NY Acad Sci; Soc Appl Spectros. *Res:* Chemistry of volatile hydrides; organotin compounds; infrared spectroscopy; radiochemistry and isotopic tracer techniques. *Mailing Add:* 735 Franklin St Westbury NY 11590-2470

DILLARD, DAVID HUGH, b Spokane, Wash, May 14, 23; m 48; c 7. SURGERY. *Educ:* Whitman Col, AB, 46; Johns Hopkins Univ, MD, 50; Am Bd Surg, dipl, 59; Bd Thoracic Surg, dipl, 61. *Hon Degrees:* Dr Sci, Witman Col, 79. *Prof Exp:* Intern, Johns Hopkins Hosp, 51; instr, 54-57, resident, 58, assoc prof, 57-69, chief, Div Cardiothoracic Surg, 72-78, PROF SURG, SCH MED, UNIV WASH, 69- *Mem:* Fel Am Col Surgeons; Am Surg Asn; Nat Psychic Sci Asn; Am Asn Thoracic Surg. *Res:* Esophageal and cardiovascular research; thoracic surgery. *Mailing Add:* Dept Surg Univ Wash Sch Med Seattle WA 98195

DILLARD, EMMETT URCEY, b Sylva, NC, Aug 12, 17; m 40; c 4. ANIMAL GENETICS. *Educ:* Berea Col, BSA, 40; NC State Col, MS, 48; Univ Mo, PhD(animal husb), 53. *Prof Exp:* From asst county supvr to county supvr, Farm Security Admin, USDA, 40-44; instr animal husb, Bur Animal Indust, 47-49, asst prof, 49-60, assoc prof, 60-79, EMER ASSOC PROF ANIMAL SCI, NC STATE UNIV, 79- *Concurrent Pos:* Livestock res adv, NC Agr Res Mission to Peru, 58-60 & 67-68. *Mem:* Am Soc Animal Sci; Sigma Xi. *Res:* Livestock, improvement through breeding and management; selection and crossbreeding in beef cattle improvement. *Mailing Add:* 1110 Dogwood Lane Raleigh NC 27607

DILLARD, GARY EUGENE, b Ridgway, Ill, Apr 26, 38; m 59; c 2. PHYCOLOGY, AQUATIC ECOLOGY. *Educ:* Southern Ill Univ, BA, 60, MS, 62; NC State Univ, PhD(bot), 66. *Prof Exp:* Asst taxon, Southern Ill Univ, 59-60, bot, 60-62; asst phycol, NC State Univ, 62-65; asst prof bot, Clemson Univ, 65-68; assoc prof biol, 68-74, assoc dean, Col Sci, 81-85, PROF BIOL, WESTERN KY UNIV, 74- *Mem:* Phycol Soc Am; Am Micros Soc; Int Phycol Soc. *Res:* Taxonomy and ecology of freshwater algae, especially of Southeastern United States. *Mailing Add:* Dept Biol Western Ky Univ Bowling Green KY 42102

DILLARD, JAMES WILLIAM, b Pueblo, Colo, Feb 26, 48; m 75; c 2. ANALYTICAL CHEMISTRY, ELECTROCHEMISTRY. *Educ:* Univ Ariz, BS, 70; NC State Univ, PhD(anal chem), 76. *Prof Exp:* Fel, Colo State Univ, 75-77; anal chemist radiochem, Tenn Valley Authority, 77-87; TECH DIR, IT CORP, 87- *Mem:* Am Chem Soc; Sigma Xi. *Res:* Finite difference simulation of electron transfer reactions; application of electrochemical techniques to environmental radiochemistry; method development of radioanalytical analysis. *Mailing Add:* PO Box 22535 Knoxville TN 37933

DILLARD, JOHN GAMMONS, b Kermit, Tex, Nov 1, 38; m 63; c 2. INORGANIC CHEMISTRY, PHYSICAL CHEMISTRY. *Educ:* Austin Col, BA, 61; Kans State Univ, PhD(chem), 66. *Prof Exp:* Asst chem, Okla State Univ, 61-63 & Kans State Univ, 63-66; fel Rice Univ, 66-67; from asst prof to assoc prof, 67-78, PROF CHEM, VA POLYTECH INST & STATE UNIV, 78- *Mem:* Am Chem Soc; Am Soc Mass Spectrometry. *Res:* Mass spectrometry, corrosion, surface chemistry oceanography. *Mailing Add:* Dept Chem Va Polytech Inst State Univ Blacksburg VA 24061-0212

DILLARD, MARGARET BLEICK, b Newark, NJ; div; c 1. MEMBRANE TRANSPORT, COMPUTER MODELS. *Educ:* Mt Holyoke Col, AB, 61; Univ NC, Chapel Hill, PhD(physics), 67; Univ Tex Med Sch, MD, 85. *Prof Exp:* Res assoc physics, Duke Univ, 67-68; instr & res assoc, Univ NC, Chapel Hill, 68-69; assoc prof physics & math, St Augustine's Col, NC, 69-71; comput programmer, NC Mem Hosp, Chapel Hill, 73-75; asst prof radiol, Sch Med, Univ NC, Chapel Hill, 75-81; med res, ECU Sch Med, NC, 85-88, renal fel, 88-89; FEL, SCH MED, UNIV NC, 89- *Concurrent Pos:* Med res, ECU Sch Med, Greenville, NC, 85-88. *Mem:* Am Phys Soc; Am Med Asn; Am Col Physicians. *Res:* Control of intracellar calcium in epithelia; epithelial chloride channels. *Mailing Add:* 3106 Buckingham Rd Durham NC 27707-7806

DILLARD, MARTIN GREGORY, b Chicago, Ill, July 7, 35; m 58; c 3. NEPHROLOGY. *Educ:* Univ Chicago, BA, 56, BS, 57; Howard Univ, MD, 65. *Prof Exp:* Chief, Hemodialysis Unit Nephrol, Freedmen's Hosp, 70-73; asst chmn postgrad educ, 73-76, from asst prof to assoc prof, 70-88, PROF MED, COL MED, HOWARD UNIV, 88-, ASSOC DEAN CLIN AFFAIRS, 76- *Concurrent Pos:* Mem, Nat Adv Coun, Regional Med Prog,

76-78; dir med educ, Howard Univ Hosp, 80- *Mem:* Am Soc Nephrol; Am Fedn Clin Res; Int Soc Nephrol; Am Col Physicians. *Res:* Clinical-pathological correlations in renal disease; protein handling by the intact and damaged nephron; insulin binding in renal disease. *Mailing Add:* Renal Div Howard Univ Hosp 2041 Georgia Ave Washington DC 20060

DILLARD, MORRIS, JR, b York, Ala, Apr 27, 27. METABOLISM, INTERNAL MEDICINE. *Educ:* Birmingham-Southern Col, AB, 49; Emory Univ, PhD(anat), 54, MD, 59. *Prof Exp:* From instr to asst prof med, 63-69, ASSOC PROF CLIN MED, SCH MED, YALE UNIV, 69- *Concurrent Pos:* Fels, Yale Univ, 61-62 & 63-64. *Mailing Add:* 4490 Whitney Ave Hamden CT 06158

DILLARD, ROBERT GARING, JR, b Clarendon, Tex, June 18, 31; m 54; c 6. CHEMICAL ENGINEERING, TECHNICAL MANAGEMENT. *Educ:* Univ Calif, LA, BSc, 56. *Prof Exp:* Process develop chemist, Shell Chem Co, 56-57, oper mgr, 63-65, plant mgr Shell Point, Calif, 66-67, mgr Geismar Plant, 69-71; oil, gen mgr, logistics, 71-73, gen mgr, Deer Park Mfg Complex, Shell Oil Co, 74-80, vpres mfg, tech, 84-87, VPRES PUB AFFAIRS, SHELL OIL CO, 87- *Mem:* Am Inst Chem Engrs; Am Chem Soc. *Mailing Add:* 315 Duncaster Dr Houston TX 77079-7015

DILLAWAY, ROBERT BEACHAM, b Washington, DC, Nov 10, 24; m 47, 71; c 6. RESOURCE MANAGEMENT, TECHNICAL MANAGEMENT. *Educ:* Univ Mich, BS(math) & BS(mech eng), 45; Univ Ill, MS, 51, PhD(mech), 53. *Prof Exp:* Develop engr, Carrier Corp, 45-46 & Eng Res Assocs, Inc, 47-48; instr mech eng, Univ Ill, 48-53; sr res engr, Rocketdyne Div, NAm Aviation, Inc, 53-54, res specialist, 54-56, supvr basic studies, 56-57, group leader basic studies & nuclear propulsion, 57-58, mgr nucleonics subdiv, Space Power Res & Develop, 58-64, corp dir res & tech mkt planning, 64-68; spec asst to secy Navy & dep dir, Off Prog Apraisal Secy Navy Staff, 68-69, Off Prog Appraisal Secy Navy Staff, 68-69; dep of labs, Lab & Res Mgt, US Army Mat Command, 69-75; sr vpres, Consol Diesel Elec Corp, 75-76; staff dir, Oversight, US House Rep Sci & Technol Comn, 76-77; consult & pres, Global Defense & Commun Corp, 77-79; pres, Dadico Systs, 83-89; exec vpres, Initiatives Inc, 83-90; PRES, DILLAWAY & ASSOCS, LTD, 79- *Concurrent Pos:* Exten lectr, Univ Calif, 53-70; indust consult, Proj Rover, Los Alamos Sci Lab, 54-55; astronaut comt del, Int Aeronaut Fedn, 60-64; chmn, Intersoc Comt on Transp, 71-73, Potomac Interface Inc, 90-; Donovan scholar. *Mem:* Fel Am Soc Mech Engrs; assoc fel Am Inst Aeronaut & Astronaut; Soc Exp Stress Anal; Brit Interplanetary Soc. *Res:* Fluid mechanics; heat transfer and thermodynamics; nuclear reactors for propulsion; transient fluid flow and boundary layer behavior; high speed temperature measurement. *Mailing Add:* 1306 Ballantrae Ct McLean VA 22101

DILLE, JOHN ROBERT, b Waynesburg, Pa, Sept 2, 31; m 55; c 2. AEROSPACE MEDICINE. *Educ:* Waynesburg Col, BS, 52; Univ Pittsburgh, MD, 56; Harvard Univ, MIH, 60; Am Bd Prev Med, dipl & cert aerospace med, 64. *Prof Exp:* Resident aerospace med, US Air Force Sch Aviation Med, Harvard Univ, 59-62, prog adv officer, Civil Aeromed Res Inst, 63-65, regional flight surgeon, Western Region, Los Angeles, 65, chief civil aeromed inst, Fed Aviation Admin, 65-87; CONSULT, 87-; COLONEL, MASTER FLIGHT SURGEON & STATE SURGEON, ARMY NAT GUARD MED DIR, OKLA DEPT CORRECTIONS, 90- *Concurrent Pos:* Assoc prof, Sch Med, Univ Okla, 62- *Honors & Awards:* Theodore C Lyster Award, Aerospace Med Asn, 78, Harry G Moseley Award, 87. *Mem:* Fel Aerospace Med Asn (pres-elect, 91); fel Am Col Prev Med; Sigma Xi; Int Acad Aviation Space Med; Asn Mil Surgeons US. *Res:* Aircraft accident investigation; pulmonary physiology; toxicity of drugs and pesticides; penetrating eye injuries. *Mailing Add:* 335 Merkle Dr Norman OK 73069

DILLE, KENNETH LEROY, b Caldwell, Idaho, May 9, 25; m 52; c 4. ORGANIC CHEMISTRY, PETROLEUM CHEMISTRY. *Educ:* Col Idaho, BA, 50; Ore State Col, MA, 52, PhD(org phys chem), 54. *Prof Exp:* Asst org chem, Ore State Col, 50-52; res chemist, Texaco Inc, 54-62, group leader chem res, 62-64, group leader fuels res, 64-68, asst supvr, 68-73, supvr, Fuels Res Sect, 73-80, coordr, Environ Affairs, 80-81 & Indust Hyg & Toxicol, 81-88, COORDR, PROD SAFETY, TEXACO, INC, BEACON, 91- *Concurrent Pos:* Texaco proj leader with US Govt, 60-62. *Res:* Synthesis and development of additives for improving motor and aviation gasolines, diesel and jet fuels and antifreezes; product development of petrochemicals. *Mailing Add:* Nine Sherrywood Rd Wappingers Falls NY 12590

DILLE, ROGER MCCORMICK, b Caldwell, Idaho, Apr 19, 23; m 55; c 1. CHEMISTRY. *Educ:* Col Idaho, BS, 44. *Prof Exp:* Chemist, Texaco Refinery, Texaco, Inc, 44-51, chemist, Montebello Res Lab, 51-58, from res chemist to sr res chemist, 58-70, supvr anal & testing, Richmond Res Lab, Va, 70-76, supvr Customer Servs Sect, Port Arthur Res Labs, 76-82; RETIRED. *Concurrent Pos:* Sr process consult, Radian Corp, 82-91; pvt consult, Dille Agency, 84- *Mem:* Am Chem Soc; Am Petrol Inst; Am Soc Testing Mat. *Res:* Petrochemicals; synthetic fuels; partial oxidation; environmental problems; coal gasification. *Mailing Add:* 4449 Lakeshore Dr Port Arthur TX 77642-1250

DILLENIUS, MARNIX FRITZ EUGEN, aeronautical engineering, aerodynamics, for more information see previous edition

DILLER, EROLD RAY, b New Stark, Ohio, May 4, 22; m 52; c 2. BIOCHEMISTRY, PHARMACOLOGY. *Educ:* Bowling Green State Univ, BA, 49; Ind Univ, MS, 51. *Prof Exp:* Res assoc biochem, Ind Univ, 49-51; sr scientist, 51-66, res scientist, 66-71, RES ASSOC, LILLY RES LABS, 71- *Mem:* NY Acad Sci; Sigma Xi; Soc Exp Biol & Med. *Res:* Intermediary metabolism and metabolic control mechanisms; lipid and sterol metabolism and metabolic control mechanisms; biochemical pharmacology; absorption. *Mailing Add:* 6640 Blossom Lane Indianapolis IN 46278

DILLER, KENNETH RAY, b Wooster, Ohio, Nov 20, 42; m 67; c 3. THERMODYNAMICS, HEAT TRANSFER. *Educ:* Ohio State Univ, BME, 66, MSc, 67; Mass Inst Technol, ScD, 72. *Prof Exp:* Res assoc biomed eng, Mass Inst Technol, 72-73; from asst prof to assoc prof mech eng, 73-84, PROF MECH & BIOMED ENG, UNIV TEX, AUSTIN, 84-, WILLIAM J MURRAY FEL, 85- *Concurrent Pos:* Alexander Von Humboldt fel, WGer, 83-84. *Mem:* AAAS; Am Soc Mech Eng; Soc Cryobiol (treas, 74-75); Microcirculation Soc; Am Ceramic Soc. *Res:* Study of energy processes in living systems involving transport of heat and mass; applications include microscopic evaluation of burn injury process and frozen preservation of living tissues and organs; computer modelling and computerized analysis of biomedical images. *Mailing Add:* 212 W 33rd St Austin TX 78705

DILLER, THOMAS EUGENE, b Orrville, Ohio, Sept 10, 50; m 77; c 3. HEAT & MASS TRANSFER, FLUID MECHANICS. *Educ:* Carnegie-Mellon Univ, BS, 72; Mass Inst Technol, SM, 74, ScD, 77. *Prof Exp:* Sr engr, Res & Develop, Polaroid Corp, 76-79; from asst prof to assoc prof, 79-90, PROF MECH ENG, VA POLYTECH INST & STATE UNIV, 90- *Concurrent Pos:* Prin investr res grants, Whitaker Found, 81-84, US Dept Energy, 82-86, NSF, 88-91, NASA, 90- *Mem:* Am Soc Mech Engrs; Am Inst Chem Engrs; AAAS; Sigma Xi. *Res:* Fundamental analysis and experimentation jet impingement heat transfer, heat transfer in unsteady and separated flows, and mass transfer in flowing blood; application of result in areas such as drying, heat exchanger design, gas turbine heat transfer and atherosclerosis. *Mailing Add:* Dept Mech Eng Va Polytech Inst & State Univ Blacksburg VA 24061

DILLER, VIOLET MARION, biophysics; deceased, see previous edition for last biography

DILLERY, DEAN GEORGE, b Fremont, Ohio, Nov 18, 28; m 51; c 4. ZOOLOGY. *Educ:* Ohio State Univ, BS, 52, MS, 55, PhD(ornith), 61. *Prof Exp:* Instr zool, Ohio Wesleyan Univ, 59-60; assoc prof, 60-73, chmn dept, 76-82, PROF BIOL, ALBION COL, 73- *Res:* Biology and behavior of arthropods; population biology and behavior of amphipods. *Mailing Add:* Dept Biol Albion Col Albion MI 49224

DILLEY, DAVID ROSS, b South Haven, Mich, Mar 10, 34; m 56; c 3. PLANT PHYSIOLOGY. *Educ:* Mich State Col, BS, 55; Mich State Univ, MS, 57; NC State Col, PhD(bot), 60. *Prof Exp:* Technician, 56-57, from asst prof to assoc prof, 60-67, PROF HORT, MICH STATE UNIV, 67- *Concurrent Pos:* Chmn Gordon Res conf Post Harvest Physiol, 72; consult, United Fruit Co & Cent Am Res Inst Indust; consult, Gen Foods Corp, 74, Grumman Allied Industs, Inc & Union Carbide Corp, 75-, Frito-Lay, 82-, Monsanto Corp, 86-; lectr, Nato Adv Studies Inst, Greece, 81. *Honors & Awards:* Dow Chem Award, Am Soc Hort Sci. *Mem:* Am Soc Plant Physiol; fel Am Soc Hort Sci; Sigma Xi; fel AAAS. *Res:* Physiology and biochemistry of fruits during growth, maturation and senescence; handling, transportation and storage of perishable commodities; instrument design for gas analysis and gas atmosphere control systems. *Mailing Add:* Dept Hort Mich State Univ East Lansing MI 48824

DILLEY, JAMES PAUL, b Athens, Ohio, Feb 19, 34. THEORETICAL PHYSICS. *Educ:* Ohio Univ, AB, 55, MS, 56; Syracuse Univ, PhD(physics), 64. *Prof Exp:* From asst prof to assoc prof, 63-76, PROF PHYSICS, OHIO UNIV, 76- *Concurrent Pos:* Res fel, Theoret Physics Inst, Univ Alta, 68-69; vis assoc prof, Univ Colo, 71-72; vis prof, Univ Ariz, 79-80. *Mem:* Am Phys Soc; Am Asn Physics Teachers; Am Geophys Union; Am Astron Soc; AAAS; Sigma Xi. *Res:* Planets. *Mailing Add:* Dept Physics Ohio Univ Athens OH 45701

DILLEY, RICHARD ALAN, b South Haven, Mich, Jan 12, 36; m 60; c 4. BIOCHEMISTRY, PLANT PHYSIOLOGY. *Educ:* Mich State Univ, BS, 58, MS, 59; Purdue Univ, PhD(biochem, plant physiol), 63. *Prof Exp:* NIH, C F Kettering Res Lab, 63-64; vis asst prof biophys, Univ Rochester, 65; staff scientist, C F Kettering Res Lab, 66-68, investr, 69-71; assoc prof, 71-75, PROF BIOL SCI, PURDUE UNIV, 75- *Honors & Awards:* Alexander von Humboldt Award, 82. *Mem:* AAAS; Am Soc Plant Physiol; Am Inst Biol Sci; Am Soc Biol Chemists. *Res:* Membrane biochemistry in energy transducing systems; ion and electron transport; photophosphorylation and membrane structure in the photosynthetic apparatus; quinones of plants. *Mailing Add:* Dept Biol Sci Purdue Univ West Lafayette IN 47907

DILLEY, WILLIAM G, b Van Nuys, Calif, Sept 21, 42; m 65; c 2. CELL BIOLOGY, ONCOLOGY. *Educ:* Univ Calif, Berkeley, AB, 65, MA, 67, PhD(endocrinol), 70. *Prof Exp:* Asst prof anat, Col Physicians & Surgeons, Columbia Univ, 70-76; asst prof exp surg, Duke Univ Med Ctr, 76-82, dir clin res, Unit Lab, 78-82; ASSOC RES PROF SURG, WASH UNIV SCH MED, 82- *Mem:* AAAS. *Res:* Breast cancer; thyroid-parathyroid endocrinology. *Mailing Add:* Dept Surg Wash Sch Med St Louis MO 63110

DILLING, WENDELL LEE, b Bluffton, Ind, June 16, 36; m 58; c 1. PHYSICAL ORGANIC CHEMISTRY. *Educ:* Manchester Col, BA, 58; Purdue Univ, PhD(org chem), 62. *Prof Exp:* res chemist, 62-68, sr res chemist, 68-72, res specialist, 72-80, RES LEADER, DOW CHEM CO, 80- *Concurrent Pos:* Counr, Am Chem Soc, 76-; comt publ, 77-84, comt on comts, 90- *Mem:* Am Chem Soc; AAAS; Inter-Am Photochem Soc; Europ Photochem Asn; Sigma Xi. *Res:* Pentacyclodecane bishomocubane chemistry; reaction mechanisms; spectroscopy; organic photochemistry; photocycloaddition reactions. *Mailing Add:* 1810 Norwood Dr Midland MI 48640

DILLMAN, LOWELL THOMAS, b Huntington, Ind, Aug 26, 31; m 54; c 4. NUCLEAR PHYSICS. *Educ:* Manchester Col, BA, 53; Univ Ill, MS, 55, PhD(physics), 58. *Prof Exp:* Asst, Univ Ill, 53-57; from asst prof to assoc prof physics, 58-68, PROF PHYSICS, OHIO WESLEYAN UNIV, 68- *Concurrent Pos:* Consult, Health and Safety Res Div, Oak Ridge Nat Lab, 67- *Mem:* Am Phys Soc; Am Asn Physics Teachers; Health Physics Soc. *Res:* Health physics internal dosimetry. *Mailing Add:* Dept Physics Ohio Wesleyan Univ Delaware OH 43015

DILLMAN, RICHARD CARL, b Ft Dodge, Iowa, Sept 4, 31; m 64; c 2. VETERINARY PATHOLOGY. *Educ:* Iowa State Univ, BS, 59, DVM, 61; Kans State Univ, MS, 64, PhD(path), 68. *Prof Exp:* Instr vet path, Kans State Univ, 64-68; from asst prof to assoc prof, Iowa State Univ, 68-75; PROF VET PATH, NC STATE UNIV, 75- *Mem:* Sigma Xi. *Res:* Clinical pathology; serum proteins; bovine respiratory diseases; reproductive disorders of domestic livestock; avian respiratory diseases. *Mailing Add:* Col Vet Med NC State Univ Raleigh NC 27606-8401

DILLMAN, ROBERT O, b Seattle, Wash, Mar 15, 47; m; c 3. EXPERIMENTAL BIOLOGY. *Educ:* Stanford Univ, BA, 70; Baylor Col Med, MD, 73; Am Bd Internal Med, dipl, 77. *Prof Exp:* Instr, Dept Med, Baylor Col Med, 78; clin instr med, Div Hemat-Oncol, Univ Calif, 80-81, asst dir, 80-84, asst prof, 81-86; assoc dir, Ida M Green Cancer Ctr, Scripps Clin & Res Found, 87; MED DIR, HOAG CANCER CTR, HOAG MEM HOSP PRESBY, 89-; CLIN PROF MED, UNIV CALIF, IRVINE SCH MED, 89- *Concurrent Pos:* Actg chief med oncol, San Diego Vet Admin Med Ctr, 83-84, chief, 84-87; dir, Exp Clin Oncol, Scripps Clin & Res Found, 86-89. *Mem:* AMA; Am Col Physicians; Am Soc Clin Oncol; Am Soc Hemat; Am Fedn Clin Res; Am Asn Cancer Res; Soc Biol Ther; AAAS; Am Asn Immunologists; Am Cancer Soc; Am Col Surgeons. *Res:* Experimental biology. *Mailing Add:* Hoag Cancer Ctr Hoag Mem Hosp Presby 301 Newport Blvd Newport Beach CA 92658

DILLMANN, WOLFGANG H, b Recklinghauser, Ger, Dec 8, 39; m 79; c 4. MEDICAL. *Educ:* Univ Munich, MD, 70; Am Bd Internal Med, cert, 72; Am Bd Endocrinol & Metab, cert, 77. *Prof Exp:* Intern, City Hosp Ctr, Elmhurst & Mt Sinai Hosp, New York, 70-71; resident, Montefiore Hosp & Albert Einstein Col Med, New York, 71-72, fel endocrinol & metab, 72-75, asst prof med, 75-76; asst prof med, Univ Minn, 76-78; from asst prof to assoc prof, 79-87, PROF MED, UNIV CALIF, SAN DIEGO, 87- *Mem:* Am Thyroid Asn; Endocrine Soc; Am Diabetes Asn; Am Heart Asn; Am Fedn Clin Res; Am Soc Clin Invest; Am Soc Biochem & Molecular Biol; Int Soc Heart Res. *Res:* Thyroid hormone action in the heart; mechanisms to induce stress proteins in the myocardium; author of numerous publications. *Mailing Add:* Med Ctr Univ Calif 225 Dickinson St H-811-C San Diego CA 92103

DILLON, HUGH C, JR, medicine; deceased, see previous edition for last biography

DILLON, JOHN JOSEPH, III, b Emmitsburg, Md, Jan 15, 47; m 72; c 1. BIOCHEMISTRY. *Educ:* Mt St Mary Col, BS, 68; Univ Pittsburgh, MS, 70; Univ Del, PhD(chem), 78. *Prof Exp:* Instr chem, Va Mil Inst, 70-73; vis prof chem, Buknell Univ, 78-79; ASST PROF CHEM, SAINT MARY'S COL, 79- *Mem:* Am Chem Soc; AAAS. *Res:* Enzymology; mechanism of RNA polymerase; gene regulation; molecular biology. *Mailing Add:* 3113 Riverside Dr South Bend IN 46628-3515

DILLON, JOHN THOMAS, structural geology; deceased, see previous edition for last biography

DILLON, JOSEPH FRANCIS, JR, b Flushing, NY, May 25, 24; m 46; c 2. MAGNETISM, MAGNETO-OPTICS. *Educ:* Univ Va, AB, 44, AM, 48, PhD(physics), 49. *Prof Exp:* Physicist, Naval Ord Lab, 48 & Brit Agr Res Coun Lab, USDA, 49-52; MEM TECH STAFF, AT&T BELL LABS, 52- *Concurrent Pos:* Vis scientist, Inst Solid State Physics, Univ Tokyo, 66-67, vis prof, 78; Guggenheim fel, 66-67. *Honors & Awards:* Distinguished lectr, Magnetic Soc, 81. *Mem:* Am Phys Soc; Magnetic Soc; Conf Magnetism & Magnetic Mat. *Res:* Ferromagnetism; ferrites; crystal growth; ferrimagnetic resonance; ferrimagnetic garnets; optical properties of solids; magnetic domains; optical study of metamagnetic transitions; critical phenomena; magnetooptical devices; magnetooptical effects in superconductors. *Mailing Add:* AT&T Bell Labs Rm 1 D-328 Murray Hill NJ 07974-2070

DILLON, LAWRENCE SAMUEL, b Reading, Pa, Apr 6, 10; m 32; c 1. EVOLUTION, MOLECULAR BIOLOGY. *Educ:* Univ Pittsburgh, BS, 33; Agr & Mech Col Tex, MS, 50, PhD(entomol), 54. *Prof Exp:* Chemist, Glidden Co, 34-37; cur entomol, Reading Pub Mus, 37-48; from instr to prof, 48-75, EMER PROF BIOL, TEX A&M UNIV, 75- *Concurrent Pos:* NSF sci fac fel, Queensland, Australia, 59-60. *Mem:* Fel AAAS; NY Acad Sci; Am Soc Zoologists; Soc Study Evolution; Soc Syst Zool. *Res:* Evolution and phylogeny of living things; Pleistocene paleoecology and bioclimatology; origin of genetic mechanism; speciation; evolution of mammalian brain; neuroanatomy. *Mailing Add:* Rockwood Park Estates 1904 Cedarwood Dr Bryan TX 77801

DILLON, MARCUS LUNSFORD, JR, b Charleston, WVa, Feb 7, 24; m 47; c 8. SURGERY. *Educ:* Duke Univ, BS & MD, 48. *Prof Exp:* Intern surg, Duke Univ Hosp, 47-49, from jr asst res to sr asst res, 49-55, resident surg, 55-56, from instr to assoc prof, 53-71; PROF SURG, UNIV KY, 71- *Concurrent Pos:* Asst chief surg serv, Vet Admin Hosp, Durham, 56-71, chief thoracic surg, 64-71; chief thoracic & cardiovasc surg, Vet Admin Hosp, Lexington, 71- *Mem:* Am Col Surg; Am Asn Thoracic Surg; Soc Thoracic Surg; AMA; Sigma Xi. *Res:* Hypothermia; extracorporeal circulation and coagulation; cardiovascular disease and hypothermia; infections; cancer. *Mailing Add:* Dept Surg Vet Admin Hosp Lexington KY 40507

DILLON, OSCAR WENDELL, JR, b Franklin, Pa, May 27, 28; m 60; c 4. ENGINEERING MECHANICS. *Educ:* Univ Cincinnati, AeroEng, 51; Columbia Univ, MS, 55, DrEngSc, 59. *Prof Exp:* Aeronaut engr, Cornell Aeronaut Lab, 53-54; asst prof mech, Johns Hopkins Univ, 58-63; lectr aerospace eng, Princeton Univ, 63-65; assoc prof, 65-67, chmn dept, 67-72, PROF ENG MECH, UNIV KY, 67- *Honors & Awards:* Res Award, Am Soc Eng Educ, 67 & Univ Ky, 68. *Mem:* Soc Natural Philos; Am Soc Mech Engrs. *Res:* Theoretical and experimental mechanics of solids, especially in the inelastic range. *Mailing Add:* Dept Eng Mech Univ Ky Lexington KY 40506

DILLON, PATRICK FRANCIS, SMOOTH MUSCLE, NUCLEAR MAGNETIC RESONANCE. *Educ:* Univ Va, PhD(physiol), 80. *Prof Exp:* ASST PROF CELL PHYSIOL, MICH STATE UNIV, 83- *Res:* Physiology. *Mailing Add:* Dept Physiol Mich State Univ 111 Giltner Hall East Lansing MI 48824

DILLON, RAYMOND DONALD, b Superior, Wis, Apr 19, 25; m 50; c 3. ZOOLOGY. *Educ:* Wis State Univ, Superior, BS, 49; Syracuse Univ, MS, 51; Univ Wis, PhD(zool, vet sci), 55. *Prof Exp:* Asst prof biol, Nebr State Teachers Col, 54-60; asst prof, Ball State Teachers Col, 60-62; asst prof zool & physiol, Univ Nebr, 62-65; prof zool & chmn dept, 65-69, PROF BIOL, UNIV SDAK, 69- *Concurrent Pos:* Partic, NSF US Antarctic Res Prog, 67-69, Arctic Biome Prog, Naval Arctic Res Lab, Alaska, 71-72; Bush grants, USD res awards, & NSF fac res, 74; vis res scholar, Ind Univ, 80-81. *Mem:* AAAS; Soc Protozool; Nat Asn Biol Teachers; Am Inst Biol Sci; Am Micros Soc. *Res:* Protozoology; research in free-living and parasitic protozoa. *Mailing Add:* Dept Biol Univ SDak Vermillion SD 57069

DILLON, RICHARD THOMAS, b Pocatello, Idaho, May 2, 28; m 51; c 4. RADIATION GENETICS, SYSTEMS ANALYSIS. *Educ:* Idaho State Col, BS, 52; Univ Ore, MS, 54, PhD(abstr harmonic anal), 58. *Prof Exp:* Staff mem systs anal, 58-61, staff mem of dir advan systs study, 61-69 & planetary quarantine, 69-72, STAFF MEM BIOSYSTS STUDIES, SANDIA LABS, 72- *Mem:* Am Microbiol Soc; Am Inst Biol Sci; AAAS; NY Acad Sci; Sigma Xi. *Res:* Experimentally determining the inactivating and mutagenic effects of heat and ionizing radiation separately and in combination on various cellular systems, recently with dry spores including the effect of relative humidity. *Mailing Add:* 5808 Pojoaque Rd NE Albuquerque NM 87110

DILLON, ROBERT MORTON, b Seattle, Wash, Oct 27, 23; m 43; c 3. CONSTRUCTION EDUCATION, BUILDING RESEARCH. *Educ:* Univ Wash, BArch, 49; Univ Fla, MAArch, 54. *Prof Exp:* Asst prof archit, Clemson Col, 49-50; instr, Univ Fla, 50-52, asst prof, 52-55; staff architect, Bldg Res Adv Bd, Nat Res Coun-Nat Acad Sci, 55-57, proj dir, 57-58, exec dir, 58-77; exec asst to pres, Nat Inst Bldg Sci, 78-81, vpres & actg controller, 83-84; exec vpres, Am Coun Construct Educ, 84; RETIRED. *Concurrent Pos:* Designer-draftsman, SC, 49-50; architect, Fla, 52-55; lectr, Cath Univ Am, 57-63; consult, Educ Facilities Labs, Inc, NY, 58-71; ed reports & publ, Bldg Res Adv Bd, 58-77; adv bd dirs, Wash Ctr Metrop Studies, 58-60; mem sub-panel housing, White House Panel Civilian Technol, 61-62; exec secy, US Nat Comt for Int Coun Bldg Res, Studies & Documentation, Nat Res Coun, Nat Acad Sci-Nat Acad Eng, 62-74; mem adv comt low income housing demonstration prog, HUD, 64-66; mem, Am Inst Archit adv comt, Voc Rehab Admin, HEW, 67-68; mem, ERIC-CEF Adv Coun, Univ Wis, 67-69; mem, Nat Adv Coun Res in Energy Conserv, 75-78; vis prof, Grad Sch Archit, Univ Utah, 78; adj assoc prof construct sci, Univ Okla, 84-; architectural consult, 84-; mem, Nat Inst Bldg Sci Consult Coun. *Mem:* Corp mem Am Inst Archit; Am Soc Civil Engrs; Am Inst Steel Construct; assoc Am Inst Constructors; Nat Acad Code Admin. *Res:* Housing, building and city planning. *Mailing Add:* PO Box 193 Angel Fire NM 87710

DILLON, ROY DEAN, b Peoria, Ill, Dec 10, 29; m 53; c 3. AGRICULTURE. *Educ:* Univ Ill, BS, 52, EdM, 58, EdD, 65. *Prof Exp:* Teacher high schs, Ill, 52-53 & 54-61; asst dir safety, Ill Agr Asn, 61-62; assoc prof agr, Morehead State Univ, 64-67; assoc prof, 67-70, prof sec educ, 70-80, PROF AGR, UNIV NEBR, LINCOLN, 70- *Res:* Knowledge and ability required by workers in off-farm agricultural occupations. *Mailing Add:* Dept Agr Educ Univ Nebr East Campus Lincoln NE 68583-0709

DILLON, WILLIAM PATRICK, b Fall River, Mass, Dec 13, 36; m 61; c 3. GEOLOGICAL OCEANOGRAPHY. *Educ:* Bates Col, BS, 58; Rensselaer Polytech Inst, MS, 61; Univ RI, PhD(geol oceanog), 69. *Prof Exp:* Res assoc marine geol, Narragansett Marine Lab, Univ RI, 61-64, instr oceanog, Sch Oceanog, 68-69; asst prof geol, San Jose State Col, 69-71; marine geologist, 71-82, chief, 82-86, RES GEOLOGIST, BR ATLANTIC MARINE GEOL, US GEOL SURV, 86- *Mem:* Geol Soc Am; Sigma Xi; Am Geophys Union. *Res:* Structure and development of continental margins; US eastern margin south of Cape Hatteras; Caribbean plate boundaries; marine gas hydrates. *Mailing Add:* Br Atlantic Marine Geol US Geol Surv Woods Hole MA 02543

DILLS, CHARLES E, b La Moure, NDak, Apr 20, 22; m 61; c 1. PHYSICAL CHEMISTRY, ORGANIC CHEMISTRY. *Educ:* NDak State Univ, BS, 49; George Washington Univ, MS, 51; Harvard Univ, PhD(chem), 56. *Prof Exp:* Res assoc org chem, Columbia Univ, 54-56; res chemist, Nat Res Corp, Mass, 56-58; instr chem, Northwest Mo State Col, 58-59, asst prof, 59-60; asst ed res, Chem & Eng News, 60-61; prof sci, Deep Springs Col, 61-63; asst prof chem, 63-67, assoc prof, 67-80, PROF CHEM, CALIF POLYTECH STATE UNIV, SAN LUIS OBISPO, 80- *Res:* Mechanisms of organic reactions. *Mailing Add:* Dept Chem Calif Polytech State Univ San Luis Obispo CA 93407

DILLS, WILLIAM L, JR, b Wilmington, Del, Jan 29, 45. METABOLISM OF CARBOHYDRATES. *Educ:* Univ Vt, PhD(biochem), 73. *Prof Exp:* ASST PROF NUTRIT & BIOCHEM, CORNELL UNIV, 76-; ASSOC PROF CHEM, SOUTHEASTERN MASS UNIV, 82- *Mailing Add:* Dept Chem 103 Violette Bldg Southeastern Mass Univ North Dartmouth MA 02747

DILLWITH, JACK W, PLANT-INSECT INTERACTION, PHEROMONE METABOLISM. *Educ:* Univ Nev, PhD(biochem), 80. *Prof Exp:* ASST PROF INSECT BIOCHEM, OKLA STATE UNIV, 86- *Mailing Add:* 1423 N Payne Stillwater OK 74075-6915

DILMINO, MICHAEL JOSEPH, OVARIAN ENDOCRINOLOGY, HORMONE ACTION. *Educ:* Rutgers Univ, PhD(reproductive physiol), 71. *Prof Exp:* Prof biochem & assoc dean admissions & acad planning, Eastern Va Med Sch, 79-89; DIR BIOMED PROG, UNIV ALASKA, ANCHORAGE, 89- *Mailing Add:* Dept Biomed Prog Univ Alaska 3211 Providence Dr Anchorage AK 99508

DILPARE, ARMAND LEON, b New York, NY, Aug 12, 32; m 52; c 1. MACHINE DESIGN, AUTOMATIC CONTROL SYSTEMS. *Educ:* City Col New York, BME, 53; Columbia Univ, MSME, 57, PhD(kinematics mechanisms), 65. *Prof Exp:* Instr mech eng, City Col New York, 53; struct engr, Repub Aviation Corp, 53-56; mem tech staff, Bell Tel Labs, NY, 56-62; div engr, Lundy Electronics & Systs Inc, 62-65; asst prof mech eng, Columbia Univ, 65-70; assoc dir res & develop, Defense Systs Div, Lundy Electronics & Systs, Inc, 70-75; dir coop engr sci progs,assoc prof, 75-78, Jacksonville Univ, 78-84; PROF MECH ENG, FLA INST TECHNOL, MELBOURNE, 84- *Concurrent Pos:* Consult, Lundy Electronics & Systs, Inc, 65-70. *Mem:* Am Soc Mech Engrs; Am Soc Eng Educ. *Res:* Kinematic analysis and synthesis of mechanisms; dynamics of machinery; computer-aided-design. *Mailing Add:* Dept Mech Eng Fla Inst Technol 150 W University Blvd Melbourne FL 32901

DILS, ROBERT JAMES, b Dayton, Ohio, Oct 2, 19; m 47; c 4. SCIENCE EDUCATION. *Educ:* Eastern Ky Univ, BS; Marshall Univ, MA, 60. *Prof Exp:* Teacher high sch, Ky, 48-51; instr chem, Ashland Jr Col, 51-57; chmn dept sci, Paul G Blazer High Sch, 57-62; coord sci, Sec Sch, Ashland, Ky, 62-63; from asst prof to assoc prof, phys sci, Marshall Univ, 64-85; PROF, OHIO UNIV, 85- *Mem:* AAAS; Am Asn Higher Educ. *Res:* Science education and curriculum development in college and secondary schools. *Mailing Add:* Dept Phys Sci Marshall Univ Huntington WV 25701

DILSAVER, STEVEN CHARLES, b Los Angeles, Calif, March 5, 53. NEUROPSYCHOPHARMACOLOGY, AFFECTIVE DISORDERS. *Educ:* Stanford Univ, BS, 75; Univ Calif, MD, 80. *Prof Exp:* Resident, Dept Psychiat, Univ Mich, 80-85, asst prof psychiat, Ment Health Res Inst, 85-87; dir, Psychopharmacol Prog & assoc prof psychiat, neurosci & nursing, Ohio State Univ, 87-90; PROF PSYCHIAT & BEHAV SCI, SCH MED, UNIV TEX, HOUSTON, 90- *Concurrent Pos:* Neuroscientist, Ohio State Univ, 87-, prin investr res & grad studies, 81-88; prin investr, NIMH, 85-, Univ Mich Med Sch, 86-87 & Bremer Found, 88; consult, Ont Ment Health Admin, 86 & 88, Vet Admin Merit Rev Comt, 85, NIMH, 85 & NC Alcohol Res Coun, 90-; travel award, Am Col Neuropsychopharmacol, 85; vis prof, Dept Psychiat, Univ Pisa, Italy, 89 & Dept Psychiat & Alcohol Res Ctr, Univ NC, Chapel Hill, 90. *Mem:* AAAS; AMA; Am Psychiat Asn; Physicians Social Responsibility; Soc Biol Psychiat; Soc Neurosci. *Res:* Neurobiology of affective disorders, stress, bright artificial light, lithium, heterocyclic antidepressants & drugs of abuse; development of paradigms useful in directing research in the fields of affective disorders and psychopharmacology. *Mailing Add:* Dept Psychiat Health Sci Ctr Univ Tex PO Box 20708 Houston TX 77225

DILTS, JOSEPH ALSTYNE, b Sommerville, NJ, Sept 12, 42. INORGANIC CHEMISTRY. *Educ:* Ohio Wesleyan Univ, BA, 64; Northwestern Univ, PhD(chem), 69. *Prof Exp:* Fel Ga Inst Technol, 68-70; asst prof, 70-78, ASSOC PROF CHEM, UNIV NC, GREENSBORO, 78- *Concurrent Pos:* Vis prof chem, Col William & Mary, 76-77. *Mem:* Am Chem Soc; The Chem Soc; NAm Thermal Anal Soc. *Res:* Organometallic and hydride chemistry of the main group elements; application of thermal analysis techniques to organometallic chemistry. *Mailing Add:* Dept Chem Univ NC 1000 Spring Garden Greensboro NC 27412

DILTS, PRESTON VINE, JR, b Louisiana, Mo, Jan 16, 34; m 57; c 3. OBSTETRICS & GYNECOLOGY. *Educ:* Wash Univ, BA, 55; Northwestern Univ, MD, 59, MS, 61; Am Bd Obstet & Gynec, dipl, 69, cert fetal & maternal med, 75. *Prof Exp:* From asst prof to assoc prof obstet & gynec, Sch Med, Univ Ky, 68-72; prof & chmn dept, Sch Med, Univ NDak, 73-75; prof obstet & gynec & chmn dept, Sch Med, Univ Tenn, 75-85; PROF OBSTET & GYNEC & CHMN DEPT, UNIV MICH, 85- *Concurrent Pos:* Examr, Am Bd Obstet & Gynec; USPHS fel reprod physiol, Sch Med, Univ Calif, Los Angeles, 67. *Mem:* Am Col Obstet & Gynec; Am Col Surg; Soc Gynec Invest; Soc Obstet Anesthesia & Perinatology. *Res:* Reproductive physiology, specifically the placental transfer of drugs and trace metals. *Mailing Add:* Dept Obstet & Gynec Univ Mich Med Sch Womens Hosp L2120 Ann Arbor MI 48109-0010

DILTS, ROBERT VOORHEES, b Plainfield, NJ, June 18, 29. ANALYTICAL CHEMISTRY. *Educ:* Wesleyan Univ, AB, 51; Princeton Univ, MA, 53, PhD(chem), 54. *Prof Exp:* From instr to asst prof chem, Williams Col, 54-60; asst prof, 60-68, dir undergrad prog chem, 67-69, ASSOC PROF CHEM, VANDERBILT UNIV, 68-, DIR UNDERGRAD PROG, 74- *Concurrent Pos:* Vis assoc prof, Univ Kans, 69-70. *Mem:* Am Chem Soc; Sigma Xi. *Res:* Coulometric titrations; atomic absorption spectroscopy; complexometric titrations. *Mailing Add:* Vanderbilt Univ Box 1590 Sta B Nashville TN 37235

DI LUZIO, NICHOLAS ROBERT, physiology; deceased, see previous edition for last biography

DILWORTH, BENJAMIN CONROY, b Monroe, La, Sept 29, 31; m 55; c 3. POULTRY NUTRITION, BIOCHEMISTRY. *Educ:* Miss State Univ, BS, 59, MS, 62, PhD(poultry nutrit), 66. *Prof Exp:* Res asst poultry nutrit, 59-66, asst prof poultry sci, 66-73, assoc prof poultry nutrit & poultry sci, 73-78, prof poultry nutrit, 78-80, PROF POULTRY SCI, MISS STATE UNIV, 78- *Mem:* Poultry Sci Asn; Sigma Xi; World Poultry Sci Asn. *Res:* Calcium, phosphorus and vitamin D3 interrelationships and requirements of poultry; protein quality of feedstuffs; sodium requirement of poultry. *Mailing Add:* Dept Poultry Sci Miss State Univ PO Box 5188 Mississippi State MS 39762

DILWORTH, ROBERT HAMILTON, III, b Bryson City, NC, Aug 6, 30; m 53; c 4. TECHNICAL VENTURE MANAGEMENT. *Educ:* Univ Tenn, BS, 52. *Prof Exp:* Chief engr, Spec Instruments Lab, Inc, Tenn, 54-55; instrument engr, Oak Ridge Inst Nuclear Studies, 55-56, develop engr, Oak Ridge Nat Lab, 56-62; chief electronics engr, Ortec, Inc, prod mgr electronic, 66-68, mgr life sci prod, 68-76; head mgt sect, Fusion Energy Div, Oak Ridge Nat Lab, 76-81, tech asst, Cent Mgt Off, 81; RETIRED. *Mem:* Inst Elec & Electronics Engrs. *Res:* Analysis and management of new technological ventures; modular nuclear instrumentation for physics research; engineering and product line management; research management. *Mailing Add:* 10033 Casa Real Cove Knoxville TN 37922

DIMAGGIO, ANTHONY, III, b New Orleans, La, Aug 17, 35; m 60; c 2. BIOCHEMISTRY. *Educ:* Loyola Univ, La, BS, 56; La State Univ, PhD(biochem), 61. *Prof Exp:* From asst prof to assoc prof, 61-85, CHMN DEPT, LOYOLA UNIV, LA, 67-, PROF CHEM, 85- *Concurrent Pos:* Consult, Photo-Dek, Inc, 64-66; dir res strontium fallout in tooth study, 64-66, co-prin investr, 66-68; chmn grad coun, Loyola Univ, 68-70; consult legal, 77-; co-dir, MST prog sci teachers, 85- *Mem:* AAAS; Am Chem Soc; Am Soc Ichthyologists & Herpetologists. *Res:* Comparative biochemistry of reptiles, especially the lizard, Anolis Carolinensis, particularly its endocrinology and metabolism; nutrition and social issues. *Mailing Add:* Dept Chem Loyola Univ New Orleans LA 70118

DI MAGGIO, FRANK LOUIS, b New York, NY, Sept 2, 29; m 63; c 2. STRUCTURAL MECHANICS. *Educ:* Columbia Univ, BS, 50, MS, 51, PhD(civil eng), 54. *Prof Exp:* PROF CIVIL ENG, COLUMBIA UNIV, 56- *Concurrent Pos:* Consult, Weidlinger Assocs, 56-; tech consult, Implements & Bal Comn, US Golf Asn, 58-66; NSF sr fel, Turin Polytech Inst, Italy, 62-63; guest scholar, Kyoto Univ, 86. *Mem:* Fel Am Soc Civil Engrs; Sigma Xi. *Res:* Structural dynamics; fluid-structure interaction; constitutive equations for geological materials. *Mailing Add:* Dept Civil Eng Columbia Univ New York NY 10027

DIMARCO, G ROBERT, b Camden, NJ, July 31, 27; m 49; c 5. FOOD SCIENCE & TECHNOLOGY. *Educ:* Rutgers Univ, BS, 54, PhD(plant pathol), 59. *Prof Exp:* From asst prof to chmn food sci dept, Rutgers Univ, 59-74; dir, Basic & Health Sci, Gen Foods Corp, 75-77, dir, Cent Res, 77-88; SR VPRES RES & DEVELOP, R J REYNOLDS TOBACCO, CO, WINSTON-SALEM, NC, 77- *Concurrent Pos:* Hon prof food sci, Cook Col, Rutgers Univ, 75-; Comt mem, Adv Bd Mil Personnel Supplies, Nat Res Coun, 78-81 & Food Indust Liaison Adv Panel, AMA, 78-; Food & Drug Law, Inst Food Update Bd Gov, 77; Adv Comt Sci Ed Res Grants, USDA, 81. *Mem:* Fel Inst Food Technologists; Sigma Xi. *Res:* Res administration in nutrition; physiology; biochemistry; physical chemistry; engineering; research from very basic to totally applied. *Mailing Add:* Bowman Gray Tech Ctr R J Reynolds Tobacco Co Winston Salem NC 27102

DI MARIA, DONELLI JOSEPH, b Waterbury, Conn, Mar 18, 46. SOLID STATE PHYSICS. *Educ:* Lehigh Univ, BS, 68, MS, 70, PhD(physics), 73. *Prof Exp:* Fel, 73-74, RES STAFF MEM INSULATOR PHYSICS, THOMAS J WATSON RES CTR, IBM CORP, 73- *Res:* Optical and electrical properties of insulators and their interfaces with other materials. *Mailing Add:* Chadeayne Rd Ossining NY 10562

DI MARIO, PATRICK JOSEPH, PROTEIN BIOCHEMISTRY, ULTRASTRUCTURAL ANALYSIS. *Educ:* Ind Univ, PhD(molecular, cellular & develop biol), 85. *Prof Exp:* RES FEL, CARNEGIE INST WASHINGTON, 85- *Res:* Cellular biology. *Mailing Add:* Dept Embryol Carnegie Inst Wash 115 W University Pkwy Baltimore MD 21210

DIMARZIO, EDMUND ARMAND, b Philadelphia, Pa, Mar 23, 32; m 56; c 3. PHYSICS. *Educ:* St Joseph's Col, Pa, BS, 55; Univ Pa, MS, 60; Cath Univ Am, PhD, 67. *Prof Exp:* Physicist, Am Viscose Co, 56-62; mem tech staff theoret chem, Bell Tel Labs, 62-63; PHYSICIST, NAT BUR STAND, 63- *Honors & Awards:* High Polymer Physics Prize, 67; Stratton Award, 71. *Mem:* Fel Am Phys Soc. *Res:* Helixcoil transition in biological macromolecules; glass transition in polymers; liquid crystal phase transitions; surface polymers; kinetics of crystallization. *Mailing Add:* 14205 Parkvale Rd Rockville MD 20853

DIMASI, GABRIEL JOSEPH, b Harrison, NJ, July 11, 36; m 62; c 5. ELECTROCHEMICAL & CHEMICAL ENGINEERING, COMPUTER PROGRAMMING. *Educ:* Newark Col Eng, BSChE, 58, MS, 63. *Prof Exp:* Chem engr primary batteries, US Army Electronics Command Labs, 58-60, engr fuel cells, 60-63, engr magnesium batteries, 63-65, res engr, 66-67, group leader electrode kinetics, 67-72, chmn, Dept Chem Educ & Training, 66-72, group leader, Electronics Technol & Devices Lab, 72-86, chmn, Dept Mat Sci & Eng, 72-87, br chief, Nondevelop Items, 87-88, BR CHIEF SATELLITE COMMAND, US ARMY ELECTRONICS COMMAND, 88- *Concurrent Pos:* Lectr & consult, Rensselaer Polytech Inst, 71; dep branch chief, tech, strategy & prog planning, US Army Electronics Command, 86-87. *Mem:* Electrochem Soc; Am Inst Chem Eng; Am Inst Chem; Int Soc Electrochem; Am Educ Res Asn; Armed Forces Commun Electronics Asn. *Res:* Design and development of lithium thionylchloride batteries; rechargeable lithium batteries; heat transfer and cooling of power supplies; computerized data collection and instrument control; algorithms for electrochemical kinetics; database development and sort routines for large amounts of data. *Mailing Add:* 24 Pal Dr Wayside Ocean NJ 07712

DIMAURO, SALVATORE, b Verona, Italy, Nov 14, 39; m 68; c 2. NEUROLOGY. *Educ:* Univ Padova, Italy, MD, 63, Specilization neurol, 67. *Prof Exp:* Inst gen path, Univ Padova, Italy, 64-68; fel neurol, Univ Pa, 68-69, instr & asst prof, 72-74; assoc prof, 74-78, PROF NEUROL, COLUMBIA UNIV, 78- *Mem:* Am Neurol Asn; Am Acad Neurol; Soc Neurosci; AAAS. *Res:* Biochemical research in neuromuscular diseases, with particular emphasis on hereditary disorders of muscle metabolism; metabolic mypathies; mitochondrial mypathies. *Mailing Add:* Dept Neurol Columbia Univ 710 W 168th St New York NY 10032

DIMEFF, JOHN, b Detroit, Mich, July 2, 21; m 44; c 2. PHYSICS. *Educ:* Harvard Univ, BS, 42. *Prof Exp:* Radio engr, Naval Res Lab, 43-44 & 45-46; physicist, Ames Res Ctr, NASA, 46, aeronaut res scientist, 46-53, chief wind tunnel instrument br, 56-59, asst chief, Instrumentation Div, 59-67, chief, 67-71, asst dir instrumentation res, 71-75; PRES, DIMEFF ASSOCS, 75- *Concurrent Pos:* Adj prof dent, Univ Calif, San Francisco, 75-; pres, Signathics, 77-84. *Res:* Sensors; instrumentation; electronics; data processing. *Mailing Add:* Dimeff Assocs 5346 Greenside Dr San Jose CA 95127

DIMENT, WILLIAM HORACE, b Oswego, NY, Oct 15, 27; m 58; c 3. GEOPHYSICS. *Educ:* Williams Col, AB, 49; Harvard Univ, AM, 51, PhD(geophys), 54. *Prof Exp:* Explor geophysicist, Calif Co, 53-56; geophysicist, US Geol Surv, 56-60, chief, Br Theoret Geophys, 60-62, res geophysicist, 62-65; prof geol, Univ Rochester, 65-73; RES GEOPHYSICIST, US GEOL SURV, 73- *Concurrent Pos:* NSF sr fel, Yale Univ, 64-65; consult, US Dept Energy, US Dept Int on nuclear reactor siting & nuclear waste disposal, 65-70. *Mem:* Fel AAAS; fel Geol Soc Am; Seismol Soc Am; Am Geophys Union; Soc Explor Geophysicists; Am Soc Limnol & Oceanog. *Res:* Terrestrial heat flow and geothermal systems; seismology; gravity; limnology and environmental effects of deicing salts. *Mailing Add:* US Geol Surv Box 25046 MS 966 Denver Fed Ctr Denver CO 80225

DIMICCO, JOSEPH ANTHONY, b New Haven, Conn, June 13, 47. PHARMACOLOGY, NEUROBIOLOGY. *Educ:* Tufts Univ, BS, 69; Georgetown Univ, PhD(pharmacol), 78. *Prof Exp:* Staff fel, Lab Clin Sci, NIMH, 78-80; ASST PROF PHARMACOL, SCH MED, IND UNIV, 80- *Mem:* AAAS; Soc Neurosci. *Res:* Central nervous system control of cardiovascular function; neuropharmacology of central and peripheral cardiovascular control mechanisms. *Mailing Add:* Dept Pharmacol Ind Univ Sch Med 635 Barnhill Dr Indianapolis IN 46223

DIMICHELE, LEONARD VINCENT, b Williamsport, Pa, Sept 12, 52; m 79; c 2. ICHTHYOLOGY. *Educ:* Villanova Univ, BA, 74; Univ Del, MS, 77, PhD(physiol), 80. *Prof Exp:* NIH fel animal physiol, 80-84, ASSOC RES SCIENTIST, CHESAPEAKE BAY INST, JOHNS HOPKINS UNIV, 84- *Mem:* AAAS; Am Soc Zoologists; Am Soc Icthyologists & Herpetologists. *Res:* Ways Teleosts adapt to their environment; mechanism of hatching in Teleosts; the relationship between polymorphic genes, development, growth and physiological adaptation. *Mailing Add:* Dept Wildlife & Fisheries Tex A&M Univ College Station TX 77893

DIMICHELE, WILLIAM ANTHONY, b Wilmington, Del, May 13, 51; m 72; c 2. PALEOBOTANY, PLANT PALEOECOLOGY. *Educ:* Drexel Univ, BS, 74; Univ Ill, Urbana, MS, 76, PhD(bot), 79. *Prof Exp:* From asst prof to assoc prof bot, Univ Wash, 79-85; ASSOC CUR PALEOBIOL, SMITHSONIAN, 85- *Concurrent Pos:* Vis lectr bot, Univ Ill, Urbana, 78; res assoc, Coal Sect, Ill State Geol Surv, 79; vis res scholar, Ind Univ, 80; assoc ed, Paleobiol, 85-87. *Honors & Awards:* Isbella C Cookson Award, Bot Soc Am, 79. *Mem:* Bot Soc Am; Soc Study Evolution; Geol Soc Am; Soc Econ Paleontologists & Mineralogists; Am Soc Naturalists; AAAS. *Res:* Terrestrial paleoecology, particularly investigation of Carboniferous plant communities and patterns of change in vegetation through time; relationships among community paleoecological parameters, species-level variability and evolutionary patterns. *Mailing Add:* Dept Paleobiol Smithsonian NMNH Washington DC 20560

DIMICK, PAUL SLAYTON, b Burlington, Vt, Sept 15, 35; m 58; c 3. FOOD SCIENCE. *Educ:* Univ Vt, BS, 58, MS, 60; Pa State Univ, PhD(dairy sci), 64. *Prof Exp:* Asst dairy sci, Univ Vt, 58-60; asst dairy sci, 61-63, res asst, 63-64, res assoc, 64-66, from asst prof to assoc prof food sci, 66-75, PROF FOOD SCI, PA STATE UNIV, 75- *Mem:* Am Dairy Sci Asn; Inst Food Technol; Sigma Xi; Am Oil Chem Soc. *Res:* Ruminant lipid metabolism, especially milk fat synthesis; food flavor research, especially lipid precursors, lipid crystallization. *Mailing Add:* Dept Food Sci Penn St Univ University Park PA 16802

DIMITMAN, JEROME EUGENE, b New York, NY, Sept 24, 20; m 85; c 3. PLANT PATHOLOGY. *Educ:* Univ Calif, BS, 43, MS, 49, PhD, 58. *Prof Exp:* Assoc plant pathologist, State Dept Agr, Calif, 47-49; chmn dept biol sci, 49-73, prof, 49-83, EMER PROF BIOL, CALIF STATE POLYTECH UNIV, 83- *Mem:* Am Phytopath Soc; Bot Soc Am; Asn Trop Biol; Sigma Xi. *Res:* Citrus diseases, viruses; Phytophthora physiology; subtropical plants; crop diseases; Phytophthora citrophthora host range studies; citrus virus studies; Exocortis virus effect on callus formation. *Mailing Add:* Dept Biol Sci Calif State Polytech Univ Pomona CA 91768

DIMITROFF, EDWARD, b Nancy, France, Feb 27, 27; US citizen; m 51; c 2. ORGANIC CHEMISTRY, PHYSICAL CHEMISTRY. *Educ:* Univ Denver, BS, 56; St Mary's Univ, Tex, MS, 65. *Prof Exp:* Res chemist, US Naval Ord Test Sta, 56-59; from assoc res chemist to sr res chemist, 59-66, staff scientist, 66-69, mgr petrol technol, 70-78, asst dir energy technol, 78-80, DIR PETROL RES, SOUTHWEST RES INST, 80- *Mem:* Am Chem Soc; fel Am Inst Chemists; Soc Automotive Eng; Am Soc Testing & Mat. *Res:* Mechanisms of dispersancy in oil media; engine sludge, varnish, rust formation; liquid fuels crystallization; development of synthetic fuels, lubricants and novel energy systems; technical management. *Mailing Add:* 8530 Pendragon San Antonio TX 78250-2052

DIMITROFF, GEORGE ERNEST, b New York, NY, July 22, 38; m 61; c 3. ALGEBRA. *Educ:* Reed Col, BA, 60; Univ Ore, MA, 62, PhD(math), 64. *Prof Exp:* Asst prof math, Knox Col, Ill, 64-73; PROF MATH, EVERGREEN STATE COL, OLYMPIA, WA, 73- *Mem:* Am Math Soc; Math Asn Am. *Res:* Partially ordered topological spaces. *Mailing Add:* Evergreen State Col Olympia WA 98505

DIMLICH, RUTH VAN WEENEN, MORPHOMETRY, CYTOCHEMISTRY. *Educ:* Univ Cincinnati, PhD(anat), 80. *Prof Exp:* ASST PROF GROSS ANAT, COL MED, UNIV CINCINNATI, 84- *Mailing Add:* 8737 Shagbark Dr Cincinnati OH 45242

DIMMEL, DONALD R, b Waseca Co, Minn, May 26, 40; m 60; c 3. ORGANIC CHEMISTRY. *Educ:* Univ Minn, Minneapolis, BS, 62; Purdue Univ, PhD(org chem), 66. *Prof Exp:* NIH fel photochem, Cornell Univ, 66-67; asst prof org chem, Marquette Univ, 67-74; from res chemist to sr res chemist, Hercules, Inc, 74-78; assoc prof, 78-82, res assoc, 78-86, PROF CHEM, INST PAPER CHEM, 82-, PRIN RES CHEM, 86- *Res:* Mechanisms for the delignification of wood; lignin chemistry; carbohydrate reactions. *Mailing Add:* Inst of Paper Sci & Technol 575 14th St NW Atlanta GA 30318-5403

DIMMICK, JOHN FREDERICK, b Thomson, Ill, Aug 7, 21; m 46; c 2. ANIMAL PHYSIOLOGY. *Educ:* Western Ill Univ, BS, 48, MS, 49; Univ Ill, PhD(mammalian physiol), 60. *Prof Exp:* Teaching asst biol, Western Ill Univ, 48-49; teacher high schs, Ill, 49-59; res asst physiol, Univ Ill, 59-60 , USDA fel, 60-61; teacher & res animal physiol & gen biol, 61-86, EMER, WAKE FOREST UNIV, 86- *Concurrent Pos:* Vis lectr, NC Acad Sci, 63-; dir training ctr, High Sch Teacher's Inserv Training Proj, NSF-NC Acad Sci, 64-68. *Mem:* Assoc mem Am Physiol Soc; Sigma Xi. *Mailing Add:* 2860 Weslyan Lane Winston-Salem NC 27106

DIMMICK, RALPH W, b Chicago, Ill, Nov 18, 34; m 81; c 2. ANIMAL ECOLOGY, WILDLIFE MANAGEMENT. *Educ:* Southern Ill Univ, BA, 57, MA, 61; Univ Wyo, PhD(zool), 64. *Prof Exp:* Asst prof animal ecol, Tenn Tech Univ, 64-66; from asst prof to assoc prof forestry, 66-78, PROF WILDLIFE FISHERIES SCI, UNIV TENN, KNOXVILLE, 79- *Concurrent Pos:* Assoc ed, J Wildlife Mgt; ed, Proc Southeastern Fish & Wildlife Conf. *Mem:* Wildlife Soc. *Res:* Ecology and management of wildlife resources. *Mailing Add:* Dept Forestry Wildlife & Fisheries Univ Tenn Plant Sci Bldg Rm 308 Knoxville TN 37916

DIMMIG, DANIEL ASHTON, b Lansdale, Pa, Mar 5, 24; m 51; c 2. ORGANIC POLYMER CHEMISTRY. *Educ:* Muhlenberg Col, BS, 45; Univ Pa, MS, 48; Univ Pittsburgh, PhD(org chem), 63. *Prof Exp:* Res chemist, Nitrogen Div, Allied Chem Corp, 48-55; jr fel, Mellon Inst, 55-63; proj leader, Pennwalt Corp, 64-85, supvr, 86-90; RETIRED. *Mem:* Am Chem Soc. *Res:* Synthesis of new organic addition and condensation polymers; photochemical synthesis of mercaptans; heterogeneous catalytic synthesis of sulfur compounds; industrial organic chemistry. *Mailing Add:* 300 Riverview Rd King of Prussia PA 19406

DIMMLER, D(IETRICH) GERD, b Karlsruhe, Ger, Nov 14, 33; m 57; c 2. SYSTEM DEVELOPMENT, COMPUTER-AIDED ELECTRONICS ENGINEERING. *Educ:* Karlsruhe Tech Univ, Elec Eng, 62. *Prof Exp:* Electronics engr logic design, Karlsruhe Nuclear Res Ctr, 62-66; assoc elec engr systs develop, Brookhaven Nat Lab, 66-69; prin mem tech staff systs planning, Xerox Corp, El Segundo, Calif, 70-71; mgr software develop, Process Control Systs Develop, Rank Xerox Corp, Munchen, Ger, 71-72; SR SCIENTIST, BROOKHAVEN NAT LAB, 72-, ASSOC DIV HEAD, 80- *Mem:* Inst Elec & Electronics Engrs; AAAS. *Res:* Architectures, hardware and software of real-time systems and computer networks in data acquisition, experiment control and process control-specific interest; computer-aided design; parallel processing. *Mailing Add:* Brookhaven Nat Lab Assoc Univ Bldg 535B Upton NY 11973-5000

DIMMOCK, JOHN O, b Garden City, NY, Nov 24, 36; m 58, 74; c 6. PHYSICS. *Educ:* Yale Univ, BS, 58, PhD(physics), 62. *Prof Exp:* Res scientist solid state physics, Raytheon Co, 62-63; mem staff, Lincoln Lab, Mass Inst Technol, 63-66, leader appl physics group, 66-71, leader appl optics group, 71-74; dir electronic & solid state sci prog, Off Naval Res, 74-81, dep dir technol progs, 81-89; tech dir, Air Force Off Sci Res, Bolling AFB, 84-89; STAFF VPRES RES, MCDONNELL DOUGLAS CORP, 89- *Mem:* Fel Am Phys Soc; sr mem Inst Elec & Electronics Engrs; assoc fel Am Inst Aeronaut & Astronaut; AAAS; Sigma Xi. *Res:* Theoretical and experimental solid state physics; magnetism and semiconductors. *Mailing Add:* McDonnell Douglas Corp PO Box 516 St Louis MO 63166-0516

DIMMOCK, JONATHAN RICHARD, b Southampton, Eng, July 6, 37; m 68; c 3. MEDICINAL CHEMISTRY. *Educ:* Univ London, BPharm, 59, PhD(pharmaceut chem), 63. *Prof Exp:* Org res chemist, Chesterford Park Res Sta, Saffron Waldon, Eng, 63-67; from asst prof to assoc prof, 67-76, PROF PHARMACEUT CHEM, UNIV SASK, 76- *Concurrent Pos:* Res grants, Can Found Advan Pharm, 68, Nat Res Coun Fund, 68-70 & Smith Kline & French Labs, Pa, 71; Med Res Coun grant, 72- *Mem:* Can Inst Chem; Can Soc Chem (bd dir, 85-88). *Res:* Design, synthesis and evaluation of novel compounds screened against cancers and respiration in mitochondria; antiparasitic agents; anti-epileptic compounds. *Mailing Add:* Dept Med Chem Col Pharm Univ Sask Saskatoon SK S7N 0W0 Can

DIMOCK, DIRCK L, b Braintree, Mass, June 23, 30; m 52; c 3. PHYSICS, SPECTROSCOPY & SPECTROMETRY. *Educ:* Antioch Col, BS, 52; Johns Hopkins Univ, PhD(physics), 57. *Prof Exp:* Mem res staff, Plasma Physics Lab, Princeton Univ, 57-62; vis scientist, Max Planck Inst Physics & Astrophys, 62-63; assoc head, Exp Div, 63-77, PRIN RES PHYSICIST, PLASMA PHYSICS LAB, PRINCETON UNIV, 76- *Concurrent Pos:* Lectr astrophys sci, 63-66; consult, Princeton Appl Res Corp, 64-65 & EMR Corp, 80. *Mem:* AAAS; fel Am Phys Soc. *Res:* Plasma physics; spectroscopic and optical plasma diagnostics, and atomic structure. *Mailing Add:* PO Box 451 Princeton NJ 08540

DIMOCK, RONALD VILROY, JR, b Melrose, Mass, Apr 11, 43; m 66; c 2. INVERTEBRATE ZOOLOGY, PHYSIOLOGICAL ECOLOGY. *Educ:* Univ NH, BA, 65; Fla State Univ, MS, 67; Univ Calif, Santa Barbara, PhD(invert zool), 70. *Prof Exp:* From asst prof to prof, 70-83, chmn dept biol, 84-90, DIR BELEWS LAKE BIOL STA, WAKE FOREST UNIV, 75- *Concurrent Pos:* Sabbatical, Univ Amsterdam, Neth, 83; consult, SC Comn Higher Ed, 88; pres, NC Acad Sci, 85-86. *Mem:* Am Soc Zool; Am Micros Soc; Sigma Xi. *Res:* Invertebrate chemical communication; physiological ecology of marine invertebrates; dynamics of host-symbiont interactions; invertebrate behavior. *Mailing Add:* Dept Biol Wake Forest Univ Winston-Salem NC 27109

DIMOND, EDMUNDS GREY, b St Louis, Mo, Dec 8, 18; m 44; c 3. CARDIOLOGY. *Educ:* Ind Univ, BS, 42, MD, 44. *Hon Degrees:* DSc, Hahnemann Grad Sch & Med Col. *Prof Exp:* Lectr, Sch Aviation Med, Randolph Field, 49; dir cardiovasc lab, Med Ctr, Univ Kans, 50-60, prof med & chmn dept, Sch Med, 53-60; dir, Inst Cardiopulmonary Dis, Scripps Clin & Res Found, 60-68; provost, 70-82, DISTINGUISHED PROF MED, UNIV MO, KANSAS CITY, 68-, EMER PROVOST HEALTH SCI, 82-

Concurrent Pos: Med consult, US Bur Labor, 55-60; Fulbright prof, Neth, 56; vis prof, Nat Heart Inst, London, 59; State Dept Am Specialists Abroad Prog, Philippines & Taiwan, 61, Columbia & Chile, 63 & Czech, Ceylon, Indonesia & Vietnam, 64; res assoc, Univ Calif, San Diego, 64-67, prof med-in-residence, 67-68; spec consult med educ to Under Secy Health, Dept HEW, 68-69; ed-in-chief, ACCEL, J in Cardiol, Am Col Cardiol, 69-77; Am specialist cardiol, People's Repub China, 71-72 & 74-85; scholar-in-residence, Rockefeller Found Ctr, Bellagio, Italy, 78; hon prof med, Shanghai Second Med Col, 80-; chmn am adv comt, Nat Heart Hosp China, 83-; mem foreign alumni comt, Chenghua Univ, Beijing, 85- *Mem:* Am Col Physicians; corresp mem Brit Cardiac Soc; Am Col Cardiol (pres, 61). *Res:* Cardiovascular physiology. *Mailing Add:* 2501 Holmes St Kansas City MO 64108

DIMOND, HAROLD LLOYD, b Paterson, NJ, May 1, 22; m 49; c 2. ORGANIC CHEMISTRY. *Educ:* NY Univ, AB, 48; Carnegie Inst Technol, MS, 52, PhD(chem), 53. *Prof Exp:* Asst chem, Carnegie Inst Technol, 48-50; res chemist, Pittsburgh Coke & Chem Co, 53-56 & Gulf Res & Develop Co, 56-66; from res chemist to sr res chemist, Hilton-Davis Chem Co, Sterling Drug, Inc, 66-69; sr res chemist, Drackett Co, Bristol-Myers Co, 69-77, sr scientist, 77-86; CONSULT, 77- *Mem:* Am Chem Soc. *Res:* Preparation of 3-pyrrolidones; industrial work on preparation of derivatives of coal tar chemicals; alkylation; petrochemicals; detergents; oxidation; general synthetic work; process development; product evaluation; new product and process development of household products. *Mailing Add:* 10629 Grimsby Lane Cincinnati OH 45241

DIMOND, JOHN BARNET, b Providence, RI, July 20, 29; m 55; c 2. ENTOMOLOGY, ECOLOGY. *Educ:* Univ RI, BS, 51, MS, 53; Ohio State Univ, PhD(entom), 57. *Prof Exp:* Sr entomologist, Maine State Forest Serv, 56-59; from asst prof to assoc prof entom, Univ Maine, 59-66, chmn dept, 74-76, PROF ENTOM, UNIV MAINE, 66-, COOP PROF FOREST RES, 77- *Mem:* AAAS; Entom Soc Am; Entom Soc Can. *Res:* Forest insects; insect ecology; populations. *Mailing Add:* Dept Entom 313 Deering Hall Univ Maine Orono ME 04473

DIMOND, MARIE THERESE, b Valdez, Alaska, Nov 13, 16. ZOOLOGY. *Educ:* Trinity Col, DC, AB, 38; Cath Univ, MS, 52, PhD(biol), 54. *Prof Exp:* Instr Ger, Trinity Col, DC, 38-39; teacher high schs, Md & Pa, 39-48; from instr to prof, 48-82, EMER PROF BIOL, TRINITY COL, DC, 82- *Concurrent Pos:* NSF sci fac fel marine biol, Marine Lab, Duke Univ, 59, Plymouth Lab, 60 & Hopkins Marine Sta, Stanford Univ, 61; res assoc, Univ BC, 70; vis prof, Bhopal Univ, India, 79-82 & Univ Bhubaneswar, India, 83, 84 & 85. *Mem:* AAAS; Am Inst Biol Sci; Am Soc Zoologists; assoc Am Physiol Soc; Sigma Xi. *Res:* Turtles: embryological development; histology of endorines and nerves; incubation and hatching of eggs; juvenile growth as related to diet, drugs, radiation and temperature; sex differentiation. *Mailing Add:* Dept Biol Trinity Col Washington DC 20017-1094

DIMOND, RANDALL LLOYD, b Salt Lake City, Utah, June 16, 46; m 68; c 8. MOLECULAR BIOLOGY, GENETICS. *Educ:* Univ Utah, BS, 70; Univ Calif, San Diego, PhD(biol), 75. *Prof Exp:* Fel molecular biol, Mass Inst Technol, 75-77; from asst prof to assoc prof microbiol, microbiol genetics & develop biol, 77-85, ADJ PROF MICROBIOL, MICROBIOL GENETICS & DEVELOP BIOL, UNIV WIS, MADISON, 85- *Concurrent Pos:* Vpres res, Promega Corp, 84; mem NIH Cell Biol & Physiol Study Sect, 82-86, chmn, 84-86; mem Govs Task Force Biotechnol, Wis, 86; adv Biotechnol, Cell Interaction & Develop Prog, Univ Calif, Irvine, 87- *Mem:* Am Soc Develop Biol; AAAS; Am Asn Clin Chemists. *Res:* Contributed to understanding of developmental gene regulation and the mechanisms responsible for the localization of proteins within cells. *Mailing Add:* 4409 Travis Terr Madison WI 53711

DIMOPOULLOS, GEORGE TAKIS, b Flushing, NY, Nov 24, 23; wid; c 1. PATHOGENESIS OF INFECTIOUS DISEASES. *Educ:* Pa State Univ, BS, 49, MS, 50; Mich State Univ, PhD(bact), 52; Am Bd Med Microbiol, dipl. *Prof Exp:* Lab asst bact, Pa State Univ, 47-48, asst, 49; asst virol, Mich State Univ, 50-52; res assoc, Univ Wis, 52-53; virologist animal dis & parasite res div, Agr Res Serv, USDA, 53-57; prof vet sci, La State Univ, 57-75; prof biol sci & chmn dept, Wright State Univ, 75-80, dean, Sch Sci Eng, 80-82; PROF BIOL SCI, UNIV ALA, HUNTSVILLE, 82- *Concurrent Pos:* Spec fel, NIH, 64-65. *Mem:* Fel AAAS; fel Am Acad Microbiol; Am Soc Microbiol; Soc Exp Biol & Med. *Res:* Finite purification of antigens for vaccines; Bordetella and Borrelia research. *Mailing Add:* Dept Biol Sci Univ Ala Huntsville AL 35899

DIMSDALE, BERNARD, b Sioux City, Iowa, Aug 3, 12; c 2. MATHEMATICS. *Educ:* Univ Minn, BCh, 33, AM, 35, PhD(math), 40. *Prof Exp:* Instr, Univ Idaho, 38-42; assoc instr, US War Dept, 42-43; instr math, Purdue Univ, 46; mathematician, Ballistic Res Lab, Aberdeen Proving Ground, 47-51 & Raytheon Mfg Co, Waltham, Mass, 51-56; sr mathematician, IBM Corp, 56-61, mgr, 61-69, indust consult, 69-77; CONSULT APPL MATH & DATA PROCESSING, 77- *Concurrent Pos:* Indust fel, IBM Corp, 74. *Mem:* Am Math Soc; Asn Comput Mach; Soc Indust & Appl Math. *Res:* Approximation theory; numerical analysis; optimization techniques; numerical geometry. *Mailing Add:* 319 Ninth St Santa Monica CA 90402

DIMUZIO, MICHAEL THOMAS, b Detroit, Mich, May 23, 49. EXPERIMENTAL BIOLOGY. *Educ:* Univ Mich, BS, 71; Northwestern Univ, PhD(biochem), 77. *Prof Exp:* Res asst, Dept Biochem, Northwestern Univ, 71-72, asst, 72-76; NIH-Fogarty fel, Pathophysiol Inst, Univ Bern, Switz, 77-79; res assoc, Dept Pediat Nephrology, Univ Minn, 79; res instr, Dept Biochem, Univ Ala, Birmingham, 79-82, res asst prof, 82-84; asst prof, Dept Biochem, Rush Med Col, Chicago, 84-88; PROJ MGR, ADD/OSTEOPOROSIS PROG, ABBOTT LABS, 87- *Concurrent Pos:* Numerous grants, foreign & US univs & insts, 77-90; vis asst prof, Dept Biochem, Rush Med Col, 88- *Mem:* AAAS; Am Soc Bone & Mineral Res; Orthop Res Soc; Sigma Xi; Am Soc Biol Chemists; Nat Osteoporosis Found. *Res:* Regulation and control of bone cell metabolism; non-collagenous proteins in matrix formation and their biosynthetic control by calcium metabolic hormones; author of numerous publications. *Mailing Add:* Div Diag Dept 90U AP-201401C Abbott Labs North Chicago IL 60064

DINA, STEPHEN JAMES, b Bronx, NY, May 2, 43; m 65; c 3. PLANT ECOLOGY, SCIENCE EDUCATION. *Educ:* Mt Union Col, BS, 65; Ohio State Univ, MS, 67; Univ Utah, PhD(biol), 70. *Prof Exp:* Asst prof, 70-75, ASSOC PROF BIOL, ST LOUIS UNIV, 75- *Concurrent Pos:* Environ Protection Agency grant, 73; Nat Endowment for the Humanities curric develop grant, 75-80. *Mem:* AAAS; Ecol Soc Am; Brit Ecol Soc; Am Inst Biol Sci; Am Soc Plant Physiol. *Res:* Implications of vascular plant distribution patterns expressed through understanding of the differential effects of the environment on plant physiology; water relations; organic energy balance; environmental stress conditions; science education. *Mailing Add:* Dept Biol St Louis Univ 3507 Laclede Ave St Louis MO 63103

DINAN, FRANK J, b Buffalo, NY, Dec 3, 33; m 54; c 2. ORGANIC CHEMISTRY, ANALYTICAL CHEMISTRY. *Educ:* State Univ NY Buffalo, BA, 59, PhD(chem), 64. *Prof Exp:* Jr chemist anal chem, Carborundum Metals Co, 53-59; chemist, Hooker Chem Corp, 59-61; res chemist org chem, E I du Pont de Nemours & Co, 63-64; res assoc anal chem, Cornell Univ, 64-66; from asst prof to assoc prof chem, 66-74, chmn dept, 70-74, PROF CHEM, CANISIUS COL, 74- *Mem:* Am Chem Soc. *Res:* Heterocyclic chemistry and instrumental methods of analysis and structure determination. *Mailing Add:* 146 Calvert Blvd Tonawanda NY 14150-4702

DINAPOLI, FREDERICK RICHARD, b Providence, RI, Nov 19, 40; m 61; c 3. ACOUSTICS. *Educ:* Univ RI, BS, 62, MS, 65, PhD(elec eng), 69. *Prof Exp:* Scientist comput, Pratt & Whitney Aircraft, 62-63; res scientist acoust, Raytheon Marine Res Lab, 67-68; res scientist acouts, Naval Underwater Systs Ctr, 70-86; SR SCIENTIST, SCI APPL INT CORP, 86- *Concurrent Pos:* Adj prof acoust, Univ RI, 71- *Mem:* Assoc Acoust Soc Am. *Res:* All aspects of the study and prediction of the transmission of underwater acoustic energy. *Mailing Add:* Box 338 RR 1 School Dr Westerly RI 02891

DINARDI, SALVATORE ROBERT, b New York City, NY, July 3, 43; m 79; c 2. INDOOR AIR POLLUTION, INDUSTRIAL HYGIENE. *Educ:* Hofstra Univ, BA, 65; Univ Mass, PhD(phys chem), 71. *Prof Exp:* Instr environ health, 70-71, asst prof, 71-75, ASSOC PROF ENVIRON HEALTH & INDUST HYGIENE, UNIV MASS, 75- *Concurrent Pos:* Vis prof, Univ NC, 79-80; indust hygiene consult, 79-; indust ventilation design continuing educ, 79-; fel, Univ NC, 79-80. *Mem:* Nat Environ Health Asn; Am Indust Hyg Asn; Am Public Health Asn; Soc Occup Environ Health; Am Chem Soc. *Res:* Characterizing the quality of indoor air and controlling indoor air pollution in private residences; impact of passive solar design on air quality; industrial hygiene. *Mailing Add:* Div Pub Health Univ Mass Amherst Campus Amherst MA 01003

DINBERGS, KORNELIUS, b Riga, Latvia, June 5, 25; nat; m 58; c 3. ORGANIC POLYMER CHEMISTRY. *Educ:* Ukrainian Tech Inst, Ger, MagPharm, 49; Baylor Univ, MA, 51; Purdue Univ, PhD(org chem), 56. *Prof Exp:* From tech man to sr tech man, 56-61, sr res chemist, B F Goodrich Co, 61-84; res chemist, Repub Powdered Metals, 84-88; CHEMIST III-RES, SHERWIN-WILLIAMS CO, 89- *Mem:* Am Chem Soc. *Res:* Polyurethanes, polymerization, structure-property relationships; diisocyanate and polyester preparation and stability; peroxide cures; vinyl chloride polymerization; paints, coatings and sealants; asphalt and aluminum roof coatings. *Mailing Add:* 3471 Ridge Park Dr Cleveland OH 44147

DINDAL, DANIEL LEE, b Findlay, Ohio, Sept 17, 36; m 58; c 3. ECOLOGY. *Educ:* Ohio State Univ, BS(wildlife mgt) & BS(sci educ), 58, MA, 61, PhD(ecol), 67. *Prof Exp:* Teacher high sch, Ohio, 58-63; teaching asst zool & ecol, Ohio State Univ, 63-64, US Fish & Wildlife Serv res assoc, 64-66; asst prof terrestrial invertebrate ecol, 67-71, assoc prof, 71-77, prof terrestrial invertebrate ecol, 77-80, PROF, DEPT ENVIRON & FOREST BIOL, COL ENVIRON SCI & FORESTRY, STATE UNIV NY, 80- *Concurrent Pos:* Acarology Inst traineeship adv acarine taxon, Ohio State Univ, 68 & 70; AEC res grant, 68-73; New Eng Interstate Water Pollution Control Comn res grant, 72-73; US Forest Serv res grant, 72-74. *Mem:* Ecol Soc Am; Soil Sci Soc Am; Am Soc Zoologists; Entom Soc Am; Acarological Soc Am. *Res:* Ecology of soil invertebrates in natural and manipulated terrestrial microcommunities; effects of dichloro-diphenyl-trichloroethane on invertebrate microcommunities and species diversity; effects of municipal wastewater disposal and urban street salting on soil invertebrates; succession and interspecific relationships of invertebrates in bird nests, carrion and soil litter. *Mailing Add:* State Univ NY Col Environ Sci & Forestry Syracuse NY 13210

DINEGAR, ROBERT HUDSON, b New York, NY, Dec 18, 21; wid; c 3. INORGANIC & ANALYTICAL CHEMISTRY. *Educ:* Cornell Univ, AB, 42; Columbia Univ, AM, 48, PhD(phys chem), 51, Col Santa Fe, AB, 76. *Prof Exp:* Asst chem, Columbia Univ, 46-50; res phys chemist, Los Alamos Nat Lab, Univ Calif, 50-87; fac mem chem, 87-89, SCI COORDR, LOS ALAMOS BR COL, UNIV NMEX, 89- *Concurrent Pos:* Adj prof chem, Univ NMex, Northern Br Col, 70-77; staff mem, Northern NMex Community Col, 77-78; mem, Shroud Turin Res Proj. *Mem:* Am Chem Soc; fel Am Inst Chem; prof mem Am Meteorol Soc; Sigma Xi. *Res:* Kinetics of phase changes and explosions; physics and chemistry of explosives; properties and reactivity of small particles. *Mailing Add:* 15 Tesuque Los Alamos NM 87544

DINER, BRUCE AARON, biophysics, biochemistry, for more information see previous edition

DINER, WILMA CANADA, b Monaville, WVa, Jan 21, 26; m 55; c 3. RADIOLOGY. *Educ:* Univ Ky, BS, 46; Duke Univ, MD, 50. *Prof Exp:* Intern med, Duke Hosp, 50; asst resident path, Mass Gen Hosp, Boston, 51, asst res resident radiol, 52-54; assoc radiologist, St Luke's Hosp, New Bedford, 55-56; from asst prof to assoc prof, 56-70, dir diag radiol residency prog, 74-89, PROF RADIOL, UNIV ARK, MED SCI, LITTLE ROCK, 70- *Concurrent Pos:* Consult, Ark Children's Hosp, Little Rock, 62-65 & Little Rock AFB Hosp, 62-65. *Honors & Awards:* Marie Curie Award, Am Asn Women

Radiologists, 88. *Mem:* Fel Am Col Radiol; Asn Univ Radiol; Am Roentgen Ray Soc; Radiol Soc NAm; Soc Gastrointestinal Radiologists. *Res:* Diagnostic radiology, especially gastrointestinal and mammography; Cronkhite-Canada syndrome. *Mailing Add:* Dept Radiol Univ Ark Med Sci Little Rock AR 72205

DINERSTEIN, ROBERT ALVIN, b New York, NY, Jan 4, 19; m 41; c 3. CHEMISTRY. *Educ:* City Col New York, BS, 39; Pa State Univ, MS, 40. *Prof Exp:* Chemist, Am Oil Co, 46-52, from group leader to sect leader, 52-62, mgr info & commun, 62-68, dir anal res, 68-70, mgr compensation & orgn, Am Oil Co, 70-73; dir orgn planning, Standard Oil Co, Ind, 73-76, orgn & domestic compensation, 76-84; RETIRED. *Concurrent Pos:* Harvey Washington Wiley lectr, Purdue Univ, 57. *Mem:* AAAS; Am Chem Soc. *Res:* Hydrocarbon reactions, properties, separation and analysis; distillation; chromatography; information science; documentation. *Mailing Add:* 345 Oakwood St Park Forest IL 60466

DINERSTEIN, ROBERT JOSEPH, b Little Rock, Ark, Mar 5, 44; m 76. CELL BIOLOGY. *Educ:* Harvard Univ, BA, 66; Univ Ore, PhD(org chem), 71; Univ Chicago, MD, 76. *Prof Exp:* Res assoc, dept pharmacol & physiol sci, Univ Chicago, 76-78, asst prof pharmacol, 79-85; AT MARION MERRELL DOW INC, 85- *Mem:* Am Chem Soc; Soc Neurosci; Am Soc Pharmacol & Exp Therapeut. *Res:* Cardiovascular diseases; acute inflammation. *Mailing Add:* Marion Merrell Dow Res Inst 2110 E Galbraith Rd Cincinnati OH 45215

DINES, ALLEN I, b Pittsburgh, Pa, Dec 16, 29; m 53; c 2. PHARMACEUTICAL CHEMISTRY, MEDICINAL CHEMISTRY. *Educ:* Univ Pittsburgh, BS, 51; Ohio State Univ, MS, 53, PhD(pharmaceut chem), 58. *Prof Exp:* Assoc dir prod develop, Flint Labs, Baxter Labs, Inc, Ill, 58-60; group leader pharmaceut res & develop, Miles Prod Div, Miles Labs, Inc, Ind, 60-64; lab dir, Warren-Teed Pharmaceut Inc, Rohm and Haas Co, 64-69; dir res, Strong Cobb Arner, Inc, 69-70; dir res & develop, Mylan Pharmaceut, Inc, 71-76; TECH MGR DEVELOP PROD, RICHARDSONS VICKS, A PROCTOR & GAMBLE CO. *Mem:* Am Chem Soc; Am Pharmaceut Asn; Acad Pharmaceut Sci. *Res:* Pharmaceutical research and development; pharmaceutics. *Mailing Add:* 27 Ironwood Dr Danbury CT 06811

DINES, MARTIN BENJAMIN, b Pittsburgh, Pa, May 9, 43; m 64; c 1. ORGANOMETALLIC CHEMISTRY, INORGANIC CHEMISTRY. *Educ:* Carnegie-Mellon Univ, BS, 64; Univ Ill, PhD(chem), 68. *Prof Exp:* Researcher chem, Exxon Res & Eng Co, Linden, NJ, 68-77; GROUP LEADER, OCCIDENTAL RES CORP, 77- *Mem:* Am Chem Soc. *Res:* New materials, especially solid-state, having application in areas of catalysis and energy storage; layered compounds, and separations. *Mailing Add:* Occidental Res Corp 10889 Wilshire Blvd Los Angeles CA 90024-4201

DINESS, ARTHUR M(ICHAEL), b New York, NY, Apr 21, 38; m 89; c 2. ENGINEERING SCIENCE, TECHNICAL MANAGEMENT. *Educ:* NY Univ, BA, 58; Pa State Univ, MS, 60, PhD(solid state sci), 66. *Prof Exp:* Asst chem, Pa State Univ, 58-60, Mat Res Lab, 60-64; res specialist mat res, Appl Res Lab, Philco Corp, 64-65, supvr, Solid State Mat Sect, 65-66; chemist, Metall Prog, Off Naval Res, 66-77, dir, Metall & Ceramics Prog, 78-80, dir, Mat Sci Div, 80, assoc dir res eng sci, 81-91, SCI DIR, OFF NAVAL RES, EUROP OFF, 91- *Concurrent Pos:* Res scientist, Ford Sci Lab, Pa, 66; guest scientist, Nat Bur Stand, 72-90; liaison scientist, Off Naval Res, Tokyo, 78. *Mem:* Fel Am Ceramic Soc; AAAS. *Res:* Materials research and engineering; ceramics and glasses. *Mailing Add:* Off Naval Res Arlington VA 22217-5000

DING, JOW-LIAN, b Taiwan, Repub China, Nov 7, 51; m 78; c 2. HIGH TEMPERATURE ENGINEERING DESIGN, HIGH TEMPERATURE MATERIAL TESTING & EVALUATION. *Educ:* Nat Taiwan Univ, BSc, 74; Brown Univ, MSc, 78, PhD(solid mech), 83. *Prof Exp:* Mech engr, China Tech Consult Inc, 76-77; teaching & res asst mech eng, Nat Taiwan Univ, 77-78; res asst high temperature creep eng mat, Brown Univ, 78-83; asst prof, 83-89, ASSOC PROF MECH ENG, WASH STATE UNIV, 89- *Concurrent Pos:* Consult, Potlatch Corp, Lewiston, Idaho, 89; vis res staff mem, Oak Ridge Nat Lab, 90-91. *Mem:* Am Soc Mech Engrs; Soc Exp Mech; Am Acad Mech. *Res:* Advanced constitutive equations for high temperature engineering analysis; high temperature fracture and damage mechanics; optical experimental techniques for high temperature applications; high temperature material testing and evaluations; numerical stress analysis based on advanced constitutive equations characterization of material damage by structural modal analysis. *Mailing Add:* Dept Mech & Mat Eng Wash State Univ Pullman WA 99164-2920

DING, VICTOR DING-HAI, b Tsing-Tao, China. BIOCHEMISTRY. *Educ:* Chang-Hsing Univ, BS, 65; Okla State Univ, MS, 72; Univ Ill, MS, 74, PhD(immunol), 78. *Prof Exp:* Res asst fungal dis plants, Okla State Univ, 70-72; teaching asst microbiol, Univ Ill, 72-73, res asst, 73-74, res asst immunol, 75-78; sr res scientist & group leader, Clin Sci Inc, Whippany, NJ, 78-81; in-chg immunoassay group, Cytogen Corp, 82-83; postdoctoral fel, 83-86, SR RES BIOCHEMIST, MERCK & CO, 86- *Mem:* Am Soc Biochem & Molecular Biol. *Res:* Protein purification, assays and screen for enzyme inhibitors; isolation, cultivation and identification of fungi and bacteria; author of numerous publications. *Mailing Add:* Dept Biochem Merck & Co Inc PO Box 2000 80Y-140 Rahway NJ 07065

DINGA, GUSTAV PAUL, b Haugen, Wis, Oct 6, 22; m 48; c 4. INORGANIC CHEMISTRY. *Educ:* St Olaf Col, BA, 47; Univ Louisville, MS, 49; Univ Wyo, PhD(inorg chem), 62. *Prof Exp:* Asst, Univ Ky, 49-50; instr chem, Ill Wesleyan Univ, 50-51 & St Cloud Teachers Col, 51-53; from asst prof to assoc prof, 53-64, chmn dept, 69-72 & 81-87, PROF CHEM, CONCORDIA COL, MOORHEAD, MINN, 64- *Concurrent Pos:* Researcher, Univ Utah, 60 & Nuclear Sci Inst, Wash State Univ, 62; resident res assoc, Argonne Nat Lab, 63; vis prof, Augustana Col, SDak, 64, 65 & 66, Calif Luthern Col, Thousand Oaks, 81, Univ Nebr, Lincoln, 84 & NDak State Univ, 87-88; fac res appointee, Pac Northwest Labs, Battelle Mem Inst, 68-69; mem grad fac, NDak State Univ, 71; Ames lab res, Iowa State Univ, 77; consult, Tex Mining Oper, 83; vis prof, polymer chem, NDak State Univ, 87-88. *Mem:* Am Chem Soc; fel Am Inst Chem; Sigma Xi; Int Asn Hydrogen Energy. *Res:* Inorganic and organic chemistry; synthesis and identification of organic foam inhibiting compounds and p-tolylmethyl ether; determination of physical constants for non-aqueous solvents at very low temperature; zirconium compounds and complexes; hydrogen as fuel of the future; inorganic polymer chemistry. *Mailing Add:* Dept Chem Concordia Col Moorhead MN 56562

DINGELL, JAMES V, b Detroit, Mich, Oct 10, 31; m 80; c 3. PHARMACOLOGY, CHEMISTRY. *Educ:* Georgetown Univ, BS, 54, MS, 57, PhD(chem), 62. *Prof Exp:* Chemist, Lab Chem Pharmacol, Nat Heart Inst, 55-62; from instr to assoc prof pharmacol, Sch Med, Vanderbilt Univ, 62-79; mem staff, Lab Chem Pharmacol, Nat Cancer Inst, 76-79; scientist adminr, Nat Heart, Lung & Blood Inst, 79-90; DIR, DIV PRE CLIN RES, NAT INST DRUG ABUSE, 91- *Mem:* Am Soc Pharmacol & Exp Therapeut; Am Col Neuropsychopharmacol. *Res:* Drug metabolism, distribution in tissues, assay and mechanism of action, especially in area of psychotherapeutic agents. *Mailing Add:* Nat Inst Drug Abuse 5600 Fishers Lane Rm 10A31 Rockville MD 20857

DINGER, ANN ST CLAIR, b Eau Claire, Wis, Oct 16, 45; m 77; c 2. COMPUTER MODELING. *Educ:* Vassar Col, AB, 67; Northwestern Univ, MS, 68, PhD(astron), 71. *Prof Exp:* Asst prof physics, Univ Wis-Eau Claire, 71-74; asst prof astron, Wellesley Col, 74-78; vis lect astron, Williams Col, 78-79; res affil, Jet Propulsion Lab, 79-82; RES AFFIL, STERLING SOFTWARE, NASA-AMES, MOFFET FIELD, CALIF, 83- *Mem:* Am Astron Soc; Royal Astron Soc; Sigma Xi; Int Soc Optical Eng. *Res:* Computer models of the interior structure and evolution of helium stars, matching the models to the observed stars; radio molecular line observations of cool stars and interstellar gas clouds. *Mailing Add:* MS-244-10 NASA-Ames Moffet Field CA 94035

DINGES, DAVID FRANCIS, b Hays, Kans, May 30, 49; m 71; c 2. PSYCHOPHYSIOLOGY, BIOLOGICAL RHYTHMS. *Educ:* St Benedicts Col,AB, 71; St Louis Univ, MS, 74, PhD(psychol), 76. *Prof Exp:* Instr psychol, Florrisant Valley Community Col, 74-75; res psychol, Nat Med Ctr, Childrens Hosp, 75-79; clin asst prof, 80-86, CLIN ASSOC PROF PSYCHIAT, UNIV PA, MED SCH, 86-; RES PSYCHOL, INST PA HOSP, 79- *Concurrent Pos:* Asst prof & lectr, George Wash Univ, 76-79; prin investr, Off Naval Res, 80-; co-dir, Unit Exp Psychiat, Pa Hosp, 82-; Nathan Lewis Hatfield Lectr, Col Physicians, Philadelphia; chmn, gov affairs, Am Sleep Dis Asn, 90- *Honors & Awards:* Roy M Dorcus Award, Soc Clin & Exp Hypnosis. *Mem:* AAAS; Am Psychol Asn; Sleep Res Soc; Aerospace Med Soc; Am Sleep Dis Asn; Human Factors Soc. *Res:* Human (infant and adult) psychophysiology, primarily in the study of sleep/wake cycles; sleep deprivation, performance and mood, napping, apnea, psychoimmunology of sleep, shiftwork, jet lag, biological rhythms. *Mailing Add:* Unit Exp Psychiat Inst Pa Hosp 111 N 49th St Philadelphia PA 19139

DINGLE, ALBERT NELSON, b Bismarck, NDak, May 22, 16; m 41; c 2. METEOROLOGY, CLOUD PHYSICS. *Educ:* Univ Minn, BS, 39; Iowa State Col, MS, 40; Mass Inst Technol, ScD, 47. *Prof Exp:* asst prof physics, Hampton Inst, 41-42; res assoc meteorol, Mass Inst Technol, 43-47; asst prof physics & meteorol, Ohio State Univ, 47-54; assoc res meteorologist, Eng Res Inst, 54-56, from assoc prof to prof, 56-81, EMER PROF METEOROL, UNIV MICH, 81- *Concurrent Pos:* Res assoc, Mapping & Charting Res Lab, Ohio State Univ Res Found, 50-54; consult & expert witness, NASA, 74-81. *Mem:* fel AAAS; Am Geophys Union. *Res:* Rain cleansing of the atmosphere; drop-size distributions in rain; study of convective storms by use of tracer indium; air and precipitation chemistry; waste disposal capacity of the atmosphere; cloud microphysics and modeling of convective rain-generating systems. *Mailing Add:* 8140 Heron River Dr Dexter MI 48130

DINGLE, ALLAN DOUGLAS, b Hamilton, Ont, May 3, 36; m 62. DEVELOPMENTAL BIOLOGY, CELL BIOLOGY. *Educ:* McMaster Univ, BSc, 58; Univ Ill, MSc, 60; Brandeis Univ, PhD(biol), 64. *Prof Exp:* NIH trainee develop biol, 64-65; asst prof, 65-75, ASSOC PROF BIOL, MCMASTER UNIV, 75- *Mem:* AAAS; Am Soc Cell Biol; Soc Develop Biol; Can Soc Cell Biol. *Res:* Organelle counting and sizing controls operating during cell differentiation; orientation of cilia and flagella by striated roottet fibres; the role of intermicrotubule linkers in maintenance of cell shape. *Mailing Add:* Dept Biol McMaster Univ 128 Main St W Hamilton ON L8S 4K1 Can

DINGLE, RAYMOND, b Perth, Australia, Sept 5, 35; m 58; c 3. DEVICE PHYSICS. *Educ:* Univ Western Australia, BSc, 55, Hons, 57, PhD(chem), 65. *Prof Exp:* Sr demonstr phys chem, Univ Western Australia, 59-62; fel chem, Univ Col, London, 62-63, amanuensis phys chem, Copenhagen Univ, 63-66 & lektor solid state chem, 64-66; mem tech staff solid state physics, Bell Labs, 66-79, suprv, explor high speed III-V device group, 79-; mem staff, Pivot III-V Corp; MEM STAFF, SEMICONDUCTOR INC. *Concurrent Pos:* Sr fel, Dept Solid State Physics, Res Sch Phys Sci, Australian Nat Univ, 72-73. *Mem:* AAAS; Am Phys Soc. *Res:* Electronic properties of solids; III-V semiconductor device studies; III-V integrated circuit technology; quantum effects in ultra thin semiconductor hetero-structures. *Mailing Add:* Pac Semiconductor Inc One Euclid Ave Summit NJ 07901

DINGLE, RICHARD DOUGLAS HUGH, b Penang, Malaya, Nov 4, 36; US citizen; m 59; c 3. GENETICS. *Educ:* Cornell Univ, BA, 58; Univ Mich, MSc, 59, PhD(zool), 62. *Prof Exp:* Instr zool, Univ Mich, 61-62; NSF fel, Cambridge Univ, 62-63; NIMH training fel, Univ Mich, 63-64; from asst prof to assoc prof zool, Univ Iowa, 64-82; PROF ENTOM, UNIV CALIF, DAVIS, 82- *Concurrent Pos:* NIH fel, Univ Nairobi, 69-70. *Mem:* AAAS; Am Soc Zool; Am Soc Naturalists; Animal Behav Soc; Ecol Soc Am; Sigma Xi; Soc Study Evolution; Europ Soc Evolutionary Biol; Int Soc Behav Ecol. *Res:* Insect behavior and populations; insect migration and life histories. *Mailing Add:* Dept Entom Univ Calif Davis CA 95616

DINGLE, RICHARD WILLIAM, b Bismarck, NDak, Jan 5, 18; m 47; c 2. SILVICULTURE. *Educ:* Univ Minn, BS, 41; Yale Univ, MF, 47, PhD(forestry), 53. *Prof Exp:* Instr forestry, Univ Mo, 48-53; from asst prof to prof, 53-83, EMER PROF FORESTRY, WASH STATE UNIV, 83- *Mem:* AAAS; Am Forestry Asn; Soc Am Foresters. *Res:* Artificial regeneration and grafting of ponderosa pine in eastern Washington; vegetative regeneration of conifers; windbreak establishment and culture in Columbia Basin and Eastern Washington drylands; Christmas tree production and culture; forest ecology; forest tree physiology; forest genetics; silviculture. *Mailing Add:* SE 625 Side Pullman WA 99163

DINGLE, THOMAS WALTER, b Burnaby, BC, Aug 6, 36; m 63; c 2. THEORETICAL CHEMISTRY. *Educ:* Univ Alta, BSc, 58, PhD(phys chem), 65. *Prof Exp:* Ramsay Mem fel chem, Math Inst, Oxford Univ, 64-66; fel, Univ Ottawa, 66-67; asst prof, 67-81, ASSOC PROF CHEM, UNIV VICTORIA, 81- *Mem:* Chem Inst Can; Can Asn Physicists. *Res:* Molecular orbital calculations. *Mailing Add:* Dept Chem Univ Victoria Box 3055 Victoria BC V8W 3P6 Can

DINGLEDINE, RAYMOND J, b Celina, Ohio, Dec 17, 48; m 71; c 2. NEUROPHYSIOLOGY, NEUROPHARMACOLOGY. *Educ:* Mich State Univ, BS, 71; Stanford Univ, PhD(pharmacol), 75. *Prof Exp:* Fel neuropharmacol, Med Res Coun Neurochem, Pharmacol Unit, Univ Cambridge, Eng, 75-77; fel neurophysiol, Neurophysiol Inst, Univ Oslo, Norway, 77-78; res assoc neurophysiol, Dept Physiol, Duke Univ, 78; asst prof neuropharmacol, 78-83, ASSOC PROF PHARMACOL, UNIV NC, CHAPEL HILL, 84- *Concurrent Pos:* Mem panel grant rev, NSF, 83-86, NIH, 87- *Mem:* Soc Neurosci; Am Soc Pharmacol & Exp Therapeut; Brit Pharmacol Soc; Europ Neurosci Asn. *Res:* Cellular mechanisms (electrophysiology) of opioid actions; synaptic pathology in experimental models of epilepsy; excitatory amino acids; patch clamp studies of drug receptors. *Mailing Add:* Dept Phamacol Univ NC CB 7365 FLOB Chapel Hill NC 27599

DINGLEDY, DAVID PETER, b Youngstown, Ohio, Mar 11, 19; m 44; c 8. PHYSICAL CHEMISTRY, PHOTOCHEMISTRY. *Educ:* John Carroll Univ, BS, 40; Marquette Univ, MS, 42; Ohio State Univ, PhD(phys chem), 62. *Prof Exp:* Res chemist, Owens-Corning Fiberglas, 43-53; res assoc cryog, Res Found, Ohio State Univ, 53-55; teacher high schs, Ohio, 55-57; instr chem, Ohio Wesleyan Univ, 57-59; ASSOC PROF CHEM, STATE UNIV NY COL FREDONIA, 62- *Mem:* Am Chem Soc; Sigma Xi. *Res:* Physical chemistry of glass; structure of inorganic glasses. *Mailing Add:* Dept Chem State Univ NY Col Fredonia NY 14063

DINGMAN, CHARLES WESLEY, II, b Springfield, Mass, June 2, 32; m 54; c 3. HOSPITAL ADMINISTRATION, CLINICAL PSYCHIATRIC RESEARCH. *Educ:* Dartmouth Col, AB, 54; Univ Rochester, MD, 59. *Prof Exp:* Res assoc neurochem, Dept Psychiat, Sch Med & Dent, Univ Rochester, 57-58; res assoc biochem, Nat Inst Neurol Dis & Blindness, 60-64; res med officer, Nat Cancer Inst, 64-75; resident in psychiat, Northern Va Ment Health Inst, 75-77; asst clin dir, 84-90, CLIN DIR, CHESTNUT LODGE HOSP, 90-, STAFF PSYCHIATRIST, CHESTNUT LODGE INC, 77- *Mem:* AAAS; Am Soc Biol Chem; Am Psychiat Asn. *Res:* Biochemical studies of nucleic acid synthesis and metabolism; effect of carcinogens on nucleic acid structure and metabolism; DNA repair mechanisms; discriminating features of patients who commit suicide. *Mailing Add:* Chestnut Lodge Hosp 500 W Montgomery Ave Rockville MD 20850

DINGMAN, DOUGLAS WAYNE, b Leon, Iowa, Sept 12, 53. MICROBIOLOGY. *Educ:* Univ Iowa, BS, 75, PhD(microbiol), 83. *Prof Exp:* Teaching asst gen microbiol, Univ Iowa, 78-83; res assoc, Dept Molecular Biol & Microbiol, Tufts Univ, 83-87; RES SCIENTIST, CONN AGR EXP STA, 87- *Mem:* Am Soc Microbiol; AAAS. *Res:* Development of molecular tools and techniques; investigation of the molecular mechanism of pathogenicity of the entomopathogen Bacillus popilliae. *Mailing Add:* Conn Agr Exp Sta 123 Huntington St New Haven CT 06504

DINGMAN, JANE VAN ZANDT, b Summit, NJ, Jan 5, 31; m 53, 74; c 2. EVOLUTIONARY BIOLOGY, PSYCHOBIOLOGY. *Educ:* Wellesley Col, BA, 53; Yale Univ, PhD(zool), 57. *Prof Exp:* NSF fel, Oxford Univ, 57-58; instr zool, Mt Holyoke Col, 58-59; res assoc biol, Amherst Col, 59-71; res dir, Mass Pub Interest Res Group, 72-73, environ res consult, 73-75; lectr biol, Univ NH, 76-78, res assoc, Family Res Lab, 79-80; RES CONSULT, 81- *Concurrent Pos:* NIH spec fel, Oxford Univ, 63-64; mem bd dirs, Conn River Ecol Action Corp; chmn, Conn River Task Force; mem sci adv group, Conn River Basin Prog-New Eng River Basins Comn; sci adv, New Eng Coal-Coalition Nuclear Pollution & For Land's Sake; spec adv, Restoration of Atlantic Salmon in Am, 70-74 & Strafford County Task Force on Family Violence, 78-79. *Mem:* Am Soc Naturalists; Lepidopterists Soc; Sigma Xi. *Res:* Origin and maintenance of mimicry in butterflies; ecological genetics of natural populations; interaction of social and biological aspects of human menstrual cycle. *Mailing Add:* HC71-Box 78 Center Strafford NH 03815

DINGMAN, STANLEY LAWRENCE, b Jersey City, NJ, Jan 31, 39; m 74; c 2. HYDROLOGY, RESOURCE MANAGEMENT. *Educ:* Dartmouth Col, AB, 60; Harvard Univ, AM, 61, PhD(geol), 70. *Prof Exp:* Res hydrologist, US Army Cold Regions Res & Eng Lab, 63-69; vis asst prof environ studies & earth sci, Dartmouth Col, 69-72; dir res water resources, Dubois & King, Inc Environ Engrs, 72-73; sr resource planner, Conn River Basin Prog, New Eng River Basins Comn, 73-75; PROF HYDROL & WATER RESOURCES, CHMN DEPT EARTH SCI, UNIV NH, 75-; VPRES, HYDROSCI ASSOCS, INC, 80- *Mem:* Am Geophys Union; AAAS; Am Water Resources Asn; Sigma Xi; Geol Soc Am; Am Inst Hydrol. *Res:* Hydrology of New England; fluvial hydrology; water quantity/quality management; hydrology of arctic and subarctic regions. *Mailing Add:* Star Rte 126 Box 78 Center Strafford NH 03815

DINGUS, RONALD SHANE, b Appleton City, Mo, Sept 17, 38; m 58; c 2. HIGH TEMPERATURE PHYSICS & CHEMISTRY. *Educ:* Univ Mo, BS, 60; Iowa State Univ, PhD(nuclear physics), 65. *Prof Exp:* Res asst, Ames Lab, Iowa State Univ, 60-65, res assoc, 65; res assoc, Physics Inst, Aarhus Univ, 65-66 & Nuclear Physics Lab, Univ Colo, Boulder, 66-68; MEM STAFF, LOS ALAMOS NAT LAB, 68- *Concurrent Pos:* Fulbright travel grant, Physics Inst, Aarhus Univ. *Mem:* Am Phys Soc; Sigma Xi; AAAS. *Res:* X-ray and gamma-ray spectroscopy; x-ray vulnerability of strategic reentry vehicle systems; dynamic response of materials to pulsed heating. *Mailing Add:* 417 E Stante Los Alamos NM 87544

DINKINS, REED LEON, b Whitehall, Ark, June 10, 38; m 60; c 1. ENTOMOLOGY. *Educ:* Ark Polytech Col, BS, 62; Univ Ark, MS, 66; Miss State Univ, PhD(entom), 69. *Prof Exp:* Surv entomologist, Miss State Univ, 66-69; assoc prof, 69-80, PROF ZOOL & ENTOM, PITTSBURG STATE UNIV, 80- *Mem:* Entom Soc Am; Cent States Entom Soc. *Res:* Insect pests of the Pecan. *Mailing Add:* Dept Biol Pittsburg State Univ Pittsburg KS 66762

DINMAN, BERTRAM DAVID, b Philadelphia, Pa, Aug 9, 25; m 50; c 4. OCCUPATIONAL MEDICINE, ENVIRONMENTAL MEDICINE. *Educ:* Temple Univ, MD, 51; Univ Cincinnati, ScD(occup med), 57; Am Bd Prev Med, dipl, 60. *Prof Exp:* From asst prof to prof prev med, Col Med, Ohio State Univ, 57-65; prof environ & indust health, Univ Mich, Ann Arbor, 65-73; med dir, 73-78, vpres health & safety, Aluminum Co Am, 78-87; CLIN PROF OCCUP MED, GRAD SCH PUB MED, UNIV PITTSBURGH, 87- *Concurrent Pos:* Mem comn environ hyg, Armed Forces Epidemiol Bd, 65-73; mem comt toxicol, Nat Res Coun, 65-70, chmn, 72-77, mem comt biol effects of atmospheric pollutants, 69-71; dir, Inst Environ & Indust Health, Univ Mich, Ann Arbor, 70-73; expert advisor, WHO, 71-; US delegation, Int Labor Org, 80-81 & 84-85; vchmn, Am Bd Prev Med, 85-86; vis fel, Green Col, Oxford Univ, 86- *Honors & Awards:* Robert A Kehoe Award, 72; William S Knudsen Award, 88. *Mem:* Fel Am Col Occup Med (pres, 73-75); fel Am Col Prev Med; Soc Toxicol. *Res:* Occupational and environmental toxicology. *Mailing Add:* Dept Indust Environ Health Sci Grad Sch Pub Med Univ Pittsburgh Pittsburgh PA 15261

DINNEEN, GERALD PAUL, b Elmhurst, NY, Oct 23, 24; m 47; c 3. MATHEMATICS. *Educ:* Queens Col, BS, 47; Univ Wis, MS, 48, PhD(math), 52. *Prof Exp:* Teaching asst math, Univ Wis, 47-51; sr develop engr, Goodyear Aircraft Corp, Ohio, 51-53; staff mem, Lincoln Lab, Mass Inst Technol, 53-58, sect leader data processing group, 58, from asst leader to leader, 58-60, assoc head info processing div, 60-63 & commun div, 63-64, head, 64-66, from asst dir to assoc dir, 66-70, dir, 70-77, prof elec eng, 71-77; asst secy defense for commun, command, control & intelligence, Dept Defense, 77-81; corp vpres sci & technol, Honeywell, Inc, 81-89; FOREIGN SECY, NAT ACAD ENG, 88- *Concurrent Pos:* Consult, Air Force Sci Adv Bd, 59-60, mem electronics panel, 60-65 & chmn info processing panel, 63-65; mem, Defense Intel Agency Sci Adv Comt, 65-66, vchmn, 66-73; mem, Defense Sci Bd Panels, 66-67; chmn, US Air Force Sci Adv Bd, 75-77; mem bd dirs, Votan & Sci Mus Minn, 81-87, Microelectronics & Comput Technol Corp, 82-89, Honeywell Found, 83-87 & Coun Nat Acad Eng, 84-, Nat Comt US-China Rel, Inc, 86-, vis comt, Dept Math, Mass Inst Technol, 82, Georgetown Univ, bd visitors, 87, Technol Assessment Comt, US Space Comm and Air Force Studies Bd, Comn Eng & Tech Syst, Nat Res Coun, 87-88; hon Trustee, Sci Mus Minn, 87; mem bd dirs, Corp Open Syst, 87-89, Honeywell-NEC Supercomputers, 87-89, bd trustees, Queens Col Found, 88. *Mem:* Nat Acad Eng; Am Math Soc. *Res:* Matrix algebra; probability theory; logical design of digital computers; military systems design; satellite communications. *Mailing Add:* 7611 Gleason Rd Edina MN 55435

DINNEEN, GERALD UEL, b Denver, Colo, May 6, 13; m 38; c 2. CHEMISTRY. *Educ:* Univ Denver, BS, 34. *Prof Exp:* Chemist, Richards Labs, Colo, 34-38; soil conserv serv, USDA, NMex, 38-42 & US Customs Serv, Calif, 42-44; chemist in charge lab invests, US Maritime Comn, 44-45; res chemist & supvr, Laramie Energy Res Ctr, US Bur Mines, 45-64, res dir, 64-75; CONSULT ENGR FOSSIL FUELS, 75- *Mem:* AAAS; Am Inst Chem Eng; Am Chem Soc; Geochem Soc. *Res:* Composition and reactions of shale oil and petroleum; chemistry of hydrocarbons and organic nitrogen, sulfur and oxygen compounds; origin of petroleum and oil shale; recovery and processing of shale oil. *Mailing Add:* 2993 S Columbine St Denver CO 80210

DINNER, ALAN, b Newton, Mass, Sept 30, 44; m 81; c 2. ORGANIC CHEMISTRY, ANALYTICAL CHEMISTRY. *Educ:* Mass Inst Technol, BS, 66; Ind Univ, Bloomington, PhD(org chem), 70. *Prof Exp:* Teaching asst org chem, Ind Univ, Bloomington, 66-67, res asst med chem, 67-69, fel dept chem, 70-71; fel sch pharm, Univ Calif, San Francisco, 71-73; sr anal chemist org & anal chem, 74-78, res scientist, 78-79, mgr dry prod develop, 79-81, head pharmaceut res, 81-83, dir prod develop, 83-86, dir prod develop & reg affairs, 86-89, EXEC DIR QUAL CONTROL, ELI LILLY & CO, 89- *Concurrent Pos:* Lectr chem, Butler Univ, 75-82; educ counr, Mass Inst Technol, 78- *Mem:* Am Chem Soc; AAAS; NY Acad Sci. *Res:* The study of drug stability and the isolation and identification of trace contaminants in new drugs; development of new drug dosage forms. *Mailing Add:* Eli Lilly & Co Lilly Corp Ctr Indianapolis IN 46285

DINNING, JAMES SMITH, b Franklin, Ky, Sept 28, 22; m 44; c 4. NUTRITION. *Educ:* Univ Ky, BS, 46; Okla State Univ, MS, 47, PhD(biochem), 48. *Hon Degrees:* DSc, Mahidol Univ, Thailand, 74, Univ Ark, 85. *Prof Exp:* Asst prof, Med Sch, Univ Ark, 48-52, assoc prof biochem, 53-58, prof & head dept, 59-63; mem staff, Rockefeller Found, 63-75, assoc dir, 75-78; res scientist, Dept Food Sci & Human Nutrit, Univ Fla, 78-85; RETIRED. *Concurrent Pos:* Asst prof, Univ Pittsburgh, 52-53; spec consult, USPHS; assoc ed, Nutrit Rev, ed, J Nutrit. *Honors & Awards:* Lederle Med Fac Award, 55; Mead-Johnson Award, Am Inst Nutrit, 64, Conrad Eluehjem Award, 84. *Mem:* Soc Exp Biol & Med; Am Soc Biol Chemists; Am Inst Nutrit. *Res:* Metabolic effects of vitamin deficiencies; blood cell formation. *Mailing Add:* 2554 SW 14th Dr Gainesville FL 32608

DINOS, NICHOLAS, b Tamaqua, Pa, Jan 15, 34; m 55; c 3. CHEMICAL ENGINEERING. *Educ:* Pa State Univ, BS, 55; Lehigh Univ, MS, 66, PhD(chem eng), 67. *Prof Exp:* Chem engr, E I du Pont de Nemours & Co, Inc, 55-57, res engr nuclear eng, Savannah River Lab, 57-62, reactor engr nuclear safety, 62-64; teaching asst chem eng, Lehigh Univ, 64-65; ceramics engr, Bethlehem Steel Corp, Pa, 65; assoc prof, 66-72, PROF CHEM ENG, OHIO UNIV, 72-, CHMN DEPT, 77- *Mem:* AAAS; Am Chem Soc; Am Inst Chem Engrs. *Res:* Applications of chemical engineering to biology and vice versa; transport phenomena, mathematics. *Mailing Add:* Ohio Univ 174 Stocker Ctr Athens OH 45701

DINSE, GREGG ERNEST, b Rochester, NY, Nov 9, 54; m 81. SURVIVAL ANALYSIS, NONPARAMETRIC STATISTICS. *Educ:* Bucknell Univ, BS, 76; State Univ NY, Buffalo, MS, 78; Harvard Univ, ScD, 81. *Prof Exp:* Res fel biostatist, Sch Pub Health, Harvard Univ, 81-82; STAFF FEL BIOMET, NAT INST ENVIRON HEALTH SCI, 82- *Mem:* Am Statist Asn; Biomet Soc; Royal Statist Soc. *Res:* Statistical analysis of incomplete observations, such as censored survival data and tumor prevalence and onset data, applied in various clinical, epidemiological and laboratory investigations. *Mailing Add:* PO Box 12233 MDB 3-02 Research Triangle Park NC 27709

DINSMORE, BRUCE HEASLEY, b Indiana, Pa, Sept 18, 15; m 39; c 3. ECOLOGY, AQUATIC BIOLOGY. *Educ:* Ind Univ Pa, BS, 37; Columbia Univ, MA, 41; Univ Pittsburgh, MS, 54, PhD(ecol), 58. *Prof Exp:* Teacher pub schs, Pa & NY, 37-47; from asst prof to assoc prof, 47-71, prof biol & chmn, Dept Biol Sci, 64-78, EMER PROF BIOL, CLARION UNIV, PA, 78- *Mem:* Ecol Soc Am; Am Inst Biol Sci; Sigma Xi. *Res:* Ecological studies of plant and animal communities of strip mine ponds in Pennsylvania; ecology of streams receiving acid mine drainage. *Mailing Add:* 88 Boundary Blvd Unit 149 Rotonda West FL 33947-9728

DINSMORE, CHARLES EARLE, b Lisbon Falls, Maine, Oct 10, 47; m 72; c 2. DEVELOPMENTAL BIOLOGY, HISTORY & PHILOSOPHY OF EIGHTEENTH CENTURY BIOLOGY. *Educ:* Bowdoin Col, AB, 70; Brown Univ, PhD(biol), 74. *Prof Exp:* Asst prof biol, Emmanuel Col, 74-76; asst prof, 76-83, ASSOC PROF ANAT, MED COL & GRAD COL, RUSH UNIV, 83- *Concurrent Pos:* Vis scholar, Univ Chicago, 90-91; chair-elect, Div Hist & Philos Biol, Am Soc Zoologists. *Mem:* AAAS; Hist Sci Soc; Sigma Xi; Am Soc Zoologists; Int Soc Hist, Philos & Social Studies Biol; Soc Develop Biol. *Res:* Urodele limb and tail regeneration; history and philosophy of regengeration research; pattern regulation in regenerating systems; history of 18th century biology. *Mailing Add:* Dept Anat Rush Med Col 600 S Paulina Chicago IL 60612

DINSMORE, HOWARD LIVINGSTONE, b Patton, Pa, May 27, 21; m 42; c 7. ANALYTICAL CHEMISTRY. *Educ:* Johns Hopkins Univ, AB, 42; Univ Minn, PhD(phys chem), 49. *Prof Exp:* Chemist, US Naval Res Lab, 42-46; res assoc phys chem, Brown Univ, 49-50; chemist, Smith Kline & French Labs, 50-51, Com Solvents Corp, 51-54 & M W Kellogg Co, 54-56; prof chem, Bethel Col, 56-66; PROF CHEM, FLA SOUTHERN COL, 66- *Concurrent Pos:* Chemist, Agr Res Corp, Lakeland, Fla, 74- *Mem:* Am Chem Soc. *Res:* Infrared and mass spectroscopy applied to chemical analysis; pesticide trace analysis by gas chromatography. *Mailing Add:* 1410 Elgin St Lakeland FL 33801

DINSMORE, JAMES JAY, b Owatonna, Minn, Feb 25, 42; m 64; c 4. ORNITHOLOGY. *Educ:* Iowa State Univ, BS, 64; Univ Wis, MS, 67; Univ Fla, PhD(zool), 70. *Prof Exp:* Asst prof zool, Univ Fla, 70-71; asst prof biol, Univ Tampa, 71-75; assoc prof, 75-81, PROF ANIMAL ECOL, IOWA STATE UNIV, 81- *Mem:* Am Ornith Union; Cooper Ornith Soc; Wilson Ornith Soc. *Res:* foraging behavior of birds; habitat requirements and habitat selection of birds; conservation biology. *Mailing Add:* Dept Animal Ecol Iowa State Univ 124 Sci Life Ames IA 50011

DINSMORE, JONATHAN H, b Marlboro, Mass, Dec 11, 61. BIOLOGY. *Educ:* Boston Col, BS, 83; Dartmouth Col, PhD(biol), 88. *Prof Exp:* ASSOC, DEPT BIOL, CTR CANCER RES, MASS INST TECHNOL, 88- *Concurrent Pos:* Sigma Xi Grant-in-Aid res award, 86; Am Cancer Soc fel, 89-91. *Honors & Awards:* Nathan Jenks Biol Prize, 84 & 85. *Mem:* Am Soc Cell Biol; NY Acad Sci; AAAS. *Res:* Biology; numerous publications. *Mailing Add:* Dept Biol MIT Bldg E17-221 400 Main St Cambridge MA 02139

DINTER BROWN, LUDMILA, b Pezen, Czech, Nov 4, 49; US citizen; m 88. MINING ENGINEERING, MINERAL PROCESSING. *Educ:* Mining Sch, Ostrava, CSSR, BS, 69, MS, 72 & PhD(mineral processing), 76. *Prof Exp:* Engr, Jesenik, Czech, 76-79; SR ENGR MINERAL PROCESSING, BOLIDEN ALLIS PROCESS RECH TEST CTR, 79- *Concurrent Pos:* Chmn, Soc Mining Eng, 88-89. *Mem:* Soc Mining Eng, Am Inst Mining, Metall & Petrol Engrs. *Mailing Add:* Boliden Allis Inc PRTC 9180 Fifth Ave Oak Creek WI 53154

DINTZIS, HOWARD MARVIN, b Chicago, Ill, May 28, 27; m 51; c 3. IMMUNOLOGY. *Educ:* Univ Calif, Los Angeles, BS, 48; Harvard Univ, PhD(med sci), 53. *Hon Degrees:* DSc, Lawrence Univ, 64. *Prof Exp:* Lilly res fel, Harvard Univ, 52-53; NSF fel, Yale Univ, 53-54; fel, Univ Cambridge, 54-56; asst prof chem, Calif Inst Technol, 56-58; sr res assoc biol, Mass Inst Technol, 58-61; prof biophys & dir dept, 61-90, PROF BIOPHYS & BIOPHYS CHEM, SCH MED, JOHNS HOPKINS UNIV, 90- *Mem:* Am Chem Soc; Biophys Soc. *Res:* Regulation of immune response by macromolecular arrays of antigen. *Mailing Add:* Dept Biophys & Biophys Chem Sch Med Johns Hopkins Univ Baltimore MD 21205-2185

DINTZIS, RENEE ZLOCHOVER, b New York, NY; m 51; c 3. CELL BIOLOGY, IMMUNOLOGY. *Educ:* Hunter Col, BA, 48; Harvard Univ, PhD(biochem), 53. *Prof Exp:* Fel physiol, Yale Med Sch, 53-54, biochem, Cambridge Univ, 54-56; instr, 71-73, asst prof, 74-84, ASSOC PROF, CELL BIOL, ANAT & BIOPHYS, MED SCH, JOHNS HOPKINS UNIV, 84- *Mem:* Am Asn Immunol; Am Soc Cell Biol; Biophys Soc. *Res:* Mechanism of stimulation of the immunoresponsive cell. *Mailing Add:* Dept Cell Biol & Anat Med Sch Johns Hopkins Univ 725 N Wolfe St Baltimore MD 21205

DINUNNO, CECIL MALMBERG, b Washington, DC, Aug 2, 49; m 76. ORGANIC CHEMISTRY. *Educ:* Gettysburg Col, BA, 71; Univ NH, MS, 74. *Prof Exp:* Res chemist, SISA Inc, 74-76; asst res scientist, 76-78, RES SCIENTIST CHEM, SOUTHWEST FOUND RES & EDUC, 78- *Mem:* Am Chem Soc; Sigma Xi; AAAS. *Res:* Organic synthesis of biologically significant compounds; steroids, analgesics and narcotic antagonists and polyfunctional macrocycles. *Mailing Add:* 6820 Glenwood Ct Glenn Dale MD 20769

DINUNZIO, JAMES E, b Buffalo, NY, Sept 26, 50; m 75; c 3. ANALYTICAL CHEMISTRY, INFORMATION SCIENCE & SYSTEMS. *Educ:* Utica Col, BS, 72; Southern Ill Univ, PhD(anal chem), 77. *Prof Exp:* Res assoc chem, Univ Ariz, 77-78; assoc prof chem, Wright State Univ, 78-85; res scientist, Bristol-Myers Co, 85-87, sr res scientist, 87-90, SR RES SCIENTIST II, BRISTOL-MYERS SQUIBB CO, 90- *Mem:* Am Chem Soc. *Res:* Separations, pharmaceutical trace analysis. *Mailing Add:* Pharmaceut Res Inst Bristol-Myers Squibb Co 100 Forest Ave Buffalo NY 14213

DINUS, RONALD JOHN, b Ford City, Pa, Feb 19, 40; m 64. FOREST GENETICS, DISEASE RESISTANCE. *Educ:* Pa State Univ, BS, 61; Univ Wash, MS, 63; Ore State Univ, PhD(forest genetics), 67. *Prof Exp:* Res plant geneticist, Inst Forest Genetics, Southern Forest Exp Sta, Forest Serv, USDA, 68-72, res leader, 72-77; mgr, Western Forest Res Ctr, Int Paper Co, 77-84, Forest & Biol Res, Corp Res Ctr, 84-86. *Concurrent Pos:* Mem task force basic res needs in forestry, USDA, 79, Res & Ext Users Adv Bd, 83- *Mem:* Am Inst Biol Sci; Soc Am Foresters; Sigma Xi; Am Phytopath Soc. *Res:* Intraspecific variation in photoperiodic responses of Douglas-fir; selection, progenytesting and breeding of slash and loblolly pines resistant to fusiform rust; Douglas-fir genetics, regeneration and silviculture, tissue culture, biochemistry and molecular biology of forest trees. *Mailing Add:* 5963 Fords Rd Alworth GA 30101

DINUSSON, WILLIAM ERLING, b Svold, NDak, Apr 28, 20; m 44; c 2. ANIMAL NUTRITION, ANIMAL HUSBANDRY. *Educ:* Okla Agr & Mech Col, BS, 41; Purdue Univ, PhD (animal nutrit), 49; Am Soc Animal Sci, cert animal nutritionist, 85. *Prof Exp:* Asst, Tex A&M Univ, 41-42; asst, Purdue Univ, 46-48; asst prof animal husb, Exp Sta, SDak State Col, 48-49; from assoc prof to prof animal husb, Exp Stat, NDak State Univ, 49-85, emer prof, 85; RETIRED. *Concurrent Pos:* Fulbright res scholar, Iceland, 60-61; mem subcomt sheep, Nat Res Coun, 55-75. *Honors & Awards:* Fac Lectr Award, 66. *Mem:* Fel AAAS; fel Am Soc Animal Sci. *Res:* Sunflower seed oil meal for livestock; protein and energy sources for beef cattle and swine; use of adjuvants in beef cattle; animal nutrition; ruminant nutrition. *Mailing Add:* 2906 Edgemont St Fargo ND 58102

DINWIDDIE, JOSEPH GRAY, JR, b Penns Grove, NJ, Oct 7, 22; m 45; c 4. ORGANIC CHEMISTRY. *Educ:* Randolph-Macon Col, BS, 42; Univ Va, PhD(chem), 49. *Prof Exp:* From asst prof to prof chem, Clemson Univ, 48-68; DEAN, AUGUSTA COL, 68- *Mem:* Fel AAAS; Am Chem Soc. *Res:* Synthetic medicinals; stereochemistry; molecular rearrangements. *Mailing Add:* 3606 St Croix Ct Augusta GA 30909

DIOKNO, ANANIAS CORNEJO, b San Luis, Batangas, Philippines, Aug 13, 42; m 67; c 4. UROLOGY. *Educ:* Univ Santo Tomas, Manila, MD, 65. *Prof Exp:* Fel, 70-71, from instr to prof, 71-84, CLIN PROF UROL, UNIV MICH, 84- *Concurrent Pos:* Consult urol, Wayne Co Gen Hosp & Vet Admin Hosp, Ann Arbor, 70-; consult, urol res Marion Labs, 72-, drugs for bladder dysfunction, AMA, 78; med adv, Mideast Region Int Asn Entro-Stomach Ther, 77-, Hydrodyn Comt, Urodyn Soc, 78-; contribr continuing med educ prog, Am Urol Asn, 77-78; chief urol, William Beaumont Hosp, Royal Oak, Mich, 84- *Honors & Awards:* Sci Exhib Award, Am Urol Asn, 75; Clin Res Award, 75; Silver Medal Award, Sci Exhib Cong Phys Med & Rehab, 75. *Mem:* Urodyn Soc; AMA; Am Col Surgeons; Philippine-Am Urol Soc (pres, 77-78); Am Cong Rehab; Am Urol Asn. *Res:* Lower urinary tract dysfunctions (urodynamics) and sexual dysfunctions; continence and incontinence in the elderly, clinical, epidemiological and social aspects. *Mailing Add:* William Beaumont Hosp 3535 W 13 Mile Rd Royal Oak MI 48072

DION, ANDRE R, b Quebec, Que, May 3, 26; m 59; c 3. PHYSICS. *Educ:* Laval Univ, BScA, 49, MSc, 50. *Hon Degrees:* DSc, Laval Univ, 53. *Prof Exp:* Defense res sci officer, Defense Res Bd, Can, 53-60; MEM STAFF, LINCOLN LAB, MASS INST TECHNOL, 60- *Mem:* Inst Elec & Electronics Eng; Sigma Xi. *Res:* Electromagnetic waves; microwaves; antennas; propagation. *Mailing Add:* 600 Hayward Mill Rd Concord MA 01742

DION, ARNOLD SILVA, b Laconia, NH, June 26, 39; m 61; c 4. BIOCHEMISTRY. *Educ:* Univ NH, BS, 64, MS, 66, PhD(biochem), 69. *Prof Exp:* NIH fel, Univ Pa, 68-70, res assoc biochem, 70-71; asst biochemist, Inst Med Res, NJ, 71-72, assoc biochemist & head molecular biol sect, 72-76, sr mem & head molecular biol dept, 76-86; MEM & DIR, INST MOLECULAR GENETICS, CTR MOLECULAR MED & IMMUNOL, NJ, 86- *Concurrent Pos:* Consult, Virus Cancer Prog, Nat Cancer Inst, NIH, 76-77; reviewer, Biochem Prog, NSF, 80-; mem, Exp Immunol Study Sect, NIH, 81-83, Breast Cancer Task Force, NIH & Nat Cancer Inst, 84, & Breast Cancer Working Group, 84-; adj assoc prof path, NY Med Col, 81-, adj prof microbiol & immunol, 89- *Mem:* AAAS; Am Soc Microbiol; Sigma Xi; NY Acad Sci; Am Chem Soc; Int Asn Breast Cancer Res. *Res:* Correlations between polyamine and nucleic acid synthesis during development in Drosophila melanogaster and in bacteriophage infected cells; biochemical characterization of oncornaviruses, including RNA-dependent DNA polymerase; retrovirus integration sites and expression; epithelial membrane antigens; synthetic peptide antigens/immunogens. *Mailing Add:* Ctr Molecular Med & Immunol One Bruce St Newark NJ 07103

DIONNE, GERALD FRANCIS, b Montreal, Que, Feb 5, 35; US citizen; m 63; c 1. MAGNETISM & SUPERCONDUCTIVITY FAR-INFRARED OPTICS & SURFACE PHYSICS. *Educ:* Concordia Univ, BSc, 56; McGill Univ, BEng, 58, PhD(physics), 64; Carnegie-Mellon Univ, MS, 59. *Prof Exp:* Engr I, Bell Tel Co Can, 58-59; jr engr, Int Bus Mach Corp, NY, 59-60; sr engr, Gen Tel & Electronics Semiconductor Div, Sylvania Elec Prod, Inc, Mass, 60-61; res asst physics, Eaton Electronics Res Lab, McGill Univ, 63-64; res assoc, Advan Mat Res & Develop Lab, Pratt & Whitney Aircraft Div, United Aircraft Corp, 64-66, sr res assoc, 66; STAFF MEM, LINCOLN LAB, MASS INST TECHNOL, 66- *Concurrent Pos:* Asst, Carnegie-Mellon Univ, 58-59. *Mem:* Am Phys Soc; sr mem Inst Elec & Electronics Eng; Sigma Xi. *Res:* Magnetism and ferrimagnetic materials; magneto-elastic phenomena, magneto-optics; superconductivity theory; electron spin resonance and relaxation; far-infrared spectroscopy and laser optics; millimeter-wave radiometry electron and ion emission; thermionic energy conversion. *Mailing Add:* 182 High St Winchester MA 01890-3366

DIONNE, JEAN-CLAUDE, b Luceville, Que, Jan 3, 35; m 66; c 3. GEOLOGY, GEOGRAPHY. *Educ:* Univ Moncton, BA, 57; Univ Montreal, MA, 61; Univ Paris, PhD(geog), 70. *Prof Exp:* Res officer geomorphol, Bur Amenagement l'Est Que, Mont-Joli, 64-66; ares officer geomorphol, Dept Forestry & Rural Develop, Can, 66-70; res scientist, Land Div, Environ Can, Que, 70-80; PROF, DEPT GEOG, UNIV LAVAL, QUE, 80- *Concurrent Pos:* Prof coastal morphol & photo-interpretation, Laval Univ, 66-67; consult, Urban Planning Off Que, 66-68. *Mem:* Geol Asn Can; Soc Econ Paleont & Mineral. *Res:* Geomorphology; physical geography; quaternary geology; sedimentology; coastal and marine geology; oceanography; photo-interpretation. *Mailing Add:* Dept Geog Univ Lavel Quebec PQ G1K 7P4 Can

DIONNE, RAYMOND A, b Providence, RI, Dec 25, 46; m 71. CLINICAL PHARMACOLOGY, PAIN. *Educ:* Univ Conn, BA, 68; Georgetown Univ, DDS, 72; Med Col Va, PhD(pharmacol), 80. *Prof Exp:* Fel, 75-78, ASST CLIN PROF ORAL SURG, MED COL VA, VA COMMONWEALTH UNIV, 78-; STAFF FEL, NAT INST DENT RES, NIH, 78- *Mem:* Int Asn Study of Pain; Int Asn Dent Res; Soc Neurosci; Am Pain Soc. *Res:* Assessment of relationship between clinical efficacy and toxicity of drugs used to control pain and apprehension in outpatients. *Mailing Add:* Bldg 10 Rm 1N103 Nat Inst Dent Res 9000 Rockville Pike Bethesda MD 20205

DIORIO, ALFRED FRANK, b Paterson, NJ, Jan 17, 33; c 2. LABORATORY MEDICINE, ENZYMOLOGY. *Educ:* Fordham Univ, BS, 54; Univ Md, MS, 62; Georgetown Univ, PhD(biochem), 67; Am Bd Clin Chem, dipl, 72. *Prof Exp:* Chemist, Nat Lead Co, 54-56 & Walter Reed Army Med Ctr, 56-58; phys chemist, Am Dent Asn, 58-59 & Nat Bur Standards, 59-63; asst dir, Georgetown Univ Med Ctr, 66-68; SECT HEAD, DEPT LAB MED, ALLEGHENY GEN HOSP, 68- *Mem:* Fel Am Asn Clin Chem; fel Nat Acad Clin Biochem; Am Chem Soc; assoc mem Am Soc Clin Pathologists. *Res:* Natural and synthetic fibrous macromolecules; enzymology as it pertains to cardiac and trauma patients. *Mailing Add:* Dept Lab Med Allegheny Gen Hosp Pittsburgh PA 15203

D'IORIO, ANTOINE, b Montreal, Que, Apr 22, 25; m 50; c 7. BIOCHEMISTRY. *Educ:* Univ Montreal, BSc, 46, PhD(biochem), 49. *Hon Degrees:* DSc, Carleton Univ, 90. *Prof Exp:* Lectr physiol, Univ Montreal, 49-51, asst prof, 52-56; fel enzymol, Univ Wis, 51-52; fel pharmacol, Oxford Univ, 56-57; assoc prof physiol, Univ Montreal, 57-61; head dept biochem, Univ Ottawa, 61-69, dean sci, 69-76, vrector acad, 76-84, rector & vchancellor, 84-90, prof, 61-90, EMER PROF BIOCHEM, UNIV OTTAWA, 90- *Concurrent Pos:* Pres, Can Mediter Inst. *Mem:* Am Chem Soc; Am Asn Biol Chemists; Soc Exp Biol & Med; Can Physiol Soc (treas, 58-61); Can Fedn Biol Soc (hon treas, 62-66). *Res:* Physiology and biochemistry of catecholamines. *Mailing Add:* 405-15 Murray Ottawa ON K1N 9M5 Can

DIOSADY, LEVENTE LASZLO, b Marosvasarhely, Hungary, Oct 2, 43; Can citizen; m 71; c 2. FOOD ENGINEERING, TRACE ORGANIC ANALYSIS. *Educ:* Univ Toronto, BSc, 66, MSc, 68, PhD, 71. *Prof Exp:* Dir res & develop, Cambrian Eng Group, Agra Indust Ltd, 72-79; assoc prof, 79-86, PROF FOOD ENG, DEPT CHEM ENG, UNIV TORONTO, 86- *Concurrent Pos:* Dir, Chem Eng Res Consults Ltd, 80-; chmn, Energy Subcomt, Can Comt Food, 84-, Food Process Eng Sect, Can Inst Food Sci & Technol, 84-85; pres, Food Biotech Corp. *Mem:* Can Inst Food Sci & Technol; Inst Food Technologists; Asn Off Anal Chemists; Am Oil Chemists Soc; Can Soc Chem Eng; Chem Inst Can. *Res:* Development of technology for edible oilseed processing, edible oil refining and hydrogenation, and oilseed protein upgrading; application of novel separation techniques, such as membrane processes and ion exchange to food processing; extrusion. *Mailing Add:* Dept Chem Eng Univ Toronto Toronto ON M5S 1A4 Can

DIOSY, ANDREW, b Szarvas, Hungary, Mar 27, 24; Can citizen; m 55; c 2. INTERNAL MEDICINE. *Educ:* Univ Szeged, MD, 50; Univ Man, MSc, 59; FRCP(C). *Prof Exp:* Asst prof physiol, Univ Szeged, 50-52; res fel, Aviation Hosp, Budapest, Hungary, 52-56; res asst, Univ Man, 57-59; jr intern med, Winnipeg Gen Hosp, Man, 59-60; sr intern, Deer Lodge Hosp, 60-61; resident med, St Michael's Hosp, Toronto, Ont, 62-63; res assoc pharmacol, Univ Toronto, 63-65; dir clin res, Warner-Lambert Can Ltd, 65-70, med dir, 70-88; RETIRED. *Concurrent Pos:* Res fel endocrinol, St Michael's Hosp, Toronto, Ont, 61-62. *Honors & Awards:* E L Drewry Mem Award, 59. *Mem:* Am Col Physicians; Can Med Asn; NY Acad Sci; Royal Soc Med. *Res:* Clinical pharmacology; new drug research. *Mailing Add:* 111 Dunuegan Rd Toronto ON M4V 2P9 Can

DIPALMA, JOSEPH RUPERT, b New York, NY, Mar 21, 16; m 48; c 5. PHARMACOLOGY, MEDICINE. *Educ:* Columbia Univ, BS, 36; Long Island Col Med, MD, 41. *Hon Degrees:* Dsc, Hahnemann Univ, 82. *Prof Exp:* Instr physiol & pharmacol, Long Island Col Med, 41-42, asst med, 44-48; assoc physiol & pharmacol, 46-48, asst prof med, 48-50; head dept, 50-67, sr vpres acad affairs & dean med col, 67- 82, prof pharmacol, 50-84, EMER PROF PHARMACOL & DEAN, HAHNEMAN MED COL, 84- *Concurrent Pos:* Fel, Harvard Med Sch, 46; consult malaria, Surgeon Gen, 67-69. *Mem:* Am Physiol Soc; Am Soc Clin Invest; Harvey Soc; AMA; Am Soc Pharmacol & Exp Therapeut; Sigma Xi. *Res:* Heart muscle; antifibrillatory drugs; chemical warfare agents; anthocyanins; gangliosides. *Mailing Add:* 100 Pembroke Ave Wayne PA 19087

DI PAOLA, JANE WALSH, b Brooklyn, NY, Sept 17, 17; m 42; c 3. COMBINATORICS, GEOMETRY. *Educ:* St Joseph's Col, NY, BA, 39; Brooklyn Col, MA, 62; City Univ New York, PhD(math), 67. *Prof Exp:* Struct engr appl math, Cornell Aeronaut Lab, Buffalo, NY, 42-45; instr math, New York City Community Col, 64-65; asst prof, NY Univ, 67-71; adj prof math, Fla Atlantic Univ, 71-83; RETIRED. *Concurrent Pos:* Assoc ed, Am Math Monthly, Math Asn Am, 71-76; vis prof, Fla Atlantic Univ, 86; found fel, Inst Combinatorics & its Applications, 90. *Mem:* Sigma Xi; Am Math Soc; Can Soc Hist & Philos Math; NY Acad Sci. *Res:* Combinatorial mathematics with particular attention to graph theory, block designs, Steiner triple systems, algebraic structures from combinatorial designs, finite geometries. *Mailing Add:* 3580 County Rd 215 Cheyenne WY 82009

DI PAOLA, ROBERT ARNOLD, b New York, NY, Nov 28, 33. COMPUTER SCIENCE. *Educ:* Fordham Col, BA, 56; Fordham Univ, MA, 57; Yeshiva Univ Grad Sch Sci, PhD(math), 64. *Prof Exp:* Mathematician, Grumman Corp, 57-59; actg asst prof math, Univ Calif, Los Angeles, 64-66; mathematician, Rand Corp, 66-72; PROF MATH & COMPUT SCI, GRAD SCH & UNIV CTR, QUEENS COL, 75- *Concurrent Pos:* Vis prof, Univ Siena, Italy, 76, 84, 85 & 89 & Oxford Univ, Eng, 89; prin investr, NSF & Nat Res Coun Italy, 83-; coun mem, Coun Sch Specialization in Math Logic, 86-90. *Mem:* Am Math Soc; Asn Symbolic Logic. *Res:* Collaborative research toward achieving category-theoretic representations of the Godel Incompleteness Theorems and the associated recursion theory; classical recursion theory, the theory of formal undecidability and increasingly the complexity of computation; application of algebra and category theory to incompleteness phenomena. *Mailing Add:* Math Dept Grad Sch & Univ Ctr 33 W 42nd St New York NY 10036

DIPAOLO, JOSEPH AMEDEO, b Bridgeport, Conn, June 13, 24; m 52; c 2. GENETICS. *Educ:* Wesleyan Univ, AB, 48; Western Reserve Univ, MS, 49; Northwestern Univ, PhD(genetics), 51. *Prof Exp:* Asst instr genetics & biol, Northwestern Univ, 49-51; asst instr, Dept Biol, Loyola Univ, 51-53; asst instr clin & exp path, Med Sch, Northwestern Univ, 53-55; sr cancer res scientist, Roswell Park Mem Inst, NY, 55-63; asst res prof biol, Grad Sch, State Univ NY Buffalo, 55-63; head cytogenetics & cytol sect, 63-77, CHIEF LAB BIOL, BIOL BR, NAT CANCER INST, 77-, ASSOC PROF LECTR ANAT, GEORGE WASHINGTON UNIV, 65- *Concurrent Pos:* Co-chmn US-USSR joint working group mammalian somatic cell genetics related to Neoplasia; consult, US-Poland Cancer Prog; chmn, US-German Carcinogenesis Prog, 79-85; bd dir, Am Asn Cancer Res, 83-86. *Mem:* Fel AAAS; Am Asn Cancer Res; Am Soc Human Genetics; Genetics Soc Am; fel NY Acad Sci; Am Asn Pathologists. *Res:* Experimental cancer research; cell biology. *Mailing Add:* 6605 Melody Lane Bethesda MD 20817

DI PAOLO, ROCCO JOHN, b Brooklyn, NY. ORTHODONTICS. *Educ:* Long Island Univ, BS, 44; Univ Mo, DDS, 49; Am Bd Orthodont, dipl, 68; Univ Mo, cert pedodont. *Prof Exp:* Asst prof, Fairleigh Dickinson Univ, 65-72, prof, chmn dept & dir orthodont, prof & dir, Oral Rehab Ctr, Sch Dent, 72-90, dean, Col Dent, 89-90; DIR, TRANSITIONAL PROGS, UNIV MED & DENT NJ, NEWARK, 90-, SPEC ASST TO SR VPRES ACAD AFFAIRS, 90- & VPRES UNIV, 90- *Concurrent Pos:* Consult, Bard-Parker Div Becton Dickinson Corp, 72-; univ grant, Sch Dent, Fairleigh Dickinson Univ, 73-74; consult & lectr, Oral Facial Rehab Ctr, St Barnabas Hosp, 74-; lectr, Sch Dent, Columbia Univ, 74-; dir, Int Dento-Facial Asn Abnormalities; chmn, adv comt orthod, Am Dent Asn, 83-86; comnr, Comn Dent Accreditation, 84-86; dir, Am Bd Orthod, 87-, secy-treas, 91-; dir, Grad Orthod Prog, Fairleigh Dickinson Univ, dir govt affairs, consult advan educ, 90- *Mem:* Am Asn Orthod; Am Cleft Palate Asn; Sigma Xi; fel Am Col Dent; fel Int Col Dent. *Res:* Cephalometric analysis; biomechanics; orthodontic development; cleft palate; oral facial rehabilitation; surgical orthodontics; birth defects; orthodontic diagnosis. *Mailing Add:* 56 Fox Hedge Rd Saddle River NJ 07458

DIPASQUALE, GENE, b New York, NY, July 17, 32; m 62; c 2. BIOLOGY, PHYSIOLOGY. *Educ:* Iona Col, BS, 54; Long Island Univ, MS, 60; NY Univ, PhD, 70. *Prof Exp:* from asst scientist to sr res assoc, Dept Pharmacodynamics, Warner-Lambert Res Inst, 57-75, assoc dir, 71-77; sect mgr immunopharmacol, 77-85, SECT MGR, SAFETY EVAL DEPT, STUART PHARMACEUT, 87- *Mem:* Endocrine Soc; NY Acad Sci; Am Physiol Soc; Am Soc Pharmacol & Exp Therapeut; Soc Exp Biol & Med; Am Pharmotism Asn. *Res:* Connective tissue metabolism in relation to inflammation, wound healing and atherosclerosis, especially effects of drugs in these areas; agents with anti-inflammatory properties, especially those useful for the treatment of osteoarthritis and rheumatoid arthritis; safety evaluation of pharmacologically active compounds. *Mailing Add:* ICI Am Inc 205 BMRL Wilmington DE 19897

DI PIETRO, DAVID LOUIS, b Philadelphia, Pa, Jan 16, 32; m 61; c 3. BIOCHEMISTRY. *Educ:* Temple Univ, BA, 55, MA, 57, PhD(biochem), 61. *Prof Exp:* Res fel med, Harvard Med Sch, 61-62; res assoc biochem, Div Res, Lankenau Hosp, Philadelphia, 62-63; res instr, Sch Med & res investr, Fels Res Inst, Temple Univ, 63-66, res asst prof, 66-68; ASST PROF OBSTET & GYNEC, SCH MED, VANDERBILT UNIV, 68- *Mem:* Am Chem Soc; Am Soc Biol Chemists; Brit Biochem Soc. *Res:* Enzymes of carbohydrate metabolism in liver and placenta; metabolism of estrogens in pregnancy. *Mailing Add:* 4421 Wayland Dr Nashville TN 37215

DI PIPPO, ASCANIO G, b Providence, RI, Jan 21, 32; m 61; c 5. ORGANIC CHEMISTRY, CLINICAL CHEMISTRY. *Educ:* Univ RI, BS, 54, MS, 58, PhD(chem), 61. *Prof Exp:* Sr res chemist, Labs, Olin Mathieson Chem Corp, 61 & Nat Labs Res & Testing, 61-62; from asst prof to assoc prof, 62-69, PROF ORG CHEM, SALVE REGINA COL, 69- *Concurrent Pos:* Consult, Newport Hosp, RI, 64- & Johns Hopkins Med Lab, 66-; chem eng, Raytheon, Submarine Signal Div, 68- *Mem:* Am Chem Soc; Nat Registry Clin Chem. *Res:* Organic mechanisms; organoboron compounds; effects of radiation on organic compounds. *Mailing Add:* Dept Chem Salve Regina Col Newport RI 02840

DIPIPPO, RONALD, b Providence, RI, June 2, 40; c 3. MECHANICAL ENGINEERING, THERMODYNAMICS. *Educ:* Brown Univ, ScB, 62, ScM, 64, PhD(eng), 66. *Prof Exp:* Mech engr res dept, US Naval Underwater Systs Ctr, 66-67; assoc prof, 67-74, PROF MECH ENG & CHMN DEPT, SOUTHEASTERN MASS UNIV, 74- *Concurrent Pos:* Mem adv comt, Mass Bd Higher Educ, 70-71; adj prof eng res, Brown Univ; consult, Stone & Webster Eng Corp, Mother Earth Indust, EG&G, Idaho, Costa Rican Elec Inst, Guatemalan Elec Authority & Los Alamos Nat Lab. *Mem:* Am Soc Eng Educ; Am Soc Mech Engrs; Geotherm Resources Coun. *Res:* Measurement and correlation of transport properties of gases; propulsion systems; geothermal energy conversion systems. *Mailing Add:* Dept Mech Eng Southeastern Mass Univ North Dartmouth MA 02747

DIPIRRO, MICHAEL JAMES, b Haverhill, Mass, July 26, 51; m 79. LOW TEMPERATURE PHYSICS. *Educ:* Clarkson Col Technol, NY, BS, 73; State Univ NY, Buffalo, PhD(physics), 79. *Prof Exp:* Res assoc, Nat Bur Standards, 79-80; PHYSICIST, GODDARD SPACE FLIGHT CTR, NASA, 80- *Mem:* Am Phys Soc; Am Inst Aeronaut & Astronaut. *Res:* Physics of liquid helium including properties of 3Helium on the surface of 4Helium; properties of 4Helium near the lambda point; porous plug containment of 4Helium in space. *Mailing Add:* Code 713 NASA Goddard Space Flight Ctr Greenbelt MD 20771

DIPNER, RANDY W, b Columbus, Ohio, July 6, 49. ASSISTIVE TECHNOLOGY FOR INDIVIDUALS WITH DISABILITIES. *Educ:* Ohio State Univ, BS, 72. *Prof Exp:* Mgr bus develop, Computer Technol Assocs Inc, 81-91; PRES, MEETING CHALLENGE INC, 89- *Concurrent Pos:* Chmn, Systs Info Group Computers & Disabilities, Asn Comput Mach. *Mem:* Asn Comput Mach; Nat Security Indust Asn; Rehab Eng Soc NAm. *Mailing Add:* 3630 Sinton Rd Suite 103 Colorado Springs CO 80907

DIPPEL, WILLIAM ALAN, b Jersey City, NJ, June 12, 26; m 52; c 5. ANALYTICAL CHEMISTRY. *Educ:* Princeton Univ, AB, 50, PhD(chem), 54. *Prof Exp:* Res chemist, Res Div, Explosives Dept, 54-59, anal supvr, 59-62, anal supt, 62-72, res supvr res & develop, Polymer Intermediates Dept, 72-78, res supvr, 78-79, RES MGR, RES & DEVELOP DIV, PETROCHEM DEPT, E I DUPONT DE NEMOURS & CO, 79- *Mem:* Am Chem Soc; Am Inst Chem. *Res:* Analytical research and development; analytical methods in process development and environmental quality gas and liquid chromatography; wet chemistry. *Mailing Add:* 203 Alapocas Dr Wilmington DE 19803-4504

DIPPELL, RUTH VIRGINIA, b Huntington, Ind, June 25, 20. CELL BIOLOGY, GENETICS. *Educ:* Ind Univ, Bloomington, PhD(zool), 49. *Prof Exp:* Asst zool, 42-45, res assoc, 45-67, ASSOC PROF ZOOL, IND UNIV, BLOOMINGTON, 67- *Concurrent Pos:* Adj prof, Ind State Univ, 66-67. *Honors & Awards:* Newcomb Prize, AAAS, 46. *Mem:* Fel AAAS; fel Am Asn Univ Women; Soc Protozool (vpres, 65-66); Am Soc Cell Biol; Int Soc Cell Biol. *Res:* Inheritance and morphogenesis of cell organelles, chromosomal organization and behavior in ciliate Protozoa. *Mailing Add:* 4800 W Ocotillo N 60 Glendale AR 85301

DIPPLE, ANTHONY, b Eng, Feb 9, 40. CANCER. *Educ:* Univ Birmingham, BSc, 61, PhD(chem), 64. *Prof Exp:* Fel oncol, McArdle Lab, Univ Wis-Madison, 64-66; lectr, Chester Beatty Res Inst, Inst Cancer Res, London, 66-75; sect head molecular aspects carcinogenesis, 75-81, ASSOC DIR LAB PHYS & CHEM CARCINOGENESIS, FREDERICK CANCER RES FACIL PROG, NAT CANCER INST, 81- *Concurrent Pos:* Ed, Carcinogenesis, 80-; mem, Chem Path Study Sect, 82-86. *Mem:* The Chem Soc; Brit Asn Cancer Res; Am Asn Cancer Res. *Res:* Studies of the mechanism of action of chemical carcinogens. *Mailing Add:* 340 Fieldpoint Blvd No A303 Frederick MD 21701

DIPPREY, DUANE F(LOYD), b Minneapolis, Minn, Dec 22, 29; m 52; c 2. MECHANICAL ENGINEERING. *Educ:* Univ Minn, BS, 51, MSME, 53; Calif Inst Technol, PhD(mech eng, physics), 61. *Prof Exp:* From develop eng to sr develop eng, 53-60, eng group supvr, 60-62, asst sect mgr liquid rockets, 62-64, sect mgr liquid propulsion, 64-72, sect mgr advan concepts, 72-74, dep div mgr propulsion, 74-76, dep div mgr control & energy conversion, 76-78, div mgr applied mech, 78-86, asst dir technol & appln, 86-91, ASSOC DIR, JET PROPULSION LAB, CALIF INST TECHNOL, 91- *Concurrent Pos:* Lectr, Univ Southern Calif, 63-65. *Honors & Awards:* Outstanding Leadership Medal, NASA. *Mem:* Assoc fel Am Inst Aeronaut & Astronaut; Sigma Xi. *Res:* Heat transfer from roughened surfaces to flowing fluids; propulsion systems analysis. *Mailing Add:* 2053 Los Amigos La Canada CA 91011

DIRCKX, JOHN H, b Dayton, Oh, Aug 1, 38; m 63; c 5. MEDICAL ETYMOLOGY & LEXICOGRAPHY. *Educ:* Univ Dayton, BS, 59; Marquette Univ, MD, 62. *Prof Exp:* Staff physician, Civilian Employee Health Sect, Wright-Patterson AFB, 64-66; emergency room physician, Kettering Mem Hosp, Dayton, 66-69; MED DIR, C H GOSIGER MEM HEALTH CTR, UNIV DAYTON, 68- *Concurrent Pos:* Consult, Gen Motors Corp, 72, ed consult, Health Prof Inst, 86- *Res:* Exploring the humanistic dimension of disease, death and the healing arts; interpretation of the literature and history of Western medicine; analysis of the origins and dynamics of scientific terminology. *Mailing Add:* Univ Dayton 300 College Pk Dayton OH 45469-0900

DIRECTOR, STEPHEN WILLIAM, b Brooklyn, NY, June 28, 43; m 65; c 4. ELECTRICAL ENGINEERING. *Educ:* State Univ NY Stony Brook, BS, 65; Univ Calif, Berkeley, MS, 67, PhD(elec eng), 68. *Prof Exp:* From asst prof to prof, Univ Fla, 68-77; PROF ELEC ENG, CARNEGIE-MELLON UNIV, 77-, U A & HELEN WITAKER PROF, ELEC & ELECTRONICS ENGR, 80-, PROF COMPUT SCI, 81-, HEAD, DEPT ELEC & COMPUT ENG, 82- *Concurrent Pos:* Consult, IBM Corp, Intel Corp, Tex Instruments, Inc, Harris Corp, Digital Equip Corp, Calma Corp & Mentor Graphics Corp; consult ed, McGraw Hill Book Co, 76-; mem bd dirs, Nextwave, Inc; dir, Res Ctr Computer-Aided Design, Pittsburgh, 82-89. *Honors & Awards:* F E Terman Award, Am Soc Eng Educ, 76; Centennial Medal, Inst Elec & Electronics Engrs, 84. *Mem:* Nat Acad Eng; Inst Elec & Electronics Circuits & Systs Soc (pres, 80); Inst Elec & Electronics Engrs Comput Soc; fel Inst Elec & Electronics Engrs. *Res:* Computer aided design; network theory. *Mailing Add:* Dept of Elec Eng Carnegie-Mellon Univ Pittsburgh PA 15213

DI RIENZI, JOSEPH, b Brooklyn, NY, Aug 12, 47; m 76; c 2. RADIATION PHYSICS, NUCLEAR ENERGY. *Educ:* Polytech Inst NY, BS, 68, MS, 72, PhD(physics), 75. *Prof Exp:* Adj instr physics, Polytech Inst NY, 72-76, NY Inst Technol, 74-76; ASSOC PROF PHYSICS & MATH, COL NOTRE DAME, 76- *Concurrent Pos:* Adj fac & mem adv comt, Sch Nuclear Med Technol, St Joseph Hosp, 81-84; sci coun, Md Acad Sci, 87- *Mem:* Am Asn Physics Teachers; Sigma Xi. *Res:* Studies in the theory of nuclear structure; application of the intermediate coupling theory in the unified nuclear model on certain odd-mass nuclei; foundations of quantum physics. *Mailing Add:* Col Notre Dame 4701 N Charles St Baltimore MD 21210

DIRIGE, OFELIA VILLA, b Manila, Philippines, July 14, 40. NUTRITION. *Educ:* Univ Philippines, BSHE, 59; Univ Hawaii, MS, 68; Univ Calif, Los Angeles, DrPH(nutrit), 72. *Prof Exp:* Res asst nutrit, Sch Pub Health, Univ Calif, Los Angeles, 68-69, teaching asst, 69-71, res asst, 71-72, res assoc, 72-74; nutritionist, Los Angeles County Community Health Serv, 74-; AT PUBLIC HEALTH DEPT, UNIV HAWAII. *Concurrent Pos:* Instr, Pepperdine Univ, 75- *Mem:* Am Dietetic Asn; Am Pub Health Asn; Am Home Econ Asn. *Res:* Biochemical assessment of nutritional status in pregnant women and malnourished children; transketolase activity in chronic uremia and in experimental uremia of rats; dietary assessment of nutritional status of pregnant women. *Mailing Add:* Div Maternal & Child Health Grad Sch Pub Health San Diego State Univ 6505 Alvarado St Rm 205 San Diego CA 92182-0505

DIRKS, BRINTON MARLO, b Newton, Kans, July 20, 20; m 44; c 2. AGRICULTURAL BIOCHEMISTRY, MICROBIOLOGY. *Educ:* Kans State Col, BS & MS, 48; Univ Minn, PhD(agr biochem), 53. *Prof Exp:* Chemist, Moundridge Milling Co, 38-40; chief operator, Tenn Eastman Corp, 45-46; head biochem res sect Pillsbury Mills, Inc, 52-56; group leader food prod develop, Procter & Gamble Co, 56-60, head flavor develop sect, 60-68, head flour technol sect, 68-74, head regulatory relat sect, 74-85; RETIRED. *Mem:* Am Chem Soc; Am Asn Cereal Chem; Inst Food Technol. *Res:* Enzymes of cereal grains, fungi and related materials; bio-chemistry of prepared baking mixes and other food products; flavor chemistry and application. *Mailing Add:* 301 Rolling Hills Dr Newton KS 67114

DIRKS, JOHN HERBERT, b Winnipeg, Man, Aug 20, 33; m 61; c 4. MEDICINE, NEPHROLOGY. *Educ:* Univ Man, BSc & MD, 57; FRCP(C), 63. *Prof Exp:* Vis scientist, NIH, 64-65; asst prof med, Royal Victoria Hosp, 65-68, assoc physician, 68-71; dir renal & electrolyte div, Univ Clin, 65-76, asst prof, 68-69, ASSOC PROF PHYSIOL, MCGILL UNIV, 70-, PROF MED, 73- *Concurrent Pos:* Med Res Coun fels, McGill Univ, 60-62, NIH fel, 62-64; grants, Med Res Coun, 65-; ed renal & electrolyte sect, Am J Physiol, 73-76; mem coun, Med Res Coun Can, 78-, Int Soc Nephrology, 84-; chmn Sci Adv Comt, Kidney Found, 82-; mem Med Adv Bd, Gairdner Award Found, 83-; assoc prof med, Royal Victoria Hosp, 68-, sr physician, 71-; Eric W Hamber prof med & head dept med, Univ BC, 76-, head dept med, Health Sci Ctr Hosp, 84-; physician-in-chief, Vancouver Gen Hosp, 76- *Honors & Awards:* Queen's Jubilee Award, 77; Med Award, Kidney Found Can, 85. *Mem:* Am Heart Asn; Am Fedn Clin Res; Am Physiol Soc; Am Soc Clin Invest; Can Soc Clin Invest (secy-treas, 72-75, pres & past pres, 75-78); fel Royal Soc Chem; Am Asn Physicians; Am Soc Nephrology; Int Soc Nephrology; Can Asn Prof Med (pres, 84-85); fel Am Col Physicians. *Res:* Renal physiology, primarily using micropuncture research; sodium, calcium and magnesium reabsorption by proximal and distal tubules in different physiological conditions; effects of diuretics on Na and K transport; clearance and micropuncture studies of magnesium transport in the renal tubule; micropuncture studies in Ca and P transport. *Mailing Add:* Univ Toronto Med Sci Bldg One Kings Col Circle Rm 2109 Toronto ON V5S 1A8 Can

DIRKS, LESLIE C, b New Ulm, Minn, Mar 7, 36; m 59; c 3. ENGINEERING. *Educ:* Mass Inst Technol, BS, 58; Oxford, BS, 60. *Prof Exp:* Instr physics, Phillips Acad, Mass, 60-61; tech staff, Cent Intel Agency, 61-76, dep dir sci & technol, 76-82; GROUP VPRES & MGR, DEFENSE INFO SYSTS DIV, HUGHES AIRCRAFT, 85- *Honors & Awards:* Annual Award, Inst Elec & Electronics Engrs, 80. *Mem:* Nat Acad Eng. *Mailing Add:* PO Box 6853 S-41 B340 PO Box 92919 Torrance CA 90504

DIRKS, RICHARD ALLEN, b Belmond, Iowa, Nov 11, 37; m 65; c 2. DYNAMIC METEOROLOGY. *Educ:* Wheaton Col, Ill, BS, 59; Drake Univ, MA, 62; Cornell Univ, MST, 65; Colo State Univ, PhD(atmospheric sci), 70. *Prof Exp:* Instr physics, Southeast Mo State Col, 62-64; jr meteorologist, Colo State Univ, 68-69; from asst prof to assoc prof atmospheric sci, Univ Wyo, 69-75; prog mgr environ res & technol, 75-77, prog dir atmospheric sci, 77-86, dir, Gale Proj Off, 84-86, MGR, OFF FIELD PROJ SUPPORT, NAT CTR ATMOSPHERIC RES, NSF, 86- *Mem:* Am Meteorol Soc; Am Geophys Union. *Res:* Observation and modeling of small scale airflow including such areas as convective circulations, thunderstorms, mountain airflow, urban circulations and boundary layer flow; airflow studies employing meteorological satellites. *Mailing Add:* Field Proj Support PO Box 3000 Boulder CO 80307

DIRKSE, THEDFORD PRESTON, b Holland, Mich, Jan 5, 15; m 42; c 4. CHEMISTRY, ELECTROCHEMISTRY. *Educ:* Calvin Col, AB, 36; Ind Univ, AM, 37, PhD(gen & phys chem), 39. *Prof Exp:* Instr chem, Iowa State Col, 39-41; asst prof, Hamline Univ, 41-42; chemist, US Naval Res Lab, Wash, 42-46; asst prof, Hamline Univ, 46-47; from assoc prof to prof, 47-80, EMER PROF CHEM, CALVIN COL, 80- *Concurrent Pos:* Co-ed, Int Union Pure & Appl Chem Solubility Data Ser, 83- *Honors & Awards:* Res Award, Electrochem Soc. *Mem:* Am Chem Soc; Electrochem Soc; Int Soc Electrochem. *Res:* Electroplating from nonaqueous media; electrical conductance of solutions; alkaline batteries; conductance of salts in monoethanolamine; electrode kinetics. *Mailing Add:* Dept Chem Calvin Col Grand Rapids MI 49546

DIRKSEN, CHRISTIAAN, b Leeuwarden, Neth, July 18, 36; US citizen; m 59; c 4. SOIL PHYSICS. *Educ:* Wageningen Agr Univ, Neth, BS, 57, Agr Eng, 59; Cornell Univ, PhD(soil physics), 64. *Prof Exp:* Res engr hydrol, Agr Univ Wageningen, Neth, 59; res assoc agron, Cornell Univ, 59-60; res physicist reservoir mech, Gulf Res & Develop Co, Pa, 63-68; soil scientist, Agr Res Serv, USDA, Madison, Wis, 68-73, soil scientist physics, Riverside, Calif, 73-78; ASSOC PROF SOIL PHYSICS, WAGENINGEN AGR UNIV, NETH, 78- *Mem:* Am Soc Agron; Soil Sci Soc Am; Int Soc Soil Sci; Europ Geophys Soc; Nederl Bodemk Ver. *Res:* Interactions of water, salts and other environmental factors with growth of crops of irrigation agriculture under arid and semi-arid conditions and in humid regions; physics of subsurface watershed hydrology; soil physical measurements. *Mailing Add:* Dept Hydrol Soil Physics & Hydraul Nieuwe Kanaal ll Wageningen 6709 PA Netherlands

DIRKSEN, ELLEN ROTER, b Lagow, Poland, May 10, 29; US citizen; m 56; c 1. CELL BIOLOGY, DEVELOPMENTAL BIOLOGY. *Educ:* Univ Ariz, BS, 49; Univ Ill, MS, 53; Univ Calif, Berkeley, PhD(zool), 61. *Prof Exp:* Res asst, Rheumatic Fever Res Inst, Sch Med, Northwestern Univ, Chicago, 49-51; assoc res physiologist, Cancer Res Inst, Univ Calif, San Francisco, 61-75, assoc prof anat, 74-75; assoc prof anat, 75-81, PROF ANAT, UNIV CALIF, LOS ANGELES, 81- *Mem:* Fel AAAS; Am Soc Cell Biol; Am Asn Anatomists; Soc Develop Biol; Soc Study Reproduction; Am Inst Biol Sci; Sigma Xi. *Res:* Origin, composition and formation of centrioles, basal bodies and cilia; ciliary activity in vertebrate epithelia; structure of oncogenic viruses and function as they relate to human cancers; computer analysis of regulation of ciliary activity; cell communication. *Mailing Add:* Dept Anat & Cell Biol Med Ctr Univ Calif (CHS) 10833 Le Conte Ave Los Angeles CA 90024-1763

DIRKSEN, THOMAS REED, b Pekin, Ill, Nov 5, 31; m 55; c 6. DENTISTRY, BIOCHEMISTRY. *Educ:* Bradley Univ, BS, 53; Univ Ill, DDS, 57; Eastman Dent Ctr, cert pedodontics, 60; Univ Rochester, MS, 61, PhD(biochem), 68. *Prof Exp:* Assoc prof oral biol, Sch Dent & assoc prof biochem, Sch Med, Med Col Ga, coordr biochem dent, 67-78, 67-71, assoc prof cell & molecular biol, Sch Med, 71-73, assoc dean biol sci, 78-90, PROF ORAL BIOL, SCH DENT, MED COL GA, 71-, PROF CELL & MOLECULAR BIOL, SCH MED, 73-, ASSOC DEAN RES, GRAD EDUC, 90- *Concurrent Pos:* Nat Inst Arthritis & Metab Dis study grant, 68-74; Nat Inst Dent Res Caries Study, 75-78; Juvenile Diabetes res grant, 76-78. *Mem:* Am Asn Dent Schs; Am Dent Asn; Int Asn Dent Res; Am Inst Nutrit; Am Asn Dent Res. *Res:* Lipid synthesis by bone and bone cell cultures; lipid constituents of calcified structure; pH of carious cavities. *Mailing Add:* Dept Oral Biol Med Col Ga Sch Dent Augusta GA 30912

DIRLAM, JOHN PHILIP, b Eugene, Ore, Oct 3, 43; m 65; c 2. NATURAL PRODUCTS CHEMISTRY, MEDICINAL CHEMISTRY. *Educ:* Pac Lutheran Univ, BS, 65; Univ Calif, Los Angeles, PhD(org chem), 69. *Prof Exp:* Fel, Univ Lund, 70-71; NSF fel, Yale Univ, 71-72; res chemist, 72-76, sr res scientist med chem, 76-81, sr res investr, 81-87, PRIN RES INVESTR, PFIZER INC, 87- *Mem:* Am Chem Soc. *Res:* Isolation and structural determination of fermentation natural products; antibiotic research; heterocyclic chemistry; synthesis of natural products. *Mailing Add:* Pfizer Cent Res Eastern Point Rd Groton CT 06340

D'ISA, FRANK ANGELO, b Youngstown, Ohio, Mar 30, 21; m 50; c 1. MECHANICAL ENGINEERING. *Educ:* Youngstown Univ, BS, 43; Carnegie Inst Technol, MS, 47; Univ Pittsburgh, PhD(mech eng), 60. *Prof Exp:* From asst prof to assoc prof, 47-60, PROF MECH ENG, YOUNGSTOWN STATE UNIV, 60-, CHMN DEPT, 56- *Mem:* Am Soc Mech Engrs; Nat Soc Prof Engrs; Am Soc Eng Educ; Sigma Xi. *Res:* Strength of materials, elasticity, plasticity, creep, impact, fatigue; engineering mechanics, dynamics of machinery, vibrations. *Mailing Add:* Dept Mech Eng Youngstown State Univ 410 Wick Ave Youngstown OH 44555

DI SABATO, GIOVANNI, immunology; deceased, see previous edition for last biography

DISALVO, ARTHUR F, b New York, NY, May 16, 32; m 58; c 1. MICROBIOLOGY. *Educ:* Univ Ariz, BS, 54, MS, 58; Med Col Ga, MD, 65; Am Bd Microbiol, dipl & cert med mycol, 71. *Prof Exp:* Jr bacteriologist, Ariz State Lab, 56-58; lab dir & bacteriologist, Milledgeville State Hosp, 58-59; res asst, Med Col Ga, 59-63; fel med microbiol, Nat Commun Dis Ctr, 66-68; chief, Bur Labs, Sci Dept Health & Environ Control, 68-90; CLIN PROF PATHOL & LAB MED, UNIV SC, 86-; DIR, NEV STATE HEALTH LAB, 91- *Concurrent Pos:* From asst clin prof to assoc clin prof microbiol, Sch Med, Univ SC, 71-80, adj prof med microbiol, 80-; consult microbiol subcomt, Food & Drug Admin, 75-81; adj prof, Univ NC, 85-; bioanal clin lab dir, Am Bd Bioanal, 78; mem, Med Lab Tech Adv Comt, Midlands Tech Col, 70-84, med dir, 84-90; ed bd, JSC Med Assoc, 81-, Diag Microbiol & Inf Disease, 82-; ed-in-chief, Mycopathologia, 89- *Mem:* AMA; Am Soc Microbiol; AAAS; Int Soc Human & Animal Mycol; Med Mycol Soc Am (secy-treas, 79-82, pres, 84); fel Am Acad Microbiol. *Res:* Medical mycology. *Mailing Add:* Nev State Health Lab 1660 N Virginia Ave Reno NV 89503

DISALVO, FRANCIS JOSEPH, JR, b Montreal, Can, July 20, 44; US citizen; m 66; c 2. SOLID STATE CHEMISTRY, PHYSICS. *Educ:* Mass Inst Technol, BS, 66; Stanford Univ, PhD(appl physics), 71. *Prof Exp:* Mem tech staff, AT&T Bell Labs, 71-78, head, chem physics res dept, 78-81, head, solid state chem res dept, 81-83, head, solid state & physics mat res dept, 83-86; PROF CHEM, CORNELL UNIV, 86- *Mem:* Nat Acad Sci; Am Chem Soc; fel Am Phys Soc. *Res:* Preparation and properties, mainly electrical and magnetic, of new materials. *Mailing Add:* Baker Lab Dept Chem Cornell Univ Ithaca NY 14853-0001

DI SALVO, JOSEPH, b Brooklyn, NY, July 1, 35; m 70; c 3. CARDIOVASCULAR PHYSIOLOGY. *Educ:* NY Univ, BA, 66; Cornell Univ, PhD(physiol), 69. *Prof Exp:* NIH fel, Mich State Univ, 69-70; res assoc pharmacol, Squibb Inst Med, Res, 70-71; asst prof physiol, Ball State Univ, 71-72; asst prof physiol & internal med, 72, assoc prof physiol, 73-78, PROF PHYSIOL, SCH MED, UNIV CINCINNATI, 78- *Concurrent Pos:* Burroughs Wellcome Vis Prof Award, Katholieke Univ, Leuven, Belg, 83; Deutscher Akademischer Austauschdieust, 83. *Mem:* Soc Exp Biol & Med; Am Physiol Soc. *Res:* Regulation of vascular smooth muscle; protein phosphates; phosphorylation-dephosphorylation of contractile proteins. *Mailing Add:* Dept Physiol Univ Minn Sch Med Ten University Dr Duluth MN 55812

DI SALVO, NICHOLAS ARMAND, b New York, NY, Nov 2, 20; m 45; c 2. ORTHODONTICS, PHYSIOLOGY. *Educ:* City Col New York, BS, 42; Columbia Univ, DDS, 45, PhD(physiol), 52; Am Bd Orthodont, dipl, 66. *Prof Exp:* Dent intern, Mem Hosp, 45; asst prof physiol, Col Physicians & Surgeons, 52-57, assoc prof & dir div orthod, 57-58, prof orthod & dir div, Sch Dent & Oral Surg, 58-87, EMER PROF DENT, COLUMBIA UNIV, 87- *Concurrent Pos:* Attend dent surgeon, Presby Hosp, New York, 72-87, consult, 87-; consult orthod, New York City & NY State Dept Health; Vet Admin Hosp, Kingsbridge Bronx, NY; consult dent educ, Proj Hope, Egypt, 76; Nat Def Med Ctr, Taipei, Taiwan ROC, 82; fel, Eighth Inst Advan Educ Dent Res, Am Col Dent. *Mem:* Fel NY Acad Sci; Am Asn Orthod; NE Soc Orthod (pres); Angle Soc Orthod (pres); Int Soc Cranio-Facial Biol (pres). *Res:* Posture and movement of jaws; growth and development of jaws and teeth. *Mailing Add:* Sch Dent & Oral Surg Columbia Univ, 630 W 168th St New York NY 10032

DISALVO, WALTER A, b Harrison, NJ, Dec 23, 20; c 2. CHEMICAL ENGINEERING. *Educ:* Newark Col Eng, BS, 47, MS, 51. *Prof Exp:* Technician, Nopco Chem Co, 39-47, head pilot lab, 47-53; sr res engr, Colgate-Palmolive Co, 53-64, sect head, 64-66, res coordr, 66-70, sr res assoc, 70-73; CONSULT, 80- *Mem:* Am Chem Soc. *Res:* Process development of pharmaceuticals such as vitamins and hormones; product and process development of detergents, paper and plastics; development of personal care products. *Mailing Add:* 237 Prospect Ave North Arlington NJ 07032

DI SANT'AGNESE, PAUL EMILIO ARTOM, b Rome, Italy, Apr 23, 14; US citizen; m 43; c 2. PEDIATRICS. *Educ:* Univ Rome, MD, 38; Columbia Univ, ScD(med), 48; Am Bd Pediat, dipl. *Hon Degrees:* DrMed, Univ Giessen, WGer, 62. *Prof Exp:* Intern, NY Post-Grad Hosp, 40-41, from asst resident to chief resident, 42-44; intern, Willard-Parker Hosp, 41-42; instr pediat, Col Physicians & Surgeons, Columbia Univ, 44-46, assoc, 46-51, assoc prof & head, Cystic Fibrosis & Celiac Prog, 51-59; chief prof pediat, Med Sch, Georgetown Univ, 60-89; chief, Pediat Metab Br, 60-83, EMER SCIENTIST & CONSULT, NIH, 83- *Concurrent Pos:* Asst pediat, Presby Hosp, New York, 44-46, asst attend pediatrician, 48-59; chief, Pediat Div, Vanderbilt Clin, 44-53; lectr pediat, Johns Hopkins Univ, 60-63; dir cystic fibrosis care res & teaching ctr, Children's Hosp, DC, 60-67, mem acad staff & consult, 60-; from vchmn to chmn & trustee, Gen Med & Sci Adv Coun, Nat Cystic Fibrosis Res Found, 62-67, mem exec comt & chmn res comt, 67-; founder trustee & mem exec comt, Int Cystic Fibrosis Asn, 65-, chmn sci med adv coun, 65-69; consult, Cystic Fibrosis Subcomt, Nat Acad Sci, 72-76; mem, Sr Exec Serv US, 79-83. *Honors & Awards:* Hon Medal, Int Cystic Fibrosis Asn, 75, Dirs Award, 83; Numerous Awards, Cystic Fibrosis Found, 59-80. *Mem:* Am Pediat Soc; Soc Pediat Res; Am Acad Pediat; AMA; Am Inst Nutrit; NY Acad Med. *Res:* Cystic fibrosis; pediatric gastroenterology; glycogen storage disease; immunology in children; author of 200 publications. *Mailing Add:* 4928 Sentinel Dr 406 Bethesda MD 20816

DI SANZO, CARMINE PASQUALINO, b Saracena, Italy, Apr 16, 33; US citizen; m 69; c 2. NEMATOLOGY. *Educ:* Univ Bari, Italy, Doctorate, 61; Univ Mass, PhD(plant path), 67. *Prof Exp:* Sr res biologist, 67-80, SR RES ASSOC, AGR CHEM DIV, FMC CORP, 80- *Concurrent Pos:* Leader task force develop stands for screening nematicides, Am Soc Testing & Mat, 72. *Mem:* Am Phytopath Soc; Soc Nematologists; Orgn Trop Am Nematologists; Europ Soc Nematologists. *Res:* Field of nematicides to discover new nematode control agents. *Mailing Add:* 14-704 Eighth Ave Flushing NY 11357

DISCH, RAYMOND L, b Lynbrook, NY, June 10, 32. PHYSICAL CHEMISTRY. *Educ:* Colgate Univ, AB, 54; Harvard Univ, AM, 56, PhD(phys chem), 59. *Prof Exp:* NIH fel phys chem, Oxford Univ & Nat Phys Lab, Eng, 59-63; asst prof chem, Columbia Univ, 63-68; ASSOC PROF CHEM, QUEENS COL, NY, 68- *Mem:* Optical Soc Am. *Res:* Experimental physical chemistry; electric, magnetic and optical properties of fluids; molecular physics. *Mailing Add:* Dept Chem CUNY Queens Col Flushing NY 11367

DISCHER, CLARENCE AUGUST, b Oshkosh, Wis, July 22, 12; m 41. PHYSICAL CHEMISTRY, INORGANIC CHEMISTRY. *Educ:* Oshkosh Teachers Col, EdB, 35; Ind Univ, MA, 45, PhD(chem), 47. *Prof Exp:* Teacher sch, Wis, 36-37; lab asst chem, Oshkosh Teachers Col, 38-39; powder inspector, Kankakee Ord Works, US War Dept, 42; chemist, Interlake Pulp & Paper Wis, 42-43; instr chem, Oshkosh Teachers Col, 43; instr, Preflight Sch, USAF, 43-44; asst chem, Ind Univ, 44-47; from asst prof to prof, 47-73, EMER PROF CHEM, COL PHARM, RUTGERS UNIV, 73- *Mem:* Am Chem Soc; Sigma Xi; Am Asn Col Pharm. *Res:* Theoretical electrochemistry; instrumental methods of analysis; inorganic pharmaceutical chemistry. *Mailing Add:* Whippoorwill Rd Birnamwood WI 54414

DISHART, KENNETH THOMAS, b Pittsburgh, Pa, July 31, 31. ORGANIC CHEMISTRY. *Educ:* Univ Pittsburgh, BSc, 53, PhD(chem), 58. *Prof Exp:* Res chemist, 58-68, supvr fuels performance group, 68-70, head chem div, Del, 70-74, tech mgr, Petrol Chem Div, 74-80, sr res assoc, Freon Prod Div, 80-87, SR RES ASSSOC, ELECTRONICS MAT DIV, E I DU PONT DE NEMOURS & CO, INC, 87- *Mem:* Am Chem Soc; Am Soc Testing & Mat. *Res:* Organic synthesis; fluorine chemistry; petroleum chemicals; automotive emissions; electronic soldering and cleaning technologies. *Mailing Add:* 319 Walden Rd Wilmington DE 19803

DISINGER, JOHN FRANKLIN, b Lockport, NY, July 7, 30; m 60; c 2. CONSERVATION. *Educ:* State Univ NY Teachers Col Brockport, 52; Univ Rochester, EdM, 60; Ohio State Univ, PhD(educ), 71. *Prof Exp:* Teacher pub schs, Rochester, NY, 56-70; from asst prof to assoc prof, 71-80, actg dir, 88-89, PROF NATURAL RESOURCES, OHIO STATE UNIV, 80- *Concurrent Pos:* Res assoc, Info Anal Ctr Sci, Math & Environ Educ, Educ Resource Info Ctr, 71-74, assoc dir, 74- *Honors & Awards:* Walter Jeske Award, NAm Asn Environ Educ, 84. *Mem:* NAm Asn Environ Educ (pres, 85-86); Nat Sci Teachers Asn. *Res:* Objectives and methodology relative to environmental management education; definitional status of environmental education; instructional methodologies for interdisciplinary teaching. *Mailing Add:* Sch Natural Resources Ohio State Univ Columbus OH 43210-1085

DISKO, MILDRED ANNE, b Athens, Ala, Mar 24, 27; m 47; c 1. MATHEMATICS, STATISTICS. *Educ:* Univ Ala, BS, 48; Ohio Univ, MS, 69, PhD(educ res math), 73. *Prof Exp:* High sch teacher sci & math, Jackson Co, Ohio, 48-49 & Athens Co, Ohio, 62-65; instr, Ohio Univ, 65-72; asst prof, Morris Harvey Col, 73-74; PROF MATH, GLENVILLE STATE COL, 74- *Mem:* Math Asn Am; Asn Women Math. *Res:* Measurement of mathematics processing skills. *Mailing Add:* 16 Briarwood Dr Athens OH 45701

DISMUKES, EDWARD BROCK, b Georgiana, Ala, Oct 9, 27; m 54; c 2. INORGANIC CHEMISTRY. *Educ:* Birmingham-Southern Col, BS, 49; Univ Wis, MS, 51, PhD(chem), 53. *Prof Exp:* Asst phys chem, Univ Wis, 49-52; chemist, Phys Div, 53-58, sr chemist, 58-59, head phys chem sect, 59-66, head chem defense sect, 66-71, head phys chem sect, 71-78, sr res adv, 78-89, PRIN CHEMIST, SOUTHERN RES INST, 89- *Concurrent Pos:* Lectr, Birmingham-Southern Col, 55-60. *Mem:* Am Chem Soc; Sigma Xi. *Res:* Analysis of pollutants in air and water; properties of molten silicates; properties of aqueous complex ion systems; electrochemistry; colloid chemistry; infrared spectrophotometry; kinetics of chemical reactions; electrostatic precipitation of fly ash. *Mailing Add:* 2321 Lane Circle Birmingham AL 35223-1713

DISMUKES, GERARD CHARLES, b Boston, Mass, July 30, 49; m; c 1. BIOCHEMISTRY. *Educ:* Lowell Technol Inst, BSc, 71; Univ Wis-Madison, PhD(phys chem), 75. *Prof Exp:* Fel biophys, Univ Calif Lab Chem Biodynamics, Lawrence Berkeley Lab, 75-77; asst prof, 77-84, ASSOC PROF CHEM, PRINCETON UNIV, 84- *Concurrent Pos:* Searle scholars award, 81; vis scientist, Serv de Biophysique, CEN-Saclay, France, 84. *Honors & Awards:* Alfred P Sloane Award, 84. *Mem:* Sigma Xi; Am Chem Soc; Biophys Soc. *Res:* Elucidation of the primary photophysical and photochemical processes involved in photosynthesis. *Mailing Add:* Dept Chem Princeton Univ Princeton NJ 08544

DISMUKES, ROBERT KEY, b Dahlonega, Ga, June 21, 43. NEUROSCIENCE. *Educ:* NGa Col, BS, 64; Vanderbilt Univ, MA, 66; Pa State Univ, PhD(biophysics), 71. *Prof Exp:* Fel, Johns Hopkins Univ Sch Med, 72-73; staff fel, NIH, 73-75; vis scientist, Free Univ, Neth, 75-76; staff scientist, Neurosci Res Prog, Mass Inst Technol, 77-79; study dir, Comt Vision, Nat Acad Sci, 79-; dir life sci, Air Force Off Sci Res Bolling AFB, Washington, DC; CHIEF, AEROSPACE HUMAN FACTOR RES DIV, NASA, 89- *Concurrent Pos:* Fel, Inst Soc, Ethics & Life Sci, 76-77. *Mem:* Soc Neurosci; Int Brain Res Orgn; AAAS. *Res:* Neurochemistry and psychopharmacology; conceptual research on brain function and behavior; writing, science and public policy. *Mailing Add:* Moffett NASA Ames Res Ctr MS 262-1 Moffett Field CA 94035-1000

DISNEY, RALPH L(YNDE), b Baltimore, Md, Feb 27, 28; m 55; c 3. PROBABILITY. *Educ:* Johns Hopkins Univ, BE, 52, MSE, 55. *Hon Degrees:* DEng, Johns Hopkins Univ, 64. *Prof Exp:* Engr, Indust Diecraft, Inc, 53-55; res analyst, Opers Res Off, 55-56; asst prof eng, Lamar State Col Technol, 56-59; assoc prof, Univ Buffalo, 59-63; vis assoc prof, Univ Mich, 63-64, assoc prof, 64-68, prof indust eng, 68-77; CHARLES O GORDON PROF ENG, VA POLYTECH INST & STATE UNIV, 77- *Concurrent Pos:* Vis lectr, Univ Mich, 62-63; Orgn Am States vis prof, Inst Aeronaut Technol, Brazil, 70-71; sr ed, Am Inst Indust Engrs, 72-; distinguished vis prof, Grad Sch, Ohio State Univ, 74-75. *Honors & Awards:* David F Baker Distinguished Res Award, Am Inst Indust Engrs, 72. *Mem:* fel Am Inst Indust Engrs; Soc Indust & Appl Math; Am Math Asn; Inst Math Statist; Opers Res Soc Am; Sigma Xi. *Res:* Stochastic networks and processes; operations research. *Mailing Add:* Dept Indust Eng Tex A&M Univ College Station TX 77843

DISPIRITO, ALAN ANGELO, b North Smithfield, RI, Jan 2, 54; m 87. MICROBIOLOGY. *Educ:* Providence Col, BS, 78; Ohio State Univ, MS, 80, PhD(microbiol), 83. *Prof Exp:* Postdoctoral res assoc, Univ Minn, 83-85 & Univ Wis-Milwaukee, 85-87; sr res assoc, Calif Inst Technol, 87-89; asst prof, Univ Tex, Arlington, 89-91; ASST PROF, DEPT MICROBIOL, IOWA STATE UNIV, 91- *Mailing Add:* Dept Microbiol 205 Sci Bldg I Iowa State Univ Ames IA 50011

DI STEFANO, HENRY SAVERIO, b Palermo, Italy, Jan 1, 20; US citizen; m 51; c 2. ANATOMY. *Educ:* Brooklyn Col, BA, 41; Columbia Univ, MA, 46, PhD(zool), 48. *Prof Exp:* Instr anat, Col Med, Syracuse Univ, 48-51; from instr to prof, 51-76, MEM ANAT FAC, STATE UNIV NY UPSTATE MED CTR, 76- *Mem:* Histochem Soc; Am Asn Anat; Am Soc Cell Biol; Electron Micros Soc Am; NY Acad Sci. *Res:* Cytochemistry and electron microscopy; viral induced avian and murine leukemias. *Mailing Add:* Dept Anat State Univ NY Sci Col Med 750 E Adams St Syracuse NY 13210

DISTEFANO, JOSEPH JOHN, III, b Brooklyn, NY, Apr 30, 38; c 2. BIOCYBERNETICS, CONTROL SYSTEMS. *Educ:* City Col New York, BEE, 61; Univ Calif, Los Angeles, MS, 64, PhD(control systs, biocybernet), 66. *Prof Exp:* PROF ENG, COMPUT SCI & MED, UNIV CALIF, LOS ANGELES, 66- *Concurrent Pos:* NATO fel clin med, Radiol Lab, Univ Rome, 67-68; Fulbright scholar, 80. *Mem:* Inst Elec & Electronics Engrs; Endocrine Soc; Am Thyroid Asn; Am Soc Clin Res; Biomed Eng Soc. *Res:* Endocrinology, particularly thyroid; modeling theory for biology and medicine; bioengineering systems. *Mailing Add:* BH 4731 Univ Calif Los Angeles CA 90024-1596

DISTEFANO, THOMAS HERMAN, b Philadelphia, Pa, Dec 21, 42; m 75; c 2. SOLID STATE PHYSICS. *Educ:* Lehigh Univ, BS, 64; Stanford Univ, MS, 65, PhD(appl physics), 70. *Prof Exp:* Staff mem, Stanford Linear Accelerator Ctr, Stanford Univ, 69-70; staff mem, T J Watson Res Ctr, 70-74, mgr interface physics, 74-77, tech planning staff res, 77-78, MGR OPTICAL STORAGE, IBM RES, IBM CORP, 78- *Concurrent Pos:* Consult, Synvar Assocs, 69-70. *Mem:* Inst Elec & Electronics Engrs; AAAS; Am Phys Soc. *Res:* Optical storage technology and materials; photoemission spectroscopy; non-destructive testing including acoustic microscopy and photovoltaic imaging; internal photoemission and interfaces. *Mailing Add:* IBM Corp PO Box 218 Yorktown Heights NY 10598

DISTEFANO, VICTOR, b Rochester, NY, Mar 17, 24; m 47; c 5. PHARMACOLOGY. *Educ:* Univ Rochester, AB, 49, PhD(pharmacol), 53. *Prof Exp:* Instr pharmacol, Sch Med, Marquette Univ, 53-54; from instr to assoc prof pharmacol, 54-72, assoc prof toxicol & radiation biol, 70-72, PROF PHARMACOL & TOXICOL, SCH MED, UNIV ROCHESTER, 72-, PROF RADIATION BIOL & BIOPHYS, 76- *Mem:* Am Soc Pharmacol; Sigma Xi. *Res:* Autonomics; antiradiation; central nervous system. *Mailing Add:* 1055 Quaker Rd Scottsville NY 14546

DISTERHOFT, JOHN FRANCIS, b Marengo, Iowa, June 9, 44. NEUROBIOLOGY. *Educ:* Loras Col, BA, 66; Fordham Univ, MA, 68, PhD(psychol), 71. *Prof Exp:* USPHS training grant, Calif Inst Technol, 70-73; asst prof anat, 73-78, assoc prof anat, 78-90, PROF CELL, MOLECULAR & STRUCT BIOL, MED SCH, NORTHWESTERN UNIV, CHICAGO, 90- *Concurrent Pos:* Res investr, Lab Biophys, Nat Inst Neurol & Commun Disorders & Stroke, Marine Biol Lab, Woods Hole, Ma, 84-86; NINCDS fel, 84-86. *Mem:* AAAS; Am Asn Anat; Soc Neurosci. *Res:* Neurophysiological and neuroanatomical foundations of plasticity in mammalian central nervous system. *Mailing Add:* Dept Cellular Molecular & Struct Biol Northwestern Univ 303 E Chicago Ave Chicago IL 60611-3008

DISTLER, JACK, JR, Pontiac, Mich, Dec 7, 28. COMPLEX CARBOHYDRATES. *Educ:* Mich State Univ, BS, 52; Univ Mich, MS, 54, PhD, 64. *Prof Exp:* Teaching asst bot, 54-55, res asst, Rackham Arthritis Res Unit, 56-59, res assoc internal med, 64-74, asst res scientist, 74-77, ASSOC RES SCIENTIST, DEPT INTERNAL MED, UNIV MICH, 77- *Concurrent Pos:* Arthritis Found fel, 68-71; instr biochem, Univ Mich, 66-71. *Mem:* Am Soc Biol Chem; Sigma Xi; AAAS. *Res:* Biochemistry of complex carbohydrates and cell membrane receptors. *Mailing Add:* 575 Scio Church St Ann Arbor MI 48103

DISTLER, RAYMOND JEWEL, b Paducah, Ky, July 3, 30; m 51; c 4. COMPUTER ENGINEERING. *Educ:* Univ Ky, BSEE, 51, MSEE, 60, PhD(math, elec eng), 64. *Prof Exp:* Mem tech staff, Bell Tel Labs, NJ, 51-58; from instr to asst prof, 58-66, ASSOC PROF ELEC ENG, UNIV KY, 66- *Mem:* Inst Elec & Electronics Engrs. *Res:* Microprocessors; logical design; digital system simulation. *Mailing Add:* Dept Elec Eng Univ Ky Lexington KY 40506

DITARANTO, ROCCO A, b Philadelphia, Pa, Aug 12, 26; m 56; c 4. ENGINEERING MECHANICS, MECHANICAL ENGINEERING. *Educ:* Drexel Inst Tech, BS, 47; Univ Pa, MS, 50, PhD(eng mech), 61. *Prof Exp:* Jr engr, Philco Corp, 48-50; sr engr, Vertol Helicopter Co, 50-53; anal engr, Aviation Gas Turbine Div, Westinghouse Elec Co, 53-55; shock & vibration consult, RCA Defense Electronics Prod, 55-59; assoc prof mech eng, Drexel Univ, 59-62; PROF ENG, WIDENER UNIV, 62- *Concurrent Pos:* Eng consult, US Navy Marine Eng Lab, 61-73; vis prof, Inst Sound & Vibration, Univ Southampton, 69-70; acoustical specialist, Environ Protection Agency, 73; mech consult, Scott Paper Co, 75-79, Boeing Helicopter, 79- *Mem:* Am Soc Mech Engrs; Am Soc Eng Educ; Am Inst Aeronaut & Astronaut; Acoust Soc Am; Sigma Xi. *Res:* Shock and vibrations; vibration damping; dynamics of structures and shells; mechanics of laminated structures; acoustic noise survey and design abatement. *Mailing Add:* Dept Eng Widener Univ Chester PA 19013

DITCHEK, BRIAN MICHAEL, b New York, NY, Jan 31, 51; m 75. MATERIALS SCIENCE, PHYSICAL METALLURGY. *Educ:* State Univ NY, Stony Brook, BE, 73; Northwestern Univ, PhD(mat sci), 77. *Prof Exp:* Res asst spinodol decomposition, Northwestern Univ, 73-77; fel, Nat Bur Standards, 77-78; res scientist ceramics, Martin Marietta Labs, 78-81; PRIN MEM TECH STAFF, GTE, 81- *Concurrent Pos:* Vchmn comt fatigue res, Am Soc Testing & Mat, 78-; assoc Nat Res Coun, Nat Bur Standards, 77-78, Indust Res, 84. *Mem:* Am Soc Metals; Am Soc Testing & Mat; Am Inst Mining, Metall & Petrol Engrs; Nat Res Soc. *Res:* Role of phase transformations in improving material properties; modulated structures; fatigue processes in metals; sintering and strengthening behavior of high temperature ceramics; directional solidification; crystol growth, silicides. *Mailing Add:* GTE Labs Inc Dept 306 40 Sylvan Rd Waltham MA 02254

DITMARS, JOHN DAVID, b Trenton, NJ, Oct 16, 43; m 66; c 4. CIVIL ENGINEERING, ENVIRONMENTAL SCIENCES. *Educ:* Princeton Univ, BSE, 65; Calif Inst Technol, MS, 66, PhD(civil eng), 71. *Prof Exp:* Vis asst prof civil eng, Mass Inst Technol, 70-72; asst prof civil eng & marine studies, Univ Del, 72-77; engr & mgr, Water Resources Sect, Argonne Nat Lab, 77-85, sr engr & mgr, Hydrol & Geol Eng Sec, 85-87, assoc dir eng geosci, Energy & Environ Systs Div, 87-88, DIR ENVIRON RESTORATION, ENVIRON ASSESSMENT & INFO SCI DIV, ARGONNE NAT LAB, 89- *Concurrent Pos:* Consult var eng & indust firms, 70-; res assoc, Energy & Environ Systs Div, Argonne Nat Lab, 76-77; mem, Study Comt Land, Sea & Air Disposal Indust & Domestic Wastes, Nat Res Coun, 83; grant review panel, Charles A Lindbergh Fund, Inc, 83-; proj dir, Roy F Weston Inc, 88-89. *Mem:* Am Soc Civil Engrs; Am Geophys Union. *Res:* Mixing and transport processes in hydrologic environments; wastewater jets and air-bubble plumes; model verification with prototype data; environmental impacts of advanced ocean energy systems; hydrologic tracer studies; performance aspects of high-level radioactive waste disposal in geologic media; waste site characterization strategies; remediation hazardous waste sites. *Mailing Add:* Environ Assessment & Info Sci Div Argonne Nat Lab Argonne IL 60439

D'ITRI, FRANK M, b Flint, Mich, Apr 25, 33; m 55; c 4. ANALYTICAL CHEMISTRY, WATER CHEMISTRY. *Educ:* Mich State Univ, BS, 55, MS, 66, PhD(anal chem), 68. *Prof Exp:* Technologist, Dow Chem Co, 60-62; asst anal chem, 62-68, from asst prof to assoc prof water chem, 68-77, PROF WATER CHEM & ASSOC DIR, INST WATER RES & DEPT FISH WILDLIFE, MICH STATE UNIV, 77- *Concurrent Pos:* NIH fel, 67-68; prin investr, Off Water Resources Res, US Dept Interior, 69-; adv environ mercury pollution, Mich House Rep, 70-72, US Environ Protection Agency, 75-77, US Dept Energy, 82-84, US House Rep oversight, 84-85 & Europ Econ Community-UN, 82-85; mem critical pollution mat comt, Mich Water Resources Comn, 71-75; adv, Revision Water Qual Criteria Publ, Nat Acad Sci, 71-73; Rockefeller scholar, 72 & 75; fel, Japan Soc Prom Sci, 80 & 87, Europ Econ Community-UN, 82-85. *Mem:* Am Chem Soc; Soc Environ Toxicol & Chem. *Res:* Analytical aspects of water and sediment chemistry with special interest on the transformation and translocation of mercury; phosphorus, nitrogen, heavy metals, and hazardous organic chemicals in the environment. *Mailing Add:* Rm 334 Natural Resources Bldg Mich State Univ East Lansing MI 48824-1222

DITTBERNER, PHILLIP LYNN, b Riverside, Calif, Feb 27, 44; m 66; c 3. BOTANY, ECOLOGY. *Educ:* WTex State Univ, BS, 67; NMex State Univ, MS, 71; Colo State Univ, PhD(range ecol), 73. *Prof Exp:* Ecologist, Nat Park Serv, 72-75; plant ecologist, US Fish & Wildlife Serv, 75-84, natural resource consult, 84; natural resource consult, Natural Resource Professionals, 84-89; PLANT ECOLOGIST, SERV CTR, BUR LAND MGT, 89- *Concurrent Pos:* Mem, Publ Comt, Soc Range Mgt, 74-76 & Colo Sect Rangeland Ref Comt, 75-76. *Mem:* Soc Range Mgt; Ecol Soc Am; Am Soc Agron; Soil Sci Soc Am; Sigma Xi. *Res:* Wildlife habitat reclamation; resource management systems; soil-plant-animal relationships; reclamation of surface disturbed areas; plant and animal species adaptation to site conditions on reclaimed mine and other disturbed lands; designing reclamation plans for specific post mining land uses. *Mailing Add:* Bur Land Mgt Serv Ctr PO Box 25047 Denver CO 80225-0047

DITTERLINE, RAYMOND LEE, b Mesquite, NMex, July 29, 41; m 63; c 3. PLANT BREEDING, AGRONOMY. *Educ:* NMex State Univ, BS, 63, MS, 70; Mont State Univ, PhD(agron), 73. *Prof Exp:* From instr to assoc prof, 73-83, PROF AGRON, MONT STATE UNIV, 83- *Mem:* Am Soc Agron; Crop Sci Soc Am; Nat Alfalfa Improv Conf. *Res:* Breeding alfalfa; nodulation of legumes; disease and insect resistance; seedling vigor; fertility; forage nutritive value. *Mailing Add:* Dept Plant & Soil Sci Mont State Univ Bozeman MT 59715

DITTERT, LEWIS WILLIAM, b Philadelphia, Pa, Jan 22, 34; m 57; c 2. PHYSICAL CHEMISTRY. *Educ:* Temple Univ, BS, 56; Univ Wis, MS, 60, PhD(pharm), 61. *Prof Exp:* Sr res pharmacist, Smith Kline & French Labs, 61-67; from asst prof to prof pharm, Col Pharm, Univ Ky, 67-78; dean, Sch Pharm, Univ Pittsburgh, 78-85; PROF PHARMACEUT & PHARMACEUT ANALYSIS DIV, COL PHARM, UNIV KY, 85- *Concurrent Pos:* Consult to pharmaceut mfrs; Centennial fel, Sch Pharm, Temple Univ, 85. *Honors & Awards:* Sprowls lectr, Sch Pharm, Temple Univ, 85. *Mem:* Am Pharmaceut Asn; fel Acad Pharmaceut Sci; Am Asn Col Pharm; Sigma Xi; fel Am Asn Pharmaceut Scientists. *Res:* Physical pharmacy; biopharmaceutics; performance of pharmaceutical products in humans, absorption of drugs from gastrointestinal tract; drug metabolism; clinical pharmacokinetics. *Mailing Add:* Col Pharm Univ Ky Lexington KY 40536-0082

DITTFACH, JOHN HARLAND, b St Paul, Minn, Apr 17, 18; m 44; c 5. MECHANICAL ENGINEERING. *Educ:* Univ Minn, BSME, 47, MSME, 48. *Prof Exp:* Instr mech eng, Univ Minn, 47-48; from asst prof to assoc prof, 48-70, PROF MECH ENG, UNIV MASS, AMHERST, 70- *Res:* Experimental testing techniques; energy conversion. *Mailing Add:* S East St Amherst MA 01002

DITTMAN, FRANK W(ILLARD), b Pittsburgh, Pa, July 22, 18; m 46; c 4. CHEMICAL ENGINEERING. *Educ:* Univ Pittsburgh, BS, 40; Cornell Univ, MChE, 43, PhD(chem eng), 44. *Prof Exp:* Asst chem eng, Cornell Univ, 40-44, instr math, 43-44; process engr, Koppers Co, Inc, 46-52; process engr, Rust Process Design Co, 52-55, chief chem process consult, 55-58; res technologist, Appl Res Lab, US Steel Corp, Pa, 58-62; PROF CHEM ENG, RUTGERS UNIV, NEW BRUNSWICK, 62- *Concurrent Pos:* Res engr, Rubber Reserve Co, 43-44; instr, Carnegie Inst Technol, 47-50; coun mem & former mayor, Bridgewater, NJ; chmn, Energy Mgt Coun, Somerset Co, NJ. *Honors & Awards:* Serv to Soc Award, Am Inst Chem Eng, 77. *Mem:* Am Inst Chem Engrs; Filtration Soc. *Res:* Mass, heat and momentum transfer, including distillation, extraction, drying, reverse osmosis, and filtration; design, construction and economics of chemical plants; distillation, extraction, drying, filtration, decomposition of residual organics in wastewater. *Mailing Add:* Dept Chem Eng Rutgers State Univ New Brunswick NJ 08903

DITTMAN, RICHARD HENRY, b Sacramento, Calif, July 5, 37; m 62; c 2. SURFACE PHYSICS. *Educ:* Univ Santa Clara, BS, 59; Univ Notre Dame, PhD, 65. *Prof Exp:* Guest scientist, Fritz-Haber-Inst, W Berlin, 65-66; from asst prof to assoc prof physics, Univ Wis-Milwaukee 66-90, chmn, Dept Physics, 72-75, assoc dean, Col Lett & Sci, 75-77, PROF PHYSICS, UNIV WIS-MILWAUKEE, 90- *Concurrent Pos:* Vis prof, Univ Fribourg, Switzer, 85-86 & 89. *Mem:* Am Asn Physics Teachers. *Res:* Physics; thermodynamics. *Mailing Add:* Dept Physics Univ Wis-Milwaukee Milwaukee WI 53201-0413

DITTMANN, JOHN PAUL, b Winfield, Kans, Nov 28, 48; m 69; c 3. OPERATIONS RESEARCH, INDUSTRIAL ENGINEERING. *Educ:* Univ Mo-Columbia, BS, 70, MS, 71, PhD(indust eng), 73. *Prof Exp:* Opers res analyst, 73-75, mgr mkt res, 75-76, MGR DISTRIB OPERS MGT SCI, WHIRLPOOL CORP, 76- *Mem:* Am Inst Indust Engrs; Nat Coun Phys Distrib Mgt. *Res:* Management science. *Mailing Add:* 2000 M63 Benton Harbor MI 49022

DITTMER, DONALD CHARLES, b Quincy, Ill, Oct 17, 27. ORGANIC CHEMISTRY. *Educ:* Univ Ill, BS, 50; Mass Inst Technol, PhD(org chem), 53. *Prof Exp:* Fel chem, Harvard Univ, 53-54; from instr to asst prof, Univ Pa, 54-61; E I du Pont de Nemours & Co fel, 61-62; assoc prof, 62-66, PROF CHEM, SYRACUSE UNIV, 66- *Mem:* Am Chem Soc; Royal Soc Chem; AAAS; Sigma Xi. *Res:* Model enzyme systems; small-ring heterocyclic chemistry; organic reaction mechanisms; organometallic chemistry; chalcoger chemistry; novel synthetic methods. *Mailing Add:* Dept Chem Syracuse Univ Bowne Hall Syracuse NY 13244

DITTMER, HOWARD JAMES, b Pekin, Ill, Jan 29, 10; m 41; c 1. BOTANY, MORPHOLOGY. *Educ:* Univ NMex, AB, 33, AM, 34; Univ Iowa, PhD(plant morphol), 38. *Prof Exp:* Prof sci, Chicago Teachers Col, 38-43; from assoc prof to prof, 43-75, asst dean, 56-70, assoc dean, Col Arts & Sci, 70-75, EMER PROF BIOL, UNIV NMEX, 75- *Concurrent Pos:* Environ consult, Kennecott Copper Co, 70-75. *Mem:* AAAS (past pres, Southwestern & Rocky Mt Div); Bot Soc Am; Ecol Soc Am. *Res:* Investigation of subterranean plant parts; quantitative study of roots and root hairs and their relation to physics of soil; flora of New Mexico; lawn problems of the southwest; phylogeny and form in the plant kingdom; root biomass; biomass of desert plants; root systems. *Mailing Add:* 60 Juniper Hill Loop NE Albuquerque NM 87122

DITTMER, JOHN EDWARD, b Colesburg, Iowa, May 10, 39; m 62; c 2. IMMUNOLOGY. *Educ:* Ariz State Univ, BS, 62; Brown Univ, MA, 66, PhD(develop biol), 70. *Prof Exp:* Instr high sch, Kans, 62-65; asst prof, 69-77, ASSOC PROF ANAT, SCH MED, BOSTON UNIV, 77-, ASST PROF PATH, 81- *Concurrent Pos:* Vis prof, Henan Med Univ, Zhengzhou, China. *Mem:* AAAS; Transplantation Soc; Am Soc Immunologists. *Res:* Transplantation immunology. *Mailing Add:* Dept Anat Sch Med Boston Univ Boston MA 02118

DITTMER, KARL, b Hebron, NDak, Jan 18, 14; m 49; c 4. CHEMISTRY. *Educ:* Jamestown Col, BA, 37; Univ Colo, MA, 39; Cornell Univ, PhD(biochem), 44. *Prof Exp:* Fel antibiotins & penicillin assay methods, Biochem Dept, Med Col, Cornell Univ, 44-45; from asst prof to assoc prof chem, Univ Colo, 45-49; prof & head dept, Fla State Univ, 49-58; head div grants & fels prog adminr petrol res fund, Am Chem Soc, 58-64; vpres acad affairs, Fla State Univ, 64-66; dean, Col Sci, Portland State Univ, 66-79, prof chem & coordr environ sci, 74-79, emer dean & prof chem, 79-90; RETIRED. *Mem:* AAAS; Am Chem Soc; Am Soc Biol Chemists; Sigma Xi. *Res:* Structural basis for antimetabolites; antivitamins; antibiotins; antiamino acids synthesis and properties; amino acid metabolism of rats and bacteria; biosynthesis of vitamins and amino acids; synthesis of nucleosides and derivatives; microbiological study of structural specificity of biotin. *Mailing Add:* 1851 S Shore Blvd Lake Oswego OR 97034

DITTNER, PETER FRED, b Vienna, Austria, Mar 24, 37; US citizen; m 80; c 2. ATOMIC PHYSICS. *Educ:* NY Univ, BS, 58, PhD(physics), 65. *Prof Exp:* Res assoc chem, 65-76, res staff mem II, 76-87, GROUP LEADER, OAK RIDGE NAT LAB, 88- *Concurrent Pos:* Humboldt sr sci fel, 72-73. *Mem:* Am Phys Soc. *Res:* Dielectronic recombination; atomic beams; atom-surface interactions; channeling. *Mailing Add:* Physics Div Oak Ridge Nat Lab PO Box 2008 Oak Ridge TN 37831-6377

DIUGUID, LINCOLN ISAIAH, b Lynchburg, Va, Feb 6, 17; m 52; c 4. ORGANIC CHEMISTRY. *Educ:* WVa State Col, BS, 38; Cornell Univ, MS, 39, PhD(org chem), 45. *Prof Exp:* Head chem dept, Ark State Col, 39-43; anal chemist, Pine Bluff Arsenal, Chem Warfare Serv, US Army, 42-43; Merrill res fel, Cornell Univ, 43-45, Off Sci Res & Develop res fel, 45- 46, res assoc, 46-47, B F Goodrich Rubber res fel, 47; prof chem, 54-74, chmn dept phys sci, 70-74, PROF PHYS SCI, HARRIS-STOWE COL, 74-, CHMN DEPT, 77-; RES DIR, DU-GOOD CHEM LAB, 47- *Concurrent Pos:* Cancer res, Jewish Hosp, St Louis, 59-; mem, Leukemia Res Proj, 61-63; vpres, Leukemia Guild, Mo & Ill, 64-; vis prof, Wash Univ, St Louis, 65-67. *Mem:* Am Chem Soc; fel Am Inst Chemists; Asn Consult Chemists & Chem Eng; Sigma Xi. *Res:* New method synthesis of primary aliphatic alcohols, large carbon ring compounds and benzothiazole derivatives; new micromethods for organic quantitative analyses; industrial development; thyroxine research (new micro methods of detection of T4 and T3) in human blood serum. *Mailing Add:* 1215 S Jefferson Ave St Louis MO 63104-1998

DIVADEENAM, MUNDRATHI, b Pulukurthy, India, June 21, 35; US citizen; m 64; c 2. NUCLEAR PHYSICS. *Educ:* Osmania Univ, India, BSc, 54, MSc, 56; Duke Univ, PhD(nuclear physics), 67. *Prof Exp:* Lectr physics, Osmania Med Col, India, 56-57; jr sci asst physics, Int Geophys Year Sponsored, 57-58; from res asst to res assoc nuclear physics, 60-69; asst prof physics, Prairie View Agr & Mech Col, 69-70; asst prof physics, NC Cent Univ & Duke Univ, 70-74; PHYSICIST, NAT NUCLEAR DATA CTR, BROOKHAVEN NAT LAB, 74- *Mem:* Am Phys Soc. *Res:* Average and high resolution neutron total cross sections; strength functions; optical model;

doorway states; intermediate structure; analog states; nuclear structure; neutron resonance parameters; thermal cross sections; neutron induced reactions. *Mailing Add:* Neutral Beam Div Brookhaven Nat Lab B-902c Upton NY 11973

DIVELEY, WILLIAM RUSSELL, b Royal Center, Ind, June 13, 21; m 49; c 1. ORGANIC CHEMISTRY. *Educ:* Manchester Col, BS, 43; Purdue Univ, MS, 50, PhD(chem), 52. *Prof Exp:* Res chemist, Hercules, Inc, 52-63, sr res chemist, 63-71, res scientist, 71-82; RETIRED. *Mem:* Am Chem Soc. *Res:* Synthetic organic chemistry; agricultural chemicals and pesticides; organophosphorous chemistry; polymer chemistry. *Mailing Add:* 3713 Valleybrook Rd Oakwood Hills Wilmington DE 19808

DIVEN, BENJAMIN CLINTON, b Chico, Calif, Jan 5, 19; m 51; c 3. NUCLEAR PHYSICS. *Educ:* Univ Calif, AB, 41; Univ Ill, MS, 48, PhD(physics), 50. *Prof Exp:* Mem staff, Los Alamos Sci Lab, Univ Calif, 43-45 & 50-77; CONSULT, 77- *Concurrent Pos:* Consult, 77- *Mem:* Fel Am Phys Soc. *Res:* Experimental nuclear physics. *Mailing Add:* 4730 Sandia Dr Los Alamos NM 87544

DIVEN, WARREN FIELD, b St Louis, Mo, Oct 10, 31; m 54; c 3. BIOCHEMISTRY. *Educ:* Hastings Col, BA, 52; Univ Nebr, MS, 60, PhD(chem), 62. *Prof Exp:* NIH fel, Univ Wis, 62-64; asst prof biochem, 64-71, ASSOC PROF PATH & BIOCHEM, SCH MED, UNIV PITTSBURGH, 71- *Mem:* NY Acad Sci; Am Asn Biol Chemists; Am Asn Clin Chem; Am Chem Soc; Nat Acad Clin Chem. *Res:* Clinical biochemistry; enzyme defects in inborn errors of metabolism; enzymology; molecular mechanisms of metabolic control. *Mailing Add:* Dept Path Sch Med Univ Pittsburgh Pittsburgh PA 15261

DIVER, RICHARD BOYER, JR, b El Paso, Tex, June 10, 51; m 75; c 2. SOLAR THERMAL ENERGY DEVELOPMENT, APPLIED THERMODYNAMICS. *Educ:* US Mil Acad, West Point, BS, 73; Univ Minn, MS, 77, PhD(mech eng), 79. *Prof Exp:* Res assoc solar energy, dept mech eng, Univ Minn, 80-84; MEM TECH STAFF SOLAR ENERGY, SANDIA NAT LABS, ALBUQUERQUE, 84- *Mem:* Sigma Xi. *Res:* Development of receivers for solar thermal energy systems; development of analytical tools and receivers for dish-electric and thermochemical applications. *Mailing Add:* Sandia Nat Labs PO Box 5800 Div 6217 Albuquerque NM 87185

DIVETT, ROBERT THOMAS, b Salt Lake City, Utah, Nov 4, 25; m 53; c 6. INFORMATION SCIENCE, SCIENCE EDUCATION. *Educ:* Brigham Young Univ, BS, 53; George Peabody Col, MA, 55; Univ Utah, EdD(educ admin & media), 68. *Prof Exp:* Teacher-librn, Idaho Jr High Sch, 53-54; first asst, Med Libr, Vanderbilt Univ, 54-56; asst prof libr sci & med librn, Univ Utah, 56-62; med sci librn, 63-77, assoc prof, 63-81, EMER PROF MED BIBLIOG, UNIV NMEX, 81- *Concurrent Pos:* Consult, Okla Med Ctr, 66 & Mayo Clin, 68; dir, Health Sci Info & Commun Ctr, 68-71; vis assoc prof, Univ Wash, 69; contribr, Third Int Cong Med Librarianship, Amsterdam, 69. *Honors & Awards:* Murray Gottlieb Prize, Med Libr Asn, 59 & 62. *Res:* Computer file structures for on-line storage and retrieval of literature citations and other biomedical data and information; history of medicine on the American frontier, particularly in the mountain West; patterns of acquisition and use of biomedical information by medical practitioners in small New Mexico communities. *Mailing Add:* 1004 Menaul NE Apt K2 Albuquerque NM 87712

DIVGI, CHAITANYA R, b Bombay, India, July 24, 53; m 78; c 1. NUCLEAR MEDICINE, CANCER RESEARCH. *Educ:* Bangalore Univ, India, MB & BS, 76; Bombay Univ, DRM, 78; NY Univ, MS, 87; Am Bd Nuclear Med, dipl, 88. *Prof Exp:* Res fel nuclear med, Jaslok Hosp & Res Ctr, Bombay, 77-81; chief, Choithram Hosp & Res Ctr, Indore, 81-82 & Al Jazzira Hosp, Abu Dhabi, UAE, 82-85; fel nuclear med, Mem Sloan-Kettering Cancer Ctr, NY, 85-87, immunol, 87-88 & nuclear med & immunol, 88-91, ASST ATTEND PHYSICIAN, MEM SLOAN-KETTERING CANCER CTR, NY, 91- *Mem:* Soc Nuclear Med; Am Col Nuclear Physicians. *Res:* Monoclonal antibodies for the diagnosis and treatment of various cancers; clinical experimental trials utilizing anti-cancer monoclonal antibodies. *Mailing Add:* Mem Sloan-Kettering Cancer Ctr 1275 York Ave Box 410 New York NY 10021

DIVILBISS, JAMES LEROY, b Parsons, Kans, Jan 17, 30; m 76; c 1. INFORMATION SCIENCE. *Educ:* Kans State Univ, BS, 52; Univ Ill, MS, 55, PhD(elec eng), 62. *Prof Exp:* Res assoc elec eng, Coord Sci Lab, Univ Ill, 55-63; mem tech staff elec eng, Bell Tel Labs, Inc, 63-65; sr res engr, dept comput sci, 65-70, ASSOC PROF, GRAD SCH LIBR SCI & SR RES ENGR, COORD SCI LAB, UNIV ILL, URBANA-CHAMPAIGN, 70- *Mem:* Inst Elec & Electronics Engrs. *Res:* Logical design of computers; work sampling; systems analysis. *Mailing Add:* Divilbliss Electronics 1912 Robert Dr Champaign IL 61821

DIVINCENZO, GEORGE D, b Rochester, NY, Sept 18, 41; m 62; c 2. BIOCHEMISTRY. *Educ:* St John Fisher Col, BS, 63; State Univ NY Buffalo, PhD(biochem), 69. *Prof Exp:* BIOCHEM TOXICOLOGIST, EASTMAN PHARMACEUT, EASTMAN KODAK CO, 68- *Mem:* AAAS; Am Chem Soc; Am Indust Hyg Asn; Soc Toxicol. *Res:* Metabolic fate of foreign compounds; biologic monitoring of industrial chemicals; mechanisms of toxicity; drug safety evaluation. *Mailing Add:* Sterling Drug Inc 25 Great Valley Pkwy Great Valley PA 19355-1304

DIVINE, JAMES R(OBERT), b Stockton, Calif, Mar 11, 39; m 74; c 2. CHEMICAL ENGINEERING. *Educ:* Univ Calif, Berkeley, BS, 61; Ore State Univ, PhD(chem eng), 65. *Prof Exp:* Chem engr food processing, Western Regional Res Labs, USDA, 61; sr res engr corrosion, Pac Northwest Labs, 65-74 & 78-83, staff engr, 83-85, mgr, Corrosion & Metall Sect, 85-91; RETIRED. *Concurrent Pos:* Mem staff, Westinghouse-Hanford, 74-78. *Mem:* Nat Asn Corrosion Engrs; Am Inst Chem Engrs; Sigma Xi; Nat Asn Prof Engrs. *Res:* Corrosion transport of corrosion products through coolant systems; decontamination; molten salt electrochemistry; corrosion management. *Mailing Add:* 1261 N 61st Ave West Richland WA 99352

DIVINSKY, NATHAN JOSEPH, b Winnipeg, Man, Oct 29, 25; m 47; c 3. MATHEMATICS. *Educ:* Univ Man, BSc, 46; Univ Chicago, MSc, 47, PhD(math), 50. *Prof Exp:* Res assoc math, Cowles Comn, Chicago, 49-50; asst prof, Ripon Col, 50-51; from asst prof to assoc prof, Univ Man, 51-59; assoc prof, 59-64, asst dean sci, 69-79, PROF MATH, UNIV BC, 64- *Concurrent Pos:* Royal Soc Can scholar, 57-58; Can Coun fel vis prof, Queen Mary Col, Univ London, 65-66 & 72-73. *Mem:* Am Math Soc; London Math Soc. *Res:* Power-associative algebras; theory of rings; radicals. *Mailing Add:* Dept Math Univ BC Vancouver BC V6T 1W5 Can

DIVIS, ALLAN FRANCIS, b Chicago, Ill, Mar 23, 46; m 67; c 3. GEOLOGY, CHEMISTRY. *Educ:* Univ Calif, BS, 68; Scripps Inst Oceanog, PhD(earth sci), 75. *Prof Exp:* Lab asst chem, Scripps Inst Oceanog, 64-68; comn officer eng, Nat Oceanic & Atmospheric Admin, 68-71; res asst geol, Scripps Inst Oceanog, 71-74; lectr, Univ Calif, Riverside, 74-75; asst prof, Colo Sch Mines, 75-77; consult geol, Resource Serv Int, Inc, 77-78; CONSULT GEOL, 78- *Concurrent Pos:* Penrose grant, Geol Soc Am, 72-73; consult, Lawrence Livermore Lab, 74-75; NSF grant, 75-77. *Mem:* Geol Soc Am; Am Geophys Union; Soc Mining Engrs; Sigma Xi. *Res:* Physical and chemical evolution of magma systems and its relationship to the formation of ore deposits; geophysical and geochemical exploration; reclamation of mine sites. *Mailing Add:* Hygienetic-Pac 615 Pllkol St No 1000 Honolulu HI 96814

DIVIS, ROY RICHARD, b Berwyn, Ill, July 23, 28; m 51; c 4. FOAM EXTRUSION. *Educ:* Morton Jr Col, BGEd, 48; St Mary's Col, Minn, BS, 50; Univ Detroit, MS, 53. *Prof Exp:* Chemist, L R Kerns Co, 52-53 & Amphenol-Borg Electronics, 53-55; chief chemist, Alkalon Corp, 55-59, head plastics & elastomers res, 59-60; mgr res, Plastic Container Div, Continental Can Co, 60-66; dir appln develop, Foster Grant Co, 66-79; vpres technol, U C Industs, 79-90; CONSULT, 90- *Mem:* Am Chem Soc; Soc Plastics Engrs. *Res:* Polymer properties and resins fabrication technology. *Mailing Add:* Roy Divis Plastics Consult 250 Trudy Ave Munroe Falls OH 44262

DIVJAK, AUGUST A, b Graz, Austria, Apr 23, 49; m 68; c 4. ANALOG CIRCUIT DESIGN, APPLICATION SPECIFIC IC DESIGN. *Educ:* Milwaukee Sch Eng, BS, 71. *Prof Exp:* Elec engr, Motorola Inc, 71-76; STAFF ENGR, JOHNSON CONTROLS INC, 76- *Concurrent Pos:* Mem staff data aquisition, Univ Wis, 87-88. *Res:* Design systems to be used in energy conversation and building controls. *Mailing Add:* 2921 Coventry Lane Waukesha WI 53188

DIWAN, BHALCHANDRA APPARAO, b Sangli, India, Apr 28, 37; m 66; c 2. BIOLOGY, MEDICAL RESEARCH. *Educ:* Univ Poona, BSc, 59, MSc, 61, PhD(zool), 64. *Prof Exp:* Asst prof biol, Vivekanand Col, 66-67; sci officer, Cancer Res Inst, 67-71; res assoc, Jackson Lab, 73-74; PRIN SCIENTIST LIFE SCI, MELOY LABS, 76- *Concurrent Pos:* Fel, Indian Coun Med Res, 64-66; fel trainee, Jackson Lab, 71-73; spec fel, Leukemia Soc Am, 74-76. *Mem:* Am Asn Cancer Res; AAAS. *Res:* Chemical carcinogenesis; viral and chemical cocarcinogenesis; embryology; teratology; genetics. *Mailing Add:* 5740 Heming Ave North Springfield VA 22151

DIWAN, JOYCE JOHNSON, b Brooklyn, NY, Dec 25, 40; m 70. CELL BIOLOGY. *Educ:* Mt Holyoke Col, AB, 62; Univ Ill, Chicago, PhD(physiol), 67. *Prof Exp:* USPHS fel, Johnson Res Found, Univ Pa, 66-69; from asst prof to assoc prof, 69-91, PROF BIOL, RENSSELAER POLYTECH INST, 91- *Concurrent Pos:* Mem exec comt, US Bioenergetics Group, 80-83 & 88-90; res grants, NSF, 70-73, NIH, 74-76, 77-81, 83-87 & 88-91. *Mem:* AAAS; Am Soc Cell Biol; Biophys Soc; Am Soc Biochem & Molecular Biol. *Res:* Physiology of intra-cellular membranes, particularly mitochondrial ion channels. *Mailing Add:* Dept Biol Rensselaer Polytech Inst Troy NY 12180-3590

DIWAN, RAVINDER MOHAN, b India, June 8, 45; US citizen; m 80; c 2. MATERIALS ENGINEERING, METALLURGICAL ENGINEERING. *Educ:* Univ Roorkee, India, BE, 67; Univ Fla, ME, 71, PhD(mat sci & eng), 73. *Prof Exp:* Asst engr, Mukand Iron & Steel Works Ltd, Bombay, India, 67-69; grad res asst, Mat Sci & Eng Dept, Univ Fla, 69-73; consult eng mat, Rayar Sci Ltd, Halifax, NS, Can, 73-74; asst prof eng mech & thermodyn measurements, St Mary's Univ, Halifax, NS, 74-78; sci adv, instr & res assoc qual control, Atomic Energy Control Bd, Ottawa, Can, Algonquin Col/Queen's Univ, Can, 78-80; assoc prof, 80-85, PROF MAT SCI, MECH, ADVAN METALL & THERMODYN, MECH ENG DEPT, SOUTHERN UNIV, BATON ROUGE, LA, 85- *Concurrent Pos:* Consult mat eng, Stone & Webster Eng Corp, Boston, Mass, 82; res assoc, Div Eng Res, La State Univ, Baton Rouge, 83; prog chmn, Am Soc Mech Engrs, 83-84; prin investr, Air Force Off Sponsored Res, 88-89 & Battelle Pac Northwest Labs, 88- *Mem:* Am Soc Metals Int; Am Soc Mech Engrs; Am Soc Eng Educ; Int Soc Stereology; Am Inst Mining, Metall & Petrol Engrs Metall Soc. *Res:* Materials and metallurgy; structure property relations; quantitative microscopy; advanced materials. *Mailing Add:* 1946 General Lee Ave Baton Rouge LA 70810

DIX, DOUGLAS EDWARD, b New Britain, Conn, Mar 12, 44; m 70; c 6. BIOLOGY, BIOCHEMISTRY. *Educ:* Fairfield Univ, BS, 66; State Univ NY, Buffalo, PhD(biochem), 71; Am Bd Clin Chem, dipl, 80. *Prof Exp:* Postdoctoral radiation biol, Roswell Park Mem Inst, 71-72; res assoc pharmacol, Sch Med, Yale Univ, 72-76; dir clin lab, Haynes Med Lab, Manchester, Conn, 76-80; asst prof, 79-85, ASSOC PROF BIOL, UNIV HARTFORD, 86- *Concurrent Pos:* Clin lab dir, Conn State Dept Health, 76; chemist, Clin Chem, Nat Registry Clin Chem, 78; co-dir, NSF Career Access Prog, Univ Hartford, 90-91, Conn Dept Educ Prog, 91-92 & US Dept Energy, Math/Sci Prog, 91-92. *Honors & Awards:* Recognition Award, Am Asn Clin Chem, 85. *Mem:* Am Asn Clin Chem; Nat Asn Biol Teachers. *Res:* Aging in cancer incidence; diagnostic tests in clinical decision analysis; science in public education. *Mailing Add:* Six Cobblestone Rd Bloomfield CT 06002

DIX, JAMES SEWARD, b Colton, Calif, Jan 6, 32; m 79; c 2. ORGANIC CHEMISTRY. *Educ:* Univ Iowa, BS, 54; Univ Ill, PhD(org chem), 57. *Prof Exp:* Res chemist, Phillips Petrol Co, 57-66, mgr fiber additives sect, Res & Develop Dept, 66-70, proj mgr, Phillips Fibers Corp, 70-77, res chemist, Phillips Petrol Co, 77-78, develop engr, 78-80, sect supvr, 80-85, TECH DIR, PHILLIPS 66 CO, 85- *Mem:* AAAS; Soc Plastics Engrs; Sigma Xi. *Res:* Polymer stabilization; fiber modifications; fiber finishes; plastics compounding. *Mailing Add:* 115 W Ninth St Bartlesville OK 74003

DIX, ROLLIN CUMMING, b New York, NY, Feb 8, 36; m 60; c 3. MECHANICAL ENGINEERING. *Educ:* Purdue Univ, BS, 57, MS, 58, PhD(mech eng), 63. *Prof Exp:* Sr engr, Bendix Mishawaka Div, 62-64; asst prof, 64-68, assoc prof, 64-80, PROF MECH ENG, ILL INST TECHNOL, 80-, ASSOC DEAN COMPUT, 81- *Concurrent Pos:* Dir res, Bronson & Braton, Inc, 84- *Mem:* Am Soc Eng Educ; fel Am Soc Mech Engrs; Soc Mfg Engrs. *Res:* Heat transfer and mechanical design; numerical computation and design optimization. *Mailing Add:* MAE Dept Ill Inst Technol Chicago IL 60616

DIXIT, AJIT SURESH, b Nadiad, India, Sept 30, 50; US citizen; m 81; c 1. CHROMATOGRAPHY, PAPER TECHNOLOGY. *Educ:* Univ Bombay, BS, 70; Univ Maine, MS, 76; Univ Miss, PhD(chem), 80. *Prof Exp:* Res asst chem, Univ Maine, 73-75; res fel chem, Univ Miss, 76-80; res assoc chem, Univ Kans, 80-81; staff res chemist, Olin Corp, 81-85; sr res chemist, Ecusta Corp, 85-86; RES ASSOC, P H GLATFELTER Co, 87- *Concurrent Pos:* Prin investr, Olin Corp, 81-85. *Mem:* Am Chem Soc; Sigma Xi; Chem Soc London; Tech Asn Pulp & Paper Indust. *Res:* Synthetic organic chemistry; amino acids and peptides; pulp and paper technology research; cellulose physics and chemistry; liquid chromatography; fiber morphology; protecting groups; paper physics and science. *Mailing Add:* 14 Gold Mine Ridge Pisgah Forest NC 28768-9774

DIXIT, BALWANT N, b Kerawade, India, Jan 7, 33; m 69; c 2. TOXICOLOGY. *Educ:* Univ Poona, BS, 54, Hons, 55, MS, 56; Univ Baroda, MS, 62; Univ Pittsburgh, PhD(pharmacol), 65. *Prof Exp:* Res asst, 63-65, from asst prof to prof pharmacol, Sch Pharm, 65-87, asst chmn dept, 68-73, actg dean, 76-78, chmn dept & assoc dean, 74-87, PROF PHARMACOL, SCH DENT MED, UNIV PITTSBURGH, 87- *Concurrent Pos:* Res asst pharmacol, Indian Coun Med Res, Med Col, Univ Baroda, 56-59, sr res fel, Coun Sci & Indust Res, 60-62. *Mem:* AAAS; Am Soc Pharmacol & Exp Therapeut; NY Acad Sci; Soc Neurosci; Sigma Xi. *Res:* Autonomic and biochemical pharmacology; neuroendocrinology; drug interactions; analytical biochemistry. *Mailing Add:* Dept Pharmacol 541-2 Salk Hall Univ Pittsburgh Pittsburgh PA 15261

DIXIT, MADHU SUDAN, b Nagpur, India, July 10, 42; Can citizen; m 71; c 2. WEAK INTERACTIONS, INTERMEDIATE ENERGY PHYSICS. *Educ:* Vikram Univ, India, BSc, 61; Univ Delhi, MSc, 63; Univ Chicago, PhD(physics), 71. *Prof Exp:* Asst res physicist, Univ Calif, Los Angeles, 71-73; res assoc, Carleton Univ, Ottawa, 73-75; sr res fel, Univ Victoria, Vancouver, 76-79; ASSOC RES OFFICER, NAT RES COUN, OTTAWA, 79- *Concurrent Pos:* Vis staff mem, Los Alamos Meson Physics Facil, Univ Calif, Los Angeles, 71-73. *Mem:* Am Phys Soc. *Res:* Pionic and muonic x-ray; elementary particle physics, weak interactions; rare and ultra-rare decays of pion and muon. *Mailing Add:* EP Div CERN Geneva Switzerland

DIXIT, PADMAKAR KASHINATH, b Calcutta, India, Aug 7, 21; m 48; c 3. ANATOMY, NUTRITIONAL BIOCHEMISTRY. *Educ:* Univ Bombay, BSc, 42, MSc, 46, PhD(nutrit biochem), 48. *Prof Exp:* Asst res officer, Nutrit Res Lab, Coonoor, India, 49-59; from res assoc to assoc prof anat, 61-70, PROF ANAT, SCH MED, UNIV MINN, MINNEAPOLIS, 70- *Concurrent Pos:* Raptakos-Brett res fel, Nutrit Res Lab, Coonoor, India, 48-49; Tech Coop Mission fel, Western Reserve Univ, 52-54; res fel, Dept Anat, Sch Med, Univ Minn, Minneapolis, 59-61. *Mem:* Am Asn Anat; Soc Biol & Exp Med. *Res:* Diabetes, especially quantitative histochemistry as applied in pancreatic islets; studies on pancreatic islet adenoma; studies in experimental rickets; studies on nutritional deficiencies; microtechnics. *Mailing Add:* 1088 18th Ave Minneapolis MN 55414

DIXIT, RAKESH, b Lucknow, India, July 1, 58; c 2. PHARMACOLOGY, BIOCHEMISTRY. *Educ:* Univ Lucknow, BS, 75, MS, 77, PhD(biochem & toxicol), 82. *Prof Exp:* Postdoctoral toxicol, Case Western Reserve Univ, Cleveland, 81-83; postdoctoral chem carcinogenesis, Med Col Ohio, Toledo, 83-85; fac assoc carcinogenesis, Univ Nebr Med Ctr, 85-87; SR TOXICOLOGIST, MIDWEST RES INST, KANSAS CITY, MO, 87- *Concurrent Pos:* Prin investr & study dir, Midwest Res Inst, 87- *Mem:* Am Asn Cancer Res; Soc Toxicol; Am Soc Pharmacol & Exp Therapeut; AAAS; NY Acad Sci; Am Soc Photobiol; Soc Investigative Dermat; Environ Mutagen Soc. *Res:* Toxicology and pharmacology of anticancer agents; carcinogenesis and its prevention; safety evaluation in vivo animal models; author of various publications. *Mailing Add:* Dept Life Sci Midwest Res Inst 425 Volker Blvd Kansas City MO 64110

DIXIT, SARYU N, b Sehud, India, Aug 8, 37; US citizen; m 56; c 2. BIOCHEMISTRY, ORGANIC CHEMISTRY. *Educ:* Agra Univ, BS, 56, MS, 58; Banaras Hindu Univ, PhD(org chem), 62. *Prof Exp:* Res assoc org chem, Univ Ill & Univ Chicago, 63-67; asst prof med chem, Banaras Hindu Univ, 67-70; res assoc biochem, Case Western Reserve Univ, 70-72; asst prof, Vet Admin Hosp & Univ Tenn, 72-78, assoc prof, 78-, res chemist, 72-; PROF & RES CHEMIST, LAKESIDE VET ADMIN MED CTR & NORTHWESTERN UNIV. *Mem:* Am Chem Soc; Sigma Xi; Am Soc Biol Chemists. *Res:* Biochemistry of interstitial and basement membrane collagens. *Mailing Add:* Dept Oral Biol Northwestern Univ 303 E Chicago Ave Chicago IL 60611

DIXIT, SUDHIR S, b Lucknow, India, June 20, 51. MULTI-MEDIA PRESENTATION & TRANSMISSION, BROADBAND NETWORKS. *Educ:* Bhopal Univ, India, BE, 72; Birla Inst Technol & Sci, ME, 74; Univ Strathclyde, Glasgow, Scotland, PhD(elec eng), 79; Fla Inst Technol, MBA, 83. *Prof Exp:* Res engr, STC Technol Ltd, Harlow, Eng, 78-80; assoc prin engr, Harris Corp, Melbourne, Fla, 80-83; prin engr, Codex Corp, Canton, Mass, 83-85; prin engr, Wang Labs Inc, Lowell, Mass, 85-87; mem tech staff, GTE Labs Inc, Waltham, Mass, 87-91; MEM TECH STAFF, NYNEX CORP, CAMBRIDGE, MASS, 91- *Mem:* Inst Elec & Electronics Engrs. *Res:* Research and development in video communication, multi-media, broadband networks, image processing and computer and telecommunication networks. *Mailing Add:* Nynex Corp 38 Sidney St Suite 180 Cambridge MA 02139

DIXON, ANDREW DERART, b Belfast, Northern Ireland, Oct 27, 25; m 48; c 3. CRANIOFACIAL GROWTH, DENTISTRY. *Educ:* Queen's Univ, Belfast, BDS, 49, MDS, 53, BSc, 54; Univ Manchester, PhD(anat), 58. *Hon Degrees:* DSc, Queen's Univ, Belfast, 65. *Prof Exp:* Demonstr dent prosthetics, Queen's Univ, Belfast, 48-49; lectr anat, Univ Manchester, 54-62, sr lectr, 62-63; vis assoc prof, Univ Iowa, 59-61; prof oral biol & anat, Sch Med, Univ NC, Chapel Hill, 63-73, asst dean & coordr dent res, 66-73, dir dent res ctr, 67-73, assoc dean res, Sch Dent, 69-73; dean, Sch Dent, 73-80, PROF ORTHOD, UNIV CALIF, LOS ANGELES, 80-, ASSOC DEAN ADMIN & FAC AFFAIRS, SCH DENT, 85- *Concurrent Pos:* Fulbright travel award, 59-61; Commonwealth Fund traveling fel, 61. *Mem:* Nat Inst Med-Nat Acad Sci; Am Soc Cell Biol; Am Asn Anat; Am Dent Asn; fel AAAS; fel Am Col Dent; fel Int Col Dent. *Res:* Development and growth of the jaws; innervation of oral tissues. *Mailing Add:* Sch Dent Univ Calif Los Angeles CA 90024

DIXON, ARTHUR EDWARD, b Woodstock, NB, Nov 16, 38; m 63; c 3. CONFOCAL SCANNING LASER MICROSCOPY, PHOTOLUMINESCENCE & FLUORESCENCE MICROSCOPY. *Educ:* Mt Allison Univ, BSc, 60; Dalhousie Univ, MSc, 62; McMaster Univ, PhD(solid state physics), 66. *Prof Exp:* Asst prof, 66-72, assoc dir corresp prog, 76-80, ASSOC PROF PHYSICS, UNIV WATERLOO, 72- *Concurrent Pos:* Vpres, Waterloo Distance Educ Inc, 79-88; chief exec officer & chmn, Waterloo Sci Inc, 83- *Mem:* Can Asn Physicists. *Res:* Confocal scanning laser microscopy; photoluminescence & fluorescence microscopy. *Mailing Add:* Dept Physics Univ Waterloo Waterloo ON N2L 3G1 Can

DIXON, BRIAN GILBERT, b Berkeley, Calif, July 6, 51; m 73; c 2. HYDROLOGY & WATER RESOURCES, SCIENCE EDUCATION. *Educ:* Pa State Univ, BS, 73; Univ Ill, PhD(chem), 81. *Prof Exp:* Tech serv scientist polymers, ICI US, Inc, 73-76; res scientist org chem, Dow Chem, Inc, 80-85; VPRES CHEM RES & DEVELOP, CAPE COD RES, INC, 85- *Concurrent Pos:* Prin investr, Dept Defense, 86-91, Ctrs Excellence, State Mass, 87-90, NSF, 88-91, Dept Energy, 89-91, Environ Protection Agency, 89-90, NASA, 90-91; lectr, Southeastern Mass Univ, 90- *Mem:* Am Chem Soc; AAAS; NY Acad Sci; NAm Membrane Soc. *Res:* Use of electrochemistry for the detection and complete degradation of trace levels of pollutants in water; environmentally benign paints and coatings. *Mailing Add:* Four Nicholas Lane Sandwich MA 02563

DIXON, CARL FRANKLIN, b La Junta, Colo, Sept 14, 26; m 60; c 3. PARASITOLOGY, ZOOLOGY. *Educ:* Univ Colo, BA, 50; Kans State Univ, PhD(parasitol), 60. *Prof Exp:* Vet parasitologist, USDA, 60-64; asst prof parasitol, 64-70, ASSOC PROF PARASITOL, AUBURN UNIV, 70- *Mem:* Am Soc Parasitol. *Res:* Interrelationship of helminth parasites and nutrition as a cause of disease in domestic animals. *Mailing Add:* Dept Zool & Wildlife Sci Auburn Univ Funchess Hall Auburn AL 36849-5414

DIXON, DABNEY WHITE, b Rochester, NY, Aug 2, 49; m 82; c 1. PHYSICAL ORGANIC CHEMISTRY. *Educ:* Brown Univ, AB, 71; Mass Inst Technol, PhD(chem), 76. *Prof Exp:* Fel, Univ Calif, San Diego, 76-79; asst prof chem, Washington Univ, 79-86; asst prof, 86-90, assoc prof, DEPT CHEM, GA STATE UNIV, 90- *Mem:* Am Chem Soc; Biophys Soc. *Res:* Mechanisms of organic and bioorganic reactions; free radical chemistry; electron transfer in heme proteins; anti-HIV therapeutic agents. *Mailing Add:* Dept Chem Ga State Univ Univ Plaza Atlanta GA 30303

DIXON, DENNIS MICHAEL, b Richmond, Va, Sept 4, 51. BIOLOGY, MEDICAL MYCOLOGY. *Educ:* Univ Richmond, BS, 73; Va Commonwealth Univ, PhD(microbiol), 78. *Prof Exp:* A D Williams fel microbiol, Va Commonwealth Univ, 74-75, teaching asst, 75-77, res asst, 77-78; ASST PROF BIOL, LOYOLA COL, 78- *Mem:* Am Soc Microbiol; Mycol Soc Am; Med Mycol Soc Am; Int Soc Human & Animal Mycol; Med Mycol Soc NY. *Res:* Distribution of zoopathogenic fungi in nature and ultrastructural studies of conidial ontogeny for use in taxonomy. *Mailing Add:* Dept Biol Loyola Col 4501 N Charles St Baltimore MD 21210

DIXON, DWIGHT R, b Fairfield, Idaho, June 16, 19; m 49; c 6. NUCLEAR PHYSICS. *Educ:* Utah State Univ, BSc, 42; Univ Calif, Berkeley, PhD(nuclear physics), 55. *Prof Exp:* Mem staff electronics, Radiation Lab, Mass Inst Technol, 42; proj engr, Sperry Co, NY, 42-46; physicist, Radiation Lab, Univ Calif, Berkeley, 48-55; sr engr missile eng, Utah Eng Lab, Sperry Gyroscope Co, 57-58, eng sect head, 58-59; from asst prof to assoc prof, 59-66, PROF PHYSICS, BRIGHAM YOUNG UNIV, 66- *Mem:* Sigma Xi; Am Asn Physics Teachers; Am Phys Soc. *Res:* Study of nuclear structure physics by means of radiative proton capture and inelastic neutron scattering reactions; study of energy loss and multiple scattering of charged particles in matter. *Mailing Add:* 524 E 2950 N Provo UT 84604

DIXON, EARL, JR, b Halifax, Va, Oct 20, 37. EXPERIMENTAL BIOLOGY. *Educ:* St Augustins Col, BS; Atlanta Univ, MS; Howard Univ, PhD(physiol & biophys), 71. *Prof Exp:* Prof physiol, blood-muscle physiol & intermediary metab, Tuskegee Inst, 71-91; COLLAB PROF, IOWA STATE UNIV, 91-; POSTDOCTORAL, NAT HEART & LUNG INST, NIH, BETHESDA, MD, 91- *Mem:* AAAS; Am Physiol Soc; Sigma Xi; Am Asn Higher Educ. *Mailing Add:* 312 Oslin Dr Tuskegee AL 36083

DIXON, EDMOND DALE, b Anacoco, La, Oct 22, 36; m 60; c 1. ALGEBRA. *Educ:* Memphis State Univ, BS, 58; Auburn Univ, MS, 60, PhD(math), 65. *Prof Exp:* Instr Math, Auburn Univ, 61-63; assoc prof, WGa Col, 63-65; PROF MATH, TENN TECHNOL UNIV, 65- *Mem:* Math Asn Am. *Res:* Algebra of matrices; linear algebra. *Mailing Add:* Dept Math Tenn Technol Univ Cookeville TN 38505

DIXON, ELISABETH ANN, b Nottingham, Eng, Jan 16, 44; Can citizen. CHEMICAL ECOLOGY. *Educ:* Univ London, BSc, 69; Univ Victoria, BC, PhD(phys org chem), 74. *Prof Exp:* Instr II, 78-83, SR INSTR CHEM, UNIV CALGARY, 83- *Mem:* Royal Soc Chem; Can Soc Chem; Int Soc Chem Ecol. *Res:* Structure activity studies in coleopteran phenomone systems; elucidation of detail of biological chemistry of host tree-bark beetle-symbiotic fungi complex. *Mailing Add:* Dept Chem Univ Calgary Calgary AB T2N 1N4 Can

DIXON, FRANK JAMES, b St Paul, Minn, Mar 9, 20; m 46; c 3. IMMUNOLOGY. *Educ:* Univ Minn, BS, 41, MB, 43, MD, 44; Am Bd Path, dipl. *Hon Degrees:* DSc, Ohio Med Col. *Prof Exp:* Intern, US Naval Hosp, Ill, 43-44; res fel path, Harvard Univ, 46-48; from instr to asst prof, Wash Univ, 48-51; prof & chmn dept, Med Sch, Univ Pittsburgh, 51-60; chmn dept exp path, 61-74, dir, 74-76, EMER DIR & INST MEM, RES INST SCRIPPS CLIN, 87- *Concurrent Pos:* Adj prof, Univ Calif, San Diego, 61-; mem expert adv panel immunol, WHO, Sci Adv Comt, Helen Hay Whitney Found, Am Cancer Soc & Comt Res Nat Found, sci adv bd, Nat Kidney Found, Nat Cancer Adv Bd & Sci Adv Bd, Irvington House Sci Avd Bd, vis comt biol div, Calif Inst Technol & Nat Arthritis Comn; chmn, biomed res depts, Res Inst Scripps, 70-74. *Honors & Awards:* Theobald Smith Award, 52; Parke-Davis Award, 57; Award Distinguished Achievement, Mod Med, 61; Martin E Rehfuss Award, 66; Von Pirquet Medal, Ann Forum Allergy, 67; Bunim Gold Medal, Am Rheumatism Asn, 68; Mayo Soley Award, West Soc Clin Res, 69; Gairdner Found Int Award, 69; Albert Lasker Basic Med Res Award; Dickson Prize in Med, Univ Pittsburgh, 75; Homer Smith Award, NY Heart Asn, 76; Rous-Whipple Award, Am Asn Path, 79; H P Smith Award, Am Soc Clin Pathologists, 85; Gold Headed Cane Award, Am Asn Pathologists, 87; Pahlavi Lectr, Iran. *Mem:* Nat Acad Sci; NY Acad Sci; Am Soc Exp Path (pres, 66); Am Asn Immunol (pres, 71); Am Asn Cancer Res; Asn Am Physicians; Sigma Xi; Scand Soc Immunol; Am Acad Allergists; AAAS; Am Soc Clin Invest; Int Acad Path; Am Asn Pathologists (pres, 66). *Res:* Immunopathology. *Mailing Add:* Dept Immunol Scripps Clin & Res Found La Jolla CA 92037

DIXON, GEORGE SUMTER, JR, b Asheville, NC, Mar 28, 38; m 66; c 2. SOLID STATE PHYSICS. *Educ:* Univ Ga, BS, 60, MS, 63, PhD(physics), 67. *Prof Exp:* Asst prof physics, Tenn Technol Univ, 67-68; AEC fel, Solid State Div, Oak Ridge Nat Lab, 68-70; from asst prof to assoc prof, 75-85, PROF PHYSICS, OKLA STATE UNIV, 85- *Concurrent Pos:* Dir, Okla State Alliance Minority Participation, 90- *Mem:* Am Phys Soc; Am Asn Phys Teachers. *Res:* Phase transitions; biophysics; phonon physics; thermal properties of laser materials. *Mailing Add:* Dept Physics Okla State Univ Stillwater OK 74078

DIXON, GORDON H, b Durban, SAfrica, Mar 25, 30; Can citizen; m 54; c 4. MOLECULAR BIOLOGY, DEVELOPMENTAL BIOLOGY. *Educ:* Cambridge Univ, MA, 51; Univ Toronto, PhD(biochem), 56. *Prof Exp:* Assoc biochem, Univ Wash, 55-56, asst prof, 56-58; Med Res Coun mem, Oxford Univ, 58-59; res assoc, Univ Toronto, 59-60, from asst prof to assoc prof, 60-63; from assoc prof to prof biochem, Univ BC, 63-72; prof & head biochem group, Univ Sussex, 72-74; PROF, DEPT MED BIOCHEM, FAC MED, UNIV CALGARY, 74-, HEAD DEPT, 83- *Concurrent Pos:* Vis scientist, Med Res Coun Lab Molecular Biol, Cambridge Univ, 70-71; Josiah Macy Jr Found fel, Univ Cambridge, UK. *Honors & Awards:* Steacie Prize, 66; Ayerst Award, Can Biochem Soc, 66; Flavelle Medal, Royal Soc Can, 80. *Mem:* Am Soc Biochem & Molecular Biol; fel Royal Soc; Can Biochem Soc (pres, 62-63); Royal Soc Can; Am Soc Cell Biol; Can Cell Biol Soc; Pan-Am Asn Biochem Soc (pres, 87-90). *Res:* Biochemistry of differentiation; structure and function of chromosomal proteins; protamines, histones high mobility group proteins and their genes; molecular biology of spermatogenesis; organization of genes expressed in testis differentiation; DNA and protein evolution. *Mailing Add:* Div Med Biochem Fac Med Health Sci Ctr Univ Calgary 3330 Hosp Dr NW Calgary AB T2N 4N1 Can

DIXON, HARRY S(TERLING), b Woodland, Calif, Nov 30, 10; m 37; c 4. FORENSIC ELECTRICAL ENGINEERING. *Educ:* Stanford Univ, BA, 31, EE, 36; Purdue Univ, PhD, 52. *Prof Exp:* Asst engr, Reclamation Dist No 108, Calif, 34-37; instr elec eng, Purdue Univ, 37-42; elec test engr, Douglas Aircraft, Calif, 42-44; elec design engr, NAm Aviation, Inc, 44-45; prof elec eng & chmn dept, Agr Col, NDak State Univ, 45-51; lectr, Univ Calif, Berkeley, 51-52; prof elec eng & chmn dept, Newark Col Eng, NJ Inst Technol, 52-56; lectr, Univ Calif, Berkeley, 56-57; CONSULT ENGR, 57- *Concurrent Pos:* UNESCO int expert, Lagos, Nigeria, 64-65. *Mem:* AAAS; Inst Elec & Electronics Engrs; Nat Soc Prof Engrs; Am Soc Eng Educ; Illum Eng Soc; Nat Acad Forensic Engrs; Nat Fire Protection Asn; Am Defense Preparedness Asn. *Res:* Engineering education; aircraft electrical design; illumination design; electrical power transmission; electrical control; gaseous electrical phenomena; safety; fires and explosions, causes, and cures. *Mailing Add:* 950 Creston Rd Berkeley CA 94708-1544

DIXON, HELEN ROBERTA, b Belvidere, Ill, Aug 13, 27. STRUCTURAL GEOLOGY. *Educ:* Carleton Col, BA, 49; Univ Calif, MA, 56; Harvard Univ, PhD(geol), 69. *Prof Exp:* GEOLOGIST, US GEOL SURV, 55- *Concurrent Pos:* Lectr, San Diego State Univ, 69. *Mem:* AAAS; Geol Soc Am; Mineral Soc Am; Am Geophys Union. *Res:* Igneous and metamorphic petrology; Precambrian of central Wyoming; metamorphics of eastern Connecticut. *Mailing Add:* 30111 Rainbow Hills Golden CO 80401

DIXON, HENRY MARSHALL, b New York, NY, June 4, 29. ELECTRONICS ENGINEERING. *Educ:* Univ Va, BA, 50, MS, 52, PhD(physics), 54. *Prof Exp:* Asst prof physics, Tulane Univ, 54-55; physicist, White Sands Signal Agency, 55-57; asst prof physics, NMex State Col, 56-57; from asst prof to assoc prof, 57-65, PROF PHYSICS, BUTLER UNIV, 65- *Mem:* AAAS; Am Phys Soc; Sigma Xi; Inst Elec & Electronics Engrs. *Res:* Engineering physics. *Mailing Add:* Dept Physics Butler Univ 4600 Sunset Ave Indianapolis IN 46208

DIXON, JACK EDWARD, b Nashville, Tenn, June 16, 43. BIOCHEMISTRY. *Educ:* Univ Calif, Los Angeles, BA, 66, Santa Barbara, PhD(chem), 71. *Prof Exp:* Res asst chem, Univ Calif, Santa Barbara, 68-71, NSF fel biochem, San Diego, 71-73; from asst prof to assoc prof, 73-82, PROF BIOCHEM, PURDUE UNIV, 82- *Concurrent Pos:* Adj asst prof biochem, Sch Med, Ind Univ, 76-78, adj assoc prof, 78-, adj prof med, 85-; career develop award, USPHS, & travel award, Am Soc Biol Chemists, 76-81; spec reviewer, Alcohol Study Sect, NIH, 83 & 84, Endocrine Study Sect, 85; consult, Mich Gastrointestinal Peptide Res Ctr Grant, Univ Mich & Wyeth, 84-; exec ed, Anal Biochem, 85- *Mem:* Am Chem Soc; AAAS; Am Soc Biol Chemist; Am Soc Neurosci; Am Physiol Soc. *Res:* Mechanisms of biosynthesis and degradation of peptide hormones and releasing factors; molecular biology of peptide hormones. *Mailing Add:* Dept Biochem Purdue Univ West Lafayette IN 47907

DIXON, JACK RICHARD, b Mich, Oct 29, 25; m 50; c 3. PHYSICS. *Educ:* Western Reserve Univ, BS, 48, MS, 50; Univ Md, PhD, 56. *Prof Exp:* Asst physics, Western Reserve Univ, 48-50; physicist chem & radiation lab, US Army Chem Corps, 50-52; res asst physics, Univ Md, 52-55; res physicist, US Naval Ord Lab, 55-75, HEAD MAT DIV, NAVAL SURFACE WEAPONS CTR, 75- *Concurrent Pos:* Assoc prof, Univ Md, 63- *Mem:* Am Phys Soc. *Res:* Theoretical weapons analysis; study of decay rates of metastable noble atoms; experimental solid state physics. *Mailing Add:* 422 Hillsboro Dr Silver Spring MD 20902

DIXON, JAMES EDWARD, b Schenectady, NY, Sept 9, 41; m 63; c 2. ORGANIC CHEMISTRY. *Educ:* St Bonaventure Univ, BS, 63, MS, 65; State Univ NY, Albany, PhD(chem), 74. *Prof Exp:* RES CHEMIST ELECTROPHOTOG, EASTMAN KODAK RES LABS, 74- *Mem:* Am Chem Soc. *Res:* Investigation of xerographic behavior of sensitized polymeric systems. *Mailing Add:* 29 Everwild Lane Rochester NY 14616

DIXON, JAMES FRANCIS PETER, b Eng; US citizen; m 68; c 1. EXPERIMENTAL PATHOLOGY. *Educ:* Calif State Univ, Los Angeles, BS, 69; Univ Southern Calif, PhD(exp path), 75. *Prof Exp:* Fel cancer res, Los Angeles County-Univ Southern Calif Cancer Ctr, 75-77, from asst prof to assoc prof, 77-91, PROF PATH, SCH MED, UNIV SOUTHERN CALIF, 91- *Mem:* AAAS; Soc Leukocyte Biol; Am Asn Pathologists; NY Acad Sci; Am Soc Clin Pathologists. *Res:* Control of cell proliferation; mechanisms of lymphocyte transformation; histochemistry of metabolic diseases. *Mailing Add:* Dept Path Sch Med Univ Southern Calif Los Angeles CA 90033

DIXON, JAMES RAY, b Houston, Tex, Aug 1, 28; m 53; c 5. HERPETOLOGY. *Educ:* Howard Payne Col, BS, 50; Tex A&M Univ, MS, 57, PhD(zool), 61. *Prof Exp:* Cur reptiles, Ross Allen Reptile Inst, 54-55; asst prof vet med, Tex A&M Univ, 56-61; asst prof wildlife mgt, NMex State Univ, 61-65; cur herpet, Life Sci Div, Los Angeles County Mus, 65-67; assoc prof wildlife sci, 67-71, PROF WILDLIFE & FISHERIES SCI, TEX A&M UNIV, 71- *Concurrent Pos:* Sigma Xi res grants, 60 & 73; consult, NMex State Dept Game & Fish, 64-65; NSF grants, 64-66, 78-79 & 80; chmn bd scientists, Chihuahuan Desert Res Inst, 81-84. *Mem:* Am Soc Ichthyologists & Herpetologists; Zool Soc London; Soc Study Amphibians & Reptiles; Herpetologists League; Sigma Xi. *Res:* Zoogeography, systematics and ecology of lizards of the family Gekkonidae; systematics of reptiles and amphibians of southwestern United States, Mexico and South America, especially Peru; snakes of the family Colubridce; general natural history of vertebrates. *Mailing Add:* Dept Wildlife Sci Tex A&M Univ College Station TX 77843

DIXON, JOE BORIS, b Clinton, Ky, Nov 15, 30; m 52; c 2. SOIL SCIENCE. *Educ:* Univ Ky, BS, 52, MS, 56; Univ Wis, PhD(soil sci), 58. *Prof Exp:* NSF fel soil sci, 58-59; assoc prof, Auburn Univ, 59-68; PROF SOIL MINERAL, TEX A&M UNIV, 68- *Concurrent Pos:* pres, Clay Minerals Soc, 82; vis researcher geol & chem, Arizona State Univ, 82-83; consult, Clay Indust, 87 & 88. *Mem:* fel AAAS; fel Am Soc Agron; fel Soil Sci Soc Am; Mineral Soc Am; Int Soil Sci Soc; Sigma Xi. *Res:* Clay mineralogy of soils; soil genesis, mineral synthesis; weathering of minerals in soils and rocks; surface properties of clays; mine spoil reclamation. *Mailing Add:* 1807 Lawyer Pl College Station TX 77840

DIXON, JOHN ALDOUS, b Provo, Utah, July 16, 23; m 44; c 3. SURGERY, PHYSIOLOGY. *Educ:* Univ Utah, BS, 44, MD, 47; Am Bd Surg, dipl, 54. *Prof Exp:* Exec vpres & prof surg, Col Med, 70-77, dean, 72-77, VPRES HEALTH SCI, UNIV UTAH, 73- *Concurrent Pos:* Chief surg, Johnson Air Force Hosp, Honshu, Japan & surg consult, Far East Air Force, 52-53. *Mem:* Fel Am Col Surg; Am Gastroenterol Asn; Soc Surg Alimentary Tract; Am Soc Gastrointestinal Endoscopy; Soc Am Gastrointestinal Endoscopic Surg. *Res:* Gastrointestinal surgery and diseases; motility of intestine; absorption and secretion; laser surgery and research. *Mailing Add:* Univ Utah Col Med Dept Surg 50 N Medical Dr Salt Lake City UT 84112

DIXON, JOHN CHARLES, b Chicago, Ill, July 21, 31; m 63; c 2. ENTOMOLOGY. *Educ:* Beloit Col, BS, 53; Univ Wis, MS, 55, PhD(entom, zool), 61. *Prof Exp:* Entomologist, Southeast Forest Exp Sta, Forest Serv, USDA, 59-64; asst prof biol, Chicago State Col, 64-68; asst prof, 68-70, ASSOC PROF BIOL, UNIV WIS-EAU CLAIRE, 70- *Mem:* AAAS; Entom Soc Am. *Res:* Ecological studies of defoliating insects; population studies and ecology of pine bark beetles. *Mailing Add:* Dept Biol Univ Wis-Eau Claire Eau Claire WI 54701

DIXON, JOHN D(OUGLAS), b Buffalo, Minn, July 29, 24; m 49; c 4. ELECTRICAL ENGINEERING. *Educ:* Univ Minn, BEE, 49; Univ Mo, MS, 52. *Prof Exp:* From instr to asst prof elec eng, Univ Mo, 49-53; from asst prof to prof elec eng, Univ NDak, 53-89, asst dean, 75-78, chmn dept, 79-89, emer prof elec eng, 89-; RETIRED. *Mem:* Am Soc Eng Educ; Inst Elec & Electronics Engrs. *Res:* Control systems; computers/logic design. *Mailing Add:* 15355 13th Ave N Plymouth MN 55446

DIXON, JOHN DOUGLAS, b Ewell, Eng, Jan 18, 37; m 61; c 3. GROUP THEORY, ANALYSIS OF ALGORITHMS. *Educ:* Univ Melbourne, BS, 57, MA, 59; McGill Univ, PhD(math), 61. *Prof Exp:* Instr math, Calif Inst Technol, 61-64; sr lectr, 64-67, assoc prof, Univ New S Wales, 68; assoc prof, 68-71, PROF MATH, CARLETON UNIV, ONT, 71- *Mem:* London Math Soc; Am Math Soc; Math Asn Am; Math Soc Can; Australian Math Soc. *Res:* Algebra and number theory; theory of linear groups and representation theory; design and analysis of algorithms in algebra and combinatorics. *Mailing Add:* Dept Math Carleton Univ Ottawa ON K1S 5B6 Can

DIXON, JOHN E(LVIN), b Roseburg, Ore, Mar 23, 27; m 48; c 2. AGRICULTURAL ENGINEERING. *Educ:* Ore State Univ, BS(agr eng) & BS(agr), 51; Univ Idaho, MS, 57; Mich State Univ, PhD(agr eng), 79. *Prof Exp:* Instr agr eng, Colo State Univ, 51-54; from instr to assoc prof, Univ Idaho, 54-67; agr eng adv, Kans State Univ at Hyderabad, India, 67-69; dir prof adv serv ctr, 69-71, assoc prof, 69-79, PROF AGR ENG, UNIV IDAHO, 79- *Concurrent Pos:* Fallout shelter analyst instr, Defense Civil Protection Agency, 70-; res asst, Mich State Univ, 72-74; dir, Am Soc Agr Engrs, 83-85. *Mem:* Am Soc Agr Engrs; Am Soc Eng Educ; Nat Soc Prof Engrs; Prof Soc Nuclear Defense. *Res:* Farm structures with emphasis on environmental control; poultry housing; crop storage; water quality; farmstead planning. *Mailing Add:* Dept Agr Eng Univ Idaho Moscow ID 83843

DIXON, JOHN KENT, b Detroit, Mich, Sept 1, 34. ROBOTICS, NEURAL NETWORKS. *Educ:* Lawrence Technol Univ, BSEE, 57; Wayne State Univ, MBA, 64, MA, 65; Univ Calif, Livermore, MS, 67, PhD(computer sci), 70. *Prof Exp:* Electronics engr, Signal Processing Dept, Bendix Res Lab, Southfield, Mich, 59-65; res asst artificial intel, Lawrence Radiation Lab, Livermore, Calif, 65-68; computer scientist artificial intel, Nat Inst Health, Bethesda, Md, 68-74, Naval Res Lab, 74-86; sr staff eng artificial intel, Martin Marietta Intel Systs Lab, 86-88; res staff neural networks, Syst Planning Corp, 88-90; CONSULT, 90- *Concurrent Pos:* Vpres, World Pop Soc, Am Univ, Washington, DC, 73-76; scientist chmn, First Ann Conf, World Pop Soc, 74. *Mem:* Inst Elec & Electronics Engrs; Asn Comput Mach; Am Asn Artificial Intel; Int Neural Network Soc. *Res:* Areas of tree searching; theorem proving, truth maintenance, uncertainty and robotics and neural networks. *Mailing Add:* 17 Arthur Dr E Ft Washington MD 20744

DIXON, JOHN MICHAEL SIDDONS, b Derby, Eng, Aug 22, 28; Can citizen; m 52; c 2. MEDICAL MICROBIOLOGY. *Educ:* Univ Wales, MB, BCh, 50, MD, 60; Univ London, dipl bact, 57; FRCPS(C), 69; FRCPath, 71. *Prof Exp:* Dir, Pub Health Lab, Ipswich, Eng, 59-66; chief bact, Wellesley Hosp, Toronto, Ont, 66-67; prof med microbiol, 67-89, EMER PROF, MED MICROBIOL & INFECTIOUS DIS, UNIV ALTA, 89- *Concurrent Pos:* Prov bacteriologist, Alta, 67-88; dir, Prov Lab Pub Health, Edmonton & Calgary, 67-88; consult bacteriologist, Univ Alta Hosp, Edmonton, 67-89; chmn, Can Nat Adv Comt Immunization, 72-89; chmn, Grants Comt Microbiol & Infectious Dis, Med Res Coun Can, 73-76. *Mem:* Can Asn Med Microbiol (pres, 70-72); emer fel Infectious Dis Soc Am; Can Asn Clin Microbiol & Infectious Dis (pres, 88-90); sr mem Path Soc Gt Brit & Ireland. *Res:* Antibiotic resistance of streptococci and pneumococci; immunization policies; diphtheria. *Mailing Add:* 8935 120th St Edmonton AB T6G 1X6 Can

DIXON, JOHN R, b Akron, Ohio, Oct 19, 30; m 51; c 1. MECHANICAL ENGINEERING. *Educ:* Mass Inst Technol, BS, 52, MS, 53; Carnegie Inst Technol, PhD(mech eng), 60. *Prof Exp:* Engr, Jarl Extrusions, Inc, NY, 55-57; proj engr, Joseph Kaye & Co, Mass, 57-58; asst prof mech eng, Carnegie Inst Technol, 60-61; assoc prof, Purdue Univ, 61-65 & Swarthmore Col, 65-66; head dept, 66-71 & 81-83, PROF MECH ENG, UNIV MASS, AMHERST, 66- *Concurrent Pos:* Pres, Dixon Energy Systs, Inc, 78-81. *Honors & Awards:* Ralph Coates Roe Award, Am Soc Eng Educ, 76. *Mem:* AAAS; fel Am Soc Mech Engrs; Am Soc Heating, Refrig & Airconditioning Engrs. *Res:* Application of artificial intelligence to mechanical design and manufacturing. *Mailing Add:* Dept Mech Eng Univ Mass Amherst Campus Amherst MA 01003

DIXON, JOSEPH ARDIFF, b Philadelphia, Pa, Nov 4, 19; m 42; c 2. ORGANIC CHEMISTRY. *Educ:* Pa State Univ, BS, 42, MS, 45, PhD(org chem), 47. *Prof Exp:* Chemist, USDA, 42-44; instr chem, Pa State Univ, 47-51; chemist, Calif Res Corp, 51-52; assoc prof chem, Lafayette Col, 52-55; assoc prof, 55-60, head dept, 71-84, PROF CHEM, PA STATE UNIV, UNIV Park, 60- *Concurrent Pos:* Gen Motors fel, 48-49; Sect ed, J Chem Eng Data, 78-; bd dirs, Am Chem Soc, 87-90, chmn bd dirs, 90- *Mem:* Am Chem Soc; Sigma Xi; fel Am Petrol Inst. *Res:* Correlation of properties with molecular structures; synthesis and properties of hydrocarbons; organo-lithium chemistry. *Mailing Add:* 152 Davey Lab Univ Park PA 16802

DIXON, JULIAN THOMAS, b Birmingham, Ala, Nov 3, 13. ELECTRONICS ENGINEERING, COMMUNICATIONS ENGINEERING. *Educ:* Ga Inst Technol, BSEE, 35. *Prof Exp:* Electronics engr radio monitoring, Fed Commun Comn, 40-44; electronics officer radar design, US Naval Res Lab, 44-46; chief, FM Broadcasting Br, Fed Commun Comn, 46-50, TV Broadcasting Br, 50-54, Tech Standards Br, 54-60, Tech Res Div, 60-74, chief, Res Standards Div, 74-80, dep chief scientist, 80-81; CONSULT COMMUN, 81- *Concurrent Pos:* Tech adv, Int Electrotech Comn, 65-; mem, Elec & Electronics Standards Mgt Bd, Am Nat Standards Inst, 68-81; Int Chmn, Study Group 1, Int Radio Consult Comt, 74-82. *Mem:* Inst Elec & Electronics Engrs. *Res:* All kinds of radio electronic systems, especially improvement of system performance by application of advanced technology, improved radio frequency spectrum utilization, new systems, and related matters. *Mailing Add:* 300 81st St S Birmingham AL 35206

DIXON, KEITH LEE, b El Centro, Calif, Jan 20, 21; m 49; c 1. VERTEBRATE ZOOLOGY. *Educ:* San Diego State Col, AB, 43; Univ Calif, MA, 48, PhD(zool), 53. *Prof Exp:* Mus technician, Mus Vert Zool, Univ Calif, 47-52, asst zool, 47-49; from asst prof to assoc prof wildlife mgt, Agr & Mech Col Tex, 52-58; asst res zoologist, Hastings Natural Hist Reservation, 58-59; from asst prof to assoc prof, 59-67, PROF ZOOL, UTAH STATE UNIV, 67- *Concurrent Pos:* Asst ed, Cooper Ornith Soc, 49-52; leader Mex expeds, Agr & Mech Col Tex, 53-54; ed bull, Wilson Ornith Soc, 55-58; mem field expeds, Calif, Mont, Ariz & Tex. *Mem:* Am Soc Mammal; fel Am Ornith Union; Cooper Ornith Soc; Wilson Ornith Soc; Ecol Soc Am. *Res:* Ecology, social behavior and distribution of birds and mammals. *Mailing Add:* Dept Biol Utah State Univ Logan UT 84322-5305

DIXON, KEITH R, b Portsmouth, Eng, Dec 4, 40; m 66. INORGANIC CHEMISTRY. *Educ:* Cambridge Univ, BA, 63; Univ Strathclyde, PhD(fluorine chem), 66. *Prof Exp:* Fel inorg chem, Univ Western Ont, 66-68; asst prof inorg chem, 68-74, ASSOC PROF CHEM, UNIV VICTORIA, 74- *Mem:* Chem Inst Can; The Chem Soc. *Res:* Chemistry of complex fluorides; organometallic chemistry of the precious metals. *Mailing Add:* Dept Chem Univ Victoria Box 1700 Victoria BC V8W 2Y2 Can

DIXON, KENNETH RANDALL, b Madison, Wis, Aug 2, 42; m 65. SYSTEMS ECOLOGY, COMPUTER SIMULATION & MODELLING. *Educ:* Univ Fla, BSF, 64, MSF, 68; Univ Mich, PhD(wildlife mgt), 74. *Prof Exp:* Res ecologist ecosyst anal, Oak Ridge Nat Lab, 73-76; asst prof wildlife ecol, Appalachian Environ Lab, Univ Md, 76-84; MGR WILDLIFE, DEPT WILDLIFE, WASHINGTON, 84- *Concurrent Pos:* Consult, Md Dept Natural Resources, 76-84. *Mem:* AAAS; Int Soc Ecol Modelling; Ecol Soc Am; Wildlife Soc. *Res:* Wildlife habitat analysis; wildlife as monitors; wildlife population dynamics. *Mailing Add:* Dept Wildlife 600 N Capitol Way GJ-11 Olympia WA 98501-1091

DIXON, LINDA KAY, b Brownsville, Pa, Aug 20, 40; c 1. BEHAVIORAL GENETICS, GENES & AGING. *Educ:* Calif State Col, Pa, BS, 62; Univ Calif, Berkeley, MS, 64; Univ Ill, Urbana, 67. *Prof Exp:* Res assoc, Inst Behav Genetics, Univ Colo, Boulder, 68; asst prof genetics, 69-76, assoc prof biol, 76-80, PROF BIOL, UNIV COLO, DENVER, 80- *Concurrent Pos:* Behav genetics training grant, Inst Behav Genetics, Colo, 68-69; vis investr, Waksman Inst NJ, 84-85. *Mem:* AAAS; Genetics Soc Am; Am Inst Biol Sci; Behav Genetics Asn; Am Aging Asn; Sigma Xi. *Res:* Developmental genetics of mouse behavior; relationships between chromosomal aberrations and behavioral differences; genetics of aging. *Mailing Add:* Dept Biol PO Box 173364 Denver CO 80217-3364

DIXON, LYLE JUNIOR, b Osborne, Kans, Feb 28, 24; m 42; c 3. MATHEMATICS. *Educ:* Okla State Univ, BS, 48, MS, 50; Univ Kans, PhD, 63. *Prof Exp:* Instr math, Okla State Univ, 47-48 & Southwestern State Col, 49-50; asst instr, Univ Kans, 50-51; asst prof, Ark State Col, 51-53, supvr sci & math areas, Gen Educ, 53-57, registr, 57-58, dir res, 58-59, assoc prof math, 59-63; from assoc prof to prof, 63-89, EMER PROF MATH, KANS STATE UNIV, 89- *Mem:* Nat Coun Teachers Math. *Res:* Matrix theory; mathematics education. *Mailing Add:* Dept Math Kans State Univ Manhattan KS 66506

DIXON, MARVIN PORTER, b Kansas City, Mo, Nov 10, 38; m 61; c 3. ORGANIC CHEMISTRY, PHYSICAL ORGANIC CHEMISTRY. *Educ:* William Jewell Col, BA, 60; Univ Ill, Urbana, MS, 63, PhD, 65. *Prof Exp:* From asst prof to assoc prof, 65-75, PROF CHEM, WILLIAM JEWELL COL, 75-, CHMN DEPT, 74- *Concurrent Pos:* Kans City Regional Coun Higher Educ Fac develop grant. *Mem:* Midwest Asn Chem; Teachers Lib Arts Cols; Am Chem Soc. *Res:* Ambident ion chemistry; kinetics; reaction mechanisms; organometallic chemistry; sympathetic inks. *Mailing Add:* 2601 Hope Ct Liberty MO 64068-9134

DIXON, MICHAEL JOHN, b Chicago, Ill, July 1, 41; m 82; c 3. COMPLEX APPROXIMATION THEORY, SIGNAL PROCESSING. *Educ:* Col St Thomas, BA, 63; Purdue Univ, MS, 65; Univ Calif, San Diego, PhD(math), 76. *Prof Exp:* Asst prof, 67-72, assoc prof, 75-80, PROF MATH, CALIF STATE UNIV, CHICO, 80- *Concurrent Pos:* NASA trainee, Purdue Univ, 63-66; vis res scientist, Univ Amsterdam, 77. *Mem:* Math Asn Am; Am Math Soc; Wiskundig Genootschap. *Res:* Complex approximation theory; signal processing. *Mailing Add:* 1186 Bonair Rd Chico CA 95926

DIXON, NORMAN REX, b Ecorse, Mich, Feb 24, 32; m 68; c 1. SPEECH SIGNAL PROCESSING. *Educ:* Western Mich Univ, BA, 58; Ind Univ, MA, 61; Stanford Univ, PhD, 66. *Prof Exp:* Instr speech sci, Cornell Univ, 60-61; linguistics, Stanford Univ, 62-63; PROF ENG & RES SPEECH PROCESSING, BROWN UNIV, 80-; PVT CONSULT, SPEECH PROCESSING TECHNOL, 89- *Concurrent Pos:* Adj prof psychol, San Jose State Univ, 64-65; vis prof speech sci, NC Central Univ, 67, 70-71; vpres, Acoustics, Speech & Signal Processing Soc, Inst Elec & Electronics Engrs, 80-81, pres, 82-83; bd dir, Inst Elec & Electronics Engrs, 86-87; vpres, Tech Activities, Inst Elec & Electronics Engrs, 88-; consult, Speech Processing Technol, IBM Res, 64-83; ed, IBM J Res & Develop, 83-89. *Mem:* Fel Inst Elec & Electronics Engrs. *Res:* Automatic document indexing, human factors, computer speech synthesis, recognition, coding, and speaker verification; numerous publications. *Mailing Add:* 1325 Pembroke Jones Dr Wilmington NC 28405

DIXON, PAUL KING, b Bridgeport, Conn, Sept 24, 61; m 87. GLASS TRANSITION, COMPLEX FLUIDS UNDER SHEAR. *Educ:* Univ Mich, BSE, 83; Univ Chicago, MS, 85, PhD(physics), 90. *Prof Exp:* Teaching asst physics, Cornell Univ, 83-84; res asst, Norden Systs, 84, Univ Chicago, 85-90; ASSOC, EXXON RES & ENG CO, 90- *Concurrent Pos:* Consult, Source Inc, 89. *Mem:* Am Phys Soc. *Res:* Study of complex fluids and the glass transition using a variety of techniques ranging from relaxation spectroscopy to light scattering. *Mailing Add:* LE378 Exxon Res Rte 22 East Annandale NJ 08801

DIXON, PEGGY A, b Cleveland, Ohio, June 9, 28; m 50; c 3. PHYSICS. *Educ:* Western Reserve Univ, AB, 50; Univ Md, MS, 54, PhD(physics), 59. *Prof Exp:* Physicist munitions res, Chem Corps, US Army, 50-52; asst physics, Univ Md, 52-58, res assoc, 58-61; from asst prof to assoc prof, 59-70, PROF PHYSICS & ENG, MONTGOMERY COL, 70- *Concurrent Pos:* NSF res fel, 62; chmn panel on physics in two-year col, Comn Col Physics, 68-69; vis prof, Univ Md, 70-71. *Mem:* AAAS; Am Asn Physics Teachers. *Res:* Solid state theory; ferromagnetism; lattice vibration theory. *Mailing Add:* Dept Math & Sci Montgomery Col Takoma Park Takoma Park MD 20912

DIXON, PETER STANLEY, b Redcar, Eng, Nov 29, 28; m 55; c 3. PHYCOLOGY. *Educ:* Univ Manchester, BSc, 49, MSc, 50, PhD(bot), 52, DSc(bot), 78. *Prof Exp:* Asst lectr bot, Univ Liverpool, 54-56, lectr, 56-64, sr lectr, 64-65; assoc prof, Univ Wash, 65-67; chmn dept pop & environ biol, 69-72, 75-76, PROF BIOL SCI, UNIV CALIF, IRVINE, 67- *Res:* Developmental morphology, ecology and taxonomy of algae, particularly Rhodophyta. *Mailing Add:* Dept Ecol Univ Calif Irvine Irvine CA 92717

DIXON, RICHARD WAYNE, b Hubbard, Ore, Sept 25, 36; m 70; c 3. SOLID STATE PHYSICS. *Educ:* Harvard Univ, AB, 58, MA, 60, PhD(appl physics), 64. *Prof Exp:* Res fel appl physics, Univ Harvard, 64-65; staff mem, 65-68, supvr, 68-79, dept head, appl physics, 79- 84, DIR, PHOTONICS DEVICES, BELL LABS, INC, 84- *Mem:* AAAS; Am Phys Soc; fel Inst Elec & Electronics Eng. *Res:* Research and development of acoustooptic, electroluminescent, and injection laser materials and devices. *Mailing Add:* Rm 2A231 AT&T Bell Labs Murray Hill NJ 07974

DIXON, ROBERT CLYDE, b Greensboro, NC, Jan 8, 32; m 55; c 3. COMMUNICATIONS & NAVIGATION SYSTEM DESIGN ANALYSIS. *Educ:* Pac States Univ, BSEE, 61; West Coast Univ, MSEE, 68. *Prof Exp:* Sr engr, Magnavox Res Labs, 59-68, sr staff engr, 71-72; staff engr, TRW Inc, 68-71; sr res engr, Northrop Corp, 72-74; sr tech staff asst, Hughes Aircraft Co, 74-75, chief scientist, 82-84; pres, Spectrack Systs Inc, 76-79, div mgr, Spectrack Systs Div, Res & Develop Assoc, 79-82; chief scientist, Spread Spectrum Sci, 84-89; CHIEF SCIENTIST, OMNIPOINT DATA CO, INC, 89- *Concurrent Pos:* Pres, R C Dixon & Assocs Consult Engrs, 75 -; lectr, Univ Calif, Los Angeles, 75- & George Washington Univ, 76 -; tech chmn, Int Telemetry Conf, 77; co-ed, Inst Elec & Electronics Engrs Commun Trans. *Mem:* Inst Elec & Electronics Engrs; Armed Forces Commun & Electronics Asn. *Res:* Development in spread spectrum, antijamming communication and navigation systems; avionics, mobile, and satellite spread spectrum systems. *Mailing Add:* PO Box 100 Palmer Lake CO 80133

DIXON, ROBERT JEROME, JR, b Seattle, Wash, Apr 26, 31; m 52; c 2. AERODYNAMICS, HYDRODYNAMICS. *Educ:* Univ Wash, BS, 58, MS, 60. *Prof Exp:* Sr prin engr aerodyn, hydrodyn & propulsion, Boeing Co, 59-90; RETIRED. *Concurrent Pos:* Consult, 90- *Honors & Awards:* Space Shuttle Atlantis Flag Award, Am Soc Mech Engrs. *Mem:* Sigma Xi; Am Inst Aeronat & Astronaut Asn; Am Soc Mech Engrs; US Naval Inst. *Res:* Viscous flows; separated flows; test simulation. *Mailing Add:* 3933 SW 109th St Seattle WA 98146

DIXON, ROBERT LELAND, b Sumter, SC, July 2, 40. RADIOLOGICAL PHYSICS. *Educ:* Univ SC, BS, 63, PhD(physics), 70. *Prof Exp:* Instr reactor physics, US Naval Nuclear Power Sch, 64-68; assoc prof, 70-80, PROF RADIOL PHYSICS, BOWMAN GRAY SCH MED, WAKE FOREST UNIV, 80- *Mem:* Am Asn Physicists Med (pres elect, 91); Am Col Radiol. *Res:* Nuclear magnetic resonance applications in biophysics; thermoluminescence phosphor research in dosimetry; computer applications in medicine. *Mailing Add:* Dept Radiol Bowman Gray Sch Med Wake Forest Univ 300 Hawthorne Rd SW Winston-Salem NC 27103

DIXON, ROBERT LOUIS, pharmacology, biochemistry; deceased, see previous edition for last biography

DIXON, ROGER L, b Pecos, NMex, Aug 14, 47; m 68; c 1. PHYSICS. *Educ:* NMex Highlands Univ, BS, 70; Purdue Univ, MS, 72, PhD(physics), 75. *Prof Exp:* Res assoc physics, Cornell Univ, 75-77; PHYSICIST, FERMILAB, 77- *Mem:* Am Phys Soc; Sigma Xi. *Res:* High energy particle accelerators, particularly extraction, beam lines and superconducting applications; high energy physics. *Mailing Add:* Fermilab MS 357 Fermilab Batavia IL 60510

DIXON, ROSS, June 6, 42; m; c 2. PHARMACOKINETICS, DRUG METABOLISM & DELIVERY. *Educ:* Sheffield Univ, Yorkshire County, UK, PhD(pharmacol), 68. *Hon Degrees:* MA, Trinity Col, Dublin, Ireland, 87. *Prof Exp:* Dir, Pharmaceut Anal Res & Develop, Schering Corp, 87-88; DIR, NEW DRUG DEVELOP, GENSIA PHARMACUET INC, 88- *Concurrent Pos:* Sect head Drug Metab, Hoffman La Roche, Inc. *Mem:* Am Soc Clin Pharmacol Therapeut; Am Asn Pharmaceut Sci. *Res:* Cardiovascular, analgesics, psychotropics. *Mailing Add:* Gensia Pharmaceut Inc 11025 Roselle St San Diego CA 92121

DIXON, SAMUEL, JR, b Six, WVa, Oct 16, 27; m 59; c 2. ELECTRONICS ENGINEERING, PHYSICS. *Educ:* WVa State Col, BS, 49; Monmouth Col, MS, 72. *Prof Exp:* Physicist components, Electron Rest Develop Command, US Army Electronics Command, Ft Monmouth, 58-80, sr electronics engr, Electron Technol & Devices Lab, 80-90; RETIRED. *Mem:* Inst Elec & Electronics Engrs. *Res:* Investigation and design of millimeter-wave components in high resistivity, high permittivity dielectrics used as waveguides; evaluate new technologies related to millimeter-wave integrated circuits. *Mailing Add:* 1018 Eaton Way Neptune NJ 07753

DIXON, THOMAS PATRICK, electronics engineering, optical physics, for more information see previous edition

DIXON, WALLACE CLARK, JR, b Winston-Salem, NC, Dec 18, 22; m 46; c 2. PHYSIOLOGY. *Educ:* Eastern Nazarene Col, AB, 46; Boston Univ, AM, 47, PhD(biol), 56. *Prof Exp:* From instr to asst prof biol, Eastern Nazarene Col, 48-54; from asst prof to prof basic studies & grad sch, Boston Univ, 56-68; prof biol & chmn, Natural Sci Dept, 68-79, ASSOC DEAN, COL NATURAL & MATH SCI, EASTERN KY, 79- *Mem:* Asn Gen & Lib Studies; Asn Integrative Studies. *Res:* Cellular physiology; endocrinology; science in general education. *Mailing Add:* 2029 Lakewood Dr Richmond KY 40475-0950

DIXON, WALTER REGINALD, ADRENERGIC & PRESYNAPTIC RECEPTORS. *Educ:* State Univ NY, PhD(pharmacol), 75. *Prof Exp:* ASSOC PROF PHARMACOL, SCH PHARM, UNIV KANS, 84- *Res:* Effects of indigenous and opioid peptides. *Mailing Add:* Dept Pharmacol & Toxicol 5044 Malott Hall Sch Pharm Univ Kans Lawrence KS 66045

DIXON, WILFRID JOSEPH, b Portland, Ore, Dec 13, 15; div; c 2. BIOMATHEMATICS. *Educ:* Ore State Col, BA, 38; Univ Wis, MA, 39; Princeton Univ, PhD(math statist), 44. *Prof Exp:* Asst, Princeton Univ, 39-41 & Nat Defense Res Comt Proj, 41-42; from instr to asst prof math, Univ Okla, 42-44; res assoc & consult, Joint Target Group, Air Corps & Air Serv, Intel & Opers Analyst, 20th Air Force, 44-45; assoc prof math, Univ Okla, 45-46; assoc prof, Univ Ore, 46-52, prof, 52-55; PROF BIOSTATIST, UNIV CALIF, LOS ANGELES, 55-, PROF BIOMATH, 67- *Concurrent Pos:* Assoc ed, Biomet, 52-66 & Annals Math Statist, 55-58; consult, Off Naval Res, Rand Corp, Vet Admin, State Dept Ment Hyg, Calif & NIH; chief statist res, Brentwood Vet Admin Hosp, 72-89; prof psychiat, 74-; chmn, BMDD Statist Software, Inc, 87-; pres, Dixon Statist Assocs, 88- *Mem:* Fel Royal Statist Soc; fel Am Statist Asn; fel Inst Math Statist; Biomet Soc; Sigma Xi. *Res:* Mathematical theory of serial correlation; sensitivity experiments; inefficient statistics; computers in medical research. *Mailing Add:* Dept Biomath Univ Calif Med Ctr Los Angeles CA 90024

DIXON, WILLIAM BRIGHTMAN, b Fall River, Mass, Dec 18, 35; m 57; c 2. PHYSICAL CHEMISTRY, COMPUTER INTERFACING. *Educ:* Wheaton Col (Ill), BS, 57; Harvard Univ, AM, 59, PhD(phys chem), 61. *Prof Exp:* NSF fel, Nat Res Coun, Ottawa, 61-62, Nat Res Coun Can fel, 62-63; from asst prof to assoc prof, Wheaton Col, Ill, 63-68; assoc prof, 68-69, PROF CHEM, STATE UNIV NY COL ONEONTA, 69- *Concurrent Pos:* Fulbright student, Copenhagen, 60-61; vis prof, Dartmouth Col, 77-78; Fulbright prof, Univ Zambia, Lusaka, 84-85 & 85-86. *Mem:* AAAS; Am Chem Soc. *Res:* Development of computer applications in chemical education. *Mailing Add:* Dept Chem State Univ NY Oneonta NY 13820-4015

DIXON, WILLIAM ROSSANDER, b Sask, July 3, 25; m 73. NUCLEAR PHYSICS. *Educ:* Univ Sask, BA, 45, MA, 47; Queen's Univ, PhD(physics), 55. *Prof Exp:* RES OFFICER X-RAYS & NUCLEAR RADIATIONS, NAT RES COUN CAN, 50- *Mem:* Can Asn Physicists; Am Phys Soc. *Res:* Gamma rays produced in charged-particle nuclear reactions to elucidate nuclear structure; radioactivity; radiation dosimetry. *Mailing Add:* 1861 Greenacre Circle Ottawa ON K1J 6S7 Can

DIXON-HOLLAND, DEBORAH ELLEN, b Passaic, NJ, May 20, 57; m 87; c 3. ANTIBODY DEVELOPMENT. *Educ:* Lafayette, BS, 79; Am Univ, MS, 83; Rutgers Univ, PhD(microbiol), 86. *Prof Exp:* Teaching asst microbiol, Am Univ, 79-81; res asst microbiol, Rutgers Univ, 81-84; res assoc food sci, Mich State Univ, 84-86; MGR DIAGNOSTIC RES & DEVELOP, IMMUNOL, NEOGEN CORP, 86- *Concurrent Pos:* Consult, Neogen Corp, 84-86; asst adj prof, Dept Food Sci, Mich State Univ, 86-; prin invest, numerous Fed & Small State Bus grants, Neogen Corp, 88-; numerous grants. *Mem:* Am Soc Microbiol; AAAS; Asn Off Analytical Chemists. *Res:* Manage and perform development work of enzyme immunoassay diagnostics for improvement of plant and animal health; authored & co-authored numerous journal articles & text chapters. *Mailing Add:* Neogen Corp, 620 Lesher Pl Lansing MI 48912

DIZENHUZ, ISRAEL MICHAEL, b Toronto, Ont, May 20, 31; m; c 2. PSYCHIATRY. *Educ:* Univ Toronto, MD, 55; Am Asn Psychiat Clin Children, cert career child psychiat, 60; Am Bd Psychiat & Neurol, dipl & cert psychiat, 62 & cert child psychiat, 65. *Prof Exp:* Rotating gen intern, E J Meyer Mem Hosp, Buffalo, NY, 55-56; resident psychiat, Ont Hosp, Hamilton, 56-57; resident, Col Med, Univ Cincinnati, 57-58, fel, 59-60, from instr to asst prof child psychiat, 61-69, from asst dir to assoc dir child psychiat, Cent Psychiat Clin, 63-70, assoc prof child psychiat, 69-78; dir psychiat, Jewish Hosp, Cincinnati, 78-87. *Concurrent Pos:* Staff child psychiatrist, Child Guide Home, Cincinnate, 61; consult, Family Serv Cincinnati, 63-67 & Jewish Family Serv, 63-68; lectr, Vol Training for Community Ment Health Serv, Cincinnati, 63-68; attend child psychiatrist & clinician child psychiat, Cincinnati Gen Hosp; mem bd trustees, Hamilton Co Diagnostic Ctr Ment Retarded, 63-78; psychiatrist, Consult Staff, Children's Hosp, Cincinnati, 66; asst examiner, Child Psychiat Exam, Am Bd Psychiat & Neurol, 67-87; mem prof adv comt, United Serv for Handicapped of Cincinnati; mem citizen's comt, Area Task Force on Ment Retardation; mem res team, Family Serv of Cincinnati, 67- *Mem:* Fel Am Psychiat Asn; Am Orthopsychiat Asn; fel Am Acad Child Psychiat. *Res:* Community psychiatry. *Mailing Add:* Dept Psychiat Jewish Hosp 3220 Burnet Ave Cincinnati OH 45229

DI ZIO, STEVEN F(RANK), b Newark, NJ, Dec 15, 38; m 58; c 1. CHEMICAL ENGINEERING. *Educ:* Ore State Univ, BS, 61; Rensselaer Polytech, PhD(chem eng), 64. *Prof Exp:* Texaco fel, Rensselaer Polytech, 61-62, Procter & Gamble fel, 62-63, Lummus Co fel, 63-64; asst prof chem eng, Rensselaer Polytech Inst, 64-68, assoc prof, 68-69, chmn dept biomed eng, 67-69; vpres, 69-72, pres & dir, Aero Vac Corp, 72-74; PRES, CHIEF EXEC OFFICER & DIR, SES INC, 74-; STAFF TOXICOL, DEPT HEALTH SERVS. *Concurrent Pos:* Adj assoc prof, Rensselaer Polytech Inst, 69- *Mem:* AAAS; Am Inst Chem Engrs; Am Chem Soc; Am Soc Eng Educ; Soc Cryobiol. *Res:* Application of engineering principles to medicine; development of a coherent academic program to train bioengineers. *Mailing Add:* Dept Health Servs 714/744 P St PO Box 942732 Sacramento CA 94234-7320

DJANG, ARTHUR H K, b Lio-Yuan, China, Feb 12, 25; US citizen; m 58; c 4. PATHOLOGY, NUCLEAR MEDICINE. *Educ:* Harbin Med Univ, China, MD, 44; Univ Minn, Minneapolis, MPH, 51; Univ Calif, Los Angeles, PhD(infectious dis), 55; Am Bd Path, dipl, 62; Am Bd Nuclear Med, dipl, 74. *Prof Exp:* Clin instr infectious dis, Sch Med, Univ Calif, Los Angeles, 50-55; asst prof microbiol, Sch Med & Sch Trop Med, Univ PR, San Juan, 56; dir div chronic dis & actg dir div communicable dis control, NMex State Dept Pub Health, 57-58; dir microbiol & virus lab, St Francis Hosp, Wichita, Kans, 58-61; chief pathologist & dir labs, Mem Gen Hosp, Las Cruces, NMex, 62-71; pres & dir, Bio-Med Sci Labs, Las Cruces & Albuquerque, NMex, 63-75; CHMN DEPT PATH & NUCLEAR MED, JAMESTOWN GEN HOSP, 75- *Concurrent Pos:* Consult prof path, NMex State Univ, University Park, 63-71; consult, Holloman AFB Hosp, NMex, 63-73; med exam-coroner, Dona Ana County, NMex, 67-70; prof biol sci, Western NMex Univ, 67-75; clin prof, Univ Fredomia, 77-; vpres & dir res, Am Heart Asn, 79- *Mem:* Col Am Pathologists; Am Soc Clin Pathologists; Soc Nuclear Med; Am Soc Cytol; Am Soc Microbiol. *Res:* Immunotherapy and chemotherapy of cancers; radioimmunoassay of hormones, isoenzymes and epidemiology of cardiovascular diseases and tumors. *Mailing Add:* 496 Hunt Rd Jamestown NY 14701-5708

DJERASSI, CARL, b Vienna, Austria, Oct 29, 23; nat US; m 85; c 1. ORGANIC CHEMISTRY. *Educ:* Kenyon Col, AB, 42; Univ Wis, PhD(org chem), 45. *Hon Degrees:* DSc, Nat Univ Mex, 53, Kenyon Col, 58, Fed Univ Rio de Janeiro, 69, Worcester Polytech Inst, 72, Wayne State Univ, 74, Columbia Univ, 75, Uppsala Univ, 77, Coe Col & Univ Geneva, 78, Univ Ghent & Univ Manitoba, 85. *Prof Exp:* Res chemist, Ciba Pharmaceut Prod, Inc, 42-43, 45-49; assoc dir chem res, Syntex, SA, Mex, 49-52; from assoc prof to prof, Wayne State Univ, 52-59; PROF CHEM, STANFORD UNIV, 59- *Concurrent Pos:* Vpres, Syntex, SA, 57-60, pres, Syntex Res Div, 68-72; pres, Zoecon Corp, 68-83, chmn bd, 83-88; chmn bd, Sci & Technol Int Develop, Nat Acad Sci, 68-75. *Honors & Awards:* Nat Medal of Sci, 73; Nat Medal of Technol, 91; Pure Chem Award, Am Chem Soc, 58, Baekeland Medal, 59, Fritzsche Medal, 60, Chem Invention Award, 73, Award Chem Contemp Technol Probs, 83, Esselen Award, 89; Intrasci Res Found Award, 70; Freedman Patent Award, Am Inst Chemists, 71 & Chem Pioneer Award, 73; Perkin Medal, Soc Chem Indust, 75; Nat Inventors Hall of Fame & Wolf Prize in Chem, 78; Roussel Prize, 88; Indust Appl Sci Award, Nat Acad Sci, 90. *Mem:* Nat Acad Sci; Inst Med-Nat Acad Sci; fel Am Acad Arts & Sci; Am Chem Soc; hon fel Royal Chem Soc; Am Acad Pharm Sci; Ger Acad Nat Sci; Swed Acad Sci; Bulgarian Acad Sci; Brazilian Acad Sci; Mex Acad Sci. *Res:* Chemistry of steroids; structure; antibiotics, terpenoids and phospholipids; synthesis of drugs, particularly antihistamines, oral contraceptives, antiinflammatory agents; optical rotatory dispersion and circular dichroism studies, organic mass spectrometry; magnetic circular dichroism of organic compounds; applications of computer artificial intelligence techniques to chemical structure elucidation. *Mailing Add:* Dept Chem Stanford Univ Stanford CA 94305-5080

DJERASSI, ISAAC, b Bulgaria, July 27, 21; nat US; m 54; c 2. HEMATOLOGY, ONCOLOGY. *Educ:* Hebrew Med Sch, Israel, MD, 51. *Hon Degrees:* DH, Villanova Univ, 77. *Prof Exp:* Resident pediat, Hadassah Hosp, Israel, 53-54; assoc path, Children's Med Ctr, Children's Cancer Res Found, 55-60; assoc dir clin labs & dir blood bank & donor ctr, Children's Hosp Philadelphia, 60-69; DIR DONOR CTR & RES HEMAT, MERCY CATH MED CTR, 69-, DIR RES, HEMAT-ONCOL. *Concurrent Pos:* Res assoc, Harvard Med Sch, 57-60; asst prof pediat, Sch Med, Univ Pa, 60-69; consult, Children's Cancer Res Found, Boston, Mass, 61-77; mem, Acute Leukemia Task Force, Nat Cancer Inst. *Honors & Awards:* Prize, Cong Hemat, 56; Lasker Award, 72; Golden Plate Award, 77; Edwin Cohn-DeLaval Award, 90. *Mem:* Soc Pediat Res; Am Asn Cancer Res; Am Soc Exp Path. *Res:* Pediatric hematology and hemostasis. *Mailing Add:* Misericordia Div Mercy Cath Med Ctr 54th St & Cedar Ave Philadelphia PA 19143

DJOKOVIC, DRAGOMIR Z, b Trap, Macedonia, Yugoslavia, Jan 1, 38; m 61; c 2. ALGEBRA, LINEAR & MULTILINEAR ALGEBRA. *Educ:* Fac Elec Eng, Belgrade, Yugoslavia, BSc, 60; Univ Belgrade, PhD(math), 63. *Prof Exp:* Instr math, Fac Elec Eng, Yugoslavia, 60-64, docent, 64-67; res assoc math, 67-68, assoc prof, 68-69, PROF MATH, UNIV WATERLOO, 69- *Concurrent Pos:* Ed, 71-72 & Elem Problem Sect, Am Math Monthly, 75-79. *Mem:* Am Math Soc; Can Math Soc. *Mailing Add:* Dept Pure Math Univ Waterloo Waterloo ON N2L 3G1 Can

DJURIC, DUSAN, b Novi Sad, Yugoslavia, Jan 17, 30; m 55; c 2. METEOROLOGY. *Educ:* Univ Belgrade, Dipl, 53, DrSc, 60. *Prof Exp:* Asst meteorol, Col Sci, Univ Belgrade, 54-60, docent, 60-65; vis scientist, Nat Ctr Atmospheric Res, Colo, 65-66; from asst prof to assoc prof, 66-82, PROF METEOROL, TEX A&M UNIV, 82- *Concurrent Pos:* Docent, Dept Meteorol, Darmstadt Tech, 62-63; prof, Free Univ Berlin, 74-75. *Mem:* Am Meteorol Soc; Meteorol Soc Serbia; Royal Meteorol Soc. *Res:* Numerical weather forecasting; mesoscale meteorology; dynamics and development of the low-level jet stream and conveyor belt in the atmosphere; numerical modeling and synoptic analysis. *Mailing Add:* Dept Meteorol Tex A&M Univ College Station TX 77843-3146

DJURIC, STEVAN WAKEFIELD, b Leicester, Eng, Jan 5, 54; m 78; c 2. MEDICINAL CHEMISTRY. *Educ:* Univ Leeds, BSc, 76, PhD(org chem), 79. *Prof Exp:* Res fel org chem, Ohio State Univ, 80-81; res investr med chem, 81-86, group leader, 86-88, GROUP LEADER II, G D SEARLE & CO, 88- *Mem:* Am Chem Soc. *Res:* Chemistry and biochemistry of arachidonic acid metabolites; asymmetric synthesis of natural and unnatural products. *Mailing Add:* Gastrointestinal Dis Res GD Searle Co 4901 Searle Pkwy Skokie IL 60077

DJURICKOVIC, DRAGINJA BRANKO, b Belgarde, Serbia, Yugoslavia, Jan 1, 40; US citizen; m 63; c 2. IMMUNOLOGY, PARASITOLOGY. *Educ:* Univ Belgrade, DVM, 63, MSc, 67. *Prof Exp:* Vet milk hyg, Vet Inst Serbia, Yugoslavia, 63-65; lectr res & develop irradiated vaccines, Inst for Appln Nuclear Energy in Agr Vet Sci & Forestry, Yugoslavia, 65-57; res asst immunol studies E Coli, Dept Immunol, Ont Vet Col, Can, 67-70, immune response gnotobiotic animals, Dept Biomed Studies, 70-71; vet clinician small animal pract, Leawood Animal Hosp, Kansas City, Mo, 72-75; SCIENTIST, IMMUNOL CANCER RES, NAT CANCER INST, FREDERICK CANCER RES FACIL, FREDERICK, MD, 76- *Concurrent Pos:* Dep State Vet, Mo. *Mem:* Am Vet Med Asn; Tissue Cult Asn. *Res:* Immunological studies on prevention of spontaneous and chemically induced cancer; development of potential vaccines; experimental surgeries; pathology; radiobiology. *Mailing Add:* 4505 Holter Ct Jefferson MD 21755

DLAB, VLASTIMIL, b Bzi, Czech, Aug 5, 32; m 59; c 2. PURE MATHEMATICS, ALGEBRA. *Educ:* Charles Univ, Prague, RNDr, 56, CSc, 59, DSc, 66; Univ Khartoum, PhD(algebra), 62. *Prof Exp:* Sr lectr math, Charles Univ, Prague, 57-59, reader, 64-65; sr lectr, Univ Khartoum, 59-64; sr res fel, Inst Adv Studies, Australian Nat Univ, 65-68, chmn dept, 71-74, PROF MATH, CARLETON UNIV, 68- *Concurrent Pos:* Nat Res Coun Can grants, 69-; Can Coun fel, 74; vis prof, Univ Paris, Brandeis Univ & Univ Bonn, 74-75, Univ Tsukuba & Univ Sao Paulo, 76, Univ Stuttgart, 77 & Univ Portiers, 78. *Honors & Awards:* Chech Acad Sci Award, 85; Sci Exchange Awards, NSERC, Univ Bielefeld, 83. *Mem:* Am Math Soc; Math Asn Am; fel Royal Soc Can; Can Math Cong; London Math Soc; Royal Soc Can. *Res:* Theory of groups; theory of rings and modules; general algebra; representation theory. *Mailing Add:* Dept Math & Statist Carleton Univ Ottawa ON K1S 5B6 Can

DLOTT, DANA D, b Los Angeles, Calif, Sept 11, 52; m. PICOSECOND LASERS. *Educ:* Columbia Univ, BA, 74; Stanford Univ, PhD(chem), 79. *Prof Exp:* Asst prof, 79-85, ASSOC PROF CHEM, UNIV ILL, URBANA-CHAMPAIGN, 85- *Concurrent Pos:* Alfred P Sloan fel, 84. *Res:* Excited state dynamics of molecular crystals, particularly vibrational relaxation; photochemistry and energy transport, using picosecond lasers. *Mailing Add:* 505 S Mathews Ave Noyes Lab Box 37 Urbana IL 61801

DMOWSKI, W PAUL, b Lodz, Poland, May 17, 37; m 67; c 1. OBSTETRICS & GYNECOLOGY, REPRODUCTIVE ENDOCRINOLOGY. *Educ:* Warsaw Acad Med, Poland, MD, 62; Med Col Ga, PhD(endocrinol), 71. *Prof Exp:* Res fel endocrinol, Med Col Ga, 67-71; asst prof obstet/gynec, Univ Chicago, 71-74, assoc prof, 74-79; prof, Univ Ark, 79-81; dir Reprod Endocrinol & Infertility, 81-88, PROF OBSTET & GYNEC, 81-, DIR INST FOR THE STUDY & TREATMENT ENDOMETRIOSIS, RUSH PRESBY ST LUKE'S MED CTR, 88- *Concurrent Pos:* Attend physician obstet/gynec, Michael Reese Hosp & Med Ctr, Chicago, 71-79, dir fertil unit reproductive endocrinol & infertil, 73-79. *Honors & Awards:* Prize Award, Am Col Obstet & Gynec, 75. *Mem:* Am Fertil Soc; Endocrine Soc; Soc Study Reproduction; Am Col Obstet & Gynec; Soc Gynec Invest. *Res:* Clinical research in the area of reproductive endocrinology and infertility; animal studies on implantation and implantation control; clinical and experimental research on the effect of synthetic steroids on reproductive process; studies on immunologic aspects of endometriosis. *Mailing Add:* Rush Presby St Luke's Med Ctr 600 S Paulina St Chicago IL 60612

DMYTRYSZYN, MYRON, b St Louis, Mo, Dec 26, 24; m 47; c 2. CHEMICAL ENGINEERING. *Educ:* Wash Univ, BS, 47, MS, 49, DSc, 57. *Prof Exp:* Res chem engr, Monsanto Co, 47-57, proj leader, Econ & Eng Eval Group, 57-58, group leader, 58-60, eng specialist, 60-64, proj mgr eng develop, 64-65, eng mgr, Cost & Eval Eng, Cent Eng Dept, 65-68, mgr eng, 68-69, dir eng sci, 69-70, dir design & construct, Corp Eng Dept, 70-75, proj dir, Major Joint Venture, 76-77, gen mgr, Technol Div, 77-82, gen mgr, Res & Develop Div, 83-87; RETIRED. *Concurrent Pos:* Adj prof, Wash Univ, St Louis, Mo. *Mem:* Fel Am Inst Chem Engrs; Am Chem Soc; Nat Soc Prof Engrs; Soc Chem Indust. *Res:* Process development. *Mailing Add:* Myron Dmytryszyn 12905 Timmor Ct St Louis MO 63131

DOAK, GEORGE OSMORE, b Prince Albert, Sask, Dec 25, 07; nat US; m 33; c 2. PHARMACEUTICAL CHEMISTRY. *Educ:* Univ Sask, BA(chem), 29, BA(pharm), 30; Univ Wis, AM, 32, PhD(chem), 34. *Prof Exp:* Asst, Univ Wis, 32-33; asst chemist, NDak Regulatory Dept, Bismarck, 34-36; res chemist, George A Breon Co, 36-38; chemist, USPHS, 38-51, from asst dir to assoc dir VD exp lab, 51-61; prof chem, 61-73, EMER PROF CHEM, NC STATE UNIV, 73- *Concurrent Pos:* Instr med, Johns Hopkins Hosp, 38-46; res assoc, Johns Hopkins Univ, 46-48; assoc prof, Sch Pub Health, Univ NC, 48-61. *Mem:* Am Chem Soc. *Res:* Synthesis of organometallic compounds, particularly those of arsenic, antimony and bismuth; preparation of organo-phosphorus compounds; biochemistry of treponemata; stereochemistry of trigonal bipyramids. *Mailing Add:* 717 Old Mill Rd Chapel Hill NC 27514

DOAK, KENNETH WORLEY, b Gallatin, Mo, Jan 27, 16; m 45; c 3. POLYMER CHEMISTRY. *Educ:* Cent Col, Mo, AB, 38; Johns Hopkins Univ, PhD(phys org chem), 42. *Prof Exp:* Res chemist, Uniroyal, 42-55; res assoc, Res Ctr, Koppers Co Inc, 55-57, mgr plastics res, 58-60; asst dir polymer res, Rexall Chem Co, NJ, 60-65, dir cent res, 65-70; mgr plastics res, Res Ctr, Koppers Co, Inc, Arco Polymers, Inc, mgr polymer res, 74-76, tech adv, 77-80; CONSULT POLYMERS, 81-; INDEPENDENT INVENTOR, 82- *Mem:* Am Chem Soc; Acad Appl Sci; fel Am Inst Chemists. *Res:* Theory of polymerization; ionic and free-radical processes; polymer structure and properties; polyethylene; polypropylene; polystyrene; foam; block and graft copolymers; polymer blends. *Mailing Add:* 3469 Burnett Dr Murrysville PA 15668-1347

DOAN, ARTHUR SUMNER, JR, b Ft Wayne, Ind, July 29, 33; m 60; c 3. PHYSICAL CHEMISTRY & METALLURGY. *Educ:* Wabash Col, AB, 55; Iowa State Univ, MS, 58; Colo Sch Mines, DSc, 63. *Prof Exp:* Chemist, Ames Lab, Atomic Energy Comn, 55-59; physicist, US Army Biol Res Lab, Md, 59-61; physicist refractory mat, Lewis Res Ctr, 62-68, PHYS CHEMIST, GODDARD SPACE FLIGHT CTR, NASA, 68- *Mem:* Am Chem Soc; Am Soc Mass Spectrometry; Microbeam Anal Soc. *Res:* Lunar and meteorite mineral and metals analysis; electron microprobe and computerized data reduction; planetary atmospheric analysis; mass spectrometry, gas chromatography, trace component enrichment. *Mailing Add:* 820 Boatswain Dr Annapolis MD 21401-6854

DOAN, DAVID BENTLEY, b State College, Pa, Jan 9, 26; m 54; c 4. EARTH SCIENCE. *Educ:* Pa State Univ, BS, 48, MS, 49. *Prof Exp:* Asst mineral, Pa State Univ, 48-49, instr geol, 49; geologist, US Geol Surv, 49-62; opers analyst, Res Anal Corp, 62-68; treas, Earth Sci Group, Inc, Washington, DC, 68-73; dir, Bobcat Properties Inc, 73-85; chief geologist, Gold Lake Mines Inc, 85-88; GEOLOGIST, US BUR MINES, 90- *Concurrent Pos:* Consult geologist, 62-; consult, US & foreign indust, govt orgn & var pvt explor progs; vis prof geol, Univ Md, College Park, 76-82; chief geol adv, Coalinga Resources, Inc, 84-90; dir, Gold Lake Mines, 82-85. *Mem:* Geol Soc Am; Am Inst Prof Geologists; Am Asn Petrol Geologists; Sigma Xi; Soc Petrol Engrs; Am Geophys Union. *Res:* Physical geology; petroleum; terrain analysis; engineering geology; mineral exploration; geology of Pacific Basin and Southeast Asia; equilibria in fluvial processes. *Mailing Add:* 5635 Bent Branch Rd Bethesda MD 20816

DOANE, BENJAMIN KNOWLES, b Philadelphia, Pa, May 9, 28; Can citizen; m 61; c 2. PSYCHIATRY. *Educ:* Princeton Univ, AB, 50; Dalhousie Univ, MA, 52, MD, 62; McGill Univ, PhD(psychol), 55; FRCP(C), 67; FACP, 76. *Prof Exp:* Res officer psychol, Defense Res Bd Can, 51-53; res assoc, Montreal Neurol Inst, McGill Univ, 55-57; McLaughlin fel, Maudsley Hosp, London, Eng, 64-65; Med Res Coun fel, Univ Montreal, 65-66; lectr physiol, 66-75, asst dean med, 73-75, from asst prof to assoc prof, 66-75, head dept, 75-81, PROF PSYCHIAT, DALHOUSIE UNIV, 75- *Mem:* Can Psychiat Asn; Can Med Asn; fel Am Col Psychiatrists. *Res:* Neurosciences. *Mailing Add:* Dept Psychiat Victoria Gen Hosp 1278 Tower Rd Halifax NS B3H 2V9 Can

DOANE, CHARLES CHESLEY, b Bradford, Ont, July 19, 25; US citizen; m 53; c 1. ENTOMOLOGY. *Educ:* Ont Agr Col, BS, 49; Univ Wis, MS, 51, PhD, 53. *Prof Exp:* Mem staff prod develop dept, Shell Chem Corp, 53-56; from asst entomologist to entomologist, Conn Agr Exp Sta, 56-77; field res mgr, Albany Int, 79-81, mgr, prod develop, 81-84; MGR PROD DEVELOP & RES, SCENTRY INC, UNITED AGR PROD CO, 84-; DIR, COM DEVELOP, 87- *Concurrent Pos:* Mem, Int Orgn Biol Control Noxious Animals & Plants. *Mem:* Soc Invert Path; Entom Soc Am; Entom Soc Can. *Res:* Sex pheromones of insects for their control; insect behavior; biological control and insect pathology. *Mailing Add:* 5934 E Calle Del Sud Phoenix AZ 85018

DOANE, DOUGLAS V, b Detroit, Mich, Nov 11, 18. METALLURGICAL ENGINEERING. *Educ:* Wayne State Univ, BS, 42; Univ Mich, MS, 43. *Prof Exp:* Assoc dir res, Climax Molybdenum Co, Amax Inc, 72-82; CONSULT, 82- *Mem:* Fel Am Soc Metals; sr mem Metall Soc; Soc Automotive Engrs; Am Foundrymen's Soc. *Mailing Add:* 2302 Vinewood Ann Arbor MI 48104

DOANE, ELLIOTT P, b Ill, July 11, 29; m; c 4. CHEMICAL ENGINEERING. *Educ:* Harvard Univ, AB, 51; Univ Ill, MS, 53, PhD(chem eng), 55. *Prof Exp:* Chem engr, Hooker Chem Co, NY, 54-57, supvr pilot plant, 57-60; res engr, Tex Butadiene & Chem Co, 61, dir res lab, 62; res engr, Phillips Petrol Co, 63-66, mgr hydrocarbon processes br, 66-69; mgr process develop dept, Western Res Ctr, Stauffer Chem Co, 69-76, dir technol planning dept, 77-88; MGR, ENG RES & DEVELOP, KERR-MCGEE CORP, 88- *Mem:* Am Chem Soc; Am Inst Chem Engrs; Sigma Xi. *Res:* High pressure studies; kinetics; absorption; catalysis. *Mailing Add:* 4117 NW 144th Terr Oklahoma City OK 73134

DOANE, FRANCES WHITMAN, b Halifax, NS, Aug 30, 28. VIROLOGY, ELECTRON MICROSCOPY. *Educ:* Dalhousie Univ, BSc, 50; Univ Toronto, MA, 62. *Prof Exp:* Res asst ophthal, Univ & Hosp for Sick Children, 53-57, res asst virol, Univ Toronto, 57-62, lectr & asst prof, 62-69, head unit, 62-80, assoc prof virol, 69-89, PROF, DEPT MICROBIOL, UNIV TORONTO, 89- *Concurrent Pos:* Consult, Hosp for Sick Children, 71-89, Toronto Gen Hosp, 83-89. *Mem:* Hon mem Micros Soc Can; Can Soc Microbiol; Am Soc Microbiol; Electron Micros Soc Am; NY Acad Sci. *Res:* Electron microscopy applied to diagnostic virology; immunoelectron microscopy; ultrastructural aspects of viral-induced cytopathology and transformation. *Mailing Add:* Fac Med Univ Toronto Dept Microbiol Toronto ON M5S 1A8 Can

DOANE, J WILLIAM, b Bayard, Nebr, Apr 26, 35; m 58; c 2. LIQUID CRYSTALS DISPLAY, NUCLEAR MAGNETIC RESONANCE. *Educ:* Univ Mo, BS, 58, MS, 62, PhD(physics), 65. *Prof Exp:* From asst prof to assoc prof, 65-74, assoc dir, Liquid Crystal Inst, 79-83, PROF PHYSICS, KENT STATE UNIV, 74-, DIR, LIQUID CRYSTAL INST, 83- *Concurrent Pos:* Res prof, Molecular Path & Biol, Northeastern Ohio Univ Col Med; assoc mem, Stefan Inst, Ljubljana, Yugoslavia, 74-; dir, Ouonic Imaging Syst, 86-; prin investr, NSF Defense Agency & Indus Grants. *Mem:* Fel Am Phys Soc; Am Asn Physics Teachers; Biophys Soc; Sigma Xi. *Res:* Liquid crystal display research and nuclear magnetic resonance studies in liquid crystals. *Mailing Add:* Liquid Crystal Inst Kent State Univ Kent OH 44242-0001

DOANE, JOHN FREDERICK, b Newmarket, Ont, Apr 14, 30; m 63; c 2. ENTOMOLOGY, ANIMAL BEHAVIOR. *Educ:* Ont Agr Col, BSA, 54; Univ Wis, MS, 56, PhD(entom), 58. *Prof Exp:* Res officer entom, 58-65, RES SCIENTIST ENTOM, RES BR, AGR CAN, 65- *Concurrent Pos:* Head, Cereals Protection Res Sta, Agr Can, Saskatoon. *Mem:* Entom Soc Can; Entom Soc Am; Coleopterists Soc; Int Orgn Biol Control. *Res:* Investigation of the biology, ecology, population dynamics and the orientation and feeding behavior of wireworms pests of grain crops in the Prairie Provinces of Canada; biology and population dynamics of the wheat midge; biological control of aphids. *Mailing Add:* Eight Simpson Cres Saskatoon SK S7H 3C6 Can

DOANE, MARSHALL GORDON, b Syracuse, NY, June 11, 37. BIOPHYSICS, BIOMEDICAL ENGINEERING. *Educ:* Univ Rochester, BS, 59, MS, 61; Univ Md, PhD(biophys), 67. *Prof Exp:* Res asst phys optics, Univ Rochester, 59-61; optical physicist, Aerojet-Gen Corp, 61; physicist, Bausch & Lomb, Inc, 61-62; res fel corneal res, Eye Res Inst, Retina Found, 67-69, res assoc, 69-72, assoc staff scientist, 72-84; instr, dept ophthal, 75-79, ASST PROF OPHTHAL, MED SCH, HARVARD UNIV, 79-; SR SCIENTIST, EYE RES INST, RETINA FOUND, 85- *Honors & Awards:* Bausch & Lomb Hon Sci Award. *Mem:* Asn Res Vision & Ophthal; Int Soc Eye Res. *Res:* Pyridine nucleotide fluorescence in nerve cells; corneal physiology and metabolism; active transport and dehydration mechanisms in the cornea; corneal prostheses; bioinstrumentation; optics; corneal wetting and lid/tear-film interactions; lid dynamics during blinking; contact lens/tear-film interactions; high-speed imaging. *Mailing Add:* 24 Highland Ave Cambridge MA 02139

DOANE, TED H, b Fairview, Okla, May 12, 30; m 54; c 1. ANIMAL BREEDING. *Educ:* Okla State Univ, BS, 52; Kans State Univ, MS, 53, PhD(animal breeding), 60. *Prof Exp:* Exten county agt, 55-56, exten livestock specialist, 56-58 & 60-64, assoc prof animal sci, Agr Exten, 62-74, assoc prof, 75-78, PROF ANIMAL SCI, UNIV NEBR-LINCOLN, 78- *Concurrent Pos:* Prof agr, Exten Admin, Univ Nebr-Ataturk Univ Contract, Turkey, 64-66; prof animal sci & adv, Kabul Univ-Univ Nebr Contract, 75-77. *Mem:* Am Soc Animal Sci. *Res:* Animal production and management; sheep breeding. *Mailing Add:* Dept Animal Sci Univ Nebr East Capmpus Lincoln NE 68583-0908

DOANE, WILLIAM M, b Covington, Ind, Sept 26, 30; m 52; c 4. ORGANIC CHEMISTRY. *Educ:* Purdue Univ, BS, 52, MS, 60, PhD(biochem), 62. *Prof Exp:* Teacher, Sec Sch, Ind, 54-55; res chemist, Northern Mkt & Nutrit Res Lab, USDA, 62-70, leader nonstarch prod invests, 70-75, res leader derivatives & polymer exploration, 75-79, lab chief, biomat conversion, 80-85, RES LEADER PLANT POLYMERS, NAT CTR AGR UTILIZATION RES, USDA, 85- *Concurrent Pos:* Instr chem, Bradley Univ, 64- *Honors & Awards:* IR-100 Award, 75 & 78. *Mem:* Am Chem Soc; AAAS; Weed Sci Soc; Am Asn Cereal Chemists. *Res:* Reaction mechanisms in carbohydrate chemistry; sulfur derivatives of carbohydrates; synthesis and characterization of starch derivatives; starch graft copolymers; encapsulation of biologically active chemicals; biodegradable plastics. *Mailing Add:* Nat Ctr Agr Utilization Res USDA 1815 N University Peoria IL 61604

DOANE, WINIFRED WALSH, b New York, NY, Jan 7, 29; m 53; c 1. DEVELOPMENTAL GENETICS, MOLECULAR GENETICS. *Educ:* Hunter Col, BA, 50; Univ Wis, MS, 52; Yale Univ, PhD(zool), 60. *Prof Exp:* Teaching asst zool, Univ Wis, 50-51, res asst, 51-53; asst prof biol, Millsaps Col, 54-55; lab asst genetics, Yale Univ, 56-58, NIH res trainee, 60-62, res assoc develop genetics, 62-75, lectr, 65-75, assoc prof biol, 75-77; PROF ZOOL, ARIZ STATE UNIV, 77- *Concurrent Pos:* Assoc ed, Develop Genetics, 79-81; mem, Genetics Study Sect, NIH, 74-77, Genetic Basis Dis Rev Comt, Nat Inst Gen Med Sci, 80-84 & Biomed Sci Study Sect, 85-88; mem, Alan T Waterman award comt, NSF, 79-81, chmn, 81; vis scholar, Univ NC, Chapel Hill, 83-84; mem, Peer Rev Comt, Vis Profs for Women, NSF, 86. *Mem:* Am Soc Cell Biol; Genetics Soc Am; Am Soc Zoologists; Int Soc Develop Biologists; Soc Develop Biol; fel AAAS. *Res:* Developmental, biochemical and molecular genetics of eukaryotes, especially Drosophila and other insects; regulatory mechanisms in cell differentiation and development with emphasis on genetic and endocrine controls; amylases in Drosophila; molecular genetics of honeybee. *Mailing Add:* Dept Zool Ariz State Univ Tempe AZ 85287-1501

DOBAY, DONALD GENE, b Cleveland, Ohio, Sept 12, 24; m 45, 66; c 5. HAZARDOUS WASTE MANAGEMENT, ENVIRONMENTAL REGULATIONS. *Educ:* Oberlin Col, AB, 44; Univ Mich, MS, 45, PhD(colloid chem), 48. *Prof Exp:* Res chemist, Union Carbide Corp, 48-51; sr res chemist, B F Goodrich Res Ctr, Ohio, 51-60, tech coordr, B F Goodrich Sponge Prod, Conn, 60-62, mgr polymer develop, 62-74; tech mgr & res dir, Grand Sheet Metal Sponge Rubber Prod, 74-75; consult, 75-78, process mgr & dir labs, Crawford & Russell, Inc, Conn, 78-84; prin consult engr, TRC Environ Consults, Conn, 84-85; PRIN, DOBAY SERV, BLOOMFIELD, CONN, 85-; PRES, CONN ENVIRON ENG SERV, INC, BLOOMFIELD, CONN, 90- *Concurrent Pos:* Staff consult, C E Maguire, Inc, New Britain, Conn, Goldberg, Zoino & Asoc, Inc, Vernon, & Crawford & Russell, Inc, Stamford; prin, Recon Assoc, Norwalk, Conn; adj prof, Civil Eng, Univ Hartford; mem, Conn Forum Regulated Environ Prof. *Mem:* Am Chem Soc; Air Pollution Control Asn; Am Indust Hyg Asn; Am Inst Chem Engrs; Sigma Xi; Water Pollution Control Fedn. *Res:* Environmental engineering; polymerization; process engineering; chemical processing; rubber and plastics production; polymer characterization; latex properties and processing; flammability; analytical test methods; pollution control; thermal and acoustic insulation; cellular properties; hazardous waste management. *Mailing Add:* Nine Kent Lane Bloomfield CT 06002-1326

DOBBELSTEIN, THOMAS NORMAN, b Wyandotte, Mich, Oct 14, 40; m 60; c 3. ANALYTICAL CHEMISTRY. *Educ:* Eastern Michigan, BS, 64; Iowa State Univ, MS, 66, PhD(anal chem), 67. *Prof Exp:* Asst prof, 67-72, assoc prof, 72-81, PROF CHEM, YOUNGSTOWN STATE UNIV, 81-, CHMN DEPT, 74- *Mem:* Am Chem Soc; Sigma Xi. *Res:* Ion-responsive membranes; high performance liquid chromatography. *Mailing Add:* Dept Chem Youngstown State Univ Youngstown OH 44555

DOBBEN, GLEN D, b Fremont, Mich, Sept 25, 28; m 54; c 5. RADIOLOGY. *Educ:* Calvin Col, BS, 50; Marquette Univ, MD, 53. *Prof Exp:* Intern med, Blodgett Mem Hosp, Grand Rapids, Mich, 53-54; res radiol, Henry Ford Hosp, Detroit, Mich, 54-55, 57-59; instr, Univ Chicago, 59-61; res fel neuro-radiol, Univ Lund, 61-62; from asst prof to assoc prof radiol, Univ Chicago, 62-72; PROF NEURORADIOL, COL MED, UNIV ILL, CHICAGO, 73-, PROF NEUROSURG, 76- *Concurrent Pos:* Dir, Div Radiol, Cook County Hosp, 69-72. *Mem:* Asn Univ Radiologists. *Res:* Diagnostic radiological sciences as applied to neuroradiology. *Mailing Add:* 6355N 600 W Michigan City IN 46360

DOBBINS, DAVID ROSS, b Indianapolis, Ind, Aug 9, 41; m 64; c 2. PLANT PHYSIOLOGY. *Educ:* Franklin Col Ind, BA, 63; Univ Mass, MA, 67, PhD(bot), 71. *Prof Exp:* Nat Res Coun Can fel, Univ Lethbridge, 70-72; asst prof bot, Dept Biol Sci, Wellesley Col, 72-77; asst prof, 77-80, assoc prof, 80-85, PROF BOT, DEPT BIOL, MILLERSVILLE UNIV, 85- *Mem:* Bot Soc Am; Am Soc Plant Physiologists; Soc Develop Biol; Sigma Xi; Int Soc Wood Anatomists. *Res:* Developmental anatomy and morphogenesis; investigations on regulation of vascular cambium, factors affecting the plane of cell division, the structural mode of parasitism in flowering plant parasites; ecological strategies of liama anatomy in the tropical rain forest. *Mailing Add:* Dept Biol Millersville Univ Millersville PA 17551

DOBBINS, JAMES GREGORY HALL, b Ashland, Ky, Jun 30, 43; m 71; c 3. COMPUTER PERFORMANCE EVALUATION, RELIABILITY. *Educ:* Univ Ky, BA, 65, MA, 66, PhD(math), 69. *Prof Exp:* Asst prof math, Marshall Univ, 69-70; from asst prof to assoc prof, Wheaton Col, 74-78; assoc prof, Mt Vernon Nazarene Col, Ohio, 76-78, prof & chmn math & comput sci, 78-81; sr prin engr, 81-87, sr performance analyst, 87-89, CONSULT PERFORMANCE/RELIABILITY ANALYSTS, NCR CORP, SC, 89- *Concurrent Pos:* Lectr, Ohio State Univ, 77-81 & Univ SC, 81-; grants, Res Corp, NSF. *Mem:* Inst Elec & Electronics Engrs. *Res:* Pure mathematics; quality control; computer performance analysis and reliability. *Mailing Add:* 743 Trafalgar Dr Columbia SC 29210

DOBBINS, JAMES TALMAGE, JR, b Chapel Hill, NC, June 13, 26; m 51; c 2. ANALYTICAL CHEMISTRY, ATOMIC SPECTROSCOPY. *Educ:* Univ NC, BS, 47, PhD(anal chem), 58. *Prof Exp:* Chief, Indust Hyg Sect, Dept Chem, 406th Med Gen Lab, US Army, 54-56, head, Dept Chem, 56; res chemist, R J Reynolds Industs, 58-66, head, Anal Instrumentation Sect, 67-72, mgr, Anal Div, 72-75, master scientist, Appl Res & Develop Dept, 75-83, master chemist, Res & Develop Tech Serv, R J Reynolds Tobacco Co, 83-91; RETIRED. *Mem:* AAAS; Soc Appl Spectros; NY Acad Sci; Am Chem Soc; fel Am Inst Chem; Sigma Xi. *Res:* Atomic absorption and emission spectroscopy by inductively coupled argon plasma (ICP); analytical chemistry of tobacco and tobacco products; automated sampling techniques using flow injection devices. *Mailing Add:* 2838 Bartram Rd Winston-Salem NC 27106

DOBBINS, JOHN POTTER, b Tianjin, Hopeh, China, July 4, 14; US citizen; m 55; c 4. GERONTOLOGY, METABOLOGY. *Educ:* Univ Calif, Berkeley, AB, 35; Saxon Inst Technol, Dresden, Germany, dipl eng, 38; Swiss Fed Inst Technol, Zurich, PhD(photophysics), 63; FAIC, 71; FICAN, 77. *Prof Exp:* Chemist, Best Foods Co, San Francisco, 36-38 & W P Fuller & Co, 38-40; chem engr, NAm Aviation, Inc, 40-44, res supvr, 45-51, res consult, 51-54; tech dir photodocumentation, Records Serv Corp, 54-57; consult engr optical anal, Switz & Conn, 57-65; sr staff scientist photophysics, Hycon Mfg Co, McDonnell-Douglas Corp, 66-68; eng consult syst design, Gen Cinetics Ltd, 68-69; staff scientist electrotechnol autonetics & electronics res div, Rockwell Int, 69-74; HEALTH CONSULT ORTHOMOLECULAR MED, 75- *Concurrent Pos:* Prin investr, US Navy Bur Aeronaut, 46-47, US Atomic Energy Comn, Sandia Corp, 52 & USAF Aerospace Med Res Lab, Wright-Patterson AFB, 71-73; instr, eng exten div, Univ Calif, Los Angeles, 50; consult, Contraves AG, Zurich, 59-60, Bolsey Assocs, Stanford, 64-65; res consult energy fuel & gen eng pract, 75-80. *Mem:* Soc Photog Scientist & Engrs; fel Am Inst Chemists; Nat Health Fedn; Cancer Control Soc; fel Int Col Appl Nutrit. *Res:* Lysine-ascorbate therapy for viral diseases; cancer causes, prevention and cures; religion and health. *Mailing Add:* 615 Allen Ave San Marino CA 91108

DOBBINS, RICHARD ANDREW, b Burlington, Mass, July 15, 25; m 53; c 2. FLUID DYNAMICS, AEROSOL SCIENCE. *Educ:* Harvard Univ, SB, 48; Northeastern Univ, MS, 58; Princeton Univ, PhD, 61. *Prof Exp:* Res engr, Arthur D Little, Inc, 50-53; sr engr, Sylvania Elec Prod Inc, 53-57; from asst prof to assoc prof eng, 60-68, chmn div, 83-88, PROF ENG, BROWN UNIV, 68- *Concurrent Pos:* Vis res assoc, Calif Inst Technol, 67-68 & Univ Essex, Eng, 71; assoc ed, Combustion Sci & Technol, 69-73; vis prof, Abadan Inst Technol, Iran, 75; consult, various govt & indust orgns; sr scientist, Nat Bur Standards, 81-82 & 88-89. *Mem:* Am Asn Aerosol Res; Am Soc Mech Engrs; Combustion Inst; Am Soc Eng Educ. *Res:* Nucleation and condensation dynamics; aerosol science and technology; high temperature effects in fluid dynamics; heat transfer; atmospheric fluid mechanics; air pollution; particle formation in combustion. *Mailing Add:* 11 President Ave Providence RI 02906

DOBBINS, ROBERT JOSEPH, b Buffalo, NY, Aug 16, 40; m 64; c 4. CELLULOSE CHEMISTRY. *Educ:* Fordham Univ, BS, 62; State Univ NY Buffalo, PhD(phys chem), 66. *Prof Exp:* Res chemist, Indust Chem Div, Am Cyanamid Co, 66-72; SR RES CHEMIST & GROUP LEADER PAPERMAKING DEVELOP, ST REGIS PAPER CO, 72-; AT CHAMPION INT TECH CTR, WNYACK, NY. *Mem:* Am Chem Soc; Tech Asn Pulp & Paper Indust. *Res:* Cellulose-water-electrolyte interactions; adsorption from aqueous solutions; chemistry of pulping and bleaching processes; mechanisms of inorganic reactions; coordination chemistry. *Mailing Add:* Betz Paperchem Inc 7510 Bay Meadows Way Jacksonville FL 32216

DOBBINS, THOMAS EDWARD, forest products; deceased, see previous edition for last biography

DOBBS, DAVID EARL, b Winnipeg, Man, Feb 4, 45; m 65; c 2. ALGEBRA, COMMUTATIVE ALGEBRA. *Educ:* Univ Man, BA, 64, MA, 65; Cornell Univ, PhD(math), 69. *Prof Exp:* Vis prof math, Univ Calif, Los Angeles, 69-70; asst prof, Rutgers Univ, 70-75; assoc prof, 75-80, PROF MATH, UNIV TENN, 80- *Concurrent Pos:* Res assoc, Off Naval Res, Univ Calif, Los Angeles, 69-70. *Mem:* Am Math Soc; Math Asn Am. *Res:* Commutative algebra; homological algebra. *Mailing Add:* Dept Math Univ Tenn Knoxville TN 37996-1300

DOBBS, FRANK W, b Chicago, Ill, Sept 8, 32; m 58; c 3. PHYSICAL CHEMISTRY. *Educ:* Univ Chicago, BA, 53, MS, 55; Mass Inst Technol, PhD(phys chem), 61. *Prof Exp:* From asst prof to assoc prof, 59-67, chmn dept chem, 71-77, dean, Col Arts & Sci, 78-89, PROF CHEM, NORTHEASTERN ILL UNIV, 67- *Mem:* AAAS; Am Chem Soc. *Res:* Molecular orbitals. *Mailing Add:* 2728 Noyes Evanston IL 60201

DOBBS, GREGORY MELVILLE, b Teaneck, NJ, Aug 19, 47; m 76; c 3. COMBUSTION, LASER DIAGNOSTICS. *Educ:* Dartmouth Col, AB, 69; Princeton Univ, MA, 71, PhD(chem), 75. *Prof Exp:* Systs programmer, Kiewit Computation Ctr, Hanover, NH, 66-69; resident res fel physics, Cent Res Dept, E I du Pont de Nemours & Co, Inc, 69; asst instr chem, Princeton Univ, 70-73; res assoc chem, Mass Inst Technol, 74-76; sr res scientist, 76-86, MGR LASER DIAG, UNITED TECHNOLS RES CTR, 86- *Concurrent Pos:* Adj instr computer sci, Hartford Grad Ctr, 79-86. *Mem:* Am Chem Soc; Am Phys Soc; Combustion Inst; Am Inst Aero & Astro. *Res:* Chemical physics; molecular spectroscopy and energy transfer; application of lasers to chemistry and combustion science; laser diagnostics of propulsion and combustion. *Mailing Add:* United Technols Res Ctr Mail Stop 129-90 411 Silver Lane East Hartford CT 06108-1010

DOBBS, HARRY DONALD, b Atlanta, Ga, Nov 23, 32; m 53; c 3. BIOLOGY. *Educ:* Emory Univ, AB, 53, MS, 54, PhD(biol), 67. *Prof Exp:* Res asst parasitol, Emory Univ, 54-55; from asst prof to assoc prof biol, 55-70, PROF BIOL, WOFFORD COL, 70- *Mem:* Am Soc Zoologists; Am Inst Biol Sci. *Res:* Parasitology, especially trichiniasis in mice; anatomical and physiological adaptations of invertebrates. *Mailing Add:* Dept Biol Wofford Col 429 N Church St Spartanburg SC 29303-3663

DOBBS, LELAND GEORGE, SURFACTANT SECRETION, ALVEOLAR CELL DIFFERENTIATION. *Educ:* Columbia Univ, MD, 69. *Prof Exp:* ASST CLIN PROF MED, CARDIOVASC RES INST, UNIV CALIF, SAN FRANCISCO, 79- *Mailing Add:* Cardiovasc Res Inst 1368 HSW Univ Calif San Francisco CA 94143-0130

DOBBS, ROBERT CURRY, b Seattle, Wash, Nov 9, 35; Can citizen; m 55, 75; c 4. SILVICULTURE. *Educ:* Univ Wash, BSc, 59, PhD(tree biol), 66; Yale Univ, MFor, 60. *Prof Exp:* Forester, US Forest Serv, Seattle, 59, res forester, Klamath, Calif, 60-61; res scientist silvicult, Can Forestry Serv, Winnipeg, 66-70, Victoria, 70-77, res mgr silvicult, Can Forestry Serv, Ottawa, 77-84; SR PROG DIR, RES, PAC FORESTRY CTR, 84- *Concurrent Pos:* Weyerhaeuser fel Univ Wash, 61-62. *Mem:* Can Inst Forestry; Sigma Xi. *Res:* Silviculture, particularly reforestation problems; research and development relating to production of forest biomass for conversion to energy; forestry research management. *Mailing Add:* Forestry Can Pac & Yukon Pac Forestry Ctr 506 W Burnside Rd Victoria BC V8Z 1M5 Can

DOBBS, THOMAS LAWRENCE, b Los Angeles, Calif, Apr 9, 43; m 64; c 3. AGRICULTURAL ECONOMICS. *Educ:* SDak State Univ, BS, 65; Univ Md, PhD(agr econ), 69. *Prof Exp:* Asst prof agr econ, Univ Wyo, 69-74; agr economics, USAID, 74-78; ASSOC PROF AGR ECON, SDAK STATE UNIV, 78- *Mem:* Am Agr Econ Asn. *Res:* Community and rural development; natural resource economics. *Mailing Add:* Dept Econ SDak State Univ Box 504a Brookings SD 57007

DOBELBOWER, RALPH RIDDALL, JR, b Bellefonte, Pa, Mar 23, 40; m 63; c 7. RADIATION THERAPY. *Educ:* Pa State Univ, BS, 62, AB, 63; Jefferson Med Col, MD, 67, PhD(radiation biol), 75; Am Bd Radiol, dipl, 75. *Prof Exp:* Instr radiation ther & nuclear med, Thomas Jefferson Univ Hosp, 74-75, asst prof & coordr, Outreach Prog Radiation Ther, 75-79, assoc prof, 79-80, dir, Gastrointestinal Radiother Serv, 75-80; PROF & CHMN RADIATION THER, MED COL OHIO, TOLEDO, 80-; PROG ADV RADIATION THER TECHNOL, MICHAEL J OWENS TECH COL, 80- *Concurrent Pos:* Jr fac clin fel, Am Cancer Soc, 74-77; asst dir, Am Col Radiol PCS Study, 74-79; asst attend physician, Radiation Ther, Bryn Mawr Hosp, 75-79; mem adv bd, Sch Radiation Ther Technol, Gwynedd-Mercy Col, 77-79. *Honors & Awards:* Medal, Univ Mont Pisllian, France; Medal, Univ Cattolica, Rome. *Mem:* Am Soc Therapeut Radiologists; Am Radium Soc; Radiol Soc NAm; Sigma Xi; Am Asn Physicists Med. *Res:* Radiation treatment of gastronintestinal malignancies; cancer of the pancreas; intraoperative radiation treatment. *Mailing Add:* Dept Radiation Oncol Med Col Ohio CA-10008 Toledo OH 43699

DOBELLE, WILLIAM HARVEY, b Pittsfield, Mass, Oct 24, 41; m 72. BIOMEDICAL ENGINEERING. *Educ:* Johns Hopkins Univ, BA, 64, MA, 67; Univ Utah, PhD(physiol), 74. *Prof Exp:* Res assoc, Dept Biophys, Johns Hopkins Univ, 61-68; dir, Neuroprostheses Prog, Inst Biomed Eng, Univ Utah, 69-76, assoc dir, 73-76; dir, Div Artificial Organs, Dept Surg, Col Physicians & Surgeons, Columbia Univ, 76-81; CHMN, INST ARTIFICIAL ORGANS, NEW YORK CITY, 82- *Concurrent Pos:* Exec dir, Med Eye Bank, Md, 64-67 & Intermountain Organ Bank, Salt Lake City, 69-70; mem bd dirs, NY Regional Transplant Prog, Inc, 78-80. *Honors & Awards:* Kusserow Award, Am Soc Artificial Internal Organs, 76 & 78. *Mem:* Int Soc Artificial Organs; Am Soc Artificial Internal Organs. *Res:* Sensory prostheses for the blind and deaf; hybrid and mechanical artificial pancreas; cardiac assist devices and artificial heart; organ banking and transplantation. *Mailing Add:* One Lincoln Plaza #37R New York NY 10023

DOBERENZ, ALEXANDER R, b Newark, NJ, Aug 17, 36; m 58; c 2. NUTRITION. *Educ:* Tusculum Col, BS, 58; Univ Ariz, MS, 60, PhD(biochem), 63. *Prof Exp:* Res assoc biophys, Univ Ariz, 63-69; vis assoc prof nutrit, Univ Hawaii, 69; assoc prof nutrit, Univ Wis-Green Bay, 69-71, asst dean col human biol, 69-71, assoc dean cols, 71-74, prof nutrit sci, 71-76, prof growth & develop, 75-76; PROF FOOD SCI & HUMAN NUTRIT & DEAN COL HUMAN RESOURCES, UNIV DEL, 76-, COORDR HOME ECON RES, DEL EXP STA, COL AGR SCI, 78- *Concurrent Pos:* Nat Inst Dent Res fel, 63-66, res career develop award, 66-69; consult, Northeastern

Wis Health Planning Coun, 74-, Bellin Hosp Obesity Clin, 75- & Gen Foods Corp, 78-; mem home econ res subcomt, Exp Sta Comt on Orgn & Policy, 78-; mem, Nat Coun Adminrs Home Econ. *Mem:* Soc Exp Biol Med; Am Home Econ Asn; Am Chem Soc; Sigma Xi; Am Inst Nutrit. *Res:* Calcified tissue; mineral metabolism. *Mailing Add:* Col Human Resources Univ Del Newark DE 19716

DOBERNECK, RAYMOND C, b Milwaukee, Wis, June 17, 32; m 57; c 6. SURGERY. *Educ:* Marquette Univ, BS, 53, MD, 56; Univ Minn, Minneapolis, PhD(surg), 65. *Prof Exp:* From asst prof to assoc prof, Sch Med, Creighton Univ, 65-70, PROF SURG & VCHMN, SCH MED, UNIV NMEX, 70- *Concurrent Pos:* John & Mary R Markle Found Scholar, 66. *Mem:* Am Col Surg; Soc Univ Surg; Asn Acad Surg; Am Gastroenterol Asn; Soc Head & Neck Surg; Am Surg Asn. *Res:* Head and neck tumors; hepatic hemosiderosis in surgical states; deglutition after oral and pharyngeal resection for cancer. *Mailing Add:* Dept Surg Univ NMex Sch Med Albuquerque NM 87106

DOBERSEN, MICHAEL J, b Bay Village, Ohio, May 13, 49; m 85; c 2. IMMUNOPATHOLOGY, AUTOIMMUNITY. *Educ:* Kent State Univ, BS, 71; Univ Miami, PhD(microbiol), 76; Univ Colo, MD, 89. *Prof Exp:* Intern Med Tech, St Thomas Hosp, 70-71; teaching asst biol, Dept Biol, Kent State Univ, 71-72; grad student & res asst, Dept Microbiol, Sch Med, Univ Miami, 72-77; sr staff fel & assoc investr, lab oral med, Nat Inst Dent Res, NIH, 77-82; sr staff fel, Nat Inst Neurol & Commun Dis & Stroke, NIH, 82-84; Univ Colo Sch Med, 85-89; asst prof, Dept Microbiol & Barbara Davis Ctr Childhood Diabetes, 84-85, RESIDENT PATHOL, UNIV COLO MED CTR, 89- *Concurrent Pos:* Consult, Developmental Psychobiol Res Group, Univ Colo, HSC, Denver, Colo, 86-88. *Mem:* AAAS; Am Soc Microbiol; AMA; Col Am Pathologists; Union Concerned Scientists. *Res:* Role of autoantibodies in the pathogenesis of insulin; dependent diabetes mellitus including the isolation and biochemical characterization of the corresponnding pancreatic antigens; anatomic and clinical pathology; forensic pathology. *Mailing Add:* Dept Path Box B216 Univ Colo Med Ctr 4200 E Ninth Ave Denver CO 80262

DOBI, JOHN STEVEN, b Passaic, NJ, Feb 13, 48; m 73; c 1. ENVIRONMENTAL SCIENCE. *Educ:* Rutgers Col, BS, 70, Rutgers Univ, MS, 72; Univ Mass, PhD(environ sci), 80. *Prof Exp:* Field div mgr & proj biologist, Quirk Lawler & Matosky Eng, 72-74; environ engr, Burns & Roe Engrs, 74; res engr, United Engrs & Constructors, 77-78; SR ENVIRON ENGR, PROJ LICENSING MGR & SCIENTIST, PUB SERV ELEC & GAS CO, 78- *Concurrent Pos:* Guest lectr, Dept Environ Sci, Rutgers Univ, Grad Sch, 81-82; consult & chmn, Union County Environ Adv Bd, 82-86; exec vpres & mem bd dirs, Malhi Pilot Asn, 84- *Mem:* Sigma Xi; Am Soc Limnol & Oceanog; Soc Power Indust Biologists. *Res:* Limnological eutrophication modeling; predictive stochastic and mechanistic math models with both steady state and dynamic scenarios. *Mailing Add:* 728 Oak Ave Westfield NJ 07090

DOBINSON, FRANK, organic chemistry, polymer chemistry, for more information see previous edition

DOBKIN, DAVID PAUL, b Pittsburgh, Pa, Feb 29, 48; m; c 2. ALGORITHMS ANALYSIS, COMPUTER GRAPHICS. *Educ:* Mass Inst Technol, BS, 70; Harvard, MS, 71, PhD(appl math), 73. *Prof Exp:* Asst prof comput sci, Yale Univ, 73-78; assoc prof, Univ Ariz, 78-81; PROF ELEC ENG & COMPUT SCI, PRINCETON UNIV, 81- *Concurrent Pos:* Consult, Bell Lab, 78; vis scientist, Xerox, Palo Alto Res Ctr, 81. *Mem:* Soc Indust & Appl Math; Asn Comput Mach. *Res:* Application of techniques from the analysis of algorithms area to problems of practical importance; computer geometry and graphics; distributed computing and computer security. *Mailing Add:* Computer Sci Dept Computer Sci Bldg Princeton Univ Princeton NJ 08544

DOBKIN, SHELDON, b New York, NY, Nov 12, 33; m 59; c 3. BIOLOGICAL OCEANOGRAPHY, INVERTEBRATE ZOOLOGY. *Educ:* City Col New York, BS, 57; Univ Miami, MS, 60, PhD(marine biol), 65. *Prof Exp:* Res aide, Rosenstiel Sch Marine & Atmospheric Sci, Univ Miami, 58-61, res instr, 61-63, instr marine biol, 63-64; from asst prof to assoc prof, 64-74, PROF ZOOL, FLA ATLANTIC UNIV, 74- *Mem:* Am Soc Zoologists; World Maricult Soc; AAAS. *Res:* Larval development of decapod crustaceans, particularly the caridean and penaeidean shrimps; taxonomy of caridean shrimps; larval development of marine invertebrates; shrimp and prawn culture. *Mailing Add:* Dept Biol Sci & Zool Fla Atlantic Univ Boca Raton FL 33431

DOBLIN, STEPHEN ALAN, b Jersey City, NJ, Feb 8, 45; m 68; c 1. MATHEMATICS EDUCATION. *Educ:* Univ Ala, BS, 67, MA, 69 & PhD(math), 72. *Prof Exp:* Asst math, dept math, Univ Ala, 68-70, fel, 70-72; asst prof, Univ Southern Miss, 72-77, assoc prof & actg chmn, 77-78, assoc prof, 78-80, assoc prof & chmn, 81-82, prof & chair math, Dept Math, 82-89, acting dean, 89-90, DEAN, COL SCI & TECHNOL, UNIV SOUTHERN MISS, 90- *Concurrent Pos:* Dir, Math Resource Ctr, Univ Southern Miss, 75-79; consult, Rand McNakky & Co, 76-78, Alcorn State Univ, 78-79; nat consult, Dekalb Community Col, 79; prin & co-investr, NSF & US Dept Educ, 78-86, adv, Off Educ Res & Improv, 87-; lect, secondary sch lectureship prog, Mat Asn Am, 82-; mem, State Miss, Math Symp Steering Comt, 87-89, chair, Math Discipline Comt, 89-; mem, Nat Joint AMS-MAA Comt Teaching Assts & Part-time Instr, 89- *Mem:* Am Math Soc; Math Asn Am; Nat Coun Teachers Math; Sigma Xi. *Res:* Complex analysis of differential equations involving complex directional derivatives; educational administration. *Mailing Add:* 3104 Southaven Dr Hattiesburg MS 39402

DOBO, EMERICK JOSEPH, b Szeged, Hungary, Oct 23, 19; US citizen; m 50; c 3. CHEMICAL ENGINEERING. *Educ:* Univ Mich, BS, 42; Univ Wash, MS, 48; Univ Tex, PhD(chem eng), 54. *Prof Exp:* Develop chem engr, Crown Zellerbach Corp, Wash, 48-51; sr develop chem engr, Chemstrand Corp, Ala, 54-57, group leader chem eng, 57-62; group leader explor eng, Monsanto Co, 62-70, eng fel, 70-76, sr fel, 76-82; RETIRED. *Mem:* Am Inst Chem Engrs; Am Chem Soc. *Res:* Engineering and chemistry related to high polymers and fibers, to nonwovens, and to gas separations. *Mailing Add:* 5048 S Rainbow Blvd No 102 Las Vegas NV 89118

DOBOSY, RONALD JOSEPH, b Cleveland, Ohio, June 15, 46; m 68; c 2. PLANETARY BOUNDARY LAYER, MODEL DEVELOPMENT. *Educ:* Ore State Univ, BA, 68; Univ Wis-Madison, MS, 72, PhD(meteorol), 77. *Prof Exp:* Postdoctoral visitor, Nat Ctr Atmospheric Res, 77-78; asst prof meteorol, Iowa State Univ, 78-84; RES METEOROLOGIST, ATDD, NAT OCEANIC & ATMOSPHERIC ADMIN, 84- *Mem:* Am Meteorol Soc. *Res:* Identifiable motion structures in atmospheric boundary layer; transport of heat, moisture, pollutants; model development from physical principles and data analysis. *Mailing Add:* Nat Oceanic & Atmospheric Admin-ATDD PO Box 2456 Oak Ridge TN 37831-2456

DOBRATZ, CARROLL J, b Park Rapids, Minn, Oct 13, 15; m 42; c 4. CHEMICAL ENGINEERING. *Educ:* Univ Minn, BChE, 38; Univ Cincinnati, PhD(chem eng), 43. *Prof Exp:* Instr chem eng, Univ Cincinnati, 41-43; sr res chemist, Shell Oil Co, Tex, 43-46; res & develop engr, Dow Chem Co, 46-74; PROCESS ENG SPECIALIST, KALAMA CHEM INC, 74- *Mem:* Am Chem Soc; Am Inst Chem Eng. *Res:* Heat and mass transfer; applied kinetics; thermodynamics. *Mailing Add:* Kalama Chem Inc PO Box 113 Kalama WA 98625-0113

DOBRIN, MILTON BURNETT, geophysics; deceased, see previous edition for last biography

DOBRIN, PHILIP BOONE, b Passaic, NJ, Dec 21, 34; div; c 3. MEDICAL PHYSIOLOGY, CARDIOVASCULAR PHYSIOLOGY. *Educ:* NY Univ, BA, 58; Conn Col, MA, 62; Loyola Univ Chicago, PhD(physiol) 68, MD, 74. *Prof Exp:* Res asst psychopharmacol, Chas Pfizer Res Labs, 59-63; asst prof med physiol, 68-74, adj asst prof med physiol & surg resident, 74-80, asst prof surg, 80-85, ASSOC PROF SURG, STRITCH SCH MED, LOYOLA UNIV, CHICAGO, 85-; ASST CHIEF-OF-STAFF, HINES VET ADMIN HOSP, 85- *Concurrent Pos:* Consult, NIH, 68-; Surg resident Loyola-Hines, Chicago, 79; adj assoc prof physiol, Stritch Sch Med, Loyola Univ, Chicago, 80-; clin investr, 80-83. *Mem:* Am Physiol Soc; Asn Acad Surg; Am Col Surg. *Res:* Biomechanics of the arterial wall; physiology of the circulation; balloon embolectomy catheters; peritonitis surgical sutives. *Mailing Add:* Dept Surg 151 Hines Vet Admin Hosp Hines IL 60141

DOBROGOSZ, WALTER JEROME, b Erie, Pa, Sept 3, 33; m 53; c 4. MICROBIOLOGY, BIOCHEMISTRY. *Educ:* Pa State Univ, BS, 55, MS, 57, PhD(bact biochem), 60. *Prof Exp:* NIH fel microbiol, Univ Ill, 60-62; from asst prof to assoc prof, 62-71, PROF MICROBIOL, NC STATE UNIV, 71- *Concurrent Pos:* NIH career develop award, 63-, res grant, 64-; NSF res grant, 64-; AEC res grant, 68-; mem ed bd, J Bacteriol, 71-76. *Mem:* Am Soc Microbiol. *Res:* Metabolic regulatory mechanisms in bacteria and higher organisms including studies on regulation of carbohydrate metabolism and nucleic acid and enzyme synthesis formation and the role of cyclic AMP in these processes. *Mailing Add:* Dept Microbiol NC State Univ Raleigh Main Campus Box 7615 Raleigh NC 27695-7615

DOBROTT, ROBERT D, b Guymon, Okla, Sept 28, 32; m 62; c 3. CRYSTALLOGRAPHY, PHYSICAL CHEMISTRY. *Educ:* Univ Wichita, BS & MS, 59; Harvard Univ, PhD(phys chem), 64. *Prof Exp:* Anal chemist, Cessna Aircraft, Kans, 54-59; mem tech staff, Tex Instruments Inc, 64-68, mgr characterization serv, 68-72, mem tech staff, 73-80; SR RES SCIENTIST, PACKAGE MAT RES, MOSTER INC, 80- *Mem:* Am Chem Soc; Am Crystallog Asn. *Res:* X-ray and electron diffraction; x-ray fluorescence; SEM, TEM and optical microscopy; auger and ion scattering spectroscopy; ion probe and electron probe microanalysis; microanalysis standards fabrication; environment testing; failure analysis. *Mailing Add:* 1429 Lamp Post Lane Richardson TX 75080-5723

DOBROV, WADIM (IVAN), b Masur, Russia, July 14, 26; nat US; m 56; c 4. SOLID STATE PHYSICS, ELECTRO-OPTICS. *Educ:* Univ Calif, PhD(physics), 56. *Prof Exp:* Res asst, Univ Calif, 54-56; res scientist, 56-65, staff scientist, 65-66, SR STAFF SCIENTIST ELECTRO-OPTICS, LOCKHEED RES LAB, 66- *Mem:* Am Phys Soc; Inst Elec & Electronics Engrs; Sigma Xi. *Res:* Nuclear moments; paramagnetic resonance; ferroelectricity; microwave ultrasonics; lasers and infrared technology. *Mailing Add:* Lockheed Res Labs 3251 Hanover St Palo Alto CA 94304

DOBROVOLNY, CHARLES GEORGE, b Budapest, Hungary, July 19, 02; nat US; m 32; c 1. VIROLOGY, PUBLIC HEALTH. *Educ:* Univ Mont, AB, 28; Kans State Univ, MS, 33; Univ Mich, PhD(parasitol, zool), 38. *Prof Exp:* High sch teacher, Idaho, 28-29; instr zool, Kans State Univ, 29-35; instr, Univ Mich, 35-40; form asst prof to assoc prof, Univ NH, 40-48, chmn div biol sci, 47-48; parasitologist, Div Trop Dis, NIH, 48-51, consult schistosomiasis control, WHO, Brazil, 51-57, consult yellow fever & trop virus invest, Guatemala, 57-59, trop virus studies, Panama, 59-60, malaria studies, Malaya & Atlanta, Ga, 61-64, consult int demonstration proj for commun dis control, Tex, 65-66, pesticide studies, 66-67, proj officer pesticides prog, 67-68, proj officer, Food & Drug Admin, 68-70, proj officer, Environ Protection Agency, 70-75; RETIRED. *Mem:* AAAS; Am Soc Parasitol; Am Micros Soc; Am Soc Trop Med & Hyg; Am Pub Health Asn; Asn Mil Surgeons US. *Res:* General and experimental parasitology; malarian chemotherapy; schistosomiasis; tropical virology; pesticides; virology; parasitology. *Mailing Add:* 1366 Vistaleaf Dr Decatur GA 30033

DOBROVOLNY, ERNEST, b Delmont, SDak, Aug 27, 12; m 40; c 4. GEOLOGY. *Educ:* Kans State Col, BS, 35; Univ Mich, MS, 40. *Prof Exp:* Chem analyst, E I du Pont de Nemours & Co, 35-37; asst geol, Univ Mich, 38-40; asst geologist, State Hwy Comn Kans, 40-42, regional geologist, 43-44; asst hwy engr, Pub Rds Admin, Alaska, 42-43; geologist, US Geol Surv, Ariz,

44, NMex, 45-46, Mont, 46-48, Colo, 48-54, asst chief eng geol br, 49-54, geologist-in-chg munic geol & eng, La Paz, Bolivia, 54-55, res geologist, Nev Test Site, 56-59, consult eng geol earthquake studies, Chilean Geol Surv, 60, geologist mapping proj, Ky, 60-61, res geologist, Sci & Eng Task Force, Fed Reconstruct & Develop Planning Comn, US Geol Surv, Alaska, 64-76; CONSULT GEOLOGIST, 76- *Concurrent Pos:* Mem comt Alaska Earthquake, Nat Acad Sci, chmn comt eng geol, Hwy Res Bd, 63- *Mem:* Geol Soc Am; Asn Eng Geol. *Res:* Stratigraphy; military geology; civil engineering; mapping geology. *Mailing Add:* 2210 S Corona Denver CO 80210

DOBROVOLNY, JERRY S(TANLEY), b Chicago, Ill, Nov 2, 22; m 47; c 2. CIVIL ENGINEERING. *Educ:* Univ Ill, BS, 43, MS, 47. *Prof Exp:* From instr to prof, 45-87, EMER PROF, GEN ENG & HEAD DEPT, UNIV ILL, URBANA, 87- *Concurrent Pos:* Civil Engr, Ill State Hwy Dept, 48-54; consult, 55-59; mem, State of Ill Adv Coun Voc Educ, 69-72 & Nat Adv Coun Voc Educ, 70-72; examr, Comn Insts Higher Educ, NCent Asn Cols & Sec Schs. *Honors & Awards:* Arthur L Williston Award, Am Soc Eng Educ. *Mem:* Fel AAAS; Am Soc Civil Engrs; Am Soc Eng Educ; Hist Sci Soc; Nat Soc Prof Engrs; Sigma Xi. *Res:* Engineering geology and technology; soil mechanics. *Mailing Add:* Trans Bldg Rm 217 Univ Ill Urbana IL 61801

DOBROWOLSKI, JAMES PHILLIP, b Los Angeles, Calif, June 2, 55; m 84; c 2. SOIL CRUSTING EFFECTS, RIPARIAN ZONE. *Educ:* Univ Calif, Davis, BS, 77; Wash State Univ, MS, 79; Tex A&M Univ, PhD(watershed sci & mgt), 85. *Prof Exp:* Systs analyst, Norac Co, Inc, 80; range scientist, Dern & Polk Resources Consults, Tex, 80-81; res asst watershed sci, dept range sci, Tex A&M Univ, 81-83, W G Mills fel hydrol, Tex Water Resources Inst, Col Sta, 82-83, Tom Slick fel agr, Col Agr, 83-84; res asst prof, 84-85, ASST PROF WATERSHED SCI, DEPT RANGE SCI, UTAH STATE UNIV, 85- *Concurrent Pos:* Prin investr, Utah Agr Exp Sta, 84- *Mem:* Sigma Xi; Soc Range Mgt; Am Soc Agr Engrs. *Res:* Process oriented approach to rangeland hydrology; soil physical barriers to infiltration in light of maximizing on-site effective precipitation; nutrient and water flux following land disturbance; hydrologic, nutrient cycling and ecologic approaches to Riparian zone. *Mailing Add:* Dept Range Sci UMC 52 Utah State Univ Logan UT 84322

DOBROWOLSKI, JERZY ADAM, b Katowice, Poland, May 9, 31; Can citizen; m 59; c 3. PHYSICAL OPTICS. *Educ:* Univ London, BSc, 53, MSc & dipl, Imp Col, 54, PhD, 55. *Prof Exp:* Nat Res Coun Can fel, 55-56; from asst res officer to assoc res officer, 56-71, sr res officer, 71-89, PRIN RES OFFICER, INST MICROSTRUCT SCI, NAT RES COUN CAN, 89- *Honors & Awards:* Joseph Fraunhofer Award, Optical Soc Am, 87; Moët Hennessy-Louis Vuitton Prize, 89. *Mem:* Optical Soc Am; Am Vacuum Soc; Int Soc Optical Eng. *Res:* Design and fabrication of optical thin film systems; optical filters. *Mailing Add:* Inst Microstruct Sci Nat Res Coun Ottawa ON K1A 0R6 Can

DOBRY, ALAN (MORA), b Chicago, Ill, Mar 19, 27; m 50; c 2. PHYSICAL CHEMISTRY. *Educ:* Univ Chicago, PhB & SB, 45, SM, 48, PhD(chem), 50; Univ Pittsburgh, MLitt, 56. *Prof Exp:* Fel calorimetry peptide reactions, Yale Univ, 51; res engr phys chem lubrication, Westinghouse Elec Co, 52-56; res chemist, Amoco Oil Co, 56-89; RETIRED. *Mem:* Am Chem Soc; Royal Soc Chem. *Res:* Emulsions; particle size distributions; adsorption from solutions; synthetic crudes; high pressure chemistry; radiochemistry; heterogeneous catalysis. *Mailing Add:* 5623 S Drexel Ave Chicago IL 60637-1417

DOBRY, REUVEN, b Bialistock, Poland, Apr 13, 30; nat US; m 60; c 2. CHEMICAL ENGINEERING, FOOD SCIENCE & TECHNOLOGY. *Educ:* Syracuse Univ, BChE, 54; Univ Ill, MS, 55; Cornell Univ, PhD(chem eng), 58. *Prof Exp:* Res chem engr, Bioferm Corp, 58-63; sr chem engr, Battelle Mem Inst, 63-67; sr proj leader, Standard Brands Inc, 67-69; eng res group leader, Beechnut Inc, 69-73; mgr res & develop, Tech Serv, Tetley Inc, 73-88; CONSULT, 88- *Mem:* Am Chem Soc; Am Inst Chem Engrs; Inst Food Technologists. *Res:* Special techniques for isolation and purification of biochemicals; engineering research process and product development, related to foods and beverages; quality control; quality assurance; consumer packaging development; microwave applications. *Mailing Add:* 87 Rolling Dr Stamford CT 06905

DOBRY, RICARDO, b Santiago, Chile, Dec 7, 37; m 66; c 2. SOIL DYNAMICS, EARTHQUAKE ENGINEERING. *Educ:* Univ Chile, BS, 63; Nat Univ Mex, MS, 64; Mass Inst Technol, ScD, 71. *Prof Exp:* Instr soil mech, Mass Inst Technol, 70-71; dir MS prog, Univ Chile, 71-73; sr proj engr earthquake eng, Woodward-Clyde Consult, 74-77; assoc prof, 77-81, PROF CIVIL ENG, RENSSELAER POLYTECH INST, 81-, DIR, GEOTECH CENTRIFUGE RES CTR, 90- *Concurrent Pos:* Prin, Solum, Chile, 69-73; mem, Comt Seismol, Nat Res Coun, 82-84, Comt Earthquake Eng, 83-88; chmn, Soil Dynamics Comt, Am Soc Civil Engrs, 82-86; vis prof, Dept Civil Eng, Univ Tex, Austin, 84-85. *Honors & Awards:* Croes Medal, Am Soc Civil Engrs, 85. *Mem:* Am Soc Civil Engrs; Earthquake Eng Res Inst. *Res:* Soil dynamics and earthquake engineering, including experimental cyclic loading of soils, constitutive relations of soils, models and engineering methods for soil failure, site amplification in earthquakes, machine foundations and geotechnical centrifuge model testing. *Mailing Add:* Dept Civil Eng Rensselaer Polytech Inst Troy NY 12180

DOBSON, ALAN, b London, Eng, Dec 20, 28; m 54; c 4. VETERINARY PHYSIOLOGY, GASTROENTEROLOGY. *Educ:* Cambridge Univ, BA, 52, MA, 70, ScD, 82; Aberdeen Univ, PhD(physiol), 56. *Prof Exp:* From sci officer to prin sci officer, Dept Physiol, Rowett Res Inst, Scotland, 52-64; assoc prof physiol, 64-70, PROF VET PHYSIOL, NY STATE COL VET MED, CORNELL UNIV, 70- *Concurrent Pos:* Vis prof, NY State Col Vet Med, Cornell Univ, 61-62; Wellcome fel, Vet Sch, Cambridge Univ, 70-71, vis worker Physiol Lab, 77-80, 82, 84-86 & 90-91; chmn, Sci Comt, Fourth Int Symp Ruminant Physiol & Metabo, 82-85; vis prof, Fac Agr & Forestry, Univ Alta, Edmonton, Can, 84; Carnegie fel, Carnegie Mellon Univ, Pittsburg, 86; chmn, Satellite Conf of XXXth Int Cong, Int Union Psychol Sci, Ruminant Digestion, 86; Quatrocentennial res fel, Emmanuel Col, Cambridge, 90. *Mem:* Am Physiol Soc; Brit Physiol Soc; Brit Biochem Soc; Sigma Xi. *Res:* Physiology of the ruminant digestive tract; acid-base physiology; control of absorption; peripheral blood flow; perfusion and ventilation in the horse. *Mailing Add:* Dept Physiol 723 VRT NY State Col Vet Med Cornell Univ Ithaca NY 14853

DOBSON, DAVID A, b Oakland, Calif, Mar 28, 37; m 57. NUCLEAR PHYSICS, ATOMIC PHYSICS. *Educ:* Univ Calif, Berkeley, BS, 59, PhD(physics), 64. *Prof Exp:* Physicist, Lawrence Radiation Lab, Livermore, 64-69; from asst prof to assoc prof, 69-78, PROF PHYSICS, BELOIT COL, 80- *Mem:* AAAS; Am Phys Soc. *Res:* Investigation of weak interactions by a study of asymmetries and angular correlations in beta decay of polarized nuclei. *Mailing Add:* Dept Physics & Astron Beloit Col Beloit WI 53511

DOBSON, DONALD C, b Central, Idaho, Sept 18, 26; m 49; c 9. POULTRY SCIENCE. *Educ:* Utah State Univ, BS, 54; Cornell Univ, MS, 55; Utah State Univ, PhD(nutrit, biochem), 61. *Prof Exp:* Res asst biochem, Univ Utah, 55-57; res asst nutrit biochem, 57-60, asst prof turkey res, 60-70, assoc prof animal sci, 70-80, ASSOC PROF ANIMAL SCI & VET, UTAH STATE UNIV, 80- *Mem:* Poultry Sci Asn. *Res:* Poultry nutrition and management, especially turkeys. *Mailing Add:* 2031 N 1600 E Logan UT 84321

DOBSON, GERARD RAMSDEN, b Lynbrook, NY, May 4, 33; m 62; c 3. ORGANOMETALLIC CHEMISTRY, CHEMICAL KINETICS. *Educ:* Fla Southern Col, BS, 55; Temple Univ, MEd, 58; Fla State Univ, PhD(chem), 64. *Prof Exp:* High sch teacher, Pa, 55-58; asst prof inorg chem, Univ Ga, 63-67; assoc prof chem, Univ SDak, 67-69; from assoc prof to prof phys chem, 69-89, REGENTS PROF INORG CHEM, UNIV NTEX, 89- *Mem:* Am Chem Soc; Royal Soc Chem; Sigma Xi; Am Inst Chemists. *Res:* Organotransition metal chemistry; kinetics and mechanism of reactions of metal carbonyls and derivatives; fast kinetics. *Mailing Add:* Dept Chem Univ NTex Denton TX 76203-5068

DOBSON, HAROLD LAWRENCE, b Liberty Co, Tex, May 10, 21; m 45; c 4. BIOCHEMISTRY, INTERNAL MEDICINE. *Educ:* Baylor Univ, BS, 43, MD, 46, MS, 56. *Prof Exp:* Intern, Salt Lake County Hosp, Utah, 46-47; instr biochem, 47-48, from instr to assoc prof internal med, 53-70, CLIN ASSOC PROF, BAYLOR COL MED, 71-; CLIN PROF MED & INTERNAL MED, UNIV TEX MED SCH HOUSTON, 73- *Concurrent Pos:* USPHS fel biochem & internal med, Col Med, Baylor Univ, 48-50; resident, Vet Admin Hosp, Houston, 50-51; fel coun on atherosclerosis, Am Heart Asn; head, Diabetic Clin, Hermann Hosp, Univ Tex Med Sch Houston, 73-80. *Mem:* AAAS; AMA; Am Col Cardiol; Am Soc Pharmacol & Exp Therapeut; Am Soc Nephrol. *Res:* Metabolic disease; diabetes; computer use for clinical records. *Mailing Add:* 7707 Fannin Suite 290 Houston TX 77054

DOBSON, JAMES GORDON, JR, b Waterbury, Conn, Jan 23, 42; m 71; c 1. PHYSIOLOGY, PHARMACOLOGY. *Educ:* Cent Conn State Univ, BS, 65; Wesleyan Univ, MA, 67; Univ Va, PhD(physiol), 71. *Prof Exp:* Res pharmacologist, Univ Calif, San Diego, 71-73; from asst prof to assoc prof, 73-84, PROF PHYSIOL, MED SCH, UNIV MASS, 85-, GRAD DIR, CELLULAR & MOLECULAR PHYSIOL GRAD PROG & PROF MED, 89- *Concurrent Pos:* Fel, Univ Calif, San Diego, 71-72, Giannini Found fel, 72-73; consult, NIH, 78- *Honors & Awards:* Career Develop Award, Nat Heart Lung & Blood Inst, 80, Merit Award, 87. *Mem:* Am Heart Asn; Am Physiol Soc; Biophys Soc; Int Soc Heart Res; AAAS; Soc Gen Physiologists. *Res:* Cardiovascular physiology; particularly the mechanisms involved in the regulation of contractile and metabolic function in the myocardium of the normal, ischemic and hypoxic heart; antiadrenergic actions of adenosine. *Mailing Add:* Dept Physiol Med Sch 55 Lake Ave N Worcester MA 01655

DOBSON, MARGARET VELMA, b Burbank, Calif, June 8, 48; m 75; c 2. VISUAL SCIENCES, CHILD DEVELOPMENT. *Educ:* Fla State Univ, BA, 70; Brown Univ, ScM, 73, PhD(psychol), 75. *Prof Exp:* Res scientist ophthal, Tufts-New Eng Med Ctr, 74-75; NIH fel, Univ Wash, 75-78, res assoc, 77-78, sr res assoc, 84-85, from res asst prof to res assoc prof psychol, 78-85; ASSOC PROF, UNIV PITTSBURGH, 84- *Concurrent Pos:* Prin investr, NIH res grant, 78- & Children's Eye Care Found grant, 80-81; affiliate, Child Develop & Mental Retardation Ctr, Univ Wash, 81-84, NIH training grant, 81-85. *Mem:* Am Res Vision & Ophthal; Soc Res Child Develop; AAAS. *Res:* Development and use of behavioral methods for the assessment of visual acuity and visual fields in infants and young children in laboratory and clinical settings. *Mailing Add:* Dept Psychiat & Psychol Univ Pittsburgh Pittsburgh PA 15260

DOBSON, PETER N, JR, b Baltimore, Md, Sept 15, 36; m 57; c 3. THEORETICAL HIGH ENERGY PHYSICS. *Educ:* Mass Inst Technol, BS, 58; Univ Md, 60-64, PhD(physics), 65. *Prof Exp:* Physicist, Raytheon Corp, 58-59; physicist, Westinghouse Elec Corp, 59-60; from asst prof to assoc prof, 65-74, actg chmn dep, 75-76, PROF PHYSICS & ASTRON, UNIV HAWAII, 74- *Mem:* Am Phys Soc. *Res:* Theory of elementary particles and their interactions at high energy. *Mailing Add:* Hong Kong Univ Sci & Tech 12/F World Shipping Ctr Seven Canton Rd Kowloon Hong Kong

DOBSON, R LOWRY, b Beijing, China, Nov 2, 19; US citizen; m 43; c 5. MEDICAL RESEARCH, RADIOBIOLOGY. *Educ:* Univ Calif, Berkeley, AB, 41, PhD(biophys), 50; Univ Calif, San Francisco, MD, 44. *Prof Exp:* Teaching asst, Univ Calif, Berkeley, 41-42; intern, Univ Calif Hosp, San Francisco, 44-45; resident physician, Permanente Found Hosp, 45-46; physician & res fel, Donner Lab, Univ Calif, 46-47, dir med servs, Lawrence Radiation Lab, 47-58; chief med officer, radiation & isotopes, WHO, Geneva, Switzerland, 58-67; sr scientist, Biomed Sci Div, Lawrence Livermore Nat Lab, Univ Calif, Livermore, 67-88; ADJ PROF RADIOL, SCH MED UNIV CALIF, DAVIS, 77- *Concurrent Pos:* Med physics physician to Univ & lectr & assoc res med physicist, Univ of Calif, 48-58, instr & res assoc med physics, Donner Lab, 47-48; consult, Int Comn Radiol Protection & Int Comn Radiol Units & Meas, 59-60, US Dept Com, 71-72 & WHO, 78; partic guest, 88- *Mem:* Radiation Res Soc; Environ Mutagen Soc; AAAS; Sci Res Soc N Am.

Res: Biological and health effects of exposure to environmental agents, radiation, radionuclides and chemicals, especially on certain mammalian cell populations during development; radiobiology; toxicology; cell biology; reproductive and developmental biology. *Mailing Add:* 2716 Cedar St Berkeley CA 94708

DOBSON, RICHARD LAWRENCE, b Boston, Mass, Apr 12, 28; m 50; c 3. DERMATOLOGY. *Educ:* Univ Chicago, MD, 53; Am Bd Dermat, dipl, 59. *Prof Exp:* From instr to asst prof dermat, Univ NC, 57-61; from assoc prof to prof, Med Sch, Univ Ore, 61-72; prof derm & chmn dept, Sch Med, State Univ NY, Buffalo, 72-79; PROF DERMAT, MED UNIV SC, 80-, CHMN DEPT. *Concurrent Pos:* NIH res fel, 56-57; attend physician, Univ Sch, Ore Hosps, 61-72; consult, Vet Admin Hosp, Portland, 61-72; scientist, Ore Regional Primate Ctr, 63-68; asst chief ed, Arch Dermat, 64-69; mem gen med study sect, NIH, 65-69, chmn, 70-72; vis prof, Univ Nijmegen, Netherlands, 69-70; head dept dermat, Buffalo Gen Hosp, E J Meyer Hosp, Buffalo. *Mem:* Am Asn Cancer Res; Am Acad Dermat; Soc Invest Dermat (pres, 76); Am Col Physicians; Am Dermat Asn. *Res:* Cutaneous carcinogenesis; physiology of eccrine sweat gland. *Mailing Add:* Dept Dermat Med Univ SC 171 Ashley Ave Charleston SC 29403

DOBY, RAYMOND, b New York, NY, Oct 9, 23; m 68. CONTINUUM & FLUID MECHANICS. *Educ:* NY Univ, BME, 47, MS, 51; Univ Pa, PhD(eng mech), 62. *Prof Exp:* Sr engr steam div, Westinghouse Elec Corp, 53-62; sr staff scientist, Avco Corp, 62-63; adv engr astronuclear labs, Westinghouse Elec Corp, 63-66; assoc prof eng, Swarthmore Col, 66-71; consult, RADO Assocs, 72-80; mgr eng systs, Conrail Corp, 78-90; DIR, R-D ASSOCS, 90- *Concurrent Pos:* Adj asst prof, Evening Col, Drexel Inst, 56-62. *Mem:* Am Soc Mech Engrs; Sigma Xi. *Res:* Gas dynamics; boundary layer analysis; heat and mass transfer; biomechanics; interaction of science and society; elastic waves in bounded media; seismic structural analysis. *Mailing Add:* 409 W Colonial Apts Cherry Hill NJ 08002

DOBY, TIBOR, b Budapest, Hungary, Aug 23, 14; US citizen; m 48. RADIOLOGY, NUCLEAR MEDICINE. *Educ:* Univ Budapest, MD, 38; Am Bd Radiol, dipl, 62. *Prof Exp:* Resident internal med, Med Sch Hosp, Univ Budapest, 38-44, asst prof, 44-48, asst prof radiol, 48-50; assoc radiologist, Trade Union & Rwy Employees Hosp, Budapest, 50-56; resident, St Raphael's Hosp, New Haven, Conn, 57-59; instr, Med Sch, Yale Univ, 59-60; assoc radiologist, 60-65, dir radiol dept, 65-80, CHIEF RADIOL & ISOTOPE SCANNING & EMER CHIEF DIAG RADIOL, MERCY HOSP, MAINE, 80- *Mem:* Fel Am Col Radiol; NY Acad Sci; Radiol Soc NAm; Soc Nuclear Med. *Res:* Circulatory shock; hemodynamics; history of medicine. *Mailing Add:* Dept Radiol Mercy Hosp 144 State St Portland ME 04101

DOBYNS, BROWN M, b Jacksonville, Ill, May 14, 13; m 40, 85; c 3. SURGERY. *Educ:* Ill Col, BA, 35; Johns Hopkins Univ, MD, 39; Univ Minn, MS, 44, PhD, 46; Am Bd Surg, dipl. *Prof Exp:* Intern surg, Johns Hopkins Hosp, 39-40; resident, Kahler Hosp, Mayo Clin, 43-45, resident, Mayo Clin, 45-46, asst to surg staff, 46; asst prof surg, Med Sch, Harvard Univ, 48-51; assoc chief surg serv, Cleveland Metrop Gen Hosp, 67-88; from assoc prof to prof, Sch Med, Case Western Reserve Unit, 58-87, emer prof, 87-; RETIRED. *Concurrent Pos:* Res fel surg, Med Sch, Harvard Univ, 46-48; asst, Mass Gen Hosp, 46-51; asst chief surg serv, Cleveland Metrop Gen Hosp, 51-66; asst surgeon, Univ Hosps, Cleveland, 51-; mem staff, Lutheran Hosp, 75-91. *Honors & Awards:* Van Meter Prize, Am Thyroid Asn, 46. *Mem:* Am Surg Asn; Soc Univ Surg; Am Col Surg; Am Soc Clin Invest; Am Thyroid Asn (pres, 56-57); Int Soc Surg. *Res:* Thyroid physiology and general oncology. *Mailing Add:* Cleveland Metrop Gen Hosp 3395 Scranton Rd Cleveland OH 44109

DOBYNS, LEONA DANETTE, b Shelby, Mont, Jan 28, 30. PHYSICAL CHEMISTRY. *Educ:* Col Great Falls, BS, 58; Univ Notre Dame, PhD(phys chem), 64. *Prof Exp:* From asst prof to assoc prof chem, Seattle Univ, 64-72; assoc prof, 72-78, PROF CHEM, CALIF STATE UNIV, DOMINGUEZ HILLS, 78- *Concurrent Pos:* Res grant, Univ Notre Dame, 68. *Mem:* Am Chem Soc; Am Phys Soc. *Res:* Microwave spectroscopy and molecular structure. *Mailing Add:* 129 W 37th St Long Beach CA 90807-2662

DOBYNS, ROY A, b Bristol, Va, Jan 31, 31; m 55; c 2. MATHEMATICS. *Educ:* Carson-Newman Col, BA, 53; Vanderbilt Univ, MA, 54; George Peabody Col, PhD(math), 63. *Prof Exp:* Asst prof math, La Col, 56-58; from asst prof to prof, McNeese State Col, 58-68; prof & chmn dept, Georgetown Col, 68-73; chmn div natural sci & math, Clayton Jr Col, 73-75; ACAD DEAN, CARSON-NEWMAN COL, 75- *Mem:* Math Asn Am. *Mailing Add:* Carson-Newman Col Jefferson City TN 37760

DOBYNS, SAMUEL WITTEN, b Norton, Va, Mar 7, 20; m 45; c 3. CONSTRUCTION MANAGEMENT, STRUCTURAL ENGINEERING. *Educ:* Va Mil Inst, BS, 41; Lehigh Univ, MS, 49. *Prof Exp:* Sr instrumentman, E I du Pont de Nemours & Co, 46; from instr to prof civil eng, Va Mil Inst, 46-85, dir, Eve Col, 67-74; RETIRED. *Concurrent Pos:* Dir, Robert A Marr Sch Surv & Comput Clin, 50-72; partner, Dobyns & Morgan Consult Engrs, 52-64; NASA fac fel, Manned Spacecraft Ctr, Univ Houston, Tex A&M Univ, 67; Off Civil Defense fel, WVa Univ, 69; consult, Thompson & Litton, 70-75; consult engr & land survr, 75-85; engr, City of Lexington, 80-, Valley Blox, Inc, 85- *Mem:* Am Soc Civil Engrs; Am Soc Photogram; Am Cong Surv & Mapping; Am Soc Eng Educ. *Res:* Bridge design; structural analysis; construction management computer programming and applications; project management; photogrammetry; solid waste; surveying; nuclear defense. *Mailing Add:* 10 Sellers Ave Lexington VA 24450-1931

DOCHERTY, JOHN JOSEPH, b Youngstown, Ohio, Dec 5, 41; m 65; c 2. VIROLOGY, MICROBIOLOGY. *Educ:* Youngstown Univ, BA, 64; Miami Univ, Ohio, MS, 66; Univ Ariz, PhD(microbiol), 70. *Prof Exp:* Fel virol, Dept Microbiol, Sch Med, Pa State Univ, 70-72, asst prof microbiol, 72-76, assoc prof microbiol, Dept Microbiol, Col Sci, 76-86; PROF & CHMN, DEPT MICROBIOL/IMMUNOL, COL MED, NORTHEASTERN OHIO UNIVS, 87- *Concurrent Pos:* Consult, Hercules Chem Co, Wilmington, Del, 73, Frederick Cancer Res Ctr, Md, 75, ENI, Columbia, Md & Carter-Wallace/Wampole Labs, East Windsor, Nev; res fel virology, Univ Glasgow, Scotland, 80; chair, Asn Med Sch Microbiol & Immunol. *Mem:* Am Soc Microbiol; Sigma Xi; AAAS; Am Soc Virol. *Res:* Molecular basis of pathogenesis of the herpes viruses; antiviral drugs and chemicals; viral DNA binding proteins. *Mailing Add:* Dept Microbiol/Immunol Col Med Northeastern Ohio Univs Rootstown OH 44272

DOCKEN, ADRIAN (MERWIN), b Holt, Minn, Nov 17, 13; m 41; c 4. SYNTHETIC ORGANIC CHEMISTRY. *Educ:* Luther Col, BA, 37; Univ Wis, PhD(org chem), 41. *Prof Exp:* Asst, Univ Wis, 37-41; res assoc, Northwestern Univ, 41-42; head dept chem, 42-70 & 77-78, prof, 42-86, RES PROF CHEM, LUTHER COL, 86- *Concurrent Pos:* Fac fel, Ford Found, Yale Univ, 54-55; fac fel, NSF, Imp Col, Univ London, 61-62; Am Chem Soc Petrol Fund fac award, Stanford Univ, 67-68; vis scholar, Univ Calif, Los Angeles, 74-75. *Mem:* AAAS; Am Chem Soc; The Chem Soc. *Res:* Organic synthesis. *Mailing Add:* Dept Chem Luther Col Decorah IA 52101

DOCKERY, JOHN T, b Cleveland, Ohio, May 2, 36; m 61; c 3. SCIENCE ADMINISTRATION, SYSTEM DESIGN & SYSTEM SCIENCE. *Educ:* John Carroll Univ, BS, 58; Fla State Univ, MS, 62, PhD(physics), 65. *Prof Exp:* Physicist, US Naval Weapons Lab, Va, 60-61; engr, Boeing Aerospace, Wash, 63-64; res physicist, Bendix Aerospace Systs, Mich, 65-66; sr scientist, Res Inst, Ill Inst Technol, 66-68; opers res analyst, Weapons-Systs Anal, Off Chief of Staff, US Army, Pentagon, 68-73, opers res analyst, US Army Concept Anal Agency, Md, 73-80, opers res br chief, Shape Tech Ctr, Hague, Neth, 80-84; sr analyst Off Joint Chief of Staff, Pentagon, Wash, DC, 84-87; fed exec fel, Brookings Inst, Wash, DC, 87-88; SR ANALYST, OFFICE JOINT CHIEFS OF STAFF, PENTAGON, WASHINGTON, DC, 89- *Concurrent Pos:* Mem Army Math Steering Comt, 68-73; mem, Land Use & Zoning, 70-80; mem, Europ Working Group Multi-Criteria Decision Theory, 75-78; mem, Exec Eng Adv Panel, Univ Pa, 88-91; vis res prof, George Mason Univ, Fairfax, Va, 91- *Mem:* Am Phys Soc; Am Geophys Union; Am Inst Aeronaut & Astronaut; Asn Comput Mach; Opers Res Soc Am; Sigma Xi; Int Fuzzy Sets Asn; Soc Indust & Appl Math. *Res:* Large-scale simulation; Fuzzy Set theory; multi-criteria decision theory; weapon systems analysis; computational physics; theory of combat with embedded command and control. *Mailing Add:* 2507 Pegasus Lane Reston VA 22901

DOCKS, EDWARD LEON, b Detroit, Mich, Jan 14, 45; m 69; c 2. NEW PRODUCT DEVELOPMENT, APPLICATION THEORY. *Educ:* Wayne State Univ, BS, 67; Univ Calif, Los Angeles, PhD(chem), 72. *Prof Exp:* NIH res fel, 72-74; res chemist, 74-79, SR SCIENTIST, US BORAX RES CORP, 79- *Mem:* Am Chem Soc. *Res:* Development of new products; finding new uses for current products; corrosion work involving testing of candidate materials and failure analysis. *Mailing Add:* US Borax Res Corp 412 Crescent Way Anaheim CA 92801-6794

DOCKTER, MICHAEL EDWARD, b Carmel, Calif, July 29, 49; m 69; c 2. BIOCHEMISTRY, BIOPHYSICS. *Educ:* Calif State Col, Sonoma, BS, 71; Wash State Univ, PhD(biochem), 75. *Prof Exp:* Res asst, Univ Basel, 75-77; asst mem, 77-82, ASSOC MEM BIOCHEM, ST JUDE CHILDREN'S RES HOSP, 82- *Concurrent Pos:* Fel, Europ Molecular Biol Orgn, 75-76; asst prof, Univ Tenn Med Units, 78-83. *Mem:* AAAS; Am Chem Soc; NY Acad Sci; Am Soc Biol Chemists. *Res:* Studies of membrane structure and function using biochemical and biophysical approaches; the protein complexes of oxidative phosphorylation; flow cytometry. *Mailing Add:* Dept Med Rm H308 Univ Tenn Ctr Health Sci 956 Court Ave Memphis TN 38163

DOCTOR, BHUPENDRA P, b Surat, India, May 1, 30; US citizen; m; c 2. RESEARCH ADMINISTRATION, SCIENCE ADMINISTRATION. *Educ:* Univ Bombay, BSc, 52; Tex A&M Univ, MS, 55; Univ Md, PhD(biochem), 59. *Prof Exp:* Res asst, Dept Biochem & Nutrit, Tex A&M Univ, 53-55; teaching & res asst, Dept Chem, Univ Tex, Austin, 55-56; grad res fel, Dept Biol Chem, Sch Med, Univ Md, Baltimore, 56-58, asst prof, 61-62; postdoctoral fel, Dept Biochem, Cornell Univ, Ithaca, NY, 59-60; dir, Div Biochem, Weizmann Inst Sci, Rehovot, Israel, 86-87; res biochemist, 60-70, chief, Dept Biol Chem, 70-77, dep dir res, 60-77, DIR, DIV BIOCHEM, WALTER REED ARMY INST RES, WASHINGTON, DC, 77- *Concurrent Pos:* Res & study fel Secy Army, Lab Molecular Biol, Med Res Coun, Cambridge Eng, 67-78; vis scholar, Div Pharmacol, Dept Med, Sch Med, Univ Calif, San Diego, 81-82; vis prof molecular biol, Salk Inst Biol Studies, San Diego, Calif, 81-82. *Honors & Awards:* Presidential Rank Meritorious Award, Pres of US, 86. *Mem:* Am Soc Biochem & Molecular Biol; Am Chem Soc; AAAS; Sigma Xi; Am Soc Microbiol; Am Soc Pharmacol & Exp Therapeut. *Res:* Use of macromolecules as scavengers of toxic substances; isolation and purification of macromolecules on large scale; mapping the topology of proteins using monoclonal antibodies. *Mailing Add:* Div Biochem Walter Reed Army Inst Res Washington DC 20307-5100

DOCTOR, NORMAN J(OSEPH), b Brooklyn, NY, Oct 5, 29; m 56; c 3. ORDNANCE ENGINEERING. *Educ:* Purdue Univ, BS, 51. *Prof Exp:* Physicist, Nat Bur Standards, 51-53; physicist, Diamond Ord Fuze Labs, US Army Lab Command, 53-56, res & develop supvr, 56-63, res & develop supvr, Harry Diamond Labs, 63-69, chief, Electron Devices Br, 69-70, Electron Timing Br, 70-81, Energy Systs & Mat Br, 81-89; RETIRED. *Concurrent Pos:* Sr engr, Advan Technol Res Corp, 89-; hands-on-sci teacher, 89- *Honors & Awards:* Arthur S Flemming Award, 62. *Mem:* Inst Elec & Electronics Engrs; Am Defense Preparedness Asn. *Res:* Dielectric and magnetic measurements; printed circuit technology; microelectronics; thin films; ammunition electronics; electronic timing; ordnance electrochemical and electromechanical power supplies. *Mailing Add:* Six Tegner Ct Rockville MD 20850-2772

DOCTOR, VASANT MANILAL, b Surat, India, Mar 19, 26; m 53; c 3. BIOCHEMISTRY. *Educ:* Royal Inst Sci, India, 46 & 48; Univ Wis, MS, 51; Tex A&M Univ, PhD, 53. *Prof Exp:* Fel biochem, Tex Agr Exp Sta, 53-54; Hite fel exp med, M D Anderson Hosp & Tumor Inst, Univ Tex, 54-55, from asst biochemist to assoc biochemist, 55-62; chief biochemist, Hindustan Antibiotics, Ltd, 62-65; assoc prof chem, Univ Houston, 65-68; PROF CHEM, PRAIRIE VIEW AGR & MECH UNIV, 68- *Concurrent Pos:* USPS res career develop award, Dent Br, Univ Tex, 59-62. *Mem:* AAAS; Am Soc Biol Chemists; Soc Exp Biol & Med. *Res:* Model enzyme systems; active site of nucleases; blood clotting mechanisms; cancer chemotherapy. *Mailing Add:* Dept Chem Prairie View Agr & Mech Univ Prairie View TX 77445-2576

DOCTROW, SUSAN R, b Philadelphia, Pa, Oct 26, 56. EXPERIMENTAL BIOLOGY. *Educ:* Univ Rochester, BS, 78; Brandeis Univ, PhD(biochem), 85. *Prof Exp:* Res fel, Children's Hosp, Boston, 84-86; res assoc, Harvard Med Sch & Children's Hosp, 86-89; res scientist, 89-90, SR RES SCIENTIST, ALKERMES, 90- *Concurrent Pos:* Postdoctoral fel, Am Cancer Soc, 84-86; instr biochem, Harvard Med Sch & Children's Hosp, 87-89; small bus innovative res, NIH, 91- *Mem:* AAAS; Am Soc Cell Biol. *Mailing Add:* Alkermes 26 Landsdowne St Cambridge MA 02139-4234

DOD, BRUCE DOUGLAS, b Englewood, NJ, Oct 23, 41; m 65; c 2. EARTH SCIENCES AND METEORITICS. *Educ:* Eastern Kentucky Univ, BS, 69; E Tex State Univ, MS, 70; Univ S Miss, PhD(earth sci), 73. *Prof Exp:* Technician, Union Carbide Nuclear Co, 65-67; metall technician, Inst Nickel Co, 67-69; teacher earth sci, Tate Creek Jr High, 69; asst instr earth sci, E Tex State Univ, 70; prof & chair physics & earth sci, Wayland Univ, 75-82; PROF PHYSICS & EARTH SCI, MERCER UNIV, 82- *Concurrent Pos:* Ed, text sect, Nat Asn Geol Teachers, 78-80; consult, Ga Site Superconducting Supercollider, 86-87; reviewer, Choice Libr Rev Serv, 84- *Mem:* Geol Soc Am; Nat Asn Geol Teachers; Meteoretical Soc. *Res:* Petrology and petrography of stony meteorites; meteorite concentration mechanisms and fall site environmental statistics; palynostratigraphy. *Mailing Add:* Dept Physics Mercer Univ Main Campus 1400 Coleman Ave Macon GA 31207

DODABALAPUR, ANANTH, b Bangalore, India, Feb 17, 63; m 91. COMPOUND SEMICONDUCTOR MATERIALS & DEVICES, MOLECULAR BEAM EPITAXY. *Educ:* Indian Inst Technol, Madras, BTech, 85; Univ Tex, Austin, MS, 87, PhD(elec eng), 90. *Prof Exp:* Grad res asst, Microelectronics Res Ctr, Univ Tex, 85-90; POSTDOCTORAL MEM TECH STAFF, PHOTONICS RES LAB, AT&T BELL LABS, HOLMDEL, NJ, 90- *Mem:* Inst Elec & Electronics Engrs; Mat Res Soc. *Res:* Molecular beam epitaxial growth and characterization of compound semiconductor materials; semiconductor optoelectronic materials, devices, and integration schemes; ion implantation and rapid thermal processing of semiconductors; photovoltaics. *Mailing Add:* AT&T Bell Labs Rm 4F-433 Crawfords Corner Rd Holmdel NJ 07733-1988

DODD, CHARLES GARDNER, b St Louis, Mo, Jan 26, 15; m 43; c 4. PHYSICAL CHEMISTRY, MATERIALS SCIENCE. *Educ:* Rice Inst, BS, 40; Univ Mich, MS, 45, PhD(phys chem), 48. *Prof Exp:* Halliburton prof petrol eng, Univ Okla, 56-62; chief advan mat res, Owens-Ill, Inc, 62-68; assoc prin scientist, Philip Morris, Inc, 68-74; sr res assoc, Warner-Lambert Co, Milford, 74-80; pres, Conn Technol Consults, Inc, 80-90; PRES, CTC TECHNOLOGIES, INC, 91- *Concurrent Pos:* Vis scientist, Dept Chem, Univ Ariz, 91- *Mem:* Fel AAAS; fel Am Inst Chemists; Am Vaccum Soc; Am Chem Soc. *Res:* Surface physics and chemistry; Langmuis-Blodgett organic film depositon; thin films; soft x-ray spectroscopy and chemical bonding. *Mailing Add:* 7925-A N Oracle Rd No 364 Tucson AZ 85704-6356

DODD, CURTIS WILSON, b Fulton, Mo, Nov 4, 39; m 63; c 1. ELECTRICAL ENGINEERING. *Educ:* Univ Mo-Rolla, BS, 63, MS, 64; Ariz State Univ, PhD(control theory), 68. *Prof Exp:* Fac assoc elec eng, Ariz State Univ, 64-66; from asst prof to assoc prof elec sci, Southern Ill Univ, Carbondale, 67-84; PRIN ENGR, UNIVERSAL DATA SYSTS, MOTOROLA, 84- *Res:* Optimal control of distributed parameter systems. *Mailing Add:* Universal Data Systs Motorola 5000 Bradford Dr Huntsville AL 35805

DODD, DAROL ENNIS, b Biloxi, Miss, Feb 9, 49; m 69; c 2. INHALATION. *Educ:* Univ Kans, BA, 71, PhD(pharmacol & toxicol), 78; Am Bd Toxicol, dipl, 83. *Prof Exp:* Fel, Chem Indust Inst Toxicol, 78-80; scientist, Union Carbide Corp, 80-81, sr scientist, 81-83, mgr, Bushy Run Res Ctr, 83-89; DEP DIR, TOXIC HAZARDS RES UNIT, MANTECH ENVIRON TECH INC, 89- *Concurrent Pos:* Prog mgr, Union Carbide Agr Prod Co, 83-87; guest lectr, dept pharmacol toxicol, Univ Pittsburgh, 81-84 Duquesne Univ, 87- *Mem:* Soc Toxicol; Am Indust Hyg Asn. *Res:* Mechanisms of pulmonary toxicity and assessments of pulmonary function, pulmonary hypersensitization and sensory irritation; carbamate toxicology, registration of pesticides and regulatory toxicology; highly dangerous gases and vapors such as methyl isocyanate. *Mailing Add:* 3912 Eagle Point Dr Beavercreek OH 45430

DODD, DAVID CEDRIC, veterinary pathology, for more information see previous edition

DODD, EDWARD ELLIOTT, b Rochester, NY, June 27, 22; wid; c 2. SPACE TELECOMMUNICATIONS SYSTEMS. *Educ:* Univ Rochester, BS, 43; Univ Calif, PhD(physics), 52. *Prof Exp:* Physicist, Nat Bur Stand, 52-53; physicist, Naval Ord Lab, 53-54; physicist, Motorola Inc, 54-63; res specialist, Lockheed Missiles & Space Co, 63-66, advan systs engr, Lockheed Eng & Sci Co, 66-87; RETIRED. *Mem:* Am Phys Soc; Audio Eng Soc; Sigma Xi. *Res:* Missile guidance systems; ranging and inertial sensors; computer information systems; space telecommunications systems design. *Mailing Add:* PO Box 57622 Webster TX 77598

DODD, GERALD DEWEY, JR, b Oaklyn, NJ, Nov 18, 22; m 46; c 7. RADIOLOGY. *Educ:* Lafayette Col, Pa, BA, 45; Jefferson Med Col, MD, 47. *Prof Exp:* Intern, Fitzgerald Mercy Hosp, 47-48; resident radiol, Jefferson Med Col Hosp, 48-50, asst radiologist, 52-55; assoc prof radiol, Univ Tex Postgrad Sch Med & assoc radiologist & head sect diag radiol, M D Anderson Hosp & Tumor Inst, 55-61; clin prof radiol & asst radiologist, Jefferson Med Col, 61-66; prof diag radiol & chmn, Dept Diag Radiol, Univ Tex Med Sch, 71-74, PROF DIAG RADIOL & HEAD DIV DIAG IMAGING, M D ANDERSON CANCER CTR, UNIV TEX, HOUSTON, 84- *Concurrent Pos:* Dir-at-large, Am Cancer Soc, 77, mem, Med & Sci Comt, 79, Nat Conf Breast Cancer, 88, chmn, Nat Adv Comt Cancer Prev & Detection, 87-; chmn, Breast Task Force, Am Col Radiol, 86-, mem, Task Force Standards Setting, 88-; Robert D Moreton chair diag radiol, 88; chmn, Med & Sci Comt, Am Cancer Soc, 87, ACS Nat Conf Breast Cancer, 88. *Honors & Awards:* Silver Medal, Am Roentgen Ray Soc, 56; Gold Medal, Radiol Soc NAm, 86 & Am Col Radiol, 89. *Mem:* Fel Am Col Radiol; Am Roentgen Ray Soc; Radiol Soc NAm; Sigma Xi. *Mailing Add:* 1749 South Blvd Houston TX 77096

DODD, JACK GORDON, (JR), b Spokane, Wash, June 19, 26; m 51; c 2. ATOMIC PHYSICS. *Educ:* Ill Inst Technol, BS, 51; Univ Ark, MS, 57, PhD, 65. *Prof Exp:* Res technician physics, Argonne Nat Lab, 51-53; pub sch teacher, 53-55, prin, 56-57; asst physics, Univ Ark, 55-56; asst prof, Drury Col, 57-60; assoc prof & chmn dept, Ark Polytech Col, 62-67; assoc prof, Univ Tenn, Knoxville, 67-71; Charles A Dana prof physics, Colgate Univ, 71-88; RETIRED. *Concurrent Pos:* Consult, McCrone Assocs, 61- & Honeywell, Inc, 64- *Mem:* Am Phys Soc; Am Asn Physics Teachers; Am Astron Soc; Sigma Xi. *Res:* Optical instrumentation and physical optics; atomic beams, shocks and detonations. *Mailing Add:* 23 Broad Hamilton NY 13346

DODD, JAMES ROBERT, b Bloomington, Ind, Mar 11, 34; m 56; c 2. CARBONATE PETROLOGY. *Educ:* Ind Univ, AB, 56, AM, 57; Calif Inst Technol, PhD(geobiol), 61. *Prof Exp:* Ford Found fel oceanog, Calif Inst Technol, 61; geologist res labs, Texaco Inc, Tex, 61-63; asst prof geol, Case Western Reserve Univ, 63-66; assoc prof, 66-73, PROF GEOL, IND UNIV, BLOOMINGTON, 73- *Concurrent Pos:* Chmn, Ind Univ, Bloomington, 87-90. *Mem:* Am Asn Petrol Geologists; Geol Soc Am; Soc Econ Paleontologists & Mineralogists; Int Asn Sedimentologists; Sigma Xi. *Res:* Carbonate petrology and sedimentology; paleoecology and geochemistry of carbonate skeletons. *Mailing Add:* Dept Geol Sci Ind Univ Bloomington IN 47405

DODD, JIMMIE DALE, b Esbon, Kans, Aug 6, 31; m 54; c 4. PLANT ECOLOGY, SOILS. *Educ:* Ft Hays Kans State Col, AB, 56, MS, 57; Univ Sask, PhD(plant ecol, soils), 60. *Prof Exp:* Asst prof forestry, Ariz State Col, 60-61; asst bot, NDak State Univ, 61-63; from asst prof to assoc prof range & forestry, 63-70, prof, 70-80, PROF RANGE SCI, TEX A&M UNIV, 80- *Concurrent Pos:* Secy range sci educ coun, Southwestern Naturalist. *Mem:* Ecol Soc Am; Bot Soc Am; Am Soc Range Mgt. *Res:* Soil-plant relationships in native vegetation; effects of prescribed fire on native vegetation; use of radionuclides in the study of ecological systems; native vegetation manipulation to increase production. *Mailing Add:* 1307 Leacrest Dr College Station TX 77840

DODD, JOHN DURRANCE, b Tarrytown, NY, Mar 15, 17; m 40; c 3. BOTANY. *Educ:* Syracuse Univ, BS, 38; Univ Vt, MS, 40; Columbia Univ, PhD(bot), 47. *Prof Exp:* Teaching asst & asst bot, Univ Vt, 38-41; asst morphol lab, Columbia Univ, 41-43 & 46-47; instr bot, Univ Wis, 47-49; from asst prof to assoc prof dept bot & plant path, 49-60, prof bot, Dept Bot & Plant Path, Iowa State Univ, 60-; RETIRED. *Mem:* Bot Soc Am. *Res:* Interspecific grafts in Viola; cell shape; plant morphology; freshwater algae; diatoms; aquatic plant biology. *Mailing Add:* RR 3 Gordonville VA 22942

DODD, RICHARD ARTHUR, b Eng, Feb 11, 22; m 47; c 3. METALLURGY. *Educ:* Univ London, BS, 44, MS, 47, DSc(metall), 74; Univ Birmingham, PhD(metall), 50. *Prof Exp:* Res metallurgist, Rolls Royce Ltd, Eng, 44-47; sr lectr, Univ Witwatersrand, SAfrica, 50-54; res metallurgist, Dept Mines, Ottawa, 54-56; asst prof metall eng, Univ Pa, 56; chmn dept, 74-80, PROF METALL ENG, UNIV WIS-MADISON, 56- *Mem:* Am Inst Mining, Metall & Petrol Engrs; Am Soc Metals; Brit Iron & Steel Inst; Brit Inst Metals; fel Royal Soc Chem. *Res:* General physical metallurgy. *Mailing Add:* 2211 Chamberlain Ave Madison WI 53705

DODD, ROBERT TAYLOR, b Bronx, NY, July 11, 36; m 58; c 3. MINERALOGY, PETROLOGY. *Educ:* Cornell Univ, AB, 58; Princeton Univ, MA, 60, PhD(geol), 62. *Prof Exp:* Spec scientist, Air Force Cambridge Res Labs, 62-65, gen phys scientist, 65; from asst prof to prof mineral, 73-77, PROF EARTH SCI, STATE UNIV NY STONY BROOK, 77- *Mem:* AAAS; Geol Soc Am; Mineral Soc Am; Am Geochem Soc; Meteoritical Soc. *Res:* Igneous and metamorphic petrology; petrology and mineralogy of meteorites. *Mailing Add:* Dept Earth & Space Sci State Univ NY Stony Brook NY 11790

DODD, ROGER YATES, b UK, Feb 23, 44; US citizen. INFECTIOUS DISEASES, BLOOD BANKING. *Educ:* Univ Sheffield, Eng, BSc, 64; George Washington Univ, PhD(microbiol), 78. *Prof Exp:* Sci officer, Microbiol Res Establishment, Porton Down, Eng, 64-70; res assoc, 70-73, res scientist, 73-80, sr res scientist, 80-81, HEAD, TRANSMISSIBLE DIS LAB, AM RED CROSS, 81- *Concurrent Pos:* Lectr, Univ Bristol, Eng, 64-65, Am Red Cross, 72-86. *Mem:* Am Asn Immunologists; Sigma Xi; Am Soc Microbiol; Am Asn Blood Banks; Int Soc Blood Transfusion. *Res:* Epidemiology of infectious disease with special reference to blood transfusion; laboratory testing for retrovirus infection; hepatitis B and non A & B; viral inactivation. *Mailing Add:* Jerome H Holland Lab Am Red Cross 15601 Crabbs Branch Way Rockville MD 20855

DODDAPANENI, NARAYAN, b 42; c 2. PHYSICAL CHEMISTRY. *Educ:* Osmania Univ, India, BSc, 62, MSc, 64; Case Inst Technol, MS, 69; Case Western Reserve Univ, PhD(phys chem), 72. *Prof Exp:* Res assoc, Case Labs Electrochem Studies, Case Western Reserve Univ, 73-77; staff scientist, Gould Labs Energy Res, 77-79; SR PRIN RES SCIENTIST, HONEYWELL POWER SOURCES CTR, HONEYWELL INC, 80- *Mem:* Electrochem Soc; Am Chem Soc. *Res:* Electrocatalysis for aqueous and non-aqueous power supplies. *Mailing Add:* Sandia Nat Labs Orgn 2523 Albuquerque NM 87185

DODDS, ALVIN FRANKLIN, b Starkville, Miss, Jan 20, 19; m 43; c 3. BIOCHEMISTRY, PHARMACEUTICAL CHEMISTRY. *Educ:* Miss State Univ, BS, 40; Northwestern Univ, MS, 42, PhD(biochem), 43; Med Col SC, BS, 49. *Prof Exp:* Asst biochem, Dent Sch, Northwestern Univ, 40-43; res chemist, Pan Am Refinery Corp, Tex, 43-45; asst prof pharmaceut chem, Loyola Univ, 45-47; assoc prof, 47-52, prof, 52-81, EMER PROF PHARMACEUT CHEM, MED UNIV SC, 81- *Mem:* Am Chem Soc; Am Pharmaceut Asn. *Res:* Synthesis of local anesthetics; toxicity of synthetic vitamin K; metabolism of mouth bacteria; medicinal chemistry. *Mailing Add:* 425 Geddes Ave Charleston SC 29407

DODDS, DONALD GILBERT, b North Rose, NY, Oct 4, 25; m 45; c 2. WILDLIFE BIOLOGY, ECOLOGY. *Educ:* Cornell Univ, BSc, 53, MSc, 55, PhD(wildlife mgt), 60. *Prof Exp:* Asst, Cornell Univ, 53-55 & 58-60; wildlife biologist, Dept Mines & Resources, Nfld, 55-58; vis prof, Acadia Univ, 61-64, prof biol, 70-75, dean sci, 75-79, from assoc prof to prof wildlife biol, 64-87; RETIRED. *Concurrent Pos:* UN Develop Prog & Food & Agr Orgn wildlife & parks policy, Zambia, 66-67, 72, Kenya, 74, Ethiopia, 81, Botswana, 83, Trinidad & Tobago, 83-84; resource consult, Eastern Can, 67-; big game biologist, Dept Land & Forests, NS, 60-63, asst dir wildlife conserv, 63-64, actg dir, 64-65. *Mem:* Wildlife Soc; Can Soc Wildlife & Fishery Biol; Can Soc Zool. *Res:* Population ecology study of snowshoe hare; ecological and management studies of white-tailed deer, moose, weasel, fox, Hungarian partridge, beaver and lynx; forest-wildlife interrelationships and population ecology; resource and environmental planning studies. *Mailing Add:* Dept Biol Acadia Univ RR 1 Site 2 Box 149 Coldbrook NS B0P 1K0 Can

DODDS, JAMES ALLAN, b Sunderland, UK, June 29, 47. PLANT VIROLOGY. *Educ:* Leeds Univ, BSc, 69; McGill Univ, MSc, 72, PhD(plant path), 74. *Prof Exp:* Nat Res Coun Can fel, Agr Can Res Sta, Vancouver, 74-76; asst plant pathologist virol, Conn Agr Exp Sta, 76-80; asst prof, 80-82, ASSOC PROF, UNIV CALIF, RIVERSIDE, 82- *Mem:* Am Phytopathological Soc; Can Phytopathological Soc; Int Orgn Citrus Virologists. *Res:* Viruses of algae and fungi and their use in biological control; consequences of mixed virus infections on viruses and plants; cross protection, induced resistance, vegetable and citrus viruses. *Mailing Add:* 735 Glenhill Dr Riverside CA 92507

DODDS, STANLEY A, b Toledo, Ore, Jan 26, 47. SOLID STATE PHYSICS. *Educ:* Harvey Mudd Col, BS, 68; Cornell Univ, PhD(physics), 75. *Prof Exp:* Asst res physicist, Univ Calif, Los Angeles, 74-77; asst prof, 77-82, ASSOC PROF PHYSICS, RICE UNIV, 82- *Mem:* Am Phys Soc; Am Asn Physics Teachers. *Res:* Muon spin rotation. *Mailing Add:* Physics Dept Rice Univ Box 1892 Houston TX 77251

DODDS, WELLESLEY JAMISON, b Faulkton, SDak, Oct 18, 15; m 37; c 4. ELECTRICAL ENGINEERING. *Educ:* SDak State Col, BS, 38; Univ Kans, MS, 41. *Prof Exp:* Asst physics, SDak State Col, 38-39, Univ Kans, 39-40 & Univ Ill, 41-42; power tube design engr, Victor Div, Radio Corp Am, Pa, 42-44, res engr, Labs Div, NJ, 45-54, mgr microwave solid state eng, 63-66, mem staff electronic components & devices, 66-70, mem tech planning staff, Picture Tube Div, 70-76, dir, prod quality, safety & reliability assurance & dir, prod qual, oper & analysis, 76-81; CONSULT QUAL ASSURANCE & RELIABILITY, 81- *Concurrent Pos:* Consult, US Dept Defense, 55- & Nat Acad Sci, 63. *Mem:* AAAS; Inst Elec & Electronics Engrs. *Res:* Invention and research in microwave radio; electrets; biophysics of nerve. *Mailing Add:* 5 Ashlea Village Conestoga St New Holland PA 17557

DODERER, GEORGE CHARLES, b Monticello, NY, Aug 3, 28; m 58; c 4. CHEMISTRY. *Educ:* Union Univ (NY), BS, 50; Univ Louisville, MS, 73. *Prof Exp:* Asst physics prog, Gen Elec Co, NY, 50-51, spectroscopist metals, Mass, 51-52, researcher vacuum insulation, Pa, 52-54, instrumental analyst, Ky, 54-55; guest worker polymers, Nat Bur Standards, 56-57; instrumental chem analyst, 58-68, mgr anal chem, 68-71, mgr chem develop, 71-77, mgr chem res & develop, 77-79, MGR ENG CHEM, GEN ELEC CO, 80- *Mem:* Soc Appl Spectros; Am Chem Soc; Sigma Xi. *Res:* Instrumental chemical analysis; chemistry of hermetic refrigerating systems; flat panel vacuum thermal insulation; vacuum technology. *Mailing Add:* 7100 Shefford Lane W Louisville KY 40242

DODGE, ALICE HRIBAL, b Fullerton, Calif; c 2. ANATOMY, CELL BIOLOGY. *Educ:* Univ Calif, Berkeley, AB, 48, MA, 51; Stanford Univ, PhD(anat), 69. *Prof Exp:* Res asst, Sch Med, Stanford Univ, 60-68, res assoc & res scientist anat, 70-77; from instr to assoc prof, 69-78, PROF, CALIF COL PODIAT MED, 78- *Concurrent Pos:* Nat Cancer Inst grant, 71-76. *Mem:* AAAS; NY Acad Sci; Am Soc Cell Biol; Histochem Soc; Am Asn Cancer Res; Am Anat Asn; Hamster Soc. *Res:* Subcellular localization of enzymes; light and electron microscopic, electrophoretic, cytogenetic and immunohistolcyto chemical study of androgen/estrogen and estrogen induced hamster tumors and hamster fetal tissues; renin immunohistochemical study of hamster vascular smooth muscle. *Mailing Add:* 2038 Maryland St Redwood City CA 94061

DODGE, AUSTIN ANDERSON, pharmacy; deceased, see previous edition for last biography

DODGE, CARROLL WILLIAM, botany; deceased, see previous edition for last biography

DODGE, CHARLES FREMONT, b Dallas, Tex, May 28, 24; m 50; c 2. GEOLOGY. *Educ:* Southern Methodist Univ, BS, 49, MS, 52; Univ NMex, PhD, 67. *Prof Exp:* Valuation engr, T Y Pickette, 47-48; instr geol, Arlington Col, 48-50; geologist, Concho Petrol Co, 50-52 & Intex Oil Co, 52-53; district geologist, Am Trading & Prod Corp, 53-57; from assoc prof to prof geol, Univ Tex, Arlington, 57-77; SR SCIENTIST & SR FEL, ISEM-SOUTHERN METH UNIV, 77-; PRES, C F DODGE & ASSOC INC, 79- *Concurrent Pos:* Consult, Core Labs, 58, McCord & Assocs & Lewis Eng, 59, Atlantic Refining Co, 59-61, Sun Oil Co, 66-67, Coastal Plains Oil, 68, Tex Steel, 69 & Sonatrach (Algeria), 69-70; NSF sci fac fel, 65-66; res proj 91A, Am Petrol Inst, 66-70; vpres explor, Arkomagas Co, 77-; chmn, dept geol, Univ Tex, 74-76. *Honors & Awards:* Distinguished Serv Award, Am Asn Petro Geol, 86. *Mem:* Am Asn Petrol Geol (pres, Southwestern Sect, 86-87); Am Inst Prof Geol; Soc Explor Geophys; Soc Independent Prof Earth Sci; Sigma Xi. *Res:* Reservoir geology; clastic sedimentology. *Mailing Add:* 1301 Briarwood Arlington TX 76013

DODGE, DONALD W(ILLIAM), b Worcester, Mass, Aug 29, 28; m 50; c 2. TECHNICAL MANAGEMENT. *Educ:* Worcester Polytech Inst, BS, 50, MS, 52; Univ Del, PhD(chem eng), 58. *Prof Exp:* Chem engr, Arthur D Little, Inc, 52-53; res engr, E I du Pont de Nemours & Co, 57-60, staff engr, 60-62, supvr res, 62-63, mgr, 63-66, tech supt, 66-67, mfg supt, 67-68, prod mgr film dept, 68-71, tech mgr, 71-75, bus mgr, Polymer Prod Dept, 75-80, mgr, New Bus Develop, 80-83, tech dir, Res & Develop Div, Polymer Prod Dept, 83-85; CONSULT, 85- *Mem:* Am Inst Chem Engrs; NY Acad Sci; Sigma Xi. *Res:* Plastics processing and technology; fluid mechanics; rheology. *Mailing Add:* 330 Brockton Rd Wilmington DE 19803

DODGE, E(LDON) R(AYMOND), civil engineering, hydraulics & hydrology; deceased, see previous edition for last biography

DODGE, FRANKLIN C W, b Oakland, Calif, Sept 18, 34; m 53; c 5. GEOLOGY. *Educ:* Univ Calif, Berkeley, BA, 59; Stanford Univ, MS, 60, PhD(geol), 63. *Prof Exp:* GEOLOGIST, US GEOL SURV, 63- *Mem:* Geol Soc Am; Mineral Soc Am; Mineral Asn Can; Norweg Geol Soc; Soc Econ Geol. *Res:* Geological study of the Sierra Nevada batholith. *Mailing Add:* 65 Linario Way Menlo Park CA 94028

DODGE, FRANKLIN TIFFANY, b Uniontown, Pa, Nov 11, 36; m 68; c 2. MECHANICAL ENGINEERING, FLUID MECHANICS. *Educ:* Univ Tenn, BS, 60; Carnegie Inst Tech, MS, 61, PhD(mech eng), 63. *Prof Exp:* Coop engr, Pittsburgh Des Moines Steel Co, 56-59; res asst fluid mech, Carnegie Inst Tech, 60-61; sr res engr dept mech sci, Southwest Res Inst, 63-70, group leader fluid mech & eng anal, 70-71; asst prof mech & aerospace eng, Univ Tenn, Knoxville, 71-73; staff engr, 73-82, INST ENGR, SOUTHWEST RES INST, 82- *Concurrent Pos:* Chmn, fluid transients comt, Am Soc Mech Engrs, 85-86. *Mem:* Assoc fel Am Inst Aeronaut & Astronaut; fel Am Soc Mech Engrs. *Res:* Hydrodynamics; nuclear reactor safety; lubrication; oil production equipment; fluid dynamics of pollution; spacecraft propellant management. *Mailing Add:* Div Mech & Fluids Eng 6220 Culebra Rd San Antonio TX 78284

DODGE, HAROLD T, b Seattle, Wash, May 20, 24; m 51; c 2. INTERNAL MEDICINE, CARDIOLOGY. *Educ:* Harvard Univ, MD, 48. *Prof Exp:* Intern med, Peter Bent Brigham Hosp, Mass, 48-49; resident, King County Hosp, Wash, 49-50; asst, Sch Med, Univ Wash, 50-51; clin investr, Nat Heart Inst, Md, 51-56; asst prof, Sch Med, Duke Univ, 56-57; from asst prof to prof, Sch Med, Univ Wash, 57-66; prof & dir cardiovasc div, Med Ctr, Univ Ala, 66-69; PROF MED & DIR CARDIOVASC RES & TRAINING CTR, SCH MED, UNIV WASH, 69-, CO-DIR DIV CARDIOL, 71- *Concurrent Pos:* Res fel med, Sch Med, Univ Wash, 50-51; fel cardiol, Emory Univ, 52-53; instr, Sch Med, Georgetown Univ, 53-56; chief cardiol, Durham Vet Admin Hosp, 56-57 & Seattle Vet Admin Hosp, 57-66. *Mem:* Am Fedn Clin Res; Am Soc Clin Invest; Asn Univ Cardiol; Asn Am Physicians. *Res:* Cardiovascular research; physiological aspects of heart disease; epidemiology of cardiovascular disease; drug research. *Mailing Add:* Dept Med Univ Wash Sch Med Seattle WA 98195

DODGE, JAMES STANLEY, b Washington, DC, June 22, 39; m 61; c 2. COLLOID CHEMISTRY, EMULSION POLYMERIZATION. *Educ:* Case Inst Technol, BS, 61, MS, 66; Case Western Reserve Univ, PhD(macromolecular sci), 69. *Prof Exp:* Res engr, TRW, Inc, 61-63; sr res chemist, Sherwin-Williams Co, 69-74, staff scientist, 74-76; res assoc, 76-80, sr res & develop assoc, 80-88, RES & DEVELOP FEL, B F GOODRICH CHEM CO, 88- *Mem:* Am Chem Soc; Soc Rheol. *Res:* Synthesis and colloidal properties of polymer latexes; rheology of polymers and colloids. *Mailing Add:* B F Goodrich Chem Co Tech Ctr PO Box 122 Avon Lake OH 44012

DODGE, PATRICK WILLIAM, b Sleepy Eye, Minn. June 2, 36; m 62; c 3. PHARMACOLOGY. *Educ:* Univ Minn, BS, 59, PhD(pharmacol), 67. *Prof Exp:* Scientist drug metab, 67-69, from group leader to head anti-inflammatory analgesic sect, 69-73, dept mgr pharmacodynamics, 73-74, dept mgr pharmacol, 74-75, dept mgr pharmacol & med chem, 75-77, DEPT MGR PHARMACOL, ABBOTT LABS, 77- *Mem:* Am Soc Pharmacol & Exp Therapeut; Sigma Xi. *Res:* Planning and administration of research associated with discovery of medicinal agents for therapeutic use in cardiovascular, gastrointestinal and central nervous system diseases. *Mailing Add:* Abbott Labs D46R AP9 Abbott Park IL 60064

DODGE, PHILIP ROGERS, b Beverly, Mass, Mar 16, 23; div; c 3. PEDIATRIC NEUROLOGY. *Educ:* Univ Rochester, MD, 48. *Prof Exp:* From instr to asst prof neurol, Harvard Med Sch, 57-67; head Edward Mallinckrodt dept pediat, 67-86, PROF PEDIAT & NEUROL, SCH MED, WASH UNIV, 67- *Concurrent Pos:* Pediat neurologist, Boston Lying-In Hosp, 61-67; investr, J P Kennedy, Jr Lab Study Ment Retardation, 63-67; consult, Fernald State Sch Retarded Children, 63-67; mem, Surgeon Gen Comt Epilepsy, USPHS, 66-70, mem child develop & ment retardation rev comt, 66-70, mem gen clin res ctrs comt, 71-74; med dir, St Louis Childrens

Hosp, 67-84; neurologist, Barnes & Allied Hosps, 67-; mem, Ment Health Comn, State of Mo, 74-78 & Nat Adv Child Health & Human Develop Coun, NIH, 74-77; spec asst to dir mental retardation, Nat Inst Children's Health & Human Develop, NIH, 87-88. *Honors & Awards:* Hower Award, Child Neurol Soc, 78. *Mem:* Am Acad Neurol; Am Neurol Asn; Soc Pediat Res; Am Pediat Soc. *Res:* Pediatric neurology; clinical and laboratory investigations of neurologic disorders of childhood, especially infectious, nutritional and metabolic diseases. *Mailing Add:* Dept Pediat Washington Univ Med Sch 400 S Kingshighway St Louis MO 63130

DODGE, RICHARD E, b Machias, Maine, Mar 4, 47; m 73; c 2. CORAL REEF ASSESSMENT. *Educ:* Univ Maine, BA, 69; Yale Univ, MPhil, 73, PhD(geol & geophys), 78. *Prof Exp:* Cur paleontol, Peabody Mus Natural Hist, Yale Univ, 78; asst prof, 78-81, dir, Inst Marine & Coastal Studies, 85, ASSOC PROF OCEAN SCI, NOVA UNIV OCEANOG CTR, 81-, ASSOC DIR, 85- *Concurrent Pos:* Geol Soc Am grant, 74-75; NSF grants, 75-77, 79-81, 84-85, 85-86 & 89-90, Environ Protection Agency grant, 80-82, Nat Oceanic Atmospheric Admin grant, 80-82, Exxon, Bermuda Biol Sta grant, 81-82, Am Petrol Inst Grants, 84-86 & White Hall Found grants, 85-86, Seagrant grant, 89-92, USCS, 89-91, Broward County, Fla, 87-94 & Biol Monitoring, various consult damage caused to coral reefs from ship groundings; subcontract, Dept Energy-Ocean Thermal Energy Conversion, Univ Miami, 80-81; field work, Red Sea-Saudia Arabia, Vicques, Puerto Rico, St Croix & US Virgin Islands; adv ed, Coral Reefs. *Mem:* Am Soc Shore & Beach; Int Soc Reef Studies; AAAS. *Res:* Ecology, paleoecology, paleoclimatology, and paleobiology of corals and coral reefs; relation of coral growth rate to the environment for recent and fossil ecology studies; geology of coral reefs to include structure, zonation, morphology, dating, sea level changes. *Mailing Add:* Nova Univ Oceanog Ctr 8000 N Ocean Dr Dania FL 33004

DODGE, RICHARD PATRICK, b Wichita, Kans, Mar 17, 32. PHYSICAL CHEMISTRY. *Educ:* Univ Wichita, BS, 54; Univ Calif, PhD, 58. *Prof Exp:* Asst chem, Univ Calif, 54-55, assoc, 55-56, asst, Lawrence Radiation Lab, 56-58; res chemist, Union Carbide Res Inst, 58-64; from asst prof to assoc prof, 64-78, PROF CHEM, UNIV PAC, 78- *Mem:* AAAS; Am Crystallog Soc. *Res:* X-ray diffraction; molecular structure; quantum chemistry; computer calculations. *Mailing Add:* Dept Chem Univ of the Pac 3601 Pacific Ave Stockton CA 95211

DODGE, THEODORE A, b Chicago, Ill, Jan 17, 11. DRILLING. *Educ:* Harvard Univ, BA, 32, PhD(geol), 36. *Prof Exp:* PRES, HOAGLAND & DODGE DRILLING INC, 54- *Mailing Add:* 1770 N Potter Pl Tucson AZ 85719

DODGE, WARREN FRANCIS, b Scottdale, Pa, May 5, 28; m 49, 82; c 3. PEDIATRICS, PHARMACOLOGY. *Educ:* Univ Tenn, BS, 53, MD, 55. *Prof Exp:* Intern, Jefferson Davis Hosp, 55-56; resident pediat, Baylor Col Med, 56-58; from asst prof to assoc prof, 60-72, PROF PEDIAT, UNIV TEX MED BR GALVESTON, 72- *Concurrent Pos:* Jessie Jones fel, 58; fel renal dis of childhood, Baylor Col Med, 58-60. *Mem:* Soc Pediat Res; Am Pediat Soc. *Res:* Clinical pharmacology; epidemiology of renal disease and hypertension; delivery of health care. *Mailing Add:* Univ Tex Med Br Galveston TX 77550

DODGE, WILLIAM HOWARD, b Moss Point, Miss, Feb 20, 43; div; c 1. ONCOLOGY. *Educ:* Millsaps Col, BS, 65; Univ Miss, MS, 67, PhD(microbiol), 70. *Prof Exp:* Asst prof microbiol, Winston-Salem State Univ, 73-75; res asst prof med, 75-81, ASSOC PROF EXP MED, BOWMAN GRAY SCH MED, WAKE FOREST UNIV, 81- *Concurrent Pos:* Fel, Vet Admin Hosp & Sch Med, Univ Fla, 70-73; vis scientist, Tumor Biol Lab, Nat Cancer Inst, 76; bd mem, Southeastern Cancer Res Asn. *Mem:* Int Asn Comp Res Leukemia; Int Soc Exp Hematol; Am Asn Cancer Res; Sigma Xi. *Res:* Factors which regulate the growth and differentiation of hematopoietic cells; how alterations in the levels of various factors contribute to leukemia progression; leukemogenic oncornaviruses. *Mailing Add:* Dept Med Bowman Gray Sch Med Winston-Salem NC 27103

DODGE, WILLIAM R, b Oregon City, Ore, Mar 21, 29; m 55. PHYSICS. *Educ:* Stanford Univ, BS, 56, PhD(physics), 62. *Prof Exp:* PHOTONUCLEAR PHYSICIST, NAT BUR STANDARDS, 61- *Mem:* Am Phys Soc; Sigma Xi. *Res:* Electron and photon induced nuclear disintegration reactions; semiconductor radiation detectors; linear accelerator physics. *Mailing Add:* 8200 Raymond Lane Potomac MD 20854

DODGEN, DURWARD F, b Winnsboro, Tex, Apr 10, 31; m 54; c 2. CHEMISTRY. *Educ:* N Tex State Univ, BS, 52; Univ Miss, MS, 57. *Prof Exp:* Res chemist, Phillips Petrol Co, 52-55; anal chemist, Am Pharmaceut Asn Lab, 57-59; med serv rep, Chas Pfizer & Co, Inc, 59-60; pub coordr, Warren-Teed Pharmaceut Div, Rohm and Haas Co, 60-61; asst dir, Food Chem Codex, Nat Acad Sci-Nat Res Coun, 61-65, assoc dir, 65-66; asst dir nat formulary & asst dir sci div, Am Pharmaceut Asn, 66-69, assoc ed, J Pharmaceut Sci, 67-69; dir, Food Chem Codex, Dir GRAS & Food Additives Surv & Staff Officer, Food & Nutrit Bd, Div Biol Sci, Nat Acad Sci-Nat Res Coun, 69-81; STAFF SCIENTIST, KELLER & HECKMAN LAW OFFICES, 81- *Concurrent Pos:* Observer, Joint Expert Comt Food Additives, Food & Agr Orgn, WHO, 64, 65, 67 & 69-72, mem, 73-; mem, Nat Formulary Bd, 66-75; mem, US Adopted Names Coun, 67-69 & 74-77; mem adv panel pharmaceut ingredients, US Pharmacopoeia, 70-75; assoc mem food additives comt, Int Union Pure & Appl Chem, 71-75, titular mem, 75-80; mem-at-large, US Pharmacopeial Conv, 80-; mem fac, Toxicol Forum, 81; mem adv comt, Int Prog Chem Safety, Food & Drug Admin, 81. *Mem:* Am Chem Soc; Inst Food Technologists. *Res:* Food additive specifications and patterns of use; pharmaceutical analysis; standards for food additives and drugs. *Mailing Add:* 5902 Mt Eagle Dr Apt 909 Alexandria VA 22303-2529

DODGEN, HAROLD WARREN, b Blue Eye, Mo, Aug 31, 21; m 45; c 3. CHEMICAL PHYSICS. *Educ:* Univ Calif, BS, 43, PhD(phys chem), 46. *Prof Exp:* Asst, Manhattan Dist Proj, Univ Calif, 43-46; Inst Nuclear Studies fel, Univ Chicago, 46-48; from asst prof to assoc prof, 48-59, prof chem, 59-66, EMER PROF CHEM & PHYSICS, WASH STATE UNIV, 66- *Concurrent Pos:* dir, nuclear reactor proj, Wash State Univ, 55-68, chmn chem physics, 68-77. *Honors & Awards:* Fel AAAS. *Mem:* Am Chem Soc; Am Phys Soc; AAAS; Sigma Xi. *Res:* Theory of ammonium chloride phase transition; complex ions of iron and thorium with fluoride; neutron scattering; radioactive exchange; quenching of fluorescence; application of radioisotopes in physical chemistry; spectrochemical analysis; nuclear magnetic and quadrupole resonance. *Mailing Add:* Dept Chem Wash State Univ Pullman WA 99163

DODINGTON, SVEN HENRY MARRIOTT, b Vancouver, BC, May 22, 12; nat US, m 40; c 3. NAVIGATION, ELECTRONICS. *Educ:* Stanford Univ, AB, 34. *Prof Exp:* From jr engr to head elec dept, Scophony Ltd, London, 35-41; dept head, Int Tel & Tel Fed Labs, 41-50, div head, 51-54, lab dir, 54-58, vpres, 58-69, asst tech dir, 69-75, avionics consult, Int Tel & Tel Corp, 75-85; RETIRED. *Honors & Awards:* Pioneer Award, Inst Elec & Electronics Engrs, 80. *Mem:* Assoc fel Am Inst Aeronaut & Astronaut; fel Inst Elec & Electronics Engrs; Am Inst Navig. *Res:* Navigation; countermeasures; television; microelectronics. *Mailing Add:* One Briarcliff Rd Mountain Lakes NJ 07046

DODSON, B C, b Magnolia, Ark, Dec 6, 24; m 46; c 3. INORGANIC CHEMISTRY, SCIENCE EDUCATION. *Educ:* Ark State Teachers Col, BSE, 48; Univ Ark, MS, 58; Kans State Teachers Col, EdS, 61; Univ Okla, EdD, 69. *Prof Exp:* High sch teacher, Ark, 48-54 & Tex, 54-55; teacher, Ark, 55-60, sci coordr, 56-60; asst prof chem, Southern Ark Univ, 60-70, actg head, Dept Chem, 66-69, chmn, Div Sci & Math, 70-73, prof chem & sci educ, 70-87, dean, Sch Sci & Technol, 77-87; RETIRED. *Concurrent Pos:* NSF sci fac fel, 65-66. *Mem:* Am Chem Soc; Nat Sci Teachers Asn; Nat Educ Asn; Asn Educ Teachers Sci (treas, 70-71). *Res:* Analysis of the objectives, materials and methods used in the introductory college chemistry course in selected colleges and universities; evaluation of individualized science modules. *Mailing Add:* 120 Columbia 223 PO Box 1397 Magnolia AR 71753

DODSON, CHARLES LEON, JR, b Knoxville, Tenn, Mar 15, 35; m 58; c 2. PHYSICAL CHEMISTRY. *Educ:* Emory & Henry Col, BS, 57; Univ Tenn, MS, 62, PhD(chem), 63. *Prof Exp:* Asst prof chem, 66-67, head dept, 68-70, assoc prof, Univ Ala, Huntsville, 67-81; tech support specialist, 81-83, prin appl chemist, 83-87, SR STAFF SCIENTIST, BECKMAN INSTRUMENTS, INC, 88- *Concurrent Pos:* European Off, US Off Aerospace Agency grant, Univ Birmingham, 63-64; Nat Res Coun fel, Can, 64-66; vis, Dept Theoret Chem, Univ Oxford, 72-73. *Mem:* AAAS; Am Chem Soc; Am Phys Soc. *Res:* Infrared, microwave and electron spin resonance spectroscopy; liquid scintillation. *Mailing Add:* Beckman Instruments Inc 2500 Harbor Blvd MS B-33-D Fullerton CA 92634-3100

DODSON, CHESTER LEE, b Eastland, Tenn, Dec 18, 21; m 47; c 2. GEOLOGY, HYDROLOGY. *Educ:* WVa Univ, BS, 50, MS, 53. *Prof Exp:* Geologist, US Geol Surv, 52-63; dir, Water Res Inst, WVa Univ, 63-85, asst prof hydrol & geol, 63-86; RETIRED. *Mem:* AAAS; Geol Soc Am; Am Geophys Union; Nat Water Well Asn; Am Water Resources Asn. *Res:* Groundwater geology; geology of mineral deposits; water resources. *Mailing Add:* 985 Hawthron Rd Cookeville TN 38501-1553

DODSON, DAVID SCOTT, b Pratt, Kans, Dec 15, 42; m 64; c 2. NUMERICAL ANALYSIS, SOFTWARE ENGINEERING. *Educ:* Kans State Univ, BS, 64; Purdue Univ, MS, 68, PhD(computer sci), 72. *Prof Exp:* Mem tech staff, TRW Space Technol Lab, 64-66; assoc mgr user serv, Comput Ctr, Purdue Univ, 70-72, mgr, 72-79; sr specialist engr, Boeing Computer Serv, 79-86; MGR MATH SOFTWARE, CONVEX COMPUTER CORP, 86- *Mem:* Asn Comput Mach; Soc Indust & Appl Math. *Res:* Vector and parallel algorithms in computational linear algebra, with a goal of improving performance on existing and planned supercomputers. *Mailing Add:* Convex Computer Corp PO Box 833851 Richardson TX 75083-3851

DODSON, EDWARD O, b Fargo, NDak, Apr 26, 16; m 40; c 6. EVOLUTIONARY BIOLOGY, GENETICS. *Educ:* Carleton Col, BA, 39; Univ Calif, PhD(zool), 47. *Prof Exp:* Asst zool, Univ Calif, 39-46; instr, Dominican Col, 46-47; from instr to assoc prof, Univ Notre Dame, 47-57; from assoc prof to prof, 57-81, EMER PROF BIOL, UNIV OTTAWA, 81- *Concurrent Pos:* Lectr, Univ Calif, Los Angeles, 45; vis prof, Univ Montreal, 62, Fondation Teilhard de Chardin, Paris, 71-72, Sta Zool, Italy, 79, Lab Arago, France, 79, Sta biol d'Arcachon, France, 79 & Sta biol de Roscoff, France, 79; vis res prof, Roswell Park Mem Inst, 64-65. *Mem:* AAAS; Genetics Soc Am; Soc Study Evolution; Genetics Soc Can. *Res:* Chromosomes of vertebrates; evolution; mutation; tissue culture. *Mailing Add:* Dept Biol Univ Ottawa Ottawa ON K1N 6N5 Can

DODSON, NORMAN ELMER, b Gregory, Ky, Oct 18, 09; m 36; c 1. MATHEMATICS. *Educ:* Berea Col, AB, 33; Univ Ala, AM, 47. *Prof Exp:* High sch teacher, Ky, 33-34 & 35-37; headmaster & teacher math, Jefferson Mil Col, 38-42; assoc prof, Newberry Col, 42-45; asst, Univ Ala, 45-46, asst prof statist, 46-47; assoc prof math, Lenoir-Rhyne Col, 47-57; from asst prof to assoc prof, Wittenberg Univ, 57-75; adj prof math, 75-77, instr math & math lab tutor, Lander Col, 79-82; RETIRED. *Concurrent Pos:* High sch teacher, Fla, 37-39. *Mem:* Math Asn Am. *Res:* Algebra and new trends in mathematical education. *Mailing Add:* Heritage Hills Retirement Ctr 1110 Marshall Rd Greenwood SC 29646

DODSON, PETER, b Ross, Calif, Aug 20, 46; m 68; c 2. PALEOBIOLOGY, VERTEBRATE ANATOMY. *Educ:* Univ Ottawa, BSc, 68; Univ Alta, MSc, 70; Yale Univ, PhD(geol), 74. *Prof Exp:* Curatorial assoc vert paleont, Peabody Mus Nat Hist, Yale Univ, 73-74; assoc anat, 74-75, ASST PROF VET ANAT, SCH VET MED, UNIV PA, 75- *Mem:* Soc Vert Paleont;

Paleont Soc; Am Soc Ichthyol & Herpet; Soc Study Evolution; Soc Syst Zool. *Res:* Relationships of morphometric anatomy to function, behavior and ecology of living and extinct vertebrates; taphonomic biases in the fossil record; dinosaurs. *Mailing Add:* 4502 Regent St Philadelphia PA 19143

DODSON, RAYMOND MONROE, b West Hazleton, Pa, July 8, 20; m 43; c 4. ORGANIC CHEMISTRY. *Educ:* Franklin & Marshall Col, BS, 42; Northwestern Univ, PhD(org chem), 47. *Prof Exp:* Asst prof org chem, Univ Minn, 47-51; chemist, G D Searle & Co, 51-55, asst to dir chem res, 55-60; PROF CHEM, UNIV MINN, MINNEAPOLIS, 60-; AT DEPT CHEM, FT HARE UNIV, SAFRICA. *Concurrent Pos:* Mem endocrinol study sect, NIH, 62-66. *Mem:* AAAS; fel Am Chem Soc; NY Acad Sci; The Chem Soc. *Res:* Heterocyclic compounds; organic sulfur-containing compounds; steroids; stereochemistry of cholesterol and its derivatives; fermentation of steroids; reactions of sulfur monoxide and disulfur monoxide. *Mailing Add:* Riverview Tower 1920 S First St Minneapolis MN 55454-1056

DODSON, RICHARD WOLFORD, b Kirksville, Mo, Jan 15, 15; m 37; c 2. INORGANIC CHEMISTRY, NUCLEAR CHEMISTRY,. *Educ:* Calif Inst Technol, BS, 36; Johns Hopkins Univ, PhD(chem), 39. *Prof Exp:* Am Can Co fel, Johns Hopkins Univ, 36-40; Nat Res fel, Calif Inst Technol, 40, mem staff, Nat Defense Res Comt Proj, Calif Inst Technol & Northwestern Univ, 40-43; group leader & later asst div leader, Los Alamos Lab, NMex, 43-45; asst prof chem, Calif Inst Technol, 46-47; chemist, Brookhaven Nat Lab, 47-68, from assoc prof to prof, 47-82, EMER PROF CHEM, COLUMBIA UNIV, 82- *Concurrent Pos:* Sr chemist, Brookhaven Nat Lab, 48-82; secy, Gen Adv Comt, AEC, 51-56. *Mem:* AAAS; Am Phys Soc; Am Chem Soc. *Res:* Reaction kinetics; radiochemistry. *Mailing Add:* 336 Oak Leaf Circle Santa Rosa CA 95409-6202

DODSON, ROBIN ALBERT, b Palmer, Alaska, Dec 15, 40; c 5. PHARMACOLOGY. *Educ:* Eastern Wash Univ, BS, 70; Wash State Univ, PhD(pharmacol sci), 78; Idaho State Univ, BS, 81. *Prof Exp:* Ward supvr, Eastern State Hosp, 65-71; bio-lab coordr, Eastern Wash Univ, 70-71; teaching asst pharmacol, Wash State Univ, 72-75; from asst prof to assoc prof, 75-86, PROF PHARMACOL, COL PHARM, IDAHO STATE UNIV, 86-; CHIEF ACAD OFF, IDAHO STATE BD EDUC, 86- *Concurrent Pos:* Consult, State Idaho, Dept Health & Welfare, Substance Abuse & Drug Abuse Prev, 76-; consult, Alcohol Rehab, Ctr I, 76-; vis scientist, Univ Wash Sch Med; Statewide Comt Drugs, AIDS, Hwy Safety, 86-; reviewer J Pharm Sci, 88- *Mem:* AAAS; Am Asn Col Pharm; Sigma Xi; Res Soc Alcoholism; Soc Neurosci; Am Soc Pharmacol Exp Ther. *Res:* Pharmacology of commonly used chemicals on various aspects of the central nervous system; how ethanol and barbiturates effect the biochemistry of the brain; how drugs affect the deformability of red blood cells. *Mailing Add:* Col Pharm Idaho State Univ Pocatello ID 83209-0009

DODSON, RONALD FRANKLIN, b Paris, Tex, Feb 14, 42; m 65. ELECTRON MICROSCOPY, CYTOLOGY. *Educ:* ETex State Univ, BA, 64, MA, 65; Tex A&M Univ, PhD(biol electron micros), 69. *Prof Exp:* Asst instr biol, ETex State Univ, 64-65; instr, Electron Micros Ctr, Tex A&M Univ, 65-69; res assoc anat, Med Sch, Univ Tex, San Antonio, 69-70; asst prof neurol, Baylor Col Med, 70-77; CHIEF, DEPT CELL BIOL & ENVIRON SCI, UNIV TEX HEALTH CTR, 77-, ASST DIR RES, 83- *Concurrent Pos:* Res fel, Electron Micros Ctr, Tex A&M Univ, 65-69; fel stroke coun, Am Heart Asn. *Mem:* AAAS; Am Chem Soc; Am Thoracic Soc; Int Acad Path; NY Acad Sci; fel Am Col Chest Physicians. *Res:* Ultrastructural studies of experimental pathology in respiratory diseases and environmental sciences and cerebrovascular diseases. *Mailing Add:* Univ Tex Health Ctr PO Box 2003 Tyler TX 75710

DODSON, VANCE HAYDEN, JR, b Oklahoma City, Okla, June 12, 22; m 53. INORGANIC CHEMISTRY. *Educ:* Univ Toledo, BChE, 44, MS, 47; Purdue Univ, PhD(chem), 52. *Prof Exp:* Mat engr, Curtiss-Wright Corp, 44-45; from instr to assoc prof chem, Univ Toledo, 45-57; asst instr, Purdue Univ, 48-50; res chemist, Elec Autolite Co, Ohio, 57-58, res mgr, Battery Div, 58-61; res mgr construct specialties, Dewey & Almy Chem Co, Cambridge, 61-63, dir res, Construct Mat Div, 63-70, tech serv dir, 70-85; CONSULT, 85- *Mem:* Am Chem Soc; Am Ceramic Soc; Am Concrete Inst; Am Soc Testing & Mat. *Res:* Silicates; elastomers; organic coatings; lead alloys; electrochemical sources of energy; portland cement; concrete. *Mailing Add:* PO Box 6 Federal Corner Rd Center Tuftonboro NH 03816-0006

DODSON, VERNON N, b Benton Harbor, Mich, Feb 19, 23; m 55; c 5. TOXICOLOGY, BIOCHEMISTRY. *Educ:* Marquette Univ, MD, 51; Univ Mich, BS, 52; Am Bd Prev Med, cert occup med, 69. *Prof Exp:* Asst pathologist, Sch Med, Johns Hopkins Univ, 53-54; from asst resident to resident internal med, Univ Hosp, Univ Mich, Ann Arbor, 54-56, jr clin instr, 56-57, instr internal med, Univ Mich, 57-61, from asst prof to assoc prof internal & industrial med & toxicol, 61-71; prof med & environ med, Med Col Wis, 71-72; vis prof prev med, Med Sch, Univ Wis, 73-74; prof prev med, Grad Sch, Univ Minn, Rochester-Minneapolis, 74-77; PROF PREV MED, MED SCH, UNIV WIS-MADISON, 77-, DIR UNIV HOSP EMPLOYEE HEALTH SERV, 77-, ACTG DIR CTR ENVIRON TOXICOL, HEALTH SCI DIV, 78- *Concurrent Pos:* Lectr, Inst Social Work, Univ Mich, 57-58, Sch Dent, 57-58, Sch Nursing, 58-60 & Dept Postgrad Med, 59, res assoc, Clin Radioisotope Unit, Med Sch, 50-60, res assoc, Inst Indust Health, 60-71; attend physician internal med, US Vet Admin Hosp, Ann Arbor, 63-71; consult, Gen Motors Corp Tech Ctr, Warren, Mich, 63-65 & 72-; mem numerous comts, Univ Mich Med Ctr, 63-71; med dir, Delco Electronics Div, Gen Motors Corp, Wis, 71-72; assoc physician, Trinity Mem Hosp, Cudahy, Wis, 71-73; mem comt toxicity determinants & lists, Bur Occup Safety & Health, USPHS, HEW, 71-72; consult, Atty Gen, Dept Natural Resources & Dept Health & Social Servs, State of Wis, 73-74, Oscar Mayer Co, Madison, 73-74, Forest Prod Div, Forest Serv, USDA, 73- & Nat Inst Occup Safety & Health, Bethesda, Md, 74-; consult plant physician, IBM Co, Rochester, Minn, 76-77; consult occup & environ health, Dept Health & Social Servs, State of Wis, 77-, mem gov's task force occup health & safety, 78-; consult prev med & internal med, Mayo Clin, 74-77, mem res comt, Div Prev Med, Dept Med, 74-77, mem educ comt, 74-77, chmn, 74-75, chmn ad hoc toxicol comt, 74-77, mem comt occup safety & health, 76-77; mem educ comt, Ctr Environ Toxicol, Health Sci Div, Univ Wis, 78-, mem infectious dis control comt, Univ Hosp, 78-; comnr, Int Comn Indust Health. *Mem:* Am Med Asn; Am Fedn Clin Res; Brit Biochem Soc; fel Am Col Physicians; fel Am Col Prev Med; Sigma Xi; fel Am Acad Occup Med; fel Am Occup Med Asn. *Res:* Auto-immunity; immunotoxicology; clinical medicine; thyroid metabolism and diseases; electrophoresis; rheumatology. *Mailing Add:* Dept Prev Med Univ Wis Med Ctr 504 N Walnut Madison WI 53706

DODZIUK, JOZEF, b Warsaw, Poland, Aug 7, 47; US citizen; m 69; c 1. MANIFOLDS, EQUATIONS. *Educ:* Columbia Univ, MA, 71, PhD(math), 73. *Prof Exp:* C L E Moore instr math, Mass Inst Technol, 73-75; asst prof math, Univ Pa, 76-81; assoc prof, 81-84, PROF MATH, QUEENS'S COL & GRAD CTR, CITY UNIV NY, 83- *Concurrent Pos:* Prin investr, NSF res grants, 73-86; vis prof, Univ Oxford, 80-81; fel, Sloan Found, 80-82. *Mem:* Am Math Soc. *Res:* Analysis on non compact manifolds; the theory of the lapace operator and analogues difference operators. *Mailing Add:* Dept Math Queens Col City Univ NY Flushing NY 11367

DOE, BRUCE R, b St Paul, Minn, Apr 24, 31; m 58, 90. GEOLOGY. *Educ:* Univ Minn, BS & BGeol E, 54; Mo Sch Mines, MS, 56; Calif Inst Technol, PhD(geol), 60. *Hon Degrees:* DSc, Univ Mo, 89. *Prof Exp:* Fel isotope chem, Geophys Lab, Carnegie Inst, 60-61; geologist, Washington, DC, US Geol Surv, 61-63 & 69-71, Colo, 63-68 & 71-81 & Switz, 68-69, chief, Br Isotope Geol, Colo, 76-81, asst chief geologist, Eastern Region, 84-86, asst res dir, 86-90, GEOLOGIST, US GEOL SURV, VA, 81- *Concurrent Pos:* Vis investr, Geophys Lab, Carnegie Inst, 61-62 & Japan-US Sci Coop Prog, 65; acad guest, Swiss Fed Inst Technol, Zurich, 68-69; staff scientist, Lunar Sample Prog, NASA, 69-71, mem lunar sample anal planning team, Houston, 70-71, vchmn, 72-73; mem, Comet & Asteroid Missions Adv Panel, 71-72 & Comet & Asteroid Sci Working Group, 72-73; coordr earth sci exhibs trailer, US Geol Surv, 72-73; reporter isotopes, US Geodynamics Comt, Nat Acad Sci, 77-81; ed, Essays & Studies, Am Geophys Union, 82-85; sr scientist degradation of mat due to acid rain, Nat Park Serv, 83-85. *Honors & Awards:* Vernadsky Mem lectr, Moscow-Leningrad, 76. *Mem:* Fel AAAS; fel Geol Soc Am; Geochem Soc (pres, 82); fel Soc Econ Geologists; Asn Women Geologists; Am Geophys Union. *Res:* Trace element and isotope geochemistry; radiogenic tracer analysis; volcanology; ore genesis. *Mailing Add:* US Geol Surv 923 Nat Ctr Reston VA 22092

DOE, FRANK JOSEPH, b Dover, NH, July 14, 37; m 73; c 2. GENETICS. *Educ:* Spring Hill Col, BS, 62; Brandeis Univ, PhD(biol), 68. *Prof Exp:* ASSOC PROF BIOL, UNIV DALLAS, 69- *Mem:* Genetics Soc Am. *Res:* Genetics of the mating-type loci in the yeast, Schizosaccharomyces pombe; general genetics of neurospora. *Mailing Add:* Dept Biol Univ Dallas 1845 E Northgate Dr Irving TX 75062

DOE, RICHARD P, b Minneapolis, Minn, July 21, 26; m 50; c 3. ENDOCRINOLOGY, INTERNAL MEDICINE. *Educ:* Univ Minn, BS, 49, MB, 51, MD, 52, PhD, 66. *Prof Exp:* Instr med, Univ Minn, Minneapolis, 52-66, assoc prof, 66-69, head endocrine sect, 69-76, prof med, Med Sch, 69-88. *Concurrent Pos:* Chief chem sect, Minneapolis Vet Admin Hosp, 55-60, chief metab & endocrine sect, 60-69, mem endocrine sect, 76-88. *Mem:* Endocrine Soc; Am Fedn Clin Res; Am Soc Clin Invest; Cent Soc Clin Res. *Res:* Role of glucocorticoid receptors in modulation of the human immune system; isolation; characterization and measurement of transcortin thyroxine-binding globulin and thyroxine-binding protein; cell mediated immunity in diabetes mellitus. *Mailing Add:* 5613 Hawkes Dr Edina MN 55436

DOEBBER, THOMAS WINFIELD, b Indianapolis, Ind, Mar 4, 46; m 72; c 2. NEUTROPHIL ACTIVATION. *Educ:* Bucknell Univ, BS, 68; Univ Rochester, PhD(biochem), 73. *Prof Exp:* Postdoctoral fel, Ames Res Ctr, NASA, Calif, 73-75; postdoctoral fel, Sch Med, Wash Univ, 75-79; SR RES FEL, MERCK SHARP & DOHME RES LAB, 79- *Mem:* AAAS; Am Soc Biochem & Molecular Biol; NY Acad Sci. *Res:* Characterization of neutrophil activation by platelet activating factor; discovery and development of receptor antagonist of platelet activating factor; discovery and development of antihyperglycemic agents. *Mailing Add:* Merck & Co PO Box 2000 Rahway NJ 07065

DOEBBLER, GERALD FRANCIS, b San Antonio, Tex, June 27, 32; m 57; c 7. BIOCHEMISTRY. *Educ:* Univ Tex, BS, 53, MA, 54, PhD(chem), 57. *Prof Exp:* Scientist, Biochem Inst Tex, 53-55; res chemist, Linde Co Div, 57-60, sr res chemist, 60-68, sr res chemist, Ocean Systs, Inc, 68-69, res assoc, 69-71, sr res scientist, Corp Res Dept, 71-76, SR RES SCIENTIST, CENT SCI LAB, UNION CARBIDE CORP, 76- *Mem:* Am Chem Soc; Am Soc Microbiol. *Res:* Biophysics, physiology of inert gases and decompression sickness; cryobiology; blood and cell preservation; analytical biochemistry; enzymology; low temperature spectroscopy; sulfur metabolism; biochemical individuality; analytical chemistry. *Mailing Add:* Dept Chem Incarnate Word Coll 4301 Broadway San Antonio TX 78209-6398

DOEDE, JOHN HENRY, b Chicago, Ill, Sept 29, 37; div; c 3. RESOURCE MANAGEMENT. *Educ:* Harvard Univ, AB, 59; Univ Chicago, MS, 62, PhD(chem), 63. *Prof Exp:* Res assoc high energy physics, Argonne Nat Lab, 63-64, asst physicist, 64-66; mgr prog, EMR Comput Div, Electro-Mech Res, Inc, 66-67; pres, Data Int Inc, Minn, 67-70; vpres, Heizer Corp, 70-72; VPRES, FIRST CAPITAL CORP, FIRST CHICAGO INVESTMENT CORP, 72-, DIR, 75- *Mem:* Am Phys Soc; Sigma Xi. *Res:* Weak interaction particle physics; laser development and applications; computer applications and data reduction; monetary investment to support economically viable and applicable research. *Mailing Add:* 785 Bangor St San Diego CA 92106

DOEDEN, GERALD ENNEN, b Onarga, Ill, Oct 16, 18; m 42; c 1. ANALYTICAL CHEMISTRY. *Educ:* Cent Normal Col (Ind), BS, 40; Ind Univ, MA, 50, PhD(anal chem), 65. *Prof Exp:* High sch teacher, Ind, 40-41; analyst, E I du Pont de Nemours & Co, Inc, 41-43; high sch teacher, Ariz, 45-46 & Ind, 46-56; from asst prof to assoc prof chem, 56-70; prof, 70-81, EMER PROF CHEM, BALL STATE UNIV, 81- *Mem:* Fel Am Inst Chem; Soc Appl Spectros. *Res:* Coulometric analysis; gas chromatography; spectrophotometry. *Mailing Add:* 3111 W Torquay Rd Muncie IN 47304-3234

DOEDENS, DAVID JAMES, b Milwaukee, Wis, July 2, 42; m 66; c 3. TOXICOLOGY. *Educ:* Univ Wis, BS, 65; Univ Ill, PhD(pharmacol), 71; Am Bd Clin Chem, cert toxicol chem. *Prof Exp:* Res assoc pharmacol, Sch Med, Ind Univ, Indianapolis, 70-71; res assoc toxicol, 71-72, lectr, 72-74, instr, 74-76, ASST PROF TOXICOL, DEPT PHARMACOL & TOXICOL, SCH MED, IND UNIV, INDIANAPOLIS, 76- *Concurrent Pos:* Asst dir, Ind State Dept Toxicol, 75- *Mem:* Am Acad Forensic Sci; Sigma Xi; Int Asn Forensic Toxicologists. *Res:* Development of analytical methods for drugs; mechanisms of drug toxicity; drug stability in biological media. *Mailing Add:* Dept Pharmacol & Toxicol Sch Med Ind Univ 1001 Walnut St MF 003 Indianapolis IN 46223

DOEDENS, ROBERT JOHN, b Milwaukee, Wis, Dec 15, 37; m 61; c 2. STRUCTURAL & COORDINATION CHEMISTRY. *Educ:* Univ Wis-Milwaukee, BS, 61; Univ Wis-Madison, PhD(chem), 65. *Prof Exp:* Res asst & instr chem, Northwestern Univ, 65-66; asst prof, 66-72, assoc prof, 72-78, PROF CHEM, UNIV CALIF, IRVINE, 78- *Mem:* Am Chem Soc; Am Crystallog Asn. *Res:* Structural inorganic chemistry; x-ray crystallography; crystal and molecular structures of coordination and organometallic compounds of the transition elements; structure and magnetism in magnetically condensed systems. *Mailing Add:* Dept Chem Univ Calif Irvine CA 92717

DOEG, KENNETH ALBERT, b Weehawken, NJ, Aug 5, 31; m 57; c 2. BIOCHEMISTRY. *Educ:* NJ State Teachers Col, AB, 52; Rutgers Univ, PhD(zool), 57. *Prof Exp:* Smith, Kline & French fel, Rutgers Univ, 57-58; fel enzyme inst, Univ Wis, 58-61; asst prof clin sci, Sch Med, Univ Pittsburgh, 61-64; asst prof zool, 64-69, assoc prof, 69-77, PROF BIOL, UNIV CONN, 77- *Concurrent Pos:* Vis prof, Cornell Univ, 73-74; vis prof, Biocenter, Univ Basel, Switz, 74. *Mem:* AAAS; Am Chem Soc; Endocrine Soc; Am Soc Biol Chem & Molecular Biol. *Res:* Mechanisms of hormone action; synthesis and turnover of cell membranes as related to secretion. *Mailing Add:* Dept Molecular & Cell Biol U-125 Univ Conn Storrs CT 06269-3125

DOEGE, THEODORE CHARLES, b Lincoln, Nebr, Dec 11, 28; m 57; c 2. EPIDEMIOLOGY, PUBLIC HEALTH. *Educ:* Oberlin Col, AB, 50; Univ Rochester, MD, 58; Univ Wash, MS, 65; Am Bd Prev Med, dipl. *Prof Exp:* Intern med, Salt Lake County Gen Hosp, Utah, 58-59; asst resident pediat, Univ Hosps, Seattle, 61-63; chief resident, King County Hosp, 63; instr prev med, Sch Med, Univ Wash, 63-65, asst prof, 65-67; assoc prof prev med, Col Med, 67-72, ASSOC PROF EPIDEMIOL, SCH PUB HEALTH, UNIV ILL, 72-; SR SCIENTIST, DEPT ENVIRON, PUB & OCCUP HEALTH, AMA, CHICAGO, 84-, DIR, DEPT RISK ASSESSMENT, 88- *Concurrent Pos:* Fel prev med, Sch Med, Univ Wash, 63-65; fel trop med, La State Univ, 66; vis assoc prof, Chiengmai Univ, Thailand, 67-70. *Mem:* Fel Am Pub Health Asn; Am Col Prev Med; Am Col Epidemiol; Sigma Xi; AMA; AAAS. *Res:* Injuries; epidemiology of public health problems. *Mailing Add:* 515 N State St Chicago IL 60610

DOEHLER, ROBERT WILLIAM, b Mt Olive, Ill, Aug 7, 29; m 54; c 2. MINERALOGY. *Educ:* Univ Ill, BS, 51, MS, 53, PhD(geol), 57. *Prof Exp:* Res geologist, Jersey Prod Res Co, 57-60; dir chem res, Am Colloid Co, 60-69; assoc prof, 69-80, PROF EARTH SCI, NORTHEASTERN ILL UNIV, 80- *Concurrent Pos:* Vis prof, Notre Dame, 73; res assoc, Argonne Nat Lab, 77 & 78. *Mem:* Mineral Soc Am; AAAS; Sigma Xi. *Res:* Production and industrial applications of non-metallic minerals; mineralogy and geochemistry of sediments; geology and geochemistry of coal. *Mailing Add:* Dept Earth Sci Northeastern Ill Univ Chicago IL 60625

DOEHRING, DONALD O, b Milwaukee, Wis, Sept 8, 35; m 58; c 2. GEOMORPHOLOGY, GEOENVIRONMENTAL SCIENCE. *Educ:* Univ Calif, Berkeley, AB, 62; Claremont Grad Sch & Univ Ctr, MA, 65; Univ Wyo, PhD(geol), 68. *Prof Exp:* Instr geol, Univ Wyo, 67-68; vis asst prof, Pomona Col, 68-69 & Washington & Lee Univ, 69-70; fel, State Univ NY, Binghamton, 70-71; asst prof, Univ Mass, Amherst, 71-75; from asst prof to assoc prof earth resources, 75-85, PROF EARTH RESOURCES, COLO STATE UNIV, 85-, DEPT HEAD, 89- *Mem:* Geol Soc Am; Am Quaternary Asn; Am Inst Prof Geologists; Am Geophys Union. *Res:* Arid region geomorphology; quantitative modeling and analysis of landforms; slope process studies; environmental geomorphology; environmental and engineering geology. *Mailing Add:* Dept Earth Resources Colo State Univ Ft Collins CO 80523

DOELL, RICHARD RAYMAN, b Oakland, Calif, June 28, 23; m 50; c 2. GEOPHYSICS. *Educ:* Univ Calif, AB, 52, PhD(geophysics), 55. *Prof Exp:* Lectr geophysics, Univ Toronto, 55-56; asst prof, Mass Inst Technol, 56-59; geophysicist, US Geol Surv, 59-79; RETIRED. *Concurrent Pos:* Co-chmn comt mineral resources & environ, mem panel environ aspects of submarine mining & appraisal of mineral resource base, mem nat mat adv bd, Nat Acad Sci. *Honors & Awards:* Vetlesen Prize, 71. *Mem:* Nat Acad Sci. *Res:* General geophysics, especially the earth's magnetic field; remnant magnetism in rocks. *Mailing Add:* 1200 Brickyard Way No 302 Pt Richmond CA 94801-4136

DOELL, RUTH GERTRUDE, b Vancouver, BC, Mar 24, 26; nat US; m 50; c 2. BIOLOGY. *Educ:* Univ Calif, AB, 52, PhD(comp biochem), 56. *Prof Exp:* Instr physiol, Med Sch, Tufts Univ, 56-59; res assoc biochem, Sch Med, Stanford Univ, 59-60, res assoc pediat, 60-65, res assoc med microbiol, 65-67; lectr, 67-69, assoc prof, 69-74, PROF BIOL, SAN FRANCISCO STATE UNIV, 74- *Res:* Viral and chemical carcinogenesis; immunology. *Mailing Add:* Dept Biol San Francisco State Univ San Francisco CA 94132

DOELLING, HELLMUT HANS, b New York, NY, July 25, 30; m 60; c 7. ECONOMIC GEOLOGY, MAPPING GEOLOGY. *Educ:* Univ Utah, BS, 56, PhD(geol), 64. *Prof Exp:* Geologist, Utah Geol & Mining Surv, 64; asst prof geol, Midwestern Univ, 64-66; econ geologist, 66-83, SR MAPPING GEOLOGIST, UTAH GEOL & MINERAL SURV, 83- *Concurrent Pos:* Consult, coal, uranium, precious & base metals. *Mem:* AAAS; Soc Econ Geologists; Sigma Xi. *Res:* Coal, uranium, salt anticlines, geology of Utah. *Mailing Add:* 270 E 10th N Centerville UT 84014

DOELP, LOUIS C(ONRAD), JR, b Philadelphia, Pa, Jan 8, 28; m 50; c 2. CHEMICAL ENGINEERING. *Educ:* Univ Pa, BS, 50; Villanova, MS, 72. *Prof Exp:* Plant engr, Am Foam Rubber Corp, 50-51; chem engr develop, Air Prod & Chem, Inc, 51-55, proj dir process develop, 55-61, sect head, 61-69, dir res & develop, Catalysts & Processes, Houdry Process & Chem Corp Div, 69-, CHIEF ENGR, AIR PROD & CHEM CORP, INC. *Res:* Hazard and risk analysis. *Mailing Add:* 7201 Hamilton Blvd Allentown PA 18195-1501

DOEMEL, WILLIAM NAYLOR, b Pittsburgh, Pa; m; c 1. MICROBIAL ECOLOGY, SAINCE POLICY. *Educ:* Heidelberg Col, BS, 66; Ind Univ, PhD(microbiol), 70. *Prof Exp:* From asst prof to assoc prof, 70-87, chmn biol, 84-88, PROF, WABASH COL, 88-, DIR, COMPUTER SERV, 91- *Concurrent Pos:* Lilly fac fel, 77-78. *Mem:* AAAS; Am Soc Microbiol. *Res:* Physiological ecology of micro-organisms existing in extreme environments including thermal effluents, acid effluents and polluted streams; scientists and the public. *Mailing Add:* Computer Serv Wabash Col Crawfordsville IN 47933

DOEMLING, DONALD BERNARD, b Chicago, Ill, Nov 20, 30; m 59, 87; c 2. PHYSIOLOGY, PHARMACOLOGY. *Educ:* St Benedict's Col, BS, 52; Univ Ill, MS, 54, PhD(physiol), 58. *Prof Exp:* Asst physiol, Univ Ill, 52-57; from instr to asst prof physiol & pharmacol, Dent Sch, Northwestern Univ, Chicago, 57-60; from instr to assoc prof, Jefferson Med Col, 61-68; PROF PHYSIOL & PHARMACOL & CHMN DEPT, SCH DENT, LOYOLA UNIV CHICAGO, 68- *Mem:* AAAS; Am Physiol Soc; Am Asn Dent Schs. *Res:* Lymph flow and composition; factors affecting saliva. *Mailing Add:* Dept Physiol & Pharmacol Loyola Univ Sch Dent Maywood IL 60153

DOEPKER, RICHARD DUMONT, b Findlay, Ohio, Jan 22, 33; m 59, 76; c 4. PHYSICAL CHEMISTRY. *Educ:* Xavier Univ, Ohio, BS, 55, MS, 57; Carnegie Inst Technol, MS, 60, PhD(chem), 61. *Prof Exp:* Nat Acad Sci-Nat Res Coun res assoc photoradiation chem, Nat Bur Standards, 63-65; from asst prof to assoc prof, 65-80, PROF CHEM, UNIV MIAMI, 80-; AT DEPT CHEM, GONZAGA UNIV, SPOKANE, WASHINGTON. *Mem:* AAAS; Am Soc Photobiol; Am Chem Soc; Sigma Xi. *Res:* Photoradiation chemistry; reaction of hot radicals; modes of decomposition of highly excited molecules produced through vacuum ultraviolet photolysis and electron impact; photo-ionization; H-atom reactions; Lyman absorption technique; environmental chemical analysis and exhaust emission methodology. *Mailing Add:* 3-USBM Spokane Res Ctr S-1501 Rockwood Blvd Spokane WA 99203

DOEPPNER, THOMAS WALTER, JR, b Augusta, Ga, Sept 5, 51. COMPUTER SCIENCE. *Educ:* Cornell Univ, BS, 73; Princeton Univ, MSE, 74, MA, 75, PhD(elec eng & comput sci), 77. *Prof Exp:* asst prof Comput Sci & Appl Math, 76-79, asst prof comput sci, 79-81, ASSOC PROF COMPUT SCI, BROWN UNIV, 81- *Concurrent Pos:* prin software eng, Text Systems, Inc, Barrington, RI, 77-; vis scientist, George Washington Univ, 80-; consult, Atlantic Res Corp, 77-78, Raytheon Corp, Missile Systs Div, 80-84, Submarine Signal Div, 80-82, Summagraphics Corp, 80-81, Brandeis Univ, 82-83, Foxboro Co, 82-84, Exxon Office Systs, 82, Compugraphics Co, 83, Gould-Modicon Div, Inst Adv Prof Study, 84. *Mem:* Inst Elec & Electronics Engrs; Asn Comput Mach; Sigma Xi. *Res:* Debugging and profiling parallel and distributed programs; efficient system support for parallel and distributed computing. *Mailing Add:* Dept Comput Sci Brown Univ Box 1910 Providence RI 02912

DOERDER, F PAUL, b Boone, Iowa, June 4, 45. GENETICS. *Educ:* Dana Col, BS, 67; Univ Ill, PhD(cell biol), 72. *Prof Exp:* Fel genetics, Univ Iowa, 72-73; asst prof biol, Univ Pittsburgh, 73-79; vis res scientist, Univ Ill, 79-81; assoc prof, 81-88, PROF BIOL, CLEVELAND STATE UNIV, 88- *Mem:* Genetics Soc Am; Soc Protozoologists; Am Genetic Asn; AAAS; Am Soc Cell Biol. *Res:* Immunogenetics of surface antigens in Tetrahymena; cytogenetics; aging; population genetics of ciliates; developmental genetics of cell cycle. *Mailing Add:* Dept Biol Cleveland State Univ 1983 E 24 St Cleveland OH 44115

DOERFLER, THOMAS EUGENE, b Dayton, Ohio, Feb 16, 37; m 68; c 2. STATISTICS. *Educ:* Univ Dayton, BS, 59; Iowa State Univ, MS, 62, PhD(stat), 65. *Prof Exp:* Analyst-programmer statist, Res Inst, Univ Dayton, 55-59; res asst, Statist Lab, Iowa State Univ, 59-65; consult, Booz-Allen Appl Res Inc, 65-70; dir mgt sci, Columbia House Div, Columbia Broadcasting Systs, 70-73; SR STATIST SECT, ARTHUR D LITTLE INC, 73- *Concurrent Pos:* Lectr, Sch Mgt, Boston Univ, 78. *Mem:* Am Stat Asn. *Res:* Statistical methods and applications; design and analysis of experiments; optimization techniques. *Mailing Add:* Arthur D Little Inc Acorn Park 35-321A Cambridge MA 02140-2390

DOERFLER, WALTER HANS, b Weissenburg-Bayern, Ger, Aug 11, 33; m 60; c 1. VIROLOGY, MOLECULAR BIOLOGY. *Educ:* Univ Munich, MD, 59. *Prof Exp:* Intern, Mercer Hosp, Trenton, NJ, 59-60; res fel, Max Planck Inst Biochem, München, Germany, 61-63 & Dept Biochem, Sch Med, Stanford Univ, 63-66; from asst prof to assoc prof virol, Rockefeller Univ, 66-71, adj prof, 71-78; PROF & DIR INST GENETICS, UNIV COLOGNE, 72- *Concurrent Pos:* Ventor Found exchangee, 59-60; City of New York Health Res Coun career scientist, 69-71; guest prof, Univ Uppsala, 71-72, guest sceintsit, Stanford Univ, 78, Princeton Univ, 86, Kawasaki Med Sch, Japan, 88 & Inst Molecular Biol Acad Sci, USSR, Moscow, USSR, 90; ed, J Gen Virol, 77-82; speaker, Spec Res Area, 74-, Ger Res Asn, 78-88, Senate, 85-91. *Honors & Awards:* Aronson Prize, 81; Robert Koch Prize, 84. *Mem:* Am Soc

Biochem & Molecular Biol; Am Soc Microbiol; AAAS; Am Soc Virol; German Soc Biol Chem; Ger Soc Hyg Microbiol; Europ Molecular Biol Orgn; Soc Virol. *Res:* Structure and function of nucleic acids; molecular biology of virus infected cells; interaction between viral (foreign) and host genomes; DNA methylation and regulation of gene expression; integrated viral genes in transformed cells; molecular biology of baculoviruses; abortive viral infections; linguistics and genetic code; gene expression in human lymphomas; mechanism of recombination in mammalian cells; patterns of DNA methylation in human genome. *Mailing Add:* Inst Genetics Univ Cologne 121 Weyertal D-5000 Cologne 41 Germany

DOERGE, DANIEL ROBERT, b Austin, Tex, Oct 22, 49. BIOCHEMICAL TOXICOLOGY, THYROID TOXICOLOGY. *Educ:* Ore State Univ, BS, 73; Univ Calif-Davis, PhD(biochem), 80. *Prof Exp:* Res assoc, Dept Marine & Atmospheric Chem, Sch Marine & Atmospheric Sci, Univ Miami, 80-81; asst environ, Dept Food Sci & Human Nutrit, 81-84; asst prof, Dept Agr Biochem, 84-89, ASSOC PROF TOXICOL, DEPT ENVIRON BIOCHEM, UNIV HAWAII, 89- *Concurrent Pos:* Prin investr, Nat Inst Environ Health Sci, 88-91. *Mem:* Am Chem Soc; Soc Toxicol; Int Soc Study Xenobiotics; Am Soc Mass Spectrometry; Sigma Xi. *Res:* Chemical and biochemical mechanisms of thyroid toxicity/carcinogenicity; mechanisms of peroxidase-catalyzed reactions; organosulfur oxidation mechanisms; anti-hyperthyroid drug mechanisms; drug metabolism; liquid chromatography/mass spectrometry. *Mailing Add:* Dept Environ Biochem Univ Hawaii 1800 East-West Rd Honolulu HI 96822

DOERGES, JOHN E, b Mar 8, 26; m; c 1. PHYSICS. *Educ:* Univ Northern Colo, BA, 59, MA, 63; NMex Highlands, MS, 64. *Prof Exp:* High sch teacher physics, chem & math, United Sch Dist No 1, Ault, Colo, 59-61, Jefferson Co Schs, Evergreen, Colo, 61-63; asst prof physics & math, Upper Iowa Univ, Fayette, 64-73; dir training computer training, Northeast Iowa Computer Network, 73-74, dir computer network, 74; radiation safety officer, 74-88, DIR SAFETY OFF, UNIV WYO, 88- *Concurrent Pos:* Wyo rep, Rocky Mountain Low Level Radioactive Waste Bd, 83-; consult, Western Radiation Consults, 87- *Mem:* Health Physics Soc; Sigma Xi. *Res:* Uranium mines. *Mailing Add:* 1702 Mill St Laramie WY 82070

DOERING, CHARLES HENRY, b Munich, WGer, Jan 7, 35; US citizen; m 61; c 4. PSYCHOENDOCRINOLOGY, HORMONE ASSAYS. *Educ:* Univ San Francisco, BSc, 56; Univ Munich, MSc, 59; Univ Calif, San Francisco, PhD(biochem), 64. *Prof Exp:* Asst biochem, Univ Calif, San Francisco, 59-63; res fel, Harvard Med Sch, 63-67; res assoc psychiat, Stanford Univ Sch Med, 67-71, sr res scientist, 71-76; chief, Chem Labs, Long Island Res Inst, 76-80, sr res scientist psychoendocrinol group, 76-83; ASSOC RES PROF PSYCHIAT, STATE UNIV NY SCH MED, STONY BROOK, 77- *Concurrent Pos:* Assoc res prof psychiat, State Univ NY Sch Med, Stony Brook, 77- *Mem:* AAAS; Endocrine Soc; Am Chem Soc. *Res:* Biosynthesis of steroid hormones and its genetic control; influence of steroid hormones on behavior; steroid hormone receptors in brain; correlation between androgens and sexual behavior/aggressiveness; developmental psychoendocrinology. *Mailing Add:* 21 Dyke Rd Setauket NY 11733-3014

DOERING, CHARLES ROGERS, b Philadelphia, Pa, Jan 7, 56; m 85; c 3. STATISTICAL PHYSICS, NONLINEAR SCIENCE. *Educ:* Antioch Col, BS, 77; Univ Cincinnati, MS, 78; Univ Tex, Austin, PhD(physics), 85. *Prof Exp:* Scientist instrument design, Measurex Corp, 78-81; postdoctoral assoc physics, Univ Tex, Austin, 86; dirs postdoctoral fel, Los Alamos Nat Lab, 86-87; asst prof, 87-91, ASSOC PROF PHYSICS, DEPT PHYSICS, CLARKSON UNIV, 91- *Concurrent Pos:* Collabr, Los Alamos Nat Lab, 87-, presidential young investr award, NSF, 89 & prin investr grant, 89- *Mem:* Am Phys Soc; Soc Indust & Appl Math. *Res:* Theoretical physics and mathematical physics concerned with methods and applications of statistical physics and nonlinear analysis to problems of fluid mechanics, condensed matter physics and quantum theory. *Mailing Add:* Dept Physics Clarkson Univ Potsdam NY 13699-5820

DOERING, EUGENE J(OHNSON), b Plankinton, SDak, Aug 28, 28; m 54; c 2. AGRICULTURAL ENGINEERING. *Educ:* SDak State Univ, BS, 52, MS, 58. *Prof Exp:* Engr, Boeing Airplane Co, Kans, 52; agr engr, Soil Conserv Serv, 54-56, agr engr, Sci & Educ Admin-Agr Res, Calif, 58-65, AGR ENGR, NORTHERN GREAT PLAINS RES CTR, AGR RES SERV, USDA, NDAK, 65- *Mem:* Am Soc Agr Engrs; Nat Soc Prof Engrs. *Res:* Salinity in agriculture; reclamation; irrigation; drainage; flow of water in saturated and unsaturated soil. *Mailing Add:* USAID/Lahore Washington DC 20523

DOERING, JEFFREY LOUIS, b Chicago, Ill, May 26, 49; m 74; c 2. GENE CLONING, GENETIC REARRANGEMENTS. *Educ:* Univ Chicago, BA, 71, PhD(develop biol), 75. *Prof Exp:* Res asst, Univ Chicago, 71-75; fel embryol, Carnegie Inst, 75-77; asst prof biol sci, Northwestern Univ, 78-83; ASSOC PROF BIOL, LOYOLA UNIV, CHICAGO, 83- *Concurrent Pos:* Res fel, NIH, 75-77; pvt invest grants, NIH, 79-81, ACS, 78, 83 & 87 & Shriners, 79-; Downs Syndrome res fund, 89-90. *Mem:* Am Soc Cell Biol; AAAS; Sigma Xi; NY Acad Sci. *Res:* Molecular genetics of human collagen disorders; genome reorganization and control of gene expression; characterization of human complex repetitive DNAs. *Mailing Add:* Dept Biol Loyola Univ 6525 N Sheridan Rd Chicago IL 60626

DOERING, ROBERT DISTLER, industrial & mechanical engineering, for more information see previous edition

DOERING, WILLIAM VON EGGERS, b Ft Worth, Tex, June 22, 17; div; c 3. PHYSICAL ORGANIC CHEMISTRY. *Educ:* Harvard Univ, BS, 38, PhD(org chem), 43. *Hon Degrees:* DSc, Tex Christian Univ, 74 & Univ Karlsruhe, 87. *Prof Exp:* Res chemist, Nat Defense Res Comt Proj, Harvard Univ, 41-42 & Plaroid Corp, Mass, 43-; from instr to assoc prof org chem, Columbia Univ, 43-52; prof, Yale Univ, 52-67, dir, div sci, 62-65; prof, 67, Mallinckrodt prof, 68-86, EMER PROF ORG CHEM, HARVARD UNIV, 86- *Concurrent Pos:* Dir res, Hickrill Chem Res Found, NY, 47-59; co-chmn, Coun Livable World, 62-78; hon prof, Fudan Univ, 80. *Honors & Awards:* Scott Award, 45; Pure Chem Award, Am Chem Soc, 53, Synthetic Chem Award, 63; Hoffman Medal, Ger Chem Soc, 62. *Mem:* Nat Acad Sci; Am Acad Arts & Sci; Am Chem Soc. *Res:* Mechanism of thermal rearrangements: Cope, Diels-Alder; vinylcyclobutane; vinylcyclopropane; methylenecyclopropane; energy flow in chemically activated molecules; conjugative interaction. *Mailing Add:* Dept Chem Harvard Univ Cambridge MA 02138-2902

DOERMANN, AUGUST HENRY, b Blue Island, Ill, Dec 7, 18; m 47; c 2. BIOLOGY, BACTERIAL VIRUS GENETICS. *Educ:* Wabash Col, AB, 40; Univ Ill, MA, 41; Stanford Univ, PhD(biol), 46. *Hon Degrees:* DSc, Wabash Col, 75. *Prof Exp:* Fel bacteriophage res, Vanderbilt Univ, 45-47, vis instr biol, 46; fel, Carnegie Inst Genetics Res Unit, 47-49; biologist, Oak Ridge Nat Lab, 49-53; from assoc prof to prof biol, Univ Rochester, 53-58; prof, Vanderbilt Univ, 58-64; prof genetics, Univ Wash, 64-85; RETIRED. *Concurrent Pos:* Vis res fel, Calif Inst Technol, 49; vis prof, Univ Cologne, 57-58; sci assoc, Basel Inst Immunol, 70-71. *Mem:* Nat Acad Sci; AAAS; Am Acad Arts & Sci. *Res:* Microbiological and biochemical genetics; viral morphogenesis; radiation genetics of bacteriophage. *Mailing Add:* Rural Rte 1 Site 20 Compart 123 Whitehorse YT Y1A 4Z6 Can

DOERNER, ROBERT CARL, b St Cloud, Minn, Sept 26, 26; m 54; c 5. FAST REACTOR PHYSICS, REACTOR SAFETY EXPERIMENTS. *Educ:* St Johns Univ, BS, 48; St Louis Univ, PhD(physics), 55. *Prof Exp:* EXPERIMENTAL REACTOR PHYSICIST, ARGONNE NAT LAB, 55- *Concurrent Pos:* Vis prof, Cornell Univ, 62-63. *Mem:* Am Nuclear Soc. *Res:* Reactor safety experiments and analysis; radiographic digital imaging. *Mailing Add:* 9700 S Cass Ave Bldg 208 Argonne IL 60439

DOERNER, WILLIAM A(LLEN), b Pullman, Wash, Feb 24, 20; m 50; c 4. CHEMICAL ENGINEERING, MATHEMATICS. *Educ:* Ore State Col, BS, 42; Univ Mich, MSE, 47, MS, 49, ScD(chem eng), 52. *Prof Exp:* Res engr cent res dept, Exp Sta, E I DuPont De Nemours & Co, 51-66, develop supvr photo prod dept, 66-69, group leader, Cent Res Dept, 69-85; RETIRED. *Mem:* Am Chem Soc; Am Inst Chem Engrs. *Res:* Thermodynamics of power fluids; process chemistry engineering research. *Mailing Add:* 19 Baynard Blvd Wilmington DE 19803

DOERR, PHILLIP DAVID, b Ft Gordon, Ga, Oct 29, 42; m 65; c 2. WILDLIFE ECOLOGY. *Educ:* Colo Col, BA, 64; Colo State Univ, MS, 68; Univ Wis, Madison, PhD(wildlife ecol), 73. *Prof Exp:* Res asst wildlife ecol, Univ Wis, 68-73; from asst prof to assoc prof zool, 73-82, PROF ZOOL & FORESTRY, NC STATE UNIV, 82- *Concurrent Pos:* Vis scientist, Nat Wildlife Fedn, 85. *Mem:* Am Ornithologists Union; Cooper Ornith Soc; Wilson Ornith Soc; Wildlife Soc. *Res:* Population studies of American woodcock in North Carolina; population studies of redcockaded woodpecker and American alligator in North Carolina; relationships of avian predator and prey populations. *Mailing Add:* Dept Fisheries NC State Univ Raleigh Main Campus Raleigh NC 27650

DOERR, ROBERT GEORGE, b Winona, Minn, Jan 26, 41; m 70. ORGANIC CHEMISTRY, ANALYTICAL CHEMISTRY. *Educ:* St Mary's Col (Minn), BA, 63; Pa State Univ, PhD(chem), 67. *Prof Exp:* Fel, Pa State Univ, 67; prof chem, Luther Col (Iowa), 67-68; CHEMIST, WATKINS PROD INC, 68- *Mem:* Am Chem Soc. *Mailing Add:* RR 2 No 131A Fountain City WI 54629

DOERSCH, RONALD ERNEST, b Cleveland, Wis, Sept 13, 35; m 57; c 4. WEED SCIENCE. *Educ:* Univ Wis, BS, 58, MS, 61, PhD(soils), 63. *Prof Exp:* From asst prof to assoc prof, 62-71, PROF WEED CONTROL, UNIV WIS-MADISON, 71-, CHAIR AGRON DEPT, 89-; EXTEN SPECIALIST, USDA, 62-, PEST MGT EDUC COORDR, 78- *Honors & Awards:* Ciba-Geigy Recognition Award, Am Soc Agron, 71; Outstanding Exten Award, Weed Sci Soc Am, 83. *Mem:* Weed Sci Soc Am; Am Soc Agron. *Res:* Weed control in field crops; the influence of soil properties upon soil-applied herbicides and their responses in terms of total crop management. *Mailing Add:* Dept Agron 356 Moore Hall Univ Wis 1575 Linden Dr Madison WI 53706-1597

DOERSHUK, CARL FREDERICK, b Warren, Ohio, Dec 24, 30; m 54; c 3. PEDIATRICS, PULMONARY DISEASE. *Educ:* Oberlin Col, BA, 52; Western Reserve Univ, MD, 56. *Prof Exp:* Instr pediat, Western Reserve Univ Sch Med, 63-65; from asst prof to assoc prof, 65-75, PROF PEDIAT, CASE WESTERN RESERVE UNIV, 75- *Concurrent Pos:* Dir pediat respiratory ther, Rainbow Babies & Childrens Hosp, 65-, dir cystic fibrosis ctr, 69-; prin investr, NIH grants, 78-86. *Mem:* Soc Pediat Res; Am Pediat Soc; Am Thoracic Soc. *Res:* Pulmonary function physiology and pathophysiology in infants and children with cystic fibrosis, other pulmonary disorders and normal children; psychosocial problems and issues in relation to pulmonary disease in children. *Mailing Add:* Dept Pediat Case Western Reserve Univ 2040 Adelbert Rd Cleveland OH 44106

DOETSCH, GERNOT SIEGMAR, b Berlin, Ger, May 25, 40; US citizen; m 66; c 2. NEUROPHYSIOLOGY, SENSORY SYSTEMS. *Educ:* DePauw Univ, BA, 62; Duke Univ, PhD(psychol), 67. *Prof Exp:* Res psychol, 6571st Aeromed Res Lab, Holloman AFB, 67-70; fel neurophysiol, Dept Physiol & Biophys, Univ Wash, 70-72; asst prof, 72-77, assoc prof surg & asst prof physiol, 77-81, ASSOC PROF SURG & PHYSIOL, MED COL GA, 81- *Concurrent Pos:* Assoc, Behav & Brain Scis, 80-; NSF grant, 85-87; Consult NIH Prog Proj, 86- *Mem:* Sigma Xi; Soc Neurosci; Int Brain Res Orgn. *Res:* Physiology of sensory systems; sensory and motor functions of cerebral cortex; information processing in neuronal populations; neural plasticity. *Mailing Add:* Sect Neurosurg Med Col Ga Augusta GA 30912-4010

DOETSCH, PAUL WILLIAM, b Cheverly, Md, Feb 14, 54; m 85. CANCER RESEARCH, MOLECULAR BIOLOGY. *Educ:* Univ Md, College Park, BS, 76; Purdue Univ, MS, 78; Temple Univ, PhD(biochem), 82. *Prof Exp:* Res fel biochem, Dana-Farber Cancer Inst & dept pathol, Sch Med, Harvard Univ,

82-85; course instr molecular biol, dept biol, Harvard Univ, 83-85; ASST PROF BIOCHEM, SCH MED, EMORY UNIV, 85- *Mem:* Am Chem Soc; AAAS; NY Acad Sci; Radiation Res Soc; Am Soc Photobiol; Sigma Xi. *Res:* The chemistry, enzymology, and genetics of DNA damage and repair; the molecular mechanisms of action of antitumor agents such as cis-dichlorodiammine platinum II and other metal coordination complexes. *Mailing Add:* 618 Stone Mountain Way Stone Mountain GA 30087-5142

DOETSCH, RAYMOND NICHOLAS, b Chicago, Ill, Dec 5, 20; m 48; c 3. BACTERIOLOGY. *Educ:* Univ Ill, BS, 42; Ind Univ, AM, 43; Univ Md, PhD(bact), 48. *Prof Exp:* Asst bacteriologist, Ind Univ, 42-43; bacteriologist, Nat Dairy Res Lab, Baltimore, 43-45; asst bact, 45-46, from instr to assoc prof, 46-60, PROF BACT, UNIV MD, COLLEGE PARK, 60- *Concurrent Pos:* Guggenheim fel, Rowett Inst, Aberdeen, 56; vis lectr, Inst Hist Med, Johns Hopkins Univ, 63. *Mem:* AAAS; Am Soc Microbiol; NY Acad Sci; Am Acad Microbiol; Am Asn Hist Med. *Res:* Bacterial cytology; general microbiology; history of science and microbiology; bacterial motility systems. *Mailing Add:* 10429 43rd Ave Beltsville MD 20705

DOETSCHMAN, DAVID CHARLES, b Aurora, Ill, Nov 24, 42; m; c 2. CHEMICAL PHYSICS. *Educ:* Northern Ill Univ, BS, 63; Univ Chicago, PhD(chem), 69. *Prof Exp:* Res fel chem, Australian Nat Univ, 69-74; res fel, Univ Leiden, 74-75; asst prof, 75-82, ASSOC PROF CHEM, STATE UNIV NY BINGHAMTON, 82- *Concurrent Pos:* Nat Inst Health prin investr, 78-83; carbon indust consult, 83-85; pulsed epr ctr coordr, 89-90. *Mem:* Sigma Xi; Am Phys Soc; Am Chem Soc. *Res:* Applications of continuous-wave and time-resolved electron paramagnetic resonance and optical-magnetic double resonance techniques to the study of chemical reaction mechanisms in solids at low temperatures; solids of interest include proteins, carbonaceous materials and polymers. *Mailing Add:* Dept Chem State Univ NY Binghamton NY 13902-6000

DOEZEMA, RYAN EDWARD, b Los Angeles, Calif, Oct 15, 42; m 67; c 2. SEMICONDUCTOR PHYSICS, SUPERCONDUCTOR PHYSICS. *Educ:* Calvin Col, BS, 64, Univ Md, PhD(physics), 71. *Prof Exp:* Res assoc, Univ Md, 71-73; res assoc, Tech Univ Munich, 73-79; assoc prof, 79-87, PROF PHYSICS, UNIV OKLA, 87-, CHMN, DEPT PHYSICS & ASTRON. *Concurrent Pos:* Vis assoc prof, Univ Md, 85-86; prin investr, NSF grants, 86, 90. *Mem:* Am Phys Soc; Sigma Xi. *Res:* Far-infrared spectroscopy of two dimensional electronic systems in semiconductors; high-T superconductors. *Mailing Add:* Dept Physics & Astron Univ Okla Norman OK 73019

DOGGER, JAMES RUSSELL, b Milwaukee, Wis, July 31, 23; m 46; c 3. ENTOMOLOGY. *Educ:* Univ Wis, BS, 47, MS, 48, PhD(entom), 50. *Prof Exp:* Instr econ entom, Univ Wis, 48-49; asst prof entom, Okla Agr & Mech Col, 50-52; from asst prof to assoc prof, NC State Col, 52-58; prof entom & chmn dept, NDak State Univ, 58-69; asst chief stored prod, Mkt Qual Res Div, Agr Res Serv, USDA, 69-72, asst area dir, Ga-SC area, 72-75, area dir, Ches-Pot area, Sci & Educ Admin, 76-78, asst to regional admin, Northeast Region, 79-84, co-dir, Doc Ctr, Beneficial Insects Introd Lab, Agr Res Serv, USDA, 84-88; RETIRED. *Concurrent Pos:* State entomologist, NDak, 58-69; Fulbright prof, Univ Nat Trujillo, Peru, 67-68. *Mem:* Entom Soc Am; Sigma Xi. *Res:* Insects affecting stored products, forage and related crops; immature insects; taxonomic, Coleoptera, Elateridae, Carabidae. *Mailing Add:* Rte 1 Box 753 Gore VA 22637

DOGGETT, LEROY ELSWORTH, b Waterloo, Iowa, Oct 22, 41; m 65. DYNAMICAL ASTRONOMY, NAVIGATION. *Educ:* Univ Mich, BS, 64; Georgetown Univ, MA, 70; NC State Univ, PhD, 81. *Prof Exp:* ASTRONR, 65-90, HEAD, NAUTICAL ALMANAC OFF, US NAVAL OBSERV, 90- *Concurrent Pos:* Assoc ed, Archaeoastron, 81. *Mem:* Am Astron Soc; Inst Navig; Int Astron Union. *Res:* Planetary theory; history of astronomy. *Mailing Add:* 7001 Barkwater Ct Bethesda MD 20817

DOGGETT, WESLEY OSBORNE, b Brown Summit, NC, Jan 24, 31; m 53; c 8. PLASMA PHYSICS. *Educ:* NC State Col, BS, 52, BS, 53; Univ Calif, Berkeley, MA, 54, PhD(physics), 56. *Prof Exp:* Res assoc physics, Radiation Lab, Univ Calif, 54-56; tech proj coordr, Air Force Nuclear Eng Test Reactor, Wright Air Develop Ctr, 56-58; from asst prof to assoc prof physics, 58-62, asst dean sch phys sci & appl math, 64-68, PROF PHYSICS, NC STATE UNIV, 62- *Concurrent Pos:* Consult, Repub Aviation Corp, 56-59, Res Triangle Inst, 62-, NASA, Langley Field, 63, Off Civil Defense, 63-65 & Becton, Dickinson & Co, 76-; dir, Troxler Electronics, Inc, 62-76. *Mem:* AAAS; Am Phys Soc; Am Soc Eng Educ; Am Inst Physics; Am Nuclear Soc; Inst Elec & Electronics Engrs. *Res:* Nuclear reactor physics; development and analysis of statistical methods for investigating half-lives; theoretical analysis of gamma ray penetration in matter; relativistic electron beam studies; relativistic magnetron studies; high power microwaves. *Mailing Add:* Dept Physics 2452 Oxford Rd NC 27608

DOGHRAMJI, KARL, b Jan 12, 54; m 87. SLEEP DISORDERS MEDICINE, HUMAN SEXUALITY. *Educ:* Muhlenberg Col, BS, 76; Jefferson Med Col, MD, 80. *Prof Exp:* Instr, dept Psychiat & Human Behavior, 84-86, ASST PROF PSYCHIAT & HUMAN BEHAVIOR, JEFFERSON MED COL, 86-, DIR SLEEP DISORDERS CTR, 84-, DIR PROG HUMAN SEXUALITY, 86- *Concurrent Pos:* Dir continuing med educ, Dept Psychiat, Jefferson Med Col, 86-; consult, Vet Admin Med Ctr, Philadelphia, 86-; co-chmn, Accreditation Comt, Am Sleep Disorders Asn. *Mem:* Am Med Asn; Am Psychiat Asn; Sleep Res Soc; Am Sleep Disorders Assoc. *Res:* Sleep and sleep disturbances; affective disorders, seasonal changes in mood, sleep apnea and cardiac rhythm disturbances during sleep. *Mailing Add:* 1015 Walnut St Suite 327D Philadelphia PA 19107

DOHANY, JULIUS EUGENE, b Banatski Karlovac, Yugoslavia, May 28, 26; nat US; m 51; c 2. POLYMER CHEMISTRY & ENGINEERING. *Educ:* Budapest Tech, MChE, 50; Swiss Fed Inst Technol, ScD, 63. *Prof Exp:* Supvr plant, Coal Chem Co, Hungary, 51-53, mgr, 53-56; supvr pilot plant, Viscose Co, Switz, 56-58, res lab, 58-59; res chemist, Cryovac Div, W R Grace & Co, 63-64; sr res chemist, Pennwalt Corp, 64-66, proj leader, 66-67, group leader, 67-81 & 81-88, dir res & develop, 81-88, asst to pres, 88-89; INDEPENDENT CONSULT, 89- *Mem:* Am Chem Soc; Soc Plastics Eng; Am Inst Chemists. *Res:* Synthesis and characterization of polymers; fluorine containing polymers, coatings, film, polymer rheology; vicose fiber process; synthetic fibers; textile chemistry; tar distillation; coal carbonization; gas desulfurization. *Mailing Add:* 480 Howellville Rd Berwyn PA 19312

DOHERTY, JAMES EDWARD, III, b Newport, Ark, Nov 22, 23; div; c 2. INTERNAL MEDICINE, CARDIOLOGY. *Educ:* Univ Ark, BS, 44, MD, 46; Am Bd Internal Med, dipl, 55; Am Bd Cardiovasc Dis, dipl, 59. *Prof Exp:* Intern, Columbus City Hosp, Ga, 46-47; resident med, 49-52, from instr to assoc prof, 52-68, PROF MED, SCH MED, UNIV ARK, LITTLE ROCK, 68-, PROF PHARM, 69- *Concurrent Pos:* Chief cardiol, Little Rock Vet Admin Hosp, 57-69; dir, Div Cardiol, Univ Ark Med Ctr-Vet Admin Hosps, 69-77. *Honors & Awards:* Wellcome lectr, 81; Casimir Funk Award, Asn Mil Surgeons, US, 75, Abernathy Award, 85. *Mem:* Am Fedn Clin Res; Soc Nuclear Med; Am Col Physicians; Am Col Cardiol; fel Am Heart Asn (assoc vpres, 88-89). *Res:* Cardiovascular disease, primarily the metabolism of radioactive digoxin and cholesterol. *Mailing Add:* 401 N Ash Little Rock AR 72205

DOHERTY, LOWELL RALPH, b San Diego, Calif, Mar 12, 30; m 65; c 2. STELLAR ATMOSPHERES. *Educ:* Univ Calif, Los Angeles, BA, 52; Univ Mich, MS, 54, PhD(astron), 62. *Prof Exp:* Lectr astron & res fel, Harvard Univ, 57-63; from asst prof to assoc prof, 63-76, PROF ASTRON, UNIV WIS-MADISON, 76- *Concurrent Pos:* Vis prof astron, Univ Hawaii, 76-77. *Mem:* Am Astron Soc; Sigma Xi; Int Astron Union; fel AAAS. *Res:* Stellar polarization; theory of stellar atmospheres. *Mailing Add:* Astron Dept Univ Wis Madison WI 53706-1582

DOHERTY, NIALL STEPHEN, b Antrim, UK, July 27, 49; m 74; c 2. INFLAMATION, AUTOIMMUNITY. *Educ:* Univ Reading, UK, BSc, 70; Univ London, UK, PhD(pharmacol), 74. *Prof Exp:* Res fel, dept med chem, Univ Turku, Finland, 74-75; sr res biologist, dept cell biol, Roche Prod Ltd, UK, 76-79; sr res pharmacologist, Dept Pharmacol, Merrell Dow Res Inst, Cincinnati, Ohio, 79-85, group leader, immunopharmacol, 85-87, dir cell biol, 87-90; MGR INFLAMMATION, PFIZER CENT RES, GROTON, CONN, 90- *Concurrent Pos:* Res award, Royal Soc, London, 74 & Farmos Group, Turku, Finland, 75. *Mem:* Brit Pharmacol Soc; Int Soc Immunopharmacol; Inflammation Res Asn. *Res:* Mechanisms of inflammatory, immunological and allergic diseases; development of novel therapeutic agents. *Mailing Add:* Pfizer Cent Res Eastern Point Rd Groton CT 06340

DOHERTY, PAUL MICHAEL, b Boston, Mass, July 7, 48. SOLID STATE PHYSICS. *Educ:* Mass Inst Technol, BS, 70, PhD(physics), 74. *Prof Exp:* Asst prof physics, Oakland Univ, 74-86. *Mem:* Am Asn Physics Teachers; Sigma Xi. *Res:* Laser intensity fluctuation spectroscopy of biological macromolecules and turbulent flow. *Mailing Add:* Dept Physics Oakland Univ Rochester MI 48065

DOHERTY, WILLIAM HUMPHREY, b Cambridge, Mass, Aug 21, 07; m 39; c 2. ELECTRONICS ENGINEERING. *Educ:* Harvard Univ, SB, 27, SM, 28. *Hon Degrees:* ScD, Cath Univ, 50. *Prof Exp:* Res assoc, radio wave propagation, Nat Bur Standards, 28-29; mem tech staff, radio equip develop, Bell Tel Labs, 29-48, dir electronic & TV res, 49-55, asst to pres, sci & technol relations, 61-70; asst vpres, admin, AT&T Co, 55-57; mgr, govt & mil contracting, Western Elec Co, 58-60, patent lic mgr, 60-61; RETIRED. *Honors & Awards:* Morris Liebmann Mem Prize, Inst Radio Engrs, 37. *Mem:* Fel Inst Elec & Electronic Engrs; fel AAAS. *Res:* Electronics and electrical communications, especially telephony, television and physical sciences; telephone science history. *Mailing Add:* 18 Belted Kingfisher Rd Hilton Head Island SC 29928

DOHLMAN, CLAES HENRIK, b Uppsala, Sweden, Sept 11, 22; m 48; c 6. OPHTHALMOLOGY. *Educ:* Univ Lund, Lic Med, 50, Med Dr & docent, 57. *Prof Exp:* Resident ophthalmol, 50-52, asst surgeon, Univ Eye Clin, Univ Lund, 55-57; dir cornea serv, Mass Eye & Ear Infirmary, 64-75, chief eye serv, 74-89; chmn dept & dir Howe Lab Ophthal, 74-89, PROF OPHTHAL, HARVARD MED SCH, 74- *Concurrent Pos:* Res fel ophthal, Johns Hopkins Hosp, 52-53; res fel, Retina Found, 53-54; fel, Mass Eye & Ear Infirmary, 58-63. *Res:* Diseases of the cornea of the eye. *Mailing Add:* Dept Opthal Harvard Med Sch 243 Charles St Boston MA 02114

DOHM, GERALD LYNIS, b Colby, Kans, Mar 1, 42; m 63; c 2. BIOCHEMISTRY, PHYSIOLOGY. *Educ:* Kans State Univ, BS, 65, MS, 66, PhD(biochem), 69. *Prof Exp:* Captain, US Army Med Res & Nutrit Lab Fitzsimons Gen Hosp, Denver, 69-72; PROF BIOCHEM, SCH MED, E CAROLINA UNIV, GREENVILLE, NC, 72- *Concurrent Pos:* Fogarty Sr Int fel. *Mem:* Am Physiol Soc; Am Diabetes Asn; Sigma Xi. *Res:* Biochemical adaptation to exercise; control of insulin sensitivity in muscle. *Mailing Add:* Dept Biochem E Carolina Univ Greenville NC 27858

DOHMS, JOHN EDWARD, b New York, NY, Apr 5, 48. AVIAN MEDICINE, AVIAN IMMUNOLOGY. *Educ:* Bowling Green State Univ, BS, 70, MS, 72; Ohio State Univ, PhD(poultry sci), 77. *Prof Exp:* Tech res assoc, Ohio Agr Res & Develop Ctr, Ohio State Univ, 72-74, res asst, Grad Sch, 74-77; asst profavian immunol, 77-83, ASSOC PROF AVIAN MICROBIOL, DEPT ANIMAL SCI & AGR BIOCHEM, UNIV DEL, 83- *Concurrent Pos:* Mem, Conf Res Workers Animal Dis. *Mem:* Am Soc Microbiologists; Poultry Sci Asn. *Res:* Poultry diseases; molecular biology of mycoplasma; immunology of the upper respiratory tract of domestic birds. *Mailing Add:* 126 E Cleveland Ave Newark DE 19711

DOHNANYI, JULIUS S, b Budapest, Hungary, Aug 29, 31; US citizen; m 60; c 1. THEORETICAL PHYSICS. *Educ:* Fla State Univ, MS, 55; Ohio State Univ, PhD(physics), 60. *Prof Exp:* Mem staff, Sandia Corp, 60-63; mem tech staff, Bell Comm, Inc, 63-72, MEM STAFF, BELL TEL LABS, INC, 72- *Mem:* Am Phys Soc; Am Geophys Union. *Res:* Physics of interplanetary debris; nuclear magnetic resonance; superconductivity; statistical models of the nucleus. *Mailing Add:* Three Orchard Ave Holmdel NJ 07733

DOHOO, ROY MCGREGOR, b Essex, Eng, Sept 3, 19; Can citizen; m 49; c 2. ELECTRICAL ENGINEERING, SATELLITE COMMUNICATIONS. *Educ:* Cambridge Univ, BA, 41, MA, 44. *Prof Exp:* Exp officer radar, Air Ministry, UK, 47-48; sci officer electronic fuzing, Armament Design Estab, 48-52; sci officer, Defence Res Telecommun Estab, Defence Res Bd Can, 52-55, sect leader radar systs, 55-57; res engr, Rand Corp, Calif, 57-59; sci officer, Defence Res Bd Can, 59-62, sect leader microwave propagation, Commun Res Ctr, 62-67, supt, Commun Lab, 67-69; dir, Nat Commun Lab, Commun Res Ctr, Dept Commun, 69-74, dir, Gen Space prog, 74-77; PRES, ROY M DOHOO LTD, 77- *Mem:* Sr mem Inst Elec & Electronics Engrs; fel Brit Inst Elec Engrs; assoc fel Aeronaut & Space Inst. *Res:* Telecommunications; radar systems; electronic fuzing; microwave propagation; satellite communications. *Mailing Add:* 2092 Woodcrest Rd Ottawa ON K1H 6H8 Can

DOI, KUNIO, b Tokyo, Japan, Sept 28, 39; m 62; c 2. MEDICAL PHYSICS, OPTICS. *Educ:* Waseda Univ, Japan, BSc, 62, PhD, 69. *Prof Exp:* Chief radiography res, Kyokko Res Labs, Dai Nippon Toryo Co, Ltd, Japan, 62-69; from asst prof to assoc prof, 69-77, PROF RADIOL & DIR KURT ROSSMANN LAB RADIOL IMAGE RES, UNIV CHICAGO, 77-, DIR GRAD PROG MED PHYSICS, 85- *Concurrent Pos:* Instr, Fac Med, Tohoku Univ, Japan, 66-69. *Honors & Awards:* Japanese Appl Physics Soc Award, 68; Japanese Soc Instrument & Control Engrs Award, 72. *Mem:* Am Asn Physicists in Med. *Mailing Add:* Dept Radiol Univ Chicago Box 420 Chicago IL 60637

DOI, ROY HIROSHI, b Sacramento, Calif, Mar 26, 33; m 58; c 2. MICROBIOLOGY. *Educ:* Univ Calif, Berkeley, AB, 53 & 57; Univ Wis, MS, 58, PhD(bact), 60. *Prof Exp:* USPHS fel, 60-63; asst prof bact & bot, Syracuse Univ, 63-65; from asst prof to assoc prof, 65-69, chmn dept biochem & biophys, 74-77, PROF BIOCHEM, UNIV CALIF, DAVIS, 69-, COORDR BIOTECHNOL, 89- *Concurrent Pos:* Mem microbiol training comt, Nat Inst Gen Med Sci, 68-71; NSF sr fel, 71-72; mem microbial chem study sect, NIH, 75-79 & 83-85; vis prof, Max Planck Inst Biochem, Munich, 78-79; sr US scientist award, Alexander von Humboldt Found, 78-79; mem, recombinant DNA Adv Comt, NIH, 90-94. *Mem:* Am Soc Microbiol; Am Soc Biol Chemists. *Res:* Isolation, cloning, and characterization of Bacillus subtilis genes; structure and function of RNA polymerase of sporulating bacteria; clostridium cellulovorans cellulase. *Mailing Add:* 1520 Lemon Lane Davis CA 95616

DOIG, MARION TILTON, III, b Charleston, SC, Sept 29, 43; m 66; c 2. BIOCHEMISTRY, ENVIRONMENTAL CHEMISTRY. *Educ:* Col Charleston, BS, 66; Univ SFla, MS, 71, PhD(chem), 73. *Prof Exp:* Anal chemist environ chem, Environ Protection Agency, 65-69; vis asst prof clin chem, Univ SFla, 73-74; asst prof, 74-79, ASSOC PROF CHEM, COL CHARLESTON, 80- *Mem:* Sigma Xi; Am Chem Soc. *Res:* Marine toxicology; folate metabolism; cancer chemotherapy; applications of flow microcalorimetry to biochemical problems. *Mailing Add:* Dept Chem Col Charleston Charleston SC 29424

DOIG, RONALD, b Montreal, Que, Feb 14, 39. GEOLOGY, GEOPHYSICS. *Educ:* McGill Univ, BSc, 60, MSc, 61, PhD(geol), 64. *Prof Exp:* Mem staff gamma-ray spectrometry, Geophys Div, Geol Surv Can, 62-64; lectr isotope geol & geophys, 64-66, from asst prof to assoc prof, 66-78, chmn dept, 75-80, PROF GEOL SCI, MCGILL UNIV, 78- *Mem:* Geol Soc Am; Geol Asn Can. *Res:* Isotope geology; geochronology; gamma-ray spectrometry. *Mailing Add:* Dept Geol Sci McGill Univ 853 Sherbrooke St W Montreal PQ H3A 2T6 Can

DOIGAN, PAUL, b Greenfield, Mass, June 8, 19; m 53; c 3. PHYSICAL CHEMISTRY. *Educ:* Univ Conn, BS, 41; NY Univ, BS, 43, PhD(chem), 50; Univ Mass, MS, 46. *Prof Exp:* Res assoc chem, Atomic Res Inst, Iowa State Col, 46; instr chem, Univ Mass, 46-47; instr chem, Univ Conn, 50-51; res chemist, Gen Elec Co, 51-55, prof personnel, 55-58, consult engr & mgr personnel placement & facility planning, 59-60, mgr eng admin, missile & space vehicle dept, 60-62 & spacecraft dept, 62-64, consult specialist, 65-66, proj mgr col placement coun, 66-67, mgr doctoral & int recruiting, 67-79, mgr entry level recruiting, 75-79, mgr tech recruiting, 79-84; CONSULT, ENG MANPOWER & IMMIGRATION, 84- *Concurrent Pos:* Mem conf elec insulation, Nat Res Coun; chmn, Eng Manpower Comn, 79-81 & 82-84; chmn eng sch fac shortage proj, Am Soc eng Educ, 80- & mem bd dir. *Mem:* AAAS; Am Chem Soc; Inst Elec & Electronics Engrs; fel Am Inst Chemists; Am Soc Eng Educ. *Res:* Engineering administration; space vehicles; manned space laboratory. *Mailing Add:* 2242 Pine Ridge Rd Schenectady NY 12309

DOIRON, THEODORE DANOS, b New Orleans, La, Aug 14, 48; m 71; c 2. PHYSICS. *Educ:* Univ New Orleans, BS, 70, MS, 72; Duke Univ, PhD(physics), 77. *Prof Exp:* Fel physics, Duke Univ, 77-78; Nat Res Coun fel, Nat Bur Standards, 78-80; asst prof physics, Va Commonwealth Univ, 80-83; MEM STAFF, NAT BUR STANDARDS, WASH, DC, 83- *Mem:* Am Phys Soc; Mach Vision Asn; Sigma Xi. *Res:* Dimensional metrology; machine vision applications. *Mailing Add:* Bldg 220 Rm A107 Nat Bur Standards Gaithersburg MD 20899

DOKU, HRISTO CHRIS, b Istanbul, Turkey, Apr 17, 28; US citizen; m 58; c 1. ORAL SURGERY. *Educ:* Univ Istanbul, DDS, 51; Tufts Univ, DMD, 58, MDS, 60. *Prof Exp:* From instr to assoc prof, 58-67, PROF ORAL SURG, SCH DENT MED, TUFTS UNIV, 67-, CHMN DEPT, 65-, ASSOC DEAN, 67- *Concurrent Pos:* USPHS trainee, 58-60; prin investr, USPHS. *Honors & Awards:* Hatton Award, Chicago, 60. *Mem:* Am Dent Asn; Am Soc Oral Surg; Int Asn Dent Res. *Res:* Blood clotting factors; thromboplastic activity and its relation to saliva; wound healing; age and dentition. *Mailing Add:* 37 Maugus Hill Rd Wellesley Hills MA 02181

DOLAK, TERENCE MARTIN, b Youngstown, Ohio, Sept 24, 51; m 77, 85; c 1. MEDICINAL & ORGANOSULFUR CHEMISTRY. *Educ:* Ohio State Univ, BS, 73; Univ SC, PhD(chem), 77. *Prof Exp:* Res assoc, Wayne State Univ, 77-78; scientist, Bristol-Myers Co, 79-81, group leader, 82-84; sect head, Ayerst Labs, 84-86; dir res, 86-88, VPRES RES & DEVELOP, BAUSCH & LOMB, 88- *Mem:* Am Chem Soc; AAAS; NY Acad Sci. *Res:* Structural design, synthesis and structure activity relationships of new compounds with useful cardiovascular properties. *Mailing Add:* PO Box 806 Canandaigua NY 14424-0340

DOLAN, CHARLES W, b Worcester, Mass, May 12, 43; m 73; c 2. STRUCTURAL ENGINEERING, CONSTRUCTION. *Educ:* Univ Mass, BS, 65; Cornell Univ, MS, 67, PhD(civil eng), 89. *Prof Exp:* Vpres, Abam Engrs Inc, 67-87; assoc prof civil eng, Univ Del, 89-91; ASSOC PROF CIVIL ENG, UNIV WYO, 91- *Concurrent Pos:* Adj assoc prof, Cornell Univ, 75; res fel, Prestressed Concrete Inst, 88. *Honors & Awards:* T Y Lin Award, Am Soc Civil Engrs, 72. *Mem:* Fel Am Concrete Inst; Prestressed Concrete Inst; Am Soc Civil Engrs; Int Asn Bridge & Struct Engrs. *Res:* Adaptation of new materials to civil engineering practice with a particular emphasis on nonmetallic prestressing systems. *Mailing Add:* Dept Civil Eng Univ Wyo Laramie WY 82071

DOLAN, DESMOND DANIEL, b Nelson-Miramichi, NB, Nov 19, 16; nat US; m 48; c 4. PLANT PATHOLOGY. *Educ:* McGill Univ, BSc, 37, MSc, 39; Cornell Univ, PhD(genetics & plant breeding), 46. *Prof Exp:* Veg pathologist, Dom Path Lab, 38-41; asst, Cath Univ Am, 41-42; asst plant breeding & veg crops, Cornell Univ, 42-46; assoc res prof hort, Univ RI, 46-53; coordr & horticulturist-in-chg plant introd, NY State Agr Exp Sta, Plant Sci Res Div, USDA, 53-87; RETIRED. *Concurrent Pos:* Assoc prof, Cornell Univ, 55-65. *Mem:* Am Soc Hort Sci; Am Soc Agron; Sigma Xi. *Res:* Fungus causing stem-streak on melons; breeding Iroquois muskmelon, Rhode Island Red watermelon and Rhode Island Early and Summer Sunrise tomatoes; plant germ plasm for the Northeast; discovery of first all-female, gynoecious cucumber. *Mailing Add:* 186 Lafayette Ave Geneva NY 14456-1550

DOLAN, JAMES F, b Kingston, NY, Jan 9, 52. LASERS, FIBER OPTICS. *Educ:* St John Fisher Col, Rochester, NY, BA, 73; Univ Conn, MS, 83, PhD(physics), 88. *Prof Exp:* Vis asst prof, Trinity College, Hartford, 85-88; ASST PROF, PHYSICS, SOUTHERN CONN STATE UNIV, 88- *Concurrent Pos:* Consult, Naval Underwater Systs Ctr, New London Lab. *Mem:* Am Phys Soc; Am Asn Physics Teachers; Optical Soc Am. *Res:* Experimental condensed matter research primarily concerned with high pressure effects and optical properties of solid state laser materials; photoluminescence spectroscopy; fiber optics and optical amplifiers. *Mailing Add:* Physics Dept Southern Conn State Univ New Haven CT 06515

DOLAN, JO ALENE, b Syracuse, NY, Dec 9, 41. BIOCHEMISTRY. *Educ:* Wellesley Col, BA, 63; Columbia Univ, PhD(pathobiol), 71. *Prof Exp:* SR RES BIOLOGIST DIABETES, MED RES DIV, LEDERLE LABS, AM CYANAMID CO, 77- *Concurrent Pos:* Res fel, Sloan-Kettering Inst Cancer Res, 71-73 & Roche Inst Molecular Biol, 73-75; cancer prog trainee, Fels Res Inst, Sch Med, Temple Univ, 75-77. *Mem:* Am Diabetes Asn; Endocrine Soc; Am Soc Pharmacol & Exp Therapeut. *Res:* Mechanism of hormone action; hormone-receptor interactions; various disease states associated with altered hormone-receptor interactions, insulin and diabetes; obesity. *Mailing Add:* Med Res Div Lederle Labs Am Cyanamid Co Middletown Rd Pearl River NY 10965

DOLAN, JOHN E, b 1923. SCIENCE ADMINISTRATION. *Prof Exp:* Vchmn & dir, Ames Elec Power Serv Corp, 50-88; CONSULT, 88- *Mem:* Nat Acad Eng; Inst Elec & Electronics Engrs. *Mailing Add:* 14448 Mark Dr Largo FL 34644

DOLAN, JOSEPH EDWARD, b Sterling, Colo, May 21, 45; m 67; c 2. ARTHROPOD ZOOLOGY, AQUATIC BIOLOGY. *Educ:* Colo State Univ, BS, 67, MS, 69; Univ Northern Colo, DA(biol sci), 78. *Prof Exp:* Instr zool, Cent Wyo Col, 69-75, chmn, Div Life & Phys Sci & instr zool, 78-86, dean extended studies, 86-88, PRES & VPRES/DEAN ACAD SERV, CENT WYO COL, 88- *Concurrent Pos:* Mem, Sci, Technol & Energy Authority Comn, Wyo. *Mem:* Assoc mem Sigma Xi; Nat Asn Biol Teachers; Am Mus Natural Hist. *Res:* Aquatic insects of running water. *Mailing Add:* Vpres & Dean Acad Serv Cent Wyo Col PO Box 1520 Riverton WY 82501

DOLAN, JOSEPH FRANCIS, b Rochester, NY, Sept 17, 39; m 71; c 2. ASTRONOMY, ASTROPHYSICS. *Educ:* St Bonaventure Univ, BS, 61; Harvard Univ, AM, 63, PhD(astrophys), 66. *Prof Exp:* Physicist, Smithsonian Astrophys Observ, 65-66; sr scientist, Jet Propulsion Lab, Calif Inst Technol, 66-68; asst prof astron, Case Western Reserve Univ, 68-75; Nat Res Coun sr res assoc, 75-77, ASTROPHYSICIST, LAB ASTRON & SOLAR PHYSICS, NASA GODDARD SPACE FLIGHT CTR, 77- *Concurrent Pos:* Vis scholar, Stanford Univ, 90-91. *Mem:* Am Astron Soc; Int Astron Union. *Res:* X-ray and gamma ray astronomy; astronomical polarization. *Mailing Add:* Code 681 Goddard Space Flight Ctr Greenbelt MD 20771-6810

DOLAN, KENNETH WILLIAM, b Yakima, Wash, July 23, 40; m 63; c 3. NUCLEAR PHYSICS. *Educ:* Univ Wash, BS, 62; Univ Ore, MS, 64, PhD(nuclear physics), 68. *Prof Exp:* Res asst, Univ Ore, 66-68; tech staff mem physics, 68-77, MEM SCI STAFF, SANDIA LABS, 77- *Mem:* Am Phys Soc. *Res:* Experimental low energy nuclear physics. *Mailing Add:* 4131 Davis Way Livermore CA 94550

DOLAN, LOUISE ANN, b Wilmington, Del, Apr 5, 50. THEORETICAL HIGH ENERGY PHYSICS. *Educ:* Wellesley Col, BA, 71; Mass Inst Technol, PhD(physics), 76. *Prof Exp:* Asst physics, Mass Inst Technol, 72-76; jr fel physics, Soc Fels, Harvard Univ, 76-79; from asst prof to assoc prof physics, Rockefeller Univ, 79-90, lab head, 90; PROF PHYSICS, UNIV NC, CHAPEL HILL, 90- *Concurrent Pos:* Vis scientist, Ecole Normale, Paris, 77;

vis fel, Princeton Univ, 78; fel, Am Phys Soc; John Simon Guggenheim fel, 88. *Honors & Awards:* Maria-Goeppert Mayer Award, 87. *Mem:* Am Phys Soc; Sigma Xi. *Res:* Coherent phenomena in quantum field theory including phase transitions, critical phenomena; spontaneous symmetry breakdown and nonabelian gauge theories; loop space; hidden symmetry; exactly integrable systems; spin systems; Kramers Wannier duality; transfer matrix; non-perturbative analysis; affine Kac-Moody Lie algebras; quantum gravity; unified dual string theory. *Mailing Add:* Dept Physics Univ NC Chapel Hill NC 27599-3255

DOLAN, THOMAS J(AMES), b Chicago, Ill, Dec 29, 06; m 29; c 2. MECHANICAL ENGINEERING. *Educ:* Univ Ill, BS, 29, MS, 32. *Prof Exp:* Asst & instr theoret & appl mech, Univ Ill, Urbana-Champaign, 29-37, from asst prof to assoc prof, 37-42, res prof, 45-52, head Dept Eng Mech, 52-70; CONSULT ENGR, 73- *Concurrent Pos:* Mem, Nat Res Develop Bd, 48; eng adv bd mem, NSF, 52-57; mem, Nat Res Coun, 55-; eng consult & mem bd dirs, Packer Eng; Am ed, Appl Mat Res, 62-73. *Honors & Awards:* Dudley Medal, Am Soc Testing & Mat, 52, Templin Award, 54. *Mem:* Fel Am Soc Mech Engrs (vpres, 58-60); fel Am Soc Testing & Mat; hon mem Soc Exp Stress Anal (pres, 51); Am Soc Eng Educ. *Res:* Fatigue of metals; photoelastic stress analysis; properties of metals; mechanical vibrations; metallurgy; failure analysis; product liability. *Mailing Add:* 401 Burwash Ave Savoy IL 61874

DOLBEAR, GEOFFREY EMERSON, b Richmond, Calif, June 1, 41; m 61; c 2. FUEL SCIENCE, SCIENCE COMMUNICATIONS. *Educ:* Univ Calif, Berkeley, BS, 62; Stanford Univ, PhD(inorg chem), 66. *Prof Exp:* Res chemist, Film Res Lab, E I du Pont de Nemours & Co, Inc, Del, 65-68, Cracking Catalyst Res Dept, 68-70, Automotive Exhaust Res Dept, 70-73 & Res Div, W R Grace & Co, 73-75; sr res chemist, Occidental Res Corp, 75-76, group leader, 76-79, mgr, 79-82; res assoc, Sci & Technol Div, Union Oil Co of Calif, 82-89; CONSULT, G E DOLBEAR & ASSOCS, 89- *Mem:* Am Chem Soc; Catalysis Soc. *Res:* Chemistry of processes for manufacturing chemical and transportation fuels from coal and heavy petroleum residue. *Mailing Add:* 23050 Aspen Knoll Dr Diamond Bar CA 91765

DOLBIER, WILLIAM READ, JR, b Elizabeth, NJ, Aug 17, 39; m 88; c 3. ORGANIC CHEMISTRY. *Educ:* Stetson Univ, BS, 61; Cornell Univ, PhD(org chem), 65. *Hon Degrees:* DSc, Stetson Univ, 90. *Prof Exp:* Fel org chem, Yale Univ, 65-66; assoc prof chem, 66-77, chmn dept, 83-88, PROF CHEM, UNIV FLA, 77- *Concurrent Pos:* A P Sloan fel, 70-72; Guggenheim fel, 73-74; SERC fel, 91. *Mem:* Am Chem Soc. *Res:* Thermal rearrangement of small ring hydrocarbons; kinetic substituent effect studies of pericycle reactions; mechanistic fluorine chemistry; synthetic fluorine chemistry. *Mailing Add:* Dept Chem Univ Fla Gainesville FL 32611-2046

DOLBY, JAMES LOUIS, applied statistics; deceased, see previous edition for last biography

DOLBY, LLOYD JAY, b Elgin, Ill, Oct 5, 35; m 56; c 3. ORGANIC CHEMISTRY, NATURAL PRODUCTS CHEMISTRY. *Educ:* Univ Ill, BS, 56; Univ Calif, Berkeley, PhD(chem), 59. *Prof Exp:* NSF fel chem, Univ Wis, 59-60; from asst prof to assoc prof chem, Univ Ore, 60-69, prof, 69-; AT ORGANIC CONSULTS, ORE. *Concurrent Pos:* Sloan Found res fel, 65-67. *Mem:* Am Chem Soc. *Res:* Synthesis of natural products; reaction mechanisms. *Mailing Add:* Dept Chem Univ Oregon Eugene OR 97403

DOLCH, WILLIAM LEE, b Kansas City, Mo, July 11, 25; m 48; c 2. CEMENT & CONCRETE. *Educ:* Purdue Univ, BSChE, 47, MS, 49, PhD(chem), 56. *Prof Exp:* Asst hwy engr, 47-56, res assoc & asst prof, 56-60, assoc prof, 60-64, PROF ENG MAT, SCH CIVIL ENG, PURDUE UNIV, WEST LAFAYETTE, 64- *Honors & Awards:* Dudley Medal, Am Soc Testing & Mat, 66; Wason Medal, Am Concrete Inst, 68, Anderson Medal, 84. *Mem:* Fel Am Soc Testing & Mat; fel Am Concrete Inst. *Res:* Physico-chemical properties of engineering materials; concrete and concrete aggregates; problems of concrete. *Mailing Add:* Sch Civil Eng Purdue Univ West Lafayette IN 47907

DOLE, HOLLIS MATHEWS, economic geology; deceased, see previous edition for last biography

DOLE, JIM, b Phoenix, Ariz, May 28, 35; m; c 3. VERTEBRATE ECOLOGY. *Educ:* Ariz State Univ, BA, 57; Univ Mich, MS, 59, PhD(zool), 63. *Prof Exp:* Instr zool, Univ Mich, 62-63; from asst prof to assoc prof biol, 63-70, PROF BIOL, CALIF STATE UNIV, NORTHRIDGE, 70- *Mem:* AAAS; Ecol Soc Am; Am Soc Ichthyologists & Herpetologists; Animal Behavior Soc. *Res:* Vertebrate ecology and behavior; population ecology. *Mailing Add:* Dept Biol Calif State Univ Northridge CA 91330

DOLE, MALCOLM, physical chemistry; deceased, see previous edition for last biography

DOLE, VINCENT PAUL, b Chicago, Ill, May 8, 13; m 42; c 3. MEDICAL RESEARCH, PHARMACOLOGY. *Educ:* Stanford Univ, AB, 34; Harvard Univ, MD, 39. *Hon Degrees:* MD, Uppsala, 88, Cagliori, 89. *Prof Exp:* Intern, Mass Gen Hosp, Boston, 40-41; from asst resident to resident, 41-46, assoc mem & assoc physician, 47-51, mem & physician, Rockefeller Inst & Hosp, 51-59, prof med, 59-83, EMER PROF MED & EMER SR PHYSICIAN, ROCKEFELLER UNIV, 83- *Concurrent Pos:* Co-ed, J Exp Med, 56-65. *Honors & Awards:* Lasker Award, 88. *Mem:* Nat Acad Sci; AAAS; Am Soc Clin Invest; Soc Exp Biol & Med; Asn Am Physicians. *Res:* Cardiovascular and metabolic research; drug addiction. *Mailing Add:* Dept Med Rockefeller Univ New York NY 10021

DOLE, WILLIAM PAUL, b Ellenville, NY, May 31, 47. CORONARY PHYSIOLOGY. *Educ:* NY Univ, MD, 73. *Prof Exp:* Asst prof med, Univ Tex, 79-85; ASSOC PROF MED, UNIV IOWA, 85- *Mailing Add:* Berlex Labs Inc 110 E Hanover Ave Cedar Knolls NJ 07927

DOLECEK, ELWYN HAYDN, b Lubbock, Tex, Mar 1, 46; m 70; c 2. RADIOLOGICAL PHYSICS. *Educ:* Valparaiso Univ, BS, 68; Rutgers Univ, MS, 71, PhD(radiol physics), 75. *Prof Exp:* Oceanogr, Naval Oceanog Off, 67-69; radiol physicist health physics, Rutgers Univ, 71-77; HEALTH PHYSICIST, ARGONNE NAT LAB, 77- *Mem:* Nat Health Physics Soc. *Res:* Thermoluminescent dosimetry with emphasis on lithium fluoride dosimeters as well as the development of DNA as a practical thermoluminescent dosimeter for measuring human exposure to ionizing radiation; current research focus on CR-39 neutron dosimetry and automated TLD personnel dosimetry systems. *Mailing Add:* OHS/HP Argonne Nat Lab 9700 Cass Ave Argonne IL 60439

DOLENKO, ALLAN JOHN, b Winnipeg, Man, June 24, 42; m 65; c 2. ORGANIC POLYMER CHEMISTRY. *Educ:* Univ Man, BSc, 64, MSc, 65; Queen's Univ, PhD(chem), 70. *Prof Exp:* Sr chemist, Chemco Ltd, 70 & Actol Chem Ltd, 71; res scientist, Eastern Forest Prod Lab, Environ Can, 71-80; DIR ALTERNATE ENERGY DIV, ENERGY MINES & RESOURCES, CAN, 80- *Mem:* Forest Prod Res Soc. *Mailing Add:* 2043 Kingsgrove Cresent Gloucester ON K1J 6E9 Can

DOLES, JOHN HENRY, III, b Youngstown, Ohio, Nov 20, 43. OCEAN ACOUSTICS, SYSTEMS ENGINEERING. *Educ:* Case Inst Technol, BS, 65; Mass Inst Technol, PhD(math), 69. *Prof Exp:* mem tech staff, 69-85, TECH SUPVR, BELL LABS, 86- *Concurrent Pos:* Chmn bd dirs, Whippany Fed Credit Union, distinguished mem tech staff, 83- *Mem:* Soc Indust & Appl Math; Am Geophys Union; Acoust Soc Am; Inst Elec Electronics Engrs. *Res:* Computational physics; ionospheric physics; ocean acoustics; phased array design. *Mailing Add:* AT&T Bell Labs 1919 S Eads St Arlington VA 22202

DOLEZAL, VACLAV J, b Ceska Trebova, Czech, May 21, 27; m 63; c 3. APPLIED MATHEMATICS. *Educ:* Tech Univ Prague, Ing, 49; Czech Acad Sci, CSc(appl math), 56, DrSc(appl math), 66. *Prof Exp:* Res assoc math, Math Inst, Czech Acad Sci, 51-56, mathematician, 56-65; vis prof appl math, Eng Col, State Univ NY Stony Brook, 65-66; mathematician, Math Inst, Czech Acad Sci, 66-68; PROF APPL MATH, STATE UNIV NY STONY BROOK, 68- *Concurrent Pos:* Vis prof, Univ Linz, Austria, 74-75; NSF grants. *Res:* Operator-theoretic analysis of nonlinear systems. *Mailing Add:* Dept Appl Math & Statist State Univ NY Stony Brook NY 11794-3600

DOLEZALEK, HANS, b Berlin, Germany, May 14, 12; US citizen; m 51; c 3. MARITIME REMOTE SENSING, ATM ELEC FOR GLOBAL CHANGE. *Educ:* Aachen Tech Univ, Diplom, 56. *Prof Exp:* Chief int rels, Dept Foreign Acad, Charlottenburg, Ger, 32-41; physicist, Aeronaut Commun Res Inst, Oberpfaffenhofen, 44-45; scientist with Dr H Israel, Aachen, 50-57; asst head, Meteorol Observ, 57-61; group leader atmospheric physics, Space Systs Div, Avco Corp, 61-66; consult, Boston Col, 66-67; physicist, 67-90, LIAISON SCIENTIST, EUROP OFF, OFF NAVAL RES, LONDON, 90- *Concurrent Pos:* Chmn subcomt I, Int Comn Atmospheric Elec, 63-75 & 84-, secy, 64-75, chmn subcomt VII, 75-84, subcomt I, 84-; chmn comt atmospheric & space elec, Am Geophys Union, 69-74; mem working group atmospheric elec, Comn Atmospheric Sci, World Meteorol Orgn, 68-78, US deleg session V, 70, rapporteur atmospheric elec, 74-78, US deleg session Caracas, 75; chmn, res study group, maritime remote sensing NATO/DRG/III, 78-80, US delegate, 80-, mem, adv bd, Fraunhofer Inst Atmospheric Environ Res, 74-86. *Honors & Awards:* Hon Mem Int Comn Atmospheric Elec. *Mem:* AAAS; Am Meteorol Soc; Am Geophys Union; NY Acad Sci; German Acad Exch Serv. *Res:* Atmospheric Electricity; maritime remote sensing; synthetic aperture radar for ocean use. *Mailing Add:* 1812 Drury Lane Alexandria VA 22307-1914

DOLFINI, JOSEPH E, b Middletown, NY, July 11, 36; m 85. ORGANIC CHEMISTRY, MEDICINAL CHEMISTRY. *Educ:* Univ Mich, BS, 57, MS, 61, PhD(chem), 62. *Prof Exp:* NIH fel org chem, Columbia Univ, 61-63; asst prof chem, Purdue Univ, West Lafayette, 63-65; res supvr, Squibb Inst, 65-73; head dept org chem, Merrell-Nat Labs, 73-78, assoc dir drug develop, 78-81; PRES & SCI CONSULT, NOBILENT INC, CINCINNATI, 81- *Mem:* Am Chem Soc; Royal Soc Chem; NY Acad Sci. *Res:* Synthetic organic chemistry; heterocyclic and carbocyclic synthesis; antibiotics; natural products synthesis; flavor chemistry; fermentation chemistry. *Mailing Add:* Suite 133 Nobilent Inc 10999 Reed Hartman Hwy Cincinnati OH 45242

DOLGIN, MARTIN, b New York, NY, Apr 12, 19; m 50; c 3. CARDIOLOGY, INTERNAL MEDICINE. *Educ:* NY Univ, AB, 40, MD, 43; Am Bd Internal Med, dipl, 51 & 74; Am Bd Cardiovasc Dis, dipl, 56. *Prof Exp:* Intern & asst resident med, Lincoln Hosp, New York, 43-45; from instr to assoc prof, Sch Med, NY Univ, 48-73; chief cardiol sect, NY Vet Admin Hosp, 54-89; PROF CLIN MED, SCH MED, NY UNIV, 73- *Concurrent Pos:* Fel internal med, Lahey Clin, Boston, 45-47; fel cardiovasc dis res, Michael Reese Hosp, Chicago, 47-48; lectr med, Columbia Univ, 48-68; consult cardiol, Will Rogers Hosp, Saranac, NY, 61-71 & Columbus Hosp, New York, 68-72; attend physician, Bellevue Hosp & Univ Hosp, New York, 73- *Mem:* Am Heart Asn; fel Am Col Physicians; fel Am Col Cardiol; Am Fedn Clin Res. *Res:* Clinical electrocardiography. *Mailing Add:* Dept Clin Med NY Univ Sch Med 550 First Ave New York NY 10016

DOLGOFF, ABRAHAM, b New Brunswick, NJ, Aug 10, 33; m 56; c 1. ENGINEERING GEOLOGY, PETROLEUM GEOLOGY. *Educ:* City Col New York, BS, 53; NY Univ, MS, 58; Rice Univ, PhD(geol), 60. *Prof Exp:* Jr civil engr, Bd Water Supply, City New York, 53-55, asst civil engr, Dept Pub Works, 55-58; res geologist, Rice Univ, 60-61 & Bellaire Labs, Texaco, Inc, 61-66; specialist eng geol of dams, Gerard Eng Inc, 66-71 & eng geol of dams & tunnels, Harza Eng Co, 71-72; sr geologist, 72-77, SR SPECIALIST GEOL & QUAL ASSURANCE ADMINR, SARGENT & LUNDY, ENGRS, 77- *Concurrent Pos:* Lectr, Hunter Col, 70; gen chmn, Annual Meeting, Asn Eng Geologists, 76-79; qual assurance expert, Sargent & Lundy, 81- *Mem:* Geol Soc Am; Asn Eng Geol. *Res:* Applied geology-geologic, seismic environmental, aspects, siting and construction engineering works (nuclear power plants, waste disposal facilities); quality assurance program writing and adminstration; regional factors, petroleum accumulation. *Mailing Add:* Sargent & Lundy Engrs 55 E Monroe St Chicago IL 60603

DOLHERT, LEONARD EDWARD, b Boston, Mass, July 16, 57. HIGH TEMPERATURE SUPERCONDUCTORS, CERAMICS RESEARCH. *Educ:* Mass Inst Technol, SB, 79, PhD(ceramics), 85. *Prof Exp:* Ceramics researcher, Raytheon Co , 79; vis scientist, Mass Inst Technol, 86; RES ENGR, W R GRACE & CO, WASH RES CTR, 87- *Mem:* Am Ceramic Soc. *Res:* Ceramic electronic packaging. *Mailing Add:* W R Grace & Co 7379 Rte 32 Columbia MD 21044

DOLHINOW, PHYLLIS CAROL, b Elgin, Ill, Nov 6, 33; m 68; c 1. PHYSICAL ANTHROPOLOGY, PRIMATOLOGY. *Educ:* Beloit Col, AB, 55; Univ Chicago, AM, 60, PhD(anthrop), 63. *Prof Exp:* Lectr anthrop, Univ Calif, Berkeley, 61-62; asst prof, Columbia Univ, 63-64; asst prof, Univ Calif, Davis, 64-66; from asst prof to assoc prof, 66-71, PROF ANTHROP, UNIV CALIF, BERKELEY, 71- *Concurrent Pos:* Fel, Ctr Advan Study Behav Sci, 62-63; asst res anthropologist, Nat Ctr Primate Biol, 64-66; Wenner-Gren Found anthrop res grant primatology in Africa, 66-68; NSF grant, Primate Behav Unit, Univ Calif, Berkeley, 67-71, NIMH grant, 73-80, Miller prof, 75-76; Smithsonian Inst grant primatology in India, 85-87. *Mem:* Fel AAAS; Sigma Xi; fel Am Anthrop Asn; Ecol Soc Am; Animal Behav Soc; Int Primatological Asn; Am Primatological Asn. *Res:* Primate social behavior and ecology in Asia and Africa; field station for behavior research colony primate development; human evolution and behavior; attachment and social bonds. *Mailing Add:* Dept Anthrop Univ Calif Berkeley CA 94720

DOLIN, MORTON IRWIN, b Brooklyn, NY, Dec 24, 20. BIOCHEMISTRY. *Educ:* City Col New York, BS, 42; Univ Ky, MS, 44; Ind Univ, PhD(bact), 50. *Prof Exp:* Res fel, Univ Ill, 50-52; biochemist, Oak Ridge Nat Lab, 52-69; vis investr, Agr Univ, Wageningen, 70-71; NIH spec fel, Dept Biochem, 71-73, RES ASSOC MICROBIOL, UNIV MICH, ANN ARBOR, 73- *Concurrent Pos:* Guggenheim fel, 59-60. *Mem:* Am Soc Microbiol; Am Soc Biol Chemists. *Res:* Electron transport reactions; pyridine nucleotides. *Mailing Add:* 3540 Greenbrier Blvd Ann Arbor MI 48105

DOLIN, RICHARD, b Brooklyn, NY, May 16, 21; m 45; c 3. ELECTRICAL ENGINEERING. *Educ:* NY Univ, BS, 47, MS, 50. *Prof Exp:* Instr elec eng, NY Univ, 47-51; engr-in-charge commun & control systs, Res Labs, Sylvania Elec Prod, NY, 51-56; prin engr reconnaissance & electronic systs, Bulova Res & Develop Labs, 56-58; sr res engr, Fairchild Camera & Instrument Co, 58-59; from assoc prof to prof, 61-83, chmn dept, 68-70 & 77-80, dir technol & pub policy, 79-83, EMER PROF ENG, HOFSTRA UNIV, 84-, CONSULT, 84- *Concurrent Pos:* Consult, Airborne Instruments Lab Div, Cutler-Hammer Inc, 59-61. *Mem:* Sr mem Inst Elec & Electronics Engrs; Am Soc Eng Educ. *Res:* Analysis and design of communications and reconnaissance systems. *Mailing Add:* 941 Alice Ct North Bellmore NY 11710

DOLL, CHARLES GEORGE, stratigraphic geology, paleontology; deceased, see previous edition for last biography

DOLL, EUGENE CARTER, b Ft Benton, Mont, Feb 15, 21; m 47; c 3. SOIL FERTILITY. *Educ:* Mont State Col, BS, 49, MS, 51; Univ Wis, PhD(soils), 53. *Prof Exp:* Asst, Mont State Col, 50-51; from asst agronomist to assoc agronomist soil fertil, Univ Ky, 53-60, assoc prof soils, 58-60; from assoc prof soils to prof soil sci, Mich State Univ, 60-74; agronomist, Int Staff, Tenn Valley Authority, 74-75; soil scientist, Int Fertilizer Develop Ctr, 75-76; int training, USDA, 78-80; supt, Land Reclamation Res Ctr, NDak State Univ Reclamation, 81-91; RETIRED. *Concurrent Pos:* Regional maize adv, Int Atomic Energy Agency, 66; regional dir, NC State Univ Proj for AID, Ecuador & Panama, 70-71. *Mem:* Fel Am Soc Agron; fel Soil Sci Soc Am. *Res:* Nitrogen, phosphorus and potash requirements of general farm crops; phosphorus availability in tropical and subtropical soils and the evaluation of various sources of phosphate fertilizers, including phosphate rock; reclamation techniques for returning stripmined lands to full agricultural productivity. *Mailing Add:* NDak State Univ Reclamation PO Box 459 Mandan ND 58554

DOLL, JERRY DENNIS, b Pocahontas, Ill, Oct 26, 43; m 73; c 4. PLANT SCIENCE, WEED SCIENCE. *Educ:* Univ Ill, BS, 65; Mich State Univ, MS, 66, PhD(crop sci), 69. *Prof Exp:* Vol, Peace Corps, 69-71; weed scientist, Int Ctr Trop Agr, 71-76; WEED SCIENTIST, UNIV WIS, MADISON, 77- *Concurrent Pos:* Secy-treas, North Cent Weed Control Conf, 81-89; bd dirs, Weed Sci Soc Am, 84-86. *Mem:* Weed Sci Soc Am; Am Soc Agron; hon mem Colombian Weed Sci Soc (vpres, 74, pres, 75); Int Weed Sci Soc; Europ Weed Res Soc. *Res:* Perennial weed research, especially on quackgrass; integrated weed management; alternative weed management. *Mailing Add:* Dept Agron Univ Wis Madison WI 53706-1597

DOLL, JIMMIE DAVE, b San Diego, Calif, Oct 19, 45; m 66; c 1. PHYSICAL CHEMISTRY. *Educ:* Univ Kans, BS, 67; Harvard Univ, PhD(phys chem), 71. *Prof Exp:* From asst prof to assoc prof chem, State Univ NY, Stony Brook, 75-80; staff mem, Los Alamos Nat Lab, 79-81, fel, 81-90; PROF CHEM, BROWN UNIV, 90- *Concurrent Pos:* NSF fel, 71-72; Sloan Found fel, 76-78; vis staff mem, Los Alamos Sci Lab, 77- *Mem:* Am Phys Soc. *Res:* Theoretical chemistry; surface chemistry. *Mailing Add:* 51 Bay Rd Barrington RI 02806-4758

DOLL, RICHARD, b Oct 28, 12. MEDICINE. *Educ:* Univ London, MB, BS, 37, MD, 34, DSc, 58; Oxford Univ, DM, 69; FRCP, 57; FRS, 66; FFOM, 87. *Hon Degrees:* DSc, Newcastle Univ, 69, Belfast Univ, 72, Reading Univ, 73, Nfld Univ, 73, Stony Brook Univ, 88, Harvard Univ, 88, London Univ, 88 & Oxon Univ, 89; DM, Tasmania Univ, 75; FRCGP, 78. *Prof Exp:* Casualty officer & house physician, St Thomas's Hosp, 38-39, jr asst, Med Unit, 45; res asst, Cent Middlesex Hosp, 46-47, hon assoc physician, 49-69; mem, Statist Res Unit, Med Res Coun, 48-69, dep dir, 56-60, dir, 61-69, dep dir, Clin Res Ctr, 66-69; Regius prof med, Univ Oxford, 69-79, hon physician, Radcliffe Infirmary, 69-79, dir, Imp Cancer Res Fund, Cancer Epidemiol & Clin Trials Unit, 78-83 & 87-89, HON MEM, CANCER STUDIES UNIT, IMP CANCER RES FUND, OXFORD, 89-, EMER PROF MED, CANCER STUDIES UNIT, UNIV OXFORD, 89- *Concurrent Pos:* House physician, Royal Postgrad Med Sch, Hammersmith, 39; William Julius Mickle fel, Univ London, 55; hon lectr epidemiol, London Sch Hyg & Trop Med, 56-62 & Univ Col Hosp Med Sch, 62-69; chmn, Adverse Reactions Subcomt, Comt on Safety of Med, Med Res Coun, 70-77, Cancer Coord Comt, 73-77, Mgt Comt, Inst Cancer Res, 78-87 & Comt Epidemiol of AIDS, Med Res Coun, 85-88; warden, Green Col, Oxford, 79-83; mem, Radiation & Cancer Subcomt, UK Coord Comt Cancer Res, 88; mem, Comn Health & Environ, WHO, 90- *Honors & Awards:* David Anderson Berry Prize, Royal Soc Edinburgh, 58; Bisset Hawkins Medal, Royal Col Physicians, 62; Award for Cancer Res, UN, 62; Buchanan Medal, Royal Soc, 72, Royal Medal, 86; Nuffield Medal, Royal Soc Med, 73; Presidential Award, NY Acad Sci, 74; Snow Award, Epidemiol Sect, Am Pub Health Asn, 76; Gold Medal, Royal Inst Pub Health, 77; Charles S Mott Prize for Cancer Res, 79; Gold Medal, Brit Med Asn, 83; Wilhelm Conrad Röntgen Prize, Acad Lincei, Rome, 84; Johann-Georg-Zimmermann Prize, 85; Alton Ochsner Award, 88. *Mem:* Hon mem Am Asn Cancer Res; hon mem Am Gastroenterol Asn; hon mem Am Epidemiol Soc; hon mem Am Acad Arts & Sci; hon mem Ital Oncol Soc; hon mem Norweg Acad Sci. *Res:* Epidemiology. *Mailing Add:* Clin Trial Serv Unit Harkness Bldg Radcliffe Infirmary Oxford OX2 6HE England

DOLLAHON, JAMES CLIFFORD, animal breeding; deceased, see previous edition for last biography

DOLLAHON, NORMAN RICHARD, b Gallup, NMex, Apr 22, 44; m 66; c 1. PROTOZOOLOGY, PARASITOLOGY. *Educ:* Univ NMex, BS, 66, MS, 68; Univ Nebr-Lincoln, PhD(zool), 71. *Prof Exp:* USPHS trainee parasitol, Rutgers Univ, 72-73; ASST PROF BIOL, VILLANOVA UNIV, 73- *Mem:* Am Soc Parasitologists; Soc Protozoologists. *Res:* Study of infections of insects, reptiles and mammals with trypanosomatid flagellates. *Mailing Add:* Dept Biol Villanova Univ Villanova PA 19085

DOLLAR, ALEXANDER M, b Vancouver, BC, Apr 7, 21; US citizen; m 44; c 1. ANIMAL NUTRITION, FOOD BIOCHEMISTRY. *Educ:* Univ Calif, Berkeley, BS, 48, MS, 49; Reading Univ, PhD(nutrit & biochem), 58. *Prof Exp:* From asst prof to assoc prof food sci, Col Fisheries, Univ Wash, 59-67; supvr, Hawaii Develop Irradiator, Hawaii Dept Agr, 67-76; prof environ health sci, Univ Hawaii, 76-85, emer prof, 85-; RETIRED. *Concurrent Pos:* UN-FAO consult, Far East Countries, 84-87. *Mem:* AAAS; Am Chem Soc; Inst Food Technologists. *Mailing Add:* 3492A Monte Hermoso Laguna Hills CA 92653

DOLLARD, JOHN D, b New Haven, Conn, Jan 19, 37. PHYSICS, MATHEMATICS. *Educ:* Yale Univ, BA, 58; Princeton Univ, MA, 60, PhD(physics), 63. *Prof Exp:* Res assoc physics, Princeton Univ, 63-65; asst prof math, Univ Rochester, 65-68; vis asst prof math & physics, Yale Univ, 68-69; assoc prof, 69-79, PROF MATH, UNIV TEX, AUSTIN, 79-, CHMN DEPT, 83- *Mem:* Am Phys Soc; Am Math Soc; Math Asn Am. *Res:* Quantum mechanical scattering theory. *Mailing Add:* Dept Math Univ Tex Austin Austin TX 78712

DOLLHOPF, WILLIAM EDWARD, b Pittsburgh, Pa, June 13, 42; m 64; c 2. NUCLEAR PHYSICS. *Educ:* Thiel Col, AB, 64; Western Reserve Univ, MS, 67; Col William & Mary, PhD(physics), 75. *Prof Exp:* Instr physics, Alliance Col, 66-69; accelerator scientist, Space Radiation Effects Lab, Newport News, 74-75; asst prof physics, Wabash Col, 75-80; ASSOC PROF PHYSICS, WITTENBERG UNIV, 80- *Concurrent Pos:* Beam transp consult, Space Radiation Effects Lab, Newport News, 74-75. *Mem:* Am Asn Physics Teachers; Am Phys Soc; Sigma Xi. *Res:* Quasi-elastic nuclear reactions at intermediate energies to determine the structure of clusters within nuclei; charged particle beam transport systems; science education for elementary and middle schools. *Mailing Add:* Dept Physics Wittenberg Univ PO Box 720 Springfield OH 45501

DOLLIMORE, DAVID, b London, Eng, Apr 3, 28; m 54; c 3. CONCENTRATED SUSPENSIONS, THERMAL ANALYSIS. *Educ:* London Univ, BSc, 49, PhD(chem), 52. *Hon Degrees:* DSc, London Univ, 76. *Prof Exp:* Res asst, Exeter Univ, 52-54; asst lectr chem, St Andrews Univ, 54-56; prin lectr chem, Royal Col Advan Technol, Salford, 56-64; sr lectr & reader chem, Univ Salford, 64-82; chmn dept, 82-84, PROF CHEM, UNIV TOLEDO, 82- *Honors & Awards:* Mettler Award, NAm Thermal Anal Soc, 79; Silver Medal, Il Cemento, 80. *Mem:* Fel Royal Soc Chem; fel Am Inst Chemists; fel Inst Ceramics; Am Chem Soc; Am Ceramics Soc; Int Confedn Thermal Anal. *Res:* Surface chemistry and the heat treatment of solids; studies of cement systems, glass technology, coal science and slurry behavior; studies of liquids and gas on ruled adsorbents. *Mailing Add:* Dept Chem Univ Toledo Toledo OH 43606

DOLLING, DAVID STANLEY, b Bournemouth, UK, Mar 21, 50; m; c 2. HIGH SPEED EXTERNAL AERODYNAMICS. *Educ:* London Univ, BSc, 71, PhD(aerospace sci), 77; von Karman Inst, Belg, dipl, 74. *Prof Exp:* Aerodynamicist, Hawker Siddeley Dynamics, UK, 71-73; mem res staff & lectr viscous flows, Mech & Aerospace Eng Dept, Princeton Univ, 76-; AT DEPT AEROSPACE ENG & MECH ENG, UNIV TEX, AUSTIN. *Mem:* Am Inst Aeronaut; Royal Aeronaut Soc. *Res:* Supersonic fluid dynamics; interactions of shock waves with turbulent boundary layers. *Mailing Add:* Dept Aerospace Eng & Mech Eng Univ Tex Austin TX 78712

DOLLING, GERALD, b Dunstable, Eng, Nov 21, 35; m 59; c 3. SOLID STATE PHYSICS. *Educ:* Cambridge Univ, BA, 57, PhD(physics), 61. *Prof Exp:* Head, Neutron & Solid State Physics Br, 79-85, dir, physics div, 85-89, RES OFFICER PHYSICS, ATOMIC ENERGY CAN LTD, 61-, VPRES, PHYS SCI, 90- *Mem:* Fel Am Phys Soc; Can Asn Physicists; French Can Asn Adv Sci; Chem Inst Can; Can Nuclear Asn. *Res:* Lattice and magnetic dynamics; neutron scattering from condensed systems; ferroelectricity; management of research. *Mailing Add:* Chalk River Labs AECL-Res Chalk River ON K0J 1J0 Can

DOLLINGER, ELWOOD JOHNSON, b Lynchburg, Ohio, Apr 20, 20; m 55; c 2. CYTOGENETICS, PLANT BREEDING. *Educ:* Ohio State Univ, BSc, 44; Pa State Univ, MSc, 47; Columbia Univ, PhD(cytogenetics, bot), 53. *Prof Exp:* Agt, US Regional Pasture Res Lab, 44-47; asst, Columbia Univ, 47-48 & Univ Ill, 49-50; res assoc, Carnegie Inst, 50-51 & Brookhaven Nat Lab, 51-53; USPHS fel, Nat Cancer Inst, 53-55; prof, 55-85, EMER PROF CYTOGENETICS, OHIO AGR RES & DEVELOP CTR, 85- *Mem:* Genetics Soc Am; Am Genetic Asn; Am Soc Agron; Sigma Xi. *Res:* Cytogenetics of maize; radiation induced mutation of maize; corn breeding. *Mailing Add:* Dept Agron Ohio State Univ Agr Res & Develop Ctr Wooster OH 44691-2614

DOLLWET, HELMAR HERMANN ADOLF, b Merzig, Ger, Jan 20, 29; US citizen; div; c 4. PLANT PHYSIOLOGY, PLANT BIOCHEMISTRY. *Educ:* Univ Mich, BS, 58; Univ Calif, Riverside, MS, 67, PhD(plant sci & physiol), 69. *Prof Exp:* From asst prof to assoc prof, 70-85, PROF BIOL, UNIV AKRON, 85- *Concurrent Pos:* Radiation safety officer, Univ Akron. *Mem:* Sigma Xi. *Res:* Antiinflamatory properties of copper in rheumatoid diseases; radon expert. *Mailing Add:* Dept Biol Univ Akron Akron OH 44304

DOLLY, EDWARD DAWSON, b Davenport, Iowa, July 29, 40; m 60; c 2. GEOLOGY. *Educ:* Univ Ill, BS, 63; Univ Okla, MS, 65, PhD(geol), 69. *Prof Exp:* Explor geologist, Shell Oil Co, 69-73; explor geologist oil & gas, Filon Explor Corp, 73-78; INDEPENDENT GEOLOGIST, 78-; OWNER/ PARTNER, JAMES ENERGY MGT, 86- *Honors & Awards:* Leverson Award, Am Asn Petrol Geologists, 80; Explorer of Yr, Rocky Mountain Asn Geologists, 80. *Mem:* Am Asn Petrol Geologists; Sigma Xi; Rocky Mountain Asn Geologists. *Res:* Oil, gas and mineral exploration worldwide. *Mailing Add:* 1580 Lincoln Denver CO 80203

DOLMAN, CLAUDE ERNEST, b Porthleven, Cornwall Eng, May 23, 06; m 31, 55; c 6. MEDICAL MICROBIOLOGY. *Educ:* Univ London, MB, BS, 30, DPH, 31, PhD(bact), 35. *Prof Exp:* House surgeon, St Mary's Hosp, 29; clin asst, Royal Chest Hosp & Hosp Sick Children, London, 30; res scholar, St Mary's Hosp Inst Path & Res, 31; clin assoc & res asst, Connaught Labs, Toronto, 31-33, res & clin assoc, 33-35, demonstr hyg & prev med, 33-35; assoc prof bact & prev med & actg head dept, 35-36, prof & head dept, 36-51, prof bact & immunol & head dept, 51-65, res prof microbiol, 65-71, EMER PROF MICROBIOL, UNIV BC, 71- *Concurrent Pos:* res mem, Connaught Labs, 35-73; actg head dept nursing & health, 35-43, prof & head dept, 43-51; dir div labs, Dept Health, BC, 35-56; hon consult bacteriologist, Vancouver Gen Hosp, 36- *Honors & Awards:* Coronation Medal, 53; Jubilee Medal, 78. *Mem:* Fel Am Pub Health Asn; hon life mem Can Pub Health Asn; fel Royal Soc Can (pres, 69-70); hon life mem, Can Asn Med Microbiologists (pres, 64-66); Am Soc Microbiologists. *Res:* Staphylococcus toxins; botulinum toxins; brucellosis; cholera; salmonellosis; bacterial food poisoning; history of bacteriology and microbiology. *Mailing Add:* 1611 Cedar Crescent Vancouver BC V6J 2P8 Can

DOLNY, GARY MARK, b Nanticoke, Pa, June 17, 55; m 85; c 1. SEMICONDUCTOR DEVICE DESIGN AND FABRICATION, INTEGRATED CIRCUIT DEVELOPMENT. *Educ:* Haverford Col, BS, 77; Univ Pittsburg, MS, 78, PhD(elec eng), 81. *Prof Exp:* Asst prof elec eng, Dept Elec Eng, Wilkes Univ-Pa, 81-85; MEM TECH STAFF, DAVID SARNOFF RES CTR CN5300-PRINCETON, NJ, 85- *Mem:* Inst Elec & Electronic Engrs. *Res:* Silicon process technology; IC design; bicmos technology; modeling, simulation and CAD; polycrystalline silicon thin film transistor development; smart power integrated circuits. *Mailing Add:* David Sarnoff Res Ctr CN5300 Princeton NJ 08543-5300

DOLOVICH, JERRY, b Winnipeg, Man, Can, Apr 2, 36; m 63; c 3. ALLERGY. *Educ:* Univ Man, MD, 59. *Prof Exp:* PROF, DEPT PEDIAT, FAC HEALTH SCI, MCMASTER UNIV, 78- *Res:* Allergy; immunology; asthma. *Mailing Add:* 391 Queen St S Hamilton ON L8P 3T8 Can

DOLPH, CHARLES LAURIE, b Ann Arbor, Mich, Aug 27, 18; m 44; c 2. APPLIED MATHEMATICS. *Educ:* Univ Mich, BA, 39; Princeton Univ, MA, 41, PhD(math), 44. *Prof Exp:* Asst, Nat Defense Res Comt, Princeton Univ, 41-42, instr math, 42; theoret physicist, Naval Res Lab, Washington, DC, 43-45; mem tech staff, Math Group, Bell Tel Labs, 45-46; from asst prof to assoc prof math, 46-59, math res, Eng Res Inst, 46-64, head theoret & comput div, Willow Run Res Ctr, 52-64, PROF MATH, UNIV MICH, ANN ARBOR, 59- *Concurrent Pos:* Consult, Ramo-Wooldridge, 54-57 & Bendix Aviation Corp, 59-; Guggenheim fel, 57; mem Nat Acad Sci-Nat Res Coun adv comt math, Off Naval Res, 64-66; vis res prof, Army Math Res Ctr, Wis, 65; vis prof, Univ Stuttgart, 72. *Honors & Awards:* Thompson Prize, Inst Elec & Electronics Engineers. *Mem:* AAAS; Am Math Soc. *Res:* Nonlinear integral equations; antenna theory; theory of compressible flow; stochastic processes; vibration theory; anomalous propagation theory; transform theory. *Mailing Add:* Dept Math Univ Mich 347 W Eng Ann Arbor MI 48104

DOLPH, GARY EDWARD, b Binghamton, NY, Oct 17, 46; m 82; c 3. PALEOBOTANY. *Educ:* State Univ NY Binghamton, BA, 68; Ind Univ, MA, 73, PhD(bot), 74. *Prof Exp:* From asst prof to assoc prof, 74-85, PROF PLANT SCI, IND UNIV, 85- *Concurrent Pos:* Ed, Int Asn Angiosperm Paleobotanists & Paleobotanical Sect, Bot Soc Am; grants, Exxon Educ Found & NSF Comprehensive Asst Undergrad Sci Educ. *Mem:* Paleont Soc; Bot Soc Am; Sigma Xi; Torrey Bot Club. *Res:* Multivariate statistics in analyzing the gross morphology and cuticular structure of modern and fossil leaves; fossil flowers from the Cretaceous-Eocene sediments of western Tennessee and Kentucky. *Mailing Add:* Ind Univ Kokomo 2300 S Washington St Kokomo IN 46904-9003

DOLPHIN, DAVID HENRY, b London, Eng, Jan 15, 40; m 63; c 3. BIO-ORGANIC CHEMISTRY, BIOINORGANIC CHEMISTRY. *Educ:* Nottingham Univ, BSc, 62, PhD(chem), 65. *Hon Degrees:* DSc, Nottingham Univ, 82. *Prof Exp:* Res fel chem, Harvard Univ, 65-66, from instr to assoc prof, 66-74; assoc prof, 74-79, actg dean sci, 88-89, PROF CHEM & ASSOC DEAN SCI, UNIV BC, 79-; VPRES TECHNOL DEVELOP, QUADRA LOGIC TECHNOLOGIES, VANCOUVER, 90- *Concurrent Pos:* Guggenheim fel, 80; vis prof, Harvard Univ, 80. *Honors & Awards:* Fel, Japanese Soc Promotion Sci, 87; Gold Medal in Health Sci, Sci Coun BC. *Mem:* Am Chem Soc; The Chem Soc; Royal Inst Chem; Chem Inst Can. *Res:* Structure, synthesis, chemistry and biochemistry of porphyrins, vitamin B12, and related macrocycles; photodynamic therapy. *Mailing Add:* Dept Chem Univ BC 2075 Wesbrook Pl Vancouver BC V6T 1W5 Can

DOLPHIN, JOHN MICHAEL, b Mahanoy City, Pa, Nov 21, 23; m 50; c 3. PATHOLOGY. *Educ:* Hahnemann Med Col, MD, 47. *Prof Exp:* Resident path, Hahnemann Med Col & Hosp, 49-50, 52-54; resident path, Pa Hosp, 54-55; from asst prof to assoc prof path, Hahnemann Med Col, 63-74; LAB DIR LAB PROCEDURES, UPJOHN CO, 74-; AT SMITH-KLINE BIO-LAB. *Mem:* AMA; Col Am Path. *Res:* Pathology of tumors; hypertensive vascular lesions. *Mailing Add:* Smith-Kline Bio-Sci Lab 400 Egypt Rd Norristown PA 19403

DOLPHIN, PETER JAMES, b Derby, Eng, Mar 5, 47; Can citizen. BIOCHEMISTRY. *Educ:* Southampton Univ, UK, BSc hons, 68, PhD(biochem), 71, DSc(biochem), 91. *Prof Exp:* Res scientist, Nat Inst Med Res, London, UK, 71-72; Can Heart Found fel biochem, McGill Univ, 72-75, res assoc, 75-76, res scholar & asst prof biochem, 76-78; from asst prof to assoc prof, 78-87, PROF BIOCHEM, DALHOUSIE UNIV, 87- *Concurrent Pos:* Mem, Grant Rev Comt, 80-83, Med Res Coun, 84-88. *Honors & Awards:* Bristol Myers Sr Award, 86; Max Forman Sr Award, 90. *Mem:* Fel Am Heart Asn; Can Soc Clin Invest; Can Biochem Soc; Am Soc Biol Chemists & Molecular Biol; Brit Biochem Soc; Can Soc Nutrit Sci. *Res:* The influence of dietary and hormonal factors on the sub- cellular assembly secretion and metabolism of plasma lipoproteins in animal models of atherosclerosis; human disorders of lipoprotein metabolism; genetic defects; lipolytic enzymes; enzyme mechanism. *Mailing Add:* Dept Biochem Dalhousie Univ Halifax NS B3H 4H7 Can

DOLPHIN, ROBERT EARL, b Worcester, Mass, Oct 4, 29; wid; c 2. ENTOMOLOGY. *Educ:* San Jose State Col, BA, 58; Purdue Univ, West Lafayette, PhD(entom), 66. *Prof Exp:* Mgr entom, Mosquito Abatement Dist, Calif, 58-61; instr agr entom, Agr Exp Sta, Purdue Univ, West Lafayette, 61-65; res entomologist, Entom Res Div, 65-68, supvr res entomologist & res leader, 68-75, assoc area dir, Mid-Great Plains Area, NCent Region, Agr Res Serv, 75-84, DIR LAB, YAKIMA AGR RES LAB, USDA, 84- *Mem:* Sigma Xi; Entom Soc Am. *Res:* Biological control and ecology of deciduous fruit insects. *Mailing Add:* 700 Cherry Rm 217 Columbia MO 65201

DOLPHIN, WARREN DEAN, b Philadelphia, Pa, Feb 17, 40; m 63; c 2. CELL PHYSIOLOGY, SCIENCE EDUCATION. *Educ:* West Chester State Col, BS, 62; Ohio State Univ, PhD(cell biol), 68. *Prof Exp:* Res fel zool, Univ Maine, 68-69, asst prof, 69-70; asst prof, 70-75, assoc prof, 75-79, PROF ZOOL, IOWA STATE UNIV, 79-, EXEC OFFICER BIOL, 77- *Concurrent Pos:* NIH grant, 68-70; NSF grants, 76-79 & 81-83; consult, Am Col Testing Prog, 78; Asst Dean of Scis and Humanities, Iowa State Univ, 87-88. *Mem:* AAAS; Iowa Acad Sci; Sigma Xi; Int Cong Individualized Instr (pres). *Res:* Nitrogen metabolism and gene regulation in Acanthamoeba; multivariate analysis of student achievement in self-paced instructional systems. *Mailing Add:* Dept Zool Iowa State Univ Ames IA 50011

DOLUISIO, JAMES THOMAS, b Bethlehem, Pa, Sept 28, 35; m 69; c 3. PHARMACEUTICS. *Educ:* Temple Univ, BS, 57, MS, 59; Purdue Univ, PhD(phys pharm), 62. *Prof Exp:* From asst to assoc prof pharm, Philadelphia Col Pharm & Sci, 61-67; prof & asst dean, Col Pharm, Univ Ky, 67-73; PROF PHARM & DEAN COL PHARM, UNIV TEX, 73- *Concurrent Pos:* NSF fel, Am Found Pharmaceut Educ; consult, Hoechst-Roussel Pharmaceut, Inc & Nat Inst Drug Abuse, 73- *Mem:* Sigma Xi; Am Found Pharmaceut Educ; Am Asn Col Pharm; AAAS; Acad Pharmaceut Sci (vpres elect, 75). *Res:* Bioequivalence as a measure of therapeutic equivalence; factors affecting drug absorption; pharmacokinetic studies in animals and in man. *Mailing Add:* Col Pharm Univ Tex Austin TX 78712

DOLYAK, FRANK, b Statford, Conn, Nov 13, 27; m 51; c 3. PHYSIOLOGY. *Educ:* Univ Conn, BA, 50; Univ Kans, PhD(zool), 55. *Prof Exp:* Asst instr zool, Univ Kans, 51-54; from instr to assoc prof physiol, Univ Conn, 54-65; assoc prof biol & chmn dept, Augusta Col, 65-66; chmn dept, 67-73 & 79-81, PROF BIOL, RI COL, 66- *Mem:* AAAS; NY Acad Sci. *Res:* Serology; radiation biology; immunogenetics. *Mailing Add:* Dept Biol RI Col Providence RI 02908

DOMAGALA, JOHN MICHAEL, b Detroit, Mich, Feb 26, 51; m 72; c 1. ANTI-INFECTIVES. *Educ:* Univ Detroit, BS, 73; Wayne State Univ, PhD(org chem), 77. *Prof Exp:* Res chemist catalysts, Chrysler Corp, 72-73; RES CHEMIST ANTI-INFECTIVES, PARKE DAVIS CO, WARNER-LAMBERT, 79- *Concurrent Pos:* Consult, BASF Wayandotte Corp, 76-77. *Mem:* Am Chem Soc. *Res:* Organic mechanism; rearrangements of epoxides; carbonyl compounds and carbonyl ions; theoretical chemistry; molecular orbital calculations; mechanism of action of anti-infectives; quinoline anti-infectives; totally synthetic anti-infectives. *Mailing Add:* 776 Georgetown Canton MI 48188-1536

DOMAGALA, ROBERT F, physical metallurgy; deceased, see previous edition for last biography

DOMAILLE, PETER JOHN, b Swan Hill, Australia, Aug 10, 48; c 2. MOLECULAR SPECTROSCOPY. *Educ:* Monash Univ, Australia, BSc, 69, PhD(chem), 73. *Prof Exp:* Fel chem, Univ Canterbury, 73-75 & Univ Calif, Santa Barbara, 75-78; fel chem, Univ Calif, Santa Barbara, 75-78; mem staff, Cent Res & Develop Dept, E I du Pont De Nemours & Co, Inc, 78-, MEM STAFF, EXP STA, DUPONT-MERCK PHARMACEUT. *Res:* Laser spectroscopy of small molecules and molecular structure determination; nuclear magnetic resonance of organometallics, paramagnetic complexes, inorganic compounds, proteins and nuclei acids; biomolecular structure determinations. *Mailing Add:* Exp Sta Bldg 328 Dupont-Merck Pharmaceut PO Box 80 Wilmington DE 19880

DOMAN, ELVIRA, b New York, NY; c 2. BIOCHEMISTRY, ENZYMOLOGY. *Educ:* Hunter Col, BA, 55; NY Univ, MS, 59; Columbia Univ, MA, 60; Rutgers Univ, PhD(physiol, biochem), 65. *Prof Exp:* Jr technician, NY Univ Hosp, 55-56; sr technician cancer chemother, Sloan Kettering Inst Cancer Res, 56-57; sr technician endocrinol, Col Physicians & Surgeons, Columbia Univ, 57, res asst, 58; res asst phys chem, Sloan Kettering Inst Cancer Res, 60-61; Pop Coun fel endocrinol, Rockefeller Univ, 65-66, res assoc, 66-68; lectr, Dept Biol Sci, Douglass Col, Rutgers Univ, 70-73; asst prof biol, Seton Hall Univ, 73-77; asst prog dir, 78-81, ASSOC PROG DIR PHYSIOL PROCESSES, NSF, 81- *Concurrent Pos:* Vis scientist, Rutgers Univ, 89. *Honors & Awards:* Dir(s) Equal Opportunity Achievement Award, NSF, 86. *Mem:* AAAS; Am Chem Soc; NY Acad Sci. *Res:* Isolation and characterization of enzymes in respiratory chain of beef heart, of enzymes in rat liver related to endocrinology, and of enzymes involved in the growth of yeast cells and callus tissue. *Mailing Add:* Nat Sci Found 1800 G St NW Rm 321 Washington DC 20550

DOMANIK, RICHARD ANTHONY, b Racine, Wis, Oct 13, 46; m 72; c 2. BIOCHEMISTRY, INSTRUMENTATION. *Educ:* Ripon Col, Wis, BA, 68; Northwestern Univ, Evanston, PhD(chem), 74. *Prof Exp:* Res assoc biophys, Univ Conn, 72-75; asst prof chem, Cent Mich Univ, 75-79; INSTRUMENTATION SCIENTIST, ABBOTT LABS, 79- *Mem:* Am Chem Soc; Biophys Soc; Sigma Xi; Soc Photo-Optical Instrumentation Engrs. *Res:* Development of instrument based systems for clinical diagnostics. *Mailing Add:* 30908 Leesley Ct Libertyville IL 60048

DOMANSKI, THADDEUS JOHN, b Jersey City, NJ, June 14, 11; m 34; c 1. CHEMICAL & PHYSICAL CARCINOGENESIS. *Educ:* NY Univ, BS, 32, MS, 36, PhD(col sci teaching), 49. *Prof Exp:* Chemist, Path Dept, Med Ctr, Jersey City, NJ, 35-43; biochemist, Lab Serv, 8th Gen Hosp, US Air Force, 44-45, chief, Bur Labs, Dept Pub Health & Welfare, Hqs, Korea, 46-47, biochemist, Sch Aviation Med, 50, chief lab, 50-54, chief lab serv & actg chief dept path, 54-57, rep, Off Air Force Surgeon Gen, US Army Biol Labs, Ft Detrick, Md, 57-58, chief epidemiol lab, Aerospace Med Ctr, Tex, 58-61, chief toxicol br, Armed Forces Inst Path, 61-66, assoc chief clin biores lab sci, Biomed Sci Corps, 65-66; scientist adminr pharmacol & toxicol, Res Grants Br, Nat Inst Gen Med Sci, 66-67; RETIRED. *Mem:* Sigma Xi. *Res:* Extramural, research grants program development in chemical carcinogenesis. *Mailing Add:* 5924 Rudyard Dr Bethesda MD 20814

DOMASK, W(ILLIAM) G(ERHARD), b Port Arthur, Tex, Mar 11, 20; m 48; c 6. CHEMICAL ENGINEERING. *Educ:* Tex A&M Univ, BS, 42, MS, 48; Univ Tex, PhD(chem eng), 53. *Prof Exp:* Res chemist, Jefferson Chem Co, 45-46; sr chem engr, Humble Oil & Refining Co, 52-60, mgr southwest region sales eng, 60-65, tech adv, Hq Mkt, 66-77; GOVT REGULATIONS ADV, MKT DEPT, EXXON CO, USA, 77- *Mem:* Am Chem Soc; Sigma Xi; Am Inst Chem Engrs. *Res:* Synthesis of chlorinated hydrocarbons; process engineering research; chemicals, plastics and petroleum products development; products research coordination, and toxicologicol aspects of products. *Mailing Add:* 11603 Starwood Dr Houston TX 77024

DOMB, ELLEN RUTH (COLMER), b Morristown, NJ, Aug 18, 46; m 68. EXPERIMENTAL SOLID STATE PHYSICS. *Educ:* Mass Inst Technol, BS, 68; Univ Pa, MA, 69; Temple Univ, PhD(physics), 74. *Prof Exp:* Res assoc physics, Univ Nebr-Lincoln, 74-76; asst prof physics, Harvey Mudd Col, 76-79; sr physicist, Pomona Div, Gen Dynamics, 79-81; tech staff specialist, 81-85, eng mgr, 85-87, DIR INFO SERV, AEROJET ELECTRO SYSTEMS CORP, 87- *Concurrent Pos:* Vis assoc, Calif Inst Technol, 77- *Mem:* Am Phys Soc. *Res:* Electrical and magnetic properties of limited dimensionality materials, including spin-glass alloys and layered compounds; properties of amorphous ferromagnetic and superconducting materials; fiber optics; infrared sensors. *Mailing Add:* Aerojet Electro Sys Co 170-1911 2443 Oceanview Dr PO Box 296 Azusa CA 91702

DOMBRO, ROY S, b Brooklyn, NY, Oct 21, 33; m 67; c 2. BIOCHEMISTRY. *Educ:* Brooklyn Col, BS, 54; Univ Wis, MS, 56, PhD(biochem), 58. *Prof Exp:* Res assoc biochem, Rockefeller Inst, 58-64; mem staff, Inst Muscle Dis, 64-65; assoc, Dept Surg, Albert Einstein Col Med, 65-67, asst prof biochem, 67-70; res scientist surg, Med Sch, 70-75, RES ASST PROF SURG, SCH MED, UNIV MIAMI, 75- *Concurrent Pos:* Res chemist, Vet Admin Hosp, Miami. *Mem:* AAAS; Am Chem Soc. *Res:* Design and synthesis of antimetabolites neurochemistry; amino acid metabolism; vasoactive amines and peptides. *Mailing Add:* 9841 SW 123 ST Miami FL 33176

DOMBROSKI, JOHN RICHARD, b Elmira, NY, June 6, 41; m 67; c 2. POLYMER & ORGANIC CHEMISTRY. *Educ:* Rochester Inst Technol, BS, 64; Clarkson Col Technol, MS, 67; Syracuse Univ & State Univ NY, Syracuse, PhD(polymer chem), 71. *Prof Exp:* Res fel polymer chem, Inst Marcomolecular Chem, Freiburg Univ, WGermany, 70-71; res chemist, 71-73, sr res chemist, 73-80, SR RES ASSOC POLYMER CHEM, RES LABS, EASTMAN CHEM DIV, 80- *Mem:* Am Chem Soc; Sigma Xi. *Res:* Vinyl and condensation polymers; sterochemistry of polymers; mechanisms of polymerization; polymer blends; network-forming polymers; reinforced plastics, adhesives. *Mailing Add:* Res Labs Eastman Chem Div PO Box 1972 Kingsport TN 37662

DOMBROWSKI, JOANNE MARIE, b Detroit, Mich, Aug 14, 46. OPERATOR THEORY. *Educ:* Marygrove Col, BS, 68; Purdue Univ, MS, 70, PhD(math), 73. *Prof Exp:* From asst prof to assoc prof, 73-90, PROF MATH, WRIGHT STATE UNIV, 90- *Mem:* Am Math Soc; Math Asn Am. *Res:* Tridiagonal matrices; commutator equations; systems of orthogonal polynomials. *Mailing Add:* Dept Math Wright State Univ Dayton OH 45435

DOMELSMITH, LINDA NELL, b Georgetown, Tex, Jan 19, 49; m 70; c 1. DETOXIFICATION BACTERIAL ENDOTOXINS, ANALYSIS COTTON. *Educ:* Univ Tex, Austin, BA, 70; Duke Univ, PhD(phys & org chem), 74. *Prof Exp:* Lectr chem, NTex State Univ, 74-75; res assoc chem, La State Univ, Baton Rouge, 75-78; asst prof org chem, Univ Nebr, Omaha, 78-81; res assoc theoret chem, Univ New Orleans, La, 81-82; RES CHEMIST, SOUTHERN REGIONAL RES CTR, ARG RES SERV, USDA, 82- *Concurrent Pos:* Lead scientist, Southern Regional Res Ctr, Agr Res Serv, USDA, 88-; co-ed, Proc of Cotton Dust Res Conf, 90- *Mem:* Am Chem Soc; Sigma Xi. *Res:* Detoxification of bacterial endotoxins; analysis of cotton and cotton dust; capillary gas chromatography of plant, bacterial and fungal extracts. *Mailing Add:* Southern Regional Res Ctr 1100 Robert E Lee Blvd New Orleans LA 70124

DOMENICONI, MICHAEL JOHN, b San Francisco, Calif, Sept 22, 46; m 74; c 2. PHYSICAL CHEMISTRY, ELECTROCHEMISTRY. *Educ:* Univ San Francisco, BS, 68; Univ Calif, Berkeley, PhD(chem), 75. *Prof Exp:* Sr chemist, Electrochimica Corp, Mt View, Calif, 75-76; mem tech staff electrochem, GTE Labs, Inc, 76-79; TECH DIR SPEC PROJS, ALTUS CORP, SAN JOSE, 79- *Mem:* Am Chem Soc; Electrochem Soc; AAAS; Sigma Xi; Int Soc Electrochem. *Res:* Exploratory physical-chemical and electrochemical investigations in non-aqueous solvents; applied research and development of advanced battery systems. *Mailing Add:* 152 Muir Ave Santa Clara CA 95051

DOMER, FLOYD RAY, b Cedar Rapids, Iowa, July 12, 31; m 65. PHARMACOLOGY. *Educ:* Univ Iowa, BS, 54, MS, 56; Tulane Univ, PhD(pharmacol), 59. *Prof Exp:* Asst pharm, Univ Iowa, 54-55, asst pharmacol, 55-56; asst, Tulane Univ, 56-59; USAF res contract, Istituto Superiore di Sanita, Italy, 60-61; asst prof pharmacol, Col Med, Cincinnati, 61-62; from asst prof to assoc prof, 63-74, PROF PHARMACOL, TULANE UNIV, 74-, ADJ PROF ANESTHESIOL, 84- *Concurrent Pos:* Life Ins Med Res Fund fel, Nat Inst Med Res, Eng, 59-60; hon res fel, Univ Col, London, 71-72; consult, Pan-Am Sanit Bur Regional Off, Arg, WHO, 71; sr res scientist, NIMH, USPHS, 84-85. *Mem:* AAAS; Soc Exp Biol & Med; Am Soc Pharmacol & Exp Therapeut; Soc Neurosci. *Res:* Transport systems, ways of affecting them, particularly the blood-brain barrier; pharmacology of hemicholiniums. *Mailing Add:* Dept Pharmacol Tulane Univ 1430 Tulane Ave New Orleans LA 70112

DOMER, JUDITH E, b Millersville, Pa, Apr 9, 39; m 65. MEDICAL MICROBIOLOGY, MYCOLOGY. *Educ:* Tusculum Col, BA, 61; Tulane Univ, PhD(microbiol), 66; Am Bd Med Microbiol, dipl, 81. *Prof Exp:* Asst prof biol, St Mary's Dominican Col, 67-68; res assoc microbiol, 68-71, from asst prof to assoc prof, 72-88, PROF MICROBIOL, SCH MED, TULANE UNIV, 88- *Concurrent Pos:* NIH fel, Tulane Univ, 66-67; Wellcome Trust fel immunol, Kennedy Inst Rheumatology, London, Eng, 71-72; mem bact & mycol study sect, NIH, 75-79, biomed sci study sect, 88-90, AIDS & related dis study sect, 90-; assoc ed, Exp Mycol J, 78-80; mem Sci Comt, Europ J Epidemiol, 84-; guest researcher, NIH, 84-85. *Mem:* Am Soc Microbiol (chmn, Med Mycol Div, 83-85); Med Mycol Soc of the Americas (pres, 87-89); Int Soc Human & Animal Mycol; Am Asn Immunol; Infectious Dis Soc Am. *Res:* Biochemistry of fungal cell walls; immunology of the systemic mycoses; host-parasite interactions in experimental marine candidiasis and cryptococcosis, with emphasis on cellular immune mechanisms; biochemistry of fungal cell walls, especially those of dimorphic organisms. *Mailing Add:* Dept Microbiol Tulane Univ Sch Med New Orleans LA 70112

DOMERMUTH, CHARLES HENRY, JR, b St Louis, Mo, Nov 16, 28; m 52; c 3. VETERINARY MICROBIOLOGY. *Educ:* Elmhurst Col, BS, 51; Univ Ky, MS, 55; Va Polytech Inst, PhD(microbiol), 62. *Prof Exp:* Asst prof, 54-70, PROF VET SCI, VA POLYTECH INST & STATE UNIV, 70- CHMN, DIV AGR & URBAN PRACT, 80- *Concurrent Pos:* Res microbiologist, Statens Serum Inst, Denmark, 62-63, EAfrican Vet Res Orgn, Kenya, 64-66 & USDA, 64- *Mem:* Am Soc Microbiol; Am Asn Avian Pathologists; Asn Am Vet Med Cols; Res Workers Animal Dis; Int Orgn Mycoplasmology. *Res:* Electron microscopy of infected tissue; studies of Mycoplasmataceae and viruses and diseases caused by them; development of vaccine for hemorrhagic enteritis of turkeys. *Mailing Add:* Dept Large Animal Clin Sci Va Polytech Inst & State Univ Blacksburg VA 24061

DOMESHEK, S(OL), b New York, NY, Dec 6, 20; m 42; c 2. DISPLAY SYSTEMS, NAVIGATION. *Educ:* City Col New York, BS, 41; NY Univ, BME, 56. *Prof Exp:* Jr engr, US Geol Surv, 42-44; proj engr, US Naval Training Device Ctr, 46-51, sr engr, 51-57, head visual systs br, 57-61, staff eng consult, 61-64, head phys sci lab, 64-66; dir instrumentation, Avionics Lab, US Army Electronics Command, 66-72, chief, Navig Div, 73-81, chief systs mgt off, US Avionics Res & Develop Activ, 81-84, dir prod eng, Prod Assurance, Installation Eng, Integrated Logistic Support, Ft Monmouth, 84-86; PVT PRACT, 86- *Mem:* Nat Soc Prof Engrs; Optic Soc Am; Am Soc Photogram; Inst Navig. *Res:* Patents in photogrammetry; display technology; positioning and navigation; radar mapping; cockpit instrumentation; geographic orientation. *Mailing Add:* 2320 Edgewood Terrace Scotch Plains NJ 07076

DOMHOLDT, LOWELL CURTIS, b Tyler, Minn, Mar 17, 34; m 55; c 4. MECHANICAL ENGINEERING. *Educ:* Univ Minn, BS, 55, MS, 57; Case Inst Technol, PhD(fluid mech), 63. *Prof Exp:* Instr mech eng, Ohio State Univ, 58; from instr to assoc prof, Case Inst Technol, 60-67; from assoc prof to prof, 67-85, EMER PROF MECH ENG, CLEVELAND STATE UNIV, 85- *Concurrent Pos:* Fac co-op coordr Eng, Lakeland Community Col, 86-87. *Honors & Awards:* Ralph R Teetor Award, Soc Automotive Engrs, 67. *Mailing Add:* 9339 Buena Vista Dr Mentor OH 44060-7060

DOMIER, KENNETH WALTER, b Norquay, Sask, Aug 30, 33; m 56; c 3. AGRICULTURAL ENGINEERING. *Educ:* Univ Sask, BE, 55, MSc, 57; Mich State Univ, PhD(agr eng), 67. *Prof Exp:* Fuels & lubricants engr, Federated Cooperatives Ltd, 55-58; asst prof agr eng, Univ Man, 58-69; chmn dept, 69-74, 82-87, PROF AGR ENG, UNIV ALTA, 69- *Concurrent Pos:* Vis prof, Swed Inst Agr Eng, 74-75; assoc prof eng, geol & geophys Alta, Alta Inst Agrologists. *Honors & Awards:* Maple Leaf Award, Can Soc Agr Engrs. *Mem:* Fel Am Soc Agr Engrs; fel Can Soc Agr Engrs (pres, 71-72); Agr Inst Can; Royal Swed Acad Agr & Forestry; Asn Prof Engrs. *Res:* Tillage and seeding; fertilizer placement; tractors; machinery; utilization of flax fibres. *Mailing Add:* Dept Agr Eng Univ Alta Edmonton AB T6G 2H1 Can

DOMINEY, RAYMOND NELSON, b Mobile, Ala, June 23, 57; m 82. INTERFACIAL CHEMISTRY, MOLECULAR RECOGNITION. *Educ:* Univ WFla, BS, 78; Mass Inst Technol, PhD(chem), 82. *Prof Exp:* Res assoc dept chem, Univ NC, Chapel Hill, 82-84; vis asst prof phys chem, Univ Va, 84-86; ASST PROF CHEM, UNIV RICHMOND, 86-; AFFIL ASST PROF, DEPT BIOCHEM & MOLECULAR BIOPHYS, MED COL VA, VA COMMONWEALTH UNIV, 88- *Concurrent Pos:* Vis lectr chem, Univ NC, Chapel Hill, 82-83. *Mem:* Am Chem Soc; Am Phys Soc; NY Acad Sci; Sigma Xi. *Res:* Chemistry within and across membranes; molecular recognition; guest-host or receptor interactions; NMR, photochemistry and electrochemistry applied to micelles, membranes and molecular binding by them; charge transfer photochemistry. *Mailing Add:* Dept Chem Univ Richmond Richmond VA 23173

DOMINGO, WAYNE ELWIN, b Weeping Water, Nebr, June 30, 16; m 37; c 4. CROP BREEDING. *Educ:* Univ Nebr, BS, 38; Utah State Col, MS, 40; Univ Ill, PhD(plant breeding), 42. *Prof Exp:* Assoc, Univ Ill, 42-43; assoc plant breeder, Bur Plant Indust, Soils & Agr Eng, USDA, 43-45, agronomist, 45-46; dir oilseeds prod div, Baker Castor Oil Co, 46-74; consult agronomist, NL Industs, 75-81; RETIRED. *Mem:* Am Soc Agron; Soc Econ Bot. *Res:* Hybridization of forage grasses; genetics of soybean; breeding condiment, insecticide and drying oil plants; domestic and foreign production of oilseeds. *Mailing Add:* 4701 Date Ave No 205 La Mesa CA 91941

DOMINGOS, HENRY, b Massena, NY, Sept 17, 34; m 58; c 3. SOLID STATE ELECTRONICS. *Educ:* Clarkson Technol Univ, BEE, 56; Univ Southern Calif, MSEE, 58; Univ Wash, Seattle, PhD(elec eng), 63. *Prof Exp:* Mem tech staff electronics, Hughes Aircraft Co, 56-58; asst prof elec eng, Univ Nev, 58-60; actg instr, Univ Wash, Seattle, 60-63; prof, 63-88, CHMN, ELEC & COMPUT ENG, CLARKSON UNIV, 88- *Mem:* Inst Elec & Electronic Engrs; Am Soc Eng Educ. *Res:* Semiconductor devices and integrated circuits. *Mailing Add:* ECE Dept Clarkson Univ Potsdam NY 13699-5720

DOMINGUE, GERALD JAMES, b Lafayette, La, Mar 2, 37; div; c 5. BACTERIOLOGY, IMMUNOLOGY. *Educ:* Univ Southwestern La, BS, 59; Tulane Univ, PhD(med microbiol, immunol), 64. *Prof Exp:* Teaching asst bact, Univ Southwestern La, 58-59; asst res instr pediat, Sch Med, State Univ NY Buffalo, 64-66; instr microbiol, Sch Med, St Louis Univ, 66-67; from asst prof to assoc prof surg, microbiol & immunol, 67-74, PROF UROL, MICROBIOL & IMMUNOL, SCH MED, TULANE UNIV, 74-, RES AFFIL, DELTA REGIONAL PRIMATE CTR, 68- *Concurrent Pos:* USPHS fel, Children's Hosp, Buffalo, NY, 64-66; dir microbiol, Snodgras Lab Path & Bact, St Louis City Hosp, 66-67; instr, Sch Med, St Louis Univ & lectr, Sch Dent, Wash Univ, 66-67; consult bacteriologist, Southern Baptist Hosp, New Orleans, 68-84; consult res scientist, Vet Admin Hosp, New Orleans, 69-72; consult & mem tech adv bd, Analytab Prod Inc, New York, 72-78; consult, Armour Pharmaceut, 75-77; consult bacteriologist, Tulane Med Ctr Hosp, 78-83; div lectr microbiol, Am Soc Microbiol, 78, found lectr, 79-80; consult, Med Technol Corp, 83-90; pres, Southwestern Asn Clin Microbiol, 86; panel distinguished profs, Tulane Univ, 83-89. *Mem:* Fel Am Acad Microbiol; NY Acad Sci; Soc Exp Biol & Med; Am Soc Microbiol; fel Infectious Dis Soc Am; AAAS; Fedn Am Scientists; Sigma Xi; Soc Basic Urol Res; affil mem Am Urol Asn. *Res:* Significance of cell wall deficient microorganisms in renal and other chronic infections; biological studies on common enterobacterial antigens; immune response in urinary tract infections; chorionic gonadotropin producing bacteria; significance of fimbriae and other bacterial cell glycocalyceal antigens in urinary tract infections; bacterial porins. *Mailing Add:* Dept Urol 1430 Tulane Ave New Orleans LA 70112-2699

DOMINGUEZ, CESAREO AUGUSTO, b Buenos Aires, Argentina, Oct 1, 42; m 68; c 2. HADRONIC SYMMETRIES, WEAK INTERACTIONS. *Educ:* Univ Buenos Aires, MS, 68, PhD(physics), 71. *Prof Exp:* Instr physics, Univ Buenos Aires, 68-71; res assoc, Stanford Linear Accelerator Ctr, 71-72; prof physics, Ctr Advan Studies, Mex, 72-78; assoc prof physics, Tex A&M Univ, 78-82; prof physics, Univ Santa Maria, Valparaiso, Chile, 83-85; Alexander von Humboldt fel, Deutsches Elektronen Synchrotron, Hamburg, Ger, 86-88; ADJOINT HON PROF PHYSICS, CATH UNIV CHILE, 85-; PROF THEORET PHYSICS, UNIV CAPE TOWN, 88- DIR, INST THEORET PHYSICS & ASTROPHYS, 90- *Concurrent Pos:* NSF exchange, Rockefeller Univ, 74-80, 83-85; vis mem staff, Los Alamos Sci Lab, 79-82; co-prin investr, US Dept Energy grant, 79-82; mem adv comt, UN Prog Develop, 84-85; vpres, Chilean Phys Soc, 85; vis scientist, Int Ctr Theoret Physics, Trieste, Italy, 85; prin investr, Found Res Develop, SAfrica, 90- *Mem:* Mex Acad Sci. *Res:* Calculation of chiral-symmetry breaking in elementary particle and nuclear physics; chirality in quantum chromodynamics. *Mailing Add:* Inst Theoret Physics Univ Cape Town Rondebosch 7700 Cape South Africa

DOMINH, THAP, b Hanoi, Vietnam, Dec 25, 38; US citizen; m 64; c 2. PHOTOCHEMISTRY, PHOTORESIST CHEMISTRY. *Educ:* Univ Chicago, MS, 62, PhD(photochem), 66. *Prof Exp:* G Swift fel photochem, Univ Chicago, 64-65; Nat Res Coun fel, Univ Alta, 66-68, res assoc, 68-69; mem tech staff chem res, AT&T Bell Labs, 69-71; res assoc, Kodak Res Labs, 71-88, TECH ASSOC, EASTMAN KODAK CO, 88- *Mem:* Am Chem Soc; AAAS; Soc Photog Scientists & Engrs. *Res:* Physical organic chemistry; catalysis involving transition metal complexes; organic and organometallic photochemistry, photochromism, photopolymerization unconventional imaging systems; photoresist and printing chemistry. *Mailing Add:* Eastman Kodak Co Kodak Colo Div Windsor CO 80551-1610

DOMINIANNI, SAMUEL JAMES, b New York, NY, Sept 21, 37; m 68. ORGANIC CHEMISTRY. *Educ:* Queens Col, NY, BS, 58; Univ Mass, MS, 60; Univ NC, PhD(org chem), 64. *Prof Exp:* Res assoc fel chem, Iowa State Univ, 64-66; SR ORG CHEMIST, ELI LILLY & CO, 66- *Mem:* AAAS; Am Chem Soc; Royal Soc Chem. *Res:* Heterocyclic synthesis; organic photochemistry. *Mailing Add:* MC 705 Bldg 882 Eli Lilly & Co Res Labs Indianapolis IN 46285-0002

DOMINICK, WAYNE DENNIS, b Chicago, Ill, Oct 19, 46; m 71; c 2. COMPUTER SCIENCE. *Educ:* Ill Inst Technol, BS, 68; Northwestern Univ, MS, 74, PhD(comput sci), 75. *Prof Exp:* Systs analyst comput sci, US Army Data Support Command, 69-70; systs programmer, Vogelback Comput Ctr, Northwestern Univ, 71-74, res asst, Dept Comput Sci, 74-75; from asst prof to assoc prof comput sci, Univ Southwestern La, 75-81; PRES, EXEC SYST INC, 81- *Concurrent Pos:* Grants, US Army Corps Engrs, 74-77; consult, US Army Corps Engrs, 75-77, La State Natural Resources & Energy Div, 75-77, Battelle Columbus Labs, 76-, OCLC, Inc, 77- & La State Dept Educ, 77-; NSF grants, 76-; vpres, Info Mgt Inc, 78-; pres, Exec Systs, Inc, 81-; NASA grants, 83- *Honors & Awards:* Royal E Cabell Sci Award, Northwestern Univ, 74. *Mem:* Am Soc Info Sci; Asn Comput Mach. *Res:* Data base management system design; information storage and retrieval system design; software monitoring; system performance measurement and evaluation; user-system interfacing; interactive graphics; computer systems analysis. *Mailing Add:* Dept Comput Study Univ SW La Box 44330 Lafayette LA 70504

DOMINO, EDWARD FELIX, b Chicago, Ill, Nov 20, 24; m 48; c 5. PHARMACOLOGY. *Educ:* Univ Ill, BS, 48, MD & MS, 51. *Prof Exp:* Rotating intern, Presby Hosp, Chicago, 51-52; instr pharmacol, Univ Ill, 52-53; from instr to assoc prof, 53-62, PROF PHARMACOL, UNIV MICH, ANN ARBOR, 62-, DIR, NEURO-PSYCHOPHARMACOL RES LAB, 66- *Concurrent Pos:* Vis prof psychiat, Wayne State Univ, 65-; mem neuropharmacol sect, Int Brain Res Orgn; rep, US Pharmacopeia, 76-89. *Honors & Awards:* Sigma Xi Prize, 51; Award, Mich Soc Neurol & Psychiat, 55; First Prize, Am Soc Anesthesiol, 63; Nikolai Pavlovich Kravkov Mem Medal, Acad Bd Inst Pharmacol & Chemother, Acad Med Sci, USSR, 68. *Mem:* Am Soc Pharmacol; NY Acad Sci; Soc Psychophysiol Res; fel Am Col Neuropsychopharmacol (vpres, 76); AMA. *Res:* Experimental and clinical neuropharmacology and psychopharmacology as a means of understanding brain function; central neural transmitters, especially cholingeric substances and interaction of various psychoactive drugs with neural transmitters; biology of mental disease, particularly schizophrenia. *Mailing Add:* Dept Pharmacol Univ Mich Ann Arbor MI 48109-0626

DOMINY, BERYL W, b Davison, Mich, Apr 2, 41; m; c 2. ORGANIC CHEMISTRY, INFORMATION SCIENCE. *Educ:* Mich State Univ, BS, 63; Univ Mich, PhD(chem), 67. *Prof Exp:* Org chemist, 67-73, info scientist, 73-82, COMPUTATIONAL CHEMIST, PFIZER INC, 83- *Mem:* Am Chem Soc. *Res:* Heterocyclic chemistry; computer utilization in pharmaceutical research and development. *Mailing Add:* 189 Seneca Dr Noank CT 06340

DOMKE, CHARLES J, b Chicago, Ill, Nov 4, 14; m 45; c 1. AIR POLLUTION. *Educ:* Loyola Univ, Ill, BS, 41. *Prof Exp:* Res chemist, Burgess Battery Co, 41; from asst chemist to chemist, Standard Oil Co, 41-48, from asst proj automotive engr to proj, 49-65; sr proj automotive engr, Am Oil Co, 66-67; procedures & standards coordr, Environ Protection Agency, 67-69, chief, Surveillance Br, 67-76, proj mgr, Characterization & Appl Br, 76-79, proj mgr inspection & maintenance staff, 79-80; RETIRED. *Concurrent Pos:* Leader vapor lock prog, Coord Res Coun, 64-65 & expression for fuel volatility panel, 64- *Mem:* Soc Automotive Eng. *Res:* Design and use of motor fuels under extreme environmental conditions, and/or in high compression engines; development of engine to run without crankcase lubrication; use of ammonia and hydrazine as fuels for spark ignition and compression ignition engines; automation and automatic data logging. *Mailing Add:* 41181 Crestwood Dr Plymouth MI 48170-2638

DOMKE, HERBERT REUBEN, b Hillsboro, Kans, Apr 6, 19; m 46; c 4. PUBLIC HEALTH. *Educ:* Univ Chicago, SB, 39, MD, 42; Harvard Univ, MPH, 48, DrPH, 59. *Hon Degrees:* DSc Cent Methodist Col, Fayette, Mo, 75. *Prof Exp:* Chief med officer, Chicago Health Dept, 44-47; health comnr, St Louis County Health Dept, Mo, 49-58; dir, Pittsburgh-Allegheny County Health Dept, Pa, 59-66; hosp comnr & dir health & hosps, City of St Louis, Mo, 66-71; dir, 71-79, DIR, SECT MED CARE, MO DIV HEALTH, 79-; PROF COMMUNITY HEALTH & MED PRACT, UNIV MO-COLUMBIA, 71- *Concurrent Pos:* Asst anat, Univ Chicago, 44-46; asst prof prev med & pub health, Wash Univ, 49-58, prof pub health in prev med & prof med, 66; clin prof prev med in internal med, St Louis Univ, 66. *Honors & Awards:* Presidential Award, Nat Med Asn, 67; St Louis City Mayors Civic Award, 70; Mo Nursing Home Asn Award, 71. *Mem:* AAAS; Asn State & Territorial Health Officers; Conf State & Prov Health Authorities NAm. *Res:* Epidemiology and public health administration. *Mailing Add:* 701 W High St Jefferson City MO 65101

DOMMERT, ARTHUR ROLAND, b Crowley, La, Apr 3, 37; m 63; c 4. VETERINARY MICROBIOLOGY. *Educ:* Tex A&M Univ, BS, 60, DVM, 61; La State Univ, MS, 63, PhD(microbiol, biochem), 66; Am Col Vet Microbiol, dipl, 69. *Prof Exp:* NIH fel microbiol & biochem, La State Univ, 61-66; assoc prof vet microbiol, Univ Mo-Columbia, 66-71; prof & head dept, 71-81, PROF VET MICROBIOL & PARASITOL & ASSOC VICE CHANCELLOR, ACAD AFFAIRS, LA STATE UNIV, BATON ROUGE, 81- *Concurrent Pos:* Mem, Exec Coun Conf Res Workers in Animal Dis, 80- *Mem:* Am Vet Med Asn; Am Soc Microbiol; Conf Res Workers Animal Dis; Asn Am Vet Med Cols. *Res:* Changes in animal tissues due to obligate anaerobic bacteria; isolation and identification of obligate anaerobic bacteria; curriculum development in veterinary medicine; academic administration. *Mailing Add:* Dept Vet Microbiol & Parasitol La State Univ Baton Rouge LA 70803

DOMNING, DARYL PAUL, b Biloxi, Miss, Mar 14, 47; m 87. EVOLUTIONARY BIOLOGY, PALEOECOLOGY. *Educ:* Tulane Univ, BS, 68; Univ Calif, Berkeley, MA, 70, PhD(paleont), 75. *Prof Exp:* Res biologist, Inst Nat Pesquisas da Amazonia, Manaus, Brazil, 76-78; asst prof, 78-85, ASSOC PROF ANAT, HOWARD UNIV, WASHINGTON, DC, 85- *Concurrent Pos:* Res assoc, Univ Calif Mus Paleont, 76-78, Smithsonian Inst, 79- & Nat Hist Mus, Los Angeles County, 85-; marine mammal specialist, Int Union Conserv Nature & Natural Resources, 79-; mem manatee tech adv coun, Dept Natural Resources, Fla, 81-; sci adv, US Marine Mammal Comn,

82-85. *Mem:* Soc Vert Paleont; Am Soc Mammalogists; Soc Syst Zool; Soc Marine Mammal. *Res:* Morphology, systematics, functional anatomy, ecology, paleoecology, fossil and recent distribution, evolution, conservation, and bibliography of the Sirenia and Desmostylia; herbivorous marine mammals. *Mailing Add:* Dept Anat Howard Univ Washington DC 20059

DOMOKOS, GABOR, b Budapest, Hungary, Mar 5, 33; US citizen; m 67; c 1. FLUIDS. *Educ:* Eotvos Lorand Univ, Budapest, dipl physics, 56; Joint Inst Nuclear Res, USSR, Dr(math sci, physics), 63. *Prof Exp:* Res assoc cosmic rays, Cent Res Inst Physics, Budapest, 56-60; mem res staff elementary particle physics, Joint Inst Nuclear Res, USSR, 61-63; vis scientist, Europ Orgn Nuclear Res, Switz, 63-64; sr res staff mem, Cent Res Inst Physics, Budapest, 64-65; lectr, Johns Hopkins Univ, 65-66; res scientist, Univ Calif, Berkeley, 66-67; sr res scientist, Cent Res Inst Physics, Budapest, 67-68; PROF PHYSICS, JOHNS HOPKINS UNIV, 68- *Concurrent Pos:* Consult, Rutherford Lab, Chilton, Eng, 73; vis sci staff mem, Europ Orgn Nuclear Res, Switz, 75-76; Alexander von Humboldt sr vis scientist award, Deutsches Electronen-Synchrotron, Hamburg, Ger, 76; vis prof, Univ Florence, Italy, 78; Fulbright fel, 83. *Honors & Awards:* Alexander von Humboldt sr vis scientist award,Deutsches Electronen-Synchrotron, Hamburg, Ger, 76. *Mem:* Am Math Soc; fel Am Phys Soc; Ital Phys Soc; Europ Phys Soc. *Res:* Interactions of elementary particles at high energies; quantum gravity; cosmology; theory of turbulence. *Mailing Add:* Dept Physics & Astron Johns Hokins Univ Charles & 34th Baltimore MD 21218

DOMOZYCH, DAVID S, PLANT BIOLOGY, CELL BIOLOGY. *Educ:* Miami Univ, Ohio, PhD(bot), 81. *Prof Exp:* ASSOC PROF BIOL, SKIDMORE COL, 83- *Res:* Ultrastructure of green algae; processes of cell division. *Mailing Add:* Dept Biol Skidmore Col Saratoga Springs NY 12866

DOMROESE, KENNETH ARTHUR, b Vincennes, Ind, May 23, 33; m 57; c 3. PHYSIOLOGY. *Educ:* Concordia Col, Ill, BS, 55; DePaul Univ, MS, 62; Northwestern Univ, PhD(biol sci), 63. *Prof Exp:* From asst prof to assoc prof, 61-64, PROF BIOL SCI, CONCORDIA COL, 77- *Concurrent Pos:* Res fel, Med Sch, Loyola Univ, 72-73. *Mem:* AAAS; Am Soc Zool; Sigma Xi. *Res:* Lipid metabolism in insect flight muscle; invertebrate endocrinology; cardiovascular physiology. *Mailing Add:* Dept Nat Sci Concordia Col 7400 Augusta River Forest IL 60305

DOMSKY, IRVING ISAAC, b Racine, Wis, Feb 3, 30; m 64; c 4. ANALYTICAL CHEMISTRY. *Educ:* Univ Wis, BS, 51, PhD(anal org chem), 59. *Prof Exp:* Anal chemist, Qm Food & Container Inst, Ill, 54; res asst chem, Yale Univ, 58-60; res assoc, Div Oncol, Chicago Med Sch, 60-64; anal chemist, Abbott Labs, 64-67; sr anal chemist, Armour Dial, Inc, 67-73; sr anal group leader, Quaker Oats Co, 73-74; sr anal res chemist, Chem Prod Div, Chemetron Corp, 75-77; PRES & LAB DIR, ALLIED LABS, LTD, 77- *Mem:* Am Water Works Asn; AAAS; Am Chem Soc; Am Oil Chemists Soc. *Res:* Gas chromatography; new analytical methods. *Mailing Add:* 6820 N Francisco Ave Chicago IL 60645

DONABEDIAN, AVEDIS, b Beirut, Lebanon, Jan 7, 19; nat US; m 45; c 3. QUALITY ASSESSMENT, HEALTH ORGANIZATIONS. *Educ:* Am Univ, Beirut, BA, 40, MD, 44; Harvard Univ, MPH, 55. *Prof Exp:* Physician & actg supt, Eng Mission Hosp, Jerusalem, 45-47; asst dermat & venerology, Am Univ, Beirut, 48-54, instr physiol, 48-50, univ physician, 49-51, dir univ health serv, 51-54; res assoc med care, Sch Pub Health, Harvard Univ, 55-57; from asst prof to assoc prof prev med, New York Med Col, 57-61; from assoc prof to prof pub health econ, Univ Mich, Ann Arbor, 61-66, prof, Med Care Orgn, 66-, Nathan Sinai distinguished prof pub health, Sch Pub Health, 79-89, EMER PROF, UNIV MICH, ANN ARBOR, 89- *Concurrent Pos:* Med assoc, Med Care Eval Studies, United Community Serv, Boston, 55-57; vis lectr, Sch Pub Health, Harvard Univ, 57-58; mem panel health serv, Nat Acad Sci, 70; ed consult, Int J Health Serv, 71; consult, Asn Am Med Col, 74; mem acad adv comt, Ctr Pub Health Res Mex, 84-; hon pres, Fundaciön Avedis Donabedian para la Mejora de la Assistencia Sanitaria, Barcelona, Spain, 90. *Honors & Awards:* Elizur Wright Award, Am Risk & Ins Asn, 78; Michael Davis lectr, Univ Chicago, 86; Baxter Am Found Health Serv Res Prize, 86. *Mem:* Nat Inst Med; fel Am Pub Health Asn; Asn Teachers Prev Med; hon fel Am Col Hosp Admin; hon fel Am Col Utilization Rev Physicians. *Res:* Medical care organization; author of 6 books and many publications. *Mailing Add:* Sch Pub Health Univ Mich 1400 Wash Heights Ann Arbor MI 48109

DONACHIE, MATTHEW J(OHN), JR, b Orange, NJ, Oct 23, 32; m 55; c 4. PHYSICAL METALLURGY. *Educ:* Rensselaer Polytech Inst, BMetE, 54; Mass Inst Technol, SM, 55, MetE, 57, ScD(x-ray & lattice strains), 58. *Prof Exp:* Metallurgist, Adv Metals Res Corp, 58; res scientist, Res Labs, United Aircraft Corp, 58-59; group leader high temperature metall, Gen Dynamics/Elec Boat, 59-61; supvr phys metall, Chase Brass & Copper Co, 61-63; gen supvr metall eng, 63-67, develop metallurgist, 67-69, STAFF PROJ ENGR, PRATT & WHITNEY GROUP, UNITED TECHNOLOGIES CORP, 69- *Concurrent Pos:* From adj asst prof to adj assoc prof, Hartford Grad Ctr, 58-68, adj prof, 68- *Mem:* Fel Am Soc Metals Int; Am Soc Testing & Mat. *Res:* Metallurgy of high temperature alloys; x-ray metallurgy; mechanical behavior of metals; environmental effects in metals; electron microscopy and metallography; creep and fatigue. *Mailing Add:* Pratt & Whitney Mail Stop 114-39 Main St East Hartford CT 06108

DONADY, J JAMES, b Hempstead, NY, June 21, 38; m 62; c 1. DEVELOPMENTAL GENETICS. *Educ:* State Univ NY Stony Brook, BS, 62; Univ Iowa, PhD(zool), 69. *Prof Exp:* NIH fel genetics, City of Hope Med Ctr, 69-70, jr res scientist develop genetics, 71-72; from asst prof to assoc prof, 73-85, PROF BIOL, WESLEYAN UNIV, 85- *Concurrent Pos:* Res grant, NIH Child Health & Human Develop & Nat Sci Found, 74, 78, 79 & 82. *Mem:* Genetics Soc; AAAS; Tissue Cult Asn; Sigma Xi. *Res:* Genetic approach to problems of cell differentiation; combines Drosophila genetics, mutants and in vitro culturing of embryonic cells; emphasis on muscle protein genes, their organization and regulation and the role of the nuclear matrix. *Mailing Add:* Dept Biol Wesleyan Univ Middletown CT 06457

DONAGHY, JAMES JOSEPH, b Cumberland, Ky, Mar 13, 35; m 60; c 2. PLASMA PHYSICS. *Educ:* Univ Fla, BS, 59; Univ NC, PhD(physics), 65. *Prof Exp:* Opers analyst, Univ NC, 64-65; asst prof physics, Va Mil Inst, 65-66; opers analyst, US Govt, 66-67; asst prof, 67-72, assoc prof, 72-79, PROF PHYSICS, WASHINGTON & LEE UNIV, 79- *Concurrent Pos:* Vis prof, 66. *Mem:* Am Asn Physics Teachers. *Res:* Positron annihilation in solids; Fermi surface in metals; negative ion sources; charged particle beams. *Mailing Add:* Dept Physics Washington & Lee Univ Lexington VA 24450

DONAGHY, RAYMOND MADIFORD PEARDON, b Eastman, PQ, Aug 18, 10; nat US; m 41; c 4. NEUROSURGERY. *Educ:* Univ Vt, BS, 33, MD, 36. *Prof Exp:* Intern neurol, Montreal Gen Hosp, Can, 36-37, asst resident internal med, 37-38; resident gen surg, Children's Mem Hosp, 38-39; asst resident neurosurg, Mass Gen Hosp, 39-40; resident psychiat, McLean Hosp, 40-41; resident, Mass Gen Hosp, 42-43; assoc prof, 46-52, PROF NEUROSURG, COL MED, UNIV VT, 52- *Concurrent Pos:* Fel neurosurg, Lahey Clin, 41-42; Dalton scholar, Mass Gen Hosp, 42. *Mem:* Fel Am Col Surgeons; Soc Neurol Surgeons; Am Asn Neurol Surgeons; Acad Neurol Surgeons; Can Neurol Soc. *Res:* Brain abscess; spastic element in cerebral thrombosis; neurovascular surgery; microsurgery; spinal cord and brain trauma. *Mailing Add:* Univ Vt Health Ctr Burlington VT 05401

DONAHOE, FRANK J, b Ashland, Pa, Mar 12, 22; m 43; c 2. SOLID STATE PHYSICS. *Educ:* La Salle Col, BA, 43; Univ Pa, PhD(physics), 54. *Prof Exp:* Instr math & phys chem, La Salle Col, 48-49; asst, Univ Pa, 49-51; res physicist, Franklin Inst, 51-64; from assoc prof to prof, 64-90, EMER PROF PHYSICS, WILKES UNIV, 90- *Mem:* Inst Elec & Electronic Engineers; Am Phys Soc; Am Asn Physics Teachers; Sigma Xi. *Res:* Low temperature physics of metals; order-disorder in ferromagnetic alloys; thermoelectricity; cosmogony. *Mailing Add:* PO Box 195 Dallas PA 18612

DONAHOE, JOHN PHILIP, b Hagerstown, Md, Dec 1, 44; m 70; c 2. VETERINARY MEDICINE. *Educ:* Univ Ga, BS, 67, DVM, 70, PhD(microbiol), 77. *Prof Exp:* Asst prof vet med, Ohio State Univ, 76-85; ASSOC DIR POULTRY HEALTH, ARBER ACRES. *Mem:* Am Vet Med Asn; AAAS; Am Asn Avian Pathologists; Asn Vet Med Col; World Vet Poultry Asn. *Res:* Mechanisms of viral immune suppression as well as molecular basis of animal oncogenic herpesviral cellular transformation. *Mailing Add:* Maine Biol Lab China Rd PO Box 255 Winslow ME 04901

DONAHOO, PAT, b Van Buren, Ark, July 22, 28; m 65; c 2. QUALITY ASSURANCE, HACCP PROGRAMS. *Educ:* Hendrix Col, BS, 50; Okla State Univ, MS, 52, PhD(chem), 55. *Prof Exp:* Chemist, Tex Co, 52-53; res chemist, Lion Oil Co Div, Monsanto Co, 55-56 & Inorg Chem Div, 56-60; res chemist, Griffith Labs, Inc, 60-62, chief chemist, 62-66, tech mkt mgr, 66-71; dir grocery prod develop, Anderson Clayton Foods, 71-77; vpres res & qual control, Rich-Seapak Corp, 77-91; RETIRED. *Concurrent Pos:* Consult, 91- *Mem:* Am Chem Soc; Inst Food Tech; Soc Adv Food Serv Res. *Res:* Seasonings for food products; hydrolyzed plant proteins; oil based grocery products; spices; frozen seafood products; frozen breaded products; food science; analytical chemistry. *Mailing Add:* 113 Augusta St St Simons Island GA 31522

DONAHUE, D JOSEPH, b Joliet, Ill, May 11, 26; m 76; c 3. MANAGEMENT. *Educ:* Univ Mich, BS, 47, MS, 48, PhD(phys chem), 51. *Prof Exp:* Engr color tubes, RCA, 51-58, mgr, adv develop semiconductors, 58-62, mgr, Indust, 62-67, chief exec officer, Semiconductors, 67-70, chief exec officer, Europe, 70-75, vpres eng, 75-77, vpres eng & mfg, 77-81, chief exec officer consumer elec, 81-87; SR VPRES HIGH DEFINITION TV, GEN ELEC & THOMSON, 87- *Concurrent Pos:* Chmn, EIA Eng Policy Comt, 88-; mem, AEA High Definition Comt, 88-, Fed Commun Comn Advan TV Adv Comt, 87- *Mem:* Fel Inst Elec & Electronics Engrs. *Res:* Developed the first color picture tubes; silicon semiconductor technology; high definition television. *Mailing Add:* 4505 Foxhall Crescents NW Washington DC 20007

DONAHUE, DOUGLAS JAMES, b Wichita, Kans, Oct 26, 24; m 48; c 5. NUCLEAR PHYSICS. *Educ:* Univ Ore, BS, 47, MS, 48; Univ Wis, PhD, 52. *Prof Exp:* Physicist, Hanford Labs, Gen Elec Co, 52-57; from asst prof to prof physics, Pa State Univ, 57-63; assoc prof, 63-64, PROF PHYSICS, UNIV ARIZ, 64- *Mem:* Fel Am Phys Soc. *Res:* Low energy nuclear physics; accelerator mass spectrometry; atomic physics; medical physics. *Mailing Add:* Dept Physics Univ Ariz Tucson AZ 85721

DONAHUE, FRANCIS M(ARTIN), b Philadelphia, Pa, May 8, 34; m 60; c 5. ELECTROCHEMICAL ENGINEERING. *Educ:* La Salle Col, BA, 56; Univ Calif, Los Angeles, PhD(eng), 65. *Prof Exp:* Res chemist, Tasty Baking Co, 56-59; group leader corrosion res, Betz Labs, Inc, 59-61; electrochemist, Stanford Res Inst, 61-63; res engr, Univ Calif, Los Angeles, 63-65; from asst prof to assoc prof, 69-79, PROF CHEM ENG, UNIV MICH, 79- *Concurrent Pos:* Vis prof, Swiss Fed Inst Technol, Zurich, 72-73. *Mem:* Electrochem Soc; Int Soc Electrochem. *Res:* Corrosion and corrosion inhibition; electrochemical energy conversion; electrocatalysis; electrosynthesis; electrodeposition; electroless plating. *Mailing Add:* Dept Chem Eng Univ Mich Col Eng Ann Arbor MI 48109

DONAHUE, HAYDEN HACKNEY, b El Reno, Okla, Dec 4, 12; m 47; c 3. PSYCHIATRY, PSYCHIATRIC EDUCATION & ADMINISTRATION. *Educ:* Univ Kans, BS, 39, MD, 41. *Prof Exp:* Intern, Univ Hosps, UNiv Ga, 41-42, res psychiat & instr med & psychiat, Sch Med, 42; res aviation med, Army Air Force Sch Aviation Med, 43; res psychiat, US Air Force Hosp, St Petersburg, Fla, 44; asst chief hosp opers, US Vet Admin, Washington, DC, 46; lectr hosp admin & psychiat, Residency Prog & asst mgr, Vet Admin Hosp, N Little Rock, Ark, 46-49; dir educ & res, Ark State Hosp, 49-51, res psychiat, 59-61; asst med dir, Tex State Bd Hosp & Spec Schs, 51-53; dir ment health, State Okla, 53-59 & 70-78; supt, Cent State Griffen Mem Hosp, 61-79; DIR, OKLA INST MENT HEALTH EDUC & TRAINING, 79- *Concurrent Pos:* Assoc prof psychiat, Sch Med, Univ Ark, 49-51 & 60-61; consult, Ark

State Dept Health, 49-51; lectr legal med, Sch Law, Univ Tex, 52; lectr, Homicide Inst, Univs Okla & Tex, 53-67; consult asst prof neurol & psychiat, Univ Okla, 54-58, assoc prof psychiat, 58-67, clin prof psychiat & behav sci, 67-88, co-chmn adv comn to chmn dept psychiat, 87-88, clin prof psychiat & behav sci, emer, 88; adv, Okla Comt, President's Comt Employ Handicapped, 57-59 & 61-71, vchmn, Okla Comt, White House Conf Children & Youth, 59; instr, Okla State Univ, 58-59; mem, Okla Gov Comn, White House Conf Aging, 59-60, Ark Gov Comn, 60-61 & nat adv comt, 59-61, chmn sect ment health & aging, 60-61; mem bd dirs, Pan Am Training Exchange Prog Psychiat, 61-63, treas, 63; mem, Am Psychiat Asn-Pan Am Exchange rep, Latin Am Sem Ment Health, WHO, Buenos Aires, 63; secy sect psychopharmacol, Am Mex Joint Ment Health Conf, Mexico City, 64; co-chmn sect pvt & pub ment hosps; Nat Conf Ment Illness & Health, Chicago, 64; chief consult, Okla State Penitentiary, 63-67; consult, Okla State Crime Bur, 63-86; consult, Base Hosp, Tinker Field, 64-70; lectr, Univ Okla, 64-70; chmn comt on juv delinq, Okla State Crime Comn, 67-79; mem state, county & local adv comts & coun ment health, Am Bar Asn comm on ment disabled exec comm, 73-80; mem, Okla Coun Juv Justice, 79; bd dir, Nat Asn Mental Health Prog Dirs, 55-59 & 70-78; pres, Okla Dist Br, Am Psychiat Asn, 63, secy, 81- *Honors & Awards:* Bowis Award, Am Col Psychiat; Health Planning Award, Okla State Planning Coun, 78; Psychiat Admin Award, Am Asn Psychiat Admin, 80; Pieperdink Award, Asn Ment Health Adminr, 88; Distinguished Serv Award, Am Psychiat Asn, 84. *Mem:* Fel Am Psychiat Asn (treas, 68-72); fel Am Geriat Soc; fel Am Col Psychiat (regent, 65, treas, 66-71); Am Col Psychiat (pres, 75); Am Asn Med Supt Ment Hosps (pres, 74-75); Nat Asn Mental Health Prog Dirs (treas, 76). *Res:* War neurosis; tuberculosis; narcosynthesis; problems of the aged; psychological selection and training of professional, technical and ancillary hospital personnel; hospital operations and management problems; use of special drugs in treatment of mental patients and rehabilitative therapies in care and treatment of institutionalized patients. *Mailing Add:* 1109 Westbrooke Terr Norman OK 73072

DONAHUE, JACK DAVID, b Chicago, Ill, Nov 21, 38; m 65; c 2. STRATIGRAPHY. *Educ:* Univ Ill, Urbana, BA, 60; Columbia Univ, PhD(geol), 67. *Prof Exp:* Lectr geol, Queens Col, NY, 64-67, asst prof, 67-70; from asst prof to assoc prof geol, 70-83, assoc prof anthropol, 78-83, PROF GEOL & ANTHROPOL, UNIV PITTSBURGH, 83- *Concurrent Pos:* Ed-in-chief, Geoarcheol, 85- *Mem:* Geol Soc Am; Soc Econ Paleontologists & Mineralogists; Int Asn Sedimentology; Sigma Xi; Soc Am Archeol. *Res:* Reconstruction of geology setting of archaeological sites and terraines; materials analysis of prehistoric ceramics. *Mailing Add:* Dept Geol & Planetary Sci Univ of Pittsburgh Pittsburgh PA 15260

DONAHUE, JAMES EDWARD, computer science, for more information see previous edition

DONAHUE, THOMAS MICHAEL, b Healdton, Okla, May 23, 21; m 50; c 3. AERONOMY, PLANETARY ATMOSPHERES. *Educ:* Rockhurst Col, AB, 42; Johns Hopkins Univ, PhD(physics), 47. *Hon Degrees:* DSc, Rockhurst Col, 81. *Prof Exp:* Res assoc & asst prof physics, Johns Hopkins Univ, 48-51; from asst prof to prof, Univ Pittsburgh, 51-59, dir, Lab Atmospheric & Space Sci, 66-74 & Space Res Coordn Ctr, 70-74; chmn dept, 74-80, prof atmospheric & oceanic sci, 80-87; EDWARD H WHITE II DISTINGUISHED UNIV PROF PLANETARY SCI, UNIV MICH, 87- *Concurrent Pos:* Guggenheim fel, Meudon Observ, France, 60; vis prof, Univ Paris, 60, 65, 79 & 80 & Harvard Univ, 71-72; chmn, Atmospheric Sci Adv Panel, NSF, 68-72; vis assoc, Dept Terrestrial Magnetism, Carnegie Inst, Washington, DC, 82-83; mem, contamination & interference space, Nat Res Coun, 58-61, Comt Polar Res, 66-69, Comt Atmospheric Sci, 69-72, climate bd, 79-82, chmn, Comt Solar Terrestrial Res, 72-75; mem, Stratospheric Res Adv Comt, 76-79, comt prog changes, Int Solar Polar Mission Panel, 81, Expendable Launch Vehicle Panel, 84, chmn, Sci Steering Group, Pioneer Venus Mission, 74-; trustee at large, Upper Atmospheric Res Corp, 68-75, chair, 71-72; trustee, secy & vchmn, Univ Corp Atmospheric Res, 77-84; trustee & chmn, Univ Space Res Asn, 77-84; mem, Space Tel Inst Vis Comn, Asn Univ Res Astron, 87-89, chair, 88-89; mem, Arecibo adv bd, 72-75 & 86-88, chmn, 88; chmn, Comn Pub Affairs, Am Geophys Union, 90- *Honors & Awards:* Henry K Arctowski Medal, Nat Acad Sci, 81; John Adam Fleming Medal, Am Geophys Union, 81; Wellock Distinguished Res Accomplishments Award, Univ Mich, 80, Henry Russel lectr, 86; Space Sci Award, Am Inst Aeronaut & Astronaut, 88. *Mem:* Nat Acad Sci; fel Am Geophys Union; fel AAAS; fel Am Phys Soc; Am Astron Soc; Int Acad Astronaut. *Res:* Atmospheric chemistry and physics; atomic and molecular physics; astrophysics. *Mailing Add:* Dept Atmospheric Oceanic & Space Sci Univ Mich Ann Arbor MI 48109

DONALD, DENNIS SCOTT, b Felicity, Ohio, May 21, 39; m 69; c 2. HETEROCYCLIC CHEMISTRY, SYNTHETIC CHEMISTRY. *Educ:* Univ Cincinnati, BS, 62; Univ Wis, PhD(org chem), 69. *Prof Exp:* RES CHEMIST, CENT RES & DEVELOP DEPT, E I DUPONT DE NEMOURS & CO, 68- *Mem:* Am Chem Soc. *Res:* Synthetic organic chemistry and physical organic chemistry supporting imaging and electro optic technology; fundamental studies in nonlinear optical phenonema in organic materials. *Mailing Add:* PO Box 321 Mendenhall PA 19357

DONALD, ELIZABETH ANN, b Edmonton, Alta, Feb 14, 26. NUTRITION. *Educ:* Univ Alta, BSc, 49; Wash State Univ, MS, 55; Cornell Univ, PhD(nutrit), 62. *Prof Exp:* Dietitian, Univ Alta, 50-51; actg jr home economist, Wash State Univ, 55-59; from asst prof to assoc prof foods & nutrit, Cornell Univ, 62-69; assoc prof to prof, 69-88, EMER PROF FOODS & NUTRIT, UNIV ALTA, 88- *Mem:* NY Acad Sci; Am Inst Nutrit; Can Soc Nutrit Sci. *Res:* Foods; nutritional status; vitamin B-6; vitamin B-6 requirement of young women using anovulatory steroids; energy balance in the elderly female; nutritional status of elderly. *Mailing Add:* Fac Home Econ Univ Alta Edmonton AB T6G 2M8 Can

DONALD, MERLIN WILFRED, b Montreal, Que; c 2. NEUROPSYCHOLOGY. *Educ:* McGill Univ, PhD(psychol), 68. *Prof Exp:* Asst prof psychol, Dept Neurol, Yale Univ Sch Med, 70-72; neuropsychologist, Vet Admin Hosp, West Haven, Conn, 70-72; assoc prof, 73-82, PROF PSYCHOL, DEPT PSYCHOL, QUEEN'S UNIV, ONT, 82- *Concurrent Pos:* Nat Res Coun fel, Neuropsychol Lab, Vet Admin Hosp, West Haven & res assoc, Dept Psychiat, Yale Univ Sch Med, 68-70; hon res fel, Univ Col, London, UK, 80-81. *Mem:* Fel Can Psychol Asn; Soc Neurosci. *Res:* Neural models of selective attention; medical applications of human psychobiology; neurolinguistics. *Mailing Add:* Dept Psychol Queen's Univ Kingston ON K7L 3N6 Can

DONALD, WILLIAM DAVID, b Donalds, SC, Apr 27, 24; m 48; c 4. MEDICINE. *Educ:* Erskine Col, AB, 45; Vanderbilt Univ, MD, 47. *Prof Exp:* Instr pediat, Sch Med, Vanderbilt Univ, 50-51; from asst prof to assoc prof, Sch Med, Univ Ala, 53-59; assoc prof, Med Col Ga, 59-60; ASSOC PROF PEDIAT, SCH MED, VANDERBILT UNIV, 60- *Mem:* AAAS; Am Acad Pediat; Am Fedn Clin Res. *Res:* Infectious diseases in children. *Mailing Add:* Dept Pediat Vanderbilt Univ Nashville TN 37232

DONALD, WILLIAM WALDIE, b Freeport, NY, Feb 16, 50; m 74; c 2. WEED SCIENCE, PLANT PHYSIOLOGY. *Educ:* State Univ NY, Stony Brook, BS, 72; Univ Minn, St Paul, MS, 74; Univ Wis, Madison, PhD(agron), 77. *Prof Exp:* Fel weed sci, USDA Metab & Radiation Lab, 77-78; asst prof weed sci, Colo State Univ, Ft Collins, 78-80; RES AGRONOMIST, METAL & RADIATION LAB, USDA, 80- *Mem:* Weed Sci Soc Am; Am Soc Agron; Am Soc Plant Physiologists; Sigma Xi. *Res:* Weed control in no-tillage cropping; applied weed control of foxtails and Canada thistle in wheat. *Mailing Add:* 244 AG Eng UMC Columbia MO 65211

DONALDSON, ALAN C, b Northampton, Mass, Oct 23, 29; m 57; c 4. GEOLOGY. *Educ:* Amherst Col, BA, 51; Univ Mass, MS, 53; Pa State Univ, PhD(geol), 59. *Prof Exp:* From asst prof to assoc prof, 57-69, PROF GEOL, WVA UNIV, 69-, CHMN DEPT, 71- *Mem:* Soc Econ Paleontologists & Mineralogists; Am Asn Petrol Geologists; fel Geol Soc Am. *Res:* Modern sediments and their depositional environments, sedimentary rocks, stratigraphy and sedimentation. *Mailing Add:* Dept Geol WVA Univ Geol White Hall Morgantown WV 26506

DONALDSON, COLEMAN DUPONT, b Philadelphia, Pa, Sept 22, 22; m 45; c 5. AERONAUTICAL ENGINEERING. *Educ:* Rensselaer Polytech Inst, BAeroE, 43; Princeton Univ, MA, 54, PhD(aeronaut eng), 57. *Prof Exp:* Mem staff, Nat Adv Comt Aeronaut, 43-44, head aerophys sect, 46-52; aeronaut engr, Bell Aircraft Corp, 46; pres & sr consult, Aeronaut Res Assocs of Princeton, Inc, 54-86; CONSULT, 86- *Concurrent Pos:* Consult, Martin-Marietta Corp, 55-72, Gen Elec Co, 56-72, Gen Precision Equip Corp, 57-67, Thompson Ramo Wooldridge Inc, 58-61 & Grumman Aerospace Corp, 64-72; mem, Res & Adv Subcomt Fluid Mech, NASA & Vehicle Response Group, Defense Atomic Support Agency; Robert H Goddard vis lectr, Princeton Univ, 70-71, mem adv coun; indust prof adv comt, Pa State Univ, 70-77; chmn, Naval Res Adv Comn, Air Warfare Bd, 72-77. *Mem:* Nat Acad Eng; Am Phys Soc; assoc fel Am Inst Aeronaut & Astronaut. *Res:* Fluid and gas dynamics; viscous and other transport phenomena; turbulence and turbulent transport phenomena; chemical aspects of high temperature gas flows. *Mailing Add:* PO Box 279 Gloucester VA 23061

DONALDSON, DAVID MILLER, b Ogden, Utah, Oct 2, 24; m 47; c 9. MICROBIOLOGY, IMMUNOLOGY. *Educ:* Univ Utah, BS, 50, MS, 52, PhD(bact), 54. *Prof Exp:* Lab fel, Univ Utah, 50-52, from asst to res instr, 52-55; prof bact, 55-68, PROF MICROBIOL, BRIGHAM YOUNG UNIV, 68- *Mailing Add:* Dept Microbiol Brigham Young Univ 775 WIDB Provo UT 84601

DONALDSON, DONALD JAY, b Toledo, Ohio, Feb 15, 40; m 60. HUMAN ANATOMY, DEVELOPMENTAL BIOLOGY. *Educ:* Univ Toledo, BEd, 62; Tulane Univ, PhD(anat), 68. *Prof Exp:* Instr histol & embryol, 68-70, from asst prof to assoc prof anat, 70-88, PROF ANAT & NEUROBIOL, CTR HEALTH SCI, UNIV TENN, MEMPHIS, 88- *Mem:* Am Asn Anatomists; Am Soc Cell Biol. *Res:* Control of growth and differentiation-regeneration of appendages in Amphibia; epidermal cell migration. *Mailing Add:* Dept Anat & Neurobiol Univ Tenn Ctr for Health Sci Memphis TN 38163

DONALDSON, EDWARD ENSLOW, b Wenatchee, Wash, Mar 7, 23; m 46; c 1. MOLECULAR PHYSICS, SURFACE PHYSICS. *Educ:* Wash State Univ, BS, 48, PhD, 53. *Prof Exp:* Physicist radiol physics, Hanford Labs, Gen Elec Co, 53-57; from asst prof to prof physics, Wash State Univ, 57-88, chmn dept, 67-74 & 80-84; RETIRED. *Concurrent Pos:* Vis prof, Univ Liverpool, 68. *Mem:* AAAS; Am Asn Physics Teachers; sr mem Am Vacuum Soc; Am Phys Soc. *Res:* Physics of surfaces; pressure sensitive adhesives. *Mailing Add:* Dept Physics Wash State Univ Pullman WA 99164-2814

DONALDSON, EDWARD MOSSOP, b Whitehaven, Eng, June 25, 39; Can citizen; m 64; c 1. COMPARATIVE ENDOCRINOLOGY, AQUACULTURE. *Educ:* Sheffield Univ, BSc, 61, DSc, 75; Univ BC, PhD(zool), 64. *Prof Exp:* USPHS fel steroid biochem, Univ Minn, Minneapolis, 64-65; scientist, Vancouver Lab, Fisheries Res Bd Can, 65-75; prog head nutrit & endocrinol, W Vancouver Lab, Dept Fisheries & Oceans, 75-81, chmn dept sci adv comt, 84-85, sect head, Fish Cult Res, Biol Sci Br, 81-89, HEAD, CTR DICIPLINE EXCELLENCE BIOTECHNOL & GENETICS AGR, W VANCOUVER LAB, DEPT FISHERIES & OCEANS, 88-, SECT-HEAD BIOTECHNOL, GENETICS & NUTRIT, 90- *Concurrent Pos:* Consult zool, Univ Calif, Berkeley, 68; mem orgn comt, Int Symp Comp Endocrinol, 69-89; vis res scientist, Oceanic Inst, Hawaii, 69; res assoc, Univ BC, 69-; consult, Int Develop Res Ctr, 73- & Can Exec Serv Overseas, 80; mem, Animal Biol Grant Selection Comt, NSERC, 83-86, Food, Agr & Aquacult Strategic Grant Comt, 90-; chmn, reproduction & stock improvement sci study group, Planning Group Aquacult, Versailles Working Group Technol, Growth & Employment, 84-; consult, CIDA,

IDRC, FAO, NATO, Bonneville Power Admin, US Dept Energy, USAID & Portuguese Nat Fisheries Inst, 85; res grant reviewer, NSF, US Sea Grant, BC Sci Coun, NSERC, Can Coun & USDA. *Mem:* Can Soc Zool; Am Soc Zool; Aquacult Asn Can; World Aquacult Soc; Royal Soc Can; Acad Sci Royal Soc Can. *Res:* Purification of gonadotropin; endocrinology and hormonal control of sex differentiation, ovulation and somatotropin induced growth in fish; stress in salmonids; diet development and testing in salmon; resource enhancement and aquaculture of Pacific salmon; effects of waste pollutants on reproduction and stress response; application of biotechnology in aquaculture. *Mailing Add:* Dept Fisheries & Oceans W Vancouver Lab 4160 Marine Dr Vancouver BC V7V 1N6 Can

DONALDSON, ERLE C, b Tela, Honduras, Dec 30, 26; m 54; c 5. CHEMICAL ENGINEERING, PETROLEUM ENGINEERING. *Educ:* The Citadel, BS, 53; Univ SC, MS, 55; Univ Houston, BS, 61; Univ Tulsa, PhD(chem eng), 75. *Prof Exp:* Anal chemist, Signal Oil & Gas Co, 55-59, chem engr, 59-61; proj leader, US Bur Mines, 61-73; proj mgr environ aspects oil recovery, 73-80, chief, reservoir evaluation sect, US Dept Energy, 80-; AT DEPT PETROL & GEOL ENG, UNIV OKLA, NORMAN, OKLA. *Mem:* Am Inst Chem Engrs; Soc Petrol Engrs. *Res:* In situ determination of oil saturation, reservoir porosity, permeability and lithology; determine the surface chemical and physical reactions of oil-water-rock systems; production of petroleum using chemical compounds to enchance recovery. *Mailing Add:* Dept Geol Eng Univ Okla Main Campus Norman OK 73019

DONALDSON, JAMES A, b Madison, Fla, Apr 17, 41. MATHEMATICS. *Educ:* Lincoln Univ, Pa, AB, 61; Univ Ill, MS, 63, PhD(math), 65. *Prof Exp:* Asst prof math, Howard Univ, 65-66; Univ Ill, Chicago Circle, 66-70; assoc prof, Univ NMex, 70-71; PROF MATH, HOWARD UNIV, 71-, CHMN DEPT, 72- *Concurrent Pos:* Vis mem, Courant Inst Math Sci, NY Univ, 76-77. *Mem:* Math Asn Am; Am Math Soc; Nat Asn Mathematicians; Soc Indust & Appl Mathematicians. *Res:* Differential equations; scattering theory; perturbation theory. *Mailing Add:* Dept Math Howard Univ Washington DC 20059

DONALDSON, JAMES ADRIAN, b St Cloud, Minn, Jan 22, 30; m 50; c 5. OTOLOGY. *Educ:* Univ Minn, BA, 50, BS, 52, MD, 54, MS, 61. *Prof Exp:* Clin instr otolaryngol, Univ Southern Calif, 60-61; from asst prof to assoc prof otolaryngol & maxillofacial surg, Col Med, Univ Iowa, 61-65; head dept, 65-75, PROF OTOLARYNGOL, HEAD & NECK SURG, SCH MED, UNIV WASH, 65- *Mem:* Am Acad Otolaryngol, Head & Neck Surg; fel Am Otol Soc; fel Am Laryngol, Rhinol & Otol Soc; fel Am Col Surg; fel Am Neurotology Soc. *Res:* Surgical anatomy of the temporal bone. *Mailing Add:* Univ Hosp Univ Wash RL-30 Seattle WA 98195-0001

DONALDSON, JOHN ALLAN, b Chatham, Ont, Oct 15, 33; m 57; c 2. GEOLOGY. *Educ:* Queen's Univ, Ont, BSc, 56; Johns Hopkins Univ, PhD(geol), 60. *Prof Exp:* Geologist, Geol Surv Can, 59-65, res scientist, 65-68; assoc prof, 68-74, PROF GEOL, CARLETON UNIV, ONT, 74- *Concurrent Pos:* Chmn dept, Carleton Univ, 80-82 & 83-84. *Mem:* Soc Econ Paleontologists & Mineralogists; Geol Asn Can; Int Asn Sedimentol; Geol Soc Am. *Res:* Precambrian sedimentology; stromatolites. *Mailing Add:* Dept Geol Carlton Univ Ottawa ON K1S 5B6 Can

DONALDSON, JOHN RILEY, b Dallas, Tex, Nov 24, 25; m 51; c 4. NUCLEAR PHYSICS. *Educ:* Rice Inst, BS, 45, MA, 47; Yale Univ, MS, PhD(physics), 51. *Prof Exp:* Nuclear physicist, Calif Res & Develop Co, 50-53; assoc prof physics, Univ Ariz, 53-54; from asst prof to assoc prof, Fresno State Col, 56-67, chmn dept, 83-91, PROF PHYSICS, CALIF STATE UNIV, FRESNO, 67- *Concurrent Pos:* Mem, Fresno County Bd Supvrs, 73-80, chmn, 77-78; vis prof ETH Zurich, 67-68, 82-83. *Mem:* Sigma Xi; AAAS; Am Asn Physics Teachers; Am Phys Soc. *Res:* General and nuclear physics. *Mailing Add:* Dept Physics Calif State Univ Fresno CA 93740-0037

DONALDSON, LAUREN RUSSELL, b Tracy, Minn, May 13, 03; m 27; c 2. BIOLOGY. *Educ:* Intermountain Union Col, AB, 26; Univ Wash, MS, 31, PhD(fisheries), 39. *Hon Degrees:* DSc, Rocky Mountain Col, 58 & Hamline Univ, 65. *Prof Exp:* Prin & teacher high sch, Mont, 26-30; from asst to prof fisheries, 32-73, dir lab radiation biol, 58-66, dir appl fisheries lab, 43-57, EMER PROF FISHERIES, COL FISHERIES, UNIV WASH, 73- *Concurrent Pos:* Consult fish & wildlife serv, US Dept Interior, 35-40, Gen Elec Co, 47- & Gen Mills Inc; biologist, State Dept Fisheries, Wash, 42-43; chief div radiobiol, Radiol Safety Sect, Oper Crossroads, Bikini, 46, chief div radiobiol, Bikini Sci Resurv, 47-48 & Bikini-Eniwetok Resurv, 49; biologist, Oper Ivy, 52; lectr, Univ Oslo, 52, Univ Helsinki, 59 & Tokyo Fish Univ, 73; rep, AEC, Japan, 54; dir radiobiol studies, Pac Weapons Testing Prog, 54- *Honors & Awards:* Pres's Medal, Univ Helsinki, 62; Lockheed Award, Marine Technol Soc, 71. *Mem:* Am Fisheries Soc (vpres, 39); Radiation Res Soc; World Maricult Soc; Marine Technol Soc. *Res:* Fisheries management; fresh water biology; radiobiology; genetics of salmonoid fishes; nutrition of the chinook salmon, particularly histological changes. *Mailing Add:* Col Fisheries Univ Wash 204 Fisheries Seattle WA 98195

DONALDSON, MERLE RICHARD, b Silverdale, Kans, Apr 7, 20; m 43; c 2. ELECTRICAL ENGINEERING. *Educ:* Ga Inst Technol, BEE, 46, MS, 47, PhD(elec eng), 59. *Prof Exp:* Instr elec eng, Ga Inst Technol, 46-50, asst prof, 50-51; engr, Oak Ridge Nat Lab, 51-55, sr engr in chg res & develop cyclotrons, 55-57; proj engr, Basic Study Projs, Norden-Ketay Corp, 57; res engr, Electronic Commun, Inc, 57-58, sr staff engr, Advan Technol Group, 58-60, prin eng scientist, 60, mgr, Advan Develop Sect, 60-61, dir, Advan Develop Lab, 61-63; acad supvr, Off-Campus Grad Ctr & assoc prof elec eng, Univ Fla, 62-64; PROF ELEC ENG & CHMN DEPT, UNIV S FLA, 64- *Concurrent Pos:* Consult, Univ Colo, 57, Am Lava Corp, 59, Capital Radio Eng Inst, 60- & Sperry Microwave Electronics Co, 63-70; newsletter ed, Inst Elec & Electronics Engrs, 58-59, ed, Transactions, 59-60. *Mem:* AAAS; fel Inst Elec & Electronics Engrs; Sigma Xi. *Res:* Microwave theory and techniques. *Mailing Add:* 1833 Almeria Way S St Petersburg FL 33712

DONALDSON, ROBERT M, JR, b Hubbardston, Mass, Aug 1, 27; m 50; c 2. INTERNAL MEDICINE, GASTROENTEROLOGY. *Educ:* Yale Univ, BS, 49; Boston Univ, MD, 52. *Prof Exp:* Asst prof med, Sch Med, Boston Univ, 59-64; assoc prof, Univ Wis, 64-67; from assoc prof to prof med, Sch Med, Boston Univ, 67-73; chief med serv, Vet Admin Hosp, West Haven, Conn, 73-82; prof med & vchmn dept internal med, 73-87, DEP DEAN, SCH MED, YALE UNIV, 87- *Concurrent Pos:* USPHS fel, Harvard Med Sch, 57-59; chmn gastroenterol res eval comt, Vet Admin, 68-71; chmn gastroenterol & nutrit training comt, NIH, 69-73; ed, J Am Gastroenterol Asn, 70-77. *Honors & Awards:* Friendenwald Medal, Am Gastroenterol Asn, 87. *Mem:* Am Gastroenterol Asn (pres, 79); Asn Am Physicians; Am Soc Clin Invest. *Res:* Gastric; secretion; intestinal absorption; gastrointestinal bacteriology. *Mailing Add:* Yale Univ Sch Med I-208 SHM 333 Cedar St New Haven CT 06510-8055

DONALDSON, ROBERT PAUL, b Berkeley, Calif, Oct 15, 41; m 64, 78; c 3. PLANT MEMBRANE, PEROXISOMES. *Educ:* Univ Tex, Austin, BA, 64; Miami Univ, Ohio, MS, 66; Mich State Univ, PhD(biochem), 71. *Prof Exp:* Res assoc, lysosome res, Rockefeller Univ, NY City, 71-73; res assoc, glyoxysomal membrane res, Univ Calif, Santa Cruz, 73-76; PROF BIOL TEACHING & RES, GEORGE WASHINGTON UNIV, 77- *Mem:* Am Soc Plant Physiologists. *Res:* Functions of proteins in the membranes of organelles such as glyoxysomes and peroxisomes; lipid metabolism in germinating seeds; tumor cell peroxisomes. *Mailing Add:* Dept Biol Sci George Washington Univ Washington DC 20052

DONALDSON, ROBERT RYMAL, b Hornell, NY, Feb 27, 17; m 42; c 3. SCIENCE EDUCATION, PHYSICS. *Educ:* Syracuse Univ, BA, 40, MA, 43; Cornell Univ, PhD(sci educ), 55. *Prof Exp:* Teacher pub sch, 40-43, 46-51 & 54-55; prof physics, Col Arts & Sci, State Univ NY Col Plattsburgh, 55-84; RETIRED. *Concurrent Pos:* Ed consult spec proj, Ford Found, Ankara, Turkey, 64-65. *Mem:* AAAS; Am Asn Physics Teachers; Nat Sci Teachers Asn. *Res:* Introductory courses in physics and physical science; meteorology; astronomy; science education. *Mailing Add:* Dept Physics Col Arts & Sci State Univ NY Col Plattsburgh NY 12901

DONALDSON, SUE KAREN, b Detroit, Mich, Sept 16, 43; c 1. MUSCLE PHYSIOLOGY. *Educ:* Wayne State Univ, BSN, 65, MSN, 66; Univ Wash, PhD(physiol & biophys), 73. *Prof Exp:* Asst assoc prof physiol & nursing, Univ Wash, 73-78; assoc prof physiol & nursing, Rush Univ, Ill, 78-84, dir clin nursing res prog, 80-84; ASSOC DEAN RES, SCH NURSING, UNIV MINN, 84- *Concurrent Pos:* Prin investr, Wash State Heart Asn grant, 73-74, NIH grants, 73-, USPHS Div Nursing grant, 80- & Muscular Dystrophy Asn grant, 81- *Mem:* Am Nurses Asn; Biophys Soc; Am Heart Asn; Coun Nurse Researchers. *Res:* Animal studies of excitation-contraction-coupling mechanisms in mammalian ventricular and skeletal muscle fibers, including alterations in muscle properties associated with ontogenetic differentiation and disease. *Mailing Add:* Dept Physiol 6-101 Unit F Univ Minn 308 Harvard St SE Minneapolis MN 55455

DONALDSON, TERRENCE LEE, b Franklin, Pa, Apr 20, 46. BIOREMEDIATION, BIOENGINEERING. *Educ:* Pa State Univ, BS, 68; Univ Pa, PhD(chem eng), 74. *Prof Exp:* Asst prof chem eng, Univ Rochester, 74-80, assoc prof, 80; coordr biotechnol, 85-86, group leader, bioprocess res & develop, 81-87, GROUP LEADER, ENVIRON BIOTECHNOL, OAK RIDGE NAT LAB, 87- *Concurrent Pos:* Adj fac chem eng, Univ Tenn, 81. *Mem:* Am Inst Chem Engrs; Am Chem Soc; Sigma Xi. *Res:* Mass transfer; carrier transport in synthetic membranes; enzyme catalysis; separations and kinetics; biochemical engineering; bioreactors; wastewater treatment; bioremediation of hazardous wastes. *Mailing Add:* Oak Ridge Nat Lab Box 2008 Oak Ridge TN 37831-6044

DONALDSON, VIRGINIA HENRIETTA, b Glen Cove, NY, Oct 3, 24. MEDICINE. *Educ:* Univ Vt, AB, 47, MD, 51. *Prof Exp:* Intern, Strong Mem Hosp, Rochester, NY, 51-53, asst res pediat, Strong Mem & Genessee Hosps, 53-54; asst res pediat, Buffalo Children's Hosp, 54-55; from instr to sr instr med, Western Reserve Univ, 62-67; assoc prof med, 67-71, PROF PEDIAT & MED, COL MED, UNIV CINCINNATI, 71- *Concurrent Pos:* Res fel pediat, Sch Med, Western Reserve Univ, 55-57, res fel med, 57-62; asst attend pediat & asst dir, Outpatient Dept, Babies' & Children's Hosp, Cleveland, Ohio, 56-57; assoc, Res Div, Cleveland Clin, 57-62, asst mem staff, 62-63; mem res staff, St Vincent Charity Hosp, 63-67, estab investr, Am Heart Asn, 64-69; dir hemat, Cincinnati Shrine Burn Inst, 67-70. *Mem:* AAAS; Am Fedn Clin Res; Am Soc Hemat; Am Soc Clin Invest; Asn Am Physicians. *Res:* Studies of hydrolytic enzymes of human blood as related to blood coagulation; fibrinolysis; action of first component of complement and relation of fibrinolytic mechanisms to complement action. *Mailing Add:* Children's Hosp Res Found Univ Cincinnati Col Med Cincinnati OH 45229-2899

DONALDSON, W(ILLIS) LYLE, b Cleburne, Tex, May 1, 15; m 38; c 4. ELECTRICAL ENGINEERING. *Educ:* Tex Tech Univ, BS, 38. *Prof Exp:* Distribution engr, Tex Elec Serv Co, 38-42, distribution supvr, 45-46; from asst prof to assoc prof elec eng, Lehigh Univ, 46-54; sr res engr, 54-55, mgr commun res, 55-59, dir elec & electronics dept, 59-63, vpres, 64-72, vpres planning & prog develop, 72-74, sr vpres planning & prog develop, 74-85, SR CONSULT, SOUTHWEST RES INST, 85- *Mem:* Sigma Xi; fel Am Soc Nondestructive Testing; fel Inst Elec & Electronics Engrs; Am Optical Soc. *Res:* Radio direction finding; communications; system engineering; electronic instrumentation and control; sonics; military physics; bioengineering; nondestructive testing of materials; new techniques for testing ferrous metals; design work on receivers, transmitters and industrial electronic devices. *Mailing Add:* PO Box 160218 San Antonio TX 78280-2418

DONALDSON, WILLIAM EMMERT, b Baltimore, Md, Dec 19, 31; m 55; c 6. POULTRY NUTRITION. *Educ:* Univ Md, BS, 53, MS, 55, PhD(poultry nutrit), 57. *Prof Exp:* Asst poultry nutrit, Univ Md, 53-57; asst prof, Univ RI, 57-62; from asst prof to assoc prof, 62-68, prof poultry nutrit, 68-80, PROF POULTRY SCI, NC STATE UNIV, 80- *Honors & Awards:* Am Feed Mfg

Nutrit Res Award, 76. *Mem:* AAAS; Am Inst Nutrit; Poultry Sci Asn; Sigma Xi. *Res:* Fat metabolism, especially fatty acid biosynthesis and interconversion. *Mailing Add:* Dept Poultry Sci NC State Univ Box 7608 Raleigh NC 27695-7608

DONALDSON, WILLIAM TWITTY, b Statesboro, Ga, June 26, 27; m 48; c 2. ANALYTICAL CHEMISTRY. *Educ:* Univ Ga, BS, 48. *Prof Exp:* Develop chemist, Ethyl Corp Develop Lab, 48-51; res supvr, Savannah River Lab, E I du Pont de Nemours & Co, 51-63, lab supvr, Martinsville Nylon Plant, 63-66; chief chem serv, US Federal Water Pollution Control Admin Southeast Water Lab, 66-69; chief anal chem, 69-78, dep dir, Environ Res Lab, 78-83, actg dir, 83-85, CHIEF ENVIRON MEASUREMENTS BRANCH, US ENVIRON PROTECTION AGENCY, 85- *Mem:* Am Chem Soc; Am Soc Testing Mat; Asn Official Anal Chemists; Sigma Xi. *Res:* Applied analytical chemistry; development of methods to measure chemical and microbial transformation rates of organic chemicals; development of spectroscopic methods and computerized data analysis techniques for elucidation of structures of organic compounds. *Mailing Add:* US Environ Protection Agency College Station Rd Athens GA 30613

DONART, GARY B, b Howard, Kans, Sept 6, 40; m 61; c 3. ECOLOGY, PLANT PHYSIOLOGY. *Educ:* Ft Hays Kans State Col, BS, 62, MS, 63; Utah State Univ, PhD(range sci), 68. *Prof Exp:* Asst prof range mgt, Humboldt State Col, 65-68; asst prof range sci, Tex A&M Univ, 68-72; assoc prof, 72-79, PROF RANGE SCI, NMEX STATE UNIV, 79- *Concurrent Pos:* Consult, Nat Parks, 65-66; mem, bd dirs, Soc Range Mgt, 86-88 & 91-93; chmn, Fac Senate, 87-88. *Mem:* Fel Soc Range Mgt (pres, 91-93); Soc Agron; Am Inst Biol Sci. *Res:* Ecology and management of range lands; poison plant problems; livestock management; forage plant physiology; soil-plant relations. *Mailing Add:* Dept Range Sci NMex State Univ Las Cruces NM 88003

DONARUMA, L GUY, b Utica, NY, Sept 6, 28; m 50; c 3. ORGANIC CHEMISTRY. *Educ:* St Lawrence Univ, BS, 49; Carnegie Inst Technol, PhD(org chem), 53. *Prof Exp:* Sr chemist explosives dept, Eastern Lab, E I du Pont de Nemours & Co, Inc, 53-55, res chemist, 55-57, sr res chemist, 57-60, res assoc & supvr, 60-62; asst prof chem, 62-64, assoc prof & exec off dept, 65-67, prof, 67-75, assoc dean grad sch, 68-73, dean grad sch & dir res, Clarkson Col Technol, 73-75; prof chem & dean math, sci & eng, Calif State Univ, Fullerton, 75-77; vpres acad affairs, NMex Inst Mining & Technol, 77-81, prof chem, 77-81; provost, Polytech Inst NY, 81-85; vpres res & grad study, Univ Ala, 85-87; vpres res & technol, Macrochem Corp, 87-90; DIR CHEM SCI, VELCRO GROUP CORP, 90- *Mem:* Am Chem Soc; NY Acad Sci; The Chem Soc; Sigma Xi. *Res:* Synthetic polymer chemistry; polymer drugs, novel polymer systems, polymers for enhanced oil recovery; adhesives. *Mailing Add:* 72 Norma Dr Nashua NH 03062

DONATH, FRED ARTHUR, b St Cloud, Minn, July 11, 31; m 52; c 2. APPLIED GEOLOGY, GEOLOGIC DEFORMATION. *Educ:* Univ Minn, BA, 54; Stanford Univ, MS, 56, PhD(geol), 58. *Prof Exp:* Asst prof geol, San Jose State Col, 57-58; from asst prof geol to prof, Columbia Univ, 58-67; head dept, Univ Ill, Urbana-Champaign, 67-77, prof geol, 67-80; pres, CGS, Inc, 80-83; PRIN & CORP VPRES RES, EARTH TECHNOL CORP, 83- *Concurrent Pos:* Vis lectr, Am Geol Inst, 66-70; lectr, continuing educ prog, Am Asn Petrol Geol, 65-78; assoc ed, Geol Soc Am, 63-73; vis scientist, Am Geophys Union, 67-72; ed, Annual Rev Earth & Planetary Sci, 70-80; consult, US Nuclear Regulatory Comn; adv, Off Sci & Tech Policy, 78-79; mem, US Nat Comt Rock Mechanics, 78-81. *Honors & Awards:* Semicentennial Medallion, Rice Univ, 62. *Mem:* Fel AAAS; fel Geol Soc Am; Am Geophys Union; Am Asn Petrol Geol; fel Geol Soc London; Sigma Xi. *Res:* Structural geology; experimental rock deformation; mechanics of earth deformation. *Mailing Add:* 1006 Las Posas San Clemente CA 92672

DONATI, EDWARD JOSEPH, b Wilkes-Barre, Pa, Sept 9, 24; m 48. ANATOMY. *Educ:* King's Col, Pa, BA, 51; Univ Md, PhD(anat), 64. *Prof Exp:* Technician, Med Labs, US Army Chem Ctr, 52-56, biologist-histologist, 56-60; sect chief microanat-cytol, Directorate Med Res, Edgewood Arsenal, 60-67, actg chief path br, Med Res Labs, 67-68; asst prof, 68-71, ASSOC PROF ANAT, SCH MED, UNIV MD, BALTIMORE CITY, 71- *Mem:* Electron Micros Soc Am; Am Asn Anat. *Res:* Electron microscopy of renal lymphatics and murine tumors. *Mailing Add:* Dept Anat Univ Md Sch Med 655 W Baltimore St Baltimore MD 21201

DONATI, ROBERT M, b Richmond Heights, Mo, Feb 28, 34. NUCLEAR MEDICINE. *Educ:* St Louis Univ, BS, 55, MD, 59. *Prof Exp:* Intern, St Louis City Hosp, 59-60; asst resident, John Cochran Hosp, St Louis, 60-62; from instr to assoc prof, 63-74, dir div nuclear med, 68-88, PROF MED, 74-, SR ASSOC DEAN, 83-, EXEC ASSOC VPRES, MED SCH, ST LOUIS UNIV, 85- *Concurrent Pos:* Fel nuclear med, Med Sch, St Louis Univ, 62-63; staff physician, St Louis Univ & John Cochran Hosp, 63-; chief nuclear med serv, St Louis Vet Admin Hosp, 68-79; consult, Walter Reed Army Inst Res, 68-75; nuclear med res prog specialist, Reg Res Adv Group, 73-74, Res Adv Comt, 73-76, Vet Admin Cent Off, Washington, DC; mem, Presidents Adv Comn Vet Admin, 72, Radiation Task Force, HEW, 78-79; actg chief staff, 76-77, chief staff, St Louis VA Hosp, 79-83. *Mem:* Am Physiol Soc; Soc Nuclear Med; NY Acad Sci; Sigma Xi; fel Int Soc Hemat. *Res:* Control of cellular proliferation in a variety of systems; clinical investigation in nuclear medicine. *Mailing Add:* 5335 Botanical Ave St Louis MO 63110

DONATO, ROSARIO FRANCESCO, b Castrovillari, Italy, Dec 1, 47; m 76; c 2. CALCIUM-BINDING PROTEINS. *Educ:* Cath Univ Sacred Heart, Rome, MD, 73, PhD(neurol), 77. *Prof Exp:* Asst prof anat, Cath Univ Sacred Heart, 73-79; assoc prof, 79-86, PROF NEUROANAT, FAC MED, UNIV PERUGIA, 86- *Concurrent Pos:* Vis prof, Dept Biol, Univ Va, Charlottesville, 89; mem, adv bd, Lab Methodology Biochem, Boca Raton, 90. *Honors & Awards:* F Nitti Award, Nat Acad Sci, Italy, 78. *Mem:* Am Soc Cell Biol; Int Soc Neurochem; Europ Soc Neurochem; Europ Neurosci Asn. *Res:* Intracellular calcium-binding proteins; immunochemistry of calcium-binding proteins. *Mailing Add:* Dept Exp Med & Biochem Sci Sect Anat Univ Perugia CP 81 Perugia Succ 3 06100 Italy

DONAWICK, WILLIAM JOSEPH, b Troy, NY, Aug 18, 40; m 61; c 2. TRANSPLANTATION BIOLOGY, VETERINARY SURGERY. *Educ:* Cornell Univ, DVM, 63; Am Col Vet Surg, dipl, 72; Univ Pa, MS, 73. *Hon Degrees:* MS, Univ Pa, 73. *Prof Exp:* Instr vet clin med & surg, 64-66, USPHS res fel transplantation biol, 66-69, from asst prof to assoc prof, 69-78, PROF SURGERY, UNIV PA, 78- *Concurrent Pos:* NIH career develop award, 73-77. *Mem:* Am Col Vet Surgeons (pres, 82-83). *Res:* Veterinary clinical surgical investigation. *Mailing Add:* Univ Pa New Bolton Ctr Kennett Square PA 19348

DONCHIN, EMANUEL, b Tel Aviv, Israel, Apr 3, 35; m 55; c 3. PHYSIOLOGICAL PSYCHOLOGY. *Educ:* Hebrew Univ, Israel, BA, 61; Univ Calif, Los Angeles, MA, 63, PhD(psychol), 64. *Prof Exp:* Res asst psychol, Univ Calif, Los Angeles, 61-64, res assoc, 64-65; res assoc neurol, Sch Med, Stanford Univ, 65-67; res assoc neurobiol, Ames Res Ctr, NASA, 66-69; assoc prof psychol, 69-73, PROF PSYCHOL, UNIV ILL, URBANA-CHAMPAIGN, 73-, HEAD, DEPT PSYCHOL, 80- *Mem:* Am Electroencephalog Soc; Psychonomic Soc; Soc Psychophysiol Res (pres, 79-80); fel Am Pschol Asn; fel AAAS. *Res:* Physiological mechanisms of attention and perception; computer analysis of brain waves. *Mailing Add:* Dept Psychol Univ Ill 603 E Daniel St Champaign IL 61820

DONDERO, NORMAN CARL, b Somerville, Mass, May 22, 18; m 52; c 1. MICROBIOLOGY, WATER POLLUTION. *Educ:* Univ Mass, BS, 41; Univ Conn, MS, 43; Cornell Univ, PhD(bact), 52. *Prof Exp:* Instr animal dis, Univ Conn, 46-48; res assoc bact, Cornell Univ, 51-53, asst prof, 53-54; assoc, Inst Microbiol, Rutgers Univ, 54-56, from asst prof to assoc prof, 56-63, prof environ sci, 63-66, chmn microbiol sect, 63-65; prof, 66-84, EMER PROF APPL MICROBIOL, CORNELL UNIV, 84- *Honors & Awards:* Fulbright Res Award, New Zealand, 82. *Mem:* Am Soc Microbiol; Am Acad Microbiol. *Res:* Aquatic microbiology; water pollution. *Mailing Add:* 864 Sheffield Rd Ithaca NY 14850

DONDERSHINE, FRANK HASKIN, b Newark, NJ, Dec 29, 31; m 58. MICROBIOLOGY. *Educ:* Seton Hall Univ, AB, 55. *Prof Exp:* Asst hematologist, Martland Med Ctr, NJ, 51-54; microbiologist, Warner-Lambert Res Inst, 57-62; microbiologist, Ethicon, Inc, 62-70; sr microbiologist, Cosan Chem Corp, 70-71; MGR, MICROBIOL DEPT, CARTER-WALLACE, INC, 71- *Concurrent Pos:* Lectr, Ctr Prof Advan. *Mem:* AAAS; Am Soc Microbiol; Soc Indust Microbiol. *Res:* Antimicrobial product evaluations; general antimetabolite and clinical evaluations; antiseptics, disinfectants and sterilizers development; ultrasonics; aerosols; antibiotics; dental research; quality control. *Mailing Add:* Carter-Wallace Inc Half Acre Rd Cranbury NJ 08512

DONDES, SEYMOUR, b New York, NY, Apr 3, 18; m 45; c 3. PHYSICAL CHEMISTRY. *Educ:* Brooklyn Col, BA, 39; Rensselaer Polytech Inst, MS, 50, PhD(chem), 54. *Prof Exp:* Res scientist radiation chem, 52-70, from assoc prof to prof 70-88, EMER PROF CHEM, RENSSELAER POLYTECH INST, 88- *Mem:* AAAS; Am Chem Soc. *Res:* Radiation chemistry; photochemistry; kinetics of gas reactions; nuclear reactor technology. *Mailing Add:* Dept Chem Rensselaer Polytech Inst Troy NY 12181

DONE, ALAN KIMBALL, b Salt Lake City, Utah, Sept 23, 26; m 47; c 10. PEDIATRICS, CLINICAL PHARMACOLOGY. *Educ:* Univ Utah, BA, 49, MD, 52. *Prof Exp:* From intern to assoc pediatrician, Salt Lake County Gen Hosp, 53-58; from res instr to res asst prof pediat, Col Med, Univ Utah, 56-58; asst prof, Stanford Univ, 58-60; from assoc res prof to prof pediat, Col Med, Univ Utah, 60-71; spec asst to dir, Bur Drugs, Food & Drug Admin, 71-74; PROF PEDIAT & PHARMACOL, COL MED, WAYNE STATE UNIV, 75-, ADJ PROF CLIN PHARM, 80- *Concurrent Pos:* Am Heart Asn fel, 54-57; adj prof pharmacol, Col Med, Univ Utah, 69-71, prof clin pharmacol, Col Pharm, 70-71. *Mem:* Am Pediat Soc; Soc Pediat Res; fel Am Acad Pediat; Soc Exp Biol & Med. *Res:* Developmental pharmacology. *Mailing Add:* 63375 S Highland Dr No 2054 Salt Lake City UT 84121

DONEFER, EUGENE, b Brooklyn, NY, Jan 2, 33; m 53; c 3. ANIMAL NUTRITION, INTERNATIONAL DEVELOPMENT. *Educ:* Cornell Univ, BS, 55, MS, 57; McGill Univ, PhD(nutrit), 61. *Prof Exp:* From asst prof to assoc prof, 61-74, PROF ANIMAL SCI, MACDONALD COL, 75-, DIR, MCGILL INT, MCGILL UNIV, 85- *Concurrent Pos:* Consult, Beef Prod Proj Caribbean, Barbados; res assoc, Int Develop Res Centre, Ottawa, 72-73; proj dir, Sugarcane Feeds Ctr, Trinidad, 76-81. *Honors & Awards:* E W Crampton Award, 82. *Mem:* Am Inst Nutrit; Can Soc Nutrit Sci. *Res:* Ruminant nutrition; nutritive evaluation of forages; improving low-quality forages; utilization of sugarcane feeds. *Mailing Add:* McGill Int McGill Univ 3550 Univ St Montreal PQ H3A 2A7 Can

DONEGAN, WILLIAM L, b Jacksonville, Fla, Nov 3, 32; m 63; c 2. SURGERY, SURGICAL ONCOLOGY. *Educ:* Yale Univ, BS, 55, MD, 59. *Hon Degrees:* FACS, Am Col Surgeons, 65. *Prof Exp:* Surgeons surg, Edis Fischel State Cancer Hosp, 64-74; PROF SURG, MED COL WIS, 74-; CHMN SURG, SINAI SAMARITAN MED CTR, MILWAUKEE, 82- *Concurrent Pos:* pres, Wis div, Am Cancer Soc, 90-91. *Mem:* Am Surg Asn; Cent Surg Asn; Soc Head & Neck Surgeons; Soc Surg Oncol; Soc Surg Alimentary Tract; Am Soc Clin Oncol. *Res:* Cancer of the breast. *Mailing Add:* Surg Dept 950 N 12th St Milwaukee WI 53201

DONELAN, MARK ANTHONY, b Grenada, West Indies, Mar 27, 42; m 67; c 2. PHYSICAL OCEANOGRAPHY, METEOROLOGY. *Educ:* McGill Univ, BEng, 64; Univ BC, PhD(oceanog), 70. *Prof Exp:* Proj engr, Procter & Gamble Can, 64-66; fel, Dept Appl Math & Theoret Physics, Cambridge Univ, 70-71; RES SCIENTIST, ENVIRON CAN, FED GOVT CAN, 71- *Concurrent Pos:* Assoc ed, Atmosphere-Ocean, 78-84; adj prof, Dept Appl Math, Univ Waterloo, 79-; prof, Dept Civil Eng, McMaster Univ, 79-; Humboldt res fel, Max Planck Inst Meteorol, Hamburg, Fed Repub Germany, 84. *Mem:* Can Meteorol & Oceanog Soc; Can Appl Math Soc; Am Meteorol Soc; Am Geophys Union; Oceanog Soc. *Res:* Air-water interaction, surface gravity waves, turbulence, gas transfer. *Mailing Add:* Can Ctr Inland Waters PO Box 5050 Burlington ON L7R 4A6 Can

DONELSON, JOHN EVERETT, b Ogden, Iowa, May 23, 43; m 66; c 4. BIOCHEMISTRY. *Educ:* Iowa State Univ, BSc, 65; Cornell Univ, PhD(biochem), 71. *Prof Exp:* Teacher math & chem, sec sch, Am Peace Corps, Ghana, WAfrica, 65-67; fel molecular biol, Med Res Ctr, Lab Molecular Biol, Cambridge, Eng, 71-73; fel biochem, Med Ctr, Stanford Univ, 73-74; from asst prof to assoc prof, 74-80, PROF BIOCHEM, UNIV IOWA, 80- *Honors & Awards:* Burroughs-Wellcome Award Molecular Parisitol, 83. *Mem:* Am Soc Advan Sci. *Res:* Mechanisms of gene expression and immune evasion in parasites causing tropical diseases. *Mailing Add:* Dept Biochem Sch Med Univ Iowa Iowa City IA 52242

DONER, HARVEY ERVIN, b Nampa, Idaho, Dec 6, 38; m 68; c 2. SOIL CHEMISTRY, SOIL MINERALOGY & SURFACE REACTIONS. *Educ:* Univ Idaho, BS, 61, MS, 64; Univ Calif, Riverside, PhD(soil sci), 67. *Prof Exp:* Res assoc, Univ Wis-Madison, 68; res assoc, Mich State Univ, 68-69; from asst prof to assoc prof, 69-83, PROF RES, UNIV CALIF, BERKELEY, 83- *Mem:* Soil Sci Soc Am; Int Soil Sci Soc; Clay Minerals Soc; Sigma Xi; AAAS. *Res:* Soil chemistry; trace element chemistry of soils and the reactions of organic compounds on mineral surfaces. *Mailing Add:* Dept Soil Sci Univ Calif Berkeley CA 94720

DONER, LANDIS WILLARD, b Winona, Minn, Aug 31, 41; m 72. CARBOHYDRATE CHEMISTRY. *Educ:* Winona State Univ, BA, 64; NDak State Univ, MS, 66; Purdue Univ, PhD(biochem), 71. *Prof Exp:* Fel biochem, Purdue Univ, 71 & 73-74; fel chem, Dundee Univ, Scotland, 71-73; RES CHEMIST BIO-ORG CHEM, EASTERN REGIONAL RES CTR, USDA, 74- *Honors & Awards:* Harvey W Wiley Award, Asn Off Anal Chemists, 90. *Mem:* Am Chem Soc; AAAS; Asn Off Anal Chemists. *Res:* Carbohydrate chemistry and biochemistry; instrumental methods of sugar analysis; food applications of natural variations in stable isotope ratios; honey composition; plant carbohydrates. *Mailing Add:* USDA Eastern Regional Res Ctr 600 E Mermaid Lane Philadelphia PA 19118

DONEY, DEVON LYLE, b Franklin, Idaho, Mar 31, 34; m 58; c 5. PLANT GENETICS. *Educ:* Utah State Univ, BS, 60, MS, 61; Cornell Univ, PhD(plant breeding, genetics), 65. *Prof Exp:* Tomato breeder, Libby, McNeil & Libby, 64-65; RES GENETICIST, AGR RES SERV, USDA, 65-, RES LEADER, SUGARBEET RES, 88- *Mem:* Crop Sci Soc Am; Am Soc Sugarbeet Technologists. *Res:* Physiological genetics of sugar beet; selection methods; alcohol fuel from beets. *Mailing Add:* Res Leader Sugar Res USDA-ARS-NSA Northern Crop Sci Lab Box 5679 Univ Sta Fargo ND 58105-5677

DONG, RICHARD GENE, b Sacramento, Calif, Mar 16, 35; m; c 2. STRUCTURAL MECHANICS & DYNAMICS, SEISMIC ENGINEERING. *Educ:* Univ Calif, Berkeley, BS, 57, MS, 59, PhD(civil eng), 64. *Prof Exp:* Develop engr, Aerojet-Gen Corp, Calif, 59-61; ENGR, LAWRENCE LIVERMORE NAT LAB, UNIV CALIF, 63- *Res:* Equipment qualification for nuclear power plants; fragilities of equipment and structures; material behavior and properties; technical bases for nuclear power plant licensing; seismic design methodology; fluid structure interaction; practical design technologies; general structural dynamics consultation. *Mailing Add:* 38 Hornet Ct Danville CA 94526

DONG, STANLEY B, b Canton, China, Apr 2, 36; US citizen; m 64; c 3. STRUCTURAL MECHANICS, STRUCTURAL DYNAMICS. *Educ:* Univ Calif, Berkeley, BS, 57, MS, 58, PhD(struct mech), 62. *Prof Exp:* Sr res engr, Aerojet-Gen Corp, 62-65; from asst prof to assoc prof eng, 65-73, PROF ENG, UNIV CALIF, LOS ANGELES, 73- *Concurrent Pos:* Off Naval Res-Am Inst Aeronaut & Astronaut res scholar struct mech, 70. *Honors & Awards:* Achievement Award, Chinese Engrs & Scientists Southern Calif. *Mem:* Am Soc Civil Engrs; Am Soc Mech Engrs; AAAS; fel Am Acad Mech. *Res:* Analysis of structural composites, structural dynamics of plates and shells. *Mailing Add:* 4514 Boelter Hall Eng & Appl Sci Univ Calif Los Angeles CA 90024

DONGARRA, JACK JOSEPH, b Jul, 18, 50; c 3. COMPUTER SCIENCE. *Educ:* Chicago State Univ, BS, 72; Ill Inst Technol, MS, 73; Univ NMex, PhD(appl math), 80. *Prof Exp:* SCIENTIST, ARGONNE NAT LAB, 80-; SCI DIR, ADV COMPUT RES FACIL, 84- *Mem:* Soc Indust Appl Math. *Mailing Add:* Dept Comp Sci Univ Tenn Knoxville TN 37996-1301

DONGIER, MAURICE HENRI, b Sorgues, France, Nov 27, 25; Can citizen; m 52; c 4. PSYCHIATRY, ALCOHOLISM. *Educ:* Univ Aix-Marseille, MD, 51; McGill Univ, dipl psychiat, 54. *Prof Exp:* Res asst psychiat, McGill Univ, 51-52; chef de clin adj neuro psychiat, Univ Aix-Marseille, 54-57, chef de clin tit psychiat, 57-60; res psychiatrist neurophysiol, Hosp Timone, Marseille, 60-63; prof psychiat & chmn dept, Univ Liege, 63-71; div, Alan Mem Inst, 71-80; chmn dept, 74-84, PROF PSYCHIAT, MCGILL UNIV, 71- *Concurrent Pos:* Res award, Med Res Sci Found, Belg, 65-70; res award, Med Res Coun Can, 73-76, mem behav sci grant comt, 76-79; mem behav & soc adv coun, Alcohol Beverage Med Res Found, Johns Hopkins, 80-86. *Mem:* Can Psychiat Asn; Am Psychiat Asn; Am Psychopath Asn. *Res:* Psychosomatics, especially coronary heart disease; alcoholism; electroencehalography in psychiatry. *Mailing Add:* McGill Dept Psychiat McGill Univ Fac Med 1033 Pine Ave W Montreal PQ H3A 1A1 Can

DONIACH, SEBASTIAN, b Paris, France, Jan 25, 34; m 55; c 4. THEORETICAL PHYSICS. *Educ:* Cambridge Univ, BA, 54; Univ Liverpool, PhD(theoret physics), 58. *Prof Exp:* Imp Chem Industs fel theoret physics, Univ Liverpool, 58-60; lectr physics, Queen Mary Col, Univ London, 60-64 & Imp Col, 64-66, reader, 67-69; PROF APPL PHYSICS, STANFORD UNIV, 69- *Concurrent Pos:* Vis scientist, Europ Orgn Nuclear Res, Geneva, 63, Inst Laue Langevin, Grenoble; consult, Atomic Energy Res Estab, Harwell, Eng, 66-69, Bell Tel Res Labs, 67, 70 & 71, Dept Physics, Univ Calif, San Diego, 68 & Argonne Nat Lab, AEC, 70-; res assoc, Dept Physics, Harvard Univ, 67-68; dir, Stanford Synchrotron Radiation Proj, 73-78, consult dir, 78-80; vis assoc prof, Univ Paris, 75-76, 78, 82 & 85, sabbatical, Univ Paris, Jussieu & Orsay, 80; Japan Soc Promotion Sci vis prof, Inst Solid State Physics, Univ Tokyo, 78; mem, Bd trustees & Corp, Aspen Ctr Physics, 78-85, Solid State Sci Panel, Nat Res Coun, 80-, adv bd, Inst Theoret Physics, Univ Calif, Santa Barbara, 86-89; vis scientist, CEA Saclay, 85 & 87; co-chair, SSRL affil fac, 89; vis fel, Advan Studies Inst, Los Alamos Nat Lab, 87-; sr vis scientist, Léon Brioullin, CEN Saclay, 90; assoc ed, Advances in P;hysics, 82- *Mem:* Brit Inst Physics & Phys Soc; fel Am Phys Soc. *Res:* Theory of cooperative phenomena: granular superconductors, heavy Fermi liquids, high temperature superconductivity, metal-semiconductor interfaces, phospholipia phases, protein organization; biophysics; x-ray absorption spectroscopy; quantum chemistry of large molecules; small angle x-ray difraction from biological materials: calcium binding proteins, acetylcholine receptor membrane, bacteriorhoolopsin. *Mailing Add:* Dept Appl Physics Stanford Univ Stanford CA 94305

DONIGER, JAY, b Brooklyn, NY, Mar 22, 44; m 67; c 2. MOLECULAR BIOLOGY. *Educ:* Brooklyn Col, BS, 65; Purdue Univ, PhD(biol), 72. *Prof Exp:* Res assoc biochem, Brandeis Univ, 72-75; asst biologist, Brookhaven Nat Lab, 75-77; EXPERT SCIENTIST, NAT CANCER INST, 77-, SR STAFF FEL, 88- *Res:* DNA repair mechanisms in Mammalian cells. *Mailing Add:* 17219 Epsilon Pl Rockville MD 20815

DONISCH, VALENTINE, b Novocherkassk, Russia, Nov 29, 19; Can citizen; m 45. BIOCHEMISTRY. *Educ:* Univ Bonn, MD, 47; Univ Western Ont, PhD(biochem), 65; FRCP(C), 77. *Prof Exp:* Physician, Extrapulmonal Tuberc Sanitorium, 47-52; physician, Lutheran Mission Hosp, Eket, Nigeria, 52-54, govt hosps, Accra & Kumasi, Ghana, 55-59 & Victoria Hosp, London, Ont, 59-61; from instr to asst prof biochem, Med Sch, Univ Western Ont, 66-73, emer prof, 73-; RETIRED. *Concurrent Pos:* Fel, Victoria Hosp, London, Ont, 65-66. *Res:* Extrapulmonal tuberculosis; phospholipids and cancer. *Mailing Add:* Box 672 Grand Bend ON N0M 1T0

DONIVAN, FRANK FORBES, JR, b Inglewood, Calif, Oct 19, 43; m 66; c 2. DEEP SPACE TELECOMMUNICATIONS, SPACECRAFT NAVIGATION. *Educ:* Univ Calif, Los Angeles, BA, 66; Univ Fla, PhD(astron), 70. *Prof Exp:* Asst dean, Grad Sch, Univ Fla, 71-73, asst prof, 70-76, assoc prof phys sci & astron, 77-79; mem tech staff, Jet Propulsion Lab, Calif Inst Technol, Pasadena, 79-84, radio sci syst engr, 84-86, mgr Voyager ground data syst eng off, 86-88, mem, Space Sta Off, 88-90; ACTG PROG MGR DSN SYSTS, OFF SPACE OPERS, GROUND NETWORKS DIV, NASA HQ, 90- *Honors & Awards:* Except Serv Award, NASA, 86. *Mem:* Am Astron Soc; Aerospace Indust Asn Am; Sigma Xi. *Res:* Radio investigation of clusters of galaxies; quasars and solar energy; developing technique for using very long baseline interferometry; navigation of interplanetary (deep space) probes; design of deep space telecommunications systems. *Mailing Add:* Off Space Opers Code OT NASA Hq Washington DC 20546

DONKER, JOHN D, b New Brunswick, NJ, Mar 2, 20; m 48; c 1. ANIMAL SCIENCE. *Educ:* Univ Calif, BS, 48; Univ Minn, PhD(reproductive physiol), 52. *Prof Exp:* Dir dairy cattle nutrit lab, Univ Ga, 54-56; from assoc prof to prof dairy husb, 56-90, EMER PROF, UNIV MINN, MINNEAPOLIS, 90- *Concurrent Pos:* Fel, Am Inst Indian Studies, 64. *Mem:* Am Dairy Sci Asn; Am Soc Animal Sci. *Res:* Nutrition of dairy cows; evaluation of energy contents of forages and/or rations; lactational physiology. *Mailing Add:* Col Agr Univ Minn St Paul MN 55108

DONLON, MILDRED A, b Quincy, Mass, Oct 1, 40; m; c 2. RADIATION BIOLOGY, RADIO PROTECTION. *Educ:* Univ Rochester, MS, 68, PhD(microbiol), 72. *Prof Exp:* SCI PROGS MGR, ARMED FORCES RES INST, NAT NAVY MED CTR, 77- *Mem:* Am Soc Cell Biol; Radiation Res Soc. *Mailing Add:* Armed Forces Radiobiol Res Inst Nat Navy Med Ctr Bethesda MD 20814-5145

DONN, BERTRAM (DAVID), b Brooklyn, NY, May 25, 19. ASTRONOMY. *Educ:* Brooklyn Col, AB, 40; Harvard Univ, PhD(astron), 53. *Prof Exp:* Physicist, Elec Testing Labs, 40-41; asst physicist, Signal Corps Labs, 41-43; res assoc radiation lab, Columbia Univ, 43-46; from instr to assoc prof physics & astron, Wayne State Univ, 50-59; head, Astrochem Br, 59-75, SR SCIENTIST, GODDARD SPACE FLIGHT CTR, NASA, 75- *Concurrent Pos:* Res assoc, Inst Nuclear Studies, Chicago, 54-56; vis prof astron, Cornell Univ, 74-75. *Mem:* AAAS; Am Phys Soc; Am Astron Soc; Am Geophys Union; Int Astron Union; Sigma Xi. *Res:* Interstellar matter; comets; cosmic chemistry. *Mailing Add:* 19 Woodland Way Greenbelt MD 20770

DONN, CHENG, b Chungking, China, Nov 24, 38; US citizen; m 64; c 1. ELECTRICAL ENGINEERING, ELECTROMAGNETISM. *Educ:* Chengkung Univ, BS, 60; Wichita State Univ, MS, 68; Univ Kans, PhD(elec eng), 73. *Prof Exp:* Res assoc antennas, Electrosci Lab, Ohio State Univ, 69-73; electromagnetic pulse analyst, Intelcom Rad Tech, 73-74; mem tech staff, TRW Defense & Space Systs Group, 74-77; sr staff engr antennas, Martin Marietta Aerospace, Denver Div, 77-80; adv systs prog engr, Gen Dynamics, 80-; DEPT MGR, RANDTRON SYSTS, CALIF. *Mem:* Inst Elec & Electronics Engrs; Sigma Xi. *Res:* Electromagnetic theory and antenna systems. *Mailing Add:* Randtron Systs 130 Constitution Menlo Park CA 94025

DONNALLEY, JAMES R, JR, b Camden, NJ, June 6, 18; m 46; c 3. CHEMICAL ENGINEERING. *Educ:* Pa State Univ, BS, 39; Cornell Univ, PhD(chem eng), 44. *Prof Exp:* Res assoc silicones, Res Lab, 43-46, group leader silicone process develop, Chem Div, 46-48, mgr, Waterford Plant, 48-52 & Mfg Silicone Prod Dept, 52-60, gen mgr, Insulating Mat Dept, 60-66 & Semiconductor Prod Dept, 66-74, mgr lighting res & tech serv oper, Lamp Bus Div, 74-80, vpres environ issues, Gen Elec Co, Fairfield, 80-; RETIRED. *Mem:* Am Chem Soc; Am Inst Chem Engrs. *Res:* Silicates for adsorption and catalyses; process research on freeze-preservation of food; chemical process research on chlorosilanes and silicones; chemical research high temperature polymers. *Mailing Add:* PO Box 936 Ponte Vedra Beach FL 32082

DONNALLY, BAILEY LEWIS, b Deatsville, Ala, June 22, 30; m 55; c 3. ATOMIC PHYSICS. *Educ:* Auburn Univ, BS, 51, MS, 52; Univ Minn, PhD(physics), 61. *Prof Exp:* Instr physics, Col St Thomas, 56-61; from asst prof to assoc prof, 61-66, chmn dept, 71-80, PROF PHYSICS, LAKE FOREST COL, 66- *Concurrent Pos:* Vis fel, Yale Univ, 66-67. *Mem:* AAAS; fel Am Phys Soc; Am Asn Physics Teachers (pres); Sigma Xi. *Res:* Mass spectrometry; polarized proton sources; nuclear scattering with polarized protons; atomic collision processes; development of experiments for advanced undergraduate laboratories. *Mailing Add:* Dept Physics Lake Forest Col Lake Forest IL 60045

DONNAN, WILLIAM W, b Keystone, Iowa, June 15, 11; m 35; c 3. ENGINEERING, AGRICULTURE. *Educ:* Iowa State Univ, BS, 34, CE, 46. *Prof Exp:* Jr engr soil conserv serv, USDA, 34-39, assoc engr, Calif, 39-41, res engr, 41-53, sr res engr agr res serv, 53-57, res invest leader, 57-61, br chief SW br soil & water conserv res, 61-71; CONSULT ENGR, 71- *Concurrent Pos:* Grant, Western Europe, 58; mem, Soil & Water Conserv Cultural Exchange Group, USSR, 58 & Tech Activ Comt, Int Comn Irrig & Drainage, 60-65; consult, Indus Basin, WPakistan, 60, Khusistan Basin, Iran, 66, Upper Ferat Basin, Turkey, 69 & Tisza River Basin, Hungary, 71; team leader consult group, FAD Tech Assistance to Egypt on Nile Delta Drainage Probs, 75-76. *Honors & Awards:* Hancock Drainage Eng Award, Am Soc Agr Engrs, 66; R J Tipton Medal, Am Soc Civil Engrs, 71. *Mem:* Fel Am Soc Civil Engrs; Am Soc Agr Engrs. *Res:* Drainage of irrigated land including the spacing and depth of drains and criteria for the design of drainage systems. *Mailing Add:* 700 S Lake Ave Pasadena CA 91106

DONNAY, GABRIELLE (HAMBURGER), crystallography; deceased, see previous edition for last biography

DONNAY, JOSEPH DESIRE HUBERT, b Grandville, Belg, June 6, 02; nat 39; m 31, 49; c 4. CRYSTALLOGRAPHY, MINERALOGY. *Educ:* Univ Liege, EM, 25; Stanford Univ, PhD(geol), 29. *Prof Exp:* Engr & geologist, Syndicat des Petroles au Maroc, French Morocco, 29-30; teaching fel mineral & res assoc geol, Stanford Univ, 30-31; assoc mineral, Johns Hopkins Univ, 31-39; prof crystallog & mineral, Laval Univ, 39-45; prof crystallog & mineral, 45-71, EMER PROF CRYSTALLOG & MINERAL, JOHNS HOPKINS UNIV, 71- *Concurrent Pos:* Res chemist, Hercules Powder Co, 42-45; prof, Univ Liege, 46-47; Fulbright lectr, Sorbonne, Paris, 58-59; guest prof, Univ Marburg, 66; lectr, Univ Montreal, 70-72, guest prof, 72-76; US deleg, Int Union Crystallog Cong, USA, 48, Sweden, 51, France, 54, Can, 57 & USSR, 66; res assoc, McGill Univ, 75- *Honors & Awards:* Roebling Medal, Mineral Soc Am, 71; Trasenster Medal, Asn Ing Univ Liege, 77; Can Silver Jubilee Medal, 77. *Mem:* Fel Geol Soc Am (vpres, 54); fel Mineral Soc Am (vpres, 49, 52, pres, 53); Am Crystallog Asn (secy-treas, Crystallog Soc Am, 44, secy, 45-46, vpres, 46, 48, 55, pres, 49, 56); fel Mineral Soc Gt Brit & Ireland; hon mem Soc Fr Mineral (vpres, 49). *Res:* Crystallography, especially crystal optics and relationships between crystal morphology and structure. *Mailing Add:* 516 Iberville St Hilaire PQ J3H 2V7

DONNELL, GEORGE NINO, b Shanghai, China, Feb 21, 19; US citizen; m 42; c 3. PEDIATRICS. *Educ:* Pomona Col, BA, 40; Wash Univ, MD, 44; Am Bd Pediat, dipl, 50. *Prof Exp:* Asst pediat, Sch Med, Wash Univ, 45-46; from instr to assoc prof, Univ Southern Calif, 45-61, Winzer prof & chmn dept pediat, pediatrician-in-chief, Children's Hosp, 71-85, co-chmn, Dept Pediat, 80-85, PROF PEDIAT, UNIV SOUTHERN CALIF, 61- *Concurrent Pos:* Mem & consult, Nat Kidney Dis Found, chmn sci adv coun, Southern Calif Chap, 58-59. *Mem:* AAAS; Am Acad Pediat; Soc Pediat Res; Am Pediat Soc; fel Royal Soc Med. *Res:* Biochemical genetics; galactose metabolism; disorders of organic acid metabolism; disorders of amino acid metabolism. *Mailing Add:* Children's Hosp Los Angeles Univ Southern Calif Sch Med 4650 Sunset Blvd Los Angeles CA 90054

DONNELL, HENRY DENNY, JR, b Vandalia, Ill, June 21, 35; m 58; c 3. EPIDEMIOLOGY. *Educ:* Greenville Col, AB, 57; Wash Univ, MD, 60; Univ Calif, Berkeley, MPH, 65; Am Bd Prev Med, dipl, 70. *Prof Exp:* Intern, Butterworth Hosp, Grand Rapids, Mich, 60-61, asst resident surg, 61-62; asst chief prev med, Ft Leonard Wood, Mo, 62-63, chief, 63-64; resident epidemiol, Calif State Dept Health, Berkeley, 65-66; asst prof, Sch Med, Univ Mo-Columbia, 66-71, clin asst prof, Dept Commun Health & Med Pract, 66-71; dir, Bur Commun Dis Control, 71-73, DIR SECT EPIDEMIOL & DIS SURVEILLANCE, DIV HEALTH OF MO, 73- *Concurrent Pos:* Chief prog methodology unit, Mo Regional Med Prog, 69-71, consult, 71-; dir grad study community health, Univ MoColumbia. 69-70. *Mem:* AAAS; Am Pub Health Asn; Asn Teachers Prev Med; Royal Soc Health. *Res:* Studies of distribution of disease, especially heart, cancer, stroke and infectious diseases; distribution of health manpower in relationship to distribution of population and various influencing variables. *Mailing Add:* 1730 E Elm PO Box 570 Jefferson City MO 65102-0570

DONNELLY, BRENDAN JAMES, b Dublin, Ireland, July 6, 37; US citizen; m 66. FOOD CHEMISTRY. *Educ:* Nat Univ Ireland, BSc, 60, PhD(chem), 65; Trinity Col, Dublin, MSc, 62; St Louis Univ, MBA, 75. *Prof Exp:* USDA fel chem, Exp Sta, Colo State Univ, 65-68; sr res chemist, Corn Prod Div, Anheuser-Busch Inc, 68-75; assoc prof cereal chem, Cereal Chem & Tech Dept, NDak State Univ, 75-79; Cereal Qual Lab mgr, North Am Plant Breeders, Colo, 79-82; DIR, NORTHERN CROPS INST, N DAK STATE UNIV, 82- *Concurrent Pos:* Consult, US Wheat Assocs, Washington, DC, 77-; mem, Nat Wheat Improv Comt, 80-; assoc ed, Cereal Chem, Am Asn Cereal Chemists, 80-; affil prof, Dept Food Sci & Nutrit, Colo State Univ, 81- *Mem:* Inst Food Technologists; Am Asn Cereal Chemists; Am Chem Soc; Phytochem Soc NAm; AAAS; Sigma Xi. *Res:* Isolation and identification of naturally occurring polyphenols (flavonoid family); precursors of haze and sediment formation in fruit juice; chemical components of cereal grain (wheat and barley) quality; high fructose syrup prroduction from corn starch. *Mailing Add:* NDak State Univ Box 5183 SU Sta Fargo ND 58105

DONNELLY, DENIS PHILIP, b Ann Arbor, Mich, Nov 27, 37; m 63; c 4. ELECTRONIC INSTRUMENTATION. *Educ:* Univ Mich, BS, 59, MS, 61, PhD(physics), 67. *Prof Exp:* Res assoc, Univ Mich, 68-70; asst prof physics, Skidmore Col, 70-76; from asst prof to assoc prof, 76-86, dean sci, 80-84, PROF PHYSICS, SIENA COL, 86- *Mem:* Am Asn Physics Teachers; AAAS; Sigma Xi. *Res:* Low energy nuclear physics; atomic beams; meteorological and electronic instrumentation; history of science; scientific pedagogy; biophysical modeling. *Mailing Add:* Dept Physics Siena Col Loudonville NY 12211

DONNELLY, EDWARD DANIEL, b Birmingham, Ala, Dec 5, 19; m 47; c 3. PLANT BREEDING. *Educ:* Auburn Univ, BS, 46, MS, 48; Cornell Univ, PhD(plant breeding), 51. *Prof Exp:* Res asst & instr forage crops, Auburn Univ, 47-48, assoc plant breeder, 51-58, prof agron & soils, 58-83; RETIRED. *Concurrent Pos:* Res asst, Cornell Univ, 48-51. *Mem:* Fel Am Soc Agron; fel Crop Sci Soc Am. *Res:* Forage crops breeding; improving for disease and root-knot nematode resistance; increased palatability and nutritive value; earlier development and larger seeded; reseeding winter annual legumes. *Mailing Add:* 751 Sherwood Dr Auburn AL 36830

DONNELLY, GRACE MARIE, b Providence, RI, Apr 5, 29. ENVIRONMENTAL SCIENCES GENERAL. *Educ:* Univ RI, BS, 52; Brown Univ, MS, 56; Univ Conn, PhD(cytol), 62. *Prof Exp:* Res librarian, Citrus Exp Sta, Univ Fla, 52-54; biol fel, Brookhaven Nat Lab, 62-64; res assoc biol, City of Hope Med Ctr, 64-67; from instr to asst prof cell biol, Univ Ky, 67-72, bioassay proj leader, 72-75, sr res biologist, Tobacco & Health Res Inst, Univ Ky, 75-80; PRES, BIOSPEC INC, PROVIDENCE, 81- *Mem:* AAAS; Am Soc Cell Biol; NY Acad Sci. *Res:* Cytology of the domestic fowl; radiosensitivity of amphibians; chromosome cytology; cellular requirements for ribonucleic acid synthesis; effect of environmental agents on cells; occupational health; environmental management. *Mailing Add:* Biospec Inc 147 Sixth St Providence RI 02906-3733

DONNELLY, JOHN, b Liverpool, Eng, June 9, 14; nat US; m 49; c 2. PSYCHIATRY. *Educ:* Univ Liverpool, MB, ChB, 38; Royal Col Physicians & Surgeons, DPM, 48; Am Bd Psychiat & Neurol, dipl, 52. *Hon Degrees:* ScD, Trinity Col, Conn, 79. *Prof Exp:* Sr registr, Cane Hill Hosp, Eng, 48-49; chief serv psychiat, Inst Living, 49-52, clin dir, 52-56, med dir, 56-65, psychiatrist-in-chief, 65-79, sr consult, 79-84; PROF PSYCHIAT, SCH MED, UNIV CONN, 77-; PSYCHIATRIST, PVT PRACTICE, 84- *Concurrent Pos:* From asst clin prof to assoc clin prof psychiat, Med Sch, Yale Univ, 52-68, lectr, 68-80; ed, Digest of Neurol & Psychiat, Inst Living, 63-79; pvt pract, 84- *Mem:* Fel Am Psychiat Asn; fel Am Col Physicians; fel Royal Soc Med; fel Royal Col Psychiat UK. *Res:* Psychodynamic and biophysical correlates; forensic aspects of psychiatry. *Mailing Add:* 400 Washington St Hartford CT 06106

DONNELLY, JOHN JAMES, III, b Philadelphia, Pa, June 26, 54; m 76; c 2. ANTIGEN PROCESSING, IMMUNOREGULATION. *Educ:* Univ Pa, BA, 75, PhD(immunol), 79. *Prof Exp:* Fel immunol, Univ Cambridge, 79-82 & Johns Hopkins Univ Sch Med, 82-83; asst prof ophthal, Sch Med, Univ Pa, 83-88; RES FEL IMMUNOL, MERCK SHARP & DOHME RES LABS, 88- *Concurrent Pos:* Adj asst prof, Sch Med, Univ Pa, 88- *Mem:* Am Asn Immunologists; Royal Soc Trop Med & Hyg; Asn Res Vision & Ophthal; NY Acad Sci; AAAS. *Res:* Antigen processing; regulation of MHC antigen expression, transplantation and tumor immunity, immunomodulators and immunoregulation; filariasis, immunity to metazoan parasites; allergy, immediate hypersensitivity. *Mailing Add:* Dept Cancer Res WP16-3 Merck Sharp & Dohme Res Labs West Point PA 19486

DONNELLY, JOSEPH P(ETER), b Brooklyn, NY, May 10, 39; m 68; c 5. ELECTRICAL ENGINEERING, SEMICONDUCTOR PHYSICS. *Educ:* Manhattan Col, BEE, 61; Carnegie Inst Tech, MS, 62, PhD(elec eng), 66. *Prof Exp:* NATO fel, Imp Col, London, 65-66; STAFF MEM APPL PHYSICS, LINCOLN LAB, MASS INST TECHNOL, 67- *Concurrent Pos:* Nat lectr, Inst Elec & Electronics Engrs, Electron Device Soc, 79. *Mem:* Fel Inst Elec & Electronics Engrs; Bohmische Physical Soc; Sigma Xi. *Res:* Ion implantation in compound semiconductors; photodiodes; lasers and LED's; integrated optics; infrared detectors; microwave devices; semiconductor heterojunctions. *Mailing Add:* PO Box 140 Carlisle MA 01741

DONNELLY, KENNETH GERALD, b Brooklyn, NY, Jan 17, 37; m 62; c 4. AUDIOLOGY, SPEECH PATHOLOGY. *Educ:* Cath Univ Am, BA, 60, MA, 61; Univ Pittsburgh, PhD(audiol), 64. *Prof Exp:* Instr speech path, Gallaudet Col, 61-62, from asst prof to assoc prof audiol, 64-66; dir speech & hearing clin, St Charles Hosp, Port Jefferson, NY, 66-67; assoc prof audiol & speech path, 67-69, prof, 69-70, head dept speech, 70-80, PROF AUDIOL & DIR GRAD STUDIES, UNIV CINCINNATI, 80- *Concurrent Pos:* Study, Off Voc Rehab, 68-69; proj dir, US Off Educ, HEW, 68-72; training grant, Vet Admin, 69-73; develop grant, Merrill Trust Fund, 69-70; secy treas, Hearing, Speech & Lang Serv Inc, 69-; pres, Hearing, Speech & Lang Assoc, Inc, 88. *Mem:* Acad Rehab Audiol; fel Am Speech-Lang-Hearing Asn; Soc Int Audiol; Am Acad Audiol. *Res:* Hearing disorders in children; methods of determining best fit for hearing aids; academic placement techniques for deaf children. *Mailing Add:* Dept Commun Sci & Disorders Mail Location 379 Cincinnati OH 45221-0379

DONNELLY, PATRICIA VRYLING, b Memphis, Tenn, July 23, 33. MICROBIOLOGY. *Educ:* Tex Woman's Univ, BA, 55; Univ Mo, MS, 63. *Prof Exp:* Bacteriologist, Dallas Pub Health Lab, 55-57; lab dir, La Rabida Hosp, 57-58; microbiologist, Walter Reed Army Hosp, 58-59 & US Air Force Aerospace Med Ctr, 59-62; biologist, Marine Lab, Fla Bd Conserv, 63-67; RES ASSOC, DEPT CELL BIOL, BAYLOR COL MED, 67- *Mem:* Am Soc Microbiol; AAAS; Am Chem Soc. *Res:* Biochemical genetics; tissue culture. *Mailing Add:* 3715 McPherson Rd Ft Worth TX 76140

DONNELLY, RUSSELL JAMES, b Hamilton, Ont, Apr 16, 30; m 56; c 1. PHYSICS, SOLID STATE PHYSICS. *Educ:* McMaster Univ, BSc, 51, MSc, 52; Yale Univ, MS, 53, PhD(physics), 56. *Prof Exp:* Instr, dept physics & James Franck Inst, Univ Chicago, 56-57, from asst prof to prof, 57-66; chmn dept, 66-72 & 82-83, PROF PHYSICS, UNIV ORE, 66- *Concurrent Pos:* Alfred P Sloan res fel, 59-63; Consult, Gen Motors Res Labs, 58-68 & NSF, 68-73; mem adv panel physics, NSF, 70-73 & 79-84, chmn, 71-72; vis prof, Niels Bohr Inst, Copenhagen, Denmark, 72, Univ Birmingham, Eng, 80 & Univ Calif, Santa Barbara, 81; sr vis fel, Sci Res Coun, Eng, 78; mem Task Group Fundamental Physics & Chem in Space, Nat Res Coun; assoc ed, Physical Review, 88-; chmn adv group, Low Temperature Facil in Space, NASA. *Honors & Awards:* Gov Northwest Sci Award, Ore Mus Sci & Indust; Otto LaPorte Mem Lectr, Am Phys Soc, 74. *Mem:* Fel AAAS; fel Am Phys Soc (secy-treas, 67-70 & 88-). *Res:* Experimental and theoretical low temperature physics, especially superfluidity; fluid dynamics; hydrodynamic stability; turbulence. *Mailing Add:* Dept Physics Univ Ore Eugene OR 97403-1274

DONNELLY, THOMAS EDWARD, JR, b Chelsea, Mass, Sept 16, 43; m 75; c 2. PHARMACOLOGY, BIOCHEMISTRY. *Educ:* Mass Col Pharm, BS, 66; Harvard Univ, MA, 68; Yale Univ, PhD(pharmacol), 72. *Prof Exp:* Pharmacologist, Leo Pharmaceut Prod, 73-74; ASSOC PROF PHARMACOL, UNIV NEBR MED CTR, 74- *Mem:* Am Soc Pharmacol & Exp Therapeut. *Res:* Role of protein Kinase C in cell proliferation; role of dietary fat in tumor promotion; phospholipids and protein kinase C in atherosclerois. *Mailing Add:* Smithkline Beecham Pharmaceut 709 Swedeland Rd L-231 PO Box 1539 King of Prussia PA 19406

DONNELLY, THOMAS HENRY, b Endicott, NY, Apr 20, 28; m 55; c 4. BIOPHYSICAL CHEMISTRY. *Educ:* Rensselaer Polytech Inst, BS, 50; Cornell Univ, PhD, 55. *Prof Exp:* Phys chemist, Swift & Co, Oak Brook, 55-67, mgr gelatin & stabilizers res, 67-71, gen mgr sci & serv, 72-77, res mgr appl chem, res & develop ctr, 77-79; vis prof, 79-84, asst chmn, 81-84, lectr chem, Dept Chem, Loyola Univ Chicago, 84-87; CHMN, ASST PROF CHEM, MUNDELEIN COL, 87- *Concurrent Pos:* Fac fel, Ill Benedictine Col, 70-81; mem educ comt, Nat Confectioners Asn, 73-75; instr, Mundelein Col, 79-87; mem, food enzym deleg Peoples' Repub China, 85. *Mem:* AAAS; Am Chem Soc; Inst Food Technologists. *Res:* Physical chemistry of proteins and polymers; ultracentrifugation, gelatin; food chemistry. *Mailing Add:* 4633 Grand Ave Western Springs IL 60558

DONNELLY, THOMAS WALLACE, b Detroit, Mich, Dec 23, 32; m 56; c 3. GEOLOGY. *Educ:* Cornell Univ, BA, 54; Calif Inst Technol, MS, 56; Princeton Univ, PhD(geol), 59. *Prof Exp:* From asst prof to assoc prof geol, Rice Univ, 59-66; assoc prof, 66-69, PROF GEOL, STATE UNIV NY, BINGHAMTON, 69- *Concurrent Pos:* Res assoc, Fla State Collection Arthropods, 66- *Mem:* Geol Soc Am. *Res:* Geology of island arcs; geochemistry of volcanic rocks; geology of Caribbean area and Guatemala; deep-sea sediment chemistry; aquatic entomology. *Mailing Add:* Dept Geol State Univ NY Binghamton NY 13901

DONNELLY, THOMAS WILLIAM, b Victoria, BC, Can, June 1, 43; m 79. NUCLEAR STRUCTURE. *Educ:* Univ BC, BSc, 64, PhD(theoret nuclear physics), 67. *Prof Exp:* Postdoctoral fel physics res, Stanford Univ, 67-69; sr postdoctoral fel physics res, Univ Toronto, 69-70, lectr physics teaching & res, 70-71, asst prof, 71-79; SR RES SCIENTIST PHYSICS TEACHING & RES, MASS INST TECHNOL, 79- *Concurrent Pos:* Consult; A P Sloan fel, 74-75; A von Humboldt award, 85-87. *Mem:* Can Asn Physicists; fel Am Phys Soc. *Res:* Theoretical nuclear physics; electron scattering; electroweak interactions; polarization studies in semi-leptonics electroweak interactions. *Mailing Add:* Ctr Theoret Physics 6-300 Mass Inst Technol Cambridge MA 02139

DONNELLY, TIMOTHY C, LASER MEDICINE, INNOVATIVE CHEMICAL EDUCATION TECHNIQUES. *Educ:* Calif State Univ, BS, 70, MS, 72; Univ Calif, Davis, PhD(chem), 79. *Prof Exp:* Lectr chem, Calif State Univ, Hayward, 71-72; res chemist, Clorox Co, 77-79; res asst, 73-75, assoc, 76-77, LECTR CHEM, UNIV CALIF, DAVIS, 79- *Concurrent Pos:* Pres, TiDon Res Inst 76- *Mem:* Am Chem Soc; Sigma Xi. *Res:* Applications of lasers to vascular disease and inoperable tumors; innovative techniques for teaching general chemistry; nature of surface active agents and practical applications of these agents. *Mailing Add:* Dept Chem Univ Calif Davis CA 95616

DONNELLY, WILLIAM H, PATHOLOGY. *Prof Exp:* Prof pediat & path, Col Med, Univ Fla, 71-91. *Mailing Add:* Dept Path Med Col Univ Fla Box J-275 JHMHC Gainesville FL 32610

DONNER, DAVID BRUCE, b New York, NY, Oct 17, 45. ENDOCRINOLOGY. *Educ:* Queens Col, BA, 66; Rensselaer Polytech Inst, PhD(chem), 72. *Prof Exp:* Fel biochem, Cornell Univ, 72-73, from instr to asst prof, 73-83; res assoc, 73-78, assoc biochem, 78-83, UNIT CHMN CELL BIOL & GENETICS & ASSOC MEM, SLOAN KETTERING INST, 83-; ASSOC PROF BIOCHEM, CORNELL UNIV, 83- *Concurrent Pos:* NIH Res Career Develop Award. *Mem:* Am Chem Soc; Endocrine Soc; Harvey Soc; NY Acad Sci. *Res:* Peptide hormone regulation of cell growth and differentiation; signal transmission in biological membranes; insulin and growth, hormone action in cells; peptide hormone-receptor binding. *Mailing Add:* Sloan-Kettering Cancer Ctr 1275 York Ave New York NY 10021-6007

DONNER, MARTIN W, b Leipzig, Ger, Sept 5, 20; US citizen; m 51; c 3. MEDICINE, RADIOLOGY. *Educ:* Univ Leipzig, MD, 45. *Prof Exp:* Intern, Leipzig Hosp, 45-46, resident med, 46-50; resident radiol, Radiol Ctr, Cologne, Ger, 50-54; resident, Mt Park Hosp, St Petersburg, Fla, 54-57; fel, Johns Hopkins Univ Hosp, 57-58, from instr to assoc prof, 58-66, prof radiol & radiologic sci & chief, Div Diag Roentgenol, 66-72, chmn dept radiol & radiol sci & radiologist-in-chief, Sch Med & Hosp, 72-87, dir, Johns Hopkins Swallowing Ctr, 81-89; RETIRED. *Concurrent Pos:* Vis investr, Carnegie Inst, 62-72; consult, Good Samaritan Hosp, Rosewood State Hosp, Baltimore City Hosps & Vet Admin Hosp, Baltimore. *Honors & Awards:* Cannon Medal, Soc Gastrointestinal Radiol. *Mem:* AAAS; AMA; fel Am Col Radiol; Asn Univ Radiol; hon mem Ger Radiol Soc. *Res:* Hematology; metabolitic joint diseases; radiotherapy and blood coagulation; cineradiography of intestinal motility; roentgen diagnosis of gastrointestinal diseases and abdominal calcifications; radioangiography of placental circulation; radiography of diabetic complications and diseases involving the hands; Roentgen tomography; swallowing and swallowing disorders. *Mailing Add:* 317 S Wind Rd Baltimore MD 21204-6737

DONOGHUE, DANIEL JAMES, b Madison, Wis, Oct 16, 52. BIOCHEMISTRY, GENETICS. *Educ:* Univ Wis, BSc, 74; Mass Inst Technol, PhD(biol), 79. *Prof Exp:* Fel tumor virol, Salk Inst, 80-82; ASST PROF CHEM, UNIV CALIF, SAN DIEGO, 82- *Concurrent Pos:* Fel, Helen Hay Whitney Found, 80-82, Searle Scholars prog, 83-86. *Mem:* AAAS. *Res:* Molecular biology of tumor viruses and growth factors, structure and function of platelet-derived growth factor (PDGF); characterization of oncogenic protein kinases; proto-oncogenes and control of the cell cycle. *Mailing Add:* Dept Chem Univ Calif San Diego La Jolla CA 92093-0322

DONOGHUE, JOHN FRANCIS, b Roslyn, NY, Nov 30, 50; m 73; c 3. THEORETICAL ELEMEMTARY PARTICLE PHYSICS. *Educ:* Univ Notre Dame, BS, 72; Univ Mass, Amherst, PhD(physics), 76. *Prof Exp:* Res assoc physics, Carnegie-Mellon Univ, 76-78; res assoc physics, Ctr Theoret Physics, Mass Inst Technol, 78-80; from asst prof to assoc prof, Univ Mass, 80-91, prof physics, 89-90; sci assoc, Cern. *Mem:* Fel Am Phys Soc. *Res:* Weak interactions, quarks; gauge theories. *Mailing Add:* Dept Physics & Astron Univ Mass Amherst MA 01003

DONOGHUE, JOHN TIMOTHY, b Holyoke, Mass, July 23, 35; c 3. RESEARCH ADMINISTRATION, ORGANIC CHEMISTRY. *Educ:* Univ Ill, PhD(chem), 63. *Prof Exp:* Chemist-supvr Chamber Works, E I DuPont de Nemours & Co, Inc, 63-67 & 69-71; lab mgr, Polaroid Corp, 71-74; area mgr, Xerox Corp, 74-78; dir res & develop, Pitney Bowes, 79-81 & Varn Prod Co, 81-85; DIR RES & DEVELOP, DELEET MERCHANDISING CORP, 85-; CONSULT, DONOGHUE, INC, 78- *Concurrent Pos:* Adj asst prof, Glassboro State Col, 69-71, Southeastern Mass Univ, 71-74 & Bridgewater State Col, 72-74; adj assoc prof, Sacred Heart Univ, 79-; adj prof, Western Conn State Univ, 83- & Ramapo Col, 82-; asst prof, Univ Ark. *Mem:* Am Chem Soc; AAAS; Asn Consult Chemists & Chem Engrs; Tech Asn Pulp & Paper Indust. *Res:* Development of products used in graphics arts industries; management of industrial research and development activity. *Mailing Add:* Printer's Oil Supply 310 Ballardvale St Wilmington MA 01887

DONOGHUE, JOSEPH F, b Philadelphia, Pa, Mar 23, 47; m 83. MARINE & COASTAL GEOLOGY, SEDIMENTOLOGY. *Educ:* Princeton Univ, BSc, 69; Univ Southern Calif, PhD(geol), 81. *Prof Exp:* Oakley fel, Dept Geol Sci, Univ Southern Calif, 76-79; Smithsonian fel & res assoc, Nat Mus Natural Hist, Washington, DC, 79-80; fel & geologist, Adv Comt Reactor Safeguards, Washington, DC, 81-82; asst prof, 83-89, ASSOC PROF, DEPT GEOL, FLA STATE UNIV, TALLAHASSEE, 89-; CONSULT, US GEOL SURVEY, 88- *Concurrent Pos:* Collabr, Sedimentology Lab, Smithsonian Inst, Washington, DC, 81-; consult, Adv Comt Nuclear Waste, Washington, DC, 83- *Mem:* Geol Soc Am; Soc Econ Paleontologists & Mineralogists; Am Geophys Union; Estuarine Res Soc; AAAS; Nat Asn Geol Teachers. *Res:* Sediment-water interactions; sea-level effects; environmental geochemistry; radiosotopic tracers; estuarine and coastal marine geology; nuclear waste disposal; coastal processes; quaternary dating methods. *Mailing Add:* Dept Geol Fla State Univ Tallahassee FL 32306-3026

DONOGHUE, TIMOTHY R, b Milton, Mass, May 3, 36; m 64; c 2. NUCLEAR PHYSICS. *Educ:* Boston Col, BS, 57; Univ Notre Dame, PhD(physics), 63. *Prof Exp:* Instr physics, Univ Notre Dame, 62-63; fel, Ohio State Univ, 63-64, from asst prof to assoc prof, 64-75, assoc dean res, Grad Sch, 79-82, prof physics, 75-88; PROF PHYSICS & VICE PROVOST RES & GRAD STUDIES, UNIV PITTSBURGH, 88- *Concurrent Pos:* Vis staff scientist, Los Alamos Sci Labs, 71-74; mem, Ind Univ Cyclotron User's Comt, 79-81; mem int prog adv comt, 5th Int Symp in Polarization Phenomena, 79-80. *Honors & Awards:* Fritz Thyssen Stiftung, 81. *Mem:* Am Phys Soc; AAAS; Am Asn Univ Prop. *Res:* Investigation of fundamental symmetries in physics including time reversal invariance, parity non-conservation and search for neutral weak currents; experimental nuclear astrophysics; polarized beam investigations of nuclear reaction mechanisms and nuclear level structure, including D-State of alpha particles. *Mailing Add:* Dept Physics Univ Pittsburgh Pittsburgh PA 15260

DONOGHUE, WILLIAM F, JR, b Rochester, NY, Sept 7, 21; m 74. MATHEMATICS. *Educ:* Univ Rochester, AB, 47, MSc, 48; Univ Wis, PhD(math), 51. *Prof Exp:* Assoc mathematician, appl physics lab, Johns Hopkins Univ, 51-52; from asst prof to assoc prof math, Univ Kans, 52-62; temp mem, Courant Inst Math Sci, NY Univ, 63-64; vis prof math, Mich State Univ, 64-66; PROF MATH, UNIV CALIF, IRVINE, 66- *Concurrent Pos:* Guggenheim fel, 58-59. *Res:* Linear topological spaces; Eigenvalue problems. *Mailing Add:* Dept Math Univ Calif Irvine CA 92717

DONOHO, ALVIN LEROY, b Humphreys, Mo, Nov 20, 36; m 64; c 3. BIOCHEMISTRY. *Educ:* Univ Mo, BS, 58, MS, 60, PhD(biochem), 65. *Prof Exp:* Res scientist, 64-80, RES ADV BIOCHEM, ELI LILLY & CO, 80- *Mem:* Am Chem Soc; AAAS. *Res:* Metabolism, degradation, and assay of drugs and agrichemicals. *Mailing Add:* 7281 Sacramento Dr Greenfield IN 46140-9695

DONOHO, CLIVE WELLINGTON, JR, b Nashville, Tenn, Jan 16, 30; m 55; c 4. HORTICULTURE, PLANT PHYSIOLOGY. *Educ:* Univ Ky, BS, 52; NC State Univ, MS, 58; Mich State Univ, PhD(hort), 60. *Prof Exp:* From asst prof to prof res hort, Ohio Agr Res & Develop Ctr & Ohio State Univ, 60-67; prof hort sci & head dept, NC State Univ, 67-73; assoc dir, Ohio Agr

Res & Develop Ctr, Ohio State Univ, 73-82, dir, 82-84, prof hort, 73-84; DIR & ASSOC DEAN, GA AGR EXP STAS, UNIV GA, 84- *Concurrent Pos:* Mem, Agr Res Inst. *Mem:* Fel Am Soc Hort Sci; Int Soc Hort Sci; Sigma Xi; Coun Agr Sci & Technol. *Res:* Fruit physiology; natural auxins; absorption, translocation and metabolism of synthetic growth regulators; influence of pesticide chemicals on fruit physiology; water relations in fruit crops. *Mailing Add:* Ga Agr Exp Stas Univ Ga 107 Conner Hall Athens GA 30602

DONOHO, DAVID LEIGH, b Pasadena, Calif, Mar 5, 57; m 84; c 1. MATHEMATICAL STATISTICS, SPECTROSCOPY & SPECTROMETRY. *Educ:* Princeton Univ, AB, 78; Harvard Univ, PhD(statist), 84. *Prof Exp:* From asst prof to assoc prof, PROF STATIST, UNIV CALIF, BERKELEY, 84-; PROF STATIST, STANFORD UNIV, 91- *Concurrent Pos:* Presidential Young Investr, NSF, 85-90; assoc ed, Probability Theory & Related Fields, 89-; fel, MacArthur Found, 91. *Mem:* Fel Inst Math Statist; Am Math Soc; Am Statist Asn; Soc Indust & Appl Math. *Res:* Nonlinear signal recovery; visualization of statistical data; robust statistical analysis; decision theory. *Mailing Add:* 2830 Buena Vista Way Berkeley CA 94708

DONOHUE, DAVID ARTHUR TIMOTHY, b Montreal, Que, Apr 11, 37; US citizen; m 61; c 1. PETROLEUM. *Educ:* Univ Okla, BS, 59; Pa State Univ, PhD(petrol eng), 63; Boston Col, JD, 71. *Prof Exp:* Prod engr, Imp Oil Ltd, Can, 58, res engr, 63-64; from asst prof to assoc prof petrol & natural gas eng, Pa State Univ, 64-69; PRES, INT HUMAN RESOURCES DEVELOP CORP, 69-; PRES, HONEOYE STORAGE CORP, 73- *Concurrent Pos:* Chmn, Pa-NY Chap, Petrol Inst, 67-68; partner, Donohue Anstem & Morrill, 76-; pres, Arlington Explor Co, 78- *Honors & Awards:* Cedric K Ferguson Medal, Soc Petrol Engrs, 68. *Mem:* Am Inst Mining, Metall & Petrol Engrs; Soc Petrol Engrs; Am Inst Chem Engrs; Am Soc Eng Educ; Am Asn Petrol Geologists. *Res:* Experimental and theoretical research dealing with developing new methods and improving old methods of producing crude oil from underground formations; analysis and development of underground gas storage reservoirs; training of international petroleum industry personnel; oil and gas exploration and development. *Mailing Add:* 17 Allen Rd Wellesley MA 02181

DONOHUE, JOHN J, b Totowa, NJ, Feb 17, 19; m 45; c 3. GEOENVIRONMENTAL SCIENCE, MINERAL RESOURCES. *Educ:* Univ Maine, BA, 44, MSc, 49; Rutgers Univ, PhD(geol sci & eng), 51. *Prof Exp:* Field geologist, State Geol Surv Maine, 47-49; proj dir & consult, Off Naval Res, 51-54; sr res & explor geologist, Arabian Am Oil Co, 54-61; dir & consult, Geo-Tek Assocs, 61-64; prof phys sci, Bennett Col, NY, 64-77, chmn, SCI & math dept, 67-77, dir, Kettering Sci Ctr, 70-77; sr mgr environ assessments, Camp Dresser & McKee, Inc, Environ Engrs, Boston, 77-79; MINERAL RESOURCES CONSULT & PRES, GEO-TEK ASSOCS, LTD, 79- *Concurrent Pos:* Asst, Univ Maine, 48-49; sr res marine geologist, head res & develop sect & assoc prof, Grad Sch Oceanog, Univ RI, 51-54; res consult, Davison Chem Div, W R Grace Co, 61-73 & Valumet Corp, 75-77; sr staff consult, United Aircraft Res Labs, 67-71; consult, Amartech, Jeddeh KSA, 81-83; vpres, Technol World Inc, 83- *Mem:* Geol Soc Am; Am Asn Petrol Geol; Am Geophys Union; Sigma Xi. *Res:* Geoenvironmental research, geotechnical and ocean engineering; environmental systems and technology; mineralogy; mining and petroleum geology; strategic mineral resources exploration and evaluation; mineral resources technology research and development. *Mailing Add:* 45 Locust St Danvers MA 01923

DONOHUE, JOYCE MORRISSEY, b Holyoke, Mass, Jan 27, 40; m 73; c 4. BIOCHEMISTRY, NUTRITION. *Educ:* Framingham State Col, BSEd, 61; Univ Mass, MS, 64; Univ NH, PhD(biochem), 72. *Prof Exp:* Teacher chem & biol, W Springfield High Sch, 62-66; instr chem & nutrit, Framingham State Col, 66-68, assoc prof biochem, 71-73; health scientist, V J Ciccone & Assocs, Inc, Woodbridge, Va, 81-89; toxicol serv mgr, Law Environ Inc, 89-90; ADJ ASSOC PROF NUTRIT, FOOD SCI & CHEM, NORTHERN VA COMMUNITY COL, 74-; TOXICOLOGIST & PROG MGR, LIFE SYSTS INC, ARLINGTON, VA, 90- *Concurrent Pos:* Hood fel, Univ Mass, 61-62; NSF fel, Yale Univ, 65-66; Nat Defense Educ Act fel, Univ NH, 68-71; adj prof human foods & nutrit, Va Polytech Inst & State Univ, 79-; adj assoc prof nutrit & chem, Northern Va Community Col, 74- *Mem:* AAAS; Am Diet Asn; Sigma Xi. *Res:* Protein structure; nutrition education; environmental health issues; biochemical nutrition; toxicology; risk assessment. *Mailing Add:* 11979 William & Mary Circle Woodbridge VA 22192

DONOHUE, MARC DAVID, b Watertown, NY, Sept 10, 51; m 74; c 3. CHEMICAL ENGINEERING. *Educ:* Clarkson Col Technol, BS, 73; Univ Calif, Berkeley, PhD(chem eng), 77. *Prof Exp:* asst prof chem eng, Clarkson Col Technol, 77-79; from asst prof to assoc prof, 79-87, CHMN DEPT, JOHNS HOPKINS UNIV, 84-, PROF CHEM ENG, 87- *Mem:* Am Inst Chem Engrs. *Res:* Thermodynamics and phase transitions. *Mailing Add:* Dept Chem Eng Johns Hopkins Univ Baltimore MD 21218

DONOHUE, ROBERT J, b Chicago, Ill, Jan 31, 34; m 57; c 4. SPECTROSCOPY, PHYSICAL OPTICS. *Educ:* DePaul Univ, BS, 55, MS, 57. *Prof Exp:* Res asst physics, DePaul Univ, 54-56; res physicist, 56-61, sr res physicist, Res Labs, 61-71, staff proj engr, automotive & safety eng, 71-78, staff proj eng, 79-83, MGR, INT REGULATIONS, GEN MOTORS ENVIRON ACTIV STAFF, 84- *Honors & Awards:* Caldwell Award, Soc Automotive Engrs. *Mem:* Soc Automotive Engrs. *Res:* Plasma spectroscopy; nuclear physics; geometrical and physical optics; illumination and lighting systems; human factors engineering; auto safety engineering. *Mailing Add:* Int Regulations Environ Activ Gen Motors Tech Ctr 30400 Mound Rd PO Box 9015 Warren MI 48090-9015

DONOIAN, HAIG CADMUS, b Ramallah, Palestine, Apr 8, 30; US citizen; m 57; c 2. PHYSICAL CHEMISTRY, COLLOID CHEMISTRY. *Educ:* Lowell Technol Inst, BS, 52; Clark Univ, MA, 54, PhD(chem), 58. *Prof Exp:* Res chemist, Am Cyanamid Co, 57-63, Cabot Corp, Mass, 63-67 & Org Chem Div, Am Cyanamid Co, 67-72; SCIENTIST, XEROX CORP, 72- *Honors & Awards:* Roon Award, Fedn Socs Paint Technol, 67. *Mem:* Am Chem Soc. *Res:* Diffusion of substances in solution; colloidal chemistry and optical properties of carbon black. *Mailing Add:* 1195 Severn Ridge Webster NY 14580

DONOVAN, ALLEN F(RANCIS), b Onondaga, NY, Apr 22, 14; m 40, 53, 74; c 3. AERONAUTICAL ENGINEERING, AEROSPACE ENGINEERING. *Educ:* Univ Mich, BSE & MS, 36. *Hon Degrees:* DEng, Univ Mich, 64. *Prof Exp:* Stress analyst, Curtiss Aeroplane & Motor Co, NY, 36-38; sr stress analyst, Glenn L Martin Co, Md, 38-39; asst chief struct, Stinson Aircraft Div, Aviation Mfg Corp, Mich, 39-40; chief struct, Nashville Div, Vultee Aircraft Corp, 40-41; asst to chief exp design engr, Airplane Div, Curtiss Wright Corp, NY, 41-42, asst head struct dept, Res Lab, 42-44, head, 44-46; head aero mech dept, Cornell Aero Lab, Cornell Univ, 46-55; dir aero res & develop staff, Space Tech Labs, Inc, 55-58, vpres, 58-60; sr vpres tech, Aerospace Corp, 60-78, RETIRED. *Concurrent Pos:* Mem, US Air Force Sci Adv Bd, 48-57 & 59-68, chmn propulsion panel, 59-60 & 63-68; US deleg, Geneva Conf Suspension of Nuclear Test, 59, consult, President's Sci Adv Comt, 57-72; consult, Sci & Technol Off, NSF, 73-76; consult aerospace eng, 78- *Honors & Awards:* Assoc Sci Award, USAF, 61, Except Civilian Serv Medal, 68. *Mem:* Nat Acad Eng; fel Am Inst Aeronaut & Astronaut; Sigma Xi. *Res:* Space and ballistic missile system design and development. *Mailing Add:* 35 Beachcomber Dr Corona del Mar CA 92625

DONOVAN, ARTHUR L, US citizen. SCIENCE IN THE ENLIGHTENMENT, HISTORY OF CHEMISTRY. *Educ:* Harvard Univ, AB, 60; Univ Wis, MS, 64; Princeton Univ, PhD(hist), 70. *Prof Exp:* Asst assoc prof hist, Univ Ill, Chicago, 68-74 & WVa Univ, 74-80; prof hist, sci & technol studies, Va Polytech & State Univ, 80-88; PROF HIST & HEAD DEPT HUMANITIES, US MERCHANT MARINE ACAD, 88- *Concurrent Pos:* Vis prof hist, Univ Chicago, 77-78; dir, Ctr Study Sci in Soc, Va tech, 80-84. *Mem:* Hist Sci Soc; Soc Hist Technol; Soc Social Study Sci. *Res:* Chemical revolution of 18th century; theories of scientific change; professionalization of engineering; history of American coal industry; history of energy in America. *Mailing Add:* Dept Humanities US Merchant Marine Acad Kings Point NY 11024-1699

DONOVAN, EDWARD FRANCIS, b Columbus, Ohio, Nov 20, 18; m 43; c 5. VETERINARY MEDICINE. *Educ:* Ohio State Univ, DVM, 49. *Prof Exp:* Pvt pract, Ohio, 49-52; assoc dir clin res, Am Cyanamid Co, 52-56; from asst prof to assoc prof vet med, 56-62, PROF VET CLIN SCI, COL VET MED & GRAD SCH, OHIO STATE UNIV, 62-, LECTR OPHTHAL, COL MED, 65- *Concurrent Pos:* Morris Animal Found res dir, 62-; mem comt prof educ, Inst Lab Animal Resources, Nat Acad Sci, 66- *Mem:* Am Vet Med Asn. *Res:* Veterinary ophthalmology; retinal diseases; animal endocrinology. *Mailing Add:* 1925 Innisbrook Ct Venice FL 34293

DONOVAN, GERALD ALTON, b Hartford, Conn, Feb 10, 25; m 48; c 3. POULTRY NUTRITION, BIOCHEMISTRY. *Educ:* Univ Conn, BS, 50, MS, 52; Iowa State Univ, PhD(poultry nutrit, biochem), 55. *Prof Exp:* Asst poultry res, Univ Conn, 51-52; res assoc, Iowa State Univ, 52-55; in-chg poultry nutrit res, Agr Res Dept, Pfizer, Inc, 55-60; assoc prof poultry sci, Univ Vt, 60-66, prof animal sci, 66-73, assoc dir agr exp sta, 66-67, assoc dir col agr & home econ, 67-73, actg dean, Col Agr & Home Econ, 70-73; PROF ANIMAL SCI, DEAN COL RESOURCE DEVELOP, DIR AGR EXP STA & DIR COOP EXTEN SERV, UNIV RI, 73- *Mem:* Sigma Xi; Asn Univ Dirs Int Agr Progs; Am Inst Nutrit. *Res:* Vitamins, particularly vitamin A; antibiotics; hormones; chemotherapeutic agents. *Mailing Add:* Woodward Hall Univ RI Kingston RI 02881

DONOVAN, JAMES, b Vancouver, Wash, Oct 26, 06; m 31; c 2. CHEMICAL ENGINEERING. *Educ:* Mass Inst Technol, BS, 28. *Prof Exp:* Instr chem eng, Mass Inst Technol, 28-29; chem engr, Hird & Connor, Inc, 29-32 & Acme Indust Equip Co, 32-34; chief engr & exec, Artisan Metal Prods, Inc, 34-61, chief engr & exec, Artisan Indust Inc, 61-76, PRES & TREAS, ARTISAN INDUST INC, 76- *Concurrent Pos:* Exec, Jet-Vac Corp, Kontro Co, Inc & Anyl-Ray Corp. *Mem:* Am Inst Chem Engrs; Am Chem Soc; Am Welding Soc; Inst Food Technol. *Res:* Chemical engineering process; evaporation, distillation and reaction costs. *Mailing Add:* 670 South St No A Waltham MA 02154

DONOVAN, JOHN FRANCIS, b St John, NB, Nov 11, 35; m 63; c 2. ECONOMIC GEOLOGY, EARTH SCIENCES. *Educ:* St Francis Xavier Univ, BSc, 57; Univ Iowa, MS, 59; Cornell Univ, PhD(geol), 63. *Prof Exp:* Resident geologist, Ont Dept Mines, 63-68; assoc prof, Winona State Col, 68-80, PROF GEOL & EARTH SCI, WINONA STATE UNIV, 80- *Mem:* Geol Soc Am; Soc Econ Geol. *Res:* Study of iron formations-sedimentary deposits of Archean age, Precambrian areas of Canadian Shield. *Mailing Add:* Dept Geol Winona State Univ Winona MN 55987

DONOVAN, JOHN JOSEPH, b Lynn, Mass; c 5. MANAGEMENT INFORMATION SYSTEMS, COMPUTER SCIENCE. *Educ:* Tufts Univ, BS, 63; Mass Inst Technol, BS, 63; Yale Univ, MEng, 64, MS, 65, MPh & PhD(comput), 66. *Prof Exp:* Assoc prof elec eng, 66-73, PROF MGT SCI, SLOAN SCH MGT, MASS INST TECHNOL, 73- *Concurrent Pos:* Ford fel, Mass Inst Technol, 66; pres, Int Comput Inc, 67-69; chmn bd, Mitrol, 70-71; clin prof pediat, Med Sch, Tufts Univ, 74-, dir financial publ, 75, dir, Ctr Birth Defects Info Systs, 77- *Res:* Computer based information systems. *Mailing Add:* Cambridge Technol Group 219 Vassar St Cambridge MA 02139

DONOVAN, JOHN LEO, b Rochester, NY, Apr 1, 29; m 53; c 4. NUCLEAR PHYSICS. *Educ:* Univ Rochester, BS, 51; Univ Pittsburgh, MS, 53; Univ Mich, PhD(nuclear sci), 64. *Prof Exp:* Instr, NY Univ, 53-56; sr res physicist, Curtiss-Wright Corp, 56-59; SR RES PHYSICIST, EASTMAN KODAK CO, 64- *Concurrent Pos:* Assoc lectr, Univ Rochester, 66-85. *Mem:* AAAS; Am Nuclear Soc; Soc Photog Sci & Eng; Sigma Xi; Am Phys Soc; Am Phys Soc. *Res:* Nonsilver image forming systems and their application to radiography, including measurement of intrinsic efficiency of phosphors irradiated with x-rays; electrophotography. *Mailing Add:* 20 Royal Oak Dr Rochester NY 14624-2814

DONOVAN, JOHN RICHARD, b St Louis, Mo, Aug 7, 17; m 40; c 3. SULFURIC ACID & INORGANIC SULFUR CHEMICALS, CATALYIS & CATALYSTS. *Educ:* Wash Univ, BS, 39, MS, 53. *Prof Exp:* Chemist & chem engr, Monsanto Co, Sauget, Ill, 40-45, sr chem engr heavy chem & intermediates, Inorg Div, 47-54, res group leader heavy chem & catalysts, 54-63, prod qual mgr copper powder, 63-64; sr chem engr nuclear reactors, Oak Ridge Nat Lab, 45-47; sr eng supvr sulfuric acid plants & catalyst, Monsanto Enviro-Chem Systs, Inc, 64-70, mgr process design sulfuric acid plants, 70-72, mgr technol res & develop, 73-82; PRES, JRD CONSULT FIRM, INC, 84- *Mem:* Fel Am Inst Chem Engrs; Am Chem Soc. *Res:* Optimization and improvement of sulfuric acid plant designs and catalyst used therin, including reduction of air pollution, economic benefits, improved safety and product quality; author of over 20 publications. *Mailing Add:* 12308 Crystal View St Louis MO 63131

DONOVAN, JOHN W, b Boston, Mass, June 7, 29; m 61; c 4. PHYSICAL CHEMISTRY. *Educ:* Boston Col, BS, 50; Col Holy Cross, MS, 53; Cornell Univ, PhD(chem), 59. *Prof Exp:* Fel chem, Univ Rochester, 59-61; trainee molecular biol, virus lab, Univ Calif, Berkeley, 61-63; res assoc chem, Univ Ore, 63-65; mem staff, Western Regional Res Lab, USDA, 65-85. *Mem:* Am Chem Soc; Am Soc Biol Chemists. *Res:* Physical chemistry of proteins and starch; absorption spectrophotometry; scanning calorimetry. *Mailing Add:* 315 Behrens St El Cerrito CA 94530-3706

DONOVAN, LEO F(RANCIS), b East Orange, NJ, Apr 12, 32; div; c 2. AERONAUTICAL ENGINEERING, CHEMICAL ENGINEERING. *Educ:* Lehigh Univ, BS, 54; Rensselaer Polytech Inst, MChE, 57, PhD(chem eng), 65. *Prof Exp:* Chem engr, Res Lab, Mobil Oil Co, 57-59; nuclear chem engr, Nuclear Power Eng Dept, Alco Prod Inc, 59-62; AEROSPACE ENGR, LEWIS RES CTR, NASA, 65- *Mem:* Am Inst Aeronaut & Astronaut. *Res:* Fluid mechanics; numerical analysis; numerical flow visualization. *Mailing Add:* 311 Parkway Dr Berea OH 44017

DONOVAN, P F, b Boston, Mass, Dec 31, 32. ENERGY POLICY & TECHNOLOGY, TECHNOLOGY ASSESSMENT. *Educ:* Northeastern Univ, BS, 55; Univ Calif, Berkeley, PhD(nuclear chem), 58. *Hon Degrees:* DSc, Northeastern Univ, 74. *Prof Exp:* Mem tech staff, Bell Labs, 58-67; actg prog dir, nuclear physics, NSF, Washington, DC, 67, actg prog dir, elem particle facil, 68, prog dir, intermediate & high energy physics, 68-69, head, physics sect, 70-71, dir, Div Advan Technol Appln, 71-73, dir, Off Energy Res & Develop Policy, 73-76; pres, DHR Inc, Washington, DC, 76-79; regents prof, Univ Calif, Santa Cruz, 80; chmn & chief exec officer, 81-86, CHMN, R&C ENTERPRISES, LTD, SANTA CRUZ, CALIF, 86-; CHMN & CHIEF EXEC OFFICER, ANADYNE, INC, SANTA CRUZ, CALIF, 86- *Concurrent Pos:* NSF predoctoral fel, 56-58; consult, Lawrence Radiation Lab, Berkeley, Calif, 58-60; guest scientist, Brookhaven Nat Lab, 60-71; grad fac assoc, Rutgers Univ, 64-71; Sigma Xi-RESA Nat lectr, 73-74; distinguished Sigma Xi lectr, 74; adj prof physics, Univ Calif, Santa Cruz, 81; chmn, Trans-Tech Marketing Co, Washington, DC, 81-86. *Mem:* Fel Am Phys Soc; Sigma Xi; AAAS. *Mailing Add:* 2099 El Rancho Dr Santa Cruz CA 95060

DONOVAN, RICHARD C, b McKees Rocks, Pa, May 15, 41; m 64; c 2. MANUFACTURING TECHNOLOGY. *Educ:* Univ Pittsburgh, BS, 62, MS, 64, PhD(mech eng), 67. *Prof Exp:* Asst prof mech eng, Univ Pittsburgh, 67-68; mem res staff, AT&T Technol, 68-71, res leader, 71-78, supvr technol planning, 78-80, dept head interconnection & semiconductor process technol, Eng Res Ctr, 80-90, DEPT HEAD, INTEGRATED PROCESS CONTROL, AT&T BELL LABS, 90- *Concurrent Pos:* Chmn, Mfg Syst Sci Tab, Semiconductor Res Corp. *Mem:* Am Soc Mech Engrs; Inst Elec & Electronics Engrs; Soc Mfg Engrs. *Res:* Semiconductor processing; computer and robotic automation of semiconductor processes; crystal and epitaxial growth of III-V semiconductor compounds; silicon materials and processing technology; polymer processes and material recycling processes; research and technical management. *Mailing Add:* AT&T Bell Labs Eng Res Ctr PO Box 900 Princeton NJ 08540

DONOVAN, SANDRA STERANKA, b Cleveland, Ohio, Sept 20, 42; m 66; c 1. PHYSICAL CHEMISTRY, ELECTROCHEMISTRY. *Educ:* Western Reserve Univ, BA, 64, MS, 66, Case Western Reserve Univ, PhD(electrochem), 69. *Prof Exp:* Res chemist mat, Hercules Inc, 69-70; sr res assoc, Horizons Res Inc, 71-73, group leader mat, 73-76, mgr mat sci & contract res & develop, 76-78; prog mgr bus develop, Standard Oil Ohio, 78-80, mgr prog, Vistron Corp, 81-82, dir planning & develop, Indust Chems Div, 82-85, vpres & gen mgr, Nitrogen Chem Div, 85-87, Nitrogen Chem Strategic Rev, 87-88; OFF EXEC VPRES, BP AM, 88- *Concurrent Pos:* mem bd overseers, Case Western Reserve, 72-78; dir, Mather Col Alumnae Asn, 73-79; pres, Com Develop Asn, 88-89; dir, Com Develop Asn, 82-84 & 87-93, Fertilizer Inst, 86-87; adv, Berkeley Exec Mgt Prog, 87- *Mem:* Electrochem Soc; Sigma Xi. *Res:* Electrochemically fabricated oxides and protective coatings; electrochemical kinetics, materials, petrochemicals and derivatives; commercial development of petrochemicals. *Mailing Add:* 246 Hawthorne Dr Chagrin Falls OH 44022

DONOVAN, TERENCE M, b Chicago, Ill, Dec 7, 31; m 55; c 3. PHYSICS. *Educ:* San Jose State Col, BS, 56; Stanford Univ, MS, 61, PhD, 70. *Prof Exp:* Physical chemist, 56-60, chem physicist, 61-65, RES PHYSICIST, NAVAL WEAPONS CTR, CHINA LAKE, 70- *Concurrent Pos:* Assoc chemist, Shockley Transistor Corp, 60-61. *Mem:* Am Phys Soc. *Res:* Optical properties of solids; optical properties and photoemission; electronic structure of amorphous and crystalline solids; laser optics. *Mailing Add:* Code 3812 Naval Weapons Ctr China Lake CA 93555

DONOVAN, TERRENCE JOHN, b Waterbury, Conn, July 27, 36; m 56; c 3. GEOLOGY, ATMOSPHERIC SCIENCES. *Educ:* Midwestern Univ, BS, 61; Univ Calif, Riverside, MA, 63; Univ Calif, Los Angeles, PhD(geol), 72. *Prof Exp:* Geologist petrol explor, Mobil Oil Corp, 63-68; asst prof geol, Midwestern Univ, 68-72; geologist & prog chief res, US Geol Surv, 72-84; geophysicist/res pilot, COMAP ExplorServ Inc, 84-86; dir spec proj & chief eng test pilot, Tessera Res Corp, 86-91; CONSULT, BROKEN WIND ASSOC, 91- *Concurrent Pos:* Geophys/Geol consult, 84- *Honors & Awards:* A I Levorsen Award, Am Asn Petrol Geologists, 71. *Mem:* Fel Geol Soc Am; Am Asn Petrol Geologists; AAAS; Soc Exp Test Pilots; Am Geophys Union; Aerospace Indust Asn Am. *Res:* Experimental and applied geophysical and aerospace sciences; airborne geophysics and remote sensing. *Mailing Add:* 469 Pine Song Trail Golden CO 80401

DONOVAN, THOMAS ARNOLD, b Galesburg, Ill, July 11, 37; m 60; c 3. INORGANIC CHEMISTRY. *Educ:* Knox Col (Ill), BA, 59; Univ Ill, PhD(inorg chem), 62. *Prof Exp:* Asst prof chem, Ind State Col, 62-64; asst prof chem, Knox Col (Ill), 64-68; from asst prof to assoc prof, 68-81, PROF CHEM, STATE UNIV NY COL BUFFALO, 81- *Mem:* Am Chem Soc. *Res:* Preparation, characterization and properties of new coordination compounds; biological applications of coordination chemistry. *Mailing Add:* Dept Chem State Univ NY Col 1300 Elmwood Ave Buffalo NY 14222

DONOVICK, PETER JOSEPH, b Champaign, Ill, Jan 14, 38; m 81; c 4. NEUROPSYCHOLOGY. *Educ:* Lafayette Col, BA, 61; Univ Wis, MS, 63, PhD(psychol), 67. *Prof Exp:* From asst prof to assoc prof psychol, Harpur Col, 66-76; chmn dept, 78-81. actg assoc dean, 82-85, PROF PSYCHOL, STATE UNIV NY, BINGHAMTON, 76- *Concurrent Pos:* Spec res fel anat & vis asst prof, Hershey Med Ctr, Pa State Univ, 70-71; clin neuropsychol trainee, Hutchings Psychiat Ctr, 83-84; staff, Neuropsychol Res Rehab Ctr, Tel-Aviv, Israel, 89-90; Fulbright fel, 89-90. *Mem:* Fel Am Psychol Asn; NY Acad Sci; Soc Neurosci; Behavior Genetics Asn; Int Neuropsychol Soc. *Res:* Genetic and environmental influences of the response to lesions of the limbic system of the brain; impact of environmental toxins, parasites and nutrition on behavior; lifespan developmental psychobiology; behavior genetics; human neuropsychology. *Mailing Add:* Dept Psychol State Univ NY Binghamton NY 13901

DONOVICK, RICHARD, bacteriology; deceased, see previous edition for last biography

DONSKER, MONROE DAVID, mathematics; deceased, see previous edition for last biography

DONTA, SAM THEODORE, b Aliquippa, Pa, Aug 9, 38; m 59; c 4. INFECTIOUS DISEASES. *Educ:* Allegheny Col, BS, 59; Albert Einstein Col Med, MD, 63. *Prof Exp:* Intern & resident, Univ Pittsburgh Hosps, 64-65; fel biochem, Brandeis Univ, 67-70; res assoc infectious dis, Univ Hosp, Boston, 70-71; from asst prof to assoc prof, 71-74, prof med, Univ Iowa, 78-82; clin investr, Vet Admin Hosp, 75-78. *Concurrent Pos:* Clin investr, Vet Admin Hosp, 75-78; mem, Bact & Mycol Study Sect, NIH, 76-80. *Mem:* Infectious Dis Soc Am; Am Fedn Clin Res; AAAS; Am Soc Microbiologists. *Res:* Mechanisms of action of cholera and Escherichia coli enterotoxins on cultured adrenal cells; toxin-membrane interactions; tissue culture models of infectious disease; bacterial adherence; viral receptors; receptor regulation. *Mailing Add:* Dept Med Univ Conn Health Ctr 263 Farmington Ave Farmington CT 06032

DOOB, JOSEPH LEO, b Cincinnati, Ohio, Feb 27, 10; m 31; c 3. POTENTIAL THEORY, PROBABILITY. *Educ:* Harvard Univ, AB, 30, AM, 31, PhD(math), 32. *Hon Degrees:* DSc, Univ Ill. *Prof Exp:* Nat Res fel math, Columbia Univ, 32-34, assoc theoret statist & Carnegie Corp fel, 34-35; from assoc math to prof, 35-78, EMER PROF MATH, UNIV ILL, URBANA-CHAMPAIGN, 78- *Honors & Awards:* Foreign assoc, Acad Sci, Paris, 75; Nat Medal Sci, 80. *Mem:* Nat Acad Sci; Am Acad Arts & Sci; Am Math Soc (pres, 63-64); fel Inst Math Statist (vpres, 45, pres, 50). *Res:* Probability and potential theory. *Mailing Add:* 208 W High St Urbana IL 61801

DOODY, JOHN EDWARD, b Chicago, Ill, Mar 30, 25. PHYSICAL CHEMISTRY. *Educ:* St Mary's Col, BS, 46; St Louis Univ, MS, 50, PhD(phys chem), 53. *Prof Exp:* Mil high sch teacher, Mo, 46-53; instr, 53-54, PROF CHEM, CHRISTIAN BROS COL, 54-, DIR GRANTS, 80- *Concurrent Pos:* Cottrell res grant, 54; NSF grant, 54, res grant, 62; AEC res grant, 55-66; Tenn heart res grant, 62 & 63; Am Chem Soc petrol res grant, 63; dir, NSF Insts Sci Teachers, 67-75; dir, Dept Health, Educ & Welfare Inst Minority Cult, 69 & 71. *Mem:* Am Chem Soc; Am Inst Chemists; AAAS. *Res:* Heavy metal ions; transport across membranes and reactions with nucleic acids and derivatives; electrochemistry. *Mailing Add:* Dept Chem Christian Bros Col 650 E Pkwy S Memphis TN 38104-5581

DOODY, MARIJO, b Chicago, Ill, Sept 14, 36. DETERGENT FORMULATION, FLOOR POLISH FORMULATION. *Educ:* Loyola Univ, BS, 58. *Prof Exp:* Jr chemist, Nallo Chem, 59-64; chemist, Armour Dial, 64-68 & Enterprise Div Valspar, 68-88; sr chemist, Hysan Corp, 88-90; TECH DIR CHEM, NYCO PROD CO, 91- *Concurrent Pos:* Interprof chair, Soc Cosmetic Chemists, 70-80. *Res:* Maintenance chemicals. *Mailing Add:* 9625 Albany Evergreen Park IL 60642

DOOLEN, GARY DEAN, b Billings, Mont, June 7, 39; m 62; c 2. NUCLEAR FUSION, PLASMA MODELLING. *Educ:* Purdue Univ, BS, 61, MS, PhD(physics), 67. *Prof Exp:* Res assoc atomic physics, Goddard Space Flight Ctr, NASA, 67-69; asst prof physics, Tex A&M Univ, 69-75; STAFF MEM, LOS ALAMOS NAT LAB, 75-, DEP DIR, CTR NONLINEAR STUDIES, 88- *Mem:* Am Phys Soc. *Res:* Nuclear fusion with special interest in pure fusion; energy transport; atomic and nuclear excited states; nuclear properties; computer modelling of plasmas; plasma radiation interactions. *Mailing Add:* Los Alamos Nat Lab MS F669 Los Alamos NM 87544

DOOLEY, DAVID MARLIN, b Tulare, Calif, May 13, 52; m 78. INORGANIC CHEMISTRY. *Educ:* Univ Calif, San Diego, BA, 74; Calif Inst Technol, PhD(chem), 79. *Prof Exp:* Res asst, Calif Inst Technol, 74-78; asst prof, 78-84, ASSOC PROF CHEM, AMHERST COL, 84- *Concurrent*

Pos: NIH fel, 74-78; vis scientist, Mass Inst Technol, 81; vis scholar, Stanford Univ, 82. *Mem:* Am Chem Soc; Am Heart Asn; Am Soc Biol Chemists. *Res:* Spectroscopic studies of the electronic structure and bonding in transition metal complexes; ligand field theory; applications to bioinorganic chemistry; copper proteins; inorganic photochemistry. *Mailing Add:* Dept Chem Amherst Col Amherst MA 01002

DOOLEY, DOUGLAS CHARLES, b Chicago, Ill, Mar 5, 46. BONE MARROW CULTURE, CELL PHYSIOLOGY. *Educ:* Univ Calif, Berkeley, BA, 67; Univ Wash, PhD(microbiol), 72. *Prof Exp:* Fel cell physiol, Worcester Found Exp Biol, 72-76; res assoc cell physiol, 76-77, HEAD, TISSUE CULT LAB, RED CROSS BLOOD RES LAB, 77-, RES DIR. *Mem:* Int Soc Exp Hematol; AAAS; Soc Cryobiol. *Res:* Long term culture of bone marrow and regulation of hematopoiesis. *Mailing Add:* Am Nat Red Cross PO Box 3200 Portland OR 97208

DOOLEY, ELMO S, b Davidson, Tenn, Feb 23, 24; m 45; c 3. PHYSIOLOGY. *Educ:* Tenn Polytech Inst, BS, 52; Univ Tenn, MS, 55, PhD, 57. *Prof Exp:* Consult microbiol, Cumberland Med Ctr, 57-58; chief microbiol br, US Army Med Res Lab, 58-61; sr scientist, US Aerospace Med Res Lab, 61-63; chmn dept biol, 64-66, PROF BIOL, TENN TECHNOL UNIV, 64- *Mem:* Aerospace Med Asn; Am Soc Microbiol; Am Fedn Clin Res; fel Royal Soc Health. *Res:* Aerospace and cardiovascular physiology. *Mailing Add:* 148 Mattson St Cookeville TN 38501

DOOLEY, GEORGE JOSEPH, III, b Greenwich, Conn, Aug 8, 41; m 63; c 3. MATERIALS SCIENCE. *Educ:* Univ Notre Dame, BS, 63; Iowa State Univ, MS, 66; Ore State Univ, PhD(mat sci), 69. *Prof Exp:* Res metallurgist, Albany Metall, Res Ctr, Bur Mines, 66-68; res scientist, Aerospace Res Labs, US Air Force Systs Command, 68-73; DIR METALL & PROCESS RES & DEVELOP, OREGON METALL CORP, 74- *Mem:* Am Soc Metals; Am Inst Mining, Metall & Petrol Eng; Am Vacuum Soc; Am Phys Soc. *Res:* Process and production metallurgy, ingot manufacture and mill products aspects of titanium metal production; surface chemistry and surface physics of both single crystal and polycrystalline materials using the techniques of low energy electron diffraction and auger electron spectroscopy. *Mailing Add:* 8804 NW Peavy Arboretum Rd Corvallis OR 97330

DOOLEY, JAMES KEITH, b Steubenville, Ohio, Aug 8, 41; div; c 2. SYSTEMATIC ICHTHYOLOGY, AQUATIC ECOLOGY. *Educ:* Univ Miami, BS, 64; Univ S Fla, MA, 70; Univ NC, Chapel Hill, PhD(zool), 74. *Prof Exp:* Res asst fisheries, Inst Marine Sci, Univ Miami, 64-67; fisheries biologist, Nat Oceanic & Atmospheric Admin, Nat Marine Fisheries Serv, Miami, 67; from asst prof to assoc prof, 73-87, PROF BIOL, ADELPHI UNIV, 87- *Concurrent Pos:* Res fisheries consult, Biol Consult Inc, 73-75, EEA Inc, 87- *Mem:* Am Soc Ichthyologists & Herpetologists; Animal Behavior Soc; AAAS; Sigma Xi; Ichthyol Soc Japan. *Res:* Systematics and biology of the tilefishes, Branchiostegidae and Malacanthidae; orientation of fishes; heavy metals in fishes; fishes of the Canary Islands. *Mailing Add:* Dept Biol Adelphi Univ Garden City NY 11530

DOOLEY, JOHN RAYMOND, JR, b Denver, Colo, Dec 12, 25; m 52; c 5. NUCLEAR PHYSICS. *Educ:* Regis Col (Colo), BS, 49; Univ Denver, MS, 51. *Prof Exp:* Engr, Colo State Hwy Dept, 49-51; instr physics, guided missile training, Lowry AFB, Colo, 51-52; staff mem nuclear physics, Sandia Corp, 52-53; physicist, Isotope Geol Br, US Geol Surv, 53-88; RETIRED. *Res:* Natural radioactivity and uranium geochemistry; uranium and thorium disequilibrium; health physics; uranium-234 fractionation; fission tracks; uranium-lead dating of uranium deposits, developed the radioluxograph for autoradiography of uranium and induced-particle autoradiography of lithium and boron. *Mailing Add:* 7475 Miller St Arvada CO 80005-3867

DOOLEY, JOSEPH FRANCIS, b New York, NY, Oct 3, 41; m 63; c 2. CLINICAL CHEMISTRY, TOXICOLOGY. *Educ:* Fordham Univ, BS, 63; Univ Minn, Minneapolis, PhD(org chem), 67. *Prof Exp:* Res chemist, E I du Pont de Nemours & Co, Inc, 67-70; staff scientist, Med Res Labs, 70-74, sr res scientist, 74-77, SR RES INVESTR, PFIZER CENT RES, PFIZER INC, 77-; ASSOC PROF TOXICOL & MED LAB SCI, QUINNIPIAC COL, HAMDEN, CONN, 77- *Concurrent Pos:* Clin assoc prof, Dept Lab Med, Sch Med, Univ Conn, 76-; sr partner, Med Lab & Consult, Biomed Assoc, 80-; nat comt chmn, Clin Lab Testing in Chem & Drug Safety Eval, Am Asn Clin Chemistry, 76-; mem comt, Effects of Drugs in Clin Lab Tests, Int Fed Clin Chemistry. *Mem:* AAAS; Am Chem Soc; Am Asn Clin Chemistry; Soc Toxicol. *Res:* Hematology; enzymology; effects of drugs on kidney function and liver metabolism; radioimmunoassay; toxicology; laboratory instrumentation. *Mailing Add:* Clin Path Lab Pfizer Med Res Lab Eastern Point Rd Groton CT 06340

DOOLEY, WALLACE T, b Conway, Ark, June 15, 17; m 39; c 4. ORTHOPEDIC SURGERY. *Educ:* Univ Kans, AB, 39, MA, 41; Meharry Med Col, MD, 47. *Prof Exp:* Rockefeller Found fel, 51-52; Nat Polio Found fel, 51-55; DIR PHYS MED, GEORGE W HUBBARD HOSP, 55-, PROF ORTHOP SURG, MEHARRY MED COL, 67-, HEAD DIV, 60- *Concurrent Pos:* Chief orthop surg, Riverside Sanitarium & Hosp, 55-; assoc prof, Meharry Med Col, 60-67, prog dir rehab med, 63-67; mem, Asn Med Rehab Dirs & Coordr & Asn Orthop Chmn. *Mem:* Am Cong Rehab Med; Nat Med Asn. *Mailing Add:* Dept Surg/Orthop Meharry Med Col Sch Med 1005 Todd Blvd Nashville TN 37208

DOOLEY, WILLIAM PAUL, b Richmond, Va, Aug 2, 15; m 41; c 2. CHEMISTRY, CHEMICAL ENGINEERING. *Educ:* Univ Richmond, BS, 38; Mass Inst Technol, SM, 40. *Prof Exp:* Tech & admin asst to gen mgr, Am Viscose Corp, 40-45, asst supt, Staple Develop Plant, 45-46, supt, 46-50, head, Develop Serv Dept, 50-55, corp prod tech supt, 55-60; TECH & VENTURE CONSULT, 78- *Concurrent Pos:* Tech & Venture Consult, 78- *Mem:* Am Chem Soc; Am Inst Chem Engrs; Chem Mkt Res Asn. *Res:* Manufacture of viscose and viscose rayon products. *Mailing Add:* Two Salette Lane The Landings Skidaway Island Savannah GA 31411

DOOLITTLE, CHARLES HERBERT, III, b Holyoke, Mass, July 18, 39; m 61, 71; c 2. CLINICAL PHARMACOLOGY, ONCOLOGY. *Educ:* Univ Mass, Amherst, BS, 62, MA, 65; George Washington Univ, PhD(pharmacol), 70; Brown Univ, MD, 75; Am Bd Internal Med, cert, 83; Am Bd Med Oncol, cert, 89. *Prof Exp:* Res asst clin pharmacol, Sch Med, Yale Univ, 64-66; USPHS res fel biochem pharmacol, Brown Univ & Roger Williams Gen Hosp, 69-75, intern, Roger Williams Gen Hosp, 75-76; resident, Roger Williams Gen Hosp, 76-77; CHIEF ONCOL SECT & PRIN INVESTR, DEPT MED & SURG, VET ADMIN HOSP, PROVIDENCE, 80- *Mem:* AMA; Am Col Physicians; Sigma Xi. *Res:* Anti-neoplastic agents. *Mailing Add:* Dept Med Brown Univ Providence RI 02912

DOOLITTLE, DONALD PRESTON, b Torrington, Conn, May 14, 33; m 57; c 2. GENETICS. *Educ:* Univ Conn, BS, 54; Cornell Univ, MS, 56, PhD(animal genetics), 59. *Prof Exp:* Asst, Cornell Univ, 54-58; fel, Roscoe B Jackson Mem Lab, 58-60; asst res prof biomet, Grad Sch Pub Health, Univ Pittsburgh, 60-65; from asst prof to assoc prof genetics, WVa Univ, 66-67; assoc prof, 67-89, EMER PROF ANIMAL SCI, PURDUE UNIV, 89-; RES ASSOC, JACKSON LAB, 87- *Concurrent Pos:* Vis prof animal sci, Cornell Univ, 85. *Mem:* Am Genetics Asn; Genetics Soc Am. *Res:* Mammalian genetics; genetic linkage; biometry. *Mailing Add:* Jackson Lab Bar Harbor ME 04609

DOOLITTLE, J(ESSE) S(EYMOUR), mechanical engineering; deceased, see previous edition for last biography

DOOLITTLE, RICHARD L, b Washington, DC, Dec 29, 53. CELLULAR BIOLOGY. *Educ:* Univ Bridgeport, BA, 75; Univ Rochester, PhD(anat & path), 81. *Prof Exp:* ASSOC PROF ANAT, UNIV NEW ENG, 82- *Mem:* Am Asn Path; Am Soc Microbiol; Reticuloendothelial Soc; NY Acad Sci; AAAS. *Mailing Add:* Dept Biol Rochester Inst Technol Rochester NY 14623-0887

DOOLITTLE, ROBERT FREDERICK, II, b Chicago, Ill, Dec 21, 25. HIGH-ENERGY ASTROPHYSICS, MICROCOMPUTER SYSTEMS SOFTWARE SPECIALIST. *Educ:* Oberlin Col, AB, 48; Univ Mich, MS, 50, PhD(physics), 58. *Prof Exp:* Asst, Univ Mich, 50-58; asst prof physics, San Diego State Col, 58-60; mem tech staff, Space Physics Dept, TRW Space Tech Labs, 60-66, mem tech staff, Space Sci Lab, TRW Systs, 66-70, staff scientist, Instrument Systs Lab, 70-78, proj scientist, High Energy Astron Observ, 78-81, proj scientist, Gamma Ray Observ, sr staff scientist, Software & Info Systs Div, TRW Defense & Space Systs Group, Redondo Beach, 82-83; SYSTS SOFTWARE SPECIALIST, VAR CO, 83- *Mem:* AAAS; Am Phys Soc; Am Astron Soc; Res & Engr Soc Am. *Res:* High energy astronomy and cosmic ray physics. *Mailing Add:* 1290 Monument St Pacific Palisades CA 90272

DOOLITTLE, RUSSELL F, b New Haven, Conn, Jan 10, 31; m 55; c 2. BIOCHEMISTRY. *Educ:* Wesleyan Univ, BA, 52; Trinity Col, Conn, MA, 57; Harvard Univ, PhD(biochem), 62. *Prof Exp:* Instr biol, Amherst Col, 61-62; Nat Heart Inst fel, 62-64; asst res biologist, 64-65, from asst prof to assoc prof chem, 67-72, PROF BIOCHEM, UNIV CALIF, SAN DIEGO, 72- *Concurrent Pos:* Career develop award, USPHS, 69-74; mem, NIH Blood Adv Comt, 74-78; Guggenheim fel, 84-85. *Mem:* Nat Acad Sci; Am Soc Biol Chemists; Am Acad Arts & Sci. *Res:* Protein chemistry and molecular evolution; structure of fibrinogen. *Mailing Add:* Ctr Molecular Genetics Univ Calif San Diego La Jolla CA 92093

DOOLITTLE, WARREN FORD, III, b Urbana, Ill, Feb 21, 42; div; c 2. MOLECULAR BIOLOGY. *Educ:* Harvard Col, BA, 63; Stanford Univ, PhD(biol sci), 69. *Prof Exp:* NIH fel, Nat Jewish Hosp, Denver, Colo, 69-70, res assoc molecular biol, 70-71; from asst prof to assoc prof, 71-82, PROF BIOCHEM, DALHOUSIE UNIV, 82-, MED RES COUN CAN SCHOLAR, 71- *Concurrent Pos:* Can Inst Advan Res fel, dir prog evoluationary biol; Guggenheim fel, 85-86. *Mem:* Am Soc Microbiol. *Res:* Molecular biology and genetics of archaebacteria algae; early cellular evoluation. *Mailing Add:* Dept Biochem Dalhousie Univ Halifax NS B3H 4H7 Can

DOOMES, EARL, b Washington, La, Feb 8, 43; m 65; c 3. ORGANIC CHEMISTRY. *Educ:* Univ Nebr, PhD(chem), 69. *Prof Exp:* Res assoc chem, Northwestern Univ, Evanston, 68-69; from asst prof to assoc prof chem, Macalester Col, 74-77; from assoc prof to prof chem, 82-87, CHMN, SOUTHERN UNIV, 87- *Concurrent Pos:* NSF sci fac fel, Fla State Univ, 75-76. *Mem:* Am Chem Soc. *Res:* Interaction of small organic molecules with nucleic acids; synthetic approaches to cyclophanes and related conjugated polyenes, nonbenzonoid and unusual bridged aromatics, and nucleophilic substitution reactions of allyl systems. *Mailing Add:* Southern Br PO Box 11513 Baton Rouge LA 70813

DOORENBOS, HAROLD E, b Morrison, Ill, Oct 4, 25; m 49; c 2. POLYMER CHEMISTRY. *Educ:* Cent Col (Iowa), BS, 49; Univ Ark, MS, 56; Univ Del, PhD(chem), 62. *Prof Exp:* Instr chem, Pa State Univ, 55-56; chemist, USDA, Md, 58 & E I du Pont de Nemours & Co, 61-62; chemist, Dow Chem Co, 62-90; RETIRED. *Mem:* AAAS; Am Chem Soc; Sigma Xi. *Res:* Organic synthesis; pharmaceutical, polymer, chlorine and fluorine chemistry. *Mailing Add:* 1908 Hollywood Blvd Iowa City IA 52240-5926

DOORENBOS, NORMAN JOHN, b Flint, Mich, May 13, 28; m 51, 79; c 7. CHEMICAL PHARMACOGNOSY. *Educ:* Univ Mich, BS, 50, MS, 51, PhD(chem), 53. *Prof Exp:* Sr res chemist sensitizing dyes, Ansco, 53-56; from asst prof to prof pharmaceut chem, Univ Md, 56-65; prof med chem, Univ Miss, 65-77, prof pharmacog & chmn dept, 67-77; prof physiol & dean col sci, Southern Ill Univ, Carbondale, 77-; VCHANCELLOR, UNIV WIS-EAU CLAIRE. *Concurrent Pos:* Vis scientist, Am Asn Cols Pharm, 63-73; consult, Malinckrodt Chem Works, 63-71; Merck Sharp & Dohme lectr, WVa Univ, 64; NSF Sci Curric Proj, Univ Ill, 64-65. *Mem:* AAAS; Am Chem Soc (treas, 49, chmn, 50); Soc Econ Bot (pres, 77-78); Acad Pharmaceut Sci; Am Soc

Pharmacog; Sigma Xi. *Res:* Medical chemistry; steroids; heterocyclic steroids; alkaloids; natural products; biopharmaceutics; pharmacology; drug abuse; marijuana and phytochemistry; food toxins and bioactive marine substances. *Mailing Add:* Univ Auburn Auburn AL 36849-3501

DOORNENBAL, HUBERT, b Utrecht, Neth, Apr 19, 27; Can citizen; m 56; c 3. PHYSIOLOGY, MEAT SCIENCE. *Educ:* Univ BC, BSA, 52, MSA, 56; Cornell Univ, PhD(physiol, meats), 61. *Prof Exp:* Res asst beef cattle, 54-56, res officer beef cattle & swine nutrit, 56-59, RES SCIENTIST PHYSIOL & MEATS, RES STA, CAN DEPT AGR, 61- *Mem:* Am Soc Animal Sci; Can Soc Animal Sci; Agr Inst Can. *Res:* Physiology; endocrinology; nutrition; radiation biology; meat science. *Mailing Add:* Res Br Exp Farm Lacombe AB T0C 1S0 Can

DOPPMAN, JOHN L, b Springfield, Mass, June 14, 28; m; c 2. RADIOLOGY. *Educ:* Holy Cross Col, AB, 49; Yale Univ, MD, 53. *Prof Exp:* Dep chief, Diag Radiol Dept, NIH, 64-70; prof, Dept Radiol, Univ Hosp, Univ Calif, San Diego, 70-72; chief, Diag Radiol Dept, 72-86, DIR RADIOL, DIAG RADIOL CLIN CTR, NIH, BETHESDA, 86-; PROF RADIOL, SCH MED, GEORGETOWN UNIV, 86- *Concurrent Pos:* Clin prof radiol, Sch Med, George Washington Univ, 72- & Uniformed Serv Univ Health Sci, 78-86; lectr, Armed Forces Inst Path, 73-85; consult, Nat Naval Med Ctr, 73-; Pub Health Serv counr, Radiol Soc NAm & Am Col Radiol, 76-86. *Honors & Awards:* Smelo Mem Lectr, Baptist Hosp, Ala; Peter Heimann Mem Lectr, Int Soc Endocrine Surgeons, WGer, 83; Dunn Mem Lectr, Houston Radiol Soc, 86; First Robert Shapiro Mem Lectr, New Haven, Conn, 86; Houghton Lectr, Royal Col Surgeons, Ireland, 87; 20th John A Campbell Lectr, Sch Med, Indiana Univ, 90. *Mem:* Am Col Radiol; Radiol Soc NAm; AMA; Asn Univ Radiologists; Am Roentgen Ray Soc; hon mem Am Asn Endocrine Surgeons; corresp mem Ger Radiol Soc; hon fel Royal Col Surgeons; hon mem Hungarian Radiol Soc. *Mailing Add:* Diag Radiol Dept NIH Bldg 10 Rm 1C-660 Bethesda MD 20892

DORAIN, PAUL BRENDEL, b New Haven, Conn, Aug 30, 26; m 50; c 2. PHYSICAL CHEMISTRY. *Educ:* Yale Univ, BS, 50; Ind Univ, PhD, 54. *Prof Exp:* Mem, Enrico Fermi Inst Nuclear Studies, Univ Chicago, 54-56; mem, Aeronaut Res Lab, Ohio, 56-58; from asst prof to prof chem, Brandeis Univ, 58-81, from co-chmn to chmn dept, 71-74; dean fac & vpres acad affairs, Colby Col, 81-82; PRES, DEPT CHEM, AMHERST COL, 82- *Concurrent Pos:* Trustee, Lowell Technol Inst, Mass, 72-74; Tallman vis prof chem & physics, Bowdoin Col, 74-75; vis fel, Yale Univ, 79-80; sr res fel appl physics, Yale Univ, 82-85. *Mem:* Am Phys Soc; Sigma Xi; AAAS; Am Chem Soc; Mat Res Soc. *Res:* Paramagnetic resonance; optical spectra of 4d and 5d transition metal ions; surface enhanced Raman scattering; surface chemistry. *Mailing Add:* Dept Chem Amherst Col Amherst MA 01002

DORAI-RAJ, DIANA GLOVER, b Rochester, NY, Sept 24, 38; wid; c 4. CHEMISTRY. *Educ:* Univ Rochester, BS, 60; Univ Ore, PhD(chem), 67. *Prof Exp:* Lectr chem, Univ Waterloo, 66-67; instr, Seneca Col Arts & Technol, 68-69; ASSOC PROF CHEM, STILLMAN COL, 80- *Mailing Add:* Dept Chem Stillman Col 3601 15th St Tuscaloosa AL 35403

DORAN, DONALD GEORGE, b Los Angeles, Calif, Oct 2, 29; m 50; c 4. SOLID STATE PHYSICS, METAL PHYSICS. *Educ:* Wash State Univ, BS, 51, MS, 55, PhD(physics), 60. *Prof Exp:* Eng asst, Hanford Works Div, Gen Elec Co, 51-53; aeronaut res scientist, Ames Aeronaut Lab, Nat Adv Comt Aeronaut, 55; physicist, Poulter Res Labs, Stanford Res Inst, 55-58 & 60-62, sr physicist, 62-64, head solid state group, 64-65, dir shock wave physics div, 65-67; res assoc mat res sect, Pac Northwest Lab, Battelle Mem Inst, 67-70; res assoc, Westinghouse-Hartford Co, 70-72, fel scientist, 72-77, mgr, irridation orgn effects, 77-85; mgr, Fusion Mat Prog, Pac Northwest Lab, 85-89; FAC, WASH STATE UNIV, 90- *Concurrent Pos:* Fel scientist, Westinghouse-Hanford Co, 72- *Honors & Awards:* Outstanding Accomplishment Award, Fusion Energy Div, Am Nuclear Soc, 86. *Mem:* Am Phys Soc; Am Nuclear Soc; Fedn Am Scientists. *Res:* Analysis of damage in metals irradiated with fission and fusion neutrons, ions and electrons; analytical and computer modeling of damage mechanisms. *Mailing Add:* 2307 Snohomish Richland WA 99352

DORAN, J CHRISTOPHER, b Dayton, Ohio, May 26, 45; m 70; c 1. BOUNDARY LAYER METEOROLOGY, MICROMETEOROLOGY. *Educ:* Fordham Univ, BS, 66; Yale Univ, PhD(physics), 71; Univ Utah, MS, 76. *Prof Exp:* Res assoc & assoc instr physics, Univ Utah, 71-73, asst res prof, 73-74; res scientist, 76-78, sr res scientist, 78-89, TECH GROUP LEADER ATMOSPHERIC SCI, BATTELLE NORTHWEST, 89- *Mem:* Am Meteorol Soc. *Res:* Boundary layer meteorology; micrometeorology; diffusion studies; numerical modeling; turbulence. *Mailing Add:* 1516 Johnston Ave Richland WA 99352

DORAN, JOHN WALSH, b Pittsburgh, Pa, Nov 12, 45; m 66; c 1. MICROBIAL ECOLOGY, AGROECOSYSTEM MANAGEMENT. *Educ:* Univ Md, BS, 67; Va Polytech Inst & State Col, MS, 70; Cornell Univ, PhD(soil sci), 76. *Prof Exp:* Med lab specialist diag bact, US Army Med Corps, 70-72; grad researcher soil microbiol, Cornell Univ, 72-75; asst prof, 75-79, ASSOC PROF AGRON, DEPT AGRON, UNIV NEBR, 79-, SOIL SCIENTIST MICROBIOL, AGR RES RES SERV, USDA, 75- *Concurrent Pos:* Res fel, Div Tropical Crops & Pastures, Commonwealth Sci & Indust Res Orgn, 83-84; assoc ed, Soil Sci Soc Am J & J Alternative Agr, Inst Alternative Agr. *Mem:* Am Soil Sci Soc; Am Soc Microbiol; Am Soc Agron; Soil & Water Conserv Soc; Int Soil Tillage Res Orgn. *Res:* Biological and chemical considerations in disposal and utilization of wastes; tillage and residue management effects on soil microbial activity and nutrient cycling; biological-physical interactions in soil as related to agricultural management. *Mailing Add:* 1501 W Park Circle Lincoln NE 68522

DORAN, PETER COBB, b Bronxville, NY, Nov 23, 36; m 64; c 3. HEALTH SCIENCES. *Educ:* Colby Col, AB, 58; Southern Ill Univ, MA, 60, PhD(health, psychol), 66. *Prof Exp:* Instr health educ, Southern Ill Univ, 61-66, supvr student teachers, Dept Health Educ, 63-66; consult psychologist & dir ment health educ, Maine Bur Ment Health, 66-67; exec dir, Maine Comn Rehab Needs, 67-70 & Maine Health Coun, 70-72; prof spec educ, 71-73, dir health educ ctr, 72-73, chmn dept, 73-80, PROF HEALTH SCI, UNIV MAINE, FARMINGTON, 73- *Concurrent Pos:* Coordr health training, Peace Corps Projs, Southern Ill Univ, 64; chmn, Nat Comt Ment Health Educ, 67, historian, 67-; res consult, Maine's Regional Med Prog, 70-71; secy, Maine's Health Educ Consortium, 73-76; consult, Nat Instrnl TV, 74-76; trustee, Maine's Health Syst Agency, 75-76. *Mem:* Am Pub Health Asn; Am Psychol Asn; Nat Rehab Asn. *Res:* Attitudinal effects of rural preceptorships on medical and dental students in a health maintenance organization; dynamics of activating patients toward preventive health practices; mental health education. *Mailing Add:* Dept Human Develop & Family Univ Maine Farmington ME 04938

DORAN, ROBERT STUART, b Winthrop, Iowa, Dec 21, 37; m 59; c 2. ANALYSIS & FUNCTIONAL ANALYSIS, ALGEBRA. *Educ:* Univ Iowa, BA, 62; MS, 64; Univ Wash, MS, 67, PhD(math), 68. *Prof Exp:* Instr math, Univ Wash, Seattle, 68; asst prof math, Univ Northern Iowa, Cedar Falls, 68-69; from asst prof to prof, 69-89, CHAIR MATH, TEX CHRISTIAN UNIV, 78-, MINNIE STEVENS PIPER PROF, 89- *Concurrent Pos:* Vis prof, Univ Tex, 79-80; mem, Inst Adv Study, 80-81; vis scholar, Mass Inst Technol, 81, Oxford Univ Math Inst, Eng, 88-; prin invest, Nat Sci Found, 88-; consult, John Wiley Publ Co, 70; Addison-Wesley Publ Co, 79; Trustee & chmn, Inst Adv Study, 84-, pres, bd trustees. *Mem:* Am Math Soc; Math Asn Am; Sigma Xi. *Res:* Banach Algebras, C*-algebras; representation Banach algebraic bundles; author, articles on Banach algebras and algebraic bundles; author, four books on Banach algebras and C*-algebras. *Mailing Add:* Dept Math Tex Christian Univ Ft Worth TX 76129

DORAN, THOMAS J, JR, b Brooklyn, NY, Dec 31, 42; m 64, 88; c 7. ORGANIC CHEMISTRY. *Educ:* Case Inst Technol, BS, 62; Western Reserve Univ, PhD(chem), 67. *Prof Exp:* Sr res chemist, 66-77, res supvr biochem, Chem Div, PPG Industs, Inc, 77-88; MGR, ENVIRON FATE RICERCA INC, 88- *Mem:* Am Chem Soc; AAAS. *Res:* Environmental chemistry; metabolism; agricultural chemicals. *Mailing Add:* 1122 Rolling Meadows Akron OH 44313

DORATO, PETER, b New York, NY, Dec 17, 32; m 56; c 4. ELECTRICAL ENGINEERING. *Educ:* City Col New York, BEE, 55; Columbia Univ, MSEE, 56; Polytech Inst Brooklyn, DEE, 61. *Prof Exp:* Instr elec eng, City Col New York, 56-57; assoc prof, Polytech Inst Brooklyn, 57-72; prof, Univ Colo, Colorado Springs, 72-76; chmn dept, 76-84, PROF ELEC & COMPUT ENG, UNIV NMEX, 76- *Concurrent Pos:* Vis prof, Univ Colo, Boulder, 69-70, Univ Cal, Santa Barbara, 84-85; hon prof, Nanjing Aeronaut Inst. *Mem:* Distinguished Mem, Inst Elec & Electronics Engrs Control Systs Soc; Fel Inst Elec & Electronics Engrs. *Res:* Control theory; stability; systems analysis; robust control. *Mailing Add:* Dept Elec Eng & Comput Sci Univ NMex Albuquerque NM 87131

D'ORAZIO, VINCENT T, b Joliet, Ill, Dec 28, 29; m 54; c 2. ORGANIC CHEMISTRY. *Educ:* Univ Ill, BS, 52; Mich State Univ, PhD(org chem), 63. *Prof Exp:* Supv chemist, US Rubber Co, 52-54; res scientist, S C Johnson & Son, Inc, 63-87; CONSULT, 87- *Mem:* AAAS; fel Am Inst Chem; Am Chem Soc; Royal Soc Chem. *Res:* Synthesis of heterocyclic organic compounds; synthesis of insecticides related to pyrethrins, pyrethrum cultivation extraction and technology; development of insecticidal products and insect repellents. *Mailing Add:* 5802 Tahoe Dr Racine WI 53406

DORCHESTER, JOHN EDMUND CARLETON, b Vancouver, BC, Aug 18, 17; nat US; m 44; c 3. PHYSIOLOGY, BIOCHEMISTRY. *Educ:* Univ BC, BA, 47, MA, 48; Univ Toronto, PhD(physiol, biochem), 52. *Prof Exp:* Asst, Univ Toronto, 49-52; from instr to asst prof physiol, Jefferson Med Col, 53-60; prof sci, West Chester Univ, 60-80, vchmn dept, 65-67, prof biol, 80-85; RETIRED. *Mem:* Am Physiol Soc. *Res:* Thermal tolerance in fish; gastrointestinal hormones; hormonal assay methods; gastrointestinal innervation and relation to structure and hormones; effects of tranquilizers on gastrointestinal motility. *Mailing Add:* 8741 Beaumaris Pl Sidney BC V8L 3Z6 Can

DORDICK, HERBERT S(HALOM), b Philadelphia, Pa, Oct 20, 25; m 48; c 2. ELECTRICAL ENGINEERING, SYSTEMS ANALYSIS. *Educ:* Swarthmore Col, BSEE, 49; Univ Pa, MSEE, 57. *Prof Exp:* Res & develop engr, Leeds & Northrup Co, Pa, 49-54; systs engr & mgr adv proj, Radio Corp Am, 54-62; dir eng, Electronic Instruments Div, Burroughs Corp, 62-64; group leader & sr mem res staff, Rand Corp, 64-68; dep dir res & spec projs, Syst Develop Corp, 68-69; pres, Info Transfer Corp, 69-71; dir, Off Telecommun, City New York, 71-73; prof commun sci & assoc dir ctr commun policy res, Annenberg Sch Commun, Univ Southern Calif, 73-85; PROF COMMUN & CHMN, DEPT RTF, SCH THEATER & COMMUN, TEMPLE UNIV, 86- *Concurrent Pos:* Contrib researcher, President's Task Force Telecommun Policy, 44-46; consult to indust, govt, educ insts & nat & int foundations, 61-; vis scholar, Ctr Int Studies, Mass Inst Technol, 78-79; Fulbright scholar, Victoria Univ, Wellington, NZ. *Mem:* AAAS; Inst Elec & Electronics Engrs; Int Commun Asn; Int Inst Commun. *Res:* Communications policy and planning; cable television; broadcast and common carrier policy; computer networks; information systems and services; communications engineering; public television; educational television; diffusion of innovation. *Mailing Add:* Sch Commun & Theater Temple Univ Philadelphia PA 19122

DORDICK, ISADORE, b Riga, Latvia, June 14, 11; US citizen; m 54. MEDICAL GEOGRAPHY. *Educ:* Univ Pa, BA, 33, MA, 37; Johns Hopkins Univ, PhD(geog), 52. *Prof Exp:* Res assoc physiol, Mt Sinai Hosp, 38-39, res ed, 40-42; librn, Libr Cong, 42-44; biologist, Off Quarter Master Gen War

Dept, 44-45; res analyst, Dept State, 45-47; climatologist, Am Meteorol Soc, 52-58; consult bioclimat, 58-62; CLIMATOLOGIST, AM METEOROL SOC, 62- *Concurrent Pos:* Consult climat, 62- *Mem:* Am Meteorol Soc; Asn Am Geog; Sigma Xi. *Res:* Bioclimatology with particular regard to the effect of climate on human work; medical climatology; applied climatology in business and engineering. *Mailing Add:* 900 N Lake Shore Dr Apt 1804 Chicago IL 60611

DORE-DUFFY, PAULA, b Hyannis, Mass, Feb 23, 48; m 72; c 3. NEUROIMMUNOLOGY. *Educ:* Simmons Col, BS, 70; La State Univ, PhD(microbiol), 76. *Prof Exp:* From asst prof to assoc prof med & neurol, Sch Med, Univ Conn, 78-88, chief, Div Neuroimmunol, & dir, Multiple Sclerosis Ctr, 82-88; PROF NEUROL, SCH MED, WAYNE STATE UNIV, 88-, CO-DIR, MULTIPLE SCLEROSIS CLIN & RES CTR & CHIEF, DIV NEUROIMMUNOL, DEPT NEUROL, 88- *Concurrent Pos:* Prin investr numerous grants, NIH & Multiple Sclerosis Soc; mem, NIH Study Sect NSP-A, 88-; vis summer scientist, Mt Desert Island Biol Labs, 85. *Mem:* Am Acad Neurol; Am Asn Immunologists; Am Fedn Clin Res; Soc Neurosci. *Res:* Role of prostaglandins and leukotrienes in immunoregulation; immunology of multiple sclerosis; monocyte activation antigens; blood brain-barrier-immune interactions. *Mailing Add:* Dept Neurol Sch Med Wayne State Univ 6E UHC 4201 St Antoine Detroit MI 48201

DOREMUS, ROBERT HEWARD, b Denver, Colo, Sept 16, 28; m 56; c 4. GLASS SCIENCE, FRACTURE. *Educ:* Univ Colo, BS, 50; Univ Ill, MS, 51, PhD(phys chem), 53; Cambridge Univ, PhD(phys chem), 56. *Prof Exp:* Phys chemist, res lab, Gen Elec Co, 55-71; NY STATE PROF GLASS & CERAMICS, RENSSELAER POLYTECH INST, 71- *Concurrent Pos:* Sabbatical, Univ Calif, Berkeley, 78-79. *Honors & Awards:* Purdy Award, Am Ceramic Soc. *Mem:* AAAS; fel Am Ceramic Soc. *Res:* Glass science; crystallization from solution; optical properties of metals; diffusion; precipitation in metals; ionic binding to polyelectrolytes; biomaterials. *Mailing Add:* Materials Sci Dept Rensselaer Polytech Inst Troy NY 12181

DOREN, DOUGLAS JAMES, b Richmond, Calif, Sept 30, 55; m 89; c 1. SURFACE SCIENCE, COMPUTATIONAL CHEMISTRY. *Educ:* Univ Calif, Berkeley, BS, 79; Harvard Univ, MS, 81, PhD(chem physics), 86. *Prof Exp:* Postdoctoral mem tech staff surface chem, AT&T Bell Labs, 86-88; ASST PROF PHYS CHEM, UNIV DEL, 88- *Mem:* Am Phys Soc; Am Chem Soc. *Res:* Theoretical and computational studies of chemical reactions at gas-solid and liquid-solid interfaces. *Mailing Add:* Dept Chem Univ Del Newark DE 19716

DORENBUSCH, WILLIAM EDWIN, b Hamilton, Ohio, Mar 14, 36; m 64. NUCLEAR PHYSICS. *Educ:* Univ Notre Dame, BS, 58, PhD(nuclear physics), 62. *Prof Exp:* Instr physics, Univ Notre Dame, 62-63; instr, Mass Inst Technol, 63-65, asst prof, 65-69; ASSOC PROF PHYSICS, WAYNE STATE UNIV, 69- *Mem:* Am Phys Soc. *Res:* Ion-solid interactions; charged particle spectroscopy with nuclear reactions. *Mailing Add:* Dept Physics Wayne State Univ Detroit MI 48202

DORENFELD, ADRIAN C, b Brooklyn, NY, Dec 16, 19; m 42; c 3. METALLURGICAL & MINING ENGINEERING. *Educ:* Columbia Univ, BS, 40, EM, 41. *Prof Exp:* Operator, Utah Copper Co, 41; metallurgist, US Vanadium Co, 41-42; mill supt, Callahan Zinc-Lead Co, 42; chief metallurgist, Mammoth-St Anthony Co, 42-43; metallurgist, Combined Metals Co, 43-45; foreman, Phelps Dodge Corp, 45-50; from asst prof to assoc prof mineral eng, Univ Ala, 51-54; sr mineral engr, C F Braun & Co, 54-56; gen mgr mines, Roberts & Assocs, 56-60; prof, 60-85, EMER PROF MINERAL ENG, UNIV MINN, MINNEAPOLIS, 85- *Concurrent Pos:* Consult, Pac Uranium Inc, 55-60, Israel Mining Industs, 56-62,Tec Trans Int Ltd, 75- & Harbridge House Inc, 77-; consult, Alcoa, Behre Dolbear & Co, Freeport McMoran Copper & various other co. *Mem:* Am Inst Mining, Metall & Petrol Engrs; Can Inst Mining & Metall. *Res:* Examination, management and design of mineral projects; application of statistics to mineral processing; mineral economics and feasibility studies. *Mailing Add:* 6120 Pine Circle Mound MN 55364

DORER, CASPER JOHN, b Cleveland, Ohio, Sept 25, 22; m 46; c 2. FUEL SCIENCE & TECHNOLOGY. *Educ:* Case Western Reserve Univ, BS, 43. *Prof Exp:* Res assoc res & develop chem, Case Western Reserve Univ, 44-48 & Carl F Prutton & Assocs, 48-50; vpres, Cleveland Indust Res, 50-72; prof mgr fuel additives, Lubrizol Corp, 72-77, proj mgr & admin asst res & develop, 77-81, mgr res fuel additives, 81-87; RETIRED. *Mem:* Am Chem Soc. *Res:* Additives for fuels and lubricants. *Mailing Add:* 4852 Fairlawn Dr Lyndhurst OH 44124-1121

DORER, FREDERIC EDMUND, b Cleveland, Ohio, Aug 24, 33; m 61; c 3. BIOCHEMISTRY, TOXICOLOGY. *Educ:* Western Reserve Univ, AB, 55, MS, 56; State Univ NY, PhD(biochem), 61. *Prof Exp:* Fel biochem, Sch Med, Univ Wis, 61-63; chemist, radioisotope serv, Vet Admin Hosp, 63-68; sr instr biochem, 69-73, ASST PROF BIOCHEM, CASE WESTERN RESERVE UNIV, 73-; RES CHEMIST, VET ADMIN HOSP, 68- *Mem:* Am Soc Biol Chemists; AAAS; Am Chem Soc; Am Asn Clin Chem. *Res:* Experimental renal hypertension; occurrence and metabolism of biologically active peptides. *Mailing Add:* Vet Admin Hosp Lab 10000 Brecksville Rd Brecksville OH 44141

DOREY, CHERYL KATHLEEN, b Wellsville, NY, Sept 7, 44; m 67. CELL BIOLOGY, HISTOCHEMISTRY. *Educ:* State Univ NY Buffalo, BA, 66; Univ Mass, MA, 68; Georgetown Univ, PhD(cell biol & histochem), 75. *Prof Exp:* Res asst electron microscopy, Litton Bionetics, Inc, 69-72, scientist histochem, 72-75; fel biochem & develop biol, Univ Southern Calif, 75-78; ASST PROF RES OPHTHAL, SCH DENT MED, SOUTHERN ILL UNIV, 78-; AT DOHENY EYE FOUND. *Mem:* Am Soc Cell Biol; Tissue Cult Asn; Am Histochem Soc; Int Asn Dent Res. *Res:* Control of the endothelial cell cycle by a cartilage factor; diabetic microangiopathy; osteogenesis. *Mailing Add:* Vitreo Retinal Biochem & Cell Biol Harvard Med Sch Eye Res Inst 20 Staniford St Boston MA 02114

DORF, MARTIN EDWARD, b Bound Brook, NJ, May 16, 44; m 76; c 2. CELLULAR IMMUNOLOGY. *Educ:* Rutgers Univ, BA, 66; Duke Univ, MA & PhD(immunogenetics), 72. *Hon Degrees:* MA, Harvard Univ, 84. *Prof Exp:* Exchange fel immunogenetics, Hosp St Louis, Paris, 71-72; res fel, 72-73, instr, 74-75, from asst prof to assoc prof, 75-83, PROF PATH, HARVARD MED SCH, 84- *Concurrent Pos:* Outstanding investr grant, Nat Cancer Inst, 85; ed, J Immunol. *Mem:* Am Asn Immunologists; Am Asn Pathologists; Transplantation Soc; AAAS. *Res:* Analysis of the mechanisms and genetic control of immune responses and immunoregulation with emphasis on the roles of the major histocompatibility and immunoglobulin gene complexes. *Mailing Add:* Dept Path Harvard Med Sch 200 Longwood Ave Boston MA 02115

DORF, RICHARD C, b New York, NY, Dec 27, 33; m 56; c 2. ELECTRICAL ENGINEERING. *Educ:* Clarkson Tech Univ, BSEE, 55; Univ Colo, MSEE, 57; US Naval Postgrad Sch, PhD(elec eng), 61. *Prof Exp:* Instr elec eng, Clarkson Tech Univ, 56-58; res assoc, Univ NMex, 58-59; instr, US Naval Postgrad Sch, 59-61; lectr, Univ Edinburgh, 61-62; assoc prof, US Naval Postgrad Sch, 62-63; from assoc prof to prof, Univ Santa Clara, 63-69, chmn dept, 63-69; dean col eng & & vpres educ serv, 69-72; dean extended learning, 72-81, PROF ELEC ENG, UNIV CALIF, DAVIS, 72-, PROF, GRAD SCH MGT, 81- *Mem:* Fel Inst Elec & Electronics Engrs; Am Soc Eng Educ; Sigma Xi. *Res:* Analysis and design of automatic control systems; design of digital control systems; engineering management. *Mailing Add:* Dept Elec & Comput Eng Univ Calif Davis CA 95616

DORFMAN, DONALD, b Bronx, NY, Mar 5, 34; m 54; c 4. FRESH WATER & MARINE BIOLOGY. *Educ:* Monmouth Col , BS, 66; Univ Conn, MS, 68; Rutgers Univ, PhD(environ sci), 70- *Prof Exp:* Res asst fisheries, Univ Conn, 66-68; res asst environ sci, Rutgers Univ, 68-70; ASSOC PROF BIOL, MONMOUTH COL, NJ, 70- *Concurrent Pos:* Chmn, Environ Assessment Coun, 73-; adj prof, Rutgers Univ, 74- *Mem:* Am Fisheries Soc; Sigma Xi. *Res:* Effect of lead on growth of brook trout; responses of anadromous fishes to increased temperature and decreased oxygen concentration; fish scrum proteins; heavy metals in fish tissues. *Mailing Add:* Dept Biol Monmouth Col West Long Branch NJ 07764

DORFMAN, HOWARD DAVID, b New York, NY, July 20, 28; m 52; c 3. PATHOLOGY. *Educ:* NY Univ, BA, 47; State Univ NY, MD, 51. *Prof Exp:* Intern, Maimonides Hosp Brooklyn, 51-52; resident path & microbiol, Mt Sinai Hosp, New York, 52-54; resident surg path, Columbia Presby Med Ctr, 54-58; pathologist & dir labs, Sharon & New Milford Hosps, Conn, 58-59; asst pathologist, Sinai Hosp Baltimore, 59-61, assoc pathologist, 62-64; dir labs & pathologist, Hosp Joint Dis & Med Ctr, 64-77; assoc prof path & orthop surg, Sch Med, Johns Hopkins Univ, 77-84, prof orthop path, 84-85; PROF ORTHOP SURG, PATH & RADIOL, ALBERT EINSTEIN COL MED, 85- *Concurrent Pos:* Asst surgery, Col Physicians & Surgeons, Columbia Univ, 57-58, asst prof, 65-67; asst prof, Med Sch, Johns Hopkins Univ, 63-64; staff pathologist, Johns Hopkins Hosp, 63-64; from assoc prof to prof clin path, Mt Sinai Sch Med, 67-74; mem staff, Sinai Hosp Baltimore, 74-; consult, Nat Cancer Inst. *Mem:* Orthop Res Soc; fel NY Acad Med; Int Acad Path; fel Am Soc Clin Path; Int Skeletal Soc (pres, 86-88). *Res:* Pathology of bone tumors and joint diseases with particular reference to rheumatoid athritis, neoplasms, secondary to pre-existing bone disease, osteoblastic tumors and vascular tumors of the bone; adamantinoma of long bones. *Mailing Add:* Montefiore Med Ctr 111 E 210th St New York NY 10467

DORFMAN, JAY ROBERT, b Pittsburgh, Pa, May 20, 37; m 84; c 3. THEORETICAL PHYSICS, STATISTICAL MECHANICS. *Educ:* Johns Hopkins Univ, BA, 57, PhD(physics), 61. *Prof Exp:* Res assoc physics, Rockefeller Inst, 61-64; from asst prof to assoc prof, dept physics & astron & Inst Fluid Dynamics & Applied Math, 64-72, dir, Inst Phys Sci & Technol, 83-85, PROF DEPT PHYSICS & ASTRON & INST PHYS SCI & TECHNOL, UNIV MD, COLLEGE PARK, 72-, DEAN, COL COMPUT, MATH & PHYS SCI, 85- *Concurrent Pos:* Vis assoc prof, Rockefeller Univ, 69-70; actg provost, Div Math, Phys Sci & Eng, Univ Md, 85-; consult, Los Alamos Sci Lab & Nat Bur Standards. *Mem:* Am Phys Soc. *Res:* Kinetic theory; theory of transport processes. *Mailing Add:* Dept Physics & Astron Univ Md College Park MD 20742

DORFMAN, LEON MONTE, b Winnipeg, Man, June 9, 22; nat US; m 48; c 3. CHEMICAL KINETICS, RADIATION CHEMISTRY. *Educ:* Univ Man, BSc, 44; Univ Toronto, MA, 45, PhD(chem), 47. *Prof Exp:* Fel, Univ Rochester, 47-48, instr chem, 48-50; res assoc, Knolls Atomic Power Lab, Gen Elec Co, 50-57; sr chemist, Argonne Nat Lab, 57-64; prof, 64-85, chmn dept, 68-77, EMER PROF CHEM, OHIO STATE UNIV, 85- *Concurrent Pos:* Consult, Argonne Nat Lab, 64-65 & Gen Dynamics Corp, 64-67; prof, Univ Toronto, 67; vis res scientist, Hebrew Univ, Jerusalem, 69; Guggenheim fel, 71-72; consult, Westinghouse Res Ctr, 74-76; mem adv rev comt, Radiation Lab, Notre Dame Univ, 79-83; mem vis comt, Chem Div, Brookhaven Nat Lab, 81-84. *Honors & Awards:* Spinks Lectr, Univ Sask, 81. *Mem:* Am Chem Soc; Am Phys Soc; Radiation Res Soc; Am Asn Univ Professors. *Res:* Chemical kinetics; fast reactions; radiation chemistry; spectroscopy of reactive transients. *Mailing Add:* 3234 Jefferson Ave Cincinnati OH 45220

DORFMAN, LESLIE JOSEPH, b Montreal Que, Can, Sept 11, 43. NEUROLOGY, CLINICAL NEUROPHYSIOLOGY. *Educ:* McGill Univ, BSc, 64; Albert Einstein Col Med, MD, 68. *Prof Exp:* Fel electromyography, Nat Hosp, Queen Square, London, 74; from instr to assoc prof, 73-90, PROF NEUROL, STANFORD UNIV SCH MED, 90- *Concurrent Pos:* Stanford dir, Lab Electromyography, Evoked Potential Lab, Multiple Sclerosis Clinic, Adult Neuromuscular Disorders Clinic, 74-; attend physician, Standford Univ Hosp, 74-; consult neurol, Palo Alto Vet Admin Med Ctr, 74-; prog med dir, Rehab Res & Develop Ctr, Palo Alto, 79-90; actg chmn, Stanford Dept Neurol & Neurol Sci, 89- *Mem:* AAAS; Am Acad Neurol; Can Neurol Soc; Am Asn Electrodiagnostic Med; Am EEG Soc. *Res:* Application of signal processing techniques to clinical electrophysiology for diagnosis of neurological disorders. *Mailing Add:* Dept Neurol Rm H3160 Stanford Univ Med Ctr Stanford CA 94305

DORFMAN, MYRON HERBERT, b Shreveport, La, July 3, 27; c 2. GEOTHERMAL ENGINEERING, PETROLEUM GEOLOGY. *Educ:* Univ Tex, Austin, BS, 50, MS, 74, PhD(petrol eng), 75. *Prof Exp:* Engr & geologist, Sklar Oil Co, 50-56, mgr prod exp, 56-57, vpres, 57-59; owner, Dorfman Oil Properties, 59-71; from asst prof to prof petrol eng, 74-80, H B Harkins prof, 80, W A Moncrief Centennial Chair, 83, DIR GEOTHERMAL STUDIES, CTR ENERGY STUDIES, UNIV TEX, AUSTIN, 74-, CHMN, DEPT PETROL ENG, 78- *Concurrent Pos:* Prin investr, Geopressured Geothermal Energy Prog, Dept Energy, 74-81; mem steering comt, Geothermal Adv Comt, Dept Energy, 75-, Geothermal Reservoir Mgt Prog, 76- & Geothermal Well Logging Prog, Sandia Nat Lab, 77- *Mem:* Nat Acad Sci; AAAS; Am Asn Petrol Geologists; fel Geol Soc Am; Soc Petrol Engrs. *Res:* Geopressured geothermal energy; well-logging in hostile environments underground. *Mailing Add:* Univ Tex PEB Austin TX 78712

DORFMAN, RONALD F, PATHOLOGY. *Prof Exp:* Prof path, 68-91, CO-DIR SURG PATH, STANFORD UNIV MED CTR, 91- *Mailing Add:* Dept Path L-229 Stanford Univ Med Ctr 300 Pasteur Dr Stanford CA 94305

DORGAN, WILLIAM JOSEPH, b Davenport, Iowa, Dec 27, 38; m 65; c 2. DEVELOPMENTAL BIOLOGY, CELL BIOLOGY. *Educ:* St Ambrose Col, BA, 61; Creighton Univ, MS, 63; Univ Colo, PhD(anat), 68. *Prof Exp:* From asst prof to assoc prof zool, 68-80, ASSOC PROF ANAT, MONT STATE UNIV, 80- *Mem:* AAAS; Soc Develop Biol; Am Asn Univ Prof. *Res:* Programmed death in embryonic development and abnormal development; programmed death in rat placental giant cells in vitro; studies on the photodynamic action of vital dyes in embryonic systems. *Mailing Add:* Dept Biol Mont State Univ Bozeman MT 59717

DORHEIM, FREDRICK HOUGE, b Bedger, Iowa, Nov 12, 12; m 39; c 1. ECONOMIC GEOLOGY. *Educ:* Iowa State Col, BS, 38, MS, 50. *Prof Exp:* Chief geologist, Iowa State Hwy Comn, 46-50 & B L Anderson, Inc, 50-56; geologist, 56-69, chief econ geol, 56-75, Chief Geologist, Iowa Geol Surv, 75-; RETIRED. *Concurrent Pos:* Consult, Indust Minerals & Sanit Landfills, 57-58; chmn, Forum Geol Indust Minerals, 72. *Mem:* Am Inst Prof Geol Sci; Am Asn Petrol Geologists. *Res:* Geophysical study of sand and gravel resources in northwest Iowa; geophysical study of area of gypsum occurrence in the Fort Dodge area; geology of Floyd County, Iowa. *Mailing Add:* 3322 Hanover Ct Iowa City IA 52240

DORIAN, WILLIAM D, b Timisoara, Romania, Mar 4, 21; Can citizen; m 44; c 2. INTERNAL MEDICINE. *Educ:* Univ Cluj, Romania, MD, 48; FRCPC(c), 70. *Prof Exp:* EXEC DIR MED RES, MERCK FROSST CAN, INC, 73- *Concurrent Pos:* Staff physician, St Mary's Hosp, Montreal, 66-86; affil, Montreal Gen Hosp, 67-86; lectr, Dept Med, McGill Univ, 70-86; staff, Wellesley Hosp, Toronto, 87- *Mem:* NY Acad Sci; Am Acad Allery. *Res:* Clinical pharmacology; cardiovascular, anti-inflammatory, antibiotic agents. *Mailing Add:* 3303 Don Mills Rd Suite 1003 Willowdale ON M2J 4T6 Can

DORITY, GUY HIRAM, b Canandaigua, NY, Jan 2, 33. ORGANIC CHEMISTRY. *Educ:* Oglethorpe Univ, BS, 54; Univ NC, MA, 59; Univ Hawaii, PhD(chem), 65. *Prof Exp:* From instr to asst prof chem, 63-71, ASSOC PROF CHEM, UNIV HAWAII, HILO, 71- *Concurrent Pos:* Vis scientist, Nat Biol Inst, Indonesia, 70-71; leader, training course natural prod chem, SE Asian Ministries of Educ Orgn, Bogor, Indonesia, 71. *Mem:* Am Chem Soc; Sigma Xi. *Res:* Aromatic and heterocyclic fluorine compounds; natural products from Polynesian and Southeast Asian medicinal plants. *Mailing Add:* 523 W Lanikaula St Hilo HI 96720

DORKO, ERNEST A, b Detroit, Mich, Sept 16, 36; m 71; c 1. ORGANIC CHEMISTRY, PHYSICAL CHEMISTRY. *Educ:* Univ Detroit, BChE, 59; Univ Chicago, MS, 61, PhD(org chem), 64. *Prof Exp:* Res chemist, Phys Sci Lab, US Army Missile Command, Redstone Arsenal, Ala, 64-67; from asst prof to assoc prof, 67-77, prof chem, Air Force Inst Technol, 77-87; RES CHEMIST, PHILLIPS LAB, 87- *Concurrent Pos:* Consult, Wright-Patterson AFB, 68- & Wright Aeronaut Labs, 75-; vis scientist, Air Force Weapons Lab, Kirtland AFB, NMex, 85-86. *Mem:* AAAS; Am Chem Soc; Sigma Xi. *Res:* Synthesis of small ring organic compounds; infrared and normal coordinate analysis; shock-tube kinetics of pollution reactions; chemiluminescence reactions in flow tubes; laser dyes; laser induced fluorescence spectroscopy; laser spectroscopy. *Mailing Add:* PL/ARDJ Phillips Lab Kirtland AFB NM 87117-6008

DORMAAR, JOHAN FREDERIK, b Djakarta, Indonesia, Feb 16, 30; Can citizen; div; c 3. SOIL CHEMISTRY, RANGE SCIENCE. *Educ:* Univ Toronto, BSA, 57, MSA, 58; Univ Alta, PhD(soil chem), 61. *Prof Exp:* Asst soil chem, Univ Alta, 58-61, lectr, 61-62; res scientist, Agr Res Sta, Can Dept Agr, 62-75, SR RES SCIENTIST, RES STA, AGR CAN, 75- *Concurrent Pos:* mem NSERC Plant Biol Grant Comt, 87-90; adj prof, Univ Sask, 86-, Univ Lethbridge, 90- *Mem:* Soil Sci Soc Am; Can Soc Soil Sci(pres, 86-87); Can Quaternary Asn; Am Quaternary Asn; Int Soil Sci Soc. *Res:* Organic matter of chernozemic soils, various grass species, and man-induced pressures; lignin chemistry as related to soil organic matter; effect of pseudo-gley on soil organic matter; palaeosols; rhizosphere chemistry; soil-root interactions; range ecology; allelo chemistry. *Mailing Add:* Soil Sci Sect Res Sta Agr Can Lethbridge AB T1J 4B1 Can

DORMAN, CLIVE EDGAR, b Granite City, Ill, Nov 30, 42; m 67; c 2. PHYSICAL OCEANOGRAPHY, METEOROLOGY. *Educ:* Univ Calif, Riverside, BA, 65; Ore State Univ, MS, 72, PhD(phys oceanog), 74. *Prof Exp:* Assoc prof oceanog, 74-80, assoc prof, 80-84, PROF GEOL, SAN DIEGO STATE UNIV, 84- *Mem:* Am Meteorol Soc; Am Geophys Union. *Res:* Air-sea interaction; climatology; coastal oceanography. *Mailing Add:* Dept Geol Sci San Diego State Univ San Diego CA 92182

DORMAN, CRAIG EMERY, b Cambridge, Mass, Aug 27, 40; m 62; c 3. OCEANOGRAPHY. *Educ:* Dartmouth Col, Hanover, NH, BA, 62; Naval Postgrad Sch, Monterey, Calif, MS, 69; Mass Inst Technol, Cambridge, PhD(phys oceanog), 72. *Prof Exp:* Rear adm, USN, 62-89; DIR, WOODS HOLE OCEANOG INST, 89- *Concurrent Pos:* Adv, Dir Naval Intel, Wash, DC, 89-; gov, Joint Oceanog Insts, Wash, DC, 89-; dir, Maritrans, Philadelphia, 91. *Honors & Awards:* Legion of Merit, USN, 74, 82 & 88. *Mailing Add:* Woods Hole Oceanog Inst Woods Hole MA 02543

DORMAN, DOUGLAS EARL, b Burbank, Calif, Mar 9, 40; m 68; c 2. SPECTROCHEMISTRY, COMPUTATIONAL CHEMISTRY. *Educ:* Univ Calif, Los Angeles, BS, 63; Brandeis Univ, MA, 65, PhD(org chem), 67. *Prof Exp:* NIH fel biogenesis, Univ Sussex, 67-68; NIH fel nuclear magnetic resonance, Calif Inst Technol, 68-71; mem tech staff, Bell Labs, 71-73; sr phys chemist nuclear magnetic resonance, 73-75, RES SCIENTIST, MOLECULAR STRUCT, LILLY RES LABS, 75- *Mem:* Am Chem Soc; The Chem Soc. *Res:* Determination of molecular structure by physical and chemical methods. *Mailing Add:* MCM 525 Lilly Res Lab Indianapolis IN 46206

DORMAN, HENRY JAMES, b Chicago, Ill, Mar 21, 28; m 57; c 2. GEOPHYSICS. *Educ:* Carleton Col, BA, 49; Northwestern Univ, MS, 51; Columbia Univ, PhD(geophys), 61. *Prof Exp:* Res scientist, Lamont-Doherty Geol Observ, Columbia Univ, 60-63, sr res assoc, 63-72, asst dir, 65-72, lectr geol, Univ, 66-72; prof geophys, Univ Tex Med Br, 72-74; res scientist, Marine Sci Inst, Univ Tex, Galveston, 74-81, prof geol sci, 75-81, prof marine sci, 76-81; sr res spec, Exxon Prod Res Co, 81-83, res assoc, 83-86; RES PROF GEOL SCI, MEMPHIS STATE UNIV, 87-, ASSOC DIR, CTR EARTHQUAKE RES & INFO, 87- *Concurrent Pos:* Mem, Panel Solid Earth Probs, Geophys Res Bd, Nat Acad Sci, 62-63; lectr geol, Univ Wis, 63; mem, Village Planning Bd, Pomona, NY, 67-72; consult, Consol Edison Co, New York, 68-; founding mem & dir, Palisades Geophys Inst, 68; mem, Mayor Lindsay's Oceanog Adv Comt, 69-73; adj prof geol, Rice Univ, 74-78. *Mem:* Soc Explor Geophysicists; AAAS; Seismol Soc Am; Am Geophys Union; fel Geol Soc Am. *Res:* Structural geology; exploration seismology; long-period surface waves and free oscillations; model seismology; seismic inverse problem; crust and mantle structure; microearthquake studies; seismicity; lunar seismology; numerical analysis and computer applications. *Mailing Add:* 142 Walnut Creek Memphis TN 38018

DORMAN, LEROY MYRON, b Virginia, Minn, Oct 15, 38; m 59; c 3. GEOPHYSICS. *Educ:* Ga Inst Technol, BS, 60; Univ Wis, MS, 69, PhD(geophys), 70. *Prof Exp:* Asst res physicist, Ga Inst Technol, 63-66; res asst geophys, Univ Wis, 66-69; Carnegie fel, Carnegie Inst, 69-71; res geophysicist, Atlantic Oceanog & Meteorol Lab, Nat Oceanic & Atmospheric Admin, 71-73; asst res geophysicist & lectr, Univ Calif, San Diego, 73-75, assoc res geophysicist & lectr, 75-77, assoc prof, 78-82, chair, geol res div, 83-85, chair grad dept, 85-88, PROF GEOPHYS, SCRIPPS INST OCEANOG, UNIV CALIF, SAN DIEGO, 82- *Mem:* Am Geophys Union; Seismol Soc Am; Soc Explor Geophysicists. *Res:* Theoretical and experimental seismology; isostasy; marine geophysics. *Mailing Add:* Scripps Inst Oceanog 0215 Univ Calif San Diego La Jolla CA 92093-0215

DORMAN, LINNEAUS CUTHBERT, b Orangeburg, SC, June 28, 35; m 58; c 2. BIOPOLYMERS, THERMOPLASTICS. *Educ:* Bradley Univ, BS, 56; Ind Univ, PhD(org chem), 61. *Hon Degrees:* DSc, Saginaw Valley State Univ, 88. *Prof Exp:* Res chemist, Cent Res-Plastics Lab, Dow Chem Co, 60-68, sr res chemist, 68-72, res specialist, 72-76, sr res specialist, 76-79, res assoc, Polymer Res Lab, 79-83, ASSOC SCIENTIST, ADVAN POLY MAT LAB, DOW CHEM CO, 83- *Honors & Awards:* Bond Award, Am Oil Chem Soc, 60. *Mem:* AAAS; Am Chem Soc; Sigma Xi; Nar Orgn Black Chemists & Chem Engrs; Soc Adv Mat & Proc Eng. *Res:* Organic synthesis; pharmaceutical chemistry; synthesis of heterocyclic compounds; diagnostic reagents; peptide synthesis; materials research. *Mailing Add:* Cent Res-Advan Polymeric Syst Lab 1702 Bldg Dow Chem Co Midland MI 48674

DORMAN, ROBERT VINCENT, b Niles, Mich, July 2, 49; m 76; c 3. BIOCHEMISTRY. *Educ:* Villanova Univ, BS, 71; Ohio State Univ, PhD(physiol chem), 76. *Prof Exp:* Res chemist neurobiol, Vet Admin, 76-78; res assoc neurochem, Ohio State Univ, 78-81; asst prof, Drexel Univ, 81-83; asst prof, 83-88, ASSOC PROF BIOL, KENT STATE UNIV, 88- *Concurrent Pos:* NIH fel, 77-78. *Mem:* Am Soc Neurochem; AAAS. *Res:* Role of membrane lipids in the functioning of excitable tissue; the involvement of phospholipid catabolism and arachidonic acid accumulation and conversion to eicosonoids in the mechanisms of neurotransmissions and trauma-induced central nervous system pathology. *Mailing Add:* Dept Biol Sci Kent State Univ Kent OH 44242

DORMANT, LEON M, enhanced oil recovery, for more information see previous edition

DORMER, KENNETH JOHN, b Ashland, Pa, Mar 10, 44; m 66; c 3. CARDIOVASCULAR PHYSIOLOGY, NEUROPHYSIOLOGY. *Educ:* Cornell Univ, BS, 66; Univ Calif, Los Angeles, MS, 69, PhD(biol), 74. *Prof Exp:* Fel cardiovasc control, Marine Biomed Inst, Univ Tex Med Br, 74-77; PROF PHYSIOL, UNIV OKLA HEALTH SCI CTR, 90-, CLIN NEUROPHYSIOLOGIST, COCHLEAR IMPLANT CLIN, VPRES, HOUGH EAR INST, 79- *Concurrent Pos:* Nat Heart, Lung & Blood Inst fel, Univ Tex Med Br, 74-77; NIH grants, 78-81 & 81-84. *Mem:* Am Physiol Soc; Am Heart Asn; Am Sci Affil; Soc Neurosci; Sigma Xi. *Res:* Neural control of the cardiovascular system; mechanisms of hypertension; auditory neuroprostheses (cochlear implant and implantable hearing devices). *Mailing Add:* Dept Physiol & Biophys PO Box 26901 Oklahoma City OK 73190

DORN, CHARLES RICHARD, b London, Ohio, June 12, 33; m 64; c 3. VETERINARY PREVENTIVE MEDICINE. *Educ:* Ohio State Univ, DVM, 57; Harvard Univ, MPH, 62; Am Col Vet Prep Med, dipl. *Prof Exp:* Staff vet, Stark Animal Hosp, Canton, Ohio, 57-58; vet inspector, Cincinnati Health

Dept, 60-61; USPHS trainee, 61-62; res specialist cancer, Calif State Dept Pub Health, 62-68; prof vet microbiol & community health & med pract, Univ Mo-Columbia, 68-75; chmn dept, 75-87, PROF, VET PREV MED, OHIO STATE UNIV, 75- *Concurrent Pos:* Lectr vet med, Univ Calif, Davis, 67-68; consult, Calif State Dept Pub Health, 68-70 & WHO, India, 75; vis scientist epidemiol, Nat Cancer Inst, 75-76; Fogarty Sr Int Fel, 87-88. *Mem:* Am Vet Med Asn; fel Am Pub Health Asn; Conf Pub Health Vet; Asn Teachers Prev Med; fel Am Col Epidemiol; Int Soc Vet Epid & Econ. *Res:* Heavy metal environmental contamination of foods; epidemiology and microbiol of Salmonella and other zoonotic infections. *Mailing Add:* Dept Vet Prev Med Ohio State Univ Col Vet Med Columbus OH 43210-1092

DORN, DAVID W, b Detroit, Mich, June 25, 30; m 54; c 4. THEORETICAL NUCLEAR PHYSICS, RESEARCH ADMINISTRATION. *Educ:* Purdue Univ, BS, 52, PhD(physics), 59. *Prof Exp:* From res physicist to group leader physicist, 59-69, staff asst to assoc dir nuclear design, 69-72, prog mgr, Environ Studies Group, 72-75, prog mgr, Technol Appln Group, 75-78, prog mgr energy & resources planning group, 78-81, PRIN DEP TREATY VERIFICATION PROJ, LAWRENCE LIVERMORE NAT LAB, 88- *Concurrent Pos:* Mem tech subcomt, explosives tagging, alcohol, tobacco & firearms, US Dept Treas, 73-75; mem solar technician training prog adv comt, Univ Calif, 78-84; sr scientist magnetic fusion, 81-84; mem US deleg conf on disarmament, Geneva, 85; tech adv, US Dept Energy, Washington, DC, 85-88. *Mem:* Am Phys Soc; Sigma Xi. *Res:* Utilization of results from scientific and engineering research; assessment of possible outcomes of adoption of technology; national and regional energy and resource problems; long range stategic planning; arms control/disarmament measures. *Mailing Add:* 5165 Lenore Ave Livermore CA 94550-2380

DORN, GORDON LEE, b Chicago, Ill, June 8, 37; m 59; c 2. MICROBIOLOGY. *Educ:* Purdue Univ, BS, 58, MS, 60, PhD(genetics), 61. *Prof Exp:* NSF fel, 61-63; univ res fel genetics, Albert Einstein Col Med, 63, asst prof, 64-68; chmn, Dept Microbiol, Baylor Univ, 68-69; CHMN, DEPT MICROBIOL & DIR, CLIN MICROBIOL LABS, WADLEY INST MOLECULAR MED, 69- *Concurrent Pos:* Adj prof, NTex State Univ, 75-; pres, Dorn Microbiol Assocs Dallas, 78- *Honors & Awards:* IR 100 Award, 82. *Mem:* AAAS; Am Genetic Asn; Genetics Soc Am; Am Soc Microbiol. *Res:* Interferon production, diagnostics test in the areas of bacteriology, mycology and virology; blood, sputum, urine, throat cultures and antibiotic susceptibility testing. *Mailing Add:* 9000 Harry Hines Dallas TX 75235

DORN, RONALD I, b New York, NY, Feb 22, 58; m 80; c 2. PHYSICAL GEOGRAPHY, CLIMATIC CHANGE. *Educ:* Univ Calif, Berkeley, BA, 80, MA, 82; Univ Calif, Los Angeles, PhD(geog), 85. *Prof Exp:* Asst prof geog, Tex Tech Univ, 85-88; asst prof, 88-90, ASSOC PROF GEOG, ARIZ STATE UNIV, 90- *Concurrent Pos:* Pres young investr award, Nat Sci Found, 87-92; mem, Am Geomorphol Field Group & Brit Geomorphol Field Group. *Honors & Awards:* Kirk Bryan Award, Geol Soc Am, 86; G K Gilbert Award, Asn Am Geogr, 88; Pres Young Investr Award, NSF, 87-92. *Mem:* AAAS; Asn Am Geogr; Am Quat Asn; Geol Soc Am; Am Geophys Union. *Res:* Geomorphology and physical geography; reconstructing how the earth's physical geography has changed over time; dating land forms in deserts and relating these age-determinations to a general theory of landscape change; dating rock art to assist in critical reconstructions. *Mailing Add:* Geog Dept Ariz State Univ Tempe AZ 85287-0104

DORN, WILLIAM S, b Pittsburgh, Pa, July 12, 28; m 52; c 3. COMPUTER EDUCATION, NUMERICAL METHODS. *Educ:* Carnegie Inst Technol, BS, 51, PhD(math), 55. *Prof Exp:* Mgr mech & eng syst comput, Gen Elec Co, 55-57; res scientist, NY Univ, 57-59; mathematician, IBM Corp, 59-65, mgr comput ctr, 65-67, asst dir math sci, 67-68; chmn dept, 74-79, PROF MATH, UNIV DENVER, 68-, DEAN, 87- *Concurrent Pos:* Fulbright scholar, Comn on Int Exchange Persons, 72-73; vis prof comput sci, Open Univ, Eng, 79-; Ed-in-chief, Comput Surveys, 68-72. *Mem:* Fel AAAS; Asn Comput Mach; Soc Indust & Appl Math (secy, 64); Math Asn Am. *Res:* Educational technology; digital computers and their application; numerical analysis; computer-assisted instruction. *Mailing Add:* Dept Math Univ Denver Denver CO 80208

DORNBUSH, RHEA L, b New York, NY. NEUROPSYCHOLOGY. *Educ:* Queen's Col, BA, 62, MA, 63; City Univ New York, PhD(exp psychol), 67; Columbia Univ, Sch Pub Health, MPH, 81. *Prof Exp:* Lectr, teaching fel & res assoc psychol, Queen's Col, City Univ New York, 63-65; asst prof psychol, Rutgers Univ, 65-68; from asst prof to assoc prof psychiat, New York Med Col, 68-76; sr res scientist, Reproductive Biol Res Found, 76-78; assoc prof psychiat, 78-80, assoc dir, Div Biol Psychiat, 78-80, PROF PSYCHIAT & DIR GRAD PROG BEHAV SCI, NY MED COL, 80- *Concurrent Pos:* Clin lectr med psychol, Dept Psychiat, Washington Univ, 76-78. *Mem:* AAAS; Soc Biol Psychiat; Am Psychol Asn; NY Acad Med; NY Acad Sci; Clin Trials Psychiat Drugs. *Res:* Closed head injury and memory deficits, neuropsychological aspects of neurological disorders; selection factors in medical school admissions; research design, methodology and statistical analysis in medical research. *Mailing Add:* NY Med Col Dept Psychiat Valhalla NY 10595

DORNE, ARTHUR, b Philadelphia, Pa, Apr 6, 17; m 39; c 1. ENGINEERING. *Educ:* Univ Pa, BA, 38, BS, 39. *Prof Exp:* Radio engr, Glenn D Gillett, consult radio engr, Washington, DC, 40-42; engr, Fed Tel & Tel, NJ, 42-43; antenna group leader & res assoc, Radio Res Lab, Harvard Univ, 43-45; supvr engr antenna group, Airborne Inst Lab, NY, 45-47; pres, Dorne & Margolin, Inc, 47-69 & Granger Asn, 69-70; PRES, DORNE CONSULT, 71- *Mem:* AAAS; fel Inst Elec & Electronics Engrs; Mineral Soc Am. *Res:* Telecommunications; radio navigation; antennas. *Mailing Add:* 1668 S Forge Mountain Dr Valley Forge PA 19481

DORNER, ROBERT WILHELM, b Bern, Switz, Oct 22, 24; nat US; div; c 2. BIOCHEMISTRY, IMMUNOLOGY. *Educ:* Univ Lausanne, dipl, 48; Univ Calif, PhD(biochem), 53. *Prof Exp:* Asst biochem, Univ Calif, 49-51, asst virus lab, 51-53; res assoc zool, Univ Southern Calif, 53-54; jr res botanist, Univ Calif, Los Angeles, 55-56, asst res botanist, 56-57; asst insect toxicol entom, Citrus Exp Sta, Univ Calif, Riverside, 57-62; from asst prof to prof, 62-89, EMER PROF BIOCHEM & INTERNAL MED, ST LOUIS UNIV, 90- *Concurrent Pos:* Spec investr, Arthritis Found, 64-67; mem bd dir, Nat Registry Clin Chem, 82-90. *Mem:* AAAS; Am Asn Clin Chem; Am Chem Soc; Am Rheumatism Asn; fel Nat Acad Clin Biochem; Am Soc Biochem & Molecular Biol. *Res:* Biochemistry of connective tissues; immunology of rheumatoid arthritis. *Mailing Add:* Rheumatol Div Sch Med St Louis Univ St Louis MO 63104

DORNETTE, WILLIAM HENRY LUEDERS, b Cincinnati, Ohio, June 22, 22; m 45, 84; c 2. ANESTHESIOLOGY. *Educ:* Univ Cincinnati, BS, 44, MD, 46, JD, 69. *Hon Degrees:* JD, Univ Cincinnati, 69. *Prof Exp:* Fel anesthesiol, Med Ctr, Georgetown Univ, 49-51; from instr to asst prof anesthesiol, Univ Wis, 52-55; asst prof, Univ Calif, Los Angeles, 55-57; prof & chmn dept, Col Med, Univ Tenn, Memphis, 58-65; chief anesthesiol, Vet Admin Hosp, Cincinnati, Ohio, 66-71; dir educ, Div Anesthesiol, Cleveland Clin, 72-76; consult anesthesia & hosp safety, 76-81; mem staff, Dept Legal Med, Armed Forces Inst Path, 82-89; LAW MED SPECIALIST, 90- *Concurrent Pos:* Anesthesiologist-in-chief, John Gaston Hosp; asst clin prof, Col Med, Univ Cincinnati, 65-71, assoc clin prof med jurisprudence, 73-80; ed, J Legal Med, 72-76; adj prof, Anesthesiol, Sch Med, Univ Md 84- *Mem:* Fel Am Col Legal Med; Nat Fire Protection Asn; Int Anesthesia Res Soc; Am Bar Asn. *Res:* Instrumentation in anesthesiology; pathology of anesthetic complications; fire and electrical safety in hospitals; legal problems related to medicine; hospital and medical risk management; HIV Infections, AIDS and the law. *Mailing Add:* 9716 Hill St Kensington MD 20895-3133

DORNFELD, DAVID ALAN, b Horicon, Wis, Aug 3, 49; m 76. INTELLIGENT SENSORS, COMPUTER AIDED MANUFACTURING. *Educ:* Univ Wis-Madison, BS, 72, MS, 73, PhD(mech eng), 76. *Prof Exp:* Asst prof systs design, Univ Wis-Milwaukee, 76-77; from asst prof to assoc prof mfg & mech eng, 77-89, vchair & instr mech eng, 87-89, PROF MFG & MECH ENG, UNIV CALIF, BERKELEY, 89-, DIR, ENG SYSTS RES CTR, 89- *Concurrent Pos:* Dir res & assoc, Ecole Nat Superieure des Mines, Paris, France, 83-84; ed, J Eng Indust, Am Soc Mech Eng Transactions, 86-91; mem, Sci Comt, NAm Mfg Res Inst, Soc Mgr Engrs, 86-, pres, 90-91; mem, Design Mfg Div Adv Bd, NSF, 90-91, Comt Unit Processes, Nat Res Bd, Nat Asn Eng, 91-94; consult, var co & law firms. *Honors & Awards:* Outstanding Young Mfg Engr Award, Soc Mfg Engrs, 82; Blackall Machine Tool & Gage Award, Am Soc Mech Engrs, 86. *Mem:* Am Soc Mech Engrs; Am Welding Soc; Soc Mfg Engrs; Japan Soc Precision Eng; Inst Elec & Electronics Engrs. *Res:* Manufacturing automation including process modelling, sensors and signal processing for intelligent manufacturing; machine tool monitoring; intelligent sensor development; sensor feedback for precision machining. *Mailing Add:* Dept Mech Eng Univ Calif Berkeley CA 94720

DORNFEST, BURTON S, b New York, NY, Oct 31, 30; m 54; c 2. HEMATOLOGY, ANATOMY. *Educ:* NY Univ, BA, 52, MS, 54, PhD(physiol), 60. *Prof Exp:* Mem staff biostatist, Mem Hosp Cancer & Allied Dis, New York, 52-53; asst biol, NY Univ, 53-54, 56-58, instr gen sci, 58-63; instr anat, NY Med Col, 63-64; from instr to asst prof, 64-73, ASSOC PROF ANAT, STATE UNIV NY HEALTH SCI CTR, 73- *Concurrent Pos:* Fel, NIH, Nat Heart Inst, 58-60, 61-63 & Leukemia Soc, 60-61; tech collabr, Dept Biol, Brookhaven Nat Lab, 59-60; res grant, NIH, Nat Inst Arthritis & Metab Dis, 64-71, Nat Cancer Inst, 73-75 & Mildred Werner League Cancer Res, 76-77; from vis assoc prof to adj prof anat, City Univ NY, 74-; from adj assoc prof to adj prof hemat, Sch Health Sci, Hunter Col, 78-82; vis prof anat, New York Med Col, 82, 83 & 85; co-prin investr, NIH res grant, Heart & Lung Inst, 82-85; adj prof anat, Touro Col Ctr Biomed Educ, 84-88; adj prof hemat, Sch Health Sci, Hunter Col, 90-; adj prof anat, Einstein Med Col, 91- *Mem:* AAAS; NY Acad Sci; Am Soc Hemat; Am Asn Anat; Reticuloendothelial Soc; Int Soc Exp Hemat; Am Asn Clin Anatomists; Harvey Soc. *Res:* Problems relating to reticuloendothelial system function and hemopoiesis; blood cell production, release and destruction in the rat; role of the liver in the production of extrarenal erythropoietin; studies on immune activation associated with phenylhydrazine-induced anemia in the rat. *Mailing Add:* Dept Anat & Cell Biol State Univ NY Health Sci Ctr Brooklyn NY 11203-2098

DORNHOFF, LARRY LEE, b Minden, Nebr, Apr 13, 42; m 68; c 2. MATHEMATICS, COMPUTER-BASED INSTRUCTION. *Educ:* Univ Nebr, Lincoln, BS, 62; Univ Chicago, MS, 63, PhD(math), 66. *Prof Exp:* Instr math, Yale Univ, 66-68; asst prof, 68-74, ASSOC PROF MATH, UNIV ILL, URBANA, 74- *Mem:* Math Asn Am. *Res:* Finite group theory; group representations; computer-aided mathematics instruction. *Mailing Add:* Dept Math Univ Ill Urbana IL 61801

DORNING, JOHN JOSEPH, b Bronx, NY, Apr 17, 38; m 63; c 3. MATHEMATICAL PHYSICS, NUCLEAR ENGINEERING. *Educ:* US Merchant Marine Acad, BS, 59; Columbia Univ, MS, 63, PhD(nuclear sci & eng), 67. *Prof Exp:* Marine engr, Merchant Marine, 60-62; fel nuclear sci & eng, AEC, 63-66; asst physicist reactor theory, Brookhaven Nat Lab, 67-69, assoc physicist & group leader, 69-70; assoc prof nuclear eng, Univ Ill, Urbana, 70-75, prof, 75-84; WHITNEY STONE PROF, NUCLEAR ENG & ENG PHYSICS, THORNTON HALL, UNIV VA, CHARLOTTESVILLE, 84- *Concurrent Pos:* Consult, Brookhaven Nat Lab, Argonne Nat Lab, Los Alamos Nat Lab & Sci Appln, Inc, Schlumberger-Doll Res, 74-; Nat Res Coun vis prof math physics, Ital Nat Res Coun, 75-76, 81, 87; physicist plasma physics, Lawrence Livermore Lab, 77-78. *Honors & Awards:* Mark Mills Award, Am Nuclear Soc, 67; Glen Murphy Award, Am Soc Engr Educ, 88; Ernest O Lawrence Award, US Dept Energy, 90. *Mem:* Fel Am Nuclear Soc; fel Am Phys Soc; Soc Indust & Appl Math; fel AAAS.

Res: Fission reactor theory, neutron transport theory, fission reactor kinetics, kinetic theory of gases, plasma physics, computational and numerical methods for fluid flow and particle transport; nonlinear dynamics and chaos. *Mailing Add:* Thornton Hall Univ Va Charlottesville VA 22903-2442

DORNY, C NELSON, b Washington, DC, Jan 20, 37; m 60; c 5. ELECTRICAL & SYSTEMS ENGINEERING. *Educ:* Brigham Young Univ, BES, 61; Stanford Univ, MSEE, 62, PhD(elec eng), 65. *Hon Degrees:* MA, Univ Pa, 76. *Prof Exp:* Engr, Westinghouse Res Labs, 63-64; from asst prof to assoc prof elec eng, 65-79, assoc dean, 79-80, chmn dept, 80-86, PROF SYSTS ENG, UNIV PA, 79- *Concurrent Pos:* White House fel, spec asst to Secy, USDA, 69-70; NSF res initiation grant, 67-69; consult, Westinghouse Res Labs, Pa, 66-, Hewlett-Packard, 75-76, BDS Technol, 78-79, & Interspec, Inc, 78-; assoc dir, Valley Forge Res Ctr, Univ Pa, 75-80. *Mem:* Sigma Xi; Systems, Man & Cybernetics Soc (vpres, 74-76); Inst Elec & Electronics Engrs; Am Soc Eng Educ. *Res:* System integration; telecommunication networks; robotics; high resolution radar and ultrasonic imaging. *Mailing Add:* 275 Towne/6315 Univ Pa Philadelphia PA 19104-6315

DOROSHOW, JAMES HALPERN, b Lynwood, Calif, Aug 19, 48; m 78; c 1. MEDICINE, PHARMACOLOGY. *Educ:* Harvard Univ, AB, 69; Harvard Med Sch, MD, 73. *Prof Exp:* Intern med, Mass Gen Hosp, 73-74, resident, 74-75; clin assoc oncol & clin pharmacol, Nat Cancer Inst, 75-78; asst prof med, Univ Southern Calif, 78-81; head, Pharmacol Sect, 81-83, DIR, DEPT MED ONCOL, CITY OF HOPE CANCER CTR, 83- *Concurrent Pos:* Chmn local arrangements, Am Soc Clin Oncol, 86. *Mem:* Am Asn Cancer Res; Am Soc Clin Oncol; fel Am Col Physicians; Am Soc Pharmacol & Exp Therapeut; Am Fed Clin Res; Am Soc Hematol. *Res:* Reactive oxygen formation in the mechanism of anti-cancer quinone-containing antibiotics. *Mailing Add:* Dept Med Oncol & Therapeut Res City of Hope Nat Med Ctr Duarte CA 91010

DOROUGH, GUS DOWNS, JR, b Los Angeles, Calif, Mar 5, 22; m 70; c 3. EXPLOSIVES CHEMISTRY, PORPHYRIN CHEMISTRY. *Educ:* Univ Calif, BS, 43, PhD(chem), 47. *Prof Exp:* Asst chem, Univ Calif, 43-44; res chemist, Manhattan Atomic Bomb Proj, 44-46; Nat Res fel, Washington, DC, 46-47; from instr to asst prof chem, Wash Univ, 47-54; chemist & div leader, Lawrence Radiation Lab, Univ Calif, 54-67, head chem dept, 67-71; dep dir res & advan technol, US Dept Defense, Washington, DC, 71-73; assoc dir, Lawrence Livermore Nat Lab, 73-85; RETIRED. *Mem:* Am Chem Soc; Sigma Xi. *Res:* Chemical explosives; material research and development. *Mailing Add:* 614 Escondido Circle Livermore CA 94550

DOROUGH, H WYMAN, b Notasulga, Ala, Dec 23, 36; m 57; c 5. TOXICOLOGY, PESTICIDE CHEMISTRY. *Educ:* Auburn Univ, BS, 59, MS, 60; Univ Wis, PhD(toxicol), 63. *Prof Exp:* Lab asst entom, Auburn Univ, 56-59, res asst, 59-61; res asst, Univ Wis, 61-63; from asst prof to assoc prof entom, Tex A&M Univ, 63-69; from assoc prof to prof toxicol & entom, Univ Ky, 69-85, dir, Grad Ctr Toxicol, 77-83; HEAD, DEPT BIOL SCI, MISS STATE UNIV, 85- *Concurrent Pos:* USPHS & Environ Protection Agency res grant, 65-81; consult, Nat Ctr Toxicol Res, 80-81; dipl, Acad Toxicol Sci, 82-; peer rev comt, Toxnet Database, Nat Libr Med, 85-88. *Honors & Awards:* Bussart Res Award, Entom Soc Am, 81. *Mem:* Soc Toxicol; Soc Environ Toxicol & Chemistry; Entom Soc Am; Am Chem Soc. *Res:* Chemistry and toxicology of pesticides. *Mailing Add:* Biol Sci Drawer GY Mississippi State MS 39762

DORR, JOHN VAN NOSTRAND, II, b New York, NY, May 16, 10; m 46; c 3. GEOLOGY. *Educ:* Harvard Univ, BS, 32; Colo Sch Mines, GeolE, 37. *Hon Degrees:* DSc, Univ Minas Gerais, 66. *Prof Exp:* Geologist, Superior Oil Co, Calif, 37-38; from jr geologist to staff geologist, 38-71, actg chief, Br Latin Am & African Geol, Off Int Geol, US Geol Surv, 71-72; CONSULT GEOLOGIST, 74- *Concurrent Pos:* Vpres working group & comn on manganese, Int Asn Genesis Ore Deposits, 70-75. *Honors & Awards:* Medal, State of Minas Gerais, Brazil, 62; Bonifacio Medal, Brazilian Geol Soc, 64. *Mem:* Fel Geol Soc Am; Soc Econ Geol; fel Brazilian Geol Soc (vpres, 56); Geol Mining & Metall Soc India. *Res:* Economic geology; ore deposits; tungsten; nickel; manganese; iron. *Mailing Add:* 4982 Sentinel Dr, Apt 304 Bethesda MD 20816

DORRANCE, WILLIAM HENRY, b Highland Park, Mich, Dec 3, 21; m 46; c 4. PHYSICAL CHEMISTRY, GAS DYNAMICS. *Educ:* Univ Mich, BS, 47, MS, 48; Occidental Univ, DSc, 78. *Prof Exp:* Res engr, Aeronaut Res Ctr, Univ Mich, 47-49, aerodyn group leader, 49-51; sr aerodyn engr, Convair Div, Gen Dynamics Corp, 51-53, aerodyn group supvr astronaut div, 53-55, asst to dir sci res, Gen Off, 55-58, sr staff scientist, 58-61; head gas dynamics dept, Aerospace Corp, 61-62, group dir adv planning div, 62-64; vpres, Conductron Corp, 64-69, group exec, 64-67, mem bd dirs, 65-67, 68-69; chmn bd dirs, Interface Systs Corp, 67-69, chmn & chief exec, 69-70; pres, Orgn Control Serv, Inc, 71-85; PRES, ENG & CONSULT CO, 70- *Concurrent Pos:* Mem bd dirs, Orgn Control Serv, Inc, 71-86. *Mem:* AAAS; Am Chem Soc. *Res:* Systems and operations analysis; high speed aerodynamics; compressible gas dynamics; hypersonic flow; nonsteady flow phenomena; thermal protection systems; high performance missile design; thermochemical cycles for producing fuels. *Mailing Add:* 385 G Kaelepulu Kailua HI 96734

DORREL, D GORDON, b Oct 14, 40; c 2. AGRICULTURE. *Educ:* Univ Bc, Bsc, 62, Univ Sack Msc, 63, Phd Mich State Univ, 68. *Prof Exp:* DIR GEN, WESTERN REGIONAL RES BR, AGR CAN. *Mailing Add:* Western Regional Res Br Agr Can Sir John Carling Bldg Ottawa ON K1A 0C5 Can

DORRELL, DOUGLAS GORDON, b Clinton, BC, Oct 14, 40; m 63; c 2. PLANT CHEMISTRY. *Educ:* Univ BC, BSA, 62; Univ Sask, MSc, 63; Mich State Univ, PhD(crop sci), 68. *Hon Degrees:* NDC, Nat Defense Col, 80. *Prof Exp:* Res officer plant breeding, 63-65; res asst crop sci, Mich State Univ, 66-68; res scientist lipid chem, 68-78, adv planning & evaluation directorate, 78-79, dir, Winnipeg Res Sta, Res Br, Can Dept Agr, 80-83; dir, Lethbridge Res Sta, Res Br, Agr Can, 83-88; DIR, GEN WESTERN REGION, RES BR, 88- *Concurrent Pos:* Mem, Can Comt Grain Qual, 80-88; adv, Nat Defence Col, 79-80; mem, Alberta Agr Res Inst Bd, 87-, Agr Develop Fund Bd, 89- *Mem:* Agr Inst Can; Can Soc Agron. *Res:* Identification and modification of the concentration of chemical components in flax and sunflower seeds to improve quality; identification of specific components in new crop plants that have commercial value. *Mailing Add:* 930 Carling Ave Rm 711 Sir John Carling Bldg Otawa ON K1A 0C5 Can

DORRENCE, SAMUEL MICHAEL, b Rock Springs, Wyo, May 21, 39; m 60; c 3. FOSSIL FUEL CHEMISTRY, ORGANIC CHEMISTRY. *Educ:* Univ Utah, BA, 61, PhD(org chem), 64. *Prof Exp:* Res chemist, Celanese Chem Co, 64-67, group leader, 67; res chemist, Laramie Petrol Res Ctr, US Bur Mines, US Dept Energy, 67-70, res chemist, 70-73, proj leader, 73-77, chem res assoc, 77-80, mgr, Physical Sci Div, Laramie Energy Technol Ctr, 80-83; dir, Off Phys Sci, 83-86, VPRES, PHYS SCIS, WESTERN RES INST, 86- *Concurrent Pos:* Adj prof fuels eng, Univ Utah, 75-88; mem adv bd, Univ Wyo Enhanced Oil Recovery Inst, 84-88; chair & bd mem, Wyo Sci Technol & Energy Authority, 87- *Mem:* Petrol & Fuel Div Am Chem Soc; Sigma Xi. *Res:* Chemical structure and physical property determinations of tar sands bitumens, shale oils and materials derived from fossil fuel raw materials during recovery experiments; analyses of nitrogen, sulfur, and metals in crudes and syncrudes; in situ recovery experiments in tar sands; asphalt chemistry and chemical-physical property relationships. *Mailing Add:* 1410 La Prele Laramie WY 82070

DORRINGTON, KEITH JOHN, b London, Eng, Oct 26, 39; m 60; c 3. BIOCHEMISTRY, IMMUNOLOGY. *Educ:* Univ Sheffield, BSc, 61, PhD(biochem), 64. *Prof Exp:* Res biochem, Univ Sheffield, 61-64, fel, 64-66; vis fel, Duke Univ Med Ctr, 66-67; sr scientist molecular pharmacol, Univ Cambridge, Eng, 67-70; assoc prof, 70-75, chmn dept, 77-82, assoc dean res, fac med, 81-82, PROF BIOCHEM, UNIV TORONTO, 75-; VPRES, RES & DEVELOP, CONNAUGHT LABS LTD, 82- *Concurrent Pos:* Mem, Immunol & Transplantation Comt, Med Res Coun Can, 75-79; sect ed, J Immunol, Am Asn Immunologists, 75-81; mem, Med Res Coun Can, 82-85. *Honors & Awards:* Ayerst Award, Can Biochem Soc, 77. *Mem:* Am Asn Immunologists; Am Soc Biol Chemists; Brit Biochem Soc; Can Biochem Soc; Can Soc Immunol. *Res:* Chemical, physical and biological properties of proteins with special reference to the immunoglobulins. *Mailing Add:* Wellcome BioTechnol Ltd Langley Court Beckenham Kent BR3 3BS England

DORRIS, GILLES MARCEL, b Varennes, Que, May 23, 52; m 77; c 2. SURFACE & COLLOID CHEMISTRY. *Educ:* Univ Que, BSc, 75; McGill Univ, PhD(chem), 79. *Prof Exp:* Fel, Pulp & Paper Inst Can, 79-81, asst scientist, 81-83, assoc scientist, 83-89, group leader, 85-86, SECT HEAD, PULP & PAPER INST CAN, 87-, SCIENTIST, 89- *Concurrent Pos:* Dow Chem fel, 90-92. *Mem:* Chem Inst Can; Can Pulp & Paper Asn; Tech Asn Pulp & Paper Indust. *Res:* Surface chemistry of polymers; physical measurements by inverse gas chromatography; flocculation and sedimentation studies on colloidal suspensions; chemical recovery; deinking. *Mailing Add:* Paprican 570 St Johns Blvd Pointe Claire PQ H9R 3J9 Can

DORRIS, KENNETH LEE, b Baytown, Tex, Sept 26, 35; m 57; c 1. PHYSICAL CHEMISTRY. *Educ:* Univ Tex, BS, 61, PhD(microwave spectros), 66. *Prof Exp:* Asst prof, 65-69, ASSOC PROF CHEM, LAMAR UNIV, 69- *Concurrent Pos:* Robert A Welch res grant, 67-70. *Mem:* Am Chem Soc. *Res:* Spectroscopic studies of heterocyclic and N-acyl compounds, especially structure, symmetry and forces in the molecule. *Mailing Add:* Dept Chem Lamar Univ St PO Box 10022 Beaumont TX 77710-0022

DORRIS, PEGGY RAE, b Holly Bluff, Miss, Feb 27, 33. ZOOLOGY. *Educ:* Miss Col, BS, 56; Univ Miss, MS, 63, PhD(zool), 67. *Prof Exp:* Asst biol, Univ Miss, 63-66; from asst prof to assoc prof, 66-74, chmn dept, 72-90, PROF BIOL, HENDERSON STATE UNIV, 74- *Res:* Spiders of Mississippi and Arkansas. *Mailing Add:* 125 Evonshire Dr Arkadelphia AR 71923

DORRIS, ROY LEE, b Choctaw, Okla, Oct 2, 32; m 56; c 3. PHARMACOLOGY. *Educ:* Southern Nazarene Univ, BS, 59; Peabody Teachers Col, MA, 60; Vanderbilt Univ, PhD(pharmacol), 69. *Prof Exp:* Asst prof pharmacol, Univ Tex Southwestern Med Sch Dallas, 69-74; from asst prof to assoc prof, 74-82, PROF PHARMACOL, BAYLOR COL DENT, 82- *Mem:* Soc Exp Biol & Med; Am Soc Exp Pharmacol & Therapeut; Sigma Xi. *Res:* Effects of drugs on uptake, storage and release of catecholamines, with particular emphasis on dopamine-containing centers in the brain. *Mailing Add:* Dept Pharmacol Baylor Col Dent Dallas TX 75246

DORRIS, TROY CLYDE, b West Frankfort, Ill, Apr 6, 18; m 43; c 2. LIMNOLOGY, WATER POLLUTION. *Educ:* Univ Ill, PhD(zool, limnol), 53. *Prof Exp:* Assoc prof biol sci, Quincy Col, 47-56; assoc prof zool, 56-61, PROF ZOOL, OKLA STATE UNIV, 61-, DIR RESERVOIR RES CTR, RES FOUND, 67- *Res:* Biological effects of oil refinery effluents; biological parameters for water quality criteria; primary productivity of tropical waters. *Mailing Add:* 112 S Melrose Dr Stillwater OK 74074

DORROH, JAMES ROBERT, b Marion, Ala, Apr 20, 37; m 59; c 1. MATHEMATICS. *Educ:* Univ Tex, BA, 58, MA, 60, PhD(math), 62. *Prof Exp:* From asst prof to assoc prof math, 67-71, PROF MATH, LA STATE UNIV, BATON ROUGE, 71- *Mem:* Am Math Soc; Math Asn Am. *Res:* Functional analysis; semigroups of linear and nonlinear transformations. *Mailing Add:* Dept Math La State Univ 319 Lockett Hall Baton Rouge LA 70803

DORROS, IRWIN, b Brooklyn, NY, Oct 3, 29; m 54; c 4. ELECTRICAL ENGINEERING. *Educ:* Mass Inst Technol, SB & SM, 56. *Hon Degrees:* DScEng, Columbia Univ, 62. *Prof Exp:* Exec dir, Bell Tel Labs, 56-78; asst vpres, AT&T, 78-83; EXEC VPRES, BELL COMMUN RES, 83- *Honors & Awards:* Inst Elec & Electronics Engrs Founders Medal, 91. *Mem:* Fel Inst Elec & Electronics Engrs. *Res:* Network planning; information networking. *Mailing Add:* Three Cobblestone Lane Morristown NJ 07960

DORSCHNER, KENNETH PETER, b Appleton, Wis, Sept 7, 21; m 51; c 2. AGRONOMY. *Educ:* Univ Wis, BS, 49, MS, 51, PhD, 54. *Prof Exp:* Res asst, Univ Wis, 49-51 & 52-54; asst plant physiologist, Miss State Col, 51-52; biologist, Niagara Chem Div, FMC Corp, 54-59, supvr biol labs, 59-62, prod mgr herbicides, 62-67; res supvr pesticides, W R Grace & Co, Md, 67-69; mgr agr chem res, Glidden-Durkee Div, SCM Corp, 69-72; chief, Plant Studies Br, Criteria & Eval Div, Environ Protection Agency, 72-; RETIRED. *Mem:* Weed Sci Soc Am. *Res:* Synthetic plant hormones as herbicidal compounds. *Mailing Add:* 2612 Lakevale Dr Vienna VA 22181

DORSCHNER, TERRY ANTHONY, b Moultrie, Ga, June 29, 43; m 65; c 2. LASER GYROSCOPES, MAGNETO-OPTICS. *Educ:* Mass Inst Technol, BS, 65; Univ Wis, MS, 67, PhD(elec & electronics), 71. *Prof Exp:* Teaching asst elec & electronics, Univ Wis, 68-69, res asst, 65-71, lectr, 74; res assoc, Inst fur Hochfrequenztechnik, Tech Univ, Braunschweig, 71-73; SR RES SCIENTIST, RAYTHEON RES DIV, 74- *Mem:* Inst Elec & Electronics Engrs; Optical Soc Am. *Res:* Development of lasers for inertial measurement applications, particularly ring laser gyroscopes; optics; magneto-optics; quantum electronics; electromagnetic field theory; thin film devices; applications of hyperbolic geometry. *Mailing Add:* 29 Glenwood Ave Newton Center MA 02159

DORSCHU, KARL E(DWARD), b Collingswood, NJ, Aug 16, 30; m 56; c 3. METALLURGICAL ENGINEERING. *Educ:* Drexel Inst Technol, BS, 53; Rensselaer Polytech Inst, MMetE, 57; Lehigh Univ, PhD(metall eng), 60. *Prof Exp:* Engr welding metall, Aviation Gas Turbine Div, Westinghouse Elec Corp, 53-54 & Frankford Arsenal, 54-55; instr & res asst, Rensselaer Polytech Inst, 55-57; instr, Lehigh Univ, 57-59; sr engr, Res Lab, Air Reduction Co, 59-63, proj supvr, 63-65, supvr, 65-67, asst dir welding res, 67-71; vpres & gen mgr, Electrochem Corp, 71-76; exec vpres, Franklin Inst Res Labs, 76-80; PRES & FOUNDER, WELDRING CO, INC, 80- *Honors & Awards:* Lincoln Gold Medal, Am Welding Soc, 69. *Mem:* Am Welding Soc. *Res:* Metallurgy and heat transfer in welding. *Mailing Add:* Weldring Co Inc PO Box 20 Broomall PA 19008

DORSET, DOUGLAS LEWIS, b York, Pa, Aug 29, 42; m 67; c 1. CRYSTALLOGRAPHY, MATERIALS SCIENCE. *Educ:* Juniata Col, BS, 64; Sch Med, Univ Md, Baltimore City, PhD(biophysics), 71. *Prof Exp:* Res assoc membrane biophysics, Dept Biol Sci, State Univ NY Albany, 70-72; NIH fel electron diffraction, Roswell Park Mem Inst, 72-73; assoc res scientist electron diffraction, Molecular Biophys Dept, 73-82, PRIN SCIENTIST & HEAD ELECTRON DIFFRACTION DEPT, MED FOUND BUFFALO, 82- *Concurrent Pos:* Adj prof chem eng, State Univ NY Buffalo, 78-80; guest prof, Microbiol Dept, Biozentrum der Universitaet Basel, Switz, 80-81; vis prof polymer sci & eng, Univ Mass, Amherst, 90; consult, Electron Diffraction Comn, Int Union Crystallog, 90- *Mem:* NY Acad Sci; Am Crystallog Asn; Biophys Soc; Electron Micros Soc Am; Sigma Xi; Am Phys Soc; AAAS. *Res:* Solid state packing of biomembrane components and linear polymers determined from electron diffraction and electron microscopy data; solid state chemistry; structure of binary solids and thermally disordered phases. *Mailing Add:* Electron Diffraction Dept Med Found Buffalo 73 High St Buffalo NY 14203

DORSEY, CLARK L(AWLER), JR, b Lakota, Va, Apr 22, 23; m 45; c 4. ENVIRONMENTAL MONITORING. *Educ:* Va Polytech Inst, BS, 45, MS, 46; Purdue Univ, PhD(chem eng), 49. *Prof Exp:* Mem staff, Textile Fibers Dept, E I Du Pont de Nemours & Co, 49-67, Org Chem Dept, 68-74, Corp Plans Dept, 75-81; PRIVATE CONSULT, STRATEGIC PLANNING, 81- *Mem:* Sigma Xi. *Res:* Venture analysis and development; corporate planning; management sciences. *Mailing Add:* Rte 1 Box 560 Lancaster VA 22503

DORSEY, GEORGE FRANCIS, b Chattanooga, Tenn, Dec 15, 42; m 66; c 3. POLYMER CHEMISTRY. *Educ:* Univ Chattanooga, AB, 64; Univ Tenn, PhD(chem), 69. *Prof Exp:* DEVELOP CHEMIST, NUCLEAR DIV, UNION CARBIDE CORP, 69- *Mem:* Am Chem Soc. *Res:* Polymers; adhesives; foamed epoxy resins; polyurethanes; protective coatings. *Mailing Add:* PO Box 2002 Bldg 9202 MS 8095 Martin Marietta Oak Ridge TN 37831

DORSEY, ROBERT T, b Cleveland, Ohio, Feb 28, 18; m 45; c 1. ELECTRICAL ENGINEERING. *Educ:* Mass Inst Technol, SB, 40. *Prof Exp:* Supvr ord, Naval Ord Lab, 40-43 & Mine Test Sta, 43-45; supvr lighting appln, Gen Elec Co, mgr lighting develop, 66-81; RETIRED. *Concurrent Pos:* Mem, Int Comn Illum, 57-, vpres, 67; chmn, US Comt Appln, 64- *Honors & Awards:* Gold Medal, Illum Eng Soc. *Mem:* Fel AAAS; fel Illum Eng Soc (pres, 72). *Res:* Illuminating engineering; new lighting techniques; design and application of new lamps; development on vision and visual performance. *Mailing Add:* 3726 Tolland Rd Cleveland OH 44122-5143

DORSEY, THOMAS EDWARD, b New York, NY, June 6, 40. ENDOCRINOLOGY, MEDICINE. *Educ:* Queens Col, BA, 63, MA, 65; City Univ New York, PhD(biochem), 75. *Prof Exp:* Mem fac, Univ Inst Oceanog, City Col New York, 74-77; sr biochemist, USV Pharmaceut COrp Revlon INc, 77-79, group leader,Relvon Health Care Group Res & Develop, 79-84 sect head, Drug Metab, 84-86; sr clin res monitor, Ayerst Labs, Am Home Prods, 86-87; GROUP PROJ DIR, CLIN RES, ORGANON INC, 87- *Concurrent Pos:* Res scientist, Lamont-Doherty Geol Observ, Columbia Univ, 75-76. *Mem:* Am Chem Soc; NY Acad Sci; Sigma Xi. *Res:* Pharmacokinetics and biopharmaceutics of drugs including enzyme kinetic studies and metabolic studies; cardiovascular drugs; clinical research in oncology and tumor research; LHRH analogs and endocrine research. *Mailing Add:* 235 E 57th St New York NY 10022

DORSKY, DAVID ISAAC, b Passaic, NJ, Aug, 1, 53; m 82. MOLECULAR VIROLOGY, ANTIVIRAL CHEMOTHERAPY. *Educ:* Brandeis Univ, AB & MA, 75; Harvard Univ MD & PhD(exp path), 82. *Prof Exp:* Intern & resident, med, 82-85, fel infectious dis, 85-88, INSTR, MED, BETH ISRAEL HOSP, 88-, INSTR, MED, HARVARD MED SCH, 88- *Concurrent Pos:* Fel infectious dis, Brigham & Women's Hosp, 85-88. *Res:* Antiviral chemotherapy; structure and function of herpes simplex virus DNA polymerase; mechanism of action of antiviral compounds as polymerase inhibitors. *Mailing Add:* 14 Farmstead Lane West Simsbury CT 06092

DORSON, WILLIAM JOHN, JR, b Nashua, NH, May 9, 36; m 58; c 2. CHEMICAL & BIOMEDICAL ENGINEERING. *Educ:* Rensselaer Polytech Inst, BChE, 58, MChE, 60; Univ Cincinnati, PhD(chem eng, biomed), 67. *Prof Exp:* Coop engr, Gen Elec Co, 56-58, develop engr, Knolls Atomic Power Lab, 58-64, res & develop engr, Space Power & Propulsion Dept, 64-65; instr chem eng, Univ Cincinnati, 65-66; assoc prof biomed eng, 66-71, PROF BIOMED ENG, ARIZ STATE UNIV, 71-, PROF CHEM ENG, 80- *Concurrent Pos:* Consult, Gen Elec Co, Randam Electronics, Inc & Hunkar Instrument Develop Labs, Inc; mem, Int Conf Med & Biol Eng, 67. *Mem:* Am Chem Soc; Am Soc Mech Engrs; Am Inst Chem Engrs; Asn Advan Med Instrumentation. *Res:* Development of medical diagnostic and prosthetic methods; body fluid rheology; transport phenomena; radioactive tracer techniques; two-phase flow and heat transfer; nucleonics. *Mailing Add:* Dept Chem & Metall Eng Ariz State Univ Tempe AZ 85287

DORST, JOHN PHILLIPS, b Cincinnati, Ohio, July 8, 26; m 50; c 4. MEDICINE, RADIOLOGY. *Educ:* Cornell Univ, MD, 53. *Prof Exp:* Intern, Univ Iowa Hosps, 53-54, resident radiol, 55-58; fel pediat radiol, Childrens Hosp, Univ Cincinnati, 58-59, from asst prof to assoc prof radiol, 59-66, asst prof pediat, 65-66; assoc prof radiol, 66-69, assoc prof pediat, 67-78, PROF RADIOL, JOHNS HOPKINS UNIV, 70-, PROF PEDIAT, 78- *Mem:* Soc Pediat Radiol; fel Am Col Radiol; Radiol Soc NAm; Am Roentgen Ray Soc. *Res:* Pediatric radiology; functional craniology; chondrodysplasias. *Mailing Add:* Dept Radiol & Pediat Johns Hopkins Hosp Baltimore MD 21205

DORT, WAKEFIELD, JR, b Keene, NH, July 16, 23; m 54, 67; c 1. QUATERNARY GEOLOGY, GEOMORPHOLOGY. *Educ:* Harvard Univ, BSc, 44; Calif Inst Technol, MSc, 48; Stanford Univ, PhD(geol), 55. *Prof Exp:* Field geologist, Bunker Hill & Sullivan, Idaho, 47 & 49 & US Geol Surv, 48; instr geomorphol & econ geol, Duke Univ, 48-50; asst prof, Pa State Univ, 52-57; assoc prof, 57-70, PROF GEOL, UNIV KANS, 70- *Concurrent Pos:* Consult engr geol, 50-52 & ground water & petrol geol, 52-57; geol ed, Petrol & Natural Gas Digest, 53-56; mem visual educ comt, Am Geol Inst, 55-63 & earth sci curric proj, 63-64; assoc ed, Monitor, 56-59, Oil & Gas Abstr, 59-61 & Mining Abstr, 59-61; mem Antarctic expeds, US, 65, 66 & 69, Japan, 67; res assoc, Idaho Mus Nat Hist, 65-; mem exec comt & ed, Inst for Tertiary-Quaternary Studies, 82-; hon lectr, Mid-Am State Univ Asn, 82-83. *Mem:* AAAS; Geol Soc Am; Am Geog Soc; Asn Am Geog; Soc Am Archaeol; Sigma Xi. *Res:* Geomorphology; glacial and Pleistocene geology; late Cenozoic paleoclimatology; geology of Idaho; geology of archaeological sites; geomorphology of Antarctica; quaternary history of central Great Plains. *Mailing Add:* Dept Geol Univ Kans Lawrence KS 66045

DORUS, ELIZABETH, b Parsons, Kans, Dec 6, 40; m 68; c 1. BEHAVIORAL GENETICS. *Educ:* Valparaiso Univ, BA, 62; Univ Chicago, PhD(personality & psychopath), 71; Univ Pa, MA, 75. *Prof Exp:* Intern clin psychol, Duke Univ Med Ctr, 64-65; trainee, Westside Vet Admin Ment Hyg Clin, 65-67; consult psychologist, Ment Health Div, Chicago Bd of Health, 67-71; instr psychol, Univ Pa, 72-73; res fel, Ill State Psychiat Inst, 74-75; RES ASSOC & ASST PROF PSYCHIAT, UNIV CHICAGO, 77- *Concurrent Pos:* Fel Found Fund Res Psychiat, 72-73; NIH fel, 75-76; NIMH fel, 77-81. *Mem:* Am Psychol Asn; Am Soc Human Genetics; AAAS; Behavior Genetics Asn; Social Biol Asn. *Res:* Genetics of psychiatric disorders; studies of lithium; human genetics. *Mailing Add:* 1515 N Astor St No 14A Chicago IL 60610

DORWARD, RALPH C(LARENCE), b Viking, Alta, July 31, 41; m 64; c 2. METALLURGY, MATERIALS SCIENCE. *Educ:* Univ Alta, BSc, 62, MSc, 64; McMaster Univ, PhD(metall), 67. *Prof Exp:* Res assoc metall, Univ Alta, 64; sr res metallurgist, Kaiser Aluminum & Chem Corp, 67-69, staff res metallurgist, 69-77, proj mgr, Wettable Cathode Develop, 78-82, sect head, advan alloy res & develop, 82-84, prog mgr, Al-Li Alloy Develop, 84-87, RES & DEVELOP MGR, AIRCRAFT-AEROSPACE ALLOY PROD, KAISER ALUMINUM & CHEM CORP, 87- *Mem:* Am Soc Metals; Am Soc Testing & Mat. *Res:* Physical metallurgy of aluminum alloys; thermodynamics of solutions, phase transformations and phase equilibria; electrochemistry of corrosion and aluminum extraction; alloy development; stress corrosion; properties of ceramics; project management. *Mailing Add:* Metals Res Div Kaiser Ctr Technol Pleasanton CA 94566

DORWARD-KING, ELAINE JAY, b Knoxville, Tenn, Dec 17, 57. AUTOMATION ROBOTICS, GAS CHROMATOGRAPHY. *Educ:* Maryville Col, BA, 79; Colo State Univ, PhD(anal chem), 84. *Prof Exp:* Res fel, Colo State Univ, 79-84; SR RES CHEMIST, NUTRIT CHEM DIV, MONSANTO CO, 84- *Mem:* Am Chem Soc; Soc Environ Toxicol & Chem; Sigma Xi. *Res:* Automated sample preparation procedures for inductively coupled plasma, gas chromatography and high performance liquid chromatography using laboratory robotics; analytical method development in capillary gas chromatography and ion chromatography. *Mailing Add:* 2515 158th Ave SE Bellevue WA 98008

DORWART, BONNIE BRICE, b Petersburg, Va, Jan 27, 42; m 63; c 2. RHEUMATOLOGY. *Educ:* Bryn Mawr Col, AB, 64; Sch Med, Temple Univ, MD, 68. *Prof Exp:* From intern to resident, Lankenau Hosp, Philadelphia, 68-72; fel rheumatology, Sch Med, Univ Pa, 72-74; instr med, 74-76, ASST PROF MED, JEFFERSON MED COL, PHILADELPHIA, 76-; ASST PHYSICIAN CONNECTIVE TISSUE DIS, LANKENAU HOSP, 74- *Mem:* Rheumatism Asn; Am Col Physicians. *Res:* Clinical correlates of cervical spondylosis; pathogenesis of edema in rheumatoid arthritis; hypothyroid arthropathy. *Mailing Add:* 124 Maple Ave Bala-Cynwyd PA 19004

DORY, ROBERT ALLAN, b St Paul, Minn, June 7, 36; m 67; c 3. MATHEMATICAL PHYSICS. *Educ:* Univ NDak, BS, 58; Univ Wis, MS, 60, PhD(physics), 62. *Prof Exp:* Res asst physics, Midwestern Univ Res Asn, 60-62; instr, Univ Wis, 62-64; physicist, 64-73, group leader, 73-75, asst sect leader, 75-79, SECT LEADER, OAK RIDGE NAT LAB, 79- *Mem:* Fel Am Phys Soc. *Res:* Thermonuclear research; plasma theory. *Mailing Add:* Oak Ridge Nat Lab PO Box 2009 Oak Ridge TN 37831

DOSANJH, DARSHAN S(INGH), b Sultanwind, India, Feb 21, 21; m 57; c 3. AERONAUTICAL ENGINEERING. *Educ:* Punjab Univ, BSc Hons, 44, MSc, 45; Univ Mich, MSE, 48; Johns Hopkins Univ, PhD(aeronaut), 53. *Prof Exp:* Jr instr mech eng, Johns Hopkins Univ, 49-50, asst, Aeronaut Dept, 50-54; res assoc, Inst Fluid Dynamics & Appl Math, Univ Md, 55-56; assoc prof mech & aerospace eng, 56-62, PROF MECH & AEROSPACE ENG, SYRACUSE UNIV, 62- *Concurrent Pos:* Vis prof, Col Aeronaut, Cranfield, Eng, 61-62; NATO sr fel, 67; Fulbright-Hays sr fac res fel & vis prof, Southampton Univ, 71-72; mem Tech Comt on Aeroacoustics, Am Inst Aeronaut & Astronaut. *Mem:* Acoust Soc Am; assoc fel Am Inst Aeronaut & Astronaut; Am Phys Soc; Am Soc Mech Engrs; Am Soc Eng Educ; Am Asn Univ Prof; Sigma Xi. *Res:* High speed gas dynamics; aerodynamics; aero acoustics; bioacoustics. *Mailing Add:* 5176 Brockway Lane Fayetteville NY 13066

DOSCH, HANS-MICHAEL, b Frankenberg, WGermany, May 20, 26. HUMAN IMMUNE SYSTEM. *Educ:* Univ Marburg, WGermany, MD, 71. *Prof Exp:* ASSOC PROF PEDIAT & IMMUNOL, HOSP SICK CHILDREN & UNIV TORONTO, 83- *Mem:* Am Asn Immunol; Can Soc Immunol; Soc Pediat Res. *Mailing Add:* Dept Pediat Immunol Hosp Sick Children 555 University Ave Toronto ON M5G 1X8 Can

DOSCHEK, GEORGE A, b Pittsburgh, Pa, Sept 3, 42; m 68. SOLAR PHYSICS, ATOMIC SPECTROSCOPY. *Educ:* Univ Pittsburgh, BS, 63, PhD(physics), 68. *Prof Exp:* Teaching asst physics, Univ Pittsburgh, 63-64, res assoc astrophys, Univ Pittsburgh at E O Hulburt Ctr Space Res, 68-70, astrophysicist, 70-79, HEAD, SOLAR TERRESTRIAL RELATIONSHIPS BR, E O HULBURT CTR SPACE RES, US NAVAL RES LAB, 79- *Mem:* Am Astron Soc; Int Astron Union; fel Optical Soc Am. *Res:* Astrophysics; ultraviolet and x-ray spectroscopy of high temperature astrophysical and laboratory plasmas. *Mailing Add:* 8221 Coach St Potomac MD 20854

DOSCHEK, WARDELLA WOLFORD, b Sewickley, Pa, May 18, 44; m 68; c 1. RESEARCH ADMINISTRATION. *Educ:* Univ Pittsburgh, BS, 63, PhD(phys chem), 68. *Prof Exp:* Res scientist photosynthesis, Res Inst Advan Studies, 68-70; asst prog dir regulatory biol prog, NSF, 70-78; RETIRED. *Mem:* Sigma Xi. *Res:* Photosynthesis. *Mailing Add:* 8221 Coach St Potomac MD 20854

DOSCHER, MARILYN SCOTT, b New York, NY, July 1, 31. BIOCHEMISTRY. *Educ:* Cornell Univ, BA, 53; Univ Wash, PhD(biochem), 59. *Prof Exp:* Chemist, Schering Corp, NJ, 53-55; asst prof, 67-72, ASSOC PROF BIOCHEM, SCH MED, WAYNE STATE UNIV, 73- *Concurrent Pos:* NIH fel, Yale Univ, 60-62, univ fel, 62-64; Am Cancer Soc res assoc grant, Brookhaven Nat Lab, 64-67. *Mem:* Am Chem Soc. *Res:* Protein structure. *Mailing Add:* Dept Biochem 374 Scott Hall Wayne State Univ Sch Med 540 E Canfield Detroit MI 48201

DOSER, DIANE IRENE, b Syracuse, NY, Dec 22, 56; m 85; c 1. SEISMOLOGY, TECTONOPHYSICS. *Educ:* Mich Technol Univ, BS, 78; Univ Utah, MS, 80, PhD(geophys), 84. *Prof Exp:* Bantrell postdoctoral fel, Calif Inst Technol, 84-85; asst prof, 86-91, ASSOC PROF GEOPHYS, UNIV TEX, EL PASO, 91-; DIR, J W KIDD SEISMIC OBSERV, 88- *Concurrent Pos:* Prin investr, Grants from NSF, US Geol Surv, Petrol Res Fund & Tex Adv Res Prog, 86-; mem rev panel, US Geol Surv, Nat Earthquake Hazards Reduction Prog, 89-91; guest ed, Spec Ed Pure & Appl Geophys Hist Earthquakes, 91-; standing comt mem, Inc Res Insts Seismol, 91- *Mem:* Geol Soc Am; Am Geophys Union; Seismol Soc Am; Asn Women Sci. *Res:* Historic earthquake studies; seismicity of continental rifts; induced seismicity associated with oil fields; uncertainty estimates associated with geophysical inversions. *Mailing Add:* Dept Geol Sci Univ Tex El Paso TX 79968

DOSHAN, HAROLD DAVID, b New York, NY, Oct 30, 41; m; c 2. CLINICAL RESEARCH, DRUG METABOLISM. *Educ:* Cornell Univ, BA, 62; Stanford Univ, PhD(org chem), 67. *Prof Exp:* Res assoc org chem, Johns Hopkins Univ, 67-68, NIH fel pharmacol, Sch Med, 68-70; sr res scientist, Pfizer Cent Res Labs, 70-80; from asst dir to assoc dir clin pharmacol, Revlon Health Care Group, 80-86; assoc dir clin pharmacol, Pharmaconsult Assocs, 87-88; DIR CLIN RES, BIO-RES LABS, LTD, 88- *Mem:* Am Chem Soc; Am Asn Pharmaceut Scientists; Am Soc Clin Pharmacol & Therapeut; Am Soc Hypertension; Drug Info Asn. *Res:* Clinical research (phases I-IV); clinical pharmacology and drug disposition; pharmacokinetics; xenobiotic metabolism. *Mailing Add:* 18 Griffith Rd Riverside CT 06878-1202

DOSHI, ANIL G, b Bombay, India, Jan 7, 52; m 79; c 2. NOVEL CAST COEXTRUDED PACKAGING FILMS, SUPERMARKET PACKAGING FILMS. *Educ:* Bombay Univ, BS, 72; Univ Lowell, MS, 75. *Prof Exp:* Prod develop chemist vinyl bldg prod, Bird Inc, 75-79; prod develop chemist calendered film, sheet, Pantasote, 79-81; res engr bldg prod, GAF Corp, 81-84; qual assurance mgr extruded sheet, Royalite, Uniroyal Plastics, 84-87; LAB MGR BLOWN & CAST FILM, BORDEN, 87- *Mem:* Soc Plastics Engrs. *Res:* Applied research and development pertaining to flexible PVC supermarket and industrial films as well as LLDPE blown and cast coextruded pallet wrap films. *Mailing Add:* Borden-Resinite One Clark St North Andover MA 01845

DOSHI, MAHENDRA R, b India, May 6, 41; US citizen; m 71; c 2. PULP AND PAPER, MEMBRANE PROCESSES. *Educ:* Bombay Univ, BS, 65; Clarkson Col, MS, 68, PhD(chem eng), 70. *Prof Exp:* Res assoc chem eng, Clarkson Col Technol, 70-71; vis asst prof, State Univ NY, Buffalo, 71-77; assoc prof eng, Inst Paper Chem, 77-86; CONSULT & INSTR, DOSHI & ASSOCS INC, 86- *Concurrent Pos:* Sr process engr, Marathon Eng/Architects/Planners. *Mem:* Am Inst Chem Engrs; Sigma Xi; Tech Asn Pulp & Paper Indust; NAm Membrane Soc. *Res:* Recycled paper processing; screening; hydrocyclones; deinking; separation processes; solid-liquid separation; membrane processes; reverse osmosis; ultrafiltration; filtration; floatation; sedimentation; fluid mechanics; heat and mass transfer; computer simulation. *Mailing Add:* Doshi & Assoc Inc 2617 N Summit St Appleton WI 54914

DOSIER, LARRY WADDELL, b Waynesboro, Va, Mar 26, 44. PLANT MORPHOGENETICS, CROP BREEDING. *Educ:* Col William & Mary, BS, 66; Univ Va, MA, 71, PhD(biol), 74. *Prof Exp:* Asst plant variety examr, 75-76, plant variety examr, Plant Variety Protection Off, LMGS Div, 76-81, COMPUT SPECIALIST, IRM DIV, AGR MKT SERV, USDA, 82- *Concurrent Pos:* Mem rev bd, Nat Grass Variety, 77-79, 79-81; mem rev bd, Nat Clover Variety, 78-79, inaugural mem, 79-81 & 81-83. *Mem:* Bot Soc Am. *Res:* Developmental phenomena of Elodea trichoblasts as they relate to plant cell differentiation. *Mailing Add:* 5211 Palco Pl College Park MD 20740

DOSKOTCH, RAYMOND WALTER, b Husiatyn, Western Ukraine, May 23, 32; Can citizen; m 55; c 2. PHARMACOGNOSY, BIOCHEMISTRY. *Educ:* McMaster Univ, BSc, 55; Univ Wis, MS, 57, PhD(biochem), 59. *Prof Exp:* Res assoc, Enzyme Inst, Univ Wis, 59-60; unit head, Mich Dept Health Labs, 60-61; asst prof res, Sch Pharm, Univ Wis, 61-63; from asst prof to assoc prof pharmacog, Col Pharm, 63-70, PROF BIOCHEM, COL BIOL SCI & PROF NATURAL PROD CHEM & PHARMACOG, COL PHARM, OHIO STATE UNIV, 70- *Mem:* Am Soc Pharmacog; Am Chem Soc; The Chem Soc. *Res:* Isolation of biologically active natural products and structure determination biosynthesis of natural products. *Mailing Add:* Ohio State Univ Col Pharm 500 W 12th Ave Columbus OH 43210-1214

DOSS, JAMES DANIEL, b Reading, Pa, Mar 9, 39; m 58; c 2. BIOMEDICAL ENGINEERING, SUPERCONDUCTIVITY. *Educ:* Ky Wesleyan Col, BS, 64; Univ NMex, MSEE, 69. *Prof Exp:* STAFF MEM, LOS ALAMOS NAT LAB, 64-; MEM SCI STAFF, BERNALILLO COUNTY MED CTR, 75- *Concurrent Pos:* Assoc prof elec eng, Univ NMex, 69-71, Sch Med, 75-76. *Honors & Awards:* IR-100 Award, 78; Spec Award Excellence Technol Transfer, Fed Lab Consortium, 86; R&D 100 Award, 89. *Mem:* Radiation Res Soc; NAm Hyperthermia Group; Inst Elec & Electronics Engrs. *Res:* Tissue hyperthermia; electrosurgery; ophthalmology; oncology, superconductivity and microwave sources; microwave dosimetry. *Mailing Add:* Group MP-8 MS-H826 Los Alamos Nat Lab Los Alamos NM 87545

DOSS, RAOUF, mathematics; deceased, see previous edition for last biography

DOSS, RICHARD COURTLAND, b Omaha, Nebr, Nov 14, 26; c 4. APPLIED CHEMISTRY. *Educ:* Creighton Univ, BSChem, 50, MS, 52. *Prof Exp:* Sr res chemist, Phillips Petrol Co, 52-77, sr patent develop chemist, Chem Appln, 77-84, sr licensing specialist, 85-88, regist patent agent, Phillips Petrol Col, 84-88; RETIRED. *Mem:* Am Chem Soc. *Res:* Petrochemicals-aimed towards organic adhesives, coatings and sealants. *Mailing Add:* 1224 SE Grandview Rd Bartlesville OK 74006

DOSSEL, WILLIAM EDWARD, b St Louis, Mo, Apr 16, 20; m 55; c 2. HISTOLOGY, EMBRYOLOGY. *Educ:* Ill Col, BA, 48; Marquette Univ, MS, 50; Johns Hopkins Univ, PhD(zool), 54. *Prof Exp:* Asst zool, Marquette Univ, 48-50; fr instr, Johns Hopkins Univ, 40-53; from instr to asst prof anat, Sch Med, Univ NC, 54-60; assoc prof, Creighton Univ, 60-69, actg chmn dept, 69-70, prof anat, Sch Med, 69-90, chmn dept, 70-90; RETIRED. *Concurrent Pos:* Asst, Marine Biol Lab, Woods Hole, Mass, 52-54. *Mem:* Am Asn Anat; Am Asn Med Col. *Res:* Functional and morphological development of chick endocrine organs; fine structure chick embryonic organs. *Mailing Add:* Dept Anat Creighton Univ Sch Med Omaha NE 68178

DOSSETOR, JOHN BEAMISH, b Bangalore, India, July 19, 25; Can citizen; m 57; c 4. TRANSPLANTATION IMMUNOLOGY, NEPHROLOGY. *Educ:* Oxford Univ, BA, 46, MA, BM, BCh, 50; McGill Univ, PhD, 61; FRCP(C), 57. *Prof Exp:* Asst prof med & exp surg, McGill Univ, 61-69; PROF MED & DIR, DIV BIOMED ETHICS & HUMANITIES, UNIV ALTA, 69- *Mem:* Fel Am Col Physicians; Can Soc Clin Invest; Am Soc Nephrology; Can Soc Immunol; Transplantation Soc. *Mailing Add:* Fac Med 222 Anri 8220 114th St Edmonton AB T6G 2J3 Can

DOSSO, HARRY WILLIAM, b Gull Lake, Sask, Jan 9, 32; m 56; c 4. GEOPHYSICS. *Educ:* Univ BC, BA, 55, MSc, 57, PhD, 67. *Prof Exp:* From instr to assoc prof physics, 57-69, head dept, 69-75, PROF PHYSICS, UNIV VICTORIA BC, 69- *Concurrent Pos:* Res tech officer, Pac Naval Lab, Defence Res Bd Can, 58-63; Nat Res Coun & Defence Res Bd Can grants, 65- *Mem:* Can Asn Physicists; Am Geophys Union; Can Geophys Union; Europ Geophys Union; Can Soc Explor Geophys. *Res:* Electromagnetic fields; geomagnetic micropulsations and magnetotelluric modelling; analytical and analogue methods of studying electromagnetic variations at the earth's surface. *Mailing Add:* Dept Physics & Astron Univ Victoria Victoria BC V8W 3P6 Can

DOST, FRANK NORMAN, b Seattle, Wash, Mar 24, 26; m 50; c 2. PHYSIOLOGY, TOXICOLOGY. *Educ:* Wash State Univ, BS & DVM, 51; Kans State Univ, MS, 59. *Prof Exp:* Pvt pract, 51-56; instr physiol & pharmacol, Kans State Univ, 56-59; Mark L Morris Found fel & jr vet, Dept Vet Physiol & Pharmacol, Wash State Univ, 59-62; asst prof chem, pharmacol & vet med, 62-67, assoc prof vet med, Environ Health Serv Ctr, 67-75, prof agr chem, Vet Med, Environ Health Sci Ctr, 75-80, PROF AGR CHEM & EXTEN TOXICOLOGIST, ORE STATE UNIV, 80- *Concurrent Pos:* Consult, Adv Comt, Environ Protection Agency, 70- & US Forest Serv, 76-; mem, ad hoc panel toxicol hydrazines, Nat Acad Sci-Nat Res Coun, 71-; mem, Ore State Bd Educ, 77-86; chmn, Gordon Res Conference Toxicol &

Safety Eval, 84; consult, Lab Toxicol, Nat Cancer Inst, NIH, 72-74. *Mem:* Soc Toxicol; Am Soc Pharmacol & Exp Therapeut; Gordon Res Conf Toxicol & Safety Eval (chmn, 84-); Acad Toxicol Scis (dipl gen). *Res:* Effects of intoxicants on biochemical and physiological mechanisms and metabolic fate of intoxicants; methods of hazard assessment. *Mailing Add:* Dept Agr Chem Ore State Univ Corvallis OR 97331

DOST, MARTIN HANS-ULRICH, b Berlin, Ger, Jan 12, 33; US citizen; m 57; c 3. AUTOMATIC CONTROL, COMPUTER SIMULATION. *Educ:* Tech Univ Berlin, BS, 55; Univ Calif Berkeley, MS, 57. *Prof Exp:* Res engr, Inst Eng Res, 56-58; instr feedback control theory, 59-60, INSTR MODELING & SIMULATION, IBM CORP, 68- *Concurrent Pos:* Assoc eng, Systs Develop Div, 58-60, staff engr, 61-67, adv engr, Math Anal & Simulation Dept, Gen Prods Div, IBM Corp, 68-; assoc ed, Simulation J, 79-86, chmn, student chap comt, 82-84, dir at large, Soc Comput Simulation, 79-, dir, Am Automatic Control Coun, 85-88; founding mem, Ger Simulation Soc of ASIM/GI, 83-87. *Res:* Analysis and design of automatic control systems; mathematical modeling and simulation of dynamic systems. *Mailing Add:* 17215 Melody Lane Los Gatos CA 95030

DOSTER, JOSEPH MICHAEL, b Chapel Hill, NC, Dec 3, 54; m 81. PARALLEL PROCESSING, RADIATION TRANSPORT. *Educ:* NC State Univ, BS, 77, PhD(nuclear eng), 82. *Prof Exp:* asst prof, 82-88, ASSOC PROF NUCLEAR ENG, NC STATE UNIV, 88- *Concurrent Pos:* Consult, Res Triangle Inst, 81- & Gen Elec Co, 85- *Mem:* Am Nuclear Soc; Sigma Xi; Soc Comput Simulation. *Res:* Applications of advanced computer architectures to the real time simulation and control of nuclear power systems; algorithm development for the parallel solution of multi-phase flow problems; modeling of high temperature liquid metal heat pipes for space reactor applications. *Mailing Add:* NC State Univ Box 7909 Raleigh NC 27695-7909

DOSWELL, CHARLES ARTHUR, III, b Elmhurst, Ill, Nov 5, 45; m 75; c 2. METEOROLOGY. *Educ:* Univ Wis, BSc, 67; Univ Okla, MSc, 69, PhD(meteorol), 76. *Prof Exp:* Res meteorologist, Nat Severe Storms Lab, 74-76; res meteorologist, Techniques Develop Unit, Nat Severe Storms Forecast Ctr, 76-82; meteorologist, Weather Res Prog, Environ Res Lab, 82-86, METEOROLOGIST, NAT SEVERE STORMS LAB, NAT OCEANIC & ATMOSPHERIC ADMIN, 86- *Concurrent Pos:* Adj assoc prof, Univ Okla, 89- *Honors & Awards:* Silver Medal, Dept Commerce. *Mem:* Am Meteorol Soc; Sigma Xi; Nat Weather Asn. *Res:* Severe thunderstorm forecasting; severe thunderstorm dynamics; objective analysis. *Mailing Add:* Nat Severe Storm Lab 1313 Halley Circle Norman OK 73069

DOTSON, ALLEN CLARK, b Badin, NC, Mar 21, 38. PHILOSOPHICAL IMPLICATIONS OF QUANTUM MECHANICS, QUANTUM MECHANICAL MEASUREMENT THEORY. *Educ:* Wake Forest Col, BS, 60; Univ NC-Chapel Hill, PhD(physics), 64. *Prof Exp:* From asst prof to assoc prof physics, Western Mich Univ, 64-81; ASSOC PROF PHYSICS, ST ANDREWS PRESBY COL, 81- *Mem:* Am Asn Physics Teachers; Sigma Xi. *Res:* Quantum field theory; elementary-particle kinematics; classical theory of fluids; philosophical implications of quantum mechanics. *Mailing Add:* Chem & Physics Dept St Andrews Presby Col Laurinburg NC 28352-5598

DOTSON, MARILYN KNIGHT, b Auburn, Ala, May 2, 52. MATHEMATICAL STATISTICS, COMPUTER SCIENCE. *Educ:* Ga Southern Col, BSEd, 75; Ga Southwestern Col, MEd, 76; EdD, Nova Univ, 89. *Prof Exp:* Teacher math, Americus High Sch, Ga, 75-78; instr math, 79-80, CHAIRWOMAN, DEPT MATH & PHYSICS, 80-, MAJOR ADV, PRE-ENG PROG, 83- *Concurrent Pos:* Consult, Sykes Enterprises, Inc, 84- *Mem:* Math Asn Am; Nat Coun Teachers Math. *Res:* Human factors: usability testing, primarily the design and implementation of testing programs that deal with human characteristics, expectations, and behavior of high tech products that people use in their work and everyday lives. *Mailing Add:* Dept Math Belmont Abbey Coll Belmont NC 28012

DOTT, ROBERT HENRY, JR, b Tulsa, Okla, June 2, 29; m 51; c 5. SEDIMENTOLOGY, TECTONICS. *Educ:* Univ Mich, BS, 50, MS, 51; Columbia Univ, PhD, 55. *Prof Exp:* Field asst, US Geol Surv, 47-48; geologist, Humble Oil & Refining Co, 54-58; from asst prof to assoc prof geol, 58-66, chmn dept geol & geophys, 74-77, PROF GEOL, UNIV WIS-MADISON, 66- *Concurrent Pos:* With Cambridge Res Ctr, US Air Force, 56-57; mem, Comn on Educ in Geol Sci, 67-69; consult, Roan Selection Trust, Zambia, 67 & Atlantic Richfield Co, 83-86; vis prof, Univ Calif, Berkeley, 69; lectr, Tulsa Univ, 69; NSF sci fac fel, Stanford Univ & US Geol Surv, Calif, 78 & Univ Colo, 79; distinguished lectr, Am Asn Petrol Geologists, 84-85; vis scientist, Oxford Univ, 85-86; Cabot Distinguished vis prof, Univ Houston, 86-87; Erskine fel, Canterbury Univ, Christchruch, NZ, 87; nat lectr, Sigma Xi, 89-90; chmn, Hist Geol Div, Geol Soc Am, 90. *Honors & Awards:* Pres Award, Am Asn Petrol Geologists, 56; Nat Lectr, Sigma Xi, 88- *Mem:* Geol Soc Am; Am Asn Petrol Geologists; Soc Econ Paleontologists & Mineralogists (secy-treas, 70-72, vpres, 72-73, pres, 81-82); Int Asn Sedimentol; Sigma Xi; Hist Earth Sci Soc (pres, 90). *Res:* Sedimentology and tectonics of orogenic belts, particularly Pacific Northwest, Southern Chile, and the Antarctic Peninsula; Paleozoic sedimentology Great Lakes region; shallow marine and eolian deposits in western United States; history of geology. *Mailing Add:* Dept Geol & Geophys Univ Wis Madison WI 53706

DOTTERWEICH, FRANK H(ENRY), b Baltimore, Md, Dec 11, 05; m 46. CHEMICAL ENGINEERING. *Educ:* Johns Hopkins Univ, BE, 28, PhD(chem eng), 37. *Prof Exp:* From cadet engr to asst gen supt gas opers, Consol Gas, Elec Light & Power Co, 28-34; instr & coach, Johns Hopkins Univ, 34-37; assoc prof eng, 37-41, dir div, 46-64, dean eng, 64-71, PROF ENG, TEX A&I UNIV, 41- *Concurrent Pos:* Tech consult, Petrol Admin War, 42-45 & Nat Gas & Gasoline. *Mem:* Am Chem Soc; Am Inst Mining, Metall & Petrol Engrs; Am Inst Chem Engrs; Am Gas Asn. *Res:* Processing of gas-condensate reservoirs; chemicals from gas. *Mailing Add:* Sch Eng Tex A&I Univ PO Box 2084 Kingsville TX 78363-8320

DOTTIN, ROBERT P, MOLECULAR GENETICS, DEVELOPMENTAL BIOLOGY. *Educ:* Univ Toronto, PhD(genetics), 72. *Prof Exp:* Assoc prof biol, Johns Hopkins Univ, 76-86; PROF BIOL, HUNTER COL, CITY COL NY, 86- *Mailing Add:* Dept Biol Sci Hunter Col 695 Park Ave New York NY 10021

DOTY, COY WILLIAM, agricultural engineering, soil conservation; deceased, see previous edition for last biography

DOTY, MAXWELL STANFORD, b Portland, Ore, Aug 11, 16; m 40; c 4. PHYCOLOGY. *Educ:* Ore State Col, BS, 40, MS, 42; Stanford Univ, PhD(biol), 45. *Prof Exp:* Asst bot, Ore State Col, 39-41; asst biol, Stanford Univ, 41-45; instr bot, Northwestern Univ, 45-46, asst prof, 46-50; chmn dept, 54-80, from assoc prof to prof, 50-87, EMER PROF BOT, UNIV HAWAII, 88- *Concurrent Pos:* Head dept bot, marine biol lab, Woods Hole Oceanog Inst, 46-51; chmn, Pac Sci Assoc, Sci Comn on Pac botany; chmn, Int Adv Comt Int Seaweed Symposia. *Mem:* AAAS; Int Phys Soc; Brit Phys Soc; Phys Soc France; Philippine Phys Soc. *Res:* Marine algae; marine algal production ecology and marine agronomy. *Mailing Add:* Dept Bot St John 614D Univ Hawaii Manoa 2500 Campus Rd Honolulu HI 96822

DOTY, MITCHELL EMERSON, b Clarksville, Iowa, Aug 18, 31; c 2. ELECTROCHEMISTRY. *Educ:* Mankato State Univ, BS; Univ Nebr, Lincoln, MS; Kansas State Univ, PhD. *Prof Exp:* Sr res scientist & proj coordr, NJ Zinc Co, Palmerton, Pa, 65-71; chemist, GAF Corp, Johnson City, NY, 73-76; HEAD, CHEM RES APPL MAT, AMETEK CORP, HARLEYSVILLE, PA, 76- *Mem:* Am Chem Soc; Electrochem Soc. *Mailing Add:* PO Box 316 Chalfont PA 18914-0316

DOTY, PAUL MEAD, b Charleston, WVa, June 1, 20; m 42, 54; c 4. BIOCHEMISTRY. *Educ:* Pa State Univ, BS, 41; Columbia Univ, MA, 43, PhD(chem), 44. *Hon Degrees:* Univ Chicago, 66. *Prof Exp:* Instr chem, Polytech Inst Brooklyn, 43-45, asst prof, 45-46; asst prof, Notre Dame Univ, 46-48; from asst prof phys chem to assoc prof chem, Harvard Univ, 48-56, prof chem, 56-68, chmn dept, 67-70, prof biochem & molecular biol, 68-80, Mallinckrodt prof, 80-88, prof pub policy, 88-90, EMER PROF BIOCHEM, HARVARD UNIV, 88-, EMER PROF PUB POLICY, 90- *Concurrent Pos:* Ed, J Polymer Sci, 45-61; Rockefeller fel, Cambridge Univ, 46-47; Harvey Lectr, 59-60; ed, J Molecular Biol, 59-63; chmn sci adv comt, Soviet-Am exchanges, Nat Acad Sci, 61-63; mem, President's Sci Adv Comt, 61-65; consult, US Arms Control & Disarmament Agency, 61-81; chmn comt rev gen ed, Harvard Univ, 62-64; mem, Coun Foreign Relations, 62-; chmn, Aspen Strategy Group, 72-82, mem, 82-; dir, Ctr Sci & Int Affairs, John F Kennedy Sch Govt Harvard, 73-85, emer dir, 85-; sr fel, Soc Fol, Harvard, 63-81, Aspen Inst, 75-87; mem bd, Aspen Inst Berlin, 74-, Int Inst Strategic Studies, London, 74-83, Mitre Corp, 75-, Dartmouth Conf Exec Comt, 76-; ed, Int Security, 75-85; mem, Comt Int Security & Arms Control, Nat Acad Sci, 80-, Comt Int Security Studies, Am Acad Arts & Sci, 81-, chmn, 88-91, chmn, Comt Int Security, Harvard Univ, 85-91. *Honors & Awards:* Pure Chem Award, Am Chem Soc, 56. *Mem:* Fel Nat Acad Sci; Am Chem Soc; Fedn Am Sci; Am Soc Biol Chem; fel Am Acad Arts & Sci; Am Philos Soc. *Res:* Molecular biology; structure, properties and function of nucleic acids and proteins. *Mailing Add:* Dept Biochem & Molecular Biol Harvard Univ Cambridge MA 02138

DOTY, RICHARD LEROY, b Boulder, Colo, Oct 14, 44. SENSORY PHYSIOLOGY, PSYCHOPHYSICS. *Educ:* Colo State Univ, BS, 66; Calif State Univ, San Jose, MA, 68; Mich State Univ, PhD(psychol), 71. *Prof Exp:* Res asst psychophys, Ames Res Ctr, NASA, 67-68; fel, Univ Calif, Berkeley, 72-73; head, Human Olfaction Sect & assoc mem, Monell Chem Senses Ctr, 73-78, FROM ASST PROF TO ASSOC PROF, DEPT OTORHINOLARYNGOL, COLON, HEAD & NECK SURG, MED SCH, UNIV PA, 73-, RES ASSOC DEPT PHYSIOL, 78-; DIR, SMELL & TASTE CTR, UNIV PA SCH MED & SCI DIR, SMELL & TASTE RES CTR, UNIV PA, 80- *Concurrent Pos:* Res assoc, Dept Oral Med, Philadelphia Gen Hosp, 73-78. *Mem:* AAAS; Am Soc Zoologists; Am Soc Mammalogists; Animal Behav Soc; Europ Chemoreception Res Orgn. *Res:* Endocrine-sensory physiology interactions; biological rhythms; chemoreception; evolutionary processes; animal behavior; trigeminal detection of chemical vapors. *Mailing Add:* 125 White Horse Pike Haddon Heights NJ 08035

DOTY, ROBERT L, b Missouri Valley, Iowa, Aug 2, 18. ELECTRICAL ENGINEERING. *Educ:* Univ Iowa, BS, 48; Iowa State Univ, PhD, 55. *Prof Exp:* From instr to asst prof elec eng, Iowa State Univ, 53-59; sect chief preliminary eng, Data Systs Div, Autonetics Div, N Am Aviation, Inc, 59-62, chief engr, 62-65, DIR ADVAN ANALYSTS, AUTONETICS DIV, ROCKWELL INT, 65- *Mem:* Sigma Xi. *Res:* Control theory; inertial navigation; digital computers. *Mailing Add:* 1506 Flippen Ct Anaheim CA 92802

DOTY, ROBERT WILLIAM, b New Rochelle, NY, Jan 10, 20; m 41; c 4. NEUROPHYSIOLOGY, NEUROPSYCHOLOGY. *Educ:* Univ Chicago, BS, 48, MS, 49, PhD(physiol), 50. *Prof Exp:* USPHS fel neurophysiol, Neuropsychiat Inst, Univ Ill, 50-51; asst prof physiol, Col Med, Univ Utah, 51-56; from asst prof to assoc prof, Med Sch, Univ Mich, 56-61; PROF PHYSIOL & PSYCHOL, CTR BRAIN RES, SCH MED, UNIV ROCHESTER, 61- *Concurrent Pos:* Mem, Nat Eye Inst Visual Sci Study Sect, 68-72, Int Brain Res Orgn, bd sci counselors, NIMH & bd sci advisors, Yerkes Primate Ctr; ed, Neurosci Transl, 63-70, Exp Neurol, 65-75, Acta Neurobiologiae Experimentalis, J Physiol (Paris) Pavlovian J Biol Sci & Behav Brain Res. *Honors & Awards:* Javits Award, 86. *Mem:* Fel AAAS; Am Physiol Soc; Soc Neurosci (pres, 76); fel Am Psychol Asn(pres, 84); Int Primatological Soc. *Res:* Visual cortex; reflex deglutition; problem of the dual brain, memory. *Mailing Add:* Dept Physiol Univ Rochester Sch Med Rochester NY 14642

DOTY, STEPHEN BRUCE, b Dekalb, Ill, Nov 20, 38; m 60. ANATOMY. *Educ:* Rice Univ, BA, 61, MA, 63, PhD(biol), 65. *Prof Exp:* Assoc prof orthop surg res, Sch Med, Johns Hopkins Univ, 65-75; res biologist, NIH, 75-78; ASSOC PROF ANAT, COL PHYSICIANS & SURGEONS, COLUMBIA UNIV, 78- *Concurrent Pos:* Fel, Dept Path, Nat Cancer Inst, NIH, 72-73. *Mem:* NY Acad Sci; Microbeam Anal Soc; Am Soc Gravitational & Space Biol. *Res:* Tissue culture, electron microscopy and cell physiology studies of mineralizing tissues; hormonal and bio-electrical controls governing bone formation; bone formation and implants. *Mailing Add:* Dept Anat & Cell Biol Columbia Univ Col Phys Surg 630 W 168th St New York NY 10032

DOTY, W(ILLIAM) D'ORVILLE, b Rochester, NY, Mar 11, 20; m 45; c 2. METALLURGY, MATERIALS ENGINEERING. *Educ:* Rensselaer Polytech Inst, BMetE, 42, MMetE, 44, PhD(metall), 46. *Prof Exp:* Welding res fel & asst, Rensselaer Polytech Inst, 42-46, supvr, 46-47; welding metallurgist, Carnegie-Ill Steel Corp, 47-52; res engr welding, US Steel Corp, 52-57, div chief bar, plate & forged prod, 58-66, res consult steel prod develop, Appl Res Lab, 66-73, chief staff engr, 73-75, chief res engr, 75-77, sr res consult, 77-85; PRIN CONSULT, DOTY & ASSOCS INC, 85- *Concurrent Pos:* With Off Sci Res & Develop, 44; chmn pressure vessel res comt, Welding Res Coun; mem heavy sect steel technol comt, Atomic Energy Comn-Oak Ridge Nat Lab; mem adv comt, Ship Hull Res Comt, Nat Res Coun; mem, main comt ASME Boiler & Pressure Vessel Comt, Welding Res Coun Comts, Am Welding Soc Comts. *Honors & Awards:* Spraragen Award, Am Welding Soc, 66; Pressure Vessel Res Comt Award, 73; J Hall Taylor Award, Am Soc Mech Engrs, 84. *Mem:* Am Inst Mining, Metall & Petrol Engrs; fel Am Soc Metals; hon mem Am Welding Soc; fel Am Soc Mech Engrs; fel Brit Welding Inst; Sigma Xi. *Res:* Development and application of new and improved carbon, high-strength low-alloy and high-yield-strength alloy steels; weldability of steels, especially quenched and tempered carbon and alloy steels; design, material selection and fabrication of pressure vessels and structures. *Mailing Add:* 1276 Earlford Dr Pittsburgh PA 15227

DOUB, WILLIAM BLAKE, b Rocky Mt, NC, Dec 10, 24; m 64. NUCLEAR PHYSICS. *Educ:* Univ Calif, Los Angeles, PhD(nuclear physics), 56. *Prof Exp:* SR SCIENTIST, BETTIS ATOMIC POWER LAB, WESTINGHOUSE ELEC CORP, WEST MIFFLIN, 57- *Mem:* Am Phys Soc; Am Nuclear Soc. *Res:* Nucleon-nucleon low energy scattering; reactor physics. *Mailing Add:* 628 Rolling Green Dr Bethel Park PA 15102

DOUBEK, DENNIS LEE, b Chicago, Ill, June 15, 44. BIOCHEMISTRY, ORGANIC CHEMISTRY. *Educ:* Univ Ill, PhD(biochem), 77. *Prof Exp:* Fac res assoc, 76-78, SR RES CHEMIST NATURAL PROD CHEM, CANCER RES INST, ARIZ STATE UNIV, 78- *Mem:* AAAS. Mem: AAAS. *Res:* Isolation and characterization of antineoplastic agents from marine organisms. *Mailing Add:* Cancer Res Inst Ariz State Univ Tempe AZ 85287-2404

DOUBLEDAY, CHARLES E, JR, b Corpus Christi, Tex, July 29, 44. PHYSICAL ORGANIC CHEMISTRY. *Educ:* Univ Kans, AB, 66; Univ Chicago, PhD(chem), 73. *Prof Exp:* Res chem, Univ Tex, Austin, 73-75; asst prof chem, State Univ NY, Buffalo, 75-; PROF CHEM, COLUMBIA UNIV. *Mem:* Am Chem Soc. *Res:* Chemically induced dynamic nuclear polarizations; potential surfaces of excited states of organic molecules. *Mailing Add:* Dept Chem Columbia Univ New York NY 10027-2399

DOUBT, THOMAS J, b Johnstown, Pa, July 29, 44. EXERCISE PHYSIOLOGY, SPORTS MEDICINE. *Educ:* Temple Univ, PhD(med physiol), 76. *Prof Exp:* HEAD CARDIOVASC RES, NAVAL MED RES INST, 80- *Mailing Add:* Dept Diving Med Nat Naval Med Res Ctr Bethesda MD 20814

DOUDNEY, CHARLES OWEN, b Dallas, Tex, Nov 5, 25; c 1. BIOCHEMISTRY, GENETICS. *Educ:* Univ Tex, PhD(genetics), 53. *Prof Exp:* Res biologist, Univ Tex, 51-53; assoc biologist, Oak Ridge Nat Lab, 53-55; res assoc, Univ Pa, 55-56; from assoc biologist to biologist, Univ Tex M D Anderson Hosp & Tumor Inst, 56-67, chief, Sect Genetics, 56-67, assoc prof biol, Post-grad Sch Med, 57-67, assoc, Grad Sch Biomed Sci, 63-67; head, Dept Genetics, Albert Einstein Med Ctr, 67-74; RES SCIENTIST, NY STATE DEPT HEALTH, 74- *Mem:* Genetics Soc Am; Am Soc Microbiol. *Res:* Biochemistry and genetics of neurospora; radiation protection and recovery of microorganisms; mutation; macromolecular synthesis; ultraviolet effects on microorganisms; investigations concern UV-induced mutation in bacteria, survival of bacteria after UV-damage and the interaction of DNA repair processes in promoting these phenomena; mutation induction in microorganisms. *Mailing Add:* 586 Pearse Rd Niskayuna NY 12309

DOUDOROFF, PETER, b Libau, Russia, June 22, 13; nat US; div; c 1. AQUATIC ECOLOGY, TOXICOLOGY. *Educ:* Stanford Univ, AB, 35; Univ Calif, PhD(zool), 41. *Prof Exp:* Asst biologist, US Bur Fisheries, 34; asst biol, Stanford Univ, 34-35; asst oceanog, Scripps Inst, Univ Calif, 35-41; biologist, La State Dept Conserv, 41-45; aquatic biologist, US Fish & Wildlife Serv, Wash & Ore, 46-47; biologist, USPHS, Ohio, 47-53, supvry fishery res biologist, Ore, 53-65; prof fisheries, 54-76, EMER PROF FISHERIES, ORE STATE UNIV, 77- *Concurrent Pos:* Consult biologist, 69-80. *Honors & Awards:* Award of Excellence, Am Fisheries Soc, 78. *Mem:* AAAS; Am Fisheries Soc. *Res:* Ecology, physiology and toxicology of fresh-water and marine fishes; inland fisheries biology; water pollution. *Mailing Add:* 3483 NW Crest Dr Corvallis OR 97330

DOUEK, MAURICE, b Cairo, Egypt, Nov 2, 48; Can citizen; m 75; c 2. PHYSICAL CHEMISTRY. *Educ:* McGill Univ, BSc, 70, PhD(chem), 75. *Prof Exp:* Fel chem, 75-77, assoc scientist, 77-85, SCIENTIST CHEM, PULP & PAPER RES INST CAN, 85- *Mem:* Can Pulp & Paper Tech Asn Pulp & Paper Indust. *Res:* Development of new methods of chemical analysis for the pulp and paper industry. *Mailing Add:* 570 St John's Blvd Pointe Claire PQ H9R 3J9 Can

DOUGAL, ARWIN A(DELBERT), b Dunlap, Iowa, Nov 22, 26; m 51; c 4. ELECTRICAL ENGINEERING, PHYSICAL ELECTRONICS. *Educ:* Iowa State Col, BS, 52; Univ Ill, MS, 55, PhD(elec eng), 57. *Prof Exp:* Engr, Collins Radio Co, 52; asst elec eng, Univ Ill, 52-56, res assoc, 56-57, from asst prof to assoc prof, 57-61; dir labs, Electronics & Rel Sci Res, 64-67, dir electronics res ctr, 71-77, PROF ELEC ENG, UNIV TEX, AUSTIN, 61- *Concurrent Pos:* Asst dir defense res & eng for res, Dept of Defense, DC, 67-69. *Honors & Awards:* Centennial Medal, Inst Elec & Electronics Engrs, 84. *Mem:* Fel Am Phys Soc; fel Inst Elec & Electronics Engrs; Soc Eng Sci; Am Soc Eng Educ; Optical Soc Am. *Res:* Electrical gaseous discharge and plasma physics; magnetohydrodynamics; controlled thermonuclear fusion; electron physics of the gaseous state; ionization phenomena and physics of the earth-earth's upper atmosphere; science and technology of lasers and coherent electro-optics; holography; scientific research administration. *Mailing Add:* Eng Sci Univ Tex B 112 Austin TX 78712

DOUGALIS, VASSILIOS, b Athens, Greece, Mar 19, 49; m 76. NUMERICAL ANALYSIS, PARTIAL DIFFERENTIATION EQUATIONS. *Educ:* Princeton Univ, BSE, 71; Harvard Univ, SM, 73, PhD(appl math), 76. *Prof Exp:* ASST PROF, DEPT MATH, UNIV TENN, 76- *Concurrent Pos:* Vis prof, Dept Math, Univ Crete, Greece, 80. *Mem:* Soc Indust & Appl Math; Am Math Soc. *Res:* Numerical methods for nonlinear partial differential equations of hyperbolic and parabolic type occuring mainly in fluid mechanics. *Mailing Add:* Inst Appl & Comp Math Res Ctr Crete PO Box 1527 Iraklion Crete Greece

DOUGALL, RICHARD S(TEPHEN), b Schenectady, NY, Apr 22, 37; m 67; c 4. HEAT TRANSFER, FLUID MECHANICS. *Educ:* Union Col, NY, BSME, 59; Mass Inst Technol, SM, 60, ME & ScD, 63. *Prof Exp:* Res asst, Heat Transfer Lab, Mass Inst Technol, 60-63; asst prof 63-68, ASSOC PROF MECH ENG, UNIV PITTSBURGH, 68-, ASSOC CHMN, MECH ENG DEPT, 83- *Concurrent Pos:* Consult, various govt agencies & indust corps. *Mem:* Am Soc Mech Engrs; Am Soc Eng Educ; Am Nuclear Soc; Soc Natural Philos; Sigma Xi; Am Soc Heating, Refrig & Air Conditioning Engrs. *Res:* Boiling heat transfer and two-phase flow in engineering systems; computational and experimental heat transfer. *Mailing Add:* Dept Mech Eng Univ Pittsburgh 648 Benedum Hall Pittsburgh PA 15261

DOUGAN, ARDEN DIANE, b Muncie, Ind, July 18, 54. NUCLEAR INSTRUMENTS & COMPUTATIONS. *Educ:* Purdue Univ, BS, 76; Univ Rochester, MS, 78, PhD(chem), 81. *Prof Exp:* CHEMIST, LAWRENCE LIVERMORE NAT LAB, 81- *Mem:* Am Chem Soc; Am Phys Soc. *Res:* Gamma spectrometry; nuclear measurements. *Mailing Add:* Lawrence Livermore Nat Lab PO Box 808 L-233 Livermore CA 94550

DOUGH, ROBERT LYLE, SR, b Ft Lauderdale, Fla, Aug 5, 31; m 53; c 5. SCIENCE EDUCATION. *Educ:* Guilford Col, BS, 53; NC State Univ, MS, 56, PhD(appl physics), 62. *Prof Exp:* Instr physics, NC State Col, 56-60; assoc nuclear engr, Astra, Inc, 60-61; asst prof physics, NC State Univ, 62-67; assoc prof sci educ, 68-77, PROF SCI EDUC, E CAROLINA UNIV, 77- *Concurrent Pos:* Vis asst prof, Ctr Res in Col Instr Sci & Math, Fla State Univ, 67-68. *Mem:* Nat Sci Teachers Asn; Am Asn Physics Teachers. *Res:* Introductory physics education; science and society; using microcomputers in teaching physical science. *Mailing Add:* Dept Sci Educ E Carolina Univ Greenville NC 27834-4353

DOUGHERTY, CHARLES MICHAEL, b Sept 16, 44; US citizen; div; c 1. ORGANIC CHEMISTRY. *Educ:* Williams Col, AB, 66; Pa State Univ, PhD(chem), 72. *Prof Exp:* Asst prof, 73-78, ASSOC PROF, DEPT CHEM, LEHMAN COL, 78- *Concurrent Pos:* NSF vis fac fel, Columbia Univ, 80-81. *Mem:* NY Acad Sci; Sci Res Soc NAm; Am Chem Soc; The Chem Soc. *Res:* Development of new synthetic methods especially as applied to systems of potential biological and/or pharmaceutical activity; computer modelling and synthetic design of enzyme inhibitors. *Mailing Add:* Dept Chem Lehman Col Bronx NY 10468

DOUGHERTY, DENNIS A, b Harrisburg, Pa, Dec 4, 52; m 73; c 2. ORGANIC CHEMISTRY. *Educ:* Bucknell Univ, BS, 74, MS, 74; Princeton Univ, PhD(chem), 78. *Prof Exp:* Fel chem, Yale Univ, 78-79; from asst prof to assoc prof, 79-89, PROF CHEM, CALIF INST TECHNOL, 89- *Concurrent Pos:* Fel, Alfred P Sloan Found, 83-85; Camille & Henry Dreyfus Teacher Scholar, 84-89. *Mem:* Am Chem Soc. *Res:* Physical organic chemistry; strained hydrocarbons and related structures; biradicals; photochemistry; molecular recognition; theoretical and computational organic chemistry; bioorganic chemistry. *Mailing Add:* Dept Chem 164-30 Calif Inst Technol Pasadena CA 91125

DOUGHERTY, EUGENE P, b Wilmington, Del, Nov 24, 53; m 75; c 3. EMULSION POLYMERIZATION, POLYMERIZATION REACTION ENGINEERING. *Educ:* St Joseph's Col, BS, 75; Princeton Univ, PhD(phys chem), 80. *Prof Exp:* SR RES SCIENTIST, ROHM & HAAS CO, 79- *Concurrent Pos:* Ed, Polymerization Reaction Eng, 91- *Mem:* Am Chem Soc; Am Phys Soc. *Res:* Polymerization reaction engineering; SCOPE model for simulation and control of emulsion polymerization; polymerization process control; polymerization kinetics; new polymer process and product development for plastics and coatings. *Mailing Add:* Rohm & Haas Co PO Box 219 Bristol PA 19007

DOUGHERTY, HARRY L, b San Jose, Costa Rica, Sept 22, 26; US citizen; m 52; c 4. ANATOMY, ORTHODONTICS. *Educ:* Univ Calif, Berkeley, AB, 51, MA, 53; Univ Calif, San Francisco, DDS, 57. *Prof Exp:* Assoc clinician, Children's Hosp, Los Angeles, 58-60; chief sci, Orthop Hosp, 60-64; ASSOC PROF ORTHOD & CHMN DEPT, SCH DENT, UNIV SOUTHERN CALIF, 60- *Mem:* AAAS; Am Asn Orthod. *Res:* Cleft palate research. *Mailing Add:* Dept Orthod Univ Southern Calif Sch Dent Los Angeles CA 90007

DOUGHERTY, HARRY W, OXYGEN METABOLISM, ENZYMOLOGY. *Educ:* Columbia Univ, PhD(biochem), 62. *Prof Exp:* SR RES FEL, MERCK INST, RAHWAY, NJ, 64- *Mailing Add:* 34 Crane Pkwy Cranford NJ 07016-3459

DOUGHERTY, JOHN A, b Shreveport, La, Jan 5, 43. BEHAVIORAL PHARMACOLOGY, BEHAVIORAL TOXICOLOGY. *Educ:* Univ Minn, BA, 69, PhD(psychol & pharmacol), 73. *Prof Exp:* CHIEF, NEUROBEHAV RES SECT TAFT LAB, NAT INST SAFETY HEALTH, CTR DIS CONTROL, 90- *Concurrent Pos:* Assoc prof, Dept Psychiat, Col Med, Univ Ky, 73-, asst prof, Dept Pharmacol, 75-, asst prof, Grad Ctr Toxicol, 82-; staff psychologist, Vet Admin Ctr, Lexington, Ky, 73- *Mem:* Soc Neurosci; Behav Pharmacol Soc; Am Psychol Asn; Am Psychol Soc. *Res:* Behavioral effects of chemicals and drugs. *Mailing Add:* Neurobehav Res Sect Nat Inst Safety Health C-24 4676 Columbia Pkwy Cincinnati OH 45226-1998

DOUGHERTY, JOHN JOSEPH, b Paterson, NJ, Oct 16, 23; m 53; c 4. ANALYSIS. *Educ:* US Naval Acad, Annapolis, Md, BS, 45, BSEE, 52, Monterey, Calif, MSEE, 53; Notre Dame Univ, MBA, 70;Ariz Col, AB, 75; Columbia Pac Univ, PhD(humane lett & Irish hist), 75. *Prof Exp:* Instr elec & electronics eng, US Naval Acad, Md, 48-49; commun eng, North Eastern Univ, Boston, 58-60; prof fac & prof, Univ Notre Dame, Ind, 69-70; sys chmn, Ctr Sys Sci, Prescott Col, Ariz, 70-71, pres, 71-72, prof comm, math & sys, 72-82; FAC ASSOC, COLUMBIA PAC UNIV, 82- *Concurrent Pos:* Invited lectr, Ireland, 88- *Mem:* Assoc fel Am Inst Aeronaut & Astronaut; Inst Elec & Electronics Engrs; Soc Naval Engrs. *Res:* Organizational analysis; radio antennae and transmissions lines; Irish history, especially the Spanish Armada. *Mailing Add:* Western Energy Mgt Consult 20 Santa Fe Ct Prescott AZ 86301

DOUGHERTY, JOHN WILLIAM, b Jerome, Ariz, Oct 13, 25; m 47; c 4. GENERATOR DESIGN. *Educ:* Univ Calif, Berkeley, BS, 47; Polytech Inst Brooklyn, PhD(syst sci), 69. *Prof Exp:* Test engr, 47-48, design engr medium steam turbine, Generator & Gear Dept, 48-55, design anal engr, 55-58, design methods engr, 58-60, consult eng ed, Corp Eng, 60-71, sr engr test facil, Turbine Div Facil Planning, 71-72, tech leader test integration, 72-76, TECH LEADER ELECTROMAGNETICS & DESIGN, LARGE STEAM TURBINE-GENERATOR DEPT, GEN ELEC CO, 77-; AT DEPT ELEC ENG, FLA POWER & LIGHT CO, JUNO BEACH. *Mem:* Inst Elec & Electronics Engrs. *Res:* Synchronous generator design; computerated design; Chebyshev approximation theory; finite element analysis. *Mailing Add:* Power Plant Eng Fla Power & Light Co Juno Beach FL 33408

DOUGHERTY, JOSEPH C, b Troy, NY, Feb 13, 34; m 59; c 4. NEPHROLOGY, MEDICAL ADMINISTRATION. *Educ:* Manhattan Col, BS, 56; Cornell Univ, MD, 60. *Prof Exp:* Intern, intenal med, Albany Med Ctr Hosp, NY, 60-61, asst res med, 61-62; sr res med, Cornell Med Div, Bellevue Hosp, NYC, 62-64, instr med, 66-67; fel cardiol, NY Hosp, 64-66; asst prof med & nephrol, Montefiore Med Ctr, Bronx, NY, 67-71; assoc prof med & nephrol, Med Sch, Univ Tex, San Antonio, 71-76, Scranton-Temple Res Prog, 76-80; DIR, WATSON WISE DIALYSIS UNIT, HARLINGEN, 80- *Concurrent Pos:* Lectr, Pfizer Pharmaceut, 81-; med dir, Valley Diag Clin, Harlington, Tex, 83-86; bd dirs, Am Heart Asn, 83- *Mem:* Am Col Physicians; Transplantation Soc; Int Soc Nephrol; Am Heart Asn. *Res:* Intermediate metabolism in isolated perfered organs and cardiovascular reactivity in transplanted organs. *Mailing Add:* Valley Diag Clin 2200 Haine Dr Harlingen TX 78550

DOUGHERTY, RALPH C, b Dillon, Mont, Jan 31, 40; m 60; c 3. PHYSICAL ORGANIC CHEMISTRY, BIOCHEMISTRY. *Educ:* Mont State Col, BS, 60; Univ Chicago, PhD(phys org chem), 63. *Prof Exp:* Resident res assoc, Argonne Nat Lab, 63-65; asst prof chem, Ohio State Univ, 65-69; assoc prof, 69-77, PROF CHEM, FLA STATE UNIV, 77- *Concurrent Pos:* Mem, Nat Res Coun Comt Analytical Chem, 75-78. *Mem:* Am Soc Andrology; Am Chem Soc; Am Soc Mass Spectrometry; Soc Toxicol. *Res:* Organic mass spectrometry; pollution chemistry; ion-molecule reactions; biochemical epidemiology; molecular orbital theory; analytical toxicology; monoclonal antibodies for analysis. *Mailing Add:* Dept Chem Fla State Univ Tallahassee FL 32306

DOUGHERTY, ROBERT MALVIN, b Long Branch, NJ, May 25, 29; m 50; c 2. MICROBIOLOGY, VIROLOGY. *Educ:* Rutgers Univ, BS, 52, MS, 54, PhD(microbiol), 57. *Prof Exp:* Instr microbiol, Sch Med & Dent, Univ Rochester, 57-60; Nat Cancer Inst spec res fel virol, Imp Cancer Res Fund, London, Eng, 60-62; assoc prof, 62-69, PROF MICROBIOL, STATE UNIV NY UPSTATE MED CTR, 69- *Mem:* AAAS; Am Soc Microbiol; Soc Exp Biol & Med; Am Asn Immunol; Brit Soc Gen Microbiol; Sigma Xi. *Res:* Virology; virus immunology; viral oncology. *Mailing Add:* Dept Microbiol 766 Irving Ave Syracuse NY 13210

DOUGHERTY, ROBERT WATSON, b Newcomerstown, Ohio, Feb 5, 04; m 48; c 2. VETERINARY MEDICINE. *Educ:* Iowa State Univ, BS, 27; Ohio State Univ, DVM, 36; Ore State Univ, MS, 41. *Prof Exp:* Instr vet med, Ore State Univ, 37-39; from asst prof to assoc prof, State Col Wash, 46-48; prof, State Univ NY Vet Col, Cornell Univ, 48-61, actg head , Dept Vet Physiol, 53-54; Leader Physiopath invests, Nat Animal Disease Lab, 61-74; RETIRED. *Concurrent Pos:* Fulbright scholar, NZ, 56-57; prof, Iowa State Univ, 62-; chmn orgn & ed comt, Int Symposium Physiol Digestion in Ruminant, 64; vis prof, Univ Wis-Madison, 78. *Honors & Awards:* Borden Award, Am Vet Med Asn, 63; Makowitz Award, Acad Exp Surg, 88. *Mem:* AAAS; NY Acad Sci; Conf Res Workers Animal Dis (pres, 71); distinguished fel Am Col Vet Pharmacol & Therapeut; Am Soc Vet Physiologists & Pharmacologists. *Res:* Reproductive physiology in ruminants; bovine spermatozoa and the vaginal pH of the female; formation and absorption of toxins from the bovine rumen; hematological changes in certain pathological conditions in ruminants; physiological studies of acute tympanites in ruminants. *Mailing Add:* Rte 2 Ames IA 50010

DOUGHERTY, THOMAS JOHN, b Buffalo, NY, Aug 2, 33. RADIOBIOLOGY. *Educ:* Canisius Col, BS, 55; Ohio State Univ, PhD(org chem), 59. *Prof Exp:* Res chemist, film dept, Yerkes Res & Develop Lab, E I du Pont de Nemours & Co, Inc, 59-67, staff scientist, 67-70; from cancer res scientist to assoc cancer res scientist, 70-75, PRIN CANCER RES SCIENTIST, ROSWELL PARK MEM INST, 75-, HEAD, DIV RADIOBIOL, DEPT RADIOL MED, 76- *Concurrent Pos:* Assoc prof, State Univ NY Buffalo, 75- *Honors & Awards:* Benjamin Franklin Award, 81; Schoelkopf Award, 88. *Mem:* AAAS; Am Chem Soc; Am Asn Cancer Res; Sigma Xi; Am Soc Photobiol. *Res:* Radiobiology, especially photochemically induced in vivo reactions; photodynamic therapy for cancer treatment. *Mailing Add:* Dept Radiation Med Roswell Park Cancer Inst 666 Elm St Buffalo NY 14263

DOUGHERTY, WILLIAM J, b Brooklyn, NY, June 3, 34; m 58; c 2. CELL BIOLOGY, HISTOLOGY. *Educ:* St Joseph's Col, Pa, BS, 56; Princeton Univ, MA, 61, PhD(biol), 63. *Prof Exp:* Biochemist, State Univ NY Downstate Med Ctr, 56-59; USPHS fel, Harvard Univ, 63-64; from asst prof to assoc prof, 64-77, PROF ANAT, MED UNIV SC, 77- *Concurrent Pos:* Vis prof anat, McGill Univ, 74-75. *Mem:* Am Soc Cell Biol; Am Soc Zoologists; Electron Micros Soc Am; Am Asn Anat; Histochem Soc. *Res:* Cytochemical and electron microscope study of cells and tissues, especially mechanisms of mitosis and cytokinetics and effects of heavy metals and other poisons on mitochondrial structure and function; muscle ultrastructure and cytochemistry; tooth mineralization. *Mailing Add:* Dept Anat & Cell Biol Med Univ SC 171 Ashley Ave Charleston SC 29425-2204

DOUGHMAN, DONALD JAMES, b Des Moines, Iowa, Sept 26, 33; m 71; c 4. OPHTHALMOLOGICAL SURGERY. *Educ:* Drake Univ, BME, 55; State Univ Iowa, MD, 61. *Prof Exp:* Instr ophthal, State Univ Iowa Hosps, 68; fel corneal dis, Mass Eye & Ear Infirmary, Boston, 68-72 & Retina Found, Boston, 70-72; asst prof, 72-75, assoc prof, 75-80, PROF & HEAD, DEPT OPHTHALMOL, UNIV MINN, 80- *Concurrent Pos:* Chief ophthal, USPHS Hosp, Boston, 68-70. *Mem:* Asn Res Vision & Ophthal; Am Acad Ophthal Otolaryngol; fel Am Col Surgeons. *Res:* Corneal transplantation; methods of corneal preservation utilizing organ culture incubation as a method of long term storage. *Mailing Add:* Dept of Ophthal Univ Minn Box 493 Mayo Mem Bldg Minneapolis MN 55455

DOUGHTY, CHARLES CARTER, b Alamosa, Colo, Dec 12, 15; m 44; c 2. HORTICULTURE. *Educ:* Kans State Univ, BS, 52; Wash State Univ, PhD(hort), 59. *Prof Exp:* Supt & asst horticulturist, Coastal Wash Res & Exten Unit, Wash State Univ, 54-65, assoc horticulturist, Western Wash Res & Exten Ctr, 67-75, horticulturist, 75-80; RETIRED. *Honors & Awards:* Dow Chem Co Award, 68. *Mem:* AAAS; Am Soc Hort Sci. *Res:* Effects of growth regulators on hardiness and freeze injury in small fruits, herbicide activity and nutrition of horticultural crops. *Mailing Add:* 7608 Orting Hwy E Sumner WA 98390

DOUGHTY, CLYDE CARL, b Hutchinson, Kans, July 21, 24; m 51; c 4. BIOCHEMISTRY. *Educ:* Univ Kans, BA, 48, MA, 50; Ill Inst Technol, PhD(microbiol biochem), 56. *Prof Exp:* Asst, Rheumatic Fever Inst, Northwestern Univ, 51-53; asst, Ill Inst Technol, 53-56; res biochemist, Charles F Kettering Found, 56-59; res assoc biol chem, 59-62, asst prof, 62-66, assoc prof biol chem & prev med, 66-76, PROF BIOL CHEM, UNIV ILL COL MED, 76- *Mem:* AAAS; Am Soc Biol Chem. *Res:* Enzymology; microbial cell wall; microbial metabolism; vision biochemistry. *Mailing Add:* Dept Biol Chem Univ Ill Med Ctr Chicago IL 60612

DOUGHTY, JULIAN O, b Tuscaloosa, Ala, June 11, 33; m 56; c 2. FLUID MECHANICS, RHEOLOGY. *Educ:* Miss State Univ, BS, 56, MS, 60; Univ Tenn, PhD(eng sci), 66. *Prof Exp:* Design engr, McDonnell Aircraft Co, 56-57; instr eng graphics, Miss State Univ, 57-60; instr & res asst basic eng & eng mech, Univ Tenn, 60-66; from asst prof to prof aerospace eng, 66-81, PROF MECH ENG, UNIV ALA, TUSCALOOSA, 81- *Concurrent Pos:* NASA res grant, 71-76. *Mem:* Am Inst Aeronaut & Astronaut; Am Soc Mech Engrs; Soc Automotive Engrs; Sigma Xi. *Res:* Low speed aerodynamics; viscoelastic fluids; two-phase flow in zero gravity; unsteady separated supersonic flow; solid fuel ramjet propulsion. *Mailing Add:* Rte 1 Box 743 Coker AL 35452

DOUGHTY, MARK, b Hull, Eng, Dec 20, 21; m 49; c 7. ORGANIC CHEMISTRY. *Educ:* Univ London, BSc, 49, PhD(org chem), 51. *Prof Exp:* Res dir org chem lab, Fothergill & Harvey, Eng, 51-54; lectr chem, Mt St Mary's Col, Eng, 54-61; lectr chem & physics, Staffordshire Col Technol, Eng, 62-63; asst prof, 63-65, ASSOC PROF CHEM & CHMN DEPT, CONCORDIA UNIV, CAN, 65-, PRIN, LONEGAN UNIV COL, 84- *Res:* Stereochemistry; science and theology; mechanism of organic reactions; philosophy of science. *Mailing Add:* Dept Chem Concordia Univ Sir G Williams 1455 Demaisonneuve Montreal PQ H3G 1M8 Can

DOUGLAS, ANDREW SHOLTO, b Pretoria, SAfrica, Feb 6, 53; US citizen; m 77; c 3. MECHANICS OF NON-LINEAR MATERIALS, CARDIAC MECHANICS. *Educ:* Univ Cape Town, BSc, 75, MSc, 77; Brown Univ, Providence, ScM, 79, PhD(eng mech), 82. *Prof Exp:* Asst prof mech eng, Rice Univ, Houston, 81-83; asst prof, 83-89, ASSOC PROF MECH ENG, JOHNS HOPKINS UNIV, BALTIMORE, 89- *Concurrent Pos:* Assoc prof biomed eng, Johns Hopkins Univ, 88- *Res:* Dynamic and ductile fracture mechanics; shear localization and stability of plastic deformation in metals; metal plasticity and viscoplasticity; mechanics of the left ventricle; cardiac mechanics. *Mailing Add:* Dept Mech Eng Johns Hopkins Univ Baltimore MD 21218

DOUGLAS, BEN HAROLD, b Wesson, Miss, Feb 20, 35; m 84; c 2. PHYSIOLOGY. *Educ:* Miss Col, BS, 56; Univ Miss, PhD(physiol), 64. *Prof Exp:* Instr basic electronics, Keesler AFB, Miss, 56-57; teacher physics, Copiah-Lincoln Jr Col, 57-60; chief instr, Sch Nursing, St Dominics Hosp, Miss, 62-63; from inst to asst prof, 64-68, ASSOC PROF PHYSIOL & BIOPHYS UNIV MISS, 68-, PROF ANAT, 77-, ASST VCHANCELLOR, 81- *Concurrent Pos:* Estab investr, Am Heart Asn, 72-; vis investr, Blood Pressure Unit, Western Infirmary, Glasgow, Scotland, 73-74; pres, Soc Study Pathophysiology Pregnancy, 90-91. *Mem:* AAAS; Am Heart Asn; Am Physiol Soc; Am Fedn Clin Res; Am Soc Study Reproduction; Sigma Xi. *Res:* Hypertension and its effects on pregnancy. *Mailing Add:* Univ Miss Med Ctr 2500 N State St Jackson MS 39216

DOUGLAS, BODIE E, b New Orleans, La, Dec 31, 24; m 45; c 4. INORGANIC CHEMISTRY. *Educ:* Tulane Univ, BS, 44, MS, 47; Univ Ill, PhD(chem), 49. *Prof Exp:* Asst, Tulane Univ, 46-47; res assoc chem, Pa State Univ, 49-50, asst prof, 50-52; from asst prof to assoc prof, 52-63, PROF CHEM, UNIV PITTSBURGH, 63- *Concurrent Pos:* Fulbright lectr, Univ Leeds, 54-55; vis prof, Osaka Univ, 70. *Mem:* fel AAAS; Am Chem Soc. *Res:* Coordination compounds; spectroscopy. *Mailing Add:* Dept Chem Univ Pittsburgh Pittsburgh PA 15260

DOUGLAS, BRUCE L, b New York, NY, July 14, 25; m 74; c 5. DENTISTRY, ORAL SURGERY. *Educ:* Princeton Univ, AB, 47; NY Univ, DDS, 48; Columbia Univ, dipl oral surg, 51, MA, 55, dipl higher educ, 57; Univ Calif, MPH, 62; Am Bd Oral & Maxillofacial Surg, dipl, 57. *Prof Exp:* Intern dent, Queens Gen Hosp, Jamaica, NY, 48-49; dent officer, US Navy, 51-53; resident oral surg, Queens Hosp Ctr, 53-54; pvt pract, 54-59; Fulbright prof oral surg, Col Med, Okayama Univ, Japan, 59-61; assoc prof oral med & coordr social aspects dent, Col Dent & assoc prof prev med, Col Med, Univ Ill Med Ctr, 63-67, prof prev med, Col Med & prof community dent, Col Dent, 67-72, PROF HEALTH ADMIN, SCH PUB HEALTH, UNIV ILL, 71-; ATTEND ORAL & MAXILLOFACIAL SURG, GRANT HOSP, CHICAGO, 81- *Concurrent Pos:* Assoc attend oral surgeon & dir oral surg training prog, New York City Hosp, Elmhurst, 56-59; attend oral surgeon, Chicago Cent Hosp, 64-80, chief, Dept Dent, 66-70, chmn, 78-80; consult var countries WHO, 64-66, 69 & 73; dir dent dept, Mile Sq Health Ctr, Presby-St Luke's Hosp, 67-68; attend oral surgeon, Rush-presby-St Luke's Med Ctr, Chicago, 67-77, dir, Dept Dent & Oral Surg, 68-75; mem, Ill House Rep, 71-74; prof dent & oral surg, Rush Med Col, 71-77; planning consult, Ill State Comprehensive Health Planning Agency, 75; dir, Off Dent Manpower Distrib, Ill Dept Pub Health, 75-76, chief, Div Dent Health, 76-77 & Health Manpower Develop, 77-78; mem bd gov, Chicago Inst Med, 70-80; consult, Med Lett Drugs & Therapeut, 78-; field dir, Water Fluoridation Res Proj, Am Dent Asn, 81-82; mem adv bd, United Blood Serv, 83- *Mem:* Am Dent Asn; Sigma Xi; fel Am Pub Health Asn; fel Am Dent Soc Anesthesiol (pres, 63-64); Am Asn Hosp Dentists (pres); fel Int Col Dentists; fel Am Med Writers Asn. *Res:* New methods of providing dental care to the chronically ill and aged; impact of water fluoridation on dental practice. *Mailing Add:* 2401 Duffy Lane Riverwoods IL 60015

DOUGLAS, BRYCE, b Glasgow, Scotland, Jan 6, 24; m 55; c 3. MEDICINAL CHEMISTRY. *Educ:* Glasgow Univ, BS, 44; Univ Edinburgh, PhD(chem), 48. *Prof Exp:* Res chemist, J F MacFarlan & Co, Scotland, 44-46; asst, Royal Col Physicians Lab, 47-49; dept biol chem, Aberdeen Univ, 49; Nat Res Coun Can fel, 49-51; fel, Harvard Med Sch, 52-53; res assoc natural prod, Ind Univ, 53-56, vis res assoc, Univ Malaya, 56-58; group leader to sect head, 58-67, dir res macrobiol, 67-71, from vpres to pres, 71-81, res & develop, Pharmaceut Prod, Smith Kline & French Labs, 71-81, vpres, Sci & Technol, Smith Kline Beckman Corp, 81-86; DIR, SCH DENT MED, UNIV PA, 80-; TRUSTEE, FRANKLIN INST, 83- *Concurrent Pos:* Mem, Study Comn on Pharm, 73-75; vpres, Royal Soc Med Found, NY, 72-; vchair bd, Beaver Col. *Mem:* AAAS; Am Chem Soc; Soc Chem Indust; fel Royal Soc Chem; Am Inst Chem; Sigma Xi. *Res:* Alkaloids; chemotherapy; heterocyclic compounds; management science. *Mailing Add:* Box 672 Kimberton PA 19442

DOUGLAS, CARROLL REECE, b Knoxville, Tenn, Sept 3, 32; m 51; c 1. POULTRY NUTRITION. *Educ:* Univ Tenn, BS, 55, PhD(animal nutrit), 66; Univ Fla, MSA, 59. *Prof Exp:* Asst poultry nutrit, Univ Tenn, 56-57; interim asst, Univ Fla, 57-59; farm mgr, Am Develop Corp, Fla, 60; asst mgr & mgr chick hatchery, Fla State Hatcheries, 60-62; nutritionist & supvr, Loret Mills, Tenn, 62; asst poultry nutrit, Univ Tenn, 62-66, poultryman & leader exten poultry, Agr Exten Serv, 66-69; asst prof poultry mgt & asst scientist, 69-73, assoc prof, 73-79, PROF POULTRY MGT & EXTEN POULTRYMAN, UNIV FLA, 79- *Mem:* AAAS; Poultry Sci Asn; World Poultry Sci Asn. *Res:* Protein, amino acid, energy and mineral requirements of broilers and laying hens; xanthophyll utilization by the chicken; management systems for growing chickens; nutrition of growing pullet. *Mailing Add:* Dept Poultry Sci Univ Fla 2A MEHR Bldg Gainesville FL 32611

DOUGLAS, CHARLES FRANCIS, b Tucson, Ariz, July 16, 30; m 52; c 2. AGRONOMY. *Educ:* Univ Ill, BS, 56, MS, 57; Purdue Univ, PhD(agron), 61. *Prof Exp:* Agr consult, Tenn Valley Authority, 61-65, supvr agr & indust educ progs on fertilizer mfr & use, Midwestern & New Eng States, 65-69; HEAD DEPT AGRON, COASTAL PLAIN EXP STA, UNIV GA, 69- *Mem:* Am Soc Agron. *Res:* Crop physiology and ecology; fertilizer education; experimental fertilizer use. *Mailing Add:* 802 Texas Dr Tifton GA 31794

DOUGLAS, CHARLES LEIGH, b Danville, Ill, Aug 2, 30; m 77; c 3. MAMMALIAN ECOLOGY. *Educ:* Antioch Col, AB, 59; Dartmouth Col, MS, 61; Univ Kans, PhD(zool), 67. *Prof Exp:* Biologist, Wetherill Mesa Archeol Proj, Mesa Verde Nat Park, 61-64; res scientist biol, Tex Mem Mus, Univ Tex, Austin, 66-69; assoc prof, Prescott Col, 69-73; assoc prof, 73-78, PROF, DEPT BIOL SCI, UNIV NEV, LAS VEGAS, 79-, RES SCIENTIST & UNIT LEADER, NAT PARK SERV COOP RESOURCES STUDIES UNIT, 73- *Concurrent Pos:* Collabr, Mesa Verde Nat Park, 65-70; chmn fac, Prescott Col, 70-71, chmn, Ctr Man & Environ, 71-73; partner, Southwestern Environ Consult, 72-74; owner-dir, Faunal Anal Lab, 73-; ed, Desert Bighorn Coun Transcripts, 76-85. *Mem:* Am Soc Mammalogists; Ecol Soc Am; Wildlife Soc; Soc Am Archeol. *Res:* Ecology of desert bighorn sheep and feral burros; analysis and interpretation of faunal remains from archeological sites; microanatomy of hair; aging parameters in southwestern mammals. *Mailing Add:* Dept Biol Sci Univ Nev Las Vegas NV 89154

DOUGLAS, CRAIG CARL, b Houston, Tex, Oct 12, 54; m 82. THEOREMS FOR ALGORITHMS, ALGORITHMS ON REAL COMPUTERS. *Educ:* Univ Chicago, AB, 77; Yale Univ, MS, 78, MPhil, 80, PhD(computer sci), 82. *Prof Exp:* Asst prof computer sci, Duke Univ, 84-86; instr, 82-84, RES AFFIL COMPUTER SCI, YALE UNIV, 88-; RES STAFF MEM MATH SCI, T J WATSON RES CTR, INT BUS MACH CORP, 86- *Concurrent Pos:* Vis prof, Univ Pavia, Italy, 88. *Mem:* Soc Indust & Appl Math; Asn Comput Mach. *Res:* New algorithms for solving partial differential equations on workstations, supercomputers and parallel computers; multigrid and domain decomposition methods. *Mailing Add:* T J Watson Res Ctr IBM PO Box 218 Yorktown Heights NY 10598

DOUGLAS, DAVID LEWIS, b Seattle, Wash, Jan 22, 20; m 48; c 2. ELECTOROCHEMISTRY. *Educ:* Calif Inst Technol, BS, 47, PhD(phys chem), 51. *Prof Exp:* Res assoc, Knolls Atomic Power Lab, Gen Elec Co, 51-55, phys chemist, Res Lab, 55-60, mgr fuel cell eng, Aircraft Accessory Turbine Dept, 60-62, tech planning, Direct Energy Conversion Oper, 62-64; dir res & develop, Gould-Nat Batteries, Inc, Gould Inc, 64-66, vpres res, 66-70, vpres & dir, Energy Tech Lab, 70-74, vpres contract res, 74-79; tech mgr, Elec Power Res Inst, 79-86; CONSULT, 86- *Concurrent Pos:* Adj prof chem eng, Univ Minn, 87- *Mem:* Fel AAAS; Am Chem Soc; Electrochem Soc; Sigma Xi. *Res:* Electrochemistry; fuel cells; batteries. *Mailing Add:* 8058 Pa Rd S Bloomington MN 55438

DOUGLAS, DEXTER RICHARD, b Benton, Ohio, Nov 14, 37; m 62; c 2. BOTANY. *Educ:* Kent State Univ, BS, 62; Univ Wyo, MS, 65; Univ Minn, PhD(plant path), 68. *Prof Exp:* Res pathologist, USDA, Res & Exten Ctr, Univ Idaho, 68-75; dir res, Chem Supply Co Inc, Twin Falls, Idaho, 76-78; pres, Hi-Alta, Inc, 78-80; area mgr, 80-81, PRES, IDAHO CROP IMPROVEMENT ASN INC. *Concurrent Pos:* Agr consult, 76- *Mem:* Am Potato Asn; Am Phytopath Soc. *Res:* Control of potato resistance and control of post-harvest pathogens; seed potato production. *Mailing Add:* Hi-Alta Inc Box 916 Moore ID 83255

DOUGLAS, DONALD WILLS, JR, b Washington, DC, July 3, 17; m 39, 50; c 2. AEROSPACE ENGINEERING. *Prof Exp:* Engr, 39-43, chief flight test group in chg testing models, 43-51, dir contact admin, 48, in chg res labs, Santa Monica Div, 49, vpres mil sales, 51-57, pres, Douglas Aircraft Co, 57-67, chmn & dir, Douglas Aircraft Co Can, Ltd, 65-74, & pres 68-71, vpres admin, 67-71, sr corp vpres, 71-72, DIR, MCDONNELL DOUGLAS CORP, 67-; PRES, DOUGLAS ENERGY CO, 81- *Concurrent Pos:* Pres, Douglas Develop Co, Calif, 72-74; sr consult mkt develop, Biphase Energy Systs, subsid Transam Deslaval. *Honors & Awards:* Chevalier, Legion of Honor, France; Officiale, Order of Merit, Repub of Italy. *Mem:* Am Inst Aeronaut & Astronaut; Aerospace Industs Asn; Nat Defense Transp Asn. *Mailing Add:* Centerior Energy Corp 6200 Oak Tree Blvd Independence OH 44131

DOUGLAS, DOROTHY ANN, b Bennington, Vt, May 13, 51. ECOLOGY, BOTANY. *Educ:* Oberlin Col, BA, 72; Univ Calif, Berkeley, PhD(bot), 77. *Prof Exp:* Asst prof biol, Earlham Col, 77-; ASSOC PROF BIOL, DEPT BIOL, BOISE STATE UNIV. *Concurrent Pos:* NSF fel, 72-75; grants, Sigma Xi, 75 & 76, Can Fac & Inst Res Prog, 85, Am Philos Soc, 86, Nat Geog Soc, 87 & US Forest Serv, 88; Prof Develop Fund grants, Earlham Col, 78-80; fac res grants, Boise St Univ, 82-84 & 86-87. *Mem:* Ecol Soc Am; Bot Soc Am; Brit Ecol Soc. *Res:* Plant ecology; reproductive ecology and demography; arctic and alpine ecology. *Mailing Add:* Dept Biol Boise State Univ 1910 University Dr Boise ID 83725

DOUGLAS, EDWARD CURTIS, b Greensburg, Pa, July 27, 40; m 73; c 2. ELECTRICAL ENGINEERING. *Educ:* Col Wooster, BA, 62; NY Univ, MEE, 64, PhD(elec eng), 69. *Prof Exp:* Mem tech staff elec eng, David Sarnoff Res Lab, 69-79, mgr, advan process technol develop, Solid State Technol Ctr, RCA Corp, 79-87; PRIN MEM TECH STAFF & CHIEF TECHNOLOGIST, GE MICROELECTRONICS CTR, 87- *Mem:* Inst Elec & Electronics Engrs; Electrochem Soc. *Res:* Technology development for very-large-scale integration in areas of ion implantation, plasma etching, direct step on wafer, photolithography, low pressure chemical vapor deposition, direct digital controlled furnaces, and advanced photo masks; managed development of 1.25 micron VHSIC silicon-on-sapphire fabrication process; managed development of submicron SOS process. *Mailing Add:* GE Microelectronics Ctr 3026 Cornwallis Rd PO Box 13049 Res Triangle Park NC 27709

DOUGLAS, GORDON WATKINS, b Midlothian, Va, June 2, 21; m 54; c 4. OBSTETRICS & GYNECOLOGY. *Educ:* Princeton Univ, AB, 42; Johns Hopkins Univ, MD, 45. *Prof Exp:* Asst, 49-52, asst prof, 53-56, PROF OBSTET & GYNEC, SCH MED, NY UNIV, 56- *Concurrent Pos:* Markle scholar, 52; dir, Am Bd Obstet & Gynec, 64- *Mem:* Am Col Obstet & Gynec; Am Col Surg; Asn Profs Gynec & Obstet; Harvey Soc; Am Obstet & Gynec Soc. *Res:* Immunology. *Mailing Add:* Dept Obstet & Gynec NY Univ Sch Med 550 First Ave New York NY 10016

DOUGLAS, HUGH, b Salisbury, SRhodesia, Oct 28, 27; m 53; c 4. NATURAL RESOURCES ECONOMICS. *Educ:* Amherst Col, AB, 49; Columbia Univ, MA, 51. *Prof Exp:* Geologist, AEC, 51 & Tex Gulf Sulphur Co, 51-55; sr geologist & asst to mgr, Am Overseas Petrol Ltd, Turkey, 55-59 & Libya, 59-62; consult, Oil Shale Corp & Swan Petrol Ltd, 62-63; secy treas, Thermonetics, Inc, 63-65; indust economist energy & natural resources & mgr mining & mineral econ, Stanford Res Inst, 65-71; mgr mineral planning, Utah Int Inc, 71-76; PRES, HUGH DOUGLAS & CO, RESOURCES ECON CONSULTS, 76- *Concurrent Pos:* Consult prof mineral econ, Stanford Univ. *Mem:* Mining & Metall Soc Am; Corp Planners Asn; Am Inst Mining, Metall & Petrol Engrs; Am Asn Petrol Geologists; Soc Econ Geologists. *Res:* Metals and minerals substitutes; economic analysis of world metals and minerals industries. *Mailing Add:* 124 16th Ave San Francisco CA 94118

DOUGLAS, J(AMES) M(ERRILL), b Aurora, Ill, July 27, 33; m; c 2. CHEMICAL ENGINEERING. *Educ:* Johns Hopkins Univ, BE, 54; Univ Del, PhD(chem eng), 60. *Prof Exp:* Res chem engr, Atlantic Refining Co, Pa, 60-62, res assoc, 62-65; asst prof chem eng, Univ Del, 65; assoc prof, Univ Rochester, 65-68; PROF CHEM ENG, UNIV MASS, AMHERST, 68- *Concurrent Pos:* Atlantic Refining Co res grant, Imp Col, Univ London, 64. *Mem:* Am Inst Chem Engrs. *Res:* Process synthesis; process dynamics and control; optimization theory; reactor design. *Mailing Add:* Dept Chem Eng Univ Mass Amherst Campus Amherst MA 01003

DOUGLAS, JAMES, b Uvalde, Tex, Oct 1, 14; m 41; c 4. CIVIL ENGINEERING. *Educ:* US Naval Acad, BS, 38; Rensselaer Polytech Inst, BCE & MCE, 43; Stanford Univ, PhD(civil eng), 63. *Prof Exp:* Civil Engr Corps, US Navy, 40-61; instr construct, 61-63, assoc prof, 63-75, PROF CIVIL ENG, STANFORD UNIV, 75- *Concurrent Pos:* Chmn comt construct mgt, Hwy Res Bd, 70. *Honors & Awards:* Thomas Fitch Rowland Prize, 69 & Construct Mgt Award, 75, Am Soc Civil Engrs. *Mem:* Fel Am Soc Civil Engrs. *Res:* Economics of construction equipment ownership, obsolescence, depreciation, standardization; simulation of construction equipment operations using statistical-computer solution. *Mailing Add:* Dept Civil Eng Sch Eng Stanford Univ Stanford CA 94305

DOUGLAS, JAMES NATHANIEL, b Dallas, Tex, Aug 14, 35; m 83; c 4. RADIO ASTRONOMY. *Educ:* Yale Univ, BS, 56, MS, 58, PhD(astron), 61. *Prof Exp:* Instr astron, Yale Univ, 60-61, from asst prof to assoc prof, 61-71, PROF ASTRON, UNIV TEX, AUSTIN, 71- *Concurrent Pos:* Mem Comn 5, Int Sci Radio Union, 63-; mem, Int Astron Union, 64-; mem, Adv Panel for Astron, NSF, 71-74. *Mem:* Fel AAAS; Am Astron Soc; Am Geophys Union. *Res:* Studies of decametric radiation from Jupiter and from discrete radio sources. *Mailing Add:* 13030 Northborough Dr No 201 Houston TX 77067

DOUGLAS, JAMES SIEVERS, RESPIRATORY PHARMACOLOGY, BRONCHIAL ASTHMA. *Educ:* Univ Wales, Cardiff, PhD(pharmacol), 67. *Prof Exp:* ASSOC FEL TOXICOL, JOHN B PIERCE FOUND LAB, 77- *Mailing Add:* John B Pierce Found Lab 290 Congress Ave New Haven CT 06519

DOUGLAS, JOCELYN FIELDING, b Delta, Utah, Jan 25, 27; m 51; c 3. BIOCHEMISTRY, ENVIRONMENTAL HEALTH. *Educ:* Univ Ill, BS, 48; Columbia Univ, MA, 50, PhD(chem), 53. *Prof Exp:* Asst chem, Columbia Univ, 48-52; res chemist, Johnson & Johnson, 52-54, proj leader, 54-58; dir biochem, Wallace Labs, 58-74; PRES, SCI SERV INC, 84- *Concurrent Pos:* Expert, Nat Cancer Inst, 76-80, dep dir, carcinogenesis testing prog & chief opers, nat toxicol prog, 79-84; consult toxicol & carcinogenesis, 74- *Honors & Awards:* Richard Neff Soc Award, 66. *Mem:* AAAS; Am Soc Pharmacol & Exp Therapeut; Am Chem Soc; Int Soc Biochem Pharmacol; NY Acad Sci; Soc Toxicol. *Res:* Scientific consulting; toxicology; chemical carcinogenesis; drug analysis; metabolism; environmental health; scientific management. *Mailing Add:* Hermitage Farm PO Box 533 Front Royal VA 22630

DOUGLAS, JOHN EDWARD, b Normal, Ill, June 29, 26; c 3. THEORETICAL CHEMISTRY, NUMERICAL METHODS IN SCIENCE EDUCATION. *Educ:* Univ Chicago, BS, 47, MS, 48; Univ Wash, PhD(chem), 52. *Prof Exp:* Instr chem, Univ Wyo, 52-56; phys chemist, Stanford Res Inst, 56-60; assoc prof, 60-66, chmn dept, 73-76, vprovost grad studies & res, 80-87, CHEM FAC, EASTERN WASH UNIV, 66- *Concurrent Pos:* Vis prof, Canterbury Univ, Eng, 68-69 & Univ Calif, San Francisco, 78-79. *Mem:* AAAS. *Res:* Gas phase kinetics; energetics of complex formation; non-linear processes; computer simulations in chemistry. *Mailing Add:* Chem Dept Eastern Wash Univ Cheney WA 99004

DOUGLAS, JOSEPH FRANCIS, b Indianapolis, Ind, Oct 31, 26; m 50; c 4. ELECTRICAL POWER CONTROL. *Educ:* Purdue Univ, BSEE, 48; Univ Mo, MSEE, 62. *Prof Exp:* Proj engr elec eng, USDA, 48-56; asst prof, Southern Univ, 56-62, assoc prof & head dept, 62-64; training coordr, Atomics Div, Am Mach & Foundry Co, 64-66; group leader elec technol & eng, Pa State Univ, York, 66-74, app assoc dean acad instr, 74-77, asst prof eng, 66-70, assoc prof elec technol, bachelor technol prog, 77-87, ASSOC PROF ENG, PA STATE UNIV, YORK, 70-; CONSULT, 85- *Concurrent Pos:* Bd dir, Inst Elec Electronic Engrs, 85 & 86. *Honors & Awards:* Lindbach Award, 72; Centennial Medal, Inst Elec & Electronics Engrs, 84. *Mem:* Sr mem Inst Elec & Electronics Engrs. *Res:* Low speed with reasonable torque output from electrical induction motors of the wound rotor class. *Mailing Add:* 2755 Trout Run Rd York PA 17402-8957

DOUGLAS, LARRY JOE, b Oklahoma City, Okla, Mar 3, 37; m 69; c 5. PHYSICAL CHEMISTRY, ELECTROCHEMISTRY. *Educ:* Univ Denver, BS, 58, PhD(phys chem), 70. *Prof Exp:* Assoc res engr, Nat Cash Register Co, Calif, 58-61, res engr, 62-64; res scientist, Denver Res Ctr, Marathon Oil Co, 70-76; sr prog mgr alcohol fuels, Solar Energy Res Inst, 78-87; PRES, ENTROPY ASSOCS, 87- *Mem:* AAAS; Sigma Xi; Electrochem Soc; Soc Petrol Engrs. *Res:* Mass transport mechanisms in oxide crystals; electrochemistry of alloy deposition; liquid/liquid and liquid/solid interfacial properties; adsorption into and from micellar solutions; electrokinetic behavior of non-electrolytic fluids; biotechnology/biofuels and energy analysis and process evaluation. *Mailing Add:* 1747 S Field Ct Denver CO 80226

DOUGLAS, LLOYD EVANS, b New York, NY, Oct 5, 51; div. SOFTWARE SYSTEMS, MATHEMATICAL STATISTICS. *Educ:* City Col NY, BS, 72; Miami Univ, MS, 74. *Prof Exp:* Teaching fel, Boston Univ, 75-76, sr teaching fel, 76; mathematician, US Naval Underwater Syst Ctr, 76-77, comput specialist, 77-79; computer specialist, Trident Command & Control Syst Maintenance Activ, 79-80; opers res analyst, US Army Commun & Electronics Command, 80- 83; comput specialist, Gen Serv Admin, 83-84; head cent syst mgr sect, 84-89, spec asst mgt, 89, DIR, POLICY & EVAL STAFF, NSF, 89- *Mem:* Math Asn Am; Am Math Soc. *Res:* Software systems; mathematical statistics; algebra; analysis and functional analysis. *Mailing Add:* 5544 Karen Elaine Dr 1521 New Carrollton MD 20784

DOUGLAS, LOWELL ARTHUR, b Durango, Colo, June 17, 26; m 53; c 2. CLAY MINERALOGY, SOIL GENESIS. *Educ:* Utah State Univ, BS, 52, MS, 59; Rutgers Univ, PhD(soils), 61. *Prof Exp:* Soil chemist, US Salinity Lab, 54-56; from instr to assoc prof, 59-69, dir grad prog in soils & crops, 76-84, PROF SOILS, RUTGERS UNIV, 69-, CHMN, SOILS & CROPS DEPT, 82- *Concurrent Pos:* Ed, Soil Sci, 79- *Honors & Awards:* Fel, Soil Sci Soc Am. *Mem:* Soil Sci Soc Am; Clay Minerals Soc; Mineral Soc Am; Int Soc Soil Sci. *Res:* Clay mineralogy, soil micromorphology; soil chemistry; soils and faulting; clay mineral genesis and alteration. *Mailing Add:* Dept Soils & Crops Rutgers The State Univ New Brunswick NJ 08903

DOUGLAS, MATTHEW M, b East Grand Rapids, Mich, Apr 3, 49; m 86; c 3. THERMOREGULATORY ECOLOGY. *Educ:* Univ Mich, Ann Arbor, BA, 72; Eastern Mich Univ, MS, 74; Univ Kans, PhD(entomol), 78. *Prof Exp:* Asst prof entom, Calif State Univ, Fresno, 77-78; asst prof, Boston Univ, 78-80; WRITER & RESEARCHER, 81- *Concurrent Pos:* Lectr, Marine Prog, Boston Univ, 79; Consult, Arthur D Little, Inc, 79, Boston Mus Sci, 80; asst prof zool, Harvard Univ, 79; adj asst prof entom & biophys ecol, Boston Univ, 81-82; adj sr res scientist, Univ Kans, 82- *Mem:* AAAS; Soc Study Evolution; Lepidopterists Soc; Entom Soc Am. *Res:* Behavioral and physiological thermoregulatory strategies of insects, especially the Lepidoptera; fern defenses, chemical and other, against insect attack. *Mailing Add:* 1503 Woodland St Jenison MI 49428

DOUGLAS, MICHAEL GILBERT, b Perth, Western Australia, June 9, 45; US citizen; m 73; c 3. BIOCHEMISTRY, MOLECULAR BIOLOGY. *Educ:* Southwestern Univ, BS, 67; St Louis Univ, PhD(biochem), 74. *Prof Exp:* Res assoc, Dept Biochem, Health Sci Ctr, Univ Tex, Dallas, 73-75; Swiss State fel, Biozentrum, Dept Biochem, Univ Basel, 75-77; asst prof to assoc prof, 77-88, PROF BIOCHEM, HEALTH SCI CTR, UNIV TEX, 88- *Concurrent Pos:* NIH fel, Univ Tex, Dallas, 74-75; NIH grant, 78-88; Burroughs Wellcome vis prof, 82, Robert A Welch prof, 85- *Mem:* Am Soc Cell Biol; Am Soc Chem; AAAS; NY Acad Sci; Am Soc Biol Chem & Molecular Biol; Am Soc Microbiol; Genetics Soc Am. *Res:* Biochemical genetics, assembly of intracellular organelle membranes and control of intracellular protein traffic. *Mailing Add:* Dept Biochem Univ NC Med Sch 405 FLOB CB No 7260 Chapel Hill NC 27599-7260

DOUGLAS, NEIL HARRISON, b Moorefield, WVa, Feb 17, 32; m 58; c 2. SYSTEMATIC ICHTHYOLOGY. *Educ:* Okla State Univ, BS, 55, MS, 59, PhD(zool), 62. *Prof Exp:* Asst zool, Okla State Univ, 57-59, asst water pollution, 59-62; from asst prof to assoc prof, 62-72, prof zool, 72-80, PROF BIOL, NORTHEAST LA UNIV, 80- *Mem:* Am Fisheries Soc; Am Soc Ichthyologists & Herpetologists. *Res:* Taxonomy of freshwater fishes; herpetology; fresh water fishes of Louisiana. *Mailing Add:* Dept Biol Northeast La Univ Monroe LA 71209

DOUGLAS, PAMELA SUSAN, b New Brunswick, NJ, Dec 2, 54. INTERNAL MEDICINE, CARDIOLOGY. *Educ:* Princeton Univ, BA, 74; Med Col Va, MD, 78; Hosp Univ Pa, Am Bd Internal Med, 81 & Am Bd Cardiovasc Dis, 83. *Prof Exp:* Asst prof med & cardiol, Univ Pa, 84-90; ASSOC PROF MED & CARDIOL, HARVARD UNIV, 90- *Concurrent Pos:* Dir, Noninvasive Lab, Beth Israel Hosp, 90- *Mem:* Fel Am Col Cardiologists; Am Heart Asn; Am Fed Clin Res; Am Med Women's Asn. *Res:* Cardiovascular health and disease especially cardiac function, hypertrophy, effects of exercise (chronic and acute), ultrasound, Doppler echocardiography; heart disease in women. *Mailing Add:* 58A Fayerweather St Cambridge MA 02138

DOUGLAS, ROBERT ALDEN, b High Point, NC, Dec 4, 25; m; c 3. MECHANICS, MATERIALS SCIENCE. *Educ:* Purdue Univ, BS, 51, MS, 52, PhD(mech), 56. *Prof Exp:* Mgr res & develop, Danly Mach Specialties, 56-58; from assoc prof to prof mech, 58-74, assoc head dept eng mech & mgr themis res prog, 67-74, PROF CIVIL ENG, NC STATE UNIV, 76- *Concurrent Pos:* Dep dir, US Eng Team, Afghanistan, 64-66. *Mem:* Soc Exp Stress Anal; Am Soc Eng Educ; Am Soc Civil Engrs; Soc Eng Sci. *Res:* Wave propagation; the behavior of solids during impact; biomechanics; wave velocities in human tissue. *Mailing Add:* Dept Civil Eng NC State Univ Raleigh Main Campus Box 7908 Raleigh NC 27695-7908

DOUGLAS, ROBERT G, b Los Angeles, Calif, Feb 20, 37. MARINE GEOLOGY. *Educ:* Univ Calif, Los Angeles, PhD(geol), 66. *Prof Exp:* DEAN, NATURAL SCI & MATH DIV, UNIV SOUTHERN CALIF, 88- *Mem:* Fel Geol Soc Am; Am Asn Petrol Geologists. *Mailing Add:* Natural Sci & Math Div Univ Southern Calif Los Angeles CA 90089-4012

DOUGLAS, ROBERT GORDON, JR, b New York, NY, Apr 17, 34; m 56; c 3. INTERNAL MEDICINE, INFECTIOUS DISEASES. *Educ:* Princeton Univ, AB, 55; Cornell Univ, MD, 59. *Prof Exp:* Intern med, NY Hosp, 59-60, asst resident, 60-61; asst resident, Johns Hopkins Hosp, 61-62; chief resident, NY Hosp, 62-63; clin assoc & clin investr, Lab Clin Invest, Nat Inst Allergy & Infectious Dis, 63-66; from instr to assoc prof microbiol & med, Bayler Col Med, 66-70; assoc prof med & microbiol, Univ Rochester, 70-74, prof, 74-82, head, Infectious Dis Unit, Sch Med & Dent, 70-82, sr assoc dean educ, 79-82; PROF MED & CHMN DEPT, MED COL, CORNELL UNIV, 82-; PHYSICIAN-IN-CHIEF, NY HOSP, 82- *Mem:* Am Soc Clin Invest; Am Soc Microbiol; Am Asn Immunol; Am Fedn Clin Res; fel Am Col Physicians. *Res:* Pathogenesis of respiratory viral infections; influenza; vaccines; antiviral chemotherapy; secretory antibody; clinical virology. *Mailing Add:* Dept Med Cornell Univ Med Col 525 E 68th St New York NY 10021

DOUGLAS, ROBERT HAZARD, b Miami, Fla, Feb 23, 47; m; c 3. REPRODUCTIVE PHYSIOLOGY. *Educ:* WVa Univ, BS, 69, MS, 72; Univ Wis, PhD(endocrinol, reproductive physiol), 75. *Prof Exp:* Asst prof reproductive physiol, Mich State Univ, 75-78; mem staff, Univ Ky, 78-80, asst prof vet sci, 80-84; PRES, BET LABS INC, 84- *Mem:* Am Asn Animal Sci. *Res:* Control of luteolysis and induction of ovulation in large domestic farm species; endocrine control of reproduction function in equids. *Mailing Add:* 305 N Broadway Lexington KY 40508

DOUGLAS, ROBERT JAMES, b Seattle, Wash, Dec 4, 37. MATHEMATICS. *Educ:* Univ Wash, BS, 61, MS, 66, PhD(math), 68. *Prof Exp:* Res assoc, Univ NC, Chapel Hill, 68-69; from asst prof to assoc prof, 69-77, PROF MATH, SAN FRANCISCO STATE UNIV, 77- *Mem:* Am Math Soc; Math Asn Am; Asn of Comput Mach. *Res:* Combinatorial analysis; graph theory; algorithms and complexity theory. *Mailing Add:* Dept Comp Sci San Francisco State Univ San Francisco CA 94132

DOUGLAS, RONALD GEORGE, b Osgood, Ind, Dec 10, 38; m 83; c 5. MATHEMATICS. *Educ:* Ill Inst Technol, BS, 60; La State Univ, PhD(math), 62. *Prof Exp:* From res instr to prof math, Univ Mich, 62-69; chmn dept, 71-73, & 81-84, dean phys sci & math, 86-90, PROF MATH, STATE UNIV NY, STONY BROOK, 69-, VPROVOST, 90- *Concurrent Pos:* Mem, Inst Advan Study, 65-66; Sloan fel, 68-74; sr res fel, Newcastle Univ, Eng, 74; coun, Exec Comt, Am Math Soc; Guggenheim fel, 80-81; sr res fel, Math Sci Res Inst, Berkeley; bd math sci, NRC & MS 2000 Comt. *Mem:* Am Math Soc; Math Asn Am; AAAS. *Res:* Operator theory; functional analysis. *Mailing Add:* Dept Math State Univ NY Stony Brook NY 11794

DOUGLAS, ROY RENE, b Chicago, Ill, Oct 8, 38; m 66, 75; c 3. TOPOLOGY. *Educ:* Northwestern Univ, BA, 60; Univ Calif, Berkeley, MA, 63, PhD(math), 65. *Prof Exp:* Res asst, Univ Calif, Berkeley, 63-65; asst prof, 65-69, assoc prof math, 69-78, PROF MATH, UNIV BC, 78- *Concurrent Pos:* Nat Res Coun Can grant, 67- *Mem:* Am Math Soc; Can Math Cong. *Res:* Algebraic topology; H-spaces and rational homotopy theory; mathematical physics; multivariate statistics; neuro-physiology. *Mailing Add:* Dept Math Univ BC 2075 Westbrook Pl Vancouver BC V6T 1W5 Can

DOUGLAS, STEVEN DANIEL, b Jamaica, NY, Feb 28, 39; m; c 1. MEDICINE, IMMUNOLOGY. *Educ:* Cornell Univ, AB, 59, MD, 63. *Hon Degrees:* MA, Univ Pa, 81. *Prof Exp:* Intern med, Mt Sinai Hosp, NY, 63-64, asst resident, 66-67; staff assoc surgeon, Lab Exp Path, Nat Inst Arthritis, Metab & Digestive Dis, 64-66; NIH fel immunol & hemat, Dept Med, Univ Calif, San Francisco, 67-69; from asst prof to assoc prof med, Mt Sinai Sch Med, 69-74; from assoc prof to prof med & microbiol, Sch Med, Univ Minn, 74-80; DIR, DIV ALLERGY-IMMUNOL, CHILDREN'S HOSP PHILADELPHIA, 80-; PROF PEDIAT & MICROBIOL, SCH MED, UNIV PA, 80- *Concurrent Pos:* Asst attend physician, Mt Sinai Med Sch, 69-73; Nat Heart & Lung Inst grant, 69-74; Nat Inst Allergy & Infectious Dis grant, 74-; rep, Am Bd Med Lab Immunol, 74- *Honors & Awards:* Redway Award, Med Soc State NY, 77; Emil Conason Mem Award, Mt Sinai Sch Med, 70; William Hammond Award, NY State J Med, 80. *Mem:* Am Soc Clin Invest; Am Acad Microbiol; Am Asn Immunologists; Soc Exp Biol & Med; Soc Cell Biol. *Res:* Cellular immunology, cell biology and genetics, arthritis; differentiation of mononuclear phagocytes; author of over 300 scientific publications. *Mailing Add:* Children's Hosp Philadelphia 34th & Civic Ctr Blvd Philadelphia PA 19104

DOUGLAS, TOMMY CHARLES, b Durant, Okla, Sept 29, 46. GENETICS, IMMUNOLOGY. *Educ:* Princeton Univ, AB, 69; Calif Inst Technol, MS, 70, PhD(immunol), 74. *Prof Exp:* Fel immunogenetics, Calif Inst Technol, 74; mem, Basel Inst Immunol, 74-76; asst prof, 76-82, ASSOC PROF MED GENETICS, UNIV TEX HEALTH SCI CTR, 82- *Concurrent Pos:* Vis asst prof biol, Calif Inst Technol, 80; dir, UTGSBS prog immunol, 86-89. *Mem:* Sigma Xi; Am Asn Immunologists; Am Genetic Asn. *Res:* Genetically determined variation in mammals, particularly with respect to cell surface antigens, immune responsiveness, immunoglobulins, the major histocompatibility complex, and apolipoproteins. *Mailing Add:* Med Genetics Ctr UTGSBS PO Box 20334 Astrodome Sta Houston TX 77225

DOUGLAS, W J MURRAY, b Briercrest, Sask, Feb 15, 27; c 1. CHEMICAL ENGINEERING. *Educ:* Queen's Univ, Ont, BSc, 48; Univ Mich, MSE, 52, PhD(chem eng), 58. *Prof Exp:* Develop engr, Polymer Corp, Ont, 48-51; from asst prof to assoc prof, 58-69, PROF CHEM ENG, MCGILL UNIV, 69-, CHMN DEPT, 77- *Concurrent Pos:* Consult, Pulp & Paper Res Inst Can, 62- *Mem:* AAAS; Am Inst Chem Engrs; Chem Inst Can. *Res:* Transport processes, including transport of heat, mass and momentum; chemical reaction and diffusion; mixing. *Mailing Add:* Dept Chem Eng McGill Univ 817 Sherbrooke St W Montreal PQ H3A 2M5 Can

DOUGLAS, WILLIAM HUGH, b Belfast, Northern Ireland, Aug 18, 38; US citizen; m 71; c 2. HYDROPHOBIC COMPOSITES, SERVOHYDRAULICS. *Educ:* Queen's Univ, Belfast, BS, 60, MS, 61, PhD(chem & biomat), 65; Guy's Hosp, London, BDS, 70. *Prof Exp:* Asst lectr chem, Queen's Univ, Belfast, 61-65; teacher clin dent, Sch Dent, Cardiff, Wales, 65-67, lectr dent mat, 71-78; assoc prof, 78-85, DIR DENT MAT, SCH DENT, UNIV MINN, 78-, H L ANDERSON ENDOWED PROF, 85- *Concurrent Pos:* Nuffield Found fel, London, UK, 65; consult Brit Standards Inst Dent Mat, 74-77, 3M Corp, St Paul, 78-, Minn Dept Health, 78-79; vis assoc prof dent mat, Univ Mich, Ann Arbor, 77-78; lectr, 3M, Continuing Dent Educ, Univ Minn & Int Asn Dent Res, 78-; prin investr, NIH grant, Develop Artificial Oral Environ, 84- *Honors & Awards:* IR-100 Award, J Res & Develop, 83. *Mem:* Int Asn Dent Res; fel Acad Dent Mat. *Res:* Development of artificial mouth; a servohydraulic model capable of reproducing the chewing cycle and environment conditions of the human mouth; aspects of restorative dentistry and occlusion; development of a computer graphics system to capture occlusal anatomy and the wear on the surface of the tooth; development of hydrophobic composites and demonstration of their properties; use of sevohydraulics in order to develop clinical measurement technology; physical biology of oral cavity. *Mailing Add:* Biomat Prog Sch Dent Moos Tower 16-212 Univ Minn 515 Delaware St SE Minneapolis MN 55455

DOUGLAS, WILLIAM J, b Saranac Lake, NY, Nov 1, 41; m 63; c 3. ANATOMY. *Educ:* State Univ NY, Plattsburgh, BS, 63; Brown Univ, MAT, 67, PhD(biomed sci), 70. *Prof Exp:* Assoc prof biol, Edinboro Col, 70-71; assoc dir & sr scientist cell culture, W Alton Jones Cell Sci Ctr, 71-78; ASSOC PROF ANAT, SCH MED, TUFTS UNIV, 78- *Concurrent Pos:* Fel, Div Biol & Med Sci, Brown Univ, 70; consult, Dept Med, Mem Hosp, Pawtucket, RI, 73-78; grants, Am Lung Asn, 73-75, NIH, 74-78 & United Cerebral Palsy Res & Educ Found, 75-78. *Mem:* Am Asn Anatomists; Histochem Soc; Am Thoracic Soc; Am Soc Cell Biol; Tissue Cult Asn. *Res:* Isolation and culture of diploid epithelial cells from lung, liver and prostate; ultrastructural and biochemical characterization of these cultures. *Mailing Add:* Dept Anat M-D 524 Tufts Univ Sch Med 136 Harrison Ave Boston MA 02111

DOUGLAS, WILLIAM KENNEDY, b Estancia, NMex, Sept 5, 22; m 46; c 1. MEDICINE, AEROSPACE MEDICINE. *Educ:* Col Mines & Metal, Tex, BS, 43; Univ Tex, MD, 48; Johns Hopkins Univ, MPH, 54; Am Bd Prev Med, dipl & cert aviation med, 56. *Prof Exp:* Command flight surgeon, Hq, Northeast Air Command, USAF, 50-52, base surgeon, Harmon AFB, Nfld, 52-53, res aviation med, Johns Hopkins Univ & Langley AFB, 53-55, chief prof serv, Hq, Mil Air Transport Serv, 55-57, chief, Aircrew Effectiveness Br, Off Surgeon Gen, 57-59, flight surgeon, Proj Mercury Astronauts, NASA, 59-62, from asst dep bioastronaut hq Air Force Missile Test Ctr to dir bioastronaut, Hq, Air ForceEastern Test Range, Patrick AFB, Fla, 62-66, asst dep chief staff bioastronaut & med, Hq, Systs Command, 66-68, chief, Cent Aeromed Serv, Hq, Europe, 68-71, vcomdr, Aerospace Med Div, Systs Command, USAF, 71-77; dir prog eng-life sci, McDonnell Douglas Astronaut Co, 77-82, sr fel, 77-88; RETIRED. *Concurrent Pos:* Mem sci adv bd, USAF, 79-85. *Honors & Awards:* Louis H Bauer Found Award, Aerospace Med Asn, 64; John Jefferies Award, Am Inst Aeronaut & Astronaut, 65; W Boynton Award, Am Astron Soc. *Mem:* Fel Am Col Physicians; AMA; fel Aerospace Med Asn; Am Inst Aeronaut & Astronaut; Int Acad Astronaut; Mercury Seven Found. *Mailing Add:* PO Box 14429 Albuquerque NM 87191

DOUGLAS, WILLIAM WILTON, b Glasgow, Scotland, Aug 15, 22; m 54; c 2. NEUROPHARMACOLOGY, NEUROENDOCRINOLOGY. *Educ:* Univ Glasgow, MB, ChB, 46, MD, 49. *Prof Exp:* Houseman, Glasgow West Infirmary, 46; houseman, Law Hosp, Carluke, 47; lectr physiol, Aberdeen Univ, 47-48; staff mem med res coun, Physiol & Pharmacol Div, Nat Inst Med Res, Eng, 50-56; vis assoc pharmacol, Col Physicians & Surg, Columbia Univ, 52-53; assoc prof, Albert Einstein Col Med, 56-58, prof, 58-68; PROF PHARMACOL, SCH MED, YALE UNIV, 68- *Mem:* Am Physiol Soc; Soc Pharmacol & Exp Therapeut; Pharmacol Soc Can; Brit Pharmacol Soc; Brit Physiol Soc. *Res:* Pharmacology and physiology of the neurohumoral transmission; cellular mechanisms of release of nervous and neuroendocrine secretions. *Mailing Add:* Dept Pharmacol Yale Univ Sch Med 333 Cedar St New Haven CT 06510

DOUGLAS-HAMILTON, DIARMAID H, b Shaftesburg, Eng, June 17, 40; m 82. BIOPHYSICS, PHYSICAL APPLICATIONS IN REPRODUCTION. *Educ:* Oxford Univ, BA, 61; Harvard, MA, 65. *Prof Exp:* Prin res scientist physics, Avco Everett Res Lab, 70-82; Staff scientist oxygen ion implantation, Ion Implant Div, Eaton Corp, 82-86; VPRES RES & DEVELOP BIOPHYS, HAMILTON THORN RES, 86- *Concurrent Pos:* Consult, Avco Everett Res Lab, 82-85, Ion Implant Div, Eaton Corp, 86-87. *Mem:* Sigma Xi; AAAS; Am Phys Soc. *Res:* Gas laser kinetics; plasma gas discharges; optics and application to biology; sperm motion analysis design; optical analysis of biological specimens; computer applications. *Mailing Add:* 729 Cabot St Beverly MA 01915

DOUGLASS, CARL DEAN, b Little Rock, Ark, Apr 27, 25; m 46; c 2. BIOCHEMISTRY. *Educ:* Hendrix Col, BS, 47; Univ Okla, MS, 49, PhD(chem), 52. *Prof Exp:* Asst, Univ Okla, 47; from instr to assoc prof biochem, Sch Med, Univ Ark, 52-59; chief nutrit res br, US Food & Drug Admin, 59-61; nutrit prog officer, Nat Inst Arthritis & Metab Dis, 61-64; chief res & training div, Nat Libr Med, 64-66, chief facil & resources div, 66-67; assoc dir prog develop, Div Res Facil & Resources, NIH, 67-69, assoc dir, Div Res Resources, 69-70, assoc dir, Statist, Anal & Res Eval, 70-71, dep dir, Div Res Grants, 71-77, dir, div res grants, 77-85; RETIRED. *Mem:* Fel AAAS; Am Chem Soc; Soc Exp Biol & Med; Am Inst Nutrit; Am Inst Biol Sci; Sigma Xi. *Res:* Intermediary metabolism of drugs; metabolism of plant pigments. *Mailing Add:* 15107 Interlachen Dr Silver Spring MD 20906

DOUGLASS, CLAUDIA BETH, b Detroit, Mich, Oct 29, 50; m 78. BIOLOGY. *Educ:* Ind Univ, Bloomington, BS, 72, MAT, 74; Purdue Univ, W Lafayette, PhD(biol educ), 76. *Prof Exp:* Res asst biol educ, Purdue Univ, 74-75, instr physiol, 75-76, vis instr inst design, 76; asst prof biol, 76-79, PROF BIOL, CENTRAL MICH UNIV, 80- *Concurrent Pos:* Consult mat develop, Univ Elem Sch, Bloomington, 73-74, Indianapolis Sickle Cell Found, 75-76; NSF grant, 77-79, book reviewer, Sci & Children, Am Biol Teacher, 73-; statist consult, Nat Inst Educ grant, 79; sci educ consult, Bilalian Child Develop Ctr, 80. *Mem:* Nat Asn Res Sci Teaching; Nat Asn Biol Teachers; Nat Sci Teachers Asn; Sch Sci & Math Asn. *Res:* Cognitive style, learning theory and motivation related to the development of instructional materials; individualized instruction including audio-tutorial, inquiry-based and mastery learning techniques for high school, especially urban disadvantaged students. *Mailing Add:* Dept Biol Cent Mich Univ Mt Pleasant MI 48859

DOUGLASS, D C, b Rochester, NY. PHYSICAL CHEMISTRY, POLYMER CHEMISTRY. *Educ:* Univ Rochester, BA, 51; Cornell Univ, PhD(phys chem), 57. *Prof Exp:* MEM STAFF, AT&T BELL LABS, 57- *Mem:* Am Chem Soc; Am Phys Soc; AAAS. *Res:* Molecular motion in polymer science; phara electric and electric optic materials by nuclear magnetic residence. *Mailing Add:* Bell Labs Rm 1A 256 Murray Hill NJ 07974

DOUGLASS, DAVID HOLMES, b Bangor, Maine, Feb 12, 32; m 53; c 3. PHYSICS, ASTROPHYSICS. *Educ:* Univ Maine, BS, 55; Mass Inst Technol, PhD(physics), 59. *Prof Exp:* Res scientist physics, Lincoln Lab, Mass Inst Technol, 59-61, instr, 61-62; from asst prof to prof, Univ Chicago, 62-68; PROF PHYSICS, UNIV ROCHESTER, 68- *Concurrent Pos:* Alfred P Sloan res fel, 64-68. *Mem:* AAAS; Am Phys Soc; Int Soc Gen Relativity & Gravitation; Am Asn Physics Teachers; Am Astron Soc. *Res:* Liquid helium; superconductivity; low temperature, solid state, condensed matter and gravitation physics; gravitation waves; elementary particles; quarks. *Mailing Add:* Dept Physics Univ Rochester Rochester NY 14627

DOUGLASS, DAVID LESLIE, b Newark, NJ, Sept 28, 31. METALLURGICAL ENGINEERING. *Educ:* Pa State Univ, 53, MS, 55; Ohio State Univ, PhD(metall eng), 58. *Prof Exp:* PROF, DEPT METALL SCI & ENG, UNIV CALIF, LOS ANGELES, 68- *Concurrent Pos:* Chmn nat sem, Am Soc Metals, 70 & Gordon Res Corrosion, 75. *Mem:* Fel Am Soc Metals; Nat Asn Corrosion Engrs. *Mailing Add:* Dept Mat Sci & Eng Univ Calif 5732 Boelter Hall Los Angeles CA 90024

DOUGLASS, IRWIN BRUCE, b Des Moines, Iowa, Sept 2, 04; m 31; c 2. ENVIRONMENTAL CHEMISTRY. *Educ:* Monmouth Col, BS, 26, DSc, 58; Univ Kans, PhD(org chem), 32. *Prof Exp:* Teacher, Ill High Sch, 26-28; teacher, Kans City Jr Col, 30-31; asst chem, Univ Kans, 29-32; asst prof, NDak Agr Col, 32-33; asst prof, Northern Mont Col, 33-37; res asst, Yale Univ, 37-38; prof, Northern Mont Col, 38-40; from asst prof to prof, 40-70, actg head dept, 41-46, head dept, 46-52, head planning off, 69-72, res prof, 72-74, EMER PROF CHEM, UNIV MAINE, 70- *Concurrent Pos:* Mem, Maine Bd Environ Protection, 74-77. *Mem:* AAAS; Am Chem Soc. *Res:* Kraft pulping odor control; organic sulfur compounds. *Mailing Add:* 904 Rte 2A Williston VT 05495

DOUGLASS, JAMES EDWARD, b Bessemer, Ala, Nov 3, 28; m 53; c 2. FORESTRY. *Educ:* Auburn Univ, BS, 51; Mich State Univ, MS, 55. *Prof Exp:* Conserv forester, Int Paper Co, 53-54; res forester, Southeastern Forest Exp Sta, US Forest Serv, Ashville, 56-84; proj leader, Coweeta Hydrological Lab, 72-84. *Mem:* Soc Am Foresters; Soil Sci Soc Am. *Res:* Soil moisture-vegetative relations; forest hydrology using control and treatment experimental watershed approach. *Mailing Add:* Coweeta Hydrologic Lab Rte 1 Box 216 95 Hillcrest Circle Franklin NC 28734

DOUGLASS, JAMES EDWARD, b Corpus Christi, Tex, May 18, 30; m 55; c 2. ORGANIC CHEMISTRY. *Educ:* Rice Univ, BA, 52; Univ Tex, PhD(chem), 59. *Prof Exp:* Hickrill Chem Res Found res fel, 58-59; asst prof, Univ Ky, 60-65; assoc prof, 65-68, dept chmn, 77-82, PROF CHEM, MARSHALL UNIV, 68- *Mem:* AAAS; Am Chem Soc; Int Soc Heterocyclic Chem. *Res:* Pyrimidinium ylide chemistry; heterocyclic synthesis. *Mailing Add:* Dept of Chem Marshall Univ Huntington WV 25701

DOUGLASS, KENNETH HARMON, b Rochester, NY. MEDICAL PHYSICS, NUCLEAR MEDICINE. *Educ:* Univ Rochester, BS, 64; Carnegie-Mellon Univ, MS, 66, PhD(physics), 69. *Prof Exp:* Asst prof physics, Pa State Univ, Schuylkill Campus, 69-71; res assoc biophysics, Johns Hopkins Univ, 71-74; from instr to asst prof nuclear med, Johns Hopkins Med Inst, 74-88; SR MED PHYSICIST & ASSOC PROF RADIOL RADIATION SAFETY, WVA UNIV, MORGANTOWN, 88- *Mem:* Soc Nuclear Med. *Res:* Non-invasive studies of cardiac function; computerized image-processing; kinetic modeling of receptor binding in the brain, using measurments from positron emission tomography. *Mailing Add:* Radiol Dept Radiation Safety WVa Hosp Box 8150 Morgantown WV 26506-8150

DOUGLASS, MATTHEW MCCARTNEY, b Port of Spain, Trinidad, Sept 21, 26; US citizen; m 54; c 5. CIVIL ENGINEERING, ENGINEERING EDUCATION. *Educ:* McGill Univ, BEng, 52; George Washington Univ, MSE, 62; Okla State Univ, PhD(civil eng), 66. *Prof Exp:* Jr engr, Kilborn Eng Co, Ont, 52; asst engr, Dept Works & Hydraul, Trinidad, 53-55, exec engr, 55-57; asst designer, E Lionel Pavlo, Consult Engrs, 57; from instr to asst prof civil eng, Howard Univ, 57-66; assoc prof, 66-87, PROF CIVIL ENG, CONCORDIA UNIV, 88-, CHMN, DEPT CIVIL ENG, 82- *Concurrent Pos:* Examr, Order Engrs of Quebec, 77- *Mem:* Fel Am Soc Civil Engrs; Am Soc Eng Educ; Can Soc Civil Engrs. *Res:* Elastic-plastic buckling of columns; application of complementary potential energy to the analysis of space frameworks; curved-girder orthotropic bridges; personalized learning systems in engineering. *Mailing Add:* Dept Civil Eng Concordia Univ Sir G Williams 1455 De Maisonneuve Montreal PQ H3G 1M8 Can

DOUGLASS, PRITCHARD CALKINS, b Jamestown, NY, Mar 22, 13; m 38; c 3. ANALYTICAL CHEMISTRY. *Educ:* Houghton Col, BS, 35. *Prof Exp:* Chemist, G C Murphy Co, 37-42; chemist, Bausch & Lomb Optical Co, 42-48; fel, Mellon Inst Indust Res, 48-51; microscopist cotton properties, Coats & Clark Res Labs, 51-53; chemist & head anal chem & controls, 53-63, chemist specialist, 63-78, CONSULT, BAUSCH & LOMB, INC, 78- *Mem:* Am Chem Soc; Am Inst Chemists. *Res:* Application of analytical and electrochemistry to chemical processes and materials; microscopy; environmental chemistry; instrumentation applied to metal finishing, glass, lubrication and polycrystalline materials. *Mailing Add:* 31 Pickford Dr Rochester NY 14618

DOUGLASS, RAYMOND CHARLES, b San Francisco, Calif, Mar 30, 23; m 48; c 2. INVERTEBRATE PALEONTOLOGY. *Educ:* Stanford Univ, BS, 50, PhD, 57; Univ Nebr, MS, 52. *Prof Exp:* Field asst, US Geol Surv, 47-48, geologist, 52-88; RETIRED. *Concurrent Pos:* Asst, Univ Nebr, 50-51; mem fac, USDA Grad Sch, 57-; dir, Cushman Found Foraminiferal Res, 61-, pres, 64-66. *Mem:* Paleont Soc; Soc Econ Paleont & Mineral. *Res:* Evolution, ecology and distribution of large foraminifera; Pennsylvanian and Permian Fusulinidae. *Mailing Add:* US Geol Surv US Nat Mus Natural Hist Washington DC 20560

DOUGLASS, ROBERT L, speech pathology, for more information see previous edition

DOUGLASS, TERRY DEAN, b Jackson, Tenn, Oct 26, 42; m 64; c 3. ELECTRONICS ENGINEERING, PHYSICS. *Educ:* Univ Tenn, BS, 65, MS, 66, PhD, 68. *Prof Exp:* Sr scientist, Life Sci Dept, EG&G, Ortec Inc, 68-76, div mgr & vpres, 76-80, pres, 80-; PRES, CTI, INC, TENN. *Mem:* Inst Elec & Electronics Engrs. *Res:* Optimization of time measurement in nuclear instrumentation; emission computed tomography; nuclear medicine. *Mailing Add:* 810 Innovation Dr PO Box 22999 Knoxville TN 37933

DOUGLIS, AVRON, b Tulsa, Okla, Mar 14, 18; m 41; c 2. MATHEMATICS. *Educ:* Univ Chicago, AB, 38; NY Univ, MS, 48, PhD(math), 49. *Prof Exp:* Res fel, Calif Inst Technol, 49-50; from instr to assoc prof math, NY Univ, 50-56; res prof, 56-58, prof, 58-89, EMER PROF MATH, UNIV MD, COLLEGE PARK, 89- *Mem:* Am Math Soc; Soc Indust & Appl Math. *Res:* Partial differential equations. *Mailing Add:* Dept Math Univ Md College Park MD 20742

DOUKAS, HARRY MICHAEL, b Washington, DC, July 30, 19; m 47; c 4. ORGANIC CHEMISTRY, BIOCHEMISTRY. *Educ:* Univ Md, BS, 42; Georgetown Univ, MS, 52, PhD(chem), 53. *Prof Exp:* Asst, Univ Md, 46-48; chemist, Bur Agr & Indust Chem, Biol Active Compounds Div, USDA, 48-52, res chemist, Eastern Utilization Res Br, 53-55; sr investr, Georgetown Univ, 52-53; head org sect, Chem Corps Biol Lab, US Army, 55-58; prog dir fels, NSF, 58-65; asst chief fels, Career Develop Rev Br, Div Res Grants, NIH, 65-72, chief, 72-73, chief, Off Res Manpower, 73-75; from asst dean to assoc dean, sponsored res, Med Ctr, Georgetown Univ, 75-91; RETIRED. *Concurrent Pos:* Consult, NIH. *Mem:* Am Chem Soc; Sigma Xi; Nat Coun Univ Res Adminr; Soc Res Adminr; Asn Am Med Col. *Res:* Natural and synthetic plant growth hormones; unsaturated naphthylenic acids; isolation and characterization of plant alkaloids; substituted pyridine compounds; health science administration. *Mailing Add:* 9920 Brixton Lane Bethesda MD 20817

DOULL, JOHN, b Baker, Mont, Sept 13, 22; m 58. PHARMACOLOGY. *Educ:* Mont State Col, BS, 44; Univ Chicago, PhD(pharmacol), 50, MD, 53. *Prof Exp:* Asst, Toxicity Lab, Univ Chicago, 46-51, res assoc, 51-53, from asst prof to assoc prof pharmacol, Univ, 53-57; res assoc, US Air Force Radiation Lab, 53-54, asst dir, 54-67; PROF PHARMACOL & TOXICOL, MED CTR, UNIV KANS, 67- *Mem:* Soc Exp Biol & Med; Am Soc Pharmacol & Exp Therapeut; Am Chem Soc; Soc Toxicol; Am Indust Hyg Asn; Sigma Xi. *Res:* Pesticides; biological aspects of ionizing radiation; toxicology. *Mailing Add:* Dept Pharmacol Univ Kans Med Ctr Kansas City KS 66103

DOUMA, JACOB H, b Hanford, Calif, May 30, 12. CIVIL ENGINEERING. *Educ:* Univ Calif, BS, 35. *Prof Exp:* At US Army Corps Engrs, 35-36 & US Bur Reclamation, 36-39; mem staff, US Army Corps Engrs, 39-55, chief hydraul engr, 55-79; independent consult hydraul engr, 79-90; RETIRED. *Concurrent Pos:* Mem, US comn large dams, Permanent Inter Cong Navig. *Mem:* Nat Acad Eng; fel Am Soc Civil Eng. *Res:* Irrigation, drainage and flood control. *Mailing Add:* 1001 Manning St Great Falls VA 22066

DOUMANI, GEORGE ALEXANDER, b Acre, Palestine, Apr 16, 29; US citizen; m 57, 85; c 3. PETROLEUM GEOLOGY, STRATIGRAPHY. *Educ:* Univ Calif, Berkeley, BA, 56, MA, 57; Pac Western Univ, PhD (environ sci), 85. *Prof Exp:* Petrol inspector, Arabian Am Oil Co, Saudi Arabia, 50-52; exploit engr, Shell Oil Co, Calif, 56; eng geologist, Hersey Inspection Bur, 57-58; geologist & glaciologist, Nat Acad Sci & NSF, Antarctic, 58- 60; res assoc geol, Inst Polar Studies, Ohio State Univ, 60-63; sect head arctic & antarctic bibliog, Cold Regions Sect, Sci & Technol Div, Libr of Cong, Washington, DC, 63-66, specialist earth sci & oceanog, Sci Policy Res Div, Cong Res Serv, 66-75; pres, Adar Corp, 75-77; pres, Tech Transfer Int Corp, 77-79; vpres, tech transfer, HRM Inc, 79-85; dir, Peace Corps, Sana'a Yemen, 85-87; dir, Off Tech Policy, Dept Energy, 88-90; PROG DIR, NSF, WASHINGTON, DC, 88- *Concurrent Pos:* Spec asst, NAfrica, Near EAsia & Pac Region, Peace Corps, Washington, DC, 88. *Mem:* Am Asn Petrol Geologists; fel Geol Soc Am; Am Polar Soc; Sigma Xi; Antarctican Soc. *Res:* Stratigraphic paleontology; earth sciences and oceanography; national science policy issues; antarctic geology; technology transfer; technology assessment; energy policy; environmental policy. *Mailing Add:* 4000 Massachusetts Ave NW Apt 1531 Washington DC 20016

DOUMAS, A(RTHUR) C(ONSTANTINOS), b Fredericksburg, Va, Jan 31, 32; div; c 5. CHEMICAL ENGINEERING. *Educ:* Va Polytech Inst, PhD(chem eng), 55. *Prof Exp:* Proj leader, 55-70, mgr, 70-77, SR PROCESS SPECIALIST, DOW CHEM CO, FREEPORT, 77- *Mem:* Am Chem Soc; Am Soc Metals; Am Soc Qual Control; Am Inst Chem Eng. *Res:* Chemical process; applications; electrochemistry; heat transmission; engineering statistics; unit operations. *Mailing Add:* Box 547 Lake Jackson TX 77566

DOUMAS, BASIL T, b Argos Orestikon, Greece, July 16, 30; US citizen; m 57; c 3. CLINICAL CHEMISTRY. *Educ:* Univ Thessaloniki, BS, 52; Univ Tenn, MS, 60, PhD(biochem), 62. *Prof Exp:* Consult path, Baptist Mem Hosp, Memphis, 60-62, head dept clin chem, 62-64; asst prof path, Med Col & asst dir clin chem, Univ Hosp, Univ Ala, 64-65, assoc dir clin chem, Univ Hosp, 65-70, assoc prof path, Med Col, 68-70; assoc prof, 70-75, PROF PATH, MED COL WIS, 75- *Concurrent Pos:* Instr, Univ Tenn, 63-64; chmn subcomt human albumin, Nat Comt Clin Lab Standards, 75- *Mem:* AAAS; Am Asn Clin Chemists; Am Chem Soc; NY Acad Sci; Acad Clin Lab Physicians & Scientists; Sigma Xi. *Res:* Electrophoresis of serum proteins and dye binding; bacterial metabolism pyrimidines; clinical chemistry methodology, standards, instrumentation and quality control. *Mailing Add:* Dept Path Med Col Wis 2430 Whipple Tree Lane Brookfield WI 53005

DOUMAUX, ARTHUR ROY, JR, b Little Neck, NY, Mar 15, 38; m 62; c 2. ORGANIC CHEMISTRY. *Educ:* Lehigh Univ, BS, 61; Yale Univ, PhD(org chem), 66. *Prof Exp:* Res & develop chemist, 66-73, proj scientist, 73-76, res scientist, 76-80, SR RES SCIENTIST, UNION CARBIDE CORP, 80- *Mem:* Am Chem Soc; Sigma Xi. *Res:* Metal-ion catalyzed oxidations with peroxidic materials; chemistry of ethylene oxide adducts; synthesis gas chemistry; polyethyleneamines, alkylamines. *Mailing Add:* 1401 Wilkie Dr Charleston WV 25341

DOUMIT, CARL JAMES, b New Iberia, La, Jan 11, 45; m 69; c 1. INORGANIC CHEMISTRY, ANALYTICAL CHEMISTRY. *Educ:* Univ Southwestern La, BS, 67; Tulane Univ, MS, 70, MAT, 71, PhD(inorg chem), 73. *Prof Exp:* Instr chem, Tulane Univ, 68-70; assoc prof phys sci, 73-80, PROF CHEM, MISS UNIV WOMEN, 80- *Mem:* Am Chem Soc; Sigma Xi. *Res:* Coordination chemistry; bioinorganic chemistry; analytical chemistry. *Mailing Add:* 703 Hemlock St Columbus MS 39702

DOUNCE, ALEXANDER LATHAM, CATALASE, LIVER PROTEINS. *Educ:* Cornell Univ, PhD(org chem), 35. *Prof Exp:* Emer prof, Sch Med & Dent, Univ Rochester; RETIRED. *Res:* Enzymology. *Mailing Add:* 615 Edgewood Ave Rochester NY 14618

DOUPLE, EVAN BARR, b Hershey, Pa, Sept 16, 43; m 66; c 2. RADIOBIOLOGY, ONCOLOGY. *Educ:* Millersville State Col, BS, 64; Kans Univ, PhD(radiation biophysics), 72. *Prof Exp:* Sci teacher biol & chem, Lebanon Sr High Sch, 64-67; lectr radiation biophysics, Kans Univ, 70-72; from instr to asst prof, 73-80, assoc prof radiobiol, 80-90, PROF MED & PHARMACOL, DARTMOUTH MED SCH, 90- *Concurrent Pos:* From adj asst prof to adj assoc prof, 76-90, adj prof, Thayer Sch Eng, Darmouth Col, 90- *Mem:* Sigma Xi; Radiation Res Soc; Am Soc Therapeutic Radiologists; Am Asn Physicists Med; AAAS. *Res:* Radiobiology of in vitro and animal tumor models; hyperthermia and radiation; radiation-platinum coordination complex interactions in vitro and in transplantable animal tumors. *Mailing Add:* Radiobiol Lab Dartmouth-Hitchcock Med Ctr Hanover NH 03756

DOUPNIK, BEN, JR, b Agenda, Kans, Aug 13, 39; m 61; c 3. FIELD CROPS, MYCOTOXINS. *Educ:* Kans Wesleyan Univ, AB, 62; Univ Nebr, MS, 64; La State Univ, PhD(plant path), 67. *Prof Exp:* Asst prof plant path, Coastal Plain Exp Sta, Univ Ga, 67-72; assoc prof, 72-79, PROF, S CENT RES & EXTEN CTR, UNIV NEBR, 79- *Concurrent Pos:* Fac develop leave, Univ Ky, 80-81. *Mem:* Am Phytopath Soc; Am Soc Microbiol; Sigma Xi. *Res:* Control of field crop diseases, grain deterioration and mycotoxin studies, and general extension plant pathology. *Mailing Add:* S Cent Res & Exten Ctr Univ Nebr Box 66 Clay Center NE 68933

DOUROS, JOHN DRENKLE, b Reading, Pa, Dec 26, 30; m 54; c 2. MICROBIOLOGY. *Educ:* Duke Univ, BA, 52; Rutgers Univ, MS, 55; Pa State Univ, PhD(bact), 58. *Prof Exp:* Assoc microbiologist, Parke, Davis & Co, 57-60 & Sun Oil Co, 60-63; sr chemist & microbiologist, Esso Res & Eng Co, 63-68; mgr microbiol res & develop, Gates Rubber Co, 68-72; chief natural prod br, Nat Cancer Inst, 72-81; asst dir licensing, Bristol Labs, 81-83, vpres licensing, Bristol Myers Co, 83-88; RETIRED. *Mem:* Am Soc Microbiol; Soc Indust Microbiol; Am Cancer Soc; Sigma Xi. *Res:* Mechanism of action of antibiotics; degradation of organic compounds by microorganisms; fermentation and natural products research. *Mailing Add:* 270 Gloucestershire Rd Winston-Salem NC 27104

DOUSA, THOMAS PATRICK, b Prague, Czech, Dec 13, 37; m 66; c 3. INTERNAL MEDICINE, BIOCHEMISTRY. *Educ:* Charles Univ, Prague, MD, 62; Czech Acad Sci, PhD(biochem), 68. *Prof Exp:* From intern to resident & res assoc, Inst Cardiovasc Dis, Prague Tech Univ, 62-65; attend physician, Community Med Ctr, 65-68; clin investr & attend physician, Dept Med I, Charles Univ, 68-69; AMA fel molecular endocrinol, Inst Biomed Res, Chicago, IL, 69-70; asst prof physiol, Med Sch, Northwestern Univ, 70-72; from asst prof to assoc prof, 72-77, PROF MED & PHYSIOL, MAYO MED SCH & MAYO GRAD SCH MED, MAYO CLIN FOUND, 77-, HEAD NEPHROL RES UNIT, 83-, HENRY & MILDRED UILEIN II PROF MED RES, 84- *Concurrent Pos:* Estab investr, Am Heart Asn, 74-79, fel high blood pressure coun; mem, Bd Sci Coun, Nat Inst Aging, NIH & rev comt, Nat Inst Diabetes & Digestive & Kidney Dis. *Honors & Awards:* Czech Acad Sci Prize, 69. *Mem:* Am Fedn Clin Res; Am Soc Clin Invest; Am Physiol Soc; Endocrine Soc; Am Soc Renal Biochem & Metab (pres, 87-89). *Res:* Cellular action of vasopressin and other hormones in the kidney; hormonal regulation of water and phosphate excretion. *Mailing Add:* Dept Med Div Nephrol Mayo Clin Mayo Med Sch Rochester MN 55901

DOUTHAT, DARYL ALLEN, b Nescopeck, Pa, Nov 3, 42; m 67. PHYSICAL CHEMISTRY, CHEMICAL PHYSICS. *Educ:* Pa State Univ, BS, 65; Univ Chicago, MS, 72, PhD(chem), 74. *Prof Exp:* Instr, Kennedy-King Col, 72-76, asst prof chem, 77-80; from asst prof to assoc prof, 81-89, PROF, DEPT PHYSICS & CHEM, UNIV ALASKA, 89- *Concurrent Pos:* Consult, Radiol & Environ Res Div, Argonne Nat Lab, 74-80; vis fel, Centre Interdisciplinary Studies in Chem Physics, Univ Western Ont, 76-77. *Mem:* Am Chem Soc; Radiation Res Soc. *Res:* Theory of electron transport; electron energy degradation in matter; calculation of the initial yields of ions and excited states produced in matter exposed to penetrating charged particles or photons. *Mailing Add:* Dept Physics & Chem Univ Alaska 3221 Providence Dr Anchorage AK 99504

DOUTHIT, HARRY ANDERSON, JR, b Raymondville, Tex, June 18, 35; m 59; c 1. MOLECULAR BIOLOGY. *Educ:* Univ Tex, BA, 61, PhD(bot), 65. *Prof Exp:* Fel molecular biol, Univ Wis, 64-67; from asst prof to assoc prof cell & molecular biol, 67-80, assoc prof, 80-85, PROF BIOL SCI, UNIV MICH, 85- *Mem:* AAAS; Am Soc Microbiol. *Res:* Enzymology, including the nature and relevance of thiaminase I; differentiation including the contributions and interactions of nucleic acids in bacterial differentiation. *Mailing Add:* Dept Biol Univ Mich Ann Arbor MI 48109

DOUTT, RICHARD LEROY, b La Verne, Calif, Dec 6, 16; m 42, 79; c 2. ENTOMOLOGY, ENVIRONMENTAL SCIENCES. *Educ:* Univ Calif, BS, 39, MS, 40, PhD(entom), 46, LLB, 59, JD, 68. *Prof Exp:* Jr entomologist, 46-48, from asst prof to prof, 48-75, chmn dept, 64-69, actg dean col agr sci & actg assoc dir agr exp sta, 69-70, EMER PROF BIOL CONTROL & ENTOM, DIV BIOL CONTROL LABS AT ALBANY, UNIV CALIF, BERKELEY, 75- *Concurrent Pos:* Prin scientist, Ecosci Div Henningson, Durham & Richardson, Inc, 75-; entomologist, Santa Barbara County, 81- *Mem:* Am Entom Soc; Ecol Soc Am. *Res:* Taxonomy of Chalcidoidea; biology of entomophagous insects. *Mailing Add:* 1781 Glen Oaks Dr Santa Barbara CA 93108

DOUTY, RICHARD T, b Williamsport, Pa, June 12, 30; m 59; c 4. STRUCTURAL ENGINEERING. *Educ:* Lehigh Univ, BS, 56; Ga Inst Technol, MS, 57; Cornell Univ, PhD(struct eng), 64. *Prof Exp:* Engr trainee, Bethlehem Steel Co, 57-58; res assoc, Cornell Univ, 61-62; from asst prof to assoc prof civil eng, 62-68, PROF CIVIL ENG, UNIV MO-COLUMBIA, 68- *Concurrent Pos:* Partic, NSF Proj Use of Comput & Math Optimization Tech in Eng Design, 65; sr fel, Univ Pa, 69-70; mem eval panel, Associateship Prog, Nat Res Coun, 71; pres, Eng Design Programmers, Inc, 85- *Mem:* Am Soc Civil Engrs; Sigma Xi. *Res:* Computer techniques in engineering design. *Mailing Add:* Dept Civil Eng Univ Mo Columbia MO 65211

DOUVILLE, PHILLIP RAOUL, b Hartford, Conn, June 24, 36; m 59; c 3. PHYSICAL CHEMISTRY. *Educ:* Univ Conn, BA, 59, PhD(phys chem), 69. *Prof Exp:* Asst instr chem, Univ Conn, 63-65; asst prof to assoc prof chem, 65-83, PROF CHEM, CENT CONN STATE UNIV, 83- *Mem:* AAAS; Am Chem Soc; NY Acad Sci. *Mailing Add:* 23 Virginia Dr Middletown CT 06457-4826

DOVE, DEREK BRIAN, b Middlesex, Eng, Jan 12, 32; m 54; c 2. PHYSICS. *Educ:* Imp Col, Univ London, BSc, 53, PhD(crystallog), 56. *Prof Exp:* Sci officer, Atomic Energy Res Estab, Eng, 55-59, sr sci officer, 59; Nat Res Coun Can fel, 59-61; mem tech staff, Bell Tel Labs, Inc, 61-67; assoc prof, Univ Fla, Gainesville, 67-71, prof mat sci eng & elec eng, 71-77; MGR, IBM WATSON RES CTR, YORKTOWN HEIGHTS, NY, 77- *Mem:* Am Vacuum Soc. *Res:* Magnetic films; structure of amorphous film using electron diffraction; surface analysis; luminescent materials. *Mailing Add:* 41 Indian Hill Rd Mt Kisco NY 10549

DOVE, JOHN EDWARD, chemical kinetics; deceased, see previous edition for last biography

DOVE, LEWIS DUNBAR, b Savannah, Ga, Oct 8, 34; m 58; c 1. PLANT PHYSIOLOGY. *Educ:* Univ Md, BS, 57, MS, 59; Duke Univ, PhD(bot), 64. *Prof Exp:* Phys sci aid, Plant Indust Sta, USDA, Md, 57; instr bot, Newcomb Col, Tulane Univ, 61-65, asst prof biol, 65-69; assoc prof, 69-74, PROF BIOL, WESTERN ILL UNIV, 74-, FAC DEVELOP ASSOC, 80- *Concurrent Pos:* Sigma Xi res grant-in-aid, 65-66; res grants, NSF, 67-69 & 81- & Ill State Acad Sci, 71-72. *Mem:* Am Soc Plant Physiol; Sigma Xi; Asn Develop Computer-Based Instrnl Syst. *Res:* Effects of environmental stress on metabolism in plants; plant proteins; computer-assisted instruction; faculty development. *Mailing Add:* Dept Biol Sci Western Ill Univ Macomb IL 61455

DOVE, RAY ALLEN, b Rockmart, Ga, Aug 17, 21; m 47; c 4. CHEMISTRY. *Educ:* Univ Akron, BS, 47, MS, 53. *Prof Exp:* Chemist, Goodyeear Tire & Rubber Co, 53-58, from jr res chemist to sr res chemist, 58- 75, sect head, 75-86; RETIRED. *Mem:* Am Chem Soc. *Res:* Polarography; ultraviolet spectrophotometry; nonaqueous titrimetry; solid-liquid-gas and ion-exchange chromatography; flame photometry; atomic absorption spectroscopy. *Mailing Add:* 906 Winton Ave Akron OH 44320-2832

DOVE, WILLIAM FRANCIS, b Washington, DC, July 30, 33; m 61; c 2. PLASMA PHYSICS. *Educ:* Pa State Univ, BS, 56; Univ Calif, Berkeley, PhD(physics), 68. *Prof Exp:* Res assoc physics & astron, Univ Md, 68-74; mem staff, Div Controlled Thermonuclear Res, 74-, AT DIV PLASMA PHYSICS, US DEPT ENERGY. *Mem:* Am Phys Soc. *Res:* High Mach number plasma flow, magnetic diffusion, shock wave propagation and turbulent ion heating; non-equilibrium plasma transport properties and measurement of thermoelectric coefficient. *Mailing Add:* Div Appl Plasma Physics US Dept Energy Washington DC 20545

DOVE, WILLIAM FRANKLIN, b Bangor, Maine, June 20, 36; m 64; c 3. GENETICS, CELL BIOLOGY. *Educ:* Amherst Col, AB, 58; Calif Inst Technol, PhD(chem), 61. *Prof Exp:* From asst prof to assoc prof, 64-73, prof oncol, 73-77, PROF ONCOL & GENETICS, MCARDLE LAB, UNIV WISMADISON, 77- *Concurrent Pos:* A A Noyes fel, Calif Inst Technol, 61-62; NSF fel, Cambridge Univ, 62-63; NIH fel, Stanford Univ, 64; Guggenheim fel, Pasteur Inst, 75-76; George Streisinger Prof, Exp Biol, Univ Wis. *Mem:* Genetics Soc Am. *Res:* Genes and the control of their replication and expression; developmental genetics and neoplasia. *Mailing Add:* McArdle Lab Univ Wis Madison WI 53706

DOVEL, WILLIAM LAWRENCE, fish biology, for more information see previous edition

DOVENMUEHLE, ROBERT HENRY, b St Louis, Mo, July 10, 24; div; c 5. PSYCHIATRY. *Educ:* St Louis Univ, MD, 48; Am Bd Psychiat & Neurol, dipl psychiat, 63. *Prof Exp:* Resident physician, Hastings State Hosp, 49-51; chief open sect psychiat, William Beaumont Army Hosp, 51-53; resident & chief resident psychiat, Duke Hosp, 54-55, chief in-patient serv, 55-57, from asst prof to assoc prof psychiat, Sch Med, Duke Univ, 57-65, res coordr, Ctr Study Aging, 57-65; dir ment health progs, Western Interstate Comn Higher Educ, 65-66; prof psychiat, Univ Mo-Kansas City, 66-69; dir geriat progs, Kansas City Gen Hosp & Med Ctr, 66-69; exec dir, Dallas County Ment Health & Ment Retardation Ctr, 69-75; clin dir, Guilford County Ment Health Ctr, 75-76; CONSULT, 76- *Concurrent Pos:* Consult, Keeley Inst, 54-65; attend, Vet Admin Hosp, Durham, NC, 55-65 & Roanoke, Va, 59-64; clin prof psychiat, Univ Tex Southwestern Med Sch, 69-75; adj prof, NTex State Univ, 69-75; spec lectr occup ther, Tex Woman's Univ, 70-75; adj prof child develop, Univ NC, Greensboro, 76-78; staff psychiatrist, Southern Va Mental Health Inst, 88- *Mem:* Geront Soc; fel Am Psychiat Asn; AMA. *Res:* Psychiatric problems of older people; criteria of effective psychiatric nursing therapy. *Mailing Add:* PO Box 2548 Danville VA 24541-0548

DOVER, CARL BELLMAN, b Milwaukee, Wis, Feb 10, 41; m 69; c 2. THEORETICAL NUCLEAR PHYSICS. *Educ:* Mass Inst Technol, BS, 63; PhD(physics), 67. *Prof Exp:* Scientific asst, Inst Theoret Phys, Univ Heidelberg, 67-70; asst res, Inst Nuclear Physics, Univ Paris, 70-71; STAFF PHYSICIST, BROOKHAVEN NAT LAB, 71- *Honors & Awards:* Humboldt Sr Res Award, 88. *Mem:* Fel Am Phys Soc. *Res:* Nuclear reaction theory; heavy ion physics; intermediate energy nuclear theory. *Mailing Add:* Dept Physics Brookhaven Nat Lab Upton NY 11973

DOVERMANN, KARL HEINZ, b Bonn, WGer, Aug 31,48. MATHEMATICS. *Educ:* Univ Bonn, Dipl, 75; Rutgers Univ, PhD(math), 78. *Prof Exp:* L E Dickson instr, Univ Chicago, 78-80; asst prof math, Purdue Univ, 80-; PROF MATH, UNIV HAWAII, MANOA. *Mem:* Am Math Soc. *Res:* Algebraic topology with emphasis on finite transformation groups on smooth manifolds and on equivariant surgery; existence and classification of actions. *Mailing Add:* Dept Math Univ Hawaii Honolulu HI 96822

DOVERSPIKE, LYNN D, b Cumberland, Okla, Mar 1, 34; m 54; c 3. ATOMIC & MOLECULAR PHYSICS. *Educ:* Okla State Univ, BS, 58; Univ Calif, Los Angeles, MS, 60; Univ Fla, PhD(physics), 66. *Prof Exp:* Asst prof, 67-74, ASSOC PROF PHYSICS, COL WILLIAM & MARY, 74- *Mem:* Am Phys Soc. *Res:* Experimental low energy atomic and molecular physics. *Mailing Add:* Dept Physics Col William & Mary Williamsburg VA 23185

DOVIAK, RICHARD J, b Passaic, NJ, Dec 24, 33; m 57; c 3. ATMOSPHERIC SCIENCE. *Educ:* Rensselaer Polytech Inst, BSEE, 56; Univ Pa, MSEE, 59, PhD(elec eng), 63. *Prof Exp:* Instr elec eng, Moore Sch Elec Eng, Univ Pa, 60-61, from assoc lectr to asst prof, 62-70, supvr, Acad Labs, 67-71; SCIENTIST, NAT SEVERE STORMS LAB, NAT OCEANIC & ATMOSPHERIC AGENCY, 71- *Concurrent Pos:* Prin investr, NSF res grant, 69-71; consult, Raytheon, 63-70 & ARK Electronics Corp, 70-71. *Mem:* AAAS; Am Geophys Union; Inst Elec & Electronics Engrs; Am Meteorol Soc; Sigma Xi. *Res:* Radar meteorology-antennas, propagation and scattering; electromagnetic interference; nonionizing radiation of biological tissues; atmospheric probing. *Mailing Add:* 1313 Halley Circle Norman OK 73069

DOW, BRUCE MACGREGOR, b Newton, Mass, Oct 30, 38. NEUROPHYSIOLOGY, VISION. *Educ:* Wesleyan Univ, AB; Univ Strasbourg, Cert; Univ Rochester, MD. *Prof Exp:* Intern med, Baltimore City Hosp, 66-67; res assoc neurophysiol, Nat Inst Dent Res, 67-70; staff fel, Nat Eye Inst, 70-75; vis scientist neurophysiol, Nat Inst Mental Health, 75-76; res assoc prof, 76-78, assoc prof, 78-86, PROF PHYSIOL, SCH MED, STATE UNIV NY, BUFFALO, 86- *Mem:* Soc Neurosci; Asn Res Vision & Ophthalmol. *Res:* Studies of color vision and mechanisms of fixation/localization in foveal cortical centers of trained monkeys, utilizing single cell neurophsiology, psychophysics, and histological reconstruction. *Mailing Add:* Physiol 124 Sherman State Univ NY Buffalo Sch Med 3435 Main St Buffalo NY 14214

DOW, DANIEL G(OULD), b Ann Arbor, Mich, Apr 26, 30; m 54; c 4. ELECTRICAL ENGINEERING. *Educ:* Univ Mich, BSE, 52, MSE, 53; Stanford Univ, PhD(elec eng), 58. *Prof Exp:* Asst prof, Calif Inst Technol, 58-61; dir, Res Tube Div, Varian Assocs, 62-65, mgr microwave semiconductor task force, Cent Res Labs, 65-68; chmn, Dept Elec Eng, 68-77, assoc dir, Appl Physics Lab, 77-79, PROF ELEC ENG, UNIV WASH, 68- *Concurrent Pos:* Mem bd dir, Wash Res Found; dir, Wash Energy Res Ctr, 79-81. *Mem:* Inst Elec & Electronics Engrs; Am Soc Engr Educ. *Res:* Microwaves; solid state devices; plasma physics; physical electronics; education. *Mailing Add:* Dept Elec Eng Univ Wash Seattle WA 98195

DOW, JOHN DAVIS, b Paterson, NJ. THEORETICAL PHYSICS. *Educ:* Univ Notre Dame, BS, 63; Univ Rochester, PhD(physics), 67. *Prof Exp:* From instr to asst prof physics, Princeton Univ, 67-72; from assoc prof to prof physics, Univ Ill, Urbana-Champaign, 72-83, res prof, Coord Sci Lab, 79-83; Freimann prof, Dept Physics, Univ Notre Dame, Ind, 83-91; PROF & CHMN, DEPT PHYSICS & ASTRON, ARIZ STATE UNIV, 90- *Concurrent Pos:* NSF fel, 67-68; consult, Western Elec Co, 68-70; vis scientist, RCA Labs, 68 & Argonne Nat Lab, 69 & 75; assoc, Ctr Advan Study, Univ Ill; hon guest prof, Tsinghua Univ, Beijing, China, 89- *Mem:* Fel Am Phys Soc. *Res:* Theory of solids, optical, electronic and transport properites. *Mailing Add:* Dept Physics & Astron Ariz State Univ Tempe AZ 85287

DOW, LOIS WEYMAN, b Cincinnati, Ohio, Mar 11, 42; div; c 2. HEMATOLOGY, ONCOLOGY. *Educ:* Cornell Univ, BA, 64; Harvard Med Sch, MD, 68. *Prof Exp:* Intern med, Bronx Munic Hosp Ctr, 68-69; asst resident, Presby Hosp, 69-70; res fel hemat, Columbia Presby Med Ctr, 70-72; from instr to asst prof med, Univ Tenn, Memphis, 72-74; res assoc hemat & oncol, St Jude Children's Res Hosp, 74-77, asst mem, 77-80, assoc mem, 80-88; assoc prof pediat, Univ Tenn, Memphis, 83-88; PROVISIONAL TO ASSOC STAFF, MED & PATH, MED CTR DEL, 88-; ASSOC PROF MED, JEFFERSON MED COL, PHILADELPHIA, PA, 88- *Concurrent Pos:* Consult, Baptist Mem Hosp, 74-88, Nat Cancer Inst, 78- *Mem:* Am Found Clin Res; fel Am Col Physicians; Am Soc Hematol; Am Asn Cancer Res; Am Soc Clin Oncol; Int Soc Exp Hemat. *Res:* Cell biology as related to hematology-oncology; normal and malignant stem cells; growth characteristics and effects of anticancer therapy. *Mailing Add:* Med Arts Pavilion Suite 129 4745 Stanton-Ogletown Rd Newark DE 19713

DOW, NORRIS F, research administration; deceased, see previous edition for last biography

DOW, PAUL C(ROWTHER), JR, b Melrose, Mass, Mar 31, 27; m 50; c 4. GUIDANCE & CONTROL. *Educ:* US Mil Acad, BS, 49; Univ Mich, MSE(aeronaut eng) & MSE(instrumentation eng), 54, PhD(aeronaut eng), 57. *Prof Exp:* Mgr guid, Control & Commun Dept, Avco Systs Div, Avco Corp, 60-68, prog dir, 68-75; dept head & FBM prog mgr, 75-88, vpres, 88-90, SR VPRES PROGS, CHARLES STARK DRAPER LAB, INC, 90- *Concurrent Pos:* Lectr, Northeastern Univ, 73-81. *Mem:* Assoc fel Am Inst Aeronaut & Astronaut. *Res:* Guidance and control. *Mailing Add:* C S Draper Lab 555 Technology Square Cambridge MA 02139

DOW, ROBERT LEE, b Royal Oak, Mich, Feb 16, 58; m 82; c 2. MEDICINAL CHEMISTRY. *Educ:* Hope Col, BA, 80; Harvard Univ, PhD(org chem), 85. *Prof Exp:* Res scientist, 85-88, SR RES SCIENTIST, MED CHEM DEPT, PFIZER CENTRAL RES, 88- *Mem:* Am Chem Soc. *Res:* Synthesis and structure activity relationships of agents for the treatment of cancer and the complications associated with obesity. *Mailing Add:* Pfizer Cent Res Groton CT 06340

DOW, ROBERT STONE, b Portland, Ore, Jan 4, 08; m 34; c 2. ANATOMY, PHYSIOLOGY. *Educ:* Linfield Col, BS, 29; Univ Ore, AM & MD, 34, PhD(anat), 35. *Hon Degrees:* DSc, Linfield Col, 63. *Prof Exp:* Asst anat, Med Sch, Univ Ore, 30-35; intern, Wis Gen Hosp, Madison, 35-36; clin instr neurol, 39-42, from asst prof to assoc prof anat, 39-46, from asst clin prof to assoc clin prof med, 46-69, CLIN PROF MED, DIV NEUROL, MED SCH, UNIV ORE, 69- *Concurrent Pos:* Nat Res Coun fel physiol, Sch Med, Yale Univ, 36-37; Belg-Am Educ Found fel, Brussels & Nat Hosp, London, 37-38; fel, Rockefeller Inst, 38-39; Fulbright scholar, Pisa, Italy, 53-54; dir lab neurophysiol, Good Samaritan Hosp, Portland, Ore, 66-74, hon mem Robert S Dow Neurol Sci Inst, 88- *Mem:* Am Physiol Soc; fel Am Col Physicians; Am Asn Neurol Surg; Am Electroencephalog Soc (pres, 57-58); Am Epilepsy Soc (pres, 64); fel Am Acad Neurol; Am Asn Anatomists; AMA; Am Neurol Asn; Am Asn Neurol Surgeons. *Res:* Neurological anatomy and physiology; innervation of the lung; comparative anatomy of the cerebellum; anatomy and physiology of the cerebellum and vestibular systems; pathology of multiple sclerosis; action potentials in central nervous system. *Mailing Add:* Neurol Sci Inst Good Samaritan Hosp Portland OR 97210

DOW, THOMAS ALVA, b Baltimore, Md, July 3, 45; m 69; c 2. PRECISION ENGINEERING, MACHINE DESIGN & CONTROLS. *Educ:* Va Polytechnic Inst, BS, 66; Case Inst Technol, MS, 68; Northwestern Univ, PhD(mech eng), 72. *Prof Exp:* Res scientist, Battelle Columbus Labs, 71-81; PROF MECH ENG, NC STATE UNIV, 82- *Concurrent Pos:* Exec dir, Am Soc Precision Eng, 86-, bd dirs, 86-89; chmn, Metalwork Lab Comt, 76-78, Tribology Div, Am Soc Mech Engrs, 86-87. *Honors & Awards:* Newkirk Award, Am Soc Mech Eng, 79. *Mem:* Fel Am Soc Mech Engrs; Soc Optical Engrs; Am Soc Precision Eng. *Res:* Machine design control and sensing for precision fabrication processes; grinding and turning of glasses and brittle ceramics; precision drive system design and characterization; control system design for machine tools; metrology; tribology. *Mailing Add:* Precision Eng Ctr NC State Univ Box 7918 Raleigh NC 27695-7918

DOW, W(ILLIAM) G(OULD), b Faribault, Minn, Sept 30, 95; m 24, 68; c 5. CONTROLLED NUCLEAR FUSION. *Educ:* Univ Minn, BS, 16, EE, 17; Univ Mich, MSE, 29. *Hon Degrees:* DSc, Univ Colo, 80. *Prof Exp:* From instr to prof elec eng, 26-66, chmn dept, 58-65, sr res space geophysicist, 66-71, EMER PROF ELEC ENG, UNIV MICH ANN ARBOR, 66- *Concurrent Pos:* Res engr, Radio Res Lab, Nat Defense Res Comt Proj, Harvard Univ, 43-45; electronics eng consult, Nat Bur Standards, 45-53; mem bd trustees & consult, Environ Res Inst Mich, 72-90, emer trustee, 90-; mem, US Rocket & Satellite Res Panel, 46-60; mem, Sci Adv Comt, Harry Diamond Labs, US Army Ordinance, 53-64; mem, Panel on Rocketry, US Nat Comt, Int Geophys Year, 56-58; mem, Sci Adv Comt, Gen Motors Res Lab, 58-63; mem staff, World Tour of Electronics Res & Electronics Educ & Space Res Facil, 69-70. *Mem:* AAAS; Inst Elec & Electronics Engrs; Am Soc Eng Educ; Am Geophys Union; Am Phys Soc; Am Welding Soc; Am Inst Aeronaut & Astronaut; Am Astronaut Soc. *Res:* Controlled nuclear fusion; space research and exploration; plasmas; physical electronics; remote sensing for earth observations; radar and radar countermeasures; microwave devices; building a thrust engine powered by nuclear fusion for space vehicles of resonable size, putting payloads into orbit with ground lift off weight 20 or more times less than is now possible. *Mailing Add:* 915 Heatherway Ann Arbor MI 48104

DOWALIBY, MARGARET SUSANNE, b Dover, NH, Mar 5, 24. OPTOMETRY. *Educ:* Los Angeles Col Optom, 50. *Prof Exp:* Clin supvr, Los Angeles Col Optom, 48-53, from assoc prof clin optom to prof optom, 53-65; PROF OPTOM, SOUTHERN CALIF COL OPTOM, 65- *Mem:* Am Acad Optom. *Res:* Cosmetic effect of lenses and frames; optics of lenses. *Mailing Add:* Southern Calif Col Optom 2575 Yorba Linda Blvd Fullerton CA 92631-1699

DOWBEN, PETER ARNOLD, b San Antonio, Tex, July 24, 55. SOLID STATE PHYSICS. *Educ:* Haverford Col, Pa, BA, 77; Cambridge Univ, PhD(physics), 81. *Prof Exp:* Staff, Fritz Haber Inst, Max Planck Soc, 80-83; asst prof, 84-90, ASSOC PROF PHYSICS, SYRACUSE UNIV, 90- *Concurrent Pos:* Vis scientist, dept physics, Univ Osnabruck, 81 & eng physics, Cornell Univ, 83. *Mem:* Sigma Xi; Math Asn Am; Inst Physics UK; Am Vacuum Soc; Am Phys Soc; Am Chem Soc. *Res:* Adsorbate band structures and surface properties using photoemission techniques. *Mailing Add:* 201 Physics Bldg Syracuse Univ Syracuse NY 13244-1130

DOWBEN, ROBERT MORRIS, b Philadelphia, Pa, Apr 6, 27; m 50; c 3. BIOPHYSICS, PHYSIOLOGY. *Educ:* Haverford Col, AB, 46; Univ Chicago, MS, 47, MD, 49. *Hon Degrees:* AM, Brown Univ, 49. *Prof Exp:* Instr, Univ Pa, 52-53; dir radioisotope unit, Vet Admin Hosp, Philadelphia, Pa, 53-55; asst prof med, Northwestern Univ, 56-62; assoc prof biol, Mass Inst Technol, 62-68; prof med sci, Brown Univ, 68-72; prof physiol & dir biophys grad prog, Univ Tex Health Sci Ctr, Dallas, 72-80; RETIRED. *Concurrent Pos:* Fel, Inst Atomenergi, Oslo, 50-51; fel, Dept Med, Johns Hopkins Univ, 51-52; Lalor fel, 60; lectr, Harvard Univ, 63-68; mem, Marine Biol Lab; adj prof physiol, Univ Tex Health Sci Ctr, Dallas, 80- *Honors & Awards:* Distinguished Serv Award, Soc Neuromuscular Dis. *Mem:* Am Chem Soc; Soc Exp Biol & Med; Am Physiol Soc; Am Soc Biol Chem; Biophys Soc; Am Soc Clin Invest. *Res:* Muscle; biological membranes; protein structure; contractile proteins; general medicine. *Mailing Add:* Baylor Res Fdn 3812 Elm St PO Box 710699 Dallas TX 75226

DOWBENKO, ROSTYSLAW, b Ukraine, Jan 8, 27; nat US; m 51; c 1. ORGANIC CHEMISTRY. *Educ:* Northwestern Univ, PhB, 54, PhD(org chem), 57. *Prof Exp:* Res chemist, Wrigley Co, 54; sr res chemist, Pittsburgh Plate Glass Co, 57-64, res assoc, 64-67, sr res assoc, 67-69, asst dir resin res, 69-73, scientist, 73-75, mgr radiation cure, 75-77, mgr polymer res & develop, 77-84, dir polymer res, 84-86, tech dir, Europe, 86-88, FEL, PPG INDUSTS INC, 89- *Mem:* Am Chem Soc; Am Inst Chemists; The Chem Soc. *Res:* Organic polymer chemistry; coatings technology. *Mailing Add:* PPG Industs Inc PO Box 9 Allison Park PA 15101

DOWD, FRANK, JR, b Omaha, Nebr, Aug 8, 39. PHARMACOLOGY. *Educ:* Creighton Univ, DDS, 69; Baylor Col Med, PhD(pharmacol), 75. *Prof Exp:* PHARMACOLOGIST, CREIGHTON UNIV SCH MED, 91- *Concurrent Pos:* Individual grant award, Nat Inst Dent Res, NIH, 91- *Mem:* Am Soc Pharmacol & Exp Therapeut; Int Asn Dent Res; AAAS. *Mailing Add:* Dept Pharmacol Creighton Univ Sch Med 24th & California St Omaha NE 68178

DOWD, JOHN P, b New Bedford, Mass, Feb 1, 38; m 61; c 2. HIGH ENERGY PHYSICS. *Educ:* Mass Inst Technol, SB, 59, PhD(physics), 66. *Prof Exp:* Res asst physics, Mass Inst Technol, 59-66; vis scientist, Ger Electron Synchrotron, Hamburg, 66-67; from asst prof to assoc prof, 67-78, chmn Dept Physics, 85-86, PROF PHYSICS, SOUTHEASTERN MASS UNIV, 78- *Concurrent Pos:* Vis scientist, Univ Bonn, 78-79. *Mem:* Am Phys Soc; Am Asn Physics Teachers. *Res:* Experimental high energy physics. *Mailing Add:* Dept Physics Southeastern Mass Univ North Dartmouth MA 02747

DOWD, PAUL, b Brockton, Mass, Apr 11, 36; m 60; c 3. ORGANIC CHEMISTRY. *Educ:* Harvard Univ, AB, 58; Columbia Univ, MA, 59, PhD(chem), 62. *Prof Exp:* Lectr chem, Harvard Univ, 63-64, instr, 64-66, from lectr to asst prof, 66-70; assoc prof, 70-77, PROF CHEM, UNIV PITTSBURGH, 77- *Concurrent Pos:* Alfred P Sloan Found fel, 70-72. *Mem:* Am Chem Soc; fel AAAS. *Res:* Reactive intermediates in organic chemistry; mechanism of action of vitamin B12; free radical rearrangements; mechanism of action of vitamin K diradicals. *Mailing Add:* Dept Chem Univ Pittsburgh Pittsburgh PA 15260

DOWD, SUSAN RAMSEYER, b Chicago, Ill, Oct 14, 36; m 60; c 3. BIOCHEMISTRY, ORGANIC CHEMISTRY. *Educ:* Reed Col, BA, 58; Columbia Univ, MA, 60, PhD, 62. *Prof Exp:* Fel, Mass Inst Technol, 62-63; res asst biol sci, Harvard Med Sch, 64-66, res assoc, 66-69; res assoc, Protein Res Inst, Univ Pittsburgh, 74-78; SR RES BIOLOGIST, DEPT BIOL SCI, CARNEGIE-MELLON UNIV, 79- *Mem:* Biophys Soc; Chem Soc. *Res:* Nuclear magnetic resonance studies of labelled phospholipids. *Mailing Add:* Dept Biol Sci Carnegie Mellon Univ 4400 Fifth Ave Pittsburgh PA 15213

DOWDA, F WILLIAM, b Cobb Co, Ga. INTERNAL MEDICINE. *Educ:* Emory Univ, MD, 49. *Prof Exp:* Intern, Peter Bent Brigham Hosp, Boston, 49-50; resident, Barnes Hosp, 50-51 & 52-54; staff, Piedmont Hosp, 53-76; clin assoc prof med, Emory Univ, 62-76; PVT PRACT INTERNAL MED, 76- *Concurrent Pos:* Lt Med Corps, US Naval Res, 50-52; path fel & mem staff, Grady Mem Hosp. *Mem:* Inst Med-Nat Acad Sci; Am Med Asn; Soc Med Admnr; Am Col Physicians. *Mailing Add:* 490 Peachtree St NE Suite 129-B Atlanta GA 30308

DOWDELL, RODGER B(IRTWELL), b Portsmouth, NH, Mar 18, 25; m 46; c 8. ENGINEERING, FLUID MECHANICS. *Educ:* Yale Univ, BE, 45; Brown Univ, MS, 52; Colo State Univ, PhD, 66. *Prof Exp:* Fluid mech engr, B-I-F Industs, RI, 52-59; assoc prof mech eng & chmn dept, Univ Bridgeport, 59-66; assoc prof mech eng & appl mech, 66-71, PROF MECH ENG & APPL MECH, UNIV RI, 71- *Concurrent Pos:* Mem & deleg, Int Stand Orgn, 59- *Mem:* Am Soc Mech Engrs. *Res:* Flow measurement and control; hydrodynamics; aerodynamics; heat transfer; propulsion. *Mailing Add:* 1439 Sea Gull Ct Punta Gorda FL 33950

DOWDLE, JOSEPH C(LYDE), b July 3, 27; US citizen; m 56; c 3. ELECTRICAL ENGINEERING. *Educ:* Ala Polytech Inst, BEE, 52, MEE, 58; NC State Col, PhD(elec eng), 62. *Prof Exp:* Instr elec eng, Ala Polytech Inst, 53-57; asst proj engr, Radiation, Inc, 57-58; instr elec eng, NC State Col, 58-62; assoc prof, 62-65, dir grad prog, 62-63, chmn dept eng, 66-69, exec asst to vpres, Huntsville Affairs, 69-77, PROF ELEC ENG, UNIV ALA, HUNTSVILLE, 65-, VPRES ADMIN, 77- *Concurrent Pos:* Consult, Troxler Elec Labs, 58-63; vis scholar, Univ Mich, 68-69. *Mem:* Am Soc Eng Educ; Inst Elec & Electronics Engrs; Am Math Soc. *Res:* High frequency electromagnetics wave phenomena; dielectric materials; powdered dielectrics at millimeter frequencies; magnetic field effect of gyros. *Mailing Add:* 1607 Alaca Pl Tuscaloosa AL 35401

DOWDLE, WALTER R, b Irvington, Ala, Dec 11, 30; m 53; c 2. VIROLOGY. *Educ:* Univ Ala, BS, 55, MS, 57; Univ Md, PhD(microbiol), 60. *Prof Exp:* Asst dir sci, 77, dir, Ctr Infectious Dis, 80, mem res staff, Virol Sect, 60-, Dir, Virol Div, 70-, DEP DIR, CTR DIS CONTROL, 87- *Concurrent Pos:* Vis prof, Univ NC; dir, Regional Reference Ctr Respiratory Dis & Int Influenza Ctr, WHO. *Mem:* Am Soc Microbiol (pres, 89); Soc Exp Biol & Med. *Res:* Bacteriophage; human respiratory disease viruses; mycoplasma agents of man. *Mailing Add:* Ctr Dis Control 1600 Clifton Rd Atlanta GA 30333

DOWDS, RICHARD E, b Cuyahoga Falls, Ohio, Feb 25, 30; m 50; c 2. MATHEMATICS. *Educ:* Kent State Univ, BS, 51; Purdue Univ, MS, 54, PhD(math), 59. *Prof Exp:* Asst prof, Purdue Univ, 58-59; from assoc prof to prof, Butler Univ, 59-65; ASSOC PROF MATH, STATE UNIV NY, FREDONIA, 65- *Mem:* Am Math Soc; Math Asn Am. *Res:* Functional analysis; topological vector spaces; measure theory. *Mailing Add:* Dept Math & Natural Sci D-Youville Col 320 Porter Ave Buffalo NY 14201

DOWDY, EDWARD JOSEPH, b San Antonio, Tex, Sept 25, 39; c 8. NUCLEAR STRUCTURE. *Educ:* St Mary's Univ, BS & BA, 61; Tex A&M Univ, MEng, 63, PhD(nuclear eng), 65. *Prof Exp:* Asst prof nuclear eng, Univ Mo-Columbia, 65-67; from asst prof to assoc prof, Tex A&M Univ, 67-73; staff mem nuclear safeguards, 72-77, alt group leader, 77-79, tech adv to dir, Off Safeguards & Security, US Dept Energy, 79-80; group leader, detection & verification, Univ Calif, 88-89, advan nuclear tech, 85-88, prog mgr, arms control verification tech, Los Alamos Nat Lab, 88-89; SCI ADV, US DELEG NUCLEAR & SPACE TREATY NEGOTIATIONS, 89- *Concurrent Pos:* Owner, Lumen Essence Stained Glass Studio. *Res:* Low energy nuclear physics; nuclear radiation detector design and development; arms control technology. *Mailing Add:* Int Technol Div MSB 234 Los Alamos Nat Lab PO Box 1663 Los Alamos NM 87545

DOWDY, ROBERT H, b Union, WVa, June 28, 37; m 58; c 2. SOIL SCIENCE. *Educ:* Berea Col, BS, 59; Univ Ky, MS, 62; Mich State Univ, PhD(soil sci), 66. *Prof Exp:* RES LEADER & PROF, SOIL & WATER MGT RES DIV, AGR RES, USDA, UNIV MINN, ST PAUL, 66-, ASSOC PROF SOIL SCI, 77- *Mem:* Am Soc Agron; Soil Sci Soc Am; Clay Minerals Soc. *Res:* Clay-chemical phenomena; clay-organic interactions and the significance of such associations on the binding together of soil masses. *Mailing Add:* Borlaug Hall/Soils Sci 439 Univ Minn St Paul 1991 Upper Buford Circle St Paul MN 55108

DOWDY, WILLIAM LOUIS, b San Antonio, Tex, Dec 3, 37; div; c 3. TECHNOLOGY MANAGEMENT, INNOVATION MANAGEMENT. *Educ:* St Mary's Univ, Tex, BS, 59; Tex A&M Univ, ME, 64. *Prof Exp:* Sr engr, Space Div, Chrysler Corp, 59-62; advan progs mgr, Space Div, Rockwell Int, 64-73, managing dir & gen mgr, Sci Ctr, 73-76; mgr prog develop, Elec Power Res inst, 76-78; dir new prod develop, BSP Div, Envirotech Corp, 78-81; dir feasibility progs, Lurgi Corp, 81-83; dir technol & innovation mgt, Stanford Res Inst, SRI Int, 83-88; vpres, PA Consult Group, 88-89; VPRES & GEN MGR, MSE ENVIRON, INC, 90- *Concurrent Pos:* Instr nuclear eng, Tex A&M Univ, 64; chmn, sensor systs tech comt, Am Inst Aeronaut & Astronaut, 73-74, prin, W L Dowdy Consult, 76- *Honors & Awards:* Technol Utilization Award, Nat Aeronaut & Space Admin, 66, Achievement Award, 66. *Mem:* Assoc fel Am Inst Aeronaut & Astronaut; assoc fel Royal Aeronaut Soc. *Res:* Management of technology assets to create strategic business advantage; interconnection of interdisciplinary research and development for formulation of new products, services and business entities; research and development management. *Mailing Add:* 97 Ramona Pl Camarillo CA 93010

DOWE, THOMAS WHITFIELD, b Eagle Pass, Tex, Jan 26, 19; m 43; c 3. ANIMAL HUSBANDRY. *Educ:* Agr & Mech Col Tex, BS, 42; Kans State Col, MS, 47, PhD(animal nutrit, biochem), 52. *Prof Exp:* Animal husbandman, SDak State Col, 47-48; asst prof animal husb, Col Anr, Univ Nebr, 48-57; DIR AGR EXP STA, UNIV VT, 57-, DEAN, COL AGR, 65-, PROF ANIMAL SCI, 80- *Concurrent Pos:* Dir, Univ Nebr Agr Mission, USAID, Columbia, SAm, 70-72. *Mem:* Am Soc Animal Sci; Am Soc Range Mgt. *Res:* Ruminant nutrition; non-protein nitrogen utilization; roughage utilization; nutrient requirements of ruminants. *Mailing Add:* 55 Crescent Rd Burlington VT 05402

DOWELL, ARMSTRONG MANLY, b Torrance, Calif, Aug 29, 21; m 50; c 3. PLASTICS. *Educ:* Univ Calif, Los Angeles, AB, 47, MA, 49. *Prof Exp:* Clin lab technician, Cedars of Lebanon Hosp, 50-55; mem tech staff, Hughes Aircraft Co, 50-85; RETIRED. *Mem:* Sigma Xi. *Res:* Properties of plastics materials including physical, mechanical, thermal, electrical and environmental; specifications expert in aerospace-military documentation. *Mailing Add:* 1919 Coronet Space No 116 Anaheim CA 92801

DOWELL, CLIFTON ENDERS, b McKinney, Tex, Dec 12, 32; c 2. MICROBIOLOGY. *Educ:* Tex Christian Univ, BA, 55, MA, 57; Univ Tex Southwestern Med Sch Dallas, PhD(microbiol), 62. *Prof Exp:* NIH res fel biophys, Calif Inst Technol, 62-64; asst prof microbiol, Sch Med, Tulane Univ, 64-66; asst prof bact, Univ Calif, Davis, 66-69; ASSOC PROF MICROBIOL, UNIV MASS, AMHERST, 69- *Concurrent Pos:* Sr int fel, John Fogarty Ctr, NIH, 79-80. *Mem:* Sigma Xi; Am Soc Microbiol. *Res:* Bacteriophage; microbial genetics. *Mailing Add:* Dept Microbiol Univ Mass Amherst Campus Amherst MA 01003

DOWELL, DOUGLAS C, b Tacoma, Wash, May 31, 24; m 45, 75; c 4. ENGINEERING MECHANICS, NUCLEAR ENGINEERING. *Educ:* Univ Iowa, BS, 49; US Air Force Inst Technol, MS, 56; Iowa State Univ, PhD(eng mech), 64. *Prof Exp:* Proj engr, Armed Forces Spec Weapons Proj, US Air Force, Sandia Base & Los Alamos Sci Lab, 50-54 & US Atomic Energy Comn, Idaho, 56-60, instr physics & math, US Air Force Acad, 60-62, from asst prof to assoc prof eng mech & dep head dept, 64-68; assoc dean, Sch Eng, 68-70, dir educ serv, 70-77, PROF MECH ENG, CALIF STATE POLYTECH UNIV, 77- *Concurrent Pos:* Pres, Dowell Eng, 77-; energy mgt consult, 77- *Mem:* Am Acad Mech; Am Soc Eng Educ; Asn Energy Engrs. *Res:* Engineering education; effect of shock waves on buried structures; energy storage systems. *Mailing Add:* 4501 Peck Rd No 64 El Monte CA 91732

DOWELL, EARL HUGH, b Macomb, Ill, Nov 16, 37; m 81; c 3. MECHANICAL ENGINEERING. *Educ:* Univ Ill, BS, 59; Mass Inst Technol, SM, 61, ScD(aeronaut eng), 64. *Prof Exp:* Res engr aerospace div, Boeing Co, 62-63; res asst aeronaut & astronaut, Mass Inst Technol, 63-64, asst prof, 64-65; from asst prof to assoc prof aerospace & mech sci, Princeton Univ, 65-72, prof, 72-, assoc chmn dept, 75-77; DEAN, SCH ENG, DUKE UNIV. *Honors & Awards:* Structural Dynamics & Materials Award, Am Inst Aeronaut & Astronaut Structures, 80. *Mem:* Am Inst Aeronaut & Astronaut; Am Soc Mech Eng; Acoust Soc Am; Am Acad Mech; US Air Force Sci Adv Bd. *Res:* Structural dynamics; aeroelasticity; unsteady aerodynamics; acoustics. *Mailing Add:* Sch Eng Duke Univ Durham NC 27706

DOWELL, FLONNIE, b Marietta, Ga, Feb 7, 47. STATISTICAL MECHANICS, LIQUID CRYSTALS & POLYMERS. *Educ:* Univ SFla, BA, 69; Tex Womans Univ, MS, 74; Georgetown Univ, PhD(phys chem), 77. *Prof Exp:* Phys scientist polymer physics, Nat Bur Standards, 77-79; phys scientist chem physics, Oak Ridge Nat Lab, 79-81; PHYS SCIENTIST THEORET PHYSICS, LOS ALAMOS NAT LAB, UNIV CALIF, 81-, PROJ LEADER, 84- *Concurrent Pos:* Sch teacher, Ocala, Fla & Carrollton, Tex, 70-71; Nat Sci Found/NATO travel award, 77; Nat Res Coun postdoctoral res assoc, 77-79; mem, Working Party Polymer Liquid Crystals, Iint Union Pure & Appl Chem. *Mem:* Am Phys Soc; Royal Soc Chem; Sigma Xi; AAAS; NY Acad Sci; Am Chem Soc; Mat Res Soc. *Res:* Statistical mechanics of chain molecule systems (including polymers) in condensed phases (liquids, liquid crystals, crystals, glasses); prediction and design of new polymers and other materials. *Mailing Add:* Theoret Div B268 Los Alamos Nat Lab Univ Calif Los Alamos NM 87545

DOWELL, JERRY TRAY, b Yreka, Calif, Feb 12, 38; m 68; c 3. ELECTRODYNAMICS OF MOVING MEDIA, DIELECTRIC PHENOMENA. *Educ:* Univ Calif, Berkeley, AB, 59, PhD(physics), 66. *Prof Exp:* Res scientist physics, Lockheed Palo Alto Res Lab, 66-70; asst prof physics, Univ Mo, Rolla, 70-75; PHYSICIST, IRT CORP, 74- *Mem:* Am Phys Soc; Inst Elec & Electronics Engrs; Sigma Xi. *Res:* Atomic and molecular physics, molecular beams, electron beams, microwave spectroscopy, quantum electronics, physical optics, astronomy, astrophysics, and electromagnetics. *Mailing Add:* 205 W Floresta Way Portola Valley CA 94025

DOWELL, MICHAEL BRENDAN, b Bronx, NY, Nov 18, 42; m 68; c 2. HIGH TEMPERATURE CHEMISTRY. *Educ:* Fordham Univ, BS, 63; Pa State Univ, PhD(inorg chem), 67. *Prof Exp:* Physicist, Pitman-Dunn Res Labs, US Army Frankford Arsenal, 67-69; res scientist, 69-74, develop mgr, 74-80, sr group leader, 80-81 mgr grafoil and boron nitride products, 82-83, MGR MARKET DEVELOP, UNION CARBIDE CORP, 83- *Concurrent Pos:* Dir, US Advan Ceramics Asn, 87- *Mem:* AAAS; Am Chem Soc; Am Phys Soc; Am Carbon Soc; Am Ceramic Soc; Am Soc Metals Int. *Res:* Kinetics and thermodynamics of phase transformations in solids; vaporization of solids; carbon fibers; intercalation of graphite; electrochemical devices; advanced ceramics. *Mailing Add:* 368 N Main St Hudson OH 44236

DOWELL, ROBERT VERNON, b San Francisco, Calif, Sept 13, 47; m 74; c 1. INSECT ECOLOGY. *Educ:* Univ Calif, Irvine, BS, 69; Calif State Univ, Hayward, MS, 72; Ohio State Univ, PhD(entom), 76. *Prof Exp:* asst res scientist entom, Univ Fla, 77-80; pest mgt specialist III, Div Pest Mgt, 80-82, economic entomologist, 82-88, primary state entomologist, 85-87, SR ECON ENTOMOLOGIST, DEPT FOOD & AGR, 84- *Honors & Awards:* Super Achievement Award, Calif Dept Food & Agr, 84. *Mem:* Lepidopterist Soc; Ecol Soc Am; AAAS; Soc Pop Ecol; Pac Coast Entom Soc. *Res:* Plant-insect interrelationships and evolution of swallowtail butterflies and whiteflies. *Mailing Add:* Dept Food & Agr-Div Plant Indust 1220 N St PO Box 94287 Sacramento CA 94271-0001

DOWELL, RUSSELL THOMAS, b Cameron, Mo, Sept 23, 41; m 63; c 4. PHYSIOLOGY, PHYSIOLOGICAL CHEMISTRY. *Educ:* Kans State Univ, BS, 64; Univ Ariz, MEd, 65; Univ Iowa, PhD(physiol), 71. *Prof Exp:* Fel med, Univ Chicago, 71-73; asst prof, Univ Tex Med Br, Galveston, 73-77; assoc prof physiol, Univ Okla Health Sci Ctr, 77-; PROF PHYSIOL, UNIV KANS MED CTR. *Concurrent Pos:* Fel, Chicago & Ill Heart Asn, 71-73; mem, Marine Biomed Inst, Galveston, 73-77. *Honors & Awards:* Jerome Frankel Found Bahr Res Award, Univ Okla, 78. *Mem:* Am Physiol Soc; Int Soc Heart Res; Am Col Sports Med; Soc Exp Biol & Med; Aerospace Med Asn. *Res:* Heart adaptation to stress; exercise physiology. *Mailing Add:* Dept Physiol/Biophys Univ Okla Health Sci Ctr Oklahoma City OK 73190

DOWELL, VIRGIL EUGENE, b Melver, Kans, June 3, 26; m 61; c 4. FISH BIOLOGY. *Educ:* Kans State Teachers Col, BS, 51, MS, 52; Univ Okla, PhD(zool), 57. *Prof Exp:* Asst, Kans State Teachers Col, 50-52; asst, Univ Okla, 52-53; asst, Okla Biol Surv, 53-56; from asst prof to assoc prof, 56-68, PROF BIOL, UNIV NORTHERN IOWA, 68- *Mem:* Am Inst Biol Sci; Am Fisheries Soc; Am Soc Ichthyologists & Herpetologists. *Res:* Aquatic biology. *Mailing Add:* 1609 Grandview Ct Cedar Falls IA 50613

DOWELL, VULUS RAYMOND, JR, b Mt Vernon, Ky, July 27, 27; m 48; c 4. MICROBIOLOGY, BACTERIOLOGY. *Educ:* Univ Ky, BS, 61; Univ Cincinnati, MS, 62, PhD(microbiol), 66. *Prof Exp:* Instr microbiol & res surg, Univ Cincinnati, 65-66; res microbiologist in chg, Anaerobis Bact Lab, 66-71, chief enterobact br, 71-81, ASST DIR LAB SCI, HOSP INFECTIONS PROG, CTR DIS CONTROL, 81-; ASST PROF, SCH MED, EMORY UNIV, ATLANTA, 77- *Concurrent Pos:* Asst prof, Ga State Univ, 71-; assoc prof, Sch Pub Health, Univ NC; mem ed bd, Appl Microbiol, 73-75; mem ed bd, J Clin Microbiol, 75- *Mem:* Am Soc Microbiol; Sigma Xi; Brit Soc Gen Microbiol; NY Acad Sci; Am Pub Health Asn. *Res:* Hospital acquired infections; clinical anaerobic bacteriology; polymicrobic infections; factors predisposing to microbiol infections; characterization of unusual bacteria, botulism, and foodborne diseases. *Mailing Add:* 2295 Clairmont Rd Atlanta GA 30329

DOWER, GORDON EWBANK, b Brit, Nov 16, 23; m 47; c 4. ELECTROCARDIOLOGY. *Educ:* St Bartholomew's Hosp Med Col, MB, BS, 49. *Prof Exp:* From instr to asst prof, 54-64, ASSOC PROF PHARMACOL, UNIV BC, 64- *Concurrent Pos:* Can Life Ins Co fel, 54-57; Heart Found Can fel, 57-; consult, Shaughnessy Hosp; res consult, Vancouver Gen Hosp. *Mem:* Fel Am Col Cardiol. *Res:* Polarcardiography; computer techniques in electrocardiographic diagnosis. *Mailing Add:* 4640 NW Marine Dr Vancouver BC V6R 1B9 Can

DOWHAN, WILLIAM, b Detroit, Mich, Dec 15, 42. BIOCHEMISTRY. *Educ:* Princeton Univ, BS, 64; Univ Calif, Berkeley, PhD(biochem), 69. *Prof Exp:* PROF & VCHAIR BIOCHEM, MED SCH, UNIV TEX, 72- *Concurrent Pos:* Guggenheim fel, 83. *Mem:* Am Soc Biochem & Molecular Biol; Am Soc Microbiologists. *Mailing Add:* Dept Biochem & Molecular Biol Med Sch Univ Tex PO Box 20708 Houston TX 77225

DOWIS, W J, b Idaho Springs, Colo, Aug 24, 08; m 32; c 3. NUCLEAR PLANT APPLICATIONS, ECONOMICS OF ENGINEERING APPLICATIONS. *Educ:* Univ Colo, BSEE, 30. *Prof Exp:* Prin engr, Gen Elec Co, 46-63, United Nuclear Co, 63-69; consult engr, Battle Inst & Exxon Corp, 69-84; RETIRED. *Concurrent Pos:* Chmn planning comn, Richland, Wash, 70-74, mem utilities adv bd, 80-84. *Mem:* Fel Inst Elec & Electronics Engrs. *Mailing Add:* 1603 Judson Ave Richland WA 99352

DOWLER, CLYDE CECIL, b Moundsville, WVa, Jan 12, 33; m 56; c 4. WEED SCIENCE, CHEMIGATION. *Educ:* Univ WVa, BS, 54, MS, 56; Ohio State Univ, PhD(agron), 58. *Prof Exp:* Res agronomist, NC, 59-63, res agronomist, PR, 63-67, RES AGRONOMIST, GA COASTAL PLAIN EXP STA, AGR RES SERV, USDA, 67- *Mem:* Weed Sci Soc Am; Coun Agr Sci & Technol; Am Soybean Asn; Plant Growth Regulator Soc; Sigma Xi. *Res:* Develop weed management systems, biological and ecological data of weeds in agronomic row crops and integrated pest management with intensive cropping sequences, chemigation, conservation tillage impact on weeds and crops. *Mailing Add:* Ga Coastal Plain Exp Sta Tifton GA 31793

DOWLER, WILLIAM MINOR, b Birch Tree, Mo, Nov 10, 32; m 58; c 3. PHYTOPATHOLOGY. *Educ:* Univ Mo, BS, 54, MS, 58; Univ Ill, PhD(plant path), 61. *Prof Exp:* Res plant pathologist, Crops Res Div, 61-75, mem, Nat Prog Staff, 75-81, RES LEADER, PLANT DIS RES LAB, AGR RES SERV, USDA, 81- *Mem:* Am Phytopath Soc; Am Soc Hort Sci; Int Soc Plant Path; Soc Nematologists. *Res:* Diseases of all crops; nematodes; physiology of pathogens and etiology of diseases. *Mailing Add:* 6519 Debolt Rd Sabillasville MD 21780

DOWLING, DIANE MARY, b Winnipeg, Man, Feb 21, 35; m 89. MATHEMATICS. *Educ:* Univ Man, BSc, 55, MSc, 56; Univ Toronto, PhD(math), 59. *Prof Exp:* Lectr, 58-62, from asst prof to assoc prof, 62-75, PROF MATH, UNIV MAN, 75- *Mem:* Am Math Soc; Math Asn Am; Can Math Soc; Soc Indust & Appl Math. *Res:* Modular representation theory; graph theory; combinatorics. *Mailing Add:* Dept Math Univ Man Winnipeg MB R3T 2N2 Can

DOWLING, EDMUND AUGUSTINE, b Waterford, Ireland, July 26, 27; US citizen; m 53; c 4. PATHOLOGY. *Educ:* Nat Univ Ireland, MB, BCh & BAO, 51; Am Bd Path, dipl, 59. *Prof Exp:* From instr to prof path, Sch Med, Univ Ala, Birmingham, 54-74; PROF PATH, SCH MED, UNIV S ALA, 74- *Concurrent Pos:* Consult, Birmingham Baptist Hosps, 62-74. *Mem:* AMA Brit Med Asn; Int Acad Path; Am Soc Cytol. *Res:* Neoplastic diseases; exfoliative cytology. *Mailing Add:* Dept Path Univ S Ala Med Col 307 University Blvd Mobile AL 36688

DOWLING, HERNDON GLENN, (JR), b Cullman, Ala, Apr 2, 21; m 43, 67; c 4. HERPETOLOGY. *Educ:* Univ Ala, BS, 42; Univ Fla, MS, 48; Univ Mich, PhD(zool), 51. *Prof Exp:* Instr biol, Univ Fla, 47-48; mus asst, Univ Mich, 48-51; instr biol, Haverford Col, 51-52; from asst prof to assoc prof zool, Univ Ark, 52-59; from assoc cur to cur reptiles, NY Zool Park, 59-67; dir herpet info search systs, Am Mus Natural Hist, 68-73; assoc prof, 73-75, PROF BIOL, NY UNIV, 75- *Concurrent Pos:* Fel, Univ Fla, 56-57; res assoc, Dept Amphibians & Reptiles, Am Mus Natural Hist, 57-78; adj prof, Univ RI, 64-76 & NY Univ, 65-73; gen ed, Catalogue Am Amphibians & Reptiles, 66-73; ed, Amphibian & Reptile Sect, Biol Abstr, 68-73 & Herpet Rev, 69-72; ed, Yearbook of Herpet, 74-75; res affil, Dept Zool, Univ, Md, 82- *Mem:* Fel AAAS; Am Inst Biol Sci; Am Soc Ichthyologists & Herpetologists; Am Soc Zoologists; Soc Study Evolution; Sigma Xi. *Res:* Systematic herpetology; taxonomic studies of colubrid snake genera; higher categories of Serpentes; evolutionary biology. *Mailing Add:* Dept Biol NY Univ New York NY 10003

DOWLING, JEROME M, b Chicago, Ill, July 9, 31; m 56; c 4. INFRARED PHENOMENOLOGY, REMOTE SENSING. *Educ:* Ill Inst Technol, BS, 53, MS, 55, PhD(physics), 57. *Prof Exp:* From asst prof to assoc prof physics, Ariz State Univ, 59-63; mem tech staff, 63-69, staff scientist, 69-79, SR SCIENTIST, AEROSPACE CORP, 79- *Concurrent Pos:* Fel pure physics, Nat Res Coun Can, 57-59. *Mem:* Fel Am Phys Soc; Am Asn Physics Teachers; Optical Soc Am; Sigma Xi. *Res:* Upper atmospheric physics and chemistry; molecular structure and dynamics; interferometry; sensor technology. *Mailing Add:* Aerospace Corp PO Box 92957-M2 251 Los Angeles CA 90009

DOWLING, JOHN, b Ashland, Ky, Sept 12, 38; m 63; c 2. PHYSICS. *Educ:* Univ Dayton, BS, 60; Ariz State Univ, MS, 62, PhD(physics), 64. *Prof Exp:* Res assoc molecular spectros, Ariz State Univ, 64; res assoc atmospheric & space physics, Univ Fla, 64-66; asst prof, Univ NH, 66-70; prof physics, 76-81, chmn physics dept, Mansfield Univ, 81-84; PROF, DEPT PHYSICS & ASTRON, MICH STATE UNIV, EAST LANSING, 84- *Concurrent Pos:* Film ed, Am J Physics, 76-82 & Bulletin Atomic Sci, 79-84. *Mem:* Am Phys Soc; Am Asn Physics Teachers; AAAS. *Mailing Add:* Physics Dept Mansfield Univ Mansfield PA 16933

DOWLING, JOHN ELLIOTT, b Pawtucket, RI, Aug 31, 35; m 75; c 3. NEUROBIOLOGY. *Educ:* Harvard Univ, AB, 57, PhD(biol), 61. *Hon Degrees:* MD, Univ Lund, Sweden, 82. *Prof Exp:* From instr to asst prof biol, Harvard Univ, 60-64; assoc prof ophthal & biophys, Sch Med, Johns Hopkins Univ, 64-71; assoc dean, 80-84, PROF BIOL, HARVARD UNIV, 71-, MASTER, LEVERETT HOUSE, 81- *Concurrent Pos:* Trustee, Marine Biol Lab, 70-77 & 86-90; chmn Comn Life Sci, Nat Acad Sci & Nat Res Coun, 86-89; trustee, Mass Eye & Ear Infirmary, 86- *Honors & Awards:* Friedenwald Award, Asn Res Vision & Ophthal, 70; Retina Res Found Award Merit, 81; Alcon Vision Res Recognition Award, 86. *Mem:* Nat Acad Sci; Am Acad Arts & Sci; Asn Res Vision & Ophthal; Soc Gen Physiol; AAAS; Soc Neurosci. *Res:* Visual physiology; chemistry and anatomy; nervous system, fine structure and function. *Mailing Add:* Biol Labs Harvard Univ 16 Divinity Ave Cambridge MA 02138

DOWLING, JOHN J, b Webster Groves, Mo, Dec 8, 34; m 58; c 5. GEOPHYSICS. *Educ:* St Louis Univ, BS, 57, PhD(geophys), 64; Univ Tulsa, MS, 60. *Prof Exp:* Res engr, Jersey Prod Res Inc, 57-60; asst to dean, Inst Technol, St Louis Univ, 60-63, instr eng, 60-64, res asst geophys, 63-64; res assoc geosci, Southwest Ctr Advan Studies, 64-67; asst prof geophys, Tex Tech Col, 67-68; MEM MARINE SCI INST, SOUTHEAST BR, UNIV CONN, 68-, ASSOC PROF GEOPHYSICS, 80- *Mem:* AAAS; Am Geophys Union; Seismol Soc Am; Soc Explor Geophys; Am Soc Eng Educ; Sigma Xi. *Res:* Interior of the earth from elastic waves; marine geophysics; crustal structure in deep oceans and at continental margins. *Mailing Add:* Marine Sci Inst Univ Conn Groton CT 06340

DOWLING, JOSEPH FRANCIS, b New York, NY, June 19, 33; m 58; c 2. FOOD CHEMISTRY. *Educ:* Adelphi Univ, BA, 55, MS, 66. *Prof Exp:* From asst chief chemist to chief chemist, 64-69, labs mgr, 69-73, tech serv mgr, Corn Prod, 73-76, TECH MGR, REFINED SUGARS INC, 76- *Concurrent Pos:* Secy, Cane Sugar Ref Res Proj, 65-76; vchmn, US Nat Comt Uniform Methods Sugar Anal, 70-74; tech adv comt, Sugar Asn, 86- *Honors & Awards:* George & Eleanor Meade Award, Sugar Indust Tech, 68. *Mem:* Sugar Indust Tech; Am Chem Soc; Am Soc Sugar Beet Technol; Inst Food Technol. *Res:* Sugar chemistry, development of analytical methods and improved means of refining sugar; applications of sugars in the food industry; gas liquid chromatography of sugar and related sugar impurities. *Mailing Add:* Refined Sugars Inc One Federal St Yonkers NY 10702

DOWLING, MARIE AUGUSTINE, b Baltimore, Md, Aug 19, 24. MATHEMATICS. *Educ:* Col Notre Dame, Md, BA, 45; Catholic Univ Am, MS, 58. *Prof Exp:* Teacher, St Marys High Sch, Md, 46-48; teacher, Notre Dame Prep Sch, 48-59; chmn dept math, 65-80 & 84-90, dir, Winterim, 72-78, ASSOC PROF MATH, COL NOTRE DAME, MD, 59- *Mem:* Math Asn Am. *Res:* Teaching mathematics on the undergraduate level; history of mathematics. *Mailing Add:* Dept Math Col Notre Dame-Md Baltimore MD 21210-2476

DOWNER, DONALD NEWSON, b Lexington, Miss, July 2, 44; m 72. MICROBIOLOGY. *Educ:* Univ Miss, BS, 66, PhD(microbiol), 71. *Prof Exp:* Fel microbiol, Med Ctr Univ Miss, 71-72; Killiam fel biochem, Univ Alta, 72-74, prof asst, 74-78; ASST PROF MICROBIOL, MISS STATE UNIV, 78- *Mem:* Am Soc Microbiol; Sigma Xi. *Res:* Virology. *Mailing Add:* Dept Biol Sci PO Drawer GY Mississippi State MS 39762

DOWNER, NANCY WUERTH, b Washington, DC, Sept 2, 43. MEMBRANE PROTEINS, VISUAL TRANSDUCTION. *Educ:* Mt Holyoke Col, AB, 65; Univ Pa, PhD(biochem), 74. *Prof Exp:* Fel molecular biol, Univ Ore, 74-76; fel biophys, Johns Hopkins Univ, 76-78; asst prof biochem, Univ Ariz, 79-84. *Concurrent Pos:* Fel, Fight for Sight, Inc, 77-78; vis prof, Mass Inst Technol, 84-85; vis scientist, Los Alamos Nat Lab, 86-87. *Mem:* Biophys Soc; AAAS. *Res:* Mechanism of visual transduction, membrane structure and function, membrane proteins; biosenor development; biosensors. *Mailing Add:* TSI Mason Res INT 57 Union St Worcester MA 01608

DOWNER, ROGER GEORGE HAMILL, b Belfast, Northern Ireland, Dec 21, 42; m 66; c 3. INSECT PHYSIOLOGY, NEUROENDOCRINOLOGY. *Educ:* Queen's Univ Belfast, 64, MS, 67; Univ Western Ont, PhD(zool), 70. *Hon Degrees:* DSc, Queen's Univ Belfast, 84. *Prof Exp:* From asst prof to assoc prof, 70-81, chmn biol, 86-89, PROF BIOL, UNIV WATERLOO, 81-, VPRES, 89- *Concurrent Pos:* Nat Res Coun Can grant, 70-; Japan-Can sr fel, 76; vpres res, Huntsman Lab; consult, Am Cyanamid; Natural Sci & Eng Res Coun Can operating, strategic, equipment & co-op grants; vpres, Biol Coun Can, 86; pres, Univ Biol chmn, Can Coun, 87-88; consult, agr chem, Several Multinat Co. *Mem:* Entom Soc Can; Can Soc Zool (pres, 86-87); Int Soc Neurochem. *Res:* Regulation of lipid and carbohydrate metabolism in insects; cyclic nucleotides; lipid absorption and transport; biogenic amines; biochemistry; insecticide toxicity and vertebrate nervous system; insecticide development. *Mailing Add:* Dept Biol Univ Waterloo Waterloo ON N2L 3G1 Can

DOWNES, JOHN ANTONY, b Wimbledon, Eng, Feb 14, 14; m 53; c 5. ENTOMOLOGY. *Educ:* Univ London, BSc, 35. *Prof Exp:* Commonwealth Fund fel, Univ Calif, Berkeley, 37-39; demonstr zool, Univ London, 39-40; from lectr to sr lectr entom, Glasgow Univ, 40-53; entomologist, Sci Serv, Can Dept Agr, 53-58, head vet & med entom, 58-59, sr entomologist, Res Br, 59-78; res assoc, Biosystematics Res Inst, Ottawa, 79-84; RES ASSOC, LYMAN MUS, MCGILL UNIV, 80-; RES ASSOC, NAT MUS NATURAL SCI, OTTAWA, 85- *Concurrent Pos:* Sr sci inspector, Ministry Food, Gt Brit, 41-45; hon sci inspector, Dept Agr, Scotland, 45-47; deleg, Med Res Coun Can, 50; secy, Int Cong Entom, 56; chmn, Sci Comt, Biol Surv Can, (Terrestrial Arthropods), 79 & 86. *Honors & Awards:* Gold Medal, Entom Soc Can, 77. *Mem:* Entom Soc Am; Soc Study Evolution; Entom Soc Can; Royal Entom Soc London. *Res:* Systematics, behavior and physiology of insects, especially Diptera and Lepidoptera; arctic insects; biogeography. *Mailing Add:* 877 Riddell Ave Ottawa ON K2A 2V8 Can

DOWNES, JOHN D, b Buckhannon, WVa, Feb 27, 19; m 40; c 4. HORTICULTURE. *Educ:* WVa Univ, BS, 42, MS, 51; Mich State Univ, PhD(hort), 55. *Prof Exp:* Instr & asst hort, WVa Univ, 43-51; asst, Mich State Univ, 51-54; asst horticulturist, Malheur Br Exp Sta, Ore State Col, 54-55; from asst prof to prof hort, Mich State Univ, 55-70; prof Hort & Dir Veg Res Prog, Tex Tech Univ, 70-84; CONSULT, 85- *Concurrent Pos:* Crop specialist, IRI Res Inst, Brazil, 67-68; tech consult veg prod, transport & mkt, Mich State Univ-Inst Nat Tech Agropecuaria, Argentina, 69-70; tech consult, IRI Res Inst, Brazil, 73 & CID, Swaziland, SAfrica, 81; res hort, Washington State Univ, USAID, NDA, Lesotho, SAfrica, 84-86, IRI Res Inst/FA, Dominican Republic, 89. *Mem:* AAAS; Sigma Xi; Am Soc Hort Sci. *Res:* Plant physiology, nutrition, breeding, and genetics; biometry; vegetable production efficiency and economics; wind and windblown soil injury on vegetables and their control. *Mailing Add:* 1231 Miami St Athens TN 37303-4554

DOWNES, WILLIAM A(RTHUR), electronics; deceased, see previous edition for last biography

DOWNEY, BERNARD JOSEPH, b Philadelphia, Pa, Jan 18, 17; m 55; c 1. PHYSICAL CHEMISTRY. *Educ:* Cath Univ, BA, 38, MS, 45, PhD(phys chem), 52. *Prof Exp:* Instr chem, De La Salle Col, 44-49; from asst prof to assoc prof, La Salle Col, 49-54; asst prof, Seton Hall Univ, 54-59; assoc prof, 59-61, chmn dept, 60-69, assoc dean grad studies, 70-74, dir res, 79-85, PROF CHEM, VILLANOVA UNIV, 61-, DEAN GRAD SCH, 74-, UNIV PATENT ADMINR, 79- *Concurrent Pos:* Vis res prof, Imp Col, Univ London, 69-70. *Mem:* Am Chem Soc; Sigma Xi. *Res:* Kinetics of metallic film oxidation; diffusion in ionic crystals. *Mailing Add:* Grad Sch Villanova Univ 1123 Windsor Dr West Chester PA 19380

DOWNEY, H FRED, b Hagerstown, Md, Aug 6, 39; div; c 2. CORONARY CIRCULATION. *Educ:* Univ Md, College Park, BS, 61, MS, 64; Univ Ill, Urbana, PhD(biophys), 68. *Prof Exp:* Asst prof vet physiol & pharmacol, Univ Ill, Urbana, 68-71, asst prof physiol & biophys, 70-71; asst prof physiol, Southwestern Med Sch, Univ Tex Health Sci Ctr, Dallas, 72-77, res assoc prof, 78-84; dir cardiovasc res, Cardiopulmonary Inst, Methodist Hosp, 72-84; PROF & VCHMN, DEPT PHYSIOL, TEX COL OSTEOP MED, 84- *Concurrent Pos:* Prin investr, Res Coronary Circulation, NIH, 77-; Coun Circulation fel, Am Heart Asn; NIH predoctoral fel. *Mem:* Am Physiol Soc; Soc Exp Biol & Med; Sigma Xi; Inst Heart Res; Microcirculatory Soc. *Res:* Coronary circulation, especially control and distribution of coronary blood flow; blood flow to ischemic myocardium; cardiovascular responses to nicotine. *Mailing Add:* Dept Physiol Tex Col Osteopath Med 3500 Camp Bowie Blvd Ft Worth TX 76107

DOWNEY, JAMES MERRITT, b Wabash, Ind, Nov 1, 44; m 67; c 2. CORONARY PHYSIOLOGY, CARDIAC PHARMACOLOGY. *Educ:* Manchester Col, Ind, BS, 67; Univ Ill, Urbana, MS, 69, PhD(physiol), 71. *Prof Exp:* Res asst, Marine Biol Lab, Woods Hole, 70; res fel, Peter Bent Brigham Hosp, Boston, 71-72; asst prof physiol, Univ SFla, 72-75; assoc prof, 75-80, PROF PHYSIOL, UNIV SALA, 80- *Concurrent Pos:* Vis asst prof, Southwestern Med Sch, Dallas, 75; vis scientist, Rayne Inst, St Thomas Hosp, London, 81. *Mem:* Fel Am Physiol Soc; fel Am Heart Asn. *Res:* Coronary artery hemodynamics and controllers of blood flow; drug and other interventions which might protect ischemic myocardium. *Mailing Add:* 3024 MSB Univ SAla Mobile AL 36688

DOWNEY, JOHN A, b Regina, Sask, Sept 16, 30; m 53; c 4. MEDICINE, PHYSIOLOGY. *Educ:* Univ Man, BSc & MD, 54; Oxford Univ, PhD(physiol), 62; Am Bd Phys Med & Rehab, dipl. *Prof Exp:* Intern, Vancouver Gen Hosp, BC, 53-54; resident phys med & rehab, Presby Hosp, New York, 54-56, 57-58, res assoc, 58-59; resident internal med, Peter Bent Brigham Hosp, Boston, Mass, 56-57, 59-60; vis worker Christ Church Col, Oxford Univ, 60-62; asst prof med & from asst prof to assoc prof phys med & rehab, Col Physicians & Surgeons, Columbia Univ, 63-70; chmn & dir rehab med, Columbia-Presby Med Ctr, 74-90; PROF REHAB MED, COL PHYSICIANS & SURGEONS, COLUMBIA UNIV, 70- *Concurrent Pos:* Consult physiatrist, Blythedale Children's Hosp, Valhalla, 62; asst attend physician, Presby Hosp, 62, assoc attend physician, 64-; vis fel phys med, Col Physicians & Surgeons, Columbia Univ, 62-63. *Mem:* Inst Med-Nat Acad Sci; Am Physiol Soc; Am Rheumatism Asn; Am Acad Phys Med & Rehab; Am Cong Rehab Med; Asn Acad Physiatrists. *Res:* Physiology of temperature regulation; control of respiration; peripheral circulation; clinical care of patients with chronic disabling illness. *Mailing Add:* Dept Rehab Med Col Physicians & Surgeons Columbia Univ New York NY 10032

DOWNEY, JOHN CHARLES, b Eureka, Utah, Apr 12, 26; m 49; c 5. ENTOMOLOGY, BEHAVIOR-ETHOLOGY. *Educ:* Univ Utah, BS, 49, MS, 50; Univ Calif, PhD(entom), 57. *Prof Exp:* Instr biol, Univ Utah, 50-52; assoc zool, Univ Calif, 52-56; from asst prof to prof, Southern Ill Univ, 56-68; prof biol & head dept, Univ Northern Iowa, 68-81, dean, Grad Col, 81-88; RETIRED. *Concurrent Pos:* Prin investr numerous grants, NSF. *Mem:* Entom Soc Am; Soc Syst Zool; Lepidop Soc (secy, 66-73); Soc Study Evolution. *Res:* General variation and evolution; taxonomy and morphology Lycaenidae; ecology and behavior of insects; ultrastructure, sound and chemical communication in Lepidoptera; eggs and immature stages of butterflies. *Mailing Add:* 5027 Kestral Park Dr Sarasota FL 34231

DOWNEY, JOSEPH ROBERT, JR, b Charleston, WVa, Nov 27, 41; m 63; c 2. PHYSICAL CHEMISTRY. *Educ:* Western Md Col, BA, 63; Fla State Univ, PhD(inorg chem), 72. *Prof Exp:* Instr chem, Fla State Univ, 72-73; assoc chem, Rensselaer Polytech Inst, 73-76; sr res chemist, 76-80, proj leader, 80-85, res leader, 85-89, RES ASSOC, DOW CHEM USA, 89- *Mem:* Am Chem Soc. *Res:* Solution interactions and structural determinations via vibrational spectroscopy; hydrogen bonding, ion pairing, complex formation; application of computerized methods to handling of spectroscopic data; curve resolution, interfacing; thermodynamics; thermochemistry; calorimetry; critical evaluation of thermodynamic and physical property data. *Mailing Add:* 1897 Building Dow Chem Co Midland MI 48667

DOWNEY, KATHLEEN MARY, DNA POLYMERASES, MUTAGENESIS. *Educ:* Univ Wash, Seattle, PhD(biochem), 65. *Prof Exp:* PROF MED & BIOCHEM, SCH MED, UNIV MIAMI, 82- *Mailing Add:* 8324 SW 168th Terr Miami FL 33157

DOWNEY, RICHARD KEITH, b Saskatoon, Sask, Jan 26, 27; m 52; c 5. PLANT BREEDING. *Educ:* Univ Sask, BSA, 50, MSc, 52; Cornell Univ, PhD, 61. *Prof Exp:* Res scientist, Exp Farm, Lethbridge, Alta, 52-57; res scientist, Crops Sect, 57-73, asst dir, 73-90, HEAD, OIL SEED SECT, AGR CAN RES STA, 90- *Concurrent Pos:* Nat dir, Agr Inst Can. *Honors & Awards:* Bond Medal, Am Oil Chemists' Soc, 63; Merit Award, Pub Serv Can, 68; Royal Bank Can Award, 75; Grindley Medal, Agr Inst Can, 73; Officer Order Can, 76; Queen's Jubilee Medal, 77; Gold Metal, Prof Inst Pub Serv, 90. *Mem:* Fel Agr Inst Can; Can Soc Agron (past pres); fel Royal Soc Can. *Res:* Oil seed and forage crop improvement. *Mailing Add:* Agr Can Res Sta 107 Sci Crescent Saskatoon SK S7N 0X2 Can

DOWNEY, RONALD J, b Manitowoc, Wis, Apr 8, 33; m 57; c 5. CELL PHYSIOLOGY. *Educ:* Regis Col, Colo, BS, 55; Creighton Univ, MS, 58; Univ Nebr, PhD(microbiol), 61. *Prof Exp:* Res assoc cell metab, USDA, 61-62; asst prof biol, Univ Notre Dame, 62-66, from assoc prof to prof microbiol, 66-72; PROF ZOOL & MICROBIOL & CHMN DEPT, OHIO UNIV, 72- *Concurrent Pos:* USPHS career develop award. *Mem:* Am Soc Microbiol; Soc Exp Biol & Med; Am Soc Cell Biol; Am Soc Exp Path. *Res:* Synthesis of nitrogen assimilatory; differentiation in the lower eucaryotes. *Mailing Add:* Dept Zool & Microbiol Ohio Univ Athens OH 45701

DOWNHOWER, JERRY F, b Indianapolis, Ind, Oct 25, 40; m 64; c 2. BEHAVIORAL ECOLOGY. *Educ:* Occidental Col, BA, 62; Univ Kans, MA, 64, PhD(zool), 68. *Prof Exp:* Res asst, Kans State Biol Surv, 62-64; res asst zool, Univ Kans, 64-68; lectr biol, Cornell Univ, 68-70; from asst prof to assoc prof, 70-84, PROF ZOOL, OHIO STATE UNIV, 85- *Concurrent Pos:* Mem fac of orgn for trop studies, 71; mem steering comt, Biol Sci Curric Studies, 71, mem bd dirs, 71-72; consult, Battelle Columbus Labs, 74-75; prog dir, Pop Biol & Physiol Ecol, NSF, 80-81. *Mem:* AAAS; Soc Syst Zool; Am Soc Mammalogists; Am Soc Naturalists. *Res:* Behavior; adaptive significance of vertebrate social organization. *Mailing Add:* Dept Zool Ohio State Univ Main Campus Bot/Zool Bldg 104 Columbus OH 43210

DOWNIE, HARRY G, b Toronto, Ont, June 11, 26; m 50; c 4. PHYSIOLOGY, EXPERIMENTAL SURGERY. *Educ:* Ont Vet Col, DVM, 48, MVSc, 52; Cornell Univ, MS, 51; Univ Western Ont, PhD(physiol), 59. *Prof Exp:* Lectr physiol, 48-49, from asst prof to assoc prof, 51-56, head dept res, 56-58, head dept physiol sci, 58-69, chmn, Dept Biomed Sci, 69-80, PROF PHYSIOL, ONT VET COL, UNIV GUELPH, 56-, PROF PHYSIOL, DEPT BIOMED SCI, 81- *Concurrent Pos:* Fel Coun Arteriosclerosis, Am Heart Asn. *Mem:* Fel AAAS; Can Vet Med Asn; Am Soc Res Workers Animal Dis; Am Soc Vet Physiol & Pharmacol; Am Physiol Soc. *Res:* Animal physiology, especially cardiovascular physiology; cardiovascular surgery; blood coagulation; blood flow and vascular disease in man and animals. *Mailing Add:* Dept Biomed Sci Ont Vet Col Univ Guelph Guelph ON N1G 2W1 Can

DOWNIE, JOHN, b Glasgow, Scotland, Dec 12, 31; Can citizen; m 59; c 3. CHEMICAL ENGINEERING. *Educ:* Univ Glasgow, BSc, 53; Univ Toronto, MASc, 56, PhD(chem eng), 59. *Prof Exp:* Res engr, Reservoir Mech, Gulf Res & Develop Co, 59-62; from asst prof to assoc prof, 62-71, head chem eng dept, 72-77, actg dean, Fac Appl Sci, 85-86, PROF CHEM ENG, QUEEN'S UNIV, ONT, 71- *Mem:* Chem Inst Can; Can Soc Chem Engrs; Sigma Xi. *Res:* Kinetics of gas/solid catalytic oxidation reaction networks using statistical analysis and design of experiments; mathematical modeling of wood gasifiers as a basis for technical assessment. *Mailing Add:* Dept Chem Eng Queen's Univ Kingston ON K7L 3N6 Can

DOWNIE, JOHN WILLIAM, b Winnipeg, Man, May 11, 45; m 87. PHARMACOLOGY, NEUROSCIENCE. *Educ:* Univ Man, BSc Hons, 67, PhD(pharmacol), 72. *Prof Exp:* Res asst urol, Queen's Univ, Ont, 72-73, lectr urol & pharmacol, 73-75, asst prof urol & pharmacol, 75-80; assoc prof, 80-85, PROF PHARMACOL & ASSOC PROF UROL, DALHOUSIE UNIV, NS, 85- *Concurrent Pos:* Med Res Coun Can scholar, 75-80; vis prof, Marine Biomed Inst, Univ Tex, Galveston, 85-86. *Honors & Awards:* Mihran & Mary Basmajian Med Res Award, Queen's Univ, Ont, 79. *Mem:* Pharmacol Soc Can; Am Soc Pharmacol & Exp Therapeut; Int Continence Soc; Urodynamic Soc; Soc Neurosci; Can Urol Asn. *Res:* Neurotransmission in urinary tract; lower urinary tract dynamics in spinal injury; autonomic nerve control of the lower urinary tract. *Mailing Add:* Dept Pharmacol Dalhousie Univ Sir Charles Tupper Med Bldg Halifax NS B3H 4H7 Can

DOWNING, DAVID ROYAL, b Lima, Ohio, July 21, 39. NAVIGATION GUIDANCE & CONTROL SYSTEMS, FLIGHT DYNAMICS. *Educ:* Univ Mich, BSE, 62, MSE, 63; Mass Inst Technol, ScD, 70. *Prof Exp:* Systs engr, Electronics Res Ctr, NASA, 66-70, aerospace engr, Langley Res Ctr, 74-81; asst prof systs eng, Boston Univ, 70-74; assoc prof, 81-86, PROF AEROSPACE ENG, UNIV KANS, 86-, CHMN AEROSPACE ENG, 88- *Concurrent Pos:* Mem, Gen Aviation Tech Comt, Am Inst Aeronaut & Astronaut, 80-, chmn, 83-85; dir, Flight Res Lab, 84-90 & Kans Space Grant Consortium, 91- *Mem:* Assoc fel Am Inst Aeronaut & Astronaut; Inst Elec & Electronics Engrs; Soc Automotive Engrs; Am Soc Eng Educ. *Res:* Navigation, guidance and control of aerospace vehicles; applied optimal control, instrumentation systems, flight dynamics and flight testing. *Mailing Add:* 2004 Learned Hall Univ Kans Lawrence KS 66044

DOWNING, DONALD LEONARD, b Willoughby, Ohio, Apr 2, 31; m 59; c 2. FOOD SCIENCE. *Educ:* Univ Ga, BSA, 57, PhD(food sci), 63. *Prof Exp:* Instr food sci, Univ Ga, 61-63; food scientist, Beech-Nut Life Savers, Inc, 63-67; asst prof, NY State Agr Sta, 67-73, mem fac, Dept Food Sci, 77-80, PROF FOOD SCI, CORNELL UNIV, 80- *Mem:* Fel Inst Food Technologists; Int Asn Milk, Food & Environ Sanit. *Res:* Science and technology related to food preservation and environmental quality. *Mailing Add:* 72 White Springs Lane Geneva NY 14456

DOWNING, DONALD TALBOT, b Perth, Western Australia, Mar 11, 29; m 52; c 6. BIOCHEMISTRY OF LIPIDS. *Educ:* Univ Western Australia, BSc, 51, PhD(org chem), 55. *Prof Exp:* Res chemist, Kiwi Polish Co Ltd, Australia, 54-55; res officer wax chem, Commonwealth Sci & Indust Res Orgn, 55-63, sr res officer, 63-64, sr res scientist, 64-66; asst res prof dermat & biochem, 66-69, from assoc prof to prof biochem & assoc res prof to res prof dermat, Sch Med, Boston Univ, 69-78; PROF DERMAT, SCH MED, UNIV IOWA, 78- *Concurrent Pos:* NIH res career develop award, 69-73. *Mem:* Am Chem Soc; Am Oil Chemists Soc; Soc Invest Dermat; AAAS; Am Soc Biol Chemists. *Res:* Chemical composition and biosynthesis of lipids from sebaceous glands, epidermis and other keratinizing tissues. *Mailing Add:* Dept Dermat Univ Iowa Col Med Med Labs Rm 270 Iowa City IA 52242

DOWNING, GEORGE V, JR, b Salem, Va, July 29, 23; m 51, 81; c 5. PHYSICAL CHEMISTRY. *Educ:* Haverford Col, BA, 47; Cornell Univ, PhD(phys chem), 52. *Prof Exp:* Chemist, Off Sci Res & Develop Malaria Res Proj, NY Univ, 44-46; res chemist natural prod develop group, Merck & Co Inc, 51-56, group leader anal methods develop, 56-62, mgr in-process controls res, 62-65, dir phys & anal res, 65-77, sr dir anal res, 77-88; RETIRED. *Mem:* Am Chem Soc. *Res:* Purity characterization of organic compounds; analytical methods development; trace methods in biological materials; chromatographic methods. *Mailing Add:* 26 Skyline Dr Warren NJ 07059-6718

DOWNING, JOHN SCOTT, b Philadelphia, Pa, July 31, 40. MATHEMATICS, TOPOLOGY. *Educ:* Princeton Univ, AB, 62; Mich State Univ, MS, 66, PhD(math), 69. *Prof Exp:* High sch teacher, Venezuela, 62-65; from asst prof to assoc prof, 69-75, PROF MATH, UNIV NEBR, OMAHA, 75- *Concurrent Pos:* Mem fac, Univ de Oriente, 74-75. *Mem:* Am Math Soc; Math Asn Am. *Res:* Topology of manifolds. *Mailing Add:* Dept Math & Comp Sci Univ Nebr Omaha NE 68182-0243

DOWNING, KENTON BENSON, b Montrose, Colo, Nov 5, 40; m 58; c 4. FORESTRY, RECREATION RESOURCE MANAGEMENT. *Educ:* Colo State Univ, BS, 62, MS, 66; Univ Mo, PhD(forestry), 73. *Prof Exp:* Dist forester, Colo State Forest Serv, Colo State Univ, 64-67; instr, Univ Mo, 67-73; asst prof, Ore State Univ, 73-77; ASSOC PROF, DEPT FOREST RESOURCES, UTAH STATE UNIV, 77- *Concurrent Pos:* Prin investr res proj, Ore State Univ & Utah State Univ, 73-; dir, Nat Outdoor Recreation Shortcourse Training Prog, Utah State Univ & US Forest Serv, 79-81; Danforth assoc; IPA assignment, Washington, DC; BLM, USDA Forest Serv, 84 & 85. *Mem:* Sigma Xi. *Res:* Social aspects of natural resource management; conflicts between outdoor recreation and other natural resource uses and outdoor recreation users and natural resource managers; land management planning. *Mailing Add:* Dept Forest Resources Utah State Univ Logan UT 84322-5215

DOWNING, MANCOURT, b Denver, Colo, May 25, 25; m 48; c 4. BIOCHEMISTRY. *Educ:* Univ Chicago, SB, 52, PhD(biochem), 55. *Prof Exp:* Res assoc, Am Meat Inst Found, 52-55; from instr to asst prof, 55-64, ASSOC PROF, UNIV COLO, BOULDER, 64-, EMER PROF CHEM. *Concurrent Pos:* NIH fel, Inst Molecular Biol, Univ Wis, 68-69. *Mem:* Fel AAAS; Am Soc Biol Chemists. *Res:* Purification of biological macromolecules; biological properties of vitamin B12 and DNA; intermediary metabolism of nucleic acids; cytokinins. *Mailing Add:* Dept Chem 2759 Park Lake Dr Boulder CO 80301

DOWNING, MARY BRIGETTA, b St Louis, Mo, Jan 28, 38; m 62; c 1. GEOCOSMIC STUDIES, HISTORY & PHILOSOPHY OF SCIENCE. *Educ:* Catholic Univ, BA, 60. *Prof Exp:* Ctr dir, Dept Defense, WGer, 60-62; ctr dir, Indust Home Blind, 63-64; publ & commun, Care, Inc, 64-68; educ & publ dir, Nat Notions Asn, 68-70; artistic dir, Lieberman Harrison Advert, 70-72; PRES, M B DOWNING ASN, 72-; EXEC DIR, NAT COUN GEOCOSMIC RES, 79-; SECY & DIR, MERCURY HOUSE PUBL, 85- *Concurrent Pos:* Lectr bus & corp timing, Nat Coun Geocosmic Res, 70-, mem credentials fac, 76-; consult financial trends, M B Downing Asn, 72-; ed, Nat Coun Geocosmic Res J & Geocosmic News, 79- *Mem:* Nat Coun Geocosmic Res; Inst Study Cycles. *Res:* Cyclic movement in financial markets; all recurrent objective phenomena and their interrelation; the interaction of measurable human activity with observable geocosmic phenomena; history of corporate cyclic activity. *Mailing Add:* 78 Hubbard Ave Stamford CT 06905

DOWNING, MICHAEL RICHARD, b North Platte, Nebr, Nov 19, 47; m 70; c 1. CLINICAL RESEARCH, HEMATOLOGY & ONCOLOGY. *Educ:* Chadron State Col, Nebr, BA, 70; Okla State Univ, PhD(biochem), 74. *Prof Exp:* Hemat researcher, Mayo Clinic, 74-77, res assoc biochem, 77-79, asst prof, Mayo Med Sch, 78-79; sr scientist, Alpha Therapeut Corp, 79-80, mgr clin proj, 80-82, dir clin res, 82-85; MGR CLIN STUDIES, AMGEN, 85- *Concurrent Pos:* Consult, protein chem, clin studies, regulatory affairs. *Mem:* Am Heart Asn; AAAS; Am Chem Soc; Regulatory Affairs Prof Soc. *Res:* Molecular biology and uses to elucidate protein structural functional relationships, proteins involved in coagulation and the inhibition of coagulation, r-DNA derived erythroporetic hormones and r-DNA derived vaccines. *Mailing Add:* 1900 Oak Terrace Lane Thousand Oaks CA 91320-1731

DOWNING, ROBERT GREGORY, b St Joseph, Mo, Dec 1, 53. NEUTRON BEAMS, NEUTRON REACTIONS & TECHNIQUES. *Educ:* Mo Western State Col, BS, 76; Univ Mo, Rolla, PhD(chem), 81. *Prof Exp:* Chemist, Farmland Industs, 75-76; res assoc, Univ Mo, Rolla, 76-80; lectr anal chem, Tex A&M Univ, 80-81; Nat Res Coun postdoctoral, Nat Res Coun/Nat Bur Standards, 81-83; RES CHEMIST, NAT INST STANDARDS & TECHNOL, 83-, DEP GROUP LEADER, 89- *Honors & Awards:* Bronze Medal, Dept Com, 90. *Mem:* Am Chem Soc; Sigma Xi; Am Soc Testing & Mat; Microbeam Anal Soc. *Res:* Application of neutron fields for analytical measurements; technique development for radiation measurement and instrument development; semiconductor, polymer and coatings materials research; boron and lithium determination in biological and nonbiological systems. *Mailing Add:* Nat Inst Standards & Technol Bldg 235 Rm B108 Gaithersburg MD 20899

DOWNING, S EVANS, b Meredith, NH, June 29, 30; m 55; c 2. PATHOLOGY, CARDIOVASCULAR PATHOPHYSIOLOGY. *Educ:* Univ NH, BS, 52; Yale Univ, MD, 56. *Prof Exp:* Intern path, Yale-New Haven Hosp, 56-57; USPHS fel, 57-58, Life Ins Med Res Fund fel, 59-60; scientist, Nat Heart Inst, 60-62; assoc physiol, George Washington Univ, 62; from asst prof to assoc prof, 62-74, PROF PATH, SCH MED, YALE UNIV, 74- *Concurrent Pos:* Keese Prize, Sch Med, Yale Univ, 56, fel, Nuffield Inst Med Res, Eng, 58-59; USPHS res career develop award, 62-72; consult, Nat Heart Lung & Blood Inst, NIH, 72-; prin investr res grants, NIH, mem, study sect. *Mem:* Am Asn Path; Soc Pediat Res; NY Acad Sci; Int Acad Path; Am Physiol Soc; Int Soc Heart Res. *Res:* Cardiovascular physiology; pathophysiology. *Mailing Add:* Dept Path Yale Univ Sch Med 310 Cedar St New Haven CT 06510

DOWNING, STEPHEN WARD, b Philadelphia, Pa, July 27, 43; m 66; c 4. CELL BIOLOGY. *Educ:* Col Wooster, BA, 65; Northwestern Univ, PhD(zool), 69. *Prof Exp:* Res assoc path, Mass Gen Hosp, Harvard Med Sch, 69-72; asst prof anat, Chicago Med Sch-Univ of Health Sci, 72-76; asst prof, 76-79, ASSOC PROF BIOMED ANAT, SCH MED, UNIV MINN, DULUTH, 79- *Mem:* Am Soc Cell Biol; Am Asn Anatomists; Sigma Xi. *Res:* Investigations of epidermal differentiation and morphogenesis in selected vertebrate models, particularly reptilian, agnathan and mammalian epidermal tissues; biology of the cytoskeleton. *Mailing Add:* 4104 Pitt St Duluth MN 55804

DOWNING, WILLIAM LAWRENCE, b Des Moines, Iowa, Oct 2, 21; m 46; c 3. FRESHWATER BIOLOGY. *Educ:* Univ Iowa, BA, 43, MS, 48, PhD(zool), 51. *Prof Exp:* Instr biol, Univ Iowa, 49-51; prof & head dept, Jamestown Col, 51-63; prof biol, Hamline Univ, 63-87; prof, 63-87, EMER PROF BIOL, HAMLINE UNIV, 88- *Concurrent Pos:* Adv pre-med, Hamline Univ; supvr, Soil & Water Conserv Dist; comnr, Metrop Waste Control. *Mem:* AAAS; Am Soc Zoologists; Soc Protozoologists; Sigma Xi. *Res:* Morphogenesis of ciliates; unionid mussels and acidic precipitation; soil and water conservation. *Mailing Add:* 1834 Simpson Ave Falcon Heights MN 55113

DOWNS, BERTRAM WILSON, JR, b St Paul, Minn, Dec 11, 25; m 56; c 3. STRONG ELEMENTARY PARTICLE INTERACTIONS. *Educ:* Calif Inst Technol, BS, 46; Univ Minn, MS, 49; Stanford Univ, PhD(physics), 56. *Prof Exp:* Res fel, Univ Birmingham, 55-56; res assoc, Lab Nuclear Studies, Cornell Univ, 56-59; from asst prof to assoc prof physics, 59-65, assoc dean grad sch & actg dir comput ctr, 65-67, PROF PHYSICS, UNIV COLO, BOULDER, 65- *Concurrent Pos:* Consult, Atomic Power Develop Assocs, Mich, 57-58; vacation consult, Atomic Energy Res Estab, Eng, 63-64; mem staff, Oxford Univ, 63-64. *Mem:* Fel Am Phys Soc. *Res:* Hyperon-nucleon interactions and hypernuclei. *Mailing Add:* Univ Colo Dept Physics Campus Box 390 Boulder CO 80309-0060

DOWNS, DAVID S, b Woodbury, NJ, Jan 4, 41; m 67; c 2. SOLID STATE PHYSICS. *Educ:* Gettysburg Col, BA, 62; Univ Del, MS, 64, PhD(physics), 69. *Prof Exp:* Nat Res Coun res assoc physics, Picatinny Arsenal, 68-69, res physicist, Feltman Res Lab, Solid State Br, 69-77, actg chief, Energy Conversion Sect, 74-75; chief, ignition & combustion sect, Appl Sci Div, 77-85, CHIEF PROPULSION BR, ENERGETICS & WARHEADS DIV, ARMAMENT RES, DEVELOP & ENG CTR, US ARMY, 85- *Honors & Awards:* Res & Develop Award, US Army, 87. *Mem:* Am Phys Soc; Am Asn Physics Teachers. *Res:* Electron spin resonance of transition metals in II-VI compounds; electrical and optical properties of metallic azides; optical and electrical properties of energetic materials; igniters, propellants and propulsion engineering. *Mailing Add:* Army Armament Res Develop & Eng Ctr Energetics & Warheads Div B382 Dover NJ 07806-5000

DOWNS, FREDERICK JON, b New York, NY, Oct 22, 39; m 63; c 2. BIOCHEMISTRY. *Educ:* Hunter Col, BA, 61; NY Med Col, MS, 65, PhD(biochem), 68. *Prof Exp:* Res scientist biochem, Union Carbide Res Inst, 68-69; teacher biol, NY City Bd Educ, 69-70; from lectr to asst prof, 70-74, assoc prof, 74-80, PROF CHEM, HERBERT LEHMAN COL, 80- *Concurrent Pos:* Res assoc, NY Med Col, 68-71, adj asst prof, 71- *Mem:* Am Chem Soc; AAAS; Soc Complex Carbohydrates. *Res:* Isolation and characterization of glycoproteins and glycolipids and their biological significance. *Mailing Add:* 213 45 29th Ave Bayside NY 11360

DOWNS, GEORGE SAMUEL, b San Antonio, Tex, Oct 4, 39; c 1. RADIO ASTRONOMY, RADAR ASTRONOMY. *Educ:* Cornell Univ, BS, 62; Stanford Univ, MS, 64, PhD(elec eng), 68. *Prof Exp:* Mem tech staff astron, Jet Propulsion Lab, 68-85; MEM TECH STAFF, LINCOLN LAB, MASS INST TECHNOL, 85- *Mem:* Am Astron Soc; Sigma Xi. *Res:* Timing measurements and analysis of pulsating radio sources including pulsars; radar probes of the terrestrial planets, particularly Mars. *Mailing Add:* Lincoln Lab Mass Inst Technol PO Box 73 Lexington MA 02173

DOWNS, JAMES JOSEPH, b St Joseph, Mo, Jan 31, 28. PHYSICAL CHEMISTRY, ORGANIC CHEMISTRY. *Educ:* St Benedicts Col, BS, 49; Univ Notre Dame, MS, 52; Fla State Univ, PhD(chem), 54. *Prof Exp:* Prin chemist, Midwest Res Inst, 56-77; dir, Chem Physics Ctr, Carnegie-Mellon Inst Res, 78-79, sr fel, 79-82; prin scientist, Franklin Inst, 82-84; vpres, US Environ, 84-89; VPRES, ROLITE, INC, 89- *Mem:* AAAS; Am Chem Soc; Am Statist Asn. *Res:* Computer data processing; mass spectroscopy; environmental science; kinetics; structural chemistry; fiberoptics. *Mailing Add:* 1600 Hagys Ford Rd Narberth PA 19072

DOWNS, MARTIN LUTHER, b Reading, Pa, March 12, 10; m; c 2. CHEMISTRY. *Educ:* Pa State Univ, BS, 31; Lawrence Col, MS, 32, PhD(pulp, paper making), 34. *Prof Exp:* Develop dept, Mead Corp, 34-37; from chief chemist to tech dir, Thilmany Pulp & Paper Co, 37-70, vpres, 64-70; OWNER & MGR, PAPER CONCEPTS CONSULT, 70- *Mem:* Am Chem Soc; Tech Asn Pulp & Paper Indust. *Res:* Paper chemistry; paper sizing; stock processing; specialty and technical papers; paper recycling. *Mailing Add:* Paper Concepts 1000 Greengrove Rd Appleton WI 54911-4224

DOWNS, ROBERT JACK, b Sapulpa, Okla, June 25, 23; m 45; c 1. BOTANY. *Educ:* George Washington Univ, BS, 50, MS, 51, PhD(bot), 54. *Prof Exp:* Student asst physics & bot, George Washington Univ, 49-50, asst bot, 50-51; phys sci, Astrophys Observ, Smithsonian Inst, 51-52; plant physiologist photoperiod proj, Plant Indust Sta, Agr Res Serv, USDA, 52-59, mem pioneering res group, Plant Physiol Lab, Agr Res Serv, Plant Indust Sta, 59-65; PROF BOT & HORT SCI & DIR PHYTOTRON, NC STATE UNIV, 65- *Concurrent Pos:* Assoc ed, Biotronics & S&E Div, Am Soc Agr Eng; consult, controlled environ facil & design. *Honors & Awards:* Alex Laurie Award, Am Soc Hort Sci; Henry Allan Gleason Award, NY Bot Garden. *Mem:* Am Soc Hort Sci; Bot Soc Am; Int Soc Bioclimateorol; Am Soc Agr Engrs; Am Soc Plant Physiologists. *Res:* Phytochrome and the regulatory effects of light on plants; taxonomy of South American plants, especially Xyridaceae and Bromeliaceae; bioengineering and environmental physiology. *Mailing Add:* Southeastern Plant Environ Labs NC State Univ Box 7618 Raleigh NC 27695

DOWNS, THEODORE, b Chicago, Ill, July 1, 19; wid; c 2. VERTEBRATE PALEONTOLOGY. *Educ:* Kans State Teachers Col, BS, 41; Univ Calif, MA, 48, PhD(vert paleont), 51. *Prof Exp:* Cur vert paleont, 52-61, chief cur, 61-80, EMER CHIEF CUR, EARTH SCI DIV, LOS ANGELES COUNTY MUS, 80- *Concurrent Pos:* Nat Res Coun fel, 51-52. *Mem:* AAAS; Geol Soc Am; Soc Vert Paleont; Am Soc Mammalogists; Soc Study Evolution; Soc Syst Zool; Paleont Soc; Sigma Xi. *Res:* Paleomammalogy; evolution; paleoecology; paleogeographic distribution of middle to late Cenozoic vertebrates; field operations in Nevada, Oregon, California and Mexico. *Mailing Add:* Los Angeles Co Mus Natural Hist 900 Exposition Blvd Los Angeles CA 90007

DOWNS, THOMAS D, b Kalamazoo, Mich, Aug 28, 33; m 62; c 1. BIOSTATISTICS. *Educ:* Western Mich Univ, BS, 60; Univ Mich, MPH, 62, PhD(biostatist), 65. *Prof Exp:* Asst prof, Case Western Reserve Univ, 65-70, assoc prof biomet, Sch Pub Health, 70-77, PROF, UNIV TEX HOUSTON, 77- *Mem:* Math Asn Am; Am Statist Asn; Am Pub Health Asn; Biomet Soc. *Res:* Applied statistics and mathematics in public health. *Mailing Add:* Univ Tex Dept Biometry Univ Tex Health & Sci Ctr PO Box 20186 Houston TX 77225

DOWNS, WILBUR GEORGE, virology; deceased, see previous edition for last biography

DOWNS, WILLIAM FREDRICK, b Santa Maria, Calif, Aug 4, 42; m 67; c 1. HAZARDOUS MATERIALS GEOCHEMISTRY, ROCK WATER INTERACTIONS. *Educ:* Univ Colo, BA, 65, MS, 74; Pa State Univ, PhD(geochem), 77. *Prof Exp:* Res asst, Pa State Univ, 71-74, proj assoc, 74-77; sci specialist, Eng Lab, EG&G Idaho, Inc, 77-88; GEOCHEMIST, JACOBS ENG GROUP & URANIUM MILL TAILINGS PROG, ALBUQUERQUE, NMEX, 88- *Concurrent Pos:* Explor geologist, Duval Mining Co, 71; consult & lectr, Univ Idaho, 79-88, Idaho State Univ, 85-88. *Mem:* Geochem Soc; Geol Soc Am. *Res:* Investigation of the solubility of scale forming minerals in synthetic geothermal brines; experimental calibration of both stable isotope, and chemical specie geothermometers; evolution of E Snake River Plain; migration of hazardous wastes through the near surface environment. *Mailing Add:* Jacobs Eng Group Inc 5301 Central Ave NE Albuquerque NM 87108

DOWS, DAVID ALAN, b San Francisco, Calif, July 25, 28; m 50; c 3. LASER CHEMISTRY. *Educ:* Univ Calif, BS, 52, PhD(chem), 54. *Prof Exp:* Instr chem, Cornell Univ, 54-56; from instr to assoc prof, 56-63, chmn dept, 66-72, PROF CHEM, UNIV SOUTHERN CALIF, 63- *Concurrent Pos:* NSF sr res fel, Oxford Univ, 62-63; NATO vis prof, Univ Florence, 70. *Mem:* Am Chem Soc; Am Phys Soc. *Res:* Molecular electronic and vibrational spectroscopy; laser photochemistry; crystal spectroscopy and intermolecular forces in crystals; chemical dynamics simulation. *Mailing Add:* Dept Chem Univ Southern Calif Los Angeles CA 90089-0482

DOWSE, HAROLD BURGESS, b Albany, NY, Jan 3, 45; m 67; c 2. GENETICS OF CARDIAC FUNCTION, ACOUSTICAL SIGNAL ANALYSIS. *Educ:* Amherst Col, BA, 66; NY Univ, PhD(biol), 71. *Prof Exp:* From instr to asst prof, 79-90, ASSOC PROF ZOOL, UNIV MAINE, 90- *Mem:* AAAS; Int Soc Chronobiol; Soc Res Biol Rhythms. *Res:* Cellular mechanism of biological oscillators using Drosophila melanogaster as the model; follow the activity of genes that control the circadian clock, as well as behavioral correlates; developed signal analysis techniques for use on behavioral data. *Mailing Add:* Dept Zool Univ Maine Orono ME 04469

DOWTY, ERIC, mineralogy, for more information see previous edition

DOXTADER, KENNETH GUY, b San Francisco, Calif, June 22, 38; m 62; c 3. SOIL MICROBIOLOGY. *Educ:* Univ Calif, Berkeley, BS, 61; Cornell Univ, MS, 63, PhD(agron), 65. *Prof Exp:* Res asst soil microbiol, Cornell Univ, 61-65; asst prof, 65-69, assoc prof agron, 69-79, PROF AGRON, COLO STATE UNIV, 79- *Mem:* AAAS; Am Soc Microbiol; Am Soc Agron; Int Soc Soil Sci; Sigma Xi. *Res:* Physiology and ecology of soil microorganisms; microbial biogeochemistry; microbial transformations of pesticides and minerals. *Mailing Add:* Dept Agron Colo State Univ Ft Collins CO 80523

DOYLE, DARRELL JOSEPH, b Allentown, Pa, July 26, 39; m 64; c 2. BIOCHEMISTRY. *Educ:* Lehigh Univ, BA, 61, MS, 63; Johns Hopkins Univ, PhD(biochem), 67. *Prof Exp:* NIH fel, Stanford Univ, 67-69; asst prof anat & cell biol, Med Sch, Univ Pittsburgh, 69-72; assoc cancer res scientist, Roswell Park Mem Inst, 72-80, dir, dept cell & tumor biol, 80-; PROF BIOL SCI, STATE UNIV NY, BUFFALO. *Concurrent Pos:* Vis prof, Univ Chile, 71. *Mem:* Am Soc Biol Chemists; AAAS. *Res:* Developmental biology; biochemical genetics; regulatory mechanisms in eukaryotic cells. *Mailing Add:* Dept Biol Sci State Univ NY Cooke Hall 109 Buffalo NY 14260

DOYLE, DARYL JOSEPH, b Devils Lake, NDak, June 1, 50; m 76; c 4. ADHESIVES, PHOTOCHEMISTRY. *Educ:* NDak State univ, BS, 71, PhD(chem), 81. *Prof Exp:* Asst prof chem & physics, Valley City State Col, 76-80; vis asst prof chem, NDak State Univ, 81-82; from asst prof to assoc prof, 82-91, PROF CHEM, GMI ENG & MGT INST, 91- *Concurrent Pos:* Chair, Sci Dept, Valley City State Col, 79-80. *Mem:* Am Chem Soc; Soc Mfg Engrs; Laser Inst Am. *Res:* Adhesives; adhesive bonding; applications and excimer laser surface modification for adhesive bonding. *Mailing Add:* GMI Eng & Mgt Inst 1700 W Third Ave Flint MI 48504-4898

DOYLE, EUGENIE F, b New York, NY, Oct 19, 21; m 44; c 5. PEDIATRIC CARDIOLOGY. *Educ:* Johns Hopkins Univ, MD, 46. *Prof Exp:* Intern pediat, Johns Hopkins Hosp, 46-47; resident, Bellevue Hosp, 47-49; from asst prof to assoc prof, 53-70, PROF PEDIAT, SCH MED, NY UNIV, 70-, DIR DEPT PEDIAT CARDIOL, MED CTR, 59- *Concurrent Pos:* Fel pediat cardiol, Med Ctr, NY Univ, 49-53. *Mem:* Am Pediat Soc; Am Heart Asn; fel Am Col Cardiol; fel Am Acad Pediat. *Res:* Treatment of acute rheumatic fever and rheumatic heart disease; natural history of congenital heart disease, especially aortic stenosis and the management of congestive heart failure. *Mailing Add:* Dept Pediat Cardiol NY Univ Med Ctr 550 First Ave New York NY 10016

DOYLE, FIONA MARY, b Newcastle-upon-Tyne, Eng, Sept 27, 56. HYDROMETALLURGY, POWDER PREPARATION. *Educ:* Univ Cambridge, BA, 78, MA, 82; Univ London, MSc, 79, PhD(metall & mat sci), 83. *Prof Exp:* Grad trainee contracting eng, Davy McKee, Metals & Minerals Div, Stockton-on-Tees, Eng, 83; asst prof, 83-88, ASSOC PROF METALL, DEPT MAT SCI & MINERAL ENG, UNIV CALIF, BERKELEY, 88- *Concurrent Pos:* Chartered engr, Gt Brit, 87-; co-ed, Mineral Processing & Extractive Metall Rev, 90- *Mem:* Electrochem Soc; Mat Res Soc; Am Inst Mining, Metall & Petrol Engrs Metall Soc; Inst Mining & Metall. *Res:* Aqueous chemistry in the processing and behavior of minerals, materials, wastes and effluents; chemical thermodynamics and kinetics; electrochemistry, leaching and transformation of minerals; hydrolysis and precipitation; solvent extraction, organic phase reactions, novel separations processes. *Mailing Add:* 372 Hearst Mining Bldg Berkeley CA 94720

DOYLE, FRANK LAWRENCE, b San Antonio, Tex, Oct 16, 26; m 62; c 1. MANAGEMENT, HYDROLOGY & GEOLOGY. *Educ:* Univ Tex, BS, 50; La State Univ, MS, 55; Univ Ill, PhD(geol), 58. *Prof Exp:* Instr geol, St Mary's Univ, San Antonio, 50-53; petrol geologist, Seeligson Eg Comt, US Geol Surv, Mont, 52-53, geologist, Fuels Br, 55; res asst, Ill State Geol Surv, 55-56, asst geologist, 56-58; from asst prof to assoc prof geol, St Mary's Univ, San Antonio, 58-62, chmn, Geol Dept, 61-62; geologist, Water Resources Div, US Geol Surv, Col & Ariz, 62-63; consult, Int Resources & Geotech, Inc, Panama, 65-68, Tex Instruments, Nicaragua, 68-70, Int Ctr Arid & Semi-Arid Land Studies, Tex Tech Univ, Algeria, 70-71; regional geologist & head, NAlabama Off, Ala Geol Surv, 71-77; consult, hydrogeologist, Fla, 77-78; chief hydrogeologist, Metcalf & Eddy, Inc, 78-79; sr hydrologist & prog mgr, US Nuclear Regulator Comn, 79-84; environ review officer, hydrologist & geologist, Off Secy, US Dept Interior, 84-87; SYSTS SCIENTIST, HYDROLOGIST & GEOLOGIST, MITRE CORP, 87- *Concurrent Pos:* Assoc geologist & res aff, Ill State Geol Surv, 58-62; consult, Johnson Environ & Energy Ctr & adj prof hydrol & chmn, Environ Sci Prog, Div Natural Sci & Math, Univ Ala, Huntsville, 71-77; mem US delg, 27th Int Geol Cong, Moscow, USSR, 84; mem, Nat Res Coun, 80-89 & US Nat Comt Geol, 88-89; mem, US comt, Int Asn Hydrogeologists, 77- & chmn, 84-89, ex officro, 89-; assoc prof geol, Univ Conn & State Univ NY, 63-65. *Mem:* Fel Geol Soc Am; Am Inst Prof Geologists; Int Asn Hydrogeologists (US comt secy-treas, 80-84, chmn, 84-89); Sigma Xi; Am Asn Petrol Geologists; Am Geophys Union; Asn Ground Water Scientist & Engrs; Am Inst Hydrol. *Res:* Groundwater; environmental geology and hydrology; Quaternary, areal and subsurface geology; applications of remotely-sensed data; hydrology and geology of volcanic, glacial and limestone terranes; geomorphology and glaciology. *Mailing Add:* Mitre Corp HSD/YAQ BAFB San Antonio TX 78235-5000

DOYLE, FREDERICK JOSEPH, b Oak Park, Ill, Apr 3, 20; m 55; c 4. PHOTOGRAMMETRY, REMOTE SENSING. *Educ:* Syracuse Univ, BCE, 51. *Hon Degrees:* Dr Eng, Hannover Tech Univ, 76, Ohio State Univ, 86, Univ Bordeaux, 87, Royal Tech Univ Stockholm, 87. *Prof Exp:* With Inter-Am Geod Surv, S Am, 46-47; from instr to assoc prof photogramm, Ohio State Univ, 52-59, chmn Dept Geod Sci, 59-60; dir intel systs, Broadview Res Corp, 60-61; res engr, Raytheon-Autometric, 61-67, chief scientist, 67-69; sr adv, Nat Mapping Div, US Geol Surv, 68-89; RETIRED. *Concurrent Pos:* Chmn, Apollo Orbital Sci Photo Team, NASA, 69-74; mem exec comt, Div Earth Sci, Nat Acad Sci, 73-74; actg dir, Earth Resources Observ Satellite, US Dept Interior, 78-80. *Honors & Awards:* Fairchild Photogram Award, Am Soc Photogram, 68, Alan Gordon Award, 85; Exceptional Sci Achievement Award, NASA, 71; Meritorious Serv Award, US Dept Interior, 77, Distinguished Serv Award; Brock Gold Medal Award, Int Soc Photogram & Remote Sensing, 84. *Mem:* Nat Acad Eng; Am Cong Surv & Mapping; Am Geophys Union; Am Soc Photogram (pres, 69-70); Int Soc Photogram (secy gen, 76-80, pres, 80-84); fel AAAS. *Res:* Aerial and space photogrammetric system design and analysis; science policy. *Mailing Add:* 1591 Forest Villa Lane McLean VA 22101

DOYLE, JOHN ROBERT, b Norwood, Mass, Dec 18, 24; m 56; c 4. ORGANOMETALLIC CHEMISTRY. *Educ:* Mass Inst Technol, BS, 49, MS, 52; Tulane Univ, PhD, 55. *Prof Exp:* Chemist, Mass Inst Technol, 49-52; from instr to assoc prof, 55-65, PROF INORG CHEM, UNIV IOWA, 65- *Mem:* AAAS; Am Chem Soc; Am Crystallog Asn. *Res:* Organometallic compounds with catalytic activity; structures of metal-olefin compounds. *Mailing Add:* Dept Chem Univ Iowa Iowa City IA 52240

DOYLE, JON, b Houston, Tex, May 12, 54; m 89; c 1. REASONING, KNOWLEDGE REPRESENTATION. *Educ:* Univ Houston, BS, 74; Mass Inst Technol, SM, 77, PhD(artifical intel), 80. *Prof Exp:* Res assoc computer sci, Stanford Univ, 80-81; res assoc computer sci, Carnegie Mellon Univ, 81-84, res scientist computer sci, 84-88; res assoc computer sci, 88-89, PRIN RES SCIENTIST COMPUTER SCI, MASS INST TECHNOL, 90- *Concurrent Pos:* Res assoc prof med, Sch Med, Tufts Univ, 88-; invited lectr, Int Symp Methodologies for Intelligent Systs, 89 & 91 & Am Asn Artificial Intel, 90; chair, Spec Interest Group Artificial Intel, Asn Comput Mach, 89-91. *Mem:* Fel Am Asn Artificial Intel; Am Math Soc; Am Econ Asn; Asn Comput Mach; Soc Indust & Appl Math. *Res:* Mathematical foundations of psychology; theories of rationality and limited rationality; rational control of reasoning, planning and knowledge representation; logic and economics of thinking. *Mailing Add:* Lab Computer Sci Mass Inst Technol 545 Technology Sq Cambridge MA 02139

DOYLE, JOSEPH THEOBALD, b Providence, RI, June 11, 18; m 44; c 2. CARDIOLOGY. *Educ:* Harvard Univ, AB, 39, MD, 43; Am Bd Internal Med, dipl, 52 & Am Bd Cardiovasc Dis, dipl, 59. *Prof Exp:* Asst med, Harvard Med Sch, 48-49; asst physiol, Sch Med, Emory Univ, 49-50; instr physiol & asst med, Grady Mem Hosp, Atlanta, Ga, 50-51, dir electrocardiographic lab, 50-52, instr med & physiol & asst coord cardiovasc training prog, 51-52; assoc med, Sch Med, Duke Univ, 52; from asst prof to assoc prof med, 52-61, head div cardiol, 60-84, dir Cardiovasc Health Ctr, 52-90, PROF MED, ALBANY MED COL, 61- *Concurrent Pos:* Whitehead res fel, Sch Med, Emory Univ, 49-50; consult, US Vet Admin Regional Off, Atlanta, 51-52; dir pvt diag clin, 57-; cardiac catheterization unit, 57-63; attend staff, Albany Med Ctr Hosp & Albany Vet Admin Hosp, 53-; fel coun arteriosclerosis, coun clin cardiol & coun epidemiol, Am Heart Asn. *Mem:* Am Fedn Clin Res; fel Am Col Physicians; fel Am Col Cardiol; Asn Univ Cardiologists. *Res:* Cardiovascular physiology and epidemiology. *Mailing Add:* Albany Med Col 47 New Scotland Ave Albany NY 12208

DOYLE, LARRY JAMES, b Denver, Colo, Jan 27, 43; m 72. MARINE GEOLOGY, SEDIMENTATION STRATIGRAPHY. *Educ:* Duke Univ, BA, 65, MA, 67; Univ Southern Calif, PhD(geol), 73. *Prof Exp:* Econ geologist, Global Marine Inc, 69; explor geologist, Gen Oceanog, 70; from asst prof to assoc prof, 72-80, PROF GEOL, DEPT MARINE SCI, 80-, DIR, CTR NEARSHORE MARINE SCI, UNIV SFLA, 87- *Concurrent Pos:* NDEA fel; AB Duke scholarship. *Mem:* Geol Soc Am; Soc Econ Paleontologists & Mineralogists; Int Asn Sedimentologists; Am Geol Inst; Am Asn Petrol Geologists; Sigma Xi. *Res:* Marine geology of continental margins, specifically sediments and sedimentary processes of the coastal zone and of continental slope of eastern North America and of the continental margin of the Gulf of Mexico; geology of the pyrenees. *Mailing Add:* Dept Marine Sci Univ SFla St Petersburg FL 33701

DOYLE, LAWRENCE EDWARD, b Cincinnati, Ohio, Mar 12, 09; m 39; c 3. MECHANICAL & INDUSTRIAL ENGINEERING. *Educ:* Yale Univ, BSME, 30; Univ Ill, ME, 50. *Prof Exp:* Engr, Cincinnati Milling Mach Co, 35-40; supvr mfg eng, Allison Div, Gen Motors Corp, 41-43; mech engr, Norman E Miller & Assoc, 43-45; from asst prof to prof mech eng, 46-75, EMER PROF MECH ENG, UNIV ILL, URBANA, 75- *Honors & Awards:* Nat Educ Award, Soc Mfg Engrs, 61. *Mem:* Am Soc Metals; Am Soc Eng Educ; fel Soc Mfg Engrs. *Res:* Establishment and development of a scientific basis for process design in manufacturing. *Mailing Add:* 2702 Perkins Rd Urbana IL 61801

DOYLE, LEE LEE, b Sacramento, Calif, Sept 22, 32; m 72. REPRODUCTIVE PHYSIOLOGY, ENDOCRINOLOGY. *Educ:* Dominican Col San Rafael, BA, 54; Stanford Univ, MA, 61; Tulane Univ, PhD(reprod physiol), 71. *Prof Exp:* Res assoc obstet & gynec, Med Sch, Stanford Univ, 58-61; asst specialist, Med Sch, Univ Calif, 61-67; from instr to asst prof, Sch Med, Tulane Univ, 67-72; assoc prof, 72-77, PROF OBSTET & GYNEC, MED SCH, UNIV ARK, LITTLE ROCK, 77- *Concurrent Pos:* NIH & inst grant, Tulane Univ & Delta Regional Primate Ctr, 68-72; NIH & Pop Coun grants, Med Sch, Univ Ark & Delta Regional Primate Ctr, 72-; adj scientist reproductive physiol, Delta Regional Primate Ctr, 68- *Honors & Awards:* Rubin Award, Am Fertil Soc, 62; Squibb Prize Paper Award, Pac Coast Fertil Soc, 64. *Mem:* Am Fertil Soc; Asn Reproductive Health Prof; Planned Parenthood; Am Pub Health Asn; Asn Prof Gynecol & Obstet. *Res:* Improved methods of contraception and delivery of services; primate reproductive physiology and endocrinology; adolescent sexuality. *Mailing Add:* Dept Obstet & Gynec Univ Ark Med Ctr Little Rock AR 72201

DOYLE, MARGARET DAVIS, b Chelsea, Okla, Sept 23, 14; m 47; c 1. NUTRITION. *Educ:* Univ Ark, BSc, 34; Univ Chicago, SM, 38, PhD(nutrit), 45; Am Bd Nutrit, dipl. *Prof Exp:* Instr foods, Univ Minn, 38-40; instr foods & nutrit, Conn Col, 40-42; from instr to asst prof, Univ Chicago, 45-53; from asst prof to prof nutrit, Univ Minn, St Paul, 60-81; RETIRED. *Mem:* Fel AAAS; Am Inst Nutrit; Am Dietetic Asn. *Res:* Protein-calorie interrelation in young adults; food habits and dietary intake patterns of obese women and of young adults. *Mailing Add:* 26 Circle Dr Dune Acres Chesterton IN 46304

DOYLE, MICHAEL P, b Minneapolis, Minn, Oct 31, 42; m 64; c 2. PHYSICAL ORGANIC CHEMISTRY. *Educ:* Col St Thomas, BS, 64; Iowa State Univ, PhD(org chem), 68. *Prof Exp:* Instr org chem, Univ Ill, Chicago Circle, 68; from asst prof to assoc prof org chem, Hope Col, 68-74, prof, 74-82; H G Herrick prof chem, 82-84, DR D R SEMMES DISTINGUISHED PROF CHEM, TRINITY UNIV, SAN ANTONIO, 84- *Concurrent Pos:* titular mem & sec, Int Union Pure & Appl Chem Comn on Physical Org Chem, 87-91; mem, Nat Res Coun Bd Chem Sci & Technol, Bd Dirs Res Corp. *Honors & Awards:* Catalyst Award, Chem Mfr Asn, 82; Am Chem Soc Award, Res Undergrad Inst, 88. *Mem:* Am Chem Soc; AAAS; Org Reactions Catalysis Soc; Am Soc Biol Chemists; Sigma Xi; Coun Undergrad Res. *Res:* Transition metal catalysis in carbenoid transformations; chiral catalysts for organic syntheses; electron transfer reactions; hydrosilylation; mechanisms of organic reactions. *Mailing Add:* Dept Chem Trinity Univ San Antonio TX 78212

DOYLE, MICHAEL PATRICK, b Madison, Wis, Oct 3, 49; m 71; c 3. FOOD MICROBIOLOGY. *Educ:* Univ Wis-Madison, BS, 73, MS, 75, PhD(food microbiol), 77. *Prof Exp:* Sr proj leader microbiol, Ralston Purina Co, 77-80; from asst prof to prof food microbiol, Food Res Inst, Univ Wis-Madison, 80-91; PROF FOOD MICROBIOL, DEPT FOOD SCI, UNIV GA, 91- *Concurrent Pos:* Assoc ed, J of Food Protection, 81-86; fel Am Acad Microbiol. *Honors & Awards:* Pound Res Award, 85; F W Tanner Lectureship, 86; Samuel Cate Prescott Res Award, 87; Fel Am Acad Microbiol, 87; James M Craig Mem Lectr, 90. *Mem:* Am Soc Microbiol; Int Asn Milk, Food & Environ Sci; Inst Food Technologists; Sigma Xi; AAAS. *Res:* Gram-negative foodborne bacterial pathogens; mechanisms of virulence, association with foods, methods of control, methods for their isolation. *Mailing Add:* Dept Food Sci Ga Exp Sta Univ Ga Griffin GA 30223

DOYLE, MILES LAWRENCE, b Ashland, Ohio, July 14, 27; m 55; c 2. BIOCHEMISTRY. *Educ:* Ashland Col, AB, 49; St Louis Univ, PhD(biochem), 55. *Prof Exp:* Instr biochem, Vanderbilt Univ, 55-58; asst prof chem, Quincy Col, 58-61; asst res biochemist, Univ Calif, Davis, 61-62; assoc prof chem, Wis State Col, Eau Claire, 62-64; assoc prof, Col St Teresa, Minn, 64-66; ASSOC PROF CHEM, ARK STATE UNIV, 66- *Concurrent Pos:* Res biochemist, Thayer Vet Admin Hosp, Tenn, 55-58. *Mem:* Fel AAAS; Am Chem Soc; Brit Biochem Soc. *Res:* Enzymology. *Mailing Add:* 3604 Marzee Ann Dr Jonesboro AR 72401

DOYLE, RICHARD ROBERT, b Camden, NJ, July 29, 37; m 63; c 2. ORGANIC CHEMISTRY, BIOCHEMISTRY. *Educ:* Drexel Univ, BS, 60; Univ Mich, MS, 63, PhD(org chem), 65. *Prof Exp:* NIH fel & res assoc biochem, Univ Mich, 65-67; from asst prof to assoc prof, 67-82, PROF ORG CHEM, DENISON UNIV, 82- *Concurrent Pos:* Vis prof, Rackham Arthritis Unit, Univ Mich, 80-81. *Mem:* AAAS; Am Chem Soc; NAm Mycological Asn. *Res:* Amino acid synthesis; mushroom chemistry. *Mailing Add:* 14 Sunset Hill Granville OH 43023

DOYLE, ROGER WHITNEY, b Halifax, NS, Mar 7, 41; m 64. ECOLOGY, GENETICS. *Educ:* Dalhousie Univ, BS, 62, MS, 63; Yale Univ, PhD(biol), 67. *Prof Exp:* Asst prof zool & NIH biol sci support grant, Duke Univ, 67-71; assoc prof, 71-80, PROF BIOL, DALHOUSIE UNIV, 80- *Mem:* Am Soc Limnol & Oceanog; Am Soc Naturalists; Ecol Soc Am; Soc Study Evolution; World Mariculture Soc. *Res:* Population ecology and ecological genetics of marine animals; aquaculture genetics. *Mailing Add:* Dept Biol Sci Dalhousie Univ Halifax NS B3H 3V1 Can

DOYLE, TERRENCE WILLIAM, b Montreal, Quebec, Aug 8, 42; m 68; c 3. CANCER RESEARCH, ANTIBIOTIC RESEARCH. *Educ:* Loyola Col, BSc, 63; Univ Notre Dame, PhD(chem), 66. *Prof Exp:* NIH fel, Cornell Univ, 66-67; sr res scientist, 67-75, ASST DIR MED CHEM, BRISTOL LABS CAN, DIV BRISTOL MEYERS CO, 75- *Mem:* Am Chem Soc; Chem Inst Can. *Res:* Natural product isolation, structural characterization, biochemistry and organic synthesis primarily in the area of antitumor antibiotics. *Mailing Add:* Five Wellsweep Lane Deep River CT 06417-1384

DOYLE, WALTER M, b Utica, NY, Sept 26, 37; m 62. ATOMIC PHYSICS. *Educ:* Syracuse Univ, BA, 59; Univ Calif, PhD(physics), 63. *Prof Exp:* Physicist, Hughes Aircraft Co, 63-64; sr scientist, Philco Corp, Ford Motor Co, 64-66, prin scientist, Aeronutronic Div, 66-69; vpres & dir res & develop, Laser Precision Corp, 69-78, pres, 78-88; PRES, AXIOM ANALYSIS, 88- *Mem:* Am Phys Soc; Am Chem Soc. *Res:* Solid state physics; optics; quantum electronics; development of optical measuring instruments and electrooptic devices. *Mailing Add:* 2875 Bernard Ct Laguna Beach CA 92651

DOYLE, WILLIAM DAVID, b Boston, Mass, June 5, 35; m 58; c 3. SOLID STATE PHYSICS. *Educ:* Boston Col, BS, 57, MS, 59; Temple Univ, PhD(physics), 64. *Prof Exp:* Sr physicist magnetics, Franklin Inst Res & Develop, 59-64; proj leader, Univac Div, Sperry Rand Corp, 64-67, mgr physics sect, 67-70, sr staff scientist, 71-80; mgr Bubble Memory Res & Develop, Motorola, Inc, 80-84; dir, Magnetic Div, Kodak Res Lab, 84-90; MINT CHAIR & PROF PHYSICS, DEPT PHYSICS & ASTRON, UNIV ALA, 90- *Concurrent Pos:* Mem adv comt magnetism, Am Inst Physics, 64-; sr res fel, Univ York, Eng, 70-71. *Mem:* Sr mem Inst Elec & Electronics Engrs; Am Phys Soc. *Res:* Magnetic properties of thin films and fine particles; magnetization processes; domain theory; bubble devices. *Mailing Add:* 3015 Yorktown Dr Tuscaloosa AL 35406

DOYLE, WILLIAM LEWIS, b Brooklyn, NY, May 19, 10; m 37; c 1. CELL BIOLOGY. *Educ:* Johns Hopkins Univ, AM, 32, PhD(zool), 34. *Prof Exp:* Bruce fel, Johns Hopkins Univ, 34-35; Rockefeller fel, Cambridge Univ, 35-36; Rockefeller fel, Carlsberg Lab, Copenhagen 36-37; asst prof biol, Bryn Mawr Col, 37-42; res assoc & asst prof pharmacol, 42-44, assoc prof, 44-45, from assoc prof to prof anat, 45-76, dir toxicity lab, 45-56, assoc dean div biol sci, 58-61, coordr basic med sci curric, 69-71, EMER PROF ANAT, UNIV CHICAGO, 76- *Concurrent Pos:* Consult, Chem Corps, US Army, 46-59; reserve officer & sci attache, US For Serv, Stockholm, 51-52; from dir to vpres, Mt Desert Island Biol Lab, 64-68, pres, 70-75. *Mem:* Am Asn Anat; Histochem Soc (pres, 61-62); Am Physiol Soc; Am Soc Zoologists; Am Soc Cell Biol. *Res:* Cellular fine structure. *Mailing Add:* 1642 E 56th St Apt 705 Chicago IL 60637-1974

DOYLE, WILLIAM T, b Coalinga, Calif, June 1, 29; m 53; c 4. BRYOLOGY. *Educ:* Univ Calif, Berkeley, BA, 57, PhD(bot), 60. *Prof Exp:* From instr to asst prof biol, Northwestern Univ, 60-65; from asst prof to assoc prof, 65-72, dean, Div Nat Sci, 80-83, PROF BIOL, UNIV CALIF, SANTA CRUZ, 72-, DIR, INST MARINE SCI, 76- *Mem:* AAAS; Am Bryol & Lichenological Soc; Bot Soc Am; Am Inst Biol Sci. *Res:* Development, morphology and cytology of bryophytes; development and evolution of land plants; algae. *Mailing Add:* Dept Biol Sci Appl Sci Bldg Univ Calif Santa Cruz CA 95064

DOYLE, WILLIAM THOMAS, b New Britain, Conn, Dec 5, 25; m 51; c 2. PHYSICS. *Educ:* Brown Univ, BSc, 51; Yale Univ, MSc, 52, PhD(physics), 55. *Prof Exp:* From instr to assoc prof, 55-64, chmn dept, 66-70, PROF PHYSICS, DARTMOUTH COL, 64- *Concurrent Pos:* NSF fel, 58-59. *Mem:* Am Phys Soc; Am Asn Physics Teachers; AAAS; Sigma Xi; Mat Res Soc. *Res:* Inhomogeneous media; solid state physics; optical and magnetic properties of defects in solids; magnetic resonance; high pressure physics. *Mailing Add:* Dept Physics Dartmouth Col Hanover NH 03755

DOYLE-FEDER, DONALD PERRY, b Rochester, NY, Feb 2, 18; m 48, 85; c 3. OPTICS. *Educ:* Univ Rochester, AB, 40. *Prof Exp:* Physicist, Geomet Optics, Univ Rochester, 41-45; supt optical design, Argus, Inc, 44-49; physicist in chg optical design, Nat Bur Standards, 49-56; optical designer, Eastman Kodak Co, 57-86; RETIRED. *Mem:* Fel Optical Soc Am. *Res:* Application of computing machinery to optical design; numerical analysis and programming digital computers. *Mailing Add:* 50 Stanford Dr Rochester NY 14610

DOYNE, THOMAS HARRY, b Pottsville, Pa, Sept 21, 27; m 55; c 3. BIOCHEMISTRY. *Educ:* Pa State Univ, BS, 50, MS, 53, PhD(biochem), 57. *Prof Exp:* Asst x-ray crystallog & chem, Pa State Univ, 50-57; from asst prof to assoc prof, 57-65, chmn dept, 70-80, PROF CHEM, VILLANOVA UNIV, 65- *Concurrent Pos:* Vis prof, US Mil Acad, 82-83; Fulbright fel, Japan, 53-54. *Mem:* Am Chem Soc; Am Crystallog Asn; The Chem Soc; Royal Soc Chem; AAAS. *Res:* Determination of the absolute configuration of molecules by means of x-ray analysis; structure of divalent cation salts of amino acids and peptides. *Mailing Add:* Dept Chem Villanova Univ Villanova PA 19085

DOYON, GILLES JOSEPH, b Troys-Rivières, Que, Feb 12, 52; m 77; c 1. LIQUID CHROMATOGRAPHY FOR CHARACTERIZATION OF FISH OR ADULTERATED FRUITS & JUICES, PACKAGING. *Educ:* Laval Univ, BScA, 77, MSc, 83, PhD(food sci), 89. *Prof Exp:* Agr officer inspection, 77-80, proj officer, 80-81 & 84, RES SCIENTIST, RES BR, FOOD RES & DEVELOP CTR, AGR CAN, ST-HYACINTHE, 89- *Mem:* Can Inst Food Sci & Technol; Inst Food Technol; Poultry Sci Asn. *Res:* Post harvest physiology and packaging material testing for prolonged shelf-life of horticultural produce. *Mailing Add:* 3600 Casavant W St-Hyacinthe PQ J2S 8E3 Can

DOYON, LEONARD ROGER, b Manchester, NH, Apr 6, 23; m 45; c 4. ASSURANCE SCIENCES, SECURITY SCIENCES. *Educ:* Northeastern Univ, BSEE, 53, MSEE, 59; Polytech Inst NY, PhD(opers res), 75. *Prof Exp:* Supvry engr, Gen Elec Co, 53-58; sr engr, Raytheon Co, 58-60; sect mgr, Sylvania Electronic Systs Inc, 60-62; sect mgr, Raytheon Co, 62-65, dept mgr, 65-69, prin engr, 69-75; assoc prof indust eng, Northeastern Univ, 75-79; staff engr, Harris Corp, 79-81; prof opers res, Fla Inst Technol, 81-90; CONSULT, 91- *Concurrent Pos:* Adj prof, Northeastern Univ, 59-69, assoc prof, 73-75, prof, Fla Inst Technol, 79-81; guest lectr, Mass Inst Technol, 66-67; consult eng, Nucle0laire de Saclay, France, 75, Simage SA France, 83-84, pvt pract, 75-79, 81- *Mem:* Sigma Xi; sr mem Inst Elec & Electronics Engrs; Am Soc Qual Control; Opers Res Soc Am. *Res:* Stochastic-network modeling techniques for analyzing the effectiveness of physical security systems against hostile and terrorist intruders; reliability modeling of complex electronic systems; author of 46 technical papers presented and published in USA and Europe. *Mailing Add:* 440 Hawthorne Ct Indian Harbor Beach FL 32937

DOZIER, LEWIS BRYANT, b Rocky Mount, NC, Oct 29, 47; m 80; c 2. ROUGH SURFACE & BOTTOM SCATTERING, STOCHASTIC ACOUSTIC WAVE PROPAGATION. *Educ:* Duke Univ, BS, 69; NY Univ, MS, 72, PhD(math), 77. *Prof Exp:* Mem tech staff radar & sonar systems, Ball Labs, Whippany, NJ, 69-74; grad res asst stochastic wave propagation, NY Univ, Courant Inst, 74-77; SR SCIENTIST UNDERWATER ACOUST, SCI APPLN INT CORP, 77- *Mem:* Acoust Soc Am; Soc Indust & Appl Math. *Res:* Underwater acoustics, including stochastic acoustic wave propagation, rough surface scattering, acoustic propagation modeling, especially coupled modes and parabolic equations, and matched field processing; large-scale computer simulations. *Mailing Add:* 8535 Aponi Rd Vienna VA 22180

DOZIER, SLATER MATHEW, biology, for more information see previous edition

DOZSA, LESLIE, b Budapest, Hungary, May 25, 24; m; c 2. VETERINARY MEDICINE. *Educ:* Univ Budapest, Hungary, DVM, 46. *Prof Exp:* Asst prof obstet sterility & clinic obstet, Vet Col, Hungary, 47-51; field veterinarian sterility, Artificial Insemination Ctr, 51-56; from assoc prof to prof animal husb & path, 57-74, PROF VET SCI & ANIMAL PATHOLOGIST, DIV ANIMAL INDUST & VET SCI, WVA UNIV, 74- *Res:* Sterility of cattle, particularly histopathology. *Mailing Add:* Dept Animal Sci WVa Univ Box 6108 Morgantown WV 26506

DRAAYER, JERRY PAUL, b Albert Lea, Minn, Aug 18, 42; m 64; c 3. NUCLEAR STRUCTURE, THEORETICAL PHYSICS. *Educ:* Iowa State Univ, BS, 64, PhD(physics), 68. *Prof Exp:* Res assoc, Univ Mich, 69-70, instr physics, 72-73; from asst prof to assoc prof ,75-83, PROF PHYSICS, LA STATE UNIV, 83-, CHMN, 85- *Concurrent Pos:* Res assoc physics, Nicls Bohr Inst, 68-69, Univ Rochester, 73-75. *Mem:* Am Phys Soc; Sigma Xi. *Res:* Group theory and the methods of statistical spectroscopy in seeking a microscopic interpretation of collective phenomena in strongly deformed nuclei; microscope models of nuclear structure; applications of group therapy in nuclear physics. *Mailing Add:* 841 Bancroft Way Baton Rouge LA 70808

DRABEK, CHARLES MARTIN, b Chicago, Ill, July 24, 42; m 65; c 2. ZOOLOGY. *Educ:* Univ Denver, BS, 64; Univ Ariz, MS, 67, PhD(zool), 70. *Prof Exp:* Asst prof biol, Cent State Univ, Okla, 70-75; from asst prof to assoc prof, 75-87, PROF BIOL, WHITMAN COL, 87- *Concurrent Pos:* Vis assoc prof biol, Harvard Univ, 80 & 81. *Mem:* Sigma Xi; AAAS; Am Soc Zoologists; Am Soc Mammalogists; Marine Mammal Soc; Am Inst Biol Sci. *Res:* Cardiovascular and respiratory morphology adaptations of diving birds and mammals. *Mailing Add:* Dept Biol Whitman Col Walla Walla WA 99362

DRACH, JOHN CHARLES, b Cincinnati, Ohio, Sept 25, 39; m 64; c 2. BIOCHEMICAL PHARMACOLOGY. *Educ:* Univ Cincinnati, BS, 61, MS, 63, PhD(biochem), 66. *Prof Exp:* Assoc res biochemist, Parke, Davis & Co, 66-68, res biochemist, 69-70; from asst prof to assoc prof dent, 70-80, assoc prof med chem, 78-80, PROF DENT, SCH DENT & PROF MED CHEM, COL PHARM, UNIV MICH, 80-, CHMN, DEPT BIOL & MAT SCI, SCH DENT. *Concurrent Pos:* Mem prof educ comt, Mich Div, Am Cancer Soc, 73-77; spec reviewer, Am Pharmaceut Asn Drug Interactions Proj, 73-78, AMA, Div Drug Eval, 82-83, US Pharmacopeia Drug Interactions Proj, 82-85; rev, Antimicrobial Agents Chemother, 82-, Antiviral Res, 84- *Mem:* Am Soc Microbiol; AAAS; Am Chem Soc; Int Asn Dent Res; Am Asn Dent Sch; Am Asn Oral Biologists. *Res:* discovery, mechanism of action and metabolism of antiviral antineoplastic and drugs. *Mailing Add:* Dept Biol & Mat Sci Sch Dent Univ Mich Ann Arbor MI 48109-1078

DRACHMAN, DANIEL BRUCE, b New York, NY, July 18, 32; m 60; c 3. NEUROLOGY, MUSCULAR PHYSIOLOGY. *Educ:* Columbia Univ, AB, 52; NY Univ, MD, 56. *Prof Exp:* Intern med, Beth Israel Hosp, Boston, Mass, 56-57; asst resident, Neurol Unit, Harvard Univ, 57-58; resident neurol, Boston City Hosp, 58-59, resident neuropath, 59-60; clin assoc neurol, NIH, 60-62, res assoc neuroembryol, 62-63; asst prof neurol, Tufts-New Eng Med Ctr, 63-69; assoc prof, 69-73, PROF NEUROL, SCH MED, JOHNS HOPKINS UNIV, 74- *Concurrent Pos:* Teaching fel, Harvard Univ, 57-60; clin instr, Georgetown Univ, 61-63. *Mem:* AAAS; fel Am Acad Neurol; fel Am Neurol Asn; NY Acad Sci; Royal Soc Med. *Res:* Neuroembryology; congenital neuromuscular defects; diseases of muscle; development and trophic relationship of nerve and muscle; histopathology; physiology; neuroimmunology; myasthenia gravis. *Mailing Add:* Dept Neurol Johns Hopkins Univ Sch Med 600 N Wolfe St Baltimore MD 21205

DRACHMAN, DAVID A, b New York, NY, July 18, 32; m 59; c 3. NEUROLOGY, PHYSIOLOGICAL PSYCHOLOGY. *Educ:* Columbia Col, AB, 52; NY Univ, MD, 56. *Prof Exp:* Intern med, Duke Univ Hosp, 56-57; resident neurol, Harvard Med Sch, 57-60; clin assoc neurol, Nat Inst Neurol Dis & Stroke, 60-63; from asst prof to prof neurol, Med Sch, Northwestern Univ, 63-77; PROF NEUROL & CHMN DEPT, MED SCH, UNIV MASS, WORCESTER, 77- *Concurrent Pos:* Teaching fel, Harvard Med Sch, 57-60; instr, Sch Med, Georgetown Univ, 61-63; consult, Vet Admin Hosp, Hines, Ill, 63-70; vis scientist, Med Sch, Georgetown Univ, 66-; lectr, Food & Drug Admin, 68- *Honors & Awards:* Sci Honoree, Nat Alzheimer's Asn, Chicago, 90. *Mem:* AAAS; fel Am Acad Neurol; Am Neurol Asn; NY Acad Sci; Sigma Xi; Nat Alzheimer's Dis & Related Disorders. *Res:* Neurology of memory; neuroophthalmology; human spatial orientation; computer diagnosis; neurobiology of aging; alzheimer's disease. *Mailing Add:* 111 Barretts Mill Rd Concord MA 01742

DRACHMAN, RICHARD JONAS, b New York, NY, June 2, 30; m 64; c 2. ASTROPHYSICS, ATOMIC PHYSICS. *Educ:* Columbia Col, AB, 51; Columbia Univ, AM, 54, PhD(physics), 58. *Prof Exp:* Asst physics, Columbia Univ, 51-53; physicist, Naval Res Lab, 56-58, Nat Res Coun res assoc, 58-59, physicist, 59-63; asst prof physics, Brandeis Univ, 59-63; MEM STAFF, GODDARD SPACE FLIGHT CTR, NASA, 63- *Concurrent Pos:* Consult, United Aircraft Res Labs, 61-62; mem fac, USDA Grad Sch, 66-73. *Mem:* Fel Am Phys Soc; AAAS. *Res:* Quantum theory; positron systems; atomic scattering. *Mailing Add:* Lab Astron & Solar Physics Code 681 Goddard Space Flight Ctr Greenbelt MD 20771

DRACUP, JOHN ALBERT, b Seattle, Wash, July 14, 34; m 72; c 5. CIVIL ENGINEERING, WATER RESOURCES. *Educ:* Univ Wash, BS, 56; Mass Inst Technol, MS, 60; Univ Calif, Berkeley, PhD(civil eng), 66. *Prof Exp:* Engr, Shell Oil Co, Tex, 56-57 & Boeing Aircraft Co, Wash, 58-59; res asst eng, Mass Inst Technol, 59-60; asst prof eng, Ore State Univ, 60-62; teaching fel, Univ Calif, Berkeley, 62-65; PROF CIVIL ENG, UNIV CALIF, LOS ANGELES, 65- *Concurrent Pos:* Univ Calif & US Govt grants, 66-; consult, US Govt Water Resources Coun, 66-67, Lockheed Calif, 67, Rocketdyne, Inc, 67, TRW Systs, Inc, 67- & Environ Dynamics, Inc, 67-; mem US deleg, Int Conf Water for Peace, 67; NSF grant, 77-; Four Nat Res Coun Comts, 77- *Mem:* Fel Am Soc Civil Engrs; Am Water Resources Asn; Am Geophys Union; Int Water Resource Asn; AAAS; Am Water Works Asn. *Res:* Water resources, hydrology. *Mailing Add:* 4532 Boelter Hall Univ Calif Los Angeles CA 90024-1593

DRACUP, KATHLEEN A, b Santa Monica, Calif, Sept 28, 42; m 72; c 5. CARDIOLOGY. *Educ:* St Xaviers Col, BS, 67; Univ Calif, Los Angeles, MN, 74; Univ Calif, San Francisco, DNSc, 82. *Prof Exp:* Admin nurse, Little Co Mary Hosp, 67-70; clin nurse, Univ Calif, Los Angeles, 71-74, asst prof, 74-78, res assoc, 79-81, assoc prof, 81-89, PROF, UNIV CALIF, LOS ANGELES, 89- *Concurrent Pos:* Ed, Heart & Lung, J Critical Care, 81- *Mem:* Am Nurses Asn; Am Heart Asn; Am Asn Critical Care Nurses; Am Thoracic Soc. *Res:* Psychosocial adaptation of patients and their families to cardiovascular disease. *Mailing Add:* 1315 Georgina Ave Santa Monica CA 90402-2121

DRAEGER, NORMAN ARTHUR, b Milwaukee, Wis. APPLIED RESEARCH FOR COATINGS, SURFACE PHYSICAL CHEMISTRY. *Educ:* Univ Wis-Milwaukee, BS, 77, PhD(phys chem), 87; Univ Wis-Madison, MS, 79. *Prof Exp:* Teaching asst, Dept Chem, Univ Wis-Madison, 77-79, lab staff, Water Chem Prog, 79-80; teaching asst, Dept Chem, Univ Wis-Milwaukee, 80-87, res asst, Lab Surface Studies, 81-87; RES DIR, MIDWEST RES TECHNOL, INC, 88- *Concurrent Pos:* Prin investr, Small Bus Innovation Res Projs, 88-; indust reviewer, Wis Small Bus Innovation Res Prog, Wis State Dept Develop, 89-; adv coun mem, Small Bus Develop Ctr, Univ Wis-Ext, 90- *Mem:* Am Vacuum Soc; Am Inst Aeronaut & Astronaut. *Res:* Thin film coatings; surface physical chemistry and analysis; solid state diffusion; high electric field effects at surfaces; aerospace coatings; vehicle exterior coatings for environmental protection; electrochromic devices; advanced power systems. *Mailing Add:* 5510 W Florist Ave Milwaukee WI 53218

DRAEGER, WILLIAM CHARLES, b San Francisco, Calif, Apr 7, 42; m 66; c 2. NATURAL RESOURCES, REMOTE SENSING. *Educ:* Univ Calif, Berkeley, BS, 64, MS, 65, PhD(wildland resource sci), 70. *Prof Exp:* Asst res forester, Univ Calif, Berkeley, 70-74; prin appln scientist, Agr & Soils, 74-78, CHIEF TRAINING & ASST SECT, EARTH RESOURCE OBSERVATION SYST DATA CTR, US GEOL SURV, 78- *Mem:* Soc Am Foresters; Am Soc Photogram; AAAS. *Res:* Applications of remote sensing techniques to natural resource management problems; specifically concerned with transfer of technology from the research community to potential operational users. *Mailing Add:* Earth Resource Observation Syst Data Ctr Sioux Falls SD 57198

DRAGO, RUSSELL STEPHEN, b Turners Falls, Mass, Nov 5, 28; m 50; c 4. INORGANIC CHEMISTRY. *Educ:* Univ Mass, BS, 50; Ohio State Univ, PhD(chem), 54. *Prof Exp:* Fel, Ohio State Univ, 54-55; from instr to prof chem, Univ Ill, Urbana-Champaign, 55-82; prof chem, 82-84, GRAD RES PROF, UNIV FLA, GAINESVILLE, 84- *Concurrent Pos:* Consult, 60-; Guggenheim fel, 73-74; distinguished vis prof, Univ Firenze, Italy, 85 & Univ Marseille, France, 88. *Honors & Awards:* Award for Res Inorg Chem, Am Chem Soc, 69. *Mem:* Am Chem Soc; Royal Soc Chem. *Res:* Lewis acid-base interactions; physical inorganic chemistry; spectroscopy; transition metals; non-aqueous solvents; binding of small molecules; catalysis. *Mailing Add:* Dept Chem Ctr Catalysis Univ Fla Gainesville FL 32601

DRAGOIN, WILLIAM BAILEY, b Dothan, Ala, Sept 15, 39; m 61; c 2. EXPERIMENTAL PSYCHOLOGY, DEVELOPMENTAL PSYCHOLOGY. *Educ:* Troy State Col, BS, 63; Auburn Univ, MS, 65; George Peabody Col, PhD(psychol), 69. *Prof Exp:* Instr psychol, Auburn Univ, 65-67, asst prof, 70-72; PROF & CHMN PSYCHOL, GA SOUTHWESTERN COL, 72-, RES DIR, CHARLES MIX MEM FOUND, 72- *Concurrent Pos:* NEH fel, Dept Philos, Stanford Univ, 79 & Univ Rochester, 84. *Mem:* Am Psychol Asn; Sigma Xi. *Res:* Biological and genetic processes which influence animal conditioning; experimental psychology; sociobiology. *Mailing Add:* Charles Mix Found Dept Psychol Ga Southwestern Col Wheatley St Americus GA 31709

DRAGON, ELIZABETH ALICE OOSTEROM, b Mineola, NY, June 12, 48; m 72; c 1. VIROLOGY. *Educ:* Univ NH, BS, 70; Yeshiva Univ, PhD(virol & cell biol), 80. *Prof Exp:* Fel, Biol Dept, Brookhaven Nat Lab, 80-82; res scientist, 82-85, SR RES SCIENTIST, CODON GENETIC ENG LAB, 85-, GROUP LEADER, 87- *Mem:* Am Soc Microbiol; AAAS; NY Acad Sci; Am Soc Parasitologists. *Res:* Regulation of expression of viral genes; cell free translation systems; genetics of adenovirus; molecular biology of Trypanosoma Cruzi; regulation & immunoglobulin expression. *Mailing Add:* 9984 Broadmoor Dr San Ramon CA 94583

DRAGONETTI, JOHN JOSEPH, b New York, NY, Nov 23, 31; m 63; c 2. SCIENCE ADMINISTRATION. *Educ:* Columbia Univ, BS, 57; Univ Southern Calif, MPA, 90. *Prof Exp:* Eng geologist, NY State Dept Transp, 58-68; sr eng geologist, NY State Conserv Dept, 68-71; chief, Bur Mineral Resources, NY State Dept Environ Conserv, 71-79; mgr, Western Region Conserv Div, 79-80, dep div chief, Onshore Minerals Conserv Div, 80-82, assoc chief, Off Earth Sci Applns, 82, asst dir, eastern region, 82-83, ASST DIR, INTERGOVT AFFAIRS, US GEOL SURV, 83- *Mem:* Geol Soc Am; Am Inst Prof Geologists; Soc Mining Engrs; Am Geol Inst. *Res:* Writings involve laws; rules and regulations concerning mining; oil and gas development. *Mailing Add:* 2018 Homer Terr Reston VA 22091

DRAGOO, ALAN LEWIS, b Grayling, Mich, Aug 29, 38; m 62; c 2. MATERIALS SCIENCE ENGINEERING. *Educ:* Univ Mich, BS, 61, MS, 63; Univ Md, PhD(physics), 75. *Prof Exp:* Chemist, US Dept Com, Nat Bur Standards, 61-65, res chemist, 65-85, supvry res chemist, Nat Inst Standards & Technol, 85-89, PROJ MGR, DIV MAT SCI, OFF BASIC ENERGY SCI, US DEPT ENERGY, 89- *Mem:* Am Chem Soc; Am Ceramic Soc. *Res:* Theory atomic bonding, structure, phenomena and properties of ceramic materials, processing and synthesis of ceramic materials, ceramic powder characterization. *Mailing Add:* Dept Energy ER-131 Mail Stop G-236 Washington DC 20545

DRAGOUN, FRANK J, b Omaha, Nebr, Sept 21, 29; m 54; c 3. HYDROLOGY, CIVIL ENGINEERING. *Educ:* Univ Nebr, BS, 53; Colo State Univ, MS, 66. *Prof Exp:* Civil engr, Soil Conserv Serv, USDA, 53-55, hydraul engr, Agr Res Serv, 55-62, res hydraul engr, 62-70; asst gen mgr, 70-82, GEN MGR, CENT NEBR PUB POWER & IRRIG DIST, 82- *Mem:* Am Soc Civil Eng; Nat Soc Prof Eng; Int Asn Sci Hydrol; Am Geophys Union; Groundwater Mgt Dists Asn; Nat Soc Prof Engrs. *Res:* Removal and transport of sediment from small agricultural watersheds; hydrologic systems of small agricultural watersheds; automated irrigation systems. *Mailing Add:* PO Box 740 Holdrege NE 68949

DRAGOVICH, ALEXANDER, b Podgorica, Yugoslavia, Feb 26, 24; US citizen; m; c 1. MARINE ECOLOGY. *Educ:* Munich Tech Univ, BS, 47; Miss State Col, MS, 50. *Prof Exp:* Lab instr aquatic biol, Miss State Col, 50 & zool, Univ Hawaii, 50-51; res asst dis pineapples, Calif Packing Corp, 51; res asst tuna ecol, Calif Fish & Game Dept, 54-55; fishery res biologist, Southeast Fisheries Ctr-Miami Lab, Nat Marine Fisheries Serv, 55-84; CONSULT, 88- *Concurrent Pos:* Consult, Fla Bd Conserv & Fla Univs; mem adj staff fac, Fla Atlantic Univ; partic spec proj, Food & Agr Orgn, UN, 85-86. *Mem:* Gulf & Caribbean Fisheries Inst. *Res:* Ecology and taxonomy of marine phytoplankton and invertebrates; ecology of tunas and penaeid shrimp; study of Guiana shrimp; fishery-biology, ecology and population dynamics of shrimp stocks; ecologic study of estuarine areas of Florida Everglades. *Mailing Add:* 1331 SW Second Terr Miami FL 33145

DRAGSDORF, RUSSELL DEAN, b Detroit, Mich, Nov 21, 22; m 48; c 2. SOLID STATE PHYSICS. *Educ:* Mass Inst Technol, SB, 44, PhD(physics), 48. *Prof Exp:* From asst prof to assoc prof, Kansas State Univ, 48-56, assoc dean & actg dean grad sch, 65-66, Prof Physics, 56-90; RETIRED. *Concurrent Pos:* Res physicist, Metals Res Lab, Union Carbide Corp, 56-57 & Lawrence Radiation Lab, 68. *Mem:* Am Phys Soc; Am Crystallog Asn; Sigma Xi. *Res:* X-ray diffraction; small angle x-ray scattering; small particle properties; crystal imperfections. *Mailing Add:* Dept Physics Kans State Univ Cardwell Hall Manhattan KS 66506

DRAGT, ALEXANDER JAMES, b Lafayette, Ind, Apr 7, 36; m 57; c 3. THEORETICAL HIGH ENERGY PHYSICS, CLASSICAL MECHANICS. *Educ:* Calvin Col, AB, 58; Univ Calif, Berkeley, PhD(physics), 64. *Prof Exp:* Sr scientist, Lockheed Missiles & Space Co, 61-62; staff scientist, Aerospace Corp, 63; mem dept physics, Inst Advan Study, 63-65; from asst prof to assoc prof, 65-69, chmn dept physics & astron, 75-78, PROF PHYSICS, UNIV MD, COLLEGE PARK, 69- *Concurrent Pos:* Vis staff mem, Los Alamos Sci Lab, 78-79 & 86-87; vis prof, Tex A & M Univ, 84; vis scientist, Tex Accelerator Ctr, 84; guest scientist, Lawrence Berkeley Lab, 85 & 87. *Mem:* Fel Am Phys Soc; Am Math Soc; Am Geophys Union; AAAS. *Res:* Theoretical elementary particle physics; space physics; accelerator design; plasma physics; optics; mathematical physics; dynamical systems. *Mailing Add:* Dept Physics & Astron Univ Md College Park MD 20742

DRAGUN, HENRY L, b Philadelphia, Pa, Feb 24, 32; m 59. PHYSICAL & ORGANIC CHEMISTRY. *Educ:* Univ Pa, AB, 53; Rutgers Univ, PhD(phys chem), 60. *Prof Exp:* Res phys chemist mech res lab, E I du Pont de Nemours & Co, 59-62; sr chemist, Elkton Div, Thiokol Chem Corp, 62-65; PROF CHEM, ANNE ARUNDEL COMMUNITY COL, 65-, CHMN DIV SCI, 69- *Mem:* AAAS; Am Chem Soc; Am Ord Asn. *Res:* High energy propellant development; investigation of relationship between microstructure of polymeric materials and their macroscopic electrical and mechanical properties. *Mailing Add:* Anne Arundel Community College 101 Col Pkwy Arnold MD 21012

DRAGUN, JAMES, b Detroit, Mich. EARTH SCIENCES. *Educ:* Wayne State Univ, BS, 71; Pa State Univ, MS, 75, PhD(soil chem), 77. *Prof Exp:* Soil chemist, US Environ Protection Agency, 78-80, sr soil chemist, 80-82; mgr, Pac Environ Lab, Kennedy/Jenks Engrs, 82-84, E C Jordan, 84-87, Stalwart, 87-88; DRAGUN CORP, 88- *Concurrent Pos:* Consult, 77-; appointed to Chess, 88- *Honors & Awards:* Bronze Medal, US Environ Protection Agency, 80; LICA Award, 90. *Mem:* Am Chem Soc; Soil Sci Soc Am; Am Soc Agron; Sigma Xi. *Res:* Test protocols to measure the adsorption and chemical reactions of organic and inorganic chemicals in sediments and soils; create estimation techniques to predict the occurrence and extent of these processes; assess the extent of human exposure and environmental fate of toxic substances. *Mailing Add:* Dragun Corp 3240 Coolidge Berkeley MI 48072-1634

DRAHOS, DAVID JOSEPH, b Oceanside, NY, Dec 28, 51; m 78; c 2. BIOTECHNOLOGY, BIOREMEDIATION. *Educ:* Manhattan Col, BS, 73; Univ Pittsburgh, PhD(molecular biol), 79. *Prof Exp:* Fel, Mcardle Lab, Univ Wis-Madison, 79-82; sr res chemist, 82-84, SR RES GROUP LEADER, MONSANTO CO, 84- *Concurrent Pos:* Adj assoc prof, Microbiol & Plant Pathology, Clemson Univ, 88- *Mem:* AAAS; Am Phytopath Soc; Am Soc Testing & Mat. *Res:* Development of microbial biological control agents active against major agronomic disease and insect pests through the use of genetic manipulation procedures. *Mailing Add:* Monsanto Co 700 Chesterfield Village Pkwy Mail Zone GG3A St Louis MO 63198

DRAHOVZAL, JAMES ALAN, b Cedar Rapids, Iowa, Feb 13, 39; m 63; c 3. GEOLOGY. *Educ:* Univ Iowa, BA, 61, MS, 63, PhD(geol), 66. *Prof Exp:* Explor geologist resources, Pan Am Petrol Corp, 65-66; res geologist & asst chief, Geol Div, Geol Surv Ala, 66-77, asst state geologist & dir tech opers, 77-78; sr res geologist, Geol & Interpretation Dept, Gulf Sci & Technol Co, Harmarville, Pa, 78-82, supvr, 82-83; sr res geologist, Gulf Research & Develop Co, Houston, Tex, 84; res assoc, Arco Oil & Gas Co, Plano, Tex, 84-88; GEOLOGIST & HEAD, PETROL & STRATIG, SEC, KY GEOL SURVEY, UNIV KY, LEXINGTON, 89- *Concurrent Pos:* NSF travelling grant, Czech, 68; lectr, Dept Geol, Univ Ala, 70; consult geologist, 74- *Mem:* Fel Geol Soc Am; AAAS; Am Asn Petrol Geologists. *Res:* Applications of remote sensing technology to regional structural geology and stratigraphy; Paleozoic invertebrate paleontology; southern Appalachian geology; geology of thrust belts; basin analysis; seismic stratigraphy; geology of vifts. *Mailing Add:* 3313 High Hope St Lexington KY 40517

DRAINE, BRUCE T, b Calcutta, India, Nov 19, 47; US citizen; m 76; c 2. THEORETICAL ASTROPHYSICS. *Educ:* Swarthmore Col, BA, 69; Cornell Univ, MS, 75, PhD(theoret physics), 78. *Prof Exp:* Ctr res fel astrophysics, Smithsonian Astrophys Observ, 77-80; mem, Inst Advan Study, Princeton, 80-82; from asst prof to assoc prof, 81-90, PROF, DEPT ASTROPHYS SCI, PRINCETON UNIV, 90- *Mem:* Am Astron Soc. *Res:* Theory of the interstellar medium, including physics of interstellar dust. *Mailing Add:* Dept Astrophys Sci Peyton Hall Princeton Univ Princeton NJ 08540

DRAKE, ALBERT ESTERN, b Stamping Ground, Ky, June 12, 27; m 52; c 4. STATISTICS, AGRICULTURAL ECONOMICS. *Educ:* Univ Ky, BS, 50, MS, 51; Univ Ill, PhD(agr econ), 58. *Prof Exp:* Res asst agr, Univ Ill, 53-55, res assoc food technol, 55-59; from assoc prof to prof biometrics, Auburn Univ, 59-63; dir comput ctr statist & comput, WVa Univ, 63-66; PROF STATIST, UNIV ALA, 66-, ASSOC DEAN UNDERGRAD PROGS, 90- *Concurrent Pos:* NSF grants, 59-63, Ford Found grants, 76 & 80; fels, NC State Univ, 59, 63 & Univ Fla, 60. *Mem:* Decision Sci Inst (secy, 73-74); Biomet Soc; Am Statist Asn; Am Agr Econ Asn. *Res:* Computer assisted instruction in statistics; application of statistical techniques to management decision making. *Mailing Add:* Col Com & BA Univ Ala PO Box 870223 Tuscaloosa AL 35487

DRAKE, ALVIN WILLIAM, b Bayonne, NJ, Sept 21, 35; m 57, 88; c 2. OPERATIONS RESEARCH. *Educ:* Mass Inst Technol, SB & SM, 58, EE, 61, ScD(elec eng), 62. *Prof Exp:* From asst prof to assoc prof, 64-73, assoc dir, Opers Res Ctr, 66-77, PROF ELEC ENG, MASS INST TECHNOL, 73- *Concurrent Pos:* Ford fel, Mass Inst Technol, 64-66. *Mem:* Opers Res Soc Am; Inst Elec & Electronics Engrs; Am Asn Blood Banks. *Res:* Probilistic applications in the delivery of public services; blood, organ, and tissue banking; decision analysis and risk assessment. *Mailing Add:* 30 F Inman St Cambridge MA 02139

DRAKE, ARTHUR EDWIN, b Elmwood Place, Ohio, Sept 17, 18; m 41; c 3. ELECTROCHEMISTRY. *Educ:* Miami Univ, AB, 40; Western Reserve Univ, MS, 42, PhD(electrochem), 43. *Prof Exp:* Res chemist, Hercules, Inc, 43-45, from sales rep to sales supvr, 45-55, sales mgr, 55-61, dir develop, 61-66, PRES, DRACO, INC, GREENVILLE, 66- *Mem:* Am Chem Soc. *Res:* Electromotive force measurements on molten binary alloys. *Mailing Add:* Two Swallow Hill Rd PO Box 4007 Wilmington DE 19807-1632

DRAKE, AVERY ALA, JR, b Kansas City, Mo, Jan 17, 27; m 63; c 2. GEOLOGY. *Educ:* Mo Sch Mines, BS, 50, MS, 52. *Prof Exp:* Asst, Mo Sch Mines, 50-52; geologist, 52-79, asst chief geologist, 79-84, RES GEOLOGIST, US GEOL SURV, 85- *Concurrent Pos:* Geologist, Nat Lead Co, 51; mem, NSF-US Navy Exped, Bellingshausen Sea, Antarctica, 61; assoc ed for geol, Am Geophys Union, Antarctic Res Series, 71-79; dir advan geol field methods course, Brazilian Dept Mines & Energy, 73-75; vis prof struct geol, Howard Univ, 74; assoc ed, Geol Soc Am Bull, 85-90. *Honors & Awards:* Antarctic Medal, 61. *Mem:* Geol Soc Am; Soc Econ Geol; Geochem Soc. *Res:* Structural geology of central Appalachians; Precambrian geology; Antarctic geology. *Mailing Add:* US Geol Surv Mailstop 928 Reston VA 22092

DRAKE, BILLY BLANDIN, b Warsaw, Mo, Dec 18, 17; wid; c 3. BIOCHEMISTRY, ENZYMOLOGY. *Educ:* Cent Methodist Col, Mo, AB, 39; Univ Pittsburgh, MS, 41, PhD(biochem), 43. *Prof Exp:* Sr scientist, Rohm and Haas, 43-68, head, Enzyme Technol Serv Lab, 68-71 & Enzyme Lab, 71-76, enzymes prod mgr, 75-76; RETIRED. *Concurrent Pos:* Consult, 76-87. *Mem:* Am Chem Soc; Sigma Xi. *Res:* Production and properties of enzymes; separation and oxidation of amino acids; microbial fermentations. *Mailing Add:* 711 Pecan Dr Philadelphia PA 19115-2817

DRAKE, CHARLES GEORGE, b Windsor, Ont, July 21, 20. NEUROLOGICAL SCIENCES. *Educ:* Univ Western Ont, London, MD, 44, MSc, 47; Univ Toronto, Ont, MS, 56; FRCS(C), 52. *Hon Degrees:* DSc, Mem Univ, St John's, Nfld, 73; LLD, Univ Toronto, 82 & Dalhousie Univ, Halifax, NS, 84; Dr, Univ Repub Montevideo, Uruguay, 84; FRCS(I), 76. *Prof Exp:* Jr rotating intern, Toronto Gen Hosp, Ont, 44-45, resident-clin asst, div neurosurg, 49-51; instr physiol, Fac Med, Univ Western Ont, 45-47; JH Brown fel, dept physiol, Yale Univ, 47-48; resident surgeon, Victoria Hosp, London, Ont, 48-49, clin prof neurosurg, 49-69; prof & chmn clin neurol sci, 69-74, head, div neurosurg, 69-75, prof surg & chmn dept, 74-84, PROF NEUROSURG, DEPT CLIN NEUROL SCI, DIV NEUROSURG, UNIV WESTERN ONT, 69- *Concurrent Pos:* Guest lectr, var org throughout US, Europe, Can, SAm & Far East; vis prof, var univs throughout US, Can, Europe & Far East; G A Routledge traveling fel, Europe, 51; fel neuropth, Banting Inst, Toronto, 51-52; McLaughlin fel clin res, Victoria Hosp, Ont, 52-53. *Mem:* Am Asn Neurol Surgeons (vpres, 66-69, pres, 77); Royal Col Physicians & Surgeons Can (vpres, 68-70, pres, 71-73); Am Acad Neurol Surgeons; Can Asn Clin Surgeons (pres, 73); World Fedn Neurosurg Soc (pres, 77-81); Soc Neurol Surgeons (pres elect, 79, pres, 80); James IV Asn Surgeons, Inc (vpres, 81); Am Col Surgeons (pres, 84-85); Can Neurol Soc; Can Med Asn. *Mailing Add:* Div Neuro Sci Univ W Ont Fac Med London ON N6A 5C1 Can

DRAKE, CHARLES HADLEY, b Waterloo, Iowa, Feb 8, 16; m 42; c 2. BACTERIOLOGY. *Educ:* Univ Minn, BA, 37, MS, 39, PhD(bact), 42. *Prof Exp:* Teaching asst bact, Univ Minn, 36-42, instr med bact, Exten Div, 42; instr, Sch Lib Arts & Sch Med, Univ Kans, 42-44; from asst prof to prof bact, Wash State Univ, 44-80; RETIRED. *Concurrent Pos:* Agent, Bur Biol Surv & Bur Plant Indust, USDA, 36-42; reserve officer, USPHS. *Mem:* Am Soc Microbiol. *Res:* Medical mycology and bacteriology; immunology; public health; pathogenicity and allergic properties of Nocardia asteroides; viral diseases, plague, Tularemia, pseudomonas aeruginosa; bacterial limnology. *Mailing Add:* 1310 Orion Dr Pullman WA 99163

DRAKE, CHARLES LUM, b Ridgewood, NJ, July 13, 24; m 50; c 3. GEOPHYSICS. *Educ:* Princeton Univ, BSE, 48; Columbia Univ, PhD(geol), 58. *Prof Exp:* Lectr, Columbia Univ, 53-55, instr, 58-59, from asst prof to assoc prof, 59-67, actg asst dir, Lamont Geol Observ, 63-65; prof & chmn dept, 67-69; dean, grad studies, Dartmouth Col, 78-81, assoc dean, Sci Div, 78-81, Albert Bradley prof earth sci, 84-89, PROF GEOL, DARTMOUTH COL, 69-, CHMN DEPT, 91- *Concurrent Pos:* NSF sr fel, Cambridge Univ, 65-66; pres, inter-union comt on geodynamics, Int Coun Sci Unions, 70-75; chmn comt on geodynamics, Nat Acad Sci, 70-79; comt adv, Environ Sci Serv Admin, 67-71, chmn, 70-71; chmn bd, Int Off Earth Sci, 72-75; mem, Assembly Math & Phys Sci, 74-76, Geol Correlation Prog, Int Union Geol Sci-UNESCO, 84-90, Films Comt, Nat Acad Sci, 86- & President's Coun Adv Sci & Technol, 90- *Honors & Awards:* G P Woollard Award, Geophys Div, Geol Soc Am. *Mem:* Seismol Soc Am; Am Asn Petrol Geol; Geol Soc Am (pres, 75-76); Royal Astron Soc; Soc Explor Geophys; Am Geophys Union (pres, 84-86); Sigma Xi; hon mem Soc Geol France. *Res:* Marine geology and geophysics; tectonics; structural geology; seismology. *Mailing Add:* RD 1 East Thetford VT 05043

DRAKE, CHARLES ROY, b Cromwell, Ky, Apr 27, 18; m 49; c 2. PLANT PATHOLOGY, PLANT PROTECTION. *Educ:* Western Ky State Col, BS, 52; Univ Wis, PhD(plant path), 56. *Prof Exp:* Asst, Univ Wis, 52-56; plant pathologist, Crops Res Div, Va Agr Exp Sta, Agr Res Serv, USDA, 56-62; from assoc prof to prof, 62-88, exten proj leader, 76-88, EMER PROF PLANT PATH, VA POLYTECH INST & STATE UNIV, 89- *Concurrent Pos:* Dept coordr, Grad Plant Protection Prog; col div coordr, Develop Innovative Capstone Course for Undergrad Plant Protection Curric; adv, Univ Acad Career; mem, SE Regional Peach Tree Short-Life Comt & Nation Peach Coun; pres, Fac Senate Va, 79-80. *Honors & Awards:* Prestigious Carroll Miller Award. *Mem:* Am Phytopath Soc; Sigma Xi. *Res:* Diseases of fruit, with special interest in apple scab and peach brown rot; delineated the cause of Golden Delicious leaf blotch as a transpirational stress; established a 43-acre peach and nectarine test with a grower 119 varieties of peaches and 29 varieties of nectarines; developed high density research and demonstration orchard for grower training. *Mailing Add:* Dept Plant Path Physiol & Weed Sci Va Polytech Inst & State Univ Blacksburg VA 24061-0331

DRAKE, CHARLES WHITNEY, b South Portland, Maine, Mar 8, 26; m 52; c 3. ATOMIC PHYSICS. *Educ:* Univ Maine, BS, 50; Wesleyan Univ, MA, 52; Yale Univ, PhD, 58. *Prof Exp:* Intermediate scientist, Westinghouse Elec Corp, 52-53; from instr to asst prof physics, Yale Univ, 57-66; assoc prof, 66-75, actg chmn dept, 76-77, chmn dept, 77-84, PROF PHYSICS, ORE STATE UNIV, 75- *Mem:* Am Phys Soc. *Res:* Atomic and molecular beams; polarized particles for nuclear reactions; beam foil spectroscopy. *Mailing Add:* Dept Physics Ore State Univ Corvallis OR 97331

DRAKE, DARRELL MELVIN, b Stillwater, Okla, Sept 5, 32; m 55; c 4. NUCLEAR PHYSICS. *Educ:* Univ Okla, BS, 54, MS, 56; Univ Wash, PhD(physics), 61. *Prof Exp:* Asst physics, Univ Ill, 62-65; STAFF MEM PHYSICS, LOS ALAMOS SCI LAB, 65- *Concurrent Pos:* Vis staff mem physics, Bruyeres le Chatel Ctr Study, Comn Atomic Energy, France, 75-76. *Res:* Radiative capture of neutrons and protons with energies of 5 to 20 MeV; pion interactions with complex nuclei; fast neutron induced gamma ray production and fast neutron scattering. *Mailing Add:* MSD438 ESS-8 Los Alamos Sci Lab Los Alamos NM 87545

DRAKE, DAVID ALLYN, b Lorain, Ohio, Sept 29, 37; m 59; c 2. MATHEMATICS. *Educ:* Harvard Univ, AB, 59; Syracuse Univ, PhD(math), 67. *Prof Exp:* Systs programmer, Int Bus Mach, 60-62; instr math, Syracuse Univ, 66-67; fel, 67-68, from asst prof to assoc prof, 68-80, PROF MATH, UNIV FLA, 80-, CHAIR, 88- *Concurrent Pos:* Alexander von Humboldt fel, 74-75 & 82; vis prof math, Free Univ, Berlin, 79; assoc ed, J Statist Planning & Inference, 79-88. *Mem:* Am Math Soc. *Res:* Combinatorics and finite geometries, particularly Hjelmslev geometries, projective planes and Latin squares. *Mailing Add:* Dept Math Univ Fla Gainesville FL 32611

DRAKE, EDGAR NATHANIEL, II, b Springfield, Mo, May 18, 37; m 59; c 3. PHYSICAL CHEMISTRY, ANALYTICAL CHEMISTRY. *Educ:* Univ Houston, BS, 60; Univ Colo, MS, 62; Tex A&M Univ, PhD(chem), 69. *Prof Exp:* Instr, Tex A&I Univ, 62-63; asst prof, Howard Payne Col, 63-65; assoc prof, 65-80, PROF CHEM, ANGELO STATE UNIV, 80- *Mem:* Am Chem Soc. *Res:* Chemistry of coordination compounds; stability of complexes; mechanisms of complex ion reactions. *Mailing Add:* Dept Chem Angelo State Univ San Angelo TX 76909

DRAKE, EDWARD LAWSON, b Charlottetown, PEI, Apr 9, 30; m 56; c 2. INVERTEBRATE PATHOLOGY, ENTOMOLOGY. *Educ:* McGill Univ, BSc, 50; Cornell Univ, MS, 59; Dalhousie Univ, PhD(biol), 69. *Prof Exp:* Entomologist, Govt of Nyasaland, 52-56; experimentalist, Cornell Univ, 56-57; from lectr to asst prof, Prince of Wales Col, 59-69, chmn dept, 69-76, ASSOC PROF BIOL, UNIV PEI, 69-, DEAN SCI, 85- *Concurrent Pos:* Mem, Sci Coun Can, 78-84. *Res:* Virus diseases of insects; insect fine structure; insect taxonomy. *Mailing Add:* Dept Biol Sci Univ PEI Charlottetown PE C1A 4P3 Can

DRAKE, ELISABETH MERTZ, b New York, NY, Dec 20, 36; m 57. CHEMICAL ENGINEERING. *Educ:* Mass Inst Technol, SB, 58, ScD(chem eng), 66. *Prof Exp:* Staff consult cryog, A D Little Inc, 58-64, sr engr, 66-70, sr engr safety & fire technol, 71-77, mgr risk assessment group, 77-78, dep head safety & fire technol sect, 78-80, vpres technol risk mgt, 80-82,; Cabot prof & chair chem eng dept, Northeastern Univ, 82-86; vpres environ health & safety pract, A D Little Inc , 86-89; ASSOC DIR, ENERGY LAB, MASS INST TECHNOL, 90- *Concurrent Pos:* Lectr, Univ Calif, Berkeley, 70-71; vis prof, Mass Inst Technol, 73-74; Sci Adv Bd, Environ Protection Agency, 76-77; dir Am Inst Chem Engrs, 88-90. *Mem:* Am Inst Chem Engrs; Am Chem Soc; AAAS; Sigma Xi. *Res:* Energy technology; risk assessment and control of hazardous materials, liquefied natural gas technology and safety, cryogenic engineering, risk management. *Mailing Add:* 30 F Inman St Cambridge MA 02139

DRAKE, ELLEN TAN, b Beijing, China, July 23, 28; US citizen; m 52; c 3. HISTORY OF EARTH SCIENCES, SCIENCE WRITING & EDITING. *Educ:* Bryn Mawr Col, BA, 49; Ore State Univ, MA, 75, PhD(oceanog), 81. *Prof Exp:* Teaching asst geol, Wesleyan Univ, 49-52; asst ed & jr res chemist, Carnegie Inst Technol, 52-53; asst ed, Alcohol Studies, Yale Univ, 53-55, ed & head publ, Peabody Mus Natural Hist, 61-66; res asst oceanog, Col Oceanog, 69-79, instr tech writing & ed, 75-81, instr, Comput Ctr, 77-79, asst prof, dept geol, 85, RES ASSOC OCEANOG, COL OCEANOG, ORE STATE UNIV, 81- *Concurrent Pos:* Mem, US Hist of Geol Comt, Nat Res Coun, 83-86; prog coordr oceanog & limnol, Pac Div, AAAS, 84- *Mem:* Geol Soc Am; Sigma Xi; Hist Sci Soc; Hist Earth Sci Soc. *Res:* Tectonic evolution of the Oregon continental margin; carbon dioxide increase in the atmosphere and oceans; history of geology; impact tectonics studies. *Mailing Add:* Col Oceanog Ore State Univ Corvallis OR 97331

DRAKE, FRANK DONALD, b Chicago, Ill, May 28, 30; m 78; c 5. RADIO ASTRONOMY, SCIENCE ADMINISTRATION. *Educ:* Cornell Univ, BEP, 52; Harvard Univ, MA, 56, PhD(astron), 58. *Prof Exp:* Field engr, Gen Elec Co, 51; electronics, eng & commun officer, USS Albany, US Navy, 52-55; dir, Astron Res Group, Ewen Knight Corp, 58; head, Telescope Oper & Sci Serv Div, Nat Radio Astron Observ, 58-63; chief, Lunar & Planetary Sci Sect, Jet Propulsion Lab, Calif Inst Technol, 63-64; from assoc prof to prof astron, Cornell Univ, 64-76, Goldwin Smith prof, 76-84; dean, Div Natural Sci, 84-88, PROF ASTRON & ASTROPHYS, UNIV CALIF, SANTA CRUZ, 84- *Concurrent Pos:* Assoc dir, Ctr Radiophysics & Space Res, Cornell Univ, 64-75, chmn, Dept Astron, 68-71, interim chmn, Dept Geol Sci, 70-71; dir, Nat Astron & Ionosphere Ctr, 71-81, sr scientist, 81-84; pres, Comn 51 & chmn, 20th Gen Assembly, Nat Organizing Comt, & US Nat Comt, Int Astron Union; chmn bd physics & astron, Nat Res Coun, 90-; pres, SETI Inst & mem sci working group, 85-; dir at large, Asn Univ Res Astron Inc; distinguished lectr, Buffalo Soc Natural Sci, 83; mem, Slipher Awards Comt, astron surv comt, Panel Astron Facil, comt employment prob astron, comt sci & pub policy, ad hoc group sci prog, Alan T Waterman Awards Comt & astron surv comt, Radio Astron Panel, Natr Acad Sci; chmn, Astron Sect, AAAS, Div Planetary Sci, Am Astron Soc, vis comt, Nat Radio Astron Observ; mem, vis comt, Kitt Peak Nat Observ & Cerno Tololo IntraAm Observ, NRC, organizing comt, First Soviet Am Symp Commun Extraterrestial Intel, Armenian SSR, comt space astron & astrophys, Nat Res Coun, adv comt very large array, Nat Radio Astron Observ, workshop interstellar commun, NASA Ames Res Ctr, workshop utilization existing radio antennas, Jet Propulsion Lab, ad hoc comt, review status sci & eng educ, Univ Nev Syst, 83. *Honors & Awards:* McNair Mem Lect, Univ NC, 64; Loeb Lect, Harvard Univ, 69; Rennie Taylor Award, Am Tentative Soc, 78; Nathan Hayward Mem Lect, Franklin Inst, 83; Winthrop Rockefeller Lectr, Univ Ark, 87; Verna & Mars Mclean Lectr, Baylor Col Med, 87. *Mem:* Nat Acad Sci; fel Am Acad Arts & Sci; Int Astron Union; Int Sci Radio Union; Am Astron Soc; fel AAAS; Brit Interplanetary Soc; Sigma Xi; Astron Soc Pac (pres, 88-90). *Res:* Solar system; 21 centimeter line research; search for extraterrestrial intelligence; radio telescope development. *Mailing Add:* Univ Calif 413 Natural Sci Bldg II Santa Cruz CA 95064

DRAKE, GEORGE M(ARSHALL), JR, b Goodlettsville, Tenn, July 25, 32; m 58; c 4. CHEMICAL ENGINEERING. *Educ:* Univ Tenn, BS, 55, MS, 57, PhD(chem eng), 61. *Prof Exp:* Res engr, Spruance Film Res & Develop Lab, 61-74, staff engr, 74-75, sr engr, Tech Sect, Tecumseh, Kans, 76-80, SR ENGR, KAPTON-TEFLON TECH, E I DU PONT DE NEMOURS & CO, INC, CIRCLEVILLE, OH, 80- *Mem:* Am Chem Soc; Am Inst Chem Engrs. *Res:* Film extrusion, process development and improvement; quality control; laboratory methods. *Mailing Add:* E I Du Pont de Nemours & Co Inc ED Dept PO Box 89 Circleville OH 43113

DRAKE, GORDON WILLIAM FREDERIC, b Regina, Sask, Aug 20, 43; m 66; c 2. ATOMIC PHYSICS. *Educ:* McGill Univ, BSc, 64; Univ Western Ont, MSc, 65; York Univ, Ont, PhD(physics), 67. *Prof Exp:* Nat Acad Sci res fel physics, Smithsonian Astrophys Observ, Mass, 67-69; from asst prof to assoc prof, 69-75, PROF PHYSICS, UNIV WINDSOR, 75- *Concurrent Pos:* Fel, Alfred P Sloan Found, 73. *Honors & Awards:* Herzberg Medal, 79; Steacie Prize, 81. *Mem:* Fel Am Phys Soc; Can Asn Physicists; fel Inst Physics; fel Royal Soc Can. *Res:* Theory of atomic processes, including relativistic effects, radiative transitions, electron-atom and atom-atom scattering; precision calculations for two-electron systems and applications to astrophysical problems. *Mailing Add:* Dept Physics Univ Windsor Windsor ON N9B 3P4 Can

DRAKE, JOHN EDWARD, b Simla, India, Apr 20, 36; m 60; c 4. MAIN GROUP CHEMISTRY, SPECTROSCOPY. *Educ:* Univ Southampton, BSc, 57, PhD(chem), 60. *Hon Degrees:* DSc, Univ Southampton, 78. *Prof Exp:* Chemist, Lawrence Radiation Lab, Univ Calif, Berkeley, 60-62; asst lectr chem, Univ Hull, 62-63; lectr, Univ Southampton, 63-69; assoc prof, 69-71, PROF CHEM, UNIV WINDSOR, 71-, HEAD DEPT, 83- *Concurrent Pos:* Vis prof, Univ Calif, Berkeley, 79 & Univ Wuppertal, 81. *Mem:* The Chem Soc; fel Chem Inst Can (secy-treas, Div Inorg Chem, 81-); Am Chem Soc. *Res:* Spectroscopic, synthetic and structural studies of main group organometalloids and hydrides; application of electrical discharges to preparative problems; vibrational, nuclear magnetic resonance, and photo-electron spectroscopy; x-ray crystallography. *Mailing Add:* Dept Chem & Biochem Univ Windsor Windsor ON N9B 3P4 Can

DRAKE, JOHN W, b Detroit, Mich, Feb 10, 32; m 60, c 2. MUTAGENESIS. *Educ:* Yale Univ, BS, 54; Calif Inst Technol, PhD(biol), 58. *Prof Exp:* Res assoc bact, Univ Ill, Urbana-Champaign, 58-59, from instr to prof microbiol, 59-77; actg chief, lab environ mutagenesis, 77-78, chief, lab molecular genetics, 79-82, HEAD, MUTAGENESIS SECT, NAT INST ENVIRON HEALTH SCI, RES TRIANGLE PARK, NC, 83- *Concurrent Pos:* Fulbright fel, Weizmann Inst, 57-58; Guggenheim fel, 64-65; mem biol sect, grad fel panel, NSF, 68-70, chmn, 71-72; NIH spec fel, dept molecular biol, Univ Edinburgh, 71-72; mem, genetics study sect, NIH, 72-76, task force health aspects energy policy proj, Am Pub Health Asn, 73-74, sci adv bd, study group mutagenicity testing, Environ Protection Agency, 76-77, int comn protection against environmental mutagens & carcinogens, 77-80, bd toxicology & environ health hazards, Nat Acad Sci-Nat Res Coun, 77-81, & bd dirs, Genetics Soc Am, 83-; prof basic med sci, Univ Ill, 72-77; assoc ed, Genetics, 75-81, ed, 82-; adj prof path, Univ NC, Chapel Hill, 78- & adj prof genetics, Duke Univ, 80-; mem, genetics curric, Univ NC, Chapel Hill, 78-, adv comt, Cancer Res Ctr, 78-81 & curric toxicol, 79-; chmn, Prog Comt, Int Congress Genetics, 87-88. *Honors & Awards:* Special Achievement Award, NIH, 78, Outstanding Performance Award, 85; Award of Excellence, Environ Mutagen Soc. *Mem:* Genetics Soc Am. *Res:* Molecular mechanisms of mutation. *Mailing Add:* Lab Molecular Genetics - E3-01 Nat Inst Environ Health Sci PO Box 12233 Research Triangle Park NC 27709

DRAKE, JUSTIN R, b Columbus, Ind, Nov 14, 21. METALLURGICAL ENGINEERING. *Educ:* Purdue Univ, BS, 50. *Prof Exp:* Proj engr, Caterpillar Tractor Co, 50-66; dir mat res develop, Cummins Indust Co, 66-83; RETIRED. *Concurrent Pos:* Mem, High Temperature Alloys Comt, Soc Automotive Engrs & var comts, Am Soc Testing & Mat & Am Soc Mech Engrs. *Mem:* Fel Am Soc Metals. *Res:* Materials research and development on diesel and gas turbines; materials problems relating to fatigue, corrosion and processing of cast irons and superalloys; cast and wrought stainless steels and superalloys. *Mailing Add:* 1905 Park Valley Dr Columbus IN 47203

DRAKE, LON DAVID, b North Tonawanda, NY, Nov 30, 39; c 4. ENVIRONMENTAL GEOLOGY. *Educ:* Univ Buffalo, BA, 61; Univ Calif, Los Angeles, MA, 65; Ohio State Univ, PhD(geol), 68. *Prof Exp:* Asst prof, 68-73, assoc prof, 73-79, PROF GEOL, UNIV IOWA, 79- *Concurrent Pos:* Consult for indust. *Mem:* Geol Soc Am; Am Asn Quaternary Res; Am Coun Reclamation Res; Sigma Xi. *Res:* Processes of glacial till production and deposition; strip mine reclamation; economic pleistocene geology; low-technology applications; landfill hydrogeology. *Mailing Add:* Dept Geol Trowbridge Hall Univ Iowa Iowa City IA 52242

DRAKE, MICHAEL CAMERON, b Spencer, Iowa, Sept 11, 48; m 70; c 3. TURBULENT COMBUSTION. *Educ:* Iowa State Univ, BS, 70; Pa State Univ, PhD(chem), 77. *Prof Exp:* Nat Res Coun assoc, Nat Bur Standards, 77-78; chemist, Corp Res & Develop, Gen Elec Co, 78-86; CHEMIST, GEN MOTORS RES LABS, 86- *Honors & Awards:* Dushman Award, Gen Elec, 81. *Res:* Chemical effect in turbulent combustion; flame stability; pollution formation; heat release; laser diagnostics in high temperature chemistry processes. *Mailing Add:* 1057 Valley Stream Dr Rochester MI 48063

DRAKE, MICHAEL JULIAN, b Bristol, England, July 8, 46; US citizen; m; c 2. GEOCHEMISTRY, PLANETARY SCIENCE. *Educ:* Univ Manchester, BSc, 67; Univ Ore, PhD(geol), 72. *Prof Exp:* Res assoc lunar sci, Smithsonian Astrophys Observ, 72-73; from asst prof to assoc prof planetary sci, Dept Planetary Sci & Lunar & Planetary Lab, Univ Ariz, 73-83, assoc dir, Lunar & Planetary Lab, 78-80, assoc dean sci, 86-87, PROF PLANETARY SCI, DEPT PLANETARY SCI & LUNAR & PLANETARY LAB, UNIV ARIZ, 83-, PROF GEOSCI, 88- *Concurrent Pos:* Mem, numerous comts & working groups, NASA, 79-; vis fel, Univ Cambridge, 81; assoc ed, J Geophys Res, 82; mem, Space Sci Working Group Steering Comt, Am Asn Univs, 85-, US/USSR Implementation Team, Mars Landing Site, 88 & Planetary Sci Coun, Univ Space Res Asn, 90-; chmn, Planetary Sci Adv Bd, Spacecause, 90- *Mem:* Am Geophys Union; fel Meteoritical Soc; Am Astron Soc Div Planetary Sci; Geochem Soc; Europ Union Geosci. *Res:* Petrology and geochemistry of lunar samples and meteorites; model calculations of evolution of planetary bodies; experimental investigations of mineral-melt equilibria. *Mailing Add:* Planet Sci & Lunar & Planet Lab Univ Ariz Tucson AZ 85721

DRAKE, MICHAEL L, b Dayton, Ohio, Feb 9, 49; m 71; c 2. STRUCTURAL DYNAMICS, VIBRATION CONTROL. *Educ:* Univ Cincinnati, BS, 72, MS, 73. *Prof Exp:* RES ENGR, UNIV DAYTON, 73- *Mem:* Am Soc Mech Engrs. *Res:* Experimental and analytical vibration analysis; vibration control techniques; passive damping material development; passive damping design, development, and implantation; reliability and maintainability assessment studies. *Mailing Add:* Univ Dayton 300 College Park JPC-36 Dayton OH 45469-0112

DRAKE, RICHARD LEE, b Columbus, Ohio, Feb 25, 50; m 72; c 2. ANATOMY, MOLECULAR ENDOCRINOLOGY. *Educ:* Mt Union Col, BS, 72; Ind Univ, Bloomington, PhD(anat), 75. *Prof Exp:* Fel cell biol, dept nutrit & food sci, Mass Inst Technol, 75-77; asst prof anat, Med Col Wis, 77-81; asst prof, 81-85, ASSOC PROF ANAT, COL MED, UNIV

CINCINNATI, 85- *Mem:* AAAS; Am Soc Cell Biol; Am Asn Anatomists; Am Diabetes Asn; Sigma Xi. *Res:* Cell and molecular biology; developmental changes which alter the structure and function of the liver cell. *Mailing Add:* Univ Cincinnati Col Med 231 Bethesda Ave ML No 521 Cincinnati OH 45267-0521

DRAKE, ROBERT E, b Los Angeles, Calif, Dec 4, 43; m 67; c 2. GEOCHRONOLOGY, ISOTOPE GEOCHEMISTRY. *Educ:* Pomona Col, BA, 65; Univ Calif, Riverside, MA, 67; Univ Calif, Berkeley, PhD(geol), 74. *Prof Exp:* Fel, 74-76, ASST RES GEOLOGIST, UNIV CALIF, BERKELEY, 76- *Res:* Geochronology in the central Chilean Andes; calibration of vertebrate; hominid fossils in East Africa. *Mailing Add:* Berkeley Geochronology Ctr Inst Human Origins 2453 Ridge Rd Berkeley CA 94709

DRAKE, ROBERT FIRTH, b Providence, RI, July 4, 47. CHEMICAL EDUCATION, TRANSITION METAL CHEMISTRY. *Educ:* Providence Col, BS, 69; Univ NC, Chapel Hill, PhD(inorg chem), 73. *Prof Exp:* Asst prof chem, Wash Col, 73-75, Metrop State Col, Colo, 75-76, Kans State Univ, 76-77, Whitman Col, 77-78, Colo Sch Mines, 78-80, Moorhead State Univ, 80-84; assoc prof chem, St Mary's Col, Minn, 84-86, Northland Col, 86-87, New Col, Fla, 87-88, Bard Col, 88-89; ASSOC PROF CHEM, NAT SCI & MATH DIV, FLA MEM COL, MIAMI, 89- *Concurrent Pos:* Res assoc, Univ NC, Chapel Hill, 74 & Wash State Univ, Pullman, 75; res chemist, Geochem & Environ Chem Res, Inc, Rapid City, SDak, 81. *Mem:* Am Chem Soc; Royal Soc Chem; Sigma Xi. *Res:* Magnetic interactions in condensed systems of transition metal ions; advances in chemical education; geochemistry; high temperature superconductors. *Mailing Add:* Div Nat Sci & Math Fla Mem Col 15800 NW 42nd Ave Miami FL 33054

DRAKE, ROBERT L, b Bradford, Pa, June 24, 26; m 50; c 4. ELECTRICAL ENGINEERING, SYSTEMS DESIGN. *Educ:* Tulane Univ, BSEE, 50, MSEE, 57; Miss State Univ, PhD(elec eng), 65. *Prof Exp:* Plant engr, Buckeye Cellulose Corp Div, Procter & Gamble Co, 50-54; from instr to asst prof elec eng, Tulane Univ, 54-59; mem tech staff, Space Technol Labs, Inc Div, Thompson-Ramo-Wooldridge, Inc, 59-61, proj engr, 61-62; from asst prof to assoc prof elec eng, 62-70, PROF ELEC ENG, TULANE UNIV, 70- *Concurrent Pos:* NSF sci fac fel, 63-64; partner, Systs Technol Inst. *Mem:* Inst Elec & Electronics Engrs; Am Soc Eng Educ. *Res:* Control, information and adaptive control systems; engineering applications of signal processing. *Mailing Add:* Dept Elec Eng Tulane Univ New Orleans LA 70118

DRAKE, ROBERT M, JR, b Eagle Cliff, Ga, Dec 13, 20; m 44; c 2. MECHANICAL ENGINEERING. *Educ:* Univ Ky, BSME, 42; Univ Calif, Berkeley, MSME, 46, PhD(eng), 50. *Prof Exp:* From instr to assoc prof heat transfer, fluid mech & thermodyn, Univ Calif, Berkeley, 47-55; prof mech eng & chmn dept, Princeton Univ, 56-63; prof, Univ Ky, 64-71, dean Col Eng, 66-71, chmn, Dept Mech Eng, 66-67; vpres res & develop, Combustion Eng, Inc, 71-75; spec asst to pres, Univ Ky, 75-77; vpres technol, Studebaker Worthington Inc, NY, 77-78; exec vpres, Univ Investment Co Inc, Ky, 78-79, pres, 79-80; consult engr, 80-90; RETIRED. *Concurrent Pos:* Engine design specialist & supvr turbine air design, Gen Elec Co, 54-56, consult, 56-57; consult, McGraw-Hill Book Co, 58-68, Rand Corp, 60-63, Air Preheater Co, 62, NSF, 62-63, Arthur D Little, Inc, 63 & Gen Tel Elec Serv Corp, 80-81; consult & dir, Intertech Corp, 62-72; sr staff consult, Arthur D Little, Inc, 63-64; dir, Magnetic Corp Am Inc, Waltham, Mass, 74-80; dir & secy-treas, Projectron, Inc, Lexington, Ky, 81-84, dir, chmn bd & chief exec officer, 84-85. *Honors & Awards:* Centennial Medal, Am Soc Mech Engrs. *Mem:* Nat Acad Eng; fel & hon mem Am Soc Mech Engrs. *Res:* Heat transfer; thermodynamics; fluid mechanics. *Mailing Add:* 648 Tally Rd Lexington KY 40502

DRAKE, STEPHEN RALPH, b Detroit, Mich, June 4, 41; m 65; c 2. FOOD SCIENCE, POST-HARVEST PHYSIOLOGY. *Educ:* Miss State Univ, BS, 64, MS, 71, PhD(food sci), 73. *Prof Exp:* Food scientist meat, US Army Natick Res Command, 73-75; food scientist fruit & veg, Wash State Univ, 75-76; res food technologist, Irrigated Agr Res to Exten Ctr, 76-84, POST-HARVEST PHYSIOLOGIST, TREE FRUIT RES LAB, AGR RES SERV, USDA, WENATCHEE, WASH, 84- *Honors & Awards:* Nat Food Processors Award, Am Soc Hort Sci, 79. *Mem:* Inst Food Technologists; Am Soc Hort Sci; Am Diet Asn; Sigma Xi. *Res:* Process calculation and varietal evaluations for fruits and vegetables; effect of orchard and field management practises on the nutritional quality of fruits and vegetables; post-harvest storage of tree fruits and associated physiological problems. *Mailing Add:* Tree Fruit Res Ctr USDA Agr Res Serv Wenatchee WA 98801

DRAKE, STEVENS STEWART, polymer chemistry, for more information see previous edition

DRAKONTIDES, ANNA BARBARA, b New York, NY, Aug 21, 33. ELECTRON MICROSCOPY. *Educ:* Hunter Col, BS, 55, MA, 60; Cornell Univ, MS, 68, PhD(anat & pharmacol), 71. *Prof Exp:* Instr physiol, Hunter Col, 62-66, lectr physiol & biol, Sch Gen Studies, 67-69, asst prof anat & physiol, 69-73; instr pharmacol, Med Col, Cornell Univ, 71-73; asst prof, 73-80, ASSOC PROF ANAT, NY MED COL, 80- *Concurrent Pos:* Instr anat & physiol, Sch Nursing, Cornell Univ, 67-68, assoc prof, 74-76; NIH fel, 71-72. *Mem:* NY Acad Sci; Am Asn Anatomists; Soc Neurosci; fel Pharmacol Mfrs Asn. *Res:* Correlated study of the physiologic, pharmacologic and morphologic aspects of the mammalian neuromuscular junction; chemicals and diseases that cause neuropathology. *Mailing Add:* Dept Cell Biol & Anat NY Med Col Valhalla NY 10595

DRANCE, S M, b Bielsko, Poland, May 22, 25; Can citizen; m 52; c 3. OPHTHALMOLOGY. *Educ:* Univ Edinburgh, MB, ChB, 48; Royal Col Surg, dipl ophthal, 53, fel, 56. *Prof Exp:* Res assoc ophthal, Oxford Univ, 55-57; assoc prof ophthal & dir glaucoma clin, Univ Sask, 57-63; assoc prof, 63-68, PROF OPHTHAL, UNIV BC, 68-, HEAD DEPT, 74-, DIR GLAUCOMA SERV, 63- *Concurrent Pos:* Mem subcomt ophthal, Can Dept Health & Welfare Res Grants, 64 & Med Res Coun Can, 67-; ed, Can J Ophthal. *Mem:* Asn Res Vision & Ophthal; Can Med Asn; Can Ophthal Soc; Brit Med Asn; Ophthal Soc UK. *Res:* Behavior of visual function under conditions of raised intraocular pressure; pharmacology of glaucoma medication; natural history of glaucoma. *Mailing Add:* Dept Ophthal Univ BC Fac Med 2194 Health Sci Mall Vancouver BC V6T 1W5 Can

DRANCHUK, PETER MICHAEL, b Poland, Sept 4, 28; Can citizen. PETROLEUM ENGINEERING. *Educ:* Univ Alta, BSc, 52, MSc, 59. *Prof Exp:* Sessional instr petrol eng, 52-53, lectr, 53-59, from asst prof to assoc prof, 59-77, PROF PETROL ENG, UNIV ALTA, 77- *Mem:* AAAS; Am Inst Mining, Metall & Petrol Engrs; Can Inst Mining & Metall; Sigma Xi; NY Acad Sci. *Res:* Petroleum production and reservoir mechanics. *Mailing Add:* 10751 86th Ave Univ Alta Edmonton AB T6E 2M8 Can

DRANE, CHARLES JOSEPH, JR, b Boston, Mass, Nov 21, 27; m 71; c 2. ANTENNA ARRAY ANALYSIS & SYNTHESIS, ELECTROMAGNETIC FIELD THEORY. *Educ:* Boston Col, BS, 50; Mass Inst Technol, SM, 54, MS, 66; Harvard Univ, PhD(appl physics), 75. *Prof Exp:* Physicist, Nat Bur Standards, 52; phys res asst, US Army, White Sands Proving Ground, 53-55; physicist, Air Force Cambridge Res Labs, 55-76; physicist, Rome Air Develop Ctr, 76-89; LECTR, DEPT ELEC & COMPUTER ENG, NORTHEASTERN UNIV, 78- *Concurrent Pos:* Lectr elec eng, Northeastern Univ, 78- *Honors & Awards:* Lord Brabazon Award, Inst Electron & Radio Engr, 71. *Mem:* sr mem Inst Elec & Electronics Engrs; assoc mem Int Sci Radio Union; Sigma Xi. *Res:* Analysis and synthesis of antenna apertures and arrays; high-resolution designs; broadband transmitting/receiving arrays; radiation pattern constraints; applied mathematics, especially optimization methods; mathematical modeling. *Mailing Add:* Five Greenwood Rd Reading MA 01867-3769

DRANE, JOHN WANZER, b Forest, La, June 26, 33. STATISTICS, BIOMETRY. *Educ:* Northwestern State Col, La, BS, 55; Univ Fla, MS, 57; Emory Univ, PhD(biomet), 67. *Prof Exp:* Nuclear engr, Newport News Shipbldg & Dry Dock Co Va, 56-61; assoc prof math, Randolph-Macon Col, 60-64; instr & spec fel statist & biomet, 63-65, assoc prof, Emory Univ, 65-68; assoc prof, Southern Methodist Univ, 68-79; prof statist, Univ SC, 79-85; vis prof biostatist, res data analysis, 85-88; PROF BIOSTATIST, AUBURN UNIV, ALA, 88- *Concurrent Pos:* Asst prof, chmn div biomath & biostatist, Univ Tex Southwestern Med Sch, 70-73; asst prof med comput sci, Univ Tex Health Sci Ctr, 73-76, adj assoc prof, 76-79; sr statist consult, Criterion Anal, Inc, 78-80; pres, Wanzer, Inc, 81-88. *Honors & Awards:* Spec fel, NIH, 63-65. *Mem:* Am Arbit Asn; Am Statist Asn; Sigma Xi; Biometric Soc; Math Asn Am; Mid South Sociol Asn; Decision Sci Inst. *Res:* Application of statistics and mathematics to problems of society, law, biology and the health sciences; modeling, nonlinear regression, contingency tables, EEOC compliance, employment practice analysis. *Mailing Add:* Epidemiol & Biostatist Sch Pub Health Univ SC Columbia SC 29208

DRANOFF, JOSHUA S(IMON), b Bridgeport, Conn, June 30, 32; m 53; c 3. CHEMICAL ENGINEERING. *Educ:* Yale Univ, BE, 54; Princeton Univ, MSE, 56, PhD(chem eng), 60. *Prof Exp:* Asst prof chem eng, Yale Univ, 57-58 & Northwestern Univ, 58-62; assoc prof, Columbia Univ, 62-63; dir, Basic Indust Res Inst, 87-88; assoc prof, 63-67, chmn dept, 71-76 & 79-82, PROF CHEM ENG, NORTHWESTERN UNIV, 67- *Mem:* Am Chem Soc; Am Inst Chem Engrs; Am Soc Eng Educ. *Res:* Chemical reactor analysis; heterogeneous catalysis; photoreaction engineering; chromatographic separations; ion exchange; adsorption on molecular sieves; applied catalysis, with emphasis on catalyst deactivation and behavior of supported liquid phase catalysts; adsorption of multicomponent gases on zeolites; analysis and behavior of photocatalytic reactors. *Mailing Add:* Dept Chem Eng Northwestern Univ Evanston IL 60208-3120

DRAPALIK, DONALD JOSEPH, b Chicago, Ill, Dec 10, 34. PLANT TAXONOMY. *Educ:* Southern Ill Univ, Carbondale, BA, 59, MA, 62; Univ NC, Chapel Hill, PhD(bot), 70. *Prof Exp:* Asst prof, 68-73, ASSOC PROF BIOL, GA SOUTHERN COL, 73-, MEM GRAD FAC, 71- *Concurrent Pos:* Fac res grant, Ga Southern Col, 70-71 & 79-80. *Mem:* Int Asn Plant Taxon; Am Soc Plant Taxon; Bot Soc Am. *Res:* Taxonomy, morphology and evolution of the North American milkweeds; floristics of vascular plants in Georgia; reproductive biology of Elliottia. *Mailing Add:* Dept Biol Ga Southern Univ Statesboro GA 30460-8042

DRAPER, ARTHUR LINCOLN, b Philadelphia, Pa, Feb 20, 23; m 46; c 3. PHYSICAL CHEMISTRY. *Educ:* Rice Univ, BA, 48, MA, 49, PhD(chem), 51. *Prof Exp:* Res chemist, Jersey Prod Res Co, 51-57; res assoc dept chem, Rice Univ, 58-59; from asst prof to assoc prof chem, Tex Tech Univ, 61-85, assoc dean, Col Arts & Sci, 76-84; ADJ PROF CHEM, WESTERN KY UNIV, 85- *Mem:* Fel AAAS; Am Chem Soc. *Res:* Physical and colloid chemistry; thermodynamics; kinetics; solid state; structure; surface chemistry; oxides and mixed oxides; adsorption; catalysis. *Mailing Add:* Dept Chem Western Ky Univ Bowling Green KY 42101

DRAPER, CARROLL ISAAC, b Maroni, Utah, Sept 27, 14; m 39; c 6. POULTRY NUTRITION. *Educ:* Utah State Univ, BS, 39; Iowa State Col, PhD, 42. *Prof Exp:* Asst prof poultry husb, State Col Wash, 41-43; head poultry dept, Univ Hawaii, 43-45; assoc prof poultry husb, 45-48, head poultry dept, 48-70, PROF ANIMAL SCI, UTAH STATE UNIV, 70-, PROF DAIRY & VET SCI & POULTRY SPECIALIST, EXTEN, 77- *Mem:* Poultry Sci Asn. *Res:* Value of protein feeds in poultry diets. *Mailing Add:* 1396 E 1700 N Logan UT 84321

DRAPER, GRENVILLE, b Leeds, UK, Sept 24, 50. CARIBBEAN GEOLOGY & TECTONICS. *Educ:* Univ Cambridge, UK, BA, 73; Imp Col, London, UK, MSc, 74; Univ WI, Jamaica, PhD(geol), 79. *Prof Exp:* Asst lectr struct geol, Univ WI, 74-75, res asst, 75-78; from asst prof to assoc prof, 78-89, PROF GEOL, FLA INT UNIV, 89- *Mem:* Geol Soc Am; Am Geophys Union; Geol Soc London; Sigma Xi; Nat Asn Geol Teachers. *Res:* Structural geology and tectonics of the Northern Caribbean and Mexico; tectonic evolution of blue schist belts; island arc structure and tectonics. *Mailing Add:* Dept Geol Fla Int Univ Tamiami Trail Miami FL 33199

DRAPER, HAROLD HUGH, b Manitoba, Can, Apr 11, 24; nat US; m 47; c 2. NUTRITION. *Educ:* Univ Manitoba, BSA, 45; Univ Alta, MSc, 48; Univ Ill, PhD(nutrit sci), 52. *Prof Exp:* Lectr, Univ Man, 45-46, Univ Alta, 48-49; nutritionist, Merck Sharpe & Dohme, Inc, 52-54; from asst prof to prof nutrit biochem, Univ Ill, Urbana, 54-75; PROF NUTRIT, UNIV GUELPH, 75- *Concurrent Pos:* Orgn European Econ Coop fel, Univ Liverpool, 61; NIH Special Fel, Agr Res Coun, Cambridge, UK, 68. *Honors & Awards:* McHenry Award, Can Soc Nutrit Sci, 85. *Mem:* Fel AAAS; Am Inst Nutrit; Can Soc Nutrit Sci. *Res:* Nutrition and osteoporosis; Vitamin E and lipid peroxides. *Mailing Add:* Dept Nutrit Sci Univ Guelph Guelph ON N1G 2W1 Can

DRAPER, JAMES EDWARD, b Kansas City, Mo, Sept 14, 24; m 48; c 4. NUCLEAR PHYSICS. *Educ:* Williams Col, BA, 44; Cornell Univ, PhD(physics), 52. *Prof Exp:* Instr & asst physics, Williams Col, 46-47; asst, Cornell Univ, 47-52, res assoc, 52; assoc physicist, Brookhaven Nat Lab, 52-56; from asst prof to assoc prof, Yale Univ, 56-62, sr res assoc, 62-63; chmn dept, 66-71, PROF PHYSICS, UNIV CALIF, DAVIS, 64- *Concurrent Pos:* Res assoc, Nuclear Physics Div, Atomic Energy Res Estab, Harwell, Eng, 71 & 79; consult nuclear physics, Lawrence Berkeley Lab, 80- *Mem:* Fel Am Phys Soc. *Res:* In-beam gamma and electron spectroscopy on high-spin states of nuclei; experimental neutron physics; neutron spectroscopy; photonuclear and particle reactions. *Mailing Add:* Dept Physics Univ Calif Davis CA 95616

DRAPER, JOHN DANIEL, b Hagerstown, Md, June 13, 19; m 46; c 4. ORGANIC CHEMISTRY. *Educ:* Franklin & Marshall Col, BS, 41; Univ Md, PhD(org chem), 48. *Hon Degrees:* ScD, Bethany Col, 90. *Prof Exp:* Asst gen chem, Univ Md, 41-43, res chemist, Comt Med Res C/ntract, 43-45, asst inorg quant anal, Univ, 45-46; sr res chemist, Phillips Petrol Co, 47-49; res chemist, J T Baker Chem Co, 49-51; from asst prof to prof, 51-84, head dept, 54-84, EMER PROF CHEM, BETHANY COL, 84- *Concurrent Pos:* Consult, Stoner-Mudge Corp, Pa, 54-63. *Mem:* Am Chem Soc; Sigma Xi. *Res:* Synthesis and analysis of antimalarials; synthetic lubricating oil additives; kinetics of organic reactions. *Mailing Add:* Dept Chem Bethany Col Bethany WV 26032

DRAPER, LAURENCE RENE, b New York, NY, Apr 14, 30; m 54; c 2. IMMUNOLOGY. *Educ:* Middlebury Col, AB, 52; Univ Chicago, PhD(microbiol), 56. *Prof Exp:* Res assoc microbiol, Univ Chicago, 56-57, asst prof, 59-60; res assoc immunol, Argonne Nat Lab, 57-59; res biologist, Physiol Lab, Nat Cancer Inst, 60-68; assoc prof, 68-73, PROF MICROBIOL, UNIV KANS, 73- *Concurrent Pos:* Logan fel microbiol, Univ Chicago, 60. *Mem:* Am Asn Immunol; Radiation Res Soc; Am Soc Microbiol. *Res:* Cellular immunology. *Mailing Add:* Dept Microbiol Univ Kans Haworth Hall 7042 Lawrence KS 66045

DRAPER, NORMAN RICHARD, b Southampton, Eng, Mar 20, 31. STATISTICS. *Educ:* Cambridge Univ, BA, 54, MA, 58; Univ NC, PhD(math statist), 58. *Prof Exp:* Statistician, Plastics Div, Imp Chem Indust, Eng, 58-60, mem, Math Res Ctr, 60-61; from asst prof to assoc prof, 61-66, chmn dept, 67-73, vis prof, Math Res Ctr, 73-74, PROF STATIST, UNIV WISMADISON, 66- *Concurrent Pos:* Res asst statist tech res group, Princeton Univ, 57; sabbatical visit, Univ Melbourne, Australian Nat Univ, Canberra & Univ NSW, Sydney, Australia, 85, Augsburg Univ, WGer, 88, 89 & Southampton Univ, UK, 90. *Mem:* Fel Inst Math Statist; fel Am Statist Asn; Biomet Soc; fel Royal Statist Soc; fel Am Soc Qual Control; Int Statist Inst. *Res:* Experimental statistics; design and analysis of experiments; statistical theory; regression analysis; nonlinear estimation; response surface methodology. *Mailing Add:* Dept Statist Univ Wis 1210 W Dayton St Madison WI 53706

DRAPER, RICHARD NOEL, b Camden, Twp, Ohio, Feb 18, 37; m 74; c 4. COMPUTER SCIENCES. *Educ:* Baldwin-Wallace Col, BS, 59; Johns Hopkins Univ, PhD(math), 66. *Prof Exp:* Instr math, Johns Hopkins Univ, 63-66; asst prof math, Notre Dame Univ, 66-72; from assoc prof to prof math, George Mason Univ, 72-86, chmn, 80-86; res staff mem algorithms, 86-89, FEL, SUPERCOMPUT RES CTR, 89- *Concurrent Pos:* Vis asst prof, Purdue Univ, 71-72; prog dir algorithms & number theory, NSF, 84-86; co ed, J Supercomput. *Mem:* Am Math Soc; Soc Indust & Appl Math; Asn Comput Mach. *Res:* Research into parallel algorithms and parallel architectures. *Mailing Add:* Supercomput Res Ctr 17100 Science Dr Bowie MD 20715-4300

DRAPER, RICHARD WILLIAM, b London, Eng, Dec 12, 42; m 72; c 2. STEROID CHEMISTRY. *Educ:* Univ London, BSc, 65, ARIC, 67, PhD(org chem), 68. *Prof Exp:* Fel, Dept Chem, Univ Rochester, 68-71; prin scientist, Natural Prod Dept, 71- 80, SR PRIN SCIENTIST, PROCESS DEVELOP DEPT, SCHERING-PLOUGH CORP, 81- *Mem:* Royal Soc Chem; Am Chem Soc. *Res:* Synthesis and reaction mechanisms. *Mailing Add:* Schering-Plough Corp 60 Orange St Bloomfield NJ 07003

DRAPER, ROCKFORD KEITH, PROTEIN TOXINS, ENDOCYTOSIS. *Educ:* Univ Calif, Los Angeles, PhD(biochem), 74. *Prof Exp:* ASSOC PROF CELL BIOL, UNIV TEX, DALLAS, 80- *Mailing Add:* Biol Prog FO3-1 Univ Tex Dallas Box 830-688 Richardson TX 75083-0688

DRAPER, ROY DOUGLAS, B Fresno, Calif, May 30, 33; m 60; c 2. BIOCHEMISTRY. *Educ:* Sacramento State Col, BS, 55, BA, 59; Univ Calif, Davis, PhD(biochem), 64. *Prof Exp:* Clin technician, Suttar Hosp, Sacramento, 55-56 & Hotel Dieu Hosp, El Paso, Tex, 57; PROF CHEM, CALIF STATE UNIV, SACRAMENTO, 64- *Concurrent Pos:* Danforth Assoc, 72-76. *Res:* Enzyme mechanism and general biochemistry. *Mailing Add:* Dept Chem Calif State Univ 6000 J St Sacramento CA 95819

DRASIN, DAVID, b Philadelphia, Pa, Nov 3, 40; m 63; c 3. MATHEMATICAL ANALYSIS. *Educ:* Temple Univ, AB, 62; Cornell Univ, PhD(math), 66. *Prof Exp:* Instr math, Rutgers Univ, 62; from asst prof to assoc prof, 66-74, PROF MATH, PURDUE UNIV, 74- *Mem:* Math Asn Am; Am Math Soc. *Res:* Functions of a complex variable; Tauberian theorems; potential theory. *Mailing Add:* Dept Math Purdue Univ West Lafayette IN 47907

DRATZ, EDWARD ALEXANDER, b Minneapolis, Minn, July 2, 40. BIOCHEMISTRY, VISION. *Educ:* Carleton Col, BA, 61; Univ Calif, Berkeley, PhD(chem), 66. *Prof Exp:* Helen Hay Whitney fel, Lab Chem Biodynamics, Univ Calif, Berkeley, 66-67 & Dept of Biol, Mass Inst Technol, 67-69; asst prof, 69-77, assoc prof chem, 77-80, ASSOC PROF BIOCHEM, UNIV CALIF, SANTA CRUZ, 77- *Mem:* Am Soc Photobiol; Asn Res Vision & Ophthal; Biophys Soc; AAAS. *Res:* Structures and mechanisms of action of biological membranes with emphasis on visual photoreceptors; preservation of membranes in vitro and the mechanisms which prevent degeneration of membranes in vivo. *Mailing Add:* Dept Chem Mont State Univ 108 Baines Hall Bozeman MT 59717

DRAUGHON, FRANCES ANN, b Tazewell, Va, April 30, 52; wid; c 2. MYCOTOXINS, FOOD BORNE DISEASES. *Educ:* Univ Tenn, BS, 73, MS, 76; Univ Ga, PhD(food sci), 79. *Prof Exp:* Grad res asst food sci, Univ Ga, 76-79; asst prof, 79-83, ASSOC PROF FOOD TECHNOL SCI, UNIV TENN, 83- *Concurrent Pos:* Fac develop res award, China, 85; Roddy res prof. *Mem:* Am Soc Microbiol; Inst Food Technologists; Int Asn Milk, Food & Environ Sanitarians; AAAS; Sigma Xi. *Res:* Monoclonal antibodies; inhibition of foodborne pathogens; diagnosis and improve methodology of foodborne pathogens. *Mailing Add:* Dept Food Technol Sci Univ Tenn Knoxville TN 37916

DRAUGLIS, EDMUND, b Philadelphia, Pa, May 21, 33; m 64; c 3. SURFACE CHEMISTRY, PHYSICS. *Educ:* Univ Pa, BS, 55; Yale Univ, MS, 59, PhD(phys chem), 61. *Prof Exp:* Phys chemist, Semiconductor Prod Dept, Gen Elec Co, 61; res scientist, United Aircraft Res Lab, 62-63; prin chemist, Battelle Mem Inst, 64-86; SR CHEMIST, MEARL CORP, 86- *Mem:* Sigma Xi. *Res:* Surface characterization by electron spectroscopy for chemical analysis; boundary lubrication; solid-liquid interface; liquid films; ellipsometry; study of adsorption by electron spectroscopy for chemical analysis; plasma polymerization; inorganic pigments; hydrothermal processing. *Mailing Add:* 217 N Highland Ave Ossining NY 10562

DRAUS, FRANK JOHN, b Dupont, Pa, Oct 30, 29; m 56; c 4. BIOCHEMISTRY. *Educ:* Alliance Col, BS, 51; Duquesne Univ, MS, 53, PhD(biochem), 57. *Prof Exp:* From asst to assoc prof, 56-65, head dept, 67-83, dir admis, 81-83, PROF BIOCHEM, SCH DENT MED, UNIV PITTSBURGH, 65-, ASSOC DEAN, 83- *Concurrent Pos:* Temporary adv, WHO, 69; indust consult. *Mem:* AAAS; assoc Am Dent Asn; fel Am Inst Chem; Int Asn Dent Res; NY Acad Sci. *Res:* Mechanism for the formation of synthetic calculus; isolation and characterization of mucoproteins and mycopolysaccharides from salivary glands. *Mailing Add:* 1024 Dale Dr Pittsburgh PA 15220

DRAWE, D LYNN, b Mercedes, Tex, Nov 3, 42; m 64; c 2. RANGE MANAGEMENT, WILDLIFE RESEARCH. *Educ:* Tex A&I Univ, BS, 64; Tex Tech Univ, MS, 67; Utah State Univ, PhD(range ecol), 70. *Prof Exp:* From asst prof to assoc prof range & wildlife mgt, Tex A&I Univ, 70-74; ASST DIR RANGE & LIVESTOCK MGT & RES, WELDER WILDLIFE FOUND, 74- *Concurrent Pos:* Consult wildlife, Winrock Int, Kenya, 82. *Mem:* Soc Range Mgt; Wildlife Soc; Animal Sci Soc. *Res:* Range livestock management, grazing systems; range animal nutrition including white-tailed deer; brush control, range plant physiology; range plant ecology; fire ecology. *Mailing Add:* Welder Wildlife Refuge PO Drawer 1400 Sinton TX 78387

DRAY, SHELDON, b Chicago, Ill, Nov 20, 20; m 53, 68; c 2. MEDICINE, IMMUNOLOGY. *Educ:* Univ Chicago, BS, 41; Univ Ill, MD, 46, MS, 47; Univ Minn, PhD(phys biochem), 54. *Prof Exp:* Intern, Univ Ill Hosps, Chicago, 46-47; med officer, Nutrit Sect, Bur State Serv, USPHS, 47-49; med officer, Phys Biol Lab, Nat Inst Arthritis & Metab Dis, 49-54, med officer, Lab Clin Invest & Immunol, Nat Inst Allergy & Infectious Dis, 55-65; head dept, 65-80, PROF MICROBIOL & IMMUNOL, UNIV ILL, CHICAGO, 80- *Concurrent Pos:* Coun mem, Int Union Immunol Socs; consult, WHO Webb-Waring Lung Inst, Denver, 72 E Tenn State Univ Med Study Comt, 73; vis prof, Weizmann Inst, Israel, 71, Osaka Univ, Japan, 78; vis lectureship, Pasteur Inst, France, 73; bd of trustees, Ill Cancer Coun, 78-; study sect, Leukemia Res Found, 78-82. *Honors & Awards:* Boris Pregel Award, NY Acad Sci. *Mem:* AAAS; Am Chem Soc; Am Soc Microbiol; Am Asn Immunol (secy-treas, 64-70); Soc Exp Biol & Med; Am Asn Cancer Res; Am Soc Biol Chemists; Reticuloendothelial Soc; Soc Biol Ther. *Res:* Physical chemistry of membranes; serum proteins; immunogenetics; maternal-fetal incompatibility; informational RNA; immunoglobulin allotypes; allotype suppression; lymphocyte biology; tumor chemotherapy and immunotherapy; cellular immunity; proteinase inhibitors. *Mailing Add:* 155 N Harbor Dr Apt 3508 Chicago IL 60601

DRAYER, DENNIS EDWARD, DRUG BIOTRANSFORMATION, PHARMACOKINETICS. *Educ:* Univ Del, PhD(org chem), 71. *Prof Exp:* ASSOC RES PROF PHARMACOL, MED COL, CORNELL UNIV, 81- *Mailing Add:* Dept Pharmacol Med Col Cornell Univ New York NY 10021

DRAYER, DENNIS EUGENE, b Frankfort, SDak, June 24, 28; m 50; c 2. CHEMICAL ENGINEERING. *Educ:* SDak Sch Mines & Technol, BS, 52; Kans State Univ, MS, 54; Univ Colo, Boulder, PhD(chem eng), 61. *Prof Exp:* Instr chem eng, Kans State Univ, 53-54; chem engr, Dow Chem Co, 54-58; chem engr, Cryogenic Eng Lab, Nat Bur Stand, 59-61; from res engr to advan res engr, 61-68, SR RES ENGR, MARATHON OIL CO, 68- *Concurrent Pos:* Adj assoc prof, Univ Denver, 66-72. *Mem:* Am Chem Soc; Am Inst Chem Engrs. *Res:* Heat transfer in cryogenic systems; liquid phase hydrocarbon oxidations; economic evaluation techniques; project evaluation; flow of fluids and slurries. *Mailing Add:* Marathon Oil Co Denver Res Ctr PO Box 269 Littleton CO 80160

DRAYSON, SYDNEY ROLAND, b Buckhurst Hill, Eng, Dec 21, 37; m 63; c 2. ATMOSPHERIC PHYSICS. *Educ:* Univ London, BSc, 60; Univ Chicago, MS, 61; Univ Mich, PhD(meteorol), 67. *Prof Exp:* Asst res mathematician, 63-66, res assoc, 66-67, lectr meteorol & assoc res engr, 68-

69, asst prof meteorol, 70-75, assoc prof, 75-80, PROF ATMOSPHERIC SCI, UNIV MICH, ANN ARBOR, 80-, INTERIM CHAIR, DEPT ATMOSPHERIC OCEANIC & SPACE SCI, 90- *Concurrent Pos:* Ed, J Atmospheric Sci. *Mem:* AAAS; Am Meteorol Soc; Am Geophys Soc. *Res:* Atmospheric radiative transfer, including radiative heating rates, remote sounding of the atmosphere and molecular spectroscopy; numerical methods and computer applications. *Mailing Add:* Dept Atmospheric Oceanic & Space Sci Univ Mich Ann Arbor MI 48109-2143

DRAZIN, MICHAEL PETER, b London, Eng, June 5, 29. MATHEMATICS. *Educ:* Cambridge Univ, BA, 50, MA, 53, PhD(math), 53. *Prof Exp:* Fel Trinity Col, Cambridge Univ, 52-53; sci off, Admiralty Res Lab, Teddington, Eng, 53-55; fel Trinity Col, Cambridge Univ, 55-57; vis lectr, Northwestern Univ, 57-58; sr scientist, Res Inst Advan Study, 58-62; ASSOC PROF MATH, PURDUE UNIV, 62- *Honors & Awards:* Smith's Prize of Univ of Cambridge, 52. *Mem:* Am Math Soc; Soc Indust & Appl Math. *Res:* Non-commutative ring theory; abstract algebra; combinatorial problems; matrix theory; applied mathematics. *Mailing Add:* Univ Math Purdue Univ West Lafayette IN 47907

DREBUS, RICHARD WILLIAM, psychology, organization analysis, for more information see previous edition

DREBY, EDWIN CHRISTIAN, III, b Haddonfield, NJ, Sept 2, 15; m 40; c 5. PHYSICAL CHEMISTRY, ORGANIC CHEMISTRY. *Educ:* Yale Univ, PhD(phys chem), 39. *Prof Exp:* Res asst, Am Soc Testing & Mat, 39-41; res chemist, 41-54, DIR LAB, SCHOLLER BROS, INC, 54- *Mem:* Am Chem Soc; Am Asn Textile Chem & Colorists; Am Soc Test & Mat; Sigma Xi. *Res:* Development of chemical products to assist in the scouring and dyeing of textile yarns, fabrics and garments and to alter their hand for ease of manufacture and consumer requirements. *Mailing Add:* 659 Medford Lane Medford NJ 08055

DRECHSEL, PAUL DAVID, b Newark, NJ, Oct 24, 25; m 50; c 4. POLYMER CHEMISTRY, TEXTILE PHYSICS. *Educ:* Rutgers Univ, BS, 46; Cornell Univ, PhD(phys chem), 51. *Prof Exp:* Sr res chemist, Allegany Ballistics Lab, Hercules Inc, 51-57, res supvr, 57-58, group Supvr, 58-61, asst dept supt, 61-62, dept supt, 62-70, res scientist, Res Ctr, 70-71, res assoc, 71-73, mgr res dept, 73-84, sr res assoc, Fibers Res & Develop Lab, 84-86; RETIRED. *Mem:* Am Chem Soc; Am Sci Affil. *Res:* Physical chemistry and mechanical properties of polymers; diffusion; textile chemistry; heat transfer; fibers technology; kinematics of the melt spinning of fibers; draw resonance phenomena; characterization of the molicular morphology of fibers from polycrystalline polymers; relation of mechanical and other physical properties to morphology; molecular weight distribution of polymeric fibers. *Mailing Add:* 22 Chipping Green Dr Arden NC 28704

DRECKTRAH, HAROLD GENE, b La Crosse, Wis, Nov 1, 38; m 62; c 2. INSECT MORPHOLOGY, AQUATIC ENTOMOLOGY. *Educ:* Wis State Univ, La Crosse, BS, 62; Iowa State Univ, MS, 64, PhD(entom), 66. *Prof Exp:* Res asst, Iowa Dept Agr, 62-66; from asst prof to assoc prof, 66-79, PROF BIOL, UNIV WIS-OSHKOSH, 79- *Mem:* Entom Soc Am; Am Entom Soc; NAm Benthological Soc; Sigma Xi. *Res:* Insect morphology, especially the internal reproductive organs of both sexes; aquatic entomology, especially Trichoptera taxonomy. *Mailing Add:* Dept Biol Univ Wis 800 Algoma Blvd Oshkosh WI 54901

DREEBEN, ARTHUR B, b New York, NY, Feb 15, 22; m 50; c 2. INORGANIC CHEMISTRY. *Educ:* Polytech Inst Brooklyn, BS, 48, MS, 50. *Prof Exp:* Asst chem, Res Lab, Gen Elec Co, 50-52; asst, Knolls Atomic Power Lab, 52-53; assoc res engr, Res Dept Lamp Div, Westinghouse Elec Corp, 53-58; mem tech staff, David Sarnoff Res Ctr, RCA Corp, 58-85; RETIRED. *Mem:* Am Chem Soc; Sigma Xi. *Res:* Solid state chemistry; luminescence; photoconduction; crystal growth and imperfections. *Mailing Add:* 75 Dodds Lane Princeton NJ 08540

DREES, DAVID T, b Dyersville, Iowa, Nov 23, 33; m 56; c 4. VETERINARY PATHOLOGY, TOXICOLOGY. *Educ:* Iowa State Univ, DVM, 57; Mich State Univ, MS, 66, PhD(path), 69; Am Bd Toxicol, dipl, 81. *Prof Exp:* Instr anat, Mich State Univ, 64-68, NIH fel path, 68-69; head animal health & path, Warren-Teed Pharmaceut, Inc, 69-76; proj leader & pathologist, Rohm & Haas Co, 76-77; dir corp biol affairs, Marion Labs, Inc, 77-89, DIR BIOL AFFAIRS, MARION MERRELL DOW INC, 89- *Mem:* Am Vet Med Asn; Sigma Xi; Soc Toxicol; Am Bd Toxicol. *Res:* Animal models; pre-clinical efficacy and safety evaluation of drugs; safety evaluation of chemicals (toxicology and pathology). *Mailing Add:* Marion Merrell Dow Inc PO Box 9627 Park B Kansas City MO 64134

DREES, JOHN ALLEN, b Chicago, Ill, Feb 6, 43; div; c 2. PHYSIOLOGY, BIOPHYSICS. *Educ:* DePauw Univ, BA, 65; Ind Univ, PhD(physiol), 72. *Prof Exp:* From asst prof to assoc prof physiol & biophys, Sch Dent, 70-86, ASSOC PROF PHYSIOL, SCH MED, TEMPLE UNIV, 86- *Mem:* AAAS; Am Physiol Soc; Am Heart Asn. *Res:* Cardiovascular physiology; control of vascular capacity; blood volume; system response to hemorrhage; molecular mechanisms of cardiac energetics in hypertrophy and heart failure. *Mailing Add:* 404 Runnymede Ave Jenkintown PA 19046

DREESMAN, GORDON RONALD, b Grundy Center, Iowa, Nov 28, 35; m; c 3. IMMUNOLOGY, VIROLOGY. *Educ:* Cent Col, BA, 57; Kans Univ, MA, 63; Univ Hawaii, PhD(microbiol), 65. *Prof Exp:* Asst prof molecular virol, Sch Med, St Louis Univ, 67-69; from asst prof to assoc prof, 69-79, PROF VIROL, BAYLOR COL MED, 79- *Concurrent Pos:* Fel, Baylor Col Med, 65-66; consult immunochemist, St John's Mercy Hosp, St Louis, Mo, 69-70. *Mem:* AAAS; Am Soc Microbiol; Am Asn Immunol. *Res:* Viral immunology; protein chemistry of viruses; humoral and cellular immune response of the host with AIDS and hepatitis. *Mailing Add:* Biotech Res Inc 12500 Network Suite 112 San Antonio TX 78249

DREESZEN, VINCENT HAROLD, b Palmyra, Nebr, July 23, 21; m 44; c 3. GEOLOGY, HYDROLOGY. *Educ:* Nebr State Teachers Col, Peru, AB, 42; Univ Nebr, MSc, 50. *Prof Exp:* Geologist, 47-59, asst dir div, 59-67, actg dir, 67-69, DIR CONSERV & SERV DIV, UNIV NEBR, LINCOLN, 69-, PROF GEOL, 70-, MEM GRAD FAC, 76- *Mem:* Am Water Works Asn; Asn Am State Geol; Geol Soc Am; Sigma Xi. *Res:* Optimum use of water resources; conservation of other natural resources; study and correlation of Pleistocene sediments and their land forms. *Mailing Add:* 3135 Prairie Hill Circle Lincoln NE 68506

DREGNE, HAROLD ERNEST, b Ladysmith, Wis, Sept 25, 16; m 43; c 4. SOIL SCIENCE. *Educ:* Wis State Univ, BS, 38; Univ Wis, MS, 40; Ore State Univ, PhD(soil chem), 42. *Prof Exp:* Asst, Ore State Univ, 40-42; jr soil scientist, Soil Conserv Serv, USDA, 42-43 & 46; asst prof agron, Univ Idaho, 46-47; asst soil scientist, Wash State Univ, 47-49; prof agron & agronomist, Exp Sta, NMex State Univ, 49-50, prof soils, 50-69; prof agron & chmn dept, 69-72, chmn dept plant & soil sci, 72-78, dir int ctr arid & semi arid land studies, 76-81, HORN DISTINGUISHED PROF SOIL SCI, TEX TECH UNIV, 72- *Concurrent Pos:* Head, Inter Col Exchange Prog, Pakistan, 55-57; mem US soil salinity deleg, Soviet Union, 60; soil fertil expert, Food & Agr Orgn, UN, Chile, 61; consult, UNESCO, Tunisia, 67, UN Environ Prog, 75-85 & Food & Agr Orgn, 79-82; mem adv comt technol innovation, Nat Acad Sci, 77-82; sr land resource adv, US Agency Int Develop, 84. *Mem:* Fel AAAS; fel Am Soc Agron; fel Am Soil Sci Soc. *Res:* Chemistry and fertility of arid region soils; saline and sodium soils; desertification of arid lands. *Mailing Add:* Dept Plant & Soil Sci Tex Tech Univ Lubbock TX 79409

DREIBELBIS, JOHN ADAM, b Emlenton, Pa, Aug 8, 28; m 52; c 3. HETEROGENOUS CATALYSIS. *Educ:* Univ Pittsburgh, BS, 50. *Prof Exp:* Quality control chemist, Pittsburgh Coke & Chem, 51-54, res chemist, 54-62; SR RES CHEMIST, CALSICAT DIV, MALLINCKRODT, 62- *Honors & Awards:* President's Award, Mallinckrodt Inc, 81. *Mem:* Am Chem Soc. *Res:* Developing new heterogenous catalysts or new processes for preparation of currently used catalysts; manipulation of the properties of aluminas and silicas; methods of incorporating active catalytic materials. *Mailing Add:* Calsicat Div 1707 Gaskell Ave Erie PA 16503-2497

DREICER, HARRY, b Bad Lausick, Ger, Oct 6, 27; US citizen; m 50; c 3. PLASMA PHYSICS. *Educ:* Mass Inst Technol, BS, 51, PhD(physics), 55. *Prof Exp:* Mem staff physics, Los Alamos Sci Lab, Univ Calif, 54-66, group leader, 66-75, asst div leader, 75-76; prof physics, astrophysics & astrogeophysics, Univ Colo, Boulder, 77; leader, Controlled Thermonuclear Res Div, 77-89, SR FEL, LOS ALAMOS NAT LAB, UNIV CALIF, 89-, ACTG ASSOC DIR AT LARGE, 91- *Concurrent Pos:* Consult, Boeing Sci Res Labs, 59- & Martin Co, 61; part-time prof, Univ NMex, 60; Europ Atomic Energy Community fel, Lab Gas Ionizzati, Frascati, Italy, 64-65 & Saclay Nuclear Res Ctr, France, 65. *Mem:* Fel Am Phys Soc; fel AAAS. *Res:* Effect of coulomb interactions and radiation field on plasma distribution functions; runaway electrons in plasmas; measurement of plasma microwave emission and transmission; anomalous high frequency electrical resistivity of plasmas. *Mailing Add:* Los Alamos Nat Lab ADAL MSA104 PO Box 1663 Los Alamos NM 87545

DREIER, WILLIAM MATTHEWS, JR, b Omaha, Nebr, Mar 20, 37; m 61; c 1. CHEMICAL ENGINEERING. *Educ:* Univ Wis, BS, 59, MS, 60; Univ Minn, PhD(chem eng), 64. *Prof Exp:* Aerospace engr, Lewis Res Ctr, NASA, 63-64; sr res chem engr, 64-68, SECT LEADER, JAMES FORD BELL RES CTR, GEN MILLS, INC, 68- *Mem:* Am Inst Chem Engrs. *Res:* Thermodynamics of cryogenic fuels; microwave heating; dehydration. *Mailing Add:* 2255 Valders Ave Minneapolis MN 55427

DREIFKE, GERALD E(DMOND), b St Louis, Mo, June 21, 18; m 51; c 4. ELECTRICAL ENGINEERING, APPLIED MECHANICS. *Educ:* Washington Univ, BS & MS, 48, DSc, 61. *Prof Exp:* Layout & drafting, Curtiss Wright Corp, 36-39, design engr, 39-44; layout engr, Douglas Aircraft, 39; from instr to prof elec eng, St Louis Univ, 48-70, secy dept, 51-54, dir dept, 54-71; mgr, res & develop, Union Elec Co, 71-77; CONSULT, 77- *Concurrent Pos:* Consult, Emerson Elec Co, 51 & Monsanto Co, 61-; mem, tech staff, Bell Tel Labs, 63; ed-in-chief, Instrument Soc Am, 67-; vis prof physics & consult engr, Univ Mo, St Louis, 79- *Mem:* Inst Elec & Electronics Engrs; Am Soc Eng Educ; Instrument Soc Am; Nat Soc Prof Engrs. *Res:* Automatic controls; electromechanical systems; system dynamics. *Mailing Add:* Six Westmoreland Pl St Louis MO 63108

DREIFUS, LEONARD S, b Philadelphia, Pa, May 27, 24. CARDIAC PACEMAKERS, CARDIAC ARRYHTHMIAS. *Educ:* Hahnemann Med Col, MD, 51. *Prof Exp:* DIR HEART STA, HAHNEMANN UNIV, PHILADELPHIA, PA, 72-, PROF MED, 89-; PROF MED, JEFFERSON MED COL, THOMAS JEFFERSON UNIV, 74- *Mailing Add:* Hahnemann Univ Broad & Vine Sts Philadelphia PA 19102

DREIFUSS, FRITZ EMANUEL, b Dresden, Ger, Jan 20, 26; US citizen; m 54; c 2. MEDICINE. *Educ:* Univ NZ, MB, ChB, 50. *Prof Exp:* House physician & res neurol officer, Nat Hosp, London, Eng, 54-57; from asst prof to assoc prof, 59-68, PROF NEUROL, SCH MED, UNIV VA, 68-, NEUROLOGIST, UNIV HOSP, 59- *Concurrent Pos:* Neurologist, Commonwealth of Va Child Neurol Prog, 59- *Mem:* AAAS; Am Acad Neurol; Asn Res Nerv & Ment Dis; AMA; NY Acad Sci; Int League Against Epilepsy (pres, 85). *Res:* Neurological sciences, especially pediatric neurology; epilepsy and mental retardation. *Mailing Add:* Dept Neurol Univ Va Hosp Charlottesville VA 22903

DREIKORN, BARRY ALLEN, b Norristown, Pa, Sept 15, 39; m 62; c 1. HETEROCYCLIC CHEMISTRY, QUINOLINE CHEMISTRY. *Educ:* Rutgers Univ, BA, 61; Univ Pa, PhD(org chem), 66. *Prof Exp:* Sr chemist, Eli Lilly Res Lab, 66-72, res scientist, 72-81, sr res scientist, 81-90, res adv, 90-91; SR RES ADV, DOWELANCO, 91- *Mem:* Am Chem Soc. *Res:* Synthesis and development of agrichemicals; fungicides, rodenticides, insecticides. *Mailing Add:* 9731 Trilobi Dr Indianapolis IN 46236

DREILING, CHARLES ERNEST, b Inglewood, Calif, Apr 6, 41; m 62; c 2. BIOCHEMISTRY, NEUROCHEMISTRY. *Educ:* Univ Wash, BA, 64; NMex State Univ, MS, 68; Ore State Univ, PhD(biochem), 70. *Prof Exp:* Teaching asst, Dept Biol, NMex State Univ, 66-67; res asst, Dept Biochem & Biophys, Ore State Univ, 68-70; asst prof endocrinol, Dept Biol, 71-75, asst prof, 75-81, ASSOC PROF BIOCHEM, SCH MED, UNIV NEV, RENO, 81- *Mem:* AAAS; Sigma Xi; NY Acad Sci. *Res:* Embryology and enzymology of the myelin sheath of the central and peripheral nervous systems; local and humoral factors which initiate normal myelination and control remyelination following injury and disease. *Mailing Add:* Biochem Div Sch Med Sci Univ Nev Reno NV 89557

DREILING, DAVID A, surgery; deceased, see previous edition for last biography

DREILING, MARK JEROME, b Kansas City, Mo, Nov 27, 40; m 68; c 2. PHYSICS. *Educ:* Kans State Univ, BS, 62, MS, 64, PhD(physics), 68. *Prof Exp:* Res physicist, 67-80, SUPVR, FUNDAMENTAL MAT SCI SECT, PHILLIPS PETROL CO, 87- *Mem:* Am Chem Soc; Am Phys Soc. *Res:* Electron spectroscopy; surface science; catalysis; x-ray diffraction; photovoltaics; composites materials. *Mailing Add:* 1301 Cherokee Hills Dr Bartlesville OK 74006

DREIMANIS, ALEKSIS, b Valmiera, Latvia, Aug 13, 14; Can citizen; m 42; c 2. QUATERNARY GEOLOGY. *Educ:* Latvian Univ, Mag rer nat, 38, habil, 41. *Hon Degrees:* DSc, Univ Waterloo, 69, Univ Western Ont, 80. *Prof Exp:* Asst geol, Inst Geol, Latvian Univ, 38-40, lectr, 41, privat-docent, 42-44; assoc prof, Baltic Univ, 46-48; lectr, 48-51, from asst prof to prof, 51-80, EMER PROF GEOL, UNIV WESTERN ONT, 80- *Concurrent Pos:* Consult, Inst Res Mineral Resources of Latvia, 42-44 & Ont Dept Planning & Develop, 50-53; consult various Can & US govt agencies, explor & eng consult co, 50-; Can deleg, Int Geol Cong, 60; Can deleg, Int Union Quaternary Res, 65, 69, 73 & 77, pres, Comn Genesis & Lithology Quaternary Deposits, 73-82; mem, Can Nat Adv Comt Res in Geol Sci & chmn subcomt quaternary geol, 67-71; foreign mem, Latrian Acad Sci, 90; corresp mem, Baltic Res Inst. *Honors & Awards:* Centennial Medal Can, Can Govt, 67; Latvian Cult Found Hon Award Tech & Natural Sci, 77; Queen Elizabeth II 25th Anniversary Medal, Can, 77; Logan Medal, Geol Asn Can, 78; Centennial Medal, Geol Surv of Finland, 87; Distinguished Career Award, Geol Soc Am, 87; Albrecht Penck Medal, Ger Quaternary Asn, 88; W A Johnston Medal, Can Quaternary Asn, 89. *Mem:* Am Quaternary Asn (pres, 81-83); Asn Advan Baltic Studies; Geol Soc Am; Geol Asn Can; fel Royal Soc Can; corresp mem Geol Soc Finland; Geol Soc Sweden; Can Quaternary Asn; Quaternary Res Asn; hon mem Int Union Quaternary Res. *Res:* Pleistocene and glacial geology; lithology and genesis of glacial deposits; indicator trains; Late Pleistocene stratigraphy of Northern Hemisphere. *Mailing Add:* Dept Geol Univ Western Ont London ON N6A 5B7 Can

DREISBACH, JOSEPH HERMAN, b Northampton, Pa, Nov 6, 49. PROTEIN BIOCHEMISTRY, MICROBIAL BIOCHEMISTRY. *Educ:* LaSalle Col, BA, 71; Lehigh Univ, MS, 74, PhD(chem), 77. *Prof Exp:* ASST PROF CHEM, UNIV SCRANTON, 78- *Mem:* Am Chem Soc; Am Soc Microbiol; Sigma Xi. *Res:* Characterization of the processes involved during bacterial enzyme induction by large extracellular molecules prior to interaction at the gene level. *Mailing Add:* Dept Chem Univ Scranton Scranton PA 18510

DREISBACH, ROBERT HASTINGS, b Baker, Ore, Mar 29, 16; m 41; c 2. PHARMACOLOGY. *Educ:* Stanford Univ, AB, 37; Univ Chicago, PhD(pharmacol) & MD, 42. *Prof Exp:* Asst pharmacol, Univ Chicago, 39-42; intern, St Mary's Hosp, 42-43; from instr to prof pharmacol, Sch Med, Stanford Univ, 43-73; CLIN PROF ENVIRON HEALTH, SCH PUB HEALTH & COMMUNITY MED, UNIV WASH, 73- *Res:* Toxicology; environmental pollution. *Mailing Add:* Sch Pub Health & Community Med Univ Wash SC-34 Seattle WA 98195

DREISS, GERARD JULIUS, b West New York, NJ, July 11, 28; m 59; c 2. THEORETICAL PHYSICS. *Educ:* Rutgers Univ, BA, 58; Univ Pa, MS, 64, PhD(physics), 68. *Prof Exp:* Mathematician, Avco-Everett Res Lab, 59-63; vis instr physics, Northeastern Univ, 68-69, asst prof, 69- 71; asst ed, 71-72, ASSOC ED, PHYS REV, 72- *Mem:* Am Phys Soc; Sigma Xi. *Res:* Theoretical nuclear physics. *Mailing Add:* Am Phys Soc PO Box 1000 Ridge NY 11961

DREIZEN, PAUL, b New York, NY, Oct 23, 29. BIOPHYSICS, MEDICINE. *Educ:* Cornell Univ, AB, 51; NY Univ, MD, 54. *Prof Exp:* Intern med, NY Univ-Bellevue Med Ctr, 54-55, asst resident, 55-56, asst med, 58-59; cardiologist, US Naval Hosp, Nat Naval Med Ctr, 56-58; res assoc biol, Mass Inst Technol, 59-62; from asst prof to assoc prof med, 62-71, PROF MED & BIOPHYS, STATE UNIV NY, BROOKLYN, 71-, DEAN SCH GRAD STUDIES, 73- *Concurrent Pos:* Nat Res fel, 59-63; career scientist, NY City Health Res Coun, 63-75. *Mem:* Biophys Soc; Soc Gen Physiol; Am Soc Cell Biol; Asn Am Physicians; Am Soc Clin Invest; Sigma Xi; Am Chem Soc. *Res:* Physical chemistry of contractile proteins; molecular mechanism of muscular contraction. *Mailing Add:* Grad Sch State Univ NY Box 41 450 Clarkson Ave Brooklyn NY 11203

DREIZEN, SAMUEL, b New York, NY, Sept 12, 18; m 56; c 1. NUTRITION. *Educ:* Brooklyn Col, BA, 41; Western Reserve Univ, DDS, 45; Northwestern Univ, MD, 58. *Prof Exp:* Res assoc, Univ Cincinnati, 45-47; from instr to assoc prof nutrit & metab, Northwestern Univ, 47-66; prof oral oncol, 76-89, EMER PROF ORAL ONCOL, INST DENT SCI, DENT BR, UNIV TEX, 89- *Concurrent Pos:* Asst sci dir, Nutrit Clin, Hillman Hosp, 48-60; Clayton Found fel, 49-54; consult nutrit, Am Dent Asn, 64-; consult, M D Anderson Hosp & Tumor Inst, Univ Tex, 67-, prof path, Inst Dent Sci, Dent Br, 66-, prof, 68-89, emer prof, Grad Sch Biomed Sci, 89- *Mem:* AAAS; Am Asn Phys Anthropologists; Int Asn Dent Res; Soc Res Child Develop; NY Acad Sci. *Res:* Dental caries; nutritional deficiency diseases; child growth and development; cancer chemotherapy; oral medicine. *Mailing Add:* Univ Tex Dent Br PO Box 20068 Houston TX 77225

DRELICH, ARTHUR (HERBERT), b Jersey City, NJ, Mar 26, 20; m 49; c 5. TEXTILE CHEMISTRY, MICROSCOPY. *Educ:* NY Univ, BA, 40; Univ Pa, MS, 42. *Prof Exp:* Jr chemist protective agents, Edgewood Arsenal, 42-43; supvr chem res, Johnson & Johnson, 43-67, sr scientist, 67-74, sr res assoc, Res Div, 74-76, sr technol counr, 76-79, consult, 79-83, res fel, 83-86, CONSULT, RES DIV, CHICOPEE MFG CO, JOHNSON & JOHNSON, 86- *Concurrent Pos:* Distinguished lectr, Polymers, Johnson & Johnson, 83. *Honors & Awards:* Johnson Medal, 76; Nonwovens Div Award, Tech Asn Pulp & Paper Indust, 89, Mark Hollingsworth Prize, 89. *Mem:* Am Chem Soc; fel Royal Micros Soc; Fiber Soc; Sigma Xi; fel NY Micros Soc. *Res:* Technology of non-woven fabrics; properties of thermoplastic polymers; absorbency; physico-chemical methods to rapidly destabilize polymer latexes; fiber and chemical microscopy and scientific photography. *Mailing Add:* 60 Parkside Rd Plainfield NJ 07060

DRELL, SIDNEY DAVID, b Atlantic City, NJ, Sept 13, 26; m 52; c 3. ARMS CONTROL. *Educ:* Princeton Univ, AB, 46; Univ Ill, MA, 47, PhD(physics), 49. *Hon Degrees:* DSc, Univ Ill, Chicago, 81. *Prof Exp:* Res assoc physics, Univ Ill, 49-50; instr, Stanford Univ, 50-52; res assoc, 52-53, asst prof, Mass Inst Technol, 53-56; from assoc prof to prof physics, Stanford Univ, 56-63, exec head theoret physics, 69-86, Lewis M Terman prof & fel, 79-84, co-dir, Stanford Ctr Int Security & Arms Control, 83-89, PROF, STANFORD LINEAR ACCELERATOR CTR, STANFORD UNIV, 63-, DEP DIR, 69- *Concurrent Pos:* Consult, Los Alamos Sci Labs, 56-64 & 68-, Off Sci & Technol, 60-73, Jason, 60-, US Arms Control & Disarmament Agency, 69-81, Nat Security Coun, 73-81, Off Technol Assessment, US Congress, 75-91, Off Sci & Technol Policy, 77-82 & Senate Select Comt, Intel, 78-82 & 89; Loeb lectr & vis prof, Harvard Univ, 62 & 70; mem, Pres Sci Adv Comt, 66-70; mem bd gov, Weizmann Inst Sci, 70-; chmn high energy physics adv panel, US Dept Energy, 74-82, mem, 74-86; mem vis comt, Dept Physics, Mass Inst Technol, 74-90; mem bd trustees, Inst Advan Study, 74-83; vis Schrodinger prof, Univ Vienna, Austria, 75; mem bd dirs, Ann Reviews, Inc, 76-; mem bd dirs, The Arms Control Asn, Washington, DC, 78-; mem, Coun Foreign Rels, NY, 80-; Guggenheim fel, 61-62 & 71-72; vis fel, All Souls Col, Oxford Univ, 79; I I Rabi vis prof, Columbia Univ & vis prof, Rockefeller Univ, 84; mem, Carnegie Comn Sci, Technol & Govt, 88-, Sci & Acad Adv Comt on Nat Labs, Univ Calif, 88-; chmn, Int Adv Comt, Inst Global Conflict & Coop, Univ Calif, 90-; John D & Catherine T MacArthur Found fel, 84-89; mem, Aspen Strategy Group, 84-; mem, Mass Inst Technol Lincoln Lab Adv Bd, 85-90; adj prof, Dept Eng & Pub Policy, Carnegie Mellon Univ, 89- *Honors & Awards:* Ernest Orlando Lawrence Mem Award, US AEC, 72; Richtmyer Mem lectr, Am Asn Physics Teachers, San Francisco, Calif, 78; Leo Szilard Award, Physics Pub Interest, Am Phys Soc, 80-; Danz Lectr, Univ Wash, 83. *Mem:* Nat Acad Sci; fel Am Phys Soc (pres, 86); Am Acad Arts & Sci; Am Philos Soc. *Res:* Quantum field theory; elementary particle physics; understanding of the structure of hadrons, particularly the quark confinement problem; arms control and national security. *Mailing Add:* Stanford Linear Accelerator Ctr PO Box 4349 Stanford CA 94309

DREN, ANTHONY THOMAS, b Chisholm, Minn, Feb 15, 36; m 61; c 4. CLINICAL PHARMACOLOGY, PHARMACY. *Educ:* Duquesne Univ, BS, 59, MS, 61; Univ Mich, PhD(pharmacol), 66. *Prof Exp:* Teaching asst pharm, Duquesne Univ, 59-61; retail pharmacist, Heckler Drug Co, 59-61; sr pharmacologist, Abbott Labs, 66-69, group leader pharmacol, 69-72, assoc res fel, 72-75, sect head neuropharmacol, 75-79; SR CLIN RES SCIENTIST, BURROUGHS WELLCOME CO, 79- *Mem:* Am Chem Soc; AAAS; Soc Pharmacol & Exp Therapeut; Soc Neurosci; Sigma Xi; Soc Clin Pharmacol Therapeut. *Res:* Neuropharmacology; cerebral vascular disease; abuse potential of drugs; electroencephalography; psychopharmacology; neurochemistry; cerebral hypoxia; neurophysiology; clinical research. *Mailing Add:* Burroughs Wellcome Co 3030 Cornwallis Rd Research Triangle Park NC 27709

DRENGLER, KEITH ALLAN, b Fremont, Ohio, Jan 6, 53; m 75; c 2. TERPENE BIOSYNTHESIS, PEPTIDE SYNTHESIS. *Educ:* Albion Col, BA, 75; Univ Ill, Urbana-Champaign, PhD(org chem), 80. *Prof Exp:* RES CHEMIST ORG CHEM, INT MINERALS & CHEM CORP, 80- *Mem:* Am Chem Soc; Sigma Xi. *Res:* Elucidating the mechanisms of enzymatic reactions involved in terpene biosynthesis; synthesis of compounds with novel biological activity for potential use as veterinary drugs; synthesis of peptide and peptide analogs for development of veterinary drugs. *Mailing Add:* 422 Teal Rd Lindenhurst IL 60046

DRENICK, ERNST JOHN, ENDOCRINOLOGY, NUTRITION. *Educ:* NY Univ, MD, 41. *Prof Exp:* PROF ENDOCRINOL & NUTRIT, DEPT MED, WADSWORTH VET ADMIN HOSP MED CTR, UNIV CALIF, LOS ANGELES, 55- *Mailing Add:* Wadsworth Vet Admin Hosp Med Ctr Wilshire & Sawtelle Blvds Los Angeles CA 90073

DRENICK, RUDOLF F, b Vienna, Austria, Aug 20, 14; nat US; m 46; c 3. APPLIED MATHEMATICS. *Educ:* Vienna Univ, PhD(theoret physics), 39. *Prof Exp:* Asst prof math & physics, Villanova Col, 39-44; engr anal, Gen Elec Co, 46-49; mgr anal group, Radio Corp Am, 49-57; res mathematician, Bell Tel Labs, Inc, 57-61; chmn dept, 66-67, PROF ELEC ENG, POLYTECH INST NY, 61-; PRES, POLYSYST ANALYSIS CORP. *Concurrent Pos:* NSF fel, 64-65; consult mathematician, Mathematica, Princeton & Res Triangle Inst, Durham, 62-65; prog dir, Syst Theory & Appln, NSF, 77-78. *Mem:* Fel Inst Elec & Electronics Engrs; Soc Indust & Appl Math; Inst Math Statist; Inst Mgt Sci. *Res:* Application of mathematical methods to engineering problems. *Mailing Add:* 35 Melody Lane Huntington NY 11743

DRENNAN, OLLIN JUNIOR, b Kirksville, Mo, Apr 11, 25; m 53; c 3. HISTORY OF SCIENCE. *Educ:* Northeast Mo State Teachers Col, AB, 49; Mo Valley Col, BS, 50; Bradley Univ, MS, 55; Univ Wis, PhD(hist sci), 61. *Prof Exp:* Prog dir & engr, Radio Sta KMMO, 49; teacher, high sch, Mo, 51-52; instr physics & chem, Mo Valley Col, 52-53; asst prof physics, Evansville Col, 53-55; instr, Northeast Mo State Teachers Col, 55-64; assoc prof physics, Western Mich Univ, 64-67, dir, 64-67, prof natural sci, 67-77, assoc dean, Col Gen Studies & Area Chmn, Gen Studies Sci, 71-77; prof physics, 77-80, REIGER-BLACK DISTINGUISHED PROF PHYSICS & NATURAL SCI, NORTHEAST MO STATE UNIV, 80- *Res:* 19th and 20th century physical science; general education science. *Mailing Add:* Dept Sci Northeast Mo State Univ Kirksville MO 63501

DREOSTI, IVOR EUSTACE, b Johannesburg, SAfrica, Feb 19, 36; Australian citizen; m 66; c 2. NUTRITION & CANCER, ANTIOXIDANTS & FREE RADICALS. *Educ:* Univ Natal, SAfrica, BSc, 60, MSc, 62, PhD(biochem), 65, DSc, 86. *Prof Exp:* Prof officer, Natal Agr Res Inst, SAfrica, 63-66; lectr biochem, Univ Witwotessrand, SAfrica, 68-69; sr lectr, Univ Natal, SAfrica, 69-75, assoc prof, 75-77; prin res scientist, 77-82, SR PRIN RES SCIENTIST & PROG COORDR, DIV HUMAN NUTRIT, COMMONWEALTH SCI & INDUST RES ORGN, 82- *Concurrent Pos:* Vis lectr nutrit, Univ Calif, Davis, 66-68, vis prof, 75, distinguished lectr, 90; mem, Comt Recommended Dieting Intakes, Nat Health & Med Res Coun, Australia, 83-89 & Comt Expert Consult Trace Requirements Humans, Food & Agr Orgn, WHO & Int Atomic Energy Agency, 83-90; lectr, Australian Asn Clin Biochemists, 91. *Mem:* Am Inst Nutrit; NY Acad Sci. *Res:* Free radicals in the initiation of cancer and the value of antioxidants as anticancer agents; highly sensitive techniques for the assessment of genotoxicity and mutagenicity. *Mailing Add:* Div Human Nutrit CSIRO PO Box 10041 Gouger St Adelaide 5000 Australia

DRESCH, FRANCIS WILLIAM, b Sharon, Pa, Sept 21, 13; m 39. ECONOMIC STATISTICS, OPERATIONS RESEARCH. *Educ:* Stanford Univ, AB, 32, AM, 34; Univ Calif, PhD(math), 37. *Prof Exp:* Asst, Univ Calif, 35-37, Florence Noble traveling fel math, Cambridge Univ & Univ Paris, 37-38; instr math, Univ Calif, 38-41; from asst dir to dir comput & ballistics, US Naval Proving Ground, 46-52; sr statistician, 52-57, mgr indust opers res & electronic data processing, 57-61, sr math economist, 62-79, opers analyst, 79-80, SR MATH ECONOMIST, CTR PLANNING & RES, STANFORD RES INST, 80- *Mem:* AAAS; Am Math Soc; Inst Math Statist; Economet Soc; Brit Oper Res Soc; Sigma Xi. *Res:* Econometrics; ballistics; statistics; computing techniques; economic impact of environmental regulations; macroeconomics; potential rate of post-attack economic recovery. *Mailing Add:* 1486 Kings Lane Palo Alto CA 94303

DRESCHER, DENNIS G, m; c 2. BIO-OTOLOGY. *Educ:* Univ Wis-Madison, BS, 63, MM, 64, PhD(biochem), 71. *Prof Exp:* Assoc prof, 78, PROF NEUROSCI, DEPT OTOLARYNGOL & BIOCHEM, SCH MED, WAYNE STATE UNIV, 87- *Concurrent Pos:* Dir basic res, Dept Otolaryngol, Wayne State Univ, 78-, Lab Bio-otol, 78-; mem, Dept Budget Comt, Wayne State Univ, 81-, Pew Found Eval Comt, 86; chmn, Res Comt, Dept Otolaryngol, Wayne State Univ, 83-, Univ-Wide Neurosci Prog, 86-89, Res Subcomt, Internal Rev Comt, Dept Otolaryngol, 88. *Res:* Bio-otol; neuroscience. *Mailing Add:* Bio-Otology Lab Wayne State Univ Med Sch 540 E Canfield Ave Detroit MI 48201

DRESCHER, WILLIAM JAMES, b Craig, Colo, Aug 20, 18; m 41; c 2. GROUNDWATER HYDROLOGY, HYDROGEOLOGY. *Educ:* Univ Colo, BSE, 40; Univ Wis, MS, 56. *Prof Exp:* Engr, Am Bridge Co Ind, 40-41; hydraul engr, US Geol Surv, Fla, 41-42, La, 42-44 & Wis, 46-51, dist engr, 51-57, br area chief, 57-62, hydraul res engr, 62-66, res hydrologist, 66-70, regional planning officer, Dept Interior, 70-74; CONSULT HYDROLOGIST, 74- *Concurrent Pos:* Lectr, Univ Wis, 74-75; US chmn steering comt, Int Field Year Great Lakes, Int Hydrol Decade, 66-77. *Mem:* Int Asn Hydrogeol; Am Geophys Union. *Mailing Add:* 322 Robin Pkwy Madison WI 53705

DRESCHHOFF, GISELA AUGUSTE-MARIE, b Monchengladbach, Ger, Sept 13, 38; US citizen. RADIATION PHYSICS, GEOPHYSICS. *Educ:* Tech Univ Braunschweig, BS, 61, MS, 65, PhD(physics), 72. *Prof Exp:* Staff scientist radiation protection, Physikalisch Tech Bundesanstalt Ger, 65-67; res assoc nuclear waste disposal, Kans Geol Surv, 71-72; dep dir, 72-84, CO-DIR, RADIATION PHYSICS LAB, SPACE TECHNOL CTR, UNIV KANS, 84- *Concurrent Pos:* Vis asst prof physics, Univ Kans, 72-74, adj asst prof, 74-; assoc prof mgr, Div Polar Prog, NSF, 78- *Mem:* Am Phys Soc; Am Geophys Union; Sigma Xi; Am Polar Soc; AAAS; Antaretican Soc; Explorers Club; US Naval Inst. *Res:* Remote sensing; nuclear waste disposal; reactor radiation protection; geophysics of the polar regions. *Mailing Add:* Space Technol Ctr 2291 Irving Hill Rd Lawrence KS 66045

DRESDEN, CARLTON F, b Dodgeville, Wis, Oct 17, 31; m 55; c 3. BIOCHEMISTRY, ORGANIC CHEMISTRY. *Educ:* Wis State Univ, Platteville, BS, 53; Univ Wis, MS, 57, PhD(org chem, biochem), 59. *Prof Exp:* Assoc prof chem, Slippery Rock State Col, 59-61; asst prof biochem, La State Univ, 61-62; chmn sci div, 62-70, actg dean, Sch Natural Sci & Math, 76-80, PROF CHEM, SLIPPERY ROCK STATE COL, 62- *Mem:* AAAS; Am Chem Soc. *Res:* Biochemistry of the myxomycetes; mechanism of action of bacterial toxins. *Mailing Add:* Dept Chem Slippery Rock State Col Slippery Rock PA 16057

DRESDEN, MARC HENRI, biochemistry, developmental biology; deceased, see previous edition for last biography

DRESDEN, MAX, b Amsterdam, Netherlands, Apr 23, 18; US citizen; m 48; c 2. THEORETICAL PHYSICS. *Educ:* Univ Mich, PhD(physics), 46. *Prof Exp:* Asst physics, Univ Mich, 41-46; from asst prof to prof, Univ Kans, 46-57; prof & chmn dept, Northwestern Univ, 57-60; prof, Univ Iowa, 60-64; PROF PHYSICS, STATE UNIV NY STONY BROOK, 64-, EXEC OFF, INST THEORET PHYSICS, 66- *Concurrent Pos:* Mem bd dirs, Midwest Univs Res Asn, 56-64; sci bd dirs, Midwest Res Inst, 62-; vis prof, Johns Hopkins Univ, 57-58; vis lectr, Am Inst Physics, 57- *Mem:* Fel Am Phys Soc. *Res:* Statistical mechanics; superconductivity; quantum field theory; behavior of positrons; parastatistics; symmetrics and S matrix theory; particle physics. *Mailing Add:* SLAC PO Box 4349 Bin 61 Stanford CA 94309

DRESDNER, RICHARD DAVID, b New York, NY, Feb 20, 18; m 50; c 2. CHEMISTRY. *Educ:* NY Univ, BA, 41; Pa State Univ, MS, 44, PhD(phys chem), 47. *Prof Exp:* Res engr, Battelle Mem Inst, 47-50; res engr, Picatinny Arsenal, 50-51; res engr, Micrometallic Corp, 51-53; assoc res prof, 53-59, assoc prof, 59-66, chmn div gen chem, 66-76, PROF CHEM, UNIV FLA, 66- *Mem:* Am Chem Soc. *Res:* Electrolytic fluorinations; viscosity; low temperature air rectification; analytical photochemistry; kinetics; pyrotechnics; plastic filters; synthetic reactions; fluorochemicals; photochemical fluorinations. *Mailing Add:* Dept Chem Univ Fla Gainesville FL 32611

DRESHFIELD, ARTHUR C(HARLES), JR, b Kalamazoo, Mich, Nov 9, 29; m 57; c 3. CHEMICAL ENGINEERING. *Educ:* Univ Ill, BS, 51; Lawrence Col, MS, 53, PhD, 56. *Prof Exp:* Res engr papermaking processes, Scott Paper Co, 55-57; from res group mgr pulping & paperboard to dir res, Fibreboard Paper Prod, 57-68, dir prod develop, 68-72; mgr res & develop, 72-81, DIR, RES & DEVELOP CTR, POTLATCH CORP, 81- *Concurrent Pos:* Chmn indust adv coun, Forest Prod Lab, Univ Calif, 64-65; inst paper chem adv, 81-84; forest prof lab adv coun, 83-86. *Honors & Awards:* Steele Medal, Inst Paper Chem, 56. *Mem:* Tech Asn Pulp & Paper Indust; Can Pulp & Paper Tech Asn; AAAS. *Res:* Alkaline pulping; papermaking; packaging. *Mailing Add:* Potlach Corp Fiber Res & Develop Ctr PO Box 503 Cloquet MN 55720

DRESHFIELD, ROBERT LEWIS, b Kalamazoo, Mich, Mar 15, 33; m 56; c 2. PHYSICAL METALLURGY. *Educ:* Univ Ill, BS, 54; Univ Mo, MS, 59; Case Western Reserve Univ, PhD(mat sci), 72. *Prof Exp:* Pratt & Whitney Aircraft Div, United Aircraft Corp, 62-64; METALLURGIST, LEWIS RES CTR, NASA, 64- *Mem:* Am Inst Mining, Metall & Petrol Engrs; Am Soc Metals; Am Soc Testing Mat. *Res:* Development of alloys for use in aircraft gas turbines and space power systems; phase relationships in superalloys. *Mailing Add:* Lewis Res Ctr 21000 Brookpark Rd Cleveland OH 44135

DRESKA, NOEL, b Kankakee, Ill, Dec 24, 28. MOLECULAR SPECTROSCOPY. *Educ:* Col St Francis, Ill, AB, 50; Ohio State Univ, PhD(physics), 64. *Prof Exp:* Instr physics, Col St Francis, Ill, 64-66; from asst prof to assoc prof, 66-84, PROF PHYSICS, LEWIS UNIV, 84-, CHMN DEPT, 70-77 & 84- *Mem:* Am Asn Physics Teachers. *Res:* High resolution, infrared and molecular spectroscopy. *Mailing Add:* Dept Physics Lewis Univ Rte 53 Romeoville IL 60441

DRESNER, JOSEPH, b Belg, Feb 11, 27; nat US; m 57. SURFACE PHYSICS, SOLID STATE PHYSICS. *Educ:* Univ Mich, BSE, 49, MS, 50; NY Univ, PhD(physics), 58. *Prof Exp:* Physicist, Hosp Joint Dis, NY, 50-53; asst, NY Univ, 54-58; SR MEM TECH STAFF, DAVID SARNOFF RES CTR, RCA CORP, 58- *Concurrent Pos:* Consult, Hosp Joint Dis, NY, 53-54; vis prof, Inst Physics Sao Carlos, Univ Sao Paulo, 71-72. *Mem:* Fel Am Phys Soc; Mat Res Soc; Soc Info Display. *Res:* Luminescence and photoconductivity of solids; electronic properties of organic and amorphous materials; thin films; electron impact phenomena; secondary emission; thermionic emission; television display and pickup devices; medical radiation physics; photovoltaic cells. *Mailing Add:* David Sarnoff Res Ctr SRI Int Princeton NJ 08543-5300

DRESS, WILLIAM JOHN, b Buffalo, NY, June 9, 18. TAXONOMIC BOTANY. *Educ:* Univ Buffalo, BA, 39; Cornell Univ, PhD(bot), 53. *Prof Exp:* Asst, 47-53, asst prof bot, 57-60, from assoc prof to prof, 60-82, EMER PROF BOT, L H BAILEY HORTORIUM, CORNELL UNIV, 82- *Mem:* Am Soc Plant Taxon; Int Asn Plant Taxon. *Res:* Plant taxonomy, especially of cultivated plants; revision of genus Chrysopsis. *Mailing Add:* L H Bailey Hortorium Cornell Univ Ithaca NY 14853

DRESSEL, FRANCIS GEORGE, b Hart, Mich, Sept 22, 04; wid; c 2. MATHEMATICS. *Educ:* Mich State Col, BS, 28; Univ Mich, MS, 29; Duke Univ, PhD(math), 33. *Prof Exp:* From instr to prof, 29-74, EMER PROF MATH, DUKE UNIV, 74- *Concurrent Pos:* Teacher, Civil Aeronaut Admin, Duke Univ, 42-43; asst, Math Sci Div, Army Res Off, 51- *Mem:* Am Math Soc; Math Asn Am. *Res:* Integral equations; Stieltjes integrals; partial differential equations. *Mailing Add:* 2502 Frances St Durham NC 27707

DRESSEL, HERMAN OTTO, electrooptics; deceased, see previous edition for last biography

DRESSEL, PAUL LEROY, statistical analysis; deceased, see previous edition for last biography

DRESSEL, RALPH WILLIAM, b Buffalo, NY, Mar 3, 22; m 45; c 4. PHYSICS, ELECTRON PHYSICS. *Educ:* Union Col, BS, 44; Univ Ill, PhD(physics), 50. *Prof Exp:* Asst instr, Union Col, 42-44; mem staff, Radiation Lab, Mass Inst Technol, 44-46; asst, Univ Ill, 46-49; from asst prof to prof physics, 50-82, from assoc physicist to physicist, Phys Sci Lab, 52-61, head, dept physics, 57-61, EMER PROF PHYSICS, NMEX STATE UNIV, 82-; PHYSICIST, NUCLEAR EFFECTS BR, WHITE SANDS MISSILE RANGE, 61- *Mem:* Am Asn Physics Teachers; Am Phys Soc. *Res:* Electromagnetic radiation; interaction of high energy quanta, electrons with matter. *Mailing Add:* 1740 Imperial Ridge Las Cruces NM 88001

DRESSELHAUS, GENE FREDERICK, b Ancon, CZ, Nov 7, 29; m 58; c 4. SOLID STATE PHYSICS. *Educ:* Univ Calif, AB, 51, PhD(physics), 55. *Prof Exp:* Instr physics, Univ Chicago, 55-56; asst prof, Cornell Univ, 56-60; mem staff, Lincoln Lab, 60-77, SR SCIENTIST, BITTER NAT MAGNET LAB, MASS INST TECHNOL, 77- *Concurrent Pos:* Consult, Gen Elec Res Lab, 56-60 & Oak Ridge Nat Lab, 58-60. *Mem:* Fel Am Phys Soc. *Res:* Electronic energy bands in solids; surface impedance of metals; excitons in insulators; surface electronic states; optical properties of solids; high temperature superconductivity. *Mailing Add:* Mass Inst Technol Lab Rm 13-3017 Cambridge MA 02139

DRESSELHAUS, MILDRED S, b Brooklyn, NY, Nov 11, 30; m 58; c 4. SOLID STATE PHYSICS. *Educ:* Hunter Col, AB, 51; Radcliffe Col, AM, 53; Univ Chicago, PhD(physics), 58. *Hon Degrees:* DEng, Worcester Polytechnic Inst, 76; DSc, Smith Col, 80, Hunter Col, 82, NJ Inst Technol, 84, State Univ NJ, Rutgers, 89, Rutgers Univ, 90; Dr, Cath Univ Louvain, Belgium, 88. *Prof Exp:* NSF fel, Cornell Univ, 58-60; staff mem, Lincoln Lab, Mass Inst Technol, 60-67, Abby Rockefeller Mauze vis prof, Dept Elec Eng & Computer Sci, 67-68, prof, 68-73, assoc head dept, 72-74, Abby Rockefeller Mauze prof elec eng, 73-85, dir, Ctr Mat Sci & Eng, 77-83, PROF, DEPT ELEC ENG & COMPUTER SCI, MASS INST TECHNOL, 68-, PROF, DEPT PHYSICS, 83-, INST PROF, 85- *Concurrent Pos:* Mem adv bd, Mat Res Div & Mat Res Labs, NSF, 74-76; mem exec comt phys & math sci, Nat Acad Sci, 75-78; chmn, Steering Comt Eval Panels, Nat Bur Standards, 78-83; mem, Sci Adv Comt, Allied Chem Corp, 80-84; mem gov bd, Nat Res Coun, 84-88, AAAS, 85; mem, Energy Res Adv Bd, 84-90; chmn, Eng Sect, Nat Acad Sci, 87-90, class chmn, Eng & Appl Sci Sect, 90- *Honors & Awards:* Achievement Award, Soc Women Engrs, 77; Graffin Lectr, Am Carbon Soc, 82; Hund-Klemm Lectr, Max Planck Inst, Stuttgart, 88; Ann Achievement Award, Eng Socs New Eng, 88; Nat Medal of Sci, 90. *Mem:* Nat Acad Sci; Nat Acad Eng; fel Am Acad Arts & Sci; corresp mem Braz Acad Sci; fel Am Phys Soc (vpres, 82, pres-elect, 83, pres, 84); fel Inst Elec & Electronics Engrs; sr mem Soc Women Engrs. *Res:* Electronic, optical and magneto-optical properties of solids; semimetals, semiconductors, magnetic semiconductors, graphite intercalation compounds. *Mailing Add:* Mass Inst Technol Rm 13-3005 77 Massachusetts Ave Cambridge MA 02139

DRESSER, HUGH W, b Utica, NY, Jan 31, 30; m 51; c 2. GEOLOGY, PALEONTOLOGY. *Educ:* Univ Cincinnati, BS, 50, MS, 51; Univ Wyo, PhD(geol), 60. *Prof Exp:* Jr geologist, Carter Oil Co, 51-52, field geologist, 52-54; geologist II, Humble Oil & Refining Co, 59-64, res geologist, Esso Prod Res Co, 64-65; from asst prof to assoc prof, 65-75, PROF GEOL, MONT COL MINERAL SCI & TECHNOL, 75- *Mem:* Am Asn Petrol Geol; Geol Soc Am; Am Inst Prof Geol; Soc Econ Paleont & Mineral; Int Asn Sedimentologists. *Res:* Palichnology, structural geology, geomorphology, stratigraphy. *Mailing Add:* Dept Geol Mont Col Mineral Sci & Technol Butte MT 59701

DRESSER, MILES JOEL, b Spokane, Wash, Dec 19, 35; m 59; c 3. SURFACE PHYSICS. *Educ:* Linfield Col, BA, 57; Iowa State Univ, PhD(physics), 64. *Prof Exp:* Asst prof, 63-70, ASSOC PROF PHYSICS, WASH STATE UNIV, 70- *Concurrent Pos:* Physicist, Nat Bur Standards, 72; sr vis res prof, Surface Sci Ctr, Univ Pittsburgh, 84-85 & 87-88. *Mem:* Am Phys Soc; Am Vacuum Soc; Am Asn Physics Teachers. *Res:* Ultra high vacuum surface physics; surface ionization; thermionic and field emission; mass spectrometry; thin film electroluminescence; surface and solid diffusion; heterogeneous catalysis; electron stimulated desorption. *Mailing Add:* Dept Physics Wash State Univ Pullman WA 99164-2814

DRESSLER, ALAN MICHAEL, b Cincinnati, Ohio, Mar 23, 48. ASTRONOMY. *Educ:* Univ Calif, Berkeley, BA; Univ Calif, Santa Cruz, PhD(astron & astrophys), 76. *Prof Exp:* Carnegie fel, Hale Observ, 76-78, Las Campanas fel, 78-81, MEM SCI STAFF, M T WILSON & LAS CAMPANAS OBSERV, CARNEGIE INST WASH, 81- *Honors & Awards:* Pierce Prize, Am Astron Soc, 83. *Mem:* Am Astron Soc; Int Astron Union. *Res:* Formation and evolution of galaxies; structure, morphology and stellar populations of galaxies as function of environment and cosmological age; cosmology;large scale structure and motions of matter in the universe. *Mailing Add:* Carnegie Observ 813 Santa Barbara St Pasadena CA 91101

DRESSLER, EDWARD THOMAS, b Pittsburgh, Pa, May 29, 43; div; c 2. PHOTONUCLEAR REACTIONS. *Educ:* Duquesne Univ, BS, 66; Am Univ, MS, 72, PhD(physics), 74. *Prof Exp:* Res fel, Univ Saskatchewan, 74-76; res assoc, Ctr Radiation Res, Nat Bur Standards, 76-78; ASST PROF PHYSICS, PA STATE UNIV, 78- *Mem:* Am Phys Soc; Am Asn Physics Teachers. *Res:* Theory of photonuclear reactions; photo disintegration of light nuclei; relativistic effects in photonuclear processes; pion photoproduction and electroproduction from nuclei. *Mailing Add:* Pa State Univ Ogontz Campus 1600 Woodland Rd Abington PA 19001

DRESSLER, HANS, b Vienna, Austria, July 21, 26; nat US; m 56; c 1. ORGANIC CHEMISTRY, BIOCHEMISTRY. *Educ:* Columbia Univ, AM, 51, PhD(chem), 54. *Prof Exp:* Asst chem, Columbia Univ, 51-52, asst, 53; fel monomer synthesis, Mellon Inst, 54-56; sr chemist, Verona Res Ctr, 56-62, group mgr, 62-67, SR GROUP MGR, MONROEVILLE RES CTR, KOPPERS CO, INC, 67- *Mem:* Am Chem Soc; NY Acad Sci; Royal Soc Chem. *Res:* Organic synthesis; product and process development. *Mailing Add:* 1236 Catalina Dr Monroeville PA 15146

DRESSLER, ROBERT EUGENE, b New York, NY, Dec 16, 44; m 67; c 3. NUMBER THEORY. *Educ:* Univ Rochester, AB, 65; Univ Ore, MA, 66, PhD(math), 69. *Prof Exp:* Asst prof math, Southern Ill Univ, Carbondale, 69-70; from asst prof to assoc prof, 71-78, PROF MATH, KANS STATE UNIV, 78- *Mem:* Am Math Soc; Math Asn Am. *Res:* Density; properties of special number theoretic functions; number theory in harmonic analysis. *Mailing Add:* Dept Math Kans State Univ Manhattan KS 66506

DRESSLER, ROBERT LOUIS, b Branson, Mo, June 2, 27; c 2. PLANT SYSTEMATICS, REPRODUCTIVE ECOLOGY. *Educ:* Univ S Calif, BA, 51; Harvard Univ, PhD(biol), 57. *Prof Exp:* Instr, Wash Univ, 58-61, asst prof 61-63; biologist, 63-87, RES ASSOC, SMITHSONIAN TROP RES INST, 87- *Concurrent Pos:* Adj cur, Fla State Mus, 72-; Co-Coordr, Orchidaceae, Flora Meso-Americana; adj res staff, Marie Selby Bot Gardens, 87-; sco adv bd, Gesneriad Res Found, 87- *Honors & Awards:* Silver Medal, Mass Hort Soc, 86. *Mem:* Fel AAAS; Am Soc Plant Taxonomists; Asn Trop Biol; Int Asn Plant Taxonomists; Soc Bot Mex. *Res:* Evolution systematics and reproductive biology of orchidaceae; systematics of euglossine bees. *Mailing Add:* Rte 2 Box 565C Micanopy FL 32667

DRETCHEN, KENNETH LEWIS, b New York, NY, Jan 15, 46; m 68; c 2. PHARMACOLOGY. *Educ:* Brooklyn Col Pharm, BS, 68; Univ Iowa, MS, 70, PhD(pharmacol), 72. *Prof Exp:* Res assoc, 72-73, instr, 73-74, from asst prof to assoc prof, 74-84, PROF PHARMACOL, SCH MED, GEORGETOWN UNIV, 84- *Concurrent Pos:* Fac develop award, Pharmaceut Mfr Asn Found, 73. *Mem:* Am Soc Pharmacol & Exp Therapeut; Soc Neurosci; Am Chem Soc; Soc Exp Biol. *Res:* Role of the cyclic nucleotides in the function of the physiology and pharmacology of the motor nerve terminal. *Mailing Add:* Dept Pharmacol Georgetown Univ Sch Med Washington DC 20006

DREVDAHL, ELMER R(ANDOLPH), b Marquette, Mich, Aug 24, 26; m 49; c 2. MINING ENGINEERING, COMPUTER SCIENCE. *Educ:* Mich Col Mining & Technol, BS, 48; Univ Wash, Seattle, MS, 51. *Prof Exp:* Mining engr, Jones & Laughlin Ore Co, Mich, 48-50; asst prof mining eng, SDak Sch Mines & Technol, 51-55; from asst prof to assoc prof, Univ Ariz, 55-63; head div tech, Ariz Western Col, 63-65; dir tech ed, 65-66, dean occup educ, 66-69, assoc prof, 69-75, PROF ENG & HEAD DEPT, CLARK COL, 75- *Concurrent Pos:* Instr & examr, US Bur Mines First Aid Training. *Mem:* Am Inst Mining, Metall & Petrol Engrs; Nat Soc Prof Engrs; Am Soc Civil Engrs; Am Soc Eng Educ. *Res:* Analysis of equipment systems and application of computers to mining operations; engineering geology. *Mailing Add:* Dept Eng Clark Col Vancouver WA 98663

DREVES, ROBERT G(EORGE), b Brooklyn, NY, Oct 24, 14; m 38; c 2. AERONAUTICAL ENGINEERING. *Prof Exp:* Coordr prod control aircraft prod, Sperry Gyroscope Co, 41-43, aircraft flight training, US Navy, 43-46; mil training specialist, US Naval Training Device Ctr, 46-52, head, Aerospace Syst Trainers Dept, 60-65, assoc tech dir maintenance eng, 65-69, dir logistics & field eng, 70-73; CONSULT, 74- *Mem:* AAAS; Soc Logistics Engrs; Am Inst Aeronaut & Astronaut. *Res:* Design, development, test and utilization of air weapon system simulators for training of pilots and air crews; logistics management. *Mailing Add:* 541 Golden Arm Rd Deltona FL 32738

DREW, BRUCE ARTHUR, b Detroit, Mich, Mar 9, 24; m 48; c 5. MATHEMATICAL MODELS, EXPERIMENT DESIGN. *Educ:* Wayne State Univ, BS, 50. *Prof Exp:* Jr chem engr, Parke, Davis & Co, Mich, 48-50; formulator surface coatings, Rinshed-Mason Co, 50-52; develop engr, Huron Milling Co, 52-57; sr chemist, Hercules Powder Co, Va, 57-59; statistician, Pillsbury Co, 59-66, dir corp qual assurance, 66-68, sr res assoc appl math, 68-80, res fel, 80-88; CONSULT IN STATIST, DESIGN EXP & INDUST APPL MATH, 88- *Concurrent Pos:* Pres, Twin Cities Sect, Am Statist Asn, 76-78; chmn subcomt ed statist, Am Soc Testing & Mat, 84- *Mem:* Soc Indust & Appl Math; Am Inst Chem Engrs; Am Statist Asn; Am Asn Cereal Chem; Royal Statist Soc; Am Soc Testing & Mat; Int Asn Math & Comput Simulation; Asn Energy Engrs. *Res:* Computer models of chemical systems. *Mailing Add:* 4425 Abbott Ave S Minneapolis MN 55410

DREW, DAN DALE, b Abilene, Tex, Sept 29, 26; m 49; c 5. COMPUTER SCIENCE, MATHEMATICS. *Educ:* NTex State Univ, BS, 50, MS, 51; Tex A&M Univ, PhD(eng), 66. *Prof Exp:* Mathematician, US Naval Ord Test Sta, Calif, 51-53; analyst digital comput, Gen Dynamics/Convair, Tex, 53-56, group engr, 56-59; appl sci rep, Int Bus Mach Corp, Tex, 59-60; comput specialist, 60-62, assoc dir, 62-68, assoc prof, 66-68, prof indust eng, dir, Comput & Info Sci Div, 68-80, PROF COMPUT SCI & DIV HEAD, TEX A&M UNIV, 80- *Mem:* Asn Comput Mach; Soc Indust & Appl Math; Sigma Xi. *Res:* Digital computing; computer languages. *Mailing Add:* 209 Sulphur Springs Bryan TX 77801

DREW, DAVID A(BBOTT), chemical engineering; deceased, see previous edition for last biography

DREW, FRANCES L, b Pittsburgh, Pa, Apr 30, 17; m 36, 49, 83; c 3. PUBLIC HEALTH. *Educ:* McGill Univ, MD, 42; Univ Pittsburgh, MPH, 61, MFA, 87. *Prof Exp:* Instr, Dept Med, McGill Univ, 44-48; clin instr, Dept Med, 49-61, clin asst prof prev med, 61-69, clin assoc prof community med, 69-74, clin assoc prof community med, Pop Div, Grad Sch Pub Health, 73-76, assoc dean student affairs, 75-84, CLIN PROF COMMUNITY MED, SCH MED, UNIV PITTSBURGH, 74- *Concurrent Pos:* Mem, Senate Univ Press Comt, Univ Pittsburgh, 70-, chmn, 72-76; mem, Chancellor's Comn Tenure, Univ Pittsburgh, 72-73, adv, Sch Nursing, Regional Med Prog grants, Nurse Practitioner Prog, 72-73, mem, Affirmative Action Comt, Sch Med, 73-, chmn, 74-, mem, Univ Ctr Int Activities, Prog Policy Comt, 73-77, chmn, Fac Asn Med Sch, 74-75, chmn comt guidelines for fac contract renewal, Sch Med, 74-75. *Mem:* AMA; Am Psychosom Soc; NY Acad Sci. *Res:* Hypertension; epidemiology of chronic diseases; social determinants of disease. *Mailing Add:* 5023 Frew Ave Pittsburgh PA 15213

DREW, HENRY D, b Buffalo, NY, Sept 8, 41; m 66; c 2. ANALYTICAL CHEMISTRY. *Educ:* Canisius Col, BS, 63; Seton Hall Univ, MS, 65, PhD(chem), 67. *Prof Exp:* Res assoc, Seton Hall Univ, 64-66; vis asst prof, Purdue Univ, 66-68; asst prof chem, Southern Ill Univ, Edwardsville, 68-72, assoc prof, 72-; AT CTR DRUG ANALYSIS, FOOD & DRUG ADMIN. *Concurrent Pos:* Vis assoc prof, State Univ NY Buffalo, 76-77. *Mem:* Am Chem Soc; Soc Appl Spectros. *Res:* Analytical instrumentation, enzyme kinetics and equilibria; computer assisted instruction. *Mailing Add:* Ctr Drug Analysis Rm 1002 Food & Drug Admin 1114 Market St St Louis MO 63101-2045

DREW, HOWARD DENNIS, b Newark, Ohio, June 7, 39; m 65. SOLID STATE PHYSICS. *Educ:* Univ Pittsburgh, BS, 62; Cornell Univ, PhD(physics), 68. *Prof Exp:* Res assoc, 67-70, asst prof, 70-75, ASSOC PROF SOLID STATE PHYSICS, UNIV MD, COLLEGE PARK, 75- *Mem:* Am Phys Soc; Am Optical Soc; AAAS. *Res:* Optical spectroscopy of metals, alloys and surfaces and far infrared spectroscopy of semimetals and semiconductors. *Mailing Add:* Dept Physics Univ Md College Park MD 20742

DREW, JAMES, b Flushing, NY, Sept 21, 30; m 56; c 3. SOIL SCIENCE. *Educ:* Rutgers Univ, BS, 52, PhD(soil sci), 57. *Prof Exp:* Asst arctic soils, Rutgers Univ, 55-57; from asst prof to prof agron, Univ Nebr, Lincoln, 57-76, asst dean grad col, 70-71, assoc dean grad studies, 71-74, dean grad studies, 74-76; PROF AGRON, DEAN SCH AGR & LAND RESOURCES MGT & DIR AGR EXP STA, UNIV ALASKA-FAIRBANKS, 76- *Mem:* Fel AAAS; fel Am Soc Agron; Int Soc Soil Sci; Soil Conservation Soc Am; Sigma Xi; fel Soil Sci Soc Am; Am Forestry Asn. *Res:* Soil genesis, classification, survey and mineralogy; remote sensing of soil resources. *Mailing Add:* 4725 Villanova Dr Fairbanks AK 99701

DREW, JOHN H, b Cleveland, Ohio, Nov 7, 43; m 69; c 2. MATHEMATICS. *Educ:* Case Inst Technol, BS, 65; Univ Minn, PhD(math), 70. *Prof Exp:* Asst prof, 70-76, chmn dept, 83-86, ASSOC PROF MATH, COL WILLIAM & MARY, 76- *Mem:* Math Asn Am. *Res:* Permanents of positive semidefinite matrices, sign nonsingularity and other strong forms of nonsingularity, positive matrices and rank; maximizing the permanent function over the set of hermetian matrices with fixed nonnegative eigenvalues. *Mailing Add:* Dept Math Col William & Mary Williamsburg VA 23185

DREW, LAWRENCE JAMES, b Astoria, NY, Dec 18, 40; m; c 1. STATISTICS, GEOLOGY. *Educ:* Univ NH, BS, 62; Pa State Univ, MS, 64, PhD, 66. *Prof Exp:* Fel, Pa State Univ, 66-67; res geologist, US Bur Mines, 67; res scientist, Geotech Div, Teledyne Inc, 67-69; sr res geologist, Cities Serv Oil Co, 69-72; GEOLOGIST, US GEOL SURV, 72- *Mem:* Int Asn Math Geol; Am Statist Asn. *Res:* Statistical and economic analysis of supply of mineral resources; analysis and modeling of the exploration process. *Mailing Add:* US Geol Surv Mail Stop 920 Reston VA 22092

DREW, LELAND OVERBEY, b Charleston, SC, June 13, 23; m 44; c 1. AGRICULTURAL ENGINEERING. *Educ:* Clemson Col, BS, 43; Iowa State Col, MS, 45; Mich State Univ, PhD(agr eng), 63. *Prof Exp:* Instr agr eng, Clemson Col, 43-44; asst agr engr, Edisto Br, SC Agr Exp Sta, 45-47; from asst prof to assoc prof, Univ Ga, 47-56; adv, Int Co-op Admin, Lebanon, 56-59 & Pakistan, 59-60; assoc prof, Clemson Univ, 63-68; assoc prof, Ohio State Univ, 68-77, prof agr eng, 77-; ENG TECHNOL DEPT, CLEMSON UNIV, SC. *Concurrent Pos:* Adv to dean, Col Technol & Agr Eng, Univ Udaipur, India. *Mem:* Am Soc Agr Engrs; Am Soc Eng Educ; Sigma Xi. *Res:* Agricultural machinery, including tillage and seedling mechanics; chemical preservation of hay; mechanization of vegetable production; instrumentation. *Mailing Add:* Dept Civil Eng Clemson Univ Main Campus Clemson SC 29634

DREW, PHILIP GARFIELD, b Dedham, Mass, Jan 25, 32; m 61, 78; c 4. MEDICAL SCIENCES, HEALTH SCIENCES. *Educ:* Carnegie Inst Technol, BS, 54; Harvard Univ, SM, 59, PhD(eng), 64. *Prof Exp:* Res asst control systs eng, Harvard Univ, 61-64; mem prof staff eng, Arthur D Little, Inc, 64-76, mem sr staff, 76-81; PRES, DREW CONSULTS, INC, 81- *Mem:* Inst Elec & Electronics Engrs; Soc Photo-Optical Instrumentation Engrs; Sigma Xi. *Res:* Systems analyses; medical imaging equipment design. *Mailing Add:* 101 Bedford Rd Carlisle MA 01741

DREW, ROBERT TAYLOR, b Red Bank, NJ, Apr 22, 36; m 62; c 3. TOXICOLOGY. *Educ:* Rensselaer Polytech Inst, BS, 58; NY Univ, MS, 62, PhD(radiation health), 68; Am Bd Toxicol, dipl. *Prof Exp:* Sanit chem trainee, NY State Dept Health, 58-59; chemist, NJ Dept Health, 60; res asst, Inst Environ Med, NY Univ Med Ctr, 60-67, assoc res scientist, 67-69, instr, 69-70; sr staff fel, Nat Inst Environ Health Sci, 70-74; supvry phys scientist, Nat Inst Environ Health Sci, 74-76; adj prof, Dept Path, State Univ NY, Stony Brook, 79-86; dir inhalation & toxicol, Brookhaven Nat Lab, 76-86; DIR, HEALTH & ENVIRON SCI DEPT, AM PETROL INST, 86- *Concurrent Pos:* Sr staff fel, Nat Inst Environ Health Sci, 70-74. *Mem:* Soc Toxicol; Health Physics Soc; Am Indust Hyg Asn; Am Conf Govt Indust Hygienists; NY Acad Sci. *Res:* Inhalation toxicology, especially effects of short and long term exposures to environmental agents and combinations of agents; environmental radiation. *Mailing Add:* Health & Environ Sci Dept Am Petrol Inst 1220 L St NW Washington DC 20005

DREW, RUSSELL COOPER, b Chicago, Ill, Aug 16, 31; m 53; c 5. ENVIRONMENTAL CHEMISTRY, INSTRUMENT DESIGN. *Educ:* Univ Colo, BS, 53; Duke Univ, PhD(physics), 61. *Prof Exp:* Sr staff mem, Off Sci & Technol, Exec Off President US, 65-72, assoc dir, 76-77; head, Off Naval Res, London, 72-73; dir, Sci & Technol Policy Off, NSF, 73-76; vpres, Wash Opers, SCI, Inc, 77-80; pres, Sci-Tech Co, 80-83; PRES, VIKING INSTRUMENTS CORP, 83- *Concurrent Pos:* Distinguished lectr, Am Inst Aeronaut & Astronaut, 68-70; consult, Off Technol Assessment, US Cong, 79-84; mem, Adv Comt Indust Sci & Technol Off, NSF, 85- *Honors & Awards:* Centennial Medal, Inst Elec & Electronics Engrs, 84. *Mem:* Fel Inst Elec & Electronics Engrs (pres, 88, 89); fel AAAS; Am Chem Soc; Am Soc Mass Spectrometry. *Res:* New instrumentation for environmental analyses using gas chromatography, mass spectrometry and advanced software for instrument control; development of tandem mass spectrometers based upon magnetic sector technology. *Mailing Add:* Viking Instruments Corp 12007 Sunrise Valley Dr Reston VA 22091

DREW, T JOHN, b Pinjarra, Australia, Oct 23, 41; m 67; c 4. FOREST MANAGEMENT, FOREST GENETICS. *Educ:* BC Inst Technol, dipl forestry, 67; NC State Univ, BSc, 70, PhD(forestry), 73; Wash State Univ, MBA, 84. *Prof Exp:* Sr cruiser & party chief, Forestal Int Ltd, Vancouver, BC, 67-68; res asst forestry, Sch Forest Resources, NC State Univ, 70-72; chief res & develop, Jari Florestal E Agropecuaria Ltd, Brazil, 72-74; scientist, Weyerhauser, Wash, 74-79, mgr forest regeneration & res, Weyerhauser, Can Ltd, 79-84; dir, Reforestation & Reclamation Div, Alberta Forest Serv, 84-89; DIR GEN, FORESTRY CAN, PAC, 89- *Mem:* Can Inst Forestry. *Res:* Promote sustainable development of British Columbia's forest resource and the international competitiveness of the forestry business. *Mailing Add:* Pac Forestry Centre 506 W Burnside Rd Victoria BC T5K 2M4 Can

DREW, WILLIAM ARTHUR, b Grosse Pointe, Mich, Apr 29, 29; m 57; c 1. ARACHNOLOGY. *Educ:* Marietta Col, AB, 51; Mich State Univ, PhD(entom), 58. *Prof Exp:* From asst prof to assoc prof, 58-68, PROF ENTOM, OKLA STATE UNIV, 68-, CUR, 58- *Mem:* Entom Soc Am; Entom Soc Can; Brit Arachnological Soc; Am Arachnological Soc. *Res:* Curator insect collection Oklahoma; taxonomy of Hemiptera and spiders. *Mailing Add:* Dept Entom Okla State Univ Main Campus Stillwater OK 74078

DREWES, HARALD D, b Ger, Nov 22, 27; nat US; m 57; c 2. GEOLOGY, APPLIED & ECONOMIC ENGINEERING. *Educ:* Wash Univ, BA, 51; Yale Univ, MA, 52, PhD, 54. *Prof Exp:* GEOLOGIST, US GEOL SURV, 54- *Concurrent Pos:* Tectonics & mineral assessment, North Punjab, Pakistan; Binny fel, 52; Stanalind fel, 53; Gilbert fel, 84-85. *Honors & Awards:* Ohle Prize, 51; Bradley Lectr, 87. *Mem:* Fel Geol Soc Am; Ger Geol Asn. *Res:* Thrust tectonics of the basin and range geologic province, especially southeastern California, eastern Nevada, southeastern Arizona, southwestern Mexico, and northern Chihuahua, related problems of faulting, gneiss dome development, magmatism, sedimentation and mineralization. *Mailing Add:* US Geol Surv Mail Stop 905 Fed Ctr Box 25046 Denver CO 80225

DREWES, LESTER RICHARD, b Deshler, Ohio, Apr 11, 43; m 66; c 3. NEUROSCIENCES. *Educ:* Capital Univ, BS, 65; Univ Minn, PhD(biochem), 70. *Prof Exp:* Fel biochem, Univ Wis, 70-73, asst scientist, 73-76; asst prof, 76-80, assoc prof, 81-87, PROF & HEAD BIOCHEM, UNIV MINN, DULUTH, 87- *Concurrent Pos:* Alexander Von Humboldt fel, 83. *Mem:* Am Soc Biol Chemists; Am Physiol Soc; Am Soc Neurochem; AAAS; Int Soc Neurochem. *Res:* Transport and utilization of substrates by cells of the central nervous system in vivo and in vitro, under normal and pathological conditions; neurochemistry; neurotoxicology. *Mailing Add:* Dept Biochem Univ Minn Duluth MN 55812

DREWES, PATRICIA ANN, b Chicago, Ill, Jan 12, 32. CLINICAL CHEMISTRY. *Educ:* Immaculate Heart Col, BA, 53; Univ Calif, Los Angeles, PhD(biol chem), 65. *Prof Exp:* Control chemist, Vitaminerals, Calif, 53-59; res biochemist, Dept Neurochem & Neuropharmaceut Res, Vet Admin Hosp, Sepulveda, Calif, 59-60; res fel med, Harvard Med Sch & Beth Israel Hosp, 65-67; res scientist, 67-74, asst dir res, 74-77, DIR GEN CHEM, BIO-SCI LABS, 77- *Mem:* AAAS; Am Chem Soc; NY Acad Sci; Am Asn Clin Chem. *Res:* Pharmaceutical analysis; serotonin metabolism; metabolic response of transplantable murine tumors and leukemia cells to radio- and chemotherapy; enzymology and radioimmunoassay. *Mailing Add:* 627 N McCadden Pl Los Angeles CA 90004-1029

DREWES, ROBERT CLIFTON, b San Francisco, Calif, Feb 14, 42; m 64; c 4. FUNCTIONAL MORPHOLOGY, AFRICAN AFFAIRS. *Educ:* San Francisco State Univ, BA, 69; Univ Calif, Los Angeles, PhD(biol), 81. *Prof Exp:* Asst dir, Nairobi Snake Park, Nat Mus Kenya, 69-70; curatorial assoc herpet, Calif Acad Sci, 70-73; teaching fel biol, Univ Calif, Los Angeles, 73-75; asst cur, 75-81, chmn dept, 82-88, ASSOC CUR, DEPT HERPET, CALIF ACAD SCI, 81-, DIR MUS PROGS, 88- *Concurrent Pos:* Res assoc, Nat Mus Kenya, 70-; bd dir, San Francisco Zool Soc, 79-81; mem world checklist comn, Asn Syst Collections, 81-; mem bd gov, Am Soc Ichthyologists & Herpetologists, 85- *Mem:* Fel Royal Geog Soc; Am Soc Ichthyologists & Herpetologists; Soc Study Amphibians & Reptiles; Herpet Asn Africa; Sigma Xi. *Res:* Systematics and ecology of African herpetofauna; comparative physiology of arid-adapted amphibians; evolutionary relationships of ranoid frogs. *Mailing Add:* Dept Herpet Calif Acad Sci Golden Gate Park San Francisco CA 94118

DREWRY, WILLIAM ALTON, b Dyess, Ark, Oct 23, 36; m 81; c 3. CIVIL ENGINEERING, ENVIRONMENTAL ENGINEERING. *Educ:* Univ Ark, BS, 59, MS, 61; Stanford Univ, PhD(environ eng), 68. *Prof Exp:* Instr civil eng, Univ Ark, 60-62; res asst environ eng, Stanford Univ, 62-65; asst prof, Univ Ark, 65-68; from assoc prof to prof, Univ Tenn, Knoxville, 68-76; prof civil eng & chmn dept, 76-84, PROF ENVIRON & SANIT ENG, OLD DOMINION UNIV, 76-, ASSOC DEAN, 88- *Concurrent Pos:* Consult, indust & munic. *Mem:* Water Pollution Control Fedn; Am Soc Eng Educ; fel Am Soc Civil Eng; Prof Engrs Educ (vpres, 90-91); Am Water Works Asn. *Res:* Water pollution control, water treatment, alum sludge dewatering; computer modeling of sanitary sewer systems. *Mailing Add:* Col Eng & Technol Old Dominion Univ Norfolk VA 23529-0236

DREWS, MICHAEL JAMES, b Milwaukee, Wis, Oct 2, 45. POLYMER CHEMISTRY, TEXTILE CHEMISTRY. *Educ:* Univ Wis-Madison, BS, 67; NTex State Univ, PhD(phys chem), 71. *Prof Exp:* Res assoc, Clemson Univ, 71-72, sr res assoc, 73-74, from vis asst prof to prof, 74-89, J E SIRRINE PROF TEXTILES, CLEMSON UNIV, 89- *Concurrent Pos:* Co-dir, Adv Eng Fibers Lab, 87- *Honors & Awards:* Educ Serv Award, Plastics Inst Am, 75. *Mem:* Am Asn Textile Chemists and Colorists; Am Chem Soc; Fiber Soc. *Res:* Flame retardants and the chemistry of flame retardant materials; fiber reinforced composite materials; surfactants and interfacial phenomena; spectroscopy as applied to polymeric materials; textile dyeing and finishing processes. *Mailing Add:* Sch Textiles Clemson Univ Clemson SC 29634-1307

DREXLER, EDWARD JAMES, b Cincinnati, Ohio, Feb 9, 38; m 63; c 5. ANALYTICAL CHEMISTRY. *Educ:* Xavier Univ, Ohio, BS, 59, MS, 62; Wayne State Univ, PhD(zirconium chem), 65. *Prof Exp:* Asst, Xavier Univ, 59-61; assoc, Wayne State Univ, 61-64; from asst prof to assoc prof, 64-81, PROF CHEM, UNIV WIS, WHITEWATER, 81- *Mem:* Am Chem Soc; Sigma Xi. *Res:* Analytical chemistry of zirconium and hafnium; specific ion electrodes. *Mailing Add:* Dept Chem Univ Wis Whitewater WI 53190

DREXLER, HENRY, b Carnegie, Pa, June 24, 27; m 57; c 2. MICROBIOLOGY. *Educ:* Pa State Univ, BS, 54; Univ Rochester, PhD(microbiol), 60. *Prof Exp:* Technician clin chem, Brookhaven Nat Lab, 54-56; instr microbiol, Sch Med, Univ Southern Calif, 60-62; USPHS fel microbial genetics, Karolinska Inst, Sweden, 62-64; from asst prof to assoc prof, 64-75, PROF MICROBIOL, BOWMAN GRAY SCH MED, 75- *Mem:* Am Soc Microbiol. *Res:* Microbial genetics, especially bacteriophage-host cell relationships and transduction. *Mailing Add:* Dept Microbiol & Immunol Bowman Gray Sch Med 300 S Hawthorne Winston-Salem NC 27103

DREYER, DAVID, organic chemistry, for more information see previous edition

DREYER, DUANE ARTHUR, b Ypsilanti, Mich, May 10, 42; m 65; c 2. NEUROPHYSIOLOGY, NEUROPHARMACOLOGY. *Educ:* Univ Cincinnati, BS, 65; Univ Pittsburgh, PhD(pharmacol), 71. *Prof Exp:* Instr pharmacol, Sch Med, Univ Pittsburgh, 71-72; res assoc physiol, Div Neurosurg, Med Ctr, Duke Univ, 72-75; from assoc prof physiol, Sch Med & asst prof oral biol, Sch Dent, Univ NC, Chapel Hill, 75-86; INSTR, DURHAM TECH COMMUNITY COL, 90- *Concurrent Pos:* Computer consult; Nat Inst Dent Res career develop award, 75-80. *Mem:* Am Physiol Soc; Soc Neurosci. *Res:* Sensory neurophysiology; psychophysics; somatic sensation. *Mailing Add:* Durham Tech Community Col 1637 Lawson St Durham NC 27703

DREYER, KIRT A, b Bemidji, Minn, Aug 10, 39; m 59; c 2. PHYSICAL INORGANIC CHEMISTRY, ANALYTICAL CHEMISTRY. *Educ:* Bemidji State Col, BS, 61; Univ Iowa, PhD(chem), 71. *Prof Exp:* PROF CHEM, BEMIDJI STATE UNIV, 65- *Mem:* Am Chem Soc. *Res:* Metal complexes of amino acids; kinetics of rapid exchange processes; cadmium content of foodstuffs, particularly sunflower seeds and an isolation of the cadmium source; levels of chromium in biological samples, particularly those which serve as sources of chromium for humans; isolation characterization of chromium complexes in biological samples. *Mailing Add:* 2514 Grange Rd NW Bemidji MN 56601-8132

DREYER, ROBERT MARX, b Chicago, Ill, Jan 6, 14. ECONOMIC GEOLOGY. *Educ:* Northwestern Univ, BS, 34; Calif Inst Technol, MS, 37, PhD(geol), 39. *Prof Exp:* Asst geol, Northwestern Univ, 34-35; soil survr, US Soil Conserv Serv, 36; geologist, US Geol Surv, 38; instr, Univ Kans, 39-41, from asst prof to prof geol & chmn dept, 42-53; chief geologist, Kaiser Aluminum & Chem Corp, 53-59; asst chief geologist, Reynolds Metals Co, 59-61; pres, Western Mineral Assocs, 62-64; mgr mineral div, Assoc Oil & Gas Co, 64-65; PRES, WESTERN MINERAL ASSOCS, 65- *Mem:* Fel Geol Soc Am; Soc Econ Geol; Am Inst Mining, Metall & Petrol Engrs; fel Mineral Soc Am; Am Asn Petrol Geologists. *Res:* Petrography; geographic geology; mineralogy; geophysics; geochemistry. *Mailing Add:* PO Box 1540 Mill Valley CA 94941

DREYER, WILLIAM J, b Kalamazoo, Mich, Aug 11, 28; m 52; c 3. MOLECULAR BIOLOGY, IMMUNOLOGY. *Educ:* Reed Col, BA, 52; Univ Wash, PhD(biochem), 56. *Prof Exp:* Fel, Polio Found, NIH, 56-57, res biochemist, 57-63; PROF BIOL, CALIF INST TECHNOL, 63- *Concurrent Pos:* Consult automated biomed anal to numerous orgn, 63-; mem adv bd, Hereditary Dis Found, 72- *Mem:* Soc Biol Chemists; Asn Advan Med Instrumentation; Biophys Soc; Genetics Soc Am. *Res:* Genetic and molecular basis of antibody formation and embryogenesis; tumor immunology; protein chemistry; development of automated biomedical instruments. *Mailing Add:* Div Biol Calif Inst Technol 156-29 Pasadena CA 91125

DREYFUS, MARC GEORGE, b Brooklyn, NY, Mar 5, 26; m 54, 78; c 3. OPTICAL PHYSICS. *Educ:* Harvard Univ, AB, 47, SM, 49. *Prof Exp:* Asst spectros, Mass Inst Technol, 49-50; physicist, Am Optical Co, 50-54, Librascope, Inc, 54-59, Barnes Eng Co, 59-63, Bulova Watch Co, 64 & Philips Labs, 65-67; chief scientist, BAI Corp, 67-75; pres, Dreyfus-Pellman Corp, 75-80; sr scientist, Am Cystoscope Makers Inc, 80-86; ASST PROF OPTICAL ENG TECHNOL, NORWALK STATE TECH COL, 87- *Concurrent Pos:* Aerial photographer, US Army Air Corps, 44-45. *Mem:* Fel Optical Soc Am; Sigma Xi. *Res:* Optics; spectroscopy; colorimetry; optical information processing; pattern recognition; electro-optical process control; fiber optics. *Mailing Add:* 337 Eastfield Dr Fairfield CT 06432

DREYFUS, PIERRE MARC, b Geneva, Switz, Oct 14, 23; US citizen; m 47; c 3. NEUROCHEMISTRY, NUTRITION. *Educ:* Tufts Col, BS, 47; Columbia Univ, MD, 51; Am Bd Psychiat & Neurol, dipl, 58. *Prof Exp:* Intern & asst resident med, NY Hosp, 51-53; asst & chief resident neurol, Harvard Med Sch, 53-55, asst neurol, 58-59, instr, 59-62, assoc, 62-66, asst prof, 66-68; PROF NEUROL & CHMN DEPT, UNIV CALIF, DAVIS, 68-; DIR, SACRAMENTO MED CTR, 68- *Concurrent Pos:* Res fel neuropath, Harvard Med Sch, 55-58; clin assoc, McLean Hosp, 62-67; adv neurosci, WHO, 74- *Mem:* Am Neurol Asn; Am Soc Clin Nutrit; Am Inst Nutrit; Int Soc Neurochem; Am Soc Neurochem. *Mailing Add:* PO Box 3248 Ketchum ID 83340

DREYFUS, RUSSELL WARREN, b Michigan City, Ind, Dec 23, 29; div; c 2. LASERS. *Educ:* Purdue Univ, BS, 51, MS, 53; Yale Univ, MS, 58, PhD(physics), 60. *Prof Exp:* Tech staff mem semiconductor res, Hughes Aircraft Co, 53-54; guest assoc scientist, Brookhaven Nat Lab, 54-56; PHYSICIST, IBM CORP, 58- *Concurrent Pos:* Guest scientist, Swiss Fed Inst Technol, 64, Kernforschwngsanlage Juelich & Max Planck Plasma Physics, WGer, 81-82. *Mem:* Fel Am Phys Soc; NY Acad Sci; fel Am Optical Soc. *Res:* Defects in ionic and semiconductor crystals by electrical, optical and mechanical properties; crystal growing; dynamical measurements using lasers; ultraviolet lasers pumped by relativistic electron beams; generation and detection of gas phase ions and atoms using tunable lasers; generation of Lyman-alpha radiation for measuring atomic hydrogen in tokamaks; photo ablation of surfaces by UV lasers; photothermal deformation of semiconductor crystal surfaces due to atomic layers. *Mailing Add:* Thomas J Watson Res Ctr IBM Corp PO Box 218 Yorktown Heights NY 10598

DREYFUS, STUART ERNEST, b Ind, Oct 19, 31. OPERATIONS RESEARCH. *Educ:* Harvard Univ, AB, 53, PhD(appl math), 64. *Prof Exp:* Actuarial clerk, Metrop Life Ins Co, 53-54; numerical analyst, Gen Elec Co, 54-55; mathematician, Rand Corp, 55-67; assoc prof, 67-72, PROF INDUST ENG & OPERS RES, UNIV CALIF, BERKELEY, 72- *Res:* Mathematical methods of optimization; operations research; computational aspects of dynamic programming and variational problems; limits of scientific decision-making. *Mailing Add:* Dept Indust Eng Univ Calif Berkeley Berkeley CA 94720

DREYFUSS, JACQUES, b St Gall, Switz, Jan 20, 37; US citizen; wid; c 2. BIOCHEMISTRY. *Educ:* Beloit Col, BS, 58; Johns Hopkins Univ, PhD(biochem), 63. *Prof Exp:* Fel biochem, Princeton Univ, 63-64; sect head drug metab, 64-77, SR RES FEL, DEPT DRUG METAB, E R SQUIBB & SONS, INC, 83- *Mem:* Am Soc Pharmacol & Exp Therapeut; NY Acad Sci; Am Soc Clin Pharmacol Therapeut. *Res:* Drug metabolism; central nervous system and cardiovascular agents. *Mailing Add:* Dept Drug Metab Squibb Inst Med Res PO Box 4000 Princeton NJ 08543-4000

DREYFUSS, M(AX) PETER, b Frankfurt, Ger, Sept 24, 32; nat US; m 54; c 2. POLYMER CHEMISTRY. *Educ:* Union Col, BS, 52; Cornell Univ, PhD(chem), 57. *Prof Exp:* From res chemist to sr res chemist, Res Ctr, Corp Res, B F Goodrich, Co, 56-67, res assoc, 66-72, sr res assoc, 72-84; res prof & sr res scientist, Mich Molecular Inst, Midland, Mich, 84-90; CONSULT, 90- *Concurrent Pos:* Imp Chem Indust fel, Univ Liverpool, 63-65. *Mem:* Am Chem Soc. *Res:* Polymerization of cyclic ethers; polymer chemistry and synthesis; polymer modification; synthesis of model compounds. *Mailing Add:* 3980 Old Pine Trail Midland MI 48640

DREYFUSS, PATRICIA, b Reading, Pa, Apr 28, 32; m 54; c 2. POLYMER CHEMISTRY. *Educ:* Univ Rochester, BS, 54; Univ Akron, PhD(polymer sci), 64. *Prof Exp:* Am Asn Univ Women Marie Curie int fel, Univ Liverpool, 64-65; res chemist, B F Goodrich Res Ctr, 65-71; res assoc chem, Case Western Reserve Univ, 71-74, NIH spec fel, 72-73; res assoc, Inst Polymer Sci, Univ Akron, 74-84; prof chem & sr res scientist, Mich Molecular Inst, 84-90; CONSULT, 90- *Mem:* Adhesion Soc (secy, 81-83); Am Chem Soc. *Res:* Mechanisms and kinetics of cationic polymerization of cyclic ethers; synthesis and characterization of related graft copolymers; chemical structure-property relationships of polymers; morphology; adhesion. *Mailing Add:* 3980 Old Pine Trail Midland MI 48640

DREYFUSS, ROBERT GEORGE, b Frankfurt-am-Main, Ger, Sept 6, 31; nat US; m 56. INORGANIC CHEMISTRY, GLASS TECHNOLOGY. *Educ:* Union Col, BS, 52. *Prof Exp:* Mgr res & develop control lab, Thatcher Glass Mfg Co, Inc, 52-61; asst dir res & eng, Glass Containers Corp, 61-65; vpres tech serv, Containers, Kraftco Corp, 66-80; vpres tech serv, Metropak Containers Corp, 66-83; vpres eng, Midland Glass Co, 80-83; RETIRED. *Mem:* Am Chem Soc. *Res:* Glass manufacturing. *Mailing Add:* 2104 NW 127th St Vancouver WA 98685

DRIBIN, DANIEL MACCABAEUS, b Chicago, Ill, Dec 10, 13; m 36; c 1. ALGEBRAIC NUMBER FIELDS. *Educ:* Univ Chicago, SB, 33, SM, 34, PhD(math), 36. *Prof Exp:* Nat Res fel, Inst Advan Study & Yale Univ, 36-38; instr math, Univ Nebr, 38-42; jr res analyst, US War Dept, 42-46; analyst, Nat Security Agency & Predecessors, 46-73; exec grad comt, Dept Math, Univ Md, 73-84; RETIRED. *Concurrent Pos:* Lectr math, George Wash Univ, 46-58, assoc prof lectr, 58-60, prof lectr, 60-81. *Honors & Awards:* Legion Merit, 45. *Mem:* Am Math Soc; Math Asn Am. *Res:* Algebraic number fields; algebraic number theory; history of modern mathematics. *Mailing Add:* 1016 Kathryn Rd Silver Spring MD 20904

DRIBIN, LORI, TROPHIC FACTORS. *Educ:* Northwestern Univ, PhD(biol), 75. *Prof Exp:* ASSOC PROF ANAT, SOUTHEAST COL OSTEOP MED, 82- *Mailing Add:* Dept Anat Southeast Col Osteop Med 1750 NE 168th St North Miami Beach FL 33162

DRICKAMER, HARRY GEORGE, b Cleveland, Ohio, Nov 19, 18; m 42; c 5. CHEMISTRY, PHYSICS. *Educ:* Univ Mich, BSE, 41, MS, 42, PhD(chem eng), 46. *Prof Exp:* Res group leader, Pan Am Refining Corp, Tex, 42-46; from asst prof to assoc prof, 46-53, PROF CHEM, PHYSICS & CHEM ENG, UNIV ILL, URBANA, 53- *Honors & Awards:* Colburn Award, Am Inst Chem Eng, 47, Alpha Chi Sigma Award, 66, Walker Award, 72; Ipatieff Prize, Am Chem Soc, 56, Langmuir Award in Chem Physics, 74; Buckley Solid State Physics Prize, Am Phys Soc, 67; Bendix Res Award, Am Soc Eng Educ, 68; P W Bridgman Award, AIRAPT, 77; Michelson-Morley Award, Case Western Reserve Univ, 78; Chem Pioneers Award, Am Inst Chemists, 83; Alexander von Humboldt Award, Fed Repub Ger, 86; Warren K Lewis Award, Am Inst Chem Eng, 86; Peter Debye Award in Phys Chem, Am Chem Soc, 87; Robert A Welch Prize in Chem, Welch Found, 87; Elliot Cresson Medal, Franklin Inst, 88; Nat Medal Sci, 89; Outstanding Mat Res-DOE, 90. *Mem:* Nat Acad Sci; Am Chem Soc; Am Inst Chem Eng; Am Phys Soc; Am Geophys Union; Am Acad Arts & Sci; Am Philos Soc. *Res:* Physical chemistry; properties of matter at high pressure. *Mailing Add:* Sch Chem Sci Univ Ill Urbana IL 61801

DRICKAMER, KURT, b Champaign, Ill, Nov 4, 52. MEMBRANE BIOCHEMISTRY, CELL BIOLOGY. *Educ:* Stanford Univ, BS, 73; Harvard Univ, PhD(biochem), 78. *Prof Exp:* res fel biochem, Duke Univ, 78-79; res fel biochem, Cold Spring Harbor Lab, 79-81, sr scientist, 81; asst prof biochem, Univ Chicago, 81-86; ASSOC PROF BIOCHEM, COLUMBIA UNIV, 86- *Mem:* Am Soc Biol Chemists; Am Soc Cell Biol; Biochem Soc UK. *Res:* Molecular mechanisms of receptor-mediated endocytosis, with emphasis on the recognition of carbohydrates by cellular receptor; pathways of membrane biosynthesis. *Mailing Add:* Dept Biochem Columbia Univ 630 W 168th St New York NY 10032

DRICKAMER, LEE CHARLES, b Ann Arbor, Mich, May 25, 46; m 79. ANIMAL BEHAVIOR, ECOLOGY. *Educ:* Oberlin Col, BA, 67; Mich State Univ, PhD(zool), 70. *Prof Exp:* Res scientist primatology, NC Found Ment Health Res, 71-72; from asst prof to prof biol, Williams Col, 72-86; PROF & CHMN DEPT ZOOL, SOUTHERN ILL UNIV, CARBONDALE, 87- *Concurrent Pos:* NSF fel, NC State Univ & NC Dept Ment Health, 70-71; NIMH res grant, 73; assoc ed, Am Midland Naturalist, 74-80; NIH res grant, Nat Inst Child Health & Develop, 75-; vice secy gen, Int Ethological Soc, 83-; prog officer, Animal Behavior Soc, 83-86; consult ed, J Comp Psychol, 85-88; ed, Animal Behaviour, 88-91; secy-gen, Int Coun Ethologists, 91-95. *Mem:* Sigma Xi; Animal Behav Soc; Ecol Soc Am; Brit Ecol Soc; Int Primatol Soc; Am Soc Mammalogists. *Res:* Development of behavior, hormones and behavior, population biology, and behavioral ecology of rodents, primates and birds. *Mailing Add:* Dept Zool Southern Ill Univ Carbondale IL 62901-6501

DRIEDZIC, WILLIAM R, b Toronto, Ont, May 27, 48; m 71; c 2. CARDIAC METABOLISM, ENERGY METABOLISM. *Educ:* Univ BC, Vancouver, PhD(zool), 75. *Prof Exp:* PROF BIOL, MT ALLISON UNIV, NB, CAN, 77- *Honors & Awards:* Fraser Gold Medal. *Mem:* Can Soc Zool; Am Physiol Soc; Int Soc Heart Res; Soc Exp Biol. *Res:* Naturally occuring differences amongst animals are exploited to study the control of oxygen delivery and energy metabolism in heart; of particular interest are fishes which are successful in a wide range of environmental conditions. *Mailing Add:* Dept Biol Mt Allison Univ Sackville NB E0A 3C0 Can

DRIES, WILLIAM CHARLES, b Milwaukee, Wis, Nov 4, 30; m 57; c 3. MECHANICAL & METALLURGICAL ENGINEERING, METALLURGY & PHYSICAL METALLURGICAL ENGINEERING. *Educ:* Univ Wis, BS, 53, MS, 56, PhD(metall eng), 62. *Prof Exp:* Consult mech engr, Lofte & Fredericksen, 56-59; res asst metall eng, 60-62, PROJ ASSOC MECH ENG, UNIV WIS-MADISON, 62-; PRES, DRIES ASSOCS, INC, 62- *Mem:* Am Soc Mech Engrs; Nat Soc Prof Engrs; Am Soc Heat, Refrig & Air Conditioning Engrs; Am Soc Metals. *Res:* Physical metallurgy of iron; protective construction for nuclear weapons; project management using network methods; computer applications to engineering problems. *Mailing Add:* 7457 Elmwood Ave Middleton WI 53562

DRIESCH, ALBERT JOHN, b Pittsburgh, Pa, Sept 2, 20. ORGANIC CHEMISTRY. *Educ:* St Francis Col, Pa, BA, 43; Univ Notre Dame, MS, 51. *Prof Exp:* From asst prof to prof chem, St Francis Col, Pa, 56-88, head dept, 52-85; ASSOC DIR, ST FRANCE XAVIER CHURCH, 88- *Mem:* Am Chem Soc. *Res:* General organic chemistry. *Mailing Add:* Dept Chem St Francis Xavier Church 912 Seventh St Moundsville WV 26041

DRIESENS, ROBERT JAMES, b Redford, Mich, Jan 12, 18; m 53; c 3. BACTERIOLOGY. *Educ:* Calvin Col, AB, 39; Mich State Col, MS, 49, PhD(bact), 52. *Prof Exp:* Bacteriologist, Labs, Mich Dept Health, 52-60; head bact prod develop dept, Corvel, Inc, Omaha, 60-64, develop assoc, 64-65; sr bacteriologist, 65-69, sr microbiologist, 69-75, MEM TECH SERV, BIOCHEM PROD, ELI LILLY & CO, 75- *Mem:* Am Soc Microbiol; AAAS; Sigma Xi. *Res:* Mass culture of microbes; microbial conversions of chemicals; microbial nutrition. *Mailing Add:* 6631 Avalon Forest Dr Indianapolis IN 46250

DRIEVER, CARL WILLIAM, b Chicago, Ill, Mar 4, 38; m 64; c 2. CLINICAL PHARMACOLOGY. *Educ:* Purdue Univ, BS, 61, MS, 63, PhD(pharmacol), 65. *Prof Exp:* Teaching asst pharm, Sch Pharm, Purdue Univ, 61-62; asst prof pharmacol, Sch Pharm, Univ Md, 65-67; asst prof, 67-68, assoc prof pharmacol, 68-73, ASSOC PROF PHARMACY PRACT, COL PHARM, UNIV HOUSTON, 73-, MEM GRAD FAC, 80- *Mem:* Am Soc Hosp Pharmacists; Am Asn Cols Pharm. *Res:* Drug dependence and tolerance; utilization review; drug interactions; pharmacokinetics; geriatrics. *Mailing Add:* Col Pharm Univ Houston Houston TX 77030

DRIGGERS, FRANK EDGAR, b El Paso, Tex, Dec 14, 19; m 45; c 2. REACTOR PHYSICS. *Educ:* Univ Calif, AB, 40; Univ Mich, AM, 48, PhD(physics), 51. *Prof Exp:* Jr astronr, Naval Observ, Washington, DC, 40-41; res physicist, Eng Res Inst, Univ Mich, 46-50; reactor physicist, 51-64, MEM STAFF ADV OPER PLANNING, E I DU PONT DE NEMOURS & CO, INC, 64- *Mem:* Fel Am Nuclear Soc; AAAS. *Res:* Isotope technology; nuclear power analysis. *Mailing Add:* 707 Winged Foot Dr Aiken SC 29803

DRILL, VICTOR ALEXANDER, b Sunderland, Eng, June 10, 16; nat US; 25; m 40; c 3. PHARMACOLOGY. *Educ:* Long Island Univ, BS, 38; Princeton Univ, PhD(physiol), 41; Yale Univ, MD, 48. *Prof Exp:* Asst chemist, Fleischmann Labs, NY, 37-38; instr pharmacol, Col Physicians & Surg, Columbia Univ, 43-44; from instr to asst prof, Med Sch, Yale Univ, 44-48; prof, Med Sch, Wayne Univ, 48-53; dir biol res, G D Searle & Co, 53-70, dir sci & prof affairs, 70-74; PROF PHARMACOL, COL MED, UNIV ILL, CHICAGO, 75- *Concurrent Pos:* Jacobus fel, Princeton Univ, 41-42; Nat Res Coun fel, Northwestern Univ, 42-43; dir educ & assoc physician, Detroit Receiving Hosp, 49-53; prof lectr pharmacol, Col Med, Univ Ill, 54-74; lectr pharmacol, Med Sch, Northwestern Univ, 54-67; mem endocrine panel, Cancer Chemother Nat Serv Ctr, 58-62; consult, Nat Cancer Inst, 65-67; mem US Comn, World Med Asn; mem, Drug Res Bd, Nat Res Coun, 71-76; mem sci adv bd, Nat Ctr Toxicol Res, Fed Drug Admin, Dept Health, Educ & Welfare, 75-76. *Honors & Awards:* Kenneth P DuBois Mem Award, Outstanding Toxicologist Midwest, Midwest Regional Chap, Soc Toxicol, 82. *Mem:* Distinguished fel Soc Toxicol (pres, 72-73); fel NY Acad Sci; fel Royal Soc Med; Am Physiol Soc; Int Family Planning Res Asn (vpres, 72-74); Am Soc Pharmacol & Exp Therapeut; hon mem Pharmacol Soc Venezuela. *Res:* Endocrinology, pharmacology, oral contraceptives. *Mailing Add:* 224 S Fulton St Princeton WI 54968

DRINKARD, WILLIAM CHARLES, JR, b Eufaula, Ala, May 11, 29. INORGANIC CHEMISTRY. *Educ:* Huntingdon Col, BA, 50; Ala Polytech Inst, MS, 52; Univ Ill, PhD(inorg chem), 56. *Prof Exp:* Chemist, Food Mach & Chem Corp, 52-53; chemist, E I du Pont de Nemours & Co, 55; asst prof chem, Univ Calif, Los Angeles, 56-60; mem staff, Exp Sta, 60-70, res supvr, Plastics Dept, 70-72, lab dir, Sabine River Lab, 72-74, ASSOC DIR, CENT RES & DEVELOP DEPT, EXP STA, E I DU PONT DE NEMOURS & CO, INC, 74- *Mem:* Am Chem Soc. *Res:* Complex inorganic compounds; ligand reactivity; homogeneous catalysis by metal ions; organometallic chemistry. *Mailing Add:* 104 Center Ct Wilmington DE 19810

DRINKER, PHILIP A, b Brookline, Mass, Apr 7, 32; m 53; c 4. BIOENGINEERING. *Educ:* Yale Univ, BS, 54; Mass Inst Technol, PhD(hydrodyn), 61. *Prof Exp:* Asst prof civil eng, Mass Inst Technol, 61-65; from res assoc to prin assoc surg & chief, Bioeng Div, Harvard Med Sch, 65-73, sr assoc surg, 73-90; chief, Dept Biomed Eng, Brigham & Women's Hosp, 80-90; RETIRED. *Concurrent Pos:* Consult surg & dir clin eng serv, Peter Bent Brigham Hosp, 65-80; from lectr to sr lectr, Mass Inst Technol, 65-; mem, Cardiol Adv Comt, Nat Heart, Lung & Blood Inst, 75-77; consult bioeng, New England Sinai Hosp, 78-90. *Mem:* Biomed Eng Soc; Am Soc Artificial Internal Organs; Rehab Eng Soc NAm. *Res:* Rehabilitation engineering; blood oxygenation; equipment design for respiratory care. *Mailing Add:* 48 Cedar Rd Belmont MA 02178

DRINKWATER, WILLIAM OTHO, b Providence, RI, Sept 29, 19; m 49; c 2. HORTICULTURE. *Educ:* Univ Mass, BVA, 47, MS, 49; Rutgers Univ, PhD(hort), 54. *Prof Exp:* Asst veg crops, Rutgers Univ, 51-52; instr, Univ Conn, 49-52, asst prof, 52-56; from asst prof to prof 56-81, EMER PROF HORT & FORESTRY, RUTGERS UNIV, 81- *Mem:* Am Soc Hort Sci; AAAS; Sigma Xi. *Res:* Undergraduate teaching; extension teaching. *Mailing Add:* 324 Midway Island Clearwater FL 34630

DRINNAN, ALAN JOHN, b Bristol, Eng, Apr 6, 32; US citizen; m 56; c 2. ORAL PATHOLOGY. *Educ:* Univ Bristol, BDS, 54, MB, ChB, 62; FDSRCS, 62; State Univ NY Buffalo, DDS, 64. *Prof Exp:* Tutor oral surg, Univ Bristol, 57-58; vis asst prof oral path, 62-65, assoc prof, 65-70, PROF ORAL MED & CHMN DEPT, SCH DENT, STATE UNIV NY BUFFALO, 70- *Concurrent Pos:* Consult dentist & chief dent serv, Buffalo Gen Hosp, 66-; WHO consult, Port Moresby Dent Col, Papua, New Guinea, 71; Fulbright prof, Univ Melbourne, 81. *Mem:* Int Asn Dent Res; fel Am Acad Oral Path (pres, 79-80); Am Dent Asn; Int Asn Oral Pathologists (pres, 90-). *Res:* Experimental carcinogenesis; dental education; medical-dental relationships. *Mailing Add:* 3435 Main St Buffalo NY 14214

DRISCOLL, BERNARD F, NEUROIMMUNOLOGY, CELL BIOLOGY. *Educ:* Cath Univ Am, PhD(cell biol), 70. *Prof Exp:* RES BIOLOGIST, NIMH, NIH, 70- *Mailing Add:* NIMH NIH Bethesda MD 20892

DRISCOLL, CHARLES F, b Tucson, Ariz, Feb 28, 50; m 72; c 2. NONNEUTRAL PLASMAS. *Educ:* Cornell Univ, BA, 69; Univ Calif, San Diego, MS, 71, PhD(physics), 76. *Prof Exp:* Res physicist, Gulf Gen Atomic Co, 69; res asst, Univ Calif, San Diego, 71-76, post grad researcher, 77-78, asst researcher, 78-84; res physicist, Molecular Biosysts, Inc, 81-82; assoc researcher, 84-90, RESEARCHER, UNIV CALIF, SAN DIEGO, 90- *Concurrent Pos:* Consult, Sci Applns, Inc, 80-81; Lectr, Univ Calif, San Diego, 85-90, sr lectr, 90- *Honors & Awards:* Excellence in Plasma Physics Res Award, Am Phys Soc, 91. *Mem:* Fel Am Phys Soc; AAAS; Math Asn Am. *Res:* Experimental and theoretical research on waves, transport, and equilibrium of nonneutral plasmas and two-dimensional inviscid fluids. *Mailing Add:* Physics Dept 0319 Univ Calif-San Diego 9500 Gilman Dr La Jolla CA 92093

DRISCOLL, DENNIS MICHAEL, b Warren, Pa, July 10, 34; m 69. METEOROLOGY. *Educ:* Pa State Univ, BS, 59, MS, 61; Univ Wis, PhD(meteorol), 71. *Prof Exp:* Res assoc meteorol, Travelers Res Ctr, 61-62; instr geog, Univ Wis-Milwaukee, 62-64, teaching asst meteorol, 65-66, instr geog, Univ Wis Ctr Syst, 64-65; ASSOC PROF METEOROL, TEX A&M UNIV, 69- *Mem:* Am Meteorol Soc; Int Soc Biometeorol. *Res:* Human biometeorology; statistical meteorology-climatology; forecast verification. *Mailing Add:* Dept Meteorol Tex A&M Univ College Station TX 77843

DRISCOLL, DOROTHY H, b Boston, Mass, May 30, 24. MEDICAL BIOPHYSICS. *Educ:* Radcliffe Col, SB, 46; Smith Col, MA, 49. *Prof Exp:* Instr sci, Northampton Sch Girls, 46-47; from asst biol sci to instr zool, Smith Col, 48-50; technician, Biophys Lab, Harvard Med Sch, 51-53, res asst clin isotopes, 53-55; res assoc & health safety officer, Atomic Bomb Casualty Comn, Nat Acad Sci, 54-56; jr tech specialist, Med Physics Div, Brookhaven Nat Lab, 56-61; res assoc radiol, 61-67, asst prof, 67-71, assoc prof med physics, 71-76, ASSOC PROF RADIOL, MED COL, THOMAS JEFFERSON UNIV, 76- *Concurrent Pos:* Investr, Marine Biol Labs, Woods Hole, 49-50; adj lectr radiol health prog, Physics Div, Manhattan Col, 65-69; consult clin radioisotopes, Sacred Heart Hosp, Allentown, 66-68, dept oncol, Children's Hosp, Philadelphia, 71- & Am Found New Affairs-New Access Routes to Med Careers, 72-; affil staff, Dept Obstet & Gynec, St Barnabas Hosp, NJ, 74- *Mem:* Am Asn Physicists in Med; Radiol Soc NAm; Radiation Res Soc; Soc Nuclear Med; Int Soc Magnetic Resonance; Sigma Xi. *Res:* Clinical isotopes; electron paramatic resonance applied to oncology. *Mailing Add:* 3856 Violet Dr Philadelphia PA 19154

DRISCOLL, EGBERT GOTZIAN, b Indianapolis, Ind, Nov 10, 29. INVERTEBRATE ECOLOGY, INVERTEBRATE PALEONTOLOGY. *Educ:* Oberlin Col, BA, 52; Univ Nebr, MS, 56; Univ Mich, PhD(geol), 62. *Prof Exp:* Asst geol, Univ Nebr, 55-56; asst, Mus Paleont, Univ Mich, 57-58, instr geol, Univ & asst cur mus, 60-63; from asst prof to assoc prof, 63-71, PROF GEOL, WAYNE STATE UNIV, 71- *Concurrent Pos:* Fulbright res fel, Finland, 75. *Mem:* Soc Econ Paleont & Mineral; Brit Palaeont Asn; Australian Geol Soc; Am Soc Limnol & Oceanog; Marine Biol Asn UK; Sigma Xi. *Res:* Benthic ecology of recent marine invertebrates, emphasis upon animal-sediment-water interactions; applications to paleoecologic interpretations; destruction of invertebrate skeletons and its geologic significance. *Mailing Add:* Dept Geol Wayne State Univ Detroit MI 48202

DRISCOLL, GEORGE C, b Mineola, NY, Jan 26, 27; div; c 2. CIVIL ENGINEERING, STRUCTURES. *Educ:* Rutgers Univ, BS, 50; Lehigh Univ, MS, 52, PhD(civil eng), 58. *Prof Exp:* Asst civil eng, Fritz Eng Lab, 50-52, asst engr tests, 52-57, assoc dir, 69-84; from res instr to res assoc prof, 57-65, PROF CIVIL ENG, LEHIGH UNIV, 65- *Honors & Awards:* Huber Res Prize, Am Soc Civil Engrs, 66; Davis Silver Medal Award, Am Welding Soc, 58, Adams Mem Award, 70. *Mem:* Am Welding Soc; Nat Soc Prof Engrs; Am Soc Civil Engrs. *Res:* Plastic analysis and design of welded continuous frames and their components; plastic design of multistory steel frames; computer-aided design. *Mailing Add:* Fritz Eng Lab 13 Lehigh Univ Bethlehem PA 18015

DRISCOLL, JOHN G, b New York, NY, Apr 17, 33. MATHEMATICS. *Educ:* Iona Col, BS, 54; St Johns Univ, MS, 57; Columbia Univ, PhD(math), 69. *Prof Exp:* Teacher, St Joseph's Acad, WI, 57-61 & Power Mem Acad, NY, 61-65; asst prof math, 65-71, asst to pres, 69-71, prof math, 71-77, PRES, IONA COL, 71- *Concurrent Pos:* Dir, Drexel Bunham Educ Fund, 83-, New Rochelle Hosp Med Ctr, 84-; vchmn, Bd Trustees, Comn Independent Col & Univ State NY 85-; dir bd, Chase/Nat Bank Westchester, 85- *Mem:* Math Asn Am. *Res:* Algebraic integration theory; mathematical education; mathematical analysis for the behavioral sciences. *Mailing Add:* 111 Bruct St No 704 Kirkland PQ H9H 3L5 Can

DRISCOLL, JOHN STANFORD, b Olean, NY, May 31, 34; m 62; c 3. MEDICINAL CHEMISTRY. *Educ:* Mich State Univ, BS, 56; Princeton Univ, MA, 58, PhD(org chem), 60. *Prof Exp:* Sr res chemist, Boston Lab, Monsanto Res Corp, 60-64, res group leader, 64-68; res chemist, Nat Cancer Inst, 68-74, actg assoc dir, Develop Therapeut Prog, 80-83, head, Drug Design & Chem Sect, 74-85, dep chief, lab med chem & biol, 77-85, CHIEF, LAB MED CHEM, NAT CANCER INST, 85- *Mem:* Am Chem Soc; Am Asn Cancer Res. *Res:* Drug design; structure-activity relationships; anticancer drugs; anti-AIDS drugs. *Mailing Add:* Four Infield Ct S Rockville MD 20854

DRISCOLL, MICHAEL JOHN, b Peekskill, NY, Sept 8, 34; wid; c 1. NUCLEAR ENGINEERING. *Educ:* Carnegie Inst Technol, BS, 55; Univ Fla, MS, 62; Mass Inst Technol, ScD(nuclear eng), 66. *Prof Exp:* Eng analyst, Phillips Petrol Co, 55-56; nuclear propulsion engr, naval reactor br, Atomic Energy Comn & nuclear propulsion div, bur ships, US Navy, 57-60; from asst prof to prof, 69-89, EMER PROF NUCLEAR ENG & RES SCIENTIST, MASS INST TECHNOL, 89- *Mem:* Am Nuclear Soc. *Res:* Nuclear reactor coolant technology; nuclear fuel management. *Mailing Add:* Nuclear Reactor Lab 138 Albany St Cambridge MA 02139

DRISCOLL, RAYMOND L, b James City County, Va, Oct 11, 05. ELECTROMAGNETISM. *Educ:* William & Mary Col, BS, 28; Univ NC, MS, 36. *Prof Exp:* Physicist, Nat Bur Standards, 34-71; RETIRED. *Mem:* Am Phys Soc. *Mailing Add:* 10139 Cedar Ln Kensington MD 20895

DRISCOLL, RICHARD JAMES, b Chicago, Ill, Aug 14, 28. ANALYTICAL MATHEMATICS. *Educ:* Loyola Univ, Ill, BS, 50, AM, 51; Northwestern Univ, PhD(math), 59. *Prof Exp:* Instr, Northwestern Univ, 57-58; from instr to asst prof, 58-66, ASSOC PROF, LOYOLA UNIV, CHICAGO, 66- *Mem:* Am Math Soc; Math Asn Am; Sigma Xi. *Res:* Ordinary differential equations; calculus of variations. *Mailing Add:* 1515 Juneway Terr Chicago IL 60626

DRISCOLL, RICHARD STARK, b Denver, Colo, Sept 16, 28; m 54; c 2. LAND USE PLANNING, REMOTE SENSING. *Educ:* Colo State Univ, BS, 51, MS, 57; Ore State Univ, PhD(range ecol & soils), 62. *Prof Exp:* Range conservationist, Pac NW Forest & Range Exp Sta, US Forest Serv, 52-56, proj leader range mgt & wildlife habitat res, 56-60, wildlife habitat res, 60-62, range mgt res, Div Watershed, Recreation & Range Res, 62-65, prin range ecologist, 65-75, mgr Resources Eval Tech Prog, 75-83; NATURAL RESOURCES CONSULT, LAND EVAL/LAND USE PLANNING, UN FOOD & AGR ORGN, 83- *Concurrent Pos:* Lectr, Ore State Univ, 58-60; fac affil, Colo State Univ, 65-; spec instr, Univ Denver, 71. *Honors & Awards:* Outstanding Achievement Award, Soc Range Mgt, 83; Pres Citation Meritorious Serv, Am Soc Photogram & Remote Sensing, 78, 86. *Mem:* Am Inst Biol Sci; Am Soc Photogram & Remote Sensing; Sigma Xi; Soc Range Mgt. *Res:* Range ecology including ecosystem classification; remote sensing of range, wildlife habitat and watershed resources; land use planning; land evaluation for natural resources. *Mailing Add:* 2217 Sheffield Dr Ft Collins CO 80526

DRISCOLL, WILLIAM, DISEASE PREVENTION. *Prof Exp:* ACTG BR MGR, DIS PREV & HEALTH PROM BR, NAT INST DENT RES, NIH, 88- *Mailing Add:* NIH Nat Inst Dent Res Dis Prev & Health Prom Br Westwood Bldg Rm 538 5333 Westbard Ave Bethesda MD 20816

DRISKA, STEVEN P, SMOOTH MUSCLE PHYSIOLOGY & BIOCHEMISTRY. *Educ:* Carnegie-Mellon Univ, PhD(chem), 76. *Prof Exp:* ASSOC PROF PHYSIOL & BIOPHYS, MED COL VA, 79- *Res:* Regulation of contraction. *Mailing Add:* Dept Physiol Temple Univ Sch Med 3420 N Broad St Philadelphia PA 19140

DRISKELL, JUDY ANNE, b Detroit, Mich, Sept 29, 43. VITAMINS. *Educ:* Univ Southern Miss, BS, 65; Purdue Univ, MS, 67, PhD(nutrit), 70. *Prof Exp:* Asst prof nutrit & foods, Auburn Univ, 70-72 & Fla State Univ, 72-74; from assoc prof to prof human nutrit & foods, Va Polytech Inst & State Univ, 74-89; PROF NUTRIT SCI & HOSP MGT, UNIV NEBR, 89- *Concurrent Pos:* USDA/CSRS Nutrition Scientist, 81-82. *Honors & Awards:* Borden Award. *Mem:* Am Inst Nutrit; Inst Food Technologists; Am Home Econ Asn; Soc Nutrit Educ; Sigma Xi; Am Diet Asn. *Res:* Vitamin B6 requirements and status; effects of massive ingestion of ascorbic acid; malnutrition and behavior and cellular alterations. *Mailing Add:* Dept Nutrit Sci & Hosp Mgt Univ Nebr Lincoln NE 68583

DRISKILL, WILLIAM DAVID, b Sharon Grove, Ky, Dec 5, 38; m 61; c 2. ATOMIC PHYSICS, SCIENCE EDUCATION. *Educ:* Murray State Univ, Ky, BS, 61; Univ of the South, MA, 66; George Peabody Col, PhD(physics educ), 75. *Prof Exp:* Teacher physics, Murray High Sch, Ky, 62-66; PROF PHYSICS, BELMONT COL, 66-, PROF SCI EDUC, 77-, COORDR PHYSICS, DEPT CHEM & PHYSICS, 80- *Concurrent Pos:* Shell merit fel, Shell Oil Co, 66. *Mem:* Am Asn Physics Teachers. *Res:* Establishing priorities in physics education at both the college level and secondary school level; use of Delphi forecasting technique. *Mailing Add:* Dept Physics Chem Belmont Col Nashville TN 37212

DRISKO, RICHARD WARREN, b San Mateo, Calif, Nov 16, 25; m 75; c 2. ORGANIC CHEMISTRY. *Educ:* Stanford Univ, BS, 47, MS, 48, PhD(chem), 50. *Prof Exp:* sr proj scientist chem, 50-78, DIR, MAT SCI DIV, CIVIL ENG LAB, 78- *Honors & Awards:* Meritorious Civilian Serv Award, US Navy, 53. *Mem:* Am Chem Soc; Nat Asn Corrosion Eng. *Res:* Isolation and identification of alkaloids; determination of amino acid sequence in proteins; analysis of creosote; paints, protective coatings and corrosion control; pollution; ecology. *Mailing Add:* 532 North N St Oxnard CA 93030

DRISTY, FORREST E, b Eakin, SDak, Oct 23, 31; m 58; c 2. GEOMETRY. *Educ:* SDak Sch Mines & Technol, BS, 53, MS, 59; Fla State Univ, PhD(math), 62. *Prof Exp:* Indust engr, Eastman Kodak Co, 54; instr math, SDak Sch Mines & Technol, 57-59; from asst prof to assoc prof, Fla Presby Col, 62-66; assoc prof, Clarkson Col Technol, 66-68; PROF MATH, STATE UNIV NY COL OSWEGO, 68- *Concurrent Pos:* Vis fel, Princeton Univ, 67-68. *Mem:* Math Asn Am. *Res:* Applications of linear algebra to geometry and topology; history of mathematics. *Mailing Add:* Dept Math State Univ NY Col Oswego NY 13126

DRITSCHILO, WILLIAM, b Villach, Austria, Nov 11, 46; m 70; c 2. ENVIRONMENTAL SCIENCE. *Educ:* Univ Pa, BA & BS, 70; Cornell Univ, PhD(ecol), 78. *Prof Exp:* Res assoc, Washington Univ, St Louis, 77-78; asst prof environ sci, Univ Calif, Los Angeles, 78-85. *Mem:* Ecol Soc Am; Entom Soc Am; AAAS. *Res:* Effects of energy supplies and sources on agriculture and environment; environmental indices; community structure in carbid beetles. *Mailing Add:* One Beaver Pond Rd Proctor VT 05765

DRIVER, CHARLES HENRY, b Orlando, Fla, Oct, 12, 21; m 43; c 3. FOREST PATHOLOGY, RANGE MANAGEMENT. *Educ:* Univ Ga, BS, 47, MS, 50; La State Univ, PhD, 54. *Prof Exp:* Res technician forest path, USDA, 46-47; instr biol, Emory Univ, 49-52; res botanist, Int Paper Co, 54-57, dir forest res, 57-65; prof forest path, Col Forest Resources, Univ Wash, 65-88; RETIRED. *Concurrent Pos:* Vis prof, protection silviculture, Forestry Sect, Univ Melbourne, Victoria, Australia, 84-85. *Mem:* Mycol Soc Am; Soc Am Foresters; Soc Range Mgt. *Res:* Pathology of intensely managed forest; forest products pathology; management of wildlands range resources. *Mailing Add:* 6254 Stemilt Loop Wenatchee WA 98801

DRIVER, GARTH EDWARD, b Waterville, Wash, Aug 18, 23; m 48; c 5. CONTROL ENGINEERING. *Educ:* Univ Kans, BS, 47, MS, 48. *Prof Exp:* Engr, Hanford Labs, 49-59, sr engr, 59-65; res assoc control systs, 65, mgr digital systs sect, 65-68, mgr, Control & Instrumentation Dept, 68-70, mgr, Comput & Control Dept, 70-80, LEADER TECH OPERATIONS, PAC NORTHWEST DIV, BATTELLE MEM INST, 80-, SR ENGR. *Mem:* Inst Elec & Electronics Engrs. *Res:* Transistor circuits; radiation telemetry; nuclear reactor instrumentation; computer process control; process simulation. *Mailing Add:* Battelle Northwest PO Box 999 Richland WA 99352

DRIVER, RICHARD D, b Aberaeron, Wales, June 23, 48; div; c 3. ATOMIC PHYSICS. *Educ:* Univ London, BSc, 70, DIC & PhD(spectros), 74. *Prof Exp:* Res fel astrophys, Harvard Col Observ, 74-76; res fel physics, Mass Inst Technol, 76-78; res physicist, Foxboro Anal Co, 78-85; scientist, Aster Corp, 86-88; DIR SENSOR DEVELOP, IRIS FIBER OPTICS, 88- *Mem:* Am Phys Soc; Int Soc Optical Eng. *Res:* Spectroscopy, laser spectroscopy, fiber optics and fiber optic sensing. *Mailing Add:* 46 Highland Ave Cambridge MA 02139

DRIVER, RODNEY D, b London, Eng, July 1, 32; US citizen; m 55; c 3. MATHEMATICAL ANALYSIS, ELECTRODYNAMICS. *Educ:* Univ Minn, BS, 53, MS, 55, PhD(math), 60. *Prof Exp:* Instr math, Univ Minn, 56-60; fel, Res Inst Advan Study, 60-61; staff mem, Math Res Ctr, US Army, Univ Wis, 61-62 & Sandia Corp, 62-69; assoc prof, 69-74, PROF MATH, UNIV RI, 74- *Mem:* Am Math Soc. *Res:* Functional-differential equations; two-body problem of classical electrodynamics. *Mailing Add:* Dept Math Univ RI Kingston RI 02881

DRLICA, KARL, b Portland, Ore, June 28, 43; m 67; c 2. MOLECULAR BIOLOGY. *Educ:* Ore State Univ, BA, 64; Univ Calif, Berkeley, PhD(molecular biol), 71. *Prof Exp:* Regents fel, Univ Calif, 65, fel plant path, Univ Calif, Davis, 71-73; fel, Am Cancer Soc, 71-73; fel biochem, Princeton Univ, 73-76; from asst prof to assoc prof biol, Univ Rochester, 77-85; fel, NIH, 79-84; assoc mem, 85-89, MEM, PUB HEALTH RES INST, NY, 89- *Concurrent Pos:* Mem, Molecular Biol Study Sect, NIH, 79, Genetic Basis Dis Sect, 83-87; fel, Europ Molecular Biol Orgn, 85; lectr, Am Soc Microbiol & Found Microbiol, 88-89. *Mem:* Am Soc Microbiol. *Res:* Bacterial chromosome structure and DNA supercoiling. *Mailing Add:* Pub Health Res Inst 455 First Ave New York NY 10016

DRNEVICH, VINCENT PAUL, b Wilkinsburg, Pa, Aug 6, 40; m 66; c 4. CIVIL ENGINEERING. *Educ:* Univ Notre Dame, BSCE, 62, MSCE, 64; Univ Mich, PhD(civil eng), 67. *Prof Exp:* Res asst soil dynamics, Univ Notre Dame, 62-64 & Univ Mich, 64-65 & 65-67; from asst prof to prof civil eng, Univ Ky, 67-91, actg dean eng, 89-90; PROF & HEAD, CIVIL ENG, PURDUE UNIV, 91- *Concurrent Pos:* NSF eng res initiation grant, 68-69; mem, Int Soc Soil Mech & Found Eng, 64-, mem soil dynamics comt & preconf pub subcomt, 68-69; chmn, Ky Geotech Eng Group, 72-73, officer, 71-76; vis engr, Bur Reclamation, Dept Interior, Denver, 74; vchmn subcomt surface & subsurface reconnaissance, 78-80, struct properties of soils, soil dynamics, exec, 76-82, 85-, chmn & tech ed, Geotech Testing J, Am Soc Testing & Mats Comt Soil & Rock Eng Purposes, 85-; vchmn, Organizing Comt, Symp Acoustic Emissions, Am Soc Testing & Mats, 81; subcomt chmn, Gov Earthquake Hazards Tech Adv-Panel Mitigation Comt; rep, publ comt, 11th Int Conf, San Francisco, 85, US Nat Soc, Int Soc Soil Mech & Found, 79-; comt soil dynamics, Am Soc Civil Engrs, 73-81, 84-89; civil engr vis, Accrediting Bd Engr & Tech, 84-89. *Honors & Awards:* Norman Medal, Am Soc Civil Engrs, 73; C A Hogentogler Award, Am Soc Testing & Mats, 79; Huber Res Prize, Am Soc Civil Engrs, 80. *Mem:* Am Soc Civil Engrs; Am Soc Testing & Mat; Am Soc Eng Educ; Nat Soc Prof Engrs; Int Soc Soil Mech & Found Eng; Earthquake Eng Res Inst. *Res:* Behavior of soils due to dynamic loading, such as caused by machines and earthquakes; improvement of soil testing techniques and equipment. *Mailing Add:* Sch Civil Eng Purdue Univ West Lafayette IN 47907

DROBECK, HANS PETER, b Bavaria, Ger, Oct 3, 23; nat US; m 48; c 4. TOXICOLOGY, EXPERIMENTAL PATHOLOGY. *Educ:* Univ Del, BS, 48; Syracuse Univ, MS, 51, PhD(zool), 53. *Prof Exp:* Asst, Syracuse Univ, 48-51; assoc res pathologist, Parke, Davis & Co, 53-56; res pathologist, Sterling Winthrop Res Inst, NY, 56-63, dir, toxicol dept, 63-87; RETIRED. *Mem:* Soc Toxicol; NY Acad Scisitol; Europ Soc Toxicol; Am Asn Lab Animal Sci. *Res:* Toxicology. *Mailing Add:* Box 149 RD 1 Rensselaer NY 12144

DROBNEY, RONALD DELOSS, b Ft Dodge, Iowa, Oct 13, 42. WATERFOWL ECOLOGY. *Educ:* Univ Northern Iowa, BA, 65; Univ Mo, MA, 73, PhD(wildlife ecol), 77. *Prof Exp:* Actg dir, Gaylord Mem Lab, Univ Mo, 76-77; asst prof wildlife mgr, Univ Mich, Ann Arbor, 78-83; assoc prof avian ecol, Univ Md, Frostburg, 83-86; ASSOC PROF WILDLIFE ECOL, UNIV MO, COLUMBIA, 86- *Honors & Awards:* Citation of Merit, Am Ornithologists Union, 77. *Mem:* Wildlife Soc; Am Ornithologists Union. *Res:* Feeding ecology, nutrition and reproductive bioenergetics of waterfowl; wetland ecology. *Mailing Add:* 112 Stephens Hall Univ Mo Columbia MO 65211

DROBNIES, SAUL ISAAC, b New York, NY, June 8, 33; div; c 1. MATHEMATICS, OPERATIONS RESEARCH. *Educ:* Univ Tex, BS, 55, MA, 58, PhD(math), 61. *Prof Exp:* Fel math, Rice Univ, 61-62; sr opers analyst, Gen Dynamics Corp, 62-63; from asst prof to assoc prof, 63-70, PROF MATH, SAN DIEGO STATE UNIV, 70- *Mem:* Am Math Soc; Math Asn Am; Soc Indust & Appl Math. *Res:* Mathematical analysis; continued fractions. *Mailing Add:* Dept Math San Diego State Univ San Diego CA 92182

DROBNYK, JOHN WENDEL, b Trenton, NJ, Sept 30, 35; m; c 2. GEOLOGY. *Educ:* Amherst Col, BA, 57; Rutgers Univ, MS, 59, PhD(geol), 62. *Prof Exp:* Asst instr geol, Rutgers Univ, 57-61, res asst, 61-62; geologist, Pan-Am Petrol Corp, 62-64; from instr to assoc prof, 65-71, PROF GEOL, SOUTHERN CONN STATE UNIV, 71-, CHMN DEPT EARTH SCI, 67- *Mem:* Am Asn Petrol Geol; Soc Econ Paleont & Mineral; Sigma Xi. *Res:* Geology of modern clastic sediments; interpretation of ancient depositional environments. *Mailing Add:* Tuttles Point Guilford CT 06437

DROBOT, STEFAN, b Cracow, Poland, Aug 7, 13; US citizen; m 41; c 3. MATHEMATICS. *Educ:* Univ Jagiellonian, MA, 37; Univ Wroclaw, PhD(math), 47. *Prof Exp:* Asst theoret mech, Lwow Polytech Inst, Poland, 39-40; lectr, Siberian Metall Inst USSR, 41-46; from asst prof to assoc prof math, Wroclaw Univ & Polytech Inst, 46-59; res assoc prof, Univ Chicago, 59-60; from assoc prof to prof, Univ Notre Dame, 60-63; PROF MATH, OHIO STATE UNIV, 63- *Concurrent Pos:* Head dept appl math, Math Inst, Polish Acad Sci, 49-58. *Mem:* Am Math Soc; Math Asn Am. *Res:* Mechanics of continua; dimensional analysis; operational calculus; variational principles. *Mailing Add:* 217 W Irving Way Columbus OH 43214

DROBOT, VLADIMIR, UNIFORM DISTRIBUTION OF SEQUENCES. *Educ:* Univ Notre Dame, BS, 63; Univ Ill, PhD(math), 67. *Prof Exp:* Asst prof math, State Univ NY, Buffalo, 67-73; ASSOC PROF MATH, UNIV SANTA CLARA, 73- *Concurrent Pos:* Ed, Problems Sect, Math Asn Am, 79- *Mem:* Am Math Soc; Math Asn Am; Sigma Xi. *Res:* Mathematical analysis. *Mailing Add:* Dept Math Univ Canta Clara Santa Clara CA 95050

DROEGE, JOHN WALTER, b Seymour, Ind, Sept 7, 21. FUEL SCIENCE, THERMOPHYSICAL PROPERTIES. *Educ:* Ind Univ, AB, 42; Ohio State Univ, PhD(chem), 53. *Prof Exp:* Chemist, Battelle Mem Inst, 53-85; RETIRED. *Res:* Thermodynamics; calorimetry; physical properties of materials; high temperature chemistry; coal chemistry; coal liquefaction; thermophysical and transport properties at elevated pressure and temperature. *Mailing Add:* RR 1 Box 31 Brownstown IN 47220

DROESSLER, EARL GEORGE, b Dubuque, Iowa, Jan 14, 20; m 44, 57; c 5. METEOROLOGY. *Educ:* Loras Col, AB, 42. *Hon Degrees:* ScD, Loras Col, 58. *Prof Exp:* Meteorologist, Off of Naval Res, 46-50, head geophys br, 50; exec dir comt geophys & geog, Res & Develop Bd, 52-53; exec secy, Coord Comt Gen Sci, Off of Asst Secy Defense for Res & Develop, 54-58; prog dir atmospheric sci, NSF, 58-66; prof atmospheric sci & vpres res, State Univ NY Albany, 66-71; prof geosci & dean res, NC State Univ, 71-79, dir univ affairs, Nat Oceanic & Atmospheric Admin, 79-83, emer prof, 79-; asst secy natural resources, NC Dept Natural Resources & Community Develop, 83-85; CONSULT, 85- *Concurrent Pos:* Mem, Adv Comt Weather Control, 53-57 & US Nat Comt for Int Geophys Year, 55-64; chmn bd trustees, Univ Corp Atmospheric Res, 69-73; mem, NC Comn Sci & Technol, 71-79; mem, Mineral Resources Comt, Nat Asn State Univs & Land-Grant Cols, 73-76; mem bd govs, Research Triangle Inst, NC, 73-79, mem exec comt, 75-79; mem, Oak Ridge Assoc Univs Coun, 74-79, mem bd dirs, 75-79, mem exec comt, 78-79; co-chmn, NSF Adv Panel Weather Modification, 75-77; mem bd trustees, Triangle Univs Ctr Advan Studies, Inc, 76-79; exec ed, Weatherwise, Helen Dwight Reid Found, 78-86. *Honors & Awards:* Charles Franklin Brooks, Am Meteorological Soc, 76. *Mem:* fel Am Geophys Union; fel Am Meteorol Soc(pres, 83-). *Res:* Cloud physics; weather modification. *Mailing Add:* 1305 Glenn Eden Dr Raleigh NC 27612-4750

DROLL, HENRY ANDREW, b New York, NY, Sept 2, 26; m 52; c 5. INORGANIC CHEMISTRY. *Educ:* George Washington Univ, BS, 52, MS, 53; Pa State Univ, PhD(chem), 56. *Prof Exp:* AEC asst, George Washington Univ, 52-53 & Pa State Univ, 54-55; sr scientist, Atomic Power Div, Westinghouse Elec Corp, Pa, 55-56; from asst prof to assoc prof, 56-71, PROF CHEM, UNIV MO-KANSAS CITY, 71- *Mem:* Am Chem Soc. *Res:* Chemistry of less familiar elements; chemical equilibrium, especially study of complex ions in aqueous systems; preparation of coordination compounds. *Mailing Add:* Dept Chem Univ Mo 5308 Oak St Kansas City MO 64112-2878

DROLSOM, PAUL NEWELL, b Martell, Wis, July 15, 25; m 50; c 2. AGRONOMY. *Educ:* Univ Wis, BS, 49, MS, 50, PhD(agron, plant path), 53. *Prof Exp:* Asst agron, Univ Wis, 49-52; pathologist, Agr Res Serv, USDA & NC State Univ, 53-58; from asst prof to prof, 58-89, EMER PROF AGRON, UNIV WIS-MADISON, 90- *Concurrent Pos:* Mem, US AID Prog, Brazil, 64-66. *Mem:* Am Soc Agron. *Res:* Corn breeding with major emphasis on development of lines with earliness, cold tolerance and disease resistance. *Mailing Add:* 310 Marinette Trail Madison WI 53705-4721

DROMGOLD, LUTHER D, b Newport, Pa, Apr 20, 25; m 50; c 4. LUBRICATION, ANALYTICAL CHEMISTRY. *Educ:* Pa State Univ, BS, 49. *Prof Exp:* Engr, Witco Chem corp, 49-54, sales engr, 55-62, mgr new prod develop, Kendall/Amalie Div, 62-87; RETIRED. *Mem:* Am Chem Soc; fel Am Soc Lubrication Engrs. *Res:* Petroleum, especially specialized and process lubricants, automotive underbody coatings; rust preventives; petroleum resins; adhesives; emulsions; hydraulic fluids. *Mailing Add:* 21 McKune Ave Bradford PA 16701-3229

DRONAMRAJU, KRISHNA RAO, evolutionary biology, population genetics, for more information see previous edition

DRONEN, NORMAN OBERT, JR, b Shelton, Wash, Oct 9, 45; m 67; c 3. PARASITOLOGY. *Educ:* Eastern Wash Univ, BA, 68, MS, 70; NMex State Univ, PhD(biol), 73. *Prof Exp:* ASST PROF BIOL, TEX A&M UNIV, 74- *Concurrent Pos:* Co-investr res grant, Environ Protection Agency, 74-75; prin investr, NIH biomed supports grants, Tex A&M Univ, 75-76; co-investr, NSF res proj estab primate colony, Col Vet Med, Tex A&M Univ, 75-76. *Mem:* Sigma Xi; Am Soc Parasitologists. *Res:* Parasites, especially the host-parasite; population dynamics of natural populations of helminths; trophic structure utilization by parasites; experimental manipulation of host-parasite systems. *Mailing Add:* Wildlife & Fisheries Tex A&M Univ Col Sta TX 77843

DROPESKY, BRUCE JOSEPH, b Philadelphia, Pa, Apr 29, 24; m 48; c 2. NUCLEAR CHEMISTRY. *Educ:* Rensselaer Polytech Inst, BS, 49; Univ Rochester, PhD(phys chem), 53. *Prof Exp:* Mem staff, 53-70, assoc group leader, 70-85, mem staff, 85-88, LAB ASSOC, LOS ALAMOS NAT LAB, 88- *Concurrent Pos:* Vis prof, Kyoto Univ, Japan, 85, lectr, Peoples Repub China, 85; chmn elect, div nuclear chem & technol, Am Chem Soc, 77, chmn, 78. *Mem:* Am Chem Soc; Am Phys Soc; Sigma Xi. *Res:* High energy nuclear reactions; nuclear decay scheme studies; nuclear spectroscopy; electromagnetic isotope separation; pi meson induced nuclear reactions. *Mailing Add:* Los Alamos Nat Lab MS-515 PO Box 1663 Los Alamos NM 87545

DROPKIN, JOHN JOSEPH, b Bobruisk, Russia, Feb 22, 10; US citizen; m 33, 57; c 4. SOLID STATE PHYSICS. *Educ:* Columbia Univ, AB, 30; Polytech Inst Brooklyn, MS, 47, PhD(physics), 48. *Prof Exp:* Asst physics, Columbia Univ, 30-35; teacher high sch, NY, 36-48; from asst prof to prof, 48-78, head dept, 57-65, EMER PROF PHYSICS & COORDR, FRESHMAN LEARNING CTR, POLYTECH INST NY, 78- *Mem:* Fel Am Phys Soc; Am Asn Physics Teachers. *Res:* Stimulated phosphorescence and photoconduction in infrared sensitive phosphors; infrared photoconduction in zinc sulfide; solid state physics, especially conduction, photoconduction and luminescence. *Mailing Add:* Freshman Learning Ctr Polytech Inst NY Brooklyn NY 11201

DROPKIN, VICTOR HARRY, b New York, NY, Mar 21, 16; m 41; c 1. ZOOLOGY. *Educ:* Cornell Univ, BA, 36; Univ Chicago, PhD(zool), 40. *Prof Exp:* Instr, Univ Chicago, 40-41 & US Air Force, 41-45; from instr to assoc prof biol, Roosevelt Univ, 46-51; fel, Naval Med Res Inst, 51-53; nematologist, plant indust sta, USDA, 53-69; prof, Univ Mo-Columbia, 69-85, chmn, 80-84, emer prof plant path, 85-; RETIRED. *Concurrent Pos:* Vis prof, Univ Wis, 66-67. *Mem:* Fel Soc Nematol (treas, 62-63, vpres, 64-65, pres, 65-66); fel Am Phytopath Soc; Europ Soc Nematol. *Res:* Nematodes parasitic on plants. *Mailing Add:* 3925 Woodview Ct St Paul MN 55127

DROPP, JOHN JEROME, b Monessen, Pa, Dec 2, 40; m 62; c 2. BIOLOGY, HISTOLOGY. *Educ:* Washington & Jefferson Col, BA, 62; Ohio Univ, MS, 64; Ore State Univ, PhD(zool), 69. *Prof Exp:* Neurophysiologist, Walter Reed Army Inst Res, Washington, DC, 68-70; asst prof biol, Wilson Col, 70-76; asst prof, 76-80, ASSOC PROF BIOL, MT ST MARY'S COL, 80- *Mem:* AAAS; Am Soc Zoologists; Am Asn Anatomists; Soc Neurosci; Sigma Xi. *Res:* Behavioral stress and its effects on various organ systems of the body-effects at the histological level; trophic effects of neurons; mast cells in the vertebrate brain. *Mailing Add:* 207 Sherwood Dr Chambersburg PA 17201

DROPPO, JAMES GARNET, JR, b Ottawa, Ont, July 24, 43; m 69; c 2. MICROMETEOROLOGY. *Educ:* Cornell Univ, BS, 65; State Univ NY Albany, MS, 68, PhD(atmospheric sci), 72. *Prof Exp:* Res assoc, Atmospheric Sci Res Ctr, Albany, 70-72; SR RES SCIENTIST ATMOSPHERIC SCI, PAC NORTHWEST LABS, BATTELLE MEM INST, 72- *Concurrent Pos:* Lectr, Joint Ctr Grad Study-Continuing Educ Prog, Richland, 75- *Mem:* Inst Elec & Electronics Engrs; Am Meteorol Soc; Can Meteorol & Oceanog Soc. *Res:* Micrometeorological processes influencing energy and pollutant fluxes in the atmospheric surface boundary layer over natural surfaces. *Mailing Add:* Battelle Pac Northwest Labs PO Box 999 Richland WA 99352

DROR, MICHAEL, biocompatible polymers, sensors, for more information see previous edition

DROSDOFF, MATTHEW, b Chicago, Ill, Dec 15, 08; m 35; c 2. SOIL SCIENCE. *Educ:* Univ Ill, BS, 30; Univ Wis, MS, 32, PhD(soil chem), 34. *Prof Exp:* Asst soils, Univ Wis, 30-35; jr soil survr, Soil Chem Div, USDA, 38-40, assoc soil technologist, Tung Lab, Bur Plant Indust, 40-42, Bur Plant Indust, Soil & Agr Eng, 42-53, soil technologist, 45-50, sr soil scientist, 50-53, Hort Crops Res Br, Agr Res Serv, 53-55, soils adv, Int Coop Admin, Peru, 55-60; food & agr officer, AID, Vietnam, 60-64; adminr, Int Agr Develop Serv, USDA, 64-66; PROF SOIL SCI, CORNELL UNIV, 66- *Concurrent Pos:* Adv, Point IV Agr Mission to Colombia, 51-53, Bolivia, 54, consult, Korea & Nigeria, 79, Colombia, Indonesia, Thailand, India, 80 & 83, Indonesia, 84 & Peru, 85 & 86. *Honors & Awards:* Int Agron Award, Am Soc Agron, 74. *Mem:* AAAS; Am Soc Agron; Soil Sci Soc Am; Soc Int Develop. *Res:* Chemical studies of colloidal clays; field and laboratory research on genesis and morphology of soils; soil chemistry and fertility related to mineral nutrition of tung trees, including minor elements; soil survey; agricultural development; tropical soils; soil fertility research in the tropics. *Mailing Add:* Dept Agron Cornell Univ Bradfield Hall Ithaca NY 14850

DROSSMAN, MELVYN MILES, b Brooklyn, NY, June 30, 37; m 58; c 3. ELECTRONICS ENGINEERING. *Educ:* Polytech Inst Brooklyn, BEE, 57, MEE, 59, PhD(elec eng), 67. *Prof Exp:* Instr elec eng, Polytech Inst Brooklyn, 58-66; res assoc psychophysiol, State Univ NY Downstate Med Ctr, 66-68; assoc prof, 68-70, chmn, Dept Elec Technol, 68-78, assoc vpres eng & res, 84, actg dean, Ctr Eng & Technol, 84-85, PROF ELEC ENG TECHNOL & COMPUT SCI, NY INST TECHNOL, 70-, DEAN SCH ENG & TECHNOL, 85- *Concurrent Pos:* Consult, Res Ctr, Rockland State Hosp, 63-66 & HZI Res Co, 74-76, NASA, 76-81, US Army, 78-81; vpres eng & develop, Appl Digital Data Systs, 83; pres, Appl Tech Solutions, Inc, 81- *Mem:* Inst Elec & Electronics Engrs; Asn Comput Mach; Am Soc Eng Educ. *Res:* Computer analysis of averaged evoked EEG responses in human subjects using visual and auditory stimuli to determine relationship between EEG and cerebral information processing. *Mailing Add:* 26 Locust Rd Old Bethpage NY 11804

DROSTE, JOHN BROWN, b Hillsboro, Ill, Nov 23, 27; m 51; c 2. SEDIMENTOLOGY. *Educ:* Univ Ill, BS, 51, MS, 53, PhD(geol), 56. *Prof Exp:* Instr geol & dir gen studies, Univ Ill, 56-57; from asst prof to assoc prof, 57-68, PROF GEOL, IND UNIV, BLOOMINGTON, 68- *Mem:* Geol Soc Am; Soc Econ Paleont & Mineral; Clay Minerals Soc. *Res:* Stratigraphy and sedimentation of Paleozoic rocks in the Midwest, particularly the patterns of deposition and interpretation of sedimentary environments. *Mailing Add:* Dept Geol Ind Univ Bloomington IN 47405

DROST-HANSEN, WALTER, b Chicago, Ill, Sept 29, 25; m 50; c 2. CHEMICAL PHYSICS. *Educ:* Univ Copenhagen, Magister scientiarum, 50. *Prof Exp:* Asst mass spectros invest, Inst Theoret Physics, Copenhagen, 46-50; Carlsberg Found fel, Inst Phys Chem, 50; spec asst, Ill Inst Technol, 50-51; res phys chemist, Bjorksten Res Labs, 51-52; assoc prof chem physics & sr chem physicist, NMex Inst Mining & Technol, 53-56; sr res engr, Pan-Am Petrol Corp, 56-61; sr res chemist, Jersey Prod Res Co, 61-64; from assoc prof to prof chem physics, Inst Marine Sci, 64-70, chmn div chem oceanog, 67-70, PROF CHEM & DIR LAB WATER RES, UNIV MIAMI, 70- *Mem:* AAAS; Am Chem Soc; NY Acad Sci; Sigma Xi. *Res:* Structure and properties of water and aqueous solutions; surface and colloid phenomena; physical chemistry of biological systems; thermal pollution. *Mailing Add:* Lab Water Res Dept Chem Univ Miami Coral Gables FL 33124

DROUILHET, PAUL R, b San Pedro, Calif, Mar 11, 33; m 54; c 3. ENGINEERING MANAGEMENT, RADAR COMMUNICATIONS. *Educ:* Mass Inst Technol, BSEE, 55, MSEE, 55, EE(elec eng), 57. *Prof Exp:* Staff, 55-80, div head, 80-85, ASST DIR, LINCOLN LAB, MASS INST TECHNOL, 85- *Mem:* Fel Inst Elec & Electronics Engrs. *Mailing Add:* Lincoln Lab Mass Inst Technol 244 Wood St Box 73 Lexington MA 02173-9108

DROZDOWICZ, ZBIGNIEW MARIAN, b Warsaw, Poland, Apr 9, 49; m; c 2. LASERS, MEDICAL SYSTEMS. *Educ:* Univ Calif, Los Angeles, BS, 72; Mass Inst Technol, PhD(physics), 78. *Prof Exp:* sr scientist lasers, res div, Raytheon Co, 78-81; supvr eng sci group, Quantronix Corp, 81-86; dir laser res & develop, Weck Syst, Squibb Co, 86-87; DIR ENG, QUANTRONIX CORP, 87- *Mem:* Inst Elec & Electronics Engrs; Optical Soc Am. *Res:* Laser pumped infrared and submillimeter lasers; electrical discharge and chemical infrared lasers; solid state lasers; pyrolytic gas decomposition. *Mailing Add:* 1363 Stony Brook Rd Stony Brook NY 11790

DRUCKER, ARNOLD, b Brooklyn, NY, Mar 18, 32; m 57; c 2. POLYMER CHEMISTRY, ORGANIC CHEMISTRY. *Educ:* City Col New York, BS, 53; Polytech Inst Brooklyn, MS, 56, PhD(polymer & org chem), 64. *Prof Exp:* Res chemist, Merck & Co, 56-58 & Am Cyanamid Co, Conn, 58-71; teacher gen chem, Ridgefield High Sch, 71-77; ASSOC PROF CHEM, UNIV CONN, 77- *Concurrent Pos:* Lectr org chem, City Col New York, 66-74 & Univ Conn, Stamford, 75-77. *Mem:* AAAS; Am Chem Soc. *Res:* Keratin chemistry; stereospecific polymerization; epoxy resins; triazine polymers; steroids; water soluble polymers; paper chemistry; polymeric flocculants for water treatment. *Mailing Add:* Six Garden Lane Westport CT 06880-5316

DRUCKER, DANIEL CHARLES, b New York, NY, June 3, 18; m 39; c 2. APPLIED MECHANICS, MATERIALS ENGINEERING. *Educ:* Columbia Univ, BS, 37, CE, 38, PhD(eng), 40. *Hon Degrees:* DEng, Lehigh Univ, 76; DSc, Technion (Israel Inst Tech), 83, Brown Univ, 84, Northwestern Univ, 85, Univ Ill, Urbana, 92. *Prof Exp:* Eng asst, Tunnel Authority, New York, 37; asst, Columbia Univ, 38-39; instr mech eng, Cornell Univ, 40-43; supvr mech solids, Armour Res Found, Ill Inst Technol, 43-45, asst prof mech, 46-47; from assoc prof to prof eng, Brown Univ, 47-64, chmn, Div Eng, 53-59, Ballou prof, 64-68, chmn phys sci coun, 61-63; dean, Col Eng, Univ Ill, Urbana, 68-84; GRAD RES PROF AEROSPACE, UNIV FLA, GAINESVILLE, 84- *Concurrent Pos:* Guggenheim fel, 60-61; NATO sr sci fel, 68; Fulbright travel grant, 68; mem gen comt, Int Coun Sci Unions, 76-86, Nat Sci Bd, 88- *Honors & Awards:* von Karman Medal, Am Soc Civil Engrs, 66; Lamme Medal, Am Soc Eng Educ, 67; M M Frocht Award, Soc Exp Stress Anal, 71; Thomas Egleston Medal, Columbia Univ Sch Eng & Appl Sci, 78; Timoshenko Medal, Am Soc Mech Engrs, 83; John Fritz Medal, Founder Eng Soc, 85; Nat Medal Sci, 88. *Mem:* Nat Acad Eng; Polish Acad Sci; fel Am Acad Arts & Sci; Int Union Theoret & Appl Mech (pres, 80-84, vpres, 84-88); Am Soc Eng Educ (pres, 81); Am Soc Mech Engrs (pres, 73); Am Acad Mech (pres, 72); AAAS. *Res:* Photoelasticity; plasticity; mechanics of metal cutting and deformation processing; stress analysis; soil mechanics; materials engineering. *Mailing Add:* Univ Fla 231 Aerospace Bldg Gainesville FL 32611

DRUCKER, E(UGENE) E(LIAS), b New York, NY, Dec 11, 24; m 46; c 3. MECHANICAL ENGINEERING. *Educ:* Mass Inst Technol, BS, 49, MS, 50. *Prof Exp:* Asst mech eng, Mass Inst Technol, 49-50; from instr to assoc prof, US Naval Postgrad Sch, 50-56; assoc prof, 56-62, PROF MECH ENG, SYRACUSE UNIV, 62- *Concurrent Pos:* Engr, Babcock & Wilcox Co, 51 & Westinghouse Atomic Power Div, 54; consult, Mat Adv Bd, Nat Acad Sci, 57-58 & Int Bus Mach, 58; Fulbright lectr, Delft Univ Technol, 65-66. *Mem:* Am Soc Mech Engrs; Am Soc Eng Educ; Am Nuclear Soc; Am Soc Heating, Refrig & Air Conditioning Engrs. *Res:* Thermodynamics, heat transfer, fluid mechanics & applications of these to nuclear power. *Mailing Add:* 228 Lockwood Rd Syracuse NY 13214

DRUCKER, HARRIS, b Brooklyn, NY, July 28, 43; m 66; c 2. ELECTRICAL ENGINEERING. *Educ:* Pa State Univ, BSEE, 64, Univ Pa, MSE, 66, PhD(elec eng), 67. *Prof Exp:* Jr engr, Philco Corp, Ford Motor Co, Pa, 64-67; engr, Radio Corp Am, NJ, 67-68; PROF & CHMN, DEPT ELECTRONIC ENG, MONMOUTH COL, NJ, 68- *Concurrent Pos:* Consult, Control Data Corp, 81-83, Perkin-Elmer, 81-83 & AT&T Bell Labs, 85-88; Bell Communications Res, 88; sabbatical, AT&T Bell Labs, 90-91. *Mem:* Sr mem Inst Elec & Electronics Engrs; Am Soc Eng Educ. *Res:* Speech recognition and speech processing for the hearing impaired; microcomputer software and hardware; computer networks; neural networks. *Mailing Add:* Dept Electronic Eng Monmouth Col West Long Branch NJ 07764

DRUCKER, HARVEY, b Chicago, Ill, Jan 1, 41; m 65; c 2. MICROBIOLOGY, BIOCHEMISTRY. *Educ:* Univ Ill, BS, 63, PhD(microbiol), 67. *Prof Exp:* Sr res investr biochem, Pac Northwest Labs, Battelle Mem Inst, 69-73, sect mgr molecular biol, 73-79, mgr, biol dept, 79-83; ASSOC LAB DIR, ENERGY ENVIRON & BIOL RES, ARGONNE NAT LAB, 83- *Concurrent Pos:* Cardiovasc Res Inst fel, Med Ctr, Univ Calif, San Francisco, 67-69; adj prof, Grad Ctr, Richland, Wash, 72- *Mem:* AAAS; Am Soc Microbiol; Am Chem Soc; Fedn Am Soc Exp Biol. *Res:* Regulation of exocellular enzymes affecting degradation of macromolecules; chemistry of proteolytic enzymes; toxicity of synthetic fuels; environmental chemistry of metals. *Mailing Add:* Argonne Nat Lab Bldg 202 Rm LA125 Argonne IL 60439

DRUCKER, WILLIAM D, b New York, NY, Mar 30, 29; div; c 2. INTERNAL MEDICINE, MEDICAL TEACHING & ADMINISTRATION. *Educ:* NY Univ, BA, 50, MD, 54; Am Bd Internal Med, dipl, 73; Am Bd Endocrinal & Metab, dipl, 75. *Prof Exp:* From instr to assoc prof med, Sch Med, NY Univ, 62-76; PROF MED, MED COL WIS, MILWAUKEE, 76-, ASST DEAN STUDENT AFFAIRS & DIR MINORITY AFFAIRS, 87-; ACAD CHIEF MED, ST JOSEPH'S HOSP, 76-; ASST DEAN STUDENT AFFAIRS & DIR MINORITY AFFAIRS, MED COL WIS, 87- *Concurrent Pos:* NIH fel, 59-62; career scientist, Health Res Coun City of New York, 62-73. *Mem:* AAAS; Endocrine Soc; fel Am Col Physicians; Am Fedn Clin Res. *Res:* Spontaneous canine Cushing's syndrome; biologic action and physiologic regulation of the adrenal androgens in man; endocrine aspects of acne, hirsutism and obesity. *Mailing Add:* St Joseph's Hosp Med Col Wis 5000 W Chambers St Milwaukee WI 53210

DRUCKER, WILLIAM RICHARD, b Chicago, Ill, Apr 5, 22; m 47; c 4. MEDICINE. *Educ:* Harvard Univ, BS, 43; Johns Hopkins Univ, MD, 46; FRCS(C). *Prof Exp:* From instr to prof surg, Western Reserve Univ, 54-66; prof surg & chmn dept, Univ Toronto, 66-72, surgeon-in-chief, Toronto Gen Hosp, 66-72; prof surg & dean, Sch Med, Univ Va, 72-77; PROF SURG & CHMN DEPT, SCH MED & DENT, UNIV ROCHESTER, 77- *Concurrent Pos:* Charles O Finley scholar, 54-57; Markle scholar, 58-63. *Mem:* Am Surg Asn; Fedn Am Socs Exp Biol; Can Soc Clin Invest; fel Am Col Surg. *Res:* Nutrition and intermediary metabolism in surgical patients; hemorrhagic shock. *Mailing Add:* Dept Surg Uniformed Serv Univ Health Sci 4301 Jones Bridge Rd Bethesda MD 20814

DRUDGE, J HAROLD, b Bremen, Ind, Feb 7, 22; m 46; c 1. VETERINARY PARASITOLOGY. *Educ:* Mich State Univ, DVM, 43; Johns Hopkins Univ, ScD(parasitol), 50. *Prof Exp:* Vet parasitologist, Agr Exp Sta, Miss State Col, 50-51; actg chmn dept, Univ Ky, 63-68, chmn dept, 68-73, parasitologist, Agr Exp Sta, 51-63, prof vet sci, 63-88, EMER PROF VET SCI, UNIV KY, 88- *Mem:* Am Vet Med Asn; Am Soc Parasitol. *Res:* Nematode parasites of cattle, sheep and horses. *Mailing Add:* Dept Vet Sci Gleck Equine Res Ctr Univ Ky Lexington KY 40546-0099

DRUEHL, LOUIS D, b San Francisco, Calif, Oct 9, 36; m 64; c 2. KELP EVOLUTION, PLANT MARICULTURE. *Educ:* Wash State Univ, BSc, 59; Univ Wash, MSc, 61; Univ BC, PhD(bot, oceanog), 65. *Prof Exp:* From asst prof assoc prof, 66-88, PROF BIOL, SIMON FRASER UNIV, 88- *Concurrent Pos:* Nat Res Coun Can grants, 66-; vis assoc prof, Hokkaido Univ, Japan, 72 & Friday Harbor Labs, Univ Wash, 82, 83, 85 & 90, Univ Alaska, 86; adv mariculture, Ministry Marines, Brazil, 75-77, Sci Coun grant, 81-83; exchange scientist, Nat Res Coun & Nat Coun Pesquinas; consult & researcher, Marine Biomass Prog, Gen Elec Gas Res Inst, 81-; Environ Can grants, 77-79; Marine Resources Br BC grants, 79-81; pres, Can Kelp Resources Ltd; dir, San Francisco Univ Inst Aquacult Res, 88-; Luigi Provasoli Award, J Phycol, 88-89. *Mem:* Western Soc Naturalists (pres, 88). *Res:* Kelp evolution and distribution studies employing classical and molecular DNA techniques; biology of the Laminariales; mariculture of seaweeds. *Mailing Add:* Dept Biol Sci Simon Fraser Univ Burnaby BC V5A 1S6 Can

DRUELINGER, MELVIN L, b South Bend, Ind, Dec 7, 40; m 59; c 2. HETEROCYCLIC CHEMISTRY, REACTIVE INTERMEDIATES. *Educ:* Ind Univ, Bloomington, BS, 62; Univ Wis-Madison, PhD(org chem), 67. *Prof Exp:* NIH res fel org chem, Iowa State Univ, 67-68; assoc prof org chem, Ind State Univ, Terre Haute, 68-81; distinguished vis prof, US Air Force Acad, 80-82; chmn & prof, Dept Chem, Millikin Univ, Decatur, Ill; PROF CHEM, UNIV SOUTHERN COLO, PUEBLO, 84-, DIR, RES & SPONSORED PROGS, 87- *Concurrent Pos:* Sabbatical leave, Colo State Univ, 74-75; Air Force Systs Command-URRP fel, US Air Force Acad, 78-80; fac res fel, Rocket Propulsion Lab, Edwards AFB, Calif, 85, fac fel, IBM, 85. *Mem:* Sigma Xi; Am Chem Soc; Royal Soc Chem; Int Soc Heterocyclic Chem. *Res:* Organic photochemistry; highly reactive organic species; organic reaction mechanisms; cycloadditions, energetic materials synthesis and heterocyclic compounds, especially small strained poly-heterocyclic systems; organofluorine compounds. *Mailing Add:* Dept Chem Univ Southern Colo Pueblo CO 81001

DRUFENBROCK, DIANE, b Evansville, Ind, Oct 7, 29. MATHEMATICS, GENERAL PHYSICS. *Educ:* Alverno Col, BA, 53; Marquette Univ, MS, 59; Univ Ill, PhD(math), 61. *Prof Exp:* From asst prof to prof math, Alverno Col, 61-74; PROF MATH, ST MARY-OF-THE-WOODS COL, IND, 74- *Concurrent Pos:* Lectr, Grad Sch, Marquette Univ; lectr, Univ Wis-Parkside. *Mem:* Am Math Soc; Math Asn Am. *Res:* Theory of prime power metabelian groups. *Mailing Add:* St Mary-of-the-Woods Col St Mary-of-the-Woods IN 47876-1099

DRUFFEL, LARRY EDWARD, b Quincy, Ill, May 11, 40; m 62; c 5. SOFTWARE ENGINEERING, PROGRAM MANAGEMENT. *Educ:* Univ Ill, BS, 63; Univ London, MSc, 67; Vanderbilt Univ, PhD(syst info sci), 75. *Prof Exp:* Develop engr comput commun integration, Air Force Commun Serv, 67-69; assoc prof comput sci, US Air Force Acad, Colo, 69-71; adj prof, Univ Colo, Colo Springs, 76-78; prog mgr, Defense Advan Res Projs Agency, 78-82; dir, Ada Joint Prog Off, Res & Eng, Off Undersecy Defence, 80-82 & Comput Software & Systs, Res & Advan Technol, 82-83; vpres, Rational, 83-86; DIR, SOFTWARE ENG INST, CARNEGIE-MELLON UNIV, 86- *Concurrent Pos:* Consult, Rome Air Develop Ctr, 77-78, Inst Defense Anal, 83-; software ed, Comput Rev, 84-86 & Defense Sci & Electronics, 85- *Mem:* Fel Inst Elec & Electronics Engrs; Asn Comput Mach. *Res:* Design automation; software development environments; computer aided education; software engineering. *Mailing Add:* 111 Wynnwood Dr Pittsburgh PA 15215-1547

DRUGAN, JAMES RICHARD, b Detroit, Mich, May 5, 38; m 57; c 3. SYSTEMS ENGINEERING, STATISTICS. *Educ:* Univ Calif, Los Angeles, BS, 60, MS, 63, PhD(eng), 69. *Prof Exp:* Engr kinematics, Singer-Librascope, 58-62; consult engr, 62-63; sr engr systs dynamics, Singer-Librascope, 63-64; res group supvr, Jet Propulsion Lab, 64-66; staff eng systs anal, Singer Librascope, 66-70, supvr appl anal, 70-74, mgr syst engr, 74-79; PRES, MEASUREMENT ANALYSIS CORP, 80-; PRES, DRUGAN ASSOC, INC, 80- *Res:* Systems analysis; pattern recognition; detection theory; experimental design; computer software systems; information theory; estimation theory; controls; kinematics; systems simulation; dynamic programming; human factors; signal processing; involutometry; dynamics and acoustics. *Mailing Add:* 522 Kenneth Rd Glendale CA 91202

DRUGER, MARVIN, b Brooklyn, NY, Feb 21, 34; m 57; c 3. GENETICS. *Educ:* Brooklyn Col, BS, 55; Columbia Univ, MA, 57, PhD(genetics), 61. *Prof Exp:* From asst prof to assoc prof zool & sci teaching, 61-71, PROF BIOL & SCI EDUC, SYRACUSE UNIV, 71- *Concurrent Pos:* NIH fel, Sydney, Australia, 61-62; NIH res grant, 62-66; Fulbright lectr, Sydney, Australia, 69-70; vis fel, Western Australian Inst Technol, 80; Danforth Assoc, 80-; chmn educ comt, Genetics Soc Am, 74-78; exec bd mem, Nat Asn Res Sci Teaching, 73-76; pres, Soc Col Sci Teachers, 81-82. *Honors & Awards:* Science Teaching Achievement Recognition Award, Nat Sci Teachers Asn, 68, Gustav-Ohaus Award, 78 & 88. *Mem:* AAAS; Genetics Soc Am; Soc Study Evolution; Am Asn Biol Teachers; Nat Sci Teachers Asn; Asn Educ Teachers in Sci (pres, 85-86). *Res:* Science teacher education; individualized instruction; scientific literacy; science teaching methods. *Mailing Add:* Lyman Hall Syracuse Univ Syracuse NY 13244

DRUGER, STEPHEN DAVID, b Brooklyn, NY, May 1, 42. CONDENSED MATTER PHYSICS, CHEMICAL PHYSICS. *Educ:* Brooklyn Col, BS, 63; Univ Rochester, MA, 66, PhD(physics), 69. *Prof Exp:* Res assoc physics, Col William & Mary, 69-70; assoc res scientist, NY Univ, 70-72; asst prof, State Univ NY, Binghamton, 72-73; res assoc physics, Univ Rochester, 73-77; sr res assoc & adj asst prof physics, Clarkson Univ, 77-83; sr res assoc, 83-87, RES SCI & ASSOC PROF, NORTHWESTERN UNIV, 87- *Mem:* Am Phys Soc; Am Chem Soc. *Res:* Charge transport in solid ionic conductors and molecular solids; electromagnetic small-particle proximity effects on molecular energy transfer and fluorescence; light scattering from small particles. *Mailing Add:* Dept Chem Northwestern Univ Evanston IL 60208-3113

DRUGG, WARREN SOWLE, b Sitka, Alaska, Jan 29, 29; m 58; c 3. GEOLOGY, PALYNOLOGY. *Educ:* Univ Wash, BS, 52, MS, 58; Claremont Grad Sch, PhD(bot), 66. *Prof Exp:* Geologist, Calif Explor Co, 58-60, assoc res geologist, Calif Res Corp, 60-66, res geologist, Chevron Res Co, 66-68, sr res geologist, 68-77, SR RES ASSOC, CHEVRON OIL FIELD RES CO, CHEVRON CORP, 77- *Mem:* Soc Econ Paleont & Mineral; Am Asn Stratig Palynologists; Am Asn Petrol Geologists; Paleont Soc; Sigma Xi. *Res:* Fossil dinoflagellates, spores and pollen of Mesozoic and Tertiary Age. *Mailing Add:* 1820 E Stearns Ave La Habra CA 90631

DRUI, ALBERT BURNELL, b St Louis, Mo, Aug 31, 26; m 49; c 2. INDUSTRIAL & MECHANICAL ENGINEERING. *Educ:* Univ Washington, St Louis, BS, 49, MS, 57. *Prof Exp:* Planning & design engr, McDonnell Aircraft Corp, 49-53; prod engr, Army Ord Corps, 53; tech economist, Olin Mathieson Chem Corp, 54; plant indust engr, Metals Div, Dow Chem Co, 55-58; sr supvr Div Planning Staff, Boeing Co, 58-60; asst prof mech eng, 60-70, ASSOC PROF MECH ENG, UNIV WASH, 70- *Concurrent Pos:* Consult indust & mech engr, 60-; US Dept HEW res grant radiol opers efficiency, 67-70, co-investr, res grant Medex doctors' asst prog, 69-71; dir, Study Pilot Prog Food Serv, US Army, Natick Labs, 71; auditor & rep in bargaining, Boeing Aerospace Co, 72; consult, Gen Tel NW Systs Anal, 73, Secy State UN, 73 & John Fluke Mfg Co, 75; rep, Teamsters Local 344, 73; supvr, eval Boeing Aerospace Co mfg plans, 74; consult, Boeing Commercial Airplane Co, 76, supvr training manual, 78; auditor, Weyerhaeuser Co, all USA plants. *Mem:* Am Inst Indust Engrs; Inst Mgt Sci. *Res:* Industrial engineering to improve operational systems; work measurement and plant layout; effective delivery of health care by examining radiologists, anesthesiologists, general practitioners and mid medical care professions. *Mailing Add:* 22035 117th Ave Kent WA 93031

DRULLINGER, ROBERT EUGENE, b Lansing, Mich, Dec 23, 44; div; c 2. CHEMICAL PHYSICS. *Educ:* Mich State Univ, BS, 67; Columbia Univ, PhD(chem), 72. *Prof Exp:* PHYSICIST, NAT BUR STANDARDS, 72-, LEADER, ATOMIC BEAMS-FREQUENCY STANDARDS GROUP, 88- *Concurrent Pos:* Nat Res Coun fel, Nat Bur Stand, 73-74. *Mem:* Am Phys Soc; AAAS. *Res:* Development of new spectroscopic techniques for the elucidation of molecular structure and kinetics; primarily concerned with the excited states of molecules with repulsive ground states; ultra-accurate spectroscopic measurements. *Mailing Add:* Div 524 Nat Inst Standards & Technol 325 Broadway Boulder CO 80303

DRUM, CHARLES MONROE, b Richmond, Va, Sept 17, 34; m 58; c 3. SOLID STATE PHYSICS. *Educ:* Washington & Lee Univ, BS, 57; Univ Va, PhD(physics), 63. *Prof Exp:* NSF fel, Atomic Energy Res Estab, Eng, 63-64; MEM TECH STAFF, BELL TEL LABS, INC, 64- *Mem:* Am Phys Soc; Inst Elec & Electronics Engrs. *Res:* Imperfections in crystals; electron diffraction; materials science engineering. *Mailing Add:* 2604 Gordon St Allentown PA 18104

DRUM, I(AN) M(ONDELET), b Ottawa, Ont, Oct 22, 13; m 43; c 2. CHEMICAL ENGINEERING. *Educ:* Royal Mil Col Can, dipl, 35; Queen's Univ, Ont, BSc, 37. *Prof Exp:* Exec vpres, Dye & Chem Co Can, Ltd, 48-51; mgr special projs, Home Oil Co, Ltd, 51-66, vpres spec projs, 66-76; PRES, DRUM CONSULT LTD, 76- *Concurrent Pos:* Consult, Chem Inst Can, 76- *Mem:* Fel Chem Inst Can; Can Inst Mining & Metall. *Res:* Processing, transporting and marketing of natural gas and crude oil. *Mailing Add:* 69 Woodmeadow Close SW Calgary AB T2W 4L8 Can

DRUM, RYAN WILLIAM, b Milwaukee, Wis, Sept 25, 39; div; c 3. HERBOLOGY, MARINE ALGAL AQUACULTURE. *Educ:* Iowa State Univ, BSc, 61, PhD(phycol), 64. *Prof Exp:* NATO fel, Univs Bonn & Leeds, 64-65; asst prof bot, Univ Ark, 66; asst prof, Univ Mass, Amherst, 66-70; asst prof, Fairhaven Col, 70-76; PRIVATE CONSULT, 76- *Concurrent Pos:* Vis asst prof, Univ Calif, Los Angeles, 69-70; lectr, Dominion Herbal Col, 77- *Mem:* Am Herb Asn. *Res:* Cytoplasmic ultrastructure of diatoms; biogenesis of silica in diatoms; grasses and sponges; single-cell ecology; synthetic petrification; silica replication of cell lumens; pre-Columbian technology; herbal medicine; seaweed harvest for food. *Mailing Add:* Waldron Island WA 98297

DRUMHELLER, DOUGLAS SCHAEFFER, b Bechtelsville, Pa, Dec 31, 42; m 65; c 2. ACOUSTIC WAVEGUIDES, SHOCK WAVES. *Educ:* Univ Southern Calif, BSME, 64, MSME, 65; Lehigh Univ, PhD(appl mech), 69. *Prof Exp:* Mem tech staff optics, Bell Tel Labs, 65-67; teaching & res asst shell theory, Lehigh Univ, 67-69; DISTINGUISHED MEM TECH STAFF SOLID MECH, SANDIA NAT LABS, 69- *Mem:* Soc Natural Philos; Am Acad Mech. *Res:* Theoretical work in mixture mechanics including bubbly liquids, fluid saturated porous solids, and composite materials; theoretical experimental, and design work in acoustic waveguides for communication in petroleum wells. *Mailing Add:* Sandia Nat Labs PO Box 5800 Orgn 6252 Albuquerque NM 87185

DRUMHELLER, JOHN EARL, b Walla Walla, Wash, Dec 19, 31; m 56; c 3. MAGNETIC RESONANCE, MAGNETIC SUSCEPTIBILITY. *Educ:* Wash State Univ, BS, 53; Univ Colo, MS, 58, PhD(physics), 62. *Prof Exp:* Engr, Douglas Aircraft Co, Calif, 53-54 & Kaiser Aluminum Co, Wash, 54; res assoc, Univ Zurich, 62-64; from asst prof to assoc prof, 64-72, PROF PHYSICS, MONT STATE UNIV, 72- *Mem:* Am Phys Soc; Am Asn Physics Teachers; Sigma Xi. *Res:* Electron paramagnetic resonance in dilute impurity crystals and in low-dimensional magnetic compounds; ferro- and antiferromagnetic resonance in low-dimensional magnets; conductivity studies; dielectric, alternating current and direct current magnetic susceptibility, and magnetization studies; studies of the exchange interaction in magnetic materials; structural and magnetic phase transitions; studies of high-temperature superconductors. *Mailing Add:* Dept Physics Mont State Univ Bozeman MT 59717

DRUMHELLER, KIRK, b Walla Walla, Wash, Jan 14, 25; m 50; c 4. SOLAR ENERGY, MECHANICAL ENGINEERING. *Educ:* Mass Inst Technol, SB, 45. *Prof Exp:* Instr math, Whitman Col, 46-47; engr, Gen Elec Co, 51-53, supvr fuel element prod, 53-55, supvr tool & equip eng, 56-57; mgr design & projs, 58-62, adv fuel develop, 62-65; adv ceramic mat develop, 65-66, mgr mat develop, 66-68, mgr, Jersey nuclear prog, 69-70; prog mgr energy-fusion, Pac Northwest Labs, Battelle Mem Inst, 71-73, mgr solar energy progs, 74-84, mgr indust res, 84-89; PRES, HELIOSTATS, INC, 89- *Concurrent Pos:* Pres, Child Safety Prod. *Mem:* Int Solar Energy Soc. *Res:* Heliostat manufacturing technology for solar energy production and toxic waste destruction. *Mailing Add:* 5015 Nicklas Pl NE Seattle WA 98105-2810

DRUMM, MANUEL FELIX, b St Louis, Mo, June 21, 22; m 42; c 6. ORGANIC CHEMISTRY. *Educ:* Monmouth Col, BS, 45; Univ Nev, MS, 48; Univ Mo, PhD(org chem), 51. *Prof Exp:* Res chemist, Plastics Div, 50-52, res group leader, 52-58, res sect leader, 58-60, MGR RES, RESIN PROD DIV, MONSANTO CO, 60- *Mem:* Am Chem Soc. *Res:* Thermosetting polymers; compounding and processing polyvinylchloride; polymer foams; coatings. *Mailing Add:* 471 Forest Hills Rd Springfield MA 01128-1215

DRUMMETER, LOUIS FRANKLIN, JR, b Minersville, Pa, Dec 27, 21; m 44; c 4. OPTICS. *Educ:* Johns Hopkins Univ, BA, 43, PhD(physics), 49. *Prof Exp:* Jr instr physics, Johns Hopkins Univ, 42-44, instr, 44, asst, 44-48; physicist, US Naval Res Lab, 48-52, supvr physicist, 52-78, assoc dir res, 78-80; INSTR, UNIV VA, 81- *Mem:* Optical Soc Am. *Res:* Research administration; atmospheric optics; radiative heat transfer; infrared. *Mailing Add:* 2011 Belfast Dr SE Ft Washington MD 20744

DRUMMOND, BOYCE ALEXANDER, III, b Denison, Tex, Feb 28, 46; m 79; c 2. ECOLOGY, ENTOMOLOGY. *Educ:* Henderson State Col, BS, 67; Univ Tex, Austin, MA, 71; Univ Fla, PhD(zool), 76. *Prof Exp:* Vis asst prof zool, Univ Fla, 76-78; asst prof ecol, Ill State Univ, 78-; DIR, PIKES PEAK RES STA, EDUC CTR, FLORISSANT, COLO. *Concurrent Pos:* Res assoc, Fla State Collection Arthropods, 74-; ed, J Lepidopterist Soc, 89-92. *Mem:* AAAS; Soc Study Evolution; Ecol Soc Am; Asn Trop Biol; Lepidopterists Soc; Sigma Xi. *Res:* Evolutionary ecology, population biology and ethology of invertebrates, especially tropical ecology, coevolution of plant-animal interactions, reproductive biology and life history evolution of insects. *Mailing Add:* Natural Perspectives PO Box 9061 Woodland Park CO 80866

DRUMMOND, CHARLES HENRY, III, b Greensboro, NC, Aug 24, 44. CERAMIC ENGINEERING, GLASS SCIENCE & TECHNOLOGY. *Educ:* Ohio State Univ, BEng Phys, 68, MS, 68; Harvard Univ, SM, 70, PhD(appl physics), 74. *Prof Exp:* asst prof, 74-80, ASSOC PROF MAT SCI & ENG, OHIO STATE UNIV, 80- *Concurrent Pos:* NSF grant, 75-77; dir, Am Conf Glass Probs, 76-; summer fel, Am Soc Elec Eng/NASA, 85- *Mem:* Am Ceramic Soc. *Res:* Structure and properties of glass; X-ray diffraction; phase transformation; physical properties; composites. *Mailing Add:* Dept Mat Sci & Eng Ohio Univ 2041 College Rd Columbus OH 43210

DRUMMOND, KENNETH HERBERT, b Riverside, Calif, Jan 19, 22; m 55; c 3. OCEANOGRAPHY. *Educ:* Univ Ariz, BS, 49. *Prof Exp:* Marine res technician, Scripps Inst, Univ Calif, 49-50; asst oceanog & chief technician, Res Found, Tex A&M Univ, 50-52, assoc oceanog & proj dir, 53-57; oceanogr, US Navy Oceanog Off, 52-53; exec officer, Satellite Tracking Prog, Smithsonian Astrophys Observ, 57-58, asst dir mgt, 59-60; asst to chancellor admin, Univ Calif, San Diego, 60-62; Washington rep, Sci Serv Div, Tex Instruments Inc, 62-69; dir prog develop, Teledyne Inc, 69-70, assoc dir, Alexandria Labs, Teledyne Geotech Corp, 71-72; asst to pres, Ensco, Inc, Springfield, 72-76; prog dir, Marine Resources Ctrs, State of NC, 76-77; asst to the dean & dir spec projs, Ctr Wetland Resources, La State Univ, Baton Rouge, 77-; RETIRED. *Concurrent Pos:* Consult, Smithsonian Inst, 60-61; partic scientist, Int Indian Ocean Exped, 62; instr, Naval Reserve Officers Sch, 63-67; exec secy indust investment panel, Comn Marine Resources & Eng Develop, Exec Off of President, 67-68; mem sci adv staff, Nat Planning Asn, 62-69; mem marine adv coun, La State Univ, 69- *Mem:* Fel AAAS; Nat Ocean Industs Asn (dir & treas, 72-76); Marine Technol Soc (treas, 65-68); fel Explorers Club. *Res:* Scientific administration in the earth sciences; satellite tracking. *Mailing Add:* Ctr Wetland Resources La State Univ Baton Rouge LA 70803

DRUMMOND, MARGARET CRAWFORD, b Tulsa, Okla, Dec 4, 22. BIOCHEMISTRY. *Educ:* Agnes Scott Col, AB, 44; Emory Univ, MS, 46, PhD(bact), 57. *Prof Exp:* Instr biochem, Nursing Sch Emory Univ, 46; biochemist, USPHS, 47-49; res biochemist tuberc res, Vet Admin Hosp, 49-53; chief res & clin biochem labs, Gravely Sanatorium, 53-54; instr bact, Med Sch & res assoc Div Basic Health Sci, 57-60, from asst prof to prof, 60-90, EMER PROF MICROBIOL, EMORY UNIV, 90- *Mem:* AAAS; Am Soc Microbiol; Sigma Xi. *Res:* Host-parasite interrelationships; host resistance factors in tuberculosis; purification and study of mechanism of action of staphylococcal coagulase. *Mailing Add:* 1351 Springdale Rd Emory Univ Atlanta GA 30306

DRUMMOND, ROGER OTTO, b Peoria, Ill, Aug 11, 31; m 53; c 2. ACAROLOGY. *Educ:* Wabash Col, AB, 53; Univ Md, PhD, 56. *Prof Exp:* Med entomologist, USDA, 56-70, invests leader, 70-72, location leader & res leader, 72-77, dir, 77-86, collabr, Knipling-Bushland US Livestock Insects Res Lab, Agr Res Serv, 86-88; CONSULT, 88- *Mem:* Fel AAAS; Am Entom Soc; Am Acarol Soc; Sigma Xi; Am Soc Parasitol. *Res:* Animal systemic insecticides; livestock parasites, expecially ticks and cattle grubs; author of 2 books. *Mailing Add:* 525 Drummond Dr Kerrville TX 78028

DRUMMOND, WILLA H, b Harrisburg, Pa, Dec 5, 43. PERINATAL CARDIOPULMONARY PHYSIOLOGY. *Educ:* Brown Univ, ScB, 66; Univ Pa, MD, 70. *Prof Exp:* Assoc prof physiol & pediat, 82-88, prof vet med, 84-88, PROF PEDIAT, PHYSIOL & VET MED, UNIV FLA, 88- *Concurrent Pos:* Resident fel cardiol, Children's Hosp, Philadelphia, fel, Cardiovasc Res Inst, San Francisco, 76-78. *Mem:* Soc Pediat Res; Am Physiol Soc; Am Heart Asn; Am Acad Pediat; Int Soc Vet Perinaiology. *Res:* Physiology and pharmacology of pulmonary vascular bed in health and diseases; focus on primary pulmonary hypertension of the new born. *Mailing Add:* Dept Pediat J296 JHMHC Univ Fla Box J296 Gainesville FL 32610

DRUMMOND, WILLIAM ECKEL, b Portland, Ore, Sept 18, 27; m 53; c 3. THEORETICAL PHYSICS. *Educ:* Stanford Univ, BS, 51, PhD(physics), 58. *Prof Exp:* Physicist, Hanford Labs, Gen Elec Co, 51-52, Calif Res & Develop, 52-54, Radiation Lab, Univ Calif, 54 & Stanford Res Inst, 55-58; prin scientist, Res Lab, Avco Mfg Co, 58-59; physicist, Gen Atomic Div, Gen Dynamics Corp, 59-65; PROF PHYSICS, UNIV TEX, AUSTIN, 65-, DIR, CTR PLASMA PHYSICS & THERMONUCLEAR RES, 66- *Concurrent Pos:* Lectr, Stanford Univ, 55. *Mem:* AAAS; fel Am Phys Soc; Sigma Xi. *Res:* Plasma physics; collective accelerators; supersonic flow and shock waves; shock waves in solids; nuclear reactor theory. *Mailing Add:* 1623 Begen Ave Mountain View CA 94040

DRUMWRIGHT, THOMAS FRANKLIN, JR, b Danville, Va, Sept 8, 28; m 58; c 2. NONDESTRUCTIVE TESTING, ELECTRICAL ENGINEERING. *Educ:* Va Mil Inst, BS, 51; Ohio State Univ, MS, 57. *Prof Exp:* Engr welding, Newport News Shipbuilding & Dry Dock Co, 53-55; engr, Alcoa Res Lab, 57-70, sr engr, 70-73, group leader non-destructive testing, 73-78, eng assoc, 78-82, sr tech specialist, 82-84, tech consult, 88-91; RETIRED. *Honors & Awards:* Comm E-7 C W Briggs Award & Award of Merit, Am Soc Testing & Mat. *Mem:* Fel Am Soc Testing & Mat; fel Am Soc Nondestructive Testing; Sigma Xi. *Res:* Development of improved methods, instrumentation, and procedures for company-wide nondestructive testing programs; consultation regarding nondestructive testing, process instrumentation and inspection use in all Alcoa plants; development of quantitative nondestructive evaluation methods. *Mailing Add:* 790 Kennedy Ave New Kensington PA 15068

DRURY, COLIN GORDON, b Skegness, UK, June 4, 41; m 62; c 2. ERGONOMICS, HUMAN FACTORS ENGINEERING. *Educ:* Univ Sheffield, UK, BSc, 62; Univ Birmingham, UK, PhD(eng prod), 68. *Prof Exp:* Res engr, Motor Indust Res Asn, UK, 62-64; lectr ergonomics, Univ Birmingham, UK, 68-69; mgr ergonomics, Pilkington Bros (Glass) Ltd, UK, 69-72; PROF INDUST ENG, STATE UNIV NY, BUFFALO, 72- *Concurrent Pos:* Chmn, Indust Div, Ergonomics Soc, 71-72; dir res & eval, Lakes Area Emergency Med Serv, 75-77; dir, Ergonomics Div, Inst Indust Engrs, 85-86; exec dir, Ctr Indust Effectiveness, TCIE, 87- *Honors & Awards:* Bartlett Medal, Ergonomics Soc, 80; IIE Engonomics Div Award, 88. *Mem:* Ergonomics Soc; Human Factors Soc; Inst Indust Engrs. *Res:* Ergonomics in theory and practice; developing models of human performance and stress in industrial quality and process control; applying ergonomics to workplace design, automation and safety in industry. *Mailing Add:* Dept Indust Eng 342 Bell Hall State Univ NY Buffalo NY 14260

DRURY, LISTON NATHANIEL, b Jacksonville, Fla, Oct 5, 24; m 52; c 2. GENERAL AGRICULTURE. *Educ:* Univ Ga, BSAE, 50, MS, 61. *Prof Exp:* Electrification adv, Suwannee Valley Elec Coop, 50-51 & Talquin Elec Coop, 51-53; instr, Univ Tenn, 53-54; res agr engr, agr res serv, USDA, 55-83; RETIRED. *Mem:* Am Soc Agr Engrs. *Res:* Poultry environmental research. *Mailing Add:* 1745 Robert Hardeman Rd Winterville GA 30683

DRURY, MALCOLM JOHN, b Eng, Oct 61, 48; UK & Can citizen. GEOTHERMICS, HYDROGEOLOGY. *Educ:* Univ Wales, BSc, 70; Dalhousie Univ, PhD(geophys), 77. *Prof Exp:* Fel, 71-79, RES SCIENTIST GEOTHERMICS, GEOL SURV CAN, ENERGY MINES & RESOURCES CAN, 79- *Concurrent Pos:* Prin investr, nuclear fuel waste mgt, Atomic Energy Can, 79- *Mem:* Am Geophys Union; Geol Asn Can; Can Geophys Union; Can Geothermal Energy Asn. *Res:* Geothermal energy; thermal evolution and tectonic development of the earth's crust; heat flow; physical properties earth materials. *Mailing Add:* 601 Booth St Ottawa ON K1A 0E4 Can

DRURY, WILLIAM HOLLAND, b Newport, RI, Mar 18, 21; m 51; c 4. ECOLOGY. *Educ:* Harvard Univ, BA, 42, MA, 48, PhD, 52. *Prof Exp:* Asst prof biol & gen educ, Harvard Univ, 52-56; dir res, Mass Audubon Soc, 56-75; MEM FAC BIOL, COL OF THE ATLANTIC, 75- *Concurrent Pos:* Lectr biol, Harvard Univ, 56-75. *Mem:* Asn Field Ornith; Am Ornith Union; Wilson Ornith Soc; Cooper Ornith Soc. *Res:* Bird ecology and behavior; plant ecology relative to geological processes. *Mailing Add:* Ten High St Bar Harbor ME 04609

DRUSE-MANTEUFFEL, MARY JEANNE, b Racine, Wis, Oct 17, 46; m 73; c 2. BIOCHEMISTRY, NEUROCHEMISTRY. *Educ:* Duke Univ, AB, 68; Univ NC, Chapel Hill, PhD(biochem), 72. *Prof Exp:* Fel neurobiol, Univ NC, Chapel Hill, 72; fel neurochem, Develop Metab Neurol Br, NIH, 72-73; from asst prof to assoc prof, 73-87, PROF BIOCHEM, STRITCH SCH MED, LOYOLA UNIV CHICAGO, 87- *Concurrent Pos:* NIMH fel, Univ NC, 69-72; Sloan-Neurobiol fel, Univ NC, 72; NIH fel, Nat Inst Neurol Commun Dis Stroke, 72-73; prin investr, Nat Found March of Dimes, 75-78, Nat Coun Alcoholism, 78-79, Nat Inst Alcohol Abuse & Alcholism, 80-; Alcohol Biomed Rev Comt; Schweppe Found career develop award, 78-81. *Mem:* AAAS; Am Soc Neurochem; Soc Neurosci; Sigma Xi; Am Chem Soc; Res Soc Alcohlism; Am Soc Biochem & Molecular Biol. *Res:* Developmental neurochemistry; influence of maternal alcoholism and maternal undernutrition on CNS neurotransmitter levels, receptors, and upstake systems, and on synaptogenesis in offspring. *Mailing Add:* 918 Lexington St Wheaton IL 60189-3845

DRUSHEL, HARRY (VERNON), b Evans City, Pa, Feb 2, 25; m 46; c 3. ANALYTICAL CHEMISTRY. *Educ:* Univ Pittsburgh, BS, 49, PhD, 56. *Prof Exp:* Chemist, Exxon Res & Develop Labs, 58-66, res assoc, 66-68, sr res assoc, 68-82, sci adv, 82-85; OWNER, DRU-TEK, 85- *Honors & Awards:* Coates Award, Am Chem Soc. *Mem:* Am Chem Soc; Soc Appl Spectros; Coblentz Soc. *Res:* Absorption spectroscopy, ultraviolet, visible, infrared; polarography; instrumental analyses; sulfur and nitrogen compounds in petroleum; luminescence spectroscopy; gas and liquid chromatography; microcoulometry; electron spectroscopy for chemical analysis. *Mailing Add:* Dru-Tek PO Box 46032 Baton Rouge LA 70895

DRUSIN, LEWIS MARTIN, b New York, NY, Sept 25, 39. PREVENTIVE MEDICINE, INFECTIOUS DISEASES. *Educ:* Union Col, BS, 60; Cornell Univ, MD, 64; Columbia Univ, MPH, 74. *Prof Exp:* Instr med, 69-70, asst prof, Pub Health, 70-77, asst prof med, 72-79, assoc prof pub health, 77-84, ASSOC PROF CLIN MED & PROF CLIN PUB HEALTH, CORNELL UNIV, 79-; DIR DEPT EPIDEMIOL, NY HOSP, 70-, ASSOC ATTEND PHYSICIAN, 79- *Concurrent Pos:* Asst attend physician, New York Hosp, 72-79 & Mem Hosp Cancer & Allied Dis, 73-81; vis assoc physician, The Rockefeller Univ, 77-87, consult, 87-; assoc attend physician, Cancer & Allied Dis, Memorial Hosp, New York, 81-84; regional dir NAm, Int Union against Venereal Dis & Treponematoses, 86- *Mem:* Fel Am Col Physicians; fel Am Col Prev Med; fel Infectious Dis Soc Am; Med Soc Study Venereal Dis; Sigma Xi; Am Venereal Dis Asn (pres, 82). *Res:* Epidemiology and clinical aspects of hospital acquired infections and sexually transmitted diseases. *Mailing Add:* 55 E End Ave No 3E New York NY 10028-7928

DRUY, MARK ARNOLD, b Jan 11, 55. ORGANIC SOLID STATE CHEMISTRY. *Educ:* Brown Univ, BA, 77; Univ Pa, PhD(chem), 81. *Prof Exp:* Sr mem tech staff, GTE Labs, 81-86; SR SCIENTIST, FOSTER-MILLER INC, 86- *Mem:* Am Chem Soc; Mat Res Soc; Soc Photo-Instrumentation Engrs; Soc Plastic Engrs. *Res:* Structure/property relationships of polymers with electrical and optical properties; principals and applications of remote fiber optic spectroscopy. *Mailing Add:* 38 Bonad Rd Arlington MA 02174

DRUYAN, MARY ELLEN, b Washington, DC, July 14, 38; m 81; c 2. BIOCHEMISTRY, PHYSICAL CHEMISTRY. *Educ:* Wellesley Col, BA, 60; Tufts Univ, MS, 62; Univ Chicago, PhD(biochem), 72; Univ Ill, MPH, 88. *Prof Exp:* Mem staff, Argonne Nat Lab, 72-74; asst prof, 74-82, ASSOC PROF BIOCHEM, LOYOLA UNIV, 82- *Concurrent Pos:* Res chemist, Hines Vet Admin Hosp, Ill, 74-76, consult, 76-; mem study sect oral biol & med, NIH, 80-84. *Mem:* Am Chem Soc; Am Crystallog Asn; Biophys Soc; Am Asn Women Sci; Am Asn Dent Schs; Am Dietitians Asn. *Res:* Structural chemistry of molecules of biological interest; biochemical nutrition; spectroscopy. *Mailing Add:* Loyola Univ Sch Dent 2160 S First Ave Maywood IL 60153

DRUZ, WALTER S, b Chicago, Ill, June 26, 17. PULMONARY PHYSIOLOGY. *Educ:* Univ Ill, PhD(physiol), 76. *Prof Exp:* ASSOC PROF PHYSIOL, STRITCH SCH MED, LOYOLA UNIV, 78- *Mem:* Am Physiol Soc; Am Thoracic Soc; Inst Elec & Electronics Engrs; Soc Motion Picture & Television Engrs. *Mailing Add:* 824 N Kalaheo Ave Kailua HI 96734-1977

DRYDEN, RICHARD LEE, b Pittsburgh, Pa, July 2, 45; m 68; c 3. MICROBIOLOGY, IMMUNOLOGY. *Educ:* Allegheny Col, BS, 67; Univ SC, MS, 69; NC State Univ, PhD(microbiol), 74. *Prof Exp:* asst prof, 73-81, ASSOC PROF BIOL, WASH & JEFFERSON COL, 81- *Concurrent Pos:* Assoc adv to health professions. *Mem:* Am Soc Microbiol. *Res:* Immunotherapy of spontaneous leukemia and lymphona in mice; Azotobacter vinelandii; encystment studies. *Mailing Add:* Dept Biol Wash & Jefferson Col Washington PA 15301

DRYER, MURRAY, b Bridgeport, Conn, Nov 4, 25; m 55; c 2. SPACE PHYSICS, FLUID MECHANICS. *Educ:* Stanford Univ, BS, 49, MS, 50; Tel-Aviv Univ, PhD(physics), 70. *Prof Exp:* Res scientist aerodyn, NASA, 50-59; assoc res scientist, Martin-Marietta Corp, 59-65; lectr gasdyn, Univ Colo, Denver, 63-76; CHIEF INTERPLANETARY STUDIES SPACE PHYSICS, NAT OCEANIC & ATMOSPHERIC ADMIN, 65-; LECTR, UNIV COLO, DENVER, 78- *Concurrent Pos:* Vis assoc prof magnetohydrodyn, Colo State Univ, 66-67; mem comt solar-terrestrial res, Nat Acad Sci, 76-79 & 84-; chmn, Scostep's proj Solar Connection to Traveling Interplanetary Phenomena, 89-; mem, comn space res, Int Astron Union, sci comn solar-terrestrial res. *Honors & Awards:* Space Sci Award, Am Inst Aeronaut & Astronaut, 75. *Mem:* Am Phys Soc; Am Astron Soc; Am Geophys Union; Am Inst Aeronaut & Astronaut; AAAS; Int Astron Union. *Res:* Theoretical research concerned with the solar wind and its interactions, under both quiet and disturbed conditions, with obstacles in the solar system; computational physics directed to various forms of solar activity and their interplanetary consequences with application to prediction of geomagnetic storms. *Mailing Add:* Space Environ Lab R/E/SE Nat Oceanic & Atmospheric Admin Boulder CO 80303-3328

DRYHURST, GLENN, b Birmingham, Eng, Sept 15, 39; m 84; c 3. ELECTROCHEMISTRY OXIDATION MECHANISMS. *Educ:* Univ Aston, UK, BSc, 62; Univ Birmingham, UK, PhD(chem), 65. *Prof Exp:* Res assoc electrochem, Univ Mich, 65-67; from asst prof to assoc prof, 67-83, CHMN DEPT CHEM, UNIV OKLA, 81-, GEORGE LYNN CROSS RES PROF CHEM, 83- *Concurrent Pos:* Fulbright sr prof, Konstanz Univ, FRG, 87-88; div ed, Jour Elec Chem Soc, 85-; prin investr, Nat Sci Foun, NIH grants, 68-; secy-treas, Electrochem Soc, 85-87, vchmn, Div Org & Biol Electrochem, 87-89, chmn, 89-91. *Honors & Awards:* Res Award, Sigma Xi; Sr Prof Award, Fulbright Comm, 87. *Mem:* Am Chem Soc. *Res:* Oxidation chemistry of Indolic neurotransmitters and neurotoxins; electrochemical and enzymatic studies of oxidation mechanisms; mechanisms of action of neurotoxins; chemistry and biochemistry of neurodegenerative diseases. *Mailing Add:* Dept Chem Univ Okla Norman OK 73019

DRYSDALE, JAMES WALLACE, b Edinburgh, Scotland, May 11, 37; div; c 2. BIOCHEMISTRY, MOLECULAR BIOLOGY. *Educ:* Univ Edinburgh, BSc, 60, MSc, 63; Univ Glasgow, PhD(biochem), 65. *Prof Exp:* Asst lectr biochem, Univ Edinburgh, 60-63; asst lectr, Univ Glasgow, 63-66; asst prof, Mass Inst Technol, 66-68; vis lectr molecular biol, Univ Edinburgh, 68-69; prin res assoc biochem, Harvard Med Sch, 69-72; assoc prof, 72-83, PROF BIOCHEM, SCH MED, TUFTS UNIV, 83- *Concurrent Pos:* Res grants, NIH, NSF & Am Cancer Soc, 69-; consult Med Res Apparatus, Immunodiagnostics, Immucell, 72- *Mem:* Brit Biochem Soc; Am Soc Biol Chemists; AAAS; Am Soc Cell Biol; Am Soc Hemat. *Res:* Molecular basis for hemochromatosis; mammalian protein synthesis, gene expression in normal and malignant cells, iron metabolism, structure, function and metabolism of isoferritins. *Mailing Add:* Dept Biochem Sch Med Tufts Univ Boston MA 02111

DRZEWIECKI, TADEUSZ MARIA, b London, Eng, Sept 16, 43; US citizen; m 67; c 2. CONTROL SYSTEMS, WEAPONS RESEARCH. *Educ:* City Col New York, BE, 66, ME, 68; Naval Postgrad Sch, DrEng, 80. *Prof Exp:* Instr mech eng, City Col New York, 66-68; Res engr, Harry Diamond Labs, US Army, 67-73, actg res & develop supvr, 73-77, res & develop supvr fluidics, 77-78, proj officer, fluid control br, 78-82; vpres, Sci & Technol Assocs, Inc, 82-84; PRES & CHIEF EXEC OFFICER, DEFENSE RES TECHNOLOGIES, INC, 84- *Concurrent Pos:* David B Steinman mem fel, City Col New York, 66-67; dir, Tadeusz M Drzewiecki, Consult Engr, 75-; assoc ed, J Dynamic Systs, Measurement & Control; mem res staff, Syst Planning Corp, 82. *Honors & Awards:* Cert Achievement, US Army, Harry Diamond Labs, 67, Hinman Award, 73, Sustained Super Performance, 75; Army Commendation Medal, US Army Corps Engrs, 71; Res & Develop Award, US Army, 77 & 82. *Mem:* Am Soc Mech Engrs; Am Inst Aeronaut & Astronaut; Instrument Soc Am; Sigma Xi; Am Defense Preparedness Asn. *Res:* Fluidics; fluid mechanics; heat transfer; control and automatic control; tactical weapons, guidance and control. *Mailing Add:* 4608 Norbeck Rd Rockville MD 20853-1709

D'SILVA, THEMISTOCLES DAMASCENO JOAQUIM, b Lira, Uganda, July 18, 32; Indian citizen; m 65; c 4. SYNTHETIC ORGANIC CHEMISTRY, PESTICIDE CHEMISTRY. *Educ:* Univ Bombay, BS, 54; Catholic Univ Am, PhD(org chem), 64. *Prof Exp:* Fel, Worcester Found Exper Biol, 64-66; sr res scientist pesticide chem, Union Carbide Corp, 67-86; PRIN SCIENTIST, RHONE-POULENC AG CO, 87- *Concurrent Pos:* Mem, MIC Incident Invest Team. *Mem:* Am Chem Soc; fel Royal Soc Chem; fel Am Inst Chem. *Res:* Synthesis and structure-activity studies of pest control agents; modifications to impart selectivity to pests and safety to mammals and the environment. *Mailing Add:* Rhone-Poulenc Ag Co 12014 T W Alexander Dr Research Triangle Park NC 27709

D'SOUZA, ANTHONY FRANK, b Bombay, India, May 9, 29; US citizen; m 65; c 2. MECHANICS, MECHANICAL & AEROSPACE ENGINEERING. *Educ:* Univ Poona, India, BE, 54; Univ Notre Dame, MS, 60; Purdue Univ, PhD(eng), 63. *Prof Exp:* Jr engr design, Mahindra & Mahindra Ltd, India, 54-55; asst supt, Air-India Int Corp, 55; consult indust mgt, Ibcon Ltd, 55-57; trainee, Ransomes & Rapier Ltd, Eng, 57-58; asst, Univ Notre Dame, 58-60 & Purdue Univ, 60-63; from asst prof to assoc prof dynamical systs & control, 63-80, PROF MECH ENG, ILL INST TECHNOL, 80- *Concurrent Pos:* NSF res grant, 63-65, lab develop grant, 64-66; consult, Argonne Nat Lab, 78-; Dept Energy res grant, 79-81; Dept Transp res grant, 79-81; dir, Ertech, Inc, 84. *Mem:* Am Soc Mech Engrs; Sigma Xi. *Res:* Vehicle dynamics, handling and stability; self-excited vibrations; stability theory; railroad engineering; control engineering; robotics; dynamic systems. *Mailing Add:* Dept Mech Eng Ill Inst Technol Chicago IL 60616

DU, DAVID HUNG-CHANG, b Kaohsiung, Taiwan, Repub China, July 1, 51; m 80; c 3. COMPUTER SCIENCE, HARDWARE SYSTEM. *Educ:* Nat Tsing-Hua Univ, BS, 74; Univ Wash, MS, 80, PhD(comput sci), 81. *Prof Exp:* Asst prof, 81-87, ASSOC PROF COMPUT SCI, UNIV MINN, 87- *Concurrent Pos:* Consult, 3M, Honeywell. *Mem:* Inst Elec & Electronics Engrs; Asn Comput Mach. *Res:* High speed communication; computer architecture; parallel systems, database design and CAD for VLSI. *Mailing Add:* Univ Minn Minneapolis MN 55455

DU, JULIE (YI-FANG) TSAI, b Tsingtao, China; m 64; c 1. BIOCHEMISTRY. *Educ:* Nat Taiwan Univ, BS, 59; Tex Tech Univ, MS, 63; Ohio State Univ, PhD(phys chem), 70. *Prof Exp:* NIH fel path, 70-71, Univ Louisville, res assoc, Biochem Dept, 71-74, toxicol res vinyl chloride, 74-80; res assoc, prostaglandin & leukotriene, dept physiol & biophys, Georgetown Univ, 81-84; TOXICOLOGIST, US ENVIRON PROTECTION AGENCY, 84- *Mem:* Soc Toxicol; Am Chem Soc; AAAS; Sigma Xi. *Res:* Chemical carcinogenesis, aging, prostglandin and leukotriene. *Mailing Add:* 1408 Colleen Lane McLean VA 22101

DU, LI-JEN, b Harbin, China, Sept 27, 35; m 64; c 1. ELECTRICAL ENGINEERING. *Educ:* Nat Taiwan Univ, BS, 58; Ohio State Univ, MS, 62, PhD(elec eng), 65. *Prof Exp:* Res engr, Electro Sci Lab, Ohio State Univ, 62-68; asst prof, Univ Louisville, 68-72, assoc prof elec eng, 72-80; NAVAL RES LAB, WASHINGTON, DC, 80- *Mem:* AAAS; Inst Elec & Electronics Engrs; Optical Soc Am; Sigma Xi. *Res:* Applied electromagnetics; microwave electronics; image processing. *Mailing Add:* Code 4230 Naval Res Lab Washington DC 20375

DUA, PREM NATH, b Bhera, India, Nov 15, 35; m 61; c 2. VETERINARY MEDICINE, PATHOLOGY. *Educ:* Punjab Univ, DVM, 56; Miss State Univ, MS, 63, PhD(nutrit, biochem), 67. *Prof Exp:* Vet surgeon, Punjab Vet Dept, India, 56-59; instr vet med, Punjab Vet Col, 59-61; res asst poultry sci, Miss State Univ, 61-67; res assoc nutrit & biochem, Vanderbilt Univ, 67-69, assoc vet, 69-70; vet, State Lab, Va Dept Agr, 70-77; VET MED OFFICER, FOOD & DRUG ADMIN, 77- *Concurrent Pos:* Ralston Purina Co res fel, 66. *Mem:* Am Inst Nutrit; Am Vet Med Asn. *Res:* Vitamins A and K; carotenoid and lipid metabolism; thiamine metabolism. *Mailing Add:* Food & Drug Admin 200 C St SW Washington DC 20204

DUANE, DAVID BIERLEIN, b Port Jervis, NY, June 4, 34; m 57; c 2. MARINE GEOLOGY, COASTAL ENGINEERING. *Educ:* Dartmouth Col, AB, 57; Univ Kans, MS, 59, PhD(geol), 63. *Prof Exp:* Explor geologist, Magnolia Petrol Co, 57 & Mobil Oil Co, 62-64; res phys scientist, US Lake Surv, Corps Engrs, US Army, 64-65, supvry res phys scientist, 65-66, chief geol br, Coastal Eng Res Ctr, 66-75; assoc dir grant mgt, Nat Sea Grant Prog, 75-87, dir, Nat Undersea Res Prog, 87-91, DEP DIR, OCEAN RES PROG, NAT OCEANIC & ATMOSPHERIC ADMIN, 91- *Concurrent Pos:* Instr, Bur Corresp Study, Univ Kans, 60-68. *Mem:* Fel Geol Soc Am; Soc Econ Paleont & Mineral; Sigma Xi; Am Geophys Union; Am Shore & Beach Preserv Asn. *Res:* Geologic processes and history as related to coastal engineering, marine mineral exploration and exploitation; impact of ocean waste disposal on the environment; shore processes, sediment transport, coastal geomorphology, regional geology. *Mailing Add:* 6202 Wilmett Rd Bethesda MD 20817

DUANE, THOMAS DAVID, b Peoria, Ill, Oct 10, 17; m 44; c 4. OPHTHALMOLOGY. *Educ:* Harvard Univ, BS, 39; Northwestern Univ, MD, 43, MS, 44; Univ Iowa, PhD(physiol), 48. *Prof Exp:* prof ophthal & chmn dept, Jefferson Med Col, Thomas Jefferson Univ, 62-81; RETIRED. *Mem:* Am Ophthal Soc; Am Acad Ophthal & Otolaryngol; Am Asn Res Vision & Ophthal; AMA; Sigma Xi. *Res:* Biophysics of the eye, especially retinal circulation. *Mailing Add:* PO Box 10 Bedminster PA 18910

DUANY, LUIS F, JR, b Santiago de Cuba, Cuba, Dec 26, 19; m 48; c 1. DENTAL EPIDEMIOLOGY, PREVENTIVE DENTISTRY. *Educ:* Univ Havana, DDS, 44; Univ NC, Chapel Hill, MPH, 65, DrPH(epidemiol), 70. *Prof Exp:* Pvt pract, Cuba, 44-61; asst prof oral cancer, Sch Dent, Univ Havana, 45-46; scientist adminr res grants, Nat Inst Dent Res, 61-63; asst prof epidemiol & prev dent, Sch Dent, Univ PR, San Juan, 66-67; asst prof oral biol & chief epidemiol sect, Inst Oral Biol, Univ Miami, 67-80; GEN DENT, 80- *Concurrent Pos:* Nat Inst Dent Res fel, Univ Miami & Univ NC, 63-66; consult, Div Oral Biol, Univ Miami, 65-67, Div Family Med, 68-69; consult, Pan-Am Health Orgn, WHO, 68-; lectr, Sch Dent, Univ Antioquia, Colombia, 68-, Sch Dent, Univ El Salvador, 70- & Nat Inst Dent Res, 71- *Honors & Awards:* Dipl, Asn Oral Surgeons Cuba, 47 & PR Pub Health Dept, 67. *Mem:* Am Dent Asn; Int Asn Dent Res; Am Pub Health Asn. *Res:* Epidemiology of dental caries, especially microbiology, dietary regimes, dental plaque extent, oral hygiene, fluoride exposure and their relationship to caries status; studies of caries, particularly free and carie active students. *Mailing Add:* 8822 SW 24th St Miami FL 33165

DUAX, WILLIAM LEO, b Chicago, Ill, Apr 18, 39; m 65; c 4. MOLECULAR BIOPHYSICS. *Educ:* St Ambrose Col, BA, 61; Univ Iowa, PhD(phys chem), 67. *Prof Exp:* Res fel inorg x-ray crystallog, Ohio Univ, 67-68; res assoc steroid struct & x-ray crystallog, 68-69, head crystallog lab, 69-72, assoc dir res, 83-87, HEAD MOLECULAR BIOPHYS DEPT, MED FOUND BUFFALO, 72-, RES DIR, 88- *Concurrent Pos:* Co-prin investr, Nat Cancer Inst grant, 71-; prin investr, Gen Med Sci Inst grant, 73-; adj assoc prof med chem, Sch Pharm, State Univ NY Buffalo, 73-; mem, Nat Comt Crystallog, 81-83, chmn, 88-90; assoc res prof, Roswell Park Div, Dept Biochem, State Univ NY, Buffalo, 81-; NIH consult & dir, US distrib Cambridge Struct Database, 83-; secy, comn small molecules, Int Union Crystallog, 85-90, chmn, 91-; pres, Am Crystallog Asn, 86, exec officer, 88-90; mem bd govs, Am Inst Physics, 87-88. *Honors & Awards:* Fulbright Scholar, Lectr Prog, Yugoslavia, 87. *Mem:* AAAS; Biophys Soc; Am Crystallog Asn (vpres, 85, pres, 86); Am Chem Soc; Am Asn Cancer Res. *Res:* Conformational analysis of crystal structure data for steroids polypeptide hormones, ion transport antibiotics and proteins; elucidation of structural-functional relationships among steroid opiates and ion transport antibiotics; development of computer based graphical techniques for rational drug design; correlations of x-ray structural data with emperial energy calculations. *Mailing Add:* Med Found Buffalo 73 High St Buffalo NY 14203

DUB, MICHAEL, b Opaka, Ukraine, Mar 11, 17; US citizen; m 49; c 1. ORGANOMETALLIC CHEMISTRY, ORGANIC CHEMISTRY. *Educ:* Univ Vienna, MS, 44; City Col New York, BS, 55. *Prof Exp:* Res chemist, NY Quinine & Chem Works, Monsanto Co, 52-55; from res chemist to sr chemist, Monsanto Chem Co, 55-82; PRES, CHEM-INFO SERV INC, 83- *Honors & Awards:* Cert Recognition, Sigma Xi, 82. *Mem:* Am Chem Soc. *Res:* Organic synthesis; animal nutrition; science communications. *Mailing Add:* 1060 Orchard Lakes Dr St Louis MO 63146

DUBACH, HAROLD WILLIAM, b St Joseph, Mo, Nov 25, 20; m 46; c 4. PHYSICAL OCEANOGRAPHY, METEOROLOGY. *Educ:* Baker Univ, AB, 42; Univ Chicago, cert meteorol, 43. *Prof Exp:* Weather officer upper air, US Air Force Weather Serv, 42-46; res meteorologist cloud physics, Thunderstorm Proj, US Weather Bur, 46-48; phys oceanog res, US Naval Oceanog Off, 48-60, staff tech asst, 68-69; actg dir & dep oceanogr info sci, Nat Oceanog Data Ctr, 60-68; asst dir oceanog & meteorol, Coastal Plains Ctr Marine Develop Serv, 69-73, chmn marine industs dept, Beaufort Tech Educ Ctr, 73-75; oceanogr & consult, Savannah River Lab, ERDA, 75-76; dir, Marine Res Ctr, NC Off Marine Affairs, 77-78; ADMINR, NAT UNDERSEA RES PROG, UNIV NC, WILMINGTON, 80- *Concurrent Pos:* Forecaster & Geophysicist, US Air Force Reserves Air Weather Serv, 46-62; mem, Oceanogr Exam Panel, US Civil Serv Comn, 54-61, Meteorologist Exam Panel, 57-60; mem, Fed Adv Comn Water Pollution, US Dept Interior, 66-69; consult oceanog, Pepperdine Univ, Beaufort, 76; coordr, Ocean Outfall Proj, NC Off Marine Affairs, 76- *Honors & Awards:* Super Accomplishment Award, US Navy Oceanog Off, 62, Patent Award, 67. *Mem:* Am Meteorol Soc; Marine Technol Soc; Int Oceanog Found; Smithsonian Assocs; Nat Geog Soc. *Res:* Marine science documentation, information processing and dissemination; coastal environment and air-sea interaction; oceanographic instrumentation design; marine climatology; cloud physics, thunderstorms and culumni forms. *Mailing Add:* 4609 Dean Dr Wilmington NC 28405

DUBACH, LELAND L, b St Joseph, Mo, Apr 15, 23; m 45; c 3. METEOROLOGY, AERONOMY. *Educ:* Univ Colo, BA, 49; Univ Calif, Los Angeles, MEd, 51, MA, 65. *Prof Exp:* Meteorologist, US Army Air Corps, 43-47; teacher math & sci, Sec Schs, Calif, 49-52; meteorologist, US Air Force, mem fac meteorol, Forecaster Supt Sch, Ill, 54-57, consult, USAF-SAC, Ohama, 59-61, chief forecaster, Lajes Field, Azores Islands, 61-63, dep sta comdr & consult, Space Systs Div, Air Force Serv Ctr, El Segundo, 63-67, coord scientist aeron, ionospheric physics & meterol, Nat Space Sci Data Ctr, Goddard Space Flight Ctr, 67-77, meteorologist, Meteorol Prog Off, 78-79, dept br head, Severe Storms Br, Goddard Space Flight Ctr, NASA, 80-87; RETIRED. *Mem:* Am Meteorol Soc. *Res:* Climate; severe storms, mesometeorology. *Mailing Add:* 3798 Aster Dr Sarasota FL 34233

DUBAR, JULES R, b Canton, Ohio, June 30, 23; m 64; c 2. INVERTEBRATE PALEONTOLOGY, BIOSTRATIGRAPHY. *Educ:* Kent State Univ, BS, 49; Ore State Univ, MS, 50; Univ Kans, PhD, 57. *Prof Exp:* Asst geol, Ore State Univ, 49-50 & Univ Ill, 50-51; instr, Southern Ill Univ, 51-53 & Univ Kans, 53-54; from instr to asst prof, Southern Ill Univ, 54-57; from asst prof to assoc prof, Univ Houston, 57-62; assoc prof, Duke Univ, 62-64; sr geologist, Res Lab, Esso Prod Res Co, 64-67; from assoc prof to prof geosci & chmn dept, Morehead State Univ, 67-81; dir explor, Int Resource Develop Corp, USA, 81-82; RES SCIENTIST & TECH ED, BUR ECON GEOL, UNIV TEX, AUSTIN, 82- *Concurrent Pos:* Consult, Fla Geol Surv, 53-59 & Int Minerals & Chem Corp, 63-64 & 76-77; NSF grants, 59-64, 68-70 & 77-81; geologist, US Geol Surv, 79-80. *Mem:* AAAS; Am Asn Petrol Geol; Geol Soc Am; Am Soc Oceanog; Am Malacol Union; Soc Econ Paleontologists & Minerologists; Paleont Soc Am; Int Paleont Union. *Res:* Stratigraphy; paleontology and paleoecology of Neogene of eastern seaboard and gulf coastal region; Neogene mollusca; stratigraphy and paleoclimatology late Cenozoic of West Texas; origin of playa lakes, West Texas. *Mailing Add:* Bur Econ Geol Univ Tex Austin TX 78712

DUBAS, LAWRENCE FRANCIS, b Evanston, Ill, Nov 28, 52; m 76; c 3. ORGANIC CHEMISTRY, DECISION ANALYSIS. *Educ:* Univ Ill, BS, 74; Stanford Univ, PhD(org chem), 78; Univ Del, MBA, 86. *Prof Exp:* Synthesis chemist, E I du Pont de Memours & Co Inc, 78-81, patent liaison, 81-83, process develop chemist, 83-86, res supvr, 87-90, DECISION ANALYST, E I DU PONT DE NEMOURS & CO INC, 90- *Mem:* Am Chem Soc. *Res:* Process development. *Mailing Add:* Du Pont B13375 E I du Pont de Nemours & Co Inc Wilmington DE 19898

DUBAY, DENIS THOMAS, b Baltimore, Md, Mar 18, 52; m 77; c 2. ANGIOSPERM SEXUAL REPRODUCTION. *Educ:* Univ Notre Dame, BS, 74; Emory Univ, MS, 79, PhD(biol), 81. *Prof Exp:* Res assoc ecol res, Dept Biol, Emory Univ, 81-82, asst prof, 82; res assoc, 82-85, RESEARCHER, DEPT BOT, NC STATE UNIV, 86-; DIR, NC SCI & MATH ALLIANCE. *Concurrent Pos:* Prin investr, USDA; NASA & ASEE summer fac fel, 88. *Mem:* Sigma Xi; Ecol Soc Am; Am Inst Biol Sci; AAAS. *Res:* Interspecific differences in the effects of natural and human-generated stress factors on the sexual reproduction of flowering plants; effects of atmospheric deposition on plants; gaseous emissions from plants in controlled environments. *Mailing Add:* NC Sci & Math Alliance 410 Oberlin Rd Suite 306 Raleigh NC 27605

DUBAY, GEORGE HENRY, b Chicago, Ill, May 26, 14; m 46; c 5. MATHEMATICS. *Educ:* Loyola Univ, Ill, BS, 36, MA, 38. *Prof Exp:* Salesman, Midcontinent Chem Corp, 38-39; purchase agent, McKesson & Robbins, Inc, 39-40; teacher pub schs, Ill, 46, 47-48; instr, Marquette Univ, 46-47; from asst prof to assoc prof, 48-69, chmn dept, 52-76, PROF MATH, UNIV ST THOMAS, 69- *Concurrent Pos:* Tech ed, J Data Mgt, Data Processing Mgt Asn. *Honors & Awards:* Minnie Piper Prof award, 58; cert, Data Processing, 64. *Mem:* Asn Comput Mach; Math Asn Am; Soc Indust & Appl Math; Data Processing Mgt Asn; Asn Educ Data Systs. *Res:* Computing. *Mailing Add:* 4114 Turnberry Circle Houston TX 77025

DUBBE, RICHARD F, b Minneapolis, Minn, Jan 9, 29; m 50; c 4. ELECTRONICS. *Educ:* Univ Minn, BEE, 53. *Prof Exp:* Tech serv engr, Minn Mining & Mfg Co, 53-58, res engr, 58-63, proj supvr electron beam recording, 63-65, proj mgr, 65-66, res mgr, 66-76, tech dir, mincom div, 76-, TECH DIR, DATA REC PROD DIV, 3M CO. *Mem:* Soc Motion Picture & TV Engrs; Inst Elec & Electronics Engrs; Soc Photog, Scientists & Engrs. *Res:* Electron beam, magnetic and sound recording; television; motion picture technology. *Mailing Add:* Electro Tech Develop 195 Nature Way Little Canada MN 55117

DUBBELDAY, PIETER STEVEN, b Surakarta, Indonesia, Dec 23, 28; nat US; m 58; c 2. HYDROACOUSTICS, MATHEMATICAL PHYSICS. *Educ:* Vrije Univ, Amsterdam, Candidatus, 50, Drs, 53, PhD(nuclear physics), 59. *Prof Exp:* Asst physics, Free Univ, Amsterdam, 51-53 & 54-56, chief asst, 56-59; res assoc & instr, Nuclear Physics Lab, Univ Wis, 59-61; engr, Missile Test Proj, RCA Serv Co, 61-63, sr engr, 63-66; instr physics, Fla Inst Technol, 61-66, from assoc prof to prof physics & oceanog, 66-81; res physicist, 81-87, SUPVRY RES PHYSICIST, UNDERWATER SOUND REF DET, NAVAL RES LAB, ORLANDO, 87- *Honors & Awards:* Alan Berman Res Publ Award, 85 & 87. *Mem:* Fel Acoust Soc Am; Am Phys Soc. *Res:* Neutron physics; problems in missile tracking; survey and error analyses of atmospheric refraction; physical oceanography; hydroacoustics; shallow water circulation; continuum mechanics. *Mailing Add:* Naval Res Lab USRD Box 568337 Orlando FL 32856-8337

DUBBS, DEL ROSE M, b Vesta, Minn, Feb 10, 28. VIROLOGY, BIOCHEMISTRY. *Educ:* Univ Minn, BA, 50, MS, 57, PhD(bact), 61. *Prof Exp:* Res asst bact, Univ Minn, 53-60; res assoc biochem, Univ Tex M D Anderson Hosp & Tumor Inst, 60-62; from asst prof to assoc prof, 62-74; PROF VIROL, BAYLOR COL MED, 74- *Concurrent Pos:* USPHS res career develop award, 65-70. *Mem:* Am Soc Microbiol. *Res:* Studies of the biochemical alterations in virus infected cells and to relate these changes to transformation of normal cells to malignant cells. *Mailing Add:* Box 1255 Redwing Dr Bandera TX 78003

DUBE, DAVID GREGORY, b Cincinnati, Ohio, Aug 28, 48; m 74; c 1. ANALYTICAL CHEMISTRY. *Educ:* Xavier Univ, BS, 70; Pa State Univ, PhD(chem), 79. *Prof Exp:* Res Engr, Chem & Metall Div, Gen Telephone & Electronics, 78-81; MGR METHODS DEVELOP & PHARMACEUT QUALITY CONTROL, MEAD JOHNSON & CO, SUBSID BRISTOL MYERS CORP, 81- *Mem:* Am Chem Soc. *Res:* Methods for the routine analysis of pharmaceuticals; computer-aided experimentation. *Mailing Add:* 7966 Owens Dr Newburgh IN 47630-2623

DUBÉ, FRANÇOIS, b Montreal, Que, Oct 4, 55; m 80; c 2. EMBRYOLOGY, BIOCHEMISTRY. *Educ:* Univ Montreal, BSc, 77, MSc, 80, PhD(biol), 83. *Prof Exp:* Postdoctoral fel cell biol, Stanford Univ, 83-85; univ res fel, Nat Sci & Eng Res Coun & Univ Que, 85-90; PROF CELL BIOL, UNIV QUE, RIMOUSKI, 90- *Concurrent Pos:* Res fel, Marine Biol Lab, Woods Hole, 86; invited prof, Ecole Normale Superieure, Lyon, 88-89. *Mem:* Am Soc Cell Biol. *Res:* Early developmental biology of marine invertebrates; ionic regulation of meiosis and mitosis; fertilization processes in echinoderms and molluscs; transduction mechanisms and biochemistry of cellular activation. *Mailing Add:* Dept Oceanog Univ Que Rimouski PQ G5L 3A1 Can

DUBE, GREGORY P, b Winchester, Mass, Feb 24, 54. PHARMACOLOGY. *Educ:* Northeastern Univ, BS, 77; Univ Cincinnati, PhD(cardiovasc pharmacol), 84. *Prof Exp:* SR PHARMACOLOGIST CARDIOVASC PHARMACOL, LILLY RES LAB, 86- *Mem:* Am Heart Asn; Am Soc Pharmacol & Exp Therapeut. *Mailing Add:* Dept Pharmacol Lilly Res Lab Mail Drop 0821 Indianapolis IN 46285

DUBE, MAURICE ANDREW, BOTANY. *Educ:* Wash State Univ, BS, 50; Ore State Univ, MS, 58, PhD(bot), 63. *Prof Exp:* Asst prof bot, 63-68, ASSOC PROF BIOL, WESTERN WASH UNIV, 68- *Res:* Life history of marine algae. *Mailing Add:* 304 Morey Ave Bellingham WA 98225

DUBE, ROGER RAYMOND, b Portland, Maine, Nov 24, 49; m 72; c 2. PHYSICS. *Educ:* Cornell Univ, AB, 72; Princeton Univ, MA, 74, PhD(physics), 76. *Prof Exp:* Astron, Kitt Peak Nat Observ, 76-77; sr staff mem physics, Jet Propulsion Lab, 77-78; asst prof physics, Univ Mich, Dearborne, 78-80; ASST RES PROF, UNIV ARIZ, 80- *Concurrent Pos:* Mem, Proposal Rev Comt, NSF, 78-80. *Mem:* AAAS; Am Astron Soc; Soc Advan Chicanos & Native Am in Sci. *Res:* Experimental relativity; cosmology; astrophysics; detector development. *Mailing Add:* Eight Circle Dr Carmel NY 10512

DUBECK, LEROY W, b Orange, NJ, Mar 1, 39. SOLID STATE PHYSICS. *Educ:* Rutgers Univ, BA, 60, MA, 62, PhD(physics), 65. *Prof Exp:* Teaching asst physics, Rutgers Univ, 60-62; asst prof, 65-69, assoc prof, 69-76, PROF PHYSICS, TEMPLE UNIV, 76- *Mem:* Am Phys Soc; Am Asn Physics Teachers; Nat Sci Teachers Asn. *Res:* Thermal conductivity; magnetization and flux motion properties of superconductors; nuclear magnetic resonance of superconductors; superconducting tunneling; science education. *Mailing Add:* Dept Physics Temple Univ Philadelphia PA 19122

DUBERG, JOHN E(DWARD), b New York, NY, Nov 30, 17; m 43; c 2. AERONAUTICAL ENGINEERING. *Educ:* Manhattan Col, BS, 38; Va Polytech Inst, MS, 40; Univ Ill, PhD(struct eng), 48. *Prof Exp:* Field engr, Caldwell Wingate Builders, 38-39; asst, Talbot Lab, Univ Ill, 40-43; aeronaut res scientist, Nat Adv Comt Aeronaut, Va, 43-46; asst chief res engr, Standard Oil Co Ind, 46-48; aeronaut res scientist, Nat Adv Comt Aeronaut, 48-52, chief, struct res div, 52-56; res engr, Aeroneutronics Systs, Inc, 56-57; prof struct eng, Univ Ill, 57-59; tech asst to chief, 59-61, tech asst to assoc dir, 61-64, asst dir, 64-68, assoc dir, 68-80, ADJ PROF, GEORGE WASHINGTON UNIV, JOINT INST FOR ADVAN FLIGHT SCI, LANGLEY RES CTR, NASA, 80- *Concurrent Pos:* Mem panel guided missiles mat, minerals & metals adv bd, Nat Acad Sci. *Mem:* Am Inst Aeronaut & Astronaut; AAAS; Sigma Xi; Am Soc Eng Educ. *Res:* Fatigue of joints in steel structures; stress analysis of stiffened shells; space environmental effects. *Mailing Add:* Four Museum Dr Newport News VA 23601

DUBES, GEORGE RICHARD, b Sioux City, Iowa, Oct 12, 26; m 64; c 6. GENETICS. *Educ:* Iowa State Univ, BS, 49; Calif Inst Technol, PhD(genetics), 53. *Prof Exp:* Res assoc inst coop res, McCollum-Pratt Inst, Johns Hopkins Univ, 53-54; res assoc, Sect Virus Res, Med Ctr, Univ Kans, 54-56, from asst prof to assoc prof, 56-64; head viral genetics sect, Eugene C Eppley Inst Res Cancer & Allied Dis, 64-68; assoc prof, 64-81, PROF MICROBIOL, COL MED, UNIV NEBR, OMAHA, 81- *Mem:* Genetics Soc Am; Am Genetic Asn; Am Soc Microbiol; Am Asn Cancer Res; NY Acad Sci; Sigma Xi; Am Asn Advan Sci; Am Inst Biol Sci; Biometric Soc; Int Soc Oncodevelop Biol & Med; Tissue Cult Asn. *Res:* Genetics of animal viruses; transfection; amoebic cysts; copper-mediated inactivation of nucleic acids; doxorubicin; opal mutants; oxygen limitation and virus multiplication; asbestos and transfection. *Mailing Add:* 7061 Starlite Dr Omaha NE 68152-2021

DUBES, RICHARD C, b Chicago, Ill, Oct 7, 34; m 59; c 1. COMPUTER SCIENCE. *Educ:* Univ Ill, BSEE, 56; Mich State Univ, MS, 58, PhD(elec eng), 62. *Prof Exp:* Mem tech staff, Hughes Aircraft Co, 56-57. *Mem:* Sigma Xi; Inst Elec & Electronics Engrs; Pattern Recognition Soc; Classification Soc NAm (pres, 90-91). *Res:* Pattern recognition; exploratory data analysis; image processing; signal processing; statistical computing. *Mailing Add:* Dept Comput Sci Mich State Univ East Lansing MI 48824-1027

DUBEY, DEVENDRA P, b Varanasi, India, Aug 10, 36. BIOCHEMICAL GENETICS. *Educ:* Banaras Univ, PhD(physics), 67. *Prof Exp:* PATH ASSOC, DANA-FARBER CANCER INST & SCH MED, HARVARD UNIV, 80- *Mem:* Inst Immunol; AAAS; Am Soc Histocompatability. *Mailing Add:* Dept Path Harvard Med Sch Dana-Farber Cancer Inst 44 Binney St Boston MA 02115

DUBEY, JITENDER PRAKASH, b India, July 15, 38; US citizen; m 64; c 2. PARASITOLOGY, PROTOZOOLOGY. *Educ:* Ujjain Univ, India, BVSc, 60; Agra Univ, India, MVSc, 63; Sheffield Univ, Eng, PhD (microbiol), 66. *Prof Exp:* Assoc prof parasitol, Col Vet Med, Ohio State Univ, Columbus, 73-78; prof parasitol, Montana State Univ, Bozeman, 78-82; MICROBIOLOGIST, ZOONETIC DIS LAB, LPSI, AGR RES SERV, USDA, 82- *Concurrent Pos:* Consult, WHO, Geneva, 87. *Mem:* Am Vet Med Asn; Am Soc Parasitologists; fel Indian Asn Advan Vet Parasitol; Indian Soc Parasitologists. *Res:* Author of 3 books; over 300 research papers in international journals; discovery of the life cycle of Toxoplasma gonchi, a parasite causing blindness in children and abortion in livestock. *Mailing Add:* 234 Lastner Lane Greenbelt MD 20770

DUBEY, RAJENDRA NARAIN, b Bihar, India, Nov 2, 38; m 67; c 4. MECHANICAL ENGINEERING, APPLIED MATHEMATICS. *Educ:* Patna Univ, BSc, 57; Ranchi Univ, BSc, 61; Univ Waterloo, PhD(civil eng), 66. *Prof Exp:* Asst lectr civil eng, Regional Inst Tech, Jamshedpur, India, 61-62; asst engr, M M Bilaney & Co, India, 63-64; fel civil eng, 66-67, from asst prof to assoc prof mech eng, 67-80, PROF DEPT ENG, UNIV WATERLOO, 80- *Mem:* Asn Am Soc Mech Engrs. *Res:* Plastic instabilities in solids; convective instabilities in fluids; boundary layer flow; dynamic instabilities. *Mailing Add:* Dept Mech Eng Univ Waterloo Waterloo ON N2L 3G1 Can

DUBEY, SATYA D(EVA), b Sakara Bajid, India, Feb 10, 30; US citizen; m 60; c 3. STATISTICS, MATHEMATICS. *Educ:* Patna Univ, BS, 51; Indian Statist Inst Calcutta, dipl, 53; Mich State Univ, PhD(statist), 60. *Prof Exp:* Tech asst statist, Indian Inst Technol, 53-56; res asst math, Carnegie Inst Technol, 56-57; instr statist & res asst, Mich State Univ, 57-60; sr math statistician, Procter & Gamble, Co, Ohio, 60-65, head statist sect, 65-66; prin statistician & group leader statist & opers res, Ford Motor Co, Mich 66-68; assoc prof indust eng & opers res, NY Univ, 68-73; actg dir, Div Biometrics, Dept Health & Human Serv, 75-76 & 89-90, chief statist eval br, Bur Drugs, Food & Drug Admin, 73-84, CHIEF STATIST EVAL & RES BR, NAT CTR DRUGS & BIOLOGICS, FOOD & DRUG ADMIN, DEPT HEALTH & HUMAN SERV, 84- *Concurrent Pos:* Consult, Mich State Budget Div, 58, HEW, 63-66, Dept Defense, 68-70; del, Int Statist Inst, 71; prin investr & dir res contract on statist, Procedures In Reliability Eng, Dept Army, 71- *Honors & Awards:* Commendable Serv award, Food & Drug Admin, 77. *Mem:* AAAS; fel Royal Statist Asn; NY Acad Sci; Int Statist Inst; sr mem Am Inst Indust Engrs; Am Soc Qual Control; Sigma Xi; fel Am Statist Asn. *Res:* Statistical inference procedures applicable to engineering, physical, social, biomedical, mathematical and computer sciences; over 60 research publications. *Mailing Add:* Bur Drugs Div Biometrics Rm 18B4T 5600 Fishers Lane Rockville MD 20857

DUBIEL, RUSSELL F, b Jamaica, NY, Dec 28, 51; m 85. CONTINENTAL SEDIMENTOLOGY, PALEOPEDOLOGY. *Educ:* Johns Hopkins Univ, BA, 74; Univ Maine, MS, 77; Univ Colo, PhD(geol), 87. *Prof Exp:* Environ scientist, Environ Protection, State Maine, 77-78; phys sci tech, 78-80, GEOLOGIST, US GEOL SURV, 80- *Mem:* Am Asn Petrol Geologists; Geol Soc Am; Int Asn Sedimentologists; Soc Econ Paleontologists & Mineralogists. *Res:* Upper Triassic rocks on the Colorado Plateau; larustrine sedimentology. *Mailing Add:* US Geol Surv MS 919 Box 25046 DFC Denver CO 80225

DUBIN, ALVIN, b Russia, Jan 23, 14; nat US; m 38; c 2. BIOCHEMISTRY. *Educ:* Brooklyn Col, BA, 40, MS, 42. *Prof Exp:* Asst dir labs, Beth-El Hosp, 38-46; chief biochemist, Hektoen Inst Med Res, Cook County Hosp, 47-53; assoc prof, 70-72, PROF BIOCHEM, RUSH MED SCH, 72-; DIR DEPT BIOCHEM, HEKTOEN INST MED RES, COOK COUNTY HOSP, 53- *Concurrent Pos:* Consult, Oak Forest Hosp, 56-, Woodlawn Hosp, 56 & MacNeal Mem Hosp, 58-; asst prof biochem, Univ Ill Col Med, 60-70 & Cook County Grad Sch, 60- *Honors & Awards:* Hektoen Gold Medal, AMA, 67 & Ames Award, 85. *Mem:* Am Asn Clin Chem; Soc Exp Biol & Med; Am Chem Soc; Am Inst Chemists. *Res:* Biochemistry and its relationship to liver and kidney disease; synthesis and degradation of proteins and the excretion of proteins as related to renal damage; relationship of uric acid and calcium metabolism to such diseases as gout and parathyroid diseases. *Mailing Add:* Kektoen Inst 629 S Wood St Chicago IL 60612

DUBIN, DONALD T, b Brooklyn, NY, Mar 23, 32; m 65; c 6. BIOCHEMISTRY, CELL BIOLOGY. *Educ:* Harvard Univ, AB, 53; Columbia Univ, MD, 56. *Prof Exp:* Intern med, Bronx Munic Hosp, NY, 56-57, resident, 59-60; sr asst surgeon biochem, NIH, 57-59; instr, Harvard Med Sch, 62-63, assoc, 63-66, asst prof, 66-67; assoc prof, 67-71, PROF MICROBIOL, RUTGERS MED SCH, COL MED & DENT, NJ, 71- *Concurrent Pos:* Res fel bact, Harvard Med Sch, 60-62. *Mem:* Am Soc Cell Biol; Am Soc Microbiol; Am Soc Biol Chem; Sigma Xi. *Res:* Nucleic acid metabolism; mitochondrial biogenesis. *Mailing Add:* One Northern Dr Bridgewater NJ 08807

DUBIN, FRED S, b Hartford, Conn, Jan 31, 14; c 2. ENERGY MANAGEMENT, HEATING & AIR CONDITIONING. *Educ:* Carnegie Inst Technol, BSME, 35; Pratt Inst, MArch, 76. *Prof Exp:* Distinguished, prof architect, prof, Columbia Univ, Univ Southern Calif, Univ Ark, fel, Carnegie Inst Technol, Mellon Univ, 36-90; PRES CONSULT ENG, DUBIN-BLOOME ASSOC, 36- *Concurrent Pos:* Bd dirs, Am Solar Energy Soc, 87; consult, energy codes, Jamaica, 81-90, Long Island Power Authority, 90-91 & Carnegie Mellon Univ, 91. *Mem:* Nat Acad Eng; fel Am Soc Heating & Vent; fel Am Consult Eng Soc; Am Consult Engrs Coun. *Res:* Interaction of lighting and air conditioning; author of various publications. *Mailing Add:* One Seaside Pl East Norwalk CT 06855

DUBIN, HENRY CHARLES, b Paterson, NJ, Sept 7, 43; m 66; c 2. OPERATIONS RESEARCH. *Educ:* St Lawrence Univ, NY, BSc, 65; Ind Univ, Bloomington, MA, 68, PhD(chem physics), 73. *Prof Exp:* Res physicist ballistics, US Army Ballistic Res Labs, 72-75; physicist electromagnetics, 75-82, tech mgr commun systs anal, Systs Anal Activ, 82-, CHIEF SCIENTIST, US ARMY OPER TEST & EVAL COMMAND. *Honors & Awards:* Systs Anal Award, US Army, 85. *Mem:* Asn Comput Mach. *Res:* Computer simulations of antenna radiation patterns and electromagnetic signal propagation. *Mailing Add:* 814 Chesney Lane Bel Air MD 21014

DUBIN, MARK WILLIAM, b New York, NY, Aug 30, 42; m 64; c 2. EDUCATION ADMINISTRATION. *Educ:* Amherst Col, BA, 64; Johns Hopkins Univ, PhD(biophys), 69. *Prof Exp:* Res fel sensory physiol, John Curtin Sch Med Res, Australian Nat Univ, 69-71, chair, 83-87; PROF, DEPT MOLECULAR, CELLULAR & DEVELOP BIOL, UNIV COLO, BOULDER, ASSOC VCHANCELLOR ACAD AFFAIRS, UNIV COLO, BOULDER, 88- *Concurrent Pos:* Nat Eye Inst grant, 72-90, NSF grant, 76-83 & March of Dimes Grant. *Mem:* AAAS; Soc Neurosci; Asn Res Vision & Ophthal. *Res:* Role of action potentials in the determination of proper synaptic connections in the vertebrate central nervous system; development of neuronal morphology. *Mailing Add:* Dept MCD Biol Box 347 Univ Colo Boulder CO 80309-0347

DUBIN, MAURICE, b Boston, Mass, Dec 29, 26; m 60; c 3. PHYSICS, ASTROPHYSICS. *Educ:* Univ Mich, BS, 48; Harvard Univ, AM, 49. *Prof Exp:* Physicist, Cornell Aeronaut Res Lab, 48-50; geophys res directorate, Air Force Cambridge Res Labs, Mass, 50-59 & Goddard Space Flight Ctr, NASA, Md, 59, head aeronomy prof, Off Space Sci, Washington, DC, 59-64, CHIEF INTERPLANETARY DUST & COMETARY PHYSICS, GODDARD SPACE FLIGHT CTR, NASA, 64- *Concurrent Pos:* Mem synoptic rocket panel working group II, Cmt Exten to Standard Atmos, 60-; mem tech panel satellites, Int Geophys Year. *Mem:* Am Phys Soc; Sci Res Soc Am; Am Inst

Aeronaut & Astronaut; Am Geophys Union. *Res:* Atmospheric physics; interplanetary dust and comets; sounding rockets and satellites for investigations of the airglow, the upper atmosphere comads and micrometeorites. *Mailing Add:* NASA Goddard Space Flight Ctr Code 616 Greenbelt Rd Greenbelt MD 20771

DUBIN, NORMAN H, b Paterson, NJ, Feb 13, 42; m 64; c 1. REPRODUCTIVE ENDOCRINOLOGY. *Educ:* Univ Rochester, BA, 63; Rutgers Univ, PhD(zool), 70. *Prof Exp:* Fel, Ohio State Univ, 69-71; asst prof obstet-gynec, Univ Md, 71-75; ASSOC PROF OBSTET-GYNEC, JOHNS HOPKINS UNIV, 75- *Honors & Awards:* Lalor Found Award, 72. *Mem:* Soc Study Reproduction; AAAS; Endocrine Soc. *Res:* Factors affecting steroid and prostaglandin production during pregnancy; fertility control. *Mailing Add:* 6210 Biltmore Ave Baltimore MD 21215

DUBIN, PAUL LEE, b New York, NY, Apr 17, 41; m 70; c 1. COLLOID CHEMISTRY, BIOANALYTICAL CHEMISTRY. *Educ:* City Col New York, BS, 62; Rutgers Univ, New Brunswick, PhD(phys chem), 70. *Prof Exp:* Res assoc, Univ Mass, Amherst, 70-71; res assoc, Univ Calif, Irvine, 71-72; sr res chemist, Dynapol, 73-78; sr staff scientist, Memorex Corp, 78-80; sr res scientist, Clairol, 80-81; from asst prof to assoc prof, 81-91, PROF, IND UNIV-PURDUE UNIV, INDIANAPOLIS, 91- *Concurrent Pos:* Consult, Reilly Indust, 81-88. *Mem:* Am Chem Soc; Am Phys Soc. *Res:* Polymer-surfactant complexes; polyelectrolyte-colloid interactions; aqueous size exclusion chromatography; protein separation by polyelectrolyte coaeruation; dynamic light scattering; mixed micelles. *Mailing Add:* Chem Dept IUPUI 1125 E 38th St Indianapolis IN 46205

DUBINS, LESTER ELI, b Washington, DC, Apr 27, 20. MATHEMATICS. *Educ:* City Col New York, BS, 42; Univ Chicago, MS, 50, PhD(math), 55. *Prof Exp:* Mathematician, Inst Air Weapons Res, Chicago, 51-55; asst prof math, Carnegie Inst Technol, 55-57; NSF fel, Inst Adv Study, 57-59; NSF fel, 59-60, from asst prof to assoc prof, 60-65, PROF MATH & STATIST, UNIV CALIF, BERKELEY, 65- *Concurrent Pos:* Consult, Inst Air Weapons Res, Chicago, 55-60. *Mem:* Am Math Soc; Math Asn Am. *Res:* Probability theory; differential geometry; game theory; functional analysis. *Mailing Add:* Univ Calif Evans Hall 869 Berkeley CA 94720

DUBINS, MORTIMER IRA, b Boston, Mass, Mar 24, 19; m 51; c 1. GEOLOGY, METEOROLOGY. *Educ:* Tufts Col, BS, 40; Univ Kans, MS, 48; Boston Univ, EdM, 49, EdD(sci educ), 53. *Hon Degrees:* Adv Study, Harvard & VA Polytech Inst State Univ. *Prof Exp:* Instr chem & sci, Boston Univ, 48-49; teacher & head, dept sci, High Sch, Foxboro, 49-53, teacher, 53-54; instr sci educ, Castleton Teachers Col, 54-55; asst prof, Northwestern Univ, 55-57; PROF GEOL & METEOROL, STATE UNIV NY COL ONEONTA, 57- *Concurrent Pos:* Instr, Boston Univ, 50-51; dir, NSF Earth Sci Inst, 61-63; vis prof, Univ NDakota, 65 & 69, Franklin & Marshall, 72. *Mem:* Fel AAAS; Am Meteorol Soc; Mineral Soc Am. *Res:* Meteorology; economic geology; earth science; mineralogy; crystallography; microscopic metallic minerals of New York and Ontario, Quebec; Grenville marble; capillary pyrite and microscopic talc. *Mailing Add:* 67 Woodside Ave Oneonta NY 13820-1380

DUBINSKY, BARRY, b Philadelphia, Pa. PHARMACOLOGY. *Educ:* Temple Univ, BS, 62; Univ Pittsburgh, MS, 65, PhD(pharmacol), 68. *Prof Exp:* Asst pharmacol, Univ Pittsburgh, 62-66; sr scientist, Dept Pharmacodyn, Warner-Lambert Res Inst, Morris Plains, 68-77; SR SCIENTIST, DIV PHARMACOL, ORTHO PHARMACEUT CORP, RARITAN, 77-, RES FEL, DIV BIOL RES, 80- *Mem:* Am Soc Pharmacol & Exp Therapeut; AAAS; Sigma Xi. *Res:* Neuropharmacology; psychopharmacology. *Mailing Add:* Ortho Pharmaceut Corp Div Pharmacol Raritan NJ 08869

DUBISCH, ROY, b Chicago, Ill, Feb 5, 17; m 39; c 3. MATHEMATICS. *Educ:* Univ Chicago, BS, 38, MS, 40, PhD(math), 43. *Prof Exp:* Asst physics, Wilson Jr Col, 38-40; instr math, Ill Inst Technol, 40-41 & Univ Mont, 42-46; asst prof & chmn dept, Triple Cities Col, 46-48; assoc prof, Fresno State Col, 48-54, prof & chmn dept, 55-61; prof, 61-87, EMER PROF MATH, UNIV WASH, 87- *Mem:* Am Math Soc; Math Asn Am; Nat Coun Teachers Math. *Res:* Mathematical education. *Mailing Add:* 210 Manito Dr Mt Vernon WA 98273

DUBISCH, RUSSELL JOHN, b Missoula, Mont, May 15, 45; m 66; c 2. GRAVITATION THEORY, PHILOSOPHY OF SPACETIME. *Educ:* Reed Col, BA, 67; Univ Wash, MS, 69; Univ Pittsburgh, PhD(physics), 78. *Prof Exp:* Vis asst prof, Univ Colo, Colorado Springs, 79-80; ASST PROF PHYSICS, SIENA COL, 80- *Mem:* Am Asn Physics Teachers. *Res:* Mathematical theory of gravitational radiation; classical-background approaches to quantization in general relativity; back-reaction problem in general relativity; philosophy of spacetime; formation of stars. *Mailing Add:* Dept Physics Siena Col Rte 9 Loudonville NY 12211

DUBISKI, STANISLAW, b Warsaw, Poland, Aug 21, 29; Can citizen; wid; c 1. IMMUNOCHEMISTRY, IMMUNOGENETICS. *Educ:* Wroclaw Univ, MD, 53; Silesian Med Acad, PhD(microbiol), 57. *Prof Exp:* Res asst microbiol, Med Sch, Wroclaw Univ, 51-54; res asst, Silesian Med Acad, 54-57, asst prof, 57-59; head dept serol, Inst Hemat, Poland, 59-61; res assoc & clin teacher, 63-67, head immunol res lab, Toronto Western Hosp, 63-78, asst prof path chem & med, 67-70, assoc prof, 70-76, mem Inst Immunol, Sch Grad Studies, 71-85 Prof med genetics 78-85, PROF CLIN BIOCHEM, UNIV TORONTO, 76-, PROF, DEPT IMMUNOL, 85- *Concurrent Pos:* Training scholar, Cambridge Univ & Lister Inst, Eng, 58; vis scientist, Basel Inst for Immunol, Switz, 73-74. *Mem:* Brit Soc Immunol; NY Acad Sci; Am Asn Immunol; Can Soc Immunol (secy, 67-68, vpres, 71-72, pres, 73-75). *Res:* Polymorphism of serum proteins; mechanism and regulation of antibody formation; genetics and immunochemistry of rabbit immunoglobulin allotypes; lymphoid cell markers (rabbit) immune cell surface markers, aging of immune system. *Mailing Add:* 4368 Med Sci Bldg Dept Immunol Univ Toronto Toronto ON M5S 1A8 Can

DUBLE, RICHARD LEE, b Galveston, Tex, May 31, 40; m 69. AGRONOMY, PLANT PHYSIOLOGY. *Educ:* Tex A&M Univ, BS, 62, MS, 65, PhD(plant & soil sci), 67. *Prof Exp:* Asst prof forage physiol, Agr Res & Exten Ctr, Overton, 67-70, asst prof turf res, 70-73, assoc prof agron, 73-74, TURFGRASS SPECIALIST, TEX AGR EXTEN SERV, TEX A&M UNIV, 75- *Mem:* Am Soc Agron; Sigma Xi. *Mailing Add:* 1003 Howe Dr College Station TX 77845

DUBLIN, THOMAS DAVID, b New York, NY, Jan 18, 12; m 39; c 2. HEALTH CARE ADMINISTRATION. *Educ:* Dartmouth Col, AB, 32; Harvard Univ, MD, 36; Johns Hopkins Univ, MPH, 40, DrPH, 41; Am Bd Prev Med, dipl, 49. *Prof Exp:* Intern, Boston City Hosp, 36-38; asst res physician, Rockefeller Inst Hosp, 38-39; epidemiologist-in-training, State Dept Health, NY, 39-40, asst dist state health officer, 40, epidemiologist, 41-42; assoc prof prev med & community health, Long Island Col Med, 42-43, prof & exec officer dept, 43-48; exec dir, Nat Health Coun, 48-53, med consult, Nat Found Infantile Paralysis, 53-55; med dir community serv prog, Off Dir, NIH, 55-60, actg chief, Geog Dis Studies, 60, chief epidemiol & biomet br, Nat Inst Arthritis & Metab Dis, NIH, 60-66; res adv health serv, Off Tech Coop & Res Agency, Int Develop, 66-68; dir, Off Health Manpower, HEW, 68-70; mem staff, Bur Health Manpower Educ, NIH, 70-76; staff consult, Am Med Asn, Coord Coun Med Educ, 76-78; consult, Res & Develop, Educ Comn, Foreign Med Grad, 78-87; CONSULT, 87- *Concurrent Pos:* Mem, Bd Dirs, Am Bd Prev Med, 61-71, vchmn prev med, 65-71. *Mem:* AAAS; fel Am Pub Health Asn; AMA; Am Epidemiol Soc; fel NY Acad Med; fel Am Col Prev Med. *Res:* Epidemiology; health policy development with special emphasis on health manpower resources. *Mailing Add:* 2938 Garfield St NW Washington DC 20008

DUBNER, RONALD, b New York, NY, Oct 12, 34; m 58; c 3. DENTISTRY. *Educ:* Columbia Univ, BA, 55, DDS, 58; Univ Mich, Ann Arbor, PhD(physiol), 64. *Prof Exp:* Intern clin dent, Pub Health Serv Hosp, Baltimore, Md, 58-59, staff dentist, Clin Ctr, NIH, 59-61, investr physiol, 61-64, investr neurophysiol, 65-68, chief neurol mechanisms sect, 68-73, CHIEF NEUROBIOL & ANESTHESIOL BR, NAT INST DENT RES, 73- *Concurrent Pos:* Vis assoc prof, Sch Dent, Howard Univ, 68-80; co-chmn, Conf Oral-Facial Mechanisms, Honolulu, 70; vis scientist, Dept Anat, Univ Col, Univ London, 70-71; sect ed, Pain, 74-89, co-chief ed, 90-; coun mem, Int Asn for the Study of Pain, 75-81; mem, Fed Interagency Comt on New Therapies for Pain and Discomfort, 78-81; assoc ed, J Neurosci, 80-; bd dir, Am Pain Soc, 80-82. *Honors & Awards:* Meritorious Serv Medal, USPHS, 75; Schlack Award, Asn Mil Surgeons, 85; Distinguished Serv Medal, USPHS, 90. *Mem:* AAAS; Am Physiol Soc; Int Asn Dent Res; Soc Neurosci; Int Asn Study Pain (vpres, 81-88, treas, 88-). *Res:* Mechanisms of oral facial sensation, especially mechanisms of pain, touch and temperature sensation; cortical and subcortical mechanisms of sensation. *Mailing Add:* Neurobiol & Anesthesiol Br Nat Inst Dent Res Bldg 30 Rm B-18 Bethesda MD 20892

DUBNICK, BERNARD, b Brooklyn, NY, May 29, 28; m 52, 66; c 2. NEUROSCIENCES, BIOCHEMISTRY & PHARMACOLOGY. *Educ:* City Col New York, BS, 49; Univ Ill, PhD(chem), 53. *Prof Exp:* Dir dept pharmacodyn, Warner-Lambert Res Inst, 53-77; DIR CARDIOVASC-CENT NERVOUS SYST RES SECT, MED RES DIV, AM CYANAMID CO, 77- *Mem:* Fel AAAS; Am Chem Soc; fel NY Acad Sci; Am Soc Pharmacol; Sigma Xi. *Res:* Biochemical pharmacology; central nervous system; allergy; drug metabolism. *Mailing Add:* Cardiovasc-CNS Res Sect Am Cyanamid Co Pearl River NY 10965

DUBOCOVICH, MARGARITA L, b Venadotuerto, Argentina, July 12, 47; m 88; c 1. NEUROPHARMACOLOGY. *Educ:* Univ Buenos Aires, PhD(pharmacol), 76. *Prof Exp:* Asst prof, 82-86, ASSOC PROF PHARMACOL, SCH MED, NORTHWESTERN UNIV, 86- *Concurrent Pos:* Sci consult, Nelson Res, 86-89, Glaxo, 90- *Mem:* Soc Neurosci; Brit Pharmacol Soc; Am Soc Pharmacol & Exp Therapeut. *Res:* Dopamine and melatonin receptors; development of melatonin agonists and antagonists. *Mailing Add:* Dept Pharmacol Northwestern Univ Med Sch 303 E Chicago Ave Chicago IL 60611

DUBOIS, ANDRE T, b Liège, Belg, Mar 16, 39; US citizen; m 78; c 2. GASTROENTEROLOGY, GASTROINTESTINAL PHYSIOLOGY. *Educ:* Free Univ Brussels, Belg, BS, 59, MD, 63, PhD(physiol), 75. *Prof Exp:* Intern, dept internal med, Univ Hosp St Pierre, Brussels, Belg, 62-63, resident, dept surg, Free Univ Brussels, 63-65, asst prof, 65-71; instr, dept med, 75-77, asst prof med, 75-78, res assoc prof, 78-85, RES PROF MED & SURG & CHIEF, LAB GASTROINTESTINAL & LIVER STUDIES, UNIFORMED SERV UNIV HEALTH SCI, BETHESDA, MD, 85-, ASST DIR, DIV DIGESTIVE DIS, 80- *Concurrent Pos:* Resident, dept surg, Mil Hosp Brussels, 64-65; instr, dept anat, Univ Brussels, 65-68, fel, lab exp surg, 66-67, staff fel, dept surg, Univ Hosp, 69-71; NATO fel, lab clin sci, NIMH, NIH, Bethesda, Md, 71-72, vis assoc & vis scientist, sect gastroenterol digestive dis br, Nat Inst Arthritis, Metab & Digestive Dis, 73-75; sr investr, Armed Forces Radiobiol Res Inst, Bethesda, Md, 83-; adj assoc prof physiol, Sch Med, Georgetown Univ, 84-; assoc ed, Prostaglandins, 84- *Honors & Awards:* Van Engelen Prize, Univ Hosp, Free Univ Brussels, Belg, 63; William Beaumont Award, Excellence Clin Res, 84. *Mem:* Am Gastroenterol Asn; Am Physiol Soc; Am Fedn Clin Res; Am Asn Acad Surg; Soc Nuclear Med. *Res:* Gastric motility and secretion in healthy men and in patients with various diseases states; effect of physical and psychological stress on gastric function; role of prostaglandins in the regulation of gastric function. *Mailing Add:* Uniformed Serv Univ Health Sci 4301 Jones Bridge Rd Bethesda MD 20814

DUBOIS, ARTHUR BROOKS, b New York, NY, Nov 21, 23; m 50; c 3. ENVIRONMENTAL HEALTH. *Educ:* Harvard Col, MD, 41-43, Cornell Univ, MD, 46. *Prof Exp:* Intern med, New York Hosp, 46-47; asst resident med, Peter Bent Brigham Hosp, 51-52; from asst prof to prof physiol, Sch Med, Univ Pa, 52-74, from assoc prof to prof med, 57-74; PROF EPIDEMIOL & PHYSIOL, MED SCH, YALE UNIV, 74- *Concurrent Pos:* Life Ins med res fel physiol, Sch Med, Univ Rochester, 49-51; NIH res career award, 63-74; investr, Am Heart Asn, 55-63; Bowditch lectr, Am Physiol Soc, 58; consult, US Naval Hosp, Philadelphia, 58-74; mem, Nat Res Coun panel rev NSF fel appln, 59-62; mem Am Inst Biol Sci adv comt physiol, Off Naval Res, 61-64; consult, Philadelphia Gen Hosp, 62-74; chmn comt gaseous environ manned spacecraft, Space Sci Bd, Nat Acad Sci, 62-63; Jackson lectr, Am Col Chest Physicians; mem cardiovasc study sect, NIH, 64-68; Nathanson lectr, Univ Southern Calif, 64; mem comt on toxicol, Nat Res Coun, 66-76, vchmn, 71-76, chmn comt on biol effects atmospheric pollutants, 68-72; chmn panel for selection of lung ctr, Nat Heart & Lung Inst, 71; mem lectr, Sch Med, Yamagata Univ, Japan, 79; dir, John B Pierce Found Lab, 74-88. *Honors & Awards:* Edward Livingston Trudeau Medal, Am Lung Asn, 89. *Mem:* Am Physiol Soc; Am Soc Clin Invest; Asn Am Physicians; Sigma Xi. *Res:* Physiology of respiration and pulmonary circulation; normal and clinical function of the lungs. *Mailing Add:* John B Pierce Lab 290 Congress Ave New Haven CT 06519

DU BOIS, DONALD FRANK, b Little Falls, NY, Jan 4, 32; m 55; c 3. THEORETICAL PHYSICS. *Educ:* Cornell Univ, BA, 54; Calif Inst Technol, PhD(physics, math), 59. *Prof Exp:* Instr physics, Calif Inst Technol, 58-59; res physicist, Rand Corp, Calif, 59-62; sr staff physicist, Hughes Aircraft Co, 62-70, sr scientist, Hughes Res Labs, Malibu, Calif, 70-72; vis prof, Univ Colo, Boulder, 72-73; GROUP LEADER STATIST & MAT PHYSICS, LOS ALAMOS SCI LAB, 73- *Concurrent Pos:* Consult, Rand Corp, 58-59 & 70-75, Lawrence Livermore Lab, 72-73 & Univ Colo, Boulder, 73-; adj prof, Univ Colo, Boulder, 79-; J S Guggenheim fel, 81-82. *Mem:* Am Phys Soc. *Res:* Many-particle theory; quantum and statistical mechanics; plasma and solid state physics; electron correlations in metals; transport in semiconductors; nonlinear interaction of radiation with plasmas and plasma turbulence. *Mailing Add:* MS B262 Los Alamos Nat Lab Los Alamos NM 87545

DUBOIS, DONALD WARD, b Oklahoma City, Okla, Jan 31, 23; c 2. SCIENCE EDUCATION. *Educ:* Univ Okla, MA, 50, PhD(math), 53. *Prof Exp:* Instr math, Univ Okla, 52-53 & Ohio State Univ, 53-55; from asst prof to assoc prof, 63-68, PROF MATH, UNIV NMEX, 68- *Concurrent Pos:* Dir & ed, proc int conf ordered fields & real algebraic geom, San Francisco, 81 & Boulder, 83. *Mem:* Am Math Soc. *Res:* Introduced analytic methods and cohesive groups in group theory, Kadison-Dubois theorem in theory of primes, real commutative algebra (including Reelnullstellen satze) in algebraic geometry, and registers in mathematics education. *Mailing Add:* Dept Math Humanities Bldg 419 811 NW 60th St Seattle WA 98107

DUBOIS, FREDERICK WILLIAMSON, b Newburgh, NY, Nov 6, 23; m 46; c 3. INORGANIC CHEMISTRY. *Educ:* St Lawrence Univ, BS, 46; Univ Mich, MS, 49, PhD(chem), 52. *Prof Exp:* Instr chem, Univ Idaho, 46-47; instr qual anal, Univ Mich, 51; asst group leader, 52-80, assoc group leader & proj leader, 80-86, LAB ASSOC, LOS ALAMOS NAT LAB, 86- *Mem:* Am Chem Soc. *Res:* Complex ions in solution; military high explosives; detonation theory; plastics and elastomers. *Mailing Add:* M-1 Los Alamos Nat Lab Los Alamos NM 87545

DUBOIS, GRANT EDWIN, b Niagara Falls, NY, May 31, 46; m 79; c 3. ORGANIC CHEMISTRY. *Educ:* Capital Univ, Columbus, BS, 67; State Univ NY, Buffalo, PhD(chem), 72. *Prof Exp:* Fel, Stanford Univ, 72-73; res chemist org chem, Dynapol, Palo Alto, 73-81; res group leader, Syva Co, Palo Alto, 81-84; DIR CHEM PROD DISCOVERY, NUTRASWEET CO, 84- *Mem:* Am Chem Soc; AAAS; Asn Chemoreception Sci; Europ Chemoreception Soc. *Res:* Synthetic organic chemistry; design and synthesis of biologically active molecules especially in the area of non-nutritive sweetener development; immunochemistry, as relates to the development of immunoassay methods for the quantitation of drugs, hormones and other biologically important compounds; gustation mechanism; complex carbohydrates as relates to food applications. *Mailing Add:* Nutrasweet Co 601 E Kensington Rd Mt Prospect IL 60056

DUBOIS, JEAN-MARIE M, b St-Jean-sur-Richelieu, Que, Dec 26, 44; m 67; c 1. REMOTE SENSING. *Educ:* Univ Sherbrooke, BA, 67, LL, 69, MA, 71; Univ Ottawa, PhD(geog), 80. *Prof Exp:* Searcher geomorphol, Fisheries Res Bd Can, 71-72; res asst field res, 69-70, lectr cartog, 69-71, lectr, geog, 72-74, asst prof, 74-79, dir, 70-82, assoc prof, 79-83, dir Mgt Inst, 82-85, admin dir, Remote Sensing Ctr, 85, PROF PHYS GEOG, UNIV SHERBROOKE, 83- *Mem:* Can Asn Geographers; Int Union Quaternary Res; Am Quaternary Asn; Int Geog Union; Can Asn Quaternary Res. *Res:* Remote sensing of biophysical environment and land use; geomorphology, surficial deposits, coasts, longshore and marine currents; algae, salmon rivers, archaeology, glaciers. *Mailing Add:* Dept Gégraphie Et Télnédnétection Sherbrook Univ 2500 University Blvd Sherbrooke PQ J1K 2R1 Can

DUBOIS, JOHN R(OGER), b Eau Claire, Wis, Aug 16, 34; m 54; c 3. ELECTRICAL ENGINEERING. *Educ:* Univ Wis, BS, 57, MS, 59, PhD(elec eng), 63. *Prof Exp:* Supvr transmitter eng, TV Wis, Inc, 57-63; sr electronics engr, North Star Res & Develop Inst, 63-65, dir electronics res, 65-71; dir commun, Hennepin County, Minn, 71-87; COMMUN CONSULT, 71- *Concurrent Pos:* Instr elec eng, Univ Wis, 57-63. *Mem:* Inst Elec & Electronics Engrs; Asn Pub Safety Commun Officers; Sigma Xi. *Res:* Solid state devices and applications; electronics technology in law enforcement; very-high frequency and ultra-high frequency communications systems; computer-aided dispatch design. *Mailing Add:* 7320 Gallagher Dr Suite 118B Edina MN 55435

DUBOIS, ROBERT DEAN, b York, Nebr, June 19, 48; m 74; c 2. IONIZATION IN ION-ATOM COLLISIONS, CHARGE TRANSFER. *Educ:* Univ Nebr, Lincoln, BS, 70, MS, 72, PhD(atomic physics), 75. *Prof Exp:* Res, Joint Inst Lab Astrophys, 75-77; lectr res, Univ Aarhus, Denmark, 77-78; res, Hahn-Meitner Inst, Berlin, 78-79; vis asst prof introd phys, Kans State Univ, 80; SR SCIENTIST RES, PAC NORTHWEST LAB, RICHLAND, WASH, 81- *Concurrent Pos:* Fulbright sr prof, J W Goethe Univ, Frankfurt, 89-90. *Mem:* Am Phys Soc. *Res:* Experimental study of ion-atom/molecule collisions; differential electron emission studies multiple ionization probability and electron transfer process; absolute cross sections. *Mailing Add:* Battelle Pac Northwest Lab PO Box 999 Richland WA 99352

DU BOIS, ROBERT LEE, b Omaha, Nebr, Jan 25, 24; m 47. GEOPHYSICS. *Educ:* Univ Wash, BS, 49, MS, 50, PhD(geol), 54. *Prof Exp:* From asst prof to prof geol, Univ Ariz, 52-67; dir, Earth Sci Observ, 67-77, KERR MCGEE PROF GEOL & GEOPHYS, UNIV OKLA, 67-, DIR, LEONARD EARTH SCI OBSERV, 67- *Concurrent Pos:* Consult econ geol, 53- *Mem:* AAAS; Am Geophys Union; fel Geol Soc Am. *Res:* Archeomagnetism; paleomagnetism; rock magnetism. *Mailing Add:* 830 S Van Vleet Oval Rm 107 Norman OK 73019

DUBOIS, RONALD JOSEPH, b Lawrence, Mass, Jan 3, 42. ORGANIC CHEMISTRY. *Educ:* Lowell Inst Technol, BS, 64; Clarkson Tech Univ, PhD(org chem), 69. *Prof Exp:* Res chemist, E I du Pont de Nemours & Co, Inc, 68-71; mem staff, Cancer Res, Microbiol Assocs, Inc, 71-75; MEM STAFF, HERCULES, INC, 75- *Mem:* Am Chem Soc. *Res:* Synthesis of agricultural chemicals for use as herbicides, fungicides and insecticides. *Mailing Add:* 13217 Superior St Rockville MD 20853-3340

DUBOIS, THOMAS DAVID, b Mexico, Ind, Nov 15, 40; m; c 2. INORGANIC CHEMISTRY. *Educ:* McMurry Col, BS, 62; Ohio State Univ, MS, 65, PhD(inorg chem), 67. *Prof Exp:* Assoc prof, 67-80, PROF CHEM, UNIV NC, CHARLOTTE, 80-, MEM GRAD FAC, 80- *Mem:* Am Chem Soc; Electrochem Soc. *Res:* Coordination compounds; magnetochemistry and superconductivity; plasma and VLSI processing chemistry; automated instrument development. *Mailing Add:* Dept Chem Univ NC Charlotte NC 28223

DUBOSE, LEO EDWIN, b Gonzales, Tex, Apr 22, 31; m 51; c 3. ANIMAL SCIENCE. *Educ:* Abilene Christian Univ, BS, 52; SDak State Univ, MS, 54; Tex A&M Univ, PhD(animal breeding), 65. *Prof Exp:* Asst prof animal sci, SDak State Univ, 54-57; teacher, San Angelo Col, 57-63; prof animal sci, 65-80, GRAD FAC PROF AGR, ABILENE CHRISTIAN UNIV, 80- *Mem:* Am Soc Animal Sci; Equine Nutrit & Physiol Soc. *Res:* Digestability of feeds by ruminants; heritability estimates of carcass traits and production traits in beef cattle; correlations between traits in beef cattle; feeding evaluation of young equines. *Mailing Add:* Dept Agr Abilene Christian Univ Box 7986 Abilene TX 79699

DUBOSE, ROBERT TRAFTON, b Dallas, Tex, Oct 17, 19; m 44; c 3. VETERINARY MEDICINE, VIROLOGY. *Educ:* Tex A&M Univ, BS & 13 DVM, 56, MS, 58. *Prof Exp:* Instr, Tex A&M Univ, 56-59; prof vet microbiol, 59-80, PROF AGR & URBAN PRACTICE, VA POLYTECH INST & STATE UNIV, 80- *Mem:* Am Vet Med Asn; Am Asn Avian Path; Am Soc Microbiol; Wildlife Dis Asn; Sigma Xi. *Res:* Turkey hemorrhagic enteritis virus in cell cultures; pathogenicity of quail bronchitis virus and chicken embryo lethal orphan virus in birds and cells; modification of infectious bronchitis virus. *Mailing Add:* 3772 Walnut Hill Dr Flint TX 75762

DUBOW, MICHAEL SCOTT, b Brooklyn, NY, May 29, 50; m 87. BIOTECHNOLOGY, GENETIC ENGINEERING. *Educ:* State Univ NY, BSc, 72; Ind Univ, MA, 75, PhD(microbiol), 77. *Prof Exp:* Instr biol, State Univ NY, 71-72; res trainee, Ind Univ, 72-77; fel, Cold Spring Harbor Lab, 77-80; asst prof, 80-86, ASSOC PROF MOLECULAR BIOL, MCGILL UNIV, 86- *Concurrent Pos:* Consult, Rutgers Univ, 77, Allelix, 87. *Honors & Awards:* Garner Award, Am Soc Microbiol, 75; Beckman prize for Innovative Res in Toxicol, 83. *Mem:* Am Soc Microbiol; Can Genetics Soc; Can Soc Microbiol; Soc Toxicol Can. *Res:* Mechanisms, regulation and manipulation of the transposition of bacteriophages Mu and D108 DNAs; isolation of cellular mobile genetic elements and characterizing their functions and use in gene cloning; molecular mechanisma of toxicity. *Mailing Add:* Dept Microbiol & Immunol McGill Univ Fac Med 3775 University St Montreal PQ H3A 2B4 Can

DUBOWSKI, KURT M(AX), b Berlin, Ger, Nov 21, 21. CLINICAL BIOCHEMISTRY, FORENSIC SCIENCES. *Educ:* NY Univ, AB, 46; Ohio State Univ, MSc, 47, PhD(chem, toxicol), 49; Am Bd Clin Chem, dipl; Am Bd Forensic Toxicol, dipl; Am Bd Bioanal, dipl. *Hon Degrees:* LLD, Capital Univ, 84. *Prof Exp:* Biochemist & asst dir labs, Norwalk Hosp, 50-53; dir chem, Iowa Methodist Hosp, 53-58; assoc prof clin chem & dir clin labs, hosp & clins, Univ Fla, 58-61; assoc prof clin chem & toxicol, 61-64, dir clin chem, 61-65, PROF BIOCHEM, CLIN CHEM, TOXICOL, MED & PATH, COL MED, UNIV OKLA, 64-, DIR TOXICOL & FORENSIC SCI LAB, UNIV OKLA, HEALTH SCI CTR, 61-, GEORGE LYNN CROSS DISTINGUISHED PROF MED, 81- *Concurrent Pos:* Toxicologist to coroner, Fairfield County, Conn, 50-53; lectr, NY Univ & Northwestern Univ, 50-56; mem, comt alcohol & other drugs, Nat Safety Coun, 50-, chmn , 69-71; chief dep coroner & toxicologist, Polk County, Iowa, 53-58; state criminalist, Iowa Dept Pub Safety, 53-58; lectr, Univ Iowa, 54-58; spec consult, USPHS, 57-62; secy-treas, Am Bd Clin Chem, 58-73, pres, 73-78; vis prof, Ind Univ, 58-; consult, US Dept Vet Affairs, 59-, Oklahoma City Police Dept, 61-, Okla State Bur Invest, 66-, Okla Med Res Found, 67-, NIH, 67-75, US Dept Transp, 68-80, Okla Dept Pub Safety, 69-, NIMH, 71-73, Okla Bur Narcotics & Dangerous Drugs Control, 71-, Exec Off of the President, 72-75, Nat Bur Standards, US Dept Com, 72-78, Ctr Dis Control, US Dept Health, Educ & Welfare, 72-82 & Chem Indust Inst Toxicol, 80-83, Nat Transp Safety Bd, 84-; dir tests alcoholic & drug influence, State Okla, 70-; prin deleg Am Asn Clin Chem & Asn Chem Soc , Nat Coun Health Lab Surv, 72-; mem toxicol resource comt, Col Am Pathologist, 72-, comt clin chem, Am Chem, 75-, adv bd, Milton Helpen Int Ctr Fornsic Sci, Wichita State Univ, 75- Nat Adv Group Drug-Use Testing, Nat Inst Drug Abuses, US Dept Health & Human Serv, 86-; pres, Am Bd Forensic Toxicol, 75-80; distinguished scientist lectr, Int Soc Clin Lab Technol, 80; vpres, Int Comt Alcohol, Drugs & Traffic Safety, 86-89; consult, Nat Res Drug Abuse, 91-, Armed Forces Inst Path, 91- *Honors & Awards:* Fisher Award, Am Asn Clinical Chem, 73; Harger Award, Am Acad Forensic Sci, 83; Widmark Award, 80. *Mem:* Fel NY Acad Sci; Am Asn Clin Chem (pres, 85); Sci Res Soc Am; AAAS; Int Asn Chiefs Police; fel Am Acad Toxiol; fel Am Acad Forensic Sci. *Res:* Effects and determination of alcohol, cannabinoids in biologic materials; clinical chemical and toxicological methodology; clinical pharmacology; evaluation of trace evidence in criminal investigations; forensic chemistry & toxicology; drug abuse. *Mailing Add:* Forensic Sci Labs Univ Okla HSC PO Box 26901 Oklahoma City OK 73190-3000

DUBOWSKY, STEVEN, b New York, NY, Jan 14, 42; m 64; c 3. MECHANICAL ENGINEERING, CONTROL SYSTEM ENGINEERING. *Educ:* Rensselaer Polytech Inst, BME, 63; Columbia Univ, MS, 64, ScD(eng), 71. *Prof Exp:* Engr, Gen Dynamics Corp, 63-64; sr engr, Optical Technol Div, Perkin Elmer Corp, 64-71; from asst prof to prof eng & appl sci, Univ Calif, Los Angeles, 71-82; PROF, DEPT MECH ENG, MASS INST TECHNOL, CAMBRIDGE, 82- *Concurrent Pos:* NSF grant & consult, 71-; assoc ed, Mechanism & Mach Theory, 75-81; assoc, Danford Found, 77-87; vis fac mem, Dept Eng & vis scholar, Queen's Col, Cambridge Univ, 77-78. *Mem:* Am Soc Mech Engrs; Inst Elec & Electronics Engrs; Sigma Xi. *Res:* Dynamics and control of mechanical and electromechanical systems, noise and vibrations in these systems; application of robotic manipulators to the industrial environment; use of microprocessor computers. *Mailing Add:* Mech Eng Dept Mass Inst Technol Cambridge MA 02139

DUBPERNELL, GEORGE, b Detroit, Mich, May 30, 01; m 27; c 3. ELECTROCHEMISTRY. *Educ:* Univ Mich, BS, 24, PhD(electrochem), 33; Columbia Univ, AM, 27. *Prof Exp:* Res engr & mgr, Waterbury lab, United Chromium, Inc, 33-55; tech adv, M&T Chem, Inc, 55-66. *Concurrent Pos:* Consult, 67- *Honors & Awards:* Hothersall Award, Brit Inst Metal Finishing, 78. *Mem:* Am Chem Soc; Electrochem Soc; hon mem Am Electroplaters & Surface Finishers Soc; Brit Inst Metal Finishing. *Res:* History, theory and technology of electrochemistry, electroplating, chromium plating and overvoltage; insulating lacquers; type IV superconductivity (room temp). *Mailing Add:* Presby Village C41 17383 Garfield Ave Redford MI 48240-2195

DUBRAVCIC, MILAN FRANE, b Gospic, Croatia, July 25, 22; US citizen; m 53; c 4. ANALYTICAL CHEMISTRY, FOOD SCIENCE. *Educ:* Univ Zagreb, BS, 48; Univ Mass, Amherst, PhD(food sci), 68. *Prof Exp:* Chemist, Dept Health, Cent Inst Hyg, Zagreb, Yugoslavia, 48-53; chief chemist, Lifeguard Milk Prod Ltd, Australia, 54-56; res chemist, Imp Chem Industs, 56-61 & Commonwealth Sci & Indust Res Orgn, 62-63; from asst prof to assoc prof, 68-79, prof chem technol, 79-86, coordr chem technol, 72-86, EMER PROF, UNIV AKRON, 87- *Concurrent Pos:* Dir NSF grant for instructional sci equip, Univ Akron, 69-71; vis scientist, Chenoweth Lab, Univ Mass, Amherst, 78-79. *Mem:* Am Chem Soc; Croatian Chem Soc; Inst Food Technol; Sigma Xi. *Res:* Food technology; methodology for analysis of foods and plastics; mechanisms of enzymatic, oxidative, thermal and radiolytic degradation of organic materials. *Mailing Add:* 2499 Hillview St Sarasota FL 34239-3034

DU BREUIL, FELIX L(EMAIGRE), b Paris, France, Feb 11, 21; nat US; m 49; c 7. MINERAL ENGINEERING. *Educ:* Nat Sch Mines, Paris, IngcMines, 45; Pa State Univ, MS, 47, PhD(elec eng), 49. *Prof Exp:* From instr to assoc prof mineral eng, Pa State Univ, 49-56; field engr, Jeffrey Mfg Co, 56-58; proj engr, 58-70, ASST MGR MINING, BITUMINOUS COAL RES, INC, 70- *Mem:* Inst Elec & Electronics Engrs; Fr Soc Mineral Indust; Instrument Soc Am. *Res:* Coal mining; health and safety; communications; automation. *Mailing Add:* 1365 Foxwood Dr Monroeville PA 15146

DUBRIDGE, LEE ALVIN, b Terre Haute, Ind, Sept 21, 01; m 25, 74; c 2. EXPERIMENTAL PHYSICS. *Educ:* Cornell Col, Iowa, AB, 22; Univ Wis, AM, 24, PhD(physics), 26. *Hon Degrees:* Twenty-eight from US cols & univs, 40-70. *Prof Exp:* Instr physics, Univ Wis, 25-26; Nat Res Coun fel, Calif Inst Technol, 26-28; from asst prof to assoc prof, Wash Univ, 28-34; prof & chmn dept, Univ Rochester, 34-46, dean fac arts & sci, 38-42; pres, 46-69, EMER PRES, CALIF INST TECHNOL, 69- *Concurrent Pos:* Assoc ed, Am Physics Teacher, 35-38, Phys Rev, 36-39 & Rev Sci Instruments, 36-42; with phys sci div, Nat Res Coun, 36-42; dir radiation lab, Mass Inst Technol, 40-45; mem gen adv comt, AEC, 46-52; trustee, Rand Corp, 48-61, Rockefeller Found, 56-67 & Mellon Inst, 58-66; mem sci adv comt, Off Defense Mobilization, 51-56, Nat Sci Bd, 52-56, 58-64, Nat Merit Scholar Corp, 63-69, Distinguished Civilian Serv Awards Bd, 63-67 & Gen Motors Corp, 70-76; mem bd dirs, Nat Educ TV, 64-69; sci adv to President, 69-70; mem, Presidents' Sci Adv Comt, 69-72. *Honors & Awards:* Brit Royal Medal, 46; Res Corp Award, 47; Lehman Award, NY Acad Sci, 69; Golden Plate Award, Am Acad Achievement, 74; Vannerar Bush Award, NSF, 82. *Mem:* Nat Acad Sci; AAAS; fel Am Phys Soc (vpres, 46, pres, 47); Am Philos Soc. *Res:* Biophysics; nuclear disintegration; photoelectric and thermionic emission; direct current amplification; energy distribution of photoelectrons; theory of photoelectric effect; radar. *Mailing Add:* 1730 Homet Rd Pasadena CA 91106

DUBROFF, LEWIS MICHAEL, b Brooklyn, NY, Jan 22, 47; m 71; c 3. NUCLEIC ACID, NUCLEIC ACID ANTIBODIES. *Educ:* Univ Pa, MD, 71, PhD(molecular biol), 75. *Prof Exp:* Assoc fel, nucleic acid, Inst Cancer Res, 72-75; fel dermat, Mayo Clin, 75-76; res asst prof, Hahnemann Med Col, 76-78; asst prof med, 78-80, RES ASST PROF PHARMACOL, UPSTATE MED CTR, STATE UNIV NY, 80- *Res:* Drug nucleic acid interactions and antibody nucleic acid interactions. *Mailing Add:* Dept Pharmacol State Univ NY Upstate Med Ctr 750 Adams St 475 Irving Ave No 314 Syracuse NY 13210-1703

DUBROFF, RICHARD EDWARD, b Chicago, Ill, June 14, 48; m 78. SIGNAL PROCESSING, WAVE PROPAGATION. *Educ:* Rensselaer Polytech Inst, BS, 70; Univ Ill, MS, 72, PhD(elec eng), 76. *Prof Exp:* Res assoc fel ionospheric measurements, Univ Ill, 76-78; sr res engr seismic explor, Phillips Petrol Co, 78-85; ASSOC PROF ELEC ENG, UNIV MO ROLLA, 85- *Concurrent Pos:* Consult, City Engr, Bartlesville, Okla, 78-80. *Mem:* Soc Indust & Appl Math; Inst Elec & Electronics Engrs; Am Geophys Union. *Res:* Geophysical applications (acoustic imaging and seismic migration) of acoustic and elastic waves; electromagnetic and atmospheric wave propagation. *Mailing Add:* Univ Mo 122 Elec Eng Bldg Rolla MO 65401

DUBROFF, WILLIAM, b Johannesburg, SAfrica, Oct 1, 37; US citizen; m 60; c 1. MATERIALS SCIENCE, METALLURGY. *Educ:* Columbia Univ, BA, 61, MS, 63, PhD(metall), 67. *Prof Exp:* Dir NASA proj metall, Columbia Univ, 65-67; res engr, Inland Steel Co, 67, sr res engr, 67-72, supv res engr, 72-76, asst dir res, 76-80, assoc dir res, 80-84, dir res, 84-85; PROG COORDR, EG&G INC, 85- *Concurrent Pos:* Ed, Ironmaking & Steelmaking, 78-85; lectr, Ironmaking, McMasters Univ, 80-85. *Mem:* Am Inst Mining, Metall & Petrol Engrs; Am Soc Metals; Sigma Xi. *Res:* Magnetic and mechanical properties, phase transformations, and recrystallization in metals; microstructure-property-processing correlations for nonmetallic materials; continuous and novel casting technologies. *Mailing Add:* 9313 Walnut Dr Munster IN 46321

DUBROVIN, KENNETH P, b Chicago, Ill, Oct 24, 31; m 54; c 5. SOIL CHEMISTRY, PLANT PHYSIOLOGY. *Educ:* Univ Ill, BS, 53; Univ Wis, PhD(soil chem), 56; Univ Mo, MBA, 66. *Prof Exp:* Pesticide res & develop, Spencer Chem Div, Gulf Oil, 56-71; dir res, Great Western Sugar Co, 71-76, vpres, Int Develop, 76-79; dir bus develop, Y-Tex Corp, 80-81; dir, Mkt & Loan Serv, Inc, 81-84; vpres, Fed Land Bank, 86-89; VPRES, FARM CREDIT BANK, 85-; VPRES, CONAGRA USSR, 90- *Concurrent Pos:* Fulbright scholar, Neth, 53-54. *Mem:* Sigma Xi. *Res:* Improved fertilizers; development of new and unique herbicides, fungicides, insecticides and nematicides; mechanism of action of herbicides. *Mailing Add:* 8600 Firethorn Dr Loveland CO 80538

DUBRUL, E LLOYD, b New York, NY, Apr 5, 09; m 56. PHYSICAL ANTHROPOLOGY. *Educ:* NY Univ, DDS, 37; Univ Ill, MS, 49, PhD, 55. *Prof Exp:* From instr to asst prof oral & maxillofacial surg, 46-55, from asst prof to prof chg oral anat, Col Dent, 47-64, asst prof post grad studies, 48-51, actg head dept oral anat, 64-66, assoc prof anat, Col Med, 58-59, PROF ANAT, COL MED, UNIV ILL MED CTR, 59-, PROF ORAL ANAT, COL DENT, 64-, HEAD DEPT, 66- *Mem:* Fel AAAS; Am Asn Phys Anthrop; NY Acad Sci. *Res:* Comparative anatomy; neuroanatomy; anthropology. *Mailing Add:* Univ Ill Sch Med 801 S Paulina St Chicago IL 60612

DUBRUL, ERNEST, b Wooster, Ohio, Mar 2, 43; m 66; c 2. BIOCHEMISTRY, DEVELOPMENTAL BIOLOGY. *Educ:* Xavier Univ, Ohio, HAB, 64; Wash Univ, PhD(biol), 69. *Prof Exp:* Fel biol, Oak Ridge Nat Lab, 69-72; res assoc, Mem Res Ctr, Univ Tenn, 72-74; ASSOC PROF BIOL, UNIV TOLEDO, 74- *Mem:* Soc Develop Biol; Am Soc Cell Biol; Soc Neurosci. *Res:* Nucleic acids and protein synthesis in development. *Mailing Add:* Dept Biol Univ Toledo 2801 W Bancroft Toledo OH 43606

DUBUC, PAUL U, b Boston, Mass, Sept 23, 38. EXPERIMENTAL BIOLOGY. *Educ:* Univ Calif, Santa Barbara, PhD(biophys), 71. *Prof Exp:* RES SCIENTIST, SANSUM, 81- *Mem:* Endocrine Soc; Am Diabetes Soc. *Mailing Add:* Sansum Med Res Found 2219 Bath S Santa Barbara CA 93105

DUBUC, SERGE, b Montreal, Que, Apr 16, 39; m 62; c 3. MATHEMATICS. *Educ:* Univ Montreal, BSc, 62, MSc, 63; Cornell Univ, PhD(math), 66. *Prof Exp:* Asst math, Cornell Univ, 64-66; asst prof, Univ Montreal, 66-71; assoc prof, Univ Sherbrooke, 71-73; assoc prof, 73-76, PROF MATH, UNIV MONTREAL, 76- *Concurrent Pos:* Ford Motor Co fel, 68-69; Killam fel, 75-76. *Mem:* Can Math Cong. *Res:* Fractal geometry; mathematical analysis, especially iteration of functions; numerical analysis, especially interpolation theory; plane geometry; nondifferentiable functions. *Mailing Add:* Dept Math Univ Montreal Montreal PQ H3C 3J7 Can

DUBY, JOHN, engineering; deceased, see previous edition for last biography

DUBY, PAUL F(RANCOIS), b Brussels, Belg, Dec 16, 33; m 59; c 3. METALLURGY. *Educ:* Univ Brussels, Mech & Elec E, 56; Columbia Univ, EngScD(mineral eng), 62. *Prof Exp:* Temp res assoc, Ctr Nuclear Sci, Royal Mil Sch, Brussels, 57-58; res assoc mineral eng, Henry Krumb Sch Mines, Columbia Univ, 61-63; asst prof metall eng, Sch Metall Eng, Univ Pa, 63-65; from asst prof to assoc prof, 65-76, PROF MINERAL ENG, HENRY KRUMB SCH MINES, COLUMBIA UNIV, 76- *Mem:* Am Inst Mining, Metall & Petrol Engrs; Electrochem Soc; Sigma Xi. *Res:* Extractive metallurgy; transport properties of fused salts; applied electrochemistry; hydrometallurgy; surface chemistry; corrosion. *Mailing Add:* Henry Krumb Sch Mines Columbia Univ New York NY 10027

DUBY, ROBERT T, b Ludlow, Mass, July 10, 40; m 61; c 2. ANIMAL PHYSIOLOGY, REPRODUCTIVE PHYSIOLOGY. *Educ:* Univ Mass, BS, 62, MS, 65, PhD(animal physiol), 67. *Prof Exp:* NIH fel, 67-68; asst prof reproductive physiol, Cornell Univ, 68-70; asst prof, 70-77, ASSOC PROF VET & ANIMAL SCI, UNIV MASS, AMHERST, 77-, ASST PROF GRAD FAC, ANIMAL SCI, 80- *Mem:* Soc Study Reproduction; Sigma Xi. *Res:* Utero ovarian relationships in domestic animals; separation of X and Y bearing sperm and ovum transfer techniques; sex differentiation in the ovine embryo. *Mailing Add:* 716 Amherst Rd Sunderland MA 01375

DUCE, ROBERT ARTHUR, b Midland, Ont, Apr 9, 35; US citizen; m; c 2. ATMOSPHERIC CHEMISTRY. *Educ:* Baylor Univ, BA, 57; Mass Inst Technol, PhD(inorg & nuclear chem), 64. *Prof Exp:* Res assoc geochem, Mass Inst Technol, 64-65; asst prof chem & asst meteorologist, Inst Geophys, Univ Hawaii, 65-68, assoc prof & assoc geochemist, 68-70; assoc prof, 70-73, PROF, UNIV RI, 73-, DIR, CTR ATMOSPHERIC CHEM STUDIES, 81- *Concurrent Pos:* Mem, Ocean Sci Comt, Nat Acad Sci, 73-75, sci adv bd, Environ Protection Agency, 76-79, World Meteorol Orgn Working Group on Interchange between Atmosphere & Oceans, 77-, chmn, 87-, Spec Comt Oceanog Res Working Group 72, 84-89; mem, Int Asn Meterol & Atmospheric Physics Comt on Atmospheric Chem & Global Pollution, 75-, secy, 79-83 & pres, 83-90; mem, World Meteorol Orgn, Comt Atmospheric Sci, Working Group on Amospheric Chem, 82-87, Nat Oceanic & Atmospheric Admin, Atmospheric Chem & Monitoring Adv Panel, 85-87; vis prof, Univ Otago, NZ, 83; mem adv commt, Ocean Sci Div, NSF, 79-82,

Atmospheric Sci Div, 84-87; mem bd trustees, Univ Corp Atmospheric Res, 86-, secy, 86-89; mem bd gov, Joint Oceanog Insts, Inc, 87-, vchmn, 90-; dean, Grad Sch Oceanog, Univ RI, 87-, vprovost marine affairs, 87-; mem, Bd Atmospheric Sci & Climate, Nat Acad Sci/Nat Res Coun, 82-86 & 89-, chmn, Comt on Raze in Nat Parks, mem, Comt on Atmospheric Chem, 87-90; sect chmn, AAAS, 87-88, coun mem, 90-, coun mem, Am Meteorol Soc, 87- *Honors & Awards:* Rosenstiel Award, 90. *Mem:* Am Meteorol Soc; Am Chem Soc; Geochem Soc; fel Am Geophys Union; fel Am Meteorol Soc; fel AAAS. *Res:* Atmospheric chemistry of the halogens, heavy metals, boron, organic carbon, phosphorus, nitrogen and mineral dust; chemical oceanography; chemical fractionation at the air-sea interface; global transport of atmospheric particulate matter and trace gases; chemistry of the air/sea interface; neutron activation analysis. *Mailing Add:* Grad Sch Oceanog Univ RI Narragansett RI 02882-1197

DUCH, DAVID S, b Bridgeport, Conn, Mar 6, 43; m. BIOCHEMISTRY. *Educ:* Univ Scranton, Pa, BS, 64; Duke Univ, Durham, NC, PhD(biochem), 69. *Prof Exp:* Postdoctoral fel, Lab Enzymol, Roswell Park Mem Inst, Buffalo, NY, 69-72; res assoc, Wistar Inst, Philadelphia, Pa, 72-75; res scientist III, Dept Med Biochem, Wellcome Res Labs, Research Triangle Park, NC, 75-78, res scientist IV, 78-82, res scientist V, 82-85, group leader, 85-87, HEAD, SECT EXP ONCOL, DIV MED BIOCHEM, WELLCOME RES LABS, RESEARCH TRIANGLE PARK, NC, 87- *Concurrent Pos:* Vis scientist, Ctr Neurochem, Strasbourg, France, 74. *Mem:* Am Asn Cancer Res; Am Soc Pharmacol & Exp Therapeut; Sigma Xi; AAAS. *Res:* Folates and antifolates; purine and pyrimidine biosynthesis; mechanism of anticancer drugs; experimental chemotherapy; author of various publications. *Mailing Add:* Div Med Biochem Wellcome Res Labs Burroughs Wellcome Co 3030 Cornwallis Rd Research Triangle Park NC 27709

DUCHAMP, DAVID JAMES, b St Martinville, La, Oct 15, 39; m 64; c 3. PHYSICAL CHEMISTRY, X-RAY CRYSTALLOGRAPHY. *Educ:* Univ Southwestern La, BS, 61; Calif Inst Technol, PhD(phys chem), 65. *Prof Exp:* res scientist, 65-87, DIR, PHYS & ANALYTICAL CHEM, UPJOHN CO, 88- *Mem:* AAAS; Am Chem Soc; Am Crystallog Asn; Asn Comput Mach; NY Acad Sci. *Res:* Molecular structure studies by x-ray diffraction; application of mathematics and computers to chemical structure problems; data acquisition and control, via computer, of physical measurement instruments. *Mailing Add:* Phys & Analytical Div Upjohn Co Kalamazoo MI 49001

DUCHAMP, THOMAS EUGENE, b St Martinville, La, Nov 12, 47; m 70. MATHEMATICS. *Educ:* Univ Ill, Urbana, BS, 69, MS, 69, PhD(math), 76. *Prof Exp:* instr math, Univ Utah, 76-79; asst prof, 79-85, ASSOC PROF MATH, UNIV WASH, 85- *Concurrent Pos:* Vis assoc prof, Tulane Univ, 83-85. *Mem:* Am Math Soc; AAAS. *Res:* Differential geometry; several complex variables. *Mailing Add:* Dept Math Univ Wash GN-50 Seattle WA 98195

DUCHARME, DONALD WALTER, b Saginaw, Mich, June 14, 37; m 58; c 3. PHARMACOLOGY. *Educ:* Cent Mich Univ, AB, 59; Univ Mich, PhD(pharmacol), 65. *Prof Exp:* Sr scientist, 65-80, res head, 80-85, assoc dir cardiovasc dis, 85-89, SR SCIENTIST, UPJOHN CO, 89- *Concurrent Pos:* Adj prof pharmacol, Med Col Ohio, 77-; fel, Coun High Blood Pressure Res, Am Heart Asn; mem, bd trustees, Am Heart Asn Mich, 72-, pres, 81-82; chmn, NCent Region, Res Adv Comt, Am Heart Asn, 82-83, mem subcomt regional & nat res. *Mem:* Am Heart Asn; Am Soc Pharmacol & Exp Therapeut. *Res:* Cardiovascular physiology and pharmacology, particularly the autonomic control of the capacity vessels and the role of these vessels in the etiology of arterial hypertension. *Mailing Add:* Safety Pharmacol Upjohn Co Kalamazoo MI 49001

DUCHARME, ERNEST PETER, b St Paul, Minn, July 15, 16; m 47; c 1. PLANT PATHOLOGY. *Educ:* St Mary's Col, Minn, BSc, 38; DePaul Univ, MS, 43; Univ Minn, PhD(plant path), 49. *Prof Exp:* Instr, high sch, 38-43; asst plant path, Univ Minn, 43-46; plant pathologist, Agr Res & Educ Ctr, 46-81, EMER PROF PATH, UNIV FLA, 81- *Concurrent Pos:* Citrus dis & citrus consult. *Mem:* Am Phytopath Soc; Bot Soc Am; Int Orgn Citrus Virologists. *Res:* Root diseases of citrus; soil microbiology in relation to the citrus root disease complex; root parasitizing nematodes; soil ecology. *Mailing Add:* 4314 Larry Lagoon Winter Haven FL 33884

DUCHARME, JACQUES R, b Montreal, Que, Jan 1, 28; m 55; c 5. PEDIATRIC ENDOCRINOLOGY, BIOCHEMISTRY. *Hon Degrees:* DSc, Université Claude Bernard, France, 87. *Prof Exp:* Univ Montreal, BA, 48, MD, 54; Univ Pa, MSc, 61. Prof Exp: Resident pediat, Children's Hosp Philadelphia, 55-57; lectr, Columbia Univ, 57-59; from lectr to assoc prof, 59-69, chmn dept, 69-75, PROF PEDIAT, UNIV MONTREAL, 69- *Concurrent Pos:* Res fel, Babies' Hosp, Columbia-Presby Med Ctr, New York, 57-59; dir, Pediat Endocrine Res Lab, L'Hopital Ste-Justine, 60-, Pediat Res Ctr, 75-78, & endocrine sect, dept biochem, 81-; chief, Endocrine Serv, Hôpital San-Justine, 88-91. *Honors & Awards:* Michel Sarrazin Award, Club Res Clin Quebec, 84. *Mem:* Am Pediat Soc; Europ Soc Pediat Endocrine; Endocrine Soc; Can Pediat Soc; NY Acad Sci. *Res:* Steroid biochemistry; steroid metabolism in newborn infants, infancy and childhood; molecular regulation of Leydig cell function and the influence of Sertoli cells on this regulation; mechanisms of pubertal development. *Mailing Add:* Dept Pediat Hosp San-Justine 3175 St Catherine Rd Montreal PQ H3T 1C5 Can

DUCHEK, JOHN ROBERT, b St Louis, Mo, Dec 6, 48; m 72; c 2. ORGANIC CHEMISTRY. *Educ:* St Louis Univ, BS, 70, PhD(chem), 75. *Prof Exp:* Teaching asst, St Louis Univ, 70-74; res assoc, Univ Ariz, 74-76; from asst prof to assoc prof org chem, 76-80, PROF NAT SCI, UNIV NFLA, 80-; RES CHEMIST, MALLINCKRODT, INC, 84- *Concurrent Pos:* Co-investr, NIH grant, 77-; mgr, Ash Stevens Inc, 80-81; troubleshooter, Sigma Chem Co, 81-84. *Mem:* Am Chem Soc; Sigma Xi. *Res:* Dye synthesis; lipid bilayer membranes; prebiotic chemistry; physiological buffers; pharmaceutical chemistry. *Mailing Add:* 351 Porchester Dr St Louis MO 63125

DUCIBELLA, TOM, b New Rochelle, NY, Nov 5, 47. TISSUE CULTURE. *Educ:* Princeton Univ, PhD(biol), 73. *Prof Exp:* Fel histol, Harvard Med Sch, 73-76, instr, 76-77; asst prof histol, Univ Tenn, 77-79; asst prof histol, Sch Med, Tufts Univ, 79-82; proj mgr, Millipore Corp, 82-88; IVF LAB DIR & ASSOC PROF OBSTET & GYNEC, TUFTS-NEW ENG MED CTR, 89- *Concurrent Pos:* NIH grant, Sch Med, Univ Tenn, 77-79; prin investr, Sch Med, Tufts Univ, 79-82; NIH grant, 88-91. *Mem:* Am Soc Cell Biol; Am Fertil Soc; Soc Study Reproduction. *Res:* Role of the cell surface in early development and organization of tumor cells; tissue culture growth systems; mammalian egg maturation. *Mailing Add:* Tufts-New Eng Eng Med Ctr 171 Harrison Ave Box 61 Boston MA 02111

DUCIS, ILZE, b Senne I, Ger, July 8, 47; US citizen; m 86; c 1. BIOCHEMISTRY, BIOPHYSICS. *Educ:* Univ Calif, Davis, BA, 70; San Jose State Univ, San Jose, MS, 74; Univ Rochester, PhD(pharmacol), 79. *Prof Exp:* Postdoctoral fel biophys, Max-Planck-Inst, Ger, 79-81 & neurochem, 81-84; postdoctoral res assoc biochem, Cornell Univ, Ithaca, 84-87; res asst prof neuropath, Sch Med, Univ Miami, 87-90; ASST PROF BIOL, TEX WOMAN'S UNIV, 90- *Mem:* Am Soc Neurochem; Soc Neurosci. *Res:* Membrane transport and binding systems; reconstituting membrane transport systems into proteoliposomes; binding properties of the peripheral-type benzodiazepine receptor in cultured astroglia under various conditions; transport and binding systems in pinealocytes and the effects of drugs and hormones on these systems. *Mailing Add:* Dept Biol Tex Woman's Univ PO Box 23971 Denton TX 76204-1971

DUCK, BOBBY NEAL, b Reagan, Tenn, Sept 6, 39; m 62; c 1. CYTOGENETICS. *Educ:* Univ Tenn, BS, 61; Auburn Univ, MS, 63, PhD(agron), 64. *Prof Exp:* Asst agronomist, Agr Exp Sta, Univ Fla, 64-66; from asst prof to assoc prof, 66-73, asst dean sch agr, 70-78, PROF AGRON, UNIV TENN, MARTIN, 73- *Concurrent Pos:* Mem, Coun Agr Sci & Technol. *Mem:* Am Soc Agron; Crop Sci Soc Am; Soil Conserv Soc Am; Wildlife Soc; Coun Agr Sci & Technol. *Res:* Instruction in plant sciences; breeding and management of forage crops. *Mailing Add:* Sch Agr Univ Tenn Martin TN 38238

DUCK, IAN MORLEY, b Kamloops, BC, Can, Oct 4, 33; m 63; c 2. NUCLEAR PHYSICS. *Educ:* Queen's Univ, Ont, BSc, 55; Calif Inst Technol, PhD(theoret nuclear physics), 61. *Prof Exp:* Res assoc, Univ Southern Calif, 61-63; res assoc, 63-65, from asst prof to assoc prof, 65-69, PROF PHYSICS, RICE UNIV, 69- *Res:* Weak interactions; nuclear reaction mechanisms; few nucleon problems. *Mailing Add:* Dept Physics Rice Univ Box 1892 Houston TX 77251

DUCK, WILLIAM N, JR, b Millheim, Pa, Feb 9, 20; m 44; c 2. BIOCHEMISTRY, PHYSICAL CHEMISTRY. *Educ:* Pa State Univ, BS, 42; Drexel Inst, dipl chem eng, 44; Franklin & Marshall Col, MS, 64. *Prof Exp:* Qual control supvr, Nat Dairy Prod, Inc, 42-50; res & develop chemist, Am Stores Co, Inc, 50-54; res dir, Pa Mfrs Confectioners Asn, 54-64; res chemist, Gen Cigar Co, 64-87; RES CHEMIST, HELME TOBACCO CO DIV, AM MAISE CO, 87- *Concurrent Pos:* Consult, flavor & confectionary indust. *Honors & Awards:* Stroud Jordon Award, Am Asn Candy Technologists, 86. *Mem:* Am Chem Soc. *Res:* Studies of moisture relations of sugar glasses and fat crystallization of chocolate and other fats. *Mailing Add:* 607 Capri Rd Lancaster PA 17603

DUCKER, THOMAS BARBEE, b Huntington, WVa, Dec 20, 37; m 65; c 3. MEDICAL SCIENCES, HEALTH SCIENCES. *Educ:* Univ Va, BA, 59, MD, 63. *Prof Exp:* Res asst, Univ Va, 62-63; intern surg, Univ Mich, 63-64, asst resident gen surg, 64-65, asst resident neurosurg, 65-66; fel neurosurg & res assoc, Walter Reed Army Med Ctr, 66-68; resident neurosurg, Univ Mich, 68-70; from asst prof to assoc prof surg & neurosurg, Col Med, Med Univ SC, 70-75, mem grad fac, Col Grad Studies, 73-75, asst dean, Col Med, 74-75; PROF SURG & NEUROSURG & HEAD DIV NEUROL SURG, SCH MED, UNIV MD, 75- *Concurrent Pos:* Attend neurosurgeon, Univ Hosp, Charleston, SC, 70-75, mem prof staff, 70-75, secy, 73-75; consult neurosurg, Roper Hosp, Vet Admin Hosp & US Naval Hosp, Charleston, 70-75; mem admis comt, Col Med, Med Univ SC, 70-75, chmn, 74-75, mem curriculum planning comt, Univ, 73-75; attend neurosurgeon, mem sch med coun & mem med staff, Univ Md Hosp, 75-; consult neurosurgeon, Mercy Hosp, Baltimore, Md, 75- *Mem:* Am Asn Neurol Surgeons; Am Col Surgeons; AMA; Am Asn Surg Trauma; Cong Neurol Surgeons. *Res:* Spinal cord injury. *Mailing Add:* 100 Cathedral St Annapolis MD 21401

DUCKETT, KERMIT EARL, b Asheville, NC, Mar 27, 36; m 65; c 2. PHYSICS, ASTRONOMY. *Educ:* Ga Inst Technol, BS, 58; Univ Colo, MS, 61; Univ Tenn, PhD(physics), 64. *Prof Exp:* Res assoc fiber physics, Am Enka Corp, 58-59; from asst prof to assoc prof, 65-85, Assoc Dean, Col Human Ecol, 86-89, PROF PHYSICS, UNIV TENN, KNOXVILLE, 85- *Concurrent Pos:* NSF res assoc, USDA, 65-66. *Mem:* Brit Textile Inst; Am Asn Physics Teachers; fel Textile Inst Eng; Sigma Xi. *Res:* Infrared spectroscopy; physical properties of cellulosic fibers; yarn mechanics. *Mailing Add:* Dept Physics Univ Tenn Knoxville TN 37996

DUCKLES, SUE PIPER, b Oakland, Calif, Mar 1, 46; m 68; c 2. NEUROSCIENCES, CARDIOVASCULAR PHYSIOLOGY. *Educ:* Univ Calif, Berkeley, BA, 68; Univ Calif, San Francisco, PhD(pharmacol), 73. *Prof Exp:* Asst prof resident pharmacol, Sch Med, Univ Calif, Los Angeles, 76-79; from asst prof to assoc prof pharmacol, Univ Ariz, Tucson, 79-85; assoc prof pharmacol, 85-88, PROF PHARMACOL, UNIV CALIF, IRVINE, 88- *Concurrent Pos:* Fac develop award, Pharmaceut Mfrs Asn Found, 76; estab investr, Am Heart Asn, 82; assoc ed, Life Sci, 80-85; specific field ed, J Pharmacol Exp Ther. *Mem:* Am Soc Pharmacol & Exp Therapeut; Soc Neurosci. *Res:* Pharmacology and physiology of vascular smooth muscle; neurogenic control of the cerebral circulation; autonomic and cardiovascular pharmacology; polypeptide transmitters and their actions on vascular smooth muscle; alterations in vascular smooth muscle responsiveness and innervation during development and aging. *Mailing Add:* Dept Pharmacol Col Med Univ Calif Irvine CA 92717

DUCKSTEIN, LUCIEN, b Paris, France, Aug 25, 32; US citizen; m 58; c 3. OPERATIONS RESEARCH. *Educ:* Univ Toulouse, MSc, 55, MSc, 56; Nat Polytech Inst Elec Eng & Hydraul, dipl, 56, PhD(fluid mech), 62. *Prof Exp:* Engr, Regie Nat Renault, France & Ger, 58-60; jr civil engr aerodyn, Colo State Univ, 61-62; PROF SYSTS ENG, UNIV ARIZ, 62- *Concurrent Pos:* Sci adv, Metra-Int. *Mem:* Inst Mgt Sci; Am Geophys Union. *Res:* Systems engineering; mathematical models water resource; optimization techniques; economic calculus; hydrology and water resources. *Mailing Add:* Dept Systs Sci Case Western Reserve Univ University Circle Cleveland OH 44106

DUCKWORTH, HENRY EDMISON, b Brandon, Man, Nov 1, 15; m 42; c 2. ATOMIC PHYSICS. *Educ:* Univ Man, BA, 35, BSc, 36; Univ Chicago, PhD(physics), 42. *Hon Degrees:* DSc, Univ Ottawa, 66, McMaster Univ, 69, Laval Univ, 71, Mt Allison Univ, 71 & Univ NB, 72; Univ Man, LLD, 78. *Prof Exp:* Lectr physics, United Col, 38-40; jr physicist, Nat Res Coun Can, 42-44, asst res chemist, 44-45; asst prof physics, Univ Man, 45-46; assoc prof, Wesleyan Univ, 46-51; prof, McMaster Univ, 51-65, chmn dept, 56-61, dean grad studies, 61-65; vpres develop, Univ Man, 65-66, acad vpres, 66-71; pres & vchancellor, 71-81, EMER PRES, UNIV WINNIPEG, 81-; EMER PROF, UNIV MANITOBA, 83-, CHANCELLOR, 86- *Concurrent Pos:* Nuffield fel, 55; ed, Can J Physics, 56-62; mem comn atomic masses, Int Union Pure & Appl Physics, secy, 60-66, chmn, 66-69; mem, Nat Res Coun Can, 61-67 & Defence Res Bd, 65-71; hon fel, United Col, 66; mem, Sci Coun Can, 73-77; mem, Can Environ Adv Coun, 73-76; mem, Nat Libr Adv Bd, 74; chmn, Coun Asn Commonwealth Univs, 77-78; mem, Natural Sci & Eng Res Coun, 78-; dir, Inst Res on Pub Policy, 78- *Honors & Awards:* Medal, Can Asn Physicists, 64; Tory Medal, Royal Soc Can, 65; Officer, Order of Can, 76. *Mem:* Fel Am Phys Soc; Royal Soc Can (pres, 71); Can Asn Physicists (pres, 60). *Res:* Mass spectroscopy; precise determination of atomic masses using mass spectrographic techniques. *Mailing Add:* Dept Physics Univ Man Winnipeg MB R3T 2N2 Can

DUCKWORTH, HENRY WILLIAM (HARRY), b Ottawa, Ont, Oct 11, 43; m 73. BIOCHEMISTRY. *Educ:* McMaster Univ, BSc, 65; Yale Univ, PhD(biochem), 70. *Prof Exp:* From asst prof to assoc prof, 72-83, PROF CHEM, UNIV MAN, 83- *Concurrent Pos:* Assoc ed, Can J Biochem, 74-80 & 81-83; mem adv comt, Atomic Energy Control Bd, 76-78. *Mem:* Can Biochem Soc; Am Soc Biochem & Molecular Biol; Biochem Soc. *Res:* Structure and function of allosteric proteins; evolution of regulatory properties, with specific reference to bacterial citrate syntheses. *Mailing Add:* Dept Chem Univ Man Winnipeg MB R3T 2N2 Can

DUCKWORTH, WALTER DONALD, b Athens, Tenn, July 19, 35; m 55; c 3. SYSTEMATIC ENTOMOLOGY. *Educ:* Middle Tenn State Univ, BSc, 57, NC State Univ, MSc, 60, PhD(entom), 62. *Prof Exp:* From assoc cur entom to cur, Smithsonian Inst, 62-84; PRES & DIR, BISHOP MUSEUM, HONOLULU, 86- *Concurrent Pos:* Spec asst, Off Dir, Nat Mus Natural Hist, Smithsonian Inst, 75-78, spec asst, Off Asst Sec Mus Progs, 78-84. *Honors & Awards:* Dir's Award, Nat Mus Natural Hist; Distinguished Serv Award, Smithsonian Inst, 80, 83, 84. *Mem:* Asn Trop Biol (secy-treas, 67-71); Entom Soc Am (pres, 83); Soc Syst Zool; Am Inst Biol Sci (secy-treas, 78-84, pres, 86); Orgn Trop Studies; Asn Syst Collections (vpres, 87-89, pres, 90-91). *Res:* Biosystematics of the Microlepidoptera, particularly tropical groups. *Mailing Add:* Bishop Mus Honolulu HI 96817

DUCKWORTH, WILLIAM C(APELL), b Jackson, Tenn, Oct 31, 19; m 51; c 3. CHEMICAL ENGINEERING. *Educ:* Univ of the South, BS, 40; Ga Inst Technol, MS, 54; Memphis State Univ, MA, 73 & 89. *Prof Exp:* Chem engr, Am Cyanamid Co, 46-47; chem technologist, Tenn Corp, 47-61; mkt res engr, Enjay Chem Co, 61-63; sr chem economist, Southern Res Inst, 63-66; sr proj engr, Gulf Res & Develop Co, 66-71; sr engr, Velsicol Chem Corp, 71-74; sr proj engr, Gulf Sci & Technol Co, 74-82; CONSULT, 82- *Mem:* Chem Mkt Res Asn. *Res:* Chemical economics; chemical technology and economics; history of technology. *Mailing Add:* 4053 Summer Fr Village Apt 4 Memphis TN 38122

DUCKWORTH, WILLIAM CLIFFORD, b Athens, Tenn, Oct 21, 41. ENDOCRINOLOGY, DIABETES. *Educ:* Univ Tenn, BS, 62; Univ Tenn, MD, 66. *Prof Exp:* Intern & resident med, City Memphis Hosps, 66-69; NIH spec fel endocrinol, Med Units, Univ Tenn, 69-71; res assoc, Vet Admin Hosp, Memphis, 71-73, clin investr, 73-76, NIH res career develop award endocrinol, 76-80; PROF MED, SCH MED, IND UNIV, 80- *Concurrent Pos:* Asst prof med, Med Units, Univ Tenn, 72-75; assoc prof med, Ctr Health Sci, Univ Tenn, 75-; vis prof, Univ Geneva, Switz, 79; med sci adv bd, Juv Diabetes Found, 80- *Mem:* Am Fedn Clin Res; Am Diabetes Asn; Soc Clin Invest; Endocrine Soc. *Res:* Diabetes and carbohydrate metabolism; mechanisms of hormonal action and hormonal metabolism. *Mailing Add:* Vet Admin Med Ctr 1481 W Tenth St Indianapolis IN 46202

DUCKWORTH, WINSTON H(OWARD), b Greenfield, Ohio, Oct 15, 18; m 41; c 2. ENGINEERING MECHANICS, TECHNICAL MANAGEMENT. *Educ:* Ohio State Univ, BChE, 40, MS, 41. *Prof Exp:* Asst supvr, 46-52, chief ceramic div, 52-66, founding dir, Defense Ceramic Info Ctr, 67-71, mem res coun, 79-85, FEL, BATTELLE MEM INST, 66- *Concurrent Pos:* Atomic Energy Comn, USN, 44; dir, Engrs Joint Coun, 75. *Honors & Awards:* Cramer Award, Am Ceramic Soc, 78, Distinguished Life Mem, 85; Greaves-Walker Award, Nat Inst Ceramic Engrs, 87. *Mem:* AAAS; fel Am Ceramic Soc (vpres, 76); Nat Inst Ceramic Engrs (pres, 75, & permanent secy, 78-); Am Soc Testing & Mat; Can Ceramic Soc. *Res:* Processing and behavior of ceramic materials; brittle behavior; aircraft and nuclear ceramics; research management; armor materials. *Mailing Add:* Battelle Mem Inst 505 King Ave Columbus OH 43201

DUCLOS, STEVEN J, b Watertown, NY, July 31, 62; m 85; c 1. HIGH-PRESSURE PHYSICS, CRYSTALLOGRAPHY. *Educ:* Wash Univ, St Louis, AB, 84; Cornell Univ, MS, 87, PhD(physics), 90. *Prof Exp:* MEM TECH STAFF, AT&T BELL LABS, 90- *Mem:* Sigma Xi; Am Phys Soc. *Res:* Characterization of high pressure phases of materials using x-ray diffraction, Raman spectroscopy, visible absorption spectroscopy and luminescence spectroscopy; design of real time data acquisition and analysis software for optics experimentation. *Mailing Add:* AT&T Bell Labs Rm 1D-148 600 Mountain Ave Murray Hill NJ 07974-2070

DUCOFF, HOWARD S, b New York, NY, May 5, 23; m 46; c 4. RADIATION BIOLOGY, EXPERIMENTAL GERONTOLOGY. *Educ:* City Col New York, BS, 42; Univ Chicago, PhD(physiol), 53. *Prof Exp:* Scientist, Argonne Nat Lab, 46-57; asst prof physiol, 57-59, assoc prof physiol & biophys, 59-66, PROF PHYSIOL & BIOPHYS, UNIV ILL, URBANA, 66-, PROF BIOENG, 73- *Concurrent Pos:* NIH fel, Univ cambridge, Eng, 64-65; staff consult, Argonne Univ Asn, 66-77; vis scientist, Lawrence Berkeley Lab, Univ Calif, 75-76; vis investr, Univ Sussex, Eng, 83-84; assoc ed, Radiation Res, 89- *Mem:* Am Soc Cell Biol; Am Soc Zoologists; Geront Soc Am; Radiation Res Soc; Soc Gen Physiologists; NAm Hyperthermia Group. *Res:* Examination of differences between proliferative and differentiated tissues in their response and ability to repair or adapt to heat or radiation injury; study of insects, germinating plant seeds and avian and mammalian erythrocytes. *Mailing Add:* Dept Physiol & Biophys Univ Ill 407 S Goodwin Urbana IL 61801

DUCSAY, CHARLES ANDREW, b Pittsburgh, Pa, Apr 5, 53; m 78. PERINATAL BIOLOGY. *Educ:* Fla State Univ, BS, 75; Univ Fla, MS, 77, PhD(reproductive physiol), 80. *Prof Exp:* Fel, Ore Regional Primate Res Ctr, 80-; AT DEPT PHYSIOL, UNIV PAC, FOREST GROVE, ORE. *Mem:* Sigma Xi; Soc Study Reproduction; Am Soc Animal Sci. *Res:* Study of uterine and endocrine factors involved in the regulation of gestational length; initiation of parturition in non-human primates. *Mailing Add:* Ore Regional Primate Res Ctr 505 NW 185th Ave Beaverton OR 97006

DUCZEK, LORNE, b Oct 12, 46; Can citizen. PLANT PATHOLOGY, CROP SCIENCE. *Educ:* Univ Sask, BSA, 69, MSc, 71; Univ Toronto, PhD(plant path), 74. *Prof Exp:* Plant path specialist, Sask Dept Agr, 74-77; CEREAL PATHOLOGIST, AGR CAN RES STA, SASKATOON, SK, 77- *Mem:* Can Phytopath Soc. *Res:* Cereal diseases, particularly those of barley and wheat; root diseases; biological control. *Mailing Add:* 30 Birch Pl Saskatoon SK S7N 2P6 Can

DUDA, EDWIN, b Donora, Pa, Oct 15, 28; m 55; c 5. MATHEMATICS. *Educ:* Washington & Jefferson Col, BA, 51; WVa Univ, MS, 53; Univ Va, PhD(math), 61. *Prof Exp:* Instr math, WVa Univ, 53, 55-57 & Univ Va, 60-61; from asst prof to assoc prof, 61-72, actg chmn dept, 67-70 & 76-78, PROF MATH, UNIV MIAMI, 72- *Mem:* Am Math Soc; Math Asn Am; Polish Math Soc. *Res:* Transformations; topological analysis. *Mailing Add:* Dept Math Univ Miami Univ Sta Coral Gables FL 33124

DUDA, EUGENE EDWARD, b Chicago, Ill, Apr 3, 40; m 73; c 1. RADIOLOGY. *Educ:* Loyola Univ, BS, 61, MD, 66. *Prof Exp:* From instr to asst prof, 72-80, dir neuroradiol, Univ Chicago Hosp & Clin, 73-85, ASSOC PROF RADIOL, PRITZER SCH MED, UNIV CHICAGO, 80-; ASSOC PROF, RUSH UNIV, 86- *Concurrent Pos:* Chmn, Dept Diag Radiol, Christ Hosp & Med Ctr, 85- *Mem:* Radiol Soc NAm; Am Soc Neuroradiol; AMA; Am Col Radiol; Asn Univ Radiologists. *Res:* Application of computed axial tomography to various diseases of the brain. *Mailing Add:* Dept Radiol Christ Hosp & Med Ctr 4440 W 95 St Oak Lawn IL 60453

DUDA, J(OHN) L(ARRY), b Donora, Pa, May 11, 36; m 62; c 4. CHEMICAL ENGINEERING. *Educ:* Case Inst Technol, BS, 58; Univ Del, MChE, 61, PhD(chem eng), 63. *Prof Exp:* Res engr, process fundamentals lab, Dow Chem Co, 63-71; PROF CHEM ENG, 71-, HEAD, CHEM ENG DEPT, PA STATE UNIV, 83- *Honors & Awards:* William H Walker Award, Am Inst Chem Engrs. *Mem:* Am Inst Chem Engrs; Am Chem Soc. *Res:* Transport phenomena; molecular diffusion; fluid mechanics; polymer processing; tribology. *Mailing Add:* Pa State Univ 160 Fenske Lab University Park PA 16802

DUDA, RICHARD OSWALD, b Evanston, Ill, Apr 27, 36; m 68; c 2. ELECTRICAL ENGINEERING. *Educ:* Univ Calif, Los Angeles, BS, 58, MS, 59; Mass Inst Technol, PhD(elec eng), 62. *Prof Exp:* sr res engr, Info Sci Lab, Stanford Res Inst, 62-; SR STAFF SCIENTIST, LAB ARTIFICIAL INTEL RES, FAIRCHILD ADVAN RES & DEVELOP, PALO ALTO, CALIF, 88-; PROF, DEPT ELEC ENG, SAN JOSE STATE UNIV. *Mem:* AAAS; Asn Comput Mach; fel Inst Elec & Electronics Engrs. *Res:* Electronic computers and artificial intelligence, especially pattern recognition and machine learning; information, control and network theory. *Mailing Add:* Dept Elec Eng San Jose State Univ One Washington Sq San Jose CA 95192

DUDAR, JOHN DOUGLAS, b Calgary, Alta, Aug 15, 40; m 62; c 2. NEUROPHYSIOLOGY. *Educ:* Univ Alta, BSc, 62, MSc, 65; Dalhousie Univ, PhD(neurophysiol), 69. *Prof Exp:* Lectr pharm, Univ Alta, 65-66; asst prof, 70-77, actg ast dean, Fac Grad Studies, 81-82, assoc prof, 82, ASSOC PROF PHYSIOL, DALHOUSIE UNIV, 77- *Concurrent Pos:* Med Res Coun Can, Univ BC, 69-70; Dalhousie Univ fel, Inst Neurophysiol, Oslo, Norway, 71-72; vis prof, Med Res Coun, 79; assoc prof Dalehousie Univ, 82; managing dir, Home Rx Serv, Halifax, 82. *Mem:* Can Physiol Soc; NS Pharmaceut Soc; Alta Pharmaceut Soc. *Res:* Cholinergic systems; septal-hippocampal relationships. *Mailing Add:* Dept Physiol & Biophys Dalhousie Univ Halifax NS B3H 4H6 Can

DUDAS, MARVIN JOSEPH, b Lethbridge, Alta, Mar 17, 43. SOIL MINERALOGY & CHEMISTRY. *Educ:* Univ Alta, BSc, 66, MSc, 68; Ore State Univ, PhD(pedology), 72. *Prof Exp:* Fel, 72-74, res assoc, 72-75, from asst prof to assoc prof, 75-85, actg chmn, 84-85, PROF PEDOLOGY, UNIV ALTA, 85- *Mem:* Am Soc Agron; Clay Minerals Soc; Sigma Xi. *Res:* Investigations on the nature of trace elements in the terrestrial ecosystem with emphasis on soils; mineralogical and chemical attributes of natural and disturbed soil bodies. *Mailing Add:* Dept Soil Sci Univ Alta Edmonton AB T6G 2M7 Can

DUDDEY, JAMES E, b Dayton, Ky, Aug 11, 41; m 65; c 3. ORGANIC CHEMISTRY, POLYMER CHEMISTRY. *Educ:* Thomas More Col, AB, 61; St Louis Univ, MS, 65, PhD(org chem), 67. *Prof Exp:* Chemist, Appl Res Lab, US Steel Corp, 66-68; sr res chemist, polyester res & develop, Goodyear Tire & Rubber Co, 68-77, sr develop engr, chem mat processes develop, 77-81, advan tire technol, 81-86, SR DEVELOP ENGR, EXTENDED MOBILITY SYSTS, TIRE TECHNOL, GOODYEAR TIRE & RUBBER CO, 86- *Mem:* Am Chem Soc; Soc Plastics Engrs; Soc Mech Engrs. *Res:* Reactions of trifluoroacetic acid with alkynes, substituted alkynes and substituted alkyl tosylates; reactions of carbon monoxide; modifications of polyesters; three United States patents. *Mailing Add:* 4548 Pineview Akron OH 44321

DUDECK, ALBERT EUGENE, b West Hazleton, Pa, Oct 16, 36; m 59; c 3. AGRONOMY. *Educ:* Pa State Univ, BS, 58, PhD(agron), 64. *Prof Exp:* Assoc prof turf res, Univ Nebr, 64-70; ASSOC PROF ORNAMENTAL HORT, INST FOOD & AGR SCI, UNIV FLA, 70- *Mem:* Am Soc Agron; Crop Sci Soc Am; Int Turfgrass Soc. *Res:* Turf grass phase of highway and fine turf research and teaching; grass breeding; turf grass breeding and genetics. *Mailing Add:* Dept Hort Univ Fla Gainesville FL 32611

DUDEK, F EDWARD, b Columbus, Nebr, Sept 12, 47; m 78; c 1. ELECTROPHYSIOLOGY. *Educ:* Univ Calif, Irvine, BSc, 69, PhD(physiol), 73. *Prof Exp:* Fel sensory physiol, Dept Ophthal Res, Col Physicians & Surgeons, Columbia Univ, 73-74; res assoc neurophysiol, Dept Psychobiol, Univ Calif, Irvine, 74; res assoc neurophysiol, Marine Biomed Inst, Med Br, Univ Tex, Galveston, 74-75; asst prof neurophysiol, Dept Zool, Erindale Col, Univ Toronto, 75-80; ASSOC PROF NEUROPHYSIOL, DEPT PHYSIOL, SCH MED, TULANE UNIV, 80- *Mem:* Am Soc Zoologists; AAAS; Soc Neurosci; Am Physiol Soc. *Res:* Local synaptic circuits in mammalian hippocampus and hypothalamus; mechanisms of synchronization of epileptiform discharges; electrophysiology of neuroendocrine cells; electrotonic coupling in mammalian brain. *Mailing Add:* Mental Retardation Res Ctr Dept Psychiat Univ Calif Los Angeles Sch Med 760 Westwood Plaza Los Angeles CA 90024

DUDEK, R(ICHARD) A(LBERT), b Clarkson, Nebr, Sept 3, 26; m 54; c 2. INDUSTRIAL ENGINEERING. *Educ:* Univ Nebr, BS, 50; Univ Iowa, MS, 51, PhD(indust eng, mgt), 56. *Prof Exp:* Plant indust engr, Fairmont Foods Co, Sioux City, Iowa, 51-52, div indust engr, Nebr, 52-53; asst instr, Univ Iowa, 53-54; asst prof mech eng, Univ Nebr, 54-56; assoc prof indust eng & res assoc schs health professions, Univ Pittsburgh, 56-58; prof indust eng & chmn dept, 58-86, dir Ctr Biotechnol & Human Performance, 69-71, horn prof, 70-88, EMER PROF, TEX TECH UNIV, 88- *Concurrent Pos:* Deleg, Am Inst Indust Engrs, 59-63, 68-70, 74, nat dir student chaps, 61-63, mem, Task Force Comt, Orgn Scholastic Dept Heads, 64, student conf chmn, 65-66, dir, Acad Affairs, 84-85; asst secy, Coun Indust Eng Acad Dept Heads, 80-81, secy, 81-82, vchmn, 82-83, chmn, 83-84. *Mem:* Am Soc Eng Educ; Am Soc Mech Engrs; fel Am Indust Engrs; Inst Mgt Sci; Nat Soc Prof Engrs; Sigma Xi. *Res:* Application of operations research; sequencing/scheduling; systems analysis; technology assessment; work analysis and design; biotechnology. *Mailing Add:* Dept Indust Eng Tex Tech Univ Lubbock TX 79409

DUDEK, THOMAS JOSEPH, b Akron, Ohio, Nov 27, 34; m 56; c 6. POLYMER SCIENCE. *Educ:* Univ Akron, BS, 56, PhD(polymer chem), 61. *Prof Exp:* Proj officer elastomers, US Air Force Mat Lab, Dayton, Ohio, 59-62; mem tech staff chem, Aerospace Corp, Calif, 62-63; aerospace technologist, Ames Res Ctr, NASA, 63-64; mgr mat res dept chem & physics, Lord Corp, Pa, 64-69; section head polymer physics, 69-75, mgr eng math & physics, 75-80, ASSOC DIR RES, RES DIV, GENCORP, 80- *Mem:* Am Chem Soc; Am Phys Soc. *Res:* Polymer structure-mechanical property relationships; failure properties of elastomers and plastics; dynamic mechanical and damping properties of polymers and polymer composites; cord/rubber composites; tire mechanics. *Mailing Add:* 517 Fairwood Dr Tallmadge OH 44278-2027

DUDERSTADT, EDWARD C(HARLES), b Kansas City, Kans, Dec 20, 28; m 50; c 2. CERAMIC ENGINEERING. *Educ:* Univ Mo-Rolla, BS, 58, MS, 59. *Prof Exp:* Tech engr, aircraft nuclear propulsion dept, 59-61, sr engr, nuclear mat & propulsion oper, 61-66, prin engr, nuclear systs progs, 66-70, engr, advan energy progs, 70-78, SR ENGR, AIRCRAFT ENGINE GROUP, GEN ELEC CO, 78- *Mem:* Am Ceramic Soc. *Res:* Physical and mechanical properties of ceramics, refractory metals and cermets; ceramic materials fabrication processes; rates and mechanisms of gas transport in solids; development of thermal barrier coatings. *Mailing Add:* 5327 Dee Alva Dr Fairfield OH 45014

DUDERSTADT, JAMES, b Ft Madison, Iowa, Dec 5, 42; m 64; c 2. APPLIED PHYSICS. *Educ:* Yale Univ, BEng, 64; Calif Inst Technol, MS, 65, PhD(eng sci, physics), 67. *Prof Exp:* From asst prof to prof nuclear eng, Univ Mich, Ann Arbor, 69-81, dean, Col Eng, 81-86, provost & vpres acad affairs, 86-88, PRES, UNIV MICH, ANN ARBOR, 88- *Concurrent Pos:* Summer res assoc, Los Alamos Scientific Lab, 64, res physicist, Lawrence Livermore Lab, 71, vis res scientist, Calif Inst Technol, 71; consult, various nat labs & co, 68-; presidential appointee, Nat Sci Bd & co-chair, Comt Educ & Human Resources; dir bd, Unisys, Indust Technol Inst, Nat Mfg Forum & Univ Mich Hosps. *Honors & Awards:* Nat Medal of Technol, 91; Mark Mills Award, Am Nuclear Soc, 68, Arthur Holly Compton Award, 85; E O Lawrence Award, US Dept Energy, 86. *Mem:* Nat Acad Eng; Am Phys Soc; Sigma Xi; fel Am Nuclear Soc; AAAS; Am Soc Eng Educ; Nat Soc Prof Engrs. *Res:* Nuclear reactor physics; statistical physics; plasma physics; applied mathematics; computer simulation; kinetic theory and statistical mechanics; author or co-author of numerous publications and articles. *Mailing Add:* 2074 Fleming Admin Bldg Univ Mich Ann Arbor MI 48109-1340

DUDEWICZ, EDWARD JOHN, b New York, NY, Apr 24, 42; m 63; c 3. STATISTICS. *Educ:* Mass Inst Technol, SB, 63; Cornell Univ, MS, 66, PhD(statist), 69. *Prof Exp:* Res asst statist, Cornell Univ, 65-66; asst prof, Univ Rochester, 67-72; assoc prof, Ohio State Univ, 72-77, grad comt chmn dept, 73-75 & 78-80, math analyst, Comput Ctr, 74-81, prof statist, 77-82; PROF MATH DEPT & STATIST COUN CHMN, SYRACUSE UNIV, 82- *Concurrent Pos:* Reviewer, Math Rev, 66- & Zentralblatt f r Mathematik, 70-; prin investr grants, Off Naval Res-Ctr Naval Anal, 67-72, 79 & Army Res Off-Durham, 72-74, Nat Cancer Inst, 79, NATO, 78-; consult, Graflex Inc, Singer Co, 69, Gleason Works, 69, Myocardial Infarction Res Unit, Strong Mem Hosp, 70-72, O M Scott & Sons Co, 76- & Ohio Dept Pub Welfare, 78; abstractor, Int Lit Dig Serv, 75-78; ed, Basic Ref on Statist Tech in Qual Control, 75-80, ed, Am J Math & Mgt Sci, 79-; vis assoc prof & vis scholar statist, Stanford Univ, 76; vis prof, Katholieke Univ, Leven, Belgium, 79. *Honors & Awards:* Res Award, Sigma Xi, 77; Jack Youden Prize, Am Soc Qual Control, 81; Jacob Wolfowitz Prize. *Mem:* Am Soc Qual Control; fel NY Acad Sci; fel Inst Math Statist; fel Am Statist Asn; Sigma Xi; Int Statist Inst. *Res:* Statistical selection and ranking procedures; estimation of ordered parameters; statistical inference with unknown and unequal variances; nonparametric techniques; simulation and Monte Carlo techniques; sequential analysis; statistical computation. *Mailing Add:* Math Dept Syracuse Univ Syracuse NY 13244

DUDGEON, DAN ED, b Elmhurst, Ill, Nov 11, 47; m 74; c 2. MULTI-DIMENSIONAL SIGNAL PROCESSING, MACHINE VISION. *Educ:* Mass Inst Technol, BS, 69, MS, 70, EE, 71, DSc, 74. *Prof Exp:* Sr staff, Bolt, Beranek & Newman, 74-78; ASST GROUP LEADER, MIT LINCOLN LAB, 79- *Concurrent Pos:* Distinguished lectr, Inst Elec & Electronics Engrs, 88. *Honors & Awards:* Browder J Thompson Mem Award, IEEE, 76. *Mem:* Fel Inst Elec & Electronic Engrs; Acoust Speech & Signal Processing Soc. *Res:* Digital signal processing; multi-dimensional digital signal processing; image processing; computed imaging; machine vision. *Mailing Add:* Lincoln Lab Mass Inst Technol Lincoln Lab 244 Wood St PO Box 73 Lexington MA 02173-0073

DUDKIEWICZ, ALAN BERNARD, b Green Bay, Wis. IMMUNOREPRODUCTION, DEVELOPMENTAL BIOLOGY. *Educ:* Univ Wis, Stevens Point, BS, 65; Univ Mass, Amherst, MA, 68; Univ Tenn, Knoxville, PhD(zool), 73. *Prof Exp:* Instr zool, Univ Tenn, Knoxville, 71-72; res assoc biochem, Reproduction Res Lab, Univ Ga, Athens, 73-76; asst prof biol, Univ Houston, 76-83; IVF LAB DIR, DEPT OBSTET-GYNEC, MT SINAI HOSP MED CTR, CHICAGO, 83-; ASSOC PROF OBSTET-GYNEC & ASST PROF BIOCHEM, RUSH-PRESBY ST LUKE'S MED CTR, CHICAGO, 84-; DIR, DIV LABS, CTR HUMAN REPRODUCTION, 90- *Mem:* AAAS; Am Andrology Soc; Soc Study Reproduction; Sigma Xi; Am Soc Cell Biol; Am Fertil Soc. *Res:* Immunoreproductive and biochemical aspects of unique gamete and embryo proteins which mediate sperm-egg interaction and early embryogenesis in mammals. *Mailing Add:* Ctr Human Reproduction 750 N Orleans St Chicago IL 60610

DUDLEY, ALDEN WOODBURY, JR, b Lynn, Mass, May 15, 37; m 59; c 3. NEUROPATHOLOGY. *Educ:* Duke Univ, AB, 58, MD, 62. *Prof Exp:* Asst prof med & path, Duke Univ, 67-68; asst prof, Univ Wis-Madison, 68-71, assoc dir, Independent Study Prog, 73-75, assoc prof path, 71-76, dir, Neuropath Training Prog, 68-76; chmn, dept path, Univ S Ala, 76-77, prof path, 76-80; dir, Neuropath Training Prog, Cleveland Clin, 80-84; clin asst prof, Case Western Res, 80-85; PROF PATH BAYLOR & CHIEF LAB SERV, V A MED CTR, HOUSTON, 85- *Concurrent Pos:* NIH training fel neuropath, Duke Univ, 65-67; res assoc, Lab Exp Neuropath, Nat Inst Neurol Dis & Stroke, 63-65; neuropathologist, Ctr Cerebrovasc Res, Durham, NC, 67-68; consult, Nat Biomed Res Found, Silver Spring, Md & Vet Admin Hosps, Madison & Tomah, Wis, 68-76; consult, Mendota & Cent Colony State Hosps, Madison, 69-76; mem adv bd, Group Res Path Educ, 74-84; mem, comt prof affairs, Am Asn Neuropathologists, 74-76 & continuing educ comt, 75-77, chmn awards comt, 75-81 & 83-84; mem, World Fedn Neurol Res Coun, 78; adj prof, Med Col Ohio, 83- *Honors & Awards:* Inst Muscles & Nerves Prize, Marseille, France, 82. *Mem:* Am Asn Path & Bact; Am Soc Exp Path; NY Acad Sci; Am Asn Neuropath; Group Res Path Educ (pres, 75-77). *Res:* Computer analysis of histologic sections of brain; drug, heavy metal and herbicide toxicology; diseases of muscle; mechanism of strokes and related diagnostic procedures; neuropathology of mental retardation, alcoholism and related areas. *Mailing Add:* 113 Lab Serv VAMC 2002 Holcombe Blvd Houston TX 77030

DUDLEY, DARLE W, q Salem, Ore, Apr 8, 17; m 41; c 2. MECHANICAL ENGINEERING. *Educ:* Ore State Univ, BS, 40. *Prof Exp:* Engr, Gen Elec Co, 40-48, sect head gear develop, 48-57, mgr advan gear eng, 57-64; mgr mech transmissions, Mech Tech Inc, 64-65, tech dir, 65-67; chief gear technol, Solar Turbines Inc, 67-78; GEAR CONSULT, DUDLEY ENG CO, INC, 78- *Honors & Awards:* Edward P Connell Award, 58; Golden Gear Award, Power Transmission Design, 66; Medaille d'Argent, French Inst Gears & Gear Transmissions, 77; Worcester Reed Warner Medal, Am Soc Mech Engrs, 79. *Mem:* Am Soc Mech Engrs; Am Gear Mfrs Asn; Int Fedn Theory of Machines & Mechanisms. *Res:* Fatigue strength; friction and wear; gear arrangements; gear lubrication; author of two publications by the American Gear Manufacturers Association. *Mailing Add:* 17777 Camino Murrillo San Diego CA 92128

DUDLEY, DONALD LARRY, b Spokane, Wash, Sept 12, 36; m 59. PSYCHOPHYSIOLOGY, PSYCHOBIOLOGY. *Educ:* Univ Puget Sound, BS, 58; Univ Wash, MD, 64. *Prof Exp:* Instr psychiat, Univ Wash, 69; chief serv psychiat, US Navy, Bremerton, Wash, 69-71; assoc prof psychiat, Univ Utah, 71-72; from assoc prof to prof, 72-75, CLIN PROF NEUROL SURG, UNIV WASH, 81-; MED DIR WASH INST NEUROSCIENCES, 83- *Concurrent Pos:* Cardiovasc res trainee, Univ Wash, 66-67; dir alcohol drug rehab ctr, Vet Admin Hosp, Salt Lake City, 71-72; staff psychiatrist, Vet Admin Hosp, Seattle, 72-75; chief behav med psychiat, Harborview Med Ctr,

Seattle, 75-81; vis prof, Univ Amsterdan, 75, vis prof Univ London, 75; Nat Inst Health, clin rev trials comt, 76, coun, Am psychosomatic Soc, 76; Nat Inst Health task force on asthma, 77, prev, control and educ in respiratory dis, 77; Am Lung Asn & Am Thoracic Soc task force on chronic obstructive pulmonary dis, 83; pres, Wash Inst Neurosci, 83. *Mem:* Am Acad Clin Res; Am Asn Advan Sci; Sigma Xi; Am Fedn Clin Res; NY Acad Sci; fel Int Col Psychosomatic Med. *Res:* Clinical and basic research in brain biology and physiology, seizure disorders and chronic medical and psychiatric disorders pulmonary diseases. *Mailing Add:* 2825 Eastlake E No 333 Seattle WA 98102

DUDLEY, FRANK MAYO, b Umatilla, Fla, Dec 27, 20; m 41; c 2. CHEMICAL EDUCATION. *Educ:* Oglethorpe Univ, AB, 44; Univ Ga, BS, 47; Ohio State Univ, MA, 50, PhD(sci educ), 62. *Prof Exp:* Clin chemist, Downey Hosp, Gainesville, Ga, 46-47 & Theresa-Holland Clin, Leesburg, Fla, 47-48; instr high sch, Ohio, 50-53, head dept sci, 53-55; asst sci educ, Ohio State Univ, 56-57; instr chem, Palm Beach Jr Col, 57-60; eval serv phys sci, Univ SFla, 60-61; clin lab off, US Army Hosp, Fort Polk, La, 61-62; from asst prof to assoc prof chem educ, Univ S Fla, 62-81; RETIRED. *Honors & Awards:* Dept Army Nat Defense Medal, 69. *Mem:* AAAS. *Res:* Improvised equipment in teaching the biological sciences; relationship of preparation and professional status to the decisions of science teachers; selected electrolytes and chromatography methods for colloidal components with various electrolytes; probable sources of ion complexes of geodes at ballast point; improvisation of materials and equipment for teaching chemistry. *Mailing Add:* PO Box 358 Oldsmar FL 34677

DUDLEY, GARY A, b Huntington, WVa, Aug 13, 52. PHYSIOLOGY. *Educ:* Ohio State Univ, MS, 75, PhD(physiol exercise), 78. *Prof Exp:* Postdoctoral res assoc, Health Sci Ctr, State Univ NY, Syracuse, 78-80; from asst prof to assoc prof zool & med sci, Dept Zool & Biomed, Ohio Univ, 80-87; prof & environ muscle physiologist, Bionetics Corp, 87-91; SR RES PHYSIOLOGIST, NASA, 91- *Concurrent Pos:* Vis scholar, Am Col Sports Med, 83. *Honors & Awards:* Pres Award, Nat Strength & Conditioning Asn, 86. *Mem:* Am Physiol Soc; Am Col Sports Med; Nat Strength & Conditioning Asn. *Mailing Add:* Biomed Oper & Res Off NASA Mail Code MDM Kennedy Space Center FL 32899

DUDLEY, HORACE CHESTER, b St Louis, Mo, June 28, 09; m 35, 54; c 4. PHYSICS, RADIOBIOLOGY. *Educ:* Mo State Teachers Col, AB, 31; Georgetown Univ, PhD, 41; Am Bd Health Physics, dipl, 60. *Prof Exp:* Lab asst, US Bur Stand, 31-32; jr chemist, Bur Chem, USDA, 33-34 & Med Res Div, Chem Warfare Serv, 34-36; biochemist, USPHS, 36-42; explosives specialist, US Navy, 42-47, head allied sci sect, Med Serv Corps, 49-52, head biochem div, Naval Med Res Inst, 47-52, head radioisotope lab, Naval Hosp, St Albans, NY, 52-62; prof physics & chmn dept, Univ Southern Miss, 62-69; prof radiation physics & chief physicist, Univ Ill Med Ctr, 69-77; RETIRED. *Concurrent Pos:* Res collabr, Brookhaven Nat Labs, 52-57; consult, Oak Ridge Inst Nuclear Studies, 48-57, Long Island Jewish Hosp, 57-61 & Vet Hosp, Hines, 70-77; Atomic Energy Comn grants, 63-64, NSF grants, 63 & 65; at Radiation Safety Asn, 76-80; mem, Comt Coop Clin Studies, 80-85, med libr staff, Vet Admin Med Ctr, Hines, Ill, 80-85. *Honors & Awards:* Nat prize, Am Chem Soc, 29. *Mem:* Fel AAAS; Health Physics Soc; Am Phys Soc; Am Asn Physics Teachers; Am Asn Physicists in Med. *Res:* Biochemistry; radioactive isotopes; nuclear theory; theory of neutrino flux as generalized sub-quantic medium. *Mailing Add:* 405 W Eighth Pl Hinsdale IL 60521

DUDLEY, J DUANE, b Twin Falls, Idaho, Sept 8, 28; m 50; c 5. PHYSICS. *Educ:* Brigham Young Univ, BS, 52; Rice Inst, MA, 53; Univ Utah, PhD(physics), 59. *Prof Exp:* Instr physics, Idaho State Col, 53-54; physicist, Hughes Aircraft Co, 54-55; instr physics, 56-58, asst prof, 58-63, assoc prof, 63-69, PROF PHYSICS, BRIGHAM YOUNG UNIV, 69- *Concurrent Pos:* Physicist, Sandia Corp, 59-61 & Kaman Nuclear Corp, 65-66. *Mem:* Am Asn Physics Teachers. *Res:* Musical acoustics. *Mailing Add:* Dept Physics Brigham Young Univ Provo UT 84602

DUDLEY, JOHN MINOT, b Boston, Mass, July 17, 26; m 60; c 3. PHYSICS. *Educ:* Mass Inst Technol, SB, 46; Univ Calif, Berkeley, PhD(physics), 60. *Prof Exp:* Instr physics, Pomona Col, 55-56; prin physicist, Aerojet-Gen Nucleonics Div, Gen Tire & Rubber Co, 60-63, staff specialist, 63-64; assoc prof, 64-85, PROF PHYSICS, COLBY COL, 86- *Concurrent Pos:* Consult, Thermonuclear Div, Oak Ridge Nat Lab, 70-71; res assoc, Univ Calif, Berkeley, 77-78 & Univ Col, London, 85-86. *Mem:* Am Phys Soc; Sigma Xi. *Res:* Nuclear physics; history and philosophy of science. *Mailing Add:* Dept Physics Colby Col Waterville ME 04901

DUDLEY, JOHN WESLEY, b Huntsville, Ind, Sept 29, 31; m 51; c 4. PLANT BREEDING, GENETICS. *Educ:* Purdue Univ, BS, 53; Iowa State Univ, MS, 55, PhD(crop breeding), 56. *Prof Exp:* Plant geneticist, Sugar Beet Sect, Agr Res Serv, USDA, 57-59; plant geneticist, Alfalfa Sect, Crop Sci Dept, NC State Univ, 59-65; assoc prof agron, 65-69, PROF AGRON, UNIV ILL, URBANA, 69- *Concurrent Pos:* Ed, Crop Sci, 71. *Mem:* Biomet Soc; fel Am Soc Agron; fel Crop Sci Soc Am; Am Genetics Asn. *Res:* Quantitative genetics; plant breeding methods. *Mailing Add:* 1802 Augusta Dr Champaign IL 61821

DUDLEY, KENNETH HARRISON, b Hagerstown, Md, Nov 12, 37; wid; c 2. PHARMACOLOGY, ORGANIC CHEMISTRY. *Educ:* Elon Col, BS, 59; Univ NC, Chapel Hill, PhD(org chem), 63. *Prof Exp:* Chemist, Chem & Life Sci Lab, Res Triangle Inst, 64-67; asst prof, Ctr Res Pharmacol & Toxicol, 67-73, assoc prof, 73-80, PROF PHARMACOL, SCH MED, UNIV NC, CHAPEL HILL, 80- *Concurrent Pos:* NSF fel, Univ Basel, 63-64. *Mem:* Am Chem Soc; Am Soc Exp Therapeut & Pharmacol. *Res:* Drug metabolism; chemical basis of the penicillin hypersensitivity reaction; organic synthesis; analytical chemistry, drug assays. *Mailing Add:* 5311 Pelham Rd Parkwood Durham NC 27713-2531

DUDLEY, MICHAEL, b Liverpool, Eng, Feb 27, 57; m 85; c 1. X-RAY TOPOGRAPHY, X-RAY CHARACTERIZATION OF MATERIALS. *Educ:* Univ Warwick, Eng, BSc, 78, PhD(eng sci), 82. *Prof Exp:* Postdoctoral, Pure & Appl Dept, Univ Strathclyde, Scotland, 81-84; asst prof, 84-90, ASSOC PROF MAT SCI & ENG, STATE UNIV NY, STONY BROOK, 90- *Concurrent Pos:* Dir, Nat Consortium Synchrotron Topograph, Beamline X-19C, Nat Synchrotron Light Source, Brookhaven Nat Lab, 87- *Mem:* Am Soc Metals; Mat Res Soc; Am Chem Soc; Metall Soc; Sigma Xi; AAAS. *Res:* X-ray topographic characterization of defects in single crystals; synchrotron topography; defects in semiconductor; superconductor and optoelectronic epilayers. *Mailing Add:* Dept Mat Sci & Eng State Univ NY Stony Brook NY 11794-2275

DUDLEY, PATRICIA, b Denver, Colo, May 22, 29. INVERTEBRATE ZOOLOGY. *Educ:* Univ Colo, BA, 51, MA, 53; Univ Wash, PhD(zool), 57. *Prof Exp:* Res assoc zool, Univ Wash, 57-59, actg instr, 59; instr, 59-62, from asst prof to assoc prof, 62-72, prof zool, 73-80, PROF BIOL, BARNARD COL, COLUMBIA UNIV, 80- *Concurrent Pos:* NSF fel, Univ Wash, 65-66; mem, Corp Marine Biol Lab, Woods Hole, Mass; chmn, Dept Biol, Barnard Col, 80-82 & 86-88. *Mem:* Am Soc Zool; Marine Biol Asn UK; Am Micros Soc. *Res:* Systematics, development and histology of copepods symbiotic in marine animals; electron microscopy of Crustacea, Mesozoa. *Mailing Add:* Dept Biol Sci Barnard Col Columbia Univ New York NY 10027

DUDLEY, PETER ANTHONY, b New York, NY, July 27, 38. BIOCHEMISTRY, VISION. *Educ:* Rutgers Univ, BS, 63; Tex A&M Univ, MS, 69; Baylor Col Med, PhD(biochem), 75. *Prof Exp:* Fel retina biochem, Baylor Col Med, 75-77, res assoc, 77-78, res instr, 78-; AT RETINAL & CHOROIDAL DIS BR, NAT EYE INST, NIH. *Mem:* Am Chem Soc; NY Acad Sci; Sigma Xi. *Res:* Visual cell biochemistry. *Mailing Add:* Nat Eye Inst NIH Bldg 31 Rm 48 Bethesda MD 20892

DUDLEY, RICHARD H(ARRISON), b Los Angeles, Calif, Mar 31, 31; m 55; c 2. PHYSICAL METALLURGY. *Educ:* Columbia Univ, BS, 54, MS, 55; Rensselaer Polytech Inst, PhD(phys metall), 62. *Prof Exp:* Res asst welding, Rensselaer Polytech Inst, 58-62; mem tech staff, 62-68, supvr silicon device technol, 68-70, supvr metals group, 70-73, supvr metals & insulators group, 73-76, SUPVR PACKAGING MAT & PROCESSES GROUP, BELL TEL LABS, INC, 76- *Mem:* Inst Elec & Electronics Engrs; Int Soc Hybrid Microelectronics. *Res:* Materials and processes for packaging of silicon integrated circuits. *Mailing Add:* 8425 Little Leaf Ct Orlando FL 32811

DUDLEY, RICHARD MANSFIELD, b E Cleveland, Ohio, July 28, 38; m 78. EMPIRICAL PROCESSES. *Educ:* Harvard Univ, AB, 59; Princeton Univ, PhD(math), 62. *Prof Exp:* From instr to asst prof math, Univ Calif, Berkeley, 62-66; assoc prof, 67-73, PROF MATH, MASS INST TECHNOL, 73- *Concurrent Pos:* A P Sloan fel, 66-68; assoc ed, Ann of Probability, Inst Math Statist, 72-78, ed, 79-81. *Mem:* Am Math Soc; Inst Math Statist; Am Statist Asn; Int Statist Inst; Am Pub Health Assoc; Asn Comput Mach. *Res:* Probability theory; mathematical statistics; analysis; convergence of empirical frequencies to underlying probability laws uniformly over classes of events (sets) or variables (functions); computation of probabilities; other areas of probability, measure theory, and mathematical statistics. *Mailing Add:* Dept Math Rm 2-245 Mass Inst Technol Cambridge MA 02139

DUDLEY, SUSAN D, b Norfolk, Va, April 8, 49; m. BEHAVIORAL ENDOCRINOLOGY, PSYCHOLOGY OF WOMEN. *Educ:* Old Dominion Univ, BS, 73; Col William & Mary, MA, 76; Univ Mass, Amherst, PhD(biopsychol), 80. *Prof Exp:* Instr psychol, Wheaton Col, Mass, 79-80; fel reprod biol, Col Med, M S Hershey Med Ctr, Pa State Univ, 80-82; Univ Md, Europ Div, New York, 82-86, Asian Div, San Francisco, 86-87; PROF PSYCHOL, AUBURN UNIV, MONTGOMERY, 87- *Mem:* Sigma Xi; Soc Neurosci; Am Psychol Asn; AAAS. *Res:* Sexual differentiation and development of central nervous system; reproductive behaviors; interactions of estradiol and insulin in reproduction; feeding and body weight maintenance; attitudes toward rape victims; psychology of women. *Mailing Add:* Dept Psychol Auburn Univ Atlanta Hwy Montgomery AL 36193-0401

DUDLEY, THEODORE R, b Boston, Mass, Dec 31, 36; m 60; c 2. PLANT GERMPLASM & HABITATS, PHYTOGEOGRAPHY. *Educ:* Univ Mass, BS, 58; Univ Edinburgh, PhD(plant taxon), 63. *Prof Exp:* Asst taxonomist, Arnold Arboretum, Harvard Univ, 59-60, hort taxonomist, 63-66; RES BOTANIST, US NAT ARBORETUM, 66- *Concurrent Pos:* Ed, Dioscorides Press. *Mem:* Int Asn Plant Taxonomists; Am Asn Plant Taxon; Am Asn Bot Gardens & Arboretum; Bot Soc Am. *Res:* Generic and specific limits of groups in the Cruciferae; monographic and revisionary studies of the Aquifoliaceae & Caprifoliaceae including cultivated plants; plant exploration in Near East, Turkey, Peruvian Andes, Tierra del Fuego, People's Republic of China, Greece and Republic of Korea, Chile and the southeastern United States. *Mailing Add:* 3501 NY Ave NE US Nat Arboretum Washington DC 20002

DUDLEY, UNDERWOOD, b New York, NY, Jan 6, 37; m 77; c 4. MATHEMATICS, HISTORY OF SCIENCE. *Educ:* Carnegie Inst Technol, BS, 57, MS, 58; Univ Mich, PhD(math), 65. *Prof Exp:* Asst prof math, Ohio State Univ, 65-67; from asst prof, to assoc prof, 67-77, PROF MATH, DEPAUW UNIV, 77- *Mem:* Am Math Soc; Math Asn Am; Soc Indust & Appl Math. *Res:* Mathematical cranks. *Mailing Add:* Dept Math DePauw Univ Greencastle IN 46135-0037

DUDNEY, CHARLES SHERMAN, b Knoxville, Tenn, June 5, 49; m 77; c 2. INDOOR AIR QUALITY. *Educ:* Va Polytech Inst, BS, 72; Mass Inst Technol, PhD(biophysics), 78. *Prof Exp:* Instr, Mass Inst Technol, 78-79; MEM RES STAFF, OAK RIDGE NAT LAB, 79- *Concurrent Pos:* Mem, Fed Ad Hoc Task Force, Indoor Air Pollution; prin investr, Radon Entry & Mitigation; mem, Radon Contractor Proficiency Prog, Environ Protection Agency. *Mem:* Health Physics Soc; AAAS. *Res:* Radon entry in houses and large buildings; impact on human health of environmental pollutants. *Mailing Add:* Health & Safety Res Div Oak Ridge Nat Lab PO Box 2008 Oak Ridge TN 37831-6113

DUDNEY, NANCY JOHNSTON, b Pittsburgh, Pa, Aug 14, 54; m 77; c 2. DEFECT CHEMISTRY, CHARGE & MASS TRANSPORT. *Educ:* Col William & Mary, BS, 75; Mass Inst Technol, PhD(ceramics), 79. *Prof Exp:* E P Wigner, fel, 79-81, RES STAFF, OAK RIDGE NAT LAB, 81- *Concurrent Pos:* Assoc ed, Am Ceramic Soc, 86- *Mem:* Am Ceramics Soc; Am Chem Soc; Sigma Xi; Am Phys Soc; Mat Res Soc. *Res:* Ceramics and solid state chemistry: defect chemistry, nonstoichiometry, electrical conductivity and diffusion; properties of solid electrolytes and mixed ionic-electronic conductors; fabrication and characterization of ceramic materials and thin film materials. *Mailing Add:* Solid State Div Oak Ridge Nat Lab PO Box 2008 Oak Ridge TN 37831-6030

DUDOCK, BERNARD S, b New York, NY, Nov 17, 39; div; c 3. GENE ORGANIZATION, RNA. *Educ:* City Col New York, BS, 61; Pa State Univ, PhD(org chem), 66. *Prof Exp:* NIH fel biochem, Cornell Univ, 66-68; from asst prof to assoc prof, 68-81, chair biochem dept, 81-84, PROF BIOCHEM, STATE UNIV NY STONY BROOK, 81- *Concurrent Pos:* Nat Cancer Inst grant, 68-88; res career develop award, NIH, 73-78. *Mem:* AAAS; Fedn Am Socs Exp Biol. *Res:* Primary structure of nucleic acids; role of nucleic acids in cellular differentiation and development. *Mailing Add:* Dept Biochem State Univ NY Stony Brook NY 11794-5215

DUDRICK, STANLEY JOHN, b Nanticoke, Pa, Apr 9, 35; m 58; c 6. REOPERATIVE SURGERY, TOTAL PARENTERAL NUTRITION. *Educ:* Franklin & Marshall Col, Lancaster, Pa, BS, 57; Univ Pa, Philadelphia, MD, 61. *Hon Degrees:* FPCS, Philippine Col Surgeons, 78. *Prof Exp:* Chief surg, Philadelphia Vet Hosp, 68-72; chief surg serv, Hermann Hosp, Houston, Tex, 72-80; prof surg, 72-82, chmn dept, 72-80, CLIN PROF SURG, SCH MED, UNIV TEX, HOUSTON, 82-; CONSULT MED & SURG, TEX INST REHAB & RES, HOUSTON, 74-; SURGEON-IN-CHIEF, HERMANN HOSP, HOUSTON, TEX, 90- *Concurrent Pos:* Asst attend physician, Philadelphia Gen Hosp, 67-72; assoc surgeon, Univ Pa Hosp, Philadelphia, 67-72; attend surgeon, Hermann Hosp, Houston, Tex, 72-, Tex Children's Hosp, 81-88 & St Luke's Episcopal Hosp, 81-88; consult surg, Univ Tex Syst Cancer Ctr & M D Anderson Hosp & Tumor Inst, 72-88, Off Pres Gen Surg, 82-88; dir, Nutrit Support Serv, St Lukes Episcopal Hosp, Houston, Tex, 81-88; chmn, Dept Surg, Pa Hosp, Philadelphia, 88-90. *Honors & Awards:* Joseph B Goldberger Award Clin Nutrit, AMA, 70; Brookdale Award Med, AMA, 75; Frank Stinchfield Award, Hip Soc Am Acad Orthopaedic Surgeons, 81; Harry M Vars Award, Am Soc Parenteral & Enteral Nutrit, 82; Grace A Goldsmith Award, Am Col Nutrit, 82; Albion O Bernstein MD Award, Med Soc NY, 86; Ladd Medal, Am Acad Pediat, 88. *Mem:* Am Col Surgeons; Am Soc Clin Invest; Am Surg Asn; Am Soc Parenteral & Enteral Nutrit (pres, 77-78); Am Soc Nutrit Support Serv (pres, 84-85); Soc Univ Surgeons; Sigma Xi. *Res:* Reoperative general surgery, inflammatory bowel disease and the management of critically ill and complex surgical patients; development of nutrient regimens specifically tailored for optimal support during various pathophysiologic conditions, particularly metabolic techniques for the arrest, reversal and prevention of atherosclerosis. *Mailing Add:* 6411 Fannin Houston TX 77030

DUDUKOVIC, MILORAD, b Belgrade, Yugoslavia, Mar 25, 44; m 69; c 2. CHEMICAL ENGINEERING, APPLIED MATHEMATICS. *Educ:* Univ Belgrade, dipl chem eng, 67; Ill Inst Technol, MS, 70, PhD(chem eng), 72. *Prof Exp:* Design engr, Inst Process Design, Belgrade, 67-68; instr, Ill Inst Technol, 70-72; asst prof, Ohio Univ, 72-74; assoc prof, 74-81, PROF CHEM ENG, WASH UNIV, 81-, DIR CHEM REACTION ENG LAB, 79- *Concurrent Pos:* Consult, Ill Environ Protection Agency, 71, Monsanto, 77-83, Exxon, 84-86, Ethyl, 85-87 & other chem co, 86-; dir, Monsanto Prof Prog Develop, 77-83. *Honors & Awards:* Cert Recognition, NASA, 87; Nat Catalyst Award, Chem Mfrs Asn, 88. *Mem:* Am Inst Chem Engrs; Am Chem Soc; Am Soc Eng Educ; Sigma Xi; AAAS; Am Asn Crystal Growth. *Res:* Chemical reaction engineering; kinetics and diffusion in gas-solid noncatalytic reactions; modelling and tracer studies of multiphase reactors of the trickle-bed, bubble columns and gas lift slurry type; reaction engineering of semiconductor materials; reaction engineering of composites; crystal growth. *Mailing Add:* Dept Chem Eng Wash Univ Box 1198 St Louis MO 63130

DUDZIAK, DONALD JOHN, b Alden, NY, Jan 6, 35; m 59; c 3. NUCLEONICS. *Educ:* US Merchant Marine Acad, BS, 56; Univ Rochester, MS, 57; Univ Pittsburgh, PhD(appl math), 63. *Prof Exp:* From assoc engr to sr engr, Bettis Atomic Power Lab, Westinghouse Elec Corp, 57-65; staff mem, 65-74, alt group leader transport & reactor theory, 75-78, sect leader Ctr Neutronics, theoret div, 74-79, group leader transport & reactor theory, 79-82, SECT LEADER HIGH TECHNOL SYSTS STUDIES, LOS ALAMOS SCI LAB, 82- *Concurrent Pos:* Adj prof, Univ NMex, 66; vis prof, Univ Va, 68-69; mem, Nat Cross Sect Eval Working Group; mem, US Nuclear Data Comt, 73-; guest scientist, Swiss Fed Inst Reactor Res, 81-82; fel, Los Alamos Lab, 88. *Mem:* Fel Am Nuclear Soc; Soc Indust & Appl Math; Health Physics Soc. *Res:* Radiation shielding analysis and nuclear safety evaluation; application of stochastic process theory to nuclear reactor kinetics; nuclear reactor analysis for development of design models; radiological physics and radiation biology; fusion reactor nuclear analysis and methods development; inertial confinement fusion systems studies; advanced technology systems studies. *Mailing Add:* MS K-575 PO Box 1663 Los Alamos Nat Lab Los Alamos NM 87545

DUDZIAK, WALTER FRANCIS, b Adams, Mass, Jan 7, 23; m 54; c 4. COMPUTER SCIENCE, NUCLEAR SCIENCE. *Educ:* Rensselaer Polytech, BS & MS, 48; Univ Calif, Berkeley, PhD(math, physics), 54. *Prof Exp:* Engr, Manhattan Dist Proj, Oak Ridge Nat Lab, 44-46; instr physics, Rensselaer Polytech, 47-48; aeronaut res scientist, Nat Adv Comt Aeronaut, Ohio Univ, 48-49; mem res staff, Lawrence Radiation Lab, Univ Calif, 49-58; tech mil planning oper, Gen Elec Co, 58-60, mgr comput sci oper, 60-64, tech mil planning oper, 64-65; dir info sci inst & exec vpres res & develop, 65-68, CONSULT & MEM BD, PAN-FAX, INC, 65-; PRES, INFO SCI, INC, 68- *Concurrent Pos:* Lectr, Univ Calif, Santa Barbara, 58-59; consult nuclear weapons simulation, Shock Physics. *Mem:* Am Phys Soc; Am Asn Comput Mach; Am Geophys Union; Sigma Xi. *Res:* Leads; application of computer sciences to various scientific disciplines; effects of nuclear detonations; propagation of electromagnetic radiation; optical scanning; digitizing techniques as applied to rapid data transmission; electro-optic systems. *Mailing Add:* Info Sci Inc 123 W Padre Santa Barbara CA 93105

DUDZINSKI, DIANE MARIE, b Erie, Pa, July 23, 46. MOLECULAR BIOLOGY. *Educ:* Villa Maria Col, BS, 68; Fordham Univ, MS, 70, PhD(phycol, limnol, biol oceanog, microbiol), 74. *Prof Exp:* Lab asst limnol & cytol, Fordham Univ, 69-72, teaching fel environ sci, gen biol & invertebrate zool, 72-74; asst prof biol, Manhattan Col, 72-78; from assoc prof to prof biol, Col Santa Fe, 78-86, chmn dept sci & math, 80-85; ASST PROF MICROBIOL, MOLECULAR & CELL BIOL, IMMUNOL, GENETICS, HUMAN BIOL & INTRODUCTORY BIOL, MERCYHURST COL, 86- *Concurrent Pos:* Instr microbiol, Pace Univ, 72-78; mem US-USSR joint oceanog exped to Bering Sea, US Dept Interior, Fish & Wildlife Serv, 77; consult, Enviro-Med Labs, La, 79-80; grants, NSF ISEP, 76-78 & 79-80, NSF URP, 77, NIH Minority Biomed Res Prog, 80-81, Sigma Xi, 87 & 90-91; NASA-Am Soc Eng Educ fel, Ames Res Ctr, Moffetfield Calif, 82-85; adv & coodr, BSN Prog Nursing, Mercyhurst Col; tech adv, Pa Dept Eviron Resources, Presque Isle State Park, 88-; adj prof microbiol, Villa Maria Col, Gannon Univ, Pa, 89- *Mem:* AAAS; Sigma Xi; Am Soc Microbiol; Ecol Soc Am; Am Inst Biol Sci; Nat Asn Biol Teachers. *Res:* Morphology, physiology and nutrition of the marine symbiotic alga, Symbodinium microadriaticum; bacterial enumeration of coliforms and pathogens at selected sample sites in the New York harbor; behavior genetics in inbred mouse strains; microbial populations in controlled environment life support systems for man in a space environment; controlled life support systems involving higher plants and human microbial interactions; space station biohabitat design; author or co-author of nine publications. *Mailing Add:* Biol Dept Mercyhurst Col Glenwood Hills Erie PA 16546

DUECK, JOHN, b Altona, Man, Aug 11, 41; m 62; c 3. PLANT PATHOLOGY. *Educ:* Univ Man, BSA, 64; Univ Minn, MSc, 66, PhD(plant path), 71. *Prof Exp:* Agronomist, Soils & Crops Br, Man Dept Agr, 66-68; pathologist bact dis, Agr Can Res Sta, Harrow, Ont, 71-73, pathologist plant quarantine, Ottawa, 73-74, plant pathologist, Saskatoon, 74-81, dir, Regina, Sask, 81-89; dep proj dir, Barani Agr Res & Develop Proj, Islamabad, Pakistan, 85-87; DIR, AGR CAN RES STA, SUMMERLAND, BC, 89- *Mem:* Can Phytopath Soc; Agr Inst Can; Am Phytopath Soc. *Res:* Diseases of rapeseed and sunflower with emphasis on biology and control of Sclerotinia sclerotiorum. *Mailing Add:* Agr Can Res Sta Summerland BC V2A 8C6 Can

DUECKER, HEYMAN CLARKE, b Seville, Ohio, July 18, 29; m 49; c 4. INORGANIC CHEMISTRY. *Educ:* Marion Col, BS, 50; Univ Toledo, MS, 56, Univ Md, PhD, 64. *Prof Exp:* Res chemist, Radio Corp Am, 51-54; res chemist, Sci Lab, Ford Motor Co, 55-58; phys chemist, Nat Bur Standards, 58-66; inorg chemist, W R Grace & Co, 66-67, mgr inorg res, 67-69, dir inorg res, 69-71, VPRES RES, CONSTRUCT PROD DIV, W R GRACE & CO, 71- *Mem:* AAAS; Am Chem Soc; Am Ceramic Soc. *Res:* Inorganic oxides, preparation, properties and identification; silicates, zeolites; catalytic materials and processes; surface and colloid chemistry; inorganic construction materials. *Mailing Add:* W R Grace & Co 62 Whittemore Ave Cambridge MA 02140

DUEDALL, IVER WARREN, b Albany, Ore, Jan 16, 39; m 61; c 2. OCEANOGRAPHY, PHYSICAL CHEMISTRY,. *Educ:* Ore State Univ, BS, 63, MS, 66; Dalhousie Univ, PhD(chem oceanog), 73. *Prof Exp:* Scientist chem oceanog, Marine Ecol Lab, Bedford Inst Oceanog, Dartmouth, NS, 66-72; from asst prof to assoc prof chem oceanog, Marine Sci Res Ctr, State Univ NY, Stony Brook, 77-82; PROF & HEAD DEPT OCEANOG/OCEAN ENG, FLA INST TECHNOL, 82- *Mem:* AAAS; Am Geophys Union; Sigma Xi; Am Soc Limnol & Oceanog. *Res:* Physical chemistry of electrolyte systems; partial molal properties; coastal and estuary processes; chemical inputs to the ocean from rivers; deep sea distribution of properties; nutrient dynamics; man's impact on the ocean; international programs in marine sciences; wastes in the ocean; waste utilization; artificial reefs; marine policy & pollution. *Mailing Add:* Dept Oceanog/Ocean Eng Fla Inst Technol Melbourne FL 32901-6988

DUEK, EDUARDO ENRIQUE, b Buenos Aires, Arg, Jan 21, 49; m; c 2. COMPUTER SIMULATION CALCULATIONS, DETECTOR DESIGN SIMULATIONS. *Educ:* Univ Buenos Aires, MS, 76; State Univ NY, Stony Brook, PhD(chem physics), 83. *Prof Exp:* Res assoc, Inst Nuclear Physics, Orsay, France & State Univ NY, 83-84; res assoc, 84-86, physics asst, 86-88, PHYSICS ASSOC, BROOKHAVEN NAT LAB, 88- *Concurrent Pos:* Consult, Nat Comn Atomic Energy, Arg, 81 & 87; vpres, Ctr Int Leadership, 89-; vpres phys sci & eng, Arg NAm Asn Advan Sci, Technol & Culture, 90-; secy, US-Latin Am Task Force Particle Physics, 90-; mem, Int Adv Bd, First & Second Symp US-Latin Am SSC Physics, Exp & Technol, 90- *Mem:* Am Phys Soc. *Res:* Computer simulation calculations for experiment design and analysis; software integration for large off-line analysis codes; software development; high-tech developments in Latin America. *Mailing Add:* Physics Dept Brookhaven Nat Lab Upton NY 11973

DUEKER, DAVID KENNETH, b Kansas City, Kans, Nov 4, 44. OPHTHALMOLOGY. *Educ:* Pomona Col, BA, 66; Yale Univ, MD, 70. *Prof Exp:* Intern, Pac Med Ctr, 70-71; resident ophthal, Yale Univ, New Haven Hosp, 71-74; fel glaucoma, Harvard Med Sch, 74-76; clin fel, Mass Eye & Ear Infirmary, 74-76, asst dir glaucoma serv & asst ophthal, 76-; AT DEPT OPHTHAL, UNIV MO MED CTR, COLUMBIA. *Concurrent Pos:* Prin investr, NIH grant, 77-; instr ophthal, Harvard Med Sch, 76-79, asst prof, 80- *Mem:* AAAS; Am Acad Ophthal & Otolaryngol; Asn Res Vision Ophthal. *Res:* Glaucoma; neovascular glaucoma; angiogenesis in the eye; fluorescein angiography. *Mailing Add:* Dept Ophthal Univ Mo Columbia MO 65212

DUELAND, RUDOLF (TASS), b Staten Island, NY, Feb 16, 33. VETERINARY MEDICINE. *Educ:* Cornell Univ, DVM, 56; Univ Minn & Mayo Grad Sch Med, MS, 70. *Prof Exp:* Assoc prof & dir small animal clin, Orthop Surg, Western Col Vet Med, Univ Saskatchewan, 71-72; guest lectr surg, Col Engr, Cornell Univ, 72-80; vis scientist, Univ Washington & Swiss Res Inst, Switz, 78-79; assoc prof, NY State Col Vet Med, Cornell Univ, 72-80; PROF & CHMN, DEPT SURG SCI, SCH VET MED, UNIV WIS, MADISON, 80- *Concurrent Pos:* Surgeon & clinician, Henry Bergh Mem Hosp, NY, 59-60; vet, NY City Dept Health, 61-67; dir, Dueland Animal Clin, Inc, Staten Island, 60-73; vis scientist, Biomech Lab, Dept Orthop, Mayo Clin, 88-89. *Mem:* Am Vet Med Asn; Am Animal Hosp Asn; Vet Orthop Soc (vpres 75, pres, 77); Orthop Res Soc; Bioelect Repair & Growth Soc. *Res:* Research in vascular and morphological effects of internal fixation of bone; biomechanical and vascular considerations of various cruciate repairs; limb salvage prosthesis. *Mailing Add:* Dept Surg Sci/Sch Vet Med Univ Wis Madison WI 53706

DUELL, ELIZABETH ANN, b Dayton, Ohio, Jan 9, 36. BIOCHEMISTRY, MOLECULAR BIOLOGY. *Educ:* Univ Dayton, BS, 58; Western Reserve Univ, PhD(biochem), 67. *Prof Exp:* Instr, 70-72, ASST PROF BIOCHEM & DERMAT, UNIV MICH, ANN ARBOR, 72- *Honors & Awards:* Taub Mem Award Psoriasis Res, 73. *Mem:* AAAS; Soc Invest Dermat; NY Acad Sci; Am Fedn Clin Res; Sigma Xi. *Res:* Role of retinoids ecosanoids and growth factors in control of cell proliferation and differentiation with special emphasis on the epidermis. *Mailing Add:* Dept Dermat Univ Mich Med Ctr R 6558 Kresge Med Res Bldg Ann Arbor MI 48109-0528

DUELL, PAUL M, b Goodland, Kans, Dec 11, 24; m 48; c 4. INORGANIC CHEMISTRY. *Educ:* Ft Hays Kans State Col, BA, 50, MS, 51; Kans State Univ, PhD(chem), 58. *Prof Exp:* Instr phys sci, Garden City Jr Col, 51-52; instr chem, Washburn Univ, 52-53 & Kans State Univ, 53-57; from asst prof to prof chem & chmn dept, 57-73, dean col lib arts, 73-77, PROF CHEM, WILLAMETTE UNIV, 77- *Mem:* AAAS; Am Chem Soc. *Res:* Physical properties of mixed aqueous salt solutions, including magnetic susceptibilities, conductivities, partial molal volumes; physical properties of binary aqueous salt solutions; ligand field effects on complex ions in aqueous and nonaqueous solutions. *Mailing Add:* 3322 Rawlins Ave Salem OR 97303

DUELLMAN, WILLIAM EDWARD, b Dayton, Ohio, Sept 6, 30; m 53; c 3. HERPETOLOGY. *Educ:* Univ Mich, BS, 52, MS, 53, PhD, 56. *Prof Exp:* Asst herpet, Mus Zool, Univ Mich, 52-56, teaching fel zool, 54-56; instr, Dept Biol, Wayne State Univ, 56-59; from asst prof to assoc prof zool, 59-68, prof, 68-70, asst curator herpet, 59-63, assoc curator, 63-70, PROF SYSTS & ECOL, UNIV KANS, 70-, CURATOR, MUS NATURAL HIST, 70- *Mem:* Am Soc Ichthyol & Herpet; Soc Study Evolution; Soc Syst Zool; Ecol Soc Am; Brit Herpet Soc; Herpetologists' League. *Res:* Systematics; zoogeographic patterns; evolution; zoogeography of New World herpetofauna; evolutionary biology of reptiles and amphibians. *Mailing Add:* Mus Nat Hist Univ Kans Lawrence KS 66045

DUELTGEN, RONALD REX, b Salem, Ore, Sept 21, 40; m 62; c 2. INFORMATION SCIENCE. *Educ:* Ore State Univ, BA, 62; Univ Mich, PhD(org chem), 67. *Prof Exp:* Res investr, 67-72, supvr chem doc, 72-74, mgr sci info, G D Searle & Co, 74-77; mgr, Tech Info Serv, 77-81, dir, Info & Admin Serv, 81-83, dir, Info Serv, Henkel Corp, 83-85; CURRENT AWARENESS INFO SPECIALIST, 3M CO, 85- *Mem:* Am Chem Soc; Sigma Xi; Am Soc Info Sci. *Res:* Synthetic organic chemistry; information transfer; computer manipulation of scientific and technical information. *Mailing Add:* 3M Co 201-25-00 3M Ctr St Paul MN 55144-1000

DUERR, FREDERICK G, b St Paul, Minn, May 17, 35; m 54; c 3. ZOOLOGY, COMPARATIVE PHYSIOLOGY. *Educ:* Univ Minn, BA, 56, PhD(zool), 65. *Prof Exp:* Assoc prof zool, Univ SDak, 61-67; assoc prof biol, Univ Sask, Regina, 67-68; assoc prof, 68-77, ADJ ASSOC PROF BIOL, UNIV NDAK, 77- *Mem:* AAAS; Am Soc Zool; Am Soc Parasitol. *Res:* Comparative physiology of digenetic trematode parasitism; physiology of respiration and nitrogen excretion. *Mailing Add:* Box 5387 Grand Forks ND 58206

DUERR, J STEPHEN, b Erie, Pa, Apr 8, 43; m 64; c 3. ELECTRON MICROSCOPY, MATERIALS FAILURE ANALYSIS. *Educ:* Mass Inst Technol, BS, 65, MS, 67, PhD(metall), 71. *Prof Exp:* Metallurgist, Battelle Mem Inst, 65-66; sr metallurgist, Bettis Atomic Power Lab, 71-74; dir anal serv, PhotoMetrics Inc, 74-77; tech dir, Struct Probe, Inc, 77-90; PRES, METUCHEN ANALYSIS, INC, 78- *Concurrent Pos:* Course dir, Ctr Prof Advan, 80-; adj prof mat & metall eng, Stevens Inst Technol, Hoboken, NJ, 84- *Mem:* Am Soc Metals Int; Microbeam Anal Soc; Am Soc Testing & Mat; Nat Soc Prof Engrs; Sigma Xi; Soc Plastics Engrs; Asn Consult Chemists & Chem Engrs. *Res:* Failure analysis and problem solving investigations in most materials systems including metals; microelectronics devices, plastics and adhesives; claims substantiation in advertising, litigation and patent development; electron microscopy/microanalysis, surface analysis and chemical analysis studies. *Mailing Add:* Nine Scenic Falls Rd Long Valley NJ 07853

DUERRE, JOHN A, b Webster, SDak, Aug 21, 30; m 57; c 3. MICROBIAL BIOCHEMISTRY, MOLECULAR BIOLOGY. *Educ:* SDak State Univ, BS, 52, MS, 56; Univ Minn, PhD(microbiol & biochem), 60. *Prof Exp:* Res bacteriologist, Rocky Mt Lab, Nat Inst Allergy & Infectious Dis, 61-63; from asst prof to assoc prof, 63-71, PROF MICROBIOL, SCH MED, UNIV NDAK, 71- *Concurrent Pos:* AEC fel, Argonne Nat Lab, 60-61; NSF reg grant, 63-71; NIH career develop award, 65-75, NIH res grant, 71-84; vis scientist, Neuropsychiat Res Unit, Med Res Coun Labs, Eng, 69-70, Walter Reed Army Inst Res, Wash, DC, 84-85; NSF res grant, 87-; USDA grant, 83-84. *Honors & Awards:* Outstanding Res Award, Sigma Xi, 77. *Mem:* Am Soc Microbiol; Am Soc Biochem & Molecular Biol; Sigma Xi. *Res:* Methylation of chromosomal proteins; relationship of methylation to cell differentiation. *Mailing Add:* Dept Microbiol Univ NDak Grand Forks ND 58202

DUERSCH, RALPH R, b Berlin, Ger, Feb 28, 26; m 57; c 2. DECISION SUPPORT SYSTEMS, SIMULATION. *Educ:* Newark Col Eng, BS, 57; Rensselaer Polytech Inst, MS, 65; Union Col, MS, 72, PhD(eng & admin), 73. *Prof Exp:* Jr engr, Kearfott Co, Paterson, NJ, 56; elec engr, Switchgear & Control Div, 57, Light Military Electronics Dept, 57-58, Heavy military Electronics Dept, 58-59, Light Military Electronics Dept, 59-60, syst anal engr, Advan Technol Lab, 60-70, INFO SYST ENGR CORP RES & DEVELOP, GEN ELEC CO, 70- *Concurrent Pos:* Adj assoc prof, Inst Admin & Mgt, Union Col, 75- *Mem:* Inst Elec & Electronics Engrs. *Res:* Improved techniques of decision support for management, factory control and scheduling using methods from artificial intelligence, optimization, simulation and computer science. *Mailing Add:* 19 Wispering Sands Dr No 1005 Sarasota FL 34242

DUERST, RICHARD WILLIAM, b Rice Lake, Wis, Aug 18, 40; m 64; c 6. PHYSICAL CHEMISTRY, ANALYTICAL CHEMISTRY. *Educ:* St Olaf Col, BA, 62; Univ Calif, Berkeley, PhD(chem), 66. *Prof Exp:* Fel phys chem, Lawrence Radiation Lab, Univ Calif, 66-67; Nat Res Coun-Nat Bur Stand res fel, Nat Bur Stand, Md, 67-69; from asst prof to assoc prof chem, 69-80; ANALYTICAL CHEMIST, 3M APRL, ST PAUL, MINN, 81- *Concurrent Pos:* NSF vis prof & res assoc, Harvard Univ, 74-76 & 78-80. *Mem:* Am Chem Soc; Royal Soc Chem; Am Phys Soc; Sigma Xi; Soc Appl Spectres. *Res:* Infrared, microwave, nmr, esr and Raman analysis characterization of hydrogen-bonded systems by spectroscopy; hydration of ionic species in aqueous solution; effect of isotopic substitution on physical properties; thin films. *Mailing Add:* 201 BS 05 3M APRL St Paul MN 55144

DUESBERG, PETER H, b Muninster, WGer, Dec 2, 36; c 3. TRANSFORMING GENES. *Educ:* Univ Frankfurt, PhD(chem), 63. *Prof Exp:* Postdoc fel, Max Planck Inst Virus Res, 63; asst res virologist, Virus Lab, Univ Calif, Berkeley, 64-66, postdoctoral fel, 66-68, asst resident prof & asst res biochemists, Dept Molecular Biol, 68-70, from asst prof to prof, 70-89, PROF, DEPT MOLECULAR & CELL BIOL, DIV BIOCHEM & MOLECULAR BIOL, UNIV CALIF, BERKELEY, 89- *Concurrent Pos:* Outstanding investr award, NIH, 86 & Fogarty scholar-in- residence, 86-87. *Honors & Awards:* Merck Award, 69; First Ann Am Med Ctr Oncol Award, 81; Litchfield Lectr, 88. *Mem:* Nat Acad Sci; Leukemia Soc Am. *Res:* Author of over 180 publications. *Mailing Add:* Dept Molecular & Cell Biol Univ Calif 229 Stanley Hall Berkeley CA 94720

DUESBERY, MICHAEL SERGE, b Hull, UK, May 20, 42; m 64; c 4. SOLID STATE PHYSICS, MATERIALS SCIENCE. *Educ:* Cambridge Univ, BA, 64, PhD(physics), 67. *Prof Exp:* Res fel metals, Dept Metall, Oxford Univ, 66-68; res officer defects, Div Physics, Nat Res Coun Can, Ottawa, 68-87; CONTRACT PHYSICIST NAVAL RES LAB, WASHINGTON LAB, 87- *Concurrent Pos:* Vis prof, George Washington Univ, 86-87. *Honors & Awards:* Canadian Metal Physics Medal, 86. *Mem:* Am Phys Soc; Metall Soc; Mat Res Soc. *Res:* Defects, particularly dislocations, in materials. *Mailing Add:* Naval Res Lab Washington DC 20350-5000

DUESER, RAYMOND D, b Ft Worth, Tex, Nov 16, 45; m 84. FISH & WILDLIFE SCIENCES. *Educ:* Univ Tex Austin, BA, 67, MS, 70; Univ Mich, PhD(wildlife & ecol), 75. *Prof Exp:* From asst prof to assoc prof environ sci, Univ Va, 74-90; PROF & DEPT HEAD FISHERIES & WILDLIFE, UTAH STATE UNIV, 90- *Concurrent Pos:* Prin investr, Va Coast Reserve Long-Term Ecol Res Prog, Univ Va, 87-90. *Honors & Awards:* Pres Stewardship Award, Nature Conservancy, 90. *Mem:* Am Soc Mammalogists; Am Soc Naturalists; Brit Ecol Soc; Ecol Soc Am; Soc Conserv Biol; Wildlife Soc. *Res:* Ecology and biogeography of mammals, the management and conservation of wildlife, and the restoration of disturbed ecosystems. *Mailing Add:* Fisheries & Wildlife Dept Utah State Univ Logan UT 84322-5210

DUESTERHOEFT, WILLIAM CHARLES, JR, b Austin, Tex, Dec 10, 21; m 49; c 1. ELECTRICAL ENGINEERING. *Educ:* Univ Tex, BS, 43, MS, 49; Calif Inst Technol, PhD(elec eng), 53. *Prof Exp:* Elec engr, Gen Elec Co, 43-46; from instr to asst prof elec eng, Univ Tex, 46-49; instr, Calif Inst Technol, 49-52; aerophysics engr, Gen Dynamics/Convair, 52-54; assoc prof, 54-60, PROF ELEC ENG, UNIV TEX, AUSTIN, 60- *Concurrent Pos:* Corp consult, 54- *Mem:* Inst Elec & Electronics Engrs; Am Soc Eng Educ; Am Phys Soc. *Res:* Plasma dynamics; geophysics; power and energy conversion systems. *Mailing Add:* Dept Elec Eng ENS 143 Univ Tex Austin TX 78712

DUEVER, THOMAS ALBERT, b Kassel, Ger, Aug 26, 58; Can citizen. APPLIED STATISTICS IN CHEMICAL ENGINEERING. *Educ:* Univ Waterloo, BASc, 82, MASc, 83, PhD(chem eng), 87. *Prof Exp:* Res engr, Abitibi-Price Inc, 87-90; ASST PROF CHEM ENG, UNIV WATERLOO, 90- *Mem:* Chem Inst Can; Can Pulp & Paper Asn; Tech Asn Pulp & Paper Indust. *Res:* Application of statistical methods in chemical engineering; modelling of polymerization reactions; quality improvement for pulping processes; environmental risk assessment. *Mailing Add:* Dept Chem Eng Univ Waterloo Waterloo ON N2L 3G1 Can

DUEWER, ELIZABETH ANN, b Champaign, Ill, June 21, 37; m 63; c 3. LICHENOLOGY. *Educ:* Univ Ill, BS, 60, MS, 62; Univ Ariz, PhD(bot), 71. *Prof Exp:* Teaching asst bot, Univ Ill, 60-62 & Duke Univ, 63; from teaching asst to teaching assoc, Univ Ariz, 63-69; consult floricult, Cushman Nurseries, 75-76 & 78; consult, 79 & 81-84, TEACHER, UNIV WIS-PLATTEVILLE, 85- *Mem:* Am Bryol & Lichenological Soc; Brit Lichen Soc; AAAS. *Res:* The taxonomy and morphology of the lichen genus Acarospora and its phycobiont Trebouxia. *Mailing Add:* 1975 Old Lancaster Rd Platteville WI 53818

DUEWER, RAYMOND GEORGE, b Auburn, Ill, Mar 28, 26; m 63; c 3. HORTICULTURE, AGRICULTURE. *Educ:* Univ Ill, BS, 61, MS, 62; Univ Ariz, PhD(hort), 69. *Prof Exp:* From asst prof to assoc prof, 69-81, PROF HORT, UNIV WIS, 81- *Mem:* Am Asn Hort Sci; Am Hort Soc; Am Soc Hort Sci. *Res:* Microsporogenesis in diploid, triploid, and tetraploid Cucumis melo L on the cytological level; breeding strawberries; germination study of lettuce; embryo culture in muskmelons. *Mailing Add:* 1975 Old Lancaster Rd Platteville WI 53818

DUFAU, MARIA LUISA, b Argentina, Sept 22, 38; US citizen; m 69; c 1. ENDOCRINOLOGY, BIOCHEMISTRY. *Educ:* Nat Univ Cuyo, MD, 62, PhD(med sci), 68. *Prof Exp:* Res assoc, Clin Res Inst Montreal, 67-68; lectr, Dept Med, Monash Univ, Australia, 68-70; sr investr endocrinol, 70-78, CHIEF SECT MOLECULAR ENDOCRINOL, ENDOCRINOL & REPROD RES BR, NAT INST CHILD HEALTH & HUMAN DEVELOP, NIH, 78- *Concurrent Pos:* Coun mem, Am Endocrine Soc, 84-88; USA del, Int Soc Endocrinol, 84-; mem, Sr Exec Serv, 90- *Honors & Awards:* Serono Award, Am Soc Andrology, 83. *Mem:* Am Endocrine Soc; Am Soc Clin Res; Am Soc Clin Invest; Am Soc Biol Chem; Asn Am Physicians. *Res:* Molecular basis of peptide hormone action with particular emphasis on control of gonadal function; molecular and gene structure, regulation, and function of gonado tropin and lactogen receptors; hormonally regulated membrane-coupling functions and intracellular events involved in modulating steroid biosynthesis in testis and ovary; cell-to-cell communication in the testis, developmental aspects of Leydig cell maturation, and induction of regulatory mechanisms in the Leydig cell and their influence on other testicular compartments; biochemical properties of luteinizing hormone in relationship to biological activity and the biological activity of stored and circulating gonadotropins, and its relevance to physiologic regulation and clinical disorders of pituitary and gonadal function. *Mailing Add:* 5673 Bent Branch Rd Bethesda MD 20816

DUFF, DALE THOMAS, b Ross Co, Ohio, Dec 26, 30; m 58; c 1. CROP PHYSIOLOGY. *Educ:* Ohio State Univ, BS, 57, MS, 64; Mich State Univ, PhD(crop sci), 67. *Prof Exp:* Trainee, Soil Conserv Serv, USDA, 55-57; agronomist, Clinton County Farm Bur Coop Asn, Ohio, 58-61; asst instr agron, Ohio State Univ, 61-64; teaching asst crop sci, Mich State Univ, 64-67; asst prof, 67-75, ASSOC PROF PLANT & SOIL SCI, UNIV RI, 75- *Mem:* AAAS; Am Soc Agron; Crop Sci Soc Am; Sigma Xi; Int Turfgrass Soc. *Res:* Stress physiology of turf grasses, especially high temperature and low temperature effects upon growth. *Mailing Add:* Dept Agron Univ RI Kingston RI 02881

DUFF, FRATIS L, medicine; deceased, see previous edition for last biography

DUFF, GEORGE FRANCIS DENTON, b Toronto, Ont, July 28, 26; div; c 5. MATHEMATICAL MODELING, NAVIERSTOKES EQUATIONS. *Educ:* Univ Toronto, BA, 48, MA, 49; Princeton Univ, PhD(math), 51. *Prof Exp:* Moore instr, Mass Inst Technol, 51-52; from asst prof to assoc prof, 52-61, chmn dept, 68-75, PROF MATH, UNIV TORONTO, 61- *Concurrent Pos:* Ed, Can J Math, 57-61 & 78-81; fel Royal Soc Can, 59. *Mem:* Am Math Soc; Can Math Soc (pres, 71-73). *Res:* Differential equations. *Mailing Add:* Dept Math Univ Toronto Toronto ON M5S 1A1 Can

DUFF, IVAN FRANCIS, b Pendleton, Ore, July 20, 15; c 2. INTERNAL MEDICINE. *Educ:* Univ Ore, AB, 38; Univ Mich, MD, 40. *Prof Exp:* Intern, Univ Hosp, 40-41, from asst resident to resident internal med, 41-46, asst physician, 48-53, from instr to assoc prof, 46-60, prof internal med charge, Rockham Arthritis Res Unit, Arthritis Div, 60-69, prof internal med, Charge Arthritis Div, 69-76, actg chief, Div Geriat Med, Dept Internal Med, 80-86, emer prof internal med, Dept Internal Med Univ Hosp, 86; RETIRED. *Concurrent Pos:* Consult, Radiation Effects Res Found, Hiroshima & Nagasaki, Japan, 64-86 Peking Union Med Col, Chinese Acad Med Sci, Beijing, China, 84; dir, Regional Arthritis Control Prog, Mich, 69-71. *Mem:* Am Fedn Clin Res; Am Rheumatism Asn; fel Am Col Physicians. *Res:* Rheumatic and thromboembolic diseases. *Mailing Add:* Dept Internal Med Univ Mich Ann Arbor MI 48109

DUFF, J(ACK) E(RROL), b Baltimore, Md, May 17, 18; m 47. ELECTRONICS. *Educ:* Case Inst Technol, BS, 39. *Prof Exp:* Jr engr, 39-41, sr engr, 41-45, dir elec sect, 45-54, res coordr, 54-55, DIR RES, HOOVER CO, 55- *Mem:* Inst Elec & Electronics Engrs. *Res:* Small series motor design; industrial electronic control and inspection equipment design; analysis of multivibrator circuits; electrostatic heating-thermal control; frequency multiplier. *Mailing Add:* 645 Glenwood SW North Canton OH 44720

DUFF, JAMES MCCONNELL, b Aviemore, Scotland, Sept 24, 40; Can citizen; m 63; c 4. ORGANIC CHEMISTRY, ORGANOMETALLIC CHEMISTRY. *Educ:* Univ Toronto, BSc, 65, PhD(org chem), 70. *Prof Exp:* Fel inorg chem, Univ Leeds, 70-72; res assoc org chem, Univ Toronto, 72-74; mem sci staff pigments, 74-79, group leader inks, 78-80, mgr polymer synthesis, 80-85, PRIN SCIENTIST, XEROX RES CTR CAN, 86- *Mem:* Am Chem Soc. *Res:* Synthesis and characterization of materials used in copying and duplicating technologies; polymer chemistry. *Mailing Add:* 2660 Speakman Dr Mississauga ON L5K 2L1 Can

DUFF, RAYMOND STANLEY, b Hodgdon, Maine, Nov 2, 23; m 45; c 3. PEDIATRICS, SOCIOLOGY. *Educ:* Univ Maine, BA, 48; Yale Univ, MD, 52, MPH, 59. *Prof Exp:* Intern & resident pediat, Yale-New Haven Hosp, 52-55; dir bur med serv, New Haven Health Dept, 55-56; instr pediat & pub health, 56-59, asst prof pediat & sociol, 59-67, assoc prof, 67-78, PROF PEDIAT, SCH MED, YALE UNIV, 78- *Concurrent Pos:* Asst dir ambulatory serv, Yale-New Haven Hosp, 56-59. *Mem:* AAAS; fel Am Pub Health Asn; fel Am Acad Pediat; Am Sociol Asn. *Res:* Behavioral aspects of health, especially health or illness care in large hospitals as influenced by patients, families, physicians, nurses, administrators and others. *Mailing Add:* Dept Pediat Yale Univ Sch Med 789 Howard Ave New Haven CT 06504

DUFF, ROBERT HODGE, b Durand, Mich, Oct 21, 29; m 58; c 3. ELECTRON MICROSCOPY, CHEMISTRY. *Educ:* Mich State Univ, BS, 51; Ind Univ, PhD(inorg chem), 61. *Prof Exp:* Control chemist, Am Agr Chem Co, Mich, 51-53; asst, Ind Univ, 53-59; electron microscopist & phys chemist, Nat Bur Standards, 59-61; staff scientist electron micros, Avco Corp, Wilmington, 61-66; res assoc, Ledgemont Lab, Kennecott Copper Corp, Lexington, 66-70, anal chemist, 70-72; consult chemist, Technol Inc, 72-73; sr physicist, Systs Res Lab, Dayton, 73-75; sr res chemist, 75-78, ANALYTICAL PROJ LECTR, STANDARD OIL OHIO, CLEVELAND, 78- *Mem:* AAAS; Am Chem Soc; fel Am Inst Chemists; Electron Micros Soc Am; Int Soc Stereology; Sigma Xi. *Res:* Use of electron microscopy in all phases of material science; use of x-rays for elemental analysis with the electron microscope. *Mailing Add:* 2201 S Overlook Cleveland Heights OH 44106

DUFF, RONALD GEORGE, b Billings, Mont, Dec 8, 36; m 62, 88; c 2. CELL BIOLOGY, VIROLOGY. *Educ:* Univ Mont, BMus, 59; Univ Colo, Phd(path), 68. *Prof Exp:* Prof microbiol, M S Hershey Med Ctr, 69-74; head, viral & cell biol, Abbott Labs, 74-83; vpres, develop, 83-84, SR VPRES, RES & DEV, DAMON BIOTECH, INC, 84- *Concurrent Pos:* Fel, Baylor Col Med, 69; prof microbiol, Chicago, 76-84; chmn, comt monoclonal antibodies, Indust Biotechnol Asn; mem, biotechnol comt, Pharmaceut Mfrs Asn, Res & Develop comt, Asn Biotechnol Co. *Mem:* AAAS; Am Soc Microbiol; Am Soc Cancer Res; Sigma Xi; Pharmaceut Mfrs Asn; Indust Biotechnol Asn; Tissue Culture Soc; Wound Healing Soc. *Res:* Molecular biology, immunology, cell biology, protein chemistry and polymer chemistry. *Mailing Add:* Sr Vpres Res & Develop Curative Tech Inc 14 Research Way South Setauket NY 11733-9052

DUFF, RUSSELL EARL, b Grand Rapids, Mich, Nov 28, 26; m 87; c 5. FLUID DYNAMICS, CHEMICAL PHYSICS. *Educ:* Univ Mich, BSE, 47, MS, 48, PhD(physics), 51. *Prof Exp:* Res assoc, Eng Res Inst, Univ Mich, 47-51; mem staff, Los Alamos Sci Lab, 51-61, Inst Defense Anal, 61-62; proj leader, Lawrence Radiation Lab, 62-64, div leader, 64-67; vpres, S-Cubed, 67-88; RETIRED. *Concurrent Pos:* Various consult tasks. *Mem:* Am Phys Soc. *Res:* Research on safety of underground nuclear testing, shock and detonation hydrodynamics, high pressure material properties; high temperature chemistry. *Mailing Add:* PO Box 1620 La Jolla CA 92038

DUFF, WILLIAM G, b Alexandria, Va, Dec 10, 36. COMMUNICATIONS & ELECTRONICS. *Educ:* Syracuse Univ, MSEE, 69; Clayton Univ, DSEE, 77. *Prof Exp:* CHIEF ENGR, DEFENSE SYSTS DIV, ATLANTIC RES CORP, 75- *Concurrent Pos:* Instr electromagnetic compatibility & interference, Interference Control Technol, 71-; pres, Electromagnetic Soc, Inst Elec & Electronics Engrs, 84-85. *Honors & Awards:* Richard Stoddard Award, Inst Elec & Electronics Engrs, 84, Lawrence G Cummings Award, 85. *Mem:* Fel Inst Elec & Electronics Engrs. *Res:* Electromagnetic compatibility. *Mailing Add:* Atlantic Res Corp 5501 Backlick Rd Springfield VA 22151

DUFFEY, DICK, b Wabash County, Ind, Aug 26, 17. NUCLEAR ENGINEERING. *Educ:* Purdue Univ, BS, 39; Univ Iowa, MS, 40; Univ Md, PhD(nuclear eng), 56. *Prof Exp:* Engr nuclear proj, Union Carbide, 40-42; nuclear engr, US Atomic Energy Comn, 47-54; head nuclear eng prog, 54-65, assoc prof, 56-59, nuclear reactor dir, 58-68, PROF NUCLEAR ENG, UNIV MD, 59- *Concurrent Pos:* Tech sect adv comt on reactor safeguards, Atomic Energy Comn, 59-66. *Mem:* Am Nuclear Soc; Am Phys Soc; Am Geophys Union; Am Inst Chem Engrs; Am Chem Soc; Health Physics Soc. *Res:* Nuclear reactor design and operation; nuclear reactor safety; neutron uses; californium 252 uses. *Mailing Add:* Nuclear Eng Dept Univ Md College Park MD 20742

DUFFEY, DONALD CREAGH, b Winchester, Va, Feb 9, 31. PHYSICAL CHEMISTRY, ORGANIC CHEMISTRY. *Educ:* Va Polytech Inst, BS, 53; Rice Inst, MA, 55; Ga Inst Technol, PhD(chem), 59. *Prof Exp:* Res fel, Pa State Univ, 59-60; asst prof, 60-61, assoc prof, 62-67, PROF CHEM, MISS STATE UNIV, 67- *Concurrent Pos:* Johnson Res Found fel, Univ Pa, 64-65; vis prof, Univ Cincinnati, 68, Univ Utah, 79 & Univ Va, 80; fac res partic, Morgantown Energy Technol, 78. *Mem:* Am Chem Soc. *Res:* Reaction mechanisms; chemical spectroscopy. *Mailing Add:* Dept Chem PO Box 35 Miss State Univ Mississippi State MS 39762

DUFFEY, GEORGE HENRY, b Manchester, Iowa, Dec 24, 20; m 45; c 3. SYMMETRY CONSIDERATIONS, IRREVERSIBLE THERMODYNAMICS. *Educ:* Cornell Col, BA, 42; Princeton Univ, AM, 44, PhD(chem physics), 45. *Prof Exp:* Asst, Princeton Univ, 42-45; from asst prof to prof chem, SDak State Col, 45-58; prof chem & physics, Univ Miss, 58-59; PROF PHYSICS, SDAK STATE UNIV, 59- *Concurrent Pos:* Western Elec Fund instruction award, Am Soc Eng Educ, 71-72; vis lectr, Univ Western Australia, Nedlands, 77; consult, Univ Utah, 78 & 79. *Honors & Awards:* F O Butler Found Award. *Mem:* Am Phys Soc; Am Chem Soc; Ital Soc Physics; AAAS; Philos Sci Asn. *Res:* Alpha-particle nuclear models; polarographic theory; valence theory; molecular-orbital calculations; detonation-wave theory; irreversible thermodynamics. *Mailing Add:* Dept Physics SDak State Univ Box 2219 Brookings SD 57007

DUFFEY, MICHAEL EUGENE, b Iowa City, Iowa, Nov 15, 45. EPITHELIAL ION TRANSPORT. *Educ:* Univ Iowa, BS, 68; Carnegie-Mellon Univ, MS, 69, PhD(bioeng), 77. *Prof Exp:* Biomed engr, Wilford Hall, US Air Force Med Ctr, 70-73; fel, Univ Pittsburgh, 77-79; ASSOC PROF PHYSIOL, STATE UNIV NY, BUFFALO, 85- *Concurrent Pos:* Vis asst prof chem eng, Carnegie-Mellon Univ, 79. *Mem:* Am Physiol Soc; Biophys Soc; Soc Gen Physiologists. *Res:* Regulation of ion channels in biological membranes, especially in epithelial tissues. *Mailing Add:* Dept Physiol Sch Med 120 Sherman Hall State Univ NY Buffalo NY 14214

DUFFEY, PAUL STEPHEN, b Oakland, Calif, Nov 24, 39. MONOCLONAL ANTIBODY TECHNIQUES, MOLECULE PROBE TECHNIQUES. *Educ:* San Jose Univ, Calif, BA, 63; Univ Mich, PhD(epidemiol sci), 73. *Prof Exp:* Microbiologist, Calif Dept Health Serv, 64-68; teaching fel cell immunol, Med Sch, Univ Mich, 73-75, from asst prof to assoc prof immunol & microbiol, 75-81; RES MICROBIOLOGIST IMMUNOL, CALIF DEPT HEALTH SERV, 81- *Concurrent Pos:* Rep, Calif Interagency Taskforce Biotechnol, 86- *Mem:* Am Asn Immonologists; NY Acad Sci. *Res:* Cellular immunology-control, cell-cell interactions; antibody or lymphokine-secreting hybridomas; recombinant nucleic acid methods to identify and classify microorganisms; immunochemical and DNA probe methods in epidemiologic studies. *Mailing Add:* Microbial Dis Lab Calif State Dept Health Serv 2151 Berkeley Way Berkeley CA 94704

DUFFEY, SEAN STEPHEN, b Toronto, Ont, Nov 28, 43; m 71; c 2. BIOCHEMICAL ECOLOGY, TOXICOLOGY. *Educ:* Univ BC, Vancouver, BSc, 68, MSc, 70, PhD(bot), 74. *Prof Exp:* fel entom, Univ Ga, Athens, 74-76; ASST PROF ENTOM, UNIV CALIF, DAVIS, 76- *Mem:* Entom Soc Am. *Res:* Mechanisms of resistance of plants against insects with emphasis of biochemical interactions between plants and insects. *Mailing Add:* Dept Entom Univ Calif Davis CA 95616

DUFFIE, JOHN A(TWATER), b White Plains, NY, Mar 31, 25; m 47; c 3. CHEMICAL ENGINEERING. *Educ:* Rensselaer Polytech Inst, BChE, 45, MChE, 48; Univ Wis, PhD(chem eng), 51. *Prof Exp:* Res engr, Electrochem Dept, E I du Pont de Nemours & Co, 51-52; sci liaison officer, Off Naval Res, 52-53; asst dir eng exp sta, Univ Wis-Madison, 57-65, dir univ-indust res prog & assoc dean grad sch, 65-76, prof chem eng, 69-88, dir Solar Energy Lab, 54-88, EMER PROF CHEM ENG, UNIV WIS-MADISON, 88- *Concurrent Pos:* Fulbright fel, Australia, 64 & 76; hon sr res fel, Univ Birmingham, UK, 84; ed, Solar Energy J. *Honors & Awards:* Charles G Abbot Award, Am Solar Energy Soc. 76; Farrington Daniels Award, Int Solar Energy Soc, 87. *Mem:* fel Am Inst Chem Engrs; Int Solar Energy Soc (pres, 72-73). *Res:* Solar energy research; solar energy thermal processes. *Mailing Add:* 5710 Dorsett Dr Madison WI 53711

DUFFIELD, DEBORAH ANN, b Boston, Mass, Nov 12, 41; m 76; c 2. CYTOGENETICS, POPULATION GENETICS. *Educ:* Pomona Col, BA, 63; Stanford Univ, MA, 66; Univ Calif, Los Angeles, PhD(genetics), 77. *Prof Exp:* Instr biol, Old Dominion Col, 68-70; teaching assoc biol, Univ Calif, Los Angeles, 72-75, lectr, genetics/marine mammal, 75-77; ASST PROF GENETICS/MARINE MAMMAL, BIOL DEPT, PORTLAND STATE UNIV, 77- *Mem:* Am Soc Mammalogists; Int Asn Aquatic Animal Med. *Res:* Genetic variability in natural populations, specifically-comparative cytogenetic and electrophoretic evaluation of cetacean species, emphasizing evolutionary implications and population structure and dynamics; species comparisons of hemoglobin and various red blood cell parameters as they relate to diving capabilities and habitat in cetaceans. *Mailing Add:* 6203 N Commercial Portland OR 97217

DUFFIELD, JACK JAY, b Long Beach, Calif, Nov 7, 33; m 58; c 2. SPECTROSCOPIC INSTRUMENTATION. *Educ:* Pomona Col, BA, 55; Mass Inst Technol, PhD(chem), 60. *Prof Exp:* Proj mgr & sr engr, Appl Physics Corp, 60-66; proj mgr, Instrument Div, Cary Prod, 66-79, res & develop, Optical Prod, 79-82, RES SCIENTIST, VARIAN RES CTR, VARIAN ASSOCS, 83- *Mem:* Sigma Xi; Am Chem Soc. *Res:* Gas chromatography; design and development of polarized light spectroscopic instrumentation (optical rotatory dispersion and circular dichroism); fluorescence and ultra violet-visual spectrophotometry-near infrared region spectroscopic instrumentation; high power x-ray tubes; development management. *Mailing Add:* 878 Helena Dr Sunnyvale CA 94087

DUFFIELD, ROGER C, b Kansas City, Kans, Apr 7, 37. MECHANICAL & AEROSPACE ENGINEERING. *Educ:* Univ Kans, BSME, 60, MSME, 64, PhD, 68. *Prof Exp:* Design engr, LFM Mfg Co Div, Rockwell Mfg Co, 60-62; asst instr eng mech, Univ Kans, 64-66; asst prof, 66-70, ASSOC PROF MECH & AEROSPACE ENG, UNIV MO-COLUMBIA, 70- *Mem:* Am Soc Mech Engrs. *Res:* Structural dynamics. *Mailing Add:* Univ Mo 2005 A Engr Columbia MO 65211

DUFFIELD, WENDELL ARTHUR, b Sisseton, SDak, May 10, 41; m 64. STRUCTURAL GEOLOGY. *Educ:* Carleton Col, BA, 63; Stanford Univ, MS, 65, PhD(geol), 67. *Prof Exp:* GEOLOGIST, US GEOL SURV, 66- *Mem:* Geol Soc Am; Am Geophys Union; Geothermal Resources Coun; Sigma Xi. *Res:* Volcanology and related geothermal resources; igneous and metamorphic petrology; mineralogy. *Mailing Add:* 2255 N Gemini Dr Flagstaff AZ 86001

DUFFIN, JOHN H, b Easton, Pa, June 18, 19; m 54. CHEMICAL ENGINEERING. *Educ:* Lehigh Univ, BS, 40; Univ Calif, Berkeley, PhD(chem eng), 59. *Prof Exp:* Lab analyst & chief chemist, Hercules Powder Co, NJ & Utah, 40-44; tech adv & prod foreman uranium isotope prod, Tenn Eastman Corp, 44-45; res engr, Allied Chem & Dye Corp, NY 45-53; res scientist, Battelle Mem Inst, 53-54; instr chem eng, Univ Calif, Berkeley, 54-59; asst prof chem eng & dir comput ctr, San Jose State Col, 59-62; assoc prof, 62-66, prof & chmn, Dept Mat Sci & Chem, 69-72, PROF CHEM & ELEC ENG, US NAVAL POSTGRAD SCH, 72- *Concurrent Pos:* Consult, Kaiser Aluminum & Chem Co, 57-58 & Int Bus Mach Corp, 60-72. *Mem:* Fel Am Inst Chemists; Am Inst Chem Engrs; Sigma Xi; NY Acad Sci. *Res:* Mathematical modeling of chemical engineering processes and use of models to investigate control problems for purposes of obtaining optimum control and designing equipment based on dynamic behavior. *Mailing Add:* 5806 W Crowley Ave Visalia CA 93291

DUFFIN, RICHARD JAMES, b Chicago, Ill, Oct 13, 09; m 47; c 2. MATHEMATICS, PHYSICS. *Educ:* Univ Ill, BS, 32, PhD(physics), 35. *Prof Exp:* Instr math, Purdue Univ, 36-40; assoc, Univ Ill, 40-42; physicist, Carnegie Inst, 42-46; prof math, 46-70, UNIV PROF MATH SCI, CARNEGIE-MELLON UNIV, 70- *Concurrent Pos:* Vis prof, Purdue Univ, 49-50 & Dublin Inst Advan Studies, 59-60; consult, Westinghouse Res Labs, 56-; dir appl math res, Duke Univ, 58-59; distinguished vis prof, State Univ NY Stony Brook, 67 & Tex A&M Univ, 68. *Honors & Awards:* Von Neumann Theory Prize, Opers Res Soc, 82; Monie A Ferst Award, Sigma Xi, 84. *Mem:* Nat Acad Sci; Am Acad Arts & Sci; Am Math Soc; Soc Natural Philos; Soc Indust & Appl Math. *Res:* Thermal and magnetic effects in conductors; quantum theory; Fourier type integrals and series; functional inequalities; navigational devices; partial difference equations; elasticity; electrical network theory; linear and non-linear programming; solar energy utilization. *Mailing Add:* Dept Math Carnegie-Mellon Univ Pittsburgh PA 15213

DUFFUS, HENRY JOHN, b Vancouver, BC, Aug 27, 25; m 48. PHYSICS. *Educ:* Univ BC, BApSci, 48, BA, 49; Oxford Univ, DPhil(physics), 53. *Prof Exp:* Lectr physics, Carleton Col, 48-51; Nat Res Coun Can spec overseas scholar, 52-53; physicist & defence sci serv officer, Defence Res Bd, 54-59; prof physics & head dept, 59-79, dean sci & eng, Royal Roads Mil Col, 79-87; RETIRED. *Res:* Paramagnetism at low temperatures; microwave spectroscopy; geomagnetism; infrared sensing; acoustics. *Mailing Add:* 139 Atkins Ave Victoria BC V9B 2Z9 Can

DUFFUS, JAMES EDWARD, b Detroit, Mich, Feb 11, 29; m 52; c 3. PLANT PATHOLOGY, PLANT VIROLOGY. *Educ:* Mich State Col, BS, 51; Univ Wis, PhD(plant path), 55. *Prof Exp:* Actg supt, Agr Res Sta, 69-70, PLANT PATHOLOGIST & RES LEADER, SUGARBEET PROD RES, USDA, 55-, LOCATION COORDR, 89- *Concurrent Pos:* Assoc, Exp Sta, Univ Calif, Davis & Berkeley, 62-; vis scientist, Tasmanian Dept Agr, Hobart, Tasmania, Australia, 80-81, adj prof, Univ Ark, Fayetteville; fel, Am Phytopath Soc, 82. *Honors & Awards:* Super Serv Award, USDA, 83; Meritorious Serv Award, Am Soc Sugar Beet Technol. *Mem:* Am Soc Sugar Beet Technol; fel Am Phytopath Soc; Int Soc Plant Path; Int Soc Hort Sci. *Res:* Virus diseases of sugar beets and vegetable crops; interrelationships of yellowing type virus diseases; virus-vector relations; insect feeding through membranes, infectivity neutralization; role of wild hosts in virus epidemiology; whitefly transmitted viruses. *Mailing Add:* US Agr Res Sta 1636 E Alisal St Salinas CA 93905

DUFFY, CLARENCE JOHN, b Pasco, Wash, Feb 5, 47; c 1. GEOLOGY, PHYSICAL CHEMISTRY. *Educ:* Mass Inst Technol, BS, 69; Univ BS, PhD(geol), 77. *Prof Exp:* STAFF MEM GEOL, LOS ALAMOS SCI LAB, 77- *Mem:* Mineral Soc Am. *Res:* Chemical aspects of nuclear waste storage; hydrologic parameters of low permeability porosity rocks. *Mailing Add:* INC-7 Mailstop J514 Los Alamos Nat Lab Box 1663 Los Alamos NM 87545

DUFFY, FRANK HOPKINS, b Honolulu, Hawaii, Jan 22, 37; m 64; c 4. NEUROPHYSIOLOGY, NEUROLOGY. *Educ:* Univ Mich, BSE(elec eng) & BSE(math), 58; Harvard Univ, MD, 63. *Prof Exp:* Internship, Yale-New Haven Hosp, 63-64; resident neurol surg, Mass Gen Hosp, 64-66; resident neurol, Peter Bent Brigham Hosp & Children's Hosp Med Ctr, Boston, 66-68; instr, 70-71, asst prof neurol, 71-79, PROF NEUROL, HARVARD MED SCH, 79-; DIR EXP NEUROPHYSIOL, CHILDREN'S HOSP, BOSTON, 74- *Concurrent Pos:* Res fel neurol, Harvard Med Sch, 66-68; mem comt on vision, Nat Acad Sci-Nat Res Coun, 69-70; neurologist, Peter Bent Brigham Hosp, 70-75, Beth Israel Hosp, 70- & Children's Hosp, Boston, 70- *Res:* Computer analysis of the EEG and evoked potentials; single cell microelectrode studies of visual and somatosensory systems of cats and primates. *Mailing Add:* 1990 Commonwealth Ave Boston MA 02135

DUFFY, JACQUES WAYNE, b Nimes, France, July 1, 22; m 50; c 3. MECHANICAL ENGINEERING. *Educ:* Columbia Univ, BA, 47, BSc, 48, MS, 49, PhD(appl mech), 57. *Hon Degrees:* DSc, Univ Nantes, France, 80. *Prof Exp:* Res engr dynamics of aircraft, Grumman Aircraft Eng Corp, 49-51; from asst prof to assoc prof eng, 53-65, chmn Ctr Biophys Sci & Biomed Eng, 69-72, PROF ENG, BROWN UNIV, 65- *Concurrent Pos:* Guggenheim fel, 64-65; Eng Found fel, 78; adj staff, Miriam Hosp, consult bio-eng, 72-76; mem, bd trustees, New Eng Tech Inst, 75-80; rep grad matters, div eng, Brown Univ, 76-81; mem comt recommendations, US Army Basic Sci Res, Nat Res Coun, 78-84; vis prof, Dept Physics Metals, Univ Nantes, France, 85. *Mem:* Soc Exp Mech; fel Am Soc Mech Engrs; Am Soc Testing & Mat; Sigma Xi; Am Soc Metals. *Res:* Dynamic plasticity; dynamic fracture; dynamic mechanical behavior of structural metals and ceramics; dynamic shear banding in metals; explosive fracture of metals, ceramics and composites; approximately 85 publications. *Mailing Add:* Dept Eng Brown Univ Providence RI 02912

DUFFY, LAWRENCE KEVIN, b Brooklyn, NY, Feb 1, 48; m; c 3. NEUROSCIENCES, TOXICOLOGY. *Educ:* Univ Alaska, PhD(biochem), 77. *Prof Exp:* asst prof biochem, Harvard Med Sch, 83-87; ASSOC PROF BIOCHEM & DIR, PROTEIN CORE FACIL, UNIV ALASKA, FAIRBANKS, 87-; CONSULT, ALCOHOL & DRUG ABUSE, ALASKA BIOTECH, 89- *Concurrent Pos:* Dir, Protein Microsequencing Lab, Ctr Neurol Dis, Brigham & Women's Hosp, Boston, 85-87. *Honors & Awards:* Fac Scholar Award, Alzheimer Soc, 87. *Mem:* Am Soc Biochem & Molecular Biol; Am Chem Soc; Sigma Xi. *Res:* Neuroscience; toxins; alcoholism. *Mailing Add:* Dept Chem Univ Alaska Fairbanks AK 99775

DUFFY, NORMAN VINCENT, JR, b Washington. DC, Nov 1, 38; m 62; c 5. INORGANIC CHEMISTRY. *Educ:* Georgetown Univ, BS, 61, PhD(chem), 66. *Prof Exp:* NATO fel, 65-66; vis asst prof chem, 66-67, asst prof, 67-70, asst dean col arts & sci, 73-75, assoc dean, 75-76, assoc prof, 70-80, chmn chem dept, 81-86, PROF CHEM, KENT STATE UNIV, 80- *Mem:* Am Chem Soc; Sigma Xi. *Res:* Dithiocarbamate complexes; metal carbonyls. *Mailing Add:* Dept Chem Kent State Univ Kent OH 44242-0001

DUFFY, PHILIP, b Nimes, France, July 22, 23; US citizen; m 49; c 2. NEUROLOGY, NEUROPATHOLOGY. *Educ:* Columbia Univ, BA, 43, MD, 47. *Prof Exp:* Instr neuropath, Columbia Univ, 54-55; asst prof neurol, State Univ NY, 55-64; assoc prof, 64-68, PROF NEUROPATH & DIR DIV, COL PHYSICIANS & SURGEONS, COLUMBIA UNIV, 68- *Mem:* Am Acad Neurol; Am Asn Neuropath; Asn Res Nerv & Ment Dis. *Res:* Central nervous system tumors. *Mailing Add:* 630 W 168th St New York NY 10032

DUFFY, REGINA MAURICE, b Jersey City, NJ. PLANT MORPHOLOGY. *Educ:* Col of New Rochelle, BA, 41; Fordham Univ, MS, 44; Columbia Univ, PhD(bot), 50. *Prof Exp:* Instr biol, NJ State Teachers' Col, 41-42 & 43-44; teacher, Pub Sch, 45-46; instr biol, NJ State Teachers' Col, 46; teaching asst bot, Columbia Univ, 47-50, res assoc, 50-55; asst prof biol, Long Island Univ, 51-52 & 53-55; asst prof, Jersey City State Col, 55-59; lectr, Hunter Col, 59; chmn dept sci, Nanuet Schs, NY, 59-67; admin head, 67-68, PRES, NORTHWESTERN CONN COMMUNITY COL, 68- *Mem:* Sigma Xi; Torrey Bot Club. *Res:* Cellular morphology; liverworts. *Mailing Add:* Four Woodruff Ct Litchfield CT 06759

DUFFY, ROBERT A, b Buck Run, Pa, Sept 9, 21; m 45; c 5. GUIDANCE, NAVIGATION. *Educ:* Ga Inst Technol, BS, 51. *Prof Exp:* Dir, Guidance & Control, Air Force Ballistic Missle Div, US Air Force, 58-63, staff officer, Dirs Staff Group, Off Secy Defense, 63-67, prog mgr, Dep Reentry Syst, Space & Missle Systs Orgn, 67-70, vcomdr, 70-71; vpres & dir, Draper Lab Div, Mass Inst Technol, 71-73, pres, chief exec officer & dir, Chas Stark Draper Lab, Inc, 73-87; RETIRED. *Concurrent Pos:* Bd dir, Draper Lab, 72-, Harvard Trust Bank, 83-; chmn, Comt Automation for Combat Aircraft, Nat Acad Eng, 81- *Honors & Awards:* Thurlow Award, Inst Navig, 64; Thomas D White Award, Nat Geog Soc, 70. *Mem:* Nat Acad Eng; Inst Navig (pres, 76-77); fel Am Inst Aeronaut & Astronaut. *Res:* Development and practical application of guidance and control systems to aircraft, missles and space craft. *Mailing Add:* C S Draper Lab Inc 555 Technology Sq Cambridge MA 02139

DUFFY, ROBERT E(DWARD), b Scranton, Pa, May 27, 30; m 53; c 3. AERONAUTICAL ENGINEERING. *Educ:* Rensselaer Polytech Inst, BAE, 51, MAE, 54, PhD, 65. *Prof Exp:* Aeronaut engr, US Govt, 51-52; from instr to asst prof, 53-65, chmn dept, 68-75, ASSOC PROF AERONAUT ENG, RENSSELAER POLYTECH INST, 65- *Concurrent Pos:* Consult, Gen Elec Co; Xerox Corp & NY State Bur Educ; dir, Panaflight Corp & Burden Lake Holding Corp. *Mem:* Assoc fel Am Inst Aeronaut & Astronaut. *Res:* Flight mechanics; experimental aerodynamics. *Mailing Add:* RD 4 Wynantskill NY 12198

DUFFY, WILLIAM THOMAS, JR, b San Francisco, Calif, June 30, 30; m 59; c 3. EXPERIMENTAL GRAVITATIONAL PHYSICS, LOW TEMPERATURE PHYSICS. *Educ:* Univ Santa Clara, BEE, 53; Stanford Univ, MS, 54, PhD(physics), 59. *Prof Exp:* From asst prof to assoc prof, 59-68, chmn dept, 74-90, PROF PHYSICS, SANTA CLARA UNIV, 68- *Concurrent Pos:* NSF fel, State Univ Leiden, 61-62; res assoc & vis scientist, Stanford Univ, 68-69 & 78. *Mem:* Am Phys Soc; Sigma Xi. *Res:* Acoustic absorption in technical materials at ultralow temperatures with applications to gravitational wave antenna design; past research on nuclear magnetic resonance, linear chain magnetism, magnetism of organic crystals. *Mailing Add:* Dept Physics Santa Clara Univ Santa Clara CA 95053

DUFLOT, LEO SCOTT, b Mayfield, Ky, June 24, 19; m 51; c 5. ANESTHESIOLOGY. *Educ:* Univ Tex, BA, 39, MD, 43; Am Bd Anesthesiol, dipl, 54. *Prof Exp:* From instr to prof anesthesiol, Univ Tex Med Br, Galveston, 51-87; dir anesthesiol, Driscoll Found Childrens Hosp, 81-87; RETIRED. *Concurrent Pos:* Res assoc, Univ Pa, 51. *Mem:* AMA; Am Soc Anesthesiologists. *Mailing Add:* 614 Bradshaw Dr Corpus Christi TX 78412-3002

DUFOUR, JACQUES JOHN, b Baie St-Paul, Que, June 10, 39; m 67; c 2. REPRODUCTIVE PHYSIOLOGY, AGRICULTURE. *Educ:* Laval Univ, BS, 64; Guelph Univ, MSc, 67; Univ Wis, PhD(reproductive physiol), 71. *Prof Exp:* Res officer physiol, 67-68, RES SCIENTIST, RES STA AGR CAN, 71- *Concurrent Pos:* Mem, Prov Eval Comt Res Subvention, 74-78 & Res Comt on Que Orientation Dairy Beef Cattle, 75-76; head, Can Mission Surv Reproductive Physiol in USSR, 76-; researcher, Folliculogenesis in sheep, France, 77-78. *Mem:* AAAS; Can Soc Animal Sci; Soc Can Zootech; Soc Study Reproduction. *Res:* Folliculogenesis in sheep and cattle to achieve twinning. *Mailing Add:* Res Sta Agr Can 2560 Blvd Hochelaja St Croix PQ G1V 2J3 Can

DUFOUR, REGINALD JAMES, b Simmesport, La, July 29, 48. ASTRONOMY. *Educ:* La State Univ, Baton Rouge, BS, 70; Univ Wis, Madison, MS, 71, PhD(astron), 74. *Prof Exp:* Nat Acad Sci-Nat Res Coun res assoc astron, L B Johnson Space Ctr, NASA, 74-75; ASST PROF SPACE PHYSICS & ASTRON, RICE UNIV, 75- *Mem:* AAAS; Royal Astron Soc; Am Astron Soc. *Res:* Observational astronomy and instrumentation; gaseous nebulae; structure and evolution of galaxies; magellanic clouds, ultraviolet spectra of stars. *Mailing Add:* Dept Space Physics & Astron Rice Univ Box 1892 Houston TX 77251

DUFRAIN, RUSSELL JEROME, b Chicago, Ill, Oct 4, 44; m 67; c 3. CYTOGENETICS, GENETIC TOXICOLOGY. *Educ:* Western Ill Univ, BSEd, 67, MS, 68; Cornell Univ, PhD(physiol), 74. *Prof Exp:* From instr to asst prof biol, Moraine Valley Community Col, 69-71; from res asst to res assoc phys biol, NY State Vet Col, Cornell Univ, 72-75; assoc scientist, Med & Health Div, Oak Ridge Assoc Univs, 75-77, scientist, 77-82, sr scientist, 83-85, prog dir, Cytogenetic Develop & Reprod Toxicol, 84-85; sr res geneticist, Biosci Lab, Corp Technol, Allied-Signal Corp, 85-88; sr res scientist Toxicol Prod, Bristol-Myers Co, 88-90, SR RES INVESTR, BRISTOL-MYERS SQUIBB CO, 90- *Concurrent Pos:* Consult, Environ Protection Agency, 88-90. *Mem:* AAAS; Am Inst Biol Sci; NY Acad Sci; Environ Mutagen Soc; Radiation Res Soc; Sigma Xi. *Res:* Use of numerical analysis of genetic, cytogenetic and morphological damage to quantify physical and/or chemical effects on mammalian somatic cells, reproductive cells and early embryos. *Mailing Add:* Dept Investigative Toxicol Bristol Myers Squibb Co PO Box 4755 Syracuse NY 13321-4755

DUFRESNE, ALBERT HERMAN, b Sault Ste Marie, Ont, May 26, 28; m 60; c 2. PHYSICS. *Educ:* Mich Inst Technol, BSc, 54. *Prof Exp:* Mgr analyt lab, Johnson & Johnson Inc, 54-70, purchasing mgr, 70-83; PRES, POLYSCI PUBLICATIONS INC, 83- *Mem:* Spectroscopy Soc Can. *Mailing Add:* Polysci Pub Inc PO Box 148 Morin Heights PQ J0R 1H0 Can

DUFRESNE, RICHARD FREDERICK, b Holyoke, Mass, Aug 4, 43. FLAVOR CHEMISTRY, TOBACCO CHEMISTRY. *Educ:* Univ Rochester, BS, 65; Carnegie-Mellon Univ, PhD(org chem), 73. *Prof Exp:* Fel, Med Sch, Johns Hopkins Univ, 72-74, Brandeis Univ, 75-77 & Univ Mass, Amherst, 79-80; MEM STAFF, LORILLARD RES CTR, 80- *Mem:* Am Chem Soc; Sigma Xi. *Res:* Synthesis of heterocycles including indoles and pyrrolidines; vinca alkaloids; benzoazanorbornadienes; flavorants; radioactively-labeled securinega and cephalotax alkaloid precursors. *Mailing Add:* 313 Spur Rd Greensboro NC 27406

DUFTY, JAMES W, b Freeport, NY, May 5, 40; m 64. STATISTICAL MECHANICS OF COMPLEX FLUIDS & AMORPHOUS SOLIDS. *Educ:* Williams Col, AB, 62; Lehigh Univ, MS, 64, PhD(physics), 67. *Prof Exp:* Fel, Armstrong-Cork Res, 66-67; postdoctoral res assoc, Lehigh Univ, 67-68; postdoctoral fel, 68-70, from asst prof to assoc prof physics, 70-78, asst prof chem eng, 70-71, PROF PHYSICS, UNIV FLA, 78- *Concurrent Pos:* Vis assoc prof physics, Inst Fluid Dynamics, Univ Md, 76-77; vis scientist, Joint Inst Lab Astrophys, Univ Colo, 79-80, Mass Inst Technol, 90; vis physicist, Nat Bur Standards, Washington, DC, 82-83; vis prof, Inst Theoret Physics, Univ Utrecht, The Neth, 88. *Mem:* AAAS; Am Phys Soc; Sigma Xi. *Res:* Theoretical studies of simple and complex fluids; development and application of classical and quantum non-equilibrium statistical mechanics; dynamical properties of plasmas, fluid type cellular automata, metastable states, and amorphous solids. *Mailing Add:* Dept Physics Univ Fla Gainesville FL 32611

DUGA, JULES JOSEPH, b Bellaire, Ohio, Mar 21, 32; m 55; c 2. RESEARCH ADMINISTRATION, SCIENCE POLICY. *Educ:* Ohio State Univ, BSc, 53, MSc, 55, PhD, 60. *Prof Exp:* From prin physicist phys chem div to sr res physicist solid state res div, Battelle Mem Inst, 56-70; tech & sci consult, 70-76; TECH & SCI CONSULT, ECON, PLANNING & POLICY ANALYSIS SECT, BATTELLE MEM INST, 76-, SR RES & DEVELOP ANALYST. *Mem:* AAAS; Am Phys Soc; Am Inst Mining, Metall & Petrol Eng; Am Ceramic Soc; Technol Transfer Soc. *Res:* Technology utilization by state and local governments; science policy in state governments; energy policy planning; technology assessment and transfer; economic input-output analysis; macroeconomic impacts of new technologies. *Mailing Add:* 505 King Ave Columbus OH 43201

DUGAICZYK, ACHILLES, b Kosztowy, Poland, Sept 9, 30; US citizen. CELL BIOLOGY. *Educ:* Jagiellonian Univ, Cracow, Poland, MSc, 55; Univ Calif, San Francisco, PhD(biochem), 63. *Prof Exp:* Res asst, Univ Calif, San Francisco, 66-67 & 72-76; vis prof, Nat Polytech, Mexico City, 67-72; asst prof biochem, Baylor Col Med, Houston, 76-81; assoc prof biochem, 83-90, PROF BIOCHEM, UNIV CALIF, RIVERSIDE, 90- *Concurrent Pos:* Vis scientist, City of Hope Res Inst, 81; res adj biochem, Med Sch, Wroclaw, Poland, 62-66. *Mem:* Am Genetic Soc; Am Soc Biol Chemists; AAAS. *Res:* Analysis of the human genome using recombinant DNA technologies; molecular basis of the function, control of expression and evolution of human genes. *Mailing Add:* Dept Biochem Univ Calif Riverside CA 92521

DUGAL, HARDEV SINGH, b Bareilly, India, Feb 1, 37; m 68. PAPER CHEMISTRY. *Educ:* Agra Univ, BSc, 55; Harcourt Butler Technol Inst, India, MS, 58; Darmstadt Tech Univ, Dr Ing(cellulose chem), 63. *Prof Exp:* Student apprentice, Distillery Dept, Daurala Sugar Works, India, 55-66; pool officer, Coun Sci & Indust Res, New Delhi, 66-68; dir new projs, Punj Sons Ltd, 68; res fel phys chem, Inst Paper Chem, 68-70, proj leader, 69-70, asst prof chem, 70-89, dir, Div Indust & Environ Systs, 74-81, dir, Info Serv Div & head fac, Dept Spec Studies, 81-89; PRES, INTEGRATED PAPER SERV, INC, 89- *Concurrent Pos:* Chemist, Gwalior Rayon Silk Mfg Co Ltd, Nagda, India, 58-59; student apprentice, Adam Opal, Russelsheim am Main, WGer, 60; fel, Inst Paper Chem, 64-66; deleg, Tech Asn Pulp & Paper Indust Conf on Air & Water Environ, Jacksonville, Fla, 69, Minneapolis, 70 & Boston, 71; consult, pulp & paper indust; mem assessment citizens adv coun, Upper Miss River Basin Comn, 75-; Int Environ Comt Meeting, Helsinki, 75, Montreal, 76, Seattle, 77 & Stockholm, 78. *Mem:* Tech Asn Pulp & Paper Indust; Am Chem Soc; Indian Tech Asn Pulp & Paper Indust. *Res:* Pulping, bleaching and paper making; wet-end additives; carbohydrate chemistry; physical and chemical aspects of aqueous environment in pulp and paper industry. *Mailing Add:* Integrated Paper Serv Inc 101 W Edison Ave Suite 250 Appleton WI 54915

DUGAL, LOUIS PAUL, b Quebec, Que, Oct 1, 11; m 37; c 3. PHYSIOLOGY. *Educ:* Laval Univ, BA, 31, MSc, 34; Univ Pa, PhD(biol), 39. *Hon Degrees:* LLD, Univ Toronto, 65, Concordia Univ, 75, Ottawa Univ, 78. *Prof Exp:* From asst prof to prof biol, Univ Montreal, 35-45; prof exp physiol, Laval Univ, 45-55; prof biol, Univ Ottawa, 55-65, chmn dept, 55-62, dean fac pure & appl sci, 62-65; dep dir, Sci Secretariat, Privy Coun Can, 65-66; vdean, Med Sch, Univ Sherbrooke, 66-67, vrector, 67-72; chmn, Comt Approval Univ Prog, 70-74; CONSULT RES, QUE PROV GOVT, 74- *Concurrent Pos:* Mem, Nat Res Coun Can, 44-50; assoc dir, Inst Hyg & Human Biol, 45-55; mem adv comt & chmn arctic med res, Defense Res Bd Can, 47-58; consult, Royal Can Air Force, 48; Guggenheim fel, Columbia Univ, 54-55. *Honors & Awards:* Officer, Order Brit Empire, 46; Laureate, Fr Acad Sci, Montyon Prize, 52. *Mem:* Can Physiol Soc (pres, 51-52); fel Royal Soc Can; Am Physiol Soc; Soc Fr Speaking Physiologists. *Res:* Resistance and acclimatization of mammal to cold environment; respiration at high altitudes. *Mailing Add:* 344 de la Corniche St St Nicolas PQ G0S 2Z0 Can

DUGAN, CHARLES HAMMOND, b Baltimore, Md, Apr 2, 31; m 54; c 5. CHEMICAL PHYSICS. *Educ:* Univ Ky, BS, 51; Univ Calif, Los Angeles, MA, 54; Harvard Univ, PhD(appl physics), 63. *Prof Exp:* Physicist, US Navy Electronics Lab, 51-52 & Smithsonian Astrophys Observ, 61-67; from asst prof to assoc prof, 67-75, PROF PHYSICS, YORK UNIV, 75- *Concurrent Pos:* Vis fel appl optics, Imperial Col, 73-74; mem staff, Physics Dept, Univ Bielefeld, 80-81 & Dept Elec Eng, Cornell Univ, 81. *Mem:* AAAS; Am Phys Soc; Can Asn Physicists. *Res:* Photo dissociation of molecules. *Mailing Add:* Dept Physics York Univ North York ON M3J 1P3 Can

DUGAN, GARY EDWIN, b Batavia, NY, Dec 9, 41; m 69; c 2. DERMATOLOGY, PRODUCT DEVELOPMENT. *Educ:* Rochester Inst Technol, BS, 64; Univ Conn, PhD(med chem), 69. *Prof Exp:* Mgr prod develop, Block Drug Co, 68-75; mgr prod develop, Gillette Co, 75-80, asst to vpres res & develop, 80-84; dir res & develop, Avon Prod Inc, 84-87 & vpres res & develop, 87-88; sr vpres res & develop, Elizabeth Arden, 88-89; VPRES RES & DEVELOP, CLAIROL, 91- *Mem:* Cosmetic, Toiletry & Fragrance Asn; Soc Cosmetic Chemists. *Res:* Developing new materials and products to treat aging skin and to prevent the aging of skin; to understand the aging process and develop products that allow people to look and feel younger. *Mailing Add:* 315 Grandview Circle Ridgewood NJ 07450

DUGAN, JOHN PHILIP, b Darby, Pa, Apr 23, 42; m 64; c 4. PHYSICAL OCEANOGRAPHY. *Educ:* Pa Mil Col, BS, 64; Northwestern Univ, MS, 66, PhD(theoret & appl mech), 67. *Prof Exp:* NSF & Off Naval Res Funds fel mech, Johns Hopkins Univ, 67-69; asst prof mech eng, Univ Toronto, 69-71; res physicist, Environ Sci Div, Naval Res Lab, 71-84; HEAD, FIELD MEASUREMENTS DEPT, ARETE ASSOC, 84- *Mem:* Inst Elec & Electronics Engrs Oceanic Eng Soc; Am Geophys Union; Sigma Xi; Am Meteorol Soc; Oceanog Soc. *Res:* Ocean instrumentation; observations of fine and microscale upper ocean structure. *Mailing Add:* 5505 Saddlebrook Ct Burke VA 22015

DUGAN, KIMIKO HATTA, b Kyoto City, Japan, Oct 21, 24; nat US; wid. MICROSCOPIC ANATOMY. *Educ:* Okla Col Women, BA, 61; Univ Okla, MS, 65, PhD(med sci), 70. *Prof Exp:* Teaching asst human histol & embryol, 64-69, from instr to asst prof, 69-78, ASSOC PROF ANAT SCI, COL MED & GRAD COL, UNIV OKLA HEALTH SCI CTR, 78- *Mem:* Am Asn Anatomists; NY Acad Sci; Am Soc Zoologists; Am Chem Soc; Int Soc Develop & Comparative Immunol; Am Inst Chemists. *Res:* Light and electron microscopies; histochemistry and cytochemistry of skin, pancreas and heart; stereological study of cardiac enlargement. *Mailing Add:* 14901 N Pa Ave No 234 Oklahoma City OK 73134-6008

DUGAN, LEROY, JR, b Petersburgh, Ind, Aug 18, 15; m 38; c 3. AGRICULTURAL & FOOD CHEMISTRY, FOOD SCIENCE & TECHNOLOGY. *Educ:* Ind Univ, BS, 37; Univ Wash, PhD(org chem), 42. *Prof Exp:* Assoc org chem, Am Meat Inst Found, 46-49, chief div org chem, 49-61; assoc prof food sci, 61-66, prof food sci & human nutrit, 66-81, asst dean, Grad Sch, 71-81, EMER PROF FOOD SCI & HUMAN NUTRIT & EMER ASST DEAN, GRAD SCH, MICH STATE UNIV, 81- *Mem:* Am Chem Soc; Am Oil Chem Soc; Soc Chem Indust; Inst Food Technol; Sigma Xi. *Res:* Autooxidation of fats; antioxidants for fats and foods; lipid composition, phospholipid studies; lipid-protein interactions; lipid browning; pesticide distribution in foods; lipid nutrition; food chemistry. *Mailing Add:* Dept Food Sci Mich State Univ East Lansing MI 48823

DUGAN, PATRICK R, b Syracuse, NY, Dec 14, 31; m 56; c 4. MICROBIOLOGY, BIOCHEMISTRY. *Educ:* Syracuse Univ, BS, 56, MS, 59, PhD(microbiol), 64. *Prof Exp:* Res asst microbiol & biochem, Syracuse Univ Res Corp, 56-58, res assoc, 58-61, assoc res scientist, 61-63; from asst prof to prof Microbiol, Ohio State Univ, 64-87, chmn dept, 70-73, actg dean, Col Biol Sci, 78, dean, 79-84; DIR CTR BIOPROCESSING, IDAHO NAT ENG LAB, 87- *Mem:* AAAS; Am Soc Microbiol; Soc Indust Microbiol; Am Chem Soc; fel Am Acad Microbiol. *Res:* Microbial physiology and ecology, particularly aquatic organisms; microbial metabolism of inorganic compounds; analytical chemistry; microbiology and biochemistry of coal mine drainage; methane and hydrocarbon oxidation; waste treatment; eutrophication; biosorption of metals and toxic waste; microbial processing of fossil fuels; gas and vapor phase bioreactors. *Mailing Add:* Dept Microbiol Idaho Nat Eng Lab EG&G PO Box 1625 Idaho Falls ID 83415

DUGAS, HERMANN, b Baie-Comeau, Que, Nov 21, 42; m 66; c 1. BIOORGANIC CHEMISTRY. *Educ:* Univ Montreal, BSc, 64; Univ NB, PhD(org synthesis), 67. *Prof Exp:* Fel org synthesis, Univ NB, 67-68; fel protein chem, Biochem Lab, Nat Res Coun Can, 68-69 & biophys, 69-70; from asst prof to prof, 70-89, PROF CHEM, UNIV MONTREAL, 89- *Mem:* Am Chem Soc; Chem Inst Can. *Res:* Mechanism of enzyme action; nuclear magnetic resonance for the study of biological systems; ion transport and crown ethers; DNA intercalants and drug receptor interactions; study on neuropeptides; molecular modeling; peptide mimic. *Mailing Add:* Dept Chem Univ Montreal Montreal PQ H3C 3J7 Can

DUGDALE, MARION, b Bellavista, Callao, Peru, Oct 7, 28; US citizen; m 55; c 2. INTERNAL MEDICINE, HEMATOLOGY. *Educ:* Bryn Mawr Col, AB, 50; Harvard Med Sch, MD, 54. *Prof Exp:* Intern & resident med, Univ NC, Chapel Hill, 54-57; resident, 58-59, from instr to assoc prof, 59-74, PROF MED, UNIV TENN, MEMPHIS, 74- *Concurrent Pos:* Fel hemat, Duke Hosp, NC, 57-58; mem coun on cerebrovasc dis & mem coun on thrombosis, Am Heart Asn; mem, Nat Heart & Lung Adv Coun, 75-79. *Mem:* Int Soc Thrombosis & Hemostasis; AMA; Am Soc Hemat; Sigma Xi. *Res:* Hemostasis. *Mailing Add:* Coleman H316 Madison Ave Memphis TN 38163

DUGDALE, RICHARD COOPER, b Madison, Wis, Feb 6, 28; m 53, 66; c 3. BIOLOGICAL OCEANOGRAPHY, LIMNOLOGY. *Educ:* Univ Wis, BS, 50, MS, 51, PhD(zool), 55. *Prof Exp:* Fel, Univ Ga, 56; instr zool, Univ Ky, 57; from instr to asst prof zool, Univ Pittsburgh, 58-62; from assoc prof to prof marine sci, Univ Alaska, 62-67; res prof oceanog, Univ Wash, 67-75; res scientist, Bigelow Lab Ocean Sci, West Boothbay Harbor, Maine, 75-80; PROF BIOL SCI & ASSOC DIR, INST MARINE COASTAL STUDIES, UNIV SOUTHERN CALIF, 80- *Concurrent Pos:* Fulbright res fel, Inst Oceanog & Fisheries Res, Athens, Greece, 72-73. *Mem:* AAAS; Am Soc Limnol & Oceanog (vpres, 74, pres, 75); Am Geophys Union. *Res:* Nitrogen cycle in the sea; biological nitrogen fixation in sea and lakes; vitamins and microorganic constituents; mechanisms of nutrient limitation; upwelling ecosystems. *Mailing Add:* Dept Biol Sci Univ Southern Calif Allan Hancock Found AHF 254 Los Angeles CA 90089

DUGGAL, SHAKTI PRAKASH, b New Delhi, India, Oct 1, 31; m 64; c 2. COSMIC RAY PHYSICS, SPACE PHYSICS. *Educ:* Univ Delhi, BSc, 51, MSc, 53; Gujrat Univ, India, PhD(physics), 59. *Prof Exp:* Res asst space physics, Phys Res Lab, India, 53-58, sr res fel, 58-60; from asst prof to assoc prof, 60-77, PROF COSMIC RAY, BARTOL RES FOUND, FRANKLIN INST, UNIV DEL, 78- *Mem:* Fel Am Phys Soc; life mem Am Geophys Union; AAAS. *Res:* Astrophysics; geophysics; solar-terrestrial physics. *Mailing Add:* 411 Dresel Pl Swarthmore PA 19081

DUGGAN, DANIEL EDWARD, b New York, NY, June 29, 26; m 56; c 3. BIOCHEMICAL PHARMACOLOGY. *Educ:* St Johns Univ, BS, 51; Univ Md, MS, 53; Georgetown Univ, PhD(biochem), 55. *Prof Exp:* Asst, Univ Md, 51-53; biochemist, Nat Heart Inst, 56-62; res assoc, Merck Inst Therapeut Res, 62-70, sr res fel, 70-73, dir biochem pharmacol, 73-82, sr dir drug metab, 82-90; CONSULT, DUGGAN ASSOCS, 90- *Concurrent Pos:* Am Instrument Co fel, Nat Heart Inst, 55-56. *Mem:* Am Chem Soc; Am Soc Pharmacol & Exp Therapeut. *Res:* Drug metabolism; pharmacokinetics; biochemistry of active transport; inflammation. *Mailing Add:* Duggan Assocs West Point PA 19486-0113

DUGGAN, DENNIS E, b Calgary, Alta, July 13, 30; m 54; c 4. BACTERIAL GENETICS. *Educ:* Univ Alta, BSc, 53; Ore State Univ, MSc, 57, PhD(bact), 61. *Prof Exp:* Sr res bacteriologist, H J Heinz Co, 58-61 & Midwest Res Inst, 61-62; fel bact, Sch Med, Univ Kans, 62; fel genetics, Oak Ridge Nat Lab, 63-64; fel, Sch Med, Yale Univ, 64-67; assoc prof bact, 67-77, ASSOC PROF MICROBIOL & CELL SCI & ASSOC MICROBIOLOGIST, UNIV FLA, 77- *Mem:* Am Soc Microbiol. *Res:* Freezing preservation of bacteria; mass culture and concentration of lactobacilli; radiation and thermal preservation of foods; radiation resistant bacteria; mechanisms of radiation resistance; gene action; transduction; mega plasmids/minichromosomes. *Mailing Add:* Dept Microbiol Univ Fla Gainesville FL 32611

DUGGAN, HELEN ANN, b Essex, Ont, Jan 11, 21; US citizen. CHEMISTRY. *Educ:* Siena Heights Col, BS, 41; Cath Univ Am, PhD(chem), 48. *Prof Exp:* Teacher, Cath Schs, Ill, 41-43; instr chem, Barry Col, 47-53, asst prof, 53-54; teacher, Cath Schs, Mich, 54-56; asst prof chem & physics, Siena Heights Col, 56-62, assoc prof chem & biol, 62-63; prof chem & biol & chmn div natural sci, St Dominic Col, 63-68; prof chem, Barry Col, 68-70; prof, 75-90, EMER PROF CHEM, SIENA HEIGHTS COL, 90- *Concurrent Pos:* Mem, Cath Round Table Sci, 49. *Mem:* Am Soc Microbiol; Am Chem Soc; Am Inst Chem; Sigma Xi. *Res:* Chemical genetics; thermal decomposition of hydrocarbons. *Mailing Add:* Siena Heights Col 1247 E Siena Heights Dr Adrian MI 49221

DUGGAN, JEROME LEWIS, b Columbus, Ohio, Aug 4, 33; m 51; c 3. PHYSICS. *Educ:* NTex State Univ, MS, 56; La State Univ, PhD(physics), 61. *Prof Exp:* Asst prof physics, Univ Ga, 61-63; mem sr staff, Spec Training Div, Oak Ridge Assoc Univs, 63-74; PROF PHYSICS, NTEX STATE UNIV, 74- *Mem:* Am Phys Soc. *Res:* Low energy nuclear physics. *Mailing Add:* Dept Physics NTex State Univ Denton TX 76203

DUGGAN, MICHAEL J, b Boulder, Colo, May 21, 31; m 58; c 2. PHYSICS. *Educ:* Mass Inst Technol, SB, 52; Ohio State Univ, MS, 53; Stanford Univ, PhD(physics), 64. *Prof Exp:* Asst prof, 64-73, ASSOC PROF PHYSICS, SAN JOSE STATE UNIV, 73- *Res:* Theoretical physics. *Mailing Add:* Dept Physics San Jose State Univ San Jose CA 95192

DUGGAR, BENJAMIN CHARLES, b Madison, Wis, July 13, 33; m 57; c 3. PROGRAM PLANNING & EVALUATION. *Educ:* Yale Univ, BS, 55; Univ Pittsburgh, MS, 56; Harvard Univ, ScD(biotechnol), 63. *Prof Exp:* Res fel, Sch Pub Health, Harvard Univ, 59-62; scientist, Mitre Corp, 62-63; vpres, Bio-Dynamics, Inc, 63-68, pres, 68-70, pres, Res Div, 70-73; dir, Off Anal Health Serv Admin, Dept Health, Educ & Welfare, 73-76; vpres, JRB Assocs, 76-86; exec vpres, 86-88, SR VPRES, CTR HEALTH POLICY STUDIES, LA JOLLA MGT CORP, 89- *Concurrent Pos:* Gen Motors res fel biotechnol, Sch Pub Health, Harvard Univ, 59-61, Guggenheim fel aviation health & safety, 61-62; dir, Carroll Rehab Ctr Visually Impaired, 69-74; preceptorial prof, Human Ecol Ctr, Antioch Col, 75-77; trustee, Howard County Gen Hosp, 78-, chmn bd, 85-87. *Mem:* Human Factors Soc; Am Pub Health Asn; Am Indust Hyg Asn; Nat Rural Health Asn. *Res:* Health services delivery systems; quality assurance; evaluation methodology for social welfare programs. *Mailing Add:* 5511 High Tor Hill Columbia MD 21045

DUGGER, WILLIE MACK, JR, b Adel, Ga, July 28, 19; m 46; c 2. PLANT PHYSIOLOGY. *Educ:* Univ Ga, BSA, 41; Univ Wis, MS, 42; NC State Col, PhD, 50. *Prof Exp:* Asst, Univ Wis, 41-42; asst prof bot, Univ Ga, 46; asst, NC State Univ, 46-50; asst prof plant physiol, Univ Md, 50-55; assoc plant physiologist, Agr Exp Sta, Univ Fla, 55-60; res plant physiologist, Agr Air Res Ctr, Univ Calif, Riverside, 60-63, chmn dept life sci, 64-68, dean, Col Natural & Agr Sci, 68-81, assoc dir, Agr Exp Sta, 70-81, prof bot, 63-90, EMER PROF BOT, UNIV CALIF, RIVERSIDE, 90- *Mem:* AAAS; Am Soc Plant Physiol; Sigma Xi. *Res:* Organic transport and minor element nutrition in plants; desert plant physiology. *Mailing Add:* Dept Bot & Plant Sci Univ Calif Riverside CA 92521

DUGGIN, MICHAEL J, b Dorking, Eng, July 30, 37; US citizen; m 79; c 2. PHYSICS OF REMOTE SENSING. *Educ:* Univ Melbourne, Australia, BSc, 59; Monash Univ, Australia, PhD(physics), 65. *Prof Exp:* Teaching fel physics, Monash Univ, Australia, 62-64 & Univ Pittsburgh, 64-66; asst prof metall eng, Univ Pittsburgh, 66; res scientist physics & metall, Commonwealth Sci & Indust Res Orgn, Australia, 67-71, sr res scientist physics & remote sensing, 71-79; assoc prof, 79-82, PROF PHYSICS & REMOTE SENSING, STATE UNIV NY, SYRACUSE, 82-, ADJ PROF GEOL, 90- *Concurrent Pos:* Prin investr, NASA, 80-85, USDA, 84-85, US Air Force, 83, 87-; assoc ed, Int J Remote Sensing, 85- *Mem:* Fel Inst Physics; fel Royal Astron Soc; assoc fel Am Inst Aeronaut & Astronaut; sr mem Inst Elec & Electronics Engrs; fel Optical Soc Am; Am Soc Photogram; Soc Photo-Optical Instrumentation Engrs; Sigma Xi. *Res:* Studies of the physics of processes controlling the level and spectral distribution of radiance recorded by remote sensing devices for terrestrial monitoring; image analysis; sensor design; radiometry; optical property measurement; infrared sensor studies; background radiation suppression; background clutter suppression; feature enhancement in image analysis; approximately 100 publications. *Mailing Add:* Div Eng CESF State Univ NY 308 Bray Hall Syracuse NY 13210

DUGGINS, WILLIAM EDGAR, b Philadelphia, Pa, Aug 3, 20; m 47; c 2. ORGANIC CHEMISTRY. *Educ:* LaSalle Col, AB, 43; Mass Inst Technol, PhD, 49. *Prof Exp:* Res chemist, Gen Aniline & Film Corp, 49-52, sr develop engr, 52-59, supvr new prod develop, 59-60, proj mgr, 61; asst to vpres res & develop, Collier Carbon & Chem Corp, 61-62, mgr mkt res & develop, 62-70, tech mgr, 70-81; TECH MGR, CHEM DIV, UNION OIL CALIF, 81- *Mem:* Am Chem Soc; Com Develop Asn; Chem Mkt Res Asn; Am Carbon Soc. *Res:* Dyes; intermediates; resins; amino acids; high-pressure acetylene chemicals; petrochemicals; fertilizers; pesticides; petroleum coke; carbon; graphite. *Mailing Add:* 3427 Corinna Dr Rancho Palos Verdes Peninsula CA 90274

DUGLE, JANET MARY ROGGE, b Pierre, SDak, June 7, 34; div. PLANT TAXONOMY. *Educ:* Carleton Col, BA, 56; Univ SDak, MA, 60; Univ Alta, PhD(bot), 64. *Prof Exp:* Secy, Hotpoint Co, Chicago, 56; teacher high sch, SDak, 57-58; instr bot, Univ SDak, 59-60, curator, 58-60; instr, Univ Alta, 64-65; lectr biol, Yale Univ, 65-67, secy biophys, 66-67; RES OFFICER BOT, ATOMIC ENERGY CAN LTD, WHITESHELL NUCLEAR RES ESTAB, 67- *Concurrent Pos:* Univ res fel biol, Yale Univ, 65-66; res assoc, Univ Man, 67, lectr, 68 & adj prof. *Mem:* AAAS; Int Asn Plant Taxon; Am Soc Plant Taxon; Can Bot Asn (secy, 76-78); Soc Study Evolution. *Res:* Plant systematics and ecology; radiation ecology; evolutionary biology. *Mailing Add:* 11 Cauchon Rd Pinawa MB R0E 1L0 Can

DUGLISS, CHARLES H(OSEA), b Staatsburg, NY, Oct 12, 21; m 46; c 2. CHEMICAL ENGINEERING. *Educ:* Pratt Inst, BChE, 43. *Prof Exp:* Chem engr, Am Cyanamid Co, 43-64, sr res scientist, Resins Res Group, 64-68, sr res scientist, Chem Eng Div, Stamford Res Labs, 68-86; RETIRED. *Mem:* Am Inst Chem Engrs; Am Chem Soc. *Res:* Thermosetting and thermoplastic polymers and processes; reinforced plastics; polymer processing; chemilumin escence. *Mailing Add:* Woodland Valley Rd HC-01 Box 102A Phoenicia NY 12464

DUGOFF, HOWARD, b Yonkers, NY, Nov 23, 36; m 58; c 3. MECHANICAL ENGINEERING, PHYSICS. *Educ:* Stevens Inst Technol, ME, 58, MS, 60. *Prof Exp:* Res asst underwater weapons, Davidson Lab, Stevens Inst Technol, 59-60, assoc res engr, 60-63, res engr, 63-65, chief vehicle res, 65-67, asst head phys factors, Hwy Safety Res Inst, Univ Mich, 67-71; chief, Anal & Simulation Br, Mobility Systs Lab, US Army Tank-Automotive Command, 71-74; chief, Handling & Stability Div, 74-75, assoc adminr planning & eval, 75-76, assoc adminr res & develop, 76-77, dep adminr, Nat Hwy Traffic Safety Admin, 77-79, adminr, res & spec prog admin, 79-84, Sci & Technol Adv to Secy Transp, US Dept Transp, 84-85; spec consult, 85-87, VPRES, ICF INC, 87- *Concurrent Pos:* Mem driving simulation comt, Hwy Res Bd, Nat Acad Sci-Nat Res Coun, 68-; mem comt F-9 on tires, Am Soc Testing & Mat; mem, fed coord comt sci, eng & technol, 79-80; mem, Exec Policy Bd, Alaska Natural Gas Pipeline Syst, 79-; mem, US Radiation Policy Coun, 79-80. *Honors & Awards:* Dept Army Res & Develop Award, 73. *Mem:* Am Soc Mech Engrs; Soc Automotive Engrs; Int Soc Terrain-Vehicle Systs. *Res:* Hydrodynamics and mechanics of submerged vehicles; hydrodynamics performance of amphibious craft; mechanics of off-road vehicles; stability and control of pneumatic tired vehicles; analysis of driver-vehicle highway systems. *Mailing Add:* 11404 Grundy Ct Potomac MD 20854

DUGOLINSKY, BRENT KERNS, b Waverly, NY, July 5, 45; m 71; c 1. MARINE GEOLOGY, MICROPALEONTOLOGY. *Educ:* Syracuse Univ, BS, 67, MS, 72; Univ Hawaii, PhD(oceanog), 76. *Prof Exp:* Res fel, Syracuse Univ, 71-72; lectr, Windward Community Col, Hawaii, 72-73; res asst, Univ Hawaii, 73-76; asst prof, Univ Wis,Oshkosh, 76-77; econ geologist, res & admin, WVa Geol Surv, 77, head, 77-80; ASSOC PROF OCEANOG, STATE UNIV COL, ONEONTA, 80- *Concurrent Pos:* Adj asst prof, WVa Univ, 77-80. *Mem:* Geol Soc Am; Sigma Xi; Soc Econ Paleontologists & Mineralogists; Int Oceanog Found; Oceanic Soc. *Res:* Description and identification of newly discovered benthic Foraminiferans from deep-sea environment and their relationship with mineral deposits on the sea floor. *Mailing Add:* 82 Maple St Oneonta NY 13820-1921

DUGRE, ROBERT, b Montreal, Que, Sept 10, 49. VIROLOGY, VACCINE. *Educ:* Univ Montreal, BSc, 73, MSc, 76; Univ Quebec, PhD(virol), 78. *Prof Exp:* Fel virol, dept microbiol & immunol, Univ Western Ont, 78-79, chief develop viral vaccines & prod Marek's Disease Vaccine, 80-85, dir, viral vaccines prod, 85-89, VPRES OPERS, UNIV WESTERN ONT, 90- *Concurrent Pos:* Assoc prof, Virol Res Ctr, Atmand-Frappeir Inst. *Res:* Mammalian cell technology; development of technologies for massive cell culture with a main interest for viral vaccines production; development of new viral vaccines. *Mailing Add:* IAF Bio Vac Inc Ville de Laval Laval-des-Rapides PQ H7N 4Z2 Can

DUGUAY, MICHEL ALBERT, b Montreal, Que, Sept 12, 39; m 63; c 2. PHYSICS. *Educ:* Univ Montreal, BS, 61; Yale Univ, PhD(physics), 66. *Prof Exp:* Mem tech staff lasers, Bell Tel Labs, 66-87; PROF ELEC ENG, LAVAL UNIV, 88- *Concurrent Pos:* Mem staff, Sandia Labs, 74-77. *Mem:* Optical Soc Am; Am Phys Soc; Inst Elec & Electronics Engrs; Sigma Xi. *Res:* Laser frequency shifting; laser pulse compression; x-ray lasers; ultrashort laser pulse displays; optical sampling; ultrafast Kerr cells; picosecond lifetime measurements; ultrahigh-speed photography; solar lighting. *Mailing Add:* Dept Elec Eng Laval Univ Quebec City PQ G1K 7P4 Can

DUGUID, JAMES OTTO, b Lusk, Wyo, May 22, 40; m 70; c 2. GEOHYDROLOGY, GEOLOGICAL ENGINEERING. *Educ:* Univ Wyo, BS, 63, MS, 64, BS, 66; Princeton Univ, ME & MA, 70, PhD (geol, civil, & geol eng), 73. *Prof Exp:* Res scientist earth sci, Oak Ridge Nat Lab, 72-76, sect head, 76-77; staff scientist earth sci, Battelle Mem Inst, 77-85; vpres & chief hydrogeologist, ICF Clement, 85-88; CONSULT HYDROGEOLOGIST, EBASCO SERV INC, 88- *Concurrent Pos:* Instr, Dept Civil Eng, Univ Wyo, 64-68; proj leader, Environ Sci Div, Oak Ridge Nat Lab, 72-77; mem, Comt for Develop of Standards for Determining Groundwater Supply, Am Nuclear Soc, 75-79, Div Waste Mgt, Dept Energy, Comt Shallow Land Burial of Radioactive Waste, 75-77, Div Biomed & Environ Res, Dept Energy, Health & Safety Comt, 76-78, Scientific Comt, Nat Coun on Radiation Protection, 80-83, Panel on Rev of Radioactive Waste Mgt Practices 83-85, Panel on Rev of Closure Plan, 85-86 & Panel on Rev of the High-Level Wastes, Nat Acad Sci, Oak Ridge Nat Lab, 86-, Peer Rev Comt, Environ Protection Agency, 84 & Div Waste Mgt, Task Force, Dept Energy, 82-83; chmn, Div Waste Mgt, Working Group, Dept Energy, 78-80 & Peer Rev Comt, 81-82; Us rep, Swed Multinational Proj, Dept Energy, 83-88; mem adv coun, Dept Hydrology, Tarleton State Univ, 84- & Dept Geol & Geol Sci, Princeton Univ, 87- *Honors & Awards:* George McJunkin Award, Inst Interamericano, 64- *Mem:* Nat Water Well Asn; Sigma Xi. *Res:* Ground water transport of radionuclides from both low-level radioactive waste disposal sites, deep geologic repositories and hazardous waste sites. *Mailing Add:* 110 13th SE Washington DC 20003

DUGUID, WILLIAM PARIS, b Barrhead, Scotland, June 25, 27; Can citizen; m 56; c 5. PATHOLOGY. *Educ:* Univ Glasgow, MB, 50; MRCP, 64; FRCP, 76. *Prof Exp:* Consult path, Western Regional Hosp Bd, 64-70; sr lectr path, Univ Glasgow, 64-70; assoc prof, McGill Univ, 70-72; sr pathologist, 70-71, CHIEF PATH, MONTREAL GEN HOSP, 71-; PROF PATH, MCGILL UNIV, 72-, PROF CANCER CTR, 78- *Concurrent Pos:* Consult path, Reddy Mem Hosp, 72-; examr, Royal Col Surgeons, 84-; dir, Bd Res Inst, Montreal Gen Hosp; guest prof, Univ Edinburg, Leeds & Western Ont. *Honors & Awards:* Queen Elizabeth II Silver Jubilee Medal, Gov Gen Can, 52-87. *Mem:* Int Acad Pathologists; Asn Clin Pathologists; Am Pancreatic Asn; Am Asn Pathologists. *Res:* Cell differentiation in the pancress with particular reference to carcinogenesis and islet cell differentiation. *Mailing Add:* 6501 Cedar Ave Rm 377 Montreal PQ H3G 1A4 Can

DUGUNDJI, JAMES, mathematics; deceased, see previous edition for last biography

DUGUNDJI, JOHN, b New York, NY, Oct 25, 25; m 65; c 2. AERONAUTICAL ENGINEERING. *Educ:* NY Univ, BAE, 44; Mass Inst Technol, ScD(aeronaut eng), 51. *Prof Exp:* Res engr aerodyn, Grumman Aircraft Eng Corp, 48-49; asst aeroelasticity, Mass Inst Technol, 50-51; prin dynamics engr, Repub Aviation Corp, 51-56; from asst prof to assoc prof, 57-70, PROF AERONAUT & ASTRONAUT, MASS INST TECHNOL, 70- *Mem:* Am Inst Aeronaut & Astronaut. *Res:* Aeroelasticity; structural dynamics. *Mailing Add:* Dept Aeronaut & Astronaut Mass Inst Technol Cambridge MA 02139

DUHAMEL, RAYMOND C, b New Bedford, Mass, Oct 4, 42; m 64; c 3. COLLAGEN BIOCHEMISTRY, VASCULAR GRAFT BIOCHEMISTRY. *Educ:* Stonehill Col, BS, 63; Boston Col, MS, 69; Univ Mass, Amherst, PhD(zool), 77. *Prof Exp:* Res assoc, dept cell & develop biol, Univ Ariz, 76-77 & dept pharmacol, 77-79, from res asst prof to res assoc prof, 85-86; MGR BIOL PROD, BARD VASCULAR SYSTS, C R BARD INC, 86- *Mem:* Am Soc Biochemists; AAAS; Sigma Xi; Am Soc Biomat; Am Soc Artificial Internal Organs; NY Acad Sci. *Res:* Biochemistry of collagen and basement membranes; biologic vascular grafts; protein coatings on synthetic vascular grafts; biology of endothelial cells. *Mailing Add:* Bard Vascular Systs Div C R Bard Inc 129 Concord Rd Box M Billerica MA 01821-4603

DUHL, DAVID M, b Paterson, NJ, Aug 12, 53. CANCER RESEARCH. *Educ:* Muhlenberg Col, BS, 75; Hahnemann Univ, MS, 77, PhD(biochem), 80. *Prof Exp:* Fel, 80-84, RES INSTR BIOCHEM, SCH MED, VANDERBILT UNIV, 84- *Mem:* AAAS; Am Chem Soc; Am Soc Cell Biol; Sigma Xi; Am Asn Cancer Res. *Res:* Nonhistone proteins of human colon cancer and their relation to the expression and maintenance of the malignant state. *Mailing Add:* Dept Biochem Vanderbilt Univ Sch Med Nashville TN 37232

DUHL, DAVID N, b Staten Island, NY, Apr 20, 39; m 63; c 2. METALLURGY. *Educ:* Rensselaer Polytech Inst, BMetE, 60; Mass Inst Technol, SM, 63, PhD(metall), 64. *Prof Exp:* Res asst metall, Mass Inst Technol, 60-64; res assoc, adv mat res & develop lab, 64-70, sr asst mat proj engr, 70-81, SR PROJ ENGR, MAT ENG & RES LAB, PRATT & WHITNEY AIRCRAFT, COM PROD DIV, UNITED TECHNOL CORP, 81- *Honors & Awards:* George W Mead Award. *Mem:* Metall Soc. *Res:* High temperature alloy development and evaluation; phase stability and equilibra. *Mailing Add:* Pratt & Whitney 400 Main St East Hartford CT 06108

DUHL, LEONARD J, b New York, NY, May 24, 26; m 80; c 5. PSYCHIATRY, PUBLIC HEALTH. *Educ:* Columbia Univ, AB, 45; Albany Med Col, MD, 48; Am Bd Psychiat & Neurol, dipl psychiat, 56. *Prof Exp:* Intern, Jewish Hosp, Brooklyn, NY, 48-49; resident psychiat, Winter Vet Admin Hosp, Topeka, Kans, 49-54; psychiatrist, Prof Serv Br, NIMH, 54-64, chief off planning, 65-66; spec asst to secy, US Dept Housing & Urban Develop, 66-68; dir dual degree prog health sci & med educ, 73-77, PROF CITY & REGIONAL PLANNING, COL ENVIRON DESIGN & PROF PUB HEALTH, SCH PUB HEALTH, UNIV CALIF, BERKELEY, 68-; PROF PSYCHIAT, MED CTR, UNIV CALIF, SAN FRANCISCO, 68- *Concurrent Pos:* Fel, Sch Psychiat, Menninger Found, 49-54; asst health officer, Contra Costa County Health Dept, Calif, 51-53; clin instr, George Washington Univ, 58-61, assoc, 61-63, asst clin prof, 63-68; mem res adv comt, Off Educ, 56-57, mem joint task force health & housing, 61-, mem comn subcomt cult deprivation-poverty, 63, liaison to asst secy health, 65-66; consult, Peace Corps, 61-65, Model Cities Prog, 66-68; mem, bd trustees, Park Forest Col, 65-68 & Robert F Kennedy Found; mem, sci & technol adv comt, Calif State Legis, 70-75; mem, sci adv coun, House Rep Comt Pub Works, 73; mem, bd trustees, Calif Sch Prof Psychol, 78; consult, Assembly Off Res, Calif State Legis, Sacramento, 81-84, Select Comts Mental Health, 83; consult, WHO, Europe, 85-; consult, UNICEF, 87-; Cal Inst Integral Studies, 91; sr assoc, Youth Policy Inst, 87-; consult mayor, Oakland, CA, 90- *Honors & Awards:* Albert Deutsch Lectr, Am Psychiat Asn, 67; Erich Lindemann Lectr, Harvard, 88. *Mem:* Am Col Psychiatrists; fel Am Psychiat

Asn; fel Am Pub Health Asn; Int Asn Child Psychiat. *Res:* Humanistic and health science education; public health; medical and health care; holistic health; community development and organization; urban planning; psychotherapy; mental health aspects of education; psychoanalytic concepts, planning and policy; healthy cities; international health. *Mailing Add:* Sch Pub Health Univ Calif Berkeley CA 94720

DUHL-EMSWILER, BARBARA ANN, b Ft Knox, Ky, Oct 13, 52. PEPTIDE CHEMISTRY. *Educ:* Mich State Univ, Bs, 74, PhD(org chem), 79. *Prof Exp:* Fel, McNeil Pharmaceut, 79-81, SCIENTIST, ORTHO PHARMACEUT CORP, JOHNSON & JOHNSON, 81- *Mem:* Am Chem Soc; AAAS. *Res:* Process development and the synthesis of small peptides; stereocontrolled organic synthesis, especially as relating to Wittig reactions; insect pheromones; leukotrenes and other natural products. *Mailing Add:* RD 1 Box 349B Pittstown NJ 08867-9729

DUHRING, JOHN LEWIS, b Plainfield, NJ, May 7, 33; c 3. OBSTETRICS & GYNECOLOGY. *Educ:* McGill Univ, BS, 55; Univ Pa, MD, 59. *Prof Exp:* Investr, Harrison Dept Surg Res, Univ Pa, 56-60; from asst prof to prof obstet & gynec, Med Ctr, Univ Ky, 63-78, dir obstet, 73-78; PROF & CHMN, DEPT OBSTET & GYNEC, MED COL OHIO, TOLEDO, 78- *Concurrent Pos:* Asst clin prof obstet & gynec, Univ Hawaii, 67-70; attend staff, Obstet & Gynec Serv, Tripler Gen Hosp, Honolulu, 67-70; vis prof, Letterman Army Med Ctr, 73 & Tripler Army Med Ctr, 74. *Mem:* Am Col Obstetricians & Gynecologists; AMA. *Res:* Continuing effort to reduce perinatal morbidity and mortality in high risk pregnancy through biochemical and biophysical means. *Mailing Add:* Dept Obstet & Gynec Med Col Ohio CS 10008 Toledo OH 43699

DUHRKOPF, RICHARD EDWARD, b Chicago, Ill, Oct 11, 49. GENETICS, ENTOMOLOGY. *Educ:* Ohio State Univ, BS, 71, MSc, 73, PhD(genetics), 77. *Prof Exp:* Vis asst prof biol, Bucknell Univ, 77-78; fel med entom, Johns Hopkins Univ, 78-80; asst prof, Grinnell Col, 80-; AT DEPT BIOL, BAYLOR UNIV, TX. *Mem:* AAAS; Am Inst Biol Sci; Genetics Soc Am; Animal Behav Soc; Soc Study Evolution; Sigma Xi. *Res:* Mosquito, behavioral, evolutionary and quantitative genetics. *Mailing Add:* Biol Dept Baylor Univ Waco TX 76798

DUICH, JOSEPH M, b Farrell, Pa, June 7, 28; m 53; c 3. AGRONOMY. *Educ:* Pa State Univ, BS, 52, PhD(agron), 57. *Prof Exp:* Instr, 54, from asst prof to assoc prof, 56-67, prof agron, 67-75, PROF TURFGRASS SCI, DEPT AGRON, PA STATE UNIV, 75- *Mem:* Fel AAAS; Am Soc Agron; Crop Sci Soc Am; Weed Sci Soc Am. *Res:* Agronomic research in turfgrass field; breeding, weed control and fertility research. *Mailing Add:* 1210 Charles St State College PA 16801

DUIGNAN, MICHAEL THOMAS, b Flushing NY, June 14, 50; m; c 1. NONLINEAR OPTICS, LASER APPLICATIONS. *Educ:* Rutgers Univ, BS, 76; Brandeis Univ, PhD(chem), 81. *Prof Exp:* Nat res coun fel, Naval Res Lab, 81-83; VPRES & SR RES SCIENTIST, POTOMAC PHOTONICS, 84- *Honors & Awards:* Naval Res Lab Publ Award, 84. *Mem:* AAAS; Am Phys Soc; Optical Soc Am; Am Chem Soc. *Res:* Infrared multiphoton absorption and decomposition in the gas phase; multiphoton ionization spectroscopy especially for free radicals; nonlinear optical scattering; laser applications to analytical chemistry. *Mailing Add:* Potomac Photonics Inc 4720-E Boston Way Lanham MD 20706

DUISMAN, JACK ARNOLD, b Ft Knox, Ky, Mar 14, 37; m 61; c 2. PHYSICAL CHEMISTRY. *Educ:* Augustana Col, BA, 59; Univ Calif, Berkeley, PhD(phys chem), 66. *Prof Exp:* NSF fel phys chem, Univ Kans, 66-67; res chemist, Union Carbide Corp, 67-70, proj scientist, 70-71, group leader, 71-73, res supvr, Linde Div, 73-76; SR RES ASSOC, CHEVRON RES CO, 76- *Mem:* Am Chem Soc. *Res:* Chemical thermodynamics; photovoltaic; solid state chemistry; zeolite catalysis; environmental control; solar energy; oxygen chemistry. *Mailing Add:* 855 Seaview El Cerrito CA 94530-3008

DUJARDIN, JEAN-PIERRE L, MENTAL IMAGERY, ANTONOMIC INTEGRATION. *Educ:* Ohio State Univ, PhD(physiol), 76. *Prof Exp:* ASSOC PROF CARDIOVASC PHYSIOL, OHIO STATE UNIV, 76- *Mailing Add:* Dept Physiol Ohio State Univ 1068 Graves Hall 333 W Tenth Ave Columbus OH 43210

DUJOVNE, CARLOS A, b Resistencia, Arg, July 9, 37; m 63; c 1. CLINICAL PHARMACOLOGY, INTERNAL MEDICINE. *Educ:* Univ Buenos Aires, MD, 61. *Prof Exp:* Instr pharmacol & med, Univ Buenos Aires, 62-63; intern, Mt Sinai Hosp, Chicago, 63-64, resident internal med, 64-66; resident, Vet Admin Hosp, Washington, DC, 66-67; from asst prof to assoc prof, 70-78, PROF MED & PHARMACOL, SCH MED, UNIV KANS, 79-, DIR, LIPID & ARTERIOSCLEROTIC PREVENTION CLINIC. *Concurrent Pos:* NIH fel liver & metab, George Washington Univ, 67-68; fel clin pharmacol, Johns Hopkins Hosp, 68-70; instr, Sch Med, Loyola Univ, Ill, 65; clin asst, Chicago Med Sch, 65-66. *Res:* Treatment and prevention of arteriosclerosis and lipid disorders; hepatotoxicity; drug toxicity and adverse reactions; clinical therapeutic trials; effects of drugs on lipoprotein regulation in man. *Mailing Add:* Dept Pharmacol Lipid & Arteriosclerosis Univ Kans Med Ctr 1347 Bell Kansas City KS 66103

DUKATZ, ERVIN L, JR, b Detroit, Mich, July 2, 52; m; c 2. CONSTRUCTION MATERIALS ANALYSIS. *Educ:* Valparaiso Univ, BSCE, 74; Pa State Univ, MSCE, 78, PhD(civil eng), 84. *Prof Exp:* Trainee engr, Pittsburgh-Des Moines Steel Co, 74-76; instr civil eng, Pa State Univ, 77-82; asst prof civil eng, Bucknell Univ, 83-84; construct mat engr, 84-86, SR MAT ENGR, VULCAN MAT CO, 86- *Concurrent Pos:* Consult engr & land surv, 80-84. *Mem:* Am Soc Civil Engrs; Nat Asphalt Pavement Asn; Am Soc Testing & Mat; Asn Asphalt Paving Technologists; Transp Res Bd. *Res:* Highway pavements and constituent materials; bituminous materials and the composition and mechanical properties of asphalt; stabilized and unstabilized base materials. *Mailing Add:* Vulcan Mat Co PO Box 530187 Birmingham AL 35253-0187

DUKE, CHARLES BRYAN, b Richmond, Va, Mar 13, 38; m 61; c 2. SOLID STATE PHYSICS, RESEARCH ADMINISTRATION. *Educ:* Duke Univ, BS, 59; Princeton Univ, PhD(physics), 63. *Prof Exp:* Staff mem physics, Gen Elec Co, 63-69; prof physics, Univ Ill, Urbana, 69-72, res prof, Coord Sci Lab, 69-72; prin scientist res fel, Xerox Corp, 72-73, mgr molecular & org mat res, 73-81, mgr theory physics & chem res, 81-88, mgr, Imaging Sci Lab, 88, dep dir, Webster Res Ctr, 88, dep dir & chief scientist, Battelle Pac Northwest Div, 88-89, SR RES FEL, WEBSTER RES CTR, XEROX CORP, 89- *Concurrent Pos:* Chmn steering comt, Int Conf Solid Surfaces, 71; consult, Gen Elec Co, 69-72; adj prof physics, Dept Physics & Astron, Univ of Rochester, 72-88; affil prof physics, Dept Physics, Univ Wash; bd dirs, Am Vacuum Soc, 72-76 & 78-81; sr res fel, Xerrox Corp, 73-88; bd gov, Am Inst Physics, 76-82 & 84-87; ed-in-chief, J Mat Res, 85-86; counr, Mat Res Soc, 88-90. *Honors & Awards:* M W Welch Award in Vacuum Sci & Technol, 77. *Mem:* Fel Am Phys Soc; Am Chem Soc; Sigma Xi; hon mem Am Vacuum Soc (pres, 79); fel Inst Elec & Electronics Engrs; Mat Res Soc (treas, 91-92). *Res:* Scattering theory; many-body theory; electron tunneling in solids; low-energy electron-solid-scattering; surface crystallography; chemisorption; catalysis; theory of electronic structure of polymers and molecular solids. *Mailing Add:* Xerox Corp Webster Res Ctr 800 Phillips Rd 0114-38D Webster NY 14580

DUKE, CHARLES LEWIS, b Asheville, NC, Sept 7, 40; m 65; c 2. NUCLEAR PHYSICS. *Educ:* NC State Univ, BS, 62; Iowa State Univ, PhD(physics), 67. *Prof Exp:* AEC fel physics, Ames Lab, Iowa State Univ, 67-68; instr, Inst Physics, Aarhus Univ, 68-69; from asst prof to assoc prof, 69-79, PROF PHYSICS, GRINNELL COL, 79- *Mem:* Am Phys Soc. *Res:* Production and study of nuclei far from stability through on-line isotope separator techniques; beta-strength function and delayed neutron measurements. *Mailing Add:* Dept Physics Grinnell Col Grinnell IA 50112

DUKE, DAVID ALLEN, b Salt Lake City, Utah, Nov 26, 35; m 55; c 4. MATERIALS SCIENCE. *Educ:* Univ Utah, BS, 57, MS, 59, PhD(geol eng, ceramics), 62. *Prof Exp:* Jr scientist, Kennecott Copper Res Ctr, 60-61; res mineralogist, 62-67, mgr adv mat res, 67-70, mgr bus develop, 70-74, gen mgr, Indust Prods Div, 75-76, bus mgr, Telecommun, 76-80; vpres telecommun, Corning Glass Works, 80-85, sr vpres res & develop, 85-87, res, develop & eng, 87-88, VCHMN TECHNOL, CORNING GLASS WORKS, 88- *Mem:* Mineral Soc Am; Am Ceramic Soc; fel Am Inst Chemists. *Res:* Geochemical and instrumental analysis of rocks and minerals; crystal chemistry; crystallization of glass and the properties of glass-ceramic materials; chemical strengthening of glass-ceramics; fiber optics. *Mailing Add:* Seven Theresa Dr Corning NY 14830

DUKE, DENNIS WAYNE, b Nashville, Tenn, Aug 29, 48; m 74; c 4. LATTICE GAUGE THEORY. *Educ:* Vanderbilt Univ, BS, 70; Iowa State Univ, PhD(physics), 74. *Prof Exp:* Teaching fel, Univ Rochester, 74-76; res fel, Fermilab, 76-78 & Rutherford Lab, 78-79; vis asst prof, 79-84, ASST PROF PHYSICS & ASSOC DIR RES, SUPERCOMPUT COMPUT RES INST, FLA STATE UNIV, 84- *Mem:* Am Phys Soc. *Res:* Computationally intensive areas of theoretical physics; parton model phenomenology. *Mailing Add:* Physics Dept Fla State Univ Tallahassee FL 32306

DUKE, DOUGLAS, b Philadelphia, Pa, Aug 7, 23; m 44; c 2. ASTRONOMY, ASTROPHYSICS. *Educ:* Univ Calif, AB, 47; Univ Chicago, PhD(astron), 50. *Prof Exp:* Asst, Yerkes Observ, Univ Wis, 47-50; asst prof astron, Univ NC, 50-51; from asst prof to assoc prof astron & phys sci, Univ Fla, 51-54; head visibility br, US Naval Electronics Lab, 54-55; res engr, Convair Astronaut, 55-57; sr engr, Atlantic Missile Range, Radio Corp Am, 57-59; mem tech staff, Inst Defense Anal, DC, 59-62; staff scientist to vpres res, Autonetics Div, NAm Aviation, Inc, 62-64; dir aerospace prog, Data Dynamics, Inc, 64-65; prof, 66-85, EMER PROF ASTRON, UNIV MIAMI, 85- *Concurrent Pos:* Asst dir, Morehead Planetarium, Chapel Hill, NC, 50-51; consult, Cmndg Gen, NAm Air Defense Command, 61-62. *Mem:* Fel AAAS; Am Astron Soc; Astron Soc Pac. *Res:* Observational astrophysics; astrometry; missile and satellite tracking instrumentation; data reduction; satellite and interplanetary orbits; celestial navigation; atmospheric optics; planetarium operation. *Mailing Add:* 919 Poppy Lane Carlsbad CA 92009

DUKE, EVERETTE LORANZA, b Goochland Co, Va, June 28, 29; m 53; c 2. SOIL SCIENCE. *Educ:* Va State Col, BS, 49; Mich State Univ, MS, 50, PhD(soil sci), 55. *Prof Exp:* Instr hort, Va State Col, 50-51; assoc prof sci, Norfolk State Col, 55-63, prof biol, 63-71, ASSOC VPRES ACAD AFFAIRS, NORFOLK STATE UNIV, 71- *Concurrent Pos:* Dir, Thirteen-College Curric Prog, Norfolk State Col, 67-71. *Mem:* Sigma Xi. *Res:* Land use planning; land utilization in relation to land character in rural-urban fringe areas in Southern Michigan. *Mailing Add:* Norfolk State Univ 2401 Corprew Ave Norfolk VA 23504

DUKE, GARY EARL, b Galesburg, Ill, Dec 16, 37; m 61. AVIAN PHYSIOLOGY, ECOLOGY. *Educ:* Knox Col, Ill, BA, 59; Mich State Univ, MS, 64, PhD, 67. *Prof Exp:* From asst prof to assoc prof, 68-76, PROF AVIAN & VET PHYSIOL, UNIV MINN, ST PAUL, 76- *Concurrent Pos:* NSF & Agr Exp Sta res grants, 67-82, 73-80. *Mem:* Am Ornith Union; Poultry Sci Asn; Am Soc Vet Physiologists & Pharmacologists; Am Physiol Soc; Comp Gastroenterol Soc. *Res:* Study of gastrointestinal motility and absorption in turkeys and birds of prey; regulation of motility in turkeys and pellet egestion in birds of prey. *Mailing Add:* Dept Vet Biol Univ Minn Col Vet Med St Paul MN 55108

DUKE, JAMES A, b Birmingham, Ala, Apr 4, 29; m 50; c 2. BOTANY. *Educ:* Univ NC, AB, 52, MA, 55, PhD(bot), 59. *Prof Exp:* Lab asst bot, Univ NC, 54-59; Chem Corps, USDA, 55-57; res asst, Mo Bot Garden, 59-61, asst cur, 61-63; asst prof bot, Univ Wash, 61-62; res botanist, USDA, 63-65; ecologist, Battelle Mem Inst, 65-71; supvry botanist, New Crops Res Br, 71-72, chief, Plant Taxon Lab, 72-77, Econ Bot Lab, 77-81 & Germplasm Lab, 81-85, ECON BOTANIST, USDA, 85- *Mem:* Asn Trop Biol; Weed Sci Soc; Int Soc Plant Taxon; Soc Econ Bot; Int Soc Trop Ecol; Am Soc Pharmacog. *Res:*

Taxonomy and ecology in Latin America; ecological amplitudes of crops and weeds; crop diversification; medicinal plants; computerized catalog of ecological, ethnomedicinal and nutritional attributes of economic plants; author of 13 books. *Mailing Add:* Bldg 001 BARC-W USDA Beltsville MD 20705

DUKE, JODIE LEE, JR, b Kennett, Mo, Aug 3, 45; m 67; c 2. CARBOHYDRATE CHEMISTRY. *Educ:* Univ Mo-St Louis, BS, 67; Univ Mich, Ann Arbor, MA, 69, PhD(biochem), 74. *Prof Exp:* Res assoc immunol, Women's Hosp, Univ Mich, 74; res chemist, A E Staley Mfg Co, 74-77; sr res scientist, R J Reynolds Foods Inc, 77-78; SR RES BIOCHEMIST, LEE SCI INC, 79- *Mem:* Am Chem Soc; Sigma Xi. *Res:* Applied research for the food industry. *Mailing Add:* Nine Clemson Ct Newark DE 19711-4301

DUKE, JOHN CHRISTIAN, JR, b Baltimore, Md, May 27, 51; m 78; c 2. NONDESTRUCTIVE EVALUATION, COMPOSITE MATERIAL. *Educ:* Johns Hopkins Univ, BES, 73, MSE, 76, PhD(mat sci eng), 78. *Prof Exp:* Physicist, Nat Bur Standards, 75-77; asst prof mech eng, US Naval Acad, 77-78; fel, Johns Hopkins Univ, 78; asst prof, 78-82, ASSOC PROF ENG MECH, VA POLYTECH INST & STATE UNIV, 82- *Concurrent Pos:* Bd dirs, Am Soc for Nondestructive Testing, 85-87; past chmn, NDE Eng Div, Am Soc Mech Engrs. *Honors & Awards:* Teetor Award, Soc Automotive Engrs, 79. *Mem:* Fel Am Soc Nondestructive Testing; Am Soc Mech Engrs; Am Soc Eng Educ. *Res:* Nondestructive methods of evaluation for understanding deformation mechanism in metal alloys, advanced composites, biomaterials, and adhesively bonded structures. *Mailing Add:* Eng Sci Mech Va Polytech Inst State Univ Blacksburg VA 24061-0219

DUKE, JOHN MURRAY, b Montreal, Que, Jan 3, 47; m 75; c 1. MAGMATIC ORES, STRATEGIC MINERALS. *Educ:* McGill Univ, BSc, 68, MSc, 71; Univ Conn, PhD(geol), 74. *Prof Exp:* Fel, Univ Toronto, 74-76; res scientist, 76-88, DIV DIR, GEOL SURV CAN, 88- *Concurrent Pos:* Mem, Can Geosci Coun Comt on Int Sci Relations. *Mem:* Geol Asn Can; Mineral Asn Can (secy, 78-83, vpres, 86-87, pres, 88-89). *Res:* The genesis of magmatic sulfide and oxide ores of nickel, copper, platinum metals and chromium and the appraisal of their resources. *Mailing Add:* Four Carr Cres Kanata ON K2K 1K4 Can

DUKE, JOHN WALTER, b Ballinger, Tex, Oct 30, 37; m 57; c 3. MATHEMATICS. *Educ:* North Tex State Univ, BA, 59; Tex Tech Univ, MS, 61; Univ Colo, PhD(math), 68. *Prof Exp:* Instr, Tex Tech Univ, 61-64, asst prof, 66-68; assoc prof, 68-77, PROF MATH, ANGELO STATE UNIV, 77- *Mem:* Am Math Soc; Math Asn Am. *Res:* Matrices over division algebras; general ring theory. *Mailing Add:* Dept Math Angelo State Univ San Angelo TX 76909

DUKE, JUNE TEMPLE, b Cambridge, Ohio, June 18, 22. ORGANIC CHEMISTRY, POLYMER CHEMISTRY. *Prof Exp:* Technician, Off Rubber Reserve, Govt Labs, Univ Akron, 44-45, lab supvr rubber res, 45-50, chemist supvr, 50-56; res chemist, Energy Div, Olin Mathieson Chem Corp, 56-59; sr chemist, Standard Oil Co, Ohio 59-60, sr res chemist & proj leader polymer res, 60-68, res assoc, 68-84; RETIRED. *Mem:* AAAS; Am Chem Soc; Fedn Socs Paint Technol. *Res:* Polymerization and structure of synthetic rubber; organo-boron polymers; synthetic coatings; emulsion polymerization; resins and plastics. *Mailing Add:* 28649 Jackson Rd Chagrin Falls OH 44022

DUKE, KENNETH LINDSAY, b Heber City, Utah, Feb 22, 12; m 34; c 3. ANATOMY. *Educ:* Brigham Young Univ, AB, 36; Duke Univ, PhD(cytol), 40. *Prof Exp:* Asst zool, Brigham Young Univ, 36-37; grad asst, from instr to assoc prof anat, 40-81, EMER PROF ANAT, SCH MED, DUKE UNIV, 81- *Concurrent Pos:* China Med Bd NY traveling fel, Malaya, 61; vis instr, Sch Med, Univ Mo, 44; vis instr, Sch Med, Univ NC, 46, vis assoc prof, 55; vis asst prof, Sch Med, Univ Tenn, 49. *Mem:* Am Asn Anat. *Res:* Histology; cytology; mammalian reproductive tract; germ cells of mammals. *Mailing Add:* Dept Anat Duke Univ Med Ctr Durham NC 27710

DUKE, MICHAEL SN, b Los Angeles, Calif, Dec 1, 35; m 58; c 4. GEOLOGY, COSMOCHEMISTRY. *Educ:* Calif Inst Technol, BS, 57, MS, 61, PhD(geochem), 63. *Prof Exp:* Geologist, Astrogeol Br, US Geol Surv, 63-70; lunar sample cur, 70-77, CHIEF LUNAR & PLANETARY SCI DIV, JOHNSON SPACE CTR, NASA, 77- *Concurrent Pos:* Adj prof, Rice Univ, 79- *Honors & Awards:* Nininger Meteorite Award, Ariz State Univ, 63. *Mem:* AAAS; Sigma Xi; Meteoritical Soc; Am Geophys Union. *Res:* Mineralogy, petrology, chemical composition of extraterrestrial material, including lunar rocks, soils, meteorites, and cosmic dust; origin of solar system; development of mineralogical, chemical techniques for laboratory; automated spacecraft applications. *Mailing Add:* 16534 Ivy Grove Houston TX 77058

DUKE, RICHARD ALTER, b Geneva, Ohio, Aug 23, 37. MATHEMATICS. *Educ:* Kenyon Col, AB, 59; Dartmouth Col, MA, 61; Univ Va, PhD(math), 65. *Prof Exp:* Instr math, Univ Wash, 65-66, asst prof, 66-72; from asst prof to assoc prof, 72-83, asst dir math, 72-79, PROF MATH, GA INST TECHNOL, 84- *Concurrent Pos:* Exam leader, Advan Placement Prog Math, Col Bd-Educ Testing Serv, 73-80; panel vis lectr, Math Asn Am, 82- *Mem:* Am Math Soc; Math Asn Am. *Res:* Graph theory and combinatorics with an emphasis on the combinatorial properties of finite complexes, Ramsey theorems and the connections with block designs. *Mailing Add:* Sch Math Ga Inst Technol Atlanta GA 30332

DUKE, ROY BURT, JR, b Houston, Tex, Sept 20, 32; m 50; c 3. PETROLEUM CHEMISTRY. *Educ:* Univ Houston, BS, 56, MS, 60; Ga Inst Technol, PhD(chem), 67. *Prof Exp:* Res chemist, Am Oil Co, 56-60 & Tex Eastman Co, 60-62; instr, Ga Inst Technol, 62-65; adv res chemist, 66-74, SR RES CHEMIST, MARATHON OIL CO, 74- *Mem:* Am Chem Soc. *Res:* Catalysis; dehydrogenation; mechanism of the Grignard reaction; kinetics and mechanism of the reaction of alkoxides with methylene halides; aldol and related condensations; pyrolysis and extraction of oil shale; identification of components in oil shale and related mineral deposits; enhanced oil recovery; surfactant flooding, emulsions and demulsification. *Mailing Add:* Marathon Oil Co Box 269 Littleton CO 80160

DUKE, STANLEY HOUSTON, b Battle Creek, Mich, Oct 9, 44; m 78. PLANT PHYSIOLOGY, PLANT BIOCHEMISTRY. *Educ:* Henderson State Univ, BS, 66; Univ Ark, MS, 69; Univ Minn, PhD(bot), 75. *Prof Exp:* Mem staff, US Army Chem Corps, 68-70; fel, 76-78, from asst prof to prof plant physiol, 78-89, PROF & ASSOC CHMN, DEPT AGRON, UNIV WIS-MADISON, 89- *Concurrent Pos:* Fel, Linnean Soc London, 75. *Mem:* Am Soc Plant Physiologists. *Res:* Carbohydrate metabolism; enzyme kinetics; low temperature effects on physiology and biochemistry; chronobiology; photomorphogenesis; plant physiological ecology; agronomy. *Mailing Add:* Dept Agron 246 Moore Hall 1575 Linden Dr Univ Wis Madison WI 53706

DUKE, STEPHEN OSCAR, b Battle Creek, Mich, Oct 9, 44; m 67; c 2. PLANT PHYSIOLOGY. *Educ:* Henderson State Univ, BS, 66; Univ Ark, MS, 69; Duke Univ, PhD(bot), 75. *Prof Exp:* Instr biol, Duke Univ, 74-75; nat res coun assoc, plant physiol, 75-76, plant physiologist, 75-85, res leader, 85-87, DIR, SOUTHERN WEED SCI LAB, AGR RES SERV, USDA, 87- *Concurrent Pos:* Assoc ed, Weed Sci, 80-85. *Mem:* Am Soc Plant Physiologists; Sigma Xi; AAAS; Weed Sci Soc Am; Scand Soc Plant Physiol; Am Chem Soc. *Res:* Photobiology of plants; secondary plant metabolism, particularly plant phenolics; greening and photosynthesis; herbicide mechanism of action. *Mailing Add:* Southern Weed Sci Lab Agr Res Serv USDA Stoneville MS 38776

DUKE, VICTOR HAL, b Kamas, Utah, Jan 15, 25; m 49; c 3. PHARMACOLOGY, PHARMACOGNOSY. *Educ:* Idaho State Univ, BS, 49, BS, 50; Univ Utah, PhD(pharm), 61. *Prof Exp:* Asst prof pharmacol, Univ NMex, 61-65; assoc prof, Col Pharm, Univ Wyo, 65-68; from assoc prof to prof, Univ Mont, 68-72; dean, Sch Health Sci, Boise State Univ, 72-85; RETIRED. *Concurrent Pos:* Dir drug abuse seminar educators, Univ Mont, 69-71; mem, Nat Adv Coun Health Manpower Educ, 71-72. *Mem:* Am Pharmaceut Asn; Am Public Health Asn; Am Soc Allied Health Prof. *Res:* Effect of drugs on analgesic threshold in animals; drug action in obesity and the evaluation of anorexigenics; neuropharmacology and behavioral pharmacology of psychotogens. *Mailing Add:* Box 576 Rexburg ID 83440

DUKELOW, DONALD ALLEN, b Ellwood City, Pa, June 13, 32; m 56; c 3. METALLURGY. *Educ:* Geneva Col, BS, 54; Carnegie Inst Technol, MS, 56, PhD(metall), 57. *Prof Exp:* Res engr, Jones & Laughlin Steel Corp, 57-64; asst prof metall eng, Univ Pittsburgh, 64-67; sr res engr, US Steel Corp, Monroeville, 67-80; sr res engr, Weirton, 80-85, MGR, PROD-PROCESSING TECHNOL, NAT STEEL CORP, PITTSBURGH, 86- *Mem:* Am Inst Mining, Metall & Petrol Engrs; Am Iron & Steel Engr. *Res:* Physical chemistry of metallurgy. *Mailing Add:* Nat Steel Corp 20 Stanwix St Pittsburgh PA 15222

DUKELOW, W RICHARD, b Princeton, Minn, Oct 23, 36; m 58; c 3. REPRODUCTIVE PHYSIOLOGY. *Educ:* Univ Minn, BS, 57, MS, 58, PhD(physiol), 62. *Prof Exp:* Instr, Univ Minn, Grand Rapids, 60-62, asst prof, 62-64; res assoc biochem, Univ Ga, 64-65, asst res prof, 65-69; assoc prof, Endocrine Res Ctr, 69-74, assoc dean res, Col Vet Med, 85-88, PROF PHYSIOL & ANIMAL HUSB & DIR ENDOCRINE RES UNIT, MICH STATE UNIV, 74- *Concurrent Pos:* NIH biomed grant, 66-67 & spec fel, 67-68; Lalor Found res grant & Pop Coun grant, 67-68; vis scientist, Ore Regional Primate Res Ctr, 67-68. *Honors & Awards:* NIH Res Career Develop Award, 70. *Mem:* Am Fertil Soc; Soc Study Fertil; Am Physiol Soc; Soc for Study Reproduction; Am Soc Pharmacol & Exp Therapeut; Sigma Xi; Am Soc Primatologists (pres, 86-88). *Res:* Biochemistry and physiology of reproduction, especially spermatozoa capacitation, intrauterine devices and embryonic mortality; primatology. *Mailing Add:* Endocrine Res Ctr Mich State Univ East Lansing MI 48824

DUKEPOO, FRANK CHARLES, b Parker, Ariz, Jan 29, 43; c 1. HUMAN GENETICS. *Educ:* Ariz State Univ, BS, 66, MS, 68, PhD(zool), 73. *Prof Exp:* Instr genetics, Mesa Community Col, Ariz, 71-72; teacher gen sci, Phoenix Indian Sch, Ariz, 72-73; asst prof biol genetics, San Diego State Univ, 73-77; prof mgr, Minority Insts Sci Improvement Prog, NSF, 77-; AT DEPT BIOL, NORTHERN ARIZ UNIV, FLAGSTAFF. *Concurrent Pos:* Consult, Ctr on Aging, Inst Minority Aging, San Diego, 73-, sec sci prog, Bur Indian Affairs, 73-75 & elderly Indian nutrit res proj, 74; mem ad hoc comt, Minority Sci Prog, NIH, 74-75. *Mem:* AAAS; Soc Advan Chicanos & Native Americans Sci. *Res:* Albinism and inbreeding among the Hopi Indians of Arizona; inter-racial variant biological response to alcohol; cross-cultural study on the minority aged. *Mailing Add:* Dept Biol Northern Ariz Univ Box 5640 Flagstaff AZ 86011-5640

DUKER, NAHUM JOHANAN, b New York, NY, Oct 27, 42; div; c 4. CHEMICAL CARCINOGENESIS. *Educ:* Univ Ill, MD, 66. *Prof Exp:* Intern, Bellevue Hosp, New York, 66-67; captain, US Army, 67-69; resident path, Med Ctr, NY Univ, 70-76, instr, Sch Med, 76-77; from asst prof to assoc prof, 77-87, PROF PATH, SCH MED, TEMPLE UNIV, 87- *Concurrent Pos:* Res career develop award, Nat Inst Aging. *Mem:* Am Asn Pathologists; Am Asn Cancer Res; Am Soc Biochem & Molecular Biol; Am Soc Photobiol; Int Acad Path; Environ Mutagen Soc. *Res:* Types of damages done to DNA by physical and chemical carcinogens; mechanisms by which cells recognize DNA damage and initiate DNA repair; physical carcinogenesis. *Mailing Add:* Dept Path Sch Med Temple Univ 3400 N Broad St Philadelphia PA 19140

DUKES, GARY RINEHART, b Lafayette, Ind, Feb 24, 39; m 65; c 2. ANALYTICAL CHEMISTRY. *Educ:* US Naval Acad, BS, 62; Purdue Univ, PhD(chem), 72. *Prof Exp:* NIH fel, Mass Inst Technol, 72-74; asst prof chem, Univ Ala, 74-78; scientist, 78-79, res head, 79-84, MGR, UPJOHN CO, 84- *Mem:* Sigma Xi; Am Chem Soc; Am Asn Pharmaceut Scientists. *Res:* Laboratory computerization; kinetics and mechanisms of ligand exchange reactions; analytical methods development; stability testing concepts. *Mailing Add:* Upjohn Co 7171 Portage Rd Kalamazoo MI 49001

DUKES, MICHAEL DENNIS, b Winter Haven, Fla, Nov 4, 52; m 73; c 2. INORGANIC CHEMISTRY. *Educ:* Univ SFla, Tampa, BA, 73; Univ SC, Columbia, PhD(inorg chem), 77. *Prof Exp:* Res chemist radioactive waste mgt, Savannah River Lab, 77-83, AREA SUPVR, LABS DEPT, SAVANNAH RIVER PLANT, E I DU PONT DE NEMOURS & CO, INC, 83- *Mem:* Am Chem Soc. *Res:* Selection and testing of solid forms for the immobilization of radioactive wastes. *Mailing Add:* Ten Belaire Terr Aiken SC 29801-2836

DUKES, PETER PAUL, b Vienna, Austria, June 27, 30; US citizen; wid; c 2. BIOCHEMISTRY, HEMATOLOGY. *Educ:* Univ Chicago, PhD(biochem), 58. *Prof Exp:* Res asst biochem, Univ Chicago, 54-55, from instr to asst prof, 58-67; from asst prof to assoc prof, 67-82, PROF BIOCHEM & PEDIAT, SCH MED, UNIV SOUTHERN CALIF, 82-; DIR RES, CHILDREN'S HOSP LOS ANGELES, 88- *Concurrent Pos:* USPHS spec fel, Physiol Chem Inst, Univ Marburg, 64-65; res assoc, Argonne Cancer Res Hosp, 58-67; mem erythropoietin comt, Nat Heart, Lung & Blood Inst, 70-74. *Mem:* Int Soc Exp Hemat (secy, 90-); AAAS; Am Soc Biochem & Molecular Biol; Am Soc Hemat; German Soc Biol Chem. *Res:* Erythropoietin; regulation of hematopoiesis; control of cell differentiation by hormones; megakaryocytopoiesis. *Mailing Add:* Mail Stop 93 Childrens Hosp 4650 Sunset Blvd Los Angeles CA 90027

DUKES, PHILIP DUSKIN, b Reevesville, SC, Jan 16, 31; m 56; c 2. PLANT PATHOLOGY, PLANT BREEDING. *Educ:* Clemson Univ, BS, 53; NC State Univ, MS, 60, PhD(plant path), 63. *Prof Exp:* Plant chief clerk, Davison Chem Corp, W R Grace & Co, 53-54; asst county agent, SC Exten Serv, 56-58; asst plant pathologist & asst prof plant path, Coastal Plain Exp Sta, Univ Ga, 62-67, assoc prof plant path, 67-70; RES PLANT PATHOLOGIST, US VEG LAB, AGR RES SERV, USDA, 70- *Concurrent Pos:* Adj prof, Clemson Univ, 80- *Honors & Awards:* L M Ware Res Award. *Mem:* Am Phytopath Soc; Int Soc Plant Path; Am Soc Hort Sci; Sigma Xi. *Res:* Physiology of soil borne phytopathogenic fungi; physiology of parasitism of root and stem pathogens; breeding disease resistant vegetables; sweet potato diseases; southern peas diseases, pepper diseases; physiology and genetics of Fusarium Oxysporum Batatas; control of vegetable diseases by breeding and other means. *Mailing Add:* US Veg Lab Agr Res Serv USDA 2875 Savannah Hwy Charleston SC 29414-5334

DUKLER, A(BRAHAM) E(MANUEL), b Newark, NJ, Jan 5, 25; c 3. CHEMICAL ENGINEERING. *Educ:* Yale Univ, BE, 45; Univ Del, MS, 50, PhD, 51. *Prof Exp:* Develop engr, Rohm and Haas Co, 45-48; res engr, Shell Oil Co, 50-52; from asst prof to assoc prof chem eng, 52-60, chmn dept, 65-73, dean eng, 76-82, PROF CHEM ENG, UNIV HOUSTON, 60- *Concurrent Pos:* Vis prof, Univ Brazil, 63; exec dir, State Tex Energy Coun, 73-75; Lady Davis vis prof, Technion, Israel, 76; consult, US Nuclear Regulatory Comn, Brookhaven Lab & Schlumberger Shell Develop Co. *Honors & Awards:* Res Lectureship Award, Am Soc Eng Educ; Alpha Chi Sigma Res Award, Am Inst Chem Engrs. *Mem:* Nat Acad Eng; Am Chem Soc; fel Am Inst Chem Engrs; Am Soc Eng Educ. *Res:* Fluid mechanics; multiphase flow; boundary layer; heat and mass transfer. *Mailing Add:* Col Eng Univ Houston Houston TX 77204

DULANEY, EUGENE LAMBERT, b Garbor, Okla, June 2, 19; m 62; c 1. MICROBIOLOGY. *Educ:* Tex Tech Col, BS, 41; Univ Tex, MA, 43; Univ Wis, PhD(bot), 46. *Prof Exp:* Asst biol, Tex Tech Col, 39-41; tutor bot, Univ Tex, 41-43; res, Off Prod Res & Develop proj, 44-45; sr microbiologist, 46-60, res assoc, 60-65, sr res fel, 65-74, sr investr, 74-79, SR SCIENTIST, MICROBIOL RES LABS, MERCK & CO, 79- *Concurrent Pos:* Can Nat Res Coun fel, 55-56. *Mem:* Fel AAAS; Bot Soc Am; Mycol Soc Am; Am Soc Microbiol; Torrey Bot Club; Sigma Xi. *Res:* Genetics and physiology of microorganisms; biology of streptomyces; mycology; industrial microbiology, cell biology. *Mailing Add:* Merck Inst PO Box 2000 Rahway NJ 07065

DULANEY, JOHN THORNTON, b Sutton, WVa, Sept 28, 37. BIOCHEMISTRY. *Educ:* WVa Wesleyan Col, BS, 58; Univ Chicago, MS, 62; Univ Ill, PhD(biochem), 67. *Prof Exp:* Asst biochemist, Eunice Kennedy Shriver Ctr, 71-76; investr, John F Kennedy Inst, Baltimore, 76-79; ASST PROF, DIV NEPHROL, DEPT MED, CTR HEALTH SCI, UNIV TENN, 79- *Concurrent Pos:* Asst biochem, Dept Neurol, Mass Gen Hosp, 72-76; res assoc neurol, Dept Neurol, Harvard Univ, 75-76; biochemist, Walter E Fernald State Sch, 75-76; asst prof, Dept Pediat, Johns Hopkins Univ, 76-79. *Mem:* Am Soc Neurochem; Soc Complex Carbohydrates; Am Asn Clin Chemists. *Res:* Lysosomal enzymes and storage diseases; structure and function of enzymes of mammalian plasma membranes; erythropoiesis; chronic renal failure. *Mailing Add:* Dept Med Univ Tenn Ctr Health Sci Memphis TN 38163

DULBECCO, RENATO, b Catanzaro, Italy, Feb 22, 14; nat US; m 40, 63; c 3. VIROLOGY. *Educ:* Univ Turin, MD, 36. *Hon Degrees:* DSc, Yale Univ, 68, Ind Univ, 84; LLD, Univ Glasgow, 70; DM, Vrije Univ, Brussels, 78. *Prof Exp:* Asst path, Inst Path, Univ Turin, 40-46, asst histol & embryol, Anat Inst, 46-47; res assoc bact, Univ Ind, 47-49; sr res fel, Calif Inst Technol, 49-52, from assoc prof to prof, 52-63; resident fel, Salk Inst, 63-72; asst dir res, Imp Cancer Res Fund, 72-74, dep dir res, 74-77; prof, Med Sch Dept Path & Med, Univ Calif, San Diego, 77-81; DISTINGUISHED RES PROF, SALK INST, 77-, SR CLAYTON FOUND INVEST, 79-, PRES, 88- *Concurrent Pos:* Guggenheim fel, 57-58; vis prof, Rockefeller Inst, 62; Distinguished vis lectr, Nat Polytech Inst, Mexico City, Mex, 70; hon founder, Hebrew Univ, 81; bd mem sci counr, dept cancer etiology, NCI, mem, Cancer Ctr, Univ Calif, San Diego. *Honors & Awards:* Nobel Prize in Med/Physiol, 75; Albert & Mary Lasker Basic Med Res Award, 64; Howard Taylor Ricketts Award, 65; Paul Ehrlich-Ludwig Darmstaedter Adj Prize, 67; Louisa Gross Horwitz Prize, Columbia Univ, 67; Selman A Waksman Award Microbiol, Nat Acad Sci, 74; Targa d'oro Villa San Giovanni, 78; Mandel Golden Medal, Czech Acad Sci, 82; Kimble Methodology Award, Pub Health Labs, 59; Gold Pub Health Medal, Italian Govt, 85; GHA Clowes mem Lect, Am Asn Cancer Res, 61; Prather Lectr, Harvard Univ, 69; Dunham Lectr, Harvard Univ, 72; Leeuwenhoek Lect, Roya Soc, 74; Culling Mem Lectr, Nat Soc Histotechnol, 83. *Mem:* Nat Acad Sci; AAAS; foreign mem Royal Soc; foreign mem Ital Acad Sci; hon mem Acad Sci & Letters; Int Physicians Prev Nuclear War; hon mem Tissue Cult Asn; Fedn Am Scientists. *Res:* Cell physiology; oncology. *Mailing Add:* PO Box 85800 San Diego CA 92186

DULGEROFF, CARL RICHARD, b Wood River, Ill, Dec 26, 29; m 57; c 2. PLASMA PHYSICS. *Educ:* Cent Methodist Col, AB, 53; Wash Univ, PhD(physics), 59. *Prof Exp:* Asst physics, Wash Univ, 53-58, res assoc, 58-59; sr physicist, Rocketdyne Div, NAm Aviation, Inc, 59-60, prin scientist, 60-63, sr tech specialist physics, 63-67; sect head, 67-71, SR STAFF PHYSICIST, HUGHES RES LABS, DIV HUGHES AIRCRAFT CO, 71- *Mem:* Am Phys Soc; Sigma Xi. *Res:* Beta and gamma ray spectroscopy; positron annihilation; electron polarization; ion propulsion; ultra high vacuum; free electron lasers. *Mailing Add:* Hughes Res Lab MS Rl 57 3011 Malibu Canyon Rd Malibu CA 90265

DULIN, WILLIAM E, zoology, for more information see previous edition

DULING, BRIAN R, b Pueblo, Colo, May 27, 37; m 56; c 3. CARDIOVASCULAR PHYSIOLOGY. *Educ:* Univ Colo, Boulder, AB, 62; Univ Iowa, PhD(cardiovasc physiol), 67. *Prof Exp:* From instr to assoc prof, 68-77, PROF PHYSIOL, SCH MED, UNIV VA, 77- *Concurrent Pos:* USPHS fel physiol, 67-68; Va Heart Asn jr physiol, 68-69; estab investr, Am Heart Asn, 74- *Mem:* AAAS; Biophys Soc; Microcirculatory Soc; Am Physiol Soc. *Res:* Investigation of the mechanisms active in controlling blood flow to tissues, especially relation between blood flow and tissue metabolic activity; delivery of oxygen to tissues. *Mailing Add:* Dept Physiol Univ Va Sch Med Box 449 Jordan Hall Charlottesville VA 22908

DULIS, EDWARD J(OHN), b Brooklyn, NY, Oct 30, 19; m 45; c 3. PHYSICAL METALLURGY. *Educ:* Univ Ala, BS, 42; Stevens Inst Technol, MS, 50. *Prof Exp:* Asst metallurgist, US Naval Air Sta, 42-45; res metallurgist, US Steel Res Lab, NJ, 45-52, supv technologist, Appl Res Lab, Pa, 52-55; res supvr, Crucible Steel Co Am, 55-59, mgr prod res, 59-64, dir prod res & develop, 64-65, dir res, 65-69, managing dir, 69-71, PRES, CRUCIBLE RES CTR, CRUCIBLE MAT CORP, 71- *Honors & Awards:* Albert Sauveur Award; EC Bain Award; W H Eisenman Medal; Robert Earll McConnell Award. *Mem:* Am Soc Metals; Am Inst Met Eng; Mat Res Soc; Am Powder Metall Inst; Magnetics Soc; Inst Elec & Electronics Engrs. *Res:* Development of new stainless steels, valves steels, high temperature alloys, powder metallurgy products and processes; technical management toward commercial application of scientific developments. *Mailing Add:* Crucible Res Ctr PO Box 88 Pittsburgh PA 15230

DULK, GEORGE A, b Denver, Colo, May 21, 30; m 58; c 1. RADIO ASTRONOMY, ASTROPHYSICS. *Educ:* US Mil Acad, BS, 55; Purdue Univ, MS, 59; Univ Colo, PhD(astro-geophys), 65. *Prof Exp:* Res assoc, 65-66, from asst prof to assoc prof, 66-77, chmn dept, 75-78, PROF ASTROPHYS, UNIV COLO, BOULDER, 77- *Concurrent Pos:* Vis fel, Commonwealth Sci & Indust Res Orgn, Australia, 69- 70, 72-73, 78-79 & 83. *Mem:* Am Geophys Union; Am Astron Soc; Int Astron Union; Astron Asn Australia; Int Sci Radio Union. *Res:* Solar-stellar physics; radio astronomy; solar radiophysics; coronal structure. *Mailing Add:* 2201 Fourth St Boulder CO 80302

DULKA, JOSEPH JOHN, b Cleveland, Ohio, Jan 12, 51; m 78. METABOLIC CHEMISTRY, ANALYTICAL CHEMISTRY. *Educ:* Miami Univ, BS, 72; Pa State Univ, PhD(chem), 78. *Prof Exp:* Res chemist pesticides, 77-87, ecol safety coordr, E I Du Pont Agr Prod, 87; CONSULT, 89- *Concurrent Pos:* Res fel, Miami Univ, 70-72. *Mem:* Sigma Xi; Am Chem Soc. *Mailing Add:* 1235 Gail Rd West Chester PA 19380

DULL, GERALD G, b Laurel, Mont, Aug 20, 30; m 53; c 2. POSTHARVEST PHYSIOLOGY, NONDESTRUCTIVE QUALITY EVALUATION. *Educ:* Mont State Col, BS, 52; Mich State Univ, PhD(chem), 56. *Prof Exp:* Head biochem dept, Pineapple Res Inst, Hawaii; 56-66; plant prod qual invest leader, Human Nutrit Res Div, 66-69, chief, 69-82, RES CHEMIST, FRUIT & VEG LAB, RICHARD B RUSSELL AGR CTR, AGR RES SERV, USDA, 82- *Mem:* Am Soc Hort Sci; Inst Food Technol. *Res:* Plant composition; post harvest fruit physiology; fruit and vegetable processing; nondestructive quality evaluation. *Mailing Add:* 115 Pioneer Ct Athens GA 30605

DULLER, NELSON M, JR, b Houston, Tex, Mar 6, 23; m 55; c 1. NUCLEAR PHYSICS. *Educ:* Tex A&M Univ, BS, 48; Rice Univ, MA, 51, PhD, 53. *Prof Exp:* Asst prof physics, Tex A&M Univ, 53-54; from asst prof to assoc prof, Univ Mo, 54-62; assoc prof, 62-72, PROF PHYSICS, TEX A&M UNIV, 72- *Concurrent Pos:* Sloan Found fel, 56-58. *Mem:* Am Phys Soc; Am Asn Physics Teachers; Ital Phys Soc. *Res:* High energy nuclear and cosmic ray physics. *Mailing Add:* Dept Physics Tex A&M Univ College Station TX 77843

DULLIEN, FRANCIS A L, b Budapest, Hungary, Dec 14, 25. CHEMICAL ENGINEERING, PHYSICAL CHEMISTRY. *Educ:* Budapest Tech Univ, ChemE, 50; Univ BC, MApplSci, 58, PhD(chem eng), 60. *Prof Exp:* Asst prof phys chem, Budapest Tech Univ, 50-56; asst prof chem eng, Okla State Univ, 60-62; sr res engr, Jersey Prod Res Co, 62-65 & Esso Prod Res Co, 65-66; PROF CHEM ENG, UNIV WATERLOO, 66- *Concurrent Pos:* Vis prof, Purdue Univ, 71-72. *Mem:* Am Inst Chem Engrs; Am Chem Soc; Soc Petrol Engrs; Chem Inst Can; Can Soc Chem; NY Acad Sci. *Res:* Diffusion and flow through porous media; determination of structure of porous media; mixing of liquids; air cleaning; diffusion in gaseous and liquid systems; surface chemistry. *Mailing Add:* Dept Chem Eng Univ Waterloo Waterloo ON N2L 3G1 Can

DULMAGE, HOWARD TAYLOR, b Bridgeport, Conn, July 13, 23; m 53; c 2. INSECT PATHOLOGY, MICROBIOLOGY. *Educ:* Univ Ill, BS, 47; Rutgers Univ, PhD(microbiol), 51. *Prof Exp:* Res microbiologist, Abbott Labs, Ill, 50-62; res microbiologist, Nutrilite Prod, Calif, 62-63, dir biol res & develop, 63-67; res microbiologist, Agr Res Serv, USDA, Brownsville, Tex, 67-89; CONSULT, 89- *Concurrent Pos:* Distinguished collabr, USDA, Weslaco, Tex, 89- *Honors & Awards:* Outstanding Res Award, USDA, 70. *Mem:* Fel Royal Soc Entomologists; Soc Invert Path. *Res:* Bacillus thuringiensis; production insecticidal agents by microorganisms; nutrition and strain studies actinomycetes; microbiological insect control; microbial fermentations; production insect viruses. *Mailing Add:* H D Assocs PO Box 4113 Brownsville TX 78520

DULMAGE, WILLIAM JAMES, b Winnipeg, Man, June 9, 19; nat US; m 42; c 3. PHYSICAL CHEMISTRY, PHOTOGRAPHIC SCIENCE & ENGINEERING. *Educ:* Univ Man, BSc, 46; Univ Minn, PhD(chem), 51. *Prof Exp:* Lectr chem, Univ Man, 46-47; res & develop chemist, Elec Reduction Co, Can, Ltd, 51-52; res assoc, 52-70, asst head photomat div, Eastman Kodak Co Res Labs, 70-83; RETIRED. *Mem:* Am Chem Soc; Soc Photog Sci & Engrs; AAAS. *Res:* X-ray crystallography; molecular structure; structure and properties of high polymers; phase rule studies; unconventional photoreduction systems; electrophotographic system. *Mailing Add:* 14115 Donat Dr Poway CA 92064-3415

DULOCK, VICTOR A, JR, b Waco, Tex, Feb 26, 39; m 59; c 5. APPLIED PHYSICS. *Educ:* Univ St Thomas, BA, 60; Univ Fla, PhD(physics), 64. *Prof Exp:* Res assoc physics, Univ Fla, 64-65; asst prof, La State Univ, New Orleans, 65-67; mem prof staff, TRW Inc, Tex, 67-69, staff eng, 70-72, proj mgr, Fla, 72-74, sr staff eng, TRW Inc, Calif, 74-78, sr proj eng, 78-82, dept mgr, 82-84, lab mgr, 84-87, ASST PROG MGR, PEACEKEEPER, 87- *Mem:* Am Phys Soc; Am Asn Physics Teachers; Sigma Xi. *Res:* Applied mechanics; aerospace sciences; space physics; systems analysis; systems engineering; dynamics. *Mailing Add:* 7568 Canyon Oak Dr Highland CA 92346-5301

DUMAN, JOHN GIRARD, b Spangler, Pa, Aug 25, 46; m 71; c 3. COMPARATIVE PHYSIOLOGY, CRYOBIOLOGY. *Educ:* Pa State Univ, BS, 68; Scripps Inst Oceanog, Univ Calif, San Diego, PhD(marine biol), 74. *Prof Exp:* From asst prof to assoc prof, 74-87, PROF BIOL, UNIV NOTRE DAME, 87- *Mem:* AAAS; Am Physiol Soc; Am Soc Zoologists; Soc Cryobiol; Sigma Xi. *Res:* Physiology and biochemistry of hemolymph proteins (antifreeze proteins and ice nucleator proteins) involved in the subzero temperature tolerance of poikilotherms (especially insects). *Mailing Add:* Dept Biol Univ Notre Dame Notre Dame IN 46556

DUMAS, H SCOTT, b Albuquerque, NMex, Aug 14, 57. APPLIED DYNAMICAL SYSTEMS. *Educ:* Rice Univ, BA, 79; Univ Colo, Boulder, MA, 81; Univ NMex, PhD(appl math), 88. *Prof Exp:* Vis asst prof physics & math, State Univ NY, Albany, 88-89; postdoctoral vis mem, Inst Math & Its Applications, 89-90; ASST PROF MATH, UNIV CINCINNATI, 90- *Mem:* Am Math Soc. *Res:* Applied dynamical systems with results applied in ergodic theory and in peturbation theory for ODE's and Hamiltonian systems to the physics of particle channeling in crystals; translation of mathematics texts and research monographs. *Mailing Add:* Dept Math Univ Cincinnati Cincinnati OH 45221-0025

DUMAS, HERBERT M, JR, b El Dorado, Ark, Dec 16, 27; m 53; c 1. SPACE PHYSICS, OPTICAL PHYSICS. *Educ:* Univ Ark, AB, 54, BS, 55, MS, 56. *Prof Exp:* Tech staff mem physics, 56-60, supvr tech sect, 60-65, seismic systs div, 65-69 & satellite sensors div, 69-72, supvr, Sensors Develop Div, 72-76, MGR SPACE SYSTS DEPT, SANDIA NAT LAB, 76- *Mem:* Optical Soc Am. *Res:* Detection instrumentation and diagnostic measurements of atomic detonations; seismic systems development; energy conversion devices; optical sensors for satellite applications. *Mailing Add:* 1304 Florida NE Albuquerque NM 87110

DUMAS, JEAN, b Montreal, Que, June 4, 25; m 53; c 5. ELECTRICAL ENGINEERING. *Educ:* McGill Univ, BEng, 48; Mass Inst Technol, SB, 51, SM, 52. *Prof Exp:* Res scientist, Can Armament Res & Develop Estab, 52-55; from asst prof to assoc prof, 55-68, asst head dept, 66-69, PROF ELEC ENG, LAVAL UNIV, 68-, CHMN DEPT. *Concurrent Pos:* Ford fel, Carnegie Inst Technol, 61-62. *Mem:* Inst Elec & Electronics Engrs. *Res:* Measurements; circuits; standards; lines; musical acoustics. *Mailing Add:* Dept Elec Eng Laval Univ St Foy Quebec PQ G1K 7P4 Can

DUMAS, KENNETH J, medicine, for more information see previous edition

DUMAS, LAWRENCE BERNARD, b Plainwell, Mich, Mar 2, 41; m 65; c 2. MOLECULAR BIOLOGY, BIOCHEMISTRY. *Educ:* Mich State Univ, BS, 63; Univ Wis, Madison, MS, 65, PhD(biochem), 68. *Prof Exp:* USPHS res fel biol, Calif Inst Technol, 68-70; from asst prof to assoc prof, 70-80, PROF BIOCHEM & MOLECULAR BIOL, NORTHWESTERN UNIV, 80-, DEAN ARTS & SCIS, 88- *Concurrent Pos:* Career develop award, USPHS, 74-79. *Mem:* AAAS; Am Soc Biol Chemists; Am Soc Microbiol. *Mailing Add:* Off Dean Northwestern Univ 1918 Sheridan Rd Evanston IL 60208

DUMAS, PHILIP CONRAD, b Wash, Apr 9, 23; m 64; c 6. VERTEBRATE ZOOLOGY. *Educ:* Ore State Univ, BS, 48, MA, 49, PhD(zool), 53. *Prof Exp:* From instr to asst prof, Univ Idaho, 53-65; from assoc prof to prof zool, 65-77, PROF BIOL, CENT WASH UNIV, 77- CHMN DEPT, 66- *Mem:* AAAS; Sigma Xi; Am Soc Ichthyologist & Herpetologist. *Res:* Zoogeography; herpetology; ecology. *Mailing Add:* 2512 Hannah Ellensburg WA 98926

DUMAS, RHETAUGH GRAVES, div; c 1. NURSING. *Educ:* Dillard Univ, BSN, 51; Yale Univ, MS, 61; Union for Experimenting Cols & Univs, PhD(social psychol), 75. *Hon Degrees:* DPS, Simmons Col, 76, Univ Cincinnati, 81; LHD, Yale Univ, 89; LLD, Dillard Univ, 90. *Prof Exp:* Res asst & instr, Sch Nursing, Yale Univ, 62-65, from asst prof to assoc prof psychiat nursing, 65-72, chairperson psychiat nursing prog, 66-72; chief, psychiat nursing educ br, NIMH, 72-75, dep dir, div manpower & training progs, 76-79, dep dir, Alcohol, Drug Abuse & Mental Health Admin, USPHS, 79-81; DEAN & PROF, SCH NURSING, UNIV MICH, ANN ARBOR, 81- *Concurrent Pos:* Group rel consult. *Mem:* Inst Med-Nat Acad Sci; fel Am Acad Nursing (pres, 87-89); Am Nurses Asn. *Res:* Nature and exercise of authority in groups and organizations; leadership, especially minority women leaders; clinical trials in nursing. *Mailing Add:* Sch Nursing Univ Mich 400 N Ingalls Ann Arbor MI 48109-0482

DUMBAUGH, WILLIAM HENRY, JR, b Butler, Pa, Dec 12, 29. GLASS CHEMISTRY. *Educ:* Univ Rochester, BS, 51; Pa State Univ, PhD(chem), 59. *Prof Exp:* Sr chemist, Corning Glass Works, 58-66, mgr glass chem res, 66-82, sr res assoc, 82-84, RES FEL, CORNING INC, 84- *Mem:* Am Chem Soc; Brit Soc Glass Technol; Am Ceramic Soc; Sigma Xi; Soc Photo-Optical Instrumentation Engrs. *Res:* Glass composition; high temperature chemistry. *Mailing Add:* Corning Inc Sullivan Park FR51 Corning NY 14831

DUMBRI, AUSTIN C, b Easton, Pa, Oct 14, 47; m 71; c 2. MICROCIRCUITS, SOLID STATE MEMORY. *Educ:* Lafayette Col, BS, 75; Lehigh Univ, MS, 78. *Prof Exp:* Tech asst, AT&T Bell Labs, 67-70, sr tech asst, 70-76, mem tech staff, 76-85, MEM TECH STAFF SUPVR, AT&T BELL LABS, 85- *Res:* Microcircuits. *Mailing Add:* 315 S Watson St Easton PA 18042-3728

DUMBROFF, ERWIN BERNARD, b Newark, NJ, Mar 20, 32; m 51; c 3. PLANT PHYSIOLOGY, ANALYTICAL BIOCHEMISTRY. *Educ:* Univ Ga, BSF, 56, MF, 58, PhD(plant physiol, biochem), 64. *Prof Exp:* Res forester, US Forest Serv, Fla, 57-60 & Ga, 60-64, plant physiologist, 64-65; asst prof biol, 65-68, assoc prof plant physiol, 68-81, PROF PLANT PHYSIOL, UNIV WATERLOO, 81- *Concurrent Pos:* Vis prof, Univ Jerusalem, 85, Israel-Technion Univ, 87. *Mem:* AAAS; Am Soc Plant Physiol; Can Soc Plant Physiol (pres, 85-86). *Res:* Physiology and biochemistry of plant stress; the effects of stress on gene expression and on the dynamics of protein synthesis; method development in bioanalytical and separation chemistry. *Mailing Add:* Dept Biol Scis Univ Waterloo Waterloo ON N2L 3G1 Can

DUMENIL, LLOYD C, b Argyle, Iowa, July 23, 20; m 44; c 1. SOIL FERTILITY. *Educ:* Iowa State Univ, BS, 42, MS, 51, PhD(soil fertil), 58. *Prof Exp:* Res assoc, 46-50, asst prof, 50-58, ASSOC PROF SOILS, IOWA STATE UNIV, 58- *Mem:* Am Soc Agron; Soil Sci Soc Am. *Res:* Influence of soil, management and climatic variables on corn yields in Iowa. *Mailing Add:* 309 N Franklin Ave Ames IA 50010

DUMIN, DAVID JOSEPH, b Westfield, Mass, Oct 6, 35; m 64; c 2. MICROELECTRONICS. *Educ:* Johns Hopkins Univ, BSEE, 57; Purdue Univ, MSEE, 61; Stanford Univ, PhD(elec eng), 64. *Prof Exp:* Engr solid state electronics, IBM, 57-61; mgr tech support microelectronics, RCA, 64-70; vpres, Inselek, 70-75; pres, DJD Sci, 75-77; tech dir, Allied Corp, 76-77; SAMUEL B RHODES PROF ELEC ENG, CLEMSON UNIV, 77- *Concurrent Pos:* Consult, US Navy, Union Carbide, Allied Chem, Siemens, Kyoto Ceramics and others, 75- *Honors & Awards:* Centennial Medal, Inst Elec & Electronics Engrs, 85. *Mem:* Am Phys Soc; Inst Elec & Electronics Engrs. *Res:* Microelectronic device material interactions including cryogenics, cryotrons, silicon-on-sapphire, silicon vidicons, epitaxial growth, radiation properties and submicron devices. *Mailing Add:* Clemson Univ Col Eng Clemson SC 29631

DUMKE, PAUL RUDOLPH, ANESTHESIOLOGY. *Educ:* Western Reserve Univ, MD, 37. *Prof Exp:* EMER ANESTHESIOLOGIST, HENRY FORD HOSP, MICH, 52- *Res:* Pharmacology of depressants; circulation and respiratory physiology. *Mailing Add:* 20081 E Ballantyne Ct Grosse Point Woods MI 48236

DUMKE, WARREN LLOYD, b Milwaukee, Wis, Oct 26, 28; m 63. CHEMICAL PHYSICS, PHYSICAL CHEMISTRY. *Educ:* Univ Wis, BS, 51; Iowa State Univ, MS, 56; Univ Nebr, PhD(phys chem), 65. *Prof Exp:* Res asst phys chem, AEC-Iowa State Univ, 53-56; instr chem physics & sci, Mankato State Col, 56-58; asst chem, Univ Nebr, 58-64; asst prof, Kent State Univ, 64-65, NASA res assoc neutron diffraction, 65-67; asst prof physics, 67-71, ASSOC PROF PHYSICS, MARSHALL UNIV, 71- *Mem:* Am Chem Soc; Am Crystallog Asn; Sigma Xi. *Res:* X-ray and neutron diffraction; quantum mechanical calculations; x-ray crystal studies of hydrazine-metal complexes. *Mailing Add:* RR1 Box 148 Chesapeake OH 45619

DUMM, MARY ELIZABETH, b Newark, NJ, Dec 9, 16. NUTRITION. *Educ:* Swarthmore Col, AB, 38; Bryn Mawr Col, MA, 40, PhD(biochem), 43. *Prof Exp:* Instr biol, Bryn Mawr Col, 42-44; instr chem, Col Med, NY Univ, 44-47, asst med, 47-50, from adj asst prof to asst prof, 50-59; lectr biochem, Christian Med Col, Vellore, India, 59-60, from assoc prof to prof, 60-71; from asst prof to prof, 71-82, EMER PROF PATH, COL MED & DENT NJ, RUTGERS MED SCH, 82- *Concurrent Pos:* Vis assoc prof, Teachers Col, Columbia Univ, 71. *Mem:* Am Physiol Soc; Endocrine Soc; Soc Exp Biol & Med; NY Acad Sci. *Res:* Endocrinology and metabolism; medical, clinical and nutritional biochemistry. *Mailing Add:* 13 Samson Ave Madison NJ 07940-2227

DUMMETT, CLIFTON ORRIN, b Georgetown, Brit Guiana, May 20, 19; nat US; m 43; c 1. PERIODONTOLOGY. *Educ:* Roosevelt Col, BS, 41; Northwestern Univ, DDS, 41, MSD, 42; Univ Mich, MPH, 47; Am Bd Periodont & Am Bd Oral Med, dipl. *Hon Degrees:* DSc, Northwestern Univ; ScD, Univ Pa, 78. *Prof Exp:* Prof periodont, oral path & pub health, dean & dir dent educ, Sch Dent, Meharry Med Col, 45-49; chief dent serv & exec secy res & educ, Vet Admin Hosp, Tuskegee, 49-64; prof dent, Sch Vet Med, Tuskegee Inst, 64-66; assoc prof periodont, Sch Dent, Univ Ala, 65-66; assoc prof dent, Northwestern Univ, 66; chief dent serv, Vet Admin Res Hosp & chmn res study group oral dis, 66; prof community dent & chmn dept & dent dir multipurpose health serv ctr, Univ Southern Calif, 66-68, dir, health ctr, 68, assoc dean extramural affairs, 68-75, prof dent, 68-89, EMER PROF DENT, UNIV SOUTHERN CALIF, 89- *Concurrent Pos:* Mem comt pub health surv dent, Am Coun Educ; deleg, Nat Citizen Comt, WHO, 53; ed, Nat Dent Asn Bull, 53-75; ed, Am Asn Hist Dent, 65-74; chmn, Dent Sect, AAAS, 75; mem, President's Comt Nat Health Ins, 77; pres, Los Angeles Dent Soc, 77. *Honors & Awards:* Nat Dent Asn Award, 52; Alfred Fones Award, 76; Pierre Fauchard Gold Medalist, 80; Hayden-Harris Award, Acad History Dent, 87; Special Salute Award, Am Col Dent, 88. *Mem:* Inst Med-Nat Acad Sci; fel Am Pub Health Asn; hon mem Am Dent Asn; Nat Dent Asn; Int Asn Dent Res (vpres, 67, pres-elect, 68, pres, 69); AAAS; Am Asn Dent Ed (pres, 76); Am Acad Hist Dent (pres, 82-83). *Res:* Oral pathology; public health dentistry; dental education; oral diagnosis; peridontology. *Mailing Add:* Sch Dent Univ Southern Calif Los Angeles CA 90007

DUMONT, ALLAN E, b New York, NY, Oct 8, 24; m 49; c 3. SURGERY, PHYSIOLOGY. *Educ:* Hobart Col, BA, 45; NY Univ, MD, 48. *Prof Exp:* From instr to prof, 55-73, J L Whitehill prof, 73-90, EMER PROF SURG, MED SCH, NY UNIV MED CTR, 90- *Concurrent Pos:* A A Berg fel exp surg, 55-59; USPHS fel, 61-62, res career develop award, 61-71; investr, NY Health Res Coun, 59-61; assoc dir surg, Bellevue Hosp, New York, 75- *Honors & Awards:* Purkinje Medal, Czech Med Soc. *Mem:* AAAS; Am Physiol Soc; Soc Exp Biol & Med; Soc Univ Surgeons; Am Surg Asn. *Res:* Function of the lymphatic system; wound healing. *Mailing Add:* 603 Steele Rd New Hartford CT 06057

DUMONT, GUY ALBERT, b Calais, France, Nov 3, 51; French & Can citizen; m 77; c 2. ADAPTIVE CONTROL, PULP & PAPER PROCESS CONTROL. *Educ:* Ecole Nationale Superieure des Arts et Metiers, Paris, Diplome d Ingenieur, 73; McGill Univ, Montreal, PhD(elec eng), 77. *Prof Exp:* Proj engr, Tioxide Sa, France, 73-74, head, Process Develop Sect, 77-79; assoc scientist, Pulp & Paper Res Inst Can, 79-81, head, Control Eng Sect, Montreal, 81-83 & Vancouver, 83-89; ASSOC PROF ELEC ENG, SR PAPRICAN & NAT SCI & ENG RES COUN CHAIR PROCESS CONTROL, UNIV BC, VANCOUVER, 89- *Concurrent Pos:* Auxiliary prof, McGill Univ, Montreal, 82-83; adj prof elec eng, Univ BC, Vancouver, 83-89. *Mem:* Sr mem Inst Elec & Electronics Engrs; Can Pulp & Paper Asn; Tech Asn Pulp & Paper Indust; Am Asn Artificial Intel; Int Neural Network Soc. *Res:* Advanced process control, particularly adaptive control algorithms; expert control, use of expert systems and neural networks in control and signal processing; sensor development, acoustic emission monitoring for process control; pulp and paper and chemical industries. *Mailing Add:* Pulp & Paper Ctr Univ BC 2385 East Mall Vancouver BC V6T 1Z4 Can

DUMONT, JAMES NICHOLAS, b Sigourney, Iowa, Sept 19, 35; m 61; c 3. ZOOLOGY, CYTOLOGY. *Educ:* State Univ Iowa, BS, 57, MS, 60; Univ Mass, PhD(zool), 64. *Prof Exp:* Instr biol, Rockford Col, 61-62; res assoc zool, Univ Mass, 64-65, lectr electron micros, 65-66; res biologist, 66-85, QUAL ASSURANCE MGR, OAK RIDGE NAT LAB, 86- *Concurrent Pos:* Prof, Univ Tenn-Oak Ridge Grad Sch Biomed Sci, 71-85. *Res:* Cell and developmental biology; reproductive biology; teratology; abnormal embryological development (teratology) in response to teratogens from environmental sources. *Mailing Add:* 111 Heritage Dr Oak Ridge TN 37830

DUMONT, KENT P, b Newburyport, Mass, July 10, 41; c 3. BOTANY. *Educ:* Gettysburg Col, AB, 63; Cornell Univ, MS, 65, PhD, 70. *Prof Exp:* Assoc cur, 69-77, cur fungi, NY Bot Garden, 77-82. *Concurrent Pos:* Adj asst prof, Lehman Col, City Univ New York; mem prog comt, Second Int Mycol Cong, 75-77; mem adv comt, Proj Flora Amazonica & sci dir cryptogams, Orgn Flora Neotropica, 76-78. *Mem:* Mycol Soc Am; Sigma Xi. *Res:* Taxonomic studies in the fungal family Sclerotiniaceae and related genera of Discomycetes. *Mailing Add:* Box 385 Brewester NY 10509

DUMONTELLE, PAUL BERTRAND, b Kankakee, Ill, June 22, 33; m 55; c 5. GEOLOGY. *Educ:* DePauw Univ, BA, 55; Lehigh Univ, MS, 57. *Prof Exp:* Geologist, Homestake Mining Co, 57-63; geologist, 63-74, coordr, Environ Geol, 74-79, geologist in chg, Earth Mat Technol Sect, Environ Geol, 79, head, Eng Geol Sect, 79-84, geologist, 79-85 head, Earth Hazards & Eng Geol Sect, 85-88, SR ENG GEOLOGIST, ILL STATE GEOL SURV, 85- *Concurrent Pos:* Dir, Ill Mine Subsidence Res Prog, 85-; bd dir, Assoc Eng Geologists, 87-89, pres, Boneyard Creek Comn, 88-; asst chief, 88-89, chief, Environ Geol & Geochem Br, 90-; mgr continuing educ, Asn Eng Geologists; chair tech prog, Asn Eng Geologists, 91. *Honors & Awards:* Autometric Award, 87. *Mem:* Fel Geol Soc Am; Am Inst Prof Geologists; Int Asn Engr Geologists; Am Cong Surv & Mapping; Sigma Xi; Asn Eng Geologists; Soc Mining Engrs. *Res:* Engineering and environmental geology; topographic and computer mapping; mine subsidence research; Illinois landslide classification and inventory; Illinois earthquake awareness program. *Mailing Add:* 2020 Burlison Dr Urbana IL 61801-5805

DUMOULIN, CHARLES LUCIAN, b Johnson AFB, Japan, June 28, 56; US citizen; m 79; c 4. PHYSICS OF RADIOLOGY, ANGIOGRAPHY. *Educ:* Fla State Univ, BS, 77, PhD(anal chem), 81. *Prof Exp:* Opers dir, NIH resource for Nuclear Magnetic Resonance, Syracuse Univ, 81-84, res asst prof chem, 82-84; RES STAFF, GEN ELEC CO, 84- *Concurrent Pos:* Assoc prof radiol, Albany Med Col, 87- *Honors & Awards:* Distinguished Inventor Award, Intellectual Property Owners Inc, 88. *Mem:* Am Chem Soc; Soc Magnetic Resonance Imaging. *Res:* Nuclear magnetic resonance applications to medicine and medical physics in general; in-vivo nuclear magnetic resonance spectroscopy; nuclear magnetic resonance flow measurement and imaging. *Mailing Add:* GE Corp Res & Develop Ctr PO Box 8 Schenectady NY 12301

DUNAGAN, TOMMY TOLSON, b Hamilton, Tex, Oct 17, 31; m 56; c 2. PHYSIOLOGY. *Educ:* Tex A&M Univ, BS, 53, MS, 55; Purdue Univ, PhD(zool), 60. *Prof Exp:* Parasitologist, Arctic Aeromed Lab, 55-57; NIH res fel, Purdue Univ, 58-62; from asst prof to assoc prof, 62-71, PROF PHYSIOL, SOUTHERN ILL UNIV, 71- *Mem:* Am Soc Parasitol; NY Acad Sci; Sigma Xi. *Res:* Physiological parasitology; biochemistry; tissue culture of helminths. *Mailing Add:* Dept Physiol Southern Ill Univ Carbondale IL 62901

DUNAHAY, TERRI GOODMAN, b San Pedro, CA, May 11, 54; m 83. ALGAL MOLECULAR BIOLOGY, CELL TRANSFORMATION. *Educ:* Univ Calif, Santa Barbara, BA, 76; Univ Colo, Boulder, PhD(molecular & cell biol), 86. *Prof Exp:* Res assoc, Univ Colo, Boulder, 86-88; res assoc, 89-90, STAFF SCIENTIST, SOLAR ENERGY RES INST, 91- *Mem:* Int Soc Plant Molecular Biol; Am Soc Plant Physiologists. *Res:* Molecular biology and genetic transformation of microalgae with the goal of algae mass culture for fuel and specialty chemicals. *Mailing Add:* Solar Energy Res Inst 1617 Cole Blvd Golden CO 80401

DUNAVIN, LEONARD SYPRET, JR, b Algood, Tenn, Dec 17, 30; m 62; c 2. AGRONOMY. *Educ:* Tenn Polytech Inst, BS, 52; Univ Fla, MSA, 54, PhD(agron), 59. *Prof Exp:* Asst agronomist, 59-67, ASSOC AGRONOMIST, AGR RES CTR, UNIV FLA, JAY, 67- *Mem:* Am Soc Agron; Crop Sci Soc Am. *Res:* Production and management of pasture and forage crops. *Mailing Add:* Agr Res Ctr Rte 3 Box 575 Jay FL 32565

DUNAWAY, GEORGE ALTON, JR, b Ironton, Mo, June 6, 41; m 67; c 2. BIOCHEMISTRY, ENZYMOLOGY DEVELOPMENT. *Educ:* Cent State Univ, Okla, BS, 65; Univ Okla, PhD(biochem), 70. *Prof Exp:* Res assoc pharmacol, Ind Univ Sch Med, 70-73; sr res assoc cell & molecular biol, State Univ NY, Buffalo, 73-75; asst prof biochem, 75-79, ASSOC PROF PHARMACOL & BIOCHEM, SCH MED, SOUTHERN ILL UNIV, 79- *Concurrent Pos:* Prin investr, Am Cancer Soc grant, 76-78 & Am Diabetes Asn grant, 76-77; Am Heart Asn grant, 79-81, 82-85 & 86-88; Nat Inst Arthritis, Metabolism & Digestive Dis grant, 79-82, Nat Inst Child Health & Human Develop grant, 83-86 & 88-92. *Mem:* AAAS; Sigma Xi; Am Soc Biol Chemists; Am Heart Asn; Am Diabetes Asn. *Res:* Enzymological changes associated with diabetes, nutrition, cancer and development; application of recombinant DNA techniques to study enzyme physiology. *Mailing Add:* Dept Pharmacol Southern Ill Univ Sch Med PO Box 19230 Springfield IL 62794-9230

DUNAWAY, MARIETTA, b Borger, Tex, Dec 17, 52. PROTEIN-DNA INTERACTIONS, EUKARYOTIC TRANSCRIPTION. *Educ:* Tex Tech Univ, BS, 75; Rice Univ, PhD(biochem), 80. *Prof Exp:* Res fel, Fred Hutchinson Cancer Res Ctr, 80-84; ASST RES MOLECULAR BIOLOGIST, VIRUS LAB, UNIV CALIF, BERKELEY, 85- *Concurrent Pos:* Adj asst prof molecular biol, Univ Calif, Berkeley, 85- *Res:* Molecular mechanisms of eukaryotic gene expression; protein-DNA interactions controlling transcription; ribosomal RNA genes of xenopus laevis; gene expression in the neurula stage of embryogenesis of xenopus. *Mailing Add:* Dept Molecular Biol Univ Calif MCB 401 Barker Hall Berkeley CA 94720

DUNAWAY-MARIANO, DEBRA, ENZYME MECHANISMS. *Educ:* Tex A&M Univ, PhD(biochem), 75. *Prof Exp:* ASSOC PROF CHEM, UNIV MD, 79- *Res:* Bioinorganic and bioorganic chemistry. *Mailing Add:* 16311 Cambridge Ct Mitchellville MD 20716

DUNBAR, BONNIE SUE, b Sterling, Colo, Feb 14, 48. REPRODUCTIVE BIOLOGY, ZOOLOGY. *Educ:* Univ Colo, BA, 70, MA, 71; Univ Tenn, PhD(zool), 77. *Prof Exp:* Scientist, Oceanog Mariculture Industs, 71-72; sr technician marine biol, Harbor Br Found, Smithsonian Inst Ft Pierce Bur, 72-74; res asst reproductive biol, Inst Molecular & Cellular Evolution, Univ Miami, 75; fel, Univ Calif, Davis, 77-78; staff scientist reproductive biol, Pop Coun, Rockefeller Univ, 78-81; ASST PROF, DEPT CELL BIOL, BAYLOR COL MED, HOUSTON, 81- *Mem:* AAAS; Soc Study Reproduction. *Res:* Reproductive biology with particular interest in problems associated with fertilization. *Mailing Add:* Dept Cell Biol Baylor Col Med 1200 Moursund Ave Houston TX 77030

DUNBAR, BURDETT SHERIDAN, b Kewanee, Ill, Dec 6, 38; m 71. MEDICINE, ANESTHESIOLOGY. *Educ:* Univ Ill, Urbana, BS, 60, MD, 63. *Prof Exp:* Intern, Springfield City Hosp, Ohio, 63-64; resident anesthesiol, Hosp Univ Pa, 64-66; clin asst prof, Med Sch, Univ Tex, San Antonio, 68-69; asst prof, Univ Chicago, 69-71; from asst prof to assoc prof, 71-74, PROF ANESTHESIOL/CHILD HEALTH & DEVELOP, GEORGE WASHINGTON UNIV, 74- *Concurrent Pos:* NIH res fel, Dept Physiol, Grad Sch Med, Univ Pa, 66-67; attend staff physician, Michael Reese Hosp & Med Ctr, 69-71; staff anesthesiologist, George Washington Univ Hosp, 71-74; sr attend anesthesiologist, Children's Hosp Nat Med Ctr, Washington, DC, 74- *Mem:* AAAS; AMA; Am Soc Anesthesiologists. *Res:* Respiratory physiology. *Mailing Add:* 111 Michigan Ave NW Washington DC 20010

DUNBAR, DENNIS MONROE, b Porterville, Calif, May 30, 45; m 67; c 3. ENTOMOLOGY. *Educ:* Fresno State Col, BA, 67; Univ Calif, Davis, PhD(entom), 71. *Prof Exp:* NSF trainee entom, Univ Calif, Davis, 68-70; assoc entomologist urban entom, Conn Agr Exp Sta, New Haven, 70-76; entomologist, Agr Chem Div, FMC Corp, Riverside, Calif, 76-77, Jackson, Miss, 78-81, Philadelphia, Pa, 82-85, Fresno, Calif, 86- 87; RES FEL, AGR RES DEPT, MERCK SHARPE & DOHME RES LABS, FRESNO, CALIF, 87- *Mem:* Entom Soc Am; Am Regist Prof Entomologists; Sigma Xi. *Res:* Biology and control of insect pests that attack turf, ornamentals, Christmas trees and shade trees; cotton insect control; insect control of nut crops and vegetables. *Mailing Add:* 631 E Utah Fresno CA 93710

DUNBAR, HOWARD STANFORD, b Jersey City, NJ, Sept 30, 19; m 44; c 3. NEUROSURGERY. *Educ:* Cornell Univ, AB, 41, MD, 44. *Prof Exp:* Asst prof neurosurg, Med Col, Cornell Univ, 52-64, from assoc prof to emer prof neurol surg, 64-80; RETIRED. *Concurrent Pos:* Ledyard fel, Med Col, Cornell Univ, 51-52; dir neurosurg, Roosevelt Hosp; consult, Montrose Vet Hosp. *Mem:* Fel Am Col Surg; Harvey Soc. *Res:* Radioisotopes in localization of brain lesions; stereotaxic surgery. *Mailing Add:* 120 Colombard Ct Ponte Verde FL 32082

DUNBAR, J(OHN) SCOTT, b Toronto, Ont, Aug 16, 21; m 49; c 3. RADIOLOGY. *Educ:* Univ Toronto, MD, 45. *Prof Exp:* Dir radiol dept, Montreal Children's Hosp, 52-71, assoc prof med, McGill Univ, 63-71, prof diag radiol & chmn dept, 68-71; prof diag radiol & chmn dept, Univ BC & dir dept diag radiol, Vancouver Gen Hosp, 71-75; DIR, DIV ROENTGENOL, CHILDREN'S HOSP MED CTR & PROF RADIOL & PEDIAT, COL MED, UNIV CINCINNATI, 75- *Mem:* Am Roentgen Ray Soc; Soc Pediat Radiol; Asn Univ Radiologists; Am Col Radiol; Radiol Soc NAm. *Res:* Diagnostic radiology. *Mailing Add:* 27 Southaven Pl Oakville CA 94402

DUNBAR, JOSEPH C, b Vicksburg, Miss, Aug 27, 44; m 67; c 2. ENDOCRINOLOGY OF THE PANCREAS, HORMONAL CONTROL OF METABOLISM. *Educ:* Alcorn State Univ, BS, 63; Tex Southern Univ, MS, 66; Wayne State Univ, PhD(physiol), 70. *Prof Exp:* Instr biol, Tex Southern Univ, 66-67; instr, Harper Hosp Sch Nursing, 70-80; from asst prof to assoc prof, 72-85, PROF PHYSIOL, WAYNE STATE UNIV, 85- *Concurrent Pos:* Rev consult, NIH, 80-84; sci counr, Nat Toxicol Bd, 80-82. *Mem:* Am Diabetes Asn; Soc Exp Biol & Med; Am Physiol Soc; Sigma Xi; AAAS. *Res:* Role of glucagon in physiological regulations; regulation of glucagon secretion, glucagon binding and glucagon's action in diabetes; role that the CNS plays in mediating the responses to altered nutrient levels in the circulation and how nutrient levels mediate other autonomic responses. *Mailing Add:* Dept Physiol Wayne State Univ 540 E Canfield Detroit MI 48201

DUNBAR, JOSEPH EDWARD, b Bristol, Conn, Feb 9, 24; m 51; c 3. PHARMACEUTICS. *Educ:* Rensselaer Polytech Inst, BS, 49; Univ Ill, MS, 52, PhD(chem), 56. *Prof Exp:* Asst org res chemist, Ill Geol Surv, 49-53; org res chemist, Dow Chem USA, 55-62, sr res chemist, 62-66, group leader, E C Britton Res Lab, 66-73, sr res specialist, Pharmaceut Res & Develop, 73-76 & Biochem Processes Lab, 76-79, res assoc, Bioprod Res Lab, 79-84, res assoc, Organic Specialities Res Lab, 84-89; PRIVATE CONSULT, 89- *Mem:* AAAS; Am Chem Soc; Am Inst Chem; Sigma Xi. *Res:* Correlation of chemical structure with biological activity; chemistry of organic sulfur compounds; cardiovascular research; design and synthesis of antithrombotic and hypolipidemic drugs; urinary tract antimicrobial research; plant growth regulators; animal growth promoters; herbicides. *Mailing Add:* 5813 Sturgeon Creek Pkwy Midland MI 48640

DUNBAR, MAXWELL JOHN, b Edinburgh, Scotland, Sept 19, 14; m 45, 59; c 6. OCEANOGRAPHY. *Educ:* Oxford Univ, BA, 37, MA, 39; McGill Univ, PhD(zool), 41. *Hon Degrees:* DSc, Mem Univ Nfld, 79. *Prof Exp:* Can Consul, Dept External Affairs, Greenland, 41-46; from asst prof to assoc, 46-59, chmn, Marine Sci Ctr, 63-77, PROF ZOOL, MCGILL UNIV, 59-, EMER PROF OCEANOG, 82- *Concurrent Pos:* Guggenheim fel, Denmark, 52-53; convenor, Int Biol Prog, 70-; fel, Arctic Inst NAm, 73. *Honors & Awards:* Bruce Medal, Royal Soc Edinburgh; Fry Medal, Can Soc Zoologists, 79; N Slope Borough (Alaska) Arctic Sci Prize, 86; Can Northern Sci Award, 87; Can Meteorol & Oceanog Soc Medal in Oceanog, 88; Officer, OC, 90. *Mem:* Fel AAAS; hon fel Am Geog Soc; fel Royal Soc Can; fel Royal Geog Soc; fel Linnean Soc. *Res:* Marine biology and oceanography; arctic regions; breeding cycles in the arctic plankton; production in arctic water; development of arctic marine resources; history of biology; evolutionary mechanisms in the ecosystem; climate change and climatic cycles. *Mailing Add:* Meteorol Dept McGill Univ 805 Sherbrooke St W Montreal PQ H3A 2K6 Can

DUNBAR, PHILIP GORDON, b Ames, Iowa, Jan 18, 55; m 80; c 1. DRUG DESIGN, DRUG SYNTHESIS. *Educ:* Wash State Univ, BPharm, 78; Univ Ariz, PhD(pharmaceut sci), 87. *Prof Exp:* Asst mgr, Low-Cost Drugs, 78-80; pharmacist, St Mary's Hosp, 81-82; teaching asst med chem, Univ Ariz, 82-85, res assoc, 85-87, postdoctoral, 87-88; ASST PROF MED CHEM, UNIV TOLEDO, 88- *Concurrent Pos:* Pharmacist, Smiths, 88, Toledo Hosp, 89-91. *Mem:* Am Chem Soc; AAAS. *Res:* Design and sythesis of neurotransmitter receptor subtype and enzymatic inhibitory agents for use in therapy of central nervous system disorders such as Alzheimers, depression and schizophrenia. *Mailing Add:* Col Pharm Univ Toledo 2801 W Bancroft St Toledo OH 43606

DUNBAR, PHYLLIS MARGUERITE, b Bronxville, NY. PHYSICAL CHEMISTRY. *Educ:* Columbia Univ, BA, 45, MA, 47, PhD(chem), 49. *Prof Exp:* Asst chem, Barnard Col, Columbia Univ, 43-45, lectr, 45-46; fel, Med Col, Cornell Univ, 49-50; assoc prof chem, Douglass Col, Rutgers Univ, 50-87; RETIRED. *Concurrent Pos:* Asst, Babies Hosp, Col Physicians & Surgeons, Columbia Univ. *Mem:* Am Chem Soc. *Res:* Reaction mechanisms; structure and reactivity. *Mailing Add:* 28 Redcliffe Ave Highland Park NJ 08904

DUNBAR, RICHARD ALAN, b Buffalo, NY, Sept 22, 40; m 64; c 3. POLYMER CHEMISTRY. *Educ:* St Bonaventure Univ, BS, 62; Univ Del, PhD(org chem), 67. *Prof Exp:* Res chemist, E I du Pont de Nemours & Co, Inc, 67-68; res specialist, 68-80, SPECIALIST PROCESS CHEM, MONSANTO CHEM INTERMEDIATES DIV, MONSANTO CO, PENSACOLA, FLA, 80- *Mem:* Am Chem Soc; Sigma Xi. *Res:* Polymer preparation; melt spinning; fabric construction; end-use testing; economic evaluation and scale-up studies. *Mailing Add:* PO Box 1042 Gonzalez FL 32560-1042

DUNBAR, ROBERT COPELAND, b Boston, Mass, June 26, 43; m 69; c 2. PHYSICAL CHEMISTRY. *Educ:* Harvard Univ, AB, 65; Stanford Univ, PhD(chem physics), 70. *Prof Exp:* Res assoc chem, Stanford Univ, 70; from asst prof to assoc prof, 70-78, PROF CHEM, CASE WESTERN RESERVE UNIV, 78- *Concurrent Pos:* Sloan fel, 73-75; Guggenheim fel, 78-79. *Mem:* Am Phys Soc; Am Chem Soc; Am Soc Mass Spectrometry; Inter-Am Photochem Soc. *Res:* Ion-molecule reaction processes in gas phase; properties of gas-phase ions; ion cyclotron resonance spectroscopy. *Mailing Add:* Dept Chem Case Western Reserve Univ Cleveland OH 44106

DUNBAR, ROBERT STANDISH, JR, b Providence, RI, Nov 30, 21; m 41; c 2. ANIMAL BREEDING. *Educ:* Univ RI, BS, 49; Cornell Univ, MS, 50, PhD(animal breeding), 52. *Prof Exp:* Assoc prof dairy husb, 52-57, assoc statistician, 57-62, chmn dept animal indust & vet sci, 62-64, dean, Col Agr & Forestry, 64-74, dir, Agr Exp Sta, 71-74, PROF ANIMAL SCI, WVA UNIV, 74- *Mem:* Am Soc Animal Sci; Am Genetic Asn. *Res:* Breeding systems and methods of selection in beef cattle; statistical analysis; biological experiments. *Mailing Add:* 313 Simpson St Morgantown WV 26505

DUNCALF, DERYCK, b York, Eng, Nov 14, 26; US citizen; m 78; c 3. MEDICINE, ANESTHESIOLOGY. *Educ:* Univ Leeds, MB & ChB, 50; Royal Col Physicians London, dipl anesthetics, 53; Royal Col Surgeons Eng, dipl, 53; fel fac anaesthetists; Am Bd Anesthesiol, dipl, 60. *Prof Exp:* House physician, St James Hosp, Leeds, Eng, 50, jr anesthetic officer, 50-51; anesthetic officer, Gen Infirmary, 51-52; res surg officer, Clayton Hosp, Wakefield, 52-53; registr anesthetist, United Leeds Hosp, 53-54; sr registr dept anesthetics, Welsh Nat Sch Med, Univ Wales, 54-56; asst prof anesthesiol, State Univ NY Downstate Med Ctr, 58-61; asst clin prof, Col Physicians & Surgeons, Columbia Univ, 62-64; from assoc prof o assoc prof, 65-71, PROF ANESTHESIOL, ALBERT EINSTEIN COL MED, 71-, UNIFIED VCHMN ANESTHESIOL, 85- *Concurrent Pos:* Exchange fel anesthesiol, Mercy Hosp, Pittsburgh, 56-57, assoc dir anesthesia res, 61-62; clin fel anesthesia, Montreal Children's Hosp, 57-58; vis instr, Sch Med, Univ Pittsburgh, 56-57; assoc attend anesthesiologist, Kings County Hosp, Brooklyn, NY, 59-61; actg chief dept anesthesia, Vet Admin Hosp, 61-62; assoc attend anesthesiologist, Montefiore Med Ctr, Bronx, NY, 62-64, attend anesthesiologist, 64-, dir anesthesiol residency training prog, 70-75, chmn dept anesthesiol, 75-85; assoc vis anesthesiologist, Morrisania City Hosp, 63-75, consult anesthesiologist, 75-; assoc vis anesthesiologist, Bronx Munic Hosp Ctr, 65-69; consult, Wyckoff Heights Hosp, Brooklyn, 66-88. *Mem:* NY Acad Med; Am Soc Anesthesiologists; Asn Anaesthetists Gt Brit & Ireland; AMA. *Res:* Pharmacology of muscle relaxants and narcotics; anesthesia in ophthalmology; physiology of respiratory insufficiency; mechanical ventilation. *Mailing Add:* Dept Anesthesiol Montefiore Med Ctr Bronx NY 10467

DUNCAN, ACHESON JOHNSTON, b Leonia, NJ, Sept 24, 04; m 59. APPLIED STATISTICS. *Educ:* Princeton Univ, BS, 25, MA, 27, PhD(econ), 36. *Prof Exp:* From instr to asst prof econ & statist, Princeton Univ, 29-45; from assoc prof to emer prof statist, Johns Hopkins Univ, 46-71; RETIRED. *Honors & Awards:* Shewhart Medal, Am Soc Qual Control, 64; Dodge Award, Am Soc Testing & Mat, 76. *Mem:* Am Statist Asn; Inst Math Statist; Am Soc Qual Control; Am Soc Testing & Mat; Biomet Soc. *Res:* Economic design of control charts; sampling of bulk material. *Mailing Add:* 830 W 40th St Apt 565 Baltimore MD 21211-2125

DUNCAN, BETTIE, b Ashland, Miss, June 21, 33. MICROBIOLOGY. *Educ:* Judson Col, AB, 55; Birmingham-Southern Col, MS, 61; Univ Ark, PhD(microbiol), 66. *Prof Exp:* Assoc biologist, Southern Res Inst, Ala, 57-62; from asst prof to assoc prof, 66-72, PROF MICROBIOL, PITTSBURG STATE UNIV, KANS, 72- *Mem:* Am Soc Microbiol; NY Acad Sci. *Res:* Rare earth metal effects on microorganisms; pigment production in fungi; microbial ecology of the strip pits of southeastern Kansas; nutrition and fat production in fungi. *Mailing Add:* Dept Biol Pittsburg State Univ Pittsburg KS 66762

DUNCAN, BRYAN LEE, b Kansas City, Kans, July 23, 42; m 66; c 2. FISHERIES. *Educ:* Kans State Univ, Pittsburg, BA, 64; Wayne State Univ, PhD(biol), 72. *Prof Exp:* Asst prof biol, Houghton Col, 70-72; US Peace Corps vol aquacult, 72-75; ASST PROF FISHERIES, AUBURN UNIV, 75- *Concurrent Pos:* Fel, Auburn Univ, 75; leader brackish water pond cult proj, Indonesia, 76-81. *Mem:* Am Fisheries Soc; World Maricult Soc. *Res:* Warm-water fish culture; parasites and diseases of cultured fish; pond construction; aquaculture information systems. *Mailing Add:* Dept Fish & Allied Aquacult Auburn Univ Auburn AL 36849

DUNCAN, BUDD LEE, b Thief River Falls, Minn, Nov 15, 36; m 58; c 3. THERMODYNAMICS, KINETICS. *Educ:* Macalester Col, BA, 58; SDak State Univ, MS, 60; Univ Tenn, PhD, 70. *Prof Exp:* Assoc prof chem, Tenn Wesleyan Col, 61-70, chmn dept, 76-79; sr res assoc, Olin Chem, 79-81; prof chem, Tenn Wesleyan Col, 70-79; GROUP LEADER, OLIN CHEM, 82- *Mem:* Am Chem Soc; Coblentz Soc. *Res:* Kinetics of Diels-Alder reactions; infra-red spectroscopy. *Mailing Add:* 1207 Towanda Trail Athens TN 37303-2353

DUNCAN, CHARLES CLIFFORD, b Overland, Mo, Feb 10, 07; m 59; c 3. SUBMARINE CABLE, COMMUNICATION MANAGEMENT. *Educ:* Washington Univ, St Louis, BS, 27, Harvard Grad Sah Bus Admin, Cert, 57. *Prof Exp:* Pres, Cuban Am Co, Transpac Communication Eastern Tel & Telegraph Co, Transoceanic Communications, Transoceanic Cable Ship Co, 70-72; vpres, AT&T, Long Island 70-72; vehmn, Group 800 NV Amsterdam, Holland, 73-86; dep chmn, Collog Serv Ltd, London, 82-84; VCHMN, SUBMARINE LIGHTWAVE CABLE CO, INC, 85-; CONSULT & MEM BD, INT 800 TELECOM CORP, 86-; DIR & CONSULT, INT TELECARD LTD, MANUET, NY, 88- *Concurrent Pos:* Consult, Int Communication Consults, 72- *Honors & Awards:* Gold Medal, Finnland Telecommunication Soc, 68. *Mem:* Fel Inst Elec & Electronics Engrs. *Res:* Telephotography; telephone service overseas, land telephone cables and underwater telephone cables; high seas services; the role of submarine cable and telephone sattlelites; role of communication in advance medicine the role of commercial telecommunication in prepardiness for world emergencies; a new link in world communication plowing cables under the seas; conservation ofmaterial protecting telephone buildings from atomic attacks; service 800; toll free development in the developing market; author of numerous, books and papers submarine cable communications. *Mailing Add:* Group 800 Nv 255 Elderfields Rd Manhasset NY 11030

DUNCAN, CHARLES DONALD, b Houston, Tex, Oct 7, 48. PHYSICAL ORGANIC CHEMISTRY, THEORETICAL CHEMISTRY. *Educ:* Rice Univ, BA, 70; Yale Univ, PhD(org chem), 74. *Prof Exp:* Instr chem, Univ Va, 74-76; ASST PROF CHEM, UNIV ALA, BIRMINGHAM, 76-; AT DIV SCI & MATH, UNIV MAINE, MACHIAS. *Mem:* Sigma Xi; Am Chem Soc. *Res:* Forbidden organic reactions; kinetics of unimolecular decompositions; singlet-triplet interconversions; chemistry of organic biradicals. *Mailing Add:* Div Sci & Math Univ Maine Machais ME 04654

DUNCAN, CHARLES LEE, b Waynesboro, Tenn, Oct 10, 39; m 68. BACTERIOLOGY. *Educ:* Univ Tenn, BS, 61; La State Univ, Baton Rouge, MS, 63; Univ Wis, Madison, PhD(bact), 67. *Prof Exp:* From asst prof to prof bact & food microbiol, Univ Wis-Madison, 68-76; DIR FOOD SAFETY & NUTRIT, CAMPBELL INST FOOD RES, 77-, VPRES, FOOD SCI & TECHNOL, CAMPBELL INST RES & TECHNOL, 79-; VPRES RES & DEVELOP, HERSHEY FOODS CORP. *Concurrent Pos:* Res Develop Award, Nat Inst Allergy & Infectious Dis, 74. *Honors & Awards:* Res Career Develop Award, NIH. *Mem:* Am Soc Microbiol; Inst Food Technologists. *Res:* Clostridium perfringens food poisoning; sporulation of anaerobic bacteria and germination of their spores; clostridial plasmids; foods and food ingredients. *Mailing Add:* Hershey Foods Corp 1025 Reese Ave PO Box 805 Hershey PA 17033-0805

DUNCAN, DENNIS ANDREW, b Edinburgh, Scotland, Sept 24, 29; US citizen; m 60; c 2. CHEMICAL ENGINEERING. *Educ:* Univ BC, BA, 49, BASc, 57. *Prof Exp:* Res & process engr oil-gas, Portland Gas & Coke Co, 57-58; sr res engr ethylene, Monsanto Co, 58-64; supvr process develop pyrolysis petrochem, Stone & Webster Eng Corp, 64-72; prof mgr coal gasification res, Am Gas Asn, 72-74; assoc dir hydrocarbon processing res, Inst Gas Technol, 74-81; PROG MGR, PROCESS DEVELOP, STONE & WEBSTER ENG CORP, 81- *Mem:* Am Inst Chem Engrs; Am Chem Soc. *Res:* High temperature, short residence time pyrolysis of hydrocarbons, including steam-cracking to ethylene and other petrochemicals; high pressure hydropyrolysis of coal to liquid and gaseous fuels. *Mailing Add:* Ten 13th Ave Texas City TX 77590-6349

DUNCAN, DON DARRYL, b Mayfield, Ky, Sept 9, 39; m 58; c 2. PHYSICS. *Educ:* Murray State Univ, BS, 61; Univ Ky, MS, 63, PhD(physics), 68. *Prof Exp:* Asst prof, 67-70, ASSOC PROF PHYSICS, MURRAY STATE UNIV, 70- *Mem:* Am Phys Soc. *Res:* Angular correlation studies in nuclear physics. *Mailing Add:* Dept Physics Murray State Univ Murray KY 42071

DUNCAN, DONALD, anatomy; deceased, see previous edition for last biography

DUNCAN, DONALD GORDON, b Lincoln, Nebr, Apr 21, 20; m 58; c 3. MATHEMATICS. *Educ:* Univ BC, BA, 42, MS, 44; Univ Mich, PhD(math), 51. *Prof Exp:* Res engr, Nat Res Coun Can, 43-44; lectr physics, McGill Univ, 44-45; lectr math, Univ BC, 45-47; asst prof, Univ Ariz, 50-54; from asst prof to assoc prof, San Jose State Col, 54-60, prof, 60-63; PROF MATH, CALIF STATE UNIV SONOMA, 63- *Concurrent Pos:* Consult, Off Naval Res, 57- *Mem:* Am Math Soc; Math Asn Am; London Math Soc. *Res:* Theory of groups; lattice theory; aerodynamics; numerical analysis. *Mailing Add:* Calif State Col Sonoma Rohnert Park CA 94828

DUNCAN, DONALD LEE, mathematics, for more information see previous edition

DUNCAN, DONALD PENDLETON, b Joliet, Ill, Feb 24, 16; m 56; c 3. FORESTRY. *Educ:* Univ Mich, BSF, 37, MS, 39; Univ Minn, PhD(forestry, bot), 51. *Prof Exp:* Shelterbelt supvr, US Forest Serv, Kans, 39-40; jr forester, Southern Forest Exp Sta, La, 40-41; instr forestry & forester, Exp Sta, Kans State Col, 41-42, asst prof exten, 45-47; from instr to prof forestry, Univ Minn, 47-65, asst dir sch, 64-65; prof & dir sch, 65-85, EMER PROF FORESTRY, UNIV MO, COLUMBIA, 85- *Concurrent Pos:* Consult, Minn Natural Resources Coun, 61-63, Ford Found Latin-Am Fel Prog, 68 & 70, Coop State Res Serv, USDA, 64, 69, 74-75, 78, 82-85, forest ecologist, 80-81 & Coun Grad Schs, 70-84; vis scientist, Nat Sci Found, 64-65, 68, 69 & 72; res assoc, Univ Minn, 86-88, Conserv Fedn Mo, 89-90. *Mem:* AAAS; Soc Am Foresters; Sigma Xi. *Res:* Forest ecology; forest influences; forest recreation; forestry education. *Mailing Add:* Sch Natural Resources Univ Mo Columbia MO 65211

DUNCAN, DOUGLAS WALLACE, b Vancouver, BC, Sept 7, 34; m 58; c 2. FOOD SCIENCE & TECHNOLOGY. *Educ:* Univ BC, BSA, 57; Mass Inst Technol, PhD(food technol), 61. *Prof Exp:* Mem res staff, Knorr Forschungs Inst, Corn Prod Co, Switz, 61-62; sr microbiologist, BC Res, 62-65, prog leader microbiol, 65- 72, assoc head, Div Appl Biol, 72-76, head, Div Chem Technol, 76- 86, assoc dir, 82-84, VPRES, BC RES, 84- *Mem:* Am Mgt Asn. *Res:* Microbiological leaching of sulfide minerals. *Mailing Add:* BC Res Corp 3650 Wesbrook Mall Vancouver BC V6S 2L2 Can

DUNCAN, GEORGE COMER, b Durham, NC, Dec 23, 41; m 67; c 2. PHYSICS. *Educ:* NC State Univ, BS, 64, MS, 66; Brandeis Univ, PhD(physics), 71. *Prof Exp:* From asst prof to assoc prof, 70-80, PROF PHYSICS, BOWLING GREEN STATE UNIV, 80- *Concurrent Pos:* Prin investr, NSF grant, 77 & 87. *Mem:* Am Phys Soc; Sigma Xi. *Res:* General relativity and gravitation; mathematical physics; computational physics. *Mailing Add:* Dept Physics Bowling Green State Univ Bowling Green OH 43403

DUNCAN, GEORGE THOMAS, b Chicago, Ill, Aug 7, 42; m 85; c 2. STATISTICS. *Educ:* Univ Chicago, BS, 63, MS, 64; Univ Minn, PhD(statist), 70. *Prof Exp:* Statistician, Texaco Res Labs, 64-65; vol, US Peace Corps, Philippines, 65-67; asst prof math, Univ Calif, Davis, 70-74; assoc prof, 74-84, PROF STATIST, CARNEGIE-MELLON UNIV, 84- *Concurrent Pos:* Dir, Decision Systs Res Inst, 88- *Mem:* Inst Math Statist; fel Am Statist Asn; Opers Res Soc; fel Royal Statist Soc; Econ Soc. *Res:* Applied statistics and decision theory. *Mailing Add:* Sch Urban & Pub Affairs Carnegie-Mellon Univ Pittsburgh PA 15213

DUNCAN, GORDON DUKE, b Clayton, NC, May 25, 26; m 51, 81; c 5. BIOCHEMISTRY. *Educ:* NC State Col, BS, 49, PhD(biochem), 53. *Prof Exp:* Asst, US Plant, Soil & Nutrit Lab, Cornell Univ, 49-53; biochemist, Biochem Res Lab, Elgin State Hosp, 53-57; biochemist, Charlotte Mem Hosp, 57-67; prof chem, Queens Col, NC, 67-82; ASSOC DIR, HEINEMAN MED RES LABS, NC, 82- *Concurrent Pos:* Chmn bd dirs, AquAir Labs Inc; consult biochemist, Cabarrus Mem Hosp, 74-80. *Res:* Clinical biochemistry; cardiovascular biochemistry. *Mailing Add:* 1320 Scott Ave Charlotte NC 28204

DUNCAN, GORDON W, b Weehawken, NJ, July 3, 32; m 55; c 4. ENDOCRINOLOGY. *Educ:* Cornell Univ, BS, 54; Iowa State Univ, MS, 55, PhD(reprod physiol), 60. *Prof Exp:* Res assoc & asst prof animal physiol, Iowa State Univ, 59-60; sr res scientist, Upjohn Co, 60-71; sr res scientist, Seattle Res Ctr, Univ Wash, 71-74, asst prof, Endocrinol, 73-80, DIR, POP STUDY CTR, BATTELLE MEM INST, 74-; VPRES PHARMACEUT RES & DEVELOP, UPJOHN CO, KALAMAZOO MICH, 80- *Concurrent Pos:* Lectr, Western Mich Univ, 60-71. *Mem:* Soc Study Reprod; Am Soc Animal Sci; Endocrine Soc; Am Physiol Soc; Brit Soc Study Fertil. *Res:* Basic and applied research in physiology of reproduction. *Mailing Add:* Upjohn Co Kalamazoo MI 49001

DUNCAN, HARRY ERNEST, b Hartford, WVa, Nov 20, 36; m 58; c 3. PLANT PATHOLOGY. *Educ:* Univ WVa, BS, 59, MS, 61, PhD(plant path), 66. *Prof Exp:* From exten instr to exten assoc prof plant path, 65-77, PROF PLANT PATH & SPECIALIST IN CHARGE, PLANT PATH EXTEN, NC STATE UNIV, 77- *Mem:* Am Phytopath Soc; Sigma Xi. *Res:* Diseases of vegetable crops and corn; pesticides in plant disease control; extension plant pathology. *Mailing Add:* NC State Univ PO Box 7616 Raleigh NC 27695

DUNCAN, I B R, b Kilmarnock, Scotland, Oct 10, 26; Can citizen; m 57; c 2. MEDICAL MICROBIOLOGY. *Educ:* Glasgow Univ, MB & ChB, 51, MD, 62; FRCP(C). *Prof Exp:* Asst lectr bact, Glasgow Univ, 52-55, lectr, 57-60; res bacteriologist, Hosp Sick Children, Toronto, Ont, 55-57; from asst prof to assoc prof bact, Univ Western Ont, 60-67; PROF MED MICROBIOL, UNIV TORONTO & DIR MICROBIOL, SUNNYBROOK MED CTR, 67- *Concurrent Pos:* Dir microbiol, St Joseph's Hosp, London, Ont, 60-67. *Mem:* Fel Royal Col Path Eng; fel Infectious Dis Soc Am; Can Soc Microbiol (vpres, 69-70); Can Asn Med Microbiol (secy-treas, 62-67, pres, 75-76); Path Soc Gt Brit & Ireland. *Res:* Echoviruses; staphylococcal epidemiology; antibiotics and gramnegative bacilli; plasmid epidemiology. *Mailing Add:* Dept Microbiol Sunnybrook Med Ctr Toronto ON M4N 3M5 Can

DUNCAN, IRMA W, b Buffalo, NY, Jan 30, 12; m 37; c 2. CLINICAL CHEMISTRY. *Educ:* Univ Buffalo, BA, 33; Univ Chicago, MA, 35, PhD(biochem), 50. *Prof Exp:* Prof sci, Colo Woman's Col, Denver, 44-48; asst prof chem, Univ Denver, 51-59; res chemist, Arctic Health Res Ctr, Dept Health, Educ & Welfare, Anchorage, Alaska, 60-67, Fairbanks, 67-74; res chemist, Ctr Dis Control, Dept Health & Human Ser, 74-82; DOCENT, MAXWELL MUS ANTHROP, UNIV NMEX, 83- & NMEX MUS NATURAL HIST & SCI, 85- *Concurrent Pos:* Pub Health Serv grant, NIH, 56-57; adj prof chem, Alaska Methodist Univ, Anchorage, 60-62; mem, Anchorage Area Subcomt, Gov Planning Comt, Mental Retardation Prog, 65-66, Gov Adv Comt Mental Retardation, Alaska, 66-72, chmn, 66-67. *Mem:* Am Asn Univ Women; Am Chem Soc; fel AAAS; Am Asn Clin Chemists; Am Pub Health Asn; Sigma Xi. *Res:* Genetic differences in enzymes of various human populations, methods and quality control for lipids, screening methods for human abnormalities. *Mailing Add:* 3100 Camino Cepillo NW Albuquerque NM 87107

DUNCAN, JAMES ALAN, b Osage, Iowa, Aug 14, 45; m 90. PHYSICAL ORGANIC CHEMISTRY. *Educ:* Luther Col, BA, 67; Univ Ore, PhD(org chem), 71. *Prof Exp:* Sub asst prof chem, Morgan State Col, 71-72; vis asst prof chem, Reed Col, 75-77; vis asst prof, 77-78, from asst prof to assoc prof, 78-88, PROF CHEM, LEWIS & CLARK COL, 88-, DEAN MATH & NATURAL SCI, 90- *Concurrent Pos:* Vis scientist, Mass Inst Technol, 83-84. *Mem:* Am Chem Soc. *Res:* Cycloaddition mechanisms and effects of molecular geometry on pericyclic reactions; silicon chemistry. *Mailing Add:* Dept Chem Lewis & Clark Col Portland OR 97219

DUNCAN, JAMES BYRON, b Georgetown, Tex, Aug 10, 47; m 84; c 3. CHEMICAL ENGINEERING. *Educ:* Univ Ark, BSc, 70; Tex Tech Univ, MSc, 72, PhD(hydraul), 85. *Prof Exp:* Scientist, US Army, 80-85; sr engr, Eastman Kodak Co, 86-91; STAFF SCIENTIST, BATTELLE-PAC NORTHWEST LABS, 91- *Concurrent Pos:* Prin investr, Med Bioeng Res & Develop Lab, US Army, 80-84 & tech consult, Mobility Energy Res & Develop Command, 81-84; mem, bd dirs, NAm Membrane Soc, 86-89. *Mem:* Sigma Xi; NAm Membrane Soc. *Res:* Isolation of hazardous materials through prevaporation and liquid-liquid separation methodologies; boundary effects of in-situ thermal input to vadose zones; filtration properties of certain inorganic polymers, polyphosphazines, in relation to hydraulic management of surface fluid regimes. *Mailing Add:* 2507 S Jefferson Ct Kennewick WA 99337

DUNCAN, JAMES LOWELL, b West Plains, Mo, Dec 14, 37; m 63; c 3. MEDICAL MICROBIOLOGY. *Educ:* Drury Col, AB, 59; St Louis Univ, DDS, 63; Univ Wash, PhD(microbiol), 67. *Prof Exp:* Asst prof, 67-74, ASSOC PROF MICROBIOL, MED & DENT SCHS, NORTHWESTERN UNIV, 74- *Mem:* AAAS; Am Soc Microbiol; Sigma Xi. *Res:* Bacterial toxins and their effects on mammalian tissues. *Mailing Add:* Dept Microbiol Sch Med Northwestern Univ Chicago IL 60611

DUNCAN, JAMES M, SOIL-STRUCTURE SYSTEMS, CIVIL ENGINEERING. *Prof Exp:* Prof civil eng, Univ Calif, Berkeley; DEPT CIVIL ENG, VA POLYTECH INST & STATE UNIV. *Mem:* Nat Acad Eng. *Mailing Add:* Dept Civil Eng 104 Patton Hall Va Polytech Inst & State Univ Blacksburg VA 24061

DUNCAN, JAMES PLAYFORD, b Adelaide, SAustralia, Nov 10, 19; m 42; c 4. MECHANICAL ENGINEERING. *Educ:* Univ Adelaide, BE, 41, ME, 54; Univ Manchester, DSc(sci publ), 64. *Prof Exp:* Develop engr, Richards Industs Ltd, 41-47; lectr mech eng, Univ Adelaide, 47-51, sr lectr, 53-54; turbine engr, Metrop Vickers Elec Co, Manchester, 52; prof mech eng & head dept, Univ Sheffield, 56-66; head dept mech eng, Univ BC, 66-78, prof, 66-84, emer prof mech eng, 85-87; ASSOC PROF MECH ENG, UNIV MD, 87- *Concurrent Pos:* Consult, Firth Brown Tools Ltd, Eng, 61-66, Perkin-Elmer Corp, 66-68 & Caterpillar Tractor Co, 67-73; vchmn adv comt, Royal Engrs Eng, 62-66; adj prof, Univ Victoria, 85-; pres, JP Duncan & Assocs, 85-; consult, 85- *Mem:* Brit Inst Mech Engrs; Brit Inst Prod Engrs; Brit Inst Physics & Phys Soc. *Res:* Design and production of automobile and aircraft sheet metal; stress analysis by optical means of steam power plant components; surface generation by numerical control; polyhedral NC method of sculpturing; computer aided design and computer aided manufacturing. *Mailing Add:* 2141 Eng Classrm Univ Md College Park MD 20742

DUNCAN, JAMES THAYER, b Chicago, Ill, Apr 15, 32; m 80; c 1. DEVELOPMENTAL BIOLOGY. *Educ:* Wabash Col, AB, 54; Stanford Univ, PhD(biol), 60. *Prof Exp:* Asst prof zool, Univ Calif, Riverside, 60-62; from asst prof to prof, 62-91, EMER PROF BIOL, SAN FRANCISCO STATE UNIV, 91- *Concurrent Pos:* Am Cancer Soc res grant, 61-63; NSF res grants, 63-65, 68-71; NSF sci fac fel, 65-66. *Mem:* AAAS; Soc Develop Biol; Sigma Xi. *Res:* Embryonic induction and the developmental control of cellular differentiation in Amphibians. *Mailing Add:* 692 B St Ashland OR 97520

DUNCAN, JAMES W, b Decatur, Ill, Sept 15, 26. ANTENNA DESIGN. *Educ:* Univ Col Boulder, BA, 50; Univ Ill Champaigne-Urbana, MS, 55 & PhD(elec eng), 58. *Prof Exp:* Sr staff, Ground Systs Group, Hughes Aircraft Co, 61-68; mgr, Antenna Lab, 81-84, STAFF MEM, TRW SPACE & DEFENSE, 68-, SPEC ASST ANTENNA TECHNOL, 84- *Mem:* Fel Inst Elec & Electronics Engrs. *Res:* Electromagnetic analysis and design of microwave antennas, especially communications satellite antennas. *Mailing Add:* TRW Space & Defense One Space Park Bldg 02/2304 Redondo Beach CA 90278

DUNCAN, JOHN L(EASK), b Adelaide, SAustralia, Dec 20, 32; m 61; c 3. MECHANICAL ENGINEERING. *Educ:* Univ Melbourne, BME, 55; Univ Manchester, MST, 63, PhD(eng), 68. *Prof Exp:* Trainee, Caterpillar Tractor Co, 56-58; planning supvr, Caterpillar of Australia Proprietary Ltd, 58-60; asst field engr, Vacuum Oil Co, 60-61; asst lectr eng educ, Inst Sci & Technol, Univ Manchester, 62-63, lectr, 63-68, sr lectr, 68-70; prof eng educ, McMaster Univ, 70-85; PROF MECH ENG, UNIV AUCKLAND, NZ, 86- *Mem:* Fel Am Soc Metals; fel Inst Prof Eng. *Res:* Engineering plasticity; sheet metal technology; manufacturing and production processes; metal forming; alloy development. *Mailing Add:* Dept Mech Eng Univ Auckland Private Bag Auckland New Zealand

DUNCAN, JOHN ROBERT, b Morden, Man, July 17, 37; m 63; c 3. IMMUNOPATHOLOGY. *Educ:* Univ Man, BSA, 59; Univ Toronto, DVM, 63; Univ Guelph, MS, 68; Cornell Univ, PhD(vet immunol), 71; Am Col Vet Pathologists, dipl, 73. *Prof Exp:* Pvt pract, Dauphin Vet Clin, 63-66; asst prof vet path, Ont Vet Col, 66-67; asst vet immunol, Cornell Univ, 68-71, asst prof vet immunol, 71-76; RES SCIENTIST, ANIMAL DISEASES RES INST, 76- *Mem:* Am Vet Med Asn; Am Col Vet Pathologists. *Res:* Immunopathology of infectious diseases of domestic animals, particularly Johne's Disease, venereal vibriosis, characterization of bovine immunoglobulins. *Mailing Add:* 32 Cramer Dr Nepean ON K2H 5X5 Can

DUNCAN, KATHERINE, b Tamaroa, Ill, Oct 14, 13; m 39; c 2. INTERNAL MEDICINE. *Educ:* Univ Ill, BS, 35, MD, 38. *Prof Exp:* Intern, Hurley Hosp, Flint, Mich, 37-38; pvt pract, Ill, 38-41; staff physician, Crab Orchard Defense Plant, Ill, 43; asst med dir, Nutrit Res Lab, 43-45; attend physician pediat & maternity, Montgomery County Health Dept, Md, 52-53; med officer pharmacol, Bur Biol & Phys Sci, Food & Drug Admin, 60-61, admin med officer, Career Develop Rev Br, 61-70, admin med officer, Off Protection from Risks, Off Dir, NIH, 70-87; RETIRED. *Res:* Rheumatoid diseases; allergy; medical administration. *Mailing Add:* 4717 Edgefield Rd Bethesda MD 20814

DUNCAN, LEONARD CLINTON, b Owensboro, Ky, Dec 28, 36; m 63; c 2. INORGANIC CHEMISTRY, ENVIRONMENTAL CHEMISTRY. *Educ:* Wabash Col, AB, 58; Wesleyan Univ, MA, 61; Univ Wash, PhD(chem), 64. *Prof Exp:* Fel inorg chem, Purdue Univ, 64-65; from asst prof to assoc prof, 65-72, chmn, Dept Chem, 66-68, PROF CHEM, CENT WASH UNIV, 72-, CHMN, CHEM DEPT, 83- *Concurrent Pos:* Res Corp & NSF res grants, 69; res grant, Wash Water Res Ctr, 80. *Mem:* Am Chem Soc. *Res:* Synthesis of highly fluorinated compounds of the lighter elements; environmental chemistry; acid precipitation studies; lake susceptibility studies. *Mailing Add:* Dept Chem Cent Wash Univ Ellensburg WA 98926

DUNCAN, LEROY EDWARD, JR, internal medicine; deceased, see previous edition for last biography

DUNCAN, LEWIS MANNAN, b Charleston, WVa, July 11, 51; m 75. NONLINEAR RADIO WAVE PROPAGATION. *Educ:* Rice Univ, BA, 73, MS, 76, PhD(space physics), 77. *Prof Exp:* MEM STAFF, LOS ALAMOS NAT LAB, 77- *Concurrent Pos:* NSF fel, Rice Univ, 77; mem, Arecibo SciAdv Comt, Nat Astron & Ionosphere Ctr, 80- *Mem:* Am Geophys Union; AAAS; Int Radio Sci Union; Sigma Xi. *Res:* High-power radio wave propagation and associated nonlinear wave-plasma interactions; ionospheric modification and active space plasma physics experiments; applications of the ionosphere as a large, natural plasma laboratory-without-walls. *Mailing Add:* 113 Inlet Reach Seneca SC 29678

DUNCAN, MARGARET CAROLINE, b Salt Lake City, Utah, June 9, 30; m 58; c 4. PEDIATRIC NEUROLOGY. *Educ:* Univ Tex, BA, 52, MD, 55; Am Bd Pediat, dipl, 62; Am Bd Psychiat & Neurol, cert neurol, 65, cert child neurol, 69. *Prof Exp:* From instr to assoc prof, 61-74, PROF PEDIAT NEUROL, SCH MED, LA STATE UNIV MED CTR, NEW ORLEANS, 74- *Concurrent Pos:* Fel pediat neurol, Johns Hopkins Univ, 60-61; consult, Handicapped Children's Prog, La State Bd Health, 64- *Mem:* Am Acad Neurol; Am Epilepsy Soc. *Res:* Reflex epilepsy; infantile neuroaxonal dystrophy; Rett syndrome. *Mailing Add:* Dept Neurol Pediat La State Univ Med Ctr 1542 Tulane Ave New Orleans LA 70112

DUNCAN, MARION M, JR, b Bloomfield, Mo, June 24, 27; m 48; c 2. THEORETICAL PHYSICS. *Educ:* Ala Polytech Inst, BS, 49, MS, 53; Duke Univ, PhD(physics), 56. *Prof Exp:* Instr, Univ NC, 56; vis asst prof, Duke Univ, 56-60; asst prof, Tex A&M Univ, 60-61; assoc prof, 61-66, PROF PHYSICS, UNIV GA, 66-, HEAD DEPT PHYSICS & ASTRON, 68- *Mem:* Am Phys Soc; Sigma Xi. *Res:* Low energy nuclear physics. *Mailing Add:* Dept Physics & Astron Univ Ga Athens GA 30602

DUNCAN, MICHAEL ROBERT, b Frederick, Okla, Oct 11, 47; m 71. CELL BIOLOGY, BIOCHEMISTRY. *Educ:* Okla Christian Col, BS, 69; Okla Univ, PhD(virol), 75. *Prof Exp:* Res assoc, 75-80, ASST SCIENTIST CELL BIOL, SAMUEL ROBERTS NOBLE FOUND, INC, 80- *Mem:* Tissue Cult Asn; Sigma Xi. *Res:* Studies of the basic molecular mechanisms of aging using human diploid fibroblasts cultured in vitro as a model system. *Mailing Add:* PO Box 2180 Ardmore OK 73401

DUNCAN, RICHARD DALE, b Alhambra, Calif, Apr 1, 41; m 70; c 1. MATHEMATICS. *Educ:* Univ Calif, Berkeley, BSc, 63, MA, 65; Univ Calif, San Diego, PhD, 70. *Prof Exp:* Asst math, Univ Calif, Berkeley, 63-65 & Univ Calif, San Diego, 65-70; asst prof, 70-75, ASSOC PROF MATH, UNIV MONTREAL, 75- *Mem:* Am Math Soc; Can Math Soc. *Res:* Probability; functional analysis; Markov processes and potential theory; probability theory. *Mailing Add:* Dept Math Univ Montreal Montreal PQ H3C 3U7 Can

DUNCAN, RICHARD H(ENRY), b St Louis, Mo, Aug 13, 22; m; c 2. PHYSICS, ELECTRICAL ENGINEERING. *Educ:* Univ Mo, BSEE, 49, MS, 51, PhD(physics), 54. *Prof Exp:* Prof elec eng & physicist, phys sci lab, NMex State Univ, 54-65, vpres res, 65-69; TECH DIR & CHIEF SCIENTIST, WHITE SANDS MISSILE RANGE, 69- *Mem:* Inst Elec & Electronics Engrs; Sigma Xi. *Res:* Integral equations occurring in electromagnetic radiation problems; antenna engineering. *Mailing Add:* 2012 Rose Ln Las Cruces NM 88005

DUNCAN, ROBERT C, b Jonesville, Va, Nov 21, 23; m 49; c 4. DEFENSE ADVANCED RESEARCH. *Educ:* US Naval Acad, BS, 45; US Navy Postgrad Sch, BS, 53; Mass Inst Technol, MS, 54, DSc, 60. *Prof Exp:* Chief Space Prog Br, Off Chief Naval Oper, Off Secy Defense, 60-61, spec asst dir Defense Res & Eng, Off Secy Defense, 61-63; asst dir, Electronics Res Ctr, Cambridge, chief Guid & Control Div, Manned Spacecraft Ctr, Houston, NASA, 64-68; corp officer, Polaroid Corp, 68-69, prog mgr, 68-75, asst vpres, 69-75, vpres eng, 75-85; dir, Defense Advan Res Proj Agency, 85-88, asst secy Defense Res & Technol, 86-88; DIR OPER TEST & EVAL, DEPT DEFENSE, 88- *Honors & Awards:* Norman P Hayes Award, Inst Navig, 67. *Mem:* Nat Acad Eng. *Res:* Author and co-author of four books and has published more than three dozen technical papers. *Mailing Add:* Off Secy Defense The Pentagon Rm 3E 318 Washington DC 20301-1711

DUNCAN, ROBERT LEON, JR, b Ayer, Mass, Nov 1, 51; m 75; c 1. CELLULAR IMMUNOLOGY, MYCOLOGY. *Educ:* Bloomsburg State Col, BA, 74; Univ Pa, MA, 77, PhD(immunol), 80. *Prof Exp:* Instr immunol, Sch Allied Med Professions, Univ Pa, 78, path, 79-80; ASSOC DERMATOL, SCH MED, EMORY UNIV, 80-, SR ASSOC MICROBIOL & IMMUNOL. *Mem:* Am Soc Microbiol; NY Acad Sci; AAAS. *Res:* Mechanisms involved in transferrin mediated and T-lymphocyte mediated host defense against a variety of opportunistic fungi. *Mailing Add:* 3573 Habersham Rd Atlanta GA 30305

DUNCAN, RONNY RUSH, b Hereford, Tex, May 21, 46; m 71; c 3. AGRONOMY, PLANT PHYSIOLOGY. *Educ:* Tex Technol Univ, BS, 69; Tex A&M Univ, MS, 74, PhD(plant breeding), 77. *Prof Exp:* PROF SORGHUM & TURF PLANT BREEDING & PHYSIOL, UNIV GA, 77- *Mem:* Crop Sci Soc Am; Am Soc Agron; Sigma Xi. *Res:* Sorghum and turf (warm season and cool season) breeding and development of improved types for environmental stress situations: acid soil tolerance, micronutrient imbalances; pest problems, anthracnose and fusarium resistance and insect pressure; and multiple cropping, minimum tillage practices; herbicide tolerance; tissue culture field evaluations. *Mailing Add:* Dept Agron Ga Exp Sta Griffin GA 30223-1797

DUNCAN, STEWART, b Danvers, Mass, Apr 18, 26; m 54; c 2. ZOOLOGY, PARASITOLOGY. *Educ:* Boston Univ, AB, 49, AM, 50, PhD(parasitol), 57. *Prof Exp:* From instr to assoc prof, 50-67, PROF BIOL, BOSTON UNIV, 67- *Concurrent Pos:* Vis prof, Univ Ceylon, 65. *Mem:* Am Soc Parasitol; Am Ornith Union; AAAS; Soc Protozool; Wildlife Dis Asn. *Res:* Parasitology, ornithology; parasites of birds and mammals, with emphasis on coccidia; morphology, taxonomy, histopathology. *Mailing Add:* Dept Biol Boston Univ Boston MA 02215

DUNCAN, THOMAS OSLER, b Cambridge, Ohio, Jan 15, 48. BOTANY, SYSTEMATICS. *Educ:* Ohio State Univ, BS, 70; Univ Mich, MS, 75, PhD(bot), 76. *Prof Exp:* Asst prof, 76-82, asst cur, 77-82, ASSOC PROF BOT, ASSOC CUR SEED PLANTS & DIR, UNIV & JEPSON HERBANA, UNIV CALIF, BERKELEY, 82- *Mem:* Am Soc Plant Taxonomists; Int Asn Plant Taxonomy; Classification Soc. *Res:* Classification and evolution of the genus Ranunculus with emphasis on the application of quantitative methods to systematic problems. *Mailing Add:* Dept Bot Univ Calif Berkeley CA 94720

DUNCAN, WALTER E(DWIN), b Red Lodge, Mont, Apr 30, 10; m 39; c 3. CHEMICAL ENGINEERING. *Educ:* Mont State Col, BS, 33; Mont Sch Mines, MS, 34, Min Dr E, 60. *Prof Exp:* Asst to A M Gaudin, Mont, 34-35; mill man, Mont Coal & Iron Co, 35; res metallurgist, Mo Sch Mines, 37-39; metallurgist, Mahoning Mining Co, Ill, 39-46 & Ozark-Mahoning Co, 46-49; from assoc prof to prof chem eng, Nat Resources Res Inst, Univ Wyo, 49-76, asst dir, 60-65, assox dir, 65-71, dir, 71-80, emer prof, 76-; consult, 76-; RETIRED. *Concurrent Pos:* Consult, Peruvian Govt, 50. *Mem:* Am Chem Soc; Am Inst Mining, Metall & Petrol Engrs. *Res:* Mineral and chemical processing; testing and beneficiation; mineral treatment and processing; submerged combustion. *Mailing Add:* 7799 SW Scholls Ferry Rd Apt 150 Beaverton OR 97005

DUNCAN, WALTER MARVIN, JR, b Front Royal, Va, Feb 25, 52; m 81; c 3. SOLID STATE CHEMISTRY, QUANTUM ELECTRONICS. *Educ:* Va Polytech Inst, BS, 74; NC State Univ, PhD(chem), 79. *Prof Exp:* SR MEM TECH STAFF, TEX INSTRUMENTS, INC, 79- *Mem:* Am Phys Soc; Am Chem Soc; Am Asn Crystal Growth. *Res:* Growth and property determinations of the III-V and related semiconductor compounds with electronic device application, including carrier transport and optical properties. *Mailing Add:* PO Box 655936 Dallas TX 75265

DUNCAN, WILBUR HOWARD, b Buffalo, NY, Oct 15, 10; m 41; c 3. TAXONOMY. *Educ:* Univ Ind, AB, 32, MA, 33; Duke Univ, PhD(bot), 38. *Prof Exp:* Instr, 38-40, from asst prof to prof, 40-78, EMER PROF BOT, UNIV GA, 78- *Mem:* Bot Soc Am; Am Soc Plant Taxon; Asn Trop Biol; Int Asn Plant Taxon; Sigma Xi. *Res:* Plant taxonomy; floristics; plant identification. *Mailing Add:* Dept Bot Univ Ga Athens GA 30602

DUNCAN, WILLIAM PERRY, b Chetopa, Kans, Mar 31, 43; m 65; c 3. SYNTHETIC ORGANIC CHEMISTRY, RADIOCHEMISTRY. *Educ:* Kans State Col, BS, 65, MS, 66; Okla State Univ, PhD(org chem), 72. *Prof Exp:* Asst prof org chem, Panhandle State Univ, 66-70 & 72-73; sr chemist org radiosynthesis, Radiochem Dept, Sci Prod Div, Mallinckrodt Chem Works, 73-74; sr radiochemist org radiosynthesis, Midwest Res Inst, 74-79, head org & radiochem synthesis sect, 79-84; VPRES & DEPT MGR, EAGLE PICHER INDUST INC, CHEMSYN SCI LABS, 84- *Concurrent Pos:* Lectr, Dept Chem, St Mary Col, Leavenworth, Kans, 79-; adj prof, Sch Basic Life Sci, Univ Mo, Kansas City. *Mem:* Am Chem Soc; AAAS. *Res:* Synthesis and analysis of isotopically labeled toxic compounds or otherwise biologically active agents; catalytic hydrogenation, hydrogenolysis and dehydrogenation; metal-amine reactions; reactions of polynuclear aromatic hydrocarbons and organophosphorus and organochlorine pesticides. *Mailing Add:* ORead Labs Inc 1501 Wakarusa Dr Lawrence KS 66046-6049

DUNCAN, WILLIAM RAYMOND, b Harlingen, Tex, Aug 14, 49; m 70; c 2. TRANSPLANTATION. *Educ:* Univ Tex, Austin, BA, 71, Dallas, PhD(immunol), 76. *Prof Exp:* Asst prof cell biol, Univ Tex Health Sci Ctr, Dallas, 78-83; assoc prof immunol, Dalhousie Univ, 83-87; spec asst to dir, Div Allergy, Immunol & Transplantation, Nat Inst Allergy & Infectious Dis, NIH, 87-88, chief, GTB, 88-90, dep dir, 89-90; DIR RES PROG, NAT CANCER INST CAN, 90- *Mem:* Fedn Am Socs Exp Biol; Am Asn Immunologists; Transplantation Soc. *Res:* Immunology. *Mailing Add:* Nat Cancer Inst Can Suite 200 Ten Alcorn Ave Toronto ON M4V 3B1 Can

DUNCKHORST, F(AUSTINO) T, b Mogollon, NMex, July 1, 31. CHEMICAL ENGINEERING. *Educ:* NMex State Univ, BS, 57; Univ Pittsburgh, MS, 59, PhD(adsorption chromatography), 64. *Prof Exp:* From jr engr to sr engr, 57-70, fel engr, 70-78, ADV ENGR, BETTIS ATOMIC POWERLAB, WESTINGHOUSE ELEC CORP, 78- *Mem:* Am Inst Chem Engrs; Am Soc Mech Engrs; Am Inst Aeronaut & Astronaut. *Res:* Heat transfer and fluid flow problems in nuclear reactor design. *Mailing Add:* Westinghouse Elec Corp PO Box 79 West Mifflin PA 15122

DUNCOMBE, E(LIOT), b Bahamas, May 30, 16; nat US; m 44; c 3. ENGINEERING MECHANICS. *Educ:* Cambridge Univ, BA, 37, MA, 42; Univ Del, MEE, 56; Univ Pittsburgh, PhD(controls), 65. *Prof Exp:* Prod engr, J Lucas Ltd, Eng, 37-46; sr scientist, Nat Gas Turbine Estab, 46-47; res engr, Nat Res Coun Can, 47-51; sect engr, aircraft power plants, 51-57, ENGR, BETTIS ATOMIC POWER LAB, WESTINGHOUSE ELEC CORP, 57- *Mem:* Am Nuclear Soc. *Res:* Gas dynamics; aviation gas turbines; nuclear power plants; fuel elements; control theory. *Mailing Add:* 551 Greenhurst Dr Pittsburgh PA 15243

DUNCOMBE, RAYNOR LOCKWOOD, b Bronxville, NY, Mar 3, 17; m 48; c 1. ASTROMETRY. *Educ:* Wesleyan Univ, BA, 40; Univ Iowa, MA, 41; Yale Univ, PhD, 56. *Prof Exp:* From jr astronr to assoc astronr, US Naval Observ, 42-48; res assoc, Yale Observ, 48-49; astronr, US Naval Observ, 49-63, dir nautical almanac off, 63-75; PROF AEROSPACE SCI, UNIV TEX, AUSTIN, 76- *Honors & Awards:* Hayes Award, Am Inst Navig. *Mem:* AAAS; Am Inst Aeronaut & Astronaut; Am Astron Soc; Royal Astron Soc Gt Brit; Am Inst Navig; Sigma Xi. *Res:* Elements of Venus; variable stars; dynamical astronomy; computing machinery. *Mailing Add:* 1804 Vance Circle Austin TX 78701

DUNDEE, HAROLD A, b Tulsa, Okla, Aug 23, 24; m 87. HERPETOLOGY. *Educ:* Univ Okla, BS, 48; Univ Mich, MS, 57, PhD, 58. *Prof Exp:* Asst prof, Montclair State Teachers Col, 56-57; from instr to assoc prof zool, 58-74, prof, 74-88, EMER PROF BIOL, TULANE UNIV, LA, 88- *Concurrent Pos:* Ed, Tulane Studies Zool, 63-68 & 76-83; dir, Meade Natural Hist Libr, 65-68; copy ed, Catalogue Am Amphibians & Reptiles, 89- *Honors & Awards:* Distinguished Tech Commun, Soc Tech Commun, 89. *Mem:* Am Soc Ichthyologists & Herpetologists; Soc Study Amphibians & Reptiles; Herpetologists League. *Res:* Aggregative behavior and habitat selection by snakes; ecology and endocrinology of neotenic salamanders; Yucatecan herpetofauna; Louisiana herpetofauna; accuracy in herpetoloqical nomenclature; higher category nomenclature in herpetology; co-author of book. *Mailing Add:* Dept Ecol Evolution & Organismal Biol Tulane Univ New Orleans LA 70118

DUNDURS, J(OHN), b Riga, Latvia, Sept 13, 22; m 52; c 3. MECHANICS. *Educ:* Northwestern Univ, BSME, 51, MS, 55, PhD, 58. *Prof Exp:* Designer diesel engines, Int Harvester Co, 52-53; res engr hydraul lab, 53-54, from instr to assoc prof, 55-66, PROF CIVIL ENG, NORTHWESTERN UNIV, 66- *Mem:* Am Soc Mech Engrs; Am Soc Civil Engrs; Am Acad Mech. *Res:* Mechanics of solids. *Mailing Add:* Dept Civil Eng Northwestern Univ Evanston IL 60208

DUNEER, ARTHUR GUSTAV, JR, b Brooklyn, NY, Aug 29, 24; m 51; c 3. RADIATION PHYSICS. *Educ:* Rensselaer Polytech Inst, BS, 49, MS, 54, PhD(physics), 59. *Prof Exp:* Scientist, Assoc Nucleonics, 56-59; prin physicist, Repub Aviation Corp, 59-64; sr tech specialist, Space & Info Div, NAm Aviation, Inc, 64-66, mem tech staff, Autonetics Div, Rockwell Int Corp, 66-67; RETIRED. *Mem:* Am Phys Soc. *Res:* Analysis of response of systems to EMP, system generated EMP, nuclear radiation; electromagnetic theory applied to antennae and shielding; application of computers to these analyses; infrared sensors. *Mailing Add:* 519 E Las Palmas Dr Fullerton CA 92635

DUNELL, BASIL ANDERSON, b Vancouver, BC, Apr 5, 23. NUCLEAR MAGNETIC RESONANCE. *Educ:* Univ BC, BASc, 45, MASc, 46; Princeton Univ, MA, 48, PhD(chem), 49. *Prof Exp:* Asst chem, 45-46, from asst prof to prof, 49-85, EMER PROF CHEM, UNIV BC, 85- *Concurrent Pos:* Univ fel, Physics Dept, Univ Nottingham, UK, 78-79. *Mem:* Fel Chem Inst Can; Royal Soc Chem. *Res:* Solid state nuclear magnetic resonance. *Mailing Add:* Dept Chem 2036 Main Mall Univ BC Vancouver BC V6T 1Y6 Can

DUNFORD, EDSEL D, m. AERONAUTICS ADMINISTRATION. *Educ:* Univ Wash, BS; Univ Calif, Los Angeles, MS. *Prof Exp:* Mem tech staff, 64-91, PRES & CHIEF OPER OFFICER, TRW INC, 91-, DIR, 91- *Concurrent Pos:* Chmn bd govs, Aerospace Indust Asn; mem, Defense Sci Bd & Defense Policy Adv Comt Trade. *Mem:* Nat Acad Eng; Inst Elec & Electronics Engrs; fel Am Inst Aeronaut & Astronaut. *Res:* Aeronautics. *Mailing Add:* TRW Inc 1900 Richmond Rd Cleveland OH 44124

DUNFORD, HUGH BRIAN, b Oyen, Alta, Oct 25, 27; m 52; c 4. MOLECULAR BIOLOGY. *Educ:* Univ Alta, MSc, 52; McGill Univ, PhD(chem), 54. *Prof Exp:* Fel chem, McMaster Univ, 54-55; asst prof, Dalhousie Univ, 55-57; from asst prof to assoc prof, 57-68, PROF CHEM, UNIV ALTA, 68- *Mem:* Am Chem Soc; fel Chem Inst Can. *Res:* Transient state kinetics; mechanisms of enzyme reactions. *Mailing Add:* Dept Chem Univ Alta Edmonton AB T6G 2G2 Can

DUNFORD, JAMES MARSHALL, b Seattle, Wash, Oct 13, 15; m 41; c 8. NUCLEAR & MARINE ENGINEERING. *Educ:* US Naval Acad, BS, 39; Mass Inst Technol, SM, 44. *Prof Exp:* Design supt marine eng, Norfolk Naval Shipyard, 54-55, dep asst chief nuclear power, Bur Ships, Dept Navy, Washington, DC, 55-61; vpres naval nuclear power, NY Shipbuilding Corp, 61-65; prof mech eng, Univ Pa, 65-67; tech dir, Naval Air Eng Ctr, 67-73; exec vpres, CDI Marine Co, 74-; RETIRED. *Concurrent Pos:* Chmn, Comt Requirements & Opportunities Develop Ocean Resources, Maritime Transp Res Bd, Nat Acad Sci; mem sr rev panel, TVA Employee Concerns Spec Prog. *Mem:* Soc Naval Architects & Marine Engrs; Am Soc Naval Engrs; Sigma Xi. *Res:* Nuclear power plants for naval submarine and surface vessels; application of nuclear power of undersea research vessels. *Mailing Add:* 5510 Rigel Ct Jacksonville FL 32250

DUNFORD, MAX PATTERSON, b Bloomington, Idaho, June 17, 30; m 54; c 6. PLANT CYTOGENETICS, BIOSYSTEMATICS. *Educ:* Brigham Young Univ, BS, 54, MS, 58; Univ Calif, Davis, PhD(genetics), 62. *Prof Exp:* Asst prof biol, Mills Col, 62-63; PROF BIOL, NMEX STATE UNIV, 63- *Mem:* AAAS; Bot Soc Am; Am Genetic Asn. *Res:* Cytogenetics and biosystematics of the genus Grindelia of the family Compositae; somatic crossing over in cotton; biology of Atriplex canescens (Fourwing Saltbush). *Mailing Add:* Dept Biol Box 3AF NMex State Univ Las Cruces NM 88003

DUNFORD, RAYMOND A, b Bristol, Eng, June 19, 14; Can citizen; m 40. ANALYTICAL CHEMISTRY. *Educ:* Univ London, BSc, 37. *Prof Exp:* Chemist, Brit Drug Houses Ltd, Canm 37-39, chief chemist, 39-53, prod mgr, 53-58, plant mgr, 58-64, dir, 64-69; vpres prod, Glaxo Can Ltd, 70-77; RETIRED. *Concurrent Pos:* Consult, pharmaceut mfg, 78- *Mem:* Fel Royal Soc Chem; fel Chem Inst Can. *Res:* Determination of steroid hormones and vitamin E. *Mailing Add:* 403 Morrison Rd Oakville ON L6J 4K3 Can

DUNFORD, ROBERT WALTER, b New York, NY, July 9, 46; m 77; c 1. ATOMIC PHYSICS. *Educ:* Univ Mich, BSE, 69, MS, 74, PhD(physics), 78. *Prof Exp:* From instr to asst prof physics, Princeton Univ, 78-86; PHYSICIST, ARGONNE NAT LAB, 86- *Mem:* Am Phys Soc. *Res:* Basic symmetries in low energy experiments; parity violation in atoms; tests of quantum electrodynamics; particular interest in testing predictions of gauge theories; ion-atom collisions; precision measurements; highly charged ions accelerator based atomic physics. *Mailing Add:* Phys 203 Argonne Nat Lab 9700 S Cass Ave Argonne IL 60439

DUNG, H C, b Pingtung, Taiwan, Repub China, Mar 7, 36; m 63; c 2. ALGOLOGY, ACUPUNCTURE. *Educ:* Nat Taiwan Univ, BSc, 60; Univ Louisville, PhD(anat), 69. *Prof Exp:* Teaching asst anat, Kaohsiung Med Col, 61-63; teaching asst anat, Univ Louisville, 63-65, res asst, 65-67, res assoc, 67-70; instr, Med Sch, 70-71, asst prof, 71-79, ASSOC PROF ANAT, UNIV TEX HEALTH SCI CTR, 79- *Concurrent Pos:* Vis prof anat, Nat Yang-Ming Med Col, 77. *Mem:* Soc Neurosci; Am Asn Clin Anatomists; Am Asn Anatomists; Soc Cell Biol; Am Pain Soc; Am Cong Rehab Med. *Res:* Experimental studies of neurological mutants of the mouse; neuropathological changes in the mouse in relation to immunodeficiency as expressed in prematural thymic involution and spleen atrophy; chronic pain management using acupuncture therapy; developing method for clinical quantification of chronic pain. *Mailing Add:* Dept Cell & Struct Biol Univ Tex Health Sci Ctr 7703 Floyd Curl Dr San Antonio TX 78284

DUNGAN, KENDRICK WEBB, b Science Hill, Ky, Jan 7, 28; m 51; c 3. PHARMACOLOGY, PHYSIOLOGY. *Educ:* Univ Ky, BS, 51. *Prof Exp:* Assoc pharmacologist, William S Merrell Co, Ohio, 51-54; prin investr pharmacol, Mead Johnson & Co, 54-; PRIN RES SCIENTIST, RES & DEVELOP DIV, BRISTOL MYERS. *Mem:* AAAS; NY Acad Sci; Am Soc Pharmacol & Exp Therapeut; Soc Exp Biol & Med. *Res:* Respiration; uterus; gut; autonomic nervous system. *Mailing Add:* 1160 Wortman Rd Evansville IN 47711

DUNGAN, WILLIAM THOMPSON, b Little Rock, Ark, June 12, 30; c 2. PEDIATRIC CARDIOLOGY. *Educ:* Vanderbilt Univ, AB, 51, MD, 54; Am Bd Pediat, dipl & cert cardiol. *Prof Exp:* Asst prof, 60-61 & 65-69, assoc prof, 69-73, PROF PEDIAT, MED CTR, UNIV ARK, LITTLE ROCK, 73- *Concurrent Pos:* Fel pediat cardiol, Univ Chicago Clin, 56-57 & Med Ctr, Univ Ark, Little Rock, 59-60. *Mem:* Am Acad Pediat; Am Col Cardiol. *Mailing Add:* Dept Pediat Cardiol Univ Ark Med Ctr Little Rock AR 72201

DUNGWORTH, DONALD L, b Hathersage, Eng, July 16, 31; nat US; m 62; c 2. VETERINARY PATHOLOGY. *Educ:* Univ Liverpool, BVSc, 56; Univ Calif, Davis, PhD(vet path), 61; Am Col Vet Pathologists, dipl, 63. *Prof Exp:* Lectr vet path, Univ Calif, Davis, 59-61 & Univ Bristol, 61-62; from asst prof to assoc prof, 62-70, assoc dean res & grad educ, 73-77, PROF VET PATH & CHMN DEPT, SCH VET MED, UNIV CALIF, DAVIS, 70- *Concurrent Pos:* WHO fel, Inst Dis of Chest, Brompton, London, Eng, 68-69; Fulbright fel, Wallaceville Res Ctr, NZ, 77-78. *Mem:* Royal Col Vet Surg; Int Acad Path; Am Asn Path; Am Thoracic Soc. *Res:* Pulmonary pathology, especially effects of air pollution; inhalation toxicology. *Mailing Add:* Dept Path Sch Vet Med Univ Calif Davis CA 95616

DUNHAM, CHARLES BURTON, b Port Alberni, BC, Jan 25, 38. MATHEMATICS. *Educ:* Univ BC, BA, 59, MA, 63; Univ Western Ont, PhD(math), 68. *Prof Exp:* Res asst comput sci, 64-65, lectr, 65-69, asst prof, 69-74, ASSOC PROF COMPUT SCI, UNIV WESTERN ONT, 74- *Mem:* Asn Comput Mach. *Res:* Approximation theory; numerical analysis; best approximation, with emphasis on Chebyshev approximation; subroutines for mathematical functions; computational arithmetic. *Mailing Add:* Dept Comput Sci Univ Western Ont London ON N6A 5B9 Can

DUNHAM, CHARLES W, b Norwich, Vt, May 9, 22; c 3. HORTICULTURE. *Educ:* Univ Mass, BS, 46; Univ Wis, MS, 48; Mich State Univ, PhD(hort), 54. *Prof Exp:* Asst hort, Univ Wis, 46-48; instr floricult, Univ Mass, 48-52; asst hort, Mich State Univ, 52-54; from asst prof to prof plant sci, Univ Del, 54-84, emer prof, 84-; RETIRED. *Concurrent Pos:* Ext specialist ornamental hort, Dept Plant Sci, Univ Del, 74- *Honors & Awards:* Alex Laurie Award, 57. *Mem:* Am Soc Hort Sci; Am Hort Soc; Sigma Xi. *Res:* Plant nutrition, especially ornamental plants; plant propagation and physiology applied to growth of ornamental plants. *Mailing Add:* PO Box 176 Odessa DE 19730

DUNHAM, DAVID WARING, b Pasadena, Calif, Sept 25, 42; m 70. ASTRONOMY, CELESTIAL MECHANICS. *Educ:* Univ Calif, Berkeley, BA, 64; Yale Univ, PhD(astron), 71. *Prof Exp:* Astronaut engr, Aeronaut Chart & Info Ctr, Mo, 69-72; sci res assoc, Dept Astron, Univ Tex, Austin, 72-75; res asst astron, Cincinnati Observ, Univ Cincinnati, 75-76; ASTRONAUT ENGR, COMPUT SCI CORP, MD, 76- *Mem:* Am Astron Soc; AAAS; Int Occultation Timing Asn (pres); Fedn Am Scientists. *Res:* Astrographic catalog plate constants; total and grazing lunar occultations of stars and solar system objects for diameters, stellar duplicity and astrometry; outer planet satellite orbits; solar eclipse astrometry; spacecraft orbital studies for geomagnetic investigations; asteroid ephemerides, occultations and satellites; astronautical engineering. *Mailing Add:* Comput Sci Corp 10110 Aerospace Rd Lanham MD 20706

DUNHAM, GLEN CURTIS, b Grand Forks, NDak, Nov 3, 56; m 79; c 4. ELECTRONICS ENGINEERING. *Educ:* Pac Lutheran Univ, BS & BA, 80; Wash State Univ, MS, 83. *Prof Exp:* Res scientist, Univ Wash, 83-89; RES SCIENTIST, WASH STATE UNIV, 89- *Concurrent Pos:* Opto-mech engr, Failure Anal Assoc, 88-90; fabrication engr, Matrix Sci, 88- *Res:* gallium arsenide photovoltaic cells and modules for use in sunlight or as lasers receivers. *Mailing Add:* Wash State Univ 100 Sprout Rd Richland WA 99352

DUNHAM, JAMES GEORGE, b Akron, Ohio, Sept 10, 50; div. INFORMATION THEORY, COMMUNICATION THEORY. *Educ:* Stanford Univ, BS & MS, 73, PhD(elec eng), 78. *Prof Exp:* From asst prof to assoc prof elec eng, Washington Univ, 78-84; ASSOC PROF ELEC ENG, SOUTHERN METHODIST UNIV, 84-, ASST DEAN UNDERGRAD & GRAD STUDIES, 90- *Concurrent Pos:* Res assoc, Biomed Comput Lab, 79-84. *Mem:* Inst Elec & Electronics Engrs; Soc Photo-Optical Instrumentation Engrs. *Res:* Information theory, in particular by data compression, channel coding, cryptography and applications of information theory to genetics. *Mailing Add:* Dept Elec Eng Southern Methodist Univ Dallas TX 75275

DUNHAM, JEWETT, b Anaheim, Calif, Feb 6, 24; m 51; c 3. ZOOLOGY, PHYSIOLOGY. *Educ:* Univ Iowa, BA, 48, MS, 52, PhD(zool), 57. *Prof Exp:* Instr biol chem, Chadwick Sch, Calif, 50-53; from asst prof to assoc prof, 57-69, PROF ZOOL, IOWA STATE UNIV, 70- *Mem:* AAAS; Am Soc Zool; Sigma Xi. *Res:* Biological barriers; electrogenesis. *Mailing Add:* 413 Lynn Ames IA 50010

DUNHAM, JOHN MALCOLM, b San Jose, Calif, June 27, 23; m 51; c 4. ANALYTICAL CHEMISTRY. *Educ:* Univ Calif, Los Angeles, BS, 48, PhD(anal chem), 57. *Prof Exp:* Chemist agr res, Dow Chem Co, Calif, 48-51; asst, Univ Calif, Los Angeles, 52-56; from instr to asst prof chem, Occidental Col, 56-58; res chemist, Sterling-Winthrop Res Inst, NY, 58-62; sr anal chemist, Pitman-Moore Div, Dow Chem Co, 62-66, proj leader, Dow Human Health Res & Develop, 66-67; SECT HEAD ANALYTICAL CHEM, SQUIBB INST MED RES, 68- *Mem:* Am Chem Soc. *Res:* Determination of drugs in pharmaceutical dosage forms and biological systems; coulometric titrations; separations; identity and purity of organic compounds. *Mailing Add:* 34 Beacon Hill Dr East Brunswick NJ 08816-3404

DUNHAM, PHILIP BIGELOW, b Columbus, Ohio, Apr 26, 37. PHYSIOLOGY. *Educ:* Swarthmore Col, BA, 58; Univ Chicago, PhD(zool), 62. *Prof Exp:* USPHS fel, 62-63; from asst prof to assoc prof zool, 63-71, PROF BIOL, SYRACUSE UNIV, 71- *Concurrent Pos:* Vis assoc prof, Sch Med, Yale Univ, 68-70; USPHS fel, 69; vis scientist, Physiol Lab, Univ Cambridge, Eng, 79. *Mem:* Soc Gen Physiol; Am Physiol Soc; Biophys Soc. *Res:* Transport of sodium and potassium in mammalian red blood cells. *Mailing Add:* Dept Biol Syracuse Univ 130 College Place Syracuse NY 13244-1220

DUNHAM, VALGENE LOREN, b Jamestown, NY, Oct 6, 40; m 62. PLANT PHYSIOLOGY, PLANT BIOCHEMISTRY. *Educ:* Houghton Col, BS, 62; Syracuse Univ, MS, 65, PhD(bot), 69. *Prof Exp:* Teacher, Pub Sch, NY, 62-63, head dept biol, 63-64; teaching asst bot, Syracuse Univ, 68-69; res assoc hort, Purdue Univ, 69-73; asst prof biol, State Univ NY Col Fredonia, 73-79, assoc prof, 79-; AT WESTERN KY UNIV. *Concurrent Pos:* Hon lectr plant biochem, Univ Col, Cardiff, UK, 80-81. *Mem:* AAAS; Soc Exp Biol; Am Soc Plant Physiol; Sigma XI. *Res:* Multiple DNA polymerases in plants; effects of iron on nitrogen metabolism in bluegreen algae; control of multiple glutamine synthetases in soyben. *Mailing Add:* Dept Biol Western Ky Univ Bowling Green KY 42101

DUNHAM, WILLIAM WADE, b Pittsburgh, Pa, Dec 8, 47; m 70; c 2. HISTORY OF MATHEMATICS, STATISTICS. *Educ:* Univ Pittsburgh, BS, 69; Ohio State Univ, MS, 70, PhD(math), 74. *Prof Exp:* Lectr math, Ohio State Univ, 75; vis assoc prof math, Ohio State Univ, 87-89; from asst prof to assoc prof, 75-87, PROF MATH, HANOVER COL, 90- *Mem:* Math Asn Am; Nat Coun Teachers Math; Am Asn Univ Professors. *Res:* History of mathematics; history of analysis in the late nineteenth and early twentieth centuries. *Mailing Add:* Dept Math Hanover Col Hanover IN 47243

DUNHAM, WOLCOTT BALESTIER, b Boston, Mass, June 15, 00; m 40; c 2. ONCOLOGY. *Educ:* Columbia Univ, AB, 24, MD, 28; Am Bd Microbiol, dipl. *Prof Exp:* Asst bacteriologist, NY Post-Grad Med Sch Hosp, Columbia Univ, 36-46; res biologist, Vet Admin Hosp, 46-61, dir gen med res lab, 49-56, asst dir prof serv res, 56-61, assoc chief of staff, 61-68; vis investr, Jackson Lab, 68-74; assoc res prof, 78-85, SR RES SCIENTIST, LINUS PAULING INST SCI & MED, 85- *Concurrent Pos:* Res fel, NY Post-Grad Med Sch Hosp, Columbia Univ, 39-40; assoc, Squibb Inst Med Res, 42-46; assoc prof, Med Col, Univ Tenn, 52-68. *Mem:* AAAS; Am Acad Microbiol; Am Asn Immunologists; Am Col Physicians; NY Acad Med; Sigma Xi. *Res:* Penicillin therapy in experimental syphilis; infectious hepatitis; virus hemagglutination; granuloma inguinale antigens; tissue cultures; iathyrism. *Mailing Add:* 440 Page Mill Rd Palo Alto CA 94306

DUNIFER, GERALD LEROY, b Silverton, Ore, Oct 21, 41; m 63. EXPERIMENTAL SOLID STATE PHYSICS. *Educ:* Walla Walla Col, BS, 63; Univ Calif, San Diego, MS, 66, PhD(physics), 68. *Prof Exp:* Asst res physicist, Univ Calif, San Diego, 68-69; mem tech staff, Bell Tel Labs, 69-71; from asst prof to assoc prof, 71-85, PROF PHYSICS, WAYNE STATE UNIV, 85- *Mem:* Am Phys Soc; Sigma Xi. *Res:* Microwave spectroscopy of electrically conducting systems at cryogenic temperatures. *Mailing Add:* 320 W Drayton Ferndale MI 48220

DUNIGAN, EDWARD P, b Marshfield, Wis, June 16, 34; m 61; c 2. SOIL MICROBIOLOGY. *Educ:* Wis State Univ, Stevens Point, BS, 58; Mich State Univ, MS, 61; Univ Ariz, PhD(agr chem, soils), 67. *Prof Exp:* Res chemist, US Rubber Co & Gen Motors Res Labs, 62-64; res assoc herbicides, Univ Ariz, 64-67; assoc prof soil microbiol, 67-74, assoc prof agron, 74-78, PROF AGRON, LA STATE UNIV, BATON ROUGE, 78-, HEAD, 85- *Mem:* Am Soc Agron; Soil Sci Soc Am; Crop Sci Soc Am. *Res:* Soil organic matter; soybean nodulation; increasing nitrogen fixation in legumes; non-symbiotic nitrogen fixation in grasses and rice; microbial-insect damage and subsequent effects on nitrogen fixation in soybeans. *Mailing Add:* Dept Agron La State Univ Baton Rouge LA 70803

DUNIKOSKI, LEONARD KAROL, JR, b Newark, NJ, Nov 5, 45; m 72. CLINICAL CHEMISTRY, CLINICAL LABORATORY ADMINISTRATION. *Educ:* Rutgers Univ, AB, 66; Pa State Univ, PhD(chem), 71; Am Bd Clin Chem, dipl, 76; Am Bd Bioanal, dipl, 79. *Prof Exp:* Fel clin chem, Georgetown Univ Hosp, 71-73; dir clin chem, Perth Amboy Gen Hosp, 73-85 & Old Bridge Regional Hosp, 79-85; ADMIN DIR LABS, RARITAN BAY MED CTR, 85- *Concurrent Pos:* Adj asst prof path, Rutgers Med Sch, 74-; adj prof med technol, Middlesex County Col, 75-; consult, NJ Dept Health, 77- & Col Am Pathologists, 77-80; mem, Acad Health Exec, NJ, 87- *Mem:* Am Asn Clin Chem; Acad Clin Lab Physicians & Scientists; Am Bd Clin Chem (secy-tres, 81-83, vpres, 83-84, pres, 84-85). *Res:* Chemistry of blood coagulation including applications of artificial substrates and advanced instrumental analysis techniques; high pressure liquid chromatography techniques for therapeutic drug level analysis in serum. *Mailing Add:* 558 Ridge Rd Fair Haven NJ 07704-3614

DUNIPACE, DONALD WILLIAM, b Bowling Green, Ohio, May 24, 07; m 28; c 2. PHYSICS. *Educ:* Ohio State Univ, BA, 29, MA, 30. *Prof Exp:* Asst physics, Ohio State Univ, 29-32, fel, 32-33; physicist, Libbey-Owens-Ford Glass Co, 35-62; staff scientist, Ball Bros Res Corp, 62-69; sr scientist, Libby-Owens-Ford Co, 69-76; RETIRED. *Mem:* AAAS; Sigma Xi; Am Phys Soc; Optical Soc Am. *Res:* Investigation of glass and glass production problems; optical instrumentation; radiative heat transfer phenomena. *Mailing Add:* 2525 E Clearview Dr Carson City NV 89701

DUNIPACE, KENNETH ROBERT, b Bowling Green, Ohio, Sept 26, 29; m 63; c 2. CONTROL SYSTEMS. *Educ:* Ohio State Univ, BS, 51; Mass Inst Technol, SB, 56; Univ Fla, ME, 65; Clemson Univ, PhD(elec eng), 68. *Prof Exp:* Staff engr, Perma-glass, Inc, 53-54; exec officer aerophys res, Mass Inst Technol, 56-58, staff engr Polaris guid, 58-60, test mgr, 60-63, sr rep Apollo guid & navig, 63-65; res asst elec eng, Univ Fla, 66 & Clemson Univ, 66-687, res assoc; from assoc prof to prof elec eng, Univ Mo-Rolla, 68-77, res assoc,

Transp Inst, 68-77; PROF ELEC ENG, ENG DIV, PURDUE UNIV, INDIANAPOLIS, IND, 77- *Concurrent Pos:* Consult, Instrumentation Lab, Mass Inst Technol, 68-71, Cummins Engine Co, 87, Naval Surface Weapons Ctr, 85-86. *Mem:* Inst Elec & Electronics Engrs; Simulation Coun. *Res:* Development of guidance and navigation systems for aerospace and highway application; application of contemporary control theory to transportation problems. *Mailing Add:* Eng Div Purdue Univ 1201 E 38th St Indianapolis IN 46205-2868

DUNIWAY, JOHN MASON, b San Francisco, Calif, Nov 6, 42; m 65; c 2. PLANT PATHOLOGY. *Educ:* Carleton Col, BA, 64; Univ Wis-Madison, PhD(plant path), 69. *Prof Exp:* NSF fel plant physiol, Res Sch Biol Sci, Australian Nat Univ, 69-70; lectr plant path & asst plant pathologist, 70-77, assoc prof, 77-82, PROF PLANT PATH, UNIV CALIF, DAVIS, 82-, DEPT CHMN, 89- *Honors & Awards:* Ciba-Geigy Award, Am Phytopath Soc, 82. *Mem:* Fel Am Phytopath Soc; Am Soc Plant Physiol. *Res:* Water relations and photosynthesis in diseased plants; water relations and ecology of soil microorganisms; epidemiology of plant diseases. *Mailing Add:* Dept Plant Path Univ Calif Davis CA 95616-8680

DUNKEL, MORRIS, b Brooklyn, NY, Dec 4, 27; m 57; c 3. ORGANIC CHEMISTRY. *Educ:* Long Island Univ, BS, 50; Brooklyn Col, MS, 54; Univ Ark, PhD(org chem), 56. *Prof Exp:* Res chemist, Norda Essential Oil & Chem Co, 51-53, group leader, 56-58; res chemist, Nopco Chem Co, 58-59, head org fine chem res, 59-61; group leader, UOP Chem Div, Universal Oil Prod Co, 61-66, dir appl res, Chem Div, 66-75; mgr, Tenneco Chem Inc, 76-81, dir org res & develop, 81-83; PRES, ACCORT LABS, INC, 83- *Mem:* Am Chem Soc; Royal Soc Chem; Sigma Xi; NY Acad Sci. *Res:* Organic synthesis; aroma and flavor chemistry; natural products; vitamins; catalytic hydrogenation; stereochemistry; reactions at elevated temperatures and pressures; flame retardancy; polymer stabilization. *Mailing Add:* W 108 Glen Ave Paramus NJ 07652

DUNKEL, VIRGINIA CATHERINE, b New York, NY, July 15, 34. VIROLOGY, ONCOLOGY. *Educ:* Col New Rochelle, BA, 56; Rutgers Univ, MS, 59, PhD(microbiol), 64. *Prof Exp:* Res scientist virol, E R Squibb & Sons, 59-61; cancer res scientist, Roswell Park Mem Inst, 64-69, asst res prof microbiol, Roswell Park Div, Grad Sch, State Univ NY Buffalo, 67-69; sr staff fel viral oncol, Nat Cancer Inst, 69-72, sr staff fel immunochem, 72-73, mgr biol & immunol segment, 73-76, coordr in vitro carcinogenesis prog, 74-79; CHIEF, GENETIC TOXICOL BR, DIV TOXICOL, CTR FOOD SAFETY & APPL NUTRIT, FOOD & DRUG ADMIN, 80- *Mem:* Am Asn Cancer Res; Sigma Xi; AAAS; Environ Mutagen Soc. *Res:* Evaluation and development of mutagenicity assays for use in determining the carceogenic/mutagenic potential of chemical compounds. *Mailing Add:* Ctr Food Safety & Appl Nutrit HFF-166 Food & Drug Admin 200 C St SW Washington DC 20204

DUNKELBERGER, TOBIAS HENRY, b Paxinos, Pa, Nov 4, 09; m 41; c 2. PHYSICAL CHEMISTRY, MICROCHEMISTRY. *Educ:* Dickinson Col, ScB, 30; Univ Pittsburgh, PhD(phys chem), 37. *Prof Exp:* Teacher high sch, Pa, 30-31; asst chem, Univ Pittsburgh, 31-36; asst prof, NTex Agr Col, 36-37, Univ Idaho, 37-38, Duquesne Univ, 38-41 & NY State Col Ceramics, Alfred Univ, 41-44; prof & head dept, Duquesne Univ, 44-52; prof chem, 52-76, assoc dean, Col Arts & Sci, 69-75, EMER PROF CHEM, UNIV PITTSBURGH, 76- *Concurrent Pos:* Prof, Univ Pittsburgh Faculties in Ecuador, Agency Int Develop Contract, 63-67; chief-of-party, Faculties in Guatemala, 67-69. *Honors & Awards:* Pittsburgh Award, Am Chem Soc, 70. *Mem:* AAAS; Am Chem Soc. *Res:* Thermodynamics of solutions; microchemistry; chemical education. *Mailing Add:* 5132 Beeler St Pittsburgh PA 15217

DUNKELMAN, LAWRENCE, b Paterson, NJ, June 28, 17; m 50; c 2. PHYSICS, ASTRONOMY. *Educ:* Cooper Union, BEE, 38, EE, 49. *Prof Exp:* Jr marine engr, US Naval Shipyard, NH, 39-41; elec engr, US Bur Ships, US Dept Navy, 41-43, electronics engr, 43-48; physicist, US Naval Res Lab, 48-58; aeronautics & space scientist & head ultraviolet detection systs & planetary optics, Goddard Space Flight Ctr, NASA, Md, 59-64; mem sr staff, Inst Defense Anal, Arlington, Va, 64-65; head planetary optics, Goddard Space Flight Ctr, NASA, 65-69, head astron systs br, 69-74; MEM STAFF, LUNAR & PLANETARY LAB, UNIV ARIZ, 74- *Concurrent Pos:* Consult var univ & indust labs; staff specialist, Astron Orbiting Observ, NASA, 59-61, mem, Comt Astronaut Training & Exp, 62-64; proj scientist, UK-US Ariel II satellite, 61-64; mem, Eclipse Exped, US Navy, Sweden, 54, Peru, 65; mem staff, Goddard Space Flight Ctr, NASA. *Mem:* AAAS; fel Optical Soc Am; Int Astron Union; Sigma Xi. *Res:* Atmospheric attenuation of ultraviolet, visible and near infrared; ultraviolet technology; electrooptics; low light level photography; spectrophotometric instrumentation for space research; planetary atmospheres; solar research; manned space sciences. *Mailing Add:* Univ Ariz PO Box 36241 Tucson AZ 85740-6241

DUNKER, ALAN KEITH, b New Orleans, La, Mar 16, 43; m 65; c 2. BIOPHYSICS. *Educ:* Univ Calif, Berkeley, BS, 65; Univ Wis-Madison, MS, 67, PhD(biophys), 69. *Prof Exp:* Fel biophys, Yale Univ, 69-70, NIH fel molecular biophys & biochem, 70-72; res asst, Sloan-Kettering Cancer Res Inst, 72-75; from asst prof to assoc prof chem, 75-83, PROF, WASH STATE UNIV, 83- *Concurrent Pos:* Investr, Am Heart Asn, 78-83; affl staff scientist, Pac Northwest Lab, 87- *Mem:* Biophys Soc; Am Chem Soc; Am Soc Microbiol; Am Soc Biochem & Molecular Biol. *Res:* Structure of alpha-helical protein bundles; alpha-helices containing proline; Raman spectroscopy of proteins and biological membranes; computer aided molecular design. *Mailing Add:* Dept Biochem & Biophys Wash State Univ Pullman WA 99163-4630

DUNKER, MELVIN FREDERICK WILLIAM, b Baltimore, Md, June 12, 13; m 41; c 3. PHARMACEUTICAL CHEMISTRY. *Educ:* Univ Md, PhG, 33, BS, 34, MS, 36, PhD(pharmaceut chem), 39. *Prof Exp:* Asst, Sch Pharm, Univ Md, 34-39, Warner fel, 39-40; Rockefeller Found assoc, Northwestern Univ, 40-41; from instr to asst prof pharm, Univ Wis, 41-45; assoc prof, Col Pharm, Wayne State Univ, 45-49, prof pharmaceut chem, 49-81; RETIRED. *Mem:* AAAS; Am Chem Soc; Am Pharmaceut Asn; Am Asn Col Pharm. *Res:* Replacement of diazonium boro-fluoride by mercury; aliphatic amines; steroids; the i-ethers of 3-hydroxy-5-cholenic acid; fluorine substituted phenols; phenobarbital. *Mailing Add:* 19748 W Kings Ct Grosse Pointe Woods MI 48236-2528

DUNKL, CHARLES FRANCIS, b Vienna, Austria, Sept 16, 41; US citizen; m 65; c 2. MATHEMATICS. *Educ:* Univ Toronto, BSc, 62, MA, 63; Univ Wis, PhD(math), 65. *Prof Exp:* Instr math, Princeton Univ, 65-67; from asst prof to assoc prof, 67-80, PROF MATH, UNIV VA, 80- *Concurrent Pos:* Vis assoc prof, Ga Inst Technol, 77-78. *Mem:* Soc Indust & Appl Math. *Res:* Harmonic analysis; special functions. *Mailing Add:* Dept Math Math Astron Bldg Rm 223 Univ Va Charlottesville VA 22903-3199

DUNKLE, LARRY D, b Helena, Okla, Feb 19, 43; m 65; c 4. PHYTOPATHOLOGY. *Educ:* Colo State Col, BA, 65; Univ Wis, MS, 68, PhD(bot), 70. *Prof Exp:* From asst prof to assoc prof plant path, Univ Nebr, 71-78; USDA-ARS RES PLANT PATHOLOGIST & PROF, PURDUE UNIV, WEST LAFAYETTE, 78- *Concurrent Pos:* NIH fel, Univ Nebr, 71. *Mem:* Am Phytopath Soc; Am Soc Plant Physiol; Mycol Soc Am. *Res:* Fungal physiology; biochemistry of host-parasite interaction and the role of microbial toxins in plant diseases. *Mailing Add:* Dept Bot & Plant Path Purdue Univ West Lafayette IN 47907

DUNKLE, SIDNEY WARREN, b Cleveland, Ohio. AQUATIC INSECTS, ODONATOLOGY. *Educ:* Baldwin-Wallace Col, BS, 62; Univ Wyo, MS, 65; Univ Fla, PhD(entom), 80. *Prof Exp:* Ranger-naturalist, Grand Teton Nat Park, 63 & 65; field biologist, Yellowstone Nat Park, 65; instr biol, Cuyahoga Community Col, 66-70; instr ecol, Orange Coast Community Col, 70-71; instr human biol, Santa Fe Community Col, 72-76; instr physiol, Fresno City Col, 77; bioacoust technician, Fla State Mus, 78-80; asst prof, Univ Fla, 84-88; MGR, INT ODONATA RES INST, 85- *Concurrent Pos:* Field biologist, Fla State Mus, 75-76; consult, environ sci & permitting, 80- *Mem:* Entom Soc Am; Am Entom Soc; Int Soc Odonatology. *Res:* Odonata biology; biology of aquatic and predatory insects. *Mailing Add:* Int Odonata Res Inst 1911 SW 34th St Gainesville FL 32608

DUNKLEY, WALTER LEWIS, b Olds, Alta, Feb 15, 18; nat US; m 43; c 3. FOOD SCIENCE, DAIRY TECHNOLOGY. *Educ:* Univ Alta, BSc, 39, MSc, 41; Univ Wis, PhD(dairy indust, biochem), 43. *Prof Exp:* Lectr & asst prof dairying, Univ Alta, 43-46; res chemist, Golden State Co, Ltd, Calif, 46-48; from asst prof to prof, 48-85, EMER PROF, FOOD SCI & TECHNOL, UNIV CALIF, DAVIS, 85- *Concurrent Pos:* Fulbright awards, Fats Res Lab, New Zealand, 56-57 & Agr Inst Ireland, 63-64; mem fac chem eng, Univ Guayaquil, 69. *Honors & Awards:* Borden Award, Am Dairy Sci Asn. *Mem:* Am Dairy Sci Asn; Inst Food Technol. *Res:* Chemistry and processing of dairy products; oxidative and hydrolytic rancidity; polyunsaturated meat and dairy products from ruminants. *Mailing Add:* Dept Food Sci & Technol Univ Calif Davis CA 95616

DUNLAP, BOBBY DAVID, b Post, Tex, Mar 21, 38; m 65; c 2. EXPERIMENTAL SOLID STATE PHYSICS. *Educ:* Tex Tech Col, BS, 59; Univ Wash, PhD(physics), 66. *Prof Exp:* Assoc mem tech staff physics, Bell Tel Labs, Inc, 62-63; resident res assoc, Argonne Nat Lab, 66-70, physicist, 70-79, asst div dir, Solid State Sci Div, 79-82, group leader, Mat Sci Div, 82-88, SR PHYSICIST, ARGONNE NAT LAB, 79-, DIV DIR, MAT SCI DIV, 88- *Mem:* Am Phys Soc; Mat Res Soc. *Res:* Electronic and magnetic properties of materials, with emphasis on rare-earth and actinide compounds and superconductors. *Mailing Add:* Mat Sci Div Argonne Nat Lab Argonne IL 60439

DUNLAP, BRETT IRVING, b Oakland, Calif, Apr 24, 47; c 2. PHYSICS, CHEMISTRY. *Educ:* Univ Iowa, BA, 69; Johns Hopkins Univ, MA, 72, PhD(physics), 76. *Prof Exp:* Assoc physics, Univ Fla, 75-77; assoc surface sci, Nat Bur Standards, 77-79; assoc chem, George Washington Univ, 79-80; RES PHYSICIST, NAVAL RES LAB, 80- *Mem:* Am Phys Soc; Sigma Xi; Am Chem Soc. *Res:* Theoretical surface science. *Mailing Add:* 3509 Swann Rd Suitland MD 20746-2114

DUNLAP, CHARLES EDWARD, b New York, NY, June 8, 08; m 37; c 4. PATHOLOGY. *Educ:* Harvard Univ, AB, 30, MD, 34. *Prof Exp:* Intern, Univ Chicago, 34-35, asst resident path, 36-37, resident & asst, 37; asst, Harvard Med Sch, 39-40, instr, 40-43; from asst prof to prof, 43-78, chmn dept, 45-75, EMER PROF PATH, SCH MED, TULANE UNIV, 78- *Concurrent Pos:* Littauer fel path, Harvard Cancer Comn, 37-39; res fel, Collis P Huntington Hosp, 39-43; asst res, Presby Hosp, Chicago, 35; vis pathologist, Charity Hosp, New Orleans, 43-46, sr vis pathologist, 46-; consult, Armed Forces Inst Path, 52-; consult path study sect, USPHS, 54-57, consult training grants comt, 57-58, mem path training comt, 58-62, mem cancer training comt, 62-64, mem clin cancer training comt, 65-67 & mem radiation biol effects adv comt, 66-70; mem comt on path effects of radiation, Nat Res Coun, 55-58; mem bd dirs, Oak Ridge Inst Nuclear Studies, 55-62 & Urban Maes Res Found, 57-61; mem adv comt, President's Comn Heart Dis & Cancer, 64. *Mem:* Am Asn Cancer Res; Am Soc Exp Path (vpres, 62-63, pres, 63-64); Soc Exp Biol & Med; Am Asn Pathologists & Bacteriologists; fel Col Am Pathologists. *Res:* Carcinogenic hydrocarbons; biologic effects of radiation; human cancer. *Mailing Add:* Tulane Univ Sch Med 1430 Tulane Ave New Orleans LA 70112

DUNLAP, CLAUD EVANS, III, b Camp Atterbury, Ind, Aug 13, 46; m 67; c 3. OPIATE PHARMACOLOGY, RECEPTOROLOGY. *Educ:* Univ Fla, BS, 71, PhD(pharmacol), 75. *Prof Exp:* NIH fel, Addiction Res Found, 76-79; ASST PROF PHARMACOL, BOWMAN GRAY SCH MED, WAKE FOREST UNIV, 79- *Mem:* AAAS; NY Acad Sci; Sigma Xi. *Res:* Mechanisms of pharmacologic receptor activity and regulation; opiate neuropharmacology; neurobiology of opioid peptides, with particular emphasis on their roles in neuroendocrine regulation. *Mailing Add:* Physiol-Pharmacol Dept Bowman-Gray Sch Med Winston-Salem NC 27103

DUNLAP, DONALD GENE, b Bend, Ore, Sept 21, 26; m 56; c 1. VERTEBRATE ZOOLOGY, ANIMAL ECOLOGY. *Educ:* Ore State Univ, BS, 49, MA, 51; Wash State Univ, PhD(zool), 55. *Prof Exp:* Asst zool, Ore State Univ, 49-51; asst, Wash State Univ, 51-55, jr physiologist, Col Vet Med, 55-56; asst prof biol, Ripon Col, 56-59; from asst prof to assoc prof zool, 59-63, actg chmn dept zool, 65-66, actg chmn dept biol, 74-75, PROF ZOOL, UNIV SDAK, VERMILLION, 64- *Concurrent Pos:* Vis prof, Coe Col, 57. *Mem:* Fel AAAS; Am Soc Ichthyol & Herpet; Am Soc Zool; Soc Study Amphibians & Reptiles; Soc Syst Zool. *Res:* Biology of the amphibians and reptiles; animal systematics; physiological ecology. *Mailing Add:* Dept Biol Univ SDak Vermillion SD 57069

DUNLAP, HENRY FRANCIS, b Ennis, Tex, Oct 4, 16; m 42; c 3. WELL LOGGING, PETROPHYSICS. *Educ:* Rice Inst, BA, 38, MA, 39, PhD(physics), 41. *Prof Exp:* Asst physicist, Dept Terrestrial Magnetism, Carnegie Inst, 41-42; physicist, Univ NMex, 42-45, lectr physics, 43-44; maj physicist, Atlantic Ref Co, 45-66, head long range res sect, 66-69, res scientist, Atlantic Richfield Co, 69-76; CONSULT, 76- *Concurrent Pos:* Adj prof, Petrol Eng Dept, Univ Tex, Austin, 76- *Mem:* Am Phys Soc; Soc Prof Well Log Analysts; Soc Explor Geophys (vpres, 64-65); Am Asn Petrol Geol; Am Inst Min, Metall & Petrol Eng. *Res:* Nuclear physics; internal and external ballistics; geophysics; instrumentation in supersonic aerodynamics; meteorites; well logging; geothermal energy; economics. *Mailing Add:* PO Box 79 Wimberley TX 78676-0079

DUNLAP, JULIAN LEE, b LaGrange, Ga, Jan 27, 32; m 60; c 2. PHYSICS. *Educ:* Ga Inst Technol, BEE, 54; Vanderbilt Univ, PhD(physics), 59. *Prof Exp:* PHYSICIST, FUSION ENERGY DIV, OAK RIDGE NAT LAB, 59- *Mem:* Am Phys Soc; Sigma Xi. *Res:* Plasma physics and controlled thermonuclear research. *Mailing Add:* Fusion Energy Div Oak Ridge Nat Lab PO Box Y-12 Oak Ridge TN 37831-2009

DUNLAP, LAWRENCE H, b Madison, Wis, Oct 23, 10; m 41; c 2. DRYING OILS, RESIN CHEMISTRY. *Educ:* Univ Mo, AB, 31, BS, 33, MA, 35; Univ Ill, PhD(chem), 39. *Prof Exp:* Chemist, Res & Develop Lab, Armstrong World Industs, 39-46, sect head, 46-52, gen mgr, Chem Div, 52-62, sr res assoc, 62-75; ASST CUR, NORTH MUS, FRANKLIN & MARSHALL COL, 75- *Concurrent Pos:* Instr, Franklin & Marshall Col, 42-43, adj prof, 47-48; contribr, Encycl Chem Technol. *Mem:* Am Chem Soc; Sigma Xi; AAAS. *Res:* Polymer chemistry; specific heats of polymers. *Mailing Add:* 1315 Quarry Lane Lancaster PA 17603

DUNLAP, RICHARD M(ORRIS), b Columbia, Mo, Sept 5, 17; m 46, 69; c 5. MECHANICAL ENGINEERING. *Educ:* Mass Inst Technol, BS & MS, 41. *Prof Exp:* Trainee engr, United Shoe Mach Corp, 38-39; jr engr, Sperry Prod Co, 41; proj engr, Res Construct Co, 41-45; teacher, Robert Col, 45-48; mech engr, US Naval Underwater Ord Sta, 48-59, head appl sci dept, 59-61; assoc dir res, US Naval Underwater Weapons Res & Eng Sta, 61-71; assoc dir long range planning, US Naval Underwater Systs Ctr, Dept Defense, 71-76, dir planning & prog, 76-78; consult, Nat Ctr Atmospheric Res, 79-; PRES, R M DUNLOP CO. *Concurrent Pos:* Adj prof mech eng & appl mech, Univ RI, 78- *Mem:* Am Inst Aeronaut & Astronaut; Inst Elec & Electronics Engrs. *Res:* Underwater weapons systems. *Mailing Add:* 452 Mitchell Lane Middletown RI 02840

DUNLAP, ROBERT BRUCE, b Elgin, Ill, Oct 14, 42; div; c 2. ENZYMOLOGY. *Educ:* Beloit Col, BS, 64; Ind Univ, PhD(biochem), 68. *Prof Exp:* NIH fel biochem, Scripps Clin & Res Found, 68-71; from asst prof to assoc prof, 71-78, PROF CHEM, UNIV SC, 78-, DR FRED M WEISSMAN PALMETTO PROF CHEM ECOL, 90-; DIR BASIC RES, CTR CANCER TREAT & RES, RICHLAND MEM HOSP, 90- *Concurrent Pos:* Consult, indus, govt, & granting agencies, 71-; prin investr grants, Am Cancer Soc, NSF & NIH, 71- *Honors & Awards:* Russell Award, Res Sci & Eng, 87. *Mem:* Am Chem Soc; Am Soc Biol Chemists; fel AAAS; Am Asn Cancer Res. *Res:* Enzyme mechanisms, coenzyme chemistry, nucleotide metabolism, application of multinuclear magnetic resonance spectroscopy to biochemical systems, model systems for enzyme catalyzed reactions. *Mailing Add:* Dept Chem Univ SC Columbia SC 29208

DUNLAP, ROBERT D, b Carbondale, Pa, Apr 5, 22; m 53; c 3. PHYSICAL CHEMISTRY. *Educ:* Colgate Univ, BA, 43; Pa State Univ, MS, 44, PhD(chem), 49. *Prof Exp:* Asst chem, Pa State Univ, 43-49; from instr to assoc prof, 49-59, chmn dept, 77-85, PROF CHEM, UNIV MAINE, ORONO, 59- *Mem:* Am Chem Soc. *Res:* Physical chemistry; fluorocarbon solutions. *Mailing Add:* Dept Chem Aubert Hall Univ Maine Orono ME 04473

DUNLAP, WILLIAM CRAWFORD, b Denver, Colo, July 21, 18; m 40; c 1. PHYSICS. *Educ:* Univ NMex, BS, 38; Univ Calif, PhD(physics), 43. *Prof Exp:* Asst physics, Univ Calif, 38-42; res physicist, USDA, 42-45; res assoc, Gen Elec Res Lab, 45-55; consult semiconductors, Gen Elec Electronics Lab, 55-56; supvr solid state res, Bendix Res Labs, 56-58; dir semiconductors res & solid state electronic res, Raytheon Res Div, 58-64; asst dir electronic components res, Electronics Res Ctr, NASA, 64-68, dir res, 68-70; sci adv, Transport Systs Ctr, US Dept Transp, Cambridge, 70-74; PRES, W C DUNLAP & CO, 74- *Concurrent Pos:* Assoc prof, evening session, Siena Col, 52-54; ed-in-chief, Solid State Electronics, 59- *Mem:* Fel Am Phys Soc; fel Inst Elec & Electronics Engrs. *Res:* Cosmic rays; dielectrics; color and spectrophotometry; semiconductors; solid state physics. *Mailing Add:* 126 Prince St West Newton MA 02165

DUNLAP, WILLIAM JOE, b Wichita Falls, Tex, Oct 9, 29; m 57; c 1. GROUND WATER POLLUTION, ENVIRONMENTAL CHEMISTRY. *Educ:* Tex A&M Univ, BS, 52; Univ Okla, PhD(chem), 61. *Prof Exp:* Food technologist, Mrs Tuckers Foods, Tex, 52-55; res assoc chem, Univ Okla, 60-62; biochemist, Kerr-McGee Oil Indust, Inc, Okla, 62-63; res chemist, Res Inst, Univ Okla, 63-67; res chemist, Robert S Kerr Environ Res Lab, Environ Protection Agency, 67-84, chief, Subsurface Processes Br, 84-91; RETIRED. *Mem:* Am Chem Soc; Asn Ground Water Scientists & Engrs. *Res:* Chemistry of water pollution; sub-surface biochemistry; bioremediation; chromatography; movement and fate of pollutants in ground water. *Mailing Add:* Rte 1 Box 124 Ada OK 74820

DUNLEAVY, JOHN M, b Omaha, Nebr, June 6, 23; m 47; c 4. PLANT PATHOLOGY. *Educ:* Univ Nebr, PhD(plant path), 53. *Prof Exp:* From asst prof to assoc prof, 53-61, PROF PLANT PATH, IOWA STATE UNIV, 61- *Concurrent Pos:* Plant pathologist, USDA, 53-59, soybean dis res coordr, 59-72; leader, Iowa Soybean Proj, 72- *Mem:* AAAS; Am Phytopath Soc; Am Soc Microbiol; Am Soybean Asn; Brit Soc Gen Microbiol. *Res:* Diseases of soybeans; the nature of pathogenicity in bacterial tan spot of soybean, a disease caused by Curtobacterium flaccumfaciens. *Mailing Add:* Dept Plant Path Iowa State Univ Ames IA 50011

DUNLOP, D L, b Medicine Hat, Alta, Feb 15, 25; m 52; c 5. OBSTETRICS & GYNECOLOGY. *Educ:* Univ Man, BSc & MD, 52; Univ Nebr, MS, 66; FRCPS(C), 63. *Prof Exp:* Asst instr, Univ Nebr, 65-66; assoc prof, 67-70, PROF OBSTET & GYNEC, FAC MED, UNIV ALTA, 70- *Concurrent Pos:* Teaching fel obstet & gynec, Univ Man, 62; NIH fel, 65-66. *Mem:* Can Med Asn; fel Am Col Obstetricians & Gynecologists. *Res:* Vasopressor substances in amniotic fluid and serum. *Mailing Add:* Dept Obstet & Gynec Univ Alta Edmonton AB T6G 2B7 Can

DUNLOP, EDWARD CLARENCE, b Center, Mo, Jan 18, 16; m 45; c 2. CHEMISTRY. *Educ:* Westminster Col, Mo, AB, 36; Univ Ill, MS, 38, PhD(chem), 42. *Prof Exp:* Instr chem, Westminster Col, Mo, 36-37; asst, Buswell & Rodebush, Ill, 38-41; asst water anal, State Ill, 41-42; chemist, E I du Pont de Nemours & Co, Del, 42-43 & Wash, 43- 45, asst head, 46-63, head phys & anal div, 63-80; RETIRED. *Concurrent Pos:* Mem sci adv comt, Winterthur Mus. *Mem:* Am Chem Soc; Optical Soc Am; Soc Appl Spectros; Am Inst Chemists. *Res:* Analytical chemistry. *Mailing Add:* 504 Cranebrook Rd Wilmington DE 19803-2924

DUNLOP, WILLIAM HENRY, b Salt Lake City, Utah, June 14, 43; m 71; c 3. NUCLEAR PHYSICS. *Educ:* Univ Pa, BA, 67; Univ Calif, Los Angeles, MS, 69, PhD(physics), 71. *Prof Exp:* Fel physics, Univ Calif, Los Angeles, 71-72; RES PHYSICIST, LAWRENCE LIVERMORE LAB, 72- *Concurrent Pos:* Mem, US Deleg Conf Disarmament, Geneva, Switz, 79; prog mgr develop W87 warhead, Peace Keeper Missile Syst, 82-85, prog mgr earth penetrator weapon develop, 85-86; div leader for thermonuclear design, 86-; mem, US deleg to nuclear test taks, Geneva, Switz, 88. *Mailing Add:* 560 Sahara Dr Carrelian Bay CA 95711

DUNLOP, WILLIAM ROBERT, avian pathology, cell biology; deceased, see previous edition for last biography

DUNN, ADRIAN JOHN, b London, Eng, June 16, 43; m 73. NEUROCHEMISTRY, BIOCHEMISTRY. *Educ:* Univ Cambridge, BA, 65, MA & PhD(biochem), 68. *Prof Exp:* From instr to asst prof biochem, Univ NC, Chapel Hill, 71-73; from asst prof to prof neurosci, Univ Fla, 73-88; PROF & DEPT HEAD PHARMACOL, LA STATE UNIV MED CTR, 88- *Concurrent Pos:* Nat Res Coun Italy fel, Int Lab Genetics & Biophys, Naples, 68-69; NIMH fel biochem, Univ NC, Chapel Hill, 70-71; Sloan Found Fel, 75-79. *Mem:* AAAS; Am Soc Neurochem; Soc Neurosci; Biochem Soc; Int Soc Neurochem; Int Soc Psychoneuroendocrinol; Am Soc Pharmacol Exp Therapeut. *Res:* Effects of stress and stress-related hormones on neurochemistry and behavior; action of corticotropic-releasing hormone on brain and behavior; nervous system immune system interactions. *Mailing Add:* Dept Pharmacol La State Univ Med Ctr Shreveport LA 71130

DUNN, ANDREW FLETCHER, b Sydney, NS, Jan 17, 22; m 43, 85; c 2. EXPERIMENTAL PHYSICS. *Educ:* Dalhousie Univ, BSc, 42, MSc, 47; Univ Toronto, PhD(physics), 50. *Prof Exp:* Jr res officer physics, Atlantic Fisheries Exp Sta, 46-47; asst res officer, Nat Res Coun Can, 50-54, assoc res officer, 54-63, sr res officer, 63-67, head elec sect, 71-87; PRES, MEASUREMENTS INT LTD, 87- *Honors & Awards:* Inst Elec & Electronics Engrs Award, I & M Soc, 86. *Mem:* Fel Inst Elec & Electronics Engrs; Can Asn Physicists. *Res:* Precision electrical measurements; national primary standards; absolute determination of electrical quantities. *Mailing Add:* Measurement Int Ltd PO Box 2359 Prescott ON K0E 1T0 Can

DUNN, ANNE ROBERTS, b Champaign, Ill, Nov 23, 40; m 67. INFRARED OPTICS. *Educ:* Beloit Col, BS, 62; Rensselaer Polytech Inst, MS, 65, PhD(astrophys), 69. *Prof Exp:* Physicist, solar physics, Teledyne Brown Eng Co, 69-76; engr, infrared optics, McDonnell Douglas Astronaut Co, 76-78; SR ENGR, INFRARED OPTICS, NICHOLS RES CORP, 78- *Res:* Analysis of data from military long-wave infrared equipment; design analysis and modeling of long-wave infrared sensors. *Mailing Add:* 1044 Joe Quick Rd Hazel Green AL 35750

DUNN, ARNOLD SAMUEL, b Rochester, NY, Jan 31, 29; m 52; c 2. PHYSIOLOGY, ENDOCRINOLOGY. *Educ:* George Washington Univ, BS, 50; Univ Pa, PhD(physiol), 55. *Prof Exp:* Res assoc metab & endocrinol, Michael Reese Hosp Res Inst, 55-56; from instr to asst prof pharmacol, Sch Med, NY Univ, 56-62; from asst prof to assoc prof, 62-70, PROF MOLECULAR BIOL, UNIV SOUTHERN CALIF, 70-, ASSOC DEAN, GRAD STUDIES, 90- *Concurrent Pos:* Vis prof, Hebrew Univ Jerusalem & vis scientist & prof, Dept of Biophys, Weizmann Inst Sci, Rehovat, Israel, 72-73; USPHS res fel, 72-73 & 84; vis prof, Dept Hormone Res, Weizmann Inst Sci, Rekovot, Israel, 83-84; Meyerhof fel, 83. *Mem:* Endocrine Soc; Am Physiol Soc; Fedn Am Socs Exp Biol; Am Soc Biol Chem. *Res:* Physiology of cell division; biochemistry; endocrine control of metabolism; carbohydrate metabolism; development of isotope techniques for metabolism; hormone mechanisms. *Mailing Add:* Dept Molecular Biol Univ Southern Calif Los Angeles CA 90007

DUNN, BEN MONROE, b Neptune, NJ, Mar 22, 45; m 67; c 2. BIO-ORGANIC CHEMISTRY, PHYSICAL BIOCHEMISTRY. *Educ:* Univ Del, BS, 67; Univ Calif, PhD(bio-org chem), 71. *Prof Exp:* Staff fel protein chem, Chem Biol Lab, NIH, 73-74; asst prof biochem, 74-80, asst prof molecular biol, 77-80, assoc prof, 80-86, PROF BIOCHEM & MOLECULAR BIOL, UNIV FLA, 86- *Mem:* Am Chem Soc; Am Soc Biol Chemists; Biophys Soc; Protein Soc; Am Peptide Soc. *Res:* The relationship between the structure and functionality of enzyme active sites and the catalytic activity studied by peptide synthesis, enzyme kinetics, protein sequencing and model systems. *Mailing Add:* Dept Biochem Univ Fla Gainesville FL 32610

DUNN, BENJAMIN ALLEN, b Etowah, Tenn, Aug 3, 41; m 65; c 3. MULTIPLE USE MANAGEMENT, POLICY. *Educ:* Univ Ga, BSF, 65, MF, 68, PhD(forestry), 71. *Prof Exp:* Res asst forestry, Univ Ga, 69-71; biologist, Tenn Valley Authority, 71-72, environ scientist, 72-73; from asst prof to assoc prof, 73-86, PROF FORESTRY, CLEMSON UNIV, 86- *Mem:* Soc Am Foresters; Sigma Xi; Am Forest Asn. *Res:* Taxonomic, biological and ecological investigations of endangered and threatened forest plant species; edaphic and vegetative relationships on forest sites impacted by recreation; socioeconomic aspects of forest management. *Mailing Add:* Dept Forestry Clemson Univ Clemson SC 29631

DUNN, BRUCE, b Chicago, Ill, Apr 22, 48; m 70; c 1. SOLID ELECTROLYTES, OPTICAL MATERIALS. *Educ:* Rutgers Univ, BS, 70; Univ Calif, Los Angeles, MS, 72, PhD(ceramic sci), 74. *Prof Exp:* Staff scientist, Res & Develop Ctr, Gen Elec Co, 76-80; assoc prof, 80-85, PROF, DEPT MAT SCI, UNIV CALIF, LOS ANGELES, 85- *Concurrent Pos:* Vis prof, Univ Paris VI, 85-86; chmn, Dept Mat Sci, Univ Calif, Los Angeles, 87-90; Fulbright Award, Coun Int Exchange Scholars, 85-86; div ed, J Electrochem Soc, 88-90. *Mem:* Am Ceramic Soc; Electrochem Soc; Mat Res Soc; AAAS. *Res:* Electrical and optical properties of inorganic materials; transport and optical properties of solid electrolytes; synthesis of tunable lasers and other materials by sol-gel methods; infra-red transmitting solids. *Mailing Add:* Dept Mat Sci Univ Calif Los Angeles CA 90049

DUNN, CHARLES NORD, b Elk River, Minn, Oct 25, 36; m 58; c 4. SOLID STATE PHYSICS. *Educ:* Univ Minn, BS, 58, MSEE, 60, PhD(elec eng), 64. *Prof Exp:* mem tech staff, 64-82, DISTINGUISHED MEM TECH STAFF, BELL LABS, INC, 82- *Mem:* Inst Elec & Electronics Engrs; Am Phys Soc. *Res:* Integrated circuit device development; development of microwave diodes; oxide-coated-cathode; charge transport in non-ionized gases. *Mailing Add:* 407 S Miller St Shillington PA 19607

DUNN, CHRISTOPHER DAVID REGINALD, b Berkeley, Eng, Nov 13, 45; m 75; c 1. PHARMACOLOGY, HEMATOLOGY. *Educ:* Univ Bradford, BPharm Hons, 67; Univ London, PhD(pharmacol), 70. *Prof Exp:* Lectr biophys, Inst Cancer Res, London, 70-72; sr sci officer, hemat, Welsh Nat Sch Med, Univ Wales, Cardiff, 72-75; asst prof med biol, Mem Res Ctr, Univ Tenn, 76-79, assoc prof, 79-80; SR SCIENTIST, NORTHROP SERV INC, 80-; RES ASSOC PROF, BAYLOR COL MED, 80- *Concurrent Pos:* Sr sci officer, Univ Hosp Wales, Cardiff, 72-75; fel, Brit Coun Younger Scientist Interchange Scheme, 73; consult, NASA, 78- *Mem:* Int Soc Exp Hemat; Am Soc Hemat; Brit Soc Hemat; Soc Exp Biol & Med; Aerospace Med Asn; Sigma Xi. *Res:* Hormonal control of hemopoiesis; hemopoietic stem cell regulation; cell kinetics and anti-tumor therapy; erythropoietic effects of space flight. *Mailing Add:* 58 Tne Street Uley Dursley Glos Gl11-5SJ England

DUNN, CLARK A(LLAN), b Stickney, SDak, Sept 9, 01; m 28; c 2. CIVIL ENGINEERING. *Educ:* Univ Wis, BS, 23; Okla State Univ, CE, 34, MS, 37; Cornell Univ, PhD, 41. *Prof Exp:* Engr, Bridge Div, SDak State Hwy Comn, 23-27; construct engr, Bridge Div, Ark State Hwy Dept, 27-29; from asst prof to prof civil eng, Okla State Univ, 29-66, head gen eng, 41-66, dir eng res & head sch gen eng, 45-66, assoc dean, Col Eng, 66-67, EMER PROF CIVIL ENG, OKLA STATE UNIV, 67- *Concurrent Pos:* Observer, Task Force Frigid Opers, Alaska; consult engr, Assoc J E Kirkham. *Mem:* Fel AAAS; Am Soc Civil Engrs; Am Soc Eng Educ; Nat Soc Prof Engrs (vpres, 55-57, pres, 58-59). *Res:* Electric welding inspection. *Mailing Add:* 3232 S Quebec Ave Tulsa OK 74135

DUNN, D(ONALD) W(ILLIAM), b Kingston, Ont, Apr 10, 23. APPLIED MATHEMATICS. *Educ:* Queen's Univ, Ont, BA, 48; Mass Inst Technol, PhD(math), 53. *Prof Exp:* Asst math, Mass Inst Technol, 52-53, res assoc, 53-54; res fel aeronaut, Johns Hopkins Univ, 54-57; res officer, Nat Res Coun Can, 57-65; math consult, 65-84; RETIRED. *Mem:* Am Math Soc; Soc Indust & Appl Math; Can Math Soc. *Res:* Fluid mechanics; laminar flow stability; turbulence; applications in engineering, geophysics and astrophysics. *Mailing Add:* 158 Campbell Cresent Kingston ON K7M 1Z5 Can

DUNN, DANNY LEROY, b Wichita, Kans, July 12, 46; m 67; c 2. ANALYTICAL CHEMISTRY, ORGANIC CHEMISTRY. *Educ:* Wichita State Univ, BS, 68, MS, 70; NTex State Univ, PhD(org chem), 76. *Prof Exp:* Fel bio-org, Health Sci Ctr, Univ Tex, 76-78; sr scientist, 78-81, head high performance liquid chromatography chem, 81-83, head methods develop, 83-84, mgr dermat, 84-86, ASSOC DIR METHODS DEVELOP/ ANALYTICAL CHEM, ALCON LABS, 86- *Mem:* Am Chem Soc; AAAS. *Res:* Applications of high performance liquid chromatography to drug analysis; optimization of analytical methods; resolution of optically active compounds using chromatography. *Mailing Add:* Alcon Labs 6201 South Freeway Ft Worth TX 76134

DUNN, DARREL EUGENE, b Clay Center, Kans, July 15, 32; m 65; c 2. HYDROGEOLOGY. *Educ:* Univ Ill, Urbana, BS, 55, PhD(geol), 67. *Prof Exp:* Geologist, Pure Oil Co, 57-61; groundwater geologist, Alta Res Coun, Edmonton, Can, 66-67; asst prof geol, Mont State Univ, 67-71; assoc prof, Univ Toledo, 71-74; HYDROGEOLOGIST, EARTH SCI SERV, INC, 74- *Mem:* Geol Soc Am; Am Asn Petrol Geol. *Res:* Computer modeling of ground water systems and watershed systems. *Mailing Add:* 3274 Woodhaven Dr Murrysville PA 15668

DUNN, DAVID BAXTER, b Mustang, Okla, Jan 10, 17; m 42; c 6. BOTANY. *Educ:* Univ Calif, Los Angeles, BA, 40, MA, 43, PhD(bot), 48. *Prof Exp:* Asst bot, Univ Calif, Los Angeles, 40-42 & 46-47; instr bot & genetics, Calif State Polytech Col, 47-48; asst bot, Atomic Energy Proj, Univ Calif, Los Angeles, 48-50; asst biol, NMex Agr & Mech Col, 50-53; vis lectr, Univ Minn, 54; vis botanist, Rancho Santa Ana Bot Garden, 53-55 & Occidental Col, 55-56; from asst prof to prof, 56-87, EMER PROF BOT & CUR HERBARIUM, UNIV MO, 87- *Concurrent Pos:* Ed, Trans, Mo Acad Sci, 66- & Mus Contrib Monographic Series, 70- *Mem:* Fel AAAS; Soc Study Evolution; Am Soc Plant Taxon; Bot Soc Am; Ecol Soc Am. *Res:* Lupinus, the taxonomy, breeding systems, genetics, ecological races, intersterility and interfertility between colonies as well as races; desert ecology. *Mailing Add:* 1306 Hinkson Columbia MO 65201

DUNN, DAVID EVAN, b Dallas, Tex, Oct 13, 35; m 58; c 2. ROCK MECHANICS, FORENSIC GEOLOGY. *Educ:* Southern Methodist Univ, BS, 57, MS, 59; Univ Tex, PhD(geol), 64. *Prof Exp:* Asst prof struct geol, Tex Tech Col, 62-63; vis asst prof, Univ NC, Chapel Hill, 63-64, from asst prof to assoc prof, 64-73, actg chmn dept geol, 67, asst chmn, 66-69, prof struct geol, 73-79; prof earth sci & dean, Col Sci, Univ New Orleans, 79-84; PROF GEOSCI & DEAN NATURAL SCI & MATH, UNIV TEX DALLAS, 84- *Concurrent Pos:* Consult geologist, 60, 61 & 80-99; first vchmn, Struct Geol & Tectonics Div, Geol Soc Am, 81 & 82, chmn, 83, counr, 85-87; counr, La Univs Marine Consortium, 79-84, chmn, 81-83; bd dirs, Dosecc Inc, 89-, chmn, 92-94. *Mem:* AAAS; Am Geophys Union; Asn Prof Geol Scientists; fel Geol Soc Am; Nat Asn Geol Teachers. *Res:* Deformation in orogenic belts; rock mechanics. *Mailing Add:* Univ Tex Dallas MS FN 3 2 PO Box 830688 Richardson TX 75083-0688

DUNN, DEAN ALAN, b Groton, Conn, Nov 11, 54; m 86. MICROPALEONTOLOGY, GEOLOGICAL OCEANOGRAPHY. *Educ:* Univ Southern Calif, BS, 76 & BSc, 77; Univ RI, PhD(oceanog), 82. *Prof Exp:* Geophys asst, Western Region Explor Off, Union Oil Co, Calif, 76; lab tech II, Dept Geol Sci, Univ Southern Calif, 76-77; teaching asst, dept geol, Fla State Univ, 77-78; res asst IV, Grad Sch Oceanog, Univ RI, 78-82, consult, Cenozoic Paleoceonog Proj, 82-83; asst prof geol, Univ Southern Miss, 83-86; staff scientist & sci ed, Deep Sea Drilling Proj, Scripps- Inst Oceanog, 83-86; ASSOC PROF GEOL, UNIV SOUTHERN MISS, 86- *Concurrent Pos:* Nat Merit Scholar, 72-77; Univ Southern Calif Trustee Scholar, 72-77; Shipboard sedimentologist, Deep Sea Drilling Proj Leg, 85-Equatorial Pac Paleoenvirons, 82; vis assoc prof, Ctr Marine Sci, Univ Southern Miss, 87; sedimentologist & shipboard sci rep, DSDP Leg 93, Western N Atlantic, 83; shipbd scientist, Oceanog Surv, US Naval Oceanog Off, 88-90, summer fac fel, 88-91. *Mem:* Geol Soc Am; Am Asn Petrol Geologists; Soc Econ Paleontologists & Mineralogists; Am Geophys Union; Sigma Xi. *Res:* Biostratigraphy of Mesozoic and Cenozoic calcareous nannofossils and Cenozoic radiolaria; paleo-climatology and paleo-oceanography of the Pacific Ocean; global geochemical and sedimentological cyclicities in the oceans; Gulf Coast stratigraphy and sedimentology. *Mailing Add:* Dept Geol Univ Southern Miss Box 5044 Hattiesburg MS 39406-5044

DUNN, DONALD A(LLEN), b Los Angeles, Calif, Dec 31, 25; m 48; c 2. COMMUNICATIONS. *Educ:* Calif Inst Technol, BS, 46; Stanford Univ, MS, 47, LLB, 51, PhD(elec eng), 56. *Prof Exp:* Res assoc elec eng, Stanford Univ, 51-59; dir res, Eitel-McCullough, Inc, 59-61; assoc prof & dir plasma physics lab elec eng, 61-68, assoc prof, 69-72, PROF ENG ECON SYSTS, STANFORD UNIV, 72- *Concurrent Pos:* Mem 20th Century Fund task force on int satellite commun; consult, Nat Acad Eng comt on telecommun; consult, Fed Commun Comn, Senate subcomt on commun & Dept Justice Antitrust Div; mem rev panels, Off Sci & Technol, NSF & Nat Acad Sci. *Mem:* Fel Inst Elec & Electronics Engrs; Asn Pub Policy & Mgt. *Res:* Telecommunications systems; public policy analysis. *Mailing Add:* Dept Eng-Econ Systs Terman 320 Stanford Univ Stanford CA 94305-4025

DUNN, DOROTHY FAY, b Sidney, Ill. INFORMATION SCIENCE. *Educ:* Univ Ill, Urbana, BS, 39; Univ NC, MSPH, 46; Purdue Univ, PhD(philos), 62. *Prof Exp:* Home economist, USDA, 39-43 & Univ Wis, 44-45; pub health educ consult, Ill State Dept Pub Health, 46-50; from instr to asst prof hyg, Univ Ill, 51-54, assoc prof hyg & pub health, 55-67; prof home econ & chmn dept, Western Ky Univ, 67-68; prof family econ & home mgt & chmn dept, Stout State Univ, 68-71; asst regional dir consumer affairs, Food & Drug Admin, US Dept Agr, Chicago, 71-80; RETIRED. *Honors & Awards:* Merit Award, US Civil Serv. *Mem:* Fel AAAS; fel Am Pub Health Asn; fel Am Col Health Asn; fel Am Sch Health Asn. *Res:* Domestic water consumption; level of tuberculosis knowledge and smoking habits of university freshmen. *Mailing Add:* 1004 E Harding Dr Apt 202 Urbana IL 61801-6344

DUNN, FLOYD, b Kansas City, Mo, Apr 14, 24; m 50; c 2. ULTRASONIC BIOPHYSICS & BIOENGINEERING. *Educ:* Univ Ill, Urbana, BS, 49, MS, 51, PhD(elec eng & biophys), 56. *Prof Exp:* Res assoc, 56-57, from asst prof to prof, 57-65, chmn, bioeng fac, 78-82, PROF ELEC ENG, BIOENG & BIOPHYS, UNIV ILL, URBANA, 65-, DIR BIOACOUST RES LAB, 77- *Concurrent Pos:* Sr res fel, NIH, vis prof, dept microbiol, Univ Col, Cardiff, 68-69; assoc ed, Bioacoustics, Acoust Soc Am; mem, Tech Electronic Prod Radiation Standards Comt, Food & Drug Admin, 74-76; Eleanor Roosevelt Int Cancer fel, Am Cancer Soc, vis sr scientist, Inst Cancer Res, Sutton, Surrey, Eng, 75-76, 82-83 & 90 & sr Fulbright fel, 82-83; mem, NIH Diag Radiol Study Sect, 76-82; Japan Soc Prom Sci fel, vis prof, Tohoku Univ, Sendai, Japan, 82 & 89-90; vis prof, Univ Nanjing, Nanjing, China, 83. *Honors & Awards:* William J Fry Mem Award, Am Inst Ultrasound Med, 84, Joseph W Holmes Basic Sci Pioneer Award, 89; Medal of Spec Merit, Acoust Soc Japan, 88; Silver Medal, Acoust Soc Am, 89. *Mem:* Nat Acad Sci; Nat Acad Eng; fel Acoust Soc Am(vpres, 81-82, pres, 85-86); fel AAAS; fel Inst Elec & Electronics Engrs; fel Am Inst Ultrasound in Med; fel Inst Acoust (UK); Biophys Soc; hon mem Japan Soc Ultrasound Med. *Res:* Ultrasonic bioengineering and biophysics; ultrasonic propagation properties of living organisms; physical mechanisms of interaction of ultrasound & biological media; ultrasonic toxicity; ultrasonic dosimetry; measurement of ultrasonic fields in liquid and liquid-like media and development of measurements and sound field production instrumentation; ultrasonic microscopy. *Mailing Add:* Bioacoust Res Lab Univ Ill 1406 W Green St Urbana IL 61801-2991

DUNN, FLOYD WARREN, b Huntington, Ark, Dec 15, 20; m 44; c 3. BIOCHEMISTRY. *Educ:* Abilene Christian Univ, BS, 44; Univ Colo, MS, 46, PhD(biochem), 50. *Prof Exp:* Prof chem, Abilene Christian Univ, 46-60; assoc prof biochem, Med Units, Univ Tenn, 60-63, prof, 63-65; prof biochem, Col Med, Univ Ill, assigned to fac med, Chiengmai Univ, Thailand, 65-68; dean Grad Sch, 74-85, PROF BIOCHEM, DEPT CHEM, ABILENE CHRISTIAN UNIV, 68- *Concurrent Pos:* Tech adv chem, US Int Co-op Admin at Chulalongkorn Univ, Bangkok, Thailand, 58-59. *Mem:* Am Chem Soc; Am Soc Biol Chem; Soc Exp Biol & Med. *Res:* Synthesis of amino acids and peptides; amino acid antagonists; proteolytic enzymes; microbiological assay. *Mailing Add:* Dept Chem 933 Washington Blvd Abilene TX 79601-4618

DUNN, FREDERICK LESTER, b Seneca Falls, NY, Dec 24, 28; m 86; c 2. MEDICAL ANTHROPOLOGY. *Educ:* Harvard Univ, AB, 51, MD, 56; Univ London, DTM&H, 60; Univ Malaya, PhD, 73. *Prof Exp:* Intern, Univ Wash, King County Hosp, Seattle, 56-57; from asst chief to chief influenza surveillance unit, Commun Dis Ctr, USPHS, Ga, 57-59; from asst res epidemiologist & asst clin prof trop med to assoc res epidemiologist & assoc clin prof, 60-67, assoc prof epidemiol, 67-69, PROF EPIDEMIOL & MED ANTHROP, UNIV CALIF, SAN FRANCISCO, 69- *Concurrent Pos:* Lectr, Sch Pub Health, Univ Calif, Berkeley, 65- *Mem:* AAAS; Soc Epidemiol Res; Am Soc Trop Med & Hyg; Soc Med Anthrop; Am Col Epidemiol. *Res:* Communicable disease epidemiology and control; international health; medical anthropology; comparative medical systems. *Mailing Add:* Dept Epidemiol & Biostatist Univ Calif San Francisco CA 94143-0560

DUNN, GEORGE LAWRENCE, b Groton, Conn, May 5, 36; m 59; c 3. ORGANIC CHEMISTRY. *Educ:* Univ Conn, BA, 58; Univ Maine, MS, 60, PhD(org chem), 62. *Prof Exp:* Sr med chemist, 62-69, sr investr, 69-75, ASST DIR CHEM, SMITH KLINE & FRENCH LABS, 75- *Mem:* Am Chem Soc; Sigma Xi; AAAS; ASM. *Res:* Steroid synthesis; heterocyclic and polycyclic cage compounds; antibiotics; semisynthetic penicillins and cephalosporins. *Mailing Add:* 690 Mallard Rd Wayne PA 19087

DUNN, GORDON HAROLD, b Montpelier, Idaho, Oct 11, 32; m 52; c 8. ATOMIC & PLASMA PHYSICS, PHYSICAL CHEMISTRY. *Educ:* Univ Wash, BS, 55, PhD(physics), 61. *Prof Exp:* Nat Bur Standards-Nat Res Coun fel atomic collisions, 61-62, physicist, joint inst lab astrophys, 62-77, chief quantum physics div, 77-85, SR SCIENTIST, NAT INST STANDARDS & TECHNOL, 85- *Concurrent Pos:* Lectr physics & astrophys, Univ Colo, 64-77, adj prof, 74-; mem gen comt, Int Conf Physics of Electronic & Atomic Collisons, 69-73, mem prog comt, 75; mem conf comt, Gaseous Electronic Conf, 66-70, secy, 68, chair, 69, chmn, 71-72; fel, Comt Sci & Technol, US House Rep, 75-76, JILA, 64-; mem prog comt, Conf Atomic Processes High Temperature Plasmas, 77-85 & 85, chair, 81; mem comt, Atomic, Molecular & Optical Sci, Nat Res Coun, 83-86, panel mem, Instruments & Facil, 83-84, panel mem, Ion Storage Ring Atomic Physics, 85-88, vchair, 89, chair, 90-92; mem pub comt & prog comt, Div Atomic, Molecular & Optical Physics, Am Phys Soc, 76-78, chairperson, 89-90; co-organizer, Advan Study Inst Physics of Electron-ion & Ion-ion Collisions, NATO, 85, US-Japan Workshop, 86; guest worker, Univ Giessen, 88 & 90. *Honors & Awards:* Gold Medal, Dept Com, 70; Davisson-Germer Prize, Am Phys Soc, 84. *Mem:* Fel Am Phys Soc. *Res:* Investigation of collisions of electrons and photons with ions and with other simple atomic and molecular systems; electron-ion collisions: recombination, excitation, dissociation and ionization; photon-ion collisions; ion-molecule reactions at ultra-low temperatures; very high accuracy atomic mass measurements; dynamics of trapped ions. *Mailing Add:* Joint Inst Lab Astrophys Univ Colo Box 440 Boulder CO 80309-0440

DUNN, HENRY GEORGE, b Leipzig, Ger, Apr 18, 17; m 54; c 2. PEDIATRIC NEUROLOGY. *Educ:* Univ Cambridge, MB & BCh, 42, MA, 43; DCH, 50; FRCP(C), 72; FRCP, 73. *Prof Exp:* Registr, Children's Dept, London Hosp, Eng, 49-51 & Hosp Sick Children, London, 51-52; asst pathologist, Babies Hosp, Columbia-Presby Med Ctr, 52-53; chief resident pediat, Vancouver Gen Hosp, 53-54; from asst prof to prof, 56-82, EMER PROF PEDIAT, UNIV BC, 82- *Concurrent Pos:* Holt fel, Columbia Univ, 52-53; fel pediat, Univ BC, 54-55; R S McLaughlin travelling fel, 66-67; res assoc, Harvard Med Sch, 59 & Children's Med Ctr, 59; consult, Woodlands Sch Retarded; vis prof, Makerere Univ, Kampala, Uganda, 84. *Honors & Awards:* Ross Award, Can Pediat Soc, 88. *Mem:* Am Asn Ment Retardation; Can Med Asn; Can Pediat Soc; Can Neurol Soc; Child Neurol Soc. *Res:* Mental retardation; nerve conduction studies in children; metabolic disorders in children; child neurology. *Mailing Add:* 4088 Maple Crescent Vancouver BC V6J 4B2 Can

DUNN, HORTON, JR, b Coleman, Tex, Sept 3, 29. DISPERSANTS, RUST INHIBITORS. *Educ:* Hardin-Simmons Univ, BA, 51; Case Western Reserve Univ, PhD(info sci), 79. *Prof Exp:* Instr chem, Hardin-Simmons Univ, 51-52; teaching fel chem, Purdue Univ, 52-53; res chemist, 53-70, SUPV RES, LUBRIZOL CORP, 70- *Concurrent Pos:* Ed, Isotopics, 61-63, chmn bd, 64-67; pres, Am Soc Info Sci, Northern Ohio Chap, 73-74; chmn, Am Chem Soc, Cleveland Sect, 87-88, dir, 90- *Mem:* Fel Am Inst Chemists; Am Chem Soc; Royal Soc Chem; AAAS; Am Soc Info Sci. *Res:* Synthesis of organosulfur and organophosphorus compounds; process for preparing phosphorous-containing acids; synthesis of maleamic acids and maleimides; polymer synthesis; additives for lubricants and fuels; industrial information systems. *Mailing Add:* Lubrizol Corp 29400 Lakeland Blvd Wickliffe OH 44092

DUNN, HOWARD EUGENE, b Kansas City, Mo, Apr 14, 38; m 61; c 2. ORGANIC CHEMISTRY. *Educ:* William Jewell Col, AB, 60; Univ Ill, PhD(org chem), 65. *Prof Exp:* Res chemist, Res Ctr, Phillips Petrol Co, 65-69; from asst prof to assoc prof, 69-78, PROF CHEM, UNIV SOUTHERN IND, 78- *Concurrent Pos:* Res grant, Southern Ill Univ, 71. *Mem:* Am Chem Soc. *Res:* Organoboron, sulfoxide and sulfone chemistry; homogeneous catalysis; trace analysis; environmental chemistry. *Mailing Add:* Dept Chem Univ Southern Ind Evansville IN 47712

DUNN, IRVING JOHN, b Berkeley, Calif, Dec 15, 38; m 62; c 3. FERMENTATION TECHNOLOGY, BIOCHEMICAL ENGINEERING. *Educ:* Univ Wash, BS, 60; Princeton Univ, PhD(chem eng), 63. *Prof Exp:* Asst prof chem eng, Univ Idaho, 64-68; assoc prof, Robert Col, 68-70; DOZENT CHEM ENG, SWISS FED INST TECHNOL, 71- *Mem:* Am Chem Soc. *Res:* Biochemical engineering including fermentation technology and biological waste water treatment; design of gas-liquid biological reactors and biological film reactors; modelling and simulation of biological processes; instrumentation and control. *Mailing Add:* Dept Chem Eng ETH Lab TCL 8092 Zurich Switzerland

DUNN, J(OHN) HOWARD, b Omaha, Nebr, Aug 29, 09; m 35; c 3. MECHANICAL ENGINEERING. *Educ:* Iowa State Univ, BS, 31. *Prof Exp:* Asst mgr, Dunn Mfg Co, 31-34; develop engr, Aluminum Co Am, 34-41, asst to mgr prod planning, 41-44, automotive develop mgr, 44-49, asst mgr, Develop Div, 49-53, mgr, 53-59, mgr, Process Develop Labs, 59-67, dir develop, 67-70, vpres res & develop, 70-74; RETIRED. *Mem:* Soc Automotive Engrs; Sigma Xi. *Res:* Applications of aluminum alloys and products, especially in the automotive field. *Mailing Add:* 1078 Wade Lane Oakmont PA 15139

DUNN, JACQUELIN, b Sidney, Ohio, Jan 24, 42; div; c 1. MEDICAL MICROBIOLOGY. *Educ:* Col Wooster, BA, 63; Univ Mo, Columbia, PhD(microbiol), 72; Univ SDak, MBA, 83. *Prof Exp:* Res asst neurol surg, Ohio State Univ Hosp, 63-64; res asst pediat immunol, Downstate Med Ctr, State Univ NY, 64-68; sr lab tech hemat, Dept Med, Univ Mo Med Ctr, 68-70; microbiologist bact, Alaska Med Labs, 73-74; asst prof, 74-80, ASSOC PROF MED TECHNOL, MT MARTY COL, 80-, HEAD, DIV HEALTH SCI, 78-; PROF MGT INFO SYSTS & ACCT, UNIV S DAK 84-, DIR BUS GRAD PROGS, 88- *Concurrent Pos:* Clin asst prof, Sch Med, Univ SD, 75-; consult, Ctr Community Orgn & Area Develop, Augustana Col, 75- *Mem:* Sigma Xi; Am Soc Microbiol; Am Inst Cert Pub Acct. *Res:* Systems analysis & design, telecommunications. *Mailing Add:* 930 Crestview Vermillion SD 57069-3611

DUNN, JAMES ELDON, b Fairbury, Nebr, Jan 8, 36; m 64; c 2. MATHEMATICAL STATISTICS. *Educ:* Univ Nebr, BSc, 57, MSc, 61; Va Polytech Inst, PhD(statist), 63. *Prof Exp:* From asst prof to assoc prof, 63-76, PROF MATH, UNIV ARK, FAYETTEVILLE, 76- *Concurrent Pos:* Statist consult, Ark State Judiciary Comn, 63-65, Bur Sport Fisheries & Wildlife, 65-71 & Sport Fishing Inst, 68-71; NSF sci fac fel, Stanford Univ, 71-72, adv contractor, USEPA, 80-90. *Mem:* Am Statist Asn; Biomet Soc. *Res:* Environmental statistics. *Mailing Add:* Dept Math Univ Ark Fayetteville AR 72701

DUNN, JAMES ROBERT, b Sacramento, Calif, Oct 18, 21; m 70; c 5. GEOLOGY. *Educ:* Univ Calif, AB, 43, PhD(geol), 51. *Prof Exp:* Asst, Univ Calif, 46-50; from assoc prof to prof, 50-73, ADJ PROF GEOL, RENSSELAER POLYTECH INST, 73- *Concurrent Pos:* Geologist, New Idria Quicksilver Mining Co, 46 & Iron Ore Co, Can, 51; indust & govt consult, 53-; pres, Dunn Geosci Corp, 60-71, chmn bd, 71-; pres, Am Inst Prof Geol Found, 81. *Honors & Awards:* Martin Van Couvering Mem Award, Am Inst Prof Geol; Hal Williams Hardinage Award, Am Inst Mining Eng. *Mem:* Am Inst Prof Geologists (vpres, 69, pres, 80); Asn Eng Geol; Soc Econ Geol; Am Soc Testing & Mat; fel Geol Soc Am; Soc Mining Engrs. *Res:* Physical and chemical characteristics of available materials for filling subsurface coal mines; potential alkali reactivity of chert in portland cement concrete; socio-economic applications of geology, including environmental geology, conservation and planning; utilization of geothermal heat; computer simulation techniques in planning for mineral resource development; asbestos mineralogy. *Mailing Add:* Dunn Geosci Corp 12 Metro Park Rd Albany NY 12205

DUNN, JOHN FREDERICK, JR, b Passaic, NJ, May 13, 30; m 53; c 5. MECHANICAL ENGINEERING. *Educ:* Mass Inst Technol, SB, 51, SM, 53, ScD(mech eng), 57. *Prof Exp:* Proj engr, Dynamic Anal & Control Lab, Mass Inst Technol, 51-57; proj engr, Res Div, Walworth Co, 57-58, asst dir valve res, 58-59, chief engr, 59-62; assoc prof mech eng, 62-63, PROF MECH ENG, NORTHEASTERN UNIV, 63- *Concurrent Pos:* Mem exec comt, Mfrs Standardization Soc Valve & Fitting Indust, 61-62. *Mem:* Am Soc Mech Engrs; Am Soc Eng Educ. *Res:* Components for high-performance electro-hydraulic and electro-pneumatic control systems. *Mailing Add:* 30 Huntington Ave Sharon MA 02067

DUNN, JOHN PATRICK JAMES, b Pottsville, Pa, May 29, 44; m 66; c 2. MOLECULAR BIOLOGY. *Educ:* West Chester State Col, AB, 66; Rutgers Univ, New Brunswick, PhD(microbiol), 70. *Prof Exp:* Fel molecular genetics, Univ Heidelberg, 70-72; asst microbiologist, 72-74, assoc microbiologist, 74-77, MICROBIOLOGIST, BROOKHAVEN NAT LAB, 77- *Concurrent Pos:* Adj asst prof microbiol, State Univ NY, Stoney Brook, 75-80, adj prof, 80- *Honors & Awards:* Ernest Orlando Lawrence Memorial Award, Dept Energy, 1984. *Mem:* AAAS; Am Soc Biol Chemists. *Res:* Transcription of DNA, processing of mRNA and translation of mRNA. *Mailing Add:* Biol Dept Brookhaven Nat Lab Upton NY 11973

DUNN, JOHN ROBERT, b Andover, Eng, May 12, 30; m 55; c 4. PHYSICAL ORGANIC CHEMISTRY, RUBBER TECHNOLOGY. *Educ:* Univ London, BSc, 51, DPhil(phys org chem), 53. *Prof Exp:* Nat Res Coun Can fel, 53-55; sr chemist, Natural Rubber Producers' Res Asn, 55-62; sr res chemist, 62-66, supvr compounding res, Polymer Corp, Ltd, 66-71, SCI ADV, POLYSAR, LTD, 71- *Mem:* Fel Chem Inst Can; Am Chem Soc; Int Orgn Standardization. *Res:* Rubber technology; oxidation and antioxidants in rubber and aldehydes; vulcanization of rubber; physical properties of polymers; photolysis of ketones; flame retardancy; polymer blends; modification of polymers. *Mailing Add:* Polysar Rubber Corp PO Box 3001 Sarnia ON N7T 7M2 Can

DUNN, JOHN THORNTON, b Washington, DC, Oct 27, 32; m 62; c 3. ENDOCRINOLOGY. *Educ:* Princeton Univ, AB, 54; Duke Univ, MD, 58. *Prof Exp:* Intern med, NY Hosp-Cornell Univ, 58-59; resident, Univ Utah Hosps, 59-61; asst prof, 66-70, assoc prof, 70-76, PROF MED, SCH MED, UNIV VA, 76- *Concurrent Pos:* Fel thyroid, Mass Gen Hosp-Harvard Univ, 61-62 & 63-64; fel, Presby Hosp-Columbia Univ, 62-63; staff mem, Div Radiol Health, USPHS, 62-64, res career develop award, 71-76; fel biochem, Harvard Med Sch, 64-66; Nat Inst Arthritis & Metab Dis res grant, 66-; consult, Pan Am Health Orgn, World Health Orgn, UNICEF & USAID; secy, Int Coun Control Iodine Deficiency Dis, 85-; ed, Iodine Deficiency Dis Newsletter, 85. *Honors & Awards:* Van Meter Prize, Am Thyroid Asn, 68. *Mem:* Am Soc Biol Chemists; Am Thyroid Asn; Endocrine Soc; Am Fedn Clin Res. *Res:* Thyroglobulin structure; iodine deficiency disorders; thyroid disease; thyroid physiology. *Mailing Add:* Univ Va Sch Med Box 511 Charlottesville VA 22908

DUNN, JON D, b La Junta, Colo, June 12, 37; m 63; c 3. NEUROENDOCRINOLOGY. *Educ:* Col of Idaho, BS, 62; Univ Kans, PhD(anat), 67. *Prof Exp:* From instr to assoc prof neuroanat, La State Univ Med Ctr, 68-74, res assoc, 74-75; assoc prof neuroanat, Stritch Sch Med, Loyola Univ Chicago, 75-78; PROF NEUROANAT, SCH MED, ORAL ROBERTS UNIV, 78- *Honors & Awards:* W B Peck Sci Res Award, Interstate Postgrad Med Asn NAm, 71. *Mem:* Soc Neurosci; Am Asn Anatomists; Am Physiol Soc; Endocrine Soc; Int Soc Neuroendocrinol; Sigma Xi. *Res:* Influence of various environmental factors on pituitary-adrenal and pituitary-gonadal function and the neural substrate involved in producing changes in these systems. *Mailing Add:* 2937 Dimrill Stair Manhattan KS 66502

DUNN, JONATHAN C, b Paterson, NJ, Jan 14, 57; m 87; c 1. INTERNAL MEDICINE, ENDOCRINOLOGY. *Educ:* Lehigh Univ, BS, 79; Vanderbilt Univ, MD, 83; Am Bd Internal Med, dipl, 86. *Prof Exp:* Researcher biochem, Mellon Found Teaching Grant, 82; med resident, Univ Pittsburgh Health Ctr, 83-86; FEL ENDOCRINOL & METAB, UNIV CONN HEALTH CTR, 86-; AT ENDOCRINOL, MEND, PROF ASSOCS. *Mem:* Am Diabetes Asn; Am Col Physicians. *Res:* Controlling the long term complications of diabetes mellitus. *Mailing Add:* 1804 Oak Tree Rd Edison NJ 08820

DUNN, JOSEPH CHARLES, b New York, NY; m 60; c 2. OPTIMIZATION, PATTERN CLASSIFICATION. *Educ:* Polytech Inst Brooklyn, BAeroE, 59, MS, 63; Adelphi Univ, PhD(math), 67. *Prof Exp:* Dynamics engr, Grumman Aerospace Corp, 59-61, res engr, 61-67, staff scientist, 67-69; asst prof theoret & appl mech, Cornell Univ, 69-76; assoc prof, 76-79, PROF MATH, NC STATE UNIV, 79- *Concurrent Pos:* Adj asst prof, Adelphi Univ, 68-69; NSF res initiation grant, 70, res grants, 78-80, 80-83, 85-86, 87-89, 90-92. *Mem:* Soc Indust & Appl Math. *Res:* Nonlinear functional analysis in optimization and control; singular optimal control problems; computational methods for optimization, control, and fixed point problems; group averaging and fuzzy partitioning techniques in pattern classification. *Mailing Add:* Dept Math NC State Univ Box 8205 Raleigh NC 27695-8205

DUNN, KENNETH ARTHUR, b St Catharines, Ont, Dec 13, 45; m 68; c 2. GENERAL RELATIVITY, COSMOLOGY. *Educ:* Univ Waterloo, BSc, 67; Univ Toronto, MSc, 68, PhD(math), 71. *Prof Exp:* Fel math, Birkbeck Col, Univ London, 71-72; asst prof, 72-78, chmn dept, 83-87, ASSOC PROF MATH, DEPT MATH, STATIST & COMPUT SCI, DALHOUSIE UNIV, 78- *Concurrent Pos:* Vis prof, Univ Waterloo, 79-80; vir prof, Univ BC, 87-88. *Mem:* Can Math Soc; Can Astron Soc. *Res:* Homogeneous and inhomogeneous cosmological models based on general relativity; group theoretical techniques for solving partial differential equations. *Mailing Add:* Dept Math Statist & Comput Sci Dalhousie Univ Halifax NS B3H 4H8 Can

DUNN, MARVIN I, b Topeka, Kans, Dec 21, 27; m 56; c 2. INTERNAL MEDICINE, CARDIOLOGY. *Educ:* Univ Kans, BA, 50, MD, 54; Am Bd Internal Med, dipl, 63, Am Bd Cardiovasc Dis, dipl, 65. *Prof Exp:* From instr to assoc prof, 58-71, PROF MED, UNIV KANS, 71-, DIR, CARDIOVASC LAB, 63-, FRANKLIN E MURPHY DISTINGUISHED PROFESSORSHIP CARDIOL, 78-, DEAN, SCH MED, 80- *Concurrent Pos:* Fel cardiovasc dis, Univ Kans, 58-60; consult, Vet Admin Hosp, Kansas City, Mo, 60-, Menorah Hosp, 61-, Bethany & Providence Hosps, Kansas City, Kans, 65- & US Air Force; fel coun clin cardiol, Am Heart Asn, 66- *Mem:* AMA; fel Am Col Cardiol; fel Am Col Physicians; Asn Univ Cardiologists; NY Acad Sci; Sigma Xi. *Res:* Clinical cardiovascular problems and hemodynamics. *Mailing Add:* Univ Kans Sch Med 39th & Rainbow Blvd Kansas City KS 66103

DUNN, MARY CATHERINE, zoology; deceased, see previous edition for last biography

DUNN, MICHAEL F, b Greeley, Colo, July 11, 39; m 58; c 4. BIOCHEMISTRY, ENZYMOLOGY. *Educ:* Colo Sch Mines, PRE, 61; Ga Inst Technol, MS, 63, PhD(phys org chem), 66. *Prof Exp:* Res assoc enzym, Inst Molecular Biol, Univ Ore, 66-69; res assoc physics & chem inst, Tech Univ Denmark, 69-70; asst prof biochem, 70-76, PROF BIOCHEM, UNIV CALIF, RIVERSIDE, 76- *Concurrent Pos:* NIH fel, 66-67; USPHS trainee, 67-69; NATO fel, 69-70; Am Cancer Soc grant, 71-72; NSF res grant, 75-78; Am Cancer Soc res grant, 73-75. *Mem:* Am Chem Soc; Am Soc Biol Chemists. *Res:* Enzyme structure, function and catalytic mechanism via rapid kinetic techniques; dehydrogenases; aldolases; growth factor proteins. *Mailing Add:* Dept Biochem Univ Calif Riverside CA 92521

DUNN, PETER EDWARD, b St Petersburg, Fla, May 26, 46; m 81; c 1. INSECT PATHOLOGY, DEVELOPMENTAL BIOLOGY. *Educ:* Fordham Univ, BS, 68; Purdue Univ, PhD(biochem), 73. *Prof Exp:* Res asst biochem, Dept Biochem, Purdue Univ, 68-73; res assoc, Dept Biochem, Univ Chicago, 73-77; from asst prof to assoc prof, 77-88, PROF ENTOM, DEPT ENTOM, PURDUE UNIV, 88-, DIR AGR BIOTECHNOL, 91- *Concurrent Pos:* Mem ed bd, J Invert Path, 87- *Mem:* AAAS; Entom Soc Am; Soc Invert Path. *Res:* Insect biochemistry; immune responses of insects; hormonal control of insect development. *Mailing Add:* Dept Entom Purdue Univ 1158 Entomol Hall W Lafayette IN 47907-1158

DUNN, RICHARD B, b Baltimore, Md, Dec 14, 27; m 51. ASTRONOMY, MECHANICAL ENGINEERING. *Educ:* Univ Minn, BME, 49, MS, 50; Harvard Univ, PhD, 61. *Prof Exp:* PHYSICIST NAT SOLAR OBSERV, 53- *Mem:* Am Astron Soc. *Res:* Solar astronomy; instrumentation. *Mailing Add:* Nat Solar Observ Sunspot NM 88349

DUNN, RICHARD LEE, b Pitt County, NC, Nov 28, 40; m 59; c 2. BIOMATERIALS, CONTROLLED RELEASE. *Educ:* Univ NC, Chapel Hill, BS, 63; Univ Fla, PhD(org chem), 67. *Prof Exp:* Res chemist, Beaunit Corp, El Paso Prod, 67-72, develop supt, 72-74, dir fiber develop, 74-77, dir viscose develop, 77-78, dir viscose tech serv, 78-79; sr chemist, Southern Res Inst, 79-80, head biomat sect, 80-82, head polymer div, 82-87; VPRES RES & DEVELOP, ATRIX LAB CORP, 87- *Mem:* Am Chem Soc; Controlled Release Soc. *Res:* Synthesis and characterization monomers and polymers for maxillofacial prostheses, biodegradable implants, and controlled-release drug delivery systems; fabrication of biomaterials by molding, casting, and melt-extrusion processes; preparation of fibers and films from synthetic polymers and ceramic precursors. *Mailing Add:* 5021 Kitchell Dr Ft Collins CO 80524

DUNN, ROBERT BRUCE, b Vancouver, BC, Can, Nov 25, 43; m 78; c 2. CORONARY ARTERY DISEASE, LEFT VENTRICULAR HYPERTROPHY. *Educ:* Univ BS, BSc, 65, MSc, 70; Univ Western Ont, PhD(physiol), 73. *Prof Exp:* Asst physiol, Univ Mo, 73-77, instr, 77-79; asst prof, 79-85, ASSOC PROF PHYSIOL, CHICAGO MED SCH, 85-, ASSOC DIR MED PHYSICS, 83- *Mem:* Am Physiol Soc; Am Heart Asn; Sigma Xi. *Res:* Regulation of coronary blood flow and heart metabolism during ischemia. *Mailing Add:* Dept Physiol Chicago Med Sch 3333 Green Bay Rd North Chicago IL 60064

DUNN, ROBERT GARVIN, b Lake Village, Ark, July 30, 17; m 50; c 3. ALTERNATE ENERGY SOURCES, ALTERNATE FUELS. *Educ:* La State Univ, BS, 42; Ohio State Univ, MS, 49, PhD(chem eng), 64. *Prof Exp:* Chem analyst, Esso Labs, Standard Oil Co La, 40-43; proj engr, Power Plant Lab, US Air Force, Ohio, 44-49, unit chief, 49-51, res engr, Propulsion Res Br, Aeronaut Res Lab, 51-54, res group chief, Fluid Dynamics Facilities Res Lab, Aerospace Res Labs, 54-61, proj scientist, Wright-Patterson AFB, 61-65, br chief, 65-67, dep lab dir, 67-74, lab dir, 74-75, exp eng br chief, Air Force Flight Dynamics Lab, 75-79; RETIRED. *Concurrent Pos:* Chmn combustion group, Coord Comt Sci, Dept Defense, 59-60; consult alternate energy sources res, 79-81. *Mem:* Combustion Inst; Am Inst Aeronaut & Astronaut; Am Chem Soc. *Res:* Fluid dynamics; propulsion; supersonic combustion; aeromechanics simulation techniques; gaseous detonation; aircraft engine testing at extreme temperatures; aerospace wind tunnel and arc tunnel testing. *Mailing Add:* 121 Redder Ave Dayton OH 45405

DUNN, ROBERT JAMES, NEUROBIOLOGY. *Educ:* Univ BC, PhD(biochem), 78. *Prof Exp:* ASST PROF MED SCI, UNIV TORONTO, 82- *Mailing Add:* Dept Neurol Montreal Gen Hosp 1650 Cedar Ave Montreal PQ H3G 1A4 Can

DUNN, SAMUEL L, b Tipton, Ind, Apr 17, 40; m 63; c 2. MATHEMATICS. *Educ:* Olivet Nazarene Col, BA, 61, BS, 62; Univ Wis-Milwaukee, MS, 64, PhD(math), 69; Univ Puget Sound, MBA, 85. *Prof Exp:* Teacher high sch, Ill, 61-62; from asst prof to assoc prof, 68-77, dir, Sch Natural & Math Sci, 78-80, PROF MATH, SEATTLE PAC UNIV, 77-, DEAN GRAD STUDIES, 80- *Mem:* World Future Soc; Math Asn Am; Am Asn Higher Educ Training & Develop. *Res:* Ring theory, particularly quasi-Frobenius quotient rings; quotient rings and their topologies; mathematical modeling, mathematics-economics; futurism. *Mailing Add:* Dept Grad Studies Seattle Pac Univ Seattle WA 98119

DUNN, STANLEY AUSTIN, b Long Beach, Calif, Nov 13, 21; m 44, 57, 75; c 4. HIGHWAY DEICING CHEMICAL SYNTHESIS, INVISCID MELT SPINNING & DRAWING. *Educ:* Calif Inst Technol, BS, 43; Johns Hopkins Univ, MA, 48, PhD(chem), 51. *Prof Exp:* Chemist, Jackson Lab, E I du Pont de Nemours & Co, 50-54; from res assoc to head anal dept, Rhodia, Inc, 54-59; from proj leader to assoc dir, Bjorksten Res Labs, Inc, 59- 73, dir, Inorg & High Temperature Div, 73-76, pres, tech dir & mem bd dirs, 76-84; VPRES & OWNER, SEDUN, INC, 84- *Concurrent Pos:* Adj prof, fac assoc & co-prin investr indust consortium develop, Univ Wis-Madison, 85- *Res:* Materials science; reaction kinetics; equilibria; thermodynamics; rheology; surface chemistry; statistical design; temperature and shear effects on alkaline fusion of lime and cellulose to form alkaline earth acetates; metastable crystalline forms in redrawn inviscid melt spun fibers. *Mailing Add:* 137 Larkin St Madison WI 53705

DUNN, THOMAS GUY, b Livingston, Mont, Jan 31, 35; m 60; c 2. REPRODUCTIVE PHYSIOLOGY, ENDOCRINOLOGY. *Educ:* Mont State Univ, BS, 62; Univ Nebr, Lincoln, MS, 65; Colo State Univ, PhD(physiol), 69. *Prof Exp:* Asst prof animal sci, Purdue Univ, 68-70; from asst prof to assoc prof, 70-78, PROF ANIMAL PHYSIOL, UNIV WYO, 78- *Concurrent Pos:* Actg dean, Col Agr, 83-84, dean, grad sch. *Mem:* AAAS; Am Soc Animal Sci; Soc Study Reproduction; Am Fertil Soc; Equine Nutrit & Physiol Soc. *Res:* Influence of nutrition on reproductive performance and endocrinology of beef cattle; influence of maternal malnutrition on growth, development and endocrinology of the ovine and bovine fetus; control of ovine fetal pituitary secretions. *Mailing Add:* Univ Sta Box 4268 Laramie WY 82071

DUNN, THOMAS M, b Sydney, Australia, Apr 25, 29; m 53; c 3. PHYSICAL CHEMISTRY. *Educ:* Univ Sydney, BSc, 49, MSc, 51; Univ London, PhD(phys chem), 57. *Prof Exp:* Teaching fel chem, Univ Sydney, 50-52; from asst lectr to lectr phys chem, Univ Col, Univ London, 54-63; PROF PHYS

CHEM, UNIV MICH, ANN ARBOR, 63-, HEAD DEPT, 74- *Concurrent Pos:* Plenary lectr, Int Conf Co-ord Chem, Stockholm, Sweden, 62. *Mem:* Am Chem Soc; fel The Chem Soc. *Res:* High resolution vapour phase spectra of organic and inorganic molecules in the visible and ultraviolet regions; electronic spectra at 4 degrees Kelvin of both organic and inorganic crystals. *Mailing Add:* Dept Chem Univ Mich Main Campus 1543 Chem Bldg Ann Arbor MI 48109-1055

DUNN, WILLIAM ARTHUR, JR, b Pittsburgh, Pa, Jan 2, 53; m 79; c 1. PROTEIN TURNOVER, RECEPTOR-MEDIATED ENDOCYTOSIS. *Educ:* Thiel Col, Greenville, Pa, BA, 74; Pa State Univ, University Park, PhD(biochem), 79. *Prof Exp:* Fel, Albert Einstein Col Med, 79-81; fel, 81-83, RES ASSOC, SCH MED, JOHNS HOPKINS UNIV, 84- *Mem:* AAAS; Am Soc Cell Biol; NY Acad Sci. *Res:* Protein degradation and targeting of exogenous and endogenous proteins to lysomes using various morphological and biochemical techniques to examine a) ligand and receptor dynamics during epidermal growth factor endocytosis and b) cellular degradation of organelle-specific membrane proteins. *Mailing Add:* Dept Cell Biol & Anat Univ Fla Sch Med Box J-235 JHMHC Gainesville FL 32610-0235

DUNN, WILLIAM HOWARD, Americus, Ga, July 16, 42; m 65; c 2. ANALYTICAL ABSORPTION SPECTROSCOPY. *Educ:* John Jay Col, City Univ NY, BS, 74; Long Island Univ, MS, 78; Polytech Inst NY, PhD(phys chem), 81. *Prof Exp:* Detective chemist, Forensic Sect, New York City Police Dept, 63-73; instr chem, Mount Moriah High Sch, New York, 74-78; SR SCIENTIST, BEECHAM PRODUCTS, 78- *Concurrent Pos:* Special investr, Bur Narcotics & Dangerous Drugs, US Dept Justice, 70-74; lectr chem, Brooklyn Col, 81- *Mem:* Am Chem Soc; Am Phys Soc; AAAS; fel Am Inst Chemists; NY Acad Sci. *Res:* Application of reaction kinetics and spectroscopy to the elucidation of atomic structure and chemical bonding; determination of the reaction mechanisms responsible for drug stability-instability; determination of molecular structures and conformations; characterization of reaction intermediates. *Mailing Add:* c/o Beecham Prod 1500 Littleton Rd Parsippany NJ 07054-3804

DUNN, WILLIAM JOSEPH, b Shreveport, La, Feb 17, 41; m 65; c 1. MEDICINAL CHEMISTRY. *Educ:* ETex Baptist Col, BS, 63; Okla State Univ, MS, 65, PhD(chem), 70. *Prof Exp:* NIH res asst, Pomona Col, 70-71; assoc prof, 71-80, PROF MED CHEM, COL PHARM, MED CTR, UNIV, 80- *Concurrent Pos:* Lectr, Am Found Pharmaceut Educ, 72. *Mem:* Am Chem Soc. *Res:* Quantitative structure-activity relationships; drug design. *Mailing Add:* Dept Med Chem Univ Ill Med Ctr Col Pharm Chicago IL 60680

DUNN, WILLIAM LAWRIE, b London, Ont, Oct 23, 27; m 56; c 3. PATHOLOGY. *Educ:* Univ Western Ont, BSc, 50, MD, 54; Univ London, PhD(exp path), 63; Royal Col Physicians & Surgeons Can, cert path, 64, FRCP(C), 72. *Prof Exp:* From asst prof to assoc prof path, Univ BC, 63-69, head dept, 69-76; dir labs, Vancouver Gen Hosp, 70-76, mem staff anat path, 76-78; prof, 69- EMER PROF PATH, UNIV BC. *Concurrent Pos:* Teaching fel path, Univ Western Ont, 57-59; teaching fel, Harvard Univ, 59-60; examr gen surg, Royal Col Physicians & Surgeons Can, 70-73 & mem credentials & anat path comts; consult, Can Tumor Reference Ctr, 71-74 & Liver Path Reference Ctr, 75-81; mem med adv bd, Can Hepatic Found, 71-76, BC Cancer Res Found; dir labs, Shaughnessy Hosp, Vancouver, 78- *Mem:* NY Acad Sci; Int Acad Path; Can Asn Pathologists. *Res:* Liver disease; tumor biology; diagnostic pathology. *Mailing Add:* Stanley Point Dr RR No 1 Tender Island BC V0N 2M0 Can

DUNNAM, FRANCIS EUGENE, JR, b Alexandria, La, Jan 29, 31; m 65; c 2. NUCLEAR PHYSICS. *Educ:* La State Univ, BS, 52, MS, 54, PhD(physics), 58. *Prof Exp:* Asst physics, La State Univ, 54-56, instr, 56-57; from asst prof to assoc prof, Univ Fla, 58-74. dep chmn dept physics, 72-74. actg chmn dept, 74-75, chmn dept 75-79, assoc dean, Col Liberal Arts & Sci, 79-89, PROF PHYSICS & ASTRON, UNIV FLA, 75- *Concurrent Pos:* Consult, Oak Ridge Nat Lab, 61-63. *Mem:* AAAS; Am Phys Soc; Am Asn Physics Teachers. *Res:* Experimental nuclear physics; nuclear astrophysics; musical acoustics; pedagogy. *Mailing Add:* Dept Physics Univ Fla Gainesville FL 32611

DUNNE, THOMAS, b Prestbury, Eng, Apr 21, 43, US & UK citizen; m 89; c 1. GEOMORPHOLOGY, HYDROLOGY. *Educ:* Cambridge Univ, BA, 64; Johns Hopkins Univ, PhD(geog), 69. *Prof Exp:* Res assoc, US Dept Agr, Danville, VT, 66-68; res hydrologist, US Geol Surv, Washington DC, 68-73; asst prof geog, McGill Univ, 69-73; from asst prof to assoc prof, 73-79, PROF GEOL SCI, UNIV WASH, 79-, CHMN GEOL SCI, 84- *Concurrent Pos:* Vis Rockefeller prof, Univ Nairobi, 69-71, consult, 72-; prin investr grants, Nat Res Coun Can, Nat Sci Found, NASA, US Dept Interior, US Forest Serv, 71-80. *Honors & Awards:* Robert E Horton Award, Am Geophys Union, 87. *Mem:* Am Geophys Union; Japanese Geomorphol Union; British Geomorphol Res Group; AAAS; Sigma Xi; Geol Soc Am; Nat Acad Sci. *Res:* Study of hillslope processes in geomorphology and hydrology with particular reference to tropical regions; erosion and sedimentation in forested regions; hydrology and sedimentation along the Amazon valley. *Mailing Add:* Dept Geol Sci Univ Wash Seattle WA 98195

DUNNE, THOMAS GREGORY, b Los Angeles, Calif, Oct 10, 30. PHYSICAL INORGANIC CHEMISTRY. *Educ:* Univ Calif, Los Angeles, BS, 52; Univ Wash, PhD(chem), 57. *Prof Exp:* Assoc chemist, Int Bus Mach Corp, 57-61; res assoc, Mass Inst Technol, 61-63; from asst prof to assoc prof, 63-81, PROF CHEM, REED COL, 81- *Mem:* Am Chem Soc; AAAS. *Res:* Coordination complex studies; oxidation-reduction mechanisms. *Mailing Add:* Dept Chem Reed Col Portland OR 97202

DUNNEBACKE-DIXON, THELMA HUDSON, b Nashville, Tenn, Dec 23, 25; m 54; c 3. PROTEIN PURIFICATION & IDENTIFICATION. *Educ:* Wash Univ, AB, 47, MA, 49, PhD(exper embryol), 54. *Prof Exp:* Instr zool, Smith Col, Northampton, Mass, 49-50; res fel, 54-56, res assoc virol, Virus Lab, Univ Calif, Berkeley, 56-76, instr electron microscope, 74-76; RES SCIENTIST, VIROL-PROTOZOA, STATE CALIF DEPT HEALTH SERV, 76- *Concurrent Pos:* Res scientist VI, State Calif Dept Health Serv, 76- *Mem:* Am Soc Cell Biol; Tissue Cult Asn; Electron Microscope Soc Am; Soc Protozoologists; Am Inst Biol Sci; Sigma Xi. *Res:* Isolation and identification of a biologically active protein from free-living ameba. *Mailing Add:* 2326 Russell St Berkeley CA 94705

DUNNER, DAVID LOUIS, b Brooklyn, NY, May 27, 40; m 64; c 2. PSYCHIATRY. *Educ:* George Wash Univ, AA, 60; Wash Univ, MD, 65. *Prof Exp:* Res psychiatrist, NY State Psychiat Inst, 71-79; from asst prof to assoc prof psychiat, Columbia Col Physicians & Surgeons, 72-79; chief psychiat, Harborview Med Ctr, 79-89; PROF PSYCHIAT, UNIV WASH, 79-, VCHMN, 89-, DIR OUTPATIENT PSYCHIAT, 89- *Concurrent Pos:* Consult, Found Depression & Manic Depression, 79-; vis scholar, Hunan Med Col, Changsha, Peoples Repub China, 82. *Honors & Awards:* Samuel Hamilton Award, Am Psychopath Asn, 87. *Mem:* Am Col Neuropsychopharm; Psychiat Res Soc (pres, 83-84); Soc Biol Psychiat (secy/treas, 87-90); Am Psychopath Asn (pres, 86); Am Psychiat Asn. *Res:* Phenomenology of mood and anxiety disorders; genetic studies of mood and anxiety disorders; treatment studies (psychopharmacology) of mood and anxiety disorders. *Mailing Add:* 4225 Roosevelt Way NE Suite 306 Seattle WA 98105

DUNNETT, CHARLES WILLIAM, b Windsor, Ont, Aug 24, 21; m 47; c 3. MATHEMATICAL STATISTICS. *Educ:* McMaster Univ, BA, 42; Univ Toronto, MA, 46; Aberdeen Univ, DSc, 60. *Prof Exp:* Instr math, Columbia Univ, 46-48 & Maritime Col, 48-49; biometrician, Can Dept Nat Health & Welfare, 49-52; res assoc statist, Cornell Univ, 52-53; statistician, Lederle Labs, Am Cyanamid Co, 53-74; prof clin epidemiol & biostatist, McMaster Univ, 74-87, prof appl math & chmn dept, 77-79, prof math & statist, 79-87, EMER PROF, MCMASTER UNIV, 87- *Honors & Awards:* Mem, Order Brit Empire. *Mem:* Fel Am Statist Asn; Biomet Soc; Royal Statist Soc; Int Statist Inst; Bernoulli Soc Math Statist & Probability. *Res:* Application of statistics to design and analysis of experiments in medical and biological sciences; statistical methodology; categorical data analysis in medical statistics. *Mailing Add:* Dept Math & Statist McMaster Univ 1280 Main St W Hamilton ON L8S 4K1 Can

DUNNICK, JUNE K, b New York, NY. TOXICOLOGY. *Educ:* Cornell Univ, BS, 65; Cornell Med Col, PhD(med sci), 69; Am Bd Toxicol, dipl. *Prof Exp:* Res assoc biochem, Univ Rochester, 69-71; hepatitis prog officer, NIH, 71-73; life sci res assoc, Med Sch, Stanford Univ, 73-76; antiviral prog officer, NIH, 76-80; CHEMIST & MGR, NAT TOXICOL PROG, NAT INST ENVIRON HEALTH SCI, 80- *Concurrent Pos:* Fel, NIH, 69-71. *Mem:* Sigma Xi; Soc Toxicol; AAAS; Am Asn Cancer Res. *Res:* Toxicology; carcinogenesis; reproductive toxicity; metal toxicity; antiviral drug development. *Mailing Add:* Nat Toxicol Prog PO Box 12233 Nat Inst Environ Health Sci Res Triangle Park NC 27709

DUNNIGAN, JACQUES, b St Jerome, Que, May 1, 35; m 63; c 4. PHYSIOLOGY. *Educ:* St Laurent Col, BA, 56; Univ Ottawa, BSc, 60, PhD, 63. *Prof Exp:* Res asst med, Laval Univ, 63-64; adj prof sci, Univ Sherbrooke, 64-68, assoc prof, 68-73, vdean res, 74-75, assoc vrector res, 75-79, dir prog res, Asbestox, 79-80, dir gen, Inst Res & Develop, Asbestos, 80-85, PROF SCI, UNIV SHERBROOKE, 73- *Res:* Toxicology of asbestos and other particles. *Mailing Add:* Univ Sherbrooke 2500 Boulevard Sherbrooke PQ J1L 1E6 Can

DUNNING, DOROTHY COVALT, b Washington, DC, Jan 1, 37; m 60. ANIMAL BEHAVIOR. *Educ:* Middlebury Col, BA, 58; Mt Holyoke Col, MA, 60; Tufts Univ, PhD(biol), 66. *Prof Exp:* Teaching fel biol, Tufts Univ, 66-67; scholar animal behav, Max Planck Inst Physiol of Behav, 67; instr zool, Duke Univ, 68-69; asst prof biol, 69-75, ASSOC PROF BIOL, WVA UNIV, 75- *Mem:* AAAS; Am Soc Zoologists; Am Soc Mammal; Sigma Xi; Animal Behav Soc. *Res:* Physiological mechanisms of animal behavior. *Mailing Add:* Dept Biol PO Box 6057 WVa Univ Morgantown WV 26506

DUNNING, ERNEST LEON, b Ky, Nov 13, 20; c 2. MECHANICAL ENGINEERING. *Educ:* Univ Rochester, BSME, 46; Univ Ky, MSME, 50; Univ Houston, PhD, 67. *Prof Exp:* Instr physics & math, Pikeville Col, 47-49; assoc prof mech eng, Evansville Col, 50-54 & La Polytech Inst, 55-57; assoc prof, Southern Ill Univ, Carbondale, 57-77, prof elec sci & systs eng, 77-; DEPT ENG TECH, UNIV TENN, MARTIN. *Concurrent Pos:* Consult, Schnacks Refrig Co, 53-54 & Christopher Unitemp Heating Co, 46-57; res engr, Sch Indust Prog, Hughes Aircraft Co, 59. *Mem:* Nat Soc Prof Engrs; Am Soc Mech Engrs; Am Soc Eng Educ. *Res:* Thermodynamics; heat transfer; refrigeration and air conditioning. *Mailing Add:* 1814 Wolff Dr Marion IL 62959

DUNNING, FRANK BARRYMORE, b Tadcaster, Eng, Apr 10, 45; m 68; c 2. ATOMIC PHYSICS, SURFACE PHYSICS. *Educ:* Univ Col, Univ London, BSc, 66, PhD(atomic physics), 69. *Prof Exp:* Imp Chem Industs fel exp atomic physics, Univ Col, Univ London, 69-71; res assoc exp atomic & laser physics, 71-74, from asst prof to assoc prof, 74-82, PROF SPACE PHYSICS & ASTRON, RICE UNIV, 82- *Concurrent Pos:* Alfred P Sloan Found Fel, 76. *Mem:* Am Phys Soc; Optical Soc Am. *Res:* Atomic collision studies; development of tunable lasers and their application to studies of photon interaction processes important in aeronomy and astrophysics; application of electron spin polarisation measurements; study of surface geometric and electronic structure. *Mailing Add:* Rice Univ PO Box 1892 Houston TX 77251

DUNNING, HERBERT NEAL, b Hazard, Nebr, June 2, 23; m 46, 74; c 2. GENERAL MEDICAL SCIENCES, CHEMISTRY. *Educ:* Nebr State Col, BS, 44; Univ Nebr, MS, 48, PhD(phys chem), 50. *Prof Exp:* Chemist, Standard Oil Co, Ind, 50-51; head, surface chem lab, US Dept Interior, 51-60; dir chem res dept, Gen Mills, Inc, 60-64, res dir, 64-70, tech dir qual control, 70-71; res dir, Delmark Foods & Doyle Pharmaceut, 71-72; dir, div consumer studies, US Food & Drug Admin, 72-74; dir, div oil & gas, US Dept Interior, 74-75; dir, div oil, gas & shale, US Energy Res & Develop Admin, 75-76; sr tech adv to asst secy, Dept Energy, 76-77, dir, field coord & univ affairs, 77-79; dir, Off Small Mfg Assistance, US Food & Drug Admin, 79-90; PRES, NEAL DUNNING ASSOC INC, 91- *Concurrent Pos:* Du Pont fel, Univ Nebr, 49-50 & Standard Oil fel, 48-49. *Honors & Awards:* Donald E Fox Lectr, Nebr State Col, Kearney, 83-84; Meritorious Serv Award, Dept of Interior, 59; Commendable Serv Award, FDA, 82; Award of Merit, USFDA, 87. *Mem:* Am Chem Soc; Am Inst Chem Engrs; fel AAAS; Explorers Club; Inst Food Technologists; Inst Chemists; NY Acad Sci; Am Soc Qual Control. *Res:* Origin of petroleum, surface chemistry of oil and gas production and porphyrins in petroleum; exploratory food processing; protein texturization; chocolate conching; cereal conversion; dietary supplements; medical devices. *Mailing Add:* 8309 Bryant Dr Bethesda MD 20817

DUNNING, JAMES MORSE, dentistry, public health; deceased, see previous edition for last biography

DUNNING, JEREMY DAVID, b Washington, DC, Feb 15, 51; m 72. ROCK MECHANICS. *Educ:* Colgate Univ, BA, 73; Rutgers Univ, MS, 75; Univ NC, PhD(geol), 78. *Prof Exp:* Res asst geol, Univ NC, 75-78; asst prof geol, Ore State Univ, 78-79; asst prof, 79-84, ASSOC PROF GEOL, IND UNIV, 84-, ASSOC DEAN RES, 85- *Concurrent Pos:* Grants, Small Bus Innovation Res Prog. *Honors & Awards:* Hearst Distinguished Lectr, Univ Calif, Berkeley; Hoosier Sherpa Award. *Mem:* Am Geophys Union; Geol Soc Am. *Res:* Low temperature rock mechanics; microseismic monitoring of crack propagation and brittle failure; microseismic modeling of fault behavior; technology transfer; SBIR grants. *Mailing Add:* Dept Geol Ind Univ Bloomington IN 47405

DUNNING, JOHN RAY, JR, b New York, NY, Nov 26, 37; m 86. PHYSICS, APPLIED NUCLEAR PHYSICS. *Educ:* Yale Univ, BS, 60, MS, 61; Harvard Univ, PhD(physics), 65. *Prof Exp:* Lectr physics, Harvard, 65-66, instr, 66-68, res fel & lectr, 68-69; from asst prof to assoc prof, 69-75, PROF PHYSICS, CALIF STATE COL, SONOMA, 75- *Mem:* AAAS; Am Phys Soc; Health Physics Soc. *Res:* Neutron activation analysis; x-ray analysis and uses of synchrotron radiation. *Mailing Add:* Dept Physics Sonoma Univ Rohnert Park CA 94928

DUNNING, JOHN WALCOTT, b Ottumwa, Iowa, Dec 20, 12; m 39; c 3. ORGANIC CHEMISTRY. *Educ:* Iowa State Col, BS, 35, PhD(biophys chem), 38. *Prof Exp:* Lab & pilot plant dir, Anderson Clayton & Co, 37-42; chemist, Northern Regional Res Lab, Bur Agr & Indust Chem, USDA, 42-44, in chg synthetic liquid fuels proj, 44-48; dir res, Anderson IBEC, 48-53, vpres, 53-64, pres, 64-75, chmn, 75-77; RES CONSULT, 78- *Mem:* Am Chem Soc; Am Oil Chem Soc. *Res:* Saccharification; industrial fermentation; pyrolysis; esterification; pulping; pilot plant developments; bacterial oxydations; vitamin C synthesis; vegetable oils; vegetable seed processing; finish drying of synthetic rubber. *Mailing Add:* 15305 Edgewater Dr Lakewood OH 44107

DUNNING, KENNETH LAVERNE, b Yale, Iowa, Sept 24, 14; m 41; c 4. NUCLEAR PHYSICS. *Educ:* Univ Minn, BEE, 38, Univ Md, MS, 50; Cath Univ Am, PhD, 68. *Prof Exp:* Commun engr, Western Union Tel Co, 38-41; electron scientist, Naval Res Lab, 45-50, physicist, 50-51, head Van de Graaff Br, Nuclear Sci Div, 51-74, consult radiation technol div, 75-80; CONSULT, 80- *Mem:* Am Phys Soc; Inst Elec & Electronics Engrs; Sigma Xi. *Res:* Centimeter and millimeter wave guide components; Cockroft-Walton accelerators; nuclear weapons tests; Van de Graaff accelerators; nuclear interactions; ion-induced x-rays; charged particle energy loss; surface analysis; computer programming; nuclear instrumentation. *Mailing Add:* Ten Foster Lane Port Ludlow WA 98365

DUNNING, RANALD G(ARDNER), b Lyndon, Ohio, Oct 6, 02; m 37; c 2. CHEMICAL ENGINEERING. *Educ:* Princeton Univ, BS, 24, AM, 25; Mass Inst Technol, SM, 27. *Prof Exp:* Chem engr, Roessler & Hasslacher Chem Co, NJ, 26-31; develop engr, Barber Asphalt Co, 31-38; chem engr, Merck & Co, Inc, 38-40, process develop head, 40-45, res proj analyst, 45-49, tech asst, off of sci dir, 49-53, tech asst, sci admin div, 53-57; mgr res & develop, Metalwash Mach Co, NJ, 57; chief engr, Chemirad Corp, 59-66; consult chem engr, 66-72; RETIRED. *Mem:* Am Chem Soc; Am Inst Chem Engrs; Chem Mkt Res Asn. *Res:* Equilibrium in the synthesis and decomposition of methanol; sulfuric acid recovery; synthesis of ethylene imine. *Mailing Add:* 227 Tuttle Pkwy Westfield NJ 07090

DUNNING, THOMAS HAROLD, JR, b Jeffersonville, Ind, Aug 3, 43; m 82; c 5. THEORETICAL CHEMISTRY. *Educ:* Univ Mo-Rolla, BS, 65; Calif Inst Technol, PhD(chem), 70. *Prof Exp:* Fel, Battelle Mem Inst, 70-71; res fel, A A Noyes Lab Chem Physics, Calif Inst Technol, 71-74; mem staff, Los Alamos Sci Lab, Univ Calif, 74-78; GROUP LEADER, THEORET CHEM GROUP, ARGONNE NAT LAB, 78-, HEAD CHEM DYNAMICS PROG, 86- *Mem:* Am Phys Soc; Am Chem Soc; Combustion Inst. *Res:* Electronic structure of atoms and molecules; theoretical chemical kinetics (molecular potential energy surfaces). *Mailing Add:* Molecular Sci Res Ctr Battelle Pac Northwest Lab Richland WA 99352

DUNNING, WILHELMINA FRANCES, b Topsham, Maine, Sept 12, 04. PATHOLOGY. *Educ:* Univ Maine, AB, 26; Columbia Univ, MA, 28, PhD(zool), 32. *Hon Degrees:* DSc, Univ Maine, 60. *Prof Exp:* Asst, Inst Cancer Res, Columbia Univ, 26-30, assoc, 30-41; instr path, Col Med, Wayne Univ, 41-48, asst prof oncol, 48-50, res assoc, 45-50; res assoc, Detroit Inst Cancer Res, 46-50; prof zool, Univ Miami, 50-52, res prof exp path, 52-65, res prof dept med, 65-75; res assoc, Papanicolaou Cancer Res Inst, 71-78, RETIRED. *Mem:* AAAS; Am Asn Cancer Res; Genetics Soc Am; Am Soc Zoologists; NY Acad Sci; Sigma Xi. *Res:* Mammalian genetics; experimental pathology; nutrition and cancer; endocrinology and genetics in experimental cancer. *Mailing Add:* 2850 Coconut Ave Miami FL 33133

DUNNY, STANLEY, b Northampton, Mass, Aug 2, 39. ORGANOMETALLIC CHEMISTRY. *Educ:* Univ Mass, BS, 61; Univ Wis, MS, 63; Purdue Univ, PhD(chem), 67. *Prof Exp:* Asst prof, 69-70, assoc prof, 70-75, PROF CHEM, HOLYOKE COMMUNITY COL, 75- *Mem:* Am Chem Soc; Sigma Xi. *Res:* Hydrosilylation of acetylenes via amine catalysis; organosilicon chemistry; synthesis of rings containing silicon atoms. *Mailing Add:* 795 A Westhampton Rd Northampton MA 01060

DUNPHY, DONAL, b Northampton, Mass, Feb 24, 17; m 44; c 3. PEDIATRICS. *Educ:* Col Holy Cross, BA, 39; Yale Univ, MD, 44. *Prof Exp:* Instr pediat, Sch Med, Yale Univ, 47-50; attend pediatrician, Bridgeport Gen Hosp, 50-53; assoc pediatrician, Sch Med, Univ Buffalo, 55-56, from asst prof to assoc prof pediat, Buffalo Children's Hosp, 56-61; prof & head dept, Col Med, Univ Iowa, 61-73; PROF PEDIAT, UNIV NC, CHAPEL HILL, 73- *Concurrent Pos:* Fel cardiol, Dept Pediat, Sch Med, Yale Univ, 50-52; dir pediat out-clin, Buffalo Children's Hosp, 55-61, dir child develop study, 55-59, dir, NIH Collab Proj, 58-61; pediat consult, Nat Insts Neurol Dis & Blindness, 60-61. *Mem:* AAAS; Am Pediat Soc; NY Acad Sci; Am Acad Pediat; Asn Am Med Cols. *Res:* Cord blood gas analysis in twins; factors affecting neurological status of children and methods for early recognition; plasmin in the therapy of hyaline membrane disease. *Mailing Add:* Dept Pediat Univ NC Sch Med Chapel Hill NC 27514

DUNPHY, JAMES FRANCIS, b Boston, Mass, May 16, 30; m 61; c 3. ENGINEERING POLYMERS. *Educ:* Boston Col, BS, 51; Univ Ill, PhD(org chem), 60. *Prof Exp:* Chemist, Nat Starch & Chem Corp, 52-57; res chemist, Film Dept, Yerkes Res & Develop Lab, 59-70, staff scientist, Tecumseh Film Plant, Topeka, Kans, 70-78, sr res chemist, Washington Lab, 78-84, RES ASSOC, WASHINGTON LAB, E I DU PONT DE NEMOURS & CO, INC, 84- *Mem:* Am Chem Soc. *Res:* Development of high performance polymer blends and grafts; organic chemistry of high polymers; preparation and characterization of addition and condensation polymers; chemistry of cellulose and the viscose process. *Mailing Add:* E I du Pont de Nemours & Co Inc Washington Lab PO Box 1217 Parkersburg WV 26102

DUNSHEE, BRYANT R, b Des Moines, Iowa, Mar 13, 21; m 49; c 3. BIOCHEMISTRY. *Educ:* Univ Mich, BS, 42; Univ Wis, PhD(biochem), 49. *Prof Exp:* Supvr prod, Hercules Powder Co, 42-45; instr physiol chem, Univ Minn, 49-52; sr biochemist food res, Cent Res Lab, Gen Mills Inc, 52-57, sect leader food develop dept, 57-62, res assoc food develop activity, James Ford Bell Res Ctr, 62-83; RETIRED. *Mem:* Am Chem Soc; Inst Food Technologists; Sigma Xi. *Res:* Use of dairy and vegetable proteins in food fortification. *Mailing Add:* 2250 N Victoria St Apt 210 St Paul MN 55113

DUNSING, MARILYN MAGDALENE, b Chicago, Ill, Feb 19, 26. HUMAN ECOLOGY. *Educ:* Univ Chicago, MBA, 48, PhD(family & consumption econ), 54. *Prof Exp:* Instr econ, Bowling Green State Univ, 48-49; asst, Univ Ill, Urbana, 49-50; instr econ & social sci, Wilson Jr Col, Chicago, 50-53; from instr to assoc prof family econ, Univ Calif, Davis, 54-62; assoc prof, 62-65, head dept, 78-79, PROF FAMILY & CONSUMPTIONS ECON, UNIV ILL, 66-, DIR, SCH HUMAN RESOURCES & FAMILY STUDIES, 79- *Mem:* Am Econ Asn; Am Home Econ Asn; Sigma Xi. *Res:* Environmental influences; quality of life; standards and levels of living; employment status of wives; volunteer work. *Mailing Add:* 2208 Stanley Rd Champaign IL 61821

DUNSON, WILLIAM ALBERT, b Cedartown, Ga, Dec 17, 41; m 63; c 3. TOXICOLOGY, MARINE SCIENCE. *Educ:* Yale Univ, BS, 62; Univ Mich, MS, 64, PhD(zool), 65. *Prof Exp:* PROF BIOL, PA STATE UNIV, UNIVERSITY PARK, 65- *Concurrent Pos:* Sr scientist, Stanford Univ, 68; NSF, US Dept Interior, Environ Protection Agency grants, 67-; chief scientist res vessel alpha helix, Scripps Inst Oceanog, 70, 72 & 75; sci collabr, Everglades Nat Park, 79-81; chief scientist, Great Barrier Reef Cruise, 81; vis res fel, Univ New England, Australia, 85; adj prof biol, Univ Miami, Coral Gables, Fla, Old Dominion Univ, Norfolk, Va & Fla State Univ, Tallahassee. *Mem:* AAAS; Am Physiol Soc; Am Soc Zool; Am Inst Biol Sci; Am Soc Ichthylogists & Herpetologists; Ecol Soc Am; Soc Study Amphibians & Reptiles; Herpetologist League; Am Soc Zoologists; Am Soc Limnol & Oceanog. *Res:* Physiological ecology and environmental physiology, ecotoxicology particularly in relation to ionic and osmotic regulation; effect of abiotic factors on communities; fish and amphibian tolerance to low pH and heavy metals; ecology of temporary pond wetlands; biology of estuarine fish and reptiles. *Mailing Add:* Dept Biol Pa State Univ 208 Mueller Bldg University Park PA 16802

DUNSTAN, WILLIAM MORGAN, b Greenville, NJ, Nov 4, 35; m 59; c 5. BIOLOGICAL OCEANOGRAPHY. *Educ:* Yale Univ, BS, 56; Fla State Univ, MS, 67, PhD(biol), 69. *Prof Exp:* Loan analyst, Int Div, Chase Manhattan Bank, 60-62; prod develop engr, Celanese Corp Am, 62-65; NSF fel, Fla State Univ, 65-69; asst scientist, Woods Hole Oceanog Inst, Mass, 69-72; assoc prof biol oceanog, Skidaway Inst Oceanog, 72-77; sci prog mgr, Interstate Electronics Corp, 77-80; CHMN & ASSOC DEAN, DEPT OCEANOG, OLD DOMINION UNIV, NORFOLK, VA, 80- *Concurrent Pos:* Adj assoc prof biol oceanog, Univ Ga, 73-77. *Mem:* Oceanog Soc; Phycol Soc Am; Am Soc Limnol & Oceanog; Estuarine Res Fedn. *Res:* Physiological ecology of marine plants involving the influence of pollutants on nearshore and continental shelf organisms. *Mailing Add:* Dept Oceanog Old Dominion Univ Hampton Blvd Norfolk VA 23529-0276

DUNWOODY, SHARON LEE, b Hamilton, Ohio, Jan 24, 47. MASS COMMUNICATION THEORY, SOCIOLOGY OF SCIENCE. *Educ:* Ind Univ, BA, 69; Temple Univ, MA, 75; Ind Univ, PhD(mass commun), 78. *Prof Exp:* From instr to asst prof jour, Ohio State Univ, 77-81; from asst prof to assoc prof, 81-89, EVJUE-BASCOM PROF JOUR, UNIV WIS-MADISON, 89-, PROF, INST ENVIRON STUDIES, 87-; HEAD, CTR ENVIRON COMMUN & EDUC STUDIES, 85- *Concurrent Pos:* Co-ed, Sciphers, 79-87; distinguished lectr, Fulbright Seminar, Brazil, 82; consult, AAAS, 80- *Mem:* AAAS; Am Asn Pub Opinion Res; Asn Educ Jour & Mass Commun; Nat Asn Sci Writers; Soc Social Studies Sci. *Res:* Aspects of the process by which scientific information finds its way via mass media into the public domain. *Mailing Add:* Sch Jour & Mass Commun Univ Wis 821 Univ Ave Madison WI 53706

DUPERON, DONALD FRANCIS, b Regina, Sask, Can, Dec 13, 37; m 60; c 2. PEDIATRIC DENTISTRY, COMPUTER PROGRAMMING. *Educ:* Univ Alta, DDS, 61; Univ Man, MSc, 70. *Hon Degrees:* MRCD, Royal Col Dentists, Can, 80. *Prof Exp:* Dentist, pvt pract, 61-67; from asst prof to assoc prof pedodontics, Univ Man, 68-74, head dept, 70-74; from asst prof to assoc prof, 74-77, PROF PEDODONTICS & DIR PEDODONTICS, AREA HEALTH EDUC CTR DOWNTOWN PEDIAT DENT CLIN, UNIV CALIF, LOS ANGELES, 83-, CHMN POSTDOCTORAL PEDODONTICS, 79-, DIR, AREA HEALTH EDUC CTR, 82- *Concurrent Pos:* Actg chief of staff pedodontics, Children's Hosp Winnipeg, 68-69, chief of staff, 70-74; attend staff hosp dent, Univ Calif, Los Angeles, 77- *Mem:* Royal Col Dentists Can; Can Dent Asn; Can Acad Pedodontics; Am Soc Clin Hypn; Am Asn Dent Schs; Am Acad Pedodontics; Am Dent Asn. *Res:* Study of dental materials science; sedation in children; computerized cephalometric analysis; ketamine sedation; treatment of immunocomprised children. *Mailing Add:* Ctr Health Sci Sch Dent Univ Calif Los Angeles CA 90024

DUPONT, ANDRE GUY, b Port-Alfred, Que. NEUROENDOCRINOLOGY. *Educ:* Laval Univ, MD, 65, PhD, 72. *Prof Exp:* Res assoc & fel, Tulane Univ, 73-75; resident surg, 66-69 & 72-73, asst prof endocrinol, Ctr Hosp & asst prof physiol, Laval Univ, 75-80, assoc prof molecular endocrinol, Laval Univ, 80- *Concurrent Pos:* Instr anat, Laval Univ, 67-68, fel physiol, 72-73; Med Res Coun Can fel, 73-75; Health Res Coun Que res scholar, 75-78. *Mem:* Can Soc Clin Invest; Int Soc Neuroendocrinol; Endocrine Soc; Am Physiol Soc. *Res:* Relationship of brain endorphins and neuroendocrine functions; isolation or characterization of prolactine inhibiting factor and prolactin releasing factor; inactivation of neuropeptides by rat plasma and tissues. *Mailing Add:* Lab Molecular Endocrinol CHUL 2705 Blvd Laurier Ste Foy PQ G1V 4G2 Can

DUPONT, BO, b Frederiksburg, Denmark, Jan 24, 41. IMMUNOGENETICS. *Educ:* Univ Aarhus, MD, 66; Univ Copenhagen, DSc, 78. *Prof Exp:* DIR & MEM, SLOAN-KETTERING INST CANCER RES, 73- *Mailing Add:* Tissue Typing Lab Sloan-Kettering Cancer Ctr 1275 York Ave New York NY 10021

DUPONT, CLAIRE HAMMEL, b Washington, DC, Apr 27, 33; m 66; c 2. BIOCHEMISTRY, PEDIATRICS. *Educ:* George Washington Univ, MD, 58; Univ Md, PhD(biochem), 64. *Prof Exp:* Intern, Philadelphia Gen Hosp, 58-59; resident pediat, Children's Hosp DC, 59-61; asst prof pediat res, Sch Med, Univ Md, 64-66; asst prof biochem, Fac Med, Univ Montreal, 66-71; asst dir biochem, Montreal Children's Hosp, 71-76; asst prof pediat, 72-78, ASSOC PROF, MCGILL UNIV, 78- *Concurrent Pos:* Nat Inst Neurol Dis & Blindness spec fel extra-mural prog, Sch Med, Univ Md, 61-64; asst, Sch Med, George Washington Univ, 60-61; dir biochem, Montreal Children's Hosp, 77- *Mem:* Can Soc Clin Investr; fel Am Acad Pediat; Can Biochem Soc; Can Soc Clin Chem (secy, 80-83); Am Asn Clin Chem. *Res:* Development of pediatric clinical chemistry methodology; medical decision-making and utilization of diagnostic tests. *Mailing Add:* Montreal Childrens Hosp 2300 Tupper St Rm C-451 Montreal PQ H3H 1P3 Can

DU PONT, FRANCES MARGUERITE, b Duluth, Minn, May 23, 44; m 69. ION TRANSPORT, ENVIRONMENTAL STRESS. *Educ:* Univ Calif, Berkeley, BA, 65, Los Angeles, MA, 70, Riverside, PhD(plant physiol), 79. *Prof Exp:* Vol biol, Peace Corps, Uganda, 65-68; teacher biol, Homer High Sch, Alaska, 71-72; lab technician, Med Genetics Dept, Univ Calif Los Angeles, Harbor Gen Hosp, 73-74; res asst, Univ Calif Riverside, 75-79; res assoc, Dept Plant Biol, Cornell Univ, 79-81 & Arco Plant Cell Res Inst, 81-83; PLANT PHYSIOLOGIST, WESTERN RES CTR, USDA, 83- *Concurrent Pos:* USDA competitive res grant, 85-88; panel mem, USDA Competitive Grants, 86 & NSF, 89-90; US-Israel Binat Agr Res & Develop Fund grant, 87-90. *Mem:* Am Soc Plant Physiologists; Biophys Soc; AAAS. *Res:* Role of ion transport in the response of plants to environmental stress; ion transport by plant ATPases; salt stress. *Mailing Add:* Western Regional Res Ctr USDA Agr Res Serv 800 Buchanan St Albany CA 94710

DUPONT, HERBERT LANCASHIRE, b Toledo, Ohio, Nov 12, 38; m 63; c 2. INTERNAL MEDICINE, INFECTIOUS DISEASES. *Educ:* Ohio Wesleyan Univ, BA, 61; Emory Univ, MD, 65. *Prof Exp:* From intern to resident internal med, Univ Minn, 65-67; from asst prof to assoc prof med, Div Infectious Dis, Univ Md, 70-73; prof med & dir Ctr Infectious Dis, 73-88, MARY W KELSEY PROF MED SCI, UNIV TEX HEALTH SCI CTR, HOUSTON, 88- *Concurrent Pos:* Fel infectious dis, Univ Md, 67-69;; assoc ed, Am J Epidemiol, 77-81 & J Infectious Dis, 83-88; counr, Infectious Dis Soc Am, 78-81, secy, 82-87; bd dir, Nat Found Infectious Dis, 81-, pres, 89-90; interim chmn, dept internal med, Univ Tex, 87-88. *Mem:* Fel Am Col Physicians; Am Fedn Clin Res; Am Soc Microbiol; Infectious Dis Soc Am; Am Soc Clin Invest; Asn Am Physicians. *Res:* Enteric diseases, particularly diarrheal disease, pathogenesis, diagnosis, treatment and vaccine development. *Mailing Add:* Ctr Infectious Dis Univ Tex Health Sci Ctr 6431 Fannin 1729 JFB Houston TX 77030

DUPONT, JACQUELINE (LOUISE), b Plant City, Fla, Mar 4, 34. NUTRITION. *Educ:* Fla State Univ, BS, 55, PhD(nutrit), 62; Iowa State Univ, MS, 59. *Prof Exp:* Home economist, Human Nutrit Res Div, USDA, 55-56, nutrit specialist, 56-62, res nutrit specialist, 62-64; asst prof biochem, Col Med, Howard Univ, 64-66; from asst prof to assoc prof food sci & nutrit, Colo State Univ, 66-73, prof, 73-78; prof & chmn dept, Iowa State Univ 78-89; consult, food & nutrit sci, 89-91; NAT PROG LEADER, HUMAN NUTRIT, AGR RES SERV, USDA, 91- *Concurrent Pos:* Consult nutrit study sect, USPHS, 72-76; res career develop award, NIH, 72-77; vis prof, Dept Pediat, Univ Colo Med Ctr, 77-78. mem coun arteriosclerosis, Am Heart Asn; vis scientist, INSERM, UI, Nutrit & Food, Hosp Bichat, Paris, France, 84; vis scientist, Inst Nutrit Cent Am & Panama, 73. *Honors & Awards:* Lotte Arnrich Lectr, Iowa State Univ, 87. *Mem:* Am Dietetic Asn; Am Oil Chem Soc; Am Inst Nutrit; Am Aging Asn; Inst Food Technol; Am Col Nutrit. *Res:* Effects of dietary fat upon metabolism of cholesterol and fatty acids, prostaglandins, atherosclerosis and aging. *Mailing Add:* Bldg 5 Rm 132 Beltsville MD 20705

DUPONT, PAUL EMILE, b Chicopee, Mass, Aug 21, 41; m 63; c 2. ORGANIC CHEMISTRY. *Educ:* Univ Mass, BS, 63; Rensselaer Polytech Inst, PhD(chem), 68. *Prof Exp:* Asst res chemist, 63-68, assoc res chemist, 68-74, PATENT AGENT, STERLING-WINTHROP RES INST, 74- *Mem:* Am Chem Soc. *Mailing Add:* Five Greenbrier Way East Greenbush NY 12061-2705

DUPONT, ROBERT L, JR, b Toledo, Ohio, Mar 25, 36; m 63; c 2. PSYCHIATRY. *Educ:* Emory Univ, BA, 58; Harvard Med Sch, MD, 63. *Prof Exp:* Res psychiatrist & actg assoc dir community serv, DC Dept Corrections, 68-70; adminr, Narcotic Treatment Admin, Dept Human Resources, DC, 70-73; dir, Spec Action Off Drug Abuse Prev, Exec Off of President, 73-75; dir, Nat Inst Drug Abuse, 73-78; PRES, INST BEHAV & HEALTH, 78-; CLIN PROF PSYCHIAT, GEORGETOWN MED SCH, 80- *Concurrent Pos:* Consult res & develop, DC Dept Corrections, 67-68, Child Res Br, NIH, 68-71 & Spec Comt Crime Prev & Control, Am Bar Asn, 71-72; mem Nat Adv Coun Drug Abuse Prev, 72-73; mem drug abuse task force, Nat Adv Comn Criminal Justice Standards & Goals, Dept Justice, 72-73, mem, Coord Coun Juv Justice & Delinq Prev, 74-78; actg adminr, Alcohol, Drug Abuse & Ment Health Admin, Dept Health, Educ & Welfare, 74. *Mem:* Fel Am Psychiat Asn; World Psychiat Asn; Pan Am Med Asn; Am Pub Health Asn; Anxiety Disorders Asn Am (pres, 80-); fel Acad Behav Sci. *Res:* Behavioral health and related policy. *Mailing Add:* 8708 Susanna Lane Chevy Chase MD 20815

DUPONT, TODD F, b Houston, Tex, Aug 29, 42; m 64; c 2. NUMERICAL ANALYSIS. *Educ:* Rice Univ, BA, 63, PhD(math), 68. *Prof Exp:* Res mathematician, Esso Prod Res Co, 67-68; from instr to assoc prof, 68-76, PROF MATH, UNIV CHICAGO, 76- *Concurrent Pos:* Co-founder & prin, Drem Inc. *Mem:* Am Math Soc; Soc Indust & Appl Math; AAAS; Asn Comput Mach. *Res:* Numerical solution of partial differential equations; nonlinear boundary-value problems; simulation of pipeline flow, leak defection. *Mailing Add:* Dept Comput Sci Univ Chicago Chicago IL 60637

DUPONT, WILLIAM DUDLEY, b Montreal, Que, Nov 6, 46; m 74; c 2. BIOSTATISTICS, EPIDEMIOLOGY. *Educ:* McGill Univ, BSc, 69, MSc, 71; Johns Hopkins Univ, PhD(biostatist), 77. *Prof Exp:* asst prof, 76-85, ASSOC PROF BIOSTATIST, VANDERBILT UNIV, 86- *Mem:* Am Statist Asn; Biomet Soc; AAAS; Soc Epidemiol Res. *Res:* Chronic disease epidemiology, breast disease and hypertension; statistical inference in medicine; data management; design, analysis and management of clinical trials and observational longitudinal studies; ecological statistics; estimation of animal abundance. *Mailing Add:* Dept Prev Med Vanderbilt Univ Sch Med Nashville TN 37232-2637

DU PRE, DONALD BATES, b Houston, Tex, Mar 17, 42; m 64; c 1. CHEMICAL PHYSICS. *Educ:* Rice Univ, BA, 64; Princeton Univ, MA, 66, PhD(chem), 68. *Prof Exp:* Fel chem physics, Sci Ctr, NAm Rockwell Corp, 68-69; asst prof chem, 69-72, assoc prof, 72-76, PROF CHEM, UNIV LOUISVILLE, 76- *Concurrent Pos:* Grants, Res Corp, Petrol Res Fund, Nat Sci Found & NIH. *Mem:* Am Phys Soc; Biophys Soc; AAAS. *Res:* Laser light scattering spectroscopy; chemical physics of polymers and liquid crystals. *Mailing Add:* Dept Chem Univ Louisville Louisville KY 40292

DUPREE, ANDREA K, b Boston, Mass, Sept 17, 39; m 61; c 2. ASTROPHYSICS. *Educ:* Wellesley Col, BA, 60; Harvard Univ, PhD(astron), 68. *Prof Exp:* Res fel, Harvard Col Observ, Harvard Univ, 68-74, res assoc, 74-75, sr res assoc astron & astrophys, 75-79; assoc dir solar & stellar, Ctr Astrophys, 80-87, SR ASTROPHYSICIST, SMITHSONIAN ASTROPHYS OBSERV, 79- *Concurrent Pos:* Mem, Space & Earth Sci Adv Comt, NASA, 82-85, Astrophys Mgt & Opers Group; Space Sci Bd, Nat Acad Sci, 82-85, 89-, Comt on Space Astron & Astrophys, 84-86, Bd Physics & Astron; Lectr, Dept Astron, Harvard Col Observ, Harvard Univ, 70-; exec comt, AAAS, 89- *Honors & Awards:* Phillips Lectr, Haverford Col, 72; Bart J Bok Prize, Harvard Col Observ, 73. *Mem:* Am Astron Soc (vpres, 86-88); Int Astron Union; Comt on Space Res; AAAS. *Res:* Stellar and solar atmospheres; H II regions and the interstellar medium; high resolution spectroscopy. *Mailing Add:* Ctr Astrophys 60 Garden St Cambridge MA 02138

DUPREE, DANIEL EDWARD, b Coushatta, La, Dec 1, 32; m 54; c 1. MATHEMATICS. *Educ:* La Polytech Inst, BS, 54; Auburn Univ, MS, 59, PhD(math), 60. *Prof Exp:* Asst prof math, Auburn Univ, 60-61; chmn dept, 61-64, PROF MATH & DEAN COL PURE & APPL SCI, NORTHEAST LA UNIV, 64- *Concurrent Pos:* Res grants, NSF, 63-64 & Sigma Xi, 64-65; consult, Marshall Space Flight Ctr, 61-62. *Mem:* Am Math Soc; Math Asn Am. *Res:* Interpolation theory and multivariable approximation. *Mailing Add:* Col Pure & Appl Sci Northeast La Univ Monroe LA 71209

DUPREE, HARRY K, FISH & WILDLIFE SCIENCES. *Prof Exp:* LAB DIR, US FISH & WILDLIFE SERV, 74- *Mailing Add:* Fish Farming Exp Lab US Fish & Wildlife Serv PO Box 860 Stuttgart AR 72160

DUPUIS, GILLES, b St Jacques, Que, Mar 11, 43; c 3. BIOCHEMISTRY. *Educ:* Univ Montreal, BSc, 64, MSc, 65; Univ Pittsburgh, PhD(biochem), 69. *Prof Exp:* Fel allergy, Univ BC, 69-72; ASSOC PROF BIOCHEM, UNIV SHERBROOKE, 72- *Mem:* Can Biochem Soc; Can Soc Immunol; NY Acad Sci; Nutrit Soc. *Res:* Study of the structure of lectins and the nature of their receptor sites; study of the mechanism of contact dermatitis using the in vitro test of lymphocytes transformation. *Mailing Add:* Biochem Dept Univ Sherbrook Sch Med De L'Universite Sherbrooke PQ J1H 5N4 Can

DUPUIS, RUSSELL DEAN, b Kankakee, Ill, July 9, 47; m 73; c 1. SEMICONDUCTOR MATERIALS & DEVICES. *Educ:* Univ Ill, Urbana, BSEE, 70, MSEE, 71, PhD(elec eng), 73. *Prof Exp:* Mem tech staff, Tex Instruments, 73-74, Rockwell Int, 75-79; mem tech staff, AT&T Bell Labs, 79-89, distinguished mem, 85-89; PROF & JUDSON S SWEARINGEN REGENTS CHAIR ENG, UNIV TEX-AUSTIN, 89- *Honors & Awards:* Liebmann Field Award, Inst Elec & Electronics Engrs, 85. *Mem:* Nat Acad

Eng; Electrochem Soc; Am Phys Soc; fel Inst Elec & Electronics Engrs. *Res:* Growth of epitaxial III-V compound semiconductor thin films by metalorganic chemical vapor deposition; studies of devices fabricated from these films. *Mailing Add:* Dept Elec & Computer Eng Univ Tex Austin TX 78712-1084

DUPUY, DAVID LORRAINE, b Asheville, NC, Mar 7, 41; m 69. ASTRONOMY. *Educ:* King Col, AB, 63; Wesleyan Univ, MA, 67; Univ Toronto, PhD(astron), 72. *Prof Exp:* Assoc prof astron & observ dir, St Mary's Univ, 72-82, chmn dept astron, 80-82; PROF ASTRON & OBSERV DIR, VA MIL INST, 82- *Concurrent Pos:* Mem, Nat Res Coun Can Assoc Comt Astron & chmn subcomt, 74-77; mem nat comt, Hist Astron in Can, 75-76; sabbatical, US Naval Observ, Kitt Peak Observ, & Lowell Observ. *Mem:* Can Astron Soc; Am Astron Soc; Royal Astron Soc Can; fel Royal Astron Soc Eng; Int Astron Union. *Res:* Observational studies of variable stars, especially RV Tauri stars; studies of young star clusters and resulting structure of galactic associations; previous research in peculiar galaxies; development of astronomical instrumentation including CCD camera. *Mailing Add:* Dept Physics & Astron Va Mil Inst Lexington VA 24450-0304

DUPUY, HAROLD PAUL, b Lockport, La, Sept 10, 22. BIOCHEMISTRY. *Educ:* La State Univ, BS, 50, MS, 53, PhD(biochem), 56. *Prof Exp:* Res chemist, 56-75, supvry res chemist, Food Flavor Res, Oilseed & Food Lab, Southern Regional Res Ctr, USDA, 75-79, Consult, 79- *Concurrent Pos:* Res prof, Va Polytech Inst & State Univ, 82- *Honors & Awards:* Super Serv Award, USDA, 67 & 77; Fed Bus Asn Sci Award, 76. *Mem:* Am Chem Soc; Am Oil Chemists' Soc; Inst Food Technol; Am Inst Chem. *Res:* Lathyrism; fats and oils; surface coatings; unconventional instrumental techniques for flavor analysis; instrumental analysis of food flavor quality. *Mailing Add:* 115 Ave B Apt K Metairie LA 70005

DUQUE, RICARDO ERNESTO, b Apr 16, 44; m; c 2. LYMPHONA, LEUKEMIA. *Educ:* Nat Univ Colombia, Bogota, MD, 73. *Prof Exp:* Asst prof path, Univ Fla, 86-89; asst prof path, Sch Med, Univ Mich, 84-86; PVT PRACT PATH, 89- *Concurrent Pos:* Exp path in-training, Am Asn Pathologists, 83. *Mem:* Am Asn Pathologists. *Res:* Flow cytometry. *Mailing Add:* Dept Path Norwood Clin 1528 N 26th St Birmingham AL 35234

DUQUESNOY, RENE J, b The Hague, Neth, May 24, 38; US citizen; m 68; c 2. IMMUNOLOGY. *Educ:* Delft Technol Univ, Ingenieur, 63; Univ Tenn, Memphis, PhD(exp path), 67. *Prof Exp:* Res assoc path, Med Units, Univ Tenn, Memphis, 63-67; asst clin prof microbiol, Med Col Wis, 70-76; dir res & develop, Milwaukee Blood Ctr, 74-82; assoc prof, Univ Wis, 76-80, clin prof health sci, 80-83; AT DEPT PATH, SCH MED, UNIV PITTSBURGH, PA, 83- *Concurrent Pos:* Res fel, Univ Minn, 68-70; assoc adj prof microbiol, Med Col Wis, 76-80; adj prof, Ned Col Wis, 80-83; adj prof, Marquette Univ, 80-84. *Mem:* Am Soc Histocompatibility & Immunogenetics; Am Asn Blood Banks; Am Asn Immunologists; Transplantation Soc; Int Soc Heart Transplantation. *Res:* Transplantation immunology; histocompatibility testing. *Mailing Add:* Dept Path Presby Univ Hosp CLSI Sch Med Univ Pittsburgh Pittsburgh PA 15261

DUQUETTE, ALFRED L, b Troy, Vt, Oct 14, 23; div. MATHEMATICS. *Educ:* Univ Mass, BS, 48; Columbia Univ, AM, 50; Univ Colo, PhD, 60. *Prof Exp:* Instr math, Mont State Univ, 50-52; asst prof, St John's Univ, 52-54; instr, Univ Colo, 54-55; off naval res asst, Univ Ill, 58-60; asst prof, Univ Ky, 60-61; sr scientist, Jet Propulsion Lab, 61-62; sr mem tech staff, ITT Fed Labs, Calif, 62-64; adv engr, Future Comput Technol, IBM Space Guidance Ctr, 64-66; PROF MATH, WGA COL, 66- *Mem:* Am Math Soc; Math Asn Am. *Res:* Advanced computer development. *Mailing Add:* Dept Math WGa Col Carrollton GA 30117

DUQUETTE, DAVID J(OSEPH), b Springfield, Mass, Nov 4, 39; m 82; c 2. MATERIALS SCIENCE, CORROSION SCIENCE & ENGINEERING. *Educ:* US Coast Guard Acad, BS, 61; Mass Inst Technol, PhD(metall), 68. *Prof Exp:* Res asst metall, corrosion lab, Mass Inst Technol, 65-68; res assoc, adv mat res & develop lab, Pratt & Whitney Div, United Aircraft Corp, 68-70; from asst prof to assoc prof metall eng, 70-76, PROF METALL ENG, RENSSELAER POLYTECH INST, 76- *Concurrent Pos:* Vis prof metall, Imperial Col, London, 73; Alcoa Found res award, 79 & 80; Centennial scholar, Case Western Reserve Univ, 81; sr vis scientist, Max Planck Inst Eisenforschung, Dusseldorf, WGer, 84. *Honors & Awards:* Alexander von Humboldt Prize, WGer, 83; Willis Rodney Whitney Award, Nat Asn Corrosion Engrs, 90. *Mem:* Fel Am Soc Metals; Am Inst Mining, Metall & Petrol Engrs; Nat Asn Corrosion Engrs; Electrochem Soc. *Res:* Corrosion science and engineering including the effect of environment on mechanical properties of crystalline materials; mechanical properties of metals and alloys; fatigue. *Mailing Add:* Dept Mat Eng Rensselaer Polytech Inst Troy NY 12180-3590

DURACK, DAVID TULLOCH, b Dec 18, 44; Australian citizen; m 70; c 4. INFECTIOUS DISEASES, INTERNAL MEDICINE. *Educ:* Univ Western Australia, BS & MB, 69; Oxford Univ, Eng, DPhil, 73. *Prof Exp:* Sr house physician, Dept Med, Hammersmith Hosp, London, Eng, 73-74; chief res med, Univ Hosp, Univ Wash, 74-75, actg instr, 74-75, actg asst prof, 75-77; assoc prof med, 77-82, assoc prof microbiol & immunol, 77-83, chief, Infectious Dis & Int Health Div, 77-, PROF MED, DUKE UNIV, 82-, PROF MICROBIOL & IMMUNOL, 83-,. *Concurrent Pos:* Rhodes scholar. *Mem:* Fel Am Col Physicians; Am Soc Microbiol; fel Royal Australasian Col Physicians; fel Royal Col Physicians; Am Fedn Clin Res; Am Soc Clin Invest. *Res:* Infectious diseases and international health. *Mailing Add:* Box 3867 Duke Univ Med Ctr Durham NC 27710

DURAI-SWAMY, KANDASWAMY, b Thondipatti, India, Sept 28, 45; US citizen; m 71; c 1. FUEL ENGINEERING, CHEMICAL ENGINEERING. *Educ:* Annamalai Univ, India, BChE, 67; Bucknell Univ, MSChE, 70; Univ Utah, MEA, 71, PhD(fuels eng), 73. *Prof Exp:* Teaching assoc chem eng, Indian Inst Technol, 67-68; res engr fuels eng, Garrett Res & Develop Co, Occidental Petrol Corp, 73-75; mem tech staff, TRW Energy Systs, 75-77; sr res engr fuels eng, Occidental Res Corp, 77-81, prin res engr & group leader, 81-; AT MGT & TECH CONSULT, INC, SANTA ANA. *Mem:* Am Chem Soc; Am Inst Chem Engrs. *Res:* Synthetic fuels from coal; pyrolysis; gasification and liquefaction of coal, oil shale, tar sand, and other carbonaceous materials; hydrotreating coal derived liquids. *Mailing Add:* Mgt & Tech Consult Inc 13080 Park St Santa Fe Springs CA 90670-4032

DURAN, BENJAMIN S, b Tularosa, NMex, Nov 25, 39; m 59; c 4. MATHEMATICS, STATISTICS. *Educ:* Albuquerque Univ, BS, 61; Colo State Univ, MS, 64, PhD(statist), 66. *Prof Exp:* Asst prof math, Eastern NMex Univ, 66-69; asst prof, Baylor Med Sch, 69-71; from asst prof to assoc prof, 71-84, PROF MATH & STATIST, TEX TECH UNIV, 84- *Concurrent Pos:* Asst prof, Div Biomath, Tex Inst Rehab & Res, 70-71; adj asst prof math sci, Rice Univ, 70-71. *Mem:* Inst Math Statist; Am Statist Asn; Math Asn Am. *Res:* Nonparametric statistics; statistical computing; mathematical statistics. *Mailing Add:* Dept Math Tex Tech Univ Lubbock TX 79409

DURAN, RUBEN, b Calif, Sept 30, 24; m 43; c 5. PLANT PATHOLOGY. *Educ:* Calif State Polytech Univ, San Luis Obispo, BS, 54; Wash State Univ, PhD(plant path), 58. *Prof Exp:* Res asst plant path, Wash State Univ, 54-58, jr instr plant path, 58-59; plant pathologist, Agr Mkt Serv, USDA, 59-61; from asst prof to assoc prof, 61-64, PROF PLANT PATH, WASH STATE UNIV, 71- *Mem:* Am Phytopath Soc; Mycol Soc Am. *Res:* Taxonomy and biology of the Ustilaginales; teaching mycology and plant pathology at the post-graduate level. *Mailing Add:* RR 2 No 303 Pullman WA 99163

DURAN, SERVET A(HMET), b Kutahya, Turkey, Jan 2, 20; nat US; m 46; c 3. PHYSICAL METALLURGY, MATERIALS SCIENCE. *Educ:* Mo Sch Mines, BS, 43; Stanford Univ, AM, 45, PhD(mat sci), 63. *Prof Exp:* Asst metallog, Stanford Univ, 46; from instr to assoc prof, 47-61, chmn dept, 59-70, PROF PHYS METALL, WASH STATE UNIV, 61- *Concurrent Pos:* Vis assoc prof, Stanford Univ, 56-58; consult, Mid E Tech Univ, Turkey, 69. *Mem:* Am Soc Metals; Am Inst Mining, Metall & Petrol Engrs; Am Soc Eng Educ. *Res:* Solid-state reactions; creep of metals; engineering education. *Mailing Add:* Dept Mat Sci & Eng Wash State Univ Pullman WA 99164-2920

DURAN, WALTER NUNEZ, b Maria Elena, Chile, Nov 1, 42; m 69; c 4. CARDIOVASCULAR PHYSIOLOGY, MICROCIRCULATION. *Educ:* Catholic Univ Chile, PhD(biol), 65; Duke Univ, PhD(physiol & pharmacol), 74. *Prof Exp:* Asst prof physiol, Catholic Univ Chile, 66-69; res assoc physiol & pharmacol, Duke Univ Med Ctr, 70-74, asst med res prof physiol & surg, 74-77; asst prof, 77-79, ASSOC PROF PHYSIOL & GRAD SCH BIOMED SCI, NJ MED SCH, UNIV MED & DENT NJ, 79- *Concurrent Pos:* Res assoc biophys, Univ Chile, 67-69; vis prof, Inst Biol Sci, Catholic Univ Chile, 80-81; mem coun, Am Heart Asn & Microcirc Soc, 84-87; investr, Am Heart Asn, 84-87; assoc ed, Microvasc Res, 84- *Mem:* Am Physiol Soc; Microcirculatory Soc; Am Heart Asn. *Res:* Regulation of microvascular transport of solutes and blood flow in heart and skeletal muscle; multiple tracer methods, intravital microscopy, computer-aided data acquisition and image analysis; elucidation of a microvascular reserve for transport in the myocardium. *Mailing Add:* Dept Physiol Univ Med & Dent NJ Med Sch 185 S Orange Ave Newark NJ 07103-2714

DURAND, BERNICE BLACK, b Clarion, Iowa, Dec 28, 42; m 70. FIELD THEORY, MATHEMATICAL PHYSICS. *Educ:* Iowa State Univ, BS, 65, PhD(physics), 71. *Prof Exp:* Lectr physics, 70-77, asst prof, 77-86, ASSOC PROF PHYSICS, UNIV WIS-MADISON, 86- *Concurrent Pos:* Vis staff mem, Los Alamos Nat Lab, 75 & 84, consult, 77-; mem, Inst Advan Study, 75-76; trustee, Aspen Ctr Physics, 80-86, mem, 90-; vis scientist, Fermi Nat Accelerator Lab, 84. *Mem:* Am Phys Soc; AAAS; Sigma Xi; Am Asn Univ Prof. *Res:* Use of algebra in theoretical physics; high energy scattering theory; gauge field theories; supersymmetry; strong interaction multiparticle production; dynamical symmetry breaking. *Mailing Add:* Dept Physics Univ Wis Madison WI 53706

DURAND, DONALD P, b New York, NY, Oct 18, 29; div; c 3. MICROBIOLOGY, VIROLOGY. *Educ:* Guilford Col, AB, 55; Kans State Univ, MS, 57, PhD(microbiol), 60. *Prof Exp:* Asst prof microbiol, Sch Med, Univ Mo, 59-64, assoc prof, 64-68; assoc prof bact, 68-72, prof bact, 72-80, PROF MICROBIOL, IOWA STATE UNIV, 80- *Concurrent Pos:* NIH grant, 60-66 & spec fel, Cambridge Univ, 66-67; mem staff, Royal Melbourne Inst Technol, Australia, 85. *Mem:* AAAS; Am Soc Microbiol; Brit Soc Gen Microbiol. *Res:* Animal and plant viruses in conjunction with their physical and biochemical properties as they relate to virus-host cell interaction. *Mailing Add:* Dept Microbiol Iowa State Univ 205 Sci Ames IA 50011

DURAND, EDWARD ALLEN, b Duluth, Minn, Dec 20, 19; m 49; c 5. METAL FINISHING, CORROSION PROTECTION. *Educ:* St Mary's Col, Minn, BS, 41; Creighton Univ, MS, 43; Univ Wis, PhD(chem), 50. *Prof Exp:* Chemist, Martin-Nebr Aircraft Co, 43-45; assoc inorg chemist, Armour Res Found, 45-46; res engr, Res Labs, Aluminum Co Am, 50; sr res chemist, Ekco Prod Co, Ill, 51-61; staff chemist, IBM Corp, 62-66; assoc ed, Metals Handbook, Am Soc Metals, 66-70, sr ed, 70-76; eng mgr, Tech Serv Co, Solon, Ohio, 76-77; mgr, Coatings Res & Develop, Empire Plating Co, Cleveland, 77-82; RETIRED. *Mem:* AAAS; Am Chem Soc; Nat Asn Corrosion Engineers; Am Electroplaters Soc. *Res:* Chemical and electrochemical surface treatment of metals; corrosion; selection, processing and fabrication of metals; writing and editing; formulation and application of corrosion-resisting coatings. *Mailing Add:* 7002 Fox Hill Dr Solon OH 44139-4465

DURAND, JAMES BLANCHARD, b Cranford, NJ, June 13, 29; m 52; c 3. BIOLOGY. *Educ:* Rutgers Univ, BSc, 51; Harvard Univ, MA, 54, PhD(biol), 55. *Prof Exp:* From instr to prof zool, Rutgers Univ, 55-88; RETIRED. *Concurrent Pos:* Prof zool, Col SJersey, Camden. *Mem:* Am Soc Zool; Sigma Xi. *Res:* Arthropods and molluscs; field work in marine biology. *Mailing Add:* 620 Maple Ct Haddonfield NJ 08033

DURAND, LOYAL, b Madison, Wis, May 19, 31; m 54, 70; c 3. THEORETICAL & ELEMENTARY PARTICLE PHYSICS, PHYSICAL MATHEMATICS. *Educ:* Yale Univ, BS, 53, MS, 54, PhD(physics), 57. *Prof Exp:* Vis mem physics, Inst Advan Study, 57-59; res assoc, Brookhav en Nat Lab, NY, 59-61; asst prof, Yale Univ, 61-65; chmn dept, 69-71, PROF PHYSICS, UNIV WIS-MADISON, 65- *Concurrent Pos:* NSF fel, 57-59; mem physics adv comt, Fermi Nat Accelerator Lab, vis staff mem, 82-83; trustee, Aspen Ctr Physics, 68-78, chmn exec comt, 68-72, pres, 72-76, hon trustee, 78-, sci mem, 90-, vis mem, Inst Advan Study, 75, chmn, 20th Int Conf High Energy Physics, 80 & Aspen Winter Physics Conf, 86-88; assoc ed, J Math Physics, 73-75; consult Theory Div, Los Alamos Nat Lab, 75-, vis staff mem, Theory Div, 76; mem, ZGS Prog Comt, Argonne Nat Lab, 77-79. *Mem:* Fel Am Phys Soc; AAAS; Am Math Soc; Am Asn Univ Professors; Sigma Xi. *Res:* Theoretical physics, mainly high energy particle physics; astrophysics; applied mathematics; theory of special functions. *Mailing Add:* Dept Physics Univ Wis Madison WI 53706

DURAND, MARC L, b Ware, Mass, Sept 24, 40; m 63; c 3. ORGANIC CHEMISTRY. *Educ:* Holy Cross Col, BS, 62; Univ NH, PhD(chem), 67. *Prof Exp:* Asst prof chem, Alliance Col, 66-68; assoc prof, 68-71, chmn dept, 74-77, PROF CHEM, WEST CHESTER STATE COL, 71-, DEAN ARTS & SCI, 81- *Mem:* Am Chem Soc. *Res:* Synthetic organic chemistry; natural products with experience in infrared, ultraviolet, nuclear magnetic resonance spectroscopy and optical rotatory dispersion. *Mailing Add:* Dept Chem West Chester State Col West Chester PA 19380

DURAND, RALPH EDWARD, b Calgary, Alta, June 16, 47; m 74. RADIOBIOLOGY. *Educ:* Univ Calgary, BSc, 69; Univ Western Ont, PhD(biophys), 73. *Prof Exp:* Asst prof radiol, Univ Wis-Madison, 73-77, from asst prof to assoc prof human oncol & radiol, 75-77; assoc prof oncol & environ health sci, Johns Hopkins Univ, 77-; PROF PATH & PHYSICS, UNIV BC, STAFF BIOPHYSICIST, MED BIOPHYS UNIT, BC CANCER RES CTR, VANCOUVER, CAN. *Mem:* Radiation Res Soc; Biophys Soc; Can Asn Physicists; Am Asn Cancer Res. *Res:* Cellular radiobiology; tumor radiobiology; tumor cell kinetics; radiation-drug interactions. *Mailing Add:* BC Cancer Res Ctr 601 W Tenth Ave Vancouver BC V5Z 1L3 Can

DURANT, JOHN ALEXANDER, III, b Lynchburg, SC, Jan 20, 39; m 58, 76; c 1. ENTOMOLOGY. *Educ:* Clemson Univ, BS, 61, MS, 63; Auburn Univ, PhD(entom), 66. *Prof Exp:* from asst prof to assoc prof, 65-79, PROF ENTOM, CLEMSON UNIV, 79- *Mem:* Entom Soc Am; Sigma Xi. *Res:* Corn and cotton insects ecology and control. *Mailing Add:* Pee Dee Res & Educ Ctr Box 271 Clemson Univ Florence SC 29503

DURANT, JOHN RIDGWAY, b Ann Arbor, Mich, July 29, 30; m 54; c 4. INTERNAL MEDICINE, ONCOLOGY. *Educ:* Swarthmore Col, BA, 52; Temple Univ, MD, 56; Am Bd Internal Med, dipl, 63. *Prof Exp:* From instr to asst prof med, Sch Med, Temple Univ, 63-67; assoc prof, Sch Med, Univ Ala, Birmingham, 68-70, dir div hemat & oncol, 69-74, prof med, 70-, dir, Comprehensive Cancer Ctr, 70-, prof radiation oncol, 80-88; PRES, FOX CHASE CANCER CTR, 88-; SR VPRES HEALTH AFFAIRS & DIR, MED CTR, UNIV ALA, BIRMINGHAM, 88- *Concurrent Pos:* Spec fel med neoplasia, Mem Hosp, New York, 62-63; Am Cancer Soc adv clin fel, 64-67; mem prof educ comt, Am Cancer Soc, 70-; consult, Vet Admin Hosp, Tuskegee, 70-; chmn, Southeastern Cancer Study Group, 75- *Mem:* AAAS; fel Am Col Physicians; Am Fedn Clin Res; Am Asn Cancer Educ; Am Asn Cancer Res. *Res:* Cancer chemotherapy; cytogenetics; immunology. *Mailing Add:* Univ Ala Birmingham UAB Sta Rm 102 MJH Birmingham AL 35294

DURANTE, ANTHONY JOSEPH, b New York, NY, Apr 8, 43. ORGANIC CHEMISTRY. *Educ:* Iona Col, BS, 64; Fordham Univ, PhD(org chem), 71. *Prof Exp:* Res & develop chemist, 68-75, proj scientist, 75-78, GROUP LEADER, UNION CARBIDE CORP, TARRYTOWN, 78- *Mem:* Am Chem Soc; Am Asn Textile Colorists & Chemists. *Res:* Synthesis of natural products; synthesis of pesticide synergists; silicone chemistry, especially synthesis and evaluation of silicone resins in high performance protective coatings; urethane coatings for carpets and textiles; silicone surfactants for urethane foam. *Mailing Add:* 390 Ridgebury Danbury CT 06817-0001

DURANTE, VINCENT ANTHONY, b Plainfield, NJ, Aug 27, 50; m 73; c 3. CATALYSIS, OXIDATION. *Educ:* Rutgers Univ, BA, 72; Univ Calif, Santa Barbara, PhD(inorg chem), 77. *Prof Exp:* Sr res chemist, Minerals & Chem Div, Engelhard Corp, 77-84; SR STAFF SCIENTIST, SUN REFINING & MKT CO, 84- *Concurrent Pos:* Consult, Bioelectrode Co. *Mem:* Am Chem Soc. *Res:* Study of heterogeneous catalytic mechanisms and reaction kinetics especially over zeolites and clay mineral supports; light olkane activation; surface anchored homogeneous catalysts; oxidation chemistry; molecular sieves. *Mailing Add:* 1055 Kerwood Rd West Chester PA 19382

DURAY, JOHN R, b Whiting, Ind, Jan 28, 40; m 66; c 3. ENVIRONMENTAL SCIENCES. *Educ:* St Procopius Col, BS, 62; Univ Notre Dame, PhD(physics), 68. *Prof Exp:* From res assoc to asst prof physics, Ohio State Univ, 68-70; instr, Princeton Univ, 70-74; mem staff, Bendix Field Eng Corp, 74-75, SR PHYSICIST, MGR SUBSURFACE SYSTS DEPT & MGR TECH MEASUREMENTS CTR, MGR TECH PROGS, CHEM-NUCLEAR GEOTECH INC, 75- *Concurrent Pos:* Asst prof, Ind Univ Northwest, 75. *Mem:* Am Phys Soc; Sigma Xi; Soc Prof Well Log Analysts. *Res:* Measurement and evaluation of radon and radon progeny in structures; calibration of exploration and assessment; all measurements associated with evaluation of low level radioactive waste and in clean up. *Mailing Add:* Chem-Nuclear Geotech Inc 2137 Banff Ct Grand Junction CO 81503

DURBECK, ROBERT C(HARLES), b Poughkeepsie, NY, Apr 26, 35; m 63; c 2. RESEARCH MANAGEMENT, ENGINEERING & PHYSICS. *Educ:* Union Col, NY, BSME, 56; Cornell Univ, MS, 58; Case Inst Technol, PhD, 65. *Prof Exp:* Assoc design engr, data systs div, 58-61, res staff mem control systs, res div, 65-68, mgr power systs studies, 68-69, mgr mech technol, 69-71, mgr appl technol, 71-73, mgr explor technol, 73-83, mgr tech staff, 83-85, MGR I/O SCI & TECHNOL, RES DIV, IBM CORP, 85- *Honors & Awards:* Outstanding contribution award, IBM Corp, 68. *Mem:* Am Soc Mech Engrs; Inst Elec & Electronics Engrs; Soc Info Display (chmn sponsor adv comt, 84-); Soc Photog Scientists & Engrs. *Res:* Design and control of large scale physical and information systems; high performance mechanical and electro-mechanical systems; display and printing technologies; magnetic and optical recording systems. *Mailing Add:* IBM Almaden Res Ctr 650 Harry Rd San Jose CA 95120

DURBETAKI, PANDELI, b Istanbul, Turkey, May 31, 28; nat US; m 54; c 3. MECHANICAL ENGINEERING. *Educ:* Robert Col, Turkey, BS, 51; Univ Rochester, MS, 54; Mich State Univ, PhD(mech eng), 64. *Prof Exp:* Asst mech eng, Univ Rochester, 51-52; asst drafting & design, Anstice Co & Rochester Button Co, NY, 52-53; from instr to asst prof mech eng, Univ Rochester, 53-60; instr, Mich State Univ, 60-61, NSF sci fac fel, 61-63, instr, 63-64; assoc prof, 64-77, coordr grad studies, 68-74, PROF MECH ENG, GA INST TECHNOL, 77- *Honors & Awards:* AT&T Found Award, Am Soc Eng Educ, 87. *Mem:* Am Soc Mech Engrs; Combustion Inst; AAUP; Sigma Xi; Am Soc Eng Educ; Instrument Soc Am. *Res:* Classical, statistical and non-equilibrium thermodynamics; combustion; particle combustion; combustion in stratified charge mixtures; stratified charge operation of spark ignition engines; flammability and fire hazard; coal and biomass pyrolysis and combustion. *Mailing Add:* Sch Mech Eng Ga Inst Technol Atlanta GA 30332

DURBIN, ENOCH JOB, b New York, NY, Sept 6, 22; m 45; c 3. AEROSPACE & MECHANICAL SCIENCES. *Educ:* City Col New York, BS, 43; Rensselaer Polytech Inst, MS, 47. *Prof Exp:* Mem res staff, Appl Physics Lab, Johns Hopkins Univ, 44-45 & A D Cardwell Mfg Co, 46; lectr transient anal linear syst, Univ Va, 47-48; head appl physics sect, Aerophys Lab, NAm Aviation, Inc, 51-53; DIR INSTRUMENT & CONTROL LAB, PRINCETON UNIV, 53-, PROF MECH & AEROSPACE ENG, 65- *Concurrent Pos:* Consult various US Corp, 50-, NATO, 53-75 & SUD Aviation, France, 59-65; dir res & labs, US Army, Washington, DC, 66-67, electronics command, Ft Monmouth, 66-75; mem exec bd, Found Instrumentation, Educ & Res; mem, Army Sci Adv Panel & Sci Adv Group for Aviation Systs; founder & dir, Alternate Fuels Lab, Univ BC. *Mem:* Fel AAAS. *Res:* Analysis of dynamic engineering data and physical transducer principles; alternate fueling of the internal combustion engine; fuel economy pollution control in the internal combustion engine. *Mailing Add:* Dept Mech & Aerospace Eng Princeton Univ Princeton NJ 08544

DURBIN, JOHN RILEY, b Elk City, Kans, Nov 18, 35; m 58; c 3. ALGEBRA. *Educ:* Univ Wichita, BA, 56, MA, 58; Univ Kans, PhD(math), 64. *Prof Exp:* Asst prof, 64-69, assoc prof, 69-79, PROF MATH, UNIV TEX, AUSTIN, 79- *Concurrent Pos:* Vis prof, Cambridge Univ, 66-67. *Mem:* Am Math Soc; Math Asn Am. *Res:* Group theory; representations and applications of groups. *Mailing Add:* Dept Math Univ Tex Austin TX 78712

DURBIN, LEONEL DAMIEN, b Riviera, Tex, Nov 13, 35; m 63; c 1. CHEMICAL ENGINEERING. *Educ:* Tex Col Arts & Indust, BS, 57; Rice Univ, PhD(chem eng), 61. *Prof Exp:* Asst prof, 61-69, PROF CHEM ENG, TEX A&M UNIV, 69- *Mem:* Am Inst Chem Engrs; Am Chem Soc. *Res:* Chemical process dynamics; analog and digital simulation with feedback, adaptive, and optimal control; dynamics of distributed flow systems; optimal design methods. *Mailing Add:* 3711 Sweetbriar Bryan TX 77802-3963

DURBIN, PAUL THOMAS, b Louisville, Ky, July 6, 33; m 84; c 6. PHILOSOPHY OF TECHNOLOGY. *Educ:* Providence Col, BA, 57; Cath Univ Am, MA, 62; Aquinas Inst Philos, PhD(philos sci), 66. *Prof Exp:* Lectr, St Stephen's Col, 62-64, asst prof, 66-68; asst prof, Lincoln Univ, 68-69; from asst prof to assoc prof, 69-83, PROF PHILOS, UNIV DEL, 83- *Concurrent Pos:* Vis lectr, Lowell Technol Inst, 68. *Mem:* Philos Sci Asn; Soc for Philos & Technol; Am Philos Asn; Humanities & Technol Asn. *Res:* Place of science and technology in contemporary society; ethics of science and technology. *Mailing Add:* Dept Philos Univ Del Newark DE 19716

DURBIN, RICHARD DUANE, b Santa Ana, Calif, Sept 6, 30; m 54; c 3. PLANT PATHOLOGY. *Educ:* Univ Calif, BS, 52, PhD, 58. *Prof Exp:* Res asst, Univ Calif, 53-54, sr lab tech, 54-57; NSF res fel, 57-58, asst prof plant path, Univ Minn, 58-62; assoc prof, 62-67, PROF PLANT PATH, UNIV WIS-MADISON, 67-; LAB CHIEF, PIONEERING RES LAB, USDA, 65- *Concurrent Pos:* Plant pathologist, Oat Invests, Pioneering Res Lab, USDA, 62-65. *Mem:* Am Phytopath Soc; Am Soc Plant Physiol. *Res:* Physiology of plant parasitism; mode of action and structure toxins; hypersensitivity. *Mailing Add:* Dept Plant Path 789A Russell Lab Univ Wis Madison WI 53706

DURBIN, RONALD PRIESTLEY, b Bement, Ill, Jan 23, 39; m 61, 87; c 5. ANALYTICAL CHEMISTRY. *Educ:* MacMurray Col, BA, 61; Univ Ill, Urbana, PhD(anal chem), 66. *Prof Exp:* Res chemist, Hercules, Inc, 65-70, sr res chemist, 70-77, res supvr, 77-78, mgr, Anal Div, 78-83, proj leader, 83-86, RES ASSOC, HERCULES, INC, 86- *Concurrent Pos:* Mem res comt, Univ Del Res Found, 81- *Mem:* Am Chem Soc; Sigma Xi. *Res:* Chromatography; thermodynamics of solute-solvent interactions; solvent effects in organic chemistry; spectrochemical methods of analysis; organic analysis via functional groups. *Mailing Add:* 1170 Doe Run Rd Newark DE 19711

DURDEN, CHRISTOPHER JOHN, b London, Eng, Feb 25, 40; Can citizen; div; c 2. PALEONTOLOGY, SYSTEMATIC ENTOMOLOGY. *Educ:* McGill Univ, BSc, 61; Yale Univ, MS, 68, PhD(geol, biol), 71. *Prof Exp:* Asst park naturalist, Algonquin Park Nature Mus, Ont Dept Lands & Forests, 55-56; asst entom, Entom Res Inst, Res Br, Can Dept Agr, 57-58; geol asst, Sudbury Basin Proj, Int Nickel Co, 59; asst reconnaissance mapping, Geol Surv Can, 60; asst party chief, Bedrock Stratig Mapping, Geol Serv, Ministry Natural Resources, Qué, 61; teaching asst geol, Yale Univ, 62-64, teaching asst biol, 64-65; res asst invert paleont, Carnegie Mus, Pa, 66-68; CUR GEOL, TEX MEM MUS, 68-, CUR ENTOM, 72- *Concurrent Pos:*

Curatorial asst invert paleont & entom, Peabody Mus Natural Hist, 62-66; dir, Tex Arch Geol Res, 70- *Mem:* Fel AAAS; Paleont Soc; Lepidop Soc; fel Geol Asn Can; Int Paleont Asn; Orthopterists' Soc. *Res:* Paleozoic insect evolution; speciation ecology in modern and fossil Lepidoptera, Orthoptera, Collembola, corals and woody plants; Paleozic coral evolution; Carboniferous and Silurian biostratigraphic correlation; biotic provinciality; ecosystem dynamics evolution; natural selection; micropaleoentomology. *Mailing Add:* 1907 Sharon Lane Austin TX 78703

DURDEN, DAVID ALAN, b Manchester, Eng, June 25, 43; m; c 2. ANALYTICAL CHEMISTRY, PHYSICAL CHEMISTRY. *Educ:* Univ BC, BSc, 63; Univ Alta, PhD(phys chem), 69. *Prof Exp:* Res fel, Chem, Univ Essex, 69-71; res assoc, Dept Psychiat & Biochem, Dept Health, Sask, 71-78, res chemist, 78-86; ADJ PROF & RES ASSOC, DEPT PSYCHIAT, UNIV SASK, 86- *Mem:* Am Soc Mass Spectrometry; Int Soc Neurochem. *Res:* Use of high resolution mass spectrometry for the quantitation of neurotransmitters amines and their metabolites and drugs in tissue and physiological fluids; study of metabolites implicated in psychiatric disorders. *Mailing Add:* Neuropsychiat Res A114 MR Bldg Univ Sask Saskatoon SK S7N 0W0 Can

DURDEN, JOHN APLING, JR, b Phoenix, Ariz, July 7, 28; m 55; c 5. PHARMACEUTICAL CHEMISTRY, ORGANIC CHEMISTRY. *Educ:* Ariz State Univ, BS, 50; Univ Miss, MS, 52; Univ Kans, PhD(pharm & org chem), 57. *Prof Exp:* Instr org & phys chem, Midwestern Univ, 54-56; res chemist, 57-67, res scientist, 67-73, sr res scientist, Chem & Plastics Operating Div, 73-76, assoc dir res & develop, Agr Prod Div, Union Carbide Corp, WVa, 76-80, assoc dir res & develop, Union Carbide Agr Prod Co, Inc, NC, 80-87; MGR DISC CHEM, RHONE-POULENE AG CO, NC, 87- *Mem:* Am Chem Soc. *Res:* Organic synthesis; reaction mechanisms; agricultural chemistry; structure-activity correlations. *Mailing Add:* 4508 Sterling Pl Raleigh NC 27612

DURE, LEON S, III, b Macon, Ga, Jan 19, 31; m 58; c 4. BIOCHEMISTRY. *Educ:* Univ Va, BA, 53, MA, 57; Univ Tex, PhD(biol), 60. *Prof Exp:* Fel biochem, 60-62, from asst prof to assoc prof, 62-69, PROF BIOCHEM, UNIV GA, 69- *Concurrent Pos:* NSF grant, 63-; AEC contract, 64-; USPHS career develop award, 67-72; consult, Biol Div, Oak Ridge Nat Lab, 63- *Mem:* Am Soc Biol Chem; Am Soc Plant Physiol. *Res:* Developmental biochemistry; nucleic acid and protein biosynthesis. *Mailing Add:* Dept Biochem Univ Ga Athens GA 30602

DUREN, PETER LARKIN, b New Orleans, La, Apr 30, 35; m 57; c 2. MATHEMATICS. *Educ:* Harvard Univ, AB, 56; Mass Inst Technol, PhD(math), 60. *Prof Exp:* Instr math, Stanford Univ, 60-62; from asst prof to assoc prof, 62-69, PROF MATH, UNIV MICH, ANN ARBOR, 69- *Concurrent Pos:* Sloan Found fel, 64-66; mem, Inst Advan Study, Princeton, NJ, 68-69; assoc ed, Proc Am Math Soc, 73-75; vis lectr,The Technion, Haifa, Israel, 75; managing ed, Mich Math J, 76-77; prin lectr, LMS/NATO conf on complex anal, Eng, 79 & Xian, China, 84; vis prof, Univ Md, 82; vis scholar, Univ Paris-Sud, Inst Mittag-Leffler, Stockholm & ETH, Zurich, 82-83; Sci & Eng Res Coun vis, Univ York, England, 85; bd govs, Math Asn Am, 79-82, mem at large coun, Am Math Soc, 82-85, chmn, Am Math Soc Comt Hist Math, 87-90, Comt Publ Prog, 85-89; assoc chmn grad studies, Dept Math, Univ Mich, 87-88; invited lectr, Am Math Soc, 76; vis scholar, Stanford Univ, 89. *Mem:* Am Math Soc; Math Asn Am; London Math Soc; Asn Women Math. *Res:* Complex analysis; univalent functions; linear spaces of analytic functions; harmonic analysis. *Mailing Add:* Dept Math Univ Mich Ann Arbor MI 48109

DURETTE, PHILIPPE LIONEL, b Manchester, NH, Aug 17, 44; m 67; c 2. MEDICINAL CHEMISTRY, PEPTIDE & CARBOHYDRATE CHEMISTRY. *Educ:* Marquette Univ, BS, 66; Ohio State Univ, PhD(org chem), 71. *Prof Exp:* Sci Res Coun fel, Dept Chem, Queen Elizabeth Col, Univ London, 71-72; Alexander von Humboldt Found fel, Org Chem & Biochem Inst, Univ Hamburg, Ger, 72-73; sr res chemist, 73-78, res fel, 78-88, SR RES FEL, MERCK SHARP & DOHME RES LAB, 88- *Mem:* Am Chem Soc; The Chem Soc; Sigma Xi. *Res:* Synthetic organic chemistry; synthetic and mechanistic carbohydrate chemistry; medicinal chemistry; conformational analysis; nuclear magnetic resonance spectroscopy; lipoxygenase inhibitors; host resistance enhancers; immuno regulants; enzyme inhibitors; complement antagonists. *Mailing Add:* Merck Sharp & Dohme Res Lab PO Box 2000 Rahway NJ 07065

DURFEE, RAPHAEL B, b Bisbee, Ariz, Apr 7, 18; m 43; c 2. OBSTETRICS & GYNECOLOGY. *Educ:* Stanford Univ, AB, 39, MD, 44. *Prof Exp:* From asst prof to prof obstet & gynec, Med Sch, Univ Ore, 57-76, clin prof, 76-88; prof, Univ Calif, San Diego, 81-88; RETIRED. *Concurrent Pos:* Regent for State of Ore, Int Col Surgeons. *Mem:* Fel Am Col Surg; fel Am Col Obstet & Gynec; Am Fertil Soc; fel Int Col Surg; Gynec & Urol Soc; Soc Gynec Surg. *Res:* Clinical research and investigation in gynecologic surgery. *Mailing Add:* 2929 Front St San Diego CA 92103

DURFEE, ROBERT LEWIS, b Farmville, Va, May 15, 36; m 60; c 2. CHEMICAL ENGINEERING. *Educ:* Va Polytech Inst, BS, 57, MS, 59, PhD(chem eng), 61. *Prof Exp:* Res engr chem systs, Atlantic Res Corp, 61-69; dir life sci, 69-71, VPRES OPERS & CHIEF RES EXEC, VERSAR INC, 71- *Res:* Solid state radiation chemistry; projects on advanced fuels, non-Newtonian flow, cryogenic systems, and boiling heat transfer; life sciences and environmental systems; new product development. *Mailing Add:* 8310 Private Lane Annandale VA 22003

DURFEE, WAYNE KING, b North Scituate, RI, Oct 1, 24; m 51; c 2. AVIAN PHYSIOLOGY, AQUACULTURE. *Educ:* Univ RI, BS, 50, MS, 53; Rutgers Univ, PhD, 63. *Prof Exp:* prof animal sci, 51-81, prof, 81-89, EMER PROF FISHERIES, ANIMAL & VET SCI, UNIV RI, 89- *Concurrent Pos:* Ombudsman, Univ RI, 74-76; res proj mgr, RI Agr Exp Sta, 76-82, prin investr oyster & quahog res. *Mem:* Nat Shellfisheries Asn; Sigma Xi; World Aquaculture Soc. *Res:* Embryology; behavior; poultry management; aquaculture; closed system culture and formulated rations for American oyster, Crassostrea virginica and hard clam, Mercenaria mercenaria. *Mailing Add:* Dept Fisheries Animal & Vet Sci Univ RI Kingston RI 02881-0804

DURFEE, WILLIAM HETHERINGTON, b Montague, Mass, Apr 12, 15; m 39; c 4. MATHEMATICS. *Educ:* Harvard Univ, AB, 36, MA, 40; Cornell Univ, PhD(math), 43. *Prof Exp:* Instr math, Cornell Univ, 40-43 & Yale Univ, 43-45; math physicist, Nat Defense Res Coun, Northwestern Univ, 45-46; from instr to asst prof math, Dartmouth Col, 46-51; mathematician, Nat Bur Stand, 51-53 & Opers Res Off, 53-55; from assoc prof to prof, 55-80, EMER PROF MATH, MT HOLYOKE COL, 80- *Mem:* Am Math Soc; Math Asn Am. *Res:* Algebra. *Mailing Add:* Dept Math Mt Holyoke Col South Hadley MA 01075

DURFLINGER, ELIZABETH WARD, b Ft Wayne, Ind, July 8, 13; div. INVERTEBRATE ZOOLOGY. *Educ:* Western Col, BA, 33; Univ Cincinnati, MA, 34, PhD(zool), 39. *Prof Exp:* From instr to prof, 40-75, dean women, 40-65, EMER PROF ZOOL, BUTLER UNIV, 75- *Concurrent Pos:* Consult & sci writer, George F Cram Co, Indianapolis, 75-84. *Mem:* Sigma Xi. *Res:* Ecology of entomostraca; aquatic invertebrates. *Mailing Add:* 1010 Oakwood Trail Indianapolis IN 46260-4021

DURGIN, HAROLD L, power supply, for more information see previous edition

DURGIN, WILLIAM W, b Framingham, Mass, Apr 26, 42; m 64; c 3. FLUID MECHANICS. *Educ:* Brown Univ, BS, 64, PhD(fluid dynamics), 70; Univ RI, MS, 66. *Prof Exp:* Asst prof eng sci & mech, Univ Fla, 70-71; asst prof mech eng & res eng, 71-75, assoc prof mech eng & head res & develop, 75-81, PROF MECH & LEAD RES ENG, WORCESTER POLYTECH INST, 81- *Mem:* Am Soc Mech Engrs; Sigma Xi. *Res:* Turbulence; flow induced vibration; physical and analytic modeling. *Mailing Add:* Worcester Polytech Inst Worcester MA 01609

DURHAM, CLARENCE ORSON, JR, b Victoria, Tex, Oct 20, 20; m 59; c 2. GEOLOGY. *Educ:* Univ Tex, BS, 42; Univ Chicago, cert meteorol, 43; Columbia Univ, PhD(geol), 57. *Prof Exp:* From lab asst to lab instr geol, Univ Tex, 46-48; asst, Bur Econ Geol, Univ Tex, 48-49 & Columbia Univ, 49-50; instr struct geol, 51-53, from asst prof to assoc prof geol, La State Univ, 53-63, prof geol, 63-77, chmn dept, 65-77, dir sch geol, 66-77, MEM STAFF, DURHAM GEOL ASSOCS, 77- *Concurrent Pos:* Res geologist, La Geol Surv, 55-57, dir res, 57-63. *Mem:* AAAS; Am Asn Petrol Geol; Soc Econ Paleont & Mineral; Am Geophys Union. *Res:* Mesozoic and Cenozoic stratigraphy; structural geology; geology of Gulf of Mexico region; sedimentary iron ores. *Mailing Add:* Durham Geol Assoc 430 Hwy 6 S Suite 208 Houston TX 77079

DURHAM, FRANK EDINGTON, b Jonesboro, La, July 12, 35; m 56, 77; c 3. NUCLEAR PHYSICS, HISTORY & PHILOSOPHY OF SCIENCE. *Educ:* La Tech Univ, BS, 56; Rice Univ, MA, 58, PhD(physics), 60. *Prof Exp:* From asst prof to assoc prof physics, 60-67, arts & science head, dept physics, 72-74, univ chmn, 74-79, PROF PHYSICS, TULANE UNIV, 67- *Concurrent Pos:* Mem La Nuclear & Space Auth, 68-74; consult, Gulf S Res Inst, 70-76; dir res & develop, Ultra Prod Syst, Inc, 75-89. *Mem:* Am Phys Soc; AAAS; Sigma Xi; Hist Sci Soc; Philos Sci Asn; Soc Lit & Sci. *Res:* Experimental studies of nuclear structure; history and philosophy of ideas; cross-disciplinary studies. *Mailing Add:* Dept Physics Tulane Univ New Orleans LA 70118

DURHAM, FRANKLIN P(ATTON), b Wiley, Colo, Dec 22, 21; m 43; c 3. AERONAUTICAL ENGINEERING. *Educ:* Univ Colo, BS, 43, MS, 49, AeroEng, 53. *Prof Exp:* Exp test engr, Pratt & Whitney Aircraft, 43-47; from instr to prof aeronaut eng, Univ Colo, 47-55, head dept, 56; group leader, Los Alamos Nat Lab, 57-61, alternate div leader, 61-76, assoc div leader, 76-78, prog mgr, 78-83, dep assoc dir, 83-85; RETIRED. *Mem:* Assoc fel Am Inst Aeronaut & Astronaut. *Res:* Thermodynamics; heat transfer; laser fusion. *Mailing Add:* 3100 Arizona Ave Los Alamos NM 87544

DURHAM, GEORGE STONE, b Portland, Ore, Dec 26, 12; m 35, 58; c 2. PHYSICAL CHEMISTRY. *Educ:* Reed Col, BA, 35; NY Univ, PhD(phys chem), 39. *Prof Exp:* Asst chem, NY Univ, 35-39; res chemist, Weyerhaeuser Timber Co, Wash, 39-40; instr chem, Ore State Col, 40-41; instr, Univ Ill, 41-43; from instr to prof, 43-77, chmn dept, 58-66 & 72-75, EMER PROF CHEM, SMITH COL, 77- *Concurrent Pos:* Vis asst prof, Univ Mass, 44-45, vis lectr, 57-62, mem grad faculty, 61-; vis asst prof, NY Univ, 45 & 46; res grants & res contracts, Sigma Xi, NSF, Off Naval Res, Air Force Off Sci Res, Off Ord Res & Army Res Off. *Mem:* Am Chem Soc. *Res:* Solid solutions of inorganic salts; solid state theory of the alkali halides. *Mailing Add:* PO Box 7 Arch Cape OR 97102

DURHAM, HARVEY RALPH, b Perry, Fla, Feb 25, 38; m 63; c 3. TOPOLOGY. *Educ:* Wake Forest Univ, BS, 59; Univ Ga, MA, 62, PhD(math), 65. *Prof Exp:* Assoc prof math, Appalachian State Univ, 65-73, chmn dept, 67-71, assoc dean fac, 71-73, assoc chancellor acad affairs, 73-79, PROF MATH, APPALACHIAN STATE UNIV, 73-, VCHANCELLOR ACAD AFFAIRS, 79- *Concurrent Pos:* Partic, Am Coun Educ Acad Admin Internship Prog, 69-70; provost, Appalachian State Univ, 89- *Res:* Combinatorial topology. *Mailing Add:* Dept Math Sci Appalachian State Univ Boone NC 28608

DURHAM, JAMES IVEY, b Alpine, Tex, Nov 6, 33; m 68; c 2. PLANT PHYSIOLOGY. *Educ:* Tex A&M Univ, BS, 55, MS, 61, PhD(biochem), 71. *Prof Exp:* plant physiologist, US Sugar Corp, 75-80; mem fac, dept plant sci, Tex A&M Univ, 80-; DIV NATURAL SCI, BLINN COL, BRENHAM, TEX. *Mem:* Am Chem Soc; Am Soc Plant Physiologists; Sigma Xi. *Res:* Growth and development of sugarcane and forage crops. *Mailing Add:* Div Natural Sci Blinn Col Brenham TX 77833

DURHAM, JOHN WYATT, b Okanogan, Wash, Aug 22, 07; m 35, 72; c 1. INVERTEBRATE PALEONTOLOGY. *Educ:* Univ Wash, BSc, 33; Univ Calif, MA, 36, PhD(paleont), 41. *Prof Exp:* Asst geol, Univ Calif, 35-36; geologist, Stand Oil Co Calif, 36-39; asst, Mus Paleont, Univ Calif, 41-42;

geologist & chief paleontologist, Tropical Oil Co, Colombia, 43-46; assoc prof paleont, Calif Inst Technol, 46-47; from assoc prof to prof, 47-75, chmn dept, 56-58, EMER PROF PALEONT, UNIV CALIF, BERKELEY, 75- *Concurrent Pos:* Guggenheim fel, 54-55 & 65-66; mem, Paleont Res Inst & US Nat Comt Geol, 66-70; Galapagos Int Sci Exped. *Honors & Awards:* Gold Medal, Paleont Soc, 88. *Mem:* Fel AAAS; Am Soc Syst Zool; fel Paleont Soc (vpres, 52-53, pres, 65-66); Soc Econ Paleontologists & Mineralogists; fel Geol Soc Am. *Res:* Tertiary Molluscan paleontology; tertiary stratigraphy; recent corals and echinoids; cretaceous ammonites; paleoclimates; paleobiogeography; lower Cambrian, Pre-Cambrian fossils. *Mailing Add:* Dept Paleontology Univ Calif Berkeley CA 94720

DURHAM, LEONARD, b Glen Carbon, Ill, Aug 27, 25; m 48; c 4. AQUATIC ECOLOGY, FISHERIES MANAGEMENT. *Educ:* Univ Ill, BS, 49, MS, 50, PhD(zool), 55. *Prof Exp:* Lab & field asst, Ill Natural Hist Surv, 47-49, tech asst, 49-50; fishery biologist, Ill Dept Conserv, 50-55; dir, div life sci, 67-82, dir, environ biol,69-86, chmn, dept zool, 75-82, PROF ZOOL, EASTERN ILL UNIV, 55-, ASSOC DEAN, COL ARTS & SCI, 82- *Concurrent Pos:* Mem, Ill Nature Preserves Comn, 70-73, Ill Dept Conserv Adv Bd, 81-82. *Mem:* AAAS; Am Soc Ichthyologists & Herpetologists; Am Fisheries Soc; Am Inst Fisheries Res Biologists. *Res:* Fishery biology; ecology of fishes and fish management; conservation; water pollution; thermal studies. *Mailing Add:* Life Sci Bldg Rm 134 Eastern Ill Univ Charleston IL 61920

DURHAM, LOIS JEAN, b Oakland, Calif, Dec 21, 31. ORGANIC CHEMISTRY. *Educ:* Univ Calif, BS, 54; Stanford Univ, PhD(org chem), 59. *Prof Exp:* Instr org chem, Stanford Univ, 59-60; sr res chemist, Stanford Res Inst, 60-61; NUCLEAR MAGNETIC RESONANCE SPECTROSCOPIST, STANFORD UNIV, 61- *Mem:* Am Chem Soc; Soc Appl Spectros. *Res:* Application of nuclear magnetic resonance spectroscopy in determination of organic structural analysis; organic reaction mechanisms. *Mailing Add:* Dept Chem Stanford Univ Stanford CA 94305-5080

DURHAM, MICHAEL DEAN, b Key West, Fla, Dec 11, 49; m 77. AEROSOL PHYSICS, AIR POLLUTION CONTROL. *Educ:* Pa State Univ, BS, 71; Univ Fla, MS, 75, PhD(environ eng), 78. *Prof Exp:* Info analyst, Nat Acad Sci, 72-73 & Am Psychol Asn, 73-74; res asst, Univ Fla, 75-78; RES ENGR, DENVER RES ENG, 78- *Concurrent Pos:* Consult, Environ Eng Consult, 75-78. *Mem:* Sigma Xi; Air Pollution Control. *Res:* Sampling and analysis of particulate matter and works in the design, development and evaluation of air pollution control equipment. *Mailing Add:* 654 S Gilpin St Denver CO 80209-4512

DURHAM, NORMAN NEVILL, b Ranger, Tex, Feb 14, 27; m 52; c 4. BACTERIOLOGY. *Educ:* North Tex State Univ, BS, 49, MS, 51; Univ Tex, PhD(bact), 54. *Prof Exp:* Lab asst bot & bact, NTex State Univ, 46-49, student instr, 49-51; res scientist, Univ Tex, 52-54; from asst prof to assoc prof bact, 54-60; dir, Environ Inst & Community Develop Inst, 73-77, PROF BACT, OKLA STATE UNIV, 60-, DEAN, GRAD COL, 68-, DIR, UNIV CTR WATER RES, 77-, ASST VPRES RES, 88- *Concurrent Pos:* Vis lectr, Sch Med, Univ Okla, 63 & Kans State Univ, 65; consult biol sci, NASA, 63-69; with div biol & med, AEC, 66-68; mem coun manpower planning, US Off Educ, 70-; mem eval team, Nat Coun Accreditation Teacher Educ, 68-; Sigma Xi regional lectr, 73-74. *Honors & Awards:* Outstanding Scientist Award, Okla, 77. *Mem:* AAAS; Am Acad Microbiol; Am Soc Microbiol; Brit Soc Gen Microbiol; Biochem Soc. *Res:* Radiations; bacterial metabolism and metabolic pathways; genetics; agricultural bacteriology; protein and enzyme synthesis; mechanism of antibiotic action; metabolic regulations, cell growth and reproduction; genetic transformation; chemotherapy, molecular interactions; cell structure, composition and conformation. *Mailing Add:* Grad Col Okla State Univ Stillwater OK 74074

DURHAM, ROSS M, b Toronto, Ont, Sept 19, 30; US citizen; m 55; c 3. NEUROPHYSIOLOGY, SPACE BIOLOGY. *Educ:* Univ Calif, Los Angeles, AB, 62, MA, 63, PhD(zool), 68. *Prof Exp:* Proj biologist, Biosatellite Project, Space Biol Labs, Brain Res Inst, Univ Calif, Los Angeles, 68-70, asst res psychologist, 70-71; from asst prof to assoc prof, 71-89, PROF BIOL, UNIV TENN, CHATTANOOGA, 89- *Concurrent Pos:* Vis prof, Univ Southern Calif, 70-71; consult, Vet Admin Hosp, Sepulveda, Calif, 70-71. *Res:* Water balance in vertebrates; thirst, its cause and control; renal physiology; physiological psychology; electrophysiological recording from units in subcortical nuclei. *Mailing Add:* Dept Biol Univ Tenn 615 McCallie Ave Chattanooga TN 37403

DURHAM, STEPHEN K, b Clinton, Okla, Feb 8, 49. PATHOLOGY, TOXIOLOGY. *Educ:* Tex A&M Univ, BS, 70, DVM, 71; Cornell Univ, PhD(path), 85. *Prof Exp:* DIR EXP PATH & ELECTRON MICROS, PHARMACEUT RES INST, BRISTOL-MYERS SQUIBB, 89- *Mem:* AAAS; Am Soc Exp Biol; Int Asn Pathologists; Soc Toxicol; Soc Toxicologists; NY Acad Sci. *Mailing Add:* Pharmaceut Res Inst Bristol-Myers Squibb Rte 206 Princeton NJ 08540

DURHAM, WILLIAM BRYAN, b Ithaca, NY, May 20, 47; m; c 2. EXPERIMENTAL ROCK MECHANICS, MATERIALS SCIENCE. *Educ:* Cornell Univ, BS, 69; Mass Inst Technol, PhD(geophys), 75. *Prof Exp:* Res assoc earth sci, Univ Paris, 75-77; PHYSICIST EARTH SCI, LAWRENCE LIVERMORE LAB, UNIV CALIF, 77- *Concurrent Pos:* Alexander von Humboldt fel, Univ Hanover, WGermany, 85-86. *Mem:* Am Geophys Union. *Res:* Rheology of upper mantle rocks and minerals; physical properties of rocks at high pressure and temperature; microstructure such as dislocations, subgrains and microcracks of rocks and minerals. *Mailing Add:* Lawrence Livermore Lab PO Box 808 Univ Calif Livermore CA 94550

DURHAM, WILLIAM FAY, b Cedartown, Ga, Apr 19, 22; m 47; c 3. BIOCHEMISTRY. *Educ:* Emory Univ, AB, 43, MS, 48, PhD(biochem), 50. *Prof Exp:* Biochemist, Toxicol Sect, Technol Br, Commun Dis Ctr, USPHS, Environ Protection Agency, 50-57, chief, Wenatachee Field Sta, 57-67, Pesticide Res Lab, 67-70, Perrine Primate Lab, 70-75, dir, Environ Toxicol Div, 75-80, sr res scientist, 80-90; CONSULT, 90- *Mem:* Fel AAAS; Soc Toxicol; Am Soc Pharmacol & Exp Therapeut; Sigma Xi; Am Chem Soc. *Res:* Toxicology of pesticides. *Mailing Add:* 6900 Bent Pine Pl Raleigh NC 27615-7002

DURICA, THOMAS EDWARD, b Cleveland, Ohio, Oct 25, 42; m 76. DEVELOPMENTAL NEUROBIOLOGY, ANATOMY. *Educ:* John Carroll Univ, BS, 66; Loyola Univ Chicago, PhD(anat), 77. *Prof Exp:* Asst prof anat, Rush Med Col, 77-; AT DEPT ANAT, COL HEALTH SCI, CHICAGO, ILL. *Mem:* AAAS; Sigma Xi; Soc Neurosci. *Res:* Development of the nervous system, specifically the axon reaction of the pyramidal cells of the cerebral cortex of the developing hamster. *Mailing Add:* Dept Anat Rush Univ Med Col 600 S Paulina Ave Chicago IL 60612

DURIG, JAMES ROBERT, b Washington Co, Pa, Apr 30, 35; m 55; c 3. PHYSICAL CHEMISTRY. *Educ:* Washington & Jefferson Col, BA, 58; Mass Inst Technol, PhD(phys chem), 62. *Prof Exp:* From asst prof to prof chem, Univ SC, 62-70, Educ Found prof, 70-73, DEAN, COL SCI & MATH, UNIV SC, 73- *Honors & Awards:* Russell Award Res, 68; Coblentz Soc Award, 70; Charles A Stone Award, Am Chem Soc, 75, Southern Chemist Award, 76; Alexander von Humbolt Sr Scientist Award, WGer, 76. *Mem:* Fel Am Phys Soc; Am Chem Soc; Soc Appl Spectros; Coblentz Soc (pres, 74-76); Sigma Xi. *Res:* Infrared, Raman and microwave spectra of polyatomic molecules, especially molecules having low frequency vibrations; torsional barriers; molecular structure of organometallic molecules. *Mailing Add:* Dept Chem Univ SC Columbia SC 29208

DURIO, WALTER O'NEAL, b Arnaudville, La, Jan 17, 38. PARASITOLOGY. *Educ:* Southwestern La Univ, BSc, 59, MS, 60; Univ Nebr, Lincoln, PhD(zool, physiol), 66. *Prof Exp:* Asst prof, 66-74, ASSOC PROF BIOL, UNIV SOUTHWESTERN LA, 74- *Mem:* Am Soc Parasitol; Am Micros Soc. *Res:* Helminthology; taxonomy of digenetic trematodes. *Mailing Add:* Dept Microbiol 41007 Univ Southwestern La Lafayette LA 70504

DURISEN, RICHARD H, b Brooklyn, NY, Nov 24, 46; m; c 1. ASTROPHYSICAL FLUID DYNAMICS, STELLAR EVOLUTION. *Educ:* Fordham Col, Bronx, NY, BS, 67; Princeton Univ, PhD(astron), 72. *Prof Exp:* Fel, Lick Observ, 71-73; res assoc, Ames Res Ctr, NASA, 74-76; chmn dept, 86-90, from asst prof to assoc prof, 76-81, PROF ASTRON, IND UNIV, 85- *Mem:* Am Astron Soc; AAAS; Int Astron Union; Union of Concerned Scientists. *Res:* Theoretical astrophysics involving fluid or particle dynamics; rotationally driven instabilities in stars and gas disks; globular cluster winds; white dwarf accretion and planetary ring dynamics; active galactic nuclei; white dwarf accretion and planetary ring dynamics. *Mailing Add:* Dept Astron Swain W 319 Ind Univ Bloomington IN 47405

DURKAN, JAMES P, b Baltimore, Md, Jan 13, 34; m 58; c 4. MEDICINE, OBSTETRICS & GYNECOLOGY. *Educ:* Loyola Col, Md, AB, 55; Univ Md, MD, 59; Am Bd Obstet & Gynec, dipl, 67. *Prof Exp:* Intern, Mercy Hosp, Baltimore, Md, 59-60; resident, 60-64, asst prof, 64-71, ASSOC PROF OBSTET & GYNEC, UNIV MD, BALTIMORE CITY, 71-; HEAD DEPT OBSTET & GYNEC, MERCY HOSP, 68- *Mem:* Fel Am Col Obstet & Gynec. *Res:* Neurologic influence on menstrual function; clinical oncology; clinical family planning. *Mailing Add:* Mercy Hosp 301 St Paul Pl Baltimore MD 21202

DURKEE, JACKSON L, b Tatanagar, British India, Sept 20, 22; US citizen; m 43; c 3. DESIGN & CONSTRUCTION OF LONGSPAN BRIDGES, STRUCTURAL ENGINEERING RESEARCH. *Educ:* Worcester Polytech Inst, BSCE, 43, CE, 51; Cornell Univ, MCE, 47. *Prof Exp:* Stress analyst, Douglas Aircraft Co, 43-44; deck officer, US Naval Reserve, 44-46; instr civil eng, Cornell Univ, 46-47; field engr, Fabricated Steel Construct Dept, Bethlehem Steel Corp, 47-52, designer, asst engr & engr, 52-65, chief bridge engr, 65-76; partner, consult engrs, Modjeski & Masters, 77-78; CONSULT STRUCT ENGR, 78- *Concurrent Pos:* Mem, US coun, Int Asn Bridge & Struct Eng, 72-; vis prof struct eng, Cornell Univ, 76; expert witness, various clients, 78-; construct indust arbitrator, Am Arbit Asn, 81- *Honors & Awards:* Ernest E Howard Award, Am Soc Civil Engrs, 82. *Mem:* Fel Am Soc Civil Engrs; fel Inst Civil Engrs; fel Inst Struct Engrs; Nat Soc Prof Engrs; Struct Stability Res Coun; Int Asn Bridge & Struct Eng. *Res:* Originated and directed development of shop-fabricated parallel-wire-strand method for construction of suspension bridge main cables, and plastic-based covering system for weather protection of such cables. *Mailing Add:* 217 Pine Top Trail Bethlehem PA 18017

DURKEE, LAVERNE H, b Darien, NY, June 23, 27; m 56; c 2. PLANT TAXONOMY. *Educ:* Syracuse Univ, BS, 51, MS, 54, PhD, 60. *Prof Exp:* Asst prof biol, Grove City Col, 58-61; asst prof, Parsons Col, 61; from asst prof to assoc prof, 62-70, PROF BIOL, GRINNELL COL, 70- *Mem:* Am Inst Biol Sci; Bot Soc Am; Am Soc Plant Taxonomists; Int Asn Plant Taxonomists. *Res:* Taxonomy of Acanthaceae. *Mailing Add:* Dept Biol Grinnell Col Grinnell IA 50112-0806

DURKEE, LENORE T, b Utica, NY, Nov 26, 32; m 56; c 2. ELECTRON MICROSCOPY. *Educ:* Syracuse Univ, AB, 54, MS, 58; Univ Iowa, PhD(bot), 77. *Prof Exp:* asst prof, 77-84, ASSOC PROF BIOL, GRINNELL, 84- *Mem:* Bot Soc Am. *Res:* Ultrastructure, physiology, and adaptiveness of floral and extra-floral nectaries in Passiflora. *Mailing Add:* Dept Biol Grinnell Col Grinnell IA 50112-0806

DURKIN, DOMINIC J, b St Johnsbury, Vt, Dec 24, 30; m 58; c 4. HORTICULTURE, PLANT PHYSIOLOGY. *Educ:* Univ NH, BS, 52; Ohio State Univ, MS, 58, PhD(hort), 60. *Prof Exp:* From asst prof to assoc prof hort, Purdue Univ, 60-69; chmn dept hort & forestry, 71-77, PROF FLORICULT, RUTGERS UNIV, 69- *Mem:* Am Soc Hort Sci; Sigma Xi. *Res:* Bud dormancy in the rose; florist crop physiology; post harvest physiology; cut flowers. *Mailing Add:* Dept Hort/Forestry Rutgers Univ New Brunswick NJ 08903

DURKIN, HELEN GERMAINE, IMMUNOLOGY. *Educ:* New York Univ, PhD(biol-immunol), 67. *Prof Exp:* ASSOC PROF PATH & IMMUNOL, DEPT PATH, STATE UNIV NY HEALTH SCI CTR, BROOKLYN, 77-, DIR CLIN IMMUNOL LAB, 85- *Res:* Lymphocyte ontogeny. *Mailing Add:* Dept Path State Univ NY Health Sci Ctr Brooklyn NY 11203

DURKIN, PATRICK RALPH, b Scranton, Pa, July 19, 46; m 91; c 1. DOSE-RESPONSE ASSESSMENT, SITE SPECIFIC RISK ASSESSMENT. *Educ:* State Univ NY, Fredonia, BA, 68; Fordham Univ, MS, 72; State Univ NY, Syracuse, PhD(environ zool), 79. *Prof Exp:* DIR, CHEM HAZARD ASSESSMENT DIV, SYRACUSE RES CORP, 72- *Concurrent Pos:* Adj asst prof, Col Environ Sci & Forestry, State Univ NY, 80-; consult, Sci Adv Bd, US Environ Protection Agency, 87- *Honors & Awards:* Award for Spec Achievement, US Environ Protection Agency, 80. *Mem:* Am Col Toxicol; Soc Risk Anal. *Res:* Toxicology and metabolism of xenobiotics in both aquatic and terrestrial systems; risk assessment methods, particularly for complex mixtures. *Mailing Add:* Syracuse Res Corp Merrill Lane Syracuse NY 13210

DURKOT, MICHAEL JOHN, METABOLISM, NUTRITION. *Educ:* Pa State Univ, PhD(physiol), 77. *Prof Exp:* RES PHYSIOLOGIST, DEPT DEFENSE, DIV HEAT RES, US ARMY RES INST ENVIRON MED, 82- *Mailing Add:* Dept Defense Div Heat Res Natick Labs US Army Res Inst Environ Med Natick MA 01760

DURKOVIC, RUSSELL GEORGE, b Cheyenne, Wyo, Jan 22, 40. NEUROPHYSIOLOGY. *Educ:* Va Polytech Inst, BS, 62; Case Western Reserve Univ, PhD(physiol), 68. *Prof Exp:* Fel, 68-70, from asst prof to assoc prof 70-84, PROF PHYSIOL, STATE UNIV NY HEALTH SCI CTR, 84- *Concurrent Pos:* Consult, Neurobiol Panel, NSF, 78-81. *Mem:* Soc Neurosci; Int Brain Res Orgn; AAAS; NY Acad Sci. *Res:* Use of a simplified model approach to identify neurophysiological mechanisms of learning, including analysis of sensory, integrative and motor systems of the cat spinal cord. *Mailing Add:* Dept Physiol State Univ NY Health Sci Ctr Syracuse NY 13210

DURLAND, JOHN R(OYDEN), b Chicago, Ill, Mar 7, 14; m 36; c 4. CHEMICAL ENGINEERING. *Educ:* Mich Col Mining & Technol, BS, 35; Univ Wis, PhD(org chem), 39. *Prof Exp:* Res chemist, 39-40, res chemist & leader group, Nitro, 41-45, develop supt, 46; plant mgr, Monsanto Co, 47-52, asst prod mgr, org chem div, 52-53, plant mgr, J F Queeny, 53-55, tech prod mgr, org chem div, 55-60, prod dir, 60-65, mgr int & interdivisional mfg, 65-67, vpres, Mitsubishi Monsanto Chem Co, 67-79; RETIRED. *Mem:* Am Chem Soc; Am Inst Chem Engrs. *Res:* Hydrogenation; process research in production of organic chemicals. *Mailing Add:* 3176 W Cumberland Ct Westlake Village CA 91362-3524

DURLEY, RICHARD CHARLES, b Hertford, Eng, Mar 11, 43; m 73. PLANT BIOCHEMISTRY. *Educ:* Bristol Univ, BSc, 65, PhD(org chem), 68. *Prof Exp:* Fel plant growth regulators, Dept Biol, Univ Sask, 69-71, res assoc, 71-74, prof res assoc, Dept Crop Sci, 74-82; asst prof, Dept Forest Sci, Ore State Univ, Corvallis, 82-84; sr res specialist, 84-89, FEL, MONSANTO CO, ST LOUIS, MO, 89- *Mem:* Int Plant Growth Substances Asn; fel Royal Soc Chem; Am Soc Plant Physiologists; Am Soc Pharmacog. *Res:* Chemical analysis, isolation and function of natural products; biosynthesis and function of the plant growth regulators, auxins, gibberellins, cytokinins, and abscisins. *Mailing Add:* Monsanto Co 700 Chesterfield Village Pkwy St Louis MO 63198

DURLING, ALLEN E(DGAR), b Summitt, NJ, Dec 21, 34; m 60; c 3. COMMUNICATIONS. *Educ:* Lafayette Col, BS, 60; Syracuse Univ, MEE, 62, PhD(elec eng), 64; Univ Fla, MFA, 80. *Prof Exp:* Elec engr, prod develop lab, Int Bus Mach Corp, 60-61; asst elec eng, Syracuse Univ, 61-62, instr, 62-64; from asst prof to prof, Univ Fla, 64-77; sr assoc, 77-80, VPRES, BOOZ, ALLEN & HAMILTON, MGT CONSULTS, 80- *Mem:* Inst Elec & Electronics Engrs. *Res:* Communications systems; digital signal processing; analog/digital and hardware/software interactions. *Mailing Add:* Booz Allen & Hamilton 4330 East West Hwy Bethesda MD 20014

DURLING, FREDERICK CHARLES, mathematical statistics, mathematics, for more information see previous edition

DURNEY, CARL H(ODSON), b Blackfoot, Idaho, Apr 22, 31; m 53; c 6. ELECTRICAL ENGINEERING. *Educ:* Utah State Univ, BS, 58; Univ Utah, MS, 61, PhD(elec eng), 64. *Prof Exp:* Assoc res engr control systs, Boeing Airplane Co, Wash, 58-59; asst res prof elec eng, 63-68, from assoc prof to res prof bioeng, 68-85, chmn, Dept Elec Eng, 77-82, PROF ELEC ENG, UNIV UTAH, 75-, PROF BIOENG, 85- *Concurrent Pos:* On leave, mem tech staff, Crawford Hill Lab, Bell Tel Labs, NJ, 65-66; comt mem, Am Nat Standards Inst, C95 subcomt III on radiation safety levels and/or tolerances with respect to personnel, 73-88; vis prof, Mass Inst Technol, 83-84; mem, Comt Man & Radiation, Inst Elec & Electronics Engrs, 79-88, chmn, 83-84; mem, Nat Coun Radiat Protection & Measurements, 90- *Mem:* Inst Elec & Electronics Engrs; Am Soc Eng Educ; Int Union Radio Sci; Bioelectromagnetics Soc (vpres, 80-81, pres, 81-82). *Res:* Electromagnetic field theory; microwave theory and devices; engineering pedagogy; interaction of electromagnetic fields and living systems. *Mailing Add:* Dept Elec Eng Univ Utah Salt Lake City UT 84112

DURNFORD, ROBERT F(RED), b Carlton, Mont, June 29, 22; m 48; c 2. ELECTRICAL ENGINEERING. *Educ:* Mont State Col, BS, 44, MS, 49; Ohio State Univ, PhD, 65. *Prof Exp:* From assoc prof to prof eng, Mont State Univ, 47-66; RETIRED. *Mem:* Am Soc Eng Educ; Inst Elec & Electronics Engrs. *Res:* Industrial electronics; control; energy conversion. *Mailing Add:* 316 N Tenth St Bozeman MT 59715

DURNICK, THOMAS JACKSON, b Ft Leavenworth, Kans, Mar 1, 46; m 75; c 2. SPECTROSCOPY, TRAINING. *Educ:* Rensselaer Polytech Inst, BS, 67, PhD(phys chem), 71. *Prof Exp:* Res assoc, Res Found, State Univ NY Binghamton, 71-72; fel, Univ Alta, 72-74; sr res chemist, Eastern Res Ctr, Stauffer Chem Co, 74-80; mgr anal chem, Am Sterilizer Co, 80-82; group leader, Arco Chem Co, 82-84; TRAINING SUPVR, KEVEX CORP, FOSTER CITY, CALIF, 84- *Concurrent Pos:* Consult. *Mem:* Am Chem Soc; Soc Appl Spectros. *Res:* Application of infrared, Raman and ultraviolet spectroscopy to problems of structure elucidation and identification; sample handling techniques for compounds of industrial interest; general industrial analytical chemistry; research and training users in x-ray fluorescence techniques and instrumentation. *Mailing Add:* 3M Co Bldg 14 Hwy 225 Hutchinson MN 55350

DURR, ALBERT MATTHEW, JR, b Nebraska City, Nebr, May 22, 23; m 45; c 2. ORGANIC CHEMISTRY. *Educ:* Okla State Univ, BS, 50, MS, 51. *Prof Exp:* Res chemist, Chem Dept, Beacon Res Labs, Tex Co, 51-55; lubricants group, 55-56, sr res chemist, 56-58, from actg res group leader to res group leader, 58-73, sr res scientist, 73-80, res assoc, 80-85, SR RES ASSOC, RES & DEVELOP DEPT, CONOCO INC, 85- *Mem:* Am Chem Soc; AAAS; Sigma Xi; Am Soc Lubrication Eng; Int Soc Gen Semantics. *Res:* Fundamental organic chemistry; petroleum lubricants. *Mailing Add:* 111 Glenside Ave Ponca City OK 74601

DURR, DAVID P, pediatric dentistry, for more information see previous edition

DURR, FRIEDRICH (E), b Poughkeepsie, NY, July 28, 33; m 57; c 4. INFECTIOUS DISEASES, THERAPEUTICS. *Educ:* St Johns Univ, BS, 55; Univ Wis, MS, 58, PhD(med microbiol), 60. *Prof Exp:* Instr microbiol, Col Med & Dent, Seton Hall Univ, 60-63; res virologist, J L Smith Mem Cancer Res, Pfizer, Inc, 63-71; group leader, 71-76, DEPT HEAD CHEMOTHER RES, LEDERLE LABS, 76- *Mem:* Am Soc Microbiol; NY Acad Sci; AAAS; Am Asn Cancer Res; Sigma Xi. *Res:* Effect of viruses on transplantable mouse tumors; bioassay of murine leukemia viruses; studies on the infectivity of EBV for human lymphoblastoid cells; anticancer chemotherapy, including studies with immunomodulating agents; viral chemotherapy, particularly effect of drugs on influenza virus and herpesvirus infections in vivo and in vitro; effect of drugs on viral antigen synthesis as determined by immunofluorescence; cancer chemotherapy; immunopharmacology. *Mailing Add:* Lederle Labs Bldg 56A Rm 201 Pearl River NY 10965

DURRANI, SAJJAD H(AIDAR), b Jalalpur, Pakistan, Aug 27, 28; US citizen; m 59; c 3. ELECTRICAL ENGINEERING. *Educ:* Govt Col, Lahore, Pakistan, BA, 46; Eng Col, Lahore, BScEng, 49; Univ Manchester, MScTech, 53; Univ NMex, Albuquerque, ScD(elec eng), 62. *Prof Exp:* Lectr elec eng, Eng Col, Lahore, 49-56, asst prof, 56-59; instr & res assoc, Univ NMex, 59-62; sr engr, Commun Prod Dept, Gen Elec Co, Va, 62-64; prof elec eng & chmn dept & dir res, Eng Univ, Lahore, 64-65; assoc prof elec eng, Kans State Univ, 65-66; sr engr, Space Ctr, RCA Corp, 66-68; mem tech staff, Commun Satellite Corp, 68-69, br mgr, Systs Anal Lab, 69-71, staff scientist, Adv Studies Lab, 71-73; sr scientist, Opers Res Inc, Md, 73-74; sr engr, Goddard Space Flight Ctr, 74-79, chief commun scientist, 79-81, syst planning mgr tracking & data relay, Satellite Syst, 81-83, res & planning mgr, Commun Div, Goddard Space Flight Ctr, 84-88, ADVAN SYST PROG MGR, OFF SPACE OPER, NASA HQ, 88- *Concurrent Pos:* Pres, Teaching Staff Asn, Eng Univ, Pakistan, 64-65; mem, US Comns C & F, Int Sci Radio Union; vis prof, Univ Maryland, 72; adj prof & lectr, George Washington Univ, 79; mem, Educ Activ Bd, Inst Elec & Electronic Engrs, 73-74, bd dirs, 84-85, Tech Activ Bd, 84-85, Publ Bd, 86-87 & 90-92. *Honors & Awards:* US Activ Citation Honor, Inst Elec & Electronics Engrs, 80, Centennial Award, 84. *Mem:* Fel Inst Elec & Electronics Engrs; assoc fel Am Inst Aeronaut & Astronaut; Aerospace & Electronics Systs Soc (pres, 82-83). *Res:* Space communications systems; antennas and propagation; adaptive multibeam phased array experiment for Spacelab. *Mailing Add:* 17513 Lafayette Dr Olney MD 20832

DURRANT, BARBARA SUSAN, b Lansing, Mich, Aug 22, 49. ANIMAL PHYSIOLOGY. *Educ:* NC State Univ, BS, 72, MS, 75, PhD(physiol), 79. *Prof Exp:* Fel, 79-81, STAFF SCIENTIST REPRODUCTIVE PHYSIOL, ZOOL SOC SAN DIEGO, 81- *Concurrent Pos:* Lectr, Univ Calif San Diego Exten Course, 81 & Zool Soc San Diego Zoo Inst Course, 82; lectr, Zoo Work Explor Student Training Prog, 79-; consult, Gifted & Talented Educ Prog, San Diego City Sch, 81; adj prof biol, SDSU, 88- *Mem:* Int Embryo Transfer Soc; Sigma Xi; Am Asn Zool Parks & Aquariums; Am Soc Andrology. *Res:* Estrus synchronization of various hoofed-stock species; artificial insemination; semen collection and freezing; interspecies embryo transfer; embryo freezing and in vitro fertilization. *Mailing Add:* 13448 Standish Dr Poway CA 92064-5140

DURRELL, WILLIAM S, b Miami, Fla, Oct 14, 31; m 53; c 5. ORGANIC CHEMISTRY, POLYMER CHEMISTRY. *Educ:* Univ Fla, BS, 53, PhD(org chem), 61. *Prof Exp:* Chemist, Peninsular Chem Res, Fla, 53-55, coordr res & develop, 61-64; res chemist, Ethyl Corp, La, 55-56; res assoc, Burke Res Co, 64-65; group leader, Geigy Chem Corp, Ala, 65-68, asst develop mgr, 68-69, develop mgr, RI, 69-70, dir res & vpres, Plastics & Additives Div, Ciba-Geigy Corp, Ardsley, 70-77; vpres technol, E F Houghton & Co, 77-79; vpres res & develop, J M Walter Corp, 79-82; dir res & develop, Agr & Performance Chem Div, PPG Indust Inc, 82-85; DIR POLYMER & FIBER SCI, ALLIED FIBERS, PETERSBURG, VA, 85- *Concurrent Pos:* Chmn, Div Chem Mkt & Econ, Am Chem Soc, 90. *Mem:* Am Chem Soc; Soc Plastic Engrs. *Res:* Fibers, polymers, lubricants and additives. *Mailing Add:* Allied Fibers PO Box 31 Petersburg VA 23804

DURRILL, PRESTON LEE, b Ft Madison, Iowa, Apr 4, 36; m 66. CHEMISTRY. *Educ:* Mass Inst Technol, SB, 57, SM, 59; Va Polytech Inst, PhD(chem eng), 66. *Prof Exp:* Chem engr, Esso Res Labs, La, 59-60; assoc prof, 65-70, PROF CHEM, RADFORD COL, 70- *Mem:* AAAS; Am Chem Soc; Am Inst Chem Eng; Soc Plastics Engr; Sigma Xi. *Res:* Diffusion of gases in solids and molten polymers. *Mailing Add:* 1309 Madison St Radford VA 24141

DURRUM, EMMETT LEIGH, b Spokane, Wash, May 4, 16; m 41; c 3. BIOCHEMISTRY. *Educ:* Harvard Univ, BS, 39; Stanford Univ, MD, 46. *Prof Exp:* Engr, Shell Develop Co, San Francisco, 39-42; med officer, Field Res Lab, Med Dept, US Dept Army, 46-51, chief biochem sect, Cardiorespiratory Dis Dept, Army Med Serv Grad Sch, 51-52, chief biochem sect, Dept Pharmacol, 52-54; assoc dir res, Spinco Div, Beckman Instruments, Inc, 55-62; chmn bd & res dir, Durrum Instrument Corp, 62-72, PRES, ELDEX LABS INC, 72- *Concurrent Pos:* Assoc clin prof, Sch Med, Stanford Univ, 54-60. *Mem:* AAAS; Am Chem Soc; Am Soc Biol Chemists; NY Acad Sci. *Res:* Electrophoresis, separations by physical methods for proteins; amino acid; atherosclerosis; computerized amino acid analyzers; automated laboratory equipment; synthetic peptides. *Mailing Add:* 170 Buckthorn Way Menlo Park CA 94025

DURSCH, FRIEDRICH, b Dresden, Ger, June 10, 30; m 51. ORGANIC CHEMISTRY. *Educ:* Dresden Tech Univ, BS, 52; Darmstadt Tech Univ, MS, 55, PhD(org chem), 56. *Prof Exp:* Res assoc natural prod, Univ Va, 57-60; sr chemist pharmaceut, Wallace Labs Div, Carter Prod NJ, 60-62, asst dir, 62-63; SR SCIENTIST, SQUIBB INST MED RES, 64-, SR RES FEL, E R SQUIBB & SONS, 81- *Mem:* Am Chem Soc; AAAS. *Res:* Synthetic organic chemistry; derivatives of hydroxylamine; alkaloids; pharmaceuticals; antibiotics; process development. *Mailing Add:* 36 Stoneybrook Rd Hopewell NJ 08525-0328

DURSO, DONALD FRANCIS, b Youngstown, Ohio, Jan 30, 25; m 48; c 5. CELLULOSE CHEMISTRY. *Educ:* Case Inst Technol, BS, 47; Purdue Univ, MS, 49, PhD(biochem), 51. *Prof Exp:* Asst, Purdue Univ, 47-51; res chemist, Buckeye Cellulose Corp, Tenn, 51-53, org group leader, 53-56, mgr res dept, 56-71; prof forest sci, Tex A&M Univ, 71-75; dir absorbent technol, 75-85, DIR DISCOVERY RES, JOHNSON & JOHNSON, CHICOPEE, 85- *Mem:* Fel Am Chem Soc; Tech Asn Pulp & Paper Indust; Forest Prod Res Soc. *Res:* Enzyme degradation and synthesis of cellulose; carbon column chromatography of sugars; structure of polysaccharides; cellulose composition; structure and preparation of cellulose derivatives; chemistry of pulping and bleaching; physics and chemistry of absorption of body fluids. *Mailing Add:* Seven Merritt Lane Rocky Hill NJ 08553

DURSO, JOHN WILLIAM, b Brooklyn, NY, Feb 1, 38; m 59; c 4. THEORETICAL PHYSICS. *Educ:* Cornell Univ, AB, 59; Pa State Univ, PhD(theoret physics), 64. *Prof Exp:* Res asst physics, Pa State Univ, 60-64; res assoc, Inst Theoret Physics, Naples, 64-65; res assoc theoret physics, Mich State Univ, 65-67; from asst prof to assoc prof physics & comput studies, 67-78, PROF PHYSICS & COMPUT STUDIES, MT HOLYOKE COL, 78- *Concurrent Pos:* Vis prof, Nordisk Inst Theoret Atomic Physics, 74-75 & NY State Univ, Stony Brook, 81-82. *Mem:* Sigma Xi; Am Phys Soc. *Res:* Nuclear theory; many-body problems; scattering theory. *Mailing Add:* Dept Physics Mt Holyoke Col South Hadley MA 01075

DURST, HAROLD EVERETT, b Morrowville, Kans, Feb 18, 24; m 49; c 1. ENVIRONMENTAL BIOLOGY. *Educ:* Kans State Univ, BS, 48; Univ Colo, MEd, 53; Ore State Univ, PhD(sci educ), 67. *Prof Exp:* Teacher, Ness City Pub Schs, Kans, 48-54, prin, 53-54; off mgr, Firestone Tire & Rubber Co, 54-57; buyer, Boeing Airplane Co, 57-58; teacher, Wichita Pub Schs, Kans, 58-61 & 62-63; consult writer, Biol-Sci Curriculum Study, 61-62; from instr to assoc prof, 63-75, PROF BIOL, EMPORIA STATE UNIV, 75-, DEAN, GRAD SCH, 74- *Concurrent Pos:* Area consult, Biol Sci Curriculum Study, 62-65; dir, Sci Workshop, Peace Corps, India, 68 & NSF In-Serv Insts for Sec Teachers of Sci & Math, 68-73. *Mem:* Fel AAAS; Am Inst Biol Sci. *Res:* Curriculum evaluation; college biology; affective behavior and assessment of postgraduate needs of high school biology teachers. *Mailing Add:* 1114 Congress St Emporia KS 66801

DURST, JACK ROWLAND, b Stow, Ohio, June 22, 26; m 48, 67; c 5. BIOCHEMISTRY, FOOD SCIENCE. *Educ:* Ohio Univ, BS, 48; Ohio State Univ, MS, 53; Purdue Univ, PhD(biochem), 56. *Prof Exp:* Chemist, Goodyear Tire & Rubber Co; res asst, Kettering Res Found, Ohio State Univ, 49-53, res assoc, 53; res chemist, Swift & Co, 56-57; sr res chemist, 57-60, sr scientist, 60-66, tech mgr, 66-68, res assoc appl res, 68-74, sr res assoc, 74-79, RES FEL, PILLSBURY CO, 79- *Mem:* Am Chem Soc; Inst Food Technol; Nutrit Today Soc; Am Asn Cereal Chemists; Sigma Xi. *Res:* Natural products research; invention of new food forms through control of structure and nutritional makeup; formulation and fabrication of foods for astronauts; source, function and nutrition of proteins; dietary studies; ions in foods; water activity in foods. *Mailing Add:* 12152 Mississipps Dr Champlin MN 55316

DURST, LINCOLN KEARNEY, b Santa Monica, Calif, Aug 5, 24; m 56; c 3. LOGIC, COMPUTATION. *Educ:* Univ Calif, Los Angeles, BA, 45; Calif Inst Technol, BS, 46, PhD(math), 52. *Prof Exp:* Instr math, Rice Univ, 51-55, from asst prof to assoc prof, 55-67; prof, Claremont Col & Claremont Grad Sch, 67-70; dep exec dir, Am Math Soc, 70-84; CONSULT, 85- *Concurrent Pos:* Exec dir, Comt on Undergrad Prog in Math, Math Asn Am, 66-67; mem bd dirs, Nat Fedn Sci Abstracting & Indexing Serv, 71-74; vis res assoc, Brown Univ, 85-87; ed, Tex Users Group, 90. *Mem:* Am Math Soc; Math Asn Am; Can Math Soc; Soc Indust & Appl Math; Soc Scholarly Publ. *Res:* Number theory, logic, computation. *Mailing Add:* 46 Walnut Rd Barrington RI 02806-2026

DURST, RICHARD ALLEN, b New Rochelle, NY, Dec 27, 37; m 64; c 3. ELECTROCHEMISTRY, BIOSENSORS. *Educ:* Univ RI, BS, 60; Mass Inst Technol, PhD(anal chem), 63. *Prof Exp:* Nat Res Coun res assoc, Nat Bur Stand, 63-64; vis asst prof, Pomona Col, 64-65; asst prof, Boston Col, 65-66; res chemist, Anal Chem Div, Nat Bur Stand, 66-70, chief, Electrochem Anal Sect, 70-73; group leader clin chem, Radiometer A/S, Copenhagen, Denmark, 73-74; asst chief, Air & Water Pollution Anal Sect, Ctr Anal Chem, Nat Bur Standards, 74-78, sci asst to dir, 78-79, group leader, Org Electrochem, 79-85, Anal Sensors, 86-88, dep dir, 85-86, dep dir, 88-90; PROF CHEM, DEPT FOOD SCI & TECHNOL, CORNELL UNIV-NY STATE AGR EXP STA, 90-, DIR, ANALYSIS CHEM LABS, 90- *Concurrent Pos:* Nat lectr, Sigma Xi; mem, comn electrochem, Int Union Pure & Appl Chem, 71-79 & comn electroanal chem, 80-; tour lectr, Am Chem Soc, 72; lectr, Swiss Romande Univ, 74; mem, expert panel, Int Fed Clin Chem, 75-82. *Honors & Awards:* Silver Medal, US Dept Com, 87. *Mem:* AAAS; Am Chem Soc; Am Asn Clin Chemists; Sigma Xi; Asn Off Analysis Chemists; Soc Electroanal Chem. *Res:* Electroanalytical chemistry; flow injection immunoanalysis; agricultural and food analyses; bioanalytical sensors; immunoassary. *Mailing Add:* Analysis Chem Labs Cornell Univ-NY State Agr Exp Sta Geneva NY 14456-0462

DURST, TONY, b San Martin, Romania, Jan 21, 38; Can citizen; m 64; c 2. ORGANIC CHEMISTRY. *Educ:* Univ Western Ont, BSc, 61, PhD(chem), 64. *Prof Exp:* From asst prof to assoc prof, 67-77, PROF CHEM, UNIV OTTAWA, 77- *Mem:* Am Chem Soc; Chem Inst Can. *Mailing Add:* Dept Chem Univ Ottawa Ottawa ON K1N 6N5 Can

DURUM, SCOTT KENNETH, b St Paul, Minn, May 8, 47. CYTOKINES, GROWTH FACTORS. *Educ:* Wake Forest Univ, BA, 69, MA, 73; Oak Ridge Assoc Univs, PhD(zool/cellular immunol), 76. *Prof Exp:* Postdoctoral fel immunochem, Nat Jewish Hosp & Res Ctr, Denver, Colo, 76-78, NIH individual postdoctoral award, 78-79; NIH individual postdoctoral award, Dept Path, Sch Med, Yale Univ, 79-80, res assoc, Howard Hughes Med Inst, Yale Univ, 80-84; sr staff fel, 84-90, SR INVESTR, LAB MOLECULAR IMMUNOREGULATION, BIOL RESPONSE MODIFIERS PROG, NAT CANCER INST, MD, 90- *Concurrent Pos:* Assoc ed, J Immunol, 87-; chairperson, Int Cong on Immune Consequences of Trauma, Shock & Sepsis, WGer, 88, 19th Int Leucocyte Cult Conf, Can, 88, Brit Soc Immunol, 88 & Int Cong Immunol, WGer, 89; chief ed, Cytokine, 88-; lectr, NIH, 88 & 89; Maxam award biomed res, 90. *Mem:* Am Asn Immunologists. *Res:* Immunology; role of cytokines in T lymphocyte development and activation: What are the cytokines involved in these processes, how are they produced, and how do they act?. *Mailing Add:* Nat Cancer Inst Bldg 560 Rm 31-45 Frederick MD 21702-1201

DURY, GEORGE H, b Hellidon, Eng, Sept 11, 16. GEOMORPHOLOGY. *Educ:* Univ London, BA, 37, MA, 44, PhD(geomorphol), 51, DSc, 71. *Prof Exp:* Lectr in chg geog & geol, Enfield Tech Col, Middlesex, Eng, 46-48; lectr geog, Birkbeck Col, Univ London, 49-62; McCaughey prof, Univ Sydney, 62-69, dean fac sci, 67-68; prof geog & geol, Univ Wis-Madision, 69-79, chmn, Dept Geog, 71-74. *Concurrent Pos:* Div staff scientist, Water Resources Div, US Geol Surv, 60-61; vis prof dept geol, Fla State Univ, 67; Fenneman professorship, Univ Wis-Madison & mem staff, Sidney Sussex Col, Univ Cambridge, Eng, 78. *Mem:* Royal Geog Soc; Geol Soc London; Am Geog Soc; Inst Brit Geographers; Sigma Xi. *Res:* General theory of meandering valleys; deep weathering and duricrusting; glacial diversions of surface drainage; pedimentation; paleoclimatology. *Mailing Add:* 46 Woodland Close Risby Bury St Edmunds Suffolk IP28 6QN England

DURYEA, WILLIAM R, b Port Jervis, NY, July 29, 38; m 61; c 5. CLINICAL MICROBIOLOGY. *Educ:* St Bernardine Siena Col, BS, 62; St Bonaventure Univ, PhD(biol), 67; St Francis Col, BS, 82. *Prof Exp:* Prof biol, St Francis Col, Pa, 66-88, assoc dir, Physician's Asst Prog, 83-88; CONSULT, 88- *Concurrent Pos:* Clin pract, Va Med Ctr, Altoona, Pa. *Mem:* Am Asn Physician's Assts; Pa Soc Phys Asst; Asn Phys Asst Progs. *Res:* Innovative teaching methods for the medical sciences, such as computer-assisted instruction; adult health maintenance, prev med; geriatrics. *Mailing Add:* Pa Prog Dept Biol St Francis Col Loretto PA 15940

DURZAN, DONALD JOHN, b Hamilton, Ont, Aug 4, 36; m 59; c 2. PLANT PHYSIOLOGY, BIOCHEMISTRY. *Educ:* McMaster Univ, BSc, 59; Cornell Univ, PhD(plant physiol), 64. *Prof Exp:* Res officer physiol, Can Dept Forestry, 59-68, res scientist & head biochem sect, Forest Ecol Res Inst, Dept Environ, 68-75, sr adv policy planning & assessment, Environ Mgt Serv, Environ Can, 75-77; prof biochem, sr res assoc & head biochem unit, Inst Paper Chem, 77-81; CHMN, DEPT POMOL, UNIV CALIF, DAVIS, 81- *Concurrent Pos:* Res asst, Cornell Univ, 60-63; Can Soc Plant Physiologists rep, Biol Coun Can, 75-78. *Mem:* AAAS; Am Soc Plant Physiol; Can Soc Plant Physiol (exec secy, 75-77); Am Soc Hort Sci; Soc Am Foresters. *Res:* Metabolism of nitrogenous compounds in relation to growth and development of fruit and forest trees. *Mailing Add:* 1013 Burr St Davis CA 95616

DUSANIC, DONALD G, b Chicago, Ill, Dec 15, 34; m 71; c 5. PARASITOLOGY. *Educ:* Univ Chicago, SB, 57, SM, 59, PhD(microbiol), 63. *Prof Exp:* Instr microbiol, Univ Chicago, 63-64; from asst prof to prof, Univ Kans, 64-72; PROF LIFE SCI, IND STATE UNIV, TERRE HAUTE, 72-, DIR INTERDISCIPLINARY CTR, CELL PRODS & TECHNOL, 86- *Concurrent Pos:* Vis asst prof parasitol, Univ Philippines, 64 & Nat Taiwan Univ, 71, adj prof microbiol, Sch Med, Ind State Univ, 72-; consult med ecol, Naval Med Res Unit 2, 76; guest, Univ Catolica de Pelotas, Brazil, 80; distinguished prof, Col Arts & Sci, Ind State Univ, 90. *Mem:* Am Soc Parasitol; Soc Protozool; Am Soc Trop Med & Hyg; NY Acad Sci; Sigma Xi; Am Asn Immunologists. *Res:* Immunology and physiology of animal parasites, trypanosomes. *Mailing Add:* Dept Life Sci Ind State Univ Terre Haute IN 47809

DUSCHA, LLOYD A, RESEARCH ADMINISTRATION. *Educ:* Univ Minn, BCE, 45. *Prof Exp:* Street & civil engr, US Army Engr Dist, Riverdale, ND; chief engr, dep dir eng & const, Off Chief Engrs, Washington, DC, 83-90; CONSULT ENG, 90- *Concurrent Pos:* Nat Res Coun; comt tunnel contracting practices, super collider, Nat Res Coun. *Honors & Awards:* Wheeler Medal, Soc Am Mil Engrs; Presidents Medal, Am Soc Civil Engrs, 89. *Mem:* Nat Acad Eng; fel Am Soc Civil Engrs; Nat Soc Prof Engrs; Soc Am Mil Engrs. *Res:* Author of over 15 articles. *Mailing Add:* 11802 Grey Birch Pl Reston VA 22091

DUSCHINSKY, ROBERT, medicinal chemistry; deceased, see previous edition for last biography

DUSEL-BACON, CYNTHIA, b San Jose, Calif, Aug 16, 46; m 77; c 1. GEOLOGY. *Educ:* Univ Calif, Santa Barbara, BA, 68; San Jose State Univ, BA, 75. *Prof Exp:* Teacher Span, Healdsburg High Sch Dist, 70-74; phys sci technician geol, 75-80, GEOLOGIST, US GEOL SURV, 80- *Mem:* AAAS; Mineral Soc Am; Am Geophys Union; Geol Soc Am. *Res:* Geology of east central Alaska; metamorphic petrology and petrography; metamorphic textures; petrology of ortho-augen gneiss and a sillimanite gneiss dome in east central Alaska; metamorphic facies map of Alaska; geochemistry of meta-igneous rocks. *Mailing Add:* Br Alaskan Geol US Geol Surv 345 Middlefield Rd Menlo Park CA 94025

DUSENBERRY, WILLIAM EARL, b Oxford, Nebr, June 11, 43; m 65; c 4. BIOMETRICS, BIOSTATISTICS. *Educ:* Univ Wyo, BA, 64, MS, 66; Va Polytech Inst & State Univ, MS, 71, PhD(statist), 73. *Prof Exp:* Statistician air pollution res, USPHS, Cincinnati, 66-70; sr statistician, Eli Lilly & Co, 73-77; statistician, Div Biostatist, Univ Utah Med Ctr, 77; SUPVRY STATISTICIAN & CHIEF RES SUPPORT BR, DENVER WILDLIFE RES CTR, USDA, APHIS, ADC, 77- *Concurrent Pos:* Instr statist, Butler Univ, 73-77. *Mem:* Am Statist Asn; Biomet Soc. *Res:* Sampling moments of moments; moment and ratio estimators; percentage points; statistical aspects of wildlife research. *Mailing Add:* Denver Fed Ctr Bldg 16 Bldg 16 Denver Fed Ctr Denver CO 80225-0266

DUSENBERY, DAVID BROCK, b Portland, Ore, Apr 30, 42; m 65. NEMATODE BEHAVIOR. *Educ:* Reed Col, BA, 64; Univ Chicago, PhD(biophys), 70. *Prof Exp:* Asst prof, 73-78, ASSOC PROF BIOL & PHYSICS, GA INST TECHNOL, 78- *Concurrent Pos:* USPHS fel, Calif Inst Technol, 70-73. *Mem:* Biophys Soc; Soc Nematologists; Soc Chemireception Soc. *Res:* Studies of behavior, especially sensory responses, nematodes and plant-parasitic nematodes; Caenorhabditis elegans using techniques such as laser microbeam, and computer tracking. *Mailing Add:* Dept Biol Ga Inst Technol Atlanta GA 30332

DUSENBERY, RUTH LILLIAN, b Chicago, Ill, May 30, 44; m 82. MOLECULAR BIOLOGY. *Educ:* Univ Chicago, BS, 66, PhD(chem), 70. *Prof Exp:* From res assoc to sr res assoc biol, Emory Univ, 75-84; ASST PROF CHEM & BIOL, SOUTHERN METHODIST UNIV, 84- *Mem:* Am Chem Soc; Genetics Soc Am; Environ Mutagen Soc. *Res:* Investigation of DNA repair pathways in the eukaryotic organism Drosophila melanogaster, using a number of repair deficient strains; correlation of unrepaired or misrepaired damage to mutagenesis and carcinogenesis. *Mailing Add:* Dept Biol Southern Methodist Univ 220 Fondren Sci Bldg Dallas TX 75275

DUSENBURY, JOSEPH HOOKER, b Troy, NY, Nov 18, 23; m 47. PHYSICAL CHEMISTRY. *Educ:* Union Col, BS, 47; Univ Calif, Berkeley, PhD(chem), 50. *Prof Exp:* Res chemist, Am Cyanamid Co, 47; asst chem, Univ Calif, 47-50; res chemist, Am Cyanamid Co, 50-53; from head phys org chem sect to assoc res dir, Textile Res Inst, NJ, 53-61; sect leader, 61-64, DEPT MGR, CHEM DEPT, MILLIKEN RES CORP, 64- *Concurrent Pos:* Mem, Textile Res Inst. *Honors & Awards:* Harold Dewitt Smith Mem Award, Am Soc Testing Mat, 80. *Mem:* Fel AAAS; Am Chem Soc; Fiber Soc; Soc Rheol; fel Am Inst Chemists. *Res:* Reactions of nitrous acid; dyeing and finishing of textiles; polymerization; physical properties and chemical modification of fibers. *Mailing Add:* 413 Overland Dr Spartanburg SC 29302

DU SHANE, JAMES WILLIAM, b Madison, Ind, Apr 17, 12; m 39; c 2. PEDIATRICS, CARDIOLOGY. *Educ:* DePauw Univ, AB, 33; Yale Univ, MD, 37; Am Bd Pediat, dipl. *Prof Exp:* From instr to prof, 47-79, EMER PROF PEDIAT, MAYO GRAD SCH MED, UNIV MINN, 79- *Concurrent Pos:* Head pediat sect, Mayo Clin, 57-69, mem bd gov, 60-72, head pediat cardiol sect, 69-73; chmn coun rheumatic fever & congenital heart dis, Am Heart Asn, 59-61; chmn sub-bd cardiol, Am Bd Pediat, 60-65; mem bd trustees, Mayo Found, 65-73. *Mem:* Am Pediat Soc; Am Heart Asn; Am Col Chest Physicians. *Res:* Congenital heart disease; pathology; symptomatology; electrocardiology. *Mailing Add:* Mayo Clin N Ten Plummer Bldg Rochester MN 55905

DUSI, JULIAN LUIGI, b Columbus, Ohio, Nov 10, 20; m 47. VERTEBRATE ZOOLOGY, ENVIRONMENTAL BIOLOGY. *Educ:* Ohio State Univ, BS, 43, MS, 46, PhD(zool), 49. *Prof Exp:* Asst, Ohio State Univ, 46-49; PROF ZOOL-ENTOM, AUBURN UNIV, 49- *Concurrent Pos:* Environ consult, Ala Power Co, 72-, Southern Eng Ga, 74, Bechtel Corp, 74-, Army Corps Engrs, 75, TVA, 77- & Burns & McDonnell, 80. *Mem:* AAAS; Am Soc Mammal; Wilson Ornith Soc; Am Ornith Union; Am Inst Biol Sci; Colonial Waterbird Soc. *Res:* Bird and mammal behavior and ecology; wading bird biology; environmental impacts on birds and mammals. *Mailing Add:* Dept Zool & Wildlife Sci Auburn Univ Auburn AL 36849-5414

DUSKY, JOAN AGATHA, b Tacoma, Wash, Aug 13, 51. WEED SCIENCE, PLANT PHYSIOLOGY. *Educ:* Baldwin-Wallace Col, BS, 73; NDak State Univ, MS, 75, PhD(bot), 78. *Prof Exp:* Res assoc plant physiol, Metab & Radiation Res Lab, Sci & Educ Admin, USDA, 78-80; asst prof, 80-86, ASSOC PROF WEED SCI, EVERGLADES RES & EDUC CTR, UNIV FLA, 86- *Mem:* Weed Sci Soc Am; Am Soc Hort Sci; Am Soc Sugar Cane Technol; Sigma Xi. *Res:* Weed control management practices in vegetables crops, sugarcane, rice; weed biology; herbicide selectivity, mode of action and metabolism; allelopathy; plant growth regulation; herbicide metabolism in relation to the environment; allelopathy and biological weed control. *Mailing Add:* Univ Fla PO Box 8003 Belle Glade FL 33430

DU SOUICH, PATRICK, b Apr 2, 44; m 68; c 2. DRUG METABOLISM, PHARMACOKINETICS. *Educ:* Univ Barcelona, Spain, MD, 68, PhD(pharmacol), 76. *Prof Exp:* PROF PHARMACOL & PHARMACOKINETICS, FAC MED, UNIV MONTREAL, 78- *Concurrent Pos:* Head Clin Pharmacol Sect, Hosp Hotel Dieu Montreal. *Honors & Awards:* Young Scientist Award, 85. *Mem:* Am Soc Pharmacol Exp Ther; Am Soc Clin Pharmacol Ther; Am Fed Clin Res; Soc Clin Pharmacol. *Res:* Effects of cardiorespiratory disease on drug kinetics and dynamics. *Mailing Add:* Dept Pharmacol Fac Med Univ Montreal PO Box 6128 Sta A Montreal PQ H3C 3J7 Can

DUSSAULT, JEAN H, b Que, Apr 6, 41. THYROID HORMONES, ENDOCRINOLOGY. *Educ:* Univ Montreal, BA, 60; Laval Univ, MD, 65; Univ Toronto, Inst Med Sci, MSc, 69; CSPQ, 71; FRCP(C), 77. *Prof Exp:* Internship, Hosp Enfant-Jesus, 64-65, chief resident, 65-67; sr reseacher endocrinol, Med Res Coun Can, Univ Toronto, Wellesley Hosp, 67-69 & UCLA Sch Med, Harbor Gen Hosp, 69-71; from asst prof to assoc prof, 71-71, PROF MED, LAVAL UNIV SCH MED, 81- & DIR RES UNIT, ONTOGENESIS & MOLECULAR GENETICS, CENT HOSP, LAVAL UNIV, 86- *Concurrent Pos:* Dir screening prog congenital hypothyroidism, Quebec Network Congenital Med, 74-; mem, Newborn Study Comt, Am Thyroid Asn, 73-78, 78-82 & 83-86, Grant Award Comt, Med Res Coun Can, Endocrinol-Nephrol Sect, 79-80; secy, Club Res Clin, Quebec, 75-78; mem coun, Can Soc Clin Invest, 77-81. *Honors & Awards:* Ross Award, Am Acad Pediat, 76; Van Meter-Armour Award, Am Thyroid Asn, 80; Sandoz lectr, Can Soc Endocrinol & Metab & Can Diabetes Asn, 87; Manning Award, 88. *Mem:* Am Thyroid Asn; Can Med Asn; Am Fedn Clin Res; Endocrine Soc; Can Soc Clin Res; Can Soc Endocrinol & Metab; Soc Pediat Res; NY Acad Sci. *Res:* Studies on all facets of thyroid-related research. *Mailing Add:* Ctr Hosp Univ Laval 2705 Boul Laurier Quebec PQ G1V 4G2 Can

DUSSEAU, JERRY WILLIAM, b Toledo, Ohio, July 30, 41; m 66. CARDIOVASCULAR PHYSIOLOGY. *Educ:* Earlham Col, BA, 63; La State Univ, Baton Rouge, MS, 66, PhD(vert physiol), 69. *Prof Exp:* Asst prof biol, Earlham Col, 69-70 & Hope Col, 70-76; res assoc, 76-79, ASST PROF RES, BOWMAN GRAY MED SCH, 79- *Mem:* AAAS; Soc Chronobiol; Am Soc Physiol. *Res:* Biological rhythms; hypertension; regulation of microvessel density; endocrine basis for biological clock mechanisms. *Mailing Add:* Dept Physiol & Pharmacol Bowman Gray Sch Med 4961 Lombandy Lane Winston-Salem NC 27103

DUSTAN, HARRIET PEARSON, b Craftsbury Common, Vt, Sept 16, 20. INTERNAL MEDICINE. *Educ:* Univ Vt, BS, 42, MD, 44; Am Bd Internal Med, dipl. *Hon Degrees:* DSc, Med Col Wis, 87. *Prof Exp:* Intern, Mary Fletcher Hosp, Burlington, Vt, 44-45; resident internal med, Royale Victoria Hosp, Montreal, 45; res fel & mem asst staff, Res Div, Cleveland Clin, 55-77, vchmn, 71-77; dir, Cardiovasc Res & Training Ctr, Med Ctr, Univ Ala, 77-87, prof med, 87-90; RETIRED. *Concurrent Pos:* Mem adv coun, Nat Heart & Lung Inst, 72-76; mem residency rev comt internal med, Am Bd Internal Med; distinguished physician, Vet Admin, 87-90. *Honors & Awards:* Sci Achievement Award, Am Med Asn, 89. *Mem:* Inst Med-Nat Acad Sci; Am Col Physicians; Am Heart Asn (pres, 76-77); Am Soc Clin Invests; Asn Am Physicians. *Mailing Add:* 28 Hagan Dr Essex VT 05452

DUSTMAN, JOHN HENRY, b Buffalo, NY, Apr 18, 40; m 63; c 2. ZOOLOGY, ENDOCRINOLOGY. *Educ:* Canisius Col, BS, 61; Ind Univ, PhD(zool), 66. *Prof Exp:* Asst prof zool & asst chmn dept, 66-70, chmn, Dept Biol, 70-80, ASSOC PROF ZOOL, IND UNIV NORTHWEST, 70-, DIR NORTHWEST CTR MED EDUC, IND UNIV SCH MED, 70-, ASSOC PROF PHYSIOL, 74- *Concurrent Pos:* Consult water pollution, Boise Cascade Corp, 70- *Mem:* Biol Photog Asn; Am Soc Zool; Soc Study Reproduction; Nat Asn Biol Teachers; Am Inst Biol Sci. *Res:* Chemical nature and biological activity of synthetic steroids on the reproductive system of fowl. *Mailing Add:* 5570 Marcella Merrillville IN 46403

DUSTO, ARTHUR RONALD, b Libertyville, Ill, Oct 22, 29; m 54; c 3. ENGINEERING MECHANICS. *Educ:* Purdue Univ, BS, 52; Univ Wash, MS, 62, PhD(aeronaut & astronaut), 63. *Prof Exp:* Stress analyst, Northrop Aircraft Inc, 52-53; jr officer, Civil Engrs Corp, US Navy, 53-58; res engr, Boeing Co, 58-60; instr aeronaut & astronaut, Univ Wash, 60-63; res specialist gas dynamics, Boeing Co, 63-64; assoc prof aerospace eng, Univ Ariz, 64-65; res specialist aerodynamics, Boeing Commerical Airplane Co, 65-80, PRIN ENGR, BOEING MILITARY AIRPLANE CO, 80- *Concurrent Pos:* Lectr fluid mech, Dept Aeronaut & Astronaut, Univ Wash, 68 & 70. *Mem:* Am Inst Astronaut & Aeronaut; Sigma Xi. *Res:* Developing analytical models and computational methods for predicting aerodynamics of flexible airplanes. *Mailing Add:* 5030 W Mercer Wy Mercer Island WA 98046

DUSWALT, ALLEN AINSWORTH, JR, b New York, NY, Nov 18, 32; m 54; c 2. ANALYTICAL CHEMISTRY, THERMAL METHODS FOR MATERIAL CHARACTERIZATION. *Educ:* Queens Col, NY, BS, 54; Purdue Univ, MS, 56, PhD(anal chem), 59. *Prof Exp:* Asst, Purdue Univ, West Lafayette, 54-57; res chemist, 58-74, sr res chemist, 74-84, RES SCIENTIST, HERCULES, INC, 84- *Res:* Differential thermal analysis; thermal gravimetric analysis; analytical instrumental design; general methods development; catalyst characterization; thermal analysis of polymers; kinetics; hazards evaluation; solution calorimetry; special polymers; software systems. *Mailing Add:* Hercules Res Ctr Lancester Pike Wilmington DE 19894

DUSZYNSKI, DONALD WALTER, b Chicago, Ill, July 28, 43; div; c 2. PARASITOLOGY, ECOLOGY. *Educ:* Wis State Univ, River Falls, BS, 66; Colo State Univ, MS, 68, PhD(zool), 70. *Prof Exp:* From asst prof to assoc prof, 70-79, PROF BIOL, UNIV NMEX, 79-, CHMN DEPT, 82- *Concurrent Pos:* Sr res scientist, dept physiol, Med Sch, Univ Tex, Houston, 76 & vis assoc prof, dept microbiol, Med Br, Galveston, 77; vis res scholar, Kyoto Univ, Japan, 80. *Mem:* Am Inst Biol Sci; Am Soc Parasitol; Soc Protozool; Wildlife Dis Asn; Am Micros Soc. *Res:* The coccidia of wild animals; host-parasite/parasite-parasite competition and interactions; gastrointestinal pathophysiology; parasite ecology. *Mailing Add:* Dept Biol Univ NMex Albuquerque NM 87131

DUTARY, BEDSY ELSIRA, arbovirology, for more information see previous edition

DUTCH, STEVEN IAN, b Milford, Conn, May 10, 47; m 75; c 2. STRUCTURAL GEOLOGY. *Educ:* Univ Calif, Berkeley, BA, 69; Columbia Univ, MPhil, 74, PhD(geol), 76. *Prof Exp:* asst prof, 76-82, ASSOC PROF GEOL, UNIV WIS, GREEN BAY, 82- *Mem:* Geol Soc Am. *Res:* Tectonics; Precambrian geology. *Mailing Add:* Natural & Appl Sci Univ Wis Green Bay WI 54311-7001

DUTCHER, CLINTON HARVEY, JR, b Vallejo, Calif, Apr 28, 32; m; c 1. PHYSICS, ELECTRICAL ENGINEERING. *Educ:* Univ Fla, BS, 59, MS, 61, PhD(physics), 68. *Prof Exp:* Mem tech staff, Bell Tel Labs, 61-63; instr physics, Univ Fla, 63-68; group leader appl res, Electronic Commun, Inc, 68-70; consult, 71-74; sect mgr, Schlumberger Well Serv, Schlumberger Tech Corp, 74-80, sr proj engr, 80-; DIR, RES & DEVELOP, G&H TECHNOL INC, CAMARILLO, CALIF. *Mem:* Sr Mem Inst Elec & Electronics Engrs; Sigma Xi. *Res:* Communications theory; electromagnetics; physical geology. *Mailing Add:* 14214 Stokesmount Dr Houston TX 77077

DUTCHER, JAMES DWIGHT, b Highland Park, Mich, Aug 4, 50; m 72; c 3. ENTOMOLOGY. *Educ:* Mich State Univ, BS, 72, BS, 74, MS, 75; Univ Ga, PhD(entom), 78. *Prof Exp:* from asst prof to assoc prof entom, Univ Ga, 78-88; CONSULT, 88- *Mem:* Entom Soc Am. *Res:* Bionomics and control of insect pests in fruit and nut crops, specifically apples, grapes and pecans; ecological energetics of phytophagous insects. *Mailing Add:* Dept Entom Univ Ga PO Box 748 Tifton GA 31793-0748

DUTCHER, RUSSELL RICHARDSON, b Brooklyn, NY, Oct 28, 27; m 52; c 2. GEOLOGY. *Educ:* Univ Conn, BA, 51; Univ Mass, MS, 53; Pa State Univ, PhD(geol), 60. *Prof Exp:* Instr geol, Univ Mass, 52-53; from res asst to res assoc, Pa State Univ, 56-63, from asst prof to assoc prof, 63-70, asst dir, coal res sect, 60-70; prof geol & chmn dept, 70-83, DEAN COL SCI, SOUTHERN ILL UNIV, 83- *Honors & Awards:* Vancouvering Award, Am Inst Prof Geologists, 85; Gordon H Wood Jr Mem Award, Am Asn Petrol Geologists, 89. *Mem:* Geol Soc Am; Am Asn Petrol Geologists; Am Inst Prof Geologists; Sigma Xi. *Res:* Coal petrology, petrography and stratigraphy; alteration of coals by igneous intrusives. *Mailing Add:* PO Box 128 Carbondale IL 62903

DUTEMPLE, OCTAVE J, b Hubbell, Mich, Dec 10, 20; m 51; c 2. NUCLEAR & CHEMICAL ENGINEERING. *Educ:* Mich Technol Univ, BS, 48, MS, 49; Northwestern Univ, MBA, 55. *Prof Exp:* Assoc chem engr fuel reprocessing, Argonne Nat Lab, 49-58; EXEC DIR, AM NUCLEAR SOC, 58- *Mem:* AAAS; Am Nuclear Soc; Am Inst Chem Engrs; Am Chem Soc; Coun Eng & Sci Soc Execs (pres, 75-76); Am Asn Eng Socs; fel Inst Nuclear Engrs, UK. *Res:* Administration of scientific society; economics of nuclear industry; archeology, prehistoric copper; agriculture. *Mailing Add:* Am Nuclear Soc Inc 555 N Kensington Ave La Grange Park IL 60525-5535

DUTHIE, HAMISH, b Aberdeen, Scotland, Aug 30, 38; m 84; c 4. FRESH WATER BIOLOGY, PHYCOLOGY. *Educ:* Univ Col NWales, BSc, 60, PhD(biol), 64. *Prof Exp:* From lectr to assoc prof, 63-85, PROF BIOL, UNIV WATERLOO, 85- *Mem:* Phycol Soc Am; Am Soc Limnol & Oceanog; Can Bot Asn. *Res:* Biology of periphyton; biology of reservoirs; algal ecology and taxonomy. *Mailing Add:* Dept Biol Univ Waterloo Waterloo ON N2L 3G1 Can

DUTKA, BERNARD J, b Ft William, Ont, Oct 5, 32; div; c 2. ECOTOXICOLOGY. *Educ:* Queen's Univ, Ont, BA, 55, MSc(microbiol & immunol), 64. *Prof Exp:* Lab supvr, Kingston Gen Hosp, Ont, 57-66; head bact labs, Pub Health Eng Div, Dept Nat Health & Welfare, 66-71; head microbiol sect, 71-87, LEADER, ECOTOXICOL PROJ, NAT WATER RES INST, CAN CTR INLAND WATERS, DEPT ENVIRON, 87- *Concurrent Pos:* Can chmn, Sub-Comt SC4 & Membrane Filter Working Group, Int Standards Orgn; co-ed, Toxicity Assessment: An Int Quart; mem, 723 Mcrobiol Probs Comt, Am Water Works Asn; mem, joint task group, Am Pub Health Asn; consult, Int Develop Res Ctr; co-chmn, symp comt, Int Toxicity Testing, 83-, & Int Water Qual, 87- *Mem:* Am Soc Microbiol; Am Water Works Asn. *Res:* Development of microbiological water pollution methodology; role of microorganisms in recycling of nutrients from sediment; evaluation of microbiological methods for assessment of toxicity and mutagen content of water and sewage; environmental impact studies. *Mailing Add:* PO Box 5050 Burlington ON L7R 4A6 Can

DUTKA, JACQUES, b New York, NY, Dec 29, 19; m 45; c 2. MATHEMATICS. *Educ:* City Col New York, BS, 39; Columbia Univ, AM, 40, PhD(math), 43. *Prof Exp:* Asst statistician, US War Dept, 42; asst res mathematician, Appl Math Panel, Columbia Univ, 43; instr math, Princeton Univ, 46-47; asst prof, Rutgers Univ, 47-53; mathematician, Norden-Ketay Corp, 53-56; sr engr, Radio Corp Am, 56-59, leader, 59-61, mgr, 61-63; staff scientist, Ford Instrument Co, 64-66, systs analyst & eval mgr, 66-67; staff scientist, Sperry-Rand Corp, 67, Riverside Res Inst, 67-70 & Am Tel & Tel Co, 71-73; CONSULT, AUDITS & SURVS CO, INC, 73- *Concurrent Pos:* Adj assoc prof elec eng, Columbia Univ, 54-59, adj prof, 59-72; consult, Opers Res Group, US Dept Navy, 44-45; consult, Off Naval Res, 46-47. *Mem:* Am Math Soc; Inst Math Statist; Sigma Xi. *Res:* Probability; statistics; systems analysis; communication theory; numerical analysis and computer applications; history of mathematics. *Mailing Add:* 39 Claremont Ave New York NY 10027

DUTRA, FRANK ROBERT, b Sacramento, Calif, Jan 7, 16; m 46. PATHOLOGY. *Educ:* Northwestern Univ, AB, 40, MS, 41, MD, 42. *Prof Exp:* Intern, City Hosp, Cleveland, Ohio, 41-42; res pathologist, Hosp, Western Reserve Univ, 42-43; from asst prof to assoc prof indust & forensic path, Kettering Lab, Cincinnati, 47-52; asst clin prof, 53-62, ASSOC CLIN PROF PATH, MED SCH, UNIV CALIF, SAN FRANCISCO, 62-, ASST CLIN PROF DERMAT, 77-; PATHOLOGIST, PATH LABS, LOS GATOS, 77- *Concurrent Pos:* Rocke Legal Med fel, Harvard Univ, 43-44; pathologist, Sutter Hosp, Sacramento, 53-54; pathologist & dir labs, Eden Hosp, 54-77; consult, San Francisco Vet Admin Hosp, 53-, Martinez Vet Admin Hosp, Oakland, 59-, US Naval Hosp, 63- & Santa Clara County Med Ctr, 63- *Mem:* Fel AMA; Am Soc Clin Pathologists; affil Am Acad Dermat; Am Soc Dermatopath. *Res:* Pathogenic aspects of certain metals; carcinogenesis; physiologic aspects of heat; pathology of the skin; toxicology of some industrial substances; surgical and clinical pathology. *Mailing Add:* Path Labs 464 Monterey Ave Los Gatos CA 95030

DUTRA, GERARD ANTHONY, b Paterson, NJ, Oct 23, 45; m 68; c 3. ORGANIC CHEMISTRY, AGRICULTURAL CHEMISTRY. *Educ:* St Louis Univ, BS, 67; Wayne State Univ, PhD(org chem), 72. *Prof Exp:* Res specialist, 72-80, sr res group leader, agr res, 80-82, mgr herbicide discovery, 82-87, MGR, FORMULATIONS, MONSANTO CO, 87- *Mem:* Am Chem Soc; Sigma Xi. *Res:* Development of formulation and manufacturing processes for agricultural chemicals, environmental chemistry and synthesis of biologically active compounds. *Mailing Add:* Monsanto Co 03F 800 N Lindbergh Blvd St Louis MO 63167

DUTRA, RAMIRO CARVALHO, b Ponta Delgada, Portugal, Sept 27, 31; US citizen; m 58; c 2. ORGANIC CHEMISTRY, FOOD CHEMISTRY. *Educ:* Univ Calif, Davis, BS, 54, MS, 56, PhD(agr chem), 59. *Prof Exp:* Jr specialist dairy indust, Calif Agr Exp Sta, 54-57, asst specialist food chem, 57-59; from asst prof to assoc prof chem, 59-69, PROF FOODS & NUTRIT & CHMN DEPT, CALIF STATE POLYTECH UNIV, 69- *Mem:* Inst Food Technol; Soc Nutrit Educ. *Res:* Isolation and identification of organoleptic compounds; nonenzymatic browning of foods; fortification of cereal proteins; technology of proteins from unconventional sources. *Mailing Add:* 948 Scripps Dr Claremont CA 91711

DUTRO, JOHN THOMAS, JR, b Columbus, Ohio, May 20, 23; m 48; c 3. GEOLOGY, PALEONTOLOGY. *Educ:* Oberlin Col, AB, 48; Yale Univ, MS, 50, PhD(geol), 53. *Prof Exp:* GEOLOGIST, US GEOL SURV, 48-; RES ASSOC, SMITHSONIAN INST, 62- *Concurrent Pos:* Mem geol panel, Bd Civil Serv Exam, 58-65; chief paleont & stratig br, US Geol Surv, 62-68, mem geol names comt, 62-68, 71-83; assoc ed, Geol Soc Am, 74-81; bd dir & field trip chmn, 9th Int Carboniferous Cong, 78-85, Paleont Res Inst, trustee, 85-; Sterling fel, Yale Univ, 49-50. *Mem:* AAAS; Geol Soc Am; Arctic Inst NAm; Palaeont Asn; Am Geol Inst (secy-treas, 66-71); Geol Soc London; Asn Earth Sci Ed; Paleont Soc; Sigma Xi; Soc Am Military Engrs; Cosmos Club. *Res:* Systematics, evolution and biostratigraphy of Devonian, Carboniferious and Permian brachiopods; late Paleozoic stratigraphy Alaska; paleogeography and paleotectonics of Paleozoic of Western North America. *Mailing Add:* Rm E-308 US Mus Natural Hist Washington DC 20560

DUTSON, THAYNE R, b Idaho Falls, Idaho, Oct 3, 42; m 62, 81; c 1. MEAT SCIENCE. *Educ:* Utah State Univ, BS, 66; Mich State Univ, MS, 69, PhD(food sci), 71. *Prof Exp:* Res asst meat sci, Mich State Univ, 66-71; fel, Univ Nottingham, 71-72; asst prof meat chem, 72-77, ASSOC PROF, TEX A&M, 77-; PROF FOOD SCI & CHMN DEPT, MICH STATE UNIV. *Honors & Awards:* Meat Res Award, Am Soc Animal Sci, 81. *Mem:* Am Soc Animal Sci; Inst Food Technologists; Am Meat Sci Asn. *Res:* Elucidation of the biochemical factors responsible for differences in meat tenderness and studies on the mechanisms responsible for biosynthesis and net accumulation of muscle tissue. *Mailing Add:* Agr Exp Sta Ore State Univ Corvallis OR 97331-2201

DUTT, GAUTAM SHANKAR, b Calcutta, India, Oct 24, 49. THERMAL PHYSICS, MECHANICAL ENGINEERING. *Educ:* Univ London, BSc, 70; Princeton Univ, MA, 75, PhD(aerospace), 77. *Prof Exp:* Res assoc, Princeton Univ, 76-83, res engr energy res, Ctr Environ Studies, 77-87; ENERGY CONSULT, 87- *Res:* Thermal performance of buildings and energy conservation strategies; development of diagnostic techniques for monitoring the energy performance of buildings; energy conservation and development in the third world; efficient wood-burning cookstoves for less developed countries. *Mailing Add:* 11715 Critton Circle Lake Ridge VA 22192

DUTT, GORDON RICHARD, b Choteau, Mont, Oct 25, 29; m 54; c 2. SOIL CHEMISTRY, WATER CHEMISTRY. *Educ:* Mont State Col, BS, 56; Purdue Univ, West Lafayette, MS, 59, PhD(soil chem), 60. *Prof Exp:* Asst res irrigationist, Univ Calif, 60-64; assoc prof, 64-68, PROF SOIL & WATER SCI, UNIV ARIZ, 68- *Mem:* Am Soc Enologists; Soil Sci Soc Am. *Res:* Physical chemistry of soil and water systems, water quality and ground water hydrology; run off farming; grape and wine production. *Mailing Add:* Dept Soils Water Sci Univ Ariz Tucson AZ 85721

DUTT, NIKIL D, b Hubli, Karnataka, India. HARDWARE DESIGN SYNTHESIS, HARDWARE DESCRIPTION LANGUAGES. *Educ:* Birla Inst Technol & Sci, Pilani, India, BE, 81; Pa State Univ, MS, 83; Univ Ill, Urbana-Champaign, PhD(computer sci), 89. *Prof Exp:* Res asst computer sci, Pa State Univ, 81-83; res asst computer sci, Univ Ill, 83-89; lectr, 89, ASST PROF COMPUTER SCI, UNIV CALIF, IRVINE, 89- *Concurrent Pos:* Prin investr, Semiconductor Res Corp & NSF, 90- *Mem:* Inst Elec & Electronics Engrs; Asn Comput Mach. *Res:* Computer systems design automation; hardware description languages; design synthesis; very large scale integration design; computer architecture; hardware-software co-design methodologies; high-level synthesis. *Mailing Add:* Dept Computer Sci Univ Calif Irvine CA 92717

DUTT, RAY HORN, b Bangor, Pa, Aug 26, 13; m 46; c 2. ANIMAL SCIENCE. *Educ:* Pa State Univ, BS, 41; Univ Wis, MS, 42, PhD(genetics), 48. *Prof Exp:* From asst to assoc animal husbandman, 48-58, prof animal sci, 58-81 & 82, EMER PROF ANIMAL SCI, UNIV KY, 82- *Concurrent Pos:* Ed, J Animal Sci, 64-66. *Mem:* Fel AAAS; Am Soc Animal Sci (pres, 67-68); Biomet Soc; Genetics Soc Am; Am Dairy Sci Asn. *Res:* Physiology of reproduction; pioneered research using progesterone to control estrons cycle in mammals. *Mailing Add:* 437 Bristol Rd Lexington KY 40502

DUTTA, HIRAN M, b Patna, India; US citizen; m 58; c 2. EFFECTS OF POLLUTANTS ON FISH, MEASUREMENT OF OPTOMOTOR BEHAVIOR. *Educ:* Patna Univ, India, MSc, 60; Lieden Univ, Neth, PhD(functional anat), 68. *Prof Exp:* Asst prof anat & physiol & chair, Biol Dept, NH Col, 66-67 & Walsh Col, 68-70; from asst prof to assoc prof, 75-90, PROF ANAT & HISTOL, KENT STATE UNIV, 91- *Concurrent Pos:* Vis asst prof anat & histol, Kent State Univ, 70-74; vis prof, Univ Jegellonian, Krakow, Poland, 87 & Polish Acad Sci, Inst Ecol, Poland, 89; Smithsonian

Inst grant, 89-93; Fulbright fel, Coun Int Exchange Scholars, 90. *Mem:* Am Soc Zoologists; Am Fisheries Soc; Am Soc Ichthyologists & Herpetologists. *Res:* Effect of pollutants on fish structures, gills, accessory respiratory organs, blood plasma, acetylcholinesterase, liver, kidneys, spleen, brain and function such as optomotor behavior and oxygen uptake. *Mailing Add:* Dept Biol Sci Kent State Univ Kent OH 44242

DUTTA, MITRA, b Patna, India, July 3, 53; m 83. SPECTROSCOPY, OPTICAL DEVICES. *Educ:* Gauhati Univ, BSc, 71; Delhi Univ, MSc, 73; Univ Cincinnati, MS, 78, PhD(physics) 81. *Prof Exp:* Lectr physics, Col Arts & Sci Technol, Kingston, Jamaica, 73-76; part time lectr physics, Univ WI, Jamaica, 73-76; res asst physics, Univ Cincinnati, 76-81; res assoc, Purdue Univ, 81-83; sr res assoc, City Col NY, 83-86; res engr, Syst Gen Corp, 86-87; res prof, Southeastern Ctr Elec Eng, 87-88; RES PHYSICIST, LAB COM, US ARMY, 88- *Mem:* Am Phys Soc; Optical Soc Am; Inst Elec & Electronics Engrs; Sigma Xi. *Res:* Use of optical and spectroscopic techniques to study the physics of semiconductor material and device structures; design, fabrication and characterization of novel optical devices with three and four semiconductor material and heterostructures. *Mailing Add:* Electronics Technol & Devices Lab US Army SLCET-ED Ft Monmouth NJ 07703-5000

DUTTA, PULAK, b Calcutta, India, Oct 1, 51. CHEMICAL PHYSICS. *Educ:* Presidency Col, Univ Calcutta, BSc, Hons, 70; Univ Delhi, MSc, 73; Univ Chicago, PhD(physics), 80. *Prof Exp:* Res fel, Argonne Nat Lab, 79-81; asst prof, 81-87, ASSOC PROF PHYSICS, NORTHWESTERN UNIV, 87- *Mem:* Am Phys Soc. *Res:* Studies of condensed matter phases and phase transitions in surfaces and overlayers; x-ray and neutron diffraction; organic monolayers, thin films and artificial structures; composition-modulated structures (superlattices). *Mailing Add:* Dept Physics & Astron Northwestern Univ 2145 Sheridan Rd Evanston IL 60208

DUTTA, PURNENDU, b Calcutta, India, Nov 10, 37; US citizen; m 63; c 2. GENERAL SURGERY, CRYOSURGERY. *Educ:* Univ Calcutta, MBBS, 60, MS, 64; FRCS, 66, FACS, 74. *Prof Exp:* Sr house surgeon, Manchester Northern Hosp, UK, 66; registrar surg, Royal Postgrad Med Sch, Hammersmith Hosp, London, 66-68; consult surgeon, BC Royal Mem Hosp Children, Calcutta India, 68-70; surg res fel, Univ Minn, 70-71; ASST PROF SURG, STATE UNIV NY & VET ADMIN MED CTR, BUFFALO, 71- *Mem:* Asn Vet Admin Surgeons; Soc Cryobiol; fel Am Col Cryosurg; fel Am Col Surgeons; Am Soc Gastrointestinal Endoscopic; AMA. *Res:* Gastrointestinal physiology and surgery; endocrine surgery with particular interest in thyroid, parathyroid and adrenal diseases. *Mailing Add:* 61 Waterford Park Buffalo NY 14221-3645

DUTTA, SARADINDU, b Dacca, Bangladesh, Jan 28, 31; m 62; c 3. PHARMACOLOGY, VETERINARY MEDICINE. *Educ:* Univ Calcutta, BSc, 52, GVSc, 53; Univ Wis, MS, 59; Ohio State Univ, PhD(pharmacol), 62. *Prof Exp:* Res asst animal physiol, Indian Vet Res Inst, 54-57; res assoc, 63-64, from instr to assoc prof pharmacol, Ohio State Univ, 70-74; PROF PHARMACOL, WAYNE STATE UNIV, 74- *Concurrent Pos:* Res grants, Nat Heart Inst, 65-68 & Cent Ohio Heart Asn, 66-68. *Mem:* AAAS; Am Soc Pharmacol & Exp Therapeut. *Res:* Isolation and characterization of digitalis receptors from the cardiac cell. *Mailing Add:* Dept of Pharmacol Wayne State Univ Sch of Med Detroit MI 48202

DUTTA, SHIB PRASAD, b Calcutta, India, Nov 27, 35; m 69; c 2. ORGANIC CHEMISTRY, MEDICINAL CHEMISTRY. *Educ:* Univ Calcutta, BSc, 55, MSc, 58, PhD(org chem), 67. *Prof Exp:* Chemist, Alkali & Chem Corp India, Ltd, 57-60; sr chemist, Union Carbide India, Ltd, 60-62; res chemist, Jadavpur Univ, 62-64; res fel, Bose Inst, Calcutta, 65-67; res assoc med chem, Sch Pharm, State Univ NY Buffalo, 67-69; CANCER RES SCIENTIST, ROSWELL PARK MEM INST, 69- *Mem:* Am Chem Soc; Am Soc Mass Spectrometry. *Res:* Chemistry and synthesis of modified nucleosides and nucleotides in transfer RNA; isolation and identification of nucleic acid metabolites in human biological fluids; toxicologic and metabolic studies on biologically active nucleoside derivatives; mass spectrometry of biological interest like peptides and nucleotides. *Mailing Add:* Dept Biophys 666 Elm St Buffalo NY 14263

DUTTA, SISIR KAMAL, b Bengal, India, Aug 28, 28; m 55; c 1. MOLECULAR GENETICS. *Educ:* Univ Dacca, BS, 49; Kans State Univ, MS, 58, PhD(genetics), 60. *Prof Exp:* Lectr biol, K N Col, Calcutta Univ, 49-50; asst plant sci, Agr Res Inst, Calcutta, 50-56; asst, Exp Sta, Kans State Univ, 56-59; res assoc bot, Chicago & Columbia Univ, 59-61; dir & chief res officer, Pineapple Res Sta, Malaya, 61-64; res assoc biol, Rice Univ, 64-65; asst prof, Tex Southern Univ, 65-66; chmn div sci & math, Jarvis Christian Col, 66-67; PROF MOLECULAR GENETICS, HOWARD UNIV, 67- *Concurrent Pos:* Res grants, US Dept Naval Res, US Environ Protection Agency, US Dept Energy, Res Corp, Anna Fuller Fund, NSF & NIH, 67- *Mem:* AAAS; Genetics Soc Am; Am Soc Microbiol; Sigma Xi; NY Acad Sci; Bioelectromagnetics Soc. *Res:* Molecular biology; microbial genetics; gene isolation; regulation; mutagenic tests; radiation genetics. *Mailing Add:* Dept Bot Howard Univ Washington DC 20001

DUTTA, SUNIL, b Naihati, India, Nov 2, 37. CERAMICS ENGINEERING. *Educ:* Univ Calcutta, BSc, 57, MScTech, 60; Univ Sheffield, MS, 62, PhD(ceramics), 65; Babson Col, MBA, 74. *Prof Exp:* Sr sci officer ceramics & refractories, Nat Metall Lab, India, 66-67; sr res ceramic engr, US Army Mat & Mech Res Ctr, 68-76; SR MAT ENGR, LEWIS RES CTR, NASA, 76- *Concurrent Pos:* Fel mat res ctr, Lehigh Univ, 67-68; tech supvr, US-Can Defense Develop Sharing Proj, 68-70, Avco Corp, 70-73 & Ceradyne, Inc, 72-73; proj mgr, Ford Motor Co. *Mem:* Fel Am Ceramic Soc; Nat Inst Ceramic Engrs; Ceramic Educ Coun; Sigma Xi; fel Brit Inst Ceramics. *Res:* Correlations among processing, microstructures and physical, mechanical, ballistic properties of carbide, oxide, and nitride ceramics; thermal shock, oxidation and phase relationships in gas turbine and nuclear ceramics; ceramic materials development for gas turbine, diesel and stirling engine and other structural applications. *Mailing Add:* 3996 Tennyson Lane Cleveland OH 44070

DUTTAAHMED, A, b India, Apr 3, 35; m 57; c 2. PHYSICAL INORGANIC CHEMISTRY, THEORETICAL CHEMISTRY. *Educ:* Univ Calcutta, BSc, 54, MSc, 57, DPhil, 63; La State Univ, New Orleans, PhD, 72. *Prof Exp:* Lectr chem, Univ Calcutta, 58-63; asst prof res inorg chem, Indian Asn Cultivation Sci, 63-70; fel chem, La State Univ, New Orleans, 68-72, res assoc spectros & theoret chem, 72-74; prof collabr, State Univ Campinas, 74; res assoc inorg chem, Tex Christian Univ, 75-76; asst prof, SDak State Univ, 76-77; assoc prof, 77-80, PROF INORG CHEM, BUCKS COUNTY COMMUNITY COL, 81-, CHMN DEPT SCI & TECH, 85- *Mem:* Am Chem Soc; Sigma Xi; fel Am Inst Chem. *Res:* Inorganic and bioinorganic syntheses-linkage isomers, complexes of DNA bases; spectroscopy-inorganic molecular complexes; molecular orbital methods and calculations-theoretical electronic structure; electrochemistry, electrochemical synthesis. *Mailing Add:* Sci & Tech Dept Bucks County Col Newton PA 18940

DUTTA-ROY, ASIM KANTI, b Gopinagar, WBengal, India, July 8, 55; US citizen; m 82. RECEPTOR PHARMACOLOGY, PROTEIN BIOCHEMISTRY. *Educ:* Univ Calcutta, India, BSc hons, 77, MSc, 79; PhD(biochem), 83. *Prof Exp:* NIH postdoctoral fel, Thrombosis Res Ctr, Temple Univ, Philadelphia, 83-85; res biochemist, Vet Admin Med Ctr, Dayton, Ohio, 85-88, prin res scientist, 88-89; dir protein biochem, Efamol Res Inst, NS, Can, 89; PRIN SCIENTIST, DIV BIOCHEM SCI, ROWETT RES INST, ABERDEEN, UK, 90- *Concurrent Pos:* Adj asst prof, Dept Biochem, Wright State Univ, Dayton, Ohio, 86-89, res asst prof, Dept Med, 87-89. *Honors & Awards:* Elisabeth Kushnir Mem Award, Am Heart Asn, 88. *Mem:* Am Soc Biochem & Molecular Biol; Biochem Soc UK. *Res:* Function of prostaglandin and insulin receptor systems of human blood platelet and red blood cell in the cardiovascular diseases and in diabetes; fatty acid-binding proteins and intracellular fatty acid metabolism. *Mailing Add:* Div Biochem Sci Rowett Res Inst Aberdeen AB2 9SB Scotland

DUTTON, ARTHUR MORLAN, b Des Moines, Iowa, July 28, 23; m 45; c 2. APPLIED STATISTICS. *Educ:* Iowa State Univ, BS, 45, PhD(statist), 51. *Prof Exp:* Instr & res assoc statist, Iowa State Univ, 47-51; instr radiation biol, Univ Rochester, 51-53, from asst prof to assoc prof, 53-68, lectr math, 51-57; chmn dept math sci, 68-74, prof, 68-76, prof statist, 76-90, EMER PROF, UNIV CENT FLA, 90- *Mem:* AAAS; Am Statist Asn; Biomet Soc. *Res:* Statistical techniques; mathematics and statistics education; application of statistical methods. *Mailing Add:* Dept Statist Univ Cent Fla Box 25000 Orlando FL 32816

DUTTON, CYNTHIA BALDWIN, b Albany, NY, July 17, 29; m 58; c 2. INTERNAL MEDICINE. *Educ:* Cornell Univ, BA, 52; Univ Rochester, MD, 56; Harvard Univ, MPH, 79. *Prof Exp:* Intern internal med, State Univ NY Upstate Med Ctr, 56-57, asst resident, 57-58, resident, 58-59; fel, Sch Med, Johns Hopkins Univ, 59-63, instr, 63-69; asst prof, 69-72, ASSOC PROF MED, ALBANY MED COL, 72- *Mem:* Am Fedn Clin Res; fel Am Col Physicians. *Res:* Methods of health care delivery; measurement of quality of health care; physician extenders, evaluation of their clinical performance. *Mailing Add:* Dept Med Albany Med Col New Scotland Ave Albany NY 12208

DUTTON, GARY ROGER, b Spokane, Wash, June 27, 38; m 62; c 2. NEUROBIOLOGY, CELL CULTURE. *Educ:* Univ Wash, BSc, 61; Ind Univ, MSc, 64, PhD(biochem), 67. *Prof Exp:* Nat Inst Mental Health fel psychiat, Albert Einstein Col Med, 68-70; asst res biochemist, Univ Calif, San Diego, 70-72; sr res fel, Open Univ, UK, 72-80; from asst prof to assoc prof, 80-84, PROF PHARMACOL, UNIV IOWA, 84- *Concurrent Pos:* Vis fel, Australian Nat Univ, Canberra, 76; vis scientist, Inst Physiol, Czechoslovak Acad Sci Prague, 77. *Mem:* Am Soc Neurochem; Soc Neurosci; Sigma Xi; Int Soc Neurochem; Am Soc Pharmacol & Exp Therapeut. *Res:* Developmental neurobiology: central nervous system cultures as models for brain development; receptor, neurotransmitter interaction and cell surface (monclonal antibodies) studies. *Mailing Add:* Dept Pharmacol Col Med Univ Iowa Iowa City IA 52242

DUTTON, GUY GORDON STUDDY, b London, Eng, Feb 26, 23; m 51; c 3. ORGANIC CHEMISTRY. *Educ:* Cambridge Univ, BA, 43, MA, 46; Univ London, MSc, 52; Univ Minn, PhD(agr biochem), 55. *Prof Exp:* Jr sci officer, UK Govt, 43-45; lectr org & inorg chem, Sir John Cass Col, 45-49; from asst prof to assoc prof, 49-64, PROF ORG CHEM, UNIV BC, 64- *Concurrent Pos:* Vis prof, Univ Grenoble, 65-66, 74-75, Rhodes Univ, 80, Univ Cape Town, 80, Tech Univ Chile, 80; NATO lectr, Tech Univ Denmark, 68 & Max Planck Inst, Freiburg, 68 & 74; chmn, 62nd Can Chem Conf, Vancouver, 79, 28th Cong, Int Union Pure & Appl Chem, Vancouver, 81 & 11th Int Carbohydrate Symp, Vancouver, 82; Univ BC fel, 56 & 57, Killam Sr fel, 73 & 84, Hugh Kelly fel, 85; mem, nat comt, Int Union Pure & Appl Chem. *Honors & Awards:* Montreal Medal, 86; Labatt Award, 89. *Mem:* Fel Royal Soc Chem; fel Chem Inst Can (pres, 87-89); Am Chem Soc. *Res:* Carbohydrate chemistry, particularly structures of bacterial polysaccharides and their reactions with bacteriophages. *Mailing Add:* Dept Chem Univ BC Univ Campus 2036 Main Mall Vancouver BC V6T 1Y6 Can

DUTTON, HERBERT JASPER, b Evansville, Wis, May 30, 14; m 37; c 3. CHEMISTRY. *Educ:* Univ Wis, BA, 36, MA, 38, PhD(plant physiol), 40. *Prof Exp:* Asst bot & chem, Univ Wis, 39-41; assoc chemist, Western Regional Res Lab, Bur Agr & Indust Chem, USDA, 41-45, head chem & phys properties invests, Northern Utilization Res & Develop Div, 53-74, chief, Oilseed Crops Lab, Northern Regional Res Lab, 74-81; HON FEL, HORMEL INST, UNIV MINN, 81- *Honors & Awards:* Super Serv Award, USDA, 56; Can Res Award, 61; Award Lipid Chem, Am Oil Chemists' Soc, 69; Hilditch Mem lectr, Liverpool, 71; Res Award, Am Soybean Asn, 79; Lewkowitsch Mem Lectr, London, 81. *Mem:* Am Chem Soc; Am Oil Chemists' Soc; AAAS. *Res:* Quantum efficiency and energy transfer in photosynthesis; conversion of soybeans to foods, feeds and industrial chemicals; selective catalytic hydrogenation of linolenic acid in soybean oil; computer-instrument aided lipid research; essential fatty acid metabolism; algal symbiosis in fresh water sponges; structure and analysis of dienoic fatty acids. *Mailing Add:* Rte 2 PO Box 197 Cable WI 54821

DUTTON, JOHN ALTNOW, b Detroit, Mich, Sept 11, 36; m 62; c 3. METEOROLOGY, APPLIED MATHEMATICS. *Educ:* Univ Wis, BS, 58, MS, 59, PhD(meteorol), 62. *Prof Exp:* From asst prof to assoc prof, 65-71, head, dept meteorol, 81-86, PROF METEOROL, PA STATE UNIV, UNIVERSITY PARK, 71-, DEAN, COL EARTH & MINERAL SCI, 86- *Concurrent Pos:* Expert, Systs Eng Group, Res & Tech Div, Air Force Systs Command, 65-71; vis scientist, Nat Lab Riso, Denmark, 71-72, 75 & 78-79; trustee, Univ Corp Atmospheric Res, 74-81, secy, 77-78, treas, 78-79, vchmn, 79-80; vis prof, Tech Univ Denmark, 78-79; ed meteorol monographs, Am Meteorol Soc, 80-84, chmn, Publ Comn, 84-; chmn, Unidata Steering comt, 82-86, Unidata Policy comt, 86-88, NSF UCAR Long-range planning comt, 86-87; mem bd atmospheric sci & climate, Nat Acad Sci, 82-83, & 88- , chmn, 89-, mem, Int Space Year planning comt, 86-89, panel experts earth sci & technol Int Space Year, 1992, 89-, space sci bd comt earth sci, 87-89, space studies bd, 87-, chmn, space studies bd task group priorities space res, 89-; mem, Am Meteorol Coun, 85-88; vpres, UCAR Found, 86-87, pres, 87- *Honors & Awards:* Zimmermann Award, US Air Weather Serv, 63. *Mem:* Fel Am Meteorol Soc; Math Asn Am; Soc Indust & Appl Math; Sigma Xi. *Res:* Theoretical meteorology; dynamics of atmospheric motion; global thermodynamics and energetics of the general circulation; spectral modeling of atmospheric and hydrodynamic flow, predictability; effects of atmospheric turbulence on structures; application of artificial intelligence methods to the atmospheric sciences. *Mailing Add:* 116 Deike Bldg Pa State Univ University Park PA 16802

DUTTON, JOHN C, b Chicago, Ill, Mar 8, 18. TRANSFORMERS & POWER SYSTEMS, COMPUTER DESIGN & APPLICATIONS. *Educ:* Swathmore Col, BS, 39; Rensselaer Polytech Inst, MS, 50. *Prof Exp:* Student test engr, Gen Elec Co, 39-41; radar officer, US Navy, 42-45; transformer engr, Gen Elec Co, Pittsfield, 45-54, supvr design, Gen Elec Co, Romega, 54-56, mgr comput appln proj, 56-58, mgr develop eng, 58-71, mgr transformer technol prog, Gen Elec Co, Schenectady, 71-73, consult engr, Indust Standards, Gen Elec Co, Romega, 74-83; RETIRED. *Mem:* Fel Inst Elec & Electronics Engrs. *Mailing Add:* John C Dutton Enterprises Inc 120 Hycliff Rd Rome GA 30161

DUTTON, JONATHAN CRAIG, b Williston, NDak, Mar 1, 51; m 73; c 1. GAS DYNAMICS, FLUID MECHANICS. *Educ:* Univ Wash, BSME, 73; Ore State Univ, MS, 75; Univ Ill, PhD(mech eng), 79. *Prof Exp:* Mech engr, Lawrence Livermore Nat Lab, 79-80; from asst prof to assoc prof mech eng, Tex A&M Univ, 80-85; assoc prof mech eng, Univ Ill, 85-91; PROF MECH ENG, UNIV ILL, 91- *Concurrent Pos:* Consult, Wright Patterson AFB, 83-86, Rocketdyne, Div Rockwell Int, 89-; prin investr, Army Res Off, 85-, Off Naval Res, 86-, Wright Patterson AFB, 87-; assoc ed, Am Soc Mech Eng, J Fluids Eng, 87-90. *Honors & Awards:* AT&T Found Award, Am Soc Eng Educ, 89. *Mem:* Assoc fel Am Inst Aeronaut & Astronaut; Am Soc Mech Engrs; Am Soc Eng Educ; Sigma Xi. *Res:* Gas dynamics; fluid mechanics; propulsion, experimental, numerical and analytical studies of high-speed separated, mixing and reacting flows; nozzle flows; ejector flows; diffuser flows; value flows. *Mailing Add:* Dept Mech & Indust Eng Univ Ill 1206 W Green St Urbana IL 61801

DUTTON, JONATHAN JOSEPH, b New York, NY, Oct 2, 42; m 73; c 1. BIOMEDICAL OPHTHALMOLOGY. *Educ:* Queens Col, NY, BA, 65; Harvard Univ, MS, 67, PhD(biol), 70; Rutgers Med Sch, MMS, 75; Wash Univ, MD, 77. *Prof Exp:* Asst prof paleobiol, Princeton Univ, 70-73; mem med staff, Univ Wash Hosp, 77-78; MEM RES STAFF OPHTHAL, SCH, WASH UNIV, 78-; ASSOC PROF, DEPT OPHTHAL, MED CTR, DUKE UNIV, DURHAM. *Concurrent Pos:* NSF fel, 70-71; Fight for Sight Found fel, 78-79; Heed found fel, 82-83. *Mem:* Am Acad Ophthalmol; Am Col Surgeons; Am Soc Plastic Reconstructive Surgeons. *Res:* Biology of vision; ophthalmic oncology. *Mailing Add:* Med Ctr Box 3802 Duke Univ Durham NC 27710

DUTTON, P LESLIE, PHOTOSYNTHESIS & RESPIRATION. *Educ:* Univ Wales, PhD(biochem), 67. *Prof Exp:* PROF BIOCHEM & BIOPHYS, UNIV PA, 68- *Res:* Redox reactions; charge separations. *Mailing Add:* Univ Pa B501 Richards Bldg Philadelphia PA 19104

DUTTON, RICHARD W, b London, Eng, May 16, 30; US citizen; m 77; c 3. CELL BIOLOGY, IMMUNOLOGY. *Educ:* Cambridge Univ, BA, 52, MA, 55; Univ London, PhD(biochem), 55. *Prof Exp:* Fel, Med Sch, Univ London, 55-56; vis lectr biochem, Med Col Va, 57-58; res instr biochem & med, Sch Med & Dent, Rochester Univ, 58-59; asst lectr chem path, Med Sch, Univ London, 59-62; assoc exp path, Scripps Clin & Res Found, 62-68; assoc prof, 68-70, PROF BIOL, SCH MED, UNIV CALIF, SAN DIEGO, 70- *Concurrent Pos:* Am Cancer Soc Dernham fel, 63-68. *Mem:* Am Soc Immunol; Fedn Am Socs Exp Biol. *Res:* Cellular immunology; molecular basis of antigen stimulation. *Mailing Add:* Dept Biol M001 Univ Calif San Diego Q-963 La Jolla CA 92093

DUTTON, ROBERT EDWARD, JR, b Milford, NH, Aug 11, 24; m 58; c 2. PHYSIOLOGY, MEDICINE. *Educ:* Med Col Va, MD, 49. *Prof Exp:* Instr med, State Univ NY Upstate Med Ctr, 56-59; from instr to asst prof environ med, Sch Hyg & Pub Health & Med, Johns Hopkins Univ, 61-68; assoc prof physiol, Albany Med Col, Union Univ, 68-74, from assoc prof to prof med, 70-86, prof physiol, 74-86; RES PROF, BIOMED ENG, RENSSELAER POLYTECH INST, 86- *Concurrent Pos:* Nat Heart Inst fel environ med, Sch Hyg & Pub Health & Med, Sch Med, Johns Hopkins Univ, 59-61; clin investr, US Vet Admin, 61-64; asst chief phys med & rehab, Baltimore City Hosp, Md, 64-68; consult pulmonary dis, Vet Admin Hosp, Baltimore, 67-68; attend physician, Albany Med Ctr Hosp & Vet Admin Hosp, Albany, 70-; prof biomed eng, Rensselaer Polytech Inst, 71-86; vis prof med, Harvard Univ, Boston, 78-79; pres, Eastern Sect, Am Thoracic Soc, 79-80; chap pres, Sigma Xi, 83-84; mem Cardiopulmonary Coun, Am Heart Asn. *Mem:* Am Fedn Clin Res; Am Thoracic Soc; Am Physiol Soc; Biomed Eng Soc; Sigma Xi. *Res:* Pulmonary physiology, particularly control of respiration; pulmonary diseases; internal medicine; biomedical engineering. *Mailing Add:* Dept Biomed Eng JEC 7049 Rensselaer Polytech Inst Troy NY 12180-3590

DUTTON, ROBERT W, ELECTRICAL ENGINEERING. *Prof Exp:* DIR RES CTR, INTEGRATED SYSTS, STANFORD UNIV, 91- *Mem:* Nat Acad Eng. *Mailing Add:* Stanford Univ AEL 201 4055 Stanford CA 94305

DUTTON, ROGER, b Tonyrefail, Gamorgan, UK, Sept 8, 41; Can citizen; m 65; c 2. MATERIALS SCIENCE, METALLURGY. *Educ:* Univ Wales, BSc, 63, PhD(metall), 66. *Prof Exp:* Res officer, 67-71, HEAD MAT SCI BR, ATOMIC ENERGY CAN LTD, 71- *Concurrent Pos:* Mem task force Can Prog Controlled Thermonuclear Fusion, Ministry State Sci & Technol, 73-74; mem social issues comt, Can Nuclear Asn. *Mem:* Can Inst Mining & Metall. *Res:* Underlying materials science research in support of the Canadian nuclear reactor system; irradiation damage, creep deformation, fracture and hydrogen embrittlememt. *Mailing Add:* 18 Lansdowne Pinawa MB R0E 1L0 Can

DUTTWEILER, DONALD LARS, b Kenmore, NY, Oct 2, 44; m 66; c 2. DIGITAL SIGNAL PROCESSING, VLSI. *Educ:* Rensselaer Polytech Inst, BEE, 66; Stanford Univ, MS, 67, PhD(elec eng), 70. *Prof Exp:* Mem tech staff, 70-78, SUPVR, BELL LAB, 79- *Concurrent Pos:* Fel, Bell Labs, 82. *Mem:* Fel Inst Elec & Electronics Engrs. *Res:* Digital signal processing and its VLSI implementation; first to realize an echo canceler as a VLSI device; image processing. *Mailing Add:* Bell Tel Labs Holmdel NJ 07733

DUTTWEILER, RUSSELL E, b Canton, Ill, May 12, 38; m 57; c 2. TITANIUM FOIL PROCESSING & DEVELOPMENT. *Educ:* Univ Ill, BS, 60. *Prof Exp:* Mgr mat eng, GE Aircraft Engines, 60-75, mgr group purchased mat qual, 75-82, mgr mat develop, 82-85, mgr technol integration, 85-88; dir mkt & technol develop, Aerospace Components, Banner Group Inc, 88-90; SR ENG SPECIALIST, UNIVERSAL TECHNOL CORP, 90- *Concurrent Pos:* Pres/dir, Univ Ill mat sci & eng alumni contingent, 79-; mem, Columbian Trade Deleg, Brazil, Am Soc Metals Int, 84, pres Cincinnati Chap, 85-86, vchair, Mat Energy Achievement Award Selection Comt, 91, chmn, 92; lectr, Int High Temperature Mat Sem, Beijing, China, 85. *Mem:* Fel Am Soc Metals Int. *Res:* Titanium alloy stress corrosion and surface effects; titanium foil and foil processing for composite material development; applications include NASP, IHPTET, HSCT, engines and airframes; application of smart systems to existing processes. *Mailing Add:* 2410 S Old Oaks Dr Beavercreek OH 45431

DUTY, ROBERT C, b Morrison, Ill, Sept 28, 31; m 58; c 2. ORGANIC CHEMISTRY, COAL CHEMISTRY. *Educ:* Univ Ill, BS, 53; St Louis Univ, cert meteorol, 54; Midwestern Univ, MEd, 56; Univ Iowa, PhD(org chem), 61. *Prof Exp:* Sr chemist, Petrol Prod Lab, Humble Oil & Refining Co, La, 60-61; assoc prof chem, Western State Col Colo, 61-63; from assoc prof to prof chem, Ill State Univ, 63-87; RETIRED. *Mem:* AAAS; Am Chem Soc. *Res:* Organic polarography; chromatography; coal chemistry. *Mailing Add:* Baylor Univ PO Box 97348 Waco TX 76798

DUTZ, WERNER, b Vienna, Austria, June 7, 28; c 2. PATHOLOGY. *Educ:* Univ Vienna, MD, 53; Am Bd Path, cert anat path, 59; FRCPath, 62; FRCPath(C), 72. *Prof Exp:* Prof path & chmn dept, Pahlavi Univ, Iran, 60-73; prof surg path, Med Col Va, 74-81; dir, Gen Hosp Planning Corp, Vienna, Austria, 81-82; prof & chmn path, King Saud Univ, Abha Br, Saudi Arabia, 84-86; PROF PATHOL, UNIV VIENNA, AUSTRIA, 86- *Concurrent Pos:* Consult, WHO, 70 & 74. *Mem:* Int Acad Path; Am Soc Clin Pathologists; Int Soc Geog Path; Col Am Pathologists. *Mailing Add:* Dept Path & Microbiol Krankenhaus Lainz 1130 Wien Vienna Austria

DUVAL, LEONARD A, b Cleveland, Ohio, Sept 17, 21; m 46; c 4. CHEMICAL & MECHANICAL PROCESS DEVELOPMENT, ENVIRONMENTAL CONTROLS. *Educ:* Univ Mich, BSE, 47. *Prof Exp:* Treas, Proj Develop Assocs, 47-50; pres, Duval Engine Co, 50-57; mgr opers, Kaiser-Nelson Co, 57-63; pres, Hess-von Bulow, Inc, 63-70 & Colerapa Industs, Inc, 68-90; CHMN, RECOTECH CORP, 90- *Concurrent Pos:* Pres-chmn, Euromed Int Co, Ltd, 68-88; consult, Recotech Corp, 90- *Mem:* AAAS; Am Chem Soc; Am Iron & Steel Engrs; Math Soc; Air & Water Pollution Control Asn. *Res:* Research, develop, and patented processes and devices for separation of liquids from liquids, and liquids from solids. *Mailing Add:* 207 Harmon Rd Aurora OH 44202

DU VAL, MERLIN KEARFOTT, b Montclair, NJ, Oct 12, 22; m 44, 84; c 3. SURGERY. *Educ:* Dartmouth Col, AB, 43; Cornell Univ, MD, 46; Nat Bd Med Exam, dipl, 47; Am Bd Surg, dipl, 56. *Hon Degrees:* DSc, NJ Col Med & Dent, Dartmouth Col, Med Col Wis, Col Osteop Med, Des Moines; LHD, Ohio Col Podiatric Med. *Prof Exp:* From instr to asst prof surg, Sch Med, State Univ NY, 54-56; assoc prof, Sch Med, Univ Okla, 57-60, prof & asst dir med ctr, 61-63; dean col med, Univ Ariz, 64-71; asst secy health & sci affairs, Dept Health, Educ & Welfare, Washington, DC, 71-73; vpres health sci & prof surg, Univ Ariz, 74-79; pres, Nat Ctr Health Educ, 79-82 & Assoc Health Systs, 82-84; pres, Am Healthcare Inst, 84-88; sr vpres med affairs, Samaritan Health Serv, Phoenix, Ariz, 88-90; HEALTH CONSULT, 90- *Concurrent Pos:* Asst attend surgeon, US Vet Admin Hosp, NY, 55-56 & US Naval Hosp, NY, 55-56; Markle scholar, 56-61. *Mem:* Inst Med-Nat Acad Sci; Soc Univ Surg; Am Med Asn; Am Col Surg; Am Surg Asn; Soc Med Adminr. *Mailing Add:* 3026 E Marlette Ave Phoenix AZ 85016

DUVALL, ARNDT JOHN, III, b St Paul, Minn, Jan 14, 31; m 56; c 4. OTOLOGY, OTOLARYNGOLOGY. *Educ:* Univ Minn, BA, 52, MD, 55, MS, 62. *Prof Exp:* From asst prof to assoc prof, 63-70, PROF OTOLARYNGOL, UNIV MINN, MINNEAPOLIS, 70-, CHMN DEPT, 85- *Concurrent Pos:* Fel, Karolinska Hosp & Karolinska Inst, Sweden, 62-63; consult, Minn Vet Hosps, 63- *Mem:* AAAS; Soc Head & Neck Surgeons; Am Col Surgeons; Asn Res Otolaryngol; Triological Soc; Am Otologic Soc; Am Laryngol Soc; Am Acad Facial Plastic & Reconstruct Surg; Am Bronchoesophrageal Soc. *Res:* Anatomy, pathology and physiology of the ear, particularly the use of electron microscopy. *Mailing Add:* Dept Otolaryngol Univ Minn Minneapolis MN 55455

DUVALL, GEORGE EVERED, b Leesville, La, Feb 6, 20; m 41; c 2. PHYSICS. *Educ:* Ore State Col, BS, 46; Mass Inst Technol, PhD(physics), 48. *Prof Exp:* Assoc physicist, underwater sound, Div War Res, Univ Calif, 41-46; res assoc, Res Lab Electron, Mass Inst Technol, 46-48; physicist, Gen Elec Co, 48-50, head theoret group, 50-54; sr physicist, Poulter Labs, Stanford Res Inst, 54-57, sci dir, 57-62, dir, 62-64; prof, 64-88, EMER PROF PHYSICS, WASH STATE UNIV, 88- *Honors & Awards:* Shock Compression Sci Award, Am Phys Soc, 89. *Mem:* AAAS; Am Phys Soc; Am Asn Physics Teachers. *Res:* Underwater sound; stochastic processes; reactor physics; shock and detonation phenomena; equations of state of solids; finite amplitude wave propagation; dislocations; kinetics of polymorphic transitions; shockwave chemistry. *Mailing Add:* Dept Physics Wash State Univ Pullman WA 99164-2814

DUVALL, HARRY MAREAN, b Lanham, Md, Oct 27, 10. ORGANIC CHEMISTRY. *Educ:* Univ Md, BS, 32, PhD(org chem), 36. *Prof Exp:* E R Squibb & Sons fel, Univ Va, 36-38; res chemist, Jackson Lab, E I du Pont de Nemours & Co, Inc, 38-50, Thiokol Chem Corp, 50-53 & Masonite Corp, 54-58; prof chem & head dept, 58-78, EMER PROF CHEM, VALDOSTA STATE COL, 79- *Mem:* Am Chem Soc. *Res:* Process development of dyestuffs; rubber chemicals; organic chemicals; synthetic rubber; liquid polymers; wood products. *Mailing Add:* Box 36 Rte 6 Valdosta GA 31601-9003

DUVALL, JACQUE L, physical chemistry, organic chemistry, for more information see previous edition

DUVALL, JOHN JOSEPH, b Sedro-Woolley, Wash, Oct 20, 36; m 56; c 6. PHYSICAL ORGANIC CHEMISTRY. *Educ:* Brigham Young Univ, BS, 58, PhD(org chem), 63. *Prof Exp:* Teaching assoc, Brigham Young Univ, 59-60; RES CHEMIST, WESTERN RES INST, US DEPT ENERGY, 63- *Concurrent Pos:* Guest chemist, Brookhaven Nat Lab, 69-70. *Mem:* Am Chem Soc; Sigma Xi. *Res:* Radiation chemistry of shale oil components; liquid fossil fuel analysis; in situ oil shale retorting; asphalt chemistry. *Mailing Add:* Western Res Inst Box 3395 Univ Sta Laramie WY 82071-3395

DUVALL, PAUL FRAZIER, JR, b Atlanta, Ga, Aug 19, 41; m 63; c 2. TOPOLOGY. *Educ:* Davidson Col, BS, 63; Univ Ga, MA, 65, PhD(math), 67. *Prof Exp:* Asst prof math, Univ Ga, 67-68; asst prof, Va Polytech Inst & State Univ, 70-71; from assoc prof to prof math, Okla State Univ, 71-86; PROF & HEAD, MATH DEPT, UNIV NC, GREENSBORO, 86- *Concurrent Pos:* Consult, Dept Defense, 70- *Mem:* Am Math Soc; Math Asn Am; Asn Comput Mach; Sigma Xi. *Res:* Geometric topology; mappings between manifolds; actions of discrete groups on manifolds; embedding of complexes in manifolds; topological dynamics. *Mailing Add:* Dept Math Univ NC Greensboro NC 27412

DUVALL, RONALD NASH, b Needham, Mass, June 17, 24; m 53; c 3. PHARMACEUTICAL CHEMISTRY. *Educ:* Mass Col Pharm, BS, 49, MS, 51, PhD(pharmaceut chem), 57. *Prof Exp:* Asst pharmaceut chem, Mass Col Pharm, 52-54, instr, 54-57, asst prof, 57-62; sect head, Corp Pharmaceut Res Lab, Miles Labs, Inc, 62-63, actg dir, 63-64, dir, 64-70, asst dir, Pharmaceut Res & Develop Lab, 70-72, dir, Appl Pharmaceut Lab, 72-82, dir prod develop-medicinals, 82-89, PROD DEVELOP CONSULT, MILE LABS, INC, 89- *Concurrent Pos:* Consult, Muro Pharmacal Labs, Mass, Chester A Baker, Inc & Hoyt Pharmaceut Co, 59-62. *Mem:* Am Pharmaceut Asn; Am Chem Soc; Acad Pharmaceut Sci; Am Assoc Pharmaceut Scientists. *Res:* metal chelates of medicinal agents; antiradiation compounds; drug dosage form stability; carbohydrate browning mechanisms; salicylamide metabolism; industrial pharmacy; powder flow. *Mailing Add:* Miles Labs Pharmaceut Res & Develop Lab 1127 Myrtle St Elkhart IN 46514

DUVALL, WILBUR IRVING, b Gaithersburg, Md, Jan 3, 15; m 45; c 4. PHYSICS. *Educ:* Univ Md, BS, 36, MS, 38. *Prof Exp:* Instr high sch, Md, 36-37; lab instr, Amherst Col, 38-39; jr phys sci aide, US Bur Mines, 39-40, from jr physicist to supvry physicist, 40-52, supvry physicist, Blasting Res, Md, 52-65, supvry res physicist, Rock Mech, 65-72; RETIRED. *Concurrent Pos:* Adj prof rock mech, Colo Sch Mines, 69-74, sr sci adv, 74-; consult rock mech, 72-82. *Res:* Experimental stress analysis; stresses in underground mining structures; generation and propagation of explosive waves; electronics; rock mechanics. *Mailing Add:* 8820 W Dover Circle Denver CO 80226

DUVARNEY, RAYMOND CHARLES, b Clinton, Mass, Oct 11, 40; m 63; c 4. SOLID STATE PHYSICS. *Educ:* Clark Univ, BA, 62, PhD(physics), 68; Univ NH, MS, 64. *Prof Exp:* Asst prof, 68-74, ASSOC PROF PHYSICS, EMORY UNIV, 74-, CHMN, 87- *Concurrent Pos:* Vis prof physics, Univ Gesamthochschole-Paderborn, Fed Repub Ger, 78-79. *Mem:* Am Phys Soc; Sigma Xi; AAAS. *Res:* Investigations of the pyroelectric and ferroelectric properties of solids using the techniques of magnetic resonance; defects and impurities in insulating solids; radiation damage. *Mailing Add:* Dept Physics Emory Univ Atlanta GA 30322

DUVICK, DONALD NELSON, b Sandwich, Ill, Dec 18, 24; m 50; c 3. GENETICS, PLANT BREEDING. *Educ:* Univ Ill, BS, 48; Wash Univ, PhD(bot), 51. *Prof Exp:* Corn breeder maize genetics & physiol, Pioneer Hi-Bred Int, Inc, 51-71, dir, Dept Corn Breeding, 71-75, dir, Div Plant Breeding, 75-84, vpres res, 84-90, dir, 82-90; AFFIL PROF PLANT BREEDING, IOWA STATE UNIV, 90- *Concurrent Pos:* Pres, Nat Coun Com Plant Breeders, 84-86, Crop Sci Soc Am, 86; trustee, CIMMYT Int Maize & Wheat Improv Ctr; mem, Nation Plant Genetics Resources Bd. *Honors & Awards:* Indust Agron Award, Am Soc Agron, 89; Genetics & Plant Breeding Award Indust, Crop Sci Soc Am. *Mem:* Fel AAAS; fel Am Soc Agron (pres elect, 90-91); Sigma Xi; fel Crop Sci Soc Am (pres, 86, 87); Coun Agr Sci & Technol. *Res:* Cytoplasmic inheritance of pollen sterility in maize; immunological identification of plant proteins; developmental morphology and anatomy of maize endosperm; measurement of genetic contributions to long-term yield gains in hybrid maize. *Mailing Add:* PO Box 446 Johnston IA 50131

DUVIVIER, JEAN FERNAND, b Rio de Janeiro, Brazil, Dec 17, 26; US citizen; m 56; c 5. AEROMECHANICS, INTERNATIONAL PROGRAMS. *Educ:* Boston Univ, BSc, 55; Mass Inst Technol, SM, 58, EAA, 66. *Prof Exp:* Teaching asst aircraft design, Tech Inst Aeronaut, Sao Paulo, Brazil, 51-53; res staff aeroelasticity, Mass Inst Technol, 55-58, proj leader aerodynamics, 58-61; sci staff opers res, Ctr Naval Anal, 61-66; sr res engr, Elec Boat Div, Gen Dynamics Corp, 66-68; mgr systs eval, Boenig Co, 68-72, mgr res & develop, 72-73, mgr Int Mkt SAm, Vertol Div, 73-82; vpres, Int Mkt Fairchild Repub, 82-85; vpres aerospace systs mkt, Lear Siegler Int, 85-87; dir, Systs Mkt, Smiths Industs, 87-90; CONSULT, 90- *Concurrent Pos:* Consult, Res Anal Corp, 66. *Mem:* Assoc fel Am Inst Aeronaut & Astronaut; Am Helicopter Soc; assoc mem US Naval Inst; Sigma Xi. *Res:* Aerodynamics; aeroelasticity; structural dynamics; operations research; systems analysis; transportation systems; risk analysis. *Mailing Add:* Duvivier Assocs PO Box 755 Georgetown CT 06829-0755

DUVOISIN, ROGER C, b Towaco, NJ, July 27, 27; m 48; c 4. NEUROLOGY. *Educ:* New York Med Col, MD, 54. *Prof Exp:* Intern, Lenox Hill Hosp, New York, 54-55, resident neurol, 55-56; resident, Presby Hosp, Columbia Presby Med Ctr, New York, 56-58; res assoc, Col Physicians & Surgeons, Columbia Univ, 62-65, from asst prof to prof neurol, 65-73; prof neurol, Mt Sinai Sch Med, City Univ New York, 73-79; CHMN & PROF NEUROL, RUTGERS MED SCH, UNIV MED & DENT, NJ, 79- *Concurrent Pos:* Parkinson's Dis Found Clin res fel, 63-64; mem, NIH Study Group Encephalitis, 63; consult, New York Bd Educ, 63-; consult adminr, Fed Aviation Admin, 65-73; mem res adv bd, Parkinson's Dis Found, 65-; mem extrapyramidal dis comn, World Fedn Neurol, 72-; vis prof, King's Col Hosp & Inst Psychiat, London, 73; mem adv panel on neurol dis ther, US Pharmacopeia, 70-75; ed trans, Am Neurol Asn, 77-81. *Mem:* AAAS; Soc Neurosci; AMA; fel Am Col Physicians; Am Neurol Asn (secy-treas, 83-85); Asn Res Nerv & Ment Dis (secy-treas, 75-79). *Res:* Neurological complications of achondroplastic dwarfs; cerebral vascular disease; syncopal mechanisms; convulsive syncope; infectious polyneuritis; clinical features, natural history, epidemiology, genetics, clinical pharmacology and treatment of Parkinson's disease; development of L-dopa therapy of Parkinsonism; clinical and genetic aspects of olivoponocerebellar atrophy and progressive supranuclear palsy; MPTP-induced parkinsonism, animal models and mechanism of toxicity. *Mailing Add:* Robert Wood Johnson Med Sch Univ Med & Dent NJ New Brunswick NJ 08903

DUWE, ARTHUR EDWARD, b Saginaw, Mich, July 17, 22; wid; c 3. IMMUNOLOGY, EMBRYOLOGY. *Educ:* Alma Col, BS, 49; Ohio State Univ, MS, 50, PhD(zool), 53. *Prof Exp:* Asst instr zool, Ohio State Univ, 52; asst prof, North State Teachers Col, 53-54; asst prof, Wis State Col, Superior, 54-59, assoc prof 59-64; prof biol, Waynesburg Col, 64-68; PROF BIOL, LAKE SUPERIOR STATE COL, 68- *Concurrent Pos:* Sigma Xi grant, 59; NIH grants, 60-65. *Mem:* AAAS; Am Soc Ichthyol & Herpet; Am Soc Parasitol; Sigma Xi. *Res:* Comparative immunology. *Mailing Add:* Lake Superior State Col Sault Ste Marie MI 49783

DUWELL, ERNEST JOHN, b Chicago, Ill, Mar 12, 29; m 50; c 4. PHYSICAL CHEMISTRY. *Educ:* Univ Iowa, BS, 50, PhD(chem), 54; Purdue Univ, West Lafayette, MS, 52. *Prof Exp:* Asst, Purdue Univ, West Lafayette, 50-52; sr res chemist, Jones & Laughlin Steel Corp, 54-56; res specialist, Minn Mining & Mfg Co, 56-66, res mgr, Abrasives Div, 66-85, CORP SCIENTIST, 3M CO, 85- *Concurrent Pos:* Chmn Abrasive Processes Comt, Soc Mfg Engrs. *Mem:* Am Chem Soc; Am Soc Metals; Soc Mfg Engrs. *Res:* Metal cutting and finishing; corrosion; ceramics; tribology (friction, lubrication, wear). *Mailing Add:* 212 Elm St Hudson WI 54016-1650

DUX, JAMES PHILIP, b New York, NY, July 15, 21; m 48; c 2. PHYSICAL CHEMISTRY. *Educ:* Queens Col, NY, BS, 42; Columbia Univ, MA, 47; Polytech Inst Brooklyn, PhD(phys chem), 55. *Prof Exp:* Anal chem, Gen Chem Co, 42-44; res chemist, Phys & Anal Chem, Merck & Co, Inc, 47-50; group leader phys chem, Cellulose Sect, Am Viscose Div, FMC Corp, 54-59, group head anal chem, 59-61, sect leader, Acetate Fibers, 61-70 & Synthetic Staple & Indust Yarns, 70-75, sect leader polyester, acetate & vinyon sect, Fibers Div, 75-76; sr tech assoc & mkt mgr, Lancaster Labs, Inc, 77-79, qual assurance dir, 79-85; CONSULT, QUALITY ASSURANCE & LAB MGT, 85- *Concurrent Pos:* Adj asst prof chem, Millersville State Univ, 85- *Mem:* Am Chem Soc. *Res:* Analytical and physical chemistry in polymers; polyelectrolytes; reaction kinetics; cellulose chemistry; statistics; radiochemistry; diffusion problems; fiber physics and chemistry; laboratory quality assurance. *Mailing Add:* Lancaster Labs Inc 2152 New Holland Pike Lancaster PA 17601-5418

DUXBURY, ALYN CRANDALL, b Olympia, Wash, Dec 1, 32; m 56; c 3. OCEANOGRAPHY. *Educ:* Univ Wash, BS, 55, MS, 56; Tex A&M Univ, PhD(phys oceanog), 63. *Prof Exp:* Res assoc & lectr phys oceanog, Bingham Oceanog Lab, Yale Univ, 60-64; ASSOC PROF, UNIV WASH, 64- *Concurrent Pos:* Asst dir, Wash Sea Grant Prog, 72-89. *Mem:* AAAS; Am Soc Limnol & Oceanog; Am Geophys Union; Explorers Club; Sigma Xi. *Res:* Descriptive physical oceanography, hydrographic survey work. *Mailing Add:* Sch Oceanog WB-10 Univ Wash Seattle WA 98195

DUXBURY, DEAN DAVID, b Tripoli, Wis, Dec 20, 34; div; c 3. FOOD SCIENCE. *Educ:* Univ Wis-Madison, BS, 56, MS, 57. *Prof Exp:* Res asst, Univ Wis-Madison, 56-57; food technologist, Swift & Co, 59-61, sect head canned meats, 61-65, group leader, 65-69, res mgr, Res & Develop Ctr, 69-78; vpres tech opers, Am Pouch Food Co, 78-80, pres, Pouch Technol Inc, 80-82, vpres, Tech Pouch Labs Inc, 82-85; ASSOC ED, FOOD PROCESSING MAG, 85- *Concurrent Pos:* Adv comt mem, Comt on Food Irradiation, Nat Res Coun, 75-78. *Mem:* Am Soc Testing & Mat; Inst Food Technol. *Res:* Storage life studies for fresh packaged cranberries; irradiated canned beef studies for sterility and flavor acceptance; canned foods new product development; development of flexible packaged foods; food sterilization and safety. *Mailing Add:* 15055 Spring Rd No 1D Villa Park IL 60181

DUXBURY, MITZI L, NURSING. *Educ:* Univ Wis, Madison, BS, 66, MA, 70 & PhD(educ admin), 72. *Prof Exp:* Staff nurse, Victory Mem Hosp, 52, US Army Hosp, 53; head nurse, Non-Acute Med Unit, War Mem Hosp, 53-54; staff nurse obstet, DC Gen Hosp, 56-59; off nurse, 60, vol off nurse, Fox Lake Med Clin, 62-63; staff nurse, Madison Gen Hosp, 63-64; instr nursing, Madison Area Tech Col, 64; asst prof & chmn, Dept Maternal & Child Health Nursing, Univ, Wis, Oshkosh, 71-72; asst prof nursing obstet, Univ Wis, Madison, 72-73; dir health personnel develop, Nat Found-March Dimes, 72-78; from assoc prof to prof, Sch Nursing, Univ Minn, Minneapolis, 77-83, dir grad studies, 77-78, asst dean grad studies, 77-81, from assoc prof to prof, Sch Pub Health, 79-83; dean, Col Nursing, head, Dept Nursing Sci & prof maternal child nursing, Col Nursing, Univ Ill, 83-88, prof health resources mgt, Sch Pub Health, 88-89; PROF, SCH NURSING, UNIV WIS-MADISON, 89- *Concurrent Pos:* Dir, Clin Nurse Scholar Prog, Robert Wood Johnson Found, 82-85; expert adv panel nursing, WHO, 87-91; guest lectr, Univ Minn, 82-83 & Univ Ill, Chicago, 84, 87-88; res grant, Nat Found-March Dimes, 77-80 & 82, Dept Health & Human Serv, Pub Health Serv, Health Serv Admin, Bur Commun Health Serv, 79-82, NIH, 81-82, Univ Minn Comput Ctr, 82-83 & Robert Wood Johnson Found, 82-84; exchange scientist, Nat Acad Sci, Poland, 87; vis prof, Sch Nursing, Univ Wis-Madison, 88-89; consult, J L Kellogg Grad Sch Mgt, Northwestern Univ, 88, Rutgers Univ, 88. *Mem:* Inst Med-Nat Acad Sci; Soc Behav Med; NY Acad Sci; fel Am Acad Nursing; Am Soc Law & Med; Asn Health Sci Res; AAAS; Am Asn Cols Nursing; Am Nurses Asn; Am Pub Health Asn. *Res:* Author of over 40 publications in nursing. *Mailing Add:* Sch Nursing Univ Wis 600 Highland Ave Madison WI 53792

DUYAN, PETER, b Pittsburgh, Pa, Apr 18, 15. AIRCRAFT, ELECTRIC ELECTRONICS. *Educ:* Univ Calif, Berkeley, BSEE, 38. *Prof Exp:* Chief, Elec Electronics Sect, Eng Dept, Douglas Aircraft, 38-73; dir, Fiber Optics Div, Deutsch Co, 73-82; RETIRED. *Mem:* Fel Inst Elec & Electronics Engrs. *Mailing Add:* 1418 Santiago Dr Newport Beach CA 92660

DUYKERS, LUDWIG RICHARD BENJAMIN, b Surabaia, Indonesia, Oct 2, 29; US citizen; m 86; c 2. PHYSICS. *Educ:* Delft Univ Technol, ME, 57. *Prof Exp:* Engr arrays, Physics Lab, Rijks Verdedigings Organisatie- Dutch Inst Appl Sci Res, Neth, 57-59; sr scientist undersea acoust, Saclant Antisubmarine Warfare Res Ctr, Italy, 59-66; physicist underwater acoust, Naval Ocean Systs Ctr, 66-86; PROG AREA MGR, MATRIX PROCESSOR, NAVAL SEA SYSTS COMMAND, PMS 412, 86- *Mem:* Acoust Soc Am. *Res:* Linear arrays, underwater acoustics, resonance of gas-filled cavities, including mammals; systems engineering; over the horizon radar (OTHR); signal processing; one US patent. *Mailing Add:* 6608 Potomac Ave No B1 Alexandria VA 22307-6660

DUYSEN, MURRAY E, b Henderson, Iowa, July 27, 36; m 56; c 2. PLANT PHYSIOLOGY. *Educ:* Univ Omaha, BA, 59; Univ Nebr, MSc, 62, PhD(bot), 66. *Prof Exp:* From asst prof to assoc prof bot, 65-75, PROF BOT, NDAK STATE UNIV, 75- *Mem:* AAAS; Am Soc Plant Physiol; Bot Soc Am; Sigma Xi. *Res:* Chloroplast development and activity; water relations and plant development. *Mailing Add:* Dept Bot NDak State Univ Main Campus Fargo ND 58102

DUZGUNES, NEJAT, CHEMOTHERAPY & IMMUNOTHERAPY. *Educ:* Middle East Tech Univ, BS, 72; State Univ NY, Buffalo, PhD(biophys sci), 78. *Prof Exp:* Asst res biochemist, Cancer Res Inst, Univ Calif, San Francisco, 81-87, assoc res biochemist, 87-91; ASSOC PROF & CHAIRPERSON, DEPT MICROBIOL, SCH DENT, UNIV PAC, 91- *Concurrent Pos:* Asst res biochemist, Children's Hosp Med Ctr, Oakland, 81-83; assoc adj prof, dept pharmaceut chem, Univ Calif, San Francisco, 87-; vis prof, Inst Cybernetics & Biophys, Genova, Italy, 86 & dept biophys, Kyoto Univ, 88; prin investr, Am Heart Asn grant-in-aid, 83-87, Univ Wide Task Force on AIDS grant, 86-90 & Nat Inst Allergy & Infectious Dis grant, 88-; prin investr, Liposome Technol, Inc grant, 89- *Mem:* Biophys Soc; Am Soc Cell Biol; AAAS; Am Soc Microbiol; Int Soc Antiviral Res; Int AIDS Soc. *Res:* Kinetics and mechanisms of fusion of HIV influenza virus and sendai virus with cell membranes; chemotherapy and immunotherapy of HIV and mycobacterial infections. *Mailing Add:* Dept Microbiol Univ Pac San Francisco CA 94115

DVONCH, WILLIAM, b Chicago, Ill, June 22, 15; m 52; c 4. BIO-ORGANIC CHEMISTRY. *Educ:* Univ Ill, BS, 39; Purdue Univ, MS, 48, PhD(biochem), 50. *Prof Exp:* Asst, Wright Jr Col, Chicago, 39-40; chemist, Starch & Dextrose Sect, Northern Regional Res Lab, Bur Agr & Indust Chem, USDA, 41-46, biochemist, Fermentation Sect, Northern Utilization Res Br, Agr Res Serv, 50-54; BIOCHEMIST, RES DIV, WYETH LABS, 54- *Mem:* Am Chem Soc. *Res:* Carbohydrate chemistry; isolation of antibiotics; enzymes; antitumor compounds; penicillins and cephalosporins. *Mailing Add:* Wyeth Labs Box 8299 Philadelphia PA 19101

DVORACEK, MARVIN JOHN, b Penelope, Tex, July 16, 32; m 57; c 2. AGRICULTURAL ENGINEERING. *Educ:* Texas A&M Univ, BS, 53 & 59; Univ Calif, Davis, MS, 62. *Prof Exp:* Soil conservationist, Soil Conserv Serv, USDA, 53 & 56; instr eng graphics, Texas A&M Univ, 57-59, instr agr eng, 59; lectr & jr specialist, Univ Calif, Davis, 60-62; asst prof, 62-67, ASSOC PROF AGR ENG, TEX TECH UNIV, 67-, CHMN AGR ENG & TECHNOL DEPT, 77- *Concurrent Pos:* NSF fac fel, Dept Hydrol & Water Resources, Univ Ariz, 70-71, lectr, 71-72; Nat Asn Col Teachers Agr fel, 76- *Mem:* Am Soc Agr Engrs; Am Soc Civil Engrs; Nat Soc Prof Engrs; Am Soc Eng Educ; Nat Asn Col Teachers Agr. *Res:* Ground water recharge; irrigation; water conservation; evaporation; ground water pollution; runoff; hydrologic cycle. *Mailing Add:* 4823 16th St Lubbock TX 79416

DVORAK, ANN MARIE-TOMPKINS, b Bangor, Maine, May 19, 38; m 62; c 3. IMMUNOPATHOLOGY. *Educ:* Univ Maine, AB, 59; Univ Vt, MD, 63. *Prof Exp:* From intern to resident pediat, Boston Floating Hosp, 63-65; resident pediat path, Children's Hosp of DC, 65-66; resident path, Georgetown Univ Sch Med & Peter Bent Brigham Hosp, 66-69; from asst prof to assoc prof path, Tufts Med Sch, 71-75; assoc pathologist, Mass Gen Hosp, 75-79; pathologist, 79-86, SR PATHOLOGIST, BETH ISRAEL HOSP, 75-; ASSOC PROF PATH, HARVARD MED SCH, 75- *Concurrent Pos:* Res fel, Harvard Med Sch, 69-71. *Mem:* Am Asn Immunologists; Am Soc Cell Biologist; Am Asn Pathologists; Internal Acad Path; Int Acad Path. *Res:* Immunopathologic reactions in which basophils predominate in order to elucidate the functions of this cell in order to understand their roles in tumors and graft rejection as well as in other disease. *Mailing Add:* Dept Path Beth Israel Hosp 330 Brookline Ave Boston MA 02215

DVORAK, FRANK ARTHUR, b Kerrobert, Sask, Aug 8, 39; US citizen; m 63. AERONAUTICAL ENGINEERING, FLUID DYNAMICS. *Educ:* Royal Mil Col Can, BASc, 62; Univ BC, MASc, 64; Cambridge Univ, PhD(aeronaut eng), 67. *Prof Exp:* Sr engr aeronaut res, Boeing Co, 67-72; div mgr comput simulation, Flow Indust Inc, 72-74; PRES, ANALYTIC METHODS INC, 74- *Concurrent Pos:* Athlone Fel, Cambridge Univ, 64-66. *Honors & Awards:* Tech Award, NASA, 76. *Mem:* Am Helicopter Soc; Can Aeronaut & Space Inst. *Res:* High lift subsonic and transonic aerodynamics; stalled flight, especially massive flow separation; boundary layer control. *Mailing Add:* Analytic Methods Inc PO Box 3786 Bellevue WA 98009

DVORAK, HAROLD FISHER, b Milwaukee, Wis, June 20, 37; m 62; c 3. PATHOLOGY. *Educ:* Princeton Univ, AB, 58; Harvard Med Sch, MD, 63. *Prof Exp:* Instr path, Harvard Med Sch, 67-69; from asst pathologist to pathologist, Mass Gen Hosp, 69-79; CHIEF, DEPT PATH, BETH ISRAEL HOSP, BOSTON, 79- *Concurrent Pos:* From asst prof to assoc prof path, Harvard Med Sch, 69-77, prof, 77-79, Mallinckrodt prof, 79-; NIH career develop awardee, 70-75. *Mem:* Am Asn Immunol; Am Asn Path & Bact; NY Acad Sci; Am Soc Exp Path; Int Acad Path. *Res:* Role of basophilic leukocytes and tissue mast cells in cell-mediated immune reactions in man and animals; tumor secreted mediators and the tumor micro environment. *Mailing Add:* 27 Mason Rd Newton Center MA 02159

DVORAK, JAN, b Jindrichuv Hradec, Czech, Dec 13, 44; Can citizen. CYTOGENETICS. *Educ:* Agr Univ, Czech, ING, 66; Univ Sask, PhD(crop sci), 72. *Prof Exp:* Fel, Dept Crop Sci, Univ Sask, 72-73, res assoc, 73-76; asst prof cytogenetics, 76-80, ASSOC PROF AGRON & RANGE SCI, DEPT AGRON, UNIV CALIF, DAVIS, 80- *Mem:* Am Soc Genetics; Crop Sci Soc Am; Can Soc Genetics. *Res:* Chromosomal evolution and homoeology in related species; evolution of polyploids; evolution of diploidizing genetic systems; interspecific transfer of chromosomes; interspecific transfer of genes. *Mailing Add:* Dept Agron Univ Calif Davis CA 95616

DVORCHIK, BARRY HOWARD, b Bridgeport, Conn, Feb 29, 44; m 66; c 3. PHARMACOKINETICS. *Educ:* Univ Conn, BS, 66; Univ Fla, PhD(pharmacol), 72. *Prof Exp:* From instr to assoc prof obstet, gynec, pharmacol & med, Hershey Med Ctr, Pa State Univ, 72-83; res fel, McNeil Pharmaceut, 83-89; DIR, BIOCLIN INC, 89- *Concurrent Pos:* Consult, Div Biopharaceutics, Fed Drug Admin, Dept Health, Commonwealth Pa; USPHS grant fetal pharmacol, Hershey Med Ctr, Pa State Univ, 75-81, Nat Found-March of Dimes grant, 75-78; USPHS grant, Hershey Med Ctr, Pa State Univ, 79-82. *Mem:* Am Col Clin Pharmacol; Am Soc Clin Pharmacol & Therapueut; Soc Toxicol; Am Soc Pharmacol & Exp Therapeut. *Res:* Drug metabolism and pharmacokinetics; clinical trials management. *Mailing Add:* BioClin 1001 E Main St Suite 808 Richmond VA 23219

DVORETZKY, ISAAC, b Houston, Tex, Jan 24, 28; m 89; c 3. ANALYTICAL & PETROLEUM CHEMISTRY. *Educ:* Rice Univ, BA, 48, MA, 50, PhD(chem), 52. *Prof Exp:* Res chemist, Houston Res Lab, Shell Oil Co, 52-56, group leader, 56-58 & 59-62, exchange scientist, Royal Dutch-Shell Lab, Amsterdam, Neth, 58-59, supvr, Petrol Chem Dept, Emeryville, Calif, 62-67, process res liaison, Head Off, Shell Oil Co, NY, 67-68, supvr, Unconventional Raw Mat Dept, 68-69, mgr, 69-72, MGR PROF RECRUITMENT & UNIV RELS, SHELL DEVELOP CO, HOUSTON, 72- *Concurrent Pos:* Coun, Gordon Res Conf, 72-; bd dir, Nat Consortium Grad Degrees for Minorities in Eng, 80-; comn on Prof in Sci & Technol, 87-, bd dirs, 91- *Mem:* Am Chem Soc; Sigma Xi. *Res:* Catalysis; refining processes; coal liquefaction and gasification; carbene chemistry. *Mailing Add:* Shell Develop Co PO Box 1380 Houston TX 77251-1380

DVORNIK, DUSHAN MICHAEL, b Mezica, Yugoslavia, Oct 23, 23; Can citizen; m 51; c 1. BIOCHEMICAL PHARMACOLOGY, PHARMACOKINETICS. *Educ:* Univ Zagreb, Chem Eng, 48, DSc(chem), 54. *Prof Exp:* Res chemist, Pliva Chem Works, Yugoslavia, 46-49, head Dept Synthetic Chem, 50-52, res scientist, 52-54; fel natural prod, Lab Org Chem, Swiss Fed Inst Tech, 54-55; fel alkaloids, Nat Res Coun Can, 55-58, synthetic chem, Univ Ottawa, 58-59; res chem, Wyeth-Ayerst Res, 59-64, dir, Biochem Dept, 64-84, prin res investr & asst vpres, 84-88; BIOMED RES CONSULT, 88- *Concurrent Pos:* Fel Coun Arteriosclerosis, Am Heart Asn. *Mem:* Am Chem Soc; Am Soc Biol Chemists; Soc Exp Biol & Med; Am Diabetes Asn; Chem Inst Can. *Res:* Diabetic complications; atherosclerosis; drug metabolism. *Mailing Add:* 174 Moore St Princeton NJ 08540

DWARAKANATH, MANCHAGONDANAHALLI H, b Bangalore, India, May 15, 43; US citizen; m 69; c 2. ELECTRICAL ENGINEERING, SOFTWARE PROJECT MANAGEMENT. *Educ:* Univ Mysore, BS, 64; Worcester Polytech Inst, MS, 67; Polytech Inst Brooklyn, PhD(elec eng), 77. *Prof Exp:* Jr engr, Mysore State Elec Bd, 64; lectr, Nat Inst Eng, Univ Mysore, 64-65; sr engr, Am Elec Power, Boeing Co, 66-78, sr specialist engr elec eng, 78-80, princ engr elec eng, Boeing Comput Serv Civ, 80-; PROG MGR, MISSION ANALYSIS, GEN DYNAMICS CONVAIR DIV, LINDBERGH FIELD, SAN DIEGO, CALIF. *Concurrent Pos:* Vis scientist, Indian Inst Sci, India, 79; tech consult, Govt of India, 81. *Mem:* Sigma Xi; sr mem Inst Elec & Electronics Engrs. *Res:* Application of digital computers, modern control theory and system identification techniques for real time monitoring and control of power system; power systems. *Mailing Add:* Gen Dynamics Convair Div PO Box 85357 MZ 84-6980 Lindbergh Field San Diego CA 92138

DWASS, MEYER, b New Haven, Conn, Apr 9, 23; m 49; c 4. MATHEMATICAL STATISTICS. *Educ:* George Washington Univ, AB, 48; Columbia Univ, AM, 49; Univ NC, PhD(math statist), 52. *Prof Exp:* Math statistician, Bur Census, 49-52; from asst prof to assoc prof math, Northwestern Univ, 52-61; prof, Univ Minn, 61-62; dir, Ctr Statistics & Probability, 80-86, PROF MATH, NORTHWESTERN UNIV, 62- *Concurrent Pos:* Prof math, Hebrew Univ Jerusalem, 71-72. *Mem:* Am Math Soc; fel Inst Math Statist; Am Statist Asn. *Res:* Nonparametric statistics; renewal theory; probability theory; computers and teaching. *Mailing Add:* Dept Math Northwestern Univ Evanston IL 60201

DWELLE, ROBERT BRUCE, b Neenah, Wis; m 70; c 2. PLANT PHYSIOLOGY. *Educ:* Carleton Col, AB, 70; Univ Mont, PhD(bot), 74. *Prof Exp:* NDEA Title IV fel plant physiol, Univ Mont, 70-73, teaching asst bot, 73-74; res assoc, 74-76, asst prof, 76-81, ASSOC PROF PLANT PHYSIOL, UNIV IDAHO, 81- *Concurrent Pos:* Lectr crop physiol, Univ Idaho, 79; prin investr, USDA Competitive Res grant, 81-84. *Mem:* Am Soc Plant Physiologists; Crop Sci Soc Am; AAAS; Potato Asn Am; Europ Asn Potato Res. *Res:* Varied rates and mechanisms of photosynthesis, partitioning of photosynthates, and the influence of plant growth regulators on sink-source relationships of potato clones. *Mailing Add:* Dept Plant Sci Moscow ID 83843

DWIGGINS, CLAUDIUS WILLIAM, JR, b Amity, Ark, May 11, 33. PHYSICAL CHEMISTRY. *Educ:* Univ Ark, BS, 54, MS, 56, PhD(chem), 58. *Prof Exp:* proj leader petrol composition res, US Bur Mines & Energy Res & Develop Admin, 58-79, res chemist, Thermodynamics, US Dept Energy, 79-84; CONSULT, 84- *Mem:* AAAS; NY Acad Sci; Am Chem Soc; Am Crystallog Asn; Sigma Xi. *Res:* X-ray diffraction and structure determination; small angle x-ray scattering; colloid physics. *Mailing Add:* 1211 S Keeler Bartlesville OK 74003

DWINGER, PHILIP, b The Hague, Netherlands, Sept 25, 14. PURE MATHEMATICS. *Educ:* Univ Leiden, PhD(math), 38. *Prof Exp:* Prof math & head dept, Univ Indonesia, 53-56; from asst prof to prof, Purdue Univ, West Lafayette, 56-62; prof, Univ Delft, 62-65; head dept, 75-79, prof math, Univ Ill, Chicago Circle, 65-85, Dean, Col Lib Arts Sci, 79-85, DEAN & EMER PROF, 85- *Mem:* Am Math Soc; corresp mem Royal Netherlands Acad Sci; Netherlands Math Soc. *Res:* Algebra, particularly ordered sets and lattices; Boolean algebras. *Mailing Add:* 111 E Chestnut Apt 27C Chicago IL 60611

DWIVEDI, ANIL MOHAN, b Ballipursani, India, Dec 15, 47; US citizen; m 75; c 2. BIOMOLECULAR SPECTROSCOPY, MOLECULAR INTERACTIONS & STRUCTURES. *Educ:* Banaras Hindu Univ, India, BS, 66; IIT Kanpur, India, MS, 69; Lucknow Univ, India, PhD(physics), 73. *Prof Exp:* Asst prof physics, G B Pant Univ, Pantnagar, India, 73-75; sr scientist molecular spectros, Johnson Wax Co, Racine, Wis, 83-86; prof biophys, Northeastern Hill Univ, India, 86-87; SR RES CHEMIST PHARM DEVELOP, DU PONT MERCK PHARMACEUT CO, 88- *Concurrent Pos:* Res assoc, Univ Ill, Urbana, 75-78; vis scientist, Swiss Fed Inst Technol, Zurich, Switz, 78-79; postdoctoral scholar, Univ Mich, Ann Arbor, 79-81, asst res scientist, 81-83, vis scientist, 87-88. *Mem:* Am Chem Soc; Am Phys Soc. *Res:* Molecular spectroscopy; vibrational; optical; mossbauer spectroscopy; biophysics; spectroscopy of polymeric materials; surfaces and thin films; structures of polymers in aqueous solutions; mechanism of drug actions; over 40 research publications. *Mailing Add:* Du Pont Merck Pharmaceut Co PO Box 80400 Wilmington DE 19880-0400

DWIVEDI, CHANDRADHAR, b India, July 1, 48. DEVELOPMENTAL MEDICINE, TOXICOLOGY. *Educ:* Univ Gorakhpur, India, BSc, 64, MSc, 66; Univ Lucknow, India, PhD(pharmacol), 72. *Prof Exp:* Lectr chem, T D Col, Jaunpur, India, 66-67; res scholar, Gorakhpur Univ, India, 62-68, lectr, 68-69; res fel pharmacol, King Georges Med Col, Lucknow, India, 69-73; fel pharmacol & biochem, Vanderbilt Univ, 73-76; chief biochemist pediat, 77-81, asst prof, 76-82, ASSOC PROF PEDIAT & BIOCHEM & DIR PEDIAT CLIN LAB, MEHARRY MED COL, 82- *Concurrent Pos:* Prin investr, Nat Sci Found, Washington, DC, 77-79; vis prof, Univ NDak, 81; prin investr, NIH grant, 81- *Mem:* Int Soc Biochem Pharmacol; Int Union Pharmacol; Am Asn Clin Chemists; Am Soc Human Genetics; Int Soc Develop Neurosci. *Res:* Study of genetic or environmental factors on the development and targe organ toxicity; diagnostic laboratory medicine for neoplastic diseases. *Mailing Add:* Dept Pharmaceut Sci SDak State Univ PO Box 2202 C Brookings SD 57007-0197

DWIVEDI, RADHEY SHYAM, Balliacuipu, India; wid; c 3. CELL BIOLOGY, BIOCHEMISTRY. *Educ:* Univ Saskatchewan, Can, PhD(cell biol), 66. *Prof Exp:* PROF CELL BIOL, HOWARD UNIV, 69- *Concurrent Pos:* Postdoctoral res, Cell Res Inst, Univ Tex, Austin, 66-69. *Mem:* Am Soc Cell Biol; AAAS; Electron Micros Soc Am; Sigma Xi; NY Acad Sci. *Mailing Add:* 103 Periwinkle Ct Greenbelt MD 20770

DWIVEDY, RAMESH C, b Etawah, India, Mar 15, 43; m 67; c 1. BIOENGINEERING, AGRICULTURAL ENGINEERING. *Educ:* Univ Allahabad, BS, 63; Univ Guelph, MS, 65; Univ Mass, Amherst, PhD(agr eng), 70. *Prof Exp:* Lectr agr eng, Univ Udaipur, India, 63-64; asst prof agr eng, Univ Del, 70-74; resources engr, Dept Natural Resources & Environ Control, State of Del, 74-78; PROJ ENGR, DELMARVA POWER & LIGHT CO, 78- *Mem:* Assoc Am Soc Agr Engrs; Water Pollution Control Fedn. *Res:* Stress analysis in grain mass inside a bin; electrophysiological research to develop a non-chemical device of insect control; aquacultural engineering. *Mailing Add:* Delmarva Power & Light Co Univ Plaza Chapman Rd PO Box 6066 Newark DE 19714

DWORETZKY, MURRAY, b New York, NY, Aug 18, 17; m 43; c 2. ALLERGY, IMMUNOLOGY. *Educ:* Univ Pa, BA, 38; Long Island Col Med, MD, 42; Univ Minn, MS, 50; Am Bd Internal Med, dipl, 52, cert allergy, 54; Am Bd Allergy & Immunol, dipl, 72. *Prof Exp:* Intern, City Hosp, New York, 42-43, asst resident path, 43, fel, 46-47; resident, Univ Chicago, 47-48; fel med, Mayo Found, 48-50; asst med, 51-52, instr, 52-56, clin asst prof, 56-61, clin asst prof pub health, 57-62, clin assoc prof med, 61-66, CLIN PROF MED, MED COL, CORNELL UNIV, 66- *Concurrent Pos:* Asst physician, Outpatient Dept, New York Hosp, 51, physician, 51-56, asst attend physician, 56-61, assoc attend physician, 61-66, attend physician, 66-, physician-in-chg, Allergy Clin, 61-88; attend physician, Manhattan Eye, Ear & Throat Hosp, 52-62; med dir-at-large asthma, Allergy Found Am, 63-78, mem med coun, 78-; examr allergy subspecialty, Am Bd Internal Med, 69-72; mem bd gov, Am Bd Allergy & Immunol, 71-74; consult med (allergy), Mem Hosp, 72. *Honors & Awards:* Distinguished Serv Award, Am Acad Allergy & Immunol, 89. *Mem:* Fel AAAS; AMA; fel Am Acad Allergy (treas, 66, pres, 68); fel Am Col Physicians. *Res:* Anaphylaxis; toxic and allergic reactions to staphylococcal fractions; management of asthma. *Mailing Add:* 115 E 61st St New York NY 10021

DWORJANYN, LEE O(LEH), b Lviv, Ukraine, Feb 18, 34; div; c 4. CHEMICAL ENGINEERING. *Educ:* Univ Sydney, BSc, 55, BE, 57; Univ London, PhD(diffusion), 62. *Prof Exp:* Leverhulme scholar, Univ London, 60-62; chem engr, Imp Chem Indust, Australia & NZ, 57-59; res engr prod develop, 62-69, sr res engr, Orlon-Lycra Tech Div, 70-72, res supvr, 72-76, process supvr, Christina Lab, 76-78, res staff engr, 79-84, RES ASSOC, E I DU PONT DE NEMOURS & CO, INC, 85- *Concurrent Pos:* Database programming. *Mem:* Am Chem Soc; Am Inst Chem Engrs. *Res:* Dyed and pigmented fibers; thermodynamic equilibria; waste incineration; electrochemical cell design; process data acquisition; microcomputer programming; professional office management systems; noise analysis; catalytic oxydation; ion exchange; pc software. *Mailing Add:* 1570 Citation Dr Aiken SC 29803

DWORKEN, HARVEY J, b Cleveland, Ohio, Aug 1, 20; m 49; c 2. INTERNAL MEDICINE, GASTROENTEROLOGY. *Educ:* Dartmouth Col, BA, 41; Case Western Reserve Univ, MD, 44. *Prof Exp:* Intern, Michael Reese Hosp, Chicago, Ill, 44-45; resident psychiat, New Eng Med Ctr, Boston, 47-48; asst resident med, Mt Sinai Hosp, Cleveland, Ohio, 48-49, chief resident, 49-50; from clin instr to asst clin prof, Case Western Univ, 52-62, physician-in-chg, Gastrointestinal Labs, 62-81, from asst prof to prof, 62-90, dir, Div Gastroenterol, Dept Med, 73-81, EMER PROF MED, CASE WESTERN UNIV, 90- *Concurrent Pos:* Fel gastroenterol, Univ Hosp Pa, 50-52; chmn curric revision comt, Case Western Reserve Univ, 66-70, coordr, Phase 2 Curric, 72-75; vis prof, Taita Jing-Fu Med Found, Taipei, 90. *Mem:* Am Col Physicians; Am Gastroenterol Asn. *Res:* Curriculum planning; clinical investigation on gastric secretion and inflammatory diseases of the intestinal tract; medical education research. *Mailing Add:* 2074 Abington Rd Cleveland OH 44106

DWORKIN, JUDITH MARCIA, b Worcester, Mass, July 14, 49; m 74. WATER RESOURCES. *Educ:* Clark Univ, MA, 75, PhD(geog), 78. *Prof Exp:* Geographer, US Army Corp Engrs, 75-76; instr geography, Univ Toronto, 76-78; asst prof Water Resources, Univ Ariz, 78-83; ATM, MYER HENDRIX LAW FIRM, 87- *Concurrent Pos:* Consult, Indianapolis Water Co, 76-77, US Water Resources Coun, 78 & US Agency Int Develop, 78- *Mem:* Am Asn Geographers; AAAS; Am Geophys Union. *Res:* Institutional arrangements for efficient and equitable groundwater management; impacts of water supply projects in developing countries; community attitudes in resource decision-making. *Mailing Add:* 2929 N Central Ave Phoenix AZ 85012

DWORKIN, MARK BRUCE, b Camden, NJ, June 21, 49; m 81. METABOLIC REGULATION. *Educ:* Univ Wis-Madison, BS, 71; Wesleyan Univ, PhD(biol), 76. *Prof Exp:* Asst prof molecular genetics, Columbia Univ, NY, 82-87; CHMN MOLECULAR BIOL, ERNST-BOEHRINGER INST, 88- *Res:* Regulation of early development in vertebrate (xenopus) embryos; regulation of carbon metabolism in early embryos and tumors; roles of malic enzyme isozymes in embryo and tumor physiology. *Mailing Add:* Bender & Co Dr Boehringergasse 5-11 Vienna A-1121 Austria

DWORKIN, MARTIN, b New York, NY, Dec 3, 27; m 57; c 2. MICROBIOLOGY. *Educ:* Ind Univ, AB, 51; Univ Tex, PhD(bact), 55. *Prof Exp:* Res scientist bact, Univ Tex, 52-53; from asst prof to assoc prof microbiol, Med Ctr, Ind Univ, 57-62; assoc prof, 62-69, PROF MICROBIOL, UNIV MINN, MINNEAPOLIS, 69-, DIR, MD/PHD PROG, 90- *Concurrent Pos:* Res fel, Univ Calif, 55-57; NIH fels, 55-57, NIH career develop award, 64 & 69; vis prof, Oxford Univ, 70-71 & Stanford Univ, 78-79; Found for Microbiol lectr, 74, 77 & 81; Guggenheim fel, 78-79; mem, cellular & molecular basis dis study sect, NIH, 81-85; Sackler fel, Tel Aviv Univ, 92. *Mem:* Am Soc Microbiol; Brit Soc Gen Microbiol. *Res:* Microbial physiology; myxobacteria; developmental microbiology. *Mailing Add:* Dept Microbiol Univ Minn Minneapolis MN 55455

DWORNIK, JULIAN JONATHAN, b Colonsay, Sask, Mar 11, 38; Can citizen; m 63; c 1. ANATOMY. *Educ:* Andrews Univ, BA, 61; Univ Man, MSc, 64, PhD(anat, neuroanat), 69. *Prof Exp:* Demonstr microanat, Univ Man, 65-66, teaching fel anat, 66-677; from instr to asst prof, Univ Louisville, 67-70; asst prof, Univ SFla, 70-73, from asst dean to dep dean admin, 72-85, interim assoc dean continuing med educ, 83-84, dir Learning Resources Ctr, 84-87, dep dean admin, Col Med, 85-86, assoc dean admis, 86-88, ASSOC PROF ANAT, UNIV SFLA, 73- *Mem:* Can Fedn Biol Soc; Can Asn Anatomists; Asn Am Med Cols. *Res:* Gross anatomy and teratology. *Mailing Add:* Med Col Univ S Fla Box 6 12901 Bruce B Downs Blvd Tampa FL 33612-4799

DWORSCHACK, ROBERT GEORGE, b Milwaukee, Wis, Feb 26, 20; m 45; c 4. INDUSTRIAL MICROBIOLOGY. *Educ:* Univ Wis, BS, 42; Bradley Univ, MS, 49. *Prof Exp:* Chemist, Kurth Malting Co, 41; chemist, Northern Regional Res Lab, Bur Agr & Indust Chem, USDA, 42-44 & 45-51, chemist, Northern Utilization Res Div, Agr Res Serv, 51-60; chemist, Columbia Malting Co, 60-64; chemist, Clinton Corn Processing Co, 64-82; CONSULT MICROBIOL, ARCHER-DANIELS-MIDLAND, 82- *Concurrent Pos:* Biol Warfare res, Navy, 44-45. *Mem:* Am Soc Microbiol. *Res:* Process and product development in microbiology; starch chemistry; enzymology. *Mailing Add:* Archer Daniels Midland Co 1251 Beaver Channel Pkwy Clinton IA 52732

DWORZECKA, MARIA, b Warsaw, Poland; c 1. NUCLEAR PHYSICS. *Educ:* Warsaw Univ, MSc, 64, PhD(physics), 69. *Prof Exp:* Asst physics, Warsaw Univ, 64-66, jr fac mem, 66-69; res assoc, Mich State Univ, 70-72; asst prof, Univ Mass, Amherst, 72-73, adj asst prof, 73-74; vis asst prof, 74-80, asst prof physics, Univ Md, College Park, 80-82; assoc prof physics, 82-88, PROF PHYSICS, GEORGE MASON UNIV, FAIRFAX, VA, 88- *Concurrent Pos:* vis scientist, Tatu Inst, Bombay, Inst Nuclear Physics, Darmstasdt, W Germany. *Mem:* Asn Women Sci; Humbolol Fel; Am Phys Soc. *Res:* Nuclear structure--Hartree-Fock theory, Strutinsky approach, semiclassical approach; nuclear dynamics--heavy ions scattering, fluid dynamical descriptions, time-dependent Hartree-Fock theory, dynamical extended Thomas-Fermi approximation, nonlinear dynamics. *Mailing Add:* Dept Physics George Mason Univ Fairfax VA 22030

DWYER, DENNIS MICHAEL, b Passaic, NJ, Feb 26, 45; m 69; c 2. PARASITOLOGY, PROTOZOOLOGY. *Educ:* Montclair State Col, BA, 67; Univ Mass, Amherst, MS, 70, PhD(zool), 71. *Prof Exp:* Nat Inst Allergy & Infectious Dis fel parasitol, Rockefeller Univ, 71-73, asst prof parasitol, 73-76; res microbiologist, Lab Parasitic Dis, 76-80, PRIN INVESTR & SUPVRY RES MICROBIOLOGIST, CELL BIOL & IMMUNOL SECT, LAB PARASITIC DIS, NAT INST ALLERGY & INFECTIOUS DIS, 80- *Concurrent Pos:* Adj assoc prof parasitol, Rockefeller Univ, 76-81; adj assoc prof zool, Univ Mass, Amherst, 77-; grant reviewer, Spec Prog Res Training Trop Dis, WHO & Cellular Biol Sect, NSF; grant reviewer, Trop Med Parasitol Study Sect, & Microbiol Infectious Dis Study Sect, Nat Inst Allergic & Infectious Dis, Nat Inst Health, AID, US Dept State, US-Israeli Bi-Nat Sci Found; mem Leishmaniasis Steering Comt, Trop Dis Res, WHO, 82-86; mem, Parasitic Dis Rev Panel, USAMDRC, Am Inst Biol Sci, 85- *Honors & Awards:* Henry Baldwin Ward Medal, Am Soc Parasitol, 80; NIH Dir Award, 87. *Mem:* Soc Protozool; Am Soc Parasitol; Am Soc Trop Med Hyg; Am Soc Cell Biol; AAAS. *Res:* Chemical and immunochemical characterization of parasite cell surfaces; intracellular parisitism; aspects of pathogenicity and immunology of various parasitic protozoan groups; physiologic mechanisms and etiology of protozoan diseases. *Mailing Add:* NIH Rm 126 Bldg 4 Bethesda MD 20892

DWYER, DON D, b Hugoton, Kans, Dec 28, 34; m 56; c 2. RANGE MANAGEMENT. *Educ:* Ft Hays Kans State Col, BS, 56, MS, 58; Tex A&M Univ, PhD(range mgt), 60. *Prof Exp:* Asst prof forestry, Ariz State Col, 59-60; asst prof agron & bot, Okla State Univ, 60-64; from assoc prof to prof range mgt, NMex State Univ, 64-71; prof range sci & head dept, Utah State Univ, 71-84; exec dir, Consortium for Int Develop, 84-87; CONSULT, 87- *Mem:* Fel, Soc Range Mgt; Ecol Soc Am; Am Soc Agron. *Res:* Grazing management on native rangeland; ecology of native range plants. *Mailing Add:* 1225 W Boutz Rd Las Cruces NM 88005

DWYER, FRANCIS GERARD, b Philadelphia, Pa, June 13, 31; m 61; c 5. CHEMICAL ENGINEERING. *Educ:* Villanova Univ, BChE, 53; Univ Pa, MS, 63, PhD(chem eng), 66. *Prof Exp:* Jr engr, Mobil Res & Develop Corp, 53-54, chem engr, 56-62, sr chem engr, 62-69, engr assoc, 69-78, sr res assoc, 78-80, mgr, catalyst synthesis & develop, 80-82, SR SCIENTIST, MOBIL RES & DEVELOP CORP, 82-, MGR, CATALYST RES & DEVELOP, 85- *Honors & Awards:* Chem Eng Pract Award, Am Inst Chem Engrs, 90. *Mem:* Am Inst Chem Engrs; Am Chem Soc; Catalysis Soc; Int Zeolite Asn. *Res:* Catalyst and process development in petroleum processing; zeolite catalysis; petrochemical processing and auto exhaust catalysis; catalytic kinetics and reaction mechanisms of complex chemical reactions. *Mailing Add:* Mobil Res & Develop Corp Paulsboro NJ 08066

DWYER, HARRY, III, b Binghamton, NY, July 30, 45; m 72; c 1. INSTRUCTION-LEVEL PARALLELISM, SUPERSCALAR SYSTEMS. *Educ:* Univ Tex El Paso, BS, 77; Univ Syracuse, MS, 83; Cornell Univ, PhD, 91. *Prof Exp:* ADV ENGR, IBM CORP, 77- *Mem:* Asn Comput Mach; Inst Elec & Electronics Engrs; Am Asn Artificial Intel. *Res:* Computer architecture; fine grain instruction parallelism; out-of-order instruction execution; superscalar processor architecture and supporting storage systems; high performance computer structures. *Mailing Add:* IBM Corp Dept F44 PO Box 8008 Endicott NY 13760

DWYER, JAMES MICHAEL, physical chemistry, forensic science; deceased, see previous edition for last biography

DWYER, JOHANNA T, b Syracuse, NY, Oct 20, 38. CLINICAL NUTRITION. *Educ:* Harvard Univ, DSc, 69. *Prof Exp:* DIR NUTRIT, FRANCES STERN NUTRIT CTR, BOSTON, 74-; PROF MED, SCH MED, TUFTS UNIV, 84- *Concurrent Pos:* Mem, Food & Nutrit Bd, Nat Acad Sci, 90- *Honors & Awards:* Robert Wood Johnson Health Policy Fel, 80-81; Harvey Wilery Award, 84. *Mem:* Am Soc Clin Nutrit; Am Soc Parenteral & Enteral Nutrit; Am Pub Health Asn; Am Dietetic Asn; Soc Nutrit Educ (pres, 76). *Res:* Diet and cancer; diet change. *Mailing Add:* Frances Stern Nutrit Ctr 750 Washington PO Box 783 Boston MA 02111

DWYER, JOHN DUNCAN, b Newark, NJ, Apr 26, 15; m 42; c 4. SYSTEMATIC BOTANY. *Educ:* St Peter's Col, AB, 36; Fordham Univ, MS, 38, PhD(bot), 41. *Prof Exp:* Instr biol, St Francis Col, NY, 42; prof, Albany Col Pharm, 42-47 & Siena Col, 47-53; from assoc prof to prof biol, St Louis Univ, 53-82, head dept, 53-63; RETIRED. *Concurrent Pos:* Nat Acad Sci res grant, Mus Natural Hist, Paris, 52; cur trop SAm Phanerogams, Mo Bot Gardens, 64-; consult, US Army Tropic Test Ctr, Panama, 65; consult, Ciba Pharm Co; consult, Am Inst Res & Indust Technol, Guatemala, 70- *Mem:* AAAS; Torrey Bot Club; Am Soc Plant Taxon; Asn Taxon Study Trop African Flora. *Res:* American species of Ochnaceae and Leguminosae; general flora of Central America and Peru. *Mailing Add:* 549 Virginia Ave St Louis MO 63119

DWYER, LAWRENCE ARTHUR, b Minneapolis, Minn, Aug 1, 47. ENZYMOLOGY, LIPID BIOCHEMISTRY. *Educ:* Univ Nev, Reno, BA, 72, PhD(biochem), 80. *Prof Exp:* Microbiol tech & lab supvr, Cetus Corp, 73-75; res fel, Univ Nev, Reno, 75-80; fel, Biochem Dept, Med Sch, Northwestern Univ, 80-82; biochemist, Abbot Park, Ill, 82-83, biochemist & group leader, South Pasadena, Calif, 85-86, BIOCHEMIST & GROUP LEADER, ABBOTT LABS, IRVING, TEX, 86- *Concurrent Pos:* Res assoc biochem, Univ Nev, 83-85. *Mem:* Am Chem Soc; AAAS. *Res:* Mechanism of action of xenobiotic detoxication and activation reactions catalyzed by purified isozymes of rabbit liver cytochrome P-450; cuticular lipid biosynthesis in the American cockroach. *Mailing Add:* Abbott Labs PO Box 152020 Irving TX 75015-2020

DWYER, ROBERT FRANCIS, b Utica, NY, Feb 20, 30; m 53; c 2. ANALYTICAL CHEMISTRY, PHYSICAL CHEMISTRY. *Educ:* Syracuse Univ, BS, 51; Pa State Univ, MS, 53. *Prof Exp:* Res chemist, Linde Div, 53-61, sr res chemist, 61-62, group leader, Low Temperature Measurement & Radiation Serv, 62-66, supvr anal servs, 66-69, mgr anal serv, Realty Div, 69-73, mgr sci serv, Gen Serv, 73-80, DIR ADMIN, UNION CARBIDE CORP, 80- *Concurrent Pos:* Mem sampling & anal panel, Group on Composition Exhaust Gases, Coord Res Coun, 61-67. *Mem:* Am Chem Soc. *Res:* Instrumental methods of analysis; research applications of radioisotopes; radiation chemistry; blood preservation; low-temperature measurements of physical properties; rheologic and thermodynamics properties of slush and solid hydrogen. *Mailing Add:* 60 Allison Rd Katonah NY 10536

DWYER, SAMUEL J, III, b San Antonio, Tex, June 8, 32; m 53; c 6. ELECTRICAL ENGINEERING, RADIOLOGY. *Educ:* Univ Tex, BS, 57, MS, 59, PhD(elec eng), 63. *Prof Exp:* Instr elec eng, Univ Tex, 57-63; from asst prof to assoc prof, 63-70, PROF ELEC ENG, UNIV MO-COLUMBIA, 70-, DIR BIOENG, 76- *Concurrent Pos:* Res engr, Defense Res Lab, 60-61; NIMH res grant, 65- *Mem:* Inst Elec & Electronics Engrs; Asn Comput Mach; Inst Math Statist; Am Statist Asn; Sigma Xi. *Res:* Application of statistics and information theory to radiology; statistical communication theory. *Mailing Add:* 5801 W 90th St Overland Park KS 66207

DWYER, SEAN G, b New York, NY, Mar 12, 45; m 68; c 3. PHYSICAL ORGANIC CHEMISTRY, POLYMER CHEMISTRY. *Educ:* Univ NDak, BS, 66, PhD(org chem), 70. *Prof Exp:* Sr chemist product res, 70-76, res & develop mgr business develop, 77-78, mgr, New Markets Prod Res Sect, 78-82, tech dir Europe, Africa & Near East, S C Johnson & Sons, Inc, 82-86,; Com Prods Res Develop Dept Mgr, 86-88, PARTNER-CORP NEW PRODS & TECHNOL, 88- *Mem:* Am Chem Soc; Sigma Xi. *Res:* Arsonium ylides and imines; attempted synthesis of stable carbenes; product development in floor care, cleaners and sanitizers; personal care products, industrial cleaners. *Mailing Add:* Corp New Prod & Technologies 1525 Howe St Racine WI 53403

DWYER, TERRY M, b Troy, NY, Nov 27, 48. MEMBRANE BIOPHYSICS. *Educ:* Univ Rochester, MD, 77, PhD(physiol), 78. *Prof Exp:* Asst prof, 80-85, ASSOC PROF PHYSIOL, MED CTR, UNIV MISS, 85- *Mem:* Biophys Soc; Soc Neurosci; Am Physiol Soc; AAAS; Soc Gen Physiol. *Mailing Add:* Dept Physiol & Biophys Med Ctr Univ Miss 2500 N State St Jackson MS 39216-4505

DWYER, THOMAS A, b New York, NY, Nov 18, 23. COMPUTER SCIENCE, APPLIED MATHEMATICS. *Educ:* Dayton Univ, BS, 45; Case Western Reserve Univ, MS, 51, PhD(math), 60. *Prof Exp:* Chmn math dept, Cathedral Latin Sch, 47-57; assoc prof comput sci, Dayton Univ, 60-67; assoc prof, 68-76, PROF COMPUT SCI, UNIV PITTSBURGH, 76- *Concurrent Pos:* NSF fel, Case Western Reserve Univ; prin investr, Proj SOLO, 69-71, NSF Soloworks, 72-75 & Solo/Net/Works, 76-80; consult ed, Addison Wesley. *Mem:* Am Math Soc; Soc Indust & Appl Math; Asn Comput Mach. *Res:* Nonlinear networks and boundary value problems; optimization; computers in education; computer assisted instruction; man-machine systems. *Mailing Add:* Dept Comput Sci 322 Alumni Hall Univ Pittsburgh Main Campus 4200 Fifth Ave Pittsburgh PA 15260

DWYER, THOMAS ALOYSIUS WALSH, III, mathematics; deceased, see previous edition for last biography

DWYER-HALLQUIST, PATRICIA, b York, Pa, June 29, 54; m 79. ORGANIC SYNTHESIS. *Educ:* Hope Col, AB, 76; Univ Chicago, PhD(biochem), 81. *Prof Exp:* Asst prof chem, Univ Wis-Oshkosh, 81-83; sr res chemist, 83-85, RES ASSOC, APPLETON PAPERS INC, BATUS, 85- *Mem:* Am Chem Soc; AAAS. *Res:* Synthesis and design of color formers for carbonless and thermal paper; mechanism of type II restriction endonucleases and methylases. *Mailing Add:* 2030 Hazel St Oshkosh WI 54901

DY, KIAN SENG, b Philippines, June 28, 40; m 67. SOLID STATE PHYSICS. *Educ:* Ohio State Univ, BSc, 61; Cornell Univ, PhD(physics), 67. *Prof Exp:* Res assoc physics, Univ Ill, Urbana, 67-68; asst prof, 68-74, ASSOC PROF PHYSICS, UNIV NC, CHAPEL HILL, 74- *Res:* Quantum statistical mechanics; transport properties of fluids; lattice dynamics. *Mailing Add:* Dept Physics Univ NC Chapel Hill NC 27514

DYAL, PALMER, b Odon, Ind, Oct 27, 33; m 55; c 2. PHYSICS, ASTROPHYSICS. *Educ:* Coe Col, BA, 55; Univ Ill, Urbana-Champaign, PhD(chem physics), 59. *Hon Degrees:* DSc, Coe Col, 78. *Prof Exp:* Proj scientist, US Air Force, Kirtland AFB, 61-66; res scientist, 66-82, ASST DIR SPACE RES, AMES RES CTR, NASA, 82- *Concurrent Pos:* Fel physics, Univ Calif, Berkeley, 73-74. *Mem:* Am Phys Soc; Am Geophys Union; Am Astron Soc; Explorers Club. *Res:* Magnetic field research on the moon in the Apollo Program; magnetic field and particle experiments on high altitude nuclear bursts; photo production of pions from complex nuclei. *Mailing Add:* 26405 Ascension Dr Los Altos Hills CA 94022

DYAR, JAMES JOSEPH, b Marietta, Ohio, Nov 1, 31; m 58; c 3. PLANT PHYSIOLOGY. *Educ:* Univ WVa, AB, 54, MS, 57; Ohio State Univ, PhD(bot), 60. *Prof Exp:* Asst prof biol, Univ WVa, 60-61; researcher tobacco, Brown & Williamson Tobacco Corp, 61-62; from asst prof to assoc prof, 62-70, PROF & AREA COORDR BIOL, BELLARMINE COL, 70-, CHMN DEPT, 69- *Concurrent Pos:* Tobacco Indust Res Comt grant, 63-64; NSF res grant, 64-65. *Mem:* AAAS; Am Soc Plant Physiol; Am Inst Biol Sci. *Res:* Organic translocation in plants; molecular research in tobacco. *Mailing Add:* Dept Biol Bellarmine Col Louisville KY 40205

DYBALL, CHRISTOPHER JOHN, b Melksham, Eng, Apr 3, 51; m 75; c 2. POLYMER CHEMISTRY. *Educ:* Lancaster Univ, PhD(polymer chem), 75. *Prof Exp:* Trainee technician, Avon Rubber Co, Eng, 68-69; assoc, Nat Res Coun Can, 75-76; group leader polymer chem, Lucidol Div, Pennwalt Co, 76-79; mgr res & develop Specialty Polymers, Duolite Int, Diamond Shamrock Corp, 79-82; VPRES OPERS/GEN MGR, LASERCARD DIV, DREXLER TECHNOL CORP, 82- *Mem:* Am Chem Soc. *Res:* Polymerisation kinetics-structure, property relationships of polymers; optical data storage media. *Mailing Add:* 241 Mirada Rd Half Moon Bay CA 94019-1309

DYBALSKI, JACK NORBERT, b Chicago, Ill, Oct 19, 24; m 51; c 3. INDUSTRIAL ORGANIC CHEMISTRY, HIGHWAY ENGINEERING. *Educ:* Univ Chicago, PhB, 51. *Prof Exp:* Chemist, Armour Indust Chem Co, 53-54, res chemist, 54-60, sect leader asphalt res, 60-64, proj mgr water conserv, 62-71, res mgr, Hwy Chem Div, 71-78, MGR, HWY CHEM DEPT, ARMAK CO, 78-, COMMERCIAL DEVELOP, 64- *Concurrent Pos:* Mem hwy res bd, Nat Acad Sci-Nat Res Coun, 60- *Mem:* Am Chem Soc; Asn Asphalt Paving Technol; Am Soc Testing & Mat; Am Soc Civil Eng. *Res:* Bituminous research and development based on cationic concept pertaining to hydrological uses for water conservation and industrial uses in building, paving and hazardous waste containment. *Mailing Add:* 102 Scotdale Rd LaGrange IL 60525

DYBAS, LINDA KATHRYN, b Chicago, Ill, Oct 15, 42; m 86. ELECTRON MICROSCOPY, INVERTEBRATES. *Educ:* Knox Col, BA, 64; Calif State Univ, San Francisco, MA, 73; Univ Ulm, WGer, Dr rer(human biol), 76. *Prof Exp:* Res asst otolaryngol, Mass Eye & Ear Infirmary, 66, path, Med Ctr, Univ Calif, San Francisco, 67-74; asst, Univ Ulm, WGer, 74-76; asst prof, 77-85, ASSOC PROF BIOL, KNOX COL, GALESBURG, ILL, 85- *Mem:* AAAS; Int Soc Develop & Comp Immunol; Sigma Xi; Am Soc Zoologists. *Res:* Cell morphology, specifically, the structure and function, including cellular defense reactions, of marine invertebrate blood cells; spermatogenesis; oogenesis in insects and marine invertebrates. *Mailing Add:* Knox Col Box K-20 Galesburg IL 61401

DYBCZAK, Z(BIGNIEW) W(LADYSLAW), b Zaleszczyki, Poland, June 27, 24; US citizen; m 57; c 2. MECHANICAL & NUCLEAR ENGINEERING. *Educ:* Univ London, BSc, 50; Univ Toronto, PhD(mech eng), 59. *Prof Exp:* Indust design engr, Eng, 49-51; grad instr & res asst mech eng, Univ Toronto, 52-54, instr, 54-56, lectr, 56-59; assoc mech engr res & develop, Argonne Nat Lab, 60; prof & dean eng, Tuskegee Inst, 60-85; Dept Mech Eng, Univ Fla, 85-86; RETIRED. *Concurrent Pos:* Consult, Can & US Govt, Industs & Founds; vis prof, Univ Fla, 85- *Honors & Awards:* Bendix Award, Am Soc Eng Educ, 81; Centennial Medal, Am Soc Mech Engrs, 79. *Mem:* Am Soc Mech Engrs; Am Nuclear Soc; Am Soc Eng Educ. *Res:* Photoelastic stress analysis; design and vibration analysis; nuclear reactor shielding; experimental reactor physics; solar energy. *Mailing Add:* 129 Arrowhead Dr Montgomery AL 36117

DYBEL, MICHAEL WAYNE, b Hammond, Ind, July 19, 46; m 77; c 3. TECHNOLOGY ASSESSMENT, MANAGEMENT CONSULTING. *Educ:* Wabash Col, BA, 68; Northwestern Univ, MS, 69; Univ Chicago, MBA, 78. *Prof Exp:* Biochemist, Int Mineralse Chem Corp, 72-73; biochemist I, Abbott Labs, 73-78; assoc, Technomic Consults, 78-81; PRES, STRATEGIC TECHNOLOGIES INT, INC, 82- *Concurrent Pos:* SBIR grant, Dept Energy, 84; vis scientist, Shanghai Inst Planned Parenthood Res, Peoples Rep China, 88, Shemyakin Inst, USSR Acad Sci, 89. *Mem:* Am Chem Soc. *Mailing Add:* 808 Paddock Lane Libertyville IL 60048

DYBING, CLIFFORD DEAN, b Deadwood, SDak, Nov 6, 31; m 53; c 4. PLANT PHYSIOLOGY. *Educ:* Colo Agr & Mech Col, BS, 53, MS, 55; Univ Calif, Davis, PhD(plant physiol), 59. *Prof Exp:* Asst bot & plant path, Colo Agr & Mech Col, 53-55; asst bot, Univ Calif, Davis, 55-58; PLANT PHYSIOLOGIST, AGR RES SERV, USDA, 60- *Concurrent Pos:* Vis scientist, Prairie Regional Lab, Nat Res Coun Can, Sask, 67-68; prof, SDak State Univ, 71-; ed-in-chief, Crop Sci Soc Am & ed, Crop Sci J, 77-79. *Mem:* Am Soc Plant Physiol; Am Soc Agron; fel Crop Sci Soc Am; Sigma Xi. *Res:* Lipid metabolism; plant growth regulators; environmental influences on plant growth; fruit and seed development; flower production. *Mailing Add:* Univ Sta Box 2207-A Brookings SD 57007

DYBOWSKI, CECIL RAY, b Yorktown, Tex, Sept 23, 46; m 79; c 1. PHYSICAL CHEMISTRY. *Educ:* Univ Tex, Austin, BS, 69, PhD(chem physics), 73. *Prof Exp:* Res fel chem eng, Calif Inst Technol, 73-76; from asst prof to assoc prof chem, 76-86, PROF CHEM, UNIV DEL, 86- *Concurrent Pos:* Vis scientist, E I Du Pont de Nemours & Co, 81-82; assoc prof, Univ Paris, 89. *Mem:* Am Chem Soc; Sigma Xi; Am Phys Soc. *Res:* Nuclear magnetic resonance of solids; surface studies; dynamics in ordered fluids and polymers; author of two books. *Mailing Add:* Dept Chem & Biochem Univ Del Newark DE 19716

DYBVIG, DOUGLAS HOWARD, b Bemidji, Minn, Feb 14, 35; m 57; c 5. ORGANIC CHEMISTRY, IMAGING SCIENCE. *Educ:* St Olaf Col, BA, 57; Univ Ill, PhD(org chem), 61. *Prof Exp:* Sr res chemist, 3M Co, 60-70, mgr lab, 70-73, res dir, Cent Res, 73-77, tech dir, Indust Graphics Div, 77-; managing dir, Minn 3M Res Ltd, Essex Harlow, Eng, DIR, I&IT LABS, 3M CO, ST PAUL, MINN. *Concurrent Pos:* Asst prof chem, St Olaf Col, 64-65. *Mem:* Am Chem Soc; Soc Imaging Sci & Tech. *Res:* Organic reaction mechanisms; synthesis and chemistry of fluorine compounds; chemistry of rocket fuels; printing; color reproduction. *Mailing Add:* 3M Co 3M Ctr St Paul MN 55144

DYBVIG, PAUL HENRY, b Dayton, Ohio, Oct 16, 55. COMMUNICATIONS FRONT END INTERFACE. *Educ:* Univ Ariz, BS, 79, BS, 87. *Prof Exp:* ELEC ENGR C3I TEST EQUIP, US ARMY ELECTRONIC PROVING GROUND, 87- *Mem:* Inst Elec & Electronics Engrs. *Res:* Development of the Army's newest command, control, communications, and intelligence test item stimulator. *Mailing Add:* Ash St Huachuca City AZ 85616

DYCE, ROLF BUCHANAN, b Guelph, Ont, Oct 12, 29; US citizen; m 59, 89; c 2. PLANETARY SCIENCES. *Educ:* Cornell Univ, BS, 51, PhD(elec eng), 55. *Prof Exp:* Res engr radio propagation, Stanford Res Inst, 57-63, staff scientist, 63-64; res assoc planetary radar, Arecibo Observ, Cornell Univ, 64-65, assoc dir, 65-79; chief scientist, Equatorial Commun Co, 79-90; DEVELOP ENG, CELLULAR DATA INC, 90- *Concurrent Pos:* Mem comn III, IV, US Nat Comt, Int Sci Radio Union, 53-; propagation ed, Trans Group Antennas & Propagation, 64-69; res assoc, Radiosci Lab, Stanford Univ, 66-68. *Mem:* Inst Elec & Electronics Engrs; Sigma Xi; AAAS. *Res:* Experimental studies of the ionosphere employing low frequency radar, satellite or cosmic signals; radar exploration of the moon and planets. *Mailing Add:* 699 Torrington Dr Sunnyvale CA 94087

DYCK, GERALD WAYNE, b Borden, Sask, July 11, 38; m 66; c 3. REPRODUCTIVE PHYSIOLOGY. *Educ:* Univ Sask, BSA, 60; Univ Man, MSc, 63; Iowa State Univ, PhD(reproductive physiol), 66. *Prof Exp:* Res scientist, 66-71, RES SCIENTIST II, AGR CAN, 71- *Concurrent Pos:* Adj prof, Univ Man, 69- *Mem:* Agr Inst Can; Brit Soc Study Fertil; Soc Study Reproduction; Am Soc Animal Sci; Can Soc Animal Sci (pres, 77-78); Sigma Xi. *Res:* Reproductive physiology in female swine: ovulation rate, embyonic survival and litter size (nutritional and environmental effects); estrus-post weaning; prepubertal development and controlled stimulation of puberty. *Mailing Add:* Agr Can Res Sta Box 610 Brandon MB R7A 5Z7 Can

DYCK, PETER LEONARD, b Manitoba, Mar 31, 29; m 58; c 2. GENETICS. *Educ:* Univ Man, BSA, 56, MSc, 57; Univ Calif, Davis, PhD(genetics), 60. *Prof Exp:* RES SCIENTIST, CAN DEPT AGR, 60- *Mem:* Genetics Soc Can; Am Soc Agron. *Res:* Genetics of wheat, particularly leaf and stem rust resistance. *Mailing Add:* Can Dept Agr 195 Dafoe Rd Winnipeg MB R3T 2M9 Can

DYCK, RUDOLPH HENRY, b Pasadena, Calif, Apr 17, 31; m 55; c 3. PHYSICAL CHEMISTRY. *Educ:* Univ Calif, BS, 52, PhD(chem), 56. *Prof Exp:* Mem tech staff, RCA Labs, 55-62; mem tech staff, 62-67, SECT MGR, OPTOELECTRONICS, FAIRCHILD RES & DEVELOP LAB, FAIRCHILD SEMICONDUCTOR, 67- *Mem:* AAAS; Am Phys Soc; Soc Photo-Optical Instrumentational Engrs. *Res:* Photosensitive devices, arrays and subsystems; light emitting diodes, arrays and subsystems; CCD color imagers, infrared imagers, and fiber optic coupled imagers; low noise CCD preamplifiers and CCD tapped delay line correlations. *Mailing Add:* 1801 McCarthy Blvd Milpitas CA 95035

DYCK, WALTER PETER, b Winkler, Man, Dec 7, 35; c 4. INTERNAL MEDICINE, GASTROENTEROLOGY. *Educ:* Bethel Col, Kans, BA, 57; Univ Kans, MD, 61. *Prof Exp:* Intern & resident med, Henry Ford Hosp, Detroit, 61-63 & 65-66; investr exp physiol, Scott & White Mem Hosp, 68-, sr staff consult gastroenterol, Scott & White Clin, 68-, DIR, DIV GASTROENTEROL, 73-; PROF INTERNAL MED & DIR, DIV GASTROENTEROL, TEX A&M UNIV, COL MED, 78- *Concurrent Pos:* Res fel gastroenterol, Univ Zurich, 63-64; res fel gastrointestinal enzyme, Univ Toronto & Hosp for Sick Children, Toronto, 64-65; NIH training fel gastroenterol, Mt Sinai Sch Med, 66-68; consult, Vet Admin Ctr, Temple, Tex, 68- & Gen Med A Study Sect, NIH, 73-77; chmn dept res, Scott & White Clin, 70-73. *Mem:* AMA; Am Fedn Clin Res; Am Gastroenterol Asn; Am Col Physicians; Soc Exp Biol & Med; Am Physiol Soc. *Res:* Hormonal control of pancreatic secretion; intestinal enzyme secretion; gastric secretion; peptic ulcer disease. *Mailing Add:* Dept Med Scott & White Clin 2401 S 31st St Temple TX 76508

DYCKES, DOUGLAS FRANZ, b New Haven, Conn, May 13, 42; m 67; c 2. BIOLOGICAL CHEMISTRY. *Educ:* Yale Univ, AB, 63; Case Western Reserve Univ, PhD(chem), 70. *Prof Exp:* Res assoc chem, Cornell Univ, 70-72; sci officer, MRC Lab Molecular Biol, 72-74; from asst prof to assoc prof, 74-88, PROF CHEM, UNIV HOUSTON, 88- *Concurrent Pos:* vis scientist, MRC Lab Molecular Biol, 81. *Mem:* AAAS; Am Chem Soc. *Res:* The synthesis, structure and basis of biological activity in naturally occurring peptides and proteins and their analogs. *Mailing Add:* Dept Chem Box 194 Univ Colo PO Box 173364 Denver CO 80217-3364

DYE, DAVID L, b Seattle, Wash, Aug 5, 25; m 52; c 3. PHYSICS. *Educ:* Univ Wash, BS, 45, PhD(physics), 52. *Prof Exp:* Res assoc, Radiol Lab, Med Ctr, Univ Calif, 54-55; chmn physics, Gordon Col, Rawalpindi, 55-58; res assoc, Radiol Lab, Med Ctr, Univ Calif, 58-59; res specialist, Aero-Space Div, Boeing Co, 59-62, chief radiation effects lab, 62-64, chief radiation effects unit, 64-68; sr scientist, Air Force Spec Weapons Ctr, 68-70; mem staff, Minuteman Physics Technol, Boeing Comput Serv, 70-72, mgr, Minuteman Hardness Data Prog, 72-74, mgr energy systs, 74-77, nuclear sci, 77-80, electronic radar counter measures, 81-90; RETIRED. *Mem:* AAAS; Am Phys Soc; Inst Elec & Electronics Engrs. *Res:* Radiation dosimetry and health physics; radiation effects on electronics components and systems; space radiation and nuclear physics; system survivability; nuclear reactor physics; fuel cycle technology; philosophy of science. *Mailing Add:* 12825 SE 45th Pl Bellevue WA 98006-2031

DYE, FRANK J, b Bronx, NY, Jan 12, 42; m 67; c 2. DEVELOPMENTAL BIOLOGY. *Educ:* Western Conn State Univ, BS, 63; Fordham Univ, MS, 66, PhD(cytol), 69. *Prof Exp:* NIH fel, Fordham Univ, 65-67; from asst prof to assoc prof, 67-78, PROF BIOL, WESTERN CONN STATE UNIV, 78- *Concurrent Pos:* Res assoc, New Eng Inst, 68-70, adj prof, 73; Nat Inst Dent Res fel, Univ Conn Health Ctr, 75-76, vis assoc prof, 76-78; vis fel, Yale Univ, 82-87; CSU-Am Asn Univ Profs grant, 85, 86 & 87. *Mem:* AAAS; Tissue Cult Asn; Am Inst Biol Scientists; Am Soc Zool; Sigma Xi; NY Acad Sci; Am Soc Cell Biol. *Res:* Origin of tissue culture populations; morphogenesis; gene activation; in vitro movement of epithelial cells; tooth germ morphogenesis; purine metabolism. *Mailing Add:* Dept Biol Western Conn State Univ Danbury CT 06810

DYE, HENRY ABEL, b Dunkirk, NY, Feb 14, 26; m 78; c 2. MATHEMATICS. *Educ:* Univ Chicago, MS, 47, PhD(math), 50. *Prof Exp:* Bateman fel, Calif Inst Technol, 50-52; instr math, Calif Inst Technol, 52-53; mem sch math, Inst Adv Study, 53-54; asst prof math, Univ Iowa, 54-56; assoc prof, Univ Southern Calif, 56-59; assoc prof, Univ Iowa, 59-60; chmn dept, 75-78, PROF MATH, UNIV CALIF, LOS ANGELES, 60- *Mem:* Am Math Soc; Sigma Xi. *Res:* Functional analysis. *Mailing Add:* Dept Math Univ Calif 405 Hilgard Ave Los Angeles CA 90024

DYE, JAMES EUGENE, b Rock Springs, Wyo, Oct 17, 39. CLOUD PHYSICS, ATMOSPHERIC SCIENCES. *Educ:* Univ Wash, BS, 62, PhD(atmospheric sci), 67. *Prof Exp:* Res assoc, Dept Atmospheric Sci, Univ Wash, 62-67; res assoc, Inst Meteorol, Univ Stockholm, 67-68; asst prof, Atmospheric Simulation Lab, Dept Mech Eng, Colo State Univ, 69-70; SCIENTIST, NAT CTR ATMOSPHERIC RES, 70- *Concurrent Pos:* Fulbright travel grant, Comt Int Exchange Persons, Fulbright-Hays prog with Sweden, 67-68; affil prof, Colo State Univ, 72-; assoc ed, J Atmospheric & Oceanic Tech, 83-88; ed, J Geophys Res Atmosphere, 88- *Mem:* Am Meteorol Soc; Am Geophys Soc. *Res:* Cloud and precipitation physics including weather modification; aerosol physics; meteorological instrumentation. *Mailing Add:* Nat Ctr Atmospheric Res Box 3000 Boulder CO 80303

DYE, JAMES LOUIS, b Soudan, Minn, July 18, 27; m 48; c 3. PHYSICAL CHEMISTRY, SOLID-STATE CHEMISTRY. *Educ:* Gustavus Adolphus Col, AB, 49; Iowa State Col, PhD(chem), 53. *Prof Exp:* Asst phys chem, Inst Atomic Res & Dept Chem, Iowa State Col, 49-53; chmn dept, 86-90, PROF PHYS CHEM, MICH STATE UNIV, 53- *Concurrent Pos:* NSF sci fac fel, Max Planck Inst, Gottingen, Ger, 61-62; vis scientist, Ohio State Univ, 68-69, AT&T Bell Labs, 82-83; chmn, Postdoctoral Fel Comt Chem, Nat Res Coun, 72-74; Fulbright res scholar & Guggenheim fel, 75-76; mem adv comt, Div Mat Res, NSF, 80-84; dir, Ctr Fundamental Mat Res, Mich State Univ, 85-86; Guggenheim fel & vis scientist, Cornell Univ, 90-91. *Honors & Awards:* Sigma Xi Jr Award, 68, Sr Award, 87. *Mem:* Nat Acad Sci; Am Chem Soc; Sigma Xi; fel AAAS; Am Acad Arts & Sci; Am Phys Soc; Mat Res Soc. *Res:* Synthesis and properties of crystalline salts of alkali metal anions (alkalides) and trapped electrons (electrides); measuremnt of solid-state optical, NMR and ESR spectra, conductivities, magnetic susceptibilities, photo and thermionic-electron emission. *Mailing Add:* Dept Chem Mich State Univ East Lansing MI 48824-1322

DYE, ROBERT F(ULTON), b Gloster, Miss, Oct 18, 20; m 47; c 3. CHEMICAL ENGINEERING. *Educ:* Miss State Univ, BS, 43; Ga Inst Tech, MS, 51, PhD(chem eng), 53. *Prof Exp:* Chem engr, Inorg & Org Res, Monsanto Chem Co, 46-49; asst, chem eng dept, Ga Inst Technol, 50-51 & 52-53; sr process & design proj engr, Process Develop Div, Phillips Petrol Co, 53-62; dir, Miss Indust & Technol Res Comn, 62-65; STAFF ENGR, SHELL OIL CO, 65- *Mem:* Fel AAAS; Am Chem Soc; fel Am Inst Chem Engrs; NY Acad Sci. *Res:* Gas diffusion; diffusional processes; chemical technology; chemical process economics. *Mailing Add:* Shell Oil Co PO Box 2099 Houston TX 77252

DYE, WILLIAM THOMSON, JR, b Chattanooga, Tenn, July 8, 18; m 44; c 2. ORGANIC CHEMISTRY. *Educ:* Univ NC, BS, 40, PhD(org chem), 44. *Prof Exp:* Res chemist, Naval Res Lab, DC, 44-46; res chemist, Monsanto Chem Co, 47-52; res chemist, Chemstrand Corp, 52-58, group leader, 58-60; group leader, Chemstrand Res Ctr, Inc, Monsanto Co, 60-70; MGR PATENT LIAISON, BURLINGTON INDUSTS CORP RES & DEVELOP, 70- *Mem:* Am Chem Soc. *Res:* Organic phosphorus compounds; synthetic fibers; textiles; patent liaison. *Mailing Add:* 379 Fearrington Post Pittsboro NC 27312-8502

DYER, ALAN RICHARD, b Seattle, Wash, Aug 19, 45. BIOSTATISTICS. *Educ:* Stanford Univ, BS, 67; Univ Chicago, MS, 68, PhD(statist), 72. *Prof Exp:* From asst prof to assoc prof, 72-85, PROF COMMUNITY HEALTH & PREV MED, MED SCH, NORTHWESTERN UNIV, 85-, ACTG CHMN, 86- *Mem:* Am Statist Asn; Biomet Soc; Am Heart Asn. *Res:* Epidemiology; cardiovascular disease. *Mailing Add:* Dept Community Health & Prev Med Northwestern Univ Med Sch Chicago IL 60611

DYER, ALLAN EDWIN, b Toronto, Ont, Aug 23, 23; m 44; c 4. PHARMACOLOGY. *Educ:* Univ Toronto, PhmB, 49, PhD(pharmacol), 55, MD, 67; Univ Buffalo, BSc, 51. *Prof Exp:* Res assoc & head bioassay dept, Connaught Med Res Labs, 56-67; intern, Toronto Western Hosp, 67-68; chief drugs & biol, 68-75, dir drugs & therapeut, 75-77, exec chmn, Area Planning Coordr, 77-81, assoc dep minister, Instnl Health, 81-84, dep minister environ, 84-85, DEP MINISTER HEALTH, ONT MINISTRY HEALTH, 85- *Concurrent Pos:* Can Life Ins Off Asn fel, Univ Toronto, 54-56, univ fel, 55-56. *Mem:* Pharmacol Soc Can; Can Fedn Biol Soc. *Res:* Methodology for screening and assessing pharmacological and toxicological activity. *Mailing Add:* Hepburn Block Ministry Health Toronto ON M7A 1R3 Can

DYER, CHARLES AUSTEN, b St Ann's Bay, Jamaica, WI, July 24, 36; US citizen; m 61; c 2. ARTIFICIAL INTELLIGENCE, DATA MANAGEMENT. *Educ:* Pratt Inst, BID, 57; Yeshiva Univ, MS, 62; City Univ NY, PhD(educ psychol), 80. *Prof Exp:* Prog instr & support, Int Bus Mach, 68-70; instr, City Univ NY, 70-73; training coordr, Digital Equip Corp, 73-76, prog analyst, 76-79, software engr, 79-84, prog mgr, 84-86, artificial int consult, 86-87, SOFTWARE ENGR, DIGITAL EQUIP CORP, 87- *Concurrent Pos:* Consult, Stuart Taylor Assoc, 82-; vis lectr, Framingham State Col, 82-83; consult, Boston Univ Med Sch, 84-, Mass Pre-Eng Prog, 87-88. *Res:* Concepts and software for large-scale data management; perceptual hyperfields (multi-dimensional, holographic, non-Euclidean fields) in cognition and data retrieval; paradigm for transfer in learning; software that generates software. *Mailing Add:* 203 Grove St Framingham MA 01701

DYER, CHARLES CHESTER, b Sutton, Que, June 8, 46. COSMOLOGY. *Educ:* Bishop's Univ, Que, BSc, 68; Univ Toronto, MSc, 69, PhD(astron), 73. *Prof Exp:* Vis astron, Kitt Peak Nat Observ, 71-72; sr visitor, Inst Astron, Univ Cambridge, Eng, 73-75; instr & res assoc, 75-78, ASST PROF DEPT ASTRON, UNIV TORONTO, 78- *Concurrent Pos:* Univ res fel, 80- *Mem:* Can Astron Soc; Int Soc Gen Relativity & Gravitation; Royal Astron Soc. *Res:* Optics in general relativity and its application to problems in observational cosmology; gravitational lenses and applications. *Mailing Add:* Dept Astron Scarborough Col Univ Toronto 1265 Military Trail Scarborough ON M1C 1A4 Can

DYER, CHARLES ROBERT, b Gilroy, Calif, Mar 14, 51; m 79. IMAGE PROCESSING, COMPUTER VISION. *Educ:* Stanford Univ, BS, 73; Univ Calif, Los Angeles, MS, 74; Univ Md, PhD(comput sci), 79. *Prof Exp:* Grad res asst, Univ Calif, Los Angeles, 74; grad res asst, Comput Sci Ctr, Univ Md, 75-76, fac res asst, 76-79, fac res assoc, 79; asst prof, Dept Elec Eng & Computer Sci, Univ Ill, Chicago, 79-82; from asst prof to assoc prof, 82-88, PROF, DEPT COMPUTER SCI, UNIV WIS-MADISON, 88-, CHAIR, 90- *Concurrent Pos:* Consult, Imtech, Inc, 76-79, NIH, 79-81, Int Harvester, 81-82 & Nat Inst Sci & Technol, 83-85. *Mem:* Asn Comput Mach; Am Asn Artificial Intel; Inst Elec & Electronics Engrs. *Res:* Parallel algorithms and computer architectures for image processing and scene analysis; hierarchial representation techniques for computer vision; texture analysis; model-based three-dimensional object recognition. *Mailing Add:* Dept Comput Sci Univ Wis Madison WI 53706

DYER, DENZEL LEROY, b McCool Jct, Nebr, Oct 12, 29; m 52; c 3. CHEMISTRY. *Educ:* York Col, BS, 50; Univ Nebr, MS, 53, PhD(chem), 55. *Prof Exp:* Chemist, Dow Chem Co, 55-59; assoc res scientist, Life Sci Dept, Martin Marietta Corp, 59-64; prin scientist, Life Sci Dept, Northrop Corp Labs, 64-69; OWNER, DYER LABS, 69- *Mem:* Am Chem Soc; Am Soc Microbiol; Soc Indust Microbiol; Consult Chemists Asn (pres, 70-75). *Res:* Microbiology; microbial physiology and chemistry. *Mailing Add:* 2531 W 237th St Suite 121 Torrance CA 90505

DYER, DONALD CHESTER, b Great Bend, Kans, July 8, 39; m 59; c 2. PHARMACOLOGY. *Educ:* Univ Kans, BS, 61, PhD(pharmacol), 65. *Prof Exp:* Asst prof pharmacol, Ore State Univ, 67-68; from asst prof to assoc prof, Sch Med, Univ Wash, 68-75; assoc prof vet physiol & pharmacol, 75-80, chmn dept, 80-81, PROF VET PHYSIOL & PHARMACOL, IOWA STATE UNIV, 80- *Concurrent Pos:* Fel pharmacol, Postdoctoral (vascular smooth muscle), Univ Man, 65-67. *Mem:* Am Soc Pharmacol & Exp Therapeut; Western Pharmacol Soc. *Res:* Polypeptides; smooth muscle stimulating lipids; hallucinogens; autonomic drugs; fetal pharmacology. *Mailing Add:* Dept Vet Physiol & Pharmacol Iowa State Univ Col VVet Med Ames IA 50011

DYER, ELIZABETH, b Haverhill, Mass, May 10, 06. ORGANIC CHEMISTRY. *Educ:* Mt Holyoke Col, AB, 27, AM, 29; Yale Univ, PhD(chem), 31. *Prof Exp:* Asst chem, Mt Holyoke Col, 27-29; asst chem, Yale Univ, 29-31, Chem Found fel, 31-33; from instr to prof chem, 33-71, EMER PROF CHEM, UNIV DEL, 71- *Mem:* Am Chem Soc. *Res:* Polymers; isocyanates; pyrimidines and purines. *Mailing Add:* 115 Cokesbury Village Hockessin DE 19707-1503

DYER, FRANK FALKONER, b Webbers Falls, Okla, Nov 18, 31; m 57; c 2. ANALYTICAL CHEMISTRY, NUCLEONICS. *Educ:* Okla State Univ, BS, 53, MS, 55; Univ Tenn, PhD(chem), 58. *Prof Exp:* Chemist, Pan Am Petrol Corp, 58-60; CHEMIST, OAK RIDGE NAT LAB, 60- *Honors & Awards:* Elected to Inventors Forum, 85. *Mem:* Am Chem Soc; Am Nuclear Soc. *Res:* Gamma-ray spectroscopy; activation analysis; radiochemistry; evaluation of condition and performance of nuclear reactor fuel; diffusion processes; fission product behavior; nuclear safety. *Mailing Add:* 4417 Crestfield Rd Knoxville TN 37921

DYER, HUBERT JEROME, botany; deceased, see previous edition for last biography

DYER, IRA, b Brooklyn, NY, June 14, 25; m 49; c 2. ARCTIC SCIENCE, ACOUSTICS. *Educ:* Mass Inst Technol, SB, 49, SM, 51, PhD(acoustics), 54. *Prof Exp:* Asst physics, Mass Inst Technol, 49-51; acoust scientist, Bolt Beranek & Newman Inc, 51-61, vpres & dir phys sci, 61-71, dir prog advan study, 64-67; dept head, 71-81 PROF OCEAN ENG, MASS INST TECHNOL, 71- *Concurrent Pos:* Assoc dir and dir sea grant prog, Mass Inst Technol, 72-75; chmn, res adv comt, US Coast Guard, 75-79; mem, Bd Visitors, Maine Maritime Acad, 73-81, mem, Marine Bd, Nat Res Coun, 78-81. *Honors & Awards:* Acoust Soc Am Biennial Award, 60. *Mem:* Nat Acad Eng; fel Acoust Soc Am (pres); fel AAAS; fel Inst Elec & Electronics Engrs. *Res:* Acoustic waves and vibrations; acoustic scattering and diffraction; noise of aerodynamic origin; structure-borne sound; underwater acoustics; Arctic Ocean acoustics; oceanographic engineering; ocean environmental acoustics; sonar engineering. *Mailing Add:* Nine Cliff St Marblehead MA 01945

DYER, JAMES ARTHUR, b San Antonio, Tex, Feb 10, 32; m 53. MATHEMATICS. *Educ:* Univ Tex, BS, 52, MA, 54, PhD(math), 60. *Prof Exp:* Res scientist physics, Defense Res Lab, Univ Tex, 54-58, asst prof math, Univ, 60-61; asst prof, Univ Ariz, 61-62; from asst prof to assoc prof, Southern Methodist Univ, 62-65; assoc prof math, 66-70, PROF, IOWA STATE UNIV, 70-; MGR CAD SYSTS, DIGITAL EQUIPMENT, MARLBOROUGH, MASS. *Mem:* Math Asn Am; Soc Indust & Appl Math; Inst Elec & Electronics Engrs. *Res:* Integration; functional analysis; integral equations; signal processing. *Mailing Add:* Polaroid Corp 565 Tech Sq 6J Cambridge MA 02139

DYER, JAMES LEE, b Long Beach, Calif, Sept 2, 34; m 60. ENGINEERING. *Educ:* Univ Calif, Los Angeles, BS, 57, MS, 60, PhD(eng), 65. *Prof Exp:* Teaching fel eng, Univ Calif, Los Angeles, 65; mem tech staff, TRW Systs, 65-66; from asst prof to assoc prof, 66-76, chmn dept, 70-74, PROF MECH ENG, CALIF STATE UNIV, LONG BEACH, 76- *Mem:* AAAS. *Res:* Thermodynamics of phase changes and irreversible processes. *Mailing Add:* Dept Mech Eng 1250 Bellflower Blvd Long Beach CA 90840

DYER, JOHN, b Beckenham, Eng, June 8, 35; nat US; m 64; c 2. POLYMER CHEMISTRY. *Educ:* ARIC, 59; Univ Manchester, PhD(chem), 64. *Prof Exp:* Lab asst, Brit Nylon Spinners, 52-57, res asst, 57-58, asst tech officer, 59; res chemist, Am Viscose Div, FMC Corp, 64-68, sr res chemist, 68-74, res assoc, Fiber Div, 74-76; res assoc, 77, Eastern Res Div, ITT Rayonier Inc, 77, mgr appl res, 78-81; STAFF SCIENTIST, JOHNSON & JOHNSON PROD, INC, 82- *Honors & Awards:* Fiber Soc Award, 73. *Mem:* Am Chem Soc; Fiber Soc; Tech Asn Pulp & Paper Indust. *Res:* Cellulose and viscose rayon; thiocarbonates, reaction mechanisms and kinetics; fibers, polymers, characterization and applications; technical-economic investigations of process and product development projects; pollution chemistry; nonwovens, personal care and wound dressings. *Mailing Add:* 57 Meadowbrook Rd Randolph NJ 07869-3800

DYER, JOHN KAYE, b Portland, Maine, Jan 25, 35; m 63; c 2. MICROBIOLOGY, IMMUNOLOGY. *Educ:* Eastern NMex Univ, BS, 63; Ore State Univ, MS, 65, PhD(microbiol), 67; Registry Med Technol, cert, 59; Nat Registry Microbiologists, cert pub health & med lab microbiol, 74. *Prof Exp:* Med technologist, Borgess Hosp, Kalamazoo, Mich, 59-63; res assoc microbiol, Mich State Univ, 67-69; from asst prof to assoc prof, 69-86, PROF MICROBIOL, DEPT ORAL BIOL, COL DENT, UNIV NEBR-LINCOLN, 86- *Concurrent Pos:* Nat Inst Dent Res spec dent res award grant, Col Dent, Univ Nebr-Lincoln, 73-76; consult, Vet Admin Hosp, Lincoln, Nebr, 72-90. *Mem:* Am Soc Microbiol; Am Soc Clin Pathologists; Int Asn Dent Res. *Res:* Metabolism in anaerobic and halotolerant bacteria; identification of anaerobic bacteria; hypersensitivity immune responses against bacterial antigens as associated with periodontal disease; isolation and characterization of the cellular components of periodontopathic and halotolerant bacteria. *Mailing Add:* Dept Oral Biol Col Dent Univ Nebr Med Ctr 40th & Holdridge Lincoln NE 68583

DYER, JOHN NORVELL, nuclear physics; deceased, see previous edition for last biography

DYER, JUDITH GRETCHEN, b New York, NY, Apr 27, 37; m 62; c 1. ENTOMOLOGY. *Educ:* Bethany Col, WVa, BA, 58; William Paterson Col NJ, MS, 71; Rutgers Univ, PhD, 75. *Prof Exp:* Ed, Int Tel & Tel, 60-62; asst prof biol, St Peter's Col, NJ, 74-78; agr tech ed/writer, Am Cyanamid Co, 79-82. *Mem:* Entom Soc Am. *Res:* Arrhenotokous reproduction in Phytoseiid mites; environmental effects on mite populations; distribution and abundance of predatory mites. *Mailing Add:* 123 Lakeshore Dr Oakland NJ 07436

DYER, LAWRENCE D, b Los Angeles, Calif, Sept 3, 30; m 52; c 3. PHYSICAL CHEMISTRY. *Educ:* Calif Inst Technol, BS, 51; Univ Va, PhD(phys chem), 57. *Prof Exp:* Eng asst, Helipot Corp, 49-50; chem asst org chem, US Naval Ord Test Sta, Calif, 51; jr chemist fused hydroxide corrosion, Oak Ridge Nat Lab, 51-53; sr res phys chemist, Res Labs, Gen Motors Corp, 57-66; mem tech staff, 66-84, SR MEM TECH STAFF, TEX INSTRUMENTS INC, 84- *Concurrent Pos:* Expert witness, glass fracture, 82- *Mem:* Electrochem Soc; Am Asn Crystal Growth; Am Chem Soc; Am Ceramic Soc. *Res:* Surface physics and chemistry; friction and wear; fused salts and hydroxides; crystal growth of copper and silicon; silicon epitaxial growth; plastic deformation in copper and silicon; crystal wafer and chip fracture; caustic etching of silicon; sterilization of di water for semiconductor processing. *Mailing Add:* Texas Instruments Inc PO Box 84 Mail Sta 884 Sherman TX 75090

DYER, MELVIN I, b Havre, Mont, Oct 18, 32; m 57; c 2. ZOOLOGY. *Educ:* Univ Idaho, BS, 55; Univ Minn, MS, 61, PhD(zool), 64. *Prof Exp:* Asst prof zool, Univ Guelph, 64-67; affil assoc prof, Ohio State Univ, 67-72; dir process studies, US Int Biol Prog Grassland Biome, Natural Resource Ecol Lab & res assoc, Dept Fisheries & Wildlife Biol, 72-74, assoc prof, 74-78, PROF DEPT FISHERY & WILDLIFE BIOL, COLO STATE UNIV, 78-; AT ENVIRON SCI DIV, OAK RIDGE NAT LAB. *Concurrent Pos:* Biologist-supvr, Patuxent Wildlife Res Ctr, US Bur Sport Fisheries & Wildlife, 67-72; dir, Ecosyst Studies Prog, Div Environ Biol, NSF, 78-; sci dir, Nat Resource Ecol Lab, 77- *Mem:* Am Ornith Union; Cooper Ornith Soc; AAAS; Ecol Soc Am; Wildlife Soc. *Res:* Systems ecology; North American grasslands; avian ecology. *Mailing Add:* Rte 2 Box 330-A Lenoir TN 37771

DYER, PEGGY LYNN, b Bryan, Tex, Sept 13, 46; m 80; c 1. NUCLEAR PHYSICS. *Educ:* Univ Tex, Austin, BS, 68; Calif Inst Technol, PhD(physics), 73. *Prof Exp:* Res fel, Kellogg Radiation Lab, Calif Inst Technol, 73-74; res assoc, Nuclear Physics Lab, Univ Wash, 74-77; asst prof physics, Cyclotron Lab, Mich State Univ, 77-81; STAFF MEM, LOS ALAMOS NAT LAB, 81- *Concurrent Pos:* Mem comt nuclear sci, Nat Res Coun, 74-77. *Mem:* Am Phys Soc. *Res:* Experimental nuclear astrophysics, heavy ion physics, weak interaction physics; laser spectroscopy; gamma ray lasers; antimatter gravity. *Mailing Add:* MS-D449 Los Alamos Nat Lab Los Alamos NM 87545

DYER, RANDOLPH H, b Ft Smith, Ark, Oct 31, 40; m 64; c 2. BIOCHEMISTRY. *Educ:* Transylvania Col, AB, 62. *Prof Exp:* Chemist, Melpar Inc, Va, 63-66; CHEMIST, ALCOHOL, TOBACCO & FIREARMS LAB, US TREASURY DEPT, 66- *Mem:* Asn Off Anal Chemists; Am Soc Brewing Chemists; Am Soc Enol; Int Food Technol. *Res:* Analytical chemistry and biochemistry, especially enzymes and fermentation processes and products. *Mailing Add:* 8504 Georgan Pl Annandale VA 22003

DYER, ROBERT FRANK, anatomy, cell biology, for more information see previous edition

DYER, ROLLA MCINTYRE, JR, b Elizabethtown, Ky, Dec 30, 22; m 47; c 3. ANALYTICAL CHEMISTRY. *Educ:* Univ Louisville, PhD, 63. *Prof Exp:* Instr chem, Campbellsville Col, 59-60; vis instr, Univ Louisville, 60 63; asst prof, Northeast La Univ, 63-67; from asst prof to assoc prof, 67-71, PROF CHEM, IND STATE UNIV, EVANSVILLE, 71- *Res:* Separation and identification of components of complex mixtures; design and preparation of instructional models especially three dimensional magnetic field maps. *Mailing Add:* Div Sci & Math Univ Southern Ind 8600 University Blvd Evansville IN 47712

DYER, WILLIAM GERALD, b Boston, Mass, Oct 27, 29; m 65; c 2. HELMINTHOLOGY. *Educ:* Boston Univ, AB, 57, AM, 58; Colo State Univ, PhD(parasitol), 65. *Prof Exp:* Res asst endocrinol, Worcester Found Exp Biol, 58-60; res asst biochem, Harvard Med Sch, 60-62; asst prof, Minot State Col, 65-66, assoc prof, 66-67, prof zool & head, Dept Biol, 67-69; from asst prof, 69-75, asst dean, Col Sci, 75-76, PROF ZOOL & ASSOC DEAN, SOUTHERN ILL UNIV, 76- *Mem:* Am Soc Parasitol; Wildlife Dis Asn; Am Inst Biol Sci. *Res:* Taxonomy; ecology; marine parasitology. *Mailing Add:* Dept Zool Southern Ill Univ Carbondale IL 62901

DYER-BENNET, JOHN, b Leicester, Eng, Apr 17, 15; nat US; m 51; c 2. ALGEBRA. *Educ:* Univ Calif, AB, 36; Harvard Univ, MA, 39, PhD(math), 40. *Prof Exp:* Asst math, Univ Calif, 36-37; instr, Vanderbilt Univ, 40-41 & 45-46; from instr to assoc prof, Purdue Univ, 46-60; assoc prof, 60-65, chmn dept math & astron, 64-66, prof, 65-80, EMER PROF MATH, CARLETON COL, 80- *Concurrent Pos:* NSF faculty fel, Switz, 58-59. *Mem:* Am Math Soc; Math Asn Am. *Res:* Abstract algebra. *Mailing Add:* Dept Math Carleton Col Northfield MN 55057

DYKE, BENNETT, US citizen. BIOLOGICAL ANTHROPOLOGY, HUMAN GENETICS. *Educ:* Trinity Col, BA, 55; Univ Mich, PhD, 68. *Prof Exp:* Instr anthrop, Bucknell Univ, 64-70; asst prof anthrop, Pa State Univ, 70-74, assoc prof, 74-81; ASSOC SCIENTIST, SOUTHWEST FOUND BIOMED RES, SAN ANTONIO, 81- *Concurrent Pos:* Pop Coun fel, 69-70. *Mem:* AAAS; Am Soc Human Genetics; Soc Study Human Biol. *Res:* Genetic demography; computer modeling; population genetics. *Mailing Add:* 98 Cliffside Pl No 512 San Antonio TX 78231

DYKE, RICHARD WARREN, b Chicago, Ill, Oct 22, 22; m 47; c 6. MEDICINE, HEMATOLOGY. *Educ:* Ind Univ, AB, 44, MD, 46; Am Bd Internal Med, dipl, 55. *Prof Exp:* Intern, Marion County Gen Hosp, Indianapolis, Ind, 46-47, resident, 49-53, dir med educ, 53-59; from instr to assoc prof, 53-72, PROF MED, SCH MED, IND UNIV, INDIANAPOLIS, 72-; CLIN PHARMACOLOGIST, LILLY LABS CLIN RES, ELI LILLY & CO, WISHARD MEM HOSP, 69- *Concurrent Pos:* Dir poison control ctr, 55-69, clin dir internal med, Marion County Gen Hosp, 59-69, chmn formulary comn, 63-68. *Mem:* Fel Am Col Physicians; AMA; Am Soc Hemat; Sigma Xi. *Res:* Clinical research in cancer chemotherapy. *Mailing Add:* 542 W 83rd St Indianapolis IN 46260

DYKE, THOMAS ROBERT, b Akron, Ohio, Dec 16, 44; m 70; c 2. PHYSICAL CHEMISTRY. *Educ:* Col Wooster, BA, 66; Harvard Univ, MA & PhD(chem), 72. *Prof Exp:* Fel chem, Univ Rochester, 71-74; from asst prof to asssoc prof, 74-86, PROF CHEM, UNIV ORE, 86- *Mem:* Am Chem Soc; Am Phys Soc. *Res:* Structure and properties of hydrogen bonded and van der Waals complexes; molecular beam spectroscopy using microwave, infrared and Raman techniques. *Mailing Add:* 2150 Monroe St Eugene OR 97403

DYKEN, MARK LEWIS, b Laramie, Wyo, Aug 26, 28; m 51; c 6. MEDICINE, NEUROLOGY. *Educ:* Ind Univ, BS, 51, MD, 54. *Prof Exp:* Resident neurol, Med Ctr, Ind Univ, 55-58; clin dir & dir res, New Castle State Hosp, 58-61; from asst to assoc prof, 61-69, prin investr & dir, Cerebral Vascular Clin Res Ctr, 66-74, PROF NEUROL, SCH MED, IND UNIV, INDIANAPOLIS, 69-, CHMN DEPT, 71- *Concurrent Pos:* Med dir, Ind Mult Sclerosis Soc Clin, 61-; asst dir, Multicategorical Clin Res Facil, Ind, 62-66; consult, Cerebral Palsy Clin, 61-70; mem adv comt, Neurol Sensory Dis Serv Prog, Bur State Serv, USPHS, 64-67; mem bd dirs, Ind Neuromuscular Res Lab, 65-; dir, Ind Regional Med Stroke Prog, 68; mem, Clin Mgt Study Group & Epidemiol Study Group, Joint Comt Stroke Facil, 70-73; fac mem, Ind Univ Chapter, 81; vchmn, Stroke Coun, Am Heart Asn, 82-84 & chmn, 84-86. *Honors & Awards:* E Ward M D Mem Lectr, Med & Chirurgical Fac, 186th Ann Meeting, Baltimore, Md, 84. *Mem:* AMA; fel Am Acad Neurol; Asn Res Nerv & Ment Dis; Am Heart Asn; Asn Univ Profs Neurol (pres-elect, 84-86, pres, 86-). *Res:* Cerebrovascular disease; epilepsy; muscle disease; changes in brain and vasculature following injury early in life; demyelinating diseases. *Mailing Add:* Dept Neurol Ind Univ Sch Med 545 Barnhill Dr Emerson Hall 125 Indianapolis IN 46223

DYKEN, PAUL RICHARD, b Casper, Wyo, Mar 14, 34; c 3. PEDIATRICS, NEUROLOGY. *Educ:* Ind Univ, BA, 56, MD, 59; Am Bd Psychiat & Neurol, dipl, 66. *Prof Exp:* Intern, Philadelphia Gen Hosp, Pa, 59-60; resident neurol, Ind Univ & Affiliated Hosps, 60-63; from asst prof to assoc prof pediat neurol, Med Ctr, Ind Univ, Indianapolis, 65-69; from assoc prof to prof neurol & pediat, Med Col Wis, 69-72, dir pediat neurol, 69-72, head dept neurol, 71-72; prof & chief pediat neutol, Med Col Talmadge Hosp, 72-; AT DEPT NEUROL, UNIV S ALA, MOBILE. *Concurrent Pos:* Fel neurophysiol, Barnes Hosp, Washington Univ, 63-64; fel pediat neurol, Univ Chicago,

64-65; chief neurologist, Milwaukee Children's Hosp, 69-72; consult, Southern Wis Colony, 69-72, Milwaukee County Gen Hosp, 69-72 & Woods Vet Hosp, 71-72; asst examr, Study Sect Develop Behav Sci, 74- *Mem:* AAAS; Asn Res Nerv & Ment Dis; Am Neurol Asn; fel Am Acad Neurol; Child Neurol Soc; Sigma Xi. *Res:* Central nervous system degenerative disease; child neurology; cerebral spinal fluid; electroencephalography. *Mailing Add:* Neurol Moorer Bldg Rm 1100 Univ SAla 2451 Fillingim St Mobile AL 36617

DYKES, ROBERT WILLIAM, b Portland, Ore, Apr 5, 43; m 66; c 3. NEUROPHYSIOLOGY, SOMESTHESIS. *Educ:* Univ Calif, Berkeley, BA, 65; Johns Hopkins Univ, PhD(physiol), 70. *Prof Exp:* From asst prof to assoc prof sensory neurophysiol, Dalhousie Univ, 71-78; from assoc prof to prof surg, physiol, neurol & neurosurg, McGill Univ & Montreal Neurol Inst, 85-89; PROF TITULAIRE, DEPT PHYSIOL, UNIV MONTREAL, 89- *Concurrent Pos:* NIH fel, New York Med Col, 70-71; for researcher, Nat Inst Health & Med Res, Paris, 75; vis prof, dept physiol, Univ Calif, 77; assoc dir, Microsurg Res Labs, Royal Victoria Hosp, 78-; Killiam res scholar, 78-80 & 81-85; res fel, Govt Quebec, 80-86. *Mem:* Can Physiol Soc; Soc Neurosci; Can Asn Neurosci. *Res:* Somatosensory neurophysiology, especially peripheral nerve injuries, and cortical somatotopic organization. *Mailing Add:* Dept Physiol Univ Montreal PO Box 6128 Sta A Montreal PQ H3C 3J7 Can

DYKHUIZEN, DANIEL EDWARD, b Muskegon, Mich, Oct 31, 42; m 68; c 3. GENETICS, EVOLUTION. *Educ:* Stanford Univ, BS, 65; Univ Chicago, PhD(biol), 71. *Prof Exp:* Fel bact genetics, Stanford Univ, 71-72; res fel genetics, Australian Nat Univ, 72-76; res scientist pop genetics, Purdue Univ, 76-81; ASSOC RES PROF, WASH UNIV, 81- *Mem:* Genetics Soc Am; Soc Study Evolution; Am Soc Microbiol; AAAS. *Res:* Experimental study of natural selection using laboratory strains of E Coli. *Mailing Add:* Dept Ecology State Univ NY Stony Brook Main Stony Brook NY 11794

DYKLA, JOHN J, b Chicago, Ill, July 15, 44; m 81; c 2. THEORETICAL ASTROPHYSICS. *Educ:* Loyola Univ Chicago, BS, 66; Calif Inst Technol, PhD(physics), 72. *Prof Exp:* Res assoc physics, Univ Tex, Austin, 71-73; asst prof, 73-77, ASSOC PROF PHYSICS, LOYOLA UNIV, CHICAGO, 77- *Concurrent Pos:* Regional sci adv, Univ Tex, Austin, 71-73; res grant, Loyola Univ Chicago, 78; res assoc, Theoret Astrophy Group, Fermilab, Batavia, Ill, 85-86. *Mem:* Am Asn Physics Teachers; AAAS; Sigma Xi. *Res:* General relativistic astrophysics, especially black holes in space and cosmology; interpretation of exact solutions of Einstein's gravitational field equations; astrophysical implications of alternate relativistic gravitational theories; quantum gravity theory and cosmology. *Mailing Add:* Dept Physics 6525 N Sheridan Rd Chicago IL 60626

DYKMAN, ROSCOE A, b Pocatello, Idaho, Mar 20, 20; m 44, 80; c 4. PSYCHOPHYSIOLOGY, BIOMATHEMATICS. *Educ:* Univ Chicago, PhD(human develop), 49. *Prof Exp:* Instr psychol, Ill Inst Technol, 47-50; USPHS fel, Johns Hopkins Hosp, 50-52, instr psychiat, 52-53; asst dir studies, Asn Am Med Cols, 53-55; assoc prof psychol, 55-61, prof & head, Div Behav Sci, 75-90, PROF PSYCHOL, MED CTR, UNIV ARK, 61-; DIR, PSYCHOPHYSIOL LAB, CARE UNIT, ARK CHILDREN'S HOSP, 90- *Concurrent Pos:* Career scientist award, NIMH, 62-72. *Mem:* Am Psychol Asn; Soc Psychophysiol Res; Pavlovian Soc Am; AAAS; Sigma Xi. *Res:* Clinical child psychology; learning. *Mailing Add:* Ark Children's Hosp Care Unit 800 Marshall St Little Rock AR 72202

DYKSTERHUIS, EDSKO JERRY, b Hospers, Iowa, Dec 27, 08; m 33; c 3. RANGE ECOLOGY. *Educ:* Iowa State Col, BS, 32; Univ Nebr, PhD(bot), 45. *Prof Exp:* Field asst, Powell Nat Forest, Utah, 30; jr forester, Nat Forests, NMex, 33-34 & Southwestern Forest & Range Exp Sta, US Forest Serv, 34; jr range exam & sr forest ranger, Crook Nat Forest, Ariz, 34-35; asst range exam, Carson Nat Forest, NMex, 35-38; assoc range exam, Southern Forest Exp Sta, Tex, 38-39; sr flood control rep, US Forest Serv, Kans & Ozarks, 39-41; range conservationist, US Soil Conserv Serv, Tex, Okla, La & Ark, 43-49, head range conservationist, Mont, Wyo, NDak, SDak & Nebr, 49-64; prof range ecol, 64-70, EMER PROF RANGE ECOL, TEX A&M UNIV, 71-; CONSULT NATURAL FORAGES, AID, US DEPT STATE, 71- *Concurrent Pos:* Vis prof, Mont State, 50; mem Nat Resources Coun, 51-52; bot assoc ed, Ecol Monogr, 59-61; vis prof, Kans State Univ, 62; exten range specialist, SDak State Univ, 64. *Honors & Awards:* Mercer Award, Ecol Soc Am, 49; Authorship Award, USDA, 57; Prof Conservationist Award, Am Motors Corp, 72; Heritage Award, Native Prairie Asn, 87. *Mem:* Fel AAAS; Soil Conserv Soc Am; Ecol Soc Am; fel Am Soc Range Mgt (pres, 68). *Res:* Prairie and savanna rangelands in relation to soils, climate, fire and grazing; quantifying ecological degeneration and predicting potential productivity by types of physical sites. *Mailing Add:* 3807 Oaklawn Bryan TX 77801-4730

DYKSTRA, CLIFFORD ELLIOT, b Chicago, Ill, Oct 30, 52; m 88. PHYSICAL CHEMISTRY, THEORETICAL CHEMISTRY. *Educ:* Univ Ill, Urbana-Champaign, BS, 73; Univ Calif, Berkeley, PhD(chem), 76. *Prof Exp:* Res assoc, Univ Calif, Berkeley, 76-77; from asst prof to assoc prof, 77-88, PROF CHEM, UNIV ILL, URBANA-CHAMPAIGN, 88- *Concurrent Pos:* Consult, Theoret Chem Group, Chem Div, Argonne Nat Lab, 78-80; dir basic sci, Ill Valley Res Ctr, 86- *Mem:* Am Chem Soc; Am Phys Soc. *Res:* Theoretical molecular electronic structure; electron pair theories; electron correlation effects in chemical systems: rearrangement reactions, organic anions, instellar molecules, and molecular clusters; hydrogen bonding magnetic effects. *Mailing Add:* Dept Chem Purdue Univ 1125 E 38th St Indianapolis IN 46205

DYKSTRA, DEWEY IRWIN, JR, b Baltimore, Md, June 13, 47; m 69; c 1. PHYSICS EDUCATION, SOLID STATE PHYSICS. *Educ:* Case Western Reserve Univ, BS, 69; Univ Tex, Austin, PhD(physics), 78. *Prof Exp:* Instr physics, Cleveland Pub Sch, 69-72, Frederick County Bd of Educ, 72-73; asst prof physics, Okla State Univ, 78-81; ASST PROF PHYSICS, BOISE STATE UNIV, 81- *Mem:* Am Asn Physics Teachers; Nat Sci Teachers Asn; Nat Asn Res Sci Teaching; Asn Educ Teachers Sci; Jean Piaget Soc. *Res:* Application of theories of cognitive development to physics education; pulsed calorimetric studies of solid-solid structural phase transition. *Mailing Add:* Dept Physics/Eng Boise State Univ 1910 University Dr Boise ID 83725

DYKSTRA, JERALD PAUL, b Holland, Mich, Aug 23, 46; m 65; c 2. ION BEAM SYSTEMS, SEMICONDUCTOR MANUFACTURING EQUIPMENT. *Educ:* Univ Tex Austin, BSEE, 69, MSE, 75. *Prof Exp:* Biomed engr, Med Monitor Systs, Inc, 69-73; chief engr, Amsco Med Electronics, Inc, 73-77; sr engr, 77-84, eng mgr, 84-86, DIR ENG, SEMICONDUCTOR EQUIP DIV, EATON CORP, 86- *Mem:* Inst Elec & Electronics Engrs. *Res:* Ion implantation systems for semiconductor doping applications; ion beam transport and optics; wafer handling and transport systems for low particulate processing of semiconductors; granted several patents. *Mailing Add:* Eaton Corp 2433 Rutland Dr Austin TX 78758

DYKSTRA, LINDA A, b Hammond, Ind, Nov 8, 44. PSYCHOLOGY & PHARMACOLOGY. *Educ:* Hope Col, BA, 66; Univ Chicago, MA, 67, PhD(psychol pharmacol), 72. *Prof Exp:* PROF BEHAV PHARMACOL & PSYCHOL PHARMACOL, UNIV NC, 73- *Concurrent Pos:* Res scientist award, Nat Inst Drug Abuse, 87- & Merit Award, 87- *Mem:* Am Soc Pharmacol; Am Psychol Asn; Behav Pharmacol Soc; AAAS. *Mailing Add:* Dept Psychol & Pharmacol Univ NC Davie Hall CB No 3270 Chapel Hill NC 27599-3270

DYKSTRA, MARK ALLAN, b Hull, Iowa, Oct 17, 46; m 71; c 2. MEDICAL MYCOLOGY, DIAGNOSTIC MICROBIOLOGY. *Educ:* Cent Col, Pella, Iowa, BA, 69; Northern Ill Univ, MS, 71; Tulane Univ, PhD(microbiol), 76. *Prof Exp:* Resident pub health & med microbiol, Hartford Hosp, 76-78; asst prof, dept med microbiol, Creighton Univ, 78-83; dir, Diag Micro Lab St Joseph Hosp, Omaha, 78-83; DIR, DIAG MICRO LABS, RES MED CTR, KANSAS CITY, MO, 83-; ASSOC CLIN PROF, UNIV MO, KANSAS CITY, 84- *Concurrent Pos:* Dir, Diag Micros Lab, St Joseph Hosp, Omaha, 83-; assoc clin prof, Univ Mo, Kansas City, 84- *Mem:* Am Soc Microbiol; Am Fedn Clin Res; Med Mycol Soc Am. *Res:* Pathogenesis of cryptococcosis, especially nonspecific resistance mechanisms in murine model. *Mailing Add:* 7575 W 106th No 303 Overland Park KS 66212

DYKSTRA, RICHARD LYNN, b Des Moines, Iowa, Oct 19, 42; m 64; c 3. ANALYTICAL STATISTICS. *Educ:* Cent Col, Iowa, BA, 65; Univ Iowa, PhD(statist), 68. *Prof Exp:* From asst prof to assoc prof, 68-81, PROF STATIST, UNIV MO-COLUMBIA, 81-; AT DEPT STATIST, UNIV IOWA. *Mem:* Am Statist Asn; Inst Math Statist. *Res:* Probability theory; stochastic processes; multivariate analysis; mathematical statistics. *Mailing Add:* Dept Statist Univ Iowa Iowa City IA 52242

DYKSTRA, STANLEY JOHN, b Eddyville, Iowa, Apr 27, 24; m 49; c 2. ORGANIC CHEMISTRY. *Educ:* Calvin Col, AB, 49; Wayne Univ, PhD(chem), 53. *Prof Exp:* Asst, Wayne Univ, 52; res assoc, Stanford Univ, 53-54; sr res chemist, 54-61, group leader, 61-68, sect leader, 68-70, dir chem develop, Mead Johnson Res Ctr, Mead Johnson & Co, 70-82; dir chem proc dev, Bristol Myers PRDD, Evansville, 82-88; RETIRED. *Mem:* Am Chem Soc; Sigma Xi. *Res:* Epoxyethers; primary aliphatic hydroperoxides; medicinals. *Mailing Add:* 2711 Hillcrest Terr Evansville IN 47712-5057

DYKSTRA, THOMAS KARL, b Grand Rapids, Mich, May 9, 35; m 59; c 2. ORGANIC CHEMISTRY, POLYMER CHEMISTRY. *Educ:* Calvin Col, BA, 57; Univ Ill, PhD(org chem), 61. *Prof Exp:* Res chemist, 61-65, lab head, 71-81, prog dir, 81-83, asst dir develop, 83-86, RES ASSOC CHEM, EASTMAN KODAK CO, 66-, MGR PERSONEL & UNIV RELS, 86- *Mem:* Am Chem Soc. *Res:* Application of organic and polymer chemistry to electrophotography. *Mailing Add:* Bldg 83 Res Lab Eastman Kodak Co Rochester NY 14650-2211

DYLLA, HENRY FREDERICK, b Atlanta, Ga, Mar 17, 49; m; c 2. PHYSICS. *Educ:* Mass Inst Technol, BS & MS, 71, PhD(physics), 75. *Prof Exp:* Res asst atomic physics, Mass Inst Technol, 70-75; res staff mem plasma physics, Princeton Univ, 75-90; ASSOC MGR, ACCELERATOR DIV, CEBAF, 90- *Concurrent Pos:* Pres, Princeton Sci Consults, Inc, 81-90; adj prof physics, Col William & Mary, Williamsburg, Va, 90-; mem, bd dirs, Am Vacuum Soc, 84-86, AIP gov bd, 90-93. *Mem:* Am Phys Soc; Am Vacuum Soc; Mat Res Soc. *Res:* Plasma-wall interactions in controlled fusion devices; surface physics; mass spectrometry; electron spectroscopy; vacuum technology; free electron lasers. *Mailing Add:* CEBAF 12000 Jefferson Ave Newport News VA 23606

DYM, CLIVE L, b Leeds, Eng, July 15, 42; US citizen; div; c 2. EXPERT SYSTEMS, APPLIED MECHANICS. *Educ:* Cooper Union BCE, 62; Polytech Inst Brooklyn, MS, 64; Stanford Univ, PhD(aeronaut), 67. *Prof Exp:* Asst prof, State Univ NY, Buffalo, 66-69; mem res staff, Inst Defense Anal, 69-70; assoc prof, Dept Civil Eng, Carnegie-Mellon Univ, 70-74; sr scientist, Bolt Beranek & Newman, Inc, Cambridge, 74-77; dept head, 77-85, PROF, DEPT CIVIL ENG, UNIV MASS, AMHERST, 77- *Concurrent Pos:* Vis assoc prof, Dept Aeronaut Eng, Israel Inst Technol, Haifa, 71; vis sr res fel, Inst Sound & Vibration Res, Univ Southampton, Eng, 73; vis scientist, Palo Alto Res Ctr, Xerox, 83-84; vis prof, Dept Civil Eng, Stanford Univ, 83-84, Dept Civil Eng, Carnegie Mellon Univ, 89; consult, Bell Aerospace Co, 67-69, Dravo Corp, 70-71, Salem Corp, 72, Gen Anal, Inc, 72, ORI, Inc, 79, Bolt Beranek & Newman, Inc, 79-81, AVCO-AERL, Inc, 81-83 & 85-86, TASC, Inc, 85-86, Amerinex Artificial Intel, 86-88. *Honors & Awards:* Huber Res Prize, Am Soc Civil Engrs, 80; Western Elec Fund Award, Am Soc Eng Educ, 83. *Mem:* Fel Am Soc Civil Engrs; fel Am Soc Mech Engrs; fel Acoust Soc Am; AAAS; Inst Noise Control Eng; Am Asn Artificial Intel; Inst Elec & Electronics Engrs Comput Soc. *Res:* Knowledge-based (expert) systems for engineering analysis and design; vibration and stability of structures; shell structures; acoustics; applied mathematics, especially variational methods; mathematical modeling. *Mailing Add:* Dept Civil Eng Univ Mass Amherst MA 01003

DYMENT, JOHN CAMERON, b Hamilton, Ont, June 7, 38; m 63; c 3. SOLID STATE PHYSICS, OPTOELECTRONIC DEVICES. *Educ:* McMaster Univ, BSc, 60; Univ BC, MSc, 62; McGill Univ, PhD(physics), 65. *Prof Exp:* Mem tech staff, Bell Labs, Murray Hill, 65-73, MEM SCI STAFF, BELL-NORTHERN RES, 73- *Concurrent Pos:* Lectr fiber optics, Carleton Univ, Ottawa, 77-85; guest ed, J Quantum Electronics, Inst Elec & Electronics Engrs, 79; Can deleg, Standards for Fiber Optic Terminals, Int Electrotech Comn, 80-85; chmn, 8th Inst Elec & Electronics Engrs Int Semiconductor Laser Conf, Ottawa, Can, 82; mem bd, Can Eng Manpower, 88- *Mem:* Am Inst Physics. *Res:* Paramagnetic resonance; optoelectronic device research; lasers, light emiting diodes, detectors, modulators; management of crystal growth, processing, device characterization and packaging; analytic techniques for materials and processes, optoelectronic integrated device structures. *Mailing Add:* Bell-Northern Res PO Box 3511 Sta C Ottawa ON K1Y 4H7 Can

DYMERSKI, PAUL PETER, b Pittsburgh, Pa, June 12, 47. PHYSICAL CHEMISTRY. *Educ:* Duquesne Univ, BS, 69; Univ Idaho, MS, 71; Case Western Reserve Univ, PhD(phys chem), 74. *Prof Exp:* Fel, Cornell Univ, 74-75; fel chem, Univ Toronto, 75-76, DIR MASS SPECTROMETRY, NY STATE HEALTH DEPT, 76- *Mem:* Sigma Xi; Am Soc Mass Spectrometry & Allied Topics; Am Chem Soc. *Res:* Applications of mass spectrometry to environmental and health chemistry. *Mailing Add:* O H Mat Co PO Box 551 Findlay OH 45840

DYMICKY, MICHAEL, b Ukraine, Oct 1, 20; nat US; m 43; c 2. ORGANIC CHEMISTRY, PHYSICAL CHEMISTRY. *Educ:* Polytech Lwiw, Ukraine, Chem Tech, 43; Innsbruck Univ, dipl chem, 47, Doctorandum, 49; Temple Univ, PhD(org chem), 60. *Prof Exp:* Fel, Univ Pa, 52-53; res chemist, Wyeth Inst, 53-56, US Agr Res Serv, 56-59 & Wyeth Inst, 59-62; assoc prof chem, Kutztown State Col, 62-65; assoc prof, Gwynedd-Mercy Col, 65-66; mem staff, Smoke Invests Tobacco Lab, Eastern Utilization Res & Develop Div, Agr Res Serv, 66-72, sr res chemist, meat, hides & leather, 72-79, SR RES CHEMIST, FOOD SAFETY LAB, EASTERN REGIONAL RES CTR, USDA, 80- *Honors & Awards:* Cert Recognition, Chem Abstr, 79; Award, USDA, 85 & 88; Award Antichlostridial Agents, US Dept Com, 87. *Mem:* Am Chem Soc; Sigma Xi; Shevchenko Sci Soc. *Res:* Chemistry of plants; physiological compounds occurring in nature; organic syntheses; spectroscopy; chemistry of the pigments of the meat; syntheses and studies of the structure of chemical compunds versus biological activity. *Mailing Add:* 9653 Dungan Rd Philadelphia PA 19115

DYMSZA, HENRY A, b Newton, NH, Jan 14, 22; m 56; c 4. NUTRITIONAL BIOCHEMISTRY, FOOD SCIENCE. *Educ:* Pa State Univ, BS, 43, PhD(agr & biochem), 54; Univ Wis, MS, 50. *Prof Exp:* Res nutritionist, Gen Foods Corp, 54-59; sr res assoc nutrit, Mass Inst Technol, 59-64; head metab sect, Food Div, US Army Natick Labs, 64-66; assoc prof, 66-70, chmn dept, 66-77, PROF FOOD & NUTRIT SCI, UNIV RI, 70- *Concurrent Pos:* Chief, Clin Nutrit Br, Bur Foods, Food & Drug Admin, Washington, DC, 79-80; mem bd sci advs, Am Coun on Sci & Health, 85- *Honors & Awards:* Commendation for Research, US Army, 66. *Mem:* AAAS; Am Chem Soc; Am Inst Nutrit; Am Dietetic Asn; Inst Food Technol. *Res:* Nutritional biochemistry; synthetic and unusual nutrients and diets; energy metabolism; food preservation and safety; fish nutrition; hyper-vitaminosis and over-nutrition; nutritional status evaluation; nutrition education and behavior; food irradiation. *Mailing Add:* Dept Food Sci & Nutrit Univ RI Woodward Hall Kingston RI 02881

DYNAN, WILLIAM SHELLEY, b Nov 4, 54. TRANSCRIPTIONAL REGULATION. *Educ:* Mass Inst Technol, SB, 75; Univ Wis, PhD(oncol), 80. *Prof Exp:* Postdoctoral fel biochem, Univ Calif, Berkeley, 80-85; ASST PROF CHEM & BIOCHEM, UNIV COLO, BOULDER, 85- *Res:* Control of eukaryotic m-RNA synthesis by sequence specific DNA binding proteins; regulation of transciption in human and animal viruses. *Mailing Add:* Dept Chem & Biochem Campus Box 215 Univ Colo Boulder CO 80309

DYNE, PETER JOHN, b London, Eng, Sept 14, 26; Can citizen; m 53; c 1. RADIATION CHEMISTRY. *Educ:* Univ London, BSc, 46, PhD(phys chem), 49. *Prof Exp:* Asst lectr phys chem, King's Col, Univ London, 49, lectr, Imp Col, 49-50; fel physics, Nat Res Coun Can, 50-52; fel chem, Jet Propulsion Lab, Calif Inst Technol, 52-53; sci officer, Atomic Energy Can, Ltd, 53-65, head, Mat Sci Br, Whiteshell Nuclear Res Estab, 65-71, dir, Chem & Mat Sci Div, 71-76; dir gen off energy res & develop, Dept Energy, Mines & Resources, 76-89; RETIRED. *Mem:* Fel Chem Inst Can; The Chem Soc. *Res:* Photochemistry; molecular spectroscopy; problems in pure and applied chemistry and in materials relevant to reactor technology; sciences technology of storing radioactive wastes; research and development in all energy technologies. *Mailing Add:* 452 Princeton Ottawa ON K2A 0N1 Can

DYNES, J ROBERT, b Miller, SDak, Oct 18, 22; m 47; c 2. ANIMAL SCIENCE. *Educ:* SDak State Col, BS, 47, MS, 49; Tex A&M Univ, PhD, 68. *Prof Exp:* Instr, High Sch, 50-51; instr, 51-56, asst prof, 56-62, ASSOC PROF MEATS, MONT STATE UNIV, 62- *Mem:* Am Soc Animal Sci; Am Meat Sci Asn; Inst Food Tech. *Res:* Meats research; carcass and red meat development in live animals. *Mailing Add:* 905 W Villard St Bozeman MT 59715

DYNES, ROBERT CARR, b London, Ont, Nov 8, 42; m 67; c 1. SOLID STATE PHYSICS. *Educ:* Univ Western Ont, BSc, 64; McMaster Univ, MSc, 65, PhD(physics), 68. *Prof Exp:* Mem tech staff physics, Bell Labs, 68-75, head, Dept Physics, 75-91; PROF PHYSICS, UNIV CALIF, SAN DIEGO, 91- *Honors & Awards:* Fifth London Prize, Inst Union of Pure & Appl Sci, 90. *Mem:* Nat Acad Sci; Am Phys Soc. *Res:* Low temperature solid state physics; details of phonons and electron transport and the phenomena of superconductivity and superfluidity. *Mailing Add:* Dept Physics Univ Calif San Diego La Jolla CA 92093

DYNKIN, EUGENE B, Leningrad, USSR, May 11, 24. PROBABILITY. *Educ:* Moscow Univ, BA, 45, PhD, 48. *Prof Exp:* From asst prof to prof, Moscow Univ, 48-68; res scholar, Acad Sci, USSR, 68-76; PROF MATH, CORNELL UNIV, 77- *Honors & Awards:* Prize of Moscow Math Soc, 51. *Mem:* Nat Acad Sci; fel Am Acad Arts & Sci; Inst Math Statist; Am Math Soc; Bernoulli Soc Math, Statist & Probability. *Res:* Lie theory and probability theory; stochastic processes; optimal control; probabilistic models of economic growth and equilibrium; sufficient statistics. *Mailing Add:* Dept Math White Hall Cornell Univ Ithaca NY 14853

DYOTT, THOMAS MICHAEL, b Rochester, NY, Aug 9, 47; m 69. RESEARCH MANAGEMENT, CHEMICALS & POLYMERS. *Educ:* Gettysburg Col, BA, 69; Princeton Univ, MA, 71, PhD(chem), 73. *Prof Exp:* sr scientist chem, 73-79, proj leader, 80-81, dept mgr, Rohm & Haas Co, 81-84; pres, Advan Genetic Sci Inc, 84-85, RES DIR, ROHM & HAAS CO, 86- *Mem:* Am Chem Soc; Asn Comput Mach. *Res:* Polymer chemistry. *Mailing Add:* 733 W Prospect Ave North Wales PA 19454

DYRBERG, THOMAS PETER, b Copenhagen, Denmark, Nov 17, 54; m 84. IMMUNOLOGY, MEDICINE. *Educ:* Univ Copenhagen, MD, 81. *Hon Degrees:* DMSc, Univ Copenhagen, 86. *Prof Exp:* Res fel autoimmunol, Hagedorn Res Lab, Denmark, 81-84; res fel, dept immunol, Scripps Clin & Res Found, 85-86; sr res fel autoimmunol, 86-88, STAFF SCIENTIST, HAGEDORN RES LAB, 88- *Res:* Etiology and pathogenesis of insulin-dependent diabetes and other autoimmune disorders; virus as etiologic factors; characterization of prediabetes state with the aim of designing prophylactic therapy. *Mailing Add:* Dept Molecular Genetic Biosci Novo-Nordisk A/S Novo Allee Bagsvaend DK-2880 Denmark

DYRKACZ, W WILLIAM, b Arnold, Pa, Apr 17, 19; m 43. METALLURGICAL ENGINEERING. *Educ:* Carnegie Inst Technol, BSc, 42. *Prof Exp:* Mat engr jet engines, Gen Elec Co, 42-46; chief metallurgist forgings, Cameron Mfg Co, 46-49; assoc dir res, chief metallurgist & mgr prod develop specialty steels, Allegheny Ludlum Steel Corp, 49-66; vpres opers titanium, Teledyne Titanium, 66-68; INDUST CONSULT METALL MFG FACIL PLANNING, METALL ENG & FAILURE ANALYSIS, 68- *Concurrent Pos:* Consult, Mat Adv Bd, Nat Acad Sci, 56-73 & Aerospace Struct Mat Handbk, Mech Properties Data Ctr, 69-79. *Honors & Awards:* F B Lounsberry Award, 55. *Mem:* Fel Am Soc Metals; sr mem Am Inst Mining & Metall Engrs; Am Vacuum Soc; Am Powder Metall Inst. *Res:* Advanced melting and metal working processes for the production of superalloys, specialty steels, titanium, zirconium, and refractory-base alloys. *Mailing Add:* 532 Connestee Trail Brevard NC 28712-9435

DYRNESS, CHRISTEN THEODORE, b Chicago, Ill, June 4, 33; m 62; c 3. FOREST SOILS. *Educ:* Wheaton Col, Ill, BS, 54; Ore State Univ, MS, 56, PhD(soil sci), 60. *Prof Exp:* Soil scientist, 59-74, PRIN SOIL SCIENTIST WATERSHED MGT RES, PAC NORTHWEST FOREST & RANGE EXP STA, US FOREST SERV, 74- *Concurrent Pos:* Assoc prof soil sci, Ore State Univ, Corvallis, 59-74; prog leader, Inst Northern Forestry, Fairbanks, Alaska, Pac Northwest Forest & Range Exp Sta, US Forest Serv, 74-85, affil prof forestry, Univ Alaska, Fairbanks, 74-90, res soil scientist, Inst Northern Forestry, Fairbanks, Alaska, 85-90. *Mem:* Fel AAAS; Soil Sci Soc Am; Am Soc Agron. *Res:* Physical and morphological properties of soil and plant soil relationships; physical properties of forest soils as affected by management practices; plant-soil relationships on pumice soils; plant-soil relationships in the taiga of interior Alaska. *Mailing Add:* 1993 Marion St SE Albany OR 97321

DYROFF, DAVID RAY, b St Louis, Mo, Feb 16, 40; m 61, 84; c 6. INDUSTRIAL CHEMISTRY. *Educ:* Univ Ill, BS, 62; Calif Inst Technol, PhD(chem), 65. *Prof Exp:* Sr res chemist, Monsanto Co, 65-67, res group leader, 67-79, sr res group leader, 79-86, SR RES SPECIALIST, DETERGENTS & PHOSPHATES DIV, MONSANTO CO, 86- *Res:* industrial process research and development; detergent ingredients. *Mailing Add:* Monsanto Co 800 N Lindbergh Blvd St Louis MO 63167

DYRUD, JARL EDVARD, b Maddock, NDak, Oct 20, 21; m 52; c 3. PSYCHIATRY. *Educ:* Concordia Col, Moorhead, Minn, AB, 42; Johns Hopkins Univ, MD, 45. *Prof Exp:* Intern, Johns Hopkins Univ, 45-46; intern, Vet Admin Ment Hyg Clin, DC, 48-49; resident psychiat, Chestnut Lodge, Inc, 49-51, staff psychiatrist, 51-56; pvt pract, 56-68; assoc chmn dept psychiat, Univ Hosp, 68-78; PROF PSYCHIAT, SCH MED, UNIV CHICAGO, 68- *Concurrent Pos:* USPHS fel & resident, Spring Grove State Hosp, 52-53; consult, Md State Hosp, 56-68; prin investr behav anal, Inst Behav Res, 63-68; dir psychiat, Chestnut Lodge Res Inst, 67-68; consult, Lab Adult Psychiat, NIMH, 67-68, mem clin projs res rev comt, 70-74; assoc dean fac affairs, Div Biol Sci & Pritzker Sch Med, Univ Chicago, 78-81. *Honors & Awards:* John Nuveen Lectr, Univ Chicago, 78. *Mem:* Am Psychiat Asn; Am Psychoanal Asn; Sigma Xi. *Res:* Ego psychology; operant analyses of behavior; psychobiology of schizophrenia; minimal brain dysfunction; normal and abnormal child development. *Mailing Add:* Dept Psychiat Univ Chicago Hosp 5841 Maryland Ave Chicago IL 60637

DYSART, BENJAMIN CLAY, III, b Columbia, Tenn, Feb 12, 40; m 60. NONPOINT SOURCE POLLUTION CONTROL, ENVIRONMENTAL IMPACT ASSESSMENT. *Educ:* Vanderbilt Univ, BEng, 61, MS, 64; Ga Tech, PhD(civil eng), 69. *Prof Exp:* Staff engr, Union Carbide Corp, 61-62 & 64-65; sci adv to asst secy Army Civil Works, Dept Army, 75-76; asst prof environ & water resources eng, 68-70, assoc prof eng & dir water resources eng grad prog, 70-75, PROF ENVIRON & WATER RESOURCES ENG, CLEMSON UNIV, 76- *Concurrent Pos:* Dir, Nat Wildlife Fedn, 74-, pres & chmn bd, 83-85; mem, Nonpoint Source Pollutant Task Force, US Environ Protection Agency, 79-81, Sci Adv Bd, 83-, Outer Continental Shelf Adv Bd, US Dept Interior, 79-82; sr fel, Conserv Found, 85-; trustee, Rene Dubos Ctr Human Environ, 85-; consult var corp, firms & fed agencies. *Mem:* Asn Environ Eng Prof (pres, 81-82); Am Geophys Union; Am Soc Civil Engrs; hon mem, Water Pollution Control Fedn; Nat Wildlife Fedn (vpres, 78-83, pres & chmn bd, 83-85). *Res:* Control of nonpoint source pollution; site evaluation; environmental impact assessment; national water resources policy; mathematical modeling; design of water quality monitoring programs; simulation of environmental protection and water resources systems. *Mailing Add:* 401 Rhodes Eng Res Ctr Clemson Univ Clemson SC 29634-0919

DYSART, RICHARD JAMES, b Chicago, Ill, May 6, 32. ENTOMOLOGY. *Educ:* Univ Ill, BS, 54, PhD(entom), 61. *Prof Exp:* assoc entomologist, Ill Natural Hist Surv, 60-65; res entomologist, European Parasite Lab, USDA, Sevres, France, 65-71, dir, Beneficial Insects Res Lab, Newark, 78-87, SUPV RES ENTOMOLOGIST, GRASSHOPPER IPM PROG, SIDNEY, MONTANA, USDA, 87- *Mem:* Entom Soc Am; Int Orgn Biol Control. *Res:* Ecology of grasshoppers on rangeland; biological control of grasshoppers (egg parasites). *Mailing Add:* USDA PO Box 1109 Sidney MT 59270

DYSINGER, PAUL WILLIAM, b Burns, Tenn, May 24, 27; m 58; c 4. TROPICAL PUBLIC HEALTH, PREVENTIVE MEDICINE. *Educ:* Southern Missionary Col, BA, 51; Loma Linda Univ, MD, 55; Harvard Univ, MPH, 62, Am Bd Prev Med dipl,65. *Prof Exp:* General Med for Blackfeet, Navajo and Hopi Indians, USPHS 56-58; Med attache, Am Embassy, Phnom Penh, Cambodia, 58-60; res assoc prev med, Sch Med, Loma Linda Univ, 61; dir pub health & trop med, Field Sta, Tanganyika, EAfrica, 62-64; asst prof pub health, 64-67, chmn, Dept Trop Health, 67-71, assoc dean acad affairs & int health, 71-78, chmn, Dept Int Health, 83-85, dir, Prev Med Res Training, 84-88, PROF INT HEALTH, LOMA LINDA UNIV, 71-; SR HEALTH CONSULT, ADVENTIST DEVELOP & RELIEF AGENCY, INT, 88- *Concurrent Pos:* WHO traveling fel, 69; consult, TRW, Inc, 67- & Voc Rehab, Calif, 67-; pub health adv, Ministry Health, Dar es Salaam, Tanzania, 78-80, Rural Health Educ Ctr, Chuhar Kana Mandi, Sheikhupura, Pakistan, 80-81, prev med chief, VA, 84-; mem, Nat Coun Inst Health. *Mem:* Fel Am Col Prev Med; fel Am Pub Health Asn; Int Health Soc; fel Royal Soc Trop Med & Hyg; Adventist Int Med Soc (pres, 83-85). *Res:* Statistical epidemiology, especially mortality studies in emphysema and accidents; manpower studies in Southeast Asia and East Africa; lifestyle intervention trials in veterans with disease and at risk. *Mailing Add:* 6840 Eastern Ave NW Washington DC 20012

DYSKEN, MAURICE WILLIAM, b Dayton, Ohio, June 16, 42; m 68; c 2. GERIATRIC PSYCHIATRY. *Educ:* Oberlin Col, AB, 64; Case Western Reserve Univ, MD, 68; Am Bd Psychiat & Neurol, cert psychiat, 76. *Prof Exp:* Internship med, Univ Hosp, Cleveland, 68-69, resident psychiat, 69-70; resident med, Univ Affil Hosp Cleveland, 70-71; med corps, US Army, 71-73, Chicago, 75-77, asst prof, 75-77, res assoc psychiat, Univ Chicago, 77-81; resident psychiat, Univ Chicago, 73-75, asst prof, 75-77, res assoc, 77-81; assoc prof psychiat, Univ Ill, 81-83; ASSOC PROF PSYCHIAT, UNIV MINN, 83- *Concurrent Pos:* Res assoc, Ill State Psychiat Inst, 77-83; lectr, Northwestern Univ, 78-83; res consult, Dept Psychiat, Hannepin County Med Ctr, 83- *Mem:* Psychiat Res Soc; Am Asn Geriatric Psychiat; Am Col Physicians; Am Psychiat Asn; Psychiat Res Soc; Soc Biol Psychiat; Gerontol Soc Am; Am Asn Psychiat Law. *Res:* Pharmacological agents for Alzheimer's dementia; the relationship between clinical improvement and neuroleptic blood levels in elderly patients; blood brain barrier integrity in patients with Alzheimer's dementia. *Mailing Add:* GRECC Prog Minneapolis Va Med Ctr One Veterans Dr Minneapolis MN 55417

DYSON, DEREK C(HARLESWORTH), b San Eduardo, Arg, Dec 23, 32; m 61; c 2. CHEMICAL ENGINEERING. *Educ:* Cambridge Univ, BA, 55; Univ London, PhD(chem eng), 66. *Prof Exp:* Develop engr, Du Pont of Can, 58-60; res engr, Pennsalt Chem Corp, 60-62; res asst chem eng, Imp Col, London, 62-65; from asst prof to assoc prof, 66-77, PROF CHEM ENG, RICE UNIV, 77- *Mem:* Am Inst Chem Engrs; Sigma Xi. *Res:* Optimization of chemical processes; theory of interfacial stability in the capillary regime; computer control of chemical processes. *Mailing Add:* Dept Chem Eng Rice Univ Box 1892 Houston TX 77251

DYSON, FREEMAN JOHN, b Crowthorne, Eng, Dec 15, 23; nat US; m 50; c 6. MATHEMATICAL PHYSICS, ASTROPHYSICS. *Educ:* Cambridge Univ, BA, 45. *Hon Degrees:* Various from US & foreign univs, 66-84. *Prof Exp:* Res fel, Trinity Col, Cambridge Univ, 46-49 & Univ Birmingham, 49-51; prof physics, Cornell Univ, 51-53; PROF PHYSICS, INST ADVAN STUDY, 53- *Honors & Awards:* Heineman Prize, Am Inst Physics, 65; Lorentz Medal, Royal Netherlands Acad, 66; Hughes Medal, Royal Soc, 68; Max Planck Medal, Ger Phys Soc, 69; J Robert Oppenheimer Mem Prize, Ctr Theoret Studies, 70; Harvey Prize, Israel Inst of Technol, 77; Wolf Prize, Israel, 81. *Mem:* Nat Acad Sci; fel Royal Soc; Am Phys Soc. *Res:* Science, technology and arms-control. *Mailing Add:* Inst Advan Study Princeton NJ 08540

DYSON, JOHN DOUGLAS, b Lemmon, SDak, Auug 9, 18; m 40; c 4. ANTENNA THEORY. *Educ:* SDak State Univ, BS, 40, BS (elec eng), 49; Univ Ill, MS, 50, PhD(elec eng), 57. *Prof Exp:* Statistician, SDak Hwy Planning Surv, 40-41; officer, US Army Separated as Lt Colonel, 41-46; engr, Res Staff, Sandia Corp, 51-52; res asst, 52-57, asst prof, 58-60, assoc prof, 60-66, PROF ELEC ENG, UNIV ILL, 66- *Mem:* Antennas & Propagation Soc; fel Inst Elec & Electronics Engrs; Int Sci Radio Union; Microwave Theory & Techniques Soc. *Res:* Log spiral frequency independent antennas; design of antenna systems; direction finding systems; measurement techniques; computer control measurements. *Mailing Add:* 1004 S Western Ave Champaign IL 61820

DYSON, ROBERT DUANE, b Minneapolis, Minn, May 18, 39; m 61, 85; c 3. CELL PHYSIOLOGY. *Educ:* Univ Ore, BS, 61; Univ Ill, Urbana, MS, 63, PhD(biophys chem), 65, Ore Health Sci Univ, MD, 77. *Prof Exp:* NIH fel & res assoc molecular biol, Univ Calif, Berkeley, 65-67; asst prof biophysics, Ore State Univ, 67-71, assoc prof biochem & biophysics, 71-73; PVT PRACT, OBSTET & GYNEC, 81- *Concurrent Pos:* Resident obstet & gynec, Univ Ore Health Sci Ctr, 78-81. *Mem:* AAAS; Am Soc Cell Biol; Am Col Obstet & Gynec; Am Fertil Soc. *Res:* Cell biology. *Mailing Add:* 10373 NE Hancock St Suite 214 Portland OR 97220

DYSON-HUDSON, V RADA, b Cold Spring Harbor, NY, July 8, 30; m 53; c 3. BIOLOGICAL ANTHROPOLOGY. *Educ:* Swarthmore Col, BA, 51; Oxford Univ, DPhil(ecol), 54. *Prof Exp:* Researcher human ecol, Karamoja, Uganda, 57-59; lectr zool, Univ Khartoum, 61-63; res assoc human ecol, Dept Social Rels, Johns Hopkins Univ, 65-71, assoc prof human ecol & res scientist, Dept Pathobiol & Dept Social Rels, 71-73; assoc prof, State Univ Ny Col Environ Sci & Forestry, 73-74; assoc prof, 74-87, SR RES ASSOC ANTHROP, CORNELL UNIV, 87- *Concurrent Pos:* J S Guggenheim & Fulbright res fels, Uganda, 54-55, Penrose Fund grants, 54-56 & Permanent Sci Found grant, 56-57; Sigma Xi grants, Uganda, 56-57, Univ Khartoum, 61-63 & Johns Hopkins Univ, 66-67; Marion Talbot fel, Oxford Univ, 58-59; NSF grant-in-aid, Johns Hopkins Univ, 65-68 & res grant, 72, Kendall Fund grant, 67 & Wenner-Gren Found Anthrop Res grant, 67-68; NSF grant-in-aid, 80-81, 84; partic, S Turkana Ecosystem Proj, Kenya, 80-; H F Guggenheim Found grant, 84-86, Wenner Gren Found grant, 83 & 84, Nat Geog grant, 84-85 & 88-89. *Mem:* AAAS; fel Am Anthrop Asn. *Res:* Study of Drosophila ecology in England; study of low-energy agricultural society of East Africa; a general synthesis of human ecology as an integration of many disciplines; ecology of nomadic pastoralists in East Africa; evolutionary theory as applied to human behavior. *Mailing Add:* Dept Anthrop Cornell Univ Col Arts & Sci Ithaca NY 14853

DZIADYK, BOHDAN, b Aschaffenburg, Ger, Mar 26, 48; US citizen; m 74; c 2. BOTANY, ECOLOGY. *Educ:* Southern Ill Univ, BA, 70, MS, 80; NDak State Univ, PhD(bot), 82. *Prof Exp:* asst prof, 80-88, ASSOC PROF BIOL, AUGUSTANA COL, 88- *Concurrent Pos:* Pew Sci Prog res grant, 88-91. *Mem:* Ecol Soc Am; Natural Areas Asn; Sigma Xi (pres, John Deere Chap, 87-88); Nature Conservancy; Int Asn Veg Sci. *Res:* Analysis of structure and function in terrestrial, especially grassland, ecosystems; dynamics of primary production; plant physiology; range sciences and management. *Mailing Add:* Dept Biol Augustana Col Rock Island IL 61201

DZIAK, ROSE MARY, b Pittston, Pa, Apr 11, 46; m 83; c 2. CELL PHYSIOLOGY. *Educ:* Col Misericordia, BS, 67; Univ Rochester, MS, 70, PhD(radiation biol), 74. *Prof Exp:* Instr biol, Col Misericordia, 69-70; NIH fel, Med Sch, Northwestern Univ, 74-75; instr physiol, Med Ctr, La State Univ, 75-76; asst res prof, 76-77, from asst prof to assoc prof, 77-89, PROF ORAL BIOL, STATE UNIV NY, BUFFALO, 90- *Concurrent Pos:* George S Clarke Mem fel, Arthritis Found, 76; Res Career Develop Award, 80-85. *Mem:* AAAS; Endocrine Soc; Am Soc Bone & Min Res; NY Acad Sci. *Res:* Calcium regulation in isolated bone cells in an attempt to understand the mechanism of action of agents that influence bone resorption and formation. *Mailing Add:* 304 Brantwood Amherst NY 14226

DZIDIC, ISMET, b Derventa, Yugoslavia, June 14, 39. PHYSICAL CHEMISTRY, ANALYTICAL BIOCHEMISTRY. *Educ:* Univ Zagreb, dipl eng, 63; Univ Alta, PhD(phys chem), 70. *Prof Exp:* Res chemist, Res Inst Org Chem Industs, 63-64; teaching asst phys chem, Univ Alta, 65-70; fel, Inst Lipid Res, Baylor Col Med, 71-74, res asst prof chem, 74-81; SR RES CHEM, SHELL DEVELOP CO, 81- *Mem:* Am Chem Soc; Am Soc Mass Spectrometry. *Res:* Mass spectrometric study of gaseous ion-molecule reactions, solvation of ions in the gas phase; proton affinities of organic molecules; application of the chemical ionization mass spectrometry for structural studies of biological compounds. *Mailing Add:* 1220 Bartlett No 2 Houston TX 77006-6408

DZIECIUCH, MATTHEW ANDREW, b Edmonton, Alta, Oct 11, 31; m 58; c 3. ENVIRONMENTAL CHEMISTRY, ELECTROCHEMISTRY. *Educ:* Univ Alta, BSc, 57, MSc, 58; Univ Ottawa, PhD(electrochem), 62. *Prof Exp:* RES SCIENTIST ELECTROCHEM, FORD MOTOR CO, 61- *Mem:* AAAS; Sigma Xi; Electrochem Soc. *Res:* Batteries; electrochemical kinetics; deposition and oxidation kinetics; pollution chemistry, particularly treatment and disposal of pollutants in water. *Mailing Add:* 1411 Plainfield Dearborn MI 48125

DZIELAK, DAVID J, b Mar 12, 54; m. CARDIOVASCULAR PHYSIOLOGY. *Educ:* Univ Miss, PhD(pharmacol), 82. *Prof Exp:* ASST PROF PHYSIOL, UNIV MISS MED CTR, 82-, ASST PROF, DEPT SURG, 87- *Concurrent Pos:* Sr res scientist, Aetria Labs, 86-87. *Mem:* Am Physiol Soc; AAAS; Sigma Xi. *Res:* Hypertension. *Mailing Add:* Dept Surg Univ Miss Med Ctr 2500 N State St Jackson MS 39216-4505

DZIERZANOWSKI, FRANK JOHN, b Plains, Pa, Aug 28, 29; m 54; c 2. PHYSICAL CHEMISTRY, MATERIAL CHARACTERIZATION. *Educ:* Rutgers Univ, BS, 56. *Prof Exp:* Jr res chemist, Minerals & Chem Corp Am, 56-60, from res chemist to sr res chemist, Minerals & Chem Phillipp Corp, 60-66; res supvr fundamental res, 66-74, group leader fundamental res, 74-76, GROUP LEADER PHYS MEASUREMENTS, ENGELHARD CORP, 76- *Mem:* Catalysis Soc; Am Chem Soc; Clay Minerals Soc; Am Soc Testing & Mat; Tech Asn Pulp & Paper Indust. *Res:* Synthesis and properties of inorganic compounds; hydothermal synthesis of minerals; clay mineralogy; adsorption and catalysis; X-ray diffraction; characterization of materials. *Mailing Add:* Eight Norfolk Rd Somerset NJ 08873

DZIEWONSKI, ADAM MARIAN, b Lwow, Poland, Nov 15, 36; m 67. SEISMOLOGY, GEOMAGNETISM. *Educ:* Univ Warsaw, MS, 60; Acad Mining & Metall, Cracow, Dr Tech Sci(appl geol), 65. *Prof Exp:* Res asst seismol, Inst Geophys, Polish Acad Sci, 61-65, res assoc, 65; res assoc, Southwest Ctr Advan Studies, Univ Tex, Dallsa, 65-69, asst prof geophysicis 69-71; assoc prof geophys & assoc, Ctr Earth & Planetary Physics, 72-76, chmn, dept geol sci, 82-86, PROF GEOL & MEM, CTR EARTH & PLANETARY PHYSICS, HARVARD UNIV, 76- *Concurrent Pos:* Mem, Polish Sci Exped to NVietnam, Int Geophys Year, 58-59. *Mem:* Seismol Soc Am; fel Am Geophys Union; Soc Explor Geophys; AAAS; fel Am Acad Arts & Sci. *Res:* Physical properties of the earth's interior from observations of seismic wave propagation; earthquake mechanism; electrical and thermal properties of the crust and upper mantle from geomagnetic deep soundings. *Mailing Add:* Dept Earth Sci Harvard Univ Cambridge MA 02138

DZIMIANSKI, JOHN W(ILLIAM), b Baltimore, Md, Dec 13, 24; m 52; c 3. ELECTRICAL ENGINEERING. *Educ:* Johns Hopkins Univ, BE, 47, Dr Eng, 52. *Prof Exp:* Asst high frequency insulation, Johns Hopkins Univ, 47-52; group leader, Elec Res Sect, Allis-Chalmers Mfg Co, 52-56; ADV ENGR, ADVAN TECHNOL DIV, WESTINGHOUSE ELEC CORP, 56- *Mem:* Inst Elec & Electronics Engrs; Electrochem Soc. *Res:* Molecular electronics; semiconductor reliability physics; MNOS semiconductor memonics; charged coupled devices. *Mailing Add:* 412 Forest Lane Catonsville MD 21228-5154

DZIUK, HAROLD EDMUND, b Foley, Minn, Apr 27, 30; m 52; c 5. ANIMAL PHYSIOLOGY. *Educ:* Univ Minn, BS, 51, DVM, 54, MS, 55, PhD(vet physiol), 60. *Prof Exp:* Instr vet physiol, Col Vet Med, Univ Minn, 51-54 & 57-60; scientist, Hanford Prod Oper, Gen Elec Co, Wash, 60-61; assoc prof, Iowa State Univ, 61; from asst prof to assoc prof, 61-69, chmn dept vet biol, 72-75, PROF VET PHYSIOL, COL VET MED, UNIV MINN, ST PAUL, 69-, CHMN DEPT VET BIOL, 76- *Mem:* Conf Res Workers Animal Dis; Am Vet Med Asn; Am Physiol Soc. *Res:* Comparative gastrointestinal physiology; pathogenesis of diseases of the gastrointestinal tract. *Mailing Add:* Dept Vet Biol Rm 335 Animal Sci Vet Bldg Col Vet Med Univ Minn St Paul MN 55108

DZIUK, PHILIP J, b Foley, Minn, Mar 24, 26; m 51; c 7. REPRODUCTIVE PHYSIOLOGY, ENDOCRINOLOGY. *Educ:* Univ Minn, BS, 50, MS, 52, PhD(dairy husb), 55. *Prof Exp:* From asst prof to assoc prof animal physiol, 55-67, PROF ANIMAL PHYSIOL, UNIV ILL, URBANA, 67- *Concurrent Pos:* Lalor fel, 58-59; vis investr, R B Jackson Lab, 59; Pig Indust Develop Auth & Lalor fels, Cambridge, Eng, 61-62; mem comt on hormones, Nat Acad Sci-Nat Res Coun, 62-; indust consult; mem, human embryol & develop study sect, NIH, 82-86; mem, coop state res serv rev team, USDA, 83 & competitive grant rev, 85. *Honors & Awards:* Upjohn Award, Am Fertil Soc, 70; Physiol & Endocrinol Award, Am Soc Animal Sci, 71; Alexander von Humboldt sr US scientist award, 80; Distinguished Serv, Soc Study Reproduction, 89. *Mem:* AAAS; Am Asn Anat; fel Am Soc Animal Sci; Soc Study Fertil; Soc Study Reproduction (exec vpres, 85, pres, 86-87). *Res:* Egg transfer in cattle and swine; control of ovulation in cattle, sheep and swine; superovulation; artificial insemination; early stages of fertilization of eggs; embryonic mortality; sperm transport in male and female; diet on reproduction. *Mailing Add:* 111 Animal Genetics Univ Ill 1301 W Taft Dr Urbana IL 61801

DZOMBAK, WILLIAM CHARLES, b McKeesport, Pa, Dec 4, 21; m 53. PHYSICAL CHEMISTRY. *Educ:* Univ Pittsburgh, BSc, 43; Purdue Univ, PhD(chem), 50. *Prof Exp:* Assoc prof, Providence Col, 50-52; assoc chemist, Argonne Nat Lab, 52-53; assoc prof, 53-64, prof, 64-86, EMER PROF PHYS CHEM, ST VINCENT COL, 86- *Mailing Add:* Dept Chem St Vincent Col Latrobe PA 15650

DZURISIN, DANIEL, b Streator, Ill, Oct 29, 51; m 73; c 1. VOLCANOLOGY. *Educ:* Univ Notre Dame, BS, 73; Calif Inst Technol, PhD(geol), 77. *Prof Exp:* Geologist, Hawaiian Volcano Observ, 77-81, GEOLOGIST, CASCADES VOLCANO OBSERV, US GEOL SURV, 81- *Concurrent Pos:* Prin investr, Planetary Geol Prog, NASA, 77-85. *Mem:* Am Geophys Union; Am Astron Soc. *Res:* Monitoring and prediction of volcanic activity; physical processes in large silicic magma bodies beneath young calderas. *Mailing Add:* 9709 NE 107th Ave Vancouver WA 98662

E

EACHUS, ALAN CAMPBELL, b Champaign, Ill, July 11, 39; m 61; c 2. SANITARY CHEMISTRY. *Educ:* Syracuse Univ, BS, 60; State Univ NY Col Forestry, Syracuse Univ, PhD(org chem), 64; Northwestern Univ, MM, 75. *Prof Exp:* Fel abstraction kinetics, State Univ NY Col Forestry, Syracuse Univ, 64; res chemist, Dow Chem Co, 66-67, develop specialist, 67-69, res chemist, 69-70; sr res chemist, Alberto-Culver Corp, Melrose Park, 70-72, group leader, 72; mkt develop specialist, Com Solvents Corp, 72-76; prod mgr, Int Minerals & Chem, 76-80, mkt mgr, 80-82; DIR TECH SERV, ANGUS CHEM CO, 82- *Mem:* Am Chem Soc; NY Acad Sci; Soc Indust Microbiol. *Res:* Disinfectant/detergent formulations; flame fuels; chemical specialty items; antimicrobial agents. *Mailing Add:* 644 S Michigan Ave Villa Park IL 60181

EACHUS, JOSEPH JACKSON, b Anderson, Ind, Nov 5, 11; m 45; c 2. MATHEMATICS. *Educ:* Miami Univ, AB, 33; Syracuse Univ, AM, 36; Univ Ill, PhD(math, physics), 39. *Prof Exp:* Asst math, Syracuse Univ, 34-36; asst, Univ Ill, 36-39; instr, Purdue Univ, 39-42; elec engr, US Dept Defense, 46-55; systs dir, DatAmatic Corp, Minneapolis-Honeywell Regulator Corp, 55-62, prin staff scientist, Electronic Data Processing Div, Honeywell Inc, 62-70, group dir, Appl Res Div, Honeywell Info Systs, 70-75, prin staff scientist, Electronic Data Processing Div, 75-76; prin engr, Equip Div, Raytheon Co, 77-81; CONSULT, 82- *Concurrent Pos:* Mem info systs panel, Comput Sci & Eng Bd, Nat Acad Sci, 70-; consult, Inst for Defense Anal, 75-85. *Mem:* Emer mem Am Math Soc; Asn Comput Mach; fel Inst Elec & Electronics Engrs; fel AAAS. *Res:* Differential equations; q-difference equations; orthogonal functions; computing machinery; communications, electronics. *Mailing Add:* 4935 Stevens Dr Sarasota FL 34234-3756

EACHUS, RAYMOND STANLEY, b Bowdon, Eng, June 11, 44; m 66. PHYSICAL INORGANIC CHEMISTRY. *Educ:* Univ Salford, BS, 66; Univ Leicester, PhD(chem), 69. *Prof Exp:* Teaching fel phys chem, Univ BC, 69-70; res assoc, 70-80, HEAD CHEM PHYS LAB, EASTMAN KODAK CO, 80- *Mem:* Sigma Xi; Royal Inst Chem; The Chem Soc; Am Chem Soc. *Res:* Study of the kinetics and mechanisms of photochemical events in inorganic solids using physical techniques, especially electron spin resonance spectroscopy. *Mailing Add:* 30 San Rafael Rochester NY 14618

EACHUS, SPENCER WILLIAM, b Plainfield, NJ, Mar 11, 44; m 66. WOOD CHEMISTRY. *Educ:* Clarkson Col Technol, BS, 65; Syracuse Univ, PhD(org chem), 72. *Prof Exp:* Res assoc wood chem, Empire State Paper Res Inst, 71-73; res scientist pulping, 73-77, group leader, 77-85, SECTION LEADER, UNION CAMP CORP, 85- *Mem:* Tech Asn Pulp & Paper Indust; Am Chem Soc. *Res:* Methods and equipment for the pulping and bleaching of wood fibers. *Mailing Add:* Union Camp Corp Res & Develop Box 3301 Princeton NJ 08543-3301

EADE, KENNETH EDGAR, b Ft William, Ont, Jan 16, 26; m 56; c 2. REGIONAL GEOLOGY. *Educ:* Queen's Univ, Ont, BSc, 48; McGill Univ, MSc, 50, PhD(geol), 55. *Prof Exp:* Geologist, Geol Surv Can, 51-53 & 55-87 & Port WAfrica Proj, 53-54; CONSULT, 88- *Mem:* Geol Soc Am; Soc Econ Geologists; Arctic Inst NAm; Can Geol Soc; Can Inst Mining & Metall. *Res:* Precambrian geology in northern Canada. *Mailing Add:* 182 Sanford Ottawa ON K2C 0E9 Can

EADES, CHARLES HUBERT, JR, b Dallas, Tex, July 19, 16; m 42; c 3. BIOCHEMISTRY. *Educ:* Southern Methodist Univ, BS, 38; Univ Tex, MA, 40; Univ Ill, PhD(biochem), 48. *Prof Exp:* Instr sci, Paris Jr Col, Tex, 40-42; supvr sect anal lab, Pan Am Refining Corp, 42-45; spec res asst, Univ Ill, 45-48; from instr to asst prof biochem, Univ Tenn, 48-55; sr chemist, Mead Johnson & Co, 55-56, group leader, 56-57, sect leader, 57-59; sr scientist, Warner Lambert Res Inst, 59-64, sr res assoc, 64-70, sr clin res assoc, 70-77, asst dir med res, 77-81; CONSULT MED RES, 81- *Concurrent Pos:* Consult, Dept Path, Booth Mem Hosp, Flushing, NY, 63-70; fel coun arteriosclerosis, Am Heart Asn, mem coun clin cardiol & mem coun high blood pressure res. *Mem:* NY Acad Sci; Soc Exp Biol & Med; Am Soc Biol Chemists; Am Inst Nutrit; Am Med Writers Asn. *Res:* Amino acid requirements of man; nutrition of lactic acid bacteria; amino acid metabolism in stress; health and disease; automatic instrumentation; human nutrition; lipid and cholesterol metabolism; clinical investigation of cardiovascular agents in angina, hypertension and allied diseases; arteriosclerosis and cardiovascular disease; use of radioisotopes in metabolic and biochemical studies. *Mailing Add:* 50 Hillcrest Rd Mountain Lakes NJ 07046

EADES, JAMES B(EVERLY), JR, b Bluefield, WVa, July 22, 23; m 50; c 3. AEROSPACE ENGINEERING, ENGINEERING MECHANICS. *Educ:* Va Polytech Inst, BS, 44, MS, 49, PhD(eng mech), 58. *Prof Exp:* Instr aeronaut eng, Va Polytech Inst, 47-48, asst prof, 48-51, 53-57, prof aerospace eng, 57-69, head dept, 61-69; sr analyst, Anal Mech Assoc, Inc, Seabrook, 69-75, sr scientist, 75-77; prin scientist, Bus & Technol Systs, Inc, 77-81; vpres & prin scientist, Eng & Sci Assoc, Inc, 81-85; prin engr, Computational Eng, Inc, 86-90; SR SCIENTIST, ANALYTICAL MECH ASSOC, INC, HAMPTON, VA, 90- *Concurrent Pos:* Dir, Conf Lunar Explor, Blacksburg, Va, 62; aeronaut res specialist, Naval Ord Lab, White Oak, Md, 63-75; Nat Acad Sci sr res assoc, Goddard Space Flight Ctr, 67-69; vpres, Celestial Mech Inst, Inc; consult, US Army Transp Corps, Naval Ord Lab, Res Anal Corp, Naval Ship Res & Develop Ctr. *Mem:* Sr mem Am Astron Soc; assoc fel Am Inst Aeronaut & Astronaut; Sigma Xi. *Res:* Transonic flow phenomenon; high speed aerodynamics, flight performance and control; wind-tunnel testing of structures and vehicles; space flight, space mechanics, astrodynamics and celestial mechanics; hydrodynamics and underwater systems; reentry dynamics and reentry systems analysis. *Mailing Add:* 1603 Peacock Ln Silver Spring MD 20904

EADES, JAMES L, b Charlottesville, Va, Apr 21, 21; m 41; c 2. INDUSTRIAL MINERALOGY. *Educ:* Univ Va, BA, 50, MA, 53; Univ Ill, PhD(geol), 62. *Prof Exp:* Soils res engr, Va Coun Hwy Invest & Res, 52-58; res asst, Univ Ill, Urbana, 58-62, res asst prof geol, 62-70; chmn dept, 73-80, ASSOC PROF GEOL, UNIV FLA, 70- *Concurrent Pos:* Mem hwy res bd, Nat Acad Sci-Nat Res Coun; consult, Lime Indust & Hwy Depts; mem staff, USAID Projs, Africa; Nat Lime Asn res grant, 58-; mem staff, USAID Proj, SAm, 72-75. *Mem:* Am Soc Testing & Mat; Clay Minerals Soc; Geol Soc Am; Am Inst Prof Geologists (pres); Sigma Xi. *Res:* Clay mineralogy; calcium silicate reactions at ambient and elevated temperatures; sulfur dioxide reactions; toxic elements in limestones. *Mailing Add:* 2233 NW 19th Lane Gainesville FL 32605

EADES, JOHN ALWYN, b Ashbourne, UK, Dec 27, 39; British. ELECTRON DIFFRACTION, CRYSTALLOGRAPHY. *Educ:* Cambridge, BA, 62, PhD(physics), 67. *Prof Exp:* Prof physics, Univ Chile, Santiago, 67-73, Univ Bristol, UK, 74-81, Univ Ill, 81-; CONSULT. *Mem:* Am Phys Soc; Electron Micros Soc Am; FOP. *Res:* Electron microscopy with electron diffraction; particularly convergent-team diffraction and diffraction from surfaces. *Mailing Add:* MRL 104 S Goodwin Urbana IL 61801

EADIE, GEORGE ROBERT, b Eldorado, Ill, Sept 24, 23; m 43; c 2. MINING ENGINEERING. *Educ:* Univ Ill, BS, 49, MS, 56, EM, 57. *Prof Exp:* From asst prof to assoc prof mining eng, Univ Ill, 54-63; mem staff, Freeman Coal Mining Corp, 63-65; assoc ed, Coal Mining & Processing, 65-68; adminr, Ill State Geol Surv, 68-76; prof mining eng technol, Univ Southern Ind, Evansville, 76-; RETIRED. *Concurrent Pos:* Prof, Univ Ill, Urbana; mining consult, 76- *Mem:* Distinguished mem Soc Mining Engrs. *Res:* Mine safety; education in minerals industry. *Mailing Add:* 1500 Roosevelt Ave Eldorado IL 62930

EADON, GEORGE ALBERT, b Islip, NY, Oct 2, 45; m 73; c 2. ORGANIC CHEMISTRY, TOXICOLOGY. *Educ:* Mass Inst Technol, BS, 67; Stanford Univ, PhD(chem), 71. *Prof Exp:* from asst prof to assoc prof chem, State Univ NY, 71-79; res scientist, NY State Dept Health, 79-81, dir Toxicol Inst, 81-83; chmn, 87-88, ASSOC PROF, DEPT ENVIRON HEALTH & TOXICOL, STATE UNIV NY, ALBANY, 84-; DIR, DIV ENVIRON SCI, NY STATE DEPT HEALTH, 83- *Mem:* Am Chem Soc; AAAS. *Res:* Chemistry of environmental contaminants; application of mass spectrometry to stereochemical biochemical and analytical problems; mechanisms of electron-impact induced, photochemical and thermal reactions especially those involving hydrogen-atom transfer. *Mailing Add:* 335 E High St Ballston Spa NY 12020

EADS, B G, b June 20, 40; m; c 3. MEASUREMENT RESEARCH & CONTROLS ENGINEERING. *Educ:* Tenn Technol Univ, BSEE, 62; Univ Va, MSEE, 71. *Prof Exp:* Elec engr, Langley Res Ctr, NASA, 62-67; DIV DIR, INSTRUMENTATION & CONTROLS DIV, OAK RIDGE NAT LAB, 67- *Mem:* Sr mem Instrumentation Soc Am; Inst Elec & Electronics Engrs Computer Soc. *Res:* Process instrumentation research and development. *Mailing Add:* Dir Instrumentation & Controls Div Oak Ridge Nat Lab PO Box 2008 Oak Ridge TN 37831-6005

EADS, EWIN ALFRED, inorganic chemistry; deceased, see previous edition for last biography

EAGAN, ROBERT JOHN, b Rochester, NY, Aug 25, 44; m 76. GLASS SCIENCE, CERAMICS ENGINEERING. *Educ:* Alfred Univ, BS, 66; Univ Ill, MS, 68, PhD(ceramic eng), 71. *Prof Exp:* Mem tech staff ceramic eng, 71-77, supvr ceramic develop div I, 77-87, DIR, MAT & PROCESS SCI, SANDIA LABS, 88- *Concurrent Pos:* Mem subcomt 7, Int Comn Glass, 77-; trustee & mem exec comt, Int Glass Cong XII, Inc. 78- *Mem:* Am Ceramic Soc (pres, 90-91); Nat Inst Ceramic Engrs. *Res:* Development and characterization of new glass and glass ceramic materials; management of materials research and development. *Mailing Add:* Box 636 Cedar Crest NM 87008

EAGAR, ROBERT GOULDMAN, JR, b Richmond, Va, Feb 12, 47; m 73; c 2. BIO-ORGANIC CHEMISTRY, MICROBIOLOGY. *Educ:* Va Polytech Inst & State Univ, BSc, 69; Calif Inst Technol, PhD(chem), 74. *Prof Exp:* Res chemist, 75-82, GROUP LEADER, UNION CARBIDE RES CORP, 82-, ASSOC DIR, UNION CARBIDE CHEM & PLASTIC CO INC. *Mem:* Am Chem Soc; AAAS; Am Soc Microbiol. *Res:* Industrial microbiocides; cosmetic chemicals. *Mailing Add:* Union Carbide Chem & Plastic Co Inc PO Box 670 Bldg 200 Boundbrook NJ 08805

EAGAR, THOMAS W, b Chattanooga, TN, Jan 09, 50. MATERIALS PROCESSING & MANUFACTURING, SELECTION OF MATERIALS & FAILURE ANALYSIS. *Educ:* Mass Inst Technol, SB, 72. *Hon Degrees:* ScD, Mass Inst Technol, 75. *Prof Exp:* Res engr, Homer Res Labs, Bethlehem Steel Corp, 74-76; from asst prof to assoc prof, 76-87, PROF MATS ENG, DEPT MATS SCI & ENG, MASS INST TECHNOL, 87- *Concurrent Pos:* Dennison K Bullens Scholar, 69-71; Found Educ Fund Scholar, 70-71; NSF grad fel, 72-74; mem awards comt, res comt & tech papers comt, Am Welding Soc; mem, Joining Coun & Int Mats Review Comt, Am Soc Metals; mem, aluminum alloys, univ res & interpretive report comt, Welding Res Coun; liaison scientist, US Off Naval Res, Far E, 84-85; mem, Am Coun Int Inst Welding. *Honors & Awards:* Metall & Mats Prize, AIME, 72; Adams Mem Mem Award, Am Welding Soc, 79-83, Charles H Jennings Mem Medal, 84; H Mathewson Gold Medal, TMS-AIME, 87; Henry Krumb Lectr, SME-AIME, 87. *Mem:* Am Welding Soc; Am Soc Metals; AAAS; Soc Automotive Engrs; Am Ceramic Soc; Japan Welding Soc; Mats Res Soc. *Res:* Studies of slag-metal reactions in welding; resistance welding of sheet steels; weldability of high alloy steels, ceramic and metal brazing; arc physics; heat flow in welding & automation; control of welding processes; author of 80 publications. *Mailing Add:* Mass Inst Technol 77 Massachusetts Ave Rm 8-106 Cambridge MA 02139

EAGEN, CHARLES FREDERICK, b Detroit, Mich, Apr 13, 46; m 70; c 2. SEMICONDUCTOR & DEVICE PHYSICS. *Educ:* Oakland Univ, BA, 67; Iowa State Univ, PhD(physics), 72. *Prof Exp:* Presidential intern, Naval Res Lab, 72-73; RES SCIENTIST PHYSICS, FORD MOTOR CO RES LABS, 73- *Mem:* Am Phys Soc. *Res:* Device and process design for integrated circuits. *Mailing Add:* 1301 Red Oak Rd Ann Arbor MI 48103

EAGER, GEORGE S, JR, b Baltimore, Md, Sept 5, 15; m 45; c 3. ELECTRICAL ENGINEERING. *Educ:* Johns Hopkins Univ, BE, 36, Dr Eng(elec eng), 41. *Prof Exp:* Physicist, Armstrong Cork Co, Pa, 45-47; asst dir res, Gen Cable Corp, 48-70, assoc dir res, 70-80; PRES, GRJ CORP, 80-; PRES, CABLE TECHNOL LABS, INC, 83- *Mem:* Fel Inst Elec & Electronics Engrs; Int Conf Large Elec Systs. *Res:* Electrical wires and cables; dielectrics; electrical transmission; high voltage testing. *Mailing Add:* 14 Bellegrove Dr Upper Montclair NJ 07043

EAGER, RICHARD LIVINGSTON, b Kenaston, Sask, Aug 27, 17; m 49; c 2. PHYSICAL CHEMISTRY. *Educ:* Univ Sask, BE, 43, MSc, 45; McGill Univ, PhD(chem), 49. *Prof Exp:* From asst prof to prof, 47-84, EMER PROF CHEM, UNIV SASK, 84- *Concurrent Pos:* Vis lectr, Univ Leeds, 64-65. *Mem:* Chem Inst Can. *Res:* Use of biomass as a source of energy. *Mailing Add:* 46 Weir Cres Saskatoon SK S7H 3A9 Can

EAGLE, DONALD FROHLICHSTEIN, b St Louis, Mo, Jan 30, 33; m 61; c 2. MEDICAL PHYSICS. *Educ:* Yale Univ, BS, 54; Ga Inst Technol, MS, 56, PhD(physics), 62. *Prof Exp:* Asst res physicist, Eng Exp Sta, Ga Inst Technol, 56-61; sr physicist, Magnetic Tape Lab, Ampex Corp, 62-67; sr res physicist, Dikewood Corp, 67-69; mem tech staff, Sandia Labs, 69-73; fel med physics, Univ Tex M D Anderson Hosp & Tumor Inst, 73-74; med physicist, St Joseph Hosp & X-Ray Assocs, Albuquerque, 74-77; CHIEF PHYSICIST, BOCA RATON COMMUNITY HOSP, BOCA RATON, FLA, 77- *Concurrent Pos:* Adj prof radiol technol, Univ Albuquerque, 74-77. *Mem:* Am Asn Physicists in Med; Am Phys Soc; Am Col Med Phys. *Res:* Microwave spectroscopy and molecular structure; magnetic materials and magnetism; systems analysis; radiological physics; physics of radiation therapy. *Mailing Add:* Dept Radiation Oncol 800 Meadows Rd Boca Raton FL 33486

EAGLE, EDWARD, physiology, toxicology; deceased, see previous edition for last biography

EAGLE, HARRY, b New York, NY, July 13, 05; m 28; c 1. MEDICINE. *Educ:* Johns Hopkins Univ, AB, 23, MD, 27. *Hon Degrees:* MS, Yale Univ, 48; DSc, Wayne State Univ, 65, Duke Univ, 81, Rockefeller Univ, 82. *Prof Exp:* Assoc bact, Sch Med, Univ Pa, 33-35, asst prof, 35-36; intern, Johns Hopkins Univ, 27-28, asst, Med Sch, 29-30, instr, 30-32, dir, Lab Exp Therapeut, 35-36; sci dir, Res Br, Nat Cancer Inst, 47-49, chief, Lab Exp Therapeut, Microbiol Inst, NIH, 49-58, chief exp ther, Nat Inst Allergy & Infectious Dis, 58-59, chief, Lab Cell Biol, 59-61; chmn, Dept Cell Biol, Albert Einstein Col Med, 61-70, prof cell biol, 61-90, assoc dean sci affairs, 70-90, dir, Cancer Res Ctr, 75-90; RETIRED. *Concurrent Pos:* Fel, Johns Hopkins Univ, 28-29, lectr, 36-47; res fel, Harvard Med Sch, 32-33; labexp therapeut, USPHS, 36-46; trustee, Microbiol Found, Rutgers Univ. *Honors & Awards:* Lilly Bronze Medal, 36; Alvarenga Prize, Col Physicians Philadelphia, 36; Borden Award, Asn Am Med Col, 64; Einstein Commemorative Award, 69; NY Acad Med Award, 70; Modern Med Distinguished Achievement Award, 72; Louisa Gross Horowitz Award, 73; Sidney Farber Med Res Award, 74; Hubert H Humphrey Cancer Res Ctr Award, 82; Waterford Int Biomed Sci Award, Scripps Clin & Res Found, 83; E B Wilson Award, Am Soc Cell Biol, 84; Nat Medal of Sci Award, 87. *Mem:* Nat Acad Sci; Am Acad Arts & Sci; Am Soc Biol Chemists; Am Soc Clin Invest; Asn Am Physicians. *Res:* Immunochemistry; antigen-antibody reaction; serodiagnosis and chemotherapy of syphilis; blood coagulation; trypanosomiasis and tropical diseases; detoxification of metal poisoning; mode of action of antibiotics; cell and tissue culture; cancer research. *Mailing Add:* Cancer Res Ctr Albert Einstein Col Med Bronx NY 10461

EAGLE, SAM, b St Anthony, Idaho, July 11, 12; m 41; c 6. CHEMICAL ENGINEERING. *Educ:* Mont State Col, BS, 34; Carnegie Inst Technol, MS, 35, DSc(chem), 38. *Prof Exp:* Night sch instr, Carnegie Inst Technol, 37-38; res chemist, Res & Develop Sect, Chevron Res Co, Richmond, 38-41, asst foreman, Cracking Div, Richmond Refinery, 41-43, foreman, 43-45, res chemist, Calif Res Corp, 46-49, sr res chemist, 49-56, tech asst chem, 56-61, res engr, 62-72; RETIRED. *Mem:* Am Inst Chem Engrs; Am Chem Soc. *Res:* Process development in oil refining; adsorption; petrochemicals; catalyst research. *Mailing Add:* 7769 Baron Ct El Cerrito CA 94530

EAGLEMAN, JOE R, b Howell Co, Mo, Oct 9, 36; m 60; c 3. ATMOSPHERIC SCIENCES. *Educ:* Univ Mo, BS, 59, MS, 61, PhD(meteorol), 63. *Prof Exp:* From asst prof to assoc prof, 63-75, PROF ATMOSPHERIC SCI, UNIV KANS, 75- *Mem:* Am Meteorol Soc; Sigma Xi. *Res:* Severe thunderstorms; methods of measuring and calculating evapotranspiration; tornadoes; microclimatology; structure of thunderstorms and tornadoes; tornado damage to buildings. *Mailing Add:* Dept Physics & Astron Univ Kans Lawrence KS 66045

EAGLES, DOUGLAS ALAN, b New Britain, Conn, Feb 22, 43; m 67; c 2. NEUROPHYSIOLOGY. *Educ:* Lake Forest Col, BA, 65; Univ Mass, MA, 68, PhD(zool), 72. *Prof Exp:* NIH fel neurobiol, Univ Iowa, 71-73; ASST PROF BIOL, GEORGETOWN UNIV, 73- *Concurrent Pos:* Grass fel, 74. *Mem:* AAAS; Am Soc Zoologists; Sigma Xi. *Res:* Physiology of mechanoreception, organization of motor systems and the roles of sensory feedback in the coordination of movement. *Mailing Add:* Dept Biol Georgetown Univ 37th & O Sts NW Washington DC 20057

EAGLESHAM, ALLAN ROBERT JAMES, plant physiology, for more information see previous edition

EAGLESON, GERALD WAYNE, b Sioux City, Iowa, Apr 24, 47; m 74; c 2. ENDOCRINOLOGY, DEVELOPMENTAL BIOLOGY. *Educ:* Univ Calif, Riverside, BA, 68; Calif State Univ, Fullerton, MA, 73; Simon Fraser Univ, Burnaby, PhD(biol), 78. *Prof Exp:* Instr biol, Toloa Col, Tonga, 70-72; teaching asst molecular biol, Calif State Univ, Fullerton, 69-70, 72-73; teaching asst biol, Simon Fraser Univ, 74-77; instr, Univ Tex, Austin, 77-78, fel develop biol, 78-79; ASST PROF, LORAS COL, IOWA, 79- *Concurrent Pos:* Fulbright res assoc; vis prof, Univ Nijmegen, Neth. *Mem:* Am Soc Zoologists; Herpetologist's League; Am Asn Scientists. *Res:* Development and maturation of the pituitary, thyroids and gonads; timing of hypothalamic control of the hypothalamo-pituitary-thyroid (and gonad) axes. *Mailing Add:* Dept Biol Loras Col Dubuque IA 52004-0178

EAGLESON, HALSON VASHON, b Bloomington, Ind, Mar 14, 03; m 32, 41; c 3. PHYSICS. *Educ:* Ind Univ, AB, 26, AM, 31, PhD(physics), 39. *Hon Degrees:* DSc, Ind Univ, 85. *Prof Exp:* Prof math & physics, Morehouse Col, 27-35, head dept physics, 35-47; prof physics, Univ DC, 47-71, RETIRED. *Concurrent Pos:* Head dept, Clark Col, 40-47. *Mem:* Acoust Soc Am; Am Asn Physics Teachers; Nat Inst Sci (pres, 48). *Res:* Architectural acoustics; sound transmission; design of optical instruments; musical acoustics; ultrasonics; shock waves. *Mailing Add:* 3818 20th St NE Washington DC 20018

EAGLESON, PETER STURGES, b Philadelphia, Pa, Feb 27, 28; m 49, 74; c 3. HYDROLOGY. *Educ:* Lehigh Univ, BS, 49, MS, 52; Mass Inst Technol, ScD(civil eng), 56. *Prof Exp:* Asst fluid mech, Lehigh Univ, 50-51; asst fluid mech, 52-54, instr, 54-55, asst prof hydraul eng, 55-61, assoc prof, 61-65, head dept, 70-75, PROF CIVIL ENG, MASS INST TECHNOL, 65- *Honors & Awards:* Res Prize, Am Soc Civil Eng, 63; Horton Award, Am Geophys Union, 79 & Horton Medal, 88. *Mem:* Nat Acad Eng; Am Soc Civil Eng; fel Am Geophys Union; Int Asn Hydrol Sci; fel Am Meteorol Soc; fel AAAS. *Res:* Hydroclimatology. *Mailing Add:* Dept Civil Eng Mass Inst Technol Rm 48-335 Cambridge MA 02139

EAGLETON, LEE C(HANDLER), chemical engineering; deceased, see previous edition for last biography

EAGLETON, ROBERT DON, b Ladonia, Tex, Aug 19, 37; m 63; c 4. SOLID STATE PHYSICS. *Educ:* Okla State Univ, 59, MS, 62, PhD(physics), 69. *Prof Exp:* Instr physics, US Naval Nuclear Power Sch, Calif, 62-65; asst prof, 68-80, PROF PHYSICS, CALIF STATE POLYTECH UNIV, POMONA, 80- *Mem:* Am Asn Physics Teachers; Sigma Xi; Am Phys Soc. *Res:* Temperature dependence of positronium annihilation in solids; thermally stimulated luminescence in stannic oxide single crystals and ceramics. *Mailing Add:* Dept Physics Calif State Polytech Univ Pomona CA 91768

EAGON, JOHN ALONZO, b Portsmouth, NH, May 5, 32; m 57; c 2. MATHEMATICS. *Educ:* Princeton Univ, BA, 54; Univ Chicago, MS, 58, PhD(math), 61. *Prof Exp:* NATO fel math, Univ Sheffield, 61-62; asst prof, Univ Ill, Urbana, 62-67; assoc prof, 67-74, PROF MATH, UNIV MINN, MINNEAPOLIS, 74- *Mem:* Am Math Soc. *Res:* Commutative rings; linear graphs. *Mailing Add:* Dept Math Grad Sch Univ Minn 127 Vincent Hall 206 Church St SE Minneapolis MN 55455-0487

EAGON, PATRICIA K, SEX HORMONE RECEPTORS, SEXUAL DIMORPHISM. *Educ:* Univ Pittsburgh, PhD(biochem), 76. *Prof Exp:* ASST PROF MED & BIOCHEM, SCH MED, UNIV PITTSBURGH, 80- *Mailing Add:* Univ Pittsburgh Sch Med 1000-E Scaife Hall Pittsburgh PA 15261

EAGON, ROBERT GARFIELD, b Salesville, Ohio, Oct 29, 27; m 52; c 1. MICROBIOLOGY. *Educ:* Ohio State Univ, BSc, 51, MSc, 52, PhD(bact), 54. *Prof Exp:* Fulbright scholar, Pasteur Inst, Paris, 54-55; asst prof bact, Univ Ga, 55-59, from assoc prof to prof microbiol, 59-87, Franklin prof, 87-90; VPRES BIOMED RES, HORIZON MICRO-ENVIRONMENTS, 90- *Honors & Awards:* P R Edwards Award, Am Soc Microbiol, 76. *Mem:* AAAS; Am Soc Microbiol; fel Am Acad Microbiol; Soc Indust Microbiol. *Res:* Bacterial metabolism; physiology. *Mailing Add:* Horizon Micro-Environments 1081 Industrial Dr PO Box 149 Watkinsville GA 30677

EAKER, CHARLES MAYFIELD, b Bonne Terre, Mo, Aug 3, 19; m 43; c 3. CHEMISTRY. *Educ:* Cent Col, AB, 41; Univ Md, PhD(org chem), 46. *Prof Exp:* Asst, Off Sci Res & Develop, Md, 43-45; res chemist, Monsanto Co, 46-52, res group leader, 52-69, res specialist, 69-82; RETIRED. *Mem:* Am Chem Soc. *Res:* Synthetic organic chemistry; rubber antioxidants and vulcanization agents; agricultural chemicals. *Mailing Add:* PO Box 241 Put In Bay OH 43456

EAKER, CHARLES WILLIAM, b St Louis, Mo, May 25, 49; m 74; c 2. THEORETICAL REACTION DYNAMICS. *Educ:* Mich State Univ, BS, 71; Univ Chicago, PhD(chem), 74. *Prof Exp:* R A Welch fel chem, Univ Tex, Dallas, 74-76; from instr to assoc prof, 76-89, PROF CHEM, UNIV DALLAS, 89- *Mem:* Am Chem Soc; Sigma Xi. *Res:* Calculation of potential energy surfaces of small molecular systems using ab initio and semiempirical techniques; determination of dynamical properties of reacting systems on calculated potential surfaces; use of computer graphics to visualize potential energy surfaces and to simulate chemical reactions. *Mailing Add:* Univ Dallas 1845 E Northgate Dr Irving TX 75062-4799

EAKIN, BERTRAM E, b Jerome, Idaho, Oct 9, 28; m 52; c 3. CHEMICAL ENGINEERING, PETROLEUM ENGINEERING. *Educ:* Mass Inst Technol, BS, 51; Ill Inst Technol, MS, 57, PhD(gas technol), 62. *Prof Exp:* Lab technician, Chem & Geol Labs, Casper, Wyo, 51-53; lab technician, Inst Gas Technol, Ill Inst Technol, 53-54, asst chem engr, 54-57, assoc chem engr, 57-62, chem engr, 62-64, sr chem engr, 64-71; dir res, P-V-T Inc, 71-75; sr res scientist, Getty Oil Co, 76-84; SR RES CONSULT, TEXACO ETHYLPHOSPHONOTHIOIC DICHLORIDE, 84- *Concurrent Pos:* From adj instr to adj asst prof, Ill Inst Technol, 53-71. *Mem:* Am Inst Chem Engrs; Soc Petrol Engrs. *Res:* Experimental measurement of thermodynamic and transport properties of gases and liquids at cryogenic temperatures and elevated pressures, enhanced oil recovery processes, rock-fluid properties; effect of water flooding on residual oil properties; oil-shale processing; publications include 62 papers, patents or books in the technical literature. *Mailing Add:* PO Box 770070 Houston TX 77215

EAKIN, RICHARD MARSHALL, b Florence, Colo, May 5, 10; m 35, 82; c 2. ZOOLOGY. *Educ:* Univ Calif, AB, 31, PhD(zool), 35. *Hon Degrees:* Berkeley Citation, Univ Calif, Berkeley, 77. *Prof Exp:* Asst zool, Univ Calif, 31-34; Nat Res Coun fel, Univ Erlangen & Univ Freiburg, 35-36; from instr to assoc prof zool, 36-49, asst dean, Col Letters & Sci, 39-42, chmn dept, 42-48 & 52-57, chmn Miller inst basic res sci, 61-67, Miller res prof, 61-62 & 69-70, prof zool, 49-77, EMER PROF ZOOL, UNIV CALIF, BERKELEY, 77- *Concurrent Pos:* Guggenheim fel, Stanford Univ, 53; NSF fel, Univ Berne, 57; assoc ed, J Exp Zool, 67-71, J Ultrastruct Res, 73-89 & Zoomorphology, 77-85; nat lectr, Sigma Xi, 74; distinguished vis prof biol, Tougaloo Col, 78; vis prof biol, Talladega Col, 79 & Fisk Univ, 81; trustee, Talladega Col, 81-85; producer, films of Darwin, Mendel, Harvey, Pasteur, Beaumont & Spemann, Univ Calif, Berkeley. *Honors & Awards:* Walker Prize, Boston Mus Sci, 76; Distinguished Invest Award, Electron Micros Soc Am, 82. *Mem:* Am Soc Zoologists (pres, 75); Sigma Xi; Soc Develop Biol; Electron Micros Soc Am. *Res:* Fine structure of photoreceptors. *Mailing Add:* Dept Integrative Biol Univ Calif Berkeley CA 94720-9978

EAKIN, RICHARD R, b New Castle, Pa, Aug 6, 38; m 60; c 2. MATHEMATICS. *Educ:* Geneva Col, BA, 60; Wash State Univ, MA, 62, PhD(math), 64. *Prof Exp:* From asst prof to assoc prof math, Bowling Green State Univ, 64-72, asst dean, Grad Sch, 69-72, vprovost student affairs, 72-78, vice provost, Instnl Planning & Student Affairs, 78-80, exec vice provost planning & budgeting, 80-83, vpres planning & budgeting, 83-87; CHANCELLOR, ECAROLINA UNIV, 87- *Mem:* Math Asn Am. *Res:* Combinatorial mathematics. *Mailing Add:* Chancellor ECarolina Univ Greenville NC 27858

EAKIN, RICHARD TIMOTHY, b Birmingham, Ala, May 25, 42. BIOPHYSICAL CHEMISTRY. *Educ:* Univ Tex, Austin, BS, 63; PhD (physics), 84, Calif Inst Technol, PhD (biochem), 68. *Prof Exp:* NIH fel, Stanford Univ, 68-71; RES ASSOC CHEM, UNIV TEX, AUSTIN, 71- *Concurrent Pos:* Vis staff mem, Los Alamos Sci Lab, 72-77; fel Nat Heart Lung & Blood Inst, Univ Wash, 85-87. *Mem:* Am Chem Soc. *Res:* Respiratory-deficient cytoplasmic mutants of Neurospora; structure and function of cytochromes; biochemical applications of carbon-13 nuclear magnetic resonance spectroscopy; kinetic modeling blood-tissue exchange processes. *Mailing Add:* 1603 Scenic Dr Austin TX 78703

EAKINS, KENNETH E, b London, Eng, July 17, 35; m 61; c 2. PHARMACOLOGY, OPHTHALMOLOGY. *Educ:* Univ London, BPharm, 58, PhD(pharmacol), 62. *Prof Exp:* From asst prof to assoc prof ophthal, Col Physicians & Surgeons, Columbia Univ, 65-72, assoc prof, 72-76, prof pharmacol, 76-80; HEAD, PHARMACOL DEPT, WELLCOME RES LABS, UK, 80-; AT DEPT ANALYTICAL PHARMACOL, RAYNE INST, KING'S COL SCH MED & DENT,. *Concurrent Pos:* Nat Coun to Combat Blindness fel, Wilmer Inst, Med Sch, Johns Hopkins Univ, 62-63; USPHS grants, Nat Eye Inst, 66-; vis prof, Inst Ophthal, Univ London, 71-72; consult, NIH, 73-80; ed, Exp Eye Res, 75- *Mem:* Am Soc Pharmacol & Exp Therapeut; Brit Pharmacol Soc; Asn Res Vision & Ophthal; Int Soc Eye Res; Brit Res Asn. *Res:* Ocular inflammation; glaucoma; prostaglandin antagonists. *Mailing Add:* Whitby Res Inc 1001 Health Science Rd W Irvine CA 92715

EAKINS, PETER RUSSELL, b Montreal, Que, May 17, 27; m 50, 62; c 3. GEOLOGY. *Educ:* McGill Univ, BSc, 48, MSc, 49, PhD(geol), 52. *Prof Exp:* Geochemist, Cerro de Pasco Corp, 52-55; explor geologist, Malartic Gold Fields, Ltd, 55-57; chief geologist, Mineral Mgt Ltd, 57-58; lectr geol, McGill Univ, 58-59, from asst prof to assoc prof, 59-90; RETIRED. *Concurrent Pos:* Prof officer & geosci consult, Expo 65; consult geologist, Mkuski Copper Mines, Zambia, Africa, 67-68; chmn exhibits comt, Georama, 72, Int Geol Cong Montreal, 72; ed, Mineral Explor Res Inst, Montreal, 77- *Mem:* Geol Soc Am; Geochem Soc fel Royal Geog Soc; fel Geol Asn Can; Can Inst Min & Metall. *Res:* Structural geology; mineral exploration; Malartic gold deposits, Quebec; structures in the Quebec Appalachians; volcanic rock types of Northwestern Quebec; geochemical prospecting techniques for copper deposits in Peru. *Mailing Add:* RR 1 452 Duhamel Rd Bedford PQ J0J 1A0 Can

EAKS, IRVING LESLIE, b Sawtelle, Calif, May 24, 23; m 48; c 3. PLANT PHYSIOLOGY, BIOCHEMISTRY. *Educ:* Colo Agr & Mech Col, BS, 48; Univ Calif, Davis, MS, 50, PhD(plant physiol), 53. *Prof Exp:* Asst, Univ Calif, Davis, 48-52; from jr plant physiologist to assoc plant physiologist, Citrus Exp Sta, 52-62, PLANT PHYSIOLOGIST, UNIV CALIF, RIVERSIDE, 62-, LECTR BIOCHEM, 73- *Mem:* Am Soc Plant Physiol; Am Soc Hort Sci. *Res:* Post-harvest physiology, handling and chemical composition of fruits; chilling injury; biogenesis of and responses to ethylene. *Mailing Add:* Dept Biochem Univ Calif Riverside CA 92521

EALES, JOHN GEOFFREY, b Wolverhampton, Eng, Sept 9, 37; m 63; c 2. THYROID FUNCTION. *Educ:* Oxford Univ, BA, 59; Univ BC, MSc, 61, PhD(zool), 63. *Prof Exp:* Asst prof biol, Univ NB, 63-67; from asst prof to prof, 67-89, DISTINGUISHED PROF ZOOL, UNIV MAN, 89- *Concurrent Pos:* Ed, Can J Zool, 89-; Killam fel, 89-91. *Mem:* Can Soc Zoologists; Am Soc Zoologists; fel Royal Soc Can. *Res:* Fish endocrinology specifically thyroid function in salmonidae,with particular reference to regulation of the peripheral metabolism of thyroid hormones and conversion of thyroxine to triodothyronine; influence of nutrition, reproduction and environmental stressors on these activities. *Mailing Add:* Dept Zool Univ Man Winnipeg MB R3T 2N2 Can

EALY, ROBERT PHILLIP, b Kay Co, Okla, July 6, 14; m 39. ORNAMENTAL HORTICULTURE, ECOLOGY. *Educ:* Okla State Univ, BS, 41; Kans State Col, MS, 46; La State Univ, PhD, 55. *Prof Exp:* Asst hort, Kans State Col, 41-42; from asst prof to prof, Okla State Univ, 46-61; prof hort & head dept, Kans State Univ, 61-63, prof hort & landscape archit & head dept, 63-66, dir landscape archit, 66-69, assoc dean, Col Archit & Design, 67-74, head dept, 69-79, prof, 66-82, emer prof landscape archit, 90; RETIRED. *Mem:* Fel Am Soc Landscape Archit. *Res:* Landscape architecture; outdoor recreation; plant ecology; environmental planning; ornamental landscape plant materials; dwarfing rootstocks; herbaceous perennial flowers; chlorosis of ornamentals; graft unions; container grown nursery stock; radioactive tracers. *Mailing Add:* 2121 Meadowlark Rd Manhattan KS 66502

EAMES, ARNOLD C, b West Paris, Maine, Feb 10, 30; m 54; c 3. PAPER CHEMISTRY. *Educ:* Rensselaer Polytech Inst, BME, 51; Inst Paper Chem, Lawrence, MS, 57, PhD(paper chem), 59. *Prof Exp:* Res engr, S D Warren Co, 53-54; proj engr, Container Corp Am, 56; res engr, 59-61, asst res dir new prod develop, 62-74, prod develop mgr, 74-77, asst res dir, Graphic Arts & Reprographic Prod, 77-84, ASST RES DIR, DIRECTED BASIC RES, S D WARREN CO, 84- *Mem:* Tech Asn Pulp & Paper Indust. *Res:* Coated paper technology for printing and other special uses. *Mailing Add:* 22 Walpham St Westbrook ME 04092

EAMES, M(ICHAEL) C(URTIS), b Birmingham, Eng, Feb 6, 31; m 56; c 2. NAVAL ARCHITECTURE. *Educ:* Univ Durham, BSc, 50 & 51; NS Tech Col, MEng, 57; Univ Newcastle-on-Tyne, DSc, 86. *Prof Exp:* Shipbldg apprentice, R & W Hawthorn-Leslie & Co, Ltd, Eng, 47-52; sci officer hydrodyn, Can Defence Res Estab Atlantic, 52-58, group leader appl math, 58-60, head spec studies team, 60-63, head fluid mech sect, 63-74, sr scientist, 74-89; ADJ PROF, CTR FOREIGN POLICY STUDIES, DALHOUSIE UNIV, HALIFAX, 90- *Concurrent Pos:* Adj prof, Naval Archit, Tech Univ NS, 79-; dir int studies naval warfare, NATO, 85-89. *Mem:* Can Aeronaut & Space Inst; Royal Inst Naval Architects. *Res:* Applied hydrodynamics and aerodynamics; naval architecture; aeronautical engineering; naval operations research; technological forecasting; maritime strategy. *Mailing Add:* 49 Murray Hill Dr Dartmouth NS B2Y 3B2 Can

EAMES, WILLIAM, b Minnedosa, Man, Sept 21, 29; m 62; c 4. MATHEMATICAL ANALYSIS. *Educ:* Brandon Univ, BSc, 50; Univ Man, BSc, 52, MSc, 53; Queen's Univ, Ont, PhD(math), 56. *Prof Exp:* Nat Res Coun fel, Univ Col, Univ London, 56-57; asst prof math, Univ NB, 57-58; from lectr to sr lectr, Sir John Cass Col, Eng, 58-66; assoc prof, 66-78, PROF MATH, LAKEHEAD UNIV, 66-, CHMN DEPT, 78- *Mem:* Am Math Soc; Can Math Cong; London Math Soc. *Res:* Measure theory. *Mailing Add:* Dept Math Lakehead Univ Thunder Bay ON P7B 5E1 Can

EAMES, WILMER B, b Kansas City, Mo, May 8, 14; m 39; c 2. OPERATIVE DENTISTRY. *Educ:* Kansas City-Western Dent Col, DDS, 39. *Prof Exp:* Pvt prac, 39-61; prof oper dent, Dent Sch, Northwestern Univ, Chicago, 61-67, assoc dean, 64-67; prof, 67-79, EMER PROF OPER DENT, SCH DENT, EMORY UNIV, 79- *Concurrent Pos:* Clin prof, Sch Dent, Colo Univ, 80- *Honors & Awards:* Hollenback Mem Prize, 78, 79; Jerome & Dorothy Schweitzer Res Award, Greater NY Acad Prosthodontics, 79; Hinman Distinguished Serv Medallion Atlanta, 81; Albert L Borish Award, Acad Gen Dent, 83; Wilmer Souder Award, Mat Group, Int Asn Dent Res, 85; William John Gies Award, Am Col Dentists, 86. *Mem:* Am Dent Asn; Int Asn Dent Res; Am Acad Restorative Dent; Acad Oper Dent; hon mem Am Dent Soc Europe; Sigma Xi; hon fel Acad Gen Dent; fel Am Col Dentists; fel Int Col Dentists. *Res:* Dental materials and operative techniques; clinical studies comparing dental materials; over 100 publications on dental materials. *Mailing Add:* 14390 E Marina Dr Apt 501 Aurora CO 80014

EANES, EDWARD DAVID, b Rochester, NY, Sept 2, 34; m 61; c 2. BIOPHYSICAL CHEMISTRY. *Educ:* Col William & Mary, BS, 57; Johns Hopkins Univ, MA, 59, PhD(crystal struct), 61. *Prof Exp:* Phys chemist, Nat Bur Standards, 60-61; comn officer, US Pub Health Serv, 61-63; asst prof phys chem, Med Col, Cornell Univ, 63-67; chief molecular struct sect, 67-80, chief mineral chem & struct sect, lab biol struct, 80-82, chief skeletal biophys sect, 82-86, CHIEF MINERAL CHEM & STRUCT SECT, BONE RES BR, NAT INST DENT RES, 86- *Honors & Awards:* Biol Mineralization Res Award, Int Asn Dent Res, 83,. *Mem:* Am Chem Soc; fel AAAS; Int Asn Dent Res; Sigma Xi. *Res:* Calcium phosphate chemistry; biological calcification. *Mailing Add:* Bone Res Br Nat Inst Dent Res Bldg 30 Rm 106 Bethesda MD 20892

EARDLEY, DIANE DOUGLAS, c 3. IMMUNO REGULATION, SUPPRESSOR T-CELLS. *Educ:* Univ Calif, Berkeley, PhD, 72. *Prof Exp:* ASSOC PROF IMMUNOBIOL & MED MICROBIOL, UNIV CALIF, SANTA BARBARA, 83- *Mem:* Am Asn Immunologists. *Mailing Add:* Dept Biol Sci Univ Calif Santa Barbara CA 93109

EARECKSON, WILLIAM MILTON, III, organic polymer chemistry; deceased, see previous edition for last biography

EARGLE, GEORGE MARVIN, b Salisbury, NC, Sept 29, 39. APPLIED MATHEMATICS. *Educ:* Univ NC, Chapel Hill, BS, 61, NC State Univ, MA, 63, PhD(math), 68. *Prof Exp:* Asst prof math, NC State Univ, 66-69; from asst prof to assoc prof, 69-79, PROF MATH SCI, APPALACHIAN STATE UNIV, 79- *Mem:* Soc Indust & Appl Math; Math Asn Am. *Res:* Applications of mathematics; elasticity theory. *Mailing Add:* 107 Owl Tree Trail Greenville TX 75401

EARGLE, JOHN MORGAN, b Tulsa, Okla, Jan 6, 31. SOUND RECORDING & REPRODUCTION. *Educ:* Eastman Sch Music, BM, 53; Univ Mich, Ann Arbor, MM, 54; Univ Tex, Austin, BS, 62; Cooper Union, ME, 70. *Prof Exp:* Mgr qual, RCA Records, 62-69; chief engr, Mercury Rec, 69-71; dir, Altec Corp, 71-74; prin, JME Assoc, 74-76; vpres, JBL Inc, 76-82; PRIN, JME CONSULT CORP, 82- *Concurrent Pos:* Lectr, Aspen Sch Music, 81- & Peabody Conserv, 85- *Honors & Awards:* Bronze Medal, Audio Eng Soc, 84. *Mem:* Fel Audio Eng Soc (pres, 75-76); Acoust Soc Am; Soc Motion Picture & TV Engrs; Inst Elec & Electronics Engrs; Nat Acad Rec Arts & Sci; Acad Motion Picture Arts & Sci. *Res:* Sound recording and reproduction; transducers, microphones and loud speakers; recording arts and sciences. *Mailing Add:* 7034 Macapa Dr Los Angeles CA 90068

EARHART, CHARLES FRANKLIN, JR, b Melrose Park, Ill, Oct 26, 41. MICROBIAL PHYSIOLOGY, BACTERIAL CELL SURFACES. *Educ:* Knox Col, Ill, AB, 62; Purdue Univ, PhD(molecular biol), 67. *Prof Exp:* NIH fel, Sch Med, Tufts Univ, 67-68; fac assoc, 68-70, from asst prof to assoc prof, 70-83, PROF MICROBIOL, UNIV TEX, AUSTIN, 83- *Concurrent Pos:* Lectr, Found Microbiol Lect Prog, Am Soc Microbiol, 88. *Mem:* AAAS; Am Soc Microbiol; Genetics Soc Am; Sigma Xi. *Res:* Modification of bacterial cell envelopes; genetics and physiology of iron assimilation in microbial systems. *Mailing Add:* Dept Microbiol Univ Tex Austin TX 78712-1095

EARHART, J RONALD, b Hershey, Pa, July 29, 41; m 63; c 2. PHYSICS. *Educ:* Lebanon Valley Col, BS, 63; Univ NH, MS, 66, PhD(physics), 69. *Prof Exp:* Sr physicist, high frequency radar-propagation, Int Tel & Tel Electro Physics Labs, 68-73; SR PHYSICIST SONAR-SYST ANALYSIS & PROJ LEADER RADAR-SYSTS ANALYSIS, APPL PHYSICS LAB, JOHNS HOPKINS UNIV, 73- *Mem:* Am Geophys Union; Sigma Xi. *Res:* Ionospheric propagation; radar; system analysis; space physics (particles); sonar; doppler navigation. *Mailing Add:* Johns Hopkins Univ Appl Physics Lab Johns Hopkins Rd Laurel MD 20707

EARHART, RICHARD WILMOT, b Columbus, Ohio, Jan 19, 40; m 73. PHYSICS. *Educ:* Middlebury Col, AB, 60; Ohio State Univ, BSc, 61, MSc, 66. *Prof Exp:* Staff asst planning, Off Dir Defense Res & Eng, 66-69; RES SCIENTIST PLANNING & ANALYSIS, COLUMBUS DIV, BATTELLE MEM INST, 71- *Mem:* AAAS; Soc Am Mil Engrs. *Res:* Economic analyses and planning of civil space systems. *Mailing Add:* 1471 Michigan Ave Columbus OH 43201

EARING, MASON HUMPHRY, b Albany, NY, Oct 23, 21; m 49; c 4. ORGANIC POLYMER CHEMISTRY. *Educ:* Rensselaer Polytech Inst, PhD(chem), 50. *Prof Exp:* Mem staff org res, Wyandotte Chem Corp, 50-60, sr res chemist, 60-63 & Ballast Dept, Gen Elec Co, 63-75; res supvr, 75-82, CONSULT SPENCER KELLOGG DIV, TEXTRON INC, 82- *Concurrent Pos:* Tutor, Empire State Col, 83-; asst prof, Daemen Col, 84 & 85. *Mem:* Fel AAAS; Am Chem Soc; fel Am Inst Chemists. *Res:* Industrial and polymer chemistry; polyethers and urethane polymers; dielectric and acoustic properties of polymers. *Mailing Add:* 94 N Union St Buffalo NY 14221

EARL, ALLAN EDWIN, b Kingston, Ont, Sept 19, 39; m 61; c 3. FOOD SCIENCE. *Educ:* Queens Univ, BS, 61, MS, 63; Univ Alta, PhD(carbohydrate chem), 68. *Prof Exp:* Sr scientist, Labatt Breweries Can Ltd, 68-74; food prod develop engr, 74-75, dir, Can Food Prod Develop Ctr, Man Res Coun, 75-77; res coordr, Rapeseed Asn Can, 77-79; EXEC DIR, CANOLA COUNCIL CAN, 79- *Mem:* Can Inst Food Sci & Technol; Inst Food Technologists; Am Asn Cereal Chemists; Chem Inst Can. *Res:* Food processing, new and improved. *Mailing Add:* BC Tree Fruits Ltd 1473 Water St Kelowna BC V1Y 1J6 Can

EARL, BOYD L, b Pa, July 20, 27; m 49; c 4. MATHEMATICS. *Educ:* Wilkes Col, BS, 52; Bucknell Univ, MS, 57. *Prof Exp:* Teacher high sch, Pa, 52-56; instr math, Bucknell Univ, 56-61; ASSOC PROF MATH, WILKES COL, 61- *Mem:* Math Asn Am; Am Math Soc. *Res:* Topology; algebra. *Mailing Add:* 764 Mercer Ave Kingston PA 18704

EARL, BOYD L, b Burley, Idaho, Aug 17, 44; m 80; c 3. PHYSICAL CHEMISTRY. *Educ:* Univ Idaho, BS, 66; Univ Calif, Berkeley, MS, 70, PhD(phys chem), 73. *Prof Exp:* Adj asst prof chem, Brooklyn Col, City Univ New York, 73-76; from asst prof to assoc prof, 76-90, PROF CHEM, UNIV NEV, LAS VEGAS, 90-, CHMN DEPT, 82-86 & 90- *Mem:* Sigma Xi; AAAS; Am Chem Soc. *Res:* Laser-induced reactions in the gas phase, using carbon dioxide lasers; product analysis; kinetic systems. *Mailing Add:* Dept Chem Univ Nev Las Vegas NV 89154

EARL, CHARLES RILEY, b San Diego, Calif, Oct 27, 33; m 60; c 2. POLYMER CHEMISTRY, PHYSICAL CHEMISTRY. *Educ:* Whittier Col, AB, 55; Polytech Inst Brooklyn, PhD(polymer chem), 70. *Prof Exp:* Chemist, US Food & Drug Admin, 55-56 & Aerojet-Gen Corp, 56-60; res fel, Jewish Hosp Brooklyn, 59-61; sr res fel, Polytech Inst Brooklyn, 62-68; sr res chemist, Milliken Res Corp, 68-72; head dept chem & textiles, Spartanburg Tech Col, 72-81; gen mgr, S & C Chem Co, 82-85; vis prof, 84-85, adj fac, 85-88, PROF & CHEM DEPT HEAD, ANDERSON COL, 88- *Concurrent Pos:* Vis assoc prof, Clemson Univ, 81-82; chem consult, 82- *Mem:* Am Chem Soc; Sigma Xi. *Res:* Adsorption, adhesion of polymers to textile fibers; relationship of polymer structure and properties; polyelectrolytes. *Mailing Add:* 440 Harrell Dr Spartanburg SC 29302-2518

EARL, FRANCIS LEE, b Jasper, Mo, Dec 12, 24; m 50; c 2. VETERINARY TOXICOLOGY. *Educ:* Mich State Col, DVM, 47. *Prof Exp:* Vet, State Vet Off, State Mo, 47-48; sta vet, Animal Husb Div, Bur Animal Indust, USDA, Md, 48-52, vet, Path Div, 52-61; vet, Div Toxicol, Food & Drug Admin, 61-63, vet med officer, Bur Foods, 64-75, facil mgr, Spec Pharmacol Animal Lab, 75-80; VET CONSULT, 80- *Mem:* Am Vet Med Asn. *Res:* Atrophic rhinitis in swine; swine erysipelas; use of swine on drug toxicity research; comparative toxicology of dogs and swine; diseases and housing of miniature swine; clinical laboratory values of dogs and swine; ocular toxicity in dogs; teratogenic effects of compounds in dogs and miniature swine; perinatal toxicology. *Mailing Add:* 2613 Hughes Rd Adelphi MD 20783

EARL, JAMES ARTHUR, b Omaha, Nebr, Aug 14, 32; m 55; c 4. PHYSICS. *Educ:* Mass Inst Technol, BS, 53, PhD(physics), 57. *Prof Exp:* Physicist, Ft Monmouth, 58; from lectr to asst prof physics, Univ Minn, 58-65; assoc prof, 65-75, PROF PHYSICS, UNIV MD, 75- *Concurrent Pos:* Vis prof, Univ Ariz, 87-88. *Honors & Awards:* Sr scientist award, Alexander von Humboldt Found, 77-78. *Mem:* Fel Am Phys Soc; Am Geophys Union; Am Astron Soc; Sigma Xi. *Res:* Cosmic ray extensive air showers; solar cosmic rays; primary cosmic ray electrons; charged particle transport theory. *Mailing Add:* Dept Physics & Astron Univ Md College Park MD 20742

EARLE, ALVIN MATHEWS, b Topeka, Kans, Mar 20, 31; m 54; c 3. ANATOMY. *Educ:* Loyola Univ, Ill, BS, 54; Univ Colo, MS, 58, PhD(zool), 62. *Prof Exp:* Assoc prof biol & chmn dept, Regis Col, Colo, 60-66; PROF & VCHMN, DEPT ANAT, UNIV NEBR MED CTR, OMAHA, 68- *Concurrent Pos:* NIH spec res fel, Med Ctr, Univ Kans, 66-68. *Mem:* Soc Neurosci; Am Asn Anatomists; Sigma Xi. *Res:* Neuroanatomy; neurophysiology; nervous control of heart rate and rhythm. *Mailing Add:* Dept Anat Univ Nebr Med Ctr Omaha NE 68105

EARLE, CLIFFORD JOHN, JR, b Racine, Wis, Nov 3, 35; m 60; c 2. MATHEMATICS. *Educ:* Swarthmore Col, BA, 57; Harvard Univ, MA, 58, PhD(math), 62. *Prof Exp:* Instr & res fel math, Harvard Univ, 62-63; mem, Inst Advan Study, 63-65; from asst prof to assoc prof, 65-69, chmn dept, 76-79, PROF MATH, CORNELL UNIV, 69- *Concurrent Pos:* Ed, Proc Am Math Soc, 89- *Mem:* Am Math Soc. *Res:* Functions of a complex variable; Riemann surfaces; quasiconformal mappings; automorphic forms. *Mailing Add:* Dept Math Cornell Univ Ithaca NY 14853

EARLE, DAVID PRINCE, JR, b Englewood, NJ, May 23, 10; m 36; c 4. MEDICINE. *Educ:* Princeton Univ, AB, 33; Columbia Univ, MD, 37, MedScD(int med), 42. *Prof Exp:* Intern, St Luke's Hosp, New York, 37-39; resident, Columbia Univ & res serv, Goldwater Hosp, 39-41; from instr to assoc prof, Sch Med, NY Univ, 41-54; chmn dept, 65-73, prof med, 54-78, EMER PROF MED, SCH MED, NORTHWESTERN UNIV, CHICAGO, 78- *Concurrent Pos:* Res assoc, NY Univ Res Serv, Goldwater Hosp, 41-46, dir, 46-47; consult, Surgeon Gen, 52-56; mem cardiovasc study sect, USPHS, 58-62; mem metab training grants comt, Nat Inst Arthritis & Metab Dis, 64-67, chmn, 66-67, mem urol res training grants comt, 67-69, chmn, 68-69, mem, Nat Adv Arthritis & Metab Dis Coun, 70-74. *Mem:* Am Soc Clin Invest; Am Clin & Climat Asn; master Am Col Physicians; Asn Am Physicians. *Res:* Clinical and experimental renal disease; chemotherapy of human malaria; streptococcal infections; hemorrhagic fever. *Mailing Add:* 303 E Chicago Ave Chicago IL 60611

EARLE, ELIZABETH DEUTSCH, b Vienna, Austria, Oct 6, 37; US citizen; m 60; c 2. PLANT TISSUE CULTURE, CELL BIOLOGY. *Educ:* Swarthmore Col, BA, 59; Radcliffe Col, MA, 60; Harvard Univ, PhD(biol), 64. *Prof Exp:* NIH fel, Princeton Univ, 64-65; res assoc biol, Harvard Univ,

68-69, res assoc & lectr floricult, 70-74, res assoc plant breeding, 75-78, sr res assoc, 78-79, assoc prof, 79-86, PROF PLANT BREEDING, CORNELL UNIV, 86- *Concurrent Pos:* Vis res assoc biol, Stanford Univ, 74-75, 86; mem, NSF develop biol rev panel, 78-82; mem, USDA competitive grants rev panel genetic mechanisms for crop improv, 82-85; ed, Plant Cell Reports, 87- *Mem:* Am Soc Plant Physiologists; Int Asn Plant Tissue Cult; Sigma Xi; Int Soc Plant Mol Biol. *Res:* Culture of plant cells and protoplasts; gene transfer and other genetic manipulations with plant cells; effects of phytotoxins on plant cells; organelle genetics; disease and insect resistance. *Mailing Add:* Dept Plant Breeding & Biomet Cornell Univ Ithaca NY 14853-1902

EARLE, ERIC DAVIS, b Carbonear, Nfld, Nov 26, 37; c 3. NUCLEAR PHYSICS. *Educ:* Mem Univ Nfld, BSc, 58; Univ BC, MSc, 60; Oxford Univ, DPhil, 64. *Prof Exp:* RES SCIENTIST NUCLEAR PHYSICS, ATOMIC ENERGY CAN, 64- *Mem:* Can Asn Physicists. *Res:* Experimental nuclear physics including neutron capture, parity violation, weak interaction, gamma-ray spectroscopy, strength functions and solar neutrino physics. *Mailing Add:* Chalk River Labs Atomic Energy Res Chalk River ON K0J 1J0 Can

EARLE, RALPH HERVEY, JR, b Cranston, RI, Apr 15, 28. SCIENCE ADMINISTRATION, INDUSTRIAL CHEMISTRY. *Educ:* Brown Univ, ScB, 49; Ga Inst Tech, MS, 50; Purdue Univ, PhD(org chem), 57. *Prof Exp:* Chemist, Hercules Powder Co, 50, 52-54, res chemist, 57-68, supvr com develop adv planning, Pine & Paper Chem Dept, Hercules Inc, 68-71, venture mgr, New Enterprise Dept, 71-74, corp mgr new technol, 74-82, dir acquisitions & divestitures, 82-89; RETIRED. *Concurrent Pos:* Fel chem, Univ Canterbury, 66-67. *Mem:* Am Chem Soc; The Chem Soc; Soc Chem Indust; World Future Soc. *Res:* Chemistry of wet strength resins; nitrogen-containing heterocycles; amine-epichlorhydrin reactions, retention and flocculation. *Mailing Add:* 1515 N Adams St Wilmington DE 19806

EARLEY, CHARLES WILLARD, b Oil City, Pa, Jan 5, 33; m 58; c 5. ANTIGRAVITY ANALOGUE, HIGH DENSITY MAGNETISM. *Educ:* Eltanin Cols Univ, BS, 50, MS, 55, PhD(philos sci), 59, BSEE, 63, BS, 72, DSc, 73. *Prof Exp:* Asst city engr, City Rouseville, Pa, 52-57; chem, bact, radiol warfare officer & liaison officer, US Army, UN Forces, 53-55; chief engr, KPAS Radio, Banning Broadcasting Co, 66-68; sci ed, Radio-TV News Serv Beaumont, 66-73; chief electrochemist & electronic engr, 57-66, field archaeologist, 72-74, DIR RES, US NATURAL RESOURCES, 74- *Concurrent Pos:* Broadcast consult, Arcsine Zero Proj, 68; Simon postdoctoral res fel philos sci, Eltanin Univ, 69-73. *Res:* Limnology with the discovery of cnidaria in tionesta Dam backwaters; magnetic effects on electronic, electrochemical, and biological systems; antigravity; application of diamond heat conductance. *Mailing Add:* US Natural Resources PO Box 9873 Ogden UT 84409-0873

EARLEY, JOSEPH EMMET, b Providence, RI, Apr 6, 32; m 56; c 3. PHYSICAL INORGANIC CHEMISTRY. *Educ:* Providence Col, BS, 54; Brown Univ, PhD(phys chem), 57. *Prof Exp:* Res assoc inorg chem, Univ Chicago, 57-58; from asst prof to assoc prof chem, 58-69, chmn dept, 84-90, PROF CHEM, GEORGETOWN UNIV, 69- *Concurrent Pos:* Consult & coordr chem res eval, US Air Force Off Sci Res, 61-87; vis assoc, Calif Inst Technol, 67-68; guest researcher, Free Univ Brussels, 76. *Honors & Awards:* Potter Prize, Brown Univ, 57. *Mem:* AAAS; Am Chem Soc; Soc Study Process Philos; Am Philos Asn. *Res:* Mechanisms of inorganic oxidation reactions in aqueous solution, especially reactions of titanium, rutheneum and rhodium complexes; ontological implications of current developments in chemical physics. *Mailing Add:* Dept Chem Georgetown Univ Washington DC 20057

EARLEY, LAURENCE E, b Ahoskie, NC, Jan 23, 31; c 2. INTERNAL MEDICINE. *Educ:* Univ NC, BS, 53, MD, 56. *Prof Exp:* From instr to asst prof, Harvard Med Sch, 63-68; from assoc prof to prof med, Univ Calif, San Francisco, 68-73; prof med & chmn dept, Univ Tex Health Sci Ctr San Antonio, 73-77; chmn dept, 77-90, PROF MED, UNIV PA, 77- *Concurrent Pos:* Boston Med Found grant, 61-63; NIH career develop award, 67-68; mem drug efficacy study, Nat Res Coun-Nat Acad Sci, 67-68; mem training grants comt, Nat Inst Arthritis & Metab Dis, 68-72; mem sci adv coun, Nat Kidney Found, 68, mem fel comt, 69-72; mem exec comt renal sect, Am Heart Asn, 68; mem, Am Bd Int Med, 80-88, chmn, 87-88. *Mem:* Inst Med-Nat Acad Sci; Am Fedn Clin Res; Am Soc Clin Invest (pres, 75-76); Asn Am Physicians (pres, 88-89); Am Physiol Soc; Am Soc Nephrology (pres, 77-78). *Res:* Physiology and pathophysiology of renal function and electrolyte physiology; physio-pharmacology of diuretic agents; physiology and pathophysiology of regulation of extracellular fluid volume; clinical renal diseases. *Mailing Add:* Med 100 Centrex Gl Univ Pa Phildadelphia PA 19104

EARLL, FRED NELSON, b Berkeley, Calif, Mar 5, 24; m 46; c 3. GEOLOGY. *Educ:* Univ Southern Calif, BS, 54; Univ Utah, PhD(geol), 57. *Prof Exp:* Geologist, Western Consult Serv, Utah, 54-57; from asst prof to assoc prof, 57-64, head dept, 58-84, PROF GEOL, MONT COL MINERAL SCI & TECHNOL, 64- *Mem:* Am Inst Mining, Metall & Petrol Engrs; Am Inst Prof Geologists; Soc Econ Geol. *Res:* Base and precious metal mining districts; geochemical prospecting; ore mineralogy. *Mailing Add:* Dept Geol Eng Mont Col Mineral Sci & Technol Butte MT 59701

EARLOUGHER, ROBERT CHARLES, JR, b Tulsa, Okla, June 26, 41; m 70; c 1. PETROLEUM ENGINEERING. *Educ:* Stanford Univ, BS, 63, MS, 64, PhD(petrol eng), 66. *Prof Exp:* Res engr, Marathon Oil Co, 66-69, adv res engr, 69-71; sr consult, Sci Software Corp, 72; advan res engr, Marathon Oil Co, 72-74, sr res engr, 74-77, mgr engr, Dept Res Div, 77-81, div reservoir engr, 81-85, prod mgr, 85-87, mgr prod coordr (UK), 88-89, MGR BRAE PROJ, MARATHON OIL CO, 90- *Honors & Awards:* Lester C Uren Award, Asn Inst Mining Engrs Soc Petrol Engrs, 79; John Franklin Carll Award, Soc Petrol Engrs, 80. *Mem:* Soc Petrol Engrs; Sigma Xi. *Res:* Thermal and miscible water flooding methods of petroleum recovery; reservoir engineering; transient testing methods; reservoir simulation technique. *Mailing Add:* Marathon Oil Co-London Box 1228 Houston TX 77251-1228

EARLS, JAMES ROE, b Summerfield, Okla, Apr 18, 43; m 70; c 2. NATURAL GAS PIPELINE. *Educ:* Okla State Univ, BS, 67; Cent State Univ, Okla, MBA, 75. *Prof Exp:* Mech engr, McDonnell Douglas Corp, 67-68; proj engr, USAF Tinker AFB, Okla, 68-80 & Fed Aviation Authority, 80-81; sr proj engr, 81-88, DIR, ENG & CONSTRUCT, OKLA GAS & ELEC CO, 88- *Mem:* Am Soc Mech Engrs; Nat Soc Prof Engrs. *Mailing Add:* 4122 Woodcutter Dr Oklahoma City OK 73150

EARLY, JAMES G(ARLAND), b Washington, DC, Nov 8, 37; m 61; c 3. METALLURGY. *Educ:* Lehigh Univ, BS, 59; Rensselaer Polytech Inst, PhD(metall), 63. *Prof Exp:* Metallurgist, Nat Bur Standards, 63-85, dep chief, Metall Div, 85-86, prog analyst, 86-87, sr prog analyst, 87-88; SCI ADV TO DIR MAT LAB, NAT INST STANDARDS & TECHNOL, 88- *Honors & Awards:* Bronze Medal, US Dept Com. *Mem:* Am Soc Metals Int; Can Metall Soc; Am Soc Testing & Mat; Planseeberichte fur Pulvermetallurgie. *Res:* Characterization of perfect metal crystals and origin of defects in metal crystals; hydrogen embrittlement of steels; fracture toughness testing; powder metallurgy; resource recovery. *Mailing Add:* Mat Sci & Eng Lab Rm B309 Bldg 223 Nat Inst Standards & Technol Gaithersburg MD 20899

EARLY, JAMES M, b Syracuse, NY, July 25, 22; m 48; c 8. PHYSICS. *Educ:* NY State Col Forestry, Syracuse, BSc, 43; Ohio State Univ, MSc, 48, PhD(elec eng), 51. *Prof Exp:* Instr & res assoc elec eng, Ohio State Univ, 46-51; mem tech staff, Bell Tel Labs, 51-56, dept head, 56-62, dir, 62-69; div vpres, Fairchild Res & Develop Div, Fairchild Camera & Instrument Corp, 69-80, mgr, Very Large Scale Integration Advan Res & Develop Lab, 80-86; CONSULT, 90- *Mem:* Fel AAAS; Am Phys Soc; Inst Elec & Electronics Engrs; Sigma Xi. *Res:* Electron devices, particularly semiconductor; technology and engineering. *Mailing Add:* 708 Holly Oak Dr Palo Alto CA 94303

EARLY, JOSEPH E, b Williamsburg, Ky, Jan 14, 40; m 63; c 1. MATHEMATICS EDUCATION. *Educ:* Cumberland Col, Ky, BS, 63; Univ Tenn, Knoxville, MMath, 66, EdD(math educ), 69. *Prof Exp:* Teacher elem sch, Ohio, 59-60 & high sch, Ky, 63-65; prof math, 69-77, chmn, Dept Math, 69-81 PROF MATH & PHYSICS, CUMBERLAND COL, KY, 77-, ACAD DEAN, 81- *Mem:* Math Asn Am. *Res:* Grade level teaching preferences of prospective elementary teachers with respect to their attitudes toward arithmetic and achievements in mathematics. *Mailing Add:* Acad Dean Cumberland Col Williamsburg KY 40769

EARNEST, ANDREW GEORGE, b York, Pa, July 14, 49; m 73. QUADRATIC FORMS. *Educ:* Elizabethtown Col, BS, 70; Ohio State Univ, MS, 72, PhD(math), 75. *Prof Exp:* Instr math, Ohio State Univ, 76; asst prof math, Univ Southern Calif, 76-81; asst prof, 81-83, ASSOC PROF MATH, SOUTHERN ILL UNIV, 83- *Mem:* Am Math Soc; Sigma Xi. *Res:* Arithmetic theory of integral quadratic forms; structural properties of class groups. *Mailing Add:* Dept Math Southern Ill Univ Carbondale IL 62901

EARNEST, SUE W, b Grand Forks, NDak, Sept 19, 07; m 28; c 2. SPEECH PATHOLOGY. *Educ:* San Diego State Col, BA, 29; Univ Southern Calif, MA, 37, PhD, 47. *Prof Exp:* Instr eng, Univ Louisville, 45-46; asst prof, 47-48, assoc prof speech, 48-54, chmn dept speech path & audiol, 54-61, prof, 54-77, EMER PROF SPEECH PATH & AUDIOL, SAN DIEGO STATE UNIV, 77- *Concurrent Pos:* Mem nat bd, western regional educ comn, United Cerebral Asn, 57-59. *Mem:* Fel Am Speech & Hearing Asn. *Res:* Aphasia; cerebral palsy; geriatrics and speech; author, producer Centennial Pageant for State of California to audience of 31,500 and cast of 150. *Mailing Add:* 1770 Ft Stockton Dr San Diego CA 92103

EARNSHAW, JOHN W, b Toronto, Ont, July 22, 39; m 65. ELECTRON PHYSICS. *Educ:* Univ Toronto, BASc, 61; Cambridge Univ, PhD(electron physics), 65. *Prof Exp:* Asst sci res officer electron physics, Nat Res Coun Can, 65-67; asst prof, 67-70, ASSOC PROF PHYSICS, TRENT UNIV, 70-, HEAD DEPT, 76- *Mem:* Am Vacuum Soc. *Res:* Ultra high vacuum electron optics. *Mailing Add:* Dept Physics Trent Univ Peterborough ON K9J 7B8 Can

EARNSHAW, WILLIAM C, b Sept 22, 50; m. CENTROMERE STRUCTURES, AUTOANTIBODIES. *Educ:* Mass Inst Technol, PhD(biol), 77. *Prof Exp:* Asst prof, 82-90, PROF CELL BIOL & ANAT, SCH MED, JOHNS HOPKINS UNIV, 90- *Concurrent Pos:* Distinguished Young Scientist, Md Acad Sci, 84. *Honors & Awards:* Devils Bag Award, Nat Arthritis Found, 85; Chromosoma Prize, 89; RR Bensley Award, Am Asn Anat, 90. *Mem:* Am Soc Cell Biol. *Mailing Add:* Dept Cell Biol & Anat Johns Hopkins Univ 720 Rutland Ave Baltimore MD 21205

EARTLY, DAVID PAUL, b Hammond, Ind, Mar 16, 42. EXPERIMENTAL HIGH ENERGY PHYSICS. *Educ:* Univ Notre Dame, BS, 63; Univ Chicago, MS, 65, PhD(physics), 69. *Prof Exp:* PHYSICIST, FERMI NAT ACCELERATOR LAB, UNIVS RES ASN, ENERGY RES & DEVELOP ADMIN, 69- *Mem:* Am Phys Soc; Sigma Xi. *Res:* Strong interaction physics via the study of fundamental nucleon-nucleon and meson-nucleon interactions. *Mailing Add:* 43W 920 Red Oaks Dr Elburn IL 60119

EASH, JOHN T(RIMBLE), b Albany, Ind, Sept 1, 06; m 27; c 2. METALLURGY. *Educ:* Purdue Univ, BS, 28; Univ Mich, MS, 29, PhD(metall), 32. *Prof Exp:* Asst, Univ Mich, 29-31; res metallurgist, Res Lab, Int Nickel Co, 31-35, sect supvr, 35-50, metall supvr, 50-54, asst mgr lab, 54-63, mgr, 63-70, mgr lab & mgr res, 70-73; METALL CONSULT, 73- *Mem:* Am Soc Metals. *Res:* Copper-nickel-tin alloys constitution and properties; platinum and palladium dental alloys; cast iron melting; high strength low alloy steels; elevated temperature alloys; maraging steels. *Mailing Add:* 43 Bradrick Lane Allendale NJ 07401

EASLEY, JAMES W, physics, for more information see previous edition

EASLEY, STEPHEN PHILLIP, b Indianapolis, Ind, Sept 4, 52; m 74; c 2. ELECTROMAGNETIC BIOEFFECTS, BIOMEDICAL QUANTITATIVE ETHOLOGY. *Educ:* Purdue Univ, BA, 73; Wash Univ, AM, 75, PhD(biol anthropology), 82. *Prof Exp:* Asst prof anthropology, Purdue Univ WLafayette, Ind, 80-86; ASST SCIENTIST, SOUTHWEST FOUND BIOMED RES, SAN ANTONIO, TEX, 86- *Concurrent Pos:* Consult, Couchman-Conant Inc, 81-86. *Mem:* Am Soc Primatologists; Sigma Xi; Am Asn Phys Anthropologists. *Res:* Quantitative ethology and statistical analyses of the social behavior of nonhuman primates to model the effects of environmental challenges on human health and disease processes. *Mailing Add:* Behav Med Lab Southwest Found Biomed Res PO Box 28147 San Antonio TX 78228-0147

EASON, ROBERT GASTON, b Bells, Tenn, May 15, 24; m 50; c 2. NEUROPSYCHOLOGY, PSYCHOPHYSIOLOGY. *Educ:* Univ Mo, BA, 50, MA, 52, PhD(psychol), 56. *Prof Exp:* Res assoc electrophysiol, Univ Calif, Los Angeles, 56-57; res psychologist, US Navy Electronics Lab, 57-60; from asst prof to prof psychol, San Diego State Col, 60-67; prof psychol & head dept, 67-80, ROSENTHAL PROF PSYCHOL, UNIV NC, GREENSBORO, 80- *Mem:* AAAS; Am Psychol Asn; Am Psychol Soc; Soc Psychophysiol Res; Soc Neurosci; Psychonomic Soc. *Res:* Electrophysiological correlates of arousal, motivation, emotion, attention and perception; event related potentials and information processing. *Mailing Add:* Dept Psychol Univ NC Greensboro NC 27412

EASSON, WILLIAM MCALPINE, b Evanston, Ill, July 3, 31; m 58; c 4. PSYCHIATRY. *Educ:* Aberdeen Univ, MB & ChB, 54, MD, 67; FRCPsychol. *Prof Exp:* Instr psychiat, Univ Sask, 59-61; staff psychiatrist, Menninger Found, 63-67; prof psychiat & chmn dept, Med Col Ohio, 67-72; prof & dir div child & adolescent psychiat, Med Sch, Univ Minn, 73-74; PROF PSYCHIAT & BEHAV SCI & HEAD DEPT, LA STATE UNIV MED CTR, NEW ORLEANS, 74- *Concurrent Pos:* Ed, J Clin Psychiat, 77-80. *Mem:* Fel Am Psychiat Asn; fel Am Orthopsychiat Asn. *Res:* Adolescent psychiatry; childhood psychosis. *Mailing Add:* Dept Psychiat La State Univ Med Ctr 1542 Tulane Ave New Orleans LA 70112

EAST, JAMES LINDSAY, b Senatobia, Miss, Nov 5, 36; m 63; c 1. VIROLOGY. *Educ:* Memphis State Univ, BS, 63, MS, 67; Univ Tenn, PhD(microbiol), 70. *Prof Exp:* Spec technologist virol, St Jude Childrens Res Hosp, 63-67; proj investr, 70-71, asst virol, 71-79, ASSOC VIROLOGIST, UNIV TEX SYST CANCER CTR, M D ANDERSON HOSP & TUMOR INST, 79- *Concurrent Pos:* USPHS trainee, St Jude Childrens Res Hosp, 67-70. *Mem:* Am Soc Microbiol; Int Asn Comp Res on Leukemia & Related Diseases; Am Asn Cancer Res. *Res:* Relatedness of retroviruses and human cancer. *Mailing Add:* 4915 Droddy St Houston TX 77091

EAST, LARRY VERNE, b Apr 16, 37; US citizen; m 59; c 1. PHYSICS. *Educ:* Univ Wichita, BS, 58, MS, 60; Case Inst Technol, PhD(physics), 65. *Prof Exp:* Part-time instr, Case Inst Technol, 62-65; staff mem physics, Los Alamos Sci Lab, Univ Calif, 67-75; systs eng mgr, Canberra Industs, 75-78, mgr software develop, 78-82, sr staff scientist, 84-89; EG&G IDAHO INC, 89- *Concurrent Pos:* Mem staff, Int Atomic Energy Agency, 82-83. *Mem:* AAAS; Am Phys Soc; Inst Nuclear Mat Mgt. *Res:* Gamma-ray and x-ray spectroscopy; fission neutrons; nuclear physics instrumentation; software systems; nuclear safeguards. *Mailing Add:* 1355 Rimline Dr Idaho Falls ID 83401

EASTER, DONALD PHILIPS, b Washington, DC, Aug 10, 19; m 50, 74; c 4. FUELS FROM BIOMASS & WASTES. *Educ:* Univ Md, BS, 42. *Prof Exp:* Res chemist, Celanese Corp Am, 42-45, Johns Hopkins Appl Physics Lab, 45-52; sr scientist chem res, Olin Mathieson Chem Corp, 52-56; chief mass spectrometry sect, Eng Res & Develop Labs, US Army, 56-62; prog scientist space res, Planetary Prog, NASA, 62-70; res mgr chem res, Inst Planetary Res, Cornell Univ, 73-74; res & develop prog analyst chem, Inst Gas Technol, 74-81; RETIRED. *Mem:* Am Chem Soc; fel AAAS. *Res:* Physical chemistry; space research; mass spectrometry; bioconversion processes; exobiology instrumentation for planetary research. *Mailing Add:* 4501 Arlington Blvd Arlington VA 22203

EASTER, ROBERT ARNOLD, b San Antonio, Tex, Oct 10, 47; m 72; c 3. SWINE. *Educ:* Tex A&M Univ, BS, 70, MS, 72; Univ Ill, PhD(animal sci), 76. *Prof Exp:* Lectr, 76, asst prof, 76-80, ASSOC PROF ANIMAL NUTRIT, UNIV ILL, 80- *Mem:* Am Soc Animal Sci. *Res:* Nutrition of the pig with particular interests in amino acid and B-vitamin nutrition of the reproducing female and young suckling piglet. *Mailing Add:* Animal Sci Univ Ill 1301 W Gregory Dr Urbana IL 61801

EASTER, STEPHEN SHERMAN, JR, b New Orleans, La, Feb 12, 38; m 63; c 2. PHYSIOLOGY, ANATOMY. *Educ:* Yale Univ, BS, 60; Johns Hopkins Univ, PhD(biophys), 67. *Prof Exp:* USPHS fel physiol, Cambridge Univ, 67-68, Miller Inst Basic Res, Univ Calif, Berkeley, 68-69 & Wilmer Ophthal Inst, Johns Hopkins Univ, 69; asst prof zool, Univ Mich, Ann Arbor, 70-74, chmn dept zool, 78-82, assoc prof biol, 74-78, dir, Neurosci Prog, 84-88, PROF BIOL, UNIV MICH, ANN ARBOR, 78- *Concurrent Pos:* Mem & chmn, Visual Sci B Study Sect, NIH, 78-82. *Mem:* Asn Res Vision & Ophthal; Soc Neurosci; Int Brain Res Orgn; Soc Develop Biol. *Res:* Development and plasticity of nervous systems, particularly visual; neurobiology. *Mailing Add:* Dept Biol Univ Mich Ann Arbor MI 48109-1048

EASTER, WILLIAM TAYLOR, b Winston-Salem, NC, Dec 19, 31; m 61; c 2. ENGINEERING. *Educ:* NC State Univ, BS, 59; Carnegie Inst Tech, Carnegie-Mellon Univ, MS, 60. *Prof Exp:* Instr, 63-66, asst prof, 66-75, ASSOC PROF ELEC ENG, NC STATE UNIV, 75-, ASSOC DEPT HEAD, 80- *Concurrent Pos:* Dir, Eng Oper Prog, NC State Univ, 70-80. *Mem:* Inst Elec & Electronics Engrs; Am Soc Eng Educ; Am Soc Eng Mgt. *Mailing Add:* Elec & Comput Eng Dept NC State Univ Box 7911 Raleigh NC 27695-7911

EASTERBROOK, DON J, b Sumas, Wash, Jan 29, 35; m 57. GLACIAL GEOLOGY. *Educ:* Univ Wash, Seattle, BS, 58, MS, 59, PhD(geol), 62. *Prof Exp:* From instr to assoc prof, 59-66, chmn dept, 65-77, PROF GEOL, WESTERN WASH UNIV, 68- *Concurrent Pos:* NSF res grants, 61-68, 71, 80-82 & 87 & Dept Interior Res Grants, 74-81. *Mem:* Am Geophys Union; Geol Soc Am; Am Quarternary Asn; Int Quarternary Asn. *Res:* Glacial geology; geomorphology; sedimentation; environmental geology; quaternary glacial chronology; paleomagnetism of sediments; genesis of sediments; depositional environments; use of remanent magnetism of sediments in Pleistocene chronology; glaciation of the Puget lowland and Columbia plateau and correlations with European countries. *Mailing Add:* Dept Geol Western Wash Univ Bellingham WA 98225

EASTERBROOK, ELIOT KNIGHTS, b Dudley, Mass, Oct 28, 27; m 57; c 2. POLYMER CHEMISTRY, ORGANIC CHEMISTRY. *Educ:* Univ NH, BS, 48, MS, 50; Ohio State Univ, PhD(phys org chem), 53. *Prof Exp:* Sr res chemist, Chem Div, US Rubber Co, 52-62, group leader stereo polymer res, 62-67, res scientist, Uniroyal Inc, 67-74, SR RES SCIENTIST, CHEM DIV, UNIROYAL INC, 74- *Mem:* Am Chem Soc. *Res:* Polymerization of butadiene and acrylonitrile; nonaqueous polymerization of ethylene and propylene elastomers. *Mailing Add:* Uniroyal Chem Co Inc Naugatuck CT 06770

EASTERBROOK, KENNETH BRIAN, b Ilford, Eng, June 4, 35; div; c 3. MICROBIOLOGY. *Educ:* Bristol Univ, BSc, 56; Australian Nat Univ, PhD(virol), 62. *Prof Exp:* Lectr virol, Univ Western Australia, 62-64; assoc prof, 67-74, PROF MICROBIOL, DALHOUSIE UNIV, 74-, HEAD DEPT, 85- *Concurrent Pos:* Fel, Australian Nat Univ, 64-65; USPHS int fel, Calif Inst Technol, 65-66; fel, Ont Cancer Inst, 66-67. *Mem:* Can Soc Microbiol; Am Soc Microbiol. *Res:* Microbial ultrastructure; structure, production and function of bacterial spinae. *Mailing Add:* Dept Microbiol Dalhousie Univ Halifax NS B3H 4H6 Can

EASTERDAY, BERNARD CARLYLE, b Hillsdale, Mich, Sept 16, 29. COMPARATIVE MEDICINE. *Educ:* Mich State Univ, DVM, 52; Univ Wis, MS, 58, PhD(vet microbiol, path), 61. *Prof Exp:* Gen practice, Mich, 52, veterinarian, Ft Detrick, Md, 55-56, 58-61; res asst, 56-58, assoc prof vet sci, 61-66, chmn dept, 68-74, PROF VET SCI, UNIV WIS-MADISON, 66-, DEAN SCH VET MED, 79- *Concurrent Pos:* Mem expert panel on zoonoses, WHO. *Mem:* Am Col Vet Microbiol; Asn Am Vet Med Cols (pres, 75-76); Am Vet Med Asn; Conf Res Workers Animal Dis. *Res:* Infectious diseases, viral; pathogenesis; influenza; epidemiology; herpesvirus latent infections. *Mailing Add:* Sch Vet Med 2015 Linden Dr W Madison WI 53706

EASTERDAY, HARRY TYSON, b Sault Ste Marie, Mich, Oct 6, 22; m 47; c 2. NUCLEAR PHYSICS. *Educ:* Univ Calif, AB, 47, PhD, 53. *Prof Exp:* Physicist, Radiation Lab, Univ Calif, 53-55; asst prof physics, Univ Ore, 55-60; assoc prof, 60-67, PROF PHYSICS, ORE STATE UNIV, 67- *Concurrent Pos:* Mem staff, Lawrence Radiation Lab, Univ Calif, Berkeley, 63-64. *Mem:* Am Phys Soc. *Res:* Low energy experimental nuclear physics. *Mailing Add:* Dept Physics Ore State Univ Corvallis OR 97331

EASTERDAY, JACK L(EROY), b Crestline, Ohio, Mar 24, 28; m 50; c 2. NUCLEAR QUALITY ASSURANCE & SOFTWARE, INFORMATION SYSTEMS. *Educ:* Univ Toledo, BS, 52. *Prof Exp:* Assoc engr, Sperry Rand Corp, 52-55; elec engr, Reliability Eng Div, 57-60, proj leader, 60-62, sr engr, 62-64, asst dir adv electronics group, 64, assoc chief adv, Electronics Div, 65-73, assoc sect mgr, Eng Physics & Electronics Sect, 73-80, qual assurance specialist, 80-84, DATA BASE MGR, SYSTS ENG, SOFTWARE DEVELOP CONTROL, BATTELLE PROJ MGT DIV, BATTELLE MEM INST, 84- *Concurrent Pos:* Mem, Nat Eng Consortium, Microcomput Archit Interfaces, Testing, Reliability and Systs Design, 78. *Mem:* Inst Elec & Electronics Eng; Proj Mgt Inst. *Res:* Reliability, system modeling, analysis and prediction of availability; maintainability, dependability, safety; design review; reliability and nuclear quality assurance; information systems and technical data base design, development and management; software development systems. *Mailing Add:* Battelle Proj Mgt Div Battelle Mem Inst 505 King Ave Columbus OH 43201

EASTERDAY, KENNETH E, b Kirksville, Ind, June 27, 33; m 59; c 1. MATHEMATICS. *Educ:* Ind Univ, BS, 55, MA, 60; Western Reserve Univ, EdD, 63. *Prof Exp:* Teacher pub sch, Ohio, 57, 59-63 & Ind, 57-59; assoc prof math, State Univ NY, 63-64; from asst prof to assoc prof, 64-72, PROF MATH IN SEC EDUC,AUBURN UNIV, 72- *Res:* Construction of mathematics programs in grades seven through twelve; methods of teaching mathematics at all levels; modern algebra. *Mailing Add:* Dept Math Educ 5064 Haley Ctr Auburn Univ Auburn AL 36830

EASTERDAY, OTHO DUNREATH, b Allen Co, Ind, Oct 3, 24; m 49; c 4. PHARMACOLOGY, RADIOBIOLOGY. *Educ:* Ball State Univ, BA, 48; Univ Iowa, MS, 50, PhD(pharmacol), 53. *Prof Exp:* Res asst pharmacol, Univ Iowa, 48-51; assoc pharmacologist, Brookhaven Nat Lab, 53-62; pharmacologist & radiobiologist, Hazleton Labs, Inc, 62-66; dir dept pharmacol & radiation, Gulf S Res Inst, La, 66-68; head toxicol & pharmacol res, Res & Develop Ctr, 68-77, VPRES & CHIEF PROD SAFETY ASSURANCE OFFICER, INT FLAVORS & FRAGRANCES, INC, 77- *Mem:* Fel AAAS; fel NY Acad Sci; Am Soc Pharmacol & Exp Therapeut; Soc Toxicol; Biomet Soc. *Res:* Chemical structure and activity; pharmacodynamics; toxicology; biometry; radiation; radioactive tracers; spectroscopy; chemical synthesis; boron and lithium pharmacology and toxicology; natural products; radio respirometry and pharmacology. *Mailing Add:* Eight Jennifer Lane Rd No Two Warren NJ 07059-6913

EASTERDAY, RICHARD LEE, b Green Bay, Wis, Apr 2, 38; m 76. BIOCHEMISTRY. *Educ:* N Cent Col, BA, 60; Va Poly Tech Inst & State Univ, MA, 62, PhD(biochem & nutrit), 67. *Prof Exp:* Res assoc biochem, NY Univ Med Sch, 64-65; supvr tech serv, Pharmacia Fine Chem, 65-67, mgr, tech serv, 67-68, mgr res tech serv & qual control, Pharmacia Inc, 68-70, dir res, 71-76, dir res & develop & mkt, 77, vpres & gen mgr, 78-79, pres, Pharmacia Biotech Group, 79-86, pres, Pharmacia PL biochem, 84-86, PRES, PHARMACIA LKB BIOTECH , INC, 87-, PRES, PHARMACIA LKB NUCLEAR, INC, 87-, CHMN, PHARMACIA PL BIOCHEM, 87-, EXEC VPRES, PHARMACIA, INC, 79- *Mem:* Am Chem Soc; AAAS; NY Acad Sci. *Res:* Separation science; liquid chromatography; gel filtration. *Mailing Add:* 1729 Merriam Dr Martinsville NJ 08836

EASTERLING, WILLIAM EWART, JR, b Raleigh, NC, Oct 8, 30; c 5. OBSTETRICS & GYNECOLOGY. *Educ:* Duke Univ, AB, 52; Univ NC, MD, 56. *Prof Exp:* From intern med to resident obstet & gynec, NC Mem Hosp, 56-61, instr, 60-61; from asst prof to assoc prof, 64-72, assoc dean, Sch Med, 74-76, vice dean, Sch Med, 76-81, PROF OBSTET & GYNEC, NC MEM HOSP, UNIV NC, 67-, CHIEF-OF-STAFF, 74- *Concurrent Pos:* Fel reprod physiol, Univ Calif, 63-64; assoc dean clin affairs, Univ NC, 81- *Honors & Awards:* Fordham Award, 61. *Mem:* AAAS; AMA; Am Col Obstetricians & Gynecologists; Endocrine Soc; Soc Gynec Invest. *Res:* Endocrinology of obstetrics and gynecology. *Mailing Add:* Dept Obstet & Gynec Univ NC Sch Med Chapel Hill NC 27515

EASTERLY, NATHAN WILLIAM, b Lewisburg, WVa, Sept 9, 27; m 52; c 3. TAXONOMIC BOTANY, BIOLOGY. *Educ:* WVa Univ, AB, 49, PhD(plant taxon), 57; State Univ Iowa, MS, 51. *Prof Exp:* Instr, 57-60, asst prof biol, 60-64, assoc prof, 64-75, PROF BOT SCI & CUR HERBARIUM, BOWLING GREEN STATE UNIV, 76- *Concurrent Pos:* Res fel, Ohio Acad Sci, Columbus, 60. *Mem:* Am Soc Plant Taxonomists; Sigma Xi. *Res:* A comparative study of flora of northwestern Ohio. *Mailing Add:* 506 Harvest Lane Bowling Green OH 43402

EASTES, FRANK ELISHA, b Wilson Co, Tenn, July 31, 24; m 53; c 6. ORGANIC CHEMISTRY. *Educ:* Tenn Polytech Inst, BS, 49; Vanderbilt Univ, MS, 51, PhD, 55. *Prof Exp:* Staff scientist, E I du Pont de Nemours & Co, 53-62; sr proj leader, Cry-o-Vac Div, W R Grace & Co, 62-66, sect head film coatings, 66-77, res assoc, 77-89, SR ANALYTICAL CHEMIST, CRY-O-VAC DIV, W R GRACE & CO, 89- *Mem:* AAAS; Am Chem Soc; Sigma Xi; NY Acad Sci. *Res:* Reactions of thiophthalic anhydride; resolutions of racemic amines; cellulose chemistry; film coating and adhesion; polymer additives and modification; polymer analysis. *Mailing Add:* 729 Otis Blvd Spartanburg SC 29302

EASTHAM, ARTHUR MIDDLETON, organic chemistry, for more information see previous edition

EASTHAM, JAMES NORMAN, b Cumberland, RI, Dec 10, 03; wid; c 3. MATHEMATICS. *Educ:* Providence Col, BS, 26; Catholic Univ, MA, 28, PhD(math), 31. *Prof Exp:* Instr math, Providence Col, 30-31; head dept math & physics, Nazareth Col, NY, 31-46; from asst prof to prof math, Cooper Union, 46-61; prof, 61-74, chmn dept, 61-69, dean open admissions serv, 71-74, EMER PROF MATH, QUEENSBOROUGH COMMUNITY COL, CITY UNIV NEW YORK, 74- *Concurrent Pos:* Instr, Univ Rochester, 40-42; consult, Rheem Mfg Co, 50-52. *Mem:* Math Asn Am; Am Soc Eng Educ. *Res:* Quartic curves; vibration theory. *Mailing Add:* 619 Stevenson Lane Towson MD 21204

EASTHAM, JEROME FIELDS, b Daytona Beach, Fla, Sept 22, 24; m 49; c 3. ORGANIC CHEMISTRY. *Educ:* Univ Ky, BS, 48; Univ Calif, PhD(chem), 51. *Prof Exp:* Asst, Univ Calif, 48-51; US AEC fel, London Univ, 51-52 & Univ Wis, 52-53; from asst prof to assoc prof, 53-62, PROF CHEM, UNIV TENN, KNOXVILLE, 62- *Mem:* AAAS; Am Chem Soc; The Chem Soc. *Res:* Chemistry of steroids and related natural products; mechanisms of organic reactions; organometallic chemistry. *Mailing Add:* 7841 Ramsgate Dr Univ Tenn Knoxville TN 37996

EASTIN, EMORY FORD, b Picayune, Miss, Nov 27, 40; m 61; c 4. WEED SCIENCE, PLANT PHYSIOLOGY. *Educ:* Miss State Univ, BS, 62, MS, 63; Auburn Univ, PhD(bot, plant physiol), 66. *Prof Exp:* Asst prof weed sci, Miss State Univ, 66-67; from asst prof to prof agron, Tex A&M Univ, 67-89; PROF AGRON & DEPT HEAD, AGRON DEPT, COASTAL PLAIN EXP STA, UNIV GA, TIFTON, 89- *Honors & Awards:* Super Performance Award for Group Res, USDA, 87. *Mem:* Weed Sci Soc Am; Am Soc Agron; Crop Sci Soc Am; Int Plant Growth Substances Asn; Int Weed Sci Soc; Asian-Pac Weed Sci Soc. *Res:* Mode of action of herbicides in plants; absorption, translocation and metabolism of herbicides by plants; plant growth regulators; weed control in rights-of-way and non-crop areas. *Mailing Add:* Agron Dept Coastal Plain Exp Sta PO Box 748 Tifton GA 31793-0748

EASTIN, JERRY DEAN, b Madrid, Nebr, Jan 18, 31; m 57; c 3. AGRONOMY. *Educ:* Univ Nebr, BSc, 53, MSc, 55; Purdue Univ, PhD(crop physiol), 60. *Prof Exp:* Nat Acad Sci-Nat Res Coun res assoc, Army Biol Labs, Ft Detrick, Md, 60-61; asst prof agron, Univ Nebr, 61-64; res plant physiologist, Crops Res Div, Agr Res Serv, USDA, 64-70; assoc prof agron, 70-72, PROF AGRON, UNIV NEBR, 72- *Concurrent Pos:* Fel, NZ Nat Climate Lab, 73. *Mem:* AAAS; fel Am Soc Agron; Crop Sci Soc Am; Am Soc Plant Physiol. *Res:* Carbon dioxide fixation in Bacillus anthracis; characterization of wheat gluten proteins by chemical and physical methods; physiology of the grain sorghum plant. *Mailing Add:* Dept Agron Univ Nebr Lincoln NE 68583-0915

EASTIN, JOHN A, b Grant, Nebr, June 13, 34; m 61. AGRONOMY, BIOTECHNOLOGY. *Educ:* Univ Nebr, BS, 58; Purdue Univ, MS, 61, PhD(crop physiol, biochem), 63. *Prof Exp:* Res asst crop physiol, Purdue Univ, 58-63; asst prof agron, Univ Wis, 63-68; res agronomist, DeKalb Agr Res Inc, 68-73; agronomic consult, 73-81; PRES, KAMTERTER INC, 81- *Concurrent Pos:* Consult, appln biotechnol to agr & soil bioremediation. *Mem:* AAAS; Crop Sci Soc Am; Am Soc Agron; Am Soc Plant Physiol; Soil Sci Soc Am. *Res:* Crop physiology and genetics; seed production technique improvement; manufacture of nitrogen fertilizer; biotechnology. *Mailing Add:* 1500 W Manor Dr Lincoln NE 68506

EASTIN, WILLIAM CLARENCE, JR, b Lorain, Ohio, Oct 22, 40; m 66; c 2. PHYSIOLOGIST, TOXICOLOGY. *Educ:* Kent State Univ, BA, 67, MS, 70; Univ Iowa, PhD(physiol), 76. *Prof Exp:* Res scientist physiol, Univ Iowa Hosp, 76-77; res physiologist, Dept Interior, US Fish & Wildlife Serv, 77-81; PHYSIOLOGIST, DEPT HEALTH HUMAN SERV, NAT INST ENVIRON HEALTH SCI, NIH, 81- *Concurrent Pos:* Proj leader physiol, Dept Interior, US Fish & Wildlife Serv, 77-81, oil team res leader, 77-79, proj officer, Nat Toxicol Prog, 81-91, toxicol disc leader, 89-91. *Mem:* Soc Exp Biol Med; Am Asn Cancer Res; Bioelectromagnetics Soc. *Res:* Animal physiology (membrane transport, initiation/promotion, and toxicology); carcinogenesis and toxicologic evaluation. *Mailing Add:* Nat Inst Environ Health Sci PO Box 12233 Res Triangle Park NC 27709

EASTLAND, DAVID MEADE, b Meridian, Miss, Nov 27, 22; m 44. THERMODYNAMICS, AIR CONDITIONING. *Educ:* Miss State Univ, BS, 44, MS, 50. *Prof Exp:* From instr to assoc prof, 46-58, PROF MECH ENG, MISS STATE UNIV, 58- *Mem:* Am Soc Mech Engrs; Nat Soc Prof Engrs; Am Soc Eng Educ. *Res:* Thermodynamics; fluid flow; instruments. *Mailing Add:* 1113 Robinhood Rd Starkville MS 39759

EASTLAND, GEORGE WARREN, JR, b Omaha, Nebr, Sept 5, 39; m 67; c 2. INORGANIC CHEMISTRY, PHYSICAL CHEMISTRY. *Educ:* Wittenberg Univ, BS, 61; SDak State Univ, PhD(inorg chem), 69. *Prof Exp:* Teacher high sch, Ohio, 61-63; lab asst chem, Wittenberg Univ, 63-64; asst, SDak State Univ, 64-69; from asst prof to assoc prof, 69-78, PROF CHEM, SAGINAW VALLEY STATE COL, 78- *Concurrent Pos:* Vis scientist, Univ Leicester, Eng, 75-76, sr res fel, 82-83. *Mem:* Am Chem Soc; Royal Soc Chem; NY Acad Sci. *Res:* Preparation and investigation of properties of novel coordination compounds of transition metals, such as rhenium and rhodium; synthesis of antitumor agents; electron spin resonance of chemical systems. *Mailing Add:* Saginaw Valley State Col 2250 Pierce Rd University Center MI 48710

EASTLER, THOMAS EDWARD, b Boston, Mass, Oct 10, 44; m 65; c 1. ENVIRONMENTAL GEOLOGY. *Educ:* Brown Univ, ScB, 66; Columbia Univ, MA, 68, PhD(geol), 71. *Prof Exp:* Asst prof geophys, USAF Inst Technol, 70-72; sr res geologist, US Army Engr Topog Labs, Washington, DC, 72-74; ASSOC PROF ENVIRON GEOL, UNIV MAINE, FARMINGTON, 74- *Concurrent Pos:* Vis lectr, Univ Dayton, & Wittenberg Col, Ohio, 70-72; fac assoc, Western Maine Energy Ctr, 77-; mem, New Eng Cong Caucus, 78. *Mem:* AAAS; Am Geophys Union; Geol Soc Am; Nat Asn Geol Teachers; Am Soc Photogram. *Res:* Environmental impact of engineering endeavors; remote sensing and environmental studies; ecosystems analysis; resources, energy and society; human ecology. *Mailing Add:* Dept Geol Univ Maine Farmington ME 04938

EASTLICK, HERBERT LEONARD, b Platteville, Wis, Apr 24, 08; m 35. ZOOLOGY. *Educ:* Univ Mont, AB, 30; Wash Univ, MS, 32, PhD(zool), 36. *Prof Exp:* Asst, Wash Univ, 31-34, instr zool, 34-35, asst, 35-36, instr, Univ Col, 35-36; instr, Stephens Col, 36-37 & Univ Mo, 37-39; Nat Res Coun fel, Univ Chicago, 39-40; from asst prof to prof zool, 40-73, chmn dept, 47-64, EMER PROF ZOOL, WASH STATE UNIV, 73- *Mem:* AAAS; fel Am Soc Naturalists; Soc Develop Biol; Am Asn Anat; Am Soc Zool. *Res:* Experimental embryology and carcinogenesis; cytology; histology; interspecies incompatibility; feather character and pathology in transplanted limbs; cytology of invertebrate and vertebrate muscle and adipose tissue; origin of avian melanoblasts. *Mailing Add:* Northeast 600 Garfield Pullman WA 99163

EASTMAN, ALAN D, b San Francisco, Calif, Oct 10, 46; m 70; c 5. INORGANIC CHEMISTRY. *Educ:* Univ Utah, BA, 71, PhD(chem), 75. *Prof Exp:* Chemist, Phillips Petrol Co, 75-81, mkt res specialist, 82-87, sr mkt specialist, 81-88, SR RES CHEMIST, PHILLIPS PETROL CO, 88- *Mem:* Am Chem Soc; Sigma Xi; Soc Appl Mat & Petrol Eng; Chem Mkt Res Asn. *Res:* Catalysis, especially the role of lattice oxygen in oxidation catalysts involving transition-metal mixed oxides. *Mailing Add:* 331 PL Phillips Petrol Co Bartlesville OK 74004

EASTMAN, CAROLINE MERRIAM, b Columbus, Ohio, Dec 25, 46; m 68. INFORMATION RETRIEVAL, DATA BASE MANAGEMENT SYSTEMS. *Educ:* Radcliffe Col, BA, 68; Univ NC, Chapel Hill, MS, 74, PhD(comput sci), 77. *Prof Exp:* Asst prof math & comput sci, Fla State Univ, 77-82; from asst prof to assoc comput sci & eng, Southern Methodist Univ, 82-85; prog dir, Info Sci, NSF, 84-85; ASSOC PROF COMPUT SCI, UNIV SC, 86- *Concurrent Pos:* Bd mem, Asn Women Comput. *Mem:* Asn Comput Mach; Am Soc Info Sci; Inst Elec & Electronic Engrs Comput Soc; Asn Women Comput; Asn Computational Ling; AAAS. *Res:* Information retrieval; data base management systems. *Mailing Add:* Dept Comput Sci Univ SC Columbia SC 29208

EASTMAN, DANIEL ROBERT PEDEN, b Semans, Sask, Jan 23, 33; m 55; c 2. PHYSICS. *Educ:* Houghton Col, BS, 55; Pa State Univ, MS, 57, PhD(physics), 61. *Prof Exp:* Optical engr, Plummer & Kershaw, Pa, 61-62; prof physics, Houghton Col, 62-65; assoc prof physics, Pa State Univ, 69-78; SR SCIENTIST, PERKIN-ELMER, 78- *Mem:* Optical Soc Am; Int Soc Optical Eng; Am Phys Soc. *Res:* Vibration-rotation near IR spectra; high precision IR spectroscopy; spontaneous brillouin spectroscopy; interferometry; optical fabrication. *Mailing Add:* 235 Poplar Ave Bridgeport CT 06606

EASTMAN, DAVID WILLARD, b Angola, NY, April 30, 39; c 6. GRAFT POLYMERIZATION, FREE RADICAL CHEMISTRY. *Educ:* Erie Co Tech Inst, AAS, 58; Canisius Col, BS, 61; State Univ NY, Buffalo, PhD(org chem), 70. *Prof Exp:* Group leader spectros, Hooker Chem & Plastics Co, 61-66, sr chemist, 69-76, res mgr, PVC Prod, 76-80, tech dir, Ruco Div, 80-81, dir technol, PVC Fabricated Prod Div, 81-84; dir res & develop, Lonza, 84-85, vpres, res & develop, 85-86, vpres mfg, 87-88, VPRES TECHNOL, LONZA, 90- *Mem:* Soc Plastics Engrs. *Res:* Polyvinyl chloride polymers; compounds and processing of vinyl polymers; graft reactions; free radical chemistry; polymer rheology; polymer alloys. *Mailing Add:* Lonza Inc 17-17 Rte 208 Fair Lawn NJ 07410

EASTMAN, DEAN ERIC, b Oxford, Wis, Jan 21, 40. SEMICONDUCTOR LOGIC. *Educ:* Mass Inst Technol, BS, 62, MS, 63, PhD(elec eng), 65. *Prof Exp:* Res staff mem, IBM, 63-71, mgr, Photoemission & Surface Physics Group, 71-81, mgr, lithography, packaging & compound semiconductor tech, 81-82, dir, Advan Packaging Lab, 83-85, dir, Prod Develop, Syst Technol Div, vpres, Logic, Memory & Packaging, 86-, VPRES SYST TECHNOL & SCI, RES DIV, IBM. *Concurrent Pos:* Vis prof, Mass Inst Technol, 72-73; chmn, Int Semiconductor Conf. *Honors & Awards:* Oliver E Buckley Prize, 80. *Mem:* Nat Acad Sci; Nat Acad Eng; fel Am Phys Soc. *Res:* Condensed matter physics, surface science and research and development management in semiconductor logic, memory and packaging technologies. *Mailing Add:* Watson Res Ctr PO Box 218 Yorktown Heights NY 10598

EASTMAN, JOHN W, b Charleston, Ill, May 10, 35; m 66; c 2. CLINICAL CHEMISTRY. *Educ:* Pa State Univ, BS, 57; Univ Calif, Berkeley, PhD(chem), 61. *Prof Exp:* Teaching asst chem, Univ Calif, Berkeley, 57-58; NSF fel, 61-62; res asst quantum chem, Univ Uppsala, 62-63; chemist, Shell Develop Co, 63-71; chemist, Med Sch, Univ Calif, 71-73; asst prof, Med Sch, George Washington Univ, 73-74; CHEMIST, US FOOD & DRUG ADMIN, 74- *Concurrent Pos:* Alexander Von Humboldt res fel phys chem, Univ Mainz, 68-69. *Mem:* AAAS; Am Chem Soc; Am Asn Clin Chemists. *Res:* Electronic structure of molecules; photochemistry; medical analysis of calcium, calcium binding, lipids; medical lab data systems. *Mailing Add:* 70 Schooner Hill Oakland CA 94618

EASTMAN, JOSEPH, b Minneapolis, Minn, Nov 2, 44; m 70; c 2. COMPARATIVE ANATOMY, HUMAN ANATOMY. *Educ:* Univ Minn, Minneapolis, BA, 66, MS, 68, PhD(zool), 70. *Prof Exp:* From instr to asst prof anat sci, Med Ctr, Univ Okla, 70-73; asst prof anat, Sch Med, Brown Univ, 73-79; assoc prof, 79-89, PROF ZOOL, OHIO UNIV, 89- *Mem:* AAAS; Am Asn Anat; Am Soc Zoologists; Am Soc Ichthyologists & Herpetologists. *Res:* Gross anatomical, histological and ultrastructural studies of fishes (especially antarctic fishes) with emphasis on morphology as related to physiology, ecology and evolution. *Mailing Add:* Dept Zool & Biomed Sci Ohio Univ Athens OH 45701-2979

EASTMAN, LESTER F(UESS), b Utica, NY, May 21, 28; m 48; c 3. ELECTRICAL ENGINEERING. *Educ:* Cornell Univ, BEE, 53, MS, 55, PhD(elec eng), 57. *Prof Exp:* Founder & dir, Joint Serv Electronic Prog, 77-87; from instr to assoc prof, 54-66, PROF ELEC ENG CORNELL UNIV, 66- *Concurrent Pos:* Consult, Westinghouse Elec Co, 55-64, Sylvania Elec Co, 56, Cornell Aero Lab, 57-64, Int Bus Mach Corp, 61-64, Raytheon Mfg Co, 61-64, MIT Lincoln Lab, 79-90, Gen Elec, 83-90 & United Technol, 83- & Boeing, 88-; vis assoc prof electronics, Chalmers Tech Inst, Sweden, 60-61; vis mem tech staff, RCA Res Labs, Princeton, 64-65; co-founder & consult, Cayuga Assoc, 67-77, pres, 70-77; mem, US Govt Adv Group Elec Devices, 77-87; co-founder, Nat Res & Resource Facil Submicron Structures, Cornell Univ, 77-; vis mem tech staff, MIT Lincoln Lab, 78-79, IBM Watson Res Lab, 85-86; John Laporte Given Prof Chair Eng, 85-; mem, Kuratorium, German Fraunmhofer Soc, Appl Phys Inst, 86-; founder & chmn bd, Northeast Semiconductor, Inc, 89- *Mem:* Nat Acad Eng; Fel Inst Elec & Electronics Engrs; Sigma Xi. *Res:* Compound semiconductor materials, microwave digital and opto-electronic devices technology and physical electronics; molecular beam epitaxy, organo-metallic vapor phase epitaxy, electron beam lithography; ballistic electrons in compound semiconductor fast transistors. *Mailing Add:* 425 Phillips Hall Cornell Univ Ithaca NY 14853

EASTMAN, MICHAEL PAUL, b Lancaster, Wis, Apr 14, 41; m 80; c 2. PHYSICAL CHEMISTRY. *Educ:* Carleton Col, BA, 63; Cornell Univ, PhD(phys chem), 68. *Prof Exp:* Fel, Los Alamos Sci Lab, 68-70; from asst prof to assoc prof, Univ Tex, El Paso, 70-80, prof chem, 80-88; PROF CHEM, NORTHERN ARIZ UNIV, 88-, CHMN CHEM, 88- *Concurrent Pos:* Asst dean sci, 81-84; asst vpres for acad affairs, 84-85. *Mem:* Am Chem Soc. *Res:* Magnetic resonance; ion pairing; geochemistry. *Mailing Add:* Dept Chem Northern Ariz Univ Flagstaff AZ 86011

EASTMAN, PHILIP CLIFFORD, b Port Hope, Ont, May 28, 32; m 56; c 2. SOLID STATE PHYSICS. *Educ:* Univ McMaster, BSc, 55, MSc, 56; Univ BC, PhD(physics), 60. *Prof Exp:* NATO fel & Rutherford Mem Award, Univ Bristol, 60-61; sr sci officer, Defence Res Telecommun Estab, Ottawa, 61-63; asst prof, 63-65, ASSOC PROF PHYSICS, UNIV WATERLOO, 65- *Mem:* Can Asn Physicists. *Res:* Galvanomagnetic and optical properties of metals; semimetals, semiconductors and insulators in single crystal, polycrystal and thin film forms. *Mailing Add:* Dept Physics Univ Waterloo Waterloo ON N2L 3G1 Can

EASTMAN, RICHARD HALLENBECK, b Erie, Pa, Oct 30, 18; m 42; c 3. ORGANIC CHEMISTRY. *Educ:* Princeton Univ, AB, 41; Harvard Univ, AM, 43, PhD(org chem), 44. *Prof Exp:* Asst, Harvard Univ, 44-46; from instr to prof, 46-83, EMER PROF ORG CHEM, STANFORD UNIV, 83- *Concurrent Pos:* NSF fel, 58-59. *Mem:* Sigma Xi. *Res:* Chemistry of natural products; ultraviolet and infrared spectroscopy; organic photochemistry. *Mailing Add:* Dept Chem Stanford Univ Stanford CA 94305

EASTMAN, ROBERT M(ERRIAM), b Dayton, Ohio, Apr 18, 18; m; c 3. INDUSTRIAL ENGINEERING. *Educ:* Antioch Col, AB, 40; Ohio State Univ, MS, 48; Pa State Univ, PhD, 55. *Prof Exp:* Asst instr eng drawing, Ohio State Univ, 47-48; instr indust eng, Pa State Univ, 48-51; assoc prof & res assoc, Eng Exp Sta, Ga Inst Technol, 51-55; chmn indust eng dept, 55-68, prof, 55-87, EMER PROF INDUST ENG, Univ Mo, Columbia, 88. *Concurrent Pos:* Fulbright lectr, Indust Univ Santander, 62-63; consult area redevelop admin, US Dept Com, 62-65; consult, US Off Educ, 68-69; vis prof, Mid East Tech Univ, Ankara, 69-71; consult, Dept Health, Educ & Welfare, 71, mem steering comt, Nat Ctr Health Serv Res & Develop, Health Serv & Ment Health Admin, 71-72; fel, NASA, 75, 78-81, NSF, 76, Navy, 82, 85-88. *Mem:* Am Soc Eng Educ; fel AAAS. *Res:* Engineering economy; technical aid to developing countries; operations research; regional development and planning; resource recovery from solid waste; robotics; Indutrial Safety. *Mailing Add:* 600 S Glenwood Ave Columbia MO 65203-2761

EASTMOND, ELBERT JOHN, b San Francisco, Calif, July 6, 15; m 37; c 4. PHYSICS. *Educ:* Brigham Young Univ, AB, 37; Univ Calif, PhD(physics), 43. *Prof Exp:* Physicist, Western Regional Res Lab, Bur Agr & Indust Chem, USDA, 42-50; from asst prof to assoc prof, 51-57, PROF PHYSICS, BRIGHAM YOUNG UNIV, 57- *Concurrent Pos:* Mem tech staff, Space Technol Labs, Los Angeles, 58 & Aerospace Corp, 63-64. *Mem:* Am Asn Physics Teachers; Optical Soc Am. *Res:* Diatomic molecular spectroscopy; spectrochemical analysis; spectrophotometry and colorimetry; rotational analysis of a band system attributed to ionized nitric oxide. *Mailing Add:* 1418 N 380 W Provo UT 84601

EASTON, DEXTER MORGAN, b Rockport, Mass, Sept 13, 21; m 53; c 4. NEUROPHYSIOLOGY. *Educ:* Clark Univ, BA, 43; Harvard Univ, MA, 44, PhD(biol), 47. *Prof Exp:* Asst, Clark Univ, 42-43 & Harvard Univ, 43-47; instr zool, Univ Wash, 47-50, res physiologist, Med Sch, 52-55; asst prof physiol, 55-58, from assoc prof to prof biol sci, 58-82, UNIV SERV PROF BIOL SCI, FLA STATE UNIV, 82- *Concurrent Pos:* Fulbright scholar, Otago Med Sch, NZ, 50-51. *Mem:* Fel AAAS; Biophys Soc; Soc Gen Physiol; Am Physiol Soc; Soc Neurosci. *Res:* Neuromuscular transmission; analysis of cell potentials and excitability; electrical activity of peripheral nerve; axonology; mathematical modelling. *Mailing Add:* Dept Biol Sci Fla State Univ Tallahassee FL 32306

EASTON, DOUGLAS P, b Minneapolis, Minn, Jan 7, 44; m 74; c 2. PHYSIOLOGICAL ECOLOGY, CELL BIOLOBY. *Educ:* Univ Iowa, PhD(zool), 72. *Prof Exp:* Res assoc microbiol, Univ Wash, 73-74, actg asst prof zool, 75-77; ASSOC PROF BIOL, BUFFALO STATE COL, 77- *Concurrent Pos:* Postdoctoral fel zool, Univ Wash, 74-76. *Mem:* Am Soc Zoologists; AAAS; Soc Develop Biol. *Mailing Add:* Dept Biol Buffalo State Col Buffalo NY 14222

EASTON, ELMER C(HARLES), b Newark, NJ, Dec 23, 09. ELECTRICAL ENGINEERING. *Educ:* Lehigh Univ, BS, 31, MS, 33; Harvard Univ, ScD(elec eng), 42. *Hon Degrees:* DEng, Lehigh Univ, 65. *Prof Exp:* Mem fac, Newark Col Eng, 35-42; mem fac, Grad Sch Eng, Harvard Univ, 42-48, asst dean, 46-48; dean col eng, 48-74, EMER DEAN COL ENG, RUTGERS UNIV, 74- *Mem:* Am Soc Eng Educ (pres, 64-65); Nat Soc Prof Engrs; Inst Elec & Electronics Engrs. *Res:* Conduction of electricity through gases. *Mailing Add:* Four Orchard Rd Piscataway NJ 08854

EASTON, IVAN G(EORGE), b Sweden, Nov 20, 16; nat US; m 41; c 4. ELECTRICAL ENGINEERING. *Educ:* Northeastern Univ, BS, 38; Harvard Univ, MS, 39. *Prof Exp:* Instr physics, Harvard Univ, 39-40; mgr eng, Gen Radio Co, 40-64, vpres, 64-68, sr vpres, 68-72; consult dir standards, Inst Elec & Electronics Engrs, 76-80; pres, US Nat Comt, Int Electrotech Comn, 80-84; RETIRED. *Mem:* Am Soc Testing & Mat; fel Inst Elec & Electronics Engrs. *Res:* National and international voluntary standards. *Mailing Add:* 56 Windsor Dr Englewood FL 34223

EASTON, NELSON ROY, b Craftsbury, Vt, Oct 8, 19; m 44; c 3. ORGANIC CHEMISTRY. *Educ:* Middlebury Col, AB, 41; Univ Ill, PhD(org chem), 46. *Prof Exp:* Chemist, Merck & Co, NJ, 41-43; asst org chem, Univ Ill, 43-44, asst 44-46; sr chemist, J T Baker Chem Co, NJ, 46-47; asst prof org chem, Lehigh Univ, 47-53; res chemist, Labs, Eli Lilly & Co, 53-62, res assoc, 62-64, asst dir chem res, 64-66, dir, 66-69, assoc dir res, 69-72, exec dir res, 72-74, vpres Lilly Res Labs, Eli Lilly & Co, 74-82; RETIRED. *Mem:* Am Chem Soc. *Res:* Hypotensive agents; reactions of acetylenes; heterocyclic compounds from acetylenic amines. *Mailing Add:* 2410 E Banta Rd Indianapolis IN 46227-4910

EASTON, RICHARD J, b Beaver, Utah, July 12, 38; m 62; c 3. COMPUTER SCIENCES. *Educ:* Univ Utah, BS, 60, MS, 63, PhD(math), 66. *Prof Exp:* Asst prof math, Weber State Col, 65-67; from asst prof to assoc prof, 67-75, PROF MATH, IND STATE UNIV, 75- *Concurrent Pos:* Lectr, Hill AFB, 66-67. *Mem:* Am Math Soc; Math Asn Am. *Res:* Vector measures; integration theory; applications software; computer sciences. *Mailing Add:* Dept Math & Comput Sci Ind State Univ Terre Haute IN 47809

EASTON, ROBERT WALTER, b Chicago, Ill, Dec 8, 41; m 65. MATHEMATICS. *Educ:* Univ Wis, BS, 63, MS, 65, PhD(math), 67. *Prof Exp:* Asst prof appl math, Brown Univ, 67-72; assoc prof, 72-80, PROF MATH, UNIV COLO, BOULDER, 80- *Mem:* Am Math Soc. *Res:* Differential equations; celestial mechanics. *Mailing Add:* Dept Math Univ Colo Campus Box 426 Boulder CO 80309

EASTON, WILLIAM HEYDEN, b Bedford, Ind, Jan 14, 16; m 40; c 2. PALEONTOLOGY. *Educ:* George Washington Univ, BS, 37, AM, 38; Univ Chicago, PhD(geol), 40. *Prof Exp:* Asst, US Nat Mus, 36-38; asst, Univ Chicago, 39-40; from asst geologist to assoc geologist, Ill Geol Surv, 40-44; from asst prof to assoc prof, 44-51, chmn dept, 63-67, prof, 51-81, EMER PROF GEOL, UNIV SOUTHERN CALIF, 81- *Concurrent Pos:* Asst, George Washington Univ, 37-38; asst, US Geol Surv, 52-53; Guggenheim fel, 59-60; actg chmn, Dept French & Italian, Univ Southern Calif, 75-76; consult geologist, 81- *Mem:* Paleont Soc (pres, 50); fel Geol Soc Am; Soc Econ Paleont & Mineral; Am Asn Petrol Geol. *Res:* Paleozoic paleontology and stratigraphy; oil and gas exploration; Carboniferous corals; radioisotopic dating coral reefs and raised shore lines. *Mailing Add:* 3818 Bowsprit Circle Westlake Village CA 91361

EASTWOOD, ABRAHAM BAGOT, b Philadelphia, Pa, Nov 8, 43; div; c 2. CYTOPATHOLOGY, ELECTRON MICROSCOPY. *Educ:* Muhlenberg Col, BS, 65; Lehigh Univ, MS, 67, PhD(biol), 71. *Prof Exp:* Staff mem cell physiol & ultrastruct, Lab Neurophysiol, Columbia Univ, 71-75, dir, Lab Muscle Morphol, 75-83, asst prof, Dept Anat, Col Physicians & Surgeons, 78-83; assoc dir res, 83-87, DIR RES INFO, MUSCULAR DYSTROPHY ASN, 87- *Mem:* AAAS; Biophys Soc; Sigma Xi; Am Soc Cell Biol. *Res:* Structural and functional substrates of excitation-contraction coupling in muscle; ultrastructural manifestations of disease in human muscle; cell membrane structure; database management. *Mailing Add:* 275 Engle St Apt E Englewood NJ 07631

EASTWOOD, BASIL R, b Argyle, Wis, Dec 17, 36; m 63; c 2. ANIMAL GENETICS, DAIRY SCIENCE. *Educ:* Wis State Univ, Platteville, BS, 58; SDak State Univ, MS, 60; Mich State Univ, PhD(dairy cattle breeding), 67. *Prof Exp:* Exten dairy specialist, Univ Mass, 63-65; from asst prof to prof animal sci, Iowa State Univ, 65-80, exten dairy specialist, 65-80; staff leader, 87-89, PROG LEADER DAIRY, LIVESTOCK & VET SCI, EXTEN SERV, USDA, 80- *Concurrent Pos:* Consult, Repub SKorea, US Feed Grains Coun, 76 & 81, Univ Costa Rica & Dairy Res Sta, 79; policy bd, Nat Coop Dairy Herd Improv Prog, 80- *Honors & Awards:* Distinguished Serv Award, Nat Dairy Herd Improv Asn, 85. *Mem:* Am Dairy Sci Asn. *Res:* Use of records in genetic improvement of dairy cattle. *Mailing Add:* USDA Exten Serv 3334 S Bldg Washington DC 20250-0900

EASTWOOD, DELYLE, b Upper Darby, Pa, Nov 19, 32. PHYSICAL & ANALYTICAL CHEMISTRY, SPECTROSCOPY. *Educ:* Univ Chicago, MS, 55 & 82, PhD(phys chem), 64. *Prof Exp:* Res asst chem res insts, Univ Chicago, 54-56, teaching asst, 57-59, phys chemist, 59-60, res asst, Inst Study Metals, 61-64; res fel phys chem & spectros, Harvard Univ, 64-66; res assoc, Univ Wash, 66-69; res assoc, Northeastern Univ, 70-71; sr chemist, Baird-Atomic, Inc, 71-72; proj chemist, Bendix Res Labs, 72-73; res chemist, US Coast Guard Res & Develop Ctr, 74-81; chemist, Dept Nuclear Energy, Brookhaven Nat Lab, 81-83; chemist, Superfund Design Ctr, US Corps Engrs, Mo River Div, 83-88; SR STAFF SCIENTIST, LOCKHEED ENG & MGT SERVS, 88- *Honors & Awards:* Silver Medal, US Dept Transp, 78. *Mem:* Am Chem Soc; Am Soc Testing & Mat; Soc Appl Spectros; Am Phys Soc. *Res:* Environmental analysis methods development; molecular and atomic spectroscopy applied to environmental pollution, especially fluorescence for oil identification, aromatics, porphyrins and other large organic molecules; plasma emission spectroscopy for trace metal analysis; geochemistry; solid state chemistry; physics applied to materials science (solid surface spectroscopy). *Mailing Add:* Lockheed Eng & Mgt Serv Co Suite 242 1050 E Flamingo Rd Las Vegas NV 89119

EASTWOOD, DOUGLAS WILLIAM, b Ellsworth, Wis, Sept 17, 18; m 43; c 4. ANESTHESIOLOGY. *Educ:* Coe Col, AB, 40; Univ Iowa, MD, 43, MS, 49. *Prof Exp:* From intern to asst resident, Receiving Hosp, Detroit, Mich, 44-45; from asst resident to resident anesthesiol, Univ Hosps, Univ Iowa, 47-49, from instr to assoc prof, Col Med, 49-55; asst prof & chief, Sch Med, Wash Univ, 50-54; prof, Sch Med, Univ Va, 55-72, chmn dept, 55-71; assoc prof med educ, 72-75, PROF ANESTHESIOL, CASE WESTERN RESERVE UNIV, 72-; dir obstet anesthesiol, Univ Hosp Cleveland, 75-84, med dir, Wright Surg Ctr, 84-88; RETIRED. *Concurrent Pos:* Instr internal med, Wayne State Univ, 44-45; dir anesthesiol prog, Vet Admin Hosp, Cleveland, 72-75 & 88- *Mem:* Am Soc Anesthesiologists; Int Anesthesia Res Soc; Sigma Xi. *Res:* Educational resources; evaluation of educational methods; obstetric anesthesia. *Mailing Add:* 3630 Mt Laurel Rd Cleveland Heights OH 44121

EASTWOOD, GREGORY LINDSAY, b Detroit, Mich, July 28, 40; m 64; c 3. GASTROENTEROLOGY. *Educ:* Albion Col, BA, 62; Case Western Reserve Univ, MD, 66. *Prof Exp:* Asst prof med, Harvard Med Sch, 74-77; assoc prof, 77-82, PROF MED, UNIV MASS MED SCH, 82-, DIR GASTROENTEROL DIV, 77- *Concurrent Pos:* Mem, Subspec Bd Gastroenterol, Am Bd Internal Med, 83-; assoc dean admissions, Univ Mass Med Sch, 86- *Mem:* Am Gastroenterol Asn (secy, 88-); Am Col Physicians. *Res:* Functional morphology of the gastrointestinal tract; epithelial renewal in peptic and preneoplastic disorders; mechanisms of gastroduodenal mucosal injury and protection. *Mailing Add:* Gastroenterol Div Univ Mass Med Sch 55 Lake Ave N Worcester MA 01605

EASTWOOD, RAYMOND L, b Pawnee City, Nebr, Aug 27, 40; m 80. GEOCHEMISTRY, PETROLOGY. *Educ:* Kans State Univ, BS, 62, MS, 65; Univ Ariz, PhD(geol), 70. *Prof Exp:* Res asst geol, Univ Ariz, 64-68; res mineralogist geochem, Res & Develop Dept, Phillips Petrol Co, Okla, 68-70; asst prof geol, Northern Ariz Univ, 70-79; SR RES GEOLOGIST, RES & DEVELOP, ARCO OIL & GAS CO, PLANO, TEX, 79- *Mem:* Am Geophys Union; Geochem Soc; Geol Soc Am. *Res:* Isotope geology; strontium isotope ratios; igneous and metamorphic petrology; bore hole geophyics; geochronology; mineralogy; geochemistry and petrology of volcanic rocks; volcanology. *Mailing Add:* 1280 Old Mill Rd McKinney TX 75069

EASTWOOD, THOMAS ALEXANDER, b London, Ont, Nov 27, 20; m 49; c 4. NUCLEAR CHEMISTRY, CHEMICAL KINETICS. *Educ:* Univ Western Ont, BA, 42, MA, 43; McGill Univ, PhD(chem), 46; Oxford Univ, DPhil(chem), 51. *Prof Exp:* Control chemist, Imperial Oil Ltd, 43-44; res officer chem, Atomic Energy Proj, Nat Res Coun Can, 47-49; res officer chem, 51-69, DIR CHEM & MAT DIV, CHALK RIVER NUCLEAR LABS, ATOMIC ENERGY CAN LTD, 69- *Concurrent Pos:* Hon lectr, McGill Univ, 46-47; Nat Res Coun Can fel, Oxford Univ, 49-50, Carnegie Res Award, 50-51; chemist, UK Atomic Energy Authority, Eng, 64-65. *Mem:* Am Phys Soc; sr mem Am Chem Soc; Can Asn Physicists; fel Chem Inst Can; Sigma Xi. *Res:* Radio chemistry; nuclear physics; kinetics of chemical reactions. *Mailing Add:* Nine Tweedsmuir Pl Deep River ON K0S 1P0 Can

EASTY, DWIGHT BUCHANAN, b Lakewood, Ohio, Mar 8, 34; m 58; c 3. ANALYTICAL CHEMISTRY. *Educ:* Ohio Wesleyan Univ, BA, 56; Lawrence Col, MS, 58, PhD(paper chem), 61. *Prof Exp:* Develop engr, Paper Sect, Res & Develop Lab, Nat Vulcanized Fibre Co, 61-62, group leader, 62-66; vis asst prof chem, Ohio Wesleyan Univ, 66-67; res assoc, Univ Wis-Madison, 67-68, lectr, 68-69; from asst prof to assoc prof chem, Inst Paper Chem, 69-89; MGR, ENVIRON ANALYTICAL LAB, JAMES RIVER CORP, 89- *Mem:* Am Chem Soc; Tech Asn Pulp & Paper Indust. *Res:* Analysis of wood, pulp, paper and pulping and bleaching liquors; environmental analysis. *Mailing Add:* 13314 SE 19th St Apt R8 Vancouver WA 98684

EATON, ALVIN RALPH, b Toledo, Ohio, Mar 13, 20; m 70; c 2. SYSTEMS ENGINEERING. *Educ:* Oberlin Col, AB, 41; Calif Inst Technol, MS, 43. *Prof Exp:* Asst aeronaut eng, Calif Inst Technol, 41-42, res supersonic aerodyn, 42-43, engr, Southern Calif Coop Wind Tunnel, 44-45; engr & supvr guided missile prog, Appl Physics Lab, Johns Hopkins Univ, 45-65, supvr, Missile Systs Div, 65-73, asst dir tactical systs, 73-79, supvr, Fleet Systs Dept, 73-83, asst dir, 79-86, assoc dir, 86-89, SR FEL & DIR SPEC PROG, JOHNS HOPKINS UNIV, 89-; RETIRED. *Concurrent Pos:* Consult to Under Secy Defense, Dept Defense, 76-83, chmn, Task Force on US Army Patriot Syst, Off Under Secy Defense, Defense Sci Bd, 76-78; consult, Asst Secy Army, 69-74 & 80-86; mem, Defense Sci Bd Task Force Countermeasures, 79-81 & Army Sci Bd, 80-86, 89-; chmn, asst secy, Res & Develop Army Independent Rev Panel for Anti-Tactical Missle Prog, 86-; mem ed bd, J Defense Res, 88- *Honors & Awards:* Distinguished Pub Serv Award, US Navy, 75. *Mem:* Cosmos Club; fel Explorers Club; fel Hudson Inst. *Res:* Defense related applied physics. *Mailing Add:* 6701 Surrey Lane Clarksville MD 21029

EATON, BARBRA L, b Abington, Pa, May 7, 41; m 73; c 2. BIOCHEMICAL EPIDEMIOLOGY. *Educ:* Smith Col, AB, 64; Reed Col, MAT, 65; Univ Pa, PhD(biochem), 72. *Prof Exp:* Staff fel, NIH, 73-76; res fel, Johns Hopkins Univ, 84-85; sr scientist & prog dir, Capital Systs Group, Inc, 85-90; SELF-EMPLOYED, SCI ANAL CONSULT, 90- *Concurrent Pos:* Proj mgr under contract, NSF, 86-90. *Mem:* Asn Women Sci; Am Soc Cell Biol; AAAS; Biophys Soc; Sigma Xi. *Res:* Epidemiological biostatistics. *Mailing Add:* 13804 Bonsal Lane Silver Spring MD 20906-3048

EATON, BRYAN THOMAS, virology, for more information see previous edition

EATON, DAVID FIELDER, b Peterborough, NH, Oct 4, 46; m 74. PHYSICAL ORGANIC CHEMISTRY. *Educ:* Wesleyan Univ, BA, 68; Calif Inst Technol, PhD(org chem), 72. *Prof Exp:* Res chemist org chem, Univ Calif, San Diego, 72-73; res chemist, Cent Res & Develop Dept, 73-79, sr res chemist, Photo Prod Dept, 79-80, group leader, 80-81, RES SUPVR, CENT RES & DEVELOP DEPT, E I DU PONT DE NEMOURS & CO INC, 81- *Concurrent Pos:* Titular mem, Comn Photochemistry, Int Union Pure & Appl Chem, mem, US Nat Comt. *Mem:* InterAm Photochem Soc; Am Chem Soc; fel Soc Photog Scientists & Engrs; Int Union Pure & Appl Chem. *Res:* Photochemistry and electron transfer reactions of organic and organometallic compounds; photoimaging systems; nonlinear optical materials. *Mailing Add:* Cent Res & Develop Dept Exp Sta E I du Pont de Nemours & Co Inc Wilmington DE 19880-0328

EATON, DAVID J, b Detroit, Mich, Dec 18, 49; m 87. ENVIRONMENTAL ENGINEERING, SYSTEMS ANALYSIS. *Educ:* Oberlin Col, BA, 71; Unit Pittsburgh, MSc, 72, MPL, 72; Johns Hopkins Univ, PhD(environ eng), 77. *Prof Exp:* Asst prof pub policy, 76-80, assoc prof pub policy & geog, Lyndon B Johnson Sch Pub Affairs, 80-85, PROF PUB AFFAIRS & GEOG, UNIV TEX, AUSTIN, 85- *Concurrent Pos:* Consult, NSF, 74-75; prin investr grants, USDA, 75-76, US Agency Int Develop, 76-79 & US Environ Protection Agency, 77-79; Fulbright res scholar, 81-82. *Mem:* AAAS; Inst Mgt Sci. *Res:* Delivery of health services; analysis of grain reserves; rural water supply and waste treatment; air and water pollution; appropriate technology. *Mailing Add:* LBJ Sch Pub Affairs Univ Tex Drawer Y Univ Sta Austin TX 78712

EATON, DAVID LEO, b Minneapolis, Minn, Jan 11, 32; m 54; c 5. INORGANIC CHEMISTRY, MATERIALS ENGINEERING. *Educ:* St Thomas Col, BS, 54. *Prof Exp:* Jr chemist ext metall, Ames, Lab, 54-56; sr design engr nuclear fuel elements, Martin Aircraft Co, 56-60; proj engr thermoelec, Gen Instruments Corp, 60-61; develop assoc chem mat eng, Corning Glass Works, 61-84; SR CHEMIST & SR RES ASSOC, ENZYME MICROBE IMMOBILIZATION CARRIERS RES & DEVELOP, MANVILLE CORP, 84- *Mem:* Am Chem Soc; Am Ceramics Soc. *Res:* Separation sciences; immobilized biological composites and their applications; material engineering. *Mailing Add:* 1349 Camellia Ct Lompoc CA 93436

EATON, DONALD REX, b Leicester, Eng, July 20, 32; m 59; c 4. PHYSICAL CHEMISTRY. *Educ:* Oxford Univ, BA, 55, MA, 59, DPhil(chem), 58. *Prof Exp:* Fel div pure physics, Nat Res Coun Can, 58-60; res chemist, Cent Res Dept, E I du Pont de Nemours & Co, 60-64, res supvr, 64-68; assoc prof chem, 68-71, PROF CHEM, MCMASTER UNIV, 71- *Mem:* Am Chem Soc; Can Inst Chem; Royal Soc Chem. *Res:* Transition metal chemistry; magnetic resonance. *Mailing Add:* Dept Chem McMaster Univ Hamilton ON L8S 4L8 Can

EATON, DOUGLAS CHARLES, b Sioux Falls, SDak, Jan 31, 45. NEUROPHYSIOLOGY, TRANSPORT. *Educ:* Calif Inst Technol, BS, 67; Scripps Inst Oceanog, MS, 69, Univ Calif, San Diego, PhD(neurosci), 71. *Prof Exp:* Res assoc neurophysiol, Med Sch, Univ Calif, Los Angeles, 71-73; vis assoc neurophysiol, Dept Biol, Calif Inst Technol, 73; PROF PHYSIOL, DEPT PHYSIOL & BIOPHYS, UNIV TEX MED BR GALVESTON, 73- *Concurrent Pos:* Coun mem, Soc Gen Physiologists. *Mem:* Biophys Soc; Soc Gen Physiologists; Am Physiol Soc. *Res:* Electrophysiology of excitable membranes and epithelial tissue. *Mailing Add:* Dept Physiol Emory Univ Med Sch Atlanta GA 30322

EATON, GARETH RICHARD, b Lockport, NY, Nov 3, 40; m 69. BIOPHYSICS, SPECTROSCOPY & SPECTROMETRY. *Educ:* Harvard Univ, BA, 62; Mass Inst Technol, PhD(chem), 72. *Prof Exp:* Dean Nat Sci, 84-88, vice provost, 88-89, from asst prof to assoc prof, 72-80, PROF CHEM, UNIV DENVER, 80- *Mem:* Am Chem Soc; Royal Soc Chem; AAAS; Soc Appl Spectros; Int Soc Magnetic Resonance; Soc Magnetic Resonance in Med; Int Electron Paramagnetic Resonance Soc; Am Phys Soc. *Res:* Synthesis and spectroscopy of inorganic complexes and organic free radicals, metal ions in biological systems with emphasis on electron paramagnetic resonance; computer simulations, instrumentation. *Mailing Add:* Dept Chem Univ Denver Denver CO 80208

EATON, GEORGE T(HOMAS), b Edmonton, Alta, Apr 18, 10; nat US; m 36; c 3. CHEMICAL ENGINEERING. *Educ:* McMaster Univ, BA, 31, MA, 33; Acadia Univ, BS, 33. *Prof Exp:* Chemist, Photog Chem, Kodak Res Labs, 37-43, staff asst, 43-46, supvr indust sales studio, Sales Div, 46-51, ed, Sales Serv Div, 51-53, staff asst, Applied Photog Div, 53-56, asst div head, 57-75, head, Photog Chem Dept, 56-75; RETIRED. *Concurrent Pos:* Lectr, Rochester Inst Technol, 52-75. *Mem:* Fel Photog Soc Am; Soc Photog Sci & Engrs (pres, 57-61); Soc Motion Picture & TV Engrs. *Res:* Chemistry of photographic processing; photoreproduction; engineering drawings; microfilming; editorial work on photographic yearbooks and encyclopedia; teaching. *Mailing Add:* 699 Heritage Rd Rochester NY 14615

EATON, GEORGE WALTER, b Upper Canard, NS, Sept 4, 33; m 56; c 2. HORTICULTURE. *Educ:* Univ Toronto, BSA, 55; Ohio State Univ, PhD(pomol), 59. *Prof Exp:* Exten specialist pomol, Ont Dept Agr, 55-58, res scientist, 58-64; from asst prof to assoc prof, 64-74, PROF, UNIV BC, 74- *Honors & Awards:* G M Darrow Award, Am Soc Hort Sci. *Mem:* Am Soc Hort Sci; Can Soc Hort Sci; Inst Hort; Agr Inst Can; Sigma Xi. *Res:* Yield component analysis; mineral nutrition; biometrics. *Mailing Add:* Dept Plant Sci Univ BC Vancouver BC V6T 2A2 Can

EATON, GORDON GRAY, b Carmangay, Alta, June 5, 41; m 75; c 2. ANIMAL BEHAVIOR. *Educ:* Univ Victoria, BA, 64; Univ Calif, PhD(psychol), 70. *Prof Exp:* Asst scientist reproductive biol & behav, Ore Regional Primate Res Ctr, 70-73; asst prof, 71-75, ASSOC PROF PSYCHOL, SCH MED, UNIV ORE HEALTH SCI CTR, 75-; assoc scientist primate behav, 73-83, SCIENTIST REPRODUCTIVE BIOL & BEHAV, ORE REGIONAL PRIMATE RES CTR, 83- *Concurrent Pos:* Mem corresp fac, Cognitive Sci Prog, Col Art & Sci, Univ Ore; assoc scientist primate behav, One Regional Primate Res Ctr, 73-83. *Mem:* AAAS; Am Soc Primatologists; Animal Behav Soc; Int Primatological Soc. *Res:* Environmental and endocrine control of primate behavior. *Mailing Add:* Ore Regional Primate Res Ctr 505 NW 185th Ave Beaverton OR 97006

EATON, GORDON PRYOR, b Dayton, Ohio, Mar 9, 29; m 51; c 2. PHYSICAL GEOLOGY. *Educ:* Wesleyan Univ, BA, 51; Calif Inst Technol, MS, 53, PhD(geol, geophys), 57. *Prof Exp:* From instr to asst prof geol, Wesleyan Univ, 55-59; from asst prof to assoc prof, Univ Calif, Riverside, 59-67; geologist, US Geol Surv, 67-72 & 74-76, dep chief, Off Geochem & Geophys, 72-74, scientist-in-chg, Hawaiian Volcano Observ, 76-78, assoc chief geologist, 78-81; dean, Col Geosci, Tex A&M Univ, 81-83, provost & vpres acad affairs, 83-86; PRES, IOWA STATE UNIV, 86- *Concurrent Pos:* Res geologist, US Geol Surv, 63-65; chmn dept geol sci, Univ Calif, Riverside, 66-67. *Honors & Awards:* US Govt Sr Exec Serv Award. *Mem:* Geol Soc Am; Am Geophys Union; Am Asn Higher Educ; Asn Am Univs; Nat Asn State Univs Land-Grant Cols. *Res:* Physical geology; regional geophysics. *Mailing Add:* Off of the Pres Iowa State Univ Ames IA 50011

EATON, HAMILTON DEAN, animal nutrition, for more information see previous edition

EATON, HARVILL CARLTON, b Nashville, Tenn, May 16, 48; m 69; c 2. MATERIAL SCIENCE. *Educ:* Tenn Technol Univ, BS, 70, MS, 72; Vanderbilt Univ, PhD(mat sci), 76. *Prof Exp:* Asst prof eng sci, Tenn Technol Univ, 78-80; asst prof eng sci, La State Univ, 76-78, from asst prof to assoc prof mech eng, 80-87, assoc dean eng, 86-89, assoc vchancellor res, 89-91, PROF MECH ENG, LA STATE UNIV, 87-, VCHANCELLOR RES & ECON DEVELOP, 91- *Concurrent Pos:* Vis scientist, Oak Ridge Nat Lab, 80 & Dept Physics, Chalmers Univ, Sweden, 81-82. *Mem:* Am Ceramic Soc; Am Soc Mech Eng; Am Mineral Soc; Mat Res Soc; Nat Inst Ceramic Engrs. *Res:* Electron and field ion microscopy of grain boundaries in metals; structure of grain boundaries; high field effects of the structure of surfaces; microstructure and microchemistry of solidified hazardous wastes; structure of calcium-silicate-hydrates. *Mailing Add:* La State Univ Baton Rouge LA 70803

EATON, J(AMES) H(OWARD), b Woodland, Calif, Nov 28, 33; m 56; c 3. ELECTRICAL ENGINEERING. *Educ:* Univ Calif, Berkeley, BS, 58, MS, 60, PhD(elec eng), 62. *Prof Exp:* Asst prof systs theory, Univ Calif, Berkeley, 62-64; Consult, Res Lab, 63-64, mgr systs dept, 64-71, lab dir, 76-78, DIR TECH PLANNING, IBM CORP, 71-, CORP TECH COMT DIR, 78-, RES STAFF MEM, 79-; RES STAFF MEM, IBM CORP. *Mem:* Inst Elec & Electronics Engrs; NY Acad Sci. *Res:* Theory of optimal control; systems theory and its application to the design and analysis of computer systems. *Mailing Add:* IBM Almaden Res Ctr K-61-802 650 Hary Rd San Jose CA 95120-6099

EATON, JEROME F, b Newark, NJ, Jan 7, 41; m 64; c 2. GEOPHYSICS. *Educ:* Lehigh Univ, BS, 63; Princeton Univ, MA, 65, PhD(geophys), 68. *Prof Exp:* Res geophysicist, Gulf Res & Develop Co, 68-73, sr staff geophysicist, Gulf Oil Co-Eastern Hemisphere, 73-76, EASTERN HEMISPHERE INTERPRETATION COORDR, GULF EXPLOR & PROD CO-INT, 76- *Mem:* Soc Explor Geophysicists. *Res:* Steady and transient states of strain in igneous rocks at high temperature; relationships between continental margins and ocean basins; seismic response of trapped fluids; exploration seismology; marine and land field data acquisition; seismic data processing and interpretation. *Mailing Add:* 5623 Bermuda Dunes Lane Houston TX 77069

EATON, JERRY PAUL, b Fresno Co, Calif, Dec 11, 26; m 47; c 4. SEISMOLOGY. *Educ:* Univ Calif, AB, 49, PhD(geophys), 53. *Prof Exp:* Asst seismol, Univ Calif, 50-53; geophysicist, Hawaiian Volcano Observ, US Geol Surv, 53-61 & crustal studies br, 61-65, res geophysicist, Off Earthquake Res & Crustal Studies, 65-70, chief, 70-75, RES GEOPHYSICIST, OFF EARTHQUAKES, VOLCANOES & ENG, US GEOL SURV, 75- *Concurrent Pos:* Lectr, Univ Calif, Berkeley, 60-61. *Mem:* Geol Soc Am; Am Geophys Union. *Res:* Mechanics of earthquake generation; detailed studies of seismicity; geophysics of volcanoes; structure of the continental crust. *Mailing Add:* Off Earthquake Res US Geol Surv 345 Middlefield Rd Menlo Park CA 94025

EATON, JOEL A, b Paducah Ky, Jan 2, 48. STELLAR ASTRONOMY. *Educ:* Auburn Univ, BS, 70; Vanderbilt Univ, MS, 71; Univ Wis, PhD(astron), 75. *Prof Exp:* Vis asst prof astronomy, Univ Ala, Tuscaloosa, 75-76; res asst, Nat Res Coun, Goddard Space Flight Ctr, NASA, 76-78; asst prof, Pa State Univ, 78-79; asst prof astron, Vanderbilt Univ, 80-; AT DEPT ASTRON, IND UNIV, BLOOMINGTON. *Concurrent Pos:* Vis astronomer, Copernicus Astron Ctr, Warsaw, 79. *Mem:* Am Astro Soc; Int Astro Union; Astron Soc Pac; Soc Mfg Engrs. *Res:* Stellar photometry and spectroscopy; untraviolet observations from spacecraft; analysis of eclipsing binary light curves; Wolf-Rayet and stars with expanding atmospheres; stellar limb and gravity darkening; stellar chromospheres and surface activity. *Mailing Add:* Tenn State Univ 330 Tenth Ave N Suite 265 Nashville TN 37203-3401

EATON, JOHN KELLY, b Camden, NJ, May 22, 54; m 84; c 2. FLUID MECHANICS, HEAT TRANSFER. *Educ:* Stanford Univ, BS, 76, MS, 77, PhD(mech eng), 80. *Prof Exp:* Engr, Hewlett Packard, 76; asst prof, 80-85, ASSOC PROF MECH ENG, STANFORD UNIV, 85- *Concurrent Pos:* Consult, var co. *Honors & Awards:* Silver Medal, Royal Soc Arts, 76; Presidential Young Investr Award, NSF, 83. *Mem:* Am Soc Mech Eng; Am Inst Aeronaut & Astronaut; Am Phys Soc. *Res:* Experimental study and large scale computation of complex turbulent flows including the effects of three dimensionality, rotation, unsteadiness or particle loading. *Mailing Add:* Mech Eng Dept Stanford Univ Stanford CA 94305

EATON, JOHN LEROY, b Decatur, Ill, Sept 21, 39; m 61; c 3. INSECT PHYSIOLOGY. *Educ:* Univ Ill, BS, 62, PhD(entom), 66. *Prof Exp:* Kettering Found teaching intern biol, Kalamazoo Col, 66-67, asst prof, 67-69; from asst prof to prof entom, 69-88, ASSOC DEAN, GRAD SCH, VA POLYTECH INST & STATE UNIV, 88- *Mem:* AAAS; Entom Soc Am; Sigma Xi. *Res:* Insect neurophysiology, morphology and behavior. *Mailing Add:* Grad Sch Va Polytech Inst & State Univ Blacksburg VA 24061-0325

EATON, JOHN WALLACE, b Ann Arbor, Mich, Mar 13, 41; div; c 1. HEMATOLOGY, PHYSIOLOGY. *Educ:* Fla State Univ, BA, 63; Univ Fla, MA, 64; Univ Mich, PhD(biol anthrop), 69. *Prof Exp:* Res asst biochem, Univ Fla, 64-65; instr anthrop, Fla State Univ, 65-66; res assoc human genetics, Univ Mich, Ann Arbor, 69-70; asst prof anthrop, Wash Univ, 70-72; from asst prof to assoc prof, 72-79, PROF MED, LAB MED & PATHOL & DIR MED GENETICS, UNIV MINN, MINNEAPOLIS, 79- *Concurrent Pos:* Consult, NIH, 73- & NSF, 74-; NSF lectureship, India, 82; vis prof, dept physiol, Univ New Eng, Armidale, NSW Australia, 85 & dept pharmaceut chem, Univ Calif San Francisco, 87; mem, Hemat-2 Study Sect, NIH, 88-93. *Honors & Awards:* Iron Bolt Award, 87. *Mem:* Am Asn Phys Anthrop; Am Fedn Clin Res; Am Soc Hemat; Am Soc Genetics; Am Asn Pathologists. *Res:* Bioligical oxidation reactions; metals in biological systems; pathogen-host interactions; physiological adaptation. *Mailing Add:* Dept Lab Med Path Univ Minn Box 198 UMHC Minneapolis MN 55455

EATON, MERRILL THOMAS, JR, b Howard Co, Ind, June 25, 20; m 42; c 3. PSYCHIATRY. *Educ:* Ind Univ, AB, 41, MD, 44; Am Bd Psychiat & Neurol, dipl. *Prof Exp:* Intern, St Elizabeth's Hosp, 44-45; resident physician, Colo State Hosp, 47-48 & Sheppard-Pratt Hosp, 48-49; assoc psychiat, Sch Med, Univ Kans, 49-51, from asst prof to asoc prof, 51-60; dir, Nebr Psychiat Inst, 68-85; assoc prof, 60-63, chmn dept, 68-85, PROF PSYCHIAT, UNIV NEBR MED CTR, OMAHA, 63- *Mem:* Am Psychiat Asn; Group Advan Psychiat; Am Col Physicians. *Res:* Psychotherapy; medical education. *Mailing Add:* 6901 N 72nd St Omaha NE 68122

EATON, MONROE DAVIS, immunology, oncology; deceased, see previous edition for last biography

EATON, MORRIS LEROY, b Sacramento, Calif, Aug 10, 39; m 64; c 1. MATHEMATICAL STATISTICS. *Educ:* Univ Wash, BS, 61; Stanford Univ, MS, 63, PhD(statist), 66. *Prof Exp:* Res assoc statist, Stanford Univ, 66; from asst prof to assoc prof, Univ Chicago, 66-71; prof, Univ Copenhagen, 71-72; from assoc prof to prof statist, 72-77, PROF THEORET STATIST & CHMN DEPT, UNIV MINN, 77- *Mem:* Fel Am Statist Asn; fel Inst Math Statist; Int Statist Inst. *Res:* Multivariate analysis; decision theory ranking procedures; invariance in statistical problems. *Mailing Add:* Dept Theoret Statist Univ Minn 206 Church St SE Minneapolis MN 55455-0485

EATON, NORMAN RAY, microbiology, for more information see previous edition

EATON, PAUL BERNARD, b Elkhart, Ind, May 21, 17. INDUSTRIAL & METALLURGICAL ENGINEERING. *Educ:* Univ Notre Dame, BS, 48; Purdue Univ, MS, 52. *Prof Exp:* Instr gen eng, Purdue Univ, 48-52, asst prof metals processing, 52-62, assoc prof, 62-; RETIRED. *Concurrent Pos:* Consult, indust orgns, 55-; mem, Foundry Educ Found, 60- *Mem:* Am Soc Metals; Am Inst Mining, Metall & Petrol Engrs; Am Soc Eng Educ; Am Foundrymen's Soc. *Res:* Physical and process metallurgy of ferrous metals. *Mailing Add:* 7240 State Rd 26E W Lafayette IN 47905

EATON, PAUL WENTLAND, b Minneapolis, Minn, Sept 30, 27; m 50; c 2. STATISTICS, ECONOMICS. *Educ:* Univ Minn, BA, 49, MA, 54, PhD(econ), 65. *Prof Exp:* Mkt analyst, Munsingwear, 53-54; instr statist, Univ Minn, 55-59; asst prof statist & econ, Mich Technol Univ, 59-75, assoc prof, 65-69, head quant anal sect, 67-69; assoc prof statist, Northern Ill Univ, 69-89; RETIRED. *Concurrent Pos:* Consult, MacKinac Bridge Surv, 62 & Western Upper Peninsula Health Coun, 67-69; actg dir, Copper Country Health Surv, 68. *Mem:* Economet Soc; Am Statist Asn. *Res:* Statistical and economic theory. *Mailing Add:* RR 1 Box 142 Kingston IL 60115

EATON, PHILIP EUGENE, b Brooklyn, NY, June 2, 36. ORGANIC CHEMISTRY. *Educ:* Princeton Univ, AB, 57; Harvard Univ, MA, 60, PhD(chem), 61. *Prof Exp:* Asst prof chem, Univ Calif, Berkeley, 60-62; from asst prof to assoc prof 62-72, PROF CHEM, UNIV CHICAGO, 72- *Concurrent Pos:* Alfred P Sloan Found res fel, 63-69; consult, E I du Pont de

Nemours & Co, Inc, 65-77, NIH, 68-72, US Army, 81- & Dow Chem, 83- *Honors & Awards:* Res Award, Rohm and Haas, 75; Alexander von Humboldt Prize, 86. *Res:* Chemistry of small ring compounds cubane and dodecahedrane; photochemistry; highfield nuclear magnetic resonance; synthesis. *Mailing Add:* Dept Chem Univ Chicago 5735 S Ellis Ave Chicago IL 60637-4931

EATON, ROBERT CHARLES, b Los Angeles, Calif, Aug 14, 46. NEUROBIOLOGY. *Educ:* Univ Calif, Riverside, BA, 68, PhD(biol), 74; Univ Ore, MS, 70. *Prof Exp:* Asst res neuroscientist, Sch Med & Scripps Inst Oceanog, Univ Calif, San Diego, 74-77; from asst prof to assoc prof, 78-90, PROF BIOL, UNIV COLO, 91- *Concurrent Pos:* NIH fel, Univ Calif, San Diego, 74-77; INSERM fel, C H U Pitie-Salpetriere, Paris, France, 77; vis scholar, Univ Calif, San Diego, 88-89. *Honors & Awards:* Res Serv Award, NIH, 74. *Mem:* Soc Neurosci; Int Soc Neuroethology; Int Brain Res Org. *Res:* Neural network properties underlying production of coordinated movement in vertebrates, specifically those acts caused by the brain stem reticulospinal system; neurophysiological, neuroanatomical and high-speed cinematic analysis techniques are used for these studies. *Mailing Add:* Dept Environ Pop & Org Biol Univ Colo Boulder CO 80309-0334

EATON, ROBERT JAMES, b Buena Vista, Colo, Feb 13, 40; m 64; c 2. AUTOMOTIVE ENGINEERING. *Educ:* Univ Kans, BSc, 63. *Prof Exp:* Engr, Gen Motors Eng Ctr, Mich, 63-69, asst mgr, 69-71, chief engr, 71-73, asst chief engr, Oldsmobile Div, 75-79, vpres, Advan Eng Staff, Gen Motors Corp, 82-86, vpres, Tech Staff, 86-88, PRES, GEN MOTORS EUROP, ZURICH, SWITZ, 88- *Concurrent Pos:* Mem bd dirs, Group Lotus, 85; chmn bd, Saab Automobile, 90. *Mem:* Nat Acad Eng; fel Soc Automotive Engrs. *Mailing Add:* Gen Motors Europ AG Steizenstasse 4 Glattbrugg 8152 Switzerland

EATON, SANDRA SHAW, b Boston, Mass, Jan 23, 46; m 69. INORGANIC CHEMISTRY, MAGNETIC RESONANCE. *Educ:* Wellesley Col, BA, 68; Mass Inst Technol, PhD(chem), 72. *Prof Exp:* From asst prof to prof chem, Univ Colo, Denver, 86-89; PROF CHEM, UNIV DENVER, 90- *Concurrent Pos:* NSF fel, 69-71; vis prof women, NSF, 84. *Mem:* Sigma Xi; Am Chem Soc. *Res:* Dynamic processes in metallorporphyrins; synthesis and spectroscopy of inorganic compounds; electron paramagnetic resonance; computer simulations. *Mailing Add:* Dept Chem Univ Denver Denver CO 80208

EATON, STEPHEN WOODMAN, b Geneva, NY, Dec 22, 18; m 46. ORNITHOLOGY. *Educ:* Cornell Univ, PhD(zool), 49. *Prof Exp:* From asst prof to prof biol, 49-84, EMER PROF BIOL, ST BONAVENTURE UNIV, 84- *Concurrent Pos:* Ed, Sci Studies, 66-78. *Mem:* Emer mem NY Acad Sci; Am Ornith Union; Wilson Ornith Soc; Cooper Ornith Soc. *Res:* Vertebrate zoology; faunal studies; biology of Parulidae and of the wild turkey; Canandaigua Lake as an ecosystem. *Mailing Add:* Ten Mile Rd Allegany NY 14706

EATON, THOMAS ELDON, b Ironton, Mo, Nov 27, 48. ELECTRICAL FIRES & FAILURES. *Educ:* Univ Mo, Rolla, BS & MS, 70; Mass Inst Technol, MS, 74, ScD(nuclear eng), 75. *Prof Exp:* Asst prof mech/nuclear eng, Univ Ky, 75-78; CONSULT ENGR, EATON ENG CO, 79- *Concurrent Pos:* Lectr, Univ Ky, 91- *Mem:* Am Soc Mech Engrs; Inst Elec & Electronics Engrs; Nat Fire Protection Asn; Soc Mining Engrs. *Res:* Electrical fires and failures; building fire cause and development; lightning damages; energy systems engineering; mechanical systems failure analysis. *Mailing Add:* 400 W Maple St PO Box 100 Nicholasville KY 40340

EATON, WILLIAM ALLEN, b Philadelphia, Pa, June 4, 38; m 62; c 2. BIOPHYSICS, PHYSICAL BIOCHEMISTRY. *Educ:* Univ Pa, AB, 59, MD, 64, PhD(molecular biol), 67. *Prof Exp:* Surgeon, Lab Phys Biol, 68-70, sr staff fel biophys & phys biochem, 70-72, sr surgeon, 72-78, MED DIR, LAB CHEM PHYSICS, NAT INST ARTHRITIS, METAB & DIGESTIVE DIS, USPHS, 79-, CHIEF, SECT MACROMOLECULAR BIOPHYSICS, 79- *Concurrent Pos:* Vis prof, Dept Biochem & Molecular Biol, Harvard Univ, 76; adj prof, Dept Chem, Univ Pa, 85- *Mem:* Biophys Soc; Found Advan Educ Sci; Am Soc Biol Chemists; Am Chem Soc. *Res:* physical chemistry of protein polymerization; molecular pathophysiology and pharmacology of sickle cell disease; time resolved optical spectroscopy of macromolecules. *Mailing Add:* Lab Chem Physics Bldg 2 Rm B1-04 NIAMDD NIH Bethesda MD 20892

EATON, WILLIAM THOMAS, b Long Beach, Calif, Feb 22, 38; m 61; c 2. MATHEMATICS. *Educ:* Univ Utah, BS, 61, MS, 63, PhD(math), 67. *Prof Exp:* Asst prof math, Univ Tenn, Knoxville, 67-70; assoc prof, 70-77, PROF MATH, UNIV TEX, AUSTIN, 77- *Concurrent Pos:* Mem, Inst Advan Study, 69-70; Alfred P Sloan fel, 69-71. *Mem:* Am Math Soc; Sigma Xi. *Res:* Topology of manifolds, particularly embeddings of manifolds in three-manifolds; piecewise linear topology and combinatories; statistics and probability theory. *Mailing Add:* 8408 Silver Ridge Dr Austin TX 78759

EATOUGH, DELBERT J, b Provo, Utah, Sept 15, 40; m 64; c 7. CALORIMETRY. *Educ:* Brigham Young Univ, BS, 64, PhD(phys chem), 67. *Prof Exp:* Res chemist, Shell Develop Co, 67-70; dir, Thermochem Inst, 70-84, PROF CHEM, BRIGHAM YOUNG UNIV, 85- *Honors & Awards:* Sunner Award, Calorimetry Conf, 80. *Mem:* Am Chem Soc; AAAS; Sigma Xi; Calorimetry Conf; Air Pollution Control Asn. *Res:* Development of solution calorimetric instrumentation and application to study of surfactant, biochemical and surface chemistry; chemical characterization of atmospheric particulate matter resulting from anthropogenic activities. *Mailing Add:* 1252 N Uinta Dr Provo UT 84601

EATOUGH, NORMAN L, b Bingham Canyon, Utah, Oct 18, 33; m 56; c 5. PHYSICAL CHEMISTRY, ATMOSPHERIC SULFUR CHEMISTRY. *Educ:* Brigham Young Univ, BS, 57, BES, 58, MS, 59, PhD(phys chem), 68; Univ Wash, MSChE, 60. *Prof Exp:* Sr develop engr, Hercules Powder Co, 60-64; asst prof chem, Dixie Jr Col, 64-65; instr chem eng, Brigham Young Univ, 65-66; from asst prof to assoc prof, 68-80, PROF CHEM, CALIF STATE POLYTECH COL, 80-, DEPT CHMN. *Concurrent Pos:* Consult, Hercules Powder Co, 64-65. *Mem:* Am Chem Soc; Sigma Xi. *Res:* Atmospheric chemistry; high pressure chemistry. *Mailing Add:* 1508 Gulf St San Luis Obispo CA 93405

EAVES, ALLEN CHARLES EDWARD, b Ottawa, Ont, Feb 19, 41; m 81; c 4. HEMATOLOGY, ONCOLOGY. *Educ:* Acadia Univ, NS, BSc, 62; Dalhousie Univ, MSc, 64, MD, 68; Univ Toronto, PhD(med biophys), 74. *Prof Exp:* Intern med, Victoria Gen Hosp, Halifax, NS, 68-69; resident internal med, Univ Toronto, Ont, 74-75; resident internal med & oncol, Univ BC, 75-79, asst prof med, 79-83, assoc prof med & path, 83-88, PROF MED & PATH, UNIV BC, VANCOUVER, 88-; DIR, TERRY FOX LAB HEMAT-ONCOL, 81- *Concurrent Pos:* Head, Div Hemat, Univ BC, Vancouver Gen Hosp & BC Cancer Agency, 85- *Mem:* Fel Royal Col Physicians & Surgeons Can; fel Am Col Physicians & Surgeons; Am Soc Hemat; Am Asn Caner Res; Am Fedn Clin Res; Int Soc Exp Hemat; Int Soc Differentiation; Int Asn Comp Res Leukemia & Related Dis. *Res:* Regulation of growth and differentiation in normal and malignant hematopoiesis with special interests in leukemia and the role of autologous bone marrow transplantation. *Mailing Add:* Terry Fox Lab Hemat-Oncol 601 W Tenth Ave Vancouver BC V5Z 1L3 Can

EAVES, BURCHET CURTIS, b Shreveport, La, Nov 25, 38; div; c 2. OPERATIONS RESEARCH. *Educ:* Carnegie Inst Technol, BS, 61; Tulane Univ, MBA, 65; Stanford Univ, MS & PhD(oper res), 69. *Prof Exp:* Asst prof bus admin & oper res, Univ Calif, Berkeley, 68-70; from asst prof to assoc prof oper res, Stanford Univ, 70-75; vis assoc prof econ & org mgt, Yale Univ, 74-75; PROF OPER RES, STANFORD UNIV, 75- *Concurrent Pos:* Guggenheim fel. *Mem:* Inst Mgt Sci; Oper Res Soc Am; Soc Indust Appl Math; Am Math Soc; Math Asn Am; Sigma Xi. *Res:* Solving equations arising in optimization, economics and game theory. *Mailing Add:* Dept Opers Res Stanford Univ Stanford CA 94305

EAVES, DAVID MAGILL, b New York, NY, Dec 7, 33; m 57, 66; c 4. APPLIED STATISTICS. *Educ:* Mass Inst Technol, BSc, 56; Univ Wash, Seattle, MSc, 63, PhD(math), 66. *Prof Exp:* ASSOC PROF MATH, SIMON FRASER UNIV, 82- *Mem:* Am Stat Asn. *Res:* Mathematical statistical interference; applications. *Mailing Add:* Dept Math Simon Fraser Univ Burnaby BC V5A 1S6 Can

EAVES, GEORGE NEWTON, b Athens, Tenn, Mar 12, 35. MEDICAL MICROBIOLOGY. *Educ:* Univ Chattanooga, BA, 57; Univ Tenn, MS, 59; Wayne State Univ, PhD(med microbiol), 62. *Prof Exp:* Asst prof biol, Washington & Jefferson Col, 62-63; grants assoc, Div Res Grants, NIH, 65-66, health scientist adminr, Div Res Facil & Resources, 66-67, exec secy molecular biol study sect, Div Res Grants, 67-73, exec secy adv coun, Nat Heart & Lung Inst, 73-74; staff mem, President's Biomed Res Panel, 74-76; asst to dir, 76-80, dep dir, Div Blood Dis & Resources, Nat Heart, Lung & Blood Inst, 80- 82, DEP DIR, DIV STROKE & TRAUMA, NAT INST NEUROL DIS & STROKE, NIH, 82- *Concurrent Pos:* Fel microbiol, Bryn Mawr Col, 63-65; guest investr, Rockefeller Univ, 70-71. *Honors & Awards:* Director's Award, NIH, 76, 86. *Mem:* Sigma Xi. *Res:* Exocellular enzymes of bacteria; metabolic effects of bacterial endotoxins. *Mailing Add:* Fed Bldg Rm 8A-13 Nat Insts Health Bethesda MD 20892

EAVES, JAMES CLIFTON, b Hillside, Ky, June 26, 12; m 38; c 2. ALGEBRA. *Educ:* Univ Ky, AB, 35, MA, 41; Univ NC, PhD(matrix algebra), 49. *Prof Exp:* Asst prof math, Univ Ala, 49-50; assoc prof, Ala Polytech Inst, 50-51, res assoc prof, 51-52; prof math & res assoc, Auburn Res Found, 52-53; prof math & admin asst, Ala Polytech Inst, 53-54; prof math & astron, Univ Ky, 54-67, head dept, 54-63; chmn dept, 67-72, centennial prof math, 67-77, EMER CENTENNIAL PROF MATH, WVA UNIV, 77- *Mem:* Am Math Soc; Math Asn Am. *Res:* Matrices; simultaneous reductions; inverse approximations; computer analysis; patents and patent law; space trajectories and transformations; higher dimensional matrices. *Mailing Add:* 557 Culpepper Rd Lexington KY 40502-2413

EAVES, REUBEN ELCO, b Baltimore, Md, Jan 20, 44. ELECTRICAL ENGINEERING. *Educ:* Johns Hopkins Univ, BES, 64; Brown Univ, ScM, 66, PhD(elec eng), 69. *Prof Exp:* Electronics engr, NASA Electronics Res Ctr, 68-70; US Dept Transp Transp Systs Ctr, 70-76; staff mem, Lincoln Lab, Mass Inst Technol, 76-81; mgr satellite design, Commun Satellite Corp, 81-84; DEPT HEAD, ADV COMMUN, MITRE CORP, 84- *Concurrent Pos:* Vis asst prof, Div Eng, Brown Univ, 73-76, sr res assoc, 76-77. *Mem:* Sigma Xi; Inst Elec & Electronics Engrs; Am Inst Aeronaut & Astronaut. *Res:* Communication systems, satellite communications, electromagnetic theory and applied mathematics; advanced communication systems employing various media including satellites and HF channels; new system concepts for HF communications. *Mailing Add:* 987 Mem Dr 672 Cambridge MA 02138

EBACH, EARL A, b Saginaw, Mich, May 13, 28; m 53; c 4. CHEMICAL ENGINEERING. *Educ:* Univ Mich, BS, 51, MS, 52, PhD(chem eng), 57. *Prof Exp:* Res & Develop, Chem Engr, Dow Chem Co, 57-72, proj leader, 62, group leader, 67, process engr, 72-84, proj mgr, 82-83, res assoc, 83, TECH SERV & DEVELOP, DOW CHEM CO, 84- *Mem:* Am Inst Chem Engrs. *Res:* Research and development in organic chemicals. *Mailing Add:* 4610 Andre Midland MI 48640

EBADI, MANUCHAIR, b Shahmirzad, Iran, Sept 6, 35; m 59; c 3. NEUROPHARMACOLOGY, NEUROCHEMISTRY. *Educ:* Park Col, BS, 60; Univ Mo, Kansas City, MS, 62; Univ Mo, Columbia, PhD(pharmacol), 66. *Prof Exp:* Res asst pediat, Sch Med, Univ Mo, Columbia, 64-66, res assoc pediat & instr pharmacol, 66-67; asst prof pediat, 67-68, from asst prof to assoc prof pharmacol, 68-71, actg chmn dept, 70-71, PROF PHARMACOL & CHMN DEPT, UNIV NEBR MED CTR, 71- *Concurrent Pos:* NIMH int prof fel, 69-70; mem, US Pharmacopoeial Conv, 70. *Honors & Awards:* AMA Golden Apple Award, 71, 72 & 75. *Mem:* AAAS; Am Chem Soc; Am Soc

Pharmacol & Exp Therapeut; Am Soc Neurochem; fel Am Soc Clin Pharmacol. *Res:* Neurochemical pharmacological aspects of central nervous system drugs. *Mailing Add:* Dept Pharmacol & Neurol Univ Nebr Col Med 600 S 42nd St Omaha NE 68198-6260

EBAUGH, FRANKLIN G, JR, medical administration, hematology; deceased, see previous edition for last biography

EBBE, SHIRLEY NADINE, PLATELET PRODUCTION, VASCULAR DISEASES. *Educ:* Univ Ore, MD, 54. *Prof Exp:* PROF LAB MED, LAWRENCE BERKELEY LAB, 77- *Mailing Add:* Lawrence Berkeley Lab Bldg 74 Berkeley CA 94720

EBBERT, ARTHUR, JR, b Wheeling, WVa, Aug 25, 22. INTERNAL MEDICINE. *Educ:* Univ Va, BA, 44, MD, 46; Yale Univ, MA, 71. *Prof Exp:* Instr & asst to dean sch med, Univ Va, 52-53; instr & asst dean med, 53-54, asst prof & asst dean, 54-63, assoc dean, 60-74, assoc prof, 63-71, dep dean, 74-88, PROF MED, SCH MED, YALE UNIV, 71- *Concurrent Pos:* Physician, Univ Va Hosp, 52-53; assoc physician, Grace-New Haven Hosp, 53-60, asst attend physician, 60-68; consult, Waterbury Hosp, 57-; attend physician, Yale-New Haven Hosp, 68-87, hon med staff, 87- *Mem:* Asn Am Med Cols. *Res:* Medical education. *Mailing Add:* 64 N Lake Dr Hamden CT 06517

EBBESEN, LYNN ROYCE, b Rapid City, SDak, Jan 27, 48; m 72; c 1. MECHANICAL ENGINEERING. *Educ:* SDak Sch Mines & Technol, BS, 70; Okla State Univ, MS, 72, PhD(mech eng), 76. *Prof Exp:* From asst prof to assoc prof, Sch Mech & Aerospace Eng, Okla State Univ, 77-81; SR ENG, SCI SIMULATION INC, ALBUQUERQUE, NMEX, 81- *Mem:* Am Soc Mech Engrs; Nat Soc Prof Engrs. *Res:* Automatic control; digital simulation of large scale systems. *Mailing Add:* Logicon RDA 2600 Yale Blvd SE Albuquerque NM 87106

EBBESMEYER, CURTIS CHARLES, b Los Angeles, Calif, Apr 24, 43; m 65; c 2. OCEANOGRAPHY. *Educ:* Calif State Univ, Northridge, BS, 66; Univ Wash, MS, 68, PhD(oceanog), 73. *Prof Exp:* Sr res engr & oceanogr, Mobil Res & Develop Corp, 69-72, assoc res engr & oceanogr, 72-74; mgr phys oceanog, 74-85, VPRES RES, EVANS-HAMILTON, INC, 85- *Concurrent Pos:* Asst adj prof, Marine Sci Res Ctr, State Univ NY Stony Brook, 74-; consult, Univ Wash, Univ BC, Mobil Oil Can, Ltd, Nat Oceanog & Atmospheric Admin, Dept Interior & City of Seattle, 74-75. *Mem:* Sigma Xi; Am Geophys Union. *Res:* Physical oceanographic studies of waves, tides, currents, icebergs, and shipwrecks diffusion processes; oil spill phytoplankton dynamics; physical oceanography of Gulf of Mexico, Gulf Stream, Puget Sound & North Atlantic Ocean. *Mailing Add:* 6306 21st Ave NE Seattle WA 98115

EBBESSON, SVEN O E, b Backaby, Sweden, Oct 14, 37; US citizen; m 62; c 3. NEUROANATOMY. *Educ:* Southwestern Col, Kans, BA, 57; Univ Md, PhD(anat), 64. *Prof Exp:* Asst anat, Tulane Univ, 58-60; neuroanatomist, Walter Reed Army Med Ctr, 62-65; asst neuroanat, Sch Med, Univ Md, 63-64, instr, 64-65; neuroanatomist, Lab Perinatal Physiol, Nat Inst Neurol Dis & Blindness, 65-69; from assoc prof to prof neurosurg & anat, Sch Med, Univ Va, 69-77; assoc dean biomed sci, Med Sci Campus, Univ PR, San Juan, 77-; DIR WANL MED EDUC PROG, UNIV ALASKA. *Concurrent Pos:* Vis asst prof, Sch Med, Univ PR & hon mem, Inst Marine Biol, 66-70. *Mem:* AAAS; Am Asn Anat; Int Soc Stereol. *Res:* Comparative neurology; stereology. *Mailing Add:* Univ Alaska PO Box 80107 Fairbanks AK 99708

EBBIN, ALLAN J, b New York, NY, May 2, 38; m; c 3. PEDIATRICS. *Educ:* NY Univ, BA, 60; State Univ NY Upstate Med Ctr, MD, 64; Univ Calif, Los Angeles, MPH, 76; Am Bd Pediat, cert, 69; Am Bd Prev Med, cert, 77. *Prof Exp:* Instr pediat, Emory Univ, 67-69; asst prof, 69-73, ASSOC PROF PEDIAT, SCH MED, UNIV SOUTHERN CALIF, 73-, ASST DIR GENETICS DIV, LOS ANGELES COUNTY-MED CTR, UNIV SOUTHERN CALIF, 69- *Concurrent Pos:* Actg chief, Congenital Malformations Unit, Leukemia Sect, Ctr Dis Control, Atlanta, Ga, 67-69; proj officer sequelae of rubella & birth defects, Nat Ctr Dis Control, 69-73; asst proj dir, Cytogenetics Lab, Child Develop & Ment Retardation Ctr, Maternal & Child Health Serv, 69-, actg dir, 77; partic, Crippled Children's Serv Panel, 70-; prin investr, Am Acad Pediat, Nat Inst Environ Health Sci, 72-75; mem adv comt, Sickle Cell Coun, Los Angeles County Health Dept, 72-; asst prog dir fetal diag high risk pregnancies, Nat Found-March Dimes, 72-, actg dir, 77; mem, Calif Tay-Sachs Dis Prev Prog, 74-; Nat Cancer Inst fel, Univ Calif, Los Angeles, 75-76; consult to var hosps. *Mem:* Am Soc Human Genetics; Soc Pediat Res; Am Acad Pediat. *Res:* Preventive medicine and genetics; epidemiology as applied to clinical genetics. *Mailing Add:* LA Co Gen Hosp 1200 N State St Los Angeles CA 90033

EBBING, DARRELL DELMAR, b Peoria, Ill, July 1, 33; m 55; c 3. CHEMISTRY. *Educ:* Bradley Univ, BS, 55; Ind Univ, PhD(phys chem), 60. *Prof Exp:* Res assoc chem, Ind Univ, 60-62; from asst prof to assoc prof, 62-69, PROF CHEM, WAYNE STATE UNIV, 69- *Mem:* Am Chem Soc; AAAS; Sigma Xi. *Res:* Quantum mechanical study of molecular properties and chemical binding; author of general chemistry book. *Mailing Add:* Dept Chem Wayne State Univ Detroit MI 48202

EBBS, JANE COTTON, b Newport, RI, May 11, 12. PHYSIOLOGY, NUTRITION. *Educ:* Univ RI, BS, 35, MS, 37. *Prof Exp:* Asst instr nutrit, Univ RI, 38-39; asst home econ, USDA, 42; nutrit adv, Off Qm Gen, Dept Army, 42-49, spec feeding & nutrit adv, 49-62; spec asst to dir, Defense Supply Agency, 62-65; adv, 65-70, CHIEF PROG PLANNING & EVAL, NUTRIT DIV, FOOD & AGR ORGN, UN, 70- *Mem:* Am Chem Soc; Am Inst Nutrit; assoc fel Am Astronaut Soc; fel Am Pub Health Asn; Am Dietetic Asn. *Res:* Vitamin A requirement of young adults; space nutrition; food processing methods; world food problems and new foods; feeding the armed forces. *Mailing Add:* 707 Gulfstream Ave 408 Sarasota FL 33536

EBDON, DAVID WILLIAM, b Detroit, Mich, Apr 9, 39; m 67; c 3. PHYSICAL CHEMISTRY. *Educ:* Univ Mich, Ann Arbor, BS, 61; Univ Md, PhD(phys chem), 67. *Prof Exp:* Lectr chem, Univ Md, 67-68; from asst prof to assoc prof phys chem, 68-80, PROF PHYS CHEM, EASTERN ILL UNIV, 80-, CHMN, DEPT CHEM, 77- *Concurrent Pos:* Sr res scientist, Nat Biomed Res Found, 67-68; res scientist, Univ Texas, Austin, 78-79. *Mem:* Am Chem Soc; Royal Soc Chem. *Res:* Thermodynamic and kinetic properties of electrolyte solutions; chemical oceanography; measurement of ionic association constants; computer modeling of natural water systems; ionic activity in multicomponent electrolyte solutions. *Mailing Add:* Dept Chem Eastern Ill Univ Charleston IL 61920

EBEL, MARVIN EMERSON, b Waterloo, Iowa, Sept 23, 30; m 60; c 4. THEORETICAL HIGH ENERGY PHYSICS. *Educ:* Iowa State Univ, BS, 50, MS, 52, PhD(physics), 53. *Prof Exp:* NSF fel, Inst Theoret Physics, Copenhagen, 53-54; from instr to asst prof, Yale Univ, 54-57; from asst prof to assoc prof, 57-64, PROF PHYSICS, UNIV WIS-MADISON, 64-, ASSOC DEAN GRAD SCH, 76- *Concurrent Pos:* Sloan Found fel, 57-62; mem bd dirs, Coun Govt Rels, 84-, chmn, 88- *Mem:* Fel Am Phys Soc; Sigma Xi. *Res:* High energy physics. *Mailing Add:* Dept Physics Univ Wis Madison WI 53706

EBEL, RICHARD E, b Milwaukee, Wis, Oct 21, 46. BIOCHEMICAL TOXICOLOGY. *Educ:* Univ Wise, PhD(biochem), 74. *Prof Exp:* Asst prof, 77-82, ASSOC PROF BIOCHEM, VA TECH, 82- *Mem:* Am Soc Biol Chemists; Am Chem Soc. *Mailing Add:* Dept Biochem & Nutrit Va Tech Blacksburg VA 24061

EBELING, ALFRED W, b Anaheim, Calif, Mar 30, 31; m 56; c 2. ZOOLOGY. *Educ:* Univ Calif, Los Angeles, BS, 54, PhD(zool), 60. *Prof Exp:* Asst prof biol, Yale Univ, 60-63; from asst prof to assoc prof zool, 63-72, contracts & grants officer, 66-67, PROF ZOOL, UNIV CALIF, SANTA BARBARA, 72- *Concurrent Pos:* NSF grants, 61-65, 66-68, 73-75 & 76-78; assoc, Los Angeles County Mus Natural Hist. *Mem:* AAAS; Am Soc Zoologists; fel Am Soc Fisheries Biologists; Am Soc Ichthyologists & Herpetologists; Am Soc Limnol & Oceanog; Am Soc Naturalists. *Res:* Ichthyology; marine ecology. *Mailing Add:* Dept Biol Sci & Marine Sci Inst Univ Calif Santa Barbara CA 93106

EBERHARD, ANATOL, b Istanbul, Turkey, Nov 13, 38; US citizen; m 64. ORGANIC CHEMISTRY, BIOCHEMISTRY. *Educ:* Univ Calif, Berkeley, BA, 59; Harvard Univ, MA, 60, PhD(chem), 64. *Prof Exp:* NIH fel, Univ Calif, Berkeley, 64-66; asst prof biol, Harvard Univ, 66-71; assoc prof chem, Fairleigh Dickinson Univ, 71-72; assoc prof, 72-84, PROF CHEM, ITHACA COL, 84- *Concurrent Pos:* Woodrow Wilson fel, 59-60. *Mem:* AAAS; Am Chem Soc. *Res:* Biochemistry of bacterial bioluminescence. *Mailing Add:* Dept Chem Ithaca Col Ithaca NY 14850

EBERHARD, EVERETT, b Topeka, Kans, Mar 15, 15; m 41; c 2. ELECTRICAL ENGINEERING. *Educ:* Univ Kans, BS, 36; Yale Univ, ME, 38. *Prof Exp:* Instr elec eng, SDak State Col, 39-40; elec engr, Hobart Bros Co, 40; instr, USAF Radio & Radar Schs, 40-42; sr engr, Victor Div, Radio Corp Am, 46-50; sect head in chg integrated circuits, Motorola Inc, 60-67, sr engr & proj leader systs develop, Western Mil Electronic Ctr, 50-80, sr elec engr, Tactical Electronics Dept, 67-80; RETIRED. *Mem:* Sr mem Inst Elec & Electronics Engrs. *Res:* Automation in field of test equipment; design of transistor oscillator circuits; application of all types of integrated circuits to military equipment. *Mailing Add:* Rte 2 Box 1711 Lakeside AZ 85929

EBERHARD, JEFFREY WAYNE, b New Braunfels, Tex, Feb 21, 50; m 79; c 2. X-RAY COMPUTED TOMOGRAPHY. *Educ:* Univ Tex, Austin, BA, 72; Univ Chicago, SM, 74, PhD(physics), 78. *Prof Exp:* PHYSICIST, GEN ELEC RES & DEVELOP CTR, 77- *Mem:* Am Phys Soc. *Res:* Nondestructive evaluation and materials characterization; x-ray computed tomography systems. *Mailing Add:* Gen Elec Co Res & Develop Ctr Bldg KW Rm 0259 Schenectady NY 12301

EBERHARD, PHILIPPE HENRI, b Lausanne, Switz, July 8, 29; m 54; c 2. PARTICLE PHYSICS. *Educ:* Polytech Sch, Univ Lausanne, Dipl, 52; Univ Paris, PhD(sci), 57. *Prof Exp:* Researcher physics, Nat Ctr Sci Res & Col France, 55-63; physicist, Lawrence Berkeley Lab, 63-71; vis scientist particle physics, Europ Orgn Nuclear Res, Geneva, Switz, 71-72; PHYSICIST, LAWRENCE BERKELEY LAB, 72- *Mem:* Am Phys Soc. *Res:* Search for magnetic monopoles; foundations of quantum theory; electron-positron physics; superconducting magnets. *Mailing Add:* Univ de Lausanne BSP Dorigny Lausanne CH 1015 Switzerland

EBERHARD, WYNN LOWELL, b Nampa, Idaho, June 10, 1944; m 68; c 3. REMOTE SENSING, TURBULENCE & DIFFUSION. *Educ:* Brigham Young Univ, BS, 69; Univ Ariz, MS, 76, PhD(atmospheric sci), 79. *Prof Exp:* PHYSICIST, ENVIRON RES LAB, WAVE PROPAGATION LAB, NAT OCEANIC & ATMOSPHERIC ADMIN, 79- *Mem:* Am Meteorol Soc; Optical Soc Am. *Res:* Development and application of optical, acoustic and radio remote sensors to the study of the motion, state and constituents of the atmosphere. *Mailing Add:* R-E-WP2 Nat Oceanic & Atmospheric Admin 325 Broadway Boulder CO 80303

EBERHARDT, ALLEN CRAIG, b Cincinnati, Ohio, Aug 30, 50; m 73; c 2. AUTOMATED SYSTEMS DESIGN, COMPUTER SCIENCE. *Educ:* NC State Univ, BS, 72, MS, 75, PhD(mech eng), 77. *Prof Exp:* Machinist, Precision Metalcrafts, Inc, 68-72; res asst, dept mech & aero eng, NC State Univ,72-77, from asst prof to assoc prof mech eng design, 77-85, dir, Integrated Mfg Systs Eng Inst, 84-86 INST, NC STATE UNIV, 84- *Concurrent Pos:* Chmn mfg systs, Nat Technol Univ, 85- *Honors & Awards:* Ralph A Teetor Award, Soc Automotive Engrs, 80. *Mem:* Soc Mfg Engrs; Am Soc Mech Engrs; Soc Automotive Engrs; Acoust Soc Am. *Res:* System dynamics, vibration and acoustics; robotic systems control and design; software for CAD, CAM and engineering analysis; tire structural dynamics and vibro-acoustic response; real time control systems; design for automation and assembly. *Mailing Add:* MAE Dept Raleigh NC 27615

EBERHARDT, KEITH RANDALL, b Los Angeles, Calif, July 10, 47; m 68; c 3. STATISTICAL APPLICATIONS IN PHYSICAL SCIENCES. *Educ:* Case Western Reserve Univ, BS, 69; Johns Hopkins Univ, PHD(statist), 75. *Prof Exp:* Instr statist, Ohio State Univ, 73-75, asst prof, 75-78; MATH STATISTICIAN, NAT INST STANDARDS & TECHNOL, 78- *Honors & Awards:* Bronze Medal, US Dept of Com, 85. *Mem:* Am Statist Asn; Inst Math Statist. *Res:* Prediction approach to survey sampling theory using linear regression models; linear models; propagation of error; general statistical inference; statistical tolerance interval estimation. *Mailing Add:* Rm A337 Admin Bldg Statist Eng Div Nat Inst Standards & Technol Gaithersburg MD 20899

EBERHARDT, LESTER LEE, b Valley City, NDak, Oct 15, 23; m 44; c 3. BIOLOGY. *Educ:* NDak State Teachers Col, BS, 46; Mich State Univ, PhD(wild life mgt), 60. *Prof Exp:* Biometrician, Game Div, State Dept Conserv, Mich, 52-61; fel, Univ Calif, Berkeley, 61-62; ECOLOGIST, PAC NORTHWEST LAB, BATTELLE MEM INST, 62- *Concurrent Pos:* Affil prof, Col Fisheries, Univ Wash, 74-; mem comt sci adv, US Marine Mammal Comn, 76- *Mem:* Fel AAAS; Wildlife Soc; Sigma Xi; Ecol Soc Am; Brit Ecol Soc. *Res:* Animal population research; statistical studies in ecology. *Mailing Add:* 2528 W Klamath Kennewick WA 99336

EBERHARDT, MANFRED KARL, b Heidenheim, Ger; Dec 5, 30. ORGANIC CHEMISTRY. *Educ:* Univ Tübingen, PhD(org chem),57. *Prof Exp:* Res assoc, Univ Chicago, 57-59, Univ Ark, 59-60 & Univ Notre Dame, 60-62; fel radiation chem & org chem, Mellon Inst, 62-64; scientist, Munich Tech Univ, 65-67; scientist, PR Nuclear Ctr, 67-76; assoc prof, Univ PR, Cayey, 76-78; investr, Sch Med Cancer Ctr & Dept Math, 78-87, ASSOC PROF, DEPT PATH, UNIV PR MED SCI CAMPUS, SAN JUAN, 87- *Concurrent Pos:* Consult, PR Dept Health, 78-79. *Mem:* Am Chem Soc. *Res:* Mechanism of organic free radical reactions; free radicals in biology and medicine; environmental carcinogenesis; radiation chemistry. *Mailing Add:* Univ PR Med Sci Campus Dept Path GPO Box 5067 San Juan PR 00936

EBERHARDT, NIKOLAI, b Rakvere, Estonia, July 2, 30; m 56; c 4. MICROWAVE ELECTRONICS, ELECTROMAGNETICS. *Educ:* Univ Munich, dipl physics, 57; Munich Inst Technol, PhD(physics), 62. *Prof Exp:* Res engr, Siemens und Halske A G, Ger, 56-62; assoc prof, 62-70, PROF ELEC ENG, LEHIGH UNIV, 70- *Concurrent Pos:* Consult, Bell Tel Labs, 63- *Res:* Physics of magnetically confined electron beams; color display tubes; theoretical and experimental investigations in the area of passive microwave devices, especially ferrite devices and filters. *Mailing Add:* Dept Elec Eng Lehigh Univ Bethlehem PA 18015

EBERHARDT, WILLIAM HENRY, b Montclair, NJ, Feb 11, 20; m 46; c 3. QUANTUM CHEMISTRY. *Educ:* Johns Hopkins Univ, AB, 41; Calif Inst Technol, PhD(phys chem), 45. *Prof Exp:* Asst, Calif Inst Technol, 41-44, instr, 44-46; assoc dean, Col Sci & Lib Studies, 70-77, from asst prof to prof 46-84, EMER PROF CHEM, GA INST TECHNOL, 84- *Concurrent Pos:* Hon fel, Univ Minn, 53; vis prof, Harvard Univ, 64, Calif Inst Technol, 78. *Mem:* AAAS; Am Chem Soc; Am Phys Soc. *Res:* Molecular structure. *Mailing Add:* Sch Chem Ga Inst Technol Atlanta GA 30332

EBERHART, H(OWARD) D(AVIS), b Lima, Ohio, Aug 16, 06; m 61; c 2. CIVIL ENGINEERING, BIOLOGICAL ENGINEERING. *Educ:* Univ Ore, BS, 29; Ore State Col, MS, 35. *Prof Exp:* Coach & instr, high sch, 29-33; jr topog engr, US Geol Surv, 34; mem staff, US Eng Off, Bonneville Dam, 35-36; from instr to prof civil eng, Univ Calif, Berkeley, 36-74, chmn dept, 59-63 & 71-74; res engr & vchmn biomech lab, Univ Hosp, San Francisco, 70-76; chmn, dept civil eng, King Abdulaziz Univ, Saudi Arabia, 76-79; EMER PROF CIVIL ENG, UNIV CALIF, BERKELEY, 74-; PROF MECH ENG, UNIV CALIF, SANTA BARBARA, 80- *Concurrent Pos:* Consult, Consol Vultee Aircraft Corp, 43-44; in-charge res concrete pavement invests, Hamilton Field, 44; consult res proj, Comt Prosthetic Res & Develop, Nat Res Coun, 45-76, mem-at-large, Div Eng & Indust Res, 60-66. *Honors & Awards:* Fulbright lectr, Univ Assiut, 64-65. *Mem:* Nat Acad Eng; fel Am Soc Civil Engrs; Am Soc Eng Educ; Am Concrete Inst; Soc Exp Stress Analysis. *Res:* Structural engineering; biomechanics; experimental stress analysis; prosthetic devices; engineering education development and evaluation. *Mailing Add:* Dept Mech Eng Univ Calif Santa Barbara CA 93106

EBERHART, JAMES GETTINS, b Columbus, Ohio, Feb 6, 36; c 2. PHYSICAL CHEMISTRY, SURFACE CHEMISTRY. *Educ:* Ohio State Univ, BSc, 57, PhD(chem), 63. *Prof Exp:* Tech staff mem mat sci, Sandia Labs, Albuquerque, NMex, 63-68; chemist, Argonne Nat Lab, Ill, 68-78; assoc prof, 78-81 PROF CHEM, AURORA UNIV, 81-; AT DEPT CHEM, UNIV COLO, COLORADO SPRINGS. *Concurrent Pos:* Consult, Argonne Nat Lab, 78-79. *Mem:* Am Chem Soc. *Res:* Wetting behavior of liquids; limit of superheat of liquids; equations of state; surface tension of liquids and solids; surface diffusion; critical properties of fluids. *Mailing Add:* Dept Chem Univ Colo PO Box 7150 Colorado Springs CO 80933-7150

EBERHART, PAUL, mathematics, for more information see previous edition

EBERHART, ROBERT CLYDE, b Oakland, Calif, Apr 17, 37; m 63; c 3. BIOMATERIALS, HEAT & MASS TRANSFER. *Educ:* Harvard Univ, AB, 58; Univ Calif, Berkeley, MS, 60, PhD(mech eng), 65. *Prof Exp:* Sr scientist biomed eng, Inst Med Sci, 64-75; assoc prof mech eng, Univ Tex, Austin, 75-76; ASSOC PROF & CHMN BIOMED ENG, SOUTHWESTERN MED SCH, UNIV TEX, ARLINGTON, 85- *Concurrent Pos:* Adj prof biomed eng, Univ Tex, Austin, 76-, Southern Methodist Univ, 78- & Univ Tex, Arlington, 79- *Mem:* Am Soc Mech Eng; Am Soc Artificial Internal Organs; Soc Critical Care Med; Inst Elec & Electronics Engrs; Am Soc Eng Educ. *Res:* Biomaterials; cardiopulmonary assist devices; ion sensing field effect transistors; computers in critical care medicine; heat and mass transfer analysis in medicine. *Mailing Add:* Dept Surg Southwestern Med Sch 5323 Harry Hines Blvd Dallas TX 75235

EBERHART, ROBERT J, b Lock Haven, Pa, Sept 9, 30; m 53; c 4. VETERINARY MEDICINE. *Educ:* Cornell Univ, AB, 52; Univ Pa, VMD, 59; Pa State Univ, PhD(physiol), 66. *Prof Exp:* Instr vet sci, Pa State Univ, 59-63; fel, Am Vet Med Asn, 63-65; from asst prof to assoc prof, 66-78, PROF VET SCI, PA STATE UNIV, 78- *Mem:* AAAS; Am Vet Med Asn. *Res:* Bovine mastitis. *Mailing Add:* 104 Mahola St Port Matilda PA 16870

EBERHART, STEVE A, b Keya Paha, SDak, Nov 11, 31; m 53; c 4. GENETICS, STATISTICS. *Educ:* Univ Nebr, BSc, 52, MSc, 58; NC State Univ, PhD(genetics, statist), 61. *Hon Degrees:* DSc, Univ Nebr, 88. *Prof Exp:* Res geneticist, Iowa State Univ, USDA, 61-64, Nat Agr Res Sta, Kenya, 64-68 & Agr Res Serv, 68-74, res geneticist, Agr Res Serv, Iowa State Univ, USDA, 74-75; assoc dir res, 75-77, vpres res, 78-85, vpres Int Tech, Funk Seeds Int, 85-87; DIR NAT SEED STORAGE LAB, USDA, 87- *Honors & Awards:* Arthur S Flemming Award, 70. *Mem:* Am Soc Agron; Am Hort Soc; CAST. *Res:* Statistical genetics of maize, including estimation of additive, dominance and epistatic variances; development of a model to study the gene action in diallels of fixed varieties and to predict variety and variety cross performance. *Mailing Add:* Nat Seed Storage Lab USDA-Agr Res Ctr Ft Collins CO 80523

EBERLE, HELEN I, b Oakland, Calif, Mar 2, 32; m 58; c 3. MOLECULAR BIOLOGY, BIOPHYSICS. *Educ:* Calif State Col, Los Angeles, BS, 56; Univ Calif, Los Angeles, PhD(microbiol), 65. *Prof Exp:* Pub Health microbiologist, Los Angeles County Health Dept, 56-60; NIH fels, Kans State Univ, 65-67, Univ Rochester, 67-68; from instr to asst prof, 68-76, ASSOC PROF RADIATION BIOL & BIOPHYS, UNIV ROCHESTER, 76- *Concurrent Pos:* Am Cancer Soc grant, 70-72; dir educ, Dept Radiation Biol & Biophys, Sch Med, Univ Rochester, 76-; res contract with Energy Res & Develop Admin-Dept Energy, 70-; mem, Nat Res Coun Comt on Pure & Appl Biophysics, 78-; consult, Chevron Corp, 80-83; NIH grant, 81-84. *Honors & Awards:* Faculty Res Award, Am Cancer Soc, 69. *Mem:* AAAS; Am Soc Microbiol; Biophys Soc; Sigma Xi. *Res:* Mechanism and regulation of DNA replication in Escherichia coli with emphasis on the characterization proteins involved in the initiation process; molecular genetics. *Mailing Add:* Dept Biophys Univ Rochester Sch Med Rochester NY 14642

EBERLE, JON WILLIAM, b Chillicothe, Ohio, Aug 28, 34; m 56; c 4. BIOMEDICAL ENGINEERING, ELECTROMAGNETICS. *Educ:* Ohio State Univ, BS, 57, MS, 60, PhD(elec eng), 64. *Prof Exp:* Res assoc phased arrays, Antenna Lab, Ohio State Univ, 57-61, assoc supvr, 61-65; mem tech staff, Tex Instruments Inc, 65-66, mgr advan radar develop br, 66, mgr surface systs dept, 66-68, mgr corp mkt, 68-69; vpres biomed, Intermed Corp, 69-70; consult to dean biomed & elec eng, 70-71, assoc prof biomed & elec eng, 71-76, ADJ PROF ELEC ENG, SOUTHERN METHODIST UNIV, 76- *Concurrent Pos:* Fel, Dept Bus Admin, Ohio State Univ, 64-65; res scientist, Div Thoracic & Cardiovasc Surg, Med Sch, Univ Tex, 70-; mem subpanel on elec safety, Am Nat Stand Inst, 70-; elec engr, Vet Hosp, Dallas, 71- *Mem:* Inst Elec & Electronics Eng; Asn Advan Med Instrumentation. *Res:* Membrane oxygenators having heparin ionically bound to their surfaces; continuous monitoring of pH of blood and partial pressures of O2 and CO2 in blood; computerized medical records; x-ray holography. *Mailing Add:* One Kingsgate Ct Dallas TX 75225

EBERLEIN, GEORGE DONALD, b New Brunswick, NJ, Nov 21, 20; m 69; c 1. GEOLOGY. *Educ:* Yale Univ, BS, 42. *Prof Exp:* Geologist, Big Sandy Mine, Inc, Ariz, 41; geologist metals sect, US Geol Surv, 42-47, Br Alaskan Mineral Resources, 47-53, staff geologist mineral deposits, 53-57, asst br chief, 57-59, br chief, 59-63, res geologist, Br Alaskan Geol, 63-81; RETIRED. *Concurrent Pos:* Teaching fel & asst, Stanford Univ; Binney fel & Penfield Prize, Yale Univ. *Mem:* Fel Geol Soc Am; Mineral Soc Am; Soc Econ Geol; Mineral Asn Can; Int Union Geol Sci; fel Geol Soc London; Sigma Xi. *Res:* Petrology of igneous and metamorphic rocks; mineral resources of Alaska; optical crystallography; Precambrian rocks of Alaska. *Mailing Add:* PO Box 1064 Los Gatos CA 95031-9998

EBERLEIN, PATRICIA JAMES, b Washington, DC, July 15, 25; m 46, 56; c 7. NUMERICAL ANALYSIS, COMPUTER SCIENCE. *Educ:* Univ Chicago, BS, 44; Mich State Univ, PhD(math), 55. *Prof Exp:* Instr math, Wayne Univ, 55-56; mathematician, Inst Advan Study, 56-57; res assoc, Comput Ctr, Univ Rochester, 57-61, asst dir anal, 61-68; assoc prof math & comput sci, 68-74, chmn dept, 81-84, PROF COMPUT SCI, STATE UNIV NY BUFFALO, 74- *Concurrent Pos:* NSF vis prof, Cornell Univ, 84-85. *Mem:* Am Math Soc; Soc Indust & Appl Math; Asn Comput Mach; AAAS; Math Asn Am; Inst Elec & Electronics Engrs Comput Soc; Asn Women Math. *Res:* Numerical linear algebra; complexity and analysis of algorithms. *Mailing Add:* State Univ NY 226 Bell Hall Buffalo NY 14260

EBERLEIN, PATRICK BARRY, b San Francisco, Calif, March 3, 44; m 68; c 2. DIFFERENTIAL GEOMETRY. *Educ:* Harvard Col, AB, 65; Univ Calif, Los Angeles, MA, 67, PhD(math), 70. *Prof Exp:* Instr & lectr math, Univ Calif, Los Angeles, 70; lectr, Univ Calif, Berkeley, 70-71; vis researcher, Univ Bonn, WGer, 71-72; lectr, Univ Calif, Berkeley, 72-73; from asst prof to assoc prof, 73-81, PROF MATH, UNIV NC, CHAPEL HILL, 81- *Concurrent Pos:* Mem, Sch Math, Inst Advan Study, 78-79; prin investr, NSF Grant, 74-78, 79-; mem, Max Planck Inst Math, Bonn, WGer, 83-84; vis prof, univ Pa, Philadelphia, PA, 88-89. *Res:* Differential geometric properties of Riemanmian manifolds of nonpositive sectional curvature; geodesic flows, isometry groups, end structure of noncompact manifolds of finite volume and geometric properties of compact manifolds that are homotopy invariants. *Mailing Add:* Dept Math Univ NC Chapel Hill NC 27514

EBERLY, JOSEPH HENRY, b Carlisle, Pa, Oct 19, 35; m 60; c 3. THEORETICAL PHYSICS, QUANTUM OPTICS. *Educ:* Pa State Univ, BS, 57; Stanford Univ, MS, 59, PhD(physics), 62. *Prof Exp:* Res physicist, Stanford Linear Accelerator Ctr, 62; resident res assoc, Nuclear Physics Div, US Naval Ord Lab, Md, 62-65; res assoc, 65-67, asst prof, 67-69, assoc prof, 69-76, prof physics, 76-78, PROF PHYSICS & OPTICS, UNIV

ROCHESTER, 79- *Concurrent Pos:* Nat Acad Sci-Nat Res Coun resident res associateship, 62-64; lectr, Univ Md, 64-65; Nat Acad Sci vis lectr, eastern Europe, 70-; consult, Dept Energy, Dept Defense; vis fel, Joint Inst Lab Astrophys, Univ Colo & Nat Bur Standards, 77-78; vis distinguished prof physics, Univ Ark, 82; vis fel, Imp Col Sci & Technol, London, 83; Humbolt sr fel, Max-Planck-Inst, Munich, 85 & 89; vis prof physics, Univ Queensland, Australia, 87; vis lectr,Auckland Found, Auckland Univ, NZ, 87; bd dir, Optical Soc Am, 87-89. *Honors & Awards:* Fulbright lectr physics, Univ Autonoma de Barcelona, Spain, 88; Alexander von Humboldt Sr Award, Alexander von Humboldt Found, WGer, 83; Marion Smoluchowski Medal, Phys Soc Poland, 87. *Mem:* Fel Am Phys Soc; fel Optical Soc Am; Am Asn Physics Teachers. *Res:* Multiphoton interactions of electrons, atoms, molecules; quantum electrodynamics; quantum optics. *Mailing Add:* Dept Physics & Astron Univ Rochester Rochester NY 14627

EBERLY, WILLIAM ROBERT, b North Manchester, Ind, Oct 4, 26; m 46; c 3. ZOOLOGY, LIMNOLOGY. *Educ:* Manchester Col, AB, 48; Ind Univ, MA, 55, PhD(zool), 58. *Prof Exp:* Sci instr pub schs, Ind, 47-52; asst zool, Ind Univ, 52-55; from asst prof to assoc prof, 55-67, PROF BIOL MANCHESTER COL, 67-, DIR ENVIRON STUDIES, 72- *Concurrent Pos:* Vis scientist, Univ Uppsala, 63-64; chair, biol dept, Manchester Col & head, Div Nat Scis; mem, Deleg Biol Educators to China, 88. *Mem:* Am Soc Limnol & Oceanog; Int Asn Theoret & Appl Limnol; Am Inst Biol Sci; Nat Asn Biol Teachers. *Res:* Eutrophication in lakes. *Mailing Add:* Manchester Col North Manchester IN 46962

EBERSOLE, JOHN FRANKLIN, b Boston, Mass, Feb 21, 46; m 68; c 2. OPTICAL PHYSICS. *Educ:* Col Holy Cross, AB, 68; Univ Fla, MS, 71, PhD(physics), 74. *Prof Exp:* Teaching asst optics, Dept Physics & Astron, Univ Fla, 68-72, instr, 71-72; from optical engr to sr optical engr, Itek Corp, 72-75, res & develop mgr optics, 74-75; sr res scientist, Appl Sci Div, Aerodyne Res, Inc, 75-; AT CREATIVE OPT INC. *Mem:* Optical Soc Am; Inst Elec & Electronics Engrs; Sigma Xi; Soc Photo-Optical Instrumentation Engrs. *Res:* Optical physics, including electro-optics, laser scattering by aerosols, optical information processing, holography, acousto-optics, integrated optics and optical materials. *Mailing Add:* 41 Mitchell Grant Way Bedford MA 01730-1264

EBERSTEIN, ARTHUR, b Chicago, Ill, Apr 23, 28; m 61; c 2. BIOPHYSICS. *Educ:* Ill Inst Technol, BS, 50; Univ Ill, MS, 51; Ohio State Univ, PhD(biophys), 57. *Prof Exp:* NSF fel, Copenhagen Univ, 57-58, NIH fel, 58-59; res scientist biophys, Inst Muscle Dis, 59-61; res scientist physics, Am Bosch Arma Corp, 61-63; head med electronics dept, Lundy Electronics & Systs, Inc, 63-64; assoc prof rehab med, 70-72, ASST PROF BIOPHYS, SCH MED, NY UNIV, 64-, PROF REHAB MED, 72-, DIR RES, 81- *Mem:* Biophys Soc; Am Physiol Soc. *Res:* Muscle physiology; electromyography. *Mailing Add:* NY Univ Med Ctr 400 E 34th St New York NY 10016

EBERT, ANDREW GABRIEL, b Brooklyn, NY, Jan 5, 36; m 61; c 3. PHARMACOLOGY, FOOD SCIENCE. *Educ:* Long Island Univ, BS, 57; Purdue Univ, MS, 61, PhD(pharmacol), 62. *Prof Exp:* Sr res scientist, Squibb Inst Med Res, 61-65; supvr pharmacol, Int Minerals & Chem Corp, 65-68, mgr pharmacol & govt registrn, 68-70, mgr prod safety eval, 70-72; dir prod safety & regulatory affairs, William Underwood Co, 72-75, vpres sci & tech affairs, 76-; VPRES, DEPT RES DEVELOP & QUAL ASSURANCE, PET, INC. *Concurrent Pos:* Guest lectr, Mass Inst Tech, 78. *Mem:* Am Pharmaceut Asn; Am Soc Pharmacol & Exp Therapeut; Soc Toxicol; Inst Food Technologists; Sigma Xi. *Res:* Quality control; product development; national and international regulation of foods; metabolic fate of foods; drugs; food additives; agricultural chemicals. *Mailing Add:* 7775 N Spalding Lake Dr Atlanta GA 30360-1054

EBERT, CHARLES H V, b Ger, June 23, 24; US citizen; m 49; c 1. PHYSICAL GEOGRAPHY, SOILS & SOIL SCIENCE. *Educ:* Univ NC, BA, 51, MA, 53, PhD(geog), 57. *Prof Exp:* Instr geog & geol, State Univ NY Buffalo, 54-56, from asst prof to assoc prof, 56-63, actg dean undergrad div, 70-71, dean undergrad div, 71-77, PROF GEOG, STATE UNIV NY BUFFALO, 63-, DISTINGUISHED TEACHING PROF, 90- *Concurrent Pos:* Adv, Peruvian Develop Found, 69-; scholar exchange, People's Rep of China, 83, Soviet Union, 90, Poland, 90. *Honors & Awards:* Distinguished Achievement Award, Acad Arts & Sci, 90. *Mem:* Asn Am Geog; Int Oceanog Found; Soil Sci Soc Am; Am Soc Agron; Int Soil Sci Soc. *Res:* Soils geography and soil morphology; environmental problems in land development; physical environmental factors in planning and development; soils field research in connection with archeological excavations at Tel Ifshar, Israel and Old Fort Niagara, New York; research and consulting on leachate diffusion from chemical dumps; natural and man-induced hazards; environmental impact. *Mailing Add:* Dept Geog State Univ NY Buffalo NY 14260

EBERT, EARL ERNEST, b Oakland, Calif, Sept 28, 31; m 57; c 3. FISH BIOLOGY. *Educ:* San Jose State Col, BA, 59. *Prof Exp:* Aquatic biologist, Marine Resources Lab, 60-63, from asst marine biologist to assoc marine biologist, Menlo Park, 63-70, SR MARINE BIOLOGIST & DIR MARINE CULT LAB, CALIF DEPT FISH & GAME, MONTEREY, 70- *Mem:* Am Inst Fishery Res Biologists; Nat Shellfisheries Asn; World Maricult Soc. *Res:* Mariculture feasibility studies; mariculture feasibility studies of selected shellfish species. *Mailing Add:* Calif Dept Fish & Game 2201 Garden Rd Monterey CA 93940

EBERT, GARY LEE, b Beaver Dam, Wis, June 30, 47; m 78; c 3. FINITE GEOMETRIES, COMBINATORIAL DESIGNS. *Educ:* Univ Wis, Madison, BS, 69, MS, 71, PhD(math), 75. *Prof Exp:* Vis lectr math, Tex Tech Univ, 75-77; from asst prof to assoc prof, 77-87, PROF MATH, UNIV DEL, 87- *Concurrent Pos:* Comput programmer, Chevrolet Eng Ctr, 69-70. *Mem:* Am Math Soc; Math Asn Am. *Res:* Finite geometries (principally affine, projective, and inversive) and their associated combinatorial structures and properties; spreads and packings in finite projective spaces. *Mailing Add:* Dept Math Sci Univ Del Newark DE 19716

EBERT, JAMES DAVID, b Bentleyville, Pa, Dec 11, 21; m 46; c 3. EMBRYOLOGY. *Educ:* Washington & Jefferson Col, AB, 42; Johns Hopkins Univ, PhD(biol), 50. *Hon Degrees:* ScD, Washington & Jefferson Col, 69; ScD, Yale Univ, 73, Ind Univ, 75; LLD, Moravian Col, 79. *Prof Exp:* Jr instr biol, Johns Hopkins Univ, 46-49; instr, Mass Inst Technol, 50-51; from asst prof to assoc prof zool, Ind Univ, 51-55; dir dept embryol, 56-76, pres Carnegie Inst Wash, 78-87; DIR CHESAPEAKE BAY INST & PROF BIOL, JOHNS HOPKINS UNIV, 87- *Concurrent Pos:* Vis scientist, Brookhaven Nat Lab, 53-54; mem adv panel, Comt Growth, Nat Res Coun, 53-55, chmn, Assembly Life Sci, 73-77; mem comt basic res aging, Am Inst Biol Sci, 55-60, pres inst, 63; mem comt genetic & develop biol, NSF, 55-56 & div comt biol & med sci, 63-67; mem cell biol study sect, USPHS, 58-62 & Comn Undergrad Educ Biol Sci, 63-67; mem vis comts, Mass Inst Technol, 59-68, Case Western Reserve Univ, 64-68, Univ Pa, 67-69, Columbia Univ, 67-69, Univ Ore, 67-72, Harvard Univ, 69-75, Princeton Univ, 70-76, Univ Chicago, 74- & Boston Univ, 77-86; Philips vis prof, Haverford Col, 60-61; Patten vis prof, Ind Univ, 62-63; dir embryol training prof, Marine Biol Lab, Woods Hole, 62-66, trustee, 64- 68, dir, 70-75 & 77-78, pres, 70-78; with Univ Sci Develop Adv Panel, NSF, 65-70; with comt Inst Progs, 71; mem bd sci coun, Nat Cancer Inst, 67-71 & Nat Inst Child Health & Human Develop, 73-76; bd sci overseers, Jackson Lab, 67-80, bd gov trustees, 74-86; bd dirs, Oak Ridge Assoc Univs, 67-71; chmn, Govt Univ Indust Res Roundtable, 87- *Honors & Awards:* First Distinguished Serv Award, Washington & Jefferson Col, 65; NIH lectr, 67; Yamagiwa mem lectr, Univ Tokyo, 68; Eminent Scientist Award, Japan Soc Prom Sci, 72; President's Medal, Am Inst Biol Sci, 72; Tamaki mem lectr, Univ Kyoto, 72; Regent's lectr, Univ Calif, 75; Storer lectr, Univ Calif, Davis, 77. *Mem:* Nat Acad Sci (vpres, 81-); Am Philos Soc; Inst Med; Soc Develop Biol (pres, 57-58); Am Soc Zool (pres, 70). *Res:* Acquisition of biological specificity; protein synthesis and interactions in development; heart development; graft versus host reactions; viruses as tools in developmental biology; melanogenesis; amino acid and vitamin metabolism in development; tumorigenic viruses; viral oncogenic sequences; cell replicating mechanisms; ionic regulation of differentiation. *Mailing Add:* Chesapeake Bay Inst Johns Hopkins Univ Suite 340 The Rotunda 711 W 40th St Baltimore MD 21211

EBERT, LAWRENCE BURTON, b Bronxville, NY, Jan 14, 49. PHYSICAL CHEMISTRY, MATERIALS SCIENCE. *Educ:* Univ Chicago, SB, 71; Stanford Univ, MS, 74, PhD(chem), 75. *Prof Exp:* Res chemist, 75-78, sr chemist res, 78-79, STAFF CHEMIST, EXXON CORP, 79- *Concurrent Pos:* Ed, Chem Engine Combustion Deposits, Plenum, NY, 85, Polynuclear Aromatic Compounds, Am Chem Soc, Advan Chem No 217, Washington, DC, 88. *Mem:* Am Chem Soc; Am Phys Soc; Sigma Xi. *Res:* Solid state chemistry and physics of carbonaceous materials; with emphasis on intercalation compounds of graphite. *Mailing Add:* Univ Chicago 1111 E 60th St Chicago IL 60637

EBERT, LYNN J, b Sandusky, Ohio, Apr 17, 20; m 43; c 4. METALLURGICAL ENGINEERING. *Educ:* Case Western Reserve Univ, BS, 41, MS, 43, PhD (metall), 54. *Prof Exp:* From asst to res assoc, 41-51, sr res assoc, 51-54, from asst prof to assoc prof metall eng, 54-65, PROF METALL ENG, CASE WESTERN RESERVE UNIV, 65- *Mem:* Am Soc Metals; NY Acad Sci; Am Inst Mining, Metall & Petrol Engrs; Sigma Xi. *Res:* Mechanical and physical behaviors of ferrous and non-ferrous metals and alloys; failure analysis; fiber composite performance. *Mailing Add:* 104 E 197th St Euclid OH 44119

EBERT, PATRICIA DOROTHY, behavioral genetics; deceased, see previous edition for last biography

EBERT, PAUL ALLEN, b Columbus, Ohio, Aug 11, 32; m 54; c 3. CARDIOVASCULAR SURGERY, PHYSIOLOGY. *Educ:* Ohio State Univ, BS, 54, MD, 58. *Prof Exp:* Intern surg, Johns Hopkins Hosp, 58-59, asst resident, 59-60; sr resident, Clin of Surg, Nat Heart Inst, 60-62; asst resident, Johns Hopkins Hosp, 62-65, chief resident, 65-66; from asst prof to assoc prof, Med Ctr, Duke Univ, 66-71; prof & chmn dept, Med Col, Cornell Univ, 71-75, surgeon in chief, NY Hosp-Cornell med Col, 71-74; prof & chmn dept, Univ Calif, San Francisco, 75-86; DIR, AM COL SURGEONS, 86- *Concurrent Pos:* Nat Cancer Inst fel, 62-63; Mead Johnson scholar, 64; Markle scholar, 67. *Mem:* Asn Acad Surg; Am Heart Asn; Am Col Surg; Soc Univ Surg; Soc Vascular Surg; Am Asn Thoracic Surg; Am Surg Asn; Soc Thoracic Surg. *Mailing Add:* 55 E Erie Chicago IL 60611

EBERT, PAUL JOSEPH, b New Orleans, La, Jan 11, 36; m 57; c 5. PHYSICS. *Educ:* La State Univ, BS, 57, MS, 59, PhD(physics), 62. *Prof Exp:* Nuclear res officer, Sch Aerospace Med, USAF, 61-64; sr physicist, L-Div, 64-69, dep group leader, L-Div, Res & Develop Group, 69-72, group leader, 72-77, assoc div leader, L-Div, 77-80, dir tech staff, 81-85, sr sci consult to off mgt & budget, 85-87, Z-DIV, LAWRENCE LIVERMORE LAB, UNIV CALIF, 87- *Mem:* Am Phys Soc; Sigma Xi; AAAS. *Res:* Applied physics. *Mailing Add:* 2148 Orion Ct Livermore CA 94550

EBERT, PHILIP E, b Milwaukee, Wis, Sept 4, 29; m 57; c 2. TEXTILE CHEMISTRY, TEXTILE ENGINEERING. *Educ:* Purdue Univ, BSChE, 51; Univ Pa, PhD(org polymer chem), 60. *Prof Exp:* Chem engr res & develop, Hercules Inc, 51-55; res chemist, 60-64, tech serv rep dyeing & finishing, 64-66, tech serv supvr, 66-78, sr res chemist, 78-87, SR MKT DEVELOP SPECIALIST, TEXTILE FIBERS DEPT, E I DUPONT DE NEMOURS & CO INC, 87- *Mem:* Sigma Xi. *Res:* Textile technology of manmade fibers. *Mailing Add:* 1711 Waterglen Lane West Chester PA 19382

EBERT, RICHARD VINCENT, b St Paul, Minn, Oct 25, 12; m 47. CLINICAL MEDICINE. *Educ:* Univ Chicago, BS, 33, MD, 37. *Prof Exp:* Intern, Boston City Hosp, Mass, 37-39; asst resident med, Peter Bent Brigham Hosp, Boston, 39-41, jr assoc, 41-42; fel, Harvard Med Sch, 40-42; chief med serv, Vet Admin Hosp, 46-52, 53-54; prof med, Sch Med, Univ Minn, 49-52, Clark prof, 52-53; prof, Northwestern Univ, 53-54; prof & chmn dept, Sch Med, Univ Ark, 54-66; prof med & chmn dept, sch med, Univ Minn,

Minneapolis, 66-78; prof med, Univ Ark, 78-91; RETIRED. *Mem:* AMA; Am Col Physicians; Am Soc Clin Invest (pres, 58); Soc Exp Biol & Med; Asn Am Physicians. *Res:* Pulmonary and cardiovascular physiology; cardiac catheterization in humans; blood volume and peripheral circulation in humans; pulmonary pathology. *Mailing Add:* Vet Admin Hosp Little Rock AR 72206

EBERT, ROBERT H, b Minneapolis, Minn, Sept 10, 14; m 39; c 3. INTERNAL MEDICINE, MEDICAL ADMINISTRATION. *Educ:* Univ Chicago, BS, 36, MD, 42; Oxford Univ, DPhil, 39; Am Bd Internal Med, dipl, 52. *Hon Degrees:* AM, Harvard Univ; DSc, Northeastern Univ, 68, Univ Md, 70; LLD, Univ Toronto, 70; LHD, Rush Univ, 74; DSc, New York Med Col, 83. *Prof Exp:* Dir med, Univ Hosps, Cleveland, 56-64; mem spec adv group, Vet Admin, Washington, DC, 59-63; chief med serv, Mass Gen Hosp, 64-65; prof med, 65-73, dean Harvard Med Sch & Fac Med & pres, Harvard Med Ctr, Harvard Univ, 65-77; pres, Milbank Mem Fund, 78-84; spec adv to the pres, 85-89, CONSULT, ROBERT WOOD JOHNSON FOUND, 89-; SPEC ADV, COMMONWEALTH FUND, 84- *Concurrent Pos:* Pres, Milbank Mem Fund, 88. *Honors & Awards:* Distinguished Serv Award, Univ Chicago, 62; Distinguished Achievement Award, Modern Med; The Abraham Flexner Award for Distinguished Serv to Med Educ, 82. *Mem:* Inst Med-Nat Acad Sci; Am Soc Clin Invest; Asn Am Physicians (vpres-pres, 71-73); Am Clin & Climat Asn; Am Col Physicians; AAAS; fel Am Public Health Asn; fel Am Acad Arts & Sci; Microcirculatory Soc; Int Epidemiol Asn; Sigma Xi. *Res:* Tuberculosis and mechanisms of inflammation. *Mailing Add:* 16 Brewster Rd Wayland MA 01778

EBERT, THOMAS A, b Appleton, Wis, July 10, 38; m 60; c 2. ECOLOGY. *Educ:* Univ Wis-Madison, BS, 61; Univ Ore, MS, 63, PhD(biol), 66. *Prof Exp:* Instr biol, Univ Ore, 66-67; asst prof zool, Univ Hawaii, 67-69; asst prof, 69-72, assoc prof, 72-75, PROF BIOL, SAN DIEGO STATE UNIV, 75- *Mem:* AAAS; Ecol Soc Am; Am Soc Limnol & Oceanog; Am Soc Naturalists; Sigma Xi. *Res:* Population biology; marine ecology; population ecology of echinoderms. *Mailing Add:* Dept Biol San Diego State Univ San Diego CA 92182

EBERT, WESLEY W, b Maple Grove Twp, Minn, Mar 22, 26; m 54; c 5. GENETICS, BOTANY. *Educ:* Univ Minn, BS, 61; Univ Calif, Davis, MS, 63, PhD(genetics), 64. *Prof Exp:* From asst prof to assoc prof, 64-71, chmn dept, 69-74, PROF BIOL, SONOMA STATE UNIV, 71- *Mem:* Bot Soc Am; Am Genetic Asn; Am Soc Agron. *Res:* Genetics and anatomical development of Carthamas tinctorius. *Mailing Add:* Dept of Biol Sonoma State Univ 1801 E Cotati Ave Rohnert Park CA 94928

EBERT, WILLIAM ROBLEY, b Philadelphia, Pa, Aug 2, 28; m 54; c 3. PHARMACEUTICAL CHEMISTRY. *Educ:* Univ Mich, BS, 52, MS, 54, PhD(pharmaceut chem), 57. *Prof Exp:* Res fel pharmaceut, Sterling-Winthrop Res Inst, 56-58; chief chemist, Vick Chem Co, 58-59; dir pharmaceut res & develop, Lemmon Pharmacal Co, 59-60; sci dir pharmaceut res, Philips Roxane Labs, 60-61, dir, 61-62, from asst dir to dir res & develop, 62-68, vpres pharmaceut res, 68-73; RETIRED. *Concurrent Pos:* Pharmaceut consult, 82- *Mem:* Am Pharmaceut Asn; Am Soc Hosp Pharmacists; fel Am Inst Chemists; Am Chem Soc; Drug Info Asn. *Res:* Technology of soft gelatin capsule dosage forms. *Mailing Add:* 4935 62nd Ave S St Petersburg FL 33715

EBERTS, FLOYD S, JR, b Easton, Pa, Dec 24, 24. DRUG METABOLISM. *Educ:* Univ Wis, PhD(biochem), 52. *Prof Exp:* SR RES SCIENTIST, DRUG METAB UNIT, UPJOHN CO, 76- *Mem:* Am Chem Soc; Am Soc Pharmacol & Exp Therapeut. *Mailing Add:* 2524 Frederick Ave Kalamazoo MI 49008

EBERTS, RAY EDWARD, b Newark, Ohio, Nov 5, 54; m 80; c 2. HUMAN FACTORS, NEURAL NETWORKS. *Educ:* Univ Calif, San Diego, BA, 77; Univ Ill, MA, 79, PhD(phychol), 83. *Prof Exp:* Asst prof, 83-88, ASSOC PROF INDUST ENG, SCH INDUST ENG, PURDUE UNIV, 88- *Concurrent Pos:* NSF presidential young investr, 87; invited prof, NTT Human Interface Labs, Japan, 90; vis assoc prof, Univ Southern Calif, 91; assoc bull ed, Human Factors Soc. *Mem:* Human Factors Soc; Psychonomics Soc. *Res:* Applies cognitive theories to the design of computer interfaces; methods to acquire information from users about how they think about tasks and then design the computer interfaces to be compatible with this thinking; statistical techniques; cognitive models; neural nets. *Mailing Add:* Dept Industrial Eng Purdue Univ West Lafayette IN 47907

EBERTS, ROBERT EUGENE, b Columbus, Ohio, May 30, 31; m 53; c 5. INORGANIC CHEMISTRY, PHYSICAL CHEMISTRY. *Educ:* Univ Dayton, BS, 53; Iowa State Univ, PhD(phys chem), 57. *Prof Exp:* Res asst phys chem, Ames Lab, USAEC, 53-57; res chemist inorg chem, Wyandotte Chem Corp, 57-62; res chemist prod develop, Nat Res Corp, 62-63, sr chemist, 63-65, sr chemist, Metals Div, Norton Co, 65-69; staff mem metall eng, Arthur D Little Inc, 69-70; sr chemist, 71-84, dir regulatory & environ affairs & staff opers, 85-88, ASST PLANT MGR, PIGMENT DIV, MEARL CORP, PEEKSKILL, NY, 89- *Mem:* Am Chem Soc; fel Am Inst Chemists. *Res:* Process development; pigments; product development. *Mailing Add:* Six Locust Hill Rd Mahopac NY 10541

EBETINO, FRANK FREDERICK, b Rye, NY, Jan 12, 27; m 50; c 2. ORGANIC CHEMISTRY. *Educ:* Ohio Univ, BS, 49; Lehigh Univ, MS, 53; Tohoku Univ, Japan, PhD(org chem), 74. *Prof Exp:* Res chemist, Eaton Labs Div, Norwich Pharmacol Co, 49-51 & Johns-Manville Corp, 53-55; sr res chemist, Chem Res Div, Morton-Norwich Prod Inc, 55-60, unit leader, 60-61, chief chem sect, 61-68, asst dir chem div, 68-69, dir chem res div, Norwich-Eaton Pharmaceut Div, 69-82, dir biol & chem res div, Norwich-Eaton Pharmaceut Div, Procter & Gamble Co, 82-83, dir, Worldwide Sci Affairs, 83-85; PHARMACEUT CONSULT, 85- *Mem:* Pharm Soc Japan; Am Chem Soc; Indust Res Inst. *Res:* Organic synthesis in field of heterocyclic chemistry; chemical structure-biological activity relationships. *Mailing Add:* 500 Park Blvd S 54 Venice FL 34285

EBIN, DAVID G, b Los Angeles, Calif, Oct 24, 42; m 71; c 4. MATHEMATICS. *Educ:* Harvard Univ, AB, 64; Mass Inst Technol, PhD(math), 67. *Prof Exp:* NSF fel, 67-68; lectr math, Univ Calif, Berkeley, 68-69; assoc prof, 69-78, PROF MATH, STATE UNIV NY STONY BROOK, 78- *Concurrent Pos:* Lectr, Ecole Polytech & Univ Paris VII, 71; mem, Courant Inst Math Sci, 76; vis prof, Univ Calif, Los Angeles, 83-84. *Mem:* Am Math Soc. *Res:* Differential geometry; infinite dimensional manifolds; nonlinear partial differential equations; mathematical theory of fluid mechanics; mathematical theory of motion of slightly compressible fluids; mathematical theory of elasto-dynamics. *Mailing Add:* Dept Math State Univ NY Stony Brook NY 11794-3651

EBLE, JOHN NELSON, b St Louis, Mo, May 19, 27; m 50; c 3. BIOCHEMISTRY. *Educ:* Univ Mo, BS, 49; Univ Wis, MS, 52, PhD(biochem), 54. *Prof Exp:* From instr to asst prof pharmacol & physiol, Kirksville Col, 54-60; pharmacologist, 60-64, assoc scientist, 64-66, ASSOC SCIENTIST, HUMAN HEALTH RES LAB, DOW CHEM CO, 66- *Mem:* AAAS; Am Physiol Soc; Soc Exp Biol & Med; Am Soc Pharmacol & Exp Therapeut; Brit Pharmacol Soc. *Res:* Warfarin; blood coagulation; interchange between somatic and autonomic nervous systems; autonomic and central nervous system pharmacology. *Mailing Add:* Dept Pathol Ind Univ Sch Med 1481 W Tenth St Indianapolis IN 46202

EBLE, THOMAS EUGENE, b Toledo, Ohio, Sept 15, 23; m 45; c 6. BIOCHEMISTRY. *Educ:* Loyola Univ, Ill, BSc, 44; Georgetown Univ, MS, 46, PhD(biochem), 48. *Prof Exp:* Chemist, Bur Standards, 45, Food & Drug Admin, 45-48 & Haris Res Labs, 48; res chemist, Upjohn Co, 48-59, sect head, 59-84; RETIRED. *Mem:* AAAS; Am Chem Soc; Am Oil Chemists Soc. *Res:* Antibiotics; chromatographic separation methods; isolation, characterization and identification of antibiotics; biosynthesis of antibiotics; structure of antibiotics; chemical and biochemical modification of antibiotics. *Mailing Add:* 7532 Woodbridge Lane Portage MI 49002

EBNER, FORD FRANCIS, b Colfax, Wash, Feb 10, 34; m 60; c 4. NEUROSCIENCES. *Educ:* Wash State Univ, BS, 54, DVM, 58; Univ Md, PhD(neuroanat), 65. *Hon Degrees:* MS, Brown Univ, 69. *Prof Exp:* Asst prof anat & physiol, Univ Md, 65-66; from asst prof to assoc prof, 66-73, PROF BIOL & MED, BROWN UNIV, 73-, CO-DIR, CTR NEURAL SCI. *Concurrent Pos:* NIH fel physiol, Johns Hopkins Univ, 60-63; spec fel anat, Univ Md, 63-65. *Mem:* AAAS; Am Asn Anat; Am Asn Neuropath; Neurosci Soc. *Res:* Anatomy and physiology of thalamus and cortex. *Mailing Add:* Dept Med Sci Brown Univ Providence RI 02912

EBNER, HERMAN GEORGE, polymer chemistry, organic chemistry, for more information see previous edition

EBNER, JERRY RUDOLPH, b LaCrosse, Wis, Nov 11, 47; m 69; c 3. INORGANIC CHEMISTRY, HETEROGENEOUS CATALYSIS. *Educ:* Univ Wis-LaCrosse, BS, 69; Purdue Univ, PhD(inorg chem), 75. *Prof Exp:* Res asst chem, AEC, Ames, Iowa, 68; instr chem, Purdue Univ, 69-75; res prof chem, 75-78, res specialist, 78-81, sr res specialist, 81-82, group leader, 82-85, SCI FEL, MONSANTO CO, 86- *Mem:* Am Chem Soc; Sigma Xi. *Res:* Chemistry of metal-metal bonded complexes, their properties and reactions, and application of electron spectroscopic chemical analysis and Raman spectroscopic techniques for the study of heterogeneous catalysts; synthesis, characterization and commercialization of heterogeneous catalysts. *Mailing Add:* 233 Bentwood Lane St Charles MO 63303

EBNER, KURT E, b New Westminster, BC, Mar 30, 31; m 57; c 4. BIOCHEMISTRY. *Educ:* Univ BC, BSA, 55, MSA, 57; Univ Ill, PhD(dairy biochem), 60. *Prof Exp:* Can Overseas Nat Res Coun fel, Nat Inst Res Dairying, Reading, Eng, 60-61; instr physiol chem & fel, Univ Minn, 61-62; from asst prof to regent prof biochem, Okla State Univ, 62-74; CHMN DEPT BIOCHEM & MOLECULAR BIOL, UNIV KANS MED CTR, 74- *Concurrent Pos:* NIH career develop award, 69; Sigma Xi lectr, 70; Nat Bd Med Examrs, 82-; pres dept biochem, Asn Med Sch, 86; adv bd, Coun Acad Soc, AAMC, 90- *Honors & Awards:* Borden Award, Am Chem Soc, 69. *Mem:* Am Soc Biol Chemists; AAAS. *Res:* Interaction of hormones with receptors; enzyme mechanisms. *Mailing Add:* Dept Biochem & Molecular Biol Univ Kans Med Ctr 39th & Rainbow Kansas City KS 66103

EBNER, STANLEY GADD, b Lincoln, Nebr, Oct 29, 33; m 56; c 2. ENGINEERING MECHANICS, AEROSPACE ENGINEERING. *Educ:* Univ Nebr, Lincoln, BS, 55; Univ Colo, Boulder, BS, 63, MS, 64, PhD(eng mech), 68. *Prof Exp:* From instr to assoc prof eng mech, USAF Acad, 64-70, dep head dept, 70-71, div chief, Space & Missile Systs Orgn, 71-76; DEAN TECHNOL, DOWNTOWN COL, UNIV HOUSTON, 76- *Concurrent Pos:* Chmn, Gulf Coast Alliance Minorities in Eng, 77-81. *Mem:* Am Soc Eng Educ; Am Soc Civil Engrs. *Res:* Dynamics and vibrations. *Mailing Add:* Dept Continuing Educ Univ Houston Downtown No One Main St Houston TX 77002

EBNER, TIMOTHY JOHN, b Minneapolis, Minn, July 15, 49; m 71; c 4. MOTOR CONTROL, CEREBELLAR PHYSIOLOGY. *Educ:* Univ Minn, BS, 71, MD, 79, PhD(physiol), 79. *Prof Exp:* Neurophysiologist, Minneapolis Vet Admin Hosp, 78-79; jr scientist, dept neurosurg, 76-78, teaching asst physiol, 76-77, res specialist, 79, asst prof, 79-85, LAB DIR, DEPT NEUROSURG, UNIV MINN, 84-, ASSOC PROF, 85- *Concurrent Pos:* Lectr, dept physiol, Univ Minn, 81-; prin investr, NSF & NIH grants. *Honors & Awards:* Hans Berger Award, Am Electroencephalographic Soc, 79. *Mem:* Am Physiol Soc; Soc Neurosci; NY Acad Sci. *Res:* Function and neurobiology of the cerebellum. *Mailing Add:* Dept Neurosurg Univ Minn 420 Delaware St SE Box 96 Minneapolis MN 55437

EBRAHIMI, FERESHTEH, b Tehran, Iran, Jan 2, 51; c 1. FRACTURE MECHANICS, MECHANICAL METALLURGY. *Educ:* Arya-Mehr Univ Technol, BS, 72; Surrey Univ, UK, MS, 76; Col Sch Mines, PhD(metall), 82. *Prof Exp:* Instr metall, Arya-Mehr Univ Technol, 72-77; consult eng, Int

Training Consultants, 77-78; teaching fel metall, Col Sch Mines, 82-83; scientist fracture & deformation, Nat Bur Standards, 83-84; asst prof, 84-89, ASSOC PROF MAT SCI, UNIV FLA, 89- *Concurrent Pos:* Sabbatical leave, Max-Planck Inst Metal Res, Stultgart, Ger, 90-91. *Mem:* Am Soc Metals; Am Soc Testing & Mat; Sigma Xi; Am Inst Mining, Metall & Petrol Engrs. *Res:* Mechanical properties of high temperature materials; fracture of materials; strengthening mechanisms in metals. *Mailing Add:* Dept Mat Sci & Eng Univ Fla 156 Rhines Hall Gainesville FL 32611

EBRAHIMI, NADER DABIR, b Tehran, Iran, June 27, 55. OPTIMIZATION, DYNAMIC SYSTEMS. *Educ:* Arya-Mehr Univ Technol, Iran, BSc, 79; Univ Wis-Madison, MSc, 81, PhD(mech eng), 83. *Prof Exp:* Teaching asst dynamic systs & theory mach, Univ Wis-Madison, 80-83; ASST PROF MACH DESIGN, THEORY MACH, NUMERICAL METHODS, DYNAMIC SYSTS, OPTIMIZATION & DYNAMICS, UNIV NMEX, 83- *Concurrent Pos:* Prin investr, Sandia Nat Labs, 84-86; consult, Liberty Mutual Ins Co, 85. *Mem:* Am Soc Mech Engrs. *Res:* Application of optimization algorithms on physical systems, robot mobility and dynamic systems. *Mailing Add:* Dept Mech Eng Univ NMex Albuquerque NM 87131

EBREY, THOMAS G, b Pittsburgh, Pa, Aug 21, 41; m 68; c 2. BIOPHYSICS. *Educ:* Univ Okla, BS, 63; Univ Chicago, PhD, 68. *Prof Exp:* Postdoctoral fel biol, Cornell Univ, 67-68; asst prof biol sci, Columbia Univ, 68-73; assoc prof, 73-78, PROF BIOPHYS, UNIV ILL, 78- *Res:* Vision; structure of visual pigments; structure of photo receptor membrane; control of permeability of photoreceptor membranes; light energy transduction in halobacteria. *Mailing Add:* Dept Physiol & Biophys Univ Ill 407 S Goodwin Urbana IL 61801

EBY, CHARLES J, b Detroit, Mich, May 29, 29; m 51; c 4. ORGANIC CHEMISTRY. *Educ:* Univ Mich, BS, 51; Dartmouth Col, MA, 53; Duke Univ, PhD(org chem), 56. *Prof Exp:* Res chemist, Monsanto Chem Co, 56-63, mem staff govt rels, 63-66, mgr res & develop mkt, 66-87; consult, Hill & Knowlton, 87-90; CONSULT, FLEISHMAN-HILLARD, 91- *Mem:* Am Chem Soc; AAAS. *Res:* Materials and systems; condensations; eliminations; substitutions; rearrangements; cyclizations. *Mailing Add:* 8408 Queen Elizabeth Annandale VA 22003

EBY, DENISE, b Baltimore, Md, Dec 8, 17. CHEMISTRY. *Educ:* St Joseph Col, Md, BS, 39; Cath Univ Am, MS, 53; Univ Md, PhD, 70. *Prof Exp:* Teacher high schs, Md, 39-41, 46-50, NY, 42-45 & WVa, 45-46; from asst prof to prof chem, St Joseph Col, Md, 50-73; res assoc, Sch Med, Univ Md, Baltimore, 74-83; ADJ PROF CHEM, LOYOLA COL, 75- *Concurrent Pos:* Consult sci educ, Sisters of Charity pvt schs, 73-; co-auth, Phys Sci. *Mem:* Nat Sci Teachers Asn; Am Chem Soc. *Res:* Chemical education; enzyme kinetics and studies of glyceraldehyde-3-phosphate dehydrogenase, including the interaction of nicotinamide adenine dinucleotide. *Mailing Add:* St Agnes Hosp 900 Caton Ave Baltimore MD 21229-5299

EBY, EDWARD STUART, SR, b Chicago, Ill, Oct 3, 34; m 65. MATHEMATICS. *Educ:* Univ Ill, BS, 56, MS, 57, PhD(math), 64. *Prof Exp:* Mathematician, US Navy Underwater Sound Lab, 57-64, res mathematician & res assoc, 64-70; RES MATHEMATICIAN, DIR INDEPENDENT RES & INDEPENDENT EXPLOR DEVELOP PROG & DIR INDUST INDEPENDENT RES & DEVELOP REVIEWS, US NAVAL UNDERWATER SYSTS CTR, 70- *Concurrent Pos:* Lectr, Dept Elec Eng, Univ Conn, 65-68. *Mem:* AAAS; Am Math Soc; Math Asn Am; Acoust Soc Am. *Res:* Underwater acoustics; signal processing. *Mailing Add:* 20 Colonial Dr Waterford CT 06385

EBY, FRANK SHILLING, b Kansas City, Mo, Apr 6, 24; m 58; c 3. NUCLEAR PHYSICS, DEFENSE SCIENCE. *Educ:* Univ Ill, BS, 49, MS, 50, PhD(physics), 54. *Prof Exp:* Res assoc physics, Univ Ill, 54; mem proj, Sherwood Res, 54-58 & Atomic Weapons Design, 58-67, div leader device design, 67-72, SR SCIENTIST, LAWRENCE LIVERMORE LAB, UNIV CALIF, 72- *Mem:* AAAS; Am Phys Soc. *Res:* Nuclear reactions; scintillation crystals; plasma physics; high explosives; weapons design. *Mailing Add:* Lawrence Livermore Lab Bldg 261 L377 Livermore CA 94550

EBY, G NELSON, b Bethlehem, Pa, Jan 6, 44; m 68; c 2. GEOCHEMISTRY, IGNEOUS PETROLOGY. *Educ:* Lehigh Univ, BA, 65, MS, 67; Boston Univ, PhD(geol), 71. *Prof Exp:* PROF, DEPT EARTH SCI, UNIV LOWELL, 70- *Concurrent Pos:* Vis sr fel, Linacre Col, Oxford Univ, 82 & 87; vis prof, Boston Univ, 85, Univ Queensland, Australia, 89 & Univ Canterbury, NZ, 90. *Mem:* Fel Geol Soc Am; Sigma Xi; Geochem Soc; Mineral Soc Am; Nat Asn Geol Teachers. *Res:* Geochemistry and petrology of alkaline rocks, granitoids and continental margin basalts; isotope geology; fission-track geochronology. *Mailing Add:* Dept Earth Sci Univ Lowell Lowell MA 01854

EBY, HAROLD HILDENBRANDT, b Platteville, Colo, Mar 3, 18; m 42; c 3. ORGANIC CHEMISTRY. *Educ:* Colo State Col, BS, 40; Univ Nebr, MA, 47, PhD(org chem), 49. *Prof Exp:* Asst, Univ Nebr, 40-42 & 46-48; res chemist, Res & Develop Dept, Conoco Inc, 48-52, sr res chemist & actg group leader, 52-54, res group leader, 54-58, supvry res chemist, 58-69, supvr, 69-72, dir, Tech Info Serv, 72- 79 & Corp Serv Sect, 79-85; RETIRED. *Mem:* AAAS; Am Soc Info Sci; Am Chem Soc. *Res:* Lubricants; fuels; waxes; specialty petroleum products; technical information; radiochemistry; environmental science. *Mailing Add:* 1416 Reveille Dr Ponca City OK 74604

EBY, JOHN EDSON, b Wabash, Ind, Mar 18, 33; m 53; c 5. SOLID STATE PHYSICS, OPTICS. *Educ:* Col Wooster, BA, 54; Univ Rochester, PhD(physics), 59. *Prof Exp:* Engr physics, Sylvania Elec Prod, Inc, 59-69, ENGR PHYSICS, GTE SYLVANIA, INC, GEN TEL & ELECTRONICS CORP, 69- *Mem:* Optical Soc Am; Am Phys Soc; Electrochem Soc. *Res:* Optical properties of solids; radiometry; high temperature interactions of gases with refractory metals. *Mailing Add:* 22 Chattanooga Rd Ipswich MA 01938

EBY, JOHN MARTIN, b Reading, Pa, Dec 8, 39; m 62; c 3. VINYL PLASTISOL & UV CURE COATINGS. *Educ:* Goshen Col, BA, 60; Univ Del, PhD(org chem), 65. *Prof Exp:* Res chemist, Res & Develop Ctr, Armstrong Cork Co, 65-75; PROJ CHEMIST, MANNINGTON MILLS, INC, 75- *Mem:* Royal Soc Chem; Soc Plastic Engrs; Radtech Int. *Res:* Organic synthesis; vinyl plastics applications; UV cure coatings. *Mailing Add:* 32 Canterbury Dr Pennsville NJ 08070

EBY, LAWRENCE THORNTON, b South Bend, Ind, May 3, 16; m 41; c 2. ORGANIC CHEMISTRY. *Educ:* Univ Notre Dame, BS, 38, MS, 39, PhD(org chem), 41. *Prof Exp:* Res chemist, Standard Oil Develop Co, 41-55; res chemist, Esso Res & Eng Co, 55-57; sr mkt develop eng, Enjay Co, Inc, 57-58, asst mgr, Mkt Develop Div, 58-64; pres, Protective Treatments, Inc, Aeroplast Corp & Dellrose Industs, Helene Curtis Industs, Inc, 64-65; res dir, Chem Div, Chrysler Corp, Mich, 65-67; mgr, Polymer Div, US Gypsum Co, 67-73, assoc dir res, 73-82, prin assoc, 82-85, dir res & develop, 85-86; RETIRED. *Honors & Awards:* Honor Scroll, Am Inst Chemists, 61. *Mem:* AAAS; Chem Mkt Res Asn; Am Chem Soc; Soc Plastics Eng; fel Am Inst Chemists; Tech Asn Pulp & Paper Indust; Soc Automotive Eng; Asn Iron & Steel Eng; Com Develop Asn; Am Indust Hyg Asn. *Res:* Adhesives; sealants; lubricating oil additives; antioxidants for synthetic rubber; vulcanization of butyl rubber; toxicity of petroleum products; synthesis of petrochemicals; polymerization; diesel fuel additives; chemicals for paper and textiles; market development of petrochemicals; paints, coatings and building materials. *Mailing Add:* 102 S Kennicott Ave Arlington Heights IL 60005

EBY, ROBERT NEWCOMER, b Pittsburgh, Pa, July 17, 31; m 55; c 2. CHEMICAL PLANT OPERATIONS, TECHNICAL MANAGEMENT. *Educ:* Princeton Univ, BSE, 52; Univ Ill, PhD(chem eng), 58. *Prof Exp:* Res engr, Plastics Div, Union Carbide Corp, 55-56, group leader, 56-62, prod supvr, Mfg Dept, 62-64, area supvr eng & qual control, 64-66, prod supt, 66-67, prod supt & mgr, Vinyl Fabrics Dept, 67-69; prod mgr film mfg, Chem Opers, Polaroid Corp, Waltham, 69-71, sr prod mgr, 71-78, plant mgr, Norwood, 78-83, plant mgr, 84-86, div mgr, 86-88; independent consult, 89-90; PRES, EBY ASSOCS, INC, LEXINGTON, MASS, 90- *Mem:* Am Chem Soc; Am Inst Chem Engrs; Sigma Xi. *Res:* Non-Newtonian flow; heat and mass transfer from high viscosity fluids; optimization of capital expenditures for process modification and expansion; diffusion transfer photographic films. *Mailing Add:* 20 Baskin Rd PO Box 248 Lexington MA 02173-0003

EBY, RONALD KRAFT, b Reading, Pa, May 7, 29; m 52; c 2. STRONG FIBERS, MOLECULAR MODELING. *Educ:* Lafayette Col, BSc, 52; Brown Univ, MS, 55, PhD(physics), 58. *Prof Exp:* Asst, Brown Univ, 52-57; physicist, Polychems Dept, Exp Sta, E I Du Pont de Nemours & Co, 57-63; physicist, Polymers Div, Nat Bur Standards, 63-67, chief polymer crystal physics sect, 67-68, chief polymers div, 68-84; prof mat sci & eng, Johns Hopkins Univ, 84-90; R C MUSSON EMINENT SCHOLAR CHAIR POLYMER SCI, UNIV AKRON, 90- *Concurrent Pos:* Lectr, Johns Hopkins Univ, 80-84; US ed, Polymer, 76- *Honors & Awards:* Medal for Meritorious Serv, US Dept Com, 80; Humboldt Sr Res Award, 89. *Mem:* Fel Acoust Soc Am; fel Am Phys Soc; Am Chem Soc; fel Soc Plastics Engrs; Soc Advan Mat & Process. *Res:* Polymer physics; molecular modeling; ultrasonic propagation; composites; fibers. *Mailing Add:* Dept & Inst Polymer Sci Univ Akron Akron OH 44325-3909

ECANOW, BERNARD, b Chicago, Ill, Nov 22, 23; m 65; c 2. PHARMACEUTICAL SCIENCES. *Educ:* Univ Minn, BChE, 47, BS, 51, PhD(pharmaceut chem), 55. *Prof Exp:* From asst prof to assoc prof mfg pharm, Butler Univ, 56-58; from asst prof to assoc prof, 58-69, PROF PHARMACEUT SCI, UNIV ILL, 70-; PRES, SYNTHETIC BLOOD CORP, DEERFIELD ILL, 85- *Concurrent Pos:* Consult, McGraw Hill-Air Force Proj, 59-60, Presby-St Lukes Hosp, 67- & Rush Med Col 70-; consult sr scientist anesthesiol, Hines Vet Admin Hosp, 73-77; consult, Ecanow & Assoc Consults Health Sci. *Mem:* Am Pharmaceut Asn. *Res:* The coacervate model of biology and theories of cancer, anesthesia, lung gas transfer, psychological states and the emergent mind; surfactants of biological interests; physical-chemical and biological aspects of pharmaceutical formulation; toxic effects of environmental pollutants; preparation of oxygen carrying synthetic blood. *Mailing Add:* Dept Pharmacol Rush Univ Med Col 600 S Paulina Chicago IL 60612

ECCLES, SAMUEL FRANKLIN, b Reno, Nev, Sept 19, 30; m 53; c 2. NUCLEAR PHYSICS. *Educ:* Univ Nev, BS, 52; La State Univ, MS, 54; Univ Wash, Seattle, PhD(physics), 58. *Prof Exp:* Res physicist, Inst Nuclear Physics, Amsterdam, Holland, 58-59; asst prof physics, Univ Nev, 59-62; res physicist, Lawrence Livermore Lab, Univ Calif, 62-90; RETIRED. *Mem:* Am Inst Physics; Am Phys Soc. *Res:* Medium energy nuclear physics including scattering and reaction experiments; nuclear structure physics; reactor and neutron physics; reactor fuel cycles; transuranium heavy element production in nuclear devices; fission processes; astrophysics. *Mailing Add:* 14 Castledown Rd Pleasanton CA 94566

ECCLES, WILLIAM J, b Owatonna, Minn, Apr 18, 32; m 67; c 2. ELECTRICAL ENGINEERING, COMPUTER SCIENCE. *Educ:* Mass Inst Technol, SB, 54, SM, 57; Purdue Univ, PhD(elec eng), 65. *Prof Exp:* Instr elec eng, Purdue Univ, 59-65; asst prof elec eng & dir comput ctr, 65-72, assoc prof comput sci, 72-79, assoc prof, 79-82, PROF ENG, UNIV SC, 82- *Mem:* Inst Elec & Electronics Eng; Am Soc Eng Educ. *Res:* Author of microprocessor texts. *Mailing Add:* Col Eng Univ SC Columbia SC 29208

ECCLESHALL, DONALD, b Warrington, UK, July 8, 27; m 57; c 2. EM TECHNOLOGY, APPLIED PHYSICS. *Educ:* Univ Liverpool, BSc, 52, PhD(physics), 56. *Prof Exp:* Prin sci officer, UK Atomic Energy Authority, Atomic Weapons Res Estab, 56-68; res physicist, radiation physics, 68-72, phys sci adminr, eng tech & physics br, 72-89, PRIN SCIENTIST, ADVAN TECHNOL, US ARMY BALLISTIC RES LAB, 89- *Concurrent Pos:* Res fel nuclear physics, Univ Pa, 66-67. *Honors & Awards:* Res & Develop

Achievement Award, US Army, 78 & 82. *Mem:* Am Phys Soc; Sigma Xi. *Res:* Mathematical modeling of physical processes; radiation and accelerator physics; EM propulsion; terminal ballistics. *Mailing Add:* Ballistic Res Lab US Army LABCOM Aberdeen Proving Ground MD 21005-5066

ECHANDI, EDDIE, b San Jose, Costa Rica, Nov 21, 26; m 52; c 2. PLANT PATHOLOGY. *Educ:* Univ Costa Rica, IngAgr, 51; IICA, Turrialba, Costa Rica, MS, 52; Univ Wis, PhD(plant path), 55. *Prof Exp:* Prof plant path, Univ Costa Rica, 55-61; head plant, Indust & Soils Dept, Inter-Am Inst Agr Sci, 62-64, basic food crops prog, 64-67; PROF PLANT PATH, NC STATE UNIV, 67- *Concurrent Pos:* Vis scientist, Univ Calif, Berkeley, 65; co-leader, Nat Bean Prog, NC State Univ Agr Mission to Peru, Lima, 67-; vis prof, Agrarian Univ, Peru, 68- *Mem:* Fel Am Phytopath Soc; Latin Am Asn Phytopath. *Res:* Diseases of tropical plants; host-parasite relations; ecology of plant disease and plant pathogens; cultivation and production of beans and pulses; bacterial diseases of vegetable crops; bacteriocins and bacteriophages as they relate to bacterial plant pathogens. *Mailing Add:* Dept Plant Path NC State Univ Raleigh NC 27607

ECHEGOYEN, LUIS ALBERTO, b Havana, Cuba, Jan 17, 51; US citizen; m 85; c 1. MAGNETIC RESONANCE, ELECTROCHEMISTRY. *Educ:* Univ PR, BS, 71, PhD(phys chem), 74. *Prof Exp:* Teaching asst phys org chem, Univ Wis-Madison, 74-75; chemist, Union Carbide Co, 75-77; from asst prof to assoc prof phys chem, Univ PR, 77-82; prog officer chem dynamics, NSF, 82-83; ASSOC PROF PHYS CHEM, UNIV MIAMI, CORAL GABLES, 83-; CONSULT, W R GRACE & CO, 84- *Concurrent Pos:* Adj assoc prof phys chem, Univ Md, College Park, 82-83. *Mem:* Am Chem Soc. *Res:* Magnetic resonance studies and electrochemistry of ions in solution; ionic interactions and dynamics. *Mailing Add:* Dept Chem Univ Miami Coral Gables FL 33124

ECHELBERGER, HERBERT EUGENE, b Buffalo, NY, Jan 6, 38; m 64; c 2. FOREST RECREATION. *Educ:* Southern Ill Univ, BS, 65, MS, 66; State Univ NY Col Environ Sci & Forestry, PhD(resource mgt), 76. *Prof Exp:* RES SOCIAL SCIENTIST, FOREST RECREATION RES, FOREST SERV, USDA, 66-, PROF LEADER, 88- *Concurrent Pos:* Adj instr, State Univ NY Col Environ Sci & Forestry, 66-69, adj asst prof, 69-77, adj assoc prof, 77-78; adj assoc prof, Univ NH, 81- *Mem:* Sigma Xi; Soc Am Foresters; Am Forestry Asn. *Res:* Identify the changing characteristics of major forest recreation trends, including participant numbers, participation styles, geographic and size distribution trends, and costs of providing forest recreation opportunities. *Mailing Add:* US Forest Serv PO Box 968 Burlington VT 05402

ECHELBERGER, WAYNE F, JR, b Pierre, SDak, Oct 23, 34; m 60; c 2. ENVIRONMENTAL HEALTH ENGINEERING. *Educ:* SDak Sch Mines & Technol, BS, 56; Univ Mich, MS, 59, MPH, 60, PhD(civil eng), 64. *Prof Exp:* Civil engr, City of Milwaukee, Wis, 56; pub health engr, State of SDak, 56-60; res asst civil eng, Univ Mich, 60 & 61, teaching fel, 60-61, instr, 64-65; asst prof, Univ Notre Dame, 65-67, assoc prof environ health eng, 67-73; prof, Sch Pub & Environ Affairs, Ind Univ, Indianapolis, 73-; AT DEPT CIVIL ENG, UNIV TEX, EL PASO. *Concurrent Pos:* NSF res grant, 66-67 & res equip grant, 67-68; Fed Water Pollution Control Admin demonstration grant, 66-69; consult & secy-treas, TenEch, Inc, 69-77; Environ Protection Agency Water Qual Off res & demonstration grant, 71-72; prin investr, HUD-Nat League of Cities, Urban Observ, 75-77, Ind Higher Educ Comn, Ind State Bd Health, Ind Dept Natural Resources, 78-79, US Environ Protection Agency Munic Environ Res Lab, 80-83 & US Environ Protection Agency Clean Lakes Prog, 81-82. *Honors & Awards:* Harrison Prescott Eddy Medal, Water Pollution Control Fedn, 73. *Mem:* Am Soc Civil Engrs; Water Pollution Control Fedn; Asn Environ Eng Prof (secy-treas, 71-73); Nat Soc Prof Engrs; Am Pub Works Asn; Sigma Xi. *Res:* Solid waste management; environmental quality planning and management; biological and chemical treatment of water and wastewater; industrial waste treatment; studies and control of freshwater eutrophication. *Mailing Add:* Dept Civil Eng & Mech Univ SFla Tampa FL 33620-5350

ECHOLS, CHARLES E(RNEST), b Alderson, WVa, Dec 5, 24; m 60, 77; c 4. CIVIL ENGINEERING. *Educ:* Univ Va, BCE, 49, MCE, 55, LLB, 54. *Prof Exp:* Instr civil eng, Mich State Univ, 55-57; ASST PROF CIVIL & APPL MECH, UNIV VA, 57- *Concurrent Pos:* Vpres construction & eng, A B Torrence & Co, Inc, 50-; vpres, Willson Finance Serv Inc, Staunton; dir, First Va Bank, Charlottesville; mem, Hwy Res Bd, Nat Acad Sci-Nat Res Coun; arbitrator, Am Arbit Asn; officer & dir, many com co. *Mem:* Am Soc Civil Engrs; Sigma Xi. *Res:* Economics and construction. *Mailing Add:* Dept Civil Eng Univ Va Charlottesville VA 22903

ECHOLS, DOROTHY JUNG, b New York, NY, Sept 9, 16; m 41; c 4. GEOLOGY. *Educ:* NY Univ, BA, 36; Columbia Univ, MA, 38. *Prof Exp:* Subsurface geologist & micropaleontologist, Am Repub Corp, 38-41; gen geologist, Foreign Div, Tex Co, 41-42; consult, Pond Fork Oil & Gas Co, 46-51; assoc prof, 51-82, EMER PROF GEOL, WASH UNIV, 82-; CONSULT. *Concurrent Pos:* Partner, Curtis & Echols Consult Geologists. *Honors & Awards:* Neil Meiner Award, 82. *Mem:* Fel AAAS; Paleont Soc; fel Geol Soc Am; Am Asn Petrol Geol; Soc Econ Paleont & Mineral; Sigma Xi; Am Inst Prof Geologists. *Res:* Micropaleontology; biological and morphological studies of microorganisms; some emphasis on Forminifera and Ostracoda; biostratigraphy; subsurface geology; paleoclimatelogy. *Mailing Add:* Curtis & Echols Geol Consults 800 Anderson Bellaire TX 77401

ECHOLS, HARRISON, b May 1, 33; m 54, 73, 77; c 5. GENETICS, BIOCHEMISTRY. *Educ:* Univ Va, BA, 55; Univ Wis, PhD(physics), 59. *Prof Exp:* From asst prof to prof biochem, Univ Wis, 60-69; PROF MOLECULAR BIOL, UNIV CALIF, BERKELEY, 69- *Mem:* Nat Acad Sci; Am Soc Biochem & Molecular Biol; Am Soc Microbiol; Genetics Soc Am. *Res:* Fidelity in DNA replication and its control; regulated initiation of DNA replication and transcription; temporal switches in viral growth. *Mailing Add:* Dept Biochem & Molecular Biol Barker Hall Univ Calif Berkeley CA 94720

ECHOLS, JOAN, b Dayton, Ohio, Jan 31, 32. VERTEBRATE PALEONTOLOGY. *Educ:* Ohio State Univ, BSc, 55; Univ Tex, Austin, MA, 59; Univ Okla, PhD(geol), 72. *Prof Exp:* Instr, 64-66 & 68-73, asst prof, 73-83, ASSOC PROF EARTH SCI, ETEX STATE UNIV, 83- *Mem:* Soc Vert Paleontology; Soc Econ Paleontologists & Mineralogists; Am Asn Petrol Geologists; Sigma Xi. *Res:* Cretaceous reptiles, invertebrates and fishes of Taylor Group, northeast Texas. *Mailing Add:* Dept Earth Sci ETex State Univ Commerce TX 75429

ECHOLS, JOSEPH TODD, JR, b Raleigh, NC, July 5, 36; m 63. PHYSICAL CHEMISTRY. *Educ:* Belhaven Col, BA, 59; Univ Miss, PhD(phys chem), 63. *Prof Exp:* Assoc prof chem, ECarolina Col, 63-64; vis asst prof & fel, La State Univ, 64-65; prof, Belhaven Col, Miss, 65-67; assoc prof, 67-80, head dept, 70-89, PROF CHEM, PFEIFFER COL, 80- *Concurrent Pos:* Consult, Radiation Disposal Syst Inc, 89-90. *Mem:* Am Chem Soc. *Res:* Kinetics; free radical reactions. *Mailing Add:* Dept Chem Pfeiffer Col Misenheimer NC 28109

ECHTENKAMP, STEPHEN FREDERICK, b Fremont, Nebr, Dec 13, 51; m 79. NEURAL CONTROL OF CIRCULATION, AUTONOMIC NERVOUS SYSTEM. *Educ:* Nebr Wesleyan Univ, BS, 74; Univ Nebr, PhD(physiol & biophys), 80. *Prof Exp:* Res assoc physiol, Sch Med, Univ Mo, 79-81; asst prof physiol, Col Med, Tex A&M Univ, 82-85; ASST PROF PHYSIOL, SCH MED, IND UNIV, 86- *Mem:* Am Physiol Soc; Soc Neurosci; Am Heart Asn. *Res:* Neural control of cardiovascular system; baroreflex modulation of renal sympathetic nerve activity; neural control of renal function; interaction of autonomic nervous system and neuropeptides. *Mailing Add:* Ind Univ Sch Med Northwest Ctr Med Educ 3400 Broadway Gary IN 46408

ECHTERNACHT, ARTHUR CHARLES, b Indianapolis, Ind, Sept 3, 39; m 62; c 2. VERTEBRATE ZOOLOGY, ECOLOGY. *Educ:* Univ Iowa, BA, 61; Ariz State Univ, MS, 64; Univ Kans, PhD(zool), 70. *Prof Exp:* Asst prof biol, Boston Univ, 68-75; from asst prof to assoc prof, 75-85, assoc dept head zool, 78-85, PROF, DEPT ZOOL & GRAD PROG ECOL, UNIV TENN, KNOXVILLE, 85-, ACTG DEPT HEAD, 85- *Concurrent Pos:* Ed, Gen Herpet, COPEDA, 86- *Mem:* Am Soc Ichthyologists & Herpetologists; Soc Study Amphibians & Reptiles; Soc Syst Zool; Asn Trop Biol; Ecol Soc Am. *Res:* Systematics and ecology of macroteiid lizards of the genera Ameiva and Cnemidophorus; systematics and ecology of tropical reptiles and amphibians. *Mailing Add:* Dept Zool Univ Tenn Knoxville TN 37996

ECK, DAVID LOWELL, b Bagely, Minn, Nov 21, 41; c 1. ORGANIC CHEMISTRY. *Educ:* Univ Mont, BA, 63; Wash State Univ, PhD(chem), 67. *Prof Exp:* Analytical chemist, Anaconda Co, 63; Petrol Res Fund res fel, Univ Calif, Santa Cruz, 67-69; Sloan Found vis asst prof chem, Reed Col, 69-70; asst prof, 70-74, ASSOC PROF CHEM, CALIF STATE COL, SONOMA, 74- *Concurrent Pos:* Cong fel, Am Chem Soc, 80-81. *Res:* Elucidation of mechanisms in biomolecular elimination reactions involving weak bases; mechanistic considerations involving the formation of small ring oxygen and sulfur heterocycles. *Mailing Add:* 7075 Adele Ave Rohnert Park CA 94928-1679

ECK, GERALD GILBERT, b Marinette, Wis, Mar 8, 42. PHYSICAL ANTHROPOLOGY, PRIMATE PALEONTOLOGY. *Educ:* Univ Chicago, BA, 64; Univ Calif, Berkeley, MA, 74, PhD(anthrop), 77. *Prof Exp:* ASST PROF ANTHROP, UNIV WASH, 75- *Mem:* AAAS; Am Asn Phys Anthropologists. *Res:* Primate paleontology and toxonomy, especially that of hominids and cercopithecoids. *Mailing Add:* Dept Anthrop Univ Wash Seattle WA 98195

ECK, HAROLD VICTOR, b Newkirk, Okla, Nov 14, 24; c 5. SOIL FERTILITY. *Educ:* Okla Agr & Mech Col, BS, 48; Ohio State Univ, PhD(agron), 50. *Prof Exp:* Asst prof agron, Okla Agr & Mech Col, 51-57; soil scientist, Agr Res Serv, USDA, 51-90; RETIRED. *Concurrent Pos:* Assoc ed, Soil Sci Soc Am J, 80-86; vis prof, Univ Ill, 69-70; collabr, Agr Res Serv, USDA, 90. *Mem:* Am Soc Agron; Soil Sci Soc Am; Soil Conserv Soc Am. *Res:* Soil management; soil fertility; plant physiology. *Mailing Add:* USDA Conserv & Prod Res Lab Bushland TX 79012

ECK, JOHN CLIFFORD, b Livingston, Mont, Dec 2, 09; m 35; c 3. FUEL TECHNOLOGY, ENVIRONMENTAL SCIENCE. *Educ:* Mont State Col, BS, 31; Univ Ill, MS, 32, PhD(org chem), 35. *Prof Exp:* Res assoc, Iowa State Univ, 35-41; res chemist, Air Reduction Co, 41-45; res engr, Allied Chem Corp, 45-72; CONSULT CONSTURCT, WILPUTTE CORP & HYDROTECHNIC CORP, 72- *Concurrent Pos:* Lectr, China & Russia. *Mem:* Fel Am Inst Chemists; Am Inst Chem Engrs; Am Chem Soc; NY Acad Sci. *Res:* Basic organic chemicals; coal gasification; oil gasification; water pollution control; incineration. *Mailing Add:* 27 Kitchell Rd Covent NJ 07961

ECK, JOHN STARK, b W Hempstead, NY, Mar 18, 41; m 64; c 3. NUCLEAR PHYSICS, SOLID STATE PHYSICS. *Educ:* Polytech Inst Brooklyn, BS, 62; Johns Hopkins Univ, PhD(physics), 67. *Prof Exp:* Jr instr physics, Johns Hopkins Univ, 62-65, res asst, 65-67; res assoc, Fla State Univ, 67-69; from asst prof to prof physics, Kans State Univ, 69-85, dept head, 81-83; prof physics & assoc dean grad studies & res, Univ Toledo, 85-89; PROF PHYSICS & ASSOC VPRES RES & GRAD STUDIES, OLD DOMINION UNIV, 89- *Concurrent Pos:* Mem, Coun Grad Sch & Coun Southern Grad Schs. *Mem:* Am Phys Soc; Sigma Xi. *Res:* Nuclear and solid state properties from coulomb excitation Mossbauer studies; nuclear heavy ion interactions from elastic and inelastic scattering of 0-16 and He-4 from medium weight nuclei; optical model interpretation of nuclear scattering of protons, alphas and heavy ions; nuclear instrumentation; nuclear interactions of the light heavy ions lithium-7 and beryllium-9; study of solids using nuclear techniques; laser annealing. *Mailing Add:* Off Res & Grad Studies Old Dominion Univ Norfolk VA 23529-0013

ECK, PAUL, b Elizabeth, NJ, Sept 3, 31; m 55; c 4. HORTICULTURE. *Educ:* Rutgers Univ, BS, 53; Univ Mass, MS, 55; Univ Wis, PhD(soils), 57. *Prof Exp:* Asst prof floricult, Univ Mass, 57-60; assoc prof, 60-70, PROF POMOL, RUTGERS UNIV, 70- *Mem:* Am Soc Hort Sci. *Res:* Physiology and nutrition of fruit crops, especially apple, blueberry and cranberry. *Mailing Add:* Dept Forestry Rutgers Univ New Brunswick NJ 08903

ECK, ROBERT EDWIN, b Ames, Iowa, Nov 28, 38; m 74; c 2. PHYSICS. *Educ:* Rutgers Univ, BA, 60; Univ Pa, MS, 62, PhD(physics), 66; Univ Calif, Santa Barbara, MA, 74(economics). *Prof Exp:* Sr scientist, Ford Sci Lab, Ford Motor Co, 66-69; mem tech staff, 69-75, mgr, IR Components Res & Develop Lab, 76-78, asst mgr IR components, 78-80, prog mgt, 80-85, DIR ENGR & TECHNOL, SANTA BARBARA RES CTR, 86- *Mem:* AAAS. *Res:* New technology in infrared detectors, including new detector materials, focal plane design concepts and computer-controlled display readout of infrared detectors. *Mailing Add:* 6539 Camino Venturoso Goleta CA 93117

ECK, THOMAS G, b Genoa, NY, Oct 19, 29; m 59. PHYSICS. *Educ:* Univ Buffalo, BA, 51; Columbia Univ, PhD(physics), 58. *Prof Exp:* From asst prof to assoc prof, 62-69, PROF PHYSICS, CASE WESTERN RESERVE UNIV, 69- *Mem:* Am Phys Soc. *Res:* Low temperature physics; solid state physics; atomic spectroscopy. *Mailing Add:* Dept Physics Case Western Reserve Univ Cleveland OH 44106

ECKARD, KATHLEEN, US citizen. FLORAL DEVELOPMENTAL MORPHOLOGY. *Educ:* Univ Calif, Riverside, BA, 69, MSc, 71. *Prof Exp:* Staff res assoc, dept plant path, 72-78, STAFF RES ASSOC BOT, DEPT BOT & PLANT SCI, UNIV CALIF, RIVERSIDE, 79- *Mem:* Sigma Xi. *Res:* Structure-function relationships; pollination events and the breeding system; computer-aided reconstruction. *Mailing Add:* Dept Bot & Plant Sci Univ Calif Riverside CA 92521

ECKARDT, MICHAEL JON, b Glendale, Calif, Apr 3, 43; m 68; c 3. MEDICAL PSYCHOLOGY, ALCOHOLISM. *Educ:* Calif State Univ, Northridge, BA 66; Univ Southern Calif, MS, 67; Univ Mich, MS, 70; Univ Ore Health Sci Ctr, PhD(psychol), 75. *Prof Exp:* Lectr, dept zool, Univ Mich, 70; res psychobiologist, Psychobiol Sect, Calif Vet Admin Hosp, Sepulveda, 75-76; fel, dept psychiat, Col Med, Univ Calif, Irvine, 76; res psychologist, clin & biobehav res, 76-78, res psychologist, Lab Preclin Studies, Div Intramural Res, 78-83, CHIEF, SECT CLIN BRAIN RES, LAB CLIN STUDIES, NAT INST ALCOHOL ABUSE & ALCOHOLISM, 83- *Mem:* AAAS; Soc Neurosci; Am Psychol Asn. *Res:* Investigating the acute and chronic effects of alcohol and other addictive substances on various anatomic and physiologic systems; studying the diagnostic utility of automated batteries of clinical laboratory tests. *Mailing Add:* Lab Clin Studies Nat Inst Alcohol Abuse & Alcoholism Bldg 10 NIH Rm 3C-102 9000 Rockville Pike Bethesda MD 20892

ECKARDT, ROBERT E, b Fanwood, NJ, May 1, 16; m 65; c 4. BIOCHEMISTRY, TOXICOLOGY. *Educ:* Antioch Col, BS, 37; Western Reserve Univ, MS, 39, PhD(biochem), 40, MD, 43; Am Bd Internal Med, dipl, 50; Am Bd Prev Med, dipl, 55. *Prof Exp:* Intern, NY Hosp, 43-44, asst resident med, 44, 47-48; spec res physician, Standard Oil Co, NJ, 48-50, asst dir med res sect, 50; dir med res div, Esso Res & Eng Co, 51-74, assoc med dir, Exxon Corp, 74-78; RETIRED. *Concurrent Pos:* Asst attend physician, Outpatient Dept, NY Hosp, 48-70, physician to outpatients, 70-78; instr med, Med Col, Cornell Univ, 43-70, assoc clin prof, 70-78; assoc clin prof, Postgrad Med Sch, NY Univ-Bellevue Med Ctr, 53-78; Distinguished Fel Toxicol Forum, 86. *Honors & Awards:* Knudsen Award, Indust Med Asn, 65; Robert A Kehoe Award, Am Acad Occup Med, 78. *Mem:* Fel Am Col Physicians; AMA; Am Indust Hyg Asn; Am Occup Med Asn (vpres, 57-59, pres, 60); NY Acad Med. *Res:* Medicine; industrial hygiene. *Mailing Add:* 6720 E Kelton Lane Scottsdale AZ 85254

ECKBLAD, JAMES WILBUR, b Minneapolis, Minn, May 13, 41; m 65; c 2. AQUATIC ECOLOGY. *Educ:* Univ Minn, BS, 64; Cornell Univ, PhD(limnol), 71. *Prof Exp:* PROF BIOL, LUTHER COL, 71- *Concurrent Pos:* Proj dir, NSF Undergrad Res Participation, 73-; consult, Interstate Power Co, 73-74; mem citizen's adv comt, Upper Miss Basin Planning Comn, 75- *Mem:* Sigma Xi; AAAS; Ecol Soc Am; Am Soc Limnol & Oceanog. *Res:* Ecology and productivity of backwater lakes of the upper Mississippi River; computer simulations and modelling. *Mailing Add:* Dept Biol Luther Col Decorah IA 52101

ECKEL, EDWIN BUTT, b Washington, DC, Jan 27, 06; wid; c 3. GEOLOGY. *Educ:* Lafayette Col, BS, 28; Univ Ariz, MS, 30. *Prof Exp:* Geologist, US Geol Surv, 30-42, in charge invests domestic quicksilver deposits, 42-43, asst chief mil geol unit, 44-45, chief eng geol br, 45-61, chief spec proj br, 62-65, res geologist, 65-68; ed, 68-71, exec secy, Geol Soc Am, 70-74; res geologist, US Geological Surv, 74-84; RETIRED. *Concurrent Pos:* Chmn, Comt Landslide Invest, Hwy Res Bd, 51-62; mem comt Alaska earthquake, Nat Acad Sci, 64-71. *Honors & Awards:* Burwell Award, Geol Soc Am, 71. *Mem:* Fel Geol Soc Am; fel Mineral Soc Am; hon mem Asn Eng Geol (pres, 65); hon fel Geol Soc London; hon mem Asn Earth Sci Editors; Soc Econ Geologists. *Res:* Engineering geology; ore deposits; German underground factories; Italian quicksilver industry; geology of Paraguay; geology of underground nuclear explosions; geology of Alaska earthquake; technical writing and minerals of Colorado; history of Geological Society of America. *Mailing Add:* PO Box 12000 Denver CO 80212-0020

ECKEL, FREDERICK MONROE, b Philadelphia, Pa, Mar 25, 39; m 63; c 2. PHARMACY. *Educ:* Philadelphia Col Pharm, BSc, 61; Ohio State Univ, MSc, 63. *Prof Exp:* Resident pharm, Ohio State Univ, 61-63, supvr pharmacists, Ohio State Univ Hosp, 63-65, asst dir pharm, 63-66; from instr to assoc prof, 66-77, PROF HOSP PHARM, UNIV NC, CHAPEL HILL, 77-, CHMN DEPT PHARM PRACT, 75- *Concurrent Pos:* Pharm consult, St Ann's Hosp for Women, 64-66; dir, Plan Pharm Assistance, 66-72, Duke Endowment & Reynolds Found Plan of Pharm Assistance grant, 66-70; dir pharm serv, NC Mem Hosp, 68-73; assoc dir pharm serv, NC Mem Hosp, 85-; Fel, Am Soc Hosp Pharmacists, 88. *Mem:* Am Soc Hosp Pharmacists (pres, 74-75); fel AAAS; Am Asn Col Pharm; Am Pharmaceut Asn. *Res:* Development, improvement and implementation of professional pharmaceutical services; drug utilization review; pharmacy practice issues; pharmacy service to small hospitals and nursing homes. *Mailing Add:* 713 Churchill Dr Chapel Hill NC 27514

ECKEL, ROBERT EDWARD, b Buffalo, NY, Mar 17, 18; m 47; c 3. MEDICINE. *Educ:* Dartmouth Col, BA, 38; Harvard Univ, MD, 42. *Prof Exp:* Fel med, Sch Med, Case Western Reserve Univ, 48-49, Am Cancer Soc fel biochem, 49-51, Nat Found Infantile Paralysis fel, 51-53; from asst prof to assoc prof, 53-83, PROF MED, SCH MED, CASE WESTERN RESERVE UNIV, 83- *Res:* Mechanism of ion transport; renal disease. *Mailing Add:* 2065 Adelbert Rd Cleveland OH 44106

ECKELBARGER, KEVIN JAY, b Goshen, Ind, May 27, 44; m 68. INVERTEBRATE ZOOLOGY. *Educ:* Calif State Univ, Long Beach, BA, 67, MA, 69; Northeastern Univ, PhD(biol), 74. *Prof Exp:* Teaching asst biol, Calif State Univ, Long Beach, 67-69, Northeastern Univ, 69-73; asst res scientist, 73-79, assoc res scientist, 79-81, SR RES SCIENTIST, HARBOR BR FOUND, INC, 81- *Mem:* Am Soc Zoologists; Sigma Xi. *Res:* Reproductive biology and development Polychaetous annelids; gametogenesis, spawning behavior, larval settlement behavior of marine invertebrates. *Mailing Add:* Harbor Br Oceanog Inst 5600 Old Dixie Hwy Ft Pierce FL 34946

ECKELMAN, CARL A, b Columbus, Ind, Feb 14, 33; m 61; c 3. WOOD SCIENCE, STRUCTURAL ENGINEERING. *Educ:* Purdue Univ, BS, 59, MS, 62, PhD(wood sci), 68. *Prof Exp:* Asst, 59-63, from instr to assoc prof, 63-79, PROF WOOD SCI, PURDUE UNIV, 79- *Res:* Furniture engineering; wood moisture relations; basic fiber science. *Mailing Add:* Dept Forestry Purdue Univ West Lafayette IN 47907

ECKELMANN, FRANK DONALD, b Englewood, NJ, May 25, 29; m 53; c 2. GEOLOGY. *Educ:* Wheaton Col, Ill, BS, 51; Columbia Univ, MS, 54, PhD, 56. *Prof Exp:* Res asst geochem, Lamont-Doherty Geol Observ, Columbia Univ, 51-52, teaching asst geol, 52-55, res assoc geochem, 56-57; from asst prof to prof geol sci, Brown Univ, 57-78, chmn dept, 61-68, dean col, 68-71; prof & head dept geol, Univ Ga, 78-81; dean, Col Arts & Sci, George Mason Univ, 81-85; DEAN, COL ARTS & SCI, OHIO UNIV, 85- *Mem:* Geol Soc Am; Geochem Soc; Mineral Soc Am; Am Geophys Union. *Res:* Nature, accurrence and geologic significance of zircon in igneous and metamorphic rocks. *Mailing Add:* Dean Col Arts & Sci George Mason Univ 4400 Univ Dr Fairfax VA 22030

ECKELMANN, WALTER R, b Englewood, May 25, 29; m 51; c 3. GEOCHEMISTRY. *Educ:* Wheaton Col, Ill, BS, 51; Columbia Univ, MA, 54, PhD(geochem), 56. *Prof Exp:* Res asst geochem, Columbia Univ, 51-55, res assoc, 55-57; res chemist, Jersey Prod Res Co, 57-59, sr res chemist, 59-62, sect head, 62-64, res mgr, European Lab, Esso Prod Res Co, Standard Oil Co, NJ, 64-66, dist prod geologist, Okla & oper mgr, New Orleans, Humble Oil & Refining Co, 66-70, gen mgr, Geol Res, 70-72, pres, 72-75, opers mgr, Explor Dept, Exxon Co USA, Houston, 75-76, dir, Esso Australia, Sydney, 76-78, dep mgr, Sci & Technol Dept, Exxon Corp, Florham Park, NY, 78-82; sr vpres, Sohio Prod Co, Houston, 83-85; PRES, RCB CO, ROANOKE, TEX, 86- *Mem:* AAAS; Geol Soc Am; Am Geochem Soc; Am Chem Soc; Am Asn Petrol Geologists; Sigma Xi. *Res:* Isotope geochemistry; oil and gas operations. *Mailing Add:* 121 Seminole Dr Roanoke TX 76262

ECKELMEYER, KENNETH HALL, b Philadelphia, Pa, June 20, 43; m 65; c 2. PHYSICAL METALLURGY. *Educ:* Lafayette Col, BS, 65; Lehigh Univ, MS, 67, PhD(metall), 71. *Prof Exp:* Mem tech staff metall, 71-79, DIV SUPVR, ELECTRON OPTICS & X-RAY ANALYSIS, SANDIA LABS, 79- *Mem:* Am Soc Metals; Inst Metallog Soc. *Res:* Phase transformations; microstructural characterization; microstructure-mechanical property relations in metals, especially uranium alloys. *Mailing Add:* 8813 Harwood Ave NE Albuquerque NM 87111

ECKELS, ARTHUR R(AYMOND), b New Haven, Conn, Nov 16, 19; m 44; c 4. ELECTRICAL ENGINEERING. *Educ:* Univ Conn, BS, 41; Harvard Univ, MS, 42; Yale Univ, DEng, 50. *Prof Exp:* Elec engr, Bur Ships, USN, 42-43; from marine engr to chief engr, US Merchant Marine, 43-46; instr elec eng, Yale Univ, 47-49; prof, NC State Col, 49-56; prof & chmn dept, Univ Vt, 56-61; PROF ELEC ENG, NC STATE UNIV, 61- *Concurrent Pos:* Res partic, Oak Ridge Inst Nuclear Studies, 53; opers analyst, USAF, 54-; Fulbright lectr, Chiao Tung Univ, 60; consult, NASA, 63-64; vis prof, Japan Nat Defense Col, 64; sr elec engr adv, Environ Protection Agency, 78-82; prog coordr, NC Cent Univ, 80-82. *Mem:* Am Soc Eng Educ; Inst Elec & Electronics Engrs; Sigma Xi. *Res:* Electrical instrumentation and control. *Mailing Add:* 1417 Dellwood Dr Raleigh NC 27607

ECKELS, KENNETH HENRY, b Baltimore, Md, Oct 11, 42; m 65, 88; c 1. VIROLOGY, VACCINE DEVELOPMENT. *Educ:* Univ Md, BS, 65, MS, 69, PhD(microbiol), 73. *Prof Exp:* Biol lab technician, Dept Hazardous Microorganisms, Walter Reed Army Inst Res, 65-66, microbiologist, 66-81, asst chief, 81-86, CHIEF, DEPT BIOL RES, WALTER REED ARMY INST RES, 86- *Concurrent Pos:* Adj assoc prof, Univ Md, 80- *Mem:* AAAS; Sigma Xi; Am Soc Microbiol; Am Soc Trop Med & Hyg. *Res:* Viral vaccine development including the development of live attenuated dengue virus vaccines; inactivated hepatitis A vaccine and other experimental vaccines. *Mailing Add:* 10607 Weymouth St Apt 102 Bethesda MD 20814

ECKENFELDER, WILLIAM WESLEY, JR, b New York, NY, Nov 15, 26; m 50; c 2. SANITARY ENGINEERING. *Educ:* Manhattan Col, BCE, 46; Pa State Univ, MS, 48; NY Univ, MCE, 56. *Hon Degrees:* DSc, Manhattan Col, 90. *Prof Exp:* Sanit engr, Atlantic Refining Co, 48-49; res assoc, NY Univ, 49-50; from asst prof to assoc prof civil eng, Manhattan Col, 55-65; prof environ health eng, Univ Tex, 65-70; distinguished prof environ & water resources eng, Vanderbilt Univ, 70-89; SR TECH DIR, ECKENFELDER INC, 89- *Concurrent Pos:* Vpres, Weston, Eckenfelder & Assocs, 52-56; consult, 56-; pres, Hydrosci Inc; Assoc Water & Air Resources Engrs, Tenn; chmn bd, Aware, Inc; exec dir, ctr indust water qual mgt. *Honors & Awards:* Gold Medal, Synthetic Organic Chem Mgrs Asn, 74; Camp Medal, Water Pollution Control Fed, 81; Imhoff-Koch Medal, Int Asn Water Pollution Res & Control, 90. *Mem:* Am Soc Civil Engrs; Am Chem Soc; Am Inst Chem Engrs; Water Pollution Control Fedn; hon mem Int Asn Water Pollution Res & Control. *Res:* Biological treatment of sewage and industrial wastes; mass transfer and aeration in waste treatment; process design of industrial waste treatment plants; water quality management. *Mailing Add:* Eckenfelder Inc 227 French Landing Dr Nashville TN 37228

ECKENHOFF, JAMES BENJAMIN, b Durham, NC, Mar 4, 43. BIOENGINEERING. *Educ:* Univ Pa, AB, 66; Northwestern Univ, MS, 73. *Prof Exp:* Jr bioengr, 73-74, sr bioengr & asst proj leader, 74-78, PROJ LEADER DEVELOP ENGR & AREA DIR BIOTECHNOL GROUP & PROG DIR RES, OSMOTIC SYSTS, ALZA RES, ALZA CORP, 78- *Mem:* Biomed Eng Soc. *Res:* Transport phenomena, reverse osmosis, optics, and instrumentation in the design of drug delivery systems. *Mailing Add:* 1080 Autumn Lane Los Altos CA 94022

ECKENHOFF, JAMES EDWARD, b Easton, Md, Apr 2, 15; m 38, 73; c 4. ANESTHESIOLOGY. *Educ:* Univ Ky, BS, 37; Univ Pa, MD, 41; Am Bd Anesthesiol, dipl. *Hon Degrees:* DSc, Transylvania Univ, 70. *Prof Exp:* Harrison fel anesthesiol, Sch Med, Univ Pa, 45-47; asst instr pharmacol, Sch Med, Univ Pa, 45-47, from asst instr to asst prof surg, 44-52, from assoc prof to prof anesthesiol, 52-55, asst dir dept, 53-65, assoc clin pharmacol, 48-65; prof, Med Sch, Northwestern Univ, 66-85, chmn dept, 66-70, dean, med sch, 70-83, pres, McGaw Med Ctr, 80-85; EMER PROF ANESTHESIA & DEAN VET ADMIN LAKESIDE MED CTR, CHICAGO, ILL, 85- *Concurrent Pos:* Asst surgeon, Children's Hosp, Pa, 49-53, consult, 53-65; consult, Valley Forge Army Hosp, Pa, 49-58, Anesthesiol Ctr, WHO, Denmark, 52, Vet Admin Hosp, Pa, 53-65 & US Naval Hosp, Pa; assoc ed, Anesthesiol, Am Soc Anesthesiol, 55-58, ed, 58-62; consult to surgeon gen, US Navy; mem surg study sect, NIH, 62-66, mem anesthesia training grants comt, 66-70; Hunterian prof, Royal Col Surg, 65; dir, Am Bd Anesthesiol, 65-73, pres, 72-73. *Mem:* Am Soc Anesthesiol; Am Physiol Soc; AMA; fel Am Col Anesthesiol; Am Col Physicians; Asn Univ Anesthetists (pres, 62); Soc Acad Anesthesiol Chmn (pres, 67-68). *Res:* Physiological and pharmacological problems pertaining to coronary circulation; effects of opiates and antagonists upon normal and anesthetized man and the effect of changing carbon dioxide tensions upon the heart and circulation; deliberate hypotension; clinical anesthesiological problems; vertebral venous plexus. *Mailing Add:* 8601 N State Rd 39 La Porte IN 46350

ECKENWALDER, JAMES EMORY, b Neuilly-sur-Seine, France, Oct 30, 49; US citizen; m 71. SYSTEMATIC BOTANY. *Educ:* Reed Col, BA, 71; Univ Calif, Berkeley, PhD(bot), 77. *Prof Exp:* Asst taxonomist, Fairchild Trop Garden, Fla, 77-78; asst prof, 78-85, ASSOC PROF BOT, UNIV TORONTO, 85- *Mem:* Am Soc Plant Taxonomists; Bot Soc Am; Int Asn Plant Taxon; Soc Study Evolution; Soc Systematic Zool. *Res:* Systematics and phylogeny of gymnosperms, Salicaceae and miscellaneous angiosperms; taxonomy of cultivated plants. *Mailing Add:* Dept Bot Univ Toronto 25 Willcocks St Toronto ON M5S 3B2 Can

ECKER, DAVID JOHN, b Pa, 12, 54; m 76; c 2. MOLECULAR BIOLOGY. *Educ:* Trenton State Col, BA, 76; Utah State Univ, PhD(biochem), 82. *Prof Exp:* Postdoctoral fel, Dept Chem, Univ Calif, Berkeley, 82-84; postdoctoral scientist, Dept Molecular Pharmacol & Med Chem, Smith Kline & French Labs, 84-86, assoc sr investr, Dept Molecular Pharmacol, 86-88, sr investr, 88, asst dir, Dept Molecular Genetics, 88-89; DIR, DEPT MOLECULAR & CELL BIOL, ISIS PHARMACEUT, INC, 89- *Mem:* Am Chem Soc; AAAS; Am Soc Biochem & Molecular Biol. *Res:* Biophysical, biochemical and cell biology experiments directed towards antisense drug discovery; coordination chemistry of natural products and the molecular biology of protein degradation; author & co-author of 37 publications. *Mailing Add:* Molecular & Cell Biology Isis Pharmaceutics 2280 Faraday Ave Carlsbad CA 92008

ECKER, EDWIN D, b Grovertown, Ind, Mar 26, 34; m 55; c 3. MATHEMATICS. *Educ:* Ball State Univ, BS, 56; Univ Ill, Urbana, MS, 59; Iowa State Univ, PhD(math), 66. *Prof Exp:* From instr to assoc prof, 59-73, PROF MATH, MACMURRAY COL, 73- *Concurrent Pos:* Vis lectr, Univ Ill, Urbana, 70-71. *Mem:* Math Asn Am; Nat Coun Teachers Math; Asn Comput Mach. *Res:* Group theory. *Mailing Add:* Dept Math MacMurray Col E College Ave Jacksonville IL 62650

ECKER, HARRY ALLEN, b Athens, Ga, Oct 22, 35; m 59; c 3. ELECTRICAL ENGINEERING. *Educ:* Ga Inst Technol, BEE, 57, MSEE, 59; Ohio State Univ, PhD(elec eng), 65. *Prof Exp:* Res asst, Ga Inst Technol, 57-59; proj engr, Navig & Guid Lab, Wright-Patterson AFB, USAF, 59-60, syst prog officer, 60-61, electronic engr opers anal br, Synthesis & Anal Div, 61-62, actg chief, 62-63, aerospace engr, directorate of synthesis, dep for studies & anal, 63-65, chief opers anal group, 65-66; sr res engr, Radar Lab, Eng Exp Sta, Ga Inst Technol, 66-69, head radar br, Electronics Div, 69-76; gen mgr, Electro-Prod Div, 77-79, vpres res & develop, 79, vpres telecommun, 79-82, DIR RES, SCI-ATLANTA, INC, 76-, VPRES TELECOMMUN, 79- *Mem:* Fel Inst Elec & Electronics Engrs; Sigma Xi. *Res:* Antennas; radar; systems analyses; bio-engineering; science communications. *Mailing Add:* 3267 Ivanhoe Dr NW Atlanta GA 30327

ECKER, JOSEPH GEORGE, b Flint, Mich, Jan 14, 42. MATHEMATICAL PROGRAMMING. *Educ:* Univ Mich, Ann Arbor, BA, 64, MS, 66, PhD(math), 68. *Prof Exp:* From asst prof to assoc prof, 68-78, PROF MATH SCI, RENSSELAER POLYTECH INST, 78-, CHMN DEPT, 84- *Concurrent Pos:* Vis prof, opers res, Ctr Opers Res & Econometrics, Belg, 75-76 & Ecole Polytechnique Federale, Lausanne, Switz, 83-84; consult, Gen Motors Res Lab, 85- & Gen Elec Corp, 88- *Mem:* Am Math Soc; Opers Res Soc Am; Math Prog Soc. *Res:* Mathematical programming, linear and nonlinear programming; multiple objective optimization; algorithm development and evaluation; geometric programming. *Mailing Add:* Dept Math Rensselaer Polytech Inst Troy NY 12180

ECKER, RICHARD EUGENE, b Waverly, Iowa, Mar 13, 30; m 53; c 5. PHYSIOLOGY. *Educ:* Iowa State Univ, BS, 58, PhD(bact), 61. *Prof Exp:* Instr bact, Iowa State Univ, 60-61; instr microbiol, Col Med, Univ Fla, 62-64; from asst biologist to assoc biologist, Argonne Nat Lab, 64-73; pres, Vitose Corp, 73-81; CONSULT PHYSIOLOGIST, 81- *Concurrent Pos:* Nat Cancer Inst fel, 61-62; NIH res grant, 62-64; distinguished vis prof, Morehouse Col, 69-70; instr physiol, Marine Biol Lab, 72. *Res:* Regulation of biological function. *Mailing Add:* 8618 Meadowbrook Dr Hinsdale IL 60521

ECKERLE, KENNETH LEE, b Jasper, Ind, Oct 18, 36. APPLIED PHYSICS. *Educ:* Ind State Univ, BS, 58; Univ Md, MS, 62. *Prof Exp:* Asst physics, Univ Md, 58-60, asst solid state physics, 60-62; atomic physicist, Nat Bur Standards, 62-67; PHYSICIST, NAT INST STANDARDS & TECHNOL, 67- *Mem:* Optical Soc Am; Am Soc Testing & Mat; Coun Optical Radiation Measurement. *Res:* Standards and measurements of transmittance and reflectance in the ultraviolet, visible and infrared spectral regions; spectrophotometry, instrument development; fluorescence; retroreflectance. *Mailing Add:* Rm B306 Metrol Bldg Nat Inst Standards & Technol Gaithersburg MD 20899

ECKERLIN, HERBERT MARTIN, b New York, NY, Oct 23, 35; m 57; c 4. ENERGY CONSERVATION, SOLAR ENERGY. *Educ:* Va Polytech Inst, BS, 58; NC State Univ, MS, 68, PhD(eng sci), 72. *Prof Exp:* Test eng, Norfolk Naval Shipyard, 58-59; efficiency engr, Va Elec & Power Co, 59-60; design engr, Combustion Eng, 60-65; sr res engr, Corning Glass Works, 65-68; exten specialist mech, 68-73, asst prof, 73-76, ASSOC PROF MECH ENG, NC STATE UNIV, 76- *Concurrent Pos:* Energy consult, Indust Com Cos, 75-; dir, NC State Univ Walk-through Prog, 77-; pres, Energy Conserve Limited, 78-; prin investr, NC State Univ Solar House, 79-; seminar leader, McGraw-Hill, 79- *Mem:* Am Soc Mech Engrs; Sigma Xi. *Res:* Evaluation of the energy conservation potential of all types of industrial processes and systems; passive and active solar thermal systems. *Mailing Add:* 4313 Azalea Dr Raleigh NC 27612

ECKERLIN, RALPH PETER, b New York, NY, Feb 10, 38; m 79; c 2. HELMINTHOLOGY, MEDICAL ENTOMOLOGY. *Educ:* Rutgers Univ, AB, 60; Univ Miami, MS, 62; Univ Conn, PhD(parasitol), 75. *Prof Exp:* Res asst, Univ Miami, 60-62; res biologist, Lederle Labs, Am Cyanamid Co, 62-66; teaching asst biol, Univ Conn, 66-70, lectr, 70-71; PROF BIOL, NORTHERN VA COMMUNITY COL, 71- *Concurrent Pos:* Prof lectr parasitol, George Washington Univ, 80-; ed, J Helminthological Soc Wash, 89- *Mem:* Am Soc Parasitologists; Am Soc Trop Med & Hyg; Wildlife Dis Asn; Entom Soc Am. *Res:* Systematics and life histories of helminth parasites and siphonapterans of wild life. *Mailing Add:* Natural Sci Div Northern Va Community Col Annandale VA 22003

ECKERMAN, JEROME, b Brooklyn, NY, Nov 18, 25; m 48; c 2. PHYSICS, ELECTRICAL ENGINEERING. *Educ:* Worcester Polytech Inst, BS, 48; Cath Univ, MS, 56, PhD(physics), 58. *Prof Exp:* Res scientist, Nat Adv Comt Aeronaut, 48-51; physicist & br chief, US Naval Ord Lab, 51-59; sr staff scientist, Avco, 59-65, assoc sect chief appl physics, Avco Corp, 65-68; physicist, NASA Electronics Res Ctr, Mass, 68-70, physicist/br chief, Microwave Sensor Br, Goddard Flight Ctr, 70-81; sr scientist, Syst Planing Corp, Arlington, VA, 81-87; ADV ENG, WESTINGHOUSE ELEC CORP, BALTIMORE, MD, 87- *Honors & Awards:* Centennial Medal & Geosci Electronics Outstanding Serv Award, Inst Elec & Electronics Engrs. *Mem:* Am Phys Soc; sr mem Inst Elec & Electronics Engrs; Am Meteorol Soc; Sigma Xi. *Res:* Ballistics range research; chemical kinetics in air and alkali metal plasmas; laminar wake transition behind hypervelocity models; turbulent wake growth; laboratory studies of flow field observables; flow analysis by interferometry; experimental and analytical development in microwave radars and radiometers for space applications; light gas gun development. *Mailing Add:* 11817 Hunting Ridge Ct Potomac MD 20854

ECKERSLEY, ALFRED, b Manchester, Eng, Dec 2, 28; nat US; m 57; c 2. ELECTRICAL ENGINEERING. *Educ:* Col Tech, Eng, BSc, 49; Univ Pa, MSEE, 54. *Prof Exp:* Assoc elec eng, Univ Pa, 49-57; lectr & res assoc, Univ NMex, 57-59; elec engr, Ark Electronics Corp, 59-61; res engr, United Control Corp, 61-64; RES ENGR, AEROSPACE GROUP, BOEING CO, 64- *Concurrent Pos:* US deleg, Spec Int Comt Radioelec Perturbations. *Mem:* Sr mem Inst Elec & Electronics Eng; assoc mem Sigma Xi. *Res:* Electronics; avionics; radio noise; interference measurement; electromagnetic compatibility. *Mailing Add:* 16745 Maplewild SW Seattle WA 98166

ECKERT, ALFRED CARL, JR, b Newark, NJ, June 12, 20; m 44; c 4. ANALYTICAL CHEMISTRY. *Educ:* Wheaton Col, BS, 41; Univ Ill, PhD(anal chem), 45. *Prof Exp:* Lab asst chem, Wheaton Col, 39-41; asst, Univ Ill, 41-44; jr chemist, Univ Chicago, 44-45; res chemist, Chem Div, Union Carbide Corp, Tenn, 45-48; res engr, Battelle Mem Inst, 48-52; tech personnel off rep, Chem Div, Union Carbide Corp, Tenn, 53-56, tech serv group coordr chem & spectros, Speedway Labs, 56-58; sr exp res chemist, Allison Div, Gen Motors Corp, 58-65; mgr, Appl Res Group, Globe-Union Inc, 65-78, MGR, MAT TESTING LAB, APPL RES GROUP, JOHNSON CONTROLS INC, 78- *Concurrent Pos:* Lectr, Univ Tenn, 47-48. *Mem:* Am Chem Soc; Soc Appl Spectros; fel Am Sci Affil; fel Am Inst Chem; Sigma Xi. *Res:* Mesomorphic state with special reference to carbon blacks; effects of magnetism on dislocation movements in nickel foil; quantative analysis of lead alloys by x-ray fluorescence; significance of blood lead analyses; working reference materials for analytical quality evaluation; trends in requirements for lead acid battery materials. *Mailing Add:* 4605 N 107th St Wauwatosa WI 53225

ECKERT, BARRY S, b Binghamton, NY, Dec 16, 49; m 71; c 2. CELL BIOLOGY, ANATOMY. *Educ:* State Univ NY Albany, BS, 71, MS, 73; Univ Miami, PhD(anat), 76. *Prof Exp:* Res assoc cell biol, Univ Colo, 76-77; asst prof, 77-83, ASSOC PROF ANAT, STATE UNIV NY BUFFALO, 83- *Concurrent Pos:* Prin investr, NIH basic Sci res grants, State Univ NY Buffalo, 77-78, NSF grant, 78-79 & 81-84, Western NY Heart Asn, 84-86. *Mem:* Am Soc Cell Biol; Sigma Xi; Soc Develop Biol. *Res:* Studying the role of microtubules, intermediate filaments and microfilaments in motility of cultured mammalian cells; studying role of phosphorylation in control of intermediate filament organization and function. *Mailing Add:* Dept Anat Sci State Univ NY at Buffalo 317 Farber Hall Buffalo NY 14214

ECKERT, CHARLES, surgery; deceased, see previous edition for last biography

ECKERT, CHARLES ALAN, b St Louis, Mo, Dec 13, 38; m 61; c 2. SANITARY & ENVIRONMENTAL ENGINEERING. *Educ:* Mass Inst Technol, BS, 60, MS, 61; Univ Calif, Berkeley, PhD(chem eng), 64. *Prof Exp:* NATO fel high pressure physics, High Pressure Lab, Nat Ctr Sci Res, Bellevue, France, 64-65; from asst prof to prof, Univ Ill, 65-89, head, Dept Chem Eng, 80-86; PROF CHEM ENG & CHEM, GA TECH, 89- *Concurrent Pos:* NATO fel, 64-65; consult various companies; Guggenheim fel, 71; vis prof, Stanford Univ, 71-72. *Honors & Awards:* Allan Colburn Award, Am Inst Chem Engrs, 73; Ipatieff Prize, Am Chem Soc, 77. *Mem:* Nat Acad Eng; Am Inst Mining, Metall & Petrol Engrs; Am Chem Soc; Am Inst Chem Engrs; Am Soc Eng Educ; Chem Soc London. *Res:* Molecular thermodynamics and applied chemical kinetics; effects of high pressure on reactions in solution; phase equilibria at high pressure and temperature. *Mailing Add:* Sch Chem Eng Ga Tech Atlanta GA 30332-0100

ECKERT, DONALD JAMES, b Akron, Ohio, Aug 30, 49; m 78. SOIL FERTILITY, CONSERVATION TILLAGE. *Educ:* Mich State Univ, BS, 71; Ohio State Univ, MA, 74, PhD(agron), 78. *Prof Exp:* ASST PROF AGRON, OHIO STATE UNIV, 78- *Mem:* Am Soc Agron; Soil Sci Soc Am; Crop Sci Soc Am; Soil Conserv Soc Am; Sigma Xi. *Res:* Energy conservation in crop production systems; conservation tillage systems, particularly fertility and weed control; crop rotations to reduce nitrogen fertilizer requirements. *Mailing Add:* Dept Agron Ohio State Univ 2021 Coffey Rd Columbus OH 43210

ECKERT, ERNST R(UDOLF) G(EORG), b Prague, Czech, Sept 13, 04; nat US; m 31; c 4. THERMODYNAMICS, HEAT & MASS TRANSFER. *Educ:* German Inst Technol, Prague, dipl ing, 27, Dr Ing, 31; Inst Technol, Danzig, Dr habil, 38. *Hon Degrees:* Dr Ing E H Munich Tech Univ, 68; Dr Eng, Purdue Univ, 68; DSc, Univ Manchester, 68, Univ Notre Dame, 72 & Polytech Inst, Romania, 73. *Prof Exp:* Asst, German Inst Technol, Prague, 28-34; lectr, Inst Technol, Danzig, 35-38; sect chief, Aeronaut Res Inst, Braunscheig, 38-45; consult power plant lab, USAF, Wright-Patterson AFB, 45-49; consult turbine & compressor div, Lewis Res Ctr, NASA, 49-51; prof mech eng, 51-73, dir, Thermodyn & Heat Transfer Div & Heat Transfer Lab, 52-73, regents prof, 66-73, EMER PROF, UNIV MINN, MINNEAPOLIS, 73- *Concurrent Pos:* Docent, Inst Technol, Braunschweig, 39-43; prof & dir inst thermodyn, German Inst Technol, Prague, 43-45; vis prof, Purdue Univ, 55-65; Fulbright Award, 62-63; past pres, Sci Coun Int Ctr Heat & Mass Transfer, Yugoslavia; US rep, Int Heat Transfer Conf; mem, Nat Comn Fire Prev & Control, 70-72; Humboldt Award, German Govt, 79. *Honors & Awards:* Max Jakob Award, 61; Gold Medal, Fr Inst Energy & Fuel, 67; Vincent Bendix Award, Am Soc Eng Educ, 72; Adams Memorial Mem Award, Am Welding Soc, 73. *Mem:* Nat Acad Eng; Fel Am Inst Aeronaut & Astronaut; hon mem Am Soc Mech Engrs; fel NY Acad Sci; Sigma Xi; Ger Soc Aeronaut & Astronaut. *Res:* Heat transfer; thermodynamics; gas turbines; jet propulsion; energy conservation. *Mailing Add:* 60 W Wentworth Ave West St Paul MN 55118

ECKERT, GEORGE FRANK, b Akron, Ohio, Feb 21, 24; m 48; c 3. CHEMICAL EDUCATION. *Educ:* Akron Univ, BS, 44; Ohio State Univ, PhD(chem), 50. *Prof Exp:* Instr phys chem, Capital Univ, 47-48; res chemist, E I du Pont de Nemours & Co, Inc, 51-54; assoc prof, 54-60, PROF CHEM, CAPITAL UNIV, 60- *Mem:* Am Chem Soc. *Mailing Add:* Dept Chem Capital Univ E Main St Columbus OH 43209

ECKERT, HANS ULRICH, b Danzig, Ger, Apr 20, 16; nat US; m 46; c 2. PHYSICS. *Educ:* Danzig Tech Univ, Cand Phys, 38; Tech Univ, Berlin, DiplEng, 41. *Prof Exp:* Test group leader, Aerodyn Inst Ger Army Ord, 43-45; task scientist, Aeronaut Res Lab, Wright Air Develop Ctr, Ohio, 46-54; sr aerodyn engr, Convair, 54-56, staff scientist, sci res lab, 56-62; staff scientist, Phys & Life Sci Lab, Lockheed-Calif Co, 62-63, head plasma physics lab, 63-67; mem tech staff chem & physics lab, Aerospace 67-78, consult, 79-83; RETIRED. *Concurrent Pos:* Jet Propulsion Lab, 81-83. *Honors & Awards:* Aerospace Inventor Yr, 78. *Mem:* Am Phys Soc. *Res:* Electrical discharges in gases; plasma flow; plasma chemistry; spectrochemical analysis. *Mailing Add:* 3901 Via Pavion Palos Verdes Estates CA 90274

ECKERT, J PRESPER, b Philadelphia, Pa, Apr 9, 19; c 4. DIGITAL COMPUTERS. *Educ:* Moore Sch, BS, 41, MS, 41. *Hon Degrees:* DSc, Univ Pa, 64. *Prof Exp:* Pres, Eckert-Mauchly Comput Corp, 45-50; dir eng, Univac Div, Sperry Rand Corp, 50-55, vpres, 55-89; CONSULT, ECKERT RES INT CORP, TOYOKO, JAPAN, 81- *Honors & Awards:* Howard N Potts Medal, Franklin Inst, 49; Medal Modern Pioneers Creative Indus, Nat Asn Mfs, 65; Harry Goode, Mem Award, 66. *Mem:* Fel Inst Elec Electronics Engrs; Sigma Xi; Nat Inventors Coun; Nat Acad Eng. *Res:* Developed and constructed ENIAC, the worlds first all electric computer. *Mailing Add:* 612 Millcreek Rd Gladwyne PA 19035

ECKERT, JOHN S, b Delta, Ohio, June 29, 10; m; c 1. PHYSICAL & BIOLOGICAL SCIENCES. *Educ:* Ohio State Univ, BChE, 33. *Prof Exp:* Jr chem engr, Goodyear Tire & Rubber Co, 33-37; area supvr, E I du Pont de Nemours & Co, 37-42; plant processing engr, B F Goodrich Co, 42-44; plant mgr, US Stoneware Co, 44-50, dir eng, Norton Co, 50-77; CONSULT ENGR, 77- *Concurrent Pos:* Adj assoc prof, Ohio State Univ, 63-75. *Mem:* Am Inst Chem Engrs; Am Soc Metals; Instrument Soc Am; Nat Asn Corrosion Engrs; Nat Soc Prof Engrs. *Res:* Mass transfer and performance of packed beds for distillation; absorption and stripping processes. *Mailing Add:* 216 Hollywood Ave Akron OH 44313

ECKERT, JOSEPH WEBSTER, b St Louis, Mo, Mar 27, 31; m 57; c 3. PLANT PATHOLOGY. *Educ:* Univ Calif, Los Angeles, BS, 52; Rutgers Univ, MS, 53; Univ Calif, Davis, PhD(plant path), 57. *Prof Exp:* Res asst, Univ Calif, Davis, 55-57, jr plant pathologist, 57-58, asst plant pathologist, 58-62, from asst prof to assoc prof, 62-70, PROF PLANT PATH, UNIV CALIF, RIVERSIDE, 70- *Concurrent Pos:* Fulbright Res Scholar, Neth-Am (Fulbright) Comn for Educ Exchange, 74-75; sr res fel, Agr Univ, Neth, 74-75. *Mem:* Am Phytopath Soc; Sigma Xi. *Res:* Post harvest fruit and vegetable diseases; physiology of fungi, fungicides. *Mailing Add:* Dept Plant Path Univ Calif Riverside CA 92521

ECKERT, JUERGEN, b Heilbronn, WGer, June 14, 47; m 79. SPECTROSCOPY, NEUTRON SCATTERING. *Educ:* Yale Univ, BS, 70; Princeton Univ, MA, 72, PhD(mat sci), 75. *Prof Exp:* Res assoc physics, 75-77, asst physicist, Brookhaven Nat Lab, 77-79; STAFF MEM, LOS ALAMOS NAT LAB, 79- *Concurrent Pos:* Vis scientist, Inst Laue-Langevin, France, 87-88; lectr, Univ Colo, 89- *Mem:* AAAS. *Res:* Neutron scattering studies on molecular solids, lattice dynamics and molecular rotations; structural and magnetic phase transitions; vibrational spectroscopy on metal hydrides and molecular crystals. *Mailing Add:* PO Box 1663 MS H805 Los Alamos NM 87545

ECKERT, RICHARD EDGAR, JR, b Kansas City, Mo, July 24, 29; m 51; c 6. RANGE SCIENCE. *Educ:* Univ Calif, BS, 52; Univ Nev, MS, 54; Ore State Univ, PhD(farm crops), 57. *Prof Exp:* Range scientist, Agr Res Serv, Mountain States Area, USDA, 57-78; RETIRED. *Mem:* Soc Range Mgt; Sigma Xi. *Res:* Range weed control and seeding; plant competition; ecological resource inventory; grazing management systems. *Mailing Add:* PO Box 7031 Incline Village NV 89450

ECKERT, RICHARD RAYMOND, b Youngstown, Ohio, July 15, 42; m 74; c 2. MICROCOMPUTERS APPLICATIONS. *Educ:* Case Inst Technol, BS, 64; Univ Kans, MS, 66, PhD(physics), 71. *Prof Exp:* Prof physics, Universidad de Oriente, Venezuela, 66-68; PROF PHYSICS & COMPUT SCI, CATH UNIV, PR, 71- *Concurrent Pos:* Prin investr, Cath Univ Biomed Res Prog, Minority Biomed Support Prog, NIH, 78- *Mem:* Am Asn Physics Teachers. *Res:* Effects of airborne pollutants on public health; microcomputers and their application to various academic tasks. *Mailing Add:* Dept Computer Sci State Univ NY Binghamton NY 13901

ECKERT, ROGER E(ARL), b Lakewood, Ohio, 26; c 3. CHEMICAL ENGINEERING. *Educ:* Princeton Univ, BSE, 48; Univ Ill, MS, 49, PhD(chem eng), 51. *Prof Exp:* Sr engr, E I du Pont de Nemours & Co, 51-64; assoc prof, 64-73, asst head, Chem Eng Sch, 70-75, PROF CHEM ENG, PURDUE UNIV, 73- *Concurrent Pos:* Consult, Glidden Co, 64-71, Packaging Corp of Am, 65-71 & Mobil Chem Co, 81-84; legal consult, 83- *Mem:* Am Inst Chem Engrs; Soc Rheology. *Res:* Design and statistical analysis of experiments; stocastic modeling of processes; rheology of viscoelastic polymer melts; organic and inorganic process development; diffusion in solids; mechanochemistry; biomedical engineering; scientific data processing and information retrieval. *Mailing Add:* 153 Indian Rock Dr West Lafayette IN 47906

ECKHARDT, CRAIG JON, b Rapid City, SDak, June 26, 40. SOLID STATE CHEMISTRY, OPTICAL SPECTROSCOPY. *Educ:* Univ Colo, BA, 62; Yale Univ, MS, 64, PhD(chem), 67. *Prof Exp:* From asst prof to assoc prof, 67-78, PROF PHYS CHEM, UNIV NEBR, LINCOLN, 78- *Concurrent Pos:* Consult, Appl Sci Knowledge, Inc, Nebr, 68-72; adv, Nebr State Dept Ed Phys Sci Proj, 68-69; mem, NSF Mat Res Lab Adv Panel, 76-79 & NSF Adv Panel for Sect Condensed Matter, 77-78. *Mem:* Am Inst Physics; Am Asn Physics Teachers; Sigma Xi; Optical Soc Am; Am Chem Soc. *Res:* Experimental and theoretical study of the electronic structure of molecules and crystals, molecular complexes and molecules of biological importance; applications of specular reflection of light; piezomodulation spectroscopy; electronic structure and spectra of dyes; electronic energy transfer in condensed phases; linear and non-linear optical properties of materials; magnetic and natural dichroism; Raman and Brillouin scattering, lattice dynamics. *Mailing Add:* Dept Chem Univ Nebr Lincoln NE 68588

ECKHARDT, DONALD HENRY, b Flushing, NY, Dec 20, 32; m 55; c 4. GEOPHYSICS. *Educ:* Mass Inst Technol, BS, 55, PhD(geophys), 61. *Prof Exp:* Geologist, Magnolia Petrol Co, Socony Mobil Oil Co, 55-56, seismic interpreter, Socony Mobil Oil Co, Venezuela, 56-58; res assoc, Ohio State Univ Res Found, 60-61, & Lunar & Planetary Lab, Univ Ariz, 61-63; res physicist, 63-75, chief, Geodesy & Gravity Br, 75-83, DIR, EARTH SCI DIV, AIR FORCE GEOPHYS LAB, 83- *Concurrent Pos:* Res assoc, Div Sponsored Res, Mass Inst Technol, 61-62. *Mem:* Am Geophys Union. *Res:* Geomagnetic induction; geodesy; selenodesy; planetary physics; lunar librations; gravity. *Mailing Add:* Air Force Geophys Lab Hanscom AFB MA 01731

ECKHARDT, EILEEN THERESA, b Passaic, NJ, May 17, 28. PHARMACOLOGY. *Educ:* Caldwell Col, BA, 49; Tulane Univ, MS, 60, PhD(pharmacol), 62. *Prof Exp:* From asst pharmacologist to assoc pharmacologist, Schering Corp, NJ, 49-58; teaching asst pharmacol, Tulane Univ, 58-62; from instr to asst prof, Univ Vt, 62-67; ASST PROF PHARMACOL, UNIV MED & DENT NJ, 67-, ASST DEAN STUDENT AFFAIRS, 73- *Res:* Liver hemodynamics; liver function; bromsulphalein excretion; role of the liver in drug metabolism. *Mailing Add:* Univ Med & Dent NJ NJ Med Sch 1855 S Orange Ave Newark NJ 07103

ECKHARDT, GISELA (MARION), b Frankfurt, Ger; m 57. PHYSICS. *Educ:* Univ Frankfurt, Ger, dipl, 52, Dr Phil nat(physics), 58. *Prof Exp:* Engr adv mat group, Semiconductor & Mat Div, Radio Corp Am, 58-60; mem tech staff, Quantum Electronics Dept, 60-66, mem plasma physics dept, 71, high voltage systems, 71-78, Optical Physics Dept, 78-80, SR MEM TECH STAFF, QUANTUM ELECTRONICS DEPT, HUGHES RES LABS, 69-, MEM TECH STAFF, CHEM PHYSICS DEPT, 80- *Concurrent Pos:* Mem, adv coun, dept physics-astron, Calif State Univ, Long Beach. *Mem:* Am Phys Soc. *Res:* Lasers and nonlinear optics; solid state physics; plasma and gas discharge physics. *Mailing Add:* 20737 Cool Oak Way Malibu CA 90265

ECKHARDT, RICHARD DALE, b De Kalb, Ill, June 24, 18; m 46; c 4. INTERNAL MEDICINE. *Educ:* Univ Ill, AB, 40; Harvard Univ, MD, 43; Am Bd Internal Med, dipl, 50. *Prof Exp:* Intern & resident, Harvard Med Serv, Boston City Hosp, 44-46, assoc, Thorndike Mem Lab, 46-49; from asst chief to chief med serv, Vet Admin Hosp, chief of staff, asst dean vet hosp affairs, 68-80; assoc internal med, 49-52, from clin asst prof to clin assoc prof, 52-60, prof, 60-80, EMER PROF INTERNAL MED, COL MED, UNIV IOWA, 80- *Concurrent Pos:* Fel, Harvard Med Sch, 44-45, res fel, 46-49, res assoc, 52; dir, Hepatitis Surv Group, Kyoto, Japan, 52; from clin asst prof to clin assoc prof internal med, Col Med, Univ Iowa, 52-60, clin prof, 60-68, prof, 68-, asst dean vet hosp affairs, 68-; chief med serv, West Side Hosp & attend physician, med serv, Res & Educ Hosps, Chicago, Ill, 57-58; assoc prof int med, Col Med, Univ Ill, 57-58. *Mem:* AAAS; Am Fedn Clin Res; fel AMA; Am Asn Study Liver Dis; Soc Exp Biol & Med. *Res:* Liver disease; protein and amino acid metabolism; nutrition. *Mailing Add:* 1675 Ridge Rd Iowa City IA 52245

ECKHARDT, ROBERT BARRY, b Jersey City, NJ, July 14, 42; m 64; c 4. BIOLOGICAL ANTHROPOLOGY, POPULATION GENETICS & EVOLUTION. *Educ:* Rutgers Univ, New Brunswick, BS, 64; Univ Mich, MS, 66, MA, 67, PhD(anthrop, human genetics), 71. *Prof Exp:* Instr anthrop, Univ Mich, 68-69, lectr, 69-71; asst prof, 71-79, ASSOC PROF ANTHROP, PA STATE UNIV, UNIVERSITY PARK, 79- *Concurrent Pos:* Dir, lab phys anthrop, Univ Mich, 68-71 & Mus Non-Western Man, 71-74; cur phys anthrop, Pa State Univ, University Park, 71-; dir, Mus Non-Western Man, 71-74; NSF grant, Pa State Univ, 72-73; book rev ed, J Human Evolution; co-ed, J Gen Educ; NIH grant, 78-79, 80-81, 84-85; Harry Frank Guggeheim Found grant, 81-82; Deutscher Akademischer Austaurchdienst Grant, 88. *Mem:* AAAS; Am Asn Phys Anthropologists; Am Asn Human Biologists; Am Anthrop Asn; NY Acad Sci; Sigma Xi. *Res:* Non-human primate and human evolution; population genetics; inheritance of quantitative characteristics, and rates of evolution, particularly in polygenic systems. *Mailing Add:* Dept Anthrop Pa State Univ State College PA 16802

ECKHARDT, RONALD A, b Baltimore, Md, Nov 18, 42. CELL & MOLECULAR BIOLOGY. *Educ:* Loyola Col, Md, BS, 64; Cath Univ Am, PhD(biol), 69. *Prof Exp:* Teaching asst biol, Cath Univ Am, 65-67; instr biol, Gallaudet Col, 64-65 & Xaverian Col, 66-67; vis investr, Oak Ridge Nat Lab, 67-69; fel biol, Yale Univ, 69-71; assoc prof, 71-80, PROF BIOL, BROOKLYN COL, CITY UNIV NEW YORK, 80- *Mem:* Am Soc Cell Biol; Soc Develop Biol; Sigma Xi. *Res:* Studies on the molecular biology of food fishes and agricultural plants with the goal being to increase food production. *Mailing Add:* 1465 E 26th St Brooklyn NY 11210

ECKHARDT, SHOHREH B, b Teheran, Iran, Oct 4, 38. PHARMACOLOGY. *Educ:* Univ Vt, BA, 60. *Prof Exp:* RES ASSOC, UNIV VT, 78- *Mem:* Sigma Xi; Am Soc Pharmacol & Exp Therapeut; Am Soc Hypertension. *Mailing Add:* Dept Pharmacol Univ Vt Burlington VT 05404

ECKHARDT, WILFRIED OTTO, b Frankfurt, Ger, Mar 30, 28; m 57. HIGH POWER SWITCHING, HIGH POWER MICROWAVE. *Educ:* Univ Frankfurt, dipl, 52, Dr phil nat(physics), 58. *Prof Exp:* Asst, Phys Inst, Univ Frankfurt, 53-57; mem staff, Microwave & Plasma Electronics Group, Radio Corp Am, 58-60; mem tech staff, Plasma Physics Dept, Hughes Res Labs, 60-63, sr staff physicist & sect head, 63-68, sr scientist & head LM cathode devices proj, 68-78, sr scientist, Explor Studies Dept & Plasma Physics Dept, 78-89; PRIN, ECKHARDT CONSULT, MALIBU, 89- *Mem:* Am Phys Soc; assoc fel Am Inst Aeronaut & Astronaut. *Res:* Plasma and gas discharge physics; high voltage and high power technology; electric propulsion; direct energy conversion; microwave and infrared physics; artificial intelligence. *Mailing Add:* 20737 Cool Oak Way Malibu CA 90265-5317

ECKHART, WALTER, b Yonkers, NY, May 22, 38; m 65. MOLECULAR BIOLOGY, VIROLOGY. *Educ:* Yale Univ, BS, 60; Univ Calif, Berkeley, PhD(molecular biol), 65. *Prof Exp:* Res assoc, 65-69, mem staff, 70-73, assoc prof, 73-79, PROF, SALK INST, 79-, CHMN, MOLECULAR BIOL & VIROL LAB, 73-, DIR, ARMAND HAMMER CTR CANCER BIOL, 76- *Concurrent Pos:* NSF fel, Salk Inst, 65-67, Am Cancer Soc fel, 67-70; assoc adj prof, Univ Calif, San Diego, 72-79, adj prof, 79-; mem rev comt, NIH Cancer Ctr, 80-85; Bd Sci Counselors, Nat Cancer Inst, 90-92. *Mem:* AAAS; Am Soc Microbiol. *Res:* Mechanisms of malignant cell transformation by tumor viruses; organization and expression of viral and cellular genes; growth regulation in mammalian cells. *Mailing Add:* Salk Inst PO Box 85800 San Diego CA 92186-5800

ECKHAUSE, MORTON, b New York, NY, May 17, 35; m 68; c 3. HIGH ENERGY PHYSICS. *Educ:* NY Univ, AB, 57; Carnegie Inst Technol, MS, 61, PhD(physics), 62. *Prof Exp:* Res assoc physics, Carnegie Inst Technol, 62; instr, Yale Univ, 62-64; from asst prof to assoc prof, 64-73, PROF PHYSICS, COL WILLIAM & MARY, 73- *Concurrent Pos:* Res visitor, Rutherford Lab, Eng, 73-74; vis prof, Swiss Inst Nuclear Res, 82-83. *Mem:* Am Phys Soc. *Res:* Experimental high-energy nuclear physics; muon lifetimes; structure of muonium; kaonic, sigma hyperonic and antiprotonic x-rays. *Mailing Add:* Dept Physics Col William & Mary Williamsburg VA 23185

ECKHOFF, NORMAN DEAN, b Meade, Kans, Apr 10, 38; m 59; c 2. NUCLEAR & INDUSTRIAL ENGINEERING. *Educ:* Kans State Univ, BS, 61, MS, 63, PhD(nuclear eng), 68. *Prof Exp:* Res engr, Boeing Co, Kans, 62-63; process engr, Litwin Eng Corp, 63; reactor engr, AEC, Tenn, 63-64; instr, 64-68, from asst prof to assoc prof, 68-76, PROF NUCLEAR ENG, KANS STATE UNIV, 76-, DEPT HEAD, 77- *Concurrent Pos:* Consult, Econ Res Serv, USDA, 69, Systs Res Co, Kans, 68-73, Comet Rice Mills, Tex, 68-69, Kemin Indust, Iowa, 71, McNally-Pittsburg, Kans, 72, Kans Gas & Elec, 73, 81-, US Atomic Energy Comn, Washington, DC, 73, Off Tech Assessment, Washington, DC, 80, Kans Corp Comn, Topeka, 85, Dynamac Corp, Washington, DC, 87- & Wolf Creek Nuclear Oper Corp, Burlington, Kans, 86- *Mem:* Am Nuclear Soc; Am Soc Eng Educ. *Res:* Neutron activation analysis; nuclear fuel management and economics; statistical models; systems analysis. *Mailing Add:* Dept Nuclear Eng Kans State Univ Manhattan KS 66502

ECKHOUSE, RICHARD HENRY, b Chicago, Ill, Jan 18, 40; m 80; c 3. GENERAL COMPUTER SCIENCES, HARDWARE SYSTEMS. *Educ:* Cornell Univ, BSEE, 62; Univ Ill, MS, 63; State Univ NY, PhD(computer sci), 71. *Prof Exp:* Instr elec engr, Bucknell Univ, 64-67; instr cc, State Univ NY, Buffalo, 67-71; sr eng mgr, Digital Equip Corp, 74-83; assoc prof computer sci, Temple Univ, 78-79; adj prof eng, Yale Univ, 84 & Dartmouth Col, 85-88; assoc prof computer sci, Amherst, 71-74, ASSOC PROF COMPUTER SCI, UNIV MASS, BOSTON, 88-; VPRES ENG, MOCO, INC, 83- *Concurrent Pos:* Adj prof, Univ Mass, Amherst, 90- *Mem:* Inst Elec & Electronics Engrs; Asn Comput Mach; Sigma Xi. *Res:* Fundamentals of computing systems; machine and assembly language programming; real-time systems; operating systems; computer architecture; microprogramming; virtual computer systems; multiprocessing; metalanguages applied to computer semantics; biomedical devices; computers in human performance and rehabilitation studies. *Mailing Add:* PO Box A Boston MA 02125

ECKLER, ALBERT ROSS, b Boston, Mass, Aug 29, 27; m 51; c 3. STATISTICS, OPERATIONS RESEARCH. *Educ:* Swarthmore Col, BA, 50; Princeton Univ, PhD(math statist), 54. *Prof Exp:* Res asst statist & opers res, James Forrestal Res Lab, Princeton Univ, 53-54; mem tech staff, Bell Tel Labs, Whippany, 54-58, supvr, 58-62, head, Mil Statist Dept, 62-72, head, Appl Math & Statist Dept, Holmdel, 72-74, head, Common Systs Anal Dept, Whippany, 74-80, head, Electronics Technol Anal Dept, Murray Hill, 80-85; RETIRED. *Concurrent Pos:* Ed & publ quart magazine, Word Ways, J Recreational Linguistics, 70- *Honors & Awards:* Best Surv Paper Award, Technometrics, 70. *Mem:* Sigma Xi. *Res:* Probability models of target coverage and missile allocation; mathematical models of telephone operations; optimal resource usage; statistical and historical evaluation of claims of extreme longevity in humans. *Mailing Add:* Spring Valley Rd Morristown NJ 07960

ECKLER, PAUL EUGENE, b Mexico, Mo, May 17, 46. SYNTHETIC ORGANIC CHEMISTRY. *Educ:* Univ Mo, Rolla, BS, 69; Univ Ore, MA, 70, PhD(chem), 75. *Prof Exp:* Clin chemist, Walter Reed Army Med Ctr, 70-72; teaching asst org chem, Univ Ore, 69-70, 73-74; res & develop chemist, 75-80, SR RES SCIENTIST, TECH SERV DEPT, PITMAN-MOORE, INC, 80- *Mem:* Am Chem Soc. *Res:* Organic synthesis and analysis, clinical chemistry; Diels-Alder reactions, natural products, monosaccharides, polyols, nitroparaffin derivatives, and heterocycles; instrumental analysis; coatings and resins; minerals processing. *Mailing Add:* Five Wheatson Ct Princeton Junction NJ 08550

ECKLES, WESLEY WEBSTER, JR, b Beaumont, Tex, Dec 15, 26; m; c 1. TECHNICAL WRITING, MICROCOMPUTER APPLICATIONS. *Educ:* Lamar Jr Col, AA, 48; Univ Tex, Austin, BS, 51. *Prof Exp:* Instr math, Pt Neches Ind Sch Dist, 51-52; consult, Sun Explor & Prod Co, 52-85; PRES, ECKLES ENTERPRISES, 60-, CONSULT, 85- *Concurrent Pos:* Pres, Gulf Coast Eng & Sci Soc, 61-68; vpres Eng, Philip C Crouse & Assocs, 87-88; consult ed, Petrol Eng Int, 86-; instr indust electronics, Lamar State Univ, Beaumont, Tex, 64-68; guest lectr petrol eng, Grad Sch, Univ Southwest La, 66-67; distinguished lectr, Soc Petrol Engrs. *Mem:* Inst Elec & Electronics Engrs; Am Inst Mining, Metall & Petrol Engrs; Soc Petrol Engrs. *Res:* Application of computers in solving oil & gas industry problems; author of numerous technical articles on production of oil and gas. *Mailing Add:* 9424 Hunters Creek Dallas TX 75243-6108

ECKLUND, EARL FRANK, JR, b Seattle, Wash, Apr 19, 45; m 76; c 1. NUMBER THEORY, OPERATING SYSTEMS. *Educ:* Pac Lutheran Univ, BS, 66; Western Wash State Col, MA, 68; Wash State Univ, PhD(math), 72. *Prof Exp:* Instr math, Northern Ill Univ, 71-72, asst prof comput sci, 72-73; fel comput sci, Univ Man, 73-74; asst prof comput sci, Ill State Univ, 74-77; asst prof, Dept Comput Sci, Ore State Univ, 77-; AT TEKTRONIX LABS, BEAVERTON, ORE. *Concurrent Pos:* Consult, 78- *Mem:* Am Math Soc; Math Asn Am; Soc Indust & Appl Math; Asn Comput Mach; Inst Elec & Electronics Engrs; Sigma Xi. *Res:* Computer science, especially database operating systems, information systems, operating systems; computational and combinatorial number theory, especially factorization and primality, distribution of residues and non-residues; computer science, especially factorization and primality testing, distribution of residues and nonresidues, factors of integers in progressions, programming methodologies and languages. *Mailing Add:* 7650 SW Belmont Dr Beaverton OR 97005-6336

ECKLUND, PAUL RICHARD, b Denver, Colo, June 20, 41; m 62; c 2. PLANT PHYSIOLOGY. *Educ:* Western State Col Colo, BA, 64; Ore State Univ, PhD(plant physiol), 68. *Prof Exp:* Asst prof biol, Vassar Col, 68-74; MEM FAC BIOL, CORNELL UNIV, 75- *Concurrent Pos:* Vis asst prof botany, Miami Univ, 74-75. *Mem:* Nat Asn Biol Teachers; Sigma Xi; Am Inst Biol Sci; Asn Biol Lab Educ. *Res:* Hormonal control of plant growth, development and senescence; biochemical and physiological changes associated with plant senescence. *Mailing Add:* 1140 Comstock Cornell Univ Ithaca NY 14853-0901

ECKLUND, STANLEY DUANE, b Minneapolis, Minn, Mar 18, 39; m 64; c 3. EXPERIMENTAL HIGH ENERGY PHYSICS. *Educ:* Univ Minn, BS, 61; Calif Inst Technol, PhD(physics), 67. *Prof Exp:* Res assoc photoprod, Stanford Linear Accelerator Ctr, 66-71; sr res assoc hyperons, Nat Accelerator Lab, Fermilab, 71-74, staff physicist elastic scattering, 74-77, head, Colliding Beams Group, 77-80; WITH STANFORD LINEAR ACCELERATOR LAB, 80- *Mem:* Am Phys Soc; AAAS. *Res:* Experimental high energy particle physics research, specializing in photoproduction, elastic and inelastic scattering, meson form factors, hyperon decays and hyperon interactions, colliding beam physics, and linear collider-accelerator physics. *Mailing Add:* Stanford Linear Accelerator Lab PO Box 4349 Stanford CA 94309

ECKMAN, MICHAEL KENT, b Denver, Colo, May 18, 42; m 63; c 2. ANIMAL PARASITOLOGY. *Educ:* Univ Northern Colo, BA, 65, MA, 66; Auburn Univ, PhD(avian dis), 70. *Prof Exp:* Sr res parasitologist, Morton-Norwich Prod, Inc, 66-68, 70-71; res specialist avian coccidiosis, Dow Chem Co, 71-77; POULTRY PATHOLOGIST, ALA COOP EXTEN SERV, AUBURN UNIV, 77- *Concurrent Pos:* Int consulting, Poultry Health & Live Prod. *Mem:* Am Soc Parasitologists; Poultry Sci Asn; Soc Protozoologists. *Res:* Characterization and profile of growth promoting, prophylactic, and medicinal feed additives in broiler chickens; quality control programs in commercial hatcheries: microbiological, true fertility, embryonic mortality, shell quality, and progeny performance. *Mailing Add:* 430 Deer Run Rd Auburn AL 36831

ECKNER, FRIEDRICH AUGUST OTTO, b Plauen, Ger, Aug 26, 26; US citizen; m 56; c 3. PATHOLOGY. *Educ:* Univ Cologne, 49-55, Dr med, 57. *Prof Exp:* Asst med, Univ Cologne, 55, 56-57, asst path, 56, 57-58; resident path, Salem Hosp, Mass, 58-60 & Univ Chicago, 60-62; pathologist in training, Congenital Heart Dis Res Ctr, Hektoen Inst Med Res, 62-64, head sect histochem, 64-70, assoc pathologist, 67-70; assoc prof path, 70-83, PROF PATH CLIN, UNIV ILL COL MED, 83- *Concurrent Pos:* Res assoc path, Univ Chicago & lectr, Univ Ill, 62-70; consult, Off Med Examr, Univ Chicago, 82-89 & Cook County, Chicago, Ill, 80- *Res:* Pathology of congenital and acquired cardiac disease by qualitative and quantitative methods at the gross and microscopic level, including study of conduction system; sudden death syndromes. *Mailing Add:* Dept Path Univ Ill Col Med M/C847 PO Box 6998 Chicago IL 60680

ECKROAT, LARRY RAYMOND, b Bloomsburg, Pa, July 18, 41; m 71; c 2. GENETICS, FISH BIOLOGY. *Educ:* Bloomsburg State Col, BS, 64; Pa State Univ, MS, 66, PhD(zool), 69. *Prof Exp:* Asst prof, 69-75, ASSOC PROF BIOL, BEHREND COL, PA STATE UNIV, 75- *Mem:* AAAS; Am Fisheries Soc; Genetics Soc Am; Am Genetics Asn. *Res:* Genetics of soluble protein polymorphisms in natural and hatchery populations of fishes and salamanders; chromosome polymorphism in natural populations of the house mouse and zebra mussel. *Mailing Add:* Dept Biol Col Sci Behrend Col Pa State Univ Erie PA 16563

ECKROTH, CHARLES ANGELO, b Mandan, NDak, May 10, 34; m 62; c 2. PHYSICS. *Educ:* St John's Univ, Minn, BA, 56; Iowa State Univ, PhD(physics), 66. *Prof Exp:* From instr to asst prof physics, Univ Mo, Columbia, 65-69; from asst prof to assoc prof physics, 69-75, dept chmn, 83-90, PROF PHYSICS, ST CLOUD STATE UNIV, 75- *Concurrent Pos:* Asst dir, Camp Uraniburg. *Mem:* Am Asn Physics Teachers. *Res:* Point symmetry groups; optics. *Mailing Add:* Dept Physics St Cloud State Univ St Cloud MN 56301

ECKROTH, DAVID RAYMOND, b Orwigsburg, Pa, Nov 20, 39; m 66. ORGANIC CHEMISTRY. *Educ:* Franklin & Marshall Col, BS, 61; Princeton Univ, MA, 63, PhD(org chem), 66. *Prof Exp:* Assoc res chemist, Sterling-Winthrop Res Inst, 65-66; asst prof chem, Wake Forest Univ, 66-69; vis asst prof, Iowa State Univ, 69-70; asst prof chem, York Col, NY, 70-75; assoc ed, Kirk-Othmer Encycl Chem Technol, John Wiley & Sons, 75-84, managing ed, Encycl Artificial Intel, Med Devices & Instrumentation & Packaging Technol, 84-88; EXEC ED, ENCYCL, MACMILLAN PUBL CO, 88- *Concurrent Pos:* Chmn Chem Sci Div, NY Acad Sci, 85. *Mem:* AAAS; Am Chem Soc; The Chem Soc; NY Acad Sci; Optical Soc Am; Sigma Xi. *Res:* Nomenclature of organic chemistry; scientific encyclopedias. *Mailing Add:* 22 Seventh Ave Brooklyn NY 11217

ECKSTEIN, BERNARD HANS, b Ulm, Germany, Dec 19, 23; nat US; m 58. PHYSICAL CHEMISTRY. *Educ:* Princeton Univ, AB, 48; Cornell Univ, PhD(phys chem), 53. *Prof Exp:* Res assoc chem, Cornell Univ, 52-54; res chemist, Textile Fibers Dept, E I du Pont de Nemours & Co, 54-57; res chemist, Parma Tech Ctr, Union Carbide Corp, 57-77, res scientist, 77-86; consult advan composites, 86-91; RETIRED. *Mem:* Am Chem Soc; Soc Advan Mat & Process Eng; Sigma Xi. *Res:* Carbon fibers; composite and high performance materials; reaction mechanisms; high temperature chemistry. *Mailing Add:* 8930 Albion Rd North Royalton OH 44133-1760

ECKSTEIN, EUGENE CHARLES, b Bucyrus, Ohio, Oct 31, 46; m 68; c 3. RHEOLOGY, ARTIFICIAL ORGANS. *Educ:* Mass Inst Technol, SB & SM, 70, PhD(mech eng), 75. *Prof Exp:* Assoc med biomed eng, Peter Bent Brigham Hosp, Harvard Sch Med, 74-75; asst prof, 75-79, assoc prof, 79-88, PROF BIOMED ENG, UNIV MIAMI, 88- *Concurrent Pos:* NIH prin investr, 78-91. *Mem:* Am Soc Mech Engrs; Am Soc Artificial Internal Organs; Int Soc Artificial Organs; Sigma Xi. *Res:* Biomechanical engineering; rheological effects in thrombus formation and development of an urinary prostheses. *Mailing Add:* 10881 SW 124th St Miami FL 33176

ECKSTEIN, JOHN WILLIAM, b Central City, Iowa, Nov 23, 23; m 47; c 5. INTERNAL MEDICINE. *Educ:* Loras Col, BS, 46; Univ Iowa, MD, 50. *Prof Exp:* Intern, Letterman Gen Hosp, 50-51; from asst resident to resident, 51-53, asst, 53-55, from instr to assoc prof, 54-65, PROF INTERNAL MED, UNIV IOWA, 65-, DEAN COL MED, 70-, ESTAB INVESTR, CARDIOVASC LAB, UNIV HOSPS, 58- *Concurrent Pos:* Rockefeller Found postdoctoral fel, 53-54; Am Heart Asn res fel, Univ Hosps, Univ Iowa, 54-55; estab investr, Am Heart Asn, 58-63; res career award, USPHS, 63-70; chmn, Cardiovasc Study Sect, NIH, 70-72 & Nat Heart, Lung & Blood adv coun, 74-78; gen res rev comt, NIH, 80-84; mem gov coun, Sect Med Schs, AMA, 85-, deleg, House Delegates, 90-; mem, Vet Admin Manpower Study, Inst Med, 88-; mem, Sci Policy Study Group, Asn Acad Health Ctrs, 88-, Study Group Info Sci, 89-90. *Honors & Awards:* Award of Merit, Am Heart Asn, Gold Heart Award. *Mem:* Inst Med-Nat Acad Sci; Am Soc Clin Invest; Asn Am Physicians; Cent Soc Clin Res (secy-treas, 65-70, pres, 73-74); Am Fedn Clin Res; Am Clin & Climatol Asn; Asn Am Med Col; AMA; Am Heart Asn (pres, 79). *Res:* Internal medicine and cardiovascular physiology. *Mailing Add:* Univ Iowa Col Med Iowa City IA 52242

ECKSTEIN, JONATHAN, b Boston, Mass. PARALLEL ALGORITHMS, CONVEX PROGRAMMING. *Educ:* Harvard Univ, SB, 80; Mass Inst Technol, SM, 86, PhD(opers res), 89. *Prof Exp:* Asst prof managerial econ, Harvard Bus Sch, 89-91; SCIENTIST, THINKING MACH CORP, 91- *Mem:* Opers Res Soc Am; Soc Indust & Appl Math. *Res:* Optimization, in particular parallel algorithms and nonlinear programming. *Mailing Add:* Thinking Mach Corp 245 First St Cambridge MA 02142

ECKSTEIN, RICHARD WALDO, b Tiro, Ohio, Oct 9, 11; m 37; c 3. MEDICINE. *Educ:* Heidelberg Col, BS, 33; Western Reserve Univ, MA, 36, MD, 38. *Prof Exp:* From intern to asst resident med, Lakeside Hosp, 39-41, sr instr physiol, Sch Med, Case Western Reserve Univ, 45-46, sr instr med, 46-49, asst prof, 49-53, asst prof physiol & med, 53-60, assoc prof, 60-; RETIRED. *Concurrent Pos:* Estab investr, Am Heart Asn, 54. *Mem:* AAAS; Am Physiol Soc; Soc Exp Biol & Med; Am Heart Asn; Am Psychiat Asn; Sigma Xi. *Res:* Coronary artery blood flow; cardiovascular; limb blood flow in dogs during shock; coronary collateral circulation; coronary blood supply of chemoreceptors. *Mailing Add:* 4360 S Hilltop Rd Chagrin Falls OH 44022-1428

ECKSTEIN, YONA, physical & analytical chemistry, for more information see previous edition

ECKSTEIN, YORAM, b Krakow, Poland, Jan 24, 38; Israel citizen; m 64; c 3. GEOTHERMICS & HYDROGEOLOGY. *Educ:* Hebrew Univ, BSc, 59, MSc, 65, PhD(geol), 77. *Prof Exp:* Hydrogeologist, 62-63; sr hydrogeologist, Geol Surv Israel, 63-74; vis scientist geol, Mass Inst Technol, 74-75; sr hydrogeologist geothermics, Hydro-Search Inc, Reno, 75-76; assoc prof, 76-81, PROF GEOL, KENT STATE UNIV, 81- *Concurrent Pos:* Sr consult, Ministry of Agr, Repub of Korea, 66-67; sr hydrogeologist, Geol Surv Israel, 74-78; sr consult, Hydro-Search Inc, 76-, Geonomics Inc, 77-78 & Int Eng Co, Inc, 78-; vis staff mem, Los Alamos Sci Lab, 77-80. *Mem:* Geol Soc Am; Am Inst Prof Geologists; Am Geophys Union; Nat Water Well Asn; Geothermal Resources Coun. *Res:* Geothermal exploration; hydro-geochemistry; terrestrial heat flow; geotectonics. *Mailing Add:* Dept Geol McGilvrey Hall Kent State Univ Kent OH 44242

ECKSTROM, DONALD JAMES, b St James, Minn. CHEMICAL PHYSICS, LASERS. *Educ:* Univ Minn, BS, 61, MS, 62; Stanford Univ, PhD(aerophysics), 71. *Prof Exp:* Aerodynamics engr, Lockheed Missiles & Space Co, 62-68; chem physicist, 70-90, DIR, MOLECULAR PHYSICS LAB, SRI INT, 90- *Mem:* Combustion Inst. *Res:* Laser development; visible chemical lasers; energy transfer; vibrational relaxation of small molecules; gas-phase reactions. *Mailing Add:* 331 Grove Dr Portola Valley CA 94025

ECOBICHON, DONALD JOHN, b Lindsay, Ont, June 21, 37; m 60; c 2. BIOCHEMICAL PHARMACOLOGY. *Educ:* Univ Toronto, BScPhm, 60, MA, 62, PhD(pharmacol), 64. *Prof Exp:* Demonstr pharmacol, Fac Med, Univ Toronto, 60-64; from asst prof to assoc prof, Ont Vet Col, Univ Guelph, 65-66, assoc prof pharmacol, Dalhousie Univ, 69-77; PROF PHARMACOL, McGILL UNIV, 77- *Concurrent Pos:* Nat Res Coun Can fel protein chem, 64-65. *Mem:* Soc Toxicol Can; Pharmacol Soc Can; NY Acad Sci; Soc Toxicol. *Res:* Study of drug hydrolysis by tissue esterases of various mammalian species; pharmacodynamics; mechanism of action; toxicology of chlorinated hydrocarbon and organophosphorus insecticides. *Mailing Add:* Dept Pharmacol & Therapeut McIntyre Bldg McGill Univ 3655 Drummond St Montreal PQ H3G 1Y6 Can

ECONOMIDES, MICHAEL JOHN, b Famagusta, Cyprus, Sept 6, 49; m 76; c 2. PETROLEUM ENGINEERING. *Educ:* Univ Kans, BS, 74, MS, 76; Stanford Univ, PhD(petrol eng), 81. *Prof Exp:* Process engr, Celanese Chem Co, 74-75; res asst, Univ Kans, 75-76; res assoc chem eng, Univ Calif, Berkeley, 76-78; res, Stanford Univ, 78-80; asst prof petrol eng, Univ Alaska, Fairbanks, 80-84; mgr tech tansfer, Dowell Schlumberger Inc, 84-89; PROF PETROL ENG & DIR, INST DRILLING & PROD, MINING UNIV LEOBEN, 89- *Concurrent Pos:* Reservoir engr, Shell Oil Co, 72; chem engr, Black & Veatch Consult Eng, 76; well testing, Ente Nazionale per l Enargia Electrica, 78; reservoir engr, Hughes Aircraft Co & Shell Oil Co, 79; distinguished lectr, Soc Petrol Engrs. *Mem:* Soc Petrol Engrs. *Res:* Well test analysis; petroleum and geothermal reservoir engineering; separation process and cold temperature rheology; fracture stimulation; production engineering. *Mailing Add:* Int Drilling & Prod Mining Univ Leoben A 8700 Leoben Austria

ECONOMOS, GEO(RGE), b Haverhill, Mass, Aug 22, 19; m 47; c 2. METALLURGY, CERAMICS & MATERIALS SCIENCE. *Educ:* Northeastern Univ, BS, 49; Mass Inst Technol, SM, 51, ScD(ceramics), 54. *Prof Exp:* Asst prof metall, Mass Inst Technol, 54-61; electronics consult res admin, Allen-Bradley Co, Milwaukee, 61-71, mgr, Mat Dept, Sprague Elec Co, Grafton, Wis, 72-77; SR PROG OFFICER, NAT ACAD SCI, NAT MAT ADV BD, WASHINGTON, 77- *Concurrent Pos:* Vis prof, Sch Appl Sci & Eng, Univ Wis, Milwaukee, 71; consult, electronic mat prep & prod, process control. *Mem:* Fel AAAS; Am Chem Soc; fel Am Ceramic Soc; Am Soc Metals; Sigma Xi; Am Inst Mining, Metall & Petrol Engrs. *Res:* Ceramic dielectrics, ferroelectrics and ferromagnetics; powder metallurgy; polymers; strategic materials availability and substitutions. *Mailing Add:* 6204 Bradley Blvd Bethesda MD 20817

ECONOMOU, DEMETRE J, b Erithrai, Attikis, Greece, Sept 25, 58. MATERIALS PROCESSING, PLASMA & CHEMICAL VAPOR DEPOSITION. *Educ:* Nat Tech Univ, Greece, BS, 81; Univ Ill Urbana, MS, 83, PhD(chem eng), 86. *Prof Exp:* ASSOC PROF CHEM ENG, UNIV HOUSTON, 86- *Concurrent Pos:* Young researcher excellence award, Halliburton Found. *Mem:* Am Vacuum Soc; Mat Res Soc; Am Inst Chem Engrs; Electrochem Soc. *Res:* Semiconductor processing; plasma etching; chemical vapor deposition; modeling of plasma processing reactors; high Tc superconducting thin films; diamond and related materials. *Mailing Add:* Dept Chem Eng Univ Houston Houston TX 77204-4792

ECONOMOU, ELEFTHERIOS NICKOLAS, b Athens, Greece, Feb 7, 40; m 66; c 1. SOLID STATE PHYSICS, SURFACE PHYSICS. *Educ:* Polytech Inst Athens, dipl, 63; Univ Chicago, MS, 67, PhD(solid state physics), 69. *Prof Exp:* Res asst physics, Univ Chicago, 66-69, res assoc, 69-70; from asst prof to prof physics, Univ Va, 73-83; DIR, RES CTR CRETE, 83-; PROF PHYSICS, UNIV CRETE, 83- *Concurrent Pos:* Expert consult, Naval Res Lab, 74-; vis assoc prof physics, Univ Chicago, 75-76; secy-gen, Res & Technol, Greece, 87-88; expert consult, Exxon Res Corp, 79-87. *Mem:* Am Phys Soc; Greek Soc Engrs; Sigma Xi. *Res:* Properties of disordered systems, mainly transport; surface plasmons in various geometries; electron-electron correlations and magnetic properties; amorphous and crystalline semiconductors. *Mailing Add:* Dept Physics Univ Crete Iraklion 71110 Greece

ECONOMOU, STEVEN GEORGE, b Chicago, Ill, July 4, 22; m 50; c 3. SURGICAL ONCOLOGY. *Educ:* Hahnemann Med Col, MD, 47. *Prof Exp:* From intern med to resident orthop, St Francis Hosp, Evanston, Ill, 47-50; resident surg, Presby Hosp, Chicago, 50-52 & 54; from asst prof to assoc prof surg, Col Med, Univ Ill, 54-68; prof surg, 71-81, JACK FRASER PROF SURG, RUSH MED COL, 81-; CLIN PROF SURG, COL MED, UNIV ILL, 68- *Concurrent Pos:* Fel path, Cook County Hosp, Chicago, 49; from asst attend surgeon to assoc attend surgeon, Presby-St Luke's Hosp, 57-65, attend surgeon, 65-, dir lab surg res, 65-69. *Mem:* AMA; Am Asn Cancer Res; Am Surg Asn; Soc Head & Neck Surgeons; Soc Surg Oncol; Am Asn Cancer Educ; Am Asn Endocrine Surgeons; Am Col Surgeons; Am Soc Clin Oncol. *Res:* Cancer surgery. *Mailing Add:* 3118 Melrose Ct Wilmette IL 60091

ECONOMY, GEORGE, b Detroit, Mich, Jan 7, 27; m 59; c 4. METALLURGY, CORROSION. *Educ:* Wayne State Univ, BS, 56; Ohio State Univ, PhD(metall eng), 60. *Prof Exp:* Engr finishes, Alcoa Res Lab, 60-64; engr corrosion, Inco Res Lab, 64-68; SR ENGR NUCLEAR MAT, WESTINGHOUSE RES & DEVELOP CTR, 68- *Mem:* Nat Asn Corrosion Engrs. *Res:* High temperature corrosion of power plant materials; high temperature chemistry relevant to power plants. *Mailing Add:* Westinghouse Sci & Tech Ctr Beulah Rd Pittsburgh PA 15235-5098

ECONOMY, JAMES, b Detroit, Mich, Mar 28, 29; m 61; c 4. ORGANIC CHEMISTRY. *Educ:* Wayne State Univ, BS, 50; Univ Md, PhD(chem), 54. *Prof Exp:* Res assoc polymer res & Marvel fel, Univ Ill, 54-56; gen res leader in charge res Semet-Solvay Petrochem Div, Allied Chem Corp, 56-60; mgr chem dept, Res Develop Div, Carborundum Co, 60-70, mgr res br, 70-74, corp scientist, 74-75; mgr polymer sci & technol, IBM Corp, 75-89; PROF & HEAD MAT SCI & ENG, UNIV ILL, 89- *Concurrent Pos:* Lectr, Canisius Col, 61-62; adj prof, State Univ NY Buffalo, 74-; mem Nat Mat Adv Bd, 84-; Off Naval Res Panel Chem Res, 85-; Univ Mat Coun, 87-89. *Honors & Awards:* Schoellkopf Gold Medal, Am Chem Soc, 72, Phillips Medal, 85; Chem Pioneer Award, Am Inst Chem, 87. *Mem:* Nat Acad Eng; Am Chem Soc; fel Am Inst Chem; NY Acad Sci; Int Union Pure & Appl Chem; AAAS. *Res:* New polymers-high temperature, ionic, thermosetting and photoconductive; new fibers-reinforcing, flame resistant, superconducting, refractory and for pollution control; ceramic composites; carbon film hyperfilter; polymer characterization; structure property relationships; polymers for microelectronic applications. *Mailing Add:* Univ Ill 1304 W Green St Urbana IL 61801

EDAMURA, FRED Y, b Vancouver, BC, Jan 25, 39; m 62; c 4. ORGANIC CHEMISTRY, ANALYTICAL CHEMISTRY. *Educ:* Univ Alta, BS, 60; Johns Hopkins Univ, MA, 62, PhD(org chem), 66. *Prof Exp:* Res chemist, Halogens Res Lab, Dow Chem Co, Midland, Mich, 65-72, sr res chemist, 72, res specialist, Ag-Org Res, Walnut Creek, Calif, 72-74, res specialist & proj mgr, Pittsburg, Ca, 74-79, group leader, 79-82, res mgr, anal lab, 82-86, mgr training & qual, Western Res & Develop, 86-90, TRAINING MGR, WESTERN DIV, DOW CHEM CO, PITTSBURG, CALIF, 90- *Mem:* Am Chem Soc; Sigma Xi; Am Soc Qual Control. *Res:* Organic synthesis; organic fluorine chemistry; synthesis of biologically active compounds; organic process research; analytical chemistry. *Mailing Add:* 2218 Lake Oaks Ct Martinez CA 94553

EDASERY, JAMES P, b Dec 11, 48; m; c 2. MASS SPECTROMETRY. *Educ:* Kerala Univ, India, BS, 69; Calicut Univ, India, MS, 71; Univ Nebr, Lincoln, PhD(chem), 79. *Prof Exp:* Res assoc, Stevens Inst Technol, 81-82; res chemist, Mt Sinai Sch Med, 82-83; res chemist, Med Sch, Cornell Univ, 83-85; res scientist, Columbia Univ, 87-89; RES SCIENTIST, TYREE ENVIRON TECH, 89- *Mem:* Am Chem Soc; Am Soc Mass Spectrometry; Sigma Xi. *Mailing Add:* 17 Raymondy Ct Garden City NY 11530

EDBERG, STEPHEN CHARLES, b New York, NY, Mar 13, 45; m 69; c 2. MICROBIOLOGY, IMMUNOLOGY. *Educ:* Lehigh Univ, BA, 67; Hofstra Univ, MA, 68; State Univ NY Buffalo, PhD(microbiol), 71; Am Acad Microbiol, dipl, 74. *Prof Exp:* Assoc dir microbiol, Montefiore Hosp Med Ctr, 71-81; asst prof microbiol & immunol, Albert Einstein Col Med, 71-75; assoc prof, Dept Lab Med, 81-89, DIR, CLIN MICROBIOL LAB, YALE-NEW HAVEN HOSP, YALE UNIV MED SCH, 81-, PROF, 89-; ASST PROF MICROBIOL, IMMUNOL & PATH, ALBERT EINSTEIN COL MED, 75- *Concurrent Pos:* Montefiore Hosp Med Ctr fel, Univ Wash, 72; adj assoc prof, City Univ New York, 74-81. *Mem:* Am Soc Microbiol; Soc Exp Biol Med; NY Acad Sci. *Res:* Theoretical and applied microbiology; immunology of tumors, transplants and dextran. *Mailing Add:* Clin Microbiol Lab Yale Univ Sch Med New Haven CT 06510

EDDE, HOWARD JASPER, b Page City, Kans, Dec 14, 37; m 61; c 3. CIVIL ENGINEERING. *Educ:* Kans State Univ, BS, 59; Univ Kans, MS, 61; Univ Tex, Austin, PhD(civil eng), 67. *Prof Exp:* Regional engr, Nat Coun Paper Indust Air & Stream Improvement, 62-64; mem staff, La State Univ, Baton Rouge, 64-66; proj engr, Roy F Weston, West Chester, Pa, 66-67; regional mgr, Nat Coun Paper Indust Air & Stream Improvement, 67-70; vpres, EKONO, Inc Consult Engrs, 70-74; AFFIL PROF, UNIV WASH, SEATTLE, 72-; PRES, HOWARD EDDE, INC ENGRS, 74- *Concurrent Pos:* Lectr, Johns Hopkins Univ, Baltimore, 67-70; chmn, Water Qual Comt, Tech Asn Pulp & Paper Asn, 78-81; chmn, Tech Coun Cold Climate Eng, Am Soc Civil Engrs; mem US-EPA contractor Selection Comt, 85-; chmn, Civil Eng Adv Bd, Seattle Univ, 86-90, prof, 87-91, mem awards comt, 88-91; mem Nat Sci & Eng Res Coun Can, 90- *Mem:* Tech Asn Pulp & Paper Asn; Water Pollution Control Asn; Am Soc Civil Engrs. *Res:* Research, design and operation of aerated stabilization basin to treat domestic sewage; development of pulp mill in-plant pollution control techniques; conduct numerous hazardous waste investigations. *Mailing Add:* Howard Edde Inc 2661 Bellevue-Redmond Rd Bellevue WA 98008

EDDINGER, CHARLES ROBERT, b Emlenton, Pa, Feb 21, 39. ORNITHOLOGY, ECOLOGY. *Educ:* Clarion State Col, BSc, 61; Univ Hawaii, MS, 67, PhD(zool), 70. *Prof Exp:* Instr biol, Kingswood Sch, Kalaw, Burma, 61-64; teacher sci, Mt Pleasant Sch Syst, Mich, 64-65; INSTR ZOOL, HONOLULU COMMUNITY COL, 71- *Concurrent Pos:* Res consult, Res Comt Environ Ctr, Univ Hawaii, 72-74; environ consult, Parsons-Brinckerhoff-Hirota Assocs; wildlife biologist, Bur Fisheries & Wildlife, US Dept Interior; researcher, Chapman Mem Fund grant & Eastern Bird-Banding Asn grant. *Mem:* Am Ornithologists' Union; Cooper Ornith Soc; Wilson Ornith Soc; Am Fedn Avicult; Avicult Soc. *Res:* Breeding biology of Hawaii's endemic birds. *Mailing Add:* Dept Biol Honolulu Community Col Honolulu HI 96817

EDDINGER, RALPH TRACY, b Wilkes-Barre, Pa, Feb 5, 22; m 45, 72; c 4. CHEMICAL ENGINEERING, FUEL TECHNOLOGY. *Educ:* Pa State Univ, BS, 42; Ohio State Univ, MS, 47, PhD(metall), 48. *Prof Exp:* Operating engr, Koppers Co, Inc, 42-46; res engr, Consolidation Coal Co, 48-51; mgr res lab, Eastern Gas & Fuel Assocs, 51-61; sr res engr, FMC Corp, 61-66, mgr proj coed, 66-72; tech/eng mgr, Cogas Develop Co, 72-81; MGR PILOT FACIL, FMC CORP, 81- *Concurrent Pos:* Chmn, Gordon Res Conf on Coal Sci, 69. *Honors & Awards:* Storch Award, Am Chem Soc, 74. *Mem:* Am Chem Soc; Am Inst Chem Engrs; Brit Inst Fuel; Am Inst Mining, Metall & Petrol Engrs. *Res:* Industrial high-temperature and fluidized-bed low-temperature carbonization of coal; coal processing and utilization; chemical process research and development. *Mailing Add:* 576 Village Rd W Princeton Junction NJ 08550

EDDINGTON, CARL LEE, b Tulsa, Okla, Dec 26, 32; m 58; c 3. BIOCHEMISTRY, SCIENCE EDUCATION. *Educ:* Univ Tulsa, BS, 55; St Louis Univ, PhD(biochem), 68. *Prof Exp:* Chemist, Indust Serv Div, Dow Chem Co, 58-62; res chemist, Samuel Roberts Noble Found, 66-71; PROF CHEM, ECENT UNIV OKLA, 71- *Concurrent Pos:* Res appointments, Environ Protection Agency, 87- *Mem:* Sigma Xi; Am Chem Soc. *Res:* Tumor-host relationships, particularly leukocyte production of humoral factors, acute phase globulins, fever, iron metabolism, enzymes, endotoxins and immunology; biochemistry of lactation, particularly hormones, nucleic acid and protein biosynthesis; ground water research-methanotropic bacteria. *Mailing Add:* Dept Chem ECent Univ Ada OK 74820

EDDLEMAN, BOBBY R, b Claude Armstrong Co, Tex, Aug 26, 37; m 56; c 3. RESEARCH EVALUATION, PRODUCTION ECONOMICS. *Educ:* Tex Tech Univ, BS, 59; NC State Univ, MS, 62, PhD(agr econ), 66. *Prof Exp:* Res instr agr econ, NC State Univ, 62-64; asst prof, Tex A&M Univ, 64-66; from asst prof to prof, Univ Fla, 66-75; prof, Miss State Univ, 75-79, dir res admin, 79-86; RESIDENT DIR AGR RES, TEX A&M UNIV, AGR RES & EXTEN CTR, CORPUS CHRISTI, 86- *Concurrent Pos:* Asst dir, Ctr Rural Develop, Univ Fla, 71-73, dir, 73-75; consult, Ministry Agr, Guyana, Venezuela & Columbia, 70-73, Fla State Legis Comt Agr & Citrus, 73-75, State Agr Exp Sta, Exp Sta Comt Orgn & Policy Res Eval Comt, 76-78, Sci & Educ Admin, USDA, 80-82 & Off Technol Assessment, 81-85. *Mem:* Am Agr Econ Asn; AAAS. *Res:* Analysis of the economics of agricultural research and the productivity of research and development investments; farm level technology adoption and impacts on productivity and economic profitability of farming operations. *Mailing Add:* TAMU Agr Res & Exten Ctr Rte Two Box 589 Corpus Christi TX 78410

EDDLEMAN, ELVIA ETHERIDGE, JR, b Birmingham, Ala, Oct 20, 22; c 1. INTERNAL MEDICINE, CARDIOLOGY. *Educ:* Howard Col, BS, 44; Emory Univ, MD, 48; Am Bd Internal Med, dipl, 56. *Prof Exp:* Intern, Grady Mem Hosp, Atlanta, Ga, 48-49; resident, Parkland Hosp, Dallas, Tex, 49-50; from instr to prof, 53-81, EMER PROF MED, SCH MED, UNIV ALA, 81- *Concurrent Pos:* Fel, Sch Med, Univ Ala & res fel, Med Col Ala, 52-53; asst chief med & chief cardiovasc sect, Vet Admin Hosp, 54-57, assoc chief of staff res, 54-, actg chief cardiovasc sect & chief med serv, 57-62. *Mem:* Fel Am Col Cardiol; fel Am Col Physicians; Am Fedn Clin Res; Am Heart Asn; Ballistocardiographic Res Soc. *Res:* Cardiovascular research. *Mailing Add:* 4209 Mountaindale Rd Birmingham AL 35213

EDDLEMAN, LEE E, b Broadus, Mont, May 8, 37; m 58; c 3. RANGE ECOLOGY. *Educ:* Colo State Univ, BS, 60, MS, 62, PhD(plant ecol), 67. *Prof Exp:* Assoc prof, 63-77, prof range sci, Sch Forestry, Univ Mont, 77-; AT DEPT RANGELAND RESOURCES, ORE STATE UNIV. *Mem:* Soc Range Mgt. *Res:* Evaluation of native plant species for reclamation of coal mine spoils; northern Rocky Mountain shrub ecology. *Mailing Add:* Dept Rangeland Resources Ore State univ Corvallis OR 97331

EDDLEMON, GERALD KIRK, b Washington, DC, Sept 16, 45; m 74; c 2. ENVIRONMENTAL IMPACT ANALYSIS. *Educ:* Univ Tenn, Knoxville, BS, 70, MS, 74. *Prof Exp:* Grad teaching asst zool & biol, Univ Tenn, 71-74; RES ASSOC, OAK RIDGE NAT LAB, 74- *Mem:* Am Fisheries Soc; Sigma Xi. *Res:* Environmental effects of conventional and unconventional energy technologies (coal conversion, oil shale, and geothermal development); transport, fate and effects of trace contaminants in aquatic ecosystems; nuclear, hazardous waste management and regulatory control; environmental auditing. *Mailing Add:* Bldg 1505 Oak Ridge Nat Lab PO Box 2008 Oak Ridge TN 37831

EDDS, GEORGE TYSON, b Heidenheimer, Tex, Jan 9, 13; m 31; c 3. PHARMACOLOGY, VETERINARY MEDICINE. *Educ:* Tex A&M Univ, BS & DVM, 36, MS, 38; Univ Minn, PhD(pharmacol), 52. *Prof Exp:* From instr to prof physiol & pharmacol, Tex A&M Univ, 35-50; vpres res, Ft Dodge Labs, Iowa, 50-62; prof, 62-83, EMER PROF VET SCI, UNIV FLA, 83- *Concurrent Pos:* Vet, Inst Food & Agr Sci; vis prof, Tex A&M, 83. *Mem:* Am Vet Med Asn; US Animal Health Asn; Sigma Xi. *Res:* Pharmacology-toxicology drug actions on animals; chemotherapy; anthelmintics; heavy metals; poisonous plants as hazards to animals; residues as hazards for mankind; aflatoxins as carcinogens. *Mailing Add:* 8601 Oakdale Waco TX 76710

EDDS, KENNETH TIFFANY, b Glen Ridge, NJ, July 17, 45; m 75; c 3. CELL MOTILITY, CYTOSKELETON. *Educ:* Univ RI, BS, 69; State Univ NY, Albany, PhD(biol), 74. *Prof Exp:* Muscular Dystrophy fel, 74-77; asst scientist, Marine Biol Lab, Woods Hole, Mass, 76-79; asst prof, 79-85, ASSOC PROF HISTOL, MED SCH, STATE UNIV NY, BUFFALO, 85- *Concurrent Pos:* Prin investr, NIH grant, 76-79 & NSF grant, 81-86 & 91-92. *Mem:* AAAS; Am Soc Cell Biol. *Res:* Structure and biochemistry of motile cells. *Mailing Add:* Dept Anat Sci State Univ NY Buffalo NY 14214

EDDY, CARLTON ANTHONY, b Boston, Mass, July 12, 42; m 68; c 2. REPRODUCTIVE PHYSIOLOGY. *Educ:* Merrimack Col, BA, 67; Univ Mass, MS, 70, PhD(animal sci), 73. *Prof Exp:* Instr obstet & gynec, 73-75, instr physiol, 74-75, asst prof obstet & gynec & asst prof physiol, 75-79, ASSOC PROF OBSTET & GYNEC, UNIV TEX HEALTH SCI CTR, SAN ANTONIO, 79- *Concurrent Pos:* Surg consult, Bexar Co Med Dist, 77- & Audey Murphy Vet Admin Hosp, 80- *Mem:* Soc Study Reproduction; Soc Gynec Invest; Am Fertil Soc; Soc Study Fertil; Sigma Xi. *Res:* Investigation of mechanisms concerned with the control of fertility; reproductive physiology and endocrinology of the non-human primate; tubal physiology; tuboplastic microsurgery. *Mailing Add:* Dept Obstet & Gynec Univ Tex Med 7703 Floyd Curl Dr San Antonio TX 78284

EDDY, DAVID MAXON, b Bridgeport, Conn, Nov 14, 41; m 81; c 3. HEALTH POLICY, DECISION SCIENCES. *Educ:* Stanford Univ, BA, 64, PhD(eng & econ systs), 78; Univ Va, MD, 68. *Prof Exp:* Chief, Bioeng Res Br, US Army Med Res & Develop Command, 71-73; res intern, Anal Res Group, Xerox Palo Alto Res Ctr, 74-76; actg asst prof, Dept Family Community & Prev Med, Sch Med, Stanford Univ, 76-78, assoc prof by courtesy, 78-81, dir, Prog for Anal of Clin Policies, Dept Eng-Econ Systs, 78-81, from assoc prof to prof, Sch Eng, 78-81; dir, Ctr Health Policy Res & Educ, 81-88, J Alexander McMahon prof health policy & mgt, 86-90, PROF COMMUNITY & FAMILY MED, DUKE UNIV, 81-, DIR, WHO COLLAB CTR RES IN CANCER POLICY, 84-, PROF HEALTH POLICY & MGT, 90- *Concurrent Pos:* Mem, Nat Comt Cancer Prev & Detection, Am Cancer Soc, 79-87, Nat Comt on Tobacco & Cancer, 81-84, Nat Comt Pub Educ, 85-86; mem bd sci counrs, Div Resources, Ctrs & Community Activ, Nat Cancer Inst, 80-83, mem, Working Group on Behav Aspects of Screening, Cancer Detection & Diagnosis of Cancer, 82, Working Group to Explore Issues in Breast Self-Exam, 82, Cancer Detection Res & Appln Comt, 86-87; consult, WHO Cancer Unit, 81-, Med Adv Panel Blue Cross & Blue Shield, 85-, Panel to Develop Guidelines for Benign Prostatic Hypertrophy, Agency for Health Care Policy & Res, 90-; mem, Expert Adv Panel on Cancer, WHO, 81-, adv, Control of Oral Cancer in Developing Countries, 83-84, consult to minister of health, India, 83-87 & Chile, 85-88, mem, Workgroup on Control of Cervical Cancer, 85, chmn, Workgroup on Cost-Effectiveness Anal for Cancer Control, 85; mem, Working Group on Mech Circulatory Support, Nat Heart Lung & Blood Inst, 83-85; mem, US-Israeli Joint Task Force on Technol Assessment, 85; adv coun, World Orgn Sci & Health, 85-; mem, Coun Res & Develop, Am Hosp Asn, 86-87, Intercoun Working Party Qual of Care, 87; mem, Panel Technol Assessment Methods, Inst Med, 86-89, Comt Clin Pract Guidelines, 90-, Comt to Advise Pub Health Serv on Med Pract Guidelines, 90-; tech adv, Nat Leadership Comn Health Care, 87-89; mem, Bd Math Sci, Nat Res Coun, 88-; chief scientist, Ctr Qual Healthcare, Blue Cross & Blue Shield, 89-90; mem, Comt Int Collab Activ, Int Union Against Cancer, 87-; mem int comt, Ctr Oncol & Biol Res Appln, 87-; mem bd sci adv, WHO Collab Ctr Prev Colorectal Cancer, 88-; columnist, JAMA, 90- *Honors & Awards:* Lanchester Prize, Opers Res Soc Am, 80. *Mem:* Inst Med-Nat Acad Sci; Int Soc Health Technol Assessment; Soc Med Decision Making. *Res:* Development and application of mathematical models and computer software for designing health policies and assessing health technologies; design of national cancer control programs and screening policies; methodology for estimating the benefits, harms, cost and cost-effectiveness of health technologies; design of system for setting research priorities. *Mailing Add:* Skyline Rte Box 32 Jackson WY 83001

EDDY, DENNIS EUGENE, b Pawnee City, Nebr, Jan 12, 35; m 59; c 1. BIOCHEMISTRY. *Educ:* Univ Omaha, BA, 66; Univ Nebr-Omaha, PhD(biochem), 73. *Prof Exp:* Asst instr chem, Univ Omaha, 66-68; res assoc biochem, Med Ctr, Univ Nebr, Omaha, 73-77; RES SCIENTIST, EXERCISE PHYSIOL LAB, QUAKER OATS-JOHN STUART RES LABS, 77- *Concurrent Pos:* Instr anal chem, Col St Mary, Omaha, 73-74, adv undergrad res proj, 73-; co-adv grad-undergrad res proj, Dept Psychol, Univ Nebr-Omaha, 74-77; instr gen chem, Metrop Tech Community Col, Omaha, 75-77. *Mem:* Am Col Sports Med; Am Aging Asn. *Res:* Aging, with emphasis on antioxidants and other dietary components as they affect biological systems. *Mailing Add:* 614 Pamela Ct Wauconda IL 60084

EDDY, EDWARD MITCHELL, b Parsons, Kans, Feb 9, 40; m 63; c 2. CELL BIOLOGY, DEVELOPMENTAL BIOLOGY. *Educ:* Kans State Univ, BS, 62, MS, 64; Univ Tex, PhD(anat), 67. *Prof Exp:* Fel anat, Harvard Med Sch, 67-69, instr, 69-70; asst prof, Univ Wash, 70-76, vchmn, 75-78, actg chmn, 78-81, assoc prof, 76-82, prof, biol struct, 82-83; HEAD, GAMETE BIOL SECT, LAB REPRODUCTION & DEVELOP TOXICOL, NAT INST ENVIRON HEALTH SCI, NIH, 83- *Concurrent Pos:* Prin investr, NSF & NIH res grants; fel, NSF & NIH. *Mem:* Am Soc Cell Biol; Am Asn Anat; Soc Develop Biol; Am Soc Zool; Int Soc Develop Biologists; Soc Study Reproduction; Am Soc Andrology; Sigma Xi. *Res:* Origin and potential of germ cell line, fertilization and initiation of development; gene expression during spermatogenesis. *Mailing Add:* Lab Reproduction & Develop Toxicol Nat Inst Environ Health Sci NIH Research Triangle Park NC 27709

EDDY, GEORGE AMOS, b Unity, Sask, June 8, 28; m 50; c 5. METEOROLOGY, CLIMATOLOGY. *Educ:* Univ BC, BASc, 50; Univ Toronto, MA, 51; McGill Univ, PhD(meteorol), 63. *Prof Exp:* From asst prof to assoc prof atmospheric sci, Univ Tex, Austin, 63-68; PROF METEOROL & ENVIRON DESIGN, UNIV OKLA, 68-; DIR, OKLA CLIMATOL SURV & STATE CLIMATOLOGIST, 78- *Concurrent Pos:* Vis assoc prof, Mass Inst Technol, 67-68; pres, Amos Eddy, Inc, 81- *Mem:* Am Asn State Climatologists. *Res:* Operational weather modification evaluation; statistical climatology; urban-rural ecosystem modeling; resource management modeling; applied climatology. *Mailing Add:* 318 Royal Oak Dr Norman OK 73069

EDDY, HUBERT ALLEN, b Boston, Mass, June 2, 30; m 59; c 2. RADIOBIOLOGY, PATHOLOGY. *Educ:* Boston Univ, BA, 52, MA, 54; Univ Rochester, PhD(radiation biol), 64. *Prof Exp:* Instr human ecol, Boston Univ, 53-54; from res asst to res assoc radiation biol, 57-63, from instr to asst prof radiol, 64-81, ASSOC PROF RADIATION ONCOL, SCH MED & DENT, UNIV ROCHESTER, 81-, ASSOC PROF PATH, 89- *Concurrent Pos:* Dir, Div Radiation Res, Dept Radiation Oncol, Univ Rochester, 83- *Mem:* AAAS; Sigma Xi; Radiation Res Soc; Am Asn Univ Professors; Int Soc Oxygen Transport to Tissue; NY Acad Sci; NAm Hyperthermia Group; Microcirculatory Soc; Am Soc Therapeut Radiol & Oncol. *Res:* Study of mechanisms of effect of ionizing radiations of mammaliam tissue and organ systems; tumor angiogenesis and the effect of ionizing radiations on tumor vasculature; comparative radiation oncology; radiation pathology. *Mailing Add:* 10810 Royal Mews Cockeysville MD 21030

EDDY, JERRY KENNETH, b Wheeling, WVa, Aug 17, 40; m 62; c 3. NUCLEAR PHYSICS. *Educ:* WLiberty State Col, AB, 62; WVa Univ, MS, 64, PhD(physics), 67. *Prof Exp:* Asst prof, 67-74, assoc prof, 74-78, PROF PHYSICS, INDIANA UNIV PA, 79- *Res:* Van de Graaff accelerators; nuclear spectroscopy; neutron induced charged particle reactions using the deuteron tritium reaction; activation analysis; charged particle induced x-ray analysis. *Mailing Add:* Stonybrook Sch Stonybrook NY 11790

EDDY, JOHN ALLEN, b Pawnee City, Nebr, Mar 25, 31; m 53; c 4. ASTROPHYSICS, SOLAR PHYSICS. *Educ:* US Naval Acad, BS, 53; Univ Colo, PhD(astrogeophys), 62. *Prof Exp:* Physicist, 62-63, SR SCIENTIST, HIGH ALTITUDE OBSERV, NAT BUR STANDARDS, 63-; SR SCIENTIST, DIR, OFF INTERDISCIPLINARY EARTH STUDIES, NAT CTR ATMOSPHERIC RES, 86- *Concurrent Pos:* Prof Adjoint, Univ Colo, 63-, res assoc, Harvard-Smithsonian Ctr Astrophys, 77-79. 67-70. *Honors & Awards:* Boulder Scientist Award, Sci Res Soc Am, 65; Arctowski Medal, Nat Acad Sci, 87. *Mem:* Am Astron Soc; fel AAAS; Am Geophys Union; Sigma Xi; Int Astron Union. *Res:* Infrared astronomy; history of astronomy; archaeo-astronomy. *Mailing Add:* 3460 Ash Boulder CO 80303

EDDY, LOWELL PERRY, b Portland, Ore, Nov 25, 20; m 46; c 3. INORGANIC CHEMISTRY. *Educ:* Ore State Col, BS, 42, MS, 48; Purdue Univ, PhD(chem), 52. *Prof Exp:* Instr chem, Univ Wyo, 50-51; res assoc & instr, Reed Col, 52-53; res chemist cellulose, Puget Sound Pulp & Timber Co, 53-57; asst prof, Western Wash State Col, 57-64, assoc prof chem, 64-85; RETIRED. *Concurrent Pos:* Am Chem Soc, Petrol Res Fund int faculty award & hon res asst, Univ Col, Univ London, 64; res assoc, Sch Chem, Univ New South Wales, 69-70 & vis UNESCO lectr, 75. *Mem:* Am Chem Soc. *Res:* Analytical methods in the sulfite pulp industry; coordination compounds of transition elements; their preparation and structure (by x-ray crystallography). *Mailing Add:* 2813 McKenzie Ave Bellingham WA 98225-6941

EDDY, LYNNE J, b New Haven, Conn, July 2, 44. CARDIOVASCULAR METABOLISM. *Educ:* Univ Ala, Birmingham, PhD(physiol), 72. *Prof Exp:* Assoc prof physiol, Univ Southern Ala, 83-86; assoc prof, Dept Phys Ther, Univ Calif, Los Angeles, 86-90; CONSULT MED WRITER, AMGEN, 91- *Mem:* AAAS; Sigma Xi; Am Physiol Soc. *Mailing Add:* 6030 Walton St Long Beach CA 90815

EDDY, NELSON WALLACE, b Burford, Ont, Jan, 15, 39; m 66; c 2. NUCLEAR PHYSICS. *Educ:* McMaster Univ, BA, 61; Univ Mass, Amherst, MS, 63; Ariz State Univ, PhD(nuclear physics), 69. *Prof Exp:* Asst prof, 68-74, ASSOC PROF PHYSICS, CONCORDIA UNIV, SIR GEORGE WILLIAMS CAMPUS, MONTREAL, 74- *Concurrent Pos:* Vis prof, Foster Radiation Lab, McGill Univ, Montreal, 75-76. *Mem:* Can Radiation Protection Asn; Can Asn Physicists. *Res:* Nuclear spectroscopy of (p,xn) reactions, fission products, fast neutron reactions and microdosimetry; Monte Carlo applications to shielding and radiation effects in biological materials. *Mailing Add:* Dept Physics Concordia Univ Montreal de Maisonneuve Blvd W PQ H3G 1M8 Can

EDDY, ROBERT DEVEREUX, inorganic chemistry; deceased, see previous edition for last biography

EDDY, THOMAS A, b Parsons, Kans, Dec 31, 34; m 64; c 1. ENTOMOLOGY, WILDLIFE MANAGEMENT. *Educ:* Kans State Univ, BS, 57, PhD(entom), 70; Univ Ariz, MS, 59. *Prof Exp:* From instr to asst prof, 60-71, ASSOC PROF BIOL, EMPORIA STATE UNIV, 71- *Res:* Hymenoptera ecology and behavior, plant ecology. *Mailing Add:* Dept Biol 1200 Commercial St Emporia KS 66801

EDDY, WILLIAM FROST, b Boston, Mass, Sept 18, 44; m 78; c 2. COMPUTATIONAL STATISTICS. *Educ:* Princeton Univ, AB, 71; Yale Univ, MA, 72, MPhil, 73, PhD(statist), 76. *Prof Exp:* From asst prof to assoc prof, 76-86, PROF STATIST, CARNEGIE-MELLON UNIV, 86- *Concurrent Pos:* Chmn bd, Interace Found NAm, 87-91. *Mem:* Fel Am Statist Asn (bd dir, 88-); fel Inst Math Statist; fel Royal Statist Soc; fel AAAS; Int Asn Statist Comput; Interface Found NAm; Asn Comput Mach; Int Statist Inst; Inst Elec & Electronics Engrs Computer Sci. *Res:* data analysis; statistical computation; applied and mathematical statistics. *Mailing Add:* Dept Statist Schenley Park Pittsburgh PA 15213-3890

EDE, ALAN WINTHROP, b Stamford, Conn, Jan 16, 33; m 57; c 3. MICROELECTRONICS. *Educ:* Worcester Polytech Inst, BS, 55; Univ Maine, Orono, MS, 63; Ore State Univ, PhD(elec eng), 68. *Prof Exp:* Engr, Raytheon Mfg Co, 55-60; from instr to assoc prof elec eng, Univ Maine, Orono, 60-71; PRES, DIRIGO ELECTRONICS ENG, 71- *Concurrent Pos:* Assoc prof indust educ, Ore State Univ, 74-81. *Mem:* Inst Elec & Electronics Engrs; Am Soc Eng Educ. *Res:* Electrofishing. *Mailing Add:* 1307 NW Buchanan Corvallis OR 97331

EDEIKEN, JACK, b Philadelphia, Pa, May 25, 23; m 42; c 5. RADIOLOGY. *Educ:* Univ Pa, MD, 47; Am Bd Radiol, dipl, 51. *Prof Exp:* Asst prof, Sch Med, Univ Pa, 51-58; assoc prof, 58-67, PROF RADIOL, THOMAS JEFFERSON UNIV, 67-, CHIEF DIAG DIV, UNIV HOSP, 69-, CHMN DEPT RADIOL, 71- *Concurrent Pos:* Consult, Vet Admin Hosps, Wilmington, Del, 55- & Philadelphia, Pa, 63-, consult, USAF. *Mem:* Sigma Xi. *Res:* Radiol Soc NAm; fel Am Col Radiol; AMA. *Mailing Add:* Diag Radiol Box 57 Anderson Hosp & Tumor Inst 6723 Bertner Ave Rm 62204 Houston TX 77030

EDELBERG, ROBERT, b NJ, Aug 2, 21; m 44; c 4. PSYCHOPHYSIOLOGY. *Educ:* Rutgers Univ, BS, 42; Univ Pa, PhD(physiol), 49. *Prof Exp:* Asst prof physiol, Long Island Univ, 49-51; from asst prof to assoc prof, Col Med, Baylor Univ, 56-63; prof psychophysiol & physiol, Med Ctr, Univ Okla, 63-70; prof psychiat, Rutgers Med Sch, 70-86, PROF PSYCHOL, GRAD SCH, RUTGERS UNIV, 70-, ADJ PROF, UNIV MED & DENT NJ, ROBERT WOOD JOHNSON MED SCH, 86- *Concurrent Pos:* Res & develop officer, Aero Med Lab, Wright Air Develop Ctr, 51-55; Sr res fel, USPHS, 58-62; assoc ed, Psychophysiol 65-83; mem, Exp Psychol Study Sect, NIH, 66-69; consult ed, J Comp & Physiol Psychol, 66-70. *Honors & Awards:* Distinguished Contrib Psychophysiol Award, Soc Psychophysiol Res, 74. *Mem:* Soc Psychophysiol Res (pres, 65-66). *Res:* Psychophysiological adaptation; electrodermal physiology and measurement; cardiovascular processes and behavior. *Mailing Add:* Dept Psychiat Univ Med & Dent NJ Robert Wood Johnson Med Sch 675 Hoes Lane Piscataway NJ 08854

EDELEN, DOMINIC GARDINER BOWLING, b Washington, DC, Jan 3, 33; m 54; c 6. GEOMETRY. *Educ:* Johns Hopkins Univ, BES, 54, MSE, 56, PhD, 65. *Prof Exp:* Jr instr math & mech, Johns Hopkins Univ, 54-56; engr, Martin Co, Md, 56-59; mem tech staff, Hughes Aircraft Co, Calif, 59-60; mem res staff, Rand Corp, 60-66; prof math, Purdue Univ, 66-69; PROF MATH, LEHIGH UNIV, 69- *Res:* Theory and applications of Lie groups and exterior differential forms; axiomatization of theoretical physics; non-local variational mechanics; gauge theory. *Mailing Add:* Dept Math Bldg 14 Lehigh Univ Bethlehem PA 18015

EDELHAUSER, HENRY F, b Dover, NJ, Sept 9, 37; m 61; c 2. PHYSIOLOGY, OPHTHALMOLOGY. *Educ:* Paterson State Col, BA, 62; Mich State Univ, MS, 64, PhD(physiol), 66. *Prof Exp:* Lab technician, Warner Lambert Pharmaceut Res Inst, 62; asst physiol, Mich State Univ, 62-65; from instr to prof physiol & ophthal, Med Col Wis, 66-89; PROF OPHTHAL & DIR RES, EMORY UNIV, 89- *Concurrent Pos:* Fel physiol, Marquette Univ, 66-67, res assoc ophthal, 67-68; grant, Nat Eye Inst, 69-92, Training Dir, 75-90; prin investr, Wis Dept Nat Res grant, 69-71, Mt Desert Island Biol Lab, 77-; sci consult, Alcon Labs, S.C. Johnson & Son & Am Cyanamid; mem, Tissue Banks Int Res Comt; Marjorie & Joseph Heil prof ophthal, 88, Ferst prof ophthal, 89; dir, Sci Rev Fight-for-Sight Inc; training grant dir, dept ophthal, Emory Univ, 90-93. *Honors & Awards:* Olga K Weiss Res Scholar; Honor Award, Am Acad Opthal. *Mem:* Am Soc Biol Sci; Am Soc Zool; Asn Res Vision & Ophthal (pres, 90-91); Am Physiol Soc; Am Acad Opthal. *Res:* Membrane physiology; pathophysiology of the eye; fish physiology and eye disease; ocular toxicology; physiological effects of vitrectomy; cellular toxicology of ophthalmic drugs. *Mailing Add:* Dept Ophthal Emory Univ Atlanta GA 30322

EDELHEIT, LEWIS S, solid state physics, general medical science, for more information see previous edition

EDELMAN, DAVID ANTHONY, b London, Eng, June 20, 38; US citizen; m 73. BIOSTATISTICS. *Educ:* Purdue Univ, BS, 61, MS, 63; Univ NC, Chapel Hill, PhD(biostatist), 72. *Prof Exp:* Staff mem exp statist, Sandia Labs, 63-69; head design & anal div med res, 72-77, ASSOC DIR RES, INT FERTIL RES PROG, 77- *Concurrent Pos:* Pres, Med Res Consult, 75- *Res:* Evaluation of newer developments in fertility control and abortion; fetal growth and development. *Mailing Add:* 130 E Longview St Chapel Hill NC 27514

EDELMAN, GERALD MAURICE, b New York, NY, July 1, 29; m 50; c 3. BIOCHEMISTRY. *Educ:* Ursinus Col, BS, 50; Univ Pa, MD, 54; Rockefeller Inst, PhD(biochem), 60. *Hon Degrees:* DSc, Univ Pa, 73; ScD, Ursinus Col, 74 & Williams Col, 79; MD, Univ Siena, Italy, 74; DSc, Gustavus Adolphus Col, Minn, 75, Univ Paris, 89, Georgetown Univ, 89; LSc, Univ Cagliari, Sardinia, 89, Univ Studi di Napoli Federico II, 90. *Prof Exp:* Med house officer, Mass Gen Hosp, 54-55; served to capt, M C AUS, 55-57; asst physician, Hosp of Rockefeller Inst, 57-60, mem fac, 60-, assoc dean grad studies, 63-66, prof, 66-74, VINCENT ASTOR PROF, ROCKEFELLER UNIV, 74- *Concurrent Pos:* Chmn adv bd, Basel Inst Immunol, 70-77, bd govs, Weizmann Inst Sci, 71-87, bd trustees, Salk Inst Biol Studies, 73-85; trustee, Rockefeller Br Fund, 72-82; mem bd overseers, Fac Arts & Sci, Univ Pa; mem adv comt, Carnegie Inst Washington, 80-87; mem bd sci overseers, Jackson Lab; sci chmn, Neurosci Res Prog, 80-, dir, Neurosci Inst, 81. *Honors & Awards:* Nobel Prize Physiol or Med, 72; Eli Lilly Award Biol Chem, 65; Rabbi Shai Shacknai Mem Prize Immunol & Cancer Res, Jerusalem, Israel, 77; C & O Vogt Award, Univ Dusseldorf, Ger, 88. *Mem:* Nat Acad Sci; Am Acad Art & Sci; fel NY Acad Sci; Am Soc Cell Biol; Genetics Soc. *Res:* Research structure antibodies; molecular embryology; neuronal group selection. *Mailing Add:* 35 E 85th St New York NY 10028

EDELMAN, ISIDORE SAMUEL, b New York, NY, July 24, 20; div; c 4. GENERAL PHYSIOLOGY. *Educ:* Ind Univ, BA, 41, MD, 44. *Prof Exp:* Intern, Greenpoint Hosp, 44-45; resident physician, Montefiore Hosp, 47-48; from asst prof med to prof med & physiol, Sch Med, Univ Calif, San Francisco, 52-67, fac res lectr, 66-67, Samuel Neider res prof med, 67-78, prof biophys, 69-78; Robert Wood Johnson Jr Prof Biochem & Chmn Dept, 78-88, EMER PROF BIOCHEM & MOLECULAR BIOPHYS, COLUMBIA UNIV, 91- *Concurrent Pos:* Dazian Found fel, Montefiore Hosp, 48-49; AEC fel, Harvard Med Sch & Peter Bent Brigham Hosp, 49-50, Am Heart Asn fel, 50-52; sr res fel chem, Calif Inst Technol, 58-59; estab investr, Am Heart Asn, 52-57; chief med serv, San Francisco Gen Hosp, Univ Calif, 56-58; mem NIH study comt, Off Sci & Technol, 64; vis scientist, Weizmann Inst Sci, 65-66; res career awards comt, Nat Inst Gen Med Sci, 69-73; Harry T Dozor vis prof biochem, Ben-Gurion Univ, Israel, 80; Deans Distinguished Lectr, Columbia Univ, 81. *Honors & Awards:* John Punnett Peters mem lectr, Yale Univ, 64; Eli Lilly Award, Endocrine Soc, 69; Gregory Pincus mem lectr, Worcester Found Exp Biol, 79; Mayo Soley Award, Western Soc Clin Res, 80; Homer W Smith Award, NY Heart Asn, 80. *Mem:* Nat Acad Sci; Am Fedn Clin Res; Endocrine Soc; Am Physiol Soc; Am Soc Clin Invest; Am Soc Biol Chem & Molecular Biol; Biophys Soc; AAAS. *Res:* Molecular biology of sodium/potassium transport; active transport across biological membranes; mechanism of action of steroid hormones and thyroid hormone. *Mailing Add:* Dept Biochem & Molecular Biophys Columbia Univ 630 W 168th St New York NY 10032

EDELMAN, JAY BARRY, b New York, NY, Oct 27, 51. PROTEIN PHYSICS, STRUCTURE ANALYSIS. *Educ:* Mass Inst Technol, MS, 73; Calif Inst Technol, PHD(biophys), 78. *Prof Exp:* Researcher biophys, Univ Calif, Irvine, 78-83; CONSULT, SCI SOFTWARE RES, 83- *Mem:* AAAS; Inst Elec & Electronics Engrs Computer Soc. *Res:* Understanding the physical basis for the structure and function of biopolymers; developing software and mathematical methods. *Mailing Add:* Dept Physiol & Biophys Univ Calif Irvine CA 92717

EDELMAN, JULIAN, b New York, NY; m 55; c 4. LOGISTICS ENGINEERING. *Educ:* City Col New York, BBA, 51; Columbia Univ, MSIE, 61. *Prof Exp:* Supvr reliability eng, Loral Electronics, 53-57; dir reliability eng, Perkin Elmer Corp, 59-63; sr eng specialist logistics & econ anal, GTE Sylvania Inc, 63-, MGR INTEGRATED LOGISTICS, GTE GOVT SYSTS. *Concurrent Pos:* Master lectr mgt, Boston Univ, 63-74 & Suffolk Univ, 68-; asst secy TC-56 & sr ed, Int Electrotech Comn. *Mem:* Fel Am Soc Qual Control; Soc Logistics Engrs; sr mem Inst Elec & Electronics Engrs. *Mailing Add:* 400 John Quincy Adams Rd Taunton MA 33062

EDELMAN, LEONARD EDWARD, polymer chemistry, for more information see previous edition

EDELMAN, MARVIN, b New York, NY, Aug 22, 39; US & Israeli citizen; m 61; c 5. PLANT MOLECULAR BIOLOGY, CHLOROPLASTS. *Educ:* Yeshiva Univ, BA & BHL, 61; Brandeis Univ, PhD(biol), 65. *Prof Exp:* Res assoc & fel, Harvard Med Sch, 66-67; res assoc & fel biochem, 67-69, intermediate scientist plant genetics, 69-71, sr scientist plant genetics, 71-80, dir grad teaching labs biol, 72-81, assoc prof, 81-84, PROF PLANT GENETICS, WEIZMANN INST SCI, 84-, HEAD, DEPT BIOL SERV, 88- *Concurrent Pos:* Vis scientist, Biol Labs, Harvard Univ, 76, Beltsville Agr Res Ctr, Md, 83-84, 86 & 87; co-organizer, French-Israeli Symp Cytoplasmic Genetics, 81 & Dynamics of Photo System II Workshop, Europ Molecular Biol Orgn, 87; mem, adv bd, Ger-Israel Biotechnol Prog, 84-; organizer, Aharon Katzir-Katchalsky Conf Plant Molecular Biol, 84, Second Cong Int Soc Plant Molecular Biol, 88; sr consult plant biotechnol & molecular biol, Argonne Nat Lab, 86-87; head, Dobrin Ctr Nutrit & Plant Biol, Weizmann Inst Sci, 87- 90. *Mem:* Israel Soc Plant Tissue Cult & Molecular Biol; Israel Biochem Soc; Am Asn Plant Physiologists; Int Soc Plant Molecular Biol. *Res:* Assembly, processing and degradation of chloroplast proteins and structure, function and photocontrol of photosynthetic reaction centers; nicotiana mutants; protein engineering and restructuring of chloroplast DNA by genetic engineering; effect of ultraviolet light on photosynthesis. *Mailing Add:* Dept Plant Genetics Weizmann Inst Sci Rehovot Israel

EDELMAN, NORMAN H, b New York, NY, May 21, 37; m 59; c 3. RESPIRATORY CONTROL PHYSIOLOGY. *Educ:* Brooklyn Col, AB, 57; NY Univ, MD, 61. *Prof Exp:* Med resident, NY Univ, 63; res assoc pulmonary physiol, genontol, Nat Heart Inst, NIH, 63-65; vis fel cardiopulmonary, Columbia Univ, 65-68; res assoc, Michael Reese Med Ctr, 67-69; asst prof pulmonary med, Univ Pa, 69-72; PROF MED & CHIEF, DIV PULMONARY MED, ROBERT WOOD JOHNSON MED SCH, 72-, DEAN, 88- *Concurrent Pos:* Vis lectr, Univ Chicago, 67-69; mem, Cardiovascular Pulmonary Study Sect, NIH, 75-79, Pulmonary Dis Adv Comt, 85-; mem grad fac physiol & biomed eng, 76-, assoc dean res, 78-, assoc chmn, dept med, Robert Wood Johnson Med Sch, 80-; chmn, Vet Admin Merit Review Bd Respiration, 80-83; consult sci affairs, Am Lung Asn, 84-;

interim dir, NJ Ctr Advan Biotech & Med, 84-86, Actg Dean, 87-88; mem, Cong Comn Sleep Res, 90- *Honors & Awards:* Merit Award, NIH, 89. *Mem:* Asn Am Physicians; Am Soc Clin Invest; Am Physiol Soc; Am Thoracic Soc; fel Am Col Physicians; fel Am Col Chest Physicians. *Res:* Metabolic and chemical environment of the central nervous system as it impacts on control of breathing; brain blood flow; brain hypoxia; endogenous opioids, central chemoreceptors; breathing during sleep. *Mailing Add:* Dept Med CN19 Robert Wood Johnson Med Sch New Brunswick NJ 08903

EDELMAN, ROBERT, b Brooklyn, NY, Apr 30, 42; m 72. CYANOACRYLATE & ANAEROBIC ADHESIVE, POLYIMIDES. *Educ:* Brooklyn Col, BS, 63; Rutgers Univ, PhD(org chem), 69. *Prof Exp:* Fel, Univ Fla, 68-69; from res chemist to sr res chemist, Celanese Res Co, 69-79, res assoc, Celanese Plastics & Specialties Co, 79-82; sr res assoc, M&T Chems, 82-89; TECHNOL DEVELOP MGR, PERMABOND DIV, NAT STARCH & CHEM CO, 89- *Mem:* Am Chem Soc. *Res:* Scyano acrylate and anaerobic adhesives, polyimides for microelectronics; resin materials for carbon fibers; polyacetals and polyesters, synthesis and evaluation. *Mailing Add:* 173 Ravenhurst Ave Staten Island NY 10310

EDELMAN, ROBERT, b St Louis, Mo, Oct 18, 36; m 58; c 4. INFECTIOUS DISEASES, ALLERGY & IMMUNOLOGY. *Educ:* Washington Univ, BA, 58, MD, 62. *Prof Exp:* Intern path, Johns Hopkins Hosp, 62-63, intern med, 63-64; resident, Barnes Hosp, St Louis, 64-65; USPHS fel prev med, Case Western Reserve Med Sch, 65-67, instr, 67-68; chief commun dis & immunol res br, Off Surg Gen, US Army, 68-70; intern virologist, SEATO Med Res Lab, Bangkok, Thailand, 70-73; chief med div, US Army Med Res Inst Infectious Dis, 73-74, virologist, 74-76; chief clin studies br, 76-81, CHIEF CLIN & EPIDEMIOL STUDIES BR, NIH, 81-, DEP DIR MICROBIOL & INFECTIOUS DIS PROG, 84- *Concurrent Pos:* Mem, Frederick Cancer Res Ctr, 75-; vchmn, NIH, Nutrit Coord Comt, 78-; mem, Life Sci Res Off Adv Comt, FASEB, 81-; enteric dis prog officer, Nat Inst Allergy & Infectious Dis, NIH, 81-; mem, Nat Dig Dis Adv Bd, 82-; clin prof med, Uniformed Serv Univ Health Sci, 84- *Mem:* Am Fedn Clin Res; Am Soc Microbiol; Am Soc Trop Med & Hyg; fel Am Col Physicians; fel Infectious Dis Soc Am; Am Asn Immunologists. *Res:* Clinical and epidemiological studies of enteric infections and vaccines, antifungal therapy, nutrition-infection immunity, rapid microbial diagnosis, Lyme disease and acquired immune deficiency disease. *Mailing Add:* Ctr Vaccine Develop Univ Md Sch Med Ten S Pine St Rm 9-34 Baltimore MD 21201

EDELMAN, WALTER E(UGENE), JR, b Oregon, Ill, July 15, 33. MECHANICAL ENGINEERING. *Educ:* Univ Minn, BME, 56, MSME, 58; Ore State Univ, PhD(mech eng), 67. *Prof Exp:* Engr, Minneapolis Honeywell, Inc, 53-56; instr mech eng, Univ Minn, 57-58; mem tech staff, Hughes Aircraft Co, 58-61; instr mech eng, Ore State Univ, 61-66; from asst prof to assoc prof, 67-75, PROF MECH ENG, CALIF STATE UNIV, LONG BEACH, 75- *Honors & Awards:* Ralph R Teetor Ed Award, Soc Automotive Engrs, 68. *Mem:* Am Soc Mech Engrs; Am Soc Eng Educ; Sigma Xi. *Res:* Design in mechanical engineering; automotive engineering. *Mailing Add:* 20822 Woodlea Lane Huntington Beach CA 92646

EDELMANN, CHESTER M, JR, b New York, NY, Dec 26, 30; m 53; c 3. PEDIATRICS. *Educ:* Columbia Univ, AB, 51; Cornell Univ, MD, 55; Am Bd Pediat, dipl & cert pediat nephrology. *Prof Exp:* From asst instr to assoc prof, 57-70, chmn dept, 73-80, PROF PEDIAT, ALBERT EINSTEIN COL MED, 70-, ASSOC DEAN, 80- *Concurrent Pos:* Res fel renal physiol, Albert Einstein Col Med, 58-59, 61-63; NIH res career develop award, 63-68; mem med adv bd, Kidney Found NY, 69-; mem sci adv bd, Nat Kidney Found, 69-75, vchmn, 73-74, chmn, 74-76; mem kidney dis & nephrology index adv comt, NIH, 70-, gen med study sect B, 71-75, chmn, 73-75; mem sub-bd pediat nephrology, Am Bd Pediat, 73-77; mem coun on circulation, Am Heart Asn; mem ed bd, Pediat Res, Pediat, Kidney Int & The Kidney. *Honors & Awards:* E Mead Johnson Award for Res in Pediat, Am Acad Pediat, 72. *Mem:* Am Acad Pediat; Am Soc Clin Invests; Am Physiol Soc; Soc Pediat Res; Am Pediat Soc; Am Soc Pediat Nephrology. *Res:* Developmental renal physiology; renal disease in infants and children. *Mailing Add:* Dept Pediat Albert Einstein Col Med 1300 Morris Park Ave Bronx NY 10461

EDELSACK, EDGAR ALLEN, b New York, NY, June 14, 24. PHYSICS. *Educ:* Univ Southern Calif, BS, 48. *Prof Exp:* Asst, Univ Southern Calif, 48-49; physicist, Emery Tumor Group, 49-53; head, Van de Graaff Accelerator Sect, US Naval Radio Defense Lab, 53-56; physicist, San Francisco Br, 56-57 & Physics Prog, Wash, 67-72, liaison scientist, London, 72-73, PHYSICIST ELECTRONIC & SOLID STATE SCI PROG, OFF NAVAL RES, WASHINGTON, DC, 73- *Concurrent Pos:* Consult, Dept Radiation Ther, Univ Md Hosp, Baltimore, 68-77. *Mem:* AAAS; Am Phys Soc; Am Asn Physicists Med; Royal Inst Gt Brit. *Res:* Superconductivity and applications; radiation therapy; accelerators and radioisotopes; biophysics; hyperthermia; microwave biological effects. *Mailing Add:* Info Ctr Georgetown Crxogenic 3530 West Pl NW Washington DC 20007

EDELSON, ALLAN L, b Los Angeles, Calif, Jan 1, 40. MATHEMATICS. *Educ:* Univ Calif, Berkeley, BSc, 62; State Univ NY Stony Brook, PhD, 68. *Prof Exp:* Instr math, State Univ NY, Stony Brook, 68-69; from asst prof to assoc prof, 69-81, PROF, UNIV CALIF, DAVIS, 81- *Mem:* AAAS; Am Math Soc. *Res:* Algebraic topology, differential equations. *Mailing Add:* Dept Math Univ Calif Davis CA 95616

EDELSON, BURTON IRVING, b New York, NY, July 31, 26; m 52; c 3. SATELLITE COMMUNICATIONS, SPACE TECHNOLOGY. *Educ:* US Naval Acad, BS, 47; Yale Univ, MS, 54, PhD(metall), 60. *Hon Degrees:* DSc, Capital Col, 86. *Prof Exp:* Mem staff, Naval Bur Ships, 59-62, Nat Aeronaut & Space Coun, 62-65 & Off Naval Res, London, 65-67; dir, Comsat Labs, Commun Satellite Corp, 68-80, sr vpres, 80-82; assoc adminr space sci & appln, NASA, 82-87; FEL, FOREIGN POLICY INST, JOHNS HOPKINS UNIV, 87-; RES PROF ELEC ENG & COMPUTER SCI, GEORGE WASHINGTON UNIV, 90- *Concurrent Pos:* Consult satellite commun, Space Systs, 82- *Honors & Awards:* Henry M Howe Award, Am Soc Metals, 63; Legion of Merit, 65; Govt Indust Award, Inst Elec & Electronics Engrs, 85; Commun Award, Am Inst Aeronaut & Astronaut, 88. *Mem:* Fel AAAS; fel Am Inst Aeronaut & Astronaut; Am Soc Metals; fel Inst Elec & Electronics Engrs; fel Brit Interplanetary Soc. *Res:* Communications satellite systems. *Mailing Add:* 116 Hesketh St Chevy Chase MD 20815

EDELSON, DAVID, b Brooklyn, NY, Nov 27, 27; m 53, 62; c 4. CHEMICAL KINETICS. *Educ:* Polytech Inst Brooklyn, BS, 46; Yale Univ, PhD(chem), 49. *Prof Exp:* Asst chem, Yale Univ, 46-49, Sterling res fel, 49-50; mem tech staff, Bell Labs, 50-85; vis prof chem, Fla State Univ, 85-86; PROF CHEM ENG, FAMU/FLA STATE UNIV COL ENG, 86- *Concurrent Pos:* Ed, Computers & Chem, 89- *Mem:* Am Chem Soc; Am Phys Soc; Combustion Inst; Am Soc Eng Educ. *Res:* Chemical kinetics; aeronomy; modeling and simulation by computer of complex chemical systems. *Mailing Add:* 1107 Kenilworth Rd Tallahassee FL 32312

EDELSON, EDWARD HAROLD, b New York, NY, Jan 28, 47; m 70; c 2. ORGANIC CHEMISTRY, ANALYTICAL CHEMISTRY. *Educ:* Lehman Col, BS, 73; Rensselaer Polytech Inst, PhD(org chem), 77. *Prof Exp:* Res assoc bio-org chem, NASA-Ames Res Ctr, 77-79; res assoc & lectr, org chem, Univ Southern Calif, 79-80; sr res chemist, synthetic fuels chem, Exxon Res & Eng Co, Baytown, Tex, 80-85; SR RES CHEMIST, PETROL COMPOS RES, MOBIL RES & DEVELOP CO, PAULSBORO, NJ, 85- *Mem:* Am Chem Soc; Soc Automotive Engrs. *Res:* Chemistry and analysis of petroleum and petroleum products; organic chemistry of coal and synthetic fuels; coal conversion processes; chemical evolution and prebiotic formation of biomacromolecules; catalysis and catalyst synthesis; clay chemistry and clay-organic reactions. *Mailing Add:* 420 Downs Dr Cherry Hill NJ 08003

EDELSON, JEROME, b New York, NY, Nov 17, 32; m 56; c 3. BIOLOGICAL CHEMISTRY, DRUG METABOLISM. *Educ:* Brooklyn Col, BS, 54; Univ Tex, MA, 57, PhD(biol chem), 60. *Prof Exp:* Asst prof chem, Univ Southwestern La, 60-63; sr biochemist, Wallace Labs, Carter-Wallace, Inc, 63-74; dir dept drug metab & disposition, Sterling-Winthrop Res Inst, 74-83, SR DIR DRUG SAFETY EVAL, STERLING RES GROUP, 83- *Mem:* AAAS; Am Chem Soc; Am Soc Pharmacol & Exp Therapeut. *Res:* Synthesis and biological activity of amino acid analogues; biochemical pharmacology; pharmacokinetics and biopharmaceutics; models for human drug metabolism. *Mailing Add:* 2906 Tuckerstown Dr Sarasota FL 34231

EDELSON, MARTIN CHARLES, b New York, NY, Nov 18, 43; m 67; c 1. SPECTROSCOPY. *Educ:* City Col NY, BS, 64; City Univ NY, MA, 67; Univ Ore, PhD (phys chem), 73. *Prof Exp:* Lectr chem, Lane Community Col, 72-73; postdoctoral teaching fel org chem, Univ BC, 73-77; postdoctoral fel phys chem, Iowa State Univ, 77-79, assoc chemist, Ames Lab US Dept Energy, 79-83, chemist, 83-86, PROG DIR NUCLEAR SAFEGUARDS, IOWA STATE UNIV, 86-, ADJ PROF MECH ENG, 89- *Mem:* Soc Appl Spectros; Optical Soc Am; Inst Nuclear Mat Mgt; Am Soc Testing & Mat; Sigma Xi. *Mailing Add:* Ames Lab US Dept Energy Ames IA 50011

EDELSON, PAUL J, Newport News, Va, Dec 5, 43; m; c 4. IMMUNOLOGY. *Educ:* Univ Rochester, AB, 64; State Univ NY Downstate Med Ctr, MD, 69. *Prof Exp:* Asst prof, Rockefeller Univ, 74-77, asst prof, Harvard Med Sch, 77- 82, ASSOC PROF PEDIAT, CORNELL UNIV, 82-; DIR DIV INFECTIOUS DIS & IMMUNOL, NY HOSP, 82- *Concurrent Pos:* Consult, NIH & NSF; mem, NY State Task Force Mat Infant Aids & Pediat Aids Comprehensive Ctrs. *Mem:* Am Asn Immunol; Am Soc Cell Biol; Soc Ped Res. *Res:* Host defense mechanisms; acquired immuno-deficiency virus; infections and immunologic diseases of children; cell biology of macrophage; history and philosophy of science. *Mailing Add:* Cornell Univ Med Col 525 E 68th St New York NY 10021

EDELSON, ROBERT ELLIS, b Camden, Tenn, Jan 20, 43; m 67; c 5. AERONAUTICAL & ASTRONAUTICAL ENGINEERING. *Educ:* Mass Inst Technol, SB, SM, 63; Univ Calif, Los Angeles, MBA, 70. *Prof Exp:* Mem tech staff, Hughes Aircraft Co, 69-70; dir, Gor Technol Ctr, GTE Labs, 80-82; sr engr, Jet Propulsion Lab, 67-69, 71-73, group supvr, 73-75, tech mgr, 75-78 & 82-87, sect mgr, 78-80, PROG MGR, JET PROPULSION LAB, 87- *Concurrent Pos:* Proj mgr, Search for Extraterrestrial Intel, 75-80, Int Astronaut Asn, 78-82. *Mem:* AAAS. *Res:* Management systems; program planning; operations; documentation management. *Mailing Add:* 4800 Oak Grove Dr Pasadena CA 91109

EDELSON, SIDNEY, b New York, NY, Aug 24, 16; m 47. MATHEMATICS, PHYSICS. *Educ:* Brooklyn Col, BA, 38; NY Univ, MA, 49; Georgetown Univ, MA, 53, PhD(solar radiation), 61. *Prof Exp:* Capt, China Waterways Transport, Shanghai, 46-47; mathematician, US Naval Observ, DC, 48-50, astronr, 50-56; astronr, US Naval Res Lab, 56-62, res astronr proj leader, 62-64; res scientist, Solar Studies, Ames Res Ctr, NASA, 64-66, res scientist optical physics & planetary atmospheres, 66-71, res scientist solar magnetic fields & non-thermal radiative processes, 71-72; consult solar physics, Ministry Sci & Res, Univ Graz, Austria, 72-74; consult staff scientist, Techno-Econ Opportunity Inst, Calif, 74-75; solar energy consult, 75-85; CONSULT, 85- *Concurrent Pos:* Vol sci adv, 19th Cong Dist, 78-81. *Mem:* Am Astron Soc; Sigma Xi; Math Asn Am; NY Acad Sci. *Res:* Optical physics; planetary atmospheres; radio astronomy; solar physics; space physics; astronomy. *Mailing Add:* PO Box 1264 Main PO Santa Barbara CA 93102

EDELSTEIN, ALAN SHANE, b St Louis, Mo, June 27, 36; m 63; c 2. SOLID STATE PHYSICS. *Educ:* Wash Univ, BS, 58; Stanford Univ, MS, 59, PhD(physics), 63. *Prof Exp:* Res assoc physics, Stanford Univ, 63-64; NSF fel, Univ Leiden, 64-65; res assoc, IBM Corp, NY, 65-68; assoc prof physics, Univ Ill, Chicago Circle, 68-80; physicist, Energy Conversion Devices, Troy, MI, 80-81; SUPVRY RES PHYSICIST, MAT SCI & TECHNOL DIV, NAVAL RES LAB, WASHINGTON, DC, 81- *Concurrent Pos:* Consult, Solid State Sci Div, Argonne Nat Lab, 69-77. *Mem:* Am Phys Soc; Mat Res

Soc. *Res:* Superconducting and normal state properties of heavy fermion, Kondo alloys, compounds and valence fluctuation systems study of small particles produced by phase separation and in an inert atmosphere; modes of investigation including resistivity, specific heat, susceptibility, sound velocity and neutron scattering measurements. *Mailing Add:* 6371 Naval Res Lab Washington DC 20375

EDELSTEIN, BARRY ALLEN, b Houston, Tex, Apr 18, 45. CLINICAL PSYCHOLOGY, BEHAVIOR THERAPY. *Educ:* Univ Tex, BA, 67, MA, 69; Memphis State Univ, PhD(psychol), 75. *Prof Exp:* PROF PSYCHOL, WVA UNIV, 74- & CHMN DEPT, 89- *Concurrent Pos:* Dir psychol, Spencer Hosp, 79-83; prin investr, Nat Instit Mental Health Clin Training Grants, 78-86; assoc ed, J Child & Adolescent Psychother, 84- & Behav Ther, 86-; adv bd, May Inst, 86-; vis fel, Univ Western Australia, 87; consult, Hopemont Hosp, 88- *Mem:* Coun Univ Dirs Clin Psychol; Coun Dir Health Psychol Training; Am Psychol Asn; Asn Advan Behav Ther; Asn Behav Anal; Am Psychol Soc. *Res:* Interpersonal competence in social and clinical settings; examining social competence in the clinical interview; clinical gerontology. *Mailing Add:* 120 Euclid Ave Morgantown WV 26505

EDELSTEIN, HAROLD, b Brooklyn, NY, Oct 26, 38; m 66; c 2. REAGENT CHEMICAL DEVELOPMENT. *Educ:* Hunter Col, AB, 59; Polytech Inst Brooklyn, PhD(chem), 65. *Prof Exp:* Chemist anal res, Cyanamid, 65-69; sr chemist org process develop, Millmaster Onyx Group, 69-72 & anal res, Beecham Ltd, 72-73; group leader anal res, Diamond Shamrock Co, 73-79; DIR RES & DEVELOP, FISHER SCI CO, 79- *Mem:* Am Chem Soc; AAAS; Sigma Xi. *Res:* Development and project management of new products; management of the introduction of new products through long term planning into company strategies. *Mailing Add:* 38-59 D'Auria Dr Fair Lawn NJ 07410

EDELSTEIN, RICHARD MALVIN, b Los Angeles, Calif, May 28, 30; m 55; c 3. PHYSICS. *Educ:* Pomona Col, BA, 51; Columbia Univ, PhD(particle physics), 60. *Prof Exp:* Res physicist, 60-62, from asst prof to assoc prof, 62-69, assoc dean, Mellon Col Sci, 78-81, PROF PHYSICS, CARNEGIE-MELLON UNIV, 69- *Concurrent Pos:* Weizmann Inst fel, 70-71; vis prof, Technion, 80; vis physicist, Stanford Linear Accelerator Ctr, 82 & High Energy Physics Lab (KEK), Japan, 83. *Mem:* Fel Am Phys Soc. *Res:* Mu meson physics; high energy proton and pi meson scattering; charmed particle production and decay. *Mailing Add:* Dept Physics Carnegie-Mellon Univ Schenley Park Pittsburgh PA 15213

EDELSTEIN, STUART J, b Perth Amboy, NJ, Sept 6, 41; m 64; c 2. BIOCHEMISTRY. *Educ:* Tufts Univ, BS, 63; Univ Calif, Berkeley, PhD(biochem), 67. *Prof Exp:* Nat Res Coun fel cellular biochem, Pasteur Inst, Paris, 67-68; asst prof biochem & molecular biol, 68-74, assoc prof biochem, molecular & cell biol, 74-77, chmn dept, 78-80, prof biochem, molecular & cell biol, Cornell Univ, 77-86; PROF & CHMN, DEPT BIOCHEM, UNIV GENEVA, 86- *Concurrent Pos:* Alfred P Sloan Found res fel, 73-75; vis scientist, Weizmann Inst Sci, 74 & Inst Pasteur, 80-81; vis prof, Univ Paris, 80-81 & 84-85; Eleanor Roosevelt Int Cancer fel, 80-81; sr fel, Fogarty Int, 84-85. *Mem:* Am Soc Biol Chemists; Am Soc Cellular Biol. *Res:* Structure of fibers of sickle cell hemoglobin and microtubules; electron microscopy, optical diffraction and image reconstruction; molecular modeling. *Mailing Add:* 30 Quai E Ansermet Univ Geneva CH-1211 Geneva 4 Switzerland

EDELSTEIN, WARREN STANLEY, b Baltimore, Md, June 11, 37; m 65; c 1. APPLIED MATHEMATICS, SOLID MECHANICS. *Educ:* Lehigh Univ, BA, 58; Duke Univ, MA, 61; Brown Univ, PhD(appl math), 64. *Prof Exp:* Fel mech, Johns Hopkins Univ, 64-65; asst prof to assoc prof, 65-80, PROF MATH, ILL INST TECHNOL, 80- *Mem:* Soc Indust & Appl Math; Soc Eng Sci; Am Acad Mech. *Res:* Boundary value problems of heat conduction viscoelasticity and nonlinear creep in metals; finite element analysis of problems in fluid mechanics and flow-induced vibrations. *Mailing Add:* Dept Math Ill Inst Technol Chicago IL 60616

EDELSTEIN, WILLIAM ALAN, b Gloversville, NY. GENERAL PHYSICS, MEDICAL & HEALTH SCIENCES. *Educ:* Univ Ill, BS, 65; Harvard Univ, AM, 68, PhD(physics), 74. *Prof Exp:* Res fel natural philos, Glasgow Univ, Scotland, 74-77; res fel med physics, Univ Aberdeen, 77-80; PHYSICIST, GEN ELEC CORP RES & DEVELOP, 80- *Concurrent Pos:* Coolidge fel, Gen Elec, 90. *Honors & Awards:* Gold Medal, Soc Magnetic Resonance Med, 90. *Mem:* Inst Physics; fel Am Phys Soc; Soc Magnetic Resonance Med. *Res:* Nuclear magnetic resonance applied to medical imaging and spectroscopy; nuclear magnetic resonance for industrial applications and example oil core analysis; environmental cleanup; contaminated soil. *Mailing Add:* Gen Elec Corp Res & Develop Bldg K-1-NMR PO Box 8 Schenectady NY 12301

EDELSTONE, DANIEL I, b Boston, Mass, Dec 6, 46. OBSTETRICS, GYNOCOLOGY. *Educ:* Univ Calif, Los Angeles, BA, 68; Univ Southern Calif, MD, 72; Am Bd Obstet & Gynec, dipl, 78. *Prof Exp:* Res fel, Cardiovasc Res Inst, San Francisco Med Ctr, Univ Calif, 76-77; resident obstet & gynec, Sch Med, Univ Pittsburgh, Magee Womens Hosp, 72-75, res assoc neonatal physiol, Dept Neonatology & Pediat, 75, chief resident obstet & gynec, 75-76, res fel maternal-fetal med, 75-76, from asst prof to assoc prof obstet & gynec, 77-87, PROF OBSTET & GYNEC, SCH MED, UNIV PITTSBURGH, MAGEE-WOMENS HOSP, 88- *Concurrent Pos:* Vis assoc prof pediat, Sch Med, Yale Univ, New Haven, Conn, 86. *Honors & Awards:* Searle-Donald F Richardson Mem Prize Award, Am Col Obstetricians & Gynecologists, 85. *Mem:* Am Gynec & Obstet Soc; Am Physiol Soc; Soc Gynec Invest; Perinatal Res Soc (secy-treas, 90-95); Am Col Obstetricians & Gynecologists; AAAS; Am Heart Asn. *Mailing Add:* Dept Obstet & Gynec Univ Pittsburgh Magee-Womens Hosp Forbes Ave & Halket St Pittsburgh PA 15213

EDEN, FRANCINE CLAIRE, biochemistry, molecular biology, for more information see previous edition

EDEN, JAMES GARY, b Washington, DC, Oct 11, 50; m 72; c 3. LASER PHYSICS, QUANTUM ELECTRONICS. *Educ:* Univ Md, BSEE, 72; Univ Ill, MS, 73, PhD, 76. *Prof Exp:* Res asst gas discharge physics, elec eng dept, Univ Ill, 72-75; res assoc laser physics, Nat Res Coun, Washington, DC, 75-76; res staff physicist laser physics, Naval Res Lab, 76-79; from asst prof to assoc prof, 79-84, PROF ELECT ENG, RES PROF, COORD SCI LAB, GRAD RES FAC PHYSICS, UNIV ILL, URBANA-CHAMPAIGN, 84- *Concurrent Pos:* Naval Res Lab publ award, 79; assoc, Ctr for Advan Study, Univ Ill, 87-88. *Honors & Awards:* Beckman Res Award, 88. *Mem:* Am Phys Soc; fel Inst Elec & Electronics Engrs; Optical Soc Am; Mat Res Soc; fel Optical Soc Am. *Res:* Atomic and molecular physics; laser-induced deposition of thin semiconductor films; molecular spectroscopy. *Mailing Add:* Everitt Lab Univ Ill 1406 W Green St Urbana IL 61801

EDEN, MURRAY, b Brooklyn, NY, Aug 17, 20; m 45, 62; c 5. INSTRUMENTATION, SIGNAL PROCESSING. *Educ:* City Col New York, BS, 39; Univ Md, MS, 44, PhD(phys chem), 51. *Prof Exp:* Phys chemist, Nat Bur Standards, 43-49; biophysicist, Nat Cancer Inst, 49-53; spec fel math biol, USPHS, Princeton Univ, 53-55; biophysicist, Nat Heart Inst, 55-59; prof elec eng, Mass Inst Technol, 59-79; CHIEF BIOMED ENG & INSTRUMENTATION BUR, DIV RES SERV, NIH, 76-; EMER PROF ELEC ENG, MASS INST TECHNOL, 79- *Concurrent Pos:* Lectr, Am Univ, 47-48 & Harvard Med Sch, 60-73; consult to dir-gen, WHO, 63-75; chmn, US Nat Comt on Eng in Med & Biol, 67-72; ed-in-chief, Info & Control, 67-84; vis prof, Swiss Fed Polytech Inst, Lausanne, 84 & 87. *Mem:* Am Physiol Soc; Biophys Soc; Inst Elec & Electronics Engrs; Soc Photo-Optical Instrumentation Engrs; Asn Advan Med Instrumentation; AAAS. *Res:* Biomedical engineering; mathematical models for biology; pattern recognition; human cognitive processes. *Mailing Add:* Div Res Serv Bldg 13 Rm 3W13 NIH Bethesda MD 20205

EDEN, RICHARD CARL, b Anamosa, Iowa, July 10, 39; div; c 2. ELECTRONIC PACKAGING & MATERIALS, SEMICONDUCTOR DEVICE PHYSICS. *Educ:* Iowa State Univ, BS, 61; Calif Inst Technol, MS, 62; Stanford Univ, PhD(solid state physics), 67. *Prof Exp:* Res asst elec eng, Calif Inst Technol, 61-62; res asst elec eng, Stanford Univ, 62-67, res assoc, 67-68; mem tech staff, NAm Rockwell Corp, 68-74; prin scientist solid state electronics, Sci Ctr, Rockwell Int, 74-80; SR VPRES & DEV, GIGABIT LOGIC INC, WESTLAKE VILLAGE, CALIF, 81- *Concurrent Pos:* Consult, electronics technol, 90- *Honors & Awards:* Centennial Medal, Inst Elec & Electronics Engrs, 85. *Mem:* Am Phys Soc; fel Inst Elec & Electronics Engrs. *Res:* Investigation of the electronic structure of solids by means of such experimental techniques as photoemission and measurements of optical properties; solid state optical detector and other semiconductor device research; development of the first Gallium Arsenide LSI integrated circuit technology. *Mailing Add:* Gigabit Logic Inc 1908 Oak Terrace Lane Newbury Park CA 91320

EDENBERG, HOWARD JOSEPH, b New York, NY, Jan 29, 48; m 78; c 3. MOLECULAR BIOLOGY, BIOCHEMISTRY. *Educ:* Queens Col, BA, 68; Stanford Univ, AM, 70, PhD(biol sci), 73. *Prof Exp:* Fel biol, Mass Inst Technol, 73-76; fel biol chem, Harvard Med Sch, 76-77; from asst prof to prof biochem, 77-86, PROF BIOCHEM & MED GENETICS, IND UNIV, 87- *Concurrent Pos:* Fel, Damon Runyon-Walter Winchell Cancer Fund, 73-75; fel, NIH, 75-77; prin investr, NIH grant, 78-, NSF grant, 84-87; vis assoc prof, Hormone Res Inst, Univ Calif, San Francisco, 85-86; mem, Alcohol Biomed Res Rev Comn, Nat Inst Alcohol Abuse & Alcoholism, 86-90. *Mem:* Am Soc Biochem & Molecular Biol; AAAS; Am Soc Microbiol; Res Soc Alcoholism; Int Soc Biomed Res Alcoholism; Sigma Xi. *Res:* Mammalian gene structure and regulation; alcohol dehydrogenase genes; protein engineering; mammalian DNA replication in vivo and in vitro; SV40 DNA replication; enzymology of replication; DNA repair. *Mailing Add:* Dept Biochem Sch Med Ind Univ 635 Barnhill Dr Indianapolis IN 46202-5122

EDENS, FRANK WESLEY, b Big Stone Gap, Va, Dec 18, 46; m 77; c 1. POULTRY PATHOLOGY, AVIAN PHYSIOLOGY. *Educ:* Va Polytech Inst, BS, 69; Va Polytech Inst & State Univ, MS, 71; Univ Ga, PhD(physiol), 74. *Prof Exp:* From asst prof physiol to assoc prof, 73-85, PROF POULTRY SCI, NC STATE UNIV, 85- *Concurrent Pos:* Consult, Embrey, Inc & United Egg Producers; owner, Edenco Consult-Sales. *Mem:* Poultry Sci Asn; Sigma Xi; AAAS; Animal Behav Soc; NY Acad Sci; Am Physiol Soc; Poultry Sci Asn; Am Asn Avian Pathol. *Res:* Body temperature regulation in birds; physiology of broodiness in turkey hens; physiological behavior in domestic fowl; respiratory pathology in turkeys and chickens; immunoregulation in poultry. *Mailing Add:* Dept Poultry Sci NC State Univ Box 7635 Raleigh NC 27695-7635

EDER, DOUGLAS JULES, b Milwaukee, Wis, Apr 26, 44; m 68; c 3. PHYSIOLOGY, BIOPHYSICS. *Educ:* Col Wooster, AB, 66; Fla State Univ, MS, 69, PhD(biophys), 73. *Prof Exp:* Res asst biophys, Fla State Univ, 70-71, psychobiol fel, 71-73; postdoc fels, Fight for Sight, 74, NIH, 74-75; asst prof, 75-80, ASSOC PROF BIOL SCI, SOUTHERN ILL UNIV, EDWARDSVILLE, 80- *Concurrent Pos:* Res grant-in-aid, Sigma Xi, Southern Ill Univ, Edwardsville, 76-77. *Mem:* Asn Res Vision & Ophthal; Soc Neurosci; Am Soc Zoologists; Sigma Xi. *Res:* Investigating molecular mechanisms whereby photoreceptor cells in retina transform light energy into neural signal, also investigating light-evoked neural activity in neuroendocrine brain centers such as pituitary, pineal, hypothalamus. *Mailing Add:* Dept Biol Sci Box 65 Southern Ill Univ Edwardsville IL 62026

EDER, HOWARD ABRAM, b Milwaukee, Wis, Sept 23, 17; m 54; c 3. MEDICAL RESEARCH. *Educ:* Univ Wis, BA, 38; Harvard Univ, MD, 42, MPH, 45; Am Bd Internal Med, cert, 52. *Hon Degrees:* Dr, Linkoping Univ, Sweden, 91. *Prof Exp:* From intern to asst resident physician, Peter Bent Brigham Hosp, Boston, 42-44; asst physician, Rockefeller Inst Hosp, 46-50; asst prof med, Med Col, Cornell Univ, 50-54; investr, Nat Heart Inst, 54-55; assoc prof, Col Med, State Univ NY, Brooklyn, 55-57; from assoc prof to prof, 57-88, EMER PROF MED, ALBERT EINSTEIN COL MED, 88-

Concurrent Pos: Adj prof, Rockefeller Univ, 74-78; mem res comt, Am Heart Asn, Coun Arteriosclerosis; chmn, Atherogenesis, Hypertension & Lipid Metab Adv Comt, Nat Heart, Lung & Blood Inst, NIH, 78-80, mem, Task Force Cholesterol Lowering Res, 85- *Honors & Awards:* Distinguished Achievement Award, Am Heart Asn. *Mem:* Inst Med-Nat Acad Sci; Am Soc Biol Chemists; Asn Am Physicians; Am Physiol Soc; Brit Biochem Soc; Am Soc Clin Invest; Am Heart Asn. *Res:* Lipid metabolism and atherosclerosis. *Mailing Add:* Dept Med Albert Einstein Col 1300 Morris Park Ave Bronx NY 10461

EDERER, FRED, b Vienna, Austria, Mar 5, 26; US citizen; m 58; c 3. EPIDEMIOLOGY, CLINICAL TRIALS. *Educ:* City Col New York, BS, 49; Am Univ, MA, 59. *Prof Exp:* statistician, Nat Cancer Inst, 57-64, Nat Heart & Lung Inst, 64-71, head, clin trials sect, Nat Eye Inst, 71-74, chief, off biomet & epidemiol, 74-84, assoc dir, 84-86, SR EPIDEMIOLOGIST, EMMES CORP, 87- *Concurrent Pos:* Lectr, Am Univ, 65-68, London Sch Hyg & Trop Med, 80; dir, Statist Ctr, Nat Diet-Heart Study, 64-67; dir, Coord Ctr, Urokinase-Pulmonary Embolism Trial, 68-71; mem, Exec Comt & Data Monitoring Comt, Diabetic Retinopathy Study, 71-79; dir, Visual Activity Impairment Survey, 81-83; assoc ed, 82-87, ed, Am J Epidemiol, 87-; co-chmn, Adv Glaucoma Intervention Study, 87-; vis prof, Johns Hopkins Univ, 87. *Honors & Awards:* David Rumbough Sci Award, Juvenile Diabetes Found, 83. *Mem:* Fel Am Col Epidemiol; Soc Clin Trials; fel Am Statist Asn; Biomet Soc; Soc Epidemiol Res. *Res:* Epidemiology; evaluation of therapeutic efficacy. *Mailing Add:* 5504 Lambeth Rd Bethesda MD 20814

EDESKUTY, F(REDERICK) J(AMES), b Minneapolis, Minn, Sept 29, 23; m 47; c 4. CRYOGENIC & CHEMICAL ENGINEERING. *Educ:* Univ Minn, BChE, 44, PhD(chem eng), 50. *Prof Exp:* Mem res staff, Los Alamos Sci Lab, Univ Calif, 50-53; consult engr air conditioning, J V Edeskuty & Assocs, 53-54; mem res staff, Los Alamos Nat Lab, Univ Calif, 54-89; RETIRED. *Mem:* Int Inst Refrig. *Res:* Cryogenic engineering, cryogenics, hydrogen energy systems; adsorption kinetics; high pressure. *Mailing Add:* 913 Tewa Loop Los Alamos NM 87544

EDGAR, ALAN D, b Glasgow, Scotland, July 5, 35; Can citizen; m 75; c 3. GEOCHEMISTRY, PETROLOGY. *Educ:* McMaster Univ, BA, 58, MSc, 61; Univ Manchester, PhD(geol), 63. *Prof Exp:* Lectr, 63-64, from asst prof to assoc prof, 64-78, PROF GEOL, UNIV WESTERN ONT, 78- *Concurrent Pos:* Res grants, Nat Res Coun, 63-88, Ont Res Found, 64-65, Geol Surv Can, 65-71, NATO, 66-67, 73-74, 81-82 & 88-89 & Ont Geol Sci Res Grant, 80-84 & 86-88; vis prof, Imp Col, Univ London, 71, Monash Univ, & Manchester Univ, 82; vis fel, Australian Nat Univ, 74; vis prof, Monash Univ, 82, Manchester Univ, 82; lectr, NATO Advan Study Inst, Rennes, France, 83; Gledden sr res fel, Univ Western Australia, 90. *Mem:* Fel Mineral Soc Am; Mineral Soc Gt Brit & Ireland; fel Geol Asn Can. *Res:* Mineralogy and crystallography of feldspathoids; experimental studies of silicate systems pertinent to alkaline igneous rocks; petrology of alkaline undersaturated rocks; geochemistry and petrology of ultrapotassic igneous rocks; petrology of the mantle. *Mailing Add:* Dept Geol Univ Western Ont London ON N6A 5B8 Can

EDGAR, ARLAN LEE, b Gratiot Co, Mich, June 3, 26; m 52; c 3. ZOOLOGY. *Educ:* Univ Mich, MA, 50, MS, 57, PhD, 60. *Prof Exp:* From instr to prof biol, Alma Col, 50-86, chmn dept, 71-77,; RETIRED. *Concurrent Pos:* Vis prof biol sta, Univ Mich, 65-76; Charles Dana prof, 80. *Mem:* Sigma Xi; Am Archeol Soc. *Res:* Proprioceptive organs of phalangids; physiological ecology and behavior of phalangids; taxonomy of phalangids in the Great Lakes region; effects of car exhaust on litter invertebrates; terrestrial litter ecology. *Mailing Add:* 4271 Riverview Dr Alma MI 48801

EDGAR, NORMAN TERENCE, b Bristol, Eng, Oct 22, 33; m 60; c 2. MARINE GEOLOGY, GEOPHYSICS. *Educ:* Middlebury Col, BA, 57; Fla State Univ, MSc, 60; Columbia Univ, PhD(marine geol), 68. *Prof Exp:* Geologist, Shell Oil Co Can Ltd, 59-63; geophysicist, Lamont Geol Observ, 63-68; coord staff geologist, Deep Sea Drilling Proj, Scripps Inst Oceanog, Univ Calif, San Diego, 68-70, chief scientist, 70-75; dep chief, Off Marine Geol, 75-79, chief, 79-83, CHIEF CARIBBEAN PROJ, OFF MARINE GEOL, US GEOL SURV, 83- *Mem:* AAAS; Geol Soc Am. *Res:* Sedimentology; stratigraphy; structural geology; marine geophysics. *Mailing Add:* US Geol Surv Nat Ctr Mail Stop 914 Reston VA 22092

EDGAR, ROBERT KENT, b New York, NY, Dec 29, 43. SYSTEMATIC BOTANY, ECOLOGY. *Educ:* Univ Va, BA, 65; Rutgers Univ, MS, 68, PhD(bot), 70. *Prof Exp:* From instr to assoc prof, 68-79, PROF BIOL, SOUTHEASTERN MASS UNIV, 79- *Concurrent Pos:* Consult, Westinghouse Environ Systs, 72-76; res assoc, Farlow Herbarium & Libr, Harvard Univ, 76- *Res:* Systematics of marine benthic diatoms; historical development of microbiology; diatom paleoecology. *Mailing Add:* Dept Biol Southeastern Mass Univ North Dartmouth MA 02747

EDGAR, ROBERT STUART, b Calgary, Alta, Sept 15, 30. DEVELOPMENTAL GENETICS. *Educ:* McGill Univ, BSc, 53; Univ Rochester, PhD(biol), 57. *Prof Exp:* Gosney & res fels, Biol Div, Calif Inst Technol, 57; fel med sci, Nat Res Coun, 57-59; from asst prof to prof biol, Calif Inst Technol, 60-70; PROF BIOL, UNIV CALIF, SANTA CRUZ, 70- *Mem:* Genetics Soc. *Mailing Add:* Dept Chem Thimann Labs Univ Calif Santa Cruz CA 95064

EDGAR, SAMUEL ALLEN, b Stafford, Kans, Feb 6, 16; m 39; c 2. MICROBIOLOGY, PATHOLOGY. *Educ:* Sterling Col, AB, 37; Kans State Univ, MS, 39; Univ Wis, PhD(zool), 44. *Hon Degrees:* ScD, Sterling Col, 62. *Prof Exp:* Asst zool, Kans State Univ, 37-38, instr, 38-41; asst, Univ Wis, 41-44; PROF POULTRY SCI & POULTRY PATHOLOGIST, AUBURN UNIV, 47- *Concurrent Pos:* Sr scientist, USPHS, Tahiti, 49-50. *Mem:* Fel AAAS; Am Soc Parasitol; Am Micros Soc; fel Poultry Sci Asn; fel NY Acad Sci. *Res:* Resistance of animals to parasitic infections; virus, bacterial and parasitic diseases of poultry; development of the protozoan parasites Eimeria in domestic poultry; immunity of poultry to coccidial infections. *Mailing Add:* Dept Poultry Sci Auburn Univ Auburn AL 36849

EDGAR, THOMAS FLYNN, b Bartlesville, Okla, Apr 17, 45; m 67; c 2. PROCESS CONTROL, OPTIMIZATION. *Educ:* Univ Kans, BS, 67; Princeton Univ, PhD(chem eng), 71. *Prof Exp:* Process engr, Conoco, 68-69; PROF CHEM ENG, UNIV TEX, AUSTIN 71-, CHMN DEPT, 85- *Concurrent Pos:* Vis prof chem eng, Univ Calif, Berkeley, 78; pres, Cache Corp, 81-84; dir, Am Automatic Control Coun, 84-86, vpres & pres, 88-91; mem, Tech Adv Comt, Fisher Controls, 85-91; Abel-Hanger chair chem engr, Univ Tex, Austin, 91; vchmn, Coun Chem Res, 91-92; dir, Am Inst Chem Engrs. *Honors & Awards:* Colburn Award, Am Inst Chem Engrs, 80, Westinghouse Award, 88; Meriam-Wiley Award, Am Soc Eng Educ, 90. *Mem:* Am Inst Chem Engrs; Am Soc Eng Educ; Instrument Soc Am; Am Chem Soc. *Res:* Mathematical modeling, optimization, and control of chemical processes, with applications to separations, chemical reactors and microelectronics. *Mailing Add:* Dept Chem Eng Univ Tex Austin TX 78712

EDGCOMB, JOHN H, PATHOLOGY. *Educ:* Univ Chicago, BS & MD; Am Bd Path, dipl. *Prof Exp:* Intern, Univ Chicago, 46-47; pathologist, Armed Forces Inst Path, Washington, DC, 52-53, Nat Cancer Inst, Bethesda, Md, 53-67 & Gorgas Mem Lab, Panama, 67-73; pathologist & dir labs, Gov Hosp NY, 73-77; asst dir toxicol & path & group leader, 77-85, SR PATHOLOGIST TOXICOL & PATH, HOFFMAN-LAROCHE, 85- *Concurrent Pos:* Mem, Libr Comt, NY Acad Med. *Mem:* AAAS; NY Acad Sci; Int Acad Path; Am Asn Path Bact; Am Asn Cancer Res; Am Soc Trop Med Hyg; Col Am Pathologists; Am Soc Clin Pathologists; Soc Invest Dermat. *Mailing Add:* Dept Toxicol & Path Bldg 100/3 Hoffmann-LaRoche Inc Kinsland Rd Nutley NJ 07110-1199

EDGE, ORLYN P, b Platteville, Wis, Mar 29, 39; m 61; c 3. MATHEMATICAL STATISTICS. *Educ:* Wis State Univ, Platteville, BS, 61, Univ Iowa, MS, 63, PhD(math), 66. *Prof Exp:* ASSOC PROF MATH, ILL STATE UNIV, 66- *Mem:* Am Statist Asn; Math Asn Am. *Mailing Add:* 1114 Sheridan Rd Normal IL 61761

EDGE, RONALD (DOVASTON), b Bolton, Eng, Feb 3, 29; m 56; c 2. NUCLEAR PHYSICS, SOLID STATE PHYSICS. *Educ:* Cambridge Univ, BA, 50, MA, 52, PhD, 56. *Prof Exp:* Res fel nuclear physics, Australian Nat Univ, 54-58; asst prof, Univ SC, 58-62; res fel, Yale Univ, 63; PROF PHYSICS, UNIV SC, 64- *Concurrent Pos:* Vis prof, Stanford Univ, 61, Calif Tech, 61-62, Univ Munich, 72, Univ Witwatersrand, 80-81 & Univ Sussex, 80; res assoc, Oak Ridge Nat Lab, 59, 65, 85 & Los Alamos Sci Lab, 73, 74. *Honors & Awards:* Russell Award; Pegram Award. *Mem:* Fel Am Phys Soc; Am Asn Physics Teachers; Sigma Xi. *Res:* Pion physics at intermediate energies; channeling in crystals; ion-solid interactions; neutron diffraction. *Mailing Add:* Dept Physics Univ SC Columbia SC 29208

EDGELL, MARSHALL HALL, b San Jose, Calif, Apr 17, 39. MOLECULAR BIOLOGY, GENETICS. *Educ:* Mass Inst Technol, BS, 61; Pa State Univ, MS, 64, PhD(biophys), 65. *Prof Exp:* From asst prof to assoc prof, 68-78, PROF BACT, MED SCH, UNIV NC, CHAPEL HILL, 78- *Concurrent Pos:* NDEA Title IV grant, Calif Inst Technol, 65-68; vis scientist, NIH, 76-77 & Lawrence Livermore Lab, 82-83; ed, J Biol Chem; dir, Prog Molecular Biol & Biotechnol, 81-88; Kenan prof. *Mem:* Am Chem Soc; Am Soc Microbiol. *Res:* Organization expression and regulation of eukaryotic genomes; mouse beta hemoglobin genes; DNA sequence evolution; Lines-1 transposons; genetic disease and gene therapy. *Mailing Add:* Microbiol Dept CB No 7290 FLOB Univ NC Chapel Hill NC 27599

EDGELL, WALTER FRANCIS, b Logansport, Ind, July 26, 16; m 37; c 4. CHEMISTRY. *Educ:* Univ Calif, BS, 39; Univ Iowa, MS, 41; Harvard Univ, PhD(phys chem), 44. *Prof Exp:* Chemist, Div Eight, Nat Defense Res Comt, Harvard Univ, 43; instr phys chem, Univ Iowa, 43-46, assoc prof, 46-49; PROF PHYS CHEM, PURDUE UNIV, 49- *Concurrent Pos:* Guggenheim fel, 56-57. *Mem:* Am Chem Soc; Soc Appl Spectros; Coblentz Soc. *Res:* Infrared and Raman spectroscopy; dynamics, structure and chemistry in electrolytic solutions; theory of symmetry in molecules and spectroscopic selection rules; metal carbonyls. *Mailing Add:* 2507 Oswego Lane West Lafayette IN 47906

EDGERLEY, DENNIS A, b Chicago, Ill, Aug 23, 48; m 81; c 2. MASS SPECTROMETRY, ENVIRONMENTAL RESEARCH. *Educ:* Tex A&M Univ, BS, 70, PhD(chem), 77. *Prof Exp:* Staff scientist chem, Rockwell Int, 79-85; ASST DIR, LAB DIV, ACZ, INC, 85- *Mem:* Sigma Xi. *Res:* Chemical analysis of liquid and gaseous synthetic fuel mixtures derived from coal and petroleum with emphasis on mass spectrometric characterization of isotopically labeled compounds produced in mechanistic studies. *Mailing Add:* 16318 Wall St Houston TX 77040

EDGERLEY, EDWARD, JR, b Lancaster, Pa, Mar 8, 31; m 54; c 4. CIVIL & SANITARY ENGINEERING. *Educ:* Pa State Univ, BS, 52; Mass Inst Technol, SM, 54; Univ Calif, Berkeley, PhD(sanit eng), 68. *Prof Exp:* From asst prof to assoc prof sanit eng, Wash Univ, 57-72, asst dean eng, 68-69; sr vpres, Envirodyne Engrs, 77-79; pres, Environ & Energy Consult, Inc, St Louis, Mo, 79-; AT SITEX CONSULT, MARYLAND HEIGHTS, MO. *Concurrent Pos:* From vpres to pres, Ryckman Edgerley Tomlinson & Assocs, Consult Engrs, 57-77; mem rev comt environ health, NIH, 69-71. *Mem:* Am Soc Civil Engrs; Am Soc Eng Educ; Air Pollution Control Asn; Am Indust Hyg Asn; Am Chem Soc; Sigma Xi. *Res:* Industrial waste water treatment; water treatment; trace organics in water; air pollution abatement; noise control; ion exchange treatment of liquid and solids waste. *Mailing Add:* Sitex Consults 2310 Creve Coeur Mill Rd PO Box 43128 Maryland Heights MO 63043

EDGERLY, CHARLES GEORGE MORGAN, b Gilmanton, NH, Nov 29, 18; m 44; c 7. DAIRY HUSBANDRY. *Educ:* Univ NH, BS, 48; Rutgers Univ, MS, 50. *Prof Exp:* Asst prof animal husb & asst animal husbandman, Agr Exp Sta, Univ Maine, 50-55; from asst prof to assoc prof, 55-84, assoc animal husbandman, Agr Exp Sta, 65-84, EMER PROF DAIRY HUSB, NDAK STATE UNIV, 84- *Mem:* Am Dairy Sci Asn; Am Fedn Mineral Socs. *Res:* Calf feeding; dairy cattle management; dairy herd waste disposal. *Mailing Add:* 1317 Eighth Ave S Fargo ND 58103-2504

EDGERTON, H(AROLD) E(UGENE), electrical engineering; deceased, see previous edition for last biography

EDGERTON, LOUIS JAMES, b Adena, Ohio, Jan 28, 14; m 46; c 3. HORTICULTURE, PLANT PHYSIOLOGY. *Educ:* Ohio State Univ, BS, 37; Cornell Univ, PhD(pomol), 41. *Prof Exp:* Asst pomol, Cornell Univ, 37-41, instr, 41; res assoc, Rutgers Univ, 42-45; prof pomol, Cornell Univ, 46-79, head dept, 70-75; RETIRED. *Concurrent Pos:* Fulbright grant, Cairo Univ, 66. *Mem:* Am Soc Plant Physiol; fel Am Soc Hort Sci. *Res:* Studies on cold hardiness of fruit plants; chemical thinning and control of preharvest apple drop with plant growth regulators; absorption, translocation and metabolism of plant growth regulators by apple and peach trees. *Mailing Add:* 110 Brandywine Rd Ithaca NY 14850

EDGERTON, MILTON THOMAS, JR, b Atlanta, Ga, July 14, 21; m 45; c 4. PLASTIC SURGERY. *Educ:* Emory Univ, AB, 41; Johns Hopkins Univ, MD, 44; Am Bd Surg, dipl, 51; Am Bd Plastic Surg, dipl, 51. *Prof Exp:* Intern surg, Barnes Hosp, St Louis, Mo, 44-45; asst resident, Johns Hopkins Hosp, 47-49, from instr to prof plastic surg, Sch Med, Johns Hopkins Univ, 49-70; prof, 70-80, alumni prof plastic surg & chmn dept, Sch Med, Univ Va, 70-88; RETIRED. *Concurrent Pos:* Surgeon in chg, Johns Hopkins Univ Hosp, 52-70; vchmn, Am Bd Plastic Surg, 69-70; trustee, Cell Sci Ctr, NY, 70-; consult, USPHS, Vet Admin, Baltimore City & Children's Hosps, Baltimore, Nat Clin Ctr, NIH & Walter Reed Hosp, Bethesda & Vet Admin Hosp, Salem, Va; chmn, Coord Coun Acad Polices in Plastic Surg, 76-79. *Honors & Awards:* Dow Corning Award, Am Soc Plastic & Reconstruct Surg, 74. *Mem:* AMA; fel Am Col Surg; Am Asn Plastic Surg (vpres, 72, pres, 74-75); Am Soc Plastic & Reconstruct Surg; Am Soc Maxillofacial Surgeons; Pan Am Med Asn (pres, 85). *Res:* Head and neck cancer surgery; reconstructive surgery; congenital defects; hand surgery; tissue transplantation research; craniomaxillo-facial surgery. *Mailing Add:* Dept Plastic Surg Univ Va Med Ctr Charlottesville VA 22908

EDGERTON, ROBERT FLINT, b Rochester, NY, Oct 16, 17; m 42; c 2. CHEMISTRY. *Educ:* Univ Rochester, BA, 40; Univ Mich, MS, 41, PhD(org chem), 44. *Prof Exp:* Fel Army specialized training prog, Univ Mich, 41-44; res chemist, Gen Elec Co, Mass, 44-45, tech sales, 46-47; tech staff, Paper Serv Div, Eastman Kodak Co, 47-53, group leader, Prof Papers, 50-52, group leader prod improv, 52-53, tech asst, Europ & Overseas Orgn, 53-55, tech asst, Int Div, 55-57, tech asst, Paris Off, France, 57-59, prod mgr graphic arts, 59-65, dir int advert planning, Int Mkt Div, 65-69, mgr advert & customer serv, Int Photog Div, 69-75, MGR INSTRNL OPERS, US & CAN MKTS DIV, EASTMAN KODAK CO, 75- *Mem:* Photog Soc Am. *Res:* Synthesis of perhydrophenanthrene derivatives and antimalarials; dehydrogenation; emulsion polymerization; purification of phenolic materials; photographic paper fixation, toning and washing. *Mailing Add:* One Rollingwood Dr Pittsford NY 14534

EDGERTON, ROBERT HOWARD, b Canton, Conn, Dec 27, 33; m 55; c 3. ENERGY, INDUSTRIAL DEVELOPMENT. *Educ:* Univ Conn, BS, 55, MS, 57; Cornell Univ, PhD(mech eng), 61. *Prof Exp:* Asst prof eng, Dartmouth Col, 62-67; head mech eng dept, Papua New Guinea Univ Tech, 86-89; ASSOC PROF ENG, OAKLAND UNIV, 67- *Concurrent Pos:* NASA fel, 64, 84, 85; Ford Found fel, IBM Corp, 66-67; vis prof, Harvey Mudd Col, 83. *Mem:* Am Soc Mech Engrs; Am Soc Eng Educ; Nat Soc Prof Engr; PNG Soc Prof Engr; PNG Nat Standards Adv Bd; PNG Univ Tech Coun; UNESCO Appropriate Tech Network Rep PNG. *Res:* Heat transfer, fluid mechanics and transport theory; flow and temperature measurements and instrumentation; biomedical engineering; blood flow; infra-red technology; applied thermodynamics; engineering education; international development; solar energy; environmental engineering. *Mailing Add:* Sch Eng & Comput Sci Oakland Univ Rochester MI 48309

EDGINGTON, DAVID NORMAN, b Oxford, Eng, Dec 18, 33. CHEMICAL LIMNOLOGY, OCEANOGRAPHY. *Educ:* Univ Oxford, Eng, BA, 57, BSc, 58, DPhil, 60. *Hon Degrees:* MA, Univ Oxford, 60. *Prof Exp:* Fel chem, Northwestern Univ, 61-62; fel, 61-62 & Radiol & Environ Res Div, Argonne Nat Lab, 62-63, staff scientist environ studies, 63-79; dir, Ctr Great Lakes Studies, Univ Wis, 79-87, PROF GEOL SCI, UNIV WIS-MULWAUKEE, 80- *Concurrent Pos:* Lectr, Northwestern Univ, 62-67. *Mem:* Royal Soc Chem; Am Soc Limnol & Oceanog; Int Asn Great Lakes Res; Geochem Soc; Am Geophys Union. *Res:* Use of natural and artificial radionuclides for studying chemical and biological processes in natural waters and sediments; long term behavior of persistent organic and inorganic pollutants in aquatic environments; ecosystem approach to environmental problems; sediment/water interactions. *Mailing Add:* 9051 N Rexleigh Bayside WI 53217

EDGINGTON, THOMAS S, b Los Angeles, Calif, Feb 10, 32; m 57; c 2. PATHOLOGY, IMMUNOLOGY. *Educ:* Stanford Univ, AB, 53, MD, 57. *Prof Exp:* Intern, Hosp Univ Pa, 57-58; resident path, Univ Calif, Los Angeles, 58-60; pathologist, Atomic Bomb Casualty Comn, Japan, 60-62; asst prof path, Univ Calif, Los Angeles, 62-65; assoc mem, 68-71, mem & prof dept path, 71-74, MEM, DEPT IMMUNOL, RES INST, SCRIPPS CLIN & RES FOUND, 74- *Concurrent Pos:* Sr res fel, Dept Exp Path, Scripps Clin & Res Found, 65-68, assoc mem, 68-71, mem, 71-74; assoc adj prof path, Univ Calif, San Diego, 68-75, adj prof, 75-; assoc ed, Hepatology, 80-85; ed, Thrombosis Res, 80-84; vis prof, Col France, 81; ann lectr, Nat Blood Club, 82; sect ed, J Immunol, 83-; counr, Am Asn Path, 83-85. *Honors & Awards:* Eli Lilly Lectr, Am Asn Study Liver Dis, 76. *Mem:* AAAS; Am Asn Immunol; Am Asn Path; Am Soc Hemat; Am Soc Clin Invest; Am Heart Asn; Am Asn Cancer Res; Reticuloendothelial Soc. *Res:* Immunology; blood coagulation; tumor immunology; inflammation; liver disease; molecular luology. *Mailing Add:* Dept Immunol Scripps Clin & Res Found 10666 N Torrey Pines Rd La Jolla CA 92037

EDGREN, RICHARD ARTHUR, b Chicago, Ill, May 28, 25; m 52; c 2. ENDOCRINOLOGY. *Educ:* Northwestern Univ, PhD(biol), 52. *Prof Exp:* Asst biol, Northwestern Univ, 49-52; sr investr, G D Searle & Co, 52-60; asst mgr, Nutrit & Endocrinol Sect, Wyeth Labs, 60-68, mgr endocrinol sect, 68-71; assoc dir clin res, Warner-Lambert Co, 71-72, dir endocrinol, 72-75; sect dir endocrinol, Parke, Davis Co, 75-78; assoc med dir, 78-81, DIR SCI AFFAIRS, SYNTEX LABS, 81- *Concurrent Pos:* Mem, Coun on Arteriosclerosis, Am Heart Asn; pres, US Int Found Res in Reproduction, Inc & Bd Sci Adv, Int Soc Reproductive Med. *Mem:* Am Soc Pharmacol & Exp Therapeut; Endocrine Soc; Royal Soc Med. *Res:* Endocrine pharmacology of steroidal and peptide hormones with special reference to reproductive physiology and oral contraceptives. *Mailing Add:* Syntex Labs Div Med Affairs 3401 Hillview Ave Palo Alto CA 94304

EDIDIN, MICHAEL AARON, b Chicago, Ill, Mar 31, 39; m 64; c 2. IMMUNOLOGY, CELL BIOLOGY. *Educ:* Univ Chicago, BS, 60; Univ London, PhD, 63. *Prof Exp:* Asst prof, 66-71, assoc prof, 71-75, PROF BIOL, JOHNS HOPKINS UNIV, 75- *Concurrent Pos:* NSF fel cell biol, Weizmann Inst, 63-64; Am Heart Asn res fel immunol, Harvard Med Sch, 64-66. *Honors & Awards:* Cole Medal, Am Biophys Soc; Merit Award, NIH. *Mem:* AAAS; Am Asn Immunol; Am Soc Cell Biol; Biophys. *Res:* Membrane structure and differentiation. *Mailing Add:* 2312 Sulgrave Ave Baltimore MD 21209

EDIGER, ROBERT I, b Hutchinson, Kans, Apr 2, 37; m 58, 81; c 4. PLANT TAXONOMY, ECOLOGY. *Educ:* Bethel Col, AB, 59; Kans State Teachers Col, MS, 64; Kans State Univ, PhD(bot), 67. *Prof Exp:* Teacher, Ford Pub Schs, 59-62; teacher, Hays Pub Schs, 62-63; from asst prof to assoc prof, 67-75, chmn dept biol sci, 74-77, PROF BOT, CALIF STATE UNIV, CHICO, 75- *Concurrent Pos:* Dir biol, Eagle Lake Biol Sta, 68-74. *Mem:* Bot Soc Am; Am Soc Plant Taxon; Int Asn Plant Taxon; Orgn Biol Field Sta (pres, 75-76). *Res:* Taxonomy of higher plants, especially the genus Senecio and Arnica in the family Compositae. *Mailing Add:* Dept Biol Sci Calif State Univ Chico CA 95929

EDIL, TUNCER BERAT, b Konya, Turkey, July 25, 45; US citizen; m 69; c 2. GEOTECHNICAL ENGINEERING, ENVIRONMENTAL GEOTECHNICS. *Educ:* Robert Col, BS, 67, MS, 69; Northwestern Univ, PhD(civil eng), 73. *Prof Exp:* Engr, Mirza Eng, Inc, 70; from asst prof to assoc prof, 73-80, PROF GEOTECH ENG, UNIV WIS-MADISON, 80- *Concurrent Pos:* Lectr geotech eng, Northwestern Univ, 73; prin investr, Univ Wis-Madison, 74-; Dow outstanding young fac, Am Soc Eng Educ, 80; ed, Am Soc Civil Engrs, 85-89, chmn Geotech Eng Div Publ Div, 85-89, mem Embankment Dams & Slopes Comn, 76- & Environ Geotech Comn, 89-; consult, Warzyn Eng, Inc, 85-89; chmn, comn D18.18, Am Soc Testing & Mat, 88- *Mem:* Am Soc Civil Engrs; Am Soc Testing & Mat; Am Soc Eng Educ; Int Soc Soil Mech & Found Eng. *Res:* Soil mechanics; geotechnical engineering; physiochemical, mechanical, and hydraulic behavior of earthen materials; stability of coastal slopes and landslides; compression of peat soils; soil-structure frictional interaction; waste geotechnics. *Mailing Add:* 37 S Eau Claire Ave Madison WI 53705

EDINGER, HENRY MILTON, b New York, NY, Feb 28, 43; m 67; c 3. PHYSIOLOGY, NEUROPHYSIOLOGY. *Educ:* City Col New York, BS, 64; Univ Pa, PhD(physiol), 69. *Prof Exp:* Asst prof physiol, 71-75, asst prof neurosci, 72-75, ASSOC PROF PHYSIOL & NEUROSCI, COL MED & DENT NJ, NJ MED SCH, 75- *Concurrent Pos:* NIH fel, Rockefeller Univ, 69-71; consult, Vet Admin Hosp, East Orange, NJ, 72- *Mem:* Soc Neurosci; Am Physiol Soc; NY Acad Sci. *Res:* Limbic system-neuroanatomy and neurophysiology. *Mailing Add:* Dept Physiol NJ Med Sch 100 Bergen St Newark NJ 07103

EDINGER, STANLEY EVAN, b Aug 9, 43. HEALTH CARE POLICY REGULATION, CLINICAL LABORATORIES. *Educ:* Brooklyn Col, BS, 64; NY Univ, MS, 69, PhD(phys chem), 70. *Prof Exp:* Teaching fel phys chem, NY Univ, 64-66, res asst, 66-70; tech translr & ed solid state chem, 70-71; clin asst chemist, tech supvr, Mt Sinai Hosp & Med Ctr, 71-76; USPHS sr scientist clin lab sci & regulations, Bur Qual Assurance, 76-77; sr scientist, 77-85, sci dir, Clin Lab Sci & Regulations, Health Care Finacing Admin, Health Standards & Qual Bur, Dept Health & Human, Serv, 85-86, sci dir, Implementation of US Pres Fed Drug Testing Prog, US Nat Inst Drug Abuse, 86-87, Med Outcome Res US Health Care Financing Admin Health Standards & Qual Bur, 87-88, SCI DIR & SR HEALTH POLICY ANALYST, USPHS, 90- *Concurrent Pos:* Consult, Prof Exam Serv, 75-76; sr scientist comdr, USPHS Comn Corps, 76-; deleg, Nat Coun Health Lab Serv, Health Care Financing Admin, 76-, nat comt, Clin Lab Studies, 80-, rep, Am Asn Blood Banks Accreditation Comt, 82-, Col Am Pathologist Inspection & Accreditation Comn, 82-, Task Force Lab Reimbursement, 83-; USPHS Task Force on Clin Labs, 78-; proj officer, Dept Health & Human Serv Clin Lab Proficiency Exam Prog; adv, Nat Comt Accreditation Chemists, 81-; fel, US Rep to US House of Rep Comt, Energy & Com, 89; comn officer USPHS Corps, sr scientist/comdr, 76-85, scientist dir/copt, 86-; chmn, Nat Coun Health Lab Serv, 86-90; comnr, Nat Comn Accreditation Chemists & Chem Engrs, 86-; mem, Lab Sect, Am Pub Health Asn; chmn, Legis Comt Coun & mem bd, US Surgeon Gen Sci Pub Adv Comt, 86-90. *Honors & Awards:* Super Performance Award, US Health Care Financing Admin, 81, Outstanding Achievement Award, 86. *Mem:* Fel Am Inst Chemists; Am Chem Soc; Am Asn Clin Chem; Am Pub Health Asn; AAAS; Sigma Xi. *Res:* Gypsum chemistry; laboratory automation and computerization; educational testing in laboratory sciences; development, writing, interpretation and implementation of quality assurance and payment regulations for clinical laboratories; quality control and proficiency testing. *Mailing Add:* 5901 Montrose Rd Apt 404N Rockville MD 20852

EDINGTON, JEFFREY WILLIAM, b Newcastle upon Tyne, Eng, May 28, 39. MATERIALS SCIENCE. *Educ:* Univ Birmingham, BSc, 60, PhD(metall), 63, DSc(metall), 75; Cambridge Univ, MA, 67. *Prof Exp:* Sr scientist metal sci, Battelle Mem Inst, 64-67; asst dir res mat, Cambridge Univ, 67-76; CHPERSON METALL & MAT & DIR MAT DURABILITY

DIV, UNIV DEL, 76- *Concurrent Pos:* Vis prof, Univ Brisbane, Australia, 74; vis prof & consult, Brown Boverie & Co, Switz, 75-76. *Mem:* Am Soc Metals; fel Inst Metallurgists. *Res:* Materials durability; erosion; stress corrosion; corrosion fatigue; electron microscopy; analytical microscopy; superplasticity. *Mailing Add:* 48 Holton Ave Westmount PQ H3Y 2G2 Can

EDISON, ALLEN RAY, b Plainview, Nebr, Sept 21, 26; m 49; c 2. ELECTRICAL ENGINEERING. *Educ:* Univ Nebr, BSc, 50, MSc, 57; Univ NMex, DSc(elec eng), 62. *Prof Exp:* Elec engr, Silas Mason Co, 50-53; from instr to assoc prof, 53-64, chmn dept, 64-70, PROF ELEC ENG, UNIV NEBR-LINCOLN, 70- *Concurrent Pos:* Res assoc, Univ NMex, 57-61; engr, Inst Telecommun, Dept Com, 65. *Mem:* Inst Elec & Electonics Engrs; Am Asn Eng Educ. *Res:* Electronic instrumentation. *Mailing Add:* 511 S 54 St Lincoln NE 68510

EDISON, LARRY ALVIN, b Aberdeen, Wash, Nov 8, 36; m 60; c 4. MATHEMATICS, COMPUTER SCIENCE EDUCATION. *Educ:* Whitman Col, BA, 58; Stanford Univ, PhD(math), 65, MS, 82. *Prof Exp:* Asst prof math, Reed Col, 64-70; prof math & chmn, Alma Col, 70-81; PROF MATH & COMPUT SCI, PAC LUTHERAN UNIV, 82- *Mem:* Am Math Soc; Math Asn Am; Asn Comput Mach. *Res:* Harmonic analysis; almost periodic functions; functional analysis; operations research; computer graphics; computer science education. *Mailing Add:* Dept Math & Comput Sci Pac Lutheran Univ Tacoma WA 98447

EDLICH, RICHARD FRENCH, b New York, NY, Jan 19, 39; m 61; c 3. PLASTIC SURGERY. *Educ:* Sch Med, NY Univ, MD, 62; Univ Minn, PhD, 73. *Prof Exp:* From instr to assoc prof, 71-76, prof plastic surg & biomed eng, Sch Med, Univ Va, 76-82, dir Emergency Med Serv & Burn Ctr, 74-85. *Concurrent Pos:* Physician tech adv, Bur Emergency Serv, HEW, 74-; consult, Div Health Manpower & Nat Ctr Health Serv Res, 77- *Mem:* Soc Univ Surgeons; Am Asn Surg Trauma; Am Burn Asn; Am Spinal Cord Injury Asn; Univ Asn Emergency Med; Am Col Surgeons; Am Soc Plastic & Reconstructive Surgeons; Soc Surgical Infection Am; Col Emergency Physicians; Am Surg Asn. *Res:* Biology of wound repair and infection. *Mailing Add:* Dept Plastic Surg Sch Med Univ Va Box 3 Charlottesville VA 22908

EDLIN, FRANK E, b Eskridge, Kans, Aug 25, 09; m 36; c 2. SOLAR ENERGY, ENGINEERING. *Educ:* Kans State Univ, BS, 31. *Prof Exp:* Field engr, Indust Eng Div, E I du Pont de Nemours & Co, 37-39, field supvr, 39-40, process engr, Design Div, 40-51, consult engr, Eng Serv Div, 51-60, engr, Develop Dept, 60-62, sr res engr, Eng Res Div, 62-65, lectr eng, Ariz State Univ, 65-71; pres, Water Appln, Inc, 71-73; CONSULT ENGR, 75- *Concurrent Pos:* Vpres, Int Plastics, Inc, Colwich, Kans, 67-75. *Mem:* AAAS; Am Inst Chem Engrs; Solar Energy Soc (exec secy, 65-67); Am Soc Mech Engrs; Nat Soc Prof Engrs. *Res:* Atomic energy; fusion; subsoil irrigation; chemical engineering unit operations. *Mailing Add:* 17826 Palo Verde Dr Sun City AZ 85373

EDLIN, GEORGE ROBERT, b Dickson Co, Tenn, Nov 17, 37; m 64; c 3. DIRECTED ENERGY WEAPON SYSTEMS, CHARGED PARTICLE ACCELERATORS. *Educ:* Austin Peay State Univ, Tenn, BS, 60; Southeastern Inst Technol, Ala, MS, 77, DSc(appl physics), 82. *Prof Exp:* Electronic engr instrumentation, Micom Test Lab, US Army, 65-69 & electromagnetics, 69-82, gen engr, Ballistic Missile Defense Advan Technol Ctr, 82-83, GEN ENGR DIRECTED ENERGY, SDC, US ARMY, 83- *Res:* Directed energy weapons office of the strategic defense commands; neutral particle beams; free electron lasers; charged particle beams; nuclear directed energy weapons. *Mailing Add:* 2608 Pentolope Dr Huntsville AL 35803

EDLIN, JOHN CHARLES, b Wilmington, Del, June 16, 43; m 69; c 3. PEDIATRICS, ADOLESCENT MEDICINE. *Educ:* Duke Univ, BS, 65; Univ Tenn, Memphis, MD, 68. *Prof Exp:* From intern to resident pediat, Duke Univ Med Ctr, 69-71; fel growth & develop, Univ London, 71-72; fel adolescent med, Harvard Med Sch, 72-73; asst prof pediat & internal med, Southwestern Med Sch, Univ Tex, 73-81; pvt pract adolescent med, 81-87; MED DIR, BRIDGEWAY DAY HOSP, DALLAS, 87- *Concurrent Pos:* Dir adolescent med prog, Dept Pediat, Univ Tex, 73-81; consult, Baylor Col Dent, 78, Timberlawn Psychiat Hosp, 78 & Western State Col, Colo. *Mem:* Am Soc Adolescent Psychiat; Soc Adolescent Med; fel Am Acad Pediat. *Res:* Growth and development; physiological changes which occur at puberty. *Mailing Add:* 12800 Hillcrest No 216 Dallas TX 75230

EDLUND, MILTON CARL, b Jamestown, NY, Dec 13, 24; m 45; c 2. PHYSICS. *Educ:* Univ Mich, MS, 48, PhD, 66. *Prof Exp:* Physicist reactor physics, Gaseous Diffusion Plant, Oak Ridge Nat Lab, 48-50, physicist & lectr, Sch Reactor Tech, 50-51, sr physicist & sect chief, 53-55; mgr, Physics & Math Dept, Babcock & Wilcox Co, 55-60, Develop Dept, 60-62 & Appln Develop Dept, 62-65, asst div mgr, 65-66; prof nuclear eng, Univ Mich, 66-67; consult, Union Carbide Corp, 67-70; prof, 70-89, EMER PROF MECH ENG, VA POLYTECH INST & STATE UNIV, 89- *Concurrent Pos:* Vis lectr, Swed Atomic Energy Comn, 53. *Honors & Awards:* Ernest Orlando Lawrence Award, 65. *Mem:* Nat Acad Eng; Am Soc Mech Engrs; Am Phys Soc; Sigma Xi; Am Nuclear Soc. *Res:* Neutron diffusion; nuclear reactor design. *Mailing Add:* 302 Neil St Blacksburg VA 24060

EDMAN, JAMES RICHARD, b Kandiyohi Co, Minn, June 5, 36; m 58; c 3. ORGANIC CHEMISTRY, POLYMER CHEMISTRY. *Educ:* Gustavus Adolphus Col, BS, 58; Univ Nebr, MS, 60, PhD(chem), 63. *Prof Exp:* Res chemist, 63-69, res supvr, 69-88, SR RES ASSOC PHOTOCHEM CATALYSIS, POLYMER SCI, E I DU PONT DE NEMOURS & CO, 88- *Mem:* Am Chem Soc; Sigma Xi. *Res:* Polymer structure property correlations. *Mailing Add:* Electronics E I du Pont de Nemours & Co Circleville OH 43113

EDMAN, JOHN DAVID, b Jan 20, 38; US citizen; m 59; c 3. MEDICAL ENTOMOLOGY. *Educ:* Gustavus Adolphus Col, BSc, 59; Univ Nebr, MSc, 61; Kans State Univ, PhD(entom), 64. *Prof Exp:* Sr scientist & asst dir, Fla Med Entom Lab, Fla State Div Health, 64-75; assoc prof entom, 75-81, PROF, UNIV MASS, AMHERST, 81-, HEAD DEPT ENTOM, 87- *Concurrent Pos:* NIH res grant, 65-; WHO travel-study fel, 75; consult, Res Resources Br, Nat Inst Allergy & Infectious Dis, 67-, Vector Biol Br, WHO, 70- & Bd Sci Technol Int Develop, Nat Acad Sci, 83-; mem trop med & parasitol study sect, NIH, 80-84; chmn, sect D, Entom Soc Am, 80; NSF coop grant, US-Australia, 81. *Mem:* Entom Soc Am; Am Mosquito Control Asn; Am Soc Trop Med & Hyg; Soc Vector Ecol. *Res:* Ecology and behavior of mosquitoes and black flies; vector-host interaction and disease transmission. *Mailing Add:* Dept Entom Univ Mass Amherst MA 01003

EDMAN, WALTER W, cosmetic chemistry; deceased, see previous edition for last biography

EDMISON, MARVIN TIPTON, b Lincoln, Nebr, July 21, 12; m 39; c 2. ORGANIC CHEMISTRY. *Educ:* Univ Nebr, AB, 33, MSc, 47; Okla Agr & Mech Col, PhD(chem), 52. *Prof Exp:* Teacher, Shattuck Mil Acad, 38-41; teacher, Wentworth Mil Acad, 47-48; from asst prof to assoc prof chem, Univ Ark, 51-55, asst to vpres & provost, 55; dir res found & prof chem, 55-78, asst vpres acad affairs, 68-78, exec dir, Okla State Univ Educ & Res Found, Inc, 72-78, EMER PROF & EMER RES COORDR, OKLA STATE UNIV, 78- *Concurrent Pos:* Proj dir, Ordark Res Proj, Univ Ark, 52-53; res adminr, Inst Sci & Technol, 54; mem Nat Coun Univ Res Adminrs & Nat Conf Advan Res; dir, Okla Water Resources Res Inst, 65-78. *Mem:* AAAS; Am Inst Chem; Nat Wildlife Fedn; Sigma Xi. *Res:* Radical substitution of aromatic nuclei; thermal decomposition of organic azides and inorganic oxidants; classified governmental research. *Mailing Add:* 1107 N Skyline Dr Stillwater OK 74074

EDMISTON, CLYDE, b Greeley, Colo, June 4, 37; m 58. PHYSICAL CHEMISTRY. *Educ:* Colo State Col, BA, 58; Iowa State Univ, PhD(chem), 63. *Prof Exp:* Nat Res Coun res assoc phys chem, Nat Bur Standards, 63-64; from asst prof to assoc prof, 64-73, PROF CHEM, UNIV WYO, 73- *Concurrent Pos:* Am Chem Soc Petrol Res Fund grant, 64-65; NSF res grant, 65-72 & NSF sr fel, Univ Fla, 70-71. *Mem:* Am Chem Soc; Am Phys Soc. *Res:* Quantum chemistry; molecular structure. *Mailing Add:* Dept Chem Univ Wyo Box 3838 Laramie WY 82071

EDMONDS, DEAN STOCKETT, JR, b Brooklyn, NY, Dec 24, 24; m 51; c 4. ATOMIC PHYSICS. *Educ:* Mass Inst Technol, BS, 50, PhD(physics), 58; Princeton Univ, MA, 52. *Prof Exp:* Res asst physics, Mass Inst Technol, 52-56, guest physicist & res fel, Cambridge Electron Accelerator, Mass Inst Technol & Harvard Univ, 59-61; from asst prof to prof, 61-89, EMER PROF PHYSICS, BOSTON UNIV, 89- *Concurrent Pos:* Dir, Tachisto, Inc, 71- & Gen Ionex, Inc, 75- *Mem:* Am Phys Soc; Am Asn Physics Teachers; Inst Elec & Electronics Engrs; Sigma Xi. *Res:* High energy accelerators; mass spectroscopy; molecular beam investigations; laser technology; communication techniques; physical electronics. *Mailing Add:* 550 Alexander Palm Rd Boca Raton FL 33432-7909

EDMONDS, ELAINE S, b Toledo, Ohio, July 14, 46; m 70, 78; c 1. NEUROANATOMY, NEUROENDOCRINOLOGY. *Educ:* Ohio Univ, BS, 68; Univ Mich, MS, 69; Univ Ariz, PhD(anat), 73. *Prof Exp:* Fel physiol, Univ Ariz, 73-74; asst prof anat, Univ NMex, 74-76; asst prof anat, Ind Univ, 77-84; MEM STAFF, DEPT NEUROL, GOOD SAMARITAN HOSP, PORTLAND, ORE, 84- *Mem:* Soc Neurosci; AAAS; Am Asn Anatomists. *Res:* Hypothalamic regulation of the endocrine system, especially reproductive system of the genetically obese rat. *Mailing Add:* 12555 SW 27th St Beaverton OR 97005

EDMONDS, FRANK NORMAN, JR, astrophysics; deceased, see previous edition for last biography

EDMONDS, HARVEY LEE, JR, b Leavenworth, Kans, Sept 23, 42; m 70. NEUROPHARMACOLOGY. *Educ:* Univ Kans, Lawrence, BA, 64, BS, 67; Univ Calif, Davis, PhD(pharmacol), 74. *Prof Exp:* Instr pharmacol, US Army Med Field Serv Sch, Ft Sam Houston, 68-70; asst prof pharmacol, Wash State Univ, 74-77; assoc prof, 77-80, PROF ANESTHESIOL, UNIV LOUISVILLE, 80- *Concurrent Pos:* Consult toxicol, 78-86; Sabbatical study, Janssen Pharmaceuts, Belgium, 83-84; pres, AMM Inc, 87- *Mem:* Int Anesthesia Res Soc; Am Soc Pharmacol & Exp Therapeut; Soc Toxicol. *Res:* Investigation of new techniques for monitoring, protecting and resuscitating the brain following head trauma or during cardiac surgery. *Mailing Add:* Dept Anesthesiol Univ Louisville Sch Med Louisville KY 40292

EDMONDS, JAMES D, JR, b Texarkana, Tex, July 28, 39; m; c 4. PHYSICS. *Educ:* San Diego State Univ, BA, 62; Cornell Univ, PhD(appl physics), 67. *Prof Exp:* Asst appl physics, Cornell Univ, 63-67, res assoc electron micros, 67; mem tech staff, Hughes Res Labs, 67-69; asst prof physics, Calif Western Col, 69-72; vis asst prof physics, Ore State Univ, 72-73; lectr physics, San Diego State Univ, 73-75; vis asst prof physics, Joint Sci Prog, Claremont Cols, 75-76; vis asst prof physics, Bucknell Univ, 76-78; assoc prof & chmn physics, Mercer Univ, 78-81; PROF, PHYSICS DEPT, MCNEESE STATE UNIV, 81- *Concurrent Pos:* Lectr, Los Angeles Valley Col, 68; lectr, Calif Luth Col, 68-69; technician & engr, Convair Astron, 59-62. *Res:* Thin film vacuum nucleation; x-ray diffractometry and structure analysis; transmission and scanning electron microscopy; ultrahigh vacuum; foundations of field theory and origins of mass, particularly quaternions and extensions to six and ten dimension space time. *Mailing Add:* Dept Physics McNeese State Univ 4100 Ryan St Lake Charles LA 70609

EDMONDS, JAMES W, b Long Beach, Calif, Oct 14, 43; m 66; c 3. CRYSTALLOGRAPHY. *Educ:* Harvey Mudd Col, Calif, BS, 65; Rice Univ, Houston, PhD(chem), 69. *Prof Exp:* Res assoc crystallog, Brookhaven Nat Lab, AEC, 68-70, sr res assoc, State Univ NY, Buffalo, 70-71 & Med Found

Buffalo, 71-74; sr res chemist, Anal Labs, Dow Chem Co, 74-81, group leader, org chem res, 81-83; mgr, mkt develop, Philips Electronic Instruments, 83-85, prod mgr, 85-88, INT SALES MGR, N V PHILIPS, 88- *Concurrent Pos:* Crystallog consult, State Univ NY, Buffalo, 71-74; mem bd dir, Joint Comt on Powder Diffraction Standards, 81-86. *Res:* X-ray and neutron powder and single crystal diffraction; optical emission spectroscopy; high resolution Guinier camera/monochromator; computer automation; aromatic monomer process development. *Mailing Add:* N V Philips Lelyweg 1 Almelo 7602 EA Netherlands

EDMONDS, MARY P, b Racine, Wis, May 7, 22. BIOCHEMISTRY. *Educ:* Milwaukee-Downer Col, BA, 43; Wellesley Col, MA, 45; Univ Pa, PhD(biochem), 51. *Hon Degrees:* DSc. Lawrence Univ, 83. *Prof Exp:* Instr chem, Wellesley Col, 45-46; fel, Univ Ill, 50-52; res assoc, Cancer Res Inst, Univ Wis, 52-55 & Montefiore Hosp Res Inst, 55-65; asst prof, Grad Sch Pub Health, 65-67, assoc res prof, 67-71, assoc prof biochem, 71-75, PROF BIOL SCI, UNIV PITTSBURGH, 76- *Mem:* Nat Acad Sci. *Res:* Molecular biology; structure and biosynthesis of nucleic acids; chemistry and enzymology of nucleic acids and nucleotides. *Mailing Add:* Dept Biol Sci A527 Langley Hall Univ Pittsburgh Pittsburgh PA 15260

EDMONDS, PETER DEREK, b Tunbridge Wells, Eng, Mar 29, 29; m 63. ULTRASONICS. *Educ:* Univ London, BSc, 52, dipl, PhD(physics), 59. *Prof Exp:* Asst physics, Phys Inst, Stuttgart Tech Hochsch, 55-56; physicist, Mullard Res Labs, Eng, 56-58, Akers Res Labs, Imp Chem Industs, Ltd, 58-61 & Plastics Div Res Labs, 61-62; res fel chem eng, Calif Inst Tech, 62-63; asst prof, 63-68, assoc prof elec eng & biomed eng, Univ Pa, 68-70; adminr tech serv, Inst Elec & Electronics Engrs, 70-76; PRIN SCIENTIST, BIOENG RES LAB, SRI INT, 76- *Concurrent Pos:* Prin investr, NSF grants, 66-70, NIH grants, 65-70 & 76-; sr res fel biomed eng, Univ Wash, 69-70; tech adv, US Nat Comt Int Electrotech Comn Tech Comt No 87, 72-; US tech adv, Int Electrotech Comn/Tech Comt No 87, 72-,; mem task group 1, Assessment of Ultrasonic Diag Instrumentation, NSF-Alliance for Eng in Med & Biol, 73-74; mem ed bd, ultrasound in med & biol, 75-; mem, Ultrasonics Tissue Signature Comt, NSF-Carnegie-Mellon Univ, 77-78, Bioeffects Comt, Am Inst Ultrasound Med, 79-81 & 86-88, Phys Acoust Comt, Acoust Soc Am, 82-85 & Standards Comt, 86-88; chmn, US Tech Adv Group, IEC/TC87, 88-; mem, Standards Comt, 89- & vice-chmn, 90-91. *Honors & Awards:* Centennial Medal, Inst Elec & Electronics Engrs, 84. *Mem:* Fel Acoust Soc Am; sr mem Inst Elec & Electronics Engrs; fel Brit Phys Soc; fel Am Inst Ultrasound Med. *Res:* Ultrasonic measurements of absorption and velocity in liquids and liquid mixtures; kinetics of fast reactions; relaxation mechanisms; mesomorphic liquid states; aqueous solutions of proteins; biomaterials; bioeffects of ultrasound; ultrasonics phantoms; biochemical physics; ultrasound tissue characterization; ultrasound hyperthermia. *Mailing Add:* SRI Int 333 Ravenswood Ave Menlo Park CA 94025

EDMONDS, RICHARD H, b Carbondale, Pa, May 10, 33; m 53; c 3. ANATOMY. *Educ:* State Univ NY Buffalo, BS, 59, PhD(anat), 65. *Prof Exp:* From asst prof to assoc prof, 65-76, from asst dean to assoc dean, 80-84, PROF ANAT, ALBANY MED COL, 76-, DIR, ANATOMICAL GIFT PROG, 76-, EXEC ASSOC DEAN, 84- *Mem:* Am Asn Anat; Am Soc Cell Biol; Am Asn Med Col; Sigma Xi; NY Acad Med. *Res:* Cell biology; developmental hematology; electron microscopy. *Mailing Add:* Dept Anat Exec Assoc Dean Albany Med Col A-34 Admin 312 Albany NY 12208

EDMONDS, ROBERT L, b Sydney, Australia, May 6, 43; m 69; c 2. FOREST PATHOLOGY, SOIL MICROBIOLOGY. *Educ:* Univ Sydney, BS, 64; Univ Wash, MS, 68, PhD(forest path), 71. *Prof Exp:* Res forestry officer, Forest Res Inst, Australia, 64-65; res asst, Dept Forestry, Australian Nat Univ, 65-66; from res asst to res assoc, Col Forest Resources, Univ Wash, 66-71; prog coordr, Bot Dept, Univ Mich, 71-73; res asst prof, 73-76, asst prof, 76-79, ASSOC PROF, COL FOREST RESOURCES, UNIV WASH, 79- *Concurrent Pos:* Mem nat comt, Nat Res Coun/Nat Acad Sci, 72-74, chmn, Aerobiol Comt, 77-80. *Mem:* Sigma Xi; Soc Am Foresters; Am Phytopath Soc; Ecol Soc Am; Am Meteorol Soc. *Res:* Forest pathology, root diseases; soil microbiology, litter decomposition, microbiology of sewage sludge; aerobiology, aerial dispersal of bacteria, fungus spores and insects. *Mailing Add:* 6311 Clearsprings Rd Baltimore MD 21212

EDMONDSON, ANDREW JOSEPH, b Leavenworth, Kans, May 11, 35; m 56; c 2. MECHANICAL ENGINEERING. *Educ:* Tex Tech Col, BS, 57; Pa State Univ, MS, 61; Tex A&M Univ, PhD(mech eng), 64. *Prof Exp:* Instr mech eng, Tex Tech Col, 57-59 & 60-61; from asst prof to assoc prof mech & aero eng, 64-76, PROF MECH & AERO ENG, UNIV TENN, KNOXVILLE, 76- *Concurrent Pos:* Consult, Oak Ridge Nat Lab, 67- *Mem:* Am Soc Eng Educ. *Res:* Theoretical and experimental studies of shell problems; experimental investigations concerning the sealing mechanisms of mechanical face seals. *Mailing Add:* Dept Mech & Aerospace Eng 410 Dougherty Univ Tenn Knoxville TN 37996

EDMONDSON, DALE EDWARD, b Morris, Ill, Oct 13, 42; m 72. BIOCHEMISTRY. *Educ:* Northern Ill Univ, BS, 64; Univ Ariz, PhD(chem), 70. *Prof Exp:* NIH fel biochem, Univ Mich, 70-72; from asst res biochemist to assoc res biochemist, Univ Calif, San Francisco, 72-80; assoc prof biochem, 80-86, PROF BIOCHEM, EMORY UNIV, 86- *Mem:* Am Chem Soc; Am Soc Biol Chemists; Sigma Xi. *Res:* Structure and function of oxidation-reduction enzymes. *Mailing Add:* Dept Biochem Emory Univ Atlanta GA 30322

EDMONDSON, FRANK KELLEY, b Milwaukee, Wis, Aug 1, 12; m 34; c 2. ASTRONOMY. *Educ:* Ind Univ, AB, 33, MA, 34; Harvard Univ, PhD(astron), 37. *Prof Exp:* Asst, Ind Univ, 29-33, Lowell Observ, 33-35 & Harvard Univ, 35-37; from instr to prof, 37-83, chmn dept, 44-78, EMER PROF ASTRON, IND UNIV BLOOMINGTON, 83- *Concurrent Pos:* Res assoc, McDonald Observ, 41-83; dir, Kirkwood Observ, 45-48 & Goethe Link Observ, 48-78; mem, Int Astron Union, 48-, chmn US Nat Comt, 62-64 & from vpres to pres Comn 20, 67-73; prog dir astron, NSF, 56-57; mem bd, Asn Univs Res Astron, 57-83, from vpres to pres, 57-65 & consult historian, 83-; cor mem, Am Mus Natural Hist, 58-; actg dir, Cerro Toldo Inter-Am Observ, 66. *Honors & Awards:* Order of Merit, Govt of Chile, 64; Meritorious Pub Serv, NSF, 83. *Mem:* AAAS (vpres, 62); Am Astron Soc (treas, 54-75); Can Astron Soc; Astron Soc Pac. *Res:* Stellar motions and distribution; radial velocities of faint stars; rediscovery and observation of asteroids on the critical list; early history of Association of Universities for Research in Astronomy, Inc, Kitt Peak National Observatory and Cerro Tololo Inter-American Observatory. *Mailing Add:* Dept Astron 319 Swain Hall W Ind Univ Bloomington IN 47405-4201

EDMONDSON, HUGH ALLEN, pathology; deceased, see previous edition for last biography

EDMONDSON, MORRIS STEPHEN, b San Antonio, Tex, Sept 9, 41; m 62; c 2. APPLIED POLYMER CHEMISTRY, ORGANIC CHEMISTRY. *Educ:* Southwest Tex State Univ, BS, 63; Univ Tex Austin, MA, 66, PhD(org chem), 70. *Prof Exp:* Res chemist, Jefferson Chem Co, 65-67; res chemist, Petro-Tex Chem Corp, 70-72, sr res chemist, 72-74, res group head, 74-77; res assoc, 77-84, assoc scientist, 84-87, ASSOC DEVELOP SCIENTIST, DOW CHEM, USA, 87- *Concurrent Pos:* Welch Found fel, 69-70. *Mem:* Am Chem Soc; TAPPI. *Res:* Liquid phase chlorination of dienes; heterogeneous catalysis; emulsian polymerization; elastomers; organometallic and Ziegler catalysis; applied polymer science and polymer chemistry. *Mailing Add:* Rte 7 130 Mohawk Dr Alvin TX 77511

EDMONDSON, W THOMAS, b Milwaukee, Wis, Apr 24, 16; m 41. LIMNOLOGY. *Educ:* Yale Univ, BS, 38, PhD(zool), 42. *Hon Degrees:* DSc, Univ Wis, 87. *Prof Exp:* Asst zool, Univ Wis, 39; asst biol, Yale Univ, 39-41; asst phys oceanog, Am Mus Natural Hist, 42-43; res assoc, Oceanog Inst, Woods Hole, 43-46; lectr biol, Harvard Univ, 46-49; from asst prof to prof, 49-86, EMER PROF ZOOL, UNIV WASH, 86- *Concurrent Pos:* NSF fel, 59-60; mem environ studies bd, Nat Acad Sci-Nat Res Coun, 74-76. *Honors & Awards:* Cottrell Award Environ Qual, Nat Acad Sci, 73; Einar Naumann-August Thienemann Medal, Int Asn Theoret & Appl Limnol, 80; Eminent Ecologist Award, Am Ecol Soc, 83; Jessie & John Danz lectr, Univ Wash, 87; Brode lectr, Whitman Col, 88; G Evelyn Hutchinson Medal Award, Am Soc Limnol & Oceanog, 90. *Mem:* Nat Acad Sci; Freshwater Biol Asn; Am Soc Limnol & Oceanog; Ecol Soc Am; Int Asn Theoret & Appl Limnol. *Res:* Ecology and taxonomy of Rotifera; population dynamics of plankton; lake productivity. *Mailing Add:* Dept Zool NJ-15 Univ Wash Seattle WA 98195

EDMONSON, DON ELTON, b Dallas, Tex, Sept 6, 25; m 51; c 3. MATHEMATICS. *Educ:* Southern Methodist Univ, BS, 45, MS, 48; Calif Inst Technol, PhD(math), 54. *Prof Exp:* Res instr math, Tulane Univ, 54-55; from asst prof to assoc prof, Southern Methodist Univ, 55-60; assoc prof, 60-64, vchmn dept, 69-70, PROF MATH, UNIV TEX, AUSTIN, 64-, PROF EDUC, 76- *Concurrent Pos:* Consult, Tex Instruments, Inc, 58-59 & Tex Ed Agency, 62-64. *Mem:* Am Math Soc; Soc Indust & Appl Math; Math Asn Am. *Res:* Abstract algebra, lattice theory; topological lattices; real and complex analysis. *Mailing Add:* Dept Math Educ Univ Tex Austin TX 78712

EDMUND, ALEXANDER GORDON, b Toronto, Ont, Aug 11, 24; m 51; c 4. VERTEBRATE PALEONTOLOGY. *Educ:* Univ Toronto, BA, 51, MA, 52; Harvard Univ, PhD, 57. *Prof Exp:* Asst cur, Royal Ont Mus, 54-64, cur, Dept Vert Paleont, 64-89; RETIRED. *Concurrent Pos:* Assoc prof, Dept Geol, Univ Toronto, 71. *Mem:* Soc Vert Paleont; Lepidopterists Soc; Sigma Xi. *Res:* Morphology; evolution; systematics and zoogeography of Pleistocene edentate mammals from North and South America. *Mailing Add:* 4228 Bardot Rd Port Charlotte FL 33953

EDMUND, RUDOLPH WILLIAM, b Lockridge, Iowa, Mar 9, 10; m 39; c 3. GEOLOGY. *Educ:* Augustana Col, AB, 34; Univ Iowa, MS, 38, PhD(struct geol), 40. *Hon Degrees:* DSc, Calif Lutheran Col, 80. *Prof Exp:* Instr geol, Coe Col, 39-40; geologist, Shell Oil Co, Inc, 40-45; asst div geologist, Globe Oil & Refining Co, Okla, 45-48; regional geologist, 51-53; from assoc prof to prof geol, Augustana Col, 48-51, prof & chmn div sci, 60-69; vpres & gen mgr, Sohio Petrol Co, Okla, 53-60; prof geol & chmn div sci, Augustana Col, 60-69; vpres acad affairs, 69-74, prof, 74-80, EMER PROF GEOL, CALIF LUTHERAN UNIV, 80- *Mem:* Fel AAAS; fel Geol Soc Am; Nat Asn Geol Teachers; Soc Explor Geophys; Soc Econ Paleontologists & Mineralogists. *Res:* Structural geology; regional stratigraphy; stratigraphic oil traps; paleogeologic maps. *Mailing Add:* Dept Geol Calif Lutheran Univ Thousand Oaks CA 91360

EDMUNDOWICZ, JOHN MICHAEL, b Nanticoke, Pa, May 18, 38; m 60; c 2. BIOCHEMISTRY. *Educ:* Philadelphia Col Pharm & Sci, BS, 60; Univ Del, MS, 63, PhD(chem), 66. *Prof Exp:* Sr biochemist, 66-75, RES SCIENTIST, ANTIBIOTIC MFG & DEVELOP DIV, ELI LILLY & CO, 75- *Mem:* Am Chem Soc; AAAS. *Res:* Isolation, purification and chemistry of the beta-lactam antibiotics, polyethers and macrolides. *Mailing Add:* Dept K418 Eli Lilly & Co Kentucky Ave Bdlg 32811 Indianapolis IN 46285-0002

EDMUNDS, GEORGE FRANCIS, JR, entomology, evolution, for more information see previous edition

EDMUNDS, LELAND NICHOLAS, JR, b Aiken, SC, Apr 21, 39; m 81; c 3. BIOLOGICAL RHYTHMS, CELL BIOLOGY. *Educ:* Davidson Col, BS, 60; Princeton Univ, MA, 62, PhD(biol), 64. *Prof Exp:* NSF-Orgn Trop Studies fel, Univ Costa Rica, 64; instr & res asst biol, Princeton Univ, 64-65; from asst prof to assoc prof, 65-76, actg provost & head div biol sci, 75-76, prof biol, 76-81, PROF ANAT SCI, HEALTH SCI CTR, STATE UNIV NY, STONY BROOK, 81- *Concurrent Pos:* Student trainee & biol aide, Insect Physiol Lab, Agr Res Ctr, USDA, Md, 56-60; vis investr, Carnegie Inst, Dept Terrestrial Magnetism, 62; NSF res grants, 65-88; vis investr, Carlsberg Found, Biol Inst, Copenhagen, 72; vis investr, Ctr Nat Res Sci, Le Phytotron, Gif-sur-Yvette, 78-79; vis prof, Sackler Sch Med, Tel Aviv Univ, 83, Univ Paris VII, Lab Biol

Cellulaire, Veg, 81, Lab Membranes Biol, 86; Int Adv Bd, Chronobiol Int, 84-; bd dirs, Int Soc Chronobiol, 87- *Mem:* Am Soc Plant Physiol; Int Soc Endocytobiol; Soc Protozoologists; Int Soc Chronobiol; Am Soc Advan Sci. *Res:* Circadian rhythms; synchrony in cell division and growth; control and regulation of the cell cycle; oscillatory enzyme systems; cellular communication. *Mailing Add:* Div Biol Sci State Univ NY Stony Brook NY 11794

EDMUNDS, LEON K, b Madison, Wis, Mar 25, 29; m 50; c 9. PLANT PATHOLOGY. *Educ:* Univ Wis, BS, 53, PhD(plant path), 58. *Prof Exp:* Res assoc plant path, Univ Wis, 58-60; RES PLANT PATHOLOGIST, USDA, AGR RES SERV, KANS STATE UNIV, 60- *Mem:* Am Phytopath Soc. *Res:* Diseases of grain sorghum in semihumid to semiarid areas of central United States. *Mailing Add:* 3630 Rocky Ford Ave Manhattan KS 66502

EDMUNDS, LOUIS HENRY, JR, b Seattle, Wash, Aug 12, 31; c 3. THORACIC SURGERY. *Educ:* Univ Wash, BS, 53; Harvard Med Sch, MD, 56; Am Bd Surg, dipl, 64; Am Bd Thoracic Surg, dipl, 65. *Prof Exp:* From intern to resident, Mass Gen Hosp, 56-63; clin asst, Mass Gen Hosp, 64; sr registrar, Leeds Gen Infirmary, Eng, 64-65; assoc, Mason Clin, Seattle, Wash, 65-66; from asst prof to assoc prof surg, Med Ctr, Univ Calif, San Francisco, 66-73, assoc, Cardiovasc Res Inst, 66-73, dir, Exp Surg Labs, 67-73; WILLIAM M MEASEY PROF SURG & CHIEF CARDIOTHORACIC SURG, UNIV PA, 73-, INSTR SURG, 80- *Concurrent Pos:* Teaching fel, Harvard Med Sch, 63-64; res fel thoracic surg, Leeds Univ, 64; investr, Va Mason Res Ctr, Seattle, Wash, 65. *Mem:* AAAS; Am Heart Asn; Am Col Cardiol; Am Soc Artificial Internal Organs; Am Asn Thoracic Surg. *Mailing Add:* Dept Surg Hosp Univ Pa 3400 Spruce St Philadelphia PA 19104

EDMUNDS, PETER JAMES, b Gloucester, Eng, Nov 28, 61. PHYSIOLOGICAL ECOLOGY, CORAL REEF BIOLOGY. *Educ:* Univ Newcastle-upon-Tyne, Eng, BSc, 83; Univ Glasgow, Scotland, PhD(physiol ecol), 86. *Prof Exp:* Sr fac trop marine biol, Sch Field Studies, 87-89; adj fac marine biol, Northeastern Univ, 89, assoc & clin lectr, 90; RES ASSOC, UNIV SOUTHERN CALIF, 91-; PROF, DEPT BIOL, CALIF STATE UNIV, NORTHRIDGE, 92- *Concurrent Pos:* adj fac, Northeastern Univ, 91- *Mem:* Sigma Xi; Am Soc Zoologists; AAAS. *Res:* Physiological ecology and molecular biology approaches to study the effects of cloniality on physiological plasticity; energy budgets and DNA fingerprints are being determined for colonial taxa including reef corals. *Mailing Add:* Dept Biol Calif State Univ Northridge CA 91330

EDMUNDSON, ALLEN B, b Flat River, Mo, June 16, 32; m 55; c 3. BIOCHEMISTRY. *Educ:* Dartmouth Col, AB, 54; Rockefeller Univ, PhD(biochem), 61. *Prof Exp:* Sr biochemist, Div Biol & Med Res, Argonne Nat Lab, 64-76; PROF BIOCHEM & BIOL, UNIV UTAH, 76- *Concurrent Pos:* USPHS fel biochem, Unit Molecular Biol, Med Res Coun, Eng, 60-64. *Mem:* Sigma Xi. *Res:* Protein chemistry; determinations of structures and genetically controlled variations of proteins; correlation of structure and functions; protein crystallography and chemistry. *Mailing Add:* Harrington Cancer Ctr 1500 Wallace Blvd Amarillo TX 79106

EDMUNDSON, HAROLD PARKINS, b Los Angeles, Calif, Dec 13, 21; m 63; c 2. MATHEMATICS. *Educ:* Univ Calif, Los Angeles, BA, 46, MA, 48, PhD(math), 53. *Prof Exp:* Asst math, Univ Calif, Los Angeles, 49-50, 51-52; mathematician, Indust Logistics Res Proj, 53, US Dept Defense, 53-54 & Rand Corp, 54-59; sr assoc, Planning Res Corp, 59-61; mem sr staff, Thompson-Ramo-Wooldridge, Inc, 61-64; sr scientist, Syst Develop Corp, 64-67; PROF COMPUT SCI & MATH, UNIV MD, 67- *Concurrent Pos:* Lectr, George Washington Univ, 53-54, Univ Southern Calif, 54-56 & Univ Calif, Los Angeles, 57-63, assoc res mathematician, 64-67. *Mem:* Am Math Soc; Math Asn Am; Soc Indust & Appl Math; Asn Comput Mach; Asn Computational Ling. *Res:* Computability theory and automata theory; mathematical statistics, logic; probability, matrix and information theories; stochastic processes; automatic translation; mathematical and computational linguistics. *Mailing Add:* Dept Comput Sci Univ Md College Park MD 20742

EDNEY, NORRIS ALLEN, b Natchez, Miss, July 17, 36; m 59; c 3. BIOLOGY. *Educ:* Tougaloo Southern Christian Col, BSc, 57; Antioch Col, MSc, 62; Mich State Univ, PhD(conserv), 69. *Prof Exp:* Prof biol, Natchez Jr Col, 57-62; from instr to asst prof, Alcorn Agr & Mech Col, 63-66; teaching asst, Mich State Univ, 66-69; PROF BIOL, ALCORN STATE UNIV, 69-, CHMN DEPT BIOL & DIR ARTS & SCI DIV, 73-, DIR GRAD STUDIES, 75- *Concurrent Pos:* Res conserv aide, Dept Natural Resources, State of Mich, 68-69; dir, Coop Col-Sch Sci Prog; dir, Microbial Conversion Proj, USDA, 73-77, dir, Pesticide Residue Res Proj, 76-, dir, Biodegradation Animal Waste Proj, 78-; dir, Instruct Implementation Improv Prog, NSF, 74-75 & 75-76, dir, Pre-Col Teacher Develop Sci Prog, 78-79, 79-80 & 80-81; dir, NIH Allied Health Grant, 75-78; mem comt, Effects of Alanap Proj, USDA Cucurbits Proj, 81. *Mem:* AAAS; Soc Protozool; Mycol Soc Am; Soc Econ Bot; Bot Soc Am. *Res:* Ecological succession of protozoa in pond and sewage water; histochemical study of certain enzyme systems in Trichomonas vaginalis; study of infections of fish by certain saprolegniaceous fungi. *Mailing Add:* Dept Biol Alcorn State Univ Box 870 Lorman MS 39096

EDNIE, NORMAN A(LEX), chemical engineering; deceased, see previous edition for last biography

EDOZIEN, JOSEPH CHIKE, b Asaba, Nigeria, July 28, 25; m 55; c 6. NUTRITION, PATHOLOGY. *Educ:* Univ London, BSc, 48; Nat Univ Ireland, BSc & MSc, 48, MD, 54; MRCP(E), 54, FRCP(E), 63; FRCPath(L), 67. *Hon Degrees:* DSc, Univ Rio de Janeiro, 63. *Prof Exp:* Prof chem path & head dept, Univ Ibadan, 61-66, dean fac med, 62-66; vis prof clin nutrit, Mass Inst Technol, 67-71, dir, Mass Nutrit Prog, 69-71; PROF NUTRIT & CHMN DEPT, SCH PUB HEALTH, UNIV NC, CHAPEL HILL, 71- *Concurrent Pos:* Leader Nigerian deleg & vpres conf, UN Con Appl Sci & Technol to Probs Less-Develop Countries, 63; mem, Expert Panel Res Immunol, WHO, 63-, chmn, Expert Comt Immunol & Parasitic Dis, 64, mem, Expert Comt Nutrit in Pregnancy & Lactation, 65 & Adv Comt Med Res, 65-70. *Mem:* Am Inst Nutrit; Am Soc Clin Nutrit; Asn Clin Pathologists; Sigma Xi. *Res:* Nutritional assessment, diet-hormone interactions and metabolic adaptation. *Mailing Add:* Dept Nutrit Univ NC Sch Pub Health Chapel Hill NC 27514

EDSALL, JOHN TILESTON, b Philadelphia, Pa, Nov 3, 02; wid; c 3. BIOCHEMISTRY & MOLECULAR BIOLOGY, PROTEIN BIOCHEMISTRY. *Educ:* Harvard Univ, AB, 23, MD, 28. *Hon Degrees:* DSc, Univ Chicago, 67, Case Western Reserve Univ, 67, New York Med Col, 67 & Univ Mich, 68; DPhil, Univ Goteborg, Sweden, 72. *Prof Exp:* Tutor biochem sci, 28-72, from instr to prof biochem, 28-73, EMER PROF BIOCHEM, HARVARD UNIV, 73- *Concurrent Pos:* Chmn, Bd Tutors Biochem Sci, 31-57; Guggenheim Mem Found fels, Calif Inst Technol, 40-41 & Harvard Univ, 54-56; mem, US Nat Comn for UNESCO, 50-56; Fulbright vis lectr, Cambridge Univ, 52 & Univ Tokyo, 64; vis prof, Col France, 55-56, Univ Calif, Los Angeles, 77, Univ Calif, Riverside, 80; ed in chief, J Biol Chem, 58-67; pres, Sixth Int Cong Biochem, NY, 64; vis fel, Australian Nat Univ, 70; scholar, Fogarty Int Ctr, NIH, 70-71; dir, Surv Sources Hist Biochem & Molecular Biol, 75-79. *Honors & Awards:* Passano Found Award, 66; Willard Gibbs Medal, Am Chem Soc, 72; Philip Hauge Abelson Award, AAAS, 89. *Mem:* Nat Acad Sci; AAAS; Am Soc Biol Chemists (pres, 57-58); Am Philos Soc; Hist Sci Soc; Am Acad Arts & Sci. *Res:* Physical chemistry of amino acids and proteins; proteins of blood and muscle; flow birefringence; Raman spectroscopy; light scattering; carbonic anhydrase; history of biochemistry and related sciences; biochemical thermodynamics; historical study of blood and hemoglobin; biochemical regulation and organization. *Mailing Add:* Dept Biochem & Molecular Biol Harvard Univ Seven Divinity Ave Cambridge MA 02138-2092

EDSBERG, ROBERT LESLIE, b Seattle, Wash, Feb 7, 22; m 48; c 3. ANALYTICAL CHEMISTRY. *Educ:* Univ Minn, BChem, 44; Univ Pa, MS, 52. *Prof Exp:* Res chemist, Gen Aniline & Film Co, 44-52; mgr chem sect, Burroughs Cent Res Labs, 52-55; mgr chem res, Burroughs Corp, Rochester, 55-66, dir res & eng, Todd Co Div, 66-77, staff engr, 77-79, mgr prod specif, Off Supplies Div, 79-82; RETIRED. *Mem:* Am Chem Soc. *Res:* Graphic arts research as related to business machines, supplies. *Mailing Add:* 141 Butler Dr Pittsford NY 14534

EDSE, RUDOLPH, b Hamburg, Ger, Dec 14, 13; nat US; m 39; c 2. AERONAUTICAL ENGINEERING. *Educ:* Univ Hamburg, Dipl Chem, 37, Dr rer nat(phys chem), 39. *Prof Exp:* Res phys chemist, Inst Aeronaut Sci, Ger, 39-44, dep head chem dept, 44-46; sci consult, Wright-Patterson AFB, USAF, 45-51; from asst prof to assoc prof aeronaut eng, 51-57, dir rocket res lab, 51-80, PROF AERONAUT & ASTRONAUT ENG, OHIO STATE UNIV, 57- *Concurrent Pos:* Consult, Allegany Ballistics Lab, Md, 57-63. *Mem:* Am Inst Aeronaut & Astronaut; Sigma Xi. *Res:* Propulsion; thermodynamics, combustion and high speed aerodynamics. *Mailing Add:* AAE 2036 Neil Ave Ohio State Univ Columbus OH 43210

EDSON, CHARLES GRANT, b West Springfield, Mass, Dec 16, 16; m 42; c 1. HYDRAULICS, MECHANICS. *Educ:* Univ Mass, BS, 38; Univ Fla, MSE, 50. *Prof Exp:* Engr, Corps Engrs, RI, 38-40, Fla, 40-42 & 46; asst prof, 46-51, ASSOC PROF HYDRAUL & MECH, UNIV FLA, 51- *Concurrent Pos:* Consult, Corps Engrs, Jacksonville, Fla, 48-49, DC, 50-52 & Brevard Eng Co, 70. *Mem:* Fel Am Soc Civil Engrs. *Res:* Open channel hydraulics; empirical relations in hydrology; three-body problem of mechanics; geometry; nomography and crystallography. *Mailing Add:* 2212 NW 15th Ave Gainesville FL 32601

EDSON, JAMES EDWARD, JR, b Fort Smith, Ark, May 26, 42. PALYNOLOGISTS. *Educ:* Ark Tech Univ, BS, 65; Univ Ark, Fayetteville, MS, 71; Tulane Univ, PhD(paleont), 76. *Prof Exp:* Sr palentologist, Mobil Oil Corp, 73-74; paleontologist, Shell Oil Co, 74-75; geologist, 75-77; ASST PROF GEOL, UNIV ARK, MONTICELLO, 77- *Mem:* Geol Soc Am; Am Asn Stratigraphic Palynologists; Nat Asn Geol Teachers. *Res:* Palyno-stratigraphy of Paleozoic and Mesozoic strata. *Mailing Add:* Dept Geol Univ Ark UAM Box 3598 Monticello AR 71655

EDSON, WILLIAM A(LDEN), b Burchard, Nebr, Oct 30, 12; m 42; c 3. ELECTRONICS ENGINEERING. *Educ:* Univ Kans, BS, 34, MS, 35; Harvard Univ, ScD(commun eng), 37. *Prof Exp:* Mem tech staff, Bell Tel Labs, 37-41, 43-44, supvr, 44-45; asst prof elec eng, Ill Inst Tech, 41-43; prof physics, Ga Inst Tech, 45-46, prof elec eng, 46-52, dir sch elec eng, 51-52; vis prof & res assoc, Stanford Univ, 52-54; consult engr, Microwave Lab, Gen Elec Co, 55-59, mgr klystron subsect, 59-61; dir res, Electromagnetic Technol Corp, 61-62, pres, 62-70; sr scientist, Vidar Corp, 70-71; dir, Wescon, 75-79,; sr res engr, 71-77, staff scientist, 77-90, SR PRIN ENGR, SRI INT, 90- *Concurrent Pos:* Consult, Nat Bur Standards, 51-64. *Mem:* Fel Inst Elec & Electronics Eng; Am Phys Soc. *Res:* Communication and radar systems; filters; electronic oscillators; cavity and quartz crystal resonators; broad band and microwave amplifiers. *Mailing Add:* SRI Int 333 Ravenswood Ave Menlo Park CA 94025

EDSTROM, RONALD DWIGHT, b Oakland, Calif, Mar 21, 36; m 87; c 2. BIOCHEMISTRY. *Educ:* Univ Calif, Berkeley, AB, 58; Univ Calif, Davis, PhD(biochem), 62. *Prof Exp:* Res assoc, Univ Mich, 62-63; asst prof, 65-71, ASSOC PROF BIOCHEM, MED SCH, UNIV MINN, MINNEAPOLIS, 71- *Concurrent Pos:* USPHS fel physiol chem, Sch Med, Johns Hopkins Univ, 63-65. *Mem:* AAAS; Am Soc Biol Chem. *Res:* Regulation of glycogen metabolism. *Mailing Add:* Dept Biochem Univ Minn Sch Med Minneapolis MN 55455

EDWALL, DENNIS DEAN, b Houston, Tex, July 26, 48; m 78; c 2. METAL-ORGANIC CHEMICAL VAPOR DEPOSITION CRYSTAL GROWTH. *Educ:* Iowa State Univ, BS, 70; Cornell Univ, MS, 72. *Prof Exp:* MEM TECH STAFF, ROCKWELL INT SCI CTR, 72- *Res:* Crystal growth and development of mercury cadmium telluride by the metalorganic chemical vapor phase epitaxy method. *Mailing Add:* 1049 Camino Dos Rios A11 Thousand Oaks CA 91360

EDWARD, DEIRDRE WALDRON, b Detroit, Mich, June 23, 23; Can citizen; m 53; c 3. BIOCHEMISTRY. *Educ:* Univ Birmingham, BSc, 44, PhD(path chem), 54. *Prof Exp:* Staff res off, Bakelite Corp, 44-46; clin biochemist, Children's Hosp, Birmingham, Eng, 46-49; res assoc, 60-65, asst prof, 65-67, ASSOC PROF EXP SURG, McGILL UNIV, 67- *Concurrent Pos:* Lasdon res fel, Trinity Col, Dublin, 54-57; consult, Div Surg Invest, St Mary's Hosp, 85- *Mem:* Can Biochem Soc; Brit Biochem Soc; NY Acad Sci. *Res:* Structure and metabolism of polysaccharides, glycoproteins; absorption of metal ions and radionuclides by gastrointestinal tract; biosynthesis of glycoproteins; acid glycosidases; structure, function of glycoproteins in the immune system. *Mailing Add:* Dept Exp Surg Donner Bldg McGill Univ Docteur Penfield Ave Montreal PQ H3A 1A4 Can

EDWARD, JOHN THOMAS, b London, Eng, Mar 23, 19; Can citizen; m 53; c 3. PHYSICAL ORGANIC CHEMISTRY. *Educ:* McGill Univ, BSc, 39, PhD(org chem), 42; Oxford Univ, DPhil, 49; Trinity Col Dublin, MA, 55, ScD, 71. *Prof Exp:* Asst res scientist, Nat Res Coun Can, 43-45 & Can Armaments Res & Develop Estab, Que, 45-; lectr, Univ Man, 45-46; Imp Chem Industs res fel, Birmingham, 49-52; lectr, Trinity Col, Dublin, 52-56; lectr, 56-57, from asst prof to prof, 57-73, Macdonald prof, 73-86, EMER PROF CHEM, MCGILL UNIV, 86- *Concurrent Pos:* Exhib of 1851 sci scholar, 46-49. *Mem:* Am Chem Soc; fel Royal Soc Can; fel Chem Inst Can. *Res:* Explosives; heterocyclic compounds; strychnine; terpenes; steroids; paper electrophoresis; amino acids; reaction mechanisms and stereochemistry; acidity functions; substituent and solvation effects on reactivities of organic molecules. *Mailing Add:* Dept Chem McGill Univ Montreal PQ H3A 2K6 Can

EDWARDS, ADRIAN L, MEDICAL EDUCATION. *Educ:* Harvard Univ, MD, 60. *Prof Exp:* PROF MED, MED SCH, CORNELL UNIV, 68- *Mem:* Inst Med-Nat Acad Sci. *Mailing Add:* One 35 E 71st St New York NY 10024

EDWARDS, ALAN KENT, b Wichita, Kans, Apr 29, 40; m 62; c 2. EXPERIMENTAL ATOMIC PHYSICS. *Educ:* Cent Methodist Col, AB, 62; Univ Nebr, Lincoln, MS, 64, PhD(physics), 68. *Prof Exp:* Res assoc physics, Univ Wash, 67-70; from asst prof, to assoc prof, 70, PROF PHYSICS, UNIV GA, 87- *Concurrent Pos:* Vis scientist, Argonne Nat Lab, 80-83. *Mem:* Am Phys Soc. *Res:* Atomic and molecular physics. *Mailing Add:* Dept Physics Univ Ga Athens GA 30602

EDWARDS, ALAN M, b Denver, Colo, Oct 8, 33; m 55; c 3. AERONAUTICS, ASTRONAUTICS. *Educ:* US Mil Acad, BS, 55; Mass Inst Technol, SM, 61; Stanford Univ, PhD(aeronaut, astronaut), 65. *Prof Exp:* From asst prof to assoc prof eng mech, USAF Acad, 61-68, dep dir academics, Aerospace Res Pilot Sch, 68-69, dir, 69-70, dir, 614th Tactical Fighter Squadron, 70-71, aerospace asst, Nat Aeronaut & Space Coun, 71-73, mil asst int progs, Off Secy Defense, 73-75; mgr syst anal, Figgie Intl Inc, 75-87; PRES, EDWARDS AFFILIATES INC, CROWNSVILLE, 87- *Concurrent Pos:* Consult, Air Force Flight Test Sch, 66; mem, USAF Sci Adv Bd, 88- *Mem:* Am Defense Preparedness Asn; Air Force Asn. *Res:* Shell structures; aeroelasticity; flight dynamics. *Mailing Add:* 203 Long Pt Rd Crownsville MD 21032

EDWARDS, ARTHUR L, b Sacramento, Calif, Feb 24, 33; m 53; c 4. CHEMICAL ENGINEERING, APPLIED MATHEMATICS. *Educ:* Univ Wash, Seattle, BS, 54; Univ Ill, MS, 59, PhD(chem eng), 61. *Prof Exp:* PHYSICS ENGR, APPL MATH & COMPUT, LAWRENCE RADIATION LAB, UNIV CALIF, 60- *Mem:* Am Inst Chem Eng; Am Chem Soc; AAAS. *Res:* High pressure solid state physics; application of computers to transient transport phenomena such as heat conduction, fluid flow and other potential flow problems; hydrodynamics. *Mailing Add:* Lawrence Livermore Lab L-298 PO box 808 Livermore CA 94550-0622

EDWARDS, BEN E, b Ross, Calif, Oct 14, 35; m 57, 66, 77; c 9. INDUSTRIAL ORGANIC CHEMISTRY. *Educ:* Mass Inst Technol, SB, 57; Ind Univ, PhD(org chem), 62. *Prof Exp:* NIH fel org chem, Columbia Univ, 62-63; res assoc, Southwest Found Res & Educ, 63-67; res chemist, Chem Sci Div, Res Inst, Ill Inst Technol, 67-68; asst prof, Dept Chem, Univ NC, Greensboro, 68-73; CHIEF CHEMIST, OLD-NORTH MFG CO, INC, 73- *Concurrent Pos:* Part-time prof, Appalachian State Univ, 81; chmn, ASTM sub-comt, Curing Mat for Concrete, 85- *Mem:* Am Chem Soc; Sigma Xi; Am Soc Testing & Mat. *Res:* Concrete curing membranes; accessory chemicals for concrete construction; organic synthesis; steroids; sulfur compounds; medicinal chemistry. *Mailing Add:* Old-North Mfg Co Inc Box 598 Lenoir NC 28645

EDWARDS, BETTY F, b Athens, Ga, Mar 13, 15; m 36; c 2. ANATOMY. *Educ:* Agnes Scott Col, AB, 35; Emory Univ, MA, 51, PhD(anat), 63. *Prof Exp:* Instr biol, Emory Univ, 43-45, instr anat, 51-55; instr biol, Ga State Col, 49-50 & 55-61; instr anat, 63-66, asst prof histol & embryol, 66-81, EMER PROF ANAT, EMORY UNIV, 81- *Mem:* Am Soc Zool; Am Inst Biol Sci; Am Asn Anat; Tissue Cult Asn; Sigma Xi. *Res:* Muscle and plant tissue culture; responses of animal and plant tissues to gravity; effects of weightlessness on growth. *Mailing Add:* 3750 Peachtree NE Atlanta GA 30319-1399

EDWARDS, BRENDA KAY, b Paducah, Ky, Oct 17, 46. BIOSTATISTICS. *Educ:* Murray State Univ, BS, 68; Vanderbilt Univ, MS, 70; Univ NC, Chapel Hill, PhD(biostatist), 75. *Prof Exp:* Asst prof biostatist, Col Med, Univ Cincinnati, 75-78; biostatistician, 78-90, ASSOC DIR, SURVEILLANCE PROG, NAT CANCER INST, 90- *Mem:* Am Statist Asn; Biomet Soc. *Res:* Analysis of survival data and other statistical methods applicable to epidemiological research; cancer surveillance. *Mailing Add:* Nat Cancer Inst EPN/Rm 343 Bethesda MD 20892

EDWARDS, BRIAN F, CRYSTALLOGRAPHY, PROTEIN STRUCTURE. *Educ:* Harvard Univ, PhD(chem), 75. *Prof Exp:* From asst prof to assoc prof, 80-90, PROF BIOCHEM, WAYNE STATE UNIV, DETROIT, 90- *Mailing Add:* Dept Biochem Wayne State Univ 540 E Canfield Ave Detroit MI 48201

EDWARDS, BRIAN RONALD, b Norwich, UK, Nov 20, 44; m 68; c 2. MEDICAL PHYSIOLOGY, RENAL PHYSIOLOGY. *Educ:* Sheffield Univ, BSc, 66, PhD(endocrinol), 69. *Prof Exp:* asst prof, 73-80, ASSOC PROF PHYSIOL, DARTMOUTH MED SCH, 73-80. *Concurrent Pos:* Can Med Res Coun res fel, McGill Univ, 69-73. *Mem:* Am Physiol Soc; Am Soc Nephrol. *Res:* Clearance studies in unanesthetized rats, emphasizing hormonal control of renal function in normal and diabetes insipidus rats. *Mailing Add:* Med Info Sci Dartmouth Med Sch Hanover NH 03756

EDWARDS, BRUCE S, b Boston, Mass, Jan 12, 49. CYTOLOGY. *Educ:* Univ Colo, PhD(biol), 79. *Prof Exp:* NIH postdoctoral res training, Univ Wisc, 79-81; DIR FLOW CYTOMETRY FACIL & SCIENTIST, RES DIV, LOVELACE MED FOUND, 81- *Concurrent Pos:* Adj assoc prof, Sch Med & Cancers, Univ NMex. *Mem:* Am Asn Immunologists; Int Soc Anal Cytol; AAAS; Sigma Xi. *Mailing Add:* Dept Biomed Res Lovelace Med Found 2425 Ridgecrest Dr SE Albuquerque NM 87108

EDWARDS, BYRON N, b Trinidad, Colo, Sept 16, 32; m 56; c 3. APPLIED PHYSICS, ELECTRICAL ENGINEERING. *Educ:* Univ Calif, Berkeley, BS, 55, MS, 57, PhD(elec eng), 60. *Prof Exp:* Assoc elec eng, Univ Calif, Berkeley, 56-58; res scientist appl physics, Aeronutronics Div, 59-62, res scientist appl physics, Philco Res Labs, 62-80, supvr Ford Aerospace & Coms Aeronutronic, Ford Motor Co, 80-90; SUPVR, LORAL AEROSPACE, 90- *Concurrent Pos:* Lectr, Univ Calif, Irvine, 70- *Mem:* Inst Elec & Electronics Engrs. *Res:* Microwave engineering and applications of gas discharges; infrared engineering and physics, communication and tracking systems; electro-optical devices. *Mailing Add:* Loral Aerospace Bldg 6 Rm 201 Newport Beach CA 92663

EDWARDS, CAROL ABE, b Hilo, Hawaii; m 62; c 1. APPLIED MATHEMATICS. *Educ:* Univ Calif, Berkeley, AB, 60; Univ Ill, Urbana, AM, 62, PhD(math), 73. *Prof Exp:* From instr to asst prof math, Hilo Col, Univ Hawaii, 62-71; asst prof, St Louis Univ, 73-75; assoc prof, 75-77, PROF MATH, ST LOUIS COMMUNITY COL, FLORISSANT VALLEY, 77- *Concurrent Pos:* Adj assoc prof math, St Louis Univ, 75, adj prof math, 80; mem, conv & conf comt, Nat Coun Teachers Math, 89-92. *Mem:* Nat Coun Teachers Math. *Res:* Subgroups of the group of permutations and complementations of the independent variables of a Boolean function; the teaching of mathematics at the undergraduate level, especially the lower division. *Mailing Add:* Dept Math St Louis Community Col-Florissant 3400 Pershall Rd St Louis MO 63135

EDWARDS, CECILE HOOVER, b East St Louis, Ill, Oct 20, 26; m 51; c 3. NUTRITION, BIOCHEMISTRY. *Educ:* Tuskegee Inst, BS, 46, MS, 47; Iowa State Univ, PhD(nutrit), 50; Am Bd Nutrit, dipl human nutrit, 63. *Prof Exp:* Res assoc nutrit, Iowa State Univ, 49-50; asst prof & res assoc foods & nutrit, Tuskegee Inst, 50-56; prof nutrit, NC A&T State Univ, 56-71, chmn dept home econ, 68-71; dean, Sch Continuing Educ, 86-87; chmn dept home econ, 71-74, dean, Sch Human Ecol, 74-86, PROF NUTRIT, HOWARD UNIV, 71- *Concurrent Pos:* Head dept foods & nutrit, Tuskegee Inst, 52-56; collabr, Bur Human Nutrit & Home Econ, Agr Res Serv, USDA, 52-55; guest scientist, Cent Food Technol Res Inst, Mysore, India, 67-68; partic distinguished scientists lect series, Bennett Col, 70; adj prof, Univ NC, Chapel Hill, 71; mem comt interpretation of recommended dietary allowances, Nat Res Coun; pres, Southeastern Col Conf Teachers of Food & Nutrit; chmn panel community nutrit educ, White House Conf Food, Nutrit & Health, 69; vpres opers, Am Home Econ Asn, 75-77; mem, Expert Panel Nitrates, Nitrites & Nitrosamines, 75-79; mem adv comt dir, NIH, 76-78; consult, Nat Inst Cancer, 72-74. *Honors & Awards:* Plaque for Contributions to Sci, Nat Coun Negro Women, 63; Scroll & Key Outstand Contribution to Sci & Educ, City East St Louis, Ill; Distinguished Sci Res Award, N Caroline Awards, Greensboro Chap for the Links Inc. *Mem:* Am Inst Nutrit; Am Home Econ Asn; Soc Nutrit Educ; Am Dietetic Asn; Nat Inst Sci. *Res:* Amino acid composition of foods; utilization of protein from vegetable sources; utilization of wheat by adult man; utilization of the amino acid methionine; utilization of amino acids in protein deficiency; pica; wheat-soy-beef mixtures; adaptation to protein deficiency; plant protein supplement from locally grown grains in Nigeria; factors affecting human pregnancy outcomes. *Mailing Add:* Sch Human Ecol Howard Univ 2400 Sixth St NW Washington DC 20059

EDWARDS, CHARLES, b Washington, DC, Sept 22, 25; m 51; c 4. PHYSIOLOGY, BIOPHYSICS. *Educ:* Johns Hopkins Univ, AB, 45, MA, 48, PhD(biophys), 53. *Prof Exp:* Hon res asst biophys, Univ Col, London, 53-54, asst lectr, 54-55; from instr to asst prof physiol optics, Johns Hopkins Univ, 55-58; asst res prof physiol, Univ Utah, 58-60; from assoc prof to prof, Univ Minn, 60-67; prof biol sci, State Univ NY, Albany, 67-84, dir Neurobiol Res Ctr, 70-84; SPEC PROJ OFFICER, NAT INST DIABETES & DIGESTIVE & KIDNEY DIS, NIH, 84- *Concurrent Pos:* Nat Found Infantile Paralysis fel, 53-54; Lalor fel, Marine Biol Lab, 57; Lederle fel, 59-60; Japan Soc Prom Sci fel; mem physiol study sect, NIH, 71-75; vis prof physiol, Czech Acad Sci, US Nat Acad & Czech Acad Exchange, 80, 82, 84; adv bd, Neurosci Res, 83-; actg assoc dean, Sch Pub Health Sci, 83-84. *Mem:* Soc Gen Physiol (secy, 71-73); Am Physiol Soc; Biophys Soc; Am Soc & Exp Pharmacol Therapeut; Soc Neurosci; fel AAAS; John Hopkins Soc Scholars, 88. *Res:* Membrane phenomena in excitable tissue; exocytosis and synaptic function. *Mailing Add:* Univ SFla Col Med 12901 Bruce B Downs Blvd Tampa FL 33612

EDWARDS, CHARLES C, b Overton, Nebr, Sept 16, 23; c 4. SURGERY. *Educ:* Univ Colo, BA, 45, MD, 48; Univ Minn, MS, 56; Am Bd Surg, dipl. *Hon Degrees:* LLD, Fla Southern Col & Philadelphia Col Pharm & Sci, 73; LHD, Pa Col Podiatry, 74. *Prof Exp:* Comnr, Food & Drug Admin, US Dept Health, Educ & Welfare, 69-73, asst secy health, 73-75; sr vpres, Becton, Dickinson & Co, 75-77, dir, 75-81; PRES & CHIEF EXEC OFFICER, SCRIPPS CLIN & RES FOUND, LA JOLLA, CALIF, 77-; PRES & CHIEF EXEC OFFICER, SCRIPPS INST MED & SCI, 91- *Concurrent Pos:* Mem, bd trustees, Scripps Clin & Res Found; mem, Priestley Soc, Mayo Clin, 89; diplomat, Am Bd Surg. *Honors & Awards:* Distinguished Serv Award, US Dept Health, Educ & Welfare, 75; Distinguished Fel, Toxicol Forum, 85. *Mem:* Inst Med-Nat Acad Sci; AMA; fel Am Col Surgeons; Soc Med Adminrs; fel Am Col Physician Execs; hon mem Am Hosp Asn. *Mailing Add:* Scripps Clin & Res Found 10666 N Torrey Pines Rd La Jolla CA 92037

EDWARDS, CHARLES HENRY, JR, b Pleasant Hill, Tenn, Sept 27, 37; m 58; c 3. MATHEMATICS. *Educ:* Univ Tenn, BS, 58, PhD(math), 60. *Prof Exp:* Asst prof math, Univ Tenn, 61; from instr to asst prof, Univ Wis, 61-64; assoc prof, 64-69, dir, Inst Res Planning, 74-76, PROF MATH, UNIV GA, 69- *Concurrent Pos:* Sloan res fel, 64-66; mem, Inst Advan Study, 65-66. *Mem:* Am Math Soc; Math Asn Am. *Res:* Topology and geometry of manifolds and applications; history of mathematics. *Mailing Add:* Dept Math Univ Ga Athens GA 30602

EDWARDS, CHARLES RICHARD, b Lubbock, Tex, Jan 22, 45; m 66; c 2. INTEGRATED PEST MANAGEMENT. *Educ:* Tex Tech Univ, BS, 68; Iowa State Univ, MS, 70, PhD(entom), 72. *Prof Exp:* PROF ENTOM, DEPT ENTOM, PURDUE UNIV, 72- *Concurrent Pos:* Chmn res soybean arthropods, Coop State Res Serv, USDA, 81-82; mem task force, Exten Comt Orgn & Policy, Integrated Pest Mgt, 77-83 & Coun Agr Sci Technol, Integrated Pest Mgt, 78-83; consult, Agri-Pace Inc, 80-89, Pesticide Users Adv Comt, Environ Protection Agency, 86-; bd mem, Consortium Int Crop Protection, 83-, Insect Pop Mgt, int experience, Asia & Africa; nat coalition on IPM, 89-; nat IPM Forum, 90- *Honors & Awards:* Distinguished Achievement Award Exten, Entom Soc Am; Researchers Recognition Award, Am Soybean Asn; Super Serv Award, USDA. *Mem:* Entom Soc Am; Am Registry Prof Entom; Am Soc Agron; Sigma Xi. *Res:* Development of management tactics for arthropod pests of soybeans, especially development of economic injury levels; host plant resistance; effect of cropping systems on arthropod pests and chemical control. *Mailing Add:* Dept Entom Purdue Univ Entom Hall West Lafayette IN 47907-1158

EDWARDS, CHRISTOPHER ANDREW, b Summit, NJ, Dec 1, 66. WAVELENGTH DIVISION MULTIPLEXERS, FIBER MICROLENSES. *Educ:* Haverford Col, BS, 88. *Prof Exp:* Res asst physics, Haverford Col, 85-88; MEM TECH STAFF, AT&T BELL LABS, 88- *Res:* Optical fiber communications, optical filters and imaging devices; nonlinear fluid dynamics; interfacial motion in porous media. *Mailing Add:* 653 Long Hill Rd Gillette NJ 07933

EDWARDS, CONSTANCE CARVER, b Baltimore, Md, June 8, 49. PURE MATHEMATICS. *Educ:* Col Notre Dame, Md, BA, 68; Univ Wis-Milwaukee, MA, 70, PhD(math), 75. *Prof Exp:* Teaching asst math, Univ Wis-Milwaukee, 68-75, instr, Univ Wis-Parkside, 74; asst prof, 75-81, ASSOC PROF MATH, IND UNIV-PURDUE UNIV, FT WAYNE, 81- *Concurrent Pos:* Res grant, Ind Univ-Purdue Univ, Ft Wayne, 78. *Mem:* Am Math Soc; Math Asn Am. *Res:* Regular semigroups and ordered semigroups. *Mailing Add:* Dept Math Ind Univ-Purdue Univ Ft Wayne IN 46805

EDWARDS, DALE IVAN, b Mattoon, Ill, Jan 12, 30; m 55; c 4. NEMATOLOGY, PLANT PATHOLOGY. *Educ:* Eastern Ill Univ, BS, 56; Univ Ill, MS, 60, PhD(plant path), 62. *Prof Exp:* Asst plant path, Univ Ill, 56-62; assoc pathologist, Tela Railroad Co, 62-65; asst prof plant path, 73-77, nematologist, 65-89, ASSOC PROF PLANT PATH, UNIV ILL, URBANA, 77-, EXTEN NEMATOLOGIST, AGR RES, USDA, UNIV ILL, 89- *Mem:* Soc Nematol; Europ Soc Nematol; Org Trop Nematol; Sigma Xi. *Res:* Host-parasite relationships and control of plant parasitic nematodes affecting soybean. *Mailing Add:* N-519 Turner Hall Univ Ill 1102 S Goodwin Ave Urbana IL 61801

EDWARDS, DAVID FRANKLIN, b Ironton, Ohio, Mar 2, 28; m 54; c 2. SOLID STATE PHYSICS. *Educ:* Miami Univ, Ohio, AB, 49; Univ Cincinnati, MS, 53, PhD(nuclear shell struct), 53. *Prof Exp:* Solid state physicist, Battelle Mem Inst, 53-55, Willow Run Lab, Univ Mich, 55-61 & Lincoln Lab, Mass Inst Technol, 61-65; prof physics, Colo State Univ, 65-69, prof physics & elec eng, 69-74; physicist, Los Alamos Nat Lab, NMex, 74-82; sr scientist, Hewlett-Packard, Santa Clara, 82-83; group leader, Lawrence Livermore Nat Lab, Calif, 83-86, chief scientist, 86-90; RETIRED. *Concurrent Pos:* Fulbright prof physics, Univ Rio Grande do Sul, Brazil, 70-71; Nat Acad Sci & USSR Acad Sci vis scientist to USSR, Lebedev Physics Inst, 75-76. *Mem:* Am Phys Soc. *Res:* Solid state spectroscopy; nonlinear optical effects; second harmonic generation; mixing at optical frequencies; laser induced damage; laser-matter interaction; laser isotope separation; effects of nuclear radiation on optical coatings harding of coatings to nuclear radiation, light scattering rough surfaces; laser interaction with window and mirror materials. *Mailing Add:* 8820 W Valpico Rd Tracy CA 95376

EDWARDS, DAVID J, b Durham, NC, Dec 5, 43; m 72; c 2. BIOGENIC AMINE METABOLISM, NEUROBIOLOGY. *Educ:* Univ NC, Chapel Hill, PhD(biochem), 71. *Prof Exp:* assoc prof, 81-88, PROF PHARMACOL, UNIV PITTSBURGH, 88- *Mem:* AAAS; Neurosci Soc. *Res:* Effects of psychotropic drugs on brain amine metabolism; trace amines. *Mailing Add:* Dept Pharmacol & Physiol Univ Pittsburgh Salk Hall Rm 542 Pittsburgh PA 15261

EDWARDS, DAVID OLAF, b Liverpool, Eng, Apr 27, 32; m 67; c 2. LOW TEMPERATURE PHYSICS. *Educ:* Oxford Univ, BA, 53, MA, PhD(physics), 57. *Prof Exp:* Pressed Steel Co res fel, Clarendon Lab, Oxford Univ, 57-58; vis asst prof, 58-60, from asst prof to assoc prof, 60-65, PROF PHYSICS, OHIO STATE UNIV, 65- *Concurrent Pos:* Vis prof, Univ Sussex, 68, Israel Inst Technol, 71 & Univ Paris, 78, 82, 86; consult, Brookhaven Nat Lab, 75-77 & Los Alamos Nat Lab, 79-81. *Honors & Awards:* Simon Mem Prize Low Temperature Phys, Brit Phys Soc, 83; Buckley Prize, Condensed Matter Physics, Am Phys Soc, 90. *Mem:* Fel Am Phys Soc; fel Royal Soc Gt Brit. *Res:* Liquid, gaseous and solid phases of He3 and He4 and the interfaces between them; superfluidity; cryogenics. *Mailing Add:* Dept Physics Ohio State Univ 174 W 18th Ave Columbus OH 43210-1106

EDWARDS, DAVID OWEN, b Buffalo, NY, Dec 15, 30; m 53; c 3. FINANCIAL CONTROL. *Educ:* Univ Mich, BS, 53; Univ Wis, PhD(chem eng), 61. *Prof Exp:* Instr chem eng, Univ Wis, 57-58; res engr, Film Dept, E I du Pont de Nemours, 60-69, consult, Cent Systs & Serv Dept, Systs & Comput Div, 69-74, auditor, Finance Dept, 74-77, planning specialist, Cent Res & Develop, 77-80, CONSULT, AUDITING DEPT, E I DU PONT DE NEMOURS & CO, INC, 80- *Res:* Plastic film manufacture and processing; applications of computer techniques; venture modeling and computer applications. *Mailing Add:* Finance Dept E I du Pont de Nemours & Co Wilmington DE 19898

EDWARDS, DONALD K, b Richmond, Calif, Oct 11, 32; m 55; c 2. THERMODYNAMICS. *Educ:* Univ Calif, Berkeley, BS, 54, MS, 56, PhD(mech eng), 59. *Prof Exp:* Thermodyn engr, Lockheed Missile & Space Div, Calif, 58-59; from asst prof to prof eng, Univ Calif, Los Angeles, 59-81, chmn, Chem Nuclear & Thermal Eng Dept, 75-78, prof mech eng & chmn dept, Univ Calif, Irvine, 82-86, prof mech eng & assoc dean eng, 86-88, PROF MECH ENG, UNIV CALIF, IRVINE, 88- *Concurrent Pos:* Fac investr, NSF grants, 59-81; consult, TRW Systs, 62-81; pres & chmn bd, Gier Dunkle Instruments, Inc, Santa Monica, 63-66; assoc ed, J Heat Transfer, 75-81, Solar Energy, 82-85, reg mech engr, 65- *Honors & Awards:* Am Soc Mech Engrs Mem Award, 73; First Thermophysics Award, Am Inst Aeronaut & Astronaut, 76. *Mem:* Fel Am Soc Mech Engrs; Optical Soc Am; Solar Energy Soc; fel Am Inst Aeronaut & Astronaut; Sigma Xi. *Res:* Heat and mass transfer; thermal radiation; radiant energy transfer between solids and through absorbing, emitting, scattering media; molecular gas radiation; thermal radiation instrumentation; natural convection; radiation and convection. *Mailing Add:* Dept Mech Aero Eng Univ Calif Irvine CA 92717

EDWARDS, DONALD M(ERVIN), b Tracy, Minn, Apr 16, 38; m 64; c 4. AGRICULTURAL ENGINEERING. *Educ:* SDak State Univ, BS, 60, MS, 61; Purdue Univ, PhD(agr & civil eng), 66. *Prof Exp:* Mem coop educ prog, Soil Conserv Serv, USDA, 56-61; asst agr eng, SDak State Univ, 60-61 & Purdue Univ, 62-66; assoc prof, 66-70, asst dean, 70-73, dir, Eng Res Ctr, 73-76, PROF AGR ENG, UNIV NEBR, LINCOLN, 70-, ASSOC DEAN COL ENG & TECHNOL, 73-, DIR ENERGY RES & DEVELOP CTR, 76-; AT DEPT AGR ENG, MICH STATE UNIV, EAST LANSING. *Concurrent Pos:* Consult & collabr to several state & fed agencies & industs. *Honors & Awards:* Outstanding Educr Award, Am Soc Agr Engrs, 76. *Mem:* AAAS; Am Soc Agr Engrs; Sigma Xi; Am Soc Eng Educ; Nat Soc Prof Engrs. *Res:* Water resources engineering, particularly irrigation; porous media; water pollution; engineering education as related to new teaching techniques and engineering educational programs; energy. *Mailing Add:* Dept Agr Eng Agr Eng Bldg Rm 215 Mich State Univ East Lansing MI 48824

EDWARDS, DOUGLAS CAMERON, b Ottawa, Ont, Oct 21, 25; m 49; c 5. RUBBER CHEMISTRY. *Educ:* Queen's Univ, Ont, BSc, 47. *Prof Exp:* Res chemist, 49-64, res assoc, 64-69, sci adv rubber chem, 69-88, mgr mat res, 84-87, SCI ADV ELASTOMERS, POLYSAR LTD, 88- *Mem:* Am Chem Soc; Chem Inst Can. *Res:* Rubber science and technology. *Mailing Add:* Elastomers Res & Develop Div Polysar Ltd Sarnia ON N7T 7M2 Can

EDWARDS, DOYLE RAY, b Dexter, Mo, Dec 22, 38; m 59; c 4. NUCLEAR ENGINEERING. *Educ:* Mo Sch Mines, BS, 59; Mass Inst Technol, SM, 61, ScD(nuclear eng), 63. *Prof Exp:* From asst prof to assoc prof, 63-73, PROF NUCLEAR ENG, UNIV MO-ROLLA, 73- *Concurrent Pos:* NSF res grant, 64-66; AEC equip grant, 65, 66 & 68, traineeship grant, 65-71; consult, Assoc Elec Coop, Inc, 75-82; Elec Power Res Inst grant, 80-84; AT&T Equip grant, 86-87. *Mem:* Am Nuclear Soc; Nat Soc Prof Engrs. *Res:* Computer methods in nuclear engineering; computer simulation of radiation damage; PRA methologics; heat transfer in fuel bundles; application of active ingredients to reactor operation. *Mailing Add:* 224 Fulton Hall Nuclear Eng Univ Mo Rolla MO 65401

EDWARDS, ERNEST PRESTON, b Landour, India, Sept 25, 19; US citizen; m 55. ECOLOGY. *Educ:* Univ Va, BA, 40; Cornell Univ, MA, 41, PhD(ornith), 49. *Prof Exp:* Instr biol, Univ Ky, 49-50; civilian biologist, US Army Chem Corps, 52-54; asst prof biol, Hanover Col, 55-56; assoc dir, Mus Natural Hist Houston, 57-60; prof biol, Univ of the Pac, 60-65; prof biol, Sweet Briar Col, 65-90; RETIRED. *Mem:* Wilson Ornith Soc; Am Ornith Union; Sigma Xi. *Res:* Ecology, distribution and taxonomy of tropical birds; ecology of the Blue Ridge Mountains; nomenclature and distribution of birds of the world. *Mailing Add:* Box AQ Sweet Briar VA 24595

EDWARDS, EUGENE H, b Birmingham, Ala, July 4, 22. MATERIALS SCIENCE ENGINEERING. *Educ:* Birmingham Southern Col, BS, 43; Univ Calif, Berkeley, MS, 50, PhD(math sci), 53. *Prof Exp:* Chief mat eng, Chevron Corp, 69-85; RETIRED. *Honors & Awards:* Mathewson Gold Medal, Am Inst Mining, Metall & Petrol Engrs, 55. *Mem:* Am Inst Mining Metall & Petrol Engrs; fel Am Soc Metals. *Mailing Add:* 69 Sleepy Hollow Lane Orinda CA 94563

EDWARDS, FREDERICK H(ORTON), b Can, June 23, 15; US citizen; m 51; c 2. ELECTRICAL ENGINEERING. *Educ:* Univ BC, BASc, 49; NS Tech Col, MSc, 55. *Prof Exp:* Design engr, Apparatus Div, Can Westinghouse, 49-55; ASSOC PROF ELEC ENG, UNIV MASS, AMHERST, 55-, ASSOC PROF ELEC & COMPUT ENG. *Concurrent Pos:* NSF fel, Cambridge Math Lab, 65-66. *Mem:* Inst Elec & Electronics Engrs. *Res:* Switching circuit theory and digital system design; energy conversion. *Mailing Add:* Dept Elec/Comput Eng Univ Mass Amherst MA 01003

EDWARDS, GAYLE DAMERON, b Alexandria, La, Mar 8, 27; m 55. PETROLEUM CHEMISTRY. *Educ:* La Col, BA, 47; Univ Tex, MA, 48, PhD(chem), 51. *Prof Exp:* Res chemist, Pan Am Refining Corp, 51-53; supvr, 53-66, mgr tech serv, Neches Plant, 66-77, SR COORDR ENVIRON CONTROL, TEXACO CHEM CO, 77- *Mem:* Am Chem Soc. *Res:* Petrochemicals; surface active agents; organometallic compounds. *Mailing Add:* Texaco Chem Co 3040 Post Oak Blvd Houston TX 77056

EDWARDS, GAYLEN LEE, b Mountain Home, Idaho, Aug 8, 56; m 80; c 2. BEHAVIORAL PHYSIOLOGY, INGESTIVE BEHAVIOR. *Educ:* Idaho State Univ, BS, 78; Univ Idaho, MS, 80; Wash State Univ, DVM, 84, PhD(vet sci), 86. *Prof Exp:* Postdoctoral fel, Univ Iowa, 86-89; ASST PROF

PHYSIOL, UNIV GA, 89- *Mem:* Soc Neurosci; Am Physiol Soc; AAAS; Soc Study Ingestive Behav. *Res:* Caudal hindbrain in the control of water and sodium balance; ingestion of water and sodium and concomitant effects on cardiovascular and renal functions. *Mailing Add:* Dept Physiol & Pharmacol Univ Ga Athens GA 30602

EDWARDS, GEORGE, b Glasgow, Scotland, June 28, 18; nat US; m 44; c 3. CHEMISTRY. *Educ:* Glasgow Univ, BSc, 41, PhD(chem), 46. *Prof Exp:* Exp officer chem eng, S W Scotland Br, Woolwich Arsenal, 41-46; lectr, Royal Col Sci & Technol, Scotland, 46-52; res assoc chem, Enrico Fermi Inst Nuclear Studies, Univ Chicago, 52-55; group leader geol age measurement, Shell Develop Co, 55-64, mgr anal chem, 64-81, res assoc, 81-83; RETIRED. *Honors & Awards:* Sir James Bielby Prize, 43. *Mem:* Brit Inst Chem Eng. *Res:* Stable isotope absolute geologic age measurement; cosmic abundance of elements; analytical chemistry. *Mailing Add:* 3100 Mid Lane Houston TX 77027

EDWARDS, GERALD ELMO, b Gretna, Va, Sept 17, 42; m 69; c 2. PLANT PHYSIOLOGY, PHOTOSYNTHESIS. *Educ:* Va Polytech Inst, BS, 65; Univ Ill, MS, 66; Univ Calif, Riverside, PhD(plant sci), 69. *Prof Exp:* Fel biochem, Univ Ga, 69-71; from asst prof to prof bot, Univ Wis, 71-80; prof & chmn dept, 81-86, PROF BOT DEPT, WASH STATE UNIV, PULLMAN, 87- *Concurrent Pos:* Vis prof bot, Univ Sheffield, 77-78, Univ Wis, 86-87. *Mem:* Am Soc Plant Physiol; Japanese Soc Plant Physiol; Scand Soc Plant Physiol; Sigma Xi. *Res:* Mechanism, regulation and efficiency of photosynthetic assimilation of carbon in plants; photosynthetic and photorespiratory metabolism studied at the whole plant, cellular and subcellular levels. *Mailing Add:* Bot Dept Wash State Univ Pullman WA 99164-4238

EDWARDS, GLEN ROBERT, b Monte Vista, Colo, July 21, 39; m 59; c 2. PHYSICAL METALLURGY, METALLURGY. *Educ:* Colo Sch Mines, BS, 61; Univ NMex, MS, 67; Stanford Univ, PhD(mat sci), 71. *Prof Exp:* Staff mem plutonium metall, Los Alamos Sci Lab, 63-67; asst mat sci, Stanford Univ, 67-71; from asst to assoc prof, Naval Postgrad Sch, 71-76; assoc prof, 76-79, PROF METALL ENG, COLO SCH MINES, 79-; DIR, CTR WELDING & JOINING RES, 87- *Honors & Awards:* Pres Award, Am Soc Metals Int, 80; Adams Mem Membership Award, Am Welding Soc, 81-85, ASM Fellow, 88; Warren Savage Award, Am Welding Soc, 87. *Mem:* fel Am Soc Metals Int; Am Inst Mining & Metall Engrs; Sigma Xi; Am Welding Soc; Int Inst Welding. *Res:* Mechanical metallurgy; deformation processes; reactive metals; welding metallurgy; advanced material processing and joining. *Mailing Add:* Dept Metal & Mat Eng Colo Sch Mines Golden CO 80401

EDWARDS, GORDON STUART, b Old Greenwich, Conn, Feb 11, 38; c 3. ENVIRONMENTAL HEALTH. *Educ:* Amherst Col, BA, 59; Harvard Univ, MA, 63; Mass Inst Technol, ScD(nutrit biochem), 70, Am Bd Toxicol, dipl. *Prof Exp:* Postdoctoral fel cell biol, Rockefeller Univ, 70-72; chief biol sect, Cancer Res Div, Thermo Electron Corp, 77-81; assoc prof pharmacol & genetics, George Washington Univ, 72-77; PRES, TOXICON ASSOCS, 81- *Concurrent Pos:* Consult, Food & Drug Admin, 74-76 & Environ Protection Agency, 77; staff mem, Med Col Va Cancer Ctr, 75-77 & Howard Univ Cancer Ctr, 75-77; consult toxicol, 81- *Mem:* Am Asn Cancer Res; Soc Toxicol; Environ Mutagen Soc; Am Chem Soc. *Res:* Toxicology consulting; occupational toxicology; environmental toxins; chemical carcinogens. *Mailing Add:* Toxicon Assocs 34 Everett St Natick MA 01760

EDWARDS, H(ERBERT) M(ARTELL), civil engineering; deceased, see previous edition for last biography

EDWARDS, HARDY MALCOLM, JR, b Ruston, La, Nov 16, 29; m 54; c 1. NUTRITIONAL BIOCHEMISTRY. *Educ:* Southwestern La Inst, BSA, 49; Univ Fla, MSA, 50; Cornell Univ, PhD(nutrit biochem), 53. *Prof Exp:* Res fel animal nutrit, Univ Fla, 49-50; res asst & nutrit chemist, Cornell Univ, 50-53; from res biochemist to sr res biochemist, Int Minerals & Chem Corp, 55-57; from asst prof to assoc prof, 57-66, dean, Grad Sch, 72-79, PROF NUTRIT BIOCHEM, UNIV GA, 66- *Concurrent Pos:* Res career develop award, NIH, 63-72; res assoc, Dept Physiol Chem, Univ Lund, 64-65; Guggenheim Mem fel & vis prof, Tours, France & Cambridge, Eng, 71-72; Nat Instit Animal Sci, Copenhagen, Denmark, 84 & 87. *Honors & Awards:* Poultry Nutrit Res Award, Am Feed Mfg Asn, 62. *Mem:* Am Soc Animal Sci; Poultry Sci Asn; Am Inst Nutrit; Soc Exp Biol & Med. *Res:* Lipid nutrition of aves; mechanism of divalent ion adsorption; etiology of leg abnormalities in poultry. *Mailing Add:* Dept Poultry Sci Univ Ga Athens GA 30602

EDWARDS, HAROLD HENRY, b Andes, NY, July 14, 32; m 62; c 3. ELECTRON MICROSCOPY. *Educ:* Cornell Univ, BEP, 55; Rensselaer Polytech Inst, PhD(biophys), 69. *Prof Exp:* Electronic engr, Res & Develop Ctr, Gen Elec Co, 55-65; asst, Rensselaer Polytech Inst, 65-69; res assoc anat, Med Ctr, Duke Univ, 69-71; asst mem biochem dept, St Jude Children's Res Hosp, 71-75, dir, Electron Micros Facil, 75-87; ASSOC PROF, PHYSICS DEPT, CHRISTIAN BROS COL, 87- *Concurrent Pos:* Adj prof biol, Memphis State Univ, 75. *Mem:* Biophys Soc; Am Soc Cell Biol; Electron Microscope Soc Am; NY Acad Sci. *Res:* Membrane structure and function; electron microscopy. *Mailing Add:* 4653 Johnson Cove Memphis TN 38117

EDWARDS, HAROLD HERBERT, b Milford, Mich, Oct 31, 37; m 62; c 2. PLANT PHYSIOLOGY, BIOCHEMISTRY. *Educ:* Albion Col, BA, 60; Univ Wis, MS, 62, PhD(bot), 65. *Prof Exp:* Wis Alumni res fel, Univ Wis, 65; NIH fel, Univ Nebr, 65-67; from asst prof to assoc prof, 67-77, PROF BIOL SCI, WESTERN ILL UNIV, 77- *Res:* Plant physiology, ultrastructure and biochemistry of the obligate parasite-host complex of powdery mildewed barley and brown spot disease of soybeans. *Mailing Add:* Dept Biol Western Ill Univ Macomb IL 61455

EDWARDS, HAROLD M, b Champaign, Ill, Aug, 6, 36; m 79. NUMBER THEORY. *Educ:* Univ Wis, BA, 56; Columbia Univ, MA, 57; Harvard Univ, PhD(math), 61. *Prof Exp:* Instr math, Harvard Univ, 61-62; res assoc math, Columbia Univ, 62-63, asst prof math, 63-66; from asst prof to assoc prof, 66-79, PROF MATH, NY UNIV, 79- *Concurrent Pos:* Vis sr lectr, Australian Nat Univ, 71; fel, John Simon Guggenheim Mem Found, 81. *Honors & Awards:* Steele Prize, Am Math Soc, 80. *Mem:* Fel NY Acad Sci; Am Math Soc; Math Asn Am. *Res:* History of mathematics and the history of number theory, particularly in the latter part of the 19th century. *Mailing Add:* Courant Inst NY Univ 251 Mercer St New York NY 10012

EDWARDS, HARRY WALLACE, b Syracuse, NY, Oct 6, 39; m 66; c 2. PHYSICAL CHEMISTRY. *Educ:* Univ Nev, BS, 62; Univ Ariz, PhD(phys chem), 66. *Prof Exp:* Asst prof, 66-70, assoc prof, 70-76, PROF MECH ENG, COLO STATE UNIV, 76- *Concurrent Pos:* Res grants, NSF, 67 77, Pub Serv Co Colo, 67-70, Environ Protection Agency, 70-73 & Resources for Scholarly progs, Colo Stat Univ, 84-86; consult, NSF, 71-72, NASA, 76-77 & Solar Energy Res Inst, 79-81; alt mem, Comt Motor Vehicle Emissions, Colo Air Pollution Control Comn, 74-76; sr vis fel, Univ Lancaster, 77-78; regional ed, Environ Technol Lett, 79-86. *Mem:* Am Chem Soc; AAAS; Air Pollution Control Asn; NY Acad Sci. *Res:* Surface chemistry; air pollution; environmental effects of trace substances; air pollution control, especially oxides of nitrogen; aerosol behavior; identification of pollution sources; indoor air pollution. *Mailing Add:* Dept Mech Eng Colo State Univ Ft Collins CO 80523

EDWARDS, HELEN THOM, b Detroit, Mich, May 27, 36. ACCELERATOR PHYSICS. *Educ:* Cornell Univ, BA, 57, PhD(physics), 66. *Prof Exp:* Res, 10 GEV Electron Synchrotron, Cornell Univ, 58-70; res, Fermi Nat Accelerator Lab, 70-87, head, Accelerator Div, 87-89; HEAD & ASSOC DIR, SUPERCONDUCTING DIV, SUPERCONDUCTING SUPERCOLLIDER LAB, DALLAS, 89- *Concurrent Pos:* MacArthur fel, 88. *Honors & Awards:* E O Lawrence Award, 86. *Mem:* Am Phys Soc. *Res:* The Booster; 400 G E V Main Ring; Superconductor tevhetron. *Mailing Add:* Superconducting Supercollider Lab 2550 Beckley Meade Ave Mail Stop 1045 Dallas TX 75237

EDWARDS, J GORDON, b Wilmington, Ohio, Aug 24, 19; m 46; c 1. MEDICAL ENTOMOLOGY, SYSTEMATIC ZOOLOGY. *Educ:* Butler Univ, BS, 42; Ohio State Univ, MSc, 46, PhD(entom), 49. *Prof Exp:* Instr entom & zool, 49-52, from asst prof to assoc prof, 52-59, PROF ENTOM, SAN JOSE STATE UNIV, 59- *Concurrent Pos:* Life fel, Calif Acad Sci; bd mem, Nat Coun Environ Balance, Consumer Alert, Am Coun Sci & Health, Entom Soc Am, Coleopterists Soc, Pac Coast Entom Soc, 85. *Mem:* Sierra Club; Entom Soc Am; Nat Audubon Soc; Am Coun Sci & Health; Nat Coun Environ Balance; Coleopterists Soc; Explorers Club. *Res:* Coleoptera biology and taxonomy; high altitude ecology and biology; medical entomology; tropical biology; pesticides and the environment. *Mailing Add:* Dept Entom San Jose State Univ Washington Sq San Jose CA 95192

EDWARDS, JAMES BURROWS, b Hawthorne, Fla, June 24, 27; m 51; c 2. DENTAL SURGERY. *Educ:* Col Charleston, SC, BS, 50; Univ Louisville, Ky, DMD, 55; Am Bd Oral & Maxillofacial Surg, dipl, 63. *Hon Degrees:* Various from cols & univs. *Prof Exp:* US Navy Dent Corps, 55-57; mem, Gov Statewide Comt Comprehensive Health Care Planning SC, 68-72, Fed Hosp Coun, 69-73; clin assoc oral surg, 70-77, clin asst prof oral surg & community dent, 77-82, PROF, ORAL & MAXILLOFACIAL SURG, COL DENT MED, MED UNIV SC, 82-, PRES. *Concurrent Pos:* Mem, Fed Hosp Coun, 69-73; mem bd, numerous found & corps, 80- *Mem:* Am Dent Asn; Am Asn Oral & Maxillofacial Surgeons; fel Int Asn Oral & Maxillofacial Surgeons; fel Int Col Dentists; hon mem Am Dent Soc Anesthesiologists. *Mailing Add:* Med Univ SC 171 Ashley Ave Charleston SC 29425

EDWARDS, JAMES MARK, b Baltimore, Md, July 26, 56; m 87. RAYNAUDS SYNDROME. *Educ:* Johns Hopkins Univ, BES, 77; Vanderbilt Univ, MD, 91. *Prof Exp:* Intern surg, Ore Health Sci Univ, 81-82, resident, 82-88, staff surgeon, 88-89; FEL VASCULAR SURG, UNIV WASH SCH MED, 89- *Concurrent Pos:* Jr eng fel, NSF, 80. *Mem:* Assoc fel Am Col Surgeons. *Res:* Small vessel vasospasm; analysis of arterial blood flow characteristics; related vascular disorders. *Mailing Add:* Seattle Vet Admin Med Ctr Surg 112 1660 S Columbian Way Seattle WA 98108

EDWARDS, JAMES WESLEY, b Evansville, Ind, Sept 1, 38; m 61; c 1. ZOOLOGY, GENETICS. *Educ:* Evansville Col, AB, 60; Utah State Univ, MS, 62, PhD(zool), 64. *Prof Exp:* Asst prof biol, St Francis Col, Pa, 64-65; assoc prof, 65-69, chmn dept, 65-80, PROF BIOL, SALEM COL, 69- *Mem:* AAAS. *Res:* Genetics of abnormal head development in Drosophila melanogaster, with special reference to eyelessness. *Mailing Add:* Dept Biol Salem Col PO Box 10548 Winston-Salem NC 27108

EDWARDS, JESSE EFREM, b Hyde Park, Mass, July 14, 11; m 52; c 2. PATHOLOGY. *Educ:* Tufts Col, BS, 32, MD, 35. *Prof Exp:* Intern, Albany Hosp, NY, 36-37; resident path, Mallory Inst Path, Boston City Hosp, 35-36, asst, 37-40, assoc pathologist, 41-53; from asst prof to prof path anat, Mayo Found, 46-60, CLIN PROF PATH, SCH MED & PROF GRAD SCH, UNIV MINN, ST PAUL, 60-; SR CONSULT ANAT PATH, UNITED HOSPS, ST PAUL, MINN, 72- *Concurrent Pos:* Fel, Nat Cancer Inst, 40-42; instr, Sch Med, Boston Univ, 38 & Tufts Col, 39-40; consult, Wash Home Incurables, 40-42, Mayo Clin, 46-60, Surgeon Gen, Dept Army, Hennepin County Hosp, Minneapolis, 64-, Dept Med, Minneapolis Vet Hosp, 66- & St Paul Ramsey Hosp, 69- *Honors & Awards:* Distinguished Serv Award, Mod Med, 64; Gold Heart Award, Am Heart Asn, 71, Res Achievement Award, 80; Distinguished Teacher Award, Minn Med Found, 74. *Mem:* Am Acad Pediat; fel AMA; Am Asn Path & Bact; Am Heart Asn (pres, 67-68); Int Acad Path (pres). *Res:* Congenital anomalies of heart and great vessels; pathology of cardiovascular diseases, congenital and acquired. *Mailing Add:* United Hosps Miller Div St Paul MN 55102

EDWARDS, JIMMIE GARVIN, b Boswell, Okla, July 27, 34; m 56; c 3. HIGH TEMPERATURE CHEMISTRY. *Educ:* Cent State Col, Okla, BS, 56; Okla State Univ, PhD(chem), 64. *Prof Exp:* Instr chem, Univ Nev, Reno, 60-61; chemist, Radiochem, Inc, 61-62; res assoc chem, Univ Kans, 64-65; asst prof chem, Univ Mo-Rolla, 65-66; res assoc chem, Univ Kans, 66-67; from asst prof to assoc prof 67-76, PROF CHEM, UNIV TOLEDO, 76-, DISTINGUISHED PROF, 87- *Concurrent Pos:* Vis prof, Justus Liebig Univ, Ger, 85. *Honors & Awards:* Outstanding Res Award, Sigma Xi, 85. *Mem:* Am Chem Soc; AAAS; Sigma Xi; Electrochem Soc; Am Asn Univ Professors; Planetary Soc. *Res:* High temperature physical and inorganic chemistry; materials research; mass spectrometry; Boron sulfide and thioboric acids; binary and ternary metal sulfides; rarefied gas dynamics. *Mailing Add:* Dept Chem Univ Toledo Toledo OH 43606

EDWARDS, JOHN ANTHONY, b Chester, Eng, Nov 9, 35; US citizen; c 2. MEDICAL GENETICS. *Educ:* Liverpool Univ, MB & ChB, 59, MD, 72. *Prof Exp:* From res asst prof to res assoc prof, 69-74, assoc prof med & clin assoc prof pharmacol, 74-79, PROF MED, STATE UNIV NY, BUFFALO, 79-, PROF FAMILY MED, 87- *Concurrent Pos:* Vis consult med genetics, West Seneca Develop Ctr, 74-; assoc cattend physician, Childrens Hosp Buffalo, 75-; asst physician, Buffalo Gen Hosp, 75-; NIH career res develop award, 72; mem staff, 78-, chief med, Sisters of Charity Hosp, 78-86; consult physicians, Buffalo Gen Hosp & Childrens Hosp, Buffalo, 80-; Hartford Found geriatric for develop award, Div Aging, Harvard Med Sch, 86, 87; med dir, Deaconess Ctr, SNF, 87-89, Episcopal Church Home, 87-, dir geriat med, Buffalo Gen Hosp, 87-89. *Mem:* Am Soc Human Genetics; Am Soc Hemat; Am Fedn Clin Res; Am Soc Clin Invest. *Res:* Genetic heterogeneity in human Mendelian traits; familial hyperlipoproteinemia; inherited anemias in laboratory animals; Alzheimer's disease and related disorders. *Mailing Add:* Deaconess Ctr 1001 Humboldt Pkwy Buffalo NY 14208

EDWARDS, JOHN AUERT, b Middletown, NY, July 2, 30; m 51; c 2. HEAT TRANSFER, FLUID MECHANICS. *Educ:* NC State Univ, BSME, 55, MS, 57; Purdue Univ, PhD(mech eng), 62. *Prof Exp:* Engr, Texaco, Inc, 57-58; res assoc heat transfer, Purdue Univ, 58-61; assoc prof eng mech, 62-70, prof eng mech & marine sci, 70-76, PROF MECH & AEROSPACE ENG, NC STATE UNIV, 76-, DIR, APPL ENERGY RES LAB, 82-, ASSOC DEPT HEAD, MECH ENG, 85- *Concurrent Pos:* NSF grants, 64-66; consult, Oak Ridge Nat Lab, 66-72; AEC res contract, 69-76; legal expert & consult, 80- *Mem:* Am Soc Eng Educ; Am Soc Mech Engrs. *Res:* Liquid metals lubrication; heat transfer with boiling alkali metals and dropwise condensation; fuels research; free convection and radiative heat transfer; turbulent lubrication; secondary turbulent flows; turbulent jets; fluidics; air source and water source heat pumps, solar. *Mailing Add:* Dept Mech & Aero Eng NC State Univ Raleigh NC 27695-7910

EDWARDS, JOHN C, b Petersburg, Va, Nov 10, 13; m 40; c 4. ANALYTICAL CHEMISTRY, ENVIRONMENTAL SCIENCE. *Educ:* Univ Richmond, BS, 36. *Prof Exp:* Asst chemist, Solvay Process Co, Va, 36-41; assoc chemist, Navy Dept, US Govt, 41-46; anal res chemist, E I du Pont de Nemours & Co, Inc, 47-50, supvr process control labs, May Plant, 50-57, anal res dir, 57-66, dir environ control, 66-78; RETIRED. *Concurrent Pos:* Co-analyst, Nat Bur Standards; mem, SC Water Resources Comn, 69-; consult, 78-88. *Mem:* Am Chem Soc. *Res:* Analytical research development of methods and techniques to identify, characterize and process acrylic fibers and related materials; application of instrumentation for analytical research; coordinator programs air and water pollution abatement. *Mailing Add:* 2000 Forest Dr Camden SC 29020

EDWARDS, JOHN D, b Hackensack, NJ, June 17, 25; m 46; c 5. PETROLEUM GEOLOGY. *Educ:* Cornell Univ, BS, 46; Columbia Univ, PhD(geol), 52. *Prof Exp:* Field geologist, US Geol Surv, 49-50; geologist spec invests, Shell Oil Co, 50-55, dist geologist, 55-62, staff geologist, 62, div explor mgr, 62-64, area explor mgr, 64-67, chief geologist, 67-68, asst to vpres explor, 68-71, struct geol teacher, 71-73, mgr explor traning, Shell Development Co, 74-79; mgr explor opers, 79-87, CONSULT, TEACHING STRUCT GEOL, PECTEIN INT, 87- *Concurrent Pos:* Lectr, Columbia Univ, 50. *Mem:* AAAS; Geol Soc Am; Am Inst Mining, Metall & Petrol Engrs; Am Asn Petrol Geologists. *Res:* Structural geology; petroleum exploration. *Mailing Add:* 148 Timberline Dr Durango CO 81301

EDWARDS, JOHN OELHAF, b Sewickley, Pa, July 21, 22; m 50; c 2. INORGANIC CHEMISTRY. *Educ:* Colgate Univ, AB, 47; Univ Wis, PhD(chem), 51. *Prof Exp:* Res assoc chem, Cornell Univ, 50-51; chemist, E I du Pont de Nemours & Co, 52; from instr to assoc prof, 52-63, PROF CHEM, BROWN UNIV, 63- *Concurrent Pos:* Consult, FMC Corp, 64-; fel, John Simon Guggenheim Mem Found, 67-68. *Mem:* AAAS; Am Chem Soc. *Res:* Chemistry of oxoanions and peroxides; kinetics and mechanisms of reactions. *Mailing Add:* Dept Chem Brown Univ Providence RI 02912

EDWARDS, JOHN R, b Streator, Ill, Feb 27, 37; m 61; c 3. BIOCHEMISTRY. *Educ:* Ill Wesleyan Univ, BS, 59; Univ Ill, PhD(biochem), 64. *Prof Exp:* NIH fel microbiol, Sch Med, Tufts Univ, 64-66; from asst prof to assoc prof, 66-79, PROF CHEM, VILLANOVA UNIV, 79-, CHMN DEPT, 80- *Concurrent Pos:* At dept microbiol, Univ Pa Dent Sch; Macromolecular Sci Dept, Smithkline Beecham, 90. *Mem:* Am Chem Soc; Am Soc Biol Chemists & Molecular Biol; Sigma Xi. *Res:* Structure and synthesis of dextrans and levans from cariogenic Streptococcus; characterization of the capsular polysaccharides from Actinomyces viscosus. *Mailing Add:* Dept Chem Villanova Univ Villanova PA 19085

EDWARDS, JOHN S, b Auckland, NZ, Nov 25, 31; m 57; c 4. ZOOLOGY, NEUROBIOLOGY. *Educ:* Univ Auckland, BSc, 54, MSc, 56; Cambridge Univ, PhD(zool), 60. *Prof Exp:* Sci officer insect physiol, Sch Agr, Cambridge Univ, 60-61; res assoc, Western Reserve Univ, 61-62, asst prof biol, 62-67; assoc prof, 67-70, dir biol prog, 82-88, PROF ZOOL, UNIV WASH, 70- *Concurrent Pos:* Vis fel, Gonville & Caius Col, Cambridge, Eng, 90. *Honors & Awards:* Alexander Von Humboldt Found Award, 81; Fel, AAAS. *Mem:* Am Soc Zool; Royal Entom Soc; Soc Neurosci; Entom Soc Am. *Res:* Neurobiology; insect nervous system, development, aging and regeneration; alpine ecology. *Mailing Add:* Dept Zool Univ Wash Seattle WA 98195

EDWARDS, JONATHAN, JR, b Richmond, Va, Jan 19, 33; m 59; c 4. STRUCTURAL GEOLOGY. *Educ:* Va Polytech Inst & State Univ, 55, MS, 60; Colo Sch Mines, DSc(geol), 66. *Prof Exp:* Field geologist, Va Dept Hwys, 59-61; GEOLOGIST, MD GEOL SURV, 66- *Mem:* Geol Soc Am. *Res:* Field investigation in the western Piedmont of Maryland to determine the stratigraphy, structure and geologic history. *Mailing Add:* Md Geol Surv Johns Hopkins Univ 2300 St Paul St Baltimore MD 21218

EDWARDS, JOSEPH D, JR, b Alexandria, La, Nov 25, 24; m 60; c 1. ORGANIC CHEMISTRY. *Educ:* La Col, BS, 44; Univ Tex, MA, 48, PhD(chem), 50. *Prof Exp:* Chemist, Oak Ridge, Tenn, 44; chemist, US Naval Res Lab, 45; fel, Univ Tex, 46-50; asst prof chem, Col Med, Baylor, 51-58; assoc prof, Clemson Univ, 59; sr res chemist, Monsanto Chem Co, Tex, 59-60; from assoc prof to prof chem, Lamar State Col, 60-67; head dept, 67-72, PROF CHEM, UNIV SOUTHWESTERN LA, 67- *Concurrent Pos:* Postdoctoral fel, Univ Ill, 50-51; prin scientist, Res Div, US Vet Admin Hosp, Houston, 51-58. *Mem:* Am Chem Soc; Royal Soc Chem London; Sigma Xi. *Res:* Alkaloids; plant pigments; synthetic organic chemistry. *Mailing Add:* Dept Chem Univ Southwestern La PO Box 44370 Lafayette LA 70504-4370

EDWARDS, JOSHUA LEROY, b Jasper, Fla, Aug 9, 18; m 53; c 3. PATHOLOGY, IMMUNOLOGY. *Educ:* Univ Fla, BS, 39; Tulane Univ, MD, 49. *Prof Exp:* Intern, Baptist Host, New Orleans, La, 43-44; resident, Touro Infirmary, 48-49; from resident to chief resident path, NE Deaconess Hosp, Boston, Mass, 49-51; instr, Duke Univ, 51-52, assoc, 52-53; asst prof, Rockefeller Inst, 53-54; prof path & chmn dept, Med Col, Univ Fla, 55-67; prof path & chmn dept, Med Col, Univ Fla, 55-67; prof path & dir combined degree prog med sci, Ind Univ, Bloomington, 67-69; prof path & chmn dept, Med Ctr, 69-80, PROF PATH, SCH MED, IND UNIV, 80- *Concurrent Pos:* Instr, Harvard Univ, 49-51. *Res:* Radiation pathology; cellular distribution of antigens; growth and development of cells and tissues; dynamics of antibody formation; cytology; immunopathology. *Mailing Add:* 9302 Robin Lane Indianapolis IN 46240

EDWARDS, KATHRYN LOUISE, b Washington, Pa, May 8, 47. PLANT PHYSIOLOGY. *Educ:* Oberlin Col, AB, 69; Univ NC, PhD(bot), 74. *Prof Exp:* Assoc plant physiol, Yale Univ, 74-76; asst prof biol, Rollins Col, 76-78; PHYSIOL, ASST PROF BIOL, KENYON COL, 78- *Mem:* Sigma Xi; Am Soc Plant Physiologists; AAAS. *Res:* Mechanism of auxin and abscesic acid uptake in root cells; investigation of the distribution, translocation, and metabolism of abscesic acid, auxin, gibberrellin, and ethylene to determine their roles in the geo response of roots. *Mailing Add:* Dept Biol Kenyon Col Gambier OH 43022

EDWARDS, KENNETH WARD, b Ann Arbor, Mich, Jan 18, 33; m 56; c 2. TRACE ANALYSIS, ENVIRONMENTAL CHEMISTRY & GEOCHEMISTRY. *Educ:* Univ Mich, BS, 54; Dartmouth Col, MA, 56; Univ Colo, PhD(kinetics, thermochem), 63. *Prof Exp:* Instr phys chem, Colo Sch Mines, 57-60; res chemist, US Geol Surv, 61-66; asst prof, 66-72, ASSOC PROF ANALYTICAL & PHYS CHEM, COLO SCH MINES, 72- *Concurrent Pos:* Chmn bd dirs, Nat Resources Lab, Inc, 70-86. *Mem:* Am Chem Soc. *Res:* Analysis of trace concentrations of radionuclides in water and the physical chemistry and geochemistry of natural radionuclides and other trace elements in water; analytical and physical chemistry of trace metals and metalloids in the environment, especially problems in geochemical and pollution analysis and pollution abatement. *Mailing Add:* 180 Estes St Lakewood CO 80226-1256

EDWARDS, KENNETH WESTBROOK, b Lansing, Mich, July 22, 34; m 60, 76; c 2. ELEMENTARY PARTICLE PHYSICS. *Educ:* Univ Mich, BSE, 56; Princeton Univ, PhD(physics), 61. *Prof Exp:* Asst physics, Eng Res Inst, Univ Mich, 54-56; asst, Princeton Univ, 56-61, instr, 61; res assoc, Univ Iowa, 61-63, asst prof, 63-67; assoc prof, 67-78, PROF PHYSICS, CARLETON UNIV, 78- *Concurrent Pos:* Mem coun, Inst Particle Physics, Can, 73-76. *Mem:* Am Phys Soc. *Res:* Scattering theory; experimental elementary particle physics, particularly electron position annihilation at ten gev with ARGUS collaboration at Deutsches Electronen-Synchrotron, W Germany. *Mailing Add:* Dept Physics Carleton Univ Ottawa ON K1S 5B6 Can

EDWARDS, LAWRENCE JAY, b Cornwall, NY, July 13, 40; m 62; c 3. PHYSIOLOGY, BIOCHEMISTRY. *Educ:* State Univ NY Albany, 62; Cornell Univ, MS, 65, PhD(insect physiol), 67. *Prof Exp:* Res asst insect physiol, Cornell Univ, 63-67; res entomologist, USDA, Ga, 67-68; res assoc biochem, O'Donnell Res Lab, NY, 68; asst prof insect physiol, toxicol & apicult, Univ Mass, Amherst, 68-73; textbook publ, C V Mosby & John Wiley & Sons, 74-78; PRES, BIO-CONTROL, 79- *Mem:* NY Acad Sci; Entom Soc Am. *Res:* Urban entomology; pest control; biological consulting; inspection services; integrated pest control; quality control. *Mailing Add:* Bio-Control 639 Bridge St Suffield CT 06078

EDWARDS, LEILA, b San Juan, PR, Apr 9, 37; US citizen; m 68; c 2. CLINICAL CHEMISTRY. *Educ:* Univ PR, BS, 58; Kans State Univ, MS, 61; State Univ NY, Buffalo, PhD(biochem), 66; Am Bd Clin Chem, dipl, 77. *Prof Exp:* Instr chem, Col Agr & Mech Arts, Univ PR, 58-59, res assoc cancer res, Sch Med, 61-62; fel biochem, Ore State Univ, 66-67; res assoc biochem, State Univ NY, Buffalo, 67-69; supvr clin chem, Erie County Lab, Med Ctr, 69-70, asst dir, 70-84, dir clin chem & toxicol, 84-90; CLIN CHEMIST, MDS LABS, 90- *Concurrent Pos:* Clin asst prof biochem, State Univ NY, Buffalo, 70-, clin asst prof med, 85-90 & pathol, 71-89, clin assoc prof pathol, 89-; consult clin chem, Buffalo Children's Hosp, 74-76; mem, Comn Radionuclides & Radioassay Clin Chem, Am Asn Clin Chem, 74-77, Nat Educ Comt, 75-77; head study group, Thyroxine Standards, 78-80; counr, Am Asn Clin Chem, 87-90, Long Ranger Planning Comn, 88-89, Legis Liaison, 85-; bd dirs, Nat Acad Clin Biochem, 88- *Honors & Awards:* Somogyi Sendror Award, Am Asn Clin Chem. *Mem:* Am Asn Clin Chem; Am Chem Soc; NY Acad Sci; Am Asn Clin Lab Physicians & Scientists; Sigma Xi; Nat Acad Clin Biochem. *Res:* clinical endocrinology; malignant hyperthermia; toxicology. *Mailing Add:* 20 Parkwood Dr Amherst NY 14226

EDWARDS, LEON ROGER, b New Ulm, Minn, May 2, 40; m 62; c 2. SOLID STATE PHYSICS. *Educ:* Univ Minn, BPhys, 62; Iowa State Univ, PhD(physics), 67. *Prof Exp:* Staff mem, orgn, 5131, 67-69, STAFF MEM, PHYSICS SOLIDS RES DIV, 5132, SANDIA CORP, 69- *Mem:* Am Phys Soc. *Res:* Low temperature physics; transport properties of rare earth metals and dilute alloys; equation of state. *Mailing Add:* Sandia Lab 2153 PO Box 5800 Albuquerque NM 87115

EDWARDS, LESLIE ERROLL, b Montesano, Wash, Dec 26, 14; m 46; c 3. GASTROINTESTINAL RESEARCH, MUSCLE METABOLISM. *Educ:* State Col Wash, BS, 37, MS, 39; Univ Rochester, PhD(physiol), 44. *Prof Exp:* Instr, Univ Rochester, 43-46 & Fels Res Found, 46-47; assoc, 47-49, from asst prof to assoc prof, 49-64, prof, 64-79, EMER PROF PHYSIOL, MED COL VA, 79- *Mem:* AAAS; Am Physiol Soc; Soc Gen Physiol; Am Chem Soc; Sigma Xi; Am Inst Biol Sci. *Res:* Lowering of ionic barriers in tissue; intermediary metabolism of carbohydrate and fat; biological value of proteins; gastric secretion; gastrointestinal hormones, insulin; glucagon; pancreatic function; muscle metabolism. *Mailing Add:* 1990 Old Hanover Rd Sandston VA 23150

EDWARDS, LEWIS HIRAM, b Frederick, Okla, Nov 6, 38; m 60; c 2. GENERAL AGRICULTURE, PHYTOPATHOLOGY. *Educ:* Okla State Univ, BS, 61; NDak State Univ, PhD(agron), 65. *Prof Exp:* Res geneticist, Agr Res Serv, USDA, 65-67; asst prof, 67-70, assoc prof, 70-73, PROF AGRON, OKLA STATE UNIV, 73-, GENETICIST-PLANT BREEDER, 67- *Concurrent Pos:* Res coordr, USAID proj, Morocco, 87-90. *Mem:* Am Soc Agron; Crop Sci Soc Am. *Res:* Genetic research in soybeans, including mutation studies, quantitative genetic studies; soybean and mungbean. *Mailing Add:* Dept Agron Okla State Univ Stillwater OK 74078

EDWARDS, LOUIS LAIRD, JR, b Bozeman, Mont, June 20, 36; m 58; c 2. CHEMICAL ENGINEERING. *Educ:* Rensselaer Polytech Inst, BChE, 58; Univ Del, MChE, 60; Univ Idaho, PhD(chem eng), 66. *Prof Exp:* From instr to asst prof chem eng, Univ Idaho, 61-66; Ford Found resident, Union Carbide Corp, 66-67; assoc prof chem eng, 67-71, PROF CHEM ENG, UNIV IDAHO, 71- *Concurrent Pos:* Guest researcher, Swed Forest Prod Lab, 71-72; consult, Pulp & Paper Indust, 74- *Mem:* Am Inst Chem Engrs; Am Soc Eng Educ; Tech Asn of Pulp & Paper Indust. *Res:* Application of chemical engineering principles to pulp and paper processes; process design and economics; mathematical modeling; optimization and computer applications; ozone technology. *Mailing Add:* Dept Chem Eng Univ Idaho Moscow ID 83843

EDWARDS, LUCY ELAINE, b Richmond, Va, Feb 28, 52. PALEONTOLOGY, BIOSTRATIGRAPHY. *Educ:* Univ Ore, BA, 72; Univ Calif, Riverside, PhD(geol sci), 77. *Prof Exp:* GEOLOGIST PALEONT, US GEOL SURV, 77- *Mem:* Am Asn Stratig Palynologists; Geol Soc Am; Paleont Soc. *Res:* Cenozoic and Mesozoic dinoflagellates; techniques of biostratigraphic correlation. *Mailing Add:* US Geol Surv 970 National Ctr Reston VA 22092

EDWARDS, MCIVER WILLIAMSON, JR, b Darlington, SC, Aug 24, 35; m 63; c 2. PHYSIOLOGY, ANESTHESIOLOGY. *Educ:* Mass Inst Technol, BS, 56; Univ Pa, MD, 62, Am Bd Anesthesiol, cert, 73. *Prof Exp:* Intern & instr med, Johns Hopkins Hosp, 62-63; instr physiol, 63-64, assoc, 64-65, 67-68, resident, 69-71, asst prof physiol, 68-80, asst prof anesthesia, 71-80, ASSOC PROF ANESTHESIA, MED SCH, UNIV PA, 80-; CHIEF, ANESTHESIA SECT, VET ADMIN MED CTR, PHILADELPHIA, 80- *Concurrent Pos:* Pa Plan fel, Univ Pa, 63-65, 67-68; USPHS fel, Middlesex Hosp Med Sch, Eng, 65-67; vis res fel physiol, Oxford Univ, 77-78. *Mem:* AAAS; Sigma Xi; Soc Cardiovasc Anesthesiologists; fel Am Col Anesthesiologists; Am Soc Anesthesiol; Asn Vet Admin Anesthesiologists (pres); Soc Technol Anesthesia; Asn Advan Med Instrumentation; Int Anesthesia Res Soc. *Res:* Respiratory physiology; neurophysiology; control of breathing; effects of anesthesia on respiration; physiological monitoring. *Mailing Add:* Anesthesia Sect 112 Vet Admin Med Ctr Philadelphia PA 19104

EDWARDS, MARC BENJAMIN, b New York, NY, Dec 19, 46; m 76; c 3. SEDIMENTARY GEOLOGY. *Educ:* City Col New York, BS, 68; Oxford Univ, DPhil(geol), 72. *Prof Exp:* Sr res scientist sedimentology, Continental Shelf Inst Norway, 72-78; res scientist, Bur Econ Geol, Univ Tex, Austin, 78-81; explor geologist, Kerr, Jain & Assoc, Houston, Tex, 81-82; CONSULT GEOLOGIST, 82- *Honors & Awards:* Sproule Award, Am Asn Petrol Geologists. *Mem:* Soc Econ & Paleont Mineralogists; Int Asn Sedimentologists; Am Asn Petrol Geologists; Geol Soc Am. *Res:* Facies analysis and reconstruction of depositional environments in ancient sedimentary rocks; interpretation of basin history, paleogeography and paleoclimatology; Spitzbergen, Arctic Norway and Texas; growth faulting and sedimentation, northern Gulf Basin. *Mailing Add:* 5430 Dumfries Houston TX 77096

EDWARDS, MARTIN HASSALL, b St Annes-on-Sea, Eng, Nov 10, 27; nat Can; m 49; c 2. PHYSICS. *Educ:* Univ BC, BA, 49, MA, 51; Univ Toronto, PhD(physics), 53. *Prof Exp:* Res assoc, Univ Toronto, 53-54; from asst prof to assoc prof, Royal Mil Col, Can, 54-61, prof physics, 61, chmn dept, 78-86; RETIRED. *Concurrent Pos:* Res assoc, Stanford Univ, 64-65; Ont Royal Comnr ducks & pesticides, 69-70; counr, Int Union Conserv Nature & Natural Resources, 75-81; mem, Environ Assessment Bd Ont, 77-86. *Mem:* Fel Am Phys Soc; Can Nature Fedn (vpres, 74-77, pres, 77-79); Can Asn Physicists. *Res:* Low temperature physics; low temperature properties of helium liquid and vapor; expansion coefficient; refractive index and density; lambda point and critical point. *Mailing Add:* Physics Dept Royal Mil Col Kingston ON K7K 5L0 Can

EDWARDS, MERRILL ARTHUR, b Amherst, NS, May 12, 32; m 56; c 2. PHYSICS, BIOPHYSICS. *Educ:* Univ NB, BSc, 53; Univ Western Ont, MSc, 56, PhD, 60. *Prof Exp:* Assoc prof, 59-74, asst dean grad studies, 74-78, PROF PHYSICS, UNIV NB, 74-, CHMN PHYSICS DEPT, 77-, ACTG ASSOC DEAN GRAD STUDIES, 78- *Concurrent Pos:* Nat Res Coun Overseas fel, 66-67. *Mem:* Biophys Soc; Can Asn Physicists. *Res:* Geophysics; peripheral circulation; use of radioactive clearance methods in determining the circulation; spectroscopy; surface physics of microspheres. *Mailing Add:* Dept Physics Univ NB PO Box 4400 Fredericton NB E3B 5A3 Can

EDWARDS, MILES JOHN, b Portland, Ore, 1929; m 56; c 4. MEDICINE. *Educ:* Willamette Univ, BA, 51; Univ Ore, MS & MD, 56; Am Bd Internal Med, dipl, 64; Am Bd Pulmonary Dis, dipl, 69. *Prof Exp:* From asst prof to assoc prof, 64-70, chief, 70-83, PROF MED, DIV PULMONARY & CRIT CARE MED, SCH MED, UNIV ORE, 70- *Concurrent Pos:* NIH fel, Cardiovasc Res Inst, Med Ctr, Univ Calif, San Francisco, 63-64. *Mem:* Am Thoracic Soc; Am Col Chest Physicians. *Res:* Respiratory physiology, alterations in blood and oxygen affinity with various types of hypoxia. *Mailing Add:* Dept Internal Med Univ Ore Portland OR 97201

EDWARDS, NANCY C, b Montgomery, Ala, Oct 16, 36. DEVELOPMENTAL NEUROBIOLOGY. *Educ:* Agnes Scott Col, BA, 58; Univ NC, Chapel Hill, MA, 66, PhD(zool), 71. *Prof Exp:* Asst dir pub rels & develop, Agnes Scott Col, 58-59, dir publicity, 59-61; from instr to assoc prof, 68-82, PROF BIOL, UNIV NC, CHARLOTTE, 82- *Mem:* AAAS; Am Soc Zool; Soc Develop Biol; Sigma Xi. *Res:* Coelenterate development; metamorphosis. *Mailing Add:* Dept Biol Univ NC Charlotte NC 28223

EDWARDS, OGDEN FRAZELLE, b Leslie, Mich, Apr 26, 09; m 32; c 2. MICROBIOLOGY. *Educ:* Mich State Col, BS, 31, MS, 33; Yale Univ, PhD(bact), 36. *Prof Exp:* Asst bact, Mich State Col, 31-33; asst bacteriologist, Yale Univ, 33-36; instr, Univ Ill, 36-42; asst prof, 46-47, assoc prof, 47-74, EMER ASSOC PROF BACT, UNIV KY, 74- *Res:* Electron microscopy of viruses; bacterial enzymes; actinomycetes. *Mailing Add:* 308 Colony Blvd Lexington KY 40502-2505

EDWARDS, OLIVER EDWARD, b Wales, Jan 8, 20; Can citizen; m 45; c 2. ORGANIC CHEMISTRY. *Educ:* Univ Alta, BSc, 41; Northwestern Univ, MS, 43, PhD(chem), 48. *Prof Exp:* CHEMIST, ORG CHEM & ALKALOIDS, NAT RES COUN CAN, 48- *Mem:* Am Chem Soc; Can Inst Chem; fel Royal Soc Can. *Res:* Reactions of dihydropyran and tetrahydropyran derivatives, electron deficient carbon and nitrogen; studies on the constitution of the Aconite alkaloids, chemistry of diterpenes; n-heterocyclics. *Mailing Add:* 678 Portage Ave Ottawa ON K1G 1T4 Can

EDWARDS, OSCAR WENDELL, b Marion, Ala, Jan 2, 16; m 43; c 1. PHYSICAL CHEMISTRY. *Educ:* Birmingham-Southern Col, BS, 36; Univ Ala, MS, 41; Emory Univ, 59. *Prof Exp:* Teacher pub schs, 36-41, 43-44; chem aide, Tenn Valley Authority, 41-42, jr anal chemist, 42-45, asst anal chemist, 45-46, res chemist III, 46-56, res chemist IV, 56-81; RETIRED. *Concurrent Pos:* Researcher ion-selective electrodes, Univ N Ala, Florence; chemist, pollution abatement, Graves Plating Co, 82-87. *Mem:* Am Chem Soc. *Res:* Determination of phosphorus in phosphatic materials; microdetermination of phosphorus by an organic reagent; diffusion of phosphates and phosphoric acid by conductimetric and optical methods; measurement of dissociation and stability constants; calculation of composition and activities in multicomponent solutions by iterative methods using computer. *Mailing Add:* 506 Riverview Dr Florence AL 35630

EDWARDS, PALMER LOWELL, b Enterprise, Ala, Mar 9, 23; m 64, 83. PHYSICS. *Educ:* La State Univ, BS, 44; Harvard Univ, SM, 47; Univ Md, PhD, 58. *Prof Exp:* Physicist, Naval Ord Lab, 44-48, res assoc, 49-55, solid state physicist, 55-60; from assoc prof to prof, Tex Christian Univ, 60-67; chmn physics fac, Univ WFla, 67-76, prof physics, 67-90; RETIRED. *Concurrent Pos:* Physicist, Naval Res Lab, 78-80, Naval Surface Weapons Ctr, 81-83. *Mem:* Am Phys Soc; Acoust Soc Am. *Res:* Magnetism; properties of materials; acoustics. *Mailing Add:* 8607 Scenic Hills Dr Pensacola FL 32514

EDWARDS, RICHARD ARCHER, paleontology; deceased, see previous edition for last biography

EDWARDS, RICHARD GLENN, b Harlan, Ky, Mar 30, 40; m 66; c 1. ENVIRONMENTAL ENGINEERING. *Educ:* Univ Ky, BS, 62, MS, 64, PhD(bioeng), 70. *Prof Exp:* Res asst, Wenner-Gren Lab, Univ Ky, 62-64 & 66-74; develop engr, Tenn Eastman Corp, 65-66; vpres AME Technol, Inc, 74-75; PRIN ENGR, WATKINS & ASSOC, INC, 75- *Mem:* Sigma Xi; Am Soc Mech Engrs; Biomed Eng Soc; Inst Environ Sci. *Res:* Response of man and animals to noise, vibration and acceleration; mobile health screening systems; audiometry, energy and environmental engineering. *Mailing Add:* 3333 Crown Crest Rd Lexington KY 40517

EDWARDS, RICHARD M(ODLIN), b Wilmington, Del, Sept 6, 20; m 43; c 2. CHEMICAL ENGINEERING. *Educ:* Purdue Univ, BS, 41; Univ Wash, MS, 48; Univ Ariz, PhD(chem eng), 64, EChem, 74. *Prof Exp:* Chem supvr, E I du Pont de Nemours & Co, 42-44; chem engr, Mallinckrodt Chem Works, 48-52, asst to tech dir, 52-54, from asst mgr to mgr process develop, 54-59; from instr to assoc prof chem eng, 59-64, asst to dean, 64-67, from asst dean to actg dean, 67-71, PROF CHEM ENG, UNIV ARIZ, 64-, VPRES STUDENT RELS, 71- *Concurrent Pos:* Consult, Am Potash & Chem Co, 66-71; assoc dean, Col Mines, Univ Ariz, 71. *Mem:* Am Inst Chem Engrs; Am Soc Eng Educ; Am Inst Mining, Metall & Petrol Engrs; Am Chem Soc. *Res:* Technology of uranium production; fluidized bed heat transfer; chemical separation processes; liquid-liquid extraction. *Mailing Add:* 8140 E Moonstone Tucson AZ 85715-9796

EDWARDS, ROBERT LEE, b Barnardsville, NC, Jan 21, 22; m 51; c 1. ANIMAL NUTRITION. *Educ:* Berea Col, BS, 46; NC State Univ, MS, 54, PhD(animal Indust), 58. *Prof Exp:* Teacher pub schs, NC, 46-52; asst agr ed, NC State Col, 52-53 & animal indust, 54-58; asst prof & asst animal husbandman, Clemson Col, 58-64, assoc prof, 64-79, prof, 79-86, EMER PROF, CLEMSON UNIV, 86- *Mem:* AAAS; Am Soc Animal Sci. *Res:* Nutrition of large animals; utilization of dietary lipids in ruminants; forage utilization by cattle and sheep; agricultural waste utilization; nitrogen requirements of equines. *Mailing Add:* Dept Animal Sci Clemson Univ Clemson SC 29634-0361

EDWARDS, ROBERT LOMAS, b Philadelphia, Pa, Aug 24, 20; m 42; c 4. RESEARCH ADMINISTRATION, ECOLOGY. *Educ:* Colgate Univ, BA, 47; Harvard Univ, AM, 49, PhD(biol), 51. *Prof Exp:* Instr, Tufts Univ, 49-50; in chg, Arctic Res Prog, 50, 53 & 54; instr, Air Staff & Command Col, Maxwell Field, Ala, 54; chief indust fishery invests, Fish & Wildlife Serv, 55-59; asst dir, Woods Hole Fisheries Res Lab, 59-69, asst dir plans & progs, Bur Com Fisheries, Washington, DC, 70-72; dir, Northeast Fisheries Ctr Hq, Woods Hole, 72-84; CONSULT, 85-; RES ASSOC, USNM. *Concurrent Pos:* Mem & coordr, Working Group US-USSR Studies Biol Productivity & Biochem World Ocean, 74-80; deleg, Int Coun Exlor Seas, 75-85; sci adv fishery negotiations USSR, Poland, Int Comn Northwest Atlantic Fisheries; lectr, Yale Univ, 80; adj prof oceanog, Univ RI, 80-85; mem, Comm Ecol Int Union Conserv of Nature, 81-87; deleg, Int Court of Justice the Hague, 85. *Honors & Awards:* Gold Medal, Dept of Com, 77; Medal of Hon, Dept State. *Mem:* Soc Am Arachnol; Am Soc Mammal; Wilson Ornith Soc; Sigma Xi; Am Soc Parasitol. *Res:* Vertebrate ecology, especially marine; systematics and evolution of bird parasites; growth of fish; ecology of spiders. *Mailing Add:* Box 505 Woods Hole MA 02543

EDWARDS, ROBERT V(ALENTINO), b Baltimore, Md, Dec 15, 40; m 62; c 2. CHEMICAL ENGINEERING. *Educ:* Johns Hopkins Univ, AB, 62, MS, 64, PhD(chem eng), 68. *Prof Exp:* From res assoc to sr res assoc, Case Western Reserve Univ, 68-70, from asst prof to assoc prof, 70-79, chmn chem eng, 85-90, PROF CHEM ENG, CASE WESTERN RESERVE UNIV, 79-, ASSOC DEAN ENG, 90- *Mem:* AAAS; Am Inst Chem Engrs; Am Chem Soc; Am Optic Soc; Am Phys Soc. *Res:* Measurement of the scalar transport properties of moving fluids and velocity measurements by light scattering; complex transport phenomena, mixing and chemical reaction; computer applications in measurements. *Mailing Add:* 2525 Edgehill Rd Cleveland OH 44106

EDWARDS, ROY LAWRENCE, b Southampton, Eng, Dec 2, 22; Can citizen; m 49; c 3. ECOLOGY. *Educ:* Oxford Univ, BA, 50, MA & PhD(entom), 52. *Prof Exp:* Lectr entom, Univ Hull, Eng, 52-57; Nat Res Coun Can fel, 57-58; res officer, Can Agr Res Lab, 58-61; asst prof biol, Univ Sask, 61-64; assoc prof, Trent Univ, 64-66, chmn dept, 66-69 & 79-83, actg vpres, 83-84, prof biol, 66-87; RETIRED. *Mem:* Brit Ecol Soc; Entom Soc Can. *Res:* Ecology and behavior of invertebrates. *Mailing Add:* Dept Biol Trent Univ Peterborough ON K9J 7B8 Can

EDWARDS, STEVE, b Quincy, Fla, June 16, 30; m 64; c 2. THEORETICAL NUCLEAR PHYSICS. *Educ:* Fla State Univ, BS, 52, MS, 54; Johns Hopkins Univ, PhD(theoret physics), 60. *Prof Exp:* From asst prof to assoc prof, 60-69, assoc chmn dept, 65-73, chmn Dept Physics, 73-79, PROF PHYSICS, FLA STATE UNIV-TALLAHASSEE, 69-, DEAN FAC & DEP PROVOST, 85- *Concurrent Pos:* Grad asst, Fla State Univ-Tallahassee, 52-55, pres fac senate, 83-85, chmn athletic bd, 80; mem adv comt, Intermidiate Sci Curriculum Study; chmn, Am Phys Soc Southeaster Sect, 82-83. *Honors & Awards:* Coyle E Moore Award, 65. *Mem:* Am Asn Physics Teachers; Am Phys Soc; Sigma Xi. *Res:* Group theoretic analysis of vibration problems; direct nuclear reaction theories, especially stripping and pick-up reactions of all types in low-energy nuclear physics; author of texbooks on general physics and also articles on physics. *Mailing Add:* Dean Faculties Fla State Univ Tallahassee FL 32306

EDWARDS, SUZAN, b Columbia, Mo, June 15, 51; m 73. ASTRONOMY. *Educ:* Dartmouth Col, BA, 73; Univ Hawaii, MS, 75, PhD(astron), 80. *Prof Exp:* ASST PROF ASTRON, SMITH COL, 80- *Mem:* Am Astron Soc; Sigma Xi; Astron Soc Pac. *Res:* Star formation. *Mailing Add:* Five Col Astron Dept Smith Col Northampton MA 01063

EDWARDS, TERRY WINSLOW, b Sheboygan, Wis, Nov 2, 35; m 58; c 1. ASTRONOMY. *Educ:* Univ Wis, BS, 58, MS, 61, PhD(astron), 68. *Prof Exp:* Satellite observer, Astrophys Observ, Smithsonian Inst, 58-59; res asst physics, Midwestern Univs Res Asn, 64-66; from instr to asst prof, 66-71, ASSOC PROF PHYSICS & ASTRON, UNIV MO, 71- *Concurrent Pos:* NASA res grant, Space Sci Res Ctr, Univ Mo, 66-69 & fel, Ames & Stanford Univs, 70-71; Aspen Ctr Physics, 77; vis assoc prof, Univ Rochester, 81-82. *Mem:* AAAS; Am Astron Soc; Int Astron Union; Sigma Xi. *Res:* Astrophysics; stellar thermodynamics; stellar structure; nucleosynthesis; digital computing; dense stars theory; general relativity. *Mailing Add:* Dept Physics & Astron Univ Mo Columbia MO 65211

EDWARDS, THOMAS CLAUDE, b San Antonio, Tex, July 10, 43; m 66; c 2. THERMODYNAMICS, MACHINE DESIGN. *Educ:* NMex State Univ, BSME, 66, MSME, 67; Purdue Univ, PhD(mech eng), 70. *Prof Exp:* Reactor engr, US AEC, 66-67; instr eng, Purdue Univ, 67-70; asst prof eng, Fla Technol Univ, 70-72; pres, 72-78, CHMN BD RES & DEVELOP MGT, ROVAC CORP, 78-, PRES ROVAC CORP; CHIEF EXEC OFFICER, INNOVATION TECHNOLOGIES, INC, 90- *Mem:* Am Soc Heating, Refrig & Air Conditioning Engrs; Am Soc Metals. *Res:* Thermodynamic cycles; invented the Edwards air vapor air conditioning cycle; positive displacement totaling compression-expansion machinery. *Mailing Add:* Innovation Technol Inc 1426 Gleneagles Way Rockledge FL 32955

EDWARDS, THOMAS F, b Pittsfield, Ill, July 17, 27; m 50; c 3. SCIENCE EDUCATION. *Educ:* Ill State Univ, BS, 51; Ariz State Univ, MA, 57; Mich State Univ, EdD(sci educ), 66. *Prof Exp:* Teacher high schs, Ill, 51-57; from instr to assoc prof chem, 57-70, prof sci educ, 72-77, PROF ELEM EDUC, ILL STATE UNIV, 70- *Concurrent Pos:* Instr, Mich State Univ, 65-66; dir NSF Insts, Ill State Univ, 71-; sci consult, Ill State Dept Educ, 71-74, Lincoln High Sch, Ill, 72-75, Ottawa Pub Schs, Ill, 75 & Rand McNally Publ Co; US Dept Educ Title III Elem & Secondary Educ Act Proj validator, 73. *Mem:* Nat Sci Teachers Asn; Nat Asn Res Sci Teaching. *Mailing Add:* 514 Snead Lane Fairfield Glade TN 38555

EDWARDS, THOMAS HARVEY, b Chilliwack, BC, Feb 12, 24; nat US; m 46; c 4. PHYSICS. *Educ:* Univ BC, BA, 47, MA, 48; Univ Mich, PhD(physics), 55. *Prof Exp:* Instr, Univ Mich, 51-53, res assoc, 53-54; from asst prof to assoc prof, 54-65, PROF PHYSICS, MICH STATE UNIV, 65- *Mem:* Optical Soc Am; Am Phys Soc. *Res:* High-resolution infrared spectroscopy; molecular structure of asymmetric and symmetric top molecules. *Mailing Add:* Dept Physics Mich State Univ East Lansing MI 48824

EDWARDS, VICTOR HENRY, b Galveston, Tex, Oct 17, 40; m 63; c 2. CHEMICAL ENGINEERING, MOLECULAR BIOLOGY. *Educ:* Rice Univ, BA, 62; Univ Calif, Berkeley, PhD(chem eng), 67. *Prof Exp:* Asst prof chem eng, Cornell Univ, 67-73; res fel, Merck & Co, Inc, 73-76; supv res eng, United Energy Resources, Inc, 76-78; process engr, 79, PROJ ENGR, ALLSTATES DESIGN & DEVELOP CO, 84- *Concurrent Pos:* Assoc prog dir eng chem prog, NSF, 71, prog mgr advan technol appln, 71-72; UNESCO lectr, Japan, 72; vis prof environ eng, Rice Univ, 79-80; consult, Edwards & Assocs, 79-, Harding-Lawson Assocs & Process Comput, Inc; sr process engr, Fluor Engrs & Constructors, 80-82; ed Southwest, Plant Serv, 82-85. *Honors & Awards:* Robert L Churchwell Award, Am Inst Chem Engrs, 81. *Mem:* Am Chem Soc; Am Inst Chem Engrs; NY Acad Sci; AAAS. *Res:* Manufacturing and environmental processes involving enzymes, microorganisms and plants; separation and purification technology; biomass resources; pollution control technology; chemical plant safety. *Mailing Add:* The Woodlands Houston TX 77387-8097

EDWARDS, W FARRELL, b Logan, Utah, Oct 5, 31; m 55. ELECTROMAGNETISM, PLASMA PHYSICS. *Educ:* Univ Utah, BS, 55; Calif Inst Technol, MS, 57, PhD(physics), 60. *Prof Exp:* From asst prof to assoc prof, 59-66, head dept, 66-72, PROF PHYSICS, UTAH STATE UNIV, 66- *Mem:* Am Phys Soc. *Res:* Electromagnetic theory of charged fluids with applications in plasmas, superconductors and quantum systems. *Mailing Add:* Dept Physics Utah State Univ Logan UT 84322

EDWARDS, W STERLING, b Birmingham, Ala, July 23, 20; m 46; c 4. SURGERY. *Educ:* Va Mil Inst, BS, 42; Univ Pa, MD, 45; Am Bd Surg, dipl, 54; Am Bd Thoracic Surg, dipl, 59. *Prof Exp:* From intern to resident surg, Mass Gen Hosp, Boston, 45-52; instr, Med Col Ala, 52-53, from asst prof to prof, 53-69; chmn cardiothoracic div, Sch Med, Univ NMex, 69-74, prof surg & chmn dept, 69-87; RETIRED. *Concurrent Pos:* USPHS res fel, Western Reserve Univ, 50-51; consult, Vet Admin Hosp, Albuquerque. *Mem:* Soc Vascular Surg; AMA; Am Col Surg. *Res:* Cardiovascular surgery and physiology; development of arterial and heart valve substitutes. *Mailing Add:* 2312 Hannett NE Albuquerque NM 87106

EDWARDS, WILLARD, b Virginia Beach, Va, Nov 4, 31. COMPUTER SCIENCE. *Educ:* Hampton Univ, BS, 61. *Prof Exp:* COMPUTER EQUIP ANALYST, NAVY COMPUTER & TELECOMMUN STA, 84-; VPRES, G&E SYSTS INC, 91- *Mem:* Asn Comput Mach; Asn Women in Comput. *Mailing Add:* Navy Computer & Telecommun Sta Wash Navy Yard Washington DC 20374

EDWARDS, WILLIAM BRUNDIGE, III, b Philadelphia, Pa, Oct 10, 42. FLAVOR CHEMISTRY, NATURAL PRODUCT CHEMISTRY. *Educ:* Lehigh Univ, BA, 64; Univ Pa, PhD(org chem), 69. *Prof Exp:* Res asst org chem, Wyeth Labs, Inc, 64-65; res chemist, Ravdin Inst, 65-66 & Arco Chem Corp, Atlantic Richfield Co, 66; spectroscopist, Univ Pa, 66-68; fel, Synvar Res Inst, 69-70; assoc prof org res, 71-73, res scientist, 73-80, ASSOC SR SCIENTIST & PROJ LEADER, RES CTR, PHILIP MORRIS INC, 80- *Concurrent Pos:* Teaching asst, Univ Pa, 65-67. *Mem:* AAAS; Am Chem Soc; The Chem Soc; Sigma Xi; Asn Chemoreception Sci. *Res:* Synthesis and study of perfume and flavor compounds; synthesis of terpenoid flavorants; factors governing structure-flavor relationships, use of computer systems for chemical information management; commercialization of chemical processes. *Mailing Add:* Res Ctr Philip Morris Inc PO Box 26583 Richmond VA 23261

EDWARDS, WILLIAM CHARLES, b Waukegan, Ill, May 17, 34; m 61; c 2. ECOLOGY. *Educ:* Carleton Col, BA, 56; Univ Wyo, MS, 58; Univ Nebr, PhD(bot), 66. *Prof Exp:* Teacher high sch, Wyo, 58-63; from asst prof to assoc prof biol, Mankato State Col, 66-70; INSTR BIOL, LARAMIE COUNTY COMMUNITY COL, 70- *Concurrent Pos:* Wyo state legislator, 74-82. *Mem:* Sigma Xi; Audubon Soc. *Res:* Reproduction in antelope; growth pattern in birch. *Mailing Add:* Dept Biol Laramie County Community Col 1400 E College Dr Cheyenne WY 82007

EDWARDS, WILLIAM DEAN, b Wichita, Kans, Nov 12, 48; m 69. CARDIOVASCULAR PATHOLOGY. *Educ:* Univ Kans, BA, 70, MD, 74; Am Bd Path, dipl, 78. *Prof Exp:* Assoc consult path, 78-80, CONSULT ANAT PATH, MAYO CLIN & FOUND, 80- *Mem:* AMA; Am Heart Asn; Am Col Cardiol; Sigma Xi. *Res:* Congenital heart disease; pulmonary hypertension; acquired heart disease. *Mailing Add:* 2205 Folwell Dr SW Rochester MN 55902-3406

EDWARDSON, JOHN RICHARD, b Kansas City, Mo, Apr 17, 23; m 48; c 3. GENETICS. *Educ:* Agr & Mech Col Tex, BS, 48, MS, 49; Harvard Univ, PhD(biol), 54. *Prof Exp:* From asst agronomist to assoc agronomist, 53-66, AGRONOMIST, UNIV FLA, 66- *Mem:* AAAS; Am Phytopath Soc; Am

Genetic Asn. *Res:* Light and electron microscopy applied to cytological information on plant virus induced inclusions used to classify viruses and to diagnose plant virus infections. *Mailing Add:* Dept Agron Bldg 193 Univ Fla Gainesville FL 32611

EDYVANE, JOHN, b Plymouth, Eng, Apr 28, 34; US citizen; m 56, 84; c 3. SENSOR & COMMUNICATION SYSTEMS. *Educ:* Plymouth Polytech, UK, HNC, 73; Royal Navy Eng Sch & Col, HNC, 75; S Dorset Tech Col, HNC, 78. *Prof Exp:* Engr officer, Brit Royal Navy Res & Develop, Admiralty Under Water Weapons Estab, Portland, 76-78; engr design combatant craft systs, Asset Inc, 78-88; DIR, ADVAN MARINE SYSTS, NKF ENG INC, 80- *Mem:* Am Soc Naval Engrs; Soc Reliability Engrs. *Res:* Emerging technology which is applicable to designs for nuclear submarine propulsion machinery; robotic systems, marine environment and ship systems applications. *Mailing Add:* Seven Knoll Ct Stafford VA 22554

EELLS, JAMES, b Cleveland, Ohio, Oct 25, 26; m 50; c 4. MATHEMATICS. *Educ:* Bowdoin Col, BA, 47; Harvard Univ, AM, 51, PhD(math), 54. *Prof Exp:* Instr math, Robert Col, Turkey, 47-48; instr, Amherst Col, 48-50; teaching fel, Harvard Univ, 51-53; instr, Tufts Col, 53-54; mem, Inst Advan Study, 54-56; asst prof, Univ Calif, Berkeley, 56-58 & Columbia Univ, 58-60, assoc prof, 60-63; overseas fel, Cambridge Univ, 63-64 & 66-67; prof math, Cornell Univ, 64-69; PROF MATH, UNIV WARWICK, 69-; DIR MATH, INT CTR THEORET PHYSICS, 86- *Concurrent Pos:* Mem, Inst Adv Study, 56-58, 62-63, 72, 77 & 82, Inst Hautes Etudes, France, 68, 72, 77, 80, 82 & 84 & Scusla Normale Sup, Pisa, Italy, 80, 83 & 86. *Mem:* Am Math Soc; Soc Math France; London Math Soc. *Res:* Global topological and differential geometric properties of analysis; calculus of variations. *Mailing Add:* Int Ctr Theoret Physics Miramare Boc 566 Trieste 34100 Italy

EER NISSE, ERROL P(ETER), b Rapid City, SDak, Feb 15, 40; m 72; c 3. ELECTRICAL ENGINEERING, PHYSICS. *Educ:* SDak State Univ, BSEE, 62; Purdue Univ, MSEE, 63, PhD(elec eng), 65; Univ NMex, MIAd, 74. *Prof Exp:* Staff mem res, Sandia Labs, 65-68, div supvr, Device Physics Res Div, 68-79; PRES, QUARTZTRONICS, INC, UTAH, 79- *Honors & Awards:* W G Cady Award, Inst Elec & Electronics Engrs, 83. *Mem:* Fel Inst Elec & Electronics Engrs; fel Am Phys Soc; sr mem Instrument Soc Am; Nat Soc Prof Engrs. *Res:* Ferroelectric, piezoelectric and semiconductor devices. *Mailing Add:* Quartztronics Inc 1020 Atherton Dr C Salt Lake City UT 84123

EESLEY, GARY, b Grove City, Ohio, May 6, 50; m 76; c 2. ULTRAFAST PHENOMENA, SOLID STATE PHYSICS. *Educ:* Case Western Reserve Univ, BSEE, 72; Univ Southern Calif, MSEE, 76, PhD(elec eng), 78. *Prof Exp:* Engr res, Westinghouse Elec Corp, 72-74; assoc sr res scientist, 78-80, staff res scientist, 80-86, SR STAFF RES SCIENTIST, GEN MOTORS CORP, 86- *Concurrent Pos:* Adj prof physics, Wayne State Univ, 88- *Honors & Awards:* Phys Sci Res Award, Sigma, Xi, 78. *Mem:* Inst Elec & Electronics Engrs; Optical Soc Am; Am Phys Soc. *Res:* Ultrashort pulsed laser measurement of transport processes in condensed matter; femtosecond time-resolved measurements. *Mailing Add:* Dept Physics Gen Motors Res Lab Warren MI 48090-9055

EFFER, W R, b Warrington, Eng, Jan 1, 27; Can citizen; m 53; c 2. PLANT PHYSIOLOGY. *Educ:* Univ Durham, BSc, 51; Univ Newcastle, PhD(plant physiol), 66. *Prof Exp:* Res chemist, Horlicks Ltd, Slough, Eng, 51-54; res scientist, Ont Res Found, Univ Toronto, 54-63 & 66-67; biologist, 67-70, supvr environ studies, 70-75, head, 75-79, MGR, ENVIRON STUDIES & ASSESSMENTS DEPT, ONT HYDRO, 80- *Mem:* Air Pollution Control Asn. *Res:* Biological changes associated with heat polution, acid rain. *Mailing Add:* Design & Develop Div Ont Hydro 700 Univ Ave Toronto ON M5G 1X6 Can

EFFROS, EDWARD GEORGE, b New York, NY, Dec 10, 35. MATHEMATICS. *Educ:* Mass Inst Technol, SB, 56; Harvard Univ, AM, 58, PhD(math), 62. *Prof Exp:* Instr math, Columbia Univ, 61-64; from asst prof to prof math, Univ Pa, 64-76, grad group chmn, 73-76, Thomas A Scott prof, 76-80; PROF MATH, UNIV CALIF, LOS ANGELES, 80- *Mem:* Am Math Soc. *Res:* Abstract analysis; representation theory of topological groups. *Mailing Add:* Dept Math Univ Calif 6356 Math Sci Los Angeles CA 90024

EFFROS, RICHARD MATTHEW, b Dec 10, 35; c 3. INTERNAL MEDICINE, PULMONARY DISEASE. *Educ:* NY Univ, MD, 61. *Prof Exp:* Prof med, Harbor Med Ctr, Univ Calif, Los Angeles, 74-89; PROF CHIEF, DIV PULMONARY & CRIT CARE MED, MED COL WIS, 89- *Mem:* Am Physiol Soc; Am Col Physicians. *Res:* Pulmonary edema; gas exchange; pulmonary physiology and disease. *Mailing Add:* Med Col Wis 8700 W Wisconsin Ave Box 134 Milwaukee WI 53226

EFFROS, RITA B, b New York, NY, Oct 8, 41. IMMUNOLOGY. *Educ:* Univ Pa, PhD(immunol), 78. *Prof Exp:* Adj asst prof, 81-86, ADJ ASSOC PROF IMMUNOL, SCH MED, UNIV CALIF, LOS ANGELES, 86- *Mailing Add:* Dept Path Health Sci 13-186 Sch Med Univ Calif Los Angeles CA 90024-1732

EFNER, HOWARD F, b Ann Arbor, Mich, Sept 23, 44; m 70; c 1. PROCESS OPTIMIZATION. *Educ:* Eastern Mich Univ, BS, 67; Univ Iowa, MS, 71; Univ NDak, PhD(chem), 75. *Prof Exp:* Nat Res Coun resident res assoc, Naval Res Lab, 75-77; res chemist, 77-82, SR RES CHEMIST, PHILLIPS PETROL CO, 82- *Mem:* Am Chem Soc; Soc Plastics Engrs; Sigma Xi. *Res:* Polymer compound development; polymer processing. *Mailing Add:* 2100 Jefferson Rd Bartlesville OK 74006

EFRON, BRADLEY, b St Paul, Minn, May 24, 38. STATISTICS. *Educ:* Calif Inst Technol, BS, 60; Stanford Univ, MS, 62, PhD(statist), 64. *Prof Exp:* From asst prof to assoc prof, 66-71, chmn dept, 76-79, PROF STATIST, STANFORD UNIV, 72-, PROF, DEPT HEALTH, RES & POLICY, 74- , CHMN, DEPT STATIST, 91- *Concurrent Pos:* Consult, var corps, labs & hosps, 62-; vis lectr, Dept Statist, Harvard Univ, 67-68; assoc ed, J Am Statist Asn, 68-69, theory & methods ed, 69-72; univ fel, Stanford Univ, 69-72, mem adv bd, 85-88, vchmn, Adv Bd Acad Coun, 86-87 & chmn, Math & Computation Sci Prog, 80-; vis scholar, Dept Math, Imp Col, London, 71-72; vis prof, Imp Col, Eng, 71-72 & Dept Statist, Univ Calif, Berkeley, 79-80; mem, Comt on Resources Math Sci, 84; chmn, External Adv Comt Statist, Univ Md, 85 & Overseas Comt, Dept Statist, Harvard Univ, 85-86; assoc dean, Humanities & Sci, 87- *Honors & Awards:* Numerous hon lectureships, 76-; Ford Prize, Math Asn Am, 78; Wilks Medal, Am Statist Asn, 90. *Mem:* Nat Acad Sci; fel Royal Statist Soc; fel Am Statist Asn; fel Inst Math Statist; AAAS; Int Statist Asn; fel Am Acad Arts & Sci; fel Int Statist Inst. *Res:* Statistics, applied math; symmetry conditions; chromosomal abnormality; sequential experiments; author of numerous articles and publications. *Mailing Add:* Dept Statist Stanford Univ Stanford CA 94305

EFRON, HERMAN YALE, b Brooklyn, NY, Apr 30, 26; m 47; c 3. HEALTH SCIENCES, RESEARCH ADMINISTRATION. *Educ:* City Col New York, BS, 47; Columbia Univ, MA, 49; NY Univ, PhD(psychol), 53. *Prof Exp:* Psychologist, Riverside Hosp, 53-54; res psychologist, Vet Hosp, Louisville, 54-56; asst chief psychol serv, Vet Hosp, Lyons, NJ, 56-68; vis fel psychometrics, Princeton Univ, 68-69; proj dir patient care eval, Vet Admin Cent Off, 69-77, health sci specialist, 77-85; CONSULT, PSYCHOLOGIST, 85- *Concurrent Pos:* James McKeen Cattell fel, 68; consult, 85- *Mem:* Am Psychol Asn; Am Statist Asn; Psychometric Soc. *Res:* Development of methodologies for assessment of quality of health care delivery; consumer satisfaction and its relationships to health care system characteristics; effectiveness of evaluation methodologies as causal agents for improvement. *Mailing Add:* 909 Brentwood Lane Silver Spring MD 20902

EFRON, ROBERT, b New York, NY, Dec 22, 27; m 67; c 3. NEUROPSYCHOLOGY, NEUROPHYSIOLOGY. *Educ:* Columbia Col, BA, 48; Harvard Univ, MD, 52. *Prof Exp:* Med house off, Peter Bent Brigham Hosp, Boston, Mass, 52-53; chief neurophysiol-biophys res unit, Vet Admin Hosp, Boston, 60-70; PROF NEUROL, MED SCH, UNIV CALIF, DAVIS, 70-; CHIEF NEUROPHYSIOL & BIOPHYS RES LAB, 70-, ASSOC CHIEF STAFF RES & DEVELOP, VET ADMIN HOSP, MARTINEZ, 75- *Concurrent Pos:* Moseley traveling fel, Harvard Univ, 53-54; Nat Found Infantile Paralysis fel, Nat Hosp, Queens Sq, London, Eng, 56-60; mem adv bd, Int Soc Study Time, 66- *Mem:* Am Acad Aphasia; AAAS; NY Acad Sci; Am Acad Neurol; fel Acoust Soc Am. *Res:* Neurophysiology and neuropsychology of perception. *Mailing Add:* Res & Develop Dept, Vet Admin Hosp 150 Muir Rd Martinez CA 94553

EFTHYMIOU, CONSTANTINE JOHN, b Athens, Greece, Apr 21, 30; m 64; c 3. APPLIED MICROBIOLOGY, IMMUNOLOGY. *Educ:* Athens Agr Col, BS, 52; Univ Md, MS, 58, PhD(microbiol), 61. *Prof Exp:* Asst dairy tech, Univ Md, 55-58, asst microbiol, 58-61; asst prof, Carnegie Inst Technol, 61-64; asst prof, 64-68, assoc prof, 68-81, PROF MICROBIOL, GRAD SCH, ST JOHNS UNIV, NY, 81- *Concurrent Pos:* Consult microbiol. *Mem:* AAAS; Am Soc Microbiol; Sigma Xi; Am Inst Biol Sci; Asn Clin Scientists; Soc Indust Microbiol. *Res:* Systematic and applied microbiology of lactic acid bacteria; biochemistry and biotechnology of fermentations; investigation of the secretory immune system and hypersensitive state as possible causes of sudden infant death syndrome. *Mailing Add:* 7861 222nd St Oakland Gardens NY 11364

EFTINK, MAURICE R, b Cape Girardeau, Mo, July 26, 51; m 72; c 2. BIOCHEMISTRY, BIOPHYSICS. *Educ:* Univ Mo-Columbia, BS, 73, PhD(biochem), 76. *Prof Exp:* Res assoc biochem, Univ Va, 76-78; ASST PROF CHEM, UNIV MISS, 78- *Concurrent Pos:* NIH fel, Univ Va, 77-78. *Mem:* Am Chem Soc; Biophys Soc. *Res:* Thermodynamics of protein, ligand interactions; protein dynamics; energetics of enzyme catalyzed reactions. *Mailing Add:* Dept Chem Univ Miss University MS 38677

EGAMI, TAKESHI, b Fukuoka, Japan, July 15, 45; m 69; c 3. MATERIALS SCIENCE, SOLID STATE PHYSICS. *Educ:* Univ Tokyo, BEng, 68; Univ Pa, PhD(metall & mat sci), 71. *Prof Exp:* Fel appl sci, Univ Sussex, 71-72; vis scientist physics, Max Planck Inst Metal Res, 72-73 & 79-80; asst prof, 73-76, assoc prof metall & mat sci, 76-80, PROF MAT SCI ENG, UNIV PA, 80- *Concurrent Pos:* Guest prof physics, Univ Tokyo, 87; assoc ed, Progress Mat Sci, 88- *Honors & Awards:* R L Hardy Gold Medal, Am Inst Mining, Metall & Petrol Engrs. *Mem:* Am Inst Mining, Metall & Petrol Engrs; Am Phys Soc; Am Soc Metals; Mat Res Soc; Inst Elec & Electronics Engrs. *Res:* Structure and magnetism of metallic glasses, quasicrystals and oxide superconductors; neutron and x-ray diffraction, in particular using pulsed neutron source and synchrotron radiation. *Mailing Add:* Dept Mat Sci & Eng Univ Pa Philadelphia PA 19104-6272

EGAN, B ZANE, b Scott Co, Va, Aug 31, 37; m 57; c 2. GENERAL CHEMISTRY, BIOCHEMISTRY. *Educ:* Berea Col, AB, 57; Ohio State Univ, MSc, 60, PhD(chem), 62. *Prof Exp:* MEM STAFF, OAK RIDGE NAT LAB, 62- *Mem:* Am Chem Soc; Am Soc Biol Chemists; AAAS; Sigma Xi. *Res:* Solvent extraction; chromatographic separations; flue gas desulfurization; metals recovery; separation and characterization of nucleic acids; photosynthetic hydrogen production; environmental and personnel monitoring. *Mailing Add:* Martin Marietta Energy Systs Oak Ridge Nat Lab Oak Ridge TN 37831

EGAN, EDMUND ALFRED, b Chicago, Ill, Apr 24, 41; m 64; c 6. NEONATAL MEDICINE, PERINATAL MEDICINE. *Educ:* Emory Univ, Atlanta, Ga, MD, 67. *Prof Exp:* Resident pediat, Col Med, Univ Fla, 67-70; chief neonatology pediat, Madigan Gen Hosp, US Army, 70-72; investr physiol, Med Sch, Univ Col Hosp, London, 72-73; asst prof pediat, Col Med, Univ Fla, 73-76, assoc prof, 76-77; chief neonatology, Children's Hosp, Buffalo, NY, 77-89; assoc prof, 77-82, PROF PEDIAT, SCH MED, STATE UNIV NY, BUFFALO, 82-, PROF PHYSIOL, 86- *Concurrent Pos:* NIH prin investr, Alveolar Permeability & Lung Dis, 74-83; mem NIH Pulmonary Technol Task Force, 80-81. *Mem:* Am Physiol Soc; Am Acad Pediat; Soc

Pediat Res; Perinatal Res Soc; Am Thoracic Soc. *Res:* Mechanisms of water and solute balance in adult and perinatal lungs; surfactant replacement in animals; clinical research in newborn infants. *Mailing Add:* Children's Hosp 219 Bryant St Buffalo NY 14222

EGAN, FRANCIS P, b New York, NY, Oct 17, 17; m 40. MATHEMATICS. *Educ:* Manhattan Col, BA, 37; Univ Notre Dame, MS, 51, PhD(math educ), 60. *Prof Exp:* From instr to prof math, Niagra Univ, 37-55; lectr, State Univ NY, Buffalo, 55-59; chmn dept, 62-73, PROF MATH, STATE UNIV NY COL ONEONTA, 59- *Mem:* Math Asn Am. *Res:* Logic; metamathematics; teacher education. *Mailing Add:* Dept Math Sci State Univ NY Oneonta NY 13820

EGAN, HOWARD L, b St Louis, Mo, May 2, 38. MATHEMATICS. *Educ:* Wash Univ, AB, 60, AM, 62, PhD(group theory), 65. *Prof Exp:* Asst prof math, Univ Md, 65-71; ASSOC PROF MATH, GALLAUDET COL, 71- *Mem:* Am Math Soc. *Res:* Infinite non-abelian group theory; algebraic coding theory. *Mailing Add:* Dept Math Gallaudet Col Washington DC 20002

EGAN, JAMES JOHN, b Oak Park, Ill, May 22, 27. PHYSICAL CHEMISTRY. *Educ:* Northwestern Univ, BS, 49; Univ Ind, PhD(phys chem), 54. *Prof Exp:* Assoc phys chemist, 53-60, chemist, 60-81, SR CHEMIST, BROOKHAVEN NAT LAB, 81- *Mem:* Am Chem Soc; Electrochem Soc; Bunsengesellschaft. *Res:* High temperature chemistry; molten salt chemistry; solid state chemistry; light scattering in aerosols. *Mailing Add:* Brookhaven Nat Lab Upton NY 11973

EGAN, JAMES JOSEPH, b Covington, Ky, Nov 7, 41; m 64; c 3. EXPERIMENTAL NUCLEAR PHYSICS. *Educ:* Thomas More Col, BA, 63; Univ Ky, MS, 66, PhD(physics), 69. *Prof Exp:* Asst prof, 69-76, ASSOC PROF PHYSICS, UNIV LOWELL, 76- *Mem:* AAAS; Am Phys Soc; Sigma Xi; NY Acad Sci. *Res:* Low energy experimental nuclear structure physics; neutron scattering cross section measurements. *Mailing Add:* 11 Charlemont Ct North Chelmsford MA 01863

EGAN, JOHN FREDERICK, b Council Bluffs, Iowa, Feb 25, 35; m 58; c 2. ELECTRONICS ENGINEERING. *Educ:* Grinnell Col, BA, 57; Northwestern Univ, MS, 58, PhD(elec eng), 61. *Prof Exp:* Proj engr, 61-64, tech dir, Air Force Electronic Systs Div, 64-67; sr staff specialist, Off Dir Defense Res & Eng, 67-71; chief scientist, Off Chief Naval Opers, 71-73; vpres, Sanders Assocs, Inc, 73-87; VPRES, LOCKHEED ELECTRONIC SYSTS GROUP, 87- *Concurrent Pos:* Mem, Computer-Security Panel, Defense Sci Bd, 69-71, Ocean Surveillence Panel, 70-73, sci adv bd panel, Nat Security Agency, 82-84, exec panel, Chief Naval Opers, 73-, Naval Studies Bd, Nat Acad Sci, 90- *Mem:* Inst Elec & Electronics Engrs; Am Inst Aeronaut & Astronaut; Asn Comput Mach; AAAS; Sigma Xi. *Mailing Add:* Sever Beverlee Dr Nashua NH 03060

EGAN, MARIANNE LOUISE, b Jersey City, NJ, June 9, 42; m 75; c 2. IMMUNOCHEMISTRY, IMMUNOBIOLOGY. *Educ:* Col St Elizabeth, AB, 64; Jefferson Med Col, PhD(biochem), 69. *Prof Exp:* Instr biochem, Jefferson Med Col, 69-70; from jr res scientist to asst res scientist, 70-72, assoc res scientist immunochem, City of Hope Nat Med Ctr, 72-76; res asst prof, 77-83, RES ASSOC PROF MICROBIOL, UNIV ALA, 84- *Concurrent Pos:* Assoc scientist, Multipurpose Arthritis Ctr, Comprehensive Cancer Ctr; chairperson, diag working group, Breast Cancer Task Force, 80-84; assoc res career scientist, Vet Admin, 88- *Mem:* Am Asn Immunologists; Am Asn Cancer Res; Int Soc Neuroimmunology. *Res:* Human natural killer cells; monoclonal antibodies against streptococci; autoimmune diseases. *Mailing Add:* Dept Microbiol Univ Ala Birmingham AL 35294

EGAN, RAYMOND D(AVIS), b Honolulu, Hawaii, Aug 22, 31; m 55; c 4. ELECTRICAL ENGINEERING. *Educ:* Stanford Univ, BS, 55, MS, 56, PhD(elec eng), 60. *Prof Exp:* Res assoc, Stanford Univ, 59-61; mgr appl res, Granger Assocs, 62-68, vpres, 68-84; vpres, Lynch Commun Systs, 84-87; VPRES & GEN MGR, NEC AM, TRANSMISSION SYSTS DIV, 87- *Concurrent Pos:* Mem, US Comn 3, Int Sci Radio Union, 61- *Mem:* Inst Elec & Electronics Engrs; Am Geophys Union; Sigma Xi. *Res:* Radio propagation; ionospheric physics; ionosphere sounding; telecommunication systems. *Mailing Add:* NEC America 14040 Park Center Rd Herndon VA 22071

EGAN, RICHARD L, b Omaha, Nebr, Dec 27, 17; m 43; c 2. MEDICINE. *Educ:* Creighton Univ, BSM, 38, MD, 40. *Prof Exp:* From instr to prof med, Sch Med, Creighton Univ, 41-71, from asst dean to dean, 54-70, asst to pres health sci, 70-71; asst dir dept undergrad med educ, AMA, 71-75, dir dept undergrad med educ & secy, Liaison Comt Med Educ, 75-76, dir, Div Med Educ & secy Coun Med Educ, 76-89, CONSULT, MED EDUC, AMA, 90- *Concurrent Pos:* Consult, Vet Admin Hosps, Omaha & Lincoln, Nebr, 59-70. *Honors & Awards:* Merit Award, Am Heart Asn, 72. *Mem:* AMA; Asn Am Med Cols; Am Asn Hist Med. *Res:* Internal medicine; medical education. *Mailing Add:* Coun Med Educ AMA 515 N State St Chicago IL 60610

EGAN, RICHARD STEPHEN, b Chicago, Ill, Aug 16, 41; m 64; c 3. ANALYTICAL CHEMISTRY, SPECTROSCOPY. *Educ:* Univ Ill Med Ctr, BS, 63, PhD(pharmaceut chem), 71. *Prof Exp:* Sect head, Abbott Labs, 66-79; dir, McNeil Pharmaceut, 79-90; DIR, MERCK, SHARP & DOHME RES LABS, 90- *Mem:* Am Chem Soc; Am Asn Pharmaceut Sci. *Res:* Analytical development of new drug substance and dosage forms including methods development, assay and stability studies. *Mailing Add:* Merck Sharp & Dohme Res Rahway NJ 07065

EGAN, ROBERT L, b Morrilton, Ark, May 9, 20; m 50; c 5. RADIOLOGY. *Educ:* Univ Pittsburgh, MD, 50; Am Bd Radiol, dipl, 56. *Prof Exp:* Asst radiologist, Univ Tex M D Anderson Hosp & Tumor Inst, 56-61, assoc prof radiol, Postgrad Sch Med, Univ Tex, 61-62; radiologist, Methodist Hosp Ind, 62-65; assoc prof, 65-68, PROF RADIOL, SCH MED, EMORY UNIV, 68-, CHIEF MAMMOGRAPHY SECT, 65- *Concurrent Pos:* Chief sect exp diag radiol & assoc radiologist, Univ Tex M D Anderson Hosp & Tumor Inst, 61-62; spec consult, cancer control prog, div chronic dis, Bur State Serv, USPHS, 61-; consult, Health Ins Plan of NY Mammography Surv Prog, 62- *Honors & Awards:* Am Cancer Soc Distinguished Service Award, 75; Lucy Wortham James Clin Award, 77. *Mem:* Am Cancer Soc; Radiol Soc NAm; Roentgen Ray Soc; fel Am Col Radiol; AMA; Soc Surg Oncol. *Res:* Cancer of the breast; teaching, evaluating and accumulating data on mammography and related procedures. *Mailing Add:* Mammography Sect Emory Univ Atlanta GA 30322

EGAN, ROBERT SHAW, b Buffalo, NY, Apr 21, 45; m 66; c 2. LICHENOLOGY. *Educ:* Univ Colo, Boulder, BA, 67, MA, 69, PhD(biol), 71. *Prof Exp:* Asst prof biol, Castleton State Col, 71-75; asst prof, Tex A&M Univ, 75-79; assoc prof & chmn dept, 79-84, PROF BIOL, UNIV NEBR, OMAHA, 85- *Mem:* Am Bryol & Lichenological Soc; Brit Lichen Soc. *Res:* Taxonomy, ecology and phytogeography of lichens; chemosystematics; scanning electron microscopy. *Mailing Add:* Dept Biol Univ Nebr Omaha NE 68182-0040

EGAN, ROBERT WHEELER, b Mineola, NY, Apr 15, 43; m 66; c 2. PULMONARY BIOLOGY. *Educ:* Brown Univ, BSc, 65; McGill Univ, PhD(chem), 72. *Prof Exp:* Chemist, Rohm & Haas Co, 65-67; NIH fel, Johns Hopkins Univ, 72-74; sr res biochemist, Merck & Co, Inc, Rahway, NJ, 74-81, assoc dir, 81-85; dir biochem, Warner-Lambert/Parke Davis, Ann Arbor, 85-86; DIR, ALLERGY DEPT, SCHERING-PLOUGH RES, BLOOMFIELD, NJ, 86-; ADJ PROF, PHYSIOL DEPT, UNIV MED & DENT NJ, NEWARK, 90- *Mem:* Am Chem Soc; Med Chem Soc; Am Soc Biol Chem; AAAS. *Res:* Enzymology and receptor binding of prostaglandins and leukotriencs free radicals from peroxidases; cytokine biology signal transduction; biochemical basis for inflammatory and immediate hypersensitivity disorders; eosinophil biochemistry; histamine biology. *Mailing Add:* 45 Virginia Ave F Manasquan NJ 08736

EGAN, THOMAS J, b Winnipeg, Man, Sept 13, 25; m 53; c 3. MEDICINE, PEDIATRICS. *Educ:* Univ BC, BA, 48; McGill Univ, MD, CM, 52. *Prof Exp:* Asst instr pediat, Univ Pittsburgh, 57-59; instr, Western Reserve Univ, 59-62; dir res lab, Children's Hosp Akron, 59-62; asst prof pediat, Sch Med, Univ Pittsburgh & dir res ctr, Children's Hosp Pittsburgh, 62-65; assoc prof pediat, Sch Med, Northwestern Univ, 65-71, prof pediat & prev & community med, 71-74; PROF PEDIAT, UNIV TORONTO, 74-; dir ambulatory serv, Hosp Sick Children, 74-88, sr attend staff, 74-91. *Concurrent Pos:* Fel res med, Univ Pittsburgh, 54-55; dir clin res ctr, Children's Mem Hosp, Chicago, 65-70, med dir ambulatory serv, 70-74. *Mem:* Soc Pediat Res; Am Acad Pediat; Can Pediat Soc; Am Cleft Palate & Craniofacial Asn. *Res:* Craniofacial anamolies. *Mailing Add:* Dept Pediat Hosp for Sick Children 555 University Ave Toronto ON M5G 1X8 Can

EGAN, WALTER GEORGE, b New York, NY, Oct 12, 23; m 63. PHYSICS & EARTH SCIENCES, SPACE SHUTTLE. *Educ:* City Col New York, BEE, 49; Columbia Univ, MA, 51; Polytech Inst Brooklyn, PhD(solid state physics), 60. *Prof Exp:* Engr, Egan Lab, NY, 46-50; nuclear physicist, Nucleonics Sect, Naval Mat Lab, 50-56; prof elec eng, City Col New York, 56-57; prin res engr, Ford Instrument Co Div, Sperry-Rand Corp, 57-58, eng proj supvr, 58-60, asst dir res, 60-62, exec asst to vpres res, 62-63; staff scientist, Grumman Aerospace Corp, 63-89; PROF PHYSICS, POLYTECH UNIV, YORK COL/CITY UNIV NY & ADELPHI UNIV, 89- *Concurrent Pos:* Mem airlines electronic eng comt, Air Lines Commun Admin Coun, 58-; adv group aeronaut res & develop, NATO, 61-; adj prof, earth sci & sr res scientist, Baruch Col/City Univ New York, 89-; res scientist space shuttle prog, NASA/Houston, 89- *Mem:* Am Phys Soc; Inst Elec & Electronics Engrs; Sigma Xi; Soc Photo-Optical Instrumentation Engrs; Res Soc Am; Am Phys Soc; Optical Soc Am; AAAS; Am Meteorol Soc. *Res:* Thin films; infrared; cryogenics; magnetics; microwaves and millimeter waves; lasers; masers; high vacuum; military and space sciences including guidance, navigation, oceanography, communications, countermeasures and antisubmarine warfare; astronomy; polarization properties of terrestrial and planetary surfaces; remote sensing from space shuttle; global warming; atmospheric properties; earth & atmospheric sciences. *Mailing Add:* 84-26 86th St Woodhaven NY 11421

EGAN, WILLIAM MICHAEL, b New York, NY, Aug 22, 44. PHYSICAL CHEMISTRY, BIOCHEMISTRY. *Educ:* Manhattan Col, BS, 66; Princeton Univ, PhD(chem), 71. *Prof Exp:* Res assoc chem, Lund Inst Technol, 71-75; staff fel, Nat Inst Child Health & Human Develop, NIH, 75-76; RES CHEMIST, BUR BIOLOGICS, FOOD & DRUG ADMIN, 76- *Concurrent Pos:* Adj assoc prof, Cath Univ Am. *Mem:* Am Chem Soc; Int Soc Magnetic Resonance; AAAS; Sigma Xi. *Res:* Nuclear magnetic resonance spectroscopy; bacterial capsular polysaccharide structure. *Mailing Add:* 5143 Eliots Oak Rd Columbia MD 21044-1824

EGAR, JOSEPH MICHAEL, b Jacksonville, Fla, Feb 2, 30; m 61; c 4. MATHEMATICS. *Educ:* Univ Okla, BS, 52; Tex A&M Univ, PhD(geol, geophys, math), 59. *Prof Exp:* Eng trainee, Savannah River Proj, E I du Pont de Nemours & Co, 52; instr geol, Marietta Col, 55; asst prof geol & math, Tex A&M Univ, 57-59, NSF sci fac fel math, 59-60; assoc prof, Ball State Teachers Col, 60-63 & Univ Akron, 63-66; ASSOC PROF MATH, CLEVELAND STATE UNIV, 66- *Concurrent Pos:* Sr lectr math, Hatfield Polytech, Hertfordshire, Eng, 74-75; hon res fel, Dept Math & Physics, Birmingham Univ, Eng, 80. *Mem:* AAAS; Math Asn Am. *Res:* Seismology, especially surface wave phenomena and underwater acoustics; applied mathematics; numerical analysis. *Mailing Add:* Dept Math Cleveland State Univ Cleveland OH 44115

EGAR, MARGARET WELLS, b Princeton, WVa, July 25, 34; m 61; c 4. REGERATION, DEVELOPMENTAL ANATOMY. *Educ:* Concord Col, BS, 56; Emory Univ, MS, 58, PhD(biol), 60. *Prof Exp:* Asst prof biol, Ball State Univ, 60-62; res assoc regeneration, Sch Med, Case Western Reserve Univ, 65-73, lectr anat, 73-77, asst prof anat, 77-85; ASSOC PROF ANAT, IND UNIV MED SCH, 85- *Honors & Awards:* Kaiser Permanente

Preclinical Teaching Award, 77; Singer Res Med, 85. *Mem:* Sigma Xi; AAAS; Am Asn Anat; Soc Develop Biol; Am Asn Clin Anat; Asn Women Sci. *Res:* Spinal cord and limb regeneration in salamanders; neural factors influencing accessory limb and supernumerary limb development; developmental pathways in axonal guidance. *Mailing Add:* 2969 Somerton Rd Cleveland Heights OH 44118

EGBERG, DAVID CURTIS, b Minneapolis, Minn, Oct 13, 42; m 70; c 1. ORGANIC CHEMISTRY. *Educ:* Macalester Col, BA, 64; Univ Minn, PhD(org chem), 72. *Prof Exp:* RES CHEMIST ANALYTICAL CHEM, GEN MILLS, INC, 72- *Concurrent Pos:* Chmn vitamin methods comt, Am Asn Cereal Chemists, 74- *Mem:* Am Chem Soc. *Res:* Development of analytical methods for the food industry; continuous flow automation and high pressure liquid chromatography. *Mailing Add:* 2845 Jewel Lane Plymouth MN 55447

EGBERT, GARY TRENT, molecular spectroscopy, for more information see previous edition

EGBERT, ROBERT B(ALDWIN), biochemistry, computer sciences; deceased, see previous edition for last biography

EGDAHL, RICHARD H, b Eau Claire, Wis, Dec 13, 26; m 53; c 4. SURGERY. *Educ:* Harvard Univ, MD, 50; Univ Minn, PhD, 57; Am Bd Surg, cert, 59. *Prof Exp:* Instr surg, Univ Minn, 57-58; dir surg res labs, Med Col Va, 58-64; chmn & Utley prof, Dept Surg, Boston Univ Med Ctr, 64-73, PROF SURG, SCH MED, BOSTON UNIV, 64-, ACAD VPRES HEALTH AFFAIRS, 73-, PROF PUB HEALTH, SCH PUB HEALTH, 83-, PROF MGT, SCH MGT, 88- *Concurrent Pos:* USPHS spec res fel, Univ Minn, 57-58; Markle scholar, 58-63; consult, Boston & Providence Vet Hosps; dir, Boston Univ Med Ctr, 73-, Boston Univ Health Policy Inst, 75-; coun mem, Inst Med-Nat Acad Sci, 81-85. *Honors & Awards:* Ciba Award, Endocrine Soc, 62. *Mem:* Inst Med-Nat Acad Sci; Am Col Surgeons; AMA; Am Soc Clin Invest; Int Asn Endocrine Surgeons (pres, 81-83); Am Physiol Soc; Am Soc Exp Path; Am Surg Asn (vpres, 80); Int Soc Surg; Soc Med Adminrs. *Res:* Experimental and clinical endocrinology; trauma; shock; pancreatitis; author or co-author of 290 publications in basic science, clinical science and health policy. *Mailing Add:* 80 E Concord St B-324 Boston MA 02118

EGE, RAIMUND K, b Hechingen, Ger, May 20, 58; m 88. OBJECT-ORIENTATION, MASSIVE PARALLELISM. *Educ:* Ore State Univ, MS, 85; Univ Stuttgart, Ger, Dipl Informatiker, 86; Ore Inst Sci & Technol, PhD(computer sci & eng), 87. *Prof Exp:* ASST PROF COMPUTER SCI, FLA INT UNIV, 87- *Mem:* Asn Comput Mach; Inst Elec & Electronics Engrs Computer Soc. *Res:* Application of object-oriented programming concepts to software engineering; interfaces and parallel computations. *Mailing Add:* Sch Computer Sci Fla Int Univ Miami FL 33199

EGE, SEYHAN NURETTIN, b Ankara, Turkey, Jan 11, 31. ORGANIC CHEMISTRY. *Educ:* Am Col for Girls, Istanbul, BS, 49; Smith Col, MA, 52; Univ Mich, PhD(org chem), 56. *Prof Exp:* Instr chem, Univ Mich, 56-57 & Am Col for Girls, Istanbul, 57-59; res assoc, Boston Univ, 59-61; asst prof, Mt Holyoke Col, 61-62; res assoc & part time lectr, Univ Toronto, 62-65; lectr, 65-67, from asst prof to prof, 67-90, ARTHUR F THURNAU PROF CHEM, UNIV MICH, 90- *Mem:* AAAS; Am Chem Soc; Inter-Am Photochem Soc; Int Soc Heterocyclic Chem. *Res:* Molecular rearrangements; organic nitrogen compounds; organic photochemistry of heterocycles; helerocyclic reactive intermediates under research. *Mailing Add:* Dept Chem Univ Mich Ann Arbor MI 48109-1055

EGEBERG, ROGER O, b Chicago, Ill, Nov 13, 03; m 29; c 4. INTERNAL MEDICINE, MEDICAL ADMINISTRATION. *Educ:* Cornell Univ, BA, 25; Northwestern Univ, MD, 29. *Prof Exp:* Chief med & prof serv, Vet Admin Hosp, 46-56; med dir, Los Angeles County Hosp, 56-58; med dir, Los Angeles County Dept Charities, 58-64; prof med & dean sch med, Univ Southern Calif, 64-69; asst secy health & sci affairs, Dept Health, Educ & Welfare, 69-71, spec asst to secy health policy & consult to pres health affairs, 71-77; SR SCHOLAR-IN-RESIDENCE, INST MED, NAT ACAD SCI, 84-; DISTINGUISHED PROF, DEPTS MED & COMMUNITY & FAMILY MED, SCH MED, GEORGETOWN UNIV, 85- *Concurrent Pos:* Med consult, Armed Forces, 46; physician-in-residence, Vet Admin Hosps, 56-; chmn, Gov Comt Study Med Aid & Health, Calif, 60; mem, President's Panel Spec Study Narcotics, 62 & Presidential Adv Comn Narcotics & Drug 63; mem & past pres, Calif State Bd Health, 63-68; mem nat adv cancer coun, Nat Cancer Inst, 64-68; mem & chmn spec med adv group, Vet Admin, 65-69; chief med officer, Medicare Bur & dir off prof & sci activ, Health Care Financing Admin, HEW, 77-83. *Honors & Awards:* Benjamin Rush Bicentennial Award, AMA. *Mem:* Fel Am Col Physicians; Am Soc Internal Med. *Res:* Ecology of Coccidioides immitis, especially the reasons for spotty distribution in the soil of endemic areas; identification of antagonists and their susceptibility to high temperatures and high salinity. *Mailing Add:* 2909 Garfield Tr NW Washington DC 20008

EGELHOFF, WILLIAM FREDERICK, JR, b Norfolk, Va, July 8, 49; m 76; c 3. CHEMICAL PHYSICS, SOLID STATE PHYSICS. *Educ:* Hampden-Sydney Col, BS, 71; Cambridge Univ, PhD(phys chem), 75. *Prof Exp:* Fel chem eng, Calif Inst Technol, 75-76; res physicist, Gen Motors Res Labs, Warren, 76-79; RES CHEMIST, SURFACE SCI DIV, NAT INST STANDARDS & TECHNOL, 79- *Mem:* Am Phys Soc; Am Vacuum Soc. *Res:* Photoelectron spectroscopy; epitaxial growth; surface physics; surface magnetism. *Mailing Add:* Surface Sci Div Nat Inst Standards & Technol Gaithersburg MD 20899

EGELSTAFF, PETER A, b London, Eng, Dec 10, 25; Can citizen; m 47; c 5. NEUTRON SCATTERING. *Educ:* London Univ, BSc, 46, PhD(physics), 54. *Prof Exp:* Metallurgist, High Duty Alloys, Ltd, 43-47; prin sci officer, UK Atomic Energy Authority, 47-57, spec merit sr appointment, 59-70; proj leader, Atomic Energy Can Ltd, 57-59; chmn dept, Univ Guelph, 70-75, prof physics, 75-90, mem bd gov, 83-86; RETIRED. *Concurrent Pos:* Vis prof, Cath Univ, Washington, DC, 67, Cornell Univ, Ithaca, NY, 68 & Tohoku Univ, Sendai, Japan, 82; mem var comts, Nat Sci & Eng Res Coun Can, 80-86. *Honors & Awards:* Achievement Medal, Can Asn Physicists, 83. *Mem:* Europ Phys Soc; Can Asn Physicists (pres, 87-88); Am Phys Soc; fel Royal Soc Can; fel Inst Physics London. *Res:* Structure and dynamics of liquids and dense gases at the atomic level, including the properties of collective modes; neutron scattering experiments on fluids. *Mailing Add:* Physics Dept Univ Guelph Guelph ON N1G 2W1 Can

EGER, EDMOND I, II, ANESTHESIA. *Prof Exp:* PROF ANESTHESIA & VCHMN RES, UNIV CALIF, 60- *Mailing Add:* Dept Anesthesia Univ Calif Box 0464 S455 San Francisco CA 94143-0464

EGER, F MARTIN, b Lwow, Poland, May 31, 36; US citizen. PHILOSOPHY OF SCIENCE. *Educ:* Mass Inst Technol, BS, 58; Brandeis Univ, PhD(physics), 63. *Prof Exp:* Res assoc theoret physics, Brandeis Univ, 63-64; fel, Lawrence Radiation Lab, Univ Calif, 65-67; asst prof physics, 67-72, ASSOC PROF PHYSICS, COL STATEN ISLAND, CITY UNIV NEW YORK, 72- *Mem:* AAAS; Am Phys Soc; Philos Sci Asn. *Res:* Statistical mechanics and the many-body-problem; history and philosophy of science; education. *Mailing Add:* 161 Stauber Rd RD2 Groton NY 13073

EGERMEIER, R(OBERT) P(AUL), b Oklahoma City, Okla, Dec 25, 27; m 52; c 2. SYSTEM ENGINEERING. *Educ:* Univ Okla, BS, 51; NMex State Univ, MS, 57; Lasalle Inst, LLB, 76. *Prof Exp:* Asst physicist, Phys Sci Lab, NMex State Univ, 53-57, asst prof mech eng, 57-62; staff engr & sect mgr, Aerospace Corp, Calif, 62-67; staff eng scientist & mgr ord systs, Radio Corp Am, 67-69; sr scientist, Hughes Aircraft Co, 69-79, scientist, 81-85; CHIEF SCIENTIST, RADAR LABS, 86- *Concurrent Pos:* Consult, 69-79; patent atty, 79-81. *Mem:* AAAS; Am Inst Aeronaut & Astronaut; Am Soc Mech Engrs; Inst Elec & Electronics Engrs. *Res:* Analog systems; thermodynamics; product liability and safety; microprocessor applications. *Mailing Add:* 22354 Malden St Canoga Park CA 91304

EGERTON, JOHN RICHARD, parasite chemotherapy, for more information see previous edition

EGERTON, RAYMOND FRANK, b Manchester, Eng, Aug 30, 44; m 76; c 1. ELECTRON ENERGY-LOSS SPECTROSCOPY, ELECTRON MICROSCOPY. *Educ:* St Johns Col, Cambridge, UK, BA, 65; Imp Col, London, UK, PhD(mat sci), 68. *Prof Exp:* Res scientist, Zenith Radio Res Lab, 68-71; res asst metall, Oxford Univ, UK, 71-75; prof mat sci, State Univ NY, 86-91; PROF PHYSICS, UNIV ALTA, CAN, 75-86 & 91- *Concurrent Pos:* Phys sci dir, Electron Micros Soc Am, 88-90. *Mem:* Micros Soc Can (pres, 87-89). *Res:* Development of techniques of analytical electron microscopy, particularly electron energy-loss spectroscopy, and their application to materials science, including thin film studies. *Mailing Add:* Physics Dept Univ Alta Edmonton AB T6G 2J1 Can

EGGAN, LAWRENCE CARL, b Fargo, NDak, Jan 10, 35; m 71; c 5. SIMULATION, GRAPH THEORY. *Educ:* Pac Lutheran Univ, BA, 56; Univ Ore, MS, 58, PhD(number theory), 60. *Prof Exp:* Teaching fel math, Univ Ore, 56-59; from instr to asst prof, Univ Mich, 60-65; assoc prof & chmn dept, Pac Lutheran Univ, 65-68; from assoc prof to prof math, 68-84, PROF & CHMN APPL COMPUT SCI, ILL STATE UNIV, 84- *Concurrent Pos:* Lectr, Imp Col, London, 63-64; vis prof, Royal Holloway Col, London, 76-77; assoc ed, Math Reviews, 79-81; consult, SID, Computer Centers & Curric, Thailand, 90. *Honors & Awards:* James Armstrong Award; Sigma Xi Res Award, Univ Ore, 60. *Mem:* Am Math Soc; Math Asn Am; London Math Soc; Sigma Xi; Asn Comput Mach; Data Processing Mgt Asn. *Res:* Number theory, especially Diophantine approximations; logic, especially automata and finite automata; combinatorics, especially graph theory; computer science, especially modeling and simulation; group decision support systems. *Mailing Add:* Appl Comput Sci Dept Ill State Univ Normal IL 61761-6901

EGGE, ALFRED SEVERIN, zoology; deceased, see previous edition for last biography

EGGEN, DONALD T(RIPP), b Hemet, Calif, Feb 11, 22; m 42; c 4. PHYSICS GENERAL. *Educ:* Whittier Col, BA, 43; Ohio State Univ, PhD(physics), 48. *Prof Exp:* Res physicist mass separation of uranium, Tenn Eastman Corp, 44-45; res asst physics, Ohio State Univ, 45-48; res engr, Exp Physics, NAm Aviation, Inc, 49-53, group engr, Component Develop, 53-57, res specialist, Exp Physics, 57-58, prog mgr, Adv Epithermal Thorium Reactor, 59-63, prog mgr fast reactor & sodium components, 63-66; prog mgr fast reactor core design, Argonne Nat Lab, 66-68; prof nuclear eng, Technol Inst, Northwestern Univ, 68-87; RETIRED. *Concurrent Pos:* Deleg, US Nuclear Regulatory Comn, Cabri Proj, Cadarache Centre E'tudes Nucleaire, France, 78-79. *Mem:* AAAS; fel Am Nuclear Soc; Sigma Xi. *Res:* Fast reactor core design; safety; liquid metal technology and heat transfer. *Mailing Add:* PO Box 7027 Thousand Oaks CA 91359-7027

EGGEN, DOUGLAS AMBROSE, b Rushford, Minn, Apr 30, 25; m 50; c 3. PATHOLOGY. *Educ:* Univ Chicago, PhB, 47, SB, 55, PhD(biophys), 57. *Prof Exp:* Res assoc biophys, Res Insts, Univ Chicago, 55-58; res assoc path, 58-61, from instr to assoc prof path, 61-70, assoc prof path & biomet, 70-72, dir, Biomed Comput Ctr, 72-75, actg head, Dept Biomet, 76, 78-79, prof path & biomet, 72-83, PROF PATH, LA STATE UNIV MED CTR, NEW ORLEANS, 83- *Concurrent Pos:* Mem coun on arteriosclerosis, Am Heart Asn. *Mem:* AAAS; Am Asn Path; Sigma Xi. *Res:* Experimental atherosclerosis and cholesterol metabolism in non-human primates; geographic pathology of atherosclerosis. *Mailing Add:* Dept Path La State Univ 1901 Perdidost New Orleans LA 70112-2262

EGGENA, PATRICK, b London, Eng, Feb 1, 38; US citizen; m 65; c 2. PHYSIOLOGY. *Educ:* Univ Cincinnati, MD, 66. *Prof Exp:* ASSOC PROF PHYSIOL, MT SINAI MED SCH, 73-, ASSOC PROF BIOPHYS, 77- *Concurrent Pos:* NIH fel, 67-70; estab investr, Am Heart Asn, 71-76, mem renal coun. *Mem:* Am Physiol Soc; Tissue Cult Soc. *Res:* Mechanism of secretion and action of the antidiuretic hormone, Vasopressin. *Mailing Add:* Dept Physiol Mt Sinai Med Sch New York NY 10029

EGGENBERGER, ANDREW JON, b Harlowton, Mont, May 8, 38. EARTHQUAKE ENGINEERING, APPLIED MECHANICS. *Educ:* Carnegie Inst Technol, BS, 61, PhD(magnetohydrodyn), 67; Ohio State Univ, ScM, 63. *Prof Exp:* Assoc res engr, Boeing Co, 61-63; fac fel magnetohydrodyn, Advan Res Inst, Lewis Res Ctr, NASA, 67-68; asst prof eng, Univ SC, 67-72; nuclear projs mgr, D'Appolonia Consult Engrs Inc, 72-84; prog dir, Earthquake Hazard Mitigation, NSF, 84-89; VCHMN, DEFENSE NUCLEAR FACIL SAFETY BD, 89- *Concurrent Pos:* Fac fel fluid turbulence, Aero-Astrodyn Lab, George C Marshall Space Flight Ctr, NASA, 69. *Honors & Awards:* Ralph Teetor Award, Soc Automotive Engrs. *Mem:* Am Inst Aeronaut & Astronaut; Am Nuclear Soc; Earthquake Eng Res Inst. *Res:* Fluid mechanics; magnetohydrodynamics; earthquake engineering. *Mailing Add:* Defense Nuclear Facil Safety Bd 625 Indiana Ave Suite 700 Washington DC 20004

EGGER, CARL THOMAS, b Monticello, Iowa, Feb 5, 37; m 57; c 5. PROCESS PLANT ENGINEERING, ALCOHOL PRODUCTION. *Educ:* Univ Iowa, BSChE, 59, MS, 60, PhD(chem eng), 62. *Prof Exp:* Reservoir engr, Shell Develop Co, Tex, 62; mgr process develop, 64-66, dir develop, 66-73, protein mgr, 73-76, dir eng, 76-80, VPRES, GRAIN PROCESSING CORP, 80- *Mem:* Am Inst Chem Engrs. *Res:* Secondary recovery; cryogenics; computer simulation; aerospace; fermentation technology; wet milling; distillation; waste treatment; enzymes; process control; solvent extraction; centrifugation; membrane separations; protein extraction and modification; energy conservation; cogeneration. *Mailing Add:* 1304 Houser Muscatine IA 52761

EGGER, M(AURICE) DAVID, b Bakersfield, Calif, June 21, 36; m 58; c 3. NEUROPHYSIOLOGY, NEUROANATOMY. *Educ:* Stanford Univ, BS, 58; Yale Univ, MS, 60, PhD(physiol psychol), 62. *Prof Exp:* From instr to assoc prof anat, Sch Med, Yale Univ, 65-74; assoc prof, 74-78, actg chmn, 86-90, PROF NEUROSCI & CELL BIOL, ROBERT WOOD JOHNSON MED SCH, UNIV MED & DENT NJ, 78- *Concurrent Pos:* USPHS fel psychiat, Sch Med, Yale Univ, 62-63; vis scientist, Med Res Coun Cerebral Function Res Group, Univ London, 69-70; NIMH res scientist develop award, 69-74, mem res scientist rev comt, 75-79. *Mem:* Fel Am Psychol Asn; Am Asn Anat; Am Physiol Soc; Soc Neurosci; fel AAAS; fel Am Psychol Soc. *Res:* Neural basis of learning; organization of spinal reflexes; anatomy of spinal cord; neurogenetics in Drosophila. *Mailing Add:* Dept Neurosci & Cell Biol Robert Wood Johnson Med Sch Piscataway NJ 08854-5635

EGGERS, A(LFRED) J(OHN), JR, b Omaha, Nebr, June 24, 22; m 50; c 2. AEROTHERMODYNAMICS, HEAT TRANSFER. *Educ:* Univ Omaha, AB, 44; Stanford Univ, MS, 49, PhD(eng mech), 56. *Prof Exp:* With NASA & predecessor, 44-71, chief vehicle environ div, Ames Res Ctr, 59-63, asst dir res & dir res & develop anal & planning, 63-64, dep assoc adminr advan res & technol, NASA Hq, 64-68, asst adminr policy, 68-71; asst dir res appln, NSF, 71-77; dir, Lockheed Palo Alto Res Lab, 77-79; PRES, RANN INC, 79- *Concurrent Pos:* Mem sci adv bd, USAF, 58-72; Hansaker Prof, Mass Inst Technol, 69-71; mem, adv bd, Solar Energy Res Inst, 85-; mem, Aerospace Eng Bd, 72-76, Long Range Planning & Develop Comt, 83-85; dir, Soc Eng Sci, 73-76, Am Inst Aeronaut & Astronaut, 62-66. *Honors & Awards:* Arthur S Flemming Award, 56; H Julian Allen Award, NASA, 69; Sylvanus Reed Award, Am Inst Aeronaut & Astronaut, 62. *Mem:* Nat Acad Eng; fel Am Inst Aeronaut & Astronaut; fel Am Astron Soc; fel AAAS. *Res:* Supersonic and hypersonic aerodynamics; aerodynamic heating, aerospace vehicles; aerospace research and development management, planning, and policy analysis and development; energy research and development and management. *Mailing Add:* Rann Inc Suite 414 260 Sheridan Ave Palo Alto CA 94306

EGGERS, DAVID FRANK, JR, b Oak Park, Ill, July 8, 22; m 45; c 3. ATOMIC & MOLECULAR SPECTROSCOPY. *Educ:* Univ Ill, BS, 43; Univ Minn, PhD(chem), 51. *Prof Exp:* Asst, Univ Minn, 43-44; chemist, Tenn Eastman Corp, 44-47; instr, 50-52, from asst prof to assoc prof, 52-63, PROF CHEM, UNIV WASH, 63- *Mem:* Am Chem Soc; Optical Soc Am. *Res:* Vibrational spectra and molecular structure; infrared and Raman spectra of solids, liquids and gases; rotational analysis of high resolution gas phase spectra. *Mailing Add:* Dept Chem BC-10 Univ Wash Seattle WA 98195

EGGERS, FRED M, b Des Moines, Iowa, Dec 8, 46; m 71; c 3. PROCESS ENGINEERING, ORGANIZATIONAL DEVELOPMENT. *Educ:* Univ Iowa, BS, 70; Univ Ala, Huntsville, MS, 73. *Prof Exp:* Process engr, 3M Co, Decatur, 70-77; plant mgr, Johnson & Johnson, 77-84; vpres & gen mgr, Consep Membranes, Inc, 84-88; prin, Fred Eggers Consult, 88-89; dir res & develop, 89-90, GEN MGR, MENTOR CORP, IRVING, TEX, 90- *Mem:* Am Soc Qual Control. *Res:* Product development activities in urology and plastic surgery, implantable products; process, product and market development in membrane science, separations and controlled release. *Mailing Add:* Mentor Corp 3041 Skyway Circle North Irving TX 75038

EGGERS, GEORGE W NORDHOLTZ, JR, b Galveston, Tex, Feb 22, 29; m 55; c 2. MEDICINE, ANESTHESIOLOGY. *Educ:* Rice Inst, BA, 49; Univ Tex, MD, 53; Am Bd Anesthesiol, dipl, 59. *Prof Exp:* From instr to asst prof, Med Br, Univ Tex, 56-61; assoc prof, 61-67, PROF ANESTHESIOL, SCH MED, UNIV MO-COLUMBIA, 67-, CHMN DEPT, 70- *Concurrent Pos:* Grants, Med Res Found Tex & Galveston Heart Asn, 59-60, Tex Heart Asn, 60-61, NIH, 61-68; vis res prof, Med Sch, Northwestern Univ, 68. *Mem:* AAAS; fel Am Col Anesthesiol; AMA; Am Soc Anesthesiol; Asn Am Med Cols. *Res:* Human pharmacology and physiology; cardiovascular dynamics; pulmonary circulation. *Mailing Add:* Univ Mo Med Ctr Columbia MO 65212

EGGERS, HANS J, b Baumholder, Ger, July 26, 27. MICROBIOLOGY, VIROLOGY. *Educ:* Univ Heidelberg, Physician, 53, Dr med biochem, 53. *Prof Exp:* Fel virol, Karolinska Inst, Stockholm, Sweden, 54; fel neurol, Univ Nervenklinik, Koeln, Ger, 55-57; fel virol, Children's Hosp Res Found, Cincinnati, Ohio, 57-59; res assoc, Rockefeller Inst, NY, 59-61, asst prof, 61-64; sect head, Max Plack Inst Virus Res, Tuebingen, Ger, 65-66; prof virol, Univ Giessen, Ger, 66-72; PROF VIROL, UNIV KOELN, GER, 72- *Concurrent Pos:* Mem bd, Ger Soc Hyg & Microbiol, 69-87; mem, Working Group AIDS, Comn Europ Communities, Brussels, 84-88; mem bd, Soc Virol, 90-; consult to several govt agencies & sci orgns; mem senate, Ger Acad Naturalists, 90. *Mem:* Am Asn Immunologists; Soc Exp Biol & Med; Am Soc Microbiol; Ger Soc Hyg & Microbiol; Soc Virol; Ger Acad Naturalists; external mem Ger Acad Sci. *Res:* Mechanism of action of selective inhibitors of virus replication; chemotherapy of virus diseases; drug-resistant virus mutants; pathogenesis of virus diseases; viral myocarditis; experimental models of virus-induced diabetes; diagnostic virology. *Mailing Add:* Univ Koeln Fuerst-Pueckler-Str 56 D-5000 Koeln 41 Germany

EGGERT, ARTHUR ARNOLD, b Shawano, Wis, Sept 9, 44; m 66; c 3. INTELLIGENT TUTORING SYSTEMS, CLINICAL LABORATORY DATA BASES. *Educ:* Univ Wis-Madison, BS, 66, PhD(anal chem), 70. *Prof Exp:* Asst prof chem, Duke Univ, 70-71; asst prof, 71-78, dir specimen processing, Clin Labs, 72-76, ASSOC PROF PATH & ASSOC DIR CLIN LABS, UNIV WIS-MADISON, 78-, DIR DATA PROCESSING, 72- *Concurrent Pos:* Dir comput programming, Lab Computing, Inc, 75-76. *Mem:* Am Chem Soc; Asn Comput Mach; Am Asn Clin Chemists; Inst Elec & Electronics Engrs. *Res:* Computer applications in quality control; clinical laboratory information systems; intelligent tutoring systems in chemistry. *Mailing Add:* Clin Labs A4 204 Univ Hosp Madison WI 53792

EGGERT, DONALD A, b Cleveland, Ohio, May 13, 34. PALEOBOTANY. *Educ:* Western Reserve Univ, BA, 56; Yale Univ, MS, 58, PhD(bot), 60. *Prof Exp:* NSF fel, Univ Ill, 60-61; asst prof bot, Southern Ill Univ, 61-65 & Univ Iowa, 65-69; assoc prof, 69-74, PROF BIOL SCI, UNIV ILL, CHICAGO CIRCLE, 74- *Concurrent Pos:* Sigma Xi grant, 62-63; NSF grants, Southern Ill Univ, 63-65 & Univ Iowa, 65-68; vis lectr, Yale Univ, 64-65. *Mem:* Sigma Xi. *Res:* Morphology; anatomy; evolution of vascular land plants with emphasis upon the fossil forms of the late Paleozoic. *Mailing Add:* 823-B S Racine Chicago IL 60607

EGGERT, FRANK MICHAEL, b Hamburg, WGermany, Apr 24, 45; Can citizen; m 76; c 1. PERIODONTAL DISEASES, SECRETORY IMMUNITY. *Educ:* Univ Toronto, DDS, 69, MSc, 71; Univ Cambridge, PhD(immunol), 78. *Prof Exp:* Res fel dent, Royal Col Surgeons Eng, 76-79; lectr oral med & periodont, London Hosp, 79-81; assoc prof, 81-83, PROF PERIODONT, UNIV ALTA, 83- *Concurrent Pos:* Pvt pract, Ala, 82-; chmn, Can Dent Asn Comt, Consumer Prods Recognition, 87- *Mem:* Royal Soc Med; Biochem Soc; Brit Soc Immunol; Brit Soc Dent Res; Int Soc Dent Res; NY Acad Sci. *Res:* Periodontal disease diagnosis and treatment; glycoproteins of secretory immune system and antibody measurement; microbiology of dental pathogens. *Mailing Add:* Fac Dent 2088 DPC Univ Alta Edmonton AB T6G 2N8 Can

EGGERT, FRANKLIN PAUL, b Buffalo, NY, May 13, 20; m 45; c 4. HORTICULTURE, GENERAL AGRICULTURE. *Educ:* Cornell Univ, PhD(pomol), 49. *Prof Exp:* Head dept, 49-63, dean grad sch, 62-75, chmn dept, 84-85, PROF HORT, UNIV MAINE, ORONO, 49- *Concurrent Pos:* Dir res, Univ Maine, Orono, 63-69, actg dir res, 70-71. *Mem:* AAAS; Am Soc Hort Sci; Am Pomol Soc. *Res:* Sustainable agriculture. *Mailing Add:* RR Two Box 365 Verona ME 04416

EGGERT, ROBERT GLENN, b Bennet, Nebr, Feb 27, 27; m 52; c 2. ANIMAL SCIENCE. *Educ:* Univ Nebr, BS, 50, MS, 52; Cornell Univ, PhD(animal nutrit), 54. *Prof Exp:* Asst, Nebr Exp Substa, 50; assoc, animal husb dept, Univ Nebr, 50-52 & Cornell Univ, 52-54; animal nutritionist, Res Div, 54-59, group leader, Agr Div, 59-68, swine prog mgr, 68-72, livestock prog mgr, 72-77, group leader, 77-80, ruminant prog mgr, Agr Res Div, 80-85, SR PROD DEVELOP MGR, AGR RES DIV, AM CYANAMID CO, 85- *Mem:* Am Soc Animal Sci; Am Dairy Sci Asn. *Res:* Amino acid and trace mineral requirements of swine; non-protein nitrogen utilization by ruminants; antibiotics; hormones for growth and reproduction; anthelmintics for swine, cattle, bovine and porcine somatotropine. *Mailing Add:* Am Cyanamid Co PO Box 400 Princeton NJ 08543-0400

EGGIMANN, WILHELM HANS, b Zurich, Switz, Apr 18, 29; m 61; c 2. ELECTRICAL ENGINEERING, SOLID STATE PHYSICS. *Educ:* Swiss Fed Inst Technol, dipl, 54; Case Inst Technol, MS, 59, PhD(elec eng), 61. *Prof Exp:* Asst microwave tech, Swiss Fed Inst Technol, 54-56; electromagnetic theory, Case Inst Technol, 56-61, asst prof elec eng, 61-64; PROF ELEC ENG, WORCESTER POLYTECH INST, 64- *Mem:* AAAS; Inst Elec & Electronics Engrs; Sigma Xi; Am Soc Eng Educ. *Res:* Electromagnetic theory; problems in diffraction theory; wave propagation and microwave techniques; plasma physics; collective interaction in solids; quantum electronics; VLSI design; computer engineering. *Mailing Add:* Dept Elec Eng Worcester Polytech Inst Worcester MA 01609

EGGLER, DAVID HEWITT, b Ashland, Wis, May 15, 40; m 74; c 2. MINERALOGY, PETROLOGY. *Educ:* Oberlin Col, AB, 62; Univ Colo, PhD(geol), 67. *Prof Exp:* Asst prof geol, Tex A&M Univ, 70-72; mem staff geochem, Geophys Lab, Carnegie Inst Wash, 72-77; res assoc, Pa State Univ, 67-70, assoc prof, 77-85, coordr grad progs, 83-88, PROF DEPT GEOSCI, PA STATE UNIV, 85-, ASSOC HEAD GRAD PROG, 88- *Concurrent Pos:* Assoc ed, Am Mineralogist, 79-82. *Honors & Awards:* L R Wager Prize, Int Asn Volcanology & Geochem of Earth's Interior, 79. *Mem:* Am Geophys Union; Geol Soc Am; Mineral Soc Am; Geochem Soc. *Res:* Geology and petrology of Kimberlites; experimental high-pressure, high temperature phase equilibria of magmatic arc rocks, mantle, carbonated melts, redox equilibria, super critical fluids. *Mailing Add:* Dept Geosci Pa State Univ University Park PA 16802

EGGLESTON, FORREST CARY, b New York, NY, Sept 28, 20; m 46; c 2. MEDICINE, SURGERY. *Educ:* Princeton Univ, AB, 42; Cornell Univ, MD, 45; Am Bd Surg & Am Bd Thoracic Surg, dipl. *Prof Exp:* Jr attend surgeon, First Surg Div & Thoracic Surg Dept, Bellevue Hosp, NY, 53; THORACIC SURGEON, CHRISTIAN MED COL, INDIA, 54-, PROF SURG, 55-, HEAD DEPT, 80- *Concurrent Pos:* Med supt & thoracic surgeon, Lady Irwin Sanatorium, 54-74. Mem, Fel Am Col Surg; Asn Surgeons India. *Mailing Add:* Box 1098 Mechanicsburg PA 17055

EGGLESTON, GLEN E, b Salt Lake City, Utah, Aug 20, 23; m 44; c 3. THERMODYNAMICS. *Educ:* Univ Utah, BS, 44; Univ Wash, MS, 49; Purdue Univ, PhD(mech eng), 53. *Prof Exp:* Supvry serv engr, Westinghouse Elec & Mfg Co, 46-48; instr mech eng, Univ Utah, 49-51; design engr, Douglas Aircraft Co, McDonnell Douglas Corp 53-58, rep air condictioning sect, Aircraft Div, 58-59, asst chief mech sect, Missles & Space Systs Div, 59-62, sect chief, 62, asst chief vehicle design br, 62-64, br chief mech sect, 64-67, asst chief engr crew systs, Manned Orbiting Lab Prog, 67-69, asst chief engr environ eng, 69-83; RETIRED. *Concurrent Pos:* Instr, night sch, Carnegie Inst Technol, 46-47; lectr, Univ Calif, Los Angeles, 54-55. *Mem:* Am Soc Mech Engrs. *Res:* Heat transfer in the specific areas of aerodynamic heating. *Mailing Add:* 27161 Fond du Lac Rd Palos Verdes Peninsula CA 90274

EGGLESTON, JANE R, b Rochester, NY, Feb 21, 47. ANTHRACITE COAL EXPLORATION, ONCOLITES & STROMATOLITES. *Educ:* Hobart & William Smith Cols, BA, 69; Syracuse Univ, MS, 72. *Prof Exp:* Consult, F F Slaney Co, Vancouver, BC, 72; environ geologist, BC Res, Vancouver & hydrogeologist-geomorphologist, Westwater Res Ctr, 73; econ geologist, WVa Geol Surv, Morgantown, 73-76; explor geologist, Skelly & Loy, Harrisburg, Pa, 76-78, sr geologist, 78-82; consult geologist, 82-83; GEOLOGIST, US GEOL SURV, RESTON VA, 83- *Concurrent Pos:* Instr, Sem Develop Corp, 81-83. *Mem:* Sigma Xi; Am Inst Prof Geologists; Geol Soc Am; Soc Econ Paleontologists & Mineralogists; Am Inst Mining, Metall & Petrol Engrs; Int Asn Sedimentologists; Am Soc Limnologists & Oceanographers. *Res:* Geologic research and related stratigraphic, structural, sedimentologic and geochemical analysis of carboniferous strata of the Eastern United States. *Mailing Add:* 8425 Idylwood Rd Vienna VA 22182

EGGLESTON, JOHN MARSHALL, solid-state lasers, nonlinear optics; deceased, see previous edition for last biography

EGGLESTON, JOSEPH C, b Memphis, Tenn, Aug 4, 36; m 62; c 3. PATHOLOGY, SURGICAL PATHOLOGY. *Educ:* Duke Univ, AB, 58; Johns Hopkins Univ, MD, 62. *Prof Exp:* PROF PATH, JOHNS HOPKINS UNIV, 85- *Mailing Add:* Dept Path Johns Hopkins Univ 600 N Wolfe St Baltimore MD 21212

EGGLESTON, PATRICK MYRON, b Panama, Apr 20, 41; US citizen; m 66; c 1. PHYCOLOGY, LIMNOLOGY. *Educ:* Mich State Univ, BS, 63; Cornell Univ, MS, 66; Ohio State Univ, PhD(zool), 75. *Prof Exp:* Teacher sci, Chenango Valley Jr High Sch, 65-67; teacher biol, Voorheesville High Sch, 67-70; teaching assoc zool, Ohio State Univ, 70-75; from asst prof to assoc prof, 75-87, PROF BIOL, KEENE STATE COL, 87- *Mem:* Phycol Soc Am. *Res:* Ecology of fresh water algae and crustacea. *Mailing Add:* 69 Timberlane Dr Keene NH 03431

EGGLETON, REGINALD CHARLES, b Hillsdale, Mich, July 6, 20; m 44; c 3. ULTRASOUND, MEDICAL SCIENCE. *Educ:* Eastern Mich Univ, BS, 61; Univ Ill, MS, 66. *Prof Exp:* Engr, Physicists Res Co, 41-47; gen mgr, Haller, Raymond & Brown, Inc, 47-50; head eng develop, Alden Prod Co, 50-53; res assoc elec eng, Univ Ill, 53-57; pres & sr res scientist, Intersci Res Inst, 57-72; assoc prof, 72-78, PROF SURG RES, SCH MED, IND UNIV, INDIANAPOLIS, 78-; DIR, FORTUNE-FRY RES LAB, INDIANAPOLIS CTR ADVAN RES, 72- *Concurrent Pos:* Res assoc, Bioacoust Res Lab, Univ Ill, 57-69; sr res scientist, Ultrasound Res Labs, 78- *Mem:* AAAS; Instrument Soc Am; Am Soc Echocardiography; Am Inst Ultrasound Med; Asn Advan Med Instrumentation. *Res:* Medical applications of ultrasound, both as diagnostic and therapeutic modalities; acoustic microscopy of viable mouse embryo hearts, ultrasonic surgery, acoustic diffraction, spectroscopy and real-time cross-sectional echocardiography. *Mailing Add:* Oral Health Res Inst Rm 145 Ind Univ Sch Dent 415 Lansing St Indianapolis IN 46202

EGLE, DAVIS MAX, b New Orleans, La, Jan 31, 39; m 63; c 2. MECHANICAL ENGINEERING, MECHANICS. *Educ:* La State Univ, BS, 60; Tulane Univ, MS, 62, PhD(mech eng), 65. *Prof Exp:* From asst prof to assoc prof, 65-73, prof, 73-81, DIR, AEROSPACE & MECH ENG, UNIV OKLA, 81- *Honors & Awards:* Okla Regents Teaching Award, 68; Am Soc Nondestructive Testing Achievement Award, 80. *Mem:* Am Soc Mech Engrs; Acoust Soc Am; Am Soc Nondestructive Testing; Am Acad Mech. *Res:* Theoretical and experimental dynamics of solids; nondestructive evaluation. *Mailing Add:* 865 Asp St Rm 212 Norman OK 73019

EGLE, JOHN LEE, JR, b Martinsburg, WVa, July 13, 39; m 61; c 2. PHARMACOLOGY, TOXICOLOGY. *Educ:* Shepherd Col, BS, 61; Univ WVa, MS, 63, PhD(pharmacol), 64. *Prof Exp:* Asst, Univ WVa, 61-64; cardiovasc training, Bowman Gray Sch Med, 64-66; asst prof pharmacol, 68-74, ASSOC PROF PHARMACOL, MED COL VA, VA COMMONWEALTH UNIV, 74- *Mem:* Am Col Pharmacists; Am Soc Pharmacol Exp Therapeut; Soc Toxicol; Sigma Xi. *Res:* Cardiovascular and respiratory pharmacology and toxicology. *Mailing Add:* Med Col Va Va Commonwealth Univ Box 613 Richmond VA 23298

EGLER, FRANK EDWIN, b New York, NY, Apr 26, 11; wid. PLANT ECOLOGY. *Educ:* Univ Chicago, BS, 32; Univ Minn, MS, 34; Yale Univ, PhD(plant ecol), 36. *Prof Exp:* Res fel, Yale & Bishop Mus, 36-37; asst prof forest bot, NY State Col Forestry, Syracuse Univ, 37-44; dir exp sta, Chicle Develop Co, 41-44; PRES ATON FOREST, INC, 45- *Concurrent Pos:* Assoc prof, Univ Conn, 47-48; tech adv, R/W Maintenance Corp, 50-54; res assoc, Dept Conserv, Am Mus Natural Hist, 51-55; Guggenheim fel, 56-58; consult vegetationist, 49-; vis prof, Wesleyan Univ, 62 & Yale Univ, 65; adv, Elec Power Res Inst, 80-86. *Honors & Awards:* Cert Mert, Ecol Soc Am, 78. *Mem:* Fel AAAS; Ecol Soc Am; fel Am Geog Soc; fel Am Mus Natural Hist. *Res:* Vegetation science and management; general ecology. *Mailing Add:* Aton Forest Norfolk CT 06058

EGLI, DENNIS B, b Ft Dodge, Iowa, Sept 4, 42. AGRONOMY, CROP PHYSIOLOGY. *Educ:* Pa State Univ, BS, 65; Univ Ill, MS, 67, PhD(agron), 69. *Prof Exp:* Asst prof, 69-74, assoc prof, 74-79, PROF AGRON, UNIV KY, 79- *Mem:* Am Soc Agron; Crop Sci Soc Am; Am Asn Plant Physiologists; Sigma Xi. *Res:* Crop production, ecology and physiology; soybean production; seed physiology, quality. *Mailing Add:* Dept Agron N-122 Agr Sci Bldg N Univ Ky Lexington KY 40546-0091

EGLI, PETER, b Zurich, Switz, Jan 23, 33; m 56; c 2. HIGH SPECIFIC ACTIVITY TRITIUM COMPOUNDS, STABLE ISOTOPE LABELED COMPOUNDS. *Educ:* Winterthur Tech Inst, BSc, 53. *Prof Exp:* Plant chemist, Swiss Soda Works, 53-56; chemist, Merck Ltd, Montreal, 56-61; radiochemist, Nuclear Res Chemicals, 61-62 & New Eng Nuclear, 62-69; RADIOCHEMIST, BRISTOL-MYERS SQUIBB, 70- *Mem:* Am Chem Soc; Am Inst Chemists. *Res:* Synthesis of carbon-14 and tritium labeled compounds for biological-medical research; basic intermediates, carbohydrates and pharmaceuticals; isotope effects. *Mailing Add:* 21 Continental Lane Titusville NJ 08560

EGLITIS, IRMA, b Riga, Latvia, Oct 13, 07; nat US; m 38. ANATOMY, DERMATOLOGY. *Educ:* State Univ Latvia, MD, 31; State Univ NY Buffalo, cert, 68. *Prof Exp:* From asst instr to instr gross anat, histol, embryol, State Univ Latvia, 31-44; instr gross anat, Ernst Moritz Arndt Univ, Ger, 44-45; from instr to prof, 52-78, EMER PROF ANAT, COL MED, OHIO STATE UNIV, 78- *Honors & Awards:* Cert Merit, AMA. *Mem:* Am Assoc Anat; Am Med Womens Asn; Sigma Xi. *Res:* Integument; visual apparatus; blood vessels. *Mailing Add:* 123 E Lane Columbus OH 43201

EGLITIS, MARTIN ALEXANDRIS, b Washington, DC, June 12, 55; m 86. GENE TRANSFER & EXPRESSION, MAMMALIAN DEVELOPMENTAL BIOLOGY. *Educ:* Univ Va, BA, 76, PhD(anat), 80. *Prof Exp:* Fel, Roche Inst Molecular Biol, 80-82; staff fel, lab molecular hemat, Nat Heart, Lung & Blood Inst, NIH sr staff fel, 86-88; LAB DIR, GENETIC THER INC, 88- *Mem:* Soc Develop Biol; AAAS. *Res:* Transfer of genes into whole animals; eukaryotic expression vectors; gene function in development. *Mailing Add:* 3022 Ferndale St Kensington MD 20895

EGLOFF, DAVID ALLEN, b Mason City, Iowa, Apr 13, 35; m 59; c 2. OCEANOGRAPHY, LIMNOLOGY. *Educ:* Amherst Col, BA, 57; Yale Univ, MS, 59; Stanford Univ, PhD(biol), 67. *Prof Exp:* From asst prof to assoc prof biol, Oberlin Col, 66-81, dir, Environ Studies Prog, 79-85, chair, Dept Biol, 88-92, PROF BIOL, OBERLIN COL, 81- *Mem:* Am Soc Limnol & Oceanog; Freshwater Biol Asn; Int Soc Limnol. *Res:* Ecology and behavior of marine rotifers. *Mailing Add:* Dept Biol Oberlin Col Oberlin OH 44074

EGLOFF, JULIUS, III, b Washington, DC, Sept 19, 46; m 80; c 4. MARINE GEOLOGY, OIL & GAS EXPLORATION. *Educ:* Univ Miami, BS, 69. *Prof Exp:* Cartographer, US Naval Oceanogr Off, 66-76; oceanogr marine geol, Seafloor Geosci Div, Naval Ocean Res & Activ, 76-91; GEOLOGIST, NAVAL RES LAB, 91- *Concurrent Pos:* Chmn, Ocean Study Group Am, 78-; owner & consult, Re-Evaluations Oil & Gas Explor, Inc, 80-; pres & chief geologist, Deep Ventures, Ltd, Oil & Gas Explor & Prod, Co, 83-86. *Mem:* Am Geophys Union; fel Explorers Club; Am Asn Petrol Geologists. *Res:* Submarine mapping and sampling; oil, gas and gold exploration. *Mailing Add:* PO Box C Pass Christian MS 39571

EGLY, RICHARD S(AMUEL), b Grabill, Ind, July 6, 14; m 49; c 4. CHEMICAL ENGINEERING. *Educ:* Purdue Univ, BS, 36; Univ Ill, MS, 38, PhD(chem eng), 40. *Prof Exp:* Asst chem, Univ Ill, 36-40; chem engr, Commercial Solvents Corp, 40-43, chief chem eng group, 43-50, dir chem res, 50-55, dir nitroparaffin develop, 55-57, assoc sci dir, 58-61, dir process develop, 62-76; CONSULT, 76- *Concurrent Pos:* With Off Sci Res & Develop, 44; pres & co-mgr, Word Christian Bookstore, Inc. *Mem:* Am Chem Soc; Soc Indust Chem; Am Inst Chem Engrs; AAAS. *Res:* High pressure properties and techniques; heat transfer; manufacture of aliphatic amines; nitration of hydrocarbons and reactions of nitroparaffins; explosives; propellants; hazard potential of chemicals. *Mailing Add:* Box 450 RR 15 West Terre Haute IN 47885-9515

EGNER, DONALD OTTO, b Cleveland, Ohio, Apr 18, 28; m 50; c 4. COMPUTER SCIENCES, APPLIED MATHEMATICS. *Educ:* Western Md Col, BA, 49. *Prof Exp:* Chief surv sect, Health Physics Group, Chem Center, US Army, 49-51, physicist, Spec Projs Div, Chem Res & Develop Lab, 51 & Nuclear Defense Lab, 53-55, chief, Theoret Physics Sect, 55-62, phys scientist, Land War Lab, 62-74, opers res analyst & head, law enforcement technol team, 74-80, directorate chief, Human Eng Lab, 80-86; INDEPENDENT TECHNOL CONSULT, 86-; INSTR MATH, ESSEX COMMUNITY COL, 86- *Concurrent Pos:* Tech dir, Tech Assocs, Inc, 68-74. *Mem:* Soc Rheol; NY Acad Sci. *Res:* Atmospheric physics; micrometeorology; air pollution; biomedical research; human engineering; technical management. *Mailing Add:* 5806 Pine Hill Dr White Marsh MD 21162

EGORIN, MERRIL JON, DRUG DEVELOPMENT, CLINICAL PHARMACOLOGY. *Educ:* Johns Hopkins Univ, MD, 73. *Prof Exp:* HEAD DEVELOP THERAPEUT, JOHNS HOPKINS UNIV, 82- *Mailing Add:* Dept Develop Therapeut Univ Md Cancer Ctr 655 W Baltimore St Baltimore MD 21201

EHLE, BYRON LEONARD, b Seattle, Wash, Jan 18, 37; m 58; c 4. NUMERICAL ANALYSIS, COMPUTER SCIENCE. *Educ:* Whitman Col, BA, 59; Stanford Univ, MSc, 61; Univ Waterloo, PhD(comput sci), 69. *Prof Exp:* Teaching asst, Stanford Univ, 59-61; from instr to asst prof, 61-77, ASSOC PROF MATH & COMPUT SCI, UNIV VICTORIA, BC, 77- *Concurrent Pos:* Lectr, Univ Waterloo, 66-69, res assoc, 68-69. *Mem:* Asn Comput Mach; Am Math Soc; Math Asn Am; Can Info Processing Soc. *Res:* Numerical methods for the solution of ordinary differential equations, particularly stiff equations; development of effective teaching tools for computer software; systems analysis. *Mailing Add:* Dept Comput Sci Univ Victoria Box 1700 Victoria BC V8W 2Y2 Can

EHLE, FRED ROBERT, b Bayside, NY, Feb 17, 52. ANIMAL SCIENCE, NUTRITION. *Educ:* Cornell Univ, BS, 74, MS, 77, PhD(animal nutrit), 80. *Prof Exp:* Res assoc dairy nutrit, Univ Ill, 79-80; RES ANIMAL SCIENTIST, AGR RES SERV, USDA, 80- *Concurrent Pos:* Adj asst prof, Dept Animal Sci, Univ Minn, 80- *Mem:* Am Soc Animal Sci; Am Dairy Sci Asn; Am Forage & Grassland Coun; Coun for Agr Sci & Technol. *Res:* Maximize the intake and efficiency of utilization of forages in ruminant diets. *Mailing Add:* 4669 Hillcrest Dr Middleton WI 53562

EHLER, ARTHUR WAYNE, b Los Angeles, Calif, May 21, 22; m 57; c 2. PLASMA PHYSICS. *Educ:* Univ Southern Calif, BE, 47, BA, 48, MS, 50, PhD(physics), 55. *Prof Exp:* Proj engr, Aerophys Develop Corp, 54-56; asst chief res sect plasma physics, Douglas Aircraft Co, 56-60; mem tech staff, Hughes Res Labs, 60-70; mem tech staff, Ames Res Ctr, 70-72; staff mem, Los Alamos Sci Lab, 72-84; RETIRED. *Mem:* AAAS; Am Phys Soc; Sigma Xi. *Res:* Absorption of radiation by plasma; production of plasma by lasers magnetically driven shocks; electric field dissociation of molecular ions; analysis of proposed methods of producing hot, dense plasma; molecular laser research. *Mailing Add:* 3641 Pebble Pl Newbury Park CA 91320-5060

EHLER, KENNETH WALTER, b Chicago, Ill, June 9, 46. BIO-ORGANIC CHEMISTRY. *Educ:* Ariz State Univ, BS, 68; Univ Ariz, PhD(chem), 72. *Prof Exp:* NIH fel chem, Swiss Fed Inst Technol, 72-73; fel, ICN Nucleic Acid Res Inst, 73-74; fel chem, Salk Inst Biol Studies, 74- *Res:* Development of methods for amide bond formation in aqueous solution; molecular basis of spiritual and psychic phenomena, cosolargy. *Mailing Add:* 3657 W 18th Ave Apt 7 Eugene OR 97402

EHLERINGER, JAMES RUSSELL, b Portland, Ore, July 2, 49. PHYSIOLOGICAL ECOLOGY, PHOTOSYNTHESIS. *Educ:* San Diego State Univ, BS, 72, MS, 73; Stanford Univ, PhD(biol), 77. *Prof Exp:* PROF BIOL, UNIV UTAH, 77- *Honors & Awards:* Murray Bell Award, Ecol Soc Am, 77; Humboldt, 84. *Mem:* Ecol Soc Am; Am Soc Plant Physiologists. *Res:* Physiology and ecology of arid land plants, with an emphasis on photosynthesis, water relations, and energy balance; stable isotopes. *Mailing Add:* Biol Dept Univ Utah Salt Lake City UT 84112

EHLERS, ERNEST GEORGE, b New York, NY, Jan 17, 27; m 51; c 2. MINERALOGY. *Educ:* Univ Chicago, MS, 50, PhD(geol), 52. *Prof Exp:* Asst geol, Univ Chicago, 50-51; geologist, NJ Zinc Co, 52-54; from asst prof to prof mineral, 54-72, prof geol & mineral, 72-84, EMER PROF GEOL & MINERAL, OHIO STATE UNIV, 84- *Concurrent Pos:* Fulbright sr lectr awards, State Univ Utrecht, 66-67 & Univs Athens, Patras & Thessaloniki, 71-72. *Mem:* Fel Mineral Soc Am; fel Geol Soc Am; Sigma Xi. *Res:* Hydrothermal and high pressure equilibria; petrology; optical mineralogy. *Mailing Add:* 402 S Church St Fairhope AL 36532

EHLERS, FRANCIS EDWARD, b Portland, Ore, Nov 5, 16; m 44; c 4. APPLIED MECHANICS, AERODYNAMICS. *Educ:* Ore State Col, BS, 41; Brown Univ, MS, 47, PhD(appl math), 49. *Prof Exp:* Asst math, Ore State Col, 41-42; staff mem, Radiation Lab, Mass Inst Technol, 42-45; from intern to assoc grad div appl math, Brown Univ, 47-50; instr math, Ore State Col, 50-51; aerodynamicist, Boeing Co, 51-54, res specialist, Math Serv Unit, Phys Res Staff, 54-58 & Math Lab, Sci Res Labs, 58-72, res specialist, Aerodyn Res Group, Boeing Com Airplane Co, 72-78, res specialist, Appl Math Group, Boeing Computer Servs, 78-84, consult, Boeing Airplane Co, 84-88; RETIRED. *Mem:* Am Math Soc; Soc Indust & Appl Math; assoc fel Am Inst Aeronaut & Astronaut; Acoust Soc Am. *Res:* Fluid mechanics; linearized theory; hodograph method; gas dynamics; mechanics; elasticity; transonic unsteady flow; supersonic flow. *Mailing Add:* 19343 Occidental Ave S Seattle WA 98148

EHLERS, KENNETH WARREN, b Dix, Nebr, Aug 3, 22; m 47; c 1. PLASMA PHYSICS, ATOMIC PHYSICS. *Educ:* Mass Inst Technol, Elec Eng, 46. *Prof Exp:* Radar technician, US Navy, 42-46; head, Electronic Landing Aids Dept, Landing Aids Exp Sta, Calif, 46-50; staff physicist, 50-78, sr physicist, Lawrence Berkeley Lab, 78-86, EMER REGENTS PROF, UNIV CALIF, 86- *Concurrent Pos:* Consult, Appl Radiation Corp, Thompson Ramo Wooldridge, Inc, Brobeck Assoc & Phys Dynamics & Sci Applns Int Corp. *Honors & Awards:* Physics Medal, USSR Acad Sci, 90. *Mem:* AAAS; Am Phys Soc; Am Vacuum Soc; Inst Elec & Electronics Engrs. *Res:* Ion beams and sources; particle accelerators; controlled thermonuclear research. *Mailing Add:* Lawrence Berkeley Lab Bldg 4 Rm 214 Univ Calif Berkeley CA 94720

EHLERS, MARIO RALF WERNER, b Pretoria, SAfrica, Feb 21, 59; Ger citizen; m 82; c 2. MOLECULAR ENZYMOLOGY, BIOCHEMICAL BASIS OF HYPERTENSION. *Educ:* Univ Cape Town, SAfrica, MBChB, 82, PhD(med), 87. *Prof Exp:* Intern, Groote Schuur Hosp, Cape Town, SAfrica, 83; med officer, Med Res Coun Liver Res Ctr, Univ Cape Town, 84-87; postdoctoral fel, Ctr Biochem & Biophys Sci & Med, 87-89, INSTR BIOCHEM, HARVARD MED SCH, 89- *Concurrent Pos:* Consult staff, Liver Clin, Groote Schuur Hosp, Cape Town, SAfrica, 84-87, tutor, Dept Internal Med, 84-87; tutor, Metab & Function Course, Harvard Med Sch, 89- *Mem:* AAAS; SAfrican Med & Dent Coun; Gen Med Coun UK. *Res:* Molecular enzymology of angiotensin-convertic enzyme, an enzyme critical to blood pressure regulation; molecular cloning; structure and function relationships; determination of the catalytic mechanism. *Mailing Add:* CBBSM Rm A-6 Brigham & Women's Hosp 75 Francis St Boston MA 02115

EHLERT, THOMAS CLARENCE, b Milwaukee, Wis, July 1, 31; div; c 4. PHYSICAL CHEMISTRY. *Educ:* Univ Wis, BS, 57, MS, 58, PhD(chem), 63. *Prof Exp:* Instr chem, Univ Wis, Milwaukee, 57-60, US Dept Defense fel, 63-64; from asst prof to assoc prof, 64-78, PROF CHEM, MARQUETTE UNIV, 78- *Mem:* Am Chem Soc; AAAS. *Res:* High temperature chemistry; mass spectroscopy; thermodynamics. *Mailing Add:* Dept Chem Marquette Univ Milwaukee WI 53233

EHLIG, PERRY LAWRENCE, b San Gabriel, Calif, May 23, 27; m 51; c 5. PETROLOGY, GEOLOGY. *Educ:* Univ Calif, Los Angeles, BA, 52, PhD(geol), 58. *Prof Exp:* Assoc prof, 56-67, PROF GEOL, CALIF STATE UNIV, LOS ANGELES, 67- *Mem:* Geol Soc Am; Mineral Soc Am; Am Asn Petrol Geol; Nat Asn Geol Teachers. *Res:* Metamorphic petrology, particularly the evolution of Pelona and similar schists in California; structural geology; history of the San Andreas fault. *Mailing Add:* Dept Geol Calif State Univ 5151 State University Dr Los Angeles CA 90032

EHLIG-ECONOMIDES, CHRISTINE ANNA, b Pasadena, Tex, June 8, 49; m 76; c 1. GEOTHERMAL RESERVOIR ENGINEERING. *Educ:* Rice Univ, BA, 71; Univ Kans, MAT, 74, MS, 76; Stanford Univ, PhD(petrol eng), 79. *Prof Exp:* Res asst petrol eng, Stanford Univ, 76-78, prog mgr Geothermal Prog, 78-80, actg asst prof petrol eng, 79-80; asst prof, 80-81, DEPT HEAD PETROL ENG, UNIV ALASKA, FAIRBANKS, 81- *Concurrent Pos:* Engr, Shell Develop Co, 77, 79 & 81. *Mem:* Soc Petrol Engrs; Sigma Xi. *Res:* Pressure transient testing of petroleum and geothermal wells; well testing; reservoir simulation; petrophysics; simulation of groundwater formations. *Mailing Add:* 2346 Underwood Houston TX 77030

EHLMANN, ARTHUR J, b St Charles, Mo, May 18, 28; m 85; c 2. MINERALOGY. *Educ:* Univ Mo, BS, 52, MA, 54; Univ Utah, PhD(mineral), 58. *Prof Exp:* Subsurface geologist, Shell Oil Co, 54-56; from asst prof to assoc prof, 58-70, chmn dept, 76-80 & 83-88, PROF GEOL, TEX CHRISTIAN UNIV, 76- *Concurrent Pos:* Hons prof, Tex Christian Univ, 78; Herndon prof geol, 81. *Mem:* Mineral Soc Am; Soc Econ Paleont & Mineral. *Res:* Mineralogy; genesis of clays; hydrothermal synthesis; trace studies; x-ray diffraction studies. *Mailing Add:* Dept Geol Tex Christian Univ Ft Worth TX 76129

EHMANN, EDWARD PAUL, b Rochester, NY, June 6, 40; m 67; c 3. FOOD SCIENCE, QUALITY ASSURANCE. *Educ:* Univ Toronto, BA, 63, MA, 65; Univ Mass, PhD(food sci & technol), 72. *Prof Exp:* Lectr, Univ Toronto, 65-68; engr, Westinghouse Res Labs, 72-74; sr food scientist, H J Heinz Co, 74-76; MGR, RES & TECH SERVS, PEPSICO, INC, 77-; MGR, QUAL ASSURANCE, PEPSI COLA CO, 85- *Mem:* Inst Food Technologists; Sigma Xi; Am Chem Soc. *Res:* Food colorimetry/appearance measurement; research management, administrative & technical; technical corporate quality assurance; quality auditing. *Mailing Add:* Pepsi Cola Co 350 Columbus Ave Valhalla NY 10595

EHMANN, WILLIAM DONALD, b Madison, Wis, Feb 7, 31; m 55; c 4. RADIOCHEMISTRY, GEOCHEMISTRY. *Educ:* Univ Wis, BS, 52, MS, 54; Carnegie Inst Technol, PhD(radiochem), 57. *Prof Exp:* Asst, Univ Wis, 52-54; AEC proj chemist, Carnegie Inst Technol, 54-57, res assoc, 57; res assoc, Nat Res Coun-Nat Acad Sci, Argonne Nat Lab, 57-58, consult, 59-67; from asst prof to assoc prof, Univ Ky, 58-66, chmn & dir grad studies, Dept Chem, 72-76, assoc dean res, Grad Sch 80-84, PROF CHEM, UNIV KY, 66-, HON ASSOC, SANDERS BROWN CTR AGING, 87- *Concurrent Pos:* proj dir, US AEC, Univ Ky, 60-71, NASA, 68-77 & NIH, 77-80, 84-91 & Dept Energy, 83-85, alumni res award, 64, univ res prof, 77-78; Fulbright res scholar & hon res fel, Inst Advan Studies, Australian Nat Univ, 64-65; distinguished prof, Col Arts & Sci, 68-69; vis prof, Ariz State Univ, 69, Fla State Univ, 72; invited lectr, Advan Study Inst, NATO, Oslo, 70; bd dir, Univ Ky Res Found, 90- *Honors & Awards:* Sturgill Award, 87. *Mem:* Fel AAAS; fel Meteoritical Soc; Am Chem Soc; Soc Environ Geochem & Health; Int Asn Geochem & Cosmochem; Sigma Xi. *Res:* Radioanalytical chemistry, activation analysis; applications of radioanalytical chemistry to problems in geochemistry, cosmochemistry, lunar and meteorite evolution, environmental chemistry and trace element's role in the aging and diseases of the human brain; over 175 research publications. *Mailing Add:* Dept Chem Univ Ky Lexington KY 40506-0055

EHRENBECK, RAYMOND, mechanical engineering, for more information see previous edition

EHRENFELD, DAVID W, b New York, NY, Jan 15, 38; m 70; c 4. ECOLOGY, CONSERVATION. *Educ:* Harvard Univ, BA, 59, MD, 63; Univ Fla, PhD(zool), 67. *Prof Exp:* Interim asst prof biol sci, Univ Fla, 67; from asst prof to assoc prof, Barnard Col, Columbia Univ, 67-74; PROF BIOL, COOK COL, RUTGERS UNIV, NEW BRUNSWICK, 74- *Concurrent Pos:* Mem, Adv Panel on Biol Diversity, Off Technol Assessment, US Congress, 85-86; Ed-in-chief, Conserv Biol, Soc Conserv Biol, 87- *Mem:* Ecol Soc Am; fel AAAS. *Res:* Orientation and navigation of sea turtles; supporting studies of visual and olfactory physiology of sea turtles; biological conservation; endangered species; philosophy of conservation, fiction. *Mailing Add:* Dept Environ Res Cook Col Rutgers Univ PO Box 231 New Brunswick NJ 08903

EHRENFELD, ELVERA, b Philadelphia, Pa, Mar 1, 42. CELL BIOLOGY, VIROLOGY. *Educ:* Brandeis Univ, BA, 62; Univ Fla, PhD(biochem), 67. *Prof Exp:* From asst prof to assoc prof cell biol, Albert Einstein Col Med, 69-74; assoc prof biochem & microbiol, 74-79, PROF BIOCHEM, CELL, VIRAL & MOLECULAR BIOL, MED CTR, UNIV UTAH, 79- *Concurrent Pos:* Fel, Albert Einstein Col Med, 67-69; NSF res grant, 70-; USPHS career develop award, 71-76; NIH res grant. *Res:* RNA virus replication and inhibition of host-cell function; regulation of gene expression in animal cells. *Mailing Add:* Dept Microbiol Univ Utah Med Ctr Salt Lake City UT 84132

EHRENFELD, JOHN R(OOS), b Chicago, Ill, May 16, 31; m 56, 83; c 3. CHEMICAL ENGINEERING, ENVIRONMENT & RESOURCES POLICY. *Educ:* Mass Inst Technol, BS, 53, ScD, 57. *Prof Exp:* Proj leader, Arthur D Little Inc, 57-61; staff engr, Prototech, Inc, 61-62; dep dir, appl sci lab, GCA Tech Div, GCA Corp, 62-65, dir, 65-66, dir appl res opers, 66-68; pres, Walden Res Corp, 68-75; vpres & tech dir, Energy Resources Co, 75-78; chmn, New England River Basin Comn, 78-81; sr staff mem, Arthur D Little Inc, 81-85; sr scientist, ABT Assoc Inc, 85-86; SR RES ASSOC & COORDR HAZARDOUS SUBSTANCE MGT PROG, MASS INST TECHNOL, 86- *Mem:* AAAS; Am Chem Soc (chmn, Indust & Eng Chem Div, 77); Am Inst Chem Engrs; Air & Waste Mgt Asn; Soc Risk Anal. *Res:* Environmental management policy; technology and the environment; hazardous substances control. *Mailing Add:* 24 Percy Rd Lexington MA 02173

EHRENFELD, ROBERT LOUIS, b New York, NY, Sept 18, 21; m 55; c 3. ORGANIC CHEMISTRY. *Educ:* Cornell Univ, AB, 42, PhD(org chem), 48. *Prof Exp:* Asst, Cornell Univ, 42-43; res chemist, Off Sci Res & Develop & Manhattan Proj, Columbia Univ, 43-46; res assoc, Mass Inst Technol, 48-49; PRES, HALOCARBON PRODS CORP, 50- *Mem:* Am Chem Soc. *Res:* Preparation, polymerization and depolymerization of organic fluorine compounds; reactions of free fluorine; photochemical chlorination of chlorofluoro ethylenes as related to their fluorination with elementary fluorine. *Mailing Add:* Halocarbon Prods Corp 82 Burlews Ct Hackensack NJ 07601

EHRENFELD, SYLVAIN, b Antwerp, Belgium, June 4, 29; nat US; m 55; c 2. MATHEMATICAL STATISTICS. *Educ:* City Col New York, BS, 50; Columbia Univ, MS, 51, PhD(math statist), 56. *Prof Exp:* Res assoc statist, Res Div, NY Univ, 57-58, assoc prof indust eng & opers res, 58-61; indust eng, Columbia Univ, 61-66; prof opers res, Sch Eng & Sci, NY Univ, 66-73; PROF STATIST, BARUCH COL, CITY UNIV NEW YORK, 73- *Mem:* Fel Am Statist Asn; Inst Math Statist; Inst Mgt Sci; Opers Res Soc Am. *Res:* Statistical methods in experimental design; sequential decision making; utility inference. *Mailing Add:* Baruch Col Dept Statist 17 Lexington Ave New York NY 10010

EHRENHAFT, JOHANN L, b Vienna, Austria, Oct 10, 15; nat US; m 53; c 1. SURGERY. *Educ:* Univ Iowa, MD, 38. *Prof Exp:* Intern surg, Hopkins Hosp, 38-39; resident, 40-42, 45-47, from instr to assoc prof, 47-53, chmn div thoracic surg, 49-86, PROF GEN SURG, COL MED, UNIV IOWA, 53- *Concurrent Pos:* Halsted fel, Johns Hopkins Hosp, 39-40; fel, Barnes Hosp, St Louis, Mo, 48-49. *Mem:* Am Surg Asn; AMA; fel Am Col Surg; Asn Thoracic Surg; Am Col Chest Physicians; Sigma Xi. *Res:* Thoracic and cardiac disease; subjects in cardiovascular surgery. *Mailing Add:* Gen Hosp Iowa State Univ Iowa City IA 52240

EHRENPREIS, LEON, b Brooklyn, NY, May 22, 30; m 61; c 8. MATHEMATICS. *Educ:* City Col, BS, 50; Columbia Univ, MA, 51, PhD(math), 53. *Prof Exp:* Instr math, Johns Hopkins Univ, 53-54; mem, Inst Adv Study, 54-57; assoc prof, Brandeis Univ, 57-59 & Yeshiva Univ, 59-61; prof, Courant Inst Math Sci, NY Univ, 62-68; prof, Yeshiva Univ, 68-; PROF MATH, TEMPLE UNIV. *Mem:* Am Math Soc. *Res:* Partial differential equations; lie groups; several complex variables; automorphic functions; theory of distributions; methodology in learning for students and researchers. *Mailing Add:* Dept Math Temple Univ Philadelphia PA 19122

EHRENPREIS, SEYMOUR, b Brooklyn, NY, June 20, 27; m 54; c 3. PHARMACOLOGY, BIOCHEMISTRY. *Educ:* City Col NY, BS; NY Univ, PhD, 54. *Prof Exp:* Res assoc biochem, Sch Med, Univ Pittsburgh, 53-55; res assoc & instr chem, Cornell Univ, 55-57; res assoc & asst prof biochem, Col Physicians & Surgeons, Columbia Univ, 57-61; assoc prof pharmacol, Sch Med, Georgetown Univ, 61-68; assoc prof pharmacol & head lab molecular pharmacol, New York Med Col, 68-70; chief pharmacol, NY State Res Inst Neurochem & Drug Addiction, 71-76; chmn, 76-85, PROF, DEPT PHARMACOL, CHICAGO MED SCH, 76- *Concurrent Pos:* Burger lectr, Univ Va, 65; mem vis sci prog, Am Pharmaceut Asn, 68-72; ed, Neurosci Res, 69-73 & Rev Neurosci, 74-76; adj prof pharmacol, Sch Pharmaceut Sci, Columbia Univ, 72-75; vis prof, Dept Pharmacol, Keio Univ, Tokyo, Japan, 74, Dept Physiol, Showa Med Sch, Tokyo, Japan, 85, Univ London, London, 88 & 89. *Honors & Awards:* Morris L Parker Award, Outstanding Res Chicago Med Sc, 81. *Mem:* Fel AAAS; fel Am Inst Chem; Am Soc Biol Chem; Am Soc Pharmacol & Exp Therapeut; fel Am Col Clin Pharmacol. *Res:* Protein interactions and modifications; kinetics and equilibria in the fibrogen-thrombin system; molecular mechanisms of nerve activity and neurotropic drug action; mechanism of opiate action; isolation of drug receptors; cholinesterase; endorphins; enkephalinase inhibitors and analgesia cardiovascular pharmacology. *Mailing Add:* Dept Pharm Chicago Med SCh 3333 Green Bay Rd North Chicago IL 60064

EHRENREICH, HENRY, b Frankfurt, Ger, May 11, 28; nat US; m 53; c 3. THEORETICAL PHYSICS. *Educ:* Cornell Univ, AB, 50, PhD(physics), 55. *Hon Degrees:* AM, Harvard Univ, 63. *Prof Exp:* Theoret physicist, Res Lab, Gen Elec Co, 55-63; Gordon McKay prof appl physics, 63-82, CLOWES PROF SCI, HARVARD UNIV, 82- *Concurrent Pos:* Vis lectr, Harvard Univ, 60-61; vis prof, Brandeis Univ & Univ Paris, 69; mem solid state comn, Int Union Pure and Appl Physics, 72-81; mem, Dept Defense, DARPA Mat Res Coun, 72-; vis prof, Univ Penn, 76; mem, White House Sr Sci Panel on Indo-US Sci & Technol Coop, 82-88; dir, Mat Res Lab, Harvard Univ, 82-90; Hund-Klemm Lectr, Max Planck Inst, Stuttgart, 84; chmn, panel pub affairs, Am Phys Soc, 91. *Honors & Awards:* Hume-Rothery Award, Metall Soc, AIME, 84. *Mem:* Fel Am Acad Sci; fel Am Phys Soc; fel Am Asn Advan Sci. *Res:* Electronic structure, optical, transport, and magnetic properties of alloys, superlattices and nanostructures; theory of chemically, structurally, and magnetically disordered semiconducting and metallic alloys. *Mailing Add:* Div Appl Scis Harvard Univ Pierce Hall Cambridge MA 02138

EHRENREICH, JOHN HELMUTH, b New London, Wis, Feb 17, 29; m 54; c 2. FORESTRY, RANGE MANAGEMENT. *Educ:* Colo State Univ, BS, 51, MS, 54; Iowa State Univ, PhD(plant ecol), 57. *Prof Exp:* Res aide, Colo State Univ, 53-54; from asst to instr gen bot, Iowa State Univ, 55-57; proj leader range & wildlife res, US Forest Serv, 57; res assoc forestry, Univ Mo, 57-60, assoc prof, 60-64; prof range mgt, Univ Ariz, 64-65, prof watershed mgt & head dept, 65-71; prof forestry, 71-77, PROF RANGE RESOURCES, UNIV IDAHO, 77-, DEAN COL FORESTRY, WILDLIFE & RANGE SCI, 71- DIR FORESTRY, WILDLIFE & RANGE EXP STA, 74- *Mem:* Ecol Soc Am; Am Soc Range Mgt; Soc Am Foresters; Wildlife Soc. *Res:* Ecological research in soil-plant relations. *Mailing Add:* Col Forestry Wildlife & Range Sci Univ Idaho Moscow ID 83843

EHRENREICH, THEODORE, b New York, NY, July 30, 13. PATHOLOGY. *Educ:* Univ Paris, MD, 39; Am Bd Path, dipl. *Prof Exp:* Pathologist, Bronx Vet Admin Hosp, 48-58; dir labs, St Francis Hosp, 58-66; dir labs, 68-81, DIR, DEPT PATH & CLIN LABS, LUTHERAN MED CTR, BROOKLYN, 81-; ASSOC CLIN PROF, COMMUNITY & ENVIRON MED, MT SINAI SCH MED, 69- *Concurrent Pos:* Littauer fel path, Harvard Univ, 42-44; assoc prof path, NY Med Col, 56-66, clin prof path, 66-; consult cancer res, Bronx Vet Admin Hosp; lectr forensic med, Sch Med, NY Univ, 67-74, asst prof, 74-; attend pathologist, Metrop Hosp 68- *Mem:* Am Soc Clin Path; fel Col Am Path; AMA; Am Asn Path; NY Acad Med. *Res:* Pathologic anatomy; cancer; nephropathology; environmental sciences. *Mailing Add:* 400 E 57th St New York NY 10022

EHRENSON, STANTON JAY, b New York, NY, Oct 13, 31; m 60; c 2. THEORETICAL & PHYSICAL ORGANIC CHEMISTRY. *Educ:* Long Island Univ, BS, 52; Univ Wis, MS, 54; Ga Inst Technol, PhD(chem), 57. *Prof Exp:* Fel chem, Pa State Univ, 57-58, instr, 58-59; res assoc, Univ Chicago, 59-62; assoc chemist, 63-66, chemist, 66-77, SR CHEMIST, BROOKHAVEN NAT LAB, 77- *Concurrent Pos:* Vis scientist, Dept Chem, Univ Calif, Irvine, 68; mem adv ed bd, Jour Org Chem, 73-78. *Mem:* Fel AAAS; Am Chem Soc; Sigma Xi. *Res:* Molecular structure; quantum chemistry; reaction mechanisms; structure-activity relationships. *Mailing Add:* 35 Hillcrest Ave Pt Jefferson NY 11777

EHRENSTEIN, GERALD, b New York, NY, Sept 27, 31; m 60; c 3. BIOPHYSICS. *Educ:* Cooper Union, BEE, 52; Columbia Univ, MA, 58, PhD(physics), 62. *Prof Exp:* Asst microwave spectros, Columbia Univ Radiation Lab, 57-62; PHYSICIST, NIH, 62-, ACTG CHIEF, LAB BIOPHYSICS, 89- *Concurrent Pos:* Mem, Sci Technol Adv Comt, Wash Tech Inst; mem corp, Marine Biol Lab, Woods Hole, Mass; vis prof, Israel Inst Technol Sch Med, 72-73. *Mem:* Am Phys Soc; Biophys Soc; Sigma Xi. *Res:* Mechanism of nervous excitation; mechanism of calcium-dependent secretion; molecular structure. *Mailing Add:* Bldg 9 Rm 1E124 NIH Bethesda MD 20892

EHRENSTORFER, SIEGLINDE K M, inorganic chemistry, physical chemistry; deceased, see previous edition for last biography

EHRENTHAL, IRVING, b New York, NY, Sept 22, 18; m 46; c 3. BIOCHEMISTRY. *Educ:* Yeshiva Univ, BA, 39; Univ Mich, MS, 40; Univ Minn, PhD(agr biochem), 50. *Prof Exp:* Org chemist, Org Med Res, Wm R Warner & Co, 46-47; sr biochemist, Maimonides Hosp, 51; sr chemist vitamins, Nopco Chem Co, 51-53; group leader carbohydrates, 54-76, RES ASSOC, CORN PROD RES DEPT, ANHEUSER-BUSCH, INC, 76- *Mem:* Am Chem Soc. *Res:* Carbohydrates; starch and starch hydrolyzates; polysaccharide structure. *Mailing Add:* 1003 Laval St Louis MO 63132

EHRET, ANNE, b Belleville, Ill, Oct 10, 37. ORGANIC CHEMISTRY. *Educ:* Univ Ill, Urbana, BS, 59; Univ Calif, Los Angeles, MS, 63, PhD(org chem), 67. *Prof Exp:* Asst prof chem, Drake Univ, 66-71; fel, Harvard Univ, 71-74, Brandeis Univ, 74-78; SCIENTIST, POLAROID CORP, 78- *Mem:* Am Chem Soc. *Res:* Investigation of mechanisms of organic reactions. *Mailing Add:* Polaroid Corp 750 Main St 4J Cambridge MA 02139-1563

EHRHART, INA C, b Yoe, Pa, Aug 30, 41; m 61; c 1. MEDICAL PHYSIOLOGY. *Educ:* Ohio State Univ, BS, 68, PhD(physiol), 72. *Prof Exp:* Res asst physiol, Ohio State Univ, 69, NIH spec nurse res fel, 69-72; NIH fel physiol, Mich State Univ, 72-74; asst prof, 74-81, ASSOC PROF PHYSIOL, MED COL GA, 81- *Mem:* Am Physiol Soc; Sigma Xi. *Res:* Lung microvessel permeability. *Mailing Add:* Dept Physiol-Endocrinol Sch Med Med Col Ga Augusta GA 30912-3000

EHRHART, L ALLEN, COLLAGEN METABOLISM, ATHEROSCLEROSIS. *Educ:* Univ Wis-Madison, PhD(biochem), 64. *Prof Exp:* MEM STAFF, BRAIN & VASCULAR RES DEPT, CLEVELAND CLINS FOUND, 66- *Mailing Add:* Res Inst Cleveland Clins Found 9500 Euclid Ave Cleveland OH 44106

EHRHART, LLEWELLYN MCDOWELL, b Dallastown, Pa, Apr 22, 42; m 64; c 2. MAMMALOGY, MARINE TURTLE BIOLOGY. *Educ:* Franklin & Marshall Col, BA, 64; Cornell Univ, PhD, 71. *Prof Exp:* Instr vertebrate zool, Cornell Univ, 67; asst prof, 69-74, ASSOC PROF BIOL SCI, UNIV CENT FLA, 74-, GORDON J BARNETT PROF ENVIRON SCI, 76- *Mem:* Am Soc Mammal; Ecol Soc Am; Soc Study Evolution; Herepetologists' League; Animal Behav Soc. *Res:* Behavioral ecology of the cricetine rodent genus, Peromyscus; population ecology of Florida rodents and carnivores; biology of loggerhead turtles and green turtles. *Mailing Add:* Dept Biol Sci Univ Cent Fla Box 25000 Orlando FL 32816

EHRHART, WENDELL A, b Dallastown, Pa, Nov 12, 34; m 58; c 2. POLYMER CHEMISTRY. *Educ:* Franklin & Marshall Col, 56; Princeton Univ, MA, 58, PhD(org chem), 61. *Prof Exp:* Chemist floor div, 60-64, chemist, 64-74, res assoc, chem div, 74-81, RES ASSOC, EXPLOR RES DIV, ARMSTRONG WORLD INDUSTS, INC, 81- *Mem:* Am Chem Soc; Sigma Xi. *Res:* Synthesis and property studies of amidines, pyrimidines, pyrimido (4,5-d) pyrimidines, polyesters and polyurethanes; formulation and processing of polymers; ultra violet curable coatings and photoresists. *Mailing Add:* 268 Windsor Rd Red Lion PA 17356-9550

EHRICH, F(ELIX) FREDERICK, b New York, NY, Oct 19, 19; m 53; c 3. ORGANIC PIGMENTS, INORGANIC PIGMENTS. *Educ:* City Univ New York, BS, 39; Univ Iowa, MS, 40; Univ Md, PhD(org chem), 42. *Prof Exp:* Asst chem, Univ Md, 40-42; org res chemist, Corn Prods Refining Co, 42; res assoc, E I du Pont de Nemours & Co, 46-58, res supvr, 58-70, tech dir, Int Dept, Colorquim SA de CV, 70-72, tech mgr pigment colors, Int Develop Div, 73-75 & Mkt Div, 75-78, tech consult, 78-82; RETIRED. *Mem:* Am Chem Soc; Sigma Xi. *Res:* Synthetic organic pigments; barbituric acids; molecular rearrangements; spiropyrimidine and heterocyclic syntheses; phthalocyanine and quinacridone pigments. *Mailing Add:* 706 Sudbury Rd Edenridge Wilmington DE 19803

EHRICH, FREDRIC F(RANKLIN), b New York, NY, Dec 17, 28; m 55; c 3. MECHANICAL ENGINEERING. *Educ:* Mass Inst Technol, BS, 47, ME, 50, ScD(mech eng), 51. *Prof Exp:* Supvr anal & mech develop & tech rep, Rolls Royce, Eng, Westinghouse Elec Corp, 51-57; mgr preliminary design, Turbomach Eng Oper & T64 Engine Design, 57-66, mgr preliminary design, Turbomach & Mech Syst Eng Oper, Aircraft Engine Group, 66-68, mgr design tech oper, 68-70, STAFF ENGR, AIRCRAFT ENGINE GROUP, GEN ELEC CO, 70- *Concurrent Pos:* Chmn, Design Eng Div, Am Soc Mech Engrs, 72-73; ed, J Acoust, Vibrations, Stress & Reliability in Design, Am Soc Mech Engrs, 79-84. *Mem:* Fel Am Soc Mech Engrs; Am Inst Aeronaut & Astronaut; Sigma Xi; Am Helicopter Soc. *Res:* Vibrations in high-speed rotating machinery; design of aircraft gas turbines. *Mailing Add:* GE Aircraft Engines 1000 Western Ave Lynn MA 01910

EHRIG, RAYMOND JOHN, b Jersey City, NJ, Dec 31, 28; m 50; c 5. POLYMER CHEMISTRY. *Educ:* Seton Hall Univ, BS, 50; Polytech Inst Brooklyn, MS, 53, PhD(chem), 57. *Prof Exp:* Anal chemist, US Testing Co, 50-52; from res chemist to sr res chemist, Shell Chem Corp, 57-62; res supvr, W R Grace & Co, 62-66; mgr polymerization res, Chemplex Co, 66-69; chief, Div Polymers, Res Lab, 69-82, res consult, Uss Chemicals, US Steel Corp, 82-86, res consult, 86-90, SR SCIENTIST, ARISTECH CHEMICAL CORP, 90- *Concurrent Pos:* Lectr, Univ Calif, Long Beach, 57-59, Col Immaculate Conception, 60-62, Col Notre Dame, Md, 63-65, Elmhurst Col, 66-69 & Duquesne Univ, 81- *Mem:* Plastics Inst Am; Am Chem Soc; Soc Plastics Engrs; Soc Plastic Industs. *Res:* Ionic and radical polymerization and copolymerization; polymer kinetics and thermal decomposition studies; polypropylene; polyethylene; unsaturated polyesters; engineering resins; plastics recycling. *Mailing Add:* 2360 Millgrove Rd Pittsburgh PA 15241-2731

EHRKE, MARY JANE, b Warren, Pa, July 10, 37; m 69; c 1. IMMUNOPHARMACOLOGY. *Educ:* Thiel Col, AB, 59; Univ Pittsburgh, MS, 63. *Prof Exp:* Jr scientist, Roswell Park Mem Inst, 63-66; res affil biochem, Univ Pittsburgh, 66-67; cancer res scientist, 68-69, cancer res scientist II, 79-82, CANCER RES SCIENTIST III IMMUNOPHARMACOL, ROSWELL PARK MEM INST, 82- *Mem:* Am Asn Immunologists; Am Asn Cancer Res; AAAS; NY Acad Sci; Int Soc Immunopharmacol; Sigma Xi. *Res:* Interactions of antineoplastic agents with host defense mechanisms, toward the ultimate goal of design of treatment protocols in which host defenses are a positive component. *Mailing Add:* Grace Cancer Drug Ctr Roswell Park Mem Inst 666 Elm St Buffalo NY 14263

EHRLICH, ALEXANDER CHARLES, b New York, NY, May 8, 35. GALVANOMAGNETIC EFFECTS, TEMPERATURE DEPENDENT ELECTRICAL RESISTIVITY. *Educ:* City Col New York, BS, 56; Carnegie Inst Technol, MS, 60, PhD(physics), 63. *Prof Exp:* Res asst, Carnegie Inst Technol, 56-63; res assoc, Lab of Physics, Univ Lausanne, Switz, 63-66; Nat Res Coun postdoctoral assoc, 66-68, res physicist, 68-77, SUPVRY RES PHYSICIST, US NAVAL RES LAB, 77- *Mem:* Am Phys Soc; Sigma Xi. *Res:* Experimental electrical transport properties and electronic structure in metallic and semi-metallic materials; highly anisotropic conductors; charge density waves; metal hydrogen alloys; transition metals; magnetic conductors; thin films; multilayers; fibers. *Mailing Add:* US Naval Res Lab Code 6341 Washington DC 20375

EHRLICH, ANNE HOWLAND, b Des Moines, Iowa, Nov 17, 33; m 54; c 1. ENVIRONMENTAL SCIENCES. *Prof Exp:* Res asst, 59-72, res assoc, 72-75, SR RES ASSOC BIOL SCI, STANFORD UNIV, 75- *Concurrent Pos:* Consult, Coun Environ Qual, 77-80; lectr, Human Biol Prog, Stanford Univ, 81-; comnr, Greater London Area War Risks Study, 84-86. *Honors & Awards:* Distinguished Serv Award, Am Humanists Soc. *Res:* Comparative morphology of Papillionoidea; reproductive strategies of Papillionoidea; world population, environmental, resource problems and their interrelationships; environmental effects of nuclear war. *Mailing Add:* Dept Biol Sci Stanford Univ Stanford CA 94305-5020

EHRLICH, EDWARD NORMAN, b Detroit, Mich, Sept 20, 28; m 61; c 3. INTERNAL MEDICINE. *Educ:* Univ Mich, BS, 48, MD, 52. *Prof Exp:* From intern to resident, Wayne County Gen Hosp, 52-54; resident, 57-60, from instr to prof med, Univ Chicago, 60-74; PROF MED, ASSOC CHMN DEPT MED & HEAD SECT ENDOCRINOL, UNIV WIS-MADISON, 74-; CHMN MED, MADISON GEN HOSP, 74- *Mem:* AAAS; Am Fedn Clin Res; Endocrine Soc; Cent Soc Clin Res; Soc Gynecol Invest. *Res:* Adrenocortical regulation of salt metabolism in humans. *Mailing Add:* 600 Highland Ave 556 Csc H4 Madison WI 53792

EHRLICH, GEORGE EDWARD, b Vienna, Austria, July 18, 28; US citizen; m 68; c 1. RHEUMATOLOGY, INTERNAL MEDICINE. *Educ:* Harvard Univ, AB, 48; Chicago Med Sch, MB, MD, 52. *Prof Exp:* Intern, Michael Reese Hosp, Chicago, Ill, 52-53; asst resident surg & path, Francis Delafield Hosp, New York, 55-56; asst resident med, Beth Israel Hosp, Boston, Mass, 56-57; resident, New Eng Ctr Hosp, 57-58; trainee, Nat Inst Arthritis & Metab Dis, Bethesda, Md, 58-59; instr med, Med Col, Cornell Univ, 59-64; asst prof med & phys med & rehab, Temple Univ, 64-66, assoc prof phys med & rehab, 66-74, assoc prof med, 67-72, prof med, sch med, 72-80, prof rehab med, 74-80; dir arthritis ctr & sect rheumatology, Albert Einstein Med Ctr & Moss Rehab Hosp, 64-80, sr attend physician, 64-80; PROF MED, DIR DIV RHEUMATOLOGY, HAHNEMANN MED COL & HOSP, 80- *Concurrent Pos:* Fel, Hosp Spec Surg NY, 59-60; fel, Sloan Kettering Inst, 60-61; spec fel, Arthritis & Rheumatism Found, 64; Squibb-Olin fel, Mem Ctr Cancer & Allied Dis, 68; consult, US Naval Hosp, Philadelphia, 64-78 & Food & Drug Admin; ed, J Albert Einstein Med Ctr, 68-72; ed, Arthritis & Rheumatic Dis Abstr, 68-70. *Honors & Awards:* Distinguished Alumnus Award, Chicago Med Sch, 69; Philip Hench Award, 71; Citation, City of Philadelphia, 69, 74; Distinguished Serv Award, Arthritis Found, 71, 72; Cavaliere, Order Star Italian Solidarity, 74. *Mem:* Fel Am Col Physicians; Am Rheumatism Asn; AMA; fel Am Col Clin Pharmacol; Am Fedn Clin Res. *Res:* Rheumatology; investigations of osteoarthritis; clinical and genetic pharmacology; pathogenesis of arthritis manifestations; Behcet's syndrome. *Mailing Add:* One Independence Pl 1101 Philadelphia PA 19106-3731

EHRLICH, GERT, b Vienna, Austria, June 22, 26, nat US; m 57. SURFACE PHYSICS. *Educ:* Columbia Univ, AB, 48; Harvard Univ, AM, 50, PhD(chem), 52. *Prof Exp:* Res assoc physics, Univ Mich, 52-53; res assoc, Metal Res Dept, Gen Elec Res Lab, 53-68; PROF PHYS METALL & RES PROF, COORD SCI LAB & MAT RES LAB, UNIV ILL, URBANA, 68- *Concurrent Pos:* Guggenheim fel, 85- *Honors & Awards:* Medard W Welch Award, Am Vacuum Soc, 79; Kendall Award, Am Chem Soc, 82. *Mem:* Nat Acad Sci; fel Am Phys Soc; Am Vacuum Soc; Am Chem Soc. *Res:* Crystal surfaces; atomic motion and interactions; surface reactions. *Mailing Add:* Coord Sci Lab Univ Ill 1101 W Springfield Ave Urbana IL 61801

EHRLICH, GERTRUDE, b Vienna, Austria, Jan 7, 23; nat US. MATHEMATICS, UNIVERSITY TEACHING. *Educ:* Ga State Col Women, BS, 43; Univ NC, MA, 45; Univ Tenn, PhD(math), 53. *Prof Exp:* Instr math, Oglethorpe Univ, 46-50; asst, Univ Tenn, 50-52, instr, 52-53; from instr to assoc prof, 53-69, PROF MATH, UNIV MD, COLLEGE PARK, 69- *Concurrent Pos:* Assoc ed, Am Math Monthly, 63-66; chmn math competition, Univ Md, 79-82, 88-90. *Mem:* Math Asn Am; Am Math Soc; Am Asn Univ Prof; Sigma Xi. *Res:* Continuous geometry; ring theory; author and co-author three textbooks. *Mailing Add:* 6702 Wells Pkwy University Park MD 20782

EHRLICH, H PAUL, b Honolulu, Hawaii, Aug 17, 41; m; c 3. BIOCHEMISTRY, BIOLOGY. *Educ:* Univ Calif, Berkeley, BS, 64; Univ Calif, San Francisco, PhD(biochem), 71. *Prof Exp:* Fel biochem, Univ Wash, 71-73 & Strangeways Res Lab, Cambridge, Eng, 73-75; asst prof path, 75-84, ASSOC PROF PATH, HARVARD MED SCH, MASS GEN HOSP & SHRINERS BURNS HOSP, 84-, ASSOC PROF PATH & SURG, 89- *Concurrent Pos:* NIH grants, 78-81, 80-83 & 84-92; estab investr, Am Heart Asn, 77-82. *Honors & Awards:* Brit Am Heart Award. *Mem:* Am Soc Cell Biol; Am Burn Asn; AAAS; NY Acad Sci; Plastic Surg Res Coun. *Res:* Collagen metabolism in biology and pathology; scar metabolism. *Mailing Add:* Shriners Burns Hosp 51 Blossom St Boston MA 02114

EHRLICH, HENRY LUTZ, b Stettin, Ger, Aug 31, 25; nat US. GEOMICROBIOLOGY. *Educ:* Harvard Univ, BS, 48; Univ Wis, MS, 49, PhD(agr bact), 51. *Prof Exp:* Alumni Res Found asst, Univ Wis, 48-50, asst, 50-51; from asst prof to assoc prof, 51-64, PROF BIOL, RENSSELAER POLYTECH INST, 64- *Concurrent Pos:* Ed-in-chief, Geomicrobiol J, 81- *Mem:* AAAS; fel Am Acad Microbiol; Am Soc Microbiol; Am Inst Biol Sci; Sigma Xi; Soc Indust Microbiol. *Res:* Investigations of geomicrobially important transformations, including oxidation of manganese, metal sulfides, arsenic; reduction of manganese, chromium; microbial role in ferromanganese nodule genesis; manganese oxidation by bacteria from deep-sea hydrothermal vents; bacterial leaching of ores; geomicrobiology. *Mailing Add:* Dept Biol Rensselaer Polytech Inst Troy NY 12180-3590

EHRLICH, HOWARD GEORGE, b Milwaukee, Wis, Nov 9, 24; m 55; c 2. CYTOLOGY. *Educ:* Marquette Univ, BS, 48; Univ Minn, PhD(cytol), 56. *Prof Exp:* Asst bot, Marquette Univ, 46-48; res asst plant path, Univ Minn, 50-52, asst bot, 52-54, 55-56, res fel, Cancer Soc, 56-58, Dight Inst, 58, instr bot, 59; from asst prof to assoc prof, 59-68, actg chmn, 70-71, PROF BIOL, DUQUESNE UNIV, 68-, CHMN DEPT BIOL SCI, 71- *Mem:* AAAS; Bot Soc Am; Mycol Soc Am. *Res:* Mycology; electron microscopy. *Mailing Add:* Dept Biol Sci Duquesne Univ Pittsburgh PA 15282

EHRLICH, I(RA) ROBERT, b Washington, DC, Sept 1, 26; m 50; c 2. AUTOMOTIVE ENGINEERING. *Educ:* US Mil Acad, BS, 50; Purdue Univ, MS, 56; Univ Mich, PhD(eng), 60. *Hon Degrees:* MEng, Stevens Inst Technol, 82. *Prof Exp:* Supvr prog, Int Elec Corp, 60-62; mgr transp res group, Stevens Inst Technol, 62-74, dean res, 74-80, head dept mech eng, 80-83, vpres res, 83-84, vpres acad affairs, 84-85; CONSULT, 85- *Concurrent Pos:* Consult, Grumman Aircraft Corp, 60-69, Midwest Appl Sci Corp, 62-65, Chrysler Corp, 65, Michelin Tire Corp, Am Motors & US Army; guest lectr, US Mil Acad, 63-73; Themis grantee. *Mem:* Fel Int Soc Terrain-Vehicle Systs (secy, 63-78, vpres, 78-81, pres, 81-84); Sigma Xi; Nat Soc Prof Engrs; Am Defense Preparedness Asn; fel Soc Auto Engrs; Am Soc Mech Engrs; Nat Safety Coun. *Res:* Automotive research; highway safety; accident reconstruction. *Mailing Add:* 859 Columbus Dr Teaneck NJ 07666

EHRLICH, JULIAN, b New York, NY. BIOPHYSICAL CHEMISTRY, BIOCHEMISTRY. *Educ:* City Col New York, BS, 42; NY Univ, DDS, 45; Stevens Inst Technol, MS, 63, PhD(chem), 71. *Prof Exp:* Pvt pract, NY, 47-60; head, Anal Div, Schwarz Bio-Res Corp, NY, 60-62; res chemist, M&T Chem, Inc, Div Am Can Co, NJ, 62-65; USPHS fel, 65-70; res assoc, Stevens Inst Technol, 70-71, res scientist, 71-73; asst prof, 73-76, ASSOC PROF, UNIV MED & DENT NJ, 76- *Concurrent Pos:* Lectr, Plastics Inst Am, 66-69; consult, Stevens Inst Technol, 73-75, adj assoc prof, 77- *Mem:* Am Chem Soc; fel Royal Soc Chem; Biophys Soc; fel Am Inst Chemists; fel Plastics & Rubber Inst, London. *Res:* Polysaccharide chemistry; biochemistry; laser light scattering; characterization of biopolymers and synthetic polymers. *Mailing Add:* Univ Med & Dent 100 Bergen St Newark NJ 07103

EHRLICH, KENNETH CRAIG, b New York, NY Sept 23, 43; m 66; c 2. FUNGAL TOXINS, DNA METHYLATION. *Educ:* Columbia Univ, AB, 65; State Univ NY, Stony Brook, PhD(chem), 69. *Prof Exp:* Res assoc biochem, Sch Med, La State Univ, 72-75; res chemist, Gulf South Res Inst, 75-79; RES CHEMIST, SOUTHERN REGIONAL RES CTR, USDA, 80- *Concurrent Pos:* Adj assoc prof, dept biochem, Sch Med, Tulane Univ, 75- *Mem:* Am Chem Soc; Am Soc Biol Chemists; Asn Off Anal Chemists; NY Acad Sci. *Res:* Mycotoxin analyses; trichothecene mycotoxins; fungal competition; sequence-specific DNA binding proteins in plant nuclei; role of DNA methylation in plant tissue. *Mailing Add:* Southern Regional Res Ctr USDA PO Box 19687 New Orleans LA 70179

EHRLICH, LOUIS WILLIAM, b Baltimore, Md, Oct 4, 27; m 59; c 3. NUMERICAL ANALYSIS. *Educ:* Univ Md, BS, 51, MA, 56; Univ Tex, PhD, 63. *Prof Exp:* Proj engr, Hercules Powder Co, 51-54; asst math, Univ Md, 54-56; numerical analyst, Space Tech Labs, Ramo-Wooldridge Corp, 56-59; res scientist, Univ Tex, 59-62; NUMERICAL ANALYST, APPL PHYSICS LAB, JOHNS HOPKINS UNIV, 62- *Mem:* Am Math Soc; Asn Comput Mach; Soc Indust & Appl Math. *Res:* Linear algebraic systems, finite difference approximations to elliptic partial differential equations and nonlinear systems. *Mailing Add:* Appl Physics Lab Johns Hopkins Rd Laurel MD 20723

EHRLICH, MARY ANN, botany, plant pathology, for more information see previous edition

EHRLICH, MELANIE, b New York, NY, June 12, 45; m 66; c 2. BIOCHEMISTRY, MICROBIOLOGY. *Educ:* Barnard Col, AB, 66; State Univ NY Stony Brook, PhD(biochem), 70. *Prof Exp:* From asst prof to assoc prof, 72-82, PROF BIOCHEM, SCH MED, TULANE UNIV, 82- *Concurrent Pos:* Jane Coffin Childs fel, Albert Einstein Col Med, 70-71; Jane Coffin Childs grant, Sch Med, Tulane Univ, 72-74, USPHS grant, 74- *Mem:* Am Soc Microbiol; Am Soc Photochem; Am Soc Biochem & Molecular Biol. *Res:* Biochemistry of nucleic acids; bacteriophage with unusually modified DNA; uptake of DNA by cultured mammalian cells; photochemistry of DNA; role of DNA methylation on transcription. *Mailing Add:* Dept Biochem Tulane Univ Sch Med 1430 Tulane Ave New Orleans LA 70112

EHRLICH, PAUL, b Vienna, Austria, Feb 26, 23; nat US; m 49; c 5. PHYSICAL CHEMISTRY. *Educ:* Queens Col, NY, BS, 44; Univ Wis, MS, 48, PhD, 51. *Prof Exp:* Phys chemist, Nat Bur Standards, 51-53; fel chem, Harvard Univ, 53-54; res chemist, Monsanto Co, 55-59, res specialist, 59-60, group leader, 60-61, scientist, 61-67; assoc prof, 67-70, PROF CHEM ENG, STATE UNIV NY, BUFFALO, 70- *Mem:* AAAS; Am Phys Soc; Am Chem Soc; Am Inst Chem Engrs. *Res:* Thermodynamic interactions of polymers; supercritical extractions; electron transport in polymers; polymerization processes. *Mailing Add:* Dept Chem Eng State Univ NY Buffalo NY 14260

EHRLICH, PAUL EWING, b Schenectady, NY, Aug 8, 48. MATHEMATICS. *Educ:* Harvard Col, BA, 70; State Univ NY Stony Brook, MA, 71, PhD(math), 74. *Prof Exp:* Vis scientist, Math Inst, Univ Bonn, 74-76; asst prof, 76-79, ASSOC PROF MATH, UNIV MO-COLUMBIA, 79- *Concurrent Pos:* Vis scholar, Sch Math, Inst Adv Study, 80. *Mem:* Am Math Soc; Math Asn Am; Sigma Xi. *Res:* Differential geometry and general relativity. *Mailing Add:* Dept Math Univ Fla 201 Walker Hall Gainesville FL 32611

EHRLICH, PAUL RALPH, b Philadelphia, Pa, May 29, 32; m 54; c 1. BIOLOGY. *Educ:* Univ Pa, AB, 53; Univ Kans, MA, 55, PhD(entom), 57. *Hon Degrees:* DHA, Univ Pac, 70. *Prof Exp:* Asst entom, Univ Kans, 53-54, assoc, 58-59; res assoc, Chicago Acad Sci, 57-58; from asst prof to assoc prof, 59-66, PROF BIOL, STANFORD UNIV, 66-, BING PROF POP STUDIES, 76- *Concurrent Pos:* Nat Sci Found postdoc fel, Univ Sydney, 65-66; pres, Zero Population Growth, 69-70, hon pres, 70-; ed bd, 74-76, sr assoc ed, Am Naturalist, 84; pres, Conservation Soc, 72-73; Bd Consult, Lizard Island Res Sta, 75- *Honors & Awards:* John Muir Award, Sierra Club, 80; First Distinguished Achievement Award, Soc Consev Biol, 87; Gold Medal, World Wildlife Fund Int, 87; First Prize, Mitchell Found, 79; Humanist Distinguished Serv Award, Am Humanist Asn, 85; Crafoord Prize, Swedish Acad Sci, 90; MacArthur Prize Fel, 90. *Mem:* fel Nat Acad Sci; fel Am Acad Arts & Sci; Soc Syst Zool; Soc Study Evolution (vpres, 70); Lepidop Soc (secy, 57-63); Ecol Soc Am; Am Soc Naturalists; fel AAAS; Soc Conservation Biol; fel Entomol Soc Am, 87; fel Am Philos Soc. *Res:* Population biology; over 500 books, papers and articles published. *Mailing Add:* Dept Biol Sci Stanford Univ Stanford CA 94305

EHRLICH, RICHARD, b New York, NY, Dec 25, 21; m 45; c 2. REACTOR PHYSICS. *Educ:* Harvard Univ, BA, 41; Cornell Univ, PhD(theoret physics), 47. *Prof Exp:* Asst physics, Cornell Univ, 41-43 & nat res fel, 46; jr scientist, Los Alamos Sci Lab, Univ Calif, 43-46; res assoc physics, Gen Elec Co, 47-54, supvr theoret physics, 54, mgr, 54-55, mgr math anal, 55-58, mgr advan develop, Schenectady, NY, 58-76, mgr nuclear safety & safeguards, 76-87; RETIRED. *Concurrent Pos:* Mem adv comt reactor physics, US AEC, 60-75 & Energy Res & Develop Admin, 75-76; chmn adv comt, Nat Nuclear Data Sect Ctr, Brookhaven Nat Lab, 67-84; bd dirs, Am Nuclear Soc, 69-72. *Mem:* Fel AAAS; Am Phys Soc; fel Am Nuclear Soc; Nat Soc Prof Engrs. *Res:* Theoretical physics work in atomic energy field; nuclear reactor design and analysis; digital computations; technical management; nuclear safety and safeguards. *Mailing Add:* 14664 Wild Berry Lane Saratoga CA 95070

EHRLICH, RICHARD, b Bedzin, Poland, Jan 19, 24; nat US; m 50; c 2. ENVIRONMENTAL & REGULATORY TOXICOLOGY. *Educ:* Munich Tech Univ, MS, 48, PhD(dairy bact), 49. *Prof Exp:* Res asst, Munich Tech Univ, 47-49; lab dir, Am Butter Inst, 49-52; assoc bacteriologist, IIT Res Inst, 52-53, res bacteriologist, 53-57, supvr biol res, 57-60, asst dir, 60-62, assoc dir, 62-63, dir life sci, 63-77, vpres res opers, 77-90; CONSULT, ENVIRON HEALTH, RES MGT, 90- *Concurrent Pos:* Mem subcomt nitrogen oxides, Div Med Sci, Nat Res Coun, Nat Acad Sci, 73-; life mem, Pan-Am Med Asn; consult, World Bank; reviewer & contribr, EPA Air Qual Criteria, reviewer, EPA, NIH, EPAI, CIAR grants; ed, Environ Res, J Am Col Toxicol, J Toxicol & Environ Health. *Mem:* AAAS; Air & Waste Mgt Asn; foreign assoc mem French Soc Tuberculosis & Respiratory Dis; Sigma Xi; Soc Occup Environ Health; Pan Am Med Asn. *Res:* respiratory infections; bacterial aerosols; infectious aerobiology; health effects of air pollution; author of more than 100 publications in fields of public health bacteriology and environmental health. *Mailing Add:* 4857 W Davis Skokie IL 60077

EHRLICH, ROBERT, b St Paul, Minn, Mar 4, 36. GEOLOGY, SEDIMENTOLOGY. *Educ:* Univ Minn, BA, 58; La State Univ, MS, 61, PhD(geol), 65. *Prof Exp:* From asst prof to assoc prof geol, Mich State Univ, 65-74; PROF GEOL, UNIV SC, 74- *Mem:* Geol Soc Am; Int Asn Sedimentol; Am Asn Petrol Geologists; Meteorit Soc; Sigma Xi. *Res:* Detrital sedimentology; geometrics; crustal dynamics as expressed in sediments; geology of North Africa; shape analysis. *Mailing Add:* 2510 Stratford Rd Columbia SC 29204

EHRLICH, ROBERT, b Brooklyn, NY, Feb 6, 38; m 61; c 2. PHYSICS. *Educ:* Brooklyn Col, BS, 59; Columbia Univ, PhD(physics), 64. *Prof Exp:* Res investr physics, Univ Pa, 63-66; asst prof, Rutgers Univ, 66-74; assoc prof physics & actg chmn dept, State Univ NY, New Paltz, 74-77; PROF PHYSICS, GEORGE MASON UNIV, FAIRFAX, VA, 77- *Mem:* Am Phys Soc. *Res:* Experimental elementary particle research. *Mailing Add:* Dept Physics George Mason Univ Fairfax VA 22030

EHRLICH, ROBERT STARK, b New York, NY, Aug 30, 40; m 67; c 1. ENZYMOLOGY. *Educ:* Columbia Col, AB, 62; Rutgers Univ, MS, 64, PhD(physics), 69. *Prof Exp:* Asst prof physics, Muskingum Col, 69-70; instr, City Col New York, 70-71; res assoc biochem, Rutgers Med Sch, 71-73; res assoc biochem,72-83, ASSOC SCIENTIST, DEPT CHEM, UNIV DEL, 83- *Mem:* Am Phys Soc; AAAS; Biophys Soc; Am Soc Biochem & Molecular Biol. *Res:* Structure and function of enzymes using techniques of fluorescence, nuclear magnetic resonance, binding and affinity labeling. *Mailing Add:* 1424 Carson Rd Wilmington DE 19803

EHRLICH, S PAUL, JR, b Minneapolis, Minn, May 4, 32; m 59; c 3. EPIDEMIOLOGY, PUBLIC HEALTH. *Educ:* Univ Minn, BA, 53, BS, 55, MD, 57; Univ Calif, MPH, 61; Am Bd Prev Med, dipl. *Prof Exp:* Intern, USPHS, Staten Island, NY, 57-58, med officer, Coast Guard, 58-59, med officer, Nat Heart Inst, 59-60, chief heart dis control prog, Field & Training Sta, 61-66, asst chief prog develop, Heart Dis Control Prog, Div Chronic Dis, 66-67, chief tech resources, Off Int Health, 67-69, dep dir off int health, 69-70, asst surgeon gen, 70-73, actg surgeon gen, 73-77, dir off int health, Off of Secy, US Dept Health, Educ & Welfare, 70-77; vpres, Am Inst Res, 78-79; dep dir, Pan Am Health Orgn, WHO, 79-83; sr adv, Am Asn World Health, 83-84; CONSULT, PUB HEALTH, 84- *Concurrent Pos:* Resident, Sch Pub Health, Univ Calif, 61-63; rep to exec bd, WHO, 69-76; assoc prof, Georgetown Univ, 69-; adj prof, Sch Pub Health, Univ Tex, Houston, 70-; fel coun epidemiol, Am Heart Asn. *Mem:* AAAS; fel Am Col Prev Med; AMA; fel Am Pub Health Asn. *Res:* Epidemiology of cardiovascular diseases; chronic disease in geriatric groups. *Mailing Add:* 6512 Lakeview Dr Falls Church VA 22041

EHRLICH, SANFORD HOWARD, b New York, NY, June 11, 31; m 55; c 2. CHEMICAL PHYSICS. *Educ:* NY Univ, BS, 53, MS, 59; Adelphi Univ, PhD(chem physics), 63. *Prof Exp:* Chem physicist, Air Prod & Chem, Inc, 63-65; sr scientist, Am Optical Co, Mass, 65-70; MEM STAFF, RES LABS, EASTMAN KODAK CO, 70- *Mem:* Am Chem Soc; Am Vacuum Soc; Am Phys Soc. *Res:* Energy transfer in gases and solids; spectroscopy of surface adsorbed molecules; photochemistry; thermodynamics of nonaqueous solutions; transference phenomena in nonaqueous media; luminescence and energy transfer in organic molecules. *Mailing Add:* 12 Courtenay Circle Pittsford NY 14534

EHRLICH, STANLEY L(EONARD), b Newark, NJ, Jan 7, 25; m 49; c 3. ACOUSTICS, SYSTEMS ENGINEERING. *Educ:* Brown Univ, ScB, 44, ScM, 45. *Prof Exp:* Physicist, US Navy Underwater Sound Lab, 48-53; sr engr transducer develop, 53-57, sr engr, Sonar Systs Develop, 57-59, sect mgr, 59-62, prin engr, 62-70, CONSULT ENGR, SONAR SYSTS DEVELOP, SUBMARINE SIGNAL DIV, RAYTHEON CO, 70- *Concurrent Pos:* Deleg, Am Nat Standards Inst, Inst Elec & Electronics Engrs, 71-82, individual expert, 83-; USA rep, Individual Syst Oper, 74-76, USA deleg, 76; chmn, Am Nat Standards Inst, 75; assoc ed, J Oceanic Eng, 75-81, ed, 82-87, assoc ed J Acoust Soc Am, 81; techn coun Acoust Soc Am, 79-81; admin comt, IEEE Coun Oceanic Eng, 81-82; admin comt, IEEE Oceanic Eng Soc, 83-; Exec Coun Acoust Soc Am, 86- *Honors & Awards:* Freeman Award for Eng Achievement, Providence Eng Soc, 76; Centennial Medal, Inst Elect & Electronics Engrs. *Mem:* AAAS; Am Phys Soc; fel Acoust Soc Am; sr mem Inst Elec & Electronics Engrs; Nat Security Indust Asn; NY Acad Sci; Am Nat Standards Inst; Sigma Xi. *Res:* Magnetostriction; electrostriction; electroacoustics; design of transducers; development of sonar systems; normal modes in solids, especially cylinders. *Mailing Add:* One Acacia Dr Middletown RI 02840

EHRLICH, WALTER, b Bosicany, Bohemia, Sept 22, 15; US citizen; m 40; c 3. CARDIOVASCULAR PHYSIOLOGY. *Educ:* Charles Univ, Prague, MD, 47; Czech Acad Sci, CSc, 61. *Prof Exp:* Secundar path, Sch Hyg, Charles Univ, Prague, 47-48, secundar internal med, 48-51; chief res group clin & exp res, Inst Cardiovasc Res, Prague, 51-63; chief physiol labor, Inst Hyg, Prague, 63-66; asst prof, 67-70, ASSOC PROF, DEPT ENVIRON PHYSIOL, SCH HYG, JOHNS HOPKINS UNIV, 70-, ASSOC PROF, DEPT PSYCHIAT & BEHAV, SCH MED, 70- *Mem:* Am Physiol Soc; Am Heart Asn; Fedn Am Scientists. *Res:* Intrinsic and reflectory regulation of circulation and respiration. *Mailing Add:* Sch Hyg Rm 7014 Johns Hopkins Univ 615 N Wolfe St Baltimore MD 21205

EHRMAN, JOACHIM BENEDICT, b Nuremberg, Ger, Nov 12, 29; nat US; m 61; c 1. PLASMA PHYSICS, APPLIED MATHEMATICS. *Educ:* Univ Pa, AB, 48; Princeton Univ, AM, 49, PhD(physics), 54. *Prof Exp:* Mem dept physics, Atomic Energy Res Dept, NAm Aviation, Inc, 51-53; instr, Sloane Physics Lab, Yale Univ, 54-55; physicist nucleonics div, US Naval Research Lab, 55-66 & plasma physics div, 66-68; PROF APPL MATH, UNIV WESTERN ONT, 68- *Concurrent Pos:* Assoc prof lectr physics, George Washington Univ, 56-57; lectr, Univ Md, 63-64; consult plasma physics div, US Naval Res Lab, 69-70; vis mem res staff, Plasma Physics Lab, Princeton Univ, 75-76. *Mem:* Am Phys Soc. *Res:* Fusion, tokamak; saturation of resistive tearing mode in toroidal plasma; electron ring accelerator; electron beams; magnetohydrodynamics; electron penetration through matter; nuclear reactors; nuclear physics. *Mailing Add:* Dept Appl Math Univ Western Ont London ON N6A 5B9 Can

EHRMAN, LEE, b New York, NY, May 25, 35; m 55; c 2. POPULATION GENETICS. *Educ:* Queens Col, NY, BS, 56; Columbia Univ, MA, 57, PhD(genetics), 59. *Hon Degrees:* DSc, Queens Col, City Univ NY, 89. *Prof Exp:* Res assoc, dept zool, Columbia Univ, 56, teaching asst, 56-57, res fel, 59-62, lectr, 62; tutor, dept biol, Queens Col, 57-58; res assoc pop genetics, Rockefeller Univ, 63-64, asst prof, 64-71; assoc prof natural sci, 71-74, PROF NATURAL SCI, STATE UNIV NY COL PURCHASE, 74- *Concurrent Pos:* Am Asn Univ Women Shirley Farr fel, 62-63; res fel, Rockefeller Univ, 62-63; Sigma Xi grant-in-aid, 63; Nat Inst Child Health & Human Develop res career develop award, 64-74; lectr, State Univ NY Col Purchase, 70-71, vis lectr, Univ PR, Rio Piedras, 87; from adj assoc prof to adj prof, Rockefeller Univ, 71-75; res assoc animal behav, Am Mus Natural Hist, 73-; State Univ NY fac exchange scholar, 74-; assoc ed, Evolution, Genetics & Cytol, Am Midland Naturalist & Am Naturalist, 77-79; NSF grant, 79; workshop coodr, behav genetics, Nat Res Coun-Nat Acad Sci, 72; mem, NSF, NIH & NIMH study sect; Dobghansky lifetime res award, Behav Genetics Asn, 88; vis distinguished prof, Univ Miami, 81; assoc ed, Evolution, co-ed, Behavior Genetics. *Honors & Awards:* Sr Res Award, Whitehall Found, 87. *Mem:* Am Soc Naturalist (pres); Behav Genetics Asn (pres, 78); Am Soc Human Genetics; Sigma Xi; Soc Study Social Biol; fel AAAS; Am Asn Univ Women; Am Soc Naturalists (treas, 72-75, vpres, 82-90, pres, 90); NY Acad Sci; Soc for Study Evolution (exec coun 86). *Res:* Reproductive isolating mechanisms, especially hybrid sterility and sexual behavioral isolation; cytoplasmic inheritance; frequency-dependent selection; author of 3 books. *Mailing Add:* Div Natural Sci State Univ NY Col Purchase NY 10577

EHRMAN, LEONARD, b New York, NY, Jan 24, 32; m 57; c 1. ELECTRICAL ENGINEERING. *Educ:* Mass Inst Technol, BS, 53, MS, 55; Northeastern Univ, PhD(elec eng), 67. *Prof Exp:* Mem tech staff, Sandia Corp, NMex, 58-61; group leader, Res & Adv Develop Div, Avco Corp, Mass, 61-63; res scientist, Signatron Inc, 63-80, sr consult scientist, 80-83, vpres, 83-88; CONSULT, 88- *Mem:* fel Inst Elec & Electronics Engrs. *Res:* Digital communications; radar analysis; signal processing; computer usage; bandwidth compression; channel simulation. *Mailing Add:* 26 Village Dr East Sandwich MA 02537

EHRMANN, RITA MAE, b Jersey City, NJ, Aug 27, 28. GEOMETRY. *Educ:* Univ Dayton, BS, 51; Villanova Univ, AM, 57; St Louis Univ, PhD(math), 69. *Prof Exp:* Teacher math-sci, Mt St Michael High Sch, 51-65; teacher math, 69-80, ASSOC PROF MATH, VILLANOVA UNIV, 80- *Concurrent Pos:* Dir, NSF In-Serv Inst, Villanova Univ, 71-73; Fulbright grant math, Univ Botswana, Africa, 82-84. *Mem:* Math Asn Am; Sigma Xi. *Res:* Geometric properties, especially convexity, of metric spaces; finite geometrics from a foundations standpoint. *Mailing Add:* Dept Math Villanova Univ Villanova PA 19085

EHRMANTRAUT, HARRY CHARLES, b Washington, DC, Nov 25, 21; m 48; c 1. INSTRUMENTATION. *Educ:* George Washington Univ, BS, 47; Georgetown Univ, MS, 48; Univ Ill, PhD(biophys), 50. *Prof Exp:* Asst photosynthesis proj, Univ Ill, 48-50; exec secy panel med aspects atomic warfare, Res & Develop Bd, Off Secy Defense, 50-51; asst to dir res, Toni Co, 51-52; res biophysicist, Armour Res Found, 52-54; sr biophysicist, Stanford Res Inst, 54-55; dir applns res dept, Spinco Div, Beckman Instruments, Inc, 56-59; pres, Mechrolab, Inc, 59-66; sci adv, Found for Nutrit & Stress Res, 66-68; pres, Gymnas Corp, 68-70; chmn, Improved Commun Inc, 70-71; partner, Bus Anal Assocs, 72-78, Pres, Cal/Quest Corp, 75-76; pres, AVM Assocs, 78-81; pres, Hollester Peak Mgt, 81-83; PRES, ALCHEM ASSOCS, 85- *Concurrent Pos:* Pres & dir, Altos Ctr, Inc, 70-71; pres, Quercine Corp, 73-74; mem, Int Oceanog Found. *Mem:* AAAS; Am Chem Soc; NY Acad Sci; Inst Elec & Electronics Engrs; Biophys Soc. *Res:* Medical and biophysical instrumentation. *Mailing Add:* 985 Northridge Ave Springfield OR 97477

EHRREICH, ALBERT LEROY, b Pipestone, Minn, June 11, 23; m 51; c 2. GEOLOGY. *Educ:* Univ Calif, Los Angeles, BA, 50, MA, 55, PhD(geol), 65. *Prof Exp:* Geologist, US Bur Reclamation, 51-53; from asst prof to assoc prof, 57-74, PROF GEOL, CALIF STATE UNIV LONG BEACH, 74- *Mem:* Geol Soc Am; Geochem Soc; Mineral Soc Am; Nat Asn Geol Teachers. *Res:* Igneous and metamorphic petrology. *Mailing Add:* PO Box 724 Willits CA 95490-0724

EHRREICH, STEWART JOEL, b Brooklyn, NY, Mar 24, 36; m 60; c 2. PHARMACOLOGY. *Educ:* Queens Col, NY, BS, 57; State Univ NY Downstate Med Ctr, MS, 61, PhD(pharmacol), 63. *Prof Exp:* Asst pharmacol, Downstate Med Ctr, 58-63; sr pharmacologist, Smith Kline & French Labs, 65-67, sr investr, 67-69; sect head, Geigy Pharmaceut Div, Ciba-Geigy Ltd, 69-71; sect leader, Scheringplough Corp, 72-77,; group leader antihypertensive drugs, Food & Drug Admin, 80-84, dep dir, Cardio- Renal Div; sr tech adv, Biometric Res Inst, 86-90; PRES, EHRREICH CONSULT, 90- *Concurrent Pos:* USPHS fel, Med Col, Cornell Univ, 63-65; lectr biol, Queens Col, NY, 63-65 & Brooklyn Col, 64-65; sr tech adv, Ernst & Young, 90- *Mem:* Am Soc Pharmacol & Exp Therapeut; Am Soc Clin Pharmacol; Sigma Xi. *Res:* Pharmacology, physiology and electro-physiology of smooth and cardiac muscle; autonomic pharmacology. *Mailing Add:* 9513 Ash Hollow Pl Gaithersburg MD 20879

EIAN, GILBERT LEE, b Fergus Falls, Minn, Apr 15, 43; m 66; c 2. ORGANIC CHEMISTRY. *Educ:* Univ Minn, Minneapolis, BChem, 65; Iowa State Univ, MS, 67, PhD(chem), 69. *Prof Exp:* Sr chemist, Imaging Res Lab 3M, 69-73, res specialist, 73-78, prod develop specialist, occup health & safety prod, Div 3M, 78-82, sr res specialist, 82-85, SR RES SPECIALIST, SPECIALTY ADHESIVES & CHEM DIV 3M, 85- *Mem:* Am Chem Soc; Sigma Xi. *Res:* Organic synthesis; organic photochemistry; free radical chemistry; nitrogen heterocycle chemistry, product development, specialty filters; polymerchemistry. *Mailing Add:* Specialty Chem Div 3M 236-3B-01 3M Ctr St Paul MN 55144

EIBECK, RICHARD ELMER, b Cincinnati, Ohio, Feb 17, 35; m 62; c 2. FLUOROCARBON DERIVATIVES. *Educ:* Univ Cincinnati, BS, 57; Univ Ill, MS, 59, PhD(chem), 61. *Prof Exp:* Res chemist, Gen Chem Div, Allied Chem Corp, 60-64, sr res chemist, 64-67, tech supvr, Indust Chem Div, 67, tech supvr, Specialty Chem Div, 67-81, mgr res & develop, Allied Chem Co, 81-87; SR RES ASSOC, ALLIED-SIGNAL CORP, 88- *Concurrent Pos:* Exec comt, ACS Div Fluorine Chem, 88-90. *Mem:* Am Chem Soc. *Res:* Chemistry of nonmetals; fluorine compounds, including sulfur fluorides; nitrogen fluorides; fluorocarbons; fluoroaromatics; chloro fluorocarbons. *Mailing Add:* 23 Pine Terrace Orchard Park NY 14127

EIBEN, GALEN J, b Monticello, Iowa, May 23, 36; m 56; c 3. ENTOMOLOGY. *Educ:* Wartburg Col, BA, 60; Iowa State Univ, MS, 62, PhD(entom), 67. *Prof Exp:* Instr biol, Tex Lutheran Col, 62-64; res assoc entom, Iowa State Univ, 64-66; asst prof biol, 67-74, ASSOC PROF BIOL, WARTBURG COL, 74- *Mem:* Am Entom Soc. *Res:* Methodology involved in screening corn for resistance to corn rootworms; corn root development and its relation to rootworm populations. *Mailing Add:* Dept Biol Wartburg Col Waverly IA 50677

EIBEN, ROBERT MICHAEL, b Cleveland, Ohio, July 12, 22; m 46; c 6. MEDICINE. *Educ:* Western Reserve Univ, BS, 44, MD, 46. *Prof Exp:* From instr to assoc prof pediat, 49-75, from asst prof to assoc prof neurol, 63-85, PROF PEDIAT, CASE WESTERN RESERVE UNIV, 75-, PROF NEUROL, 85- *Concurrent Pos:* Fel, Univ Wash, 60-63; asst med dir, Dept Contagious Dis, Cleveland City Hosp, 49-50, actg dir, 50-52, asst dir dept pediat & contagious dis, 54-60, med dir, Respirator Care & Rehab Ctr, 54-60; asst pediatrist, Univ Hosps, 49-51, assoc pediatrist, 51-; pediat neurologist, Cleveland Metrop Gen Hosp, 63- *Mem:* Am Acad Neurol; Am Soc Human Genetics; Am Acad Pediat; Child Neurol Soc (pres, 83-85); Am Pediat Soc. *Res:* Biochemical studies of heredofamilial disorders of the nervous system. *Mailing Add:* Two Oakshore Dr Bratenahl OH 44108

EIBERT, JOHN, JR, b St Louis, Mo, Sept 18, 18; c 4. PHARMACEUTICAL CHEMISTRY. *Educ:* Washington Univ, BS, 40, MS, 42, PhD(chem), 44. *Prof Exp:* Chemist, Scullin Steel Co, St Louis, 42-43; lectr physics, Washington Univ, 43-44; res chemist, Pan Am Refining Corp, Tex, 44-45 & Anheuser-Busch, Inc, Mo, 45-46; consult chemist & secy-treas, Sci Assocs, 46-61, pres, 61-87; RETIRED. *Mem:* AAAS; Am Chem Soc; NY Acad Sci; Am Inst Chemists; assoc Inst Food Technol. *Res:* Calorimetry; measurement of vapor pressure; catalysis; protein hydrolysis; thermodynamic properties of certain aromatic hydrocarbons. *Mailing Add:* 1423 Jamnica Ct St Louis MO 63122

EIBLING, JAMES A(LEXANDER), b Marion, Ohio, Nov 22, 17; m 40; c 2. SOLAR ENERGY, REFRIGERATION. *Educ:* Ohio State Univ, BME, 39. *Prof Exp:* Mech engr, B F Goodrich Co, Akron, 40-42; res engr, Battelle Mem Inst, 46-56, asst div chief, 56-60, chief, Thermal Systs Div, 60-73, prog mgr solar energy res, 73-79; HEAD CONSULT, EIBLING THERMAL SYSTS TECHNOL, 79- *Concurrent Pos:* Div engr, US Army Corps Engrs, 42-46; mem adv panels, US Dept Energy & NSF. *Honors & Awards:* Am Soc Heat, Refrig & Air-Conditioning Engrs Award, 74; Solar Hall Fame, 86. *Mem:* Fel Am Soc Mech Engrs; Am Soc Heat, Refrig & Air-Conditioning Engrs; Int Solar Energy Soc (pres, Am Sect, 73-74, int vpres, 75-78); Int Inst Refrig; Sigma Xi. *Res:* Refrigeration; air conditioning; thermal systems; sea water conversion; energy conversion devices; heat transfer; thermal and flow models. *Mailing Add:* Eibling Thermal Systs 1380 Camelot Dr Columbus OH 43220

EICH, ROBERT, b Ann Arbor, Mich, Oct 4, 23; m 51; c 3. MEDICINE, CARDIOLOGY. *Educ:* Univ Mich, AB, 47, MD, 51. *Prof Exp:* From asst prof to assoc prof, 57-71, PROF MED, STATE UNIV NY UPSTATE MED CTR, 71- *Res:* Cardiovascular disease; internal medicine. *Mailing Add:* Dept Med State Univ NY Upstate Med Ctr 155 750 E Adams Syracuse NY 13210

EICHBAUM, BARLANE RONALD, b New Brunswick, NJ, Sept 1, 26; m 50; c 3. SOLID STATE PHYSICS, MATERIAL SCIENCE ENGINEERING. *Educ:* Rutgers Univ, BS, 51, PhD(ceramic eng), 56; Univ Tex, MS, 53; Temple Univ, BA, 65. *Prof Exp:* Prod & control engr, Hercules Powder Co, 50-52; res & develop engr, Am Rock Wool Corp, 52; Tex State scholar & Edward Orton Jr fel, 52-53; proj engr, US Signal Corps, Rutgers Univ, 53-56; res engr, Int Bus Mach Corp, 56-57, develop engr & tech consult, Prod Develop Labs, 57-59; mgr magnetic tech, Aeronutronic Comput Div, Ford Motor Co, 59-60, mgr solid state devices, Aeronutronic Res Labs, 60, mgr molecular eng, 60-62, asst dir, Philco Res Labs, 62-63, dir phys electronics, Philco Corp, 63-65; dir corp develop & tech asst to vpres eng & res, AMP, Inc, Pa, 65-67; dir res & asst to vpres eng, Gulton Industs, Inc, NJ, 67-68; dir res & develop & staff scientist, Lear Motors Corp, 68-70; consult engr, P & S Domestique, 72-73; TECH CONSULT, 71-; CHEM ENGR, US BUR MINES, 75- *Mem:* Electrochem Soc; Am Ceramic Soc; sr mem Inst Elec & Electronics Eng; Sigma Xi; fel Am Inst Chemists. *Res:* Chemistry; ceramics; crystallography; metallurgy; mineral processing; general, electronic and mechanical engineering; materials in advanced equipment; electronic systems technology. *Mailing Add:* 12065 Stoney Brook Dr Reno NV 89511

EICHBERG, JORG WILHELM, b Stuttgart, Ger, Nov, 14, 39; m 63; c 2. VETERINARY MEDICINE, IMMUNOLOGY. *Educ:* Univ Munich, DVM, 67; Univ Tex, San Antonio, PhD, 74. *Prof Exp:* Asst prof, Free Univ Berlin, 67-70; assoc found scientist, Southwestern Found Res & Educ, 74-86, SCIENTIST, SOUTHWEST FOUND BIOMED RES, 86- *Mem:* Am Soc Microbiol; AAAS; Am Asn Lab Animal Sci; Am Asn Immunologists. *Res:* Primatoloty; virology; immunology. *Mailing Add:* Southwest Found Biomed Res PO Box 28147 8848 W Commerce San Antonio TX 78284

EICHBERG, JOSEPH, b Oct 5, 35; US Citizen; m 64; c 3. BIOCHEMISTRY. *Educ:* Mass Inst Technol, BS, 57; Harvard Univ, PhD(biochem), 62. *Prof Exp:* Res assoc, Harvard Med Sch, 64-68, assoc, 68-69, from asst prof to assoc prof biol chem, 69-75; assoc prof biophys sci, 75-80, PROF BIOCHEM & BIOPHYS SCI, UNIV HOUSTON, 80- *Concurrent Pos:* USPHS fel, 62-65; estab investr, Am Heart Asn, 68-73; tutor biochem sci, Harvard Univ, 69-74. *Mem:* Am Soc Neurochem; Int Soc Neurochem; Brit Biochem Soc; Soc Neurosci; Am Soc Biol Chem. *Res:* Lipid biochemistry and metabolism; neurochemistry; membrane biochemistry. *Mailing Add:* Dept Biochem & Biophys Sci Univ Houston 4800 Calhoun Houston TX 77204-5500

EICHBERGER, LE ROY CARL, b Chicago, Ill, Oct 26, 27; m 55; c 3. MECHANICAL ENGINEERING. *Educ:* Univ Ill, BS, 51, MS, 55, PhD(theoret & appl mech), 59. *Prof Exp:* Layout draftsman, McCormick Works, Int Harvester Corp, 51-53; res asst theot & appl mech, Univ Ill, 53-55, res assoc, 55-57, instr, 57-59; from asst prof to assoc prof mech eng, Univ Houston, 59-77; mgr eng & tech serv, Weatherford/Lamb, 77-80; mgr res & develop, NL Atlas Bradford, 80-89; CONSULT, 89- *Concurrent Pos:* Tech consult, Reed Roller Bit Co, Houston, 59-61, Houston Eng Res Co, 61-62 & Humble Oil & Refining Co, 68-77; vis prof, Univ Mich, 61-62. *Mem:* Soc Exp Stress Anal; Am Soc Mech Engrs; Sigma Xi (secy-treas, 64-65, pres-elect, 65-67). *Res:* Applied mechanics, specifically in areas of elasticity, shell analysis, vibration, dynamics, mechanics of materials, photoelasticity and experimental stress analysis. *Mailing Add:* 5310 Dumfries Dr Houston TX 77096

EICHEL, HERMAN JOSEPH, b Toledo, Ohio, Aug 2, 24; m 49; c 5. BIOCHEMISTRY, ORGANIC CHEMISTRY. *Educ:* Univ Dayton, BS, 48; DePaul Univ, MS, 56; Univ Cincinnati, PhD(biochem), 66. *Prof Exp:* Res chemist, Abbott Labs, 48-53 & C F Kettering Found, 53-57; sr res chemist, Diamond Labs, 57-58; proj leader microencapsulation, Nat Cash Register Co, 58-62; asst res dir, Hoechst Pharmaceut Co, 62-64, dir pharmaceut res, 64-66, vpres pharmaceut res & prod, 66-68, exec vpres, 68-74; pres, Adria Labs Inc, 74-84; PRES, CHARTWELL TECHNOL INC, 84- *Mem:* Am Chem Soc; Am Inst Chem (pres, 86-87); NY Acad Sci. *Res:* Synthesis of photosynthetic intermediates, coacervation studies; molecular heterogeniety of D-amino acid oxidase; synthetic studies of pyran chemistry; research in areas of tropical disease, tuberculosis, thyroid disease, CNS and cardiovascular agents and controlled drug release, antiviral and antineoplastic drugs; diagnostics and veterinary products. *Mailing Add:* Chartwell Technologies Inc 1445 Summit St Columbus OH 43201-2105

EICHELBERGER, EDWARD B, b Norfolk, VA, Feb 8, 34. ELECTRICAL ENGINEERING. *Educ:* Lehigh Univ, BS, 56; Princeton Univ, MA & PhD(elec eng), 63. *Prof Exp:* assoc engr, 56-59, staff engr, 59-64, develop engr, 64-65, sr engr, 68-82, FEL, IBM CORP, 82- *Mem:* fel Inst Elec & Electronics Engrs. *Res:* Computer design; logic circuit testing; semiconductor chip design. *Mailing Add:* IBM Corp Neighborhood Rd Kingston NY 12401

EICHELBERGER, JOHN CHARLES, b Syracuse, NY, Oct 3, 48; m 69; c 2. VOLCANOLOGY. *Educ:* Mass Inst Technol, BS & MS, 71; Stanford Univ, PhD(geol), 74. *Prof Exp:* Staff mem geol, Los Alamos Sci Lab, 74-79; staff mem, 79-89, distinguished staff mem, 89-90, DIV SUPVR, SANDIA NAT LABS, 90- *Concurrent Pos:* Chief scientist, Inyo Domes Sci Drilling Proj, 83-88, Katmai Drilling Proj, 88-; adj prof geochem, NMex Inst of Mines & Technol, 88- *Mem:* Am Geophys Union; Geol Soc Am. *Res:* Origin and evolution of intermediate and silicic magmas; volatiles in magmas; eruption processes. *Mailing Add:* Org 6233 Sandia Nat Labs Albuquerque NM 87185-5800

EICHELBERGER, ROBERT JOHN, b Washington, Pa, Apr 10, 21; m 43; c 4. PHYSICS. *Educ:* Washington & Jefferson Col, AB, 42; Carnegie Inst Technol, MS, 48, PhD(physics), 54. *Prof Exp:* Instr physics, Washington & Jefferson Col, 42; res physicist, Carnegie Inst Technol, 43-45, res supvr, 45-55; supvry physicist, US Army Ballistic Res Labs, 55-62, assoc tech dir, 62-67, dir, 67-86, assoc tech dir res & technol, US Armament Res & Develop Command, 77-81; CONSULT, 86. *Honors & Awards:* Presidential Meritorious Exec Award, Dept Defense, 82; Crozier Award, 84; Clifford R Gross Award, 85. *Mem:* AAAS; Am Phys Soc; Am Inst Aeronaut & Astronaut. *Res:* Effects of extremely high temperatures and pressures on solids; non-steady fluid dynamics at extreme pressures; hypervelocity ballistics; shock phenomena; brittle fracture in metals. *Mailing Add:* 409 Catherine Bel Air MD 21014

EICHELBERGER, ROBERT LESLIE, b Wichita, Kans, Jan 18, 26; m 47; c 3. PHYSICAL CHEMISTRY. *Educ:* Univ Calif, Los Angeles, BS, 48, MS, 49; Univ Wyo, PhD(chem), 57. *Prof Exp:* Supvr anal chem, Truesdail Labs, Calif, 49-52; instr col eng, Univ Wy, 52-54; res chemist liquid metals, NAm Aviation, Inc, 54-57, supvr liquid metal chem, 57-66; rep, US Dept Energy, Japan, 85-89; sr tech specialist, Atomics Int Div, NAm Rockwell Corp, 66-73, SR STAFF SCIENTIST, ENERGY TECHNOL ENG CTR, ROCKWELL INT, 73-85, 89- *Mem:* Am Chem Soc; Sigma Xi. *Res:* Liquid metal chemistry; metal-gas reactions; analytical methods; chemical engineering. *Mailing Add:* 937 Sandpiper Cir Thousand Oaks CA 91361

EICHELBERGER, W(ILLIAM) H, b Wichita, Kans, Dec 5, 21; m 43; c 4. ELECTRICAL ENGINEERING. *Educ:* Univ Colo, BS, 43. *Prof Exp:* Elec design engr, Radio Corp Am, 43-46; asst chief engr, Hathaway Instrument Co, 46-52; res engr, Res Inst, Univ Denver, 52-62, mgr comput ctr, 62-88, dir systs develop, 70-88, sr systs anal, 76-88; RETIRED. *Mem:* Sr mem Inst Elec & Electronics Engrs; Asn Comput Mach. *Res:* Applications of digital computers; digital computer logic and circuit design; pulse and digital instrumentation; audio engineering; radio receivers; electronic instrumentation. *Mailing Add:* 570 S Franflin Univ Denver Denver CO 80208

EICHELMAN, BURR S, JR, b Hinsdale, Ill, Mar 20, 43. BEHAVIORAL NEUROCHEMISTRY, AGGRESSION. *Educ:* Univ Chicago, MD, 68, PhD(biopsychol), 70. *Prof Exp:* Intern pediat, Univ Calif Med Ctr, San Francisco, 69 70; staff assoc biopsychol, NIH & USPHS, Bethesda, 70-72; resident fel psychiat, Stanford Univ Med Ctr, 72-75, Kennedy fel med, law & ethics, 75-76; asst prof, Univ Wis, 76-79; assoc prof, 79-84, PROF PSYCHIAT, UNIV WIS-MADISON, 84-; CHIEF PSYCHIAT SERV, WILLIAM S MIDDLETON MEM VET ADMIN HOSP, 76- *Concurrent Pos:* Prin investr, Vet Admin grant, 77-; dir, Lab Behav Neurochem, Vet Admin Hosp, Univ Wis, 76- *Honors & Awards:* A B Bennett Award, Soc Biol Psychiat, 72. *Mem:* Fel Am Psychiat Asn; fel Am Psychol Asn; Soc Neurosci; Am Col Neuropsychopharmacol; Int Soc Res Aggression. *Res:* Behavioral neurochemistry and pharmacology focusing on animal models of aggression and human violent behavior; concerned with ethical issues in aggression research. *Mailing Add:* Clin Dir Dorothea Dix Hosp S Boylan Ave Raleigh NC 27611

EICHENBERGER, HANS P, b Fribourg, Switz, Dec 29, 21; US citizen; m 66; c 4. ARTIFICIAL INTELLIGENCE. *Educ:* Swiss Fed Inst Technol, Dipl, 46; Pa State Univ, MS, 49; Mass Inst Technol, ScD(mech eng), 52. *Prof Exp:* Asst internal combustion eng, Swiss Fed Inst Technol, 46-48; res asst diesel eng, Pa State Univ, 48-49; instr gas turbines fluid flow, Mass Inst Technol, 50-52; asst chief aerodyn, Garret Corp, Calif & Ariz, 52-56; chief eng sci, TRW Inc, 57-62; dir res, Res Ctr, Ingersoll Rand Co, 63-68; dir res lab, IBM, Switz, 68-71, asst to corp dir res, 71-76; sci rep Europ, Gen Elec Ctr Res & Develop, 76-85; PRES EHI, EICHENBERGER INST HOCHSCHUL INNOVATIONS, 88- *Concurrent Pos:* Lectr, Case Inst Technol, 57-59; pvt consult, 87- *Mem:* Inst Elec & Electronics Engrs. *Res:* Internal flow in machinery; external flow of bodies; drag reduction. *Mailing Add:* Johannisburgstr 10 Kusnacht CH 8700 Switzerland

EICHENBERGER, RUDOLPH JOHN, b Pawnee City, Nebr, June 10, 41; m 69. EDUCATIONAL ADMINISTRATION, MECHANICS. *Educ:* Peru State Col, BSE, 64; Kans State Teachers Col, MS, 66; Univ Northern Colo, EdD, 72. *Prof Exp:* Teaching asst physics/phys sci, N Tex State Univ, 64-65; asst prof, Eureka Col, 66-75, assoc prof physics/phys sci/math, 75-80, prof physics/math/chem, 80-82; assoc prof physics/phys sci, 82-88, PROF PHYSICS/SCI, SOUTHERN ARK UNIV, 88- *Concurrent Pos:* Chmn sci-math, Eureka Col, 77-78, dir, Title III Proj, 77-80, Freshman Studies, 78-80 & Inst Res, 80-81; actg head, Phys Eng, Southern Ark Univ, 84-87, chmn Phys Sci, 89-91; consult, Amfuel Coated Fabrics, 86, Depper Legal Serv & Indust Elec, 87, El Dorado Pub Sch, 91. *Mem:* Sigma Xi; Am Asn Physics Teachers; Nat Sci Teachers Asn; Asn Educ Teachers Sci; Coun Undergrad Res. *Res:* Solar energy collection; microcomputer data collection; publications in creativity measurement in physics and microwave application to agriculture products and/or operations. *Mailing Add:* Southern Ark Univ Box 1176 Magnolia AR 71753

EICHENHOLZ, ALFRED, b Dabrowa, Poland, Apr 5, 27; US citizen; m 52; c 1. INTERNAL MEDICINE, METABOLISM. *Educ:* Univ Munich, MD, 51; Univ Minn, Minneapolis, MSc, 64; Am Bd Internal Med, dipl, 60. *Prof Exp:* Staff physician pulmonary dis serv, Vet Admin Hosp, Minneapolis, 57-59, asst chief, 59-61, chief clin radioisotope sect, 61-67; chief radioisotope serv, 67-70, CHIEF MED SERV, VET ADMIN HOSP, 70-; ASSOC PROF MED, SCH MED, UNIV PITTSBURGH, 67- *Concurrent Pos:* Instr, Univ Minn, Minneapolis, 57-64, asst prof, 64-67. *Mem:* AMA; Am Fedn Clin Res; NY Acad Sci; Soc Nuclear Med. *Res:* Acid Base, fluid and electrolyte balance; renal physiology; pulmonary disease. *Mailing Add:* Vet Admin Outpatient Clin Los Angeles CA 90013

EICHENWALD, HEINZ FELIX, b Ger, Mar 3, 26; nat US; m 51; c 3. PEDIATRICS, MICROBIOLOGY. *Educ:* Harvard Univ, AB, 46, Cornell Univ, MD, 50. *Prof Exp:* From instr to prof, Cornell Univ, 55-64; prof & chmn dept, 64-82, WILLIAM BUCHANAN PROF PEDIAT, UNIV TEX HEALTH SCI CTR, DALLAS, 82- *Concurrent Pos:* USPHS career res investr award, 63; consult, USPHS, 55-; chmn antibiotics panel I, Nat Drug Study, Nat Res Coun-Nat Acad Sci, 67-70; vis prof, Fac Med, Univ Saigon, 69-; dir pediat prog, AMA-Vietnam Med Sch Proj, 69-; mem nat adv coun, Nat Inst Child Health & Human Develop, 69-; consult, Food & Drug Admin, 70-; mem res comt, United Cerebral Palsy Found, 70-; mem, Bd Maternal, Child & Family Health, Nat Res Coun-Nat Acad Sci, 73- & Bd Dir, Lamplighter Sch, Winston Sch, Children's Develop Ctr; chief of staff, Children's Med Ctr, Dallas; pediatrician in chief, Parkland Hosp, Dallas; consult, var hosps. *Honors & Awards:* USPHS career res investr award, 63; Markle Award, 53; von Humboldt Prize, 78; Weinstein-Goldensen Award, WGermany, 79; Med Res Award, United Cerebral Palsy Res Found, 79. *Mem:* Am Pediat Soc; Infectious Dis Soc Am; Harvey Soc; Soc Pediat Res; Sci Res Soc Am; hon mem Pediat Soc Costa Rica; hon mem Pediat Soc Columbia; hon mem Pediat Soc Venezuela; hon mem Pediat Soc Panama; hon mem Pediat Soc El Salvador. *Res:* Infectious diseases of children; host-parasite interaction. *Mailing Add:* Dept Pediat Univ Tex Health Sci Ctr Dallas TX 75235

EICHER, DON LAUREN, b Lincoln, Nebr, Dec 12, 30; m 54. GEOLOGY. *Educ:* Univ Colo, BA, 54, MS, 55; Yale Univ, PhD(geol), 58. *Prof Exp:* From asst prof to assoc prof, 58-70, PROF GEOL, UNIV COLO, BOULDER, 70- *Mem:* Geol Soc Am; Paleont Soc. *Res:* Cretaceous micropaleontology; stratigraphy; marine paleocology. *Mailing Add:* Dept Geol Sci Univ Colo Campus Box 250 Boulder CO 80309-0250

EICHER, EVA MAE, b Kalamazoo, Mich, Sept 26, 39. GENETICS. *Educ:* Kalamazoo Col, BA, 61; Univ Rochester, MS, 63, PhD(genetics), 67. *Prof Exp:* Res assoc, Univ Rochester, 67-70; from assoc staff scientist to staff scientist, 71-80, SR STAFF SCIENTIST, JACKSON LAB, 80- *Concurrent Pos:* Lectr, Univ Rochester, 68, Univ Maine, 75-81 & coop prof zool, 81-; investr, Oak Ridge Nat Lab, 70-71; mem adv cont Animal Resources, NIH, 76-80, Mammalian Genetics Study Sect, 87- *Honors & Awards:* Donald R Charles Mem Award, Univ Rochester, 64. *Mem:* AAAS; Genetics Soc Am; Soc Develop Biol; Am Soc Human Genetics; Am Genetic Asn. *Res:* Mammalian genetics; mouse cytogenetics; evolution of mammation X and Y chromosome; genetics of sex determination. *Mailing Add:* Jackson Lab 600 Main St Bar Harbor ME 04609

EICHER, GEORGE J, b Bremerton, Wash, Aug 27, 16; m 51; c 2. FISH BIOLOGY. *Educ:* Ore State Univ, BS, 41. *Prof Exp:* Field party leader, US Bur Fisheries, 39-41; free lance writer, 41-43; chief fisheries biologist, Ariz Game & Wildlife Comn, 43-47; proj leader, US Fish & Wildlife Serv, 47-56; aquatic biologist, Portland Gen Elec Co, 56-71, mgr dept environ serv, 71-78; PRES, EICHER ASSOCS, INC, 78- *Concurrent Pos:* Consult indust & govt orgn, US, Australia, SAm & Can, 57- *Mem:* Am Fisheries Soc (1st & 2nd vpres, 62-64, pres, 64); Wildlife Soc; Am Soc Limnol & Oceanog; Am Inst Fishery Res Biol; fel Int Acad Fishery Sci. *Res:* Fish behavior; fish passage at dams; red salmon in Alaska; aquatic weed control; fish screen patent US and Canada. *Mailing Add:* 8787 SW Becker Dr Portland OR 97223

EICHER, JOHN HAROLD, b Dayton, Ohio, Mar 30, 21; m 57; c 2. ORGANIC CHEMISTRY. *Educ:* Purdue Univ, BS, 42, PhD(chem), 52. *Prof Exp:* Asst mineral, Purdue Univ, 42, org chem, 45-48 & asst instr, 48-51; Am Petrol Inst asst, Ohio State Univ, 43; chemist, Manhattan Proj, Carbide & Carbon Chem Corp, SAM Labs, Columbia Univ, 43-45; res consult, Tungston Plantation, Newport Ships, 51-52; from asst prof to prof, 52-89, EMER PROF CHEM, MIAMI UNIV, 89- *Concurrent Pos:* Faculty res fel, Miami Univ, 58. *Mem:* AAAS; Am Chem Soc. *Res:* Gas viscosities; natural products; organic nitrogen compounds; stereochemistry. *Mailing Add:* Dept Chem Hughes Labs Miami Univ Oxford OH 45056

EICHHOLZ, ALEXANDER, b Zagreb, Yugoslavia, Dec 12, 26; US citizen; m 55. BIOCHEMISTRY. *Educ:* Blackburn Col, BA, 54; Univ Ill, MS, 60, PhD(biochem), 62. *Prof Exp:* From instr to asst prof biochem, Chicago Med Sch, 62-66; from asst prof to assoc prof physiol, 66-78, prof physiol & biophys, 78-88, EMER PROF, ROBERT WOOD JOHNSON MED SCH, 88- *Concurrent Pos:* Schweppe Found fel, 63-; Chicago Med Sch bd trustees res award, 65. *Mem:* AAAS; Fedn Am Soc Exp Biol; Am Soc Cell Biol; Brit Biochem Soc. *Res:* Membrane structure and transport; intestinal transport; cancer. *Mailing Add:* Rte 1 Box 369 Hollywood MD 20363

EICHHOLZ, GEOFFREY G(UNTHER), b Hamburg, Ger, June 29, 20; US citizen. NUCLEAR ENGINEERING, HEALTH PHYSICS WASTE MANAGEMENT. *Educ:* Univ Leeds, BSc, 42, PhD(physics), 47, DSc, 79. *Prof Exp:* Exp officer radar develop, Brit Admiralty, 42-46; demonstr physics, Univ Leeds, 46-47; asst prof, Univ BC, 47-51; head radiation lab & physics & radiotracer subdiv, Can Bur Mines, 51-63; regents prof, 63-89, EMER REGENTS PROF NUCLEAR ENG, GA INST TECHNOL, 89- *Concurrent Pos:* Lectr, Univ Ottawa, 56-58; Int Atomic Energy Agency regional adv, Southeast Asia, 68, 78; vis prof, Mex, 72, 74 & Iran, 78. *Mem:* Am Phys Soc; fel Health Physics Soc; fel Am Nuclear Soc; Can Asn Physicists; fel Brit Phys Soc. *Res:* Industrial applications of radioisotopes; radiation effects; radioactive waste management; nuclear radiation detectors; environmental aspects of nuclear technology; architectural acoustics; activation analysis. *Mailing Add:* 1784 Noble Dr NE Atlanta GA 30306

EICHHORN, GUNTHER LOUIS, b Frankfurt am Main, Ger, Feb 8, 27; nat US; m 64; c 2. INORGANIC BIOCHEMISTRY. *Educ:* Univ Louisville, AB, 47; Univ Ill, MS, 48, PhD(chem), 50. *Prof Exp:* Asst inorg chem, Univ Ill, 47-49, asst, General Aniline stipend, 49-50; from asst prof to assoc prof inorg chem, La State Univ, 50-57; assoc prof, Georgetown Univ, 57-58; chief, Molecular Biol Sect, Gerontol Res Ctr, Nat Heart Inst, Md, 58-66; chief, Molecular Biol Sect, Nat Inst Child Health & Human Develop, 66-78; CHIEF, LAB CELLULAR MOLECULAR BIOL & HEAD, INORG BIOCHEM SECT, GERONTOL RES CTR & ACTG SCI DIR, NAT INST ON AGING, NIH, 78- *Concurrent Pos:* Sr asst scientist, Nat Inst Mental Health, USPHS, 54-57; distinguished lectr, Mich State Univ, 71 & 72; mem, Panel on Nickel, Nat Res Coun, 73-75; co-ed, Advan Inorg & Biochem, 79- *Honors & Awards:* Watkins lectr, Wichita State Univ, 83; Henry Lardy Lectership, SDak State Univ, 88. *Mem:* Fel AAAS; fel Geront Soc; fel Am Inst Chem; Am Chem Soc; Am Soc Biol Chem & Molecular Biol. *Res:* Function of metals in biological processes; hemoproteins; metal ion catalysis; coordination chemistry; nucleic acids; gerontology; metal ions and the transfer of genetic information; metal ions and aging. *Mailing Add:* Ger Res Ctr Nat Inst Aging NIH 4940 Eastern Ave Baltimore MD 21224

EICHHORN, HEINRICH KARL, b Vienna, Austria, Nov 30, 27; nat US; m 52, 77; c 4. ASTRONOMY, STATISTICAL ANALYSIS. *Educ:* Univ Vienna, DrPhil(astron), 49. *Prof Exp:* From instr to asst prof, Univ Vienna, 50-56; asst prof astron, Georgetown Univ, 56-69; assoc prof, Wesleyan Univ, 59-64; prof & chmn, Dept Astron, Univ South Fla, 64-79; prof & chmn, Dept Astron, 75-85, prof astron & civil eng, 85-88, PROF ASTRON, UNIV FLA, 88- *Concurrent Pos:* Brit Coun scholar, Univ Glasgow, 51-52; Int Coop Admin fel, McCormick Observ, 54-56; sr consult, Geonautics, Inc, 57-70; consult, Radio Corp Am Serv Co, 59, Perkin-Elmer Co, 62, Smithsonian Astrophys Observ, 61-63, Yale Univ Observ, 63-64 & 74-75, Minneapolis Honeywell Regulator Co, 64, Geo-Space Co, 65, US Army Map Serv, 65-70, Lockheed Missiles & Space Co, 80-83 & Space Telescope Sci Inst, 84; vis prof, Univ Vienna, 71 & 90, Univ Graz, 71 & 76, Yale Univ, 76, Fed Univ Rio de Janeiro, 84, Univ Innsbruk, 87, Fulbright, Univ Prague, 88; grants, NSF, NASA, US Army, US Air Force & US Naval Res Lab; partic & lectr, nat & int conf astron; hon prof, Univ Graz, Austria, 76-, Univ Vienna, 90-; chmn, Div Dynamic Astron, Am Astron Soc; distinguished scholar, Univ SFla, 79; pres,Comn 24, Int Astron Union, 80-83; prof astron, Paris Observ, 88, 90. *Mem:* Am Astron Soc; Int Astron Union; foreign corresp mem, Austrian Acad Sci. *Res:* Photographic astrometry; positional astronomy; celestial mechanics; stellar dynamics; statistical adjustment of data. *Mailing Add:* Dept Astron Univ Fla BRT 202 Gainesville FL 32611-2015

EICHHORN, J(ACOB), b Sheboygan, Wis, Sept 14, 24; m 59; c 3. CHEMICAL ENGINEERING. *Educ:* Univ Mich, BS, 46, MS, 47, PhD(chem eng), 50. *Prof Exp:* Mem staff, 50-56, div leader, 56-61, mgr spec proj, Plastics Dept, 61-62, develop mgr, Packaging Dept, 62-66, mgr spec proj, 66-68, exec asst, 68-71, ventures mgr, Plastics Dept, 71-79, lab dir, 79-82, SR PROJ MGR, DOW CHEM CO, 82- *Concurrent Pos:* Instr, Univ Mich, 52-54; mem prog mgt develop, Harvard Univ, 68. *Mem:* Am Chem Soc; Am Inst Chem Engrs; Tech Asn Pulp & Paper Indust. *Res:* Research, development and venture operations. *Mailing Add:* 4501 Arbor Dr Midland MI 48640-2645

EICHHORN, ROGER, b Slayton, Minn, Apr 1, 31; m 52; c 5. HEAT TRANSFER, FLUID MECHANICS. *Educ:* Univ Minn, BEE, 53, MSME, 55, PhD(mech eng), 59. *Prof Exp:* Instr mech eng, Univ Minn, 55-59; from asst prof to assoc prof, Princeton Univ, 59-63, assoc prof aerospace & mech sci, 63-67; prof mech eng & chmn dept, Univ Ky, 67-75, actg dean, Col Eng, 75-76, assoc dean res, Grad Sch, 76-78, dean, Col Eng, 79-; DEAN, CULLEN COL ENG, UNIV HOUSTON, 82- *Concurrent Pos:* NSF fel, Imp Col, Univ London, 63-64. *Honors & Awards:* Heat Transfer Dir Mem Award, Am Soc Mech Eng, 83. *Mem:* Fel AAAS; fel Am Soc Mech Engrs; Am Inst Aeronaut & Astronaut; Am Soc Eng Educ; Nat Soc Prof Engrs. *Res:* Heat Transfer in natural convection; channel and boundary layer flows; multiphase flows. *Mailing Add:* Cullen Col Eng Univ Houston Houston TX 77204-4814

EICHINGER, BRUCE EDWARD, b Canby, Minn, Oct 25, 41; m 62; c 3. POLYMER CHEMISTRY. *Educ:* Univ Minn, BCh, 63; Stanford Univ, PhD(polymer solutions), 67. *Prof Exp:* Fel chem, Yale Univ, 67-68; from asst prof to assoc prof, 68-80, PROF PHYS CHEM, UNIV WASH, 80- *Mem:* Am Chem Soc. *Res:* Thermodynamics of polymer solutions; theory of elasticity. *Mailing Add:* Biosym Technologies Inc 10065 Barnes Canyon Rd Suite A San Diego CA 92121-2777

EICHINGER, JACK WALDO, JR, b Ottumwa, Iowa, Sept 11, 04; m 26; c 2. INORGANIC CHEMISTRY. *Educ:* Iowa State Col, BS, 26, PhD(chem), 31. *Prof Exp:* From asst to instr, Iowa State Col, 31-34; asst prof chem, Univ Detroit, 34-37, assoc prof, 37-41; assoc prof, Williams Col, 46-47; prof & coordr nuclear sci prog, Fla State Univ, 48-62; prof, Univ Baghdad, 62-64; from prof to emer prof chem, Fla State Univ, 64-72; RETIRED. *Concurrent Pos:* Asst dir chem, US Mil Acad, 44-46. *Mem:* Am Chem Soc. *Res:* Levulose sugar; corrosion of metals; caramels; sour taste of acids; phosphorescence spectroscopy; electron configurations; electron chart. *Mailing Add:* 555 N Pantano Rd Tucson AZ 85710

EICHLER, DUANE CURTIS, b Miami, Fla, Jul 11, 46; m 69; c 1. NUCLEIC ENZYMOLOGY. *Educ:* Univ Calif, Los Angeles, PhD(biochem). *Prof Exp:* From asst prof to assoc prof, 77-88, PROF BIOCHEM, COL MED, UNIV SFLA, 88- *Concurrent Pos:* Fel Jane Cofkin Childs Mem Fund Med Res. *Mem:* Sigma Xi; Am Soc Biol Chemists; Am Soc Microbiol; Nat Bd Med Exam. *Res:* Nucleic acid enzymology; types of activities responsible for processing of ribosomal RNA in mammalian cells. *Mailing Add:* Dept Biochem Col Med Univ SFla 12901 N 30th St Tampa FL 33612

EICHLER, VICTOR B, b Dixon, Ill, July 13, 41; m 65; c 2. EXPERIMENTAL EMBRYOLOGY. *Educ:* Univ Ill, BS, 63, MS, 64; Univ Iowa, PhD(zool, embryol), 69. *Prof Exp:* NIH fels anat, Univ Chicago, 69-71; asst prof biol sci, 71-77, ASSOC PROF BIOL SIC, WICHITA STATE UNIV, 77- *Concurrent Pos:* Danforth tutorship, 70; NASA Am Soc Eng Educ Fel, 75, 76; Danforth Assoc, 78-; guest investr & vis scholar, Mus Comparative Zool, Univ Calif, Berkeley, 79; vis res assoc prof, Dept Biol Sci, Dartmouth Col, 80; dir educ prog, Fetzer Found, 87- *Mem:* AAAS; Am Soc Zoologists; Soc Study Amphibians & Reptiles; Am Asn Anatomists; Soc Neurosci. *Res:* Development, morphology, physiology of vertebrate visual systems and epithalamic structures; influence of environmental lighting on body growth and development; evolution of pineal structures and role in body functions. *Mailing Add:* 6413 Angling Rd Kalamazoo MI 49002

EICHMAN, PETER L, b Philadelphia, Pa, Nov 18, 25; c 4. NEUROLOGY, MEDICINE. *Educ:* St Joseph's Col, BS, 45; Jefferson Med Col, MD, 49. *Prof Exp:* Intern, Fitzgerald-Mercy Hosp, 49-50; resident, Walter Reed Army Hosp, 50-51; Jefferson Med Col, 51-52, Mayo Found, 52-54 & Univ Wis, 54-55; from instr neuropsychiat to prof neurol & med, Univ Wis-Madison, 55-71, dir student health, 62-65, asst dean clin affairs, 65, dean med sch & dir med ctr, 65-71; dep dir, Bur Health Manpower Educ, NIH, 72-73; PROF NEUROL, UNIV WIS-MADISON, 73- *Concurrent Pos:* Fel, Walter Reed Army Hosp, 50-51, Jefferson Med Col, 51-52, Mayo Found, 52-54 & Univ Wis, 54-55; consult, Vet Admin Hosp, 60-65; State Dept Pub Welfare, 60-65. *Mem:* Am Acad Neurol; Am Col Physicians; AMA; Am Col Health Asn. *Res:* Relationship of perinatal injury and central nervous system defects; trace mineral excretion in porphyria; immunological responses in liver disease; development of a university health service. *Mailing Add:* Univ Wisc Ctr Health Sci 600 Highland Ave Madison WI 53792

EICHMANN, GEORGE, b Budapest, Hungary, Nov 3, 36; US citizen; m 62; c 1. ELECTRICAL ENGINEERING. *Educ:* City Col New York, BEE, 61, MEE, 63; City Univ New York, PhD(eng), 68. *Prof Exp:* Jr engr elec eng, Data Syst Div, IBM, 61-62; lectr, 63-68, from asst prof to prof, 68-79, chmn dept, 82-88, H G KAYSER PROF ELEC ENG, CITY COL NEW YORK, 88- *Concurrent Pos:* Nat Sci Found grant, 69-71; City Univ New York Res Found grant, 71-72 & 77-78 & 85-; Air Force Off Sci Res grant, 77-; consult, Naval Res Lab, Washington, DC, 78-80 & NAm Philips, NY, 79-80. *Mem:* Am Soc Eng Educ; Inst Elec & Electronics Engrs; Optical Soc Am; Soc Photog Instrumentation Engrs. *Res:* Electromagnetic theory; quantum electronics; fiber optics and image processing; optical computing; neural networks. *Mailing Add:* Dept Elec Eng Steinman Hall No 405t City Col NY 139 St & Convent Ave New York NY 10031

EICHNA, LUDWIG WALDEMAR, b Tallin, Estonia, May 9, 08; nat US. MEDICINE. *Educ:* Univ Pa, AB, 29, MD, 32. *Prof Exp:* Asst instr, Sch Med, Univ Pa, 34-35; instr med, Sch Med, Johns Hopkins fel, 36-40; instr med, Col Med, NY Univ, 40; from asst prof to prof, 41-60; prof med & chmn dept, State Univ NY Downstate Med Ctr, 60-74; RETIRED. *Concurrent Pos:* Fel, Sch Med, Johns Hopkins Univ, 36-40, Commonwealth fel & asst physician, Hosp, 36-40, asst vis physician, 47-; vis physician & dir med serv, Kings County Hosp, Brooklyn, 60-74; mem fel sect, Comt on Growth, Nat Res Coun, 53, Comt Cardiovasc Syst, 54, chmn, 57, med fel bd, 56; mem res coun, Pub Health Res Inst, 57; mem comt clin invest, Nat Found, 58; med scientist training comt, NIH, 63; panel & review comt, Health Res Coun, NY City. *Mem:* Am Soc Clin Invest; Am Physiol Soc; Soc Exp Biol & Med; Asn Prof Med (pres, 69-70); Asn Am Physicians (pres, 70-71); Sigma Xi. *Res:* Cardiovascular investigation involving hemodynamics of congestive heart failure. *Mailing Add:* 210 Columbia Heights Brooklyn NY 11201

EICHNER, EDUARD, b Cleveland, Ohio, Nov 11, 05; m 31; c 2. OBSTETRICS, GYNECOLOGY. *Educ:* Western Reserve Univ, AB, 25, MD, 29; Am Bd Obstet & Gynec, dipl. *Prof Exp:* Jr vis obstetrician & gynecologist, 36-41, dir, Family Planning Clin, 69-78, ASSOC, MT SINAI HOSP, 41-; ASSOC CLIN PROF REPRODUCTIVE BIOL, SCH MED, CASE WESTERN RESERVE UNIV, 73- *Concurrent Pos:* Dir educ, St Ann Hosp, 49-56; consult, Cleveland State Hosp, 49-74; asst clin prof obstet & gynec, Sch Med, Case Western Reserve Univ, 54-73; dir res obstet & gynec, Mt Sinai Hosp, 58-, asst dir, Div Obstet-Gynec, 69-73; med dir, Preterm-Cleveland, 73-78, consult, 78- *Honors & Awards:* Gold Award, Ohio State Med Asn, 55. *Mem:* AAAS; Endocrine Soc; Soc Exp Biol & Med; Am Col Surg; Am Col Obstet & Gynec; Sigma Xi. *Res:* Medicine; anatomy; pathology; physiology; pharmacology; gynecic lymphatics; physiology fertility, pregnancy and labor; internal and external genitalia; effects of drugs on mother and fetus in pregnancy, labor and the neonatal period. *Mailing Add:* 3333 Daleford Rd Cleveland OH 44120-3437

EICHNER, EDWARD RANDOLPH, b Fort Wayne, Ind, Mar 7, 38; m 62; c 3. HEMATOLOGY. *Educ:* Baylor Univ, BA, 59; Johns Hopkins Univ, MD, 63. *Prof Exp:* Fel hematol, Univ Wash, 68-70, instr med, 70-71; from asst prof to prof med, Sch Med, La State Univ, Shreveport, 71-76, chief hematol & oncol, 71-76; PROF MED & CLIN PROF PATH, UNIV OKLA SCH MED, 77-, CHIEF HEMATOL & ONCOL, 77- *Mem:* Am Fedn Clin Res; Am Soc Hematol; Soc Exp Biol & Med; Sigma Xi. *Res:* Effect of ethanol on hematologic and nutritional status of man; effect of ethanol and other drugs on human folate and other vitamin metabolism. *Mailing Add:* PO Box 33932 Shreveport LA 71130

EICHTEN, ESTIA JOSEPH, b Stillwater, Minn, Oct 12, 46. FIELD THEORY. *Educ:* Mass Inst Technol, BS, 68, PhD(physics), 72. *Prof Exp:* Res asst, Stanford Linear Accelerator Ctr, 72-74, Cornell Univ, 74-77; asst prof physics, Harvard Univ, 77-81; SCIENTIST II, FERMI NAT ACCELERATOR CTR, 81- *Concurrent Pos:* Mem, Inst Advan Study, 75-76; vis scientist, Enrico Fermi Inst, Univ Chicago, 81-84. *Mem:* Am Phys Soc; AAAS. *Res:* Theoretical high energy particle physics; dynamical symmetry; breaking; lattice gauge theories; phenomenology. *Mailing Add:* Theory Group Fermi Nat Accelerator Lab PO Box 500 Batavia IL 60510

EICHWALD, ERNEST J, b Ger, Dec 13, 13; nat US; m 41; c 3. PATHOLOGY. *Educ:* Univ Freiburg, MD, 38; Univ Utah, MD, 53. *Prof Exp:* Instr path, Med Sch, Harvard Univ, 46-48; from asst prof to assoc prof, Univ Utah, 48-54; dir lab exp med, Mont Deaconess Hosp, 54-66; prof microbiol, Mont State Univ, 66-70; PROF PATH & SURG & CHMN DEPT PATH, COL MED, UNIV UTAH, 70- *Concurrent Pos:* Ed, Transplantation Bull, 53-61; ed adv, Cancer Res, 54-57; chmn comt tissue transplantation, Nat Acad Sci-Nat Res Coun, 57-69; ed, Transplantation, 61-; dir, McLaughlin Res Inst, Great Falls, Mont, 66-70. *Mem:* Am Soc Exp Path; Soc Exp Biol & Med; Am Asn Cancer Res; Am Asn Path & Bact. *Res:* Transplantation immunity. *Mailing Add:* Dept Path Univ Utah Med Ctr 50 N Medical Dr Salt Lake City UT 84132

EICK, HARRY ARTHUR, b Rock Island, Ill, Dec 9, 29; m 54; c 8. SOLID STATE CHEMISTRY, RARE EARTH CHEMISTRY. *Educ:* St Ambrose Col, BS, 50; Univ Iowa, PhD(chem), 56. *Prof Exp:* Asst prof, Univ Ky, 56-57; res assoc, Univ Kans, 57-58; from asst prof to assoc prof, 58-67, assoc chmn dept, 70-71, interim dir comput lab, 71-72, assoc dean, Col Natural Sci, 74-81, PROF CHEM, MICH SATE UNIV, 67- *Mem:* Am Chem Soc; Am Crystallog Asn. *Res:* Solid state preparatory and x-ray diffraction studies on lanthanide halide compounds. *Mailing Add:* Dept Chem Mich State Univ East Lansing MI 48824-1322

EICKELBERG, W WARREN B, b New York, NY, Jan 19, 25; m 52; c 4. BIOMECHANICS. *Educ:* Hope Col, AB, 49; Wesleyan Univ, MA, 51. *Prof Exp:* Assoc prof, 52-69, PROF BIOL, ADELPHI UNIV, 69- *Concurrent Pos:* Dir develop, Adelphi Univ, 58-60, vpres, 60-66; res consult, Human Resources Ctr; mem biochem consult group, President's Comt Employ the Handicapped, 70-71; chmn premed curriculum, Adelphi Univ, 70. *Mem:* AAAS; Int Soc Biomech; Sigma Xi; NY Acad Sci; Nat Asn Biol Teachers. *Res:* Cell physiology; protein metabolism; neurophysiology; industrial physiology and ergonomics. *Mailing Add:* Amityville 38 Ungua Pl Amityville NY 11701

EICKHOFF, THEODORE C, b Cleveland, Ohio, Sept 13, 31; m 52; c 3. MEDICINE. *Educ:* Valparaiso Univ, AB, 53; Western Reserve Univ, MD, 57; Am Bd Internal Med, dipl, 66. *Prof Exp:* Dept chief invests sect, Epidemiol Br, Commun Dis Ctr, USPHS, 64-66, chief bact dis sect, 66-67; from asst prof med to assoc prof med, Med Ctr, Univ Colo, 67-75, head, Div Infectious Dis, 68-80, vchmn, Dept Med, 76-81, PROF MED, MED CTR, UNIV COLO, 75- *Concurrent Pos:* Mem comt meningococcal infections, Armed Forces Epidemiol Bd, 64-73, assoc mem comn acute respiratory dis, 65-73; assoc mem comn influenza, 69-73; clin asst prof prev med, Sch Med, Emory Univ, 64-67; consult, Fitzsimons Army Hosp, 68-; consult to med dir, NASA, 69-72; mem adv comt immunization practices, USPHS, 70-74; mem infectious dis test comt, Am Bd Internal Med, 73-80; mem ed bd, Antimicrobial Agents & Chemother, 74-88; consult & chmn, vaccines adv comt, Nat Cent for Drugs & Biologics, Food & Drug Admin, 80-85; mem comt on infections within hosps & chmn, Am Hosp Asn, 78-87; dir, Dept Med, Denver Gen Hosp, 78-81; dir internal med, Presbyterian-St Lukes Med Ctr, 81-; mem, Nat Comn Orphan Dis, Dept Health & Human Serv, 86-90 & Antiviral Drugs Adv Comt, Food & Drug Admin, HHS, 90- *Mem:* AAAS; Infectious Dis Soc Am (secy, 77-82 & pres, 83-84); Am Epidemiol Soc (secy-treas, 77-82 & pres, 85-86); Am Soc Microbiol; Am Fedn Clin Res; Am Soc Clin Invest; Asn Am Physicians; Am Col Physicians. *Res:* Internal medicine; infectious diseases. *Mailing Add:* Dept Internal Med Presbyterian-St Lukes Med Ctr Denver CO 80203

EICKHOLT, THEODORE HENRY, pharmacology; deceased, see previous edition for last biography

EICKSTAEDT, LAWRENCE LEE, b Davenport, Iowa, July 20, 39; m 60; c 2. ECOLOGY, MARINE BIOLOGY. *Educ:* Buena Vista Col, BS, 61; Univ Iowa, MS, 64; Stanford Univ, PhD(biol), 69. *Prof Exp:* Asst prof biol, Calif State Col, Hayward & Moss Landing Marine Labs, 68-69; asst prof, State Univ NY Col Old Westbury, 69-70; MEM FAC BIOL, EVERGREEN STATE COL, 70- *Mem:* AAAS. *Res:* Ecological physiology; reproductive biology of marine invertebrates; environmental design. *Mailing Add:* Dept Biol Evergreen State Col 3095 Yerba Buena Rd Olympia WA 98505

EICKWORT, GEORGE CAMPBELL, b New York, NY, June 8, 40; m 65; c 3. SYSTEMATICS. *Educ:* Mich State Univ, BS, 62, MS, 63; Univ Kans, PhD(entom), 67. *Prof Exp:* From asst prof to assoc prof, 67-78, PROF ENTOM, STATE UNIV NY COL AGR & LIFE SCI, CORNELL UNIV, 78- *Concurrent Pos:* Consult, Harvard Univ Mus Comp Zool, 70 & Time-Life Books, 74; res assoc entom div, Univ Calif, Berkeley, 75-76; vis prof entom, Univ Calif, Davis, 79; fac, Rocky Mt Biol Lab, 83-88; ed, Series Insect Biol, Cornell Univ Press, 88. *Mem:* AAAS; Asn Trop Biol; Entom Soc Am; Int Union Study Social Insects; Animal Behavior Soc; Acarological Soc Am. *Res:* Systematics, morphology, behavior and ecology of wild bees; systematics and biology of mites associated with insects. *Mailing Add:* Dept Entom Comstock Hall St Univ NY Col Agr Life Sci Cornell Ithaca NY 14853

EIDELBERG, EDUARDO, b Lima, Peru, Apr 30, 30; m 56; c 2. ANIMAL PHYSIOLOGY. *Educ:* San Marcos Univ, Lima, MD, 55. *Prof Exp:* Assoc res prof anat, Univ Calif, Los Angeles, 57-61; res prof neurophysiol, Ariz State Univ, 61-69; chmn div neurobiol, Barrow Neurol Inst, 61-77; adj prof neurol, Col Med, Univ Ariz, 71-77; PROF NEUROSURG & PHYSIOL, UNIV TEX HEALTH SCI CTR, SAN ANTONIO, 77- *Concurrent Pos:* Med investr, Vet Admin. *Res:* Physiology of the nervous system; physiology of movement control; spinal cord injury. *Mailing Add:* Vet Admin Hosp 112-D 7400 Merton Minter Blvd San Antonio TX 78284

EIDELS, LEON, b Jersey City, NJ, May 25, 42; m 73; c 2. BIOCHEMISTRY. *Educ:* Univ Calif, Davis, BS, 64, MS, 66, PhD(biochem), 69. *Prof Exp:* Teaching asst biochem, Univ Calif, Davis, 66-69; res assoc microbiol, Health Ctr, Univ Conn, 70-74; from asst prof to assoc prof, Univ Tex Health Sci Ctr, Dallas, 74-89, PROF MICROBIOL, UNIV TEX SOUTHWESTERN MED CTR, DALLAS, 89- *Concurrent Pos:* Arthritis Found fel, 70-73; mem US-India Exchange of Scientists Prog, NSF, 83; found lectr, Am Soc Microbiol, 85-86. *Mem:* AAAS; Am Soc Microbiol; Sigma Xi; Am Soc Biol Chemists; Am Asn Immunologists. *Res:* Biosynthesis and translocation; biochemistry of diphtheria toxin and pseudomonas exotoxin A and its receptor. *Mailing Add:* Dept Microbiol Univ Tex Southwestern Med Ctr Dallas TX 75235-9048

EIDER, NORMAN GEORGE, b Brooklyn, NY, May 3, 30; m 54; c 4. ANALYTICAL CHEMISTRY. *Educ:* City Univ NY, BS, 52. *Prof Exp:* Jr chemist, Ames Lab, US AEC, Iowa State Univ, 52-54; sr chemist, Uniroyal Chem Div, US Rubber Co, 56-67; group leader anal chem, 67-76, lab mgr chem anal, 76-90, CHEM HYG OFFICER, HULS AM, 91- *Mem:* Am Chem Soc; Am Soc Testing Mat. *Res:* Analytical method research and development in wet chemical analysis; atomic absorption spectrophotometry, gas chromatography; infrared analysis; emission and plasma spectroscopy. *Mailing Add:* Huls Am PO Box 365 Piscataway NJ 08854

EIDINGER, DAVID, b Montreal, Que, Jan 4, 31; m 57; c 2. MICROBIOLOGY, IMMUNOLOGY. *Educ:* McGill Univ, BSc, 52, PhD(anat), 58; Columbia Univ, MD, 59. *Prof Exp:* Demonstr anat, Columbia Univ, 55-56; res assoc allergy, Royal Victoria Hosp, Montreal, Que, 61-64; from asst prof to prof microbiol, 73-77; head, Dept Microbiol, Univ Sask, 77-89, prof, 77-; ALLERGIST, PVT PRACT. *Concurrent Pos:* Fel, Banting Res Found, Can, 53-54; Ont Heart Found grant, 64-; Nat Cancer Inst grant, 67-; Med Res Coun grant, 67-; res assoc, Can Heart Found, 61-64; dir, Clin Microbiol & Immunol Labs, Univ Hosp, 77- *Mem:* Brit Soc Immunol; Can Soc Immunol; Can Soc Microbiol. *Res:* Mechanism of antigenic competition; cell cooperation in delayed hypersensitivity; effect of hormonal changes on immune response. *Mailing Add:* 20700 Bond Rd NE Poulsbo WA 98370

EIDINOFF, MAXWELL LEIGH, b New York, NY, Feb 16, 15; m 38; c 2. PHYSICAL CHEMISTRY, BIOCHEMISTRY. *Educ:* Brooklyn Col, BA, 34; Pa State Col, PhD(phys chem), 38. *Prof Exp:* Asst phys chem, Pa State Col, 34-38; instr chem, Queens Col (NY), 38-42; res supvr div war res, Columbia Univ, 42-43; res group leader, Metall Lab, Univ Chicago, 43-44; from asst prof to prof, 45-82, EMER PROF CHEM, QUEENS COL (NY), 83- *Concurrent Pos:* Assoc mem, Sloan-Kettering Inst Cancer Res, NY, 49-66, assoc scientist, 66-70. *Mem:* Am Chem Soc; Am Asn Cancer Res; Am Asn Biol Chem; Sigma Xi. *Res:* Biochemistry of viruses; low temperature calorimetry; statistical mechanics; isotope exchange and separation; isotope mass effects in chemical reaction rates; radiochemical measurements; application of radioactive tracers in medical research; intermediary metabolism studies. *Mailing Add:* 34 Joseph St New Hyde Park NY 11040

EIDMAN, RICHARD AUGUST LOUIS, b Belleville, Ill, Sept 19, 36; m 59; c 3. CHEMICAL ENGINEERING. *Educ:* Washington Univ, St Louis, BS, 58, ScD(chem eng), 63. *Prof Exp:* Engr polyolefins res, Sabine River Works, 63-66, tech rep, Polyolefins Div, Plastics Dept, 66-77, MGR BARRIAR MAT & ENVIRON PACKAGING, E I DU PONT DE NEMOURS & CO, INC, 77- *Mem:* Tech Asn Pulp & Paper Indust. *Res:* Physical and chemical treatment of surfaces of high polymers; physical properties of polymers and relationships with molecular parameters; extrusion coatings. *Mailing Add:* Dupont Polymers PO Box 80011 Wilmington DE 19880-0011

EIDSON, WILLIAM WHELAN, b Indianapolis, Ind, July 22, 35; m 60; c 3. NUCLEAR PHYSICS, ATOMIC PHYSICS. *Educ:* Tulane Univ, BS, 57; Ind Univ, MS, 59, PhD(physics), 61. *Prof Exp:* Teaching asst physics, Ind Univ, 57-59, from instr to assoc prof, 61-67; prof & chmn dept, Univ Mo-St Louis, 67-72; prof physics & atmospheric sci & head dept, Drexel Univ, 72-83, prof physics, 83-85; DEAN ARTS & SCI, LOYOLA UNIV, NEW ORLEANS, 85- *Concurrent Pos:* Vis prof, Univ Wash, 67, Kansas State Univ, 72. *Mem:* AAAS; Am Asn Physics Teachers; Am Physics Soc; Inst Elec & Electronics Engrs; NY Acad Sci; Soc Physics Students (pres, 78-84). *Res:* Accelerator studies of nuclear structure and reaction mechanisms; nuclear instrumentation; solid state detectors; ion-electron recombination; corona discharge signature analysis; world energy impact studies; non-linear optics. *Mailing Add:* Col Arts & Sci Loyala Univ Box 3 New Orleans LA 70118

EIDT, DOUGLAS CONRAD, b Fergus, Ont, July 2, 28; m 51; c 5. ECOLOGY, TOXICOLOGY. *Educ:* Univ Guelph, BSA, 50; Univ Toronto, MSA, 52; Univ Sask, PhD(biol), 55. *Prof Exp:* Res officer entom, Agr Can, 51-56; res scientist, 56-88, RES SCIENTIST ENTOM, CAN FORESTRY SERV, 88- *Concurrent Pos:* Forest entom, Food & Agr Org of UN, 61-63; sci ed, Entom Soc Can, 78-83. *Mem:* Fel Entom Soc Can (pres, 88-89); Acadian Entomol Soc. *Res:* Effects of forest spraying on aquatic invertebrates; entomopathogenic nematodes for insect control. *Mailing Add:* Forestry Can Maritimes Region PO Box 4000 Fredericton NB E3B 5P7 Can

EIDT, ROBERT C, soils analysis; deceased, see previous edition for last biography

EIDUSON, SAMUEL, b Buffalo, NY, Dec 15, 18; c 1. NEUROCHEMISTRY. *Educ:* Univ Calif, Los Angeles, BS, 47, PhD(biochem), 52. *Prof Exp:* Res asst chem, Univ Calif, Los Angeles, 48-50, res asst physiol chem, 51-52; res biochemist, Vet Admin Ctr, Los Angeles, 52-54; asst clin prof physiol chem, 57-63, assoc prof psychiat & biol chem, 63-73, PROF BIOL CHEM & PSYCHIATRY, UNIV CALIF, LOS ANGELES, 73- *Concurrent Pos:* USPHS grant, Brain Res Inst, Univ Calif, Los Angeles, 74-75, Ralph L Smith Found grant, 74-; Prin scientist & chief neurobiochem res, Brentwood Neuro-psychiat Hosp Los Angeles, Vet Admin Ctr, 52-61; consult neurobiochem lab, 61-; chief res biochemist, Neuropsychiat Inst, Univ Calif, Los Angeles, 62-70, dep dir training & educ, Brain Res Inst, 74-; consult, Career Res Scientist Develop Award Prog, NIMH, 67-71, Biol Sci Training Rev Comt, 74-78. *Mem:* AAAS; Am Soc Biol Chem; Am Soc Neurochem; Int Soc Neurochem; Int Soc Develop Psychobiol (treas, 70-71, pres, 73-74). *Res:* Investigation of monoamine oxidase; blood platelets of human schizophrenic and control subjects - glutothione perozidase related to anti-oxidant status of blood and brain organelles in health and disease. *Mailing Add:* 941 Stonehill Lane Los Angeles CA 90049

EIFLER, GUS KEARNEY, JR, b Taylor, Tex, Feb 14, 08; m 42; c 2. EXPLORATION SHALLOW LIGNITE, OIL & GAS PROSPECTS. *Educ:* Univ Tex, BA, 29, MA, 30; Yale Univ, PhD(geol), 41. *Prof Exp:* Instr, 29-38, asst prof, geol, Univ Tex, 38-50; consult, Sch Land Bd, State Tex, 51-58, pvt consult, 59-64; res geol, field mapping, Bur Econ Geo Univ Tex, 64-73; CONSULT, 73- *Concurrent Pos:* Lab teaching asst, geol, Yale Univ, 38-39. *Mem:* Geol Soc Am; Am Asn Petrol Geologists; Am Instit Prof Geologists; Am Geol Instit; Sigma Xi. *Res:* Stratigraphic and structural geology in Trans-Pecos Texas; mapping the surface geology of the Texas Panhandle and West Texas. *Mailing Add:* 1901 Meadowbrook Austin TX 78703

EIFRIG, DAVID ERIC, b Oak Park, Ill, Jan 4, 35; m 57; c 4. OPHTHALMOLOGY. *Educ:* Carleton Col, BA, 56; Johns Hopkins Univ, MD, 60; Am Bd Ophthal, dipl. *Prof Exp:* Asst prof ophthal, Univ Ky, 68-70; assoc prof ophthal, Univ Minn, Minneapolis, 70-77; PROF & CHMN DEPT OPHTHAL, UNIV NC, CHAPEL HILL, 77- *Mem:* Retina Soc; Am Acad Ophthal; Am Col Surgeons. *Res:* Intraocular lenses; choriocapillaris. *Mailing Add:* Dept Ophthal Univ NC Chapel Hill NC 27514

EIGEL, EDWIN GEORGE, JR, b St Louis, Mo, June 4, 32; m 59; c 2. MATHEMATICS. *Educ:* Mass Inst Technol, BS, 54; St Louis Univ, PhD(math), 61. *Prof Exp:* From asst prof to prof math, St Louis Univ, 61-79, acad vpres, 72-79; PROF MATH, PROVOST & ACAD VPRES, UNIV BRIDGEPORT, 79- *Concurrent Pos:* from asst to dean to dean grad sch, St Louis Univ, 65-71, assoc acad vpres, 71-72; Danforth assoc, 64-; comnr, McDonnell Planetarium, St Louis, 72-79. *Mem:* Am Math Soc; Sigma Xi; Math Asn Am. *Res:* Numerical applications of functional analysis; theory of approximation; analytic theory of numbers. *Mailing Add:* 33 Pepperbush Lane Fairfield CT 06430

EIGEN, EDWARD, b New York, NY, June 29, 23; m 45; c 2. CHEMISTRY, MICROBIOLOGY. *Educ:* Brooklyn Col, BA, 44, MA, 55. *Prof Exp:* Bacteriologist, Food Res Labs, 44-46, microbiologist, 47-52; microbiologist, US Vitamin & Pharmaceut Corp, 52-53, asst supvr anal labs, 53-59; sr res biochemist, Colgate Palmolive Co, 59-62, sect head biochem, 62-70, sect head household prod res, 70-75, sr res assoc, 75-79, assoc res fel, 85, sr scientist, 79-89; RETIRED. *Mem:* Am Chem Soc; Int Asn Dent Res. *Res:* Isolation of materials from natural products; microbiology; chromatography; oral health and skin research; body odor research; cyclodextrin complexes. *Mailing Add:* Eight Glenside Ct East Brunswick NJ 08816

EIGHME, LLOYD ELWYN, b Wenatchee, Wash, Jan 15, 27; m 51. ENTOMOLOGY, HORTICULTURE. *Educ:* Pac Union Col, BA, 51, MA, 53; Ore State Univ, PhD(entom), 65. *Prof Exp:* Instr biol, Pac Union Col, 58-62; res asst entom, Ore State Univ, 62-65; from asst prof to prof, 65-89, EMER PROF BIOL & AGR, PAC UNION COL, 89- *Mem:* Entom Soc Am; emer mem Am Registry Prof Entomologists; NAm Heather Soc (pres, 87-89). *Res:* Applied entomology; insects in stored grain; taxonomy of Hymenoptera, Sphecidae; plant propagation-Ericaceous shrubs. *Mailing Add:* PO Box 1366 Lyman WA 98263

EIGNER, JOSEPH, b Swampscott, Mass, Dec 13, 33; m 63; c 2. ENVIRONMENTAL MANAGEMENT, HAZARDOUS WASTES. *Educ:* Dartmouth Col, AB, 55; Harvard Univ, AM, 58, PhD(phys chem), 60. *Prof Exp:* Neth Orgn Health Res fel, State Univ Leiden, 60-62; NSF fel biol chem, Univ Mich, 62-64; asst prof microbiol, Sch Med, Wash Univ, 64-74; chief, Hazardous Waste Proj, Mo Dept Natural Resources, 75-77; dir solid waste proj mgt, Bi-State Develop Agency, 78-80; PRES, JOSEPH EIGNER & CO INC, 81- *Concurrent Pos:* Pres, Mo Waste Control Coalition, 84. *Res:* Solid waste resource recovery; hazardous waste management. *Mailing Add:* 6802 Waterman Ave St Louis MO 63130

EIKENBARY, RAYMOND DARRELL, b Quay, Okla, Nov 2, 29; m 58; c 5. ENTOMOLOGY, FORESTRY. *Educ:* Okla State Univ, BA, 57; Clemson Univ, MS, 63, PhD(entom), 64. *Prof Exp:* Dist forester, Bur Land Mgt, 57-59; exten agent educ, 59-61, from asst prof to assoc prof, 64-73, PROF ENTOM, OKLA STATE UNIV, 73- *Mem:* Entom Soc Am. *Res:* Ecology; biological control of insects. *Mailing Add:* Dept Entom Okla State Univ Stillwater OK 74078

EIKENBERRY, JON NATHAN, b Oelwein, Iowa, Jan 15, 42; m 63; c 2. PHYSICAL ORGANIC CHEMISTRY. *Educ:* Iowa State Univ, BS, 63; Tex A&M Univ, MS, 66; Univ Wis, PhD(org chem), 72. *Prof Exp:* Fel bio-org chem, Mass Inst Technol, 72-73; SR RES CHEMIST PHYS ORG CHEM, EASTMAN KODAK CO RES LABS, 73- *Mem:* AAAS; Am Chem Soc; Am Asn Clin Chem. *Res:* Photographic film. *Mailing Add:* 168 Pinecrest Dr Rochester NY 14617

EIK-NES, KRISTEN BORGER, b Sparbu, N Trondelag, Sept 28, 22; US citizen. BIOCHEMISTRY & PHYSIOLOGY TESTICULAR FUNCTION, CONSTRUCTION & FUNCTION ARTIFICIAL ORGANS IN MAN. *Educ:* Univ Oslo, Norway, MD, 51. *Prof Exp:* Postdoctoral biochem, Univ Utah, 51-54, instr biochem, 54-56, from asst prof to assoc prof biochem, 56-67; prof physiol, Univ Southern Calif, 68-73; PROF & CHMN BIOPHYS, UNIV TRONDHEIM, NORWAY, 73- *Concurrent Pos:* Training prog dir endocrin, Univ Utah, 58-67; mem & chmn, US Pub Health Study Sect Biol Reprod, 68-72; mem, NAVF Team Med Res & Offshore Activ, 80-82; vis prof, Inst Biomed Eng, Univ Utah, 84-85. *Honors & Awards:* Brown-Mendeles, Soc Argentina Biol, 71; Med Award, Trondelag Med Soc, 80; Distinguished Andrologist, Am Soc Andrology, 83; Gunnerius Award, Royal Norweg Res Soc, 89. *Res:* Biochemistry and physiology of testicular function; construction and function of artificial organs and their use in man. *Mailing Add:* Presthytta Nedre Bakken Steinkjer 7700 Norway

EIKREM, LYNWOOD OLAF, b Lansing, Mich, June 11, 19; m 46; c 4. SPECTROCHEMISTRY, INSTRUMENTATION. *Educ:* Mich State Col, BS, 41; Mass Inst Technol, SM, 48. *Prof Exp:* Spectrochemist, Diamond Alkali Co, 41-42; chief spectrochemist, Chrysler Evansville Ord Plant, 42-44; field engr spectros, Harry Dietert Co, 44; assoc prof chem, La Polytech Inst, 46-47; adv fel, Mass Inst Technol, 47-49; tech dir spectros, Jarrell-Ash Co, 49-53; proj engr emission spectros, Baird-Atomic Inc, 53-59; staff engr, Geophys Corp Am, Inc, 59-60; prod develop mgr, David W Mann Co, 60-63, dir mkt, 64-65; vpres, Appl Res Labs, Inc, 65-72; vpres & dir res progs, 73-79, PRES, DARLING, PATERSON & SALZER MGT CONSULTS, 80- *Concurrent Pos:* Chmn, Strategic Directions Int, 81- *Mem:* Optical Soc Am; Am Soc Testing & Mat; NY Acad Sci; fel, Am Inst Chemists; Soc Appl Spectros. *Res:* Design, testing and applications of spectroscopic analytical instrumentation; interferometry; meteorological instrumentation; x-ray; microprocessor applications to analytical instruments; strategic planning; new market evaluations - analytical instruments. *Mailing Add:* 6242 Weschester Parkway Suite 100 Los Angeles CA 90045

EIL, CHARLES, b Minneapolis, Minn, Dec 15, 46; m 78; c 3. ENDOCRINOLOGY, INTERNAL MEDICINE. *Educ:* Univ Rochester, BA, 68; Univ Chicago, PhD(biochem), 72, MD, 74; Am Bd Internal Med, cert internal med, 77, cert, endocrinol, 79. *Prof Exp:* Intern internal med, Univ Mich, 74-75, resident, 75-76; fel endocrinol, NIH, 76-79, med officer, 79-80; staff physician, Bethesda Naval Hosp, 80-82, asst br chief, 82-86; from asst prof to assoc prof med, Uniformed Serv Univ Health Sci, 80-86; ASSOC PROF MED, BROWN UNIV, 86-; DIR, CLIN RES CTR, BROWN UNIV, ROGER WILLIAMS GEN HOSP, 86- *Honors & Awards:* Peter Forsham Award, Soc Uniformed Endocrinologists, 85. *Mem:* Am Soc Clin Invest; Endocrine Soc; Am Fedn Clin Res; Am Soc Bone & Mineral Res; Am Thyroid Asn. *Res:* Disorders of hormone resistance; mechanism of hormone action for intracellular hormones such as androgens, vitamin D, and thyroid hormone; calcium and vitamin D metabolism; hormone antagonists. *Mailing Add:* Dept Med Roger Williams Med Ctr 825 Chalkstone Ave Providence RI 02908

EILBER, FREDERICK RICHARD, b Detroit, Mich, Aug 17, 40; m 65; c 4. SURGERY, ONCOLOGIC SURGERY. *Educ:* Univ Mich Med Sch, MD, 65. *Prof Exp:* Clin assoc surg br, Nat Cancer Inst, NIH, 67-70; asst & chief surg resident, Univ Md Hosp, 71-72; M D Anderson Hosp & Tumor Inst, 72-73; assoc prof, 75-79, PROF SURG, DIV ONCOL, SCH MED, UNIV CALIF, LOS ANGELES, 79- *Concurrent Pos:* Staff surgeon, Vet Admin Hosp, Sepulveda, 73- *Honors & Awards:* Ewing Award, Soc Surg Oncol, 70. *Mem:* Soc Head & Neck Surg; Soc Surg Oncol; Soc Univ Surg; Am Col Surg; Pac Coast Surg Asn. *Res:* Tumor immunology; viral oncology; cancer chemotherapy. *Mailing Add:* 9th Floor Factor Bldg Sch Med John Wayne Cancer Clin Univ Calif 10833 Lee Conte Ave Los Angeles CA 90024

EILENBERG, SAMUEL, b Warsaw, Poland, Sept 30, 13; nat US. MATHEMATICS. *Educ:* Univ Warsaw, MA, 34, PhD(math), 36. *Hon Degrees:* PhD, Brandeis Univ, 80, Univ Pa, 85, Columbia Univ, 87, Univ Cathlique de Louvain, 87, Univ Paris VII, 88. *Prof Exp:* From instr to assoc prof math, Univ Mich, 40-46; prof, Ind Univ, 46-47; prof, 47-74, univ prof, 74-82, spec prof, 82-83, EMER UNIV PROF MATH, COLUMBIA UNIV, 82- *Concurrent Pos:* Vis lectr, Princeton Univ, 45-46; vis prof, Fulbright & Guggenheim fel, Paris, 50-51; vis prof, Tata Inst, Bombay, 53-54, 56-57, Hebrew Univ, 54 & Univ Paris, 66-67. *Honors & Awards:* Wolf Prize, 86; Steele Prize, Am Math Soc, 87. *Mem:* Nat Acad Sci; Am Acad Arts & Sci; Am Math Soc; Math Asn Am; hon SE Asian Math Soc; hon NY Acad Sci. *Res:* Topology, algebra and computer mathematics. *Mailing Add:* Dept Math 522 Math Bldg Columbia Univ New York NY 10027

EILER, HUGO, b Santiago, Chile, Nov 4, 35; m 62; c 3. ENDOCRINOLOGY, REPRODUCTION. *Educ:* Univ Chile, DVM, 60; Univ Ga, MS, 74; Univ Ill, PhD(physiol), 76. *Prof Exp:* Asst prof physiol, Col Vet Med, Univ Chile, 62-66, assoc prof, 67-71, asst prof, Sch Human Med, 63-67, prof, Sch Nursing, 64-71; res assoc endocrinol, Dept Animal Sci, Univ Ill, 71-73, teaching assoc physiol, Col Vet Med, 73-76; ASSOC PROF PHYSIOL, COL VET MED, UNIV TENN, 76- *Mem:* Soc Study Reproduction. *Res:* Development of endocrine testing procedures in animals; endocrine factors affecting the physiology of the uterus. *Mailing Add:* Dept Animal Sci Univ Tenn Knoxville TN 37996

EILER, JOHN JOSEPH, b Jacksonville, Fla, Jan 25, 10; m 44; c 1. PHARMACEUTICAL CHEMISTRY. *Educ:* Univ Calif, AB, 33, PhD(biochem), 37. *Prof Exp:* Mem enzymes res staff, Cutter Labs, 37-38; from instr to assoc prof biochem & pharm, 38-51, prof, 51-72, prof biochem & biophys, 72-76, asst dean, Col Pharm, 48-56, assoc dean, 56-72, chmn, Dept Pharmaceut Chem, 58-72, EMER PROF BIOCHEM & BIOPHYS, SCH MED, UNIV CALIF, SAN FRANCISCO, 76- *Concurrent Pos:* Consult, USPHS, 47 & US Army, 47-; mem, Int Union Physiol Sci. *Mem:* Am Chem Soc; Soc Exp Biol & Med; Am Pharmaceut Asn; Am Soc Biol Chemists; NY Acad Sci. *Res:* Chemistry and metabolism of purines and nucleic acids; metabolism of carbohydrates; thyroid hormone and renal and intestinal functions; action of drugs on aerobic phosphorylation; biological action of narcotics and stimulants; mode of action of antimitotic agents. *Mailing Add:* 4315 Paradise Dr Tiburon CA 94920

EILERS, CARL G, b Fairbury, Ill, Mar 21, 25. ELECTRICAL ENGINEERING. *Educ:* Purdue Univ, BEE, 48; Northwestern Univ, MEE, 56. *Prof Exp:* Proj engr, pay TV res, Zenith Radio Corp, 48-61, div chief, circuit res group, 61-78, MGR, ELEC SYSTS RES & DEVELOP, ZENITH RADIO CORP, 78- *Mem:* Fel Inst Elec Electronics Engrs; Audio Eng Soc; Soc Motion Picture & TV Engrs. *Res:* Television; stereophonic sound for television; pay television technology. *Mailing Add:* Zenith Radio Corp 1000 Milwaukee Ave Glenview IL 60025

EILERS, FREDERICK IRVING, b Milwaukee, Wis, July 5, 38; m 61; c 2. GENETICS, MYCOLOGY. *Educ:* Univ Wis, Milwaukee, BS, 61; Univ Mich, MS, 63, PhD(bot), 68. *Prof Exp:* Res asst, Univ Mich, 61-64, teaching fel bot, 64-66; instr microbiol, Ohio Wesleyan Univ, 66-67; instr gen biol & plant physiol, Oberlin Col, 67-68; ASSOC PROF MICROBIOL, GENETICS & MYCOL, UNIV SFLA, 68-, ASST CHMN DEPT, 79- *Concurrent Pos:* Vis prof, Carnegie-Mellon Univ, 75-76. *Res:* Physiology involving fungi such as: mushroom growth, spore germination, mushroom toxins and mushroom cap digestion in the genus Coprinus. *Mailing Add:* Dept Biol Univ SFla 4202 Fowler Ave Tampa FL 33620

EILERS, LAWRENCE JOHN, b Ireton, Iowa, May 21, 27; m 49; c 4. PLANT TAXONOMY, PHYTOGEOGRAPHY. *Educ:* State Col Iowa, BS, 49, MA, 60; Univ Iowa, PhD(bot), 64. *Prof Exp:* Instr high schs, Iowa, 49-51; elec engr, Collins Radio Co, 52-56, environ engr, 56-57; environ engr, Admiral Radio Corp, Ill, 57-58; instr sci, Charles City Consol Schs, Iowa, 58-59; instr forest bot, State Univ NY Col Forestry, Syracuse, 64-65; asst prof life sci, Ind State Univ, 65-68; assoc prof, 68-77, PROF BIOL, UNIV NORTHERN IOWA, 77- AAAS; Soc Study Evol; Am Soc Plant Taxon; Bot Soc Am; Am Inst Biol Sci. *Res:* Biosystematics of the genus Sullivantia; flora and phytogeography of the Midwest. *Mailing Add:* Dept Biol Univ Northern Iowa Cedar Falls IA 50613

EIME, LESTER OSCAR, b Sappington, Mo, June 22, 22; div. INDUSTRIAL CHEMISTRY. *Educ:* Univ Mo, AB, 44, AM, 47; Ohio State Univ, cert, 69. *Prof Exp:* Instr chem, Christian Col, 44-47; res chemist, Aluminum Co Am res labs, 47-57; Petreco Div, Petrolite Corp, 57-62; sr res chemist, Space & Electronics Div, Emerson Elec Co, 62-65; engr, McDonnell Aircraft Div, McDonnell-Douglas Corp, 65-73; res chemist, Brown Shoe Co, Brown Group, Inc, 73-75; RES CHEMIST, WESTERN LITHOPLATE & SUPPLY CO, DIV BEMIS CO, 76- *Concurrent Pos:* Asst chem, Univ Mo, 46-47 & res fel, 46. *Mem:* Am Chem Soc. *Res:* Purification of hydrocarbons and other chemicals through electrical and catalytic processes; aluminum organic compounds; pigments for paint, plastics, rubber and paper coating industries; petroleum catalyst preparations; ablative materials and high temperature polymers; electrochemical corrosion studies of metals; printed circuits; urethane adhesives; long wearing lithographic coatings, lithographic developers and chemicals; granted US patent on extending life of lithographic image on printing plates. *Mailing Add:* 111 Pebble Acres Ct St Louis MO 63141

EIMERL, DAVID, b Eng. PHYSICS. *Educ:* Oxford Univ, Eng, BA, 69; Northwestern Univ, PhD(physics), 73. *Prof Exp:* Res physicist physics, Univ Calif, San Diego, 73-75; staff scientist, Phys Dynamics, Inc, 75-76; STAFF SCIENTIST LASER FUSION, LAWRENCE LIVERMORE LAB, UNIV CALIF, 76- *Concurrent Pos:* Consult, Phys Dynamics, Inc, 76-77. *Mem:* Am Phys Soc. *Res:* Raman scattering; Raman compressors; amplified spontaneous emission and parasitics control in large lasers; laser system design for laser fusion applications; theoretical physics. *Mailing Add:* 4953 Black Ave Pleasanton CA 94566

EIN, DANIEL, b Nov 26, 38; m; c 2. ALLERGIES. *Educ:* Albert Einstein Col Med, MD, 64. *Prof Exp:* CLIN PROF MED & ALLERGIES, COL MED, GEORGE WASHINGTON UNIV, 84- *Mem:* Asn Clin Path; Am Asn Immunologists; AAAS; fel Am Acad Allergy; AMA. *Res:* Immunoglobulin structure and genetics and disorders of immunity; author of articles on clinical allergic disorders. *Mailing Add:* 908 New Hampshire Ave NW Washington DC 20037

EINARSSON, ALFRED W, b Berkeley, Calif, Apr 13, 15; m 47; c 2. PHYSICS. *Educ:* Univ Calif, PhD(physics), 46. *Prof Exp:* Asst prof physics, Univ Southern Calif, 46-50; from asst prof to prof, San Jose State Univ, 57-85; RETIRED. *Mem:* Am Phys Soc; Am Solar Energy Soc. *Res:* General physics; physical optics; electrical conduction in gases. *Mailing Add:* 18851 Overlood Rd Los Gatos CA 95030

EINAUDI, FRANCO, b Turin, Italy, Oct 31, 37; m 66; c 2. ATMOSPHERIC PHYSICS. *Educ:* Turin Polytech Inst, BSc, 61; Cornell Univ, MSc, 65, PhD(atmospheric & plasma physics), 67. *Prof Exp:* Instr elec eng, Turin Polytech Inst, 61-62; res asst microwaves, Cornell Univ, 62-64, res asst plasma & atmospheric physics, 64-67; fel atmospheric physics, Univ Toronto, 67-69; vis fel, Univ Colo, Nat Oceanic Atmospheric Admin, 69-71, fel, 71-74, physicist, Aeronomy Lab, 75-76, physicist, Wave Propagation Lab, 76-79; prof geophys, Sch Geophys Sci, Ga Inst Technol, 79-87; CHIEF LAB ATMOSPHERES, GODDARD SPACE FLIGHT CTR, NASA, 87- *Mem:* Am Meteorol Soc; Am Geophys Union; Royal Meterol Soc. *Res:* Atmospheric physics; atmospheric dynamics. *Mailing Add:* 11006 Willowbottom Dr Columbia MD 21044

EINAUDI, MARCO TULLIO, b New York, NY, Dec 24, 39; m 63; c 3. ECONOMIC GEOLOGY. *Educ:* Cornell Univ, BA, 61; Harvard Univ, MA, 65, PhD(geol), 69. *Prof Exp:* Geologist, Anaconda Co, 69-75; from asst prof to assoc prof, 75-80, chmn, dept appl earth sci, 82-86, WJ & ML CROOK PROF APPL EARTH SCI & GEOL, STANFORD UNIV, 80- *Concurrent Pos:* Geol consult, Anaconda Co, 75-76, SRI Int, 76-79, Conoco Minerals, 78-81, Noranda Inc, 80, Hunt, Ware, & Proffett, 83-84, & Homestake Mining Co, 87; ed, spec issues Econ Geol, 78, 82, 86; chmn, Gordon Res Conf Geochem, 81; vis investr, Wash Geophys Lab, Carnegie Inst, 81-82; Thayer Lindley vis lectr, Soc Econ Geologists, 82-83; assoc dean res, Sch Earth Sci, Stanford Univ, 86- *Honors & Awards:* Hugh E McKinstry Mem Lectr, Harvard Univ, 86. *Mem:* Geol Soc Am; Soc Econ Geologists. *Res:* Genesis of porphyry copper molybdenum deposits, cordilleran base, precious metal vein and skarn deposits, and epithermal gold-silver deposits; development of empirical and theoretical ore deposit models for use in exploration. *Mailing Add:* Dept Appl Earth Sci Stanford Univ Stanford CA 94305-2225

EINERT, ALFRED ERWIN, b Kearney, NJ, Feb 6, 39; c 2. ORNAMENTAL HORTICULTURE, PLANT PHYSIOLOGY. *Educ:* Ark State Univ, BSA, 64; Miss State Univ, MS, 66, PhD(ornamental hort), 69. *Prof Exp:* Res assoc, Mich State Univ, 69-70; from asst prof to assoc prof, 70-78, PROF ORNAMENTAL HORT, UNIV ARK, FAYETTEVILLE, 78- *Concurrent Pos:* Netherlands Flower-Bulb Inst trainee, Mich State Univ, 68-70; vis prof, Hort & Landscape Arch, Miss State Univ, 78. *Mem:* Am Soc Hort Sci; Am Soc Landscape Archit. *Res:* Growth and development of bulb flower crops; role of environmental factors on landscape plants; marketing of ornamentals. *Mailing Add:* Dept Hort & Forestry Univ Ark Fayetteville AR 72701

EINHELLIG, FRANK ARNOLD, b Independence, Mo, July 7, 38; m 61; c 2. BOTANY, PHYSIOLOGICAL ECOLOGY. *Educ:* Kans State Univ, BS, 60; Univ Kans, BS, 61; Univ Okla, MS, 64, PhD(bot), 69. *Prof Exp:* Sci teacher, Shawnee Mission High Sch Dist, Kans, 61-67; from asst prof to assoc prof biol, 69-78, PROF BIOL, UNIV SDAK, 78-, CHAIR BIOL, 89- *Concurrent Pos:* Prin investr, eight sci educ and res grants, 72-82 & six res grants, allelopathy, 79-90; crop scientist, Weed Sci Res Lab, USDA, 83 & 88. *Mem:* Am Soc Plant Physiologists; Sigma Xi; Plant Growth Regulator Soc Am; Int Soc Chem Ecol; Weed Sci Soc Am. *Res:* Allelopathy; mechanism of inhibition caused by weedy species and specific inhibitors such as effects on photosynthesis, stomatal aperture, plant water status and respiration; interaction among allelochemicals and other stresses of the plant environment; author of chapters and more than 35 journal articles. *Mailing Add:* Dept Biol Univ SDak Vermillion SD 57069

EINHORN, MARTIN B, b Dayton, Ohio, Aug 14, 42; m 67; c 2. THEORETICAL HIGH ENERGY PHYSICS. *Educ:* Calif Inst Technol, BS, 65; Princeton Univ, PhD(physics), 68. *Prof Exp:* Res physicist, Stanford Linear Accelerator Ctr, 68-70 & Lawrence Berkeley Lab, 70-72; res physicist, Fermi Nat Accelerator Lab, 72-76; assoc res scientist, Univ Mich, 76-79, from assoc prof to prof physics, 79-90; dep dir, Inst Theoret Physics, Univ Calif Santa Barbara, 90-92. *Mem:* Fel Am Phys Soc; AAAS; Am Asn Univ Professors. *Res:* Quantum chromodynamics, electroweak interactions, especially the Higgs sector; grand unification and cosmology; elementary particle physics; field theory. *Mailing Add:* Inst Theoret Physics Univ Calif Santa Barbara Santa Barbara CA 93106

EINHORN, PHILIP A, b Philadelphia, Pa, Jan 27, 42; m 76; c 1. TECHNICAL MANAGEMENT, ANALYTICAL CHEMISTRY. *Educ:* Univ Pa, BS, 63; St Joseph's Col, MS, 70. *Prof Exp:* Res anal chemist, Sadtler Res Lab, 63-70; admin chemist, Rollins Purle, Inc, 70-71; tech mgr, Rollins Environ Serv, 71-73; mgr process develop, Chem Waste Mgt, 73-76; tech dir, Waste Conversion, Inc, 77-84; consult & prin, SRE Consult, 84-86, PRES, SRE ANALYTICS, INC, 86- *Concurrent Pos:* Consult, 76-77. *Mem:* Am Chem Soc; fel Am Inst Chemists. *Mailing Add:* 24 Jack Ladder Circle Horsham PA 19044-1904

EINOLF, WILLIAM NOEL, b Baltimore, Md, July 5, 43; m 65; c 2. ANALYTICAL CHEMISTRY, ORGANIC CHEMISTRY. *Educ:* Johns Hopkins Univ, BA, 64; Univ Del, PhD(chem), 71. *Prof Exp:* Teacher chem, Great Valley High Sch, 64-66; chemist, Wyeth Labs, 66-67; instr chem,

Lincoln Univ, 67-68; chemist, Johns Hopkins Univ Sch Med, 71-72; CHEMIST, PHILIP MORRIS RES CTR, 72- *Mem:* Am Chem Soc; Am Soc Mass Spectrometry. *Res:* Tobacco and smoke chemistry; natural products; mass spectrometry. *Mailing Add:* Philip Morris Res Ctr PO Box 26583 Richmond VA 23261

EINSET, EYSTEIN, b Geneva, NY, Mar 19, 25; m 54; c 3. FOOD SCIENCE. *Educ:* Cornell Univ, BS, 50, MS, 51, PhD(biochem), 56. *Prof Exp:* Chemist, Bur Com Fisheries, US Fish & Wildlife Serv, 55-57, lab dir, 57-60; asst prof biochem, Agr Exp Sta, Cornell Univ, 60-62; mgr food res, Tectrol Div, Whirlpool Corp, 62-64, sr res biochemist, Res & Eng Div, 64-69, staff scientist, 69-88; RETIRED. *Mem:* Am Chem Soc; Inst Food Technol; Sigma Xi; Int Inst Refrig. *Res:* Food chemistry and processing; biochemistry of storage life extension of fresh plant, animal and fishery food products by control of environmental variables. *Mailing Add:* 1916 Sunset Dr St Joseph MI 49085

EINSET, JOHN WILLIAM, b Waterloo, NY, Nov 14, 47; m 81. PLANT PHYSIOLOGY. *Educ:* Cornell Univ, BS, 69; Univ Wis, PhD(bot), 74. *Prof Exp:* Res assoc plant physiol, Univ Wis, 74-77; from asst prof to assoc prof plant physiol, Univ Calif, 77-83; ASSOC PROF PLANT PHYSIOL, HARVARD UNIV, 83- *Concurrent Pos:* Lectr plant physiol, Univ Wis, 74. *Mem:* Sigma Xi; Am Soc Plant Physiologists; Tissue Culture Asn; Am Soc Hort Sci; Int Asn Plant Tissue Culture; Int Asn Plant Growth Substances; AAAS. *Res:* Regulation of plant growth and development by hormones in tissue cultures; cytokinin physiology and biochemistry. *Mailing Add:* Plant Sci Dept Univ Calif Riverside CA 92521

EINSPAHR, DEAN WILLIAM, b Sioux City, Iowa, May 24, 23; m 46; c 2. FOREST GENETICS. *Educ:* Iowa State Univ, BS, 49, MS, 50, PhD(soils, silvicult), 55. *Hon Degrees:* MS, Lawrence Univ, 86. *Prof Exp:* Asst wood technol, Gamble Bros, Inc, Ky, 50-51; res assoc soils & silvicult, Iowa Agr Exp Sta, 52-55; res asst forest genetics, Inst Paper Chem, 55-58, res aide, 59-62, res assoc & chief genetics & physiol group, 63-70, sr res assoc forest genetics & group coordr, Div Natural Mat & Systs, 70-77, sr res assoc & dir, Forest Biol Sect, 77-85, head dept Advan Studies, 80-83; PRES, FIBER RESOURCES, INC, 85- *Mem:* Tech Asn Pulp & Paper Industs; Soil Sci Soc Am; Soc Am Foresters. *Res:* Forest soils; silviculture; wood quality-paper quality relationship. *Mailing Add:* Fiber Resources, Inc 2808 Crestview Dr Appleton WI 54915

EINSPAHR, HOWARD MARTIN, b Beaumont, Tex, Feb 7, 43; m 75; c 2. PROTEIN CRYSTALLOGRAPHY. *Educ:* Rice Univ, BA, 64; Univ Pa, PhD(chem), 70. *Prof Exp:* Res fel crystallog, Calif Inst Technol, 70-72; res assoc, Inst Dent Res, Univ Ala, Birmingham, 72-77, investr, 77-84, instr biochem, 77-81, assoc scientist, Comp Cancer Ctr, 77-84, res asst prof, 81-84; SR RES SCIENTIST MACROMOLECULAR CRYSTALLOG, UPJOHN CO, 84- *Honors & Awards:* Res Career Develop award, USPHS, 81-84. *Mem:* AAAS; Am Crystallog Asn; Sigma Xi; Biophys Soc; Am Soc Biochem & Molecular Biol; Protein Soc. *Res:* Structural studies of macromolecules by X-ray diffraction; the geometry of calcium binding by organic ligands. *Mailing Add:* Phys & Analytical Chem Upjohn Co Kalamazoo MI 49001

EINSPRUCH, NORMAN G(ERALD), b Brooklyn, NY, June 27, 32; m 53; c 3. SOLID STATE SCIENCE. *Educ:* Rice Inst, BA, 53; Univ Colo, MS, 55; Brown Univ, PhD(appl math), 59. *Prof Exp:* Asst, Univ Colo, 53-55; asst, Metals Res Lab, Brown Univ, 56-59, res assoc, 59; mem tech staff, Tex Instruments Inc, 59-62, mgr, Electron Transport Physics Br, Physics Res Lab, 62-68, actg head, Thin Film Physics Br, 64-65, dir advan technol lab, Cent Res Labs, 68-69, dir technol, Chem Mat Div, Tech Ctr, 69-72, dir, Cent Res Labs, 72-75, asst vpres, 75-77; dean, 77-90, PROF ELEC & COMPUT ENG, COL ENG, UNIV MIAMI, 77-, SR FEL SCI & TECHNOL, 90- *Concurrent Pos:* chmn panel film microstructure sci & technol, Nat Res Ctr, 78-79; dir, Ogden Corp, Image Data Corp, Exact Sci, Inc; mgr corp develop, Tex Instruments Inc, 75-76, mgr consumer prod technol & planning, 76-77. *Honors & Awards:* George Washington Medal of Freedoms Found. *Mem:* Fel AAAS; fel Am Phys Soc; fel Acoust Soc Am; fel Inst Elec & Electronics Engrs; Sigma Xi; Am Soc Eng Educ. *Res:* Transport in solids; physical acoustics; ultrasonic wave propagation in solids; management technology; microstructure science. *Mailing Add:* Col Eng Univ Miami PO Box 248581 Coral Gables FL 33124-8581

EINSTEIN, ELIZABETH ROBOZ, b Szaszvaros, Hungary; nat US; m 59. NEUROCHEMISTRY. *Educ:* Univ Budapest, PHD(biochem), 38. *Prof Exp:* Res asst bioorg chem, Calif Inst Technol, 41-45; assoc prof chem, Univ Wyo, 45-48; res assoc enzyme & carbohydrate res, Stanford Univ, 48-52; assoc prof biochem, Sch Med, Georgetown Univ, 52-58; assoc prof, neurochem, Sch Med, Stanford Univ, 58-59; PROF NEUROCHEM & LECTR BIOCHEM, SCH MED, UNIV CALIF, SAN FRANCISCO, 59- *Concurrent Pos:* NIH, Multiple Sclerosis Soc & Hartford Found grants; SEATO scholar, Univ Bangkok, 61-62; res consult, Sugar Res Found NY, 45-48; lectr orient & advan researcher, Univ Bangkok, 61-62. *Honors & Awards:* Raskob Award, Georgetown Univ, 56; Medaglia d'oro di Milano, Int Cong Neurochem, Milan, Italy, 69; mem, Inst Human Develop, Univ Calif, Berkeley. *Mem:* Am Chem Soc; Soc Exp Biol & Med; Am Acad Neurol. *Res:* Neurochemical investigations; chemistry of demyelinating diseases; developing brain; cerebrospinal fluid; author or coauthor of over ninety publications. *Mailing Add:* 1090 Creston Rd Univ Calif Berkeley CA 94708-1504

EINSTEIN, FREDERICK W B, b Auckland, NZ, Nov 7, 40; m 65; c 2. INORGANIC CHEMISTRY. *Educ:* Univ NZ, BSc, 62; Univ Canterbury, MSc, 63, PhD(chem), 65. *Prof Exp:* Fel chem, Univ BC, 65-67; from asst prof to assoc prof, 67-75, PROF CHEM, SIMON FRASER UNIV, 75- *Mem:* Am Crystallog Asn; fel Chem Inst Can. *Res:* Inorganic structural chemistry; crystal structure analysis; computing. *Mailing Add:* Dept Chem Simon Fraser Univ Burnaby BC V5A 1S6 Can

EINSTEIN, THEODORE LEE, b Cleveland, Ohio, Jan 20, 47; m 82; c 2. THEORETICAL SOLID STATE PHYSICS, SURFACE PHYSICS. *Educ:* Harvard Univ, BA, MA, 69; Univ Pa, PhD(physics), 73. *Prof Exp:* Res investr physics, Univ Pa, 73-74; vis asst prof, 75-77, from asst prof to assoc prof, 77-87, PROF PHYSICS, UNIV MD, 87- *Concurrent Pos:* Guest researcher, Chalmers Univ Technol, Goteborg, Sweden, 85; guest researcher & vis prof, Univ Padova, Padua, Italy, 86 & 87; physicist & guest worker, Nat Bur Standards, 86-; expert consult, patent case, Spensley, Horn, Jubas, Lubitz, 88-89; expert/part-time prog dir, NSF, 89-90. *Mem:* Am Phys Soc; Am Vacuum Soc; Fedn Am Scientist. *Res:* Theory of gas adsorption onto materials: energetics, electronic distributions, experimental probes, order-disorder transitions, 2-d critical phenomena, multi-adatom effects and mechanical changes; extended appearance; potential fine structure; properties of vicinal (stepped) surfaces. *Mailing Add:* Dept Physics Univ Md College Park MD 20742-4111

EINWICH, ANNA MARIA, b Baltimore, Md, Feb 7, 17; c 4. OCEANOGRAPHY, MARINE GEOLOGY. *Educ:* Johns Hopkins Univ, BA, 67. *Prof Exp:* Civil eng technologist mapping, Forest Serv, USDA, 67-68; oceanogr magnetics, US Naval Oceanog Off, 69-76; OCEANOGR MARINE GEOL, SEA FLOOR DIV, NAVAL OCEAN RES & DEVELOP ACTIV, 76- *Res:* Histories of ocean basins, especially in the Caribbean, Western Atlantic and Gulf of Mexico; velocities in sediments. *Mailing Add:* 112 Palm Ave Pass Christian MS 39571

EINZIG, STANLEY, b Brooklyn, NY, July 25, 42; m 66; c 3. MYOCARDIAL BLOOD FLOW, HEART FAILURE. *Educ:* Univ Calif, Los Angeles, BA, 64, MD, 67; Univ Minn, PhD(physiol), 77. *Prof Exp:* From asst prof to assoc prof, 77-90, PROF PEDIAT CARDIOL, MED SCH, UNIV MINN, 90- *Mem:* Am Physiol Soc; Soc Pediat Res; Soc Exp Biol & Med. *Res:* Pathophysiology of congestive cardiomyopathy; noninvesive diagnosis of congenital heart disease (echocardiography). *Mailing Add:* Pediatrics WVa Univ Morgantown WV 26506

EINZIGER, ROBERT E, b Asbury Park, NJ, Feb 19, 45; m 70; c 1. NUCLEAR WASTE MANAGEMENT. *Educ:* Ga Inst Technol, BS, 67; Rensselaer Polytech Inst, MS, 72, PhD(physics), 73. *Prof Exp:* Asst, Argonne Nat Lab, 74-76, asst scientist, 76-79; sr scientist, Westinghouse Hanford Co, 79-87; SECT MGR, BATELLE PAC NORTHWEST LABS, 87- *Honors & Awards:* Sam Tour Award, Am Soc Testing Mat, 86. *Mem:* Am Phys Soc; Am Nuclear Soc; Metall Soc. *Res:* The behavior of spent fuel, particularly its oxidation and degradation mechanisms; determination of its suitability as a waste form for storage and disposal. *Mailing Add:* 2363 Davison Ave Richland WA 99352

EIPPER, BETTY ANNE, b Elmira, NY, Nov 11, 45; m 68; c 2. ENDOCRINOLOGY, PEPTIDES. *Educ:* Brown Univ, ScB & MS, 68; Harvard Univ, PhD(biophys), 73. *Prof Exp:* Am Cancer Soc fel, Anna Fuller Fund, Univ Ore, 73-75; asst prof, 76-79, ASSOC PROF PHYSIOL, UNIV COLO HEALTH SCI CTR, 79- *Concurrent Pos:* Prin investr, NIH, 76- *Mem:* Endocrine Soc; Am Soc Biol Chemists. *Res:* Biosynthesis of bioactive peptides (especially ACTH and endorphin); post-translational processing of peptide hormones (proteolysis, glycosylation, acetylation, amidation, phosphorylation); tissue culture of peptide synthesizing endocrine and neural tissues. *Mailing Add:* Dept Neurosci John Hopkins Univ Sch Med 725 N Wolfe St Baltimore MD 21205

EIPPER, EUGENE B(RETHERTON), mechanical engineering, for more information see previous edition

EIRICH, FREDERICK ROLAND, b Vienna, Austria, May 23, 05; m 36; c 2. POLYMER CHEMISTRY. *Educ:* Univ Vienna, PhD(phys chem), 29, DSc(phys chem), 38. *Hon Degrees:* MA, Cambridge Univ, 39. *Prof Exp:* Res & assoc colloid chem, Univ Vienna, 28-32, first chem inst, 33-38; res assoc colloid sci, Cambridge Univ, 38-40, res assoc & lectr phys chem, 44-46; sr res officer, Univ Melbourne, 41-43; from assoc prof to prof polymer chem, 47-69, dean res, 67-70, DISTINGUISHED PROF POLYMER CHEM, POLYTECH INST BROOKLYN, 69- *Concurrent Pos:* Vis prof, Univ Upsala, 52 & Bristol Univ, 64-65; dir colloid sci & rubber, Am Chem Soc; ed, J Soc Rheol, 52-56. *Honors & Awards:* Bingham Medal; Higgins Award; Humboldt Prize, Germany, 81-82. *Mem:* Fel NY Acad Sci; Am Chem Soc; Soc Rheol (from vpres to pres, 70-73); Sigma Xi. *Res:* Constitution of colloid gold; colloidal metals; colloidal solutions; serum proteins; rheology; biopolymers; liquid explosives; polymer chemistry; strength of materials; surface chemistry; chemical evolution. *Mailing Add:* Polytech Inst Brooklyn 333 Jay St Brooklyn NY 11201

EISA, HAMDY MAHMOUD, b Hihia, Egypt, Aug 4, 38; m 64; c 1. PLANT BREEDING, VEGETABLE CROPS. *Educ:* Cairo Univ, BSc, 59; Cornell Univ, MSc, 66, PhD(plant breeding), 69. *Prof Exp:* Asst prof hort, Univ Nebr, Lincoln, 69-70; plant breeder, Environ Res Lab, Univ Ariz, 70-71, res assoc, 71-75; HORTICULTURIST, WORLD BANK, WASHINGTON, DC, 75-88, SR AGRICULTURIST, 88- *Concurrent Pos:* Plant breeder, Univ Ariz team, Arid Lands Res Ctr, Abu Dhabi, Arabian Gulf, 71- *Honors & Awards:* Asgrow Award, Am Soc Hort Sci, 69. *Mem:* Am Soc Hort Sci. *Res:* Breeding of vegetable crops adapted to warm, humid conditions. *Mailing Add:* World Bank 1818 H St NW Washington DC 20433

EISBERG, ROBERT MARTIN, b Kansas City, Mo, July 1, 28; m 51. NUCLEAR PHYSICS. *Educ:* Univ Ill, BS, 49; Univ Calif, PhD(physics), 53. *Prof Exp:* Res asst physics, Univ Calif Radiation Lab, 51-53; res assoc, Brookhaven Nat Labs, 53-55 & Univ Minn, 55-56; physicist, Cavendish Lab, Eng, 56-57; from asst prof to assoc prof physics, Univ Minn, 57-60; Fulbright-Guggenheim fel, Univ Tokyo, 60-61; assoc prof, 61-62, PROF PHYSICS, UNIV CALIF, SANTA BARBARA, 63- *Concurrent Pos:* Physicist, Cyclotron Lab, Arg, 62 & Rutherford Lab, Eng, 65; visitor, Europ Coun Nuclear Res, Switz, 69; Fulbright-Hays fel, Univ Peireira, Colombia, 70; vis prof, Flinders Univ, Australia, 72, Univ Surrey, Eng, 78 & Univ Zaragoza,

Spain, 86. *Mem:* Fel Am Phys Soc; Am Asn Physics Teachers. *Res:* Experimental research in nuclear scattering and reactions; passage of particles through matter particle physics; textbook writing and computer assisted instruction. *Mailing Add:* 1400 Dover Rd Santa Barbara CA 93103

EISCH, JOHN JOSEPH, b Milwaukee, Wis, Nov 5, 30; m 53; c 5. ORGANIC CHEMISTRY. *Educ:* Marquette Univ, BS, 52; Iowa State Univ, PhD(chem), 56. *Prof Exp:* Union Carbide Corp fel, Ger, 56-57; res assoc, Europ Res Assocs, Belg, 57; asst prof chem, St Louis Univ, 57-59 & Univ Mich, 59-63; from assoc prof to prof & chmn dept, Cath Univ Am, 63-72; prof & chmn dept, 72-83, DISTINGUISHED PROF CHEM, STATE UNIV NY BINGHAMTON, 83- *Concurrent Pos:* Consult, var corp. *Mem:* Am Chem Soc; Am Inst Chem. *Res:* Synthesis and properties of organometallic compounds; reactive intermediates, particularly anions, radical-anions and charge transfer complexes; mechanisms of organic reactions; stereoselectivity and regioselectivity of carbon-metal and hydrogen-metal bond additions; non-benzenoid aromatic rings, polymerization and desulfurization; ceramics and electronic devises. *Mailing Add:* Dept Chem State Univ NY Binghamton Binghamton NY 13902-6000

EISDORFER, CARL, b Bronx, NY, June 20, 30; c 5. PSYCHIATRY, PSYCHOBIOLOGY. *Educ:* NY Univ, BA, 51, MA, 53, PhD(psychol), 59; Duke Univ, MD, 64; Am Bd Psychiat & Neurol, dipl, 74. *Prof Exp:* Social investr, Children's Placement Serv, Bur Child Welfare, City of New York, 52-53, psychologist-in-training, Bur Child Guid, Bd Educ Intern Prog, 53-54; clin psychol specialist, Ment Hyg Consult Serv, US Army, Ft Dix, NJ, 54-55, neuropsychiat serv, Ryukyus Army Hosp, Okinawa, Japan, 55-56; res asst, Duke Univ, 56-58, from instr to prof med psychol & head div, 58-72, from assoc prof to prof psychiat, 68-72; prof psychiat & chmn, Dept Psychiat & Behav Sci, Sch Med, Univ Wash, Seattle, 72-81; pres, Montefiore Hosp & Med Ctr, 81-85; prof, dept psychiat & neurosci, Albert Einstein Col Med, Bronx, NY, 81-85; PROF & CHMN, DEPT PSYCHIAT, UNIV MIAMI, FLA, 86-, DIR, CTR ADULT DEVELOP & AGING, PROF PSYCHOL & DIR, CTR BIOPSYCHOSOCIAL STUDIES ON AIDS, 86- *Concurrent Pos:* Lectr, Dept Psychol, Duke Univ, 59-72, intern, Med Ctr, 65, dir training & res coordr, Ctr Study Aging & Human Develop, 65-70, dir ctr, 70-72, dir med studies, Dept Psychiat, 68-71, dir behav sci prof, Sch Med, 68-72; coordr community ment health serv, Halifax County Health Dept, NC, 59-69, prog dir, 69-71, prin consult, 71-72; vis prof, Dept Archit, Sch Environ Design, Univ Calif, Berkeley, 69-70; vis prof psychiat, Dept Psychiat, Univ Calif Med Ctr, San Franciso, 69-70; adj prof psychol, Univ Wash, 72-81; spec consult, White House Comt Aging, 61, mem, 71-73; NIMH spec fel, 62-64; consult adult develop & aging, Res & Training Rev Comt, Nat Inst Child Health & Human Develop, 69-71; mem primary care study comt, Coun Acad Socs & Asn Am Med Cols, 71; mem panel death with dignity, Inst Med of Nat Acad Sci, 71-72; mem adv comt older Americans, Dept Health, Educ & Welfare, 71-73; consult, Psychiat Educ Br, NIMH, 71, 72 & 74-; mem, Fed Coun Aging & chmn, Res & Manpower Subcomt, 74-; soc sci award & Kesten award, Ethel Percy Andrus Geront Ctr, Univ Southern Calif, 75-76; consult, Vet Admin Health Care Resources Comt, Div Med Sci, Nat Res Coun, 75-; H T Dozor Distinguished Vis Prof Geriat & Psychiat, Ben Gurion Univ, Negev, 80. *Honors & Awards:* Robert W Kleemeier Award, Geront Soc, 69; Edward B Allen Award, Am Geriat Soc, 74; Joseph Freeman Award, Geront Soc, 79; Jack Weinberg Mem Award, Am Psychiat Asn, 84; Henderson Award, Am Geriat Soc, 88; William C Menninger Mem Award, Am Col Physicians, 90. *Mem:* Inst Med Nat Acad Sci; fel Am Psychiat Asn; fel Am Psychol Asn; fel Geront Soc (pres, 71-72); fel Am Geriat Soc; fel AAAS; fel Am col Psychiat; fel Soc Behav Med; Am Fedn Aging Res (pres-elect). *Res:* Psychobiology of aging, learning, memory & stress capacity. *Mailing Add:* 1425 NW Tenth Ave SW No 200 Miami FL 33136

EISELE, CAROLYN, b New York, NY. MATHEMATICS. *Educ:* Hunter Col, AB, 23; Columbia Univ, AM, 25. *Hon Degrees:* DHH, Tex Tech Univ, 80; DSc, Lehigh Univ, 82. *Prof Exp:* From instr to prof, 23-72, EMER PROF MATH, HUNTER COL, 72- *Concurrent Pos:* Res grants, Am Philos Soc, 52-54 & 64-67 & NSF, 64-67 & 78-81; Hunter Col deleg, Int Cong Math, Hist of Sci, Logic, Methodology & Philos of Sci Semiotics, 54-; Am Coun Learned Socs travel grants, 58, 59, 83; chmn adv screening comt, Fulbright & Smith-Mundt Awards, 60-68; John Dewey Found publ grants, 72 & 79; mem staff, Inst Studies in Pragmaticism, Tex Tech Univ, 76-; assoc, seminars philos, Columbia Univ 83; Am Res Coun grants, 77 & 78. *Honors & Awards:* Award in Behav Sci, NY Acad Sci, 85. *Mem:* Fel AAAS; fel NY Acad Sci; Charles S Peirce Soc; Am Math Soc; Math Asn Am; Sigma Xi. *Res:* History and philosophy of mathematics and science of the late nineteenth century; historical perspective on Charles S Peirce's logic of science. *Mailing Add:* Apt 27E 215 E 68th St New York NY 10021

EISELE, CHARLES WESLEY, b New Albany, Ind, Apr 6, 06; m 33, 63; c 2. MEDICINE. *Educ:* NCent Col, BA, 28; Northwestern Univ, MS, 31, MB, 32, MD, 33. *Prof Exp:* Asst med, Univ Chicago, 34-35, from instr to assoc prof, 35-51, secy dept med, 41-47, chief gen med clin, 41-51; from asst dean to assoc dean, Postgrad Med Educ, 51-72, from assoc prof to prof med, 51-74, prof prev med & comprehensive health care, 69-74, EMER PROF MED, MED SCH, UNIV COLO, DENVER, 81- *Concurrent Pos:* Mem staff & consult, various hosps; consult surgeon gen, USPHS, 60-; prog dir & pres, Estes Park Inst, 74- *Mem:* Fel Am Col Physicians; fel AMA; Soc Exp Biol & Med; hon fel Am Col Hosp Adminr. *Res:* Brucellosis; toxoplasmosis; salmonellosis; evaluation and control of quality of medical care. *Mailing Add:* Estes Park Inst PO Box 400 Englewood CO 80151

EISELSTEIN, HERBERT LOUIS, b Pomeroy, Ohio, Mar 15, 19; c 3. METALLURGY. *Educ:* Univ Cincinnati, ChemE, 41. *Prof Exp:* Coop student metall, Huntington Alloys, Inc, 37-41, metallurgist, 41-54, chief testing engr lab, 54-55, sect head metall lab, Res & Develop Lab, 55-67, prod develop mgr, 67-71, asst vpres & res & develop mgr, 71-76, vpres technol, 76-85; RETIRED. *Mem:* Am Inst Mining Metall & Petrol Engrs; fel Am Soc Metals; Am Soc Testing & Mat. *Mailing Add:* Six Elwood Ave Huntington WV 25705

EISEMAN, BEN, b St Louis, Mo, Nov 2, 17; m 46; c 4. SURGERY. *Educ:* Yale Univ, BA, 39; Harvard Univ, MD, 43; Am Bd Surg, dipl, 51, Am Bd Thoracic Surg, dipl, 58. *Prof Exp:* Instr & asst prof surg, Sch Med, Wash Univ, 50-53, asst dean, 50-52; assoc prof & prof, Sch Med, Univ Colo, 53-61; prof surg & chmn dept, Col Med, Univ Ky, 61-67; PROF SURGERY, UNIV COLO MED CTR & DIR DEPT SURGERY, DENVER GEN HOSP, 67- *Concurrent Pos:* Chief surg serv, Vet Admin Hosp, Denver, 53-61; chief surg, Denver Gen Hosp, 67-71; mem comt trauma, Nat Res Coun, 60-69; surg study sect, NIH, 61-65; exec coun cardiovasc surg, Am Heart Asn, 62-68; mem bd, Am Bd Surg, 64-70; chief surgery, Rose Med Ctr, 77- *Mem:* Am Col Surg; Am Surg Asn; Soc Clin Surg; Soc Univ Surg (pres, 62); Soc Vascular Surg. *Res:* General surgery; tracheostomy and trauma; experimental coronary arterial surgery; peptic ulcer; histamine metabolism; gastric hypersecretion; role of ammonia in production of hepatic coma; treatment of hepatic coma with extracorporeal liver. *Mailing Add:* Three Village Rd Englewood CO 80110

EISEMANN, KURT, b Nuremberg, Ger, June 22, 23; US citizen; div; c 2. APPLIED MATHEMATICS, OPERATIONS RESEARCH. *Educ:* Yeshiva Univ, BA, 50; Mass Inst Technol, MS, 52; Harvard Univ, PhD(appl math), 62. *Prof Exp:* Sr mathematician, Int Bus Mach Corp, NY, 52-56, res mathematician, NY & Mass, 56-61; mgr math res, Univac Div, Sperry Rand Corp, DC, 61-63; assoc prof, Sch Eng & dir comput ctr, Cath Univ Am, 63-66; tech dir, Comput Usage Develop Corp, Mass, 66-68; dir acad comput serv & prof comput sci, Northeastern Univ, 68-74; dir comput serv & prof math & comput sci, Univ Mo-Kansas City, 74-82; dir, Univ Comput Ctr, 82-87, PROF MATH & COMPUT SCI, SAN DIEGO STATE UNIV, 82- *Concurrent Pos:* Lectr, Yeshiva Univ, 53-55 & Cath Univ Am, 62-63. *Res:* Applications of mathematics to concrete problems; linear programming; numerical analysis; effective use of computers; physical and engineering problems. *Mailing Add:* Dept Math San Diego State Univ 5300 Campanile Dr San Diego CA 92182

EISEN, EUGENE J, b New York, NY, May 14, 38; m 60; c 3. GENETICS, STATISTICS. *Educ:* Univ Ga, BSA, 59; Purdue Univ, MS, 62, PhD(genetics), 65. *Prof Exp:* From asst prof to assoc prof, 64-73, PROF ANIMAL SCI & GENETICS, NC STATE UNIV, 73- *Concurrent Pos:* Vis prof, Univ Edinburgh, 79-80; William Neal Reynolds prof, NC State Univ, 88. *Honors & Awards:* Rockefeller-Prentice Animal Breeding Award, Am Soc Animal Sci, 86. *Mem:* AAAS; Genetics Soc Am; Am Soc Animal Sci. *Res:* Experimental quantitative genetical studies with mice, involving genetical aspects of dynamics of growth and maternal influences on quantitative traits, and effects on inbreeding and selection on these traits. *Mailing Add:* Dept Animal Sci Box 7621 NC State Univ Raleigh NC 27695-7621

EISEN, FRED HENRY, b Tulsa, Okla, June 2, 29; m 54; c 3. PHYSICS. *Educ:* Calif Inst Technol, BS, 51; Princeton Univ, MA, 53, PhD(physics), 56. *Prof Exp:* Asst, Princeton Univ, 51-56; from sr physicist to res specialist, Atomics Int Div, 56-65, mem tech staff, 65-78, group leader, 78-80, DIR GALLIUM ARSENIDE ELECTRONIC DEVELOP RES, SCI CTR, ROCKWELL INT CORP, INC, 80- *Concurrent Pos:* Vis scientist, Inst Physics, Aarhus Univ, 70-71. *Mem:* Am Phys Soc; Sigma Xi; Inst Elec & Electronics Engrs. *Res:* Diffusion in solids; semiconductors; radiation damage; ion implantation in semiconductors; channeling. *Mailing Add:* 5410 Ellenvale Ave Woodland Hills CA 91367

EISEN, HENRY, b Brooklyn, NY, Dec 18, 21. PHARMACEUTICAL CHEMISTRY. *Educ:* St John's Univ (NY), BS, 49; Rutgers Univ, MS, 51; Univ Conn, PhD(pharmaceut chem), 54. *Prof Exp:* From asst prof to assoc prof pharm, 54-61, chmn dept, 61-76 & 79-82, PROF PHARMACEUT, ST JOHN'S UNIV, NY, 61- *Mem:* Am Pharmaceut Asn; Am Asn Univ Prof. *Res:* Pharmacy research and development; dosage form design and evaluation. *Mailing Add:* Col Pharm & Allied Hlth Profns St John's Univ Jamaica NY 11439-9989

EISEN, HERMAN NATHANIEL, b Brooklyn, NY, Oct 15, 18; m 48; c 5. IMMUNOLOGY. *Educ:* NY Univ, AB, 39, MD, 43. *Prof Exp:* Asst path, Col Physicians & Surgeons, Columbia Univ, 44-46; NIH fel, Col Med, NY Univ, 47-48, fel chem, 48-49, asst prof indust med, 49-53, assoc prof, 53-55; prof med, 55-61, prof microbiol & head dept, Sch Med, Wash Univ, St Louis, 61-73; prof, 73-82, Whitehead Inst prof, 82-89, EMER PROF IMMUNOL, 89- *Concurrent Pos:* Consult to Surgeon Gen, USPHS & US Army; mem comn immunization, Armed Forces Epidemiol Bd, 60; mem allergy & immunol study sect, NIH, 56-60, 61-66, chmn, 63-66; mem bd sci adv, Children's Hosp, Boston, Merck Sharp & Dohme Res Labs & Roche Inst Molecular Biol; mem, Howard Hughes Med Inst, 78-88; mem comt med affairs, Yale Univ, 83-86; mem dept molecular biol, Mass Gen Hosp, 83-85; mem bd trustees, Found Microbiol, 84-; Burroughs-Wellcome vis prof, Med Col SC, Charleston, 79; mem, Bd Toxicol & Environ Health Hazards, Nat Res Coun, 84-; outstanding investr award, Nat Cancer Inst, NIH, 86-; Nat Acad Sci, Inst Med Comt Study AIDS Res Prog, NIH, 89- *Honors & Awards:* J Howard Mueller Mem Lectr, Harvard Med Sch, 64; Solomon Berson Award, New York Univ Sch Med, 68; Jules Freund Mem Lectr, NIH, 71; Damon Lectr, Univ Ill, 72; Samit Lectr, Univ Pa, 77; Carl Prausnit Mem Lectr, Col Int Allergologicom, New Orleans, 78; Burroughs-Wellcome Lectr, Med Col SC, 79; Lowry Lectr, Wash Univ, 89. *Mem:* Inst Med-Nat Acad Sci; Am Asn Immunol (pres, 68-69); Am Soc Biol Chem; fel Am Acad Arts & Sci; Am Soc Clin Invest (vpres, 65). *Res:* Antigen recognition; antibody structure and function; molecular and cellular biology of cytotoxic T lymphocytes; antigen-specific T-cell receptors and cytolytic molecules released by T-cells; author or co-author of 4 books and over 190 publications. *Mailing Add:* Ctr Cancer Res E17-128 Mass Inst Technol Cambridge MA 02139

EISEN, JAMES DAVID, human genetics, cytogenetics, for more information see previous edition

EISENBARTH, GEORGE STEPHEN, b Brooklyn, NY, Sept 17, 47; m 69; c 2. ENDOCRINOLOGY, CELLULAR BIOLOGY. *Educ:* Columbia Col, NY, BA, 69; Duke Univ, PhD(physiol, pharmacol) & MD, 74. *Prof Exp:* Intern, Duke Univ, 75, jr resident med, 76; fel endocrinol, Lab Biochem Genetics, Nat Heart, Lung & Blood Inst, 77, res assoc cellular biol, 77-79; asst prof med, Duke Univ, 79-89; CHIEF, SECT IMMUNOL, JOSLIN DIABETES CTR, HARVARD MED SCH, 89- *Honors & Awards:* Lilly Award, Am Diabetes Asn. *Mem:* Endocrine Soc; Am Fedn Clin Res; Am Diabetes Asn. *Res:* Immunoendocrinology and developmental cellular biology; studies of the immunogenetics of polyglandular failure and cell surface membrane antigens using immunologic hybridoma techniques. *Mailing Add:* Joslin Diabetes Ctr One Joslin Pl Boston MA 02215

EISENBERG, ADI, b Breslau, Ger, Feb 18, 35; US citizen; div; c 1. PHYSICAL CHEMISTRY OF IONOMERS. *Educ:* Worcester Polytech Inst, BS, 57; Princeton Univ, MA, 59, PhD(phys chem), 60. *Prof Exp:* Res assoc polymer chem, Princeton Univ, 60-61; NATO fel, Univ Basel, 61-62; asst prof chem, Univ Calif, Los Angeles, 62-67; assoc prof, 67-74, prof chem, McGill Univ, 75-86; CONSULT, 86- *Concurrent Pos:* Consult, Jet Propulsion Lab, 62-67, Owens-Ill, 64-68 & Energy Conversion Devices, 70-86. *Honors & Awards:* Killam fel, Can Coun, 87-89; Dunlop Award, CIC, 88. *Mem:* Am Chem Soc; Soc Rheol; fel Am Phys Soc; Sigma Xi; Chem Inst Can. *Res:* Viscoelastic properties and relaxation mechanisms in organic and inorganic polymers and glasses; glass transition phenomena in amorphous materials; properties of ionic polymers; polymer chemistry. *Mailing Add:* Dept Chem Otto Maass Bldg 801 Sherbrooke St W Montreal PQ H3A 2K6 Can

EISENBERG, BENNETT, b Washington, DC, Oct 9, 42; m 70; c 1. SEQUENTIAL ANALYSIS, STOCHASTIC PROCESSES. *Educ:* Dartmouth Col, AB, 64; Mass Inst Technol, PhD(math), 68. *Prof Exp:* Instr math, Cornell Univ, 67-70; vis asst prof, Univ NMex, 70-72; ASSOC PROF MATH, LEHIGH UNIV, 72- *Concurrent Pos:* Assoc ed, Commun Statist, 81- *Mem:* Indust Math Soc; Math Asn Am. *Res:* Properties of sequential statistical tests. *Mailing Add:* Dept Math Lehigh Univ Bethlehem PA 18015

EISENBERG, BRENDA RUSSELL, b Eng, Oct 11, 42; m 64; c 4. ANATOMY. *Educ:* Univ London, BSc, 65, PhD(physiol), 71. *Prof Exp:* Teaching fel, Univ Calif, Los Angeles, 71-72, asst res biochemist, 72-74, adj asst prof med, 74-76; from asst prof to assoc prof, 76-83, PROF PHYSIOL, RUSH MED COL, 83-, DIR, CELL BIOL DEPT, 78- *Concurrent Pos:* Vis prof, Univ Chicago, 85; assoc ed, Am J Anat, 85- *Mem:* Int Stereol Soc; Am Soc Cell Biol; Biophys Soc; Am Asn Antomists; Am Heart Asn; Soc Gen Physiol. *Res:* Adaption of muscle structure to function. *Mailing Add:* Dept Physiol Univ Ill M/C 901 Box 6998 Chicago IL 60680

EISENBERG, CAROLA, b Buenos Aires, Arg; US citizen. PSYCHIATRY. *Educ:* Univ Buenos Aires, MD, 43. *Prof Exp:* Psychiat resident, Mercedes Hosp, Buenos Aires, 43-45; fel child psychiat, Johns Hopkins Hosp, 45-47, psychiatrist outpatient dept, 47-50; consult psychiat, Dept Educ, City of Baltimore, 51-53; instr, Univ Md, 55-59; instr psychiat & pediat, Johns Hopkins Univ, 58-66, asst prof, 66-67; staff psychiatrist, Mass Inst Technol, 68-72, dean student affairs, 72-78; DEAN STUDENT AFFAIRS, HARVARD MED SCH, 78- *Concurrent Pos:* Pvt pract child & adolescent psychiat, Balitmore, 55-67; consult psychiat, Park Sch Baltimore, 57-67, Sheppard Pratt Hosp, 60-67, Mass Gen Hosp, 68- & McLean Hosp, 69-; lectr, Harvard Med Sch, 68- *Mem:* Fel Am Psychiat Asn; fel Am Orthopsychiat Asn; Asn Adolescent Psychiat; Am Asn Univ Prof; Am Women's Med Soc. *Res:* Psychiatric disturbances in adolescence; pediatric psychiatry; psychotherapy. *Mailing Add:* Dept Phychiat Harvard Med Sch 25 Shattuck St Boston MA 02115

EISENBERG, DAVID, b Chicago, Ill, Mar 15, 39; m 63; c 2. BIOPHYSICAL CHEMISTRY. *Educ:* Harvard Univ, AB, 61; Oxford Univ, DPhil(theoret chem), 64. *Prof Exp:* NSF fel chem, Princeton Univ, 64-66; res fel, Calif Inst Technol, 66-69; from asst prof to assoc prof, 69-76, PROF CHEM & BIOCHEM, UNIV CALIF, LOS ANGELES, 76- *Concurrent Pos:* Alfred P Sloan fel, 69-71; USPHS career develop award, 72-77; mem, Biophys Chem Study Sect, NIH, 74-77; counr, Biophys Soc, 76-80; co-chair, Gordon Conf Diffraction Methods in Molecular Biol, 80; mem, Adv Coun Chem, Princeton Univ, 83-89; Guggenheim fel, 85; ed, Advan in Protein Chem, 88-; fac res lectr, Univ Calif, Los Angeles, 89. *Honors & Awards:* Lawrence J Onclay Lectr, Univ Mich, 86; Henry Bull Lectr, Univ Iowa, 87; Pfizer Lectr, Harvard Univ, 88; Karling Lectr, Purdue Univ, 90. *Res:* Study of biological macromolecules by x-ray diffraction; structure and properties of water. *Mailing Add:* Molecular Biol Inst 201 Univ Calif 405 Hilgard Ave Los Angeles CA 90024

EISENBERG, EVAN, b Cleveland, Ohio, Oct 31, 38. MUSCLE BIOCHEMISTRY. *Educ:* State Univ NY, Buffalo, PhD(biophys), 70. *Prof Exp:* HEAD, SECT CELL PHYSIOL, NAT HEART, LUNG & BLOOD INST, NIH, 69- *Mem:* AAAS; Biophys Soc; Am Soc Biol Chemists. *Mailing Add:* Lab Cell Biol Bldg 3 Rm B1-20 Nat Heart Lung & Blood Inst NIH Bldg 3 Rm B1-22 Bethesda MD 20892

EISENBERG, FRANK, JR, b Philadelphia, Pa, Apr 14, 20; m 48; c 4. BIOCHEMISTRY. *Educ:* Univ Pa, BS, 41, PhD, 51. *Prof Exp:* Chemist, Synthetic Fiber Res, Celanese Corp Am, 41-42; chemist org chem, Gen Foods Corp, 43-44; instr biochem, Univ Pa, 50-51; biochemist, Gen Med Res, Vet Admin, 51-52; asst intermediary metab, Pub Health Res Inst NY, 52-54; BIOCHEMIST, NAT INSTS HEALTH, 54- *Mem:* Sigma Xi; NY Acad Sci; Am Soc Biol Chemists. *Res:* Mechanism of action of dextransucrase and levansucrase; biosynthesis of inositol; metabolism of glucuronic acid; biosynthesis of glucuronic acid; gas chromatography of sugars and sugar phosphates; glucuronic acid pathway. *Mailing Add:* 6028 Avon Dr Bethesda MD 20814

EISENBERG, JOHN FREDERICK, b Everett, Wash, June 20, 35; m 57; c 2. ETHOLOGY. *Educ:* Wash State Univ, BS, 57; Univ Calif, Berkeley, MA, 59, PhD(zool), 62. *Prof Exp:* Asst prof zool, Univ BC, 62-64; asst prof, 64-65, res assoc prof, 65-72, RES PROF, UNIV MD, COLLEGE PARK, 72-; resident scientist, 65-79, asst dir, Nat Zool Park, Smithsonian Inst, 79-82; ORDWAY PROF, FLA STATE MUSEUM, UNIV FLA, 82- *Concurrent Pos:* Adj prof zool, Univ Md, 72-; assoc dept ment hyg, Johns Hopkins Univ, 73-78. *Honors & Awards:* C Hart Merriam Award, Am Soc Mammal, 81. *Mem:* Am Soc Zoologists; Am Soc Mammal; Animal Behav Soc (pres, 73); Ecol Soc Am. *Res:* Mammalian social behavior; analysis of social structure; determination of factors responsible for limiting population growth; philosophy of science. *Mailing Add:* Dept Forestry Univ Fla Gainesville FL 32611

EISENBERG, JOHN MEYER, b Atlanta, Ga, Sept 24, 46; m 69; c 2. INTERNAL MEDICINE, HEALTH CARE ADMINISTRATION. *Educ:* Princeton Univ, AB, 68; Wash Univ, MD, 72; Univ Pa, MBA, 76; Am Bd Internal Med, cert, 75. *Prof Exp:* Asst prof, 76-78, from Sol Katz asst prof to assoc prof, 78-86, SOL KATZ PROF GEN INTERAL MED, DEPT MED, UNIV PA, 86-, INTERIM CHMN, 90- *Concurrent Pos:* Assoc dir med affairs, Nat Health Care Mgt Ctr, Leonard Davis Inst Health Econ, Univ Pa, 76-78, sr fel, 76-, co-dir, residency gen med & primary care, Dept Med, 76-82, sr scholar, Clin Epidemiol Unit, 86-, mem, Cancer Ctr, 87-; consult, Nat Prof Standard Rev Orgn, 78- & Bur Radiol Health, 79-; mem, Health Technol Study Sect, Nat Ctr Health Sources Res, 81-85; vis assoc prof, Stanford Univ, Palo Alto, Calif, 84-; mem, Comt Utilization Mgt, Inst Med, 87-89; mem sci coun, Agence Nationale pour le Developpement de l'Evaluation Medicale, French Ministry Health, 90- *Mem:* Inst Med-Nat Acad Sci; Am Fedn Clin Res; Soc Med Decision Making (vpres, 80-81); fel Am Col Physicians; Soc Res & Educ Primary Care Internal Med (secy-treas, 78-80, pres, 81-82); Asn Am Physicians; Am Soc Clin Invest; AMA; Am Pub Health Asn; Am Soc Internal Med. *Res:* Cost containment by physicians; use of diagnostic tests; economic analysis of clinical practice. *Mailing Add:* Silverstein Pavilion 3 3400 Spruce St Philadelphia PA 19104

EISENBERG, JUDAH MOSHE, b Cincinnati, Ohio, Dec 17, 38; m 88; c 3. THEORETICAL PHYSICS. *Educ:* Columbia Univ, AB, 58; Mass Inst Technol, PhD(physics), 63. *Hon Degrees:* Dr, Univ Frankfurt, 85. *Prof Exp:* From asst prof to prof physics, Univ Va, 62-75, chmn dept, 70-74, Francis H Smith prof, 75-76; YUVAL NE'EMAN PROF THEORET NUCLEAR PHYSICS, TEL-AVIV UNIV, 75- *Concurrent Pos:* Mem prog adv comt, Los Alamos Meson Physics Facil, 72-75; Giulio Racah vis prof physics, Hebrew Univ Jerusalem, 74-75; Humboldt fel, 90. *Mem:* Am Phys Soc; Israel Phys Soc. *Res:* Medium-energy nuclear physics; quark substructure in nuclei. *Mailing Add:* Dept Physics & Astron Tel-Aviv Univ Ramat-Aviv Tel-Aviv 69978 Israel

EISENBERG, LAWRENCE, b New York, NY, June 9, 33; m 58; c 5. SYSTEMS ENGINEERING, CONTROL SYSTEMS. *Educ:* Fairleigh Dickinson Univ, BSEE, 60; NY Univ, MS, 61; Newark Col Eng, DEngSc, 66. *Hon Degrees:* MA, Univ Pa, 73. *Prof Exp:* Elec engr, Syst Develop Corp, 60-61; from instr to assoc prof elec eng, Newark Col Eng, 61-68; from asst prof to assoc prof, Univ Pa, 68-76, dir, Energy Ctr, Univ, 75, assoc dean, 80-85, PROF ELEC ENG, MOORE SCH ELEC ENG, UNIV PA, 76- *Concurrent Pos:* Deleg, Am Automatic Control Coun, 69-78; consult, Elec Safety Comt, Univ Pa Hosp, 70; mem, Energy Div, City Philadelphia, 73-78 & Energy Adv Coun, Camden, NJ, 78-80. *Mem:* Inst Elec & Electronics Engrs; Instrument Soc Am; Am Soc Eng Educ; Franklin Inst; AAUP; Sigma Xi. *Res:* Linear and nonlinear automatic controls; lumped and distributed circuit theory; system theory applied to transportation problems; power system analysis; energy system analysis; environmental impact of energy. *Mailing Add:* Dept Elec Eng Univ Pa Moore Sch Eng Philadelphia PA 19174

EISENBERG, LAWRENCE, b New York, NY, Dec 21, 19; m 50; c 2. ELECTRONICS ENGINEERING. *Educ:* City Col New York, BS, 40, BEE, 44; Polytech Inst Brooklyn, MEE, 52, PhD(elec eng), 66. *Prof Exp:* Sr instr electronics, Sch Indust Technol, 50-52; proj engr, Polytech Res & Develop Corp, 52-56; sr logician, Digitronics Corp, LI, 56-58; lectr elec eng, City Col New York, 58; res assoc electronics, 58-66, ASST PROF, ROCKEFELLER UNIV, 66-, CO-HEAD DEPTS ELECTRONICS & COMPUT SCI & SR RES ASSOC, 70- *Concurrent Pos:* Instr in charge, Grad Dept Elec Eng, Polytech Inst Brooklyn, 56- *Mem:* Sigma Xi. *Res:* Electrical stimulation of tissue by radiofrequency methods, particularly the heart, bladder and phrenic nerve; control of scholiosis by electronic bioconditioning. *Mailing Add:* Depts Electronics & Comput Sci Rockefeller Univ New York NY 10021

EISENBERG, LEON, b Philadelphia, Pa, Aug 8, 22; m 67; c 2. SOCIAL MEDICINE. *Educ:* Univ Pa, AB, 44, MD, 46. *Hon Degrees:* AM, Harvard Univ, 67; ScD, Univ Manchester, 73. *Prof Exp:* Instr physiol, Med Sch, Univ Pa, 47-48; asst instr neurophysiol, Basic Sci Course, Army Med Dept Res & Grad Sch, 48-50; res physician psychiat, Sheppard Pratt Hosp, Md, 50-52; asst psychiatrist, Children's Psychiat Serv, Johns Hopkins Hosp, 52-53, from instr psychiat & pediat to prof child psychiat, 53-67; prof psychiat, 67-74; Pressley prof psychiat, 74-80, PRESSLEY PROF & CHMN SOCIAL MED, HARVARD MED SCH, 80-; SR ASSOC PSYCHIAT, CHILDRENS HOSP, 74- *Concurrent Pos:* Dir, Glen Burnie Ment Hyg Clin, 53-54; psychiat ed, Crownsville State Hosp, 53-57; psychiatrist, Johns Hopkins Hosp, 54-57, asst psychiatrist-in-chg, Children's Psychiat Serv, 59-; consult, Rosewood State Training Sch, 56-58, Baltimore City Hosp, 58 & Sinai Hosp, 64-67; psychiatrist-in-chief, Mass Gen Hosp, 67-74; ed, J Orthopsychiat, 62-73; consult ed, Social Psychiat, J Pediat & J Child Psychol & Psychiat; Vis prof, Royal Soc Med, 83; mental health adv panel, World Health Orgn. *Honors & Awards:* Morris prize, Med Sch Univ Pa, 46; Theobald Smith Award, Albany Med Col, 79; Aldrich Award, Am Acad Pediat, 80; Orton Award, Orton Soc, 80; Goldbloom Lecturer, Montreal Childrens Hosp, 81; Lilly Lecturer, Royal Col Psychiatrists, 86; Queen Elizabeth II Lectr, Can Pediat Soc, 87. *Mem:* Inst Med, Nat Acad Sci; Am Pediat Soc; Am Acad Pediat; Am Psychiat Asn; Asn Res Nerv & Ment Dis; AAAS; fel, Royal Col Psychiatrists; Greek Soc

Neurol & Psychiat; Am Acad Arts & Scis. *Res:* Child psychiatry, especially early infantile autism, school phobia, psychopharmacology and studies in the development of cognition; health policy; sociology of medicine; aids. *Mailing Add:* Dept Social Med Harvard Med Sch Boston MA 02115

EISENBERG, M MICHAEL, b New York, NY, Jan 27, 31; m 53; c 3. SURGERY, GASTROENTEROLOGY. *Educ:* NY Univ, AB, 52; Harvard Univ, MD, 56. *Prof Exp:* From instr exp surg to assoc prof surg, Col Med Univ Fla, 62-68; prof surg, Col Med, Univ Minn, Minneapolis, 68-81; PROF SURG & VCHMN DEPT, STATE UNIV NY DOWNSTATE MED CTR, 81-; PROF, DEPT SURG, LONG ISLAND COL HOSP. *Concurrent Pos:* Res fel, Col Med, Univ Fla, 62-63 & Univ Calif, 65-66; sr investr, NIH res projs, 66-; chief surg, Mt Sinai Hosp, Minneapolis, 68-75; attend surgeon, Univ Minn Hosps, 68-; consult, Minneapolis Vet Admin Hosp, 69-; dir surg, Long Island Col Hosp, 81- *Mem:* Am Col Surg; Soc Univ Surg; Am Gastroenterol Asn; Soc Exp Biol & Med; Am Physiol Soc. *Res:* Physiology of secretory and motor mechanisms in the pancreas, biliary tract and stomach and duodenum. *Mailing Add:* Dept Surg State Univ NY Downstate Med Ctr LICH 340 Henry St Brooklyn NY 11201

EISENBERG, MARTIN A(LLAN), b Brooklyn, NY, Mar 8, 40; m. SOLID MECHANICS. *Educ:* NY Univ, BAeroE, 60, MS, 62; Yale Univ, ME, 64, DEng(solid mech), 67. *Prof Exp:* Asst res scientist, Eng Res Div, NY Univ, 60-61; struct engr, Sikorsky Aircraft Div, United Aircraft Corp, 61-64; asst instr eng & appl sci, Yale Univ, 65-66; from asst prof to assoc prof, 66-75, assoc chmn eng sci, 80-86, PROF ENG SCI, UNIV FLA, 75-, CHMN AEROSPACE ENG, MECH & ENG SCI, 86- *Concurrent Pos:* Prod liability consult, 80-; chmn, Am Soc Mech Engrs Comt Constitutive Equations, 85-90; dir, Fla Space Grant Consortium, 89- *Mem:* Fel Am Soc Mech Engrs; Am Acad Mech; Am Soc Eng Educ; Soc Eng Sci; Am Inst Aeronauts & Astronauts. *Res:* Theory of plasticity; stress wave propagation in solids; continuum mechanics. *Mailing Add:* Dept Aerospace Eng Mech & Eng Sci Univ Fla Gainesville FL 32611

EISENBERG, MAX, b Oct 29, 41; US citizen; c 2. INORGANIC CHEMISTRY, PHYSICAL CHEMISTRY. *Educ:* City Col NY, BS, 65; Univ Mass, MS, 68; Northeastern Univ, PhD(inorg chem), 71. *Prof Exp:* Teaching asst chem, Univ Mass, 65-67; res asst, Northeastern Univ, 67-71; lab liaison, 71-74; ASST DIR, ENVIRON HEALTH ADMIN, MD DEPT HEALTH, 74-, DIR SCI & ENVIRON HEALTH, DEPT HEALTH & MENT HYG, 81-; ASSOC PROF CHEM, TOWSON STATE UNIV, 75- *Concurrent Pos:* Mem, Gov Md Hazardous Substances Adv Coun, Comn Atomic Energy, 78-, Gov Toxic Substances Adv Coun; teaching asst, Univ Mass, 65-67; adj prof, Med Sch, Univ Md, Baltimore, 84- *Mem:* Sigma Xi: Am Chem Soc. *Res:* Behavior of paramagnetic metal ions, such as chronium (III), in solution utilizing nuclear magnetic resonance relaxation techniques; utilizing vacuum line techniques; chemistry of certain environmental pollutants; new phosphates. *Mailing Add:* Dept Biochem Columbia Univ 630 W 168th St New YOrk NY 10032

EISENBERG, MURRAY, b Philadelphia, Pa, May 23, 39; m 61; c 2. TOPOLOGICAL DYNAMICAL SYSTEMS, APL PROGRAMMING LANGUAGE. *Educ:* Univ Pa, BA, 60, MA, 62; Wesleyan Univ, PhD(math), 65. *Prof Exp:* From asst prof to assoc prof, 65-80, PROF MATH, UNIV MASS, AMHERST, 81- *Concurrent Pos:* Prin investr, NSF Instrnl Sci Equip Prog grant, 80-81 & Local Course Improvement grant, 81-83; mem, Spec Interest Group APL, Asn Comput Mach. *Mem:* Am Math Soc; Math Asn Am; Sigma Xi; Asn Comput Mach. *Res:* Dynamical systems; uses of computers, especially with APL programming language, in teaching mathematics and statistics. *Mailing Add:* Dept Math & Statist Univ Mass Amherst MA 01003

EISENBERG, RICHARD, b New York, NY, Feb 12, 43; m 66; c 2. INORGANIC CHEMISTRY. *Educ:* Columbia Col, AB, 63; Columbia Univ, MA, 64, PhD(chem), 67. *Prof Exp:* From asst prof to assoc prof chem, Brown Univ, 67-73; assoc prof, 73-76; PROF CHEM, UNIV ROCHESTER, 76- *Concurrent Pos:* Alfred P Sloan fel, Brown Univ, 72; Guggenheim fel, 77-78; vis scientist, Calif Inst Technol, 77; vis prof, Cambridge Univ, 78 & Columbia Univ, 85; chmn, Organometallic Gordon Conf, 88; mem, Adv Bd Petrol Res Fund, 88- *Mem:* Am Chem Soc; Am Crystallog Asn; The Chem Soc. *Res:* Synthetic and structural studies of transition metal complexes; organo-transition metal chemistry; systems of catalytic interest. *Mailing Add:* Dept Chem Univ Rochester Rochester NY 14627

EISENBERG, RICHARD MARTIN, b Weehawken, NJ, May 15, 42; m 66; c 3. PHARMACOLOGY, NEUROENDOCRINOLOGY. *Educ:* Univ Calif, Los Angeles, BA, 63, MS, 67, PhD(pharmacol), 70. *Prof Exp:* Res fel anat, Univ Rochester, 70-71; from asst prof to assoc prof, 71-85, PROF PHARMACOL, UNIV MINN, DULUTH, 85-, ACTG CHMN DEPT, 77-, DEPT HEAD, 80- *Mem:* Endocrine Soc; Am Soc Pharmacol. *Res:* Mechanisms underlying development of tolerance and physical dependence to narcotics; effects of drugs of abuse on ACTH release; influence of neurotransmitter systems on the adenohypophysis; pituitary-adrenal feedback relationships. *Mailing Add:* Dept Pharmacol Univ Minn Sch Med Duluth MN 55812

EISENBERG, ROBERT, b Aug 25, 44. AUTOIMMUNITY. *Educ:* Stanford Univ, MD, 71. *Prof Exp:* ASST PROF MED, UNIV NC, CHAPEL HILL, 78- *Mem:* Am Asn Immunologists; Am Arthritis Asn; Am Soc Clin Invest. *Mailing Add:* Dept Med Div Rheumatology & Immunol Univ NC 932 FLO Bldg 231H Chapel Hill NC 27599-7280

EISENBERG, ROBERT C, b Denison, Tex, Aug 5, 38; div; c 2. MOLECULAR BIOLOGY. *Educ:* Northwest Mo State Col, BS, 60; NC State Univ, MS, 62, PhD(microbiol), 66. *Prof Exp:* Res assoc dairy sci, Univ Ill, Urbana, 66-67; from asst prof to assoc prof, 67-80, PROF BIOMED SCI, WESTERN MICH UNIV, 80- *Mem:* AAAS; Am Soc Microbiol. *Res:* Pseudomonas genetics and carbohydrate metabolism; bioreclamation of ground water. *Mailing Add:* Dept Biomed Sci Western Mich Univ Kalamazoo MI 49008

EISENBERG, ROBERT MICHAEL, b Chicago, Ill, Mar 11, 38; m 62; c 3. ECOLOGY. *Educ:* Univ Chattanooga, BA, 61; Univ Mich, MS, 64, PhD(zool), 65. *Prof Exp:* Asst prof biol, Rice Univ, 65-73; asst prof, 73-77, ASSOC PROF BIOL, UNIV DEL, 77- *Mem:* Ecol Soc Am; AAAS. *Res:* Factors determining population size and structure. *Mailing Add:* Dept Ecol & Org Biol Univ Del Newark DE 19711

EISENBERG, ROBERT S, b Brooklyn, NY, Apr 25, 42; m 64; c 4. ELECTROPHYSIOLOGY, BIOPHYSICS. *Educ:* Harvard Univ, AB, 62; Univ Col, London, PhD(biophys), 65. *Prof Exp:* Post-doc physiol, Duke Univ, 65-68; from asst prof to prof physiol & biomath, Univ Calif, Los Angeles, 68-76; CHMN DEPT PHYSIOL, RUSH PRESBY ST LUKES'S MED CTR, 76- *Concurrent Pos:* Chmn, Physiol Study Sect, NIH. *Mem:* Am Physiol Soc; Biophys Soc; Soc Gen Physiologists; Am Soc Cell Biol; Inst Elec & Electronics Engrs. *Res:* Ionic channels of biological membranes; impedance measurements; three dimensional electrical field problems. *Mailing Add:* 7320 Lake St No 5 River Forest IL 60305-2231

EISENBERG, RONALD LEE, b Philadelphia, Pa. DIAGNOSTIC RADIOLOGY. *Educ:* Univ Pa, AB, 65, MD, 69. *Prof Exp:* Intern Mount Zion Med Ctr, San Francisco, 69-70; resident radiol, Mass Gen Hosp, 70-71; resident, Univ Calif, San Francisco, 73-75, asst prof, 75-80; staff radiologist, Vet Admin Med Ctr, 75-80; PROF & CHMN DEPT RADIOL, MED CTR, LA STATE UNIV, 80- *Concurrent Pos:* Var vis prof, US, Europe & Israel, 77-; prin investr, Vet Admin grant, 78 & NIH grant, 80; consult, Berlex Labs, 82-85. *Mem:* Radiol Soc NAm; Am Roentgen Ray Soc; Asn Univ Radiologists; Am Col Radiol; Soc Gastrointestinal Radiologists. *Res:* Efficacy of radiographic procedures and the designing of referral criteria and diagnostic algorithms. *Mailing Add:* Dept Radiol La State Univ Med Ctr PO Box 33932 Shreveport LA 71130

EISENBERG, SHELDON MERVEN, b Philadelphia, Pa, May 14, 42; m 71. MATHEMATICS. *Educ:* Temple Univ, AB, 63; Lehigh Univ, MS, 65, PhD(math), 68. *Prof Exp:* Instr math, Temple Univ, 65-68; from asst prof to assoc prof, 68-78, PROF MATH, UNIV HARTFORD, 78- *Mem:* Am Math Soc; Math Asn Am. *Res:* Approximation theory. *Mailing Add:* Dept Math Univ Hartford 200 Bloomfield Ave West Hartford CT 06117

EISENBERG, SIDNEY EDWIN, b New Britain, Conn, Jan 15, 13; m 46; c 2. MEDICINE. *Educ:* Wesleyan Univ, AB, 35; Univ Rochester, MD, 39; certified, Am Bd Internal Med, 50. *Prof Exp:* Clin instr internal med, Sch Med, Yale Univ, 45-54, asst clin prof, 54-60; dep chief med, 64-67, ASSOC CHIEF MED, NEW BRITAIN GEN HOSP, 67- *Concurrent Pos:* Sr attend physician & cardiologist, Hosps, 45- *Mem:* Fel Am Col Physicians. *Res:* Anemia, cardiology. *Mailing Add:* 41 Brookside Rd New Britain CT 06052

EISENBERG, SYLVAN, b New York, NY, Aug 30, 13; div; c 3. THERMODYNAMICS. *Educ:* Univ Pa, BA, 34, MS, 35; Stanford Univ, PhD(chem), 43. *Prof Exp:* Dir lab & corp consult, West Foods Lab, Lactol Corp, Calif, 36-41; DIR & OWNER, ANRESCO, 41-; PRES, MICRO TRACERS, INC, 61- *Concurrent Pos:* Consult, Vacudry Corp, 43-44; tech dir & co-owner, Desiccated Foods Co, NY, 43-47; asst prof, Univ Santa Clara, 46-48; lectr, Univ San Francisco, 50-56. *Mem:* Am Chem Soc; Am Asn Cereal Chemists; Inst Food Technol; Nat Soc Prof Engrs; Am Soc Testing & Mat. *Res:* Mixing of solids; foods; cleaning materials; corrosion. *Mailing Add:* Anresco, Inc 1370 Van Dye Ave San Francisco CA 94124

EISENBERGER, PETER MICHAEL, b New York, NY, July 20, 41; m 65; c 1. X-RAY PHYSICS. *Educ:* Princeton Univ, BA, 63; Harvard Univ, MA, 65, PhD(appl physics), 67. *Prof Exp:* Fel, Harvard Univ, 67-68; mem staff, Bell Labs, 68-76, dept head, 76-81; DIR, EXXON RES & ENG CO, 81- *Concurrent Pos:* Adj prof appl physics, Stanford Univ, 76. *Mem:* Am Phys Soc; AAAS; Optical Soc Am; Am Chem Soc. *Res:* X-ray techniques which use the capabilities presented by synchrotron radiation to determine microscopic properties of materials of interest to the physicist, chemist and biologist. *Mailing Add:* Exxon Res & Eng Co Clinton Township Rte 22 E Annadale NJ 08801

EISENBRANDT, DAVID LEE, b Kansas City, Mo, July 14, 45; m 65; c 2. COMPARATIVE PATHOLOGY, TOXICOLOGY. *Educ:* Kans State Univ, BS, 67, DVM, 69, MS, 73; Colo State Univ, PhD(path), 76; Am Col Vet Pathologists, dipl, 80. *Prof Exp:* Chief vet serv, pub health, Grand Forks AFB, 69-71; instr path, Kans State Univ, 71-73; chief vet serv, pub health, Edwards AFB, 73-74; head comp path, Naval Med Res Inst, 76-79; head anat path, Sch Aerospace Med, Brooks AFB, Tex, 79-81; res assoc, Dow Chem USA, Midland, Mich, 81-89, RES ASSOC DOW CHEM EUROPE, HORGEN, SWITZ, 89- *Concurrent Pos:* Lectr & contribr, Armed Forces Inst Path, 78-; adj prof, Mich State Univ, 81- *Mem:* Int Acad Path; Soc Toxicol Pathologists; Sigma Xi; Am Col Vet Pathologists; Electron Micros Soc Am. *Res:* Mechanisms and pathogenesis of disease, especially renal pathology and laboratory animal pathology; applications of transmission and scanning electron microscopy to pathology; utilization of morphometry and stereology in pathology; toxicology. *Mailing Add:* Health & Environ Sci Dow Chem Co 1803 Bldg Midland MI 48674

EISENBRANDT, LESLIE LEE, b Chanute, Kans, June 23, 08; m 36, 52; c 4. PHARMACOLOGY. *Educ:* Col Emporia, AB, 32; Kans State Univ, MS, 34; Rutgers Univ, PhD(zool), 36. *Prof Exp:* Asst, Kans State Univ, 32-34; asst, Rutgers Univ, 34-36; from instr to asst prof biol, Univ Mo-Kansas City, 36-42, from asst prof to assoc prof physiol, Sch Dent, 42-47, from assoc prof to prof pharmacol, Sch Pharm, 47-66, dean sch, 53-66; prof, 67-73, chmn, 72-73, EMER PROF PHARMACOL, SCH MED, UNIV MO-COLUMBIA, 73- *Concurrent Pos:* Res assoc, Sch Med Univ Calif, 48-49; consult, Midwest Res Inst, 55-62; bd trustees, US Pharmacopeia Convention Inc, 70-75. *Mem:* Am Soc Pharmacol & Exp Therapeut; Am Asn Hist Med; Am Soc Clin Pharmacol & Therapeut; Am Soc Trop Med & Hyg; Sigma Xi. *Res:* Drug metabolism and biliary excretion. *Mailing Add:* 9705 Monrovia St No 304 Lenexa KS 66215-1564

EISENBRAUN, ALLAN ALFRED, b Lodz, Poland, Nov 7, 28; m 57; c 2. ORGANIC CHEMISTRY. *Educ:* Univ Innsbruck, BSc, 52; McGill Univ, PhD(org chem), 59. *Prof Exp:* Res chemist, Ogilvie Flour Mills Co, Ltd, 52-59, res assoc, Diamond Labs, 59-60; sr res chemist, Nitrogen Div, Allied Chem Corp, 60-66; from polymer chemist to sr polymer chemist, 66-72, res assoc, 72-89, RES ADV, ETHYL CORP, 89- *Mem:* Am Chem Soc. *Res:* Carbohydrates; amino acids; heterocyclic chemistry; polymerization kinetics; polymer stabilization; new polymers for packaging applications; syntheses of new polymers. *Mailing Add:* Ethyl Corp Res & Develop Dept Gulfstate Rd POB 341 Baton Rouge LA 70821

EISENBRAUN, EDMUND JULIUS, b Wewela, SDak, Dec 10, 20; m 49; c 3. ORGANIC CHEMISTRY. *Educ:* Univ Wis, BS, 50, MS, 51, PhD, 55. *Prof Exp:* Res chemist, Monsanto Chem Co, Ohio, 55-56; res fel, Wayne State Univ, 56-59; sr res assoc chem, Stanford Univ, 59-61; res dir, Aldrich Chem Co, Wis, 61-62; assoc prof, 62-68, prof chem, 68-75, REGENTS PROF CHEM, OKLA STATE UNIV, 75- *Concurrent Pos:* Dir res proj 58A, Am Petrol Inst, 62-68. *Honors & Awards:* Sigma Xi lectr, Okla State Univ, 80. *Mem:* Am Chem Soc; The Chem Soc. *Res:* Synthesis structure proof and reaction of hydrocarbons and methylcyclopentane monoterpenoids, metal-amine reactions; Favorskii reaction; catalytic hydrogenation, hydrogenolysis and dehydrogenation; effect of steric crowding on the regiochemistry and spectroscopy of aromatic methoxy groups. *Mailing Add:* Dept Chem Okla State Univ Stillwater OK 74078

EISENBUD, DAVID, b New York, NY, Apr 8, 47; m 70; c 2. ALGEBRA. *Educ:* Univ Chicago, BS, 66, MS, 67, PhD(math), 70. *Prof Exp:* Lectr & res assoc math, 70-72, asst prof, 72-77, assoc prof, 77-80, PROF MATH, BRANDEIS UNIV, 80- *Concurrent Pos:* Fel, Alfred P Sloan Found, 74-76; vis prof, Bonn, 79-80, MSRI Berkley, 85-86, Harvard Univ, 86-87. *Mem:* Am Math Soc; Math Asn Am. *Res:* Commutative algebra; algebraic geometry. *Mailing Add:* Dept Math Brandeis Univ Waltham MA 02154

EISENBUD, LEONARD, b Elizabeth, NJ, Aug 3, 13; m 46; c 1. THEORETICAL PHYSICS. *Educ:* Union Univ, NY, BS, 35; Princeton Univ, PhD(theoret physics), 48. *Prof Exp:* Physicist, Bartol Res Found, Pa, 48-58; chmn dept, 58-62 & 68-69, prof, 58-83, EMER PROF PHYSICS, STATE UNIV NY STONYBROOK, 83- *Concurrent Pos:* Mem, Inst Adv Study, 41; physicist, radiation lab, Mass Inst Technol, 43-46. *Mem:* Fel Am Phys Soc; Sigma Xi. *Res:* Nuclear physics; quantum mechanics. *Mailing Add:* Dept Physics State Univ NY Stony Brook NY 11794

EISENBUD, MERRIL, b New York, NY, Mar 18, 15; m 39; c 3. ENVIRONMENTAL HEALTH. *Educ:* NY Univ, BSEE, 36. *Hon Degrees:* ScD, Fairleigh Dickinson Univ, 60; DHC, Cath Univ Rio de Janeiro, 70. *Prof Exp:* Indust hygienist, Liberty Mutual Ins Co, 36-47; from assoc prof to prof, 45-84, EMER PROF ENVIRON MED, NY UNIV MED CTR, 84- *Concurrent Pos:* Dir health & safety lab, US Atomic Energy Comn, 47-57, mgr, NY Opers Off, 54-59; mem bd on radioactive waste mgt, Nat Acad Sci, 56-82; mem expert adv panel radiation, WHO, 57-85; mem, Nat Coun Radiation Protection, 64-; adminr, Environ Protection Admin, New York, 68-70; mem, NY State Health Adv Coun, 69-81; mem, Nat Adv Coun, Electric Power Res Inst; adj prof, Dept Environ Sci & Eng, Univ NC, 84- *Honors & Awards:* Arthur Holly Compton Award, Am Nuclear Soc; Power-Life Award, Inst Elec & Electronics Engrs; Gold Medal, US Atomic Energy Comn. *Mem:* Nat Acad Eng; fel AAAS; Radiation Soc; fel Am Nuclear Soc; fel Health Physics Soc (pres, 64-66). *Res:* Environmental radioactivity; urban pollution; environmental effects of power generation; human ecology. *Mailing Add:* 711 Bayberry Dr Chapel Hill NC 27514

EISENFELD, ARNOLD JOEL, b Pittsburgh, Pa, July 26, 36; m 60; c 2. PHARMACOLOGY, INTERNAL MEDICINE. *Educ:* Washington & Jefferson Col, AB, 58; Yale Univ, MD, 62. *Prof Exp:* Intern & resident med, Yale-New Haven Hosp Ctr, Conn, 62-64; res assoc, NIH, 64-66; from asst prof to assoc prof pharmacol & internal med, Yale Univ, 67-77, sr res scientist obstet & gynec, 77-89; CONSULT, 89- *Concurrent Pos:* Fel pharmacol, Sch Med, Yale Univ, 66-67. *Mem:* Am Soc Pharmacol & Exp Therapeut. *Res:* Interaction of estrogens, androgens and progestins with target organs; birth control; clinical pharmacology; hypertension. *Mailing Add:* 65 Forest Hill Rd North Haven CT 06473

EISENFELD, JEROME, b New York, NY, Oct 13, 38; m 62; c 2. APPLIED MATHEMATICS, BIOMATH. *Educ:* City Col New York, BS, 60; Univ Chicago, MS, 64, PhD, 66. *Prof Exp:* Res assoc math, Univ Chicago, 66; asst prof, Rensselaer Polytech Inst, 66-72; assoc prof, 72-76, PROF MATH, UNIV TEX, ARLINGTON, 76- *Concurrent Pos:* Vis prof, Univ Tex Health Sci Ctr, Dallas, 75-81, adj prof, 73-83; IPA, NIH, 84. *Res:* System identification; mathematic modeling in medicine; differential equations; linear algebra. *Mailing Add:* Dept Math Univ Tex Arlington TX 76019

EISENHARDT, WILLIAM ANTHONY, JR, b Lorain, Ohio, Nov 15, 42; m 69; c 1. FOOD CHEMISTRY, BIOPHYSICS. *Educ:* Case Western Reserve Univ, AB, 65; State Univ NY Buffalo, PhD(chem), 70. *Prof Exp:* Res assoc chem, Univ Chicago, 70-71; res scientist chem, Union Carbide Corp, 71-76, group leader, 76-78, sr group leader immunochem & clin chem, 78-79; mgr, Phys Chem Lab, Gen Foods Corp, 79-81; dir, anal res & develop, 81-83, dir, prod res & develop, 83-85, VPRES RES & DEVELOP, ROSS LABS, DIV ABBOTT LABS, 85- *Mem:* Am Chem Soc; Sigma Xi. *Res:* Organic and bio-organic chemistry related to novel analytical and diagnostic reagents; analytical biochemistry related to design of new clinical analytical systems; surface chemistry-physics and chemistry-physics of water at low temperatures; synthesis of bioactive and radioisotopically-labeled molecules. *Mailing Add:* Ross Labs 625 Cleveland Ave Columbus OH 43216

EISENHART, CHURCHILL, b Rochester, NY, Mar 11, 13; m 39; c 2. MATHEMATICAL STATISTICS. *Educ:* Princeton Univ, AB, 34, AM, 35; Univ London, PhD(math statist), 37. *Prof Exp:* From instr to assoc prof math, 37-47, statistician & biometrician, Exp Sta, 37-47; chief, Statist Eng Lab, 46-63, sr res fel, 63-83, GUEST WORKER, NAT BUR STANDARDS, 83- *Concurrent Pos:* Res assoc, Tufts Col, 43; res mathematician, Appl Math Group, Columbia Univ, 43-44, prin math statistician, Statist Res Group, 44-45. *Honors & Awards:* Rockefeller Pub Serv Award, 58; Samuel S Wilks Mem Medal, Am Statist Asn, 77; William A Wildhack Award, Nat Conf Standards Labs, 82. *Mem:* Fel AAAS; Hist Sci Soc; fel Inst Math Statist (vpres, 48); Math Asn Am; fel Am Statist Asn (vpres, 58-59, pres, 71). *Res:* Mathematical statistics and its applications in the biological and physical sciences and in engineering and industry; history of statistical methodology. *Mailing Add:* 9629 Elrod Rd Kensington MD 20895-3116

EISENHAUER, CHARLES MARTIN, b New York, NY, Feb 6, 30; m 58; c 2. RADIATION SHIELDING, NEUTRON STANDARDS. *Educ:* Queens Col, NY, BS, 51. *Prof Exp:* Jr mathematician reactor physics, Brookhaven Nat Lab, 51-52; nuclear physicist radiation penetration, Armed Forces Spec Weapons Proj, 53-54; vis scientist cold neutron exp, Brookhaven Nat Lab, 56-57; RADIATION PHYSICIST, NAT BUR STANDARDS, 58- *Concurrent Pos:* Consult, Oak Ridge Nat Lab, 64; adj prof, Cath Univ Am, 68-72; president's comn on accident at Three Mile Island, 79; guest lectr, Georgetown Univ, 86-88. *Honors & Awards:* Silver Medal, US Dept Com, 62 & 82. *Mem:* Fel Am Nuclear Soc; Health Physics Soc. *Res:* Penetration of nuclear radiation; experimental use of cold neutrons to study the dynamics of solids and liquids; neutron spectra and detector responses in reactors; energy deposition by charged particles; calibration of neutron personnel monitors. *Mailing Add:* Nat Inst Standards & Technol Gaithersburg MD 20899

EISENHAUER, HUGH ROSS, b Lethbridge, Alta, Oct 11, 27; m 71; c 5. ENVIRONMENTAL SCIENCES. *Educ:* Univ Saskatchewan, BA, 49, MA, 50; Univ Wis, PhD(org chem), 53. *Prof Exp:* Res chemist, Can Indust Ltd, 53-54; res chemist, Du Pont Co Can, 54-62, sr res chemist, 62-66; res scientist, Pub Health Eng Div, Dept Environ, Gout Can, 66-70, head water sci subdiv, Hydrol Sci Div, 70-72, chief, Water Qual Res Div, 72-74, sr res adv, 74-78, chief res planning & coordr, Inland Waters Directorate, 78-84, spec adv, Toxic Chem Prog, 84-89, SPEC ADV TECHNOL DEVELOP PROGS, CONSERV & PROTECTION, DEPT ENVIRON, GOVT CAN, 89- *Mem:* Fel Chem Inst Can; Int Asn Water Pollution Res & Control; Can Asn Water Pollution Res & Control. *Res:* Water chemistry. *Mailing Add:* Conserv & Protection Dept Environ Place Vincent Massey Ottawa ON K1A 0H3 Can

EISENLOHR, W(ILLIAM) S(TEWART), JR, b Philadelphia, Pa, Nov 16, 07; m 32; c 3. ENGINEERING. *Educ:* Univ Pa, BS in CE, 28. *Prof Exp:* Hydraul engr, US Geol Surv, Mass, 28, Va, 28-29, Ala, 29-31, Ariz, 31-33, Washington, DC, 33-61, Colo, 61-70; RETIRED. *Mem:* Fel Am Soc Civil Engrs; Am Geophys Union. *Res:* Hydraulics of natural channels; coefficients for velocity distribution in open channel flow; effect of water temperature on flow of natural streams; floods of North Central Pennsylvania; hydrology of prairie potholes. *Mailing Add:* 350 Ponca Pl Boulder CO 80303-3828

EISENMAN, GEORGE, b New York, NY, May 6, 29; m 52; c 2. BIOPHYSICS. *Educ:* Harvard Univ, AB, 49, MD, 53. *Hon Degrees:* Dr, Univ Uppsala, Sweden. *Prof Exp:* Res assoc, Harvard Univ, 54-55; sr staff scientist, Dept Basic Res, Eastern Pa Psychiat Inst, 56-62; assoc prof physiol, Col Med, Univ Utah, 62-65; prof physiol, Univ Chicago, 65-69, prof biophys, 67-69; PROF PHYSIOL, SCH MED, UNIV CALIF, LOS ANGELES, 69- *Concurrent Pos:* Res fel, Harvard Univ, 53-54; consult, Corning Glass Works, 62-72; mem biophys & biophys chem study sect, NIH, 67-71; mem US nat comt biophys, Nat Res Coun, 82-85, chmn, 85- *Honors & Awards:* Elizabeth Robert Cole Award, Biophys Soc. *Mem:* Am Physiol Soc; Biophys Soc (pres, 78-79); fel Royal Soc Arts; Soc Gen Physiologists. *Res:* Membrane biophysics; molecular biology; physical chemistry. *Mailing Add:* Dept Physiol Univ Calif Med Ctr 405 Hilgard Ave Los Angeles CA 90024

EISENMAN, LEONARD MAX, b Brooklyn, NY, July 7, 42; m 66; c 4. ANATOMY. *Educ:* Brooklyn Col, BA, 64; Miami Univ, MA, 65; Duke Univ, PhD(physiol psychol), 74. *Prof Exp:* Fel anat, Col Physicians & Surgeons, Columbia Univ, 73-76; from asst prof to assoc prof, 76-87, PROF ANAT, JEFFERSON MED COL, 87- *Concurrent Pos:* Individual res fel, Nat Inst Neurol Dis & Stroke, 74, R01 res grant, 83-86, 86- *Mem:* Soc Neurosci; AAAS; Sigma Xi; Am Asn Anatomists. *Res:* Anatomy physiology and development of the cerebellum. *Mailing Add:* Dept Anat Thomas Jefferson Med Col Philadelphia PA 19107

EISENMAN, RICHARD LEO, b Bridgeport, Conn, July 12, 28; m 52; c 3. INTELLIGENCE SYSTEMS, POLICY ANALYSIS. *Educ:* Col Holy Cross, AB, 49; Univ Conn, MA, 50; Univ Mich, PhD(math), 64. *Prof Exp:* Instr math, Fairfield Univ, 49, Univ Md, 50-52 & US Air Force, 52-72; res mathematician, Wright Patterson AFB, 53-56, US Air Force Acad, 58-68; chief tactical anal, Tan Son Nhut AFB, Vietnam, 68-69; chief systems anal personnel plans, Hq, US Air Force, DC, 69-72; prog dir, Human Resources Res Orgn, 72-76; specialist in nat defense, Libr of Cong, 77-78; LECTR, UNIV MARYLAND, 78- *Concurrent Pos:* Consult, Air Battle Anal Hq, US Air Force, 65, Kaman Nuclear Corp, 65-66 & Holly Sugar Corp, 68 & Rand Corp, 85-; staff asst systs anal, Off Secy Defense, DC, 66-67; vpres, Ferry Landing Woods, Inc, 74-, Chaney Station, Inc, 78- & Rand, 83- *Mem:* Math Asn Am; Sigma Xi. *Res:* Theory and application of mathematical models. *Mailing Add:* PO Box 143 PO Box 143 Dunkirk MD 20754-0002

EISENREICH, STEVEN JOHN, b Eau Clarie, Wis, Sept 9, 47; m. AQUATIC & ENVIRONMENTAL CHEMISTRY. *Educ:* Univ Wis-Eau Claire, BS, 69; Univ Wis-Milwaukee, MS, 73; Univ Wis-Madison, PhD(water chem), 75. *Prof Exp:* From asst prof to assoc prof, 75-83, PROF ENVIRON ENG, DEPT CIVIL & MINERAL ENG, UNIV MINN, MINNEAPOLIS, 83- *Concurrent Pos:* Consult, Environ Protection Agency, Int Joint Comn, Dept Justice/EPA on Superfuncl Cases. *Mem:* Am Chem Soc; Int Asn Great Lakes Res; Am Soc Limnol & Oceanog; Am Geophys Union; Geochem Soc. *Res:* Environmental chemistry of inorganic and organic compounds in surface and subsurface. *Mailing Add:* Univ Minn 122 CME Bldg Minneapolis MN 55455-0100

EISENSON, JON, b New York, NY, Dec 17, 07; m 31, 77; c 2. SPEECH & HEARING SCIENCES, PSYCHOLOGY OF AGING. *Educ:* City Col New York, BSS, 28; Columbia Univ, MA, 30, PhD(educ psychol), 35; Am Bd Prof Psychol, dipl. *Prof Exp:* Instr, NY Schs, 28-35; instr speech, Brooklyn Col, 35-42; from asst prof to prof & dir speech clin, Queens Col, NY, 46-62; prof hearing & speech sci & dir, Inst Childhood Aphasia, 62-73, EMER PROF HEARING & SPEECH SCI, SCH MED, STANFORD UNIV, 73- *Concurrent Pos:* Consult, US Vet Admin Hosps, Calif, 62-; chmn spec educ adv comt & mem med & sci comt, United Cerebral Palsy Asn, 64-69; lectr, Col Physicians & Surgeons, Columbia Univ; distinguished prof spec educ, San Francisco State Univ, 75-81. *Mem:* Fel AAAS; fel Am Speech & Hearing Asn (pres, 58-59); fel Am Psychol Asn; Speech Commun Asn; World Fedn Neurol; Acad Aphasia; Int Neuropsychol Asn. *Res:* Language; speech pathology with emphasis on stuttering and aphasia; psychology of speech; communication; psycholinguistics; child language and language delay; confirmation and information in rewards and punishments; psychology of aging. *Mailing Add:* 82 Pearce Mitchell Place Stanford CA 94305

EISENSTADT, ARTHUR A, b New York, NY, Jan 23, 18; m 42; c 3. SPEECH PATHOLOGY. *Educ:* Brooklyn Col, AB, 38, MA, 46; NY Univ, PhD(speech path), 54. *Prof Exp:* Instr speech, Brooklyn Col, 45-46; asst, Cornell Univ, 46-48; asst prof, Rutgers Univ, 48-55; speech pathologist, Newark Div Speech Educ, 55-63; Prof Speech & Dir Speech Educ, St John's Univ, NY, 63-87; CONSULT, 87- *Concurrent Pos:* Fulbright award, US Dept HEW, 59-60; Shell Oil Co res grant, 72. *Mem:* Fel Am Speech & Hearing Asn; Am Asn Univ Prof. *Res:* Geriatric speech pathology; pediatric speech pathology; language acquisition disorders. *Mailing Add:* 40 Oxford St Montclair NJ 07042

EISENSTADT, BERTRAM JOSEPH, b New York, NY, Mar 28, 23; m 58; c 1. MATHEMATICS. *Educ:* City Col New York, BS, 43; Brown Univ ScM, 46; Univ Mich, PhD(math), 51. *Prof Exp:* Res assoc physics, Nat Adv Comt Aeronaut, 44-46; from asst prof to assoc prof, 49-61, PROF MATH, WAYNE STATE UNIV, 61-, CHMN DEPT, 75- *Mem:* Am Math Soc; Math Asn Am; Sigma Xi. *Res:* Functional analysis. *Mailing Add:* Dept Math Wayne State Univ Detroit MI 48202

EISENSTADT, JEROME MELVIN, b Chicago, Ill, June 11, 26; m 60. BIOCHEMISTRY, HUMAN GENETICS. *Educ:* Roosevelt Univ, BS, 52; Brandeis Univ, MA, 59, PhD(biol), 60. *Prof Exp:* From asst prof to assoc prof microbiol, 62-74, assoc prof, 74-77, PROF HUMAN GENETICS, SCH MED, YALE UNIV, 77- *Concurrent Pos:* Fel biochem, Oak Ridge Nat Labs, 60-62; NIH fel, 61-62; consult, 62- *Mem:* Am Soc Cell Biol; Am Soc Biol Chem; Am Soc Microbiol; Sigma Xi. *Res:* genetics of mitochondria; control mechanisms of protein and nucleic acid synthesis; somatic cell genetics; gene expression in transgenic mice. *Mailing Add:* Dept Human Genetics Yale Univ Sch Med New Haven CT 06510

EISENSTADT, MAURICE, b New York, NY, June 10, 31; m 61. PHYSICS. *Educ:* City Col New York, BCE, 52; Columbia Univ, AM, 53, PhD(physics), 58. *Prof Exp:* Asst physics, Columbia Univ, 54-58; physicist, Watson Labs, Int Bus Mach Corp, 58-63; physicist, Hudson Labs, Columbia Univ, 63-69; asst prof, 69-80, PHYSICIST, ALBERT EINSTEIN COL MED, 69-, ASSOC PROF MED, 80- *Mem:* Am Phys Soc. *Res:* Nuclear magnetic resonance; dielectric studies of macromolecules. *Mailing Add:* Albert Einstein Col Med Ullman 517 1300 Morris Park Ave Bronx NY 10461

EISENSTADT, RAYMOND, b Brooklyn, NY, May 13, 21; m 57; c 4. MECHANICAL ENGINEERING. *Educ:* City Col New York, BME, 41; Columbia Univ, MSME, 43, PhD(mech eng), 53. *Prof Exp:* Construct engr, Mediter Theater, US Govt War Dept, 46; tech consult war surplus mat, Italy, 46-47; design engr heat, vent & air conditioning, P M Gussow, 48 & Corgett-Tinghir, 48-49; engr res lubrication, Atomic Energy Comn, Columbia Univ, 52-53; engr wind tunnel proj inst res, Lehigh Univ, 53-54, asst prof mech eng, 53-54; assoc prof, 54-69, PROF MECH ENG, UNION COL, NY, 69-, DIR, SUMMER INST, 70- *Concurrent Pos:* NSF fac fels, Mass Inst Technol & Univ Mich, 60-61, Inst Mech & Dynamics, Yale Univ, 63 & Smith Inst Solid State Sci Res, 64; NASA Lewis Struct Div res grant, 68-69; Gen Elec Co res fel, 69-78; mem subcomt plastic fatigue strength, pressure vessel res comt, Welding Res Coun; consult, Gen Elec, Am Locomotive Co, Watervliet Arsenal; mem, Comt Crack Growth & Fatigue, Am Soc Testing Mat. *Mem:* Fel Am Soc Mech Engrs; Am Soc Eng Educ; Am Soc Metals; Soc Exp Stress Anal; Sigma Xi. *Res:* Mechanical behavior of materials; fatigue of metals; author of over 25 publications. *Mailing Add:* Dept Mech Eng Union Col Schenectady NY 12308

EISENSTARK, ABRAHAM, b Warsaw, Poland, Sept 5, 19; nat US; m 48; c 3. BACTERIOLOGY. *Educ:* Univ Ill, AB, 41, AM, 42, PhD(bact), 48. *Prof Exp:* Electron microscopist, Univ Ill, 46-48; asst prof bact, Okla Agr & Mech Col, 48-51; from assoc prof to prof, Kans State Univ, 51-71; DIR & PROF DIV BIOL SCI, UNIV MO-COLUMBIA, 71- *Concurrent Pos:* Guggenheim fel, Inst Microbiol, Copenhagen, Denmark, 59; NSF sr fel, Univ Leicester, 66; sect head & prog dir molecular biol sect, NSF, Washington, DC, 69-70. *Mem:* AAAS; Am Soc Microbiol; Am Soc Photobiol. *Res:* Microbial genetics and photobiology. *Mailing Add:* Div Biol Sci 105 Tucker Hall Univ Mo Columbia MO 65201

EISENSTATT, PHILLIP, b Omaha, Nebr, Oct 17, 22; m 52; c 3. GEOLOGY. *Educ:* Univ Nebr, BSc, 43. *Prof Exp:* Geologist, 43-50, div geologist, 50-62, staff geologist, Explor Dept, Shell Oil Co, 62-81; mgr geol, 81-84, CONSULT, GRAHAM RESOURCES, 84- *Mem:* Fel Geol Soc Am; Am Asn Petrol Geologists. *Res:* Subsurface and field geology assignments which could result in the discovery of oil and gas reserves; geological data processing applications. *Mailing Add:* 4625 Neyrey Dr Metairie LA 70002

EISENSTEIN, ALBERT BERNARD, internal medicine, endocrinology, for more information see previous edition

EISENSTEIN, BARRY I, b Brooklyn, NY, Feb 14, 48; m 69; c 2. MOLECULAR PATHOGENESIS OF BACTERIAL INFECTIONS, CLINICAL INFECTIOUS DISEASES. *Educ:* Kenyon Col, AB, 68; Columbia Univ, MD, 72. *Prof Exp:* Intern, NC Mem Hosp, Chapel Hill, 72-73, resident, 73-75; fel infectious dis, Univ NC, 75-77; from asst prof to assoc prof med, Univ Tenn, 77-82; assoc prof med, microbiol & immunol, Univ Tex Health Sci Ctr, San Antonio, 82-86; PROF & CHMN MICROBIOL & IMMUNOL, MED CTR, UNIV MICH, 86-, PROF INTERNAL MED, 86- *Concurrent Pos:* Mem, Bact-Mycol I Study Sect, NIH, 84-86, Nat Acad Sci Comt on Subtherapeut Antibiotics in Animal Feed, 87-88, Microbiol Test Comt, Nat Bd Med Examr, 90-; chair, Div B, Am Soc Microbiol, 89, Gordon Res Conf, Bact Adherence, 89; dir, Molecular Pathogenesis Training Prog, 89-; ed, Infection & Immunity, 89; Res Career Develop Award, Nat Inst Allergy & Infectious Dis, USPHS, 82. *Honors & Awards:* Clin Invest Award, Nat Inst Arthritis, Metab & Digestive Dis, USPHS, 79. *Mem:* Asn Am Physicians; Am Soc Clin Invest; Am Soc Microbiol; fel Infectious Dis Soc Am; fel Am Col Physicians; Cent Soc Clin Res. *Res:* Pathogenic mechanisms, at the molecular level, of two bacterial systems, Eschericia coli, the major causative agent of urinary tract infections and gastroenteritis, and Legionella pneumophila, the causative agent of Legionnaires' disease. *Mailing Add:* Dept Microbiol & Immunol Univ Mich Med Ctr 6643 Med Sci II Ann Arbor MI 48109-0620

EISENSTEIN, BOB I, b New York, NY, Feb 4, 39; m 64. HIGH ENERGY PHYSICS. *Educ:* Columbia Univ, AB, 59, AM, 61, PhD(physics), 64. *Prof Exp:* Res assoc physics, Columbia Univ, 64; res fel, Harvard Univ, 64-67; asst prof, 67-70, assoc prof, 70-78, PROF PHYSICS, UNIV ILL, URBANA, 78- *Mem:* Am Phys Soc. *Res:* Electron position annihilation; charm particles; data analysis. *Mailing Add:* Dept Physics Univ Ill 1110 W Green St Urbana IL 61801

EISENSTEIN, BRUCE A, b Philadelphia, Pa, Sept 10, 41; m 63; c 3. PATTERN RECOGNITION. *Educ:* Mass Inst Technol, BSEE, 63; Drexel Inst Tech, MSEE, 65; Univ Pa, PhD(elec eng), 70. *Prof Exp:* HEAD, DEPT ELEC ENG, DREXEL UNIV, 80- *Concurrent Pos:* Vis prof, Princeton Univ, 70-71; chair, Philadelphia Sect, Inst Elec & Electronics Engrs, 89-90. *Mem:* Fel Inst Elec & Electronics Engrs. *Res:* Pattern recognition by digital signal processing to achieve compaction of feature set. *Mailing Add:* Elec Eng Dept Col Eng Drexel Univ Philadelphia PA 19104

EISENSTEIN, JULIAN (CALVERT), b Warrenton, Mo, Apr 3, 21; m 48; c 3. THEORETICAL PHYSICS. *Educ:* Harvard Univ, BS, 41, PhD(physics), 48. *Prof Exp:* Res assoc acoust, Harvard Univ, 42-45; instr physics, Univ Wis, 48-52; from asst prof to assoc prof, Pa State Univ, 53-57; physicist, Nat Bur Standards, 57-66; prof physics, George Washington Univ, 66-87; RETIRED. *Concurrent Pos:* Nat Res fel, 52-53. *Mem:* Fel Am Phys Soc; Brit Inst Physics. *Res:* Low temperature physics; paramagnetism; absorption spectra of complex ions. *Mailing Add:* 82 Kalorama Circle NW Washington DC 20008

EISENSTEIN, REUBEN, b Brooklyn, NY, May 3, 29; m 58; c 4. ARTERIOSCLEROSIS & CANCER RESEARCH. *Educ:* Tulane Univ, BS, 49; LA State Univ, MD, 53. *Prof Exp:* Intern, Mt Sinai Hosp NY, 53-54; resident path, Presby-St Lukes Hosp, Chicago, 57-60, pathologist, 60-75; from asst prof to prof, Med Sch, Univ Ill, 60-69; prof, Northwestern Univ, 75-80; chief, Va Lakeside Hosp, Chicago, 76-80; chief path, Mt Sinai Med Ctr, Milwaukee, 80-91; PROF PATH & CHMN DEPT, UNIV WIS-MILWAUKEE, 80- *Concurrent Pos:* Mem, study sect, NIH, 67-71; clin prof, Med Col Wis, 81- *Mem:* Am Asn Pathologists; Am Heart Asn; Int Acad Path; Orthop Res Soc; Int Skeletal Soc. *Res:* Skeletal diseases; surgical pathology; arteriosclerosis; connective tissue biology; growth factors; immunology; ophthalmic diseases. *Mailing Add:* Sinai-Samaritan Med Ctr 950 N 12th St Milwaukee WI 53201

EISENSTEIN, ROBERT ALAN, b St Louis, Mo, July 17, 42; m 67; c 2. NUCLEAR PHYSICS, PARTICLE PHYSICS. *Educ:* Oberlin Col, AB, 64; Yale Univ, MS, 66, PhD(physics), 68. *Prof Exp:* Res fel physics, Weizmann Inst Sci, Israel, 68-70; from asst prof to prof, Carnegie-Mellon Univ, 70-84; DIR NUCLEAR PHYSICS LAB, UNIV ILL, 85- *Concurrent Pos:* Weizmann fel, Weizmann Inst, Rehovot, Israel, 68-70; mem, bd dirs Los Alamos Meson Physics Facil User's Group, 78-80, chmn, 80, div rev comt, 88, Dept Energy Rev Comt, 88; mem, adv comt physics, NSF, 80-82, Physics Surv Comt, NAS, 83-84, Nuclear Sci Adv Comt, 87-90, SURA Rev Comt for CEBAF, 88-; chmn Am Phys Soc Div Nuclear Physics, 89-90, vchair, 88-89. *Mem:* Fel Am Phys Soc; Sigma Xi. *Res:* Experimental studies of the interaction of elementary particles with nuclei. *Mailing Add:* Loomis Lab Physics Univ Ill Urbana IL 61801

EISENSTEIN, TOBY K, b Philadelphia, Pa, Sept 15, 42; m 63; c 3. MICROBIOLOGY, IMMUNOLOGY. *Educ:* Wellesley Col, BA, 64; Bryn Mawr Col, PhD(microbiol), 69. *Prof Exp:* Instr microbiol, 69-71, from asst prof to assoc prof, 71-84, PROF MICROBIOL & IMMUNOL, SCH MED, TEMPLE UNIV, 84-, ACTG CHAIR, 90- *Concurrent Pos:* USPHS res grants & contracts, 72-; bact & mycol study sect, NIH, 76-80 & 88- *Mem:* AAAS; Am Soc Microbiol; Sigma Xi; Reticuloendothelial Soc; Am Asn Immunologists; Int Endotoxin Soc; fel Am Acad Microbiol. *Res:* Immunity to bacterial infections, salmonella, group B streptococci, Legionella; vaccines; macrophages and cellular immunity. *Mailing Add:* Dept Microbiol & Immunol Temple Univ Sch Med 3400 N Broad St Philadelphia PA 19140

EISENTRAUT, KENT JAMES, b Troy, NY, July 31, 38; m 64; c 2. ANALYTICAL CHEMISTRY. *Educ:* St Michael's Col, AB, 60; Rensselaer Polytech Inst, PhD(anal chem), 64. *Prof Exp:* Res scientist, Aerospace Res Labs, 64-75, res chemist, Air Force Mat Lab, Wright-Patterson AFB, 75-79, RES CHEMIST, AIR FORCE WRIGHT AERONAUT LABS, 79- *Concurrent Pos:* US Air Force Res & Develop award, 66. *Mem:* Am Chem Soc; fel Am Inst Chemists; NY Acad Sci; Sigma Xi; Soc Appl Spectros. *Res:* Metal coordination chemistry; gas chromatography of volatile rare earth, transition and alkali metal chelates; atomic absorption spectroscopy;

synthesis of volatile chelates; infrared and nuclear magnetic resonance spectroscopy; differential thermal and thermal gravimetric analysis; metal analysis of Apollo 11, 12, 14, 15 & 16 lunar samples; research support of Air Force Oil Analysis Program; fluids, lubricants and lubrication technology research and development. *Mailing Add:* 1827 Trebein Rd Xenia OH 45385-9558

EISER, ARTHUR L, b Geneva, Ill, Apr 16, 28; m 55; c 3. PLANT TAXONOMY, PLANT ECOLOGY. *Educ:* Univ Denver, BA, 50; Iowa State Univ, MS, 52; Va Polytech Inst, PhD(plant taxon, ecol), 61. *Prof Exp:* Asst biol, Univ Denver, 47-50; asst bot, Iowa State Univ, 50-52; instr hort, Va Polytech Inst, 54-56, asst, 56-58; asst prof sci, 58-63, assoc prof biol, 63-69, PROF BIOL, BALL STATE UNIV, 69- *Mem:* Ecol Soc Am; Sigma Xi. *Res:* Plant taxonomy of flowering and ornamental plants; uses in landscaping. *Mailing Add:* 1121 Yale Ave Muncie IN 47304

EISERLING, FREDERICK A, b San Diego, Calif, May 8, 38. MICROBIOLOGY. *Educ:* Univ Calif, Los Angeles, BA, 59, PhD(microbiol), 64. *Prof Exp:* USPHS fel biophys, Univ Geneva, 64-66; from asst prof to assoc prof, 66-74, chmn dept, 81, PROF MICROBIOL, UNIV CALIF, LOS ANGELES, 74-, DEAN, LIFE SCIENCES, 87- *Concurrent Pos:* Prog dir, Res Training Molecular Biol; assoc ed, Virol; ed, Fedn Europ Microbiol Socs Lett. *Mem:* Am Soc Microbiol; AAAS; Soc Gen Microbiol; Sigma Xi. *Res:* Structure of bacteria and bacterial viruses; HIV protein structures. *Mailing Add:* Dept Microbiol Univ Calif Los Angeles CA 90024-1489

EISINGER, JOSEF, b Vienna, Austria, Mar 19, 24; US citizen; m 63; c 2. BIOPHYSICS, FUORESCENCE SPECTROSCOPY. *Educ:* Univ Toronto, BA, 47, MA, 48; Mass Inst Technol, PhD(physics), 51. *Prof Exp:* Res assoc, Mass Inst Technol, 51-52, Nat Res Coun Can, 52-53 & Rice Univ, 53-54; mem tech staff, Bell Labs, 54-86; PROF DEPT PHYSIOL & BIOPHYS, MT SINAI SCH MED, 86- *Concurrent Pos:* Adj assoc prof physics, NY Univ, 60-63; Guggenheim fels, Switz, 63-64 & Eng, 78-79. *Mem:* Fel Am Phys Soc; Biophys Soc; Am Soc Biol Chemists; Am Soc Photobiol. *Res:* Excited states of biological molecules; energy transfer and emission spectroscopy; diagnostic screening tests in environmental medicine; history of medicine; membrane biology. *Mailing Add:* Dept Physiol Mt Sinai Sch Med Box 1218 Fifth Ave & 100th St New York NY 10029

EISINGER, ROBERT PETER, b New York, NY, Oct 29, 29; m 56; c 2. NEPHROLOGY, DIALYSIS. *Educ:* Swarthmore Col, AB, 51; Columbia Univ, MD, 55. *Prof Exp:* From instr to assoc prof nephrology, NY Univ Sch Med, 61-73; PROF MED-NEPHROLOGY, UNIV MED & DENT NJ, ROBERT WOOD JOHNSON MED SCH, 73-, CHIEF NEPHROLOGY, 73- *Concurrent Pos:* Chief nephrology, NY Vet Admin Med Ctr, 66-73, Robert Wood Johnson Univ Hosp, 73- *Mem:* Am Soc Nephrology; Int Soc Nephrology; Am Phys Soc. *Res:* Dialysis; clinical nephrology. *Mailing Add:* Dept Med Div Nephrology R W Johnson Med Sch UMDNJ CN-19 One Robert Wood Johnson Pl New Brunswick NJ 08903-0019

EISLER, RONALD, b Brooklyn, NY, Feb 23, 32; m 63; c 2. MARINE BIOLOGY, POLLUTION BIOLOGY. *Educ:* NY Univ, BA, 52; Univ Wash, MS, 57, PhD(fisheries biol), 61. *Prof Exp:* Asst marine lab, Univ Miami, 52-53; biol aide, US Army Med Nutrit Lab, Colo, 53-55; asst, Col Fisheries, Univ Wash, 56; aquatic biologist, NY State Conserv Dept, 57-58; asst, Lab Radiation Biol, Univ Wash, 58-61; fishery res biologist, Sandy Hook Marine Lab, US Fish & Wildlife Serv, 61-66, res aquatic biologist, Nat Marine Water Qual Lab, US Environ Protection Agency, 66-79, biosci adv, Fish Wildlife Serv, 79-85, SR RES BIOLOGIST PATUXENT WILDLIFE RES CTR, US FISH WILDLIFE SERV, US DEPT INTERIOR, 85- *Concurrent Pos:* Adj prof, Grad Sch Oceanog, Univ RI, 70-83; vis prof, Hebrew Univ Jerusalem, Israel & resident dir, Marine Biol Lab, Israel, 72-73; assoc ed, Transactions of the Am Fisheries soc, 75-77, Marine Ecol Progress Series, 80-; adj prof, Biol Dept, Am Univ, 80-89, Biol Dept, George Mason Univ, 88- *Mem:* Am Fish Soc; Marine Biol Asn UK; Israel Ecol Soc; Nat Shellfish Asn; Am Soc Zool. *Res:* Ecological aspects of coastal pollution; physiological ichthyology; aquatic toxicology; environmental contaminants evaluation. *Mailing Add:* Patuxent Wildlife Res Ctr US Fish & Wildlife Serv Laurel MD 20708

EISLEY, JOE G(RIFFIN), b Auglaize Co, Ohio, Apr 7, 28; m 56; c 2. AEROSPACE ENGINEERING. *Educ:* St Louis Univ, BS, 51; Calif Inst Technol, MS, 52, PhD(aeronaut, physics), 56. *Prof Exp:* From asst prof to assoc prof, 56-65, PROF AERONAUT & ASTRONAUT ENG, UNIV MICH, 65-, ASSOC DEAN COL ENG, 67- *Concurrent Pos:* Consult, Bendix Systs Div, 59-60, Boeing Co, 61 & Conductron Corp, 64; NSF fac fel, 62-63. *Mem:* Am Inst Aeronaut & Astronaut; Am Soc Eng Educ; Am Soc Mech Engrs; Soc Hist Technol. *Res:* Structural dynamics; nonlinear vibrations; aeroelasticity; stress analysis; engineering curriculum development; history of technology; graduate and continuing education. *Mailing Add:* Aerospace & Engr 203 AEB 2140 Univ Mich Ann Arbor MI 48109-2140

EISNER, ALAN MARK, b Brooklyn, NY, June 3, 43. EXPERIMENTAL HIGH ENERGY PHYSICS. *Educ:* Harvard Univ, BA, 64, MA, 65, PhD(physics), 71. *Prof Exp:* Asst res physicist, 71-73, asst prof physics, Univ Calif, Santa Barbara, 73-80; assoc res physicist, 81-87, RES PHYSICIST, UNIV CALIF INTERCAMPUS INST RES AT PARTICLE ACECLERATORS, 87- *Concurrent Pos:* Adj assoc prof physics, Univ Calif, San Diego, 81-87, adj prof, 87- *Mem:* Am Phys Soc. *Res:* Studies of hadronic structure using electromagnetic probes; interactions of high energy photons with matter and with each other; physics with electron-positron colliding beams. *Mailing Add:* Dept Physics B019 Univ Calif San Diego La Jolla CA 92093

EISNER, EDWARD, renewable energy technology, for more information see previous edition

EISNER, ELMER, b Poughkeepsie, NY, Mar 8, 19; m 43; c 3. EXPLORATION PHYSICS. *Educ:* Brooklyn Col, BA, 39; Johns Hopkins Univ, PhD(physics), 44. *Prof Exp:* Asst physicist, Nat Bur Standards, 43-44; asst prof physics, Rutgers Univ, 44-47; physicist, Argonne Nat Lab, 47-50; physicist, 51-60, res assoc, 60-69, sr res assoc, 69-86, EMER SR SCIENTIST, TEXACO INC, 86- *Concurrent Pos:* Adj prof math sci, Rice Univ, 88- *Mem:* AAAS; Am Phys Soc; Am Math Soc; Soc Explor Geophysicists. *Res:* Analysis of nuclear scattering; design of electronic proximity fuses; analysis of pile behavior; geophysical exploration methods; numerical analysis; inverse problems in wave equation. *Mailing Add:* 4401 Laurel Dr Houston TX 77021

EISNER, GILBERT MARTIN, b Nov 22, 32; m 70; c 2. NEPHROLOGY, HYPERTENSION. *Educ:* Harvard Col, AB, 54; Yale Univ, MD, 56. *Prof Exp:* CLIN PROF MED, & ADJ PROF PHYSIOL & BIOPHYS, SCH MED, GEORGETOWN UNIV, 79- *Concurrent Pos:* Asst med ed, Am Fam Physician; clin prof med, George Washington Sch Med, 90. *Mem:* Fel Am Col Physicians; Am Soc Hypertension; Am Fedn Clin Res; Am Soc Nephrol; Am Physiol Soc; Int Soc Nephrol. *Mailing Add:* Dept Physiol Med Sch Georgetown Univ 3900 Reservoir Rd NW Washington DC 20007

EISNER, HOWARD, b New York, NY, Aug 8, 35; m 57; c 3. ELECTRICAL ENGINEERING, OPERATIONS RESEARCH. *Educ:* City Col New York, BEE, 57; Columbia Univ, MS, 58; George Washington Univ, DSc(eng appl sci), 66. *Prof Exp:* Teaching asst elec eng, Columbia Univ, 57; substitute teacher physics, Brooklyn Col, 57-59; res engr, ORI, Inc, Rockville, MD, 59-63, prog dir eng & opers, 64-66, assoc dir eng anal div, 68-, vpres, 68-71, exec vpres, 71-85, corp exec vpres, 85- 87, pres, ORI/Intercon Systs Corp, 85-89, pres C3I div & exec vpres, ARC Prof Serv Group, 89; DISTINGUISHED RES PROF & PROF ENG MGT DEPT, GEORGE WASHINGTON UNIV, 89- *Concurrent Pos:* Lectr physics, Sch Gen Studies, Brooklyn Col, 58-59; lectr, Sch Eng & Appl Sci, George Washington, 60-64, asst prof lectr, 64-66, Col Gen Studies, 66-67; vpres, Atlantic Res Corp, 87-89. *Mem:* Fel Inst Elec & Electronic Engrs; Am Inst Aeronaut & Astronaut; Opers Res Soc Am; Inst Mgt Sci; fel NY Acad Sci. *Res:* Communications; information theory; systems. *Mailing Add:* 10837 Deborah Dr Potomac MD 20854

EISNER, MARK JOSEPH, b Poughkeepsie, NY, July 18, 38; m 62; c 2. OPERATIONS RESEARCH. *Educ:* Harvard Univ, BA, 60; Cornell Univ, PhD(opers res), 70. *Prof Exp:* Opers analyst, Res Analysis Corp, 60-65; asst prof opers res, Cornell Univ, 69-75; sect head, Math, Computers & Systs Dept, 75-79, sect head, 79-83, planning mgr, Exxon Res & Eng Co, Exxon Corp, 84-87; LEAD TELECOMMUNICATION PLANNER, EXXON CO INT, 87- *Concurrent Pos:* NSF grant, 71-74; assoc ed, Transactions, 75 & Opers Res, 78; coun mem, Inst Mgt Sci, 82-85. *Mem:* Opers Res Soc Am; Inst Mgt Sci; Am Inst Indust Engrs. *Res:* Computer project management; game theory; traffic control theory; military operations research; mathematical programming. *Mailing Add:* Exxon Co Int 200 Park Ave Florham Park NJ 07932

EISNER, PHILIP NATHAN, b Springfield, Mass, Mar 7, 34; m 60; c 1. OPERATIONS RESEARCH. *Educ:* Mass Inst Technol, BS, 55; NY Univ, PhD(physics), 69. *Prof Exp:* Sr engr, ITT Labs, 57-61; res assoc, Dewey Electronics Corp, 61-65, assoc dir, Physics Res Lab, 68-72; proj mgr laser isotope separation, Corp Res Labs, Exxon Res & Eng Co, 72-80, proj mgr, Res & Develop Strategic Planning & Technol Dept, 81-86; EISNER CONSULT, 87- *Concurrent Pos:* Instr, NY Univ, 69-70. *Mem:* AAAS; Am Phys Soc. *Res:* Ionospheric physics; ion-molecule interactions; atomic and molecular scattering of electrons; laser isotope separation; sub-micron particles; chemistry of the upper atmosphere; decision analysis, probability. *Mailing Add:* 25 Newcomb Dr New Providence NJ 07974

EISNER, ROBERT LINDEN, b Brooklyn, NY, June 21, 27; m 53; c 3. APPLIED PHYSICS, FORENSIC SCIENCE. *Educ:* Brooklyn Col, BA, 48; Univ Iowa, PhD(physics), 54. *Prof Exp:* Asst physics, Univ Iowa, 48-51, res asst, 51-53; res engr, Westinghouse Res Labs, 53-61, sr engr, 61-66, semiconductor div, 66-67, sr reliability engr, Westinghouse Astronuclear Lab, 67-71 & Westinghouse Transportation Div, 71-72; MGR TECH CONSULT, BASIC TECHNOL, INC, 73-, MGR PROD ASSURANCE SERV, 75- *Mem:* Am Inst Mining, Metall & Petrol Engrs; Am Phys Soc; Electrochem Soc. *Res:* Reliability physics in special materials applications; systems safety and reliability analyses; physics of failure; accident reconstruction; failure analysis; hazards anticipation; nondestructive testing. *Mailing Add:* Basic Technol Inc 7125 Saltsburg Rd Pittsburgh PA 15235

EISNER, THOMAS, b Berlin, Ger, June 25, 29; nat US; m 52; c 3. ZOOLOGY. *Educ:* Harvard Univ, BA, 51, PhD, 55. *Hon Degrees:* DSc, Univ Wurzburg, 82, Univ Zurich, 83. *Prof Exp:* Res assoc, Harvard Univ, 55-57; from asst prof to prof, 57-76, JACOB GOULD SCHURMAN PROF BIOL, CORNELL UNIV, 76- *Concurrent Pos:* USPHS & NSF grants, 55-; Guggenheim fels, 64 & 72; mem bd dirs, Zero Pop Growth, 69-70 & Nat Audubon Soc, 70-75; mem, Nat Coun Nature Conservancy, 69-74; vis res prof, Univ Fla, 77-78; vis scientist, Dept Entom, Sch Agr, Neth, 64-65 & Smithsonian Trop Res Lab, Barro Colo Island, 68; consult, World Environ & Resources Prog, MacArthur Found, 87- & Ctr Conserv Biol, Stanford Univ, 83-; res assoc, Archbold Biol Sta, 73-; vis prof, Hopkins Marine Lab, Stanford Univ, 79-80 & Univ Zurich, 80-81; writer & presenter, BBC Film Secret Weapons, 83; mem steering comt, Ctr Consequences Nuclear War, 83-; mem comt res opportunities biol, Nat Acad Sci, 85-, film comt, 86-, comt human rights, 87-; mem Nat Adv Coun, Monell Chem Senses Ctr, 88-; nat sponsor, Comt Concerned Scientists, 88-; mem adv coun, World Resources Inst, 88-; mem external sci adv comt, Marine Biol Lab, Woods Hole, 89-; mem task force for 90's, Am Inst Biol Sci, 90-; mem Nat Adv Coun, Survival Int, USA, 90-; mem sci adv comt, Xerces Soc, 90- *Honors & Awards:* Newcomb-Cleveland Prize, AAAS, 67; Founder's Mem Award, Entom Soc Am, 69; Archie Carr Medal, 83; Procter Prize, Sigma Xi, 86; Kenneth Roeder Mem lectr, Tufts Univ, 86; Anson L Clark Mem lectr, Univ Tex, 87; Franklin lect, Auburn Univ, 87; John Henry lect, Princeton Univ, 88; Karl Ritter von Frisch Medal, Ger Zool Soc, 88; Tyler Prize, 90. *Mem:* Nat Acad Sci; fel Am

Acad Arts & Sci; fel Royal Soc Arts; fel AAAS; fel Animal Behav Soc; fel Entom Soc Am; Am Soc Naturalists (pres, 88-89). *Res:* Insect physiology; chemical ecology; comparative behavior; biocommunication; pheromones; defensive secretions; animal behavior and evolution; the chemical language of insects; author of 5 books and over 250 technical articles. *Mailing Add:* NB&B 347 Mudd Bldg Cornell Univ Ithaca NY 14853

EISS, ABRAHAM L(OUIS), b New York, NY, Dec 28, 34; m 57; c 2. MATERIALS ENGINEERING, NUCLEAR ENGINEERING. *Educ:* Purdue Univ, BS, 55, MS, 56; Drexel Inst Technol, MS, 64. *Prof Exp:* Engr, Sylcor Div, Sylvania Elec Co, 56-59; eng specialist, Nuclear Div, Martin Co, 59-63; proj engr, Hittman Assocs, Inc, 63-65, chief mat eng & anal, 65-70; vpres eng, Mat Resources, Inc, Cockeysville, 70-72; sr mat engr, US Atomic Energy Comn, 72-76; TECH ASST TO DIR ENG, OFF RES, US NUCLEAR REGULATORY COMN, 76- *Mem:* Sci Res Soc Am; Am Inst Mining, Metall & Petrol Engrs; Am Soc Metals; Am Soc Testing & Mat; Am Soc Mech Engr. *Res:* Powder metallurgy; thermoelectrics; thermionics; nuclear power plant engineering; nuclear materials including structural, fuels, radioisotopes. *Mailing Add:* 32 River Oaks Circle Baltimore MD 21208

EISS, ALBERT FRANK, science education; deceased, see previous edition for last biography

EISS, NORMAN SMITH, JR, b Buffalo, NY, Mar 13, 31; m 75; c 3. TRIBOLOGY, SURFACE ROUGHNESS. *Educ:* Rensselaer Polytech Inst, BME, 53; Cornell Univ, MS, 59, PhD(mech eng), 61. *Prof Exp:* Process engr, E I du Pont de Nemours & Co, 53-54; res engr, Cornell Aeronaut Lab, Inc, 56-66; assoc prof, 66-77, GEORGE R GOODSON PROF MECH ENG, VA POLYTECH INST & STATE UNIV, 77- *Concurrent Pos:* Nat Sci Found sci fac fel, Imp Col, London, 70-71. *Honors & Awards:* Nat Award, Soc Tribologists & Lubrication Engrs, 91. *Mem:* Am Soc Mech Engrs; fel Soc Tribologists Lubrication Eng; Am Soc Eng Educ; Am Soc Testing & Mat. *Res:* Friction; wear; polymer tribology; friction induced noise; surface topography characterization. *Mailing Add:* Dept Mech Eng Va Polytech Inst & State Univ Blacksburg VA 24061

EISSENBERG, DAVID M(ARTIN), b Brooklyn, NY, Aug 5, 29; m 53; c 5. ENGINEERING. *Educ:* Col William & Mary, BS, 50; Mass Inst Technol, BS, 52; Univ Tenn, MS, 63, PhD(chem eng), 72. *Prof Exp:* Mem staff reactor opers, Homogeneous Reactor Exp, Oak Ridge Nat Lab, 52-53, mem staff eng res & develop, 53-72 & long-range planning, 72-76, sr develop eng proj mgt, 76-84, RES & DEVELOP PROG MGR, OAK RIDGE NAT LAB, 85- *Res:* Fluid mechanics and heat transfer of non-Newtonian suspensions; boiling heat transfer; condensation heat transfer; seawater desalination technology; energy technology; nuclear plant aging research; diagnostic engineering research. *Mailing Add:* 101 Carlton Lane Oak Ridge TN 37830

EISSENBERG, JOEL CARTER, b Johnstown, Pa, Feb 25, 55; m 77; c 1. CHROMATIN BIOCHEMISTRY. *Educ:* Univ Tenn, Knoxville, BA, 77; Univ NC, Chapel Hill, PhD(genetics), 82. *Prof Exp:* NIH trainee, Univ NC, Chapel Hill, 77-81, res asst, 81-82; res assoc, Wash Univ, St Louis, 82-83, NIH fel, 83-86; ASST PROF, ST LOUIS UNIV MED SCH, 87- *Mem:* Genetics Soc Am; Am Genetics Asn; Am Soc Cell Biol; AAAS. *Res:* Chromatin structure and chromosome organization in Drosophila; role of chromosome architecture in promoting gene expression; nucleic acid biochemistry. *Mailing Add:* Dept Biochem & Molecular Biol St Louis Univ Med Sch 1402 S Grand Blvd St Louis MO 63104

EISSLER, ROBERT L, b Evansville, Ind, Feb 8, 21; m 48; c 4. PHYSICAL CHEMISTRY. *Educ:* Evansville Col, BS, 49; Univ Ill, MS, 56, PhD(chem), 60. *Prof Exp:* Asst chem engr, Ill State Geol Surv, 52-60, assoc chemist, 60-61; mgr graphic arts res, Ball Bros Res Corp, 61-65; PRIN CHEMIST, NORTHERN REGIONAL LAB, AGR RES SERV, USDA, 65- *Concurrent Pos:* Vis prof, Ball State Teachers Col, 62-63. *Mem:* Am Chem Soc; Sigma Xi. *Res:* Surface chemistry; coal and petroleum; photoengraving and lithographing processes; organic and inorganic coatings; emulsions; photosynthesis; supercritical fluid extraction. *Mailing Add:* 1033 Brookview Lane Peoria IL 61615

EISSNER, ROBERT M, b Newark, Del, Nov 10, 26; m 57; c 3. STATISTICS. *Educ:* Univ Del, BA, 48, MA, 53. *Prof Exp:* Mathematician, Comput Lab, Aberdeen Proving Ground, Md, 48-50, supvry math statistician, Surveillance & Reliability Lab, 50-70, supvry phys scientist, US Army Mat Systs Anal Activity, 70-85; consult, Test & Eval Int, Inc, 85-87; RETIRED. *Res:* Design and evaluation of experiments to study the reliability, availability and maintainability characteristics of developmental and fielded US Army weapons, munitions, communications and electronic equipment. *Mailing Add:* 49 Kells Ave Newark DE 19711

EISZNER, JAMES RICHARD, chemistry; deceased, see previous edition for last biography

EITEN, GEORGE, b Morristown, NJ, Nov 20, 23; m. VEGETATION SCIENCE, ANGIOSPERM TAXONOMY. *Educ:* Cornell Univ, AB, 49; Columbia Univ, MS, 58, PhD(bot), 59. *Prof Exp:* Prof biol, Willimantic Teachers Col, 57-58; prof bot genetics, Hofstra Col, 58-59; botanist, Inst de Bot, Sao Paulo, 59-75; PROF BOT & ECOL, UNIV BRASILIA, 75- *Concurrent Pos:* Plant collecting for herbarium in many countries in var continents; consult for biomass study in area in Mato Grosso to be inundated by dam, Engevix s a, Brasilia. *Mem:* Asn Trop Biol; Soc Bot do Brasil. *Res:* Description of vegetation of the world; phytosociology of Brazilian vegetation, principally the cerrado; phytosociological concepts; taxonomy of the genus Oxalis. *Mailing Add:* Univ de Brasilia BOT-IB Caixa Postal 153081 Brasilia DF 70910 Brazil

EITENMILLER, RONALD RAY, b Pekin, Ill, Nov 23, 44; m 66; c 2. FOOD SCIENCE. *Educ:* Univ Ill, BS, 66; Univ Nebr, MS, 68, PhD(food sci), 71. *Prof Exp:* From asst prof to assoc prof, 71-81, PROF FOOD SCI, UNIV GA, 81- *Mem:* Inst Food Technologists; Am Dairy Sci Asn; Sigma Xi. *Res:* Role of enzymes in quality, storage stability and safety of foods; nutritional evaluation of foods; vitamin assay methodology. *Mailing Add:* Dept Food Sci Univ Ga Athens GA 30602

EITTREIM, STEPHEN L, b St Petersburg, Fla, Mar 2, 41; m 64; c 3. MARINE GEOPHYSICS, MARINE GEOLOGY. *Educ:* Colby Col, AB, 63; Columbia Univ, PhD(marine geophys), 70. *Prof Exp:* Res asst, Lamont-Doherty Geol Observ, 63-70, res assoc, 70-75; GEOLOGIST, US GEOL SURV, 75- *Mem:* Am Geophys Union; AAAS; Geol Soc Am. *Res:* Marine geology and geophysics; marine seismic reflection and tectonics; oceanography; suspended particulate matter studies. *Mailing Add:* US Geol Surv MS 999 345 Middlefield Rd Menlo Park CA 94025

EITZEN, DONALD GENE, b Hillsboro, Kans, Oct 21, 42; m 64; c 2. ENGINEERING MECHANICS, ACOUSTICS. *Educ:* Univ NMex, BS, 64, MSc, 66; Univ Wash, PhD(eng mech), 71. *Prof Exp:* Staff mem shock testing, Sandia Labs, 64-67; mech engr, 71-75, LEADER ULTRASONICS, NAT BUR STANDARDS, 75- *Concurrent Pos:* Teaching assoc, Univ Wash, 70-71. *Mem:* Am Soc Mech Engrs; Am Soc Testing & Mat. *Res:* Research on methods for characterizing ultrasonic transducers and on reference artifacts for ultrasonic measurement; basic research on acoustic emission. *Mailing Add:* Sound A147 Nat Inst Standards & Technol Gaithersburg MD 20899

EITZMAN, DONALD V, b Madison, Wis, June 6, 27; m 54; c 5. MEDICINE, PHYSIOLOGY. *Educ:* Northeastern Univ Ill, BS, 50; Univ Iowa, MD, 54. *Prof Exp:* Instr pediat, Univ Minn, 56-57; instr, Univ Tex Health Sci Ctr, Dallas, 57-58; from asst to assoc prof, 58-68, PROF PEDIAT, UNIV FLA, 68- *Concurrent Pos:* NIH trainee immunol, 57-58, fel, 58-60; Daland fel clin med, 60-64; spec fel with Dr K Cross, 67-68. *Mem:* Soc Pediat Res; Am Pediat Soc; Am Physiol Soc. *Res:* Newborn physiology; ontogeny of the immune response in infants; comparative immunology in primates; perinatal physiology; acid base control; control of pulmonary blood flow. *Mailing Add:* Dept Pediat Univ Fla Med Col Box J-296 JHMHC Gainesville FL 32610

EK, ALAN RYAN, b Minneapolis, Minn, Sept 5, 42; m 64; c 2. FORESTRY, RESOURCE MANAGEMENT. *Educ:* Univ Minn, BS, 64, MS, 65; Ore State Univ, PhD(forestry), 69. *Prof Exp:* Res off, Can Dept Forestry & Rural Develop, 66-69; from asst prof to assoc prof forestry, Univ Wis-Madison, 75-77; from assoc prof to prof, Col Forestry, 77-84, PROF & HEAD, DEPT FOREST RESOURCES, UNIV MINN, ST PAUL, 84- *Concurrent Pos:* Consult, forest indust & govt agencies; prin, Forestronics, Inc. *Mem:* Biomet Soc; Soc Am Foresters; Am Soc Photogram & Remote Sensing; Int Union Forest Res Orgn; Sigma Xi; Am Statist Asn. *Res:* Mensuration; sampling; forest inventory and survey design; biomathematical modeling; quantitative silviculture; forest growth modelling; research planning. *Mailing Add:* Col Nat Resources Univ of Minn St Paul MN 55108

EKBERG, CARL E(DWIN), JR, b Minneapolis, Minn, Oct 28, 20; m 44; c 4. CIVIL ENGINEERING. *Educ:* Univ Minn, BCE, 43, MS, 47, PhD, 54. *Prof Exp:* Instr math & mech, Univ Minn, 46-51; asst prof civil eng, NDak Agr Col, 51-53; from asst prof to assoc prof, Lehigh Univ, 53-59; head dept, 59-85, prof, 59-88, EMER PROF CIVIL ENG, IOWA STATE UNIV, 88- *Mem:* Am Soc Civil Engrs; hon mem Am Concrete Inst; Am Soc Eng Educ; Am Rwy Eng Asn; Nat Soc Prof Engrs. *Res:* Structural engineering with particular emphasis on prestressed concrete; continuing study of composite floor system utilizing light-gage steel. *Mailing Add:* 111 Lynn Ave Apt 902 Ames IA 50010

EKBERG, DONALD ROY, b Hinsdale, Ill, Dec 23, 28; m 61; c 2. ENVIRONMENTAL PHYSIOLOGY. *Educ:* Univ Ill, BS, 50, PhD(physiol), 57; Univ Chicago, MS, 52. *Prof Exp:* Instr physiol, Univ Ill, 55-58; physiologist, Gen Elec Co, 58-65, mgr life sci, 65-71, mgr appl sci, 71-76; chief, Environ & Tech Serv Div, Southeastern Region, 76-85, southeastern regional sci coordr, 85-90, CHIEF, COOP PROGS DIV, NAT MARINE FISHERIES SERV, NAT OCEANIC & ATMOSPHERIC ADMIN, 90- *Concurrent Pos:* Fel, US Pub Health, Ger, 59-60; adj prof, Drexel Univ, 69-71. *Mem:* Am Physiol Soc; Soc Gen Physiol; Am Soc Zool; Sigma Xi. *Res:* Marine environmental biology; environmental assessment; aviation and space physiology. *Mailing Add:* Duval Bldg 9450 Koger Blvd St Petersburg FL 33702

EKDALE, ALLAN ANTON, b Burlington, Iowa, Aug 30, 46; m 69; c 2. PALEOECOLOGY. *Educ:* Augustana Col, BA, 68; Rice Univ, MA, 72, PhD(geol), 74. *Prof Exp:* from asst prof to assoc prof, 74-85, PROF GEOL, UNIV UTAH, 85- *Mem:* Soc Econ Paleontologists & Mineralogists; Paleont Soc; AAAS; Geol Soc Am; Int Asn Sedimentologists. *Res:* Trace fossils and bioturbation of sediments; ecology and paleoecology of marine mollusks. *Mailing Add:* Dept Geol & Geophys Univ Utah Salt Lake City UT 84112-1183

EKENES, J MARTIN, b Holt, Norway, Dec 5, 50; US citizen; m 75; c 8. CONTINUOUS CASTING, PROCESS METALLURGY. *Educ:* Univ Wash, BS, 77. *Prof Exp:* Ingot plant metallurgist, Aluminum Co Am, 77-80, sr metallurgist, 80-81; concentrator engr, Phelps Dodge Corp, 81-82, financial anal & engr, 82-83; res & develop scientist, 84-88, DIR TECHNOL, WAGSTAFF ENG, INC, 88- *Mem:* Am Inst Mech Engrs; Am Soc Metals. *Res:* Development of continuous and semi-continuous casting processes for aluminum and other non-ferrous metals and their alloys; air-cushion casting techniques applied to develop superior metallurgical structures in as-cast products. *Mailing Add:* N 3910 Flora Rd Spokane WA 99216

EKERDT, JOHN GILBERT, b Port Washington, Wis, May 17, 52; m 81; c 2. CATALYSIS, CHEMICAL VAPOR DEPOSITION. *Educ:* Univ Wis, Madison, BS, 74; Univ Calif, Berkeley, PhD(chem eng), 79. *Prof Exp:* Asst prof chem eng, Univ Tex, Austin, 79-85, assoc prof, 85-88, GRAD ADVISOR, 85-, PROF CHEM ENG, 88- *Mem:* Am Inst Chem Eng; Am Chem Soc; Am Soc Eng Educ; N Am Catalysis Soc. *Mailing Add:* Dept Chem Eng Univ Tex Austin TX 78712-1104

EKERN, PAUL CHESTER, b Ardmore, Okla, July 2, 20; m 50, 56; c 4. SOIL PHYSICS & HYDROLOGY, EVAPOTRANSPIRATIONS. *Educ:* Westminster Col, Mo, BA, 42; Univ Wis, PhD, 50. *Prof Exp:* Instr soils & meteorol, Univ Wis, 50-52, asst prof, 52-55; soil physicist, Pineapple Res Inst, 55-63; prof soils & agron, Univ Hawaii & Hydrologist, Water Resources Res Ctr, 64-87; RETIRED. *Mem:* AAAS; Soil Sci Soc Am; Am Meteorol Soc; Soil Conserv Soc Am; Am Geophys Union; Am Water Resources Asn. *Res:* Consumptive use of moisture in evapotranspiration; micrometeorology of mulches and tillage; soil erosion. *Mailing Add:* 3133 Huelani Pl Honolulu HI 96822

EKERN, RONALD JAMES, b Rice Lake, Wis, Dec 22, 38; m 67; c 2. SURFACE PHYSICS, ELECTROCHEMICAL SYSTEMS. *Educ:* Univ Wis-Milwaukee, BS, 68, MS, 70; Clarkson Col Technol, PhD(physics), 75. *Prof Exp:* Fel surface sci, Argonne Nat Lab, 74-76; proj scientist, Gould Labs, 76-79; MGR, RAYOVAC CORP, 79- *Mem:* Am Phys Soc; Am Vacuum Soc; Sigma Xi. *Res:* Ultra-high vacuum technology; ion accelerators for surface physics research; surface electrochemistry; electrochemical systems research and development; problem definition and analysis for active, inactive and envelope materials; corrosion, compatibility, performance domain of materials for batteries. *Mailing Add:* Rayovac Corp 601 Rayovac Dr Madison WI 53711-0960

EKLOF, PAUL CHRISTIAN, b Brooklyn, NY, Dec 28, 42; m 72; c 2. ALGEBRA, LOGIC. *Educ:* Columbia Col, AB, 64; Cornell Univ, PhD(math), 68. *Prof Exp:* J W Gibbs instr math, Yale Univ, 68-70; asst prof, Stanford Univ, 70-73; assoc prof, 73-78, PROF MATH, UNIV CALIF, IRVINE, 78- *Concurrent Pos:* Vis prof, Simon Fraser Univ, 85. *Mem:* Am Math Soc; Asn Symbolic Logic. *Res:* Logic and algebra; model theoretic and set theoretic algebra; co-author of monograph on applications of set theory to algebra. *Mailing Add:* Dept Math Univ Calif Irvine CA 92717

EKLUND, CURTIS EINAR, microbiology; deceased, see previous edition for last biography

EKLUND, DARREL LEE, b Ottawa Co, Kans, July 28, 42; m 65; c 3. APPLIED STATISTICS. *Educ:* Kans State Univ, BS, 64, MS, 66, PhD(statist), 71. *Prof Exp:* Surv statistician, Nat Ctr Health Statist, 66-68; asst prof agron & statist, Univ Mo-Columbia, 72-77; chief, Res & Anal Sect, Kans Dept Health & Environ, 77-84, MGR WATER RESOURCE, RES & DATA SECT, KANS WATER OFF, 84- *Mem:* Am Statist Asn; Am Asn Vital Records & Pub Health Statist; Sigma Xi. *Res:* Statistical computation; survey sampling; recreational use surveys; human nutrition; perinatal mortality. *Mailing Add:* 6410 SW 26th Ct Topeka KS 66614

EKLUND, KARL E, b New York, NY, July 3, 29; m 67, 81; c 2. APPLIED PHYSICS. *Educ:* Mass Inst Technol, BS, 50; Columbia Univ, MA, 56, PhD(physics), 60. *Prof Exp:* Physicist, Army Nuclear Defense Labs, 51-54; dir res, Radiation Dynamics, Inc, 60-61; res assoc physics, Columbia Univ, 62; asst dir nuclear struct lab, Yale Univ, 63-65; cur phys labs, State Univ NY Stony Brook, 65-66, dir, 66-68, asst to exec vpres systs, 68-69, dir budget, 70-71, asst to exec vpres, 71; PRIN ASSOC, EKLUND ASSOCS, 71- *Concurrent Pos:* Consult, Parameters, Inc, 54-60; lectr, Stratos Div, Fairchild, 56-58; assoc prof, US Merchant Marine Acad, 62; aerospace consult, 62-63; energy coordr, Southeast Mass Regional Planning and Econ Develop Dist, 80-82; consult, Div Hazardous Waste, Dept Environ Qual Eng, Mass; pres, Environ Restoration Eng Inc, 87-89; vpres, Ashland Indust Fuel Corp, 86-89, Am Reclamation Corp, 87-89. *Res:* Nuclear structure physics; research management and administration; management systems; energy management; educational theory; matrix theory of human behavior with applications to collective behavior; solid and hazardous waste management; environmental remediation; recycling. *Mailing Add:* Eklund Assocs 76 Myricks St Berkley MA 02779

EKLUND, MELVIN WESLEY, b Saco, Mont, July 16, 33; m 60; c 2. MICROBIOLOGY, FOOD MICROBIOLOGY. *Educ:* Wash State Univ, BS, 55, MS, 57; Purdue Univ, PhD(food sci & microbiol), 62. *Prof Exp:* Supvry res microbiologist, Utilization Div, 61-88, DIR RES, NORTHWEST & ALASKA FISHERIES CTR, US DEPT COM, 88- *Concurrent Pos:* Dept of Com rep, US-Japan Coop Prog Dev & Utilization of Nat Res, 75-; consult nutrit comt, Nat Acad Sci, 81; affil assoc prof, Univ Wash, 80- *Honors & Awards:* Gold Medal, US Dept Com, 83. *Mem:* Sigma Xi; Am Acad Microbiol; Am Soc Microbiol; Inst Food Technol; Soc Appl Bact. *Res:* Clostridium botulinum incidence in marine environment; heat destruction and factors affecting toxin production; yeasts in foods; bacteriophages and plasmids and toxicity of C botulinum, C novyi and other Clostridia; fish diseases, Vibrio parahaemolyticus; sudden infant death; infant botulism; botulism as major cause of juvenile fish mortality in US and development of methods for preventing and controlling botulism in fish; efficacy of nitrites as inhibitor of C botulinum and search for replacements for nitrites; smoke components as C botulinum inhibitor; bacteriocin production in Clostridia and relationship to plasmids; transformation studies with C botulinum and related species; listeria monocytogenes research-incidence; inactivation and inhabition in foods; development of new isolation procedures. *Mailing Add:* 18727 35th Ave NE Seattle WA 98155

EKMAN, CARL FREDERICK W, b Caribou, Maine, Feb 13, 32; m 59; c 1. INORGANIC CHEMISTRY, SOLID STATE CHEMISTRY. *Educ:* Northeastern Univ, BS, 55; Mass Inst Technol, PhD(inorg chem), 61. *Prof Exp:* Sect head inorg chem res, Itek Corp, 61-65, mgr res lab, 65-66; vpres & dir res & develop, Carter's Ink Co, 66-77; mem staff, 77-80, vpres prod, 81-88, VPRES & DIR RES, COULTER SYSTS CORP, 80-; VPRES MFG, STORK BEDFORD BV. *Mem:* AAAS; Am Chem Soc; fel Am Inst Chem; Soc Photog Scientists & Engrs; NY Acad Sci; Sigma Xi. *Res:* Physical inorganic chemistry of divalent silver; solid state photochemistry; chemical and physical consequences of actinic light on solids, particularly related to image forming systems; physics; radio frequency sputtering. *Mailing Add:* Stork Bedford BV 35 Wiggins Ave Bedford MA 01730

EKMAN, FRANK, b London, Eng, Mar 26, 17; US citizen; m 44; c 3. POLLUTION ENGINEERING, CHEMICAL ENGINEERING. *Educ:* Univ BC, BASc, 44, MASc, 46; Univ Ill, PhD(chem eng), 50. *Prof Exp:* Mem staff, MacMillian Indust, Ltd, 44-45; asst, Univ Ill, 46-48; mem staff, Howard Smith Paper Mills, Ltd, 49-51; Nat Lead Co, 51-58; asst mgr res & develop, Glidden Co, 58-59; mem staff, Pittsburgh Plate Glass Co, 59-64; res eng, Babcock & Wilcox Co, 64-66; tech dir, Environeering Inc, 66-74; sr educ consult, Inst Gas Technol, 74-77; sr develop eng, Barber-Greene Co, 77-79; consult, 79-81; regional partic & sulfur dioxide expert, US Environ Protection Agency, 81-87; RETIRED. *Honors & Awards:* President's Award, Indust Gas Cleaning Inst, 74; Am Foundrymen's Soc Award, 70. *Mem:* Am Chem Soc; Air Pollution Control Asn. *Res:* Industrial air pollution abatement. *Mailing Add:* RR5 Tamarack Lane Barrington IL 60010-9805

EKNOYAN, GARABED, MEDICAL. *Prof Exp:* PROF MED NEPHROLOGY, BAYLOR COL MED, 68- *Mailing Add:* Dept Med Baylor Col Med One Baylor Plaza Houston TX 77030

EKNOYAN, OHANNES, b Jan 15, 44; US citizen. SEMICONDUCTOR DEVICES, ELECTROOPTICS. *Educ:* Tex A&M Univ, BS, 69, MS, 70, Columbia Univ, MPhil & PhD(elec eng), 75. *Prof Exp:* Mem tech staff, Microwave Semiconductor Devices, Bell Telephone Lab, Murray Hill, NJ, 72-73; fac res high temperature semiconductor, Sandia Nat Lab, Albuquerque, 79-80; from asst prof to assoc prof, 75-87, PROF ELEC ENG, TEX A&M UNIV, 88- *Concurrent Pos:* Vis scientist, Naval Res Lab, Washington, DC, 83-85; sr fel, Tex Eng Exp Sta. *Mem:* Inst Elec & Electronic Engrs; Sigma Xi; Optical Soc Am; Am Phys Soc. *Res:* Semiconductor devices, fabrication and characterization; all phases of processes from initial material preparation to the final characterization of packaged devices; integrated optics design, fabrication and characterization. *Mailing Add:* Elec Eng Dept Tex A&M Univ College Station TX 77843-3128

EKRAMODDOULLAH, ABUL KALAM M, b Sendra, Bangladesh, Sept 9, 39; Can citizen; m 70; c 3. FOREST BIOTECHNOLOGY, ALLERGY. *Educ:* Dhaka Univ, BS, 59, MS, 61; McGill Univ, PhD(immunol), 68. *Prof Exp:* Asst prof biochem, Dhaka Univ, Bangladesh, 68-70; head qual control pharm, Squibb Corp, Bangladesh, 70-72; sr res assoc immunol, Univ Man, Can, 72-88; RES SCIENTIST FOREST BIOTECHNOL, PAC FORESTRY CTR, FORESTRY CAN, VICTORIA, 88- *Concurrent Pos:* Mem comt-at-large, Allergen Nonmenclature, Int Union Immunol Socs, 86-, Allergen Standardization, 86-; mem, Biotechnol Grant Selection Comt Sci Coun, BC, 89-; adj assoc prof forest biol, Univ Victoria, 90- *Mem:* Am Asn Immunologists; NY Acad Sci; Am Electrophoresis Soc; Can Soc Immunol; Can Soc Plant Path; Int Electrophoresis Soc; Can Phytopath Soc. *Res:* Molecular analysis of fungal disease resistance of coniferous trees with a view to developing biochemical marker for screening resistant tree during breeding and possible genetic engineering of disease resistant tree. *Mailing Add:* Pac Forestry Ctr Forestry Can 506 W Burnside Rd Victoria BC V8Z 1M5 Can

EKSTEDT, RICHARD DEAN, b East St Louis, Ill, Nov 13, 25; m 55; c 1. MICROBIOLOGY, IMMUNOLOGY. *Educ:* Wash Univ, AB, 49; Univ Mich, MS, 51, PhD(bact), 55. *Prof Exp:* Res assoc biochem, Sch Med, Univ Ill, 54-55; res assoc med, 55-56, from instr to prof microbiol, 56-87, EMER PROF, MED SCH, NORTHWESTERN UNIV, 87- *Concurrent Pos:* Helen Hay Whitney Found fel, 58-61. *Mem:* Am Soc Microbiol; Am Asn Immunol. *Res:* Mechanisms of microbial pathogenicity; nonspecific host defense mechanisms; bacteriocidal activity of blood; immunity to staphylococci; lectins in tumor and immunobiology; host parasite relationships. *Mailing Add:* 2111 Lincoln St Evanston IL 60201-2281

EKSTRAND, KENNETH ERIC, b Chicopee, Mass, Sept 3, 42; m 70; c 1. MEDICAL PHYSICS. *Educ:* Mass Inst Technol, BS, 64; Cornell Univ, PhD(physics), 71; Am Bd Radiol, dipl, 76. *Prof Exp:* Res assoc physics, Nat Comn Nuclear Energy, Frascati, Italy, 71-72; res assoc, 72-73, from instr to asst prof, 73-78, ASSOC PROF RADIOL, BOWMAN GRAY SCH MED, WAKE FOREST UNIV, 78- *Mem:* Am Asn Physicists Med; Am Col Radiol; Am Soc Therapeut Radiol; Sigma Xi. *Res:* Radiological physics; use of computers in radiology and radiation therapy; medical applications of nuclear magnetic resonance. *Mailing Add:* Bowman Gray Sch Med Wake Forest Univ Winston-Salem NC 27103

EKSTROM, LINCOLN, b Providence, RI, Aug 21, 32; m 57. PHYSICAL CHEMISTRY. *Educ:* Brown Univ, ScB, 53; Mass Inst Technol, PhD(chem), 57. *Prof Exp:* MEM TECH STAFF, LABS, RCA CORP, 57- *Honors & Awards:* David Sarnoff Outstanding Achievement Award, 63. *Mem:* AAAS; Am Chem Soc; Am Phys Soc; fel Am Inst Chem. *Res:* Thermoelectric materials; compound semiconductors; physical chemistry of compound semiconductors; absorption spectrophotometry of ionic solutions; magnetic recording materials. *Mailing Add:* 78 Westerly Rd Princeton NJ 08540

EKSTROM, MICHAEL P, b Nov 10, 39; m; c 3. TWO DIMENSIONAL DIGITAL SIGNAL PROCESSING, IMAGE PROCESSING. *Educ:* Ariz State Univ, BS, 62; Univ Calif, Davis, PhD, 71. *Prof Exp:* Engr, Field Test Div, Lawrence Livermore Lab, 62-72, group & sect leader, Eng Res Div, 72-76, assoc div leader, 76-78; Humboldt scientist, Inst Tele Commun Eng, Univ Erlangen, Nurnberg, 78-79; lectr extension short course, Univ Calif, Los Angeles, 79-80; sci consult, Schlumberger Doll Res, Ridgefield, Conn, 80-82, dir, Mech & Elec Dept, 82-84; vpres eng, Schlumberger Prod Design, Clamart, France, 84-87; vpres res & eng, Wireline, Seismics & Testing, 87-90,

VPRES ENGR, SCHLUMBERGER WELL SERVICES, 90- *Concurrent Pos:* Lectr, Univ Extension Short Course, Univ Calif, Berkeley, 76-77, & co-organizer, 77, Eng Extension Short Course, Rensselaer Polytech Inst, 77; assoc eng, Dept Elec Eng, Univ Calif, Davis, 69-70; guest lectr, Dept Elec Eng & Comput Sci, Univ Santa Clara, 76-78; tech prog chmn, Ninth Asilomar Conf Circuits, Systs & Comput, 75, gen chmn, Tenth Asilomar Conf, 76; organizer & chmn, Lawrence Symp Systs & Decision Sci, 77, First ASSP Workshop Two-Dimensional Digital Signal Processing, 79 & tech comt, Multidimensional Signal Processing, 81-84, Second Workshop Two-Dimensional Signal Processing, Mohonk Mountain, 81; mem, tech prog comt digital signal processing, Circuits & Systs Soc, Inst Elec & Electronic Engrs, 76-78. *Honors & Awards:* Alexander von Humboldt Sr Sci Award, 78; Centennial Medal, Inst Elec & Electronic Engrs, 84, Sr Award, ASSP, 76, 82. *Mem:* Fel Inst Elec & Electronic Engrs; Speech & Signal Processing Soc (secy-treas, 84). *Res:* Two-dimensional digital signal processing; applied estimation theory; applications of signal processing geology and geophysics; technical management; author of over 20 publications. *Mailing Add:* Schlumberger Well Serv PO Box 2175 Houston TX 77252-2175

EKVALL, SHIRLEY W, b Minden, Nebr, July 25, 34; c 2. HUMAN NUTRITION, NUTRITONAL BIOCHEMISTRY. *Educ:* Univ Nebr, Lincoln, BS, 56; Univ Calif, Berkeley, MS, 66; Univ Cincinnati, PhD(physiol & nutrit educ), 78. *Prof Exp:* Dietetic intern, Univ Wis-Madison, 56-57; res dietician, Univ Calif Med Ctr, Los Angeles, 57-59, pediat dietician, San Francisco, 59-63; res dietician, Children's Hosp, Oakland, 63-64; res asst, Univ Calif, Berkeley, 64-65; nutrit researcher, 66-67, from instr to prof nutrit, Univ Cincinnati, 69-85; CHIEF NUTRIT SERV, UNIV AFFIL CINCINNATI CTR DEVELOP DISORDERS, 69- *Concurrent Pos:* Interdisciplinary coun deleg nutrit, Nat Univ Affil Fac Develop Disabilities Res Comt, 71-73 & 74-76; chmn, Nat Nutrit Workshop, Cincinnati, 74, 84, 85, 86, 87 & 88; Am Asn Mental Deficiency (nat chmn, nutrit dietetics, 84-86); Nat Nutrit Comn, Nat Nutrit Workshop, 80; regional nutrit rep, Am Dietary Asn, Am Soc Allied Health Prof, 81. *Honors & Awards:* Am Asn Mental Deficiency Award, 86. *Mem:* AAAS; Am Dietetic Asn; fel Am Asn Mental Deficiency; Soc Nutrit Educ; Univ Affil Fac Handicapped; fel Am Col Nutrit. *Res:* General pediatrics; developmental disorders ranging from hereditary metabolic disorders to nutrient deficiencies; human nutritional biochemistry of the pediatric age range. *Mailing Add:* Univ Affil Cincinnati Ctr Develop Disorders 3300 Elland Ave Cincinnati OH 45229

EL-AASSER, MOHAMED S, b Shobrakhit, Egypt, Feb 10, 43; m 72; c 2. EMULSION POLYMERS, COLLOID SCIENCE. *Educ:* Alexandria Univ, BSc, 62, MSc, 66; McGill Univ, PhD(chem), 72. *Prof Exp:* from asst prof, 74-82, PROF CHEM ENG, LEHIGH UNIV & LEHIGH UNIV & CO-DIR, EMULSION POLYMERS INST, 82- *Concurrent Pos:* Consult, Du Pont, 85-, S C Johnson, 86- *Mem:* Am Chem Soc; Am Inst Chem Engrs; Sigma Xi; AAAS. *Res:* Emulsion polymers and latex technology; preparation of homo and co-polymer latexes; colloidal stability; surface characterization; kinetics of emulsion polymerization; emulsions; film formation and drying of latexes; electrodeposition. *Mailing Add:* Dept Chem Eng Emulsion Polymers Inst Lehigh Univ Mountaintop Campus Bldg A Bethlehem PA 18015

EL-ABIAD, AHMED H(ANAFI), electrical engineering; deceased, see previous edition for last biography

ELACHI, CHARLES, b Apr 18, 47. AERONAUTICS ADMINISTRATION. *Educ:* Univ Grenoble, France, BSc, 68; Inst Polytech Grenoble, France, dipl ing, 68; Calif Inst Technol, MS, 69, PhD(elec sci), 71; Univ Southern Calif, MBA, 78; Univ Calif, Los Angeles, MS, 83. *Prof Exp:* Res fel, Calif Inst Technol, 71-74, sr scientist, Jet Propulsion Lab, 71-74, leader, Radar Remote Sensing Team, 74-80, SR RES SCIENTIST, JET PROPULSION LAB, CALIF INST TECHNOL, 81-, ASST LAB DIR SPACE SCI & INSTRUMENTS, 87- *Concurrent Pos:* Mgr radar develop, Jet Propulsion Lab, Calif Inst Technol, 80-87, lectr, 82-, mgr, Earth & Space Sci Div, 84-87; prin investr var res studies & develop, NASA, 73-87; proj scientist, Spaceborne Imaging Radar Proj, 85-89; mem, Solar Syst Explor Comt Coun, NASA, 88-, Astrophys Coun, 88-; mem, Electromagnetic Acad, 90-95. *Honors & Awards:* Prof R W P King Award for Outstanding Contrib in Field of Electromagnetics, 73; Autometric Award, Am Soc Photogram, 80; Except Sci Achievement Medal, NASA, 82; Caltech Watson Lectr, 83; William T Pecora Award, 85; Geosci & Remote Sensing Distinguished Achievement Award, Inst Elec & Electronics Engrs, 87. *Mem:* Nat Acad Eng; fel Inst Elec & Electronics Engrs; Am Astronaut Soc; Am Phys Soc; Am Geophys Union; Sigma Xi; Electromagnetic Soc; Planetary Soc; Am Inst Aeronaut & Astronaut. *Res:* Use of spaceborne active microwave instruments and remote sensing of planetary surfaces spheres and subsurfaces; analysis and interpretation of radar imagery of planetary surfaces; wave propagation and scattering; electromagnetic theory; lasers and integrated optics; author of 200 publications; granted 4 patents. *Mailing Add:* Jet Propulsion Lab Calif Inst Technol 4800 Oak Grove Dr MS 180-704 Pasadena CA 91109

ELAKOVICH, STELLA DAISY, b Texarkana, Tex, Jan 15, 45. ORGANIC CHEMISTRY. *Educ:* Tex Christian Univ, BS, 66; La State Univ, PhD(chem), 71. *Prof Exp:* Instr chem, La State Univ, 71; fel, Univ Saarland, Ger, 71-72 & Univ Tex Med Br, Galveston, 72-74; asst prof, 74-78, assoc prof, 78-87, PROF CHEM, UNIV SOUTHERN MISS, 87- *Concurrent Pos:* Int asst vpres acad affairs, Univ Southern Miss, 88-89. *Mem:* Am Chem Soc; Sigma Xi. *Res:* Chemical education; organic natural products. *Mailing Add:* Dept Chem Univ Southern Miss Box 5043 Hattiesburg MS 39406-5043

ELAM, JACK GORDON, b Glendale, Calif, Aug 25, 21; m 74; c 4. GEOLOGY. *Educ:* Univ Calif, Los Angeles, AB, 43, MA, 48; Rensselaer Polytech Inst, PhD(geol), 60. *Prof Exp:* Geologist, Stanley & Stolz, Calif, 46-47, Richfield Oil Corp, 47-49 & Cameron Oil Co, Calif & Tex, 49-51; consult geologist, 51-56; asst prof petrol geol, Rensselaer Polytech Inst, 56-60; CONSULT GEOLOGIST, 60-; PRES, JACK G ELAM, INC, 83- *Concurrent Pos:* Independent oil producer, 64-; pres, Permian Basin Grad Ctr, 70-79, chmn bd, 79-81; gen partner, Explor Ltd, 73-78; dir & vpres, Keba Oil & Gas Co, 75-81; adj prof, Univ Tex, Arlington, 78-80. *Honors & Awards:* Pub Serv Award, Am Asn Petrol Geologists, 86. *Mem:* Am Asn Petrol Geologists; Geol Soc Am. *Res:* Petroleum geology; sedimentology, particularly carbonate sedimentology; structural geology and global tectonics; tectonic evolution of the Permian Basin. *Mailing Add:* 219 N Main St PO Box 195 Midland TX 79702

ELAM, LLOYD C, b Little Rock, Ark, Oct 27, 28; m 57; c 2. MEDICINE. *Educ:* Roosevelt Univ, BS, 50; Univ Wash, MD, 57; Am Bd Neurol & Psychiat, cert, 65. *Hon Degrees:* LLD, Harvard Univ, 73; DSc, St Lawrence Univ, 74; LHD, Roosevelt Univ, 74. *Prof Exp:* Intern, Univ Ill Hosp, Chicago, 57-58; resident psychiat, Univ Chicago, 58-61; asst prof & chmn dept psychiat, 61-63, prof & chmn dept, George W Hubbard Hosp, 63-68, interim dean, Sch Med, 66-68, pres, 68-81, chancellor, 81-82, DISTINGUISHED SERV PROF, DEPT PSYCHIAT, MEHARRY MED COL, NASHVILLE, TENN, 82- *Concurrent Pos:* NIMH fel, Riverside Hosp, Nashville, 58-61, consult, 62-67; lectr, Univ Tenn, 63-67. *Mem:* Inst Med-Nat Acad Sci; Am Col Psychiatrists; Am Psychiat Asn; Nat Med Asn; AMA. *Res:* Comprehensive health care; author of numerous technical publications. *Mailing Add:* 710 Ledford Dr Nashville TN 37207

ELAM, WILLIAM WARREN, b Alamo, Ga, May 1, 29; m 54; c 3. FOREST PHYSIOLOGY. *Educ:* Univ Ga, BS, 61, PhD(plant sci), 65. *Prof Exp:* Asst prof bot & forestry, 65-70, assoc prof forestry, 70-85, PROF FORESTRY, MISS STATE UNIV, 85- *Concurrent Pos:* Fire specialist, US Forest Serv, 72. *Mem:* Soc Am Foresters; Bot Soc Am. *Res:* Tree physiology, especially growth and development in Southern conifers; hardwood regeneration; prescribed burning; physiology of seeds and seedlings. *Mailing Add:* Dept Forestry Miss State Univ PO Drawer 5328 Mississippi State MS 39762-5726

ELAND, JOHN HUGH DAVID, b Salop, Eng, Aug 6, 41; m 67; c 3. PHYSICAL CHEMISTRY, PHYSICS. *Prof Exp:* Res lectr chem, Christ Church, 68-73; fel, Univ Freiburg, Ger, 73-74; assoc prof, Univ Paris, 74-76; physics, Argonne Nat Lab, 76-80; lectr, Queen's Col, Oxford, England, 80-83; FEL, WORCESTER COL, OXFORD, ENGLAND, 83- *Mem:* Am Phys Soc; Am Soc Mass Spectrometry. *Res:* Molecular ions. *Mailing Add:* Seven Beaumont St Oxford 0X1 2LP England

ELANDER, RICHARD PAUL, b Worcester, Mass, Sept 17, 32; m 58; c 3. MICROBIOLOGY. *Educ:* Univ Detroit, BS, 55, MS, 56; Univ Wis, PhD(bot, bacteriol), 60. *Prof Exp:* Asst bot, Univ Wis, 56-60; sr microbiologist, Antibiotic Mfg & Develop Div, Eli Lilly & Co, 60-65, res scientist, Res Labs, 65-67; mgr process develop, Wyeth Antibiotic Labs, 67-68, mgr bulk antibiotic prod & process develop, Wyeth Labs, Inc, 68-71, assoc dir prod & process develop, Bulk Prod, 71-73; assoc dir microbiol res, Smith Kline & French Labs, 73-75; dir fermentation develop, Bristol-Myers Co, 75-79, sr dir fermentation res & develop, Indust Div, 79-82; VPRES BIOTECHNOL, BRISTOL-MYERS SQUIBB CO, 83- *Concurrent Pos:* Fel, Univ Minn, 66; res prof, Syracuse Univ, 82-; lectr, Rensselaer Polytech Inst, 83- *Honors & Awards:* Charles Thom Award, 84. *Mem:* Sigma Xi; Am Chem Soc; fel Soc Indust Microbiol (secy, 66-67, pres, 74-75); NY Acad Sci; fel Am Acad Microbiol; fel Am Inst Chemists. *Res:* Genetics of industrial microorganisms; antibiotic fermentation process development; screening for microbial metabolites; genetic engineering; biochemical engineering; mammalian cell culture. *Mailing Add:* Indust Div Bristol-Myers Squibb Co PO Box 4755 Syracuse NY 13221-4755

ELASHOFF, JANET DIXON, b Princeton, NJ, Mar 22, 42; div; c 2. APPLIED STATISTICS. *Educ:* Stanford Univ, BS, 62; Harvard Univ, PhD(statist), 66. *Prof Exp:* Asst prof statist, Sch Educ, Stanford Univ, 67-74; statist adv, Ctr Advan Study Behav Sci, 74-75; from assoc res statistician to res statistician, 75-81, ADJ PROF, DEPT BIOMATH, CTR ULCER RES, UNIV CALIF, LOS ANGELES, 81-; DIR, DIV BIOSTATIST, CEDARS-SINAI MED CTR, 89- *Concurrent Pos:* Consult, Nat Assessment of Educ Progress, 73-83; mem biomet & epidemiol contract rev comt, Nat Cancer Inst, 73-76; assoc ed, Am Statist Asn, 76-79; assoc ed, Biometrics, 84-86; mem, Adv Comt Gastrointestinal Drugs, Food & Drug Admin, 82- 85; consult & vpres, Dixon Statist Assoc. *Honors & Awards:* Palmer O Johnson Mem Award, Am Educ Res Asn, 71. *Mem:* Fel Am Statist Asn; Soc Clin Trials; Biomet Soc; fel AAAS; fel Royal Statist Soc. *Res:* Improvement of design and data analysis techniques for problems in the biological and behavioral sciences, specifically robustness of statistical methods to violation of assumptions, and sample size estimation. *Mailing Add:* Sci & Clin Comput Serv Cedars-Sinai Med Ctr Los Angeles CA 90048-1869

EL-ASHRY, MOHAMED T, b Cairo, Egypt, Jan 21, 40; c 2. GEOLOGY. *Educ:* Cairo Univ, BS, 59; Univ Ill, MS, 63, PhD(geol), 66. *Prof Exp:* Teaching asst geol, Cairo Univ, 59-61, asst prof, 66-69; from asst prof to prof environ sci, Wilkes Col, 69-75, chmn dept environ sci, 72-75; sr scientist & co-chmn water & land resources prog, Environ Defense Fund, 75-79; dir environ qual, Tenn Valley Authority, 79-83; vpres, 88-90, SR VPRES, WORLD RESOURCES INST, WASHINGTON, DC, 90- *Mem:* Fel Geol Soc Am; fel AAAS; Third World Acad Sci; Sigma Xi. *Res:* Environmental management; water resource management; arid lands development; energy policy; coastal geology. *Mailing Add:* World Resources Inst 1709 New York Ave NW Washington DC 20006

ELATTAR, TAWFIK MOHAMMED ALI, b Cairo, Egypt, Nov 6, 25; m 56; c 1. BIOCHEMISTRY, ENDOCRINOLOGY. *Educ:* Cairo Univ, BSc, 47, MSc, 55; Univ Mo, PhD(biochem), 57. *Prof Exp:* Biochemist nutrit, Agr Chem Dept, Ministry Agr, Egypt, 47-55; asst prof physiol chem, Univ Bonn, 61-64; asst res biochemist med, Med Sch, Univ Calif, San Francisco, 64-68; assoc prof, 68-70, PROF BIOCHEM & DIR HORMONE RES, SCH DENT & MED, UNIV MO KANSAS CITY, 70- *Concurrent Pos:* Fel biophys, Univ Calif, Berkeley, 57-58; fel biochem, Worcester Found Exp Biol, Mass, 58-59; fel endocrinol, Med Sch, Univ Miami, 59-60; metabolism and mode of action of steroid hormones; biochemistry of steroid hormones and prostaglandins. *Mailing Add:* Dept Hormone Res Univ Mo Sch Dent 650 E 25th St Kansas City MO 64108

EL-AWADY, ABBAS ABBAS, b Dakahlia, Egypt, Jan 2, 39; m 65; c 4. PHYSICAL INORGANIC CHEMISTRY. *Educ:* Univ Cairo, BSc, 58; Univ Minn, PhD(phys inorg chem), 65. *Prof Exp:* Lab instr chem, fac sci, Univ Cairo, 58-59; res assoc, Univ Calif, Los Angeles, 65-66; from asst prof to prof chem, 66-84, PROF, WESTERN ILL UNIV, 84-; AT KING ABDULAZIZ UNIV, JEDDAH, SAUDI ARABIA. *Concurrent Pos:* With Environ Protection Agency, 74-75; vis scholar, Univ Cambridge, UK, 71 & Univ Venice, Italy, 72; vis res prof, State Univ NY, Buffalo, 79-80. *Mem:* Am Chem Soc. *Res:* Applications of physical methods to the study of inorganic reactions in solution; kinetics and mechanisms of transition metal complexes; photochemistry; environmental chemistry. *Mailing Add:* Dept Chem King Abdulaziz Univ Jeddah Box 9028 Saudi Arabia

ELBADAWI, AHMAD, MORPHOLOGY & ULTRASTRUCTURE. *Educ:* Alexandria Univ, Egypt, MCh, 63. *Prof Exp:* PROF PATH & UROL, STATE UNIV NY HEALTH SCI CTR, 81- *Mailing Add:* Dept Path State Univ NY Health Sci Ctr 750 E Adams St Syracuse NY 13210

ELBAUM, CHARLES, b May 15, 26; US citizen; m 56; c 3. SOLID STATE SCIENCE, NEUROSCIENCES. *Educ:* Univ Toronto, MASc, 51, PhD(appl sci), 54. *Hon Degrees:* MA, Brown Univ, 61. *Prof Exp:* Res fel metal physics, Univ Toronto, 54-57 & Harvard Univ, 57-59; asst prof appl physics, Brown Univ, 59-61, assoc prof physics, 61-63, chmn, Dept Physics, 80-86, prof, 63-91, HAZARD PROF PHYSICS, BROWN UNIV, 91- *Concurrent Pos:* Indust consult. *Mem:* Fel Am Phys Soc; Am Inst Mining, Metall & Petrol Engrs; Soc Neurosci; AAAS. *Res:* Crystal defects; mechanical properties of solids; ultrasonic wave propagation; phonon and electron transport and interactions; phase transitions; liquid and solid helium; biophysics. *Mailing Add:* Dept Physics Brown Univ Providence RI 02912

ELBAUM, DANEK, biophysics, for more information see previous edition

ELBAUM, MAREK, b Kovel, USSR, May 8, 41; US citizen; m 68; c 1. ELECTRICAL ENGINEERING, QUANTUM ELECTRONICS. *Educ:* Warsaw Tech Univ, MSc, 66; Columbia Univ, PhD(elec eng), 77. *Prof Exp:* Res assoc, Microwave Lab, Polish Acad Sci, 66-68; mem res staff, Electro-Optics Lab, Riverside Res Inst, 69-79, mgr, Electro-Optical Sensor Syst, 80-82; res dir, advan optics concepts, Riverside Res Inst, 82-89; PRES, ELECTRO-OPTICAL SCI, INC, 90- *Concurrent Pos:* Adj prof, Columbia Univ, 78- *Mem:* Optical Soc Am. *Res:* Application of lasers; theory of coherence; concepts for and feasibility of electro-optical sensors for space applications; author or coauthor of over 55 technical papers; several United States patents. *Mailing Add:* 79 Beechdale Road Dobbs Ferry NY 10522

EL-BAYOUMI, MOHAMED ASHRAF, b Cairo, Egypt, Dec 7, 34; m 56; c 3. PHYSICAL CHEMISTRY, ELECTRONIC SPECTROSCOPY. *Educ:* Univ Alexandria, BSc, 54; Fla State Univ, MSc, 57, PhD(phys chem), 61. *Prof Exp:* Res assoc electronic spectra, Fla State Univ, 61 & Mass Inst Technol, 61-62; lectr phys chem, fac sci, Univ Alexandria, 62-67; res assoc electronic spectra, Fla State Univ, 67-68; from assoc prof to prof electronic spectra, Mich State Univ, 68-81, adj prof chem, 82-83. *Mem:* AAAS; Am Phys Soc; Biophys Soc; Am Chem Soc; NY Acad Sci. *Res:* Relaxation processes in excited molecular systems; nanosecond time resolved spectroscopy; fluorescence probes of biomolecular conformation; electronic spectra of organic and biological molecules. *Mailing Add:* Fac Sci Alexandria Univ Moharrum Alexandria Egypt

EL-BAZ, FAROUK, b Zagazig, Egypt, Jan 1, 38; US citizen; m 63; c 4. GEOLOGY, SPACE SCIENCE. *Educ:* Ain Shams Univ, Cairo, BSc, 58; Mo Sch Mines, MS, 61; Univ Mo, PhD(geol), 64. *Hon Degrees:* DSc, New England Col. *Prof Exp:* Demonstr micropaleontol, Assiut Univ, 58-60; lectr econ geol, Univ Heidelberg, 64-65; explor geologist petrol, Pan Am UAR Oil Co, Egypt, 65-66; supvr lunar sci planning geol, Bellcomm Inc, 67-72 & lunar explor syst eng, Bell Tell Labs, 72-73; res dir earth & planet studies, Nat Air & Space Mus, Smithsonian Inst, 73-82; vpres sci & technol, Itek Optical Systs, Lexington,Mass, 82-86; DIR, CTR FOR REMOTE SENSING, BOSTON UNIV, MASS, 86- *Concurrent Pos:* Mem Apollo Orbital Sci Photographic Team NASA, 70-72, Crew Training Group, 70-72, Lunar Sci Review Panel, 74-75; mem, US/USSR joint lunar cartographic activ, 73-75 & Int Astron Union, 74-78; prin investr, Earth Observ Apollo-Soyuz Mission, 74-78; adj prof geol & geophys, Univ Utah, 74-76; adj prof geol, Ain Shams Univ, 76-78. *Honors & Awards:* Exceptional Sci Achievement Medal, NASA, 71. *Mem:* AAAS; fel Geol Soc Am; Am Soc Photogram; Am Geophys Union; Explorers Club. *Res:* Comparative planetology with emphasis on interpretation of space-born photographs of surface features of planets and their moons, applications of space technology to desert studies and applications of remote sensing to the fields of archaeology, geography and geology. *Mailing Add:* Boston Univ 725 Commonwealth Ave Boston MA 02215

ELBEIN, ALAN D, b Lynn, Mass, Mar 20, 33; m 54; c 3. MICROBIOLOGY, BIOCHEMISTRY. *Educ:* Clark Univ, AB, 54; Univ Ariz, MS, 56; Purdue Univ, PhD(microbiol), 60. *Prof Exp:* Res assoc microbiol, Purdue Univ, 60-61 & Univ Mich, 61-63; asst res biochemist, Univ Calif, Berkeley, 63-64; from asst prof to assoc prof biol, Rice Univ, 64-69; assoc prof biochem, Univ Tex Health Sci Ctr, San Antonio, 69-70, prof, 70-91; PROF & CHMN, BIOCHEM, UNIV ARK MED SCI, 91- *Concurrent Pos:* Consult, NSF, Vet Admin & NIH; Career Develop Award, NIH, 65-69. *Mem:* Am Soc Biol Chem; Am Soc Plant Physiol; Am Chem Soc; Am Soc Microbiol; Am Acad Microbiol. *Res:* Biosynthesis of microbial and plant cell walls; trehalose metabolism and biosynthesis; glycoprotein biosynthesis in aorta tissue; effects of polyelectrolytes on enzymes; effects of inhibitors on glycoprotein biosynthesis. *Mailing Add:* Dept Biochem & Molecular Biol Univ Ark Med Sci Little Rock AR 72205

ELBEL, ROBERT E, b Hannibal, Mo, July 8, 25; m 60; c 3. ZOOLOGY & PARASITOLOGY, PUBLIC HEALTH & EPIDEMIOLOGY. *Educ:* Univ Kans, BA, 48, MA, 50; Univ Okla, PhD(zool), 64. *Prof Exp:* Med biol technician typhus-dysentary studies, USPHS, Thomasville, Ga, 50-51; plague-malaria adv, US Opers Mission Thailand, 51-55 & 61-63; grad asst zool, Univ Okla, 56-59; med zoologist & entomologist, Ecol & Epidemiol, US Army, Dugway, Utah, 63-75; res assoc prof, 80-83, RES PROF BIOL, UNIV UTAH, 74- *Concurrent Pos:* Consult fleas of NAm, Brigham Young Univ, Provo, 76; consult, vec-borne dis, US Army, Dugway, Utah, 75-77, 79-80 & 83-85; res assoc entomol, Bishop Mus, Honolulu, Hawaii, 82- *Mem:* Sigma Xi; Am Mosquito Control Asn; Soc Syst Zool. *Res:* Bird Mallophaga; flea larvae; arboviruses; nearly 90 publications. *Mailing Add:* 1518 Evergreen Salt Lake City UT 84106

ELBERGER, ANDREA JUNE, b New York, NY, Feb 11, 52; m 78. DEVELOPMENTAL NEUROBIOLOGY, NEUROSCIENCES. *Educ:* State Univ NY, Albany, BA, 72; State Univ, Stony Brook, PhD(psychobiol), 77. *Prof Exp:* Teaching asst psychol, State Univ NY, Stony Brook, 73-74, res asst, 74, res asst psychiat, Med Sch, 75-76; trainee anat, Sch Med, Univ Pa, 77-80; ASST PROF ANAT, MED SCH, UNIV TEX, HOUSTON, 80- *Mem:* Soc Neurosci; Int Soc Develop Neurosci. *Res:* Developmental neurobiology of the corpus callosum and the visual system; using the behavioral anatomical and electrophysiological techniques to investigate the functional interactions of interhemispheric communication and visual development. *Mailing Add:* Dept Anat Univ Tenn Col Med 800 Madison Ave Memphis TN 38163

EL-BERMANI, AL-WALID I, anatomy, for more information see previous edition

ELBERT, JEROME WILLIAM, b Whittemore, Iowa, Oct 24, 42; m 75. ULTRA HIGH ENERGY X-RAY ASTRONOMY,. *Educ:* Iowa State Univ, BS, 64; Univ Wis-Madison, MS, 66, PhD(physics), 71. *Prof Exp:* Teaching asst physics, Univ Wis-Madison, 64-65, res asst, 65-71; res assoc-assoc instr, 71-73, from res asst to res assoc prof, 74-86, RES PROF, PHYSICS DEPT, UNIV UTAH, 86- *Concurrent Pos:* Prin investr, NSF grants, 84-86; vis prof, Univ Naples, 89-90. *Mem:* Am Phys Soc; AAAS. *Res:* High energy nuclear physics; cosmic ray physics and astrophysics; high energy particle production; underground muons; ultra high energy cosmic ray air showers; gamma rays. *Mailing Add:* Dept Physics Univ Utah Salt Lake City UT 84112

ELBERTY, WILLIAM TURNER, JR, b East Orange, NJ, Mar 8, 30; m 52; c 3. GEOLOGY. *Educ:* St Lawrence Univ, BS, 53; Dartmouth Col, MA, 55; Ind Univ, PhD(geol), 60. *Prof Exp:* Instr geol, 58-60, from asst prof to assoc prof geol & geog, 60-74, PROF GEOL, ST LAWRENCE UNIV, 74- *Concurrent Pos:* Dir prof in Nairobi, St Lawrence Univ, 77-78. *Mem:* AAAS; Geochem Soc. *Res:* Mineralogy of Pleistocene sands and gravels; conservation and property zoning; mineralogy; geochemistry; economic geology. *Mailing Add:* PO Box 476 Canton NY 13617

ELBLE, RODGER JACOB, b Alton, Ill, Aug 10, 48; m 71; c 3. NEUROLOGY, NEUROPHYSIOLOGY. *Educ:* Purdue Univ, BSAE, 71; Ind Univ, PhD(physiol), 75, MD, 77. *Prof Exp:* Intern med, Ind Univ Med Ctr, 77-78; resident neurol, Barnes Hosp, Wash Univ, 78-81; ASSOC PROF NEUROL, SCH MED, SOUTHERN ILL UNIV, 81-, DIR, CTR ALZHEIMER DIS & RELATED DISORDERS. *Mem:* Am Acad Neurol; Soc Neurosci; Am Physiol Soc; AMA; NY Acad Sci. *Res:* Neuromuscular control in man and the mechanisms of physiologic and pathologic tremors; Alzheimer disease; gait disturbances in the elderly. *Mailing Add:* 707 N Rutledge Springfield IL 62708

ELBLING, IRVING NELSON, b Salem, Mass, July 30, 20; m 46; c 2. ORGANIC CHEMISTRY, RESEARCH ADMINISTRATION. *Educ:* Northeastern Univ, BSc, 43. *Prof Exp:* From lab asst to instr, Northeastern Univ, 40-43; student & engr, Westinghouse Elec Co, 43-44, res chemist plastics & resins, res lab, 44-50, chemist, org coatings, paint & varnish dept, 50-52, res chemist, res lab, 52-53, supvr insulation dept, 53-66, mgr specialty coatings, 66-78, opers mgr, Chem Sci Div, Res Labs, 78-85; RETIRED. *Honors & Awards:* Roon Award Winner, 59. *Mem:* Am Chem Soc; Fedn Socs Paint Technol. *Res:* Organic coatings; phenolic alkyd and epoxy resins; electrical insulating varnishes; insulating tapes; adhesives; fluidized bed powders; financial management. *Mailing Add:* 1437 Severn St Pittsburgh PA 15217

ELCE, JOHN SHACKLETON, b Aug 7, 38; m; c 3. IMMUNOCHEMISTRY. *Educ:* Univ London, Eng, PhD(biochem), 66. *Prof Exp:* PROF BIOCHEM, QUEEN'S UNIV, CAN, 68- *Mem:* Am Soc Biol Chem. *Res:* Protein turnover; calpain. *Mailing Add:* Dept Biochem Queen's Univ Kingston ON K7L 3N6 Can

ELCHLEPP, JANE G, b St Louis, Mo, May 26, 21. PATHOLOGY. *Educ:* Harris Teachers Col, BA, 43; Univ Iowa, MS, 46, PhD(zool), 48; Univ Chicago, MD, 55. *Prof Exp:* Teacher, Pub Sch, Mo, 43-44; asst zool, Univ Iowa, 45-48; instr biol, Roosevelt Univ, 48-51; asst anat, Univ Chicago, 51-53; asst pharmacol, Wash Univ, 54; asst med, Univ Chicago, 54-55; field investr, Nat Cancer Inst, 55-62; asst prof path, 62-67, asst to vpres health affairs, 69-71, assoc prof path, 67-80, ASST VPRES HEALTH AFFAIRS, PLANNING & ANALYSIS, DUKE UNIV, 71- *Mailing Add:* Dept Path Duke Univ Sch Med Box 3322 Med Ctr Durham NC 27710

ELCRAT, ALAN ROSS, b Chicago, Ill, Jan 13, 42. APPLIED MATHEMATICS. *Educ:* Univ NMex, BS, 64; Ind Univ, MA, 65, PhD, 67. *Prof Exp:* Asst prof, 67-70, assoc prof, 70-77, PROF MATH, WICHITA STATE UNIV, 77- *Mailing Add:* Dept Math Wichita State Univ Wichita KS 67208-1595

EL DAREER, SALAH, ANTI-CANCER AGENTS, XENOBIOTICS. *Educ:* Cairo Univ, DVM, 47; Mich State Univ, MS, 51. *Prof Exp:* SR PHARMACOLOGIST, SOUTHERN RESOURCE INST, 65- *Mailing Add:* Southern Res Inst PO Box 55305 Birmingham AL 35255-5305

ELDE, ROBERT PHILIP, b Mt Vernon, Wash, March 7, 47; m 69; c 2. NEUROSCIENCE, NEUROHISTOLOGY. *Educ:* NPark Col, Chicago, BA, 69; Univ Minn, PhD(anat), 74. *Prof Exp:* Instr anat, Univ Minn, Minneapolis, 74-75; guest scientist histol, Karolinska Inst, Stockholm, 75-76; from asst prof to assoc prof, 77-84, PROF ANAT, UNIV MINN, MINNEAPOLIS, 84- *Honors & Awards:* McKnight Scholar in Neurosci. *Mem:* Soc Neurosci; Am Asn Anatomists; Histochem Soc. *Res:* Distribution and function of transmitter-coded circuits in the mammalian central nervous system, including opioid peptides and neuropepetide hormones. *Mailing Add:* Cell Biol 4-135 Jackson Hall Univ Minn 321 Church St Southeast Minneapolis MN 55455

ELDEFRAWI, AMIRA T, b Giza, Egypt, Feb 10, 37; US citizen; m 57; c 3. NEUROPHARMACOLOGY, NEUROTOXICOLOGY. *Educ:* Univ Alexandria, BSc, 57; Univ Calif, Berkeley, PhD(entom, toxicol), 60. *Prof Exp:* Asst prof toxicol, Univ Alexandria, 63-68; res assoc neurobiol, Cornell Univ, 68-72, sr res assoc, 72-76; assoc res prof, 76-79, res prof pharmacol, Univ MD, 79-88, PROF PHARMACOL, SCH MED, UNIV MD, BALTIMORE, 88- *Concurrent Pos:* Consult, Lab Neuromuscular Physiol, Dept Neurol, Sch Med, Univ Va, 77-79; NIH grants; mem, NIH Environ Health Sci Rev Comt. *Mem:* Am Soc Pharmacol & Exp Therapeut; Entom Soc Am; Soc Neurosci; Soc Toxicol. *Res:* Neuropharmacology; n-acetylcholine receptor; neurotoxicology; insecticides; GABA receptor. *Mailing Add:* Dept Pharmacol & Exp Therapeut 660 W Redwood St Baltimore MD 21201

ELDEFRAWI, MOHYEE E, b Egypt, Oct 15, 32; US citizen; m 57; c 3. NEUROPHARMACOLOGY, TOXICOLOGY. *Educ:* Univ Alexandria, BSc, 53; Univ Calif, Berkeley, PhD(entom), 60. *Prof Exp:* From asst prof to assoc prof toxicol, Univ Alexandria, 60-68; sr res assoc neurobiol, Cornell Univ, 68-72, assoc prof, 72-76; PROF PHARMACOL, SCH MED, UNIV MD, BALTIMORE, 76- *Concurrent Pos:* Grants, NSF, NIH, Muscular Dystrophy Asn Am, US Army & Nat Inst Drug Abuse; consult, Nat Sci Found, 77, World BAnk, 80, United Nations, 81, US State Dept Agency Int Develop, 82-83, Biomed Study Sect, NIH, 89-90, Drug Abuse Study Sect, 90- *Mem:* Am Soc Pharmacol & Exp Therapeut; Am Chem Soc; NY Acad Sci; Entom Soc Am; fel AAAS; Soc Toxicol; fel Nat Asn Advan Sci. *Res:* Toxicology mechanisms of actions of insecticides in the nervous systems of mammals and insects; cocaine receptor and addiction; biotechnology; toxicological and pharmacological biosensors; author of over 200 papers and chapters and 80 abstracts in three areas of science. *Mailing Add:* Dept Pharmacol & Exp Therapeut Univ Md Sch Med 655 W Baltimore St Baltimore MD 21201

ELDEN, RICHARD EDWARD, b Seneca Falls, NY, Feb 25, 23; m 55; c 4. INDUSTRIAL CHEMISTRY. *Educ:* Mass Inst Technol, SB, 44; Univ Wash, MS, 52, Seton Hall Univ, JD, 81. *Prof Exp:* Chemist, Carbide & Carbon Chem Co, Columbia, 44; from chemist to chief chemist, Inorg Chem Div, FMC Corp, 46-56, prod mgr, 56-59, resident mgr, Vancouver Plant, 59-63, mgr spec res proj, 63-66, asst to dir, Indust Chem Div, 67-77, patent liaison, 77-81, PATENT ATTY & LAWYER, FMC CORP, 81- *Mem:* Am Chem Soc. *Res:* Analytic instrumentation; administration. *Mailing Add:* 357 Dodds Lane Princeton NJ 08540

ELDER, ALEXANDER STOWELL, b Medford, Mass, July 29, 15; m 47; c 2. MATHEMATICS. *Educ:* Harvard Univ, BA, 38; Boston Univ, MEd, 40; Univ Del, MA, 56. *Prof Exp:* Elec technician, Niagara, Lockport & Ont Power Co, NY, 40-41; physicist, Watertown Arsenal, Mass, 46-49; from mech engr to chief, Dynamics Sect, 50-59, chief, Solid Properties Group, 69-76, chief analyst, mech & structures br, Propulsion div, Army Ballistic Res Lab, 76-88; RETIRED. *Concurrent Pos:* Eve instr math, Harford Jr Col, 59-61. *Honors & Awards:* Kent Award, Army Ballistic Res Lab, 83. *Mem:* Am Acad Mech; Math Asn Am; Am Math Soc. *Res:* Mechanics of solids, including vibration theory, elasticity and viscoelasticity; applied mathematics; calculation of special functions. *Mailing Add:* 814 Matthews Ave Aberdeen MD 21001

ELDER, CURTIS HAROLD, b Laramie, Wyo, March 30, 21; m 48; c 4. GEOLOGY. *Educ:* Univ Mo, BS, 51. *Prof Exp:* Jr res engr, Pan Am Petrol Co, Standard Oil Ind, 52-56, geologist, 56-63; independent consult, 63-65; res geologist, US Bur Mines, Pittsburgh Mining Res Ctr, 65- 81 & Off Surface Mining, 81-88; CONSULT, 88- *Honors & Awards:* Honor Award for Super Serv, US Dept Interior, 88. *Mem:* Am Inst Prof Geologists; sr fel Geol Soc Am; Soc Econ Paleont & Mineraliogists. *Res:* Methane gas as it relates to coal; removal of methane gas & mine safety; mine subsidence, landslides & gas problems related to abandoned coal mines. *Mailing Add:* 33172 Lynx Lane Rte 2 Evergreen CO 80439

ELDER, FRED A, b Carrollton, Ohio, Dec 4, 29; m 55; c 4. PHYSICAL CHEMISTRY, XEROGRAPHY. *Educ:* Muskingum Col, BS, 51; Univ Chicago, MS, 62, PhD(chem phys), 68. *Prof Exp:* Chemist, Nat Bur Standards, 54-55; lab researcher, Calif Res Corp, 57-59; res asst mass spectros, Univ Chicago, 62-69; asst prof chem, Rochester Inst Technol, 69-72; scientist, 72-80, SR TECH SPECIALIST, XEROX CORP, 80- *Mem:* Am Soc Mass Spectrometry. *Res:* Mass spectroscopy. *Mailing Add:* 800 Phillips Rd, Bldg 139 Webster NY 14580

ELDER, FRED F B, b Los Angeles, Calif, Aug 10, 47. CHROMOSOME STUDY. *Educ:* Univ Ill, PhD(zool), 78. *Prof Exp:* ASST PROF PEDIAT & GENETICS, MED BR, UNIV TEX, 81- *Mem:* AAAS; Am Soc Cell Biol; Soc Study Evolution. *Mailing Add:* Dept Pediat Univ Tex Health Sci Ctr PO Box 20708 Houston TX 77225

ELDER, FRED KINGSLEY, JR, b Coronado, Calif, Oct 19, 21; m 47; c 8. PHYSICS. *Educ:* Univ NC, SB, 41; Yale Univ, MS, 43, PhD(physics), 47. *Prof Exp:* Student asst, Nat Bur Standards, 41; lab asst physics, Yale Univ, 41-43, res asst, 43, instr, 43-44, asst instruction, 46-47; physicist, Naval Res Lab, Washington, DC, 44-46; instr physics, Univ Pa, 47-49; asst prof, Univ Wyo, 49-50; sr physicist, Appl Physics Lab, Johns Hopkins Univ, 50-53; assoc prof, Wabash Col, 53-55, prof & chmn dept, Belhaven Col, 55-59; physicist, Antisubmarine Warfare Lab, Naval Air Develop Ctr, 59-60, head res br, 60-65; chmn dept, 65-72, PROF PHYSICS, ROCHESTER INST TECHNOL, 65- *Concurrent Pos:* Physicist, Nat Bur Standards, 49 & Naval Ord Lab, Md, 57-59; lectr, vis sci prog, 58-59; scientist, Physics Res Labs, Eastman Kodak Co, 82; hon vis lectr physics, Aston Univ, Birmingham, Eng, 85-86; vis res fel, Lanchester Polytech, Coventry, Eng, 85-86. *Mem:* Am Phys Soc; Am Asn Physics Teachers; Netherlands Phys Soc; Am Geophys Union; Sigma Xi; Inst Physics Gt Brit. *Res:* Isotope separation by thermal diffusion; nuclear and atomic physics; separation and transmutation of neon isotopes; fluid dynamics; physics and geophysics of submarine detection; undergraduate physics curriculum. *Mailing Add:* PO Box 9848 Rochester NY 14623-0848

ELDER, H E, b Eldorado, Ill, Nov 5, 24; m 48; c 2. ELECTRONICS ENGINEERING. *Educ:* Univ Ill, BS, 48; Newark Col Eng, MS, 54. *Prof Exp:* Engr, Bell Tel Labs, Inc, Am Tel & Tel Co, 48-51, mem tech staff, 51-57, supvr, 57-89; RETIRED. *Res:* Microwave device development; magnetrons, traveling wave tubes, mixer and detector diodes, masers, klystrons and varactor diodes; solid state light-emitting diode and laser development. *Mailing Add:* 180 Oakmont Ct Reading PA 19607

ELDER, HARVEY LYNN, b Mayfield, Ky, April 23, 34. MATHEMATICS EDUCATION. *Educ:* Murray State Univ, BA, 55, MAEd, 57; Univ Ill, MA, 61, PhD(math educ), 68. *Prof Exp:* From instr to assoc prof, 57-77, PROF MATH, MURRAY STATE UNIV, 77- *Mem:* Am Math Asn. *Res:* Motivation in mathematics; error analysis in mathematics. *Mailing Add:* Dept Math Murray State Univ Murray KY 42071

ELDER, JAMES FRANKLIN, JR, b Mount Airy, NC, Aug 10, 49; m 71; c 2. MASS SPECTROMETRY, LABORATORY DATA SYSTEMS. *Educ:* Univ NC, Chapel Hill, 71; Purdue Univ, PhD(anal chem), 76. *Prof Exp:* Sr anal chemist, Dow Chem Co, 76-77; sr res chemist, 77-80, res & develop prog mgr, 80-87, RES & DEVELOP LAB MGR, R J REYNOLDS TOBACCO CO, 87- *Mem:* Am Soc Mass Spectrometry. *Res:* Applications of mass spectrometry to chemical and physical properties of gaseous ions; applications of analytical chemistry to production problems, especially to applications of computers to spectrometry. *Mailing Add:* Res & Develop Dept Bowman Gray Tech Ctr R J Reynolds Tobacco Co Winston-Salem NC 27102

ELDER, JOE ALLEN, b Richland, Ga, Mar 9, 41; m 63; c 2. CELL BIOLOGY, RADIO FREQUENCY RADIATION. *Educ:* Berry Col, BA, 63; Vanderbilt Univ, MS, 65; Pa State Univ, PhD(biophys), 70. *Prof Exp:* Fel physiol chem, Johns Hopkins Sch Med, 70-73; cell biologist, US Environ Protection Agency, 73-74, chief molecular path sect, 74-76, chief neurobiol br, 76-79, chief cellular biophys br, 79-82, actg dir exp biol div, 82-84, chief cellular biophys br, 84-87, chief cell biology br, 87-88, SPEC ASST, OFF DIR, HEALTH EFFECTS RES LAB, US ENVIRON PROTECTION AGENCY, 88- *Concurrent Pos:* Mem, Nat Coun Radiation Protection & Measurements, 83-89; Environ Protection Agency proj leader, US/USSR Coop Prog Environ Health, Biol Effects Microwave Radiation, 75-85; mem work group, Radio Frequency Radiation, World Health Orgn Regional Off Europ, 84-87; mem work group, Am Nat Standards Inst Comt Radio-Frequency Radiation Hazards, 77-82; chmn, Work Group Nat Res Agenda Electromagnetic Fields, 90-91. *Mem:* Bioelectromagnetics Soc; Biophys Soc. *Res:* Biological effects of radio frequency radiation. *Mailing Add:* Health Effects Res Lab MD-51 US Environ Protection Agency Research Triangle Park NC 27711

ELDER, JOHN PHILIP, b London, UK, Jan 30, 31; US citizen; m 75; c 2. THERMAL ANALYSIS, SOLID STATE REACTION KINETICS THEORY. *Educ:* Liverpool Univ, UK, BSc, 53, Hons, 54, MSc, 56, PhD(electrochem), 61. *Prof Exp:* Sr res electrochemist, Tex Instruments, Inc, 65-68; sr scientist, ESB, Inc Technol Ctr, 70-72; mgr, Mettler Instrument Corp, 72-79; mgr res, Inst Mining & Mineral Res, Univ Ky, 79-85; RES FEL, MERCK & CO INC, 85- *Mem:* Am Chem Soc; NAm Thermal Anal Soc; Int Confederation Thermal Anal; Am Soc Testing & Mat. *Res:* Study of thermal behavior of solid drugs to examine stability, oxidative stability and environmental impact; theoretical study of the kinetics of multiple solid state thermally stimulated chemical and physical reactions. *Mailing Add:* 411A Newport Way Jamesburg NJ 08831

ELDER, JOHN THOMPSON, JR, b Fall River, Mass, June 30, 27; m 58; c 2. PHARMACOLOGY. *Educ:* Mass Col Pharm, BS, 53, MS, 55; Univ Wash, PhD(pharmacol), 59. *Prof Exp:* From instr to asst prof pharmacol, Univ Wash, 57-65; from asst prof to assoc prof, 65-74, PROF PHARMACOL, CREIGHTON UNIV, 74- *Mem:* Am Soc Pharmacol & Exp Therapeut. *Res:* Psychopharmacology and autonomic pharmacology, especially as it applies to central nervous system function. *Mailing Add:* Dept Pharmacol Creighton Univ Omaha NE 68178-0225

ELDER, JOHN WILLIAM, b Ann Arbor, Mich, June 27, 33. ORGANIC CHEMISTRY. *Educ:* Spring Hill Col, BS, 58; Loyola Univ, Ill, MS, 60, PhD(org chem), 62. *Prof Exp:* Asst prof chem, Regis Col, Colo, 67-69; from asst prof to assoc prof, 69-78, PROF CHEM, FAIRFIELD UNIV, 78- *Mem:* Am Chem Soc; Sigma Xi. *Res:* Organic chemistry of natural products. *Mailing Add:* Dept Chem Fairfield Univ Fairfield CT 06430

ELDER, REX ALFRED, b Laquin, Pa, 17. ENGINEERING, HYDRAULICS. *Educ:* Carnegie Inst Tech, BS, 40; Oregon State Col, MS, 42. *Prof Exp:* Dir Eng Lab, Tenn Valley Authority, 61-73; eng mgr hydraulics, Bechtel Co, 73-85; CONSULT, 85- *Honors & Awards:* James Laurie Prize, 58, Hunter Rouse Lectr 84, & Hydraul Struct Award, Am Soc Civil Eng, 91. *Mem:* Nat Acad Eng; Am Soc Civil Eng; Am Soc Chem Eng. *Res:* Broad application of hydrollc model studies, design and operation of dams, power stations and mining facilities. *Mailing Add:* 2180 Vistazo E Tiburon CA 94920

ELDER, RICHARD CHARLES, b Ann Arbor, Mich, June 9, 39; m 79; c 2. INORGANIC CHEMISTRY, STRUCTURAL CHEMISTRY. *Educ:* St Louis Univ, BS, 61; Mass Inst Technol, PhD(chem), 64. *Prof Exp:* Res assoc chem, Mass Inst Technol, 64-65; from instr to asst prof, Univ Chicago, 65-70; assoc prof, 70-78, PROF CHEM, UNIV CINCINNATI, 78- *Concurrent Pos:* Vis res fel, Res Sch Chem, Australian Nat Univ, 76-77. *Mem:* Am Crystallog Asn; Am Chem Soc. *Res:* Structural chemistry of transition metal complexes; gold-based anti-arthritic drugs; X-ray absorption spectroscopy; technetium complexes; single crystal X-ray diffraction; extended X-ray absorption fine structure spectroelectrochemistry; X-ray scattering. *Mailing Add:* Dept Chem Univ Cincinnati Cincinnati OH 45221-0172

ELDER, ROBERT LEE, b Louisville, Ky, Apr 5, 31; m 55; c 2. RADIOLOGICAL PHYSICS. *Educ:* Ind Univ, BS, 53; Univ NC, MS, 55; Ore State Univ, BS, 58; Johns Hopkins Univ, ScD(biophys), 64. *Prof Exp:* Asst sanit eng, Univ NC, 53-55; engr, USPHS, Nev, 58-60, sr engr, 60-61, sr scientist, Ala, 64-65, sr scientist, Div Radiol Health, 65-67, assoc dir, Bur Radiol Health, 67-70, dir div electronic prod, 70-72, dep dir, Bur Radiol Health, 72-76, dep assoc comnr sci, 76-79, DIR COSMETIC INGREDIENT REV, FOOD & DRUG ADMIN, USPHS, 79- *Concurrent Pos:* Res lectr, Auburn Univ, 64-65; assoc prof, Ore State Univ, 65-68; asst surgeon gen, USPHS. *Mem:* AAAS; Am Soc Civil Eng; Am Soc Microbiol. *Res:* Response of cells and organ systems to toxic substances. *Mailing Add:* 8610 Buckhnon Dr Potomac MD 20854

ELDER, SAMUEL ADAMS, b Baltimore, Md, July 13, 29; m 55; c 5. PHYSICS. *Educ:* Hampden-Sydney Col, BS, 50; Brown Univ, ScM, 53, PhD(physics), 56. *Prof Exp:* Asst physics, Brown Univ, 50-55; physicist assoc staff, appl physics lab, Johns Hopkins Univ, 56, sr staff, 56-64; assoc prof physics, 64-68, PROF PHYSICS, US NAVAL ACAD, 68- *Mem:* Acoust Soc Am; Catgut Acoust Soc; Am Asn Physics Teachers; Sigma Xi. *Res:* Nonlinear acoustics; fluid dynamics; computer science; musical acoustics. *Mailing Add:* Dept Physics US Naval Acad Annapolis MD 21402

ELDER, VINCENT ALLEN, b Yankton, SDak, May 3, 48; m 86; c 1. ENVIRONMENTAL CHEMISTRY, SOIL SCIENCE. *Educ:* Mich State Univ, BS, 70; Univ Hawaii, MS, 75, PhD(soil sci), 78. *Prof Exp:* Volunteer qual control anal, Am Peace Corps, India, 70-72; fel chem eng, Mass Inst Technol, 78-79; res assoc, Sch Pub & Environ Affairs, Ind Univ, 79-80; group leader, Tech Servs, Mary Kay Cosmetics, 80-87; PRIN SCIENTIST ANALYTICAL SERV, FRITO-LAY, INC, 87- *Mem:* Am Chem Soc; Am Oil Chemists Soc; Asn Off Anal Chemists; Am Soc Mass Spectroscopists. *Res:* Analysis of foods and cosmetics; organic mass spectroscopy; environmental chemistry; gas and liquid chromatography; soil science. *Mailing Add:* 900 N Loop 12 Irving TX 75061

ELDER, WILLIAM HANNA, b Oak Park, Ill, Dec 24, 13; m 41; c 2. WILDLIFE CONSERVATION. *Educ:* Univ Wis, BS, 36, PhM, 38, PhD(zool), 42. *Prof Exp:* Asst zool, Univ Wis, 36-41; game technician, Nat Hist Surv, Ill, 41-43; asst pharmacol, Univ Chicago, 43-45; from asst prof to prof zool, 45-54, Rucker Prof Fisheries & Wildlife, 54-83, EMER PROF ZOOL, UNIV MO-COLUMBIA, 83- *Concurrent Pos:* Mem staff, toxicol lab, Univ Chicago, 43-45; Guggenheim fel, 56-57; Fulbright fel, 65-66; NSF grant, Africa, 67-68. *Mem:* Am Soc Mammalogists; Wildlife Soc; Wilson Ornith Soc; Wilderness Soc; Nature Conservancy. *Res:* Physiology of reproduction; biology of the Canada goose; measures of productivity in wild populations; lead poisoning; biology of bats; avian chemosterilants; African elephant; natural area inventories. *Mailing Add:* Stephens Hall Univ Mo Columbia MO 65211

ELDERKIN, CHARLES EDWIN, b Seattle, Wash, Aug 6, 30; m 59; c 1. ATMOSPHERIC SCIENCES. *Educ:* Univ Wash, BS, 53, PhD(atmospheric sci), 66. *Prof Exp:* Sr res scientist atmospheric sci, 65-66, sect mgr atmospheric physics, 66-72, assoc mgr dept atmospheric sci, 72-79, proj mgr wind characteristics prog element, Fed Wind Energy Prog, 76-79, MGR DEPT ATMOSPHERIC SCI, PAC NORTHWEST LAB, BATTELLE MEM INST, 79- *Concurrent Pos:* Mem, Field Observ Fac Adv Panel, Nat Ctr Atmospheric Res, 72-75. *Honors & Awards:* Ernest Orlando Lawrence Mem, US Dept Energy, 76. *Mem:* Am Meteorol; Sigma Xi. *Res:* Air pollution; physics and chemistry of clouds and precipitation; atmospheric turbulence and its effects on structures and siting of facilities; wind and solar energy; climatic effects from energy developments; evaluation of air sampling methodologies and pollution control techniques; astronomy and weather forecasting services. *Mailing Add:* 531 Holly St Richland WA 99352

ELDERKIN, RICHARD HOWARD, b Butte, Mont, May 4, 45; m 70; c 2. FINITE ELASTICITY, POPULATION BIOLOGY. *Educ:* Whitman Col, BA, 67; Univ Colo, Boulder, MA, 68, PhD(math), 71. *Prof Exp:* Asst prof math, State Univ NY, Albany, 72-73; asst prof & res fel appl math, Brown Univ, 73-74; from asst prof to assoc prof, 74-88, PROF MATH, POMONA COL, 88- *Concurrent Pos:* Sr fel, Inst Math & Appln, Univ Minn, 84-85. *Mem:* Asn Comput Mach; Soc Indust & Appl Math; Math Asn Am. *Res:* Nonlinear elastic plate theory; application of differential equations to population biology; mathematical modelling. *Mailing Add:* Dept Math Pomona Col Claremont CA 91711

ELDERS, MINNIE JOYCELYN, b Schaal, Ark, Aug 13, 33; m 60; c 2. PEDIATRICS, ENDOCRINOLOGY. *Educ:* Philander Smith Col, BA, 52; Brooke Army Med Sch, cert phys ther, 54; Univ Ark, Little Rock, MD, 60; Am Bd Pediat, dipl, 64. *Prof Exp:* Intern pediat, Univ Minn Hosp, 60-61; resident, 61-64, from instr to asst prof, 64-71, assoc prof, 71-74, PROF PEDIAT, MED CTR, UNIV ARK, LITTLE ROCK, 74- *Concurrent Pos:* Nat Inst Child Health & Human Develop res fel, Med Ctr, Univ Ark, Little Rock, 64-67, career develop award, 67- *Mem:* Soc Pediat Res; Endocrinol Soc; Am Fedn Clin Res. *Res:* Metabolism; effect of glucocorticoids on growth and maturation; control of mucopolysaccharide synthesis and degradation; growth hormone and somatomedin in acute leukemia. *Mailing Add:* Dept Pediat Univ Ark Med Ctr 4301 W Markham St Little Rock AR 72205

ELDERS, WILFRED ALLAN, b Sunderland, Eng, Mar 25, 33; m 61. GEOLOGY. *Educ:* Univ Durham, BSc, 57, PhD(geol), 61. *Prof Exp:* Demonstr petrol, Univ Durham, 59-61; from instr to asst prof geol, Univ Chicago, 61-68, assoc prof, Univ Ill, Chicago Circle, 68-69; assoc prof, 69-73, PROF GEOL, UNIV CALIF, RIVERSIDE, 73- *Concurrent Pos:* Louis Block Fund res grant, Univ Chicago, 62-63; NSF grants, 64-69 & 72-78; res geologist, Inst of Geophys & Planetary Physics, 73- *Mem:* AAAS; fel Brit Geol Soc; Am Mineral Soc; Am Geophys Union; Geol Soc Am. *Res:* Geology, geochemistry and geophysics of geothermal areas; water/rock interaction in geothermal reservoirs; geology of Salton Trough. *Mailing Add:* Dept Geol Scis Univ Calif 900 University Ave Riverside CA 92521

ELDIN, HAMED KAMAL, b Cairo, Egypt, Dec 31, 24; US citizen; m 51; c 3. INDUSTRIAL ENGINEERING, ENGINEERING MANAGEMENT. *Educ:* Cairo Univ, BSc, 45; Calif Inst Technol, MSc, 48; Univ Iowa, PhD(indust eng), 51. *Prof Exp:* Asst prof indust eng, Cairo Univ, 51-52; employee & pub rels mgr, Esso Standard Near East Inc, 52-57; admin mgr, Mobil Oil Egypt, 57-61; mem bd dirs, Nat Inst Mgr Develop, 61-63; consult comput systs, Mobil Oil Corp, 63-65 & Mobil Int, 65-66, rels dept, 66-67; PROF INDUST ENG & MGT, OKLA STATE UNIV, 68- *Concurrent Pos:* Consult, Ford Found, 71-72, UN Indust Develop Orgn, 71-73 & Ministry Pub Works, Saudi Arabia, 80-82; spec adv to Prime Minister of Egypt, 77-81; eng adv, Am Univ Cairo, 82-85; ed, Computer & Indust Eng, 76- *Honors & Awards:* Arab Petrol Conf Award, 61. *Mem:* Int Mgt Systs Asn; Eng Mgt Asn Egypt; sr mem Inst Indust Engrs; Am Soc Eng Educ. *Res:* Management science; software package library; computer oriented management information systems; decision support system; metric systems. *Mailing Add:* Boston Univ Overseas Prog APO New York NY 09102

ELDIS, GEORGE THOMAS, b Detroit, Mich, Apr 19, 44. METALLURGY. *Educ:* Mass Inst Technol, BS, 66, PhD(metall), 71. *Prof Exp:* Vis scientist, Max Planck Inst Metall, 71-73; sr res assoc, Amax Mat Res Ctr, Div Amax, Inc, 73-77, res supvr, 77-79, res mgr, 79-87; METALL ENGR, CLIMAX RES SERV, INC, 87- *Mem:* Am Inst Mining, Metall & Petrol Eng; Am Soc Metals; Am Foundrymen's Soc; Soc Automotive Engrs. *Res:* Development of wrought and cast ferrous alloys: rail steels, carburizing steels, low-alloy constructional and abrasion resistant steels in the wrought category; various graphitic and abrasion-resistant white irons in the cast category. *Mailing Add:* Climax Res Serv Inc 39205 Country Club Dr Farmington Hills MI 48331

EL-DOMEIRI, ALI A H, b Cairo, Egypt, Aug 23, 35; US citizen. SURGERY. *Educ:* Cairo Univ, MD, 58; Univ Ill, MS, 72. *Prof Exp:* Asst prof surg, Al-Azhar Univ, Cairo, Egypt, 71-72; from asst prof to assoc prof, Abraham Lincoln Sch Med, 73-78; PROF SURG, TEX TECH HEALTH SCI CTR, 78- *Concurrent Pos:* Lectr gen surg, Al-Azhar Univ, Cairo, Egypt, 65-66. *Mem:* Am Col Surgeons; Soc Surg Oncol; Soc Univ Surgeons; Royal Col Surgeons. *Mailing Add:* 2201 Oxford Ave Suite 103 Lubbock TX 79410

ELDON, CHARLES A, b Tacoma, Wash, Oct 15, 26. PROCESS ENGINEERING. *Educ:* Stanford Univ, BS, 48, MBA, 50. *Prof Exp:* Assoc, Hewlett-Packard Co, 51-69; pres, Ness Pac Div, Ness Indust, 69-72; MGR, CAPITOL EQUIP, HEWLETT-PACKARD CO, 72- *Concurrent Pos:* Consult, Process Eng, 69- *Mem:* Fel Inst Elec & Electronic Engrs (treas, 81-82, pres, 85). *Mailing Add:* Capital Equip 3500 Deer Creek Rd Bldg 26413 Palo Alto CA 94304

ELDRED, EARL, b Tacoma, Wash, Feb 27, 19; m 44; c 3. NEUROPHYSIOLOGY. *Educ:* Univ Wash, BS, 39; Northwestern Univ, MD & MS, 50. *Prof Exp:* Intern, Va Mason Hosp, Seattle, Wash, 50-51; from instr to assoc prof, 51-62, PROF ANAT, MED SCH, UNIV CALIF, LOS ANGELES, 62- *Concurrent Pos:* Mem staff, Karolinska Inst, Sweden, 52-53; Markle fel, 51-56; vis prof, Univ Fribourg, Switz, 72, Univ Lausanne, 81, Univ Otago, NZ, 87. *Mem:* Inst Brain Res Orgn; Soc Neurosci; Am Asn Anat. *Res:* Sensory receptors in muscle; motor control. *Mailing Add:* Dept Anat Univ Calif Los Angeles CA 90024

ELDRED, KENNETH M, b Springfield, Mass, Nov 25, 29; m 57; c 1. ACOUSTICS, ENVIRONMENTAL ENGINEERING. *Educ:* Mass Inst Technol, BS, 50. *Prof Exp:* Supvry engr, Boston Naval Shipyard, 51-54; supvry physicist & chief, Phys Acoust Sect, US Air Force, Wright Field, Ohio, 56-57; vpres & consult acoust, Western-Electro Acoust Labs, Los Angeles, 57-63; tech dir sci serv & systs group, Wyle Labs, El Segundo, Calif, 63-73; vpres & dir, Archit Technol & Noise Control Div, Bolt Beranek & Newman Inc, 73-77, prin consult, 77-81; DIR, KEN ELDRED ENG, 81- *Concurrent Pos:* Chmn, Peer Group Gen Eng, Nat Acad Eng, 77-78; mem comt hearing, bioacoust & diomedics, Nat Res Coun, 63-; mem comt aircraft noise, Soc Automotive Engrs, 62-; vchmn, Exec Standards Coun, Am Nat Standards Inst, 82-84, chmn, 85-87; standard adv, Acoustic Soc Am, 87. *Mem:* Nat Acad Eng; Inst Noise Control Eng (pres, 76); fel Acoust Soc Am; Soc Naval Architects & Marine Engrs; Am Nat Standards Inst; Soc Automotive Engrs. *Res:* Aircraft acoustics; response of humans to noise; modeling of national noise impact relative to sources. *Mailing Add:* PO Box 1037 Concord MA 01742

ELDRED, NELSON RICHARDS, b Oberlin, Ohio, Mar 6, 21; m 45, 68; c 5. ORGANIC CHEMISTRY. *Educ:* Oberlin Col, AB, 43; Wayne State Univ, MS, 47; Pa State Univ, PhD (chem), 51. *Prof Exp:* Asst, Parke Davis & Co, 43-46; res chemist, Union Carbide Co, 50-68; asst mgr develop, Buckman Labs, Inc, 69; supvr chem div, 70-75, MGR TECHNO-ECON FORECASTING, GRAPHIC ARTS TECH FOUND, 75- *Mem:* AAAS; Tech Asn Pulp & Paper Indust; Am Chem Soc; fel Am Inst Chemists. *Res:* Synthetic resins and fibers; chemistry of papermaking; chemistry of paper and ink; technological forecasting; graphic arts industries. *Mailing Add:* 2719 Bellemede Blvd New Port Richey FL 34655-1441

ELDRED, NORMAN ORVILLE, b Vicksburg, Mich, Sept 13, 16. NUCLEAR POWER STATION DESIGN. *Educ:* Univ Mich, Ann Arbor, BS, 38, MS, 47. *Prof Exp:* Dist gales & eng mgr, Permutit Co, New York, 38-43; line eng officer, USN, 43-46; feature engr, Grinnell Corp, US AEC Nuclear Separations Plant, Ohio, 54; sr design engr, Leonard Construct Co, Monsanto Chem Co, Chicago, 55-57; proj engr, steel mills & blast furnace designs, US Steel Corp Cent Eng, Chicago, 65-68; proj engr, Utility Power Sta Pipe Hangers, Cent Iron Co, Long Island, 73-77; proj engr, Nuclear Reactor seismic piping, Long Island Lighting Co, Shorehan Nuclear Power Sta, Courter & Co, 77-83; CONSULT ENGR, NORMAN O ELDRED & ASSOCS CONSULT ENGRS, 83- *Mem:* Nat Soc Prof Engrs; Am Soc Mech Engrs; Inst Elec & Electronics Engrs; Am Inst Chem Engrs; Am Soc Testing & Mat; US Naval Inst. *Res:* All phases of nuclear power station design including reactors, steam cycle, turbo-generators, nuclear waste treatment; seismic design of nuclear piping, hangers and supports, both large and small bore. *Mailing Add:* 508 Draper St Vicksburg MI 49097

ELDRED, WILLIAM D, b Denver, Colo, June 16, 50; m 76. RETINAL NEUROCHEMISTRY, RETINAL NEUROANATOMY. *Educ:* Univ Colo Boulder, BS, 72; Univ Colo Health Sci Ctr, PhD(anat), 79. *Prof Exp:* Postdoctoral fel, Dept Anat, State Univ NY Stony Brook, 79-80 & neurobiol, Dept Neurobiol & Behav, 80-81, res asst prof neurobiol, 81-82; asst prof, 82-88, ASSOC PROF DEPT BIOL, BOSTON UNIV, 88- *Concurrent Pos:* Prin investr, Nat Eye Inst Grants, NIH, 79- *Mem:* Soc Neurosci; Asn Res Vision & Ophthal. *Res:* Anatomical, biochemical, and quantitative studies of retinal neurons and their synaptic interactions, using immunocytochemical methods and computer analysis at the light and electron microscopic levels; regional differences in the retina. *Mailing Add:* Dept Biol Boston Univ Five Cummington St Boston MA 02215

ELDREDGE, DONALD HERBERT, b South Bend, Ind, July 5, 21; m 47; c 4. PHYSIOLOGY, BIOPHYSICS. *Educ:* Harvard Univ, SB, 43; Harvard Med Sch, MD, 46. *Prof Exp:* Intern, Boston City Hosp, 46-47; asst resident, Barnes Hosp, St Louis, 52-53; res asst, 54-58, res instr, 58-60, res asst prof, 60-61, res assoc prof, 61-64, RES PROF OTOLARYNGOL, SCH MED, WASH UNIV, 64-; ASSOC DIR RES, CENT INST DEAF, 78- *Concurrent Pos:* Mem, Comt Hearing & Bioacoustics, Armed Forces-Nat Res Coun, 53-70; res assoc, Cent Inst Deaf, 53-70, asst dir, 70-78. *Mem:* Acoust Soc Am; Soc Neurosci; Asn Res Otolaryngol; Sigma Xi. *Res:* Normal and pathologic physiology and mechanical function of the ear. *Mailing Add:* 6819 Waterman Ave St Louis MO 63130

ELDREDGE, KELLY HUSBANDS, b Salt Lake City, Utah, Apr 5, 21; m 45, 54; c 4. VIROLOGY. *Educ:* Univ Utah, BS, 43, MS, 45; Stanford Univ, PhD(virol), 49. *Prof Exp:* Asst, Univ Utah, 42-45; instr life sci, San Francisco State Col, 48-49; asst prof bact, Ariz State Univ, 49-52; lab technologist, Mem Med Ctr, 53-54; teacher, Pub Sch, 54-55; from asst prof to assoc prof life sci, 55-70, prof 70-85, EMER PROF BIOL SCI, CALIF STATE UNIV, SACRAMENTO, 85- *Concurrent Pos:* Consult, Friedlanders Labs, Sacramento County Hosp, Eskaton Health Care Ctr & Am River Hosp. *Mem:* Am Soc Microbiol; Am Pub Health Asn. *Res:* Effect of certain compounds on bacteriophage production; pathogenic fungi. *Mailing Add:* 2904 Crescent Ct Sacramento CA 95825

ELDREDGE, LUCIUS G, b East Greenwich, RI, Mar 1, 38; m 58; c 4. MARINE SCIENCE. *Educ:* Univ RI, BS, 59; Univ Hawaii, PhD(zool), 65. *Prof Exp:* Assoc prof biol, Univ Guam, 65-67, prof & chmn dept, 67-71, dir marine lab, 71-73, prof biol, 73-87; RES ASSOC MARINE ZOOL, BERNICE P BISHOP MUS, HONOLULU, 64- *Concurrent Pos:* Prog officer, Food & Agr Orgn of the UN, Rome, 87-89; exec secy, Pac Sci Asn, 89- *Mem:* Explorers Club; Soc Syst Zool; Crustacean Soc; Brit Soc Bibliog Natural Hist. *Res:* Taxonomy and ecology of ascidians; taxonomy of Indo-Pacific intertidal gastropods, crustaceans and lancelets; Indo-Pacific marine invertebrates; zoogeography; taxonomy; ecology. *Mailing Add:* Pac Sci Asn PO Box 17801 Honolulu HI 96817

ELDREDGE, NILES, b Brooklyn, NY, Aug 25, 43; m 64; c 2. PALEOBIOLOGY. *Educ:* Columbia Univ, AB, 65, PhD(geol), 69. *Prof Exp:* Asst cur, 69-74, assoc cur invert paleont, 74-79, CUR INVERTEBRATES, AM MUS NATURAL HIST, 79- *Concurrent Pos:* Adj prof biol, City Univ NY, 72-80; adj assoc prof geol, Columbia Univ, 75-; co-ed, Syst Zool, 74-77. *Honors & Awards:* Schuchert Award, Paleont Soc, 79. *Mem:* Paleont Soc; Brit Palaeont Asn; Soc Study Evolution; Soc Syst Zool. *Res:* Trilobite systematics and evolutionary theory. *Mailing Add:* Dept Invert Am Mus Natural Hist Cenral Park W & 79th St New York NY 10024

ELDRIDGE, BRUCE FREDERICK, b San Jose, Calif, Mar 26, 33; m 57; c 3. MEDICAL ENTOMOLOGY. *Educ:* San Jose State Univ, AB, 54; Wash State Univ, MS, 56; Purdue Univ, PhD(entom), 65. *Prof Exp:* Instr, Army Med Serv Sch, Tex, US Army, 57-58, med entomologist, Walter Reed Army Inst Res,Washington, DC, 58-60 & 61-63, instr med entom, Med Field Serv Sch, 65-66, med entomologist, Atlantic-Pac Interoceanic Canal Study Comn, 68-69,entomologist, 68-69, chief dept entom, Walter Reed Army Inst Res, 69-77; consult med entom, US Army Surg Gen, 77; prof entom & chmn dept, 78-86, PROF ENTOM & DIR UNIVERSITYWIDE MOSQUITO RES PROG, UNIV CALIF, 86- *Concurrent Pos:* Actg chief entom res br, US Army Med Res & Develop Command, 69-70; chmn, Armed Forces Pest Control Bd, 71-73. *Mem:* Entom Soc Am; fel AAAS; Am Mosquito Control Asn; Am Soc Trop Med & Hyg. *Res:* Ecology and physiology of mosquitos. *Mailing Add:* Dept Entom Univ Calif Davis CA 95616

ELDRIDGE, CHARLES A, b Sacramento, Calif, Mar 15, 49; m 76; c 1. COMPUTER NETWORKING. *Educ:* Pomona Col, BA, 71; Yale Univ, MPhil, 74, PhD(biophys), 76. *Prof Exp:* Res fel biophys, Cornell Univ, 76-79; mem tech staff, InterSysts Inc, 79-80; mem tech staff, Syst Develop Corp, 80-82; mem tech staff, MITRE Corp, 82-83; PRIN INVESTR, SPARTA, INC, 83- *Concurrent Pos:* Assoc prof lectr, George Washington Univ, 84- *Mem:* Inst Elec & Electronics Engrs; Asn Comput Mach; Soc Indust & Appl Math. *Res:* Machine techniques for communication, including protocols, signal processing and resource-sharing; practical application and design mechanisms. *Mailing Add:* 999 Seaton Lane Falls Church VA 22046

ELDRIDGE, DAVID WYATT, b Chattanooga, Tenn, Oct 31, 40; m 62; c 2. MICROBIAL PHYSIOLOGY. *Educ:* Tenn Polytech Inst, BS, 62; Auburn Univ, MS, 64, PhD(bot), 69. *Prof Exp:* Instr bot, Auburn Univ, 64-66, res asst, 66-68; ASSOC PROF BIOL, BAYLOR UNIV, 68- *Mem:* Am Soc Microbiol. *Res:* Biochemistry and physiology of microbial toxins, especially aflatoxins produced by Aspergillus flavus. *Mailing Add:* Dept Biol Baylor Univ Col Arts & Sci Waco TX 76798

ELDRIDGE, FREDERIC L, b Kansas City, Mo, July 8, 24; m 51; c 2. RESPIRATORY PHYSIOLOGY. *Educ:* Stanford Univ, AB, 45, MD, 48. *Prof Exp:* Intern med, Stanford Univ Hosps, 47-48; resident pediat, Univ Tex Med Br, 48-50; resident med, hosps, 50-51, from instr to prof, Sch Med, Stanford Univ, 54-73, head div respiratory med, 64-73; PROF MED & PHYSIOL, SCH MED, UNIV NC, CHAPEL HILL, 73- *Concurrent Pos:* Chief med serv, Palo Alto Vet Hosp, 60-68; Irving fel, Sch Med, Stanford Univ, 53, Neizer fel, 53-54; USPHS spec res fel pulmonary physiol, St Bartholomews Hosp, Univ London, 54-55 & neurophysiol, Col Med, Univ Utah, 68-69; Bank Am-Giannini Found fel, 55-56; Markle Found scholar med sci, 56-61. *Mem:* Am Fedn Clin Res; Am Soc Clin Invest; Soc Neurosci; Am Physiol Soc; Am Thoracic Soc. *Res:* Cardiopulmonary physiology; mechanics of respiration; metabolism of lactic acid; respiratory control mechanisms. *Mailing Add:* Dept Physiol Univ NC Chapel Hill NC 27599-7545

ELDRIDGE, JOHN CHARLES, b Chicago, Ill, June 7, 42; m 86. REPRODUCTIVE PHYSIOLOGY, MOLECULAR ENDOCRINOLOGY. *Educ:* NCent Col, BA, 65; Northern Ill Univ, MS, 67; Med Col Ga, PhD(endocrinol), 72. *Prof Exp:* Instr, Dept Biol & Health Sci, Orange County Comn Col, NY, 67-68; fel, NIH & Med Res, Bordeaux, France, 71-72; res assoc, Dept Endocrinol, Med Col Ga, 73; asst prof lab med, Med Univ SC, 73-78; asst prof, 78-88, ASSOC PROF PHYSIOL & PHARMACOL, 88-, MEM GRAD FAC, BOWMAN GRAY SCH MED, WAKE FOREST UNIV, 78- *Concurrent Pos:* Consult non-med prof staff, Med Univ Hosp & mem joint fac, Col Allied Health & Col Grad Studies, Med Univ SC, 74-78; consult, endocrine toxicol, Nida & HIH PI, 84-, Ciba Geigy Corp, 88- *Mem:* Endocrine Soc; Soc Study Reprod; AAAS; Am Fertility Soc; Am Soc Androl; Soc Neurosci. *Res:* Reproductive physiology and toxicology; substance abuse and endocrine function; hormone receptor activity; pathophysiology of aging; medical education. *Mailing Add:* Dept Physiol Pharmacol Bowman Gray Sch Med Winston-Salem NC 27103

ELDRIDGE, JOHN W(ILLIAM), b Nashua, NH, Aug 22, 21; m 42; c 3. CHEMICAL ENGINEERING. *Educ:* Univ Maine, BS, 42; Syracuse Univ, MS, 45; Univ Minn, PhD(chem eng), 49. *Prof Exp:* Chem engr, Semet-Solvay Co, 42-46 & Barrett Div, Allied Chem & Dye Corp, 49-50; from asst prof to assoc prof chem eng, Univ Va, 50-62; head dept, 62-76, PROF CHEM ENG, UNIV MASS, AMHERST, 76- *Concurrent Pos:* Consult, Albemarle Paper Mfg Co, 51-63 & Holyoke Water Power Co, 62-64; vpres, Gen Aerosols Corp, 55-, KSE, Inc & Southern Res Inst, 86- *Mem:* Am Chem Soc; Am Inst Chem Engrs; Am Soc Eng Educ. *Res:* Continuous flow chemical reactor systems; thermodynamics; kinetics; polymerization and catalysis. *Mailing Add:* Dept Chem Eng Univ Mass Amherst MA 01003

ELDRIDGE, KLAUS EMIL, b Breslau, Germany, June 5, 38; US citizen; m 60; c 2. MATHEMATICS, COMPUTER SCIENCE. *Educ:* Hardin-Simmons Univ, BA, 60; Okla State Univ, MS, 62; Univ Colo, PhD(math), 65. *Prof Exp:* Instr math & physics, Hardin-Simmons Univ, 62; asst prof math, 65-70, ASSOC PROF MATH, OHIO UNIV, 70-, CHAIR DEPT, COMPUT SCI, 81- *Mem:* Inst Elec & Electronics Engrs; Am Asn Univ Prof; Math Asn Am; Am Math Soc; Asn Comput Mach; Comput Soc. *Res:* Structures of fields, rings and algebras and their relations to the structure of groups; computer sciences; intelligent systems. *Mailing Add:* Dept Comput Sci Ohio Univ Morton Hall Rm 423 Athens OH 45701

ELDRIDGE, MAXWELL BRUCE, b Chicago, Ill, Dec 15, 42; m 65; c 2. FISH BIOLOGY, ECOLOGICAL PHYSIOLOGY. *Educ:* Univ Calif, Santa Barbara, BA, 65; Humboldt State Univ, MS, 70. *Prof Exp:* Marine biologist aquacult, Inmont Corp, 70; fishery biologist res, Bur Sport Fisheries, US Fish & Wildlife Serv, 70-71; SUPVR FISHERY BIOLOGIST POLLUTION PHYSIOL, NAT MARINE FISHERIES SERV, DEPT COM, 71- *Mem:* Am Fisheries Soc; Am Inst Fishery Res Biologists; Sigma Xi. *Res:* Physiological ecology of marine aquatic organisms with special interest in physiology of early life stages of fishes; pollution physiology and pollutant effects on bioenergetic processes. *Mailing Add:* Tiburon Lab NOAA/NMFS 3150 Paradise Dr Tiburon CA 94920

ELDRIDGE, ROSWELL, b Great Neck, NY, Jan 1, 34; m 68; c 5. MEDICAL GENETICS, NEUROLOGY. *Educ:* Haverford Col, BS, 55; Univ Rochester, MD, 60. *Prof Exp:* Sr surg genetics, Epidemiol Br, Nat Inst Neurol & Dis, NIH, 66-68, head sect genetics & epidemiol, 68-75; head sect neurogenetics, 75-76, HEAD SECT GENETICS IN EPIDEMIOL, INFECT DIS BR, INTRAMURAL PROG, NAT INST NEUROL & COMMUNICATIVE DIS & STROKE, NIH, 76- *Concurrent Pos:* Mem med adv bd, Dystonia Found, 74- & Nat Found Jewish Genetic Dis, 75- *Mem:* Am Soc Human Genetics; Am Fedn Clin Res. *Res:* Application of Mendelian genetics to study of general neurological syndromes in effort to define discreet hereditary disease; role of genetic predisposition in disorders of immune response such as multiple sclerosis. *Mailing Add:* Bldg Fed Rm 8704 Nat Inst Health Bethesda MD 20892

ELDUMIATI, ISMAIL IBRAHIM, b Damanhour, Egypt, Jan 19, 40; m 64; c 3. ELECTRICAL ENGINEERING. *Educ:* Univ Alexandria, BScEE, 62; Univ Mich, Ann Arbor, MS, 66 & 68, PhD(elec eng), 70. *Prof Exp:* Instr elec eng, Univ Alexandria, 62-65; assoc dir biophys, Sensors Inc, 70-72; mem tech staff, 72-81, SUPVR, AT&T BELL LABS, 85-, HEAD DEPT, 85- *Mem:* AAAS; Inst Elec & Electronics Engrs. *Res:* Solid state materials and devices; integrated circuits; digital communications; signal processing; microcomputer design. *Mailing Add:* Bell Labs Mountain Ave Rm 2A422 Murray Hill NJ 07974

ELESPURU, ROSALIE K, b Memphis, Tenn, May 3, 44; m 73; c 1. CANCER RESEARCH. *Educ:* Univ Rochester, AB, 66; Univ Tenn, Oak Ridge, PhD(biomed sci), 76. *Prof Exp:* Fel, Cancer Biol Prog, 76-78, scientist, Biol Carcinogenesis Prog, 79-81, group leader, Fermentation Prog, 81-84, SCIENTIST II, LAB CHEM & PHYS CARCINOGENESIS, FREDERICK CANCER RES FACIL & LEO BAECK INST, BASIC RES PROG, NAT CANCER INST, 84- *Mem:* Environ Mutagen Soc; AAAS; Am Soc Microbiol. *Res:* Bacterial assays for use in the screening of chemical carcinogens and in the search for new cancer treatment drugs. *Mailing Add:* FDA HFZ-113 5600 Fishers Lane Rockville MD 20857

ELEUTERIO, HERBERT SOUSA, b New Bedford, Mass, Nov 23, 27; m 51; c 6. ORGANIC CHEMISTRY. *Educ:* Tufts Col, BS, 49; Mich State Univ, PhD(chem), 53. *Prof Exp:* Res chemist, Polychem Dept, Exp Sta, 54-58, supvr, 58-59, supvr, Indust & Biochem Dept, 59-62, res sect head, Eastern Lab, 62-64 & Exp Sta Lab, 64-68, dir, 68-70, dir, Eastern Lab, Explosives Dept, 70-72 & Exp Sta Lab, 72, asst dir, Res & Develop Div, Explosives Dept, 72-76, prod mgr, Nylon Intermediates Div, Polymer Intermediates Dept, 76-77, res dir, Polymer Intermediates Dept, 77-78, res dir, Petrochem Dept, 78-85, dir, Tech Div AED, 85-89, DIR, NEW TECHNOL, E I DUPONT DE NEMOURS & CO, 89- *Concurrent Pos:* Chair gov bd, Coun Chem Res. *Honors & Awards:* Chem Pioneer Award, 87. *Mem:* Am Chem Soc; Am Nuclear Soc; Chem Soc; Sigma Xi; AAAS; fel Am Inst Chemists. *Res:* Reaction mechanisms; stereochemistry; polymer chemistry; nuclear chemistry. *Mailing Add:* E I du Pont de Nemours & Co Petrochem Dept Wilmington DE 19898

ELEUTERIO, MARIANNE KINGSBURY, b Cassopolis, Mich, Aug 7, 29; m 51; c 6. GENE REGULATION, BIODEGRADATION. *Educ:* Mich State Univ, BS, 50; Univ Del, PhD(biol), 71. *Prof Exp:* Postdoctoral, Skin & Cancer Hosp Mycol, Temple Univ Med Sch, 71-72 & Biol Dept, Univ Del, 72-73; asst prof genetics & microbiol, 73-76, assoc prof genetics, 76-88, PROF GENETICS, WEST CHESTER UNIV, PA, 88- *Concurrent Pos:* Vis scientist, Cent Res Dept, E I Du Pont de Nemours & Co, 86-87; vis fac biol, Indiana Univ, Bloomington, 90-91; deleg, People to People Ambassador Prog, Biotechnol Deleg to China, 90- *Mem:* Am Soc Microbiol; Am Soc Human Genetics; Genetics Soc Am; AAAS; Sigma Xi; Am Inst Biol Sci. *Res:* Isolation and characterization of photosynthetic bacteria from unusual environments; genetic studies of oxygen regulation of photosynthetic apparatus in bacteria and of catabolite repression in escherichia coli-L-arabinose operon. *Mailing Add:* 513 Ivydale Rd Wilmington DE 19803

ELEUTERIUS, LIONEL NUMA, b Biloxi, Miss, Dec 25, 36; m 69; c 2. MARINE BOTANY. *Educ:* Univ Southern Miss, BS, 66, MS, 68; Miss State Univ, PhD(bot), 74. *Prof Exp:* Biol technician plant path, US Forest Serv, 61-65; res asst bot, Univ Southern Miss, 65-68; botanist marine bot, 68-70, HEAD BOT SECT, GULF COAST RES LAB, 70- *Concurrent Pos:* Instr night classes bot & gen biol, Univ Southern Miss, 68-72; adj prof, Univ Miss, Miss State Univ & Univ Southern Miss; mem, Nat Wetland Tech Coun, Washington, DC. *Mem:* Am Bot Soc; Ecol Soc; Am Soc Naturalists; Sigma Xi. *Res:* Botanical aspects of marine communities to include ecology of salt marsh plants, seagrasses and marine algae, especially the rush, Juncus Roemerianus; island plants and silica content of coastal plants; transplanting sea oats and other plants for erosion control and dune building; halophyte morphology, salt tolerance and genetic adaptation; author of several publications. *Mailing Add:* Dept Botany Gulf Coast Res Lab 703 E Beach Drive PO Box 7000 Ocean Springs MS 39564-7000

ELEY, JAMES H, b Montgomery, Ala, July 16, 40; m 62; c 2. BIOCHEMISTRY, PLANT PHYSIOLOGY. *Educ:* Univ Tex, Austin, BA, 62, MA, 64, PhD(physiol), 67. *Prof Exp:* NIH fel biochem of photosynthesis, Brandeis Univ, 67-68; asst prof bot, Univ Ky, 68-74; ASSOC PROF BIOL, UNIV COLO, 74- *Mem:* AAAS; Am Soc Plant Physiol; Japanese Soc Plant Physiol. *Res:* Photosynthesis; plant physiology. *Mailing Add:* Dept Biol Univ Colo-Austin Bluffs Pkwy Box 7150 Colorado Springs CO 80933

ELEY, RICHARD ROBERT, b Lima, Ohio; m 66; c 5. RHEOLOGY. *Educ:* Kent State Univ, BS, 66, PhD(inorg chem), 73. *Prof Exp:* Asst prof chem, US Naval Acad, Annapolis, 73-75; SR SCIENTIST, GLIDDEN CO, 75- *Concurrent Pos:* Lectr, Dept Continuing Educ, Kent State Univ, 81-, Miteck Corp, 87- *Honors & Awards:* First Place Roon Award, Fedn Socs Coatings Technol, 86 & 88. *Mem:* Am Chem Soc; Soc Rheology; Fedn Socs Coatings Technol. *Res:* Applied rheology of coatings, polymer melts, solutions, colloids and pigment dispersions. *Mailing Add:* Glidden Corp SCM Corp 16651 Sprague Rd Strongsville OH 44136

ELFANT, ROBERT F, b New York, NY, Mar 12, 36. SEMICONDUCTOR & PACKAGING COMPONENTS, MAGNETIC STORAGE. *Educ:* Southern Methodist Univ, BS, 57, NY Univ, MEE, 59; Purdue Univ, PhD(elec eng), 61. *Prof Exp:* Vpres, systs, Omex Corp, 80-82; vpres, Microcomponents Group, Burroughs Corp, 82-85; dir assembly & parts, Intel, 85-86, gen mgr cellular prod, 86-89; VPRES ENG, 89-91, SR VPRES ENG, WESTERN DIGITAL CORP, 91- *Mem:* Fel Inst Elec & Electronics Engrs. *Mailing Add:* Western Digital Corp 8105 Irvine Ctr Dr Irvine CA 92718

ELFBAUM, STANLEY GOODMAN, b Boston, Mass, Sept 24, 38; m 70; c 3. CLINICAL CHEMISTRY. *Educ:* Northeastern Univ, BS, 61; Northwestern Univ, PhD(biochem), 66; Am Bd Clin Chem, dipl. *Prof Exp:* Clin chemist, Boston Med Lab, Mass, 57-61; sr biochemist, Gillette Med Res Inst, 65-67, proj supvr biomed div, 67-70; dept supvr, 70-73, tech dir, Boston Med Lab, Inc, 73-77; PRES & TECH DIR, CLIN SCI LAB, INC, 77- *Concurrent Pos:* Clin chem instr, Northeastern Univ, 73-83. *Mem:* Am Chem Soc; Am Asn Clin Chem; Am Asn Bioanalysts. *Res:* Radioimmunoassay of thyroid hormones; detection of abnormal hemoglobins; physical biochemistry of proteins. *Mailing Add:* 66 Cocaset St Unit 11 Foxboro MA 02035

ELFENBEIN, GERALD JAY, b Norristown, Pa, Mar 4, 45; c 2. BONE MARROW TRANSPLANTATION, MEDICAL ONCOLOGY. *Educ:* Harvard Col, AB, 66; Johns Hopkins Univ, MD, 70. *Prof Exp:* Intern med, Johns Hopkins Hosp, 70-71; res assoc immunol, NIH, 71-73; resident med, Johns Hopkins Hosp, 73-74, fel oncol, Oncol Ctr, 74-76; asst prof, Johns Hopkins Univ, 76-81; ASSOC PROF MED, UNIV FLA, 81-; MED DIR, BONE MARROW TRANSPLANT UNIT, SHANDS HOSP, 81- *Concurrent Pos:* Investr immunol, Howard Hughes Med Inst, 76-80; vis prof, Shaare Zedek Hosp, Jerusalem, Israel, 85; chmn, Bone Marrow Transplant Comt, Southeastern Cancer Study Group, 85-86. *Mem:* Int Soc Exp Hemat; Transplantation Soc; Am Asn Immunologists; Am Soc Hemat; Am Soc Clin Oncol; Am Asn Cancer Res; Am Fedn Clin Res. *Res:* Regulation of human T cell differentiation in vitro; human bone marrow transplant biology; marrow transplantation as a curative therapy for patients with leukemia and related disorders. *Mailing Add:* Dir Bone Marrow Transplant Prog H Lee Moffitt Cancer Ctr Univ SFla 12902 Magnolia Dr Tampa FL 33612

ELFNER, LLOYD F, b Manitowoc, Wis, Sept 13, 23; m 63; c 3. PSYCHOACOUSTICS, EXPERIMENTAL PSYCHOLOGY. *Educ:* Univ Wis, BS, 58, MS, 60, PhD(exp psychol), 62. *Prof Exp:* From asst prof to assoc prof psychol, Kent State Univ, 62-67; assoc prof, 67-70, PROF PSYCHOL, FLA STATE UNIV, 70- *Concurrent Pos:* Grants, NSF, 64-66 & 67-69, NIH, 64-66 & 69-72; mem, Evoked Audiometry Study Group. *Mem:* Am Psychol Asn; Acoust Soc Am; Psychonomic Soc. *Res:* Temporal, intensive and spectral resolving powers of the human auditory system; auditory evoked potentials; effects of noise on behavior. *Mailing Add:* Dept Psychol Fla State Univ Tallahassee FL 32306

ELFNER, LYNN EDWARD, b Springfield, Ohio, Aug 29, 44; m 68; c 3. PLANT ECOLOGY, ZOOLOGY. *Educ:* Ohio State Univ, BS, 67, MS, 71. *Prof Exp:* Sci teacher, Mount Orab Local Schs, 67-70; teaching asst botany, Ohio State Univ, 70-71; executive dir admin, Ohio Environ Coun, 71-73; budget mgt analyst admin, Ohio Off Budget & Mgt, 73-74; EXEC OFFICER ADMIN, OHIO ACAD SCI, 75- *Mem:* AAAS; Asn Academies Sci; Sigma Xi. *Res:* Plant ecology; post glacial plant migration; history of science; science policy. *Mailing Add:* 445 King Ave Columbus OH 43201

ELFONT, EDNA A, ULTRASTRUCTURE, ONCOLOGY. *Educ:* Univ Md, PhD(histol & embryol), 70. *Prof Exp:* DIR ELECTRON MICROS LAB, PATH SERVICES, SINAI HOSP, 72- *Mailing Add:* Dept Path Electron Micros Lab Sinai Hosp 6767 W Outer Dr Detroit MI 48235

ELFORD, HOWARD LEE, b Chicago, Ill, Sept 2, 35; m 62; c 2. BIOCHEMISTRY, CANCER. *Educ:* Univ Ill, BS, 58; Cornell Univ, PhD(biochem), 62. *Prof Exp:* Res asst biochem, Cornell Univ, 62; asst prof, Med Sch, Univ Mich, Ann Arbor, 64-69; asst prof exp med & pharmacol, Med Sch, Duke Univ, 69-75; assoc prof biochem, Med Col Va, Va Commonwealth Univ, 75-80; sr biochemist, Va Assoc Res Campus, Col William & Mary, 80-84; PRES & RES DIR, MOLECULES HEALTH, INC, 83- *Concurrent Pos:* NIH fel, Mass Inst Technol, 62-64. *Mem:* Am Soc Biol Chem; AAAS; Am Asn Cancer Res; Sigma Xi; Free Radical Soc. *Res:* Deoxyribonucleotide and DNA biosynthesis; biochemistry and pharmacology of cancer and cancer drug development; mechanisms of vitamin B-12 and its role in mammalian metabolism. *Mailing Add:* Molecules for Health Inc 3313 Gloucester Rd Richmond VA 23227

ELFSTROM, GARY MACDONALD, b Vancouver, BC, Aug 14, 44; m 69; c 2. AERODYNAMICS. *Educ:* Univ BC, BASc, 68; Univ London, PhD(aeronaut eng), 71. *Prof Exp:* Asst prof aeronaut eng, Space Inst, Univ Tenn, 71-73; assoc res officer aeronaut eng, Nat Res Coun Can, 73-81; CHIEF AERODYNAMICIST, DSMA INT, INC, MISSISSAUGA, ONT, 81- *Concurrent Pos:* Res fel, Space Inst, Univ Tenn, 71-72; sessional lectr aeronaut eng, Carleton Univ, Ottawa, Ont, 79-81. *Mem:* Am Inst Aeronaut & Astronaut; assoc fel Can Aeronaut & Space Inst; Sigma Xi. *Res:* Fluid mechanics, specifically turbulent boundary layer flow at high speeds; experimental facility development. *Mailing Add:* 1032 Deer Run Mississauga ON L5C 3N4 Can

ELFTMAN, ALICE G, anatomy, for more information see previous edition

ELFTMAN, HERBERT (OLIVER), anatomy, for more information see previous edition

ELFVING, DONALD CARL, b Albany, Calif, June 20, 41; m 85. APPLE PRODUCTION, APPLE PHYSIOLOGY ADMINISTRATION. *Educ:* Univ Calif, Davis, BS, 64, MS, 66; Univ Calif, Riverside, PhD(plant physiol), 71. *Prof Exp:* Asst prof pomol, Cornell Univ, 72-77, assoc prof, 77-79; sr res pomologist, 79-91, CHIEF RES SCIENTIST, HORT RES INST ONT, VINELAND STATION, ONT, 91- *Concurrent Pos:* Consult, US Agency for Int Develop, 77, Int Agr Develop Serv, NY, 81-82, Mountain Agr Proj, Cornell Univ, 78, govt of Ont in China, 83; vis prof, Dept Entom, Mich State Univ, 78. *Mem:* Am Soc Hort Sci; Int Dwarf Fruit Tree Asn; Sigma Xi. *Res:* Integration of cultivar-rootstock interactions, pruning and training, growth regulators, irrigation and soil management into orchard-management systems for improved apple production; relevant tree-physiology problems. *Mailing Add:* Hort Res Inst Ont Vineland Station ON L0R 2E0 Can

ELGAVISH, ADA S, b Cluj, Romania, Jan 23, 46; Israeli & US citizen; m 68; c 2. MEMBRANE TRANSPORT, CYSTIC FIBROSIS. *Educ:* Tel Aviv Univ, Israel, BSc, 69, MSc, 72; Weizmann Inst Sci, Rehovoth, PhD(biochem), 78. *Prof Exp:* NIH vis fel, NIA, Baltimore, MD, 79-81; instr, Lab Membrane Biol, Univ Ala, 81-82, res assoc, 82-84, res asst prof, dept pharmacol, 84-89, ASSOC SCIENTIST, DEPT PHARMACOL, CYSTIC FIBROSIS CTR, UNIV ALA, 84-, ASST PROF, DEPT COMP MED, 89- *Concurrent Pos:* Grantee, Am Lung Asn, 84-86, Cystis Fibrosis Found, 86-90, NIH, 89. *Honors & Awards:* Career Investr Award, Am Lung Asn, 87. *Mem:* AAAS; NY Acad Sci; Sigma Xi; Am Phys Soc; Am Thoracic Soc. *Res:* Membrane function; mechanisims of action and molecular structure of anion exchange

(Chinese hamster ovary cells); effects of bacterial toxins, infectious agents and cytokines on ion transport (epithelial cells and fibroblast cell cultures isolated from the human urinary bladder); cystic fibrosis; sulfate transport and glycoprotein sulfation (pancreas duct epithelial cells). *Mailing Add:* Dept Comp Med Volker Hall 402C Univ Ala Birmingham AL 35294

ELGEE, NEIL J, b Nova Scotia, Can, 26; m; c 5. ENDOCRINOLOGY. *Educ:* Univ Rochester, MD, 50. *Prof Exp:* Intern, Peter Bent Brigham Hosp, 50-51; asst res, Strong Mem Hosp, 51-52; res fel, Univ Washington, 52-54; resident, Harborview Med Ctr, 54-55; CLIN PROF, DEPT MED, UNIV WASH, 55- *Mem:* Inst Med Nat Acad Sci. *Mailing Add:* 1229 Madison Ave Seattle WA 98104

EL-GENK, MOHAMED SHAFIK, b Tanta, Egypt, Feb 27, 47; US citizen; m 76; c 1. THERMAL MANAGEMENT, SPACE POWER. *Educ:* Univ Alexandria, Egypt, BS, 68, MS, 75; Univ New Mex, PhD (nuclear eng), 78. *Prof Exp:* Engr, Egyptian Gen Organ Indust, 68-73; res engr, Egyptian Atomic Energy Authority, 73-78; eng specialist, Idaho Nat Eng Lab, 78-81; from asst prof to assoc prof, 81-88, PROF NUCLEAR ENG, UNIV N MEX, 88-, DIR INST SPACE NUCLEAR POWER STUDIES, 84- *Concurrent Pos:* Consult, Los Alamos Nat Lab, 82-, Sci & Eng Assoc Inc, 82-83, BDM Corp, 82-83 & Sci Appl Int Corp, 86- *Mem:* Am Soc Mech Engrs; Am Nuclear Soc; Am Inst Chem Engrs; Am Soc Eng Educ. *Res:* Nuclear reactors; thermal hydraulics and safety analysis; design and modelling of space based nuclear power sources; two-phased flow in microgravity; survivability safety and integration of space power systems; transient modeling of heat pipes. *Mailing Add:* Inst Space Nuclear Power Studies Dept Chem & Nuclear Eng Dept Univ New Mex Albuquerque NM 87131

ELGERD, O(LLE) I(NGEMAR), b Oxberg, Sweden, Mar 31, 25; nat US; m 48; c 3. ELECTRICAL ENGINEERING. *Educ:* Orebro Tech Col, Sweden, 45; Royal Inst Technol, Sweden, Dipl, 50; Washington Univ, St Louis, DSc(elec eng), 56. *Prof Exp:* Designer relay & control of hydroplants, Swed Elec Co, Sweden, 48-51; asst chief engr, Utility Co, 51-52; designer control windtunnels & steamplants, Sverdrup & Parcel, Consult Engrs, Mo, 52-53; instr elec eng, Washington Univ, St Louis, 53-56; PROF ELEC ENG, UNIV FLA, 56- *Concurrent Pos:* Vis prof, Univ Colo, 64-65; consult, Maloney Transformer Co, Gen Elec Co, Va, Hughes Aircraft Co, Calif, Martin Co, Fla, St Regis Paper Co & Aerospace Corp. *Mem:* Sr mem Inst Elec & Electronics Engrs; Swed Soc Eng & Archit. *Res:* Electromechanical componentry; general control theory with computer applications; electric energy conversion and automatic control. *Mailing Add:* Dept Elec Eng Univ Fla 1602 NW 35th Way Gainesville FL 32605

ELGERT, KLAUS DIETER, b Schwarmstedt, Ger, Mar 5, 48; US citizen; m 69; c 2. CELLULAR IMMUNOLOGY, TUMOR IMMUNOLOGY. *Educ:* Evangel Col, BS, 70; Univ Mo-Columbia, Sch Med, PhD(immunol), 73. *Prof Exp:* Teaching asst med microbiol, Med Sch, Univ Mo-Columbia, 71-72, from res asst to res assoc immunol, 73-74; asst prof, 74-80, ASSOC PROF MICROBIOL, VA POLYTECH INST & STATE UNIV, 80- *Concurrent Pos:* Prin investr, Nat Cancer Inst, NIH, 81-84 & numerous other grants. *Mem:* Reticuloendothelial Soc; AAAS; Am Soc Microbiol; Am Asn Immunologists. *Res:* Cell-mediated immunity during tumor growth; the phenotypic and functional differences between normal and tumor-bearing host macrophages at the cellular level and to establish the molecular origins for these tumor-induced alterations. *Mailing Add:* Dept Biol Va Polytech Inst & State Univ Blacksburg VA 24061

EL-GHAZALY, SAMIR M, b Luxor, Egypt, July 1, 59; m 85; c 3. MICROWAVE ACTIVE & PASSIVE DEVICES, SUPERCONDUCTING MICROWAVE DEVICES. *Educ:* Cairo Univ, BSc, 81, MSc, 84; Univ Tex, Austin, PhD(elec eng), 88. *Prof Exp:* Asst lectr electronics eng, Fac Eng, Cairo Univ, 81-84, lectr, 84-88; res asst elec eng, Univ Tex Austin, 86-88, postdoctoral fel, 88; ASST PROF ELEC ENG, ARIZ STATE UNIV, 88- *Concurrent Pos:* Researcher, Univ Lille, Francc, 82-83; res & teaching asst, Univ Ottawa, Ont, Can, 84-85; publicity chmn, Waves & Devices Group, Phoenix Sect, Inst Elec & Electronic Engrs, vchmn, 90- *Honors & Awards:* Young Scientist Award, Int Union Radio Sci, 90. *Mem:* Sr mem Inst Elec & Electronics Engrs; Int Union Radio Sci. *Res:* Microwave and millimeter-wave semiconductor devices and passive circuits; semiconductor device stimulation; ultra-short pulse propagation; superconductor microwave transmission lines; electromagnetics; wave-device interactions; numerical techniques applied to monolithic microwave integrated circuits. *Mailing Add:* Dept Elec Eng Ariz State Univ Tempe AZ 85287

ELGIN, JAMES H, JR, b Washington, DC, Feb 2, 42; m 62; c 2. GENETICS. *Educ:* Univ Md, BS, 64, MS, 66; Pa State Univ, PhD(plant breeding), 69. *Prof Exp:* Res agronomist, Agr Res Serv, USDA, Wash, 69-74, res geneticist, 74-81, supvry res geneticist, 81-86, nat prog leader, forage pasture & turf, 86-87, dep area dir, Plant Sci Inst, Beltsville, 88-89, nat prog leader, Forage & Pasture, 89-91, NAT PROG STAFF, AGR RES SERV, USDA, BELTSVILLE, 91- *Concurrent Pos:* Exec secy, NAm Alfalfa Improvement Conf; US rep seed cert schemes, OECD. *Mem:* Am Soc Agron; Crop Sci Soc Am. *Res:* Genetics and plant breeding of alfalfa; inheritance of disease and insect resistance; investigation of new breeding strategies; development of enhanced germplasm for release to private breeders. *Mailing Add:* Nat Prog Staff Agr Res Serv-USDA Bldg 005 Rm 113 Beltsville MD 20705

ELGIN, JOSEPH C(LIFTON), chemical engineering; deceased, see previous edition for last biography

ELGIN, ROBERT LAWRENCE, b Pasadena, Calif, Feb 14, 44; m 67; c 2. ECONOMIC ANALYSIS. *Educ:* Pomona Col, BA, 66; Calif Inst Technol, PhD(physics), 73. *Prof Exp:* Res affil, Energy Lab, Mass Inst Technol, 74; sr physicist, Energy Resources Co, Inc, 75-77; staff physicist, Mass Gov Comn Cogeneration, 78; energy policy analyst IV, NY State Energy Off, 79-81; econ consult, Market Opaque Debt, 82-90; CONSULT, 90- *Mem:* Am Phys Soc; Am Real Estate & Urban Econ Asn. *Res:* Design and demonstration of a home mortgage loan that gives the lender current market earnings and full amortization while keeping borrower payments smooth and affordable: FAIR, fixed payment rate adjustable interest rate. *Mailing Add:* 7261 Kingsbury Blvd University City MO 63130-4141

ELGIN, SARAH CARLISLE ROBERTS, b Washington, DC, July 16, 45; m 67; c 2. MOLECULAR GENETICS. *Educ:* Pomona Col, BA, 67; Calif Inst Technol, PhD(biochem), 71. *Prof Exp:* Teaching asst biochem, Calif Inst Technol, 67-71, Jane Coffin Childs Mem Fund res fel, 71-73; asst prof, 73-77, assoc prof biochem & molecular biol, Harvard Univ, 77-81; assoc prof biol, 81-84, PROF BIOL & BIOCHEM-MOLECULAR BIOPHYS, WASHINGTON UNIV, 84- *Concurrent Pos:* Res career develop award, Nat Inst Gen Med Sci, 76-81. *Mem:* Am Soc Cell Biol; fel AAAS; Am Soc Biochem & Molecular Biol; Genetics Soc Am. *Res:* Chromosomal proteins; chromatin structure in relation to the control of gene expression; euchromatin/heterochromatin structural determinants. *Mailing Add:* Dept Biol Box 1137 Washington Univ St Louis MO 63130

EL GUINDY, MAHMOUD ISMAIL, b Cairo, Egypt. CHEMICAL METALLURGY. *Educ:* Ain Shams Univ, Cairo, BSc, 60; Rensselaer Polytech Inst, MSc, 66, PhD(inorg chem), 68. *Prof Exp:* Instr chem, Assiut Univ, Egypt, 60-63; lectr, Siena Col, NY, 66-67; res assoc metall, McGill Univ, 67-70; res group leader, Refinery Dept, Engelhard Industs, 70-72, mgr refining develop, Minerals & Chem 72-74, tech mgr refining, Engelhard Minerals & Chem Corp, Newark, 74-78; VPRES, SABIN METALS & PRES, PLATINUM GROUP REFINERY INC, 78- *Mem:* Am Inst Mining & Metall Eng; Int Precious Metals Inst; Am Chem Soc; Sigma Xi. *Res:* Recovery and purification of precious metals from ores; solvent extraction of metals, hydrometallurgy; electrochemistry; inorganic compounds of Pt metals; recovery of metals from spent catalyst; pyrometallurgical refining of Pt metals and Re chemistry and metallurgy. *Mailing Add:* 5308 Honeywood Ln Anaheim CA 92807

ELHAMMER, AKE P, b Falun, Sweden, Oct 18, 48. POLYMER CHEMISTRY. *Educ:* Univ Stockholm, PhD(biochem), 82. *Prof Exp:* Postdoctoral fel, NAH, Wash Univ, St Louis, 83-85; res assoc path, Huddinge Hosp, Sweden, 85-86, Kabigen, 86-87; assoc prof, Karolinska Inst, Stockholm, 87-88; RES SCIENTIST BIOCHEM, BIOPOLYMERS DEPT, UPJOHN CO, 87- *Mem:* Am Soc Cell Biol. *Mailing Add:* Biopolymers Dept Upjohn Co Kalamazoo MI 49001

ELHARRAR, VICTOR, b Apr 18, 44; Can citizen; m 74; c 1. MYOCARDIAL ISCHEMIA. *Educ:* Univ Montreal, Can, PhD(physiol), 74; Indiana Univ Sch Bus, MBA 89. *Prof Exp:* ASSOC PROF PHARMACOL, MED & BIOPHYS, SCH MED, IND UNIV, 81- *Mem:* Am Soc Pharmacol & Exp Therapeut; Acad Pharmacol Soc; Am Fed Clin Res. *Res:* Electrophysiology and pharmacology of cardiac arrhythmias; modelling of physiological systems. *Mailing Add:* Dept Pharmacol Sch Med Ind Univ Indianapolis IN 46202

EL-HAWARY, MOHAMED EL-AREF, b Sohag, Egypt, Feb 3, 43; Can citizen; m 66; c 3. SYSTEMS ENGINEERING. *Educ:* Univ Alexandria, Egypt, BSc, 65; Univ Alberta, Can, PhD(elec eng), 72. *Prof Exp:* Instr elec eng, dept elec eng, Univ Alexandria, 65-68; teaching & res asst, dept elec eng, Univ Alberta, 68-72, res assoc, 72; assoc prof, Grad Sch, Fed Univ Rio de Janeiro, 72-73; from asst prof to prof, Mem Univ, Nfld, 74-81, chmn dept, 76-81; PROF ELEC ENG, TECH UNIV, NOVA SCOTIA, 81- *Concurrent Pos:* Consult, Newfoundland & Labrador Hydro, 74-; mem, Natural Sci(s) & Eng Res Coun Can, Grant Selection Comt Elec Eng, 81-; assoc ed, Can Elec Eng J, 81-87. *Mem:* Inst Elec & Electronics Engrs; Can Soc Elec Eng (pres, 86-88); Can Elec Asn; fel Eng Inst Can. *Res:* Modeling, simulation and optimization in electric power systems; planning, operation and control. *Mailing Add:* Dept Elec Eng Tech Univ Nova Scotia PO Box 1000 Halifax NS B3J 2X4 CAN

ELHILALI, MOSTAFA M, b Minia, UAR, Nov 3, 37; m 69; c 5. CANCER. *Educ:* Univ Cairo, MD, 59, DS, 62, DU, 63, MCh, 64; McGill Univ, PhD(exp surg), 69; FRCS(C), 69. *Prof Exp:* Intern med, Univ Cairo Hosp, 60-61, resident urol, 61-63, clin demonstr, 63-65; clin fel urol, Royal Victoria Hosp, Montreal, 65-69, resident surgeon, 67-68; from asst prof to prof urol, fac med, Sherbrooke Univ, 69-83, chmn dept, 75-83; PROF UROL & CHMN DEPT, MCGILL UNIV, 83-; UROLOGIST-IN-CHIEF, ROYAL VICTORIA HOSP, MONTREAL, 83- *Concurrent Pos:* Resident, Ottawa Civic Hosp, 69; Nat Cancer Inst Can fel, 66-68 & grant, 72-77; Med Res Coun Can grant, 69-72. *Mem:* Can Med Asn; Am Urol Asn; Can Urol Asn; Can Soc Immunol. *Res:* Isoenzyme changes in prostatic cancer and of lactate dehydrogenase in bladder cancer; urodynamics; immunological aspects of genitourinary cancer. *Mailing Add:* Royal Victoria Hosp-Urol Montreal PQ H3G 1Y6 Can

ELIA, VICTOR JOHN, b Portland, Ore, May 11, 42; m 66; c 4. ENVIRONMENTAL HEALTH, ANALYTICAL CHEMISTRY. *Educ:* Portland State Univ, BS, 65; Univ Nebr, Lincoln, PhD(chem), 70. *Prof Exp:* Fel mech org chem, Notre Dame Univ, 69-71; fel environ res, Kettering Lab, Sch Med, Univ Cincinnati, 71-73; asst prof, Defiance Col, 73-74; res assoc, Kettering Lab, Dept Environ Health, Univ Cincinnati, 74-76, asst prof, 76-81; RES CHEMIST & INDUST HYGIENIST, NAT COUN PAPER INDUST FOR AIR & STREAM IMPROV, CORVALLIS, 81- *Honors & Awards:* Merck, Sharp & Dohme, Inc Outstanding Res Award, 69. *Mem:* Am Chem Soc; Am Indust Hyg Asn; Am Acad Indust Hyg. *Res:* Synthesis, reaction and mechanistic studies of organic nitrogen compounds; alkaline decomposition of organic disulfides; trace metals; isolation and biological studies of potential toxic environmental pollutants; development and evaluation of air sampling and analysis methods; analysis of organics in air, water, wastewater and biological samples. *Mailing Add:* West Coast Regional Ctr PO Box 458 Corvallis OR 97339

ELIADES, THEO I, b Cyprus, Dec 30, 38; Can citizen; m 76; c 2. LUBRICANT ADDITIVES, MICELLAR STRUCTURES. *Educ:* Univ London, Eng, BSc, 64; Univ Toronto, Can, MA, 66, PhD(inorg chem), 68. *Prof Exp:* Teaching asst chem, Univ Toronto, 68-70; RES DIR, SURPASS CHEMICALS, DIV WITCO CO, 70- *Concurrent Pos:* Consult, Toronto Res Labs, 79-81. *Mem:* Chem Inst Can. *Res:* New products and processes for the manufacture of lubricating oil additives, especially overbased sulfonates and phenates; rust inhibitive coatings, mainly for the automotive industry. *Mailing Add:* Surpass Chemicals Ltd Ten Chemical Ct West Hill ON M1E 3X7 Can

ELIAS, ANTONIO L, SPACE TECHNOLOGY. *Educ:* Mass Inst Technol, BS, BS, EAA & PhD(aeronaut & astronaut). *Prof Exp:* Res asst & staff mem, Space Guidance & Navigation Div, C S Draper Lab, 72-80; asst prof aeronaut & astronaut, Mass Inst Technol, 80-86; SR VPRES ENG, ORBITAL SCI CORP, FAIRFAX, VA, 86- *Concurrent Pos:* Tech dir, Pegasus Air-Launched Orbital Booster Prog, Orbital Sci Corp. *Honors & Awards:* Nat Medal of Technol, 91; Aircraft Design Award, Am Inst Aeronaut & Astronaut, 91. *Mem:* Am Inst Aeronaut & Astronaut; Inst Navigation. *Res:* Author of over 25 technical publications and patents. *Mailing Add:* Orbital Sci Corp 12500 Fair Lakes Circle Fairfax VA 22033

ELIAS, HANS GEORG, b Bochum, Ger, Mar 29, 28; m 56; c 2. POLYMER CHEMISTRY. *Educ:* Tech Univ Hannover, dipl-chem, 54; Tech Univ Munich, Dr rer nat, 57; Swiss Fed Inst Technol, Zurich, pvtdozent, 61. *Prof Exp:* Sci asst chem technol, Tech Univ Munich, 56-59; from sci head asst chem technol to assoc prof polymer chem, Swiss Fed Inst Technol, Zurich, 60-71; pres, Mich Molecular Inst, Midland, 71-83; consult, 83-88; RETIRED. *Concurrent Pos:* Adj prof, Mich Technol Univ, 72-83, Case Western Reserve Univ, 78-84 & Cent Mich Univ, 81-84; exec dir, Mich Found Advan Res, 74-79. *Mem:* Am Chem Soc; Am Phys Soc; Sigma Xi; Soc Ger Chem; Swiss Chem Asn. *Res:* Synthesis and solution properties of polymers. *Mailing Add:* 4009 Linden Dr Midland MI 48640

ELIAS, LAURENCE, b Brooklyn, NY, Aug 20, 46. MEDICAL. *Educ:* Princeton Univ, AB, 67; Stanford Univ, MD, 72; Am Bd Internal Med, dipl, 77. *Prof Exp:* Resident med, Med Ctr, Stanford Univ, 73-74, Nat Cancer Inst spec fel, Div Hemat, 74-76, sr clin fel, Div Oncol, 76-77; asst prof med, Sch Med, Univ NMex, 77-82, actg chief med oncol, Cancer Res Treatment Ctr, 79-86, assoc prof med, 82-87, PROF MED, DIV HEMAT & ONCOL, SCH MED, UNIV NMEX, 87-, CHIEF, 91- *Concurrent Pos:* Med oncologist, Cancer Res & Treatment Ctr, Univ NMex, 77- *Mem:* Am Fedn Clin Res; Am Soc Hemat; Am Sox Clin Oncol; Sigma Xi; Am Asn Cancer Res; AAAS; Int Soc Exp Hemat; Am Soc Biochem & Molecular Biol; Int Soc Hemat. *Res:* Medical; one patent. *Mailing Add:* Dept Med Cancer Ctr Univ NMex 900 Camino de Salud NE Albuquerque NM 87131

ELIAS, LORNE, b Ottawa, Ont, Feb 2, 30; m 57; c 4. PHYSICAL CHEMISTRY. *Educ:* Carleton Univ, BSc, 52; McGill Univ, PhD(chem), 56. *Prof Exp:* Res assoc chem, McGill Univ, 56-60; PRIN RES OFFICER, INST AEROSPACE RES, NAT RES COUN CAN, 60- *Res:* Trace vapour detection; detection of explosives vapours; atmospheric monitoring of pesticides; detection of narcotics; trace vapour calibration sources. *Mailing Add:* Nat Res Coun Bldg M-10 Montreal Rd Ottawa ON K1A 0R6 Can

ELIAS, PETER, b New Brunswick, NJ, Nov 26, 23; m 50; c 3. INFORMATION THEORY. *Educ:* Mass Inst Technol, SB, 44; Harvard Univ, MA, 48, ME, 49, PhD(appl sci), 50. *Prof Exp:* Asst appl sci, Harvard Univ, 48 & 49; from asst prof to prof elec eng, 53-69, head dept, 60-66, Cecil H Green Prof, Elec Eng, 70-72, assoc head, dept elec eng & comp sci, 81-83, EDWIN S WEBSTER PROF ELEC ENG, MASS INST TECHNOL, 74- *Concurrent Pos:* Jr fel, Harvard Univ, 50-53, vis prof, 67-68 & 83-84; vis assoc prof, Univ Calif, Berkeley, 58; vis scientist, Imperial Col Sci & Technol, 75-76. *Honors & Awards:* Shannon Lectr, Inst Elec Engr & Electronics, Info Theory Soc, 77. *Mem:* Nat Acad Sci; Nat Acad Eng; fel Inst Elec & Electronics Engrs; Am Acad Arts & Sci; Asn Comput Mach; Inst Math Statist. *Res:* Information theory; reliable communication over unreliable channels; reliable computation with unreliable components; data compression; economical storage and retrieval of information; complexity of data processing. *Mailing Add:* Dept Elec Eng Mass Inst of Technol Cambridge MA 02139

ELIAS, PETER M, b Waltham, Mass, Apr 1, 41. DERMATOLOGY. *Educ:* Univ Calif, San Francisco, MD, 67. *Prof Exp:* CHIEF DERMAT, VET ADMIN MED CTR, SAN FRANCISCO, CALIF, 74-; CLIN PROF DERMAT, UNIV CALIF, SAN FRANCISCO, 78- *Mailing Add:* Vet Admin Med Ctr 4150 Clement St San Francisco CA 94121

ELIAS, ROBERT WILLIAM, b Canton, Ill, Apr 18, 42; m 65; c 2. ENVIRONMENTAL SCIENCES, GENERAL. *Educ:* Univ Ill, Urbana, BS, 64, MS, 70; Univ Tex, Austin, PhD(bot), 73. *Prof Exp:* Res fel geochem, Calif Inst Technol, 73-76; asst prof bot, Va Polytech Inst & State Univ, 76-82; ASST PROF BOT, US ENVIRON PROTECTION AGENCY, 82- *Mem:* Am Inst Biol Sci; Ecol Soc Am; Am Chem Soc; Soc Environ Geochem & Health. *Res:* Biogeochemistry of trace metals in natural ecosystems and humans. *Mailing Add:* 5317 Newhall Rd Durham NC 27713-2521

ELIAS, THOMAS ITTAN, b S Piramadom, Kerala, India, Mar 16, 47; m 79; c 2. THERMAL-FLUID SCIENCES. *Educ:* Univ Kerala, India, BS, 70, MS, 72; Univ Cincinnati, MS, 77; Univ Minn, PhD(eng mech), 81. *Prof Exp:* Develop engr, Vikram Sarabhai Space Ctr, Trivandrum, India, 72-75; res asst aerospace eng, Univ Cincinnati, 76-77; teaching assoc aerospace eng, Univ Minn, 77-78, res asst, 78-80; asst prof aerospace eng & eng mech, Ind Inst Technol, Fort Wayne, 80-82; from asst prof to assoc prof mech eng, Youngstown Univ, Ohio, 87-88; LEAD ENGR, MCDONNELL DOUGLAS CORP, ST LOUIS, MO, 88- *Concurrent Pos:* Reviewer, J Heat Transfer, Am Soc Mech Engrs & Am Inst Chem Engrs J, 84- *Mem:* Am Soc Mech Engrs; Am Inst Aeronaut & Astronaut. *Res:* Thermal-fluid sciences; boundary layer transition; energy management; energy conservation. *Mailing Add:* 368 Littany Lane Chesterfield MO 63017-2308

ELIAS, THOMAS S, b Cairo, Ill, Dec 30, 42; wid; c 2. SYSTEMATIC BOTANY, MORPHOLOGY. *Educ:* Southern Ill Univ, BA, 64, MA, 66; St Louis Univ, PhD(biol), 69. *Prof Exp:* Teaching asst, Southern Ill Univ, 64-66; asst curator bot, Arnold Arboretum, Harvard Univ, 69-72; adminr & dendrologist, Cary Arboretum, New York Bot Garden, 72-74, asst dir, 74-84; DIR, RANCHO SANTA ANA BOT GARDEN, 84- *Concurrent Pos:* Dir, Rancho Santa Ana Bot Garden; chmn, Dept Bot, Claremont Grad Sch; coord, US/USSR Bot Exchange Prog. *Honors & Awards:* Cooley Award, Am Soc Plant Taxonomists. *Mem:* Am Soc Plant Taxon; Int Asn Plant Taxon; AAAS; Bot Soc Am; Sigma Xi. *Res:* Trees of North America; floristics and phytogeography of woody plants of temperate Asia and Eastern Europe. *Mailing Add:* Rancho Santa Ana Bot Garden 2447 San Mateo Ct Claremont CA 91711

ELIASON, MORTON A, b Fargo, NDak, Apr 26, 32; m 56; c 4. PHYSICAL CHEMISTRY. *Educ:* Concordia Col, Moorhead, Minn, BA, 54; Univ Wis, PhD(phys chem), 59. *Prof Exp:* From asst prof to assoc prof, 58-69, PROF CHEM, AUGUSTANA COL, 69- *Concurrent Pos:* Vis assoc prof chem, Theoret Chem Inst, Univ Wis, 66-67; vis scholar, CTNS Grad Theol Union, Berkeley, CA, 86-87. *Mem:* Am Chem Soc; Sigma Xi. *Res:* Quantum theory of small molecules; theory of reaction rates, liquids; solubility of inert gases in fused salts; collision processes and energy transfer in gases; theory of charge transfer complexes; oscillating chemical reactions; relation of theology and science. *Mailing Add:* Dept Chem Augustana Col Rock Island IL 61201

ELIASON, STANLEY B, b McVille, NDak, Aug 31, 39; m 64. MATHEMATICAL ANALYSIS. *Educ:* Concordia Col, Moorhead, Minn, BA, 61; Univ Nebr, Lincoln, MA, 63, PhD(math), 67. *Prof Exp:* Instr math, Univ Nebr, Lincoln, 63-67; from asst prof to assoc prof math, Univ Okla, 67-83, interim chmn, 85-86, chmn, 86-89, ASSOC CHMN, MATH DEPT, UNIV OKLA, 81-, PROF MATH, 83- *Concurrent Pos:* US Air Force res grants, 71-72 & US Army Res Off-Durham, grants, 70-71 & 72-73; Danforth Assoc, 78-84. *Mem:* Am Math Soc; Sigma Xi; Math Asn Am. *Res:* Mathematical analysis; ordinary differential equations; distance between zeros; comparison theorems; second order linear and nonlinear differential equations; differential equations with deviating arguments; second order elliptic linear partial differential equations. *Mailing Add:* Dept Math Rm 423 Univ Okla 601 Elm Ave Norman OK 73019-0315

ELIASSEN, ROLF, b New York, NY, Feb 22, 11; m 41; c 2. ENVIRONMENTAL ENGINEERING. *Educ:* Mass Inst Technol, BS, 32, MS, 33, ScD(sanit eng), 35. *Prof Exp:* Sanit engr, Dorr Co, Inc, Chicago & Los Angeles, 36-40; prof sanit eng, NY Univ, 40-49 & Mass Inst Technol, 49-61; partner & sr vpres, Metcalf & Eddy, 61-73, chmn bd, 73-88; prof, civil eng, 61-73, EMER PROF ENG, STANFORD UNIV, 73- *Concurrent Pos:* Dir, Millipore Filter Corp, Mass, 58-62; consult, Int Atomic Energy Agency, 57-62, Off Sci & Tech, Exec Off President, 61-73 & Calif Dept Water Resources, 64-68; mem gen adv comt, Atomic Energy Comn, 70-75. *Honors & Awards:* George Westinghouse Award, 50. *Mem:* Nat Acad Eng; hon mem Am Soc Civil Engrs; Am Water Works Asn; Water Pollution Control Fedn. *Res:* Methods of water and sewage treatment; industrial and radioactive waste treatment processes. *Mailing Add:* 850 Webster Palo Alto CA 94301

ELIASSON, SVEN GUSTAV, b Malmo, Sweden, Apr 16, 28; m 51; c 3. NEUROLOGY. *Educ:* Univ Lund, PhD(physiol), 52; Royal Carolina Univ, Lund, MD, 54. *Prof Exp:* From asst prof to assoc prof physiol, Univ Lund, 49-52, instr neurol, 53; jr res anatomist, Univ Calif, Los Angeles, 54-55, asst res anatomist, 55-56; asst prof neurol, Univ Tex Health Sci Ctr, Dallas, 56-63; assoc prof, 63-67, PROF NEUROL, SCH MED, WASH UNIV, 67- *Concurrent Pos:* Rotary Int fel, 54-55. *Mem:* Am Physiol Soc; Am Acad Neurol; Am Neurol Asn; Am Soc Clin Invest; Am Fedn Clin Res; Sigma Xi. *Res:* Gastrointestinal physiology; neurochemical and neurophysiological disturbances in peripheral nerves. *Mailing Add:* Dept Neurol 660 S Euclid Ave St Louis MO 63110

EL-IBIARY, MOHAMED YOUSIF, b Alexandria, Egypt, Sept 9, 28; m; c 2. ELECTRICAL ENGINEERING, NUCLEAR ENGINEERING. *Educ:* Cairo Univ, BSc, 49; Univ London, PhD(elec eng), 54. *Prof Exp:* Lectr elec eng, Cairo Univ, 55-61, prof, 61-73; vis prof, 73-80, PROF ELEC ENG, UNIV OKLA, 80- *Concurrent Pos:* Res assoc, Joint Estab Nuclear Energy Res, Norway, 55-56; res assoc nuclear eng, Atomic Energy Can, 60-61; consult, Intertrade Corp, 61-67; vis prof elec eng, Univ Mosul, 68-69 & Univ Tripoli, 70-71. *Honors & Awards:* State Merit Award, Egyptian Acad Sci, 59; Merit Medal Sci & Arts, President of Egypt, 59. *Mem:* Sr mem Inst Elec & Electronics Engrs; Am Soc Eng Educ. *Res:* Nuclear electronics, especially semiconductor detectors; microwaves, particularly communications; fiber optics. *Mailing Add:* Sch Elec Eng Univ Okla Norman OK 73019

ELICEIRI, GEORGE LOUIS, b Buenos Aires, Arg, Oct 27, 39; US citizen; m 66; c 3. MOLECULAR CELL BIOLOGY, BIOCHEMISTRY. *Educ:* Univ Buenos Aires, MD, 60; Univ Okla, PhD(biochem), 65. *Prof Exp:* Instr cell biol, Sch Med, NY Univ, 68-69; from asst prof to assoc prof, 69-76, PROF PATH, SCH MED, SCH MED, ST LOUIS UNIV, 76- *Concurrent Pos:* Damon Runyon Mem Fund fel, Univ Chicago, 65-67; USPHS spec fel, Sch Med, NY Univ, 67-69, USPHS career develop award, Sch Med, St Louis Univ, 72-77; mem, biomed sci study sect, NIH, 85-86. *Mem:* Am Soc Cell Biol; Am Soc Biol Chem; Am Asn Pathologists. *Res:* Structure, function and biosynthesis of mammalian small nucleolar RNAs; mechanism of the ultraviolet light-induced inhibitions of the synthesis of small nuclear RNA species U1 through U5. *Mailing Add:* Dept Path St Louis Univ Sch Med 1402 S Grand Blvd St Louis MO 63104

ELICH, JOE, b Tooele, Utah, Sept 28, 18. MATHEMATICS. *Educ:* Utah State Agr Col, BS, 40; Univ Calif, MA, 42. *Prof Exp:* Asst prof, 46-58, PROF MATH, UTAH STATE UNIV, 58-, ASST HEAD DEPT, 77- *Mem:* Am Math Soc; Math Asn Am. *Mailing Add:* 1710 E 1140 N Logan UT 84321

ELIEL, ERNEST LUDWIG, b Cologne, Ger, Dec 28, 21; US citizen; m 49; c 2. ORGANIC CHEMISTRY. *Educ:* Univ Havana, Dr phys-chem Sci, 46; Univ Ill, PhD(org chem), 48. *Hon Degrees:* DSc, Duke Univ, 83. *Prof Exp:* From instr to prof chem, Univ Notre Dame, 48-72, head dept, 64-66; W R KENAN, JR PROF CHEM, UNIV NC, CHAPEL HILL, 72- *Concurrent Pos:* NSF sr fels, 58-59, 67-68; Guggenheim fel, Stanford & Princeton Univs, 75-76, Duke Univ, 83-84; Benjamin Rush lectr, Univ Pa, 78; Sir C V Raman vis prof, Univ Madras, India, 81; Hill lectr, Duke Univ, 85; chmn & mem bd dirs, Am Chem Soc, 87-89; chmn, chemsect, AAAS, 91-92. *Honors & Awards:* Morley Medal, 65; Laurent Lavoisier Medal, 68; Harry & Carol Mosher Award, 82; Coates Lectr, Univ Wyo, 89; Smith, Klyne & French Lectr, Univ Ill, 90; Herty Medal, 91. *Mem:* Nat Acad Sci; Am Acad Arts & Sci; Am Chem Soc (pres elect, 91); Royal Soc Chem; fel AAAS; Sigma Xi. *Res:* Stereochemistry; conformational analysis; asymmetric synthesis; heterocyclic chemistry; organosulfur chemistry; carbanion chemistry; nuclear magnetic resonance. *Mailing Add:* Dept Chem Univ NC Chapel Hill NC 27599-3290

ELIEZER, ISAAC, b Sofia, Bulgaria; US citizen; c 3. PHYSICAL CHEMISTRY, INORGANIC CHEMISTRY. *Educ:* Hebrew Univ Jerusalem, MSc, 56, PhD(phys & inorg chem), 60; FRIC. *Prof Exp:* Instr anal, inorg & phys chem, Hebrew Univ Jerusalem, 54-60; sr scientist dept inorg radiochem, Atomic Energy Comn Labs, 60-65; res assoc theoret chem, Univ Southern Calif, 65-67; assoc prof chem, Tel Aviv Univ, 67-72; dir, Col Pract Engrs, 72-75; adj prof chem & mgr, MHD Energy Res, Mont State Univ, 75-79; PROF CHEM & ASSOC DEAN, COL ARTS & SCI, OAKLAND UNIV, 79- *Concurrent Pos:* Mem, Comt Data Sci & Technol, Int Coun Sci Unions, 71-81; ed, Israel Jour Chem, 71-75; mem, Gen Comt, Int Conf Physics Electronic & Atomic Collisions, 71-75; vis prof, Weizmann Inst Sci, 72-75; mem, Subcommission Solubility, Comn Equilibria, Anal Div, Int Union Pure & Appl Chem, 74-82; mem, Adv Comt, Mont Energy Res & Develop Inst, 76-79; consult, Old West Regional Comn, 77-79, mem ed bd, Pergamon Press Solubility Series, 77-80, mem, Mich Energy & Resources Res Asn, 80-, mem, Europe Acad Sci, Arts & Humanities, 81-, mem, bd gov, William Beaumont Hosp Res Inst, 81-, bd mem, Am Biograph Inst, 84-, charter mem, Coun Liberal Learning, Asn Am Col, 85- , evaluator, Comn Higher Educ, 85-, mem biotechnol comt, Mich Technol Coun, 86-; mem, bd gov, William Beaumont Hosp Res Inst, 81- *Mem:* Am Chem Soc (chmn Mont sect, 79); Sigma Xi; AAAS; fel London Chem Soc; Int Studies Asn; Int Union Pure & Appl Chem; NY Acad Sci; Nat Comt Int Studies & Progs Adminr; Am Asn Col Teacher Educ; Europe Acad Sci, Arts & Humanities. *Res:* Physical inorganic and analytical chemistry; theoretical chemistry; high-temperature inorganic materials; ion exchange; thermodynamics; solution chemistry; energy and the environment; computer modeling; science education; science administration; research administration. *Mailing Add:* Dean Col Arts & Sci Oakland Univ Rochester MI 48309

ELIEZER, NAOMI, b Tel Aviv, Israel; US citizen; c 3. BIOCHEMISTRY. *Educ:* Hebrew Univ, MSc, 60; Weizmann Inst, PhD(chem), 65. *Prof Exp:* Scientist polymer sci, Weizmann Inst, 65-75; res scientist chem, Mont State Univ, 75-79; RES SCIENTIST & ADJ ASST PROF CHEM & BIOL, OAKLAND UNIV, 79- *Res:* Physical chemistry; biopolymers; biochemical endocrinology; hormone receptors. *Mailing Add:* Dept Biol Sci Oakland Univ Rochester MI 48309

ELIEZER, ZWY, b Buhusi, Romania, Sept 21, 33; US citizen; m 66; c 1. METALLURGY. *Educ:* Univ Bucharest, dipl physics, 59; Israel Inst Technol, MS, 69, DSc 72. *Prof Exp:* Engr operator, Geophys & Geol Prospecting Co, Bucharest, 58-63; res assoc mat eng, 72-74, from asst prof to assoc prof, 74-84, PROF MECH ENG, UNIV TEX, AUSTIN, 84- *Mem:* Am Inst Mining, Metall & Petrol Engrs; Am Soc Metals; Am Soc Testing & Mat; Sigma Xi; Nat Asn Corrosion Engrs. *Res:* Tribological properties of metals and graphite fiber-metal matrix composites; corrosion of metals in geothermal environments; sensitization of welded austenitic stainless steels. *Mailing Add:* Dept Mech Eng Univ Tex Austin TX 78712

ELIMELECH, MENACHEM, b Hadera, Israel, July 19, 55; Israel & US citizen; m 85; c 2. COLLOID SCIENCE. *Educ:* Hebrew Univ, Israel, BSc, 83, MSc, 85; Johns Hopkins Univ, PhD(environ eng), 89. *Prof Exp:* ASST PROF ENVIRON ENG, UNIV CALIF, LOS ANGELES, 89- *Mem:* Am Chem Soc; Am Water Works Asn. *Res:* Physical, chemical and colloidal processes in water and wastewater treatment. *Mailing Add:* Civil Eng Dept Univ Calif Los Angeles CA 90024-1593

ELIN, RONALD JOHN, b Minneapolis, Minn, Apr 14, 39; m 69; c 2. PATHOLOGY, CLINICAL CHEMISTRY. *Educ:* Univ Minn, BA, 60, BS, 62, MD, 66, PhD(biochem), 69. *Prof Exp:* Resident anat path, Univ Minn, 66-69; intern, Univ Hosp, San Diego County, 69-70; res assoc infectious dis, Nat Inst Allergy & Infectious Dis, 70-73; resident clin path, 73-74, CHIEF, DEPT CLIN PATH, NIH, 75-, CHIEF CLIN CHEM SERV, 77- *Concurrent Pos:* Clin prof, Uniformed Serv Univ Health Sci, 78-; mem, Nat Comt Clin Lab Standards. *Mem:* Am Soc Clin Pathologists; Am Asn Clin Chem; Am Col Nutrit; Acad Clin Lab Physicians & Scientists. *Res:* Magnesium metabolism; host defense mechanisms; endotoxin; development of better and new laboratory tests for the diagnosis of disease; factors which affect the production of acute phase reactants in humans; clinical chemistry. *Mailing Add:* Clin Path Dept NIH 9000 Rockville Pike Bldg Ten Rm 2C-306 Bethesda MD 20892

ELING, THOMAS EDWARD, b Cincinnati, Ohio, Oct 26, 41. PHARMACOLOGY, BIOCHEMISTRY. *Educ:* Univ Cincinnati, MS, 64; Univ Ala, PhD(biochem), 68. *Prof Exp:* RES CHEMIST, NAT INST ENVIRON HEALTH SCI, 69- *Concurrent Pos:* Fel drug metab, Univ Iowa, 68-69. *Mem:* Am Soc Pharmacol & Exp Therapeut; Sigma Xi. *Res:* Synthesis and metabolism of labeled drugs; effect of drugs on microsomal drug metabolizing enzymes, factors controlling the development of these enzymes in newborn animals and man; pharmacokinetics of environment agents; non-respiratory lung function; prostaglandin biosynthesis and metabolism; oxidation of chemicals by prostaglandin synthetase by lungs; mechanism of chemical uptake processes of the lung. *Mailing Add:* 8909 Willow Wood Ct Raleigh NC 27613

ELINGS, VIRGIL BRUCE, b Des Moines, Iowa, May 9, 39; m 62; c 1. PHYSICS, INSTRUMENTATION. *Educ:* Iowa State Univ, SB, 61; Mass Inst Technol, PhD(physics), 65. *Prof Exp:* Res assoc, Mass Inst Technol, 66; asst prof, 66-72, ASSOC PROF PHYSICS, UNIV CALIF, SANTA BARBARA, 72- *Concurrent Pos:* AEC grant, 66- *Mem:* Am Phys Soc. *Res:* Elementary particle physics; cardiac modeling; medical instrumentation. *Mailing Add:* Physics Dept Univ Calif Santa Barbara CA 93106

ELINSON, JACK, b New York, NY, June 30, 17; m 41; c 4. SOCIOMEDICAL SCIENCES. *Educ:* City Univ NY, BS, 37; George Washington Univ, MA, 46, PhD, 54. *Prof Exp:* Sr study dir, Nat Opinion Res Ctr, Univ Chicago, 51-56; from assoc prof to prof, 56-86, EMER PROF SOCIOMED SCI, SCH PUB HEALTH, COLUMBIA UNIV, 86- *Concurrent Pos:* Bd dir, Pub Health Asn New York City, 76-77, Med & Health Res Asn New York City, 77-89; chair, Med Soc Sect, Am Sociol Asn, 77-78; sr serv fel, Nat Ctr Health Statist, US Dept Health & Human Serv, 77-81; pres, Am Asn Pub Opinion Res, 79-80; res consult, Med & Health Res Asn New York City, 85-; vis prof, Robert Wood Johnson Med Sch, 86-; distinguished vis prof, Inst Health Care Policy, Rutgers Univ, 86-89; prin investr, Pediat Resource Ctr Outcome Study, 87- *Honors & Awards:* Leo G Reeder Award, Am Sociol Asn. *Mem:* Inst Med-Nat Acad Sci; fel Am Sociol Asn; fel Am Pub Health Asn. *Res:* Evaluation of health action programs; primary medical care & prevention; social epidemiology; measurement of health status, quality of health care, lifestyles & health. *Mailing Add:* 1181 E Laurelton Pkwy Teaneck NJ 07666

ELINSON, RICHARD PAUL, DEVELOPMENTAL BIOLOGY, FERTILIZATION. *Educ:* Yale Univ, PhD(biol), 70. *Prof Exp:* Assoc prof, 76-85, PROF DEVELOP BIOL, UNIV TORONTO, 85- *Res:* Embryo pattern formation. *Mailing Add:* Dept Zool Univ Toronto 25 Harbord St Toronto ON M5S 1A1 Can

ELIOFF, THOMAS, b Monroe, La, Dec 11, 33; m 56; c 1. HIGH ENERGY PHYSICS. *Educ:* La Polytech Inst, BS, 54; Univ Calif, Berkeley, PhD(physics), 60. *Prof Exp:* PHYSICIST, LAWRENCE BERKELEY LAB, UNIV CALIF, 60- *Mem:* Am Phys Soc. *Res:* Experimental high energy physics concerning interactions of elementary particles; accelerator development. *Mailing Add:* Bldg 50 Lawrence Berkeley Lab Univ of Calif Berkeley CA 94720

ELION, GERTRUDE BELLE, b New York, NY, Jan 23, 18. BIOCHEMISTRY, PHARMACOLOGY. *Educ:* Hunter Col, AB, 37; NY Univ, MS, 41. *Hon Degrees:* Numerous from US cols & univs, 69-89. *Prof Exp:* Lab asst biochem, sch nursing, NY Hosp, 37; asst org chem, Denver Chem Co, 38-39; teacher chem & physics, New York, 41-42; analyst food chem, Quaker Maid Co, 42-43; res chemist org chem, Johnson & Johnson, 43-44; sr res chemist, Wellcome Res Labs, 44-67, asst to dir, chemother div, 63-67, head exp ther, 67-83, EMER SCIENTIST, BURROUGHS WELLCOME CO, 83- *Concurrent Pos:* Consult, chemother study sect, USPHS, 60-64; adj prof pharmacol & exp med, Duke Univ, 71-83, res prof, 83-; adj prof pharmacol, Univ NC, 73-; mem, Nat Cancer Adv Bd, 84-90; hon staff mem, Arthur G James Cancer Hosp & Res Inst, Ohio State Univ, 90. *Honors & Awards:* Nobel Prize in Physiol/Med, 88; Nat Medal of Sci, 91; Garvan Medal, Am Chem Soc, 68, Distinguished Chemist Award, 85; Judd Award, Sloan-Kettering Inst, 83; Cain Award, Am Asn Cancer Res, 85; Medal of Honor, Am Cancer Soc, 90. *Mem:* Nat Acad Sci; Inst Med-Nat Acad Sci; Am Chem Soc; NY Acad Sci; Chem Soc; Am Soc Biol Chemists; Am Soc Pharmacol & Exp Therapeut; hon mem Acad Pharmaceut Sci; fel Am Acad Pharmaceut Scientists; hon mem Am Asn Cancer Res (pres, 83-84). *Res:* Chemistry of Purines, Pyrimidines and Pteridines; bacterial metabolism; metabolism of radioactive purines in bacteria and animals; chemotherapy; immunosuppression; antiparasitic drugs. *Mailing Add:* Burroughs Wellcome Co 3030 Cornwallis Rd Research Triangle Park NC 27709

ELIOT, ROBERT S, b Oak Park, Ill, Mar 8, 29; m 57; c 2. CARDIOLOGY, CHEMISTRY. *Educ:* Univ NMex, BS, 51; Univ Colo, MD, 55. *Prof Exp:* Intern, Evanston Hosp, Ill, 55-56; resident internal med, Univ Colo, 56-58; from instr to asst prof med, Med Ctr, Univ Minn, 63-67; assoc prof med & cardiologist, Sch Med, Univ Fla, 67-69, prof med, 69-72; prof med, dir Div Cardiovasc Med & dir Cardiovasc Ctr, Univ Nebr Med Ctr, Omaha, 72-88; DIR, INST STRESS MED INT, 88- *Concurrent Pos:* Actg chief cardiol, Vet Admin Hosps, Denver, Colo, 59, consult physician, Minneapolis, Minn, 65-67 & actg chief med serv, Gainesville, Fla, 69-70, chief cardiol sect, 69-72; creator educ TV ser heart dis prev, Heartline to Health; res fel cardiol, Sch Med, Univ Colo, 58-60, med fel specialist cardiac path, Univ Minn, 62-63; Nat Heart Inst fel & res trainee cardiovasc path, Charles T Miller Hosp, St Paul, Minn, 62-63; pres elect, Interstate Postgrad Med Assembly, 81; chmn, Bethesda Conf Comt on Prevention of Coronary Dis in Occupational Setting, 80-81; consult central nervous system & sudden death, Nat Acad Sci, 80-; bd dir, Am Inst Stress, 80-; mem adv bd, Stress & Cardiovascular Res Ctr, Eckerd Col, St Petersburg, 80- *Honors & Awards:* Kent award, 56. *Mem:* Am Med Asn; fel Am Col Physicians; Am Heart Asn; Biophys Soc; fel Am Col Cardiol. *Res:* Abnormal hemoglobin oxygen affinity in smokers and patients having signs of coronary insufficiency; myocardial proteins; cardiac pathology; effects of changes in blood-oxygen transport and their role in producing or in treating heart disease; mechanisms causing heart attacks; role of stress in heart disease; vectorcardiography; electrocardiography. *Mailing Add:* Inst Stress Med Int 4600 S Ulster St Suite 700 Denver CO 80237

ELISBERG, BENNETT LA DOLCE, b New York, NY, Nov 11, 25; m 64. INFECTIOUS DISEASES, INTERNAL MEDICINE. *Educ:* NY Univ, BA, 44; Tulane Univ, MS, 48, MD, 50; Am Bd Med Microbiol, dipl. *Prof Exp:* Intern, St Joseph's Mercy Hosp, 50-51; from jr to sr resident internal med, Kern Co Gen Hosp, 51-52, chief resident, 54-55; sr med officer res infectious dis, US Army Med Res Unit, Malaya, 55-58, dep dir, 59-61; med officer res, Dept Virus Dis, Walter Reed Army Inst Res, 61-62, med officer res, Dept Rickettsial Dis, 62-63, chief, 63-72; dir div path, Bur Biologics, Food & Drug Admin, 73-80, dir, Div Prod Qual Control, 80-85; RETIRED. *Mem:* AAAS; Am Soc Microbiol; Am Soc Trop Med & Hyg; Infectious Dis Soc Am; Am Asn Immunol. *Res:* Virus and rickettsial diseases of man. *Mailing Add:* 4008 Queen Mary Dr Olney MD 20832

ELITZUR, MOSHE, b Borzchow, Poland, April 29, 44; US & Israeli citizen; m 70; c 3. ASTROPHYSICS. *Educ:* Hebrew Univ, Jerusalem, BSc, 64; Weizmann Inst, Israel, MSc, 66, PhD(physics), 71. *Prof Exp:* Res assoc physics, Rockefeller Univ, 70-72; res fel physics, Calif Inst Technol, Pasadena, 72-74; scientist physics, Weizman Inst, Israel, 74-75, sr scientist, 75-79; assoc prof, 80-86, PROF PHYSICS, UNIV KY, LEXINGTON, 86- *Concurrent Pos:* Vis res asst prof physics, Univ Ill, Urbana, 77-80. *Mem:* Am Astron Soc; Int Astron Union. *Res:* Theoretical astrophysics, mainly problems related to the interstellar medium and active galactic nuclei. *Mailing Add:* Dept Physics & Astron Univ Ky Lexington KY 40506

ELIZAN, TERESITA S, b Naga City, Philippines, Dec 12, 31. NEUROLOGY, NEUROVIROLOGY. *Educ:* Univ Philippines, MD, 55; Am Bd Psychiat & Neurol, dipl, 63 & 75. *Prof Exp:* Asst resident gen path, St Mary's Hosp, Waterbury, Conn, 55-56, asst resident neurol, Sch Med, Yale Univ & Grace-New Haven Hosp, Conn, 56-58; clin fel neurol & neuropath, Inst Neurol, Nat Hosp, Univ London, 58-59; chief resident neurol, Montreal Neurol Inst, McGill Univ, 59-60; res asst, Mt Sinai Hosp, New York, 60-61, res assoc, 61-62; asst prof & head neurol sect, Col Med, Univ Philippines, 62; vis scientist & neurologist-in-chg, Res Ctr, Nat Inst Neurol Dis & Blindness, Guam, 63-65, vis scientist, Epidemiol Br, Md, 65-66, vis scientist, res neurologist & neuropathologist, Sect Infectious Dis, Perinatal Res Br, Nat Inst Neurol & Commun Dis & Stroke, NIH, 66-68; asst prof neurol & asst attend neurologist, 68-71, assoc prof neurol & assoc attend neurologist, 71-77, PROF NEUROL & ATTEND NEUROLOGIST, MT SINAI SCH MED, 77-, HEAD, LAB NEUROVIROL, 68- *Concurrent Pos:* USPHS clin fel neurol, Sch Med, Yale, 56-58; Dazian Found Med Res fel, Mt Sinai Hosp, New York, 60-62; consult neurologist, US Naval Hosp & US Air Force Base, Guam, 63-65; clin asst prof, Sch Med, Georgetown Univ, 65-68; mem prog proj rev comn (neurol B), Nat Inst Neurol Dis & Commun Dis & Stroke, NIH, 72-76; consult neurologist, Bronx Vet Admin Hosp, 75- *Mem:* Fel Am Acad Neurol; Am Asn Neuropath; Am Neurol Asn; Soc Neurosci; Am Soc Virol; Int Brain Res Orgn. *Res:* Neurobiology of central nervous system degenerations; Parkinson's disease; Alzheimer's disease; role of viruses in CNS degenerations. *Mailing Add:* Dept Neurol Mt Sinai Med Ctr Sch Med New York NY 10029

ELIZONDO, REYNALDO S, b Sept 9, 40; m 63; c 2. PHYSIOLOGY, PATHOLOGY. *Educ:* Tulane Univ, Phd(physiol), 67. *Prof Exp:* prof physiol & head dept, Ind Univ, 79-87; DEAN SCI, UNIV TEX EL PASO, 88- *Concurrent Pos:* Career Develop Award, NIH. *Mem:* Am Physiol Soc; Sigma Xi. *Res:* Physiology of temperature; regulation and adaptation; physiology of fever and stress hormones. *Mailing Add:* Dept Anat-Phys Univ Ind Bloomington IN 47401

ELK, SEYMOUR B, b Passaic, NJ, Sept 4, 32; m 59; c 2. GEOMETRICAL CHEMISTRY, MATH AND COMPUTER SCIENCE EDUCATION. *Educ:* Yale Univ, BS, 54; Purdue Univ, MS, 57; Univ Zurich, ABD, 60; Washington Univ, MS, 70; Eurotechnical Res Univ, DSc(math chem), 89. *Prof Exp:* R & D Missile Eng, McDonnell Aircraft Corp, St Louis, Mo, 54-55; design data coordr, Raytheon, Bedford & Waltham, Mass, 57-59; R & D Missile eng, Martin Co, Baltimore, 55-56; eng writing supvr, Anco Tech Serv, Boston, Mass, 60-61; staff math, Northrop Corp, Norwood, Mass, 61-62; eng res specialist, Lockheed Missile & Space Co, Sunnyvale, Calif, 62-65; chmn, Gen Sci Dept, Parks Col Aero Tech, St Louis Univ, Cahokia, Ill, 65-68; math teacher, Albrecht Thaer Gymnasium, Hamburg, Ger, 71-73; math teacher, Fieldston Sch, NY, 73-75; specialist lectr math, NJ Inst Technol, 77-81; asst prof, Comp Sci, Union County Col, 82-83; asst prof, William Paterson Col, NJ, 83-84; vis assoc prof, Computer Sci, 86-87; CONSULT & WRITER, ELK TECH ASSOCS, NEW MILFORD, NJ, 84-; PROF, CHEM & MATH, UNIV BRIDGEPORT, 90- *Concurrent Pos:* Consult, Math, 62 & Comp Sci Ed, 70-71. *Mem:* Am Chem Soc; Int Soc Math Chem; World Asn Theoret Org Chemists. *Res:* Orismology and taxonomy of chemical structure and nonmenclature from a geometry-graph theory-topology perspective; canonical graph-theory based "nomenclating" of geometrical structures; postulation of new chemical moieties based on geometrical ideals; chemistry education; author of textbooks. *Mailing Add:* 321 Harris Pl New Milford NJ 07646-1203

ELKAN, GERALD HUGH, b Berlin, Ger, Aug 3, 29. BACTERIOLOGY. *Educ:* Brigham Young Univ, AB, 51; Pa State Univ, MS, 55; Va Polytech Inst & State Univ, PhD(bact), 59. *Prof Exp:* From asst prof to assoc prof, 58-70, asst univ dean res, 77-79, PROF MICROBIOL, NC STATE UNIV, 70- *Concurrent Pos:* Fulbright res fel, Inst Microbiol, Sweden, 63-64; mem, State of NC Water Control Comn, 70-76; Fulbright res fel, Nat Univ East Monagas, Venezuela, 80-81; res grants, NSF, NIH, AID, USDA & Nat Soybean Asn. *Mem:* AAAS; Am Soc Microbiol; fel Am Acad Microbiol; Can Soc Microbiol; Sigma Xi. *Res:* Microbial physiology and metabolism; function and physiology of symbiotic nitrogen fixing bacteria. *Mailing Add:* Dept Microbiol NC State Univ Box 7615 Raleigh NC 27695

EL-KAREH, AUGUSTE BADIH, b Baabda, Lebanon, July 9, 32; US citizen; m 58; c 2. ELECTRICAL ENGINEERING, APPLIED PHYSICS. *Educ:* Delft Univ Technol, DipIng, 56, DSc, 62. *Prof Exp:* Asst, res labs, Europe, 53-55; asst engr, Delft Univ Technol, 55-58; instr elec eng, Univ NMex, 58-59; instr, Univ Pa, 59-60; mem tech staff, RCA Labs, NJ, 60-63; assoc prof elec eng & head electron physics lab, Pa State Univ, 63-66; prof elec eng, Clarkson Col Technol, 66-67; prof, Syracuse Univ, 67-71; prof elec eng & dir, Electron Beam Lab, 71-81, ASSOC DEAN GRAD PROGS & RES, COL ENG, UNIV HOUSTON, 76-; PRES, ABEK, INC, COLORADO SPRINGS, 81- *Concurrent Pos:* Chmn, Int Electron Beam Symp, 64-65. *Mem:* Sr mem Inst Elec & Electronics Engrs. *Res:* High power microwave tubes; electron beam techniques; millimeter wave generation; electron optics; electron physics. *Mailing Add:* Abek Inc 815 N Wooten Rd Colorado Springs CO 80915

ELKES, JOEL, b Germany, Nov 12, 13; m 43; c 1. PSYCHIATRY, PSYCHOPHARMACOLOGY. *Educ:* Univ Birmingham, MB, ChB, 47, MD, 49. *Prof Exp:* Lectr, Univ Birmingham, 45-48, sr lectr & actg dir dept pharmacol, 48-50, prof exp psychiat & chmn dept, 51-57; clin prof psychiat, Sch Med, George Washington Univ, 57-63; Henry Phipps prof & dir, dept psychiat, John Hopkins Univ & psychiatrist-in-chief, John Hopkins Hosp, 63-74, distinguished serv prof, 75; Samuel McLaughlin prof-in-residence, McMaster Univ, Hamilton, 75-80; DIR, DEPT PSYCHIAT, DIV BEHAV MED, UNIV LOUISVILLE, 80- *Concurrent Pos:* Consult psychiatrist, Birmingham United Hosp & Birmingham Regional Hosp Bd, 53-57; sci dir, Birmingham Regional Psychiat Early Treat Ctr, 53- 57; examr, Univ London, 53-56; chief, Clin Neuropharmacol Res Ctr, NIMH, 57-63, mem, Psychopharmacol Study Sect, 57-64; dir behav & clin studies ctr, St Elizabeths Hosp, DC, 57-63; mem adv comt biol sci, Air Force Off Sci Res, 64; dir, Found Fund for Res in Psychiat, 67-; Distinguished Serv Prog, Johns Hopkins Univ, 74; Samuel McLaughlin Prof-in-Residence, McMaster Univ, Hamilton, Ont, 75. *Mem:* Fel Am Psychiat Asn; Am Soc Pharmacol & Exp Therapeut; fel Am Psychopath Asn (pres, 69); Am Col Neuropsychopharmacol (1st pres, 61); NY Acad Sci; Am Col Psychiat. *Res:* Psychopharmacology; behavioral medicine. *Mailing Add:* Div Behav Med Univ Louisville Genesis Ctr 310 E Broadway Louisville KY 40202

EL KHADEM, HASSAN S, b Cairo, Egypt, Mar 24, 23, US citizen; m 51; c 2. ORGANIC & CARBOHYDRATE CHEMISTRY. *Educ:* Cairo Univ, BSc, 46; Swiss Fed Inst Technol, DSc Tech, 49; Univ London, PhD(org chem) & DIC, 52. *Hon Degrees:* DSc, Univ Alexandria, 63 & Univ London, 67. *Prof Exp:* From lectr to prof chem, Univ Alexandria, 52-71; prof chem, Mich Technol Univ, 71-81, head, dept chem & chem eng, 75-80, presidential prof chem, 81-84; ISBELL PROF CHEM, AM UNIV, 84- *Concurrent Pos:* Fulbright scholar, Ohio State Univ, 63-67. *Honors & Awards:* Nat Sci Award, Egypt, 61; M L Wolfrom Award, Am Chem Soc, 89. *Mem:* Am Chem Soc; Sigma Xi; AAAS. *Res:* Carbohydrates; nitrogen heterocycles; metal chelates; polysaccharides; glycosides; natural product chemistry; published 260 research papers and books. *Mailing Add:* Dept Chem Am Univ 4400 Massachusetts Ave NW Washington DC 20016

EL-KHATIB, SHUKRI M, b Ein Yabroud, Palestine, Aug 20, 33; US citizen; m 60; c 4. BIOCHEMISTRY. *Educ:* Okla City Univ, BS, 57; ETex State Univ, MS, 65; Tex A&M Univ, PhD(biochem), 68. *Prof Exp:* Instr pharm, Baylor Col Med, 68-70; asst prof biochem, Univ Tenn Med Sch, 70-74; assoc prof, Univ PR Med Sch, 74-76; CHMN & PROF BIOCHEM, MED SCH, UNIV CENTRAL DEL CARIBE, 76- *Mem:* Am Asn Biochem & Molecular Biol; NY Acad Sci; AAAS; Am Soc Exp Biol & Med; Am Chem Soc. *Res:* Nutrition and colon cancer; relation to dietary fat and its effect on colon cancer. *Mailing Add:* Dept Biochem & Nutrit Univ Central del Caribe Med Sch Box 60-327 Bayamon PR 00621-6032

ELKHOLY, HUSSEIN A, b Elmansoura, Egypt, Oct 30, 33; m 61; c 3. PHYSICS. *Educ:* Cairo Univ, BSc, 57; Hungarian Acad Sci, Kandidat, 61; Eotvos Lorand Univ, Budapest, Dr rer nat(physics), 61. *Prof Exp:* Fel, Eotvos Lorand Univ, Budapest, 61-62; asst prof physics, Cairo Univ, 62-63 & Univ Khartoum, 63-64; asst prof physics, 64-65, chmn dept, 65-67, chmn dept math & physics, 67-81, PROF PHYSICS & COMPUT SCI, FAIRLEIGH DICKINSON UNIV, 81- *Mem:* AAAS; Am Phys Soc; Am Asn Physics Teachers; Am Math Soc; Math Asn Am. *Res:* Phase transformation; lattice defects; radiation effects in solids; microelectronics; local area networks. *Mailing Add:* Dept Physics Fairleigh Dickinson Univ Madison NJ 07940

ELKIN, LYNNE OSMAN, b New York, NY, June 10, 46; m 67. PLANT PHYSIOLOGY, PHOTOMICROSCOPY. *Educ:* Univ Rochester, AB, 67; Univ Calif, Berkeley, PhD(bot), 73. *Prof Exp:* Lectr, 71-72, asst prof, 72-76, ASSOC PROF BIOL SCI, CALIF STATE UNIV, HAYWARD, 76- *Concurrent Pos:* Asst specialist instrnl biol, Univ Calif, Berkeley, 75-78; grant-in-aid res, Sigma Xi, 75. *Mem:* Am Soc Plant Physiol; Sigma Xi; Bot Soc Am. *Res:* Light reactions of photosynthesis; C4 pathway of photosynthesis; fluorescence photomicroscopy; scientific photography; photorespiration; crassulacean acid metabolism. *Mailing Add:* Dept Biol Sci Calif State Univ Hayward CA 94542

ELKIN, MILTON, b Boston, Mass, Feb 24, 16; m 43; c 3. RADIOLOGY. *Educ:* Harvard Univ, AB, 37, MD, 41. *Prof Exp:* Assoc radiologist, Peter Bent Brigham Hosp, Boston, 51-52; asst radiologist, New Eng Med Ctr, Boston, 52-53; chmn radiol dept, Albert Einstein Col Med, Yeshiva Univ, 54- 86; dir radiol, Bronx Munic Hosp Ctr, 54-86, ATTEND RADIOLOGIST, BRONX MUNIC HOSP CTR, 54-; PROF RADIOL, ALBERT EINSTEIN COL MED, YESHIVA UNIV, 54- *Concurrent Pos:* Assoc radiologist, Cedars of Lebanon Hosp, Los Angeles, 53-54; Knox lectr, London, Eng & Holmes lectr, Boston, Mass, 70; Rigler lectr, Tel Aviv, Israel, 71; prog dir, Nat Inst Gen Med Sci training grant diag radiol, 66-77; prin investr, Nat Inst Gen Med Sci res grant, 68-71; spec consult, USPHS-Nat Inst Gen Med Sci Res Prog-Proj Comt, 70-72. *Honors & Awards:* Gold Medal, Am Col Radiol, 77; Gold Medal, Radiol Soc NAm, 85. *Mem:* Fel Am Col Radiol; Radiol Soc NAm (pres elect, 79-80, pres, 80-81); Am Roentgen Ray Soc; AMA; Asn Univ Radiol; Soc Uroradiol. *Res:* Renal physiology; effects of radiation on tissue; uroradiology. *Mailing Add:* Albert Einstein Col Med Yeshiva Univ Bronx NY 10461

ELKIN, ROBERT GLENN, b Passaic, NJ, May 7, 53; m 81. AMINO ACID NUTRITION, AMINO ACID METABOLISM. *Educ:* Pa State Univ, BS, 75; Purdue Univ, MS, 77, PhD(animal nutrit), 81. *Prof Exp:* asst prof, 81-86, ASSOC PROF ANIMAL SCI, PURDUE UNIV, 86- *Concurrent Pos:* Assoc ed, Poultry Sci, 85-90. *Mem:* Poultry Sci Asn; AAAS; World's Poultry Sci Asn; Asn Off Anal Chemists; Am Chem Soc; Am Inst Nutrit. *Res:* Amino acid nutrition; amino acid metabolism in poultry; amino acid analysis by high performance liquid chromatography; lipid and bile acid metabolism in poultry; lipid nutrition; reduction of egg cholesterol content. *Mailing Add:* Dept Animal Sci Lilly Hall Purdue Univ West Lafayette IN 47907

ELKIN, WILLIAM FUTTER, b Atlantic City, NJ, May 5, 16; m 44; c 1. BIOSTATISTICS, PUBLIC HEALTH. *Educ:* Harvard Univ, BS, 37; Univ Mich, MS, 41, MSPH, 42. *Prof Exp:* Asst statistician, Res Div, Off Price Admin, 42-43; res secy, Philadelphia Tuberc & Health Asn, 43-45; statistician, Oak Ridge Dept Health, Tenn, 45-48 & Health Physics Div, Oak Ridge Nat Lab, 48-49; res staff mem, Dept Biostatist, Sch Pub Health, Univ NC, 50-54; biostatist consult, Fife-Hamill Mem Health Ctr, 54-58; biostatistician, Periodontal Res Proj, Sch Dent, Temple Univ, 58-60; res assoc, Henry Phipps Inst, Sch Med, Univ Pa, 60-62, guest lectr, Sch Nursing, 59-62; statistician & statist consult, US Dept Health & Human Serv, 62-82. *Mem:* Fel AAAS; fel Am Pub Health Asn; Am Statist Asn. *Res:* Public health. *Mailing Add:* 6005 McKinley St Bethesda MD 20817

ELKIND, JEROME I, b New York, NY, Aug 30, 29; m 59; c 3. COMPUTER SCIENCES. *Educ:* Mass Inst Technol, SB & SM, 52, ScD(elec eng), 56. *Prof Exp:* Staff mem psychol, Lincoln Lab, 54-56; head human eng, Airborne Systs Lab, Radio Corp Am, 56-58; sr scientist eng psychol, Bolt Beranek & Newman, 58-61, dept head, 61-64, vpres info sci, 64-66, sr vpres, 66-70; vis prof mgt, Mass Inst Technol, 70-71; mgr comput sci lab, Xerox Corp, 71-77, vpres advan systs dept, 77-79, vpres info systs, 79-85, vpres syst integration, 85-90; CHMN, LEKIA INST, 90-; PRIN, ELKIND ASSOCS, 90- *Concurrent Pos:* Ed, Trans Inst Elec & Electronics Engrs, 59-63; mem res adv comt guid, control & navig, NASA, 60-65; mem, Nat Acad Sci-NSF Comt Human Factors, 85-90, Comt Ballistic Acoust, 85-88. *Honors & Awards:* Franklin Taylor Award, Inst Elec & Electronics Engrs, 68. *Mem:* Inst Elec & Electronic Engrs; Asn Comput Mach; fel Human Factors Soc; Sigma Xi. *Res:* Interactive man-computer systems; manual control; office information systems. *Mailing Add:* 2040 Tasso St Palo Alto CA 94301

ELKIND, MICHAEL JOHN, b Detroit, Mich, July 23, 22; m 52; c 4. INORGANIC CHEMISTRY. *Educ:* Univ Detroit, BS, 43, MS, 48; Wayne Univ, PhD(chem), 51. *Prof Exp:* Develop chemist pharmaceut prods, R P Scherer Corp, 43-44; fel chem, Univ Detroit, 46-48 & Wayne Univ, 48-50, instr, 50-51; sr res & develop chemist photo prods dept, E I du Pont de Nemours & Co, 51-52, Wyandotte Chem Corp, 52-53 & J T Baker Chem Co, 53-56; mem tech staff, AT&T Bell Labs, 56-62, supvr, 62-87; RETIRED. *Mem:* Sigma Xi. *Res:* Inorganic fine and heavy chemicals; photographic chemistry; phase studies in non-aqueous media; chemistry of electron device materials and processing. *Mailing Add:* 1543 Rose Virginia Rd Wyomissing PA 19610-2842

ELKIND, MORTIMER M, b Brooklyn, NY, Oct 25, 22; m 60; c 3. BIOPHYSICS, RADIOBIOLOGY. *Educ:* Cooper Union, BME, 43; Polytech Inst Brooklyn, MME, 49; Mass Inst Technol, MS, 51, PhD, 53. *Prof Exp:* Asst proj engr, Wyssmont Co, 43-44; proj engr, Safe Flight Instrument Corp, 46-47; head instrumentation, Sloan-Kettering Inst Cancer Res, 47-49; biophysicist, Nat Cancer Inst, 49-69, sr scientist, Biol Dept, Brookhaven Nat Lab, 69-73; mem staff, Exp Radiopath Res Unit, Med Res Coun, Hammersmith Hosp, London, Eng, 71-73; prof radiol, Univ Chicago, 73-82; CHMN, DEPT RADIOL & RADIATION BIOL, COLO STATE UNIV, FT COLLINS, 81-; SR BIOPHYSICIST, DIV BIOL & MED RES, ARGONNE NAT LAB, 73- *Concurrent Pos:* Mem, Radiation Study Sect, NIH, 61-65 & Molecular Biol Study Sect, 69-71; sr fel, Nat Cancer Inst, 71-73. *Honors & Awards:* E O Lawrence Award, US AEC, 67; L H Gray Award, Int Comn on Radiation Units & Measurements, 77; E W Bertmer Award, 79; Arthur W Erskine Award, Radiol Soc NAm, 80; Gold Medal, Am Soc Therapeut Radiologists, 83; Albert Soiland Found Mem Award, 84; G Failla Mem Award, Radiation Res Soc, 84. *Mem:* Biophys Soc; Radiation Res Soc (pres, 81-82); AAAS. *Res:* Radiobiology of microorganisms and mammalian cells in tissue culture; DNA damage and repair. *Mailing Add:* Dept Radiol & Radiation Biol Colo State Univ Ft Collins CO 80523

ELKINS, DONALD MARCUM, b Woodville, Ala, Sept 15, 40; m 63; c 2. AGRONOMY. *Educ:* Tenn Polytech Inst, BS, 62; Auburn Univ, MS, 64, PhD(agron), 67. *Prof Exp:* From asst prof to prof agron, ASSOC DEAN INSTR, COL AGR, SOUTHERN ILL UNIV, 85- *Concurrent Pos:* Agron resident educ award, Am Soc Agron, 81. *Mem:* fel Am Soc Agron; Nat Asn Cols & Teachers Agr; Plant Growth Regulator Soc Am. *Res:* No-tillage or conservation tillage methods; intercropping; row cropping in living or suppressed forage sods; herbicides; growth regulators. *Mailing Add:* Col Agr Dean's Off Southern Ill Univ Carbondale IL 62901-4416

ELKINS, EARLEEN FELDMAN, b South Bend, Ind, Mar 20, 33; m 54; c 3. NEUROSCIENCES. *Educ:* Univ Md, BA, 54, MA, 56, PhD(audiol), 67. *Prof Exp:* Instr speech, Univ Md, 54-56; rehab audiologist, Walter Reed Army Med Ctr, 56-57; res assoc, Electronic Teaching Labs, DC, 60-61; asst speech & hearing, Univ Md, College Park, 63-67, res assoc, Biocommun Lab, 67-70, res asst prof, 70-76; res audiologist, Vet Admin Hosp, Washington, DC, 67-76; prog adminr, Nat Inst Neurol & Commun Disorders & Stroke, Bethesda, 76-88, dept health and human services, 88-89, CHIEF, NEUROSCI & BEHAV BR, NAT INST ALCOHOL ABUSE & ALCOHOLISM, ROCKVILLE, 88-, CHIEF, SCI REV BR, NIH, 89- *Concurrent Pos:* Res asst, Nat Inst Child Health & Human Develop, 64; consult audiol, Nat Ctr Health Statistics, 74-86; mem, Fed Panel Noise Res, 76-89. *Honors & Awards:* Merit Award, NIH, 87; Outstanding Serv Award, Asn Res Otolaryngol, 89. *Mem:* Fel Am Speech & Hearing Asn; Acoust Soc Am; Am Auditory Soc; AAAS; Asn Res Otolaryngol. *Res:* Speech recognition and perception. *Mailing Add:* Nat Inst Deafness & Other Commun Disorders 6120 Executive Blvd Exec Plaza S No 750 Rockville MD 20892

ELKINS, JOHN RUSH, b Beckley, WVa, Nov 16, 41; wid; c 2. ORGANIC CHEMISTRY. *Educ:* WVa Inst Technol, BS, 63; WVa Univ, PhD(org chem), 66. *Prof Exp:* Instr chem, WVa Inst Technol, 66-67; fel, Univ Ky, 67-68; from asst prof to assoc prof chem, Bluefield State Col, 68-78, chmn dept, 69-74; assoc prof, 75-78, PROF CHEM, CONCORD COL, 78- *Concurrent Pos:* Vis prof plant path, Va Polytech Inst & State Univ, 74-75, adj prof, 75- *Mem:* Am Phytopath Soc; Sigma Xi; Am Chem Soc. *Res:* Chestnut blight-chemical basis for host-parasite interaction between the American chestnut tree and the blight fungus with emphasis on developing a method for screening chestnut progeny for blight resistance. *Mailing Add:* Dept Chem Concord Col Athens WV 24712

ELKINS, JUDITH MOLINAR, b Stamford, Conn, Apr 20, 35; c 3. APPLIED MATHEMATICS. *Educ:* Wellesley Col, BA, 56; Harvard Univ, MA, 59; Univ Wis-Madison, PhD(math), 66. *Prof Exp:* Instr math, Mt Holyoke Col, 60-63; asst prof, Calif State Univ, San Diego, 66-67, Rutgers Univ, 67-68 & Ohio State Univ, 68-75; assoc prof, 75-79, PROF MATH, SWEET BRIAR COL, 79- *Concurrent Pos:* NSF-sci fac prof develop, Comput Sci, Univ Md, 81-82. *Mem:* Am Math Soc; Math Asn Am; Asn Women Math; Asn Comput Mach. *Res:* Best approximation real-valued functions on the positive real axis; algebraic structures. *Mailing Add:* Dept Math Sci Sweet Briar Col Sweet Briar VA 24595

ELKINS, L(LOYD) E(DWIN), b Golden, Colo, Apr 1, 12; m 34; c 3. PETROLEUM ENGINEERING. *Educ:* Colo Sch Mines, PPE, 34. *Hon Degrees:* ScD, Col of Ozarks, 62. *Prof Exp:* From roustabout to prod res dir, Pan Am Petrol Corp, 34-71; prod res dir, Amoco Prod Co, 71-77; RETIRED. *Concurrent Pos:* Petrol consult, 77- *Mem:* Nat Acad Eng; Am Petrol Inst; hon mem Am Inst Mining, Metall & Petrol Engrs (pres, 62); Am Asn Petrol Geol. *Res:* Oil field reservoir engineering; oil field drilling and well completion; oil field appraisals. *Mailing Add:* PO Box 4758 Tulsa OK 74159

ELKINS, LINCOLN F, b Denver, Colo, Feb 12, 18; m 42; c 3. PETROLEUM ENGINEERING. *Educ:* Colo Sch Mines, BS. *Prof Exp:* Res Engr, Am Prod Co, 41-45; prod engr, Conoco, 45-77; chief engr, sr engr & spec projs engr, Sohio, 47-83; PETROL CONSULT, 83- *Honors & Awards:* DeGolyer Medal, Soc Petrol Engr, 71. *Mem:* Nat Acad Eng; Soc Petrol Eng (pres, 65); hon mem Am Inst Mining Metall & Petrol Eng. *Mailing Add:* 2615 Oak Dr No 28 Lakewood CO 80215

ELKINS, ROBERT HIATT, b Marion, Ind, Oct 2, 18; m 50; c 5. CHEMISTRY. *Educ:* DePauw Univ, AB, 40; Western Reserve Univ, PhD(chem), 45. *Prof Exp:* Res chemist, Standard Oil Co, Ohio, 40-43, res chemist, Great Lakes Carbon Corp, 44-49; res chemist, Sinclair Res Lab, Inc, 49-56; mgr, Org & Polymer Chem, Borg Warner Res Ctr, 56-58, assoc dir, Chem Dept, 58-60; tech mgr com develop & cent res, Nalco Chem Co, 60-66; SR PROG ADV, INST GAS TECHNOL, ILL INST TECHNOL, 66- *Mem:* AAAS; Am Chem Soc; Sigma Xi. *Res:* Polymer chemistry; water soluble polymer applications; hydrocarbon processes; plastics; catalysis; petrochemicals; sulfur recovery from hydrogen sulfide; reactive carbon; indoor environmental control: air quality, energy conservation. *Mailing Add:* 119 N Grant St Hinsdale IL 60521

ELKINS, RONALD C, b Lubbock, Tex, July 6, 36. CARDIOVASCULAR SURGERY, THORACIC SURGERY. *Educ:* Univ Okla, MD, 62. *Prof Exp:* CHIEF, CARDIOVASC & THORACIC SECT, HEALTH SCI CTR, UNIV OKLA, 75- *Mem:* Soc Thoracic Surgeons; Soc Univ Surgeons; Am Surg Asn. *Mailing Add:* Health Sci Ctr Univ Okla PO Box 26901 Oklahoma City OK 73190

ELKINS, WILLIAM L, b Boston, Mass, Aug 2, 32. TRANSPLANTATION IMMUNOLOGY. *Educ:* Princeton Univ, AB, 54; Harvard Univ, MD, 58. *Prof Exp:* Intern surg, St Vincents Hosp, New York, 58-59; resident, Univ Hosp, 59-61, from instr to assoc prof path, Sch Med, Univ Pa, 66-87; RES ASSOC, CHILDREN'S HOSP, PHILADELPHIA, 78- *Concurrent Pos:* Fel Wistar Inst, Univ Pa & fel path, Sch Med, 65-66. *Mem:* AAAS; Am Soc Exp Path; Am Asn Immunologists; Transplantation Soc. *Res:* Genetic control of muscling in beef cattle. *Mailing Add:* RD 8 Box 432 Coatesville PA 19320

ELKINTON, J(OSEPH) RUSSELL, b Pa, Oct 12, 10; m 40; c 2. MEDICINE. *Educ:* Haverford Col, AB, 32; Harvard Univ, MD, 37; Am Bd Internal Med, dipl, 45. *Prof Exp:* Intern, Pa Hosp, 37-39, resident physician, 39-40, asst instr, Sch Med, 39-40; from instr to asst prof med, Yale Univ, 42-48; assoc physician, New Haven Hosp, 42-48; ward physician, Univ Hosp, 48-72, from asst prof to prof, 48-72, EMER PROF MED, UNIV PA, 72- *Concurrent Pos:* Fel med, Univ Pa, 39-40; Nat Res Coun fel electrolyte physiol, Sch Med, Yale Univ, 40-42; estab investr, Am Heart Asn, 49-59; consult surg gen, USPHS, 54-58; ed, Ann Internal Med, Am Col Physicians, 60-71, consult ed, 71-72, emer ed, 72-; hon sr res fel, Univ Birmingham, 73-85. *Mem:* Am Soc Clin Invest; Am Physiol Soc; fel & master Am Col Physicians; Asn Am Physicians; fel Royal Col Physicians London. *Res:* Electrolyte physiology; cardiovascular science; metabolic and renal diseases. *Mailing Add:* Four Andover Ct Bedford MA 01730

ELL, WILLIAM M, b Minot, NDak, Oct 13, 39; m 70; c 1. ENERGY CONSERVATION, SELECTOR-CALCULATOR DESIGN. *Educ:* NDak State Univ, BS, 63. *Prof Exp:* Appln engr, Honeywell Inc, 67-69; proj engr, Airflow Co, 69-71; prog mgr, Deleuw Cather & Co, 74-76; energy conserv engr, Smithsonian Inst, 77-78; mech engr, Naval Facil Eng Command, 78-80; prog mgr, USMC Hq, 80-84; mech engr, Facil Eng Support Agency, 84-88, ENERGY ENGR, ENG & HOUSING SUPPORT CTR, US ARMY, 88- *Concurrent Pos:* Owner, W Michael L Co, 85- *Mem:* Am Soc Heating Refrig & Air Conditioning Engrs; United Inventors & Scientists Am. *Res:* Air conditioner selectors; fire protection; energy conservation; mechanical engineering; author of 17 publications; awarded two US patents. *Mailing Add:* 1266 Everett Ave Woodbridge VA 22191

ELLEFSEN, PAUL, b Oak Park, Ill, June 20, 39; m 62; c 4. ANALYTICAL CHEMISTRY. *Educ:* Monmouth Col, BA, 61; Case Inst Technol, PhD(chem), 65. *Prof Exp:* ASSOC PROF CHEM, HANOVER COL, 65- *Mem:* Am Chem Soc. *Res:* Reaction mechanisms and analytical chemistry in nonaqueous solvents. *Mailing Add:* 610 Ball Dr Hanover IN 47243-9673

ELLEFSON, RALPH DONALD, b Glenwood, Minn, Jan 25, 31; m 55; c 4. MEDICAL BIOCHEMISTRY, LABORATORY MEDICINE. *Educ:* Luther Col, Iowa, BA, 53; Univ Iowa, MS, 56, PhD(chem), 58. *Prof Exp:* Asst chem, Luther Col, Iowa, 52-53; asst, Univ Iowa, 53-57, Vet Admin Hosp Iowa City, 57; instr org chem, Iowa Wesleyan Col, 57-58; from instr to asst prof biochem, 63-76, assoc prof biochem & lab med, Mayo Grad Sch Med, Univ Minn, 76-83, ASSOC PROF BIOCHEM, MOLECULAR BIOL & LAB MED, MAYO MED SCH, 83- *Concurrent Pos:* Res fel biochem, Mayo Clin & Found, 58-60, res assoc, 60, asst to staff, 60-63; consult in clin chem, NIH; mem staff, Mayo Clin, 63-, dir, Lipids, Lipoproteins, Porphyrins & Occult Blood Labs, 63- *Mem:* AAAS; Am Chem Soc; Sigma Xi; Am Oil Chem Soc; Am Asn Clin Chem. *Res:* Nitrogen heterocycles; lipid and lipoprotein chemistry and metabolism; lipoproteins components; porphyrinogens, porphyrins and porphyrias; medical laboratory systems; specialized medical laboratory software. *Mailing Add:* Mayo Med Ctr 210 Hilton Bldg Rochester MN 55905

ELLEMAN, DANIEL DRAUDT, physics; deceased, see previous edition for last biography

ELLEMAN, THOMAS SMITH, b Dayton, Ohio, June 19, 31; m 54; c 3. PHYSICAL CHEMISTRY. *Educ:* Denison Univ, BS, 53; Iowa State Col, PhD(chem), 57. *Prof Exp:* Chemist inst atomic res, Ames Lab, 53-57; radiochemist, Battelle Mem Inst, 57-64; assoc prof, 64-67, head dept, 74-79, PROF NUCLEAR ENG, NC STATE UNIV, 67-; vpres, Nuclear Safety & Res Dept, Carolina Power & Light Co, 79-85; assoc dean, 85-89, PROF, NC STATE UNIV, 89- *Mem:* Am Chem Soc; Am Nuclear Soc. *Res:* Radioisotopes applications; radiation effects; reactor chemistry. *Mailing Add:* 704 Davidson St Raleigh NC 27609

ELLENBERG, JONAS HAROLD, b Long Beach, NY, Apr 26, 42; m 69; c 2. BIOSTATISTICS. *Educ:* Univ Pa, BSc, 63; Harvard Univ, AM, 64, PhD(statist), 70. *Prof Exp:* CHIEF BIOMET & FIELD STUDIES BR, NAT INST NEUROL DIS & STROKE, NIH, 69- *Concurrent Pos:* Assoc ed, Am Statistician. *Mem:* Fel Am Statist Asn; Inst Math Statist; Biometrics Soc; Int Biomet Soc (pres, 88-89); Int Statist Inst. *Res:* Applied: epilepsy, cerebral palsy, Parkinson's disease; theoretical: general linear model, outlier theory, database utilization. *Mailing Add:* Nat Inst Neurol Dis & Stroke NIH 7550 Wisconsin Ave Rm 7A12 Bethesda MD 20892

ELLENBERG, SUSAN SMITH, b Tucson, Ariz, May 27, 46; m; c 3. BIOSTATISTICS RESEASRCH. *Educ:* Radcliffe Col, AB, 67; Harvard Univ, MAT, 68; George Washington Univ, MPhil, 76, PhD(statist), 80. *Prof Exp:* Math programmer, George Washington Univ, 71-73, statistician, 73-79; statistician, Emmes Corp, 79-82; biostatistician, Biomet Res Br, Div Cancer Treatment, Nat Cancer Inst, 82-88, CHIEF, BIOSTATIST RES BR, DIV AIDS, NAT INST ALLERGY & INFECTIOUS DIS, NIH, 88- *Concurrent Pos:* Secy, Biomet Sect, Am Statist Asn, 87-89 & mem, ad hoc Comt AIDS, 89; mem sci adv comt, Am Found AIDS Res, 90-91; consult, Food & Drug Admin, 90 & 91; bd dirs, Soc Clin Trials, 90-93. *Mem:* Am Statist Asn; Biomet Soc; Soc Clin Trials; AAAS. *Mailing Add:* Div AIDS Nat Inst Allergy & Infectious Dis NIH 6003 Executive Blvd Rm 214P Rockville MD 20892

ELLENBERGER, HERMAN ALBERT, b Annville, Pa, Mar 24, 16; m 50; c 3. CLINICAL CHEMISTRY. *Educ:* Lebanon Valley Col, BS, 38; Pa State Univ, MS, 46, PhD(biochem, org chem), 48. *Prof Exp:* Technician, Pa Dept Agr, 39-41; chemist, Whitmoyer Labs, Inc, 41-43; res asst & fel, Pa State Univ, 43-48; biochemist, Limestone Prod Corp, 48-63; toxicologist, Conn State Dept Health, 63-68; toxicologist, NS Dept Pub Health, 68-75; toxicologist, Victoria Gen Hosp, Halifax, NS, 75-82; asst prof path, Dalhousie Univ, 69-82; RETIRED. *Mem:* Can Soc Clin Chem; Chem Inst Can; Am Asn Clin Chem; Can Soc Forensic Sci; Am Chem Soc; Sigma Xi. *Res:* Analytical methodology related to fields of toxicology; analysis of street drugs received through medical sources. *Mailing Add:* 216 Donegal Dr Dartmouth NS B2V 1P3 Can

ELLENBOGEN, LEON, b Brooklyn, NY, May 3, 27; m 51; c 3. NUTRITION, HEMATOLOGY. *Educ:* City Col, BS, 49; NY Univ, MS, 51; Ind Univ, PhD(chem), 54. *Prof Exp:* Anal chemist, Novocol Chem Co, 45; res technician 1st res div, Columbia, Goldwater Mem Hosp, 49-51; from asst to res asst, Ind Univ, 51-53; chemist, 53-59, sr res chemist & group leader, Lederle Labs Div, 59-77, SR ASSOC DIR, MED DEVELOP & CHIEF NUTRIT SCI, LEDERLE LABS CONSUMER HEALTH PROD DIV, AM CYANAMID CO, 77- *Concurrent Pos:* Adj prof nutrit med, Cornell Univ Med Col, 78-; adj prof community & preventive med, NY Med Col, 81- *Mem:* Am Chem Soc; Am Inst Nutrit; Am Soc Hemat; Am Soc Clin Nutrit; Am Soc Biol Chemists; fel Am Col Nutrit. *Res:* Cardiovascular biochemistry; thrombosis; vitamin B-12 and intrinsic factor; brain biochemistry; protein fractionation; absorption and metabolism of vitamin B-12 and iron; gastrointestinal absorption; nutrition; vitamins; biochemical pharmacology; biogenic amines; catecholamine metabolism; coenzymes. *Mailing Add:* Lederle Labs Am Cyanamid Co 697 Rte 46 Clifton NJ 07015

ELLENBURG, JANUS YENTSCH, b Linthicum, Md, Jan 14, 22; m 43. SPECTROCHEMISTRY. *Educ:* Western Md Col, BS, 42, ScD(chem), 68. *Prof Exp:* Asst dir res, Crown Cork & Seal Co, Md, 44-47; spectroscopist, Fairchild Engine & Aircraft Corp, Tenn, 49-52; sr chemist, Oak Ridge Nat Lab, 52-57; chemist, Southern Res Inst, Ala, 57-58; spectroscopist, Hayes Int Corp, 58-60 & McWane Cast Iron Pipe Co, 60-61; sr chemist, Hayes Intr Corp, 61-68, sr scientist, 68-85; RETIRED. *Concurrent Pos:* Adv, Bessemer State Technol Inst, Ala, 71-72. *Mem:* Am Inst Aeronaut & Astronaut; Am Chem Soc; Optical Soc Am; Am Inst Phys; Soc Appl Spectros. *Res:* Trace elements in matrices by emission spectroscopy; laminar sublayer effects in circulating fluids; dust particle blocking mechanisms; surface cleanliness of aerospace fluid systems; additives for dispersing fine particle clouds. *Mailing Add:* 1133 Lido Dr Birmingham AL 35226

ELLENSON, JAMES L, b Fort Dodge, Iowa, Apr 25, 46; m 78; c 2. PLANT STRESS PHYSIOLOGY. *Educ:* Oberlin Col, BA, 68; Univ Calif, Berkeley, PhD(chem), 73. *Prof Exp:* Res fel, Harvard Univ, 73-79; res assoc, Boyce Thompson Inst Plant Res, 79-86; SR AGR RES CHEMIST, BASF CORP, 87- *Mem:* Biophys Soc; AAAS; Am Chem Soc; Am Soc Plant Physiologists. *Res:* Biophysical aspects of photosynthesis; plant stress physiology; air pollution effects on plants; environmental fate of pesticides. *Mailing Add:* BASF Corp PO Box 13528 Research Triangle Park NC 27709

ELLER, ARTHUR L, JR, b Chilhowie, Va, Nov 5, 33; m 60; c 4. ANIMAL SCIENCE. *Educ:* Va Polytech Inst & State Univ, BS, 55, MS, 66; Univ Tenn, PhD(animal sci), 72. *Prof Exp:* Asst county agent agr, Coop Exten Serv, 55-56, county exten agent, 56-60; asst prof, 60-61, exten specialist, 61-80, PROF ANIMAL SCI, VA POLYTECH INST & STATE UNIV, 80- *Mem:* Am Soc Animal Sci. *Res:* Beef cattle breeding and genetics. *Mailing Add:* Dept Animal Sci Va Polytech Inst Blacksburg VA 24060

ELLER, CHARLES HOWE, b Bloomington, Ind, June 5, 04; m 33; c 2. EPIDEMIOLOGY, PUBLIC HEALTH ADMINISTRATION. *Educ:* Stanford Univ, AB, 27; Univ Colo, MD, 30; Johns Hopkins Univ, DrPH, 34. *Prof Exp:* Health officer, NMex & Va, 32-36; assoc prof prev med, Sch Med, Univ Va, 35-36; dir rural health, State Dept Health, Va, 36-37; dir eastern health dist, Baltimore, 37-46; assoc prof pub health, Med Col Va, 46-49; prof community health, Med Sch, Univ Louisville, 49-59; prof, 59-73, EMER PROF PUB HEALTH, SCH MED, WASH UNIV, 73- *Concurrent Pos:* Assoc prof, Sch Hyg & Pub Health & lectr, Sch Med, Johns Hopkins Univ, 37-40; dir pub health, Univ Richmond, 46 & Louisville & Jefferson County Health Dept, Ky, 49-55; consult, Community Res Assoc, NY, 55- & USPHS; comnr health, St Louis County, 59-73; exec dir, Health Delivery Systs, Inc, 73-75. *Mem:* Fel Am Pub Health Asn. *Res:* Diphtheria; medical care. *Mailing Add:* Apt 515 250 E Alameda Santa Fe NM 87501

ELLER, PHILLIP GARY, b New Martinsville, WVa, Aug 18, 47. ENVIRONMENTAL SCIENCES. *Educ:* WVa Univ, BS, 67; Ohio State Univ, PhD(chem), 71. *Prof Exp:* Fel inorg chem, Ga Inst Technol, 71-73; MEM STAFF, LOS ALAMOS SCI LAB, 73- *Concurrent Pos:* Fel, Univ Melbourn 1988. *Mem:* Sigma Xi; Am Chem Soc; Am Crystallog Soc. *Res:* Structure and bonding in transition metal and actinide compunds; environmental chemistry. *Mailing Add:* MS J519 Canc Los Alamos NM 87545

ELLERS, ERICH WERNER, b Berlin, Ger, Sept 11, 28; m 56; c 2. PURE MATHEMATICS. *Educ:* Univ Hamburg, Staatsexamen, 56, Dr rer nat, 59. *Prof Exp:* Asst prof math, Univ Hamburg, 58-63 & Univ Braunschweig, 63-64; Reader, Flinders Univ SAustralia, 66-68; assoc prof, Univ NB, 68-69; assoc prof, 69-75, PROF MATH, UNIV TORONTO, 75- *Mem:* Ger Math Asn; Am Math Soc; Can Math Soc. *Res:* Classical groups; geometry; factorization of members of classical groups into simple isometries or involutions; length problem; geometric and algebraic characterizations of classical groups. *Mailing Add:* Dept Math Univ Toronto Toronto ON M5S 1A1 Can

ELLERSICK, FRED W(ILLIAM), b Jersey City, NJ, May 12, 33; m 58; c 4. ELECTRICAL ENGINEERING. *Educ:* Rensselaer Polytech Inst, BEE, 54; Syracuse Univ, MEE, 61; Univ Md, PhD(elec eng), 67. *Prof Exp:* Engr, Int Bus Mach Corp, 54-56, assoc engr, Mil Prods Div, 56-58, staff engr, Fed Systs Div, 58-63, develop engr, 63-69, sr engr, 69; mem tech staff, 69-70, group leader, 70-76, ASSOC DEPT HEAD, COMMUN SYSTS DIV, MITRE CORP, BEDFORD, 76- *Concurrent Pos:* Ed, Inst Elec & Electronics Engrs Trans on Commun, 81, ed-in-chief, Commun Mag, 82- *Mem:* AAAS; sr mem Inst Elec & Electronics Engrs; Sigma Xi. *Res:* System engineering for computer-communications systems; communication theory; operations research. *Mailing Add:* The Rotonda 8370 Greensboro Dr No 301 McLean VA 22102

ELLERT, FREDERICK J, SYSTEMS DEVELOPMENT. *Educ:* Rensselaer Polytech Inst, BEE, MEE & PhD(elec eng). *Prof Exp:* Develop engr, Gen Eng Labs & Corp Res & Develop, Gen Elec Co, 52-64, mgr & dir, Tech Resources Oper, Transmission Div, 64-77, mgr, AC Transmission Eng, 77-80, gen mgr, Systs Develop & Eng Dept, 80-89; CONSULT, 89- *Concurrent Pos:* Mem, US-USSR Working Groups on UHV power transmission; Ellert Consult Group Inc, 89- *Mem:* Nat Acad Eng; fel Inst Elec & Electronics Engrs. *Res:* Author of 20 scientific articles. *Mailing Add:* 37 Via Maria Dr Scotia NY 12302

ELLERT, MARTHA SCHWANDT, b Jersey City, NJ, Nov 27, 40; m 62, 72; c 2. PHYSIOLOGY. *Educ:* Barry Col, BS, 62; Univ Miami, PhD(physiol), 67. *Prof Exp:* Lab technician physiol, Sch Med, Univ Miami, 64-66; instr, 67-70, asst prof physiol & dir summer refresher prog, Sch Med, St Louis Univ, 70-75; res grant, 68-78, ASSOC PROF PHYSIOL SCH MED, SOUTHERN ILL UNIV, 75-, ASSOC PROF MED EDUC, 87- *Concurrent Pos:* Asst dean curric, Sch Med, 81- *Mem:* Soc Sigma Xi; Am Physiol Soc. *Res:* Intestinal transport, particularly of sugars; effects of intestinal resection and dietary glucose on transport rates; mechanism of gastric acid secretion; effects of drugs. *Mailing Add:* Sch Med Southern Ill Univ Carbondale IL 62901

ELLESTAD, GEORGE A, b Coalinga, Calif, Dec 8, 34; m 60; c 3. ORGANIC CHEMISTRY. *Educ:* Ore State Univ, BS, 56 & 57, MS, 58; Univ Calif, Los Angeles, PhD(chem), 62. *Prof Exp:* NIH fel with Prof W B Whalley, Sch Pharm, Univ London, 62-64; CHEMIST, LEDERLE LABS, AM CYANAMID CO, 64- *Mem:* Am Chem Soc. *Res:* Organic chemistry of natural products. *Mailing Add:* Med Res Div Am Cyanamid Co Pearl River NY 10965

ELLESTAD, MYRVIN H, b Santa Maria, Calif, Aug 17, 21; m 65; c 8. CARDIOLOGY, PHYSIOLOGY. *Educ:* Univ Calif, Berkeley, BA, 43; Univ Louisville, Ky, MD, 46. *Prof Exp:* Asst clin prof med cardiol, Sch Med, Univ Calif Los Angeles, 55-63, chief of staff, Harbor Hosp, 61; chief dept med, 62-63, chief of staff, 77-80, CHIEF DIV CARDIOL, MEM HOSP MED CTR,

LONG BEACH, 62-; MED DIR,MEM HEART INST. *Concurrent Pos:* Sr attend physician, Munic Hosp Med Ctr, 54-; clin prof med, Sch Med, Univ Calif, Irvine, 73-; consult, Fed Air Surgeon, Fed Aviation Authority, 76-87. *Mem:* Fel Am Col Cardiol; fel Royal Soc Med. *Res:* Cardiology; diagnosis and treatment of ischemic heart disease and heart failure. *Mailing Add:* Mem Hosp Med Ctr 2801 Atlantic Ave Long Beach CA 90806

ELLESTAD, REUBEN B, b Lanesboro, Minn, May 10, 00; m 34; c 3. CHEMISTRY. *Educ:* Univ Minn, BS, 22, MS, 24, PhD(anal chem), 29. *Prof Exp:* instr chem, Tufts Col, 28-29 & Univ Minn, 29-42; res dir, Lithium Corp Am, 42-67, sr scientist, 67-74, vpres, 56-74; RETIRED. *Concurrent Pos:* Chemist, Rock Anal Lab, 29-42. *Mem:* Am Chem Soc; Sigma Xi. *Res:* Inorganic and analytical chemistry. *Mailing Add:* 620 Holiday Rd Gastonia NC 28054

ELLETT, CLAYTON WAYNE, b Northfield, Ohio, Nov 12, 16; m 54; c 1. PLANT PATHOLOGY. *Educ:* Kent State Univ, BS, 38; Ohio State Univ, MS, 40, PhD, 55. *Prof Exp:* Asst bot, 39-42, asst plant path, 42-44, instr bot & plant path, 46-55, from asst prof to prof, 56-81, EMER PROF PLANT PATH, OHIO STATE UNIV, 81-; consult, Chem Lawn Corp, 81-87; RETIRED. *Concurrent Pos:* Res asst, Univ Minn, 48; dir, Plant Dis Clin, Col Agr, Ohio State Univ, 70-81. *Mem:* Am Phytopath Soc; Mycol Soc Am. *Res:* Diseases of ornamentals and corn; parasitic fungi of Ohio; teaching; mycology. *Mailing Add:* Dept Plant Path Ohio State Univ 2021 Coffey Rd Columbus OH 43210-1087

ELLETT, D MAXWELL, b Richmond, Va, July 1, 22; m 58; c 2. MECHANICAL ENGINEERING, APPLIED MECHANICS. *Educ:* Univ Va, BME, 43; Yale Univ, MEng, 50, DEng, 52. *Prof Exp:* Mem staff mech eng, Sandia Lab, 46-49; asst, Yale Univ, 49-52; mem eng staff, Sandia Corp, 52-86; RETIRED. *Mem:* Am Nuclear Soc; Soc Am Mil Engrs. *Res:* Vibration and shock theory and analysis; operation of steady state and pulsed reactors; weapon system vulnerability and associated test methods; ground motion and building response from underground nuclear explosions. *Mailing Add:* PO Box 5062 Albuquerque NM 87185

ELLETT, EDWIN WILLARD, b Midlothian, Va, May 21, 25; wid; c 1. VETERINARY MEDICINE & SURGERY, COLLEGE FUND RAISING. *Educ:* Univ Ga, DVM, 53; Va Polytech Inst, BSc, 54; Tex A&M Col, MS, 61. *Prof Exp:* Vet in pvt pract, 53-56; asst prof vet med & surg, Okla State Univ, 56-58; from assoc prof to prof vet med & surg, Tex A&M Univ, 58-86, chief, Small Animal Clin, 61-76, proj coordr, Clin Facil Bldg Prog, 76-81, develop coordr, Col Vet Med, 80-86, EMER PROF VET MED & SURG, TEX A&M UNIV, 86-; DIR, COMPANION ANIMAL GERIAT CTR, 87- *Concurrent Pos:* Consult, M D Anderson Hosp & Tumor Inst, 61-71 & Alcon Labs, 64-68; trainee comp ophthal, Sch Med, Stanford Univ, 70; pres & dir, Pet Care, Inc, 84-87. *Mem:* Am Vet Med Asn; Am Asn Vet Clinicians; Am Soc Vet Ophthal; Am Animal Hosp Asn; Sigma Xi. *Res:* Surgery, especially ophthalmic and nonsuture techniques; cataracts; evaluation of new drugs; ocular diseases of animals and their comparison with diseases of man; geriatrics of animals. *Mailing Add:* Dean's Off Col Vet Med Tex A&M Univ College Station TX 77843

ELLETT, WILLIAM H, b Alliance, Ohio, Oct 10, 29; m 65; c 3. RADIOLOGICAL HEALTH. *Educ:* Rensselaer Polytech Inst, BS, 53; NY Univ, MS, 58; Univ London, PhD(radiation physics), 68. *Prof Exp:* Res assoc radiol physics, Sloan-Kettering Inst, 53-55; physicist, Hosp for Joint Dis, 55-57; asst physicist, Biophys, Mass Gen Hosp, Boston, 57-64; phys sci adminr, US Naval Radiol Defense Lab, 64-66; hon res asst, Royal Post Grad Med Sch, Univ London, 66-68; assoc prof radiation biophys, Radiation Ctr, Ore State Univ, 68-71; sr scientist, Off Res & Monitoring, US Environ Protection Agency, 71-72, chief bioeffects anal br, Off Radiation Progs, 73-85; SR STAFF OFFICER, COMN LIFE SCI, NAT RES COUN, NAT ACAD SCI, WASHINGTON, DC, 85- *Concurrent Pos:* Attend physicist, Manhattan Vet Admin Hosp, 56-57; fel, Harvard Med Sch, 59-61, res assoc, 61-64. *Mem:* AAAS; Radiation Res Soc; Soc Risk Anal. *Res:* Radiation physics; radiation dosimetry; environmental health physics and radiation risk analyses. *Mailing Add:* Comn Life Sci Nat Res Coun 2101 Constitution Ave Washington DC 20418

ELLGAARD, ERIK G, b Des Moines, Iowa, June 5, 39; m 62; c 3. GENETICS, DEVELOPMENTAL BIOLOGY. *Educ:* Drake Univ, BA, 61; Univ Iowa, PhD(zool, genetics), 68. *Prof Exp:* USPHS fel, Purdue Univ, 68-70; asst prof, 70-76, ASSOC PROF BIOL, TULANE UNIV, 76- *Mem:* AAAS; Genetics Soc Am. *Res:* Chromosomal puffing and its relationship to RNA and protein metabolism; developmental control of gene action; effects of aquatic pollutants on locomotor activities of fish; fisheries biology; effects of acid precipitation on the environment. *Mailing Add:* Dept Biol Tulane Univ New Orleans LA 70118

ELLIAS, LORETTA CHRISTINE, b Jacksonville, Fla, Sept 3, 19. BACTERIAL METABOLISM. *Educ:* Fla State Col Women, BS, 43; Univ Ky, MS, 46; Univ Mich, PhD(bact), 58. *Prof Exp:* From instr to prof, 46-87, EMER PROF BACT, FLA STATE UNIV, 87- *Concurrent Pos:* Dir, Tallahassee Regional Lab, State Bd Health, 48-49. *Mem:* AAAS; emer mem Am Soc Microbiol; Sigma Xi. *Res:* Terminal oxidation and redox potential; development of audio and video learning tools for courses in microbiology; purification of the streptococcus camp factor. *Mailing Add:* 1547 San Luis Rd Tallahassee FL 32304

ELLIKER, PAUL R, b La Crosse, Wis, Feb 12, 11; div; c 3. MICROBIOLOGY. *Educ:* Univ Wis, BS, 34, MS, 35, PhD(dairy bact), 37. *Prof Exp:* Asst, Univ Wis, 34-37, instr dairy & food bact, 37-38; Nat Elec Mfrs Asn fel, Univ Md, 38-39; instr agr bact, Univ Wis, 39-40; from asst prof to assoc prof dairy bact, Purdue Univ, 40-47; prof microbiol & microbiologist in charge, Agr Exp Sta, 47-76, chmn, Dept Microbiol, 52-76, EMER PROF MICROBIOL, ORE STATE UNIV, 76- *Concurrent Pos:* Tech dir, Dairy Soc Int-US exhibit, Madrid Int Trade Fair, 59; US State Dept off deleg, Int Dairy Cong, Denmark, 62, Munich, 66; consult to food indust on sanit. *Honors & Awards:* Borden Award, Am Dairy Sci Asn, 54, Kraftco Award, 74. *Mem:* Int Asn Milk, Food & Environ Sanit (pres, 66-67); Am Soc Microbiol; Am Dairy Sci Asn; Am Acad Microbiol; Inst Food Technol. *Res:* Microbiology of dairy products; dairy farm and plant sanitation; germicides used in the food and institutional industries and medicine. *Mailing Add:* Dept Microbiol Ore State Univ Corvallis OR 97331

ELLIN, ROBERT ISADORE, b Poland, Nov 25, 25; nat US; m 50; c 2. PHARMACEUTICAL CHEMISTRY. *Educ:* Johns Hopkins Univ, AB, 46; Univ Md, PhD(pharmaceut chem), 50. *Prof Exp:* Asst chem, Col Pharm, Univ Md, 48-50; pharmaceut chemist, Med Div, Army Chem Ctr, 50-51; prof pharmaceut chem & chmn dept chem, RI Col Pharm, 51-56; med dir pharmaceut chem, Clin Lab Br, Med Res Lab, Edgewood Arsenal, 56-65, supvry res chemist, Clin Res Dept, 65-66, chief, Clin Lab Br, 66-77; RES CHEMIST, BIOSCI DIV, ABERDEEN PROVING GROUND, 77- *Mem:* AAAS; Am Asn Clin Chem; Am Pharmaceut Asn; Am Chem Soc. *Res:* Organophosphorus and pesticide toxicology and chemistry; pesticide and nerve gas antidotes; dimercaprol toxicology; skin protection; pharmacokinetics and bioavailability; trace drug analysis; human effluent detection; gas and liquid chromatography and mass spectrometry. *Mailing Add:* Nine Oak Hollow Ctr Baltimore MD 21208

ELLING, LADDIE JOE, b Lawton, Okla, June 18, 17; m 42, 78; c 7. PLANT BREEDING. *Educ:* Okla Agr & Mech Col, BS, 41; Univ Minn, MS, 48, PhD(plant genetics), 50. *Prof Exp:* Res assoc alfalfa breeding & seed prod, 50-53, from asst prof to prof, 53-85, EMER PROF AGRON & PLANT GENETICS, UNIV MINN, ST PAUL, 86- *Mem:* Fel Am Soc Agron; Crop Sci Soc Am. *Res:* Forage and turf seed production; undergraduate teaching. *Mailing Add:* Agron-Plant Path 212 A6R Univ Minn St Paul MN 55108

ELLINGBOE, ALBERT HARLAN, b Lakeville, Minn, Apr 3, 31; m 58; c 4. PLANT PATHOLOGY, GENETICS. *Educ:* Univ Minn, BS, 53, MS, 55, PhD(plant path), 57. *Prof Exp:* Asst plant path, Univ Minn, 54-57, res fel, 57-58; res fel biol, Harvard Univ, 58-60; from asst prof to assoc prof bot & plant path, Mich State Univ, 60-70, prof biol, Bot & Plant Path, 70-80, dir, Genetics Prog, 73-80; res dir, Int Plant Res Inst, San Carlos, Ca, 80-83; PROF PLANT PATH & GENETICS, UNIV WIS-MADISON, 83- *Concurrent Pos:* NIH spec fel & vis assoc prof genetics, Univ Wash, 66-67; vis prof, Univ Sydney, 75. *Mem:* AAAS; Genetics Soc Am; Bot Soc Am; fel Am Phytopath Soc. *Res:* Genetics of host-parasite relationships; recombination in somatic tissues in fungi; genetics and breeding of mushrooms. *Mailing Add:* Plant Path Univ Wis Madison WI 53706

ELLINGBOE, J(ULES) K, b Tucson, Ariz, Mar 18, 27; m 48, 73; c 2. ELECTRICAL ENGINEERING. *Educ:* Univ Ariz, BSEE, 50. *Prof Exp:* Electronics engr, Radio Corp Am, 50-52; mem tech staff, Hughes Aircraft Co, 52-55; sr proj engr, Am Electronics, 55-56; mem tech staff, Space Tech Labs, 56-66; proj mgr, Defense Space Systs Div, TRW Defense Systems Groups, 66-71, mgr orbital opers, Space Vehicles Div, 71-74, asst mgr mat, 74-77, proj mgr, TDRSS Ground Sta, 77-83, spec projs mgr, 83-86; RETIRED. *Mem:* Inst Elec & Electronics Engrs. *Res:* Ballistic missile ground support equipment systems engineering; aerospace support equipment; spacecraft and ground station systems engineering; project management; spacecraft orbital operations. *Mailing Add:* 1420 E Sycamore Ave El Segundo CA 90245

ELLINGBOE, JAMES, b Wilmington, Del, June 10, 37; m 67; c 3. BIOCHEMISTRY, PHARMACOLOGY. *Educ:* Oberlin Col, AB, 59; Harvard Univ, PhD(biochem), 66. *Prof Exp:* Fel biochem, Karolinska Inst, Stockholm, 66-68; asst res biochemist, Sch Med, Univ Calif, San Diego, 68-70; assoc, 70-72, prin res assoc psychiat, 72-84, ASSOC PROF PSYCHIAT NEUROSCI, HARVARD MED SCH, 84-; ASSOC BIOCHEMIST, MCLEAN HOSP, BELMONT, MASS, 73-; CHIEF BIOCHEM LAB, ALCOHOL & DRUG ABUSE RES CTR, MCLEAN HOSP & HARVARD MED SCH, 73- *Concurrent Pos:* NIH fel, 66-67; dir biochem, Drug Surveillance & Biochem Lab, Boston City Hosp, 70-73; sci ed, Eaton Publ, Natick, Mass, 89- *Mem:* Am Chem Soc; Endocrine Soc; Am Soc Pharmacol Exp Therapeuts; Soc Neurosci; Int Soc Biomed Res Alcoholism; Res Soc Alcoholism; Int Union Pure & Appl Chem; Int Brain Res Orgn; World Fedn Neuroscientists; Coun Biol Ed. *Res:* Neuroendocrine regulatory mechanisms; biochemical pharmacology; lipid biochemistry; analytical biochemistry; endocrinology. *Mailing Add:* Biochem Lab Alcohol & Drug Abuse McLean Hosp 115 Mill St Belmont MA 02178

ELLINGER, MARK STEPHEN, b Crookston, Minn, Apr 12, 49; m 71. DEVELOPMENTAL BIOLOGY, EMBRYOLOGY. *Educ:* Augsburg Col, BA, 71; Univ Minn, Minneapolis, PhD(zool), 76. *Prof Exp:* Vis asst prof biol, Eastern Wash Univ, 76-77; asst prof, 77-81, ASSOC PROF ZOOL, SOUTHERN ILL UNIV, CARBONDALE, 81- *Concurrent Pos:* Am Cancer Soc res grant, 79; vis scientist, Merck Sharp & Dohme Res Labs, 83-84; res grant, Am Cancer Soc, 85. *Mem:* Am Soc Zoologists; AAAS; Int Soc Differentiation; Sigma Xi; Soc Develop Biol. *Res:* Gene transfer in lower vertebrates, with emphasis on amphibians; chemical carcinogenesis and developmental potentials of embryonic nuclei, as revealed by egg microinjection and nuclear transplantation; role of oncogenes in eary embryogenesis; evolution of long interspersed repetitive DNA's in lower vertebrates. *Mailing Add:* 14902 64th Pl Maple Grove MN 55367-8500

ELLINGHAUSEN, HERMAN CHARLES, JR, b Annapolis, Md, Nov 3, 26; m 51; c 3. BACTERIOLOGY. *Educ:* Univ Md, BS, 50, PhD(bact), 55; Univ NC, MSc, 52. *Prof Exp:* Asst bact, Univ Md, 52-55; PRIN RES MICROBIOLOGIST, PROJ LEADER LEPTOSPIROSIS RES, AGR RES SERV, ANIMAL DIS & PARASITE RES DIV, BACTERIAL & MYCOTIC DIS INVESTS, NAT ANIMAL DIS LAB, 55- *Concurrent Pos:* Mem staff, Grad Sch, USDA & lectr, NIH, 55-60; adv to grad studies, Univ NC & Colo State Univ; mem staff, Dept Biol, Drake Univ. *Mem:* Am Soc Microbiol; Am Acad Microbiol; Conf Res Workers Animal Dis; Am Asn Vet Lab Diag; US Animal Health Asn. *Res:* Microbial nutrition; bacterial metabolism; enzymes of pathogenic bacteria, visible and infrared spectrophotometry; electron microscopy; pathogenesis and immunological aspect of leptospirosis; serological characteristics of leptospires. *Mailing Add:* 105 Spa View Ave Annapolis MD 21401

ELLINGSON, JACK A, GEOLOGY. *Educ:* Univ Wash, BS, 58, MS, 59; Wash State Univ, PhD(geol), 68. *Prof Exp:* Cur, Ginkgo Petrified Forest State Mus, Vantage, Wash, 59-60; instr phys sci, Pac Lutheran Univ, Tacoma, Wash, 63-66; asst prof geol, Univ Northern Iowa, Cedar Falls, 68-70; chmn dept, 73-79, PROF GEOL, FT LEWIS COL, COLO, 70- *Concurrent Pos:* Interim dean, Sch Arts & Sci, Ft Lewis Col, 88-; consult, Creamland Dairies, Pagosa Develop, Standard Metals. *Mem:* Fel Geol Soc Am; assoc mem Sigma Xi. *Res:* Mineralogy and optical mineralogy; igneous and metamorphic petrology and petrography; author of several technical articles. *Mailing Add:* Geol Dept Ft Lewis Col Durango CO 81301

ELLINGSON, JOHN S, b Rockford, Ill, Mar 25, 40. BIOCHEMISTRY. *Educ:* Univ Ill, Urbana, BS, 62; Univ Mich, Ann Arbor, MS, 64, PhD(biochem), 67. *Prof Exp:* Asst prof, 70-77, ASSOC PROF BIOCHEM, MED SCH, WVA UNIV, 77- Assoc Prof Dept Path. Thomas Jefferson Univ Med Sch. *Concurrent Pos:* Trainee, Brandeis Univ, 67-70. *Res:* Phospholipids and membranes; regulation of development in cellular slime molds. *Mailing Add:* Dept Path Thomas Jefferson Univ Med Sch 1020 Locust St Philadelphia PA 19107

ELLINGSON, ROBERT GEORGE, b Chicago, Ill, Dec 30, 45; m 69; c 2. ATMOSPHERIC RADIATION, CLIMATE MODELING. *Educ:* Fla State Univ, BS, 67, MS, 68, PhD(meteorol), 72. *Prof Exp:* Comput programmer, Nat Ctr Atmospheric Res, 67; res asst, Fla State Univ, 68-72; fel, Nat Ctr Atmospheric Res, 72-73; asst prof, 73-78, ASSOC PROF METEOROL, UNIV MD, 78- *Concurrent Pos:* aircraft scientist, Atlantic Trop Exp, Global Atmospheric Res Prog, 74, Monsoon Exp, 77-79; consult, Lawrence Livermore Lab, 78- *Mem:* Am Meteorol Soc. *Res:* Theoretical and computer modeling of atmospheric radiative transfer, airborne observations of electromagnetic radiation, remote sensing of climate parameters and testing of numerical weather/climate prediction models. *Mailing Add:* Dept Meteorol Univ Md College Park MD 20742-2425

ELLINGSON, ROBERT JAMES, b Chicago, Ill, June 7, 23; m 48; c 2. ELECTROENCEPHALOGRAPHY. *Educ:* Northwestern Univ, BS, 47, MA, 49, PhD(psychol), 50; Univ Nebr, MD, 63. *Prof Exp:* Res assoc psychol, Mooseheart Lab Child Res, 48-50; DEPT NEUROL, CREIGHTON UNIV; ASSOC DIR, NEBR PSYCHIAT INST, 63- *Concurrent Pos:* Chief EEG lab, Nebr Psychiat Inst, 50-76; mem adv comt prog sci, technol & human values, Nat Endowment for the Humanities, 74-76. *Mem:* Fel AAAS; Int Soc Develop Psychobiol; Am Electroencephalog Soc (secy, 64-67, pres, 68-69); Int Fedn Soc Electroencephalog & Clin Neurophysiol (secy, 69-81). *Res:* Clinical neurophysiology. *Mailing Add:* Univ Nebr Med Ctr 600 S 42nd St Omaha NE 68198

ELLINGSON, RUDOLPH CONRAD, b Madison, Wis, Mar 8, 11; m 39; c 2. ORGANIC CHEMISTRY. *Educ:* St Olaf Col, BA, 33; Johns Hopkins Univ, PhD(org chem), 38. *Prof Exp:* Lab technician, Mass Inst Technol, 33-34; asst chem, Johns Hopkins Univ, 34-38, instr, 38-39; res chemist, Mead Johnson & Co, 39-50, chief res div, 50-53, asst dir res, 53-54, dir nutrit res & prod develop, 54-59, sr res scientist, Prod Develop, 59-60, dir sci admin, 60-65, exec dir admin, 65-66, dir clin & med res admin, 66-68, dir planning & coord, Res Ctr, 68-75; RETIRED. *Mem:* Am Chem Soc; Am Inst Chem; NY Acad Sci. *Res:* Vitamins; pyrrole and porphyrin chemistry; pyrazine chemistry; chemotherapeutics; carbohydrates; nutrition. *Mailing Add:* 6921 Arcadian Hwy Evansville IN 47715-5136

ELLINGTON, EARL FRANKLIN, b Salt Lick, Ky, Nov 15, 33; m 53; c 2. ANIMAL PHYSIOLOGY. *Educ:* Univ Ky, BS, 55, MS, 56; Univ Calif, PhD(animal physiol), 62. *Prof Exp:* Asst prof, Ore State Univ, 62-68; from assoc prof to prof animal physiol, 68-78, asst dean, 78-83, ASSOC DEAN, COL AGR, UNIV NEBR, LINCOLN, 83- *Mem:* Am Dairy Sci Asn; Am Soc Animal Sci; Soc Study Reproduction; Nat Asn Col & Teachers Agr. *Res:* Endocrinology and reproduction. *Mailing Add:* CT03 Animal Sci Univ Nebr Lincoln NE 68583

ELLINGTON, JAMES JACKSON, b Conyers, Ga, June 11, 44; m 70; c 2. MEDICINAL CHEMISTRY. *Educ:* Berry Col, BA, 66; Univ Ga, PhD(med chem), 74. *Prof Exp:* Res chemist, Russell Res Ctr, USDA, 74-78; RES CHEMIST, US ENVIRON PROTECTION AGENCY, 78- *Honors & Awards:* Gold Medal for Except Serv, Environ Protection Agency. *Mem:* Am Chem Soc. *Res:* Development of methods for the identification and quantitative analysis of organic compounds sorbed to soils and sediments. *Mailing Add:* US Environ Protection Agency College Station Rd Athens GA 30613

ELLINGTON, JOE J, b Yuma, Ariz, Dec 3, 34. ENTOMOLOGY. *Educ:* Univ Ariz, BS, 56; Cornell Univ, MS, 58, PhD(entom), 63. *Prof Exp:* Field res specialist, Calif Chem Co, 63-65; from asst prof to assoc prof, 65-76, PROF ENTOM, N MEX STATE UNIV, 76- *Mem:* Entom Soc Am; Sigma Xi. *Res:* Insects of economic importance, specifically alfalfa and cotton insects; beneficial insects; host plant resistance; sampling. *Mailing Add:* Dept Bot & Entom NMex State Univ University Park NM 88003

ELLINGTON, WILLIAM ROSS, b Asheville, NC, July 2, 49; m 71; c 2. COMPARATIVE PHYSIOLOGY. *Educ:* Univ Fla, BS, 71; Univ SFla, MA, 71; Univ RI, PhD(biol sci), 76. *Prof Exp:* Fel biochem dept chem, Pomona Col, 76-77; vis asst prof zool, Univ Vt, 77-78; asst prof biol, Univ Southwestern La, 78-81; from asst prof to assoc prof, Biol Sci, 81-86, dir, Mar Lab, 86-89, PROF BIOL SCI, FLA STATE UNIV, 90- *Mem:* Am Soc Zoologists; AAAS; Am Soc Cell Biol; Am Physiol Soc; Biophys Soc. *Res:* Physiological and biochemical adaptions of invertebrates to life in extreme environments; comparative and evolutionary biochemistry; in vivo NMR spectroscopy. *Mailing Add:* Dept Biol Sci B-157 Fla State Univ Tallahassee FL 32306

ELLINGWOOD, BRUCE RUSSELL, b Evanston, Ill, Oct 11, 44; m 69; c 1. STRUCTURAL RELIABILITY, RANDOM VIBRATION. *Educ:* Univ Ill, Urbana-Champaign, BS, 68, MS, 69, PhD(civil eng), 72. *Prof Exp:* Res asst, dept civil eng, Univ Ill, 68-72; res engr, US Naval Ship Res & Develop Ctr, 72-75; res struct engr, Ctr Bldg Technol, Nat Bur Standards, 75-82, leader, struct eng group, 82-86; PROF CIVIL ENG, JOHNS HOPKINS UNIV, 86-, CHMN, DEPT CIVIL ENG, 90- *Concurrent Pos:* Adj prof, dept civil eng, Johns Hopkins Univ, 83-86. *Honors & Awards:* W L Huber Res Prize, Am Soc Civil Engrs, 80, Norman Medal, 83; Silver Medal, US Dept Com, 80; Engr of th Yr, Dept Com, Nat Soc Prof Engrs, 86; Higgins Lectureship, Am Inst Steel Construct; Moissieff Award, Am Soc Civil Engrs, 88. *Mem:* Am Soc Civil Engrs; Am Soc Testing & Mat; Am Inst Steel Construct; Am Nat Standards Inst; Sigma Xi. *Res:* Structural loads and the application of methods of probability and statistics to structural engineering; analysis of structural loads; statistical studies of performance; development of safety and serviceability criteria for design; studies of progressive collapse of structures; analysis of response of structural systems to fires. *Mailing Add:* Dept Civil Eng Johns Hopkins Univ Baltimore MD 21218

ELLINWOOD, EVERETT HEWS, JR, b Wilmington, NC, June 27, 34; c 3. BEHAVIORAL, NEUROPHARMACOLOGY. *Educ:* Univ NC, BS, 56, MD, 59. *Prof Exp:* Clin instr psychiat, Univ Ky, 64-65; assoc, 66-67, from asst prof to assoc prof, 67-72, DIR BEHAV NEUROPHARMACOL SECT & DIR COGNITIVE NEUROMOTOR ASSESSMENT FAC, MED CTR, DUKE UNIV, 72-, PROF PSYCH & PHARMACOL, 73- *Concurrent Pos:* Res fel, Med Ctr, Duke Univ, 65-66; res coun, Am Psychiat Asn; Psychopharmacol Adv Comt, Food & Drug Admin. *Mem:* AAAS; Soc Neurosci; Am Psychiat Asn; Am Psychopath Asn; Soc Biol Psychiat; Am Col Neuropsycho Pharmacol. *Res:* Behavioral neuropharmacology; clinical and animal models of psychoses and movement disorders; nigrostriatal and mesolimbic systems; the relationship of pharmokinetics to pharmacodynamics of psychotropic drugs; neuropsycho pharmacology of eating disorders. *Mailing Add:* Dept Psychiat Box 3870 Duke Univ Med Ctr Durham NC 27710

ELLINWOOD, HOWARD LYMAN, b Davenport, Iowa, Feb 25, 26; m 51; c 5. GEOLOGY. *Educ:* Univ Minn, BA, 49, PhD(geol), 53. *Prof Exp:* Geologist, Bur Econ Geol, Tex, 49-50 & Calif Co, 52-58, dist geologist, 58-61, div geologist, 61-68, div supt, 68-70, chief geologist, 70-73, geol consult, 73-74, CHIEF GEOLOGIST, STANDARD OIL CO CALIF, 74- *Mem:* Geol Soc Am; Am Asn Petrol Geol. *Res:* Petroleum geology. *Mailing Add:* 984 Tyner PO Box 8870 Incline Village NV 89450

ELLINWOOD, WILLIAM EDWARD, b Chicago, Ill, Aug 7, 50; m 73; c 2. REPRODUCTIVE BIOLOGY, ENDOCRINOLOGY. *Educ:* Univ Colo, BA, 72; Colo State Univ, PhD(physiol), 78. *Prof Exp:* Fel, Colo State Univ, 73-78; fel, 78-80, ASST SCIENTIST, ORE REGIONAL PRIMATE RES CTR, 80- *Mem:* Endocrine Soc; Soc Study Reproduction; AAAS. *Res:* Endocrine function of the nonhuman primate ovary; nonhuman primate fetal endocrinology; relationships between gonadal secretions and pituitary-hypothalamic function. *Mailing Add:* Dept Obstet/Gynec Ore Health Sci Univ 3181 SW Sam Jackson Park Rd Portland OR 97201

ELLION, M EDMUND, b Boston, Mass, Jan 20, 23; m 54; c 2. MECHANICAL ENGINEERING, PHYSICS. *Educ:* Northeastern Univ, BS, 44; Harvard Univ, MS, 47; Calif Inst Technol, PhD(physics, mech eng), 53. *Prof Exp:* Res engr, Bell Aerospace Corp, 47-50, consult aerospace indust, 53-60; exec dir appl mech & aerodyn, Nat Eng Sci Co, 60-62; pres, Dynamic Sci Corp, 62-65; mgr propulsion & elec power systs, Hughes Aircraft Co, Los Angeles, 65-78, asst div mgr space & commun & technol develop, 78-87, dir technol, 87-88; PRES, SCI INDUST, ARCADIA, CALIF, 88- *Concurrent Pos:* Lectr, Univ Calif, Los Angeles, 55-56, Calif Inst Technol, 75-84, Stanford Univ, 85. *Mem:* Assoc fel Am Inst Aeronaut & Astronaut; sr mem Inst Elec & Electronics Engrs; Sigma Xi. *Res:* Propulsion, flight control, structures and thermal control of missile and space craft systems; fourteen patents in communications satellites and propulsion. *Mailing Add:* PO Box 645 Santa Ynez CA 93460

ELLIOT, ALFRED JOHNSTON, b Calgary, Alta, Aug 16, 11; m 42; c 4. OPHTHALMOLOGY. *Educ:* Univ BC, BA, 32; Univ Toronto, MD, 37; Columbia Univ, MedScD, 41; Am Bd Ophthal, dipl, 41; Royal Col Physicians & Surgeons, Univ London, dipl ophthalmic med & surg, 45; FRCS(C), 57. *Prof Exp:* Clin teacher ophthal, Univ Toronto, 45-46, prof & head dept, 46-61; prof ophthal & head dept, 61-73, EMER PROF OPHTHAL, UNIV BC, 78- *Concurrent Pos:* Chief serv, Toronto Gen Hosp & Sunnybrook Vet Admin Hosp, 46-61; consult, Hosp for Sick Children, 46-61; guest lectr, Univ Mich, 55; head ophthal, Vancouver Gen Hosp, 61-73 & Shaughnessy Hosp, 61-73; adv to dir gen, Can Dept Vet Affairs, 66-78; chmn med adv bd, Shaughnessy Hosp DVA, 68-70 & Vancouver Gen Hosp, 69-72; sen active, Vancouver Gen Hosp; hon consult, Shaughnessy, Vancouver & Toronto Gen Hosps; chmn, Planned Giving & Endowment Comt, Can Nat Inst Blind, 90. *Honors & Awards:* Golden Jubilee Award, Can Nat Inst Blind, 68; Alfred J Elliot Lectr, 74; Ophthal Soc Award, Can, 80; Arthur Napier Magill Award, Can Nat Inst, 88. *Mem:* Am Acad Ophthal & Otolaryngol; Am Ophthal Soc; sr mem Can Med Asn; Can Ophthal Soc; Ophthal Soc UK; Can Nat Inst Blind. *Res:* Recurrent intraocular hemorrhage; carotid-cavernous fistulae; keratoconjunctivitis sicca. *Mailing Add:* 6600 Lucas Rd No 47 Richmond BC V7C 4T1 Can

ELLIOT, ARTHUR MCAULEY, b Minneaplis, Minn, May 13, 28; m 54; c 4. PHYTOPATHOLOGY, MYCOLOGY. *Educ:* Univ Minn, BSc, 53, MSc, 60, PhD(plant path), 61. *Prof Exp:* Asst prof, 61-66, ASSOC PROF BIOL, TEX TECH UNIV, 66- *Mem:* AAAS; Am Phytopath Soc; Mycol Soc Am; Am Inst Biol Sci; Int Soc Plant Path; Sigma Xi. *Res:* Field and vegetable crop diseases; fruit and ornamental tree diseases; vector transmission of plant pathogens; plant and insect phenology. *Mailing Add:* Dept Biol Tex Tech Univ Lubbock TX 79409-3131

ELLIOT, DAVID HAWKSLEY, b Eng, May 22, 36. GEOLOGY, PETROLOGY. *Educ:* Cambridge Univ, BA, 59; Univ London, DIC, 60; Univ Birmingham, PhD(geol), 65. *Prof Exp:* Geologist, Brit Antarctic Surv, 60-66; fel, Dept Geol & Mineral, Ohio State Univ, 66-67, res assoc, 67-69, from asst prof to assoc prof, 69-79, dir, Byrd Polar Res Ctr, 73-89, PROF, DEPT GEOL & MINERAL, OHIO STATE UNIV, 79- *Mem:* Geol Soc London; Geol Soc Am; Am Geophys Union; Sigma Xi. *Res:* Antarctic geology, including Beacon stratigraphy; petrology and geochemistry of Jurassic igneous rocks of the Transantarctic Mountains; stratigraphy and petrology of Mesozoic and Cenozoic strata and evolution of the Antarctic Peninsula. *Mailing Add:* Byrd Polar Res Ctr Ohio State Univ 125 S Oval Mall Columbus OH 43210

ELLIOT, JAMES I, b Toronto, Ont, Aug 21, 38; c 2. ANIMAL NUTRITION, ANIMAL MANAGEMENT. *Educ:* Univ Toronto, BSA, 62; Univ Alta, MSc, 65, PhD(animal nutrit), 69. *Prof Exp:* Lectr animal sci, Univ Alta, 68-69; asst prof, Macdonald Col, McGill Univ, 69-72; RES SCIENTIST, AGR CAN ANIMAL RES INST, 72- *Concurrent Pos:* Dir, Mgt Accountability, Agr Can. *Mem:* Agr Inst Can; Can Soc Animal Sci; Am Soc Animal Sci. *Res:* Investigation of copper as a growth promotant in swine rations and its involvement in the synthesis of unsaturated fatty acids in the pig; neonatal and prenatal metabolism of methionine in the pig; survival of colostrum-deprived neonatal piglets reared artificially; nutrient requirements of the neonatal pig; nutrition of the breeding female. *Mailing Add:* Dean Fac Agr Univ Man Agr Bldg No 300 Winnipeg MB R3T 2N2 Can

ELLIOT, JAMES LUDLOW, b Columbus, Ohio, June 17, 43; m 67; c 2. ASTRONOMY. *Educ:* Mass Inst Technol, SB & SM, 65; Harvard Univ, AM, 67, PhD(astron), 72. *Prof Exp:* Fel astron, Smithsonian Astrophys Observ, 72; res assoc, Cornell Univ, 72-74, sr res assoc, 74-77, asst prof astron, 77-78; assoc prof astron & physics & dir, 78-85, PROF ASTRON & PHYSICS & DIR, WALLACE ASTROPHYS OBSERV, MASS INST TECHNOL, 85- *Mem:* Am Astron Soc; Int Astron Union; Astron Soc Pac; Am Asn Variable Star Observers. *Res:* Astronomical instrumentation; optical observations; lunar and planetary occulatations observations. *Mailing Add:* Mass Inst Technol 77 Massachusetts Ave Bldg 54-422A Cambridge MA 02139

ELLIOT, JOE OLIVER, b Ames, Iowa, Feb 8, 23; m 50; c 2. OCEANOGRAPHY. *Educ:* Iowa State Col, BS, 43; Columbia Univ, AM, 47; Univ Md, PhD(physics), 55. *Prof Exp:* Res physicist, Div War Res, Iowa State Univ, 44; dept terrestrial magnetism, Carnegie Inst, 45; lectr & asst, Columbia Univ, 46-49; res nuclear physicist, Naval Res Lab, 49-65, phys sci adminr, 65-82; ASSOC PROF, PRINCE GEORGE COMMUNITY COL, 83- *Concurrent Pos:* Sabbatical fel, Inst Oceanog, Univ BC, 68-69. *Mem:* Am Phys Soc; Am Geophys Union. *Res:* Parameters of excited states of nuclei; neutron scattering; decay time of luminescence of phosphors; scintillation spectrometry; reactor technology; physical oceanography. *Mailing Add:* 5123 Temple Hills Rd Temple Hills MD 20748

ELLIOT, JOHN MURRAY, dairy cattle nutrition, for more information see previous edition

ELLIOTT, ALFRED MARLYN, zoology; deceased, see previous edition for last biography

ELLIOTT, ALICE, b Reece, Kans, Oct 7, 19. ZOOLOGY, PARASITOLOGY. *Educ:* Kans State Teachers Col, Emporia, BS, 42; Kans State Univ, MS, 47, PhD(parasitol), 50. *Prof Exp:* Asst zool, Kans State Univ, 46-47, instr, 47-50; asst prof biol, Kans State Teachers Col, Pittsburg, 50-51 & Hope Col, 52-54, assoc prof, 54-55; asst prof sci, Ball State Teachers Col, 55-59; prof biol, Hope Col, 59-62; from assoc prof to prof biol, Cent Mo State Col, 62-81; RETIRED. *Concurrent Pos:* Res assoc, Inst Cellular Res, Univ Nebr, 58-59. *Mem:* AAAS; Am Soc Cell Biol; Am Soc Parasitol; Am Micros Soc. *Res:* Taxonomy and ecology of fishes; helminthology; physiology of parasites; tissue culture. *Mailing Add:* 1110 Turnberry Ln Maryville TN 37801

ELLIOTT, ARTHUR YORK, b Tyler, Tex, Feb 8, 36; m 56; c 2. MICROBIOLOGY, VIROLOGY. *Educ:* NTex State Univ, BA, 57, MS, 58; Purdue Univ, PhD(virol), 69. *Prof Exp:* Res assoc virol, M D Anderson Tumor Inst, 60-61; res assoc, Baylor Col Med, 61-62; sr res staff & sr virologist, Pitman Moore Div Dow Chem Co, 62-70; asst prof microbiol, Med Sch, Univ Minn, Minneapolis, 70-72, assoc prof microbiol & urol surg, 72-78; MGR BIOL MFG, MERCK SHARP & DOHME, WEST POINT, PA, 78-, EXEC DIR, BIOL OPER, MERCK PHARMACEUT MFG DIV, 89- *Mem:* Am Soc Microbiol; Tissue Cult Asn. *Res:* Human urologic tumors; tissue cultrue and virology. *Mailing Add:* Box 4 West Point PA 19486-0004

ELLIOTT, BERNARD BURTON, b Ottawa, Ont, Can, Nov 10, 21; US citizen; m 45; c 4. ENZYMOLOGY. *Educ:* McGill Univ, BS, 49, MS, 50; Purdue Univ, PhD, 52. *Prof Exp:* Demonstr plant physiol, McGill Univ, 49-50; enzymologist, R J Reynolds Tobacco Co, 52-59; asst dir res & develop, Froedtert Malt Corp, 59-62; pres & managing dir, Brewing & Malting Res Inst, 62-64; dir res & develop, Froedtert Malt Corp, 64-84; RETIRED. *Mem:* AAAS; Am Chem Soc; Am Soc Brewing Chem; Master Brewers Asn Am; Am Asn Cereal Chem. *Res:* Fermentations; isolation of natural products. *Mailing Add:* 1309 Lynne Dr Waukesha WI 53186

ELLIOTT, CECIL MICHAEL, b Ga, Aug 1, 49. ANALYTICAL CHEMISTRY, BIO-INORGANIC CHEMISTRY. *Educ:* Davidson Col, BS, 71; Univ NC, Chapel Hill, PhD(anal chem), 75. *Prof Exp:* Asst prof chem, Univ Vt, 77-80, assoc prof, 80-; AT DEPT CHEM, COLO STATE UNIV. *Mem:* Am Chem Soc. *Res:* Electrochemistry. *Mailing Add:* Dept Chem Colo State Univ Ft Collins CO 80521

ELLIOTT, CHARLES HAROLD, b Kans City, Mo, Dec 30, 48; c 1. CLINICAL PSYCHOLOGY, PEDIATRIC PSYCHOLOGY. *Educ:* Univ Kans, BA, 71, MA, 74, PhD(clin psychol), 76. *Prof Exp:* Asst prof psychol, ECent Univ, 76-79; from asst prof to assoc prof, dept psychiat & behav sci, Univ Okla Health Sci Ctr, 79-85; coordr, Behav Med Training, 85-87, MEM FAC, FIELDING INST, UNIV NMEX SCH MED, 87- *Concurrent Pos:* Co-prin investr, Nat Cancer Inst grant, 82-85, consult, 85- *Mem:* Am Psychol Asn; Asn Advan Behav Ther; Soc Pediat Psychol. *Res:* Behavioral medicine, especially acute and chronic pain and children coping with medical procedures. *Mailing Add:* 801 Encino Pl NE B-10 Albuquerque NM 87102

ELLIOTT, DAN WHITACRE, b Greenville, Ohio, Aug 5, 22; m 62; c 2. SURGERY. *Educ:* Yale Univ, MD, 49; Ohio State Univ, MSc, 56. *Prof Exp:* From asst prof to prof surg, Ohio State Univ, 57-64; prof surg, Univ Pittsburgh, 64-76; prof & chmn dept surg, Wright State Univ, 76-88; RETIRED. *Concurrent Pos:* Consult, Vet Admin Hosps, Dayton, Ohio, 57-64, 76- & Pittsburgh, Pa, 64-70, chief surg, 70-76. *Mem:* Am Surg Asn; Soc Univ Surg; Am Gastroenterol Asn; Am Burn Asn; Int Soc Surg; Cent Surg Assoc; Soc Surg Alimentary Tract; Soc Univ Surgeons; Western Surg Asn. *Res:* Diseases of the pancreas; gastric acid secretion; biliary tract surgery and infection. *Mailing Add:* 701 Murrell Dr Kettering OH 45429-1321

ELLIOTT, DANA EDGAR, b New Vienna, Ohio, Aug 23, 23; m 48; c 3. NONDESTRUCTIVE TESTING, ELECTRICAL ENGINEERING. *Educ:* Univ Louisville, BS, 48. *Prof Exp:* Betatron engr, Allis Chalmers Mfg Co, 49-54; STAFF MEM NONDESTRUCTIVE TESTING, LOS ALAMOS SCI LAB, 54- *Mem:* Fel Am Soc Nondestructive Testing; Am Soc Metals. *Res:* Nondestructive testing development, primarily industrial radiography with emphasis on materials involved in the nuclear energy field. *Mailing Add:* 2060 46th St Los Alamos NM 87544

ELLIOTT, DANA RAY, b Grain Valley, Mo, Feb 7, 45; m 78; c 3. BIOLOGY. *Educ:* William Jewell Col, Liberty, Mo, BA, 67; Cent Mo State Univ, MS, 71; Univ Mo, Columbia, PhD(entom), 81. *Prof Exp:* PROF BIOL, CENT METHODIST COL, FAYETTE, MO, 74- *Mem:* Ecol Soc Am; Entom Soc Am. *Res:* Correlation of the species diversity of leafhoppers and vascular plants as indicators of the stage development of old-field succession in central Missouri. *Mailing Add:* Dept Biol Cent Methodist Col Fayette MO 65248

ELLIOTT, DAVID DUNCAN, b Los Angeles, Calif, Aug 4, 30; m 62; c 1. RESEARCH MANAGEMENT. *Educ:* Stanford Univ, BS, 51; Calif Inst Technol, MS, 53, PhD(physics), 59. *Prof Exp:* Res scientist space physics, Lockheed Missiles & Space Co, 59-61; mem tech staff, Aerospace Corp, 61-66, staff scientist, 66-67, head, Space Radiation & Atmospheric Dept, 67-70, sci adv, Nat Aeronaut & Space Coun, 70-72; sr staff mem for sci affairs, Nat Security Coun, 72-77; exec dir, 77-81, vpres & dir, Res & Anal Div, SRI Int, 81-86; SR VPRES, SAIC, 86- *Concurrent Pos:* Consult, Off Sci & Tech Policy, Exec Off of the President, Cent Intel Agency, Army Sci Bd, SDI Adv Comn. *Mem:* AAAS; Am Phys Soc; Am Geophys Union; Am Inst of Aeronaut & Astronaut; Armed Forces Commun & Electronics Asn. *Res:* Scientific aspects of national security and foreign policy. *Mailing Add:* SAIC 10260 Campus Point Dr San Diego CA 92121

ELLIOTT, DAVID LEROY, b Cleveland, Ohio, May 29, 32; m 84; c 2. SYSTEMS THEORY, APPLIED MATHEMATICS. *Educ:* Pomona Col, BA, 53; Univ Southern Calif, MA, 59; Univ Calif, Los Angeles, PhD(systs sci), 69. *Prof Exp:* Mathematician, US Naval Ord Test Sta, 55-69; instr & lectr systs sci, Univ Calif, Los Angeles, 68-71; from asst prof to assoc prof, 71-80, PROF MATH THEORY SYSTS, WASH UNIV, 80- *Concurrent Pos:* Consult, Electro-Optical Systs, Inc, 60-69; NSF grants, 71-75, 77-81, 81-83, 83-85 & 85-87; Nat Inst Heart, Lung & Blood Dis grant, 76-81; assoc ed, Math Systs Theory, 76-87, Soc Indust & Appl Math Rev, 79-83 & Systs & Control Letters, 90-; vis assoc prof, Brown Univ, Providence, RI, 79; vis prof elec eng, Univ Calif, Los Angeles, 86-87; prog dir systs theory, NSF, 87-89. *Mem:* Soc Indust & Appl Math; Am Math Soc; fel Inst Elec & Electronics Engrs; Math Asn Am; Sigma Xi. *Res:* Control systems on manifolds; bilinear systems. *Mailing Add:* Dept Systs Sci & Math Wash Univ St Louis MO 63130-4899

ELLIOTT, DENIS ANTHONY, b Los Angeles, Calif, Dec 2, 46. IMAGE PROCESSING, ASTRONOMY. *Educ:* Calif Inst Technol, BS, 68; Univ Calif, Los Angeles, MA, 70, PhD(astron), 74. *Prof Exp:* MEM TECH STAFF IMAGE PROCESSING, JET PROPULSION LAB, CALIF INST TECHNOL, 73- *Mem:* Am Astron Soc; AAAS; Soc Photoptical Indust Engrs; Sigma Xi. *Res:* Image processing, especially analysis of multispectral images, galactic structure and evolution; behavior of infrared detectors; infrared detector calibration. *Mailing Add:* 637 Santa Rosa Rd Arcadia CA 91007-2748

ELLIOTT, DOUGLAS FLOYD, b Rapid City, SDak, May 21, 32; m 56; c 3. ELECTRICAL ENGINEERING. *Educ:* Univ Hawaii, BA, 54; Univ Wash, MS, 56; Univ Southern Calif, PhD(elec eng), 69. *Prof Exp:* Elec engr, Pac Tel & Tel, 56-59; MEM TECH STAFF ELEC ENG, ROCKWELL INT, 59- *Concurrent Pos:* Instr, Rockwell Int, 74-, Univ Southern Calif, 75-76 & Inst Elec & Electronics Engrs, Orange County, 76-78. *Mem:* Sigma Xi; sr mem Inst Elec & Electronics Engrs. *Res:* Digital signal processing, including digital filters, fast transforms and sample-data control systems; continuous control systems; sonar and radar system design and analysis. *Mailing Add:* 5562 Mountain View Pl Yorba Linda CA 92686

ELLIOTT, EUGENE WILLIS, b Longmont, Colo, May 29, 16; m 46; c 2. INORGANIC CHEMISTRY, BOTANY. *Educ:* Univ Mont, BA, 41; Univ Iowa, MS, 47, PhD(mycol), 48. *Prof Exp:* Res asst plant physiol, Univ Wis, 41-42; dir microbiol res lab, Monsanto Chem Co, Mo, 48-51; eng draftsman, US Bur Reclamation, 51-52; sr draftsman, Shell Oil Co, Mont, 52-58; from instr to asst prof chem & bot, 58-62, assoc prof chem & plant physiol, 62-66, prof chem & bot, 66-81, EMER PROF CHEM, EASTERN MONT COL, 81- *Concurrent Pos:* Consult, Nat Parks Asn, 62- *Mem:* Sigma Xi. *Res:* Effects of paradichlorobenzene on fungi; swarm cells of Myxomycetes; flora of the Beartooth Mountains and adjacent plains. *Mailing Add:* 641 Ave C Billings MT 59102

ELLIOTT, GEORGE ALGIMON, b Trappe, Md, June 6, 25; m 49; c 3. PATHOLOGY. *Educ:* Va Mil Inst, 47-48; Univ Md, 48-49; Univ Ga, DVM, 53; Univ Pa, MS, 57; Am Col Vet Path, dipl. *Prof Exp:* Instr vet path, Vet Sch, Univ Pa, 55-58, assoc, 58-59, asst prof, 59-60; asst prof, Sch Med, Vanderbilt Univ, 60-62; res assoc path & toxicol res, Upjohn Co, 62-79, sr res vet pathologist & toxicologist, 79-90; RETIRED. *Mem:* Int Acad Path; NY Acad Sci. *Res:* Neuropathology; toxicology; immunopathology. *Mailing Add:* 4430 Romence Rd Portage MI 49002

ELLIOTT, GEORGE ARTHUR, b Montreal, Que, Jan 30, 45; m 74. OPERATOR ALGEBRAS, K-THEORY. *Educ:* Queen's Univ, Kingston, BSc, 65, MSc, 66; Univ Toronto, PhD(math), 69. *Prof Exp:* Res fel math, Univ BC, 69-70 & Queen's Univ, Kingston, 70-71; mem res, Inst Advan Study, 71-72; ASSOC PROF MATH, UNIV COPENHAGEN, 72- *Concurrent Pos:* Res assoc math, Nat Ctr Sci Res, France, 72-73 & 84-85; sr res fel, Univ Newcastle, 74; guest scholar, Res Inst Math Sci, Japan, 74 & 77; assoc prof, Univ Ottawa, 78-79, adj prof, 79-81 & 82-85; vis fel, Univ Warwick, 82, 87 & Australian Nat Univ, 84; vis prof, Univ New South Wales, 79 & Univ Toronto, 86, 88; mem res, Math Sci Res Inst, Berkeley, 85; adj prof, Univ Toronto, 84- *Mem:* Fel Royal Soc Can; Int Asn Math Physics; Am Math Soc; Can Math Soc; Danish Math Soc; London Math Soc. *Res:* Theory of rings of operators (von Neumann algebras) in Hilbert space; von Neumann algebras as related to other mathematical fields, quantum field theory and quantum statistical mechanics; closed operator algebras; non-commutative topology. *Mailing Add:* Dept Math Univ Toronto Toronto ON M5S 1A1 Can

ELLIOTT, H(ELEN) MARGARET, b Galveston, Tex, Aug 16, 25; wid. MATHEMATICAL ANALYSIS. *Educ:* Rice Univ, BA, 45; Univ Calif Berkeley, MA, 46; Harvard Univ, PhD(math), 48. *Prof Exp:* Asst, Univ Calif, 46; Am Asn Univ Women Hill fel, Harvard Univ, 48-49; instr, Washington Univ, St Louis, 49-51, from asst prof to assoc prof, 51-64; actg prof, Col William & Mary, 64-65; chmn dept, 69-73, PROF MATH, UNIV BRIDGEPORT, 68- *Concurrent Pos:* NSF faculty fel, Mass Inst Technol & Harvard Univ, 57-58. *Mem:* Am Math Soc; Math Asn Am; Nat Coun Teachers Math; Sigma Xi; Am Asn Univ Prof. *Res:* Analysis; approximation theory. *Mailing Add:* Dept Math Univ Bridgeport Bridgeport CT 06601

ELLIOTT, HOWARD CLYDE, b Birmingham, Ala, Sept 21, 24; m 58; c 5. BIOCHEMISTRY. *Educ:* Birmingham-Southern Col, BS, 48; Univ Ala, MS, 51, PhD(biochem), 56. *Prof Exp:* Chemist, Cancer Res Dept, 49-50, from asst instr biochem to prof chem, 51-69, RES PROF CHEM, MED COL, UNIV ALA, BIRMINGHAM, 70-, ASST PROF BIOCHEM MED, 66-; BIOCHEMIST, BAPTIST MED CTR, 70- *Mem:* Soc Exp Biol & Med; Am Chem Soc; fel Nat Acad Clin Biochem; Am Asn Clin Chem; Sigma Xi; fel Asn Clin Scientists. *Res:* Biochemistry of benzoate congeners; renal transport mechanisms; electrolyte metabolism; cholesterol metabolism in man. *Mailing Add:* 4260 Sharpsburg Dr Birmingham AL 35213

ELLIOTT, IRVIN WESLEY, b Newton, Kans, Oct 21, 25; m 52; c 2. ORGANIC CHEMISTRY. *Educ:* Univ Kans, BS, 47, MS, 49, PhD(chem), 52. *Prof Exp:* Instr chem, Southern Univ, 49-50; assoc prof, Fla Agr & Mech Col, 52-53, prof, 53-57; fel, Harvard Univ, 57-58; prof chem, 58-, RESEARCHER, DIV NATURAL SCI & MATH, FISK UNIV, NASHVILLE. *Concurrent Pos:* Vis prof, Howard Univ, 64-65; fel, Orsted Inst, Univ Copenhagen, 74. *Mem:* Am Chem Soc; Nat Inst Sci; Royal Soc Chem; Sigma Xi. *Res:* Synthetic organic chemistry; alkaloids. *Mailing Add:* Div Nat Sci & Math Box 14 Fisk Univ Nashville TN 37203

ELLIOTT, J LELL, physical organic chemistry, for more information see previous edition

ELLIOTT, JAMES ARTHUR, b Pierceland, Sask, Feb 24, 41; m 67; c 2. AIR-SEA INTERACTION, OCEANIC MIXING. *Educ:* Univ Sask, BSc, 62; Univ BC, MSc, 65, PhD(physics), 70. *Prof Exp:* Res scientist, 62-85, head, Ocean Circulation Div, 79-85, DIR, ATLANTIC OCEANOG LAB, BEDFORD INST OCEANOG, 85- *Concurrent Pos:* Dir, A G Huntsman Found, 82-; hon res assoc, Dalhousie Univ, 82-; mem, Sci Comt, Can Meteorol & Oceanog Soc, 83- *Mem:* Am Meteorol Soc; Can Meteorol & Oceanog Soc; Am Geophys Union; AAAS; Arctic Inst NAm. *Res:* Air/sea interaction; buoy technology; oceanic microstructure and mixing; internal waves and surface waves; mesoscale meteorology. *Mailing Add:* Bedford Inst Oceanog PO Box 1006 Dartmouth NS B2Y 4A2 Can

ELLIOTT, JAMES GARY, b Atlanta, Ga, Mar 23, 45; m 72; c 3. NUTRITION, FOOD SCIENCE. *Educ:* Univ Ga, BS, 68, MS, 71; Rutgers Univ, PhD(nutrit), 77. *Prof Exp:* Food technologist, Great Western United Corp, 69-70; food chemist, Res Ctr, Hercules, Inc, 70-71; res scientist, Thomas J Lipton, Inc, 71-74; res asst, Rutgers Univ, 74-77; mgr nutrit sci, Campbell Soup Co, 77-80; from assoc scientist to sr res scientist, Ralston Purina Co, 80-88; VPRES ANIMAL NUTRIT MKT, DEPT PROTEIN SPECIALTIES, ARCHER DANIELS MIDLAND CO, 88- *Mem:* Am Inst Nutrit; Inst Food Technologists; Am Chem Soc; Nutrit Today Soc; NY Acad Sci. *Res:* Protein quality methodology; dietary fiber and cholesterol metabolism; bioavailability of essential nutrients and toxic elements; effects of food processing on nutrient retention; nutritional toxicology of intentional and non-intentional food additives. *Mailing Add:* Protein Specialties Dept Archer Midland Co 1001 Brush College Rd Decatur IL 62526

ELLIOTT, JAMES H, b Hastings, Nebr, Jan 15, 27; m 78. OPHTHALMOLOGY. *Educ:* Phillips Univ, BA, 49; Univ Okla, MD, 52. *Prof Exp:* Intern, Mercy Hosp, Oklahoma City, Okla, 52-53; res ophthal, Med Ctr, Univ Okla, 60-62; instr, Harvard Med Sch, 65-66; assoc prof, 66-68, PROF OPHTHAL & CHMN DEPT, SCH MED, VANDERBILT UNIV, 68-, OPHTHALMOLOGIST-IN-CHIEF, UNIV HOSP, 72- *Concurrent Pos:* Consult, Vet Admin Hosp, Nashville, 66-; res fel, Harvard Med Sch, 62-65. *Mem:* Asn Res Vision & Ophthal (vpres, 78); Am Acad Ophthal & Larcyncol; AMA; fel Am Col Surg; assoc mem Am Ophthal Soc. *Res:* Immunology as applied to ophthamology, specifically immunosuppression of experimental corneal hypersensitivity and corneal graft rejection reactions. *Mailing Add:* Dept Ophthal Vanderbilt Univ Med Ctr Nashville TN 37232

ELLIOTT, JANE ELIZABETH INCH, micropaleontology; deceased, see previous edition for last biography

ELLIOTT, JERRY CHRIS, b Oklahoma City, Okla, Feb 6, 43. PHYSICS, SPACE SCIENCES. *Educ:* Univ Okla, BS, 66. *Prof Exp:* Guid engr, Gemini Prog, 66-67, guid engr, Apollo Prog, 67-68, trajectory engr, 68-72, optics sensor engr earth resources, 72-74, prog engr, Apollo-Soyuz, 74-75, proj engr space shuttle, 75-85, mgr, Configuration Design, Space Sta, 85-87, TECH MGR, SPACE STA PROG SUPPORT EQUIP, NASA, 87- *Honors & Awards:* Bausch & Lomb Nat Sci Award, 61. *Mem:* Am Indian Sci & Eng Soc; AAAS; Nat Mgt Asn; Native Am Sci Educator's Asn. *Res:* Slow-scan televideo applications; wide and narrow-band telecommunications; general technology transfer; energy technology. *Mailing Add:* PO Box 58182 Houston TX 77258

ELLIOTT, JOANNE, b Providence, RI, Dec 5, 25. MATHEMATICS. *Educ:* Brown Univ, BA, 47; Cornell Univ, MA, 49, PhD(math), 50. *Prof Exp:* Asst, Cornell Univ, 47-50; instr math, Swarthmore Col, 50-52; asst prof, Mt Holyoke Col, 52-55; from asst prof to assoc prof, Barnard Col, Columbia Univ, 55-64; PROF MATH, RUTGERS UNIV, 64- *Concurrent Pos:* Vis asst prof, Brown Univ, 54-55; NSF sr fel, 61-62. *Mem:* Am Math Soc; Math Asn Am. *Res:* Integral equations; applications of semigroups to integro-differential equations; differential equations. *Mailing Add:* Dept Math Rutgers Univ New Brunswick NJ 08903

ELLIOTT, JOHN FRANK, b St Paul, Minn, July 31, 20; m 46; c 2. CERAMICS ENGINEERING, PHYSICAL CHEMISTRY. *Educ:* Univ Minn, BS, 42; Mass Inst Technol, ScD(metall), 49. *Prof Exp:* Phys chemist, Res Lab, US Steel Corp, 49-51; res metallurgist, Inland Steel Co, 51-54, asst supt qual control, 54-55; assoc prof, 55-60, PROF METALL, DEPT MAT SCI & ENG, MASS INST TECHNOL, 60-, AM IRON & STEEL INST DISTINGUISHED PROF, 81-, DIR MINING & MINERAL RESOURCES RES INST, 78- *Honors & Awards:* Howe Mem Lectr, Am Inst Mining, Metall & Petrol Engrs, 63, Douglas Gold Medal, 76; Sr Julius Wehrner Lectr, Inst Mining & Metall, 85. *Mem:* Nat Acad Eng; fel Metall Soc; fel AAAS; fel Am Inst Chem Engrs; hon mem Am Inst Mining, Metall & Petrol Engrs; hon mem Iron & Steel Inst Japan; hon mem Inst Metall Japan; Iron & Steel Soc. *Res:* Physical chemical studies of phases and reactions at high temperatures; extractive metallurgy, steelmaking methods. *Mailing Add:* Dept Mat Sci & Eng Rm 4-138 Mass Inst Technol Cambridge MA 02139

ELLIOTT, JOHN HABERSHAM, b Baltimore, Md, July 20, 13; m 41; c 2. RHEOLOGY, POLYMER CHEMISTRY. *Educ:* Haverford Col, AB, 35; Univ Pa, MS, 37, PhD(phys chem), 40. *Prof Exp:* Res chemist, Philadelphia Lab, E I Du Pont de Nemours & Co, Inc, 35-37; chemist, J E Rhodes & Sons, 37-42; res chemist, Res Ctr, Hercules, Inc, 42-43, supvr, 43-73, res scientist, 73-78; RETIRED. *Concurrent Pos:* Consult, 78- *Mem:* Am Chem Soc; Soc Rheology. *Res:* Acid strength studies; polymer rheology; effect of substituents on the acid strength of benzoic acid; rheology of polyelectrolytes in aqueous solution. *Mailing Add:* 305 Wilson Rd Newark DE 19711

ELLIOTT, JOHN RAYMOND, organic polymer chemistry, for more information see previous edition

ELLIOTT, JOSEPH ROBERT, b Kansas City, Kans, Dec 11, 23; m 53; c 2. CLINICAL CHEMISTRY. *Educ:* Univ Kans, AB, 49; Univ Mo, MS, 51, PhD(biochem), 53. *Prof Exp:* Fel biochem, Med Sch, Northwestern Univ, 53-55; res assoc biol, Rice Univ, 55-58; instr biochem, Col Med, Baylor Univ, 58-67; assoc prof lab med & dir clin chem lab, Med Sch, Univ Okla, 67-70; CLIN CHEMIST, ST LUKE'S HOSP, 70- *Concurrent Pos:* Mem staff, Jefferson Davis Hosp, 57-59; biochemist, St Luke's Episcopal Hosp, 59-67. *Mem:* AAAS; Am Physiol Soc; Am Asn Clin Chemists; Soc Exp Biol & Med. *Res:* Clinical chemistry; hormone assay. *Mailing Add:* Dept Path St Luke's Hosp 44th & Wornall Rd Kansas City MO 64111

ELLIOTT, LARRY P, b Fleming, Mo, Sept 27, 38; m 61; c 3. MEDICAL MICROBIOLOGY. *Educ:* William Jewell Col, BA, 60; Univ Wis, MS, 62, PhD(bact), 65. *Prof Exp:* From asst prof to assoc prof, 65-78, PROF BIOL, WESTERN KY UNIV, 78-, MED TECHNOL COORDR, 81- *Concurrent Pos:* Clin microbiologist, Greenview Hosp, 74-86. *Mem:* AAAS; Am Sci Affil; Am Soc Microbiol; Sigma Xi; Asn Southeastern Biologists. *Res:* Environmental, food and clinical microbiology; microbiology of sludge applied to land. *Mailing Add:* Dept Biol Western Ky Univ Bowling Green KY 42101

ELLIOTT, LARRY PAUL, b Manhattan, Kans, Oct 16, 31; m 56; c 2. RADIOLOGY. *Educ:* Univ Fla, BS, 54; Univ Tenn, MD, 57. *Prof Exp:* Assoc prof radiol, Sch Med, Wash Univ, 65-67; prof radiol, Shands Teaching Hosp, Col Med, Univ Fla, 67-76; PROF RADIOL, ALA MED CTR, 76- *Res:* Diagnostic radiology; congenital heart disease, especially roentgenographic and pathologic correlation. *Mailing Add:* Dept Radiol/CG 300 Georgetown Univ 37th & O Sts NW Washington DC 20057

ELLIOTT, LLOYD FLOREN, b Clear Lake, SDak, July 7, 37; m 58; c 2. SOIL MICROBIOLOGY, SOIL BIOCHEMISTRY. *Educ:* SDak State Univ, BS, 59; Kansas State Univ, MS, 61; Ore State Univ, PhD(microbiol), 65. *Prof Exp:* Res microbiologist, Soil & Water Conserv Res Div, Northern Plains Br, Agr Res Serv, 65-75, microbiologist soil, water & air sci, Land Mgt & Water Conserv Res Unit, Western Region, 75-88, RES LEADER, COTTON RES UNIT, USDA-AGR RES SERV, PAC AREA, 88- *Honors & Awards:* Underwood Award, 82. *Mem:* Am Soc Microbiol; Soil Conserv Soc Am; fel Am Soc Agron; fel Soil Sci Soc Am; AAAS. *Res:* Soil microbiology; energetics of bacterial oxidations; research on beef feedlot wastes; pollution control; waste disposal; crop residue management for improved crop growth and soil conservation; plant rhizosphere microbiology; inhibitory pseudomonads. *Mailing Add:* Nat Forage Seed Res Ctr USDA-ARS Ore State Univ 3450 SW Campus Way Corvallis OR 97331-7102

ELLIOTT, LOIS LAWRENCE, b Cincinnati, Ohio, July 3, 31; wid. PSYCHOACOUSTICS, DEVELOPMENT OF AUDIO PERCEPTION. *Educ:* Bryn Mawr Col, AB, 53; Cornell Univ, PhD, 56. *Prof Exp:* Res psychologist, Oper Lab, US Air Force Personnel & Training Res Lab, 56-58, res psychologist, Personnel Lab, Lackland AFB, 58-60; res psychologist, Audiol Lab, US Air Force Sch Aerospace Med, 60-63; res psychologist, Cent Inst for Deaf, 63-70, res psychologist, Bur Educ for Handicapped, US Off Educ, 70-73, head directed res in communicative dis, Nat Inst Neurol & Communicative Dis & Stroke, NIH, 73-75, head prog audiol & hearing impairment, 76-78, PROF AUDIOL, NORTHWESTERN UNIV, 76-, PROF OTOLARYNGOL, 77- *Concurrent Pos:* Assoc prof psychol, Washington Univ, 66-70; coordr, Human Commun Sci Prog, Northwestern Univ, 80-84 & 89. *Mem:* Fel Am Psychol Asn; fel AAAS; fel Acoust Soc Am; Psychonomic Soc; Am Speech & Hearing Asn; Int Soc Audiol; Am Pshychol Soc. *Res:* Psychoacoustics; audiology and experimental design; speech perception; auditory development. *Mailing Add:* 2299 Sheridan Rd Northwestern Univ Evanston IL 60208-3550

ELLIOTT, PAUL M, b Kingsville, Tex, Oct 20, 22; m 46; c 3. PHYSICS. *Educ:* US Naval Acad, BS, 44; Tex Col Arts & Indust, MS, 60; Tex A&M Univ, PhD(physics), 64. *Prof Exp:* From instr to assoc prof physics, Tex A&I Univ, 60-69, prof physics, 69-87; RETIRED. *Mem:* AAAS; Am Phys Soc; Am Asn Physics Teachers. *Res:* Atmospheric physics. *Mailing Add:* 930 E Lincoln Cobbs NM 88240

ELLIOTT, PAUL RUSSELL, b Pueblo, Colo, Aug 26, 33; div; c 2. BIOCHEMISTRY, MEDICAL EDUCATION. *Educ:* Phillips Univ, BA, 55; Univ Mich, MS, 57, PhD(zool), 60. *Prof Exp:* From asst prof to assoc prof zool, Univ Fla, 63-71, asst dean preprof educ, Cols Arts & Sci, Dent & Med, 69-71, asst dean Tallahassee progs, Col Med, 71-78; dir prog med sci, Fla State Univ, Fla A&M Univ & Univ Fla, 71-78; asst vpres, 78-80, assoc vpres acad affiars, 80-85, PROF BIOL SCI, FLA STATE UNIV, 71- *Concurrent Pos:* NIH fel biochem, Johns Hopkins Univ, 60-63; NIH res grant, 64-67; dir, Div Student Prog, Asn Am Med Col, 85. *Res:* Cytochemistry and physiology of the epididymis; bioluminescence in bacteria and ctenophorans; medical education research; minority student access to graduate and professional education. *Mailing Add:* 234 Conradi Hall Fla State Univ Tallahassee FL 32306-2049

ELLIOTT, RALPH BENJAMIN, physical chemistry, for more information see previous edition

ELLIOTT, RICHARD AMOS, optical scattering, semiconductor lasers; deceased, see previous edition for last biography

ELLIOTT, ROBERT A, b Darke Co, Ohio, Dec 29, 24; m 46; c 4. BACTERIOLOGY. *Educ:* Miami Univ, AB, 49; Ohio State Univ, MS, 51, PhD(bact), 53. *Prof Exp:* Asst & fel dairy sci, Ohio State Univ, 54-55; bacteriologist, Biol Develop Dept, 55-64, sr bacteriologist, Biol Assay Develop Dept, 64-69, sr bacteriologist, biol tech serv & qual control, 69-76, sr mirobiologist, 76-78, res scientist, vet res, Eli Lilly & Co, Greenfield, 78-86; RETIRED. *Mem:* Am Soc Microbiol; NY Acad Sci; Sigma Xi. *Res:* Bacterial resistance. *Mailing Add:* 9817 E Michigan St Indianapolis IN 46229

ELLIOTT, ROBERT DARYL, b Nashville, Tenn, June 4, 35; m 68; c 4. ORGANIC CHEMISTRY. *Educ:* Vanderbilt Univ, BA, 57; Wayne State Univ, PhD(org chem), 61. *Prof Exp:* Sr asst scientist, Tech Develop Labs, USPHS, Ga, 60-63; SR STAFF CHEMIST, SOUTHERN RES INST, 63- *Mem:* Am Chem Soc. *Res:* Synthesis of folic acid analogs and nucleosides as anticancer drugs. *Mailing Add:* Southern Res Inst PO Box 55305 Birmingham AL 35255-5305

ELLIOTT, ROBERT JAMES, b Swanwick, Derby, Eng, June 12, 40; Can citizen; m 62; c 2. SIGNAL PROCESSING, MATHEMATICAL FINANCE. *Educ:* Oxford Univ, BA, 61, MA, 64; Cambridge Univ, PhD(math), 65, ScD, 85. *Prof Exp:* Lectr math, Univ Newcastle, UK, 64-65; instr math, Yale Univ, Conn, 65-66; lectr math, Oxford Univ, 66-69; sr res fel math, Warwick Univ, 69-73; prof math, Hull Univ, UK, 73-86, G F Grant prof & head dept, 76-86; PROF PROBABILITY THEORY, UNIV ALTA, 86- *Concurrent Pos:* Fel, Berkeley Col, Yale, 65-66 & Oriel Col, Oxford, 66-69; assoc prof, Northwestern Univ, 72-73; vis prof, Inst Hautes Etudes Sci, 76-84, Univ Alta, 77, Univ Ky, 78, Tech Univ Denmark, 83, Australian Nat Univ, 83-91. *Mem:* Am Math Soc; Can Math Soc; Soc Indust & Appl Math; Math Asn Am. *Res:* Theory of stochastic processes and applications to filtering, signal processing; stochastic control and mathematical finance. *Mailing Add:* Dept Statist & Appl Probability Univ Alta CAB 434 Edmonton AB T6G 2G1 Can

ELLIOTT, ROBERT S(TRATMAN), b New York, NY, Mar 9, 21; m 51; c 4. ELECTRICAL ENGINEERING, ENGINEERING ECONOMICS. *Educ:* Columbia Univ, AB, 42, BS, 43; Univ Ill, MS, 47, PhD(elec eng), 52; Univ Calif, Santa Barbara, MA, 71. *Prof Exp:* Jr engr radar, Appl Physics Lab, 43-46; asst prof elec eng, Univ Ill, 46-52; res physicist electromagnetic probs, Hughes Aircraft Co, 53-56; tech dir & vpres, Rantec Corp, 56-59; asst dean grad studies eng, 66-69, chmn dept, 69-86, PROF ENG, UNIV CALIF, LOS ANGELES, 59-; HUGHES ENDOWED CHAIR ELECTRO MAGNETICS, 90- *Mem:* Nat Acad Engrs; fel Inst Elec & Electronics Engrs. *Res:* Electromagnetics; microwave tubes and antennas; electrical properties of materials; engineering economics. *Mailing Add:* Sch Eng & Appl Sci Univ Calif 405 Hillgard Los Angeles CA 90024

ELLIOTT, ROSEMARY WAITE, b Great Yarmouth, Eng, June 16, 34; wid; c 6. MOLECULAR GENETICS, BIOCHEMISTRY. *Educ:* Univ Birmingham, BSc, 55; Univ Buffalo, MA, 62; State Univ NY Buffalo, PhD(biochem), 64. *Prof Exp:* Cancer res scientist, 64-66, sr cancer res scientist, 66-79, cancer res scientist IV, 79-82, CANCER RES SCIENTIST V & CHMN, CELL & MOLECULAR BIOL DEPT, ROSWELL PARK MEM INST, 82- *Mem:* AAAS; Genetic Soc Am; Am Soc Microbiol. *Res:* Regulation of levels of lysomal enzymes in mice; host-controlled variation in bacteriophages; lysogenic association between bacteriophages and host cells; genetic determination of processing of lysosomal proteins; detection and mapping mouse polymorphisms detectable by 2D-electrophoresis; detection and mapping of mouse DNA polymorphisms; insertional mutagenesis by retroviruses. *Mailing Add:* Dept Molecular Biol Roswell Park Mem Inst Buffalo NY 14263

ELLIOTT, SHELDEN DOUGLESS, JR, b Anaheim, Calif, Feb 3, 31; m 54; c 2. PHYSICS, SOLAR ENERGY. *Educ:* Yale Univ, BS, 53, MS, 54, PhD, 59. *Prof Exp:* Physicist, US Naval Weapons Ctr, 58-77; PHYSICIST, US DEPT ENERGY, 77- *Mem:* Am Phys Soc; Int Solar Energy Soc; Weather Modification Asn (pres, 74); Sigma Xi. *Res:* Low temperature physics; second sound in liquid helium isotope mixtures; atmospheric physics; astronomy; meteor physics; weather modification; solar energy. *Mailing Add:* 1801 NWY 128 Philo CA 95466

ELLIOTT, SHELDON ELLWOOD, b Asuncion, Paraguay, July 9, 25; US citizen; m 48; c 3. EXPLORATION GEOPHYSICS, APPLIED MATHEMATICS. *Educ:* Phillips Univ, AB, 48; Univ Mich, MA, 49. *Prof Exp:* Fel, Univ Mich, 50-51 & 55-56, jr instr, 56-57; mathematician, Phillips Petrol Co, 51-55 & 57-64, mgr geophys res, 64-77, math geophysicist, 69-77, chief geophysicist, 77-80, mgr geophys, 80-85; RETIRED. *Mem:* Am Math Soc; Soc Explor Geophysicists; Inst Elec & Electronics Engrs; Sigma Xi; Am Asn Petrol Geologists. *Res:* Seismology; operational calculus; differential equations; numerical methods; topology. *Mailing Add:* 1512 Macklyn Lane Bartlesville OK 74006

ELLIOTT, STUART BRUCE, b Oakland, Calif, July 11, 27; m 53; c 2. PHYSICS. *Educ:* Stanford Univ, BS, 49, MS, 51, PhD(physics), 60. *Prof Exp:* Asst prof physics, Kenyon Col, 55-60; from asst prof to assoc prof physics, 60-86, chmn dept, 71-74, 86-89, PROF PHYSICS, OCCIDENTAL COL, 86- *Concurrent Pos:* Sci ed, Optics & Spectros, Optical Soc Am, 68- *Mem:* Am Asn Physics Teachers; fel Optical Soc Am. *Res:* Holography; holographic contouring. *Mailing Add:* Dept Physics Occidental Col Los Angeles CA 90041

ELLIOTT, WILLIAM H, b St Louis, Mo, June 4, 18; m 49; c 4. BIOCHEMISTRY, ORGANIC CHEMISTRY. *Educ:* St Louis Univ, BSChem, 39, MS, 41, PhD(org chem), 44. *Prof Exp:* From instr to assoc prof, St Louis Univ, 44-59, actg chmn dept, 70-71, prof biochem, 59-88, prof chem, 77-88, EMER PROF BIOCHEM & CHEM, SCH MED, ST LOUIS UNIV, 88- *Concurrent Pos:* Consult radioisotopes, Vet Admin, 55-58, consult biochem, 69-71 & 72-79; consult, Res Career Award Comt, Nat Inst Gen Med Sci, 62-65; mem ad hoc study sects, NIH, 65-; Nat Defense Res Comt fel, Ind Univ, Bloomington, 44; St Louis Univ-NIH-Med Res Coun fel & vis prof, Dept Biochem, Univ Edinburgh, 74. *Mem:* Fel AAAS; Am Chem Soc; Am Soc Biol Chem; Am Soc Mass Spectrometry; Soc Exp Biol & Med; Am Soc Study Liver Dis; Sigma Xi. *Res:* Chemistry and metabolism of sterols, bile acids and fat-soluble vitamins; mass spectrometry. *Mailing Add:* Dept Biochem & Molecular Biol St Louis Univ Sch Med St Louis MO 63104

ELLIOTT, WILLIAM J, b San Diego, CA, July 29, 39. GEOLOGY. *Educ:* San Diego State Univ, BA, 64, MS, 66. *Prof Exp:* Exp petrol geologist, Standard Oil, CA, Bakersfield, 66-70; eng geologist, Southern CA Testing Lab, 70-71, Benton Eng, 71-74, sr eng geologist, Frugro Inc, 74-78,; CONSULT ENG GEOLOGIST, 78- *Mem:* Fel Geol Soc Am; Asn Eng Geologists. *Res:* Private consulting practice in geological engineering. *Mailing Add:* PO Box 541 Solana Beach CA 92075

ELLIOTT, WILLIAM JOHN, b St Louis, Mo, Jan 27, 51; m 81. CLINICAL PHARMACOLOGY, HYPERTENSION. *Educ:* Univ Notre Dame, BS, 73; Univ Chicago, PhD(chem), 76, MD, 79. *Prof Exp:* Resident med, Barnes Hosp, St Louis, 79-82; fel pharmacol, Wash Univ, St Louis, 82-85; ASST PROF MED & PHARMACOL, UNIV CHICAGO, 85- *Mem:* Am Fedn Clin Res; Am Soc Clin Pharmacol & Therpeut; Am Col Clin Pharmacol; AAAS; Am Soc Hypertension. *Res:* Clinical pharmacology of hypertension and anti hypertensive drugs; non-metabolic roles of essential fatty acids and other natural products, especially columbinic acid, in the regulation of both skin homeostasis and blood pressure and cholesterol levels; preventive cardiology. *Mailing Add:* Univ Chicago 947 E 58th St Chicago IL 60637

ELLIOTT, WILLIAM PAUL, b Geneva, Ill, June 16, 28; m 52; c 2. CLIMATOLOGY, METEOROLOGY. *Educ:* St John's Col, Md, AB, 47; Univ Chicago, MS, 52; Tex A&M Univ, PhD(phys oceanog), 58. *Prof Exp:* Res assoc meteorol, Tex A&M Univ, 52-57, from instr to asst prof, 53-56; atmospheric physicist micrometeorol, US Air Force Cambridge Res Lab, 57-68; res assoc prof, Sch of Oceanog, Ore State Univ, 68-74; SUPVRY METEOROLOGIST, AIR RESOURCES LAB, NAT OCEANIC & ATMOSPHERIC ADMIN, 74- *Concurrent Pos:* Assoc ed, J Climate & Appl Meteorol, 83-89, J Geophys Res; chmn, World Meteorol Orgn, 83-87, Rapporteurs on Carbon Dioxide, 84-88. *Mem:* AAAS; Am Meteorol Soc; Am Geophys Union; Wilson Ornith Soc. *Res:* Climatology of water vapor; micrometeorology; long-range transport of pollutants, ornithology. *Mailing Add:* Air Resources Labs 1325 East-West Hwy Silver Spring MD 20910

ELLIS, ALBERT, b Pittsburgh, Pa, Sept 27, 13. PSYCHOTHERAPY, SEX & MARITAL THERAPY. *Educ:* City Col, New York, BBA, 34; Columbia Univ, MA, 43, PhD(clin psychol), 47. *Prof Exp:* Clin psychologist, Northern NJ Mental Hyg Clin, 48-49; chief psychologist, NJ State Diag Ctr, 49-50; chief psychologist, NJ Dept Insts & Agencies, 50-52; pvt pract psychotherapy, 52-68; exec dir, 68-90, PRES, INST RATIONAL-EMOTIVE THER, 90- *Concurrent Pos:* Instr psychol, Rutgers Univ & NY Univ, 49-50; adj prof psychol, Rutgers Univ, 71-80 & Pittsburg State Univ, Kans, 78-; dir consult psychol, Am Psychol Asn, 61-62. *Mem:* Soc Sci Study of Sex (pres, 60-62); Am Psychol Asn (pres, 61-62); Am Acad Psychotherapists (vpres, 62-64); Am Asn Sex Educators, Counr & Therapists; Am Asn Marriage & Family Therapists; Am Orthopsychiat Asn. *Res:* Many articles and books on sex, love and marriage, and rational-emotive therapy. *Mailing Add:* 45 E 65th St New York NY 10021-6508

ELLIS, ALBERT TROMLY, b Atwater, Calif, Apr 22, 17; m 54. APPLIED MECHANICS. *Educ:* Calif Inst Technol, BS, 43, MS, 47, PhD(mech eng, physics), 53. *Prof Exp:* Asst physiol, Calif Inst Technol, 42-43; res engr electronics & instrumentation, Columbia Univ, 44-46; electronics engr & asst head electronics hydrodyn lab, Calif Inst Technol, 46-47; physicist theory controls & servomechanisms, US Naval Ord Test Sta, 47-49; res engr hydrodyn, Calif Inst Technol, 49-54, sr res fel eng, 54-57, assoc prof appl mech, 58-67; prof, 67-85, EMER PROF APPL MECH, UNIV CALIF, SAN DIEGO, 85- *Concurrent Pos:* Consult, Space Technol Labs, 56- & Naval Undersea Ctr, San Diego; sr vis, Dept Appl Math & Theoret Physics, Cambridge Univ, 64-65; sr postdoctoral fel, Nat Sci Found, 64-65; vis fel math, Wolfson Col, Oxford Univ, 84-85. *Mem:* AAAS; Inst Elec & Electronics Engrs; Am Soc Mech Engrs; Am Phys Soc; Acoust Soc Am. *Res:* Cavitation and drag reduction in dilute polymer flows; laser radiation interactions with solids and liquids; wave propagation in composite materials; high speed holography; ultrasonics; chaos; theory of bubble dynamics. *Mailing Add:* Dept Appl Mech 6120 A Urey Hall La Jolla CA 92093

ELLIS, BOYD G, b Havensville, Kans, Nov 3, 32; m 55; c 3. SOIL CHEMISTRY. *Educ:* Kans State Univ, BS, 54, MS, 55; Mich State Univ, PhD(soil chem), 61. *Prof Exp:* From asst prof to assoc prof soil sci, 61-68, PROF SOIL SCI, MICH STATE UNIV, 68- *Mem:* Fel Am Soc Agron; Clay Minerals Soc; fel Soil Sci Soc Am; Int Soil Sci Soc; fel AAAS. *Res:* Soil fertility; chemistry of nutrients in the soil and their effect on plant growth; factors affecting availability of potassium, phosphorus, magnesium and zinc; Soil chemistry related to environmental quality. *Mailing Add:* Dept Crop & Soil Sci Mich State Univ East Lansing MI 48823

ELLIS, C N, b Hillsdale, Mich, Oct 2, 52; m 76; c 1. DERMATOLOGY, CLINICAL DRUG DEVELOPMENT. *Educ:* Univ Mich, BS, 75, MD, 77. *Prof Exp:* From asst prof to assoc prof, 81-90, PROF & ASSOC CHMN DERMAT, UNIV MICH MED CTR, 90- *Concurrent Pos:* Consult, Plymouth Ctr Human Develop, 80-81, Northville Regional Psychiat Ctr, 80-81; chief dermat serv , 82-86, consult, Vet Admin Med Ctr, Ann Arbor, 86-; vis prof, Univ Louisville, Ky, 85, Univ Miami & Harvard Univ Med Sch, 88. *Mem:* Am Acad Dermat; Am Dermat Asn; Soc Investigative Dermat. *Res:* Drug development for new dermatologicals; new therapies for dermatological disorders; cyclosporine and retinoids in the treatment of psoriasis and acne. *Mailing Add:* Dept Dermat Univ Mich Med Ctr 1910 TC 1500 E Med Ctr Dr Ann Arbor MI 48109-0314

ELLIS, CLIFFORD ROY, b Nova Scotia, Can, Jan 7, 41; c 2. ECONOMIC ENTOMOLOGY. *Educ:* McGill Univ, BS, 63, MS, 65; Univ Alta, PhD(entom), 70. *Prof Exp:* Exten entomologist, NS Dept Agr & Mkt, 65-67; from asst prof to assoc prof, 70-87, PROF ECON ENTOM, DEPT ENVIRON BIOL, UNIV GUELPH, 87- *Mem:* Entom Soc Can; Entom Soc Am; Can Pest Mgt Soc. *Res:* Pests of forage and field crops with particular emphasis on the biology, control and economic thresholds of various pests. *Mailing Add:* Dept Environ Biol Univ Guelph Guelph ON N1G 2W1 Can

ELLIS, DANIEL B(ENSON), b Rochdale, Eng, May 15, 37; US citizen; m 63; c 2. BIOCHEMISTRY, BIOCHEMICAL PHARMACOLOGY. *Educ:* Univ Sheffield, BSc, 58; McGill Univ, PhD(biochem), 61. *Prof Exp:* Res assoc microbiol, Wistar Inst, Univ Pa, 62-64; biochemist cancer res, Stanford Res Inst, 64-65; sr res biochemist, Smith Kline & French Labs, 65-70, sr investr, 70-74; sect head biochem, Betz Labs, 74-76; mgr biochem, 76-90, assoc dir biol res, 78-90, ASSOC DIR, PRECLIN DEVELOP, WORLDWIDE NEUROSCI GROUP, HOECHST-ROUSSEL PHARMACEUT INC, 90- *Concurrent Pos:* Radiation safety officer, Hoechst-Roussel Pharmaceut Inc, 77- *Mem:* Am Soc Biochem & Molecular Biol; AAAS; Am Chem Soc; Biochem Soc; Am Soc Pharmacol Exp Therapeut. *Res:* Drug research and development; neurochemistry; regulatory control mechanisms; drug metabolism; influence of drugs on central nervous system; psychotropic and anti-hypertensive agents; Alzheimer's disease; cancer chemotherapy. *Mailing Add:* Hoechst-Roussel Pharmaceut Inc Rte 202-206 PO Box 2500 Somerville NJ 08876-1258

ELLIS, DAVID ALLEN, b Seattle, Wash, Mar 25, 17; m 43; c 1. PHYSICAL CHEMISTRY, INORGANIC CHEMISTRY. *Educ:* Univ Wash, BS, 39; Univ Southern Calif, MS, 48, PhD(chem), 50. *Prof Exp:* Chemist, Puget Sound Pulp & Timber Co, 40; observer, US Weather Bur, 40-41; asst, Univ Southern Calif, 45-48; res chemist, Dow Chem Co, 49-67; prof chem, 67-82, chmn, Div Sci & Math, 71-78, EMER PROF CHEM, AZUSA PAC COL, 82- *Mem:* Am Chem Soc. *Res:* Solvent extraction; inorganic chemicals; ion exchange. *Mailing Add:* 15914 D Village Green Dr Mill Creek WA 98012

ELLIS, DAVID GREENHILL, b Marietta, Ohio, Mar 9, 36. THEORETICAL PHYSICS. *Educ:* Marietta Col, AB, 58; Cornell Univ, PhD(theoret physics), 64. *Prof Exp:* Res assoc theoret physics, Ind Univ, 63-65; from asst prof to assoc prof physics & astron, 65-75, chmn dept, 74-79, PROF PHYSICS & ASTRON, UNIV TOLEDO, 75-, CHMN DEPT, 85- *Concurrent Pos:* vis prof, Univ Lund, Sweden, 76-77. *Mem:* Am Phys Soc. *Res:* Theoretical atomic physics; quantum mechanics; beam-foil spectroscopy; astrophysics. *Mailing Add:* 2326 Orchard Rd Toledo OH 43606

ELLIS, DAVID M, b Ithaca, NY, Nov 15, 37; m 63; c 3. MEDICAL ELECTRONICS. *Educ:* Pa State Univ, BS, 59; Univ Wash, MS, 65, PhD(elec eng), 68. *Prof Exp:* Res engr, Burroughs Corp, Pa, 59-63; asst prof elec eng, Univ Vt, 68-72; develop engr, Hewlett-Packard Co, 72-86; PRES/OWNER, CAMBRIDGE AERO INSTRUMENTS, INC, 86- *Concurrent Pos:* Consult components div, IBM Corp, Vt, 68-70 & Vertek Inc, 70-; consult & mem bd dirs, Cambridge Aero Instruments. *Mem:* Inst Elec & Electronics Engrs; Sigma Xi; Soc Critical Care Med; AAAS. *Res:* Measurement and processing of physiological blood pressures; measurement and processing of pressure signals related to soaring aircraft; nuclear magnetic resonance; semiconductor electronics. *Mailing Add:* Rural Box 122-2 Warren VT 05674

ELLIS, DAVID WERTZ, b Huntingdon, Pa, Feb 8, 36; m 61; c 3. ANALYTICAL CHEMISTRY. *Educ:* Haverford Col, AB, 58; Mass Inst Technol, PhD(analytical chem), 62. *Hon Degrees:* LLD, Lehigh Univ, 79; DSc, Susquehanna Univ, 82, Ursinus Col, 85. *Prof Exp:* Asst prof, Univ NH, 62-67, asst dean, Col Technol, 67-68, assoc off acad vpres, 68-69, actg acad vpres, 69, assoc acad vpres, 70-71, assoc prof chem, 67-68, vice provost acad affairs, 71-77, vpres, 77-78; PRES, LAFAYETTE COL, EASTON, PA, 78- *Mem:* AAAS; Am Chem Soc; Sigma Xi. *Res:* Excited-state chemical reactions using fluorescence and phosphorescence; development of new methods of chemical analysis; analysis of water pollutants. *Mailing Add:* Graves Landing No 710 Six Canal Park Cambridge MA 02141

ELLIS, DEMETRIUS, b Greece, Aug 23, 46; US citizen; m 70; c 2. PEDIATRICS, PEDIATRIC NEPHROLOGY. *Educ:* New York Univ, BA, 69; State Univ New York, Buffalo, MD, 73. *Prof Exp:* Asst prof, 77-83, ASSOC PROF PEDIAT, UNIV PITTSBURGH SCH MED, 84-; DIR PEDIAT NEPHROLOGY, CHILDREN'S HOSP PITTSBURGH, 77- *Concurrent Pos:* Chmn med adv bd, Nat Kidney Found, 79-80. *Mem:* Am Soc Nephrol; Int Soc Nephrol; Am Soc Pediat Nephrol; Am Pediat Soc; Am Diabetes Asn; Soc Pediat Res. *Res:* Evaluation of microalbuminuria at an early marker of renal disease in patients with insulin-dependent diabetes mellitus; development of a cell culture model for the investigation of the pathogenesis of complications in diabetes mellitus; investigation of mechanisms of gluco-corticoid-induced modulation of Nx-K ATPase in polycystic kidney disease. *Mailing Add:* Children's Hosp Pittsburgh 125 DeSoto St Pittsburgh PA 15213

ELLIS, DEREK V, b Windsor, Eng, July 26, 30; Can citizen; m 57; c 3. BIOLOGY. *Educ:* Univ Edinburgh, BSc, 51, Hons, 52; McGill Univ, MSc, 54, PhD(zool), 57. *Prof Exp:* From asst scientist to assoc scientist, Fisheries Res Bd Can, 57-63; asst prof zool, Univ Man, 63-64; assoc prof biol, 64-81, PROF BIOL, UNIV VICTORIA, BC, 81- *Res:* Ethology and ecology of aquatic animals; marine environmental impact assessment. *Mailing Add:* Dept Biol Univ Victoria Box 1700 Victoria BC V8W 2Y2 Can

ELLIS, DON EDWIN, b Ames, Iowa, Apr 8, 08; m 28; c 1. PLANT PATHOLOGY. *Educ:* Nebr Cent Col, AB, 28, BS, 29; La State Univ, MS, 32; Univ NC, PhD(plant path), 45. *Prof Exp:* Teacher high sch, Juniata, Nebr, 29-31; asst plant path, Exp Sta, La State Univ, 31-33; supv technician forest path, State Emergency Conserv Work, Iowa State Col, 33; asst pathologist, Div Forest Path, Bur Plant Indust, USDA, 34-40; asst plant pathologist, Exp Sta, 40-44, from assoc prof to prof, 44-73, head dept, 54-73, EMER PROF PLANT PATH, NC STATE UNIV, 73- *Mem:* Am Phytopath Soc (pres, 70); Mycol Soc Am; Asn Trop Biol; Am Inst Biol Sci; AAAS. *Res:* Diseases of vegetable crops. *Mailing Add:* 244 Springmoor Dr Raleigh NC 27695-7616

ELLIS, DONALD EDWIN, b San Diego, Calif, Feb 20, 39; m 65. SOLID STATE PHYSICS, MOLECULAR PHYSICS. *Educ:* Mass Inst Technol, SB, 61, MS, 64, PhD(physics), 66. *Prof Exp:* Asst prof physics, Univ Fla, 66-68; from asst prof to assoc prof, 68-78, PROF PHYSICS & CHEM, NORTHWESTERN UNIV, EVANSTON, 78- *Mem:* Am Phys Soc; Am Chem Soc. *Res:* Electronic structure of molecules; transition metal, rare earth, and actinide complexes; band structure of solids. *Mailing Add:* Dept Chem Northwestern Univ Evanston IL 60201

ELLIS, DONALD GRIFFITH, b Colorado Springs, Colo, Aug 10, 40; m 77. BIOINSTRUMENTATION, DESIGN. *Educ:* Univ Colo, BS, 62; Univ Mich, MS, 64, PhD(bioeng), 70. *Prof Exp:* Res asst metall, Denver Res Inst, 62-63 & mech eng, Univ Colo, 63; res engr phys med & rehab, Univ Mich, 64-68, res assoc, 68-70; fel, Webb-Waring Inst, 70-72; res assoc phys med & rehab, Univ Mich, 72-75, consult child measurement study, 72-73; res assoc chem eng, Univ Colo, 75-76, elec eng, 76-78, res assoc, Health Sci Ctr, Univ Colo, 78-87; DESIGNER, SPIDERWORT DESIGN, 76- *Mem:* Am Soc Mech Engrs; AAAS; Soc Plastics Eng. *Res:* Design and development of biomedical instrumentation and equipment; Doppler ultrasound instrumentation; respiratory mass spectroscopy; laboratory computer systems; equipment for studies in exercise physiology, medication monitoring and connective tissue biomechanics. *Mailing Add:* Geneva Park Boulder CO 80302-7162

ELLIS, EDWIN M, b Watertown, SDak, Mar 11, 14; m 42; c 1. MICROBIOLOGY, VETERINARY MEDICINE. *Educ:* Adrian Col, BS; Wayne State Univ, MS; Mich State Univ, PhD(microbiol) & DVM. *Prof Exp:* Asst microbiol, Mich State Univ, 51-56; animal pathologist, Ft Detrick, Md, 53-54; immunologist, Ga Coastal Plain Exp Sta, Dept Animal Dis, USDA, 56-59; bacteriologist, Fla State Dept Agr, 59-60; CHIEF MICROBIOL, LAB SERV, NAT ANIMAL DIS CTR AGR RES SERV, USDA, 60- *Honors & Awards:* Res Award, Sigma Xi. *Mem:* Am Soc Microbiol; Am Vet Med Asn; fel Am Acad Microbiol. *Res:* Fluorescent antibody techniques applied to animal viruses; animal disease; tuberculosis; isolation and immunology. *Mailing Add:* 6912 South Shore Dr South Pasadena FL 33707

ELLIS, EFFIE O'NEAL, b Hawkinsville, Ga, June 15, 13; m 53; c 1. PUBLIC HEALTH, EPIDEMIOLOGY. *Educ:* Spelman Col, AB, 33; Atlanta Univ, MS, 35; Univ Ill, MD, 50. *Prof Exp:* Dir med educ, Provident Hosp, Baltimore, Md, 53-61; pediat consult & dir maternal & child health, Ohio Dept Health, Columbus, 61-65; regional med dir, US Children's Bur, HEW, Chicago, 61-67, regional comnr, Social & Rehab Serv, Chicago, 67-70; SPEC ASST FOR HEALTH SERV, AMA, 70- *Concurrent Pos:* Lectr, 53-; consult & co-dir, Quality of Life Ctr, Chicago; consult community health progs; mem, Ohio Planning Comt Health Educ, 61-65 & Ohio Comn Ment Health & Ment Retardation Planning, 63-65; chmn panel group, White House Conf Food & Nutrit, 69; vchmn panel group, White House Conf on Children, 70. *Honors & Awards:* Golden Plate Award, Am Acad Achievement, 70; Distinguished Serv Award, Nat Med Asn. *Mem:* Inst Med-Nat Acad Sci; hon fel Am Sch Health Asn; Am Pub Health Asn; Am Pub Welfare Asn; Am Asn Ment Deficiency. *Mailing Add:* 300 N State St Apt 4605 Chicago IL 60610

ELLIS, ELIZABETH CAROL, b Columbus, Ga. ATMOSPHERIC GAS MEASUREMENT. *Educ:* Auburn Univ, BS, 68; Duke Univ, PhD(phys chem), 75. *Prof Exp:* Res chemist, US Environ Protection Agency, 73-77; SUPVR RES SCIENTIST, SOUTHERN CALIF EDISON CO, 77- *Concurrent Pos:* Lectr, Univ Calif, Los Angeles, 79- *Mem:* Am Chem Soc; Sigma Xi; Am Asn Aerosol Res. *Res:* Environmental effects from gaseous; particulate and thermal emission from fossil-fueled electric power generating stations on the properties of the atmosphere; ambient air quality; long range transport; deposition; visibility. *Mailing Add:* 1926 Sierra Madre Villa Pasadena CA 91107-1235

ELLIS, ELLIOT F, b Englewood, NJ, Apr 7, 29; m 55; c 4. PEDIATRICS, ALLERGY. *Educ:* Kenyon Col, AB, 50; Case Western Reserve Univ, MD, 54; Am Bd Pediat, dipl; Am Bd Pediat Allergy, dipl; Am Bd Allergy & Immunol, dipl. *Prof Exp:* Intern, Lenox Hill Hosp, New York, 54-55; resident, Babies Hosp, Columbia Presby Med Ctr, 57-59; instr pediat, Col Med, Univ Fla, 63-66; from asst prof to assoc prof, Univ Colo, Denver, 66-74; PROF PEDIAT, STATE UNIV NY, BUFFALO, 74- *Concurrent Pos:* Chief pediat, Nat Jewish Hosp & Res Ctr, Denver, 70-74; dir div allergy & immunol, Children's Hosp Buffalo, 74-75; chmn pediat dept, State Univ NY Buffalo, 75-84, co-dir, div allergy & immunol, 76-; fel, Children's Asthma Res Inst & Hosp, Denver, Colo, 62-63; fel allergy & immunol, Col Med, Univ Fla, 63-66. *Honors & Awards:* Bela Schick Award, Am Col Allergists, 64. *Mem:* Am Acad Pediat; Am Acad Allergy & Immunol; Am Col Allergists; Am Asn Immunologists. *Res:* Pediatric allergy and clinical immunology. *Mailing Add:* Dept Allergy & Immunol Nemours Childrens Clin PO Box 5720 Jacksonville FL 32247

ELLIS, ERIC HANS, b Mannheim, Ger, Aug 25, 35; US citizen; m 58; c 3. PHYSICS. *Educ:* Syracuse Univ, BS, 56, PhD(physics), 65. *Prof Exp:* Asst physics, Syracuse Univ, 56-64; from instr to assoc prof, 64-80, PROF PHYSICS, UNIV OF THE SOUTH, 80- *Mem:* Am Asn Physics Teachers; Sigma Xi. *Res:* Infrared background radiation; atmospheric optical noise; fluorescence and spectroscopic analysis of bone. *Mailing Add:* Dept Physics Univ of the South Sewanee TN 37375

ELLIS, EVERETT LINCOLN, b Kent, Wash, May 13, 19; m 43; c 4. WOOD SCIENCE & TECHNOLOGY. *Educ:* Univ Wash, BS, 41, PhD(wood prod), 56; Mich State Univ, MS, 43. *Prof Exp:* Asst, Mich State Univ, 41-42; res wood technologist, Chem Div, Borden Co, 43-46; from asst to assoc prof wood utilization, Univ Idaho, 46-56; assoc prof wood technol, Univ Mich, 56-65; prof forest prod, Forest Res Lab, Ore State Univ, 65-71, head dept, 65-70; Forest Prod Ltd prof wood sci, Sch Forestry, Univ Canterbury, 71-85, emer NZ Forest Prod Ltd prof, 85-90; prof wood sci, 85-90, AFFIL PROF, UNIV WASH, 91- *Concurrent Pos:* Vis prof, Univ Ariz, 81 & Univ Wash, 82; adj prof wood sci, Univ Wash, 85- *Mem:* Forest Prod Res Soc; fel Brit Inst Wood Sci; Soc Wood Sci & Technol; Tech Asn Pulp & Paper Indust; NZ Inst Foresters. *Res:* Wood and fiber anatomy and structure, including chemistry and mineral composition; factors of growth as related to wood structure and properties; education in wood science and technology; energy and forest biomass; forest products marketing. *Mailing Add:* 19614 24th Ave NW Seattle WA 98177

ELLIS, FRANK RUSSELL, b Celina, Ohio, Oct 19, 15; m 41; c 4. CLINICAL PATHOLOGY. *Educ:* Univ Mich, MD, 43; Am Bd Path, dipl, 53. *Prof Exp:* Resident asst path, Sch Med, Univ Utah, 47-49; pathologist, St Anthony Hosp, Wenatchee, Wash, 50-52; instr path, Sch Med, Univ Colo, 53-54; clin pathologist, Wayne Co Gen Hosp, Eloise, Mich, 54-66; dir, Southeastern Mich Red Cross Blood Ctr, 66-72; assoc med dir, Am Nat Red Cross Blood Prog, 73-76; med dir, Mo-Ill Regional Red Cross Blood Serv, 76-83; CONSULT, 84- *Concurrent Pos:* Pathologist, DePaul Hosp, Wyo, 52-54; consult, Warren AFB, Wyo, 54; tech adv blood transfusion res div, Army Med Res Lab, Ky, 65-74; res fel, Univ Wash, 49-50. *Mem:* Fel AAAS; Am Acad Forensic Sci; fel Am Soc Clin Path; AMA; Int Soc Blood Transfusion; fel Am Asn Blood Banks. *Res:* Detection and surveillance of the healthy carrier of hepatitis among volunteer blood donors; blood banking; transfusion therapy. *Mailing Add:* 7185 Routt St Arvada CO 80004

ELLIS, FRANKLIN HENRY, JR, b Washington, DC, Sept 20, 20. THORACIC & CARDIOVASCULAR SURGERY. *Educ:* Yale Univ, AB, 41; Columbia Univ, MD, 44; Univ Minn, PhD(surg), 51; Am Bd Surg, dipl, 53; Am Bd Thoracic Surg, dipl, 54. *Prof Exp:* Instr surg, Mayo Grad Sch Med, Univ Minn, 52-56, from asst prof to prof, 56-70; CHMN, DEPT THORACIC & CARDIOVASC SURG, LAHEY CLIN FOUND & NEW ENGLAND DEACONESS HOSP, 71- *Concurrent Pos:* Lectr surg, Harvard Med Sch, 70-74, assoc clin prof, 74-80, clin prof, 80- *Honors & Awards:* Billings Gold Medal, AMA, 55. *Mem:* Am Asn Thoracic Surg; Int Soc Surg; Int Cardiovasc Soc; Am Heart Asn; Am Surg Asn. *Res:* Thoracic and cardiovascular surgery. *Mailing Add:* Dept Thoracic & Cardiovasc 110 Francis St Suite 2C Boston MA 02215

ELLIS, FRED E, b Hutchinson, Kans, Apr 22, 26; m 58; c 2. PHYSICS, ASTRONOMY. *Educ:* ETex State Col, BS, 54, MS, 55; La State Univ, PhD(physics), 65. *Prof Exp:* Instr physics, ETex State Col, 55-58; asst prof, Southern Miss Univ, 64-65; res collabr, Nat Radio Astron Observ, 65; assoc prof physics, Pan Am Univ, 66-80, prof astron, 80-; LAURENT GILES & PARTNER, LTD. *Res:* Astrometry of extended radio sources; radio galactic spur. *Mailing Add:* Laurent Giles & Partner Ltd Four Quay Hill Shaw Island WA 98286

ELLIS, FRED WILSON, b Heath Springs, SC, Apr 24, 14; m 40; c 4. PHARMACOLOGY, BIOCHEMISTRY. *Educ:* Univ SC, BS, 36; Univ Fla, MS, 38; Univ Md, PhD(pharmacol), 41; Univ NC, Cert(med), 48; Duke Univ, MD, 51. *Prof Exp:* Supt biochem lab, Univ Md Hosp, 41-42; assoc pharmacol, Jefferson Med Col, 42-43; from asst prof to prof, 43-80, emer prof pharmacol, Sch Med, Univ NC, Chapel Hill, 80-; RETIRED. *Concurrent Pos:* Res fel pharmacol, Univ Md, 41-42; mem NIH Psychopharmacol Study Sect, 58-60, res grantee, 64- 80; mem res grants rev bd, Nat Coun Alcoholism, 73-78, chmn, Fetal Alcohol Study Group, 83-84; mem, NC Alcoholism Res Authority, 74-85, vchmn & mem exec comt, 78-83. *Mem:* AAAS; AMA; Am Soc Pharmacol; Res Soc Alcoholism; Soc Exp Biol & Med. *Res:* Adrenal function in experimental alcoholism; experimentally-induced physical dependence on ethanol in animals; development of animal models of alcoholism in monkeys and dogs; development of Beagle model of the fetal alcohol syndrome; development of an Auto Analyzer Blood Analysis for ethanol using an enzymic method in human subjects. *Mailing Add:* 805 Old Mill Rd Chapel Hill NC 27514

ELLIS, HAROLD BERNARD, b Havre, Mont, Dec 31, 17; m 44; c 4. CIVIL ENGINEERING. *Educ:* Washington State Univ, BSEE, 41; Mass Inst Technol, MSCE, 47; Iowa State Univ, PhD(civil eng), 63. *Prof Exp:* Chief of schs & training sect, Engr Off, Hq Cent Pac Area, US Army, Hawaii, 44, commandant, Hq Amphibious Training Ctr, 44-45, instr & chief, Opers & Construct Sect, Engrs S Ft Belvoir, Va, 45-46, unit instr, 118th Engr Combat Battalion, Nat Guard, RI, 47-50, asst dist engr, Dist Engr Off, Washington, DC, 50-52; exec off & commander, Hq 931st Engr Group, 52-53, staff off, Aviation Engr Force, Wolters AFB Tex, 53-55, assoc prof mil sci & tactics, Army ROTC, Iowa State Univ, 55-59, dep comdr depot oper, Army Gen Depot, France, 59-62; from asst prof to prof civil eng, Iowa State Univ, 62-73, head tech inst, 62-73; dir personnel develop, Black & Veatch Consult Engrs, Kansas City, 73-79; assoc prof civil eng, Univ Mo, Kansas City, 79-80; RETIRED. *Mem:* Am Soc Civil Engrs. *Mailing Add:* 8501 Ensley Place Leawood KS 66206

ELLIS, HAROLD HAL, b New York, NY, Apr 6, 18. SPECIALTY CHEMICALS FORMULATION, FIREPROOFING. *Educ:* Cornell Univ, BS, 38; Harvard Univ, MA, 40; Univ Minn, PhD(entom & agr chem), 42. *Prof Exp:* Bacteriologist, Naval Med Res Inst, 42-43; entomologist, Off Quartermaster Gen, US War Dept, 43-44 & Pan-Am Sanit Bur, Guatemala, 44-45; med entomologist, Standard Oil Co, NJ & SAm, 45-47; consult engr, Venezuela, 47-50; asst prof parisitol & microbiol, Chicago Med Sch, 50-55; res dir, Colombian Inst Technol Invest, Bogota, 55-57; sr engr & head, 1st Stage Minuteman, Intercontinental Ballistic Missile, Aerojet-Gen Corp, Sacramento, 57-62; mgr reliability & qual control & eng standards, Marquardt Corp, Van Nuys, 62-65; pres & tech dir, Pan-Am Indust & Mining Corp, Bogota, 65-70; vpres & tech dir, Pyrotite Corp, Miami, 83-89; PRES & TECH DIR, DELPHIC RES LABS, MIAMI, 70- *Res:* Specialty coatings (electro-conductive, fire barrier); chemical specialties and industrial chemicals formulation; cementitions binders; fireproofing; utilization of waste products; infra-red technology; nuclear radiation protection; thermal insulation; medical geography; research and development management; technology transfer; reliability and quality control engineering. *Mailing Add:* Box 2158 Ocean View Br Miami Beach FL 33140

ELLIS, HOMER GODSEY, b Paris, Tex, Sept 29, 33; m 57; c 2. MATHEMATICS, PHYSICS. *Educ:* Univ Tex, BA, 55, MA, 58, PhD(math), 61. *Prof Exp:* Asst prof math, Univ Utah, 61-62 & Univ Wash, 62-65; vis asst prof appl math, 65-67, asst prof, 67-68, ASSOC PROF MATH, UNIV COLO, BOULDER, 68- *Concurrent Pos:* Fac fel, Univ Colo, 74-75; guest scientist, Int Ctr Theoret Physics, Italy, 74-75. *Mem:* Am Math Soc; Math Asn Am; Int Soc Gen Relativity & Gravitation. *Res:* Relativity theory; differential geometry; mathematical physics. *Mailing Add:* Dept Math Box 426 Univ Colo Boulder CO 80309-0426

ELLIS, J S, b Kingston, Ont, Jan 10, 27; m 64. CIVIL ENGINEERING. *Educ:* Queen's Univ, Ont, BSc, 48; McGill Univ, MEng, 49; Cambridge Univ, PhD(civil eng), 57. *Prof Exp:* Design engr, H G Acres & Co Ltd, 49-54; struct engr, J D Lee & Co Ltd, 57-59; prof struct eng, 59-80, head dept, 69-77, PROF CIVIL ENG, ROYAL MIL COL, 77- *Concurrent Pos:* Defence Res Bd Can res grants, 62-64 & 65; chmn task group III, US Column Res Coun; trustee, Rd Safety Res Fund, Eng Inst Can; vis prof, Royal Mil Col 71. *Honors & Awards:* Duggan Medal, Eng Inst Can, 65. *Mem:* Eng Inst Can. *Res:* Ultimate capacity of steel columns; automation and traffic control. *Mailing Add:* Dept Civil Eng Royal Mil Col Kingston ON K7L 2W3 Can

ELLIS, JACK BARRY, b Toronto, Ont, Jan 4, 36; m 61; c 2. ELECTRICAL ENGINEERING. *Educ:* Univ Toronto, BASc, 58; Univ London, MSc & 61; Mich State Univ, PhD(elec eng), 65. *Prof Exp:* Lectr elec eng, Univ Waterloo, 61-62; asst, Mich State Univ, 62-63, asst instr resource develop, 64-65; asst prof elec eng, Univ Waterloo, 65-66, assoc prof, 66-70, dir continuing res prog probs urbanization, 68-70; PROF FAC ENVIRON STUDIES, YORK UNIV, 70- *Concurrent Pos:* Consult, res contract, Dept Conserv, Mich, 65-66; Nat Res Coun Can oper grant, 65-; Dept Hwy, Ont res grant, 65- *Mem:* Inst Elec & Electronics Engrs. *Res:* Nonlinear control systems; optimal control and systems theories; applications of systems theory and control theory to transportation networks, planning, economic and sociological systems. *Mailing Add:* Downsview York Univ 4700 Keele St Toronto ON M3J 1P6 Can

ELLIS, JAMES WATSON, b Uruguaiana, Brazil, Aug 16, 27; US citizen; m 51; c 3. PURE MATHEMATICS. *Educ:* Wofford Col, AB, 48; Tulane Univ, MS, 51, PhD(math), 52. *Prof Exp:* Asst prof math, Fla State Univ, 52-57, assoc prof, 57-58; assoc prof, 58-61, prof & chmn dept, 61-64, dean jr div, 64-88, ASSOC PROVOST, UNIV NEW ORLEANS, 89- *Mem:* Am Math Soc; Math Asn Am. *Res:* Linear topological spaces and topological algebras. *Mailing Add:* 2328 Lark St New Orleans LA 70122

ELLIS, JASON ARUNDEL, b Newell, SDak, Dec 22, 18; m 52; c 2. PHYSICS. *Educ:* SDak Sch Mines & Technol, BS, 42; Univ Iowa, MS, 52, PhD(physics), 62. *Prof Exp:* Engr, Gen Elec Co, 42-45; physicist, US Bur Standards, 47 & Radiation Res Lab, Univ Iowa, 51-53; asst prof physics, North Tex State Univ, 58-63; from asst prof to assoc prof physics, Univ Tex, Arlington, 63-85; vpres res & develop, 85-89, CONSULT, CURTIS AEROSPACE, 89- *Concurrent Pos:* Co-ed, Spacetime. *Mem:* Am Phys Soc; Am Asn Physics Teachers; Int Soc Gen Semantics. *Res:* Theoretical physics; mathematical biophysics. *Mailing Add:* 1714 Sesco Arlington TX 76013-3928

ELLIS, JEFFREY RAYMOND, b London, Eng, Feb 21, 44; US citizen; m 85. MEDICAL MATERIALS & PACKAGING, ELECTRONICS MATERIALS. *Educ:* Columbia Univ, BA, 64; Purdue Univ, PhD(org chem), 69; Xavier Univ, MBA, 77. *Prof Exp:* Res chemist, Verona Chem Corp, 69-70; sr scientist, Olivetti Corp Am, 71-74 & Sun Chem Corp, 74-75; sr scientist & res mgr, Tile Coun Am, Inc, 75-78; consult scientist, Princeton Polymer Labs, Inc, 78-87; PRES, J R ELLIS, INC, TECH & ECON SERV, 83- *Mem:* Am Chem Soc; Soc Plastics Engrs; Chem Mgt & Resources Asn. *Res:* Medical uses of materials; packaging of medical and food products; electronics and optics applications of materials; assessment of technologies; formulation and measurement of properties of polymeric composites; radiation processing of materials; solution of problems in polymer synthesis and processing. *Mailing Add:* 35 Teaberry Lane Newtown PA 18940

ELLIS, JERRY WILLIAM, b Pittsburg, Kans, Aug 22, 37; m 58; c 2. ORGANIC CHEMISTRY, CARBOHYDRATE CHEMISTRY. *Educ:* Kans State Col, BS, 59, MS, 61; Okla State Univ, PhD(org chem), 65. *Prof Exp:* Instr chem, Kans State Col, Pittsburg, 61 & Okla State Univ, 61-62; asst prof org chem, Wis State, Whitewater, 65-66; from asst prof to assoc prof, 66-76, PROF ORG CHEM, EASTERN ILL UNIV, 76- *Concurrent Pos:* Vis prof chem, Johns Hopkins Univ, 79-80. *Mem:* Am Chem Soc. *Res:* organic syntheses; carbohydrate chemistry. *Mailing Add:* Dept Chem Eastern Ill Univ Charleston IL 61920-3099

ELLIS, JOHN EMMETT, b San Pedro, Calif, May 26, 43; m 66. ORGANOMETALLIC CHEMISTRY, INORGANIC CHEMISTRY. *Educ:* Univ Southern Calif, BSc, 66; Mass Inst Technol, PhD(chem), 71. *Prof Exp:* Res fel inorg chem, Mass Inst Technol, 66-71; from asst prof to assoc prof, 71-84, PROF CHEM, UNIV MINN, MINNEAPOLIS, 84- *Concurrent Pos:* Dir, NSF Undergrad Res Partic Prog, 74-82, DuPont Young fac award, 76; Bush sabbatical fel, 90-91. *Mem:* Am Chem Soc; Royal Soc Chem; Sigma Xi; Am Sci Glassblowers Soc. *Res:* Synthesis, characterization and chemical properties of highly reduced organometallic and carbonyl compounds of the transition elements; chemistry of coordinated carbon monoxide and related substances. *Mailing Add:* Dept Chem 207 Pleasant St SE Minneapolis MN 55455

ELLIS, JOHN FLETCHER, b Laurel, Del, Apr 26, 37; m 59; c 2. WEED SCIENCE. *Educ:* Univ Del, BS, 59; Rutgers Univ, MS, 61, PhD(weed sci), 64. *Prof Exp:* Res rep, 66-67, res specialist herbicides, 67-69, group leader, 69-71, prod planner all prod, 71-74, mgr biol res herbicides & plant growth regulators, Agr Div, 74-78, dir, Regist & Toxicol, 78-80, DIR BIOL RES, AGR DIV, CIBA-GEIGY CORP, 80- *Mem:* Am Soc Agron; Weed Sci Soc Am; Sigma Xi; AAAS. *Res:* Evaluation of chemical products for herbicidal plant growth regulator insecticidal and fungicidal properties that might lead to valuable tools for food production. *Mailing Add:* 3605 Mossborough Dr Greensboro NC 27410

ELLIS, JOHN FRANCIS, b Torrington, Conn, July 29, 22; m 46; c 3. BIOLOGY. *Educ:* Amherst Col, BA, 48, MA, 50; Univ Edinburgh, PhD(animal genetics), 55. *Prof Exp:* Teaching asst, Amherst Col, 48-50, from instr to asst prof biol, 53-58, res assoc, 58-59; from asst prof to assoc prof, 59-68, PROF BIOL, HAMILTON COL, 68- *Concurrent Pos:* USPHS trainee, State Univ NY Upstate Med Ctr, 66-67; consult, NY State Dept Ment Health, Marcy State Hosp. *Mem:* Fel AAAS; Genetics Soc Am; Am Inst Biol Sci; Sigma Xi. *Res:* Human cytogenetics; genetics of mental retardation and mental illness; tissue culture and cell hybridization; Drosophila biochemical genetics. *Mailing Add:* 77 College St Clinton NY 13323

ELLIS, JOHN TAYLOR, b Lufkin, Tex, Dec 27, 20; m 42; c 3. PATHOLOGY. *Educ:* Univ Tex, BA, 42; Northwestern Univ, MD, 45; Am Bd Path, dipl, 51. *Prof Exp:* Rotating intern, St Luke's Hosp, 45-46; asst, William Buchanan Blood Ctr, Baylor Hosp, 48; asst path, Col Med, Cornell Univ, 48-59, from instr to assoc prof, 49-62; prof path & chmn dept, Sch Med, Emory Univ, 62-68; PROF PATH & CHMN DEPT, COL MED, CORNELL UNIV, 68-,. *Concurrent Pos:* From asst resident pathologist to asst attend pathologist, New York Hosp, 50-55, assoc attend pathologist, 55-; mem comt blood, health resources adv comt, Off Emergency Planning, Exec Off Pres, 64-71; mem path study sect, NIH, 65-69, chmn, 69-70; mem bd ed, Am J Path, 66-; mem sci adv bd consult, Armed Forces Inst Path, 70-75; chmn histopath comt, Polycythemia Vera Study Group, 72-88; attend pathologist, Mem Hosp Cancer & Allied Dis, 73-; pathologist-in-chief, New York Hosp, 68- *Mem:* Am Soc Exp Path; Soc Exp Biol & Med; Am Asn Path & Bact; Col Am Path; Int Acad Path. *Res:* Nephrosis and experimental proteinuria; muscular dystrophy and experimentally induced diseases of muscle; iron metabolism; polycythemia vera. *Mailing Add:* Dept Path Cornell Univ Med Col New York NY 10021

ELLIS, KEITH OSBORNE, b Albany, NY, Oct 18, 41; m 63; c 2. PHARMACOLOGY, PHYSIOLOGY. *Educ:* Heidelberg Col, BA, 63; Univ Cincinnati, PhD(pharmacol), 69. *Prof Exp:* High sch teacher biol, chem & physics, Ohio Bd Educ, North Baltimore, 63-64; sr res pharmacologist, 70-73, unit leader pharmacol, 74-77, proj mgr, 78-81, SECT CHIEF PHARMACOL, NORWICH-EATON PHARMACEUT, 77- *Concurrent Pos:* Fel neurophysiol, Grass Found, Woods Hole, Mass, 69; fel, Baylor Col Med, 69-70; adj instr, State Univ NY Agr & Tech Col Morrisville, 73-78, Upper Div Col, State Univ NY Utica-Rome, 75-77, Sch Gen Studies, State Univ NY Binghamton, 77-79 & Dept Pharmacol, Upstate Med Ctr, State Univ NY, 79- *Mem:* NY Acad Sci; Soc Exp Biol & Med; Am Soc Pharmacol & Therapeut. *Res:* Skeletal muscle pharmacology; central nervous system pharmacology; cardiovascular pharmacology. *Mailing Add:* Norwich-Eaton Pharmaceut Inc 17 Eaton Ave Norwich NY 13815

ELLIS, KENNETH JOSEPH, b Terre Haute, Ind, July 14, 44; m 71; c 2. MEDICAL PHYSICS, HEALTH PHYSICS. *Educ:* Univ Tex-El Paso, BS, 66; Vanderbilt Univ, PhD(physics), 72. *Prof Exp:* Med assoc physics, Med Dept, Brookhaven Nat Lab, 72-73, res assoc, 73-74, asst scientist, 74-76, assoc scientist, 76-79, scientist, 79-86; ASSOC PROF, CHILDREN'S NUTRIT RES CTR, 86- *Concurrent Pos:* Atomic Energy Comn Health Physics fel, 80. *Mem:* Health Physics Soc. *Res:* Medical physics; in vivo neutron activation; whole body counting; tracer kinetics; compartmental modeling; photon absorptionetry; neutron shielding; health physics. *Mailing Add:* Children's Nutrit Res Ctr Med Towers Bldg 1100 Bates St Houston TX 77030

ELLIS, LEGRANDE CLARK, b Farmington, Utah, June 20, 32; m 54; c 7. REPRODUCTIVE ENDOCRINOLOGY. *Educ:* Utah State Univ, BS, 54, MS, 56; Okla State Univ, PhD(physiol, endocrinol), 61. *Prof Exp:* From instr to asst prof cellular physiol, Okla State Univ, 57-62; fel quant biol, Univ Utah, 62-64; asst prof physiol & endocrinol, 64-66, assoc prof, 66-71, PROF PHYSIOL & BIOCHEM, UTAH STATE UNIV, 71- *Mem:* Am Physiol Soc; Endocrine Soc; Soc Study Reproduction; Am Soc Zoologists; Sigma Xi; Soc Exp Biol Med; Am Soc Andrology. *Res:* Environmental influences on endocrinology and reproduction; male infertility; hormonal control of fur replacement and pigmentation; prostaglandins and pineal gland. *Mailing Add:* Dept Biol UMC-53 Utah State Univ Logan UT 84322-5305

ELLIS, LEONARD CULBERTH, b Portsmouth, Va, Dec 13, 34; m 54; c 3. INORGANIC CHEMISTRY. *Educ:* Col William & Mary, BS, 56; Univ Va, PhD(org chem), 62. *Prof Exp:* Res chemist, 61-76, assoc dir res & develop, 76-82, dir Chem Res, 82-88, DIR RES & DEVELOP, VA CHEM CO, 88. *Res:* Stabilization of dithionites; groundwood pulp bleaching; sodium dithionite process development; alkyl and allyl amine products and processes; amidation products/processes. *Mailing Add:* Hoechst-Celanese Va Chem Co 3340 W Norfolk Rd Portsmouth VA 23703-2433

ELLIS, LESLIE LEE, JR, b Norfolk, Va, Sept 13, 25; m 49; c 4. ZOOLOGY. *Educ:* Tulane Univ, BS, 48, MS, 49; Univ Okla, PhD(zool), 52. *Prof Exp:* Asst zool, Tulane Univ, 48; asst, Univ Okla, 49-50, 51-52, spec instr & asst, Univ Biol Sta & State Biol Surv, 50-51; res aide, USPHS, 52; from asst prof to prof zool & entom, Miss State Univ, 52-68, head dept zool, 62-68; prof biol sci, Univ Cent Fla, 68, chmn dept, 68-69, dir grad studies & res, 69-70, dean grad studies & res, 70-74, actg pres, 78, vpres acad affairs, 79-81, provost & vpres acad affairs, Univ Cent Fla, 81-87, interim dean, Col Health & Prof Studies, 87-90, dir, off res park affairs, 87-90; RETIRED. *Concurrent Pos:* NSF panelist; indust consult; chmn, Orange County Res & Develop Authority. *Mem:* Entom Soc Am; Sigma Xi; AAAS. *Res:* Aquatic biology; medical entomology and parasitology; bee diseases. *Mailing Add:* 250 Nottoway Trail Maitland FL 32751

ELLIS, LYNDA BETTY, b Los Angeles, Calif, Nov 25, 45; m 66. BIOMEDICAL COMPUTING. *Educ:* Univ Southern Calif, BS, 65; Brandeis Univ, PhD(biochem), 71. *Prof Exp:* Res specialist biochem, 71-73, asst prof, 73-80, ASSOC PROF HEALTH COMPUTER SCI, UNIV MINN, 80- *Mem:* Sigma Xi. *Res:* Computer-aided instruction; health information systems; biochemical simulation. *Mailing Add:* Univ Minn 420 SE Delaware St Minneapolis MN 55455

ELLIS, LYNN W, b San Mateo, Calif, Feb 27, 28. TELECOMMUNICATIONS, MANAGEMENT. *Educ:* Cornell Univ, BEE, 48; Stevens Inst Technol, MS, 54; Pace Univ, DPS(telecommun mgt), 79. *Prof Exp:* Dir res, ITT Co, NY, 75-79; vpres, eng, Bristol-Babcock Inc, 80-82; Consult, Int Bus, 82-85; PROF MGT, UNIV NEW HAVEN, 85- *Mem:* Fel AAAS; fel Inst Elec & Electronics Engrs. *Mailing Add:* Univ New Haven 300 Orange Ave West Haven CT 06516

ELLIS, PATRICIA MENCH, b Rochester, NY, Mar 6, 49. GEOLOGY, EARTH SCIENCES. *Educ:* Univ Rochester, BA, 71; Duke Univ, MA, 73. *Prof Exp:* Teaching asst geol, Duke Univ, 71-73; teaching asst, Univ Tex, 73-76, res asst, Bur Econ Geol, 76-78; res geologist, Explor Res Div, 78-81, PROJ GEOLOGIST, CONOCO, INC, 81- *Concurrent Pos:* Geologist, Phillip Petrol Co, 74 & Texaco, Inc, 75. *Mem:* Soc Econ Paleontologists & Mineralogists; Geol Soc Am; Am Asn Petrol Geologists; Can Asn Petrol Geologists. *Res:* Geological research in carbonate petrology, diagenesis, depositional environments and geochemistry; exploration geology, Montana and Idaho overthrust belt. *Mailing Add:* 210 Harris Circle Newark DE 19711

ELLIS, PAUL JOHN, b Northampton, Eng, May 25, 41; m 73; c 1. NUCLEAR PHYSICS. *Educ:* Bristol Univ, BSc, 62; Univ Manchester, PhD(nuclear physics theory), 66. *Prof Exp:* Res assoc nuclear theory, Univ Mich, 66-68 & Rutgers Univ, 68-70; res officer nuclear theory, Nuclear Physics Lab, Oxford Univ, 70-73; from asst prof to assoc prof, 73-82, PROF NUCLEAR THEORY, UNIV MINN, 82- *Mem:* Am Phys Soc. *Res:* Theoretical nuclear physics. *Mailing Add:* Sch Physics & Astron Univ Minn Minneapolis MN 55455

ELLIS, PHILIP PAUL, b Saginaw, Mich, Oct 30, 23. OPHTHALMOLOGY. *Educ:* Baylor Univ, MD, 48; Am Bd Ophthal, dipl, 55. *Prof Exp:* From instr to asst prof ophthal, Col Med, Univ Iowa, 54-58; assoc prof & head dept, Sch Med, Univ Ark, 58-60; assoc prof, 60-67, PROF OPHTHAL, SCH MED, UNIV COLO, DENVER, 67-, HEAD DEPT, 60- *Mem:* AMA; Am Acad Ophthal & Otolaryngol; Asn Res Vision & Ophthal; Am Ophthal Soc. *Res:* Ocular pharmacology and toxicology; therapeutics; immunology. *Mailing Add:* Dept Ophthal Univ Colo Sch Med 4200 E Ninth Ave Denver CO 80262

ELLIS, REX, BIOCHEMISTRY. *Educ:* Am Univ, MS, 67. *Prof Exp:* RES CHEMIST, US DEPT AGR, 58- *Res:* Bioavailabilty of zinc. *Mailing Add:* 11800 Ellington Dr Beltsville MD 20705-1309

ELLIS, RICHARD AKERS, b Brewster, Mass, May 5, 28. BIOLOGICAL STRUCTURE. *Educ:* Univ Mass, AB, 49; Harvard Univ, AM, 51, PhD, 54; Brown Univ, AM, 62. *Prof Exp:* Instr anat, Sch Med, Harvard Univ, 54; res assoc, Bermuda Biol Sta, 56; from instr to assoc prof, 56-67, PROF BIOL, BROWN UNIV, 67- *Mem:* AAAS; Am Asn Anatomists; Histochem Soc; Am Soc Zoologists; Am Soc Cell Biol; fel AAAS. *Res:* Histochemistry and electron microscopy of salt-secreting epithelia and skin; innervation and vascularization of taste buds; histochemistry of the metrial gland in pregnancy and pseudopregnancy. *Mailing Add:* Div Biol & Med Brown Univ Providence RI 02912

ELLIS, RICHARD BASSETT, b Abilene, Tex, May 12, 15; m 40; c 3. PHYSICAL CHEMISTRY. *Educ:* Vanderbilt Univ, BA, 36, MS, 37, PhD(anal & phys chem), 40. *Prof Exp:* Supvr anal div, Res & Develop Dept, Joseph E Seagram & Sons, 40-42; res chemist, Corning Glass Works, 42-46; instr chem, Univ Fla, 46-47; asst prof, Univ Miami, 47-51; from sr chemist to head inorg sect, Southern Res Inst, 51-68; prof physics & head dept, Huntingdon Col, 68-70; prof sci, Troy State Univ, Montgomery, 70-80, head dept sci, 71-76; RETIRED. *Concurrent Pos:* Lectr, Birmingham-Southern Col, 54-66 & Univ Ala, Birmingham, 66-68. *Res:* Analysis distillery products; surface chemistry; electrochemistry; fused salt technology; high-temperature materials. *Mailing Add:* 3115 Partridge Rd Montgomery AL 36111-3023

ELLIS, RICHARD JOHN, b New Castle, Ind, Apr 11, 39; m 62, 86; c 2. PLANT PHYSIOLOGY. *Educ:* Univ Calif, Santa Barbara, BA, 60; Univ Calif, Berkeley, PhD(bot), 67. *Prof Exp:* Asst prof biol, City Col New York, 67-68; from asst prof to assoc prof, 68-81, PROF BIOL, BUCKNELL UNIV, 81- *Concurrent Pos:* NSF fac fel sci, 75. *Mem:* Am Soc Plant Physiol; Japanese Soc Plant Physiologists; Am Soc Photobiol. *Res:* Regulation of chlorophyll synthesis; phototropism in higher plants; photomorphogenesis. *Mailing Add:* Dept Biol Bucknell Univ Lewisburg PA 17837

ELLIS, RICHARD STEVEN, b Brookline, Mass, May 15, 47; m 69; c 2. MATHEMATICS. *Educ:* Harvard Univ, BA, 69; NY Univ, MS, 71, PhD(math), 72. *Prof Exp:* Mem tech staff appl probability, Bell Tel Labs, 69-72; asst prof math, Northwestern Univ, 72-75; from asst prof to assoc prof, 75-81, PROF MATH, UNIV MASS, AMHERST, 81- *Concurrent Pos:* Mem, Comt Concerned Scientists; Alfred P Sloan res fel, 77-81; Lady Davis fel, 82. *Mem:* Am Math Soc; Int Asn Math Physics. *Res:* Applications of probability theory to statistical mechanics; correlation inequalities and limit theorems for Ising models of ferromagnetism; large deviations and asymptotic problems in probability theory and statistical mechanics. *Mailing Add:* Dept Math & Statist Univ Mass Amherst MA 01003

ELLIS, ROBERT ANDERSON, JR, physics; deceased, see previous edition for last biography

ELLIS, ROBERT L, b Richmond, Ind, July 26, 38; m 62; c 1. MATHEMATICS. *Educ:* Miami Univ, AB, 60; Duke Univ, PhD(math), 66. *Prof Exp:* From instr to asst prof math, Duke Univ, 64-66; asst prof, 66-71, ASSOC PROF MATH, UNIV MD, COLLEGE PARK, 71- *Concurrent Pos:* NSF grant, 67-69, Air Force grant, 87- *Honors & Awards:* Sr US Scientist Award, WGer Govt-Alexander von Humboldt Found, 72. *Mem:* Am Math Soc; Math Asn Am. *Res:* Functional analysis, operator theory and topology. *Mailing Add:* Dept Math Univ Md Col Arts & Sci College Park MD 20742

ELLIS, ROBERT MALCOLM, b Meaford, Ont, Mar 16, 36; m 89; c 2. GEOPHYSICS. *Educ:* Univ Western Ont, BA, 57, MSc, 58; Univ Alta, PhD(physics), 64. *Prof Exp:* Instr math, Univ Western Ont, 59-60; from asst prof to assoc prof, 64-74, PROF GEOPHYS, UNIV BC, 74-, DEPT HEAD, 89- *Concurrent Pos:* Prin fel, Seismic Proj, Arctic Inst NAm, 66; vis scientist, Univ Alta, 71-72; vis prof, Ahmadu Bello Univ, Nigeria, 74-76; vis res scholar, Macquarie Univ, 81; vis fel, Australian Nat Univ, 82-84. *Mem:* Am Geophys Union; Geol Asn Can; Can Geophys Union. *Res:* Seismology: refraction and earthquake studies. *Mailing Add:* Dept Geophys & Astron Univ BC Vancouver BC V6T 1W5 Can

ELLIS, ROBERT WILLIAM, b Wendell, Idaho, Aug 26, 40; m 63; c 2. BIOCHEMISTRY, TOXICOLOGY. *Educ:* Col Idaho, BS, 63; Ore State Univ, MS, 65, PhD(biochem), 70. *Prof Exp:* Fel biochem, Kans State Univ Med Ctr, 69-70; instr chem, Lorado Jr Col, 70-71; PROF BIOCHEM, BOISE STATE UNIV, 71- *Concurrent Pos:* NIH grant, 69-70; consult, Valley Trout Co, 78-79. *Res:* Toxicology of heavy metals; water quality in aquaculture; geothermal aquaculture; aquaculture nutrition. *Mailing Add:* Dept Chem Boise State Univ Boise ID 83725

ELLIS, ROBERT WILLIAM, b Richmond, Va, Oct 16, 39; m 60; c 4. ENGINEERING MECHANICS. *Educ:* Va Polytech Inst, BS, 62, MS, 63, PhD(eng), 66. *Prof Exp:* From asst prof to assoc prof eng, Univ SFla, 65-69, asst dean acad affairs grad studies, 71-72, asst vpres acad affairs & prof eng, 71-72; dean, Sch Technol, Fla Int Univ, 72-78, dean, Sch Bus & Orgn Sci, 72-74, provost, North Miami Campus, 77; exec vpres & chief acad officer, Detroit Inst Technol, 78-80, pres, 80-81; mech engr, US Army Tank Automotive Command, Warren, Mich, 81-84; dean eng, 84-89, PROVOST, LAWRENCE TECHNOL UNIV, 89- *Concurrent Pos:* Metall engr, Polysci Div, Litton Indust, Inc, 62-63; instr, Va Polytech Inst, 64-65; lectr mechanical eng, Lawrence Inst Technol, 81-84. *Mem:* Soc Exp Stress Analysis; Nat Soc Prof Engrs; Am Soc Metals; Am Soc Eng Educ; Soc Automotive Engrs. *Res:* Engineering and composite materials; materials. *Mailing Add:* Provost Lawrence Technol Univ 21000 W Ten Mile Rd Southfield MI 48075

ELLIS, ROSEMARY, nursing; deceased, see previous edition for last biography

ELLIS, ROSS COURTLAND, b McKinney, Tex, Feb 23, 29; m 51; c 3. GEOLOGY. *Educ:* Occidental Col, BA, 53; Univ Wash, Seattle, PhD, 59. *Prof Exp:* Asst prof geol, Univ Wash, 57-62; from asst prof to assoc prof, 62-74, PROF GEOL, WESTERN WASH STATE COL, 74- *Res:* Structural geology; geomorphology; petrology. *Mailing Add:* Dept Geol Western Wash State Univ 516 High St Bellingham WA 98225

ELLIS, SAMUEL BENJAMIN, b Reardan, Wash, 04; m 29; c 2. ELECTROANALYTICAL CHEMISTRY. *Educ:* Univ Wash, BS, 26; Lafayette Col, MS, 27; Columbia Univ, PhD(phys inorg chem), 33. *Prof Exp:* Asst, Columbia Univ, 27-30, instr, 30-32; res chemist, US Rubber Prod, Inc, 33-37, Hellige, Inc, 37-49 & New York Lab Supply Co, 49-66; prof chem, Dutchess Community Col, 66-71; mem staff res & develop, Delta Sci Corp, 71-75; RETIRED. *Mem:* Am Chem Soc; Instrument Soc Am. *Res:* Role of electrostatics in certain basic concepts of physics and chemistry; rheology; electrical measurements; determination of hydrogen-ion concentration; general colorimetric analysis; laboratory equipment; electrostatics; automatic analyzers, water and sewage; the experimental development of a space field-force hypothesis explaining electrostatic forces solely as attractions, negating any possibility of repulsion. *Mailing Add:* Box 775 Goshen NH 03752

ELLIS, STANLEY, b California, Pa, Sept 2, 23; m 43; c 2. ENDOCRINOLOGY, BIOCHEMISTRY. *Educ:* Wayne State Univ, BS, 47, MS, 49, PhD(biochem), 51. *Prof Exp:* Res biochemist hormones, Inst Exp Biol, Univ Calif, Berkeley, 51-55; asst prof biochem, Emory Univ, 56-60; sr protein chemist, Cutter Labs, Berkeley, 60-62; br chief endocrine biochem, 62-76, RES SCIENTIST BIOCHEM, AMES RES CTR, NASA, 77- *Concurrent Pos:* Career Develop Award, NIH, 59-60, mem, Endocrinol Study Sect, 69-73; mem, med adv bd, Nat Pituitary Agency, 77-81. *Honors & Awards:* Except Sci Achievement Award, NASA, 71. *Mem:* AAAS; Am Soc Biol Chemists; Endocrine Soc. *Res:* Anterior pituitary hormones; intracellular peptidases; muscle biochemistry; space biology. *Mailing Add:* 5770 Arboretum Dr Los Altos CA 94022

ELLIS, STEPHEN DEAN, b Detroit, Mich, June 7, 43; m 66, 81; c 1. ELEMENTARY PARTICLE PHYSICS. *Educ:* Univ Mich, Ann Arbor, BSE(eng physics) & BSE(eng math), 65; Calif Inst Technol, PhD(theoret physics), 71. *Prof Exp:* Res physicist, Fermi Nat Accelerator Lab, 71-73; vis scientist, Europ Orgn Nuclear Res, Geneva, Switz, 73-74; res physicist, Fermi Nat Accelerator Lab, 74-75; res asst prof high energy physics, 75-80, assoc prof, 80-85, PROF PHYSICS, UNIV WASH, 85- *Concurrent Pos:* Vis lectr, Dept Appl Math & Theoret Physics, Cambridge Univ, 75 & 82; vis assoc, Europ Orgn Nuclear Res, Geneva, Switz, 84-85. *Mem:* AAAS; fel Am Phys Soc; Sigma Xi. *Res:* Studies of the theoretical problems of gauge field theories of the strong interactions and phenomenological application of these ideas to hadronic processes, especially large transverse momentum reactions, and lepton induced reactions, especially ete-annihilation. *Mailing Add:* Dept Physics FM-15 Univ Wash Seattle WA 98105

ELLIS, SYDNEY, b Boston, Mass, Apr 20, 17; m 42; c 2. PHARMACOLOGY, BIOCHEMISTRY. *Educ:* Boston Univ, SB, 38, AM, 39, PhD(med sci), 41. *Prof Exp:* Asst biochem, Boston Univ, 39-41; res fel pharmacol, Harvard Med Sch, 41-42, asst, 42-44; asst prof, Sch Med, Duke Univ, 46-49; assoc prof, Sch Med, Temple Univ, 49-57; prof pharmacol & toxicol & chmn dept, Woman's Med Col Pa, 57-67; prof pharmacol & toxicol & dept chmn, Univ Tex Med Br, Galveston, 67-80; DEP DIR, DIV DRUG BIOL, FOOD & DRUG ADMIN, WASHINGTON, DC, 80- *Concurrent Pos:* Toxicologist, Nat Defence Res Comt, NY Univ, 42; consult, Smith Kline & French Labs, 57-79, NIH Study Sects, Pharmacol & Exp Therapeut, 60-64 & Med Chem B, 64-68, Nat Bd Med Exam, 64-68, Astra Pharmaceut Prod, 73-78; vis investr, Univ Paris, 72; vis scientist, Intergovt Personnel Act, Food & Drug Admin, Washington, DC, 79-80. *Honors & Awards:* Lindback Found Award, 64. *Mem:* Fel AAAS; Am Soc Pharmacol; Am Chem Soc; Soc Exp Biol & Med; NY Acad Sci. *Res:* Enzymes; effects of enzyme inhibitors on tissues; chemistry and biochemistry of drug decomposition and metabolism; toxins and toxicological mechanisms; biochemical mechanisms of action of autonomic agents; autonomic pharmacology; catecholamines on metabolism; marine toxins. *Mailing Add:* HFD 470 Div Res & Testing Food & Drug Admin 200 C St SW Washington DC 20204

ELLIS, THOMAS STEPHEN, b Manchester, Eng, Feb 29, 52; m 74; c 2. THERMAL ANALYSIS OF POLYMERS. *Educ:* Univ Manchester Inst Sci & Technol, Eng, BSc, 73, PhD(polymer sci), 78. *Prof Exp:* Lectr polymer sci, Univ Strathclyde, Scotland, 79-80; postdoctoral res assoc, Univ Mass, 80-85; STAFF RES SCIENTIST, POLYMERS DEPT, GEN MOTORS RES LABS, 85- *Mem:* Am Phys Soc; Am Chem Soc; NAm Thermal Analysis Soc. *Res:* Thermodynamics and chemical structure as an influence of phase behavior of multi-component polymer systems; structure-property relationships in multi-component polymer systems. *Mailing Add:* Polymers Dept Gen Motors Res Labs Warren MI 48090

ELLIS, WADE, b Chandler, Okla, June 9, 09; m 32; c 2. MATHEMATICS. *Educ:* Wilberforce Col, BS, 28; Univ NMex, MS, 38; Univ Mich, PhD(math), 44. *Hon Degrees:* DHumL, Marygrove Col, 80. *Prof Exp:* Instr math, Fisk Univ, 38-40 & Univ Mich, 43-45; staff mem radiation lab, Mass Inst Technol, 45-46; physicist, US Air Force Lab, Cambridge, 46-48; from asst prof to prof math, Oberlin Col, 48-67; prof & assoc dean grad sch, 67-77, EMER PROF MATH, UNIV MICH, 77- *Concurrent Pos:* Lectr math, Boston Univ, 47-48; fac fel, India & France, 54-55; mem, Entebbe Workshop, Africa, 62; vis writing panels, Sch Math Study Group, 63 & 65, bd adv, 64-67; vis prof, Nat Univ Eng, Peru, 64; mem bd trustees, Marygrove Col, 69-71, chmn acad affairs comt, 70-71; comt int educ, Asn Grad Schs, 70-72; mem exec bd, Comn Insts Higher Educ, NCent Asn Cols & Sec Schs, 70-75; mem bd trustees, Inst Man & Sci, 70-; chmn deleg, US-Japan Bi-Nat Conf Math Educ, Tokyo, 71; vpres, Col Placement Servs, Inc, 72-81; vchancellor, Acad Affairs, Univ Md, Eastern Shore, 78-79; interim pres, Marygrove Col, Detroit, 79-80; consult mathemetician microcomputers, 80- *Honors & Awards:* Comdr, Orden de las Palmas Magisteriales del Peru, 64. *Mem:* Math Asn Am; Am Math Soc. *Res:* Computer science; analytic mechanics; curriculum development; finite fields. *Mailing Add:* Dept Math West Valley Col 14000 Fruitvale Ave Saratoga CA 95070-5698

ELLIS, WALTON P, b Mammoth Spring, Ark, Aug 25, 31; m 56; c 2. SURFACE CHEMISTRY. *Educ:* Univ Calif, Berkeley, BS, 53; Univ Chicago, PhD(chem), 57. *Prof Exp:* STAFF MEM CHEM RES, LOS ALAMOS NAT LAB, 57- *Mem:* AAAS; Am Chem Soc; Am Vacuum Soc; Mat Res Soc. *Res:* Gas-solid reaction kinetics; electronic, chemical and physical properties of surfaces and thin films; low energy electron diffraction; Auger, loss and photoelectron spectra; synchrotron radiation. *Mailing Add:* Los Alamos Nat Lab CLS-2 MS-G738 Los Alamos NM 87545

ELLIS, WILLIAM C, b Clay, La, Apr 23, 31; m 55; c 1. ANIMAL NUTRITION. *Educ:* La Polytech Inst, BS, 53; Univ Mo, MS, 55, PhD, 59. *Prof Exp:* Asst, Univ Mo, 53-55, from instr to asst prof animal husb, 55-61; from asst prof to assoc prof, 61-70, PROF ANIMAL SCI, TEX A&M UNIV, 70- *Concurrent Pos:* N Atlantic fel, 59-60. *Mem:* Fed Am Socs Exp Biol; Am Soc Animal Sci. *Res:* Ruminant physiology; nutrition and metabolism; protein and energy metabolism by rumen microorganisms and the ruminants tissues; forage utilization by ruminants. *Mailing Add:* Dept Animal Sci Tex A&M Univ College Station TX 77843

ELLIS, WILLIAM HAYNES, b Cedar Hill, Tenn, Dec 4, 31; m 52; c 1. SYSTEMATIC BOTANY, PLANT ECOLOGY. *Educ:* Austin Peay State Col, BS, 53, MA, 56; Univ Tenn, PhD(bot), 63. *Prof Exp:* Teacher high sch, Tenn, 55-56; from instr to asst prof biol, Austin Peay State Col, 56-60; investr plant taxon, Oak Ridge Inst Nuclear Studies, 61 & Highlands Biol Sta, 62; instr bot, Univ Tenn, 62-63, vis asst prof, 62; assoc prof biol, Austin Peay State Univ, 63-68, dir grad studies, 66-67, from assoc dean to dean fac, 67-68, vpres acad affairs, 68-72, dir instnl res, 72-77, dean grad sch, 77-86, PROF BIOL, AUSTIN PEAY STATE UNIV, 68-, DEAN, COL GRAD & PROF PROGS, 86- *Concurrent Pos:* Vis scientist, Tenn Acad Sci, 64-65. *Mem:* AAAS; Am Soc Plant Taxon; Bot Soc Am; Nat Asn Biol Teachers; Sigma Xi. *Res:* Systematic revision of the genus Acer, section rubra, with emphasis on cytogenetics; pigment studies by chromatography, ecological considerations and morphological study; ecology of the woody flora of Kentucky and Tennessee. *Mailing Add:* 323 Fairway Dr Clarksville TN 37043

ELLIS, WILLIAM HOBERT, b Albany, Ga, Dec 28, 28; m 50; c 4. NUCLEAR CHEMISTRY & ENGINEERING. *Educ:* NGa Col, BS, 57; State Univ, PhD(nuclear & inorg chem), 63. *Prof Exp:* Asst prof, 62-66, ASSOC PROF NUCLEAR ENG & CHEM, UNIV FLA, 66- *Mem:* AAAS; Am Nuclear Soc; Am Chem Soc. *Res:* Nuclear and radio chemistry; nuclear instrumentation and spectrometry; activation analysis and direct energy conversion; inorganic chemistry. *Mailing Add:* Dept Nuclear Eng Sci Univ Fla Gainesville FL 32611

ELLIS, WILLIAM RUFUS, b Greenville, SC, Jan 22, 40; m 66; c 2. MAGNETIC FUSION RESEARCH, HIGH POWER MICROWAVE RESEARCH. *Educ:* Clemson Univ, BS, 62; Prince Univ, MS, 64 & PhD(aerospace & mech sci), 67. *Prof Exp:* Vis scientist, Culham Lab, Abingdom, Eng, 67-70; res staff mem, Los Alamos Nat Lab, 70-74, assoc group leader, 74-76; chief, Open Systs Br, Energy Res & Develop Admin, 76-79; dir, Mirror Syst Div, Dept Energy, 80-83; ASSOC DIR RES, US NAVAL RES LAB, 83- *Concurrent Pos:* Fusion power coord comt, Energy Res & Develop Admin, US Dept Energy, 78-81; steering comt, 3rd Nat Conf High Power Microwaves, 65-86 & 4th Nat Conf High Power Microwaves, 87-88; res adv comt, Naval Res Lab, 83- *Mem:* Am Phys Soc; Am Nuclear Soc; AAAS; Am Geophys Union. *Res:* Plasma physics; magnetic fusion technology; high power microwave and radio frequency weapons research; author of over 100 technical publications. *Mailing Add:* Code 400 Naval Res Lab 4555 Overlook Ave SW Washington DC 20375-5000

ELLIS-MENAGHAN, JOHN J, b Mass. SILICON DEVICE PHYSICS, VERY-LARGE SCALE INTEGRATION. *Educ:* Univ Notre Dame, BS, 81; Univ Vt, MS, 88. *Prof Exp:* Staff engr, 81-88, RESIDENT PROF, MICROELECTRONICS CTR NC, IBM, 89- *Concurrent Pos:* Doctoral res, NC State Univ, 89- *Mem:* Inst Elec & Electronics Engrs; Am Phys Soc. *Res:* Silicon and silicon derivative devices; sub quarter micron field effect transistors and advanced numerical analysis of such using Monte-Carlo models. *Mailing Add:* Microelectronics Ctr NC Rm 309 PO Box 12889 Research Triangle Park NC 27709

ELLISON, ALFRED HARRIS, b Quincy, Mass, Dec 23, 23; m 51; c 5. SURFACE CHEMISTRY, AIR POLLUTION. *Educ:* Boston Col, BS, 50; Tufts Univ, MS, 51; Georgetown Univ, PhD(surface chem), 56. *Prof Exp:* Chemist, US Naval Res Lab, 51-56; res chemist, Texaco Res Ctr, 56-65; res chemist, Harris Res Labs, 65; res chemist, Gillette Res Inst, Inc, 65-69; dep dir, Environ Protection Agency, 69-79, actg dir, 79-82, dir, Environ Sci Res Lab, 82-84, dir Atmospheric Sci Res Lab, Off Res & Develop, 84-88; RETIRED. *Mem:* Am Chem Soc. *Res:* Management of chemistry and physics research and development programs on atomspheric processes and on methods and instrumentation for measuring air pollutants in the air and in the emissions from sources. *Mailing Add:* 4708 Stiller St Raleigh NC 27609

ELLISON, BART T, b San Diego, Calif, Apr 21, 42; m 65; c 4. CHEMICAL ENGINEERING, CORROSION. *Educ:* Univ Calif, Berkeley, BS, 64, MS, 66, PhD(mech eng), 69. *Prof Exp:* engr, 69-70, res supvr, 80-82, petrol eng mgt, 82-86, RES MGR, SHELL DEVELOP CO, 86- *Mem:* AAAS; Soc Petrol Engrs; Sigma Xi; Nat Asn Corrosion Engrs. *Res:* Corrosion. *Mailing Add:* 3111 Newcastle Houston TX 77027

ELLISON, FRANK OSCAR, b Omaha, Nebr, June 18, 26; m 59; c 2. QUANTUM CHEMISTRY. *Educ:* Creighton Univ, BS, 49; Iowa State Univ, PhD(phys chem), 53. *Prof Exp:* Res asst spectros & theoret chem, Inst Atomic Res, Iowa State Univ, 50-53; from instr to asst prof phys chem, Carnegie Inst Technol, 53-65; from assoc prof to prof, 68-88, EMER PROF PHYS CHEM, UNIV PITTSBURGH, 88- *Mem:* AAAS; Am Phys Soc; Am Chem Soc; Am Inst Chemists; NY Acad Sci; Int Union Pure Appl Chem. *Res:* Theory of molecular spectra and electronic structure of molecules; chemical physics. *Mailing Add:* Dept Chem Univ Pittsburgh Pittsburgh PA 15260

ELLISON, GAYFREE BARNEY, b Brownwood, Tex, Feb 2, 43. CHEMISTRY, ORGANIC CHEMISTRY. *Educ:* Trinity Col, Conn, BS, 65; Yale Univ, PhD(chem), 74. *Prof Exp:* Appointment chem physics, Joint Inst Lab Astrophys, 75-77; ASST PROF CHEM, UNIV COLO, 77- *Mem:* Am Chem Soc; Am Phys Soc; Am Soc Mass Spectrometry. *Res:* Chemical physics of organic molecules. *Mailing Add:* Chem Dept Campus Box 215 Univ Colo Boulder CO 80309-0215

ELLISON, JOHN VOGELSANGER, b Cape Girardeau, Mo, Aug 7, 19; m 49; c 4. ELECTRONICS, UNDERWATER ACOUSTICS. *Educ:* Southeast Mo State Col, AB, 39. *Prof Exp:* Instr high sch, 39-41; instr physics, Ill Inst Technol, 41; instr & dean physics & electronics, Am TV Labs, 41-43; staff mem underwater sound, Div War Res, Columbia Univ, 43-45; sect head appl electronics, Sound Div, Naval Res Lab, 45-59; assoc scientist, Res Div, McDonnell Aircraft Co, McDonnell Douglas Corp, 59-62, proj electronics engr, Electronics Systs Eng Dept, 62-65, mgr, Reconnaissance Lab, 65-73, prin staff engr, electronic systs technol, 73-81; CONSULT ACOUST DESIGN, 81- *Mem:* Acoust Soc Am. *Res:* Propagation and scattering of electromagnetic waves; remote sensing; electro-optics; underwater sound; communications. *Mailing Add:* Two Douglass Lane St Louis MO 63122-4406

ELLISON, LOIS TAYLOR, b Ft Valley, Ga, Oct 28, 23; m 45; c 5. MEDICAL & HEALTH SCIENCES. *Educ:* Univ Ga, BS, 43; Med Col Ga, MD, 50. *Prof Exp:* Asst res prof physiol, 51-65, asst res prof surg, 60-65, assoc res prof physiol & surg, 65-68, assoc prof, 68-71, assoc dean curric, 74-75, PROF MED & SURG, MED COL GA, 71-, DIR, CARDIOPULMONARY LAB, 56-, PROVOST, 75- *Concurrent Pos:* NIH res career develop award, 63-68. *Mem:* Am Physiol Soc; fel Col Chest Physicians; Am Thoracic Soc; Am Heart Asn; Asn Am Med Col; Sigma Xi. *Res:* Pulmonary disease; preoperative and postoperative pulmonary function; heart and lung transplantation; lung surfactant; oxygen transport. *Mailing Add:* Med Col Ga 1120 15th St BA-A-202 Augusta GA 30912

ELLISON, MARLON L, b Woodbine, Iowa, Dec 18, 16; m 49. BOTANY. *Educ:* Iowa State Univ, BS, 40; Trinity Univ, MS, 61; Univ Kans, PhD(bot), 64. *Prof Exp:* Instr bot, Stephen F Austin State Col, 63-64; from asst prof to prof biol, Univ Tampa, 64-87; RETIRED. *Res:* Marine algae of Tampa Bay. *Mailing Add:* 5125 W San Jose St Tampa FL 33629

ELLISON, ROBERT G, b Millen, Ga, Dec 4, 16; m 45; c 5. CARDIOVASCULAR SURGERY, THORACIC SURGERY. *Educ:* Vanderbilt Univ, AB, 39; Med Col Ga, MD, 43; Am Bd Surg, dipl; Am Bd Thoracic Surg, dipl. *Prof Exp:* From instr to assoc prof thoracic surg, 47-59, res assoc physiol & asst res prof, 53, PROF SURG, MED COL GA, 59-, CHIEF, DIV THORACIC SURG, 55- *Concurrent Pos:* Mem surg study sect, NIH, 69-73 & Am Bd Thoracic Surg, 71-; consult, Crippled Children Serv, Atlanta, Vet Admin Hosp; chmn, Am Bd Thoracic Surg, 79-81. *Mem:* Am Surg Asn; Soc Univ Surgeons; fel Am Col Chest Physicians; Am Physiol Soc; Am Col Surgeons; Sigma Xi. *Mailing Add:* 606 Wellesley Dr Augusta GA 30909

ELLISON, ROBERT HARDY, b Temple, Tex, June 22, 50; m 72; c 3. PROCESS CHEMISTRY. *Educ:* Univ Tex, BS, 72, Northwestern Univ, PhD(chem), 76. *Prof Exp:* Fel org chem, Syntex Res, 76-77; staff chemist org chem, Gen Elec Res & Develop Ctr, 77-80; RES CHEMIST, SHELL DEVELOP CO, 80- *Mem:* Am Chem Soc. *Res:* Synthesis of natural products and compounds relating to new monomer-polymer systems; process chemistry. *Mailing Add:* PO Box 1380 Houston TX 77251

ELLISON, ROBERT L, b Williamsport, Pa, Jan 14, 30; m; c 1. GEOLOGY, MICROPALEONTOLOGY. *Educ:* Cornell Univ, AB, 52; Pa State Univ, PhD(geol), 61. *Prof Exp:* Lab asst, Pa State Univ, 55-58, instr geol, 58-59; actg asst prof, 59-61, asst prof, 61-67, chmn dept environ sci, 69-70 & 71-72, assoc prof geol, 67-80, ASSOC PROF ENVIRON SCI, UNIV VA, 80- *Concurrent Pos:* Ed, J Foraminiferal Res, 76-80; bd dirs, Cushman Found Foraminiferal Res, 80- *Mem:* Geol Soc Am; Nat Asn Geol Teachers. *Res:* Distribution and ecology of Foraminifera; structure of benthic marine communities; estaurine geology and ecology; ecology and paleoecology of testate amoebae. *Mailing Add:* Dept Environ Sci Brooks Hall Univ Va Charlottesville VA 22903

ELLISON, ROSE RUTH, b New York, NY, June 5, 23; m 46; c 2. ONCOLOGY. *Educ:* Columbia Univ, BA, 43, MD, 48. *Prof Exp:* From intern to resident med, Maimonides Hosp, Brooklyn, NY, 48-50; asst, State Univ NY Downstate Med Ctr, 50-51; res fel hemat, Sloan-Kettering Inst, New York, 51-54, spec fel, asst & assoc mem, 51-62; assoc chief med A, Roswell Park Mem Inst, 62-72; from asst res prof to assoc res prof med, State Univ NY Buffalo, 64-72, from assoc prof to prof, 72-78, res prof pharmacol, Roswell Park Div, Grad Sch, 70-78; Am Cancer Soc Enid A Haupt prof clin oncol, 79-84, prof med, 78-89, EMER PROF MED, COL PHYSICIANS & SURGEONS, COLUMBIA UNIV, 89- *Concurrent Pos:* Spec fel med, Mem Hosp, New York, 51-53; mem, Pharmacol & Exp Therapeut B Study Sect, NIH, 66-70, exec officer, Acute Leukemia Group B & chmn, Acute Leukemia Comt, 66-76, chmn, Chemother Comt, 77-79; consult, Dept Med A & Dept Exp Therapeut, Roswell Park Mem Inst, 72-79; mem, Cancer Clin Invest Rev Comt, Nat Cancer Inst, 72-76. *Mem:* Am Soc Hemat; Am Fedn Clin Oncol Socs; Am Asn Cancer Res; Am Soc Clin Oncol (secy-treas, 70-73, pres elect, 73-74, pres, 74-75); Leukemia Soc Am (vpres med & sci affairs, 75-79). *Res:* Cancer chemotherapy and natural history of neoplastic diseases. *Mailing Add:* 231 N Woodland St Englewood NJ 07631

ELLISON, SAMUEL PORTER, JR, b Kansas City, Mo, July 1, 14; m 40; c 3. GEOLOGY. *Educ:* Univ Kansas City, AB, 36; Univ Mo, AM, 38, PhD(geol), 40. *Prof Exp:* From instr to asst prof geol, Mo Sch Mines, 39-44; from geologist to dist geologist, Stanolind Oil & Gas Co, 44-48; prof geol sci, 48-71, chmn, Dept Geol, 52-56, actg dean, Col Arts & Sci, 70-71, dean, Col Natural Sci, 71-73, Deussen prof energy resources, 73-79, EMER DEUSSEN PROF ENERGY RESOURCES IN GEOL SCI, UNIV TEX, AUSTIN, 79-; CONSULT, 79- *Concurrent Pos:* Asst geologist, US Geol Surv, 42-44;

consult, Humble Oil & Refining Co, 58-71, Repub Gypsum Co, 78-81, Alpine Resources Ltd, 80-81, Ashton Resources Co, 81, Dresser Industs, 81, Univ Tex Syst, 84-85 & Flowers Drilling Co, 85; Fulbright sr res fel, Ger, 70. *Honors & Awards:* Conodont Res Medal Award, C H Pander Soc, 77. *Mem:* Am Inst Prof Geologists; Soc Petrol Engrs; hon mem Am Asn Petrol Geologists (vpres, 72-73); hon mem Soc Econ Paleont & Mineral (secy-treas, 53-58, pres, 59-60); Nat Asn Geol Teachers (vpres, 63-64, pres, 64-65); fel Geol Soc Am; Paleont Soc. *Res:* Micropaleontology; stratigraphy; petroleum geology; sedimentation; structural geology. *Mailing Add:* Dept Geol Sci Univ Tex Austin TX 78712

ELLISON, SOLON ARTHUR, b New York, NY, July 13, 22; m 46; c 2. MICROBIOLOGY. *Educ:* City Col New York, BS, 42; Columbia Univ, DDS, 46, PhD(microbiol), 58. *Prof Exp:* Instr microbiol, Col Physicians & Surgeons, Columbia Univ, 51-52, assoc, 52-58, asst prof, 59-62; assoc prof, Sch Dent & Oral Surg, State Univ NY Buffalo, 62-64, prof oral biol & chmn dept, 64-78; PROF MICROBIOL, COL PHYSICIANS & SURGEONS, COLUMBIA UNIV, 78-, PROF DENT, 80- *Mem:* Am Soc Microbiol; Am Asn Immunol; Int Asn Dent Res. *Res:* Dentistry; immunology; salivary physiology. *Mailing Add:* 231 N Woodland St Englewood NJ 07632

ELLISON, THEODORE, b Milwaukee, Wis, July 15, 30; m 53; c 3. TOXICOLOGY. *Educ:* Univ Wis, BS, 52, MS, 56, PhD(biochem, pharmacol), 59; Iona Col, MBA, 75. *Prof Exp:* Asst vet sci, Univ Wis, 52-59, sr res biochemist, Smith Kline & French Labs, 59-65; sr biochemist & head drug metab group, Riker Labs, 65, leader pharmacokinetics group, 65-69, sect head pharmacokinetics, 69-71; head bioavailability sect, Vick Divs Res, 71-74; biomed specialist, Gen Foods Corp, 74-76; sr toxicologist, Mobil Oil Corp, 76-79; PRES, T ELLISON ASSOC, INC, 79- *Concurrent Pos:* Adj assoc prof pharmacol, NY Med Col, 75-84 & adj prof toxicol, Drexel Univ, 81-84. *Mem:* Am Chem Soc; NY Acad Sci; Am Pharmaceut Asn; Am Soc Pharmacol & Exp Therapeut; Soc Toxicol. *Res:* Drug metabolism; radioisotopes; pharmacokinetics; biopharmaceutics; bioanalytical research; biochemical pharmacology; bioavailability; toxicology of industrial chemicals; risk assessment. *Mailing Add:* 1216 Yardley Rd Yardley PA 19067-7402

ELLISON, WILLIAM THEODORE, b Wilmington, NC, Nov 30, 41; m 68, 87; c 2. UNDERWATER ACOUSTICS, NAVAL ARCHITECTURE. *Educ:* US Naval Acad, BS, 63; Mass Inst Technol, MS & NavEng, 68, PhD(acoust), 70. *Prof Exp:* Dep proj mgr & tech dir, AN/SQS-26 Sonar Proj, US Navy, 70-72, exec asst to comdr, Naval Sea Systs Command, 72-73, asst qual assurance officer to supvr of shipbldg, Groton, Conn, 73-74; scientist, 74-76, VPRES & SR SCIENTIST, CAMBRIDGE ACOUST ASSOCS INC, 76-; PRES, MARINE ACOUST, 88- *Mem:* Acoust Soc Am; fel Explorers Club. *Res:* Underwater acoustics, particularly detection and localization of objects; bioacoustics, spectral analysis of marine mammal vocalizations; arctic underwater acoustics and ice dynamic modeling. *Mailing Add:* Marine Acoust Inc 15 Clay St Newport RI 02840

ELLISTON, JOHN E, b Palmerton, Pa, Feb 10, 44; m 68; c 1. PLANT PATHOLOGY. *Educ:* Muhlenberg Col, BS, 67; Purdue Univ, PhD(plant pathol), 75. *Prof Exp:* Anal chemist, Shell Chem Co, 67-69; asst, Purdue Univ, 69-75; asst plant pathologist, Conn Agr Exp Sta, 75-; RETIRED. *Concurrent Pos:* fel, Univ Ky, 75. *Mem:* Am Pathol Soc; Mycol Soc Am. *Res:* Hypovirulence agents for the biological control of chestnut blight. *Mailing Add:* Chestnut Mountain HCR 74 Box 57B Hinton WV 25951

ELLMAN, GEORGE LEON, b Chicago, Ill, Dec 27, 23; m 48; c 1. BIOCHEMISTRY. *Educ:* Univ Ill, BS, 48; State Col Wash, MS, 49; Calif Inst Technol, PhD(chem), 52. *Prof Exp:* Res biochemist, Dow Chem Co, 52-59; chief res biochemist, Langley-Porter Inst, San Francisco, 59-74; ASSOC PROF BIOCHEM, DEPT PSYCHIAT, UNIV CALIF, SAN FRANCISCO, 74-; MEM STAFF, BRAIN BEHAV RES CTR, SONOMA DEVELOP CTR, ELDRIDGE, 80- *Concurrent Pos:* Res assoc pharmacol & psychiat & lectr biochem, Med Ctr, Univ Calif, 59. *Mem:* Soc Pharmacol & Exp Therapeut. *Res:* Drug action; cell growth; methodology. *Mailing Add:* Brain Behav Res Ctr Sonoma Develop Ctr Eldridge CA 95431

ELLNER, PAUL DANIEL, b New York, NY, May 2, 25; m 48, 65; c 3. CLINICAL MICROBIOLOGY. *Educ:* Long Island Univ, BS, 49; Univ Southern Calif, MS, 52; Univ Md, PhD(microbiol), 56; Am Bd Med Microbiol, dipl. *Prof Exp:* Clin bacteriologist, Los Angeles Hosp, 48-52; res asst, Mt Sinai Hosp, New York, 52-53; teaching & res asst, Ind Univ, 53-54; US Navy res fel, Col Med, Univ Md, 54-56; instr microbiol, Col Med, Univ Fla, 56-60; asst prof, Col Med, Univ Vt, 60-63; from asst prof to assoc prof, Col Physicians & Surgeons, 63-70, asst microbiologist & co-dir, clin microbiol serv, Presby Hosp, 63-66, assoc microbiologist, 66-70, PROF MICROBIOL, COL PHYSICIANS & SURGEONS & DIR, CLIN MICROBIOL SERV, PRESBY HOSP, COLUMBIA UNIV, 71-, PROF PATH, 78- *Concurrent Pos:* Dir, med, USPHS, 56-, postdoctoral training prog clin microbiol, 63-84 & master's degree prog clin microbiol; consult, MetPath Labs, Inc, 67-83, Med Technol Corp, 75-83 & Scott Labs, Inc, 72-; vis prof, NY Med Col, Valhalla, 79, Am Soc Microbiol Latin Am, Medillin, Columbia, 82 & Am Bur Med Advan China, Taiwan, 82; lectr, med & dent students & Found Microbiol, Am Soc Microbiol, 81; co-owner, Meris Labs. *Mem:* Am Soc Microbiol; fel Am Acad Microbiol; Acad Clin Lab Physicians & Scientists; fel Asn Clin Scientists; assoc fel NY Acad Med; spec affil AMA; fel Infectious Dis Soc Am; Am Venereal Dis Asn; Sigma Xi. *Res:* Clinical bacteriology; infectious diseases; clostridia. *Mailing Add:* Clin Microbiol Serv Columbia-Presby Med Ctr 622 W 168th St New York NY 10032

ELLS, CHARLES EDWARD, b Canard, NS, Jan 20, 23; m 62; c 3. PHYSICAL METALLURGY. *Educ:* Univ Toronto, BASc, 50, MA, 51; Univ Birmingham, PhD(metall), 57. *Prof Exp:* Res officer metall, Can Dept Mines & Tech Survs, 51-53; engr, Westinghouse Elec Corp, Can, 56-57; RES OFFICER METALL, ATOMIC ENERGY CAN LTD, 57- *Mem:* Am Soc Metals. *Res:* Gases in metals; beryllium metallurgy; metallurgy of zirconium alloys; properties of fuel channels of the Candu power reactors. *Mailing Add:* Atomic Energy Can Ltd Chalk River ON K0J 1J0 Can

ELLS, FREDERICK RICHARD, b Norwalk, Conn, Apr 3, 34; m 68. POLYMER CHEMISTRY. *Educ:* Lehigh Univ, BS, 56; Univ Akron, MS, 60, PhD(polymer chem), 63. *Prof Exp:* Res chemist polymers, Firestone Tire & Rubber Co, 56-58; sr res chemist, Chem Div, PPG Industs, 63-67 & J T Baker Chem Co, Richardson-Merrell, 67-68; develop scientist, B F Goodrich Chem Co, 68-75; sr polymer chemist, Tremco, Inc, 76-87; CONSULT, 88- *Mem:* Am Chem Soc. *Res:* Synthesis and mechanism of polymers, acrylic and urethane sealants and water proofing materials. *Mailing Add:* 7606 Kenton Ave Cleveland OH 44129-4826

ELLS, JAMES E, b Cambridge, Mass, June 15, 31; m 58; c 1. HORTICULTURE. *Educ:* Univ Mass, BS, 57; Mich State Univ, MS, 58, PhD(hort), 61. *Prof Exp:* Asst prof, 61-69, ASSOC PROF HORT, EXP STA, COLO STATE UNIV, 69-, PROCESSING CROPS SPECIALIST, EXTEN SERV, 61- *Res:* Culture and processing of horticultural crops. *Mailing Add:* Dept Hort Colo State Univ Ft Collins CO 80523

ELLS, VICTOR RAYMOND, b Benton Harbor, Mich, Dec 7, 14. PHYSICAL CHEMISTRY. *Educ:* Kalamazoo Col, AB, 35; Brown Univ, MS, 38; Univ Rochester, PhD(phys chem, photochem), 39. *Prof Exp:* Instr chem & spectros, Univ Mo, 39-43; res phys chemist, Method Develop Res Norwich Pharmacol Co, Eaton Labs Div, 43-53, head, Phys & Anal Res Lab, 53-61, sr phys anal chemist, 61-77; RETIRED. *Mem:* AAAS; Am Phys Soc; Optical Soc; Am Chem Soc; Coblentz Soc; Sigma Xi. *Res:* Spectrographic analysis; spectrophotometry and absorption spectra; photochemistry; application of physical chemical methods to drug development, characterization and analysis. *Mailing Add:* 119 S Broad St Norwich NY 13815-1796

ELLSON, ROBERT A, b New York, NY, Dec 15, 34; m 58; c 3. MECHANICAL ENGINEERING. *Educ:* City Col New York, BS, 57; Univ Rochester, MS, 63, PhD(mech eng), 66. *Prof Exp:* Mech engr, Gen Elec Co, NY, 57-58; jr engr, New York Authority, 58-59; instr mech eng, Univ Rochester, 59-64; engr, NASA, Lewis Res Ctr, ASSOC PROF MECH ENG, ROCHESTER INST TECHNOL, 68- *Honors & Awards:* Dow Chem Co Award, 70; Award, Am Soc Eng Educ, 70; Centennial Medallion, Am Soc Mech Engrs, 80. *Mem:* Am Soc Mech Engrs; Am Soc Eng Educ. *Res:* Educational television; computer assisted instruction; engineering mechanics and thermodynamics. *Mailing Add:* 20 Dartford Rd Rochester NY 14618

ELLSTRAND, NORMAN CARL, b Elmhurst, Ill, Jan 1, 52; m 83; c 1. PLANT EVOLUTIONARY GENETICS, GENETICS OF ANNONA. *Educ:* Univ Ill, Urbana, BS, 74; Univ Tex, Austin, PhD(biol), 78. *Prof Exp:* Res assoc, Duke Univ, 78-79; asst prof plant ecol, 79-86, assoc prof genetics, 86-91, PROF GENETICS, UNIV CALIF, RIVERSIDE, 91- *Concurrent Pos:* Vis res assoc, Univ Canterbury, Christchurch, New Zealand, 81, Univ Calif, Berkeley, 87-88; assoc ed, J Heredity, 87- *Honors & Awards:* Eminent Ecologist, WK Kellogg Biol Sta, 88. *Mem:* Soc Study Evolution; Ecol Soc Am; Am Soc Naturalists; Am Genetics Asn; Soc Conserv Biol. *Res:* Evolutionary genetics; plant breeding systems; population genetic structure; plant mating patterns; horticulture and genetics of cherimoya. *Mailing Add:* Bot & Plant Sci Univ Calif Riverside CA 92521-0124

ELLSWORTH, LOUIS DANIEL, b Hamler, Ohio, Apr 27, 17; m 41; c 2. PHYSICS. *Educ:* Case Inst Technol, BS, 37; Ohio State Univ, MSc, 38, PhD(physics), 41. *Prof Exp:* Asst physics, Ohio State Univ, 37-40; instr, Haverford Col, 41-42; staff mem nat defense res comt, Radiation Lab, Mass Inst Technol, 42-45; assoc prof, 45-58, PROF PHYSICS, KANS STATE UNIV, 58- *Concurrent Pos:* Res physicist, Rauland Corp, Ill, 45-46. *Mem:* Am Phys Soc; Am Asn Physics Teachers. *Res:* X-ray diffraction analysis; mutations in bacteria by x-rays; electronics; nuclear reactor instrumentation; low energy nuclear physics; interactions of medium energy charged particles with matter. *Mailing Add:* Dept Physics Kans State Univ Manhattan KS 66506

ELLSWORTH, MARY LITCHFIELD, MICROCIRCULATION. *Educ:* Albany Med Col, PhD(physiol), 81. *Prof Exp:* Res assoc respiratory physiol, Med Col Va, 82-88; ASST PROF, DEPT PHARMACOL & PHYSIOL SCI, SCH MED, ST LOUIS UNIV, 88- *Res:* Oxygen transport; skeletal muscle. *Mailing Add:* St Louis Univ Sch Med 1402 S Grand Blvd St Louis MO 63104

ELLSWORTH, ROBERT KING, b Plattsburgh, NY, Nov 22, 41; m 63; c 2. PLANT BIOCHEMISTRY. *Educ:* State Univ NY Col Plattsburgh, BS, 63, MS, 66; Iowa State Univ, PhD(biochem), 68. *Prof Exp:* Teacher, Lake Placid Cent Sch, NY, 63-64 & Beekmantown Cent Sch, 64-65; instr chem, State Univ NY Col Plattsburgh, 65-66; from asst prof to assoc prof, 68-73, PROF BIOCHEM, STATE UNIV NY COL PLATTSBURGH, 73- *Mem:* Am Soc Biol Chemists; AAAS; Am Chem Soc; Am Soc Plant Physiol. *Res:* Chlorophyll biosynthesis and enzymology; chloroplast physiology; analytical biochemistry; radio biochemical techniques. *Mailing Add:* 41 Colligan Point Rd Plattsburgh NY 12901

ELLWEIN, LEON BURNELL, b Roscoe, SDak, Dec 21, 42; m 65; c 2. OPERATIONS RESEARCH, INDUSTRIAL ENGINEERING. *Educ:* SDak State Univ, BSME, 64, MS, 66; Stanford Univ, PhD(opers res & indust eng), 70. *Prof Exp:* Assoc engr design, Int Bus Mach Corp, 64-65; mech engr, NIH, 66-67, systs anal officer planning, Nat Cancer Inst, 70-72; sr scientist appl res, Sci Applns, Inc, 72-83, CONSULT, SCI APPLNS INT CORP, 83-; PROF & ASSOC DEAN, UNIV NEBR MED CTR, 83- *Concurrent Pos:* Rev comt mem, Diag Res Adv Group, Nat Cancer Inst, 72-74; consult, Nat Bladder Cancer Proj, 73-84, Bur Drugs, Food & Drug Admin, 74-76, Sch Pub Health, Univ Calif, Los Angeles, 75-77 & Nat Eye Inst, 83-; working cadre mem, Nat Pancreatic Cancer Proj, 74; trustee, Urol Res & Educ Found, San Diego, 78-; consult, Nat Inst Ophthal, Lima, Peru, 86-, Ciencias Medicas de Unicamp, Campinas, Brazil, 86-, Aravind Eye Hosp, Madurai, India, 85- *Mem:* Opers Res Soc Am; AAAS; Inst Mgt Sci; Am Mgt Asn; Inst Elec & Electronics Engrs Comput Soc; Soc Med Decision Making; Soc Risk Analysis; Soc Info Mgt. *Res:* Application of quantitative methodology to management and decision making under uncertainty, operations and process modeling, and risk and cost-benefit assessments. *Mailing Add:* Univ Nebr Med Ctr 600 S 42nd St Omaha NE 68198-6545

ELLWOOD, BROOKS B, US citizen; c 3. PALEOMAGNETISM, ARCHEOMAGNETISM. *Educ:* Fla State Univ, BS, 70; Grad Sch, Univ RI, MS, 74, PhD(oceanog), 76. *Prof Exp:* Res assoc, Ohio State Univ, 76-77; from asst prof to assoc prof geophys, Univ Ga, 77-83; assoc prof, 83-88, PROF GEOPHYS, UNIV TEX-ARLINGTON, 88-, ACTG CHMN, 89- *Mem:* Sigma Xi; Soc Explor Geophysicists; Am Geophys Union; Geol Soc Am; Soc Hist Archeol. *Res:* Paleomagnetism of igneous, metamorphic, and sedimentary (including deep-sea sediments) rocks; geophysical archeology. *Mailing Add:* Dept Geol Univ Tex-Arlington PO Box 19049 Arlington TX 76019

ELLWOOD, ERIC LOUIS, b Melbourne, Australia, Sept 8, 22; US citizen; m 47; c 3. FOREST PRODUCTS. *Educ:* Univ Melbourne, BSc, 44, MSc, 51; Yale Univ, PhD(wood technol), 53. *Prof Exp:* Asst forester, Victorian Forests Comn, Australia, 45-47; res officer wood technol, Commonwealth Sci & Indust Res Orgn, 47-51, prin res officer, 53-57; wood technologist & lectr, Forest Prod Lab, Univ Calif, 57-61; prof wood & paper sci & head dept, 61-71, DEAN SCH FOREST RESOURCES, NC STATE UNIV, 71-, PROF WOOD & PAPER SCI, 80- *Mem:* Forest Prod Res Soc; fel Tech Asn Pulp & Paper Indust; Soc Wood Sci & Technol (pres, 65-66); Australian Inst Foresters (secy-treas, 46); Int Acad Wood Sci; Soc Am Foresters. *Res:* Wood physics, especially wood-fluid relations and its application to drying and treating processes; relation between anatomy, wood and fiber properties. *Mailing Add:* Sch Forest Resources NC State Univ Raleigh NC 27650

ELLWOOD, PAUL M, JR, b San Francisco, Calif, July 16, 26; m 49; c 3. REHABILITATION MEDICINE. *Educ:* Stanford Univ, BA, 49, MD, 53. *Prof Exp:* Dir, Kenny Rehab Inst, 62-63; exec dir, Am Rehab Found, 63-73,; pres, 73-85, CHMN BD, INTERSTUDY, 87-; PRES, PAUL ELLWOOD & ASSOCS, 85- *Concurrent Pos:* Fel pediat, Univ Minn, 53-55, fel neurol, 55-57; consult, Arg Ministry Pub Health, 56 & US Dept State Int Coop Admin, Nicaragua, 58; clin prof neurol & pediat, Univ Minn, 58-, clin prof phys med & rehab, 62-; pres, mem bd dirs & res comt, Asn Rehab Ctrs, 60-62; mem, Surgeon Gen Nat Adv Health Serv Coun, 66; mem sci & prof adv bd, Nat Ctr Health Serv Res & Develop, 68-69; consult, Off Asst Secy Health & Sci Affairs, Dept Health, Educ & Welfare, 69, consult, Off Secy, 69; mem bd dirs, Nat Health Coun, 71-76; mem bd dirs, Group Health Asn Am, Inc, 75-76; consult, AMA, 77-; mem, Nat Indust Coun HMOs-OHMO, 87-; mem, Comt New Am Realities, Nat Planning Asn, 81- *Honors & Awards:* Award, Arg Ministry Pub Health, 57 & Am Acad Neurol, 58; Cert of Merit, AMA; Gold Key Award, Am Acad Phys Med & Rehab. *Mem:* Inst Med-Nat Acad Sci; hon fel Am Col Healthcare Execs. *Res:* Human resources policy research; health services research; pediatric neurology. *Mailing Add:* Paul Ellwood & Assocs PO Box 600 Excelsior MN 55331

ELLYIN, FERNAND, b Urmia, Azerbaijan, Iran, Aug 27, 38; Can citizen; m 66; c 2. STRUCTURAL & SOLID MECHANICS. *Educ:* Univ Teheran, MSc, 62; Univ Waterloo, PhD(civil eng), 66. *Prof Exp:* Engr, Beta Co, Iran, 62-63; res asst civil eng, Univ Waterloo, 63-66; asst prof, 66-69, assoc prof civil eng, 69-73, head struct & solid mech sect, 70-81, PROF CIVIL ENG, UNIV SHERBROOKE, 73- *Concurrent Pos:* Sr res officer, Dept Eng, Oxford Univ, 68-69; vis Div Theoret Studies & Res, Ctr Exp Res Bldg & Pub Works, Paris, France, 69; proj dir, subcomt reinforced openings & external loads, Pressure Vessel Res Comt, New York, 66-76; consult, Ingersoll Rand Co, S W Hooper & Co, Unit Cast Div, Mildland-Ross Can, Lynn & MacLoad Metall; vis prof, Carleton Univ, 77-78; spec consult, Atomic Energy Control Bd, 77-78. *Mem:* Am Soc Civil Engrs; Can Standards Asn; Am Soc Mech Engrs; Eng Inst Can; Am Acad Mech. *Res:* Stress analysis and design of pressure vessels and piping systems; inelastic material behaviour under multiaxial states of stress; stress concentration in plates and shells; reliability of structures and mechanical systems; low cycle fatigue and fracture; dynamic response of mechanical systems including interaction effects; developing models to predict material response under multiaxial states of stress and varying environments. *Mailing Add:* Dept Mech Eng Univ Alberta Edmonton AB T6G 2M7 Can

ELLZEY, JOANNE TONTZ, b Baltimore, Md, Mar 23, 37; m 69. MYCOLOGY, ULTRASTRUCTURE. *Educ:* Randolph-Macon Woman's Col, BA, 59; Univ NC, Chapel Hill, MA, 63; Univ Tex, Austin, PhD(bot), 69. *Prof Exp:* From teaching asst to instr biol, Univ NC, Greensboro, 62-64; asst prof, 69-75, ASSOC PROF BIOL, UNIV TEX, EL PASO, 75-, DIR, ULTRASTRUCT LAB, 73- *Mem:* AAAS; Mycol Soc Am; Electronic Micros Soc Am; Am Soc Microbiol; Sigma Xi. *Res:* Utilization of electron microscopy and cytochemistry to study the ultrastructure of gametogenesis and cell wall formation in saprolegniaceous fungi; ultrastructure of diabetic mice and the chestnut blight fungus, endothia parasitica. *Mailing Add:* Dept Biol Sci Univ Tex El Paso TX 79968

ELLZEY, MARION LAWRENCE, JR, b Shattuck, Okla, Apr 13, 39; m 69. CHEMISTRY, QUANTUM CHEMISTRY. *Educ:* Rice Univ, BA, 61; Univ Tex, Austin, PhD(physics), 66. *Prof Exp:* R A Welch fel, Univ Tex, Austin, 66-68; asst prof chem, 68-74, chmn fac senate, 79-81, ASSOC PROF CHEM, UNIV TEX, EL PASO, 74- *Concurrent Pos:* Robert A Welch Found grant, Tex, 71-79. *Mem:* Am Phys Soc; Am Chem Soc; Sigma Xi. *Res:* Molecular quantum mechanics with emphasis on complexed transition metal ions and their electronic and magnetic properties; group theory and linear algebra; graph theory applications on chemistry. *Mailing Add:* 310 Olivia El Paso TX 79912

ELLZEY, SAMUEL EDWARD, JR, b Mobile, Ala, May 16, 31; m 60; c 3. ORGANIC CHEMISTRY. *Educ:* Spring Hill Col, BS, 54; Tulane Univ, MS, 57, PhD(chem), 59. *Prof Exp:* CHEMIST, SOUTHERN REGIONAL RES CTR, USDA, 59- *Mem:* Am Chem Soc. *Res:* Organic fluorine compounds; chemical modification of cotton; nuclear magnetic resonance; fire-retardant textiles; phosphorus compounds; natural products. *Mailing Add:* 4809 Craig Ave Metairie LA 70003-7610

ELMADJIAN, FRED, b Aleppo, Syria, Oct 5, 15; nat US; m 52. PHYSIOLOGY. *Educ:* Mass Col Pharm, BS, 40, MS, 42; Clark Univ, MA, 47; Tufts Col, PhD(physiol), 49. *Prof Exp:* Staff mem, Worcester Found, 44-62, sr scientist, 55-62, res assoc, Worcester State Hosp, 47-58, dir biol res, 58-62, dir labs, 55-62; chief biol sci sect, Training & Manpower Resources Br, 62-66, chief biol sci sect, Behav Sci Training Br, NIMH, 66-75, ASSOC DIR RES TRAINING, DIV MANPOWER & TRAINING PROGS, NAT INST MENT HEALTH, ALCOHOL, DRUG ABUSE & MENT HEALTH ADMIN, DHEW, 75- *Concurrent Pos:* Asst chemist, Mass State Dept Ment Health, 45; physiologist, Mem Found Neuro-Endocrine Res, 46; res assoc, Med Sch, Tufts Col, 50-51; prof lectr & res assoc, Mass Col Pharm, 50-53; physiologist, field sta, NIMH, Mass, 51-53; neurophysiologist, Mass State Dept Ment Health, 57-62; mem revision comt, US Pharmacopoeia, 55-60; consult, opers res off, Johns Hopkins Univ, 52-61, dept psychiat, New Eng Med Ctr, 59-62 & Worcester Found, 62-63. *Mem:* AAAS; Am Physiol Soc; Am Pharmaceut Asn; Endocrine Soc; NY Acad Sci. *Res:* Stress physiology; adrenal cortical physiology; adrenal medulla; adrenaline and nonadrenaline in normal and mental diseases. *Mailing Add:* 402 Banyan Way Melbourne Beach FL 32951

ELMAGHRABY, SALAH ELDIN, b Fayoum, Egypt, Oct 21, 27; US citizen; m 64; c 3. OPERATIONS RESEARCH, SYSTEMS ENGINEERING. *Educ:* Cairo Univ, BSc, 48; Ohio State Univ, MSc, 55; Cornell Univ, PhD(indust eng), 58. *Prof Exp:* Tutor mech eng, Sch Eng, Cairo, Egypt, 49; engr, Foreign Inspection Off Egyptian State Rwys, London, Brussels & Budapest, 48-54; res asst opers res, Cornell Univ, 55-58; res leader systs anal, Western Elec Co Res Ctr, Princeton, NJ, 58-62; assoc prof opers res, Yale Univ, 62-67; prof opers res & indust eng & dir opers res prog, NC State Univ, 67-90; RETIRED. *Honors & Awards:* Distinguished Res Award, Am Inst Indust Engrs, 70; Res Div Award, Am Inst Indust Engrs, 80. *Mem:* Opers Res Soc Am; Inst Mgt Sci; fel Am Inst Indust Engrs; Am Prod & Inventory Control Soc; Sigma Xi. *Res:* Activity networks, scheduling theory, production control and operations research applications in quality control and enhancement. *Mailing Add:* NC State Univ PO Box 7913 Raleigh NC 27695-7913

EL-MASRY, EZZ ISMAIL, b Alexandria, Egypt; Can citizen; m 74; c 2. CIRCUIT THEORY & DESIGN. *Educ:* Univ Alexandria, BSc, 67, MSc, 72; Univ Man, PhD(elec eng), 77. *Prof Exp:* Instr & res engr elec eng, Univ Alexandria, 67-72; res & teaching asst, Univ Man, 72-77; res assoc, Nat Res Coun Can, 78; asst prof elec eng, elec eng dept, res asst prof, Coord Sci Lab, Univ Ill, Urbana-Champaign, 78-83; assoc prof, 83-87, PROF ELEC ENG, TECH UNIV NS, 87-; PRES, EEC ENG CONSULT, DARTMOUTH, 86- *Concurrent Pos:* Prin investr, Defence Res Estab Atlantic. *Mem:* Inst Elec & Electronics Engrs; Asn Prof Engrs. *Res:* Analytical and computer aided design of low-sensitivity structures for analog (active), digital and switched-capacitor filters. *Mailing Add:* Elec Eng Dept Tech Univ NS PO Box 1000 Halifax NS B3J 2X4 Can

ELMEGREEN, DEBRA MELOY, b South Bend, Ind, Nov 23, 52; m 76; c 2. ASTRONOMY. *Educ:* Princeton Univ, AB, 75; Harvard Univ, AM, 77, PhD(astron), 79. *Prof Exp:* Res asst, Thermophysics Div, Goddard Space Flight Ctr, 69, Lab Cosmic Ray Physics, Naval Res Lab, 71-72, Spectros Div, Nat Bur Standards, 73, Kitt Peak Nat Observ, 74, Arecibo Observ, 75; teaching fel, Harvard Univ, 77; Carnegie fel, Mt Wilson & Las Campanas Observ, 79-81; vis astronomer, Royal Greenwich Observ & Inst Astron, Cambridge Univ, Eng, 81; vis scientist, T J Watson Res Ctr, IBM Corp, 82-88; asst prof, 85-89, ASSOC PROF, VASSAR COL, 90- *Mem:* Am Astron Soc; Sigma Xi; Royal Astron Soc; Int Astron Union. *Res:* optical and millimeter observations of external galaxies to study sites of star formation and the origins of spiral structure. *Mailing Add:* Vassar Col Observ Poughkeepsie NY 12601

ELMENDORF, CHARLES HALSEY, III, b Los Angeles, Calif, July 1, 13; m 45; c 7. COMMUNITY ENGINEERING. *Educ:* Calif Inst Technol, BS, 35, MS, 36. *Prof Exp:* Mem staff, Bell Tel Labs, 36-55, asst dir, Submarine Cable Systs Dept, Am Tel & Tel Co, 55-59, dir, 59-61, assoc exec domestic systs, Mass, 61-66, asst vpres eng dept, 66-78; RETIRED. *Mem:* Nat Acad Eng; fel Inst Elec & Electronics Engrs. *Mailing Add:* 34 Cross Gate Rd Madison NJ 07940

ELMER, OTTO CHARLES, b Vienna, Austria, Jan 8, 18; nat US; m 45; c 7. ORGANIC CHEMISTRY. *Educ:* Bluffton Col, BA, 43; Univ Minn, PhD(org chem), 48. *Prof Exp:* Asst, Carleton Col, 43-44; asst, Univ Minn, 44-45 & 46-48; chemist, Tex Co, 48-53; chemist, Gen Tire & Rubber Co, 53-71, group leader, 71-79, res scientist, 79-82; CONSULT, 82- *Concurrent Pos:* Chemist, Rubber Reserve Co, 44-45, asst scientist, 46. *Mem:* Am Chem Soc; Am Inst Chemists. *Res:* Synthetic glycerides; autooxidation; reaction mechanisms; heterocycles; synthetic lubricants; lubricant additives; polymer chemistry; urethane elastomers and coatings; tire cord adhesives. *Mailing Add:* 720 Hillsdale Ave Akron OH 44303

ELMER, WILLIAM ARTHUR, b Bridgeton, NJ, May 12, 38; m 62; c 2. DEVELOPMENTAL GENETICS. *Educ:* Susquehanna Univ, AB, 60; NMex Highlands Univ, MS, 63; Univ Conn, PhD, 67. *Prof Exp:* Fel biol, Oak Ridge Nat Lab, 67-69; asst prof, 69-75, ASSOC PROF BIOL, EMORY UNIV, 75- *Concurrent Pos:* vis prof, Univ Glasgow, 79; Minna-James Heineman fel, NATO, 79. *Mem:* AAAS; Teratology Soc; Am Soc Zoologists; Int Soc Develop Biol; Sigma Xi. *Res:* Genetic control of limb development; molecular aspects of hereditary skeletal anomalies; in vitro growth of genetically abnormal chondrocytes. *Mailing Add:* Dept Biol Emory Univ Atlanta GA 30322

ELMER, WILLIAM B, b New York, NY, Jan 22, 01; m; c 3. REFLECTOR DESIGN. *Educ:* Mass Inst Technol, BSEE, 22. *Prof Exp:* Head tech div, Outside Lines Dept, Boston Edison Co, 22-42, Street Lighting Design Sect, Westinghouse Co, 42-46; managing dir, Hart Mfg, 46-49; CONSULT REFLECTOR DESIGN, 49- *Mem:* Fel Optical Soc Am; fel Inst Elec & Electronics Engrs; fel Illum Engrs Soc. *Res:* Design and development of assymetric reflectors; pendulum design; author of one book. *Mailing Add:* Two Chestnut St Andover MA 01810

ELMES, GREGORY ARTHUR, b Shoreham, Sussex, Eng, Apr 26, 50. QUANTITATIVE SPATIAL ANALYSIS. *Educ:* Univ Newcastle Upon Tyre, BSc, 68; Pa State Univ, MS, 74, PhD(geog), 79. *Prof Exp:* ASST PROF GEOG, WVA UNIV & RES ASSOC, REGIONAL RES INST, 78- *Mem:* Asn Am Geographers; AAAS. *Res:* Energy and transportation modeling; investigation of fly ash as a source of strategic minerals from location and economic perspectives; analysis of United States transportation trends. *Mailing Add:* Dept Geog WVa Univ Morgantown WV 26506

ELMORE, CARROLL DENNIS, b Pheba, Miss, Apr 3, 40; m 77; c 3. PLANT PHYSIOLOGY. *Educ:* Miss State Univ, BS, 62; Univ Ariz, MS, 66; Univ Ill, PhD(agron), 70. *Prof Exp:* PLANT PHYSIOLOGIST, DELTA STATES AGR RES CTR, USDA, STONEVILLE, MISS, 70- *Mem:* Am Soc Agron; Crop Sci Soc Am; AAAS; Am Inst Biol Sci; Am Soc Plant Physiol; Sigma Xi; Weed Sci Soc Am. *Res:* Weed control aspects of conservation tillage; weed identification and weed communities; crop production with conservation tillage concepts. *Mailing Add:* PO Box 36 Stoneville MS 38776

ELMORE, DAVID, b Los Alamos, NMex, Dec 19, 45; m 68; c 2. SOLAR VARIABILITY. *Educ:* Case Inst Technol, BS, 68; Univ Rochester, PhD(physics), 74. *Prof Exp:* Res assoc, Univ Rochester, 74-80, sr res assoc, Nuclear Structure Res Lab, 80-89; ASSOC PROF PHYSICS, PURDUE UNIV, 89- *Concurrent Pos:* Prin investr, NSF grants. *Mem:* Am Phys Soc; Am Geophys Union; AAAS. *Res:* Measure and interpret very low concentration of long lived radioisotopes in natural samples using the new technique of tandem accelerator mass spectrometry. *Mailing Add:* Dept Physics Purdue Univ West Lafayette IN 47907

ELMORE, GLENN VAN NESS, b Topeka, Kans, Apr 2, 16; m 48; c 1. ELECTROCHEMISTRY. *Educ:* Washburn Univ, BS, 38; Ga Inst Technol, MS, 40. *Prof Exp:* Instr chem, Ga Inst Technol, 40-41; res chemist, Tenn Valley Auth, 41-47; res chemist, Oak Ridge Nat Labs, 47-48; res chemist, Tenn Valley Auth, 48-53; electrochemist, Kettering Found, 53-62; ELECTROCHEMIST, SYSTS PROD DIV, IBM CORP, 62- *Mem:* Am Chem Soc; Electrochem Soc. *Res:* Chemistry of phosphorus nitrides; radiation chemistry of decomposition of water; endothermic photochemical reactions; electrodeposition; fuel cells; adhesion of electroless metals. *Mailing Add:* 3133 Briarcliff Ave Vestal NY 13850-2859

ELMORE, JAMES LEWIS, b Chattanooga, Tenn, May 28, 48; m 72; c 2. AQUATIC ECOLOGY. *Educ:* Univ Tenn at Chattanooga, BA, 71, Knoxville, MS, 73; Univ SFla, PhD(biol), 80. *Prof Exp:* Teaching asst limnol & gen biol, Univ Tenn, 71-73; teaching asst ecol, sex & reproduction, Univ SFla, 74-76; res assoc, environ sci div, Oak Ridge Nat Lab, 80-85, environ scientist, Oak Ridge Inst, Tenn, 86-87; PEER CONSULT, OAK RIDGE, TENN, 87- *Concurrent Pos:* Consult, Biol Res Assoc, 74-80, Dames & Moore, 79; prin investr, Dept Health & Rehab Serv, State Fla, 79-80. *Mem:* Am Soc Limnol & Oceanog; Soc Int Limnol; Sigma Xi; Ecol Soc Am. *Res:* Factors regulating distributions of freshwater copepods of the genus Diaptomus; longterm changes in zooplankton community composition; effects of synthetic oils on zooplankton communities in experimental ponds; environmental assessment of US Department of Energy projects; interaction of pollutants, microbes, and macroinvertebrates in sediments and soils; environmental regulatory documents. *Mailing Add:* 324 Watt Rd Knoxville TN 37922

ELMORE, JOHN JESSE, JR, b Spokane, Wash, Jan 3, 36. PHYSICAL BIOCHEMISTRY, MOLECULAR BIOLOGY. *Educ:* Reed Col, BA, 57; Hofstra Univ, MA, 67; State Univ NY Stony Brook, PhD(biochem), 72. *Prof Exp:* Res assoc, Dept Med, Brookhaven Nat Lab, 72-74, asst scientist, 74-76, assoc scientist, 76-78, SCIENTIST, DEPT MED, BROOKHAVEN NAT LAB, 78- *Mem:* AAAS; Am Chem Soc; NY Acad Sci. *Res:* Reaction mechanisms involved in oxidative cytotoxicity and damage to biological membranes, proteins and nucleic acids and in biological actions of antioxidants, drugs, hormones and carcinogens and mutagens. *Mailing Add:* PO Box 703 Bayview ID 83803

ELMORE, KIMBERLY LAURENCE, b Tulsa, Okla, Apr 9, 56; m 86. MESO-SCALE METEOROLOGY, RADAR METEOROLOGY. *Educ:* Univ Okla, BS, 78, MS, 82. *Prof Exp:* ASSOC SCIENTIST, NAT CTR ATMOSPHERIC RES, 82- *Mem:* Am Meterol Soc. *Res:* Doppler radar meteorology; effects of wind shear on aircraft flight; storm morphology, kinematic and dynamic structure; aircraft icing; aviation weather. *Mailing Add:* Res Applns Prog Nat Ctr Atmospheric Res PO Box 3000 Boulder CO 80307

ELMORE, STANLEY MCDOWELL, b Raleigh, NC, Dec 17, 33; m 58; c 4. MEDICINE, ORTHOPEDIC SURGERY. *Educ:* Vanderbilt Univ, AB, 55, MD, 58; Am Bd Orthop Surg, dipl, 69. *Prof Exp:* Assoc prof orthop surg & chmn div, Med Col Va, 66-73; ATTEND ORTHOP SURGEON, CHIPPENHAM HOSP, 73- & JOHNSTON-WILLIS HOSP, 73- *Concurrent Pos:* USPHS grant orthop surg, Sch Med, Vanderbilt Univ, 65-66; consult, McGuire Vet Admin Hosp, Richmond, Va, 66-70. *Honors & Awards:* Borden Award, 58. *Mem:* Am Acad Orthop Surg; Orthop Res Soc; Continental Orthop Soc. *Res:* Physical properties of articular cartilage; genetics of orthopedic diseases; bone changes in renal transplant patients; total joint replacement. *Mailing Add:* Chippenham Med Bldg 7135 Jahnke Rd Richmond VA 23225

ELMS, JAMES CORNELIUS, b East Orange, NJ, May 16, 16; m 42; c 4. PHYSICS. *Educ:* Calif Inst Technol, BS, 48; Univ Calif, Los Angeles, MA, 50. *Prof Exp:* Jr stress analyst, Consol Vultee Aircraft Corp, 40-41; chief develop engr, G M Giannani Co, 48; res assoc geophys, Univ Calif, Los Angeles, 49-50; res engr electronics, NAm Aviation Corp, 50-52, proj engr electromech eng, 52, asst sect chief electronics, 53-55, mgr fire control, 55-57; mgr avionics dept, Martin Co, 57-59; vpres electronics systs, Crosley Div, AVCO Corp, 59, exec vpres, 59-60; gen opers mgr, Aeronutronic Div, Ford Motor Co, 60-63; dep dir, Manned Spacecraft Ctr, NASA, Tex, 63-64; vpres & div gen mgr, Space & Info Systs Div, Raytheon Co, Mass, 64-65; dep assoc adminr for Manned Space Flight, Hq, NASA, 65-66, dir, Electronics Res Ctr, 66-70; dir, Transp Systs Ctr, US Dept Transp, 70-74; consult govt & indust, 75-81; adminr, NASA, 81-87, dir, SDIO, 84-88; RETIRED. *Concurrent Pos:* Mem space systs comt, Space Prog Adv Coun, NASA, 70-77; mem comt fed labs, 70-76; dep dir, Space Shuttle Oper Mgt Assessment Team, 75; consult to adminr, Energy Res & Develop Admin, 75-77; SDI adv comt, 84-88. *Mem:* Nat Acad Eng; Am Phys Soc; fel Am Inst Aeronaut & Astronaut; fel Inst Elec & Electronics Engrs. *Res:* Seismic investigation of earth's crustal structure; ordnance mechanisms; armament control; radar; missile guidance and control; spacecraft design; transportation analysis; energy policy; strategic defense policy. *Mailing Add:* 112 Kings Pl Newport Beach CA 92663

ELMSLIE, JAMES STEWART, b Quincy, Ill, Dec 30, 30; m 54; c 2. ORGANIC POLYMER CHEMISTRY. *Educ:* Grinnell Col, AB, 52; Univ Del, PhD(org chem), 59. *Prof Exp:* Anal chemist, Hercules Powder Co, 52-53 & 55, sr res chemist, Allegany Ballistics Lab, Hercules, Inc, 58-69, TECH SPECIALIST, HERCULES, INC, MAGNA, 69- *Mem:* Am Chem Soc. *Res:* Encapsulation of solid particles; new ablative insulators; new sprayable insulation material; synthesis of high energy compounds for use in solid propellants; new binders for solid propellants. *Mailing Add:* 4825 Wander Lane Salt Lake City UT 84117-6410

ELMSTROM, GARY WILLIAM, b Chicago, Ill, Jan 10, 39; m 67; c 3. PLANT BREEDING, PLANT NUTRITION. *Educ:* Southern Ill Univ, BS, 63, MS, 64; Univ Calif, Davis, PhD(plant physiol), 69. *Prof Exp:* From asst prof to assoc prof, 69-81, PROF HORT, UNIV FLA, 81-; asst ctr dir, 86-89, PROF HORT, CENT FLA RES & EDUC CTR, 89- *Concurrent Pos:* Consult, Japan, Dominican Repub, Honduras, Guatemala, Costa Rica. *Mem:* Am Soc Hort Sci; Coun Agr Sci & Technol. *Res:* Development of cultural practices including weed control, nutrition, and growth regulators to maximize cucurbit production; evaluation of cucurbit varieties; watermelon, cantaloupe and squash breeding. *Mailing Add:* Cent Fla Res & Educ Ctr Univ Fla 5336 Univ Ave Leesburg FL 34748-8203

EL-NAGGAR, MOHAMMED ISMAIL, b El-Mahala, Egypt, Aug 26, 51; US citizen. MICROELECTRONICS, INTEGRATED CIRCUITS & SYSTEMS. *Educ:* Cairo Univ, Egypt, BS, 74, MS, 78; Univ Manitoba, Can, PhD(elec eng), 83. *Prof Exp:* Res eng, Elec Res & Develop Ctr, Cairo, Egypt, 74-78; telecommunications engr, develop & design, Arab Orgn Ind, Cairo, Egypt, 78-80; res & teaching asst, Dept Elec Eng, Univ Manitoba, Can, 80-83; asst prof elec eng, Univ Nebr, Lincoln, 83-86, assoc prof elec eng, 86-88; ASSOC PROF ELEC ENG, OHIO STATE UNIV, COLUMBUS, 88- *Concurrent Pos:* Chmn, Nat Comt Anal Signal Processing, Inst Elec & Electronic Engrs; ed-in-chief, Int J Annlog Integrated Circuits & Signal Processing. *Honors & Awards:* US Presidential Young Investr Award, The White House, 85. *Mem:* Inst Elec & Electronics Engrs. *Res:* Analog and digital very large scale integrated circuits; analog and digital signal processing; circuits for telecommunications, instrumentation and neural information processing. *Mailing Add:* Dept Elec Engr Ohio State Univ 205 Dreese Lab 2015 Neil Ave Columbus OH 43210-1272

EL-NEGOUMY, ABDUL MONEM, b Cairo, Egypt, May 23, 20; US citizen; m 51; c 2. AGRICULTURAL BIOCHEMISTRY. *Educ:* Cairo Univ, BS, 43; Univ Wis, MS, 48, PhD(dairy chem), 51. *Prof Exp:* Instr bot, Univ Alexandria, 43-46, from lectr to assoc prof food biochem, 51-58; res assoc & fel dairy chem, Iowa State Univ, 58-62; from asst prof to assoc prof, 62-69, PROF AGR BIOCHEM, MONT STATE UNIV, 69- *Mem:* AAAS; fel Am Inst Chemists; Am Dairy Sci Asn; Inst Food Technologists; Am Asn Cereal Chemists. *Res:* Dairy and food chemistry; autooxidation of food fats and oils; genetic variants and interactions of food proteins. *Mailing Add:* 1308 S Black Ave Bozeman MT 59715

ELOFSON, RICHARD MACLEOD, b Ponoka, Alta, June 14, 19; m 41; c 4. ORGANIC CHEMISTRY. *Educ:* Univ Alta, BSc, 41; Univ Wis, PhD(org chem), 44. *Prof Exp:* Chemist, Defence Indust, Ltd, Can, 41; res chemist, F W Horner, Ltd, 44-45, dir res, 45-46; res chemist, Gen Aniline & Film Corp, 47-49; self employed, 49-53; sr res chemist, Alta Res Coun, 53-84; RETIRED. *Concurrent Pos:* Vis scientist, Univ Alberta. *Mem:* Fel Can Inst Chem; Am Chem Soc. *Res:* Polarography and oxidation reduction potentials of organic compounds; protein hydrolysates; chemistry of acetylene; coalification; electron spin resonance spectroscopy; nitrogen fixation. *Mailing Add:* 11579 University Ave Edmonton AB T6G 1Z4 Can

ELOWE, LOUIS N, b Baghdad, Iraq, Apr 2, 22; US citizen; m 53; c 3. PHARMACEUTICAL SCIENCE, CHEMISTRY. *Educ:* Col Pharm & Chem, Iraq, PhC, 44; Univ Wis, MS, 52, PhD(pharm, chem), 54. *Prof Exp:* Managing dir, Eastern Wholesale Drug Store, Iraq, 46-50; asst, Univ Wis, 51-54; asst prof pharm, Univ Toronto, 54-57; assoc prof, Fordham Univ, 57-59; sr pharmaceut chemist, Schering Corp, 59; develop chemist, Lederle Labs, Am Cyanamid Co, 59-60, group leader, 60-64, mgr pharmaceut prod develop, Cyanamid Int, 64-66; mgr int res & control, Chesebrough-Pond's, Inc, 66-69, dir int res & control, 69-74; dir int res & develop, Consumer Prod, Schering-Plough Inc, 74-79; dir int regulatory affairs, G D Searle Co, 79-81; res prof, Univ Toronto, 81-83. *Concurrent Pos:* Int Res Coord; consult, 83- *Mem:* Am Pharmaceut Asn; Am Chem Soc; Can Pharmaceut Asn; Soc Cosmetic Chem. *Res:* Development of analytical methods in drug analysis; development pharmaceutical, cosmetic, and personal care products; academic/industrial pharmaceutical research coordination. *Mailing Add:* 6440 Gillham Dr Memphis TN 38134-7560

EL-REFAI, MAHMOUD F, HORMONE RECEPTOR, GLUCOSE METABOLISM. *Educ:* Univ Southern Calif, PhD(biomed eng), 78. *Prof Exp:* ASST RES PROF TOXICOL, PHYSIOL & BIOPHYS, UNIV SOUTHERN CALIF, 81- *Mailing Add:* 17503 Jeffrey Ave Artesia CA 90701

ELRICK, DAVID EMERSON, b Toronto, Ont, Sept 6, 31; m 58; c 4. SOIL PHYSICS. *Educ:* Ont Agr Col, BSA, 53; Univ Wis, MS, 55, PhD(soils), 57. *Prof Exp:* Asst prof physics, Ont Agr Col, 57-60; res officer, Div Plant Indust, Commonwealth Sci & Indust Res Orgn, Australia, 60-62; assoc prof, Univ Guelph, 62-65, chmn dept, 71-75, actg dean grad studies, 75-76, actg dean res, 78, PROF SOIL PHYSICS, UNIV GUELPH, 65- *Concurrent Pos:* Mem hydrol subcomt, Nat Res Coun Can, 66-73; sr fel, Grenoble, France, 68-69; ed, Can J Soil Sci, 74-77. *Mem:* AAAS; Am Geophys Union; Soil Conserv Soc Am; Agr Inst Can; fel Soil Sci Soc Am; fel Am Soc Agron; fel Can Soc Soil Sci. *Res:* Fluid flow in unsaturated soil; solute dispersion in soils; field measurement of saturated hydraulic conductivity; percolation test for septic tank disposal. *Mailing Add:* Dept Land Resource Sci Univ Guelph Guelph ON N1G 2W1 Can

ELROD, ALVON CREIGHTON, b Walhalla, SC, Dec 28, 28; m 51; c 6. MECHANICAL ENGINEERING. *Educ:* Clemson Univ, BME, 49, MME, 51; Purdue Univ, PhD(mech eng), 59. *Prof Exp:* From instr to assoc prof mech eng, Clemson Univ, 53-88; RETIRED. *Concurrent Pos:* NSF res grant, 60-62; Ford Found indust residency, Gen Elec Flight Propulsion Div, 67-68; consult, Nat Coun of Eng Examr. *Mem:* Am Soc Mech Engrs; Soc Automotive Engrs. *Res:* Internal combustion. *Mailing Add:* 300 Ripple View Dr Clemson SC 29631

ELROD, DAVID WAYNE, b Niles, Mich, Sept 25, 52; m 74; c 1. COMPUTATIONAL CHEMISTRY, NEURAL NETWORKS. *Educ:* Kalamazoo Col, BA, 74; Western Mich Univ, MA, 82. *Prof Exp:* Res chemist med chem, 74-85, res chemist computational chem, 85-91, INFO SCIENTIST COMPUTATIONAL CHEM, UPJOHN CO, 91- *Mem:* Am Chem Soc. *Res:* Prediction of structure, properties and reactivity of organic molecules using artificial neurol network methods; isolation, synthesis, modification and biosynthesis of anticancer compounds. *Mailing Add:* Upjohn Co 301 Henrietta St Kalamazoo MI 49001

ELROD, HAROLD G(LENN), JR, b Manchester, NH, Nov 19, 18; m 42; c 3. ENGINEERING. *Educ:* Mass Inst Technol, BSc, 42; Harvard Univ, PhD(eng sci), 49. *Prof Exp:* Instr marine eng, US Naval Acad, 42-45; res engr refrig, Clayton & Lambert Mfg Co, 46; heat transfer & fluid flow, Babcock & Wilcox Co, 49-51; asst prof mech eng, Case Inst Technol, 51-55; from assoc prof to prof, Columbia Univ, 55-62; prof, Mich State Univ, 62-63; prof, 63-85, EMER PROF ENG SCI, COLUMBIA UNIV, 85- *Concurrent Pos:* Vis prof, Univ Southampton England, 69-70; liaison scientist, US Off Naval Res, London, 74; vis prof, Tech Univ Denmark & Inst Nat des Sci Appl, France, 77-78. *Mem:* Am Soc Mech Engrs; Am Inst Aeronaut & Astronaut. *Res:* Heat transfer and fluid flow, with emphasis on fundamental lubrication theory for granular flow, thermohydrodynamics, cavitation, surface roughness, gas and foil bearings. *Mailing Add:* 14 Cromwell Ct Old Saybrook CT 06475

ELROD, JOSEPH HARRISON, b Boone, NC, Nov 14, 41; m 69. FISHERIES. *Educ:* Appalachian State Teachers Col, BS, 63; Auburn Univ, PhD(zool), 66. *Prof Exp:* Res fishery biologist, Nat Reservoir Res Prog, Pierre SDak, 66-72, Clemson, SC, 72-76, supvry res fishery biologist, Great Lakes Fishery Lab, Ann Arbor, Mich, 76-77, SUPVRY RES FISHERY BIOLOGIST, GREAT LAKES FISHERY LAB, US FISH & WILDLIFE SERV, OSWEGO, NY, 77- *Mem:* Am Fisheries Soc; Am Inst Fishery Res Biologists; Int Asn Great Lakes Res. *Res:* Dynamics and interrelationships of major species of Lake Ontario's fish community; evaluation of sea lamprey control and lake trout restoration programs in Lake Ontario. *Mailing Add:* US Fish & Wildlife Serv 17 Lake St Oswego NY 13126

EL-SADEN, MUNIR RIDHA, b Baghdad, Iraq, Aug a 16, 28; US citizen; m 50; c 3. THERMODYNAMICS. *Educ:* Univ Denver, BSc, 51; Univ Mich, MS, 53, PhD(mech eng), 57. *Prof Exp:* Instr eng mech, Univ Mich, 55-57; unit engr, Daura Ref, Iraq, 57-58; lectr mech eng, Univ Baghdad, 58-59; asst prof, Univ Tex, 59-61; from assoc prof to prof, NC State Uni 61-66; PROF ENG, CALIF STATE UNIV, FULLERTON, 66- *Concurrent Pos:* Lectr, Ga Inst Technol, 62, E I du Pont de Nemours & Co, NC, 64 & Catholic Univ, 65; US Air Force Off Sci Res grant, 65-67. *Mem:* Am Soc Mech Engrs; Am Inst Aeronaut & Astronaut; Am Soc Eng Educ. *Res:* Heat transfer; fluid mechanics; magneto-hydrodynamics; energy conversion; nonequilibrium thermodynamics; thermomagnetic and galvanomagnetic devices. *Mailing Add:* Mech Eng Dept Calif State Univ Fullerton CA 92634

EL SAFFAR, ZUHAIR M, b Baghdad, Iraq, Sept 23, 34; m 65; c 5. CHEMICAL PHYSICS. *Educ:* Univ Wales, BS, 57, PhD(physics), 60. *Prof Exp:* Asst prof physics, Mich State Univ, 60-62; from asst prof to assoc prof, Univ Baghdad, 62-67; res assoc solid state physics, Oak Ridge Nat Lab, 65-66; res assoc chem, Johns Hopkins Univ, 67-68; from asst prof to assoc prof physics, 68-76, PROF PHYSICS, DEPAUL UNIV, 76-, CHMN DEPT, 79- *Concurrent Pos:* Consult, Argonne Nat Lab, 69. *Mem:* Am Phys Soc; Am Crystallog Asn; Brit Inst Physics & Phys Soc. *Res:* Nuclear magnetic resonance in nonmetallic solids, especially hydrocarbons, hydrates and ferroelectrics; neutron diffraction in solids. *Mailing Add:* 7811 Greenfield River Forest IL 60305

ELSAS, LOUIS JACOB, II, b Atlanta, Ga, Feb 10, 37; m 60; c 3. MEDICAL GENETICS. *Educ:* Harvard Univ, BA, 58; Univ Va, MD, 62; Am Bd Internal Med, dipl, 72. *Prof Exp:* From intern to sr resident, Yale-New Haven Hosp, 62-65; clin fel metab, Sch Med, Yale Univ, 65-66, res fel genetics, 66-68, from instr to asst prof pediat & med, 68-70; from asst prof to prof pediat & biochem, 70-77, dir, Med Sci Prog, 77-80, DIR, GENETICS DIV, SCH MED, EMORY UNIV, 70-; DIR, COMPREHENSIVE GENETIC SYST, STATE OF GA, 74- *Concurrent Pos:* Res career develop award, Nat Inst Child Health & Develop, 73-78; consult, Ga Dept Human Resources, 74-; dir, Tay Sachs Dis Prev Prog, 75-; vis res prof, Japan Soc Prom Sci, 76-77; founding pres, Soc Inherited Metabolic Dis, 78; chmn Southeastern Regional Genetics Group, 82; Professore a contralto, Italy, 84. *Honors & Awards:* John Horsely Mem Award, Sch Med, Univ Va, 72. *Mem:* Sigma Xi; Am Soc Human Genetics; Soc Pediat Res; Endocrine Soc; Am Soc Clin Invest; Asn Am Physicians. *Res:* Genetic, biochemical and developmental control of aminoacids and hexose transport by plasma membrane; genetic control of human enzyme complexes and the mechanism of action of active co-factors. *Mailing Add:* Dept Pediat & Genetics Emory Univ 2040 Ridgewood Dr Atlanta GA 30322

ELSASSER, WALTER M, b Mannheim, Ger, Mar 20, 04; nat US; m 37, 64; c 2. PHYSICS, GEOPHYSICS. *Educ:* Univ Gottingen, PhD(physics), 27. *Prof Exp:* Asst, Tech Univ, Berlin, 28-30; instr physics, Univ Frankfurt, 30-33; res fel, Inst Henri Poincare, Sorbonne, 33-36; asst meteorol, Calif Inst Technol, 36-41; res assoc, Blue Hill Meteorol Observ, Harvard Univ, 41-42; war res, Signal Corps Labs, 42-44; mem radio wave propagation comt, Nat Defense Res Comt, 44-45; indust res on electronics labs, Radio Corp of Am, NJ, 45-47; assoc prof physics, Univ Pa, 47-50; prof, Univ Utah, 50-56 & Univ Calif, San Diego, 56-62; prof geophys, Princeton Univ, 62-68; res prof geophys, Inst Fluid Dynamics & Appl Math, Univ Md, College Park, 68-74; HOMEWOOD PROF, JOHNS HOPKINS UNIV, 75- *Concurrent Pos:* Lectr, Sorbonne, 35 & 36; lectr, Mass Inst Technol, 38; actg head dept physics, Univ NMex, 60-61. *Honors & Awards:* German Phys Soc Prize, 32; Bowie Medal, Am Geophys Union, 59, Fleming Medal, 71; Gauss Medal, Ger, 77. *Mem:* Nat Acad Sci; fel Am Phys Soc; Am Geophys Union; Nat Medal Sci, US, 87. *Res:* Theoretical physics; quantum theory; physics of the earth and atmosphere; geomagnetism; theoretical biology. *Mailing Add:* Dept Earth & Planetary Sci Johns Hopkins Univ Baltimore MD 21218

ELSAYED, ELSAYED ABDELRAZIK, b Egypt, Dec 29, 47; m 75; c 4. INDUSTRIAL ENGINEERING, PRODUCTION SYSTEMS & RELIABILITY ENGINEERING. *Educ:* Cairo Univ, BSc, 69, MSc, 73; Univ Windsor, PhD(indust eng), 76. *Prof Exp:* Lectr mech eng, Cairo Univ, 69-73; teaching asst indust eng, Univ Windsor, 73-76; teaching & res assoc, Univ Utah, 76-77; from asst prof to assoc prof, 77-85, PROF INDUST ENG, RUTGERS UNIV, 85-, CHMN, 83- *Concurrent Pos:* Prin investr, Rutgers Res Coun, 78, Sea-Land Serv Inc res grant, 78-81, NSF, 80-81, 91 & Environ Protection Agency, 80-85; Fed Aviation Admin, 82-83; Naval Air Engr Ctr, 83-84; NJ Dept Higher Educ, 86-87, FAA, 89-91, DOD, 89-91. *Mem:* Am Inst Indust Engrs; Am Soc Mech Engrs; Am Soc Eng Educ; Sigma Xi; Soc Mfg Eng. *Res:* Production planning and control; stochastic processes with specialization in reliability engineering; computer aided manufacturing. *Mailing Add:* Dept Indust Eng PO Box 909 Piscataway NJ 08855

EL-SAYED, MOSTAFA AMR, b Zifta, Egypt, May 8, 33; US citizen; m 57; c 5. PHYSICAL CHEMISTRY. *Educ:* Ain Shams Univ, Cairo, BSc, 53; Fla State Univ, PhD(phys chem), 59. *Prof Exp:* Instr chem, Ain Shams Univ, Cairo, 53-54; res asst spectros, Fla State Univ, 54-57, res assoc, 58-59; res fel, Harvard Univ, 59-60; res assoc, Calif Inst Technol, 60-61; from asst prof to assoc prof, 61-67, PROF CHEM, UNIV CALIF, LOS ANGELES, 67- *Concurrent Pos:* Alfred P Sloan fel, 65-67; John Simon Guggenheim Mem Found fel, 67-68; vis prof, Am Univ Beirut, 67-68 & Univ Paris-Sud, Orsay, 90; ed, J Phys Chem, Am Chem Soc, 80- & Int Rev Phys Chem, 84-; Sherman Fairchild distinguished scholar, Calif Inst Technol, 80; Alexander von Humboldt sr US scientist award, WGermany, 82; mem-at-large, US Nat Res Coun Comt, Int Union Pure & Appl Chem, 85-, vchmn, 88-90, chmn, 90-; mem Steering comt, Int Ctr Pure & Appl Chem, Trieste, Italy, 88-; bd trustees, Assoc Univs, Inc, 88; mem adv comt, Chem Div, NSF, 90; fac res lectr, Univ Calif, Los Angeles, 90-91; ed-in-chief, J Phys Chem. *Honors & Awards:* Fresenius Nat Award in Pure & Appl Chem, 67; McCoy Award, 69; Gold Medal Award, Am Chem Soc, 71 & Tolman Award, 90; King Faisal Int Prize in Sci, 90. *Mem:* Nat Acad Sci; Am Chem Soc; Am Phys Soc; Am Asn Univ Prof; NY Acad Sci; AAAS; fel Am Acad Arts & Sci; Sigma Xi; Third World Acad Sci. *Res:* Molecular spectroscopy; interaction of intense laser beam of short pulse duration with molecules; time resolved studies on photochemical and photobiological systems; energy transport in disordered materials. *Mailing Add:* Dept Chem & Biochem Univ Calif Los Angeles CA 90024

ELSAYED, NABIL M, b Cairo, Egypt, Jan 5, 46. TOXIOLOGY. *Educ:* Cairo Univ, BS, 67; Univ Calif, Los Angeles, MS, 81, PhD(environ health & toxicol), 83. *Prof Exp:* Asst dir, Instrument Anal Res Develop, Indola Cosmetics, 70-72; mgr air pollution, Pub Health, Univ Calif, Los Angeles, 77-79, adj asst prof med & pub health, 85-87; SR TOXICOLOGIST & GROUP LEADER, DIV MILITARY RES & COMBAT CASUALTIES, LETTERMAN ARMY INST RES, 87- *Concurrent Pos:* Adj asst prof environ health, Univ Calif, Los Angeles; adj asst prof nutrit, Univ Nev, Reno. *Mem:* Soc Toxicologists; Am Soc Biochem & Molecular Biol; Oxygen Soc; Am Thoracic Soc; NY Acad Sci; Am Conf Govt & Indust Hygienists. *Mailing Add:* Mil Trauma Res Letterman Army Inst Res Presido San Francisco San Francisco CA 94129-6800

ELSBACH, PETER, b Zeist, Neth, Nov 9, 24; m 59; c 2. MEDICINE, CELL BIOLOGY. *Educ:* Univ Amsterdam, MD, 50; State Univ Leiden, Dr, 64. *Hon Degrees:* Dr, State Univ Leiden, 64. *Prof Exp:* Asst resident med, Bellevue Hosp, Sch Med, NY Univ, 53-55, resident, 55-56; res assoc & asst physician, Rockefeller Inst, 56-59; from instr to assoc prof, 59-72, PROF MED, SCH MED, NY UNIV, 72- *Concurrent Pos:* NY Heart Asn sr res fel, 59-64; Health Res Coun career scientist award, NY, 64-75; Josiah Macy Jr Found fac scholar award, 75-76; mem, Coun Atheroschlerosis, Am Heart Asn, 63-81. *Honors & Awards:* Merit Award, NIH, 87. *Mem:* NY Acad Sci; Am Physiol Soc; Am Soc Clin Invest; Am Asn Physicans; Royal Dutch Acad Sci. *Res:* Biochemical and clinical investigation pertaining to leukocyte function and bactericidal mechanisms; regulation of phospholipase action. *Mailing Add:* Dept Med NY Univ Sch Med New York NY 10016

ELSBERND, HELEN, b Calmar, Iowa, Jan 15, 38. INORGANIC CHEMISTRY. *Educ:* Viterbo Col, BA, 65; Univ Ill, Urbana, MA, 67, PhD(chem), 69. *Prof Exp:* ASSOC PROF CHEM, VITERBO COL, 69-, ACAD DEAN, 76- *Mem:* AAAS; Asn Advan Health Educ. *Res:* Electron exchange reaction rates; studies of tris (ethylenediamine) ruthenium complexes; kinetic studies of ligand substitution reactions in complexes. *Mailing Add:* Viterbo Col LaCrosse WI 54601

ELSBERRY, RUSSELL LEONARD, b Audubon, Iowa, Sept 26, 41; m 63; c 4. TROPICAL METEOROLOGY, AIR-SEA INTERACTION. *Educ:* Colo State Univ, BS, 63, PhD(atmospheric sci), 68. *Prof Exp:* Jr meteorologist, Colo State Univ, 66-68; from asst prof to assoc prof, 68-79, PROF METEOROL, NAVAL POSTGRAD SCH, 79- *Mem:* Fel Am Meteorol Soc; Sigma Xi. *Res:* Diagnosis and prediction of tropical weather disturbances; large-scale air-sea interaction; extratropical cyclogenesis. *Mailing Add:* 501 Country Club Dr Carmel Valley CA 93924

ELSENBAUMER, RONALD LEE, b Allentown, Pa, July 18, 51; m 73; c 3. ORGANIC CHEMISTRY, CONDUCTING POLYMERS. *Educ:* Purdue Univ, BS, 73; Stanford Univ, PhD(org chem), 77. *Prof Exp:* Res chemist, Allied Chem Co, 77-79, sr res chemist, 79-83, RES ASSOC, ALLIED CHEM CORP, 83- *Mem:* Am Chem Soc. *Res:* Asymmetric organic reactions; reaction mechanisms; industrial chemistry; electrically conducting polymers; lithium, polymer batteries; organic batteries. *Mailing Add:* Allied-Signal Inc PO Box 1021R Morristown NJ 07962-1001

ELSEVIER, ERNEST, b Amsterdam, Holland, Dec 26, 14; nat US; m 44; c 2. MECHANICAL ENGINEERING. *Educ:* Ala Polytech Inst, BS, 49; Ga Inst Technol, MS, 50. *Prof Exp:* Instr mech eng, Ala Polytech Inst, 49; grad asst mech eng, Ga Inst Technol, 50; from instr to asst prof, 50-57, ASSOC PROF MECH ENG, COL ENG, DUKE UNIV, 57- *Concurrent Pos:* Res assoc, John D Latimre & Assoc, 58-; consult, Pub Serv Co, NC & Burlington Industs, 64-; partner, Gardner, Elsevier & Kline; mem adv bd, Fayetteville Technol Inst; mem adv bd, NC State Bd Prof Engrs & Land Surveyors, 67-70, secy, 70. *Honors & Awards:* Award, Am Soc Mech Engrs, 62; Gov Award, 71. *Mem:* AAAS; Am Soc Mech Engrs. *Res:* Environmental sciences and fuels; thermodynamics; internal combustion; temperature; instantaneous temperature and pressures; heating, ventilating and air conditioning. *Mailing Add:* Col Eng Duke Univ Durham NC 27706

ELSEVIER, SUSAN MARIA, b Mobile, Ala, Mar 25, 45; m 72. CELL GENETICS. *Educ:* Vanderbilt Univ, BA, 67; Univ Wis, MS, 70; Yale Univ, PhD(biol), 76. *Prof Exp:* Fel, Inst Molecularbiol II, Univ Zurich, 76-77; asst prof biol sci, Univ Pittsburgh, 78-80; AT UNIV PARIS VII. *Mem:* Sigma Xi. *Res:* Mammalian somatic cell genetics. *Mailing Add:* Univ Paris VII Two Pl Jussieu Tour 43 Paris 75005 France

ELSEY, JOHN C(HARLES), b Salt Lake City, Utah, May 27, 35; m 55; c 4. ELECTRICAL ENGINEERING. *Educ:* Univ Utah, BS, 56; Mass Inst Technol, MS, 60; Univ Ill, PhD(elec eng), 63. *Prof Exp:* MEM TECH STAFF, ROCKWELL INT CORP, 63- *Mem:* Inst Elec & Electronics Engrs. *Res:* Digital computer technology; digital image processing; switching circuit theory; programing; numerical techniques; guidance and control of aerospace vehicles; computer applications. *Mailing Add:* Rockwell Int Corp 3370 Miraloma Ave Anaheim CA 92803

ELSEY, KENT D, b Seattle, Wash, Sept 20, 41; m 82; c 1. ENTOMOLOGY. *Educ:* Wash State Univ, BS, 63; NC State Univ, MS, 66, PhD(entom), 69. *Prof Exp:* RES ENTOMOLOGIST, US VEGETABLE LAB, AGR RES SERV, USDA, 69- *Mem:* AAAS; Entom Soc Am. *Res:* Integrated and biological control of vegetable insect pests. *Mailing Add:* USDA-ARS US Veg Lab 2875 Savannah Hwy Charleston SC 29414

ELSEY, MARGARET GRACE, b St Louis, Mo, Aug 16, 29. MATHEMATICS. *Educ:* Webster Col, AB, 51; Cath Univ, MA, 60, PhD(math), 63. *Prof Exp:* Instr, Loretto High Sch, 54-58; from instr to assoc prof, 62-73, chmn dept, 62-71, PROF MATH & PHYSICS, LORETTO HEIGHTS COL, 73- *Mem:* Am Math Soc; Math Asn Am; Am Asn Physics Teachers; Nat Coun Teachers Math; Asn Educ Data Systs; Sigma Xi. *Res:* Finitely compact spaces and convergence of normal Ritt series and also computer extended calculus. *Mailing Add:* 4000 S Wadsworth Blvd Littleton CO 80123

EL-SHARKAWI, MOHAMED A, b Cairo, Egypt, Apr 28, 48; m 73; c 2. POWER SYSTEMS. *Educ:* Cairo High Inst Technol, BSc, 71; Univ Brit Columbia, MASc, 77, PhD(elec eng), 80. *Prof Exp:* Res asstelec eng, Univ BC, 75-80; asst prof, 80-85, ASSOC PROF ELEC ENG, UNIV WASH, 85- *Concurrent Pos:* Consult, Puget Sound Power & Light Co, 81-84; prin investr, Bonneville Power Admin, 82-, Boeing Military Airplane Co, 83-, Southern Calif Edison Co, 85-, Elec Power Res Inst, 85-86. *Mem:* Inst Elec & Electronics Engrs; Eng Inst Can; Can Soc Elec Eng; Sigma Xi. *Res:* Modeling, computer aided design and control of electric power systems; power electronics and electric drives systems; solid state motor control design. *Mailing Add:* Dept Elec Eng FT-10 Univ Wash Seattle WA 98195

EL-SHARKAWY, TAHER YOUSSEF, gastroenterology, electrophysiology, for more information see previous edition

ELSHERBENI, ATEF ZAKARIA, b Cairo, Egypt, Jan 8, 54; c 3. MICROWAVES, ELECTROMAGNETICS. *Educ:* Cairo Univ, BSc, 76, BSc, 79, MEng, 82; Univ Man, Can, PhD(elc eng), 87. *Prof Exp:* Teaching asst physics & math, Fac Eng, Cairo Univ, 76-82; res asst elec eng, Univ Man, 83-86, postdoctoral fel, 87; asst prof, 87-91, ASSOC PROF ELEC ENG, UNIV MISS, 91- *Concurrent Pos:* Commun syst engr, United Group Co, Cairo, Egypt, 77-80; software & syst design engr, Automated Data Syst Ctr, Cairo, Egypt, 80-82. *Mem:* Inst Elec & Electronics Engrs; Sigma Xi; Electromagnetics Acad. *Res:* Scattering and diffraction of electromagnetic waves; antenna analysis and design; numerical methods; computer aided design; computer application for under graduate education; microwaves; published over 23 journal articles. *Mailing Add:* Elec Eng Dept Univ Miss University MS 38677

EL-SHERIF, NABIL, b Cairo, Egypt, Apr 15, 38; US citizen; m; c 4. PHYSIOLOGY. *Educ:* Cairo Univ, Egypt, MB & BCh, 59, dipl internal med, 62, dipl cardiovasc dis, 63, MD, 65; Am Bd Internal Med, dipl, 75; Am Bd Cardiovasc Dis, dipl, 75. *Prof Exp:* Intern, Cairo Univ Hosp, 60-61, resident, Dept Med, 61-63, staff cardiologist & asst dir, Hemodynamic Lab, Cardiol Dept, 63-72, asst prof, Fac Med, 72-74; res assoc, Dept Cardiol, Mt Sinai Med Ctr, Miami Beach, Fla, 73-74; from asst prof to assoc prof, Dept Med, Sch Med, Univ Miami, 74-78; assoc prof, 79-83, PROF, DEPT MED, STATE UNIV NY, DOWNSTATE MED CTR, 78-, PROF, DEPT PHYSIOL, GRAD SCH, 83-; CHIEF, CARDIOL DIV, VET ADMIN MED CTR, 78- *Concurrent Pos:* Clin demonstr, Cardiol Dept, Fac Med, Cairo Univ, 63-67, lectr, 67-72; fel cardiol, Mt Sinai Med Ctr, Miami Beach, Fla, 72-73; consult, Nat Heart, Lung & Blood Insts, 74-76; dir, Coronary Care Unit, Vet Admin Hosp, Miami, Fla, 74-78; chmn, Res & Develop Comt, Vet Admin Med Ctr, Brooklyn, NY, 80-82, 84-88, 91-; cardiovasc prog specialist, Vet Admin Cent Off, Res Prog, 81-84; chief, Cardiol Div, State Univ NY, Downstate Med Ctr, 84-86, dir electrophysiol, Health Sci Ctr, Brooklyn, 86-, assoc vchmn res, 86-; vis prof, Cardiac Electrophysiol Labs, Univ Chicago, 90-91. *Mem:* Am Col Physicians; Am Col Cardiol; Am Heart Asn; Am Col Chest Physicians; Am Physiol Soc; Am Fedn Clin Res; Sigma Xi; NY Acad Sci; AAAS; Nat Asn Vet Admin Physicians. *Mailing Add:* Dept Med & Physiol State Univ NY Health Sci Ctr 450 Clarkson Ave Brooklyn NY 11203

EL-SHIEKH, ALY H, b Rahmania, Egypt, Apr 27, 31; m 58; c 2. POLYMER SCIENCE, MECHANICAL ENGINEERING. *Educ:* Univ Alexandria, BSc, 56; Mass Inst Technol, MS, 61, MechE, 64, DSc(mech eng), 65. *Prof Exp:* Instr workshop technol, Univ Alexandria, 56-58, lectr textile technol, 65-68; vis lectr, 68-70, assoc prof, 70-74, PROF TEXTILE TECHNOL, NC STATE UNIV, 74- *Mem:* Fiber Soc; fel Brit Textile Inst; Sigma Xi. *Res:* Mechanics of textile structures; fiber crimp; processing dynamics; textured yarns; spindle vibration; journal bearings; carpet mechanics; fiber migration; snagging of knitted fabrics; dynamic properties of wool yarns; yarn forming systems. *Mailing Add:* 2225 Lash Ave Raleigh NC 27607

EL-SHIMI, AHMED FAYEZ, physical chemistry, for more information see previous edition

ELSHOFF, JAMES L(ESTER), b Sidney, Ohio, Jan 3, 44; m 67; c 2. SOFTWARE ENGINEERING, PROGRAMMING ENVIRONMENTS. *Educ:* Miami Univ, BA, 66; Pa State Univ, MS, 69, PhD(comput sci) 70. *Prof Exp:* Asst comput sci, Pa State Univ, 66-69, asst elec eng, 69-70; RES SCIENTIST, GEN MOTORS RES LABS, 70- *Concurrent Pos:* Vis prof eng, Oakland Univ, 73-79; lectr, Prof Develop Sem, Am Chem Soc, 74-82, tech chmn, Nat Conf, 79; adj prof, Univ Detroit, 75-77. *Honors & Awards:* McKuen Award, 77; John M Campbell Award, 83. *Mem:* Asn Comput Mach; Inst Elec & Electronics Engrs. *Res:* Software engineering; software metrics; programming environments; microprocessor software. *Mailing Add:* Gen Motors Res Labs 12 Mile & Mound Rd Warren MI 48090

ELSIK, WILLIAM CLINTON, b Snook, Tex, Oct 8, 35; m 57; c 3. PALYNOLOGY. *Educ:* Tex A&M Univ, BS, 57, MS, 60, PhD(geol), 65. *Prof Exp:* Geologist, Exxon Co, USA, 62-90; CONSULT STRATIGRAPHER, 90- *Mem:* Am Asn Stratig Palynologists (vpres, 74-75, pres, 77-78); Mycol Soc Am; Geol Soc Am; Sigma Xi. *Res:* Fossil fungal spores and their classification and taxonomy; identification of fossil microscopic angiosperm remains other than pollen. *Mailing Add:* 12410 Stafford Springs Dr Houston TX 77077

ELSLAGER, EDWARD FAITH, b Decatur, Ill, Oct 7, 24; m 46; c 3. CANCER CHEMOTHERAPY, PARASITE CHEMOTHERAPY. *Educ:* James Millikin Univ, BS, 47; Univ Ill, MS, 48, PhD(chem), 51. *Prof Exp:* Assoc res chemist, Parke-Davis, 51-52, res chemist, 52-54, res leader, 54-58, lab dir, 58-63, group dir, 63-70, sect dir, antiinfectives, 70-75, dir basic res, 75-76, vpres res, Warner-Lambert-Parke-Davis, 76-82, asst legal secy, 83-86, vpres chemotherapy, 82-86; PRES, ELSLAGER ASSOCS, 86- *Concurrent Pos:* Asst ed, J Heterocyclic Chem, 68-70, assoc ed, 71-; vpres, Int Cong Heterocyclic Chem, 75-77; mem, China deleg, Nat Acad Sci, 75. *Mem:* Am Chem Soc; Am Asn Cancer Res; Int Soc Heterocyclic Chem (pres-elect, 73-75, pres, 76-77); hon mem Pharmaceut Soc Egypt. *Res:* Synthesis of heterocyclic compounds, quinones, amines and dyes; discovery and development of new chemotherapeutic and cardiovascular drugs; antineoplastics, antibacterials, antivirals, anthelmintics, antimalarials, amebicides, antischistosomals, antifilarial, antitrydanosomal, antileprotic, antithrombotic and hypocholestermic agents. *Mailing Add:* 4081 Thornoaks Dr Ann Arbor MI 48104-4253

ELSNER, NORBERT BERNARD, b Queens Village, NY, Oct 25, 33; m 58; c 2. METALLURGICAL ENGINEERING. *Educ:* Va Polytech Inst, BS, 55; Pepperdine Univ, MBA, 76. *Prof Exp:* Res assoc, Solar Div, Int Harvester, 59-61; staff assoc thermionic power conversion, 61-64, mgr Thermoelec Prog, Gen Atomic, Inc, 64-85, dir oper, ETC, 86-88; PRES, HI-Z TECHNOL, 88- *Mem:* Am Soc Metals. *Res:* Thermoelectric materials; alloy development; joining of semiconductors to conductors; high temperature testing; refractory metals; vapor deposition of W; mechanical forming; specialty braze alloys; diffusion bonding and welding; reentry Oxide fuel capsules; behavior of organic insulations in inert atmospheres. *Mailing Add:* 5656 Soledad Rd La Jolla CA 92037

ELSNER, ROBERT, b Boston, Mass, June 3, 20; m 46; c 3. COMPARATIVE PHYSIOLOGY, MARINE BIOLOGY. *Educ:* NY Univ, BA, 50; Univ Wash, MS, 55, PhD(physiol), 59. *Prof Exp:* Res physiologist, Arctic Aeromed Lab, 53-56; physiologist, Inst Andean Biol, Peru, 59-61; assoc res physiologist, Scripps Inst Oceanog, Univ Calif, San Diego, 63-70, assoc prof, 70-73; PROF MARINE PHYSIOL, INST MARINE SCI, UNIV ALASKA, 73- *Concurrent Pos:* USPHS res career develop award, 65-70; mem biol & med sci comt polar res, Nat Acad Sci-Nat Res Coun, 71-74; US deleg mem, US-USSR Marine Mammal Proj, Environ Protection Agreement, 73- *Mem:* AAAS; Am Physiol Soc; Arctic Inst NAm; Microcirculatory Soc; Undersea Med Soc; Sigma Xi. *Res:* Marine physiology; marine mammal biology and ecology; comparative physiology of diving. *Mailing Add:* Inst Marine Sci Univ Alaska Fairbanks AK 99775-1080

ELSOHLY, MAHMOUD AHMED, b Egypt, Dec 31, 45; US citizen; m 72; c 4. DRUGS OF ABUSE, ANALYTICAL TESTING. *Educ:* Cairo Univ, BSc, 66, MSc, 71; Univ Pittsburgh, PhD(pharmacog), 75. *Prof Exp:* Instr, Dept Pharm, Cairo Univ, 66-72; teaching fel, Dept Pharmacog, Univ Pittsburgh, 72-75; res assoc, 75-76, actg asst dir, 76-79, from res asst prof to res assoc prof, 78-84, RES PROF & ASST DIR, RES INST PHARMACEUT SCI, SCH PHARM, UNIV MISS, UNIVERSITY, 84 -; PRES & LAB DIR, ELSOHLY LABS, INC, OXFORD, MISS, 85 - *Concurrent Pos:* Pharmacist, Egyptian Co Pharmaceut Trading, Cairo, 66; consult, Eradication of Cannabis & Coca, Dept State, 82-, UN Indust Orgn to set up natural prod res facilities in Tanzania & Oman, 82, Air Force Drug Testing Lab, Brooks AFB, Tex, 84- & Army Urinalysis Testing Prog, Ft Meade, Md, 84. *Mem:* Sigma Xi; Am Chem Soc; fel Am Inst Chemists; Am Pharmaceut Asn; Am Acad Forensic Sci; Soc Forensic Toxicologists. *Res:* Isolation and structure elucidation of naturally occuring substances with emphasis on those substances with potential pharmacological activity; development of analytical procedures for natural products and drugs of abuse. *Mailing Add:* PO Box 373 University MS 38677

ELSON, CHARLES, b Des Moines, Iowa, July 15, 34; m 71; c 2. NUTRITION. *Educ:* Iowa State Univ, BS, 56, MS, 61; Mich State Univ, PhD(food sci), 64. *Prof Exp:* Res fel nutrit, Sch Pub Health, Harvard Univ, 64-66; from asst prof to assoc prof, 66-77, PROF NUTRIT SCI, UNIV WIS, MADISON, 77- *Mem:* AAAS; Am Inst Nutrit; Am Soc Clin Nutrit; Sigma Xi. *Res:* Nutrient induced alterations in lipid metabolism. *Mailing Add:* Dept Nutrit Sci Univ Wis Madison WI 53706

ELSON, CHARLES O, III, b Chicago, Ill, Aug 21, 42. GASTROENTEROLOGY. *Educ:* Wash Univ, MD, 68. *Prof Exp:* ASSOC PROF MED, MED COL VA, 80-, ASSOC PROF MICROBIOL-IMMUNOL, 82- *Mem:* Am Asn Immunologists; Am Fedn Clin Res; Am Gastroenterol Asn; Am Col Physicians; Gastroenterol Res Group. *Mailing Add:* Univ Ala Birmingham Div Gastro Univ Sta Birmingham AL 35294

ELSON, ELLIOT, b St Louis, Mo, June 15, 37; m; c 2. BIOCHEMISTRY. *Educ:* Harvard Univ, AB, 59; Stanford Univ, PhD(biochem), 65. *Prof Exp:* Fel chem, Univ Calif, San Diego, 65-68; from asst prof to prof chem, Cornell Univ, 68-78, prof, 78-79; PROF BIOL CHEM, SCH MED, WASHINGTON UNIV, 79- *Mem:* Sigma Xi; Am Soc Biol Chemists; Am Soc Cell Biol. *Res:* Physical chemistry of nucleic acids and proteins; interactions between cell surface and cytoskeleton; cell surface phenomena; cell mechanics. *Mailing Add:* Dept Biol Chem Sch Med Washington Univ St Louis MO 63110

ELSON, HANNAH FRIEDMAN, b Poland, July 10, 43; US citizen; m 73; c 2. BIOCHEMISTRY, CELL BIOLOGY. *Educ:* Vassar Col, BA, 64; Mass Inst Technol, PhD(biophys), 70. *Prof Exp:* Fel cell biol, MRC Lab Molecular Biol, Cambridge, Eng, 70-72; asst prof biol, Univ Calif, San Diego, 72-77, asst res biologist, 77-79; res pathologist, Vet Admin Med Ctr, La Jolla, 79-82; Nat Res Coun-Nat Acad Sci, sr res assoc, Walter Reed Army Inst Res, Washington, DC, 88-90; SR STAFF FEL, LAB MOLECULAR CARDIOL, NAT HEART & LUNG INST, NIH, BETHESDA, MD, 90- *Concurrent Pos:* Fel, Arthritis Found, 70-72; dir res, Medsat Res Co, Bethesda, 82-88. *Mem:* Am Chem Soc; Sigma Xi; Am Soc Cell Biol. *Res:* Membrane proteins of differentiating skeletal muscle cells in cell culture; modulation of development by components of the extracellular matrix; identification and isolation of membrane receptors; biosynthesis of membrane proteins; muscular dystrophy; developmental biology; gene therapy. *Mailing Add:* 6320 Lenox Rd Bethesda MD 20817-6024

ELSON, JESSE, b Brooklyn, NY, Apr 6, 10; m 40; c 2. PHYSICAL CHEMISTRY. *Educ:* Rutgers Univ, BS, 37, PhD, 53; NC State Col, MS, 39; Va Polytech Inst, BS, 43. *Prof Exp:* Soil surveyor, Soil Conserv Serv, 38-39; soil technologist, Exp Sta, Va Polytech Inst, 39-41; proj supvr, Soil Conserv Serv, USDA, 41-42; soil technologist, Exp Sta, Va Polytech Inst, 42-43; PROF CHEM, DELAWARE VALLEY COL, 46- *Concurrent Pos:* Res fel, NC State Col & Soil Conserv Serv. *Mem:* Fel Am Inst Chem; Am Chem Soc. *Res:* Relationship between atom size and ion size of element and its position in the periodic table, calculation of bonding parameters; derivation of empirical bond energy equations. *Mailing Add:* Box 484 Doylestown PA 18901-0484

ELSON, JOHN ALBERT, b Kaiting, China, Mar 2, 23; m 57; c 2. QUATERNARY GEOLOGY. *Educ:* Univ Western Ont, BSc, 45; McMaster Univ, MSc, 47; Yale Univ, MS, 50, PhD, 56. *Prof Exp:* Instr geog, McMaster Univ, 45-46; geologist, Geol Surv, Can, 46-56; from asst prof to prof, McGill Univ, 56-87, dir grad studies, 69-87, chmn dept, 74-75; RETIRED. *Concurrent Pos:* Assoc mem, comt quaternary res, Nat Res Coun Can, 71-74; mem, Comn Genesis & Lithol Quaternary Deposits, Int Asn Quaternary Res, 74-78. *Mem:* Fel Geol Soc Am; fel Geol Asn Can; Glaciol Soc; Am Quaternary Asn; Am Soc Photogram; Can Quaternary Asn. *Res:* Surficial geology in Southwestern Manitoba; Quaternary geology; Glacial Lake Agassiz; Champlain Sea; freeze-thaw processes; glacial lakes. *Mailing Add:* Dept Geol Sci McGill Univ 3450 Univ St Montreal PQ H3A 2A7 Can

ELSON, LEE STEPHEN, b Chicago, Ill, June 26, 47; m 72; c 2. ATMOSPHERIC SCIENCES. *Educ:* Univ Calif, Berkeley, BS, 69; Univ Wash, MS, 70, PhD(atmospheric sci), 75. *Prof Exp:* Resident res assoc planetary atmospheres, NASA-Nat Res Coun, 76-78; sr scientist planetary atmospheres, 78-80, MEM TECH STAFF, JET PROPULSION LAB, CALIF INST TECHNOL, 80- *Mem:* Am Meteorol Soc; Am Astron Soc. *Res:* Winds and thermal structure of the upper atmospheres of Mars and Venus from remote sensing; normal modes of planetary atmospheres. *Mailing Add:* 183/301 Jet Propulsion Lab Cal Tech 4800 Oak Grove Dr Pasadena CA 91109

ELSON, ROBERT EMANUEL, inorganic chemistry, for more information see previous edition

ELSPAS, B(ERNARD), b New York, NY, July 26, 25; m 51; c 2. COMPUTER SCIENCES. *Educ:* City Col New York, BEE, 46; NY Univ, MEE, 48; Stanford Univ PhD(elec eng), 55. *Prof Exp:* Instr, City Col New York, 46-49; from res asst to res assoc, Electronics Lab, Stanford Univ, 51-55; sr res engr, Comput Tech Lab, 56-58, staff scientist, SRI Int, 68-86; RETIRED. *Concurrent Pos:* Vis lectr, Dept Elec Eng, Stanford Univ & Univ Hawaii. *Mem:* Sigma Xi; fel Inst Elec & Electronics Engrs; Asn Comput Mach. *Res:* Expert systems; software technology; information theory; electronic computers. *Mailing Add:* 3464 Janice Way Palo Alto CA 94303

ELSTON, CHARLES WILLIAM, power generation; deceased, see previous edition for last biography

ELSTON, DONALD (PARKER), b Chicago, Ill, June 17, 26; m 47; c 4. PALEOMAGNETISM. *Educ:* Syracuse Univ, AB, 50, MS, 51; Univ Ariz, PhD, 68. *Prof Exp:* Geologist, Mineral Deposits & Fuels Br, 53-62, Br Astrogeol & Br Petrophys & Remote Sensing, 62-73, GEOLOGIST, REGIONAL GEOPHYS, US GEOL SURV, 73- *Concurrent Pos:* Res assoc geol, Univ Ariz, 67-69. *Honors & Awards:* Nininger Meteorite Award, Ctr Meteorite Studies, Ariz State Univ, 68; Antarctica Serv Medal, NSF, 81. *Mem:* AAAS; Geol Soc Am; Am Asn Petrol Geol; Am Geophys Union; Meteoritical Soc. *Res:* Geologic field investigations of salt structures and uranium deposits, Colorado Plateau; lunar geologic mapping; manned lunar exploration studies; Apollo 16 geologic support, Earth Resources Technology Satellite and Skylab studies; paleomagnetism and magnetostratigraphy of Precambrian and Phanerozoic rocks; paleomagnetism of glaciogenic sediment, Dry Valleys and McMurdo Sound, Antarctica; climate change studies. *Mailing Add:* US Geol Surv 2255 N Gemini Dr Flagstaff AZ 86001

ELSTON, STUART B, b Elmira, NY, Mar 11, 46. ATOMIC PHYSICS. *Educ:* Rochester Inst Technol, BS, 68; Univ Mass, MS, 72, PhD(physics), 75. *Prof Exp:* res assoc, 75-79, asst prof, 79-83, ASSOC PROF ATOMIC PHYSICS, UNIV TENN, KNOXVILLE, 84- *Mem:* Am Phys Soc. *Res:* Heavy ion collision studies, emphasis on auger and autoionization electron spectroscopy. *Mailing Add:* 1517 Whitower Rd Knoxville TN 37919

ELSTON, WOLFGANG EUGENE, b Berlin, Ger, Aug 13, 28; nat US; m 52; c 2. VOLCANOLOGY, ECONOMIC GEOLOGY & PLANETARY GEOLOGY. *Educ:* City Col New York, BS, 49; Columbia Univ, AM & PhD(geol), 53. *Prof Exp:* Lectr geol, City Col New York, 49-51; lectr, Columbia Univ, 51-52; asst prof, Tex Tech Col, 55-57; from asst prof to assoc prof geol, 57-67, actg chmn dept chem, 82, PROF GEOL, UNIV NMEX, 67-, COORDR, VOLCANOLOGY PROG, 90- *Concurrent Pos:* NASA res grants, 64-; vis scientist, Am Geophys Union, 68 & 71; vis geol scientist, Am Geol Inst, 68; NSF res grants, 78-; exchange scientist, Univ Queensland, Australia, 79; vis scientist, US Geol Surv, Hawaii Volcano Observ, 78-79; found vis, Univ Auckland, NZ, 85-86; guest res fel, Royal Soc, UK, 86. *Mem:* Fel AAAS; fel Geol Soc Am; Int Astron Union; Am Inst Prof Geologists; fel Meteoritical Soc; Sigma Xi; Soc Econ Geologists. *Res:* Volcanology, economic geology, planetology. *Mailing Add:* Dept Geol Univ NMex Albuquerque NM 87131

EL-SWAIFY, SAMIR ALY, b Port Said, Egypt, July 14, 37; m 61; c 3. SOIL SCIENCE, SOIL & WATER CONSERVATION. *Educ:* Univ Alexandria, BS, 57; Univ Calif, Davis, PhD(soil sci), 64. *Prof Exp:* Asst researcher, Soil Salinity Lab, Alexandria, Egypt, 58-59; fel soil chem, Univ Calif, Riverside, 64-65; from asst prof to assoc prof soil sci, 65-75, PROF SOIL SCI, UNIV HAWAII, 75-, CHMN, DEPT AGRON & SOILS, 85- *Concurrent Pos:* Vis scientist, Commonwealth Sci & Indust Res Orgn, Australia, 73; adj res fel, E-W-Ctr, Environ & Policy Inst, Honolulu, 80; prin scientist, Int Crops Res Semi-acid Trop, Hyderabad, India, 82-84. *Mem:* Fel Am Soc Agron; Soil Sci Soc Am; Int Soc Soil Sci; Soil Conserv Soc Am. *Res:* Soil and water conservation; soil salinity; irrigation water quality; physicochemical properties of tropical soils; clay colloidal properties and rheology; swelling properties of clays; soil structural stability in relation to sesquioxides; soil erosion. *Mailing Add:* Dept Soil Sci Hig 410 Univ Hawaii at Manoa 2500 Campus Rd Honolulu HI 96822

ELSWORTH, DEREK, b Glasgow, Scotland. GEOMECHANICS. *Educ:* Portsmouth Polytechnic, Eng, BSc, 79; Imperial Col, London Univ, MSc, DIC, 80; Univ Calif, Berkeley, PhD(eng), 84. *Prof Exp:* Asst prof civil eng, Univ Toronto, 84-85; res assoc prof, 88-89, ADJ PROF, UNIV WATERLOO, 90-; ASST PROF MINERAL ENG, PA STATE UNIV, 85- *Concurrent Pos:* Komex Consult Ltd, 80-82; Lawrence Berkeley Lab, 84. *Honors & Awards:* Manuel Rocha Medal, Int Soc Rock Mech, 87- *Mem:* Am Geophys Union; Am Soc Civil Engrs; Am Acad Mech; Can Geotech Soc; Int Soc Rock Mech; Mat Res Soc. *Res:* Research into the mechanical, hydraulic and transport behavior of fractured geologic media. *Mailing Add:* 119 Mineral Sci Bldg University Park PA 16802

ELTERICH, G JOACHIM, b Dresden, Ger, May 22, 30; US citizen; m 61; c 3. AGRICULTURAL ECONOMICS. *Educ:* Rheinische Fredrich-Wilhelms Univ Bonn, dipl sci agr, 56; Univ Ky, MS, 60; Mich State Univ, PhD(agr econ), 64. *Prof Exp:* Res assoc, Rheinische Friedrich-Wilhelms Univ Bonn, 64-67; PROF AGR ECON, UNIV DEL, 67- *Concurrent Pos:* Vis prof, Univ Kiel, Germany, 74; innovative teaching grant comp assisted instr, Univ Del, 75; consult, US Agency Int Develop, 77-, & Minimum Wage Study Comn, 80-81; Fulbright grant, 85-86; chair, Opers Res Comt, 86- *Mem:* Am Agr Econ Asn; Am Econ Asn; Sigma Xi; Int Asn Agr Econ; Western Econ Asn Int; Oper Res Soc Am. *Res:* Quantitative supply analysis (milk); economic analysis of impact of Manpower programs; unemployment insurance, minimum wage, programming of agricultural production and risk analysis; evaluation and prescription of development projects; agricultural policy. *Mailing Add:* 145 Timberline Dr Newark DE 19711

ELTGROTH, PETER GEORGE, b Baltimore, Md, Sept 17, 40; m 67. PLASMA PHYSICS. *Educ:* Calif Inst Technol, BS, 62; Harvard Univ, AM, 63, PhD(physics), 66. *Prof Exp:* Nat Res Coun-Nat Acad Sci res fel astrophysics, NASA, 66-67; PHYSICIST, LAWRENCE LIVERMORE LAB, UNIV CALIF, 67- *Res:* Hydrodynamics; astrophysics; general relativity. *Mailing Add:* 1245 Lillian Livermore CA 94550

ELTHERINGTON, LORNE, b Calgary, Alta, June 2, 33; m 60; c 4. ANESTHESIOLOGY, PHARMACOLOGY. *Educ:* Univ BC, BA, 57; Univ Wash, MSc, 59, PhD(pharmacol), 61; Univ Calif, San Francisco, MD, 67. *Prof Exp:* Instr pharmacol, Univ Wash, 61-62; intern San Francisco Gen Hosp, 67-68; resident, Dept Anesthesiol, Med Ctr, Univ Calif, San Francisco, 68-70; vis prof pharmacol, Mahidol Univ, Bangkok, 70-74; ASSOC PROF ANESTHESIA, STANFORD UNIV, 74- *Concurrent Pos:* NIH grant, 61-63. *Mem:* Sigma Xi. *Res:* Rational use of drugs, local anesthesia for outpatient surgery; methods of chronic pain relief. *Mailing Add:* Dept Anesthesia Sch Med Stanford Univ Stanford CA 94305

ELTHON, DONALD L, b Forest City, Iowa, Dec 15, 52. IGNEOUS PETROLOGY, GEOCHEMISTRY. *Educ:* Univ Iowa, BS, 76; Columbia Univ, MA, 78, MPhil, 80, PhD(geochem), 80. *Prof Exp:* Fel, Carnegie Inst, Wash, 79-80; res assoc geol, Columbia Univ, 80-81; asst prof, 81-87, ASSOC PROF PETROL, UNIV HOUSTON, 87- *Concurrent Pos:* Vis res assoc, Smithsonian Inst, 81-83; mem, Deep Sea Drilling Prof, Ocean Crust Panel, 82-83; vis scientist, Lunar & Planetary Inst, 85-86; prin investr, Ctr Anal Facil, Tex Ctr Superconductivity, 88- *Mem:* Mineral Soc Am; Geochem Soc; Am Geophys Union. *Res:* Petrology and geochemistry of basalts; evolution of Earth's mantle. *Mailing Add:* Dept Chem Univ Houston Houston TX 77204-5641

ELTHON, THOMAS EUGENE, b Forest City, Iowa, Feb 7, 54. PHYSIOLOGY, BIOCHEMISTRY. *Educ:* Ariz State Univ, BS, 77; Iowa State Univ, MS, 80, PhD(bot), 83. *Prof Exp:* Postdoctoral fel mitochondrial res, Univ Pa, 83-84; postdoctoral fel, Mich State Univ-Dept Energy Plant Res Lab, 84-85, NSF fel, 85-87; ASST PROF MITOCHONDRIAL RES, UNIV NEBR, LINCOLN, 89- *Concurrent Pos:* Vis asst prof mitochondrial res, Dept Biol Sci, Univ Md, Baltimore County, 89- *Mem:* Am Soc Plant Physiologists. *Res:* Physiology, biochemistry and molecular biology of plant mitochondria. *Mailing Add:* Sch Biol Sci Manter Hall Rm 348 Univ Nebr Lincoln NE 68588-0118

ELTIMSAHY, ADEL H, b Damanhoor, Egypt, June 10, 36; m 67. ELECTRICAL ENGINEERING. *Educ:* Cairo Univ, BS, 58; Univ Mich, Ann Arbor, MS, 61, PhD(elec eng), 67. *Prof Exp:* Res asst elec eng, Nat Res Ctr, Egypt, 58-59 & 62; consult temperature control, Maxitrol Co, Mich, 66; asst prof elec eng, Univ Tenn, 67-68; from asst prof to assoc prof, 68-78, chmn elec eng dept, 80-88, PROF ELEC ENG, UNIV TOLEDO, 78- *Concurrent Pos:* Consult, Nat Bur Standards, 73-74 & DANA Corp, 75. *Honors & Awards:* Paper Award, Indust Appln Soc, Inst Elec & Electronics Engrs, 77. *Mem:* Inst Elec & Electronics Engrs; Simulation Coun; Int Solar Energy Soc; Am Soc Eng Educ. *Res:* The optimal control of solar heating and photovoltaic systems; optimal control of robotic systems; system simulation. *Mailing Add:* Dept Elec Eng Univ Toledo Toledo OH 43606

ELTINGE, LAMONT, b Chicago, Ill, May 9, 26; m 53; c 2. RESEARCH ADMINISTRATION, ENGINEERING. *Educ:* Purdue Univ, BS, 47; Ill Inst Technol, MS, 56, PhD(mech eng), 65. *Prof Exp:* Trainee/foreman, Electro-Motive Div, Gen Motors Co, 47-51; sect leader, Am Oil, Standard Oil, Co, Inc, 51-62; res asst, Inst Gas Technol, Ill Inst Technol, 63-64; dir, Automotive Res, Ethyl Corp, 64-68; vpres res & technol, Cummins Engine Co, Inc, 68-73; DIR RES, EATON CORP, 73- *Concurrent Pos:* Soc Automotive Engrs fel, Office Sci & Technol Policy, US White House, 89-90. *Honors & Awards:* Horning Award, Soc Automotive Engrs, 73. *Mem:* Soc Automotive Engrs (pres, 91); Indust Res Inst; Am Soc Mech Engrs; Sigma Xi. *Res:* Energy; engines; fuels and lubricants; identification and application of emerging technology to transporation and industrial systems and components. *Mailing Add:* Eaton Corp 26201 Northwest Hwy PO Box 766 Southfield MI 48037

ELTON, EDWARD FRANCIS, b Teaneck, NJ, Dec 3, 35; m 57; c 5. CHEMICAL ENGINEERING, PULP & PAPER TECHNOLOGY. *Educ:* Stevens Inst Technol, ME, 57; Lawrence Col, MS, 59, PhD(chem kinetics), 62. *Prof Exp:* From asst prof to assoc prof chem eng, Univ Maine, 62-70; supvr pulp processes res & develop, Am Can Co, 70-76; corp coordr pulp & paper, Air Prod & Chem Inc, 76-82; tech dir, Herty Found, 82-84; PRES, PAPER RES ASSOC, 84- *Concurrent Pos:* NSF eng res initiation grant, 64-66. *Mem:* Tech Asn Pulp & Paper Indust; Am Chem Soc. *Res:* Chemical kinetics; kinetics of delignification reaction; black liquor oxidation; oxygen bleaching; oxygen pulping; bleach process development; chemical reactor design; biochemical engineering. *Mailing Add:* PO Box 1139 North Highlands CA 95660-1139

ELTON, RAYMOND CARTER, b Baltimore, Md; m; c 3. ATOMIC PHYSICS, PLASMA PHYSICS. *Educ:* Va Polytech Inst, BS, 53; Univ Md, MS, 56, PhD(physics), 63. *Prof Exp:* Physicist ballistics, US Naval Weapons Ctr, Dahlgren, Va, 51-52; electronic engr radar, Bendix Corp, Md, 53-54; asst upper atmosphere physics, Univ Md, 54-58; PHYSICIST ATOMIC, LASER & PLASMA PHYSICS, NAVAL RES LAB, 58- *Concurrent Pos:* Physicist missile guid, Appl Physics Lab, Johns Hopkins Univ, 55; teaching, Univ Md, Am Univ Washington DC & Catholic Univ of Am. *Mem:* Fel Am Phys Soc; fel Optical Soc Am; AAAS; Sigma Xi (pres). *Res:* Research on x-ray lasers; spectroscopy on high temperature plasmas in vacuum ultraviolet region; solar and astrophysical spectroscopy; atomic physics as related to plasmas. *Mailing Add:* Naval Res Lab Code 4733 Washington DC 20375-5000

ELTZ, ROBERT WALTER, b Callicoon, NY, June 22, 32; m 62; c 3. BIOTECHNOLOGY. *Educ:* Rensselaer Polytech Inst, BS, 53; Cornell Univ, PhD(bact), 58. *Prof Exp:* Asst Cornell Univ, 53-57; res microbiologist, Chas Pfizer & Co, Inc, 57-61; res microbiologist, Sun Oil Co, Pa, 61-67, chief appl microbiol sect, 67-68, asst to mgr, basic res div, 69-70, tech planning analyst, 70; dir, biol process develop, E R Squibb & Sons, Inc, 70-80; tech dir, Krause Milling Co, 80-82; dir, bioprocess develop, 82-89, DIR, BIOPROCESS TECHNOL, CORP RES, MONSANTO CO, 89- *Concurrent Pos:* Secy-treas, Microbiol & Biochem Technol Div, Am Chem Soc, 77-80, chmn-elect, 80-81, chmn, 81-82. *Mem:* Am Soc Microbiol; Sigma Xi; Am Chem Soc. *Res:* Microbial metabolism; biochemical production of antibiotics and organic chemicals; hydrocarbon processing; grain processing. *Mailing Add:* Corp Res GG3K Monsanto Co 700 Chesterfield Village Pkwy St Louis MO 63198-0001

ELTZE, ERVIN MARVIN, b Crete, Nebr, May 9, 38; m 63; c 3. MATHEMATICAL ANALYSIS. *Educ:* Doane Col, BA, 60; Univ SDak, MA, 62; Iowa State Univ, PhD(math), 70. *Prof Exp:* Instr math, Creighton Univ, 62-65 & Iowa State Univ, 68-70; from asst prof to assoc prof, 70-79, PROF MATH, FT HAYS STATE UNIV, 79- *Mem:* Am Math Soc; Math Asn Am; Soc Indust & Appl Math; Asn Comput Mach. *Res:* Integration. *Mailing Add:* Dept Math Ft Hays State Univ 600 Park St Hays KS 67601

ELVEBACK, LILLIAN ROSE, b Sidney, Mont, Dec 5, 15. BIOSTATISTICS. *Educ:* Univ Minn, BA, 41, PhD(statist), 55; Columbia Univ, MA, 48. *Prof Exp:* Instr math, Univ Minn, Minneapolis, 43-44; tech aide, Nat Defense Res Coun, 44-45; instr biostatist, Sch Pub Health, Columbia Univ, 46-50; lectr, Univ Minn, 55-59; prof statist, Tulane Univ, 59; head sect, Pub Health Res Inst, New York, 60-65; PROF BIOSTATIST, MAYO MED SCH, UNIV MINN & MAYO CLIN, 65- *Concurrent Pos:* Consult epilepsy, Nat Inst Neurol Dis & Stroke, 72-; clin appln & prev adv comt, Nat Heart, Lung & Blood Inst, 75-80. *Mem:* Am Epidemiol Soc; fel Am Pub Health Asn; Biomet Soc; Inst Math Statist; fel Am Statist Asn. *Res:* Statistical methods in the design, execution and evaluation of experimental research in biological science, medicine and public health. *Mailing Add:* Mayo Clin Rochester MN 55902

ELVERUM, GERARD WILLIAM, JR, b Minneapolis, Minn, Sept 29, 27; m 48; c 2. ROCKET PROPULSION, DIRECTED ENERGY. *Educ:* Univ Minn, BS, 49. *Prof Exp:* Group supvr combustion, Jet Propulsion Lab, Calif Inst Technol, 49-59; dept mgr propulsion, Space Technol Labs, TRW, 59-63, lab dir, TRW Systs Group, 63-66, opers mgr, 66-81, vpres & gen mgr, Appl Technol Div, TRW Space & Defense, 81-90; RETIRED. *Concurrent Pos:* Chmn, Propulsion Subcomt, Nat Res Coun Space Shuttle Rev Bd, 74-75, Propulsion Group, NASA Space Systs & Technol Adv Comt, 78- 81 & Propulsion, Power, Hi-Power Dir Energy Group, Star Comt, 88-90; mem, Space Power & Propulsion, NASA Res & Tech Adv Comt, 75-78, Liquid Rocket Rev Comt, Nat Acad Eng, 80-81, NASA Aerospace Safety Adv Panel, 82-90 & Nat Res Coun Shuttle Criticality & Hazard Audit Comt, 86-88; comnr, Nat Res Coun Comn Eng & Tech Systs, 91- *Honors & Awards:* Spec Achievement Award, Am Soc Mech Engrs, 71; James H Wyld Propulsion Award, Am Inst Aeronaut & Astronaut, 73. *Mem:* Nat Acad Eng; Sigma Xi; Am Inst Aeronaut & Astronaut. *Res:* High and low thrust rocket propulsion; high-power directed energy devices; electro-optics; space science instruments. *Mailing Add:* 2220 Potrillo Rd Rolling Hills Estates CA 90274

ELVING, PHILIP JULIBER, b Brooklyn, NY, Mar 14, 13; m 37; c 2. ANALYTICAL CHEMISTRY. *Educ:* Princeton Univ, AB, 34, AM, 35, PhD(chem), 37. *Prof Exp:* Instr chem, Pa State Col, 37-39; from instr to asst prof analytical chem, Purdue Univ, 39-43; asst dir chem res, Publicker Indust, Inc, 43-47; assoc prof chem, Purdue Univ, 47-49; prof analytical chem, Pa State Col, 49-52; PROF CHEM, UNIV MICH, ANN ARBOR, 52-, H H WILLARD PROF CHEM, 81- *Concurrent Pos:* Vis lectr, Harvard Univ, 51-52 & Hebrew Univ, Jerusalem, 66, 73 & 81. *Honors & Awards:* Anachem Award, Am Chem Soc, 57, Fisher Award, 60. *Mem:* Am Chem Soc; Electrochem Soc. *Res:* Polarography and electrochemistry of organic compounds; methods of inorganic and organic analysis. *Mailing Add:* 2309 Devonshire Rd Ann Arbor MI 48104

ELVIN-LEWIS, MEMORY P F, b Vancouver, BC, May 20, 33; m 57; c 2. VIROLOGY, BACTERIOLOGY. *Educ:* Univ BC, BA, 52; Univ Pa, MSc, 57; Baylor Univ, MSc, 60; Univ Leeds, PhD(microbiol), 66. *Prof Exp:* Med technologist trainee, Shaughnessy Mil Hosp, Vancouver, BC, 52-53; bacteriologist, Pearson Tuberc Hosp, 54-55; med technician, Am Soc Clin Path, 55; from asst prof to assoc prof, 67-81, PROF MICROBIOL, SCH DENT, WASH UNIV, 81- *Concurrent Pos:* Asst clin prof oral path, Sch Dent, St Louis Univ, 68; adj assoc prof biol, Wash Univ, 74-75, adj prof, 81-91. *Mem:* Am Asn Dent Socs; Am Soc Microbiol; Int Asn Dent Res; Nat Coun Int Health. *Res:* Medical botany; oral cavity bacteriology; recurrent aphthous stomatitis; epidemiology, oral virus diseases; epidemiology; phytochemicals that affect dental and medical disease; ethnobotany; ethnopharmagognosy. *Mailing Add:* Dept Biol Sch Arts & Sci St Louis MO 63130

EL-WAKIL, M(OHAMED), b Alexandria, Egypt, Mar 9, 21; m 50; c 2. MECHANICAL & NUCLEAR ENGINEERING. *Educ:* Cairo Univ, BS, 43; Univ Wis, MS, 47, PhD(mech eng), 49. *Prof Exp:* Lectr mech eng, Univ Alexandria, 50-52; res assoc, Univ Wis, 52-54; asst prof, Univ Minn, 54-55; from assoc prof to prof mech eng, 56-64, PROF MECH & NUCLEAR ENG, UNIV WIS-MADISON, 64- *Concurrent Pos:* Fulbright fel, 66 & 78. *Honors & Awards:* Am Soc Mech Engrs Award, 51; Western Elec Award, Am Soc Eng Educ, 69, Benjamin Smith Reynolds Award, 70, Nuclear Eng Award, 71; Arthur Holly Compton Award, Am Nuclear Soc, 79. *Mem:* Am Soc Mech Engrs; fel Am Nuclear Soc; Am Soc Eng Educ. *Res:* Heat and mass transfer; fuel vaporization studies; two-phase flow; nuclear power; nuclear heat transport; nuclear energy conversion. *Mailing Add:* 1513 University Ave Madison WI 53706

EL WARDANI, SAYED ALY, b Alexandria, Egypt, Feb 26, 27; m 56; c 3. OCEANOGRAPHY, ENVIRONMENTAL SCIENCES. *Educ:* Univ Alexandria, BS, 48; Univ Calif, Scripps Inst Oceanog, La Jolla, MS, 52, PhD(chem, oceanog), 56. *Prof Exp:* Sr oceanogr, Univ Wash, 56-57, res asst prof, 57-59; asst prof, Portland State Col, 59-60 & San Jose State Col, 60-63; res oceanogr, US Naval Radiol Defense Lab, 63-64; staff scientist, Lockheed Ocean Lab, San Diego, 64-68, chief scientist, Gen Ocean Sci & Resources, San Diego, 68-73; CONSULT, 73- *Concurrent Pos:* Assoc prof, US Int Univ, 64-68; dir, Middle East Prog, Nat Educ Corp, Newport Beach Calif, 83-84. *Mem:* AAAS; Am Chem Soc; Geochem Soc. *Res:* Oceanography; marine geochemistry and biogeochemistry; marine environmental pollution; impacts of water resources projects; impacts of municipal and industrial waste disposal; land impacts of wastewater reclamation; planning and management of technical assistance and training programs in developing countries. *Mailing Add:* 1830 Avenida Del Mundo Coronado CA 92118

ELWARD-BERRY, JULIANNE, b Chicago, Ill, Nov 25, 46; div; c 2. DRILLING FLUID CHEMISTRY, COLLOIDAL CLAY CHEMISTRY. *Educ:* Univ Calif, Berkeley, BS, 68, MS, 69; Univ Wis-Madison, PhD(phys chem), 78. *Prof Exp:* Chemist, Allied Chem Corp, 76-78; sr res chemist, Merck & Co, 78-80; sr res chemist, 80-83, RES SPECIALIST, EXXON PROD RES CO, 83- *Mem:* Am Chem Soc; Soc Petrol Engrs; Clay Minerals Soc; Am Asn Drilling Engrs. *Res:* Chemistry of polymer-based and high temperature water-based drilling fluids; surface and colloid chemistry of clay dispersions; molecular reaction dynamics; raman spectroscopy; rheological characterization; flocculation of colloidal clays; mechanisms of heterogeneous catalysis; polymer absorption; infrared spectroscopy; lasers. *Mailing Add:* Exxon Prod Res Co PO Box 2189 Houston TX 77252-2189

ELWELL, DAVID LESLIE, b Newton, NJ, Oct 6, 40; m 65; c 3. LOW TEMPERATURE PHYSICS, SOLAR ENERGY. *Educ:* Amherst Col, BA, 62; Duke Univ, PhD(low temperature physics), 67. *Prof Exp:* Res asst liquid helium, Duke Univ, 62-67; asst prof physics, Col Wooster, 67-76; res assoc, 77-80, res scientist, Ohio Agr Res & Develop Ctr, 80-83; SR RESEARCHER, OHIO STATE UNIV, 83- *Mem:* Am Phys Soc. *Res:* Properties of liquid helium and the nature of the superfluid transition; energy conservation; properties of solar ponds; waste heat utilization; computer modeling of soybean. *Mailing Add:* 938 N Bever St Wooster OH 44691

ELWOOD, JAMES KENNETH, b Ladysmith, Wis, Apr 21, 36; m 63; c 2. ORGANIC CHEMISTRY. *Educ:* Wis State Col, Eau Claire, BA, 58; Mich State Univ, PhD(org chem), 63. *Prof Exp:* Sr res chemist, 63-70, RES ASSOC, EASTMAN KODAK CO, 70- *Mem:* Am Chem Soc. *Res:* Heterocyclic chemistry; spectroscopy; organic dyes; reaction mechanisms; relationship between structure and color of dyes. *Mailing Add:* PO Box 281 Victor NY 14564

ELWOOD, JERRY WILLIAM, b Kalispell, Mont, Dec 18, 40. ECOLOGY. *Educ:* Mont State Univ, BS, 63; Univ Minn, PhD(aquatic ecol), 68. *Prof Exp:* Fisheries biologist, Mont Fish & Game Dept, 61-63; res asst, Univ Minn, 63-68; res assoc, 68-72, RES STAFF MEM, OAK RIDGE NAT LAB, 72- *Concurrent Pos:* Prin investr, US Dept Energy res grant, 71-73; adj prof, Univ Tenn, 78-; prin investr, NSF, 78-, adv & consult, 85-; consult, Elec Power Res Inst, 81-84, prins investr, 83- *Mem:* Am Soc Limnol & Oceanog; Ecol Soc Am; Int Asn Theoret & Appl Limnol; Sigma Xi; AAAS. *Res:* Nutrient and carbon dynamics in stream ecosystems; production biology of benthic invertebrates and fishes in streams; behavior of toxic materials and radionuclides in aquatic food chains. *Mailing Add:* 3242 Arrowhead Circle No 1 Fairfax VA 22030

ELWOOD, JOHN CLINT, b Beatrice, Nebr, Mar 5, 30; m 53; c 2. BIOCHEMISTRY. *Educ:* Willamette Univ, BS, 56; Univ Ore, Med Sch, MS, 58, PhD(biochem), 60. *Prof Exp:* From instr to prof, 61-81, prof & actg chmn biochem dept, 80-82, vchmn biochem dept, State Univ NY Upstate Med Ctr, 82-88. *Concurrent Pos:* USPHS res fel, Cancer Res Inst, Philadelphia, Pa, 60-61. *Mem:* AAAS; Am Asn Clin Chemists; fel Nat Acad Clin Biochemists; fel Am Col Nutrit. *Res:* Lipoprotein metabolism in hyperlipidemic and diabetic humans; phospholipid metabolism in relation to plasma membrane; structure; trace mineral metabolism-chromium; clinical chemistry. *Mailing Add:* Dept Biochem & Molecular Biol State Univ NY Upstate Med Ctr Syracuse NY 13210

ELWOOD, WILLIAM K, b Ashtabula, Ohio, Oct 15, 28; m 56; c 3. DENTISTRY. *Educ:* Ohio Wesleyan Univ, BA, 50; Ohio State Univ, MSc, 53, DDS, 57; Wayne State Univ, PhD(anat), 65. *Prof Exp:* Res assoc dent, Henry Ford Hosp, Detroit, 57-65; guest investr, Rockefeller Inst, 65-66; asst prof restorative dent, 66-70, asst prof anat, 66-72, asst prof oral biol, 70-72, ASSOC PROF ORAL BIOL & ANAT, MED CTR, UNIV KY, 72- *Mem:* Am Asn Anat; Am Soc Cell Biol; Int Asn Dent Res; Electron Micros Soc Am; Am Dent Asn. *Res:* Bioelectrical aspects of mineralized tissues. *Mailing Add:* Dept Anat Univ Ky Med Ctr Lexington KY 40536

ELWYN, ALEXANDER JOSEPH, b New York, NY, May 14, 27; m 52; c 3. EXPERIMENTAL NUCLEAR PHYSICS, HEALTH & RADIATION PHYSICS. *Educ:* Grinnell Col, AB, 51; Washington Univ, PhD(physics), 57. *Prof Exp:* Res assoc physics, Brookhaven Nat Lab, 56-59; from asst scientist to scientist, Argonne Nat Lab, 59-79, sr scientist, physics div, 79-82; ENG PHYSICIST, FERMI NAT ACCELERATOR LAB, 82- *Mem:* Am Phys Soc; Sigma Xi; Health Physics Soc. *Res:* Nuclear reaction mechanisms; neutron scattering and reactions; nuclear spectroscopy; neutron polarization in scattering and reactions; charged particle induced reactions; nuclear astrophysics; accelerator health physics; radiation field measurements. *Mailing Add:* Fermilab PO Box 500 Batavia IL 60510

ELWYN, DAVID HUNTER, b NY, Jan 9, 20; m 41; c 4. BIOCHEMISTRY. *Educ:* Columbia Univ, AB, 41, PhD(biochem), 50. *Prof Exp:* Instr biochem, Harvard Med Sch, 53-54, assoc, 54-57, asst prof, 57-60; res assoc & asst dir dept surg res, Michael Reese Hosp & Med Ctr, 60-63; assoc dir dept surg res, Hektoen Inst Med Res, Cook County Hosp, 63-68; asst prof biol chem, Med Sch, Univ Ill, 64-68; assoc prof surg, Mt Sinai Sch Med, 68-74; res scientist, Col Physicians & Surgeons, Columbia Univ, 75-87; RES SCIENTIST, ALBERT EINSTEIN COL MED & MONTEFIORE MED CTR, 87-; RES ASSOC, ST LUKES-ROOSEVELT HOSP CTR, 90- *Concurrent Pos:* Life Ins Med Res fel, Columbia Univ, 50-53. *Mem:* AAAS; Am Soc Biol Chemists; AAAS. *Res:* Amino acid metabolism; metabolism of trauma and sepsis; parenteral nutrition; in vivo reaction rates. *Mailing Add:* 329 Spring St Ossining NY 10562

ELY, BERTEN E, III, b Newark, NJ, Nov 26, 48; m 70; c 2. MICROBIAL GENETICS. *Educ:* Tufts Univ, BS, 69; Johns Hopkins Univ, PhD(biol), 73. *Prof Exp:* From asst prof to assoc prof, 73-85, PROF BIOL, UNIV SC, 85-, DIR, INST BIOL RES & TECHNOL, 90- *Concurrent Pos:* Vis prof, Calif Tech, 85. *Mem:* Am Soc Microbiol. *Res:* Genetics of differentiation in Caulobacter crescentus; genetic regulation of Salmonella histidine operon; cloning the Caulobacter crescentus genome; pulsed field gel electrophoresis of large DNA molecules; genetic analysis of striped bass. *Mailing Add:* Dept Biol Univ SC Columbia SC 29208

ELY, CHARLES A, anatomy, for more information see previous edition

ELY, CHARLES ADELBERT, b Wellsboro, Pa, Jan 8, 33; m 57; c 2. VERTEBRATE ZOOLOGY. *Educ:* Pa State Univ, BS, 55; Univ Okla, MS, 57, PhD(zool), 60. *Prof Exp:* From asst prof to assoc prof, 60-67, PROF ZOOL, FT HAYS KANS STATE COL, 67- *Concurrent Pos:* Field dir Pac proj biol res, Div Birds, Smithsonian Inst, 63-66. *Mem:* Am Ornith Union; Cooper Ornith Soc; Wilson Ornith Soc. *Res:* Avian distribution and taxonomy; speciation; birds of Mexico, High Plains area, United States and central Pacific. *Mailing Add:* Dept Zool Ft Hays State Univ 600 Park St Hayes KS 67601

ELY, DANIEL LEE, b Dayton, Ohio, Feb 6, 45; m 67; c 2. MEDICAL PHYSIOLOGY, NEUROENDOCRINOLOGY. *Educ:* Univ Southern Calif, BA, 67, MS, 69, PhD(physiol), 71. *Prof Exp:* NIH fel physiol, Univ Southern Calif, 71-74; lectr physiol & behavior, Univ Calif, Riverside, 74-76; from asst prof to assoc prof, 76-86, PROF BIOL & BIOMED ENG, UNIV AKRON, 86- *Concurrent Pos:* Vis prof, Swed Med Coun, Univ Gothenburg, Sweden, 83-84; mem, High Blood Pressure Coun, Am Heart Asn; vpres, med sci, Ohio Acad Sci, 90. *Mem:* Sigma Xi; Am Heart Asn; Am Physiol Soc; Int Soc Hypertension. *Res:* Neuroendocrine and autonomic mechanisms of hypertension; behavioral medicine; stress physiology; environmental physiology; risk factor assessment for coronary heart disease; myocardial preservation for heart transplant. *Mailing Add:* Dept Biol Univ Akron Akron OH 44325

ELY, DONALD GENE, b Hastings, Okla, Dec 15, 37; m 60; c 2. ANIMAL SCIENCE. *Educ:* Okla State Univ, BS, 61, MS, 65; Univ Ky, PhD(animal sci), 66. *Prof Exp:* Asst, Okla State Univ, 61-63, res asst, 63-66; asst prof animal sci, Kans State Univ, 66-68; from asst prof to assoc prof, 68-76, PROF ANIMAL SCI, UNIV KY, 76- *Mem:* Am Soc Animal Sci; Sigma Xi. *Res:* Digestion and metabolism of protein and nonprotein nitrogen sources by ruminant animals. *Mailing Add:* 2007 Camperdown Ct Lexington KY 40504

ELY, JAMES FRANK, b Indianapolis, Ind, Dec 25, 45; m 65; c 2. FLUID PROPERTIES, COMPUTER MODELLING. *Educ:* Butler Univ, BS, 68; Ind Univ, PhD(chem physics), 71. *Prof Exp:* Res assoc, Nat Bur Standards, 71-73; res assoc, Dept Chem Eng, Rice Univ, 73-75; sr res chemist, Shell Develop Co, 75-79; CHEMIST, NAT BUR STANDARDS, 79-; ADJ PROF CHEM ENG, COLO SCH MINES, 80- *Concurrent Pos:* Mem, Prog Comt, Am Inst Chem Engrs, 81-86. *Honors & Awards:* Dept Com Silver Medal, 86. *Mem:* Am Inst Chem Engrs; Sigma Xi. *Res:* Theoretical and experimental studies of the thermophysical properties of pure fluids and fluid mixtures; phase equilibria and transport phenomena. *Mailing Add:* 3470 Longwood Ave Boulder CO 80303

ELY, JOHN FREDERICK, b Chicago, Ill, Mar 20, 30; m 52; c 5. SOLID MECHANICS. *Educ:* Purdue Univ, BSCE, 54; Northwestern Univ, MS, 58, PhD(mech), 63. *Prof Exp:* Struct draftsman, Am Bridge Div, US Steel Corp, Ind, 47-50; instr civil eng, Northwestern Univ, 56-58, lectr, 58-62, asst prof civil eng & dir truss bridge res proj, 62-63; from asst prof to assoc prof civil eng & eng mech, 63-76, assoc dean acad affairs, 76-80, asst dean undergrad prog, sch eng, 80-84, PROF CIVIL ENG, NC STATE UNIV, 76- *Concurrent Pos:* NSF res grants, & 70-72. *Mem:* Am Soc Civil Engrs; Sigma Xi. *Res:* Theory of elasticity; optimization of structural systems; nonlinear stability analysis of trussed domes. *Mailing Add:* Dept Civil Eng NC State Univ Mann Hall Raleigh NC 27650

ELY, JOHN THOMAS ANDERSON, b San Francisco, Calif, June 24, 23; div; c 2. COSMIC RADIATION, GLOBAL CHANGE. *Educ:* Eastern Wash State Col, BA, 52; Univ Wash, MS, 59, PhD(physics), 69. *Prof Exp:* Physicist, Cambridge Res Labs, US Air Force, 60-66; sr res assoc physics, 69-90, RES ASSOC PROF, UNIV WASH, 91- *Concurrent Pos:* Lectr physics, Grad Prog, Northeastern Univ, 62-64; consult, space sci, Off Naval Res, 75-; teacher, radiation physics, first astronaut sch, Edwards Air Force Base. *Mem:* Am Phys Soc; Sigma Xi; AAAS; Am Geophys Union. *Res:* Discovered four modulations of cosmic rays; theory to explain: solar activity influence on weather by cosmic ray modulation of atmospheric ionization near 10 kilometer altitude and of cirrus cloud cover (crucial to understanding global warming), contributions of cosmic radiation to aging and cancer; contributions of cosmic radiation to aging and cancer; theory and experimental verification of glycemic modulation of ascorbate metabolism producing fetal anomalies, variations in immune response. *Mailing Add:* Physics Dept FM-15 Univ Wash Seattle WA 98195

ELY, KATHRYN R, b Omaha, Nebr, Oct 2, 44; m 66; c 1. BIOCHEMISTRY. *Educ:* Clarke Col, Dubuque, Iowa, BA, 66; Univ Utah, Salt Lake City, PhD(biol), 81. *Prof Exp:* Sci asst, Argonne Nat Lab, 66-74, sci assoc, 74-76; res instr biochem, 76-81, RES ASSOC PROF BIOL & ADJ ASSOC PROF BIOENG, UNIV UTAH, 81- *Mem:* Am Soc Biol Chemists. *Res:* X-ray crystallographic investigations of the three-dimensional structures of proteins that interact with nucleic acids; crystallization and characterization of proteins; structural analysis using computer graphics. *Mailing Add:* Dept Biol Univ Utah Salt Lake City UT 84112

ELY, RALPH LAWRENCE, JR, b Roney's Point, WVa, Nov 26, 17; m 48; c 4. NUCLEAR PHYSICS. *Educ:* Washington & Jefferson Col, BS, 40; Univ Colo, MS, 44; Univ Pittsburgh, PhD(physics), 51. *Prof Exp:* Asst physics, Univ Colo, 40-42, instr, 42-44; assoc physicist, US Navy Radio & Sound Lab, 44-46; instr physics, Univ Pittsburgh, 46-48; res assoc, Sarah Mellon Scaife Radiation Lab, 48-51; sr scientist, Westinghouse Atomic Power Div, 51-54; tech dir & vpres, Nuclear Sci & Eng Corp, 54-59; dir measurement & controls lab, 59-65, dir off indust serv, 65-69, assoc for res, Off VPres, 69-75, dir, Univ Rels, Res Triangle Inst, 75-88; CONSULT, 88- *Mem:* Am Phys Soc; Am Nuclear Soc; Am Soc Testing & Mat. *Res:* Application nuclear techniques to industry; nuclear instrumentation; isotope tracing; interdisciplinary research liaison, research administration, university liaison. *Mailing Add:* 5004 Timmons Dr Durham NC 27713

ELY, RAYMOND LLOYD, b Warren, Ohio, Sept 12, 19; m 42; c 3. MISSILE SYSTEMS ENGINEERING. *Educ:* Carnegie Inst Technol, BS, 40, DSc(math), 51; Calif Inst Technol, MS, 44. *Prof Exp:* Aeronaut engr, Grumman Aircraft Eng Corp, 40-42 & Aircraft Lab, Wright-Patterson AFB, 46-47; teaching asst mech, Carnegie Inst Technol, 47-50; engr struct res, Pittsburgh-Des Moines Steel Co, 50-51; engr missile design, Appl Physics Lab, Johns Hopkins Univ, 53-86; RETIRED. *Concurrent Pos:* US Air Force Officer, Aircraft Lab, Wright-Patterson AFB, 42-46 & 51-53. *Mem:* Am Inst Aeronaut & Astronaut; Am Defense Preparedness Asn; Math Asn Am; Sigma Xi. *Res:* Guided missile system design; warhead and fuze systems; structural dynamics; aerodynamics. *Mailing Add:* 3102 Homewood Pkwy Kensington MD 20895-2757

ELY, ROBERT P, JR, b Freeport, Ill, Apr 2, 30; m 52; c 4. PHYSICS. *Educ:* Mass Inst Technol, BS, 52, MS, 53, PhD(physics), 60. *Prof Exp:* Res assoc, Lawrence Berkeley Lab, 59-62, from asst prof to assoc prof, 62-75, PROF PHYSICS, UNIV CALIF, BERKELEY, 75- *Mem:* Fel Am Phys Soc. *Res:* Elementary particle physics; quark and glvon structure at proton-antiproton colliders. *Mailing Add:* Lawrence Berkeley Lab Univ Calif Berkeley CA 94720

ELY, THOMAS HARRISON, b Roanoke, Va, Mar 24, 42; m 63; c 1. ANIMAL PHYSIOLOGY, CELL PHYSIOLOGY. *Educ:* Emory & Henry Col, BS, 63; Vanderbilt Univ, MS, 65; Univ Ala, PhD(biol), 71. *Prof Exp:* Asst prof biol, Longwood Col, 69-75; asst prof biol, Univ North Ala, 75-79; assoc prof,Capital Univ, 79-84, Prof Biol, 84-89; ASSOC PROF BIOL, LIVINGSTON UNIV, 89- *Mem:* Sigma Xi; Am Soc Cell Biol; Am Soc Zoologists; Am Inst Biol Sci; NY Acad Sci. *Res:* Genetic controls of the induction of males in Volvox aureus. *Mailing Add:* PO Box 97 Livingston LA 35470

ELY, THOMAS SHARPLESS, b Philadelphia, Pa, Sept 26, 24; m 45; c 5. NONIONIZING RADIATION HAZARD EVALUATION. *Educ:* Georgetown Univ, MD, 48; Univ Rochester, MS, 63. *Prof Exp:* Asst chief, Med Br, US Atomic Energy Comn, 56-58, chief, Health Protection Br, 58-61; asst prof occup med, Univ Rochester, 62-64; staff physician, Med Dept, Eastman Kodak Co, 64-66 & Lab Indust Med, 66-69, supvr, Environ Health Serv, 69-72, asst dir, Health, Safety & Human Factors Lab, 72-84, dir, Occup Health Lab, 84-86; RETIRED. *Concurrent Pos:* Mem, Comt Indust Hyg, Nat Acad Sci/Nat Res Coun, 55-58, Comt Biol Effects of Extremely Low Frequency Radiation, 76-77 & Chem Hazards in the Lab Comt, 82-83; mem, C-95 Comt RF Protection & Measurements, Am Nat Standards Inst, 58-79, Z-136 Comt Safe Use of Lasers & Masers, 72-83 & Z-311 Comt Photobiol Safety of Lamps & Lighting Systs, 77-; mem, Tech Electronic Problem Radiation Safety Standards Comt, Food & Drug Admin, 75-78; dir, Nat Coun Radiation Protection & Measurements, 75-80. *Mem:* Am Acad Occup Med; Am Col Occup Med (pres, 73-74); Nat Coun Radiation Protection & Measurements; Am Indust Hyg Asn; Health Physics Soc. *Res:* Biological effects of microwave radiation; hazards of oil mist exposure; ocular and skin hazards of visible and infrared radiation. *Mailing Add:* Box 17 West Bloomfield NY 14585-0017

ELZAY, RICHARD PAUL, b Lima, Ohio, Dec 6, 31; m 51; c 1. DENTISTRY, ORAL PATHOLOGY. *Educ:* Ind Univ, BS, 57, DDS, 60, MSD, 62. *Prof Exp:* From instr to assoc prof oral path, 62-74, asst dean acad affairs, 70-74, PROF ORAL PATH, MED COL VA, 74-, CHMN DEPT, 66- *Concurrent Pos:* Consult, US Navy Hosp, Portsmouth, Va, 62- & Vet Admin Hosp, Richmond, Va, 62-; USPHS fel, Ind Univ, 60-62. *Mem:* AAAS; Am Acad Oral Path (pres, 74); Am Soc Clin Path; Am Dent Asn; Int Asn Dent Res. *Res:* Radiation effects on oral structure and oral carcinogenesis. *Mailing Add:* Univ Minn 515 Delaware St SE 15-209 Moos Tower Minneapolis MN 55455

ELZERMAN, ALAN WILLIAM, b Ann Arbor, Mich, April 2, 49; m 70; c 2. ENVIRONMENTAL CHEMISTRY. *Educ:* Williams Col, BA, 71; Univ Wis, Madison, PhD(water chem), 76. *Prof Exp:* Res asst water chem, Univ Wis, 73-76; fel scholar, Woods Hole Oceanog Inst, Mass, 76-78; fel, Atomic Energy Res Estab, Eng, 78; from asst prof to assoc prof, 78-88, PROF ENVIRON CHEM, ENVIRON SYSTS ENG, CLEMSON UNIV, 88- *Mem:* Am Chem Soc; Water Pollution Cont Fedn; Asn Environ Eng Prof; Am Geophys Union. *Res:* Environmental chemistry; environmental engineering chemistry; analytical chemistry; sources, fate, distribution and control of chemicals in the environment; air-water interfacial processes. *Mailing Add:* Environ Systs Eng Clemson Univ 401 Rhodes Clemson SC 29634-0919

ELZINGA, D(ONALD) JACK, b Coupeville, Wash, Jan 16, 39; m 81; c 3. EDUCATIONAL ADMINISTRATION. *Educ:* Univ Wash, BE, 60; Northwestern Univ, MS, 65, PhD(chem eng), 68. *Prof Exp:* Teaching asst chem eng, Northwestern Univ, 63-65; asst prof chem eng, Johns Hopkins Univ, 67-68, asst prof opers res, 68-73, assoc prof, 73-78, res scientist & assoc prof, Ctr Metrop Planning & Res & Geog & Environ Eng, 78-79; PROF & CHMN OPERS RES, INDUST & SYSTS ENG DEPT, UNIV FLA, 79-, DIR, MFG SYSTS ENG PROG, 84- *Concurrent Pos:* Engr, Shell Develop Co, 60-61; volunteer, Peace Corps, 61-63; mathematician, Bur Health Manpower, Health Educ & Welfare & Grad Med Educ Nat Adv Comt, 77-80. *Mem:* Opers Res Soc Am; Math Prog Soc; Inst Indust Engrs; Sigma Xi; Soc Mfg Engrs; Inst Mgt Sci. *Res:* Mathematical programming-theoretical aspects and its application to real-world problems including facility location and manufacturing; multi-criterion decision making. *Mailing Add:* 1819 SW 81st St Gainesville FL 32607

ELZINGA, MARSHALL, b Hudsonville, Mich, Mar 25, 38; m 60; c 3. BIOCHEMISTRY. *Educ:* Hope Col, AB, 60; Univ Ill, Urbana, MS, 63, PhD(physiol), 64. *Prof Exp:* Res assoc biol, Brookhaven Nat Lab, 64-66, asst biochemist, 66-67; res assoc biochem, Retina Found, 67-69; staff scientist, Boston Biomed Res Inst, 69-73, sr scientist biochem, 73-76; biochemist dept biol, Brookhaven Nat Lab, 77-88; CHMN, DEPT PATH BIOCHEM, INST BASIC RES, 88- *Concurrent Pos:* Estab investr, Am Heart Asn, 68-73; assoc, Harvard Med Sch, 69-76; vis scientist, Max Planck Inst, Heidelberg, Ger, 72-73. *Mem:* Biophys Soc; Am Soc Biol Chemists. *Res:* Structure and function of proteins; amino acid sequence of myosin, actin, digestive enzymes; microtubule protein; chemical modification of proteins; sequence techniques. *Mailing Add:* Dept Pharmacol NY Inst Basic Res 1050 Forest Hill Rd Staten Island NY 10314

ELZINGA, RICHARD JOHN, b Salt Lake City, Utah, Apr 23, 31; m 57; c 5. ACAROLOGY, MORPHOLOGY. *Educ:* Univ Utah, BS, 55, MS, 56, PhD(entom), 60. *Prof Exp:* Nat Acad Sci resident res assoc, US Army Biol Warfare Lab, Md, 60-61; from asst prof to assoc prof, 61-73, PROF ENTOM, KANS STATE UNIV, 73- *Concurrent Pos:* Vis prof, Univ Minn, 81, 82 & 83; lectr, Mid Am State Univs Asn, 78-79. *Mem:* Entom Soc Am; Acarological Soc Am; AAAS. *Res:* Acarology; mites associated with army ants; fly mouthpart structure. *Mailing Add:* Dept Entom Kans State Univ Manhattan KS 66506

EMAMI, BAHMAN, b Tabriz, Iran, Feb 11, 43; US citizen; m 77; c 2. RADIATION ONCOLOGY, HYPERTHERMIA. *Educ:* Tehran Univ, MD, 68; FACR. *Prof Exp:* Asst prof radiation oncol, Tufts Univ, 77-80; from asst prof to assoc prof, Wash Univ, St Louis, 81-86; RADIATION ONCOLOGIST, MALLINCKRODT INST RADIOL, 81-; PROF RADIATION ONCOL, WASH UNIV, ST LOUIS, 86- *Concurrent Pos:* Radiation oncologist, New Eng Med Ctr, Boston, 77-80, dir exp protocols, 79-80; asst clin chief, Radiation Oncol Ctr, Wash Univ, 87-, chief hyperthermia sect, 87-90, assoc dir res, 90-; assoc ed, Int J Radiation Oncol Biol Physics, 87-, Endocurither/Hyperthermia Oncol J, 86-; fel, Tufts-New Eng Med Ctr, 75-77; mem, Radiation Ther Oncol Group, vchmn, Lung Comt & Hyperthermia Comt. *Mem:* AMA; Am Soc Therapeut Radiol & Oncol; Am Col Radiol; Am Radium Soc; Int Asn Study Lung Cancer; Pan Am Med Asn; Am Endocuriether Soc. *Res:* Three dimension treatment planning in radiation oncology; basic and clinical research in hyperthermia for the treatment of cancer; radiotherapy of lung cancer; radiotherapy of HVN cancer. *Mailing Add:* Radiation Oncol Ctr 4939 Audubon Ave Suite 5500 St Louis MO 63110

EMANUEL, ALEXANDER EIGELES, b Bucuresti, Romania, Mar 8, 37; US citizen; m 60; c 1. ENGINEERING. *Educ:* Israel Inst Technol, BSc, 63, MSc, 65, DSc(elec eng), 69. *Prof Exp:* Instr elec eng, Israel Inst Technol, 63-65, lectr, 68-69; sr res & develop engr, High Voltage Eng Corp, 69-74; from asst prof to prof, 74-82, PROF ELEC ENG, WORCESTER POLYTECH INST, 82- *Concurrent Pos:* Investr, Elec Power Res Inst, 73-74, consult, 75-77 & Allied Chem Res Div, 74-; prin investr, New England Elec res grant, 75- *Mem:* Inst Elec & Electronics Engrs; Soc Royale Belg Electriciens; Sigma Xi. *Res:* Power electronics; dielectrics; electro-mechanical energy conversion; harmonics in power systems; electrocution, electric fire. *Mailing Add:* Dept Elec Eng Worcester Polytech Inst Worcester MA 01609

EMANUEL, GEORGE, b New York, NY, Apr 3, 31; m 58; c 2. ENGINEERING, LASERS. *Educ:* Univ Calif, Los Angeles, BA, 52; Univ Southern Calif, MS, 56; Stanford Univ, PhD(gas dynamics), 62. *Prof Exp:* Res assoc gas dynamics, Stanford Univ, 62-63; mem tech staff, Aerospace Corp, 63-72; mem tech staff, TRW, 72-76; mem tech staff, Los Alamos Nat Lab, 76-80; PROF AERO, MECH & NUCLEAR ENG, UNIV OKLA, 80- *Mem:* AAAS; Am Inst Aeronaut & Astronaut; Am Phys Soc; Sigma Xi. *Res:* Gas dynamics; chemical nonequilibrium; thermal radiation; chemical lasers; photochemistry. *Mailing Add:* Dept Eng Univ Okla 865 Asp Norman OK 73019

EMANUEL, IRVIN, b Baltimore, Md, Oct 9, 26; m 89; c 3. EPIDEMIOLOGY. *Educ:* Rutgers Univ, BS, 51; Univ Ariz, MA, 56; Univ Rochester, MD, 60; Univ Wash, MS, 66. *Prof Exp:* Asst prof anthrop & asst dir, US Air Force Anthrop Proj, Antioch Col, 53-55; phys anthropologist, US Dept Air Force, Aerospace Med Lab, 55-56, consult anthrop, 57-60; intern pediat, Cleveland Metrop Gen Hosp, Ohio, 60-61; dir, Child Develop & Ment Retardation Ctr, 73-83, dir, Maternal & Child Health Prog, Sch Pub Health & community Health, 83-88, PROF EPIDEMIOL & PEDIAT, UNIV WASH, 74-, CO-DIR, MATERNAL & CHILD HEALTH PROG, SCH PUB HEALTH & COMMUNITY HEALTH, 88- *Concurrent Pos:* Guest investr, US Naval Med Res Univ 2, Taipei, Taiwan, 64-66; mem, Harvard-Peabody Mus, Solomon Islands Exped, 66 & Wash State Develop Disabilities Planning Coun, 75-79; bd dirs, Am Asn Univ Affil Progs for Developmentally Disabled, 77-80; sr fel prev med & pediat, Univ Wash, 62-66; USPHS res career develop award, Nat Inst Child Health & Human Develop, 66-71; mem Task Force Chems in the Environ & Reproductive Probs in Humans, Dalhousie Univ, 82-85; chmn, Comt Epidemiol, Forward Plan for Ment Retardation/Develop Disabilities Br, Nat Inst Child Health & Human Develop, 83-84; vis prof, Dept Epidemiol, Univ Ala, Birmingham, 87-88, Dept Clin Epidemiol, London Hosp Med Col, Eng, 87-88; fel, Fogarty Int Ctr, NIH, 87-88; vis lectr, Fac Dent, Chiang Mai Univ, Thailand, 88. *Mem:* Am Pub Health Asn; Int Epidemiol Asn; Teratology Soc; Soc Epidemiol Res; Am Epidemiol Soc. *Res:* Epidemiology of child health problems; maternal and international health; public health; pediatrics. *Mailing Add:* Dept Epidemiol SC-36 Univ Wash Seattle WA 98195

EMANUEL, JACK HOWARD, b Centerville, Iowa, Sept 26, 21; m 46; c 2. STRUCTURAL ENGINEERING, CIVIL ENGINEERING. *Educ:* Iowa State Univ, BS, 43, MS, 60, PhD, 65. *Prof Exp:* Weight control engr, Curtiss-Wright Corp, NY, 43-44; archit designer, Early Lumber Store, Iowa, 47-51; mgr & part-owner, Kingsley Lumber Co, 51-54; asst to pres & owner, H F Phelps, Oltmann and Phelps Bank, 54-58; from instr to asst prof civil eng, Iowa State Univ, 58-65; from asst prof to assoc prof, Univ NDak, 65-68; assoc prof, 68-77, PROF CIVIL ENG, UNIV MO-ROLLA, 77- *Concurrent Pos:* From pvt to second lieutenant, US Army, 44-47; mem comt A2C01 on general struct, Transp Res Bd, Nat res Coun, 76-85. *Mem:* Am Soc Civil Engrs; Am Concrete Inst; Sigma Xi; Nat Soc Prof Engrs; Prestressed Concrete Inst. *Res:* Temperature distribution and thermal stresses and movements in bridges; bridge inspection, maintenance, repair and upgrading; model analysis of structures; concrete growth; bridge supporting and expansion devices; stability of frames. *Mailing Add:* PO Box 261 Rolla MO 65401

EMANUEL, RODICA L, b Bucharest, Romania, May 4, 39; US citizen; m 60; c 1. ENDOCRINOLOGY, ANALYTICAL CHEMISTRY. *Educ:* Technion, Haifa, Israel, BSc, 64, MSc, 67; Northeastern Univ, Boston, PhD(med chem), 85. *Prof Exp:* Res asst teaching, Technion, Haifa, Israel, 65-67, sr res asst, 67-69; res asst clin lab, Brigham & Women's Hosp, Boston, 70-71, tech assoc supv clin lab, 71-73, tech dir superv & res, 73-84; res assoc, Howard Hughes Med Inst, Boston, 84-88; res assoc, 89-90, SR RES ASSOC REGULATION NEUROPEPTIDES, CHILDREN'S HOSP, BOSTON, 90- *Mem:* Endocrine Soc; Am Asn Clin Chem. *Res:* Role of signal transduction in the regulation of hypothalanic-pituitary-adrenal axis; emphasis on regulation of neuropeptides vasopression and corticotropin releasing factor. *Mailing Add:* Endocrine Div Enders 4 Children's Hosp Med Ctr 300 Longwood Ave Boston MA 02115

EMANUEL, WILLIAM ROBERT, b Denver, Colo, Sept 16, 49; m 71; c 3. SYSTEMS ECOLOGY, GLOBAL ELEMENT CYCLING. *Educ:* Okla State Univ, BS, 71, MS, 73, PhD(elec eng), 75. *Prof Exp:* Res asst, Okla State Univ, 73-75; res assoc, 75-80, RES STAFF MEM, ENVIRON SCI DIV, OAK RIDGE NAT LAB, 80- *Concurrent Pos:* Res Scholar, Int Inst Appl Systs Anal, Vienna, Austria. *Honors & Awards:* Martin Marietta Energy Systs Award, Pupl Excellence, 85. *Mem:* Ecol Soc Am; Sigma Xi; Am Geophys Union. *Res:* Application of systems analysis in ecology and geochemistry with particular interest in global element cycling. *Mailing Add:* Environ Sci Div Oak Ridge Nat Lab PO Box 2008 MS-6335 Oak Ridge TN 37831-6335

EMBERTON, KENNETH C, b Lexington, Ky, Oct 3, 48. MALACOLOGY, SYSTEMATICS. *Educ:* Ohio Univ, Athens, BS, 71, MS, 79; Univ Chicago, PhD(evolutionary biol), 86. *Prof Exp:* Vis asst cur, Field Mus Natural Hist, Chicago, 84-86; postdoctoral fel environ biol, NSF, 87-89; ASST CUR, ACAD NATURAL SCI, 89- *Mem:* Am Malacological Union; Asn Systs Collections; Coun Syst Malacologists; Soc Study Evolution; Soc Syst Zoologists. *Res:* Land snails: evolution, ecology and conservation. *Mailing Add:* Acad Natural Sci 1900 Benjamin Franklin Pkwy Philadelphia PA 19103-1195

EMBLETON, TOM WILLIAM, b Guthrie, Okla, Jan 3, 18; m 43; c 5. PLANT NUTRITION. *Educ:* Univ Ariz, BS, 41; Cornell Univ, PhD(pomol), 49. *Prof Exp:* Jr sci aide bur plant indust, soils & agr eng, USDA, US Date Garden, Indio, Calif, 42, sci aide to horticulturist P-1, Fruit & Veg Crops & Dis, 46; asst horticulturist irrig exp sta, State Col Wash, 49-50; from asst horticulturist to horticulturist, Citrus Res Ctr, Univ Calif, Riverside, 50-86, emer prof hort sci, 86-; RETIRED. *Concurrent Pos:* Hort consult, numerous corp, govt & int agencies, 72- *Honors & Awards:* Citrograph Res Award, 65; POPEOE Award, Am Soc Hort Sci, 85. *Mem:* Fel AAAS; fel Am Soc Hort Sci; Soil Sci Soc Am; Am Soc Agron; Coun Soil Testing & Plant Analysis; Coun Agr Sci & Technol. *Res:* Education; nitrogen fertilizer management programs for subtropical fruit trees versus nitrate pollution of ground waters; nitrate pollution of groundwater. *Mailing Add:* Dept Bot & Plant Sci Univ Calif Riverside CA 92521-1024

EMBLETON, TONY FREDERICK WALLACE, b Hornchurch, Eng, Oct 1, 29; m 53; c 1. ACOUSTICS. *Educ:* Univ London, BSc, 50, PhD(physics), 52, DSc, 64. *Prof Exp:* Fel, Nat Res Coun Can, 52-53, from assoc res officer to sr res officer, 54-74, prin res officer, 74-90; RETIRED. *Concurrent Pos:* Vis lectr, Univ Ottawa, 59-69 & Mass Inst Technol, 64, 67 & 72; adj prof, Carleton Univ, 78- *Honors & Awards:* Award, Acoust Soc Am, 64 & 86; Award, Soc Automotive Engrs, 74. *Mem:* Nat Acad Eng; Acoust Soc Am; fel Royal Soc Can; Inst Noise Control Eng. *Res:* Acoustic radiation forces, standards; shock and explosion waves; industrial noise control; sound propagation. *Mailing Add:* 80 Sheardown Dr Box 786 Nobleton ON L0G 1N0 Can

EMBLEY, DAVID WAYNE, b Salt Lake City, Utah, Oct 30, 46; m 70; c 8. COMPUTER SCIENCE. *Educ:* Univ Utah, BA, 70, MS, 72; Univ Ill, PhD(computer sci), 76. *Prof Exp:* Asst prof, 76-81, assoc prof computer sci, Univ Nebr, 81-; AT DEPT COMPUTER SCI, BRIGHAM YOUNG UNIV. *Mem:* Asn Comput Mach. *Res:* Very high level programming languages; interactive computing; database query languages; forms-based programming and query language; behavioral aspects of computer text editors. *Mailing Add:* Dept Computer Sci 286 Tmcb Brigham Young Univ Provo UT 84602

EMBODEN, WILLIAM ALLEN, JR, b South Bend, Ind, Feb 24, 35. SYSTEMATIC BOTANY, ETHNOBOTANY. *Educ:* Purdue Univ, BA, 57; Ind Univ, MA, 60; Univ Calif, Los Angeles, PhD(bot), 65. *Prof Exp:* From asst prof to assoc prof, 65-74, PROF BIOL, CALIF STATE UNIV, NORTHRIDGE, 74- *Concurrent Pos:* Mem hon fac, Los Angeles County Mus Natural Hist, 67-70; res fel, Harvard Univ, 80-85; post doctoral fel, Univ Calif Los Angeles, 89- *Mem:* Am Soc Plant Taxon; Sigma Xi; Int Asn Plant Taxon; fel Linnean Soc London. *Res:* Chemotaxonomy; cytogeography and chemogeography; chemical and cytological bases for distribution patterns in populations of Salvia, Bursera, Solanum and Cannabis. *Mailing Add:* Dept Biol Calif State Univ Northridge CA 91330

EMBODY, DANIEL ROBERT, b Ithaca, NY, July 10, 14; wid; c 3. STATISTICS. *Educ:* Cornell Univ, BS, 38, MS, 39. *Prof Exp:* Sr math statistician, Arnold Bernhard & Co, Inc, 47-48; res statistician, Washington Water Power Co, 49-52; from head statist sect to coordr electronic data processing, E R Squibb & Sons, Olin Mathieson Chem Corp, 53-65; math statistician, Bur Ships, Dept Navy, 65-67; biometrician, Biomet Serv Staff & Plant Protection Div, 67-72, biometrician, 72-88, CONSULT BIOMETRICIAN, ANIMAL & PLANT HEALTH INSPECTION SERV, USDA, 81- *Concurrent Pos:* Consult, State Dept Fish & Game, Idaho, 49-60, NJ, 55-65 & US Geol Surv, 52-58. *Mem:* Am Statist Asn; Biomet Soc; Entom Soc Am; Am Fisheries Soc. *Res:* Application of statistical and biometric science to problems of medical research; fishery biology; entomology; pesticide monitoring; dynamics of insect trap operations. *Mailing Add:* 5025 Edgewood Rd College Park MD 20740

EMBREE, CHARLES GORDON, b NS; m 64; c 3. POMOLOGY. *Educ:* Univ Guelph, BSc, 63; Univ BC, MSc, 68. *Prof Exp:* Exten specialist tree fruits, Horticult & Biol, NS Dept Agr & Mkt, 63-78, dir, 78-82; RES SCIENTIST APPLE PROD, AGR CAN, 82- *Concurrent Pos:* Vis scientist, NY State Agr Exp Sta, Cornell Univ, Geneva, 89. *Mem:* Can Soc Horticult Sci; Am Soc Horticult Sci; Am Pomol Soc. *Res:* Orchard management studies: factors influencing production and quality, characterization of genetic diversity in apple rootstocks, production systems and development of expert systems for management choices in orcharding. *Mailing Add:* Res Sta Agr Can Kentville NS B4N 1J5

EMBREE, EARL OWEN, b Alton, Ill, Feb 17, 24; m 62. MATHEMATICS. *Educ:* Morgan State Col, BS, 50; Univ Ill, MS, 52, PhD(math), 63. *Prof Exp:* Mathematician, Ballistics Res Labs, Md, 53-55; teacher, 55-58, from asst prof to assoc prof, 60-74, PROF MATH, MORGAN STATE COL, 74- *Mem:* Am Math Soc; Math Asn Am. *Res:* Ordinary linear differential equations involving distributions; symbolic logic; algebra. *Mailing Add:* 700 Camberley Circle B-4 Towson Br Baltimore MD 21204

EMBREE, HARLAND DUMOND, b Monmouth, Ill, May 8, 23; m 47; c 4. ORGANIC CHEMISTRY. *Educ:* Univ Calif, BS, 48; Univ Minn, PhD(chem), 52. *Prof Exp:* Res chemist, Charles Pfizer & Co, 52-53; asst prof chem, Hamline Univ, 53-57; assoc prof, 57-65, PROF CHEM, SAN JOSE STATE UNIV, 65- *Mem:* Am Chem Soc. *Res:* Chemical education; author of college textbooks on general chemistry and organic chemistry. *Mailing Add:* 7547 Arden Way Aptos CA 95003-3808

EMBREE, JAMES WILLARD, JR, b Tacoma, Wash, June 22, 48. TOXICOLOGY. *Educ:* Univ Wash, BS, 70, MS, 72; Univ Calif-San Francisco, PhD(toxicol), 76; Am Bd Toxicol, dipl. *Prof Exp:* Res toxicologist, Univ Calif, San Francisco, 76-77, asst res toxicologist, 77, instr toxicol, 77-78, asst prof toxicol, 78; EXP TOXICOLOGIST, CHEVRON ENVIRON HEALTH CTR, 78- *Concurrent Pos:* Mgr, Toxicol Res Lab, Univ Calif, San Francisco, 76-77, dir 77-78; lectr, Univ Calif, San Francisco, 78- *Res:* Environmental and industrial toxicology; genetic toxicology. *Mailing Add:* 700 Arlington Circle Novato CA 94947

EMBREE, M(ILTON) L(UTHER), b Marceline, Mo, May 27, 24; m 50; c 2. ELECTRICAL ENGINEERING, ENGINEERING PHYSICS. *Educ:* Univ Ill, BS, 49, MS, 50; Lehigh Univ, MS, 57. *Prof Exp:* Mem res staff, Univ Ill, 59-51; MEM TECH STAFF, SEMICONDUCTOR DEVICE & INTEGRATED CIRCUIT DEVELOP, BELL TEL LABS, INC, 51- *Mem:* Inst Elec & Electronics Engrs; Sigma Xi. *Mailing Add:* 3225 Eisenbrown Rd Riverview Park Reading PA 19605

EMBREE, NORRIS DEAN, b Kemmerer, Wyo, Nov 29, 11; m 37, 80; c 3. APPLIED CHEMISTRY. *Educ:* Univ Wyo, BA, 31; Yale Univ, PhD(phys chem), 34. *Prof Exp:* Res chemist, Distillation Prod Industs, Eastman Kodak Co, 34-48, dir res, 48-60, vpres in charge tech opers, 60-68, asst dir res, Res Labs, Tenn Eastman Co, 68-74; consult, 75-85; RETIRED. *Concurrent Pos:* Mem oil & fat sect, Int Union Pure & Appl Chem, 61-83; chmn comt fats & oils, Assembly Math & Phys Sci, Nat Res Coun, 68-76. *Mem:* Am Chem Soc; Am Oil Chemists' Soc (pres, 59); Am Inst Chem Engrs; Am Soc Biol Chemists. *Res:* High vacuum equipment and distillation; chemistry and technology of fats and oils; chemical products for use in nutrition and health care. *Mailing Add:* 89 Crown Colony Kingsport TN 37660-9703

EMBREE, ROBERT WILLIAM, b Elliott, Iowa, Dec 9, 32; m 59; c 3. BOTANY. *Educ:* Simpson Col, BA, 54; Univ Nebr, MS, 56; Univ Calif, Berkeley, PhD(bot), 62. *Prof Exp:* NSF fel, Birkbeck Col, London, 61-63; vis lectr bot, Univ Calif, Berkeley, 63-64; asst prof, Brown Univ, 64-68; ASSOC PROF BOT, UNIV IOWA, 68- *Mem:* AAAS; Bot Soc Am; Mycol Soc Am; Phycol Soc Am; Brit Mycol Soc. *Res:* Growth and development of fungi; ecology of coprophilous fungi; biology of mucoraceous fungi. *Mailing Add:* Dept Bot Univ Iowa Iowa City IA 52242

EMBRY, BERTIS L(LOYD), b Drummonds, Tenn, Nov 23, 14; m 41; c 5. ELECTRICAL ENGINEERING. *Educ:* Utah State Univ, BS, 41, MS, 49; Stanford Univ, EE, 54; Univ Mo, PhD(elec eng), 66. *Prof Exp:* Jr engr, Rural Electrification Admin, 41-42 & 46; electronics technician, Radiation Lab, Univ Calif, 42-43; from asst prof to assoc prof agr eng, 46-56, prof elec eng, 56-80, EMER PROF ELEC ENG & AGR & IRRIG ENG, UTAH STATE UNIV, 80- *Concurrent Pos:* Adv elec & agr eng, Utah State Contract, Iran, 60-62; consult drainage, Colombia & Cent Am, 71; irrig consult & adv, Guatemala, 77-81. *Mem:* Inst Elec & Electronics Engrs; Am Soc Eng Educ; Nat Soc Prof Engrs. *Res:* Electrical engineering in power and machinery; electronics; agricultural engineering; irrigation and direct energy conversion. *Mailing Add:* Utah State Univ Logan UT 84322

EMBRY, LAWRENCE BRYAN, b Morgantown, Ky, June 25, 18; wid. ANIMAL HUSBANDRY. *Educ:* Univ Ky, BSA, 42; Cornell Univ, MSA, 48, PhD(animal husb), 50. *Prof Exp:* From assoc prof to prof, 50-83, EMER PROF ANIMAL SCI, SDAK STATE UNIV, 83- *Mem:* AAAS; fel Am Soc Animal Sci; Am Inst Nutrit; NY Acad Sci. *Res:* Nutritive requirements of livestock; compositon and nutritive value of feeds; feed additives. *Mailing Add:* 2004 Derdall Dr Brookings SD 57006

EMBRY-WARDROP, MARY RODRIGUEZ, b Monroe, La, Aug 22, 33. MATHEMATICAL ANALYSIS. *Educ:* Southwestern at Memphis, BS, 55; Univ Va, MA, 58; Univ NC, Chapel Hill, PhD(math), 64. *Prof Exp:* From asst prof to prof math, Univ NC, Charlotte, 64-77; PROF MATH, CENT MICH UNIV, MT PLEASANT, 77- *Concurrent Pos:* Res prof, Cent Mich Univ, 80-81. *Mem:* Am Math Soc; Math Asn Am. *Res:* Operators on Hilbert space; semigroups of operators, the numerical range of an operator, invariant subspaces of operators. *Mailing Add:* Dept Math Central Mich Univ 901 Canal Rd Mt Pleasant MI 48858

EMBURY, JANON FREDERICK, JR, b Baltimore, Md, Sept, 9, 45; m 69. MILLIMETER-CENTIMETER WAVE MEASUREMENTS. *Educ:* Johns Hopkins Univ, BS, 67; Drexel Univ, MS, 73, PhD(physics), 77. *Prof Exp:* Physicist, Rohm & Haas Co, 68-77, consult, 77-78; PHYSICIST, US ARMY CHEM RES & DEVELOP ENG CTR, ABERDEEN PROVING GROUND, 78- *Concurrent Pos:* Adj fac, Camden County Community Col, 75-76. *Mem:* Sigma Xi; Optical Soc Am. *Res:* Aerosol physics; radiative transfer; imaging; electromagnetic radiation propagation through aerosols from ultraviolet to microwave; powder spectroscopy. *Mailing Add:* US Army Chem Res & Develop Eng Ctr SMCCR-RSP-B Aberdeen Proving Ground MD 21010

EMBURY, JOHN DAVID, b Grantham, Eng, July 12, 39; m 63. PHYSICAL METALLURGY. *Educ:* Univ Manchester, BSc, 60; Cambridge Univ, PhD(metall), 63. *Prof Exp:* Res scientist, US Steel Res Ctr, Pa, 63-65; sr res assoc metall, Univ Newcastle, 65-66; from asst prof to assoc prof, 66-77, PROF METALL & MAT SCI, MCMASTER UNIV, 77- *Concurrent Pos:* Vis fel, Battelle Mem Inst, 71. *Mem:* Am Soc Metals; Am Inst Mining, Metall & Petrol Engrs; Brit Inst Metals. *Res:* Microstructure of deformed materials; mechanisms of deformation and fracture in metals; stress corrosion failure; mechanism of nucleation and growth processes in solids. *Mailing Add:* Dept Metall & Mat Sci McMaster Univ 1280 Main St W Hamilton ON L8S 4L8 Can

EMCH, GEORGE FREDERICK, b Washington, DC, June 17, 25; m 50; c 2. PHYSICS. *Educ:* Trinity Col, BS, 47. *Prof Exp:* Assoc physicist, 48-56, physicist, 56-62, PRIN PROF STAFF, APPL PHYSICS LAB, JOHNS HOPKINS UNIV, 62- *Mem:* Am Phys Soc; Inst Elec & Electronics Engrs. *Res:* Simulation; radar and weapon control system design and analysis; automation and artificial intelligence. *Mailing Add:* 14616 Peach Orchard Rd Silver Spring MD 20904

EMCH, GERARD G, b Geneva, Switz, July 21, 36; m 59; c 2. MATHEMATICAL PHYSICS. *Educ:* Col Geneva, Switz, Maturite Sci, 55; Univ Geneva, Physics Dipl, 59, PhD(quant mech, spec relativity), 63. *Prof Exp:* Asst exp physics, Exp Physics Res Lab, Univ Geneva, 59-60, asst theoret physics, Univ, 59-63, chief, 63-64; res assoc, Princeton Univ, 64-65 & Univ Md, 65-66; asst prof physics, 66-71, assoc prof math & physics, 71-78, PROF MATH & PHYSICS, UNIV ROCHESTER, 78- *Concurrent Pos:* Vis prof, Roman Cath Univ Nijmegen, 70-71; mem, Ctr Interdisciplinary Res, Univ Bielefeld, 75-76 & 84; Gauss prof, Akad f Wissenschaffen zu Göttingen, 85. *Mem:* Am Phys Soc; Math Asn Am; Am Math Soc. *Res:* Operator algebras and differential geometry for quantum theory and general relativity. *Mailing Add:* Univ Fla Gainesville FL 32611

EMELE, JANE FRANCES, b Phillipsburg, NJ, Nov 14, 25. MEDICAL SCIENCES, MEDICAL WRITING. *Educ:* Upsala Col, BS, 47; Univ Ill, MS, 49; Yale Univ, PhD(pharmacol), 54. *Prof Exp:* Res asst, Biol Div, Schering Corp, 47-48; res asst physiol, Univ Ill, 49-50; microanal chemist, Bell Tel Labs, 51; chief sect pharmacodyn, Div Pharmacol, Eaton Labs, Norwich Pharmacol Co, 54-55; sr res assoc, div pharmacol, Warner Lambert Co, 55-65, mgr, div proprietary pharmacol, 65-66, dir dept pharmacol, Consumer Prod Res Div, 66-70, assoc dir biol res, proprietaries & toiletries, Warner-Lambert Co, 70-72, dir biol res, Consumer prod groups, 72-74, dir biol res, Am Chicle Div, 74-77, dir biol & clin affairs, Am Chicle Div, 77-81, dir clin res & toxicol & prod safety, Consumer Prod Res & Develop, 81-85; EXEC DIR, RESOURCES INT, 85- *Concurrent Pos:* Nat Heart Inst res fel, 52-54; mem, Morris County Asn Ment Health, 56-69, chmn educ comt, chmn indust adv comt, vpres, bd dirs, secy-exec comt & bd dirs; mem, Morris County Bd Ment Health, 59-65; mem bd dirs, dirs & educ comt, NJ Asn Ment Health, 62-64; mem, Morris County Asn Health & Welfare Agencies, 64-66; lectr & vis scientist, Rutgers Univ, 64-67; mem bd trustees, Upsala Col, 78-; mem invest subcomt, Morristown Mem Hosp, 78-80. *Mem:* AAAS; Am Soc Pharmacol & Exp Therapeut; Am Soc Clin Pharmacol & Ther; Am Col Toxicol. *Res:* Neuropharmacology; analgesia, gastrointestinal and respiratory; cardiovascular neurophysiology; clinical and laboratory evaluation of cough-cold, gastrointestinal, analgesics, breath, skin, shaving products, confections and food products. *Mailing Add:* 19 Kathleen Pl Morris Plains NJ 07950

EMERICH, DONALD WARREN, b Schuylkill Haven, Pa, July 12, 20; m 43; c 3. ANALYTICAL CHEMISTRY. *Educ:* Pa State Univ, BS, 42; Ohio State Univ, PhD, 51. *Prof Exp:* Chemist, Hercules Powder Co, 42-45; chem engr, Badger Ord Works, 45-47; res chemist, Niacet Chem Div, US Vanadium Corp, 47-49; asst, Ohio State Univ, 49-51; asst prof chem, Kans State Univ, 51-54; prof, Centenary Col, 54-60; actg head dept, 64-66 & 80-81, head dept, 66-76, prof analytical chem, 60-85, EMER PROF ANALYTICAL CHEM, MISS STATE UNIV, 85- *Concurrent Pos:* Lectr, Univ Rhode Island, 79-80, Calif Polytech, San Luis Obispo, 85-86, Utah State Univ, 86-87; admin asst, Miss State Univ, 88. *Mem:* Am Chem Soc; Sigma Xi. *Res:* Electrochemistry; nonaqueous titrimetry. *Mailing Add:* 2007 Pin Oak Dr Starkville MS 39759

EMERICK, HAROLD B(URTON), b New Brighton, Pa, July 6, 13; m 38; c 2. METALLURGICAL ENGINEERING. *Educ:* Carnegie Inst Technol, 32-38. *Prof Exp:* Supvr, Jones & Laughlin Steel Corp, 35-55, dir tech serv, 55-69, vpres res & technol, 69-72; CONSULT, 72- *Concurrent Pos:* Mem, Int Exec Serv Corps, Philippines, S Korea, Mex & Peru, 74-81. *Honors & Awards:* McKune Mem Award, Am Inst Mining, Metall & Petrol Engrs, 42. *Mem:* Am Inst Mining, Metall & Petrol Engrs; fel Am Soc Metals; fel Metall Soc (pres, 65); Asn Iron & Steel Engr. *Res:* Process metallurgy; iron and steel production technology. *Mailing Add:* 479 Salem Dr Pittsburgh PA 15243-2075

EMERICK, ROYCE JASPER, b Tulsa, Okla, Jan 1, 31; m 53; c 3. ANIMAL NUTRITION. *Educ:* Okla State Univ, BS, 52; Univ Wis, MS, 55, PhD(biochem, animal husb), 57. *Prof Exp:* PROF CHEM & ANIMAL SCI, SDAK STATE UNIV, 57- *Concurrent Pos:* Res fel, Univ Wis, 65-66. *Mem:* Am Inst Nutrit; Am Soc Animal Sci. *Res:* Urinary calculi; nitrate and mineral metabolism. *Mailing Add:* Animal Sci Complex Box 2170 Brookings SD 57007

EMERMAN, JOANNE TANNIS, b Kenora, Ont; Can citizen; m 63; c 3. ONCOLOGY, MAMMARY CELL BIOLOGY. *Educ:* Hofstra Univ, Hempstead, BA, 65; Univ Calif, Berkeley, MA, 67, PhD(zool), 77. *Prof Exp:* Fel, Lab Chem Biodynamics, Univ Calif, Berkeley, 77-80; asst prof obstet & gynec, fac med, Univ BC, Vancouver, 80-82, assoc mem, dept anat, 81-82, asst prof, 82-86, ASSOC PROF, DEPT ANAT, UNIV BC, VANCOUVER, 86-, ASSOC MEM OBSTET & GYNEC, 82- *Concurrent Pos:* Vis prof, Dept Cancer Endocrinol, Cancer Control Agency, 80-82; res scholar, Nat Cancer Inst Can, 81-87; prin investr, Nat Cancer Inst Can grant, 81- *Mem:* Can Fedn Biol Soc; Can Soc Clin Invest; Am Soc Cell Biol; Tissue Culture Asn; Sigma Xi; Am Asn Cancer Res; Int Asn Breast Cancer Res. *Res:* Effect of hormones and growth factors on growth and differentiation of normal and malignant mammary epithelial cells; hormone and drug interactions in mammary carcinoma therapy. *Mailing Add:* Dept Anat Univ BC 2177 Wesbrook Mall Vancouver BC V6T 1W5 Can

EMERSON, DAVID EDWIN, b Checotah, Okla, May 15, 32; m 53; c 3. ANALYTICAL CHEMISTRY. *Educ:* Southeastern State Col, BS, 55. *Prof Exp:* From chemist to supvry chemist, 58-63, chief br lab serv, Helium Res Ctr, 63-71, CHIEF SECT RES & ANALYTICAL SERV, HELIUM OPERS, BUR MINES, 71- *Res:* Development of gas analysis apparatus, especially helium and the impurities in helium; development methods in isotopic analysis, preparation of primary standards and sample preparation; analysis of helium-3 in natural gas. *Mailing Add:* Helium Opers Bur Mines 1100 S Fillmore St Amarillo TX 79101

EMERSON, DAVID WINTHROP, b Littleton, Mass, Mar 13, 28; m 54; c 3. ORGANIC CHEMISTRY. *Educ:* Dartmouth Col, AB, 52; Univ Mich, MS, 54, PhD(chem), 58. *Prof Exp:* Res chemist, Shell Oil Co, 57-63; from asst prof to assoc prof chem, Univ Mich-Dearborn, 63-69, chmn, Div Lit, Sci & Arts, 67-69 & dept natural sci, 73-75, prof chem, 69-81, dean, Col Arts, Sci & Letters, 79-81; dean, Col Sci, Math & Eng, 81-88, dean, Col Sci & Math, 88-89, PROF CHEM, UNIV NEV, LAS VEGAS, 81- *Mem:* Am Chem Soc; Royal Soc Chem; AAAS. *Res:* Mechanisms of organic reactions; reactive intermediates; organosulfur chemistry; solid phase reagents. *Mailing Add:* Dept Chem Univ Nev Las Vegas Las Vegas NV 89154

EMERSON, DONALD ORVILLE, b Long Beach, Calif, July 19, 31; m 57; c 3. GEOLOGY, PETROLOGY & GEOCHEMISTRY. *Educ:* Calif Inst Technol, BS, 53; Pa State Univ, MS, 55, PhD(mineral, petrol), 59. *Prof Exp:* From asst prof to assoc prof geol, Univ Calif, Davis, 57-69; GEOLOGIST, LAWRENCE LIVERMORE NAT LAB, 69- *Res:* Applied geology; aqueous modeling. *Mailing Add:* 1122 Avenida De Las Palmas Livermore CA 94550

EMERSON, FRANK HENRY, b Kansas City, Mo, June 22, 21; m 44; c 3. HORTICULTURE, PLANT PATHOLOGY. *Educ:* Univ Kans, AB, 47, MA, 48; Cornell Univ, PhD, 51. *Prof Exp:* Asst bot, Univ Kans, 47-48; asst plant path, Cornell Univ, 48-51; asst tech dir agr chem, Stauffer Chem Co, NY, 51-55; asst prof exten horticulturist, Purdue Univ, 55-58, from assoc prof to prof, 58-87, EMER PROF HORT RES, PURDUE UNIV, 87- *Mem:* Am Soc Hort Sci; Am Pomol Soc; Am Phytopath Soc; Sigma Xi. *Res:* Tree fruits; chemical growth regulators; winter hardiness; frost control; population density; training systems; chemical thinning, breeding and selection of new varieties resistant to apple scab. *Mailing Add:* 3106 Carriage Rd West Lafayette IN 47906

EMERSON, FREDERICK BEAUREGARD, JR, b Wellsville, NY, Nov 21, 35; m 58; c 3. MEDICINE. *Educ:* Alfred Univ, BA, 57; Cornell Univ, PhD(wildlife mgt), 61; Vanderbilt Univ, MD, 70. *Prof Exp:* Asst wildlife mgt, Cornell Univ, 57-61; biologist wildlife mgt, Tenn Valley Auth, 62-65; asst prof forestry, Univ Tenn, 65-66; intern, Med Ctr, Univ Colo, Denver, 70-71, resident med, 71-73, instr med, 73-75, asst prof med, 75-77; physician, Emergency Dept, 77-88, MEM STAFF, GOLETA VALLEY COMMUNITY HOSP, 77-; PHYSICIAN, UNIV CALIF, SANTA BARBARA, 77- *Concurrent Pos:* Staff physician, Dept Emergency Med Servs, Denver Gen Hosp, 73-77; NIH fel marine biol, Univ Miami, 61-62. *Res:* Emergency medicine; ecology; conservation of natural resources. *Mailing Add:* 2409 Vista Del Campo Santa Barbara CA 93101

EMERSON, GERALDINE MARIELLEN, b Dec, 30, 25; wid; c 1. MEDICAL SCIENCE, GERONTOLOGY. *Educ:* Univ Miami, BA, 49; Univ Ala Med Ctr, PhD(physiol, biochem, pharmacol & neuroanat), 60. *Prof Exp:* From asst prof to prof biochem, Med Ctr, Sch Med, Univ Ala, Birmingham, 64-87; RETIRED. *Concurrent Pos:* Partic, Max Planck Inst Brain Res, Frankfort, 77-78; consult, Ctr Aging, 81- *Mem:* AAAS; Gerontol Soc; Am Physiol Soc; Sigma Xi. *Res:* Endocrine factors in growth; aging in the Long-Evans rat; comparison of enzyme activities from different organs of the rat; parameters of human longevity; gerontology. *Mailing Add:* PO Box 1330 Umatilla FL 32784

EMERSON, JAMES L, b Garrett, Ind, Jan 23, 38; m 62; c 3. VETERINARY PATHOLOGY, TOXICOLOGY. *Educ:* Ohio State Univ, DVM, 62; Purdue Univ, MS, 64, PhD, 66. *Prof Exp:* From instr to asst prof path, Sch Vet Sci & Med, Purdue Univ, 62-66; res assoc pathologist, Norwich Pharmacal Co, 66-69; pathologist, Human Health Res & Develop Labs, Dow Chem Co, 69-75, sr res specialist, Health & Consumer Prod Dept, Dow Chem USA, 75-76; mgr dept path, Abbott Labs, North Chicago, 76-79; mgr life sci, 79-80, assoc dir, 80-81, DIR, EXTERNAL TECH AFFAIRS DEPT, COCA COLA CO, 81- *Concurrent Pos:* Assoc mem fac, Ind Univ-Purdue Univ, Indianapolis, 69-73; mem path/toxicol expert comt, Int Life Sci Inst, pres chmn, Saccharin Tech Comt, 84-; pres bd dirs, Jacquelyn McClure Lupus Found, Atlanta, Ga, 84-; mem bd gov, Flavor Extract Mfrs Asn, 88-, chmn, Safety Eval Comt. *Mem:* Am Vet Med Asn; Am Col Vet Pathologists; Soc Toxicol; Int Acad Path. *Res:* Drug safety evaluation; carcinogenesis bioassay; food safety. *Mailing Add:* 290 Landfall Rd NW Atlanta GA 30328

EMERSON, JOHN DAVID, b Oswego, NY, July 18, 46; m 68; c 2. SURVIVAL THEORY, EXPLORATORY DATA ANALYSIS. *Educ:* Univ Rochester, BA, 68; Cornell Univ, MS, 71, PhD(math), 73, cert statist, 76-77. *Prof Exp:* Teaching fel math, Cornell Univ, 68-73; asst prof, 73-81, assoc prof, 81-86, PROF MATH & DEAN COL, MIDDLEBURY COL, 86- *Concurrent Pos:* Appl math, Eastman Kodak, 68; adj prof, Tompkins-Cortland Community Col, 69-70 & Cornell Univ, 73; asst to dean, Middlebury Col, 76 & asst to acad vpres, 76-77; teaching fel, Univ Minn, 78; vis res fel biostatist, Harvard Univ, 78-79 & res fel, 80-81; vis res assoc statist, Harvard Univ, 82-83. *Mem:* Am Statist Asn; Biomet Soc; Math Asn Am. *Res:* Robust exploratory statistical methods; data analytical techniques; medical statistics; least-squares fitting methodology. *Mailing Add:* Dept Math Middlebury Col Middlebury VT 05753

EMERSON, JOHN WILFORD, b Bloomington, Ind, Dec 27, 33; m 59; c 3. GEOLOGY. *Educ:* Univ NMex, BS, 59, MS, 61; Fla State Univ, PhD(geol), 66. *Prof Exp:* Geologist, Pan Am Petrol Corp, Colo, 66-67; from asst prof to assoc prof, 67-76, PROF EARTH SCI, CENT MO STATE UNIV, 76-HEAD DEPT, 68- *Mem:* Geol Soc Am; Am Quaternary Asn; Sigma Xi. *Res:* Sedimentology; sedimentary petrology; stratigraphy. *Mailing Add:* Dept Earth Sci Cent Mo State Univ Warrensburg OH 64093

EMERSON, KARY CADMUS, b Sasakwa, Okla, Mar 13, 18; m 39; c 3. MEDICAL ENTOMOLOGY, PARASITOLOGY. *Educ:* Okla State Univ, BS, 39, MS, 40, PhD(entom), 49. *Prof Exp:* US Army, 40-88, med entomologist, Philippines, 40-42, asst prof, Okla State Univ, 46-49, tech liaison, Off Chief Res & Develop, US Army, 59-60, spec asst for res, 60-74, dep for sci & technol & actg asst secy army for res & develop, 75-79, mem Army Sci Bd, 79-83; pres, Care & Rehab Wildlife Inc, 83-85; CONSULT, 85- *Concurrent Pos:* Consult, Univs & US Depts Agr & Interior, 49-; res assoc, Smithsonian Inst, 59-; res assoc, Seminole Indian Nation Mus; res assoc, Mus, Okla State Univ & adj prof, 71-; US mem, NATO Long-Term Sci Study Panel; mem, Defense Comt Res, White House Panel Systs & Taxon & White House Comt Environ Qual; res assoc, Fla Dept Agr & Consumer Affairs & Bernice P Bishop Mus; mem bd, Int Osprey Found, 86-88. *Honors & Awards:* Spec Award, US Fish & Wildlife Serv. *Mem:* Soc Syst Zool; fel Entom Soc Am; Am Soc Trop Med & Hyg; Am Soc Parasitol; Wildlife Dis Asn. *Res:* Mallophaga and Anoplura; ectoparasites; arthropod-borne diseases; research of external parasites on wild birds and mammals, world-wide. *Mailing Add:* 560 Boulder Dr Sanibel FL 33957

EMERSON, KENNETH, b Pasadena, Calif, Nov 9, 31; m 56; c 3. PHYSICAL CHEMISTRY, INORGANIC CHEMISTRY. *Educ:* Harvard Univ, BA, 53; Univ Ore, MA, 58; Univ Minn, PhD(chem), 61. *Prof Exp:* Noyes fel chem, Calif Inst Technol, 61-62; from asst prof to assoc prof, 62-70, PROF CHEM, MONT STATE UNIV, 70- *Concurrent Pos:* Fulbright res fel, Univ Canterbury, 68-69; vis prof, Univ of the Andes, Venezuela, 72-73, Univ Canterbury, 79 & Univ del Pis Yasco, Bilbao, Spain, 88-89; exchange lectr, Jilin Univ, Chang Chun, China, 81. *Mem:* AAAS; Am Chem Soc; Am Crystallog Asn; Sigma Xi. *Res:* Inorganic structure and its relation to the theory of the chemical bond; solid state chemistry of low-dimensional structures. *Mailing Add:* Dept Chem Mont State Univ Bozeman MT 59717

EMERSON, MARION PRESTON, b Washburn, Mo, Feb 24, 18; m 47; c 3. ALGEBRA. *Educ:* Southwest Mo State Univ, BS, 38; Univ Wis, MS, 48; Univ Ill, PhD(math), 52. *Prof Exp:* Asst prof math, Harpur Col, 52-56; assoc prof, Southwest Mo State Col, 56-61; head dept, 61-79, prof, 79-87, EMER PROF MATH, EMPORIA STATE UNIV, 87- *Mem:* Am Math Soc; Math Asn Am; Nat Coun Teachers Math; Sigma Xi. *Res:* Modular lattices. *Mailing Add:* 1425 Luther St Emporia KS 66801-6040

EMERSON, MERLE T, b Spokane, Wash, Aug 19, 30; m 54; c 2. PHYSICAL CHEMISTRY, ANALYTICAL CHEMISTRY. *Educ:* Whitworth Col, Wash, BS, 52; Wash State Col, MS, 58; Univ Wash, PhD(phys chem), 58. *Prof Exp:* Res assoc chem, Fla State Univ, 58-60, asst, Inst Molecular Biophys, 60-62; asst prof, Wayne State Univ, 62-64; asst prof, Fla State Univ, 64-69; actg dir natural sci & math div, 71, ASSOC PROF CHEM, UNIV ALA, HUNTSVILLE, 69- *Concurrent Pos:* NIH res grant, 64-67; vis assoc prof, Univ Hawaii, Hilo Campus, 68-69, Univ Hawaii, 78-79. *Mem:* Am Crystallog Soc. *Res:* Molecular structure using nuclear magnetic resonance spectroscopy; x-ray determination of crystal and molecular structures and instrumentation. *Mailing Add:* Dept Chem Univ Ala Huntsville AL 35899

EMERSON, THOMAS EDWARD, JR, b Wilson, Okla, Feb 3, 35; m 55, 85; c 3. CARDIOVASCULAR PHYSIOLOGY. *Educ:* Univ Okla, BS, 58, PhD(med physiol), 64; Univ Alta, MSc, 61. *Prof Exp:* Res physiologist, Civil Aeromed Res Inst, Okla, 61-65; asst prof physiol & res asst prof med surg, Med Ctr, Univ Okla, 65-66; from assoc prof to prof physiol, Mich State Univ, 66-83; prin scientist, 83-90, RES FEL, DEPT PHYSIOL & EXP THERAPEUT, CUTTER BIOL, MILES INC, BERKELEY, 90- *Concurrent Pos:* Fel, Coun Circulation, Am Heart Asn, 75- *Mem:* Soc Exp Biol & Med; Am Physiol Soc; Am Fedn Clin Res; Soc Critical Care Med. *Res:* Cardiovascular physiology; cardiovascular mechanisms during endotoxin and hemorrhagic shock; regulation of vasoactive agents and effects on peripheral blood flow; effects of vasoactive hormones on arteries and veins; metabolic abnormalities in muscle and adipose tissue during circulatory shock; cerebral blood flow regulation; development of therapeutic regimens for inflammatory disease states such as septicemia/endotoxemia. *Mailing Add:* Dept Physiol & Exp Therapeut Cutter Biol Miles Inc Fourth & Parker Sts Berkeley CA 94710

EMERSON, THOMAS JAMES, b St Paul, Minn, Sept 4, 47; c 2. SOFTWARE SYSTEM TESTING, SOFTWARE METRICS. *Educ:* Univ Minn, BS, 79; NY Univ, MS, 88. *Prof Exp:* Mem tech staff, AT&T Bell Labs, 80-91; SR SOFTWARE ENGR, AMDAHL CORP, 91- *Mem:* Asn Comput Mach; Soc Indust & Appl Math; Inst Elec & Electronics Engrs Computer Soc; Am Math Soc. *Res:* Theories of computation; non-monotonic logic; software reliability and failure models. *Mailing Add:* Amdahl Corp 1250 E Arques Ave MS 317 Sunnyvale CA 94088

EMERSON, WILLIAM KEITH, b San Diego, Calif, May 1, 25. MALACOLOGY, PALEONTOLOGY. *Educ:* San Diego State Col, AB, 48; Univ Southern Calif, MS, 50; Univ Calif, Berkeley, PhD(paleont), 56. *Prof Exp:* Mus paleontologist, Univ Calif, Berkeley, 51-55; from asst cur to assoc cur invert, 55-66, chmn dept living invert, 60-74, CUR INVERT, AM MUS NATURAL HIST, 66- *Concurrent Pos:* Leader, Puritan-Am Mus Natural Hist Exped, 57 & mem, Belvedere Exped, 62 & Western Mex; res assoc, San Diego Natural Hist Mus, 62-80; mem, bd trustees, Del Mus Natural Hist, 89- *Honors & Awards:* Dorothy K Palmer Award for Res, Univ Calif, 54. *Mem:* Fel AAAS; Paleont Soc; Am Malacol Union (pres, 62); Soc Syst Zool; hon mem Am Malacologial Union Coun Syst Malacologists (pres, 89-91). *Res:* General invertebrate zoology; systematic malacology of New World marine faunas, especially Gastropoda and Scaphopoda; geographical distribution and ecology of Cenozoic marine mollusks. *Mailing Add:* Am Mus Natural Hist New York NY 10024

EMERT, JACK ISAAC, b Brooklyn, NY, Oct 22, 48; m 69; c 2. ORGANIC CHEMISTRY. *Educ:* Brooklyn Col, BS, 70; Columbia Univ, MS, 72, PhD(chem), 74. *Prof Exp:* ASST PROF ORG CHEM, POLYTECH INST NEW YORK, 74- *Mem:* Am Chem Soc; Sigma Xi; Royal Chem Soc. *Res:* Rates and mechanism of heme oxidation in aqueous solution in the presence and absence of micelle-forming surfactants; novel intramolecular excimer forming; microviscometric probes of the interior of micelles. *Mailing Add:* PO Box 536 Linden NJ 07036-0536

EMERY, ALAN ROY, b Feb 21, 21, 39; Canadian citizen; m 62; c 2. ICHTHYOLOGY, MARINE SCIENCES. *Educ:* Univ Toronto, BSc, 62; McGill Univ, MSc, 64; Univ Miami, PhD(marine sci, ichthyol), 68. *Prof Exp:* Scientist in chg resource mgt, Fisheries Res Bd Can, 64-65; res scientist, Ont Ministry Nat Resources, 68-73; cur ichthyol, Royal Ont Mus, 73-80; assoc prof, Univ Toronto, 78-83; DIR, CAN MUS NATURE, 83- *Concurrent Pos:* Sr scientist, Sublimnos Proj, J A MacInnis Found, 69-72; mem, Oil Pollution Working Group, Int Joint Comn, Great Lakes, 71; sci coord, Royal Ont Mus, 75-78; ed, Am Soc Ichthyologists & Herpetologists, 82-87. *Honors & Awards:* Conservation Award, Found Ocean Res, 85. *Mem:* Am Soc Ichthyologists & Herpetologists; Japanese Ichthyol Soc; Am Fisheries Soc; Sigma Xi; Royal Can Inst (vpres, 82-83, pres, 83-84); Can Mus Asn; Am Asn Sci Mus Dir; Asn Systematic Collections (pres). *Res:* Systematics of coral reef fishes; field and experimental ethology of fishes, particularly communication; theoretical and applied ecology; particularly as applied to resource management and human society; systematics; behavior. *Mailing Add:* Victoria Mem Mus Bldg Can Mus Nature PO Box 3443 Sta D Ottawa ON K1P 6P4 Can

EMERY, ALDEN H(AYES), JR, b Pittsburgh, Pa, May 2, 25; m 52; c 2. CHEMICAL ENGINEERING. *Educ:* Pa State Univ, BS, 47; Mass Inst Technol, SM, 49; Univ Ill, PhD, 55. *Prof Exp:* Chem engr, E I du Pont de Nemours & Co, 49-52; from asst prof to assoc prof, 54-64, PROF CHEM ENG, PURDUE UNIV, 64- *Concurrent Pos:* Fulbright fel, 67-68. *Mem:* AAAS; Am Chem Soc; Am Inst Chem Engrs; Tissue Cult Asn; Soc Indust Microbiol; Am Asn Univ Profs. *Res:* Immobilized enzyme technology; reactor design; biochemical engineering; plant cell culture. *Mailing Add:* 4231 Black Forest Ln West Lafayette IN 47906

EMERY, ASHLEY F, b San Francisco, Calif, Oct 16, 34; m 59; c 2. MECHANICAL ENGINEERING. *Educ:* Univ Calif, Berkeley, BS, 56, MS, 58, PhD(mech eng), 61. *Prof Exp:* Res engr, Univ Calif, Berkeley, 55-60; assoc prof, 61-69, PROF MECH ENG, UNIV WASH, 69- *Mem:* Am Soc Mech Engrs. *Res:* Heat transfer; gas dynamics; thermal stresses; fracture mechanics; building energy conservation. *Mailing Add:* Dept Mech Eng Univ Wash Seattle WA 98195

EMERY, DONALD ALLEN, b South Berwick, Maine, Dec 22, 28; m 56; c 2. PLANT BREEDING. *Educ:* Univ NH, BS, 50, MS, 55; Univ Wis, PhD(agron), 58. *Prof Exp:* Asst agron, Univ NH, 53-55 & Univ Wis, 55-58; from asst prof to assoc prof, 58-66, PROF CROP SCI, NC STATE UNIV, 66-, ASSOC DEAN GRAD SCH, 86- *Concurrent Pos:* Vis prof agron, Univ Fla, 75. *Honors & Awards:* Ensminger Award, Nat Asn Col & Teachers Agr, 74; Agron Educ Award, Am Soc Agron, 75; Golden Peanut Res Award, Nat Peanut Coun, 76. *Mem:* Fel Am Soc Agron; Sigma Xi. *Res:* Radiation genetics; breeding of peanuts; cytoplasmic inheritance and physiological genetics of cultivated peanut. *Mailing Add:* Grad Sch NC State Univ Raleigh NC 27650

EMERY, DONALD F, b Amboy, Ill, Dec 19, 28; m 52; c 2. FOOD CHEMISTRY & SAFETY. *Educ:* Knox Col, BA, 50; Purdue Univ, MS, 52, PhD(biol chem), 55. *Prof Exp:* Proj leader food res, 55-62, head mix develop dept, 62-68, tech dir qual control, 68-70, dir tech & qual control serv, 70-87, CONSULT LAB MGT, FOOD PROD DEVELOP & QUAL ASSURANCE & WATER QUAL, GENERAL MILLS, INC, 87- *Mem:* Am Asn Cereal Chemists; Inst Food Technologists. *Res:* Food product analysis; microbiology; cereal product development; food preservation methods; quality control; nutrition. *Mailing Add:* Emery & Assocs 64471 E Canyon Shadows Lane Tucson AZ 85737

EMERY, EDWARD MORTIMER, b New York, NY, Jan 23, 26; m 49; c 4. PHYSICAL CHEMISTRY. *Educ:* Univ Colo, BS, 48, PhD(chem), 52. *Prof Exp:* From res engr to sr res engr, Res Dept, Servel, Inc, 52-55; res chemist, Org Chem Div, Monsanto Co, 55-60, res proj leader, 60-64, res group leader, 64-70, sr res group leader, 70-80, sr res group leader, Animal Sci Div, 83-90; RETIRED. *Concurrent Pos:* VChmn comt E-19 on chromatog, Am Soc Testing & Mat, 65-68, chmn, 69-70; pres, Mo Acad Sci, 82-83. *Mem:* Am Chem Soc; Sigma Xi; Chromatographic Soc (UK). *Res:* Gas and liquid chromatography; organic spectroscopy; physical analytical chemistry; absorption refrigeration; calorimetry; aliphatic fluorine chemistry; instrument development. *Mailing Add:* 55 York Dr Brentwood MO 63144

EMERY, GUY TRASK, b Manchester, NH, May 22, 31; m 55; c 2. PHYSICS. *Educ:* Bowdoin Col, AB, 53; Harvard Univ, AM, 54, PhD(physics), 59. *Prof Exp:* Res assoc physics, Brookhaven Nat Lab, 59-61, from asst physicist to assoc physicist, 61-66; assoc prof, 66-69, prof physics, Ind Univ, 69-88; PROF PHYSICS, BOWDOIN COL, 88- *Concurrent Pos:* Vis assoc prof, State Univ NY Stony Brook, 65-66; guest scientist, Kernphysisch Versneller Inst, Univ Groningen, 78-79, 86; fel Japan Soc Prom Sci, Osaka Univ, 85. *Mem:* Fel Am Phys Soc; Am Asn Physics Teachers; Nederlandse Natuurkundige Ver; Hist Sci Soc. *Res:* Intermediate-energy nuclear physics; radioactive decay; neutron-capture gamma rays; atomic effects on nuclear properties; nuclear structure and spectroscopy; history of physics. *Mailing Add:* Dept Physics Bowdoin Col Brunswick ME 04011

EMERY, KEITH ALLEN, b Lansing, Mich, Oct 31, 54; m 80; c 2. SEMICONDUCTOR DEVICES, PHOTOVOLTAICS. *Educ:* Mich State Univ, BS, 76, MS, 79. *Prof Exp:* Res assoc semiconductor processing, Colo State Univ, 78-79, 81-82; SR SCIENTIST PHOTOVOLTAIC CHARACTERIZATION, SOLAR ENERGY RES INST, 79- *Concurrent Pos:* Scientist & lectr, Inst Elec & Electronics Engrs, 83-; mem working group two, Int Electrotech Comn, 81- *Mem:* Inst Elec & Electronics Engrs; Am Soc Testing & Mat; Am Vacuum Soc. *Res:* Rating photovoltaic devices from cells to arrays for the United States and international photovoltaic community; author of over 85 publications; maintain United States national center for terrestrial photovoltaic calibrations; modeling photovoltaic devices, solar spectra and chemical lasers; fabricating a variety of thin film devices and solar cells. *Mailing Add:* Solar Energy Res Inst 1617 Cole Blvd Golden CO 80401-3393

EMERY, KENNETH ORRIS, b Swift Current, Sask, June 6, 14; US citizen; m 41; c 2. MARINE GEOLOGY. *Educ:* Univ Ill, BS, 37, MS, 39, PhD(geol), 41. *Hon Degrees:* DSc, Univ Southern Calif, 90. *Prof Exp:* Assoc geologist, Ill State Geol Surv, 41-43; marine geologist, Div War Res, Univ Calif, 43-45; from asst prof to prof geol, Univ Southern Calif, 45-62; marine geologist, Woods Hole Oceanog Inst, 62-79; RETIRED. *Concurrent Pos:* Geologist, US Geol Surv, 46-60; mem, Navy Res & Develop Bd; mem comt paleoecol, Nat Res Coun Del, Pac Sci Cong, NZ, 49 & Philippines, 53; Guggenheim fel, Mid East, 59; oceanog adv to govts of US, Israel & Rep of China; spec adv comt coord offshore prospecting, Econ Coun Asia & Far East, 66-71; mem, Nat Acad Sci Comt Oceanog, 71-72. *Honors & Awards:* F P Shepard Award Marine Geol, Soc Econ Paleontologists & Mineralogists, 69, W H Tulenhofel Award Sedimentation, 89; Compass Distinguished Serv Award, Marine Technol Soc, 74; Maurice Ewing Award Geophys, 85. *Mem:* Nat Acad Sci; Am Acad Arts & Sci; fel Geol Soc Am; Soc Econ Paleont & Mineral; Am Asn Petrol Geologists. *Res:* Physiography, sediments and lithology of sea floor off California; general marine geology; marine geology of Bikini and nearby atolls, Guam, Persian Gulf; Dead Sea; eastern Mediterranean Sea; geological history of Atlantic continental shelf and slope; oil regions of continental margin off eastern Asia; structure of continental margin off western Africa; geology of Atlantic Ocean; sea levels; land levels; tide gauges of the world. *Mailing Add:* 74 Ransom Rd Falmouth MA 02540

EMERY, PHILIP ANTHONY, b Neodesha, Kans, Oct 20, 34; m 60; c 2. HYDROGEOLOGY, GEOLOGY. *Educ:* Univ Kans, BS, 60, MS, 62. *Prof Exp:* Hydrologist, 62-75, dist chief, Water Resources Div, 75-89, CONSULT GEOHYDROLOGIST, US GEOL SURV, 89- *Mem:* Geol Soc Am; Int Asn Hydrogeologists; Nat Water Well Asn; Sigma Xi; AAAS; Am Inst Prof Geologists. *Res:* Hydrologic modeling; hydrology of valley-fill aquifers; evapotranspiration of ground water. *Mailing Add:* 2910 Star Dr Lake Havasu City AZ 86403

EMERY, RICHARD MEYER, b Toledo, Ohio, Mar 19, 39; c 2. LIMNOLOGY, RADIATION ECOLOGY. *Educ:* Univ Toledo, BEd, 62; Ohio State Univ, MS, 66; Univ Wash, PhD(appl limnol), 72. *Prof Exp:* Teacher biol, Clay Sr High Sch, Ohio, 63-66; biologist, Tenn Valley Authority, 66-69; Environ Protection Agency res fel appl limnol, Univ Wash, 69-72; res scientist appl limnol, Battelle Northwest Lab, 72-81; PRIN SCIENTIST, ENVIROSPHERE CO, 84- *Mem:* Am Soc Limnol & Oceanog; NAm Benthological Soc; Int Soc Theoret & Appl Limnol; Sigma Xi. *Res:* Systems ecology pertaining to holistic response patterns and cumulative impacts, environmental regulatory analysis and design/compliance of hazardous waste disposal facilities; characterization and measurement of the ecological behavior of radionuclides in freshwater ecosystems. *Mailing Add:* PO Box 10562 Winslow WA 98110-0562

EMERY, ROY SALTSMAN, b Ill, Sept 22, 28; m 52; c 4. ANIMAL NUTRITION, BIOCHEMISTRY. *Educ:* Colo Agr & Mech Col, BS, 50, MS, 52; Mich State Univ, PhD(nutrit, biochem), 55. *Prof Exp:* Asst, 52-55, from asst prof to assoc prof, 55-67, PROF DAIRY NUTRIT, MICH STATE UNIV, 67- *Honors & Awards:* Am Feed Mfrs Award, 61; Sigma Xi Jr Res Award, 69. *Mem:* Am Soc Microbiol; Am Inst Nutrit; Am Dairy Sci Asn; Am Soc Animal Sci; fel AAAS. *Res:* Biochemistry and fermentation; digestion and nutrition in ruminants; intermediate and microbial metabolism. *Mailing Add:* Dept Animal Sci Mich State Univ Anthony Hall East Lansing MI 48824

EMERY, THOMAS FRED, b Ross, Calif, July 31, 31; m 62; c 1. BIOCHEMISTRY. *Educ:* Calif Inst Technol, BS, 53; Univ Calif, Berkeley, PhD(biochem), 60. *Prof Exp:* NSF fel, Nat Ctr Sci Res, France, 60-61; from instr to assoc prof biochem, Sch Med, Yale Univ, 61-70; PROF CHEM, UTAH STATE UNIV, 70- *Concurrent Pos:* Brown Mem grant, 61-62; prin investr, USPHS res grant, 62-88. *Mem:* Am Soc Microbiol; Fedn Am Soc Exp Biol. *Res:* Isolation, structure, function and biosynthesis of naturally occurring iron chelates or siderophores; microbial iron transport. *Mailing Add:* Dept Chem & Biochem Utah State Univ Logan UT 84322-0300

EMERY, VICTOR JOHN, b Boston, Eng, May 16, 34; m 59; c 3. THEORETICAL PHYSICS. *Educ:* Univ London, BSc, 54; Univ Manchester, PhD(theoret physics), 57. *Prof Exp:* Res assoc, Cambridge Univ, 57-59; Harkness fel, Commonwealth Fund, Univ Calif, Berkeley, 59-60; lectr, Birmingham Univ, 60-63; vis asst prof, Univ Calif, Berkeley, 63-64; SR PHYSICIST, BROOKHAVEN NAT LAB, 64- *Concurrent Pos:* Vis prof, Nordita, Copenhagen, 71-72 & Univ Paris, France, 76 & 81. *Mem:* Fel Am Phys Soc. *Res:* Theoretical solid state; low temperature and statistical physics. *Mailing Add:* Dept Physics Brookhaven Nat Lab Upton NY 11973

EMERY, W(ILLIS) L(AURENS), b Salt Lake City, Utah, Nov 23, 15; m 41; c 3. ELECTRICAL ENGINEERING. *Educ:* Univ Utah, BS, 36; Iowa State Col, MS, 40, PhD(elec eng), 47. *Prof Exp:* Instr elec eng, Univ Utah, 36-37; sales engr, Campbell-Elsey Co, 37-38; instr elec eng, Univ Utah, 38-39; instr, Iowa State Col, 41-43; radio engr, Naval Res Lab, Washington, DC, 43-45; asst prof, Iowa State Col, 46-47; assoc prof elec eng, Univ Utah, 47-50; assoc prof, 50-53, PROF ELEC ENG, UNIV ILL, URBANA, 53- *Concurrent Pos:* Vis prof, Indian Inst Technol, Kharagpur, 60-62. *Mem:* Inst Elec & Electronics Engrs; Am Phys Soc. *Res:* Microwaves; electrooptics; displays. *Mailing Add:* Univ Ill Urbana IL 61801

EMERY, WILLIAM HENRY PERRY, b Wickford, RI, Feb 10, 24; m 46; c 2. CYTOLOGY, TAXONOMY. *Educ:* RI State Col, BS, 48; Univ Conn, MS, 50; Univ Tex, PhD(cyto-taxon), 56. *Prof Exp:* Fel cyto-taxon, Southwest Tex State Univ, 57, from instr to prof biol, 57-90; RETIRED. *Concurrent Pos:* Plant taxonomist, Espey Huston Assoc & Southwest Res Inst, 74- *Mem:* AAAS; Am Bot Soc. *Res:* Cyto-taxonomy and breeding of aquatic grasses. *Mailing Add:* Dept Biol Southwest Tex State Univ San Marcos TX 78666

EMERY, WILLIAM JACKSON, b Honolulu, Hawaii, Apr 15, 46; m 70; c 3. PHYSICAL OCEANOGRAPHY, SATELLITE REMOTE SENSING. *Educ:* Brigham Young Univ, BS, 71; Univ Hawaii, PhD(phys oceanog), 75. *Prof Exp:* Res assoc phys oceanog, Univ Hawaii, 75-76; res assoc phys oceanog, Tex A&M Univ, 76-78; asst prof, 78-84, ASSOC PROF PHYS OCEANOG, UNIV BC, 84- *Concurrent Pos:* Vis prof, Inst F Meereskunde, Univ Kiel, W Ger, 84-85. *Mem:* Am Meteorol Soc; Am Geophys Union; Can Meteorol Oceanog Soc. *Res:* Large scale ocean and ocean-atmosphere problems; temperature and salinity structures of the Pacific ocean; mesoscale temperature studies; satellite remote sensing. *Mailing Add:* Dept Aerospace Eng CCAR Univ Colo Box 431 Boulder CO 80309

EMESON, EUGENE EDWARD, b Apr 10, 34; m; c 2. IMMUNOPATHOLOGY, AUTOIMMUNITY. *Educ:* Univ Colo, MD, 59. *Prof Exp:* PROF PATH & DIR DIAG IMMUNOL, COL MED, UNIV ILL, 83- *Mem:* Am Asn Pathologists; Am Asn Immunol Transplantation Soc. *Res:* Growth factors; molecular biology of neoplasia; athero sclerosis. *Mailing Add:* Dept Path Col Med M/C 847 Univ Ill 1853 W Polk St Chicago IL 60612

EMIGH, CHARLES ROBERT, b Seattle, Wash, Apr 7, 20; m 46; c 3. PHYSICS. *Educ:* Univ Colo, BS, 42; Univ Ill, MS, 48, PhD(physics), 51. *Prof Exp:* Jr engr, Res & Develop, Westinghouse Elec Corp, 42-44; asst physics, Univ Ill, 46-51; mem staff, Los Alamos Nat Lab, 51-72, dir intense neutron source facil, 72-78, assoc div leader energy technol, 78-85; prof physics & tech dir, Univ NMex Los Alamos, 85-88; PRES, CRESS COMPUTER SYSTS, 82- *Concurrent Pos:* Ford Found sr fel, Europ Ctr Nuclear Res, Geneva, 57; adj prof physics, Univ NMex, 67-73 & 80-; computer sci consult, 88-; dir, Soc Nondestructive Testing, 54-56. *Mem:* Am Phys Soc; Soc Nondestructive Testing; Inst Elec & Electronics Engrs; Am Nuclear Soc. *Res:* Experimental and theoretical physics; design and development of apparatus used in experimental physics; solid state physics; accelerator physics and engineering; computer programming techniques. *Mailing Add:* 215 Barranca Rd Los Alamos NM 87544

EMIGH, G DONALD, b Burley, Idaho, Jan 21, 11; m 38; c 2. MINING, METALLURGY. *Educ:* Univ Idaho, BS, 32, MS, 34; Univ Ariz, PhD(geol), 56. *Prof Exp:* Mining engr, Gen Elec Co, 36-37 & US Vanadium Corp, 37-48; dir mining, Monsanto Co, St Louis, 49-76; CONSULT, 76- *Mem:* Distinguished mem Am Inst Mining, Metall & Petrol Engrs; Soc Econ Geol; fel Geol Soc Am; Am Inst Prof Geologists; Can Inst Mining & Metall. *Res:* Mineral industry and exploration. *Mailing Add:* 202 Churchill Dr Burley ID 83318

EMILIANI, CESARE, b Bologna, Italy, Dec 8, 22; m 51; c 2. CLIMATOLOGY, MARINE GEOLOGY. *Educ:* Univ Bologna, Doctorate, 45; Univ Chicago, PhD(geol), 50. *Prof Exp:* Geologist petrol, Nat Soc Hydrocarbons, Italy, 46-48; res assoc geochem, Univ Chicago, 50-56; assoc prof, 57-63, PROF MARINE GEOL, UNIV MIAMI, 63- *Honors & Awards:* Vega Medal, Sweden, 83; Agassiz Medal, Nat Acad Sci, 89. *Mem:* Fel AAAS; fel Am Geophys Union. *Res:* Isotope paleoclimatology. *Mailing Add:* 151 Edgewater Dr Miami FL 33133

EMIN, DAVID, b New York, NY, Oct 2, 41; m 63. SOLID STATE PHYSICS. *Educ:* Fla State Univ, BA, 62; Univ Pittsburgh, PhD(physics), 68. *Prof Exp:* Asst res physicist, Univ Calif, Los Angeles, 68-69; mem tech staff theoret physics, 69-83, DISTINGUISHED MEM TECH STAFF, SANDIA NAT LABS, 83- *Mem:* Fel Am Phys Soc. *Res:* Low-mobility electrical transport theory; small-polaron motion; polaron theory; atomic diffusion; amorphous semiconductors; magnetic semiconductors; boron-rich solids; high-temperature superconductivity. *Mailing Add:* Solid State Theory Div 1152 Sandia Nat Labs Albuquerque NM 87185

EMINO, EVERETT RAYMOND, b Milford, Mass, Feb 8, 42; m 67; c 2. HORTICULTURE, FLORICULTURE. *Educ:* Univ Mass, BS, 65; Mich State Univ, MS, 67, PhD(hort), 72. *Prof Exp:* Grad asst hort, Mich State Univ, 65-67, instr, 67-72; asst prof plant sci, Univ Mass, 72-75; assoc prof floricult, Tex A&M Univ, 75-82; prof hort & head dept plant sci, Univ Conn, 82-87; ASST DEAN RES, INST FOOD & AGR SCI, UNIV FLA, GAINESVILLE, 87- *Mem:* Am Soc Hort Sci; Am Soc Agron; Bot Soc Am; Int Plant Propagators Soc; Nat Asn Col & Teachers Agr; Crop Sci Soc Am. *Res:* Investigations on the morphology and physiology of floricultural crops with special emphasis on the soil plant relationship and the environmental control of flowering. *Mailing Add:* Inst Food & Agr Sci Univ Fla Gainesville FL 32611-0000

EMKEN, EDWARD ALLEN, b Yates City, Ill, Aug 12, 40; m 64; c 2. LIPID CHEMISTRY, ORGANIC BIOCHEMISTRY. *Educ:* Bradley Univ, BS, 63; Univ Iowa, MS, 68, PhD(org chem), 69. *Prof Exp:* Sci trainee, 59-63, asst res chemist, 63-64, assoc res chemist, 64-69, RES LEADER, NORTHERN REGIONAL RES LABS, USDA, 69- *Concurrent Pos:* Teaching asst, Univ Iowa, 66-69; adj prof, Peoria Sch Med; assoc ed, Lipids, 82- *Honors & Awards:* Bond Award, Am Oil Chemist's Soc, 72; USDA Distinguished Res Award, 89. *Mem:* Am Chem Soc; Am Oil Chemist's Soc. *Res:* Enzyme reactions of lipids; synthesis of radioisotope and deuterium-labeled compounds and dietary fatty acids; methods developed for lipid and fatty acid isomer analysis, separation and determination of physical properties; biochemistry and nutrition of dietary fats and oils; clinical studies on lipid related diseases and disorders. *Mailing Add:* USDA Northern Regional Res Ctr 1815 N University St Peoria IL 61604

EMLEMDI, HASAN BASHIR, b Misurata, Libya, Oct 12, 59; div; c 1. PROTECTIVE COATING, GLASS FORMULATIONS. *Educ:* Univ Pittsburgh, BS, 82, MS, 86. *Prof Exp:* Proj engr, Bar & Rod Mill, Iron & Steel Corp, 82-83; res asst exp work & thesis, Univ Pittsburgh, 83-86, teaching asst mat eng, 86; develop engr electrostatic application of enamel powder, 86-90, SR DEVELOP ENGR COATING DEVELOPMENT FOR STEEL, CLAYWARE & GLASS SUBSTRATES, O HOMMEL CO, 90- *Concurrent Pos:* Chmn elect, Am Ceramic Soc, Pittsburgh Sect, 90- *Mem:* Assoc mem Nat Inst Ceramic Engrs; assoc mem Metall Soc; Am Soc Metals; Am Ceramic Soc. *Res:* Development of a wide range of ceramic coating products for steel, claybody, and glass substrates; porcelain enamel coating application both wet system and dry electrostatic powder system. *Mailing Add:* 1257 Meadowlark Dr Pittsburgh PA 15243

EMLEN, JOHN MERRITT, b Sacramento, Calif, Jan 15, 38; m; c 1. FISH & WILDLIFE SCIENCES. *Educ:* Univ Wis, BA, 61; Univ Wash, PhD(zool), 66. *Prof Exp:* Asst prof biol, Univ Colo, Boulder, 66-68; asst prof, State Univ NY Stony Brook, 68-71; from asst prof to assoc prof biol, Ind Univ, Bloomington, 71-83; Fisheries & Wildlife Dept, Mark O Hatfield Marine Sci Ctr, Ore State Univ, 83-86; US Environ Protection Agency, Corvallis, Ore, 84-86; US FISH & WILDLIFE SERVICE, SEATTLE, WA, 87- *Concurrent Pos:* NSF grant, 67-69. *Mem:* AAAS; Ecol Soc Am; Soc Conserv Biol; Sigma Xi. *Res:* Theoretical and behavioral ecology; natural selection; population genetics; applied ecology. *Mailing Add:* Nat Fishers Res Ctr US Fish & Wildlife Serv Sand Point Way Seattle WA 98115

EMLEN, JOHN THOMPSON, JR, b Philadelphia, Pa, Dec 28, 08; m 34; c 3. ZOOLOGY. *Educ:* Haverford Col, BS, 31; Cornell Univ, PhD(ornith), 34. *Hon Degrees:* DSc, Haverford Col, 70. *Prof Exp:* Jr biologist, Bur Biol Surv, USDA, 34-35; from instr zool & jr biologist to asst prof zool & asst zoologist, Exp Sta, Univ Calif, 35-43; res assoc, Rockefeller Inst, 43-46; from assoc prof to prof, 46-73, chmn dept, 51-53 & 54-55, EMER PROF ZOOL, UNIV WIS-MADISON, 73- *Concurrent Pos:* Guggenheim fel, Cent Africa, 53-54; NSF res fel, Africa, 59 & Antarctica, 62-64. *Mem:* AAAS; Am Ornith Union (pres, 75); Am Soc Mammalogists; Wilson Ornith Soc (pres, 60); Ecol Soc Am. *Res:* Population and behavior studies of birds and mammals. *Mailing Add:* Dept Zool Univ Wis Madison WI 53706

EMLEN, STEPHEN THOMPSON, b Sacramento, Calif, Aug 21, 40; m 73; c 2. ANIMAL BEHAVIOR, ECOLOGY. *Educ:* Swarthmore Col, BA, 62; Univ Mich, MS, 64, PhD(zool), 66. *Prof Exp:* From asst prof to assoc prof, 66-75, PROF ZOOL, CORNELL UNIV, 75- *Concurrent Pos:* John Simon Guggenheim Found fel, 73, Ctr Adv Behav Sci, 80. *Honors & Awards:* William Brewster Award, Am Ornithologists Union, 84. *Mem:* AAAS; Animal Behav Soc; Ecol Soc Am; Am Ornith Union; Cooper Ornith Soc; Wilson Ornith Soc; Am Soc Naturalists; Soc Ger Ornith. *Res:* Orientation and navigation behavior; visual and acoustical communication systems; evolution of social organization among vertebrates. *Mailing Add:* Div Biol Sci Cornell Univ Seeley & Mudd Hall Ithaca NY 14853

EMLET, HARRY ELSWORTH, JR, b New Oxford, Pa, Sept 21, 27; m 51; c 2. SYSTEMS ANALYSIS, HEALTH SCIENCES. *Educ:* Princeton Univ, AB, 52. *Prof Exp:* Systs reviewer automatic data processing, Prudential Ins Co, 55-56; aeronaut engr, Martin Co, 56-57; res analyst weapons systs analysis, Melpar, Inc, 57-58; aeronaut engr & proj leader, Analytical Serv, Inc, 58-65, chief, Plans Br, 65-67, chief, Tactical Br, 67-70, mgr, Tactical Div & Health Serv Studies, 70-74, mgr, Analytical Serv, Inc, 74-76 , vpres health systs, 76-83, dir spec projs, 83-91; RETIRED. *Concurrent Pos:* Mem, US AF Keese Comt Air Force Space Plan, 61 & Holzapple Comt Air Force Space Prog, 62; mem adv comt health model develop, Mil Health Care Study, Off Mgt & Budget, 74-75; chmn, Health Appln Sect, Opers Res Soc Am, 76-77; mem bd dirs, Symp Comput-Appln in Med Care, Inc, 79-87, gen chmn, 80; treas, Alliance Eng in Med & Biol, 82-86; mem bd dirs, Am Asn Med Systs & Informatics, 81-84. *Mem:* Opers Res Soc Am; Soc Advan Med Systs (vpres, 75-76, pres-elect, 76-77, pres, 77-78); Am Inst Aeronaut & Astronaut; Am Asn Med Systs & Informatics (vpres, 85-86); AAAS. *Res:* Military operational requirements analysis; research and development planning; weapons system analysis; planning techniques; philosophy; health systems analysis and evaluation. *Mailing Add:* 3302 Clearwood Falls Church VA 22042

EMLING, BERTIN LEO, b Erie, Pa, July 9, 05. ORGANIC CHEMISTRY. *Educ:* St Vincent Col, AB, 31; Johns Hopkins Univ, MA, 38; Univ Notre Dame, PhD(chem), 41. *Prof Exp:* From instr to prof org chem, 37-80, mem bd dirs, 57-65, EMER PROF ORG CHEM, ST VINCENT COL, 80- *Concurrent Pos:* Proj dir, Nat Coop Undergrad Chem Res Prog, 48-57. *Res:* Sulfonic acid esters; furyl amines; acetylenes; olefins; Schiff bases; polyester resins; autoxidation. *Mailing Add:* Dept Chem St Vincent Col Latrobe PA 15650-9121

EMMANOUILIDES, GEORGE CHRISTOS, b Drama, Greece, Dec 17, 26; US citizen; m 59; c 5. PEDIATRICS, CARDIOLOGY. *Educ:* Univ Thessaloniki, MD, 51; Univ Calif, Los Angeles, MS, 63. *Prof Exp:* From asst prof to assoc prof, 63-73, PROF PEDIAT, SCH MED, UNIV CALIF, LOS ANGELES, 73- *Concurrent Pos:* Chief div pediat cardiol & neonatology, Harbor Gen Hosp, Torrance, Calif, 63-69 & div pediat cardiol, 69-; mem coun cardiovasc dis of young, clin cardiol & cardiopulmonary dis, Am Heart Asn; fel pediat hemat, Children's Hosp of DC, 59-60; Ont Heart Asn fel, Hosp for Sick Children, Toronto, 60-61; USPHS trainee pediat cardiol, Med Ctr, Univ Calif, Los Angeles, 61-63; chmn exec comt, Sect Cardiol, Am Acad Pediat, 78-80. *Mem:* AAAS; Am Pediat Soc; fel Am Col Cardiol; Am Soc Pathologists; Am Acad Pediat; Am Heart Asn. *Res:* Cardiorespiratory adjustments of the newborn; fetal and neonatal physiology; pediatric cardiology. *Mailing Add:* Harbor-UCLA Med Ctr 1000 W Carson St Torrance CA 90509

EMMANUEL, GEORGE, b Tanta, Egypt, Sept 19, 25; US citizen; m 59; c 4. CARDIOPULMONARY PHYSIOLOGY. *Educ:* Nat Univ Athens, MD, 52. *Prof Exp:* Mem fac cardiopulmonary physiol, Belleview Hosp, Columbia Univ, 57-59; from instr to asst prof, 59-67, ASSOC PROF MED, STATE UNIV NY DOWNSTATE MED CTR, 67- *Mem:* Am Physiol Soc; Am Soc Clin Invest; Am Heart Asn; Harvey Soc. *Res:* Cardiopulmonary physiology; teachings of medicine. *Mailing Add:* Dept Med VA Med Ctr Bay Pines FL 33504

EMMEL, THOMAS C, b Inglewood, Calif, May 8, 41. POPULATION BIOLOGY, GENETICS. *Educ:* Reed Col, BA, 63; Stanford Univ, PhD(pop biol), 67. *Prof Exp:* Lectr entom, San Jose State Col, 65-66; course coordr, Orgn Trop Studies, Inc, Costa Rica & prof trop biol, Univ Costa Rica, 67-69; asst prof biol sci & zool, 68-73, assoc prof zool, 73-75, PROF ZOOL & CHMN DEPT, UNIV FLA, 75- *Concurrent Pos:* NIH fel genetics, Univ Tex, Austin, 67-68; Guggenheim fel, 86-87. *Honors & Awards:* John Abbott Award, 90. *Mem:* AAAS; Soc Study Evolution; Asn Trop Biol; Ecol Soc Am; Lepidop Soc; fel Royal Entom Soc London. *Res:* Population biology of tropical and Nearctic organisms; ecological genetics of natural populations, especially satyrid and nymphalid butterflies and land snails; territorial behavior. *Mailing Add:* Dept Zool Univ Fla Gainesville FL 32611

EMMEL, VICTOR MEYER, b St Louis, Mo, Mar 22, 13; m 43; c 4. HISTOLOGY. *Educ:* Brown Univ, AB, 35, MS, 37, PhD(biol), 39; Univ Rochester, MD, 47. *Prof Exp:* From instr to prof, 40-78, EMER PROF ANAT, SCH MED & DENT, UNIV ROCHESTER, 78- *Concurrent Pos:* Intern, Strong Mem Hosp, Rochester, 47-48; trustee & secy, Biol Stain Comn, 55-80; Nat Res Coun fel med sci, Sch Med, Yale Univ, 39-40; assoc ed, Stain Technol. *Mem:* Am Asn Anat; AAAS; Soc Exp Biol & Med; Nutrit Today Soc; Histochem Soc. *Res:* Chemistry of sea-water; menstruation in the monkey; cytology and cytochemistry of the kidney and intestine; histopathology of kidney and intestine in vitamin E deficiency. *Mailing Add:* Peckins Wayland NY 14620

EMMERICH, WERNER SIGMUND, b Dusseldorf, Ger, June 3, 21; nat US; m 53; c 3. PLASMA PHYSICS. *Educ:* Ohio State Univ, BS, 49, MS, 50, PhD(physics), 53. *Prof Exp:* Res engr nuclear physics, Westinghouse Res Ctr, 53-57, adv physicist, 57-64, mgr arc & plasma res, 64-75, dir corp res & develop, 75-86; RETIRED. *Mem:* AAAS; fel Am Phys Soc. *Res:* Optical model of atomic nucleus; beta, gamma and neutron spectroscopy; power circuit interruption; research planning. *Mailing Add:* 1883 Beulah Rd Pittsburgh PA 15235-5004

EMMERLING, DAVID ALAN, b E Liverpool, Oh, Apr 11, 48; m 76; c 2. OTHER MEDICAL & HEALTH SCIENCE. *Educ:* Univ Va, BA, 71, Med, 73, EdD, 79. *Prof Exp:* Psychologist, Ctr Psychol Serv, 75-77, & Cardiac rehab, Wake Forest Univ, 76-78; coun psychol, 78-81 & assoc dir, Coun & Study Develop Ctr, James Madison Univ, 81-84; dir, count & study develop ctr, Northern Ill Univ, 84-86; EXEC DIR, NAT WELLNESS INST INC, 86- *Concurrent Pos:* Chairperson comn VIII Wellness, Am Col Personnel Asn; vis prof, Calif State Univ, 86. *Mem:* Am Col Personnel Asn Wellness; Am Asn Coun Develop; Asn Fitness Bus; Am Col Personnel Asn. *Res:* Legal and ethical issues in counseling and health centers; evaluation of the effectiveness of wellness programs. *Mailing Add:* Nat Wellness Inst S Hall 1319 Fremont Stevens Point WI 54481

EMMERLING, EDWARD J, b Norwood, Ohio, Oct 20, 08. PROTECTIVE RELAYING, INDUSTRIAL POWER SYSTEMS. *Educ:* Univ Cincinnati, BS, 32, BA, 53. *Prof Exp:* Staff engr, Cincinnati Gas & Elec Co, 32-72; consult engr, A M Kenny Co, 72-77; PROJ ENGR, CAMARGO ASSOCS, 77- *Mem:* Fel Inst Elec & Electronics Engrs. *Mailing Add:* Camargo Assocs 1329 E Kemper Rd Cincinnati OH 45241

EMMERS, RAIMOND, b Liepaja, Latvia, Apr 19, 24; US citizen; m 56. MEDICAL PHYSIOLOGY, NEUROPHYSIOLOGY. *Educ:* ETex Baptist Col, BA, 53; Univ NC, MA, 55; Syracuse Univ, PhD(neurophysiol), 58. *Prof Exp:* Asst prof, 61-71, ASSOC PROF PHYSIOL, COL PHYSICIANS & SURGEONS, COLUMBIA UNIV, 71- *Concurrent Pos:* Dir, Nat Inst Neurol Dis & Stroke Res Proj, 61; Nat Inst Neurol Dis & Blindness res fel, 58-60; res fel neurophysiol, Univ Wis, 59-61. *Mem:* Am Physiol Soc; Am Asn Anat. *Res:* Neural mechanisms of taste and somesthesia; significance of taste in nutrition; sensory coding in the central nervous systems. *Mailing Add:* Dept Physiol Columbia Univ Col Physicians & Surgeons 630 W 168 St New York NY 10032

EMMERSON, JOHN LYNN, b Princeton, Ind, Nov 21, 33; m 57; c 2. PHARMACOLOGY, TOXICOLOGY. *Educ:* Purdue Univ, BS, 58, MS, 59, PhD(pharmacol), 61. *Prof Exp:* Sr pharmacologist, Eli Lilly & Co, 61-65; assoc prof toxicol, Purdue Univ, 65-66; sr toxicologist, Eli Lilly & Co, 66-67, head metab & path, 67-77, dir toxicol studies, 77-84, Lilly res fel & dir toxicol, 84-90, DISTINGUISHED LILLY RES SCHOLAR, ELI LILLY & CO, 91- *Mem:* AAAS; Am Soc Pharmacol & Exp Therapeut; Soc Toxicol. *Res:* Biochemical mechanisms and metabolic aspects of drug toxicity. *Mailing Add:* Lilly Res Labs Box 708 Greenfield IN 46140

EMMERT, GILBERT A, b Merced, Calif, June 2, 38; m 64; c 2. PLASMA PHYSICS. *Educ:* Univ Calif, Berkeley, BS, 61; Rensselaer Polytech Inst, MS, 64; Stevens Inst Technol, PhD(physics), 68. *Prof Exp:* Anal engr, Energy Conversion Systs, United Aircraft Corp, 61-64; asst prof, 68-72, assoc prof, 72-79, PROF NUCLEAR ENG, UNIV WIS-MADISON, 79- *Concurrent Pos:* Guest scientist, Max Planck Inst Plasma Physics, 76-77. *Mem:* Am Phys Soc. *Res:* Theoretical plasma physics; waves and instabilities in magnetically confined plasmas; plasma-wall interactions; systems studies of controlled thermonuclear fusion. *Mailing Add:* Dept Nuclear Eng Res Bldg Univ Wis Madison WI 53706

EMMERT, R(ICHARD) E(UGENE), b Iowa City, Iowa, Feb 23, 29; m 49; c 3. CHEMICAL ENGINEERING. *Educ:* Univ Iowa, BS, 51; Univ Del, MChE, 52, PhD(chem eng), 54. *Prof Exp:* Res engr chem eng, 54-58, res proj supvr, 58-61, sr res engr, 61, res supvr, 61-63, mgr indust develop, 63-64, area supvr mfg, 64-66, mfg supt, 66-67, asst plant mgr, 67-69, mgr eng technol & mat res, 69-72, dir res & develop, Pigments dept, 72-74, dir instrument prod, 75-76, dir electronic prod, 77-78, gen mgr textile fibers, 79-80, vpres corp plans, 81-83, vpres, photosysts & electronic prod dept, 83-86, vpres, Electronics Dept, 86-87, E I Du Pont de Nemours & Co, Inc, 86-87; EXEC DIR, AM INST CHEM ENGRS, 88- *Mem:* Nat Acad Eng; Am Inst Chem Engrs; Am Chem Soc. *Res:* Reaction kinetics; polymerization technology; mass transfer; gas absorption. *Mailing Add:* Exec Dir Am Inst Chem Engrs 345 E 47 St New York NY 10017

EMMETT, EDWARD ANTHONY, b Sydney, Australia, Feb 29, 40. ENVIRONMENTAL HEALTH, OCCUPATIONAL MEDICINE. *Educ:* Univ Sydney, MB, BS, 64; FRACP, 74; Univ Cincinnati, MS, 75. *Prof Exp:* Med intern, Royal Prince Alfred Hosp, Sydney, 64, resident physician, 65; med resident, Repatriation Gen Hosp, Concord, Australia, 66-69; fel environ health, Univ Cincinnati, 70, asst prof environ health & internal med, 71-75, assoc prof environ health, med, & dermat, 75-78; prof environ health sci & dir div occup med, Johns Hopkins Univ, 78-; CHIEF EXEC OFFICER, WORKSAFE, AUSTRALIA. *Mem:* Am Fedn Clin Res; Soc Investigative Dermat; Am Soc Photobiol; NY Acad Sci; Am Occup Med Asn. *Res:* Epidemiology, pathogenesis and prevention of occupational and environmental disease; photobiology. *Mailing Add:* 92 Parramatta Rd Camperdown 2050 Australia

EMMETT, JOHN L, b Rochester, Minn, July 20, 39. SOLID STATE LASERS. *Educ:* Calif Inst Technol, BS, 61; Stanford Univ, PhD(physics), 67. *Prof Exp:* Assoc dir prog develop, Lawrence Livermore Nat Lab, 88-89; PRES, JLE ASSOC, 89- *Concurrent Pos:* Mem, bd dir, Hoya Corp, 89-; pres, Crystal Res, 89- *Honors & Awards:* E O Lawrence Award, US Dept Energy, 78. *Mem:* Sigma Xi; fel Am Phys Soc. *Mailing Add:* JLE Assoc 4309 Hacienda Dr Suite 420 Pleasanton CA 94588

EMMETT, WILLIAM W, b Miami, Fla, Feb 12, 37; m 70; c 2. HYDROLOGY, WATER RESOURCES. *Educ:* Ga Inst Technol, BCE, 59, MSCE, 61; Johns Hopkins Univ, PhD(geog), 68. *Prof Exp:* RES HYDROLOGIST, US GEOL SURV, 59- *Concurrent Pos:* Org mem, Am Geomorphol Field Group. *Honors & Awards:* Meritorious Serv Award, US Dept Interior. *Mem:* Am Soc Civil Engrs; fel Geol Soc Am; Am Inst Hydrol; Am Geophys Union; Brit Geomorphol Res Group; Geol Soc Wash. *Res:* Fluvial processes and geomorphology; river channel behavior: sediment transport, especially bedload transport; hydraulics of overland flow on hill slopes. *Mailing Add:* 5960 S Wolff Ct Littleton CO 80123-6734

EMMETT-OGLESBY, MICHAEL WAYNE, b Portland, Ore, Sept 12, 47; m 75. PHARMACOLOGY. *Educ:* Univ Chicago, BA, 69; State Univ NY Buffalo, PhD(pharmacol), 73. *Prof Exp:* NIH fel pharmacol, Univ Chicago, 73-75; asst prof pharmacol, 75-80, assoc prof pharmacol, 80-88, PROF, TEX COL OSTEOP MED, 88- *Honors & Awards:* Humboldt Fel, Munich, WGer. *Mem:* AAAS; Soc Neurosci; Am Soc Pharmacol & Exp Therapeut; Soc Stimulus Properties Drug (secy/treas, 87-88, pres, 88-). *Res:* Behavioral and neurochemical studies of mechanisms mediating the effects of drugs on behavior. *Mailing Add:* Dept Pharmacol Tex Col Osteop Med 3516 Camp Bowie Ft Worth TX 76107-2690

EMMICK, THOMAS LYNN, b Indianapolis, Ind, Aug 1, 40; m 62; c 2. ORGANIC CHEMISTRY. *Educ:* Wabash Col, AB, 62; Northwestern Univ, PhD(org chem), 67. *Prof Exp:* Sr org chemist res, Eli Lilly & Co, 66-71, head agr prod develop, 71-72, HEAD AGR ORG CHEM, LILLY RES LABS, DIV ELI LILLY & CO, 72- *Mem:* Am Chem Soc. *Res:* Synthetic organic chemicals useful in the agricultural area as pesticides, as agents to stimulate growth in meat producing animals, or as agents effective against diseases in animals. *Mailing Add:* Lilly Corp Ctr Eli Lilly & Co Indianapolis IN 46285-0002

EMMONS, DOUGLAS BYRON, agriculture, dairy industry, for more information see previous edition

EMMONS, HAMILTON, b London, Eng, Dec 30, 30; US citizen; m 59; c 3. OPERATIONS RESEARCH. *Educ:* Harvard Col, AB, 52; Univ Minn, MS, 58; NY Univ, MS, 62; Johns Hopkins Univ, PhD(opers res), 68. *Prof Exp:* Mem tech staff, Bell Tel Labs, 58-64; asst prof opers res, Cornell Univ, 68-73; assoc prof, 73-87, PROF OPERS RES, CASE WESTERN RESERVE UNIV, 87-, CHMN DEPT, 76-79 & 87- *Mem:* AAAS; Inst Mgt Sci; Opers Res Soc Am. *Res:* Management of health care; scheduling theory; stochastic processes; semi-Markov decision processes. *Mailing Add:* Dept Opers Res Enterprise Hall Case Western Reserve Univ Cleveland OH 44106

EMMONS, HOWARD W(ILSON), b Morristown, NJ, Aug 30, 12; m 38; c 3. MECHANICAL ENGINEERING, FIRE SAFETY ENGINEERING. *Educ:* Stevens Inst Technol, ME, 33, MS, 35; Harvard Univ, ScD(eng), 38. *Hon Degrees:* DEng, Stevens Inst Technol, 63; Dsc, Worcester Polytech Inst. *Prof Exp:* Res engr, Westinghouse Elec Co, 37-39; asst prof mech eng, Univ Pa, 39-40; from asst prof to prof, 40-83, EMER PROF MECH ENG, HARVARD UNIV, 83- *Concurrent Pos:* Fulbright-Guggenheim fel, Eng, 52-53; Hunsaker vis prof, Mass Inst Technol, 57-58; consult, Pratt & Whitney Aircraft, 40-78, Army Ord Ballistics Res Lab, Aberdeen Proving Ground, Md, 40-55 & Naval Ord Lab, 46-52; mem space sci technol panel, Off Sci & Technol, 60-71; mem Govt adv comt on sci & Technol, 65-71; chmn Mass Sci & Technol Found, 70-74; mem Mass Nuclear Safety Comn, 74-75; with Nat Adv Comt Aeronaut, 44 & Adv Bd Nav Ord Test Sta, 49-55; mem, Fire Res Comt, Nat Acad Sci, 56-73; mem, Gas Centrifuge Theory Consult Group, 79-85; chmn, Nat Eng Lab, Nat Bur Standards, 81-83. *Honors & Awards:* Edgerton Gold Medal, Int Combust Inst, 68; Timoshenko Medal, Am Soc Mech Engrs, 71; Fluid Dynamics Prize, Fluid Dynamics Div, Am Phys Soc, 82; Man of the Year Award, Soc Fire Protection Engr, 82. *Mem:* Nat Acad Sci; Nat Acad Eng; Am Phys Soc; Am Soc Mech Engrs; Am Acad Arts & Sci. *Res:* Aerodynamics of combustion; supersonic aerodynamics; numerical solution of differential equations; fundamentals of gas dynamics; fire science. *Mailing Add:* 317 Pierce Hall Harvard Univ Cambridge MA 02138

EMMONS, LARRIMORE BROWNELLER, b Dover, NJ, Oct 6, 35; m 59; c 2. COLORIMETRY, IMAGE ANALYSIS. *Educ:* Lehigh Univ, BS, 57; Univ Rochester, PhD(physics), 66. *Prof Exp:* Advan develop engr physics, Lighting Group, GTE-Sylvania, 66-71; instr physics, Salem State Col, 71; RES SCIENTIST, ARMSTRONG WORLD INDUST, 72- *Mem:* Optical Soc Am; Am Asn Physics Teachers. *Res:* Optical properties of polymers; light scattering by particles and fibers; analysis and synthesis of images by computer for research and process control; illuminating engineering. *Mailing Add:* 974 Hermosa Ave Lancaster PA 17601

EMMONS, LYMAN RANDLETT, b Lawrence, Mass, June 14, 27; m 71; c 2. BIOLOGY. *Educ:* Trinity Col, Conn, BS, 51; Univ Va, MA, 59, PhD(biol), 61. *Prof Exp:* Master, Episcopal High Sch, Va, 51-57; from asst prof to assoc prof, Washington & Lee Univ, 61-69, prof biol, 69-89; RETIRED. *Mem:* Sigma Xi. *Res:* Mammalian cytogenetics; biochemical and microbial genetics; immunology. *Mailing Add:* Blaesiring No 11 Basel CH-4057 Switzerland

EMMONS, RICHARD CONRAD, b Winnipeg, Man, Aug 28, 98; nat US; m 26; c 1. GEOLOGY. *Educ:* Univ BC, BA, 19, MA, 20; Univ Wis, PhD(geol), 24. *Prof Exp:* Instr geol, Univ Chicago, 24; from instr to prof, Univ Wis-Madison 25-69; RETIRED. *Concurrent Pos:* Geologist, Geol Surv Can, 20-28. *Mem:* Fel Geol Soc Am (vpres, 45); fel Mineral Soc Am (pres, 44). *Res:* Mineralogy; optical mineralogy; petrology; geology of the original Huronian area; the Ontario Pre-Cambrian; five axis universal stage; optical properties of feldspars; silicosis; geology of central Wisconsin; selected petrogenic aspects of plagioclase; steel penetration in foundry sand; genesis of geosynclinal granites; granites by recrystallization; gem stones. *Mailing Add:* 5013 Bayfield Terr Madison WI 53705-4810

EMMONS, RICHARD WILLIAM, b New York, NY, Oct 21, 31; m 59; c 7. PUBLIC HEALTH, VIROLOGY. *Educ:* Earlham Col, BA, 53; Univ Pa, MD, 57; London Sch Hyg & Trop Med, DTM&H, 61; Univ Calif, Berkeley, MPH, 62, PhD(epidemiol), 65; Am Bd Prev Med, cert, 69. *Prof Exp:* Intern, Cincinnati Gen Hosp, 57-58; med officer, Div Indian Health, USPHS, 58-60; pub health med officer, 65-78, CHIEF, VIRAL & RICKETTSIAL DIS LAB, CALIF STATE DEPT HEALTH SERV, 78- *Concurrent Pos:* Lectr, Sch Pub Health, Univ Calif, Berkeley, 71-; consult arbovirus dis, WHO, 71- *Mem:* Am Soc Trop Med & Hyg; Am Pub Health Asn; Am Soc Microbiol. *Res:* Investigation by laboratory research, field studies, and epidemiological research, of viral, rickettsial, and bacterial zoonotic infectious diseases. *Mailing Add:* Viral & Rickettsial Dis Lab 2151 Berkeley Way Berkeley CA 94704

EMMONS, SCOTT W, b Boston, Mass, July 14, 45. EUKARYOTE GENE EXPRESSION, DEVELOPMENTAL BIOLOGY. *Educ:* Harvard Univ, AB, 67; Stanford Univ, PhD(biochem), 74. *Prof Exp:* Fel molecular biol, Carnegie Inst of Washington, Baltimore, 74-76, Univ Colo, Boulder, 76-79; ASST PROF, ALBERT EINSTEIN COL MED, 79- *Res:* Genome organization and control of gene expression on the nematode caenorhabditis elegans. *Mailing Add:* 583 Olmstead Ave Bronx NY 10473

EMMONS, WILLIAM DAVID, b Minneapolis, Minn, Nov 18, 24; m 49; c 3. ORGANIC CHEMISTRY. *Educ:* Univ Minn, BS, 47; Univ Ill, PhD(chem), 51. *Prof Exp:* Sr chemist, 51-52, group leader org chem, 52-57, lab head, 57-61, res supvr, 61-72, DIR PIONEERING RES, ROHM & HAAS CO, 73- *Concurrent Pos:* Ed, Org Syntheses, 61-69. *Mem:* Am Chem Soc. *Res:* Peracids; small ring heterocycles; organophosphorous chemistry; polymers and surface coatings. *Mailing Add:* Rohm Haas Co Res Lab Spring House PA 19477-1100

EMOND, GEORGE T, b New Britain, Conn, May 26, 43. ENVIRONMENTAL SCIENCES, POLYMER CHEMISTRY. *Educ:* Univ Conn, BA, 65; St Joseph Col, MA, 69; RPI, MS, 82. *Prof Exp:* Process engr, Pratt & Whitney, 65-66; coatings chemist, Stanley Works, 66-68, chief chemist, 71-79; biochem researcher, Yale Univ, 69-71; mgr chem develop, AMF-CUNO, 79-82; res engr, Hamilton Std, 82-84; DIR RES & DEVELOP/QUAL CONTROL, NORTON CO, 84- *Concurrent Pos:* Anal res chemist, UniRoyal, 68-69; consult, process & prod develop, 85-90. *Mem:* Am Chem Soc; Am Soc Mat; Soc Plastic Engrs; Soc Tribology & Lubrication Engrs; Am Soc Qual Control. *Res:* Elutriation and sedimentation of submicronic particles frm aqueous solutions; colloid science, superabrasive slurries and compounds; rheology of pastes and fluids containing particlate; platiny process research; surface chemistry of materials; author of four publications; recipient of three patents. *Mailing Add:* 156 Meander Lane Southington CT 06489

EMPEN, JOSEPH A, b Ashton, Ill, Mar 6, 40; m 62; c 3. ORGANIC POLYMER CHEMISTRY. *Educ:* Knox Col, Ill, BA, 62; Univ Iowa, MS, 64, PhD(org chem), 66, Univ Ill, MBA, 86. *Prof Exp:* Fel polymers, Univ Ariz, 66-67; res chemist, Plastics Dept, E I du Pont de Nemours & Co, 67-71; sr res chemist, 71-75, group leader, paper & paper converting, 75-78, group leader, paper, 75-78, lab mgr, New Prod Div, 78-80, MGR, GUNTHER PROD DIV, A E STALEY MFG CO, 80- *Mem:* Am Chem Soc; Tech Asn Pulp & Paper Indust. *Res:* Modification of starches to fit the ever changing bonding needs of the paper industry; specialty chemicals from carbohydrates; development of functional proteins. *Mailing Add:* A E Staley Mfg Co Decatur IL 62525

EMR, SCOTT DAVID, b NJ, Feb 8, 54; m 77; c 2. CELL BIOLOGY, ORGANELLE STRUCTURE & FUNCTION. *Educ:* Univ RI, BS, 76; Harvard Univ, PhD(microbiol & molecular genetics), 81. *Prof Exp:* NIH predoctoral trainee, Harvard Med Sch, 76-79, teaching fel med microbiol, 79; traveling scholar, Frederick Cancer Res Facil, Cancer Biol Prog, Nat Cancer Inst, Frederick, Md, 79-81; Miller Res Inst fel, Biochem Dept, Univ Calif, Berkeley, 81-83; asst prof, 83-88, ASSOC PROF, DIV BIOL, CALIF INST TECHNOL, 89-; PROF, DIV CELLULAR & MOLECULAR MED, SCH MED, UNIV CALIF SAN DIEGO, LA JOLLA, CALIF, 91- *Concurrent Pos:* Vis scholar, Inst Pasteur, Paris, France, 78; Searle scholar award, 84-87; presidential young investr award, NSF, 85-90; mem, Biol Sci Study Sect, NIH, 90-; assoc investr, Howard Hughes Med Inst, Sch Med, Univ Calif San Diego, La Jolla, 91-; ad hoc reviewer, Microbial Physiol & Genetics Study Sect, NIH. *Mem:* Am Soc Cell Biol; Am Soc Microbiol; Genetics Soc Am; AAAS. *Res:* Selective recognition, sorting and transport of proteins from one intracellular organelle to another via vesicular carries represents an essential feature of eukaryotic cells; characterize the role(s) VPS genes and their products play in the sorting and delivery of vacuolar enzymes from the Golgi complex to the vacuole in yeast. *Mailing Add:* Div Cellular & Molecular Med Sch Med Univ Calif San Diego La Jolla CA 92093-0668

EMRAGHY, HODA ABDEL-KADER, b Cairo, Egypt, Aug 15, 45; Can citizen; m 68; c 2. ASSEMBLY & FLEXIBLE MANUFACTURING, ARTIFICIAL INTELLIGENCE & EXPERT SYSTEMS. *Educ:* Cairo Univ, BSc Hons, 67; McMaster Univ, MEng, 72, PhD(mech eng), 76. *Prof Exp:* DIR, FMS CTR, ENG FAC, MCMASTER UNIV, 84-, ASSOC MEM, COMPUTER SCI & SYSTS DEPT, 86-, PROF MECH ENG, 86- *Concurrent Pos:* Consult, Westinghouse Can Inc, 82-83 & UN, 82; prin investr, Mfg Res Corp Ont, 84-, Inst Robotics & Intel Systs & Ont Technol Fund & Baden-Wurtemberg, Ger, 90-; mem, Indust Eng Grants Selection Comt, Natural Sci & Eng Res Coun Can, 90-, Strategic Grants Selection Comt, 90-93; assoc, Can Inst Advan Res, 91- *Mem:* Am Soc Mech Engrs; Soc Mfg Engrs. *Res:* Manufacturing and design automation, with particular emphasis on aspects of flexible manufacturing, assembly, robotics and artificial intelligence; expert tolerancing; hardware and software to improve effectiveness and facilitate implementation. *Mailing Add:* McMaster Univ 1280 Main St W Hamilton ON L8S 4L7

EMRICH, GROVER HARRY, b Englewood, NJ, Apr 9, 29; m 52, 72; c 6. HYDROLOGY & WATER RESOURCES. *Educ:* Franklin & Marshall Col, BS, 52; Fla State Univ, MS, 57; Univ Ill, PhD(geol), 62. *Prof Exp:* Asst geol, Fla State Univ, 54-56; field surveyor, Fla Geol Surv, 55; asst, Ill Geol Surv, 56-58, asst geologist, 58-63; ground water geologist, State Dept Health, Pa, 63-71; mgr, Environ Resources Dept, 71-74, vpres, A W Martin Assocs, 74-79, PRES, SMC MARTIN INC, 80- *Concurrent Pos:* Vis scientist, AIG, 71. *Mem:* Fel Geol Soc Am; Nat Water Well Asn; Am Water Resources Asn; Water Pollution Control Fedn. *Res:* Ground water pollution and geology; stratigraphy and sedimentation; areal and ground water geology of Pennsylvania, Illinois and the Upper Mississippi Valley; development and management of programs for land disposal of wastes; ground water development. *Mailing Add:* SMC Martin Inc PO Box 859 Valley Forge PA 19482-0859

EMRICH, LAWRENCE JAMES, cancer research, clinical trials; deceased, see previous edition for last biography

EMRICH, RAYMOND JAY, b Denver, Colo, Nov 30, 17; m 42; c 2. FLUID DYNAMICS. *Educ:* Princeton Univ, AB, 38, AM & PhD(physics), 46. *Prof Exp:* Asst Nat Defense Res Comt, Princeton Univ, 41-45; from asst prof to prof, 46-87, EMER PROF PHYSICS, LEHIGH UNIV, 87- *Concurrent Pos:* Vis scientist, Ernst Mach Inst, Ger, 68; Nat Acad Sci exchange vis, USSR, 70-71 & 79; mem comt sci & arts, Franklin Inst; guest prof, Ruhr Univ, Bochum, WGer, 80; Fulbright lectr, Istanbul Tech Univ, Turkey, 87-88; chmn, physics dept, Lehigh Univ, 56-68. *Mem:* Fel AAAS; fel Am Phys Soc; Am Asn Physics Teachers. *Res:* Small scale and short time fluid motions; fluctuations in non-equilibrium processes; small particle deposit and transport; shock tube; explosive boiling. *Mailing Add:* 517 Seventh Ave Bethlehem PA 18018

EMRICK, DONALD DAY, b Waynesfield, Ohio, Apr 3, 29. ORGANIC CHEMISTRY, POLYMER CHEMISTRY. *Educ:* Miami Univ, BS, 51; Purdue Univ, MS, 54, PhD(org chem), 56. *Prof Exp:* Asst, Purdue Univ, 51-55; sr chemist, Standard Oil Co of Ohio, 55-56, tech specialist, 56-61, res assoc, 61-64; sr res chemist, Nat Cash Register Co, 65-73; sr res chemist, Monsanto Co, 73-74; CHEM CONSULT, 74- *Mem:* Am Chem Soc; AAAS; Sigma Xi. *Res:* Organic sulfur chemistry; stereochemistry of ring compounds; lubricants; polymers; electronic absorption spectra; rare earths; polyolefins; borate esters; encapsulation. *Mailing Add:* 4240 Lesher Dr Kettering OH 45429

EMRICK, EDWIN ROY, b Pittsburgh, Pa, Mar 1, 29; m 54; c 2. ANALYTICAL CHEMISTRY. *Educ:* Duquesne Univ, BS, 51; Univ Pittsburgh, PhD (analytical chem), 59. *Prof Exp:* Sr chemist, Pratt & Whitney Aircraft Div, United Aircraft Corp, 59-61, proj chemist, 61-63; analytical chemist, Nalco Chem Co, 63-77; analytical supvr, 78-85, RES MGR, C H PATRICK & CO, 85- *Mem:* Am Chem Soc; fel Am Inst Chemists; Am Asn Textile Chemists & Colorists; Sigma Xi. *Res:* Polymer characterization; gel, thin-layer, gas and liquid chromatography; infra-red and nuclear magnetic resonance spectroscopy. *Mailing Add:* C H Patrick & Co PO Box 2526 Greenville SC 29602

EMRICK, ROY M, b Akron, Ohio, May 6, 32; div; c 3. SOLID STATE PHYSICS. *Educ:* Cornell Univ, AB, 54; Univ Ill, MS, 58, PhD(physics), 60. *Prof Exp:* Res assoc physics, Univ Ill, 60; assoc prof, 60-72, PROF PHYSICS, UNIV ARIZ, 72- *Mem:* Fel AAAS; Am Phys Soc. *Res:* Magnetic properties of artificial metallic superlattices and amorphous thin films and particles; study of lattice defects in metals. *Mailing Add:* Dept Physics Univ Ariz Tucson AZ 85721

EMRY, ROBERT JOHN, b Ainsworth, Nebr, Nov 4, 40; c 2. VERTEBRATE PALEONTOLOGY, GEOLOGY. *Educ:* Colo State Univ, BS, 66; Columbia Univ, PhD(vert paleont), 70. *Prof Exp:* RES CUR VERT PALEONT, SMITHSONIAN INST, 71- *Mem:* Soc Vert Paleont; Paleont Soc; Am Soc Mammal. *Res:* North American Tertiary fossil mammals and the geology Tertiary deposits of North America; relationships of North American fossil faunas to those of other continents. *Mailing Add:* Dept Paleobiol Smithsonian Inst Nat Mus Natural Hist Washington DC 20560

EMSHWILLER, MACLELLAN, b Grand Rapids, Mich, Oct 27, 27; m 59. PHYSICS, COMMUNICATIONS ENGINEERING. *Educ:* Univ Mich, BS, 52, MS, 53; Univ Calif, Berkeley, PhD(physics), 59. *Prof Exp:* Mem tech staff physics, Bell Tel Labs, 69-89; RETIRED. *Mem:* AAAS; Am Phys Soc; Inst Elec & Electronics Engrs; Sigma Xi; Audio Eng Soc. *Res:* Signal processing techniques using optical techniques; nuclear magnetic resonance. *Mailing Add:* 15 Nutmeg Lane Andover MA 01810

EMSLEY, JAMES ALAN BURNS, b Oban, Scotland, May 12, 43; Can citizen; m 66; c 4. ANIMAL GENETICS, QUANTITATIVE ANALYSES. *Educ:* Carleton Univ, Ont, BSc, 68; Univ Nebr-Lincoln, PhD(genetics), 73. *Prof Exp:* Res scientist dairy breeding, Agr Can, 73-77, res scientist poultry breeding, 77-78; geneticist, H&N Inc, 78-82; DIR RES, ISA BABCOCK, 82- *Mem:* Am Soc Animal Sci; Am Genetics Asn; Can Soc Animal Sci; Poultry Sci Asn; Coun Agr Sci & Technol. *Res:* Layer breeding program including primary breeding stock and experimental lines; evaluation of alternative breeding schemes; production systems in poultry. *Mailing Add:* ISA Breeders Babcock Div PO Box 280 Ithaca NY 14851-0280

EMSLEY, MICHAEL GORDON, b Bedford, Eng, May 2, 30; US citizen; m 72; c 2. ENTOMOLOGY, HERPETOLOGY. *Educ:* Univ London, BS, 53; Royal Col Sci, London, ARCS, 53; Univ London, PhD(zool), 64. *Prof Exp:* Entomologist, Empire Cotton Growing Corp, Nigeria, 54-57; asst lectr zool, Univ West Indies, 57-65; resident dir, William Beebe Tropical Field Sta, NY Zool Soc, Trinidad, 65-66; assoc cur, Acad Natural Sci, Philadelphia, 66-69; chmn dept biol, 69-74, PROF BIOL, GEORGE MASON UNIV, 69- *Concurrent Pos:* Ed, Biotropica, Asn Trop Biol, Inc, 72-82. *Mem:* Asn Trop Biol; AAAS; Inst Biol London. *Res:* Systematics of the tettigoniidae; taxonomy of the Schizopteridae; Caribbean snakes. *Mailing Add:* Dept Biol George Mason Univ Fairfax VA 22030

EMSLIE, ALFRED GEORGE, b Aberdeen, Scotland, Nov 28, 07; nat US; m 33; c 2. APPLIED PHYSICS. *Educ:* Aberdeen Univ, MA, 28; Cornell Univ, PhD(physics), 33; Cambridge Univ, PhD(physics), 38. *Prof Exp:* Asst physics, Aberdeen Univ, 28-30; from instr to asst prof, Williams Col, 37-43; staff mem, Radiation Lab, Mass Inst Technol, 43-46; res lectr electronics, Harvard Univ, 46-47; from assoc prof to prof physics, Williams Col, 47-51; head physics group, Arthur D Little, Inc, 51-61, staff assoc, 61-72; CONSULT IN APPL PHYSICS, 72- *Concurrent Pos:* Consult, Arthur D Little, Inc, 46-50. *Res:* Classical theoretical physics; electromagnetic waves; underwater sound; physical optics; heat radiation and transmission; exotic inertial sensing; hydrodynamics of viscous fluids; infrared radiative transfer. *Mailing Add:* 14 Prospect Ave Scituate MA 02066

EMSLIE, RONALD FRANK, b Winnipeg, Man, Feb 27, 32; m 60; c 3. GEOLOGY, PETROLOGY. *Educ:* Univ Man, BSc, 56, MSc, 58; Northwestern Univ, PhD, 61. *Prof Exp:* Geologist, Geol Surv Can, 60-65; vis asst prof geol, Queen's Univ, Ont, 65-66, geologist, Geol Surv Can, 66-69; guest investr, Geophys Lab, Carnegie Inst Washington, 69-70; RES SCIENTIST, GEOL SURV CAN, 70- *Mem:* Fel Geol Soc Am; fel Geol Asn Can; Mineral Asn Can; fel Mineral Soc Am. *Res:* Anorthositic and related rocks of the Eastern Canadian shield; high temperature-high pressure mineral equilibria; igneous and metamorphic petrology. *Mailing Add:* Geol Surv Can 601 Booth St Ottawa ON K1A 0E8 Can

EMSON, HARRY EDMUND, b Swinton, Eng, Nov 16, 27; m 53; c 2. PATHOLOGY & FORENSIC PATHOLOGY, BIOETHICS. *Educ:* Oxford Univ, BA, 48, BM, BCh, 52, MA, 53; Royal Col Physicians & Surgeons Can, dipl, 58, FRCPS(C), 72; Univ Sask, MD, 59. *Prof Exp:* Intern, Manchester Royal Infirmary, Eng, 52, resident clin pathologist, 52-53; pathologist, Brit Mil Hosp, Ger, 53-55; registr path, Birmingham Accident Hosp, 55-56; resident, St Paul's Hosp, 56-57; asst resident, Univ Hosp, 57-58, asst pathologist, 58-60, dir labs, St Paul's Hosp, Saskatoon, 60-75; head, Dept Path, Col Med, Univ Sask, 75-90, Univ Hosp, Saskatoon, 75-90; PROF PATH, UNIV SASK, 75- *Mem:* Can Med Asn; Can Asn Path (past pres); Can Soc Forensic Sci (past pres); Brit Asn Clin Path; Am Acad Forensic Sci. *Res:* Diagnostic human pathology; forensic pathology; bioethics. *Mailing Add:* Royal Univ Hosp Saskatoon SK S7N 0X0 Can

EMTAGE, PETER ROESCH, b London, Eng, Jan 1, 35; UK citizen; m 60; c 3. SOLID STATE PHYSICS. *Educ:* Oxford Univ, BA, 56, MA & PhD(physics), 59. *Prof Exp:* Fel physics, Northwestern Univ, Evanston, 59-60; physicist, Electronics Lab, Gen Elec Co, 60-63; PHYSICIST, WESTINGHOUSE RES LAB, WESTINGHOUSE ELEC CO, 63- *Res:* Theoretical solid state physics, including electron structure and transport in solids, properties of interfaces, and magnetic media. *Mailing Add:* 1411 Old Beulah Rd Pittsburgh PA 15235

ENDAHL, GERALD LEROY, b Lane, SDak, Dec 16, 24; m 54; c 2. BIOCHEMISTRY. *Educ:* Augustana Col, BA, 49; Univ SDak, MA, 53; Univ Okla, PhD(biochem), 59. *Prof Exp:* Res assoc, Okla Med Res Found, 53-57; res fel, Med Sch, Univ Ala, 57-60; from asst prof to assoc prof physiol chem & surg, Ohio State Univ, 60-70; ASSOC PROF PATH & BIOCHEM, UNIV SOUTHERN CALIF, 70- *Mem:* AAAS; AMA; Am Chem Soc; Am Asn Clin Chemists; NY Acad Sci; Sigma Xi. *Res:* Enzymes of carbohydrate metabolism; metabolism of steroid hormones; hormones of gastric acid secretion; aminotransferases. *Mailing Add:* 29222 Beachside Dr Rancho Palos Verdes CA 90274

ENDAL, ANDREW SAMSON, b Brooklyn, NY, Sept 1, 49; m 73; c 3. THEORETICAL ASTROPHYSICS, CLIMATE PHYSICS. *Educ:* Univ Rochester, BS, 71; Univ Fla, PhD(astrophys), 74. *Prof Exp:* Res assoc astrophys, NASA Goddard Space Flight Ctr, 74-76; vis asst prof physics, Kans State Univ, 76-78; from asst prof to assoc prof physics & astron, La State Univ, 78-81; vpres res & develop, 83-88, SR SCIENTIST, APPL RES CORP, 82-, SR VPRES RES & DEVELOP, 88- *Mem:* Am Astron Soc; Int Astron Union; Am Geophys Union. *Res:* Stellar evolution and rotation; nucleosynthesis; climate modeling; astrophysics involving computer modeling of stellar interiors and solar processes; climate physics emphasizing simple models for perturbed climates, including increased atmospheric carbon dioxide and changes in radiation balance due to dense smoke clouds. *Mailing Add:* Appl Res Corp 8301 Corp Dr Suite 920 Landover MD 20785

ENDE, NORMAN, b Petersburg, Va, Apr 5, 24; m 48; c 1. IMMUNOLOGY. *Educ:* Univ Richmond, BS, 45; Med Col Va, MD, 47; Am Bd Clin Path, dipl, 53, cert anat path, 54. *Prof Exp:* Intern, Bronx Hosp, NY, 48; resident path & surg, Vet Admin Hosp, New Orleans, La, 49-52, pathologist, Houston, Tex, 54-55, chief path, Fresno, Calif, 55-58; asst clin prof, Vanderbilt Univ, 58-61, from asst prof to assoc prof, 61-67; chief path, Grady Mem Hosp, 67-70; actg chmn dept, 74-76, PROF PATH, COL MED & DENT NJ, 70- *Concurrent Pos:* Instr, Col Med, Baylor Univ, 54-55, asst prof, 55; chief lab serv path, Vet Admin Hosp, Nashville, 58-67; prof path, Emory Univ, 67-69; chief clin path, Martland Hosp, 70-74, chief path, 74-77. *Mem:* Fel Am Col Physicians; fel Col Am Path; Am Soc Exp Path; Am Asn Path & Bact; fel Am Soc Clin Path. *Res:* Mast cell, fibrinolysis and the hyper-coagulable state; carcinoma of the prostate and thromboangiitis obliterans; starvation; transplantation; circulating antibodies. *Mailing Add:* Dept Pathol UMDNJ New Jersey Med Sch 185 S Orange Ave Newark NJ 07103

ENDERBY, CHARLES ELDRED, b Chicago, Ill, Nov 15, 34; m 57; c 3. PHYSICS. *Educ:* Univ Ill, BS, 57, MS, 58, PhD(elec eng), 61. *Prof Exp:* Asst prof elec eng, Univ Ill, 60-61; mem tech staff, Gen Elec Co, 61-65; from vpres to pres, Electro Optics Assocs, Inc, 66-71; exec vpres, Optics Technol, Inc, 71-76; vpres, Molectron Corp, 77-83 & Cooper Laser Sonics, 83-88; VPRES, LASER PHOTONICS, 88- *Mem:* Inst Elec & Electronics Engrs; Optical Soc Am; Am Mktg Asn; fel Am Soc Lasers Med. *Res:* Optical modulation; millimeter wave generation; gas laser design; medical lasers. *Mailing Add:* 1852 Edgewood Dr Palo Alto CA 94303-3015

ENDERS, ALLEN COFFIN, b Wooster, Ohio, Aug 5, 28; m 50, 76; c 4. HUMAN ANATOMY, REPRODUCTIVE BIOLOGY. *Educ:* Swarthmore Col, AB, 50; Harvard Univ, AM, 52, PhD, 55. *Prof Exp:* Teaching fel biol, Harvard Univ, 52-53 & Brandeis Univ, 53-54; res assoc, Rice Univ, 54-55, from asst prof to assoc prof, 55-63; from assoc prof to prof anat, Sch Med, Washington Univ, 63-75; chmn dept, 76-86, PROF HUMAN ANAT, UNIV CALIF, DAVIS, 76- *Concurrent Pos:* Consult, RCDA comt, NIH, 64-67, HED comt, 75-79 & REB comt, 83-87; mem anat test comt, Nat Bd Med Examr, 69-73. *Honors & Awards:* Amorose lectr, Soc Study Fertil & Steril, 87. *Mem:* Am Asn Anat (vpres, 79-81, pres, 83-84); Soc Study Reproduction; Am Soc Cell Biol; Perinatal Res Soc (pres, 82). *Res:* Fine structure of placenta and female reproductive tract; mechanisms of implantation; development of the blastocyst. *Mailing Add:* Dept Human Anat Univ Calif Davis CA 95616-8643

ENDERS, GEORGE CRANDELL, CELL BASEMENT MEMBRANE ADHESION. *Educ:* Univ Calif, San Francisco, PhD(anat), 83. *Prof Exp:* RES ASSOC, MED SCH, HARVARD UNIV, 83- *Mailing Add:* Dept Anat & Cell Biol Univ Kans Med Ctr 39th & Rainbow Blvd Kansas City KS 66103

ENDERS, GEORGE LEONHARD, JR, b Glendale, NY, Nov 13, 45; m 69; c 3. CLINICAL MICROBIOLOGY. *Educ:* Rutgers Univ, BA, 67; Immaculate Heart Col, Los Angeles, MA, 71; Univ Kans, PhD(microbiol), 74. *Prof Exp:* Res assoc microbiol, Food Res Inst, Univ Wis, 74-77; res scientist, 77-80, SR RES SCIENTIST MICROBIOL, BIOTECHNOL GROUP, AMES CO, MILES LABS, INC, 80- *Concurrent Pos:* Nat Inst Environ Health Sci fel, Food Res Inst, Univ Wis, 75- *Mem:* Am Soc Microbiol; Sigma Xi; Inst Food Technologists; Am Soc Animal Sci. *Res:* Food and agricultural microbiology fermentations; characterization of the enterotoxin elaborated by Clostridium perfringens: studies dealing with the structure, biological and serological activities and mechanism of action of this enterotoxin; anaerobes, especially physiological aspects and toxins; effects of bacteriophage infection on the metabolism of Escherichia coli envelope components; lipids, proteins and Lipopolysaccharides; characterization of the lipids of thermophiles. *Mailing Add:* 22797 Bainbridge Dr Elkhart IN 46514

ENDERSON, JAMES H, b Sioux City, Iowa, Nov 3, 36; m 57; c 1. ZOOLOGY. *Educ:* Univ Ill, BS & MS, 59; Univ Wyo, PhD(zool), 62. *Prof Exp:* From asst prof to assoc prof, 62-74, PROF BIOL, COLO COL, 74- *Concurrent Pos:* NSF fac fel systs & ecol, Cornell Univ, 69-70. *Mem:* Am Inst Biol Sci; Am Ornith Union; Wilson Ornith Soc; Cooper Ornith Soc. *Res:* Raptor ecology. *Mailing Add:* Dept Biol Colo Col Colorado Springs CO 80903

ENDERTON, HERBERT BRUCE, b Hawaii, Apr 15, 36; m 61; c 2. MATHEMATICAL LOGIC. *Educ:* Stanford Univ, BS, 58; Harvard Univ, MA, 59, PhD(math), 62. *Prof Exp:* Instr math, Mass Inst Technol, 62-64; asst prof, Univ Calif, Berkeley, 64-68; LECTR MATH, UNIV CALIF, LOS ANGELES 68- *Concurrent Pos:* Ed, J Symbolic Logic, Univ Calif, Los Angeles, 68- *Mem:* Am Math Soc; Asn Symbolic Logic. *Res:* Recursive function theory; definability theory; models of analysis; computational complexity; history of logic. *Mailing Add:* Dept Math Univ Calif Los Angeles CA 90024-1555

ENDICOTT, JOHN F, b Eugene, Ore, Aug 1, 32; m 65; c 3. CHEMICAL DYNAMICS, PHOTOPHYSICS OF COORDINATION COMPOUNDS. *Educ:* Reed Col, BA, 57; Johns Hopkins Univ, PhD(phys chem), 61. *Prof Exp:* Res assoc inorg chem, Stanford Univ, 61-63; from asst prof to assoc prof, Boston Univ, 63-69; assoc prof, 69-72, PROF CHEM, WAYNE STATE UNIV, 72- *Concurrent Pos:* Res collabr, Dept Chem, Brookhaven Nat Lab, 77-78; consult, UNESCO/UNDP, India, 77; vis prof, Ohio State Univ, 86. *Mem:* Am Chem Soc; Am Phys Soc; AAAS. *Res:* Mechanisms of inorganic reactions; photochemistry of coordination complexes; reactivity of macrocyclic complexes. *Mailing Add:* Dept Chem Wayne State Univ Detroit MI 48202

ENDLER, JOHN ARTHUR, b Montreal, Que, Oct 8, 47; US citizen; m 83. BEHAVIORAL ECOLOGY, EVOLUTIONARY BIOLOGY. *Educ:* Univ Calif, Berkeley, BA, 69; Univ Edinburgh, PhD(zool), 73. *Prof Exp:* Fel pop biol, Princeton Univ, 72-73, asst prof biol, 73-79; from assoc prof to prof biol, Univ Utah, 79-86; PROF BIOL, UNIV CALIF, SANTA BARBARA, 86- *Concurrent Pos:* John Simon Guggenheim fel, 85-86; ed, Evolution, 89-92. *Mem:* Am Soc Naturalists; Am Soc Icthyologists & Herpetologists; Ecol Soc Am; Soc Study Evolution; Asn Trop Biol. *Res:* Evolution and ecology of animal color patterns and vision; the adaptive nature, causes and mechanisms of geographic variation and natural selection; ecological genetics and genetical biogeography of tropical freshwater fishes; biogeography. *Mailing Add:* Dept Biol Sci Univ Calif Santa Barbara CA 93106

ENDO, BURTON YOSHIAKI, b Castroville, Calif, Feb 5, 26; m 52; c 2. NEMATOLOGY. *Educ:* Iowa State Univ, BS, 51; NC State Univ, MS, 55, PhD(plant path), 58. *Prof Exp:* Asst hort, NC State Univ, 53-55; asst nematologist, Nematol Sect, US Dept Agr, 55-58, nematologist, WTenn Exp Sta, 58-63, res nematologist, Beltsville Agr Res Ctr-West, 63-74, chief nematol lab, 74-75, chmn, Plant Protection Inst, 74-88, RES PLANT PATHOLOGIST, US DEPT AGR, 88- *Mem:* AAAS; Am Phytopath Soc; fel Soc Nematol (secy, 68-71, vpres, 72, pres, 73); Am Inst Biol Sci; Sigma Xi. *Res:* Host-parasite relations of nematode infected plants; plant disease resistance; video enhanced light microscopy and ultra structure. *Mailing Add:* Nematol Lab BARC-W Biosci Bldg Rm 138 Beltsville MD 20705

ENDO, ROBERT MINORU, b Mountain View, Calif, Mar 30, 25; m 50; c 3. PLANT PATHOLOGY. *Educ:* Rutgers Univ, BS, 50; Univ Ill, MS, 52, PhD(plant path), 54. *Prof Exp:* Res asst plant path, Dept Hort, Univ Ill, 51-54, agent sect cereal crops & dis, Field Crops Res Br, Agr Res Serv, USDA, 54-56, plant pathologist, 56-58; asst prof plant path & asst plant pathologist, Univ Calif, Los Angeles, 59-61, from asst prof & asst plant pathologist to assoc prof & assoc plant pathologist, 61-71, PROF PLANT PATH & PLANT PATHOLOGIST, UNIV CALIF, RIVERSIDE, 71- *Mem:* Am Phytopath Soc; Mycol Soc Am; Am Inst Biol Scientists. *Res:* Diseases of turf grass and vegetables; yellow dwarf disease of cereals. *Mailing Add:* Dept Plant Path Univ Calif 900 University Ave Riverside CA 92521

ENDO, YOUICHI, b Tokyo, Japan, Nov 45, 49; m 85. ALGEBRAIC GEOMETRY & MODULAR FORM. *Educ:* Saitama Univ, BS, 74; Rutgers Univ, MS, 79; Temple Univ, PHD(math), 85. *Prof Exp:* MANAGING DIR, PLAN DIV, PERMA-STONE NIPPON CO LTD, 85- *Mem:* Am Math Soc; Math Soc Japan. *Res:* Insulation; prevention of condensation in super-insulated buildings. *Mailing Add:* 20 Weil Pl Cresskill NJ 07626

ENDOW, SHARYN ANNE, b Hood River, Ore, May 22, 48. RIBOSOMAL GENE GENETICS, MEIOTIC MUTANTS. *Educ:* Stanford Univ, BA, 70; Yale Univ, MPhil, 72, PHD(cell & molecular biol), 75. *Prof Exp:* Fel, Cold Spring Harbor Lab, 74-76 & Mammmalian Genome Unit, Med Res Coun, Edinburgh, Scotland, 76-78; asst prof, 78-84, ASSOC PROF GENETICS, DUKE UNIV, 84- *Concurrent Pos:* Staff scientist, Mammalian Genome Unit, Med Res Coun, Edinburgh, Scotland, 78; NIH res career develop award, 82-87, genetics study section, 87, 89 & 91. *Mem:* Genetics Soc Am; Sigma Xi; Am Soc Cell Biol. *Res:* Eukaryotic gene amplicication; regulation of gene copy number in eukaryotes; genes that control chromosome segregation; chromosome segregation. *Mailing Add:* Microbiol Med Ctr Duke Univ Box 3020 Durham NC 27710

ENDRENYI, JANOS, b Budapest, Hungary, Nov 9, 27; m 56. ELECTRIC POWER, SYSTEMS RELIABILITY FIELD. *Prof Exp:* Lectr, TV Budapest, 49-52; res engr, 52-56 & 59-79, SECT HEAD, RELIABILITY & STATIST, RES DIV, ONT HYDRO, 79- *Concurrent Pos:* adj lectr, 72-82, adj prof elec eng, Univ Toronto, 82- *Mem:* Fel Inst Elec & Electronics Engrs. *Res:* Power transmission and station system reliability, especially Grounding of Transmission Lines and Stations; reliability models for complex components; methods for optimizing preventive maintenance. *Mailing Add:* Ont Hydro Res Div 800 Kipling Ave Toronto ON M8Z 5S4 Can

ENDRENYI, LASZLO, b Budapest, Hungary, May 6, 33; Can citizen; m 56. PHARMACOKINETICS, MODEL BUILDING. *Educ:* Tech Univ Budapest, Hungary, Dipl Eng, 56; Univ Toronto, PhD(chem), 65. *Prof Exp:* Res assoc, 66-69, from asst prof to assoc prof, 69-76, PROF PHARMACOL & BIOSTATS, DEPT PHARMACOL, PREV MED & BIOSTATIST, UNIV TORONTO, 76-, ASSOC DEAN GRAD STUDIES, 88- *Mem:* Am Statist Asn; Statist Soc Can; Phamacol Soc Can; Biometric Soc. *Res:* Design and statistical analysis of enzyme and pharmacokinetic experiments and of clinical trials; kinetic modeling; model identification; parameter estimation; robust procedures. *Mailing Add:* Dept Pharmacol Univ Toronto Toronto ON M5S 1A8 Can

ENDRES, JOSEPH GEORGE, b Chicago, Ill, Aug 15, 32; m 59; c 3. FOOD SCIENCE. *Educ:* Univ Ill, BS, 55, PhD(food chem), 61. *Prof Exp:* Res chemist fat & oil chem, Food Res Div, Armour & Co, 61-62, sect head, 62-64, asst mgr, 64-70; vpres res & develop, CFS Continental, 70-72; dir food prod develop, 72-78, dir food res, 78-84, DIR CORP RES PROJS, CENT SOYA-FOOD RES, 84- *Concurrent Pos:* Prof food sci dept, Univ Ill, 74- *Mem:* Am Oil Chemists' Soc; Inst Food Technologists. *Res:* Food product and process development for institutional, food service and retail; technology development, animal nutrition and biotechnology. *Mailing Add:* 4806 W Hamilton Rd S Ft Wayne IN 46804

ENDRES, LELAND SANDER, b Akron, Ohio, Mar 31, 36; m 59; c 4. ORGANIC CHEMISTRY, PHYSICAL CHEMISTRY. *Educ:* Middlebury Col, AB, 58; Univ Ore, MA, 63; Univ Ariz, PhD(chem), 67. *Prof Exp:* Instr & res assoc chem, Univ Nebr, 66-67; sr res chemist, Minn Mining & Mfg Co, 67-69; from asst prof to assoc prof, 69-78, PROF CHEM, CALIF POLYTECH STATE UNIV, SAN LUIS OBISPO, 79- *Mem:* Am Chem Soc; Sigma Xi. *Res:* Reaction kinetics; carbonium ions; small ring heterocyclics; fluorocarbons. *Mailing Add:* 790 Islay San Luis Obispo CA 93401

ENDRES, PAUL FRANK, b Peoria, Ill, Feb 10, 42. PHYSICAL CHEMISTRY. *Educ:* Bradley Univ, BS, 63; Univ Rochester, PhD(chem), 67. *Prof Exp:* Fel chem, Univ Rochester, 67-69; asst prof, 69-74, from assoc prof chem to PROF, BOWLING GREEN STATE UNIV, 74- *Mem:* Am Phys Soc. *Res:* Energy transfer in molecular collisions; molecular dynamics; molecular beam spectroscopy. *Mailing Add:* Dept Chem Bowling Green State Univ Bowling Green OH 43403

ENDRIZZI, JOHN EDWIN, b Wilburton, Okla, July 28, 23; m 55; c 5. CYTOGENETICS. *Educ:* Tex A&M Univ, BS, 49, MS, 51; Univ Md, PhD(bot), 55. *Prof Exp:* Asst prof cytogenetics, Tex Agr Exp Sta, 55-63; head dept, Univ Ariz, 63-71, prof plant genetics, 63-85; RETIRED. *Honors & Awards:* Cotton Genetics Res Award, Nat Cotton Coun Am, 69. *Mem:* AAAS; Am Genetic Asn; Genetics Soc Am; Genetics Soc Can; Am Inst Biol Sci. *Res:* Cytogenetics of Gossypium. *Mailing Add:* 2335 E Ninth St Tucson AZ 85719

ENDSLEY, L(OUIS) E(UGENE), JR, b Lafayette, Ind, May 24, 12; m 40, 84; c 3. MECHANICAL ENGINEERING. *Educ:* Purdue Univ, BS, 34, MS, 36. *Prof Exp:* Asst instr mech eng, Purdue Univ, 34-36; mech engr, Texaco, Inc, 36-54, asst to mgr tech servs, 54-56, dir, 56-58, asst mgr, 58-60, planning dir eng, 60-69, mgr sci planning, 69-75, mgr planning & admin, Res & Technol Dept, 75-77; RETIRED. *Concurrent Pos:* Mech engr, Wright Field, US Army Air Force, 39; consult to Asst Secy Defense, 50-54 & 55-63. *Mem:* Sigma Xi; Soc Automotive Engrs. *Res:* Petroleum fuels and lubricants; research and development. *Mailing Add:* Six St Johns Pkwy Poughkeepsie NY 12601-5116

ENEA, VINCENZO, b Palermo, Italy, Jan 11, 46; m 71; c 1. MOLECULAR BIOLOGY OF PLASMODIA. *Educ:* Univ Palermo, Laurea, 70; Rockefeller Univ, NY, PhD(genetics), 76. *Prof Exp:* Res Asst, Univ Palermo, 68-70; teaching fel, Mass Inst Technol, 76-79; asst prof, Rockefeller Univ, 79-83; ASSOC PROF GENETICS, NY UNIV MED CTR, 84- *Concurrent Pos:* Vis scientist, Int Lab Genetics & Biophys, Naples, 68; lectr, Univ Palermo, 82; adj prof, Rockefeller Univ, 84-85; H T Hirschl Career Develop Award, 85. *Mem:* AAAS; Am Soc Microbiol; NY Acad Sci. *Res:* Molecular genetics of microorganisms; molecular biology of parasites; developmental biology; mechanisms of genetic recombinations; molecular evolution. *Mailing Add:* Dept Med & Molecular Parasitol NY Univ Med Ctr 550 First Ave New York NY 10016

ENELL, JOHN WARREN, applied statistics, for more information see previous edition

ENEMARK, JOHN HENRY, b Springfield, Minn, Aug 24, 40; m 62; c 3. INORGANIC CHEMISTRY. *Educ:* St Olaf Col, BA, 62; Harvard Univ, AM, 64, PhD(chem), 66. *Prof Exp:* Res assoc chem, Northwestern Univ, 66-68; from asst prof to assoc prof, 68-72, PROF CHEM, UNIV ARIZ, 72- *Concurrent Pos:* NSF res fel, Northwestern Univ, 66-67; Fulbright sr scholar, 89. *Honors & Awards:* Alexander von Humboldt Sr Scientist Award, 81. *Mem:* AAAS; Am Chem Soc; Am Crystallog Asn. *Res:* Transition metal compounds; bioinorganic chemistry; x-ray crystallography. *Mailing Add:* Dept Chem Univ Ariz Tucson AZ 85721

ENENSTEIN, NORMAN H(ARRY), b Los Angeles, Calif, Nov 2, 23; m 43; c 2. ELECTRICAL ENGINEERING. *Educ:* Univ Calif, Los Angeles, AB, 46; Calif Inst Technol, MS, 47, PhD(elec eng, physics), 49. *Prof Exp:* Res physicist, Hughes Aircraft Co, 49-54, proj anti-airborne defense tactical data syst, US Army, 54-56; vpres & dir engr, Electro-Pulse, Inc, 56-58; dir tactical systs lab, Litton Industs, 58-62; mgr systs div, 62-70, MGR DATA PROCESSING PRODS DIV, GROUND SYSTS GROUP, HUGHES AIRCRAFT CO, 70- *Mem:* Inst Elec & Electronics Engrs. *Res:* Management of large scale military programs in the fields of data processing, radar, communications and weapon systems. *Mailing Add:* Process Improv Assoc 3009 Milagro Way Fullerton CA 92635

ENESCO, HILDEGARD ESPER, b Seattle, Wash, June 16, 36; m 64. CELL BIOLOGY. *Educ:* Reed Col, BA, 58; Columbia Univ, MA, 59, PhD(zool), 62. *Prof Exp:* USPHS fel anat, McGill Univ, 62-63; Shirley Farr fel biol, Univ Montreal, 63-64; res assoc biochem, Allan Mem Inst, McGill Univ, 64-68; from asst prof to assoc prof, 68-78, actg chmn dept, 74-75, chmn dept, 75-77, PROF BIOL, SIR GEORGE WILLIAMS CAMPUS, CONCORDIA UNIV, 78- *Mem:* Geront Soc; Can Soc Cell Biol; Am Soc Cell Biol; Geront Soc Can; Soc Develop Biol. *Res:* Cell function and aging. *Mailing Add:* Dept Biol Sci Concordia Univ SGW 1455 DeMaisonneuve W Montreal PQ H3G 1M8 Can

ENFIELD, CARL GEORGE, b Indianapolis, Ind, Nov 13, 42; m 66; c 3. SOIL PHYSICS, SOIL CHEMISTRY. *Educ:* Purdue Univ, BS, 65, MS, 68; Univ Ariz, PhD(agr chem & soils), 73. *Prof Exp:* Environ hydrologist instrumentation, Burr Brown Res Corp, 68-69; sr res scientist, Battelle Northwest, Battelle Mem Inst, 69-71; res assoc soil physics, Univ Ariz, 71-72; civil engr, 72-80, SOIL SCIENTIST & DIR GROUND WATER RES, ROBERT S KERR ENVIRON RES LAB, US ENVIRON PROTECTION AGENCY, 80- *Concurrent Pos:* Vis scientist, Univ Lund, Lund, Sweden, 85-86. *Mem:* Fel Am Soc Agron; Soil Sci Soc Am; Sigma Xi; Am Soc Agr Eng; Int Soil Sci Soc. *Res:* Chemical interactions with soil and geologic materials; primary emphasis on the reactions of phosphorus, nitrogenous and organic compounds; mathematical prediction of the movement of chemical pollutants through soil profiles. *Mailing Add:* Rte 1 Box 112 Ada OK 74820

ENFIELD, FRANKLIN D, b Woolstock, Iowa, Dec 26, 33; m 55; c 3. GENETICS. *Educ:* Iowa State Univ, BS, 55; Okla State Univ, MS, 57; Univ Minn, Minneapolis, PhD(animal breeding), 60. *Prof Exp:* Asst prof animal breeding, Univ Minn, Minneapolis, 60-65, assoc prof animal sci, 65-66; assoc prof, 66-70, dir grad studies in genetics, 71-75, PROF GENETICS, UNIV MINN, ST PAUL, 70- *Concurrent Pos:* NSF Panel, Population Biol & Physiol Ecol, 87-88 & 89, vchmn, Gordon conf on Quantitative Genetics & Biotechnol, 91, chmn, 93. *Mem:* Genetics Soc Am; AAAS. *Res:* Population and quantitative genetics of Tribolium; computer simulation of molecular marker assisted selection in plant breeding. *Mailing Add:* Dept Genetics & Cell Biol Col Biol Sci Univ Minn St Paul MN 55108

ENG, GENGHMUN, b New York, NY, Oct 28, 51; m 79; c 1. SOLID STATE PHYSICS, APPLIED MATHEMATICS. *Educ:* Cooper Union, BS, 72; Univ Ill, Urbana-Champaign, PhD(physics), 78. *Prof Exp:* Mem tech staff, 78-87, RES SCIENTIST, AEROSPACE CORP, 87- *Mem:* Am Phys Soc; Sigma Xi. *Res:* Studies of diffusion; surface science; electron transport in support of defense systems and materials development. *Mailing Add:* Aerospace Corp Mail Sta M2-272 PO Box 92957 Los Angeles CA 90009-2957

ENG, LAWRENCE F, b Spokane, Wash, Feb 19, 31; m 58; c 4. BIOCHEMISTRY, NEUROCHEMISTRY. *Educ:* Wash State Univ, BS, 52; Stanford Univ, MS, 54, PhD(chem), 62. *Prof Exp:* Res scientist biochem toxicol, Aero Med Lab, Wright Air Develop Ctr, 54-57; asst biochem, 58-61, res assoc pathol, 66-70, adj prof, 75- 82, PROF PATHOL, SCH MED, STANFORD UNIV, 82-; BIOCHEMIST, PALO ALTO VET ADMIN HOSP, 61- *Concurrent Pos:* Sr scientist path, Sch Med, Stanford Univ, 70-75, adj prof, 75-; mem, ed bd, Neurobiol, 70-75, J Neurochem, 78-85, J Neuroimmunol, 80-83, Molec & Chem Path, 82-, Glia, 87-; mem neurol sci study sect, Nat Inst Neurol Dis & Stroke, 79- 83; mem adv bd, Vet Admin Off Regeneration Res Progs, 85-89; mem merit rev bd neurobiol, Vet Admin Res Serv, 87-90, PI, Nat Inst Neurol Dis & Stroke, 70-; Jacob Javits Neurosci Investr Award, 87-94. *Honors & Awards:* William S Middleton Res Award, Vet Admin, 88. *Mem:* Am Chem Soc; Int Soc Neurochem; Am Soc Neurochem; AAAS; Am Soc Biol Chemists; Sigma Xi; Soc Neurosci. *Res:* Clinical chemistry; biochemistry, development, and protein chemistry of the brain; investigation of myelinating diseases and regeneration of the central nervous system. *Mailing Add:* 1445 Harker Ave Palo Alto CA 94301

ENG, LESLIE, b Baltimore, Md, June 9, 41; m 69; c 1. PHYSICAL ORGANIC CHEMISTRY. *Educ:* Johns Hopkins Univ, BA, 65; Pa State Univ, PhD(chem), 70. *Prof Exp:* Org chemist, 70-73, res chemist, 73-77, CHEMIST, US ARMY TOXIC & HAZARDOUS MAT AGENCY, EDGEWOOD ARSENAL, 77- *Mem:* Am Chem Soc; Sigma Xi. *Res:* Kinetics and mechanisms of organic reactions; thermal decomposition of organic and inorganic salts; decontamination of toxic chemicals; methods of analysis for trace quantities of toxic chemicals; analytical quality assurance and quality control. *Mailing Add:* 3623 Ady Rd Street MD 21154-1433

ENG, ROBERT H K, b Sept 17, 48. INFECTIOUS DISEASES, MICROBIOLOGY. *Educ:* Univ Wash, BS, 70; Harvard Univ, MA, 72; Einstein Col Med Yeshiva, MD, 75. *Prof Exp:* Asst prof, 80-85, ASSOC PROF MED, NJ MED SCH, NEWARK, 85- *Concurrent Pos:* Attend staff med serv, Vet Admin Med Ctr, E Orange, NJ, 80- *Mem:* Fel Am Col Physicians; fel Infectious Dis Soc Am; Soc Exp Biol & Med. *Res:* Pathogenesis of candida and cryptococcal infections; mechanisms of bacterial resistance. *Mailing Add:* Med Serv Vet Admin Med Ctr East Orange NJ 07019

ENG, SVERRE T(HORSTEIN), b Skaanland, Norway, July 30, 28; US citizen; m 57; c 3. SOLID STATE ELECTRONICS, ELECTROOPTICS. *Educ:* Chalmers Univ Technol, Sweden, MS, 53, PhD(appl physics), 67. *Prof Exp:* Res engr, Res Lab Electronics, Chalmers Univ Technol, Sweden, 53-56; mem tech staff, Semiconductor Div, Hughes Aircraft Co, 56-57; asst electronics, Stanford Univ, 57-58; sect head microwave & optical semiconductor physics & electronics, Res Labs, Hughes Aircraft Co, 58-62, dept head, 62-67; staff scientist & mem tech staff, Autonetics Div, N Am Rockwell Corp, 67-71; head, Inst Elec Measurements, 71-80, PROF MICROWAVE & ELECTROOPTIC ELECTRONICS, CHALMERS UNIV TECNOL, SWEDEN, 71-, DIR, INST ELEC MEASUREMENTS, 80- *Mem:* Am Phys Soc; Inst Elec & Electronics Engrs; Sigma Xi; Optical Soc Am. *Res:* Microwave semiconductor devices, especially high frequency transistors, parametric diodes, mixers, tunnel and backward diodes, integrated electronics, measurements and application studies; infrared detection; semiconductor and carbon dioxide lasers; superheterodyne instrumentation; laser radar. *Mailing Add:* Elec Measurements Chalmers Univ Gothenburg 41296 Sweden

ENGBRETSON, GUSTAV ALAN, b Fargo, NDak, Nov 28, 43; c 2. SENSORY BIOLOGY, NEUROBIOLOGY. *Educ:* Calif State Univ, Sacramento, BA, 68, MA, 71; Univ Okla, PhD(zool), 76. *Prof Exp:* Lectr anat sci, State Univ NY Stony Brook, Health Sci Ctr, 75-76, fel, 76-77, res asst prof, 78-79; res asst prof sensory sci, Inst Sensory Res, 79-85, asst prof, 85-87, ASSOC PROF BIOENG, SYRACUSE UNIV, 87- *Concurrent Pos:* Res asst prof, dept neurobiol & behav, Cornell Univ, 78-81, res asso prof, 87-; res asst prof, dept anat, State Univ NY, Upstate Med Ctr, 79-87. *Mem:* Asn Res Vision & Ophthal; AAAS; Am Soc Zoologists; Sigma Xi; Soc Neurosci. *Res:* Information processing in sensory organs, especially in the visual system, using anatomical, electrophysiological and immunological techniques. *Mailing Add:* 132 Cambridge Syracuse NY 13210

ENGBRING, NORMAN H, b Milwaukee, Wis, Mar 30, 25; m 50; c 3. MEDICINE. *Educ:* Marquette Univ, MD, 51. *Prof Exp:* From intern to resident, Milwaukee Co Gen Hosp, 51-55; from instr to assoc prof med, Sch Med, Marquette Univ, 55-72; asst dean, Med Col Wis, 72-80, assoc dean, 78-87, prof, 72-88, emer prof med, Grad Med Educ, 88-; RETIRED. *Concurrent Pos:* Dir radioisotope lab, Milwaukee Co Gen Hosp, 58-67, chief metab serv, 65-74; fel metab, Milwaukee Co Gen Hosp, 55-58; exec dir, MCW Affiliated Hosps, Inc, 80-88. *Honors & Awards:* Distinguished serv award, Med Col Wis, 83. *Mem:* AAAS; Am Fedn Clin Res; Am Diabetes Asn; Endocrine Soc; Sigma Xi; Ctr Soc Clin Res. *Res:* Endocrine and metabolic disorders. *Mailing Add:* 8701 Watertown Plank Rd Milwaukee WI 53226

ENGDAHL, ERIC ROBERT, b Worcester, Mass, June 11, 37; m 57; c 3. GEOPHYSICS. *Educ:* Rensselaer Polytech Inst, BS, 58; St Louis Univ, PhD(geophys), 68. *Prof Exp:* Explor geophysicist, Delta Explor Co, 58-60; res geophysicist, Theoret Studies Group, Environ Res Lab, Nat Oceanic & Atmospheric Admin, 60-77; SUPVRY GEOPHYSICIST, BR GLOBAL SEISMOL, OFF EARTHQUAKE STUDIES, US GEOL SURV, 77- *Concurrent Pos:* Mem Stand Earth Model Interas Comt & Chmn Int Asn Seismol & Physics Earths Interior Working Group Optimization Algorithms Determination Earthquake Parameters, Int Union Geol & Geophys, 71-; Adj assoc prof geophys, Univ Colo, Boulder, 72-, fel geophys, Coop Inst Res Environ Sci, 72-77; assoc ed, J Geophys Res, 74-76; liaison res, Adv Panel Earthquake Progs, US Geol Surv, 74-; mem, Comt Seismol, Nat Res Coun-Nat Acad Sci, 74-77, chmn, Panel Seismog Networks. *Mem:* Fel Am Geophys Union; Seismol Soc Am; AAAS. *Res:* Plate tectonics and earthquake prediction in an active subduction zone; application of seismic ray tracing to the study of plate structure and lateral heterogeneities in the Earth. *Mailing Add:* US Geog Surv Denver Fed Ctr Box 25046 Stop 967 Denver CO 80225

ENGDAHL, RICHARD BOTT, b Elgin, Ill, Apr 16, 14; m 40; c 2. MECHANICAL ENGINEERING. *Educ:* Bucknell Univ, BS, 36; Univ Ill, MS, 38; Am Acad Environ Engrs, dipl. *Prof Exp:* Asst, Univ Ill, 36-39, instr mech eng, 39-40; res engr, Battelle Mem Inst, 41-45, asst supvr, 45-46, supvr fuels, 47-50, chief, Fuels & Air Pollution Div, 50-57, chief, Thermal Eng Div, 57-58, staff engr, 58-65, fel environ res, 65-76, res consult, Battelle-Columbus Labs, 76-90; RETIRED. *Concurrent Pos:* environ consult, Malaysia, WHO, 79-80; combustion consult, Turkey, Int Exec Serv Corps, 80, UN Develop Prog, China, 81. *Mem:* Fel AAAS; Am Soc Mech Engrs; Am Soc Heat, Refrig & Air-Conditioning Engrs. *Res:* Combustion of pulverized coal; steam ejector performance; gas turbine locomotive; meter for flow of pulverized coal suspended in air; air pollution; heat pump; environmental control; incineration. *Mailing Add:* 1750 W First Ave Columbus OH 43212-3251

ENGE, HARALD ANTON, b Fauske, Norway, Sept 28, 20; nat US; m 47, 89; c 3. ION OPTICS. *Educ:* Tech Univ Norway, Eng Dipl, 47; Univ Bergen, Dr Philos, 54. *Prof Exp:* Lab engr, Tech Univ Norway, 47; res assoc & lectr, Univ Bergen, 48-55; from instr to assoc prof, Mass Inst Technol, 55-63, prof physics, 63-66; RETIRED. *Concurrent Pos:* Co-founder & chmn, Deltaray Corp, 69-73 & Gammaray Corp, 81. *Honors & Awards:* Tom W Bonner Prize Nuclear Physics, 84. *Mem:* Fel Am Phys Soc. *Res:* Low energy nuclear physics; nuclear instrumentation. *Mailing Add:* 25 Hillside Rd Lincoln MA 01773-4106

ENGEBRECHT, RONALD HENRY, b Oregon City, Ore, Jan 18, 34; m 54; c 3. SYNTHETIC ORGANIC & NATURAL PRODUCTS CHEMISTRY, ORGANIC SYNTHESIS. *Educ:* Ore State Univ, BS, 56, PhD(org chem), 64; Mich State Univ, MA, 59. *Prof Exp:* Instr, Ore High Schs, 56-59; asst prof cancer res, Ore State Univ, 63-65; SR RES CHEMIST, EASTMAN KODAK CO, 65- *Mem:* Am Chem Soc. *Res:* Organic photochemistry; synthesis of light-sensitive polymers; application of light-sensitive polymers for photoresists. *Mailing Add:* Res Lab Eastman Kodak Co Rochester NY 14650

ENGEL, ADOLPH JAMES, b Erie, Pa, Oct 13, 29; m 52; c 3. ELECTROANALYTICAL CHEMISTRY. *Educ:* Columbia Union Col, BA, 51; Univ Md, MS, 55; Univ Md, PhD, 72. *Prof Exp:* Lab asst, Columbia Union Col, 48-51; asst, Univ Md, 51-54; chemist fertilizer sect, USDA, 54-56; instr chem, Union Col, 56-58 & Wis State Univ, Eau Claire, 58-64; asst prof, Columbia Union Col, 64-74, North Adams State Col, 74-75 & Chattanooga State Tech Community Col, 80-86; analysis lab supvr, McKee Baking Co, 76-80; QUAL CONTROL MGR, EUREKA FOUNDRY CO, CHATTANOOGA TENN, 87- *Mem:* Am Chem Soc; Am Foundrymens Soc. *Res:* Micro methods of determination of metals; instrumental methods, including polarography & chromopotentiometry. *Mailing Add:* PO Box 1025 Collegedale TN 37315-1025

ENGEL, ALBERT EDWARD JOHN, b St Louis, Mo, June 16, 16; m 44; c 2. GEOLOGY, GEOCHEMISTRY. *Educ:* Univ Mo, BA, 38, MA, 39; Princeton Univ, MA, 41, PhD(geol), 42. *Prof Exp:* From instr to asst prof geol, Univ Mo, 38-42; from asst prof to prof, Calif Inst Technol, 48-58; prof, 58-86, EMER PROF GEOL, UNIV CALIF, SAN DIEGO & SCRIPPS INST OCEANOG, 86-; GEOLOGIST, US GEOL SURV, 42. *Mem:* Nat Acad Sci; fel AAAS; Am Acad Arts & Sci; Geol Soc Am; Am Geophys Union. *Res:* Crustal evolution. *Mailing Add:* 847 Indian Prairie Loop Victor MT 59875

ENGEL, ALFRED J, b Munich, Ger, Mar 30, 27; US citizen; m 53; c 4. CHEMICAL ENGINEERING. *Educ:* Cornell Univ, BChE, 52; Univ Wis, PhD(chem eng), 61. *Prof Exp:* Engr petrochem, Calif Res Corp, 52-55; instr chem eng, Univ Wis, 57-59; from asst prof to assoc prof, 59-71, PROF CHEM ENG, PA STATE UNIV, 71- *Concurrent Pos:* Consult, Socony Mobil Oil Co, NY, 64; Fulbright sr lectr, Ben Gurion Univ, Israel, 74-75; consult, Union Carbide AG Chem Co, 84-85. *Mem:* Fel Am Inst Chem Engrs; Am Chem Soc. *Res:* Chemical reaction kinetics and mass transfer; air pollution control and administration. *Mailing Add:* Dept Chem Eng Pa State Univ University Park PA 16802

ENGEL, ANDREW G, b Budapest, Hungary, July 12, 30; US citizen; m 58; c 2. NEUROPATHOLOGY, BIOCHEMISTRY. *Educ:* McGill Univ, BS, 53, MD, 55. *Prof Exp:* Resident internal med & neurol, Mayo Clin, 56-57; clin assoc neurol, Nat Inst Neurol Dis & Blindness, 58-59; resident internal med & neurol, Mayo Clin, 60-62; instr neurol, 66-67, assoc prof, 67-73, PROF NEUROL, MAYO MED SCH, 73-, W L MCKNIGHT 3M PROF NEUROSCI, 84- *Concurrent Pos:* Consult, Mayo Clin, 65-; spec fel neuropath, Col Physicians & Surgeons, Columbia Univ, 62-65. *Honors & Awards:* Javitts Neurosci Investr Award, NIH. *Mem:* Am Asn Neuropath; Am Soc Cell Biol; Soc Neurosci. *Res:* Experimental neuropathology; muscle biochemistry; biochemical and ultrastructural studies of myopathies. *Mailing Add:* Mayo Clin Rochester MN 55901

ENGEL, BERNARD THEODORE, b Chicago, Ill, Apr 18, 28; m 51; c 3. PHYSIOLOGICAL PSYCHOLOGY, BEHAVIORAL MEDICINE. *Educ:* Univ Calif, Los Angeles, BA, 54, PhD(physiol psychol), 56. *Prof Exp:* Jr res psychologist, Univ Calif, Los Angeles, 56; res psychologist, Inst Psychosom & Psychiat Res & Training, Michael Reese Hosp, 57-58; lectr med psychol & mem sr staff, Cardiovasc Res Inst, Sch Med, Univ Calif, 59-67; CHIEF, LAB BEHAV SCI & BEHAV PHYSIOL SECT, GERONT RES CTR, NAT INST AGING, 67- *Concurrent Pos:* Assoc prof, 70-82, prof behav biol, Sch Med, Johns Hopkins Univ, 82- *Honors & Awards:* Pavlovian Soc Award, 79. *Mem:* Fel AAAS; Soc Psychophysiol Res (pres, 70-71); Am Psychosom Soc (secy-treas, 81-83, pres, 85-86); Biofeedback Soc Am (pres, 81-82); Sigma Xi; fel Gerontol Soc; Soc Behav Med; fel Acad Behav Med Res. *Res:* Psychophysiological mechanisms underlying the learned control of autonomic responses; application of operant conditioning techniques to the clinical control of cardiovascular responses; psychophysiological changes associated with aging; behavioral treatment of incontinence in the elderly. *Mailing Add:* Gerontol Res Ctr NIA Nat Inst Aging Baltimore MD 21224

ENGEL, CHARLES ROBERT, b Vienna, Austria, Jan 28, 22; Can & French citizen; m 51; c 4. ORGANIC CHEMISTRY. *Educ:* Univ Grenoble, BS, 41; Swiss Fed Inst Technol, Chem Eng, 47, DSc(org chem), 51; Univ Paris, DSc, 70. *Prof Exp:* From asst prof to assoc prof med res & hon spec lectr chem, Univ Western Ont, 51-58; prof chem, Laval Univ, 58-90; RETIRED. *Concurrent Pos:* Can Life Ins Off Asn Med fel, 52-58; ed, Steroids, 63-, Can J Chem, 87-91; vis prof, Inst Natural Chem Substances, Nat Ctr Sci Res, France, 66-67; consult. *Honors & Awards:* Legion of Honour, France. *Mem:* Am Chem Soc; NY Acad Sci; Fr Chem Soc; Swiss Chem Soc; fel Royal Soc Chem; fel Chem Inst Can. *Res:* Synthetic organic chemistry; steroids and related products; carbanionic rearrangements; chemical endocrinology; biologically active natural products; stereochemistry; chiroptical properties. *Mailing Add:* Dept Chem Laval Univ Quebec PQ G1K 7P4 Can

ENGEL, ERIC, b Geneva, Switz, Oct 12, 25; m 50; c 3. MEDICINE. *Educ:* Univ Geneva, BS, 47, MD, 51, PhD, 58. *Prof Exp:* Instr internal med, Geneva Univ Hosp, 58-60; instr med, Harvard Med Sch, 60-63; from asst prof to prof med, Sch Med, Vanderbilt Univ, 63-78, head, Div Gen Med, 72-78; PROF, UNIV INST MED GENETICS, GENEVA, 78- *Concurrent Pos:* Clin & res fel cytogenetics, Mass Gen Hosp, 60-63; Prof Genetics & Dir Genetics Inst, Cantonal Hosp, Univ of Geneva, Switz, 79- *Mem:* Am Soc Clin Invest; Am Soc Human Genetics. *Mailing Add:* Univ Inst Med Genetics CMU 9 Ave de Champel Geneva 4 1211 Switzerland

ENGEL, GEORGE LIBMAN, b New York, NY, Dec 10, 13; m 38; c 2. MEDICINE, PSYCHIATRY. *Educ:* Dartmouth Col, BA, 34; Johns Hopkins Univ, MD, 38. *Hon Degrees:* Univ Bern, Switz, 80; DSc Med Col Ohio, 86. *Prof Exp:* From instr to asst prof med & psychiat, Col Med, Univ Cincinnati, 42-46; psychiatrist & physician, Strong Mem Hosp, 57-; from asst prof to prof, 46-83, EMER PROF PSYCHIAT MED, SCH MED & DENT, UNIV ROCHESTER, 83- *Concurrent Pos:* Clinician, Med Serv, Cincinnati Gen Hosp, 42-44, asst attend psychiatrist, 43-44; consult, Off Surgeon Gen; consult, Fitzsimons Gen Hosp, 48 & res studies sect, Nat Adv Ment Health Coun, USPHS, 49-53; fel med, Harvard Med Sch, 41-42; USPHS career res award, 62; George L Engel teaching fel, Sch Med, Univ Rochester; Albert M Biele vis prof psychiat, Med Col, Thomas Jefferson Univ, 90. *Honors & Awards:* Robert Glaser Award, Soc Gen Internal Med, 91. *Mem:* Inst Med-Nat Acad Sci; AAAS; Am Soc Clin Invest; Am Psychosom Soc; Am Psychiat Asn; Am Psychoanalysis Asn. *Res:* Physiology of respiration and circulation; electroencephalography; syncope; delirium; migraine; decompression sickness; problems of clinical and psychosomatic medicine; psychoanalysis; medical education. *Mailing Add:* 91 San Gabriel Dr Rochester NY 14610

ENGEL, GERALD LAWRENCE, b Cleveland, Ohio, July 5, 42; m 64. COMPUTER SCIENCE EDUCATION, SOCIAL IMPACT OF COMPUTING. *Educ:* Hampden Sydney Col, BS, 64; La State Univ, MA, 65, Pa State Univ, DEd, 74. *Prof Exp:* Dir comput & statistics, Va Inst Marine Sci, 73-77; PROF COMPUTER SCI & ENG, UNIV CONN, STAMFORD REGIONAL CAMPUS, 84- *Concurrent Pos:* Chairperson var comts, 78-; vis prof, Univ Ore, 83 & 84; consult, NSF, 79-, UN, 85-86; alt dir, Accreditation Bd Eng & Technol, 88- *Mem:* Math Asn Am; Nat Coun Teachers Math; Am Asn Univ Profs; Inst Elec & Electronics Engrs Comput Soc (vpres educ, 88-); Asn Comput Mach. *Res:* Computer science education. *Mailing Add:* 15 Avon Ct Naugatuck CT 06770

ENGEL, JAMES DOUGLAS, b Staten Island, NY, Apr 6, 47; m 68; c 2. MOLECULAR BIOLOGY. *Educ:* Univ Calif, San Diego, BA & BS, 70; Univ Ore, PhD(chem), 75. *Prof Exp:* ASST PROF MOLECULAR BIOL, DEPT BIOCHEM & MOLECULAR CELL BIOL, NORTHWESTERN UNIV, 78- *Concurrent Pos:* Consult, Centaur Genetics, 81- *Res:* Eucaryotic gene expression. *Mailing Add:* Dept Biochem Hogan 4-120 Northwestern Univ 2153 Sheridan Rd Evanston IL 60208

ENGEL, JAMES FRANCIS, b Kansas City, Mo, Jan 25, 41; m 61; c 4. SYNTHETIC ORGANIC CHEMISTRY. *Educ:* Rockhurst Col, AB, 61. *Prof Exp:* Instr chem, Rockhurst High Sch, Mo, 63-64; from asst chemist to assoc chemist org synthesis, 64-72, sr chemist, 72-76, head org & radiochem synthesis sect, 76-77, assoc dir Chem Sci Div, 77-79, DIR BIORGANIC CHEM DEPT, MIDWEST RES INST, 79- *Concurrent Pos:* Instr chem, Donnelly Jr Col, 65-68. *Mem:* Am Chem Soc. *Res:* Research and development synthesis and analysis of toxic materials, metabolites and/or degradation products with or without isotopic label, mass or radioactive, especially polynuclear aromatic hydrocarbon derivatives. *Mailing Add:* 124 W 78th Terr Kansas City MO 64114-1832

ENGEL, JAN MARCIN, b Gdansk, Danzig, May 1, 24; US citizen; wid; c 2. SOLID STATE PHYSICS, COMPUTER SCIENCES. *Educ:* Univ London, BSc, 46. *Prof Exp:* Res physicist, Socony-Vacuum Oil Co, NJ, 50-51; res physicist, Electronics Lab, Gen Elec Co, NY, 51-53; sr proj engr, Motorola Inc, Ariz, 53-54; res physicist, Pac Semiconductors Inc, 54-57; assoc physicist, Res Lab, IBM Corp, 58-59, staff physicist, Adv Systs Div, 59-60, adv physicist thin film devices, Components Lab, NY, 60-63, adv physicist electron devices, Gen Prods Div, 64-65, Systs Develop Div, 65-71, adv physicist, Develop Lab, Gen Prods Div 71-84; RETIRED. *Concurrent Pos:* Consult, Electro-Optical Systs Inc, Calif, 57-58; lectr, Dept Liberal Arts, Univ Exten, Univ Calif, Los Angeles, 57-58 & Univ Calif, Berkeley, 58-60; vol ed progs, IBM Corp, 62-84; pres, Genealogical Data Systs, San Jose, Calif, 80- *Mem:* Am Inst Physics; Am Phys Soc; Sigma Xi; fel Brit Inst Physics & Phys Soc; sr mem Inst Elec & Electronics Engrs; Int Soc Optical Eng. *Res:* Development of novel devices utilizing physical phenomena that are usually studied in the fields of solid state physics and utilizing semiconductor, thin film and/or electron beam technologies. *Mailing Add:* 2980 Cambridge Dr San Jose CA 95125

ENGEL, JEROME, JR, b Albany, NY, May 11, 38; m 67; c 3. NEUROPHYSIOLOGY, NEUROLOGY. *Educ:* Cornell Univ, BA, 60; Stanford Univ, MD, 65, PhD(physiol), 66; Am Bd Qualification EEG, cert; Am Bd Psychiat & Neurol, cert. *Prof Exp:* Intern med, Ind Univ Med Ctr, 66-67; resident neurol, Albert Einstein Col Med, 67-68; NIH, NINCDS, Lab Perinatal Physiol, 68-69; staff assoc, Lab Neural Control, 69-70; resident neurol, 70-72, asst prof neurol & neurosci, Albert Einstein Col Med, 72-76; assoc prof, 76-79, PROF NEUROL & ANAT, SCH MED, UNIV CALIF, LOS ANGELES, 80- *Concurrent Pos:* Vis asst prof, Sch Med, Univ PR, 68-69; NIH fel, Stanford Univ & Nat Ctr Sci Res, France, 65-66; Fulbright scholar, Inst Neurol & Psychiat, London, 71-72; NIH career develop award, 72-76; Guggenheim fel, 83-84; vis prof, dept anat, Sydney Univ, 84. *Honors & Awards:* Michael Prize, 82. *Mem:* Soc Neurosci; Am Neurol Asn; Am Acad Neurol; Am EEG Soc; Am Epilepsy Soc (pres, 84-85). *Res:* Clinical neurophysiology; pathophysiology of epilepsy; positron emission tomography. *Mailing Add:* Reed Neurol Res Ctr Sch Med Univ Calif Los Angeles Los Angeles CA 90024

ENGEL, JOEL S, b New York, NY, Feb 4, 36; m 59; c 3. TELECOMMUNICATIONS. *Educ:* City Col NY, BEE, 57; Mass Inst Technol, MSEE, 59; Polytech Inst, PhD (elec eng), 64. *Prof Exp:* Mem tech staff, Mass Inst Technol Instrumentation Lab, 57-59; mem tech staff, Bell Labs, 59-65, dept head, 67-83; vpres eng res & develop, Satellite Bus Systs/MCI Commun, 83-87; VPRES TECHNOL MGT, AMERITECH, 87- *Honors & Awards:* Alexander Graham Bell Medal, Inst Elec & Electronics Engrs, 87. *Mem:* Fel Inst Elec & Electronics Engrs. *Res:* Research and development for Ameritech, one of the seven regional bell operating companies, encompassing all technologies in support of telecommunications. *Mailing Add:* Amertech Servs 2820 W Golf Rd Rolling Meadows IL 60008

ENGEL, JOHN FRANCIS, b Cincinnati, Ohio, Sept 15, 42; m 65; c 2. ORGANIC CHEMISTRY. *Educ:* Univ Cincinnati, BS, 64; Duke Univ, MS, 70, PhD(org chem), 71. *Prof Exp:* Res chemist org chem, Cincinnati Milacron, Inc, 64-67; RES CHEMIST ORG CHEM, AGR CHEM DIV, FMC CORP, 71- *Mem:* Am Chem Soc. *Res:* Organic pesticides. *Mailing Add:* FMC Corp - Lithium Div PO Box 795 Bessemar City NC 28016

ENGEL, JOHN HAL, JR, b Detroit, Mich, Dec 12, 30; m 57; c 3. ORGANIC CHEMISTRY, POLYMER CHEMISTRY. *Educ:* Univ Detroit, BS, 56, MS, 58; Mich State Univ, MBA, 75. *Prof Exp:* Res chemist, Gen Motors Res Labs, 58-62; sr res chemist, R P Scherer Corp, 62-63; from res scientist to sr res scientist, Chrysler Corp Res Labs, Detroit, 63-67, group leader polymer res, 67-70, group leader emissions, 70-74, res staff scientist, 74-75, sr res staff scientist emissions & govt rels & mgr high temperature mat res, 75-79, mgr rubber & plastics eng, 79-81; assoc mat engr, 81-82, mgr, Basic Coatings Lab, Fisher Body, 82-85, MGR, PAINT SYSTS LAB, CPC GROUP, GEN MOTORS CORP, 85- *Concurrent Pos:* Adj prof chem, Marygrove Col, 68-; adj prof mat eng, Detroit Inst Technol, 80-81; adj prof chem, Macomb Col, 82- *Mem:* Am Chem Soc; Soc Automotive Engrs. *Res:* Solventless and aqueous paint and adhesive systems; polymer synthesis; infrared analysis of polymer systems; polyurethanes; conversion of emissions from engines and factories, superalloys, ceramics for structural applications; plastics for automotive application, automotive coatings. *Mailing Add:* 705 Washington Rd Grosse Pointe MI 48230

ENGEL, JOHN JAY, b Milwaukee, Wis, July 27, 41; m 62; c 2. BRYOLOGY. *Educ:* Univ Wis-Milwaukee, BS, 65, MS, 67; Mich State Univ, PhD(bot), 72. *Prof Exp:* Asst cur, 72-77, assoc dur bryol, 77-86, CUR BRYOL, FIELD MUS NATURAL HIST, 86- *Mem:* Am Bryol Soc; Int Asn Bryologists. *Res:* The taxonomy and phytogeography of south temperate and subantarctic liverworts; evolution, relationships, ecology and distribution of several liverwort families. *Mailing Add:* Dept Bot Field Mus Natural Hist Roosevelt Rd at Lake Shore Dr Chicago IL 60605

ENGEL, JOSEPH H(ENRY), b New York, NY, May 15, 22; m 43; c 3. MATHEMATICS, OPERATIONS RESEARCH. *Educ:* City Col New York, BS, 42; Univ Wis, MA, 47, PhD(math), 49. *Prof Exp:* Opers analyst opers eval group, Mass Inst Technol, 49-57, dep dir, 57-62; dir opers eval group, ctr naval anal, Franklin Inst, 62-65, asst chief scientist, 65-67; dir planning res & serv, Commun Satellite Corp, 67-70; head, Systs Eng Dept, Univ Ill, Chicago Circle, 70-76, prof systs eng, 70-79; chief, Math Anal Div, Nat Ctr Statist & Anal, Nat Hwy Traffic Safety Admin, Dept Transp, 79-82; chief, Opers Res Div, Ctr Appl Math, Nat Eng Lab, Nat Bur Standards, 82-86; RETIRED. *Concurrent Pos:* Chmn, adv panel oper res, NATO, 70-73, chmn, spec prog panel systs sci, NATO, 73; prin investr, Chicago Area Transp Study, 73 & Dupage County Regional Planning Comn, 75-. MGT CONSULT. *Honors & Awards:* Jacinto Steinhardt Mem Award, 86. *Mem:* Fel AAAS; Am Math Soc; Opers Res Soc Am (secy, pres, 65-69); Int Fedn Opers Res Soc (vpres, 74); Am Inst Indust Engrs. *Res:* Operations research; stochastic processes; decision problems; transportation and land use; equations of combat; safety standards for television receivers; establishment of manufacturing facilities in space; mathematical modelling of complex systems. *Mailing Add:* 7311 Broxburn Ct Bethesda MD 20817

ENGEL, LAWRENCE J, b St Louis, Ill, May 10, 29; m 52; c 4. CHEMICAL ENGINEERING. *Educ:* Ga Inst Technol, BChE, 51, MSChE, 54, PhD(chem eng), 56. *Prof Exp:* Asst instr chem eng, Ga Inst Technol, 53-56; engr, Esso Res & Eng Co, NJ, 56-59, sr engr, Spec Projs Unit, 59-63, mem chem staff, 63-65; staff adv, Enjay Polymer Labs, 65-67, res assoc, Enjay Additives Labs, 67-68, proj coordr, Paramins Div, Enjay Chem Co, NY, 68-70; sr assoc, Exxon Res & Eng Co, 70-76, MGR, EXXON CHEM CO, 76- *Res:* Petroleum and chemical process research from lab bench sales through design and operation of commercial unit; economics and staff planning and analysis of projects. *Mailing Add:* Exxon Chem Co PO Box 45 Linden NJ 07036

ENGEL, MILTON BAER, b Chicago, Ill, Aug 7, 16; m 42, 87; c 2. HISTOCHEMISTRY, EXPERIMENTAL PATHOLOGY. *Educ:* Univ Ill, DDS, 38, MS, 40. *Prof Exp:* Prof orthod & histol, 57-86, RES ASSOC, COL DENT, 45-, EMER PROF ORTHOD & HISTOL, UNIV ILL MED CTR, 86- *Concurrent Pos:* Practicing orthodontist; Carnegie fel orthod, Col Dent, Univ Ill, Med Ctr, 41-42; dental study sect, NIH, 65-69. *Mem:* AAAS; Am Dent Asn; Am Asn Orthodont; Am Soc Exp Path; Int Asn Dent Res; Sigma Xi. *Res:* Experimental pathology; histochemistry; growth; orthodontics. *Mailing Add:* Dept Histol Col Dent Univ Ill 801 S Paulina Chicago IL 60612

ENGEL, NIELS N(IKOLAJ), b Bern, Switzerland, Nov 21, 04; US citizen; m 34; c 3. METALLURGY, THEORETICAL PHYSICS. *Educ:* Tech Univ, Denmark, BS, 25, MS, 28; Univ Copenhagen, Cand Phil, 26;. *Hon Degrees:* Dr Ing, Aachen Tech Univ, 31. *Prof Exp:* Assoc prof, Tech Univ, Denmark, 36-51; assoc prof metall eng, Univ Ala, 51-59; prof, Ga Inst Technol, 59-72; prof metall eng, NMex Inst Technol, 73-79; RETIRED. *Concurrent Pos:* Consult, Oak Ridge Nat Lab, 64-70, Southern Saw, Ga, 66-80 & Southwire, Ga, 68-80. *Honors & Awards:* Adolf Martens Medal, Ger, 66. *Mem:* Am Soc Metals. *Res:* Basic conception of physical world; bonding between atoms; metallic properties; applied physical metallurgy. *Mailing Add:* 3902 Commander Dr Chamblee GA 30341-1831

ENGEL, PAUL SANFORD, b Pittsburgh, Pa, July 19, 42; m 87. PHYSICAL ORGANIC CHEMISTRY. *Educ:* Univ Calif, Los Angeles, BS, 64; Harvard Univ, PhD(chem), 68. *Prof Exp:* Sr asst scientist, NIH, 68-70; from asst prof to assoc prof, 70-80, PROF CHEM, RICE UNIV, 80- *Concurrent Pos:* Fel Alfred P Sloan, 75-78. *Mem:* Am Chem Soc. *Res:* Organic photochemistry; azo compounds; extrusion reactions; energy transfer; free radical chemistry. *Mailing Add:* Dept Chem Rice Univ PO Box 1892 Houston TX 77251-1892

ENGEL, PETER ANDRAS, b Kassa, Czeck, July 10, 35; US citizen; m 86; c 2. APPLIED MECHANICS, TRIBOLOGY. *Educ:* Vanderbilt Univ, BE, 58; Lehigh Univ, MS, 60; Cornell Univ, PhD(theoret & appl mech), 68. *Prof Exp:* Struct civil engr, Praeger-Kavanagh-Waterbury Engrs, NY, 60-62; res engr, Boeing Co, New Orleans, La, 62-65; RES ENGR MAT SCI, ENDICOTT LAB, IBM CORP, 68- *Honors & Awards:* Charles Russ Richards Mem Award, Am Soc Mech Engrs, 83. *Mem:* Fel Am Soc Mech Engrs; sr mem Inst Elec & Electronics Engrs; Am Acad Mech; Int Electronics Packaging Soc. *Res:* Contact of solids, with applications to sliding, rolling, and impact; wear studies; structural dynamics; electronic packaging; author of over 70 papers. *Mailing Add:* Dept Mech & Indust Eng State Univ NY Binghamton NY 13902-6000

ENGEL, ROBERT DAVID, b Los Angeles, Calif, Nov 22, 32; m 57; c 2. COMPUTER SCIENCE. *Educ:* Univ Calif, Los Angeles, BS, 58, MS, 59, PhD(eng), 63. *Prof Exp:* Res engr plasma properties, Univ Calif, Los Angeles, 61-63, asst prof eng, 63-66; mem staff comput, opers dept, Beckman Instrument Corp, 66-67; assoc provost, 67-80, PROF ENG, UNIV REDLANDS, 67-, CHMN DEPT, 80- *Concurrent Pos:* Consult, electronics div, Rand Corp, 60-66. *Mem:* Inst Mgt Sci; Inst Elec & Electronics Engrs. *Res:* Applied electromagnetic theory; microwaves; electronics; computers; simulation; systems. *Mailing Add:* Dept Eng Univ Redlands Redlands CA 92373

ENGEL, ROBERT RALPH, b Pittsburgh, Pa, Aug 30, 42; m 66; c 2. SYNTHETIC ORGANIC CHEMISTRY. *Educ:* Carnegie Inst Technol, BS, 63; Pa State Univ, PhD(chem), 66. *Prof Exp:* Asst prof chem, 68-71, assoc prof & chmn, Dept Chem, 77-79, dep exec officer PhD prog biochem, 77-84, PROF CHEM & BIOCHEM, QUEENS COL, CITY UNIV NY, 76- *Concurrent Pos:* NATO sr res fel, McGill Univ, 75; vis scientist, Rohm & Haas Co, 85-86. *Mem:* Am Chem Soc; NY Acad Sci; The Chem Soc; Sigma Xi. *Res:* Mechanism studies on reactions of organophosphorus compounds; organophosphorus synthesis; preparation of analogues of natural phosphates; hydrogenolysis mechanisms; aromatic substitution reactions. *Mailing Add:* Dept Chem Queens Col Flushing NY 11367

ENGEL, RUBEN WILLIAM, b Shawano, Wis, July 10, 12; m 39; c 3. BIOCHEMISTRY. *Educ:* Univ Wis, PhB, 36, PhD(biochem), 39. *Prof Exp:* Asst, Univ Wis, 36-39; assoc animal nutritionist, Ala Polytech Inst, 39-43, animal nutritionist, 46-52; head dept, 52-66, assoc dean res, Col Agr, 66-78, prof, 68-78, EMER PROF BIOCHEM & NUTRIT, VA POLYTECH INST & STATE UNIV, 78- *Concurrent Pos:* Mem panel on nutrit, Comt on Growth, Nat Res Coun, mem food & nutrit bd & mem US nat comt nutrit; nutrit adv, US Agency Int Develop, Manila, Philippines, 68-80. *Honors & Awards:* Conrad A Elvehjem Award, Am Inst Nutrit, 74; Gamma Sigma Delta Award for Distinguished Serv to Int Agr. *Mem:* Am Asn Cancer Res; Am Chem Soc; Am Inst Nutrit (pres, 64); Am Dairy Sci Asn; hon mem Philippine Asn Nutrit. *Res:* Chemical and pathological changes associated with B vitamin deficiencies; choline metabolism; relation of nutrition to cancer; minor elements in animal nutrition; socio-economic characteristics of malnutrition in infants in developing countries. *Mailing Add:* 1726 Donlee Dr Blacksburg VA 24060

ENGEL, RUDOLF, b Bonn, Ger, Aug 28, 04; US nat; m 33; c 4. MEDICINE. *Educ:* Univ Bonn, MD, 29; Univ Berlin, Dr med habil, 35; Univ Minn, MSc & MD, 49. *Prof Exp:* Resident, Univ Berlin Hosp, 29-30; asst pediat, Univ Minn, 31; resident, Univ Heidelberg Hosp, 32-34 & Univ Berlin Hosp, 34-36; asst prof, Univ Hamburg, 36-38; assoc prof internal med, Univ Berlin, 38-45; head dept, Luisen Hosp, Aachen, Ger, 46-48; from clin asst prof to assoc prof pediat, Univ Minn, 50-52; clin instr neurol, 52-57, from asst prof to prof pediat, 57-73, EMER PROF PEDIAT, MED SCH, UNIV ORE, 73- *Concurrent Pos:* Fulbright lectr, Univ Ceylon, 55-56, Univ Hamburg, 62 & Univ Munich, 74. *Mem:* AAAS; Am EEG Soc; Am Acad Cerebral Palsy; fel Am Col Physicians. *Res:* Causes of cerebral palsy; mental retardation and other malformations; biological sciences; neonatal electroencephalography; evoked potentials. *Mailing Add:* Dept Pediat Ore Health Sci Univ Portland OR 97201

ENGEL, THOMAS WALTER, b Yokohama, Japan, Apr 2, 42; US citizen; m 79; c 1. PHYSICAL CHEMISTRY, SURFACE SCIENCE. *Educ:* Johns Hopkins Univ, BA, 63, MA, 64; Univ Chicago, PhD(chem), 69. *Prof Exp:* Res assoc surface sci, Clausthal Tech Univ, 69-75 & Univ Munich, 75-78; res assoc surface sci, IBM Res Lab Zurich, 78-80; assoc prof, 80-84, chmn, Dept Chem, 87-90, PROF CHEM, UNIV WASH, 84- *Mem:* Am Chem Soc; Am Vacuum Soc. *Res:* Surface physics and chemistry; structure of solid surfaces; oxidation; molecular beam scattering from surfaces. *Mailing Add:* Dept Chem Univ Wash Seattle WA 98195

ENGEL, TOBY ROSS, b New York, NY, Mar 6, 42; m 65; c 3. CARDIOLOGY. *Educ:* Univ Col, NY Univ, BA, 62, Sch Med, MD, 66. *Prof Exp:* Resident, Univ Pa, 66-68; resident, Ohio State Univ, 70-71, fel cardiol, 71-73; from asst prof to prof med, Med Col Pa, 73-85; PROF MED, UNIV NEBR, 85- *Concurrent Pos:* Assoc ed, Annals of Internal Med, 77-85. *Mem:* Am Col Physicians; Am Col Cardiol; Am Heart Asn; Am Col Clin Phamacol; Am Soc Clin Pharmacol & Therapeut. *Res:* Clinical electrophysiology, concentrating on sinus node dysfunction; ventricular tochycardia; atrial fibrillation and flutter; pharmacologic approach to arrthymias; signal averaged ECG. *Mailing Add:* Dept Internal Med Univ Nebr Med Ctr 42nd & Dewey Sts Omaha NE 68105

ENGEL, WILLIAM KING, b St Louis, Mo, Nov 19, 30; m 54; c 3. NEUROLOGY. *Educ:* Johns Hopkins Univ, BA, 51; McGill Univ, MD, 55; Am Bd Neurol & Psychiat, dipl, 62. *Prof Exp:* Intern neurol, Univ Mich Hosp, 55-56; clin assoc, Med Neurol Br, Nat Inst Neurol Dis & Blindness, 56-59; clin clerk, Nat Hosp, London, Eng, 59-60; assoc neurologist, 60-62, actg chief, 62-63, CHIEF MED NEUROL BR, NAT INST NEUROL DIS & STROKE, 63- *Concurrent Pos:* Nat Inst Neurol Dis & Blindness trainee, 59-60; mem med bd, NIH, 68-69; clin prof, Sch Med, George Washington Univ, 69-; mem med adv bd, St Jude's Children's Res Hosp, Memphis, 70-; mem med adv bd, Myasthenia Gravis Found, 70-; mem exec comt, Res Group Neuromuscular Disorders, World Fedn Neurol, 70-; mem adv coun, Amyotrophic Lateral Sclerosis Found, 71-; assoc exam, Am Bd Neurol & Psychiat. *Honors & Awards:* S Weir Mitchell Award, Am Acad Neurol, 62; Meritorious Serv Medal, USPHS, 71. *Mem:* Fel Am Acad Neurol; Histochem Soc; Am Soc Cell Biol; Am Asn Neuropath; AMA. *Res:* Clinical and investigative neurology; neuromuscular diseases. *Mailing Add:* Dept Neurol & Path Univ Southern Calif 637 S Lucas Ave Los Angeles CA 90017

ENGELAND, WILLIAM CHARLES, NEUROENDOCRINOLOGY. *Educ:* Univ Calif, San Francisco, PhD(endocrinol), 76. *Prof Exp:* ASSOC PROF SURG & NEUROBIOL, RI HOSP, BROWN UNIV, 79- *Mailing Add:* Dept Surg Brown Univ 593 Eddy St Providence RI 02905

ENGELBART, DOUGLAS C(ARL), b Portland, Ore, Jan 30, 25; m 51; c 4. ELECTRICAL ENGINEERING. *Educ:* Ore State Col, BS, 48; Univ Calif, EE, 53, PhD(elec eng), 55. *Prof Exp:* Elec engr, Nat Adv Comt Aeronaut, Ames Aero Lab, Calif, 48-51; assoc elec eng, Univ Calif, 54-55, asst prof, 55-56; pres & tech dir, Digital Tech, Inc, 56-57; res engr, Man-Mach Info Systs, SRI Int, 57-65, prog head, 65-78; sr scientist, Tymshare, Inc, 78-84; sr scientist, Info Systs Group, McDonnell Douglas Corp, 84-89; DIR, BOOTSTRAP INST, 89- *Concurrent Pos:* Consult, Marchant Res, Inc, Calif, 55-56. *Mem:* Inst Elec & Electronics Engrs. *Res:* Digital computers; new device research with gas discharge and magnetics; man-computer on-line problem solving. *Mailing Add:* 89 Catalpa Dr Atherton CA 94025

ENGELBERG, JOSEPH, b Vienna, Austria, June 2, 28; nat US; m 54; c 2. THEORY OF LIVING SYSTEMS. *Educ:* Cooper Union, BME, 50; Univ Pa, MS, 53, PhD(physics), 58. *Prof Exp:* Res engr surg instrumentation, Univ Pa, 50-53; res engr bioeng, Franklin Inst, 53-54; instr biophys, Univ Colo, 58-60; Am Cancer Soc fel, Univ Calif, Berkeley, 60-61; from asst prof to assoc prof, 61-70, PROF PHYSIOL & BIOPHYS, SCH MED, UNIV KY, 70-, DIR, OFF INTEGRATIVE STUDIES, 82- *Concurrent Pos:* USPHS fel, 58-60. *Honors & Awards:* Lederle Award, 63-66. *Mem:* Am Physiol Soc. *Res:* Design of artificial kidney and artificial heart-lung machines; mechanics of pulmonary circulation; mammalian cells, especially physiology and effects of radiation; development of health delivery systems; theory of living systems; integrative study. *Mailing Add:* Dept Physiol & Biophys Univ Ky Med Ctr Lexington KY 40536-0840

ENGELBERGER, JOSEPH FREDERICK, b New York, NY, July 26, 25; m 54; c 2. ROBOTICS MANUFACTURING. *Educ:* Columbia Univ, BS, 46, MS, 49. *Prof Exp:* Engr, Manning, Maxwell & Moore, 46-53, chief engr, 53-56, gen mgr div, 56-57; founder & gen mgr, Consol Controls Corp, 57-77; DIR, FOUNDER & PRES, UNIMATION, INC, 62-; VPRES, CONDEC CORP, 65-; CHMN, TRANSITION RES CORP, 84. *Mem:* Nat Acad Eng. *Mailing Add:* Transition Res Corp 15 Pasture Rd Danbury CT 06810

ENGELBORGHS, YVES, b Diegem, Belg, Mar 18, 46; m 69; c 3. ENZYMOLOGY, PROTEIN PHYSICAL CHEMISTRY. *Educ:* Univ Louvain, PMD(protein chem), 72. *Prof Exp:* Postdoctoral enzym, Univ Bristol, UK, 74-75; researcher, 76-85, PROF BIOCHEM, UNIV LOUVAIN, 85- *Mem:* Am Soc Cell Biol. *Res:* Assembly of microtubules; fast kinetics of proteins; multi frequency phase flodrometry; tubulin-cytostatic interactions. *Mailing Add:* Univ Louvain Celestynenlaan 200D Louvain B-3001 Belgium

ENGELBRECHT, R(ICHARD) S(TEVENS), b Ft Wayne, Ind, Mar 11, 26; m 48; c 2. WATER QUALITY CONTROL, WATER & WASTEWATER TREATMENT. *Educ:* Univ Ind, AB, 48; Mass Inst Technol, MS, 52, ScD, 54. *Prof Exp:* Asst microbiol, sch med, Univ Ind, 48-50; asst civil & sanit eng, Mass Inst Technol, 50-52, instr, 52-54; from asst prof to assoc prof sanit eng, 54-59, PROF ENVIRON ENG, UNIV ILL, URBANA, 59-, DIR, ADVAN ENVIRON CONTROL TECH RES CTR, 79-, IVAN RACHEFF PROF, 87- *Concurrent Pos:* Mem, Ohio River Valley Water Sanit Comn, 78, chmn, 80-82. *Honors & Awards:* Harrison P Eddy Medal, Water Pollution Control Fedn, 66, Arthur Sidney Bedell Award, 73 & Gordon Maskew Fair Medal, 87; George W Fuller Award, Am Water Works Asn, 74, Publ Award, 75; Ernest Victor Balsom Commemmoration Lect, Inst Pub Health Engrs, London, 78; Eric H Vick Award, Inst Pub Health Engrs, 79; Benjamin Garver Lamme Award, Am Soc Eng Educ, 85. *Mem:* Nat Acad Eng; Am Soc Microbiol; Am Water Works Asn; Water Pollution Control Fedn (pres, 77-78); AAAS; Int Asn Water Pollution Res & Control (pres, 80-86). *Res:* Water quality; water and wastewater treatment; environmental science. *Mailing Add:* Dept Civil Eng Univ Ill Urbana-Champaign 205 N Mathews St Urbana IL 61801

ENGELBRECHT, RUDOLF S, b Atlanta, Ga, Apr 18, 28; m 50; c 3. ELECTRICAL ENGINEERING. *Educ:* Ga Tech, BS, 51, MS, 53; Ore State Univ, PhD(elec eng), 79. *Prof Exp:* MTS, Bell Labs, Whippany NJ, 53-61; Supv, Bell Labs Murray Hill, NJ, 61-64, dept head, 64-69; dept head, Bell Labs, Holmdel, NJ, 69-70; dir, RCA Solid State Tech Ctr, 70-71; group leader, RCA Labs, Zurich, Switz, 72-77; ASSOC PROF ELEC ENG, ORE STATE UnIV, 77- *Honors & Awards:* Centennial Medal, Inst Elec & Electronics Engrs, 84. *Mem:* Fel Inst Elec Electronics Engrs; Sigma Xi. *Res:* Microwave devices and circuits; communications theory; electromagnetics; wave propagation. *Mailing Add:* Dept Elec & Computer Eng Ore State Univ Corvallis OR 97331

ENGELDER, THEODORE CARL, b Detroit, Mich, Aug 31, 27; m 52; c 2. NUCLEONICS. *Educ:* Univ Mich, BS, 49; Yale Univ, MS, 50, PhD(physics), 53. *Prof Exp:* Proj engr, Dow Chem Co, 52-56; physicist, Chrysler Corp, 56; group supvr, Atomic Energy Div, Babcock & Wilcox Co, 56-60, chief exp physics sect, Res & Develop Div, 60-67, mgr physics labs, 67-69, asst dir nuclear develop ctr, 69-71, dir, Lynchburg Res Ctr, 71-87; RETIRED. *Mem:* Am Phys Soc; Am Nuclear Soc. *Res:* Nuclear reactor design and development. *Mailing Add:* 2236 Taylor Farm Rd Lynchburg VA 24503

ENGELER, WILLIAM E, b Brooklyn, NY, Nov 13, 28; m 55; c 4. SOLID STATE PHYSICS, SEMICONDUCTOR DEVICES. *Educ:* Polytech Inst Brooklyn, BS, 51; Syracuse Univ, MS, 58, PhD(physics), 61. *Prof Exp:* Physicist, 51-52, physicist semiconductor prod dept, 52-55, PHYSICIST, RES LAB, GEN ELEC CO, 61- *Mem:* Inst Elec & Electronics Engrs; Electrochem Soc; Am Phys Soc. *Res:* Semiconductors; infrared optical properties; electronics; metal-oxide semiconductor devices; device physics; integrated circuit electronics. *Mailing Add:* Gen Elec Res & Develop Ctr Bldg K1 PO Box 8 Schenectady NY 12301

ENGELHARD, ARTHUR WILLIAM, b Dayton, Ohio, Apr 9, 28; c 3. PHYTOPATHOLOGY, BOTANY. *Educ:* Ohio Univ, BS, 50; Yale Univ, MS, 52; Iowa State Univ, PhD(plant path), 55. *Prof Exp:* Asst plant path, Iowa State Univ, 52-55; asst plant pathologist, Ill State Natural Hist Surv, 55-56; res biologist, E I Du Pont de Nemours & Co, Inc, 56-64, sr sales res biologist, 64-65, sr res biologist, 66; assoc prof plant path & assoc plant pathologist, 66-77, PROF & PLANT PATHOLOGIST, RES & EDUC CTR, UNIV FLA, 77- *Concurrent Pos:* Int consult to floral indust, 70- *Mem:* Am Phytopath Soc; Int Soc Plant Path. *Res:* Cause and control of diseases of floral and ornamental crops; foliage and soil fungicides; integrated control of Fusarium wilt; integrated pest management of ornamental crops; disease control of vegetable crops. *Mailing Add:* Gulf Coast Res & Educ Ctr 5007 60th St E Bradenton FL 34203

ENGELHARD, ROBERT J, b Milwaukee, Wis, May 16, 27; m 60; c 2. FORESTRY. *Educ:* Utah State Univ, BS, 50; Univ Denver, MS, 52; Mich State Univ, PhD(forestry), 69. *Prof Exp:* Forester, US Forest Serv, 52-56 & Trees for Tomorrow, Inc, 56-65; from instr to assoc prof, 65-77, PROF FORESTRY, UNIV WIS-STEVENS POINT, 77- *Mem:* Soc Am Foresters. *Res:* Resource economics; forest policy; forest products acquisition and marketing. *Mailing Add:* 4309 Janick Circle N Stevens Point WI 54481

ENGELHARD, VICTOR H, b Louisville, Ky, Feb 10, 51. HISTOCOMPATABILITY. *Educ:* Univ Ill, PhD(biochem), 77. *Prof Exp:* From asst prof to assoc prof, 80-89, PROF MICROBIOL, UNIV VA, 89- *Concurrent Pos:* Exp Immunol Study Sect, 86-90; sect ed, J Immunol, 90-

Mem: Am Asn Immunol. *Res:* Structure of histocompatibility antigens and their recognition by cytotoxic T lymphocytes; signals involved in cytotoxic T lymphocyte activation. *Mailing Add:* Dept Microbiol Jordan Hall Univ Va Box 441 Charlottesville VA 22908

ENGELHARDT, ALBERT GEORGE, b Toronto, Ont, Mar 17, 35; m 60; c 3. COMPUTER SCIENCES GENERAL, COMPUTER ENGINEERING PHYSICS. *Educ:* Univ Toronto, BASc, 58; Univ Ill, MS, 59, PhD(elec eng, math), 61. *Prof Exp:* Asst elec eng, Univ Ill, 60-61; res engr physics & elec eng, Westinghouse Elec Corp, 61-62, sr engr, 62-66, mgr advan plasma concepts, 66-70; sr staff mem, Inst Res Hydro-Que, 70-74; staff mem, Los Alamos Sci Lab, 74-86; PRES, ENFITEK INC, 86- *Concurrent Pos:* Vis prof, Univ Que, 71-77; adj prof elec eng, Tex Tech Univ, 76-; vis indust prof computer sci, Tex Tech Univ, 87-88. *Mem:* Am Phys Soc; Inst Elec & Electronics Engrs. *Res:* Plasma physics; collision phenomena in atomic and molecular gases; laser interaction with matter; electromagnetic wave propagation. *Mailing Add:* Enfitek Inc 2306 Ave Q Lubbock TX 79405

ENGELHARDT, DAVID M, b Austria, June 15, 12; nat US; m 52; c 3. PSYCHIATRY. *Educ:* City Col New York, BS, 32; Univ Vienna, MD, 37; Am Bd Psychiat & Neurol, dipl. *Prof Exp:* Assoc, Long Island Col Med & State Univ NY Col Med, 48-49, assoc prof clin psychiat, 49-50, clin prof psychiat, 50-53, assoc prof, 53-62, actg chmn dept, 58-60, exec officer, Dept, 60-64, dir psychopharmacol treat & res unit, 56-80, prof psychiat, 62-80, EMER PROF PSYCHIAT, STATE UNIV NY HEALTH SCI CTR, 80- *Concurrent Pos:* Asst dir, assoc dir & actg dir, Kings County Psychiat Hosp, Brooklyn, 47-60, vis psychiatrist hosp ctr, 59-, dir clin servs, 60-64; mem comt clin drug eval, NIMH, 61-65, chmn adv coun childhood ment illness, 62-64, mem psychopharmacol study sect, 65-67, mem clin psychopharmacol res rev comt, 67-69, mem clin projs res rev comt, 70-74, chmn, 72-74; mem panel on drugs used in psychiat, Drug Efficacy Study, Nat Res Coun-Nat Acad Sci, 66-68. *Mem:* Fel AAAS; fel Acad Psychoanalysis; fel Am Col Neuropsychopharmacol; fel Am Psychiat Asn; Am Psychopath Asn. *Res:* Psychopharmacology; psychiatric treatment research. *Mailing Add:* 442 Montauk Hwy Southampton NY 11968

ENGELHARDT, DEAN LEE, b Oak Park, Ill, Jan 15, 40; m 70; c 2. CELL BIOLOGY, MOLECULAR BIOLOGY. *Educ:* Amherst Col, BA, 61, MA, 63; Rockefeller Univ, PhD(molecular biol), 67. *Prof Exp:* Asst prof biol, Univ Conn, 69-73; asst prof microbiol, Col Physicians & Surgeons, Columbia Univ, 73-80, assoc prof, 80-81; vpres res, 81-88, SR VPRES, ENZO BIOCHEM INC, 88- *Concurrent Pos:* Am Cancer Soc fels, Salk Inst, La Jolla, Calif, 68 & Albert Einstein Col Med, 68-69; adj assoc prof microbiol, Col Physicians & Surgeons, Columbica Univ; dir, Biotech Ctr, State Univ NY & Univ Conn. *Res:* Growth controls of cultured animal cells; in vitro protein synthesis; mechanisms in cellular aging; immunology. *Mailing Add:* Enzo Biochem Inc 325 Hudson St New York NY 10013

ENGELHARDT, DONALD WAYNE, b Blue Island, Ill, Feb 25, 35; m 58; c 3. PALEOBOTANY. *Educ:* Wabash Col, AB, 57; Ind Univ, MA, 61, PhD(paleobot), 62. *Prof Exp:* Res scientist palynology, 61-67, staff paleontologist of palynology, 67-78, regional paleontologist Houston Region, 78-85, DIV OPER GEOL, AMOCO PROD CO RES CTR, 85- *Mem:* Inst Org Paleobot; Soc Econ Paleontologists & Mineralogists; Am Asn Stratig Palynologists (vpres, 71, pres, 73); Am Asn Petrol Geologists. *Res:* Pleistocene geology and palynology; palynology of Gulf Coast Tertiary; Tertiary and Mesozoic sediments of Alaska. *Mailing Add:* 13975 Hollowgreen Dr Houston TX 77082

ENGELHARDT, EDWARD LOUIS, b Paramaribo, Suriname, Aug 22, 19; m 47; c 1. ORGANIC CHEMISTRY. *Educ:* Haverford Col, BS, 41; Univ Wis, PhD(org chem), 44. *Prof Exp:* Asst org chem, Univ Wis, 42-44; res assoc, Sharp & Dohme, Inc, 44-53, res assoc, Sharp & Dohme Div, Merck & Co, Inc, 53-56, res assoc, 56-60, asst dir, Med Chem Dept, 60-73, dir med chem, 73-75, sr dir med chem, 75-79, distinguished sr scientist, 79-88; CONSULT, 88- *Concurrent Pos:* Vol, Chase Cancer Ctr, 89- *Mem:* Fel AAAS; Am Chem Soc; NY Acad Sci; Royal Soc Chem; Sigma Xi. *Res:* Chemistry of synthetic drugs; cardiovascular diseases; nervous and mental disorders; allergy; cancer. *Mailing Add:* 2302 Terwood Rd Huntingdon Valley PA 19006-5509

ENGELHARDT, HUGO TRISTRAM, b Houston, Tex, Jan 17, 12; m 39; c 2. INTERNAL MEDICINE. *Educ:* Tulane Univ, MD, 37; Am Bd Internal Med, dipl, 45. *Prof Exp:* Asst internal med, Sch Med, Tulane Univ, 40-41, instr, 41-45; from instr to assoc prof clin med, Col Med, Baylor Univ, 45-64; lectr, Sch Med, Tulane Univ, 64-70; CLIN PROF PHYSIOL & MED, MED SCH, UNIV TEX HEALTH SCI CTR, SAN ANTONIO, 67- *Concurrent Pos:* Vis physician, Charity Hosp, New Orleans & chief, White Diabetes Clin, Tulane Serv, 43-45; chief internist, Humble Oil & Refinery Co, 45-64; chief diabetes clin, Jefferson Davis Hosp, Houston, 47-57, assoc physician, 48-57; adj physician, Brady Hosp, San Antonio, 64- *Mem:* Fel Am Col Physicians; AMA. *Res:* Diabetes mellitus. *Mailing Add:* Ctr Ethics Med & Pub Issue Baylor Col Med Houston TX 77030

ENGELHARDT, VAUGHN ARTHUR, b Chicago, Ill, Apr 14, 18; m 47; c 3. ORGANIC CHEMISTRY. *Educ:* Northwestern Univ, BS, 40; Univ Minn, MS, 44, PhD(org chem), 48. *Prof Exp:* Jr res chemist, Com Solvents Corp, Ind, 40, 41; res chemist, Nat Defense Res Comt, Columbia Univ, 42; res chemist, E I Du Pont de Nemours & Co, Inc, 48-57, assoc dir, 57-81, dir, 82-83; RETIRED. *Mem:* Am Chem Soc. *Res:* Synthesis of organic compounds; cyclopropanes; chloracetone; acetylene; cyanocarbons; agrichemicals. *Mailing Add:* 119 Canterbury Dr Wilmington DE 19803

ENGELKE, CHARLES EDWARD, b New York, NY, July 26, 30; m 55; c 3. FOUNDATIONS OF QUANTUM THEORY, ATOMIC PHYSICS. *Educ:* Queens Col, NY, BS, 51; Columbia Univ, MA, 53, PhD(physics), 61. *Prof Exp:* Asst prof, Hunter Col, 61-66, ASSOC PROF, PHYSICS DEPT, LEHMAN COL, CITY UNIV NEW YORK, 66-, ASSOC PROF GRAD FAC, 66- *Mem:* Am Phys Soc; Sigma Xi. *Res:* Interference effects involving photons whose wave functions can be reduced by observations on correlated systems; community total energy systems exploiting interseasonal thermal storage; neutron-proton interaction. *Mailing Add:* Dept Physics & Astron Herbert H Lehman Col Bronx NY 10468

ENGELKE, JOHN LELAND, b Ancon, CZ, Sept 5, 30; m 57, 70, 78; c 1. PHYSICAL CHEMISTRY. *Educ:* Mich Col Mining & Technol, BS, 52, MS, 54; Univ Calif, PhD(chem & phys chem), 59. *Prof Exp:* Resident student assoc, Argonne Nat Lab, 52-54; asst, Univ Calif, 54-55, asst, Radiation Lab, 57-59; solid state chemist, Stanford Res Inst, 59-62; mem staff, Arthur D Little, Inc, 62-68; assoc prof, 68-73, PROF CHEM, SALEM STATE COL, 73-, CHMN, DEPT CHEM & PHYSICS, 86- *Mem:* Am Chem Soc; Sigma Xi. *Res:* High temperature chemistry; thermodynamics; molecular spectroscopy; solid state chemistry; refractory materials; meteorites. *Mailing Add:* Nine West St Marblehead MA 01945

ENGELKE, RAYMOND PIERCE, b New York, NY, June 22, 38. DETONATION PHYSICS, QUANTUM MECHANICS. *Educ:* Long Beach State Col, BS, 60, MA, 62; Univ NMex, PhD(physics), 70. *Prof Exp:* Teacher physics & math, US Peace Corps, Nigeria, 64-66; res assoc ozone distrib, Univ NMex, 70-71; STAFF MEM DETONATION PHYSICS, LOS ALAMOS SCI LAB, 71- *Mem:* Am Phys Soc; Sigma Xi; Am Chem Soc. *Res:* Molecular quantum mechanics; chemistry in shocked materials. *Mailing Add:* LANL MS P952 PO Box 1663 Los Alamos NM 87545

ENGELKING, PAUL CRAIG, b Glendale, Calif, May 11, 48; m 75; c 2. PHYSICAL CHEMISTRY. *Educ:* Calif Inst Technol, BS, 71; Yale Univ, MPhil, 74, PhD(chem), 76. *Prof Exp:* Res assoc chem, Joint Inst Lab Astrophys, Univ Colo & Nat Bur Standards, 75-78; ASST PROF CHEM, UNIV ORE, 78- *Honors & Awards:* Sloan Fel. *Mem:* Am Phys Soc; Am Chem Soc. *Res:* Radicals, ions and reaction intermediates; laser spectroscopy; flowing afterglow; photodetachment of negative ions; vibronic mixing and Jahn-Teller effects in molecules; photoionization. *Mailing Add:* Dept Chem Univ Ore Eugene OR 97403

ENGELMAN, ARTHUR, b New York, NY, Mar 17, 30; m 55; c 3. ATMOSPHERIC PHYSICS, METEOROLOGY. *Educ:* City Col New York, BS, 50; NY Univ, MS, 51. *Prof Exp:* Asst meteorol, NY Univ, 51-52 & 54-55; physicist, Rome Air Develop Ctr, 55-57; sect head radio propagation, Missile Div, Raytheon Co, 57-59; dept mgr atmospheric physics, 59-70, div vpres & tech dir, Technol Div, 70-74, vpres & mgr Aerospace Sci Lab, 74-85, vpres & mgr contracts, Technol Div & Dir, Govt Contracts, GCA Corp, 85-87; dep prog mgr, Comput Technol Assoc, 87-90, PROG MGR, PEACE SHIELD, CTA INC, 90-, DEP DIR SERV DIV, 91- *Mem:* Am Geophys Union. *Res:* Micrometeorology; tropospheric and ionospheric radio propagation and military defense system analysis; upper atmosphere physics; nuclear weapons effects; physics of detonation-atmosphere interaction; software for air defense systems. *Mailing Add:* 15 Woodward Rd Framingham MA 01701

ENGELMAN, DONALD MAX, b Los Angeles, Calif, Jan 25, 41; div; c 2. BIOPHYSICS, BIOCHEMISTRY. *Educ:* Reed Col, BA, 62; Yale Univ, MS, 64, PhD(biophys), 67. *Prof Exp:* From asst prof to assoc prof, 70-78, PROF MOLECULAR BIOPHYS & BIOCHEM, YALE UNIV, 78-, CHMN DEPT, 87- *Concurrent Pos:* Guest biophysicist, Brookhaven Nat Lab, 78-; Bay area res fel, Cardiovasc Res Inst, Med Ctr, Univ Calif, San Francisco, 67-68; NIH fel, King's Col, Univ London, 68-70, Guggenheim fel, Inst Laue-Langevin, Grenoble, France & Lab Molecular Biol, Med Res Coun, Cambridge, Eng, 78-79; ed-in-chief, Ann Rev Biophysics & Biophys Chem; vis prof, dept biochem, Stanford Univ, 84; mem bd dirs, Stryker Corp, 89- *Mem:* Biophys Soc; Am Soc Biol Chem; Am Chem Soc. *Res:* Studies of the molecular organization of biological membranes amd proteins folding using biophysical methods including spectroscopy and scattering. *Mailing Add:* Dept Molecular Biophys & Biochem Yale Univ New Haven CT 06520

ENGELMANN, FRANZ, b Kenzingen, Ger, Dec 19, 28; US citizen; m 54; c 2. ENDOCRINOLOGY. *Educ:* Univ Berne, PhD(zool), 57. *Prof Exp:* Mem staff, Albert Einstein Col Med, 58-60; asst zool, Univ Mainz, 60-63; from asst prof to prof zool, 63-69, PROF BIOL, UNIV CALIF, LOS ANGELES, 69- *Concurrent Pos:* Privatdozent, Univ Mainz, 62. *Honors & Awards:* Alexander von Humboldt Award, 79. *Mem:* Am Soc Zool; Entom Soc Am. *Res:* Endocrinology of invertebrates; reproduction in insects, insect physiology. *Mailing Add:* Dept Biol Univ Calif 2203 Life Sci Los Angeles CA 90024

ENGELMANN, MANFRED DAVID, b Chicago, Ill, June 1, 30; m 54; c 2. ANIMAL ECOLOGY. *Educ:* Northwestern Univ, BS, 53; Univ Ill, MS, 55; Univ Mich, PhD(ecol), 60. *Prof Exp:* Instr zool, Univ Mich, 59-60; from instr to assoc prof natural sci, 60-70, PROF NATURAL SCI, MICH STATE UNIV, 70- *Concurrent Pos:* Res consult, Univ Mich, 61; NSF grants, 63-65. *Mem:* Fel AAAS; Ecol Soc Am; Am Soc Zool; Sigma Xi. *Res:* Ecology of arthropod fauna found in field or grassland; soils and physiology of soil arthropods, particularly oribatid mites. *Mailing Add:* Dept Natural Sci Mich State Univ East Lansing MI 48823

ENGELMANN, REINHART WOLFGANG H, b Berlin, Germany, Aug 21, 34; m 67; c 4. ELECTRONIC PHYSICS. *Educ:* Tech Univ, Munich, dipl-physics, 58, Dr rer nat, 61. *Prof Exp:* Mem tech staff semicon devices, CBS Labs, Stamford, Conn, 61-63, Hewlett-Packard Labs, Palo Alto, 63-66 & AEG-Telefunken, Ulm, Germany, 67-73; mem tech staff & lab proj mgr, Hewlett-Packard Labs, 73-85; SR SCI ADV & GROUP LEADER, OPTOELECTRONIC DEVICES, SIEMENS RES & TECHNOL LABS, 85- *Concurrent Pos:* Instr optoelectronics, Univ Santa Clara, Calif, 84-85. *Honors & Awards:* Study fel, German People, 54. *Mem:* Am Phys Soc; sr mem Inst Elec & Electronics Engrs. *Res:* Microwave and optoelectronic semiconductor device physics, optical waveguides. *Mailing Add:* Ore Grad Inst Sci Technol 19600 NW Von Neumann Dr Beaverton OR 97006

ENGELMANN, RICHARD H(ENRY), b Cincinnati, Ohio, Jan 6, 23; m 47; c 2. ELECTRICAL ENGINEERING. *Educ:* US Naval Acad, BS, 44; Univ Cincinnati, MS, 49. *Prof Exp:* From instr to prof elec eng, Univ Cincinnati, 48-88, head dept, 67-79, assoc dean, 81-86; RETIRED. *Concurrent Pos:* Guest prof, Bengal Eng Col, India, 63-64; consult, Planet Prod Corp, Midland Discount Co, Binns Mach Prod; Avco Corp, Welco Indust, Inc & Cartridge TV, Inc, 54-73; contract rev, NIH, 71-80. *Mem:* Am Soc Eng Educ; Inst Elec & Electronics Engrs; US Naval Inst. *Res:* Electromagnetic devices; feedback control systems. *Mailing Add:* Dept Elec & Computer Eng Mail Location 30 Cincinnati OH 45221-0030

ENGELSTAD, ORVIS P, b Fertile, Minn, Feb 19, 28; m 52; c 3. RESEARCH ADMINISTRATION. *Educ:* Univ Minn, BS, 52, MS, 54; Iowa State Univ, PhD(soils), 60. *Prof Exp:* Res assoc soil fertil, Iowa State Univ, 55-60; agronomist, 60-77, chief, Soils & Fertilizer Res Br, 78-81, asst dir, Div Agr Develop, 81-83, dir, 83-85, DIR, DIV RES, TENN VALLEY AUTHORITY, 85- *Mem:* Fel Am Soc Agron; fel Soil Sci Soc Am. *Res:* Agronomic and economic interpretation of crop responses to fertilizer; study of residual effectiveness of fertilizers; evaluation of fertilizers for temperate and tropical soils. *Mailing Add:* 306 Roxie Dr Florence AL 35633

ENGELSTAD, ROXANN LOUISE, b Stoughton, Wis, Oct 3, 54. VIBRATIONS & MACHINE DESIGN. *Educ:* Univ Wis-Madison, BS, 77, MS, 79, PhD(eng mech), 88. *Prof Exp:* ASST PROF VIBRATIONS & MACH DESIGN, MECH ENG DEPT, UNIV WIS-MADISON, 88- *Concurrent Pos:* Mem, Tech Comt Vibration &Sound, Am Soc Mech Engrs, 89-92 & Adv Comt Mech & Struct systs, NSF, 90-91; NSF presidential young investr award, 90. *Mem:* Am Soc Mech Engrs; Am Nuclear Soc; Am Inst Aeronaut & Astronaut; Am Soc Eng Educ. *Res:* Fusion reaction chamber design; mechanical modeling of x-ray lithography masks; fluid-structure interaction problems in fission core-melt accidents. *Mailing Add:* Dept Mech Eng Univ Wis 1513 University Ave Madison WI 53706

ENGEMANN, JOSEPH GEORGE, b Belding, Mich, Nov 27, 28; m 64; c 3. INVERTEBRATE ZOOLOGY. *Educ:* Aquinas Col, BA, 50; Mich State Univ, MS, 56, PhD(zool), 63. *Prof Exp:* From instr to assoc prof, 60-80, PROF BIOL, WESTERN MICH UNIV, 80- *Concurrent Pos:* Consult freshwater ecol. *Mem:* AAAS; NAm Benthol Soc; Am Micros Soc; Soc Syst Zool; Am Soc Zool. *Res:* Invertebrate zoology; aquatic ecology; evolution; creativity. *Mailing Add:* Dept Biol Western Mich Univ Kalamazoo MI 49008

ENGEN, GLENN FORREST, b Battle Creek, Mich, Apr 26, 25; m 52; c 3. PHYSICS, ELECTRICAL ENGINEERING. *Educ:* Andrews Univ, BA; Univ Colo, PhD, 69. *Prof Exp:* Physicist instrumentation, US Naval Ord Lab, 50-52; electronics engr, Johns Hopkins Univ, 52-54; sr res scientist, Nat Bur Standards, 54-86; CONSULT, 86- *Honors & Awards:* US Dept Com Silver Medal for meritorious serv, 61, Gold Medal, 76. *Mem:* Int Sci Radio Union; fel Inst Elec & Electronics Engrs. *Res:* Microwave circuit theory; microwave measurements; measurements via six-port methods. *Mailing Add:* 333 Sunrise Lane Boulder CO 80302

ENGEN, IVAR A, b Everett, Wash, Nov 13, 32; m 53; c 1. RADIOACTIVE WASTE MANAGEMENT, PROGRAM PROJECT MANAGEMENT. *Educ:* Idaho State Col, BS, 62, Univ Idaho, MS, 70. *Prof Exp:* Psychiat technician, Idaho Dept Health, 54-62; teacher math, Blackfoot High Sch, 62; sci asst, Argonne Nat Lab, 63-72; mathematician, EG & G Idaho, Inc, 73-77, sr engr, 78-79, prog specialist, 80-81, proj engr, 82-87, sr prog specialist, 87-90; ENGR III & CONSULT, WASTREN, INC, 90- *Concurrent Pos:* Resource person, AAAS, 78- *Mem:* Am Nuclear Soc. *Res:* Research and development project manager, safety, environmental, and project planning documents for transport, receipt, storage and disposition of rad-waste and commercial nuclear reactor spent fuel; project engineer for receipt of spent fuel shipments; preparation of proposals; geothermal applications, feasibility and economics. *Mailing Add:* Wastren Inc 477 Shoup Ave Suite 209 Idaho Falls ID 83402

ENGEN, RICHARD LEE, b Irene, SDak, Oct 30, 32; m 55; c 3. PHYSIOLOGY, BIOMEDICAL ENGINEERING. *Educ:* Iowa State Univ, BS, 54, PhD(physiol), 65; Colo State Univ, MS, 58. *Prof Exp:* Res nutritionist, Morris Res Lab, Univ Kans, 58-62; from asst prof to assoc prof, 65-74, PROF PHYSIOL, COL VET MED, IOWA STATE UNIV, 74- *Mem:* Am Physiol Soc; assoc mem Soc Exp Biol & Med; Am Soc Vet Physiol & Pharmacol. *Res:* Pulmonary physiology; cardiopulmonary vascular dynamics. *Mailing Add:* Dept Physiol & Pharmacol Col Vet Med Iowa State Univ Ames IA 50011

ENGER, CARL CHRISTIAN, b Chicago, Ill, Oct 26, 29. BIOMEDICAL ENGINEERING, IMPLANTABLE ELECTRONICS. *Prof Exp:* Consult biomed engr, Biol Powered Pacemakers, 60-67; clin instr biophys, Case Western Reserve Univ, 66-67, sr instr, 67-72; asst prof surg bioeng, Med Univ SC, 72-73; prin investr, heart-powered pacemaker proj, Cleveland Clin Found, 74-77; sr res scientist, Electrophysiol Found, 77-79; consult, Biomed Eng, 80-83; PRES, BIOTELEMETRICS, INC, 84- *Mem:* Soc Biomat; Am Soc Artificial Internal Organs; AAAS; sr mem Instrument Soc Am; Asn Advan Med Instrumentation. *Res:* Principal investigator for the development of a biologically energized cardiac pacemaker utilizing piezoelectric energy converters; research and development of impermeable biomaterials for encapsulated electronic implants; research and development of implantable sub-miniature biotelemetry devices. *Mailing Add:* 6520 Contempo Ln Boca Raton FL 33433-6476

ENGER, MERLIN DUANE, b Williston, NDak, Dec 8, 37; m 58; c 3. BIOCHEMISTRY. *Educ:* NDak State Univ, BS, 59, MS, 61; Univ Wis-Madison, PhD(biochem), 64. *Prof Exp:* Staff mem group H-9, 64-80, GROUP LEADER, GENETICS GROUP, LOS ALAMOS NAT LAB, UNIV CALIF, 80- *Concurrent Pos:* Adj asst prof microbiol, Univ NMex, 74- *Mem:* Fel Am Inst Chemists; fel Am Soc Biol Chemists; Am Chem Soc; AAAS; Am Soc Cell Biol. *Res:* RNA and protein metabolism in cultured animal cells. *Mailing Add:* Dept Zool Rm 339 Sci II Iowa State Univ Ames IA 50011

ENGERMAN, RONALD LESTER, b Chicago, Ill, May 4, 29; m 51; c 5. OPHTHALMOLOGY. *Educ:* Univ Wis, BS, 51, MS, 58, PhD(zool, biochem), 64. *Prof Exp:* Technician, Bjorksten Res Labs, 51; rubber technologist, Army Engr Res & Develop Labs, Va, 53-54; fel, Dept Surg, Univ Wis, 60-61, from instr to asst prof, 61-68, assoc prof, 68-77, PROF OPHTHAL, SCH MED, UNIV WIS-MADISON, 77- *Honors & Awards:* Fight for Sight Citation, Nat Coun Combat Blindness & Asn Res Ophthal, 64. *Mem:* Asn Res Vision & Ophthal; AAAS; Am Diabetes Asn; Soc Exp Biol & Med; Europ Asn Study Diabetes. *Res:* Diabetic retinopathy and microangiopathy; biology of the microvasculature. *Mailing Add:* Dept Ophthal Univ Wis Sch Med Madison WI 53706

ENGERT, MARTIN, b Chicago, Ill, Nov 6, 38; m 71. MATHEMATICAL ANALYSIS. *Educ:* Carleton Col, BA, 60; Stanford Univ, MS, 62, PhD(math), 65. *Prof Exp:* Asst prof math, Univ NC, Chapel Hill, 65-67; staff mem, Aarhus Univ, 67-69; ASSOC PROF MATH, UNIV WIS-WHITEWATER, 69- *Mailing Add:* Dept Math Univ Wis 800 W Main St Whitewater WI 53190

ENGH, HELMER A, JR, b Litchfield, Ill, May 21, 35; m 60. GENETICS. *Educ:* Wash Univ, AB, 57; Southern Ill Univ, MS, 59; Univ Md, PhD(poultry sci), 66. *Prof Exp:* Asst prof biol, Mankato State Univ, 66-70, assoc prof, 70-75, prof biol sci, 75-85. *Mem:* Genetics Soc Am; Am Genetic Asn; Soc Study Evolution; Nat Asn Adv Health Proffessions; Sigma Xi. *Res:* Biochemical genetics of serum enzymes. *Mailing Add:* Dept Biol Sci Mankato State Univ Box 34 Mankato MN 56001

ENGHETA, NADER, b Tehran, Iran, Oct 8, 55; US citizen; m 83; c 1. ELECTROMAGNETICS, OPTICS. *Educ:* Univ Tehran, Iran, BS, 78; Calif Inst Technol, MS, 79, PhD(elec eng & physics), 82. *Prof Exp:* Grad res asst elec eng, Calif Inst Technol, 79-82, postdoctoral res fel, 82-83; sr res scientist, Kaman Sci Corp, 83-87; asst prof, 87-90, ASSOC PROF ELEC ENG, UNIV PA, 90- *Concurrent Pos:* Vis assoc, Calif Inst Technol, 83-87; vis lectr, Univ Calif, Los Angeles, 86; NSF presidential young investr, 89. *Mem:* Sr mem Inst Elec & Electronics Engrs; Am Phys Soc; Optical Soc Am; AAAS. *Res:* Applied electromagnetics and microwave; optics; chiral materials and chiral electrodynamics; radar absorbing materials and chiral shields; waveguides; integrated optics; printed-circuit antennas; nonlinear guided-wave phenomena; optical characteristics of visual systems; wave propagation in complex and chiral media. *Mailing Add:* Moore Sch Elec Eng 6314 Univ Pa 200 S 33rd St Philadelphia PA 19104-6390

ENGIBOUS, JAMES CHARLES, b Norway, Mich, Aug 12, 23; m 47; c 4. SOIL BIOCHEMISTRY. *Educ:* Northern Mich Col, BS, 47; Ore State Col, MS, 50; Ohio State Univ, PhD(biochem), 52. *Prof Exp:* Res asst, Ore State Col, 48-50 & Ohio State Univ, 50-52; soil biochemist, Monsanto Chem Co, 52-54; group leader agron res, Int Minerals & Chem Corp, 54-55, supvr soils & plant nutrit res, 55-59, mgr agr prod res, 59-62, agr serv, Mat Dept, 62-63, Tech Serv Dept, 63-70; CHMN, DEPT AGRON & SOILS, WASH STATE UNIV, 71- *Mem:* Fel Am Inst Chem; fel Am Soc Agron; fel Soil Sci Soc Am; Soil Cons Soc Am. *Res:* Soil chemistry, bacteriology and physics. *Mailing Add:* 103 NW Lancer Lane Pullman WA 99163

ENGIN, ALI ERKAN, b Samsun, Turkey, Feb 23, 43; US citizen; m 71, 83; c 4. MECHANICS, BIOMECHANICS. *Educ:* Mich State Univ, BS, 65, Univ Mich, MS, 66, PhD(eng mech), 68. *Prof Exp:* Asst prof eng sci, Mid East Tech Univ, Ankara, 69-70; assoc res engr biomech, Hwy Safety Res Inst, Univ Mich, 70-71; from asst prof to assoc prof, 71-77, PROF MECH & DIR BIOMECH LAB, OHIO STATE UNIV, 77- *Concurrent Pos:* Consult, UN Develop Prog, 77, 82 & UNESCO, 79; expert witness, personal injury & prod liability. *Honors & Awards:* Ervin G Bailey Award, 74; Borelli Award, Am Soc Biomech, 85. *Mem:* Am Acad Mech; Am Soc Mech Engrs; Am Soc Biomech; Int Soc Biomech. *Res:* Mechanics, especially fluid-solid interaction problems and analysis of shells; biomechanics, particularly head injury, biological material properties, experimental and theoretical biomechanics of the major human joints and biodynamic modeling of various parts of the human body; published over 120 scientific papers. *Mailing Add:* Dept Eng Mech 155 W Woodruff Ave Columbus OH 43210

ENGLAND, ALAN COULTER, b Belleville, Ill, Mar 1, 32; c 3. PHYSICS. *Educ:* Univ Ill, BS, 54; Univ Rochester, PhD(physics), 61. *Prof Exp:* PHYSICIST, FUSION ENERGY DIV, OAK RIDGE NAT LAB, 60- *Concurrent Pos:* Vis scientist, Inst Plasma Physics, Munich, Ger, 67-68. *Mem:* Sigma Xi; Am Phys Soc. *Res:* High-energy nuclear physics; tokamak plasmas; hot-electron plasmas. *Mailing Add:* Oak Ridge Nat Lab PO Box 2009 Oak Ridge TN 37831

ENGLAND, ANTHONY W, b Indianapolis, Ind, May 15, 42; m 62; c 2. GEOPHYSICS, ASTROGEOLOGY. *Educ:* Mass Inst Technol, SB & SM, 65, PhD(geophys), 70. *Prof Exp:* Scientist-Astronaut, Manned Spacecraft Ctr, NASA, 67-72; geophysicist, US Geol Surv, 72-79; astronaut, Manned Spacecraft Ctr, NASA, 79-88; PROF ELEC ENG, UNIV MICH, ANN ARBOR, 88- *Honors & Awards:* Outstanding Sci Achievement Medal, NASA, 73 & Space Flight Medal, 85; US Antarctic Medal, 79; Flight Achievement Award, Am Inst Aeronaut & Astronaut, 85; Except Achievement Medal, NASA, 88. *Mem:* Am Geophys Union; Soc Explor Geophys; Sigma Xi. *Res:* Remote sensing geophysics. *Mailing Add:* Univ Mich Col Eng 3229 EECS Bldg Ann Arbor MI 48109-2122

ENGLAND, BARRY GRANT, b Tooele, Utah, Oct 13, 40; m 66; c 3. ENDOCRINOLOGY, REPRODUCTIVE PHYSIOLOGY. *Educ:* Utah State Univ, BS, 65, MS, 67; Univ Wis, PhD(endocrinol, reprod physiol), 71. *Prof Exp:* Instr, 73-74, asst prof, 74-78, ASSOC PROF BIOL REPRODUCTION, UNIV MICH, ANN ARBOR, 78- *Concurrent Pos:* Ford Found, Univ Mich, Ann Arbor, 71-73. *Mem:* AAAS; Soc Study Reproduction; Am Soc Animal Sci. *Res:* Study of reproductive endocrinology in the mammalian female; development of new radio-ligand assay techniques. *Mailing Add:* Dept Path Univ Mich Ann Arbor MI 48109

ENGLAND, DAVID CHARLES, b Myrtle, Mo, Jan 4, 22; m 46, 73; c 5. ANIMAL BREEDING. *Educ:* Wash State Univ, BS, 49; Univ Minn, MS, 50, PhD, 52. *Prof Exp:* Res asst, Univ Minn, 49-51, res fel, 51-52, asst prof animal husb, 52-55; from asst prof to assoc prof, 55-69, PROF ANIMAL HUSB, ORE STATE UNIV, 69- *Concurrent Pos:* Mem coop state res serv, USDA, 68-69; consult, Ore Regional Primate Ctr, 75-, Livestock Indust, 72-, Kroc Found Med Res, 76-, Battelle Labs, 78-, Elec Power Res Inst, 81- & Noti Ranches, 81- *Mem:* Am Soc Animal Sci; Sigma Xi; Am Soc Animal Sci (secy-treas, 77-80, pres-elect, 80-81, pres, 81-82). *Res:* Genetics; swine production. *Mailing Add:* Dept Animal Sci Ore State Univ Corvallis OR 97331

ENGLAND, JAMES DONALD, b Lyles, Tenn, Feb 4, 37; m 61; c 2. ORGANIC BIOCHEMISTRY. *Educ:* Austin Peay Univ, BS, 58; Univ Ark, MS, 60; Univ Miss, PhD(med chem), 66. *Prof Exp:* from asst prof to assoc prof, 60-71, DISTINGUISHED PROF CHEM, HARDING UNIV, 71- *Concurrent Pos:* Fel, Miss Heart Asn, 63. *Mem:* Am Chem Soc; Sigma Xi. *Mailing Add:* Dept Med Chem Box 903 Harding Univ Searcy AR 72143

ENGLAND, JAMES WALTON, b Newton, Kans, July 20, 38; m 61; c 1. MATHEMATICS. *Educ:* Kans State Col Pittsburg, AB, 60; Univ Mo, MA, 61, PhD(math), 64. *Prof Exp:* Asst prof math, Univ Va, 64-68; assoc prof math, Swarthmore Col, 69-74, prof, 74-81, chmn, Dept Math, 77-81; dean fac & acad vpres, Occidental Col, 81-; PROVOST, SWARTHMORE COL. *Mem:* Am Math Soc; Math Asn Am; Sigma Xi. *Res:* Topological dynamics, ergodic theory. *Mailing Add:* Provost Off Swarthmore Col Swarthmore PA 19081

ENGLAND, RICHARD JAY, b Springfield, Ill, Aug 5, 26; m 51; c 3. POLYMER CHEMISTRY, ANALYTICAL CHEMISTRY. *Educ:* Bradley Univ, BS, 51. *Prof Exp:* Lieutenant, Res & Develop Command, Ohio, US AF, 51-53; chemist, Explosives Dept, NJ, 53-55, process chemist, 55-56, res chemist, WVa, 56-57, Textile Fibers Dept, Va, 57 & NC, 57-59, Film Dept, Ohio, 59-82, SR RES CHEMIST, POLYMER PROD DEPT, E I DU PONT DE NEMOURS & CO, 82- *Res:* Polyamide and polyester intermediates; high explosives; polyester fibers and films; supervision; petroleum products. *Mailing Add:* Polymer Prod Dept Circleville Res Lab E I Du Pont de Nemours & Co Inc PO Box 89 Circleville OH 43113

ENGLAND, SANDRA J, b N Tonawanda, NY, June 23, 49. RESPIRATORY PHYSIOLOGY. *Educ:* Dartmouth Col, PhD(physiol), 80. *Prof Exp:* Asst prof, 81-87, ASSOC PROF PEDIAT, HOSP SICK CHILDREN, 87- *Mem:* Am Physiol Soc. *Mailing Add:* Hosp Sick Children 555 University Ave Toronto ON M5G 1X8

ENGLAND, TALMADGE RAY, b Bonham, Tex, Dec 22, 29; m 49; c 4. NUCLEAR DATA, REACTOR THEORY. *Educ:* Lincoln Mem Univ, BS, 56; Univ Pittsburgh, MS, 62; Univ Wis, PhD(nuclear eng), 69. *Prof Exp:* Jr scientist, Bettis Atomic Power, Westinghouse, 50-57, sr scientist, 57-72; STAFF MGR, THEORET DIV, LOS ALAMOS NAT LAB, 72- *Concurrent Pos:* Vis prof, Univ Wis, 66; chmn, Yields & Decay Data, US Eval Nuclear Data, Brooklyn Nat Lab, 73-, Yields Standard Comt, Am Nuclear Soc, 80- & Decay Delayed Neutron Comt, 89-; mem, Decay Heat Data Task Force, Nuclear Energy Agency, 87-90 & Coord Res Comn Yields, Int Atomic Energy Agency, Vienna, 88-91 & Delayed Neutron Comn, NEANDS/NEACRP, Paris, 89- *Mem:* Am Nuclear Soc; Inst Elec & Electronics Engrs; AAAS. *Res:* Author of 149 publications on nuclear data and reactor theory. *Mailing Add:* 613 Meadow Lane Los Alamos NM 87544

ENGLAND, WALTER BERNARD, b Hallettsville, Tex, Aug 23, 42. COMPUTATIONAL CHEMISTRY, PHYSICAL CHEMISTRY. *Educ:* Purdue Univ, BS, 65; Iowa State Univ, PhD(phys chem), 73. *Prof Exp:* Fel physics, Colo State Univ, 73-74; res assoc chem, Argonne Nat Lab, 74-78; ASST PROF CHEM, UNIV WIS, MILWAUKEE, 78- *Mem:* Am Phys Soc; Am Chem Soc; Am Vacuum Soc; NY Acad Sci. *Res:* Adaptation of ordinary many-body field-theoretic methods to chemical bonds and surfaces; application of computational quantum chemical methods to atoms, molecules and surfaces. *Mailing Add:* Dept Chem Lab Surface Studies Univ Wis PO Box 413 Milwaukee WI 53201

ENGLANDE, ANDREW JOSEPH, b New Orleans, La, July 31, 44; m 66; c 3. ENVIRONMENTAL SCIENCES. *Educ:* Tulane Univ, BS, 67, MS, 69; Vanderbilt Univ, PhD(environ eng), 74. *Prof Exp:* Asst prof, 72-76, ASSOC PROF ENVIRON HEALTH SCI & LAB DIR RIVERSIDE RES LABS, SCH PUB HEALTH, TULANE UNIV, 74- *Concurrent Pos:* Assoc environ div, Gulf SRes Inst, 74- *Honors & Awards:* Delta Omega Key, Pub Health Nat Honor Soc, 74. *Mem:* Int Asn Water Pollution Res; Water Pollution Control Fedn; Am Water Works Asn; Am Chem Soc; Sigma Xi. *Res:* Biological and physical chemical wastewater treatment methods for industrial wastes; general water quality management techniques; and trace contaminant accumulation and translocation. *Mailing Add:* Dept Environ Health Tulane Univ Med Sch 1430 Tulane Ave New Orleans LA 70118

ENGLANDER, HAROLD ROBERT, b New York, NY, Dec 11, 23; m 49; c 1. DENTISTRY. *Educ:* City Col New York, BS, 45; Columbia Univ, DDS, 48, MPH, 51; Am Bd Dent Pub Health, dipl. *Prof Exp:* Dir, US Naval Dent Inst, Great Lakes, Ill, 53-58; prof dent, Dent Col, Univ Ill, 59-62; mem staff, Nat Inst Dent Res, 62-67, dent dir & chief field trials, Epidemiol & Biomet Br, 67-75; mem fac, Univ Tex Health Sci Ctr, 75; prof dent pub health, Sch Pub Health, Houston, 78; MEM STAFF, DENT SCH, UNIV TEX HEALTH SCI CTR, 79- *Concurrent Pos:* Vis prof, Howard Univ; vis lectr, Johns Hopkins Univ Sch Pub Health & Univ Md; past consult prev dent, Surgeon Gen, US Army. *Mem:* AAAS; Am Dent Asn; fel Am Pub Health Asn; fel Am Col Dent; Sigma Xi. *Res:* Epidemiology; caries research in hamsters; clinical trials of anticaries agents; studies of peridontal disease; studies on waterborne fluoride. *Mailing Add:* 11502 Whisper Bluff San Antonio TX 78230

ENGLANDER, SOL WALTER, b Baltimore, Md, Jan 25, 30; m 54; c 3. BIOPHYSICS. *Educ:* Univ Md, BS, 51; Univ Pittsburgh, MS, 54, PhD(biophys), 59. *Prof Exp:* Biophysicist, NIH, 59-61; from instr to asst prof biochem, Dartmouth Med Sch, 61-67; assoc prof, 67-74, assoc chmn dept, 78-82, PROF BIOCHEM & BIOPHYS, SCH MED, UNIV PA, 74-, JACOB GERSHON-COHEN PROF MED SCI, 90- *Concurrent Pos:* Am Cancer Soc fel, 61-63; vis scientist, Danish AEC Res Estab, 65. *Res:* Physical biochemistry; protein and nucleic acid structure and function; hydrogen exchange. *Mailing Add:* Dept Biochem & Biophys Univ Pa Sch Med Philadelphia PA 19104

ENGLARD, SASHA, b Antwerp, Belg, June 28, 29; nat US; m 51; c 3. BIOCHEMISTRY. *Educ:* City Col New York, BS, 49; Western Reserve Univ, PhD, 53. *Prof Exp:* Assoc prof, 61-68, asst dean admis, 69-72, PROF BIOCHEM, ALBERT EINSTEIN COL MED, YESHIVA UNIV, 68- *Concurrent Pos:* NIH fel & vis prof, Hebrew Univ, Israel, 61-62 & 69; Am Heart Asn fel, Pratt Inst, Johns Hopkins Univ, 53-55. *Mem:* Am Soc Biol Chem; Brit Biochem Soc. *Res:* Mechanism of enzyme reaction; stereospecificity of enzymatically catalyzed reactions; carbohydrate metabolism; enzymology of carnitine biosynthetic and degradative pathways. *Mailing Add:* Dept Biochem Albert Einstein Col Med 1300 Morris Park Ave Bronx NY 10461

ENGLE, DAMON LAWSON, b Troy, WVa, June 22, 19; m 43; c 4. ANALYTICAL CHEMISTRY. *Educ:* Marshall Univ, BS, 41. *Prof Exp:* Analytical chemist, Chem Div, 42-51, group leader synthetic fibers & polymers, 52-59, asst dir res & develop, 59-66, dir, 66-69, asst plant mgr, Chem & Plastics Div, 69-72, plant mgr, Union Carbide Corp, 72-73; EXEC DIR, HOUSTON BUS ROUNDTABLE, 83- *Mem:* Am Chem Soc; Soc Rheology; Am Soc Eng Educ. *Res:* Synthetic polymer technology. *Mailing Add:* 2718 Frostwood Circle Dickinson TX 77539

ENGLE, IRENE MAY, b Harrisburg, Pa. THEORETICAL PHYSICS. *Educ:* Pa State Univ, BS, 63, MS, 66, PhD(physics), 70. *Prof Exp:* Instr physics, Ripon Col, 65-66; asst prof physics, Juniata Col, 69-79, chmn dept, 75-77; ASSOC PROF, US NAVAL ACAD, 79- *Concurrent Pos:* Vis physicist, Argonne Nat Lab, 70; adj fac mem & sr res assoc, Univ Kans, 77-79. *Mem:* Am Phys Soc; Am Geophys Union; Am Asn Physics Teachers; Astron Soc Pac. *Res:* Electromagnetic structure of the nucleon; symmetries and symmetry breaking; planetary magnetospheric physics; electronic properties of metals, especially thermionic-field emission phenomena; high energy charged particle kinematics in the solar system; cometary atmospheres and imaging. *Mailing Add:* Physics Dept US Naval Acad Annapolis MD 21402-5026

ENGLE, JESSIE ANN, b Chicago, Ill, Sept 17, 18; m 51; c 3. ANALYSIS, FUNCTIONAL ANALYSIS. *Educ:* Bennington Col, BA, 40; Ohio State Univ, MS, 64, PhD(math), 71. *Prof Exp:* Teacher music, Black Mountain Col, 40-42; engr, Columbia Broadcasting Corp, 42-44; res assoc, Radio Res Lab, Harvard Univ, 44-46; engr, Airborne Instruments Lab, 46-68; recording engr, Juilliard Sch Music, 49-53; violist, Columbus Symphony, 58-62; vis asst prof, 71-72, from asst prof to assoc prof, 72-85, EMER ASSOC PROF MATH, OHIO STATE UNIV,85- *Concurrent Pos:* Instr math, ITM/Midwest Univs Consortium Int Activ, Malaysia, 86. *Mem:* Am Math Soc; Math Asn Am; Asn Women Math; Nat Coun Teachers Math. *Res:* Generalization of hear measure; design and testing of microwave antennas and components for aircraft. *Mailing Add:* 1112 E 17th St Tulsa OK 74120

ENGLE, JOHN FRANKLIN, b Shoshone, Idaho, Aug 15, 21; m 44; c 5. ELECTRICAL ENGINEERING. *Educ:* Ore State Univ, BS, 47, MS, 51. *Prof Exp:* From instr to assoc prof, 47-69, PROF ELEC ENG, ORE STATE UNIV, 69- *Concurrent Pos:* Consult, hydroelec design br, N Pac Div, US Army Corps Engrs, 62- *Mem:* Asn Comput Mach; Am Soc Eng Educ; Inst Elec & Electronics Engrs. *Res:* Electric power systems; digital simulation; component modeling; on-line digital control. *Mailing Add:* 1015 NW Fernwood Pl Corvallis OR 97330

ENGLE, MARY ALLEN ENGLISH, b Madill, Okla, Jan 26, 22; m 45; c 2. PEDIATRICS, CARDIOLOGY. *Educ:* Baylor Univ, AB, 42; Johns Hopkins Univ, MD, 45; Am Bd Pediat, dipl & cert cardiol. *Prof Exp:* Intern pediat, Johns Hopkins Hosp, 45-46, asst dir outpatient dept, 46-47, asst physician, Cardiac Clin, 47-48, instr Sch Med, 46-48; asst pediat, 48-49, asst pharmacol, 49-50, from instr to assoc prof pediat, 50-69, PROF PEDIAT, MED COL, CORNELL UNIV, 69-; DIR PEDIAT CARDIOL, NEW YORK HOSP, 62-, STAVROS S NIARCHOS PROF PEDIAT CARDIOL, 79- *Concurrent Pos:* Asst resident, Sydenham Hosp, Md, 46; from asst resident to sr asst resident, New York Hosp, 48-49, from asst attend pediatrician to assoc attend pediatrician, 52-62, attend pediatrician, 62-; med dir, Inst Care of Premature Infants, 52-55; mem cardiac surg adv comt, New York City Dept Health; mem, NY State Cardiac Adv Comt; assoc ed, J Cardiol. *Honors & Awards:* Spence-Chapin Award, 58; Pres Panel Heart Dis Award, 72; Cummings Humanitarian Award, 73; 76; Award Merit, Am Heart Asn, 75; Helen B Taussig Award, 76; Philoptochos Award of Merit, 78. *Mem:* Am Acad Pediat; Soc Pediat Res; Am Heart Asn; Am Pediat Soc; Am Col Cardiol. *Res:* Pediatric cardiology, especially congenital malformations of heart and great vessels. *Mailing Add:* Dept Pediat New York Hosp Cornell Univ Med Col 1300 York Ave New York NY 10021

ENGLE, MICHAEL JEAN, CELL BIOLOGY, ENDOCRINOLOGY. *Educ:* St Louis Univ, PhD(biochem), 76. *Prof Exp:* ASSOC SCIENTIST, DEPT PEDIAT, UNIV WIS, 78- *Mailing Add:* Dept Gastroenterol Wash Univ 660 S Euclid Ave Box 8124 St Louis MO 63110

ENGLE, PAUL RANDAL, b Newton, Iowa, Oct 16, 19; m 45; c 1. ASTRONOMY, PHYSICS. *Educ:* Pan Am Col, BA, 58; Mich State Univ, MAT, 72. *Prof Exp:* Jr engr & instr astron, Phys Sci Lab, NMex State Univ, 48-51; flight instr & asst to dir, Calif Eastern Aviation, Tex, 51-58; dir observ, Pan Am Univ, 56-74, dir, Geod Satellite Prog, 66-74; dir planetarium, 75-80,

ASST PROF PLANETARIUM EDUC ASTRON, UNIV ARK, LITTLE ROCK, 75- *Concurrent Pos:* Prin investr, NSF grants, 62-65; dir, NASA-Goddard Minitrack Optical Tracking Sta, 66- *Mem:* Asn Lunar & Planetary Observers; Am Astronaut Soc; Am Asn Physics Teachers; Pan Am Soc Astrophys Res (pres, 61-); Int Soc Planetarium Educr (pres, 71 & 72). *Res:* Astronomical instrumentation; development of course programs in astro-science and astronomy for college and high school levels. *Mailing Add:* Dept Astron Univ Ark 33rd & Univ Ave Little Rock AR 72204

ENGLE, RALPH LANDIS, JR, b Philadelphia, Pa, June 11, 20; m 45; c 2. MEDICINE, HEMATOLOGY. *Educ:* Univ Fla, BS, 42; Johns Hopkins Univ, MD, 45; Am Bd Internal Med, dipl. *Prof Exp:* Intern path, NY Hosp, 45-46 & internal med, 48-49, asst resident, 49-51, resident hemat, 50-51; asst, Cornell Univ, 49-51, from asst prof to assoc prof, 52-69, prof med, 69-90, prof pub health, 73-90, assoc dir res & sponsored progs, 75-90, EMER PROF MED & PUB HEALTH, CORNELL UNIV, 90- *Concurrent Pos:* From asst attend physician to assoc attend physician, NY Hosp, 54-69, attend physician, 69-90, chief hemat div, Dept Med, 61-67, dir dept med systs & comput serv, 68-72, emer staff, 90-; chief div med systs & comput sci, Dept Med, NY Hosp-Cornell Univ, 67-74; mem, Cancer Clin Invest Rev Comt, Nat Cancer Inst, 68-72 & Comt Sci & Tech Commun, Nat Acad Sci-Nat Acad Eng, 67-70; res fel, Med Col, Cornell Univ, 50; Am Cancer Soc fel, Wash Univ, 51-52; Markle scholar, 52-57. *Mem:* AAAS; Am Soc Hemat; Am Fedn Clin Res; Soc Exp Biol & Med; Sigma Xi; Am Clin Climat Asn. *Res:* Hematology; pathology; computer applications to medicine; public health. *Mailing Add:* One Country Club Lane New Rochelle NY 10803

ENGLE, ROBERT RUFUS, b Sullivan Co, Ind, Jan 29, 30; m 54; c 6. ORGANIC CHEMISTRY, MEDICINAL CHEMISTRY. *Educ:* DePauw Univ, BA, 51; Wayne State Univ, MS, 53, PhD(chem), 58. *Prof Exp:* Teaching asst, Wayne State Univ, 51-52, 55; sr res chemist, Riker Labs, Inc, 58-65; head chem & drug procurement sect, Nat Cancer Inst, 65-76, head chem resources sect, 76-84; DEP DIR DIV EXP THERAPEUT & CHIEF DEPT MED CHEM, WALTER REED ARMY INST RES, US ARMY MED RES & DEVELOP COMMAND, US ARMY, 85- *Mem:* Am Chem Soc. *Res:* Synthesis; chemistry of natural products; structure determination; pharmaceuticals; cancer chemotherapy; chemical defense, antiparasitic, antiradiation; drug development. *Mailing Add:* 8305 Tuckerman Lane Potomac MD 20854

ENGLE, THOMAS WILLIAM, b Southbend, Ind, June 5, 51; m 78; c 2. AGRICULTURAL CHEMISTRY. *Educ:* Rose-Hulman Inst Tech, BS, 73; State Univ NY Buffalo, PhD(med chem), 79. *Prof Exp:* Technician, Pfizer Co, 72, Anaconda Aluminum Co, 73; fel, Catholic Univ, 78-81; MEM STAFF, UNION CARBIDE AGR PRODS CO, INC, 81- *Concurrent Pos:* Lectr, Univ DC, 81. *Mem:* Am Chem Soc. *Res:* Development of systems to measure physical chemical parameters used in quantitative structure activity relationships; investigation of such relationships for insecticides and herbicides. *Mailing Add:* 104 Alnick Ct Durham NC 27712-1106

ENGLEHART, EDWIN THOMAS, JR, b Johnstown, Pa, Aug 7, 21; m 51; c 1. CHEMICAL METALLURGY. *Educ:* Pa State Univ, BS, 43. *Prof Exp:* Res engr, Corrosion Res Labs, Aluminum Co Am, 43-59, sect head chem metall div, 59-77, sect head alloy technol div, Alcoa Labs, 77-83; RETIRED. *Mem:* Sigma Xi. *Res:* Solution of fundamental and practical corrosion problems dealing with aluminum, its alloys and other metals. *Mailing Add:* 450 Dakota Dr Lower Burrell PA 15068

ENGLEHART, RICHARD W(ILSON), b Sept 2, 38; m 64; c 2. NUCLEAR ENGINEERING, ENVIRONMENTAL SCIENCES. *Educ:* Carnegie-Mellon Univ, BS, 60; Pa State Univ, MS, 63, PhD(nuclear eng), 69. *Prof Exp:* Proj engr, US Army reactors group, AEC, 64-66; asst prof nuclear eng, Univ Fla, 68-73; mgr Radiol prog dept, Nus Corp, 73-79, sr exec consult & mgr Defense & Aerospace prog, 80-90; DIR, OFF ENVIRON, NEW PROD REACTORS, US DEPT ENERGY, 90- *Concurrent Pos:* Reactor supvr, Univ Fla, 69-71; Am Soc Eng Educ-Ford Found resident fel prog consult engr, Gilbert Assocs, Inc, 71-72; Nat Sci Found Coop Fels, 61-63; Atomic Energy Com Reactor Utilization Fels, 66-67. *Mem:* Am Nuclear Soc; Sigma Xi. *Res:* Radiological analyses and environmental impact assessments related to the nuclear fuel cycle; safety and environmental assessments for applications of nuclear power in outer space. *Mailing Add:* 13830 Dufief Mill Rd Gaithersburg MD 20878

ENGLEMAN, CHRISTIAN L, MICROCOMPUTER SYSTEM INTEGRATION & MARKETING. *Educ:* US Naval Acad, BS, 30; Harvard Univ, MS, 39. *Prof Exp:* Exec Officer & Chief Tech Officer, Submarine Forces, Atlantic Fleet, US Navy, 39, Cmndg officer, 40, staff squadron comdr & asst dir Underwater Sound Sch, Destroyer Squadron Fifty, Pac Fleet, 41-42, off-in-charge, elec installation & maintenance br, Bureau of Ships, Navy Dept, 42-44, comdr & elec officer, Pac Ocean Areas, US Navy, 44-45; elec coord officer, Joint Task Force One (Oper Crossroads), Dept Defense, 45-46; organizer & dir, Scientific Resurvey Bikini Atoll, Dept Defense, 47-48; staff, Joint Task Force Seven, Dept Defense, 48-49; staff, comt electronics, Res & Develop Bd, off Secy Defense, Pentagon Bldg, Washington, DC, 49-50; dir, res & develop, electronic equip for naval vessels & shore facilities, 50-51; dep dir, US Navy Shipbuilding Scheduling Agency, Philadelphia, 51-52; pres, Engleman & Co Inc, 53-60; vpres, Corp Econ & Industr Res, 60-61; exec vpres, Wachtel & Co Inc, 62-64; dir & treas, Data Dynamics Inc, 63-78; pres & chmn, Metro Lists, Ltd, 79-84; PRES & CHMN, US COMPUT CTR LTD, 85- *Concurrent Pos:* Pres, Am Eng Asn, Inc, 54-60. *Mem:* Fel Inst Elec & Electronics Engrs (founder & chmn, Inst Elec & Electronics Engrs Prof Group Commun Systs, 60, founder, Inst Elec & Electronics Engrs Aerospace & Electronics Systs Prof Group, 64). *Res:* Designed equipment to locate lost ships at sea, 23-24, suggested the name "Sonar" be used to identify underwater sound techniques used in anti-submarine warfare, 42-44; established electronic programs and communications facilities for testing nuclear devices at Bikini Atoll and directed study of the flora and fauna as it related to the nuclear fallout, 45-46; developed underwater television equipment, 47-48; produced information on the structure and history of the atolls, discovered a new species of ocean fish, named Synodus Englemani by Smithsonian Institute, 47-48; developed transistor, the basic element of advanced computer chips and development of the radar (microwave) range, 50-51. *Mailing Add:* USCC Inc MP O 82R Skamania Mines Rd Washougal WA 98671

ENGLEMAN, EPHRAIM PHILIP, b San Jose, Calif, Mar 24, 11; m 41; c 3. MEDICINE. *Educ:* Stanford Univ, AB, 33; Columbia Univ, MD, 37. *Prof Exp:* From asst clin prof to assoc clin prof, 48-64, chief rheumatic dis group, Med Ctr, 58-77, PROF MED, SCH MED, UNIV CALIF, SAN FRANCISCO, 64-, DIR, ROSALIND RUSSELL ARTHRITIS CTR, MED CTR, 77- *Concurrent Pos:* Consult, Regional Off, US Vet Admin, San Francisco, 50-77; San Francisco Army Hosp, 60-77; pres-elect, Int League Against Rheumatism, 61-81, pres, 81-85; chmn, Nat Comn Arthritis, 75-77; mem, Nat Arthritis Adv Bd, 77-80. *Mem:* Am Rheumatism Asn (pres, 62-63); Am Col Physicians; Am Fedn Clin Res. *Res:* Rheumatic diseases. *Mailing Add:* Univ Calif 400 Parnassus Ave San Francisco CA 94122

ENGLEMAN, KARL, b New York, NY, June 23, 33; m 56; c 3. MEDICINE, PHARMACOLOGY. *Educ:* Rutgers Univ, BS, 55; Harvard Univ, MD, 59; Univ Pa, MS, 71. *Prof Exp:* Intern & asst resident med, Mass Gen Hosp, Boston, 59-61, sr resident, 64; sr investr, Nat Heart Inst, NIH, Bethesda, Md, 65-70; ASSOC PROF MED & PHARMACOL, SCH MED, UNIV PA, 70- *Concurrent Pos:* Clin assoc, Nat Heart & Lung Inst, NIH, Bethesda, Md, 61-66; consult, Vet Admin Hosp, Philadelphia, 70- & Children's Hosp, Philadelphia, 72-; dir, Clin Res Ctr, Hosp Univ Pa, 72-; mem, Hypertension Res Coun, Am Heart Asn; mem adv panel, US Pharmacopia & Nat Formulary. *Mem:* Fel Am Col Physicians; Am Heart Asn; Am Soc Pharmacol & Exp Therapeut. *Res:* Hypertension; catecholamines and clinical pharmacology. *Mailing Add:* Hosp Univ Pa 3400 Spruce St Three Silverstein Suite A Philadelphia PA 19104

ENGLEMAN, ROLF, JR, b Norman, Okla, Mar 16, 34; m 56; c 5. ATOMIC & MOLECULAR SPECTROSCOPY. *Educ:* Univ Okla, BS, 55; Calif Inst Technol, PhD(chem), 59. *Prof Exp:* staff mem chemist, Los Alamos Sci Lab, 59-88; ADJ PROF CHEM, UNIV ARIZ, 88- *Mem:* Am Chem Soc; fel Am Optical Soc; Sigma Xi. *Res:* Spectroscopy and kinetics of atoms and simple molecules studied by high resolution optical spectroscopy. *Mailing Add:* Dept Chem Univ Ariz Tucson AZ 85721

ENGLEMAN, VICTOR SOLOMON, b Brooklyn, NY, Dec 31, 40; m 65; c 2. CHEMICAL ENGINEERING. *Educ:* Calif Inst Technol, BS, 62; Univ Calif, MS, 64, PhD(chem eng), 67. *Prof Exp:* Proj engr, Air Force Rocket Propulsion Lab, 67-70; res engr, Exxon Res & Eng Co, 70-77; asst div mgr, 76-80, DIV MGR, SCI APPLN INC, 80- *Concurrent Pos:* Adj assoc prof, Columbia Univ, 76. *Mem:* Am Inst Chem Engrs; Combustion Inst; Am Chem Soc. *Res:* Fossil fuel conversion and utilization; combustion; chemical kinetics; air pollution, hazardous waste incineration. *Mailing Add:* Sci Appln Int Corp 6791 Edmonton Ave San Diego CA 92122-2518

ENGLER, ARNOLD, b Czernovitz, Romania, July 19, 27; m 61; c 2. PHYSICS. *Educ:* Univ Berne, PhD(physics), 53. *Prof Exp:* Res assoc physics, Univ Berne, 53-54, Bristol Univ, 54-56 & Univ Rochester, 56-58; sr res officer, Oxford Univ, 58-60; res assoc & assoc prof, Duke Univ, 60-61; assoc prof, Northwestern Univ, 61-62; assoc prof, Carnegie Inst Technol, 62-66, PROF PHYSICS, CARNEGIE-MELLON UNIV, 66- *Concurrent Pos:* Fel, St Cross Col, Oxford Univ, 66-67. *Mem:* Fel Am Phys Soc; Ital Phys Soc. *Res:* Elementary particle physics; cosmic ray physics. *Mailing Add:* Dept Physics Carnegie-Mellon Univ Schenley Park 5000 Forbes Ave Pittsburgh PA 15213

ENGLER, CADY ROY, b Topeka, Kans, May 13, 47; m 70; c 2. BIOCHEMICAL ENGINEERING. *Educ:* Kans State Univ, BS, 69, MS, 74; Univ Waterloo, PhD(chem eng), 80. *Prof Exp:* Process engr, Exxon, 69-72; asst prof bioeng, 78-81, asst prof agr eng, 81-84, ASSOC PROF AGR ENG, TEX A&M UNIV, 84- *Concurrent Pos:* Engr res assoc, Food Protein Res & Develop Ctr, Tex Eng Exp Sta, 80-84, assoc res engr, 84-86. *Mem:* Am Inst Chem Engrs; Am Chem Soc; Am Soc Agr Engrs. *Res:* Production of chemicals from biomass via fermentation processes; bioreactor systems development; plant tissue culture systems. *Mailing Add:* Dept Agr Eng Tex A&M Univ College Station TX 77843

ENGLER, EDWARD MARTIN, b Brooklyn, NY, Aug 22, 47; m 67; c 3. PHYSICAL ORGANIC CHEMISTRY, SOLID STATE CHEMISTRY. *Educ:* Providence Col, BS, 69; Princeton Univ, MA, 71, PhD(chem), 73. *Prof Exp:* RES STAFF ORG CHEM, THOMAS J WATSON RES CTR, IBM CORP, 73- *Mem:* Am Chem Soc. *Res:* Design and synthesis of organic molecules with unusual solid state properties; chemistry and properties of organochalogens. *Mailing Add:* 467 Curie Dr San Jose CA 95123-4925

ENGLER, HAROLD S, b Augusta, Ga, Jan 10, 23; c 5. MEDICINE, SURGERY. *Educ:* Med Col Ga, MD, 50; Am Bd Surg, dipl, 58. *Prof Exp:* From instr to assoc prof, 50-66, PROF SURG, MED COL GA, 66- *Mem:* Am Col Surgeons; James Ewing Soc. *Res:* Cancer and surgical research. *Mailing Add:* 1503 Winter St Augusta GA 30902

ENGLER, RETO ARNOLD, b Zurich, Switz, Nov 20, 31; US citizen; c 1. ORGANIC CHEMISTRY. *Educ:* Swiss Fed Inst Technol, dipl chem eng, 54; Univ Tübingen, PhD(chem), 58. *Prof Exp:* Asst virol, Max Planck Inst Virus Res, 58-60; lectr pediat, Med Ctr, Univ Kans, 60-61, asst prof, 61-66, asst biochem, 66-68; virologist div microbiol, Fed Food & Drug Admin, 68, res chemist, 68-71; toxicologist & chemist, Hazard Eval Div, 71-79, chief, Disinfectants Br, Regist Div, 79-83, CHIEF, SCI ANALYST & COORD BR, HEALTH EFFECTS DIV, ENVIRON PROTECTION AGENCY, 83- *Concurrent Pos:* Lectr, Univ Mo-Kansas City, 63-67. *Mem:* AAAS; NY Acad Sci; Am Soc Microbiol; Soc Invert Path. *Res:* Biosynthesis of viruses in plants and animals, especially the synthesis of viral nucleic acid; health hazard of viruses in food; virology; microbiological control of pests; toxicology of pesticide chemicals. *Mailing Add:* 1329 F St NE Washington DC 20002

ENGLERT, DU WAYNE CLEVELAND, b WaKeeney, Kans, Dec 1, 32; m 53; c 2. GENETICS. *Educ:* Univ Kans, BS, 54; Purdue Univ, MS, 61, PhD(genetics), 64. *Prof Exp:* From asst prof to assoc prof zool, 63-77, PROF ZOOL, SOUTHERN ILL UNIV, CARBONDALE, 77-, DIR, BIOL SCI PROG, 89- *Concurrent Pos:* Vis prof dept animal sci, Purdue Univ, 68-69; consult, UN Develop Prog, Food & Agr Orgn, Nat Dairy Res Inst, Karnal, India, 79. *Mem:* Am Genetics Asn; Genetics Soc Am; Biomet Soc; Genetics Soc Can; Sigma Xi. *Res:* Genotype by environment interactions in Drosophila and Tribolium; selection and interrelationship of growth traits; genetic differences in growth curves; genetic recombination in Tribolium castaneum; population regulatory mechanisms in Tribolium. *Mailing Add:* Dept Zool Southern Ill Univ Carbondale IL 62901

ENGLERT, EDWIN, JR, internal medicine, gastroenterology; deceased, see previous edition for last biography

ENGLERT, MARY ELIZABETH, chemistry; deceased, see previous edition for last biography

ENGLERT, ROBERT D, b Portland, Ore, Feb 11, 20; m 54; c 3. TECHNICAL MANAGEMENT, RESEARCH ADMINISTRATION. *Educ:* Univ Portland, BS, 42; Ore State Univ, BS & MS, 44; Univ Colo, PhD(org chem), 49. *Prof Exp:* Asst, Ore State Univ, 43; biochemist, Naval Med Res Inst, 46; sr org chemist, Stanford Res Inst, 49-55, mgr phys sci res, Southern Calif Labs, 55-59, chmn, 59-62, dir, 62-68, exec dir, 68-70; vpres & gen mgr, Advan Technol Ctr, Dresser Indust Inc, 70-84; RETIRED. *Concurrent Pos:* Consult, RDE Consult Group, 84- *Mem:* Am Chem Soc; Sigma Xi. *Res:* Fats and oils; chemistry of boron and antimony; herbicides and insecticides; pollution control hardware; new equipment in energy field; new materials; ecology; environmental science. *Mailing Add:* 3741 Geranium Ave Corona Del Mar CA 92625

ENGLESBERG, ELLIS, b New York, NY, Oct 19, 21; m 87; c 3. GENE REGULATION, AMINO ACID TRANSPORT. *Educ:* Brooklyn Col, BA, 45; Univ Calif, MA, 48, PhD(bact), 50. *Prof Exp:* Teaching asst bact, Univ Calif, 46-49, res asst, 49-50, asst res bacteriologist, Hooper Found Med Res, 50-54; microbiologist, Long Island Biol Asn, 54-58; prof bact, Univ Pittsburgh, 58-65; chmn dept biol sci, 66-69, PROF MICROBIOL, UNIV CALIF, SANTA BARBARA, 65- *Concurrent Pos:* Mem adv panel genetic biol, NSF, 62-65; Guggenheim fel, 71-72. *Mem:* Nat Acad Sci; Genetics Soc Am; AAAS; Soc Am Microbiol. *Res:* A regulation transport of amino acids by mammalian cells in culture; insulin as a mitogen and growth control. *Mailing Add:* Dept Biol Sci Univ Calif Santa Barbara CA 93106

ENGLEY, FRANK B, JR, b Wallingford, Conn, Oct 26, 19; m 48; c 4. MICROBIOLOGY. *Educ:* Univ Conn, BS, 41; Univ Pa, MS, 44, PhD(bact), 49; Am Bd Microbiol, dipl. *Prof Exp:* Asst instr bact, Sch Med, Univ Pa, 41-44; bacteriologist, Chem Corps, US Dept Army, Camp Detrick, 46-50; assoc prof bact & parasitol & consult bacteriologist, Med Br Hosps, Univ Tex, 50-55; asst dean sch med, 56-60, prof prev med & chmn dept, 60-61, chmn dept, 55-77, PROF MICROBIOL, SCH MED, UNIV MO-COLUMBIA, 55- *Concurrent Pos:* Consult bacteriologist, Vet Admin Hosp, 54-56; mem, Am Inst Biol Sci Adv Comt to NASA on Sterilization of Spacecraft, 66-68; vis prof microbiol, Nigeria, 73, Libya, 79; mem & consult, Food & Drug Admin OTC Panel on Antimicrobials, 73-81; consult various agencies, indust & com groups. *Mem:* Fel AAAS; Am Soc Microbiol; Soc Exp Biol & Med; fel Am Pub Health Asn; Asn Am Med Cols; Sigma Xi; NY Acad Sci; Asn Advan Med Instrumentation; Res Soc Am; Royal Soc Health. *Res:* Bacterial toxins; antiseptics and disinfectants; plastics in microbiology; ethylene oxide sterilization; survival of microorganisms; author of over 100 publications, several laboratory manuals and 2 reference textbooks. *Mailing Add:* Dept Microbiol Univ Mo Sch Med M642 Med Sci Bldg Columbia MO 65212

ENGLISH, ALAN DALE, b San Diego, Calif, June, 8, 47; m 69; c 2. MACROMOLECULAR STRUCTURE DYNAMICS, INTRACTABLE MATERIALS. *Educ:* Univ Calif, Los Angeles, BS, 69, Santa Barbara, PhD(chem), 73. *Prof Exp:* res chemist, Cent Res & Develop Dept, 73-79, sr res chemist, Plastic Prod & Resins Dept, 79-80, PRIN SCIENTIST, CENT RES & DEVELOP DEPT, E I DU PONT DE NEMOURS & CO, INC, 80- *Mem:* Fel Am Phys Soc; Am Chem Soc. *Res:* Using modern solid state nuclear magnetic resonance techniques used to examine both structure and dynamics of macromolecular systems; using spectroscopic and mechanical testing methods to establish structure/property relationships of intractable materials. *Mailing Add:* Cent Res & Develop Du Pont Wilmington DE 19880-0356

ENGLISH, ALAN TAYLOR, b Los Angeles, Calif, Mar 14, 34; m 55; c 4. METALLURGY. *Educ:* Stanford Univ, BS, 56; Mass Inst Technol, MS, 60, PhD(metall), 63. *Prof Exp:* Mem tech staff, 63-73, supvr, Reliability Studies, 73-82, DEPT HEAD, TECH EXCHANGE & INT VISITS, AT&T BELL LABS, MURRAY HILL, 82- *Mem:* Am Vacuum Soc. *Res:* Structure and properties of magnetic metals and alloys; thin film materials; reliability physics in electronic devices. *Mailing Add:* Four Drum Hill Dr Summit NJ 07901

ENGLISH, ARTHUR WILLIAM, b Ft Hueneme, Calif, Oct 20, 45. NEUROBIOLOGY, MOTOR SYSTEMS NEUROBIOLOGY. *Educ:* Univ Ore, BS, 67; Univ Ill, BS, 70, PhD(anat), 74. *Prof Exp:* Instr, 74-76, asst prof, 76-81, ASSOC PROF ANAT, COL MED, EMORY UNIV, 81- *Concurrent Pos:* Affil scientist neurobiol, Yerkes Regional Primate Res Ctr, 82- *Mem:* Am Soc Zoologists; Am Asn Anatomists; Soc Neurosci. *Res:* Neural mechanisms used in the control of movement and in particular the structural and functional features of the neural control of locomotion. *Mailing Add:* Dept Anat Emory Univ 1364 Clifton Rd NE Atlanta GA 30322

ENGLISH, BRUCE VAUGHAN, b Richmond, Va, Aug 6, 21; m 49. ENVIRONMENTAL PHYSICS, ENVIRONMENTAL SCIENCES. *Educ:* Randolph-Macon Col, BS, 42; Ind Univ, MS, 43; Univ Va, PhD(physics), 58. *Prof Exp:* Asst physics, Ind Univ, 42-43; assoc prof, Randolph-Macon Col, 43-44; physicist, US Naval Res Lab, Washington, DC, 44-45, NJ, 45-46 & Fla, 46-48; assoc prof physics, Randolph-Macon Col, 48-58, prof, 58-64, actg head dept, 52-58, head dept, 58-64; consult physicist, 64-67; pres, Pollution Control Assocs, 67-71; PHYSICIST & CONSULT, 71- *Concurrent Pos:* Ford fel, Pa State Col, 51-52; Danforth fel, Univ Va, 56-57, duPont fel, 57-58; Herald-Progress columnist, Impact on Environ, 71- *Mem:* AAAS; Am Phys Soc; Brit Soc Clean Air; Air & Waste Mgr Asn; Sigma Xi. *Res:* Ultracentrifuge; gravity; clean air; pollution control. *Mailing Add:* PO Box 267 Ashland VA 23005

ENGLISH, DARREL STARR, b Newton, Kans, Sept 6, 36; m 60; c 3. GENETICS. *Educ:* Southwestern Col, BA, 59; La State Univ, Baton Rouge, MS, 61; Iowa State Univ, PhD(genetics), 68. *Prof Exp:* Asst zool, La State Univ, 59-61; instr biol, Millsaps Col, 61-64; from asst to instr genetics, Iowa State Univ, 64-67; from asst prof to assoc prof, 67-85, PROF GENETICS, NORTHERN ARIZ UNIV, 85- *Concurrent Pos:* Danforth Assoc, 69; NSF partic histochem, Vanderbilt Univ, 71; sabbatical leave, M D Anderson Hosp & Tumor Inst, Houston, Tex, 78-; instr basic methods in tissue cult, Calif State Univ, Long Beach, 81; minority biomed res support grant, 82-85. *Mem:* Am Genetics Asn; Sigma Xi; Am Cytogenetics Asn; Am Soc Mammalogists. *Res:* cytogenetics of humans and small mammals. *Mailing Add:* Box 5640 Dept Biol Northern Ariz Univ Flagstaff AZ 86011

ENGLISH, FLOYD L, b East Nicolaus, Calif, June 10, 34; m 55; c 3. SOLID STATE PHYSICS. *Educ:* Calif State Univ, Chico, AB, 59; Ariz State Univ, MS, 62, PhD(physics), 65. *Prof Exp:* Staff mem tech, Sandia Lab, 65-69, div supvr microelectronics, 69-73; dir mkt, MOS & Components Div, Collins Radio Group, Rockwell Inst, 73-74, gen mgr,74-75; pres, Darcom Inc, 75-79; from vpres to pres US opers, 80-83, PRES & CHIEF EXEC OFFICER, ANDREW CORP, 83- *Mem:* Inst Elec & Electronics Engrs. *Res:* Electrical characteristics of rectifying junctions; surface effects on ferroelectric and piezoelectric materials. *Mailing Add:* Andrew Corp 10500 W 153rd St Orland Park IL 60460

ENGLISH, GERALD ALAN, b Chester, Pa, Sept 17, 46. NUCLEAR CHEMISTRY, ANALYTICAL CHEMISTRY. *Educ:* LaSalle Col, BA, 68; Purdue Univ, West Lafayette, MS, 72, PhD(nuclear chem), 74. *Prof Exp:* Teaching asst chem, Purdue Univ, 68-70, res asst nuclear chem, 70-74; coal & analytical chemist, Energy Systs Group, Rockwell Int, 74-80; NUCLEAR CHEMIST, PAC GAS & ELEC CO, SAN FRANCISCO, CALIF, 81- *Mem:* Am Chem Soc; Am Phys Soc; Sigma Xi. *Res:* Energy-related problems concerned with breeder reactor fuel, advanced reactor fuels, coal gasification and liquefaction, upgrading of fossil fuels, pollution control and radioactive waste treatment; post-accident sampling and analysis (Diablo Canyon). *Mailing Add:* 3785 Highland Rd Lafayette CA 94549

ENGLISH, JACKSON POLLARD, b Richmond, Va, Jan 25, 15; m 39; c 2. ORGANIC CHEMISTRY. *Educ:* Va Mil Inst, BS, 35; Johns Hopkins Univ, PhD(chem), 40. *Prof Exp:* Chemist, Am Cyanamid Co, 39-42; sr group leader, Chemother Div, 42-54, asst to dir, 54-55, unit leader org chem sect, Pearl River Labs, 55-56, asst dir exp therapeut sect, Lederle Labs, 56-60, dir chem res & develop, Agr Div, 60-69; dir res admin, Polaroid Corp, 69-71; ADJ PROF CHEM, DARTMOUTH COL, 75- *Mem:* Fel AAAS; Am Chem Soc; fel NY Acad Sci; The Chem Soc. *Res:* Synthesis of chemotherapeutic agents; natural products; experimental therapeutics; pesticides; agricultural chemicals. *Mailing Add:* Dept Chem Dartmouth Col Hanover NH 03755

ENGLISH, JOSEPH T, b Philadelphia, Pa, May 21, 33; m 69; c 3. PSYCHIATRY. *Educ:* St Joseph's Col, AB, 54; Jefferson Med Col, MD, 58. *Prof Exp:* Intern, Jefferson Med Col Hosp, Philadelphia, 58-59; resident psychiat, Inst Pa Hosp, Philadelphia, 59-61 & NIMH, 61-62; psychiatrist, Off Dir, NIMH, 64-65, asst chief policy & prog coord, 65-66, dept chief off interagency liaison, 66; dep asst dir health affairs, Off Econ Opportunity, 66, asst dir, 66-68 & 69; adminr, Health Serv & Ment Health Admin, HEW, 69-70; pres, NY City Health & Hosps Corp, 70-73; CHMN, DEPT PSYCHIAT, ST VINCENT'S HOSP & MED CTR, 73-; PROF PSYCHIAT, NY MED COL, 79-, ASSOC DEAN, 80- *Concurrent Pos:* Pvt pract psychiat, 62-; chief psychiatrist, Med Prog Div, Peace Corps, 62-66; adj prof psychiat, Cornell Univ; chmn interagency task force emergency food & med prog for US, Off Econ Opportunity-HEW, USDA, 68-69; chmn Alaska subcomt fed health progs, President's Rev Comn Alaska, 69-; chmn adv comt accessible environ for disabled, Bldg Res Adv Bd, 74-; chmn mental health & substance abuse servs comt, Greater NY Hosp Asn, 74-; exec coordr panels on ment health serv delivery, President's Comn Ment Health, 77; vis fel, Woodrow Wilson Nat Fel Found, 79- *Honors & Awards:* John XXIII Medal, Col New Rochelle, 66; Meritorious Award Exemplary Achievement Pub Admin, William A Jump Mem Found, 66. *Mem:* Inst Med Nat Acad Sci; fel Am Psychiat Asn; NY Acad Med; Am Col Psychiatrists; AMA; NY Psychiat Soc. *Mailing Add:* St Vincent's Hosp & Med Ctr 203 W 12th St New York NY 10011

ENGLISH, LEIGH HOWARD, b Cleveland, Ohio, July 24, 55; m 78; c 2. MEMBRANE BIOCHEMISTRY, INSECT BIOCHEMISTRY & PHYSIOLOGY. *Educ:* Cornell Univ, BS, 76; NDak State Univ, PhD(insect biochem), 80; Harvard Univ, MTS, 82. *Prof Exp:* Fel biochem & molecular biol, Harvard Univ, 81-85; res assoc physiol, Sch Med, Tufts Univ, 85-86; res scientist, 86-88, sr res scientist, 88-90, PRIN RES SCIENTIST BIOCHEM & PHYSIOL, ECOGEN INC, 90- *Concurrent Pos:* Res team leader, Ecogen Inc, 88-, radiation safety officer, 90- *Mem:* AAAS; Entom Soc Am; Am Soc Biochem & Molecular Biol. *Res:* Structure and function of membrane-bound proteins as well as the structure and function of toxins that purturb membrane physiology; commercial development of economically valuable proteins. *Mailing Add:* Ecogen Inc 2005 Cabot Blvd W Langhorne PA 19047

ENGLISH, LEONARD STANLEY, b Hull, UK, July 2, 37; m 60; c 3. IMMUNOLOGY. *Educ:* Univ Wales, BSc, 71; Australian Nat Univ, PhD(immunol), 75. *Prof Exp:* Lectr, Mem Univ Nfld, 75-76, asst prof immunol, 76-77; asst prof, 77-81, ASSOC PROF IMMUNOL, SCH MED, E CAROLINA UNIV, 81- *Concurrent Pos:* Med Res Coun Can res grant, 75-77; NIH grant, 79-82. *Res:* Examination of the in vivo production of soluble factors by lymphoid cells during the immune response in sheep. *Mailing Add:* Bristol Polytech Sci Dept Cold Harbor Lane Frenchay Bristol BS 161 QY England

ENGLISH, ROBERT E, b Paris, Ill, July 9, 53; m 77; c 2. INTERACTIVE VIDEO DELIVERY. *Educ:* Ind State Univ, BS, 76, MS, 81. *Prof Exp:* Mfg engr, Zenith Radio Corp, 76-77, mfg eng mgr, 77-82; ASSOC PROF ELEC, IND STATE UNIV, 82- *Concurrent Pos:* Tech res consult, Beeco, 85-86, Micro Link Corp, 85-87 & Hane Indust Training, 87-89. *Mem:* Soc Mfg Engrs; Nat Asn Indust Technol. *Res:* Interactive video delivery and development systems including hardware and software development; design and development of industrial training courseware. *Mailing Add:* RR 1 Box 68A Lewis IN 47858

ENGLISH, THOMAS SAUNDERS, biological oceanography; deceased, see previous edition for last biography

ENGLISH, WILLIAM HARLEY, b LaCrosse, Wash, Apr 12, 11; m 36; c 3. PLANT PATHOLOGY. *Educ:* State Col Wash, BS, 35, PhD(plant path), 40. *Prof Exp:* Asst plant path, State Col Wash, 35-37, instr, 37-39; jr pathologist, USDA, 39-43, asst pathologist, 43-46; assoc prof plant path, Ore State Col, 46-47; from asst prof to assoc prof, 47-56, PROF PLANT PATH, UNIV CALIF, DAVIS, 56- *Mem:* AAAS; Am Phytopath Soc. *Res:* Bacterial and fungus diseases of deciduous fruit trees; mycology. *Mailing Add:* Univ Calif Davis CA 95616

ENGLISH, WILLIAM JOSEPH, b Oil City, Pa, Nov 29, 41; m 78; c 3. SPACECRAFT ENGINEERING, ANTENNAS & PROPAGATION. *Educ:* St Vincent Col, BA, 63; Carnegie Inst Technol, BSEE, 64, MS, 65; Carnegie-Mellon Univ, PhD(space sci), 69. *Prof Exp:* Mem tech staff, Commun Satellite Corp, 70-77; sect chief antennas & propagation, 77-87, MGR, SPACECRAFT ENG DEPT, INTELSAT, 87- *Concurrent Pos:* Asst prof & lectr, George Washington Univ, 72-76. *Mem:* Inst Elec & Electronics Engrs; Am Inst Aeronaut & Astronaut; Int Union Radio Sci. *Mailing Add:* Intelsat 3400 International Dr NW Washington DC 20008-3098

ENGLISH, WILLIAM KIRK, b Lexington, Ky, Jan 27, 29. INFORMATION SCIENCE. *Educ:* Univ Ky, BS, 50; Stanford Univ, MS, 64. *Prof Exp:* Staff mem, Sandia Corp, 50-52; staff engr, Univ Chicago, 52-54; sr res mem, Stanford Res Inst, 58-71; prin scientist, 71-80, MGR, INT BUS PLANNING, XEROX PALO ALTO RES CTR, 80- *Mem:* Inst Elec & Electronics Engrs; Soc Info Display. *Res:* Computer-based information processing and information retrieval. *Mailing Add:* Advance Mfg Syst Oper Hewlett Packard 1266 Kifer Rd Sunnyvale CA 94086-4709

ENGLUND, CHARLES R, b Oak Park, Ill, Feb 20, 36; m 56; c 3. ORGANIC CHEMISTRY. *Educ:* Wheaton Col, Ill, BS, 58; Southern Ill Univ, MA, 63, PhD(chem), 68. *Prof Exp:* Teacher, High Sch, 58-60; asst chem, Southern Ill Univ, 60-62; from instr to asst prof, Concordia Teachers Col, Ill, 62-65; instr, Southern Ill Univ, 65-67; assoc prof phys sci & math, 68-80, PROF CHEM, BETHANY COL, KANS, 80- *Mem:* Am Chem Soc. *Res:* Preparation and structure determination of steroidal derivatives. *Mailing Add:* 415 N Chestnut Lindsborg KS 67456-1705

ENGLUND, JOHN ARTHUR, b Omaha, Nebr, June 4, 26; m 52; c 5. OPERATIONS RESEARCH. *Educ:* Creighton Univ, BS, 49; Mass Inst Technol, SM, 51. *Prof Exp:* From instr to asst prof math, Creighton Univ, 51-56; opers analyst, Hq Strategic Air Command, 56-62; mil systs analyst, US Arms Control & Disarmament Agency, 62-63; mathematician, 63-64, chief, Strategic Br, 64-70, mgr, Strategic Div, 70-76, mgr intel studies, 71-80, exec vpres, 76-81, PRES, ANALYTICAL SERV, INC, 81- *Concurrent Pos:* Vpres admin, Mil Opers Res Soc, 78-79, pres, 79-80, past pres, 80-81. *Res:* Algebraic and analytic number theory; military systems analysis. *Mailing Add:* Analytical Serv Inc 1215 Jefferson Davis Hwy Suite 800 Arlington VA 22202

ENGLUND, JOHN CALDWELL, b Birmingham, Ala, Sept 16, 52. QUANTUM & NONLINEAR OPTICS. *Educ:* Vanderbilt Univ, BA, 74; Univ Tex, Austin, PhD(physics), 85. *Prof Exp:* Nat Res Coun res assoc, US Army Missile Command, 85-86; asst prof physics, Southern Methodist Univ, 86-89; ASST PROF PHYSICS, UNIV TEXAS, DALLAS, 89- *Concurrent Pos:* Subcontractor, Battelle Columbus Div, 87- *Mem:* Am Phys Soc; Optical Soc Am; Sigma Xi. *Res:* Quantum theory of soliton generation in stimulated raman scattering; quantum theory of lasers and lasers containing a saturable absorber; nonlinear dynamics and instabilities of lasers. *Mailing Add:* Ctr Appl Optics Univ Tex-Dallas PO Box 830688 Richardson TX 75083-0688

ENGLUND, PAUL THEODORE, b Worcester, Mass, Mar 25, 38; m 90; c 2. BIOCHEMISTRY, PARASITOLOGY. *Educ:* Hamilton Col, BA, 60; Rockefeller Univ, PhD(biochem), 66. *Prof Exp:* From asst prof to assoc prof, 68-80, PROF BIOCHEM, SCH MED, JOHNS HOPKINS UNIV, 80- *Concurrent Pos:* Fel, Sch Med, Stanford Univ, 66-68; Burrougs-Wellcome scholar molecular parasitol & molecular biol; mem bd dirs, Int Lab Res Animal Dis, Nairobi, Kenya, 88-; co-dir biol paratism course, Marine Biol Lab, Woods Hole, Mass, 85-88. *Mem:* Am Soc Biochem & Molecular Biol; Am Chem Soc. *Res:* Protein chemistry; enzymology of nucleic acids; biochemistry of parasites. *Mailing Add:* Dept Biol Chem Johns Hopkins Univ Sch Med 725 N Wolfe St Baltimore MD 21205

ENGSTER, HENRY MARTIN, b Troy, NY, May 6, 49; m 84; c 3. ANIMAL SCIENCE & NUTRITION, ANIMAL HUSBANDRY. *Educ:* St Lawrence Univ, BS, 71; Univ Vt, MS, 74, PhD(nutrit), 77. *Prof Exp:* Res fel nutrit & biochem, Hormel Inst, Univ Minn, Austin, 77; nutritionist, Purina Mills, Inc, 77-79, sr nutritionist, 79- 81, res mgr, 81-85, tech mgr, 85-89; DIR RES & NUTRIT, PERDUE FARMS, INC, 90- *Concurrent Pos:* Head, Subcomt Determination Biol Availiability Amino Acids Poultry, Animal Nutrit Res Coun, 81-82, Biokyawa Amino Acid Coun, 84 & 85; gen prog chair, Poutry Sci Asn, 91. *Mem:* Sigma Xi; Poultry Sci Asn; AAAS; Am Oil Chemists Soc; World's Poultry Sci Asn. *Res:* Dietary fats and lipids; effects of fats and essential fatty acids on endocrine physiology; nutritional and management research on pullets, commercial layers, roosters, turkeys, energy and amino acid content of feedstuffs, nutrient requirements and product and program development for poultry. *Mailing Add:* RR 5 Box 675 Cobblestone Ct Salisbury MD 21801

ENGSTROM, HERBERT LEONARD, b San Francisco, Calif, Dec 16, 41; m 72. ELECTRONICS ENGINEERING. *Educ:* Univ Calif, Berkeley, AB, 64, PhD(physics), 72. *Prof Exp:* Solid state physicist, Univ Paris, Orsay, 72-74; solid state physicist, Brookhaven Nat Lab, 75-77; solid state physicist, Oak Ridge Nat Lab, 77-81; instrumentation engr, Optical Storage Int, Santa Clara, 81-86; OPTICAL RECORDING PHYSICIST, VERBATION CORP, 86- *Mem:* Inst Elec & Electronics Engrs; Am Phys Soc. *Res:* Optical properties and light scattering from defects and impurities in semiconductors and insulators. *Mailing Add:* Verbatim Corp M5 2 29 435 Indio Way Sunnyvale CO 94086

ENGSTROM, LEE EDWARD, b Rock Island, Ill, Sept 30, 41; c 1. DEVELOPMENTAL GENETICS. *Educ:* Iowa Wesleyan Col, BS, 65; Univ Ill, Urbana, MS, 67, PhD(develop genetics), 71. *Prof Exp:* From asst prof to assoc prof biol, Ball State Univ, 70-79; assoc prof anat, 75-79, PROF MERD EDUC IND UNIV, MUNCIE CTR MED EDUC, 79-; PROF BIOL, BALL STATE UNIV, 79-; assoc prof anat, 75-79, PROF MED EDUC, IND UNIV, MUNCIE CTR MED EDUC, 79- *Concurrent Pos:* Vis prof develop genetics & anat, Case Western Reserve Univ Sch Med, 82-84. *Mem:* Genetics Soc Am; Soc Develop Biol. *Res:* Developmental genetics of Drosophila melanogaster. *Mailing Add:* Ctr Med Educ Ball State Univ Muncie IN 47306

ENGSTROM, NORMAN ARDELL, b DeKalb, Ill, July 2, 45; m 67. MARINE ECOLOGY, ETHOLOGY. *Educ:* Cornell Col, AB, 67; Univ Miami, MS, 70; Univ Wash, PhD(invert ecol), 74. *Prof Exp:* Asst prof marine sci, Univ PR, 74-77; ASST PROF BIOL, NORTHERN ILL UNIV, 77- *Mem:* AAAS; Ecol Soc Am; Am Soc Zoologists. *Res:* Escape responses of invertebrates to marine gastropod predators; prey-predator relations between cassis tuberosa and sea urchins in seagrass beds; agonistic behavior of brachyuran crabs. *Mailing Add:* 1015 Garden Rd DeKalb IL 60115

ENGSTROM, RALPH WARREN, b Grinnell, Iowa, Oct 24, 14; m 37; c 2. PHYSICS. *Educ:* St Olaf Col, BA, 35; Northwestern Univ, MS, 37, PhD(physics), 39. *Prof Exp:* Instr physics & math, St Cloud State Teachers Col, Minn, 39-41; res physicist, Nat Defense Res Comt, 41; res physicist, RCA Mfg Co, NJ, 41-43, tech adv, Electro-optics Prod, RCA Corp, 43-80; CONSULT, 80- *Mem:* Fel Am Phys Soc; Optical Soc Am; Sigma Xi. *Res:* Multiplier phototubes; television camera tubes; photoconductors; image converter tubes. *Mailing Add:* 3481 Dawn View Dr Lancaster PA 17601

ENGVALL, EVA SUSANNA, b Stockholm, Sweden, Mar 11, 40. IMMUNOCHEMISTRY. *Educ:* Univ Stockholm, BSc, 64, PhD(immunol), 75. *Prof Exp:* Res assoc biochem, Res Lab LKB, Stockholm, 65-66 & KABI AB, Stockholm, 66-69; jr res scientist immunol, Univ Stockholm, 69-75; fel immunol, Univ Helsinki, 75-76; fel, City Hope Med Ctr, 76-77, asst res scientist immunol, 77-79; SCIENTIST, LA JOLLA CANCER RES FOUND, 79- *Concurrent Pos:* Fel, Europ Molecular Biol Orgn, 75-77. *Honors & Awards:* Biochem Analysis Award, Ger Soc Clin Chem, 76. *Mem:* Am Asn Cancer Res; Am Asn Immunologists. *Res:* Molecular interactions of extracellular matrix components. *Mailing Add:* La Jolla Cancer Res Found 10901 N Torrey Pines Rd La Jolla CA 92037

ENICK, ROBERT M, b McKees Rocks, Pa, Nov 15, 58; m 82; c 2. CHEMICAL ENGINEERING. *Educ:* Univ Pittsburgh, BS, 80, MS, 83, PhD(chem eng), 85. *Prof Exp:* ASST PROF CHEM & PETROL ENG, UNIV PITTSBURGH, 85- *Concurrent Pos:* Oak Ridge fac partic, Lilly teaching fel. *Mem:* Soc Petrol Engrs; Am Inst Chem Engrs; Am Chem Soc. *Res:* High pressure themodynamics; multiple phase equilibrium; modelling unsteady state displacement in stratified porous media; viscosity enhancement of supercritical fluids. *Mailing Add:* Dept Chem & Petrol Eng Univ Pittsburgh 1249 Benedum Hall Pittsburgh PA 15260

ENKE, CHRISTIE GEORGE, b Minneapolis, Minn, July 8, 33; m; c 2. INSTRUMENTATION, MASS SPECTROMETRY. *Educ:* Principia Col, BS, 55; Univ Ill, MS, 57, PhD(chem), 59. *Prof Exp:* From instr to asst prof chem, Princeton Univ, 59-66; assoc prof, 66-74, PROF CHEM, MICH STATE UNIV, 74- *Concurrent Pos:* Alfred P Sloan res fel, 64-67. *Honors & Awards:* Chem Instrumentation Award, Am Chem Soc, 74, Computers Chem Award, 89. *Mem:* AAAS; Am Chem Soc; Am Soc Mass Spectrometry. *Res:* Electrochemistry, spectrometry, mass spectrometry; chemical instrumentation, intelligent systems for chemical measurement and instrument control; instrumentation education. *Mailing Add:* Dept Chem Mich State Univ E Lansing MI 48824

ENKE, GLENN L, b Oakland, Calif, Jan 8, 09; m 34; c 4. CIVIL ENGINEERING. *Educ:* Univ Calif, Berkeley, BS, 28; Utah State Univ, CE, 72. *Prof Exp:* Detailer, Am Bridge Co, Ind, 28-29; designer, Indust Plant, Giffels & Vallet, Inc, Mich, 29-31; engr design, Calif Bridge Dept, 31-41; struct engr, D R Warren Co, 41-42; asst chief engr, Utah-Pomeroy-Morrison, 42-43; dist engr, Morrison-Knudsen, Inc, Idaho, 43-47; struct engr, Caldwell, Richards & Sorensen, Utah, 47-48; dist engr, Utah Construct Co, 48-52; prof civil & mech eng & chmn dept, Brigham Young Univ, 52-53; chief engr, Church Jesus Christ Latter Day Saints, 53-55; gen supvr design eng, US Steel Corp, 55-62; prof civil eng sci, 62-74, EMER PROF CIVIL ENG SCI, BRIGHAM YOUNG UNIV, 74-; OWNER, GLENN L ENKE, CONSULT CIVIL & STRUCT ENGR, 74- *Concurrent Pos:* Consult struct engr, 36-;

partner, Enke & Long, Consult Engrs, Calif, 54-; vpres, Van Sickle Assocs, Consult Engrs, Colo, 56-; dir, Western Zone, Nat Coun State Bd Eng Exam, 60-62. *Honors & Awards:* Lincoln Arc Welding Found Award, 37, 42, 47 & 59. *Mem:* Fel Am Soc Civil Engrs. *Res:* Structural analysis methods for indeterminate structures; arc welding design; dynamics of long-span suspension systems; seismic force effects on multi-story buildings; engineering economics and law. *Mailing Add:* 1920 S 140th E 1455 W 820 N Roem UT 84058

ENLOE, LOUIS HENRY, b Eldorado Springs, Mo, Mar 4, 33; m 56; c 3. ELECTRICAL ENGINEERING, COMMUNICATIONS. *Educ:* Univ Ariz, BS, 55, MS, 56, PhD(elec eng), 59. *Prof Exp:* Instr elec eng, Univ Ariz, 56-59; mem tech staff commun res, 59-66, head visual systs res dept, Commun Systs Div, 66-67, head opto-electronics res dept, Commun Systs Div, Bell Tel Labs, 67-79, HEAD SYSTS DEVELOP DEPT, ATT-BELL LABS, 79- *Mem:* Inst Elec & Electronics Engrs. *Res:* Noise and modulation theory, particularly problems associated with space communications; visual systems research. *Mailing Add:* ATT-Bell Lab Syst Develop Dept 11900 N Pecos St Denver CO 80234

ENLOW, DONALD HUGH, b Mosquero, NMex, Jan 22, 27; m 45; c 1. ANATOMY. *Educ:* Univ Houston, BS, 49, MS, 50; Agr & Mech Univ, Tex, PhD, 55. *Prof Exp:* Instr zool, Univ Houston, 49-52; asst, Agr & Mech Col, Tex, 52-53, cur natural hist & anthrop, Witte Mus, 54-55; asst prof biol, WTex State Col, 55-56; instr anat, Med Col, Univ SC, 56-57; from instr to prof anat, Sch Med, Univ Mich, Ann Arbor, 57-72, dir phys growth prog, Ctr Human Growth & Develop, 68-72; prof anat & chmn dept, Sch Med, WVa Univ, 72-77; prof orthod chmn dept, asst dean grad studies, res & Thomas Hill distinguished prof, sch dent, 77-86, actg dean, 83-86, EMER PROF ORTHOD, CASE WESTERN UNIV, 89- *Honors & Awards:* Mershon Mem Lectr & Churchill Mem Lectr, Am Orthod Asn. *Mem:* Am Asn Anat; fel Royal Soc Med; Int Asn Dent Res. *Res:* Histology; embryology; gross and comparative anatomy; comparative histology of bone tissue; bone remodeling; facial growth. *Mailing Add:* Case Western Reserve Univ 2040 Adelbert Rd Cleveland OH 44106

ENNA, SALVATORE JOSEPH, b Kansas City, Mo, Dec 19, 44; m 69; c 3. PHARMACOLOGY, NEUROBIOLOGY. *Educ:* Rockhurst Col, BA, 65; Univ Mo-Kansas City, MS, 67, PhD(pharmacol), 70. *Prof Exp:* Fel pharmacol, Univ Tex Med Sch, Dallas, 70-72, Hoffmann La Roche & Co, Basel, Switzerland, 73-74, & Sch Med, Johns Hopkins Univ, 74-76; from asst prof to assoc prof, 76-80, PROF PHARMACOL & NEUROBIOL, UNIV TEX MED SCH HOUSTON, 80- *Concurrent Pos:* Consult pharmacol, ICI-USA, Wilmington, Del, 75-77, Merck Sharp & Dohme Res Lab, West Point, Pa, 76-81 & Mead Johnson Pharmaceut Div, Evansville, Ind, 81-, Panlabs Inc, Seattle, Wash, 85-; Nat Inst Neurol & Commun Dis & Stroke res career develop award, Univ Tex Med Sch Houston, 78-83, NIMH res scientist develop award, Univ Tex Med Sch, Houston, 84-89. *Honors & Awards:* John J Abel Award, Am Soc Pharmacol & Exp Therapeut. *Mem:* Soc Neurosci; Am Soc Neurochem; Am Soc Pharmacol & Exp Therapeut; Am Chem Soc; AAAS; Am Col Neuropsychopharmacol. *Res:* Central nervous system pharmacology, with particular emphasis on the interaction of drugs with neurotransmitter receptors; neurotransmitter receptors. *Mailing Add:* Nova Pharm Corp 6200 Freeport Ctr Baltimore MD 21224-2788

ENNEKING, EUGENE A, b Idaho Co, Idaho, Jan 17, 40; div; c 2. MATHEMATICAL STATISTICS. *Educ:* St Martins Col, BS, 62; Wash State Univ, MA, 64, PhD(math), 66. *Prof Exp:* Asst math, Wash State Univ, 62-66; asst prof, St Louis Univ, 66-68; from asst prof to assoc prof, 68-80, HEAD, DEPT MATH, PORTLAND STATE UNIV, 78-, PROF MATH, 80- *Mem:* Am Math Soc; Inst Math Statist; Math Asn Am; Am Statist Asn. *Res:* Combinatorial theory; probability theory. *Mailing Add:* Dept Math Box 751 Portland State Univ Portland OR 97207

ENNEKING, MARJORIE, b Eugene, Ore, June 21, 41; m 65; c 2. MATHEMATICS, EDUCATION. *Educ:* Willamette Univ, BA, 62; Wash State Univ, MA, 64, PhD(math), 66. *Prof Exp:* Teaching asst math, Wash State Univ, 62-66; asst prof, Univ Mo-St Louis, 66-68; from asst prof to assoc prof, 68-81, PROF MATH, PORTLAND STATE UNIV, 81- *Mem:* Am Math Soc; Math Asn Am; Nat Coun Teachers Math. *Mailing Add:* Dept Math Portland State Univ Box 751 Portland OR 97207-0751

ENNEKING, WILLIAM FISHER, b Madison, Wis, May 9, 26; m 47; c 7. MEDICINE. *Educ:* Univ Wis, BS, 45, MD, 49. *Prof Exp:* Intern, Med Ctr, Univ Colo, 49-50; prof orthop surg & dir div, Med Ctr, Univ Miss, 56-59; assoc prof surg & path & dir div orthop surg, 59-62, prof & chief div, 62-74, chmn dept, 74-80, PROF ORTHOP, COL MED, UNIV FLA, 74-, DISTINGUISHED SERV PROF, 80- *Mem:* Orthop Res Soc; AMA; NY Acad Sci. *Res:* Clinical orthopedic pathology; immunological aspects of bone transplantation. *Mailing Add:* Dept Orthop Surg Univ Fla Med Col J Hills Miller Health Ctr Gainesville FL 32610

ENNEVER, JOHN JOSEPH, b Ossining, NY, June 7, 20; m 46; c 2. DENTISTRY, MICROBIOLOGY. *Educ:* Wash Univ, DMD, 47; Ohio State Univ, MSc, 50. *Prof Exp:* Res assoc, Ohio State Univ, 47-50; asst prof periodont, Univ Kans City, 50-56; mem res staff, Procter & Gamble Co, 56-68; PROF DENT, DENT SCI INST, UNIV TEX, HOUSTON, 68- *Mem:* Sigma Xi; Soc Exp Biol Med; NY Acad Sci. *Res:* Microbiologic calcification; biologic calcification model systems. *Mailing Add:* 4329 Hazelton Dr Houston TX 77035

ENNIS, ELLA GRAY WILSON, b Sampson Co, NC, May 2, 25; m 62. PHYSIOLOGY. *Educ:* Univ NC, Greensboro, AB, 45, Univ NC, Chapel Hill, MA, 48, PhD(physiol), 64. *Prof Exp:* Teacher high sch, NC, 45-47; dir health & phys educ, St Mary's Jr Col, Md, 48-49; asst prof, Furman Univ, 49-56; res assoc pharmacol, 65, instr physiol, 65-67, asst prof, Sch Med, 67-69, asst prof, Sch Med & Sch Nursing, 69-74, lectr, Dept Med, 74-75, assoc prof, Dept Med Allied Health Prof & Lectr, Dept Physiol, Sch Med, Univ NC, Chapel Hill, 75-82; RETIRED. *Res:* Effect of various enzymes on in vitro blood coagulation tests; factor V and thrombin in the intrinsic and extrinsic clotting systems; preparation of self-instructional materials; colloidal aspects of blood clotting; nursing research with physiological implications; development of videotape study guide instructional program on cardiovascular physiology. *Mailing Add:* PO Box 2245 Chapel Hill NC 27515-2245

ENNIS, FRANCIS A, MOLECULAR GENETICS. *Prof Exp:* PROF MED MOLECULAR GENETICS & MICROBIOL, UNIV MASS MED CTR, 83- *Mailing Add:* Dept Med & Molecular Genetics Univ Mass Med Ctr 55 Lake Ave N Worcester MA 01605

ENNIS, HERBERT LEO, b Brooklyn, NY, Jan 6, 32; m 60; c 2. DEVELOPMENTAL BIOLOGY. *Educ:* Brooklyn Col, BS, 53; Northwestern Univ, MS, 54, PhD, 57. *Prof Exp:* USPHS fel, Northwestern Univ, 57-58; res fel bact & immunol & USPHS fel, Harvard Univ, 58-59; fel, Brandeis Univ, 59-60; instr pharmacol, Harvard Med Sch, 60-64; from asst prof to assoc prof biochem, St Jude Hosp & Col Med, Univ Tenn, 64-69; assoc mem, 69-77, MEM, ROCHE INST MOLECULAR BIOL, 77- *Concurrent Pos:* Ed, Antimicrobial Agents & Chemother, 77-88; mem, Sci Adv Comt on Nucleic Acids & Protein Synthesis, Am Cancer Soc, 90- *Mem:* AAAS; Am Soc Microbiol; Am Soc Biochem & Molecular Biol; NY Acad Sci; Sigma Xi. *Res:* Developmental biology; molecular genetics; nucleic acids. *Mailing Add:* Roche Inst Molecular Biol Nutley NJ 07110-1199

ENNIST, DAVID L, b Maracay, Venezuela, Jan 24, 57; m 78; c 2. MOLECULAR IMMUNOLOGY. *Educ:* Univ Rochester, BA, 78; Ohio State Univ, PhD(anat), 84. *Prof Exp:* Staff fel immunol, Nat Inst Allergy & Infectious Dis, 84-88; ADJ SCIENTIST MOLECULAR IMMUNOL, NAT INST CHILD HEALTH & HUMAN DEVELOP, 88- *Concurrent Pos:* Dissertation reader, George Wash Univ, 89-91; reviewer, Cytokine, 90- *Mem:* AAAS; NY Acad Sci; Am Asn Immunologists; Am Soc Microbiol. *Res:* Transcriptional regulation of major histo compatibility complex class I genes; mechanisms of gamma-IFN action on this gene. *Mailing Add:* Nat Inst Child Health & Human Develop Bldg 6 Rm 145 NIH Bethesda MD 20892

ENNOR, KENNETH STAFFORD, b Wadebridge, Eng, May 15, 33; m 65; c 2. ORGANIC CHEMISTRY. *Educ:* Univ London, BSc, 54, PhD(org chem), 57. *Prof Exp:* Fels, Ohio State Univ, 57-58 & Boston Univ, 58-59; tech officer plastics res, Imp Chem Industs Ltd, Eng, 59-62; scientist appln res resins, Esso Res Ltd, Eng, 62-65; sr scientist appln res tall oil prod, Brit Oxygen Chem Ltd, 65-68; SUPVR INFO & CHEM CONTROL, CHEM PROD DIV, UNION CAMP CORP, 68- *Mem:* Brit Oil & Colour Chem Asn; Am Chem Soc. *Res:* Wood-based fatty acids and resins for surface coatings, printing inks, ore flotation and adhesives; tall oil products; dimer acids; polyamide resins. *Mailing Add:* Chem Prod Div Union Camp Corp PO Box 2668 Savannah GA 31402

ENNS, ERNEST GERHARD, b Alta, June 13, 40; m 64, 77. MATHEMATICAL STATISTICS, OPERATIONS RESEARCH. *Educ:* Univ BC, BSc, 61, PhD(statist mech), 65. *Prof Exp:* Mem sci staff appl math, Northern Elec Res & Develop Labs, Ottawa, 65-67; consult teletraffic, Australian PMG, Melbourne, 67; lectr statist, Univ Queensland, 68; mem sci staff teletraffic, Siemens AG, Munich, 69; assoc prof, 69-77, assoc dean sci, 86-89, PROF STATIST, UNIV CALGARY, 77- *Mem:* Statist Soc Can. *Res:* Decision theory, queueing theory and geometrical probability. *Mailing Add:* Dept Math & Statist Univ Calgary Calgary AB T2N 1N4 Can

ENNS, JOHN HERMANN, b Schonau, Russia, July 18, 07; nat US; m 38; c 3. PHYSICS. *Educ:* Kans State Col, BS, 32; Univ Mich, AM, 35, PhD(physics), 41. *Prof Exp:* Instr physics, Detroit Inst Technol, 36-39; res physicist, Diamond Chain & Mfg Co, Indianapolis, 41-42; asst prof physics, Mich State Col, 42-44; res physicist, 44-58, from assoc prof to prof eng mech, 58-73, EMER PROF ENG MECH, UNIV MICH, ANN ARBOR, 73- *Mem:* Am Phys Soc; Optical Soc Am. *Res:* Sound and vibrations; emission spectroscopy; air interrupter type stabilized control gap for spark and alternating current arc source spectroscopy; solid state studies of photographic latent image formation; lattice dynamics and micromechanics of solids. *Mailing Add:* 12546 Nacido Dr Rancho Bernardo CA 92128

ENNS, MARK K, ELECTRICAL ENGINEERING. *Educ:* Kans State Univ, BS, 53; Univ Pittsburg, MS, 60, PhD, 67. *Prof Exp:* Radar maintenance officer, US Air Force, 53-55; Staff, Elec Utility Eng Dept, Westinghouse Elec Corp, Pittsburgh, Pa, 56-64; Staff, Systs Technol Dept, Westinghouse Res Labs, 64-67; asst prof, Elec Eng, Carnegie-Mellon Univ, Pittsburgh, Pa, 67-69; from assoc prof to prof, dept elect & Comput Eng, Univ Mich, Ann Arbor, 69-77, dir, Power Syst Lab, 72-77; engr mgr, vpres & pres, Shared Appln, Inc, Ann Arbor, 72-78, dir, 77-81; PRES, ELECTROCON INT INC, ANN ARBOR, 81- *Concurrent Pos:* Consult, Westinghouse Elec Corp, Res Labs, Pa Graphic Arts Tech Found, Pa, Pac Pumps Div, Dresser Industs, Detroit Edison Co, Control Data Corp, Minn, Gilbert Assocs Inc, Pa, New Eng Power Exchange, West Springfield, Bonneville Power Admin, Portland Comput Sci Corp, Va, Los Angeles Dept Water & Power, Calif, Ky Power Co, Ashland, Ky, Assoc Elec Corp, Springfield, Mo, Nucor Steel Co, Norfolk, NB; mem, Power Syst Eng Comt, USAB Manpower Comt; chmn, Fourth Biennial Workshop on Real-Time Monitoring & Control Power Syst, 76; assoc ed, Automatica, 68-72. *Mem:* Fel Inst Elec & Electronics Engrs (pres, 73-78); Power Eng Soc; Sigma Xi. *Res:* Author of 34 publications. *Mailing Add:* Electrocon Int Inc 611 Church St Ann Arbor MI 48104

ENNS, RICHARD HARVEY, b Winnipeg, Man, Nov 5, 38; m 67; c 3. THEORETICAL PHYSICS. *Educ:* Univ Alta, BSc, 60, PhD(theoret physics), 64. *Prof Exp:* Asst prof, 65-70, assoc prof, 70-76, PROF PHYSICS, SIMON FRASER UNIV, 76- *Concurrent Pos:* Nat Res Coun Can fel, Univ Liverpool, 64-65, res grant, 65-; dir, Theoret Sci Inst, 79-81 & 84-85; fel, Australian Nat Univ, 81-82. *Mem:* Can Asn Physicists. *Res:* Transport theory in solids; absorption and dispersion of sound in gases, nonlinear optics; other nonlinear problems. *Mailing Add:* Dept Physics Simon Fraser Univ Burnaby BC V5A 1S6 Can

ENNS, WILBUR RONALD, b Henderson, Nebr, Feb 26, 13; m 46; c 1. ENTOMOLOGY. *Educ:* Univ Mo, BS, 41, AM, 46; Univ Kans, PhD(entom), 55. *Prof Exp:* Asst, 42, from instr to prof, 48-78, dir entom mus, 52-78, EMER PROF ENTOM, UNIV MO-COLUMBIA, 78- *Honors & Awards:* Mus named in honor, Wilbur R Enns Entom Mus. *Mem:* Entom Soc Am; Soc Syst Zool; Am Entom Soc; Am Inst Biol Sci; Entom Soc Can. *Res:* Insect taxonomy; control of insects and mites. *Mailing Add:* 506 Bourn Ave Columbia MO 65203-1367

ENO, CHARLES FRANKLIN, b Atwater, Ohio, May 21, 20; m 48; c 2. SOIL SCIENCE. *Educ:* Ohio State Univ, BS, 42, MS, 48; Purdue Univ, PhD(soil microbiol), 51. *Prof Exp:* Soil microbiologist, 50-65, chmn dept soil sci, 65-82, ASST DIR, INT PROGS, UNIV FLA, 82- *Concurrent Pos:* Mem Coun Agr Sci & Technol. *Mem:* Fel Am Soc Agron (pres, 83); fel Soil Sci Soc Am (pres, 75). *Res:* Soil microbiology and related research in soil fertility. *Mailing Add:* 3964 NW 27th Lane Gainesville FL 32606

ENOCH, JACOB, b Berlin, Ger, Feb 17, 27; nat US; m 55; c 3. PLASMA PHYSICS. *Educ:* Brooklyn Col, BS, 52; Univ Wis, MS, 54, PhD(physics), 56. *Prof Exp:* Asst physics, Univ Wis, 52-56; mem staff, Midwestern Univs Res Asn, 56-57, Los Alamos Sci Lab, Univ Calif, 57-60 & Space Sci Lab, Gen Elec Co, 60-62; asst prof, 62-64, assoc prof physics, 64-80, ASSOC PROF PHYSICS & ASTRON, UNIV KANS, 80- *Concurrent Pos:* Vis staff mem, Los Alamos Sci Lab, 68-70, 73-; guest scientist, Max Plank Inst Plasma Physics, Ger, 70-71; Fulbright sr res fel, 70-71; vis prof, Ben Gurian Univ of the Negev, Israel, 71-73. *Mem:* Am Phys Soc. *Res:* Statistical mechanics; kinetic theory; equilibrium and stability of high temperature plasmas. *Mailing Add:* Dept Physics & Astron Univ Kansas Lawrence KS 66045

ENOCH, JAY MARTIN, b New York, NY, Apr 20, 29; m 51; c 3. PHYSIOLOGICAL OPTICS, VISION. *Educ:* Columbia Univ, BS, 50; Ohio State Univ, PhD(physiol optics), 56. *Prof Exp:* Asst sect head contact lenses, Army Med Res Lab, Ft Knox, Ky, 51-52; asst physiol optics, Ohio State Univ, 53-55, fel, 55-56, asst prof, 57-58, res assoc mapping & charting res lab, 56-57, assoc supvr, 57-58; from res instr to res asst prof physiol optics & ophthal, Med Sch, Wash Univ, 58-64, res assoc prof, 65-70, assoc prof psychol, 67-70, res prof ophthal & psychol, 70-74; grad res prof ophthal, psychol & physics, Col Med, Univ Fla, 74-80; PROF & DEAN, SCH OPTOM, UNIV CALIF, BERKELEY, 84-, PROF OPHTHAL, SCH MED, SAN FRANCISCO, 84- *Concurrent Pos:* Vis fel, Nat Phys Lab, Eng, 59-60; fel, Barnes Hosp, St Louis, 59-74, consult ophthal, 70-74; mem comt vision, Nat Acad Sci-Nat Res Coun, 59-, mem exec coun, 74-76; USPHS career develop award, 63-73; exec secy subcomt vision & its dis, Nat Inst Neurol Dis & Blindness, 65-69; assoc ed, Invest Ophthal, 66-72 & 82-88, Sight Saving Rev, Nat Soc Prevent Blindness, 74-84, Sensory Processes, 76-80, Int Ophthal, 80-, Binocular Vision, 85- & Clin Visual Sci, 86-; chmn subcomt contact lens standards, Am Nat Standards Inst, 69-79; mem adv coun, Aniseikonia Forum, 72; chmn comt standards, 74-82 & mem bd dirs, Int Perimetric Soc, 74-90; mem nat adv eye coun, Nat Eye Inst, NIH, 75-77 & 80-84; mem, Nat Comt Int Comn Optics, Nat Res Coun, 76-79; mem bd dirs, Friends of Eye Res, Rehab & Treatment, 76-88; vis prof, Waseda Univ, Tokyo, 78; chmn, comt res, Am Acad Optom, 80-84; chmn, Task Force, Res & Develop, Am Optom Asn, 83-85; chmn, Grad Group Vision Sci, Univ Calif, Berkeley, 84-; mem, Nat Soc Prevent Blindness, Fight-For-Sight Grants Comt, 88-; mem, NY Asn Visually Impaired Pisart Award Comt, 91- *Honors & Awards:* Glenn A Fry Award, Am Acad Optom, 71 & Charles F Prentice Medal, 74; Past Pres Award, Asn Res Vision & Ophthal, 74 & Francis Proctor Medal, 77; Honor Award, Am Acad Ophthal, 85; Wichtere Medal, Int Contact Lens Asn, 86; Everett Kinsey Award, Contact Lens Asn Opthal, 91. *Mem:* Fel AAAS; fel Optical Soc Am; Asn Res Vision & Ophthal (pres, 73); fel Am Acad Optom; Biophys Soc. *Res:* Retinal receptor optics and function; visual search; experimental perimetry; contact lenses; science administration; science policy. *Mailing Add:* Dept Optom Univ Calif Berkeley CA 94720

ENOCHS, EDGAR EARLE, b McComb, Miss, Sept 13, 32; m 58; c 7. MATHEMATICS, ALGEBRA. *Educ:* La State Univ, BS, 58; Univ Notre Dame, PhD(math), 58. *Prof Exp:* Instr math, Univ Chicago, 58-60; from asst prof to assoc prof, Univ SC, 60-67; PROF MATH, UNIV KY, 67- *Concurrent Pos:* NSF res grant, 63-64. *Res:* Abelian group theory; modules over integral domains; general topology; homological algebra. *Mailing Add:* Dept Math Univ Ky Lexington KY 40506

ENOS, PAUL (PORTENIER), b Topeka, Kans, July 25, 34; m 58; c 4. GEOLOGY. *Educ:* Univ Kans, BSc, 56; Stanford Univ, MSc, 61; Yale Univ, PhD(geol), 65. *Prof Exp:* Asst instr geol, Yale Univ, 62-64; geologist, Shell Develop Co, 64-65, res geologist, 65-70; assoc prof, State Univ NY, Binghamton, 70-76, prof geol, 76-82; HAAS PROF, DEPT GEOL, UNIV KANS, 82- *Concurrent Pos:* consult, Petrol Co; assoc ed, J Sedimentary Petrol, 77-80, 84-87. *Mem:* Soc Econ Paleontologists & Mineralogists; Asn Sedimentologists; Am Asn Petrol Geologists. *Res:* Sedimentology of flysch deposits; recent carbonates, Florida and Bahamas; Cretaceous carbonates, Mexico; deep sea sediments, Western North Atlantic; carbonate diagenesis; Permian and Triassic carbonates, SChina. *Mailing Add:* Dept Geol Univ Kans 120 Lindley Lawrence KS 66045

ENQUIST, IRVING FRITIOF, b Superior, Wis, June 25, 20; m 44; c 3. SURGERY. *Educ:* Univ Minn, BS, 42, MD, 44, MS, 51. *Prof Exp:* From instr to assoc prof, 52-60, PROF SURG, STATE UNIV NY DOWNSTATE MED CTR, 60-, ASSOC DEAN, 77-; DIR SURG, METHODIST HOSP BROOKLYN, 65- *Concurrent Pos:* Consult surg, US Vet Hosp, Brooklyn & St John's Episcopal Hosp, Brooklyn. *Mem:* AAAS; Am Col Surg; Am Surg Asn; Soc Surg Alimentary Tract; Int Soc Surg. *Res:* Wound healing; gastrointestinal physiology. *Mailing Add:* Methodist Hosp 506 Sixth St Brooklyn NY 11215

ENQUIST, LYNN WILLIAM, b Denver, Colo, Oct 23, 45; m 68; c 1. MOLECULAR BIOLOGY, MICROBIOLOGY. *Educ:* SDak State Univ, BS, 67; Med Col Va, PhD(microbiol), 71. *Prof Exp:* NSF fel, Med Col Va, 71; fel, Roche Inst Molecular Biol, 71-73; staff fel, Nat Inst Child Health & Human Develop, NIH, 73-77, staff scientist molecular biol, NIH, 77-81; exec scientist, Molecular Genetics, Inc, Minnetonka, Minn, 81-84; res leader, E I du Pont de Nemours, Co, Inc, 84-90; RES LEADER, DUPONT-MERCK PHARMACEUT, 91- *Mem:* Am Soc Microbiol; AAAS; Sigma Xi; Am Soc Virol. *Res:* Genetics and molecular biology of DNA viruses; structure and function of virus membrane proteins; herpes virus molecular biology; recombinant DNA technology. *Mailing Add:* 920 Fairthorne Ave Greenville DE 19807-2266

ENRIETTO, JOSEPH FRANCIS, b Spring Valley, Ill, May 7, 31; m 52; c 4. METALLURGY. *Educ:* Univ Ill, BS, 56, MS, 57, PhD(metall), 60. *Prof Exp:* Sr res engr, Res Lab, Jones & Laughlin Steel Corp, 60-63, supvr physics of metals group, 63-66, asst dir phys metall, 66-71; mgr mat eng, 71-77, mgr metall & nondestructive eval anal, 77-78, MGR MAT TECHNOL, NUCLEAR TECHNOL DIV, WESTINGHOUSE ELEC CORP, 78- *Mem:* Am Inst Mining, Metall & Petrol Engrs; Am Soc Metals; Welding Res Coun. *Res:* Internal friction in ferrous alloys; strain ageing; precipitation in ferrous base alloys; deep drawing; fatigue; nuclear pressure vessel materials; nondestructive examination; welding; stainless steel corrosion; fracture mechanics. *Mailing Add:* 916 Essex Ave Greensburg PA 15601

ENRIGHT, JAMES THOMAS, b Baker, Ore, Nov 23, 32; m 68; c 4. BEHAVIORAL PHYSIOLOGY. *Educ:* Univ Calif, Los Angeles, AB, 57, MA, 59, PhD(zool), 61. *Prof Exp:* NSF fel, Max-Planck Inst Physiol of Behav, Ger, 61-63; asst prof zool, Univ Calif, Los Angeles, 63-66; from asst prof to assoc prof oceanog, 66-73, PROF BEHAV PHYSIOL, UNIV CALIF, SAN DIEGO, 73- *Concurrent Pos:* Fulbright fel, 78; Alexander von Humboldt Prize, 81. *Mem:* Fel AAAS; Sigma Xi; Am Sec Nat; Asn Res Vision & Ophthal. *Res:* Behavioral physiology; biological rhythms; orientation; photoperiodism; marine ecology; human eye movements. *Mailing Add:* 7561 Cabrillo Ave La Jolla CA 92037

ENROTH-CUGELL, CHRISTINA, b Helsingfors, Finland, Aug 27, 19; US citizen; m 55. VISION, NEUROPHYSIOLOGY. *Educ:* Karolinska Inst, Sweden, Med lic, 48, Med dr(neurophysiol of vision), 52, Ophthal Specialist Cert, 57. *Prof Exp:* Resident ophthal, Sabbatsberg Hosp, Stockholm, Sweden, 48-49; res fel, Karolinska Inst, Sweden, 50-53, resident, Karolinska Hosp, 53 & 54-56; from asst prof to assoc prof physiol & biol sci, 62-72, from assoc prof to prof biol & elec eng, 72-81, PROF DEPTS BIOMED ENG, NEUROBIOL & PHYSIOL, NORTHWESTERN TECHNOL INST, EVANSTON, 81- *Concurrent Pos:* NIH res fel biol, Harvard Univ, 53-54; NIH spec trainee, Med Sch, Northwestern Univ, 58-61 & career develop award, 62-72; gen ed, Invest Ophthal, 79-83 & Clin Vision Sci, 86-; mem, Training & Ctr Grant study sect, Nat Eye Inst, 74-78, Nat Adv Coun, NIH, 79-83 & prog comt electrophysiol, Asn Res Vision & Ophthal Inc, 75-77, chmn, 78; chmn dept neurobiol & physiol, Northwestern Univ, 84-86; pres, Soc Neurosci, local chap, 84-85. *Honors & Awards:* von Sallmann Prize Vision & Ophthal, Fifth Inst Cong Eye Res Meeting, Veldhoven, 82; Friedenwald Award, Asn Res Vision & Ophthal, 83. *Mem:* Am Physiol Soc; Physiol Soc UK; fel Am Acad Arts & Sci. *Res:* Visual physiology, particularly retinal neurophysiology. *Mailing Add:* N U McCormick Sch Engr Bio-Med 2145 Sheridan Rd Evanston IL 60208-3107

ENS, E(RICH) WERNER, b Altona, Man, Sept 14, 56; m 82; c 3. LASER DESORPTION, TIME-OF-FLIGHT MASS SPECTROMETRY. *Educ:* Univ Winnipeg, BSc, 77; Univ Man, PhD(physics), 84. *Prof Exp:* Vis scholar physics, Univ Uppsala, Sweden, 86-87; postdoctoral scholar physics, 84-85, res assoc, 85-86, ASST PROF PHYSICS, UNIV MAN, 87- *Concurrent Pos:* Lectr, Univ Man, 84-86; consult, Bio Ion Nordic, Uppsala, Sweden, 86-89, Waters Div, Millipore, 90- *Mem:* Am Soc Mass Spectrometry; Can Asn Physicists. *Res:* Mass spectrometry of biological molecules and clusters using particle and laser desorption with time-of-flight; understanding the fundamentals of the desorption processes. *Mailing Add:* Physics Dept Univ Man Winnipeg MB R3T 2N2 Can

ENSIGN, PAUL ROSELLE, b Shantung, China, Aug 27, 06; US citizen; m 39; c 1. PUBLIC HEALTH, PEDIATRICS. *Educ:* Univ Kans, BA, 27; Northwestern Univ, MD, 36; Johns Hopkins Univ, MPH, 42. *Prof Exp:* Pediat consult, Ga State Health Dept, 43-45; div dir maternal & child health, Kans State Bd Health, 45-50, dept state health officer, 50-51; div dir maternal & child health, Mont State Health Dept, 51-55, dep state health officer, 55-57; health officer, City-Co Health Dept, Great Falls, 57-62; dir div ment health, Utah State Health Dept, 62-64; field consult, Ford Found, 64-69; DIR PREV DIS & ENVIRON HEALTH, ACTG STATE DIR HEALTH & DIR CHILD HEALTH, UTAH STATE DIV HEALTH, 69- *Concurrent Pos:* Assoc prof prev med & instr pediat, Univ Kans, 46-51; assoc prof prev med, Univ Utah, 62-64; consult health & family planning, Govt of India, 64-; pres, Asn State Maternal & Child Health Dirs, 54; NIMH grant community ment health, Great Falls, Mont, 60-65. *Mem:* Am Pub Health Asn. *Res:* Maternal and child health, particularly nutrition, prevention of otitis media in Indian children; mental health and hospital nursery infections. *Mailing Add:* 4725 S Bron Breck Dr Salt Lake City UT 84117

ENSIGN, RONALD D, b Cameron, Mo, Apr 10, 22; m 47; c 2. AGRONOMY, CROP BREEDING. *Educ:* Northwestern Mo State Col, BS, 47; Colo State Univ, MS, 49; Cornell Univ, PhD(plant breeding), 52. *Prof Exp:* Asst agron, Colo State Univ, 47-49; asst plant breeding, Cornell Univ, 49-52; supt, 52-55, assoc dir, Agr Exp Sta, 55-71, prof, 71-, EMER PROF PLANT SCI, AGR EXP STA, UNIV IDAHO. *Mem:* Am Soc Agron; AAAS; Sigma Xi. *Res:* Improvement of biological performance of Idaho fescue, Kentucky bluegrass and white clover by various plant breeding techniques; studies of various cultural treatments in production of Kentucky bluegrass seed; cultural treatments in turf production. *Mailing Add:* 1816 Eastmont Ave East Wenatchee WA 98802-4201

ENSIGN, STEWART ELLERY, b Waterloo, Iowa, Nov 25, 25; m 46; c 3. GENETICS. *Educ:* Bob Jones Univ, BA, 50; Univ Wyo, MS, 54; Univ Nebr, PhD(genetics), 59. *Prof Exp:* Instr biol, Bob Jones Univ, 52-55; res fel, Yale Univ, 59-61 & Univ Calif, San Diego, 61-63; asst prof, 63-70, PROF BIOL, WESTMONT COL, 70-, CHMN DEPT, 73- *Mem:* AAAS; Am Sci Affil; Genetics Soc Am; Asn Am Med Cols. *Res:* Reproductive isolation in the Affinis Subgroup of the genus Drosophila; gene-enzyme relations in the tryptophan synthetase system of Neurospora crassa; production of ovarian proteins in Blattella germanica. *Mailing Add:* 729 Circle Dr Santa Barbara CA 93108

ENSIGN, THOMAS CHARLES, b Minneapolis, Minn, Mar 6, 41; m 62; c 3. SOLID STATE PHYSICS. *Educ:* Macalester Col, BA, 63; Univ Wyo, MS, 65, PhD(physics), 68. *Prof Exp:* Lab asst physics, Macalester Col, 60-63; teaching asst, Univ Wyo, 63-64, res asst, 64-65, teaching asst, 65-66; Nat Res Coun res assoc, Nat Bur Standards, 68-69; sr res scientist, Res Inst Advan Studies, Martin Marietta Corp, 69-74; res specialist, Cent Res Labs, 74-78, LAB MGR, SOLID STATE PROCESS DEVELOP, 3M, 78- *Mem:* Am Phys Soc. *Res:* Solid state physics; microelectronics; lasers; thin films. *Mailing Add:* 3M Fiber Opt Ten Industrial Way E Eatontown NJ 07724

ENSINCK, JOHN WILLIAM, b Montreal, Que, Feb 19, 31; m 60; c 1. ENDOCRINOLOGY. *Educ:* McGill Univ, BSc, 52, MD, 56. *Prof Exp:* Resident med, Royal Victoria Hosp, 56-58; res assoc & asst physician, Rockefeller Inst, 58-60; asst med, Sch Med, Univ Wash, 60-61, instr & asst dir, Clin Res Ctr, Univ Hosp, 61-62; vis lectr, dept med, Univ Newcastle, 62-64; asst dir, Clin Res Ctr, Univ Hosp, 64-72, from asst prof to assoc prof med, Sch Med, 64-72, PROF MED, SCH MED, UNIV WASH, 72-, DIR CLIN RES CTR, UNIV HOSP, 72-, PROG DIR, 70- *Mem:* AAAS; Am Diabetes Asn; Am Fedn Clin Res; Am Soc Clin Invest; Endocrine Soc. *Res:* Endocrinological investigation with application of protein chemistry in relationship of insulin to carbohydrate metabolism. *Mailing Add:* Dept Med Univ Wash Seattle WA 98195

ENSLEY, HARRY EUGENE, b Charleston, SC, Aug 18, 45; m 66; c 2. ORGANIC CHEMISTRY. *Educ:* Vanderbilt Univ, BS, 70; Harvard Univ, PhD(chem), 75. *Prof Exp:* asst prof, 75-80, ASSOC PROF, DEPT CHEM, TULANE UNIV, 81- *Mem:* AAAS; Am Chem Soc; Sigma Xi. *Res:* Mechanism of reactions of singlet oxygen; synthesis of natural products and the development of new synthetic methodology. *Mailing Add:* Dept Chem 15010 Sci Tulane Univ 6823 St Charles Ave New Orleans LA 70118

ENSMINGER, DALE, b Mt Perry, Ohio, Sept 26, 23; m 48; c 6. ULTRASONICS. *Educ:* Ohio State Univ, BME, 50, BEE, 50. *Prof Exp:* Res engr ultrasonics, Battelle Mem Inst, 50-70, sr elec engr, 70-78, sr res scientist, 78-81, mgr ultrasonics res progs, 81-87; RETIRED. *Concurrent Pos:* Sr res, scientist, Battelle Mem Inst, 87-; consult, ultrasonics, 87- *Mem:* Acoust Soc Am; Soc Nondestructive Testing. *Res:* Low and high intensity applications of ultrasonics; all applications of ultrasonics. *Mailing Add:* 198 E Longview Ave Columbus OH 43202

ENSMINGER, LEONARD ELROY, b Stover, Mo, Sept 25, 12; m 41; c 1. AGRONOMY. *Educ:* Univ Mo, BS, 35; Univ Ill, PhD(soil chem), 40. *Prof Exp:* Asst prof agr chem, Univ Idaho, 39-42; soil chemist, Exp Sta, Univ Fla, 42-44; assoc prof, 44-53, prof, 53-79, head, Dept Agron & Soils, 66-79, EMER PROF AGRON, AUBURN UNIV, 79- *Mem:* Fel Am Soc Agron. *Res:* Factors affecting the availability to plants of native and added phosphorus in soils; identification of clay minerals in soils; sulfur in relation to soil fertility. *Mailing Add:* Rte 1 Box 49 Waverly AL 36879

ENSOR, DAVID SAMUEL, b Spokane, Wash, Aug 22, 41; m 70. AIR POLLUTION. *Educ:* Wash State Univ, BS, 63; Univ Wash, MS, 68, PhD(eng), 72. *Prof Exp:* Chem engr, E I Du Pont de Nemours & Co, Inc, 63-65; res scientist aerosols, Meteorol Res Inc, 72-79; HEAD CENT TECHNOL RES, RES TRIANGLE INST, 79- *Concurrent Pos:* Adj prof chem eng, NC State Univ; air pollution special fel. *Mem:* Am Chem Soc; Air Pollution Control Asn; Am Optical Soc; Am Inst Chem Engrs; Sigma Xi. *Res:* Applied aerosol science; light scattering, particle size distribution and chemical composition measurement; control technology; chemical engineering; fabric filtration; economic analysis of gas cleaning equipment; plume opacity and visibility. *Mailing Add:* 12 Sweetbriar Lane Chapel Hill NC 27514

ENSOR, PHYLLIS GAIL, b Baltimore, Md, Nov 11, 38. COMMUNITY HEALTH, MEDICAL ADMINISTRATION. *Educ:* Taylor Univ, BA, 61; Univ Md, MA, 68; NY Univ, PhD(community health), 77. *Prof Exp:* Dir health educ, Nat Found-March Dimes, White Plains, NY, 69-73; ASSOC PROF COMMUNITY HEALTH, TOWSON STATE UNIV, 73- *Concurrent Pos:* Eval grant, Acad Dean's Off, Towson State Univ, 78. *Mem:* Soc Pub Health Educ; Am Pub Health Asn. *Res:* Alcoholism, relationship of drinking patterns husbands, wives and socioeconomic status research in progress. *Mailing Add:* Dept Health Sci Towson State Univ Towson MD 21204

ENSSLIN, NORBERT, b Nov 30, 44; US citizen. NUCLEAR PHYSICS. *Educ:* Calif Inst Technol, BS, 67; Mass Inst Technol, PhD(physics), 72. *Prof Exp:* Fel nuclear physics, Naval Res Lab, Washington, DC, 72-73; fel, Meson Physics Facil, 73-75, MEM STAFF NUCLEAR PHYSICS, LOS ALAMOS SCI LAB, 75- *Mem:* Am Phys Soc. *Res:* Nuclear safeguards. *Mailing Add:* Los Alamos Sci Lab MS E 540 Los Alamos NM 87545

ENSTROM, JAMES EUGENE, b Alhambra, Calif, June 20, 43; m 78. EPIDEMIOLOGY, PHYSICS. *Educ:* Harvey Mudd Col, BS, 65; Stanford Univ, MS, 67, PhD(physics), 70; Univ Calif, Los Angeles, MPH, 76. *Prof Exp:* Res assoc, Stanford Linear Accelerator Ctr, Stanford Univ, 70-71; res physicist & consult, Lawrence Berkeley Lab, Univ Calif, 71-75; C D Rogers cancer res fel, 73-75, Nat Cancer Inst fel, 75-76, cancer epidemiol researcher, 76-81, ASSOC RES PROF, SCH PUB HEALTH, UNIV CALIF, LOS ANGELES, 81-; RES EPIDEMIOL, JONSSON COMPREHENSIVE CANCER CTR, 78- *Concurrent Pos:* Consult physicist, Rand Corp, Santa Monica, Calif, 69-73 & R & D Assocs, Marina Del Ray, 71-75; consult epidemiologist, Linus Pauling Inst Sci & Med, 76-; Preventive Oncology Academic Award, Nat Cancer Inst, 81-87. *Mem:* AAAS; Am Pub Health Asn; Am Phys Soc; Soc Epidemiol Res; Am Heart Asn. *Res:* Epidemiology of cancer and other diseases, especially among low-risk populations such as Mormons; experimental nuclear particle physics. *Mailing Add:* Sch Pub Health Univ Calif Los Angeles CA 90024-1772

ENSTROM, RONALD EDWARD, b New York, NY, Mar 22, 35; m 58; c 2. ELECTRONICS ENGINEERING, SOLID STATE PHYSICS. *Educ:* Mass Inst Technol, SB, 57, SM, 62, ScD, 63. *Prof Exp:* Asst metall, metals res lab, Union Carbide Corp, 57-58; mat engr, Nuclear Metals, Inc, Div, Textron, Inc, 58-60; asst metall, Mass Inst Technol, 60-63; MEM TECH STAFF, DAVID SARNOFF RES CTR, 63- *Concurrent Pos:* Vis scientist, Swiss Fed Inst Technol, 73-74; bd dirs & pres, Electrochem Soc, 86-87. *Honors & Awards:* RCA Labs Outstanding Res Awards, 66, 72 & 83. *Mem:* Electrochem Soc (vpres, 83-86); Am Phys Soc; Am Inst Mining, Metall & Petrol Engrs; sr mem Inst Elec & Electronics Engrs; Am Soc Crystal Growth; Sigma Xi. *Res:* Organometallic chemical vapor deposit synthesis and characterization of III-V compounds for microwave, power and opto-electronic applications; finite element modeling; glass and plastics technology; preparation and properties of superconducting materials. *Mailing Add:* 67 Colfax Rd Skillman NJ 08558

ENTEKHABI, DARA, b Tehran, Iran, Jan 16, 61, US citizen. HYDROLOGY & WATER RESOURCES. *Educ:* Clark Univ, BA, 83, MA, 84, MA, 87; Mass Inst Technol, PhD(civil eng), 89. *Prof Exp:* Postdoctoral civil eng, Mass Inst Technol, 89-90; ASST PROF HYDROL, UNIV ARIZ, 90- *Concurrent Pos:* Vis asst prof, Mass Inst Technol 90-92; Presidential young investr, NSF, 91. *Mem:* Am Geophys Union; Am Meteorol Soc. *Res:* Global hydrology and hydroclimatology. *Mailing Add:* Mass Inst Technol Br PO Box 193 Cambridge MA 02139

ENTIN, MARTIN A, b Simferopole, Crimea, Oct 19, 12; Can citizen; c 3. PLASTIC SURGERY, RECONSTRUCTIVE SURGERY. *Educ:* Temple Univ, BA, 41; McGill Univ, MSc, 42, MD & CM, 45. *Prof Exp:* Resident surg, Montreal Gen, Royal Victoria & Montreal Children's Hosps, 46-49; clin asst plastic surg, Royal Victoria Hosp, 50-55; asst lectr, McGill Univ, 57-62, asst prof, 64-71; asst surgeon, Royal Victoria Hosp, 63-70, actg surgeon-in-chief, Sub-Dept Plastic Surg, 70-71, surgeon-in-chg, Subdept plastic surg, 71-77, SR SURGEON, DEPT SURG, ROYAL VICTORIA HOSP, 77-; ASSOC PROF SURG, MCGILL UNIV, 71- *Concurrent Pos:* Chmn, Plastic Surg Res Coun, 59; Nat Res Coun Can fel, McGill Univ, 46-47, Med Res Coun Exp Work & Nat Res Coun grants, 58-65; Nat Res Coun Can res fel surg hand, Stanford Univ Hosp, 49-50; Defense Res Bd Can grants, 51-54. *Mem:* Am Soc Plastic & Reconstruct Surg; fel Am Col Surg; Am Soc Surg Hand (vpres, 71, pres-elect, 72-73, pres, 73-74); Can Soc Plastic Surg; Brit Asn Plastic Surg. *Res:* Experimental and clinical investigation and reconstruction of congenital anomalies of upper extremities; experimental production of rheumatoid arthritis; wound healing evolve toward pathogenesis of thermal injury; investigation of feasibility of autogenous whole joint transplantation. *Mailing Add:* Royal Victoria Hosp Montreal PQ H3A 1A1 Can

ENTMAN, MARK LAWRENCE, b New York, NY, Dec 24, 38; m 68; c 2. CELL PHYSIOLOGY. *Educ:* Duke Univ, MD, 63. *Prof Exp:* Res fel cardiol, Med Ctr, Duke Univ, 62-63; intern osler serv, Johns Hopkins Hosp, Baltimore, 63-64; fel, Med Ctr, Duke Univ, 64-65, fel, Res Training Prog, 65-66, from asst resident to assoc, 67-68; res physician & actg chief muscle metab, Armed Forces Inst Path, 68-70; asst prof med & myocardial biol, Baylor Col Med, 70-73, asst prof cell biophys, 72-73, from assoc prof to prof med & cell biophys, 73-77; investr cardiol & biophys, Howard Hughes Med Inst, 72-78; PROF & CHIEF SECT CARDIOVASC SCI, DEPT MED, BAYLOR COL MED, 77-; DIR DIV RES, DEBAKEY HEART CTR, 84- *Concurrent Pos:* Mem, Pharmacol Study Sect, Div Res Grants, NIH, 75-79; mem, Vet Admin Cardiovasc Res Rev Comt, 87-90. *Honors & Awards:* Young Investr Award, Am Col Cardiol, 67; Award for Outstanding Res Accomplishment, Int Soc Heart Res, 85; Res Merit Award, Nat Heart, Liver, Blood Inst, 88. *Mem:* Am Soc Pharmacol & Exp Therapeut; Am Physiol Soc; Am Soc Biol Chem; Biophys Soc; Am Soc Clin Invest; Asn Am Physicians. *Res:* Cardiac and skeletal muscle cell biology, in particular, the role of ion flux, sarcoplasmic reticulum, in excitation-contraction coupling; basic mechanisms of myocardial ischemic damage and the cellular and molecular basis for cardiovascular inflammation. *Mailing Add:* Baylor Col Med 6516 Bertner Dr Houston TX 77030

ENTREKIN, DURWARD NEAL, b Ga, Nov 25, 26; m 54. PHARMACY. *Educ:* Univ Ga, BS, 50; Univ Fla, MS, 51, PhD(pharm), 53. *Prof Exp:* Res assoc pharm, E R Squibb & Sons, 53-57; from asst prof to assoc prof, 57-65, PROF PHARM, SCH PHARM, UNIV GA, 65-, ASSOC DEAN, 68- *Mem:* Am Pharmaceut Asn. *Res:* Use of imitation flavors for masking distasteful drugs. *Mailing Add:* 455 Ponderosa Dr Athens GA 30605

ENTRINGER, ROGER CHARLES, b Iowa City, Iowa, May 17, 31; m 55; c 4. NUMBER THEORY. *Educ:* Iowa State Univ, BS, 52; Univ NMex, MS, 57, PhD(math), 63. *Prof Exp:* From instr to assoc prof, 58-74, PROF MATH, UNIV NMEX, 74- *Mem:* Am Math Soc; Math Asn Am; Soc Indust Appl Math; Sigma Xi. *Res:* Graph and combinatorial theory. *Mailing Add:* 1505 Dakota NE Albuquerque NM 87110

ENTZMINGER, JOHN N, b Memphis, Tenn, Dec 17, 36; m 61; c 2. ELECTRICAL ENGINEERING, ELECTRONICS ENGINEERING. *Educ:* Univ SC, BS, 59; Syracuse Univ, MS, 68. *Prof Exp:* Proj eng, Bell Tel Labs, Winston, Salem, NC, 59-60; proj eng, US Air Force, Rome Air Develop Ctr, 60-63 & 63-73, tech dir, Intel & Reconnaissance Directorate, 81-83; DIR, TACTICAL TECHNOL OFF, DEFENSE ADVAN RES PROJ AGENCY, DEPT DEFENSE, 83- *Concurrent Pos:* Chief, Location & Control, Rome Air

Develop Ctr, 73-81. *Mem:* Fel Inst Elec & Electronics Engrs; AAAS. *Res:* Air defense systems; surveillance, target acquisition and target recognition sensor systems; smart weapon systems; signal and data processing; communications, command and control and intelligence systems. *Mailing Add:* 3203 Dominy Court Oakton VA 22124

ENY, DESIRE M(ARC), b Algiers, France, Feb 8, 15; nat US; m 44; c 2. ENVIRONMENT, PUBLIC WORKS. *Educ:* Univ Algiers, BS, 35; Breguet Inst, Paris, MSE, 38; Cornell Univ, PhD(chem, physiol), 48. *Prof Exp:* Mgr electrochem dept, Precision Metal Prod Co, 40-43; res assoc, Univ Calif, 44-46; prof, Univ Fla, 48-49; bio-engr, Firestone Tire & Rubber Co, 49-51; coordr for Latin Am, USDA, 51-53; chief biol warfare br, US Army Chem Ctr, Md, 53-54, chem biol & nuclear protection div, 54-57, dir chem, biol & nuclear protection directorate, 57-62; dir planning, US Naval Exp Sta, 62-63; mgr eng & prod div, chem group, Glidden Co, 63-66; environ opers div, Spindletop Ctr, Ky, 66-67; pres, D Marc Eny Assocs, 67-82; CONSULT WATER & POLLUTION CONTROL, 82- *Concurrent Pos:* Consult, Army Res Off & Off Civil Defense & Mobilization, 56-63, States of KY & Md, US Econ Develop Admin, 66-68, City of Baltimore, 68-71 & US Small Bus Admin, 68-74; mem, State of Ky Sci Comn, 66-68; tech adv, Baltimore Harbor Pollution Comt, 68- *Honors & Awards:* Commendation, US Small Bus Admin, 70; Wisdom Award, 70. *Mem:* Fel Am Inst Chemists; Am Soc Civil Engrs; Am Inst Chem Engrs; Inst Elec & Electronics Engrs; Am Water Works Asn; Sigma Xi. *Res:* Environmental engineering; water; waste-water; air; solid waste studies; design; master-plans. *Mailing Add:* PO Box 257 Kingsville MD 21087

ENZ, JOHN WALTER, US citizen. AGRICULTURAL CLIMATOLOGY, MICROCLIMATOLOGY. *Educ:* Univ Wis-Stevens Point, BS, 68; Univ Minn, MS, 71, PhD(soil sci), 76. *Prof Exp:* Res asst soil sci, Univ Minn, 69-76; res assoc, 76-77; from asst prof to assoc prof, 77-89, PROF, AGR CLIMAT, DEPT SOIL SCI, NDAK STATE UNIV, 89- *Mem:* Am Meteorol Soc; Am Soc Agron; Sigma Xi. *Res:* Meso and micro climatology and meteorology; solar radiation and energy budget studies; crop or yield modeling. *Mailing Add:* Soil Sci Dept NDak State Univ Fargo ND 58105

ENZER, NORBERT BEVERLEY, b Milwaukee, Wis, Nov 26, 30; m 56; c 3. CHILD PSYCHIATRY, PEDIATRICS. *Educ:* Yale Univ, BA, 52; McGill Univ, MD, 56. *Prof Exp:* Intern pediat, Med Ctr, Duke Univ, 56-57, resident, 57-58 & 60-61, resident psychiat, 61-64, asst prof psychiat & assoc pediat, 65-68; from assoc prof to prof psychiat & pediat, Sch Med, Univ New Orleans, 68-73, head dept psychiat & biobehav sci, 71-73; chmn dept psychiat, 73-81, assoc dean acad affairs, Col Human Med, 81-84, PROF PSYCHIAT, COL MED, MICH STATE UNIV, 73- *Concurrent Pos:* Fel child psychiat, Duke Univ, 63-65. *Mem:* Am Acad Pediat; Am Psychiat Asn; Am Acad Child Psychiat; Soc Res Child Develop. *Res:* Child psychiatry and development. *Mailing Add:* Dept Psychiat Mich State Univ East Lansing MI 48824-1316

ENZINGER, FRANZ MICHAEL, b Rohrbach, Austria, Feb 17, 23; US citizen; m 62; c 1. PATHOLOGY. *Educ:* Innsbruck Univ, MD, 50. *Prof Exp:* Asst anat & histol, Innsbruck Univ, 50-51, asst forensic med, 53-54; intern, Westchester Hosp, Mt Kisco, NY, 51-52; resident & instr path, Univ Iowa, 52-53, 54-57; assoc pathologist, 57-59, CHIEF, SOFT TISSUE DIV, ARMED FORCES INST PATH, 60- *Concurrent Pos:* Chief, Int Ctr Soft Tissue Tumors, WHO; clin prof path, Uniformed Serv Univ, Bethesda, MD. *Mem:* Am Soc Clin Path; Int Acad Path. *Res:* Neoplastic diseases, especially soft tissue tumors; diagnostic pathology. *Mailing Add:* 6825 16th St Washington DC 20305

ENZMANN, ROBERT D, b Peking, China, Nov 5, 30; US citizen; m 58; c 6. GEOLOGY, ELECTRICAL ENGINEERING. *Educ:* Harvard Univ, AB, 49; Univ Witwatersrand, SAfrica, BS & MS, 53; Mass Inst Technol, PhD, 56; Uppsula, Swed, PhD, 78. *Prof Exp:* Consult geol & geophys mining co, Africa, Mediterranean Basin & Greenland, 50-57; consult, Radio Corp Am, 58-59, Convair rep & design specialist, 59-60, sr engr, Alaska & Greenland, 60-62; consult plans & projects, Avco Corp, 62-65; sr engr, Raytheon, 65-73; PRES, NORTHEAST UNIV, 72-; PROF, TRANSKEI INT UNIV, 80- *Concurrent Pos:* Asst, Mass Inst Technol, 54-55; res asst prof, Radiation Lab, Univ Mich, 62; asst prof, Northeastern Univ, 63-; asst prof, Univ Boston. *Mem:* Am Geophys Union; Am Inst Aeronaut & Astronaut; Geol Soc SAfrica; Swedish Geol Soc; fel NY Acad Sci. *Res:* Field geology; weapons systems design; space mission planning and planetology; use of instruments and engineering values in planetary orbital space, electrospheres, atmospheres, hydrospheres and endospheres. *Mailing Add:* 29 Adams St Lexington MA 02173

EOFF, KAY M, b Refugio, Tex, Sept 20, 32; m 65. PHYSICS. *Educ:* Tex Col Arts & Sci, BS, 53, MS, 55; Univ Fla, PhD(physics), 63. *Prof Exp:* Nuclear engr, Convair Div, Gen Dynamics Corp, 54-55; asst prof physics & astron, 56-77, asst prof phys sci, 77-80, ASST PROF PHYS SCI & GEOG, UNIV FLA, 80- *Mem:* Am Phys Soc. *Mailing Add:* Dept Phys Sci Univ Fla Gainesville FL 32611

EOLL, JOHN GORDON, b Detroit, Mich, Feb, 9, 43; m 65; c 3. STELLAR NUCLEOSYNTHESIS, THERMONUCLEAR FUSION. *Educ:* Wayne State Univ, BS, 64; Ind Univ, MA, 71, PhD (astrophys), 76. *Prof Exp:* Staff mem, Los Alamos Nat Lab, 76-80; asst prof physics, Lenoir-Rhyne Col, 80-84; sr scientist, Sci & Eng Assoc, 84-86; mem tech staff, Alphatech, Inc, 86-87; SR SCIENTIST, GEN RES CORP, 87- *Concurrent Pos:* Consult, Naval Surface Weapons Ctr, White Oak Lab, 81-83 & Air Force Weapons Lab, 83-84. *Mem:* Sigma Xi. *Res:* Numerical modeling of physical systems; hydrodynamics; combustion; radiation transport; atmospheric chemistry; nuclear physics processes. *Mailing Add:* 12 Boden St Beverly MA 01915-3805

EPAND, RICHARD MAYER, b New York, NY, Dec 31, 37; m 65; c 2. BIOPHYSICAL CHEMISTRY. *Educ:* Johns Hopkins Univ, AB, 59; Columbia Univ, PhD(biochem), 64. *Prof Exp:* Fel biophys chem, Cornell Univ, 65-68; vis scientist, Inst Biochem Res, Buenos Aires, Arg, 68-69; asst prof chem, Univ Guelph, 69-74; assoc prof, 74-78, PROF BIOCHEM, MCMASTER UNIV, 78-, ASSOC MEM, DEPT CHEM, 87- *Concurrent Pos:* vis prof molecular biochem & biophys, Yale Univ, 80-81. *Mem:* Am Soc Biol Chemists; AAAS; Am Chem Soc; Can Biochem Soc; Biophys Soc; Am Peptide Soc. *Res:* Biophysical properties and membrane function; viral membrane fusion, protein kinase C, multidrug resistance protein-lipid interactions; peptide hormone mechanism of action. *Mailing Add:* Dept Biochem McMaster Univ Health Sci Centre Hamilton ON L8N 3Z5 Can

EPEL, DAVID, b Detroit, Mich, Mar 26, 37; m 60; c 3. DEVELOPMENTAL BIOLOGY, CELL BIOLOGY. *Educ:* Wayne State Univ, AB, 58; Univ Calif, Berkeley, PhD(zool), 63. *Prof Exp:* Asst prof biol, Hopkins Marine Sta, Stanford Univ, 65-70; from assoc prof to prof marine biol, Univ Calif, San Diego, 70-77; actg dir, 84-88, PROF BIOL SCI, HOPKINS MARINE STA, STANFORD UNIV, 77- *Concurrent Pos:* Johnson Res Found fel, Sch Med, Univ Pa, 63-65; assoc ed, Develop Biol, 72-83; mem adv panel, Develop Biol Sect, NSF, 74-75, 87-; co-dir embryol, Marine Biol Lab, Woods Hole, 75-78; Guggenheim fel, 76-77; overseas fel, Churchill Col, Univ Cambridge, 76-77; coun mem, Am Soc Cell Biol, 79-81; mem-at-large, Sect G Biol Sci, AAAS, 79-84; trustee, Monterey Bay Aquarium, 84-89, bd dirs, Res Inst, 87-89; vis prof, Toho Univ, 89; chairperson, Develop Biol, Am Soc Zoologists, 90- *Mem:* Fel AAAS; Am Soc Cell Biol; Soc Develop Biol; Int Soc Develop Biol; Am Soc Zoologists. *Res:* Cell biology; biochemistry of fertilization and early development; comparative embryology. *Mailing Add:* Hopkins Marine Sta Stanford Univ Pacific Grove CA 93950

EPHREMIDES, ANTHONY, b Athens, Greece, Sept 19, 43; US citizen; m 74. COMMUNICATION SYSTEMS, COMMUNICATION NETWORKS. *Educ:* Nat Tech Univ, Athens, BS, 67; Princeton Univ, MA, 69, PhD(elec eng), 71. *Prof Exp:* Res asst elec eng, Princeton Univ, 67-71; asst prof, 71-74, assoc prof, 74-81, PROF ELEC ENG, UNIV MD, 81- *Concurrent Pos:* Prin investr res grants, Univ Md, 71-; invited lectr, numerous int agencies & univs, 71-; consult, Naval Res Lab, 77-; ed, Transactions Automatic Control, Inst Elec & Electronics Engrs, 78-80; pres, Pontos Inc, 81- *Mem:* Inst Elec & Electronics Engrs; AAAS. *Res:* Statistical communications; computer communications. *Mailing Add:* Elec Eng Dept Univ Md College Park MD 20742

EPIFANIO, CHARLES EDWARD, b New York, NY, Aug 28, 44; m 66; c 2. MARINE BIOLOGY. *Educ:* Lafayette Col, AB, 66; Duke Univ, PhD(zool), 71. *Prof Exp:* From asst prof to assoc prof, 71-85, dir, marine biol prog, 85-90, PROF MARINE STUDIES, UNIV DEL, 85- *Mem:* AAAS; Am Soc Zoologists; Nat Shellfisheries Asn; Estuarian Res Fedn; Crustacean Soc; Am Soc Limnol & Oceanog. *Res:* Biology and ecology of marine invertebrate larvae; reproductive biology of tropical invertebrates. *Mailing Add:* Col Marine Studies Univ Del Lewes DE 19958

EPLER, JAMES L, b Lancaster, Pa, Aug 10, 37; m 75; c 5. GENETICS, BIOCHEMISTRY. *Educ:* Millersville State Col, BS, 59; Fla State Univ, MS, 62, PhD(zool), 63. *Prof Exp:* Instr radiation biol, Fla State Univ, 64; USPHS fel, 64-66, GENETICIST, BIOL DIV, OAK RIDGE NAT LAB, 66-; LECTR ZOOL ENTOM, UNIV TENN, 69-, SECT HEAD MUTAGENESIS, 80- *Mem:* Genetics Soc Am; Environ Mutagen Soc; AAAS. *Res:* Human genetics; chemical mutagenesis; nucleic acids. *Mailing Add:* 614 Greenwood Dr Clinton TN 37716

EPLEY, RICHARD JESS, b Pana, Ill, Aug 31, 42. MEAT SCIENCE, FOOD SCIENCE. *Educ:* Univ Ill, Urbana, BS, 65; Univ Mo-Columbia, MS, 67, PhD(animal husb), 70. *Prof Exp:* Instr food sci, Univ Mo-Columbia, 69-70; assoc prof, 70-74, exten specialist, 74-80, PROF ANIMAL SCI, AGR EXTEN, UNIV MINN, ST PAUL, 80- *Mem:* AAAS; Am Soc Animal Sci; Am Meat Sci Asn; Inst Food Technologists. *Res:* Extension meat science. *Mailing Add:* Univ Minn St Paul Campus 136 Bl Meat Sci St Paul MN 55108

EPLING, GARY ARNOLD, b Elkhorn City, Ky, June 22, 45; m 66, 85; c 2. PHYSICAL ORGANIC CHEMISTRY, PHOTOCHEMISTRY. *Educ:* Mass Inst Technol, SB, 67; Univ Wis, PhD(org chem), 72. *Prof Exp:* Res assoc chem, Yale Univ, 72-73; asst prof, Fordham Univ, 73-78; from asst prof to assoc prof, 78-87, PROF CHEM, UNIV CONN, 87- *Concurrent Pos:* Asst chemist (dermat), Mass Gen Hosp, 85- *Mem:* Am Chem Soc; Am Soc Photobiol. *Res:* Mechanistic organic photochemistry and exploratory photochemistry of biologically important compounds; chemical reactions associated with phototoxicity; environmental chemistry. *Mailing Add:* Dept Chem Univ Conn Storrs CT 06269-3060

EPP, CHIROLD DELAIN, b Fairview, Okla, Mar 31, 39; m 61; c 3. PHYSICS. *Educ:* Northwestern State Col, Okla, BS, 61; Univ Okla, MS, 65; Univ Tex, Austin, PhD(physics), 69. *Prof Exp:* Instr physics, Northwestern State Col, Okla, 63-65; from asst prof to assoc prof physics, Midwestern Univ, 69-80; space shuttle trajectory flight controller, 80-85, SECT HEAD, NASA, 85- *Mem:* Am Phys Soc; Am Asn Physics Teachers. *Res:* Orbital mechanics and space shuttle trajectory. *Mailing Add:* 16383 Havenpark Houston TX 77059

EPP, DONALD JAMES, b Hastings, Nebr, June 23, 39; m 61; c 2. AGRICULTURAL ECONOMICS, RESOURCE ECONOMICS. *Educ:* Univ Nebr, BS, 61; Mich State Univ, MS, 64, PhD(agr econ), 67. *Prof Exp:* Instr agr econ, Mich State Univ, 65-67; from asst prof to assoc prof, 67-81, PROF AGR ECON & ASST DIR, ENVIRON RESOURCES RES INST, PA STATE UNIV, 81- *Concurrent Pos:* Consult, Govt Comt Preserv Agr Land, Pa, 68-69, US Cong, 78-79 & US Army Corps of Engrs, 80. *Mem:* Am Agr Econ Asn; Am Econ Asn; Asn Environ & Resource Economists. *Res:* Economic aspects of environmental quality and resource policy, especially the effects of land use and water quality policies; interaction between economic growth and environmental quality; hazardous waste management. *Mailing Add:* Dept Agr Econ & Rural Sociol Pa State Univ University Park PA 16802

EPP, EDWARD RUDOLPH, b Saskatoon, Sask, July 21, 29; m 57. MEDICAL PHYSICS, RADIATION PHYSICS. *Educ:* Univ Sask, BA, 50, MA, 52; McGill Univ, PhD(physics), 55; Am Bd Health Physics, dipl, 61. *Prof Exp:* Asst physics, Nat Res Coun Can, 52-53; physicist radiation physics, Dept Radiol, Montreal Gen Hosp, 55-57; asst biophys, Sloan-Kettering Div, Cornell Univ Med Col, 57-58, assoc, 58-60, from asst prof to prof biophysics, 60-74, chmn dept, 66-74, mem & chief div phys biol, Sloan-Kettering Inst Cancer Res, 68-74; HEAD, DIV RADIATION BIOPHYSICS, DEPT RADIATION MED, MASS GEN HOSP, BOSTON & PROF RADIATION THER, HARVARD MED SCH, 74- *Concurrent Pos:* Consult, Reddy Mem Hosp, 55-57 & Montreal Children's Hosp, 56-57; asst, Sloan-Kettering Inst Cancer Res, 57-60, assoc, 60-64, assoc mem, 64-68; mem task group, Int Comn Radiol Units & Measurements, 65-70; assoc attend physicist, Dept Med Physics, Mem Hosp for Cancer & Allied Dis, 67-74; mem radiation study sect, NIH, 71-75; mem ad hoc comt hot particles, Adv Comt Biol Effects Ionizing Radiations, Nat Acad Sci, 74-76 & comt review use ionizing radiations for treatment of benign dis, 75-77; mem, Clin Cancer Prog Proj Rev Comt, Nat Cancer Inst, 77-81; Dem, Comt Dept Energy Res Health Effects Ionizing Radiation, Nat Acad Sci, 78-79, Comt Fed Res Biol & Health Effects Ionizing Radiation, 79-; assoc ed, Int J Radiation Oncol Biol Physics, 79- *Mem:* AAAS; Am Phys Soc; Health Physics Soc; Radiation Res Soc (secy-treas, 81-); Am Asn Physicists in Med. *Res:* Radiobiology, especially cellular radiobiology; biophysics; health physics; effects of ionizing radiation of ultra-high intensity on living cells. *Mailing Add:* Dept Radiation Mass Gen Hosp Boston MA 02114

EPP, LEONARD G, b Neptune, NJ, Aug 14, 44; m 69; c 2. DEVELOPMENTAL BIOLOGY. *Educ:* Gettysburg Col, BA, 66; Pa State Univ, MS, 68, PhD(zool), 70. *Prof Exp:* From asst prof to assoc prof, 70-85, PROF BIOL, MT UNION COL, 85- *Concurrent Pos:* Vis scientist, Indiana Univ, 77 & 80 & Zool Inst, Univ Zurich, 78-79 & 84-85, Univ Kansas, 81 & Kansas State Univ, 83. *Mem:* AAAS; Am Soc Zoologists; Sigma Xi. *Res:* Biology of hydra; pigmentation in amphibia; effect of microgravity on cell function. *Mailing Add:* Dept Biol Mt Union Col Alliance OH 44601

EPP, MELVIN DAVID, b Newton, Kans, June 16, 42; m 64; c 2. PLANT TISSUE CULTURE, OSTRICH BREEDING. *Educ:* Wheaton Col, BS, 64; Univ Conn, MS, 67; Cornell Univ, PhD(genetics), 72. *Prof Exp:* Damon Runyan fel, Brookhaven Nat Lab, 72-74; sr res biologist, Monsanto Co, 74-77; res supt, Philippine Packing Corp, 78-82, mgr, plant propagation & tissue cult res, Del Monte Corp, 82-84; prin scientist, Plant Cell Res Inst, 84-89; OWNER/MGR, PRIMROSE BIRD FARM, 90- *Concurrent Pos:* Consult, United Fruit Co, 87 & SRI Int, 87, 88. *Mem:* Bot Soc Am; Genetics Soc Am; AAAS; Int Plant Tissue Cult Soc. *Res:* Select and breed bloodlines of African Black and Blue Neck ostriches. *Mailing Add:* Primrose Bird Farm RR 1 Box 164 Whitewater KS 67154

EPPENSTEIN, WALTER, b Berlin, Ger, Dec 14, 20; nat US; m 44; c 3. PHYSICS. *Educ:* Robert Col, Istanbul, BS, 42; Rensselaer Polytech Inst, MS, 52. *Prof Exp:* Instr physics, Robert Col, Istanbul, 42-46; from instr to assoc prof, Rensselaer Polytechnic Inst, 46-73, exec officer, 69-81, prof physics, 73-91, assoc chmn, 81-91, EMER PROF PHYSICS, RENSSELAER POLYTECH INST, 91- *Concurrent Pos:* Hon res assoc, Harvard Univ, 64-65. *Honors & Awards:* David M Darrin Award, 84. *Mem:* Am Soc Eng Educ; Am Asn Physics Teachers. *Res:* Educational developments in physics including new demonstration and laboratory experiments; uses of microcomputers and microprocessors. *Mailing Add:* Dept Physics Rensselaer Polytech Inst Troy NY 12180-3590

EPPERLY, W ROBERT, b Christiansburg, Va, Mar 17, 35; m 57; c 3. RESEARCH & DEVELOPMENT MANAGEMENT, ORGANIZATION DEVELOPMENT. *Educ:* Va Polytech Inst, BS, 56, MS, 58. *Prof Exp:* Dir, Fuels Res Lab, Exxon Res & Eng Co, 68-71, sr adv, Dept Mkt, Exxon Corp, 71-73, mgr, res & develop div, Exxon Res & Eng Co, Baytown, Tex, 73-76, mgr proj develop synfuels res, 76-77, gen mgr coal liquefaction, 77-79, gen mgr synthetic fuels, 80-83, sr prog mgr petrol & synfuels, 83-84, gen mgr corp res, Exxon Res & Eng Co, 84-86; exec vpres, Fuel Tech Inc, 86-89, chief exec officer, 89-90; PRES, EPPERLY ASSOCS INC, 90- *Concurrent Pos:* Mem, air pollution res adv comt, Coord Res Coun, 69-71; mem adv comt, Fossil Energy Prog, Oak Ridge Nat Labs, 78-81; mem, comt synthetic fuels safety, Nat Res Coun, 82, chmn, Comt Coop Govt Indust Res, 83. *Honors & Awards:* Pract Award, Am Inst Chem Engrs, 84. *Mem:* AAAS; Am Inst Chem Engrs; Am Petrol Inst. *Res:* Integration of scientific and engineering excellence with advanced management concepts in organization; career development, employee motivation and creativity. *Mailing Add:* PO Box 871 New Canaan CT 06840-0871

EPPERS, WILLIAM C, b Balitmore, MD, Jan 17, 30. LASER TECHNOLOGY, MICROWAVES. *Educ:* Johns Hopkins Univ, BEng, 50, PhD(eng), 62. *Prof Exp:* dir electronic technol div, 72-74, chief scientist, 74-75, dir, Air Force Avionic Lab, 75-77; deputy dir, Avionics Lab, Air Force Wright Aeronautical Labs, 77-85; ASST PROF ELEC ENG, WRIGHT STATE UNIV, 85- *Mem:* Fel Inst Elec & Electronics Engrs. *Res:* laser technology. *Mailing Add:* Dept Elec Systs Eng Wright State Univ Dayton OH 45435

EPPERSON, EDWARD ROY, b Burnsville, Miss, Oct 14, 32; m 60; c 2. INORGANIC CHEMISTRY. *Educ:* Millsaps Col, BS, 54; Univ NC, MA, 57; Univ of the Pac, PhD, 65. *Prof Exp:* From asst prof to prof chem, Elon Col, 57-66; PROF CHEM, 66-, VPRES ADMIN AFFAIRS, HIGH POINT COL, 82- *Concurrent Pos:* NSF fac fel, 64-65. *Mem:* Am Chem Soc. *Res:* Synthesis of the anhydrous metal halides; lower oxidation states of molybdenum and tungsten. *Mailing Add:* Admin Affairs High Point Col High Point NC 27262

EPPINK, RICHARD THEODORE, b Cleveland, Ohio, May 7, 31; m 75; c 2. STRUCTURAL ENGINEERING, APPLIED MECHANICS. *Educ:* Case Inst Technol, BS, 53; Univ Ill, MS, 56, PhD(civil eng), 60. *Prof Exp:* Struct engr, Glenn L Martin Co, 53; instr, Univ Ill, 59-60; mem tech staff, Nat Eng Sci Co, 60-62; MEM FAC CIVIL ENG, UNIV VA, 62- *Mem:* Am Soc Civil Engrs. *Res:* Structural dynamics and vibrations; blast effects of nuclear weapons; application and accuracy of finite element methods in structural analysis; numerical methods of structural analysis; stress analysis of arteriol bifurcations. *Mailing Add:* Dept Civil Eng Univ Va Charlottesville VA 22903

EPPLER, RICHARD A, b Lynn, Mass, Apr 30, 34; m 59; c 5. CERAMICS, INORGANIC CHEMISTRY. *Educ:* Carnegie-Mellon Univ, BS, 56; Univ Ill, MS, 58, PhD(chem eng & physics), 60. *Prof Exp:* Res chemist, Corning Glass Works, 59-65; sr scientist, Pemco Ceramics Group, Mobay Chem Corp, 65-84; mgr ceramics, Olin Corp, 84-86; PRES & OWNER, EPPLER ASSOCS, 86-; ASSOC PROF, UNIV LOWELL, 86- *Concurrent Pos:* Instr math, Elmira Col, 64. *Honors & Awards:* Dwight Joyce Award, SCM Corp, 69; John Marquis Award, Am Ceramic Soc, 74; Fel & Award of Merit, Am Soc Testing & Mat, 85. *Mem:* Fel Am Ceramic Soc; Nat Soc Prof Engrs; Am Chem Soc; Electrochem Soc; Sigma Xi; fel Am Soc Testing & Mat. *Res:* Crystallization phenomena, particularly from vitreous media; glass; glass-ceramics; glaze and enamel compositions and properties; solid state chemistry, reaction kinetics; high pressure research on solid materials; inorganic pigments; luminescent and photochromic materials. *Mailing Add:* 400 Cedar Lane Cheshire CT 06410

EPPLEY, RICHARD WAYNE, b Puyallup, Wash, Oct 12, 31; m 53; c 2. BIOLOGICAL OCEANOGRAPHY. *Educ:* State Col Wash, BS, 53; Stanford Univ, PhD(biol), 57. *Prof Exp:* From instr to asst prof biol, Univ Southern Calif, 57-60; plant physiologist, Northrop Corp, 60-63; assoc res biologist, Univ Calif, San Diego, 63-70, assoc dir, Inst Marine Resources, 75-87, lectr & res biologist, Scripps Inst Oceanog, 70-90; RETIRED. *Concurrent Pos:* Adj assoc prof biol, Univ Southern Calif, 61-63; consult, NSF, 70-74; marine biologist, US AEC, 72-73. *Honors & Awards:* Darbaker Prize Phycol, Bot Soc Am, 71; Hutchinson Medal, Am Soc Limnol Oceang, 84. *Mem:* Am Soc Limnol & Oceanog (pres, 81); fel AAAS; Phycol Soc Am; fel Am Geophys Union; Oceanogr Soc. *Res:* Physiology of marine phytoplankton; ocean productivity. *Mailing Add:* 1969 Loring St San Diego CA 92109

EPPRIGHT, MARGARET, b Manor, Tex, Apr 21, 13. BIOCHEMISTRY, NUTRITION. *Educ:* Univ Tex, BA, 33, MA, 35, PhD(chem), 45. *Prof Exp:* Instr, Pub Sch, Tex, 36-40; asst, Univ Tex, 41-44; assoc prof chem, Sam Houston State Col, 45-46; assoc prof nutrit & res assoc, Iowa State Univ, 46-48; prof chem & head dept, Sam Houston State Col, 48-49; assoc prof home econ, 49-54, head div nutrit, 54-64, chmn dept home econ, 61-71, PROF HOME ECON, UNIV TEX, AUSTIN, 54- *Mem:* Am Dietetic Asn; Am Chem Soc; fel Am Inst Chem; Am Home Econ Asn. *Res:* Synthesis of hydantoins; vitamin assay methods; nutritonal status of population groups; mineral nutrition of microorganisms; metabolic patterns in health and disease. *Mailing Add:* 2517 Hartford Rd Austin TX 78703

EPPS, ANNA CHERRIE, b New Orleans, La, July 8, 30; wid. IMMUNOLOGY. *Educ:* Howard Univ, BS, 51, PhD(zool), 66; Loyola Univ La, MS, 59. *Prof Exp:* Technologist, Clin Labs, Our Lady of Mercy Hosp, Cincinnati, Ohio, 53-54 & Clin Labs, Flint-Goodridge Hosp, New Orleans, 54-55; asst prof microbiol, Sch Med, Howard Univ, 61-69; USPHS fac fel & asst prof dept med, Johns Hopkins Univ Sch Med, 69; USPHS fac fel & asst prof, 69-71, assoc prof, 71-75, asst dean student servs, 80-86, DIR MED REP, MED CTR, TULANE UNIV, 69-, PROF MED, SCH MED, 75-, ASSOC DEAN STUDENT SERVS, 86- *Concurrent Pos:* Technologist, Clin Labs, Mercy Hosp, Hamilton, Ohio, 54 & Sch Med, La State Univ, 59-60; asst prof med technol & actg chmn dept, Xavier Univ La, 54-60. *Mem:* Am Soc Clin Pathologists; Am Soc Med Technologists; Am Soc Bacteriologists; Sigma Xi; Am Soc Trop Med & Hyg; Am Asn Blood Banks; Am Asn Univ Prof. *Res:* Immunological studies in autoimmune diseases; immunological embryology; transplantation immunology; hepatitis immunology. *Mailing Add:* 7200 Thornley Dr New Orleans LA 70126

EPPS, HARLAND WARREN, b Hawthorne, Calif, July 29, 36; div; c 2. ASTRONOMY, OPTICAL DESIGN. *Educ:* Pomona Col, BA, 59; Univ Wis, MS, 61, PhD(astron), 64. *Prof Exp:* Asst prof astron, San Diego State Col, 64-65; from asst prof to prof astron, 76-89, PROF ASTRON & ASTROPHYS & ASTRONR, UNIV CALIF, LOS ANGELES, 89- *Concurrent Pos:* Consult optical design; res grants, Air Force Res Lab & Regents of Univ Calif, NSF; mem sci adv bd, USAF, 88- *Mem:* Int Astron Union; Am Astron Soc; Soc Photo-Optical Instrumentation Engr. *Res:* Astronomical instrumentation; optical design; spectroscopy of planetary nebulae and quasi stellar sources; spectroscopy of peculiar stars. *Mailing Add:* Lick Observ Univ Calif Santa Cruz CA 95064

EPPS, JON ALBERT, b Merced, Calif, April 27, 42; m 64; c 2. PAVEMENT MATERIALS, PAVEMENT DESIGN. *Educ:* Univ Calif, Berkeley, BS, 65, MS, 66, PhD(eng), 68. *Prof Exp:* Asst prof, Tex A&M Univ, 68; prog mgr, 75-81, HEAD, DIV II & MAT ENG DIV, TEX TRANSP INST, TEX A&M UNIV, 81- *Mem:* Am Soc Testing & Mat; Asn Asphalt Paving Technologists; Sigma Xi; Nat Soc Prof Engrs. *Res:* Transportation materials and pavement rehabilitation maintenance design. *Mailing Add:* 3345 Markridge Reno NV 89509

EPPS, JOYCE E, drug metabolism, for more information see previous edition

EPPS, WILLIAM MONROE, b Latta, SC, Oct 31, 16; m 42; c 2. PLANT BREEDING, HORTICULTURE. *Educ:* Clemson Univ, BS, 37; Cornell Univ, PhD(plant path), 42. *Prof Exp:* Asst, NY State Col Agr, Cornell, 38-42; assoc plant pathologist, Exp Sta, Clemson Univ, 45-56, head dept bot & bact, 56-69, head dept plant path & physiol, 69-78; RETIRED. *Concurrent Pos:* Vis prof plant path, Clemson Univ, 83-85. *Mem:* Am Phytopath Soc. *Res:* Vegetable diseases and breeding. *Mailing Add:* 211 Wyatt Ave Clemson SC 29631

EPPSTEIN, DEBORAH ANNE, b Kalamazoo, Mich, Oct 16, 48; m 87; c 1. VIROLOGY, BIOCHEMISTRY. *Educ:* Grinnell Col, BA, 70; Univ Ark, PhD(biochem), 75. *Prof Exp:* NIH fel virol, Univ Calif, Santa Barbara, 76-78; staff researcher, vaccine develop, 78-79, sect leader, Inst Bio-Org Chem, 80-84, head biochem dept, 84-87, head tumor biol, 87-88, DIR BUS DEVELOP, SYNTEX CORP, 88- *Mem:* Am Asn Cancer Res; AAAS; Int Soc Interferon Res; Am Mgt Asn. *Res:* Protein delivery systems; mechanisms of interferon action; liposomes; tumor invasion and metastasis. *Mailing Add:* Syntex 3401 Hillview Palo Alto CA 94304

EPREMIAN, E(DWARD), b Schenectady, NY, Sept 3, 21; m 48; c 2. METALLURGY. *Educ:* Mass Inst Technol, BS, 43; Rensselaer Polytech Inst, MS, 47; Carnegie Inst Technol, DSc(metall), 51. *Prof Exp:* Res assoc metall, res lab, Gen Elec Co, 43-46; res assoc, metals res lab, Carnegie Inst Technol, 50-51; sci liaison officer, Off Naval Res, London Br, 51-52, asst sci dir, 52-53, dept sci dir, 53-54; chief metals & mat br, res div, Atomic Energy Comn, 54-57; sr metallurgist, res lab, metals div, Union Carbide Corp, 57-59, tech coordr, tech dept, 60-61, mgr, new prod mkt, 61-63, asst dir res, carbon prod div, 63-65, gen mgr, aerospace mat dept, 65-68, gen mgr, advan mat dept, 68-70, mgr tantalum & columbium prod, mining & metals div, 70-71, mgr, spec prod, 71-72, dir new ventures, 73-76; EXEC DIR, COMN SOCIOTECH SYSTS, NAT ACAD SCI, NAT RES COUN, 76-; CONSULT. *Concurrent Pos:* Mem adv bd, Int Symposium High Temperature Technol, 56, 59 & 63; US sci secy, Atoms for Peace Conf, Geneva, 56; mem mat adv bd, Nat Acad Sci, chmn, refractory metals comt, 57-59, mem panel solid propellant rocket motors, 59-60 & mem comt standing rev of Dept Defense Mat Prog, 61-63; mem, adv comt, Univ Pa, Sch Metall & Mat Sci, 69-73; mem bd dir, Acta Metallurgica, 71-73; mem bd trustees, Webb Inst Naval Archit & Marine Eng, 76- *Mem:* Fel Am Soc Metals; Am Inst Mining, Metall & Petrol Engrs; fel AAAS; Sigma Xi. *Res:* Fatigue of metals; refractory metals; nuclear and high temperature metallurgy; physical chemistry; graphite technology. *Mailing Add:* Ten Maple Ct PO Box 504 Sunapee NH 03782

EPSTEIN, ALAN NEIL, b New York, NY, July 29, 32; m 57; c 3. NEUROPSYCHOLOGY. *Educ:* Johns Hopkins Univ, BA & MA, 54, MD, 58. *Prof Exp:* Nat Found fel, Inst Neurol Sci, Sch Med, 58-61, asst prof zool, 61-64, assoc prof biol, 64-69, prof, 69-77, mem, Inst Neurol Sci, 63-80, PROF NEUROPSYCHOL, UNIV PA, 77- *Concurrent Pos:* Mem neuropsychol res rev comt, NIMH, 69-73, biol sci fel rev comt, 75-; NIMH spec fel, Col of France, Paris, 69-70; Overseas fel, Churchill Col, Univ Cambridge, 75-76. *Mem:* Fel Am Psychol Asn; AAAS; Soc Neurosci; Am Physiol Soc. *Res:* Neurological basis of behavior; feeding and drinking and the specific hungers; behavioral thermoregulation. *Mailing Add:* Leidy Lab Dept Biol Univ Pa Philadelphia PA 19104-6018

EPSTEIN, ARTHUR JOSEPH, b Brooklyn, NY, June 2, 45; m 69; c 2. CONDUCTING POLYMERS, MOLECULAR MAGNETS. *Educ:* Polytech Inst Brooklyn, BS, 66; Univ Pa, MS, 67, PhD(physics), 71. *Prof Exp:* Tech staff technol assessment, Mitre Corp, 71-72; sr scientist, Xerox Corp, 72-80, prin scientist, Webster Res Ctr, 80-85; PROF PHYSICS & CHEM, OHIO STATE UNIV, 85-, DIR, CTR MAT RES, 89- *Concurrent Pos:* Mem organizing comt, Conf Synthesis & Properties Low-Dimensional Mats, NY Acad Sci, 77; vis physicist, dept physics & Mats Res Lab, Penn State Univ, 78, dept physics, Univ Md, 79 & City Col, City Univ New York, 80; vis scientist, Dept Physics, Univ Calif, Los Angeles Calif, 77, 78 & 79; guest scientist, Francis Bitter Nat Magnet Lab, Mass Inst Technol, Cambridge, 78; lectr, Advanced Study Inst Physics & Chem Low Dimensional Solids, Tomar, Portugal, 79, 86, Spetses, Greece, 89 & Mons, Belgium, 89; vis prof, Lab Physique Solides, Univ Paris-Sud, Orsay, France, 80, 88 & 90; co-chmn, Int Conf Low Dimensional Conductors, Boulder, Colo, 81; co-ed, Proceedings Int Conf Low-Dimensional Conductors, Molecular Crystals & Liquid Crystals; mem prog comt, Int Conf on Synthetic Metals, France, 82, Italy, 84 & Japan, 86; regional ed, J Synthetic Metals, 83-; mem, Comt on Appl, Am Physics Soc; vis prof, Technion, Haifa, Israel, 84-85; adj prof, Dept Physics, Univ Fla, 83-; mem exec comt, Int Conf, Synthetic Metal, Santa Fe, NM, 88. *Mem:* Fel Am Phys Soc; Am Chem Soc. *Res:* Charge transport mechanisms; electronic and magnetic phenomena; optical and nonlinear optical phenomena of molecular and polymeric magnet, conducting polymers, organic charge transfer salts and high Tc ceramic superconductors. *Mailing Add:* Dept Physics Ohio State Univ 174 W 18th Ave Columbus OH 43210-1106

EPSTEIN, ARTHUR WILLIAM, b New York, NY, May 15, 23; m 55; c 4. PSYCHIATRY, NEUROLOGY. *Educ:* Columbia Univ, AB, 44, MD, 47. *Prof Exp:* Intern, Mt Sinai Hosp, NY, 47-48; clin asst psychiat, State Hosp, Norristown, Pa, 48; resident neurol, Mt Sinai Hosp, NY, 49-50; asst, 50-52, instr psychiat & neurol, 52-54, asst prof neurol, 54-58, assoc prof psychiat & neurol, 58-64, PROF PSYCHIAT & NEUROL, SCH MED, TULANE UNIV, 64- *Honors & Awards:* Silverberg Award, Am Acad Psychoanalysis, 85. *Mem:* AAAS; Am Acad Neurol; Soc Biol Psychiat (asst secy, 73-76, secy-treas, 76-79, pres, 81); Am Psychiat Asn; Am Epilepsy Soc; Am Acad Psychoanalysis (pres, 87-88). *Res:* Brain behavior relationships, epileptic and dream mechanisms. *Mailing Add:* 1430 Tulane Ave New Orleans LA 70112

EPSTEIN, AUBREY, b Detroit, Mich, June 4, 22; m 50; c 2. DIAGNOSTIC AUDIOLOGY, HEARING AIDS. *Educ:* Ind Univ, BA, 46; Western Reserve Univ, MA, 47; Univ Iowa, PhD(speech path, audiol), 53. *Prof Exp:* From asst prof to assoc prof speech path & audiol, Univ Pittsburgh, 53-63; assoc prof speech path & audiol, 63-68, PROF AUDIOL, IND UNIV, BLOOMINGTON, 69- *Concurrent Pos:* Consult, Vet Admin, 55-, Dept of Justice, 80- *Mem:* Fel Am Speech & Hearing Asn; Acoust Soc Am; assoc fel Am Acad Ophthal & Otolaryngol. *Res:* Auditory phenomena relative to aural pathology; noise induced hearing loss. *Mailing Add:* 2938 Ramble Rd E Bloomington IN 47408

EPSTEIN, BARRY D, b New York, NY, Mar 30, 42; m 66, 81; c 2. ELECTROCHEMISTRY, ANALYTICAL CHEMISTRY. *Educ:* City Col New York, BS, 62; Univ Calif, Riverside, PhD(anal chem), 66. *Prof Exp:* Res fel chem, Calif Inst Technol, 66-67; staff scientist, 67-80, prin staff scientist, Gen Atomic Co, 80-; GEN PROBE. *Honors & Awards:* Samuel Goldman Award, 62. *Mem:* Am Chem Soc; Electrochem Soc; Am Soc Testing & Mat. *Res:* Instrumentation; pollution analysis; batteries; biomaterials; nuclear fuels; fission product chemistry; gas chromatography. *Mailing Add:* 9880 Campus Point Dr San Diego CA 92121-1514

EPSTEIN, BERNARD, b Harrison, NJ, Aug 10, 20; m 47; c 6. MATHEMATICS. *Educ:* NY Univ, BA, 40, MS, 42; Brown Univ, PhD(appl math), 47. *Prof Exp:* Jr physicist, Nat Bur Stand, Washington, DC, 41-43; from asst physicist to assoc physicist, Manhattan Proj, 43-45; asst div appl math & instr eng, Brown Univ, 45-46; res assoc aeronaut eng, Grad Sch, Harvard Univ, 46-47; from instr to assoc prof math, Univ Pa, 47-60; prof, Yeshiva Univ, 60-63; prof math, UNIV NMEX, 63-84; GEORGE MASON UNIV, 84- *Concurrent Pos:* Vis res assoc inst math sci, NY Univ, 53; vis assoc prof, Stanford Univ, 57-58; liaison scientist, US Off Naval Res, London, Eng, 64-66; vis prof math, Israel Inst Technol, 71-72. *Mem:* Am Math Soc; Sigma Xi. *Res:* Study of motion of compressible fluid by hodograph method; conformal mapping; boundary value problems of potential theory; extremal problems relating to analytic functions. *Mailing Add:* 3176 Summit Square Pl Apt C-2 Oakton VA 22124

EPSTEIN, CHARLES JOSEPH, b Philadelphia, Pa, Sept 3, 33; m 56; c 4. MEDICAL GENETICS, DEVELOPMENTAL BIOLOGY. *Educ:* Harvard Univ, AB, 55, MD, 59. *Prof Exp:* Intern med, Peter Bent Brigham Hosp, Boston, Mass, 59-60, asst resident, 60-61; res assoc, Nat Heart Inst, 61-63, med officer, Nat Inst Arthritis & Metab Dis, 63-66, chief sect genetics & develop, Lab Chem Biol, 66-67; assoc prof pediat, 67-72, assoc prof biochem, 70-72, PROF PEDIAT & BIOCHEM, SCH MED, UNIV CALIF, SAN FRANCISCO, 72- *Concurrent Pos:* Asst med & res fel med genetics, Sch Med, Univ Wash, 63-64; prof lectr inherited metab dis, Sch Med, George Washington Univ, 65-67; asst med, Johns Hopkins Univ, 65-67; ed, Human Genetics, 84-; ed, Am J Human Genetics, 87-; mem, Recombinant DNA Adv Comt, NIH, 85-90; pres, Am Bd Med Genetics, 90- *Honors & Awards:* Nancy & Daniel Weisman Award. *Mem:* Am Pediat Soc; Asn Am Physicians; Am Soc Biochem & Molecular Biol; Am Soc Clin Invest; fel AAAS; Am Soc Human Genetics. *Res:* Hereditary diseases; biochemistry of early mammalian development; developmental genetics. *Mailing Add:* Dept Pediat Box 0748 Sch Med Univ Calif San Francisco CA 94143

EPSTEIN, DAVID AARON, interferon, gene expression, for more information see previous edition

EPSTEIN, DAVID L(EE), b Chicago, Ill, June 23, 44; m 68; c 1. OPHTHALMOLOGY. *Educ:* Johns Hopkins Univ, BA, 65, MD, 68. *Prof Exp:* Resident ophthal, 73-75, fel glaucoma, 75-76, instr, 76-77, asst prof, 78-81, ASSOC PROF OPHTHAL, HARVARD MED SCH, 81-; DIR GLAUCOMA SERV, MASS EYE & EAR INFIRMARY, 82- *Concurrent Pos:* Asst dir glaucoma serv, Mass Eye & Ear Infirmary, 76-81. *Honors & Awards:* Alcoa Res Inst Award, Am Acad Ophthal. *Mem:* Asn Res Vision & Ophthal; AAAS. *Res:* Glaucoma. *Mailing Add:* Dept Ophthal Mass Eye & Ear Infirmary 243 Charles St Boston MA 02114

EPSTEIN, EDWARD SELIG, b New York, NY, Apr 29, 31; m 54; c 4. METEOROLOGY. *Educ:* Harvard Univ, AB, 51; Columbia Univ, MBA, 53; Pa State Univ, MS, 54, PhD(meteorol), 60. *Prof Exp:* Res assoc & lectr meteorol, Univ Mich, 59-61, asst prof, 61-63; consult to asst secy com for sci & tech, US Dept Com, 63-64; from assoc prof to prof meteorol, Univ Mich, 63-73, from actg chmn to chmn dept meteorol & oceanog, 69-73; assoc adminr for environ monitoring & prediction, 73-77, dir, US Climate Prog Off, 78-81, dir, Climate & Earth Sci Lab, Nat Earth Satellite Serv, 81-83, PRIN SCIENTIST, CLIMATE ANALYSIS CTR, NAT WEATHER SERV, NAT OCEANIC & ATMOSPHERIC ADMIN, 84- *Concurrent Pos:* NSF grants, 61-64, 65-68, 70-72; vis prof, Int Meteorol Inst, Stockholm, 68-69; Univ Corp Atmospheric Res rep, 69-73; mem goal & eval comt, 69-71; ed, J Appl Meteorol, 71-73; mem adv panel atmospheric sci, NSF, 71-73; fed coordr for meteorol, 75-78. *Mem:* Fel Am Meteorol Soc; fel AAAS. *Res:* Probability and statistics in meteorology; stochastic dynamic prediction. *Mailing Add:* Nat Oceanic & Atmospheric Admin World Weather Bldg Washington DC 20233

EPSTEIN, EMANUEL, b Detroit, Mich, July 1, 22; m 50; c 3. CLINICAL CHEMISTRY. *Educ:* Wayne State Univ, BS, 45, MS, 53, PhD(biochem), 57. *Prof Exp:* Chemist, Frederick Stearns & Co, 45-46 & Fund Crippling Dis, 46-47; asst, Wayne State Univ, 48-50, res assoc, 50-57; prof plant nutrit & plant physiologist, 65-87, prof bot, 74-87, EMER PROF BOT, UNIV CALIF, DAVIS, 87-, EMER PROF NUTRIT & PLANT PHYSIOLOGIST, 87-; CLIN CHEMIST, WILLIAM BEAUMONT HOSP, 68- *Concurrent Pos:* Asst prof, Wayne State Univ, 63-77, clin assoc prof, 77- *Mem:* Am Asn Clin Chem; Endocrine Soc; Am Chem Soc; Sigma Xi. *Res:* Steroids; analysis; electrophoretically separated proteins, analysis; chromatography; automation. *Mailing Add:* William Beaumont Hosp 3601 W 13 Mile Road Royal Oak MI 48072

EPSTEIN, EMANUEL, b Ger, Nov 5, 16; nat US; m 43; c 1. PLANT NUTRITION. *Educ:* Univ Calif, BS, 40, MS, 41, PhD(plant physiol), 50. *Prof Exp:* Asst bot, Univ Calif, 43, asst plant nutrit, 46-49; assoc plant physiologist, USDA, 49-51, plant physiologist, 51-58; lectr plant nutrit & assoc plant physiologist, 58-65, prof plant nutrit & plant physiologist, 65-87, prof bot, 74-87, EMER PROF PLANT NUTRIT, PLANT PHYSIOLOGIST & BOT, UNIV CALIF, DAVIS, 87- *Concurrent Pos:* Guest investr biophys lab, Carnegie Inst, 50; adv, US Deleg, Conf Peaceful Uses Atomic Energy, 55; Guggenheim fel, Calif Inst Technol, 58; partic, Nat Acad Sci-Nat Res Coun Desalination Res Conf, Mass, 61; Fulbright res grant, Australia, 65-66; Fulbright sr res scholar, Sci & Indust Res, NZ, 74-75. *Honors & Awards:* Cherubim Gold Medal, Univ Pisa, 62; Charles Reid Barnes Hon Life Mem

Award, Am Soc Plant Physiol, 86. *Mem:* Nat Acad Sci; AAAS (pres pac div, 90-91); Am Soc Plant Physiol; Am Inst Biol Sci; Scand Soc Plant Physiol; Australian Soc Plant Physiol; Sigma Xi. *Res:* Mechanisms of ion transport in plants; salt relations of plants; development of salt tolerant crops; genetic and ecological aspects of mineral plant nutrition. *Mailing Add:* Dept Land Air & Water Resources Univ Calif Davis CA 95616-8627

EPSTEIN, ERVIN HAROLD, b Vallejo, Calif, May 17, 09; m 36; c 2. DERMATOLOGY. *Educ:* Univ Calif, AB, 31, MD, 35. *Prof Exp:* ASSOC CLIN PROF DERMAT MED, SCH MED, UNIV CALIF, SAN FRANCISCO, 62- *Concurrent Pos:* Consult var Calif hosps; assoc clin prof dermat med, Sch Med, Stanford Univ, 50-64. *Mem:* Am Dermat Asn; Soc Invest Dermat; Am Acad Dermat; Int Soc Dermat Surg; Am Soc Dermat Surg; Dermat Radiother Soc. *Res:* Disease of the skin; skin surgery; regional dermatologic diagnosis; radio-dermatitis; techniques in skin surgery; controversies in dermat. *Mailing Add:* Five Soteld Ave Piedmont CA 94611

EPSTEIN, EUGENE ETHAN, astronomy, for more information see previous edition

EPSTEIN, FRANKLIN HAROLD, b Brooklyn, NY, May 5, 24; m 51; c 4. INTERNAL MEDICINE, PHYSIOLOGY. *Educ:* Brooklyn Col, BA, 44; Yale Univ, MD, 47. *Prof Exp:* House officer, Yale Med Ctr, 47-49; res fel, Sch Med, Boston Univ, 49-50; res assoc physiol, Walter Reed Army Med Ctr, 50-52; from asst prof to prof med, Sch Med, Yale Univ, 54-72, chief metab div, 65-72; head dept med, 72-80, PROF MED, HARVARD MED SCH, 72- *Concurrent Pos:* Estab investr, Am Heart Asn, 56-61; consult, Off Surgeon Gen, US Army, 64-69; career investr, USPHS, 64-72; chmn nephrol test comt, Am Bd Internal Med, 70-74; trustee, Mt Desert Island Biol Lab, 70-, vpres, 83-; dir, Harvard Med Unit, Boston City Hosp & Thorndike Mem Lab, 72-73; Herrman Blumgart prof & physician-in-chief, 73-80, William Applebaum prof med & dir, Renal Div, Beth Israel Hosp, Boston, 81-; assoc ed, New Eng Jour Med, 82-, Quart J Med, 82- *Honors & Awards:* John P Peters Award, Am Soc Nephrol, 85. *Mem:* Am Soc Clin Invest (vpres, 69-70); Am Physiol Soc; Asn Am Physicians. *Res:* Renal physiology and disease. *Mailing Add:* Dept Med Beth Israel Hosp 330 Brookline Ave Boston MA 02215

EPSTEIN, GABRIEL LEO, b Manhattan, NY, Apr 8, 41; m 70. SOLAR PHYSICS, ATOMIC PHYSICS. *Educ:* City Col New York, BChE, 62; Univ Calif, Berkeley, PhD(physics), 69. *Prof Exp:* Teaching asst elem physics, Univ Calif, Berkeley, 63-65, res asst atomic physics & spectros, 65-69; Nat Acad Sci-Nat Res Coun Res assoc, Nat Bur Standards, Dept Com, 69-70; STAFF SCIENTIST OPTICAL SYSTS, NASA GODDARD SPACE FLIGHT CTR, 70- *Mem:* AAAS; Optical Soc Am; Am Astron Soc. *Res:* Fabry-Perot interferometry; optical isotope shifts; production and analysis of the spectra of moderately ionized atoms; design of equipment for solar observations; deduction of solar atmospheric conditions from such observations. *Mailing Add:* 14307 Long Green Dr Silver Spring MD 20906

EPSTEIN, GARY MARTIN, b Los Angeles, Calif, Jun 9, 42; m 65; c 2. PARTICLE PHYSICS. *Educ:* Univ Calif, Riverside, BA, 64, PhD(physics), 69. *Prof Exp:* Assoc prof, 69-80, PROF MATH, CALIF POLYTECH STATE UNIV, SAN LUIS OBISPO, 80- *Mem:* Am Phys Soc. *Res:* Construction of dispersion relations for Regge parameters in high energy scattering theory; mathematical modelling of physiological systems at the cellular level. *Mailing Add:* 1764 Jami Lee Ct San Luis Obispo CA 93401

EPSTEIN, GEORGE, b Boston, Mass, Nov 9, 26; m 51; c 2. PLASTICS CHEMISTRY, MATERIALS SCIENCE. *Educ:* Univ Mass, BS, 48; Mass Inst Technol, MS, 51. *Prof Exp:* Asst adhesives lab, Mass Inst Technol, 50-52; res engr, NAm Aviation, Inc, 52-55; prin engr & asst chief engr struct mat div, Aerojet-Gen Corp Div, Gen Tire & Rubber Co, 55-61; prin engr & staff scientist, Aeronutronic Div, Ford Motor Co, 61-63; proj scientist, Aerospace Res Assocs, Inc, 63-66; sr staff engr, Mat Sci Lab, 66-85, DIR, MFG ENG OFF, AEROSPACE CORP, 85- *Concurrent Pos:* Instr, Univ Calif, Los Angeles, 54-74; mem, Fed Steering Comt Adhesives Technol, 71-74; ed, Pac Coast Plastics & Rubber, 75-80; pres, Technol Conf Asn, 75-; ed, Composites & Adhesives Newslett, 84-; fel, Soc Aerospace Mat & Eng, 87. *Honors & Awards:* Meritorious Serv Award, Soc Aerospace Mat & Process Eng, 63; Distinguished Serv Award, Western Sect, Soc Plastics Indust, 74; Willard Lundberg Mem Award, Soc Plastics Engrs, 75 & Int Award Eng/Technol, 90. *Mem:* Soc Plastics Engrs; Am Chem Soc; Am Soc Testing & Mat; Soc Plastics Indust; fel Soc Aerospace Mat & Process Eng (vpres, 63-64). *Res:* Nonmetallic materials, plastics, resins, adhesives, sealants and coatings for high-performance applications; energy absorbing materials; advanced test and inspection methods; major applications to rockets, missiles and space vehicles. *Mailing Add:* Mfg Eng Dept Aerospace Corp PO Box 92957 Los Angeles CA 90009

EPSTEIN, GEORGE, b Bayonne, NJ, July 4, 34; m 56; c 3. MATHEMATICS, COMPUTER SCIENCE. *Educ:* Calif Inst Technol, BS, 55; Univ Ill, MS, 57; Univ Calif, Los Angeles, PhD(math), 59. *Prof Exp:* Mem tech staff computer sci, Hughes Aircraft, Calif, 57-59; sr staff scientist, ITT Gilfillan, Inc, Van Nuys, 59-72; prof computer sci, Ind Univ, Bloomington, 73-85; PROF COMPUTER SCI, UNIV NC, CHARLOTTE, 85- *Mem:* Am Math Soc; Asn Symbolic Logic; Asn Comput Mach; Inst Elec & Electronics Engrs. *Res:* Cybernetics; philosophy; psychology; literature; electrical circuits and systems; linguistics and education; algebra and logic. *Mailing Add:* Dept Computer Sci Univ NC Charlotte NC 28223

EPSTEIN, GERALD LEWIS, b Wash, DC, Dec 13, 56; m 85; c 2. TECHNOLOGY POLICY. *Educ:* Mass Inst Technol, SB(physics) & SB(elec eng), 78; Univ Calif, Berkeley, MA, 80, PhD(physics), 84. *Prof Exp:* Sr analyst, US Cong Off Technol Assessment, 83-89; prof dir, Ctr Sci & Int Affairs, Kennedy Sch Govt, Harvard Univ, 89-91; PROJ DIR, OFF TECHNOL ASSESSMENT, US CONG, 91- *Mem:* Am Phys Soc; AAAS. *Res:* Defense technology; relationship between military and commercial technologies; national science and technology policy, especially concerning the federal role in promoting the technological competence of US industry; proliferation; arms control. *Mailing Add:* Off Technol Assessment US Cong Washington DC 20510

EPSTEIN, HARVEY IRWIN, b Brooklyn, NY, Mar 29, 46. SOFTWARE SYSTEMS. *Educ:* Brooklyn Col, BA, 67; Univ Wis, Madison, MA, 69, PhD(math), 75. *Prof Exp:* Instr math, Viterbo Col, 69-71; asst prof math & comput sci, St Cloud State Univ, 74-75; asst prof math, Boston Col, 75-78; sr eng, Raytheon Co, 78-81; tech staff, MItre Corp, 81-; MEM TECH STAFF, AT&T BELL LABS. *Mem:* Asn Comput Mach; Am Math Soc; Am Math Assoc; NY Acad Sci. *Res:* Computer security, especially operating system and database. *Mailing Add:* Mast Rd Lee NH 03824

EPSTEIN, HENRY, b Frankfurt, Ger, Apr 5, 27; m 52; c 3. ELECTRICAL ENGINEERING. *Educ:* Brown Univ, SCB, 48; Harvard Univ, MS, 50. *Prof Exp:* PRES & CHMN, PENRIL CORP, 87- *Concurrent Pos:* Sr group vpres, Loral Corp, 77-86; vpres, Tex Instruments Inc. *Mem:* Inst Elec & Electronics Engrs; Sigma Xi. *Res:* Reduction and suppression of pollutant gases from automotive internal combustion engines; electrical and electronic controls. *Mailing Add:* 1300 Quince Orchard Blvd Gaithersburg MD 20878

EPSTEIN, HENRY F, b Bronx, NY, 1944; m 69, 82; c 4. BIOCHEMISTRY GENETICS. *Educ:* Columbia Univ, AB, 64; Stanford Univ, MD, 68; Nat Bd Med Exam, dipl, 70. *Prof Exp:* Res asst biochem, Stanford Univ, 65-67, res fel, 68-69; res staff molecular biophysicist, Yale Univ, 68-69; staff assoc chem biol, NIH, 69-71; Nat Found fel genetics cell biol, Med Res Coun Lab Molecular Biol, Cambridge, Eng, 71-73; asst prof pharmacol, Stanford Univ, 78-79; assoc prof neurol & biochem, 78-79, assoc prof, neurol biochem med, 79-81, PROF NEUROL, CELL BIOL & BIOCHEM, BAYLOR COL MED, 81- *Concurrent Pos:* Scientific adv comt, Muscular Dystrophy Asn, 79-, Task Force Genetics, 83-; Nat Aging Review Comt, Nat Inst Aging & NIH, 81. *Honors & Awards:* Borden Award, 68. *Mem:* AAAS; Am Soc Biol Chemists; Soc Neurosci; Am Soc Cell Biol; Am Chem Soc; Am Soc Clin Invest. *Res:* Molecular biology of muscle development; muscle structure, genetics of muscle disease, biochemistry of protein structure, function and regulation. *Mailing Add:* Dept Neurol Baylor Col Med One Baylor Plaza Houston TX 77030

EPSTEIN, HERMAN THEODORE, b Portland, Maine, Apr 13, 20; m 47; c 4. BIOPHYSICS. *Educ:* Univ Mich, BA, 41, MA, 43, PhD(physics), 49. *Prof Exp:* Physicist, Nat Adv Comt Aeronaut, 44-46; asst prof biophys & physics, Univ Pittsburgh, 49-53; asst prof physics, Brandeis Univ, 53-55, from assoc prof to prof biophys, 55-87, chmn dept biol, 71-74, EMER PROF BIOPHYS, BRANDEIS UNIV, 87- *Concurrent Pos:* NSF sr fel, 59-60; Guggenheim fel, 69-70; vis prof, Tel Aviv Univ, 70-71. *Mem:* Biophys Soc; Soc Neurosci; Int Soc Develop Psychobiol; NY Acad Sci; AAAS. *Res:* Brain and intelligence development in humans and in mice; educational implications of brain and mind growth stages. *Mailing Add:* Dept Biol Brandeis Univ South St Waltham MA 02154

EPSTEIN, HOWARD I, b Brooklyn, NY, Dec 16, 41; c 2. STRUCTURAL ENGINEERING, STRUCTURAL DESIGN. *Educ:* Cooper Union, BSCE, 63; Northwestern Univ, Evanston, MS, 65, PhD(appl mech), 67. *Prof Exp:* Res engr, IIT Res Inst, 67-69; asst prof, Univ Minn, 69-76; from asst prof to assoc prof, 76-88, PROF, DEPT CIVIL ENG, UNIV CONN, 88- *Concurrent Pos:* Consult, var co, 69-; engr, Brown Minneapolis Tank, 73-76; ed, J Prof Issues Eng, 85-86; sr engr, Torello Engrs, 88- *Mem:* Fel Am Soc Civil Engrs; Am Asn Univ Prof; Am Soc Eng Educ; Am Inst Steel Construct. *Res:* Structural engineering including wave propagation, steel codes, soil-structure interaction; specialized structures such as floating roofs, storage tanks and transmission poles; author of over 70 technical publications. *Mailing Add:* Dept Civil Eng Univ Conn Storrs CT 06269-3037

EPSTEIN, IRVING ROBERT, b Brooklyn, NY, Aug 9, 45; m 71; c 2. MATHEMATICAL MODELING, NONLINEAR DYNAMICS. *Educ:* Harvard Univ, AB, 66; Oxford Univ, dipl math, 67; Harvard Univ, MA, 68, PhD(chem physics), 71. *Prof Exp:* NATO fel, Cavendish Lab, Cambridge Univ, 71; from asst prof to prof chem, 71-89, chmn dept, 83-87, HELENA RUBINSTEIN PROF CHEM, BRANDEIS UNIV, 89- *Concurrent Pos:* NSF fel, Max Planck Inst, Gottingen, 77-78; mem adv bd, J Phys Chem, 81-89; Guggenheim fel, 87-88; consult, Du Pont, 84-, Bolt, Beranek & Newman, 91-; ed, Chaos, 91- *Honors & Awards:* Liebmann Award, Am Chem Soc. *Mem:* Am Chem Soc. *Res:* Experimental and theoretical studies of oscillating chemical reactions, chaos and pattern formation; mathematical modeling of phenomena in biochemical kinetics and neurobiology. *Mailing Add:* Dept Chem Brandeis Univ Waltham MA 02254-9110

EPSTEIN, ISADORE, astronomy, astrophysics; deceased, see previous edition for last biography

EPSTEIN, JACK BURTON, b New York, NY, Dec 27, 35; m 58; c 1. GEOLOGY. *Educ:* Brooklyn Col, BS, 56; Univ Wyo, MA, 58; Ohio State Univ, PhD, 70. *Prof Exp:* Field asst geol, 56, GEOLOGIST, US GEOL SURV, 57, 58-60, 64- *Concurrent Pos:* Instr, Ohio State Univ, 64. *Mem:* Geol Soc Am; Am Asn Petrol Geol; Soc Econ Paleont & Mineral. *Res:* Areal geology of western United States and eastern Pennsylvania; groundwater investigations in Louisiana; principals of environmental geology in United States. *Mailing Add:* US Geol Surv Nat Ctr 926 National Ctr Reston VA 22092

EPSTEIN, JOHN HOWARD, b San Francisco, Calif, Dec 29, 26; m 53; c 3. PHOTOBIOLOGY. *Educ:* Univ Calif, Berkeley, BA, 48; Univ Calif, San Francisco, MD, 52; Univ Minn, MS, 56. *Prof Exp:* Chief dermat, San Francisco Gen Hosp, 59-69 & Mt Zion Med Ctr, San Francisco, 71-81; PVT PRACT DERMAT, EPSTEIN, EPSTEIN & TUFFANELLI, MDS, INC, 56-; CLIN INVESTR PHOTOCARCINOGENESIS, DEPT DERMAT, UNIV CALIF, SAN FRANCISCO, 57-, CLIN PROF DERMAT, 72-

Concurrent Pos: Consult & vis prof, US Army, 60-86; chief ed, Archives Dermat, AMA, 73-78; mem, US Nat Comt Photobiol, Nat Res Coun, 76-81; Essex prof, Australasian Col Dermatologists, 77; permanent vis prof, Zagazig Univ, Egypt, 82- *Honors & Awards:* Gold Award, Am Acad Dermat, 69; Westwood Lectr, NAm Clin Dermat Soc, 80 & 84; Nomland-Carney Mem Lectr, Iowa Dermat Soc, 83; Herbert Luscombie Lectr, Jefferson Med Col, Pa, 85; Marion B Sulzberger Int Lectr, Miles Pharmaceut, Inc & Am Acad Dermat, 85; Clark W Finnerud Award, Dematol Found, 90. *Mem:* Am Acad Dermat (pres, 81-82); Soc Invest Dermat (vpres, 79-80); Am Fedn Clin Res; Am Bd Dermat (pres, 81-82); Am Soc Photobiol; fel Am Col Physicians & Surgeons; Pac Dermat Asn (pres, 85-86); hon mem in Denmark, Polish, French, Mex, Brit & Span Socs of Dermat; Am Dermat Asn (pres, 90-91). *Res:* Cutaneous photobiology including human diseases such as photoallergy, phototoxicity, porhyrias, and lupus erythematosus, etc.; experimental photocarcinogenesis especially as related to chemo-therapeutic agents; DNA damage and repair; oncogene activation. *Mailing Add:* 450 Sutter St Apt 1306 San Francisco CA 94108

EPSTEIN, JOSEPH, b Philadelphia, Pa, June 30, 18; m 45; c 4. PHYSICAL CHEMISTRY, ORGANIC CHEMISTRY. *Educ:* Temple Univ, AB, 38; Univ Pa, MS, 40; Univ Del, PhD, 66. *Prof Exp:* Plant chemist, Chem Corps Med Labs, Army Chem Ctr, 40-44, res assoc toxicol, Edgewood Arsenal, 44-45, chief analytical sect, Gassing Br, 45-47, Sanit Chem Br, 47-58 & Protection Res Br, 58-62, chief defense res div, Chem Res & Develop Labs, 62-66, chief defense res dept, Res Labs, 66-71, chief defense res br, Chem Labs, 71-73, chief, Environ Res Div, 73-80; CHEM CONSULT, 80- *Concurrent Pos:* Consult, USPHS, 54. *Honors & Awards:* Res & Develop Award, Off Res & Develop, US Army, 74 & 77. *Mem:* AAAS; Sigma Xi; Am Chem Soc; Am Ord Asn; NY Acad Sci. *Res:* Protection against and decontamination, detection and identification of chemical warfare agents; properties of chemical warfare materials in aqueous and nonaqueous media; reaction mechanisms and kinetics of reactions; development of analytical methods, especially micro methods; environmental problems, pollution abatement and control. *Mailing Add:* 4020 Essex Rd Baltimore MD 21207-4620

EPSTEIN, JOSEPH WILLIAM, b Brooklyn, NY, May 9, 38; m 64; c 2. ORGANIC CHEMISTRY. *Educ:* Cooper Union, BChE, 60; NY Univ, MS, 64, PhD(org chem), 65. *Prof Exp:* Fel org chem, Technion-Israel Inst Technol, 65-66; SR RES CHEMIST & GROUP LEADER, LEDERLE LABS, MED RES DIV, AM CYANAMID CO, 66- *Concurrent Pos:* Adj assoc prof, Dept Chem, Hunter Col, City Univ New York, 68-76. *Mem:* Am Chem Soc. *Res:* Synthesis of compounds related to the functions of the central nervous system and the cardiovascular-renal system. *Mailing Add:* 19 Briarwood Ave Monroe NY 10950-3105

EPSTEIN, JOSHUA, b Israel, May 10, 40; US citizen; m 83; c 1. CELL BIOLOGY. *Educ:* Bar-Ilan Univ, Israel, BSc, 67, MSc 69; Technion, Israel, DSc(biol), 72. *Prof Exp:* Res fel pediat & genetics, Harvard Med Sch & Mass Gen Hosp, 72-74; from instr to asst prof pediat, Johns Hopkins Univ Sch Med, 74-77; asst prof exp oncol, Ind Univ Sch Med, 77-78; cancer res scientist, Roswell Park Mem Inst, Buffalo, NY, 78-86; asst res prof exp path, Grad Sch, State Univ NY, Buffalo, 85-86; assoc cell biologist, Anderson Cancer Ctr, Univ Tex, Houston, 86-89; ASSOC PROF, UNIV ARK MED SCI, LITTLE ROCK, ARK, 89- *Honors & Awards:* First Ann Prize, Israeli Geront Soc, 72. *Mem:* Am Asn Cancer Res; Int Soc Exp Hemat; Asn Gnotobiotics; Int Asn Gnotobiotics; NY Acad Sci; AAAS. *Res:* Nature and biology of hematological malignancies and prediction of response to treatment; prevention of GVHD in allogeneic bone marrow transplantation. *Mailing Add:* Div Hemat/Oncol Ark Cancer Res Ctr Univ Ark Med Sci 4301 W Markham Slot 508 Little Rock AR 72205

EPSTEIN, L(UDWIG) IVAN, b Duisburg, Ger, Nov 25, 18; US citizen; m 55. OPTICS, COMPUTER SCIENCES. *Educ:* Calif Inst Technol, BS, 40, MS, 41; Ohio State Univ, PhD(physics), 67. *Prof Exp:* Res staff mem rockets, Calif Inst Technol, 43-46; optical engr, Bausch & Lomb Optical Co, 47-54; engr math physics, Martin Co, Baltimore, 54-58; asst prof physics, Lowell Technol Inst, 58-63; assoc prof, Marietta Col, 63-64; assoc prof biophys, Med Col Va, Va Commonwealth Univ, 67-82; RETIRED. *Concurrent Pos:* NIH grants, Va Commonwealth Univ, 73-74 & 79-81; vis scholar, Optical Sci Ctr, Univ Ariz, 82- *Mem:* Optical Soc Am; Asn Res Vision & Ophthal. *Res:* Geometrical, physical, and physiological optics; molecular orbital theory; nomography; physics applied to anesthesiology. *Mailing Add:* 8121 E Renaud Lane Tucson AZ 85710-8521

EPSTEIN, LAWRENCE MELVIN, b Brooklyn, NY, Apr 23, 23; m 46; c 4. PHYSICAL CHEMISTRY. *Educ:* Cooper Union, BChE, 43; Polytech Inst Brooklyn, MS, 52, PhD(phys chem), 55. *Prof Exp:* Chemist, Am Aniline Prod, 46-47; instr analytical chem, New York Community Col, 47-55; fel polymer properties, Mellon Inst, 55-56; sr scientist radiation res, Westinghouse Res Labs, 56-63, supvry scientist, 63-67; ASSOC PROF CHEM, UNIV PITTSBURGH, 67- *Mem:* Am Chem Soc. *Res:* Radiation chemistry and processing; radiation damage to materials; Mossbauer effect; chemical kinetics; polymer properties. *Mailing Add:* 4332 Saline St Pittsburgh PA 15217-2912

EPSTEIN, LEO FRANCIS, b New York, NY, Dec 9, 13; m 42; c 2. PHYSICAL CHEMISTRY. *Educ:* Mass Inst Technol, BS, 35, PhD(phys chem), 39. *Prof Exp:* Asst photochem, Solar Energy Res Comt, Mass Inst Technol, 39-41; asst chemist, Nat Defense Res Comt, High Explosives Res Div, US Bur Mines, 41-42; res engr, Crystal Res Labs, Inc, Conn, 45-47; res assoc, Knolls Atomic Power Lab, Gen Elec Co, 47-57, phys chemist, Vallecitos Atomic Lab, 57-68; sr chemist, Proj Mgt Support Div, Argonne Nat Lab, 68-80; RETIRED. *Concurrent Pos:* Sci adv, US Deleg Geneva Conf Peaceful Uses Atomic Energy, 55 & 71. *Mem:* Am Chem Soc; Am Phys Soc; Am Nuclear Soc; Math Asn Am; Sigma Xi. *Res:* Physical chemistry of solutions such as nonaqueous, electrolytes and liquid metals; dyestuff solutions; high explosives; mathematical analysis; ultrasonics; applications of physical chemistry to materials problems in nuclear systems; liquid sodium; fast breeder reactors. *Mailing Add:* 1100 Gough St 12A San Francisco CA 94109

EPSTEIN, LEON J, b Jersey City, NJ, June 7, 17; c 2. PSYCHIATRY. *Educ:* Vanderbilt Univ, AB, 37, MA, 38; George Peabody Col, PhD, 41; Univ Tenn, MD, 49. *Prof Exp:* Staff psychiatrist, St Elizabeths Hosp, Washington, DC, 54-56; dep dir res, Calif State Dept Ment Hyg, 56-61; from asst prof to prof psychiat, 61-87, vchmn dept, 69-87, EMER PROF, DEPT PSYCHIAT, SCH MED, UNIV CALIF, SAN FRANCISCO, 87- *Concurrent Pos:* Assoc med dir, Langley Porter Neuropsychiat Inst, 61-80. *Mem:* Am Psychiat Asn; Am Col Neuropsychopharmacol; Geront Soc; Am Col Psychiat; The Geriatric Soc. *Res:* Psychopharmacology; emotional disorders in the elderly. *Mailing Add:* Dept Psychiat Univ Calif San Francisco CA 94143

EPSTEIN, LOIS BARTH, b Cambridge, Mass, Dec 29, 33; m 56; c 4. MEDICAL SCIENCE, HEALTH SCIENCE. *Educ:* Radcliffe Col, AB, 55; Harvard Med Sch, MD, 59. *Prof Exp:* Resident path, Peter Bent Brigham Hosp, 59-60; intern med, New Eng Ctr Hosp, 60-61; res med officer, Nat Inst Arthritis & Metab Dis, 62-63; res fel, Med Sch, Univ Wash, 63-64; NIH spec res fel, Nat Inst Arthritis & Metab Dis & Nat Inst Allergy & Infectious Dis, 64-66, res med officer, Nat Inst Allergy & Infectious Dis, 66-69; from asst res physician to assoc res physician, 69-74, assoc dir cancer res inst, 74-77, assoc prof pediat, 74-80, PROF PEDIAT, UNIV CALIF, SAN FRANCISCO, 80- *Concurrent Pos:* Mem allergy & immunol training grants adv comt, Nat Inst Allergy & Infectious Dis, 72-73, mem allergy & immunol res adv comt, 73-76; vis scientist, Univ Col, Univ London, 73-74; Nat Cancer Inst res grants, 73-; Nat Inst Allergy & Infectious Dis res grant, 74-79; mem grad comt immunol, Univ Calif, San Francisco, 75-82; mem immunol sci study sect, NIH, 77-81; res assoc, Cancer Res Inst, 77-; March Dimes Birth Defect Found grant, 79-85; mem support rev comt, Cancer Ctr, Nat Cancer Inst, 84-87, bd sci couns, Div Cancer Biol, Diag & Ctrs, 91-; mem, Women Cancer Res, 88-; mem, Pediat Coun, Univ Calif, San Francisco, 88- & elec mem, Fac Coun, 89-; assoc ed, Cancer Res, 89. *Mem:* Am Soc Clin Invest; Am Asn Immunologists; Am Soc Hemat; Am Fedn Clin Res; Asn Am Physicians. *Res:* Cellular and tumor immunology and the role of interferon in each; the production, action, genetic control and delivery of interferon; mechanism of action of interleukin-1, interleukins 4 and 9 and tumor necrosis factor and immune defects in Down Syndrome; author of more than 100 scientific papers. *Mailing Add:* Univ Calif Med Sch Cancer Res Inst Moffitt 1282 San Francisco CA 94143

EPSTEIN, MARCELO, b Buenos Aires, Argentina, Dec 3, 44; div; c 3. GEOMETRY. *Educ:* Univ Buenos Aires, BSc, 67; Technion Israel, MSc, 70 & PhD(civil eng), 72. *Prof Exp:* Sr lectr civil eng, Technion Israel Inst Technol, 72-73; res assoc, Univ Alberta, 74-76; from asst prof to assoc prof, 76-83, PROF MECH ENG, UNIV CALGARY, 83- *Concurrent Pos:* Various consultancies. *Mem:* Am Soc Civil Engrs; Am Acad Mech; Soc Nat Phil. *Res:* Continuum mechanics; shell theory; wave propagation; material uniformity; modern mathematical tools; numerical techniques. *Mailing Add:* Dept Mech Eng Univ Calgary Calgary AB T2N 1N4 Can

EPSTEIN, MARVIN PHELPS, mathematics, computer science; deceased, see previous edition for last biography

EPSTEIN, MARY A FARRELL, b New York, NY, May 9, 39. RESPIRATORY SYSTEM MECHANICS & GAS EXCHANGE, BIOFLUID MECHANICS. *Educ:* Cooper Union Sch Eng, BS, 59; Columbia Univ, MS, 62, DEngSc, 67. *Prof Exp:* Asst prof, 83-84, ASSOC PROF PHARMACOL, MED SCH, UNIV CONN, 84- *Concurrent Pos:* Adj assoc prof aerospace mech & nuclear eng, Univ Okla, Norman, 85-; mem bd dirs, Biomed Eng Soc, 89-92; ed, Modeling & Physiol, Am Physiol Soc, 91- *Mem:* Am Physiol Soc; Am Inst Chem Engrs; Am Soc Eng Educ; Soc Women Engrs; AAAS; Biomed Eng Soc. *Res:* Mathematical modeling of physiological systems. *Mailing Add:* Dept Pharmacol Med Sch Univ Conn Med Sch 263 Farmington Ave Farmington CT 06032

EPSTEIN, MAX, b Lodz, Poland, Feb 5, 25; US citizen; m 63; c 3. ELECTRICAL ENGINEERING, SOLID STATE PHYSICS. *Educ:* Israel Inst Technol, BS, 52; Ill Inst Technol, MS, 55, PhD(elec eng), 63. *Prof Exp:* Instr elec eng, Ill Inst Technol, 54-58, res engr, ITT Res Inst, 58-64, sr res engr, 64-67; PROF ELEC ENG, NORTHWESTERN UNIV, 67- *Concurrent Pos:* Consult, IIT Res Inst Israel Defense Army, 48-49. *Mem:* AAAS; Am Phys Soc; Inst Elec & Electronics Engrs. *Res:* Fiber optics; lasers in medicine. *Mailing Add:* Dept Elec & Eng & Computer Sci Northwestern Univ Evanston IL 60201

EPSTEIN, MORTON BATLAN, b New York, NY, June 8, 17; m 42; c 2. CLINICAL CHEMISTRY. *Educ:* NY Univ, BS, 37; Univ Ill, PhD(chem), 42. *Prof Exp:* Asst, Univ Ill, 41-42; chemist, Picatinny Arsenal, NJ, 42-45; Am Petrol Inst res assoc, Nat Bur Stand, 46-49; res assoc, Colgate Palmolive Co, 49-54, Onyx Oil & Chem Co, 54-57, Colgate Palmolive Co, 57-63 & div adult health & aging, Chicago Bd Health, 63-69; CLIN CHEMIST, CHRIST HOSP, 69- *Concurrent Pos:* Assoc clin path, Chicago Med Sch, 70-74; asst prof, Med Sch, Northwestern Univ, Chicago, 74- *Mem:* Am Asn Clin Chemists; Am Chem Soc; Am Oil Chem Soc. *Res:* Physical and thermodynamic properties of hydrocarbons; analysis of petroleum; physical chemistry of detergents and foams; atherosclerosis and hypertension. *Mailing Add:* 5039 S Ellis Ave Chicago IL 60615-2711

EPSTEIN, MURRAY, b Tel Aviv, Israel, Aug 11, 37; US citizen; m 78; c 3. NEPHROLOGY. *Educ:* Columbia Univ, BA, 59 & MD, 63. *Prof Exp:* Res fel nephrology, Harvard Med Sch, 66-68; from asst prof to assoc prof, 70-78, PROF MED, SCH MED, UNIV MIAMI, 78 - *Concurrent Pos:* Investr, Howard Hughes Med Inst, 72-76. *Honors & Awards:* Distinguished Scientist Award, Nat Kidney Found, 90. *Mem:* Fel Am Col Physicians; Am Soc Clin Invest; Am Physiol Soc; Soc Exp Bio & Med; Int Soc Nephrology; Am Soc Nephrology; Am Soc Hypertension; Int Soc Hypertension; Ital Soc Study Liver Dis; Am Asn Study Liver Dis. *Res:* Kidney function; diseases chacterized by abnormal volume regulation; role of the renin-aldosterone system (vasoactive hormones); causes of hypertension. *Mailing Add:* Nephrology Sect (111C1) Vet Admin Med Ctr 1201 NW 16th St Miami FL 33125

EPSTEIN, NATHAN BERNIC, b New Waterford, NS, Mar 3, 24; m 51; c 3. PSYCHIATRY. *Educ:* Dalhousie Univ, MD & CM, 48; McGill Univ, dipl psychiat, 52; Columbia Univ, cert psychoanal, 55; Am Bd Psychiat & Neurol, dipl, 53. *Prof Exp:* Sr intern & asst resident psychiat, Allan Mem Inst & Royal Victoria Hosp, Montreal, 48-50; resident & jr physician, Boston State Hosp, Mass, 50-51; vol clin asst child psychiat, Outpatient Dept, Mt Sinai Hosp, New York, 51-53; actg dir, Ment Health Ctr, Paterson, NJ, 53-54; mem staff psychiat, Jewish Gen Hosp, 55-58, head sub-dept child & family psychiat, 58-59, asst psychiatrist, 59-60, psychiatrist-in-chief, 60-67; prof psychiat & chmn dept, McMaster Univ, 67-75; PROF & CHMN, SECT PSYCHIAT & HUMAN BEHAVIOR, BROWN UNIV, 78-; MED DIR, BUTLER HOSP, PROVIDENCE, RI, 78- *Concurrent Pos:* Res assoc, McGill Univ, 55-58, lectr & res asst, 58-61, assoc prof, 61-67, co-dir human develop study, 58, training analyst comt psychoanal, 59; vis prof, McMaster Univ, 78. *Mem:* Fel Acad Psychoanalysis; fel Am Psychiat Asn; Can Psychiat Asn; Can Med Asn; Can Psychoanalysis Soc. *Res:* Family structure, organization and transaction dynamics and application to family therapy groups. *Mailing Add:* Parkwood Hosp 4499 Acushnet Ave New Bedford MA 02745

EPSTEIN, NORMAN, b Montreal, Can, Dec 6, 23; m 47; c 3. CHEMICAL ENGINEERING. *Educ:* McGill Univ, BE, 45, ME, 46; NY Univ, EngScD(chem eng), 53. *Prof Exp:* Lectr chem eng, McGill Univ, 46-48; instr, NY Univ, 49-51; from instr to prof, 51-88, HON PROF, UNIV BC, 89- *Concurrent Pos:* Consult, Heat Transfer Res Inc, 71-; Killam sr fel, 75-76; hon res fel, Univ Col, London, 76; chmn, Can Nat Comt Heat Transfer, 79-84; res assoc, Atomic Energy Res Estab, Harwell, 81; ed, Can J Chem Eng, 85-90. *Honors & Awards:* Jules Stachiewicz Medal, Can Soc Chem Eng & Can Soc Mech Eng, 88; R S Jane Mem Lecture Award, Can Soc Chem Eng, 89. *Mem:* Am Chem Soc; fel Am Inst Chem Engrs; fel Chem Inst Can; Can Soc Chem Eng (vpres, 78-79, pres, 79-80). *Res:* Momentum; heat; mass transfer; fouling of heat exchangers; liquid- and three-phase fluidization; spouted beds; colloid mechanics. *Mailing Add:* Dept Chem Eng Univ BC Vancouver BC V6T 1W5 Can

EPSTEIN, PAUL MARK, b Brooklyn, NY, June 24, 46; m 75; c 3. ENZYMOLOGY, PHARMACOLOGY. *Educ:* Columbia Univ, AB, 67; Albert Einstein Col Med, Yeshiva Univ, PhD(molecular biol), 75. *Prof Exp:* Res assoc pharmacol, Univ Tex Med Sch, Houston, 75-78, instr, 78-79; asst prof, 79-84, ASSOC PROF PHARMACOL, UNIV CONN HEALTH CTR, 84- *Concurrent Pos:* Proj investr immunol, Dept Develop Theapeut, M D Anderson Hosp & Tumor Inst, Univ Tex & Syst Cancer Ctr, 75-78. *Mem:* Sigma Xi; AAAS; NY Acad Sci; Soc Neurosci; Am Asn Pharmacol & Exp Therapeut. *Res:* Cyclic nucleotide metabolism, calcium and calmodulin in the control of cell growth and development; purification, characterization and hormonal regulation of cyclic nucleotide phosphodiesterases; mediation of second messenger actions by protein phosphorylation. *Mailing Add:* Dept Pharmacol Univ Conn Health Ctr Farmington CT 06030

EPSTEIN, ROBERT B, b Chicago, Ill, June 9, 34; m 57; c 3. ONCOLOGY, HEMATOLOGY. *Educ:* Univ Ill, BS, 57, MD, 59; Am Bd Internal Med, cert, 67. *Prof Exp:* Intern, Res & Educ Hosps, Univ Ill, Chicago, 59-60; med resident, Vet Admin West Side Hosp, 60-63; res fel hemat, Univ Wash, Seattle, 63-65, from instr to asst prof med, 65-70; from assoc prof to prof med, Abraham Lincoln Sch Med, Univ Ill, Chicago, 71-82, prof exp med, Biol Resources Lab, 74-82; chief, Oncol Sect, Dept Med, 82-90, EASON PROF MED, SCH MED, UNIV OKLA, OKLAHOMA CITY, 82-; CHIEF, ONCOL SECT, VET ADMIN MED CTR, OKLA, 82- *Concurrent Pos:* Asst chief med, USPHS Hosp, Seattle, 66-70; consult, Children's Orthop Hosp & Med Ctr, 68-70, Dept Internal Med, Rush Presby St Luke's Hosp, 76-82 & Presby Hosp, Oklahoma City, 82-; Boerhaave prof med, Univ Leiden, Neth, 70-71; clin investr, Vet Admin West Side Hosp, Chicago, 71-74, chief, Med Oncol Sect, 74-75, med investr, 75-82; dir clin oncol, Okla Mem Res Found, 82-85; adj prof path, Sch Med, Univ Okla, 82- *Mem:* Transplantation Soc; Am Soc Hemat; Am Soc Clin Oncol; Am Asn Immunologists; Am Asn Cancer Res; Sigma Xi; Int Soc Exp Hemat; fel Am Col Physicians; AAAS; Am Fedn Clin Res. *Res:* Transplantation biology, specifically bone marrow transplantation. *Mailing Add:* PO Box 26901/4SP-100 Oklahoma City OK 73190

EPSTEIN, ROBERT MARVIN, b New York, NY, Mar 10, 28; m 50; c 3. ANESTHESIOLOGY. *Educ:* Univ Mich, BS, 47, MD, 51; FRCS(E), 81. *Prof Exp:* Intern, Univ Mich Hosp, 51-52; asst resident anesthesiol, Presby Hosp, NY, 52-53 & 55-56; from instr to prof, Col Physicians & Surgeons, Columbia Univ, 56-72; PROF ANESTHESIOL & CHMN DEPT, SCH MED, UNIV VA, 72- *Concurrent Pos:* NY Heart Asn fel med, Columbia Univ, 56-57, Nat Inst Gen Med Sci grant, 59-72; asst anesthesiol, Presby Hosp, New York, 56-58, from asst attend anesthesiologist to assoc attend anesthesiologist, 58-70, attend anesthesiologist, 70-72; vis scientist & Guggenheim Mem Found fel, Dept Pharmacol, Oxford Univ, 66-67; mem anesthesiol training comt, Nat Inst Gen Med Sci, 66-69 & comt anesthesia, Nat Res Coun, 70-71; dir, Am Bd Anesthesiol, 72-, pres, 79-80; ed, Anesthesiol, 74-; Am Bd Anesthesiol rep, Am Bd Med Specialties, 74- *Mem:* AAAS; Am Soc Anesthesiol; Asn Univ Anesthetists (secy, 69-72, pres, 74-75); Am Soc Pharmacol & Exp Therapeut; Am Physiol Soc; Sigma Xi. *Res:* Circulatory physiology; effects of anesthetics on splanchnic circulation and its neural control; effects of anesthetics and neuromuscular blocking agents on neuromuscular transmission and the electromyogram; effects of carbon dioxide; pharmacokinetics of anesthetics. *Mailing Add:* Univ Va Med Ctr Box 238 Charlottesville VA 22908

EPSTEIN, SAMUEL, b Poland, Dec 9, 19; US citizen; m 46; c 2. GEOCHEMISTRY. *Educ:* Univ Man, BSc, 41, MSc, 42; McGill Univ, PhD(phys chem), 44. *Hon Degrees:* LLD, Univ Manitoba, Winnipeg, Can, 80. *Prof Exp:* Res chemist, Nat Res Coun Can, 44-47; res assoc, Inst Nuclear Studies, Univ Chicago, 48-52; res fel, Calif Inst Technol, Pasadena, 52-53, sr res fel, 53-54, assoc prof, 54-59, prof geochem, 59-84, William H Leonhard prof geol, 84-90, EMER WILLIAM H LEONHARD PROF GEOL, CALIF INST TECHNOL, PASADENA, 90- *Concurrent Pos:* Hon foreign fel, Europ Union Geol Sci, 80. *Honors & Awards:* Goldschmidt Medal, Geochem Soc, 77; Arthur L Day Medal, Geol Soc Am, 78. *Mem:* Nat Acad Sci; fel AAAS; Geol Soc Am; Geochem Soc (pres, 78-79); fel Am Geophys Union; fel Am Acad Arts & Sci; hon foreign fel Europ Union Geosci. *Res:* Stable isotope geochemistry; application of isotope measurments to the problems related meteorology, hydrology, glaciology, petrology, biochemistry, plant physiology, climatology, paleontology, meteoritics, and lunar samples. *Mailing Add:* Div Geol & Planetary Sci Calif Inst Technol Pasadena CA 91125

EPSTEIN, SAMUEL DAVID, b Brooklyn, NY, Aug 4, 46; m 68; c 2. COMPUTER SCIENCE, SYSTEMS ENGINEERING. *Educ:* Drexel Univ, BS, 68. *Prof Exp:* Analyst comput sci, Auerbach Corp, 65-68; proj leader, Analytics Inc, 68-69, dir, Philadelphia Div, 69-70, vpres, 70-84; pres, ACS Commun Systs Inc, 83-87; PRES & DIR, RACAL-GUARDATA INC, 87- *Concurrent Pos:* NSF fel, 67-68. *Mem:* Am Inst Physics; Armed Forces Commun & Electronics Asn; Tech Mkt Soc Am. *Res:* Multi-user interactive data storage and retrieval concentrating on free form English and semantic interpretation; one patent for encrypting transponder. *Mailing Add:* Racal-Guardata Inc 480 Spring Park Pl Suite 900 Herndon VA 22070

EPSTEIN, SAMUEL STANLEY, b Middlesborough, Eng, Apr 13, 26; US citizen; m 59; c 3. PATHOLOGY, ENVIRONMENTAL SCIENCES. *Educ:* Univ London, BSc, 47, MB & BS, 50, dipl trop med & surg, 52, dipl path, 54, MD, 58, dipl microbiol & pub health, 63; Am Bd Microbiol, dipl, 63. *Prof Exp:* Demonstr morbid anat, Guy's Hosp, Univ London, 50; house physician med, St Johns's Hosp, Eng, 51; lectr path & bact, Inst Laryngol & Otol, Eng, 55-58; tumor pathologist & Brit Empire Cancer Campaign fel, Hosp Sick Children & Chester Beatty Cancer Res Inst, Eng, 58-60; consult path, Mem Hosp Peterborough, Eng, 60; chief labs environ toxicol & carcinogenesis, Children's Cancer Res Found, Inc, Boston, Mass, 61-71; prof pharmacol & Swetland Prof environ health & human ecol, Med Sch, Case Western Reserve Univ, 71-76; PROF ENVIRON & OCCUP MED, SCH PUB HEALTH, UNIV ILL MED CTR, 76- *Concurrent Pos:* Res assoc path, Harvard Med Sch, 62-71; consult, US Senate Comt Pub Works, 70-, Ctr Studies Narcotic & Drug Abuse, NIMH, 70- & Environ Health Progs, Inc, 70-; mem panel on polycyclic org matter, Nat Acad Sci, 70-; mem pesticide bd, Commonwealth of Mass, 70-; mem, US Senate Subcomt Exec Reorgn & Govt Res; pres, Rachel Carson Trust, Inc, DC; chairperson, Comn Advan Pub Interest Orgns, DC; dir environ health resource ctr, State Ill, 78- *Honors & Awards:* Montefiore Gold Medal Trop Med, Royal Army Med Corps, 53, Montefiore Prize Trop Hyg, 53, Ranald Martin Prize Mil Surg, 53; Achievement Award, Soc Toxicol, 69. *Mem:* AAAS; Soc Occup & Environ Health (pres); fel Royal Soc Health; Environ Mutagen Soc (secy, 69); Air Pollution Control Asn. *Res:* Toxicology; carcinogenesis; mutagenesis; preventive medicine; bacteriology and protozoology; biological hazards, including carcinogenesis, mutagenesis, due to chemical pollution of the environment, including food additives, pesticides, fertilizers, industrial chemicals and drugs; ecological effects of chemical pollutants. *Mailing Add:* Sch Pub Health M-C922 Box 6998 Med Ctr Univ Ill Chicago IL 60680

EPSTEIN, SAUL THEODORE, b Southampton, NY, June 14, 24; m 48; c 3. THEORETICAL PHYSICS. *Educ:* Mass Inst Technol, SB, 44, PhD(physics), 48. *Prof Exp:* With Inst Advan Study, 47-48; instr physics, Columbia Univ, 48-51 & Boston Univ, 52-53; from asst prof to prof, Univ Nebr, 54-63; PROF PHYSICS & CHEM & MEM THEORET CHEM INST, UNIV WIS-MADISON, 63- *Mem:* Am Phys Soc; Am Asn Physics Teachers. *Res:* Basic quantum theory; atomic and molecular structure. *Mailing Add:* Dept Physics Univ Wis 15289 Chamberlin Hall 1150 University Ave Madison WI 53706

EPSTEIN, SEYMOUR, b New York, NY, Mar 19, 21; m 55; c 4. PHYSICS, ENGINEERING. *Educ:* City Col New York, BME, 43; Polytech Inst Brooklyn, MS, 51, PhD(physics), 61. *Prof Exp:* Instr physics & math, Univ Akron, 43-44; physics, Assoc Cols Upper NY, 46-47; descriptive geometry, Brooklyn Col, 47-50; eng writer consult, CBS-Columbia, Inc, 50-51; PHYS SCIENTIST, LASER TECH AREA, COMBAT SURVEILLANCE & TARGET ACQUISITION LAB, US ARMY ELECTRONICS COMMAND, FT MONMOUTH, 52- *Concurrent Pos:* Tool engr, Goodyear Aircraft Corp, 43-44. *Mem:* AAAS; Am Phys Soc; Sigma Xi. *Res:* Electrooptics; solid state physics. *Mailing Add:* Ten Lady Bess Dr Deal NJ 07723

EPSTEIN, STEPHEN EDWARD, b Brooklyn, NY, Dec 23, 35; m 57; c 3. CARDIOLOGY. *Educ:* Columbia Col, BA, 57; Cornell Univ, Med Sch, MD, 61. *Prof Exp:* Intern, New York Hosp, 61-62, resident, 62-63; clin assoc, Nat Heart Inst, 63-66, cardiologist to surg serv, 66-68; clin instr med, Sch Med, Georgetown Univ, 67-68; actg chief, Cardiol Br, Nat Heart Inst, 68-69, chief, sect Circulatory Physiol, Nat Heart & Lung Inst, 68-73, CHIEF, CARDIOL BR, 69- *Concurrent Pos:* Clin assoc prof med, Sch Med, Georgetown Univ, 71-74, clin prof, 74-; mem, Coun Circulation & Coun Basic Sci, Am Heart Asn. *Honors & Awards:* William C Thro Prize; William Mecklenburg Polk Prize; Harold H Bix Lectr, 78; James B Herrick Mem Lectr, Chicago Heart Asn, 79; George R Herrman Lectureship, Univ Tex Med Br, Galveston, 80; Ramon L Langes Mem Lectr, Med Col Wisc, 83; Hammond Lectr, St John's Mercy Med Ctr, St Louis, 83; Kenneth Rosen Mem Lectr, Univ Ill, Chicago, 84; William L MacDonald Lectr, Heart & Stroke Found, Ont, 85; Hugo Roesler Mem Lectr, Temple Univ, 86. *Mem:* Am Fedn Clin Res; Am Col Cardiol; Am Soc Clin Invest; Am Heart Asn; hon mem Am Col Chest Physicians. *Res:* Biological mechanisms controlling cardiac function and those that lead to cardiac dysfunction; to develop new diagnostic and therapeutic approaches to cardiac disease. *Mailing Add:* NIH Bldg Ten Rm 7B15 Bethesda MD 20892

EPSTEIN, SUZANNE LOUISE, IDIOTYPE ANALYSIS, TRANSPLANTATION BIOLOGY. *Educ:* Mass Inst Technol, PhD(biol), 79. *Prof Exp:* SR STAFF FEL IMMUNOL, FOOD & DRUG ADMIN, NIH, 85- *Mailing Add:* Div Biochem & Biophys Off Biol Res & Rev NIH 8800 Rockville Pike Bldg 29 Rm 522 Bethesda MD 20892

EPSTEIN, WALLACE VICTOR, b New York, NY, Dec 10, 26; m 49; c 3. INTERNAL MEDICINE. *Educ:* City Col New York, BS, 48; Columbia Univ, MD, 52. *Prof Exp:* Res fel internal med, Columbia Univ, 55-56; res fel, 56-58, from asst prof to assoc prof med, 58-73, prof community med, 73-77, PROF MED, UNIV CALIF, SAN FRANCISCO, 77- *Mem:* AMA; Am Rheumatism Asn. *Res:* Arthritis; clinical and experimental immunology. *Mailing Add:* Dept Med Univ Calif 350 Parnassus St Suite 407 San Francisco CA 94117

EPSTEIN, WILLIAM L, b Cleveland, Ohio, Sept 6, 25; m 54; c 2. MEDICINE, DERMATOLOGY. *Educ:* Univ Calif, Berkeley, AB, 49; Univ Calif, San Francisco, MD, 52. *Prof Exp:* Instr dermat, Univ Pa, 56-57; from asst prof to assoc prof, 57-69, actg chmn div, 66-67, chmn div, 67-70, actg chmn dept, 70-74, chmn dept, 74-85, DIR DERMAT RES, UNIV CALIF, SAN FRANCISCO, 57-, PROF DERMAT, 69- *Concurrent Pos:* Consult dermatologist, 57- *Honors & Awards:* Dohi Lectr, Tokyo, 82. *Mem:* AAAS; APCR; Am Acad Dermat; Soc Invest Dermat; Am Dermat Asn; Sigma Xi. *Res:* Immunology, especially delayed and granulomatous hypersensitivity with plant-induced dermatitis and viral and cancer immunology; skin anatomy and epidermal cell turnover. *Mailing Add:* Sch Med Univ Calif Box 0536 San Francisco CA 94143-0536

EPSTEIN, WILLIAM WARREN, b Kremmling, Colo, Sept 10, 31; m 53, 80; c 3. BIO-ORGANIC CHEMISTRY. *Educ:* Univ Denver, BS, 53; Univ Calif, PhD(chem), 59. *Prof Exp:* Res scientist, Weyerhaeuser Co, 58-60; fel, Univ Ill, 60-61; from asst prof to assoc prof, 61-73, PROF CHEM, UNIV UTAH, 73-; VPRES, CHEM BIOCHEM RES, INC, 80- *Mem:* Am Chem Soc; Chem Soc London; Sigma Xi. *Res:* Biosynthesis of terpenoids; natural products. *Mailing Add:* Dept Chem Univ Utah Salt Lake City UT 84112

EPSTEIN, WOLFGANG, b Breslau, Ger, May 7, 31; US citizen; m 61; c 3. BIOCHEMISTRY, GENETICS. *Educ:* Swarthmore Col, BA, 51; Univ Minn, MD, 55. *Prof Exp:* Fel biophys, Harvard Med Sch, 61-63; guest investr molecular biol, Pasteur Inst, Paris, 63-65; res assoc biophys, Harvard Med Sch, 65-67; from asst prof to prof biochem, 67-84, PROF MOLECULAR GENETICS & CELL BIOL, UNIV CHICAGO, 84- *Concurrent Pos:* Mem, NIH Microbial Chem Study Sect, 73-77. *Mem:* Am Soc Biol Chemists; Am Soc Microbiol; Biophys Soc; AAAS. *Res:* Transport, regulation and genetic mechanisms in bacteria with major reliance on genetic analysis of these processes. *Mailing Add:* Dept Molecular Genetics & Cell Biol Univ Chicago Chicago IL 60637

ERASMUS, BETH DE WET (FLEMING), b Niagara Falls, NY, Oct 19, 35; m 65; c 2. PHYSIOLOGY. *Educ:* Wellesley Col, AB, 57; State Univ NY Buffalo, PhD(physiol), 67, MArch, 86. *Prof Exp:* Res asst pharmacol, Chas Pfizer Co Inc, 58-60; from instr to asst prof, 67-72, CLIN ASST PROF PHYSIOL, STATE UNIV NY BUFFALO, 72- *Mem:* Assoc Am Phys Soc. *Res:* Gas exchange and acid-base balance in the avian embryo; fall risk assessment and preventation. *Mailing Add:* Dept Physiol Main St Campus State Univ NY Sch Med Buffalo NY 14214

ERATH, LOUIS W, b Abbeville, La, June 10, 17. GEOPHYSICS. *Educ:* La State Univ, BS, 39. *Prof Exp:* Assoc engr, United Gas Pipeline Co, 40-42; head, Vt Fuse Eng Dept, US Navy Silver Spring, Md, 42-47; res engr, Schulmberger Well Logging Co, 47-48; vpres & tech dir, Southwestern Indust Electronics, 48-61, vpres & partner, Test Equip Corp, 61-66, vpres & tech dir, Geo Space Corp, 66-69, pres & proprietor, Erath Sound Co, 69-80; INDEPENDENT ELECTRONIC CONSULT, 80- *Mem:* Fel Inst Elec & Electronics Engrs. *Res:* Electronic instrumentation for geophysical exploration; magnetometers; sound, feedback loudspeakers. *Mailing Add:* 209 N State St Box 177 Abbeville LA 70510

ERB, DENNIS J, b Philadelphia, Pa, Apr 19, 52; m 82; c 3. MEDICINAL CHEMISTRY. *Educ:* E Stroudsburg State Col, BA, 73; State Univ NY, Buffalo, PhD(med chem), 78. *Prof Exp:* Res asst prof chem, Kalamazoo Col, 78-79; from asst prof to assoc prof, 79-90, PROF & CHMN CHEM, EAST STROUDSBURG UNIV, 90- *Concurrent Pos:* Vis prof chem, Lehigh Univ, 85, 86. *Mem:* Am Chem Soc; NY Acad Sci; Sigma Xi. *Res:* Design and synthesis of new compounds whose most profound activity involves a direct interaction with nucleic acids and elicit biological responses through those interactions. *Mailing Add:* Dept Chem East Stroudsburg Univ East Stroudsburg PA 18301

ERB, KENNETH, b Souderton, Pa, Apr 28, 39; m 64. HORTICULTURE. *Educ:* Goshen Col, BA, 61; WVa Univ, MS, 64, PhD(agr microbiol), 66. *Prof Exp:* Lectr biol, Vassar Col, 66-67, asst prof, 67-69; asst prof, 69-77, ASSOC PROF BIOL, HOFSTRA UNIV, 77- *Mem:* Am Hort Soc; Phycol Soc Am; Bot Soc Am. *Res:* field botany (ecosystem flora determination). *Mailing Add:* Dept Biol Hofstra Univ 1000 Fulton Ave Hempstead NY 11550

ERB, ROBERT ALLAN, b Ridley Park, Pa, Jan 30, 32; m 53; c 3. PHYSICAL CHEMISTRY. *Educ:* Univ Pa, BS, 53; Drexel Inst, MS, 59; Temple Univ, PhD(phys chem), 65. *Prof Exp:* Chemist, Gates Eng Co, Del, 53-54; res asst, 54-56, res engr, 56-61, sr res chemist, 61-65, sr staff chemist, 65-68, prin scientist, 68-81, inst fel, 81-84, STAFF SCIENTIST, DIV ARVIN-CALSPAN, FRANKLIN RES CTR, 84- *Mem:* AAAS; Am Chem Soc; Soc Plastics Engrs; Am Anaplastology Asn; Sigma Xi. *Res:* External prosthetics; human simulators; adhesion; rheology; heterogeneous nucleation; desalination; medical applications of physical science; environmental science; waste management; solar energy applications. *Mailing Add:* Div Arvin-Calspan Franklin Res Ctr 2600 Monroe Blvd Norristown PA 19403

ERBACHER, JOHN KORNEL, b Washington, DC, Aug 4, 42; m 66; c 3. TECHNOLOGY ASSESSMENT, CHEMICAL POWER SOURCES. *Educ:* Cath Univ Am, BA, 64, MS, 67; Colo State Univ, PhD(phys chem), 73. *Prof Exp:* Res chemist, Directorate Chem Sci, Frank J Seiler Res Lab, 73-75, dep dir, 75-77; chief thermal battery res, USAF, 77-78, asst chief metals br, Air Force Mat Lab, 78, chief electronics br, Man Tech Div, Air Force Wright Aeronaut Labs, 79-82, technol transfer mgr, Tech Div, FTD Wright-Patterson Air Force Base, USAF, 82-86; Dayton Opers Mgr, Strategic Res & Mgt Servs Inc, 87-88; PROF MGR & SR PROG MGR, UNIVERSAL TECHNOL CORP, 88- *Concurrent Pos:* Mem, Subgroup Chromates, Joint Deps Labs Comts Subpanel Thermal Batteries, 75-82; mem, export control subcomt, Technol Transfer Intel Comt, 82-86; assoc mem, AFLC Indust Technol Exchange Group Electronics & Avionics, 88- *Mem:* Sigma Xi; Am Defense Prep Asn. *Res:* Plan, advocate and design implementation of new manufacturing and repair technologies for electronics and avionics for air logistic repair depots; evaluate and recommend new US Air Force manufacturing programs in solar cells, batteries, and other electronic technologies. *Mailing Add:* 368 Bowman Dr Fairborn OH 45324

ERBAR, JOHN HAROLD, chemical engineering; deceased, see previous edition for last biography

ERBE, LAWRENCE WAYNE, b Ancon, CZ, June 30, 24; m 55; c 5. BOTANY. *Educ:* Univ Vt, BS, 53, MS, 55; Univ Tex, PhD(bot), 60. *Prof Exp:* Asst prof, 60-66, ASSOC PROF BIOL, UNIV SOUTHWESTERN LA, 66- *Mem:* Bot Soc Am. *Res:* Hybridization of Lotus tenuis and Lotus corniculatus; biosystematics of annual phloxes. *Mailing Add:* Dept Biol Univ Southwestern La Lafayette LA 70504

ERBE, RICHARD W, b Maquoketa, Iowa, July 18, 39. GENETICS. *Educ:* Univ Mich, MD, 64. *Prof Exp:* CHIEF, GENETICS UNIT, MASS GEN HOSP, 74-; PROF GENETICS & PEDIAT, SCH MED, HARVARD UNIV, 84- *Mailing Add:* Genetics Unit Mass Gen Hosp Boston MA 02114

ERBER, THOMAS, b Vienna, Austria, Dec 6, 30; nat US; m 57. PHYSICS, ELECTROMAGNETISM. *Educ:* Mass Inst Technol, BSc, 51; Univ Chicago, MS, 53, PhD(physics), 57. *Hon Degrees:* Univ Graz, Dr, 71. *Prof Exp:* From asst prof to assoc prof, 57-69, PROF PHYSICS, ILL INST TECHNOL, 69-, PROF MATH, 86- *Concurrent Pos:* Ill Inst Technol fac fel, 58-59; res fel, Univ Brussels, 63-64; vis scientist, Stanford Linear Accelerator Ctr, 70; vis prof physics, Univ Calif, Los Angeles, 78, 84 & 87, Univ Grenoble & Univ Graz, 82. *Mem:* Sr mem Inst Elec & Electronics Engrs; fel Am Phys Soc; Am Math Soc; Austrian Physics Soc; Europ Phys Soc; Magnetics Soc. *Res:* Classical and quantum electrodynamics; ultra high magnetic fields; hysteresis systems; statistical physics. *Mailing Add:* Dept Physics Ill Inst Technol Chicago IL 60616

ERBISCH, FREDERIC H, b Sebewaing, Mich, June 24, 37; m 57; c 2. BOTANY, CYTOLOGY. *Educ:* Mich State Univ, BS, 59; Univ Mich, MS, 61, PhD(bot), 66. *Prof Exp:* prof bot, Mich Technol Univ, 63-80, actg dir res, 80-82, dir, res serv, 82-88, EXEC DIR, INTELLECTUAL PROPERTIES OFF, MICH TECHNOL UNIV, 89- *Mem:* Sigma Xi; Asn Univ Tech Managers; Bot Soc Am; Am Bryol & Lichenological Soc; Mycol Soc Am; Soc Res Adminr; Nat Coun Univ Res Adminr. *Res:* Cytodevelopment of lichen asci and ascospores; effects of gamma irradiation on lichens and lichen-forming fungi and algae; enzyme systems of wood rotting fungi; using vegetation as means of controlling infrared signatures. *Mailing Add:* 1117 E Sixth Ave Houghton MI 49931

ERBY, WILLIAM ARTHUR, b Lebanon, Pa, Apr 4, 33; m 57; c 2. ORGANIC CHEMISTRY. *Educ:* Lebanon Valley Col, BS, 55; Bucknell Univ, MS, 57; State Univ NY Col Forestry, PhD(chem), 61. *Prof Exp:* Chemist, Kimberly Clark Corp, 60-63; chemist, Air Prod & Chem Inc, 63-64, group leader, 64-68; mgr res, 68-70, VPRES RES, DAUBERT CHEM CO, 70- *Mem:* Am Chem Soc. *Res:* Wood, polymer and textile chemistry. *Mailing Add:* 1060 Black Oak Dr Downers Grove IL 60515

ERCK, ROBERT ALAN, b Chicago, Ill, Oct 11, 54. ION-BEAM ASSISTED COATINGS, TRIBOLOGY. *Educ:* Goshen Col, BA, 76; Univ Ill-Urbana, MS, 79, PhD(metall), 88. *Prof Exp:* Postdoctoral, 87-89, ASST METALLURGIST, TRIBOLOGY SECT, ARGONNE NAT LAB, 89- *Concurrent Pos:* Assoc ed, Lubrication Eng, 89-; fac mem, Col DuPage, 90- *Mem:* Mat Res Soc; Am Vacuum Soc; Am Soc Metals Int; Minerals, Metals & Mat Soc; Soc Tribologists & Lubrication Engrs; AAAS. *Res:* Study of ion-beam-assisted methods for producing thin films for tribological application; study of kinetics of high-energy radiation effects in alloys. *Mailing Add:* Argonne Nat Lab MCT-212 9700 S Cass Ave Argonne IL 60439

ERCKEL, RUEDIGER JOSEF, research & development management, organic chemistry, for more information see previous edition

ERDAL, BRUCE ROBERT, b Albuquerque, NMex, June 15, 39; m 70. INORGANIC CHEMISTRY, HYDROLOGY & WATER RESOURCES. *Educ:* Univ NMex, BS, 61; Washington Univ, St Louis, PhD(nuclear chem), 66. *Prof Exp:* Asst chem, Washington Univ, 61-63, asst nuclear chem, 63-66, res assoc, 66-67; res assoc, Brookhaven Nat Lab, NY, 67-69; NSF fel, Europ Orgn Nuclear Res, 69-70, vis scientist, 70-71; asst physicist, Ames Lab, US AEC, 71-72; staff mem, 72-80, dep group leader, Isotope Geochem Group, 80-83, assoc div leader, 83-87, TECH COORDR, INC DIV, LOS ALAMOS NAT LAB, 87- *Mem:* Am Chem Soc; Sigma Xi; AAAS; Am Inst Chemists; Sigma Xi; Am Geophys Union. *Res:* Nuclear structure, fission and reactions; radiochemistry; nuclear waste management; geochemistry; environmental chemistry. *Mailing Add:* INC-DO-MSJ519 Los Alamos Nat Lab Los Alamos NM 87545

ERDELYI, IVAN NICHOLAS, b Timisoara, Romania, Apr 14, 26; US citizen; m 50. MATHEMATICAL ANALYSIS. *Educ:* Univ Cluj, grad, 51; Univ Rome, Docent, 68. *Prof Exp:* Asst prof physics & math, Polytech Inst Timisoara, 51-59; math analyst, Olivetti Gen Elec, Italy, 62-67; assoc prof computer sic, Kans State Univ, 67-69; assoc prof, 69-72, PROF MATH, TEMPLE UNIV, 72- *Mem:* Am Math Soc; NY Acad Sci. *Res:* Abstract and functional analysis. *Mailing Add:* Dept Math Temple Univ Philadelphia PA 19122

ERDLEY, HAROLD F(REDERICK), b Los Angeles, Calif, Nov 27, 25; m 52; c 5. ELECTRONICS ENGINEERING. *Educ:* Univ Calif, BS, 48, MS, 50. *Prof Exp:* Systs engr, NAm Aviation, Inc, 50-54; electromech sect head, Litton Industs, 54-57, mgr guid & control dept, 57-59, div guid systs lab, 59-60, vpres & dir eng, Litton Systs, Inc, 60-61, electromech eng, 61-63, instrument res, 63-68; dir navig systs progs, Teledyne Systs Co, 68-70, vpres navig systs progs, 70-80, vpres dir adv systs, 80-85, VPRES DIR ADV SYSTS, TELEDYNE CONTROLS, 85- *Mem:* Inst Elec & Electronics Engrs; Instrument Soc Am. *Res:* Inertial navigation systems and associated devices. *Mailing Add:* Teledyne Controls Co Adv Systs 12333 Olympic Bl Los Angeles CA 90064

ERDMAN, ANNE MARIE, b Voorburg, Neth, June 10, 16; nat US. NUTRITION. *Educ:* Col Home Econ, Neth, BS, 39, dipl, 42; Fla State Univ, MS, 53, PhD, 56. *Prof Exp:* Instr foods & nutrit, Col Home Econ, Neth, 42-52; res assoc, Inst Rural Home Econ Res, 56-57; assoc prof, 57-71, PROF FOOD & NUTRIT, FLA STATE UNIV, 71- *Concurrent Pos:* Govt dietitian, Pub Health Serv, Indonesia. *Mem:* AAAS; Am Home Econ Asn; Am Dietetic Asn; Inst Food Technol. *Res:* Food consumption patterns, factors affecting them. *Mailing Add:* Lake Mich Col 2755 E Napier Benton Harbor MI 49022

ERDMAN, ARTHUR GUY, b Hackensack, NJ, July 7, 45; m; c 3. COMPUTER AIDED DESIGN MECHANISMS. *Educ:* Rutgers Univ, BS, 67; Rensselaer Polytech Inst, MS, 68, PHD(mech eng), 71. *Prof Exp:* Asst prof, 71-75, assoc prof, 75-80, PROF MECH ENG, UNIV MINN, 80- *Concurrent Pos:* Consult, Truth Inc, 78-, Yamaha Corp, 80-88, 3M Co, 83-, CDC, 83-, pres, MINTT Inc, 84-; co-prin investr, CDC, 81-88, prin investr, NSF, 82-90; ed, Computer-Aided Design, Mechanism & Machine Theory, 82-; mem & chmn, numerous nat tech comts, ASME. *Honors & Awards:* Ralph R Teetor Eng Educ Award, Soc Automotive Engrs, 77 & 81; Gustus L Larson Mem Award, Am Soc Mech Engrs, 80. *Mem:* Am Soc Mech Engrs; Am Soc Eng Educ; Sigma Xi; Soc Automotive Engrs; fel Am Soc Mech Engrs. *Res:* Mechanical design; computer aided design; bioengineering; kinematics; dynamics and computer graphics. *Mailing Add:* Dept Mech Eng Univ Minn 111 Church St SE Minneapolis MN 55455

ERDMAN, HOWARD E, b Hazleton, Pa, May 18, 30; m 59; c 2. GENETICS, RADIATION ECOLOGY. *Educ:* Muhlenberg Col, BS, 53; Lehigh Univ, MS, 55; NC State Col, PhD(genetics), 59. *Prof Exp:* Scientist ecol, Gen Elec Co, 59-65, sr res scientist, Battelle Northwest Labs, 65-70; ASSOC PROF BIOL, TEX WOMAN'S UNIV, 71-, RADIATION SAFETY OFFICER, 75- *Concurrent Pos:* Prog officer, Int Atomic Energy Agency, Vienna, Austria, 66-69. *Mem:* AAAS; Ecol Soc Am; Entom Soc Am; Entom Soc Can; Sigma Xi. *Res:* Genetics, morphology, cytology and histopathology of radiation effects; population ecology. *Mailing Add:* Tex Woman's Univ Univ Sta Box 23971 Denton TX 76204

ERDMAN, JOHN GORDON, b Baltimore, Md, Apr 12, 19; m 48. PETROLEUM, ORGANIC GEOCHEMISTRY. *Educ:* Johns Hopkins Univ, BA, 40, PhD(org chem), 43. *Prof Exp:* Jr instr org chem, Johns Hopkins Univ, 42-43; res chemist, Nat Defense Res Comt, 43-45; res chemist & fel, Off Sci Res & Develop, Mellon Inst, 45-56, sr fel, 56-65; mgr, Geochem Br, Phillips Petrol Co, 65-74, sr scientist explor & prod, 74-87; RETIRED. *Mem:* Am Asn Petrol Geologists; Am Chem Soc; Geol Soc Am; Geochem Soc; Am Geophys Union. *Res:* Genesis of petroleum; physics and electrochemistry of primary and secondary migration; applied exploration and production methods; integrated computerized systems utilizing geological, geochemical and geophysical data for evaluation of economic potential of basins; biotechnology, biochemical products. *Mailing Add:* 6045 Olohena Rd Kapaa Kauai HI 96746

ERDMAN, JOHN WILSON, JR, b Hackensack, NJ, July 7, 45; m 74; c 2. NUTRITION. *Educ:* Rutgers Univ, BS, 68, MS, 73, MPhil, 74, PhD(food sci), 75. *Prof Exp:* Flavor chemist prod res, Pepsico Corp Res, 68; from res asst to res intern food sci, Rutgers Univ, 70-75; from asst prof to assoc prof, 75-85, PROF FOOD SCI, UNIV ILL, 85-, PROF NUTRIT INTERNAL MED, 87- *Concurrent Pos:* Mem, subcomt on the uses of the Recommended Daily Allowance, NAS, 81-85; vis prof Biochem & Biophysics Dept, Iowa State Univ, 86-87. *Honors & Awards:* Samuel Cate Prescott Res Award, Inst Food Technologists, 80. *Mem:* Inst Food Technologists; Am Inst Nutrit; Am Asn Cereal Chemists; Sigma Xi; Am Oil Chem Soc; AAAS. *Res:* The effects of food processing upon nutrient retention with emphasis on soybean products; the metabolic role of vitamin A and B carotene; bioavailability of minerals from foods. *Mailing Add:* 501 E Evergreen Ct Urbana IL 61801

ERDMAN, KIMBALL S, b Salt Lake City, Utah, June 13, 37; m 67; c 3. BOTANY. *Educ:* Brigham Young Univ, BA, 59, MS, 61; Iowa State Univ, PhD(bot), 64. *Prof Exp:* Asst prof bot, Weber State Col, 64-67; assoc prof biol, 67-71, PROF BIOL, SLIPPERY ROCK STATE COL, 71- *Concurrent Pos:* Field consult, Natural Nat Landmark Prog, Dept Interior, 72-80. *Mem:* Sigma Xi. *Res:* Distribution of the native trees of Utah; monograph of the genus Spenopholis; natural areas of western Pennsylvania; forest ecology; alternative evolution models; alternative ancient history models. *Mailing Add:* 1612 N Meridian Newberg OR 97132

ERDMAN, TIMOTHY ROBERT, b Eau Claire, Wis, July 16, 44; m 68; c 2. ORGANIC CHEMISTRY. *Educ:* Univ Chicago, BS, 66; Stanford Univ, PhD(chem), 70. *Prof Exp:* Nuffield Found fel chem, Univ Aberdeen, 70-71; NIH fel, Univ Hawaii, 71-73; res chemist, 73-80, sr res chemist, 80-83, RES SCIENTIST & GROUP LEADER, CHEVRON RES & TECHNOL CO, STANDARD OIL CO, CALIF, 83- *Honors & Awards:* Arch T Colwell Award, Soc Automotive Engrs. *Mem:* Am Chem Soc. *Res:* Lubricating oil additives. *Mailing Add:* Chevron Res & Technol Co PO Box 1627 Richmond CA 94802

ERDMAN, WILLIAM JAMES, II, medicine; deceased, see previous edition for last biography

ERDMANN, DAVID E, b St Charles, Minn, July 15, 39; m 69. ANALYTICAL CHEMISTRY, INORGANIC CHEMISTRY. *Educ:* Winona State Col, BS, 60; Univ Nebr, Lincoln, MS, 66, PhD(chem), 68. *Prof Exp:* Chemist, 68-71, res chemist, 71-78, LAB DIR, WATER RESOURCES DIV, US GEOL SURV, 78- *Mem:* Am Chem Soc. *Res:* Investigation of spectroscopic properties of some copper II beta-ketoamine chelates; improvement and development of analytical methods for water analysis; automation of water analysis methods. *Mailing Add:* 7072 Ellis St Arvada CO 80004-1004

ERDMANN, DUANE JOHN, b Rochester, Minn, Apr 21, 46; m 68. PHOTOGRAPHIC CHEMISTRY. *Educ:* Iowa State Univ, BS, 68; Univ Minn, PhD(chem), 74. *Prof Exp:* Res chemist, 74-80, RES SUPVR, PHOTO PROD DEPT, E I DU PONT DE NEMOURS & CO, INC, 80-, LAB DIR RES & DEVELOP. *Mem:* Am Chem Soc; Soc Photog Scientists & Engrs. *Res:* Research and development of photographic emulsion with emphasis on those having monodisperse grains; products for graphic arts applications; related raw materials testing. *Mailing Add:* Seven Blackfoot Dr Manville NJ 08835

ERDMANN, JOACHIM CHRISTIAN, b Danzig, June 5, 28; m 57; c 3. ELECTROOPTICS, HYDRODYNAMICS. *Educ:* Tech Univ Braunschweig, Ger, Dipl Phys, 55, Dr rer nat(physics), 58. *Prof Exp:* Res specialist, Boeing Sci Res Labs, Boeing Com Airplane Co, 60-72, prin engr, 73-84, SR PRIN ENGR, BOEING AEROSPACE & ELECTRONICS, 85- *Concurrent Pos:* Vis prof, Max Planck Inst Metal Res, 68-69. *Mem:* Am Phys Soc; Optical Soc Am; Soc Photo-Optical Instrumentation Engrs. *Res:* Low-temperature experimental physics; laser applications; statistical optics; ocean hydrodynamics. *Mailing Add:* 14300 Trillium Blvd SE Mill Creek WA 98012

ERDMANN, JOHN HUGO, b Ravenna, Ohio, June 22, 64; m 88; c 1. PHYSICS OF LIQUID CRYSTALS, APPLIED PHYSICS DISPLAYS. *Educ:* Oberlin Col, BA, 85; Kent State Univ, MA, 87, PhD(physics), 90. *Prof Exp:* Teaching asst physics, Kent State Univ, 86-87, mem res staff, Liquid Crystal Inst, 87-90; MEM TECH STAFF, HUGHES AIRCRAFT CO RES LABS, 90- *Mem:* Am Phys Soc; Soc Info Display; Soc Photo-Optical Instrumentation Engrs. *Res:* Electro-optic properties of nematic liquid crystals and polymer dispersed liquid crystal technology for display application. *Mailing Add:* Hughes Res Labs RL 70 3011 Malibu Canyon Rd Malibu CA 90265

ERDMANN, ROBERT CHARLES, b Paterson, NJ, Jan 3, 39; m 69; c 4. NUCLEAR ENGINEERING. *Educ:* Newark Col Eng, BS, 60; Univ Calif, Los Angeles, MS, 62; Calif Inst Technol, PhD(appl mech & physics), 65. *Prof Exp:* Mem tech staff, Hughes Aircraft Co, 60-62; from asst prof to prof eng, Univ Calif, Los Angeles, 65-75; mgr, Sci Appln, Inc, 73-86; ATTY, BRASIER & ERDMANN, 90- *Concurrent Pos:* Consult to industs, labs & comns, 65-; mem, Atomic Indust Forum Comt Reactor Safety & Licensing, 73- *Mem:* Fel Am Nuclear Soc; Am Phys Soc; Sigma Xi. *Res:* Nuclear reactor safety and reliability analysis; risk analysis in engineering; fast reactor accident physics; chemical industry safety analysis. *Mailing Add:* Brasier & Erdmann 915 21st St Sacramento CA 95949

ERDOGAN, FAZIL, b Kars, Turkey, Feb 5, 25; m 61; c 2. APPLIED MECHANICS. *Educ:* Tech Univ Istanbul, MS, 48; Lehigh Univ, PhD(mech eng), 55. *Prof Exp:* Instr eng, Tech Univ Istanbul, 48-52; asst, 52-55, from asst prof to assoc prof mech eng, 57-63, PROF MECH, LEHIGH UNIV, 63-, CHMN DEPT, 83- *Concurrent Pos:* Interim dean eng, Lehigh Univ, 90- *Honors & Awards:* Lehigh Univ Res Award, 82; Alexander von Humboldt Sr US Scientist Award, 83. *Mem:* Am Soc Mech Engr; Soc Eng Sci; Soc Indust & Appl Math; Am Math Soc; Turkish Soc Pure & Appl Math. *Res:* Mechanics of non-homogeneous media; thermoelasticity; viscoelasticity; fracture mechanics; integral equations. *Mailing Add:* Dept Mech Eng & Mech Lehigh Univ Bethlehem PA 18015

ERDS, ERVIN GEORGE, b Budapest, Hungary, Oct 16, 22; US citizen; m 66; c 3. PHARMACOLOGY. *Educ:* Univ Budapest & Univ Munich, MD, 50. *Prof Exp:* Asst, Inst Pathophysiol, Budapest, 47-50; res assoc biochem, Res Lab, Surg Clin, Munich, 52-54; res fel, Sch Med, Univ Pittsburgh, 54-55; res assoc anesthesia, Mercy Hosp, Pa, 55-58; from clin asst prof to clin assoc prof pharmacol, Univ Pittsburgh, 58-63; prof, Sch Med, Univ Okla, 63-70, George Lynn Cross med res, 70-73; prof pharmacol & internal med, Univ Tex Health Sci Ctr, Dallas, 73-85; PROF PHARMACOL, DEPTS PHARMACOL & ANESTHESIOL, COL MED, UNIV ILL, CHICAGO, 85-, DIR, LAB PEPTIDE RES, 89- *Concurrent Pos:* Fel biochem, Mellon Inst, 58-63; mem, Coun High Blood Pressure Res; distinguished fulbright prof, 75- *Honors & Awards:* Merit Award, NIH, 88; Gold Medal, Frey-Werle Found, Munich, 88. *Mem:* AAAS; Am Soc Pharmacol & Exp Therapeut; Am Heart Asn; Int Soc Hypertension; Am Soc Biochem & Molecular Biol; Biochem Soc Gt Brit. *Res:* Biochemical pharmacology; cardiovascular diseases and hypertension; enzymology; peptide metabolism; proteins. *Mailing Add:* Lab Peptide Dept Pharmacol Univ Ill Col Med 835 S Wolcott Ave M/C 868 Chicago IL 60612

ERDOS, GREGORY WILLIAM, b Akron, Ohio, Sept 21, 45. MYCOLOGY. *Educ:* Ohio State Univ, BS, 67; Univ NC, PhD(bot), 71. *Prof Exp:* Res assoc mycol, Univ Wis-Madison, 71-77; MEM FAC MYCOL, UNIV FLA, 77- *Mem:* Mycol Soc Am; Am Soc Microbiol; AAAS; Am Soc Cell Biol. *Res:* Developmental ultrastructure and genetics of cellular slime molds and related organisms. *Mailing Add:* Dept Microbiol & Cell Sci Univ Fla 1059 McCarty Hall Gainesville FL 32611

ERDOS, MARIANNE E, software configuration management, for more information see previous edition

ERDTMANN, BERND DIETRICH, b Breslau, Ger, Aug 17, 39. PALEONTOLOGY, PALEOECOLOGY. *Educ:* Univ Hamburg, MS, 62, DSc(geol), 75; Univ Oslo, PhD(geol), 65. *Prof Exp:* Fel geol, Laval Univ, 66; Can Nat Res Coun fel, Carleton Univ, 66-68; from asst prof to assoc prof, 68-76, chmn dept, 76-80, PROF GEOL, IND UNIV, FT WAYNE, 76- *Concurrent Pos:* Attend, Int Conf Continental Drift, Nfld, 67; consult, Can Geol Surv, 68-71; mem, Cambrian-Ordovician Boundary Comt, Int Union Geol Sci, 74- & Graptolite Res Comt, Int Asn Planetology, 77-; vis prof, Ariz State Univ, 78- *Mem:* Ger Geol Soc; Norweg Geol Soc; Swedish Geol Soc; Geol Soc Am; Soc Econ Paleontologists & Mineralogists. *Res:* Paleontology, biostratigraphy, and numerical taxonomy of Ordovician graptolites; taphonomy and fossilization of recent invertebrate marine biota; paleoichnology (trace fossils); carbonate sedimentary systems analysis. *Mailing Add:* Inst Geol & Paleontol EB-10 Tech Univ Berlin Ernst Reuter Plaiz 1 Berlin 12 D-1000 Germany

ERECINSKA, MARIA, b Warsaw, Poland, Aug 17, 39. MEDICINE, BIOCHEMISTRY. *Educ:* Med Sch, Gdansk, MD, 61; Polish Acad Sci, PhD(biochem), 67. *Prof Exp:* Res assoc biochem, Univ Tex Med Br Galveston, 61-63; res assoc, Polish Acad Sci, 64-67, asst prof, 67-69; fel biophys, Univ Pa, 69-71; asst prof biophys, 71-78, assoc prof, 78-84, PROF PHARMACOL, UNIV PA, 84- *Concurrent Pos:* Travel fel, Nat Ctr Sci Res, France, 68; Pa Plan scholar, 70; WC Stroud Established Investr, Am Heart Asn, 76-81. *Honors & Awards:* Merck Award, 71. *Mem:* Am Soc Biol Chem. *Res:* Mitochondrial structure and function; bioenergetics; hemoproteins; ion transport; neurotransmitter metabolism. *Mailing Add:* Dept Pharmacol Univ Pa 37th & Hamilton Walk Philadelphia PA 19104-6084

ERENRICH, ERIC HOWARD, b West Palm Beach, Fla, Jan 10, 44; m 67; c 2. ORGANIC COATINGS, WAXES. *Educ:* Carnegie Inst Technol, BS, 65; Cornell Univ, PhD(phys chem), 71. *Prof Exp:* Res fel, Cornell Univ, 71-72; sr chemist, Rohm and Haas Co, 72-76; group leader, NL Indust, 76-79; appl res, Rheometrics, Inc, 80; tech mgr, Allied Corp, 80-90; MKT MGR, ALLIED-SIGNAL CORP, 90- *Mem:* Am Chem Soc; Sigma Xi; Fedn Soc Coating Technol. *Res:* Physical chemistry of polymers and coatings; applied rheology; modification of coatings, polishes and inks by low moleculer weight polyethylene and waxes. *Mailing Add:* Allied-Signal Corp Box 2332 Morristown NJ 07960

ERENRICH, EVELYN SCHWARTZ, b New York, NY, Dec 16, 46; m 67; c 2. ENZYMOLOGY, PHYSICAL BIOCHEMISTRY. *Educ:* Cornell Univ, BS, 67, MS, 69, PhD(biophys chem), 71. *Prof Exp:* Lectr chem, Cornell Univ, 71-72; from scientist to prin scientist, Leeds & Northrup Co, 72-78; CHEM CONSULT, 78-; LECTR CHEM, RUTGERS UNIV, 90- *Mem:* Am Chem Soc. *Res:* Structure, function and application of enzymes and biopolymers; analytical applications of biochemicals; enzyme immobilization. *Mailing Add:* Nine Constitution Ct East Brunswick NJ 08816

ERF, ROBERT K, b Cleveland, Ohio, Oct 29, 31; m 54; c 4. LASER METROLOGY. *Educ:* Univ Mich, BSEE, 53; Harvard Univ, MS, 54. *Prof Exp:* Res engr, 54-61, supvr gen instrumentation, 61-68, chief, optics & acoust, 68-85, MGR ENG SYST, UNITED TECHNOL RES CTR, 85- *Concurrent Pos:* Ed, Laser Appl Series, Acad Press. *Mem:* Optical Soc Am; Acoust Soc Am; Am Soc Nondestructive Testing; Soc Photo-Optical Instrumentation Engrs. *Res:* Interferometry; ultrasonics; applications of lasers to optical instruments; holographic technology for flow visualization; nondestructive testing; strain measurement and vibration analysis; laser machining systems development; acoustic emission. *Mailing Add:* 127 Carriage Dr Glastonbury CT 06033

ERHAN, SEMIH M, b Bursa, Turkey; US citizen; m 58; c 3. CHEMICAL ENGINEERING, MOLECULAR BIOLOGY. *Educ:* Univ Ankara, MS, 53; Okla State Univ, PhD(biochem), 65. *Prof Exp:* Res chem engr, Mineral Explor Inst Turkey, 54-56, consulting, Union Chamber Com & Indust, 56-59; admin, Div Raw Mat, Cement Corp Turkey, 59-61; res chem engr chem technol, Lurgi Ges Waermetechnik, 61-62; postdoctoral fel biochem, Univ Pittsburgh, 65-66; res investr biochem, Univ Pa, 66-70, asst prof, 70-76; consult chem, 76-81; DIR RES, DEPT DENT MED, ALBERT EINSTEIN MED CTR, 81- *Mem:* Am Chem Soc; AAAS; Int Asn Dental Res; Mat Res Soc. *Res:* Development of water displacing polymers; protein-based adhesives and coatings; preparation of icephobic coatings. *Mailing Add:* A Einstein Med Ctr York & Tabor Rds Philadelphia PA 19141

ERHARDT, PAUL WILLIAM, b Minneapolis, Minn, Oct 31, 47; m 71; c 5. MEDICINAL CHEMISTRY. *Educ:* Univ Minn, BA, 69, PhD(med chem), 74. *Prof Exp:* Res assoc, Drug Dynamics Inst, Col Pharm, Univ Tex, Austin, 74-75; asst prof med chem, Col Pharm, Northeastern Univ, 75-76; res investr, 76-79, sr res investr & group leader, Med-Organic Chem Dept, Am Critical Care, McGaw Park, Ill, 76-83; SECT HEAD & ASST DIR, MED-ORGANIC CHEM DEPT, BERLEX LABS, CEDAR KNOLLS, NJ, 83- *Mem:* Am Chem Soc. *Res:* Cardiovascular and dopaminergic medicinal chemistry; design of prodrugs and soft drugs; peptide mimetics. *Mailing Add:* Berlex Labs 110 E Hanover Ave Ceder Knolls NJ 07927

ERHARDT, PETER FRANKLIN, b Grand Rapids, Mich, June 26, 33; m 58; c 4. POLYMER PHYSICS. *Educ:* Aquinas Col, BS, 55; Univ Mich, MS, 57; Univ Mass, PhD(phys chem), 68. *Prof Exp:* Instr chem, Aquinas Col, 57-60; res trainee polymer physics, Gen Elec Res Lab, NY, 60-63; res asst, Polymer Res Inst, Univ Mass, Amherst, 63-66; res chemist, Gen Elec Co, 66-67; from scientist to sr scientist, 67-73, prin scientist, 73-80, MGR, MKT MAT AREA, XEROX CORP, 80- *Mem:* Am Chem Soc; Am Phys Soc. *Res:* Rheology; rheo-optics and optical properties of synthetic polymers; mechanical, thermal property measurements on polymers and structure-property correlations; xerographic developer materials; toners and carriers; inks; color. *Mailing Add:* 706 Balsam Lane Webster NY 14580

ERHART, RAINER R, b Monstab, Ger, June 25, 35; US citizen; m 61; c 2. PHYSICAL GEOGRAPHY. *Educ:* Eastern Mich Univ, BA, 59; Univ Ill, MA, 61, PhD(geog), 67. *Prof Exp:* Asst prof, 65-68, ASSOC PROF GEOG, WESTERN MICH UNIV, 68- *Mem:* AAAS; Asn Am Geogr; Am Geog Soc; Nat Coun Geog Educ; Am Soc Photogram. *Res:* New media development in geography; remote sensing; agricultural geography. *Mailing Add:* Dept Geog Western Mich Univ Kalamazoo MI 49008

ERICKSEN, GEORGE EDWARD, b Butte, Mont, Mar 17, 20; m 48. ECONOMIC GEOLOGY, MINERALOGY. *Educ:* Mont State Univ, BA, 46; Ind Univ, MA, 49; Columbia Univ, PhD(geol), 54. *Prof Exp:* Geologist, US Geol Surv, 42-45 & 46; geologist, Ind Geol Surv, 47; GEOLOGIST, US GEOL SURV, 48- *Concurrent Pos:* Instr, Ind Univ, 47-49. *Mem:* AAAS; Geol Soc Am; Soc Econ Geol; Am Mineral Soc; Geochem Soc; hon mem Col Geol Chile; Soc Geol Peru; Soc Mining Engrs; Am Inst Mining Metall & Petrol Engrs; Soc Mining Engrs Chile; Soc Geol Chile. *Res:* Study of metalliferous and saline deposits of the Andes; geology and mineralogy of Chilean nitrate deposits; engineering geology related to earthquakes. *Mailing Add:* US Geol Surv Nat Ctr MS 954 Reston VA 22092

ERICKSEN, JERALD LAVERNE, b Portland, Ore, Dec 20, 24; m 46; c 2. LIQUID CRYSTAL THEORY, THEORY OF MARTENSITIC TRANSFORMATION. *Educ:* Univ Wash, Seattle, BS, 47; Ore State Univ, MA, 49; Univ Ind, PhD(math), 51. *Hon Degrees:* DSc, Nat Univ Ireland, 84, Heriot-Watt Univ, 88. *Prof Exp:* Mathematician & solid state physicist, Naval Res Lab, 51-57; from asst prof to assoc prof mech, Johns Hopkins Univ, 57-82; prof mech & math, Univ Minn, 82-90; RETIRED. *Honors & Awards:* Bingham Medal, Soc Rheology, 68; Timoshenko Medal, Am Soc Mech Engrs, 79. *Mem:* Soc Natural Philos (treas, 63-64); Soc Rheology; Soc Interaction Mech & Math. *Res:* Research on non-linear continuum theories of elasticity, viscoelasticity and liquid crystals; work aimed at finding better constitutive equations for newer materials. *Mailing Add:* 5378 Buckskin Bob Lane Florence MN 55455

ERICKSEN, MARY FRANCES, b Fortville, Ind, Aug 25, 25; m 48. PHYSICAL ANTHROPOLOGY, HUMAN ANATOMY. *Educ:* Ind Univ, AB, 47; Columbia Univ, MA, 57; George Washington Univ, PhD(anat), 73. *Prof Exp:* Cur paleont, Univ Ky, 47-49; teaching fel, 67-73, spec lectr, 73-78, asst prof, 78-81, ASSOC PROF LECTR ANAT, GEORGE WASHINGTON UNIV, 81- *Concurrent Pos:* Guest researcher, Mus Nac Argueologia y Antropologia, Lima, 49-52; guest researcher, Mus de La Serena, Mus Nac Hist Nat, Chile, 55-62; vis prof summer anat prog, Sch Med, Univ Md, 76-80; vis prof, St George Univ Med Sch, 81, 84. *Mem:* Am Asn Phys Anthropologists; Am Asn Anatomists; Am Anthrop Asn; Human Biol Coun. *Res:* Aging changes in the skeleton; archaeological and contemporary populations; physical anthropology of archaeological populations; paleopathology. *Mailing Add:* Dept Anat George Washington Univ Med Ctr Washington DC 20037

ERICKSEN, RICHARD HAROLD, b Seattle, Wash, Aug 17, 38; m 67; c 3. COMPOSITE MATERIALS, FABRICS. *Educ:* Whitman Col, BA, 61; Columbia Univ, BS, 61, MS, 63; Univ Wash, PhD(metall eng), 67. *Prof Exp:* STAFF MEM METALL & MAT ENG, SANDIA NAT LABS, 67- *Mem:* Am Soc Metals; Am Inst Aeronaut & Astronaut. *Res:* Mechanical behavior of composite materials; reinforcing filaments and materials for parachute applications. *Mailing Add:* Org 9113 Sandia Nat Labs Albuquerque NM 87185-5800

ERICKSEN, WILHELM SKJETSTAD, b Green Bay, Wis, May 3, 12; m 39; c 3. MATHEMATICS. *Educ:* St Olaf Col, BA, 36; Univ Wis, MA, 38, PhD(math), 43. *Prof Exp:* Asst prof math, St Olaf Col, 42-43; prof, NDak State Teachers Col, Minot, 43-44; fel mech, Brown Univ, 44; aerodynamicist, Bell Aircraft Corp, Buffalo, 45-46; mathematician, Forest Prod Lab, US Forest Serv, 46-53; PROF MATH, AIR FORCE INST TECHNOL, 53- *Mem:* Am Math Soc; Math Asn Am; Sigma Xi. *Res:* Sandwich construction for aircraft; stresses in wood structural members. *Mailing Add:* 3598 Eastern Dr Dayton OH 45432

ERICKSON, ALAN ERIC, b Boston, Mass, Feb 6, 28; m 51; c 4. SCIENCE COMMUNICATIONS. *Educ:* Middlebury Col, AB, 49; Boston Univ, MA, 55, PhD(biol), 60; Simmons Col, MLS, 68. *Prof Exp:* Asst instr biol, Boston Univ, 54-55, instr, 55-60; scientist embryol, Worcester Found Exp Biol, 60-66; actg assoc librn for admin, Harvard Col Libr, 70-72, SCI SPECIALIST, HARVARD UNIV LIBR, 66-, LIBRN, GODFREY LOWELL CABOT SCI LIBR, 73-, ASSOC LIBRN, COL SCI, HARVARD UNIV, 84- *Mem:* Am Libr Asn; Sigma Xi; AAAS. *Res:* Hormones in development of the embryonic reproductive system. *Mailing Add:* Cabot Sci Libr Harvard Univ Cambridge MA 02138

ERICKSON, ANTON EARL, b Chicago, Ill, June 5, 19; m 43; c 3. SOIL SCIENCE. *Educ:* Univ Ill, BS, 41, PhD(agron), 48. *Prof Exp:* Asst soil surv, Univ Ill, 41-48; from asst prof to assoc prof soil sci, 48-59, PROF SOIL SCI, MICH STATE UNIV, 59- *Mem:* Fel AAAS; fel Am Soc Agron; fel Soil Sci Soc Am. *Res:* Soil physics; physical soil-plant relations; soil aeration; soil water conservation; land treatment and disposal of waste. *Mailing Add:* 4594 Comanche Okemos MI 48864

ERICKSON, BRUCE WAYNE, b New Haven, Conn, Oct 19, 42; m 69; c 2. BIOCHEMISTRY. *Educ:* Ohio State Univ, BS, 63; Harvard Univ, AM, 65, PhD(org chem), 70. *Prof Exp:* Res assoc, Rockefeller Univ, 69-73, from asst prof to prof biochem, 73-86; PROF CHEM, UNIV NC, CHAPEL HILL, 86- *Mem:* Am Soc Biol Chemists; Am Chem Soc; Am Peptide Soc; NY Acad Sci; AAAS. *Res:* Solid-phase peptide synthesis; biochemistry of synthetic peptides; protein engineering. *Mailing Add:* 1703 Audubon Rd Chapel Hill NC 27514-2607

ERICKSON, CARLTON KUEHL, b Manistee, Mich, Apr 6, 39; m 65; c 4. PHARMACOLOGY. *Educ:* Ferris State Col, BS, 61; Purdue Univ, MS, 63, PhD(pharmacol), 65. *Prof Exp:* From asst prof to prof pharmacol, Sch Pharm, Univ Kans, 65-77; PROF PHARMACOL, COL PHARM, UNIV TEX, 78- *Concurrent Pos:* Vis researcher toxicol, Karolinska Inst, Sweden, 73-74. *Mem:* AAAS; Am Soc Pharmacol & Exp Therapeut; Soc Neurosci; Res Soc Alcoholism (vpres, 81-82). *Res:* Effects of ethanol on central neurotransmitters; mechanisms of drugs of abuse; central cholinergic mechanisms; development of sustained release forms of abused drugs. *Mailing Add:* Div Pharmacol & Toxicol Univ Tex Col Pharm Austin TX 78712

ERICKSON, CYRUS CONRAD, b Alexandria, Minn, Aug 18, 09; wid; c 3. PATHOLOGY. *Educ:* Univ Minn, BS, 30, BM, 32, MD, 33; Am Bd Path, dipl. *Prof Exp:* Asst resident & res assoc path, Med Sch, Univ Rochester, 35-37; instr, Sch Med, Duke Univ, 37-39, assoc, 39-46, assoc prof, 46-50, assoc pathologist, Univ Hosp, 39-42, 46-50; prof path, 50-77, EMER PROF PATH, MED SCH, UNIV TENN, 77-, HEAD CLIN LAB SCI & DIR SCH MED TECHNOL, 71- *Concurrent Pos:* Fel pediat, Univ Minn, 34-35; consult, US Vet Admin Bur, 55-; actg chmn dept path, Univ Tenn, 68-71. *Honors & Awards:* Award, Am Soc Cytol, 65; Distinguished Serv Award, Am Cancer Soc, 69. *Mem:* AMA; Am Soc Exp Path (secy, 53-56, vpres, 58-59, pres, 59-60); Am Soc Clin Path; Am Soc Cytol (vpres, 57-59, pres, 60-61); fel Col Am Path (bd dir, 54-56); hon mem Int Acad Cytol. *Res:* Choline deficiency and tumor incidence in rats; histogenesis and incidence of intraepithelial carcinoma of cervix in humans; factors influencing development of carcinoma of cervix in mice with prolonged sex steroid administration; investigations on uterine cancer by epidemiological study following genital cytology population screening. *Mailing Add:* 267 Kimbrough St Memphis TN 38104

ERICKSON, DAVID EDWARD, b Grand Island, Nebr, July 15, 31; m 56; c 3. PHYSICAL CHEMISTRY, COLLOID CHEMISTRY. *Educ:* SDak Sch Mines & Technol, BS, 52; Ohio State Univ, PhD(chem), 56. *Prof Exp:* Res chemist, E I du Pont de Nemours & Co, 56-64; sr res chemist, Res Div, Gen Tire Co, 64-66, group leader, 66-69, develop assoc, Chem-Plastics Div, 69-71, res scientist, 71-79, group leader, res & develop div, Gencorp, 79-87; CONSULT, 87- *Mem:* Am Chem Soc; Sigma Xi. *Res:* Polymer chemistry; colloids; adhesion; surface science. *Mailing Add:* 3390 Charring Cross Dr Stow OH 44224-4607

ERICKSON, DAVID R, b Portland, Ore, Oct 26, 29; m 51; c 3. AGRICULTURAL CHEMISTRY. *Educ:* Ore State Univ, BS, 57, MS, 58; Univ Calif, Davis, PhD(agr chem), 63. *Prof Exp:* Res lipid chemist, Swift & Co, 63-67, head edible fats & margarine res div, 67-70, gen mgr indust prod res, 70-75; dir res, Unitech Chem Inc, 75-78; DIR TECH SERVS, AM SOYBEAN ASN, 78- *Honors & Awards:* Bailey Award, 89. *Mem:* Sigma Xi; Am Chem Soc; Inst Food Technol; Am Oil Chemists Soc (pres, 90-91). *Res:* Oxidation of milk and antioxidants in milk and milk products; chemistry of processing oil seeds and edible fats and oils; vegetable proteins, emulsifiers, stabilizers, adhesives. *Mailing Add:* 9059 Monmouth Dr St Louis MO 63117

ERICKSON, DENNIS JOHN, b Minneapolis, Minn, June 9, 42; m 65; c 3. SOLID STATE PHYSICS. *Educ:* Augsberg Col, BA, 64; Univ Tenn, Knoxville, PhD(physics), 71. *Prof Exp:* Res asst physics, Univ NC, Chapel Hill, 68-70, res assoc, 71; Fel physics, 72-73, STAFF MEM, DYNAMIC TESTING DIV, LOS ALAMOS NAT LAB, 74-, DEP GROUP LEADER, SHOCK WAVE PHYSICS, 78- *Mem:* Am Phys Soc. *Res:* Explosive driven pulse power and applications of magnetic flux compression material behavior in ultrahigh magnetic fields, low temperature properties of materials; magnetism, lattice dynamics, super conductivity. *Mailing Add:* Los Alamos Nat Lab MS-P915 Los Alamos NM 87545

ERICKSON, DUANE GORDON, b Vinton, Iowa, Jan 30, 31; m 52; c 1. ENTOMOLOGIST, BIOCHEMIST. *Educ:* Univ Minn, BA, 53, MS, 57, PhD, 64. *Prof Exp:* US Army, 56-85, parasitologist, Sixth Army Med Lab, Calif, 56, parasitologist, Second Army Med Lab, Md, 57-58, chief helminth dept, Army Trop Res Med Lab, PR, 59-61, asst chief dept med zool, Walter Reed Army Inst Res, 64-65, consult parasitol, US Army & chief parasitol br, 9th Med Lab, Vietnam, 66-67, coordr schistosomiasis res & chief schistosomiasis unit, Dept Med Zool, Walter Reed Army Inst Res, 67-71, chief dept med zool, 406th Med Lab, Japan, 71-74, exec secy, Armed Forces Epidemiol Bd, Washington, DC, 74-78, exec officer, US Army Med Bioeng Res & Develop Lab, Ft Detrick, Md, 78-83, consult, parasitol to surgeon gen, US Army, 83-85; ADJ ASSOC PROF, UNIFORMED SERV HEALTH SCI, 83- *Concurrent Pos:* Adj lectr, Montgomery Col & adj prof, Prince Georges Community Col, 87- *Mem:* Am Soc Parasitol; Am Soc Trop Med & Hyg; fel Royal Soc Trop Med & Hyg; Sigma Xi. *Res:* Schistosomatoidea; immunity and pathology of parasitic infections; medical entomology; electron microscopy; immunity and pathology of schistosomiasis; chemotherapy of schistosomiasis; ultrastructural pathology; science education. *Mailing Add:* 12917 Neola Rd Wheaton MD 20906

ERICKSON, DUANE OTTO, b Fargo, NDak, Mar 26, 30; m 54; c 3. ANIMAL NUTRITION. *Educ:* NDak State Univ, BS, 57, MS, 60, PhD(animal sci, biochem), 65. *Prof Exp:* Instr & asst animal sci, 59-65, from asst prof to assoc prof, 65-77, PROF ANIMAL SCI, NDAK STATE UNIV, 77- *Mem:* AAAS; Am Soc Animal Sci. *Res:* Development of methods for forage evaluation; metabolism within the rumen of a ruminant; nutritional requirements of feeder lambs. *Mailing Add:* Dept Animal & Range Sci NDak State Univ Fargo ND 58105

ERICKSON, EDWARD HERBERT, b Oakland, Calif, Feb 16, 44; m 66; c 3. MEDICINAL CHEMISTRY. *Educ:* Univ Calif, Santa Barbara, BA, 65, PhD(org chem), 68. *Prof Exp:* Res chemist, Riker Labs, Calif, 68-71; res chemist, 71-78, proj leader pulmonary allergy, 78-81, mgr new molecule res, 81-84, MGR DRUG DISCOVERY, RIKER LABS, 3M CO, 84- *Mem:* AAAS; Am Chem Soc. *Res:* Enzyme inhibition and rational drug design; preparation of antiinflammatory and antiallergic drugs. *Mailing Add:* 270-3A-01 3M Ctr St Paul MN 55144-1000

ERICKSON, EDWIN FRANCIS, b Seattle, Wash, July 13, 34; m 63; c 4. ASTROPHYSICS. *Educ:* Stanford Univ, BS, 57, MS, 59, PhD(physics), 66. *Prof Exp:* Res scientist physics, US Army, 63-65 & NASA, 65-66; res assoc physics, Univ Strasbourg, 66-70; RES SCIENTIST INFRARED ASTRON, AMES RES CTR, NASA, 70- *Concurrent Pos:* Sr res assoc, Nat Res Coun, 70-72. *Mem:* Am Phys Soc; Am Astron Soc; Sigma Xi; AAAS; Am Fedn Scientists. *Res:* Observational infrared astronomy from aircraft and ground-based telescopes; infrared emissions from stars, planets and nebulae. *Mailing Add:* Astrophys Exp Br NASA Ames Res Ctr MS 245-6 Moffett Field CA 94035

ERICKSON, EDWIN SYLVESTER, JR, b Brooklyn, NY, July 9, 28; m 53; c 4. MINERALOGY, PETROGRAPHY. *Educ:* City Col New York, BS, 51; Pa State Univ, PhD(mineral), 63. *Prof Exp:* Geologist, US Geol Surv, 52-53; res asst geochem & mineral, Pa State Univ, 53-58; prin Mineralogist, Res Dept Bethlehem Steel Corp, 58-85; DIR-MINERALS, APPL TECHNOL GROUP, 87-; ADJ PROF GEOL, NORTHAMPTON CO COMMUNITY COL, 87- *Concurrent Pos:* Assoc dept geol, Lehigh Univ, 66- *Mem:* Sigma Xi. *Res:* Mineralogical investigation of raw materials and refractories for iron and steelmaking; agglomerated iron and manganese ores; mineralogy of refractory clays and bentonites; evaluation of ores and raw materials and interpretation of beneficiation test results. *Mailing Add:* 306 Prospect Ave Bethlehem PA 18018

ERICKSON, ERIC HERMAN, JR, b Denver, Colo, Apr 26, 40; c 4. APICULTURE, CROP POLLINATION. *Educ:* Colo State Univ, BS, 63, MS, 65; Univ Ariz, PhD(entom), 70. *Prof Exp:* Instr med entom, Army Med Field Serv Sch, San Antonio, Tex, 65-66; med entomologist, US Army, Repub Vietnam, 66-67; res entomologist, USDA, 70-78; from asst prof to assoc prof, 73-81, PROF ENTOM, UNIV WIS-MADISON, 81-; RES LEADER, CARL HAYDEN BEE RES CTR, AGR RES SERV, USDA, 78- *Concurrent Pos:* Mem Fac, Biomed Eng Ctr, Univ Wis-Madison, 76- *Honors & Awards:* Commendation Medal, US Army. *Mem:* Sigma Xi; Entom Soc Am; Am Soc Agron; Int Bee Res Asn; Int Comn Bee Bot. *Res:* Honey bee biology, behavior and crop pollination: physical, chemical, genetical and environmental interactions as well as factors associated with integrated crop management practices. *Mailing Add:* Carl Hayden Bee Res Ctr USDA Agr Res Serv 2000 E Allen Rd Tucson AZ 85719

ERICKSON, EUGENE E, b Fargo, NDak, Sept 15, 23; m 46; c 5. CHEMICAL ENGINEERING. *Educ:* Univ Minn, BChE, 44; NC State Col, PhD(chem eng), 57. *Prof Exp:* Res chem engr, Russell-Miller Milling Co, Minn, 44-50, Minn Mining & Mfg Co, 50-52 & NC State Col, 52-57; res chem engr, Atomic Energy Div, Phillips Petrol Co, 57-60, group leader, 60-62, staff engr reactor safety, 62-63; sr chem engr, N Star Res Inst, 63-64, assoc dir phys sci & eng div, 64-66, dir div, 66-74, mgr environ sci, Midwest Res Inst, 75-77; pres, Filmtec Corp, 77-85, chmn, 85-86; pres, 89-90, CHMN, AERAS WATER SYSTS, 90- *Mem:* Am Chem Soc; Am Inst Chem Engrs; Water Supply Improvement Asn. *Res:* Nuclear fuels processing; research administration; environmental systems; water treatment; reverse osmosis. *Mailing Add:* 3020 St Albans Mill Rd No 108 Minnetonka MN 55343-1207

ERICKSON, GLEN WALTER, b St Paul, Minn, Aug 1, 34; m 55; c 4. THEORETICAL PHYSICS. *Educ:* Univ Minn, BS, 55, PhD(physics), 60. *Prof Exp:* Res assoc, Inst Field Physics, Univ NC, 60-62, vis asst prof physics, 61-62; adj asst prof, NY Univ, 62-64; assoc prof, 64-74, PROF PHYSICS, UNIV CALIF, DAVIS, 74- *Concurrent Pos:* Mem comt fundamental constants, Nat Res Coun-Nat Acad Sci, 69-71; sr res fel, Sci Res Coun Eng, 70-71. *Mem:* Am Phys Soc. *Res:* Quantum field theory, especially quantum electrodynamics. *Mailing Add:* Dept Physics Univ Calif Davis CA 95616

ERICKSON, HAROLD PAUL, b Chattanooga, Tenn, Jan 16, 40; div; c 1. CELL BIOLOGY, ELECTRON MICROSCOPY. *Educ:* Carnegie-Mellon Univ, BS, 62; Johns Hopkins Univ, PhD(biophys), 68. *Prof Exp:* NIH res fel, Med Res Coun Lab Molecular Biol, Cambridge, Eng, 68-70; from asst prof to assoc prof, Duke Univ, 70-81, prof anat, 82-88, prof & chmn, cell biol, 88-90; CONSULT, 90- *Concurrent Pos:* NIH res career develop award, 72-76; mem staff biophys chem study sect, NIH, 80-83. *Mem:* Am Soc Cell Biol; Biophys Soc. *Res:* Microtubules and cytoskeleton; fibronectin and extracellular matrix; structure and self-assembly of protein complexes. *Mailing Add:* Dept Cell Biol Duke Univ Med Ctr Durham NC 27710

ERICKSON, HOWARD HUGH, b Wahoo, Nebr, Mar 16, 36; m 59; c 2. CARDIOPULMONARY PHYSIOLOGY, EQUINE EXERCISE PHYSIOLOGY. *Educ:* Kans State Univ, BS & DVM, 59; Iowa State Univ, PhD(physiol biomed eng), 66. *Prof Exp:* Area vet, 59th Vet Inspection Flight, USAF, UK, 60-63, res vet officer, Biodynamics Br, 66-70, vet scientist, Appl Physiol Br, 70-73, asst chief, Biodynamics Br, USAF Sch Aerospace Med, 73-74, chief, Tech Plans & Analytical Div, 76-79, chief, Mech Forces Div, Aerospace Med Div, 79-81; PROF PHYSIOL, DEPT ANAT & PHYSIOL, COL VET MED, KANS STATE UNIV, 81- *Concurrent Pos:* Vis mem grad fac, Tex A&M Univ, 68-81; spec mem grad fac, Colo State Univ, 70-75; clin asst prof, Health Sci Ctr, Univ Tex, 72-81; sci adv, Nat Res Coun-Air Force Systs Command Associateship Prog, 71-73; Res Coun, Am Vet Med Asn, 79-85, chmn, 83-84. *Mem:* Fel AAAS; Am Vet Med Asn; Inst Elec & Electronics Engrs; assoc fel Aerospace Med Asn; Am Physiol Soc. *Res:* cardiovascular instrumentation and control; comparative exercise physiology; equine exercise physiology and sports medicine. *Mailing Add:* 2017 Arthur Dr Manhattan KS 66502

ERICKSON, HOWARD RALPH, b Indiana, Pa, Nov 23, 19; m 55; c 3. VERTEBRATE ZOOLOGY. *Educ:* Indiana Univ Pa, BS, 52; Pa State Univ, MS, 56; Cornell Univ, PhD(vert zool), 59. *Prof Exp:* Res asst zool, Pa State Univ, 54-56; res asst vert zool, Cornell Univ, 56-59; instr biol & zool, 59-62, assoc prof vert zool, 62-66, prof vert zool, 66-77, PROF BIOL, TOWSON STATE UNIV, 77-, CHMN DEPT, 66- *Concurrent Pos:* Annual res grant, Towson State Col, 66-68. *Mem:* AAAS; Am Soc Mammal; Ecol Soc Am; Wildlife Soc. *Res:* Environmental conservation; vertebrate ecology; muskrat growth; reproduction; population dynamics; movements; control methods and procedures; ecology of fresh water piscine species. *Mailing Add:* Dept Biol Towson State Univ Towson MD 21204

ERICKSON, JAMES C, III, b Philadelphia, Pa, Oct 7, 27; m 56; c 1. ANESTHESIOLOGY. *Educ:* Univ Pa, BA, 49; Temple Univ, MD, 53, MSc, 58; Am Bd Anesthesiol, dipl, 60. *Prof Exp:* Instr anesthesiol, Med Sch, Temple Univ, 57-58; assoc, Guthrie Clin, 58-61; assoc, Med Sch, Temple Univ, 61-64, asst prof, 64-67; prof, Woman's Med Col Pa, 67-69; prof anesthesiol, Jefferson Med Col, Thomas Jefferson Univ, 69-80; PROF CLIN ANESTHESIA, NORTHWESTERN UNIV MED SCH, 80- *Concurrent Pos:* Consult, Rehab Inst Chicago & Lakeside Vet Admin Hosp & Children's Mem Med Ctr, 80-; dir, pain control serv, Northwestern Mem Hosp. *Mem:* AMA; Am Soc Regional Anesthesiol; Am Soc Clin Hypn; Soc Clin & Exp Hypn; Int Soc Hypn. *Res:* Vasopressor effect of indigo carmine; clinical evaluation of local anesthetic drugs; use of hypnosis in anethesia and pain control; pain control and diagnostic evaluation; evaluation of neurolytic agents; intraspinal narcotics; semantics of anesthesia; metaphors of pain. *Mailing Add:* Dept Anesthesiol Northwestern Univ Med Sch 303 E Superior St Chicago IL 60611

ERICKSON, JAMES ELDRED, b Stambaugh, Mich, Dec 1, 49; m 72; c 2. VERTEBRATE ZOOLOGY, ICHTHYOLOGY. *Educ:* North Park Col, BA, 71; Univ Minn, MS, 74, PhD(zool), 77. *Prof Exp:* Asst prof zool, Western Mich Univ, 77-82; lectr biol, Augsburg Col, 83; asst prof biol, St Olaf Col, 83-88; asst prof biol, Univ St Thomas, 88-90; INSTR BIOL, NORMANDALE COMMUNITY COL, 90- *Concurrent Pos:* Vis prof, Au Sable Inst Environ Studies, 87- *Mem:* Am Fisheries Soc; Am Soc Ichthyologists & Herpetologists; Sigma Xi; AAAS. *Res:* Zoogeography and ecological life histories of fresh water fishes, especially minnows and darters; winter ecology of stream fishes; parasites of freshwater fishes. *Mailing Add:* 7008 Bristol Blvd Edina MN 55435

ERICKSON, JAMES GEORGE, b Sioux City, Iowa, May 18, 29; m 50; c 3. ZOOLOGY. *Educ:* Doane Col, BA, 50; Iowa State Univ, MS, 51; Univ Wyo, PhD(zool & physiol), 64. *Prof Exp:* Fisheries biologist, Div Wildlife, Ohio Dept Natural Resources, 53-56; fisheries biologist, Wyo Game & Fish Comn, 56-61; from asst prof to assoc prof biol, 64-74, PROF BIOL, FT LEWIS COL, 74- *Mem:* Am Soc Zoologists; Soc Sci Study Sex. *Res:* Behavior and endocrinology of fishes. *Mailing Add:* Dept Biol Ft Lewis Col Durango CO 81303

ERICKSON, JEANNE MARIE, b Bryan, Tex. PHOTOSYNTHESIS, PROTEIN STRUCTURE-FUNCTION. *Educ:* Ohio State Univ, BA, 70, MS, 72; Univ Mich, MS, 79, PhD(human genetics), 81. *Prof Exp:* Teaching asst microbiol, Ohio State Univ, 71-72; instr bact, Biol Dept, Kent State Univ, 72-73; staff res assoc, Univ Calif, San Diego, 73-77; teaching asst genetics, Univ Mich, 78-79, trainee, 77-81; postdoctoral teaching asst physiol, Univ Geneva, Switz, 82-86; ASST PROF GENETICS & PLANT MOLECULAR BIOL, DEPT BIOL, UNIV CALIF, LOS ANGELES, 86- *Mem:* Genetics Soc Am; Int Soc Plant Molecular Biol; Am Soc Plant Physiologists; AAAS. *Res:* Molecular genetics; photosynthesis; structure-function relationships of membrane proteins; chloroplast gene expression; RNA splicing; herbicide-resistance; electron transfer. *Mailing Add:* Dept Biol Univ Calif Los Angeles CA 90024

ERICKSON, JOHN (ELMER), b Sioux City, Iowa, June 17, 23; m 46; c 3. CYTOGENETICS. *Educ:* Univ Omaha, BA, 48; Ind Univ, MA, 50; Univ Ore, PhD(genetics), 64. *Prof Exp:* Instr biol, McCook Jr Col, 54-59 & Univ Ore, 59-62; asst prof, 64-66, assoc prof, 66-89, EMER ASSOC PROF BIOL, WESTERN WASH UNIV, 89- *Mem:* Genetics Soc Am; Am Genetic Asn. *Res:* Cytogenetics of meiotic drive in Drosophila males; fragmentation of Y chromosome; mating, insemination and fertility problems; programming for computer-assisted instruction. *Mailing Add:* Dept Biol Western Wash Univ Bellingham WA 98225-9060

ERICKSON, JOHN GERHARD, b Northfield, Minn, July 14, 17; m 54; c 4. ORGANIC CHEMISTRY. *Educ:* St Olaf Col, BA, 38; NDak State Col, MS, 40; Univ Minn, PhD(org chem), 44. *Prof Exp:* Asst, NDak State Col, 38-40; instr, St Olaf Col, 40-41; chemist, Minn Valley Canning Co, 41; res asst soils, Univ Minn, 41, asst, 42-44; res chemist, Am Cyanamid Co, 44-50; sr res chemist, Gen Mills, Inc, 51-56; sr res chemist, Minn Mining & Mfg Co, 56-59, supvr, 59-64, mgr synthesis res, 64-66, dir sci & tech commun, 66-76, dir technol anal, 76-79; RETIRED. *Concurrent Pos:* Dir & treas, Avochem, Inc, 53-61, pres, 59-61. *Res:* Organic nitrogen compounds; heterocycles; sugar derivatives; acrylic derivatives; hydrogen cyanide; orthoformic esters; glycidyl esters; polymers; fluorine chemistry; textile agents; greases; rocket propellants; birds of the Labrador coast; flight behavior of the Procellariiformes. *Mailing Add:* 1344 S Second St Stillwater MN 55082

ERICKSON, JOHN M, b Curtiss, Wis, Apr 28, 18; m 48; c 4. PHYSICAL CHEMISTRY. *Educ:* Univ Wis, BS, 40; SDak State Col, MS, 53; Iowa State Univ, PhD(phys chem), 56. *Prof Exp:* From prod dept foreman to cost control engr, Procter & Gamble Mfg Co, Ill, 40-43; instr chem & math, SDak State Col, 47-51; asst, Iowa State Univ, 51-56; assoc prof phys chem, SDak State Col, 56-60; assoc prof, St Cloud State Univ, 60-66, prof chem, 66-82; RETIRED. *Mem:* Am Chem Soc. *Res:* Complex ions in mixed solvents; ion-exchange equilibria; water pollution. *Mailing Add:* 821 SE Tenth St St Cloud MN 56304

ERICKSON, JOHN MARK, b Orange, NJ, Dec 21, 43. INVERTEBRATE PALEONTOLOGY. *Educ:* Tufts Univ, BS, 65; Univ NDak, MS, 68, PhD, 71. *Prof Exp:* From asst prof to assoc prof invert paleont, 71-85, chmn dept geol, 81-89, PROF INVERT PALEONT, ST LAWRENCE UNIV, 85-, DIR , ST LAWRENCE AQUARIUM & ECOL CTR, 89- *Concurrent Pos:* Pres, Randolph Co, 83-85. *Mem:* Fel AAAS; Geol Soc Am; Paleont Soc; Sigma Xi; Soc Econ Paleontologists & Mineralogists; Paleont Asn. *Res:* Upper Cretaceous gastropod taxonomy, paleoecology and zoogeography; molluscan paleontology and paleoecology; sedimentation, paleolimnology and post-Pleistocene history of temperate lakes; paleosynecological relationships of invertebrates; applications of fossil oribatid mites to problems of Pleistocene and Holocene paleoecology and paleoclimate. *Mailing Add:* Dept Geol St Lawrence Univ Canton NY 13617

ERICKSON, JOHN ROBERT, b Flaxton, NDak, Sept 4, 39. PLANT BREEDING. *Educ:* Wash State Univ, BS, 63; NDak State Univ, PhD(agron), 67. *Prof Exp:* Res geneticist, Crops Res Div, USDA, 67-69; from asst prof to assoc prof agron, NDak State Univ, 69-78; hybrid wheat breeder, Dekalb Hybrid Wheat, Inc, 78-81, RES DIR, HYBRITECH SEED, 82- *Mem:* Am Soc Agron; Crop Sci Soc. *Res:* Winter wheat breeding; genetics. *Mailing Add:* HybriTech Seed 5912 N Meridian Wichita KS 67204

ERICKSON, JOHN WILLIAM, b Chicago, Ill, June 13, 25; m 47; c 8. EXPLORATION GEOLOGY. *Educ:* Augustana Col, AB, 50. *Prof Exp:* Jr comput geophys, Atlantic Refining Co, 51-52; stratigr, Chem & Geol Labs, 52-54; geologist, Gulf Oil Corp, 54-58, rep, Denver Div, Gulf Res & Develop Co, 58-60, sr geologist, 60-74; dist geologist, Mich Wis Pipeline Co, 74-77; explor mgr mid-continent, Tex Int Petrol Corp, 77-79; chief geologist, Walter Duncan Oil Properties, 79-90; RETIRED. *Mem:* Am Asn Petrol Geol; Geol Soc Am; Am Inst Prof Geologists. *Res:* Stratigraphy; subsurface and petroleum geology; geological data processing techniques. *Mailing Add:* 4715 Larissa Lane Oklahoma City OK 73112

ERICKSON, JON JAY, b Minot, NDak, Sept 16, 41; m 65; c 2. MEDICAL PHYSICS. *Educ:* St John's Univ, Minn, BS, 63; Vanderbilt Univ, MS, 65, PhD(physics), 72. *Prof Exp:* Res asst physics, Vanderbilt Univ, 64-67; res assoc, 67-73, from instr to asst prof radiol, Div Nuclear Med, Vanderbilt Hosp, 75-81, ASSOC PROF RADIOL, VANDERBILT MED CTR, 81-; HEALTH PHYSICIST, NASHVILLE VETERANS ADMIN MED CTR, 85- *Concurrent Pos:* Tech expert comput appln nuclear med, Int Atomic Energy Agency, Bolivia, Columbia, Peru, Uruguay & Brazil, 79- *Mem:* NY Acad Sci; Soc Nuclear Med; Am Asn Physicists in Med; Am Col Med Physics. *Mailing Add:* 103 Virginian Ct Franklin TN 37004

ERICKSON, KAREN LOUISE, b Covington, Mich, Aug 4, 39. ORGANIC CHEMISTRY, CHEMICAL DYNAMICS NATURAL PRODUCTS CHEMISTRY. *Educ:* Siena Heights Col, BS, 60; Purdue Univ, PhD(org chem), 65. *Hon Degrees:* Siena Heights Col, 81. *Prof Exp:* NIH fel org chem, Cornell Univ, 64-65; from asst prof to assoc prof, 65-79, PROF ORG CHEM, CLARK UNIV, 79- *Concurrent Pos:* NIH spec fel, Univ Hawaii, 72-73; vis lectr, Univ Canterbury, NZ, 75; res fel, Roche Res Inst Marine Pharmacol, Sydney, Australia, 79-80; vis prof, Ariz State Univ, 86-87. *Mem:* Am Chem Soc; Sigma Xi; Am Asn Women Sci. *Res:* Natural products; rearrangement reactions. *Mailing Add:* Dept Chem Clark Univ Worcester MA 01610

ERICKSON, KENNETH LYNN, b Rapid City, SDak, Apr 13, 46; m 69; c 1. TRANSPORT PHENOMENA, SURFACE PHENOMENA. *Educ:* Univ Ariz, BS, 68, MS, 73; Univ Tex, Austin, PhD(chem eng), 77. *Prof Exp:* Res engr, Tex Div, Dow Chem Co, 74; MEM TECH STAFF, SANDIA NAT LABS, 76- *Mem:* Am Inst Chem Engrs; Am Chem Soc; Mat Res Soc. *Res:* Experimental and theoretical investigations of chemical reactions and ignition phenom, associated with pirotechniques & explosives. *Mailing Add:* 1168 Laurel Loop NE Albuquerque NM 87122-1109

ERICKSON, KENNETH NEIL, b Minneapolis, Minn, Oct 1, 40; m 65; c 2. ELEMENTARY PARTICLE PHYSICS, SPACE PHYSICS. *Educ:* Augsburg Col, BA, 62; Mich State Univ, MS, 64; Colo State Univ, PhD(physics), 70. *Prof Exp:* Instr, 64-65, assoc prof, 70-81, PROF PHYSICS, AUGSBURG COL, 81- *Concurrent Pos:* Assoc prof, Univ Minn, Minneapolis, 70-81; adj prof physics, Univ Minn, 81- *Mem:* Am Geophys Union. *Res:* Cosmic rays; magnetospheric physics; ionospheric physics. *Mailing Add:* Dept Physics Augsburg Col 731 21st Ave S Minneapolis MN 55454

ERICKSON, KENT L, HUMAN ANATOMY. *Prof Exp:* ACTG CHMN, DEPT CELL BIOL, SCH MED, UNIV CALIF, 74- *Mailing Add:* Dept Human Anat Sch Med Univ Calif Davis CA 95616

ERICKSON, LARRY EUGENE, b Wahoo, Nebr, Oct 8, 38; m 62. BIOCHEMICAL & CHEMICAL ENGINEERING. *Educ:* Kans State Univ, BS, 60, PhD(chem eng), 64. *Prof Exp:* From instr to assoc prof, 64-72, PROF CHEM ENG, KANS STATE UNIV, 72- *Concurrent Pos:* US Pub Health Serv spec res fel, 67-68, career develop award, 70-75; dir, Ctr Hazardous Substance Res, Kans State Univ, 89- *Mem:* Am Inst Chem Engrs; Am Chem Soc; Am Soc Microbiol; Am Soc Eng Educ; Inst Food Technologists. *Res:* Optimum process design; biochemical engineering; food engineering; environmental engineering. *Mailing Add:* Dept Chem Eng Kans State Univ Manhattan KS 66506

ERICKSON, LAURENCE A, b July 16, 53; m 75; c 4. THROMBOSIS, CELL BIOLOGY. *Educ:* Univ Ill, BS, 75; Western Ill Univ, MS, 79; Univ Kans, PhD(cell biol), 82. *Prof Exp:* Sr postdoctoral res fel, Scripps Clinic, 82-85; RES SCIENTIST, UPJOHN CO, 85- *Mem:* Am Soc Hemat; Int Soc Thrombosis & Hemostasis; Am Heart Asn. *Res:* Basic mechanisms of thrombosis/thrombolysis regulation; control of cellular proliferation. *Mailing Add:* 7243-209-4 Cardiovasc Dis Res Upjohn Co Kalamazoo MI 49001

ERICKSON, LOUIS CARL, b Wilmington, Calif, Feb 13, 14; m 41; c 2. PLANT PHYSIOLOGY. *Educ:* Univ Calif, Los Angeles, AB, 37, MA, 39; Univ Calif, Berkeley, PhD(bot), 46. *Prof Exp:* Asst plant physiologist, Bur Plant Indust, USDA, 43-45; plant physiologist, Thompson Hort Chem Corp, 46-48; asst plant physiologist, Citrus Res Ctr & Agr Exp Sta, 48-54, assoc plant physiologist, 54-60, plant physiologist, 60-67, prof bot, 67-77, dir botanic gardens, 73-81, EMER PROF BOT, UNIV CALIF, RIVERSIDE, 77- *Mem:* Am Asn Bot Gardens & Arboretums; Int Lilac Soc; Royal Hort Soc. *Res:* Lilac hybridizing. *Mailing Add:* Dept Bot & Plant Sci Univ Calif Riverside CA 92521

ERICKSON, LUTHER E, b Pulaski, Wis, June 30, 33; m 57; c 2. PHYSICAL CHEMISTRY. *Educ:* St Olaf Col, BA, 55; Univ Wis, PhD(phys chem), 59. *Prof Exp:* Asst prof chem, Dickinson Col, 59-62; asst prof, 62-63, from assoc prof to prof, 64-75, DODGE PROF CHEM, GRINNELL COL, 75-

Concurrent Pos: NSF sci fac fel, Univ NC, 68-69. *Honors & Awards:* Catalyst Award, Chem Mfrs Asn, 83. *Mem:* AAAS; Am Chem Soc. *Res:* Nuclear magnetic resonance spectroscopy; complex ions and coordination compounds; conformational analysis of metal chelates. *Mailing Add:* Dept Chem Grinnell Col Grinnell IA 50112

ERICKSON, LYNDEN EDWIN, b Ft William, Ont, June 6, 38; m 65; c 3. MAGNETIC RESONANCE, OPTICAL SPECTROSCOPY. *Educ:* Queen's Univ, Ont, BSc, 59; Univ Chicago, SM, 61, PhD(physics), 66. *Prof Exp:* RES OFFICER PHYSICS, NAT RES COUN, 65- *Mem:* Am Phys Soc. *Res:* High resolution spectra of rare earth ions in solids at low temperature using fluorescence line narrowing and saturated and enhanced absorption techniques; optically detected nuclear magnetic resonance of rare earth ions in solids. *Mailing Add:* Nat Res Coun M50 Ottawa ON K1A 0R6 Can

ERICKSON, MITCHELL DRAKE, b Chicago, Ill, Aug 31, 50; m 76; c 3. ENVIRONMENTAL CHARACTERIZATION, ENVIRONMENTAL REMEDIATION. *Educ:* Grinell Col, AB, 72; Univ Iowa, PhD(anal chem), 76. *Prof Exp:* Chemist, Res Triangle Inst, 76-81; prin chemist, Midwest Res Inst, 81-87; group leader org anal, 87-89, ASSOC DIR, RES & DEVELOP PROG, COORD OFF, ARGONNE NAT LAB, 89- *Concurrent Pos:* Consult, 87- *Mem:* Am Soc Mass Spectrometry; Am Chem Soc; Soc Appl Spectros; Sigma Xi. *Res:* Analysis, chemistry, and destruction of PCBs and dioxins; general environmental analysis sensing and characterization techniques; infrared spectrometry applied to gas and air monitoring. *Mailing Add:* Argonne Nat Lab Bldg 205 Argonne IL 60439

ERICKSON, RALPH O, b Duluth, Minn, Oct 27, 14; m 45; c 2. BOTANY, DEVELOPMENTAL BIOLOGY. *Educ:* Gustavus Adolphus Col, BA, 35; Wash Univ, MS, 41, PhD(bot), 44. *Prof Exp:* Instr biol, Gustavus Adolphus Col, 35-39; asst bot, Wash Univ, 40-41; asst chemist, Western Cartridge Co, Ill, 42-44; instr bot, Univ Rochester, 44-47; res assoc, 47-49, from assoc prof to prof bot, 49-85, actg chmn dept biol, 61-63, 76-78, chmn biol grad group, 68-76, EMER PROF BOT, UNIV PA, 85- *Concurrent Pos:* Guggenheim fel, 54-55. *Mem:* Bot Soc Am; Soc Develop Biol (secy, 57-58, pres, 59); Am Soc Plant Physiologists; AAAS; Am Inst Biol Sci. *Res:* Analysis of plant growth and morphogenesis; phyllotaxis; packing of structural proteins. *Mailing Add:* Dept Biol Univ Pa Philadelphia PA 19104-6018

ERICKSON, RANDALL L, b Harris, Minn, Apr 24, 39; m 61; c 1. PHYSICAL CHEMISTRY, POLYMER CHEMISTRY. *Educ:* Concordia Col, Moorhead, Minn, BA, 61; NDak State Univ, PhD(phys chem), 65. *Prof Exp:* Res chemist, Textile Fibers Dept, E I du Pont de Nemours & Co, Va, 65-67; sr res chemist, Adhesives, Coatings & Sealers Div, 67-68, supvr, Bldg Serv & Cleaning Prod Div, 68-73, MGR SAFETY SYSTS DIV, 3M CO, 73- *Mem:* Am Chem Soc. *Res:* Solvation of extracted metal complexes by organic solvents; physical chemistry of polymers-structure property relationships. *Mailing Add:* Eight Evergreen Lane North Oaks St Paul MN 55127

ERICKSON, RAY CHARLES, b St Peter, Minn, Jan 30, 18; m 53; c 4. WILDLIFE RESEARCH. *Educ:* Gustavus Adolphus Col, AB, 41; Iowa State Col, MS, 42, PhD(econ zool), 48. *Prof Exp:* Collabr, US Fish & Wildlife Serv, 39-41, jr biologist, Patuxent Res Refuge, Md, 41, asst, Malheur Nat Wildlife Refuge, 42 & 46-47, refuge mgr, 48, wildlife mgt biologist, 48-55, head habitat improv sect, Wildlife Refuges Br, 55-58, res staff specialist, Wildlife Res Div, 58-65, asst dir in-chg endangered wildlife res prog, Patuxent Wildlife Res Ctr, 65-80; CONSULT, 80- *Honors & Awards:* Spec Conserv Award, Nat Wildlife Fedn, 76; Honor Award, Whooping Crane Conserv Asn, 80; Spec Recognition Serv Award, Wildlife Soc, 81. *Mem:* Am Ornith Union; assoc Wildlife Soc; Ecol Soc Am. *Res:* Waterfowl and waterfowl habitat ecology and management studies; preservation of rare and endangered wildlife species; endangered wildlife research supervision. *Mailing Add:* 1943 37th Ave NW Salem OR 97304

ERICKSON, RICHARD AMES, b Bryant, SDak, Sept 12, 23; m 43; c 4. PHYSICS. *Educ:* SDak Sch Mines & Technol, BS, 44; Agr & Mech Col, Tex, PhD(physics), 52. *Prof Exp:* Asst prof physics, Univ Tenn, 51-53; from asst prof to prof, 54-79, EMER PROF PHYSICS, OHIO STATE UNIV, 79- *Res:* Neutron diffraction; solid state physics; chemical physics; low temperature physics. *Mailing Add:* 1981 Indianola Ave Columbus OH 43201

ERICKSON, ROBERT PORTER, b South Bend, Ind, Feb 13, 30; c 3. PHYSIOLOGICAL PSYCHOLOGY. *Educ:* Northwestern Univ, BS, 51; Brown Univ, MSc, 56, PhD(psychol), 58. *Prof Exp:* Res assoc psychol, Brown Univ, 58-59, instr, 59; fel physiol & biophys, Univ Wash, 59-61; PROF PSYCHOL, DUKE UNIV, 61-, ASSOC PROF PHYSIOL, 77- *Mem:* Soc Neurosci. *Res:* Neurophysiology of sensory systems and behavior processes. *Mailing Add:* Dept Psychol Duke Univ Durham NC 27706

ERICKSON, ROBERT PORTER, b June, 27, 39; m 64; c 6. DEVELOPMENTAL GENETICS, CLINICAL GENETICS SYNDROMES. *Educ:* Stanford Univ, MD, 65. *Prof Exp:* John Simon Guggenheim fel, 75; dir pediat, Genetics Div, 76-86, PROF HUMAN GENETICS, SCH MED, UNIV MICH, 80-, PROF PEDIAT 82- *Concurrent Pos:* Eleanor Roosevelt Int Cancer Res fel, 84; ed bd, J Reproductive Immunol; Fulbright grant, 84; chmn, Gordon Res Conf Mammalian Gametogenesis & Embryogenesis; adv comn, NIH Recombinant DNA, 87. *Res:* Sex determination. *Mailing Add:* 1137 E Catherine St Ann Arbor MI 48109-0618

ERICKSON, ROBERT W, b McIntosh, Minn, Jan 31, 29; m 56; c 8. FOREST PRODUCTS. *Educ:* Univ Minn, St Paul, BS, 58, MS, 63, PhD(forest prod eng), 66. *Prof Exp:* Lab technician, Forest Prod Lab, Univ Calif, 58-61; asst forest prod, 61-62, instr wood seasoning, 62-66, from asst prof to assoc prof wood-fluid rels, 66-75, PROF FOREST PROD, COL FORESTRY, UNIV MINN, ST PAUL, 75- *Concurrent Pos:* Vis scientist, Soc Wood Sci & Technol, NC, 71; vis seminarist, Can Forest Prod Labs, BC, 70; sabbatical res in wood drying, Tech Ctr, Weyerhaeuser Co, Wash, 73-74. *Mem:* Soc Wood Sci & Technol (pres, 78). *Res:* Effect of surface tension upon the permeability of wood to distilled water; effects of pre-freezing upon the subsequent accelerated drying behavior of redwood and other collapse susceptible species; factors that influence the flexual creep behavior of wood during dehydration from the green condition; drying of South East Asian hardwoods; nonisothermal moisture movement in wood. *Mailing Add:* Col Forestry Univ Minn 2004 Folwell Ave St Paul MN 55108

ERICKSON, RONALD E, b Peoria, Ill, Apr 20, 33; m 58; c 2. ENVIRONMENTAL CHEMISTRY. *Educ:* Bradley Univ, BS, 55; Univ Iowa, PhD(org chem), 59. *Prof Exp:* Welch fel, Univ Tex, 58-59; NATO fel, Karlsruhe Tech Univ, 60-61; Rosalie B Hite fel, Univ Tex, 61; asst prof chem, Canisius Col, 61-65; from assoc prof to prof chem, 65-84, dir environ study prog, 76-84, PROF ENVIRON STUDIES, UNIV MONT, 81. *Res:* environmental problems of coal gasification and liquefaction. *Mailing Add:* Environ Studies Prog Univ Mont Missoula MT 59812

ERICKSON, STANLEY ARVID, US citizen. SIMULATION, SYSTEMS ANALYSIS. *Educ:* Mass Inst Technol, SB(physics) & SB(math), 64, PhD(math), 74. *Prof Exp:* Proj leader res, Naval Underwater Systs Ctr, 65-77; proj leader, 77-80, GROUP LEADER PHYSICS, LAWRENCE LIVERMORE LAB, 80- *Concurrent Pos:* Consult, RI Dept Educ, 75-77. *Mem:* Opers Res Soc Am; Inst Elec & Electronics Engrs. *Res:* Simulation and modeling of physical and social systems. *Mailing Add:* L-85, PO Box 808 Lawrence Livermore Nat Lab Livermore CA 94551

ERICKSON, WAYNE DOUGLAS, b Lansing, Mich, Feb 19, 32; m 59; c 1. AEROTHERMOCHEMISTRY, COMBUSTION. *Educ:* Mich State Univ, BS, 54, MS, 55; Mass Inst Technol, SM, 58, ScD(chem eng), 62. *Prof Exp:* Res engr, Langley Res Ctr, NASA, 55-59, head, thermal proj sect, 59-64, res engr, 64-65, head aerothermochem br, 65-70, head life support br, 70-72, sr scientist, off dir, 72-82, chief scientist, low-speed Aerodynamic Div, 83-87, CHIEF SCIENTIST, HIGH SPEED AERODYN DIV, 87-; PROF LECTR ENG, GEORGE WASHINGTON UNIV, 80- *Concurrent Pos:* Mem res staff, phys chem dept, Cambridge Univ, 70-71; vis assoc prof, Dept Chem Eng, Mass Inst Technol, 80-82; mem sci adv bd, USAF, 81- *Mem:* Combustion Inst; Am Inst Aeronaut & Astronaut; Am Inst Chem Engrs. *Res:* Chemical kinetics of high temperature systems; combustion research; thermodynamics; aerothermochemistry; chemical process systems; molecular computations; turbulent flow with combustion. *Mailing Add:* 241 William Barksdale Williamsburg VA 23185-6568

ERICKSON, WAYNE FRANCIS, b Kingston, NY, July 1, 46. ORGANIC CHEMISTRY, MEDICAL PRODUCTS. *Educ:* Univ Mass, BS, 68; Mass Inst Technol, MS, 70; Pa State Univ, PhD(org chem), 72. *Prof Exp:* Res chemist photog sci, 74-80, RES CHEMIST MED PROD, EASTMAN KODAK CO RES LABS, 80- *Res:* High speed color negative films; development medical product systems. *Mailing Add:* Eastman Kodak Co 343 State St Rochester NY 14650

ERICKSON, WILLIAM CLARENCE, b Chicago, Ill, Nov 21, 30; m 52; c 4. ASTRONOMY. *Educ:* Univ Minn, BA, 51, MA, 55, PhD(physics), 56. *Prof Exp:* Lectr physics, Univ Minn, 55-56; fel, Carnegie Inst, 56-57; sr staff scientist, Convair Sci Res Lab Div, Gen Dynamic Corp, 57-62 & Benelux Cross Antenna Proj, 62-63; from assoc prof to prof astron, 63-76, prof, 76-88, EMER PROF PHYSICS & ASTRON, UNIV MD, COLLEGE PARK, 88- *Mem:* Int Sci Radio Union; Int Astron Union; Am Astron Soc; Royal Astron Soc. *Res:* Theory and observations in radio astronomy. *Mailing Add:* Dept Physics & Astron Univ Md College Park MD 20742

ERICKSON, WILLIAM HARRY, b McKeesport, Pa, Apr 4, 16; m 41; c 2. ELECTRICAL ENGINEERING. *Educ:* Univ Pittsburgh, BS, 38; Carnegie Inst Technol, MS, 46. *Prof Exp:* Jr engr, Duquesne Light Co, Pa, 38-42; from instr to assoc prof elec eng, Naval Training Prog, 42-53, asst dir elec eng, 59-65, assoc dean col eng, 65-71, PROF ELEC ENG, CORNELL UNIV, 53- *Mem:* Fel Inst Elec & Electronics Engrs. *Mailing Add:* Cornell Univ Ithaca NY 14853

ERICSON, ALFRED (THEODORE), b Quincy, Kans, Oct 8, 28; m 48; c 2. BIOCHEMISTRY. *Educ:* Kans State Teachers Col, BSEd, 50; Kans State Univ, MS, 53, PhD(chem), 56. *Prof Exp:* Asst instr chem, Kans State Univ, 53-56; from asst prof to assoc prof, 56-63, PROF CHEM, EMPORIA STATE UNIV, 63- *Mem:* AAAS; Am Chem Soc; Sigma Xi. *Res:* Chemical education; tracer techniques; protein chemistry. *Mailing Add:* 1406 Col Dr Emporia KS 66801

ERICSON, AVIS J, b Chicago, Ill, Feb 20, 47; m 78. TERATOLOGY, CLINICAL PHARMACY. *Educ:* Ferris State Univ, BS, 71; Univ Ky, Pharm D, 73. *Prof Exp:* Asst prof clin pharm, Col Pharm, Univ Ky, 74, asst prof obstet & gynecol, Col Med, 74-75; from asst prof to assoc prof clin pharm, Mass Col Pharm & Allied Health Sci, 76-85, chair, dept clin pract, 82-85; dir div clin pharm, 85-90, ASSOC PROF PHARM PRACT, ST LOUIS COL PHARM, 85- *Concurrent Pos:* Mem consult bds, Am Found Maternal-Child Health & Int Childbirth Educ Asn, 74-82; asst dir clin pharm serv, Boston Hosp Women, 76-80; clin pharmacist reviewer, Bur Drugs, Food & Drug Admin, 80-82; mem adv comt, Omeprazole-Astra Pharmaceut Prod, Inc, 84-; chair, Astra Clin Pharm Res Award, 83- *Mem:* Am Soc Hosp Pharmacists; Am Pharmaceut Asn; Am Asn Col Pharm; Int Childbirth Educ Asn. *Res:* Drug use in the obstetric and gynecologic populations; neonatal outcome; endocrinology. *Mailing Add:* Div Pharm Pract St Louis Col Pharm 4588 Parkview Pl St Louis MO 63110

ERICSON, GROVER CHARLES, b Oak Park, Ill, Feb 17, 41; m 64; c 2. HUMAN ANATOMY, MUSCLE BIOLOGY. *Educ:* N Cent Col, BA, 64; Loyola Univ, Chicago, MS, 68, PhD(anat), 72. *Prof Exp:* Asst prof anat, Univ Tex Southwestern Med Sch, Dallas, 72-78; ASSOC PROF ANAT, COL MED & ASSOC PROF ZOOL, OHIO UNIV, 78-, PROF, DEPT ANAT, SOUTHEASTERN UNIV HEALTH SCI, 83- *Concurrent Pos:* Anat consult,

Med Plastics Lab Inc, 76- *Mem:* Am Asn Clin Anatomists. *Res:* Exercise biology; response of skeletal muscle to weight-lifting exercise. *Mailing Add:* Dept Anat Southeastern Univ Health Sci 1750 NE 168th St North Miami Beach FL 33162

ERICSON, WILLIAM ARNOLD, b Larchmont, NY, Jan 9, 34; c 2. STATISTICS. *Educ:* Univ Pa, BS, 55, MA, 58; Harvard Univ, PhD(statist), 63. *Prof Exp:* Res asst statist, Harvard Univ, 60-62; from asst prof to assoc prof math, Univ 62-69, assoc prof statist, 69-70, prof statist & chmn dept, 70-75, DIR, STATIST RES LAB, UNIV MICH, ANN ARBOR, 71- *Concurrent Pos:* Assoc ed, J Am Statist Asn, 67-75; vis lectr, Inst Math Statist, 71- *Mem:* Biomet Soc; Inst Math Statist; fel Am Statist Asn. *Res:* Sampling theory; Bayesian inference. *Mailing Add:* Dept Statist 1447 Mason Hall Univ Mich Ann Arbor MI 48109

ERICSSON, LOWELL HAROLD, b St Paul, Minn, July 30, 28; m 53; c 2. PROTEIN STRUCTURE, AMINO ACID COMPOSITION & SEQUENCE OF PROTEINS. *Educ:* Beloit Col, BS, 50. *Prof Exp:* Metallurgist, Barber Coleman Co, 50-52; technician dept biochem, Univ Chicago, 52-56; chemist, Boeing Airplane Co, 56-60; technician biochem, 60-67, res assoc, 67-74, SR RES ASSOC BIOCHEM, UNIV WASHINGTON, 74-; PARTNER, AAA LAB, 72- *Mem:* AAAS; Am Chem Soc; Am Soc Biol Chemists; Sigma Xi; Protein Soc. *Res:* Structure of proteins, specifically their amino acid composition and sequence. *Mailing Add:* 6206 89th SE Mercer Island WA 98040

ERICSSON, RONALD JAMES, b Belle Fourche, SDak, July 17, 35; m 56; c 2. REPRODUCTIVE PHYSIOLOGY, ANDROLOGY. *Educ:* Colo State Univ, BS, 57; Univ Ky, MS, 61, PhD(physiol), 64. *Prof Exp:* Res asst, Univ Ky, Lexington, 60-63; sr res scientist, Upjohn Co, Kalamazoo, 64-71; spec consult, Schering Ag, Berlin, 71-73; investr, Inst Res Hum Reproduction, Iran, 73-74; pres, Sausalito, Calif, 74-85, PRES, GAMETRICS LTD, COLONY, WYO & LAS VEGAS, NEV, 85- *Concurrent Pos:* Adj assoc prof, Western Mich Univ, Kalamazoo, 64-71; dir, Lonesome Country Ltd, 69-, Gametrics Ltd Europ, 76- & Pac Export, 78-, Ericsson Corp, 81- & X-Y Genetics Ltd, 82- *Mem:* AAAS; Am Asn Tissue Banks; Am Fertil Soc; Am Soc Andrology; Asn Study Animal Behav; Endocrine Soc; Soc Advan Contraception; Soc Exp Biol & Med; Soc Study Fertil; Soc Study Reproduction; Sigma Xi. *Res:* Human reproduction; infertility; in vitro fertilization; sex selection; chemosterilants; male fertility; author of 71 publications and co-author of one book; holder of ten patents. *Mailing Add:* Gametrics Ltd HC 69 Box 50 Alzada MT 59311

ERIKS, KLAAS, b Alkmaar, Neth, June 16, 22; nat US; m 49; c 3. STRUCTURAL CHEMISTRY. *Educ:* Univ Amsterdam, Chem Cand, 43, Chem Drs, 48, PhD(chem), 52. *Prof Exp:* Res assoc physics, Pa State Univ, 52-53; fel chem, Univ Minn, 53-54; from asst prof to assoc prof, 54-65, PROF CHEM, BOSTON UNIV, 65- *Concurrent Pos:* Fulbright res scholar, Copenhagen Univ, 63-64; mem dent training comt, Nat Inst Dent Res, 65-69; vis res scholar, Neth Reactor Ctr, Petten, 71-72; mem, Nat Comt on Crystallog; vis res fac, Argonne Nat Lab, 83-84. *Mem:* Am Chem Soc; Am Crystallog Asn. *Res:* X-ray structure determination in fields of calcium salts of amino acids; transition metal ion complexes, especially with amino acids; organic and inorganic sulphates, phosphates and fluorides; organometallics. *Mailing Add:* 145 Washington Ave Needham MA 02192

ERIKSEN, CLYDE HEDMAN, b Santa Barbara, Calif, May 1, 33; m 90; c 2. PHYSIOLOGICAL ECOLOGY, LIMNOLOGY. *Educ:* Univ Calif, Santa Barbara, BA, 55; Univ Ill, MS, 56; Univ Mich, PhD(zool), 61. *Prof Exp:* Instr biol, Exten Serv, Univ Calif, 54-55; teaching fel, Univ Ill, 55-57; field res asst limnol, Great Lakes Res Inst, 57; field specialist, Nat Sanit Found, 58-59; asst prof zool, Los Angeles State Col, 60-63 & NSF Inst Biol Sci, 61; from asst prof to assoc prof, Univ Toronto, 63-67; assoc prof, 67-72, chmn joint sci dept, 68-72, PROF BIOL, CLAREMONT COLS, 72-, DIR, BERNARD BIOL STA, 77- *Concurrent Pos:* Consult, Allen Hancock Found, Univ Southern Calif, 61-63; mem, Environ Resources Task Force, The City of Claremont, 70-71; ecol & admin specialist, US Forest Serv, Beaverhead Nat Forest, Mont, 73, consult, 74; vis res entomologist, Div Biol Control, Univ Calif, Berkeley, 81. *Mem:* Ecol Soc Am; Int Asn Theoret & Appl Limnol; Can Soc Zool; NAm Benthological Soc; Crustacean Soc. *Res:* Respiratory physiological ecology of aquatic invertebrates, ecology of playa lakes and rain pools; limnology of crater beds; public land management. *Mailing Add:* Joint Sci Dept Joint Sci Ctr Claremont Cols Claremont CA 91711

ERIKSEN, STUART P, b San Francisco, Calif, Nov 13, 30; m 55; c 3. OPERATIONS RESEARCH. *Educ:* Univ Calif, BSc, 52, MSc, 54, PhD(pharmaceut chem), 56, PhD(admin), 90. *Prof Exp:* Sr res scientist, Smith, Kline & French Labs, 55-60; assoc prof pharm, Univ Wis-Madison, 60-65; dir med res, Allergan Pharmaceut, 65-70, dir res opers, 70-72, vpres res & develop, 72-77, vpres contact lens prod res, 77-85; opers res, 85-90, CONSULT DECISION ANALYSIS, UNIV CALIF GRAD SCH MGT, 90- *Concurrent Pos:* Instr, Hahnemann Med Col, 58-60; vis lectr, Sch Pharm, Temple Univ, 59-60; adj prof, Sch Pharm, Univ Southern Calif, 82- *Mem:* AAAS; Am Chem Soc; Am Pharmaceut Asn; Asn Res Vision & Ophthal; Soc Invest Dermat; Sigma Xi. *Res:* Physical pharmacy; drug molecules in solution; rheology; drug degradation; lipid chemistry; ophthalmic and derm clinical investigations; surface phenomena on contact lenses; decision theory. *Mailing Add:* 13341 Eton PL Santa Ana CA 92705

ERIKSON, GEORGE EMIL, b Palmer, Mass, May 3, 20; m 50; c 4. ANATOMY. *Educ:* Mass State Col, BS, 41; Harvard Univ, MA, 46, PhD(biol), 48. *Prof Exp:* Reader hist of sci & learning, Harvard Univ, 43-45; instr anat, Harvard Med Sch, 47-49, asst prof biol, 49-52, assoc anat, 52-55, asst prof anat, 55-65; prof, 65-90, EMER PROF MED SCI, DIV BIOL & MED, BROWN UNIV, 90-; ANATOMIST, DEPT SURG, MASS GEN HOSP, 90-; PRES, ERIKSON BIOG INST, INC, 90- *Concurrent Pos:* Guggenheim fel, 49; asst prof gen educ biol, Harvard Univ, 49-52; consult med & pub health, Rockefeller Found, 59; assoc cur, Warren Anat Mus, Harvard Med Sch, 61-65; state dept specialist, Brazil, 62; anatomist, Dept Surg, RI Hosp, 65-, orthropedies & rehab, 65-, neurosurg, 87-; chmn, Sect Morphol, Div Biol & Med, Brown Univ, 68-85, anat, 85-89, co-chmn, Sect Pop Biol, Morphol & Genetics, 85-89 & 89-90; adj mem fac, RI Sch Design, 70-; coun Am Asn Hist Med, 71-72; archivist, Am Asn Anatomists, 72-; archivist & co-hist, Am Asn Phys Anthropologists, 80-; vis prof, Dept Anat & Cell Biol, Harvard Med Sch, 89-90; vis lectr surg, Harvard Med Sch, 90- *Honors & Awards:* Kate Hurd Mead Lectr, Col Physicians Philadelphia, 77; Seventh Raymond C Truex Distinguished Lectr, Hahnemann Univ Sch Med, 85. *Mem:* AAAS; Am Soc Zool; Am Asn Anat; Am Asn Hist Med; Am Soc Mammalogists; Anat Gesell Hist Sci Soc; Hist Sci Soc; Am Asn Anatomists; Am Asn Phys Anthropologists; Am Soc Zoologists; Oral Hist Asn; Sigma Xi. *Res:* Comparative biology of new world primates; history of biology and medicine; biographical studies; archives. *Mailing Add:* Dept Morphol Prog Med Brown Univ 97 Waterman St Providence RI 02912

ERIKSON, J ALDEN, b Milwaukee, Wis, Mar 3, 26; m 55; c 3. ORGANIC CHEMISTRY. *Educ:* Univ Wis, BS, 50; Mass Inst Technol, PhD(chem), 53. *Prof Exp:* Chemist, Paint & Brush Div, Pittsburgh Plate Glass Co, 53-59, sr res chemist, Coatings & Resins Div, 59-68, proj leader, 68-83, sr res assoc, PPG Industs, Inc, 83-88; RETIRED. *Mem:* Am Chem Soc; Fedn Soc Coatings Technol. *Res:* Alkyds; melamine and urea formaldehyde resins; thermosetting free radically polymerised copolymers; oil free polyesters; polyurethanes; silicone modified resins and aqueous dispersions. *Mailing Add:* 4212 E Ewalt Rd Gibsonia PA 15044

ERIKSON, JAY ARTHUR, b Seattle, Wash, May 2, 22; m 55; c 3. SURFACE CHEMISTRY. *Educ:* Univ Wash, BS, 48, MS, 49, PhD(phys chem), 54. *Prof Exp:* Instr petrol ref eng, Colo Sch Mines, 54-55; chemist, Agr Div, Shell Develop Co, 55-61, Santa Barbara Res Ctr, 61-63 & US Polymeric Div, HITCO, 64-67; mat & process engr autonetics, NAm Rockwell Corp, 67-73, mat & process engr, Space Div, Rockwell Int Corp, 73-78; RETIRED. *Mem:* Am Chem Soc; Sigma Xi. *Res:* Reaction kinetics of pesticide decompositions and lead sulfide film depositions; surface treatments of carbon fibers and powders; surface preparation for adhesive bonding. *Mailing Add:* 3404 Vereda Baja Santa Fe NM 87505-9232

ERIKSON, MARY JANE, b Grove City, Pa, Mar 29, 30; m 55; c 3. ENVIRONMENTAL HEALTH. *Educ:* Univ Mich, BS, 52. *Prof Exp:* Geologist, US Geol Surv, Denver, 52-56 & Doug Moran Inc, Tustin, Calif, 77-78; ed, Bendix Field Eng, Dept Eng Compound, Grand Junction, 79-80; geologist, Dept Eng, US Geol Surv & Bur Land Mgt, Colo & geologist, Bio-Resources, US Govt Bur Land Mgt, NMex, 80-90; RETIRED. *Mailing Add:* 3404 Vereda Baja Santa Fe NM 87505-9232

ERIKSON, RAYMOND LEO, b Eagle, Wis, Jan 24, 36; m 58. MOLECULAR BIOLOGY. *Educ:* Univ Wis, BS, 58, MS, 61, PhD(molecular biol), 63. *Prof Exp:* USPHS fel, Sch Med, Univ Colo, 63-65, from asst prof to prof path, 65-82; PROF CELLULAR & DEVELOP BIOL, HARVARD UNIV, 82-, AM CANCER SOC RES PROF, 83- *Concurrent Pos:* Am Cancer Soc scholar, Imp Cancer Res Fund, London, Eng, 72-73; mem, Exp Virol Study Sect, NIH, 75-78; Am Bus Cancer Res Found, Inc Award, 81-86; mem, Adv Comt Microbiol & Virol, Am Cancer Soc, 83-86 & Adv Comt Personnel for Res, 86-90. *Honors & Awards:* Papanicolaou Award, Papanicolaou Inst, 80; Robert Koch Prize, Robert Koch Found, 82; Albert Lasker Basic Med Res Award, Albert & Mary Lasker Found, 82; Distinguished Res Biomed Sci Award, Asn Am Med Cols, 82; Alfred P Solon Jr Prize, Gen Motors Cancer Res Found, 83; Hammer Prize Cancer Res, 84. *Mem:* Nat Acad Sci; Am Soc Biol Chemists; Am Soc Microbiol; Am Soc Virol; Am Acad Arts & Sci. *Res:* Tumor viruses; protein kinases; cell proliferation. *Mailing Add:* Biol Labs Harvard Univ 16 Divinity Ave Cambridge MA 02138

ERIKSSON, KARL-ERIK LENNART, b Bohus-Malmön, Sweden, May 27, 32; m 58; c 3. ENZYMES DEGRADING WOOD & WOOD COMPONENTS, PULP & PAPER MILL OPERATION. *Educ:* Univ Uppsala, Sweden, BS, 58, PhD fil lic(biochem), 63; Univ Stockholm, Sweden, DrSci(biochem), 67. *Prof Exp:* Res asst, Swed Forest Prods Lab, 58-64, div head, 64-88; PROF & EMINENT SCHOLAR, UNIV GA, 88- *Concurrent Pos:* Consult to FAD, UNIDO & pulp & paper indust worldwide, 64-; Fulbright fel, Calif Inst Technol, 68-69; mem bd, Royal Swed Acad Eng Sci, 82-85; adj prof, Inst Paper Sci & Technol, 90- *Honors & Awards:* Marcus Wallenburg Prize, Marcus Wallenburg Found, 85. *Mem:* Royal Swed Acad Eng Sci; Int Acad Wood Sci; World Acad Arts & Sci; Swed Chem Biochem Microbial & Pulp & Paper Asn; Tech Asn Pulp & Paper Indust; AAAS; Am Soc Microbiol. *Res:* Microbial and enzymatic degradation of wood and wood components. *Mailing Add:* Dept Biochem Life Sci Bldg Univ Ga Athens GA 30602

ERIKSSON, KENNETH ANDREW, b SAfrica, June 6, 46; m 75. GEOLOGY, SEDIMENTOLOGY. *Educ:* Univ Witwatersrand, BSc Hons, 68, MSc, 72, PhD(geol), 77. *Prof Exp:* Lectr geol, Univ Witwatersrand, 69-77, sr lectr geol, 78; asst prof geol, Univ Tex, Dallas, 78-79, assoc prof, 80-81; PROF GEOL, VA POLYTECH INST & STATE UNIV, 81- *Mem:* Geol Soc SAfrica; Geol Soc Am; Soc Econ Paleontologists & Mineralogists; Int Asn Sedimentologists; Sigma Xi. *Res:* Archean paleoenvironments and crustal evolution; proterozoic and paleozoic shelf sedimentation; sedimentology of Appalachian clastic wedges; atmospheric evolution. *Mailing Add:* Dept Geol Sci Va Polytech Inst & State Univ Blacksburg VA 24061-0219

ERIKSSON, LARRY JOHN, b Milwaukee, Wis, Feb 12, 45; m 67; c 2. ACOUSTIC NOISE CONTROL, ADAPTIVE FILTERS. *Educ:* Northwestern Univ, BSEE, 67; Univ Minn, MSEE, 69; Univ Wis, PhD(elec eng), 85. *Prof Exp:* Res scientist, Honeywell Corp Res Ctr, 68-71; pres, Sonotek, 71-72; acoust engr, AMF-Harley Davidson Motor Co, 72-73; VPRES RES, NELSON INDUSTS, INC, 73- *Concurrent Pos:* Adj asst prof, Univ Wis-Madison, 86-89. *Mem:* Inst Elec & Electronics Engrs; Am Soc Mech Engrs; Acoust Soc Am; Soc Automotive Engrs. *Res:* Acoustic noise control products and systems related to wave propagation in ducts; acoustic measurement apparatus; active sound attenuation using adaptive digital signal processing techniques. *Mailing Add:* 5301 Greenbriar Lane Madison WI 53714

ERINGEN, AHMED CEMAL, b Kayseri, Turkey, Feb 15, 21; m 49; c 4. APPLIED MECHANICS, CONTINUUM PHYSICS. *Educ:* Advan Eng Sch Istanbul, MS, 43; Polytech Inst Brooklyn, PhD(appl mech), 48. *Hon Degrees:* DSc, Univ Glasgow, Scotland, 81. *Prof Exp:* Res engr, Turkish Aircraft Co, 43-44; int trainee, Glenn L Martin Co, 44-45; group head fuselage sect & head struct sect, Turkish Air League Co, 45-47; from res asst prof to res assoc prof mech, Ill Inst Technol, 48-53; assoc prof eng mech, Purdue Univ, 53-55, prof eng sci, 55-66; prof aerospace & mech sci & chmn solid mech prog, 66-74, PROF CONTINUUM PHYSICS, CIVIL ENG & APPL MATH, PRINCETON UNIV, 74- *Concurrent Pos:* Consult, Armour Res Found, 48-50, Gen Motors, Picatinny Arsenal & Gen Tech Corp; ed-in-chief, Int J Eng Sci, Soc Eng Sci, 63- *Honors & Awards:* Outstanding Researcher of Yr Award, Sigma Xi, 63; Distinguished Serv Award, Soc Eng Sci, 73 & Certs of Appreciation, 74 & 75, A C Eringen Award, 76. *Mem:* Soc Eng Sci (pres, 63-73); Am Soc Mech Engrs; Am Soc Eng Educ; Soc Rheology; Soc Natural Philos; Sigma Xi. *Res:* Continuum physics; development of theories of micropolar and micromorphic continua; nonlocal continuum physics, solids, fluids, electromagnetism; material theories. *Mailing Add:* Dept Civil Eng Princeton Univ Eng Quadrangle E 307 Princeton NJ 08544

ERIQUEZ, LOUIS ANTHONY, b Danbury, Conn, Sept 12, 49; m 70; c 1. BIOCHEMISTRY, MICROBIOLOGY. *Educ:* Fairfield Univ, BS, 71; St John's Univ, MS, 73, PhD(biochem), 76. *Prof Exp:* Assoc dir res & develop, Analytab Prod Inc, Div Ayerst Labs, 76-80; with Carr Scarborough, 80-; INNOVATIVE DIAGNOSTIC SYSTS, ATLANTA. *Mem:* Am Soc Microbiol; Soc Indust Microbiol; Am Chem Soc; Am Asn Clin Chem. *Res:* Microbial physiology, especially enzymolgy, regulatory processes, fermentations and antibiotic biosynthesis; deliniation of metabolic pathways and the interplay of primary and secondary metabolism. *Mailing Add:* Innovative Diagnostic Systs Inc 3404 Oakcliff Rd STE C-1 Atlanta GA 30340-3011

ERK, FRANK CHRIS, b Evansville, Ind, Dec 17, 24; m 48; c 3. GENETICS. *Educ:* Univ Evansville, AB, 48; Johns Hopkins Univ, PhD(genetics), 52. *Prof Exp:* Jr instr biol, Johns Hopkins Univ, 48-51; assoc prof & head dept, Wash Col, 52-57; prof natural sci, 57-61 & biol sci, 62-74, chmn, dept biol, 58-61, div sci & math, 59-60 & dept biol sci, 62-67, prof cellular & comp biol, 74-76, chmn, dept biol, 76-78, prof biol, 76-81, PROF BIOCHEM, STATE UNIV NY, STONY BROOK, 81- *Concurrent Pos:* Vis assoc prof, Univ Chicago, 54-55; consult, Biol Sci Curriculum Study, 60-70, 85-; vis investr, Poultry Res Ctr, Agr Res Coun, Scotland, 64-65, Genetics Inst, Milan, 65, Masonic Med Res Lab, 68-71, Univ Sussex, 71-72, 85-86, Galton Lab, Univ Col London, 78-79 & Univ Edinburgh, 79; exec ed, Quart Rev Biol, 66-69, ed, 69-; examr advan placement biol, Col Entrance Exam Bd, 67-71; asst examr, Int Baccalaureate Prog, Geneva, 77-83; vis prof, Univ Essex, Eng, 78-79. *Mem:* Genetics Soc Am; Am Genetic Asn; Soc Study Evolution; Nat Asn Biol Teachers; Genetics Soc Can; Sigma Xi; Am Soc Zoologists. *Res:* Developmental genetics of drosophila; mutagenesis; dermatoglyphics; insect nutrition; aging. *Mailing Add:* Dept Biochem SUNY at Stony Brook Life Sci Bldg Stony Brook NY 11794-5215

ERKE, KEITH HOWARD, b Rogers City, Mich, Dec 16, 38. MEDICAL MYCOLOGY. *Educ:* Mich State Univ, BS, 64, MS, 66; Tulane Univ, PhD(med mycol), 72. *Prof Exp:* RES MICROBIOLOGIST, WILLIAM BEAUMONT ARMY MED CTR, 73- *Mem:* Sigma Xi; Am Soc Microbiol. *Res:* DNA of the pathogenic yeast organism Cryptococcus neoformans and study of hypha-forming strains. *Mailing Add:* AVC Box 400 Plymouth Montserrat Leeward Islands British West Indies

ERKILETIAN, DICKRAN HAGOP, JR, b Mayfield, Ky, Sept 22, 13; m 40; c 3. MATHEMATICS. *Educ:* Western Ky State Col, AB, 36; Univ Ill, AM, 38. *Prof Exp:* Asst math, Univ Ill, 40-41; instr, Fenn Col, 41-42; from instr to prof, 42-78, actg chmn dept, 62-63, chmn, 63-64, prof-in-chg-freshman-sophomore math, 65-67, EMER PROF MATH, UNIV MO-ROLLA, 78- *Mem:* Am Math Soc; Am Soc Eng Educ; Math Asn Am. *Res:* Differential equations. *Mailing Add:* PO Box 46 Rolla MO 65401-0046

ERLANDSEN, STANLEY L, b Chicago, Ill, May 21, 41; m 62; c 3. HISTOLOGY, MICROSCOPIC ANATOMY. *Educ:* Dana Col, BS, 63; Univ Minn, Minneapolis, PhD(anat), 67. *Prof Exp:* USPHS sr fel electron micros, Univ Wash, 67-69; asst prof anat, Univ Iowa, 69-74; assoc prof, 74-78, PROF ANAT, UNIV MINN, MINNEAPOLIS, 78- *Concurrent Pos:* counsult, Diabetes Ctr, Univ Wash, Seattle, 78- *Mem:* Am Soc Microbiol; Histochem Soc; Am Asn Anatomists; Am Soc Cell Biol. *Res:* Ultrastructure of intestinal microorganisms and their interaction with intestinal mucosa; fine structure and function of the Paneth cell; ultrastructural immunocytochemistry; biology of the intestinal protozoan, Giardia, and its host response. *Mailing Add:* Dept Cell Biol & Neuroanat Univ Minn Med Sch 4-135 Jackson Hall Minneapolis MN 55455

ERLANDSON, ARVID LEONARD, b Norway, Mich, Sept 26, 29; m 55; c 5. MICROBIOLOGY. *Educ:* Univ Mich, BS, 51, MS, 52, PhD, 54. *Prof Exp:* City bacteriologist, Ann Arbor Health Dept, 51-53; asst bact, Univ Mich, 53-54; res bacteriologist, Parke, Davis & Co, 58-64; asst prof microbiol & immunol, Sch Med, Marquette Univ, 64-71; DIR MICROBIOL & INFECTION CONTROL OFFICER, BRONSON METHODIST HOSP, 71- *Concurrent Pos:* Guest prof, Sch Med & Inst Trop Med, Sao Paulo Univ, Brazil, 75; guest lectr, Pan Am Surg Cong, Santo Domingo, Dominican Repub, 78; consult, Upjohn Co, 71. *Mem:* Am Soc Microbiol; Am Practrs Infection Control; Sigma Xi. *Res:* Bacterial physiology; bacillary dysentery; experimental pyelonephritis; pathogenesis of experimental infections; host resistance mechanisms. *Mailing Add:* Bronson Methodist Hosp 252 E Lovell St Kalamazoo MI 49006

ERLANDSON, PAUL M(CKILLOP), b Washington, DC, Oct 27, 20; m 41; c 5. ENGINEERING PHYSICS. *Educ:* Mass Inst Technol, BS, 41; Univ Tex, MA, 49, PhD(physics), 50. *Prof Exp:* Test design engr radio mfg, Crosley Corp, 37-39; cost control engr, electronics mfg, Radio Corp Am, 41-42; res physicist, Defense Res Lab, Univ Tex, 46-50; chmn dept physics, Southwest Res Inst, 50-56, asst vpres, 55-56; dir res physics, Continental Can Co, Inc, 56-59, dir eng res, 61-67, dir corp res, 67-77, gen mgr, 77-85; dir res, Schlumberger Well Surv Corp, 59-61; partner, Conn Venture Mgt Corp, 85-87; PRIN, OMNIVENTURE GROUP INC, 88- *Concurrent Pos:* Vpres, C T Venture Group. *Honors & Awards:* Eli Whitney Award. *Mem:* Am Phys Soc; Sigma Xi; Am Soc Mech Engrs; Inst Elec & Electronics Engrs; Indust Res Inst. *Res:* Information systems; high energy rate metalworking; energy conversion devices; instrumentation and physical measurements; industrial inspection systems; welding methods; research management; packaging systems. *Mailing Add:* 181 Red Fox Rd Stamford CT 06903

ERLANGER, BERNARD FERDINAND, b New York, NY, July 13, 23; m 46; c 3. BIOCHEMISTRY, IMMUNOLOGY. *Educ:* City Col New York, BS, 43; NY Univ, MS, 49; Columbia Univ, PhD(biochem), 51. *Prof Exp:* Assoc biochem, 51-52, assoc microbiol, 52-55, from asst prof to assoc prof, 55-66, PROF MICROBIOL, COLUMBIA UNIV, 66- *Concurrent Pos:* Vis scientist, Instituto Superiore di Sanita, Univ Rome, 61-62; mem comt Fulbright-Hays Acts Awards, Nat Acad Sci, 66-72; Fulbright scholar, Univ Repub, Uruguay, 67; Guggenheim fel, Inst Biophys & Biochem, Univ Paris, 69-70; adv ed, Immunochemistry, 76-80; vis scientist, Inst Cell Biol, Shanghai, People's Repub of China, 78; Scholar, Am Cancer Soc, Inst Pasteur, Paris, 79. *Honors & Awards:* 600th Anniversary Medal, Copernican Med Acad, Jagiellonian Univ, Cracow, Poland, 79. *Mem:* AAAS; Am Chem Soc; Am Soc Biol Chemists; Am Soc Cell Biol; Am Soc Photobiol; Am Asn Immunol. *Res:* Protein chemistry; peptide synthesis and enzymology; immunochemistry of nucleic acids; receptor biochemistry. *Mailing Add:* Dept Microbiol Columbia Univ 701 W 168th St New York NY 10032

ERLBACH, ERICH, b Wuerzburg, Ger, Dec 17, 33; US citizen; m 57; c 3. PHYSICS. *Educ:* Columbia Univ, AB, 55, AM, 57, PhD(physics), 60. *Prof Exp:* Asst physics, Columbia Col, 55-57; asst, Watson Lab, Int Bus Mach, 57-60, physicist, Res Ctr, 60-62; from asst prof to assoc prof physics, 62-73, chmn dept, 74-80, PROF PHYSICS, CITY COL NEW YORK, 73-, DIR HONS PROG, 81- *Concurrent Pos:* Lectr, City Col New York, 57-59; NSF res partic, Univ Md, 68-69. *Mem:* Am Phys Soc; Am Asn Physics Teachers; Asn Orthodox Jewish Scientists. *Res:* Cryogenics, especially superconductivity; solid state physics, particularly semiconductors and metals. *Mailing Add:* Dept Physics 138 St & Covent Ave New York NY 10031

ERLENMEYER-KIMLING, L, US citizen. MEDICAL GENETICS. *Educ:* Columbia Univ, BS, 57, PhD(psychol & genetics), 61. *Prof Exp:* Asst behav genetics & lectr psychol, Col Physicians & Surgeons, Columbia Univ), 58-60, asst, 62-66, res assoc, 66-69, asst prof, dept psychiat, 69-74, assoc prof, dept psychiat & dept human genetics, 74-78; from res scientist to prin res scientist, 60-78, RES SCIENTIST VII, MED GENETICS, NY STATE PSYCHIAT INST, 78-, DIR, DIV DEVELOP BEHAV STUDIES, 78-; PROF PSYCHIAT & GENETICS, COL PHYSICIANS & SURGEONS, COLUMBIA UNIV, 78- *Concurrent Pos:* Genetics Soc Am travel grant, Int Cong Genetics, Neth, 63, USSR, 78; Am Psychol Asn travel grant, Int Cong Psychol, USSR, 66; NIMH res grants, 67-70, 71- & 90-93; res grant, Scottish Rite Comt Schizophrenia, 71-75, 84-86 & 89-91, W T Grant Found, 77-84, McArthur Found, 81; vis prof, Dept Psychol, New Sch Soc Res, 72-; mem peer rev grant appln, NIH Study Sect, 76-80, NIMH Study Sect, 81-85; mem task panel, Cong Comt Huntington's Dis, 76-77 & President's Comn Mental Health, 77-78; mem, bd dirs, Soc Study Social Biol, 69-84,. *Honors & Awards:* Dobzhansky Award, 85. *Mem:* Behav Genetics Asn; AAAS; Scientists Ctr Animal Welfare; Soc Life Hist Res Psychopath; Soc Study Soc Biol (secy, 72-75, pres, 75-78); fel Am Psychopath Asn; Am Soc Human Genetics; Sigma Xi; fel Am Psychol Asn; Am Psychol Soc. *Res:* Experimental behavior genetics; medical and psychiatric genetics; population-genetic and demographic aspects of schizophrenia and other mental disabilities. *Mailing Add:* Dept Med Genetics NY State Psychiat Inst New York NY 10032

ERLICH, DAVID C, b Brooklyn, NY; m 69; c 2. HIGH PRESSURE PHYSICS, SHOCK WAVE PHYSICS. *Educ:* Calif Inst Technol, BS, 68. *Prof Exp:* RES PHYSICIST HIGH PRESSURE PHYSICS, SRI INT, 68- *Res:* Shock-wave propagation in and fracture of condensed media; response of condensed media to high pressure transient loading. *Mailing Add:* SRI Int 333 Ravenswood Ave Menlo Park CA 94025

ERLICH, RONALD HARVEY, b Detroit, Mich, May 14, 45; m 74. ANALYTICAL CHEMISTRY. *Educ:* Wayne State Univ, BS, 67; Mich State Univ, PhD(analytical chem), 71. *Prof Exp:* Res fel chem pharmacol, Heart & Lung Inst, NIH, 71-72; sr scientist pharmaceut analysis, McNeil Labs, Subsid Johnson & Johnson, 73-77; mem staff, Bristol Labs, 77-89; HEAD ANALYTICAL CHEM DEPT, MEDA-VENTURES INC, 89- *Mem:* Am Chem Soc. *Res:* Development of analytical methods for the determination of pharmaceuticals; application of instrumental methods of analysis to the specific assay of drugs in dosage forms. *Mailing Add:* 3677 Tanglewood Ct Ann Arbor MI 48105-9563

ERLICHMAN, JACK, HORMONE ACTION. *Educ:* Albert Einstein Col, PhD(molecular biol), 74. *Prof Exp:* ASSOC PROF BIOCHEM & MED, ALBERT EINSTEIN COL, 77- *Mailing Add:* Dept Med & Biochem Albert Einstein Col 1300 Morris Park Ave Bronx NY 10461

ERLICHSON, HERMAN, b Brooklyn, NY, Mar 22, 31; c 4. PHYSICS. *Educ:* City Col New York, BS, 53; Harvard Univ, AM, 54; Columbia Univ, MA, 61, PhD(philos), 68; Rutgers Univ, MPh, 76, PhD(physics), 80. *Prof Exp:* Res & develop engr, Bell Aircraft Corp, 53; res develop & sales engr, Gen Elec Co, 54-56; res & develop engr, Kollsman Instrument Corp, 56-60; from asst prof to assoc prof, 60-70, PROF PHYSICS, COL STATEN ISLAND, 70- *Concurrent Pos:* NSF grant, 84-86. *Mem:* Am Asn Physics Teachers; Am Phys Soc; Brit Soc Hist Sci. *Res:* History of physics. *Mailing Add:* Dept Appl Sci Col Staten Island Staten Island NY 10301

ERLIJ, DAVID, b Mexico City, Mex, Mar 13, 38. MEMBRANE TRANSPORT. *Educ:* Univ Mexico, MD, 65; Nat Polytech Inst, Mex, PhD(physiol), 68. *Prof Exp:* PROF PHYSIOL, DOWNSTATE MED CTR, STATE UNIV NY, 76- *Mailing Add:* 219 Congress St Brooklyn NY 11203

ERMAN, DON COUTRE, b Richmond, Ind, Mar 7, 40; m 65. AQUATIC ECOLOGY, FISHERIES. *Educ:* DePauw Univ, AB, 62; Purdue Univ, MS, 65; Utah State Univ, PhD(fisheries, aquatic ecol), 69. *Prof Exp:* from asst prof to assoc prof, 69-81, vchmn, Dept Forestry & Conserv, 75-77 & 78-81, co-chmn, Dept Conserv & Resources Studies, 81, PROF FISHERIES & AQUATIC ECOL, UNIV CALIF, BERKELEY, 81- *Concurrent Pos:* Mem coord bd, Calif Water Resources Ctr, 76-82. *Mem:* Am Fisheries Soc; Ecol Soc Am; Am Inst Biol Sci; Am Soc Limnol & Oceanog; Sigma Xi. *Res:* Freshwater benthic invertebrate communities; stream ecology; secondary production; wetland ecology. *Mailing Add:* Dept Forestry & Conserv Univ Calif Berkeley CA 94720

ERMAN, JAMES EDWIN, b Lodi, Calif, Dec 16, 40; m 71. PHYSICAL BIOCHEMISTRY. *Educ:* Univ Calif, Berkeley, BS, 62; Mass Inst Technol, PhD(phys chem), 66. *Prof Exp:* Res chemist, Chevron Res Co, 66-68; fel, Johnson Res Found, Univ Pa, 68-70; from asst prof to assoc prof, 70-81, PROF CHEM, NORTHERN ILL UNIV, 81- *Mem:* AAAS; Am Soc Biochem & Molecular Biol; Sigma Xi; Am Chem Soc; Biophys Soc. *Res:* Enzyme catalyzed oxidation-reduction reactions; rapid reaction techniques; NMR studies of proteins. *Mailing Add:* Dept Chem Northern Ill Univ DeKalb IL 60115

ERMAN, LEE DANIEL, b Chicago, Ill, 1944. ARTIFICIAL INTELLIGENCE. *Educ:* Univ Mich, BS, 66; Stanford Univ, MS, 68, PhD(comput sci), 74. *Prof Exp:* Res assoc comput sci, Carnegie-Mellon Univ, 71-74, res comput scientist, 74-76, asst prof, 76-78; res scientist, Info Sci Inst, Univ Southern Calif, 78-82; PRIN SCIENTIST, CIMFLEX TEKNOWLEDGE CORP, 82- *Mem:* Asn Comput Mach; Am Asn Artificial Intel; Inst Elec & Electronics Engrs; Computer Soc. *Res:* Structures of and tools for building knowledge based artificial intelligence expert systems; control architectures for such systems. *Mailing Add:* Cimflex Teknowledge Corp PO Box 10119 Palo Alto CA 94303-3308

ERMAN, WILLIAM F, b Butler, Mo, May 22, 31; m 60; c 4. DRUG DESIGN, TERPENE CHEMISTRY. *Educ:* Univ Notre Dame, BS, 53; Mass Inst Technol, PhD(org chem), 57. *Prof Exp:* Res chemist & group leader natural prod chem, 57-66, sect head natural prod chem, 66-71, sect head med chem, 71-77, ASSOC DIR, CORP RES DIV, PROCTER & GAMBLE CO, 77- *Concurrent Pos:* Lectr, Xavier Univ, Ohio, 63. *Honors & Awards:* Cincinnati Chemist Award, Am Chem Soc, 73. *Mem:* AAAS; Am Chem Soc; NY Acad Sci. *Res:* Natural products chemistry, particularly isolation, structure determination, reactions and synthesis of mono- and sesquiterpenes; organic photochemical transformations; medicinal chemistry; pesticide chemistry; perfume chemistry. *Mailing Add:* Res Dept Miami Valley Labs Procter & Gamble Co Cincinnati OH 45239-8707

ERMENC, EUGENE D, b Milwaukee, Wis, Mar 29, 19; m 43; c 1. CHEMICAL ENGINEERING. *Educ:* Univ Wis, BS, 40; Ga Inst Technol, MS, 42. *Prof Exp:* Jr engr, Monsanto Chem Co, 42-46, sr engr, 46-50, plant engr, 50; proj engr, plaskon div, Libbey-Owens-Ford, 50-51; dir eng, Wis Alumni Res Found, 51-55; staff engr, Food Mach & Chem Corp, 55-56; coordr pilot opers, 56-58, mgr process eng, 59-61; dir res, Philip Cary Mfg Co, 61-69; dir, Air Pollution Control, City of Cincinnati, 69-80; prof, Raymond Walters Br, Univ Cincinnati, 76-81; asst county adminr, Hamilton County, 80-81; CONSULT, 81- *Mem:* Fel Am Inst Chem Engrs; Air Pollution Control Asn; dipl Am Acad Environ Engrs. *Res:* Engineering economics; process evaluation; oxides of nitrogen; simplification of complex engineering calculations. *Mailing Add:* 543 Princewood Dr DeLand FL 32724

ERMENC, JOSEPH JOHN, b Milwaukee, Wis, Nov 11, 12; m 51; c 3. HISTORY OF TECHNOLOGY. *Educ:* Univ Wis, BS, 34; Univ Mich, MS, 40. *Hon Degrees:* MA, Dartmouth Col, 45. *Prof Exp:* Cadet gas engr, Milwaukee Gas Light Co, 34-36; engr-draftsman, Badger Meter Mfg Co, Milwaukee, 36-37; instr practical mech, Purdue Univ, 37-38; instr mech eng, Rensselaer Polytech Inst, 38-42; asst prof, 42-45, prof, 45-78, EMER PROF MECH ENG, THAYER SCH ENG, DARTMOUTH COL, 78- *Concurrent Pos:* Mem adv comt to Selective Serv Dir, NH, 55-62; hon res assoc, Univ Col, Univ London, 62-63; Nat Acad Sci panel renewable energy resources, 75-76; chmn, Natural Hist & Heritage Comn, Am Soc Mech Engrs, 78- *Honors & Awards:* NSF Sci Fac Fel, 62-63. *Mem:* Am Soc Mech Engrs. *Res:* History and philosophy of technological innovation. *Mailing Add:* 77 E Wheelock St Hanover NH 03755

ERMENTROUT, GEORGE BARD, b 1954; m; c 1. PHYSIOLOGICAL OSCILLATIONS, INTEGRO-DIFFERENTIAL EQUATIONS. *Educ:* Johns Hopkins Univ, BA & MA, 75; Univ Chicago, PhD(biophys), 79. *Prof Exp:* Fel, NIH, 79-82; from asst prof to assoc prof, 82-89, PROF MATH, UNIV PITTSBURG, 89- *Concurrent Pos:* Sloan fel, 82. *Mem:* Soc Indust & Appl Math; Soc Math Biol. *Res:* Nonlinear differential equation; models of physiology; integral equations; animal locomotion; nonlinear oscillations; graphics software. *Mailing Add:* Dept Math Univ Pittsburgh Pittsburgh PA 15260

ERMLER, WALTER CARL, b Chicago, Ill, July 22, 46; m 67; c 2. THEORETICAL CHEMISTRY. *Educ:* Northern Ill Univ, BS, 69; Ohio State Univ, MSc, 70, PhD(phys chem), 72. *Hon Degrees:* ME, Stevens Inst Technol, 89. *Prof Exp:* Fel theoret chem, Ohio State Univ, 72-73; res assoc theoret chem, Univ Chicago, 73-76 & Univ Calif, Berkeley, 76-78; assoc prof, 78-83, PROF CHEM, STEVENS INST TECHNOL, 84- *Concurrent Pos:* Vis scientist, Argonne Nat Lab, 74-76, 82-; joint study partic, IBM Corp, 74-76; vis prof chem, Ohio State Univ, 90; visitor, Ohio Supercomputer Ctr, 90. *Mem:* Am Phys Soc; AAAS; Am Chem Soc; Sigma Xi. *Res:* Application of ab initio quantum mechanical methods for studying potential energy surfaces of ground and excited electronic states of molecules, properties of large molecules and solids; intra-molecular nuclear motion; relativistic effects on the chemical bond. *Mailing Add:* Dept Chem & Chem Eng Stevens Inst Technol Hoboken NJ 07030

ERMUTLU, ILHAN M, b Istanbul, Turkey, June 24, 27; US citizen; m 56; c 2. PSYCHIATRY. *Educ:* Univ Ankara, MD, 52. *Prof Exp:* Intern, Knickerbocker Hosp, NY, 54-55; resident psychiat, Bellevue Hosp, 55-56; resident, Hillside Hosp, 56-58; resident neurol, Goldwater Mem Hosp, 58-59; chief of serv psychiat, Eastern State Hosp, Va, 59-61; dir, Tidewater Ment Health Clin, 61-63; pvt pract, Richmond, Va, 63-64; asst to dir div ment health, Ga Dept Pub Health, 64-65, dir community serv br, 65-70, supt, Ga Regional Hosp, Savannah, 70-73; chief outpatient serv, William S Hall Psychiat Inst, 73-78; assoc prof psychiat, Sch Med, Univ SC, 77-78; dir, DeKalb County Ment Health & Ment Retardation Div, 78-80; dir forensic serv, Ga Ment Health Inst, 80-81; dir, Div Mental Health & Retardation, Ga Dept Human Resources, 81-83; consult psychiat, Ment Hosp, Taif, Saudi Arabia, 84-85; consult psychiat, 86-90, PSYCHIATRIST, OBRA DETERMINATION UNIT, N DEKALB MENT HEALTH CTR, ATLANTA, GA, 90- *Concurrent Pos:* Assoc in psychiat, Sch Med, Emory Univ, 64-73; clin assoc prof psychiat, Sch Med, Emory Univ, 78-84. *Mem:* Fel Am Psychiat Asn. *Res:* Mental health. *Mailing Add:* 7011 Somerset Circle Alpharetta GA 30201-3847

ERN, ERNEST HENRY, b Irvington, NJ, Apr 27, 33; m 56; c 3. PETROLOGY, GEOLOGY. *Educ:* Bates Col, BS, 55; Lehigh Univ, MS, 57, PhD(geol), 59. *Prof Exp:* Asst geol, Lehigh Univ, 55-59; asst prof, Marshall Univ, 59-62; asst prof, 62-68, asst dean col arts & sci, 66-67, dean admis, 67-73, PROF ENVIRON SCI, UNIV VA, 75-, VPRES STUDENT AFFAIRS, 73- *Concurrent Pos:* Consult, Vt Geol Surv, 56-59 & Va Div Mineral Resources, 62-68. *Mem:* AAAS; fel Geol Soc Am; Am Inst Prof Geol Scientists. *Res:* Metamorphic petrology; geotechnics; metamorphism and structural evolution of the Appalachian Piedmont; foundation investigation studies for engineering structures. *Mailing Add:* Rotunda Univ Va Charlottesville VA 22903

ERNEST, J TERRY, b Sycamore, Ill, June 26, 35; m 65; c 2. MEDICAL RETINA, OCULAR PHYSIOLOGY. *Educ:* Northwestern Univ, BA, 57; Univ Chicago, MD, 61, PhD(visual sci), 67. *Prof Exp:* Prof ophthal, Univ Wis, 77-79; prof & chmn ophthal, Ind Univ, 80-81; prof ophthal, Univ Ill, 81-85; PROF & CHMN OPHTHAL, UNIV CHICAGO, 85- *Concurrent Pos:* Mem visual sci A study sect, NIH, 75-78, chmn, 78-79; chmn, visual disorders study sect, NIH, 79-80; res prof, Res to Prevent Blindness, Inc, 81-84; mem, Vision Res Prog Comt, 82-84; ed, Investigative Ophthal & Visual Sci, 88- *Honors & Awards:* Res Career Develop Award, NIH, 72; Honor Award, Am Acad Ophthal, 82. *Mem:* Am Ophthal Soc; Asn Res Vision & Ophthal; Am Acad Ophthal; AAAS. *Res:* Study of the ocular circulation with special emphasis on glaucoma and diabetic retinopathy using various methods of in vivo blood flow measurements. *Mailing Add:* Visual Sci Ctr Univ Chicago 939 E 57th St Chicago IL 60637-1454

ERNEST, JOHN ARTHUR, b New York, NY, Dec 12, 35. MATHEMATICS. *Educ:* Drew Univ, BA, 57; Univ Ill, MS, 58, PhD(math), 60. *Prof Exp:* Lectr math, Franklin & Marshall Col, 60; mem, Inst Advan Study, 60-62; asst prof, Univ Rochester, 62-66; vis assoc prof, Tulane Univ, 66-67; assoc prof, 67-70, chmn dept, 80-83, PROF MATH, UNIV CALIF, SANTA BARBARA, 70- *Concurrent Pos:* NSF fel, 60-62; vis asst prof, Univ Calif, Berkeley, 65-66; sci fel, Ctr Int Security & Arms Control, Stanford Univ, 84-85; chair global peace and security prog, Univ Calif, Santa Barbara, 86- *Mem:* Am Math Soc; Int Peace Res Asn. *Res:* Functional analysis; infinite dimensional representations of topological groups; operator theory; arms control; environmental effects of nuclear war; peace research. *Mailing Add:* Dept Math Univ Calif Santa Barbara CA 93106

ERNEST, MICHAEL JEFFREY, b New York, NY, Mar 10, 48. MOLECULAR BIOLOGY, ENDOCRINOLOGY. *Educ:* Cornell Univ, BS, 68; Purdue Univ, PhD(biochem), 74. *Prof Exp:* Fel, Inst Cancer Res, Columbia Univ, 74-77; ASST PROF BIOL, YALE UNIV, 77- *Mem:* NY Acad Sci; Sigma Xi; AAAS. *Res:* Gene expression in normal and neoplastic cells. *Mailing Add:* 210 Sleeper Ave Mountain View CA 94040

ERNSBERGER, FRED MARTIN, b Ada, Ohio, Sept 20, 19; c 4. GLASS SCIENCE, NUCLEAR WASTE DISPOSAL. *Educ:* Ohio Northern Univ, AB, 41; Ohio State Univ, PhD(phys chem), 46. *Prof Exp:* Res chemist, US Naval Ord Test Sta, 47-54, Southwest Res Inst, 54-56 & Mellon Inst of Indust Res, 57; res chemist, Glass Res Ctr, PPG Industs, Inc, 58-82; RETIRED. *Concurrent Pos:* Adj prof, dept mat sci & eng, Univ Fla, Gainesville, 82- *Honors & Awards:* Frank Forrest Award, Am Ceramic Soc, 64, Toledo Glass & Ceramic Award, 70, & G W Morey Award, 74; IR-100 Award, 81; Scholes Award, 89. *Mem:* Am Chem Soc; fel Am Ceramic Soc; Soc Glass Technol. *Res:* Surface chemistry, surface structure, mechanical properties of glass and glass-ceramics; chemistry of float glass. *Mailing Add:* 1325 NW Tenth Ave Gainesville FL 32605

ERNST, CARL HENRY, b Lancaster, Pa, Sept 28, 38; m 69; c 2. HERPETOLOGY, MAMMALOGY. *Educ:* Millersville Univ, BS, 60; Westchester Univ, MEd, 63; Univ Ky, PhD(biol), 69. *Prof Exp:* Asst prof biol, Elizabethtown Col, 66-67; cur vert, Univ Ky, 67-69; asst prof biol, Southwest Minn Univ, 69-72; assoc prof, 72-78, PROF BIOL, GEORGE MASON UNIV, 78- *Concurrent Pos:* Sigma Xi res grant-in-aid, 69 & 73; Minn State Col grants, 70, 71; res assoc, US Nat Mus Natural Hist, 74-; Am Philos Soc grant, 76, 81 & 89; Ala Coal Asn grant, 83. *Honors & Awards:* Wildlife Publication Award, Wildlife Soc, 73. *Mem:* Am Soc Ichthyologists & Herpetologists; Soc Study Amphibians & Reptiles; Am Soc Mammal; Soc Trop Biol; Sigma Xi; Herpetologist's League. *Res:* Ethology, ecology and taxonomy of turtles and snakes. *Mailing Add:* Dept Biol George Mason Univ Fairfax VA 22030-4444

ERNST, DAVID JOHN, b St Paul, Minn, May 29, 43; m 67; c 2. NUCLEAR PHYSICS. *Educ:* Mass Inst Technol, SB, 65, PhD(physics), 70. *Prof Exp:* From asst prof to prof physics, Ctr Advan Studies, Nat Polytech Inst, 70-72; res assoc, Case Western Reserve Univ, 72-74, res assoc & instr, 74-75; from asst prof to assoc prof, 75-85, PROF PHYSICS, TEXAS A&M UNIV, 85-, ASSOC DIR, CTR THEORET PHYSICS, 88- & INTERIM DIR, INT INST THEORET PHYSICS, 89- *Concurrent Pos:* Vis staff scientist, Los Alamos Nat Lab, 72-85, prog adv comt, Los Alamos Meson Physics Facil, 85-89; prin investr, NSF grant, 76-85 & 87-; vis asst prof phys, Univ Washington, 79-80; consult, Oak Ridge Nat Lab, 84- & Los Alamos Nat Lab, 85-; prog comt, Nuclear Physics Div, Am Physical Soc, 84-85, chmn elect, 90, chmn, 91; bd dirs, Los Alamos Meson Physics Facil Users Group, Inc, Los Alamos Nat Lab; prof, Univ Frankfurt, 89- *Mem:* Am Phys Soc; Sigma Xi. *Res:* Theoretical nuclear physics; reactions of nucleons with finite nuclei, particularly at intermediate energies; pion-nucleon interaction; pion and kaon induced reactions and elastic scattering with finite nuclei; heavy ion reactions. *Mailing Add:* Dept Physics Tex A&M Univ College Station TX 77843

ERNST, DAVID N, IMMUNOLOGY. *Prof Exp:* SR RES ASSOC, DEPT IMMUNOL, IMM 9 RES INST, SCRIPPS CLIN, 87- *Mailing Add:* Dept Immunol Imm9 Res Inst Scripps Clin 10666 N Torrey Pines Rd La Jolla CA 92037

ERNST, EDWARD W, b Great Falls, Mont, Aug 28, 24; m 50, 75; c 5. ELECTRICAL ENGINEERING. *Educ:* Univ Ill, BS, 49, MS, 50, PhD(elec eng), 55. *Prof Exp:* Res assoc elec eng, Univ Ill, 46-55; res engr, Gen Elec Co, NY, 55 & Stewart-Warner Electronics Corp, 55-58; assoc prof elec eng, 58-68, assoc head dept, 71-85, assoc dean eng, 85-87, PROF ELEC ENG, UNIV ILL, URBANA, 68- *Concurrent Pos:* Pres, Nat Electronics Conf, 64; bd dirs, Accreditation Bd Eng & Technol, Inst Elec & Electronics Engrs; mem, Eng Accreditation Comn, 79-87, vchmn, 81-85, chmn, 85-86; prog dir, undergrad eng educ, NSF, Univ Ill, Urbana, 87-89. *Mem:* Fel AAAS; fel Inst Elec & Electronics Engrs; fel Am Soc Eng Educ; Sigma Xi. *Res:* Radiolocation; electronic systems; digital systems; experimentation. *Mailing Add:* Swearingen Eng Univ SC Columbia SC 29208

ERNST, GEORGE W, b St Marys, Pa, May 25, 39; c 3. COMPUTER SCIENCE, ELECTRICAL ENGINEERING. *Educ:* Carnegie Inst Technol, BS, 61, MS, 62, PhD(elec eng), 66. *Prof Exp:* Asst prof eng, 66-70, assoc prof comput & info sci, 70-80, ASSOC PROF ARTIFICIAL INTEL, CASE WESTERN RESERVE UNIV, 80- *Mem:* Asn Comput Mach. *Res:* Artificial intelligence. *Mailing Add:* 3164 Warrington Rd Cleveland OH 44120-2429

ERNST, MARTIN L, b New York, NY, Mar 28, 20; m 53; c 2. PHYSICS. *Educ:* Mass Inst Technol, BS, 41. *Prof Exp:* Physicist, Naval Ord Lab, 41 & Bur Ord, 42; opers analyst, US Air Force, 43-46, electronics engr, Cambridge Res Ctr, 46-48; opers analyst, Opers Eval Group, Off Chief Naval Opers, 48-53, assoc dir, 53-59; sr staff mem & vpres opers res sect, Arthur D Little, Inc, 59-80, vpres mgt sci div, 59-85. *Mem:* Opers Res Soc Am (secy, 55-58, vpres, 59-60, pres, 60-61). *Mailing Add:* 16 Williams Terr Swampscott MA 01907

ERNST, RALPH AMBROSE, b Saline, Mich, July 5, 38; m 67, 88; c 1. POULTRY SCIENCE, PHYSIOLOGY. *Educ:* Mich State Univ, BS, 59, MS, 63, PhD(avian physiol), 66. *Prof Exp:* Teacher, Milan, Mich, 59-60 & Carson City, Mich, 60-61; EXTEN POULTRY SPECIALIST, UNIV CALIF, DAVIS, 66- *Honors & Awards:* Extension Award, Poultry Sci Asn, 78. *Mem:* Poultry Sci Asn; World Poultry Sci Asn; Sigma Xi. *Res:* Poultry physiology with emphasis on production. *Mailing Add:* Dept Avian Sci Univ Calif Davis CA 95616

ERNST, RICHARD DALE, b Long Beach, Calif, Oct 23, 51; m 73; c 3. STRUCTURAL CHEMISTRY. *Educ:* Univ Calif, Berkeley, BS, 73; Northwestern Univ, PhD(chem), 77. *Prof Exp:* Asst prof, 77-84, assoc prof, 84-87, PROF CHEM, UNIV UTAH, 87- *Concurrent Pos:* Consult, Phillips Petroleum Co, 79- *Mem:* Am Chem Soc; AAAS. *Res:* Synthesis, characterization, chemical and structural studies of inorganic and organometallic compounds, particularly those containing allyl or pentadienyl ligands. *Mailing Add:* Dept Chem Univ Utah Salt Lake City UT 84112

ERNST, RICHARD EDWARD, b Elgin, Ill, July 2, 42; m 67; c 3. INDUSTRIAL CHEMISTRY. *Educ:* Grinnell Col, BA, 64; Univ Wis-Madison, PhD(phys chem), 68. *Prof Exp:* RES CHEMIST, E I DU PONT DE NEMOURS & CO, INC, 68-, RES FEL, 80- *Mem:* Am Chem Soc; Sigma Xi. *Res:* Stereochemistry of transition metal complexes by nuclear magnetic resonance; radiation chemistry of polymers; pollution control; industrial process development. *Mailing Add:* Unionville-Lenape Rd Kennett Square PA 19348

ERNST, RICHARD R, NUCLEAR MAGNETIC RESONANCE SPECTROSCOPY. *Educ:* Eidgenossische Techniche Hochscule, PhD(chem),62. *ProfExp:* Res scientist, Varian Assocs, 63-68; Lecturer, 68-70, Asst Prof 70-72, Assoc Prof, 72-76, PROF, EIDGENOSSISCHE TECHNISCHE HOCHSCULE, ZURICH, 76- *Honors & Awards:* Nobel Prize Chem, 91; Loursa Gross Horwitz Prize, Columbia Univ, 91; Wolf Prize Chem, 91; Benoist Prize, 86; Ruzicka Prize, 69. *Mem:* Nat Acad Sci. *Mailing Add:* Dept Chem Eidgenossische Technische Hochscule Ramistrasse 101 ETH-Zentrum 8092 Zurich Switzerland

ERNST, ROBERTA DOROTHEA, b St Louis, Mo, Oct 12, 44. IMMUNOASSAYS. *Educ:* Univ Ill, Urbana, BS, 66. *Prof Exp:* Res chemist, 72-74, prod specialist, 74-77, supvr mfg & develop, 77-81, res scientist, 81-84, SR CHEMIST RES & DEVELOP, SYVA CO, 84- *Mem:* Am Asn Clin Chemists. *Mailing Add:* 902 Rockefeller Apt 15B Sunnyvale CA 94087-2132

ERNST, STEPHEN ARNOLD, b St Louis, Mo, Mar 31, 40; m 71, 81; c 2. CELL BIOLOGY & PHYSIOLOGY. *Educ:* Brown Univ, AB, 62, PhD(cell biol), 68; Syracuse Univ, MS, 64. *Prof Exp:* Fel cell biol, Brown Univ, 68; res assoc cell biol, Rice Univ, 68-71; from asst prof to assoc prof anat, Sch Med, Temple Univ, 71-78; assoc prof, 78-81, PROF ANAT & CELL BIOL, MED SCH, UNIV MICH, ANN ARBOR, 81- *Concurrent Pos:* Nat Found Cystic Fibrosis res fel, 68-71; Moody Found res grant, 70-71; Am Heart Asn grant-in-aid, 74-75; NSF res grant, 74-76; NIH res career develop award, 75-80, NIH res grants, 77- *Mem:* Histochem Soc; Am Physiol Soc; Am Asn Anat; Am Soc Cell Biol. *Res:* Physiology; cytochemistry of electrolyte transporting tissues; histochemistry of the cell surface; electron microscopy. *Mailing Add:* Dept Anat & Cell Biol Med Sci II Sch Med Univ Mich Ann Arbor MI 48109

ERNST, SUSAN GWENN, b New York, NY, Sept 21, 46. MOLECULAR BIOLOGY. *Educ:* La State Univ, BS, 68; Univ Mass, PhD(zool), 75. *Prof Exp:* Instr, Div Radiation Biol, Case Western Univ, 74-75; res fel, Kerckhoff Marine Lab, Calif Inst Technol, 75-78, res fel biol, 78-79; asst prof, 79-85, ASSOC PROF BIOL, TUFTS UNIV, 85- *Concurrent Pos:* Assoc prof Biotechnol Eng Ctr, Tufts Univ; assoc Inst Teaching Math & Sci Am Adolescent. *Mem:* AAAS; Soc Develop Biol; Am Soc Cell Biol; Sigma Xi; Int Soc Develop Biologists. *Res:* Transcriptional and translational expression and regulation of gene activity in developing systems; cytoplasmic localization of maternal RNAs in developing embryos. *Mailing Add:* Dept Biol Tufts Univ Medford MA 02155

ERNST, WALLACE GARY, b St Louis, Mo, Dec 14, 31; m 56; c 4. PETROLOGY, GEOCHEMISTRY. *Educ:* Carleton Col, BA, 53; Univ Minn, MS, 55; Johns Hopkins Univ, PhD(geol), 59. *Prof Exp:* Geologist, Petrol Br, US Geol Surv, 55-56; fel, Geophys Lab, Johns Hopkins Univ, 58-60; from asst prof to assoc prof, Stanford Univ, 60-68, chmn dept geol, 70-74, chmn dept earth & space sci, 78-82, prof geol & geophys, Univ Calif, Los Angeles, 68-89, dir, Inst Geophys & Planetary Physics, 86-89, PROF GEOL & GEOPHYSICS & DEAN, SCH EARTH SCIS, STANFORD UNIV, 89- *Concurrent Pos:* Fulbright fel, Univ Tokyo, 63; NSF sr fel, Univ Basel, 70-71; Guggenheim fel, Swiss Fed Inst Technol, 75-76; William Evans vis prof, Univ Otago, 82-83; chmn bd earth sci, Nat Res Coun, 84-87. *Honors & Awards:* Mineral Soc Am Award, 69. *Mem:* Nat Acad Sci; AAAS; fel Mineral Soc Am (pres, 80-81); fel Am Acad Arts Sci; fel Am Geophys Union; fel Geol Soc Am (pres, 84-85). *Res:* Igneous and metamorphic petrology; application of theoretical and experimental phase equilibria to geologic problems; plate tectonics. *Mailing Add:* Sch Earth Scis Stanford Univ Stanford CA 94305-2210

ERNST, WALTER, mechanical engineering; deceased, see previous edition for last biography

ERNST, WILLIAM ROBERT, b Hanover, Pa, June 3, 43; m 66; c 3. CHEMICAL ENGINEERING. *Educ:* Pa State Univ, BS, 65, MS, 71; Univ Del, PhD(chem eng), 74. *Prof Exp:* Proj engr, Allied Chem Corp, 65-69; from asst prof to assoc prof, chmn eng dept, 80-90, 73-90, PROF CHEM ENG, GA INST TECHNOL, 90- *Mem:* Sigma Xi; Am Inst Chem Engrs; Am Chem Soc. *Res:* Catalysis; chemical reactor design; science writing. *Mailing Add:* Sch Chem Eng Ga Inst Technol 225 North Ave NW Atlanta GA 30332

ERNST, WOLFGANG E, b Minden, Fed Repub Ger, May 31, 51. MOLECULAR BEAMS, NONLINEAR LASER SPECTROSCOPY. *Educ:* Univ Hannover, Fed Repub Ger, dipl physicist, 75, Dr rer nat, 77; Free Univ Berlin, Fed Repub Ger, Habilitation physics, 83. *Prof Exp:* Res assoc quantum electronics, Elec Eng Dept, Rice Univ, 78-79; asst prof physics, Free Univ Berlin, Fed Repub Ger, 79-83, privat-dozent, 83-88, prof, 88-89; PROF PHYSICS, PA STATE UNIV, 90- *Concurrent Pos:* Vis scholar, Dept Chem, Stanford Univ, 86; vis prof, Max Planck Inst Fluid Dynamics, Göttingen, Fed Repub Ger, 91. *Honors & Awards:* Heisenberg Award, Ger Res Asn, 85; Physics Prize, Ger Phys Soc, 87. *Mem:* Ger Phys Soc; Am Phys Soc. *Res:* High resolution spectroscopy of free radicals and small clusters; generation and application of coherent vacuum ultraviolet and extreme ultraviolet radiation as well as laser-microwave double resonance methods. *Mailing Add:* Dept Physics Pa State Univ 104 Davey Lab University Park PA 16802

ERNST-FONBERG, MARYLOU, b Harrisburg, Pa, Jan 18, 37; m 69; c 1. BIOCHEMISTRY. *Educ:* Susquehanna Univ, BA, 58; Temple Univ, MD, 62; Yale Univ, PhD(biochem), 67. *Prof Exp:* Vis scientist, Dept Chem Immunol, Weizmann Inst Sci, 66-67; res fel chem, Harvard Univ, 68-69; from asst prof to assoc prof biol, Yale Univ, 69-78; assoc prof, 78-82, PROF BIOCHEM, COL MED, EAST TENN STATE UNIV, 82-, CHMN, DEPT BIOCHEM, 87- *Res:* Biochemical characterization of the enzymes of lipid biosynthesis and investigation of alterations in this enzymology in conjunction with the development of organelles. *Mailing Add:* Dept Biochem Col Med Box 19930A E Tenn State Univ Johnson City TN 37601

EROR, NICHOLAS GEORGE, JR, b Bingham Canyon, Utah, Apr 9, 37; m 59; c 7. SOLID STATE CHEMISTRY, MATERIALS SCIENCE. *Educ:* Yale Univ, BS, 59, MS, 62; Northwestern Univ, PhD(mat sci), 65. *Prof Exp:* Sr scientist solid state chem, Sprague Elec Co, 64-70; assoc prof, 70-83, PROF MAT SCI & ENG, ORE GRAD CTR, 83- *Concurrent Pos:* Vis prof math, Williams Col, 66, vis prof chem, 68-70. *Mem:* Am Ceramic Soc; Nat Inst Ceramic Engrs; Sigma Xi; Electrochem Soc; AAAS. *Res:* Point defects in pure and doped nonmetallic compounds; nonstoichiometric compounds; oxidation-reduction kinetics; electrical properties; Raman spectroscopy; oxide catalysts; synthesis of thermodynamically defined multicomponent compounds; dielectrics; electroluminescent materials; compound semiconductors; photolysis of water (solid state chemistry). *Mailing Add:* Dept Mat Sci-Eng Univ Pittsburgh 848 Benedum Hall Pittsburgh PA 15261

EROSCHENKO, VICTOR PAUL, b May 15, 38; US citizen; m 64; c 4. HUMAN ANATOMY. *Educ:* Univ Calif, Davis, BA, 61, MS, 70, PhD(anat), 73. *Prof Exp:* PROF ZOOL, DEPT BIOL SCI, UNIV IDAHO, 73- *Mem:* Soc Study Reproduction; Soc Toxicol; AAAS; Am Asn Anatomists. *Res:* Morphological studies on reproductive organs in birds and mammals as altered or effected by environmental pollutants especially insecticide Kepone. *Mailing Add:* Dept Biol Univ Idaho Moscow ID 83843

ERPENBECK, JEROME JOHN, b Ft Thomas, Ky, Aug 22, 33; m 56; c 8. THEORETICAL CHEMISTRY. *Educ:* Villa Madonna Col, BS, 53; Univ Louisville, MS, 55; Univ Ill, PhD(chem), 57. *Prof Exp:* Asst, Sterling Chem Lab, Yale Univ, 57-59; STAFF MEM, LOS ALAMOS SCI LAB, 59- *Res:* Transport and detonation theories; reactive hydrodynamics; statistical mechanics of fluids. *Mailing Add:* Three Kiowa Lane White Rock Los Alamos NM 87544

ERPINO, MICHAEL JAMES, b Schenectady, NY, May 30, 39; m 61; c 3. ENDOCRINOLOGY, HISTOLOGY. *Educ:* Pa State Univ, BS, 62; Univ Wyo, MS, 64, PhD(physiol), 67. *Prof Exp:* NIH fel endocrinol, Cornell Univ, 67-68; from asst prof to assoc prof zool, 68-77, PROF BIOL SCI, CALIF STATE UNIV, CHICO, 77- *Mem:* AAAS; Am Soc Zool; Am Ornith Union; fel Int Soc Res Aggression; Sigma Xi. *Res:* Steroid hormones and behavior; histology of vertebrate endocrines. *Mailing Add:* Dept Biol Sci Calif State Univ Chico CA 95929

ERREDE, BEVERLY JEAN, b New Britain, Conn, Aug 4, 49. YEAST GENETICS, GENE EXPRESSION. *Educ:* Conn Col, BA, 71; Univ Calif, San Diego, PhD(chem), 76. *Prof Exp:* Fel, Dept Chem, Univ Calif, San Diego, 76-77 & Univ Rochester, 77-81; ASST PROF CHEM, UNIV NC, CHAPEL HILL, 81- *Mem:* Am Chem Soc; Am Soc Microbiol; NY Acad Sci. *Res:* Identify and define molecular mechanisms utilized by eukaryotic organisms for regulation gene expression; involving genetic and biochemical characterization of defined regulatory mutations in Saccharomyces cerevisiae. *Mailing Add:* Dept Chem Univ NC Chapel Hill NC 27514

ERREDE, LOUIS A, b New Britain, Conn, May 26, 23; m 46; c 4. POLYMER SWELLING & DRYING, MEMBRANE DIFFUSION. *Educ:* Univ Mich, BS, 47; Univ Minn, PhD(org & phys chem), 51. *Prof Exp:* Res scientist, M W Kellogg Co, 51-57; instr physics & thermodyn, Newark Col Eng, 54-56; res scientist, 3M Co, St Paul, Minn, 57-63, mgr physics chem sect, 63-68, dir explor res, Harlow, Eng, 68-73, CORP SCIENTIST, 3M CO, ST PAUL MINN, 73- *Mem:* Am Chem Soc. *Res:* Fast flow pyrolysis; bond dissociation energies; p-xylylene and heterocyclic chemistry; effect of surface chemistry on physics of water flow through microporous membranes; polymer swelling and polymer drying; membrane chemistry; sorption in polymers. *Mailing Add:* 3M Co 3M Corp Res Labs 3M Ctr Bldg 201-2N-22 St Paul MN 55133

ERREDE, STEVEN MICHAEL, b Newark, NJ, Dec 24, 52; m 79; c 1. ELEMENTARY PARTICLE PHYSICS. *Educ:* Univ Minn, BSc, 75; Ohio State Univ, MSc, 78, PhD(physics), 81. *Prof Exp:* Grad teaching asst physics, Ohio State Univ, 75-78, grad res asst, 78-81; scholar physics, Univ Mich, 81-84; asst prof, 84-88, ASSOC PROF, UNIV ILL, 88- *Concurrent Pos:* Sloan res fel, 85-89. *Honors & Awards:* Beckman Res Award, 85; Bruno Rossi Prize, High Energy Astrophys Div, Am Astron Soc, 89. *Mem:* Sigma Xi; Am Phys Soc; AAAS. *Res:* Lifetimes of charmed particles (produced in neutrino interactions); IBM nucleon decay experiment, high resolution particle spectrometry; high energy physics detector hardware and data analysis; 1.8 TeV proton-antiproton collider physics; axion search experiments; detection of supernova neutrinos; SSC detector design and R and D, 40 TEV proton-antiproton collider physics. *Mailing Add:* Loomis Lab Physics Univ Ill 110 W Green St Urbana IL 61801

ERRERA, SAMUEL J(OSEPH), b Hammonton, NJ, Jan 7, 26; m 49; c 2. STRUCTURAL ENGINEERING. *Educ:* Rutgers Univ, BS, 49; Univ Ill, MS, 51; Cornell Univ, PhD(struct), 65. *Prof Exp:* Engr of tests, Fritz Eng Lab, Lehigh Univ, 51-62, from instr to assoc prof, 51-62; mgr struct lab, Cornell Univ, 62-65, mgr struct res, 65-70, assoc prof civil eng, 68-70; sr engr, 70-76, CONSULT ENGR, BETHLEHEM STEEL CORP, 76- *Concurrent Pos:* Consult, Allegheny-Ludlum Steel Corp & R C Mahon Co, 66-70; methods engr, adv composites group, Grumman Aircraft Co, 68-69. *Mem:* Am Soc Civil Engrs. *Res:* Research management; light gauge steel structures; diaphragm bracing; structural materials; composite design; advanced structural composites. *Mailing Add:* 1730 Maple St Bethlehem PA 18017

ERRETT, DARYL DALE, b Gridley, Kans, Dec 2, 22; m 53; c 1. OPTICS. *Educ:* Kans State Teachers Col, BA, 44; Purdue Univ, MS, 48, PhD(physics), 51. *Prof Exp:* Teaching asst, Kans State Teachers Col, 41-44; res asst, Purdue Univ, 46-51; res specialist, Electro Mech Dept, NAm Rockwell Corp, 51-54; res physicist, Santa Barbara Res Ctr, Goleta, 54-67, sr staff physicist, 67-80; CONSULT PHYSICIST, 80- *Concurrent Pos:* Asst, Purdue Res Found, 46-51. *Res:* Electrical discharge through gases; linear electron accelerators; inertial guidance systems; infrared detection systems; infrared optical design; computer programs for optical design and analysis; evaporated thermopile detectors; solid state laser communication systems; infrared proximity detectors; radiometric calibration instrumentation. *Mailing Add:* 363 Moreton Bay Lane No 2 Goleta CA 93117-0363

ERSHOFF, BENJAMIN H, nutrition; deceased, see previous edition for last biography

ERSKINE, ANTHONY J, b Whinfield, Eng, June 25, 31; Can citizen; m 55; c 3. WILDLIFE BIOLOGY. *Educ:* Acadia Univ, BSc, 52; Queen's Univ, Ont, MA, 55, PhD(org chem), 57; Univ BC, MA, 60. *Prof Exp:* Tech officer animal chem, Sci Serv, Can Dept Agr, 52-53; Nat Res Coun Can fel, Atlantic Regional Lab, 56-57; biologist, Dept Environ, 60-67, sci mgr, Can Wildlife Serv, 77-91; RETIRED. *Concurrent Pos:* Ed, Can Soc Environ Biologists, 69-74; ed assoc, Can Field Naturalist, 75- *Honors & Awards:* E P Edwards Prize, Wilson Ornith Soc, 72. *Mem:* Am Ornith Union; Brit Ornith Union; Int Ornith Cong. *Res:* Structures of plant polysaccharides; bird populations, densities, trends, habitats and management. *Mailing Add:* PO Box 1327 Sackville NB E0A 3C0 Can

ERSKINE, CHRISTOPHER FORBES, b Worcester, Mass, Apr 30, 27; m 52; c 3. GEOLOGY. *Educ:* Harvard Univ, AB, 49. *Prof Exp:* Geologist, Eng Geol Br, US Geol Surv, 50-57; consult eng geologist, E B Waggoner, 57-60; proj geologist, Woodward-Clyde-Sherard & Assocs, 60-62; water geologist, Kennecott Copper Corp, 62-70; ground water geologist, Amax Explor, Inc, 70-82; HYDROLOGIST, PINTO VALLEY COPPER DIV, MAGMA COPPER CO, 85- *Concurrent Pos:* Groundwater consult, 82-85. *Mem:* Geol Soc Am; Am Inst Mining, Metall & Petrol Engrs; Asn Eng Geol; Am Inst Prof Geol; Am Water Resources Asn; Nat Water Well Asn; Am Inst Hydrologists. *Res:* Engineering geology; groundwater geology. *Mailing Add:* Magma Copper Co Pinto Valley Div Box 100 Miami AZ 85539

ERSKINE, DONALD B, b Utica, NY, Jan 18, 23; m 53; c 2. CHEMICAL ENGINEERING. *Educ:* Cornell Univ, BCE, 48. *Prof Exp:* Prod engr, Calumet & Hecla Copper Co, 48-49, res engr, 49-50; develop engr, Chem Construct Corp, 50-53; proj engr, Titanium Metals Corp, 53-59, Pittsburgh Chem Co, 59-61 & Fluor Corp, 61-63; systs engr, 63-64, DIR ENG, CALGON CORP, 64- *Mem:* Am Chem Soc; Am Inst Chem Engrs. *Res:* Hydrometallurgy of copper; application of activated carbon. *Mailing Add:* 110 Lang Dr Coraopolis PA 15108-2643

ERSKINE, JAMES LORENZO, b Seattle, Wash, Oct 25, 42; m 66; c 2. SOLID STATE PHYSICS. *Educ:* Univ Wash, BS, 64, MS, 66, PhD(physics), 73. *Prof Exp:* Res engr nuclear weapons effects, Boeing Co, 66-73; res assoc physics, Univ Wash, 73-74; res asst prof physics, Univ Ill, Urbana, 74-77; from asst prof to assoc prof, 77-86, TRULL CENTENNIAL PROF, UNIV TEX, AUSTIN, 86- *Concurrent Pos:* Sr res engr consult, Boeing Co, 72-74. *Mem:* Fel Am Phys Soc; Am Vacuum Soc. *Res:* Optical and magneto-optical properties of solids; physics and chemistry of surfaces and adsorbed atoms; electron spectroscopy applied to the study of bulk and surface electronic and structural properties; application of synchrotron radiation. *Mailing Add:* Dept Physics Univ Tex Austin TX 78712

ERSKINE, JOHN ROBERT, b Milwaukee, Wis, Mar 18, 31; m 56; c 6. NUCLEAR PHYSICS. *Educ:* Univ Rochester, BS, 53; Univ Notre Dame, PhD(physics), 60. *Prof Exp:* Res assoc nuclear physics, Mass Inst Technol, 60-61, instr physics, 61-62; res assoc, Argonne Nat Lab, 62-63, from asst physicist to assoc physicist, 63-73, sr physicist, 73-80; PROG MGR, US DEPT ENERGY, WASHINGTON, DC, 80- *Concurrent Pos:* Vis assoc prof, Univ Minn, 71-72; prog officer, US Dept Energy, Washington, DC, 77-79. *Mem:* Fel Am Phys Soc. *Res:* Experimental nuclear physics; heavy ion reactions. *Mailing Add:* US Dept Energy ER-23 GTN Washington DC 20545

ERSLEV, ALLAN JACOB, b Copenhagen, Denmark, Apr 20, 19; nat US; m 47; c 4. MEDICINE. *Educ:* Copenhagen Univ, MD, 45. *Prof Exp:* Rosenstock Mem fel, Sloan-Kettering Inst, 46-47; asst res med, Sch Med, Yale Univ, 48-50, instr, 51-53; assoc, Harvard Med Sch, 55-58, asst prof, 58-59; assoc prof, 59-63, Cardeza res prof med & dir Cardeza Found Hemat Res, 63-65, DISTINGUISHED PROF, JEFFERSON MED COL, THOMAS JEFFERSON UNIV, 65- *Concurrent Pos:* Runyon res fel, 50-51; res assoc, Thorndike Mem Lab, Mass, 55-59; consult, US Army, 53-55. *Mem:* Am Soc Clin Invest; Am Fedn Clin Res; Asn Am Physicians; Am Soc Hemat; Int Soc Hemat. *Res:* Hematology. *Mailing Add:* Dept Med Thomas Jefferson Univ 1025 Walnut St Philadelphia PA 19107

ERSLEV, ERIC ALLAN, b Harvard, Mass, Jan 30, 54; m 76; c 2. FOLD-FAULT RELATIONSHIPS, STRAIN ANALYSIS. *Educ:* Wesleyan Univ, BA, 76; Harvard Univ, AM, 78, PhD(geol), 81. *Prof Exp:* Asst prof geol, Lafayette Col, 81-83; ASSOC PROF GEOL, COLO STATE UNIV, 83- *Mem:* Geol Soc Am; Am Asn Petrol Geologists; Sigma Xi. *Res:* Geometry, fabrics and element flux during the deformation of the crust; regional tectonics of the western United States. *Mailing Add:* Dept Earth Resources Colo State Univ FT Collins CO 80523

ERSOY, OKAN KADRI, b Istanbul, Turkey, Sept 4, 45; Norweg citizen; m 72; c 2. MATHEMATICS. *Educ:* Robert Col, BSEE, 67; Univ Calif-Los Angeles, MS, 68, MS, 72, PhD(elec eng), 72. *Prof Exp:* Asst prof elec eng, Bosphorus Univ, Istanbul, 72-73; researcher comput eng & appl math, Ctr Indust Res, Oslo, Norway, 73-85; ASSOC PROF ELEC ENG, PURDUE UNIV, 85- *Concurrent Pos:* Teaching & res asst elec eng, Univ Calif, Los Angeles, 67-72; assoc prof, Bosphorus Univ, Istanbul, 76-80; vis scientist, Univ Calif, San Diego, 81-82; consult, Schlumberger-Doll Res, 88-, Amoco Res Ctr, 87- *Mem:* Inst Elec & Electronics Engrs; Optical Soc Am; Asn Comput Mach; Math Asn Am; Norweg Soc Scientists & Engrs. *Res:* Digital signal & image processing & recognition; neural computing & information processing; optical information processing; holography; fast-parallel algorithms & architecture related to the above categories; applied mathematics. *Mailing Add:* Sch Elec Eng Purdue Univ West Lafayette IN 47907

ERSOY, UGUR, b Mersin, Turkey, Mar 28, 32; m 59; c 2. REINFORCED CONCRETE, EARTHQUAKE ENGINEERING. *Educ:* Robert Col, Instanbul, BS, 55; Univ Tex, Austin, MS, 56, PhD(struct), 65. *Prof Exp:* Res engr, Univ Tex, 63-65; asst prof reinforced concrete, 59-63, assoc prof, 65-72, PROF STRUCT, MID EAST TECH UNIV, 72- *Concurrent Pos:* Design engr, R C Reese & Assoc, Toledo, 56-58; vis prof, Univ Toronto, 80-81. *Honors & Awards:* Wason Award, Am Concrete Inst, 69; Prof M Parlar Sci Award, 88. *Mem:* Am Concrete Inst. *Res:* Reinforced concrete; seismic behavior of reinforced concrete members; behavior of repaired and strengthened reinforced concrete members and structures. *Mailing Add:* Civil Eng Dept Mid East Tech Univ Ankara Turkey

ERSPAMER, JACK LAVERNE, b Chehalis, Wash, Apr 9, 18; m 45; c 1. PLANT MORPHOLOGY, PLANT ANATOMY. *Educ:* Univ Wash, BS, 41; Univ Calif, PhD(bot), 53. *Prof Exp:* Teaching asst, Univ Calif, 48-51, res asst plant path, Citrus Exp Sta, 53-56; PROF BIOL, CALIF STATE POLYTECH UNIV, POMONA, 56- *Mem:* Bot Soc Am. *Res:* Morphology and anatomy of gymnosperms. *Mailing Add:* 815 Hillcrest Dr Pomona CA 91768

ERSTFELD, THOMAS EWALD, b Erie, Pa, Mar 5, 51; m 79. INORGANIC CHEMISTRY. *Educ:* Gannon Col, BS, 73; Brown Univ, PhD(chem), 77. *Prof Exp:* Fel chem, Lunar & Planetary Inst, 77-78; sr scientist chem, Lockheed Electronics Co, Inc, 78-80; electronic mat res chemist, 80-83, lab chief, 83-86, assoc prof, 86-91, PROG MGR, USAF, 91- *Mem:* Am Chem Soc. *Res:* Vapor phase epitaxial growth of III-V quaternary compounds. *Mailing Add:* AFOSR/NC Bolling AFB DC 20332-6448

ERTEKIN, TURGAY, b Ayvalik, Turkey, July 9, 47; m 75; c 2. SURFACE CHEMISTRY, TRANSPORT PROBLEMS IN POROUS MEDIA. *Educ:* Mid East Tech Univ, Ankara, Turkey, BSc, 69, MSc, 71; Pa State Univ, PhD(petrol & natural gas eng), 78. *Prof Exp:* Res assoc petrol eng, Mid East Tech Univ, Ankara, Turkey, 70-74, instr, 74-75; res asst, 75-78, from asst prof to assoc prof, 78-87, PROF PETROL ENG, PA STATE UNIV, 87-, CHMN DEPT, 84- *Concurrent Pos:* Prin investr, Enhance Oil Recovery Consortium, Pa State, 80- & Coal Seam Degasification Projs, US Steel, UEG & Gas Res Inst, Dept Energy, Gas Res Inst; consult, UN Develop Prog, 83-84 & 85. *Mem:* Soc Petrol Engrs; Soc Indust & Appl Math; NY Acad Sci; Chamber Petrol Engrs Turkey (treas, 72-75). *Res:* Numerical modeling of fluid flow dynamics in porous media; well test analysis and interpretation; coal seam degasification process; enhanced oil recovery techniques; conventional engineering analysis for hydrocarbon reservoirs. *Mailing Add:* Petrol & Natural Gas Eng Dept Pa State Univ 102 Mineral Sci Bldg University Park PA 16802

ERTEL, NORMAN H, b Brooklyn, NY, Nov 15, 32; m 67; c 3. INTERNAL MEDICINE, ENDOCRINOLOGY. *Educ:* Harvard Univ, AB, 53; Columbia Univ, MD, 57. *Prof Exp:* Intern med, Bronx Munic Hosp, Albert Einstein Col Med, 57-58, asst resident, 58-59; NY Heart Asn fel steroid biochem, Col Physicians & Surgeons, Columbia Univ, 59-60; chief, Endocrinol Div, Andrews AFB, Washington, DC, 60-62; resident med, Bronx Munic Hosp Ctr, Albert Einstein Col Med, 62-63; NIH fel endocrinol, Med Sch, Cornell Univ, 63-65; dir, Steroid Res Lab, Jewish Hosp & Med Ctr, Brooklyn, 65-71; from instr to asst prof med, State Univ NY Downstate Med Ctr, 67-71; prof dept med, NJ Med Sch, 71, dir div enocrinol & metab, 71-82, VCHMN, DEPT MED, NJ MED SCH, 85-; CHIEF MED SERV, VET ADMIN HOSP, 71- *Concurrent Pos:* Nat Inst Arthritis, Metab & Digestive Dis res grant, Jewish Hosp & Med Ctr, Brooklyn, 68-71, Nat Cancer Inst res grant, 70-71; vis prof, Univ Guadalajara, 74; vis prof, Pa Col Med, 77 & 86. *Honors & Awards:* Award, Surgeon Gen, USAF, 62; Life Dir Award, Am Diabetes Asn, 83; Spec Affil Award, Am Diabetes Asn, 87; Brandman Award, Am Diabetes Asn, 88. *Mem:* Am Fedn Clin Res; fel Am Col Physicians; Endocrine Soc; Am Diabetes Asn; fel NY Acad Sci. *Res:* Clinical endocrinology; steroid biochemistry; metabolic disease. *Mailing Add:* Med Serv 111 Vet Admin Hosp East Orange NJ 07019

ERTEL, ROBERT JAMES, b Buffalo, Minn, May 8, 32; m 58; c 4. PHARMACOLOGY, ENDOCRINOLOGY. *Educ:* Col St Thomas, BS, 54; Univ Minn, Minneapolis, PhD(pharmacol), 65. *Prof Exp:* Jr scientist, Med Sch, Univ Minn, 58-65; res fel biochem, Univ Minn, 65-66; res assoc, Nat Heart Inst, 66-68; asst prof, 68-71, assoc prof pharmacol, Sch Pharm, 71-87, ASSOC PROF PHARMACOL, SCH DENT MED, UNIV PITTSBURGH, 87- *Mem:* Am Soc Pharmacol & Exp Therapeut. *Res:* Actions of drugs on the spontaneous activity of heart cells grown in tissue culture; control of cytoplasmic calcium in isolated heart cells; neuroendocrinology and protein biosynthesis; effect of dietary sodium on cardiovascular and autonomic nervous function. *Mailing Add:* Dept Pharmacol Physiol 541 Salk Hall Univ Pittsburgh Pittsburgh PA 15261

ERTELT, HENRY ROBINSON, b New Haven, Conn, Apr 29, 24; m 48; c 3. ORGANIC CHEMISTRY. *Educ:* Yale Univ, BS, 49, PhD(org chem), 52; NY Univ, JD, 70. *Prof Exp:* Res chemist, Esso Res & Eng Co, 52-66; patent atty, FMC Corp, 66-74, group counsel, Chem Patent & Licensing, 74-90; CONSULT, PATENT LAW, 91- *Mem:* Am Chem Soc. *Res:* Organic synthesis; agricultural pesticides; patent law. *Mailing Add:* 184 Berwind Circle Radnor PA 19087

ERTEZA, AHMED, b Rajbari, East Pakistan, Aug 1, 24; nat US; m 57; c 2. ELECTRICAL ENGINEERING. *Educ:* Univ Calcutta, BSc, 45, MS, 47; Stanford Univ, MS, 51, EE, 52; Carnegie Inst Technol, PhD(elec eng), 54. *Prof Exp:* Engr, Dacca Broadcasting Sta, Radio Pakistan, 47-48; sr lectr electronics & radio eng, Univ Dacca, 48-50; elec engr, nuclear instrumentation, Nuclear Res Ctr, Pa, 53-54; asst prof elec eng, Ahsanullah Eng Col, East Pakistan, 55-58 & Bradley Univ, 58; assoc prof, 58-63, PROF ELEC ENG, UNIV NMEX, 63- *Mem:* Inst Elec & Electronics Engrs; assoc mem Inst Eng Pakistan. *Res:* Electromagnetics and microwave; magnetohydrodynamics; nuclear instrumentation; analog and digital computation; energy conversion; controls. *Mailing Add:* Dept Elec Eng & Comput Sci Univ NMex Albuquerque NM 87106

ERULKAR, SOLOMON DAVID, b Calcutta, India, Aug 18, 24; m 50; c 2. NEUROPHYSIOLOGY. *Educ:* Univ Toronto, BA, 48, MA, 49; Johns Hopkins Univ, PhD(physiol), 52; Oxford Univ, PhD, 57. *Prof Exp:* USPHS fel, Johns Hopkins Univ, 52-54; Brit Med Res Coun grant, Oxford Univ, 55-58; asst prof otol & physiol, dir dept & dir res otol, Temple Univ, 59-60; dept demonstr physiol, Oxford Univ, 55-58; from asst prof to assoc prof, 60-67, PROF PHARMACOL, UNIV PA, 67- *Concurrent Pos:* Hon res assoc, Univ Col, Univ London, 67-68; Guggenheim fel, Hadassah Med Sch, Hebrew Univ, Israel, 74-75, vis prof physiol, 74-75; Josiah Macy Found, École Normale Supérieure, Paris; vis prof, Pierre & Marie Curie Univ, Paris, 81-82; sr int fel, Fogarty Int Ctr, 88-89; Biozentrum, Univ Basel, Basel, Switz; mem, Inst Neurol Sci, Univ Pa, 61-, assoc dir, 80-81. *Mem:* Soc Neurosci; Am Physiol Soc; fel AAAS. *Res:* Mechanisms of central synaptic transmission. *Mailing Add:* Dept Pharmacol Med Sch Univ Pa Philadelphia PA 19104

ERVE, PETER RAYMOND, b Buffalo, NY, June 25, 26; m 61; c 1. CLINICAL CHEMISTRY, TOXICOLOGY. *Educ:* Univ Toronto, BA, 51, MA, 52; Ill Inst Technol, PhD(biochem), 68; Am Bd Clin Chem, dipl, 85. *Prof Exp:* Res chemist, Vet Admin Med Ctr, Westside, 68-75; CLIN CHEMIST, VET ADMIN MED CTR, NORTH CHICAGO, 75-; ASSOC PROF, DEPT PATH, CHICAGO MED SCH, 77- *Honors & Awards:* Cert Appreciation, Am Col Surgeons, 71; Cert of Merit, AMA, 72. *Mem:* Am Chem Soc; Am Asn Clin Chem; AAAS; Nat Acad Clin Biochem. *Res:* Pathophysiology of septic shock; bacterial toxins; metabolic derrangements in endotoxic shock; geriatric nutrition. *Mailing Add:* 4100 Marine Dr Univ Health Sci-Chicago Med 3333 Greenbay Rd Chicago IL 60613-2358

ERVEN, BERNARD LEE, b Bowling Green, Ohio, Aug 25, 38; m 61; c 5. AGRICULTURAL ECONOMICS. *Educ:* Ohio State Univ, BSAgr, 60, MS, 63; Univ Wis, PhD(agr econ), 67. *Prof Exp:* Asst prof agr econ, Univ Wis, 67-69; from asst prof to prof, 69-79, PROF AGR ECON, OHIO STATE UNIV, 79- *Mem:* Am Agr Econ Asn; Nat Asn Col Teachers Agr. *Res:* Labor management problems of agricultural employers; dairy farm management problems. *Mailing Add:* 2120 Fyffe Rd Columbus OH 43210

ERVIN, C PATRICK, b Danville, Ill, Aug 5, 43. GRAVITY, MAGNETICS. *Educ:* Wash Univ, St Louis, BS, 65, MA, 68; Univ Wis, Madison, PhD(geophysics), 72. *Prof Exp:* Res assoc, 72-75, asst prof, 75-81, ASSOC PROF GEOPHYS, NORTHERN ILL UNIV, 81- *Concurrent Pos:* Geophysicist, Ill State Geol Surv, 72-73; assoc ed geophysics, Geosci Wis, Wis Geol Surv, 76-85; assoc scientist, Dry Valley Drilling Proj, 75-76; US Geol Surv, 77-78. *Mem:* Soc Explor Geophysicists; Am Geophys Union; Europ Asn Explor Geophysicists; Geol Soc Am; Am Asn Petrol Geol; Sigma Xi; Seismol Soc Am. *Res:* Applied geophysics, potential fields; computer applications in the earth sciences; geophysics of continental interiors. *Mailing Add:* Dept Geol Northern Ill Univ DeKalb IL 60115

ERVIN, FRANK (RAYMOND), b Little Rock, Ark, Nov 3, 26; m 47; c 5. PSYCHIATRY, GENETICS. *Educ:* Tulane Univ, MD, 51; Am Bd Psychiat & Neurol, dipl. *Prof Exp:* Asst psychiat & neurol, Med Sch, Tulane Univ, 52-57; from instr to asst prof psychiat, Harvard Med Sch, 57-68, assoc clin prof, 68-69, assoc prof, 69-72; prof psychiat, Univ Calif, Los Angeles, 72-89; PROF PSYCHIAT, MCGILL UNIV, MONTREAL, 80- *Concurrent Pos:* Dir, Monroe Area Guid Ctr, La, 53-57; consult psychiatrist, State Training Inst, La, 53-57; asst, Mass Gen Hosp, 57-63, psychiatrist, 63-; NIMH career res fel, 62-; adj prof, Hampshire Col & Sch Med, Univ PR; pres, Behav Sci Found; dir, Stanley Cobb Labs Psychol Res, Mass Gen Hosp, 62-72; mem, Brain Res Inst, Univ Calif Los Angeles, Ctr Human Genetics, McGill Univ. *Res:* Neurophysiologic aspects of behavior; pathological aggression; behavior-ethology. *Mailing Add:* Dept Psychiat McGill Univ Sch Med Drummond St Montreal PQ H3G 1Y6 Can

ERVIN, GUY, JR, physical chemistry, ceramics, for more information see previous edition

ERVIN, HOLLIS EDWARD, b Delhi, LA, Apr 17, 52; m 52; c 2. WOOD PRESERVATION, ENVIRONMENTAL REMEDIATION. *Educ:* La Technol Univ, BS, 74; Miss State Univ, MS, 76. *Prof Exp:* Res asst, Miss Forest Prod Lab, 74-76; tech dir, Weyehaesur Co, 77-84; MGR, TECH SRVC, INT PAPER CO, 84- *Concurrent Pos:* Vchmn, Info & Tech Develop Comt, Am Wood Preservation Asn. *Mem:* Am Wood Preservation Asn; Forest Prod Res Soc. *Res:* Major prod develop; wood deterioration; waste minimization. *Mailing Add:* 2139 Goldfinch Dr Lewisville TX 75067

ERWAY, LAWRENCE CLIFTON, JR, b Lawrenceville, Pa, Apr 27, 38; m 60; c 4. DEVELOPMENTAL GENETICS. *Educ:* Barrington Col, BA, 60; Brown Univ, MA, 63; Univ Calif, Davis, PhD(genetics), 68. *Prof Exp:* Instr biol, Barrington Col, 61-64; trainee genetics, Univ Calif, Davis, 64-66, res technician, 66-68; asst prof, 68-74, ASSOC PROF BIOL, UNIV CINCINNATI, 74- *Concurrent Pos:* NIH fel, Pompeiano's Lab, Pisa, Italy, 76-77. *Mem:* Asn Res Otolaryngol. *Res:* Effects of trace elements and genes on birth defects, including pigmentary and neurological defects in mice and man, with particular emphasis on prevention of certain hereditary disorders, specifically otolith defects and deafness. *Mailing Add:* Dept Biol Sci Univ Cincinnati Cincinnati OH 45221

ERWIN, ALBERT R, b Charlotte, NC, May 1, 31; m 54; c 1. PHYSICS. *Educ:* Duke Univ, BS, 53; Harvard Univ, MA, 57, PhD(physics), 59. *Prof Exp:* Jr res assoc, Brookhaven Nat Lab, 57-58; from instr to assoc prof, 58-65, PROF PHYSICS, UNIV WIS-MADISON, 65- *Res:* High energy particle physics. *Mailing Add:* Dept Physics Chamberlin Hall Univ Wis Madison WI 53706

ERWIN, CHESLEY PARA, b Okla, June 5, 20; m 49; c 2. MEDICINE, PATHOLOGY. *Educ:* Univ Okla, BA, 42, MD, 51; Am Bd Path, dipl, 56, cert forensic path. *Prof Exp:* Resident path, Milwaukee County Hosp, 52-55; resident, Milwaukee Hosp, 55-56; from instr to assoc prof, 56-74, PROF PATH, MED COL WIS, 74- *Concurrent Pos:* Chief med examr, Milwaukee County, 74. *Mem:* Int Acad Path; NY Acad Sci; Am Acad Forensic Sci; AMA; Col Am Path. *Res:* Study of animal tumors; oncogenic and oncolytic factors; Rickettsiae; viruses; medical education; drug deaths. *Mailing Add:* 8700 W Wisconsin Ave Milwaukee WI 53226-3595

ERWIN, DAVID B(ISHOP), SR, b Flushing, NY, May 30, 24; m 45; c 5. ELECTRICAL ENGINEERING. *Educ:* Purdue Univ, BS, 48. *Prof Exp:* Engr test set develop, Hawthorne Works, 48-53 & St Paul Shops, 53-55, dept chief electronic switching systs develop, Hawthorne Works, 55-60 & Columbus Works, 60-62, asst supt develop eng, 62-66, mgr electronic switching syst eng, Northern Ill Works, 66-72, mgr regional eng, Cent Region, 72-78, mgr eng servs, 78-80, MGR DIST SERV, WESTERN ELEC CO, 80- *Mem:* Sr mem Inst Elec & Electronics Engrs. *Res:* Manufacturing techniques for production of electronic switching system. *Mailing Add:* 514 Hamilton St Geneva IL 60134-2139

ERWIN, DONALD C, b Concord, Nebr, Nov 24, 20; m 48; c 2. MYCOLOGY. *Educ:* Univ Nebr, BS, 49, MA, 50; Univ Calif, PhD(plant path), 53. *Prof Exp:* Asst plant path, Univ Nebr, 49-50 & Univ Calif, Davis, 50-53; from jr plant pathologist to asst plant pathologist, 53-60, assoc prof plant path, 61-66, chmn dept, 77-80, ASSOC PLANT PATHOLOGIST, UNIV CALIF, RIVERSIDE, 60-, PROF PLANT PATH, 66- *Concurrent Pos:* Mem comt cotton study team, Study Problems Pest Control & Technol Assessment, Nat Acad Sci, 73-75; chmn, Cotton Dis Coun, 73-75. *Honors & Awards:* Guggenheim Res Fel, 59; Fel, Am Phytopath Soc. *Mem:* Am Phytopath Soc; Mycol Soc Am; Am Inst Biol Sci; Sigma Xi. *Res:* Study of etiology and control of diseases of alfalfa and cotton; biology and reproduction of Phytophthora. *Mailing Add:* Dept Plant Path Univ Calif Riverside CA 92521

ERWIN, JAMES V, b Horton, Kans, Sept 28, 25; m 44; c 4. CHEMICAL ENGINEERING. *Educ:* Univ Nebr, Lincoln, BSc, 50. *Prof Exp:* Res & develop engr, Minn Mining & Mfg Co, 50-54, proj leader reflective prod, 54-56, supvr res & develop thermal & reflective prod, 56-59, mgr thermal, reflective & decorative prod, 59-61, tech mgr reflective & decorative prod, 61-64, tech mgr, decorative prod dept, 64-67, tech dir, visual prod div, 67-72, tech dir com tape, 72-76, proj mgr personal care prod, 76-88, mgr consumer mkt, 85-88; RETIRED. *Concurrent Pos:* Consult, New Prod Develop & Mkt, 88. *Mem:* Am Inst Chem Engrs. *Res:* Radiation physics; surface chemistry and physics; imaging chemistry; surface coatings technology; environmental technology; optics; adhesion physics. *Mailing Add:* 2641 Natchez Ave S St Louis Park MN 55416

ERWIN, LEWIS, b Ind, Apr 2, 51; m 73; c 3. MANUFACTURING ENGINEERING, POLYMER ENGINEERING. *Educ:* Mass Inst Technol, SB, 72, SM, 74, PhD(mech eng), 77. *Prof Exp:* Asst prof mech eng, Univ Wis-Madison, 76-79; from asst prof to assoc prof mech eng, Mass Inst Technol, 79-85; DIR PACKAGING TECHNOL, KRAFT, INC, 85- *Mem:* Am Soc Mech Engrs; Soc Mfg Engrs; Soc Rheology; Soc Plastics Engrs. *Res:* Manufacturing issues for plastic and rubber components; mixing of highly viscous liquids; innovative processes in plastics molding; food packaging. *Mailing Add:* 511 Ash St Winnetka IL 60093

ERWIN, ROBERT BRUCE, b Burlington, Vt, Mar 19, 28; m 56; c 4. GEOLOGY. *Educ:* Univ Vt, BA, 52; Brown Univ, ScM, 55; Cornell Univ, PhD, 59. *Prof Exp:* Geologist, Vt Geol Surv, 55 & Texaco, Inc, 59-61; asst prof geol, St Lawrence Univ, 61-64; geologist, 64-65, dir res, 65-66, asst state geologist, 66-69, DIR & STATE GEOLOGIST, WVA GEOL & ECON SURV, 69-; PROF GEOL, WVA UNIV, 74-; PROF GEOL, MARSHALL UNIV, 78- *Mem:* AAAS; Am Asn Petrol Geol; Soc Econ Paleont & Mineral; Geol Soc Am; Am Inst Prof Geol. *Res:* Paleontology and stratigraphy of lower Paleozoic rocks; Gulf Coast stratigraphy and structure; lower Paleozoic time-stratigraphic relationships; paleontology of West Virginia. *Mailing Add:* RR 8 No 204 Box 879 Morgantown WV 26505

ERWIN, VIRGIL GENE, b Cahone, Colo, Nov 1, 37; m 59; c 3. PHARMACOLOGY, BIOCHEMISTRY. *Educ:* Univ Colo, BS, 60, MS, 62, PhD(pharmacol & biochem), 65. *Prof Exp:* Fel, Sch Med, Johns Hopkins Univ, 65-67; from asst prof to assoc prof, 67-77, PROF PHARM, UNIV COLO, BOULDER, 77-, DEAN SCH PHARM, 74- *Mem:* Am Soc Neurochem; Am Soc Pharmacol & Exp Therapeut. *Res:* Biochemical pharmacology; mechanism of action of therapeutic agents on various cellular processes. *Mailing Add:* Pharm Dept Univ Colo Campus Box 2797 Boulder CO 80309-0297

ERZURUMLU, H CHIK, b Istanbul, Turkey, Mar 7, 34; m 63. STRUCTURAL & CIVIL ENGINEERING. *Educ:* Tech Univ Istanbul, Prof degree civil eng, 57; Univ Tex, Austin, MS, 62, PhD(civil eng), 70. *Prof Exp:* Instr eng, 62-65, from asst prof to assoc prof struct eng, 65-72, head, Civil & Struct Eng Dept, 76-79, PROF CIVIL ENG, PORTLAND STATE UNIV, 72-, DEAN, SCH ENG & APPL SCI, 79- *Concurrent Pos:* Consult, 66- *Mem:* Am Soc Civil Engrs; Sigma Xi; Am Soc Eng Educ; Nat Soc Prof Engrs; Am Soc Engr Mgt. *Res:* Static and fatigue investigations of orthotropic plate bridge decks and of tubular joints; ultimate strength considerations of welded steel tubular members. *Mailing Add:* Sch Eng & Appl Sci Portland State Univ Portland OR 97207

ESAKI, LEO, b Osaka, Japan, Mar 12, 25; m 59, 86; c 3. SEMICONDUCTOR PHYSICS. *Educ:* Univ Tokyo, BS, 47, PhD(physics), 59. *Hon Degrees:* Various from US & foreign univs. *Prof Exp:* RESEARCHER, IBM THOMAS J WATSON RES CTR, YORKTOWN HEIGHTS, NY, 60- *Concurrent Pos:* Adj prof, Univ Pa & Indust Sci Inst, Univ Tokyo. *Honors & Awards:* Nobel Prize in Physics, 73; Morris N Liebman Mem Prize; Stuart Ballantine Medal; Medal of Honor, Inst Elec & Electronics Engrs, 91; Int Prize New Mat, Am Phys Soc, 85. *Mem:* Foreign assoc Nat Acad Sci; foreign assoc Nat Acad Eng; fel Am Acad Arts & Sci; fel Am Phys Soc; fel Inst Elec & Electronics Engrs; Phys Soc Japan; Inst Elec Commun Engrs Japan. *Res:* Artificial semiconductor superlattices in search of predicted quantum mechanical effects. *Mailing Add:* IBM Thomas J Watson Res Ctr PO Box 218 Yorktown Heights NY 10598

ESARY, JAMES DANIEL, b Seattle, Wash, Apr 8, 26; m 47; c 2. MATHEMATICAL STATISTICS. *Educ:* Whitman Col, AB, 48; Univ Calif, MA, 51, PhD(statist), 57. *Prof Exp:* Asst statist, Univ Calif, 50-51 & 53-57; math statistician, Boeing Airplane Co, 57-58 & Boeing Sci Res Labs, 59-70; assoc prof, 70-74, PROF OPER RES, NAVAL POSTGRAD SCH, 74- *Concurrent Pos:* Vis lectr, Univ Calif, Berkeley, 67; vis lectr prog statist, 68-71. *Mem:* fel Am Statist Asn; fel Inst Math Statist; Opers Res Soc Am. *Res:* Reliability theory; probability; military applications. *Mailing Add:* Dept Opers Res Naval Postgrad Sch OR-EY Monterey CA 93943

ESAU, KATHERINE, b Ekaterinoslav, Russia, Apr 3, 98; nat US. BOTANY. *Educ:* Univ Calif, PhD, 31. *Hon Degrees:* DSc, Mills Col, 62; LLD, Univ Calif, 66. *Prof Exp:* Instr bot & jr botanist, Univ Calif, Davis, 31-37, from asst prof & asst botanist to prof & botanist, 37-63; prof, 63-65, EMER PROF BOT, UNIV CALIF, SANTA BARBARA, 65- *Concurrent Pos:* Guggenheim fel, 40; spec lectr, Bot & Plant Res Inst, Univ Tex, 56; nat lectr, Sigma Xi, 65-66; lectr, J C Walker conf plant path, Univ Wis, 68. *Mem:* Prather Lectr, Harvard Univ, 60; John Wesley Powell Lectr, AAAS, 73; NBat Medal Sci, 89 Sci; Am Philos Soc; Bot Soc Am (pres, 51); Am Acad Art & Sci; Sigma Xi; Int Soc Plant Morphologists; Am Soc Plant Physiol. *Res:* Anatomy and ultrastructure of healthy and virus diseased plants; author of 160 publications. *Mailing Add:* Dept Biol Sci Univ Calif Santa Barbara CA 93106

ESBER, HENRY JEMIL, b El-Mina, Lebanon, Aug 28, 38; US citizen; c 2. IMMUNOLOGY, MICROBIOLOGY. *Educ:* Col William & Mary, BS, 61; Univ NC, MS, 63; WVa Univ, PhD(microbiol), 67. *Prof Exp:* Bacteriologist-in-chg, Portsmouth Gen Hosp, 63-64; fel diag virol, Los Angeles County Health Dept, 67; SR SCIENTIST & IMMUNOLOGIST, MASON RES INST, 67- *Concurrent Pos:* Instr immunol, Clark Univ, 69-; clin lab consult, Hahnemann Hosp, Worcester, 69-; affil, Grad Sch, Anna Maria Col, 75-; clin microbiol & immunol lectr, Worcester City Hosp, 79- *Mem:* AAAS; Am Soc Microbiol; Am Pub Health Asn; Am Asn Immunologists; Am Asn Cancer Res. *Res:* Lung antibodies and autoimmunity; silicosis and environmental health; immunosuppressive activity of inhalents; immune enhancement by bacterial fractions; tumor immunology; inflammation; radioimmunoassays of protein hormones and steroids; competitive protein binding of steroids. *Mailing Add:* EG&G Mason Res Inst 57 Union St Worcester MA 01608

ESCH, GERALD WISLER, b Wichita, Kans, June 22, 36; m 58; c 3. ANIMAL PARASITOLOGY, ECOLOGY. *Educ:* Colo Col, BS, 58; Univ Okla, MS, 61, PhD(zool), 63. *Prof Exp:* NIH trainee, Univ NC, 63-65; from asst prof to assoc prof biol, 65-75, dean grad sch, 84-90, PROF BIOL, WAKE FOREST UNIV, 75-, CHMN DEPT, 75-84 & 90- *Concurrent Pos:* Vis prof, W K Kellogg Biol Sta, 65-74; WHO res fel, Univ London, 70-71; res assoc, Savannah River Ecol Lab, 74-75; mem coun, Am Soc Parasitol, 84-88; mem, NC Bd Sci & Technol, 88-90. *Mem:* Am Inst Biol Sci; Am Soc Parasitol. *Res:* Ecology of animal parasites. *Mailing Add:* Dept Biol Wake Forest Univ Winston-Salem NC 27109

ESCH, HARALD ERICH, b Düsseldorf, Ger, Dec 22, 31; m 55; c 3. ANIMAL BEHAVIOR. *Educ:* Univ Wurzburg, Dr rer nat(zool), 60. *Prof Exp:* Res scientist, Ger Res Asn, 60-62; sci asst radiation res, Univ Munich, 62-64; res scientist, 64-65, from asst prof to assoc prof, 65-69, PROF BIOL, UNIV NOTRE DAME, 69- *Concurrent Pos:* Vis prof bee commun, Univ Sao Paulo, 64; mem comt African Honey Bee, Nat Res Coun; vis prof animal behav, Univ Innsbruck, 83-84. *Mem:* AAAS; Int Union Study Social Insects; Sigma Xi. *Res:* Communication in bees; electrophysiology; radiation effects on cellular level; sensory physiology; computers in biology. *Mailing Add:* Dept Biol Univ Notre Dame Notre Dame IN 46556

ESCH, LOUIS JAMES, b Grand Rapids, Mich, Apr 11, 32; m 52; c 6. DATA ANALYSIS SYSTEMS DEVELOPMENT. *Educ:* Aquinas Col, BS, 59; Rensselaer Polytech Inst, MS, 62, PhD(nuclear eng & sci), 71. *Prof Exp:* From physicist to sr physicist, 59-84, supvr software develop, 84-89, PROG MGR, ADVAN TECHNOL, KNOLLS ATOMIC POWER LAB, GEN ELEC CO, 89- *Res:* Integral and differential measurements of cross sections of fissile and nonfissile nuclides to neutrons of low and intermediate energies; proton recoil neutron spectroscopy; reactor neutronic noise and vibration analysis; data analysis methods development, computer applications, user interface development. *Mailing Add:* Three Woodcrest Dr Scotia NY 12302

ESCH, ROBIN E, b Md, Feb 25, 30; m 66; c 3. DATA ANALYSIS, EXPERIMENTAL DESIGN. *Educ:* Harvard Univ, BA, 51, AM, 53, PhD(appl math), 57. *Prof Exp:* Asst prof appl math, Harvard Univ, 58-62; head appl mech dept, Sperry Rand Res Ctr, Sudbury, 62-66; chmn dept, 68-79, PROF MATH, BOSTON UNIV, 66- *Res:* Numerical analysis; applied mechanics; data analysis. *Mailing Add:* 371 Plainfield Rd Concord MA 01742

ESCHBACH, CHARLES SCOTT, b St Paul, Minn, July 29, 46; m 68; c 2. ORGANOSILICON CHEMISTRY, APPLICATIONS RESEARCH. *Educ:* Macalester Col, BA, 68; Mass Inst Tech, PhD(chem), 75. *Prof Exp:* Spec prof 5, US Army, 69-71; chemist, 75-80, group mgr, 80-84, ASSOC DIR RES & DEVELOP, UNION CARBIDE CORP, 84- *Mem:* Sigma Xi; Am Asn Textile Chemists & Colorists. *Res:* Product and applications research and development in the field organosilican chemistry andprimary silicone polymers; applications areas of primary interest include textile finishing, foam control, surfactants and cosmetics and toiletries. *Mailing Add:* Ethan Allen Dr Rte 2 Box 691 Stormville NY 12582

ESCHBACH, JOSEPH W, INTERNAL MEDICINE. *Prof Exp:* CLIN PROF, DEPT NEPHROLOGY & DEPT MED, UNIV WASH, 75- *Mem:* Inst Med-Nat Acad Sci. *Mailing Add:* Dept Med Univ Wash 515 Minor Ave Seattle WA 98104

ESCHENBERG, KATHRYN (MARCELLA), b St Louis, Mo, Dec 12, 23. ZOOLOGY, EMBRYOLOGY. *Educ:* Miami Univ, BA, 46; Univ Colo, MA, 50; Univ Wash, PhD(zool), 57. *Prof Exp:* Asst biol & vert physiol, Univ Colo, 48-49, instr, 49-51; asst zool, embryol & cell physiol, Univ Wash, 51-56; Nat Cancer Inst res fel, Princeton Univ, 57; asst prof zool, cell biol & embryol, 58-64, assoc prof biol sci, 64-70, prof biol sci & chmn dept, 70-80, IDA & MARION VAN NATTA PROF BIOL SCI, MT HOLYOKE COL, 80- *Mem:* AAAS; Am Inst Biol Sci; Am Soc Zool; Soc Develop Biol; Sigma Xi. *Res:* Developmental biology; cytology; oogenesis. *Mailing Add:* Dept Biol Sci Mt Holyoke Col South Hadley MA 01075

ESCHENFELDER, ANDREW HERBERT, b Newark, NJ, June 13, 25; m 49; c 2. PHYSICS. *Educ:* Rutgers Univ, BS, 49, PhD(physics), 52. *Prof Exp:* Ultrasonic res, Aberdeen Proving Ground, 50; physicist, IBM Res Ctr & IBM Corp, 52-55, mgr magnetics dept, 57-60, dir solid state sci, 60-63 & components, 63-66, consult to dir res, Res Div, 66-67, Dir Lab, 67-73, SR PHYSICIST, IBM CORP, SAN JOSE RES LAB, 56-, CONSULT, 73- *Mem:* Am Phys Soc; sr mem Inst Elec & Electronics Engrs. *Res:* Magnetism; cryogenics; paramagnetic resonance; solid state. *Mailing Add:* 19580 Moray Ct Saratoga CA 95070

ESCHENROEDER, ALAN QUADE, b St Louis, Mo, Feb 22, 33; m 84; c 5. HEALTH RISK ASSESSMENT, ENVIRONMENTAL SIMULATION MODELING. *Educ:* Cornell Univ, BME, 55, PhD(eng), 59. *Prof Exp:* Res engr, US Army Ballistic Res Labs, 55-57 & Cornell Aeronaut Lab, 59-62; sect leader, Gen Motors Res Labs, 62-67; assoc dept head, Gen Res Corp, 67-75; div dir, Environ Res & Technol, 75-78; sr consult, Arthur D Little, Inc, 78-83; PRIN, ALANOVA, INC, 83- *Concurrent Pos:* Mem, Emissions Air Qual Comn, Nat Acad Sci, 74-75, Ozone & Photochem Oxidant Comn, 75-76, TRB Air Qual Comt, 78-, Diesel Impact Study Comn, 80-82, chair, NBS

Environ Eval Panel, 75-81; consult, Lincoln Lab, Mass Inst Technol, 84-89; dir, Gradient Corp, 86-90. *Mem:* Air & Waste Mgt Asn; Soc Risk Anal. *Res:* Computer simulations of environmental fate and transport of chemicals in air, water, and soil; aerodynamic analyses of fugitive dust resuspension, transport and deposition; health risk research and policy analysis for hazardous waste sites, waste disposal facilities and power plants. *Mailing Add:* 76 Todd Pond Rd Lincoln MA 01773

ESCHER, DORIS JANE WOLF, b New York, NY, July 1, 17; m 38; c 2. CARDIOLOGY. *Educ:* Barnard Col, BA, 38; NY Univ, MD, 42. *Prof Exp:* Resident med, Jewish Hosp of Brooklyn, 43-44; asst physician, Cardiac Clin, Bellevue Hosp, 45-48; asst, Med Div, 48-49, head, Cardiac Catheterization Unit, 50- 84, EMER DIR & SR CONSULT, CARDIAC CATHERTERIZATION LAB, MONTEFIORE HOSP & MED CTR, 84-, EMER PROF, 87- *Concurrent Pos:* Fel, Col Med, NY Univ, 45-46; fel, Med Div, Montefiore Hosp, 46; clin asst, Mt Sinai Hosp, 46-48; Rosenstock Mem Found fel, 47-48; lectr, Columbia Univ, 57-64; assoc attend radiol, Montefiore Hosp & Med Ctr, 57-, attend physician med, 67-, from asst prof to prof med, 66-75; consult cardiol, Lawrence Hosp, 73-; lectr bioeng, Polytech Inst NY 74. *Mem:* Am Heart Asn; Am Fedn Clin Res; NY Cardiol Soc (pres, 83-84); fel Am Col Cardiol; Am Soc Artificial Internal Organs; NAm Soc Pacing & Electrophysiol. *Res:* Diagnosis and clinical research in cardiovascular disease; artificial cardiac pacing; methods, materials, physiology; artificial organs. *Mailing Add:* Cardiac Catheterization Lab Montefiore Med Ctr 111 E 210 St Bronx NY 10467

ESCHER, EMANUEL, b Basel, Switz, Dec 16, 43; Can citizen; m 80; c 4. PEPTIDE PHARMACOLOGY, RECEPTORS. *Educ:* Freies Gym, Basel, Matura, 64; Fed Polytechnicum, Zurich, Dipl chem, 71, Doctorate(chem), 74, postgrad dipl molecular biol, 76. *Prof Exp:* Practicant org chem, Sandoz Inc, Basel, 70; res asst peptide chem, Inst Molecular Biol & Biophys, Swiss Fed Inst Technol, Zurich, 71-76; from asst prof to assoc prof, 76-87, PROF PHARMACOL, UNIV SHERBROOKE, 87- *Concurrent Pos:* Res dir, Univ Sherbrooke, 76-, mem bd gov, 90-; vis prof, Nat Ctr Sci Res, Nat Inst Health & Med Res, 84; mem, Scholar Comt, Med Res Coun Can & Adv Comt, Que Heart Found, 90- *Honors & Awards:* Jonathan Ballon Award, Can Heart Found, 78. *Mem:* Am Chem Soc; Endocrine Soc; Am Soc Biol Chem; Am Soc Pharmacol & Exp Therapeut; Peptide Soc. *Res:* Central activity in the production of receptor-specific tools with synthetic organic chemistry; peptide hormone receptor identification and receptor isolation; receptor function and potential pharmaceuticals. *Mailing Add:* Dept Pharmacol Fac Med Sherbrooke Univ Sherbrooke PQ J1H 5N4 Can

ESCHLE, JAMES LEE, b Groom, Tex, Jan 20, 37; m 58; c 2. ENTOMOLOGY. *Educ:* Tex Tech Col, BS, 60; Univ Wis, MS, 62, PhD(entom), 64. *Prof Exp:* Asst entom, Univ Wis, 60-64; entomologist, Entom Res Div, USDA, 64-; PRES, RAINBOW EXTERMINATORS. *Mem:* Entom Soc Am. *Res:* Biology and control of flies affecting livestock; aquaculture. *Mailing Add:* Rainbow Exterminators 860 Halekauwila Honolulu HI 96813

ESCHMAN, DONALD FRAZIER, b Granville, Ohio, Oct 22, 23; m 46; c 4. GEOMORPHOLOGY. *Educ:* Denison Univ, AB, 47; Harvard Univ, MA, 50, PhD(geol), 53. *Prof Exp:* Fel geol, Harvard Univ, 47-51; instr Tufts Col, 51-53, actg head dept, 52-53; from instr to assoc prof geomorphol, 53-64, chmn dept geol & mineral, 61-66, PROF GEOMORPHOL, UNIV MICH, 64- *Concurrent Pos:* Geologist, US Geol Surv, 48-70; chmn fac counsr, Col Lit, Sci & Arts, Univ Mich, 58-59 & Dir, Environ Studies Prog, 71-78. *Mem:* AAAS; Sigma Xi; Geol Soc Am; Nat Asn Geol Teachers. *Res:* Glacial geology; engineering geology; Pleistocene of Michigan; geomorphology and Cenozoic history of Rocky Mountains; geology and land use planning; slope processes. *Mailing Add:* 3035 Fairlane Ann Arbor MI 48104

ESCHMEYER, PAUL HENRY, b New Bremen, Ohio, June 7, 16. BIOLOGY. *Educ:* Univ Mich, BSF, 38, MS, 39, PhD(zool), 49. *Prof Exp:* Fish mgt ag, State Div Conserv & Nat Resources, Ohio, 39-40; dist fisheries biologist, Inst Fisheries Res, Mich, 41-42, asst fisheries biologist, 47-49; aquatic biologist, State Conserv Comn, Mo, 49-50; fishery res biologist, US Fish & Wildlife Serv, 50-56; asst dir, Inst Fisheries Res, Mich, 56-61; asst dir, Great Lakes Biol Lab, US Bur Com Fisheries, 61-64, biol ed, Div Biol Res, 65-70; fishery biologist, Great Lakes Fishery Lab, US Fish & Wildlife Serv, 70-74, fishery ed, 74-90; RETIRED. *Mem:* Fel Am Inst Fishery Res Biol; Am Fisheries Soc; Am Soc Zoologists; Am Soc Ichthyologists & Herpetologists; Wildlife Soc. *Res:* Fishery biology; natural history of freshwater fishes. *Mailing Add:* 1628 Herber Dr Ft Collins CO 80524

ESCHMEYER, WILLIAM NEIL, b Knoxville, Tenn, Feb 11, 39; div; c 3. ICHTHYOLOGY. *Educ:* Univ Mich, BS, 62; Univ Miami, MS, 64, PhD(marine biol), 67. *Prof Exp:* Res asst ichthyol, Inst Marine Sci, Miami, 66-67; asst cur, 67-69, chmn ichthyol & assoc cur, 69-73, cur, 73-83, dir res, 77-83, SR CUR, CALIF ACAD SCI, 83- *Concurrent Pos:* Supv ichthyologist, Vanderbilt Proj, Calif Acad Sci, 67-69; prin investr, NSF & Nat Oceanic & Atmospheic Admin Grants. *Mem:* Am Soc Ichthyologists & Herpetologists. *Res:* Systematics, zoogeography and biology of fishes. *Mailing Add:* 400 Oak Crest San Anselmo CA 94960

ESCHNER, ARTHUR RICHARD, b Buffalo, NY, Sept 29, 25; m 51; c 3. FORESTRY, HYDROLOGY. *Educ:* State Univ NY Col Forestry, Syracuse, BS, 50, PhD(silvicult), 65; Iowa State Col, MS, 52. *Prof Exp:* Res forester, Cent States Forest Exp Sta, US Forest Serv, 53-54, Northeastern Forest Exp Sta, Pa, 54-58 & WVa, 59-61, res forester & proj leader, NY, 61-64; from asst prof to assoc prof, 64-70, PROF FOREST INFLUENCES, STATE UNIV NY COL ENVIRON SCI & FORESTRY, 70- *Concurrent Pos:* Consult, Guatemala, 74, Burma, 82, NY, NH, Austria, Ger. *Mem:* Soc Am Foresters; Soil Sci Soc Am; Am Geophys Union; Am Water Resources Asn; Sigma Xi. *Res:* Effect of forest conditions on the disposition of precipitation and energy; watershed management, especially soil moisture, evaporation, transpiration and snow accumulation and dissipation. *Mailing Add:* State Univ NY Col Environ Sci & Forestry Syracuse NY 13210

ESCHNER, EDWARD GEORGE, b NY, June 3, 13; m 36; c 5. RADIOLOGY. *Educ:* Univ Buffalo, MD, 36. *Prof Exp:* Resident orthop, Buffalo Gen Hosp, 37-38 & radiol, 38-40, assoc, 40-42; pvt pract, 46-47; actg head dept, 47-72, CLIN PROF RADIOL, SCH MED, STATE UNIV NY, BUFFALO, 54- *Concurrent Pos:* Dir radiol, E J Meyer Mem Hosp, 47-72. *Mem:* Radiol Soc NAm; Am Col Radiol; AMA; Pan-Am Med Asn. *Mailing Add:* 755 Ochard Park Rd Buffalo NY 14224-3320

ESCOBAR, JAVIER I, b Medellin, Colombia, July 26, 43; m 67; c 2. PSYCHIATRY, PSYCHOPHARMACOLOGY. *Educ:* Col San Ignacio, Medellin, Colombia, BS, 60; Univ Antioquia, Colombia, MD, 67; Univ Minn, Minneapolis, MSc, 73. *Prof Exp:* Resident psychiat, Complutense Univ, Madrid, Spain, 68-69; resident, Univ Minn Hosps, 69-72, res fel psychiat & genetics, 72-73, asst prof psychiat, Univ Minn, 73-76; assoc prof psychiat & pharmacol, Univ Tenn, Memphis, 76-79; assoc prof, 79-85, PROF PSYCHIAT, UNIV CALIF, LOS ANGELES, 85- *Concurrent Pos:* Proj psychiatrist, NIMH-PRB collab study, St Paul Ramsey Hosp, 73-76; staff psychiatrist, St Paul Ramsey Hosp & Med Ctr, 73-76, Memphis Ment Health Inst, 76-78 & Brentwood Vet Admin Med Ctr, 79-; consult psychopharmacol, Ment Health Ctr, Univ Tenn, 76-78, chmn res comt, Dept Psychiat & dir clin res unit, 78; mem task force psychoactive drugs, State Tenn, 78; dir, Neighborhood Ctr, Vet Admin, Los Angeles, 79-, chief Affective Dis CRC, Med Ctr, 84; co-prin investr, Los Angeles Epidemiol Catchment Area Prog, 80-; prin investr, Behav Family Therapy Schizophrenics, 84; founding fel, Pac Rim Col Psychiat. *Mem:* Am Psychiat Asn; fel Am Col Clin Pharmacol; Soc Biol Psychiat. *Res:* Clinical projects in schizophrenia and affective disorders, including testing of new treatments and assessment of diagnostic tests; cross-cultural psychiatry as it concerns somatization traits across different cultures; psychiatric epidemiology, cross-cultural prevalence of psychiatric disorders. *Mailing Add:* 232 S Main St West Hartford CT 06107

ESCOBAR, MARIO R, b Lima, Peru, Jan 31, 31; US citizen; m 59; c 5. VIROLOGY, IMMUNOLOGY. *Educ:* Univ Louisville, BA, 54; Georgetown Univ, MS, 60; Ind Univ, PhD(microbiol, biochem), 63; Am Bd Med Microbiol, cert med virol, 72, med immunol, 77; Am Bd Med Lab Immunol, 79 health & med virol, 72. *Prof Exp:* Med technologist, Clin Path Lab, St Mary & Elizabeth Hosp, Louisville, Ky, 52-54; supvr, Miller Clin, Morgantown, 54-55; res asst hemat, Walter Reed Army Inst Res, 55-57; serologist & supvr, Path & Serol Clin Labs, Children's Hosp, Washington, DC, 57-61; teaching asst bact & med microbiol, Ind Univ, 61-63; Nat Acad Sci-Nat Res Coun res assoc enteric bact, Commun Dis Ctr, USPHS, Ga, 63-64, dir clin path, Clin Pub Health Labs, Ohio, 64-65; Nat Cancer Inst fel cancer & virol, Univ Miami, 65-67; asst prof clin path, 67-71, assoc prof, 71-78, PROF PATH & SCI DIR, CLIN IMMUNOPATH & VIROL SECT, MED COL VA, VA COMMONWEALTH UNIV, 78- *Concurrent Pos:* Head diag, Res Virus & Immunol Labs, 67-76;consult, Vet Admin Med Ctr, Richmond, 74-; adj prof biol sci, Old Dominion Univ, 78- *Mem:* Reticuloendothelial Soc; Am Soc Microbiol; Am Soc Clin Path; Clin Immunol Soc; Am Asn Path; Sigma Xi; Am Asn Immunol. *Res:* Preservation of human erythrocytes; serology of histoplasma capsulatum; serology of Serratia; bacteriophages and immunogenetics of Salmonella; laboratory diagnosis of viral and immune diseases; virus-host interactions; oncogenic viruses; role of mononuclear cells in cancer immunity; immunopathology of viral hepatitis and AIDS. *Mailing Add:* Med Col Va Va Commonwealth Univ Univ Sta-Box 106 Richmond VA 23298-0106

ESCUE, RICHARD BYRD, JR, b Denton, Tex, Apr 24, 19; wid; c 3. CHEMISTRY. *Educ:* N Tex State Univ, BA, 39, MA, 40; Calif Inst Technol, Ph(chem & physics), 44. *Prof Exp:* Lab asst chem, N Tex State Univ, 40; Physicist, Neches Butane Prod Co, Tex, 43-45; asst prof chem, N Tex State Univ, 45-47, assoc prof chem & physics, 47-53, prof chem, 53-85; CONSULT, 85- *Concurrent Pos:* Res grants-in-aid, Res Corp, NY, 45-48, Robert A Welch Found, Tex, 55-64, Am Acad Arts & Sci, 56 & NIH, 57-58; fel Oak Ridge Inst Nuclear Studies, 51, counsr, 58-85. *Mem:* AAAS; Am Chem Soc. *Res:* Molten Salts; radiochemistry; phase systems; microscopy. *Mailing Add:* 707 Ridgecrest Denton TX 76205

ESDERS, THEODORE WALTER, b Buffalo, NY, May 6, 45; m 67; c 3. BIOCHEMISTRY. *Educ:* Gannon Col, BA, 67; Fla State Univ, PhD(chem), 71. *Prof Exp:* NIH fel fac chem, Harvard Univ, 71-73; sr res chemist, 73-80, RES ASSOC, RES LABS, EASTMAN KODAK CO, 80- *Mem:* Am Soc Microbiol. *Res:* Lipid metabolism and the enzymes involved, as well as processes by which lipid metabolism is controlled; isolation and characterization of microbial enzymes. *Mailing Add:* 741 Sugar Creek Trail West Webster NY 14580

ESEN, ASIM, b Fethiye, Turkey, Nov 10, 38; m 66; c 2. BIOCHEMISTRY, AGRONOMY. *Educ:* Univ Ankara, Turkey, Dipl, 65; Univ Calif, Riverside, MS, 68, PhD(plant genetics), 71. *Prof Exp:* Fel plant genetics, Univ Calif, Riverside, 72-74, biochem, 74-75; asst prof, 75-81, ASSOC PROF BOT, VA POLYTECH INST & STATE UNIV, 81- *Mem:* Am Genetic Asn; AAAS; Am Chem Soc. *Res:* Molecular biology of crop plants; structure, function and evolution of maize beta-glucosidases; immunochemistry of prolamins in maize and other grasses. *Mailing Add:* Dept Biol Va Polytech Inst & State Univ Blacksburg VA 24061

ESFAHANI, MOJTABA, b Broujerd, Iran, May 7, 39; m 66; c 2. BIOCHEMISTRY. *Educ:* Am Univ Beirut, BS, 63, MS, 65; Duke Univ, PhD(biochem), 70. *Prof Exp:* Res assoc biochem, Duke Univ, 70-71; from instr to asst prof biochem, Baylor Col Med, 71-73; asst biochemist, M D Anderson Hosp & Tumor Res Inst, 73-75; assoc prof biol sci, Drexel Univ, 75-78; assoc prof 78-89, PROF BIOL CHEM, HAHNEMANN MED COL, 89- *Res:* Membrane structure and function; lipid metabolism. *Mailing Add:* Dept Biol Chem Hahnemann Med Col 85230 N Broad St Philadelphia PA 19102

ESHBACH, JOHN ROBERT, b Bethlehem, Pa, Oct 7, 22; m 44; c 4. SOLID STATE PHYSICS. *Educ:* Northwestern Univ, BS, 44, BS, 46, MS, 47; Mass Inst Technol, PhD(physics), 51. *Prof Exp:* Res assoc photoconductive cells, Northwestern Univ, 46-47; res asst microwave spectros, Mass Inst Technol, 48-50; mgr light prod studies, Gen Elec Res & Develop Ctr, 62-68, mgr microwave br, 68-74, res assoc, 51-85; RETIRED. *Concurrent Pos:* Consult, 85- *Mem:* Fel Am Phys Soc; Inst Elec & Electronics Engrs; Sigma Xi. *Res:* Microwave spectroscopy; magnetism; ferrite devices; luminescence; microwave electronics. *Mailing Add:* 2755 Rosendale Rd Schenectady NY 12309-1305

ESHBAUGH, WILLIAM HARDY, b Glen Ridge, NJ, May 1, 36; m 58; c 4. PLANT TAXONOMY, ETHNOBOTANY. *Educ:* Cornell Univ, AB, 59; Ind Univ, MA, 61, PhD(bot), 64. *Prof Exp:* Lectr bot, Ind Univ, 62; asst prof bot & cur herbarium, Southern Ill Univ, 65-67; from asst prof to assoc prof, 67-77, chmn, 83-88, PROF BOT, MIAMI UNIV, 77- *Concurrent Pos:* Prog dir syst biol, NSF, 82-83; mem coun, Am Soc Plant Taxon, 84-86. *Mem:* Am Soc Plant Taxon; Bot Soc Am (pres-elect, 87-88, pres, 88-89, past pres, 89-90); Am Inst Biol Sci; Int Asn Plant Taxon; Soc Econ Bot (secy, 78-81, vpres, 82-83, pres, 83-84); fel AAAS (pres elect, 90-91, pres, 91-92). *Res:* Biosystematic and phytogeographic studies in the Solanaceae, especially the genus Capsicum and allied genera; Flora of the Bahamas; medicinal plants-ethnobotany. *Mailing Add:* Dept Bot Miami Univ Oxford OH 45056

ESHLEMAN, RONALD L, b Shellsville, Pa, Aug 24, 33; m 59. MECHANICAL & AEROSPACE ENGINEERING. *Educ:* Lafayette Col, BSME, 59; Lehigh Univ, MS, 61; Ill Inst Technol, PhD(mech & aerospace eng), 67. *Prof Exp:* Teaching asst mech eng, Lehigh Univ, 59-61; instr, Ill Inst Technol, 61-64; from asst res engr to res engr, IIT Res Inst, 64-69, sr res engr, 69-73, sci adv, 73-75, DIR & PRES, VIBRATION INST, 75- *Concurrent Pos:* Consult, Am Nat Standards Inst, 68-, chmn comt S2; tech ed, Shock & Vibration Digest, 69- *Mem:* Am Soc Mech Engrs; Am Sci Affil; Inst Environ Sci; Soc Automotive Engrs; Sigma Xi; Vibration Inst; Aerospace Industs Asn Am. *Res:* Engineering analysis; rotor dynamics; shock and vibration isolation; dynamics of machines; digital simulation of machines; tape dynamics; vehicle dynamics; torsional vibrations; machine condition and fault analysis; modal analysis. *Mailing Add:* 333 Ridge Ave Clarendon Hills IL 60514

ESHLEMAN, VON R(USSEL), b Darke Co, Ohio, Sept 17, 24; m 47; c 4. PLANETARY EXPLORATION, ELECTRICAL ENGINEERING. *Educ:* George Washington Univ, BEE, 49; Stanford Univ, MS, 50, PhD(elec eng), 52. *Prof Exp:* Res assoc, Radio Propagation Lab, Stanford Univ, 52-56, from instr to assoc prof elec eng, 56-61, dir, Radiosci Lab, 74-83, PROF ELEC ENG & CO-DIR, CTR RADAR ASTRON, STANFORD UNIV, 61- *Concurrent Pos:* Consult, Nat Acad Sci, Nat Bur Standards, SRI Int & Jet Propulsion Lab; mem, Int Astronaut Cong, Int Astron Union & Int Sci Radio Union; dir, Watkins-Johnson Co; radio sci team leader, Voyager missions to Jupiter & Saturn, 76-80; radio sci team mem, Galileo Mission to Jupiter 79-80. *Mem:* Nat Acad Eng; Fel AAAS; fel Inst Elec & Electronics Engrs; Am Geophys Union; fel Royal Astron Soc. *Res:* Radar astronomy; planetary exploration; ionospheric and plasma physics; radio wave propagation; astronautics. *Mailing Add:* Ctr Radar Astron Stanford Univ Stanford CA 94305

ESKELSON, CLEAMOND D, b American Fork, Utah, Sept 27, 27; m 46; c 1. BIOCHEMISTRY, ORGANIC CHEMISTRY. *Educ:* Univ Utah, BS, 50; Univ Louisville, MS, 57; Univ Nebr, Omaha, PhD(biochem), 67. *Prof Exp:* Res technician, Dept Internal Med, Univ Utah, 50-51; analyst, Geneva Steel Co, 51-52; biochemist, Army Med Res Lab, Ft Knox, Ky, 52-55; clin chemist, Vet Admin Hosp, Omaha, Nebr, 57-59, biochemist, Vet Admin Radioisotope Lab, 59-67; chemist, Vet Admin Hosp, Tucson, 67-76, RES ASSOC & ADJ ASSOC PROF BIOCHEM, SCH PHARM, UNIV ARIZ, 76- *Concurrent Pos:* Consult, Creighton Univ, 58-59; instr, Col Med, Univ Nebr, 67-68; Licensed Beverage Indust res grant, 68-69; res assoc, Sch Pharm, Univ Ariz, 72-76. *Mem:* AAAS; NY Acad Sci; Am Chem Soc; Soc Nuclear Med; Asn Clin Med; Sigma Xi. *Res:* Thyroid physiology; effects of x-irradiation on biological systems; mechanisms for controlling cholesterolgenesis; vitaminology; development of radiometric procedures for diagnosing cancer; biochemistry of alcoholism. *Mailing Add:* 7402 Calle Toluca Tucson AZ 85710

ESKELUND, KENNETH H, b Waterville, Maine, Feb 13, 24; m 50; c 3. VETERINARY MEDICINE. *Educ:* Mich State Univ, DVM, 51. *Prof Exp:* Diagnostician, SJersey Diag Lab, 51-52; vet-in-chg poultry, State of Ind, 52-53; vet mgr, Ft Halifax Poultry Co, 53-57; pres & dir, Maine Biol Labs, Inc, 57-66, gen mgr, Maine Biol Lab Div, Morton-Norwich Prod Inc, 66-73; pres, Northeast Lab Serv, 71-88; PRES & MGR, MAINE BIOL LABS, INC, 75- *Concurrent Pos:* Pres, Maine Poultry Serv, 58-72 & Maine Poultry Consults, 59- *Mem:* Am Vet Med Asn; Am Asn Avian Path; Am Asn Indust Vet; US Animal Health Asn; Asn Avian Vets. *Res:* Development of inactivated avian vaccines. *Mailing Add:* Maine Biol Labs PO Box 255 Waterville ME 04093-0255

ESKEW, CLETIS THEODORE, b Cloud Chief, Okla, July 10, 04; m 36; c 1. BIOLOGY. *Educ:* Southwestern State Univ, Okla, BS, 31; Univ Okla, MS, 37; NTex State Univ, EdD, 60. *Prof Exp:* Prof biol, Mangum Jr Col, 37-39 & Southwest State Univ, Okla, 39-42; prof, 42-69, dean admis, 48-56, dean instr, 56-59, dean lib arts, 59-61, dir div sci & math, 61-68, assoc dean instr, 61-69, dean div grad studies, 68-69, EMER PROF BIOL, MIDWESTERN STATE UNIV, 75- *Res:* Taxonomy and ecology; plant physiology; vegetation of Oklahoma and Texas. *Mailing Add:* 2718 Chase Dr Wichita Falls TX 76308-5255

ESKEW, DAVID LEWIS, b Lebanon, Tenn, Nov 18, 50. PLANT PHYSIOLOGY, AGRONOMY. *Educ:* Univ Tenn, BS, 71; Univ Wis, MS, 73, PhD(agron & bot), 75. *Prof Exp:* Res biologist, Univ Calif, Riverside, 75-77; assoc, Boyce Thompson Inst, 77-80, ASSOC, US PLANT, SOIL & NUTRIT LAB, CORNELL UNIV, 80- *Mem:* Am Soc Plant Physiologists; Crop Sci Soc Am; Am Soc Agron. *Res:* Biochemistry and physiology of nitrogen fixation; nitrogen fixing organisms living in association with C-4 plants and rice. *Mailing Add:* Plant Molecular Genetics Univ Tenn 269 Ellington Places Sci Bldg Knoxville TN 37901

ESKIN, NEASON AKIVA MICHAEL, b Birmingham, Eng, May 4, 41; m 70; c 4. FOOD CHEMISTRY, BIOCHEMISTRY. *Educ:* Univ Birmingham, BSc, 63, PhD(physiol chem), 66. *Prof Exp:* Lectr biochem, Borough Polytech, Eng, 66-68; head, Dept Foods & Nutrit, 80-85, PROF FOOD CHEM, UNIV MAN, 68- *Concurrent Pos:* Lady Davis fel, 83-84. *Mem:* Can Inst Food Sci & Technol; Brit Inst Food Sci & Technol; Inst Food Technologists; Am Oil Chem Soc. *Res:* Food enzymes, rancidity in foods; cereal chemistry; phytate and mineral availability; vegetable oil quality; titanium chloride properties and applications to food analysis. *Mailing Add:* Dept Foods & Nutrit Univ Man Fac Human Ecol Winnipeg MB R3B 2E9 Can

ESKINAZI, SALAMON, mechanics, for more information see previous edition

ESKINS, KENNETH, b Beckley, WVa. PHOTOSYNTHESIS, GENETICS. *Educ:* Wheaton Col, BS, 62; Southern Ill Univ, PhD(org chem), 66. *Prof Exp:* Fel, Roswell Park Cancer Res Inst, 66-67; RES CHEM, NORTHERN REGIONAL RES CTR, AGR RES SERV, USDA, 67- *Mem:* Am Soc Plant Physiol. *Res:* Analysis of chloroplast pigments; pigment protein complexes and control of chloroplast development by light, aging and genetic factors; light control of greening and photomorphogenesis. *Mailing Add:* Northern Regional Res Ctr 1815 N University St Peoria IL 61604

ESLICK, ROBERT FREEMAN, plant breeding, agronomy; deceased, see previous edition for last biography

ESLYN, WALLACE EUGENE, b Lawrenceville, Ill, Nov 13, 24; m 47; c 5. WOOD PRODUCTS PATHOLOGY. *Educ:* Univ Mont, BS, 50, MS, 53; Iowa State Col, PhD(plant path), 56. *Prof Exp:* Plant pathologist, Forest Prod Lab, US Forest Serv, 57 & Forest Insect & Dis Lab, NMex, 57-60, plant pathologist, Forest Prod Lab, 60-69, supv res plant pathologist, biodegradation of wood res work unit, 69-84; consult Shiitake res, NC State Univ, 85-86; CONSULT, 87- *Mem:* Soc Am Foresters; Am Phytopath Soc; Mycol Soc Am; Int Union Forestry Res Orgn; Int Res Group Wood Preservation. *Res:* Forest products pathology and mycology. *Mailing Add:* Rte 1 Box 579-A Lenoir NC 28645

ESMAIL, MOHAMED NABIL, b Port-Said, Egypt, Jan 22, 42; m 64; c 2. UNIT OPERATIONS, MODELLING. *Educ:* Moscow State Univ, BSc & MSc, 64, PhD(appl math), 72. *Prof Exp:* Lectr mech, Ein-Shams Univ, Egypt, 65-72, asst prof, 72-73; res assoc chem eng, Univ Toronto, Can, 73-76; from asst prof to assoc prof, 77-82, PROF CHEM ENG, UNIV SASK, CAN, 82-, HEAD CHEM ENG, 82- *Mem:* Can Soc Chem Eng; Am Inst Chem Eng; Asn Prof Engrs Ont; Asn Prof Engrs Sask. *Res:* Fluid mechanics and engineering applications, flows of thin liquid films, stability and analysis; liquid coating process; extrudate swell of newtonian and nonnewtonian liquids; numerical methods in fluid mechanics and modelling. *Mailing Add:* Univ Sask Rm 141 Thorvaldson Bldg Saskatoon SK S7N 0W0 Can

ESMAY, DONALD LEVERN, b Murdo, SDak, Nov 1, 17; m 45; c 6. POLYMER CHEMISTRY, ADHESIVES. *Educ:* Dakota Wesleyan Univ, BA, 46; Iowa State Univ, PhD(chem), 51. *Prof Exp:* Res chemist, Standard Oil Co, Ind, 52-55; tech dir org prod div, Lithium Corp Am, 56-60; sr res specialist, Minn Mining & Mfg Co, 61-82; RETIRED. *Mem:* Am Chem Soc; Sigma Xi. *Res:* Organometallic and fluorine compounds; heterocycles; hydrocarbon conversions; petroleum, lithium, propellant and fluorine chemistry; polymers; organic syntheses; adhesives; polyurethanes; foams. *Mailing Add:* 1237 98th Lane NW Coon Rapids MN 55433

ESMAY, MERLE L(INDEN), b Greene Co, Iowa, Dec 27, 20; m 42; c 2. AGRICULTURAL ENGINEERING. *Educ:* SDak State Univ, BSAE, 42; Iowa State Univ, MSAE, 47, PhD(agr eng & struct eng), 51. *Prof Exp:* Exten agr engr, SDak State Univ, 46; asst prof teaching & res, Iowa State Univ, 47-51; prof, Univ Mo, 51-55; chief party, univ contract adv team Taiwan, 62-64, prof agr eng, Mich State Univ, 55-86; PRES, AGR TECHNOL CONSULT CORP, 86- *Concurrent Pos:* Partic in various int symposia & meetings, 62-; agr eng consult in various develop countries, 62- *Honors & Awards:* Metal Bldg Mfrs Asn Award, Am Soc Agr Engrs, 66; Int Kishida Award, Am Soc Agr Engrs, 82. *Mem:* Fel Am Soc Agr Engrs; Am Soc Eng Educ. *Res:* Environmental requirements of livestock shelters; rice drying, storage and handling; mechanization in developing countries; institution building in developing countries. *Mailing Add:* Agr Technol Consult Corp 1272 Scott Dr East Lansing MI 48823

ESMEN, NURTAN A, b Ankara, Turkey, Jan 1, 40; m 69; c 1. INDUSTRIAL HEALTH ENGINEERING, AEROSOL PHYSICS. *Educ:* Northeastern Univ, BS, 65; Univ Pittsburgh, MS, 66, PhD(chem eng), 70. *Prof Exp:* Sr researcher aerosol physics, Univ Pittsburgh, 65-70; asst prof environ eng, Univ Del, 70-74; prin technol dir, Owens-Corning Fiberglas Co, 74-75; assoc prof, Univ Pittsburgh, 75-80, prof indust hyg, Grad Sch Pub Health, 80-90; CONSULT, ESMEN RES & ENG, 90- *Concurrent Pos:* Mem study sect, Nat Inst Occupational Safety & Health, HEW, 80-84. *Mem:* Am Inst Chem Engrs; Am Indust Hyg Asn; Am Chem Soc; Sigma Xi; Am Conf Gov Indust Hyg. *Res:* Application of aerosol physics and chemical engineering to the investigation and solution of industrial health problems; development of risk assesment techniques and methodology for occupational exposures to potentially hazardous agents. *Mailing Add:* Esem Res & Eng 2531 Wickline Rd Gibsonia PA 15044

ESMON, CHARLES THOMAS, b Centralia, Ill, Apr 3, 47; m 75. BIOCHEMISTRY. *Educ:* Univ Ill, BS, 69; Univ Wash, PhD(biochem), 73. *Prof Exp:* Asst prof path, Health Sci Ctr, Univ Okla, 76-80, adj asst biochem, 76-81; assoc mem, Thrombosis/Hemat Res Prog, 82-85, MEM, C V BIOL

RES PROG, OKLA MED RES FOUND, 85-; ASSOC PROF PATH, HEALTH SCI CTR, UNIV OKLA, 80-, ASSOC PROF BIOCHEM, 81- *Concurrent Pos:* Prin investr var grants, 76-91; mem, Microbiol & Immunol Rev Comt, Am Heart Asn, 79-82, Thrombosis & Haemostasis Int Adv Comt, 80-83, established investr, 80-85, pres, Okla Affil, 84, mem, Fel Rev Comt, 87-90; mem, Hemat St Sect, NIH, 87-91, ad hoc reviewer; Distinguished Investr Award for Contrib to Hemostasis, Ninth Int Cong on Thrombosis & Hemostasis, Stockholm, Sweden, 83. *Honors & Awards:* John L Dickenson Mem Award, Am Heart Asn, 78 & 79, Sol Sherry lectr, Thrombosis, 90. *Mem:* Am Soc Biol Chemists; AAAS. *Res:* Control of blood coagulation and fibrinolysis; structure and function of enzymes; cell surface receptors; role of the endothelial cell in regulation of hemostasis. *Mailing Add:* Okla Med Res Found 825 NE 13th St Oklahoma City OK 73104

ESOGBUE, AUGUSTINE O, b Kaduna, Nigeria, Dec 25, 40. SYSTEMS & ELECTRICAL ENGINEERING. *Educ:* Univ Calif, Los Angeles, BS, 64; Columbia Univ, MS, 65; Univ Southern Calif, PhD(eng & opers res), 68. *Prof Exp:* Res assoc eng & med, Univ Southern Calif, 65-68; asst prof opers res, Case Western Reserve Univ, 68-72; assoc prof eng, 72-77, PROF ENG, GA INST TECHNOL, 77- *Concurrent Pos:* Instr sci, Lagos Ctr Higher Studies, Lagos, Nigeria, 60-61; adj prof community med, Morehouse Sch Med, Morehouse Col, 79-; reader & instr, Dept Math, Columbia Univ, 65, assoc fac mem, 70-; consult, Environ Dynamics, Calif, 68-70; prin & consult, Univ Assocs, Inc, Ohio, 69-72; prof in residence, Sch Eng, Howard Univ, 72; vpres, Atlantic Systs Inc, Ga, 72-80, pres, AESO Systs Inc, 80-; mem numerous panels, Nat Acad Sci-Nat Res Coun, 73-; adv ed, Int J Fuzzy Sets & Systs & US rep, Int Ctr Cybernetics & Systs, Romania, 76-; assoc ed, J Math Anal & Appln & assoc ed at large, Opers Res Soc Am Health News, 77-82; chmn Opers Res Soc Am & Inst Mgt Sci Vis Lectr Prog, 77-80. *Mem:* Opers Res Soc Am; Sigma Xi; fel AAAS. *Res:* Dynamic programming and optimal control theory; fuzzy sets and decision making in fuzzy environments; large scale systems analysis; operations research; theory and applications to health care, water resources and pollution; urban systems and transportation; computer sciences; software systems design. *Mailing Add:* Sch Indust & Systs Eng Ga Inst Technol 225 North Ave NW Atlanta GA 30332

ESPACH, RALPH H, JR, b Bartlesville, Okla, May 10, 32. EXPLORATION GEOLOGY, PETROLEUM GEOLOGY. *Educ:* Columbia Col, BA, 54; Univ Wyo, MA, 58. *Prof Exp:* Jr engr & jr geologist, Ohio Oil Co, 50-54; explor geologist, Standard Oil Calif, US Army CEngr, 54- 54; proj explor geologist, Tenneco Oil Co, 64-70; CONSULT, RALPH ESPACH INC, 70- *Mem:* Fel Geol Soc Am; Am Asn Petrol Geologists; Sigma Xi. *Mailing Add:* 5024 NW 61st Pl Oklahoma City OK 73122

ESPANA, CARLOS, b Mexico City, Mex, Mar 1, 19; m 49. INFECTIOUS DISEASES. *Educ:* Nat Polytech Inst, Mex, MS, 43; Univ Calif, Berkeley, PhD, 49. *Prof Exp:* Head sect virol, E R Squibb & Sons, 49-51, res assoc microbiol, Squibb Inst Med Res, 51-52, head dept chemother, 52-54; res assoc vet med, Univ Pa, 54-57, res asst prof, 57-63; specialist, Nat Ctr Primate Biol, Univ Calif, Davis, 64-68, res assoc virol, 68-78; RETIRED. *Concurrent Pos:* Prin investr grants, E R Squibb & Sons of Mex Div, Mathieson Corp, 54-60 & Nat Inst Allergy & Infectious Dis, 60-63; rep, Nat Anaplasmosis Conf, 57 & 62; mem ad hoc comt of consult on simian viruses, Diag Virol Sect, Viral Carcinogenesis Br, Nat Cancer Inst, 65-67; partic, Workshops on Virus Dis of Non-Human Primates, 68 & 71 & Int Conf Exp Med & Surg in Primates, 69; mem working team, Prog Compar Virol & Simian Viruses, WHO, Food & Agr Orgn UN, 73- *Mem:* Am Soc Microbiol; fel NY Acad Sci; fel Am Acad Microbiol. *Res:* Diagnosis, pathogenesis, prevention and control of viral diseases of non-human primates; studies on the nature of slow virus diseases also known as spongiform encephalopathies or transmissible virus dementias; Kuru, Creutzfeldt-Jakob disease. *Mailing Add:* 809 Cherry Lane Davis CA 95616

ESPELIE, KARL EDWARD, b New London, Conn, Apr 20, 46; m 74; c 1. LIPID CHEMISTRY, ECOLOGY. *Educ:* Augustana Col, AB, 67; Univ Wis-Madison, MSc, 70, PhD(biochem), 72. *Prof Exp:* Fel biol, Calif Inst Technol, 72-74, protein chem, Imp Cancer Res Fund, London, 74-76; res scientist plant biochem, Inst Biol Chem, 76-85, assoc prof, dept entom, Wash State Univ, 85-86; assoc entomologist, 86-90, ASSOC PROF, DEPT ENTOM, UNIV GA, 90- *Mem:* Am Soc Plant Physiologists; Entom Soc Am; Int Soc Chem Ecol. *Res:* Plant/insect interactions; chemistry of insect integument; chemistry and biosynthesis of the plant polymers cutin and suberin which serve as protective barriers on the aerial and subterranean surfaces of all plants. *Mailing Add:* Dept Entom Univ Ga Athens GA 30601

ESPELIE, (MARY) SOLVEIG, mathematical functional analysis, topology; deceased, see previous edition for last biography

ESPENSHADE, EDWARD BOWMAN, JR, b Chicago, Ill, Oct 23, 10; m 39; c 2. CARTOGRAPHY. *Educ:* Univ Chicago, BS, 30, MS, 32, PhD(geog), 44. *Prof Exp:* Instr geog, Univ Chicago, 37-44; chmn dept, 57-77, EMER PROF, NORTHWESTERN UNIV, 77- *Concurrent Pos:* Cartog consult, Rand McNally & Co, 45-, adv geog ed, 54-; chmn, Earth Sci Div, Nat Acad Sci-Nat Res Coun, 60-62; mem, Comn Higher Educ, NCent Asn, 70-75, mem exec bd, 72-77. *Mem:* Asn Am Geog (pres, 64-65); Am Cong Surv & Mapping; Can Cortographic Soc. *Res:* Design and development of thematic maps. *Mailing Add:* 1440 Sheridan Rd Apt 605 Wilmette IL 60091

ESPENSON, JAMES HENRY, b Los Angeles, Calif, Apr 1, 37; m 60; c 1. INORGANIC CHEMISTRY, ORGANOMETALLIC CHEMISTRY. *Educ:* Calif Inst Technol, BS, 58; Univ Wis, PhD(inorg chem), 62. *Prof Exp:* Res assoc, Stanford Univ, 62-63; from instr to assoc prof, 63-71, PROF CHEM, IOWA STATE UNIV, 71-, DISTINGUISHED PROF, 88- *Concurrent Pos:* Fel, Alfred P Sloan Found, 68-70. *Mem:* Fel AAAS; Am Chem Soc; Sigma Xi. *Res:* Inorganic and organometallic reaction mechanisms; kinetics of reactions of metal complexes; formation and cleavage of metal-carbon bonds; electron transfer and substitution reactions; homogeneous catalysis; free radical reactions. *Mailing Add:* Dept Chem Iowa State Univ Ames IA 50011

ESPERSEN, GEORGE A, b Jersey City, NJ, May 17, 06. MICROWAVES, ELECTRON OPTICS. *Educ:* NY Univ, BS, 31. *Prof Exp:* Engr, Hygrade-Sylvania Corp, 32-39, Nat Union Radio Corp, 39-40, Sperry Gyroscope Co, 42-45; res engr, 45-59, section chief, Microwave Tube & Electron Optic Sect, North Am Phillips Labs, 59-71; RETIRED. *Concurrent Pos:* Andrew Carnegie Scholar, Inst Theoret Physics, Copenhagen, 29-30; fel Inst Res Engrs, 56. *Honors & Awards:* Samuel Morse Medal for Excellence in Physics, NY Univ, 29. *Mem:* Inst Res Engrs; Am Phys Soc; Inst Elec & Electronics Engrs. *Res:* Electron tube development; microwave tube development; transmitting tube development. *Mailing Add:* 65 Bellewood Ave Dobbs Ferry NY 10522

ESPEY, LAWRENCE LEE, b Mercedes, Tex, Sept 5, 35; m 78; c 4. REPRODUCTIVE PHYSIOLOGY, ENDOCRINOLOGY. *Educ:* Univ Tex, BA, 58, MA, 61; Fla State Univ, PhD(physiol), 64. *Prof Exp:* Res technologist physiol, Dent Br, Univ Tex, 59-60; NIH fel & res assoc, Univ Mich, 64-66; from asst prof to assoc prof, 66-72, prof physiol & chmn, dept environ studies, 73-80, Cowles prof life sci, 79-82, PROF BIOL, TRINITY UNIV, 80- *Concurrent Pos:* Nat Inst Child Health & Human Develop grants, 67-; Morrison Trust grant, 67-68; adj prof, dept obstet & gynec, Univ Tex Health Sci Ctr, San Antonio, 75-; scholar pop affairs, US Dept State, 76; Fulbright-Hays sr res award, Romania, 77-79; guest res scholar, Fac Med, Kyoto Univ, Japan, 82; vis prof, dept pharmacol med, Univ SC, 84. *Mem:* AAAS; Am Physiol Soc; Endocrine Soc; Soc Study Reproduction; Am Fertil Soc. *Res:* Physical and chemical mechanisms of mammalian ovulation. *Mailing Add:* Dept Biol Trinity Univ San Antonio TX 78212

ESPINO, RAMON LUIS, chemical engineering, for more information see previous edition

ESPINOSA, ENRIQUE, b Quillota, Chile. LABORATORY CLINICAL IMMUNOLOGY, IMMUNOPATHOLOGY. *Educ:* Univ Chile, BSc, 47, MD, 55. *Prof Exp:* Asst prof physiopath, Sch Med, Univ Chile, Santiago, 56-64; fel med, Sch Med, Case Western Reserve Univ, Cleveland, 65-67, asst prof med, 67-70; assoc prof, 70-76, PROF PATH, SCH MED, UNIV LOUISVILLE, 76- *Concurrent Pos:* Chief, Sect Immunopath & Serol, Humana Hosp Univ, Louisville, Ky, 81- *Mem:* Am Asn Pathologists; Am Asn Immunologists; Int Acad Path; AAAS. *Res:* Antigenic components of normal and abnormal tissues in their relationship to pathogenesis and diagnosis of disease; role of autoantibodies in diagnosis and pathogenesis of autoimmune disorders. *Mailing Add:* Dept Path Sch Med Univ Louisville Louisville KY 40292

ESPINOSA, JAMES MANUEL, b Mexico City, Mex, Nov, 7, 42; US citizen; m 66; c 2. THEORETICAL PARTICLE PHYSICS. *Educ:* Calif Inst Technol, BS, 65; Univ Calif, Los Angeles, MS, 70, PhD(physics), 76. *Prof Exp:* Asst prof physics, Loyola Marymount Univ, 72-78; ASST PROF PHYSICS, TEX WOMAN'S UNIV, 78- *Mem:* Am Phys Soc; Am Asn Physics Teachers; Hist Sci Soc; Sigma Xi. *Res:* General relativity; astrophysics; mathematics and history of science. *Mailing Add:* 3505 Dunes Denton TX 76201

ESPINOZA, LUIS ROLAN, CLINICAL IMMUNOLOGY, RHEUMATOLOGY. *Educ:* Cayetano Heredia Univ, Lima, Peru, MD, 69. *Prof Exp:* PROF INTERNAL MED, LA STATE UNIV SCH MED, NEW ORLEANS, 90- *Res:* Infectious diseases. *Mailing Add:* Dept Internal Med La State Univ Sch Med 1542 Tulane Ave New Orleans LA 70112-2822

ESPLIN, BARBARA, NEURO-PHARMACOLOGY, ANTI-EPILEPTIC DRUGS. *Educ:* Med Acad, Warsaw, Poland, MD, 52, PhD(pharmacol), 61. *Prof Exp:* ASSOC PROF NEURO-PHARMACOL, MCGILL UNIV, 77- *Mailing Add:* Dept Pharmacol & Therapeut McGill Univ Montreal PQ H3G 1Y6 Can

ESPOSITO, F PAUL, b Newark, NJ, Aug 24, 44. THEORETICAL PHYSICS. *Educ:* Cornell Univ, AB, 66; Univ Chicago, PhD(physics), 71. *Prof Exp:* Asst prof, 71-77, ASSOC PROF PHYSICS & MATH, UNIV CINCINNATI, 77- *Mem:* Am Phys Asn; Am Math Asn; Am Astron Soc; Int Astron Union; Sigma Xi. *Res:* General relativity; differential geometry. *Mailing Add:* Dept Physics Univ Cincinnati 316A Braun Cincinnati OH 45221

ESPOSITO, JOHN NICHOLAS, b Youngstown, Ohio, July 23, 38; m 63, 87; c 3. INORGANIC CHEMISTRY. *Educ:* Youngstown Univ, BS, 60; Case Inst Technol, PhD(chem), 66. *Prof Exp:* Sr res scientist, Westinghouse Elec Corp, 66-73, mgr corrosion & solution chem, Res Lab, 74-76, mgr chem technol, Nuclear Steam Generation Div, 76-81, mgr steam generator opers, Nuclear Technol Div, 81-89, ADV ENGR, NUCLEAR SERV DIV, WESTINGHOUSE ELEC CORP, 89- *Mem:* Nat Asn Corrosion Engrs; Am Chem Soc. *Res:* Combustion of metals; inorganic chemistry of hexacoordinated silicon and germanium; surface chemistry; high temperature chemistry; corrosion; water chemistry; electrodeposition; nuclear steam generators. *Mailing Add:* 8014 Westmoreland Ave Pittsburgh PA 15218-1721

ESPOSITO, LARRY WAYNE, b Schenectady, NY, Apr 15, 51; m 75; c 2. ATMOSPHERIC SCIENCES. *Educ:* Mass Inst Technol, SB, 73; Univ Mass, Amherst, PhD(astron), 77. *Prof Exp:* RES ASSOC, ATMOSPHERIC & SPACE PHYSICS LAB, PLANETARY ATMOSPHERIC SCI, UNIV COLO, 77-, ASSOC PROF, DEPT ASTROPHYS, 84- *Concurrent Pos:* Lectr, Dept Astro-Geophys, Univ Colo, 79-84; appointee, Mgt Oper Working Group Planetary Atmospheres, NASA, 81-84; exec mem COSPAR Comn B; mem Comt Planetary & Lunar explor, Space Sci Bd, Nat Acad Sci, 82-86, chmn, 89-, dep chmn, Task Group Planetary Explor, 84-86, mem Space Sci Bd, 86-; comt mem, Dir Planetary Sci, Am Astron Soc, 83-86. *Honors & Awards:* Harold C Urey Prize, Am Astron Soc, 85; Except Sci Achievement Medal, NASA, 86; Richtmyer Mem Lectr Award, Amer Phys Soc & Am Asn Phys Teachers, 91. *Mem:* Am Astron Soc; Am Geophys Union; Int Astron Union. *Res:* Spectroscopy, photometry and polarimetry of solar system objects; development of associated radiative transfer methods; theoretical studies of planetary atmospheres; dynamics of planetary rings; investigations of Pioneer Venus, Pioneer Saturn, Voyager, Galileo; Mars Observer; USSR Phobos; USSR Mars 94; Cassini. *Mailing Add:* Lab Atmospheric & Space Physics Univ Colo Campus Box 392 Boulder CO 80309-0392

ESPOSITO, MICHAEL SALVATORE, b Brooklyn, NY, Nov 30, 40; c 1. GENETICS, BIOCHEMISTRY. *Educ:* Brooklyn Col, BS, 61; Univ Wash, PhD(genetics), 67. *Prof Exp:* NIH fel, Lab Molecular Biol, Univ Wis-Madison, 67-69; asst prof, Univ Chicago, 69-74, actg chmn dept, 78, assoc prof biol, 75-80; div fel biol & med, 80-81, sr staff scientist, Biol & Med Div, 81-88, DEP DIR, CELL & MOLECULAR BIOL DIV, LAWRENCE BERKELEY LABS, UNIV CALIF, 88- *Concurrent Pos:* NSF grants, 69-70, 71-73, 73-74 & 75-; mem genetics study sect, NIH, 76-78; mem, Gov Comt Life Sci, Nat Ctr Sci Res, France; assoc ed, Current Genetics, 79- *Mem:* AAAS; Genetics Soc Am; Am Soc Microbiol. *Res:* Yeast genetics and physiology; biochemical and genetic regulation of gene mutation, recombination, meiosis and sporulation. *Mailing Add:* Cell & Molecular Biol Div Lawrence Berkeley Lab Univ Calif M/S 934-47A Berkeley CA 94720

ESPOSITO, PASQUALE BERNARD, b New York, NY, Feb 28, 40; m 64; c 3. CELESTIAL MECHANICS, GEOPHYSICS. *Educ:* Manhattan Col, BS, 64; Univ Pa, MS, 66; Yale Univ, PhD(celestial mech), 69. *Prof Exp:* MEM TECH STAFF EARTH & SPACE SCI, JET PROPULSION, CALIF INST TECHNOL, 70- *Mem:* Am Geophys Union; Am Astron Soc. *Res:* Planetary science, such as gravity fields of planets; fundamental geodetic constants such as goecentric gravitational constant; solar physics such as solar corona electron density distribution. *Mailing Add:* 4800 Oak Grove Pasadena CA 91103

ESPOSITO, RAFFAELE, b Rome, Italy, Aug 11, 32; m 62; c 2. COMMUNICATION SYSTEMS. *Educ:* Univ Rome, DrIng, 56. *Prof Exp:* Tech dir, Selenia Spa, 75-81, dep gen mgr, 81-84, gen mgr, 84-90, GEN MGR, ALENIA-AERITALIA & SELENIA, SPA, 90- *Mem:* Fel Inst Elec & Electronics Engrs. *Res:* Communications theory; systems analysis. *Mailing Add:* Via Tiburtina Km 12 400 Rome 00131 Italy

ESPOSITO, ROCHELLE E, b Brooklyn, NY, June 28, 41; c 2. GENETICS. *Educ:* Brooklyn Col, BS, 62; Univ Wash, PhD(genetics) 67. *Prof Exp:* Asst prof, 69-75, ASSOC PROF BIOL, UNIV CHICAGO, 75- *Mem:* Genetics Soc Am; Am Soc Microbiol. *Res:* Genetic recombination; genetic and biochemical control of meiosis. *Mailing Add:* Dept Biol Ebc 308A Molecular Genetics Cell Biol Univ Chicago 920 E 58th St Chicago IL 60637

ESPOSITO, VITO MICHAEL, b Logan, WVa, Sept 11, 40; m 60; c 2. IMMUNOLOGY, MICROBIOLOGY. *Educ:* Marshall Univ, BS, 62; WVa Univ, MS, 65, PhD(biochem genetics), 66. *Prof Exp:* Res asst biochem genetics, WVa Univ, 62-66; scientist, Commissioned Corp, NIH, 67-69, sr staff fel & immunologist, Lab Blood Prod, Div Biologics Standards, 69-71; dir biol res & develop, Dade Div Am Hosp Supply Corp, 71-74; dir, Becton, Dickinson & Co, 74-75, vpres res & develop, Qual Assurance & Regulatory Affairs, Bio Quest Div, 75-78, vpres BBL Instruments, 78-79; pres anal prods, Am Home Prod Corp, 79-81; pres, Miles Sci, 81-84; pres, Genex Corp, 84-85; PRES, THERACEL CORP, 86- *Concurrent Pos:* Dir, Genex Corp, Theracel Corp, Karyon Technol Corp & Am Biotechnol Corp. *Mem:* AAAS; Int Soc Blood Transfusion; Am Soc Clin Path; Genetics Soc Am; Am Soc Microbiol; Sigma Xi. *Res:* Genetic basis of antibody synthesis; biochemistry of host-pathogen relationship; immunohematology biochemistry of development and aging, especially enzymes and subcellular organelles; laboratory instruments and automated systems; mechanisms of immunity; diagnostic reagents; genetics; immunotherapeutic. *Mailing Add:* 9404 Tobin Circle Potomac MD 20854

ESPOY, HENRY MARTI, b San Francisco, Calif, Oct 22, 17; m 45; c 2. ANALYTICAL CHEMISTRY. *Educ:* Loyola Univ, Calif, BS, 39; Univ Calif, Los Angeles, AM, 42. *Prof Exp:* Res chemist, Van Camp Labs, 41-49; from chemist to chief chemist, Barnett Labs, 59-57; lab dir, Terminal Testing Labs, Inc, 57-59; lab dir, Daylin Labs Inc, 69-77; VPRES, ASSOC LABS, 77- *Concurrent Pos:* Referee chemist, Am Oil Chem Soc, 65- *Mem:* Am Chem Soc; Am Oil Chem Soc; Inst Food Technol. *Res:* Pesticide residues; fats and oils; food and agricultural chemistry; environmental analysis. *Mailing Add:* 16 St Tropez Laguna Niguel CA 92677-2700

ESPY, HERBERT HASTINGS, b Rochester, NY, June 4, 31; m 53; c 2. PHYSICAL ORGANIC CHEMISTRY, PAPER CHEMISTRY. *Educ:* Harvard Univ, AB, 52; Univ Wis, PhD(phys org chem), 56. *Prof Exp:* Res chemist, Hercules, Inc, 56-70, sr res chemist, 71-78, res scientist, 78-82 & 85-90, mgr, 83-85, RES ASSOC, HERCULES, INC, 90- *Mem:* Am Chem Soc; Sigma Xi; Tech Asn Pulp & Paper Indust. *Res:* Vinyl and condensation polymers; paper chemicals; polyelectrolyte interactions. *Mailing Add:* 35 Marsh Woods Lane Wilmington DE 19810-3942

ESQUIVEL, AGERICO LIWAG, b Manila, Phillippines, June 5, 32. PHYSICS, ELECTRONICS. *Educ:* Berchmans Col, Philippines, AB, 55, MA, 56; St Louis Univ, PhD(physics), 63. *Prof Exp:* Res physicist, Res Inst Advan Studies, Md, 63-64 & Mat Res Lab, Martin Co, 64-66; sr res engr, Boeing Co, 66-71; res assoc, Univ Southern Calif, 71-73; mem tech staff, Hughes Aircraft Co, 73-76; MEM TECH STAFF, TEX INSTRUMENTS, 76- *Mem:* Sr mem Inst Elec & Electronics Engrs; Electrochem Soc; Am Phys Soc. *Res:* Device physics; materials science; residual stress analysis; electron diffraction; electron microscopy; x-ray diffraction; x-ray topography; radiation effects; cathodoluminescence in GaAs; deep-level transient spectroscopy; x-ray lithography; high density nonvolatile memories; trench-isolated electrically programmable read-only memories; high density FLASH electrically eraseable and programmable read-only memories; submicrometer complementary metal oxide semiconductor fabrication technology; process integration; device characterization; author of thirty-eight technical publications; twenty papers presented at national and international conferences; eight US patents. *Mailing Add:* 13912 Waterfall Way Dallas TX 75240-3840

ESRIG, MELVIN I, b Brooklyn, NY, Mar 15, 30; m 54; c 3. CIVIL ENGINEERING, SOIL MECHANICS. *Educ:* City Col New York, BBA, 51; Polytech Inst Brooklyn, BCE, 54; Univ Ill, MS, 59, PhD(civil eng), 61. *Prof Exp:* Trainee, Corps Engrs, US Army, 54; civil engr, Lockwood, Kessler & Bartlett, Inc, 54-56 & H G Holzmacher & Assocs, 56-57; soils engr, Warzyn Eng, 61-62; from asst prof to assoc prof civil eng, Cornell Univ, 62-70; assoc, Woodward-Moorhouse & Assoc, Inc, Clifton, 70-73; PRIN & VPRES, WOODWARD-CLYDE CONSULTS, 73- *Concurrent Pos:* Mem, US Comt Large Dams. *Honors & Awards:* Hogentogler Award, Am Soc Testing & Mat, 80. *Mem:* Am Soc Testing & Mat; Am Soc Civil Engrs; Sigma Xi. *Res:* Foundation engineering; capacity of piles; properties of soils. *Mailing Add:* 43 Royden Rd Tenafly NJ 07670

ESSE, ROBERT CARLYLE, b Walnut Grove, Minn, May 17, 32; m 54; c 2. ORGANIC CHEMISTRY. *Educ:* St Olaf Col, BA, 54; Ore State Univ, PhD(org chem), 59. *Prof Exp:* Group leader antibiotics, Pharmaceut Prod Develop Sect, Lederle Labs Div, 59-66, head chem process develop, 66-74, TECH DIR FINE CHEMICALS DEPT, AM CYANAMID CO, 74- *Mem:* Am Chem Soc. *Res:* Synthesis of coumarins and furocoumarins; synthetic modification of tetracycline antibiotics. *Mailing Add:* 64 Mountain View Ave Pearl River NY 10965

ESSELEN, WILLIAM B, b Boston, Mass, July 31, 12; m 47; c 1. FOOD SCIENCE. *Educ:* Mass State Col, BS, 34, MS, 35, PhD(food tech), 38. *Prof Exp:* Asst nutrit, Mass State Col, 36-38; food technologist, Owens-Ill Glass Co, 39-41; from asst res prof to res prof, 41-57, head dept, 57-71, prof, 71-75, EMER PROF FOOD TECHNOL, UNIV MASS, AMHERST, 76- *Concurrent Pos:* Consult, War Food Admin, USDA, 42-45; food technician, Qm Corps, US Dept Army, 45; vis prof, Hokkaido Univ, 60-61 & Univ West Indies, Trinidad, 71-72. *Mem:* Am Soc Microbiologists; Am Chem Soc; Inst Food Technologists. *Res:* Nutritive fruits; effect of canning, freezing and dehydration on nutrition of foods; determination of process times for canned foods; use of glass containers for foods; apple products. *Mailing Add:* 55 Hills Rd Amherst MA 01002

ESSELMAN, W(ALTER) H(ENRY), b Hoboken, NJ, Mar 19, 17; m 43; c 4. ELECTRICAL ENGINEERING, NUCLEAR RNGINEERING. *Educ:* Newark Col Eng, BS, 38; Stevens Inst Technol, MS, 44; Polytech Inst Brooklyn, DrE(elec eng), 53. *Prof Exp:* Elec engr regulating & control systs, Westinghouse Elec Corp, 38-40, elec engr servo & comput systs Navy fire control equip, 40-45, develop engr control & servo-systs, 45-50, mgr power plant syst, Bettis Atomic Power Div, 50-52, mgr syst subdiv atomic power div, 52-53, tech asst to mgr, Test & Develop Nuclear Power Plants, Submarine Thermal Reactor Test Facility, 53-58, mgr adv develop & planning, Bettis Atomic Power Div, 58-59, sr dept mgr, Astronuclear Lab, 59-61, mgr eng develop, Nerva Proj, 61-64, dep mgr, Pa, 64-68, proj mgr, 68-69, exec asst to gen mgr, 69-70; dir, Hanford Eng Develop Lab, 70-72; dir strategic planning, Elec Power Res Inst, 74-80, dir eng assessment, 81-86; RETIRED. *Concurrent Pos:* Instr, Polytech Inst Brooklyn, 48. *Honors & Awards:* E Weston Award & A Cullimore Award, NJ Inst Technol. *Mem:* Fel Am Nuclear Soc; fel Inst Elec & Electronics Engrs; Am Inst Aeronaut & Astronaut; AAAS. *Res:* Energy research and development; nuclear power systems; electric utility. *Mailing Add:* 1141 Buckingham Dr Los Altos CA 94024

ESSENBERG, MARGARET KOTTKE, b Troy, NY, Apr 21, 43; m 67; c 2. BIOCHEMISTRY, BOTANY & PHYTOPATHOLOGY. *Educ:* Oberlin Col, AB, 65; Brandeis Univ, PhD(biochem), 71. *Prof Exp:* Res assoc, Univ Leicester, Eng, 71-73; res assoc, 73-74, from asst prof to assoc prof, 74-84, PROF BIOCHEM, OKLA STATE UNIV, 84- *Concurrent Pos:* NSF fel, 71; NIH fel, 72-73; vis assoc prof bot, Univ Toronto, 84. *Mem:* Am Phytopath Soc; Sigma Xi; Am Soc Biol Chemists; Phytochem Soc NAm. *Res:* Biochemistry of interactions between plants and plant pathogenic bacteria; biosynthesis; mechanism of toxicity and role in disease resistance of antibiotics produced by plants in response to infection; biosynthesis of antibacterial sesquiterpenes in cotton plants; photoactivated toxicity of the sesquiterpenes. *Mailing Add:* Dept Biochem Okla State Univ Stillwater OK 74078-0454

ESSENBERG, RICHARD CHARLES, b Santa Monica, Calif, Dec 10, 43; m 67; c 2. TRANSPORT PHYSIOLOGY, GENETIC ENGINEERING. *Educ:* Calif Inst Technol, BS, 65; Harvard Univ, PhD(chem), 71. *Prof Exp:* Res fel biochem, Univ Leicester, 71-73; from asst prof to assoc prof, 73-84, PROF BIOCHEM, OKLA STATE UNIV, 84- *Concurrent Pos:* Vis prof med genetics, Univ Toronto, 84. *Mem:* Am Inst Biol Sci; Am Soc Microbiol; AAAS; Am Soc Biochem & Molecular Biol. *Res:* Mechanism of energy transduction in membrane transport; gene cloning in studying mechanisms of pathogenicity; intracellular signalling mechanisms. *Mailing Add:* Dept Biochem Okla State Univ Stillwater OK 74078-0454

ESSENBURG, F(RANKLIN), b Holland, Mich, Aug 2, 24; m 46; c 2. MECHANICS. *Educ:* Univ Mich, BSE, 45, LLB, 48, MS, 49, MSE, 50, PhD(mech), 56. *Prof Exp:* Patent atty, Bell Tel Labs, 50-51; pvt construct bus, 51-53; from instr to asst prof mech, Univ Mich, 53-58; from assoc prof to prof, Ill Inst Technol, 58-62; chmn dept mech eng, 62-70, chmn dept aerospece eng sci, 76-80, PROF MECH, UNIV COLO, BOULDER, 62-, GRAD SCH FAC, 80- *Mem:* Am Soc Mech Engrs. *Res:* Continuum mechanics; plate and shell theory; dynamics and vibration. *Mailing Add:* Dept Mech Eng Univ Colo Boulder CO 80309

ESSENE, ERIC J, b Berkeley, Calif, Apr 26, 39; m 85; c 4. PETROLOGY, MINERALOGY. *Educ:* Mass Inst Technol, BS, 61; Univ Calif, Berkeley, PhD(geol), 66. *Prof Exp:* NSF fel geol, Cambridge Univ, 66-68; res fel, Australian Nat Univ, 68-70; from asst prof to assoc prof, 70-80, PROF GEOL, UNIV MICH, 80- *Res:* Petrology of rocks from the lower crust and upper mantle; thermodynamics and phase equilibria of inorganic solids; petrologic mineralogy of rock forming minerals. *Mailing Add:* Dept Geol Sci Univ Mich Ann Arbor MI 48109

ESSENFELD, AMY, b Stamford, Conn, Sept 30, 59; m 90. CHEMILUMINESCENCE. *Educ:* Johns Hopkins Univ, BA, 81; Mass Inst Technol, PhD(org chem), 85. *Prof Exp:* Res chemist chem light, 85-88, sr res chemist chem light, 89-90, GROUP LEADER CHEM LIGHT, AM CYANAMID CO, 90- *Mem:* Am Chem Soc. *Res:* Charge of the research and development group that works in the area of chemiluminescence; research activities include organic synthesis, formulations, packaging design, trouble-shooting for the plant and product support. *Mailing Add:* 1937 W Main St PO Box 60 Stamford CT 06904-0060

ESSENWANGER, OSKAR M, b Munich, Ger, Aug 25, 20; US citizen; m 47; c 2. ATMOSPHERIC PHYSICS, APPLIED STATISTICS. *Educ:* Univ Vienna, Dipl, 43; Univ Würzburg, Dr rer nat, 50. *Prof Exp:* Instr meteorol, Ger Air Force, 44-45; res meteorologist, Ger Weather Serv, 46-57; proj assoc, Univ Wis, 56-57; res meteorologist, Ger Weather Serv, 57 & Nat Weather Records Ctr, 57-61; chief aerophys br, US Army Missile Command, 61-89; RES PROF, UNIV ALA, HUNTSVILLE, 89- *Concurrent Pos:* Ger rep comt statist methods, World Meteorol Org, 57; affil prof, Dept Atmospheric Physics, Colo State Univ, 68-73; adj prof, Univ Ala, Huntsville, 71-89; ed, Vol 1, World Survey of Climat, 85, Elements of Statistical Analysis, 86. *Honors & Awards:* Outstanding Res Award, Sigma Xi, 77; Hammond Oberth Award, Am Inst Aeronaut & Astronaut. *Mem:* Fel Am Meteorol Soc; assoc fel Am Inst Aeronaut & Astronaut; Am Soc Qual Control; Sigma Xi; Ger Meteorol Soc; Am Statist Asn; fel Am Biog Res Inst. *Res:* Physical structure of the atmosphere, especially mathematical analysis and statistical representation of atmospheric parameters for missile design and trajectory analysis; development of statistical methods in climatology. *Mailing Add:* 610 Mountain Gap Dr Huntsville AL 35803

ESSER, ALFRED F, b Lauf, Ger, Feb 11, 40; m 70. BIOCHEMISTRY. *Educ:* Univ Frankfurt, Diplomchemiker, 66, PhD(phys biochem), 69. *Prof Exp:* Fel biochem, Univ Calif, Santa Barbara, 69-71; res assoc biophys, NASA-Ames Res Ctr, Moffett Field, 71-73; asst prof biochem, Calif State Univ, Fullerton, 73-75; assoc mem molecular immunol, Scripps Clin & Res Found, 75-81; PROF COMP PATH & BIOCHEM, UNIV FLA, GAINESVILLE, 81- *Concurrent Pos:* Estab investr, Am Heart Asn, 76-81. *Mem:* Am Soc Biol Chemists; Am Chem Soc; Biophys Soc; Am Asn Immunologists. *Res:* Structure-function relationships in biological membranes; complement-immunochemistry. *Mailing Add:* Univ Fla Box J-145 JHMHC Gainesville FL 32610

ESSER, ARISTIDE HENRI, b Padalarang, Indonesia, May 11, 30; US citizen; m 56; c 2. PSYCHIATRY, HUMAN ECOLOGY. *Educ:* Univ Amsterdam, MD, 55. *Prof Exp:* Intern, Amsterdam Univ Hosp, 55-56; resident psychiat, Wolfheze Ment Hosp, Neth, 56-57; staff resident, Endegeest Psychiat Clin, 58-61; Lederle Inst res fel, Yale Univ, 61-62; med dir res ward, Rockland State Hosp, 62-68, dir psychiat res, Letchworth Village, 69-71; dir, Cent Bergen Community Ment Health Ctr, 71-77; chief psychiat servs, Mission of the Immaculate Virgin, 77-80; DIR QUAL ASSURANCE, BRONX PSYCHIATRIC CTR, 80-; CONSULT, 85- *Concurrent Pos:* City of Leyden travel grant to Ger, Austria & Switz, 60; supvry psychiatrist, Endegeest Psychiat Clin, 64; ed, Man-Environ Systs, 69-; dir social biol labs, Rockland State Hosp, 69-; attend psychiatrist, Col Physicians & Surgeons, Columbia Univ, 70-; assoc prof psychiat, Albert Einstein Col Med, 80-85; Consult, 85- *Mem:* Fel AAAS; Soc Gen Syst Res; fel Am Psychiat Asn; Animal Behav Soc. *Res:* Methodology for clinical and behavioral evaluations in psychiatry, mental retardation and animal studies; mental health and man-environment relations; social pollution. *Mailing Add:* 435 S Mountain Rd New York NY 10956

ESSERY, JOHN M, b Plymouth, Eng, June 15, 36; m 62; c 2. ORGANIC CHEMISTRY, MEDICINAL CHEMISTRY. *Educ:* Univ Exeter, BSc, 57, PhD(chem), 60. *Prof Exp:* Fel, Nat Res Coun Can, 60-62; sr res scientist, Bristol-Meyers Co, Syracuse, 62-79, mgr licensing, Int Div, 79-; AT ORGANON INC, WEST ORANGE. *Mem:* Am Chem Soc; Royal Soc Chem. *Res:* Chemistry of betalactam antibiotics; chemistry of heterocyclic compounds; biosynthesis of natural products. *Mailing Add:* Ortho Pharm Corp Rte 202 POB 300 Raritan NJ 08869-0602

ESSEX, MYRON, b Coventry, RI, Aug 17, 39; m 66; c 2. VIROLOGY, IMMUNOLOGY. *Educ:* Univ RI, BS, 62; Mich State Univ, MS, 67, DVM, 67; Univ Calif, Davis, PhD(microbiol), 70. *Hon Degrees:* MA, Harvard Univ, 80. *Prof Exp:* Vis scientist cancer biol, Karolinska Inst, Stockholm, 70-72; from asst prof to prof microbiol, 72-82, dept head, 78-82, PROF & HEAD DEPT CANCER BIOL, HARVARD SCH PUB HEALTH, 82-, MARY WOODARD LASKER PROF HEALTH SCI, 89- *Concurrent Pos:* Leukemia Soc Am scholar, 72-77; vis prof, Univ Tehran, 75; lectr, Harvard Med Sch, 76-; mem immunotherapy comt, NIH, 77-; mem med & sci adv bd, Leukemia Soc Am, 78-; mem res comt, Am Cancer Soc, Mass, 76- & bd sci coun, Div Cancer Etiology, Nat Cancer Inst. *Honors & Awards:* Bronze Medal, Am Cancer Soc, 80; Res Award, Ralston Purina, 85; Outstanding Investr Award, Nat Cancer Inst, 85. *Mem:* Am Soc Microbiol; Am Asn Immunologists; Am Asn Cancer Res; Am Vet Med Asn; Am Asn Vet Oncologists. *Res:* Viral oncology; tumor immunology; cancer biology; cell membranes; viral genetics; analysis of human, feline and primate viruses that cause cancer and immunosuppression; link between retrovirus and human AIDS; antigens used for blood screening and vaccine development with human leukemia and AIDS viruses. *Mailing Add:* Dept Cancer Biol Harvard Sch Pub Health 655 Huntington Ave Boston MA 02115

ESSIEN, FRANCINE B, b Augusta, Ga, Mar 7, 43; c 3. DEVELOPMENTAL GENETICS. *Educ:* Temple Univ, BA, 65; Albert Einstein Col Med, PhD(genetics), 70. *Prof Exp:* Fel cell biol & genetics, Univ Conn, Storrs, 70-71; asst prof, 71-76, ASSOC PROF BIOL, DEPT BIOL SCI, DOUGLASS CAMPUS, RUTGERS UNIV, 76- *Concurrent Pos:* Consult, NJ Dept Higher Educ, 73, Robert Wood Johnson Found, 76, Nyerere Educ Inst, 81-; panelist, Am Coun Educ, NSF, 76-81; mem, NJ Task Force Genetics, 78-; fac fel cell biol & genetics, Univ Tenn, Oak Ridge Biomed Sci Grad Sch, 73-75; co-dir Biomed Careers Prog, Rutgers Univ Med Sch, 75-; Fulbright Award, Comparative & Int Educ Soc, 82. *Mem:* AAAS; Am Soc Cell Biol; Am Soc Zoologists; NY Acad Sci; Am Women Sci. *Res:* Analysis of the genetic control of developmental processes, using mutations in the mouse as tools to investigate the cellular and molecular bases of birth defects. *Mailing Add:* Dept Biol Sci Douglass Campus Rutgers Univ PO Box 1059 Piscataway NJ 08854-1059

ESSIG, ALVIN, b Canton, Ohio, Feb 16, 23. PHYSIOLOGY. *Educ:* Harvard Univ, SB, 44; Ohio State Univ, MD, 48. *Prof Exp:* Asst resident med, Vet Admin Hosp, Cleveland, 50-51; asst resident med, Montefiore Hosp, 53-54, res fel, Cardio- vascular & Res Serv, 54-56; res fel, Col Physicians & Surgeons, Columbia Univ & Presby Hosp, 56-59; clin & res fel, Harvard Med Sch & Mass Gen Hosp, 59-61, instr med, 61-64, assoc med, 64-65; from asst prof to assoc prof med, Sch Med, Tufts Univ, 66-73, assoc prof physiol, 71-73; PROF PHYSIOL, SCH MED, BOSTON UNIV, 73-, RES PROF MED, 73- *Concurrent Pos:* Lectr physiol, Sch Med, Tufts Univ, 69-71, biophys, Harvard Med Sch, 70-75. *Mem:* AAAS; Am Physiol Soc; Am Soc Nephrol; Biophys Soc; Int Soc Nephrology; Nat Kidney Found. *Res:* Physiology. *Mailing Add:* Dept Physiol & Med Boston Univ Sch Med 80 E Concord St Boston MA 02118

ESSIG, CARL FOHL, b Canton, Ohio, July 31, 19; m 74; c 3. EXPERIMENTAL NEUROLOGY, ELECTROENCEPHALOGRAPHY. *Educ:* Kent State Univ, BS, 42; Case Western Reserve Univ, MD, 47. *Prof Exp:* Neurophysiol res, NIMH, 49-52; clin fel neurol, Mass Gen Hosp, 52-54, drug res, Addiction Res Ctr, Lexington, Ky, 54-71; med dir, USPHS, 62-71; RETIRED. *Mem:* Am Phys Soc. *Res:* Drug addiction; neuropharmacology. *Mailing Add:* 1455 Forbes Rd Lexington KY 40511

ESSIG, FREDERICK BURT, b Los Angeles, Calif, Aug 25, 47; m 74. SYSTEMATIC BOTANY. *Educ:* Univ Calif, Riverside, AB, 69; Cornell Univ, PhD(plant taxon), 75. *Prof Exp:* Lectr biol, Cornell Univ, 74-75; asst prof bot & dir, Bot Gardens, 75-80, ASSOC PROF BIOL, UNIV SFLA, 80- *Mem:* Int Asn Plant Taxon; Soc Econ Bot; Asn Trop Biol. *Res:* Systematics of the palms and clematis (Ranunculacese). *Mailing Add:* Dept Biol Univ SFla Tampa FL 33620

ESSIG, GUSTAVE ALFRED, b Philadelphia, Pa, June 11, 15; wid; c 2. PHYSICS. *Educ:* Washington & Lee Univ, BA, 42. *Prof Exp:* Asst radio engr Sig C, US Army, 42, officer in chg radar training, 42-46; asst physicist nucleonics, Monsanto Chem Co, 46-48, supvr phys chem processes & radiog, Mound Lab, Monsanto Res Corp, 48-55, gen supvr radiochem & appl nuclear & phys chem, 55-62, asst to vpres & plant mgr, 62-63, mgr planning & reporting, 63-65, asst to vpres & plant mgr, 65-69, tech specialist, Mound Lab, Monsanto Res Corp, 69-83; CONSULT. *Mem:* Am Phys Soc. *Res:* Research and development through pilot plant to production; radio-chemistry; radiography; nuclear physics; physical chemistry; vacuum techniques; electronics; instrumentation. *Mailing Add:* 1215 Meadowview Dr Miamisburg OH 45342

ESSIG, HENRY J, b Eppstein, Ger, Apr 26, 26; US citizen; m 49; c 4. ORGANIC POLYMER CHEMISTRY. *Educ:* Western Reserve Univ, BS, 51, MS, 52, PhD(org chem), 60. *Prof Exp:* Chemist, Grant Photo Prod, 51-52; chemist, BF Goodrich Chem Co, 52-60, from assoc develop scientist to develop scientist, 60-68, sr develop scientist, 68-77, supvr res & develop, 77-79, mgr res & develop, 79-84; RETIRED. *Mem:* Am Chem Soc. *Res:* Kinetic studies of heterogeneous catalysis; monomer synthesis and polymerization. *Mailing Add:* 31480 Detroit Rd Westlake OH 44145

ESSIG, HENRY WERNER, b Paragould, Ark, Dec 9, 30; m 53; c 2. ANIMAL NUTRITION. *Educ:* Univ Ark, BSA, 53, MS, 56; Univ Ill, PhD(animal nutrit), 59. *Prof Exp:* From asst prof to assoc prof, 59-66, PROF NUTRIT, DEPT ANIMAL SCI, MISS STATE UNIV, 66- *Concurrent Pos:* Nutrit consult; chmn, Southern Pasture & Forage Crop Improv Conf, 89. *Honors & Awards:* Gamma Sigma Res Award, Miss State Univ, 73, Fac Award Outstanding Res, 76. *Mem:* Am Soc Animal Sci; Am Dairy Sci Asn; Animal Nutrit Res Coun; Am Forage & Grasslands Coun. *Res:* Ruminant nutrition and forage evaluation research with cattle and sheep. *Mailing Add:* Dept Animal Sci Miss State Univ Box 5228 Mississippi State MS 39762

ESSIGMANN, JOHN MARTIN, b Woburn, Mass, Apr 26, 47; m 70. TOXICOLOGY, BIOCHEMISTRY. *Educ:* Northeastern Univ, BA, 70; Mass Inst Technol, MS, 72, PhD(toxicol), 76. *Prof Exp:* Nat Inst Environ Health Sci fel, 76-77, res assoc toxicol, 77-80, ASST PROF TOXICOL, MASS INST TECHNOL, 80- *Mem:* Am Chem Soc. *Res:* Chemical carcinogenesis research, especially carcinogen-macromolecule interactions. *Mailing Add:* 18 Vassr St Bldg 20 Rm 56-243 Cambridge MA 02139-4309

ESSIGMANN, MARTIN W(HITE), b Bethel, Vt, Jan 14, 17; m 43; c 3. ELECTRICAL ENGINEERING. *Educ:* Tufts Col, BS, 38; Mass Inst Technol, SM, 47. *Prof Exp:* Instr elec eng, Northeastern Univ, 38-44; vis instr, Mass Inst Technol, 44-47; asst prof, 47-50, prof & coordr electronics res, 50-61, head dept elec eng, 54-61, DEAN RES, NORTHEASTERN UNIV, 61-, PROF ELEC ENG, 80- *Concurrent Pos:* Vis asst prof, Mass Inst Technol, 47-48. *Mem:* AAAS; Am Soc Eng Educ; Inst Elec & Electronics Engrs. *Res:* Digital computers; nonlinear devices; speech analysis; principles of radar; information theory. *Mailing Add:* Northeastern Univ 360 Huntington Ave Boston MA 01015

ESSINGTON, EDWARD HERBERT, b Santa Barbara, Calif, Feb 19, 37; m 57; c 2. SOIL SCIENCE. *Educ:* Calif State Polytech Col, BS, 58; Univ Calif, Los Angeles, MS, 64. *Prof Exp:* Lab technician, Dept Nuclear Med & Radiation Biol, Univ Calif, Los Angeles, 57-65; sr assoc soil scientist, Hazleton-Nuclear Sci Corp Div, Isotopes Inc, 65-66; from sr assoc soil scientist to soil scientist, Palo Alto Labs, Teledyne Isotopes, Calif, 66-71, soil scientist, Nev, 71-73; soil scientist, H-7 Indust Waste, Los Alamos Nat Lab, 73-75. H-8 Environ Studies, 75-77, LS-6 Environ Sci, Life Sci Div, 78-81,

HSE-12 Environ Sci, Health Safety & Environ Div, 82-89, SOIL SCIENTIST, EES-15 ENVIRON SCI, EARTH & ENVIRON SCI DIV, LOS ALAMOS NAT LAB, 89- *Mem:* AAAS; Health Physics Soc; Soil Sci Soc Am. *Res:* Soil-plant relations; radionuclide uptake by plants; radionuclide and pollutant migration and chemistry in soils and groundwater systems; industrial radioactive and toxic waste management with emphasis on transuranic nuclides. *Mailing Add:* 118 Balboa Dr White Rock NM 87544

ESSLER, WARREN O(RVEL), b Davenport, Iowa, Apr 22, 24; m 44; c 3. ELECTRICAL ENGINEERING. *Educ:* Univ Iowa, BS, 53, MS, 55, PhD(elec eng, physiol), 60. *Prof Exp:* Res assoc audiol, Univ Iowa, 53-54; res assoc elec eng, Collins Radio Corp, 54-55; from instr to asst prof, SDak State Col, 55-61; prof & dean, Col Technol, Univ Vt, 61-72; chmn dept, 72-80, PROF ELEC ENG, TENN TECHNOL UNIV, 72- *Mem:* Am Soc Eng Educ; Nat Soc Prof Engrs; Inst Elec & Electronics Engrs; Sigma Xi. *Res:* Medical electronics. *Mailing Add:* 1385 Sherwood Lane Cookeville TN 38501

ESSLINGER, JACK HOUSTON, b Ponca City, Okla, July 19, 31; m 55; c 3. PARASITOLOGY. *Educ:* Univ Okla, BS, 53; Rice Inst, MA, 55, PhD(biol), 58. *Prof Exp:* From instr to asst prof, Sch Med, 62-67, ASSOC PROF PARASITOL, SCH PUB HEALTH & TROP MED, TULANE UNIV, 67- *Mem:* Am Soc Parasitologists; Am Soc Trop Med & Hyg; Soc Syst Zool. *Res:* Filarial systematics; filariasis of mammals. *Mailing Add:* PO Box 1110 Mountainview AK 72560

ESSLINGER, THEODORE LEE, b Spokane, Wash, Dec 13, 44; m 67; c 1. LICHENOLOGY. *Educ:* Eastern Wash State Col, BA, 68; Duke Univ, PhD(bot), 74. *Prof Exp:* Res fel bot, Smithsonian Inst, 74-75; asst prof, 75-80, assoc prof, 80-89, PROF BOT, NDAK STATE UNIV, 90- *Mem:* Am Bryol & Lichenological Asn; Mycol Soc Am; Sigma Xi. *Mailing Add:* Dept Bot NDak State Univ State Univ Sta Fargo ND 58105

ESSLINGER, WILLIAM GLENN, b Huntsville, Ala, Oct 21, 37; m 58; c 3. ORGANIC CHEMISTRY. *Educ:* Univ Ala, BS, 62, MS, 64, PhD(org chem), 66. *Prof Exp:* Asst prof chem, Union Univ, Tenn, 66-68; from asst prof to assoc prof, 73-77, chmn dept, 73-78, PROF CHEM, WGA COL, 77- *Concurrent Pos:* NSF grant dir, Ga Sci Teacher Proj, 69-; brief org exam comt, Am Chem Soc, 75- *Mem:* Am Chem Soc; Sigma Xi. *Res:* Conformational effects on reactions at exocyclic positions of cycloalkylcarbinyl derivatives; thermal decomposition of esters; conversion of biomass to alcohol and other organics. *Mailing Add:* Dept Chem WGa Col Carrollton GA 30117

ESSMAN, JOSEPH EDWARD, b Portsmouth, Ohio, Nov 15, 35; m 57; c 5. COMMUNICATIONS ENGINEERING. *Educ:* Ohio Univ, BSEE, 57, MS, 61; Purdue Univ, PhD(elec eng), 72. *Prof Exp:* Actg instr elec eng, 57-60, from instr to prof 60-79, asst chmn, 72-79, asst dean, 79-81, ASSOC DEAN, COL ENG, OHIO UNIV, 81- *Concurrent Pos:* Res engr & proj dir, Wright-Patterson AFB, 78-; Nat Sci fel; consult, North Elec; campus coordr, Ohio Aerospace Inst. *Mem:* Inst Elec & Electronics Engrs; Sigma Xi. *Res:* Communications and digital signal processing; data compression with applications to images. *Mailing Add:* Col Eng & Technol Ohio Univ Athens OH 45701

ESSMAN, WALTER BERNARD, b New York, NY, Dec 25, 33; m 62; c 1. PSYCHOPHYSIOLOGY. *Educ:* NY Univ, BA, 54; Univ NDak, MA, 55, PhD(psychol), 57; Univ Milan, MD, 72. *Prof Exp:* Asst psychol, Univ NDak, 54-55; asst, Univ Nebr, 55-56; asst, Univ NDak, 56-57; fel neurophysiol, Albert Einstein Col Med, 59-61, res assoc physiol, 61-62; from asst prof to assoc prof, 62-67, PROF PSYCHOL, QUEENS COL, NY, 67- *Concurrent Pos:* Fel neurochem, Mt Sinai Hosp, 64-68. *Mem:* AAAS; Fedn Am Socs Exp Biol; Am Psychol Asn; Am Acad Neurol; fel Am Col Nutrit. *Res:* Neural and chemical basis of learning and memory; physiological stress. *Mailing Add:* Dept Psychol Queens Col 65-30 Kissena Blvd Flushing NY 11367

ESSNER, EDWARD STANLEY, b New York, NY, Mar 31, 27; m 58; c 2. CELL BIOLOGY, ELECTRON MICROSCOPY. *Educ:* Long Island Univ, BS, 47; Univ Pa, PhD(zool), 51. *Prof Exp:* Sr asst scientist, Lab Chem Pharm, Nat Cancer Inst, 52-56; Nat Cancer Inst spec fel path, Albert Einstein Col Med, 56-58; res asst prof, Albert Einstein Col Med, 58-64; assoc mem, Div Cytol, Sloan-Kettering Inst Cancer Res, 64-75; PROF OPHTHAL, SCH MED, WAYNE STATE UNIV, 75- *Mem:* Am Soc Cell Biol; Electron Micros Soc Am; Int Soc Cell Biol; Am Asn Cancer Res; Histochem Soc. *Res:* Cell ultrastructure; enzyme localization; cell organelles. *Mailing Add:* Dept Ophthal Krsq Eye Inst Wayne State Univ Sch Med 4717 St Antoine Detroit MI 48201

ESTABROOK, FRANK BEHLE, b Nampa, Idaho, June 22, 22; m 50; c 1. PHYSICS. *Educ:* Miami Univ, Ohio, BA, 43; Calif Inst Technol, MA, 47, PhD(physics), 50. *Prof Exp:* From asst prof to assoc prof physics, Miami Univ, Ohio, 50-52; sr engr reactor physics, NAm Aviation, 52-55; physicist, US Army Off Ord Res, 55-60; sr mem tech staff, 60-80, SR RES SCIENTIST, JET PROPULSION LAB, CALIF INST TECHNOL, 80- *Mem:* AAAS; Am Phys Soc; Sigma Xi. *Res:* Relativity and gravitation; applied mathematics; cosmology. *Mailing Add:* 853 Lyndon St South Pasadena CA 91030

ESTABROOK, GEORGE FREDERICK, b Carlisle Bks, Pa, Nov 1, 42; m 76; c 3. SYSTEMATIC BOTANY, EVOLUTIONARY PLANT ECOLOGY. *Educ:* Darmouth Col, Hanover, NH, AB, 64; Univ Colo, Boulder, MA, 67. *Prof Exp:* Res assoc, NY Bot Garden, 64-65; botanist, Dept Bot & Plant Path, Colo State Univ, 65-67; res mathematician, Dept Biol, Univ Col, 67-70; asst prof bot & zool, 70-76, RES SCIENTIST HERBARIUM, UNIV MICH, ANN ARBOR, 73-, ASSOC PROF DIV BIOL SCI, 76- *Concurrent Pos:* Assoc ed, Math Biosci, 76- *Mem:* Am Soc Plant Taxonomists; Am Soc Naturalists; Ecol Soc Am; Soc Study Evolution; Int Asn Plant Taxon. *Res:* Reconstruct evolutionary relationships among kinds of organisms; explain in evolutionary terms when seeds germinate, fruits disperse, etc. *Mailing Add:* Dept Bot Univ Mich Herbarium Ann Arbor MI 48109-1048

ESTABROOK, KENT GORDON, b Astoria, Ore, Oct 12, 43; m 68; c 2. PLASMA PHYSICS, COMPUTER SCIENCES. *Educ:* Vanderbilt Univ, AB, 65; Univ Tenn, PhD(physics), 71. *Prof Exp:* Fel physics, Oak Ridge Assoc Univs, 69-71, Ecole Polytech, Paris, 71 & Univ Tenn, 72; PHYSICIST, LAWRENCE LIVERMORE LAB, 72- *Mem:* Fel Am Phys Soc. *Res:* Computer simulation of laser-plasma interaction for laser fusion and basic plasma physics. *Mailing Add:* 707 Geraldine Livermore CA 94550

ESTABROOK, RONALD (WINFIELD), b Albany, NY, Jan 3, 26; m 47; c 4. BIOCHEMISTRY. *Educ:* Rensselaer Polytech Inst, BS, 50; Univ Rochester, PhD(biochem), 54. *Hon Degrees:* DSc, Univ Rochester, 80; Dr, Karsinska Inst, Stockholm, 80. *Prof Exp:* Fel, Johnson Found Med Physics, Univ Pa, 54-57 res assoc, 57-58, from asst prof to prof phys biochem, 59-68; res assoc, Univ Pa, 57-58, from asst prof to prof phys biochem, 59-68; dean grad sch biomed sci, 73-76, VIRGINIA LAZENBY O'HARA PROF BIOCHEM & CHMN DEPT, UNIV TEX HEALTH SCI CTR, DALLAS, 68-, CECIL H & IDA M GREEN, CHAIR BIOMED SCI, 90- *Concurrent Pos:* Am Heart Asn fel, 57-58; USPHS fel, 58-60. *Mem:* Nat Acad Sci; Inst Med-Nat Acad Sci; Fedn Am Socs Exp Biol; Am Soc Biol Chem; Am Chem Soc; Am Soc Pharmacol & Exp Therapeut. *Res:* Application of physical methods to study of intracellular biochemical processes. *Mailing Add:* Dept Biochem Southwest Med Ctr 5325 Harry Hines Blvd Dallas TX 75235

ESTEE, CHARLES REMINGTON, b Hecla, SDak, Oct 7, 21; m 43; c 3. PHYSICAL CHEMISTRY. *Educ:* Jamestown Col, BS, 42; Univ Iowa, MS, 44, PhD(phys chem), 47. *Prof Exp:* Asst chem, Univ Iowa, 42, asst, 43, instr, 43-44; jr chemist, Tenn Eastman Corp, Oak Ridge, 44-45; instr, Univ Iowa, 46-47; actg prof, 47-48, head dept, 52-84, interim dean, Col Arts & Sci, 87-88, PROF CHEM, UNIV SDAK, 48- *Mem:* AAAS; Am Chem Soc; Nat Sci Teachers Asn. *Res:* Electrical properties of colloids; science education. *Mailing Add:* Dept Chem Univ SDak Vermillion SD 57069

ESTENSEN, RICHARD D, b Minneapolis, Minn, May 17, 34. MEDICAL & PATHOLOGY. *Educ:* Science Univ, MD, 61. *Prof Exp:* Res path, Chicago W Mem Hosp, 62-66, NIH trainee path; maj, Walter Reed Army Hosp, 66-69; PROF PATH, UNIV MINN HOSP, 69- *Mem:* Am Asn Cancer Res; AAAS; Am Asn Pathologists; Am Soc Cell Biol. *Mailing Add:* Dept Lab Med & Path Univ Minn Hosps Box 609 Mayo Minneapolis MN 55455

ESTEP, CHARLES BLACKBURN, b Middlesboro, Ky, Aug 9, 23; m 49; c 3. ENTOMOLOGY, PHARMACOLOGY. *Educ:* Univ Tenn, BA, 48, MS, 53, PhD(entom), 75. *Prof Exp:* Res asst radio-chem, Oak Ridge Inst Nuclear Studies, 48-50 & Univ Tenn, AEC, 50-52; prod supvr radio-pharmaceut, Abbott Labs, Oak Ridge, Tenn, 52-65, CHEM PHARMACOLOGIST, ABBOTT LABS, NORTH CHICAGO, 66- *Mem:* Am Chem Soc; NY Acad Sci; AAAS; Entom Soc Am; Biol Res Inst Am; Sigma Xi. *Res:* Radiotracer-labeled drugs in absorption, distribution and excretion studies of candidate drugs in vivo in various animal species; wholebody autoradiography for drug distribution studies in different animal species; environmental entomology. *Mailing Add:* Abbot Labs D-463 AP-9 14th St Sheridan Rd North Chicago IL 60064

ESTEP, HERSCHEL LEONARD, b Dunbar, Va, Nov 29, 29; m 52; c 6. MEDICINE, ENDOCRINOLOGY. *Educ:* King Col, AB, 52; Johns Hopkins Univ, MD, 56. *Prof Exp:* Fel endocrinol, Vanderbilt Univ, 61-62; dir endocrinol res lab, 62-76, ASSOC PROF MED, HEALTH SCI DIV, VA COMMONWEALTH UNIV, 62- *Concurrent Pos:* USPHS res grant, 62-65; Am Cancer Soc grant, 68-70; consult, McGuire Vet Admin Hosp, 68-71. *Mem:* AAAS; fel Am Col Physicians; Endocrine Soc; Am Fedn Clin Res; Sigma Xi. *Res:* Neuroendocrine regulation of adrenocorticotropic and gonadotropic hormones; mechanisms of control of parathyroid hormone secretion. *Mailing Add:* DePaul Hosp 150 Kinglsley Lane Norfolk VA 23505

ESTERGREEN, VICTOR LINÉ, b Lynden, Wash, Dec 15, 25; m 50; c 1. ANIMAL PHYSIOLOGY, ENDOCRINOLOGY. *Educ:* Wash State Univ, BS, 50, MS, 56; Univ Ill, PhD(dairy sci), 60. *Prof Exp:* Asst herdsman, Wash State Univ, 50-51; instr, Wash High Schs, 52-54; trainee biol chem, Steroid Training Inst, Univ Utah, 60-61; res assoc animal physiol, 61-62, from asst prof to assoc prof dairy sci, 62-72, PROF DAIRY SCI, WASH STATE UNIV, 72- *Mem:* Endocrine Soc; Am Soc Animal Sci. *Res:* Bovine reproduction and corticosteroid hormones; endocrinology of reproduction in farm animals; progesterone metabolism in cattle; milk progesterone assay; loteal progesterone synthesis. *Mailing Add:* Dept Animal Sci Wash State Univ Pullman WA 99164-6310

ESTERLING, DONALD M, b Chicago, Ill, Nov 18, 42; m; c 3. THEORETICAL SOLID STATE PHYSICS, MATERIALS SCIENCE. *Educ:* Univ Notre Dame, BS, 64; Brandeis Univ, MA, 66, PhD(physics), 68. *Prof Exp:* From asst prof to assoc prof physics, Ind Univ, Bloomington, 68-75; assoc res prof eng, George Wash Univ, 75-78, assoc prof eng & appl sci, 78-80, prof, Joint Inst Advan Flight Sci, NASA Langley Res Ctr, 80-83, PROF ENG & APPL SCI, GEORGE WASHINGTON UNIV, WASHINGTON, DC, 83- *Concurrent Pos:* Pres, Microcompatibles, Inc. *Mem:* Am Phys Soc; Inst Mech Engrs. *Res:* Solid model/computer graphics simulation of numerical control machining; intelligent manufacturing. *Mailing Add:* CMEE Dept Sch Eng & Appl Sci George Washington Univ 801 22 St NW Suite T733 Washington DC 20052

ESTERLY, NANCY BURTON, b New York, NY, Apr 14, 35; m 57; c 4. PEDIATRICS, DERMATOLOGY. *Educ:* Smith Col, BS, 56; Johns Hopkins Univ, MD, 60; Am Bd Pediat, dipl, 66; Am Bd Dermat, dipl, 70. *Prof Exp:* USPHS fel dermat, Sch Med, Johns Hopkins Univ, 64-67; instr pediat, Johns Hopkins Univ, 67-68; from instr to asst prof, Univ Chicago, 68-70; asst prof dermat, Col Med, Univ Ill Med Ctr, 70-72, assoc prof dermat & pediat, 72-73; assoc prof pediat, Pritzker Med Sch, Univ Chicago, 73-78; prof pediat & dermat, Sch Med, Northwestern Univ, 78-87; PROF PEDIAT & DERMAT MED, MED COL WIS, 87- *Concurrent Pos:* Attend physician & dir pediat dermat, Michael Reese Hosp & Med Ctr, 73-78; head, Dermat Div, Children's

Mem Hosp, 78-87 & Div Dermat, Dept Pediat, Children's Hosp Wis, 87- *Mem:* Soc Invest Dermat; Soc Pediat Res; Soc Pediat Dermat; Am Acad Dermat; Am Dermat Asn; Am Acad Pediat. *Res:* Neonatal skin disorders; skin infections; genetics disorders of the skin. *Mailing Add:* Med Col Wis 8701 Watertown Plank Rd Milwaukee WI 53226

ESTERMANN, EVA FRANCES, b San Francisco, Calif, Feb 26, 32. PLANT PHYSIOLOGY. *Educ:* Univ Calif, BS, 53, PhD(plant nutrit soils), 58. *Prof Exp:* Jr res biochemist, Univ Calif, Berkeley, 58-60; from asst prof to assoc prof, 60-69, PROF BIOL, SAN FRANCISCO STATE UNIV, 69- *Res:* Bacterial nutrition at surfaces; physiological ecology; spore metabolism. *Mailing Add:* Box 249 Graton CA 05444

ESTERSON, GERALD L(EE), b Baltimore, Md, June 29, 27; m 52; c 2. PROCESS COMPUTER SIMULATION, CHEMICAL PROCESS SCALE-UP. *Educ:* Johns Hopkins Univ, BEng, 51, DEng(elec eng), 56. *Prof Exp:* Res assoc oceanog instrumentation, Inst Co-op Res, Johns Hopkins Univ, 54-56; sr engr air arm div, Westinghouse Elec Co, 56-58; asst prof chem eng, Wash Univ, 58-61, assoc prof appl math, 61-65, dir inst continuing educ, 65-71, from assoc prof to prof eng, 65-76; PROF APPL CHEM, HEBREW UNIV JERUSALEM, 76- *Concurrent Pos:* Consult, Compumatix, Inc, 59, Monsanto Co, 59-61 & 64-68 & McGraw-Hill Bk Co, Inc, 62-63. *Mem:* Inst Elec & Electronics Engrs; Am Inst Chem Engrs; Instrument Soc Am. *Res:* Simulation and modeling of industrial and environmental systems; automatic control; computer control; pilot plant operations and process development; crystallizer systems; energy and cement from oil shale. *Mailing Add:* Casali Inst Appl Chem Hebrew Univ Jerusalem 91904 Israel

ESTERVIG, DAVID NELS, b Madison, Wis, 1952; m; c 3. CANCER, CELL DIFFERENTIATION. *Educ:* Univ Mo, PhD(genetics), 86. *Prof Exp:* Fel, Mayo Clin, Rochester, Minn, 84-87, res assoc, 87-89, lectr, 90; ASST PROF, UNIV WIS, EAU CLAIRE, 91- *Concurrent Pos:* John M Dalton fel, 80. *Mem:* Am Soc Cell Biol; Int Cell Cycle Soc. *Res:* Induction of cancer suppressor activity by cellular differentiation; general analysis of cell cycle and differentiation; control of viral gene promoters by cellular differentiation; mitotic centrosome mutants. *Mailing Add:* Biol Dept 360 Phillips Hall Univ Wis Eau Claire WI 54702-4004

ESTES, CARROLL L, b Ft Worth, Tex, May 30, 38. NURSING. *Educ:* Stanford Univ, AB, 59; Southern Methodist Univ, MA, 61; Univ Calif, San Diego, PhD, 72. *Hon Degrees:* LHD, Russel Sage Col, 86. *Prof Exp:* Res asst & assoc, Brandeis Univ, 62-67; res dir, Simmon's Col, 63-64; asst prof, social work, San Diego State Col, 67-72; asst prof, Dept Psychiat, 72-75, assoc prof, Dept Social & Behav Sci, 75-79, DIR, INST HEALTH & AGING, UNIV CALIF, SAN FRANCISCO, 79-, PROF, DEPT SOCIAL & BEHAV SCI, 79-, CHMN, 81- *Concurrent Pos:* Co-investr or prin investr grants, 76- *Honors & Awards:* Award Leadership in Field of Aging, Am Soc Aging, 88. *Mem:* Inst Med-Nat Acad Sci; Am Soc Aging; Asn Geront Higher Educ (pres, 81-82). *Res:* Author or co-author of over 80 articles and 7 books. *Mailing Add:* Dept Social & Behav Sci Box 0612 6th Fl N6314 Sch Nursing Univ Calif San Francisco CA 94143

ESTES, DENNIS RAY, b Stradford, Okla, June 18, 41; m 67; c 2. QUADRATIC FORMS, COMMUTATIVE RING THEORY. *Educ:* ECent State Col, Okla, BS, 61; La State Univ, MS, 63, PhD(math), 65. *Prof Exp:* Asst prof math, La State Univ, 65-66; fel, Calif Inst Technol, 66-68; from asst prof to assoc prof, 68-85, PROF MATH, UNIV SOUTHERN CALIF, 85- *Concurrent Pos:* Vis assoc prof, La State Univ, 74-75; vis prof, Ohio State Univ, 82. *Mem:* Math Asn Am; Am Math Soc. *Res:* Arithmetic theory of quadratic lattices and matrix theory over commutative rings. *Mailing Add:* Dept Math Univ Southern Calif Los Angeles CA 90007

ESTES, EDNA E, b Jasper, Ala, Nov 23, 21. BOTANY, MICROBIOLOGY. *Educ:* Univ Ala, BS, 48, MS, 49, PhD(bot), 57. *Prof Exp:* Asst prof biol, Flora Macdonald Col, 49-53; instr, Mobile Ctr, Univ Ala, 53-54; instr sci & biol, St Mary's Sem-Jr Col, 57-59; asst prof biol, Del Mar Col, 59-60; from assoc prof to prof biol, Salisbury State Col, 60-78; RETIRED. *Mem:* AAAS; Bot Soc Am; Am Inst Biol Sci; Sigma Xi. *Res:* Plant physiology, particularly relation of phosphorus nutrition to photosynthesis; correlating the uptake and distribution of phosphorus-32 in higher plants with certain photosynthetic factors, particularly chlorophyll pattern, light, and carbon dioxide supply. *Mailing Add:* PO Box 2305 Jasper AL 35502-2305

ESTES, EDWARD HARVEY, JR, b Gay, Ga, May 1, 25; m 48; c 5. MEDICINE. *Educ:* Emory Univ, BS, 44, MD, 47. *Prof Exp:* Intern med, Grady Mem Hosp, Atlanta, Ga, 47-48, asst resident, 49-50; sr asst resident med, Duke Univ Hosp, 52-53; fel med, Med Sch, Duke Univ, 53-54; chief cardiovasc sect, Vet Admin Hosp, 54-55; chief cardiol dept, Duke Univ Hosp, 55-58; chief med serv, Vet Admin Hosp, Durham, 58-63; chmn, Dept Community Health Sci, Duke Univ, 66-85, prof med, 63-90, dir, Duke Watts Family Med Prog, 85-89, EMER PROF MED, DUKE UNIV, 90-, EMER DIR, DUKE WATTS FAMILY MED PROG, 89-; DIR KATE B REYNOLDS, COMMUNITY PRACTR PROG, NC MED SOC FOUND, 90- *Concurrent Pos:* Fel physiol, Emory Univ, 48-49 & cardiovasc physiol, 50. *Mem:* Inst Med-Nat Acad Sci. *Res:* Cardiovascular physiology; electrocardiography. *Mailing Add:* PO Box 27167 Raleigh NC 27611

ESTES, EDWARD RICHARD, b Richmond, Va, Mar 2, 25; m 50; c 5. STRUCTURAL STEEL CONNECTIONS, COLD FORMED STEEL. *Educ:* Tulane Univ, BE, 45; Va Polytech Inst, MS, 48. *Prof Exp:* Fel civil eng, Va Polytech Inst, 46-47; asst prof, Univ Va, 48-55; res engr, Am Inst Steel Construct, 55-60; dir eng, Fla Steel Corp, 60-66; chief res engr, Am Iron & Steel Inst, 67-69; engr mgr, Repub Steel Corp, 69-74; consult engr, Estes & Assocs, 74-78; chmn, 79-84, assoc dean, 84-88, PROF, OLD DOMINION UNIV, 79- *Honors & Awards:* A F Davis Silver Medal, Am Welding Soc, 64. *Mem:* Am Soc Testing & Mat; Am Welding Soc; Am Soc Civil Engrs; Am Railway Eng Asn; Am Soc Eng Educ; Sigma Xi. *Res:* Structural analysis and design concentrating on both hot rolled and cold formed steel; testing methods; bolted and welded connections. *Mailing Add:* Col Eng & Technol Old Dominion Univ Norfolk VA 23529-0240

ESTES, FRANCES LORRAINE, b Mendon, Mich, Dec 25, 15. ENVIRONMENTAL CHEMISTRY. *Educ:* Kalamazoo Col, AB, 40; Univ Chicago, MS, 48; Rutgers Univ, PhD, 53. *Prof Exp:* Control chemist, Johnson & Johnson Surg Supplies, 41-42; asst catalysis, Inst Gas Tech, Ill Inst Technol, 43-47; from instr to asst prof, Douglass Col, Rutgers Univ, 48-56; res assoc biochem, Col Med, Baylor Univ, 56-61; res asst prof, Med Br, Univ Tex, 61-67; biochemist, Vet Admin Hosp, Houston, Tex, 67-69; dir environ chem, Gulf South Res Inst, 69-71; CONSULTANT, 71- *Concurrent Pos:* Mem air pollution chem & physics adv comt, US Environ Protection Agency, 72-75. *Mem:* AAAS; Am Chem Soc; NY Acad Sci; fel Am Inst Chem; Soc Appl Spectros. *Res:* Gas phase reactions and biological interactions of environmental concern. *Mailing Add:* 777 N Post Oak Rd Houston TX 77024

ESTES, JAMES ALLEN, b Sacramento, Calif, Oct 2, 45; m 81. ECOLOGY, VERTEBRATE BIOLOGY. *Educ:* Univ Minn, BA, 67; Wash State Univ, MS, 70; Univ Ariz, PhD(biol), 74. *Prof Exp:* RES BIOLOGIST, NAT ECOL RES CTR, US FISH & WILDLIFE SERV, 74- *Concurrent Pos:* Affil asst prof, Ctr Quant Sci, Univ Wash, 76-83; res assoc, Ctr Coastal Marine Studies, Univ Calif, Santa Cruz, 78-; adj prof biol, Univ Calif, Santa Cruz, 80- *Mem:* Am Soc Mammalogists; Ecol Soc Am; Am Soc Naturalists; Soc Conserv Biol. *Res:* Sea otter, community interactions in rocky systems; algae-herbivore interactions; ecology and evolution of marine macroalgae; marine vertebrates. *Mailing Add:* Inst Marine Sci Univ Calif Santa Cruz CA 95064

ESTES, JAMES RUSSELL, b Burkburnett, Tex, Aug 28, 37; m 62; c 2. SYSTEMATIC BOTANY, POLLINATION BIOLOGY. *Educ:* Midwestern State Univ, BS, 59; Ore State Univ, PhD(bot), 67. *Prof Exp:* From asst prof to assoc prof, 67-82, asst cur, 71-79, PROF BOT, DEPT BOT & MICROBIOL, UNIV OKLA, 82-, CUR, ROBERT BEBB HERBARIUM, 79- *Concurrent Pos:* Vis res scientist, dept entom, Univ Calif, 74; assoc ed bot Southwestern Naturalist, 80-82; prog chmn, Am Soc Plant Taxonomists, 81-83; dir, Okla Natural Heritage Prog, 82-83; asst ed, Flora Okla Proj, 84- *Mem:* Am Soc Plant Taxon (secy, 81-83, pres-elect, 84-85, pres, 85-86); Bot Soc Am; Sigma Xi. *Res:* Floristics of Oklahoma, Great Plains, the Southwest, and Southeast; taxonomy of the Poaceae; systematics of Asteraceae including biosystematics, cytotaxonomy, and pollination biology; autopolyploidy as an evolutionary process. *Mailing Add:* 770 Van Vleet Oval Norman OK 73019

ESTES, JOHN H, b Youngstown, Ohio, Jan 10, 16; m 46; c 4. ORGANIC CHEMISTRY, METALLURGY. *Educ:* Youngstown Univ, BS, 40; Wash State Univ, MS, 48, PhD(org chem), 52. *Prof Exp:* Metallurgist, Carnegie Ill Steel Co, 40 & Mullins Mfg Corp, 40-42; res chemist, 52-72, SR RES ASSOC, TEXACO RES CTR, 72- *Mem:* Am Chem Soc; Sigma Xi. *Res:* Zeolite synthesis; commercial process; catalysis in reforming field; gasoline additive studies. *Mailing Add:* 201 Cedar Hill Rd Wappingers Falls NY 12590-2332

ESTES, RICHARD, b San Rafael, Calif, May 9, 32; m 55; c 1. VERTEBRATE PALEONTOLOGY, HERPETOLOGY. *Educ:* Univ Calif, Berkeley, BA, 55, MA, 57, PhD(paleont), 60. *Prof Exp:* Mus preparator vert paleont, Univ Calif, Berkeley, 57-58, asst, 58-59, mus preparator, 59-60, asst vert zool, 60; from asst prof to prof biol, Boston Univ, 60-73; PROF ZOOL, SAN DIEGO STATE UNIV, 73- *Concurrent Pos:* Res assoc vert paleont, Mus Comp Zool, Harvard Univ Mus Nat Hist, San Diego, Mus Paleo Univ Calif, Berkeley, 60-73; NSF res grants, 61-67, 68-70, 73-76, 77-79, 80-83 & 86-; Am Philos Soc res grant, 64-65; Sigma Xi res grant, 65; Nat Acad Sci, Marsh Fund res grant, 70; Nat Geog Soc res grant, 73-74; ed, J Vert Paleont, 84-87. *Mem:* Am Soc Ichthyol & Herpet; Soc Vert Paleont; fel Herpetologists League; Soc Study Reptiles & Amphibians; Soc Study Evolution; AAAS. *Res:* Paleoecological and evolutionary phenomena in fossil lower vertebrate faunas; anatomy and relationships of fossil and recent Amphibia and Reptilia. *Mailing Add:* Dept Biol San Diego State Univ San Diego CA 92182

ESTES, TIMOTHY KING, b Kalamazoo, Mich, Oct 1, 40; m 61; c 2. PAPER CHEMISTRY. *Educ:* Western Mich Univ, BS, 62; Lawrence Univ, MS, 64, PhD(paper chem), 67. *Prof Exp:* SR RES SPECIALIST PAPERBOARD, PACKAGING CORP AM, 66- *Mem:* Am Chem Soc; Tech Asn Pulp & Paper Indust. *Res:* Secondary fibers; semi-chemical pulping; computer modeling; paperboard physics. *Mailing Add:* Packaging Corp Am 5401 Old Orchard Rd Tech Ctr Skokie IL 60077-1031

ESTEVEZ, ENRIQUE GONZALO, b Havana, Cuba, Nov 25, 48; US citizen; m 76. CLINICAL MICROBIOLOGY. *Educ:* Fla State Univ, BA, 71; Univ Miami, PhD(microbiol), 76. *Prof Exp:* Med technologist clin chem, Lahuis Clin Labs, 72-74; teaching asst microbiol, Univ Miami, 73-76; trainee, 76-78, develop specialist, 78-79, ASST DIR, CLIN MICROBIOL LAB, DUKE UNIV MED CTR, 79- *Mem:* Am Soc Microbiol; Am Pub Health Asn. *Res:* Rapid methods of laboratory diagnosis of infectious diseases; clinical significance of microbiological culture results; lab diagnosis of parasitic infections. *Mailing Add:* Clin Microbiol Lab Duke Univ Med Ctr PO Box 3322 Durham NC 27710

ESTEY, RALPH HOWARD, b Millville, NB, Dec 9, 16; m 44; c 2. PLANT PATHOLOGY, HISTORY OF CANADIAN AGRICULTURE. *Educ:* McGill Univ, BSc, 51, PhD(plant path), 56; Univ Maine, MS, 54; Univ NB, BEd, 60; Univ London, DIC, 65. *Prof Exp:* Instr voc sch, NB, Can, 56-57; instr plant path, Univ Conn, 56-57; from asst prof to prof, 57-82, chmn dept, 70-76, EMER PROF PLANT SCI, MACDONALD COL, MCGILL UNIV, 82- *Mem:* Soc Nematol; Mycol Soc Am; Agr Inst Can; Brit Mycol Soc; fel Can Phytopath Soc (pres, 78-79); Can Hist Asn. *Res:* History of Canadian agriculture with emphasis on agricultural education; early plant pathology and mycology in Canada. *Mailing Add:* Dept Plant Sci Macdonald Col McGill Univ Ste Anne de Bellevue PQ H9X 1C0 Can

ESTIENNE, MARK JOSEPH, b Norfolk, Va, Apr 16, 60; m 84; c 2. ANIMAL SCIENCE & NUTRITION ENDOCRINOLOGY. *Educ:* Va Polytech Inst & State Univ, BS, 82, MS, 84; Univ Ga, PhD(animal & dairy sci), 87. *Prof Exp:* Postdoctoral scholar, Dept Animal Sci, Univ Ky, 87-90; ASST PROF SWINE PROD & MGT, DEPT AGR, UNIV MD EASTERN

SHORE, 90- *Mem:* Am Soc Animal Sci; Soc Study Reproduction. *Res:* Neuroendocrine control of adenohypophysial hormone secretion, particularly luteinizing hormone and growth hormone; mechanisms controlling the onset of puberty in domestic animals. *Mailing Add:* Dept Agr Univ Md Eastern Shore Princess Anne MD 21853

ESTILAI, ALI, b Iran, Sept 29, 40; m 70; c 2. PLANT BREEDING, CYTOGENETICS. *Educ:* Univ Tehran, BS, 64; Univ Calif, Davis, MS, 68, PhD(genetics), 71. *Prof Exp:* Asst prof teaching & res fac sci, Univ Tehran, 71-77, assoc prof, 77-80; asst res agronomist,81-87, ASSOC RES AGRONOMIST, UNIV CALIF, DAVIS, 87- *Concurrent Pos:* Mem ed coun, Ministry Educ, Iran, 72-78; dep dir, Inst Biochem & Biophys, Univ Tehran, 77-78, mem res coun, 78-79; vis prof, Univ Calif, Davis, 78-79; prin invest, Guayule Bree ding & Develop Proj, 83- *Honors & Awards:* Outstanding Basic Res Award, Univ Tehran, 78. *Mem:* Am Soc Agron; Genetics Soc Am; Guayule Rubber Soc (secy, 82-83, pres-elect, 83-84, pres, 84-85, 85-86); Genetic Soc Iran (vpres, 76-80). *Res:* Evolution, cytogenetics, and breeding of various crops including safflower and saffron; breeding for increased rubber production in Guayule. *Mailing Add:* Dept Bot & Plant Sci Univ Calif Riverside CA 92521

ESTILL, WESLEY BOYD, b Enid, Okla, Mar 24, 24; m 48; c 7. ANALYTICAL CHEMISTRY, MATERIAL SCIENCE. *Educ:* Coe Col, BS, 49; Okla State Univ, MS, 51. *Prof Exp:* Mem staff res lab, Armour & Co, 51-52 & res lab, Ozark-Mahoning Co, 52-54; emission spectroscopist, Oak Ridge Nat Lab, 54-57; ELECTRON MICROSCOPIST & CHEM METALLURGIST, SANDIA CORP, 57- *Mem:* Electron Micros Soc Am; Electron Probe Anal Soc Am. *Res:* Complex ions; distillation of fluoride and ruthenium; adhesion of thin films; shock loaded metals; microanalysis; electron microprobe; electron diffraction; electron scanning; transmission microscopy; x-ray computer imaging; hydrogen effects on materials; laser interaction with materials; welding; design and fabrication of equipment for toxic material and gas transfer systems (hydrogen); chemical engineering. *Mailing Add:* Div 8441 Sandia Nat Labs Sandia Corp Livermore CA 94550

ESTIN, ARTHUR JOHN, b Feb 15, 27; US citizen; wid; c 4. SATELLITE COMMUNICATIONS, ANTENNA THEORY. *Educ:* Cooper Union, BEE, 49; Univ Colo, MS, 58, PhD(elec eng), 66. *Prof Exp:* Elec engr radio propagation, Nat Bureau Standards, 48-54, elec scientist microwave spectros, 54-58, physicist microwave & plasma physics, 58-66, asst div chief electromagnetics, 66-72, sect chief microwave eng, 72-76, supv physicist satellite commun eng, 76-80; consult engr, 80-83, CHIEF SCIENTIST, CYBERLINK CONSULT ENGR CORP, 83- *Concurrent Pos:* Sci & technol fel, Dept Com, 70-71. *Mem:* AAAS; Inst Elec & Electronics Engrs; Soc Instrumentation & Measurements; Sigma Xi. *Res:* Microwave antenna design and measurement; critical measurements in satellite communications systems. *Mailing Add:* Cyberlink Corp 1790 30th St Suite 300 Boulder CO 80301

ESTIN, ROBERT WILLIAM, physics, science education; deceased, see previous edition for last biography

ESTLE, THOMAS LEO, b Columbus Junction, Iowa, Jan 8, 31; m 53; c 4. MAGNETIC RESONANCE, DEFECTS IN CRYSTALS. *Educ:* Rice Inst, BA, 53; Univ Ill, MS, 54, PhD(physics), 57. *Prof Exp:* Fulbright scholar, 57-58; mem tech staff, Tex Instruments, Inc, 58-62, head defect physics sect, 62-66, sr res physicist, 66-67; chmn dept, 82-86, PROF PHYSICS, RICE UNIV, 67- *Mem:* Am Phys Soc. *Res:* Magnetic resonance; point imperfections in nonmetals; muon spin rotation. *Mailing Add:* Dept Physics Rice Univ PO Box 1892 Houston TX 77251-1892

ESTLER, RON CARTER, b Boonton, NJ, Dec 11, 49; m 79; c 1. CHEMICAL PHYSICS, CHEMISTRY. *Educ:* Drew Univ, BA, 72; Johns Hopkins Univ, MA, 74, PhD(chem), 76. *Prof Exp:* Res assoc chem, Columbia Univ, 76-77 & Stanford Univ, 77-78; asst prof chem, Univ Southern Calif, 78-82; AT FT LEWIS COL, 82- *Concurrent Pos:* Vis staff mem, Los Alamos Nat Lab, 84-; Coun Undergrad Res, Am Phys Soc. *Mem:* Am Chem Soc; Am Phys Soc. *Res:* Laser-induced chemical reactions; molecular reaction dynamics; laser spectroscopy; resonance ionization mass spectrometry. *Mailing Add:* Dept Chem Ft Lewis Col Durango CO 81301

ESTRADA, HERBERT, b Philadelphia, Pa, July 25, 03. POWER & TRANSMISSION, LINES. *Educ:* Univ Pa, BS, 26, MS, 27. *Prof Exp:* mgr, elec prod, steam & hydro, Philadelphia Elec, 26-63; proj mgr, Consort PA Utilities for Construct of Keystone Plant, 63-68; RETIRED. *Mem:* fel Am Soc Mech Engrs; fel Inst Elec & Electronics Engrs. *Mailing Add:* 203 62 Ave Avalon NJ 08202

ESTRADA, NORMA RUTH, b Oakland, Calif, Oct 29, 26; m 59; c 2. PSYCHOPHYSIOLOGY, PSYCHONEUROIMMUNOLOGY. *Educ:* Antioch Col West, BA, 75; Union Grad Sch West, Univ Pac, PhD(psychol), 78. *Prof Exp:* Mgt, Res & Develop, Everett A Gladman Mem Hosp, Telecare Corp, 66-88; CO-THERAPIST, PSYCHOTHER & BIOFEEDBACK, ARTHUR E GLADMAN, GLADMAN CTR, 73- *Concurrent Pos:* Bd mem, Alt Care Res Found, 77-86; mem, Adm Comt, Union Grad Sch West, 78-79; co-chmn, Coun Grove Conf, Menninger Found, 80; bd mem, Biofeedback Soc Calif, 79-80; bd mem, Calif Grad Sch Family & Marital Ther, 86-87; adj prof, Calif State Univ Hayward, 82-, William Lyon Univ, 87. *Mem:* Am Psychol Asn; Menninger Found fel; Biofeedback Soc Am; Am Assoc Adv Sci; Am Bd Med Psychotherapists. *Res:* The use of biofeedback, behavior therapy, and supportive psychotherapeutic counseling as an alternative to pharmacologic treatment of attention deficit disorder; using entrainment of specific immuno modulatory EEG frequencies for long term stimulation of the immune system. *Mailing Add:* 572 Contada Circle Danville CA 94526

ESTRIN, GERALD, b New York, NY, Sept 9, 21; m 41; c 3. COMPUTER SCIENCE. *Educ:* Univ Wis, BS, 48, MS, 49, PhD(elec eng), 51. *Prof Exp:* Res engr, Inst Advan Study, Princeton Univ, 50-53 & 55-56; dir electronic comput proj, Weizmann Inst Sci, Israel, 53-55; assoc prof eng, 56-58, chmn, Comput Sci Dept, 79-82, 85-88, PROF ENG, UNIV CALIF, LOS ANGELES, 58- *Concurrent Pos:* Lipsky fel, 54; consult, Nat Cash Register Co, 57, Telemeter Magnetics, Inc, 58-60 & Ampex Corp, 61-63; Guggenheim fel, 63 & 67; mem adv bd, appl math div, Argonne Nat Lab, 66-68, 74-80 mem, assoc univs rev comt for chmn, 76-77; dir, Comput Communs, Inc, 66-67; mem int prog comt, Int Fedn Info Processing Cong, 68; int prog chmn, Jerusalem Conf Info Technol, 71; mem bd gov, Weizmann Inst Sci, Israel, 71; Asn Comput Mach lectr; Inst Elec & Electronics Engrs distinguished speaker; mem math & comput sci res adv comt, AEC; dir, Systs Eng Labs, 77-80; mem, Sci Adv Comt & Operating Bd, Gould Inc, Rolling Meadows, Ill, 81-85. *Mem:* Fel Inst Elec & Electronics Engrs; Asn Comput Mach; Am Soc Eng Educ; NY Acad Sci; Sigma Xi. *Res:* Digital computer systems. *Mailing Add:* 500 Warner Ave Los Angeles CA 90024

ESTRIN, NORMAN FREDERICK, b Brooklyn, NY, Apr 1, 39; m 61; c 3. SCIENCE POLICY, TECHNOLOGY TRANSFER. *Educ:* Brooklyn Col, BS, 59; NY Univ, MS, 62; Fla State Univ, PhD(phys org chem), 68. *Prof Exp:* Chemist, Clairol Res Labs, Conn, 62-64; res asst, Fla State Univ, 64-68; dir sci, 68-72, vpres sci, 72-81, sr vpres sci, Cosmetic, Toiletry & Fragrance Asn Inc, DC, 81-85; vpres sci & technol, Health Indust Mfg Asn Inc, DC, 85-90; PRES, ESTRIN CONSULT GROUP INC, 90- *Concurrent Pos:* Ed, Cosmetic Ingredient Dict, Cosmetics, Toiletry & Fragrance Asn, 73, 76 & 82, Tech Guidelines, 74, 82 & Compendium of Cosmetic Ingredient Compos, 71 & 82; consult, Nat Cancer Inst, 81-83; mem bd, sci adv, Cath Univ, 81-83; ed, Sci & Regulatory Founds of the Cosmetic Indust, 84; mem, Nursing & Med Devices Steering Comt, Food & Drug Admin, 86-; mem bd dir, Nat Comt Clin Lab Standards, 88-91, Nat Mus Health & Med Found, 89-, Inst Advan Med Commun, 89- *Honors & Awards:* M deNavarre Medal Award, 85. *Mem:* Soc Cosmetic Chem; fel Royal Soc Health; Technol Transfer Soc; Am Soc Asn Exec. *Res:* Consultant for the medical device, food, drug, and cosmetic industries helping clients meet FDA regulatory requirements and helping them access new government and university developed technologies. *Mailing Add:* 9109 Copenhaver Dr Potomac MD 20854

ESTRIN, THELMA A, b New York, NY, Feb 21, 24; m 41; c 3. COMPUTERS IN MEDICINE, ENGINEERING EDUCATION. *Educ:* Univ Wis, BS, 48, MS, 49, PhD(elec eng), 51. *Prof Exp:* Res engr, Health Sci Ctr, Univ Calif, Los Angeles, 60-70, dir, Data Processing Lab, Brain Res Inst, 70-80; dir, div elect, comput & systs eng, NSF, 82-84; PROF ENG, COMP SCI DEPT, UNIV CALIF, LOS ANGELES, 80-, ASST DEAN, SCH ENG & APPL SCI, 84-, DIR, DEPT ENG & SCI, EXT, 84- *Concurrent Pos:* Fulbright fel, Weizmann Inst Sci, Rehovot, Israel, 63; prin investr, USPHS grant, Data Processing Lab, Brain Res Inst, Univ Calif, Los Angeles, 70-80, adj prof anat & comput sci, 78-80; bd trustees, Aerospace Corp, 79-82; Army Sci Bd, 80-82; NIH Biotechnol Resources Rev Comt, 81-86; NRC Bd Telecommun & Comput Applns, 82; NRC Energy Eng Bd, 84- *Mem:* Alliance for Eng in Med & Biol (vpres, 78); fel Inst Elec & Electronics Engrs (exec vpres, 82); Biomed Eng Soc; fel Inst Advan Eng; Asn Advan Med Instrumentation; fel AAAS; Asn Comput Mach; Soc Women Engrs. *Res:* Application of technology and computers to health care delivery; computer methods in neuroscience; electrical activity of the nervous system; engineering education. *Mailing Add:* Sch Eng & Appl Sci Univ Calif Los Angeles CA 90024

ESTRUP, FAIZA FAWAZ, b Joun, Lebanon, Apr 15, 33; US citizen; m 60. BIOPHYSICS, MOLECULAR BIOLOGY. *Educ:* Boston Univ, AB, 53; Yale Univ, MS, 60, PhD(biophys), 61; Brown Univ, MD, 75; Am Bd Internal Med, dipl, 80, Am Bd Rheumatology, dipl, 82. *Prof Exp:* Res asst spectros, Huntington Res Labs, Harvard Univ, 53-55; mem tech staff biophys, Bell Tel Labs, Inc, 62-63; vis asst prof chem, Haverford Col, 64-65, res assoc biol, 65-68; res assoc biol & med sci, Brown Univ, 68-75, resident internal med, RI Hosp, 75-76, resident clin pathol, 76-77, resident internal med, Mem Hosp, 77-78, chief resident internal med, 78, fel rheumatology, Brown Univ Prog, Roger Williams Gen Hosp, 78-80, clin asst prof, 83-91, CLIN ASSOC PROF MED, BROWN UNIV, 91-, STAFF RHEUMATOLOGIST, PROG MED, ASSOC HOSPS, 80-; CHIEF DEPT RHEUMATOLOGY, MEM HOSP, RI, 83- *Concurrent Pos:* Res fel, Inst Biophys, Geneva, Switz, 61-62; asst prof biol, RI Jr Col, 69-71; rheumatologist, pvt pract, 80. *Mem:* AAAS; fel Am Col Physicians; Biophys Soc; Sigma Xi; Am Rheumatism Asn; AMA; Am Soc Internal Med; Am Med Women's Asn; fel Am Col Rheumatology. *Res:* Research on ribosomal proteins using immunochemical techniques; research on host-induced modification of phage deoxyribonucleic acid; clinical research in rheumatology. *Mailing Add:* 15 Adelphi Ave Providence RI 02906

ESTRUP, PEDER JAN Z, b Copenhagen, Denmark, July 15, 31; m 60. PHYSICAL CHEMISTRY, SOLID STATE PHYSICS. *Educ:* Royal Polytech Inst, Denmark, MSc, 54; Yale Univ, PhD(phys chem), 59. *Prof Exp:* Res assoc nuclear chem, European Ctr Nuclear Res, Switz, 59-61; res scientist phys chem, Bell Tel Labs, NJ, 61-64 & Bartol Res Found, Franklin Inst, 64-67; assoc prof, 67-70, PROF PHYSICS & CHEM, BROWN UNIV, 70- & CHMN CHEM, 89- *Concurrent Pos:* Sr ed, J Phys Chem, 90- *Mem:* Am Phys Soc; Am Chem Soc; Am Vacuum Soc; fel Am Phys Soc. *Res:* Physics and chemistry of solid surfaces; low energy electron diffraction; electron spectroscopy; adsorption phenomena. *Mailing Add:* Dept Physics Brown Univ Providence RI 02912

ETCHES, ROBERT J, b Vancouver, BC, Nov 27, 48; m 70; c 3. AVIAN REPRODUCTION. *Educ:* Univ BC, BSc, 70; McGill Univ, MSc, 72; Reading Univ, UK, PhD(physiol & biochem), 75, DSc, 85. *Prof Exp:* From asst prof to assoc prof, 75-86, PROF PHYSIOL, DEPT ANIMAL & POULTRY SCI, UNIV GUELPH, 86-, PROF ZOOL, 89- *Concurrent Pos:* Young scientist res award, Poultry Sci Asn, 80; vis scientist, INRA Ctr Tours-Nouzilly, France, 81-82; pres, Pintade Farms Ltd, 83-87; vis prof physiol,

Dept Physiol, Nagoya Univ, Japan, 88, Dept Avian Sci, Univ Calif-Davis, 91; ed, Manipulation Avian Genome, 91. *Honors & Awards:* Sigma Xi Res Award, 90. *Mem:* Poultry Sci Asn; World's Poultry Sci Asn; Sigma Xi; Can Soc Animal Soc. *Res:* Developing methods for the production of transgenic chickens; developing embryonic stem cell lines that are suitable for homologous recombination; producing germline chimeras that can incorporate changes made in vitro into selected lines of chickens. *Mailing Add:* Dept Animal & Poultry Sci Univ Guelph Guelph ON N1G 2W1 Can

ETESON, DONALD CALVERT, electrical engineering; deceased, see previous edition for last biography

ETGEN, GARRET JAY, b Hackensack, NJ, Aug 20, 37; m 60; c 3. MATHEMATICS. *Educ:* Col William & Mary, BS, 59; Univ Wis, MS, 61; Univ NC, PhD(math), 64. *Prof Exp:* Asst chief appl math br, HQ, NASA, 64-67; from asst prof to assoc prof, 67-75, PROF MATH, UNIV HOUSTON, 75-, CHMN DEPT, 78- *Concurrent Pos:* Asst prof lectr, George Washington Univ, 65-67. *Mem:* Soc Indust & Appl Math; Am Math Soc; Math Asn Am; Sigma Xi. *Res:* Differential equations; matrix theory. *Mailing Add:* 3607 Thunderbird St Missouri City TX 77459

ETGEN, WILLIAM M, b Toledo, Ohio, May 7, 29; m 50; c 6. DAIRY SCIENCE, ANIMAL SCIENCE. *Educ:* Ohio State Univ, BS, 51, MSc, 55, PhD(dairy sci), 58. *Prof Exp:* Res assoc dairy sci, Ohio State Univ, 54-55; dairy husbandman, USDA, 55-58; from asst prof to assoc prof dairy sci, Univ RI, 59-64, assoc prof animal sci, 64-68; PROF DAIRY SCI, VA POLYTECH INST & STATE UNIV, 68- *Concurrent Pos:* Chmn, Dept Animal Sci, Univ RI, 67-68. *Mem:* Am Dairy Sci Asn. *Res:* Dairy management and production. *Mailing Add:* Dept Dairy Sci Va Polytech Inst Blacksburg VA 24061

ETGES, FRANK JOSEPH, b Chicago, Ill, June 18, 24; m 85; c 5. PARASITOLOGY, MALACOLOGY. *Educ:* Univ Ill, AB, 48, MS, 49; NY Univ, PhD(invert zool), 53. *Prof Exp:* Asst biol, NY Univ, 49-53; asst prof zool, Univ Ark, 53-54; from asst prof to assoc prof, 54-65, dir grad studies, 67-80, PROF ZOOL, UNIV CINCINNATI, 65- *Concurrent Pos:* Interam res fel, 62-63; fel grad sch, Univ Cincinnati, 71; NIH fel, London Sch Hyg & Trop Med, 71-72; WHO fel, Africa, 75. *Honors & Awards:* Distinguished Res Award, Sigma Xi, 66. *Mem:* Am Soc Trop Med & Hyg; Am Soc Parasitol; Soc Protozool; Am Micros Soc; Royal Soc Trop Med & Hyg. *Res:* Orientation, behavior, growth and reproduction of snail hosts of schistosomes; morphology, life history, taxonomy and physiology of animal parasites. *Mailing Add:* Dept Biol Sci 006 Univ Cincinnati Cincinnati OH 45221

ETHEREDGE, EDWARD EZEKIEL, b Jacksonville, Fla, May 22, 39; m 61; c 2. SURGERY, TRANSPLANTATION IMMUNOLOGY. *Educ:* Yale Univ, BA, 61, MD, 65; Univ Minn, PhD(surg), 74. *Prof Exp:* Asst resident surg, Univ Minn Hosps, 66-68, fel genetics & cell biol, Univ, 68-69, res fel transplant surg, Hosps, 69-71, chief resident surg, 71-73; staff surgeon, Walter Reed Army Med Ctr, 73-75; asst prof surg, Sch Med, Wash Univ, 75-79, assoc prof, 79-84; PROF SURG & DIR, DIV ORGAN TRANSPLANTATION, SCH MED, TULANE UNIV, 84- *Mem:* Transplantation Soc; Asn Acad Surg; Am Soc Transplant Surgeons; Soc Univ Surgeons; Am Asn Immunologists; Cent Surg Asn. *Res:* Classes of antibodies against transplantation antigens; leukocyte immunobiology. *Mailing Add:* Div Organ Transplantation Sch Med Tulane Univ 1430 Tulane Ave New Orleans LA 70112

ETHERIDGE, ALBERT LOUIS, b Wilmar, Ark, Aug 9, 40; m 62. ZOOLOGY, DEVELOPMENTAL BIOLOGY. *Educ:* Ark Agr & Mech Col, BS, 64; Univ Miss, MS, 65; Univ Tex, Austin, PhD(zool), 68. *Prof Exp:* Asst prof zool, La State Univ, Baton Rouge, 68-71; assoc prof, 71-77, head, Dept Natural Sci, 78-81, PROF BIOL, UNIV ARK, MONTICELLO, 77-, VCHANCELLOR ACAD AFFAIRS, 81- *Mem:* AAAS; Am Soc Zool; Soc Develop Biol; Sigma Xi. *Res:* Experimental embryology; embryonic induction of the mesonephric kidney in amphibians. *Mailing Add:* Nicholls State Univ PO Box 2002 Univ Sta Thibodaux LA 70310

ETHERIDGE, DAVID ELLIOTT, b Montreal, Que, July 1, 18; m 47; c 4. FOREST PATHOLOGY. *Educ:* Univ NB, BSc, 50; McGill Univ, MSc, 53; Univ London, PhD(plant path) & DIC, 56. *Prof Exp:* Asst forest pathologist, Forest Entom & Path Br, Can Dept Forestry, NB, 50-52, forest pathologist, Alta, 52-58 & Que, 58-67, res scientist, Pac Forest Res Ctr, Can Dept Environ, Can Forestry Serv, 57-75; RETIRED. *Concurrent Pos:* Forest pathologist, Food & Agr Orgn, UN, Govt Tanganyika, 63-64; sr res fel, Forest Res Inst, Rotorua, NZ, 66-67; consult trop forest path, UN Develop Prog Forestry Proj, Dominican Repub, 69-70 & Forest Protection, Forestry Dept, Dominica, 78; consult, 76-; vis scientist, Cent Plantation Crop Res Inst, Kayangulam, Kerala India & Inst Pertanian, Bogor, Indonesia, 75-76. *Res:* Temperate and tropical forest pest and disease control problems; tropical plantation crop protection; coconut palm; timber identification. *Mailing Add:* 3941 Oakdale Pl Victoria BC V8N 3B6 Can

ETHERIDGE, RICHARD EMMETT, b Houston, Tex, Sept 16, 29. HERPETOLOGY, PALEONTOLOGY. *Educ:* Tulane Univ, BS, 51; Univ Mich, MS, 52, PhD(zool), 59. *Prof Exp:* Lectr zool, Univ Southern Calif, 59-61; from asst prof to assoc prof, 61-70, chmn dept, 69-72, PROF BIOL, SAN DIEGO STATE UNIV, 70- *Concurrent Pos:* NSF fel, 60-61; cur herpet San Diego Natural Hist Mus & res assoc, Los Angeles County Mus, 61-73; res assoc, Harvard Univ, 87-91. *Mem:* AAAS; Am Soc Ichthyologists & Herpetologists; Soc Vert Paleont; Soc Study Amphibians & Reptiles; Herpetologists League. *Res:* Comparative osteology; systematics and evolution of lizards, especially the iguania; late Cenozoic lizard fossils of North America and the West Indies. *Mailing Add:* Dept Biol San Diego State Univ San Diego CA 92182-0057

ETHERINGTON, HAROLD, b London, Eng, Jan 7, 1900; nat US; m 28; c 1. NUCLEAR ENGINEERING. *Educ:* Univ London, BSc, 21. *Prof Exp:* Supt steel plant, Lena Goldfields, Ltd, 26-30; res engr, A O Smith Corp, 30-32; instr, Milwaukee Voc Sch, 32-36; asst engr, Allis-Chalmers Mfg Co, 37-42, mech engr, Eng Develop Div, 42-46; sect leader, Oak Ridge Nat Lab, 46-47, dir power pile div, 47-48; from dir naval reactor div to dir reactor eng div, Argonne Nat Lab, 48-53; asst to vpres mfg, ACF Industs, Inc, 53-56, vpres, Nuclear Prod, Erco Div, 56-59; mgr, Atomic Energy Dept, Allis-Chalmers Mfg Co, 59-61, gen mgr, Atomic Energy Div, 61-63; ATOMIC ENERGY CONSULT, 63- *Concurrent Pos:* Ed, Nuclear Eng Handbk; mem adv comt reactor safeguards, US Nuclear Regulatory Comn, 64-74 & 76-80, emer mem, 80- *Mem:* Nat Acad Eng; Am Soc Mech Engrs. *Mailing Add:* 84 Lighthouse Dr Jupiter FL 33458

ETHERTON, BUD, b Wardner, Idaho, Nov 16, 30; m 57; c 2. BOTANY. *Educ:* Wash State Univ, BS, 56, PhD(bot), 62. *Prof Exp:* Res assoc bot, Wash State Univ, 61-62; NSF fel biophys, Edinburgh, 62-63; lectr plant sci, Vassar Col, 63-64, asst prof biol, 64-67; vis scientist biol & med res, Argonne Nat Lab, 67-68; assoc prof bot, 68-80, PROF BOT, UNIV VT, 80- *Mem:* Am Soc Plant Physiol. *Res:* Electrical potentials and ion uptake in plant cells; electrophysiology of aluminum toxicity. *Mailing Add:* 42 Elsom Pkwy South Burlington VT 05403

ETHERTON, TERRY D, b Springfield, Ill, Oct 30, 49. ENDOCRINOLOGY, NUTRITION. *Educ:* Univ Minn, PhD(animal sci), 78. *Prof Exp:* ASSOC PROF DAIRY & ANIMAL SCI, PA STATE UNIV, 79- *Mailing Add:* Dept Dairy & Animal Sci Pa State Univ 301 W L Henning Bldg University Park PA 16802

ETHINGTON, ROBERT LOREN, b State Center, Iowa, Feb 13, 32; m 54; c 2. WOOD SCIENCE & TECHNOLOGY, ENGINEERING MECHANICS. *Educ:* Iowa State Univ, BS, 57, MS, 59, PhD(wood technol), 63. *Prof Exp:* Instr, Iowa State Univ, 59-63; technologist, US Forest Prod Lab, 63-64, proj leader fundamental properties, 64-74, asst dir, US Forest Prod Lab, Madison, Wis, 74-76, dir, Forest Prod & Eng Res, 76-79, DIR, PAC NORTHWEST FOREST & RANGE EXP STA, FOREST SERV, USDA, 79- *Honors & Awards:* L J Markwardt Award, Am Soc Testing & Mat, 76. *Mem:* Forest Prod Res Soc; Soc Wood Sci & Technol; Am Soc Testing & Mat. *Res:* Fundamental physical and mechanical properties of wood; stress grading; development of allowable stresses for wood; sampling methods for wood property evaluation. *Mailing Add:* 5266 NE 38th St Portland OR 97211

ETHINTON, RAYMOND LINDSAY, b State Center, Iowa, Aug 28, 29; m 55; c 2. GEOLOGY, PALEONTOLOGY. *Educ:* Iowa State Col, BS, 51, MS, 55; Univ Iowa, PHD(geol), 58. *Prof Exp:* from asst prof to assoc prof, 62-68, PROF GEOL, UNIV MO-COLUMBIA, 68- *Concurrent Pos:* Co-ed, J Paleontol, 69-74; ed, Spec Publ Soc Econ Paleontologists & Mineralologists, 81-85; pres, Soc Econ Paleontologists & Mineralologists, 89-90; chief panderer, Pander Soc, 90- *Honors & Awards:* Hartley Vis Fel, Univ Southampton, Eng, 87. *Mem:* Paleont Soc; Geol Soc Am; Soc Econ Paleontologists & Mineralogists; Am Asn Petrol Geol; Int Palaeont Union. *Res:* Ordovivian conodonts of North America. *Mailing Add:* Dept Geol Univ Mo Columbia MO 65211

ETHRIDGE, FRANK GULDE, b Meridian, Miss, Dec 21, 38; m 64; c 3. SEDIMENTOLOGY. *Educ:* Miss State Univ, BS, 56; La State Univ, MS, 66; Tex A&M Univ, PhD(geol), 70. *Prof Exp:* Prod geologist, Chevron Oil Co, 65-67; from asst prof to assoc prof geol, Southern Ill Univ, 70-75; assoc prof, 75-81, PROF EARTH RESOURCES, COLO STATE UNIV, 81-, ACTG HEAD DEPT, 89- *Concurrent Pos:* Consult petrol & uranium geol; prin investr res, State & Fed agencies & pvt indust. *Mem:* Soc Econ Paleontologists & Mineralogists; Int Asn Sedimentologists; Int Asn Math Geol; Sigma Xi. *Res:* Sedimentology; sandstone petrology; depositional models of Holocene environments; sequence stratigraphy, interpretation of environments of deposition and diagenesis of ancient detrital sedimentary rocks with application to exploration of petroleum, coal, and uranium. *Mailing Add:* Dept Earth Resources Colo State Univ Ft Collins CO 80523

ETHRIDGE, NOEL HAROLD, b Plains, Ga, Aug 7, 27; m 49; c 3. BLAST PHENOMENA, INSTRUMENTATION DEVELOPMENT. *Educ:* Ga Inst Technol, BS, 48. *Prof Exp:* Physicist, US Army Ballistic Res Lab, Aberdeen Proving Ground, 50-58, supvry physicist, 60-85; physicist, Oak Ridge Nat Lab, 58-60; res physicist, Tech Reps, Inc, 85-89; PRIN SCIENTIST, APPL RES ASSOCS, INC, 89- *Mem:* Am Phys Soc; NY Acad Sci; AAAS; Instrument Soc Am. *Res:* Blast phenomena and effects from nuclear and high explosives explosions involving military equipment; dust sweep up by blast; blast instrumentation. *Mailing Add:* 503 E Lee Way Bel Air MD 21014-3318

ETKIN, ASHER, b New York, NY, Mar 12, 43; m 67; c 2. HIGH ENERGY PHYSICS, PARTICLE DETECTOR DEVELOPMENT. *Educ:* City Col New York, BS, 64; Yale Univ, MS, 66, MPh, 69, PhD(physics), 71. *Prof Exp:* Res staff physicist, Yale Univ, 71-73, res assoc, 73-74; sr res assoc, City Col New York, 74-75; from asst physicist to assoc physicist, 75-78, PHYSICIST, BROOKHAVEN NAT LAB, 78- *Mem:* Am Phys Soc; Inst Elec & Electronics Engrs; AAAS; Sigma Xi. *Res:* Study of strong interaction physics, particularly relitivistic heavy ion interactions and multi-particle final states; development of new state of art particle detector systems; study of spin dependence utilizing polarized targets. *Mailing Add:* Bldg 510A Brookhaven Nat Lab Upton NY 11973

ETKIN, BERNARD, b Toronto, Ont, May 07, 18; m 42; c 2. AEROSPACE ENGINEERING. *Educ:* Univ Toronto, BASc, 41, MASc, 47. *Hon Degrees:* DEng, Carleton Univ, Ont, 71. *Prof Exp:* Lectr aerospace eng, 42-48, from asst prof to prof, 48-83, chmn div eng sci, Inst Aerospace Studies, 67-72, dean, Fac Eng, 73-79, EMER PROF AEROSPACE ENG, UNIV TORONTO, 83- *Concurrent Pos:* Indust consult, 40-; mem aerodyn subcomt, Adv Comt Aeronaut Res, Nat Res Coun Can, 44-49, assoc comt aerodyn, 61-, chmn, 62; aerodynamicist, Nat Res Coun Can, 45; mem aerodyn dept, Royal Aircraft

Estab, Eng, 58-59. *Honors & Awards:* Centennial Medal of Can, 67; McCurdy Award, Can Aeronaut & Space Inst, 67; Mech & Control Flight Award, Am Inst Aeronaut & Astronaut, 75, Wright Brothers Lectureship, 80; Thomas W Eadie Medal, Royal Soc Can, 80. *Mem:* Fel Am Inst Aeronaut & Astronaut; fel Can Aeronaut & Space Inst; fel Royal Soc Can; fel Can Acad Eng. *Res:* Subsonic aerodynamics; wing theory; turbulence; dynamics of atmospheric flight; air classification of particles; architectural aerodynamics; university government. *Mailing Add:* Inst Aerospace Studies Univ Toronto Toronto ON M5S 1A1 Can

ETLINGER, JOSEPH DAVID, b Albany, NY, Feb 23, 46; m 70; c 3. CELL BIOLOGY, BIOCHEMISTRY. *Educ:* Rensselaer Polytech Inst, BS, 68; Univ Chicago, PhD(biophys), 74. *Prof Exp:* Res assoc physiol, Harvard Med Sch, 74-76; from asst prof to prof anat & cell biol, State Univ NY Downstate Med Ctr, 77-88, vchmn dept, 87-88; PROF & CHMN CELL BIOL & ANAT, NY MED COL, 88- *Concurrent Pos:* Muscular Dystrophy Asn fel, 74-76; prin investr, Nat Heart Lung & Blood Inst grants, Muscular Dystrophy Asn, NASA, Nat Ins Arthritis & Musculoskelatal & Skin Dis, 78-; estab investr, Am Heart Asn, 80-85. *Honors & Awards:* Irma T Hirsch Award, 80-85. *Mem:* Am Physiol Soc; Am Soc Cell Biol; AAAS; Am Heart Asn; Sigma Xi. *Res:* Mechanisms and physiological control of protein synthesis and degradation in skeletal muscle, cardiac muscle and erythroid cells; muscle hypertrophy and atrophy; myofibrillar assembly; turnover of abnormal proteins. *Mailing Add:* Dept Cell Biol & Anat NY Med Col Valhalla NY 10595

ETNIER, DAVID ALLEN, b St Cloud, Minn, Dec 2, 38; m 64; c 3. ICHTHYOLOGY. *Educ:* Univ Minn, BS, 61, PhD(zool), 66. *Prof Exp:* Asst prof, 66-72, assoc prof, 72-78, PROF ZOOL, UNIV TENN, KNOXVILLE, 78- *Mem:* Am Soc Ichthyologists & Herpetologists. *Res:* Taxonomy and ecology of freshwater fishes of eastern United States; biology of aquatic insects, especially Trichoptera taxonomy. *Mailing Add:* Dept Zool Univ Tenn Knoxville TN 37916

ETTEL, VICTOR ALEXANDER, b Prague, Czech, Feb 26, 37; Can citizen; m 60; c 2. HYDROMETALLURGY, ELECTROMETALLURGY. *Educ:* Univ Chem Technol, dipl, 60; Prague Univ, PhD(inorg chem), 66. *Prof Exp:* Res engr & scientist inorg chem, Inst Inorg Chem, Czech Acad Sci, 60-68; res chemist electro, 68-82, SECT HEAD HYDRO-METALL, J ROY GORDON RES LAB, MANITOBA DIV, INCO METALS, 68-, MGR PROCESS TECHNOL, 82-, SECT HEAD CARBONYL TECHNOL, 88- *Concurrent Pos:* Fel inorg chem, Univ Toronto, 66-67. *Honors & Awards:* Sherritt Hydrometall Award, 83. *Mem:* Can Inst Metall; Can Inst Chem; Can Soc Chem Eng; Metall Soc-Am Inst Mining, Metall & Petrol Engrs. *Res:* Hydrometallurgy and electrometallurgy of base metals, mainly copper, nickel and cobalt, and precious metals. *Mailing Add:* Inco Metals Process Res Lab Sheridan Park Mississauga ON L5K 1Z9 Can

ETTENBERG, M(ORRIS), electrical engineering; deceased, see previous edition for last biography

ETTENBERG, MICHAEL, b Brooklyn, NY, Mar 26, 43. LASER & FIBER OPTICS, OPTOELECTRONICS. *Educ:* Polytechnic Inst NY, BS, 64; NY Univ, MS, 67, PhD(mat sci), 69. *Prof Exp:* Mem tech staff, RCA Labs, 69-78, head, laser diode res, 78-86; DIR OPTOELECTRONICS RES, DAVID SARNOFF RES CTR, 86- *Mem:* Fel Inst Elec & Electronics Engrs; fel Optical Soc Am. *Mailing Add:* David Sarnoff Res Ctr CN 5300 Princeton NJ 08543

ETTENSOHN, FRANCIS ROBERT, b Cincinnati, Ohio, Feb 6, 47; m 78; c 2. PALENTOLOGY, STRATIGRAPHY SEDIMENTATION. *Educ:* Univ Cincinnati, BS, 69, MS, 70; Univ Ill, Urbana-Champaign, PhD(geol), 75. *Prof Exp:* Combat engr, US Army CEngr, 71; asst prof, 75-81, actg chmn dept, 85-86, assoc prof geol, 81-87, PROF GEOL, UNIV KY, 87- *Concurrent Pos:* Prin investr, US Dept Energy, 76-80; NSF, 85-88, US Bur Mines, 89-90. *Honors & Awards:* Fulbright lectureship, Soviet Union, 89. *Mem:* Fel Geol Soc Am; AAAS; Paleont Soc; Int Paleont Asn; Sigma Xi; Paleont Res Inst. *Res:* Paleoenvironments and paleoecology of Mississippian carbonates in eastern Kentucky; paleoecology of Carboniferous echinoderms; stratigraphy and paleoenvironments of Devonian black gas shales of eastern Kentucky; tectonics and sedimentation. *Mailing Add:* Dept Geol Sci Univ Ky Lexington KY 40506

ETTER, DELORES MARIA, b Denver, Colo, Sept, 25, 47; m 67; c 1. DIGITAL SIGNAL PROCESSING. *Educ:* Wright State Univ, BS, 70, MS, 72; Univ NMex, PhD(elec eng), 79. *Prof Exp:* Fac assoc comput sci, Math Dept, Wright State Univ, 72-73; lectr elec eng, Univ NMex, 73-78, from asst prof to assoc prof elec eng, 79-90. *Concurrent Pos:* Prin investr grants from, Sandia Nat Labs, 79-81, Southwest Resource Ctr, 80-81 & NSF, 81-83. *Mem:* Sigma Xi; Inst Elec & Electronics Engrs. *Res:* Adaptive digital signal processing; algorithms for adaptive time-delay estimation and adaptive recursive filter coefficients. *Mailing Add:* 3167 Nelson Rd Longmont CO 80501-9003

ETTER, MARGARET CAIRNS, b Wilmington, Del, Sept 12, 43; m 72; c 2. SOLID STATE ORGANIC CHEMISTRY. *Educ:* Univ Pa, BA, 65; Univ Del, MS, 71; Univ Minn, PhD(org chem), 74. *Prof Exp:* Chemist, Du Pont Co, 65-67; fel solid state chem, Univ Minn, 74-75; asst prof org chem, Augsburg Col, 75-76; res chemist, 3M Co, 76-83; fel solid state NMR, 83-84, from asst prof to assoc prof, 84-90, PROF, UNIV MINN, 90- *Mem:* Am Chem Soc; Am Crystallog Soc. *Res:* Solid state chemistry and packing patterns of hydrogen bonded organics; structure property relations in organic materials; use of x-ray crystallography and solid state NMR as complementary tools; organic nuclear optical materials. *Mailing Add:* 181 Woodlawn Ave St Paul MN 55105

ETTER, PAUL COURTNEY, b Philadelphia, Pa, Oct 27, 47; m 69; c 2. UNDERWATER ACOUSTICS, PHYSICAL OCEANOGRAPHY. *Educ:* Tex A&M Univ, BS, 69, MS, 75. *Prof Exp:* Technician, Technitrol, Inc, 69; res asst, Tex A&M Univ, 73-76; sr engr, MAR, Inc, 76-82; tech dir, Syntek Eng & Comp Systs Inc, 82-89; SR SCIENTIST, RADIX SYSTS, INC, 89- *Concurrent Pos:* Speaker, Soc Physics Students, 80-; lectr, Technol Serv Corp, 82- *Mem:* Am Meteorol Soc; Acoust Soc Am; Am Geophys Union; Inst Elec & Electronics Engrs. *Res:* Numerical simulations of underwater acoustic phenomena for application to naval undersea warfare system development and operation; ocean-atmosphere interactions and their application to climatic assessments and modeling. *Mailing Add:* 16609 Bethayres Rd Rockville MD 20855

ETTER, RAYMOND LEWIS, JR, b Sherman, Tex, Aug 10, 31; m 57; c 2. ORGANIC POLYMER CHEMISTRY. *Educ:* Univ Tex, BS, 52, PhD(org chem), 57. *Prof Exp:* Asst org chem, Univ Tex, 52-56; res chemist, 56-62, SR CHEMIST, TEX EASTMAN CO, 62-, SUPVR QUAL CONTROL, 72- *Mem:* Am Chem Soc. *Res:* Low molecular weight polymers; heterocyclic nitrogen compounds; polyolefins; synthetic resins; chlorinated polyolefins; paints; high pressure polymerization. *Mailing Add:* Rte 9 Box 178-A Maxey Rd Longview TX 75601

ETTER, ROBERT MILLER, b Chambersburg, Pa, July 13, 32; m 57; c 3. ORGANIC CHEMISTRY. *Educ:* Gettysburg Col, AB, 54; Pa State Univ, PhD(org chem), 59. *Prof Exp:* Res chemist dyes, Am Cyanamid Co, NJ, 58-62, res chemist explosives, Pa, 62-63; sr res chemist org synthesis, 63-65, res supvr, 65-71, prod res mgr, dir res & develop Europe, 72-78, dir res & develop, 78-80, vpres corp res, 81-82, vpres US Consumer Prod Res & Develop, 82-88, VPRES RES & DEVELOP, WORLDWIDE INDUST PROD SC JOHNSON & SON, INC, 81- & VPRES EXTERNAL AFFAIRS RES & DEVELOP, 88- *Mem:* AAAS; Am Chem Soc; NY Acad Sci. *Res:* Carbenes; reaction mechanisms; dyes; fiber finishes; explosives; insecticides; insect repellents; plant biochemistry; consumer chemical specialties. *Mailing Add:* 544 Blue Bird Ridge Asheville NC 28804

ETTINGER, GEORGE HAROLD, PHYSIOLOGY. *Educ:* Queens Univ, Ont, MD, 20. *Prof Exp:* Dean med, Queens Univ, Ont, 49-62; RETIRED. *Mailing Add:* Cartwright's Pt Kingston ON K7K 5E2 Can

ETTINGER, HARRY JOSEPH, b New York, NY, July 20, 34; m 58; c 3. INDUSTRIAL HYGIENE. *Educ:* City Col New York, BCE, 56; NY Univ, MCE, 58. *Prof Exp:* Lectr civil eng, City Col New York, 56-58; sanitary engr environ health, NIH, USPHS, 58-61; staff mem indust hyg, Los Alamos Nat Lab, Univ Calif, 74-80, proj mgr indust hyg studies, 81-87, proj dir, Occup Safety & Health Admin, 87-89, TECH RES COORDR, LOS ALAMOS NAT LAB, UNIV CALIF, 89- *Concurrent Pos:* Adj prof radiation sci, Sch Pharm, Univ Ark, 69-; consult, Div Reactor Licensing, AEC, 70-71 & Environ Protection Agency, 72-74 & Dept Energy Adv Comt on Nuclear Facil Safety; vis mem grad fac, Tex A&M, 81-; adj fac mem, Div Occup & Environ Health, San Diego State Univ, 81-86; fac affil, Colo State Univ, 83-; bd dirs, Am Indust Hyg Asn, 87-90 & Int Soc Respiratory Protection, 85-87; chmn, Am Bd Indust Hyg. *Honors & Awards:* Meritorious Achievement Award, Am Conf Govt Indust Hygienists, 85; Edward Baler Award, Am Indust Hyg Asn, 90. *Mem:* Am Indust Hyg Asn; Am Acad Indust Hyg; Am Conf Govt Indust Hygienists; Int Soc Respiratory Protection; Am Asn Aerosol Res; Am Bd Indust Hyg. *Res:* Properties of fine particles as related to the performance of air cleaning systems and inhalation health hazards; engineering control of toxic material and physical agents in the work environment; evaluation and application of respiratory protection in occupational health. *Mailing Add:* 55 Navajo Los Alamos NM 87544

ETTINGER, MILTON G, b La Crosse, Wis, Aug 3, 30; m 57; c 2. NEUROLOGY. *Educ:* Univ Minn, BA, 51, BS, 52, MD, 54. *Prof Exp:* Intern Internal med, Long Beach Vet Admin Hosp, Calif, 54-55; CHIEF NEUROL, HENNEPIN COUNTY MED CTR, 63-; PROF NEUROL, UNIV MINN, MINNEAPOLIS, 71- *Concurrent Pos:* Fel neurol, Univ Minn & Affil Hosps, 55-58; staff neurologist, Univ Minn, 60- & Hennepin County Gen Hosp, 60-; consult, Kenney Rehab Inst, Minneapolis, 62-68; chmn bd dirs, Hennepin Fac Assocs, 87-91. *Mem:* Am Acad Neurol (asst secy-treas, 73-77); Minn Soc Neurol Sci (past pres); Acad Aphasia; Int Neuropsychol Asn; Pan-Am Neurol Soc. *Res:* Coagulation, lysis and platelet abnormalities in cerebrovascular disease; cerebrovascular disease, including catecholamines and drug therapy; sleep disorders; headache; Huntington's disease. *Mailing Add:* Dept Neurol Univ Minn Box 295 Mayo Minneapolis MN 55455

ETTINGER, MURRAY J, ENZYME MECHANISMS, CELLULAR BIOCHEMISTRY. *Educ:* Hahnemann Univ, PhD(pharmacol), 65. *Prof Exp:* PROF BIOCHEM, STATE UNIV NY, BUFFALO, 84- *Mailing Add:* Dept Biochem State Univ NY Buffalo NY 14214

ETTRE, LESLIE STEPHEN, b Szombathely, Hungary, Sept 16, 22; US citizen; m 53; c 1. ANALYTICAL CHEMISTRY. *Educ:* Budapest Tech Univ, MS, 45, DSc(anal chem), 69; Am Inst Chem, cert. *Prof Exp:* Process chemist, G Richter Pharmaceut Co, Hungary, 46-49; res assoc, Hungarian Res Inst-Heavy Chem Industs, 49-51, head tech off, 51-53; mgr indust dept, Hungarian Plastics Indust Res Inst, 53-56; chemist, Lurgi Labs, Ger, 57-58; appln chemist, Perkin-Elmer Corp, 58-60, chief appln chemist, 62-68; exec ed, Encycl Indust Chem Anal, John Wiley & Sons, Publ, 68-74; sr staff scientist, Perkin-Elmer Corp, 72-87, sr scientist, 88-90; ADJ PROF, DEPT CHEM ENG, YALE UNIV, 90-; CONSULT, 90- *Concurrent Pos:* Ed, J Chromatographia, 71-; res assoc, Dept Eng & Appl Sci, Yale Univ, 77-78; adj prof, Col Natural Sci & Math, Univ Houston, 78-80; assoc mem, Comn Nomenclature Anal Chem, Int Union Pure & Appl Chem, 81-91. *Honors & Awards:* M S Tswett Chromatography Award, 78; Chromatography Mem Medal, USSR, 79; L S Palmer Award, Minn Chromatogr, Forum, 80; A J P Martin Award, Chromatography Soc, 82; Chromatography Award, Am Chem Soc, 85. *Mem:* Am Chem Soc; fel Am Inst Chem; Chromatography Soc. *Res:* Theory, practice and application of chromatography; analytical instrumentation; scientific editing. *Mailing Add:* PO Box 2175 Norwalk CT 06852

ETZEL, HOWARD WESLEY, b Brooklyn, NY, Aug 5, 22; m 44; c 2. SCIENCE ADMINISTRATION, SOLID STATE PHYSICS. *Educ:* Carnegie Inst Technol, BS, 44, MS & DSc(physics), 49. *Prof Exp:* Res physicist, Naval Res Lab, 50-56, head radiation effects sect, 56-62; assoc prog dir physics, NSF, Washington, DC, 62-63, prog dir solid state & low temperature physics, 63-71, dep dir, Div Mat Res, 71-79; assoc dean res & vis prof elec & computer eng, NC State Univ, 79-90; RETIRED. *Concurrent Pos:* Fulbright res scholar, France, 49-50. *Honors & Awards:* Meritorious Serv Award, NSF, 76. *Mem:* Fel Am Phys Soc; Sigma Xi; AAAS. *Res:* Electronic and optical properties of solids, lasers. *Mailing Add:* 406 Annandale Dr Cary NC 27511

ETZEL, JAMES EDWARD, b Reading, Pa, Nov 9, 29; m 50; c 5. LIQUID INDUSTRIAL WASTE TREATMENT. *Educ:* Pa State Univ, BSSE, 51; Purdue Univ, MSCE, 55, PhD(sanit eng), 57. *Prof Exp:* Jr engr, Capitol Eng, Inc, 51; construct engr, US Army CEngrs, 51-53; serv engr indust wastes, E I du Pont de Nemours & Co, Inc, 57-58; dir res & develop, Environ Eng, R F Weston, Inc, 58-59; chaired prof, 79-81, from asst prof to prof, 59-90, EMER PROF ENVIRON ENG, PURDUE UNIV, 90- *Concurrent Pos:* Consult, Gen Motors Corp, Monsanto Inc, US Environ Protection Agency, Colgate-Palmolive Corp, Gen Foods Corp & Sime Darby, Malaysia, 61- *Mem:* Water Pollution Control Fedn. *Res:* Treatment of water to remove impurities to make it suitable for industrial use and development of wastewater treatment processes for specific process industries. *Mailing Add:* 710 Cardinal Dr Lafayette IN 47905

ETZLER, DORR HOMER, b Westboro, Wis, Apr 30, 15; wid; c 2. CHEMISTRY. *Educ:* Univ Wis, BS, 35; Univ Calif, PhD(chem), 38. *Prof Exp:* Res chemist, Standard Oil Co Calif, 38-42 & Calif Res Corp, 45-46, admin asst, 46-50, asst to gen mgr, 50-55, mgr gen serv, 55-63, admin & lab serv, Chevron Res Co, 63-67 & orgn planning, 67-70, mgr res serv dept, 70-78, gen mgr res serv dept, 78-80; RETIRED. *Concurrent Pos:* US Army Chem Corps Officer, 36-75. *Mem:* Indust Res Inst; Am Chem Soc. *Res:* Photochemistry of acetyl halides; compounded lubricating oils; research administration and management. *Mailing Add:* 130 Miramonte Dr Moraga CA 94556

ETZLER, FRANK M, b Detroit, Mich, Mar, 20, 52; m 74. PHYSICAL CHEMISTRY. *Educ:* Central Mich Univ, BS, 73; Univ Miami, PhD(phys chem), 78. *Prof Exp:* Res specialist, Univ Minn, 78-80; asst prof phys chem, ECarolina Univ, 80-85; prof phys chem, Univ Ky, 85-87; PROF PHYS CHEM, INST PAPER SCI & TECHNOL, ATLANTA, 87- *Mem:* Am Chem Soc; NY Acad Sci; Sigma Xi. *Res:* Measurement and structural interpretation of the properties of water near interfaces; relations between structure of interfacial water and biological function. *Mailing Add:* 1280 Winesap Dr Austell GA 30001

ETZLER, MARILYNN EDITH, b Detroit, Mich, Oct 30, 40. LECTINS, BIOCHEMISTRY. *Educ:* Otterbein Col, BS(biol) & BA(chem), 62; Wash Univ, PhD(biol), 67. *Prof Exp:* Res assoc develop biol, Wash Univ, 66-67; NIH fel immunochem, Dept Microbiol, Col Physicians & Surgeons, Columbia Univ, 67-69; asst prof, 69-75, assoc prof, 75-80, PROF BIOCHEM, UNIV CALIF, DAVIS, 80- *Concurrent Pos:* NIH grants, 71-, NSF grant, 83- *Mem:* Am Soc Cell Biol; Am Soc Biol Chemists; Soc Complex Carbohydrates; Am Soc Plant Physiologists. *Res:* Structure, specificity and function of plant lectins. *Mailing Add:* Dept Biochem & Biophys Univ Calif Davis CA 95616

ETZWEILER, GEORGE ARTHUR, b Lewistown, Pa, Mar 14, 20; m 42; c 3. ELECTRICAL ENGINEERING. *Educ:* Pa State Univ, BS, 49, MS, 50, PhD(elec eng), 64. *Prof Exp:* Develop engr, Ahrendt Instrument Co, Litton Industs, Inc, 50-55, chief develop engr, 55-57; from instr to asst prof elec eng, 57-67, ASSOC PROF ELEC ENG, PA STATE UNIV, 67- *Concurrent Pos:* Lectr, Univ Md, 56-57; consult, Bausch & Lomb, Inc, NY, P R Hoffman Co, Pa & Carborundum Co, Pa; chmn tech comt components & awards comt, Am Automatic Control Coun; mem tech comt components, Int Fedn Automatic Control, US paper selection comt, Fifth Cong. *Mem:* Sr mem Inst Elec & Electronics Engrs; Am Soc Eng Educ. *Res:* Stability and performance of feedback control systems and control system components. *Mailing Add:* Dept Elec Eng Pa State Univ University Park PA 16802

ETZWILER, DONNELL D, b Mansfield, Ohio, Mar 29, 27; m 52; c 4. PEDIATRICS, PEDIATRIC DIABETES. *Educ:* Ind Univ, BS, 50; Yale Univ, MD, 53. *Prof Exp:* Instr pediat, Cornell Med Ctr-NY Univ, 56-57; from asst clin prof to assoc clin prof, 57-77, CLIN PROF PEDIAT, UNIV MINN, 77-; PRES, INT DIABETES CTR, PARK NICOLET MED CTR, 67- *Concurrent Pos:* Mem staff, Hope Ship, Peru, 62; mem, Nat Comn Diabetes, 76-77; dir, WHO Collaborating Ctr Diabetes Educ & Training, 85; dir, Diabetes Joint Proj, Int Diabetes Ctr-Cent Inst Advan Med Studies, Moscow, 88- *Honors & Awards:* Outstanding Serv to Diabetic Youth, Am Diabetes Asn, 76, Outstanding Serv Award, 77, Banting Award, 77, Upjohn Award, 83; Becton-Dickinson Award, 79. *Mem:* Inst Med-Nat Acad Sci; Am Diabetes Asn (pres, 76-77); fel Am Acad Pediat; Int Diabetes Fedn (vpres, 79-85); Am Asn Pub Health; Soc Pub Health Educrs; hon mem Am Dietetic Asn; Am Med Asn. *Res:* Diabetes management and health care delivery. *Mailing Add:* Int Diabetes Ctr 5000 W 39th St Minneapolis MN 55416

EU, BYUNG CHAN, b Seoul, Korea, July 7, 35; m 64; c 2. THEORETICAL CHEMISTRY. *Educ:* Seoul Nat Univ, BS, 59; Brown Univ, PhD(chem), 65. *Prof Exp:* Res assoc chem, Brown Univ, 65-66; res fel, Harvard Univ, 66-67; from asst prof to assoc prof, 67-72, PROF CHEM, McGILL UNIV, 75- *Concurrent Pos:* A P Sloan Found fel, 72-74. *Mem:* Am Phys Soc; NY Acad Sci; Korean Chem Soc. *Res:* Nonequilibrium statistical mechanics; theory of transport processes; nonlinear irreversible thermodynamics; fluid dynamics; rheology. *Mailing Add:* Dept Chem McGill Univ Montreal PQ H3A 2K6 Can

EUBANK, HAROLD PORTER, b Baltimore, Md, Oct 23, 24; m 48; c 3. PHYSICS. *Educ:* Col of William & Mary, BS, 48; Syracuse Univ, MS, 50; Brown Univ, PhD(physics), 53. *Prof Exp:* Asst physics, Syracuse Univ, 48-50; asst physics, Brown Univ, 50-52, res assoc, 52-54, asst prof, 54-59; res staff, Princeton Univ, 59-72, res physicist, 72-73, prin res physicist, Plasma Physics Lab, 73-86; RETIRED. *Concurrent Pos:* Chmn, div plasma physics, Am Phys Soc, 76-77. *Honors & Awards:* Elliot Cresson Award, Franklin Inst, 82; Distinguished Assoc Award, Dept Energy, 81. *Mem:* Am Phys Soc. *Res:* Experimental nuclear and plasma physics. *Mailing Add:* Rte 1 Box 823 Kilmarnock VA 22482

EUBANK, PHILIP TOBY, b Greenup, Ill, May 12, 36; m 60; c 2. CHEMICAL ENGINEERING. *Educ:* Rose Polytech Inst, BS, 58; Northwestern Univ, PhD(chem eng), 61. *Prof Exp:* From asst prof to prof, 61-68, PROF CHEM ENG, TEX A&M UNIV, 68- *Concurrent Pos:* NSF grant, 83-90. *Mem:* Fel Am Inst Chem Engrs; Am Chem Soc; Am Soc Eng Educ; Sigma Xi. *Res:* Thermophysical properties of fluids; mixtures; phase equilibria; electrical discharge machining. *Mailing Add:* Dept Chem Eng Tex A&M Univ College Station TX 77843-3122

EUBANK, RANDALL LESTER, b Dallas, Tex, Jan 1, 52. STOCHASTIC PROESSES, APPROXIMATION THEORY. *Educ:* NM State Univ, BS, 74, MS, 75, Tex A&M Univ, MS, 76, PhD(statist), 79. *Prof Exp:* Asst prof, Ariz State Univ, 79-80; ASST PROF STATIST, SOUTHERN METHODIST UNIV, 80- *Mem:* Inst Math Statist; Am Statist Asn; Soc Indust & Appl Math; Sigma Xi. *Res:* Regression analysis and design in the presence of correlated error, splines and stochastic processes; use of quantiles in data analysis; survival data analysis. *Mailing Add:* Dept Statist Southern Methodist Univ Dallas TX 75275

EUBANK, WILLIAM RODERICK, b Cynthiana, Ky, Jan 21, 19; m 45; c 2. PHYSICAL CHEMISTRY. *Educ:* Univ Ky, BS, 40, MS, 41; Johns Hopkins Univ, PhD(phys chem), 47. *Prof Exp:* Asst phys chem, Univ Ky, 40-41; asst ceramic lab, Pa State Univ, 41-42; res phys chemist, Keasbey & Mattison, Pa, 42-43; res assoc, Nat Bur Stand, 44-48; consult, US Naval Ord Test Sta, Calif, 48-51; res chemist, Edgar Bros Co, Ga, 51-52; gen mgr, Ind Hone Mfg Co, Mich, 52-53; sr res chemist, Cent Res Lab, 3M Co, 53-61, proj supvr, Magnetic Prod Lab, 61-64, mgr mat res, Revere-Mincom Div, 64-68, mgr anal res serv, Magnetic Prod Div, 68-81; RETIRED. *Mem:* AAAS; Am Chem Soc; Am Ceramic Soc; Electrochem Soc; Electron Micros Soc Am. *Res:* Phase equilibrium; temperature control; microscopy; refractories; enamels; cements; calcination; flame photometry; explosives; electrical ceramics; paint extenders; abrasive honing stones; low-melting glasses; metalloids and intermetallic compounds; ferrites; magnetic metals; electron microscopy; semiconductors; analytical chemistry. *Mailing Add:* 188 Lewis St Edgewater FL 32032-7306

EUBANKS, ELIZABETH RUBERTA, b Jacksonville, Fla. MICROBIOLOGY. *Educ:* NTex State Univ, 68, MA, 69; La State Univ, PhD(microbiol), 73. *Prof Exp:* Res assoc microbiol, Univ Tex, Austin, 73-75; asst prof microbiol, Ariz State Univ, 75-79; SR BACTERIOLOGIST, MASS PUB HEALTH BIOLOGIC LABS, 79-; ASST PROF MED, TUFTS UNIV, BOSTON, 79- *Mem:* Am Soc Microbiol; AAAS. *Res:* Pathogenic microbiology; developments of and immune response to subcellular vaccines; pathogenic mechanisms; bacterial flagella; chemistry, location and function of surface components of gram negative bacteria. *Mailing Add:* State Lab Inst 305 South St Jamaica Plain MA 02130

EUBANKS, ISAAC DWAINE, b San Angelo, Tex, Sept 22, 38; m 59; c 3. INORGANIC CHEMISTRY. *Educ:* Univ Tex, BS, 60, PhD(inorg chem), 63. *Prof Exp:* Chemist, Savannah River Lab, E I Du Pont de Nemours & Co, Inc, 63-67; from asst prof to assoc prof, 67-78, assoc chmn dept, 78-81, PROF CHEM, OKLA STATE UNIV, 78-, DIR, CTR EFFECTIVE INSTR, 82- *Concurrent Pos:* Vis prof, York Univ, 81. *Mem:* Am Chem Soc; Royal Soc Chem; AAAS; Nat Sci Teachers Asn; Sigma Xi. *Res:* Chemical education writing and consulting. *Mailing Add:* Dept Chem Okla State Univ Stillwater OK 74075

EUBANKS, L(LOYD) STANLEY, b San Antonio, Tex, Sept 24, 31; m 51; c 3. CHEMICAL ENGINEERING. *Educ:* Rice Univ, BA, 52, BS, 53, PhD(chem eng), 57. *Prof Exp:* Res engr, 57-59, sr res engr, 59-61, sr chem engr, 61-64, process specialist, 64-73, fel, 73-80, SR FEL, MONSANTO CO, 80- *Honors & Awards:* Edgar M Queeny Medal, 83. *Mem:* Am Inst Chem Engrs. *Res:* Phase equilibria. *Mailing Add:* 1143 Sunset Lane Texas City TX 77590

EUBANKS, ROBERT ALONZO, b Chicago, Ill, June 3, 26. THERORETICAL & APPLIED MECHANICS. *Educ:* Ill Inst Technol, BS, 50, MS, 51, PhD, 53. *Prof Exp:* From instr to asst prof mech, Ill Inst Technol, 50-54; sr engr, Bulova Res Lab, NY, 54-55; res engr, Am Mach & Foundry Co, Ill, 55-56; scientist, Borg Warner Res Ctr, 56-60; sr scientist, Armour Res Found, 60-62, mgr vibrations, 62-64; sci adv mech & struct eng, IIT Res Inst, 64-65; George A Miller vis prof, 64-65, PROF CIVIL ENG & THEORET & APPL MECH, UNIV ILL, URBANA, 65- *Concurrent Pos:* Adj prof, Ill Inst Technol, 62-65; consult, Continental Can Co, 68-75 & various govt agencies, 73-; vis distinguished prof civil eng, mech & aerospace eng & math, Univ Del, Newark, 73-74; mem exec comt Nat Consortium for grad degress for minorities in eng, 76- *Mem:* Am Soc Mech; Am Math Soc; Soc Indust & Appl Math; Am Soc Civil Engrs; Acoust Soc Am; Sigma Xi. *Res:* Mathematical theory of elasticity; rotor stability; elastic wave propagation; protective construction; vibrations and shock; terminal ballistics. *Mailing Add:* 1806 S Vine St Urbana IL 61801

EUBANKS, WILLIAM HUNTER, b Columbus, Miss, Dec 13, 21; m 44; c 3. ENGINEERING GRAPHICS. *Educ:* Miss State Univ, BS, 47, MS, 53. *Prof Exp:* Draftsman, Mobile Dist Corp Eng Design & Construct, Columbus AFB, 41 & 42; from instr to assoc prof, 60-73, PROF ENG GRAPHICS, MISS STATE UNIV, 73- *Mem:* Am Soc Eng Educ; Nat Soc Prof Engrs. *Res:*

Interpretation and graphical analysis of research data; use of photography in presenting graphical research data; methods of graphic presentation; graphic analysis; creative projects for freshman engineering students. *Mailing Add:* PO Box 926 Mississippi State MS 39762

EUBIG, CASIMIR, b Poland, Feb 21, 40; US citizen; m 65; c 2. MEDICAL PHYSICS. *Educ:* Fordham Univ, BS, 62; Univ Ariz, MS, 65, PhD(physics), 70. *Prof Exp:* Instr & res assoc, Dept Physics, Univ Ariz, 70-72, health physicist, Dept Hazards Control, 72-75; asst prof, 75-80, ASSOC PROF RADIOL, MED COL GA, 80- *Mem:* Am Asn Physicists Med; Soc Nuclear Med; Health Physics Soc; Am Inst Ultrasound Med. *Res:* Nuclear medicine; pediatric nuclear cardiology; radiation safety in medical institutions; quality control in radiology imaging equipment. *Mailing Add:* Dept Radiol Med Med Col Ga Augusta GA 30912

EUDY, WILLIAM WAYNE, b Oakboro, NC, Sept 1, 39; c 1. MICROBIOLOGY, BIOCHEMISTRY. *Educ:* Wake Forest Univ, BS, 61, MA, 69; NC State Univ, PhD(microbiol), 69. *Prof Exp:* Sr res microbiologist, Norwich Pharmacal Co, 68-71; sr res microbiologist, 71-76, mgr forest sci, 76-81, ASSOC DIR RES, INT PAPER CO, 81- *Mem:* AAAS; Am Chem Soc; Am Soc Microbiol; Tech Asn Pulp & Paper Indust. *Res:* Developmental biology; enzymology; lignin catabolism. *Mailing Add:* 24 Ave A Cornwall NY 12520-1003

EUGERE, EDWARD JOSEPH, b New Orleans, La, May 26, 30; m 54; c 4. PHARMACY, PHARMACOLOGY. *Educ:* Xavier Univ, BS, 51; Wayne State Univ, MS, 53; Univ Conn, PhD(pharmacol), 56. *Prof Exp:* Asst prof pharmacol, Detroit Inst Technol, 56-58; dean, Sch Pharm, 58-70, PROF PHARMACOL, TEX SOUTHERN UNIV, 58 - *Concurrent Pos:* Mem pharm rev comt, Pub Adv Group, NIH, 69-72, reviewer consult, Minority Biomed Support Prog & Bur Health Resources Develop; consult, Regional Off Bur Health Resources Develop, USPHS. *Mem:* Am Pharmaceut Asn; Am Asn Cols Pharm; AAAS. *Res:* Cardiovascular research; environmental toxicity. *Mailing Add:* Sch Pharm Tex Southern Univ Houston TX 77004

EUGSTER, A KONRAD, b Langenegg, Austria, Dec 10, 38; US citizen; m 65; c 2. VETERINARY VIROLOGY. *Educ:* Vienna Vet Col, Dr med vet, 63; Colo State Univ, PhD(microbiol), 70; Am Col Vet Microbiol, dipl, 71. *Prof Exp:* Res assoc virol, Southwest Res Ctr, Tex, 64-68; head diag microbiol, 70-80, EXEC DIR, TEX VET MED DIAG LAB, TEX A&M UNIV, 80- *Mem:* Am Vet Med Asn; Am Asn Vet Lab Diagnosticians; Conf Res Workers Animal Dis. *Res:* Improvements of diagnostic techniques in veterinary microbiology and virology; pathogenesis of hitherto unrecognized viruses. *Mailing Add:* PO Drawer 3040 College Station TX 77841-3040

EUKEL, WARREN W(ENZL), b Plummer, Minn, Mar 4, 21; m 46; c 3. ENGINEERING PHYSICS. *Educ:* Univ Calif, BS, 50. *Prof Exp:* Physicist, Radiation Lab, Univ Calif, 50-53, Chromatic TV, Calif, 53-54 & Appl Radiation Corp, 54-64; mem staff, W M Brobeck & Assocs, Berkeley, 64-66, opers mgr, 66-67, vpres, 67-89; SR ENGR, KMI ENERGY SERV, 89- *Concurrent Pos:* Mem comt high level dosimetry, Nat Acad Sci; mem adv bd, qm res & develop, Nat Res Coun. *Mem:* AAAS; Inst Elec & Electronics Engrs. *Res:* Ion sources and gaseous discharge; electron linear accelerators; peaceful uses of radiation. *Mailing Add:* 315 Crest Ave Walnut Creek CA 94595

EULER, KENNETH L, b Natrona Heights, Pa, July 25, 37. PHARMACOGNOSY. *Educ:* Univ Pittsburgh, BS, 49, MS, 62; Univ Wash, PhD(pharmacog), 65. *Prof Exp:* Asst prof pharmacog, Univ Md, 65-67; asst prof, 67-68, ASSOC PROF PHARMACOG, UNIV HOUSTON, 68- *Mem:* Am Pharmaceut Asn; Am Acad Pharmaceut Sci; Am Soc Pharmacog; Soc Econ Biol. *Res:* Plant chemistry and biochemistry; isolation and identification of plant constituents having physiological activity and study of their biosynthetic pathways. *Mailing Add:* Dept Pharm Univ Houston 4800 Calhoun Rd Houston TX 77204

EURE, HERMAN EDWARD, b Corapeake, NC, Jan 7, 47; m 69; c 2. PARASITOLOGY. *Educ:* Md State Col, BS, 69; Wake Forest Univ, PhD(biol, parasitol), 74. *Prof Exp:* Asst prof, 74-80, ASSOC PROF BIOL, WAKE FOREST UNIV, 80- *Mem:* Am Soc Parasitol; Sigma Xi; Brit Soc Parasitol. *Res:* Parasites of bass and the effects of thermal pollution on their population dynamics. *Mailing Add:* Box 7325 Reynolda Sta Wake Forest Univ Winston-Salem NC 27109

EUSTICE, DAVID CHRISTOPHER, b Wharton, NJ, Sept 26, 52; m 84; c 2. BIOCHEMISTRY. *Educ:* State Univ NY, Geneseo, BS, 74, MA, 77 & Binghamton, PhD(biol), 80. *Prof Exp:* Fel, cancer res, Dartmouth Med Sch, 80-81 & biochem, Univ Rochester, 81-85; res biochemist, E I du Pont de Nemours & Co Inc, 85-91; SR RES INVESTR, BRISTOL-MYERS SQUIBB RES INST, WALLINGFORD, CONN, 91- *Mem:* Am Soc Microbiol. *Res:* The mode of action of antibiotics; protein synthesis inhibitors and antiviral agents; enzymology of gene regulation at both the transcriptional and translational levels. *Mailing Add:* Bristol-Myers Squibb Res Inst Wallingford CT 06492

EUSTIS, ROBERT H(ENRY), b Minneapolis, Minn, Apr 18, 20; m 43; c 2. MECHANICAL ENGINEERING. *Educ:* Univ Minn, BMechEng, 42, MS, 44; Mass Inst Technol, ScD(mech eng), 53. *Prof Exp:* Instr mech eng, Univ Minn, 43-44; aeronaut res scientist, Nat Adv Comt Aeronaut, 44-47; from instr to asst prof mech eng, Mass Inst Technol, 48-51; chief engr & asst to pres, Thermal Res & Eng Corp, 51-53; sr mech engr, Stanford Res Inst, 53-56; assoc prof mech eng, 55-62, dir, High Temp Gasodynamics Lab, 61-80, prof mech eng, 61-81, prof, 81-90, EMER WOODARD PROF, STANFORD UNIV, 90- *Concurrent Pos:* Chmn tech adv coun, Emerson Elec Corp, 66-90. *Honors & Awards:* High Temp Inst Medal, USSR Sci Acad. *Mem:* Fel Am Soc Mech Engrs; Am Soc Eng Educ; Combustion Inst; fel Am Inst Aeronaut & Astronaut; fel AAAS. *Res:* Energy systems; heat transfer; fluid mechanics. *Mailing Add:* Dept Mech Eng Stanford Univ Stanford CA 94305

EUSTIS, WILLIAM HENRY, b Coeur D'Alene, Idaho, Dec 26, 21; m 49; c 1. SCIENCE EDUCATION. *Educ:* Univ Calif, PhD(chem), 51. *Prof Exp:* Technologist, Shell Oil Co, 51-60, sr engr, Shell Chem Co Div, 60-65; instr, Yakima Valley Col, 65-83; RETIRED. *Mem:* Am Chem Soc; Sigma Xi. *Res:* Analytical petroleum; organic reaction mechanisms. *Mailing Add:* 703 N 51st Ave Yakima WA 98908

EUTENEUER, URSULA BRIGITTE, protein motility, mitosis, for more information see previous edition

EVALDSON, RUNE L, b Okelbo, Sweden, Nov 21, 18; nat US; m 42; c 5. ENGINEERING MECHANICS. *Educ:* Univ Ill, BS, 41; Stanford Univ, PhD(eng mech), 50. *Prof Exp:* Sr anal engr, Hamilton Standard Propellers, United Aircraft Corp, 41-47; asst dynamics, elasticity, Stanford Univ, 47-50; consult, Booz-Allen & Hamilton, 50-53; assoc prof mech eng, 53-56, assoc dir, Inst Sci & Technol, 58-70, managing dir, Willow Run Labs, 65-67, dir, 67-70, PROF MECH ENG, UNIV MICH, 56- *Res:* Dynamics; elasticity; fatigue of metals; operations research; research administration. *Mailing Add:* 2950 Hickory Lane Ann Arbor MI 48104

EVANEGA, GEORGE R, b Cementon, Pa, Feb 6, 36; m 63; c 2. ORGANIC CHEMISTRY. *Educ:* Lehigh Univ, BS, 57; Yale Univ, MS, 58, PhD(org chem), 60. *Prof Exp:* NIH fel, Univ Freiburg, 60-61; res chemist, Union Carbide Res Inst, NY, 62-69, mgr biomed instrumentation, 69; res chemist, Med Res Labs, Pfizer Inc, 69-70, proj leader diabetes, 70, mgr diag res, 71-73, mgr immunol, Cent Res, 73-75; dir chem res & develop, Boehringer Mannheim Corp, Am, 75-78, head div, Boehringer Mannheim Res Tutzing, WGer, 78-79, vpres, Biodynamics, Boehringer Mannheim Diagnostics Inc, Indianapolis, 79-80, Houston, vpres new prod develop, Indianapolis, 80-82, vpres mkt & sales, 82-84, vpres technol, 85-88; VPRES & CHIEF ACCT OFFICER, MILES INC, 88- *Mem:* Am Chem Soc; Royal Soc Chem; Am Asn Clin Chemists; NY Acad Sci; Clin Radioassay Soc. *Res:* Photochemistry; medicinal chemistry; clinical chemistry; immunochemistry; microbiology; instrumentation. *Mailing Add:* 1220 E Jackson Blvd Elkhart IN 46516-4419

EVANOCHKO, WILLIAM THOMAS, b Trenton, NJ, June 6, 51; m 73; c 3. NUCLEAR MAGNETIC RESONANCE, CARDIOVASCULAR DISEASE. *Educ:* Trenton State Col, BA, 73; Auburn Univ, PhD (chem), 79. *Prof Exp:* Jr res chemist, Carter Wallace, Inc, 74-75; post doc fel, Argonne Nat Lab, 79-80; mgr, Cancer Resonance Facil, 80-84; asst prof med, dept med, div cardiovasc 84-90, ASSOC PROF MED, DEPT MED, UNIV ALA, BIRMINGHAM, 90- *Concurrent Pos:* Mgr, Cancer Resonance Facil, 80-84, Cardiovasc, Nuclear & Magnetic Resonance, Univ Ala, Birmingham, 84- *Honors & Awards:* Gold Award, Soc Magnetic Resonance Med, 82. *Mem:* Soc Magnetic Resonance Med; AAAS; Am Chem Soc. *Res:* Nuclear magnetic resonance imaging and spectroscopy for application to biomedical purposes specifically cardiovascular disease. *Mailing Add:* Div Cardiovasc Dis Univ Alabama 828 Cts Birmingham AL 35294

EVANS, ALAN G, b Upland, Pa, June 8, 42; m 65; c 2. SATELLITE APPLICATIONS. *Educ:* Widener Univ, BS, 64; Drexel Univ, MS, 67, PhD(elec eng), 72. *Prof Exp:* Asst engr, Philadelphia Elec Co, 64-65; teaching asst elec eng, Drexel Univ, 65-71; assoc engr, Calspan Corp, 72-74; ELECTRONIC ENGR, NAVAL SURFACE WARFARE CTR, 74- *Concurrent Pos:* Asst prof elec eng, US Naval Acad, 83-84; mem, spec study group, Int Asn Geod, 88-92; US Defense Mapping Agency res & develop award, 88. *Mem:* Sigma Xi; Inst Navig; Am Geophys Union; Inst Elec & Electronics Engrs. *Res:* Global positioning system satellite static and dynamic relative positioning and orientation; signal multipath analyses; determination of geodetic quantities and receiver development. *Mailing Add:* 5036 Woodhaven Dr SR-5 La Plata MD 20646

EVANS, ALBERT EDWIN, JR, b Tarrytown, NY, Apr 21, 30; m 56; c 4. NUCLEAR PHYSICS. *Educ:* Yale Univ, BS, 52; Ohio State Univ, MS, 53; Univ Md, PhD(nuclear physics), 65. *Prof Exp:* Engr, Nuclear Div, Martin-Marietta Co, 57-58, sr engr, 58; physicist, Radiation Physics Div, US Naval Ord Lab, 58-67; staff physicist, Nuclear Assay Res Group, Los Alamos Nat Lab, Univ Calif, 67-75, staff physicist, Critical Exp & Diag Group, 75-86; phys scientist, Off Weapons Res, Develop & Testing, 86-90, PHYSICIST, OFF BASIC ENERGY SCI, US DEPT ENERGY, WASHINGTON, DC, 90- *Mem:* Am Phys Soc; Am Nuclear Soc. *Res:* Neutron and gamma ray spectroscopy of nuclear reactions; low-energy particle accelerators, construction and renovation; measurement of flux and power distributions in nuclear reactors; nondestructive assay of fissionable materials, physics of delayed neutrons; LMFBR fuel-motion diagnostics instrumentation; nuclear reactor critical experiments; research administration. *Mailing Add:* 15005 Carry Back Dr Gaithersburg MD 20878

EVANS, ALFRED SPRING, b Buffalo, NY, Aug 21, 17; m 50; c 3. EPIDEMIOLOGY, INTERNAL MEDICINE. *Educ:* Univ Mich, AB, 39, MPH, 60; Univ Buffalo, MD, 43; Am Bd Internal Med, dipl, 51. *Hon Degrees:* MA, Yale Univ, 66. *Prof Exp:* Asst prof prev med, Yale Univ, 46-52; assoc prof prev med & med microbiol, Univ Wis, 52-59, prof prev med & chmn dept, 59-66; dir div int epidemiol, 66-77, PROF EPIDEMIOL & DIR WHO SERUM REF BANK, DEPT EPIDEMIOL & PUB HEALTH, SCH MED, YALE UNIV, 66- *Concurrent Pos:* Ed-in-chief, Yale J Biol & Med; vpres, Soc Med Consult to Armed Forces, 79-80, pres, 80-81; col, US Army, retired. *Mem:* AMA; Am Epidemiol Soc (secy-treas, 68-73, pres, 73-74); Soc Epidemiol Res; Int Soc Epidemiol; Infectious Dis Soc Am; Am Col Epidemiol (pres, 89-90); Am Asn Hist Med. *Res:* Infectious mononucleosis; E viruses; respiratory viruses; serological surveys; viruses and cancer. *Mailing Add:* Dept Epidemiol Yale Univ 501 Leph 60 College St Box 3330 New Haven CT 06510

EVANS, ALLISON BICKLE, industrial chemistry; deceased, see previous edition for last biography

EVANS, ANTHONY GLYN, b Porthcawl, Britain, Dec 4, 42; m 64; c 3. MATERIALS SCIENCE, CERAMICS. *Educ:* Univ London, BSc, 64, PhD(metall), 67. *Prof Exp:* Proj leader ceramics, Atomic Energy Res Estab, NY, 67-71 & Nat Bur Standards, 71-74; group leader, Rockwell Sci Ctr, 74-78; PROF CERAMICS, UNIV CALIF, BERKELEY, 78- *Concurrent Pos:* Consult & mem, Mat Res Coun, 74-; mem, Nat Mat Adv Bd, 76- *Honors & Awards:* Ross Coffin Purdy Award, Am Ceramic Soc, 74. *Mem:* Am Ceramic Soc. *Res:* Mechanical properties of brittle materials, particularly fracture of ceramics under conditions of impact, thermal and mechanical stress; failure prediction based on non-destructive evaluation. *Mailing Add:* Dept Mat Sci & Mineral Eng Univ Calif Santa Barbara CA 93106

EVANS, ARTHUR T, b Huron, SDak, Nov 26, 19; m 42; c 4. UROLOGY, SURGERY. *Educ:* Miami Univ, AB, 41; Univ Chicago, MD, 44. *Prof Exp:* PROF UROL, MED CTR, UNIV CINCINNATI, 69-, DIR, 61-; DIR, CHRISTIAN R HOLMES HOSP, 69- *Concurrent Pos:* Mem staff, Div Urol, Cincinnati Gen Hosp, 61-; mem staff, Cincinnati Children's Hosp; mem, Residency Rev Comn Urol, 77-; mem coun med educ, Am Urol Asn. *Mem:* AMA; Am Col Surgeons; Am Urol Asn; Soc Genito Urinary Surg; Soc Univ Urologists. *Res:* Translumbar arteriography as presently used in renal angiography and in studying renal circulation. *Mailing Add:* Ocean Reef Club 11 Channel Cay Rd Key Largo FL 33037

EVANS, AUDREY ELIZABETH, b York, Eng, Mar 6, 25; US citizen. PEDIATRICS, BIOCHEMISTRY. *Educ:* LRCPS(E), 50; Am Bd Pediat, dipl, 57. *Prof Exp:* Resident med & surg, Royal Infirmary, Edinburgh, 51-52; clin fel pediat, Children's Med Ctr, Boston, 53-54, resident tumor ther, 57-58, asst physician, 58-62, assoc med, 62-65; asst prof, Univ Chicago, 65-69; assoc prof, 69-74, PROF PEDIAT, UNIV PA, 74- *Concurrent Pos:* Resident pediat, Johns Hopkins Hosp, 54-56; instr, Harvard Med Sch, 61-65; Fulbright scholar & USPHS spec fel, 63-65. *Mem:* AAAS; Am Acad Pediat; Am Asn Cancer Res; Am Pediat Soc; Royal Soc Med. *Res:* Pediatric hematology and oncology; biochemistry and enzymology of leukemia. *Mailing Add:* Div Oncol 4th Floor Wood Bldg 34th & Civic Center Blvd Philadelphia PA 19104

EVANS, B J, b Macon, Ga, Aug 18, 42; m 63; c 3. MAGNETIC MATERIALS, SOLID STATE CHEMISTRY. *Educ:* Morehouse Col, BS, 63; Univ Chicago, PhD(chem), 68. *Prof Exp:* Nat Res Coun Can fel physics, Univ Man, 68-69; asst prof chem, Howard Univ, 69-70; sr scientist, Scientific Labs, Ford Motor Co, 76-77; prof & chmn, Dept Chem, Atlanta Univ, 86-87; from asst prof mineral to assoc prof solid state chem, 70-78, PROF INORG/SOLID STATE CHEM, UNIV MICH, ANN ARBOR, 78- *Concurrent Pos:* Res assoc Univ Chicago, 68; consult, Nat Bur Standards, 70-78, US Geol Survey, 80-83; guest prof, Philipps Univ Marburg, 77-78; Alfred P Sloan res fel, 72-74; Alexander von Humboldt found fel, 77-78; assoc Danforth Found, 76-86; distinguished fel, Am Coun Eng Educ, 88. *Mem:* Am Chem Soc; Am Phys Soc; Am Mineral Soc; Mineral Asn Can; Sigma Xi. *Res:* Mossbauer spectroscopy; order-disorder in oxides; crystal growth; magnetism and magnetic materials; processing of ceramic materials; mixed valence compounds. *Mailing Add:* Dept Chem Univ Mich Ann Arbor MI 48109-1055

EVANS, BEN EDWARD, b Wilkes-Barre, Pa, Apr 9, 44. SYNTHETIC ORGANIC & MEDICINAL CHEMISTRY. *Educ:* Kings Col, Pa, BS, 66; Princeton Univ, MA, 69, PhD(org chem), 71. *Prof Exp:* Instr biochem, Univ NC, Chapel Hill, 71-73; sr res chemist, 73-79, res fel, 78-87, SR INVESTR, MERCK SHARP & DOHME RES LABS, MERCK INC, 91- *Mem:* Am Chem Soc; Sigma Xi; AAAS. *Res:* Organic synthetic methodology; synthesis of pharmacologically active compounds; design and synthesis of enzyme inhibitors; peptide synthesis. *Mailing Add:* 501 Perkiomen Ave Lansdale PA 19446

EVANS, BERNARD WILLIAM, b London, Eng, July 16, 34; m 62. PETROLOGY. *Educ:* Univ London, BSc, 55; Oxford Univ, DPhil(geol), 59. *Prof Exp:* Asst geol, Glasgow Univ, 58-59; demonstr mineral, Oxford Univ, 59-61; asst res geologist, Univ Calif, Berkeley, 61-66, assoc prof geol, 66-69; chmn dept, 74-79, PROF GEOL, UNIV WASH, 69- *Honors & Awards:* Award, Mineral Soc Am, 70; US Sr Scientist Award, Humboldt Found, 88. *Mem:* AAAS; Mineral Soc Am; Am Geophys Union; Geol Soc Am; Brit Geol Soc. *Res:* Petrology and mineralogy of metamorphic and igneous rocks; electron probe microanalysis. *Mailing Add:* Dept Geol Sci AJ-20 Univ Wash Seattle WA 98195

EVANS, BOB OVERTON, b Grand Island, Nebr, Aug 19, 27; m 49; c 4. COMPUTER DESIGN. *Educ:* Iowa State Univ, BEE, 49. *Prof Exp:* Elec operating engr, Northern Ind Pub Serv Co, Hammond, 49-51; mem staff, IBM Corp, 51-62, vpres develop, Data Systs Div, 62-64, pres, Fed Systs Div, 65-69, Systs Develop Div, 69-75 & Systs Commun Div, 75-77, vpres eng, prog & technol, 77-84; gen partner, Hambrecht & Quist, 84-88; EXEC VPRES & PARTNER, TECHNOL STRATEGIES & ALLIANCES, 88- *Concurrent Pos:* Mem, Stark Draper Labs, Inc; consult govt agencies; mem, Defense Sci Bd; mem elec eng vis comt, Mass Inst Technol; mem bd gov, Aerospace Industs Asn, 65-69, Gov Sci Adv Comn, State of Md, 65-69, Telecommun Policy Comt, Inst Elec & Electronics Engrs, 78-80 & Sci Adv Comt, Commun Satellite Corp, 85-87; nat dir, Armed Forces Commun & Electronics Asn, 66-69; trustee, Nat Security Indust Asn, 66-69, Rensselaer Polytechnic Inst, 74-84 & NY Pub Libr, 81-84; chmn, Eng, Sci Manpower Study, Nat Acad Eng, 73. *Honors & Awards:* Distinguished Pub Serv Award, NASA, 69; Edwin Armstrong Award, Inst Elec & Electronics Engrs, 84; Nat Medal Technol, President US, 85. *Mem:* Nat Acad Eng; fel Inst Elec & Electronics Engrs; AAAS; Asn Comput Mach. *Mailing Add:* Technol Strategies & Alliances 3000 Sand Hill Rd Bldg 2 Suite 235 Menlo Park CA 94025

EVANS, BURTON ROBERT, b Harvey, Ill, Sept 26, 29; m 59; c 4. MEDICAL ENTOMOLOGY, VEGETABLE GARDEN INSECTS. *Educ:* Millikin Univ, BA, 51; Univ Md, MEd, 55, PhD(med entom), 58; Tulane Univ, MPH, 65. *Prof Exp:* Sta entomologist, Foreign Quarantine Div, USPHS, 58-67; entomologist, Aedes Aegyptic Eradication Proj, Commun Dis Ctr, 67-69; proj officer, Pesticides Prog, Div Pesticide Community Studies, Environ Protection Agency, 69-73; mem staff pesticide training, 73-78, PROF ENTOM, GA COOP EXTEN SERV, UNIV GA, 78- *Honors & Awards:* D W Brooks Award, Univ Ga, 85. *Mem:* Am Entom Soc; Sigma Xi. *Res:* Transmission of filariae by mosquitoes; evaluation of insecticides for mosquito control; surveys for the presence of various insects of medical importance. *Mailing Add:* Ga Coop Exten Serv Univ Ga Athens GA 30602

EVANS, CHARLES ANDREW, JR, b Harriman, Tenn, Feb 19, 42; m 78; c 2. ANALYTICAL CHEMISTRY. *Educ:* Cornell Univ, BA, 64, PhD(chem), 68. *Prof Exp:* Anal chemist mass spectros, Ledgemont Lab, Kennecott Copper, 68-70; sr res chemist, Mat Res Lab, Univ Ill, 70-76, prin res chemist, 76-78, assoc prof chem, Dept Chem, 75-78; PRES, CHARLES EVANS & ASSOCS, 78- *Mem:* Am Chem Soc; Am Soc Mass Spectrometry; Microbeam Anal Soc. *Res:* Materials analysis using microanalytical techniques such as secondary ion mass spectrometry; Rutherford backscattering spectrometry and Auger electron spectrometry. *Mailing Add:* Charles Evans Assoc 301 Chesapeake Dr Redwood City CA 94063-1393

EVANS, CHARLES HAWES, b Orange, NJ, Apr 16, 40; m 65; c 1. IMMUNOLOGY, CANCER. *Educ:* Union Col NY, BS, 62; Univ Va, MD & PhD(microbiol, immunol), 69. *Prof Exp:* Intern pediat, Med Ctr, Univ Va, 69-70, resident, 70-71; res assoc, 71-73, sr scientist, 73-76, CHIEF TUMOR BIOL SECT, NAT CANCER INST, NIH, 76- *Concurrent Pos:* Assoc ed, J Nat Cancer Inst, 81- & J Cell Biochem, 89-; ed Azalean, J Azalea Soc Am, 83-; mem, Arts & Sci Coun, Univ Va, 87-; bd Trustees, Suburban Hosp, Bethesda, MD, 88- *Honors & Awards:* John Horsley Mem Prize Med Res, Univ Va, 82; Citation, USPHS, 80, Commendation Medal, 85; Distinguished Serv Award, Azalea Soc Am, 89; Sir Henry Wellcome Medal, 90. *Mem:* Am Asn Immunologists; Am Asn Cancer Res; AAAS; Asn Mil Surgeons US; Int Soc Anal Cytol; fel Am Inst Chemists; Am Acad Med Adminr. *Res:* Immunobiology of carcinogenesis and mammalian cell model systems for the evaluation of lymphokines in the prevention and control of cancer. *Mailing Add:* Rm 2A17 Bldg 37 Nat Cancer Inst NIH Bethesda MD 20892

EVANS, CHARLES P, b New York, NY, May 8, 30; m 57; c 3. ORGANIC CHEMISTRY. *Educ:* Univ Bridgeport, BA, 59. *Prof Exp:* From anal chemist to res chemist, Olin Mathieson Chem Corp, 57-62; res chemist, Escambia Chem Corp, 62-68; vpres res & opers, 68-72, PRES, VITEK RES CORP, STAMFORD, 72- *Mem:* Am Chem Soc; Sigma Xi. *Res:* Suspension and emulsion polymerization reaction. *Mailing Add:* 12 Glenbrook Rd Trumbull CT 06611

EVANS, CLAUDIA T, b New York, NY, Aug 1, 52. EXPERIMENTAL BIOLOGY. *Educ:* Cornell Univ, BS, 74, MS, 75; Case Western Reserve Univ, PhD(biochem), 81. *Prof Exp:* ASST PROF BIOCHEM, DEPT VET AFFAIRS, SOUTHWEST MED CTR, UNIV TEX, 86- *Mem:* Protein Soc; Am Soc Biochem & Molecular Biol; AAAS. *Mailing Add:* Res Div 151B Vet Admin Med Ctr 4500 S Lancaster Rd Dallas TX 75216

EVANS, CLYDE EDSEL, b Arley, Ala, Dec 29, 27; m 51; c 3. SOIL FERTILITY. *Educ:* Abilene Christian Col, BS, 55; Auburn Univ, MS, 57; NC State Univ, PhD(soil sci), 68. *Prof Exp:* PROF SOIL SCI, AUBURN UNIV, 57- *Mem:* Am Soc Agron; Soil Sci Soc Am. *Res:* Soil fertility research with phosphorus and potassium, particularly soil testing for fertilizer requirements; fertility requirements for certain vegetable crops. *Mailing Add:* Dept Agron Auburn Univ Auburn AL 36849-5412

EVANS, DANIEL DONALD, b Oak Hill, Ohio, Aug 13, 20; m 46; c 4. HYDROLOGY. *Educ:* Ohio State Univ, BS, 47; Iowa State Univ, MS, 50, PhD(soil physics), 52. *Prof Exp:* Assoc soil physics, Iowa State Univ, 50-52, asst prof, 52-53; from assoc prof to prof soils, Ore State Univ, 53-63; prof agr chem & soils, 63-73, head dept hydrol & water resources, 67-74, PROF HYDROL & WATER RESOURCES, UNIV ARIZ, 63- *Concurrent Pos:* Adv to Kenya Ministry Agr, 60-62. *Mem:* AAAS; fel Soil Sci Soc Am; fel Am Soc Agron; Am Geophys Union; fel Am Water Resources Asn. *Res:* Soil physics and hydrology. *Mailing Add:* Dept Hydrol & Water Resources Univ Ariz Tucson AZ 85721

EVANS, DAVID A, b Washington, DC, Jan 11, 41. SYNTHETIC ORGANIC CHEMISTRY. *Educ:* Oberlin Col, AB, 64; Calif Inst Technol, PhD(chem), 67. *Hon Degrees:* MA, Harvard Univ, 83. *Prof Exp:* From asst prof to prof org chem, Univ Calif, Los Angeles, 67-74; prof chem, Calif Inst Technol, Pasadena, 74-83; PROF CHEM, HARVARD UNIV, 83-, ABBOTT & JAMES LAWRENCE PROF CHEM, 90- *Concurrent Pos:* Camille & Henry Dreyfus teacher-scholar award, 71-76; Alfred P Sloan Found fel, 72-74; consult, Upjohn Co, 72-74, Eli Lilly Co, 74-89 & Merck Sharp & Dohme, 89-; mem adv bd, Off Chem & Chem Technol, Nat Res Coun, 81-84 & Comt Chem Sci, 81-85; Arthur C Cope scholar award, Am Chem Soc, 88. *Honors & Awards:* Numerous hon lectureships, 80-90; Award for Creative work in Synthetic Org Chem, Am Chem Soc, 82. *Mem:* Nat Acad Sci; Am Chem Soc; Am Acad Arts & Sci. *Mailing Add:* Dept Chem Harvard Univ 12 Oxford St Cambridge MA 02138

EVANS, DAVID ARNOLD, b San Mateo, Calif, Sept 24, 38; m 63; c 2. ENTOMOLOGY, ENVIRONMENTAL SCIENCE. *Educ:* Carleton Col, BA, 60; Univ Wis, MS, 62, PhD(entom), 65. *Prof Exp:* Res asst entom, Univ Wis, 60-64, instr, 64-65; from asst prof to assoc prof, 65-82, PROF BIOL, KALAMAZOO COL, 82- *Concurrent Pos:* Res assoc entom, Univ Ga, 73-74; Fulbright Lectureship, Sierra Leone, 82-83; AAAS diplomacy fel, AID/Off Foreign Disaster Assistance, 90-91. *Mem:* AAAS; Entom Soc Am; Sigma Xi. *Res:* Migratory behavior of the corn leaf aphid; taxonomy and bionomics of the velvet ants; chalcidoid reproductive behavior. *Mailing Add:* Dept Biol Kalamazoo Col Kalamazoo MI 49007

EVANS, DAVID ARTHUR, b Gloucester, Eng, Aug 5, 39; m 62; c 2. STATISTICAL ANALYSIS, MARINE SCIENCES. *Educ:* Cambridge Univ, BA, 60, MA, 64; Oxford Univ, PhD(particle physics), 64. *Prof Exp:* Dept Sci & Indust Res fel particle physics, Nuclear Physics Lab, Oxford Univ,

63-64, dept res asst, 64-65; res assoc, Univ Calif, Riverside, 65-67; asst prof particle physics, State Univ NY, Buffalo, 67-74; marine scientist, Deepsea Ventures, Inc, 75-79; ASSOC PROF MARINE SCI, COL WILLIAM & MARY, VA INST MARINE SCI, 79- *Concurrent Pos:* Res assoc, Rutherford High Energy Lab, Eng, 71. *Res:* Statistical techniques in particle physics data analysis; design of online measuring systems; development of analysis systems for bathymetric and other charting of the deep ocean; particle shape analysis; time series analysis. *Mailing Add:* Va Inst Marine Sci Gloucester Point VA 23062

EVANS, DAVID C(ANNON), b Salt Lake City, Utah, Feb 24, 24; m 47; c 7. ELECTRICAL ENGINEERING. *Educ:* Univ Utah, BS, 49, PhD(physics), 53. *Hon Degrees:* Comput Sci, Univ Utah, 87. *Prof Exp:* Dir eng, Comput Div, Bendix Corp, 53-62; prof elec eng & assoc dir comput ctr, Univ Calif, Berkeley, 62-65; prof elec eng & dir comput sci, Univ Utah, 66-76; MEM STAFF, EVANS & SUTHERLAND COMPUT CORP, 76- *Concurrent Pos:* Mem comn on educ, Nat Acad Eng. *Honors & Awards:* Emanual Piore Award, 86. *Mem:* Nat Acad Eng; Am Phys Soc; Asn Comput Mach; fel Inst Elec & Electronics Engrs. *Res:* Computing and information processing systems. *Mailing Add:* Evans & Sutherland Comput Corp 580 Arapeen Dr Salt Lake City UT 84108

EVANS, DAVID HUDSON, b Chicago, Ill, June 9, 40; m 62; c 2. COMPARATIVE PHYSIOLOGY, ICHTHYOLOGY. *Educ:* DePauw Univ, BA, 62; Stanford Univ, PhD(biol), 67. *Prof Exp:* NIH fel, Lancaster Univ, 67-69; from asst prof to assoc prof biol & marine sci, Univ Miami, 69-78, prof marine sci, 78-80, prof biol & chmn dept, 78-81; chmn, 82-85, PROF ZOOL, 81-, COORD DIV BIOL SCI, UNIV FLA, 88- *Concurrent Pos:* Dir, Mt Desert Island Biol Lab, Salisbury Cove Maine, 83- *Mem:* Am Soc Zoologists; Soc Exp Biol; Am Phys Soc. *Res:* Ion and water balance of fish, crustacea and reptiles. *Mailing Add:* Dept Zool Univ Fla Gainesville FL 32611

EVANS, DAVID HUNDEN, b Philadelphia, Pa, Apr 16, 24; m 76; c 2. APPLIED MATHEMATICS. *Educ:* Lehigh Univ, BS, 48; Brown Univ, PhD(appl math), 53. *Prof Exp:* Mem tech staff, Bell Tel Labs, 53-61; sr res mathematician, Res Labs, Gen Motors Corp, 61-68; lectr & res engr, Univ Mich, 68-69; vis prof eng, 69-71, PROF ENG, OAKLAND UNIV, 71- *Mem:* Opers Res Soc Am; Am Soc Qual Control; Am Statist Asn; Sigma Xi; Inst Mgt Sci. *Res:* Applied probability tolerancing; operations research; quality control. *Mailing Add:* Dept Elec & Systs Eng Oakland Univ Rochester MI 48309-4401

EVANS, DAVID L, b Chester, Pa, Apr 28, 46; m 80. PHYSICAL OCEANOGRAPHY, SMALL SCALE DYNAMICS. *Educ:* Univ Pa, BA, 68; Univ RI, PhD(oceanog), 75. *Prof Exp:* Teacher math, Rose Tree Media Sch Dist, 68-71; res assoc, 75-78, asst prof phys oceanog, 78-84, ASSOC PROF OCEANOG, UNIV RI, 84- *Concurrent Pos:* Consult, SAI, Inc, 80- & ASA, Inc, 83- *Mem:* Am Geophys Union; AAAS. *Res:* Small scale vertical mixing processes; instrumentation development; large scale circulation in the South Atlantic; remote sensing. *Mailing Add:* Grad Sch Oceanog Univ RI Kingston RI 02881

EVANS, DAVID L, b San Francisco, Calif, Mar 15, 06; m 37; c 3. MINING EXPLORATION, OIL WELL COMPLETIONS. *Educ:* Stanford Univ, AB, 27, MA, 28. *Prof Exp:* Asst geologist, Cananea Copper Mining Co, 28-32; ranger naturalist, US Nat Park Serv, Ore, 32-33; chief geologist, Cia Unificada del Potosi, Bolivia, 33-35; resident geologist, Climax Molybdenum Co, 36-40; field geologist western states explor, Freeport Sulphur Co, 40-42; sr mineral specialist nickle prod, Bd Econ Warfare, Washington, DC, 42-43; shift boss mining, Climax Molybdenum Co, 43-45; geologist oil prod, Ohio Oil Co, 45-48, geophysicist oil explor, Okla, 48-51; consult geologist, mining & petrol, Wichita, Kans, 51-64; CONSULT GEOLOGIST MINING & PETROL, RENO, NEV, 64- *Mem:* Am Inst Mining Engrs; Am Asn Petrol Geologists; Soc Econ Geologists; Geol Soc Am; Soc Independent Prof Earth Scientist. *Res:* Global plate tectonics and how it can be applied to mining and oil exploration, especially in the Americas. *Mailing Add:* 1700 Royal Dr Reno NV 89503

EVANS, DAVID LANE, b Denver, Colo, Sept 16, 54; m 80, 84; c 1. PROCESS RESEARCH & DEVELOPMENT, COMMERCIAL DEVELOPMENT. *Educ:* Univ Denver, BS, 75; Mass Inst Technol, PhD(org chem), 79. *Prof Exp:* sr res chemist, Monsanto Co, 79-84; Synthetech, Inc, 84-86; Great Western Inorg, 87-89; PRES, METRE-GEN, INC, 90- *Mem:* Am Chem Soc; Sigma Xi. *Res:* Removal of toxic and heavy metals from wastewater; synthesis of immobilized ligands. *Mailing Add:* 7161 Saulsbury Circle Arvada CO 80003-3557

EVANS, DAVID R, b Chicago, Ill, Mar 20, 41. PROTEIN STRUCTURE, MULTIDOMAIN PROTEINS. *Educ:* Wayne State Univ, PhD(biochem). *Prof Exp:* NIH fel, dept chem, 68-72, res fel chem, 72-74, consult, Mass Gen Hosp & asst prof, Sch Med, Harvard Univ, 75; from asst prof to assoc prof, 76-87, PROF BIOCHEM, SCH MED, WAYNE STATE UNIV, 87- *Mem:* Am Soc Biol Chemists; Protein Soc. *Res:* Structure, catalytic and regulatory mechanisms and inter domain interactions of multifunctional proteins, especially mammalian pyrimidine biosynthetic complexes; cloning, expression and mutagenesis of protein domains and construction of chimeric multidomain proteins. *Mailing Add:* Dept Biochem Wayne State Univ Sch Med Detroit MI 48201

EVANS, DAVID STANLEY, b Cardiff, Wales, Jan 28, 16; m 49; c 2. ASTRONOMY. *Educ:* Cambridge Univ, BA, 37, MA & PhD(astron), 41, ScD, 71. *Prof Exp:* Res asst astron, Univ Observ, Oxford Univ, 38-46; second asst, Radcliffe Observ, Pretoria, SAfrica, 46-51; chief asst, Royal Observ, Cape, SAfrica, 51-68; prof, 68-86, EMER PROF ASTRON, UNIV TEX, AUSTIN, 86 - *Concurrent Pos:* NSF sr vis scientist, Univ Tex, 65-66; Jack S Josey Centennial prof astron, 84. *Honors & Awards:* Gill Medal, Astron Soc SAfrica, 88. *Mem:* Fel Royal Astron Soc; fel Royal Soc SAfrica; Astron Soc Southern Africa (past pres & vpres); fel Brit Inst Physics & Phys Soc; Am Astron Soc. *Res:* Observational astronomy; history of astronomy. *Mailing Add:* Dept Astron Univ Tex Austin TX 78712

EVANS, DAVID W, b Erie, Pa, Oct 6, 33; m 63; c 2. AGRONOMY, PLANT PHYSIOLOGY. *Educ:* Yale Univ, BS, 55; Cornell Univ, MS, 58, PhD(agron), 62. *Prof Exp:* Res assoc, Univ Mich, Ann Arbor, 61-63; assoc agronomist, 63-80, AGRONOMIST, IRRIGATED AGR RES & EXTEN CTR, WASH STATE UNIV, 80- *Mem:* Am Inst Biol Sci; Am Soc Agron; Crop Sci Soc Am; Am Soc Plant Physiologists; Sigma Xi. *Res:* Plant root-oxygen relationships; plant-water relationships; forage crop production and management; grass seed production and physiology; alfalfa physiology and forage quality; alfalfa-stem nematode relationships. *Mailing Add:* Irrigated Agr Res & Exten Ctr Wash State Univ PO Box 30 Prosser WA 99350

EVANS, DAVID WESLEY, b Philadelphia, Pa, Aug 10, 54; m 81; c 2. HETEROCYCLIC SYNTHESIS. *Educ:* Dickinson Col, BS, 76; La State Univ, PhD(org chem), 87. *Prof Exp:* Res asst biochem, La State Univ, 86-87; ASST PROF ORG CHEM & BIOCHEM, PRESBY COL, 87- *Mem:* Am Chem Soc. *Res:* Synthesis and characterization of organo-palladium and platinum compounds, both singly and doubly-bonded; 1,2,4-Triazine synthesis. *Mailing Add:* PO Box 975 Clinton SC 29325

EVANS, DENNIS HYDE, b Grinnell, Iowa, Mar 28, 39; m 58; c 3. ANALYTICAL CHEMISTRY, ELECTROCHEMISTRY. *Educ:* Ottawa Univ, BS, 60; Harvard Univ, AM, 61, PhD(chem), 64. *Prof Exp:* Instr chem, Harvard Univ, 64-66; from asst prof to prof, Univ Wis-Madison, 66-84, Meloche-Bascom prof chem, 84-86; PROF CHEM, UNIV DEL, 86- *Mem:* Am Chem Soc; Int Soc Electrochem; Soc Electroanal Chem; Electrochem Soc. *Res:* Characterization and analytical application of electrode reactions; organic electrochemistry. *Mailing Add:* Dept Chem & Biochem Univ Del Newark DE 19716-0002

EVANS, DONALD B, b Cleveland, Ohio, Oct 11, 33. METALLURGICAL ENGINEERING. *Educ:* Mass Inst Technol, BS, 55; Univ Mich, MS, 59, PhD(metall eng), 63. *Prof Exp:* Develop engr, Mallinckrodt Chem Works, 58; sr engr, Martin Marietta Corp, 63-69; STAFF ENGR, TRW SYSTS GROUP, 69- *Mem:* Am Inst Petrol Engrs; Am Soc Metals. *Res:* New thermoelectric power generation materials and devices; thermodynamics of chemical reactions involved in steel making. *Mailing Add:* 708 Camino Real Redondo Beach CA 90277

EVANS, DONALD LEE, b Clinton, Mo, May 22, 43; m 63; c 3. MONOCLONAL TECHNOLOGY, FLOW CYTOMETRY. *Educ:* Univ Mo, Kansas City, BS, 65, MS, 67; Univ Ark, Fayetteville, PhD(microbiol & immunol), 71. *Prof Exp:* Post-doctoral, M D Anderson Hosp & Tumor Inst Univ Tex Syst Cancer Ctr, 71-72, asst prof, 72-73; asst prof immunol, Sch Med, Tex Tech Univ, 73-75; assoc prof immunol, Bowman Gray Sch Med, Wake Forest Univ, 75-82; assoc prof, 82-84, PROF, COL VET MED, UNIV GA, 84- *Concurrent Pos:* Prin investr, NIH grants, 76-80 & 87- & Am Cancer Soc grant, 79-82; comt mem, Rosalie B Hite Fel, M D Anderson Hosp & Tumor Inst Univ Tex Syst Cancer Ctr, 81-83, dir, Flow Cytometry & Monoclonal Antibody Facil. *Mem:* Sigma Xi; Am Soc Microbiol; Am Asn Immunologists; Int Soc Develop & Comp Immunol. *Res:* Characterization of the mechanisms of lymphoid cell and natural killer cell lysis of target cells; mechanisms of lymphocyte regulation by soluble factors; nonspecific cytotoxic cells in teleosts; natural killer cell receptors and target antigen recognition. *Mailing Add:* Col Vet Med Univ Ga Athens GA 30602

EVANS, DONOVAN LEE, b Verona, Ohio, Mar 14, 39; m 60; c 3. MECHANICAL ENGINEERING. *Educ:* Univ Cincinnati, BSME, 62; Northwestern Univ, PhD(mech eng), 67. *Prof Exp:* Res asst mech eng, Northwestern Univ, 65-66; from asst prof to assoc prof, 66-75, PROF MECH ENG, ARIZ STATE UNIV, 75- *Mem:* Am Soc Mech Engrs; Int Solar Energy Soc; Am Soc Heat, Refrig & Air Conditioning Engrs; Am Soc Eng Educ. *Res:* Thermosciences; high temperature gas dynamics; radiation from gasses; solar energy systems; building heating and cooling. *Mailing Add:* Dept Mech & Aero Eng Ariz State Univ Tempe AZ 85281-6106

EVANS, DOUGLAS FENNELL, b Carlsbad, NMex, Mar 16, 37; m 63; c 1. PHYSICAL CHEMISTRY, BIOPHYSICAL CHEMISTRY. *Educ:* Pomona Col, BA, 59; Mass Inst Technol, PhD(chem), 63. *Prof Exp:* Fel, Mellon Inst, 63-66; from asst prof to assoc prof chem, Case Western Reserve Univ, 66-72; PROF CHEM & CHEM ENG, DIV COLLOID, POLYMER & SURFACE PROG, CARNEGIE-MELLON UNIV, 72- *Mem:* Am Inst Chem Engrs; Am Chem Soc. *Res:* Electrolyte solutions; physical properties of bile; membrane transport. *Mailing Add:* Dept Chem Eng Univ Minn Minneapolis MN 55455

EVANS, E GRAHAM, JR, US citizen. MATHEMATICS. *Educ:* Dartmouth Col, AB, 64; Univ Chicago, MS, 65, PhD(math), 69. *Prof Exp:* Asst prof math, Univ Calif, Los Angeles, 69-70; instr, Mass Inst Technol, 70-72; from asst prof to assoc prof, 72-81, PROF MATH, UNIV ILL, URBANA, 81- *Res:* Commutative ring theory; homological algebra; algebraic geometry. *Mailing Add:* Dept Math 130 IColde Univ Ill 1409 W Green St Urbana IL 61801

EVANS, EDWARD WILLIAM, b Frackville, Pa, Sept 1, 32; m 54; c 4. MATHEMATICS. *Educ:* Kutztown State Col, BS, 54; Temple Univ, MEd, 58; Univ Mich, MA, 61, PhD(math, math educ), 64. *Prof Exp:* Teacher high sch, Pa, 55-61; NSF fel, Univ Mich, 61-63, univ & teaching fels, 64; PROF MATH, KUTZTOWN STATE COL, 64-, CHMN DEPT, 65- *Concurrent Pos:* Nat Defense Educ Act vis lectr high schs, 65-; consult high schs math progs, 64- *Mem:* Math Asn Am; Am Math Soc. *Res:* Teaching and learning of mathematics; abstract algebra; foundations of geometry. *Mailing Add:* Dept Math & Computer Sci Kutztown Univ Kutztown PA 19530

EVANS, EDWIN CURTIS, b Milledgeville, Ga, June 30, 17; m 45; c 6. GERIATRICS. *Educ:* Univ Ga, BS, 36; Johns Hopkins Univ, MD, 40; Am Bd Internal Med, cert, 51 & 77. *Prof Exp:* Intern, Hartford Hosp, Conn, 40-42; first lt to major, Med Corps, US Army, Southwest Pac Area, 42-45; chief res med, Baltimore City Hosps, 46-47; res path, Univ Pa Hosp, 47-48; pvt pract internal med, Atlanta, 48-87; DIR CLIN GERIAT, GA BAPTIST MED CTR, ATLANTA, 87- *Concurrent Pos:* Asst med, Johns Hopkins Sch

Med, 46-47; from clin instr to assoc clin prof med, Emory Univ, 48-87, clin emer prof, 87-; pres staff, Ga Baptist Med Ctr, Atlanta, 67-68, chief staff, 73-79. *Mem:* Inst Med-Nat Acad Sci; Am Soc Internal Med (pres, 72-73); fel Am Col Physicians (govenor, GA, 72-76); AMA; Am Geriat Soc. *Res:* Hospital-based patient care; teaching and clinical research in geriatrics. *Mailing Add:* 500 Westover Dr NW Atlanta GA 30305

EVANS, EDWIN VICTOR, nutrition; deceased, see previous edition for last biography

EVANS, ERNEST EDWARD, JR, b Parkersburg, WVa, Dec 14, 22; m 47. MICROBIOLOGY, IMMUNOLOGY. *Educ:* Ohio Univ, AB, 45; Ohio State Univ, MS, 47; Univ Southern Calif, PhD(microbiol), 50. *Prof Exp:* Asst prof bact, Univ Mich, 50-55; from assoc prof to prof, 55-74, chmn dept, 61-70, EMER PROF MICROBIOL, SCH MED & DENT, UNIV ALA, BIRMINGHAM, 74- *Concurrent Pos:* Rackham Fund fel, Univ Mich, 51-52; vis investr, Lerner Marine Lab, 66-72; vis res biologist, Univ Calif, Santa Barbara, 70; sr res microbiologist, Mote Marine Lab, Sarasota, Fla, 70-74. *Mem:* Fel AAAS; emer fel Am Acad Microbiol. *Res:* Antigenic structure of pathogenic fungi; evolution of immunity. *Mailing Add:* PO Box 28355 San Diego CA 92198

EVANS, ERSEL ARTHUR, b Trenton, Nebr, July 17, 22; m 45; c 2. NUCLEAR ENGINEERING, METALLURGY. *Educ:* Reed Col, BA, 45; Ore State Univ, PhD(chem), 47. *Prof Exp:* Sr engr nuclear fuel develop, Gen Elec Co, Richland, Wash, 51-55, mgr ceramic fuels, 56-64, mgr Pu fuels develop, Vallecitos, Calif, 64-67; mgr, Fuels & Mat Dept, Battelle Mem Inst, 67-70; mgr, Mat Technol Dept, Hanford Eng Develop Lab, Westinghouse Hanford Co, 70-71, assoc dir, 72-76, vpres co, 72-87, mgr technol dept, 73-76, tech dir, 76-84, labtech dir, Hanford Eng Develop Lab, 84-87; RETIRED. *Concurrent Pos:* Res fel, Res Corp, 49; Du Pont Corp fel, 50-51; consult, 87- *Mem:* Nat Acad Eng; fel Am Inst Chem; fel Am Ceramic Soc; fel Am Soc Metals; fel Am Nuclear Soc. *Res:* High temperature ceramic and metallurgical research, development and engineering involving a variety of nuclear reactors, isotopic heat sources, volcanology. *Mailing Add:* 1910 S Lyle Kennewick WA 99337

EVANS, ESSI H, b Bad-Schwalbach, WGer, Jan 12, 50; US citizen; m 74. ANIMAL NUTRITION & HEALTH, ANIMAL PHYSIOLOGY. *Educ:* Univ Md, BS, 72; Univ Guelph, MSc, 74, PhD(animal sci), 76. *Prof Exp:* Res & teaching asst animal sci, Univ Guelph, 72-76; proj leader nutrit, 76-85, tech mgr, 86-89, NUTRIT & RES MGR, SHUR-GAIN DIV, CAN PACKERS INC, 89- *Concurrent Pos:* Nat Res Coun Can indust fel, 76-79; adj prof, Univ Guelph, 87- *Mem:* Am Soc Animal Sci; Am Dairy Sci Asn; Am Asn Vet Nutrit; Coun Agr Sci Technol. *Res:* Initiating and directing research projects in animal nutrition; developing rumen function models and computer programs; developing new animal health products; managing research related to all species. *Mailing Add:* Shur-Gain Div 2700 Matheson Blvd E Suite 600 E Tower Mississauga ON L4W 4V9 Can

EVANS, EVAN CYFEILIOG, III, b San Francisco, Calif, Nov 19, 22; m 45; c 3. BIOPHYSICS. *Educ:* Univ Calif, Berkeley, AB, 48, PhD(biophys), 63. *Prof Exp:* Head, Weapon Capabilities Br, US Naval Radiol Defense Lab, 64-66 & Weapon Effects Br, 66-69; dir marine environ mgt off, Naval Ocean Systs Ctr, Hawaii Lab, 69-87; RETIRED. *Concurrent Pos:* Naval Radiol Defense Lab fel, 61; lectr, Univ Calif, Berkeley, 64-65; affil grad fac, Univ Hawaii, 71- *Mem:* Sigma Xi; Nature Conservancy. *Res:* Absorption and translocation of radionuclides in higher plants; effects of ionizing radiation on plants; atomic physics; micromeritics; radioecology of Pacific Ocean basin; underwater acoustics; bioacoustics of marine mammals; environmental survey of harbors and estuaries. *Mailing Add:* Two Arroyo Dr Kentfield CA 94904

EVANS, EVAN FRANKLIN, b Kenesaw, Nebr, Mar 17, 18; m 44; c 5. POLYMER CHEMISTRY, COMPUTER SCIENCE. *Educ:* Univ Nebr, BA, 39, MA, 40; Ohio State Univ, PhD(chem), 43. *Prof Exp:* Res chemist, Hercules Powder Co, 44-50; res chemist, E I du Pont de Nemours & Co, Inc, 51-52, res supvr, 52-59, res mgr, 59-71, res assoc, 71-85; RETIRED. *Mem:* AAAS; Am Chem Soc; Sigma Xi. *Res:* Synthesis acyclic sugar derivatives; fundamental and applied research on cellulose and cellulose derivatives; polymers; synthetic fibers; computer system analysis and programming; textiles. *Mailing Add:* 1702 Cambridge Dr Kinston NC 28501

EVANS, FOSTER, b Salt Lake City, Utah, Jan 16, 15; m 78; c 1. THEORETICAL PHYSICS. *Educ:* Brigham Young Univ, BS, 36; Univ Chicago, PhD(physics), 41. *Prof Exp:* Asst physics, Brigham Young Univ, 35-36; instr optics, Northern Ill Univ Col Optom, 40; actg instr physics, Univ Wis, 41; from instr to asst prof, Univ Colo, 41-46; staff mem, Los Alamos Nat Lab, 46-82; CONSULT PHYSICS, 82- *Concurrent Pos:* Res fel, Radiation Lab, Univ Calif, 42. *Mem:* Am Phys Soc; Am Asn Physics Teachers. *Res:* Cosmic rays; hydrodynamics; nuclear physics; transport theory in plasmas. *Mailing Add:* 196 Paseo Penasco Los Alamos NM 87544

EVANS, FRANCIS COPE, b Germantown, Pa, Dec 2, 14; m 42; c 4. ECOLOGY. *Educ:* Haverford Col, BS, 36; Oxford Univ, PhD(animal ecol), 40. *Prof Exp:* Asst, Hooper Found, Univ Calif, 39-41, jr zoologist, Exp Sta & Col Agr, 42-43; from instr to asst prof biol, Haverford Col, 43-58, actg dean, 45; from asst prof to prof, 48-82, EMER PROF ZOOL, UNIV MICH, ANN ARBOR, 82- *Concurrent Pos:* Mem Oxford Univ expeds, Faeroe Islands, 37 & Iceland, 39; lectr, Bryn Mawr Col, 48; asst biologist, Lab Vert Biol, Univ Mich, Ann Arbor, 48-52; ed, Ecol Monogr, 56-62; dir, E S George Reserve, 59-81; Guggenheim fel, 62. *Mem:* Fel AAAS; Am Soc Mammalogists; Ecol Soc Am (pres, 83-84); Brit Ecol Soc. *Res:* Ecology of natural communities; dynamics of vertebrate populations; patterns of spatial distribution; animal epidemiology. *Mailing Add:* Dept Biol Univ Mich Ann Arbor MI 48109

EVANS, FRANCIS EUGENE, b Olney, Ill, June 18, 28; m 50; c 4. ORGANIC CHEMISTRY. *Educ:* DePauw Univ, AB, 50; Mich State Univ, PhD(org chem), 55. *Prof Exp:* Res chemist, Nat Aniline Div, 55-65, sr scientist, Indust Chem Div, 65-70, TECH SUPVR, SPECIALTY CHEM DIV, ALLIED CHEM CORP, 70- *Mem:* Am Chem Soc; Sigma Xi. *Res:* Organometallics; surfactants; anhydrides; Friedel-Crafts reactions; catalytic chemistry. *Mailing Add:* 4768 Woodside Ave Hamburg NY 14075

EVANS, FRANCIS GAYNOR, b LeMars, Iowa, Dec 7, 07; m 38. ANATOMY, BIOMECHANICS. *Educ:* Coe Col, BA, 31; Columbia Univ, MA, 32, PhD(zool), 39. *Prof Exp:* Instr biol, City Col New York, 35-36; lectr zool, Columbia Univ, 36; instr, Univ NH, 38-41; instr, Duke Univ, 41-43; asst prof human gross anat, Sch Med, Univ Md, 43-45; from asst prof to prof, Col Med, Wayne State Univ, 45-59; prof, 59-77, EMER PROF ANAT, UNIV MICH, ANN ARBOR, 77- *Concurrent Pos:* Fulbright res scholar, Italy, 56-57; vis prof, Gothenburg Univ, 62-63 & Kyoto Prefectural Univ Med, 68. *Honors & Awards:* Morrison Prize, NY Acad Sci, 38. *Mem:* Fel AAAS; Am Asn Anat; Am Physiol Soc; Am Asn Phys Anthrop; Am Soc Biomech. *Res:* Comparative osteology; biomechanics of the human skeleton; stress and strain in bones; mechanical properties and structure of bone. *Mailing Add:* RWC Cottage 626 Lancaster Dr Irvington VA 22480

EVANS, FRANKLIN JAMES, JR, b Hazelton, Pa, June 1, 21; m 47; c 2. ORGANIC CHEMISTRY. *Educ:* Lafayette Col, BS, 42; Pa State Col, MS, 49; Ohio State Univ, PhD(chem), 52. *Prof Exp:* Engr prod explosives, E I du Pont de Nemours & Co, 42-45, engr rayon res, 45-47, from res chemist to sr res chemist, 53-66, res assoc, Textile Fibers, 66-77; CONSULT, 78- *Mem:* Am Chem Soc; Am Inst Chem Eng; Sigma Xi. *Res:* Textile fibers; organo silicones; steric hindrance. *Mailing Add:* 406 Garland Rd Northwood Wilmington DE 19803

EVANS, FREDERICK EARL, b Springfield, Mass, Nov 11, 48; m 78. BIOCHEMISTRY, PHYSICAL CHEMISTRY. *Educ:* Univ Mass, Amherst, BS, 70; State Univ NY, Albany, PhD(chem), 74. *Prof Exp:* Fel biochem, Univ Calif, San Diego, 75-78; SR RES CHEMIST, NAT CTR TOXICOL RES, 78- *Mem:* Am Asn Cancer Res; Am Chem Soc. *Res:* Nuclear magnetic resonance spectroscopy; chemical carcinogenesis; nucleotide and oligonucleotide conformation; phosphorus-31 nuclear magnetic resonance studies of cells and tissues; structure elucidation; clinical NMR spectroscopy. *Mailing Add:* Div Biochem Toxicol HFT-110 Nat Ctr Toxicol Res Jefferson AR 72079

EVANS, FREDERICK READ, b Salt Lake City, Utah, Sept 9, 13; m 36; c 4. PROTOZOOLOGY. *Educ:* Univ Utah, BA, 34, MA, 36; Stanford Univ, PhD(protozool), 41. *Prof Exp:* Instr biol, Stanford Univ, 40-45; from asst prof to assoc prof, Univ Utah, 45-60, prob biol, 60-; RETIRED. *Mem:* Am Inst Biol Sci; fel Am Micros Soc; fel Soc Protozoologists. *Res:* Nutrition of free-living protozoa; cystment in protozoa; protozoan populations; nuclear reorganization in the ciliates; uptake by protozoa of radioactive substances; parasitic protozoa; morphogenesis of ciliates. *Mailing Add:* 2020 Herbert Ave Salt Lake City UT 84108

EVANS, GARY R, b Twin Falls, Idaho, Aug 2, 41. ECOLOGY. *Educ:* Univ Idaho, Moscow, BS, 64, MS, 67; Colo State Univ, PhD, 75. *Prof Exp:* Dep adminr, Agr Res Serv, 88-89, SPEC ASST, GLOBAL CHANGE ISSUES, USDA, 89- *Mem:* Am Inst Biol Sci; Ecol Soc Am; Soc Range Mgt; Soil Water Conserv Soc; Sigma Xi. *Res:* Natural resources; decision theory; ecosystem modeling. *Mailing Add:* Off Asst Secy Sci & Educ USDA Washington DC 20250

EVANS, GARY WILLIAM, b Greybull, Wyo, Jan 5, 40; m 60; c 2. BIOCHEMISTRY, NUTRITION. *Educ:* Eastern Mont State Col, BS, 62; Univ NDak, PhD(biochem), 70. *Prof Exp:* Res chemist, Agr Res Serv, Beltsville, Md, 71, res chemist, Human Nutrit Lab, Sci & Educ Admin-Agr Res, Grand Forks, 71-82; res instr biochem, Univ NDak, 71-82; assoc prof, 82-86, PROF CHEM, BEMIDJI STATE UNIV, 86- *Mem:* Soc Exp Biol & Med; Am Physiol Soc; Sigma Xi. *Res:* Absorption and metabolism of trace elements. *Mailing Add:* Dept Chem Sattgast Hall Bemidji State Univ Bemidji MN 56601

EVANS, GEOFFREY, b Mountain Ash, Wales, Jan 25, 35; m 59; c 3. CARDIOVASCULAR SURGERY. *Educ:* Univ London, MB & BS, 58; FRCS, 62, 78. *Prof Exp:* House surgeon, St Mary's Hosp, London, 58; house physician, Paddington Gen Hosp, 58-59; casualty surgeon, St Mary's Hosp, London, 59, tutor anat & physiol, 59-60; sr house officer, Royal Nat Orthop Hosp, 60; surg registr, Southlands Hosp, Sussex, 61-62; lectr surg, St Mary's Hosp, London, 62-66, sr registr, 64-67; ASSOC PROF SURG, MCMASTER UNIV, 68- *Concurrent Pos:* Can Heart Found fel path, McMaster Univ, 67-69; consult, Hamilton Civic, St Joseph's, Chedoke & Joseph Brant Hosps, 69; mem, Am Heart Found. *Mem:* Asn Acad Surg; Am Heart Asn; Soc Univ Surg. *Res:* Importance of platelet interaction with surfaces in determining the duration of arterial prosthetic replacements and with formed complexes such as antigen antibody complexes in the etiology of disseminated intravascular thrombosis. *Mailing Add:* McMaster Univ Hamilton ON L8S 4L8 Can

EVANS, GEORGE EDWARD, b Great Falls, Mont, Aug 31, 32; m 55; c 2. ORNAMENTAL HORTICULTURE. *Educ:* Mont State Univ, BS, 57; Mich State Univ, MS, 58, PhD(ornamental hort), 69. *Prof Exp:* From instr to assoc prof, 62-75, PROF HORT, MONT STATE UNIV, 75- *Mem:* Am Hort Soc; Int Plant Propagators Soc; Sigma Xi. *Res:* Graft compatibility studies in intergeneric grafts of members of the rose family and in the genus Juniperus with major emphasis on anatomical aspects; ornamental plant hardiness and adaptability; turfgrass investigations. *Mailing Add:* 7100 S 19th Rd Bozeman MT 59715

EVANS, GEORGE LEONARD, b Wilkes-Barre, Pa, Aug 3, 31; m 58; c 3. MICROBIOLOGY. *Educ:* King's Col, Pa, BS, 54; Fordham Univ, MS, 57; Temple Univ, PhD(microbiol), 62. *Prof Exp:* Dir microbiol, 75-84, RES FEL, BD MICROBIOL SYSTS, 84- *Concurrent Pos:* Mem nat comt clin lab standards & subcomt cult media, 81-; specialist microbiologist, Am Acad Microbiol. *Mem:* Am Soc Microbiol; Am Asn Clin Chem. *Res:* Development of in vitro diagnostic products; medical microbiology; immunology; clinical chemistry. *Mailing Add:* BD Microbiol Systs PO Box 243 Cockeysville MD 21030

EVANS, GERALD WILLIAM, b New Albany, Ind, Sept 29, 50; m 74; c 2. SIMULATION MODELING, MULTI CRITERIA OPTIMIZATION. *Educ:* Purdue Univ, BS, 72, MS, 74, PhD(oper res & indust eng), 79. *Prof Exp:* Indust engr, Rock Island Arsenal, US Army, 74-75; sr res engr, res lab, Gen Motors Corp, 78-81; asst prof, dept mgt, 81-83, asst prof, 83-87, ASSOC PROF OPERS RES, DEPT INDUST ENG, UNIV LOUISVILLE, 87- *Concurrent Pos:* Consult, Naval Ord Sta, USN, 82 & Appliance Park, Gen Elec Corp, 83; NASA fac fel, Langley Res Ctr, Hampton, Va, 87. *Mem:* Opers Res Soc Am; Inst Mgt Sci; Inst Indust Engrs; Decision Sci Inst. *Res:* Operations research and management science as applied to problems arising in systems design and production and operation management, including production planning and scheduling, quality control, energy planning and public policy determination. *Mailing Add:* Dept Indust Eng Univ Louisville Louisville KY 40292

EVANS, GLENN THOMAS, b Elizabeth, NJ, July 31, 46; m 69; c 2. CHEMICAL PHYSICS. *Educ:* Seton Hall Univ, BS, 68; Brown Univ, PhD(chem), 73. *Prof Exp:* Fel, Phys Chem Labs, Oxford Univ, 73-74; fel, Sterling Chem Labs, Yale Univ, 74-77; asst prof chem, 77-81, assoc prof, 82-86, PROF, ORE STATE UNIV, 86- *Mem:* Am Chem Soc; Am Phys Soc. *Res:* Theory of transport and equilibrium properties of polyatomic fluids. *Mailing Add:* Dept Chem Ore State Univ Corvallis OR 97331-4003

EVANS, GREGORY HERBERT, b Quincy, Ill, July 19, 49; m 72; c 1. NUMERICAL FLUID MECHANICS, HEAT & MASS TRANSFER. *Educ:* Purdue Univ, BS, 71; Wash State Univ, MS, 78, PhD(mech eng), 81. *Prof Exp:* MEM TECH STAFF, SANDIA NAT LABS, 81- *Mem:* Am Soc Mech Engrs. *Res:* Numerical solution of multi-dimensional fluid flow problems with heat transfer(buoyancy) and mass transfer; rotating disks, chemical reactions, multi-phase flows, mixed convection. *Mailing Add:* Div 8245 Sandia Nat Labs PO Box 550 Livermore CA 94566

EVANS, HAROLD J, b Woodburn, Ky, Feb 19, 21; m 46; c 2. PLANT NUTRITION, PLANT BIOCHEMISTRY. *Educ:* Univ Ky, BS, 46, MS, 48; Rutgers Univ, PhD(soil chem & plant physiol), 50. *Prof Exp:* Asst prof bot, NC State Univ, 50-51; fel, Johns Hopkins Univ, 51-52; from assoc prof to prof bot, NC State Univ, 51-61; dir, Lab Nitrogen Fixation, 78-88, prof plant physiol, 61-88, distinguished prof, 88-89, EMER DISTINGUISHED PROF, ORE STATE UNIV, 89- *Concurrent Pos:* Consult, NSF, 64; Rockefeller Found vis prof, Univ Sussex, 69; George A Miller vis prof, Univ Ill, 73. *Honors & Awards:* Hoblitzelle Nat Award, 65; Charles Reid Barnes Award, Am Soc Plant Physiologists, 85. *Mem:* Nat Acad Sci; Am Soc Plant Physiologists (pres-elect, 70, pres, 71); Am Soc Biol Chemists; Brit Biochem Soc; Am Soc Microbiol. *Res:* Biochemical role of cations; mechanism of nitrogen fixation; molecular biology of hydrogen oxidation. *Mailing Add:* Lab Nitrogen Fixation Dept Bot & Plant Path Ore State Univ Corvallis OR 97331-2902

EVANS, HARRISON SILAS, b Monroe, Iowa, Aug 4, 11; m 34; c 2. MEDICINE, PSYCHIATRY. *Educ:* Col Med Evangelists, MD, 36; Am Bd Psychiat & Neurol, dipl, 46. *Prof Exp:* Resident psychiat, Harding Hosp, Worthington, Ohio, 36-39, staff psychiatrist, 39-42, co-dir, 46-62; dean sch med, 75-77, vpres med affairs, 77-80, PROF PSYCHIAT & CHMN DEPT, SCH MED, LOMA LINDA UNIV, 62- *Concurrent Pos:* Clin assoc prof, Ohio State Univ, 46-62, asst prof, Sch Social Admin, 58-62. *Mem:* Life fel Am Psychiat Asn; Am Acad Neurol. *Res:* Medical student education in psychiatry and psychotherapy; community mental health. *Mailing Add:* Dept Psychiat Loma Linda Univ Loma Linda CA 92350

EVANS, HELEN HARRINGTON, b Cleveland, Ohio, May 11, 24; m 66; c 1. RADIATION BIOLOGY. *Educ:* Purdue Univ, BS, 46; Western Reserve Univ, PhD(biochem), 53. *Prof Exp:* Sr instr biochem, 56-58, from asst prof to assoc prof, 58-75, PROF BIOCHEM & RADIOL, CASE WESTERN RESERVE UNIV, 76- *Concurrent Pos:* Vis scientist, Scripps Clin & Res Found, Calif, 65-66, McCardle Lab, Univ Wis, 73-74 & Radiation Study Sect, NIH, 73-77. *Mem:* Tissue Cult Asn; Radiation Res Soc; Am Soc Biol Chemists; Am Soc Microbiol. *Res:* Effect of ionizing radiation on DNA; relationship of mutagenesis, carcinogenesis and DNA repair; control of macromolecular synthesis and the mitotic cycle; DNA Replication and repair. *Mailing Add:* Div Radiation Biol Wearn Res Bldg Case Western Reserve Univ 2074 Abington Rd Cleveland OH 44106

EVANS, HERBERT JOHN, Franklin, Pa, Jan 22, 37; m 65; c 2. BIOCHEMISTRY OF HEMOSTASIS. *Educ:* Pa State Univ, BS, 63; Case Western Reserve Univ, PhD(biochem), 69. *Prof Exp:* NIH fel biochem, Duke Univ Med Ctr, 69-72; NIH staff fel biochem, Nat Inst Dent Res, 72-74; asst prof, 74-85, ASSOC PROF BIOCHEM, MED COL VA, VA COMMONWEALTH UNIV, 85- *Concurrent Pos:* Vis scientist, Strangeways Res Lab, Cambridge, Eng, 85-86. *Mem:* Int Soc Toxinology; Am Soc Biol Chemists. *Res:* Effects of snake venoms on hemostasis, including anticoagulant effects, fibrinolytic effects, antiplatelet effects and hemorrhagic effects of cobra venoms. *Mailing Add:* Dept Biochem Med Col Va Box 614 MCV Sta Richmond VA 23298-0614

EVANS, HIRAM JOHN, b Granville, NY, May 13, 16; m 44; c 4. EMBRYOLOGY. *Educ:* Hamilton Col, BA, 37; Williams Col, MA, 39; Harvard Univ, AM, 41, PhD(biol), 42. *Prof Exp:* Asst, Williams Col, 37-39; res assoc zool, Swarthmore Col, 46; from asst prof to assoc prof, Syracuse Univ, 47-64; prof biol, New Col, 64-65; prof biol & dean, Curry Col, 65-79, vpres res & planning, 73-79; RETIRED. *Concurrent Pos:* Vchmn dept zool, Syracuse Univ, 47-54 & 59-64; secy bd trustees, Biol Abstracts, 50-65. *Mem:* Soc Develop Biol. *Res:* Development and innervation of the chick ear; analysis of microquantities of respiratory gases; adrenal steroids and embryonic development. *Mailing Add:* Pawlet VT 05761-0177

EVANS, HOWARD EDWARD, b New York, NY, Sept 22, 22; m 49; c 2. COMPARATIVE ANATOMY. *Educ:* Cornell Univ, BS, 44, PhD(zool), 50. *Prof Exp:* Asst herpet, Mus Natural Hist, 38-40; technician zool, Cornell Univ, 40-42, asst, 46-50, from asst prof to assoc prof anat, NY State Vet Col, 50-60, secy Vet Col, 60-72, prof anat, 60-86, chmn dept, 76-86, EMER PROF, NY STATE VET COL, CORNELL UNIV, 86- *Concurrent Pos:* NSF fel & vis prof, Univ Calif, 57; vis prof, Phipps Inst, Med Sch, Univ Pa, 64; vis prof, Marine Inst, Univ Ga, 73-74; lectr, marine inverts, Shoals Marine Lab, Maine, 73-75; co-ed & reviewer, Avian Anat Nomenclature, 74-76; consult, NIH, 75; vis prof, Univ Hawaii, 79 & Vet Col, Univ Pretoria, 81; assoc ed, J Morph & assoc ed, Am J Anat, 86-; vis prof, Beijing Agr Univ, 87-88; consult, Vet Inst, Taiwan, 88; ext examr, Univ Zimbabwe Vet Col, 86 & 87; lectr, Envirovet, Wis Superior, 91. *Honors & Awards:* Anat Award, Am Asn Vet Anat. *Mem:* Hon mem Am Vet Med Asn; Am Asn Anat; Am Soc Zool; Am Asn Vet Anat; Am Soc Mammal; Am Soc Ichthyol & Herpet. *Res:* Anatomy of birds; cyclopia in sheep; anatomy and fetal skeletal development in dogs; anatomy of fishes; author of four books. *Mailing Add:* Dept Anat NY State Vet Col Cornell Univ Ithaca NY 14853

EVANS, HOWARD ENSIGN, b East Hartford, Conn, Feb 23, 19; m 54; c 3. ENTOMOLOGY. *Educ:* Univ Conn, BA, 40; Cornell Univ, MS, 41, PhD(entom), 49. *Hon Degrees:* MA, Harvard Univ, 69. *Prof Exp:* Asst entom, Cornell Univ, 47-49; asst prof, Kans State Col, 49-52; from asst prof to assoc prof insect taxon, Cornell Univ, 52-60; assoc cur, Mus Comp Zool, Harvard Univ, 60-64, cur insects, 64-69, Alexander Agassiz prof zool, 69-73; prof, 73-85, EMER PROF ENTOM, COLO STATE UNIV, 85- *Concurrent Pos:* Guggenheim fel, Nat Univ Mex, 59 & Commonwealth Sci & Indust Res Orgn, Canberra, Australia, 69. *Honors & Awards:* Daniel Giraud Elliot Medal, Nat Acad Sci, 76. *Mem:* Nat Acad Sci; Animal Behav Soc. *Res:* Taxonomy of Pompilidae and Bethylidae; comparative ethology of solitary wasps. *Mailing Add:* 79 Mckenna Ct Livermore CO 80536

EVANS, HOWARD TASKER, JR, b Ancon, CZ, Sept 9, 19; m 42, 66; c 1. CRYSTAL STRUCTURE, CRYSTAL CHEMISTRY. *Educ:* Mass Inst Technol, SB, 42, PhD(inorg chem), 48. *Prof Exp:* Mem res staff, Div Indust Coop, Mass Inst Technol, 43-44 & 47-49, instr sect graphics, 45-48; res physicist, Philips Labs, Inc, 49-52; RES CHEMIST, US GEOL SURV, 52- *Concurrent Pos:* Guggenheim Found vis res scientist, Royal Inst Technol, Stockholm, Sweden, 60-61. *Mem:* Mineral Soc Am; Am Crystallog Asn (secy, 50-51, vpres, 63, pres, 64); Am Chem Soc; AAAS; Sigma Xi. *Res:* Crystal chemistry; x-ray crystallography of inorganic compounds and minerals; x-ray diffraction and crystal structure determination. *Mailing Add:* US Geol Surv Nat Ctr 959 Reston VA 22092

EVANS, HUGH E, b New York, NY, July 6, 34; m 60; c 2. PEDIATRICS, INFECTIOUS DISEASE. *Educ:* Columbia Univ, BA, 54; State Univ NY Downstate Med Ctr, MD, 58; Am Bd Pediat, dipl, 63. *Prof Exp:* Intern pediat, Johns Hopkins Hosp, 58-59, asst resident, 59-60; clin assoc infectious dis, NIH, 60-62; from sr asst resident to chief resident pediat, 62-63; pvt pract, Ohio, 63-65; res assoc, Mt Sinai Hosp, 65-66; from asst prof to assoc prof clin pediat, Columbia Univ, 66-73; PROF PEDIAT, STATE UNIV NY DOWNSTATE MED CTR, BROOKLYN, 73-; DIR, DEPT PEDIAT, JEWISH HOSP & MED CTR BROOKLYN, 73- *Concurrent Pos:* Asst attend pediatrician, Harlem Hosp, NY, 66-69, attend pediatrician, 69-70, vis pediatrician, 70-73, assoc dir pediat, 66-73, consult pediatrician, 73-; assoc prof pediat, Columbia Univ, 73; vis physician, Dept Pediat, Kings County Hosp Ctr, Brooklyn, 73-; consult, St Johns Episcopal Hosp, Brooklyn, Cath Med Ctr, Brooklyn & Health Ins Prog, New York, 73-; mem, Hosp Respiratory Care Consult Team Serv Prog, Am Thoracic Soc; mem, Comt Fac, Downstate Med Ctr, State Univ New York, 81-83; mem comt, Foreign Med Grads, Health & Hosp Corp; grant reviewer, Birth Defect Found, March of Dimes, 81. *Mem:* Am Pediat Soc; NY Acad Sci; Soc Pediat Res; Soc Exp Biol & Med; Am Lung Asn. *Res:* Alpha 1-antitrypsin levels in Respiratory Distress Syndrome; ethnic influence on alpha-1-antitrypsin phenotypes; effect of various drugs on chemotaxis; infection in the neonatal period; bacterial flora of newborns; enzyme inhibitor levels in the Respiratory Distress Syndrome; clinical and laboratory aspects of new respiratory viruses. *Mailing Add:* UMDNJ Med Sch 185 S Orange Ave Newark NJ 07103-2757

EVANS, HUGH LLOYD, b Brownsville, Pa, Mar 7, 41; m 68; c 1. PSYCHOPHARMACOLOGY, TOXICOLOGY. *Educ:* Rutgers Univ, BA, 63; Temple Univ, MA, 65; Univ Pittsburgh, PhD(psychobiol), 69. *Prof Exp:* NIMH fel toxicol, 70, instr radiation biol, 71-73, asst prof radiation biol & environ health sci, Univ Rochester, 73-77; assoc prof, 77-88, PROF ENVIRON MED, NY UNIV, 89- *Concurrent Pos:* Mem, Health Effects Res Rev Panel, US Environ Protection Agency, Comt Animal Res Ethics, Am Psychol Asn, 91-93. *Mem:* Am Soc Pharmacol & Exp Therapeut; Soc Neurosci; Am Psychol Asn; Soc Toxicol; Behav Toxicol Soc (pres, 90-92). *Res:* Behavioral effects of drugs and toxins; brain-behavior relationships. *Mailing Add:* Dept Environ Med 550 First Ave New York NY 10016

EVANS, IRENE M, b Feb 25, 43; c 3. TISSUE CULTURE, PHARMACOLOGY. *Educ:* Univ Rochester, PhD(biophysics), 74. *Prof Exp:* ASSOC PROF BIOL, ROCHESTER INST TECHNOL, 83- *Mem:* Am Soc Pharmacol & Exp Therapeut; Asn Res Vision & Opthal; Genetic Toxicol; Tissue Culture. *Mailing Add:* Dept Biol Rochester Inst Technol One Lomb Memorial Dr Rochester NY 14623

EVANS, JAMES BOWEN, radiochemistry; deceased, see previous edition for last biography

EVANS, JAMES BRAINERD, bacteriology; deceased, see previous edition for last biography

EVANS, JAMES ERIC LLOYD, b Miniota, Man, May 25, 14; m 40; c 2. EXPLORATION GEOLOGY. *Educ:* Univ Man, BSc, 36; Queens Univ, Ont, MA, 42; Columbia Univ, PhD, 44. *Prof Exp:* Res geologist, Falconbridge Nickel Mines, Ltd, 42-45; field geologist, Frobisher, Ltd, 45-50; mgr, Amco

Explor, Inc, 50-54; field mgr, Tech Mines Consult, Ltd, 54-56; chief geologist, Rio Tinto Can Explor, Ltd, 56-70; dir explor, Denison Mines, LTD, 70-79; CONSULT GEOL, 80- *Concurrent Pos:* Adj prof, Univ Toronto, 80-84. *Honors & Awards:* Can Centennial Medal. *Mem:* Geol Soc Am; Soc Econ Geol; Geol Asn Can (secy, 63-65, pres, 67-68); Can Inst Mining & Metall. *Res:* Petrography and petrology of the mine; identification and metallurgy of various mine, mill and smelter products; mineral exploration in Canada, United States, Chile, Brazil, Costa Rica, Algeria, Italy, Greece, Jordan, Australia, New Zealand and the Philippines. *Mailing Add:* 1375 Stavebank Rd Mississauga ON L5G 2V4 Can

EVANS, JAMES ORNETTE, b Roanoke, Tex, July 27, 20; m 59; c 1. AGRICULTURAL ENGINEERING, SOIL SCIENCE. *Educ:* Univ Wis, BS, 47, MS, 52. *Prof Exp:* Land classification specialist, Bur Reclamation, US Dept Interior, 48-53; conservationist, Div Lands & Soil, Ohio Dept Natural Resources, 53-56 & 57-64; soil scientist, Kuljian Corp Philadelphia, opers in Iraq, 56-57; proj soil scientist, HEW, 64-67; res soil scientist, Cincinnati Water Res Lab, Fed Water Pollution Control Admin, US Dept Interior, 67-68; RES HYDROLOGIST, DIV FOREST ENVIRON RES, FOREST SERV, USDA, 68- *Mem:* Am Soc Agron; Soil Conserv Soc Am; Int Soc Soil Sci; Am Geophys Union. *Res:* Soil and water relationships-infiltration, hydraulic conductivity, drainage; irrigation, salinity; oxidation and assimilation of organic and inorganic sludges and effluents; soil fertility and conditioning; erosion, sedimentation and pollution abatement; water yields; recycling wastes; energy from biomass; reclamation of disturbed lands; assessment of atmospheric deposition on forest watersheds. *Mailing Add:* Forest Prod Lab Stat Dept One Gifford Pinchot Dr Madison WI 53705-2398

EVANS, JAMES R, b La Habra, Calif, June 8, 31. GEOLOGY. *Educ:* Whittier Col, BA, 56, Univ Southern Calif, MS, 58. *Prof Exp:* Mining geologist, Calif Div Mines, 58-77; mineral commodity specialist, US Bur Mines, 77-78; geologist, Geol Surv, Reston, Va, 79-80, dist supvr, Resource & Eval Off, Western Region, 81-82; SR TECH MINERAL SPECIALIST, BUR LAND MGT, SACRAMENTO, 82- *Concurrent Pos:* Independent consult & pvt explorer, 58-77. *Mem:* Fel Geol Soc Am; fel Mineralogic Soc Am. *Res:* Examination & appraisal of mines & mineralized areas. *Mailing Add:* 2800 Cottage Way Sacramento CA 85825

EVANS, JAMES SPURGEON, b Big Sandy, Tenn, Aug 19, 31; m 56; c 2. ANATOMY. *Educ:* Univ Tenn, BS, 58; La State Univ, MS, 60; Univ Ky, PhD(reproductive physiol), 64. *Prof Exp:* NIH fel anat, Med Sch, Univ Ky, 64-65; instr, 65-69, ASST PROF ANAT, UNIV TENN CTR HEALTH SCI, 69- *Mem:* AAAS; Sigma Xi. *Res:* Neuroendocrinology; general endocrinology and reproductive physiology. *Mailing Add:* Dept Anat Univ Tenn Ctr for Health Sci Memphis TN 38103

EVANS, JAMES STUART, b Bridgton, Maine, Jan 16, 41. EDUCATIONAL COMPUTING, INORGANIC CHEMISTRY. *Educ:* Bates Col, BS, 62; Princeton Univ, MA, 64, PhD(chem), 66. *Prof Exp:* Res assoc chem, Princeton Univ, 66; from asst prof to assoc prof, 66-85, chmn dept, 68-71, DIR COMPUT SERV, 79-, PROF CHEM, LAWRENCE UNIV, 85- *Concurrent Pos:* Vis asst prof, Univ Ore, 72-73; sr vis, Univ Oxford, Eng, 78-79, 86, 88. *Mem:* AAAS; Am Chem Soc; Am Phys Soc; NY Acad Sci. *Res:* Computer applications in chemistry, including simulation programs and computer-controlled instrumentation; inorganic biochemistry; coordination compounds, especially sulfato complexes. *Mailing Add:* Dept Chem Lawrence Univ Appleton WI 54912-0599

EVANS, JAMES WARREN, b Edna, Tex, Oct 31, 38; m 88; c 1. ANIMAL PHYSIOLOGY. *Educ:* Colo State Univ, BS, 64; Univ Calif, Davis, PhD(physiol), 68. *Prof Exp:* Assoc dean, Col Agr & Environ Sci, Univ Calif, Davis, 82-85, prof animal sci & animal physiologist, 78-85; PROF ANIMAL SCI, TEX A&M UNIV, 85- *Mem:* Equine Nutrit & Physiol Soc (vpres, 75-77, pres, 77-79); Am Soc Animal Sci; Am Physiol Soc; Endocrine Soc; Sigma Xi; Soc Study Reproduction; NAm Riding Handicapped Asn. *Res:* Physiology of reproduction in the equine. *Mailing Add:* Dept Animal Sci Tex A&M Univ College Station TX 77843-2471

EVANS, JAMES WILLIAM, b Chilhowee, Mo, Sept 21, 08; m 30; c 1. AGRICULTURAL BIOCHEMISTRY. *Educ:* Cent Mo State Teachers Col, BS, 28; Univ Minn, PhD(agr biochem), 40. *Prof Exp:* Spec analyst, Union Starch & Refining Co, 30-32, res chemist, 32-37; asst biochem, Univ Minn, 38-39, instr, 39-40; res chemist, Union Starch & Refining Co, 40-43; from res chemist to sect leader, Gen Mills, Inc, Minn, 43-50; dir res, Am Maize Prod Co, 50-59, vpres res & develop, 59-64, pres & chief exec officer, 64-75; RES & DEVELOP CONSULT, 75- *Concurrent Pos:* Mem bd & treas, Central Mo State Univ Found, Inc, 79- *Mem:* Am Chem Soc; Am Asn Cereal Chem; Am Asn Textile Chem & Colorists; Inst Food Tech. *Res:* Corn and wheat starches; corn syrups and sugars; caramel coloring; moisture methods for syrups and sugars; composition of starch hydrolysates; development of packaged foods as soups, cake, pie crust and biscuit mixes, breakfast cerals and instant puddings. *Mailing Add:* Rte 2 Windsor MO 65360-9802

EVANS, JAMES WILLIAM, b Dobcross, Eng, Aug 22, 43; US citizen; m 85; c 3. EXTRACTIVE METALLURGY, CHEMICAL ENGINEERING. *Educ:* Univ London, BS, 64; State Univ NY, Buffalo, PhD(chem eng), 70. *Prof Exp:* Tech adv prog, Int Comput Ltd, 64-65; chemist, Cyanamid Can Ltd, 65-67; chem engr, Ethyl Corp, 70-72; from asst prof to assoc prof, 72-80, chmn dept, 86-90, PROF METALL, UNIV CALIF, BERKELEY, 80- *Concurrent Pos:* Consult, San Louis Mining Co, 74, Summer Chem Co, 75, Razor Assocs, 76 & 77, Repub Steel Corp, 78, Leach & Garner, 85, Dept Energy, Kaiser Aluminum, EG & G Idaho, ISET & Comalco. *Honors & Awards:* Extractive Metall Sci Award, Am Inst Mining Metall & Petrol Engrs, 73 & 83, Mathewson Gold Medal, 84. *Mem:* Am Inst Mining, Metall & Petrol Engrs; Iron & Steel Inst Japan; Electrochem Soc. *Res:* Fluid flow; heat transport; mass transport and chemical reaction kinetics in metallurgical process; fluidized bed electrodes; electrochemistry of metals production and refining, diffusion of gases in porous solids; electromagnetic casting. *Mailing Add:* Dept Mat Sci & Mineral Eng Univ Calif Berkeley CA 94720

EVANS, JOE SMITH, b Wilson County, Tenn, Sept 1, 33; m 62; c 3. MATHEMATICS EDUCATION. *Educ:* Middle Tenn State Univ, BS, 54; Vanderbilt Univ, MS, 59; George Peabody Col, Vanderbilt Univ, PhD(math), 72. *Prof Exp:* Instr math, Martin Jr Col, 57-59; PROF MATH, MID TENN STATE UNIV, 59- *Concurrent Pos:* Nat Coun Teachers Math. *Mem:* Math Asn Am; Am Asn Univ Prof; Nat Educ Asn. *Res:* Consultant and reviewer for mathematics textbooks. *Mailing Add:* Mid Tenn State Univ Box 355 Murfreesboro TN 37132

EVANS, JOHN C, b Oklahoma City, Okla, Jan 21, 38; m 59; c 3. ASTROPHYSICS, STELLAR ATMOSPHERES. *Educ:* Univ Okla, BS, 60; Rensselaer Polytech Inst, MS, 62; Univ Mich, MS, 64, PhD(astron), 66. *Prof Exp:* From asst prof to assoc prof physics, Kans State Univ, 66-76; ASSOC PROF PHYSICS & ASSOC DEAN GRAD SCH, GEORGE MASON UNIV, 76-, ASST DIR, GEORGE MASON INST, 80- *Concurrent Pos:* Vis prof, Univ Western Ont, 72. *Mem:* Nat Coun Univ Res Admin; Am Astron Soc; Astron Soc of the Pac; Sigma Xi. *Res:* Stellar structure, specifically stellar atmospheres, thermal and dynamic structures of stellar atmospheres, line formation, chemical abundances, stellar and solar magnetic fields. *Mailing Add:* 5309 Kaywood Ct Fairfax VA 22032

EVANS, JOHN CHARLES, JR, b Jamaica, NY, Dec 19, 44; m 65. RADIOCHEMISTRY. *Educ:* Fla State Univ, BS, 66; Univ Calif, San Diego, PhD(chem), 71. *Prof Exp:* Res assoc chem, Brookhaven Nat Lab, 71-73, assoc chemist, 73-76, chemist, Dept Chem, 76-77; SR RES SCIENTIST, DEPT PHYS SCI, PAC NORTHWEST DIV, BATTELLE MEM INST, 77- *Res:* Cosmic ray interactions in matter; environmental analytical chemistry; high sensitivity nuclear counting techniques; solar neutrino detection. *Mailing Add:* 1620 SE Oxford Richland WA 99352

EVANS, JOHN EDWARD, b Sisseton, SDak, July 15, 25; m 50; c 3. BACTERIOLOGY. *Educ:* Luther Col, Iowa, BA, 49; Univ SDak, MA, 51; Univ London, PhD(bact), 58. *Prof Exp:* Asst, Univ Wis, 51-52 & Rheumatic Fever Res Inst, Northwestern Univ, 52-54; vis asst prof, 58-59, from asst prof to assoc prof, 59-68, PROF MICROBIOL, UNIV HOUSTON, 68- *Concurrent Pos:* Sr scientist, NATO; Rotary fel & Fogarty fel. *Mem:* AAAS; Am Soc Microbiol; Soc Indust Microbiol. *Res:* Regulation of growth and cell division in microorganisms; DNA technology; biodegradation. *Mailing Add:* Dept Biol Univ Houston Houston TX 77004

EVANS, JOHN ELLIS, b Oak Hill, Ohio, Oct 2, 14; m 48. PHYSICS. *Educ:* Ohio State Univ, BSc & BA, 36, MA, 37; Rice Univ, PhD(physics), 47. *Hon Degrees:* DSc, Rio Grande Univ, Ohio, 83. *Prof Exp:* Tutor math, Ohio State Univ, 36-37; teacher high sch, Ohio, 37-38; prof physics & math, Civilian Pilot Training Prog, Rio Grande Univ, 38-41; asst physics, Ohio State Univ, 41-42; staff mem, Radiation Lab, Mass Inst Technol, 42-45; fel physics, Rice Inst, 45-48; staff mem, Los Alamos Sci Lab, 48-52; group leader, Atomic Energy Div, Phillips Petrol Co, 52-54, sect head, 54-56, dir nuclear physics res, 56-61; SR CONSULT SCIENTIST, LOCKHEED MISSILES & SPACE CO, 61-, SR MEM, RES LAB, 63- *Mem:* AAAS; fel Am Phys Soc; Am Geophys Union; Am Nuclear Soc; assoc fel Am Inst Aeronaut & Astronaut; Sigma Xi. *Res:* Auroral, atmospheric and nuclear physics; homogeneous reactor development; neutron crystal spectrometry. *Mailing Add:* 615 Joandra Ct Los Altos CA 94024-5336

EVANS, JOHN FENTON, b Sewickley, Pa, Mar 20, 49; m 71. ANALYTICAL & SURFACE CHEMISTRY. *Educ:* Washington & Jefferson Col, BA, 71; Univ Del, PhD(chem), 77. *Prof Exp:* Teaching & res asst, Dept Chem, Univ Del, 71-75; res assoc, Dept Chem, Ohio State Univ, 75-77; ASST PROF DEPT CHEM, UNIV MINN, 77- *Mem:* Am Chem Soc; Sigma Xi; Am Vacuum Soc. *Res:* Surface modification using plasma chemistry; photochemistry and conventional chemistry; surface analysis using particle and light spectroscopies. *Mailing Add:* Dept Chem Univ Minn 207 Pleasant St SE Minneapolis MN 55455

EVANS, JOHN N, b 1948; m; c 1. PULMONARY RESEARCH. *Educ:* Clark Univ, BA, 70; Univ Fla, PhD(physiol), 76. *Prof Exp:* Postdoctoral fel, Univ Vt, 76-77, res asst prof, 77-82, from asst prof to assoc prof, 82-88, PROF PHYSIOL & BIOPHYSICS, COL MED, UNIV VT, 88- *Mem:* Am Thoracic Soc; Am Physiol Soc; Am Heart Asn. *Res:* Smooth muscle; vascular remodelling. *Mailing Add:* Dept Physiol Univ Vt Col Med Burlington VT 05405

EVANS, JOHN R, b Toronto, Ont, Oct 1, 29; m 54; c 6. MEDICAL RESEARCH ADMINISTRATION. *Educ:* Univ Toronto, MD, 52; Oxford Univ, DPhil, 55; FRCP(C), 58, FRCP, 86. *Hon Degrees:* Several from Can univs, Yale, Johns Hopkins, US & Maastricht, Neth. *Prof Exp:* Postdoctoral fel myocardial metab, Harvard Med Sch, 60; dean, Fac Med, McMaster Univ, Hamilton, 65-72, vpres, Health Sci, 67-72; pres, Univ Toronto, 72-78; dir, Pop, Health & Nutrit Dept, World Bank, Washington, DC, 79-83; CHMN BD, ALLELIX INC, MISSISSAUGA, ONT, 83- *Concurrent Pos:* Mem coun, Royal Col Physicians & Surgeons, 72-78; mem, Adv Comt Med Res, WHO, 76-80, Expert Adv Panel Cardiovasc Dis, 72-; mem, Pepin-Robarts Comn Nat Unity, 77-78; chmn, Nat Biotechnol Adv Comt, Can, 83-88, Comn Health Res Develop, 87-; chmn, bd trustees, Rockefeller Found, 87- *Honors & Awards:* Lilly Lectr, 86. *Mem:* Inst Med-Nat Acad Sci; master Am Col Physicians. *Mailing Add:* Allelix Inc 6850 Goreway Dr Mississauga ON L4V 1P1 Can

EVANS, JOHN STANTON, b Camilla, Ga, Oct 12, 21; m 48; c 3. FLUID DYNAMICS, PHYSICAL CHEMISTRY. *Educ:* Berry Col, BS, 42; Emory Univ, MS, 48; Univ Tenn, Knoxville, PhD(physics), 59. *Prof Exp:* Instr physics, Univ Tenn, Martin, 52; aerospace technologist, Langley Res Ctr, NASA, 52-83; RETIRED. *Mem:* Am Phys Soc. *Res:* Computational fluid dynamics. *Mailing Add:* 243 E Queens Dr Williamsburg VA 23185

EVANS, JOHN V, b Manchester, Eng, July 5, 33; m 58; c 3. RADIO PHYSICS. *Educ:* Univ Manchester, BSc, 54, PhD(physics), 57. *Prof Exp:* Leverhulme res fel radio astron, Jodrell Bank Exp Sta, Univ Manchester, 57-60; staff mem, Lincoln Lab, Mass Inst Technol, 60-66; George A Miller vis prof elec eng, Univ Ill, 66-67; staff mem, Lincoln Lab, Mass Inst Technol, 67-70, from assoc group leader to group leader, 70-75, assoc dir head, 75-77, asst dir, 77-83, prof meteorol, Dept Meteorol & Phys Oceanog, 80-83; dir, Northeast Radio Observ Corp, 80-83; DIR, COMSAT LABS, COMSAT DRIVE, 83 - *Concurrent Pos:* Mem, Int Union Radio Sci, 63-, mem, US Nat Comt, 68-70, secy, 70-72, vchmn, 73-75, chmn, 75-78. *Honors & Awards:* Appleton Prize, Royal Soc, 75. *Mem:* Nat Acad Eng; AAAS; fel Inst Elec & Electronics Engrs; Am Geophys Union. *Res:* Radar studies of the moon, Venus, meteors and the ionosphere. *Mailing Add:* Comsat Labs 22300 Comsat Dr Clarksburg MD 20871-9475

EVANS, JOHN W, b Mt Vernon, NY, Jan 20, 35; div; c 2. MATHEMATICAL BIOLOGY. *Educ:* Cornell Univ, MD, 58; Univ Calif, Los Angeles, PhD(math), 66. *Prof Exp:* Sr surgeon, Math Res Br, Nat Inst Arthritis & Metab Dis, 66-68; assoc prof, 68-73, PROF MATH, UNIV CALIF, SAN DIEGO, 73- *Mem:* Am Math Soc. *Res:* Mathematical models of nerve impulse conduction; mathematical models in pulmonary physiology. *Mailing Add:* Dept Math Univ Calif San Diego C-012 La Jolla CA 92093

EVANS, JOHN WAINWRIGHT, JR, b New York, NY, May 14, 09; m 32; c 3. OPTICS. *Educ:* Swarthmore Col, AB, 32; Harvard Univ, AM, 36, PhD(astron), 38. *Hon Degrees:* ScD, Univ NMex, 67 & Swarthmore Col, 70. *Prof Exp:* Instr astron, Univ Minn, 37-38; from instr to asst prof astron & math, Mills Col, 38-42; optical res worker, Nat Defense Res Comt, Univ Rochester, 42-46, asst prof optics, 45-46; astronomer, High Altitude Observ Harvard-Colo, 46-52; dir, Sacramento Peak Observ, 52-75, sr scientist, 75-79; CONSULT, 79- *Honors & Awards:* Cleveland Prize, AAAS, 57; Rockefeller Pub Serv Award Sci, Technol & Eng, 69; George Ellery Hale Prize, Am Astron Soc, 82; David Richardson Medal, Optical Soc Am, 87. *Mem:* AAAS; Am Astron Soc; fel Am Acad Arts & Sci; fel Optical Soc Am. *Res:* Solar physics; solar terrestrial effects; optical design. *Mailing Add:* One Baya Rd Eldorado Santa Fe NM 87505

EVANS, JOSEPH LISTON, b Lebanon, Ky, June 27, 30; m 55; c 2. NUTRITION, ANIMAL NUTRITION. *Educ:* Univ Ky, BS, 52, MS, 55; Univ Fla, PhD(nutrit, biochem), 59. *Prof Exp:* Asst animal sci, Univ Ky, 54-55; asst nutrit, Univ Fla, 55-59; from asst prof to assoc prof, 59-69, PROF NUTRIT, RUTGERS UNIV, 69- *Mem:* Am Soc Animal Sci; Am Inst Nutrit; Am Dairy Sci Asn. *Res:* Nutritional biochemical mechanisms involving utilization of minerals in rats, man and cattle and nitrogen utilization cattle. *Mailing Add:* Bartlett Hall Box 231 Cook Col Rutgers Univ New Brunswick NJ 08903

EVANS, KENNETH, JR, b Decatur, Ill, Jan 25, 41; c 3. COMPUTATIONAL PHYSICS, CONTROLLED FUSION. *Educ:* Univ Ill, BS, 63, MS, 64, PhD(physics), 70. *Prof Exp:* Res teaching asst physics, Univ Ill, 63-70; res asst, Coord Sci Lab, 68-70; res assoc & instr, Univ Wis, 70-74; PHYSICIST, ARGONNE NAT LAB, 74- *Concurrent Pos:* Vis scientist, Princeton Plasma Physics Lab, Princeton Univ, 83. *Mem:* Am Phys Soc. *Res:* Computational physics, mathematical modeling and theoretical research in plasma physics and controlled thermonuclear fusion. *Mailing Add:* Argonne Nat Lab 9700 S Cass Ave Argonne IL 60439

EVANS, KENNETH JACK, b Chickasha, Okla, July 8, 29; m 55; c 4. VERTEBRATE ZOOLOGY, ECOLOGY. *Educ:* Univ Okla, BS, 57, MS, 58; Univ Calif, Riverside, PhD(zool), 64. *Prof Exp:* Instr zool, Univ Redlands, 60-61; assoc biol, Univ Calif, Riverside, 63-64; from asst prof to assoc prof, 64-73, PROF BIOL, CALIF STATE UNIV, CHICO, 73- *Mem:* Ecol Soc Am. *Res:* Ecology of amphibians and reptiles. *Mailing Add:* Dept Biol Sci Calif State Univ Chico CA 95929

EVANS, LANCE SAYLOR, b Philadelphia, Pa, Sept 29, 44; m 65; c 3. PLANT PHYSIOLOGY, CELL CYCLE KINETICS. *Educ:* Calif State Polytech Col, Kellogg-Voorhis, BS, 67; Univ Calif, Riverside, PhD(plant sci & plant physiol), 70. *Prof Exp:* Nat Inst Environ Health Sci res fel plant sci, Univ Calif, Riverside, 70-71, res biologist, 71-72, Nat Cancer Inst fel, 72-73; res biologist, Brookhaven Nat Lab, 72-75; PROF BIOL, LAB PLANT MORPHOGENESIS, MANHATTAN COL, 75- *Concurrent Pos:* Consult, Acid Rain Res Prog, Brookhaven Nat Lab, Upton, NY, 75- *Mem:* Am Soc Plant Physiologists; Bot Soc Am; Sigma Xi. *Res:* Cell cycle regulation; air pollution; plant morphogenesis and anatomy; cell cycle kinetics and analysis of cell proliferation and cell arrest; effects of air pollutants on crop growth and yield; physiological response of cacti to environmental conditions. *Mailing Add:* Lab Plant Morphogenesis Manhattan Col Bronx NY 10471

EVANS, LARRY GERALD, b Chicago, Ill, July 15, 43; m 79; c 1. PLANETARY SCIENCE, SPACE SCIENCE. *Educ:* Purdue Univ, BS, 65, MS, 67; Northwestern Univ, PhD(mech eng & astron), 71. *Prof Exp:* Resident res assoc, Goddard Space Flight Ctr, 71-73; sr mem tech staff & prin scientist, Systs Sci Div, 73-84, MGR ADVAN ASTRON PROGS, ASTRON OPERS, COMPUT SCI CORP, 84- *Concurrent Pos:* Adj prof, dept geol, Univ Md, 86-; discipline scientist planetary instruments, Solar Syst Explor Div, NASA, 86-88. *Mem:* Am Phys Soc; Am Inst Aeronaut & Astronaut; AAAS; Planetary Soc; Sigma Xi. *Res:* Planetary geochemistry from orbiting spacecraft using x-ray and gamma-ray spectroscopy; preservation and restoration of historic buildings and monuments. *Mailing Add:* Goddard Space Flight Ctr Comput Sci Corp Code 682 Greenbelt MD 20771

EVANS, LATIMER RICHARD, b Washington, DC, Nov 4, 18; m 42; c 4. CHEMISTRY. *Educ:* Am Univ, BS, 41; Purdue Univ, PhD(org chem), 45. *Prof Exp:* Asst chem, Purdue Univ, 41-42, Manhattan Proj fel, 45; res chemist, E I du Pont de Nemours & Co, Inc, 46-50; from asst prof to assoc prof, NMex State Univ, 50-61, prof chem, 61-86; RETIRED. *Honors & Awards:* Am Inst Chem Medal, 41. *Mem:* Am Chem Soc; Sigma Xi. *Res:* Chlorination; fluorination of chloro compounds; azeotrope distillation as separation method; fluorination with antimony pentafluoride. *Mailing Add:* 4155 Tellbrook Rd Las Cruces NM 88001

EVANS, LAURIE EDWARD, b Unity, Sask, Oct 14, 33; m 60; c 3. CYTOGENETICS, PLANT BREEDING. *Educ:* Univ Sask, BSA, 54; Univ Man, MSc, 56, PhD(plant sci, cytogenetics), 59. *Prof Exp:* From res asst to res assoc, Univ Man, 59-63, from asst prof to assoc prof, 63-73, dept head, 80-88, PROF PLANT SCI, UNIV MAN, 73- *Concurrent Pos:* Plant breeder, Kenya, 68-69; vis prof, Sydney Australia, 76. *Mem:* Genetics Soc Can. *Res:* Wheat breeding. *Mailing Add:* Dept Plant Sci Univ Man 76 DaFoe Rd Winnipeg MB R3T 2N2 Can

EVANS, LAWRENCE B(OYD), b Ft Sumner, NMex, Oct 27, 34; m 63; c 2. CHEMICAL ENGINEERING. *Educ:* Univ Okla, BS, 56; Univ Mich, MSE, 57, PhD(chem eng), 62. *Prof Exp:* From asst prof to assoc prof, 62-76, PROF CHEM ENG, MASS INST TECHNOL, 76- *Concurrent Pos:* Pres, Aspen Technol, Inc, 81-83; trustee, Cache Corp, 78-83; chmn, Aspen Technole, Inc, 84- *Honors & Awards:* Donald L Katz Lectr, Univ Mich, 80; Comput & Systs Technol Div Award, Am Inst Chem Engrs, 82. *Mem:* Am Inst Chem Engrs; Am Chem Soc; Am Soc Mech Engrs; Asn Comput Mach. *Res:* Computer aided chemical process analysis; process dynamics and control; applied mathematics. *Mailing Add:* Dept Chem Eng Mass Inst Technol Cambridge MA 02139

EVANS, LAWRENCE EUGENE, b San Antonio, Tex, Sept 18, 32; m 56; c 2. PHYSICS. *Educ:* Birmingham-Southern Col, BS, 53; Johns Hopkins Univ, PhD(physics), 60. *Prof Exp:* Res assoc physics, Univ Wis, 60-62, instr, 62-63; from asst prof to assoc prof, 63-80, PROF PHYSICS, DUKE UNIV, 80- *Mem:* Am Phys Soc. *Res:* Quantum field theory; quantum electrodynamics; theory of elementary particles. *Mailing Add:* Dept Physics Duke Univ Durham NC 27706

EVANS, LEE E, b Newton, Miss, May 27, 22; m 48; c 3. ANIMAL HUSBANDRY. *Educ:* Alcorn Agr & Mech Col, BS, 43; Iowa State Univ, MS, 47; Univ Ill, PhD(animal sci), 56. *Prof Exp:* Teacher voc agr, Newton Voc Sch, 47-51; dir dept agr, Alcorn Agr & Mech Col, 51-53; asst, Univ Ill, 53-55; head, Dept Animal Husb, Fla A&M Univ, 55-68, prof, 68-74, prof rural develop, 74-90; RETIRED. *Mem:* Am Soc Animal Sci; Genetics Soc Am. *Res:* Improvement of farm livestock through better methods and techniques of breeding; experimenting with cattle feed crops fertilized with sewage as part of recycling of human waste. *Mailing Add:* 208 Osceola St Tallahassee FL 32301

EVANS, LEONARD, b London, Eng, Feb 21, 39; m 66; c 3. TRAFFIC SAFETY RESEARCH, HUMAN FACTORS. *Educ:* Queen's Univ, Belfast, BSc, 60; Oxford Univ, DPhil(physics), 65. *Prof Exp:* Fel physics, Div Pure Physics, Nat Res Coun Can, 65-67; PRIN RES SCIENTIST, OPER SCI DEPT, GEN MOTORS RES LABS, 67- *Honors & Awards:* J M Campbell Award, Gen Motors Res Labs, 84; A R Lauer Award, Human Factors Soc, 85. *Mem:* Asn Advan Automotive Med; fel Human Factors Soc; Soc Automotive Engrs; AAAS; Sigma Xi; Soc Risk Anal. *Res:* Traffic safety; author of book. *Mailing Add:* Oper Sci Dept Gen Motors Res Labs Warren MI 48090

EVANS, MARLENE SANDRA, b London, Ont, Jan 26, 46. LIMNOLOGY, OCEANOGRAPHY. *Educ:* Carleton Univ, BSc, 69; Univ BC, PhD(zool, oceanog), 73. *Prof Exp:* Res investr, Univ Mich, 74-76, asst scientist, 76-81, assoc scientist limnol, Great Lakes Res Div, 81-88; RES SCIENTIST, NAT HYDROL RES INST, 88- *Honors & Awards:* Anderson-Everett Award. *Mem:* Can Soc Zool; Am Soc Limnol & Oceanog; Int Asn Great Lakes Res (pres, 81-82); AAAS; Am Fish Soc. *Res:* Zooplankton ecology; limnology and biological oceanography; water quality; saline lakes; lipids. *Mailing Add:* Aquatic Ecol Div Nat Hydrol Res Inst Saskatoon SK S7N 3H5

EVANS, MARY JO, b Maysville, Mo, Nov 28, 35; m 68; c 2. MOLECULAR BIOLOGY, CANCER. *Educ:* William Jewell Col, BA, 57; Univ Mo, MS, 65; Univ Tenn, PhD(microbiol), 68. *Prof Exp:* Res asst virol, Univ Mo, 57-58, res fel, 64-65; teaching fel microbiol, Univ Tenn, 65-66; trainee virol, St Jude Children's Res Hosp, 66-68; cancer res scientist, 68-69, dir grad studies microbiol, 73-75, sr cancer res scientist & asst res prof, 69-78, CANCER RES SCIENTIST IV, ROSWELL PARK MEM INST & ASSOC RES PROF, ROSWELL PARK DIV, STATE UNIV NY BUFFALO, 78- *Mem:* Am Soc Microbiol; Am Soc Biol Chemists; Am Soc Cancer Res. *Res:* Oncogenes in human cancer. *Mailing Add:* Dept Biol Resources Roswell Park Mem Inst 666 Elm St Buffalo NY 14263

EVANS, MICHAEL ALLEN, b New Albany, Ind, Oct 19, 43; m 73; c 3. TOXICOLOGY. *Educ:* St Joseph Col, BS, 67; Ind Univ, PhD(toxicol), 74. *Prof Exp:* Fel, Vanderbilt Med Ctr, 74-76; asst prof, 76-82, ASSOC PROF PHARMACOL, UNIV ILL MED CTR, 82- *Concurrent Pos:* Lectr toxicol, Cook County Grad Sch Med, 78; consult, Naval Res Labs, US Dept Defense, 84 & Am Inst Drug Detection, 84; mem, Sci Rev Panel Health Res, US Environ Protection Agency, 84; reviewer, Off Orphans Prod Develop, US Food & Drug Admin, 85; hon prof, Kunming Med Col, Kunming, China, 85- *Mem:* NY Acad Sci; AAAS; Am Acad Forensic Sci; Soc Toxicol; Teratology Soc. *Res:* Mechanism of chemical hepatotoxicity; drug disposition during development; relationship between drug metabolism and drug toxicity. *Mailing Add:* Am Inst Toxicol 2345 S Lynhurst Dr No 210 Indianapolis IN 46241

EVANS, MICHAEL DOUGLAS, b Pittsburgh, Pa, Dec 13, 59; m 83; c 1. ELECTROMETALLURGY, CORROSION SCIENCE. *Educ:* Franklin & Marshall Col, AB, 81; Columbia Univ, MS, 83, DESc, 86. *Prof Exp:* Res engr, Asarco, Inc, 81; group leader, Castle Technol Corp, 85-87; MEM TECH STAFF, AT&T BELL LABS, 87- *Mem:* Am Inst Mining, Metall & Petrol Engrs; Nat Asn Corrosion Engrs; Electrochem Soc; Am Chem Soc. *Res:* Energy systems (battery and fuel cell), corrosion, electrometallurgy and materials development; high temperature; electrochemical and hydrometallurgical process analysis; battery and energy systems; advanced coating technology; high temperature service; hostile environments. *Mailing Add:* Ten Apple Tree Lane Wilmington MA 01887-3915

EVANS, MICHAEL LEIGH, b Detroit, Mich, July 26, 41; m 62; c 3. BOTANY. *Educ:* Univ Mich, BA, 63, MS, 65; Univ Calif, Santa Cruz, PhD(biol), 67. *Prof Exp:* Teaching asst biol, Univ Calif, Santa Cruz, 65-67; asst prof, Kalamazoo Col, 67-70; from asst prof to assoc prof, 71-78, PROF BOT, OHIO STATE UNIV, 78- *Concurrent Pos:* NATO fel, Univ Freiburg, WGer, 70-71. *Mem:* Am Soc Plant Physiol; Japanese Soc Plant Physiol. *Res:* Plant growth hormones, especially short-term effects. *Mailing Add:* Dept Bot Ohio State Univ Columbus OH 43210

EVANS, NANCY REMAGE, b Taunton, Mass, May 19, 44; m 68; c 2. ASTRONOMY. *Educ:* Wellesley Col, BA, 66; Univ Toronto, MSc, 69, PhD(astron), 74. *Prof Exp:* Fel astron, Univ Toronto, 74-76, res assoc, 76-82, asst prof, 82-83; resident astronr, IUE Satellite, Comput Sci Corp, Goddard Space Flight Ctr, 83-86; res assoc, Univ Toronto, 86-88; ASSOC SCIENTIST, INST SPACE & TERRESTRIAL SCI, YORK UNIV, 88- *Mem:* Am Astron Soc; Int Astron Union; Can Astron Soc. *Res:* Masses, luminosities, radii companions and temperatures of Cepheid variable stars; extragalactic distance; star formation. *Mailing Add:* c/o CRESS Inst Space & Terrestrial Sci York Univ North York ON M3J 1P3 Can

EVANS, NEAL JOHN, II, b San Antonio, Tex, Sept 22, 46; div; c 1. INTERSTELLAR MATTER, STAR FORMATION. *Educ:* Univ Calif, Berkeley, AB, 68, PhD(physics), 73. *Prof Exp:* Res fel astron, Owens Valley Radio Observ, Calif Inst Technol, 73-75; res scientist assoc IV astron, 75-76, from asst prof to assoc prof, 76-86, PROF ASTRON, UNIV TEX, AUSTIN, 87- *Concurrent Pos:* Consult, NASA, 77-78 & 84-86; mem, Infrared Telescope Facil Proposal Rev Comt, millimeter array adv comt. *Mem:* Am Astron Soc; Int Astron Union; Int Union Radio Sci. *Res:* Molecular line studies of dense interstellar clouds and infrared studies of objects embedded in the clouds to elucidate star formation. *Mailing Add:* Dept Astron Univ Tex Austin TX 78712

EVANS, NORMAN A(LLEN), b SDak, Dec 3, 22; m 44; c 4. WATER RESOURCES, IRRIGATION ENGINEERING. *Educ:* SDak State Col, BS, 44; Utah State Univ, MS, 47; Colo State Univ, PhD, 63. *Prof Exp:* Asst, Col Eng, Utah State Univ, 46-47; asst prof agr eng, NDak Agr Col, 47-51; asst civil eng, 51-52, from asst prof to assoc prof, 52-57, assoc prof agr eng, 57-59, head dept, 57-69, dir environ resources ctr, 67-78, assoc dir agr exp sta, 69-70, dir, Off Gen Univ Res, 70-72, PROF AGR ENG, COLO STATE UNIV, 59-; DIR, COLO WATER RES INST, 67- *Concurrent Pos:* Mem, Colo Water Pollution Control Comn, 66-80; mem bd dirs, Engr Coun Prof Develop, 70-76; vpres, Ft Collins City Water Bd, 63-83, pres, 83- *Mem:* Fel AAAS; Am Soc Agr Engrs (vpres, 68-70); Am Soc Eng Educ; Soil Sci Soc Am; Am Soc Civil Engrs. *Res:* Fluid mechanics of porous media; drainage; irrigation practices. *Mailing Add:* 1847 Michael Lane Ft Collins CO 80526

EVANS, RAEFORD G, b Coleman, Tex, Aug 20, 19; m 42; c 3. AGRONOMY, GENETICS. *Educ:* Tex A&M Univ, BS, 41; Univ Wyo, MS, 63, PhD(agron), 69. *Prof Exp:* Agronomist, Tex A&M Univ, 41-44; farm mgr, 55-58, from instr to assoc prof, 59-73, PROF AGRON, TARLETON STATE UNIV, 73- *Concurrent Pos:* Asst, Univ Wyo, 66-67. *Mem:* Am Soc Agron; Crop Sci Soc Am; Am Genetic Asn; Soil Conser- Soc Am. *Res:* Plant breeding; crop production; weed and range science. *Mailing Add:* Rte 2 Box 189 Stephenville TX 76401

EVANS, RALPH AIKEN, b Oak Park, Ill, Feb 2, 24; m 67; c 1. PHYSICS. *Educ:* Lehigh Univ, BS, 44; Univ Calif, PhD(physics), 54. *Prof Exp:* Radio engr, Centimeter Wave Sect, Naval Res Lab, 44-46; physicist, Inst Eng Res, Univ Calif, 47-54; res physicist, Power Transmission & Mat Handling, Linkbelt Res Lab, 54-59, dir res lab, 59-61; sr physicist, Res Triangle Inst, 61-72; PROD ASSURANCE CONSULT, EVANS ASSOCS, 72- *Concurrent Pos:* Managing ed, Inst Elec & Electronics Engrs Trans on Reliability; founding ed, J Reliability Rev, Am Soc Qual Control. *Honors & Awards:* R A Evans Award, Am Soc Qual Control. *Mem:* fel Am Soc Qual Control; fel Inst Elec & Electronic Engrs. *Res:* Engineering statistics; reliability; quality control. *Mailing Add:* 804 Vickers Ave Durham NC 27701

EVANS, RALPH H, JR, b Teaneck, NJ, July 5, 29; m 50; c 1. ANALYTICAL CHEMISTRY. *Educ:* Fairleigh Dickinson Univ, AA, 49, BS, 58; State Univ NY, AS, 52. *Prof Exp:* Res scientist, 58-68, SR RES SCIENTIST, LEDERLE LABS, HOFFMANN-LA ROCHE, INC, 69- *Mem:* Am Chem Soc; Sigma Xi. *Res:* Isolation, assay and structural elucidation of antibiotics and natural products from microbial sources. *Mailing Add:* Chem Res Dept Hoffmann-La Roche Inc Nutley NJ 07110

EVANS, RAYMOND ARTHUR, b Albuquerque, NMex, Mar 31, 25; m 50; c 1. RANGE SCIENCE, WEED SCIENCE. *Educ:* Univ Redlands, AB, 50; Univ Calif, PhD, 56. *Prof Exp:* Asst specialist rangeland soils & plants, Univ Calif, 54-58; range scientist, Pasture & Range Mgt, Agr Res Serv, USDA, 58-86, res leader, 72-86; RETIRED. *Concurrent Pos:* Consult range mgt, Soc Range Mgt, 85- *Mem:* AAAS; Ecol Soc Am; Soc Range Mgt; Weed Sci Soc Am. *Res:* Range weed control and revegetation; competition studies involving range weeds and forage species emphasizing factors of soil moisture, temperature and nutrients; utilization of field, greenhouse and laboratory techniques; employment of microenvironmental monitoring. *Mailing Add:* 1560 California Ave Reno NV 89509

EVANS, RICHARD TODD, b Evanston, Ill, Oct 2, 32; m 69; c 3. MICROBIOLOGY, IMMUNOLOGY. *Educ:* Cent Methodist Col, AB, 54; Univ Mo, MS, 59, PhD(microbiol), 63. *Prof Exp:* Asst microbiol, Univ Mo, 57-59, asst instr, 59-62; Am Dent Asn res assoc, NIH, 63-66; asst prof oral biol, Sch Dent, 66-73, asst prof microbiol, Sch Med, 72-75, dir grad studies oral biol, 70-76, assoc prof oral biol & microbiol & assoc chmn, Dept Oral Biol, 75-87, ASSOC PROF ORAL BIOL & MICROBIOL, SCH DENT, STATE UNIV NY BUFFALO, 87- *Mem:* Am Asn Immunologists; Int Asn Dent Res; Am Soc Microbiol; NY Acad Sci; Soc Mucosal Immunol. *Res:* Immunochemistry of bacterial antigens; host-parasite relationships of periodontal disease; preventive (chemotherapeutic) measures in oral disease; microbiology of dental caries. *Mailing Add:* Dept Oral Biol State Univ NY 219 Foster Hall Buffalo NY 14214

EVANS, ROBERT, MACROPHAGE IMMUNOBIOLOGY, IMMUNOLOGICAL NETWORK. *Educ:* London Univ, Eng, DSc, 79. *Prof Exp:* SR STAFF SCIENTIST, JACKSON LAB, 80- *Mailing Add:* The Jackson Lab Bar Harbor ME 04609

EVANS, ROBERT JOHN, b Osage City, Kans, Mar 18, 28; m 51; c 2. ORGANIC CHEMISTRY. *Educ:* Univ Nebr, BSc, 51; Univ Wash, PhD(chem), 59. *Prof Exp:* Org chemist, Merck & Co, NJ, 51-54; org chemist, Hydrocarbons Div, Monsanto Co, 59-66; from asst prof to assoc prof, 66-73, PROF CHEM, ILL COL, 73- *Mem:* Am Chem Soc. *Res:* Oxidation of organic compounds. *Mailing Add:* Dept Chem Ill Col Jacksonville IL 62650

EVANS, ROBERT JOHN, b Logan, Utah, Mar 18, 09; m 41; c 2. BIOCHEMISTRY, NUTRITION. *Educ:* Utah State Univ, BS, 34, MS, 36; Univ Wis, PhD(biochem), 39. *Prof Exp:* Grad asst biochem, Utah State Univ, 34-36; instr chem, Carbon Col, 39-40; assoc chemist, Wash Agr Exp Sta, 40-47; prof biochem, 47-77, EMER PROF BIOCHEM, MICH STATE UNIV, 77- *Concurrent Pos:* USPHS grant seed proteins in nutrit, Mich State Univ, 63-65, USPHS grant lipoproteins, 66-74. *Honors & Awards:* Poultry & Egg Nat Bd of USA Res Achievement Award, 58. *Mem:* AAAS; Am Chem Soc; Am Inst Nutrit; Poultry Sci Asn. *Res:* Lipid-protein binding in egg yolk lipoproteins which includes the study of lipoprotein structure and the structure of proteins and lipids; nutritive availability of the methionine in dry beans. *Mailing Add:* 760 Polk Ave Ogden UT 84404

EVANS, ROBERT MORTON, b Cleveland, Ohio, Oct 28, 17; m 42, 65; c 4. POLYMER CHEMISTRY. *Educ:* Antioch Col, BS, 41; Case Western Reserve Univ, PhD(chem), 59. *Prof Exp:* VPRES RES & ENG, MAMECO INT, 45-; CONSULT & EXEC OFFICER, CTR FOR COATINGS, ADHESIVES & SEALANTS, CARE WRU. *Concurrent Pos:* Res assoc, Case Inst Technol, 60-61; pres, Prog Design, Inc, 67-70; pres & founder, Isonetics, Inc, 70-75; mem res adv comt, Fedn Socs Paint Technol. *Honors & Awards:* Roon Award, 64 & 77; Fel, Am Soc Testing & Mat, 85, Hall of Fame, 87. *Mem:* Am Soc Testing & Mat; Fedn Socs Paint Technol; NY Acad Sci; fel Am Inst Chem; Sigma Xi; Am Chem Soc. *Res:* Specialty coatings and adhesives; sealants and flooring materials; all polymeric and some organic-inorganic alloys; insulated glass; solar heating; abrasion resistance; coauthor or author of over 15 publications. *Mailing Add:* 1365 Forest Hills Blvd Cleveland Heights OH 44118

EVANS, ROBLEY D(UNGLISON), b University Place, Nebr, May 18, 07; m 28, 90; c 3. PHYSICS. *Educ:* Calif Inst Technol, BS, 28, MS, 29, PhD(physics), 32; Am Bd Health Physics, dipl, 61. *Prof Exp:* Asst engl, Calif Inst Technol, 28, asst hist, 27-30 & 31-32, teaching fel physics, 29-32; Nat Res Coun fel, Univ Calif, 32-34; from asst prof to prof, 34-72, dir radioactivity ctr, 35-72, EMER PROF PHYSICS & EMER DIR RADIOACTIVITY CTR, MASS INST TECHNOL, 72- *Concurrent Pos:* Dir res lab, C F Braun Co, Calif, 29-31; chmn comt standards radioactivity, Nat Res Coun, 38-46, chmn subcomt shipment radioactive sunstances, 46-59 & vchmn comt nuclear sci, 46-72; chmn, Int Conf Appl Nuclear Physics, 40; mem adv comt safe handling radioactive luminous compounds, Nat Bur Stand, 41, consult, 60, mem Adv Panel on Radiation Physics, 63-66, chmn, 64; consult, Off Sci Res & Develop, 44-46, US Dept State, 46-47, US Dept Army, 47-49 & Surgeon Gen, 62-69, Los Alamos Sci Lab, 48-64, US Secy Defense, 49-54, Walter Reed Army Med Ctr, 49-56, US Naval Radiol Defense Lab, 52-69, USPHS, 61-71, Fed Radiation Coun, 65-69; consult, Peter Bent Brigham Hosp, 45-72, mem isotopes comt, 64-75, Roger Williams Hosp, 65-; consult, Brookhaven Nat Lab, 47-55, mem vis comt med dept, 65-68; consult, Mass Gen Hosp, 48-73, mem comt isotopes, 50-75; mem mixed comn radiobiol, Int Union Pure & Appl Physics, 47-53; mem joint comn standards, units & constants radioactivity, Int Coun Sci Unions, 48-51, mem joint comn radioactivity, 51-55; consult, Biol & Med Div & Biomed & Environ Res Div, AEC, ERDA, DOE, 50-81, mem, Adv Comt Isotope Distribution, 48-53, chmn, 52-53; mem, Aircraft Nuclear Propulsion Med Adv Group, 53-55; mem comt radiation protection, Mass Inst Technol, 55-72, mem comt radioisotope utilization & adv med dept, 59-72, mem clin res ctr policy comt, 64-72; mem adv comt rules & regulations radiation protection, Mass Dept Labor & Industs, 57; mem subcomt rel hazard factors, Nat Coun Radiation Protection & Measurements, 57-63; ed in physics, Radiation Res, 59-62; mem subcomt symbols, units & nomenclature, Comt Nuclear Sci, Nat Acad Sci-Nat Res Coun, 62-67, panel adv to Nat Bur Stand, 63-66, chmn, 64; mem ad hoc adv comt radiation path, Armed Forces Inst Path, 62-64; mem sci adv bd, Cancer Res Inst, New Eng Deaconess Hosp, 63-69; chmn task group high energy & space radiation dosimetry, Int Comn Radiol Units & Measurements, 64-67; adv, Univ Chicago & mem res comt, Radiol Physics Div, Argonne Nat Lab, 64-68, chmn, 67-68, chmn adv comt, Ctr Human Radiobiol, 72-75; coun mem, Nat Coun Radiation Protection & Measurements, 65-71, hon life mem, 75-; sr US deleg, Int Asn Radiation Res, 66; vis prof, Ariz State Univ, 66-67; consult, Fed Aviation Agency, 67, chmn standing comt radiation biol aspects of supersonic transport, 67, mem, comt radioactive waste mgt, Nat Acad Sci, 68-70, vchmn adv comt, US Transuranium Registry, 68-86, mem tech adv comt, Ariz AEC, 71-72, spec proj assoc, Mayo Clin, 73-81. *Honors & Awards:* Theobald Smith Medal, AAAS, 37; US Presidential Cert Merit, 48; Hull Gold Medal, AMA, 63; Silvanus Thompson Award & Medal, Brit Inst Radiol, 66; Disting Achievement Award, Health Physics Soc, 81; William D Coolidge Award & Gold Medal, Am Asn Physicists Med, 84; Enrico Fermi Award & Gold Medal, US Dept Energy, 90. *Mem:* Fel AAAS; fel Am Phys Soc; fel Am Acad Arts & Sci; fel Health Physics Soc (pres-elect, 71, pres, 72-73); Radiation Res Soc (vpres, 65-66, pres, 66-67); fel Am Asn Physicists Med; Am Asn Physics Teachers; Am Nuclear Soc; assoc Am Roentgen Ray Soc; hon mem Soc Nuclear Med. *Res:* Radioactivity; radioactive tracers in engineering and biology; geological age measurement by radioactivity; instrumentation; biological effects of radiation; nuclear medicine; health physics; pure and applied nuclear physics. *Mailing Add:* 4621 E Crystal Lane Scottsdale AZ 85253-2939

EVANS, ROGER JAMES, b Oxford, Eng. STRUCTURAL ENGINEERING, SOLID MECHANICS. *Educ:* Univ Birmingham, BSc, 55; Brown Univ, ScM, 59; Univ Calif, Berkeley, PhD(struct eng), 65. *Prof Exp:* Lectr civil eng, Univ Birmingham, 59-61; preceptor, Civil Eng Dept, Columbia Univ, 65-66; asst prof, 66-77, PROF CIVIL ENG, UNIV WASH, 77- *Res:* Theory of elasticity; elastic wave propagation; geophysical problems including rheological behavior of ice and seismic phenomena. *Mailing Add:* Dept Civil Eng Univ Wash Seattle WA 98195

EVANS, ROGER LYNWOOD, b Ipswich, Eng, June 25, 28; m 54; c 3. INORGANIC CHEMISTRY. *Educ:* Oxford Univ, BA, 52, MA, 55, DPhil(natural sci), 58; Univ Minn, MS, 55. *Prof Exp:* Sr chemist, Cent Res Labs, 58-67, sr chemist, Nuclear Prod Lab, 67-78, patent liaison, 78-87, CORP PATENT LIAISON, MED PROD LAB, 3M CO, 87- *Concurrent Pos:* Instrnl videos, Intellectual Property Mgt, 87- *Res:* Elements of periodic groups III and IV; polymer chemistry; radiopharmaceuticals; intellectual property management. *Mailing Add:* Corp Patent Liaison 3M Ctr Minn Mining & Mfg Co St Paul MN 55144

EVANS, ROGER MALCOLM, b Coronation, Alta, May 27, 35; m; c 2. ANIMAL BEHAVIOR. *Educ:* Univ Alta, BSc, 60, MSc, 61; Univ Wis, PhD(behav of gulls), 66. *Prof Exp:* From asst prof to assoc prof, 66-78, PROF ZOOL, UNIV MAN, 78- *Mem:* Am Ornith Union; Animal Behav Soc; Can Soc Zool. *Res:* Behavioral ecology of colonial water birds. *Mailing Add:* Dept Zool Univ Man Winnipeg MB R3T 2N2 Can

EVANS, RONALD M, MEDICAL GENETICS. *Educ:* Univ Calif, Los Angeles, BA, 70, PhD(microbiol), 74. *Prof Exp:* Assoc, Dept Molecular Cell Biol, Rockefeller Univ, 75-78; asst res prof, Tumor Virol Lab, Salk Inst, 78-83, assoc prof, Molecular Biol & Virol Lab, 83-84, sr mem, 84-86, PROF, GENE EXPRESSION LAB, SALK INST, LA JOLLA, CALIF, 86-; INVESTR, HOWARD HUGHES MED INST, 85- *Concurrent Pos:* Am Cancer Soc fel, NIH, 75-78; mem, molecular biol study sect, NIH, 83-86, molecular neurobiol study sect, 84-85; assoc ed, J Neurosci, 85-90, Molecular Brain Res, 85-; adj prof, Dept Biol, Univ Calif, 85-, Dept Biomed Sci, 89-; Pew scholars prog biomed sci, Nat Adv Comt, 87-; Searle scholars prog comt, 89- *Honors & Awards:* Leslie L Bennett Lectr, Univ Calif, 87; Louis S Goodman & Alfred Gilman Award, Am Soc Pharmacol & Exp Therapeut, 88; Van Meter/Rorer Pharmaceut Prize, Am Thyroid Asn, 89; C P Rhoads Mem Award, Am Asn Cancer Res, 90; McGinnis Mem Lectr, Duke Univ, 91; Pfizer Lectr, Harvard Univ, 91; Gregory Pincus Mem Lectr, Worcester Found Exp Biol, 91; Mortimer B Lipsett Mem Lectr, NIH, 91. *Mem:* Nat Acad Sci. *Res:* Author or co-author of over 90 publications. *Mailing Add:* Howard Hughes Med Inst Salk Inst 10010 N Torrey Pines Rd La Jolla CA 92037

EVANS, RUSSELL STUART, wood chemistry, for more information see previous edition

EVANS, T(HOMAS) H(AYHURST), b Los Angeles, Calif, Apr 8, 06; m 45; c 3. CIVIL ENGINEERING. *Educ:* Calif Inst Technol, BS, 29, MS, 30. *Prof Exp:* Instr eng mech, Eng Sch, Yale Univ, 30-35; from asst prof to assoc prof civil eng & mech, Univ Va, 35-42, dir eng sci, mgt & war training, 40-42; prof civil eng & dir, Sch Eng, Ga Tech Univ, 45-49; dean, Colo State Univ, 49-63; dean, Sch Eng, Fresno State Univ, 63-73; RETIRED. *Concurrent Pos:* Active duty, US Army Corps of Engrs, 42-45; first dean, Asian Inst Technol, Bangkok, Thailand, 59-61. *Mem:* Am Soc Civil Engrs; Am Soc Eng Educ; Nat Soc Prof Engrs; Am Soc Mech Engrs; Inst Elec & Electronics Engrs. *Res:* Mechanics of plates; structural stresses; city planning; administration of engineering education and research; artificial rainmaking. *Mailing Add:* 922 La Tierra Dr Lake San Marcos CA 92069

EVANS, TAYLOR HERBERT, organic chemistry; deceased, see previous edition for last biography

EVANS, THOMAS EDWARD, b Springfield, Vt, July 22, 39; m 66; c 1. MOLECULAR BIOLOGY, RESEARCH ADMINISTRATION. *Educ:* DePauw Univ, BA, 61; Case Western Reserve Univ, PhD(biol), 67; Case Western Reserve Univ, MBA, 86. *Prof Exp:* asst prof radiol & microbiol, 67-81, DIR ADMIN SERV, SCH MED, CASE WESTERN RESERVE UNIV, 81- , DEPT RADIATION BIOL. *Mem:* Radiation Res Soc. *Res:* Nucleic acid metabolism in eukaryotes, especially as related to nuclear division cycles; molecular genetics of DNA replication in Physarum polycephalum; nonnuclear DNA metabolism. *Mailing Add:* Sch Med Case Western Reserve Univ 2109 Adelbert Rd Cleveland OH 44106

EVANS, THOMAS F(REDERICK), b New York, NY, Oct 18, 24; m 59. RESEARCH ADMINISTRATION. *Educ:* Univ Wash, BS, 45; Princeton Univ, PhD(chem eng), 50. *Prof Exp:* Res engr, Textile Res Inst, 50-53; develop engr, Gen Elec Co, 53-63; assoc prof, Univ Columbia, 63-64; asst prof chem eng, Pa State Univ, 65-71; chem design engr, 71-76, assoc sr res engr, 76-80, sr res specialist, Niagara Mohawk Power Corp, 80-86; CONSULT ENGR, 86- *Honors & Awards:* Fulbright lectr, Univ Seville, 64-65. *Mem:* Am Chem Soc; Am Inst Chem Engrs. *Res:* project management for power generation from fossil fuels using current and developmental technologies; development of environmental control processes related to power generation. *Mailing Add:* 15815 35th Ave NE Seattle WA 98155-6659

EVANS, THOMAS GEORGE, b Taylor, Pa, Feb 16, 34. SOFTWARE SYSTEMS. *Educ:* Princeton Univ, BA, 55; Mass Inst Technol, PhD(math), 63. *Prof Exp:* Res mathematician, US Air Force Cambridge Res Labs, 62-72; PRES, EVANS GRIFFITHS & HART, 72- *Mem:* Asn Comput Mach; Am Math Soc; Inst Elec & Electronics Engrs Comput Soc; Am Asn Artifical Intel. *Res:* Heuristic programming approach to artificial intelligence, emphasizing description and processing of complex patterns; development of facilities for convenient conversational use of computers, especially for program debugging. *Mailing Add:* Evans Griffiths & Hart Inc 55 Waltham St Lexington MA 02173

EVANS, THOMAS P, b West Grove, Pa, Aug 19, 21; m 47; c 4. TECHNOLOGY TRANSFER, PATENTS & LICENSING. *Educ:* Swarthmore Col, BS, 42; Yale Univ, MEng, 48. *Prof Exp:* Engr, Atomic Power Div, Westinghouse Elec Corp, Pa, 48-51; dir res & develop, AMF, Inc, NY, 51-60; dir res, O M Scott & Sons Co, Ohio, 60-62; vpres res & develop, W A Sheaffer Pen Co, 62-67; dir res, Mich Technol Univ, 67-80; dir res & prof bus admin, 80-86, MGT & PROD LIC CONSULT, BERRY COL, 86- *Concurrent Pos:* Mem, Mich Energy & Resource Res Asn, mem bd trustees, 75-80, mem exec comn, 77-80. *Mem:* Inst Elec & Electronics Engrs; Licensing Execs Soc; Am Phys Soc; Am Forestry Asn; Sigma Xi; AAAS; Soc Plastics Engrs; Nat Coun Univ Res Admin. *Res:* Management of research and development, teaching organization theory and management, new product/process and management counseling, nuclear power plants and reactor shielding; solar energy; water conversion; power generation, publication in field; patents. *Mailing Add:* 25 Wellington Way SE Rome GA 30161-9417

EVANS, THOMAS WALTER, b Tioga, NDak, May 27, 23; m 45; c 2. TECHNOLOGICAL ANALYSIS & FORECASTING. *Educ:* NDak State Univ, BS, 47; Univ Wis, PhD(phys chem), 52. *Prof Exp:* Metallurgist, Hanford Works, Gen Elec Co, 52-56, sr engr, Hanford Atomic Prods Oper, 56-67; res assoc, Pac Northwest Labs, Battelle Mem Inst, 67-68, prog consult, 68, assoc dept mgr, Fast Flux Test Facility Div, 68-70; staff consult, Wadco Corp, Westinghouse Elec Corp, 70-72, adv scientist, 72-76, mgr prod planning, 76-82, fel scientist, Westinghouse Hanford Co, 82-87; RETIRED. *Mem:* Am Soc Metals; Am Nuclear Soc; Sigma Xi. *Res:* X-ray crystallography; physical metallurgy of uranium; irradiation damage; volcanology; design, development and testing of reactor fuel elements; analysis of energy facility costs; analysis of trends in energy research, development, production and usage. *Mailing Add:* 2466 Pershing Ave Richland WA 99352

EVANS, TODD EDWIN, b Jackson, Mich, June 3, 47; m 81; c 4. TOKAMAK STABILITY & CONFINEMENT PLASMA PHYSICS. *Educ:* Wright State Univ, BS, 78; Univ Tex, Dallas, MS, 79; Univ Tex, Austin, PhD(elec eng & physics), 84. *Prof Exp:* Div prof eng, Jabsco Div, Int Telephone & Telegraph, 72-75; res asst, physics dept, Wright State Univ, 75-78; teaching asst physics, Univ Tex, 78-79; teaching asst elec eng, Univ Tex, Austin, 79-80, res assoc plasma, 80-83, res fel, 84-85; sr scientist, GA Technol Inc, 85-87; assoc staff scientist, 87-90, STAFF SCIENTIST, GEN ATOMICS INC, 90- *Concurrent Pos:* Consult, Aeropropulsion Lab, Wright Patterson Air Force Base, 78, Desktop Resources Inc, 87-90; instr physics, dept physics, Austin Community Col, 80-81, lectr, dept physics, Univ Tex Austin, 84-85; prin investr, US Dept Energy & CEA, France, 87- *Mem:* NY Acad Sci; Am Phys Soc; Inst Elec & Electronic Engr; Sigma Xi. *Res:* Plasma surface interactions in tokamaks and confinement physics in toroidal systems with magnetic perturbations which produce resonant helical divertor and stochastic boundary configuration; Plasma heating at radio and microwave frequencies. *Mailing Add:* Gen Atomics Inc PO Box 85608 San Diego CA 92186-9784

EVANS, TOMMY NICHOLAS, b Batesville, Ark, Apr 12, 22; m 45; c 1. MEDICINE. *Educ:* Baylor Univ, AB, 42; Vanderbilt Univ, MD, 45; Am Bd Obstet & Gynec, dipl. *Prof Exp:* From instr to prof obstet & gynec, Univ Mich, 49-65; dean, Sch Med, Wayne State Univ, 70-72, dir, C S Mott Ctr Human Growth & Develop, 72-83, prof obstet & gynec & chmn dept, 65-83; prof obstet & gynec, chief gynec & vchmn, 83-89, EMER PROF OBSTET & GYNEC, UNIV COLO, DENVER, 89- *Concurrent Pos:* Consult, Vet Admin Hosp, 56-65. *Mem:* AMA; Am Col Surg; Am Col Obstet & Gynec; Am Asn Obstet & Gynec; Am Gynec Soc. *Res:* Human reproduction; gynecologic endocrinology; obstetrics and gynecology. *Mailing Add:* 8146 E Whispering Wind Dr Scottsdale AZ 85255-2840

EVANS, TREVOR, b Wolverhampton, Eng, Dec 22, 25; m 53; c 4. MATHEMATICS. *Educ:* Oxford Univ, BA, 46, MA, 50, DSc, 60; Manchester Univ, MSc, 48. *Prof Exp:* Asst lectr pure math, Manchester Univ, 46-50; instr math, Univ Wis, 50-51; mem, Inst Advan Study, 52-53; res assoc, Univ Chicago, 53-54; from asst prof to prof, 54-80, head dept, 63-78, FULLER E CALLAWAY PROF MATH, EMORY UNIV, 80- *Concurrent Pos:* Vis prof, Univ Nebr, 59-60; mem comt exam, Math Achievement Test, Col Entrance Exam Bd, 64-69, chmn, 69-; vis prof, Calif Inst Technol, 68 & Technische Hochschule, Darmstadt, WGer, 75. *Mem:* Am Math Soc; Math Asn Am; Sigma Xi; London Math Soc. *Res:* Algebraic aspects of combinatorics; decision problems in algebra; varieties of algebras. *Mailing Add:* Dept Math & Comput Sci Emory Univ Atlanta GA 30322

EVANS, VIRGINIA JOHN, b Baltimore, Md, Mar 19, 13. CELL BIOLOGY, CANCER. *Educ:* Goucher Col, AB, 35; Johns Hopkins Univ, MSc, 40, ScD(biochem), 43. *Prof Exp:* Chem technician, Blood Chem Lab, Johns Hopkins Univ, 38-39; asst tissue culturist, Tissue Cult Lab, Dept Surg, Johns Hopkins Hosp, 40-41; asst to dermatologist, Med Sch, Johns Hopkins Univ, 41-42, instr biochem, Sch Hyg & Nurses Sch, 43-44; fel, Lab Biol, Nat Cancer Inst, 44-46, biologist, 46-64, head, Tissue Cult Sect, 64-73; RETIRED. *Concurrent Pos:* Mem bd gov & chmn exec comt, W Alton Jones Cell Sci Ctr, 70; guest scientist, Am Found Biol Res, 73- & Biomed Res Inst, Am Fedn Biol Res; rep, Tissue Cult Asn, Am Type Cult Collection, 80- *Mem:* Soc Develop Biol; Tissue Cult Asn (vpres, 68-72, pres, 72-74); Am Asn Cancer Res; Am Soc Exp Pathologists; Am Soc Cell Biol. *Res:* Nutritional dermatoses of rats; cell physiology with special reference to tissue culture in cancer; nutrition and endocrinology of tissue cultures; carcinogenesis studies in mammalian tissue culture. *Mailing Add:* 5824 Bradley Blvd Bethesda MD 20814-1128

EVANS, W E, b Moreland, Ga, Oct 20, 38. POWER. *Educ:* Georgia Tech, BEE, 62. *Prof Exp:* Elec engr, 67-72, proj eng, 72-83, group leader, air conditioning & elec eng, 83-88, MGR CONSTRUCTION & ELEC ENG, WEST POINT PEPPEREL INC, 88- *Mem:* Inst Elec & Electronics Engrs. *Mailing Add:* West Point Pepperel Inc Po Box 71 West Point GA 31833

EVANS, WARREN WILLIAM, b Wis, Nov 23, 21. PHYSICAL CHEMISTRY. *Educ:* Univ Wis, BS, 43, PhD(chem), 52. *Prof Exp:* Chemist res & develop, Carbide & Carbon Co, 47-49; sr res chemist photog, Photo Prod Dept, E I Du Pont de Nemours & Co, 52-70, res assoc, 70-75, res fel, 75-82; RETIRED. *Mem:* emer mem Am Chem Soc. *Res:* Photographic chemistry; mechanical properties of polymers. *Mailing Add:* 25 S Second Ave Highland Park New Brunswick NJ 08904-2238

EVANS, WAYNE ERROL, b Indianapolis, Ind, June 1, 51. CATALYST DEVELOPMENT. *Educ:* Butler Univ, BS(chem) & BS(physics), 73; Univ Calif, Los Angeles, PhD(chem), 79. *Prof Exp:* Lectr, Univ Calif, Los Angeles, 77-79; RES CHEMIST, SHELL DEVELOP CO, 81- *Concurrent Pos:* Vis res assoc, Ohio State Univ, 79-81. *Mem:* Am Chem Soc. *Res:* Organometallic chemistry, especially in the areas of organometallic photochemistry and bioinorganic chemistry (synthetic macrocyclic complex synthesis and mimicking of enzyme activity); development of novel catalyst systems for industrial application and petrochemical process development. *Mailing Add:* Shell Develop Co Westhollow Res Ctr PO Box 1380 Houston TX 77001

EVANS, WAYNE RUSSELL, physics, engineering, for more information see previous edition

EVANS, WILLIAM BUELL, b Monticello, Miss, June 5, 18; m 45; c 3. MATHEMATICS, METEOROLOGY. *Educ:* Southern Miss Univ, BS, 39; La State Univ, MS, 41; Mass Inst Technol, MS, 44; Univ Ill, PhD(math), 50. *Prof Exp:* Assoc prof math, Ga Inst Technol, 50-60; assoc prof eng, Univ Calif, Los Angeles & with Eng Gadjah Mada Proj, Indonesia, 60-64; assoc prof, Emory Univ, 65-68, dir comput ctr, 65-82, prof, 68-88, EMER PROF MATH & BIOMET, EMORY UNIV, 88- *Concurrent Pos:* Vis prof, Fed Univ Pernambuco, Recife, Brazil, 73-74. *Mem:* Am Math Soc; Math Asn Am; Soc Indust & Appl Math; Biomet Soc; Sigma Xi. *Res:* Numerical analysis; general methods of approximation; differential equation of potential distribution in a biological cell. *Mailing Add:* Math Dept Emory Univ Atlanta GA 30322

EVANS, WILLIAM GEORGE, b Swansea, Wales, Aug 11, 23; nat US; m 56; c 2. INSECT ECOLOGY. *Educ:* Cornell Univ, BS, 52, MS, 54, PhD(entom), 56. *Prof Exp:* Asst prof entom, Va Polytech Inst, 56-58; from asst prof to prof, 59-88, EMER PROF ENTOM, UNIV ALTA, 88- *Mem:* Entom Soc Am; Ecol Soc Am; Sigma Xi; Entom Soc Can; Int Soc Chem Ecol. *Res:* Insect ecology; insect behavior; rhythmic activities; marine insects; chemical ecology; habitat selection, chemosensory orientation. *Mailing Add:* Dept Entom Univ Alta Edmonton AB T6G 2E8 Can

EVANS, WILLIAM HARRINGTON, physical chemistry, for more information see previous edition

EVANS, WILLIAM JOHN, b Madison, Wis, Oct 14, 47. SYNTHETIC INORGANIC & ORGANOMETALLIC CHEMISTRY. *Educ:* Univ Wis, Madison, BS, 69; Univ Calif, Los Angeles, PhD(chem), 73. *Prof Exp:* Res assoc chem, Cornell Univ, 73-75; asst prof chem, Univ Chicago, 75-81, assoc prof, 82; ASSOC PROF, UNIV CALIF, IRVINE, PROF, DEPT CHEM, 83- *Mem:* Am Chem Soc. *Res:* Exploratory synthesis and systematic reaction chemistry of lanthanide metal complexes; metal alkoxide and oxide chemistry. *Mailing Add:* Dept Chem Univ Calif Irvine CA 92717

EVANS, WILLIAM L, b Calvert, Tex, Aug 28, 24; m 48; c 3. CYTOLOGY. *Educ:* Univ Tex, BA, 49, MA, 50, PhD(zool), 55. *Prof Exp:* From instr to assoc prof, 55-68, chmn biol dept, 68-72, PROF ZOOL, UNIV ARK, FAYETTEVILLE, 68- *Mem:* Am Genetic Asn; Sigma Xi; AAAS. *Res:* Chromosomal analysis of a montane population of Circotettix in which numerous chromosomal anomalies are present in intergeneric hybrids due to accidental introgression by a second species. *Mailing Add:* 111 Nolan Ave Fayetteville AR 72703

EVANS, WILLIAM PAUL, b Peoria, Ill, July 19, 22; m 49, 81; c 3. PHYSICS. *Educ:* Univ Ill, BS & MS, 47. *Prof Exp:* Instr math, Evening Sch, Bradley Univ, 47-78; mem staff eng, Res Dept, Caterpillar Tractor Co, 48-84; INSTR MATH, ILL CENT COL, 87- *Mem:* Am Phys Soc; Soc Automotive Eng; Am Soc Metals. *Res:* Fatigue, crack propagation; fracture mechanics; residual stress; radioisotope techniques; publications. *Mailing Add:* 2127 W Laura Peoria IL 61604

EVANS, WILLIAM R, SYMBIOSIS. *Educ:* Purdue Univ, PhD(biochem), 61. *Prof Exp:* PRIN SCIENTIST, BATTELLE KETTERING LAB, 63- *Mailing Add:* 1431 Meadow Lane Yellow Springs OH 45387

EVANS, WINIFRED DOYLE, b Logansport, La, Sept 10, 34; m 56; c 2. PHYSICS. *Educ:* La Polytech Inst, BS, 56; Univ Calif, Los Angeles, MS, 58; Univ NMex, PhD(physics), 67. *Prof Exp:* Asst prof physics, La Polytech Inst, 58-60; staff mem, Solid State Physics Dept, Langley Res Ctr, NASA, 60-61; staff mem, Physics Div, 61-79, group leader space sci group, 79-81, DEP DIV LEADER, EARTH & SPACE SCI DIV, LOS ALAMOS NAT LAB, 81- *Mem:* Am Geophys Union; Am Astron Soc. *Res:* X-ray emission from the solar corona; stellar x-ray sources; ultra-soft x-ray spectroscopy; gamma-ray astronomy. *Mailing Add:* Lanl Group ET-AC F PO Box 1663 Los Alamos NM 87545

EVANSON, ROBERT VERNE, b Hammond, Ind, Nov 3, 20; m 47; c 2. PHARMACY. *Educ:* Purdue Univ, BS, 47, MS, 49, PhD(pharm admin), 53. *Prof Exp:* Retail sales clerk, E C Minas Co, 40-41; apprentice pharmacist, Physician's Supply Co, 46; asst instr pharm, 47-48, Purdue Univ, from instr pharm admin to prof, 48-63, assoc head dept pharm pract, 81-86, head dept, 66-72, EMER PROF, PHARM ADMIN, PURDUE UNIV, 86- *Concurrent Pos:* Am Found Pharmaceut Educ fel; consult pharm mgt, Phadman. *Mem:* Fel Acad Pharmaceut Sci; Am Pharmaceut Asn; Am Asn Col Pharm; Nat Asn Retail Druggists. *Res:* Disintegration of compressed tablets; economic study of drug store operation; pharmacy management and administration. *Mailing Add:* 400 Lindberg Ave West Lafayette IN 47906

EVARD, RENE, biochemistry; deceased, see previous edition for last biography

EVARTS, RITVA POUKKA, b Vesilahti, Finland, Jan 27, 32; US citizen; m 71. EXPERIMENTAL PATHOLOGY. *Educ:* Vet Col, Finland, DVM, 60, PhD(muscular dystrophy), 65. *Prof Exp:* Instr biochem, Col Vet Med, 60-71; vis assoc nutrit biochem, Nat Inst Arthritis & Metab Dis, NIH, 71-75; vis scientist path, Carcinogen Metab & Toxicol Br, 75-81, VET MED OFFICER, NAT CANCER INST, NIH, 81- *Mem:* Am Asn Can Res. *Res:* experimental pathology. *Mailing Add:* Lab Exp Carcinogenis Bldg 37 Rm 3B17 Nat Cancer Inst Bethesda MD 20892

EVATT, BRUCE LEE, b Wayne, Okla, June 4, 39; m 60; c 1. MOLECULAR CELL BIOLOGY, MECHANISMS OF THROMBOSIS & HEMOSTASIS. *Educ:* Univ Okla, MD, 64. *Prof Exp:* Osler intern & res med, Sch Med, Johns Hopkins Univ, 64-66, fel hematol, 66-68, fel med, Med & Hematol, 70-72, instr med & chief resident med, 72-73, asst prof med, 73-76; clin instr med hematol, Sch Med, Emory Univ, 68-70; EIS officer cancer, 68-70, DIR HEMATOL DIV & DIR IMMUNOL, ONCOL & HEMATOL DIV, CTRS DIS CONTROL, 76- *Concurrent Pos:* Asst physician, Johns Hopkins Hosp, 73-76; investr, Howard Hughes Res Ctr, 73-76; assoc prof med, Emory Univ, 76-; mem, Nat Comt Clin & Lab Standards, 77-, Subcomt Clin Lab Standards, Am Soc Hemat, 79-83 & Nat Blood Resource Educ Prog, 88-; chmn, AIDS Educ Comt, World Fedn Hemophilia, 88-, med secy, 88-; mem, Subcomt Lab Standards & Subcomt Coagulation, Am Soc Hemat. *Honors & Awards:* Murray Thelin Award Distinguished Res, Nat Hemophilia Found, 85, L Michael Kuhn Award Gov Sci, 85. *Mem:* Am Soc Hematol; Am Fedn Clin Res. *Res:* Control mechanism of cell growth and differentiation; pathogenesis and molecular biology of cancer; prevention of thrombotic and hemorrhagic diseases; prevention of HIV infection transmitted by blood and blood products. *Mailing Add:* Immunol Oncol & Hematol Dis Ctrs Dis Control 1600 Clifton Rd NE Atlanta GA 30333

EVCES, CHARLES RICHARD, b East Liverpool, Ohio, Dec 31, 38; m 62; c 2. ENGINEERING MECHANICS, MECHANICAL ENGINEERING. *Educ:* Univ Notre Dame, BSME, 60, MSME, 62; Univ WVa, PhD(eng), 67. *Prof Exp:* Asst prof, 67-74, ASSOC PROF MECH ENG, UNIV ALA, TUSCALOOSA, 74- *Mem:* Am Soc Mech Engrs. *Res:* Dynamics and vibrations; acoustical noise control. *Mailing Add:* Dept Mech Eng Univ Ala Tuscaloosa AL 35487

EVELAND, HARMON EDWIN, b Urbana, Ill, Feb 9, 24; m 44; c 4. GEOLOGY. *Educ:* Univ Ill, BS, 47, MS, 48, PhD(geol), 50. *Prof Exp:* Asst prof geol, Univ Tenn, 50-51; PROF GEOL & HEAD DEPT, LAMAR UNIV, 51- *Mem:* Geol Soc Am; Soc Econ Paleontologists & Mineralogists; Sigma Xi. *Res:* Pleistocene stratigraphy; geomorphology; physiography. *Mailing Add:* 4650 Baywood Lane Beaumont TX 77706

EVELAND, WARREN C, medical bacteriology; deceased, see previous edition for last biography

EVELEIGH, DOUGLAS EDWARD, b Croydon, Eng, Dec 6, 33; m 62; c 2. MICROBIOLOGY. *Educ:* Univ London, BSc, 56; Univ Exeter, PhD(mycol), 59. *Prof Exp:* Fel, Nat Res Coun, Halifax, 59-61; Nat Acad Sci-Nat Res Coun vis scientist, US Dept Army, Natick, Mass, 61-63; res assoc bact, Univ Wis, 63-65; assoc res officer, Nat Res Coun, Sask, 65-70; PROF, DEPT BIOCHEM & MICROBIOL, RUTGERS UNIV, 70- *Concurrent Pos:* Assoc ed, Can Soc Microbiol, 70-73 & Europ J Appl Microbiol, 78- *Mem:* Am Soc Microbiol; Brit Soc Gen Microbiol; Can Soc Microbiol; Mycol Soc Am; Sigma Xi. *Res:* Microbial polysaccharases; non-microbial polysaccharides; leguminous symbiotic nitrogen fixation; fungal ecology. *Mailing Add:* Dept Biochem & Microbiol Cook Col Rutgers Univ New Brunswick NJ 08903

EVELEIGH, ELDON SPENCER, b Deer Lake, Nfld, Nov 14, 50; m 73; c 2. INSECT PARASITOLOGY, INSECT PATHOLOGY. *Educ:* Mem Univ Nfld, BSc, 71, MSc, 74; Univ Toronto, PhD(ecol & acarology), 79. *Prof Exp:* Postdoctoral fel ecol, Univ BC, 79-80; res scientist, Agr Can, 80-82; RES SCIENTIST, FORESTRY CAN, 82- *Concurrent Pos:* Res assoc, Dept Forest Resources, Univ NB, 88- *Res:* Population dynamics of insect population, particularly the eastern spruce budworm; host-parasite and predator-prey interactions; insect pathology; integrated pest management systems; taxonomy and nematology. *Mailing Add:* Forestry Can-Maritime Region Fredericton NB E3B 5P7 Can

EVELEIGH, VIRGIL W(ILLIAM), b Dexter, NY, Aug 20, 31; m 56; c 3. ELECTRICAL ENGINEERING. *Educ:* Purdue Univ, BS, 57, MS, 58, PhD(elec eng), 61. *Prof Exp:* Technician commun systs, Gen Elec Co, NY, 53-54, field serv engr, radar systs, 54, engr control systs, 61-64; chmn dept, 79-83, PROF ELEC & COMPUT ENG, SYRACUSE UNIV, 64-, DIR CONTINUING EDUC ENG, 86- *Concurrent Pos:* Consult, JDR Syst Corp, 80- *Mem:* Inst Elec & Electronics Engrs; Am Asn Univ Prof; Sigma Xi. *Res:* Control systems; computational methods for optimization; adaptive control; radar; signal processing. *Mailing Add:* 2-220 Sci Tech CASE Ctr Syracuse Univ Syracuse NY 13244-4100

EVEN, WILLIAM ROY, JR, b Great Falls, NH, Jan 3, 52; m 75; c 3. POLYMER CHEMISTRY. *Educ:* Univ Utah, BS(chem) & BS(mat sci), 74; Northwestern Univ, PhD(mat sci), 79. *Prof Exp:* SR MEM TECH STAFF POLYMERS, SANDIA NAT LABS, 79- *Honors & Awards:* Hovor Award, Dept Educ, 90. *Mem:* Am Chem Soc; Sigma Xi. *Res:* Implications of molecular conformation and organization on macroscopic polymer properties. *Mailing Add:* Sandia Nat Labs PO Box 969 Livermore CA 94550

EVENS, F MONTE, b Herculaneum, Mo, Jan 21, 32; m 52; c 5. ANALYTICAL CHEMISTRY, SPECTROSCOPY. *Educ:* Southeast Mo State Col, BS, 55; Iowa State Univ, MS, 59, PhD(anal chem), 62. *Prof Exp:* Chem technician, Mallinckrodt Chem Co, 53; jr res assoc, Ames Lab, AEC, Iowa State Univ, 54-62; res chemist, Procter & Gamble Co, 62-63; res scientist, 63-68, ASSOC DIR, CONOCO INC, 68- *Mem:* Am Chem Soc; Soc Appl Spectros. *Res:* Instrumental methods of chemical analysis; atomic and molecular spectroscopy; gas chromatography; chemical separations. *Mailing Add:* 2716 Larchmont Ponca City OK 74604

EVENS, LEONARD, b Brooklyn, NY, June 28, 33; m 58; c 3. MATHEMATICS. *Educ:* Cornell Univ, AB, 55; Harvard Univ, AM, 56, PhD(math), 60. *Prof Exp:* Instr math, Univ Chicago, 60-61; asst prof, Univ Calif, Berkeley, 61-64; assoc prof, 64-69, PROF MATH, NORTHWESTERN UNIV, 69- *Mem:* Am Math Soc; Math Asn Am. *Res:* Homological algebra; group theory. *Mailing Add:* Dept Math Northwestern Univ Evanston IL 60208

EVENS, MARTHA WALTON, b Boston, Mass, Jan 1, 35; m 58; c 3. COMPUTER SCIENCE, COMPUTATIONAL LINGUISTICS. *Educ:* Bryn Mawr Col, AB, 55; Radcliffe Col, AM, 57; Northwestern Univ, PhD(comput sci), 75. *Prof Exp:* Instr math, Calif State Univ, Hayward, 61-64 & Nat Col Educ, Evanston, 66-68; lectr, Northwestern Univ, 65-66, instr Dept Comput Sci, 72-74; asst prof, 75-81, assoc prof, 81-86, PROF COMPUT SCI ILL INST TECHNOL, 86- *Concurrent Pos:* Assoc ed, Am Math Monthly & Am J Computational Linguistics; Fulbright fel, Univ Paris, 55-56; prin investr, NSF Awards, Info Sci Div & Math & Comp Sci Directorate; co-ed, Proc Nat Comput Conf, 81. *Mem:* Asn Comput Mach; Asn Comp Ling (vpres, 83, pres, 84); Math Asn Am; Inst Elec & Electronics Engrs Comput Soc; Am Asn Artificial Intel. *Res:* Artificial intelligence; natural language processing; programming languages; compilers; computational lexicography. *Mailing Add:* Dept Comput Sci Ill Inst Technol Chicago IL 60616

EVENSEN, JAMES MILLARD, geology; deceased, see previous edition for last biography

EVENSEN, KATHLEEN BROWN, b Tupper Lake, NY; m; c 2. POSTHARVEST PHYSIOLOGY, PLANT SENESCENCE. *Educ:* State Univ NY Col Potsdam, BA, 74; Univ NH, MS, 76; Univ Fla, PhD(hort), 78. *Prof Exp:* Res assoc fel, Univ Mo, 79-80; asst prof, 80-88, ASSOC PROF, POSTHARVEST PHYSIOL, PA STATE UNIV, 88- *Mem:* Am Soc Hort Sci; Am Soc Plant Physiologists; Japanese Soc Plant Physiologists; Scand Soc Plant Physiol. *Res:* Ethylene sensitivity; quality and storage of fruits and vegetables; flower senescence and abscission. *Mailing Add:* Dept Hort Pa State Univ University Park PA 16802

EVENSEN, THOMAS JAMES, b Menominee, Mich, Jan 21, 33; m 55; c 2. ORGANIC CHEMISTRY. *Educ:* Augustana Col, Ill, AB, 55; Univ Minn, Minneapolis, PhD(org chem), 59. *Prof Exp:* Teaching asst org chem, Univ Minn, Minneapolis, 55-57; sr chemist, 59-67, res specialist, 68-70, res supvr, 70-73, MGR MAT DEVELOP, COPYING PRODS DIV, 3M CO, 73- *Mem:* Am Chem Soc. *Res:* New product designs; imaging systems. *Mailing Add:* 708 Harriet Dr Stillwater MN 55082

EVENSON, DONALD PAUL, b Story City, Iowa, Oct 30, 40; m; c 3. CELL BIOLOGY, ENVIRONMENTAL HEALTH. *Educ:* Augustana Col, BA, 64; Univ Colo, PhD(cell biol), 68. *Prof Exp:* Fel, Inst Molecular Biophys, Fla State Univ, 68-70; staff scientist electron micros & virol, Union Carbide Res Inst, NY, 70-72; assoc automated cytol, Sloan Kettering Inst Cancer Res, 72-83; asst prof biol, Grad Sch Med Sci, Cornell Univ, 72-83; PROF CHEM, SDAK STATE UNIV, 83- *Mem:* Am Soc Cell Biol; Int Soc Anal Cytol; Am Soc Andrology; Am Fertility Soc; Soc Study Reproduction; Environ Mutagen Soc. *Res:* Controls of DNA synthesis and cell division; chromatin structure, male fertility, flow cytometry, sperm biochemistry. *Mailing Add:* Dept Chem SDak State Univ Box 2170 Brookings SD 57007

EVENSON, EDWARD B, b Milwaukee, Wis, Dec 30, 42; m 63; c 1. GLACIAL GEOLOGY, GEOMORPHOLOGY. *Educ:* Univ Wis, Milwaukee, BS, 65, MS, 69; Univ Mich, PhD(geol), 72. *Prof Exp:* Sr res geologist, Exxon Prod Res Lab, Exxon Corp, 72-73; from asst prof to assoc prof, 73-85, PROF GEOL, LEHIGH UNIV, 85- *Concurrent Pos:* Dir, Environ Sci & Resource Mgt, Lehigh Univ, 73-; res fel, Univ Western Ont, 75-76. *Mem:* Geol Soc Am; Sigma Xi; Am Asn Quaternary Geologists; Asn Geol Argentina. *Res:* Glacial geology of the Great Lakes Region and northeast Pennsylvania; deglaciation of Idaho Rockies; sedimentology and fabric of glacial deposits; glacial geomorphology; glacial history of Argentina; mineral exploration in gluciated areas. *Mailing Add:* Dept Geol Sci Lehigh Univ Bethlehem PA 18015

EVENSON, KENNETH MELVIN, b Waukesha, Wis, June 5, 32; m 55; c 4. ATOMIC PHYSICS, MOLECULAR PHYSICS. *Educ:* Mont State Col, BS, 55; Ore State Univ, MS, 60, PhD(physics), 64. *Prof Exp:* PHYSICIST, NAT INST STANDARDS & TECHNOL, 63- *Mem:* Am Phys Soc. *Res:* Quantum electronics; atomic and molecular structure; chemical kinetics; electron paramagnetic resonance; microwave and optical spectroscopy. *Mailing Add:* Nat Inst Standards & Technol 325 Broadway Boulder CO 80303

EVENSON, MERLE ARMIN, b La Crosse, Wis, July 27, 34; m 57; c 2. ANALYTICAL CHEMISTRY, TOXICOLOGY. *Educ:* Univ Wis-La Crosse, BS, 56; Univ Wis-Madison, MS(guid) & MS(sci educ), 60, PhD(anal chem), 66; Am Bd Clin Chemists, dipl. *Prof Exp:* From instr to asst prof med, Univ Wis-Madison, 65-69, asst dir clin labs, Univ Hosps, 65-67, dir clin chem, 67-69; vis lectr biol chem & NIH spec res fel, Harvard Med Sch, 69-71; assoc prof, 71-75, dir, Toxicol Labs, Univ Hosps, 71-85, PROF MED, UNIV WIS-MADISON, 75- *Concurrent Pos:* Consult, Instrument Prod Div, E I du Pont de Nemours & Co, Inc, 67-71, Nat Inst Gen Med Sci, NIH, 68-72, Anal Div, Oak Ridge Nat Lab, AEC, 69-73, Med Devices Sect, Fed Drug Admin, 75-81 & Millipore Corp, 79-82; nat res coun eval panel anal chem, Nat Bur Standards, 81-84. *Mem:* AAAS; Am Chem Soc; Am Asn Clin Chem; Acad Clin Lab Physicians & Sci; Sigma Xi. *Res:* Development of analytical procedures for drugs and trace elements and the relationship of these results to human health and disease; physical-chemistry studies of structure-function relationships in enzymes and changes of metalloproteins and metalloenzymes in human disease. *Mailing Add:* Dept Med & Path Univ Wis 600 Highland Ave Madison WI 53792

EVENSON, PAUL ARTHUR, b Chicago, Ill, Jan 27, 46; m 68; c 3. COSMIC RAY PHYSICS, SPACE PLASMAPHYSICS. *Educ:* Univ Chicago, BS, 67, MS, 68, PhD(physics), 72. *Prof Exp:* Res assoc, Enrico Fermi Inst, Univ Chicago, 72-76, sr res assoc cosmic ray physics, 76-83; assoc prof, 83-90, PROF, BARTOL RES INST, UNIV DEL, 90- *Concurrent Pos:* NATO fel, Danish Space Res Inst, 73-74; vis scientist, Max Planck Inst, Garching, 81. *Mem:* Am Phys Soc; Am Geophys Union; Am Astron Soc. *Res:* Experimental study cosmic radiation using high altitude balloon and satellite instrumentation. *Mailing Add:* Bartol Res Inst Univ Del Newark DE 19716

EVENSON, WILLIAM EDWIN, b Martinez, Calif, Oct 12, 41; m 64; c 5. THEORETICAL SOLID STATE PHYSICS. *Educ:* Brigham Young Univ, BS, 65; Iowa State Univ, PhD(physics), 69. *Prof Exp:* Res assoc physics, Univ D[illegible], 69-70, from asst prof to assoc prof, Brigham Young Univ, 70-79, assoc dir gen educ, 80-81, dir, 81-82, dean, 82-84, assoc acad vpres, 85-89, PROF PHYSICS, BRIGHAM YOUNG UNIV, 79- *Concurrent Pos:* NSF fel, 68-69; vis colleague, Univ Hawaii, 77-78; vis prof, Ore State Univ, 89-90. *Mem:* Am Phys Soc; Am Asn Physics Teachers; Sigma Xi. *Res:* Theory of magnetism in metals; theory of dilute magnetic alloys; theory of melting; applications of physics in ecology; theory of defects in solids; theory of dynamic effects in hypertine interactions. *Mailing Add:* Dept Physics Brigham Young Univ Provo UT 84602

EVERARD, NOEL JAMES, b New Orleans, La, Dec 24, 23; m 50; c 2. CIVIL ENGINEERING, MECHANICS. *Educ:* La State Univ, BS, 48, MS, 57; Tex A&M Univ, PhD(civil eng), 62. *Prof Exp:* From instr to asst prof civil eng, La State Univ, 48-60; design engr, David W Godat & Assocs, Consult Engrs, La, 49-53, chief engr, 53-56; from assoc prof to prof eng mech, 60-72, PROF CIVIL ENG & CHMN DEPT, UNIV TEX, ARLINGTON, 72- *Concurrent Pos:* Consult, William Dawson, Civil Engr, La, 56-60, J Weldon Hunnicut, Consult Engr, Tex, 60-, Freese, Nichols & Endress, 63-, Young-Hadawi Assocs, 70-, Welton-Becket Assocs, 75- & Young-Hadawi, 74- *Mem:* Am Soc Civil Engrs; Am Concrete Inst; fel Sigma Xi. *Res:* Theoretical and applied mechanics; column design and torsion in beams of reinforced concrete; computer methods in structural engineering. *Mailing Add:* Dept Civil Eng Univ Tex Arlington TX 76010

EVEREST, F(REDERICK) ALTON, b Gaston, Ore, Nov 22, 09; m 34; c 3. ELECTRICAL ENGINEERING, ACOUSTICS. *Educ:* Ore State Col, BS, 32; Stanford Univ, EE, 36. *Hon Degrees:* DSc, Wheaton Col, 59. *Prof Exp:* TV engr, Don Lee Broadcasting Co, 36; asst prof elec eng, Ore State Col, 36-41; engr & sect chief, Univ Calif Div War Res, US Navy Radio & Sound Lab, San Diego, 41-45; assoc dir, Moody Inst Sci, Los Angeles, 45-53, dir sci & prod, 53-70; sr lectr & head div radio, TV & cinematog, Dept Commun Hong Kong Baptist Col, 70-73; acoust consult, 73-88; RETIRED. *Concurrent Pos:* Tech writing, 73- *Mem:* Audio Eng Soc; Acoust Soc Am; sr mem Inst Elec & Electronics Engrs; fel Am Sci Affiliation; fel Soc Motion Picture & TV Engrs. *Res:* Television video amplifiers; high efficiency radio-telephone transmitters; electric fence controllers; directional broadcast antennae; propagation of underwater sound; underwater sounds of biological origin; scientific films; studio acoustics. *Mailing Add:* 2661 Tallant Rd No 619 Santa Barbara CA 93105

EVERETT, ALLEN EDWARD, b Kansas City, Mo, July 8, 33; m 66. ELEMENTARY PARTICLE PHYSICS. *Educ:* Princeton Univ, AB, 55; Harvard Univ, AM, 56, PhD(physics), 61. *Prof Exp:* From asst prof to assoc prof, 60-76, chmn dept, 77-80, PROF PHYSICS, TUFTS UNIV, 76- *Mem:* Am Phys Soc. *Res:* Theory of elementary particles, cosmology. *Mailing Add:* Dept Physics & Astron Tufts Univ Medford MA 02155

EVERETT, ARDELL GORDON, b Cambridge, Mass, July 27, 37; m 60; c 3. GEOLOGY, HYDROGEOLOGY. *Educ:* Cornell Univ, AB, 59; Univ Okla, MS, 62; Univ Tex, Austin, PhD(geol & geochem), 68. *Prof Exp:* Jr geologist, Shell Oil Co, Colo, 60 & Texaco Inc, 62; teaching asst geol & geophys, Univ Okla, 61-62; from instr to asst prof geol, Ohio State Univ, 67-69; staff asst water qual & res, Dept Interior, Washington, DC, 69-70, actg dep asst secy appl sci & eng, 70, dep asst secy appl sci, 70, dir, Off Tech Anal, Environ Protection Agency, 70-74; tech adv & dir, Regulatory Litigation Dept, Am Petroleum Inst, 74-77; PRES, EVERETT & ASSOC, 78- *Concurrent Pos:* Consult, Am Petrol Inst, Dept Mineral & Energy, Papua New Guinea & Rocky Nat Oil & Gas Asn; vis assoc prof petrol geol, Dept of Geol, Univ Md, 85. *Honors & Awards:* Franklin Gillian Prize, Univ Tex Libr, 67. *Mem:* Fel Geol Soc Am; Am Inst Prof Geol; Geochem Soc; Am Asn of Petrol Geol Health; Soc Min Eng; fel AAAS. *Res:* Economic geology; applied geochemistry; environmental geology and resources management; sedimentology; geochemistry of petroleum and economic mineral formation; environmental geochemistry. *Mailing Add:* 203 Dale Dr Rockville MD 20850

EVERETT, GLEN EXNER, b St George, Utah, Oct 3, 34; m 58; c 3. SOLID STATE PHYSICS. *Educ:* Univ Utah, BA, 56, Univ Chicago, MS, 57, PhD(physics), 61. *Prof Exp:* Actg asst prof, 60-62, from asst prof to assoc prof, 62-74, PROF PHYSICS, UNIV CALIF, RIVERSIDE, 74- *Concurrent Pos:* Consult, US Naval Weapons Ctr, Calif, 63- *Mem:* Am Phys Soc. *Res:* Cyclotron resonance in metals; ferro and antiferromagnetic resonance in binary rare earth compounds. *Mailing Add:* Dept Physics Univ Calif Riverside CA 92521

EVERETT, GUY M, b Missouri Valley, Iowa, Feb 6, 15. NEUROPHARMACOLOGY, PSYCHOPHARMACOLOGY. *Educ:* Univ Iowa, BA, 37; Univ Md, PhD(physiol), 43. *Prof Exp:* Sect head neuropharmacol, Abbott Labs, 43-68, res scientist, 68-71; LECTR PHARMACOL, SCH MED, UNIV CALIF, SAN FRANCISCO, 72- *Concurrent Pos:* Consult, Res Div, Abbott Labs, 71-76. *Mem:* Am Soc Pharmacol & Exp Therapeut; Am Col Neuropsychopharmacol; Int Col Neuropsychopharmacol. *Res:* Neuropharmacology, biogenic amines, dopamine in parkinsonism and behavior; interrelation of brain biogenic amines; antiepileptic drugs. *Mailing Add:* Dept Pharmacol Sch Med Univ Calif 281 Castro St San Francisco CA 94114

EVERETT, HERBERT LYMAN, b New Haven, Conn, Aug 9, 22; m 44; c 2. PLANT BREEDING, GENETICS. *Educ:* Yale Univ, BA, 44, MS, 47, PhD(genetics), 49. *Prof Exp:* Res asst, Dept Plant Breeding, Conn Agr Exp Sta, New Haven, 49-52; from asst prof to assoc prof, 52-64, dir resident instr, Col Agr, 66-77, PROF PLANT BREEDING, CORNELL UNIV, 64- *Concurrent Pos:* Proj leader & vis prof, Cornell Univ Grad Educ Prog, Col Agr, Univ Philippines, 64-65. *Mem:* AAAS; Genetics Soc Am; Am Soc Agron; Sigma Xi. *Res:* Plant breeding research and genetics research in corn. *Mailing Add:* 520 Bradfield Hall Col Agr Cornell Univ Ithaca NY 14850

EVERETT, JAMES LEGRAND, III, b Charlotte, NC, July 24, 26. MECHANICAL ENGINEERING. *Educ:* Pa State Univ, BS, 48, MS, 49; Mass Inst Technol, MS, 59. *Prof Exp:* Instr mech eng, Pa State Univ, 48-50; exec vpres, 68-71, pres, 71-82, CHMN & CHIEF EXEC OFFICER, PHILADELPHIA ELEC CO, 82- *Mem:* Nat Acad Sci; fel Am Soc Mech Eng; Am Nuclear Soc; Inst Elec & Electronics Engrs. *Mailing Add:* 743 Mancill Rd Wayne PA 19087

EVERETT, JOHN WENDELL, b Ovid, Mich, Mar 5, 06; m 32; c 2. ANATOMY, NEUROENDOCRINOLOGY. *Educ:* Olivet Col, AB, 28; Yale Univ, PhD(zool), 32. *Hon Degrees:* DSc, Olivet Col, 84. *Prof Exp:* Instr biol, Goucher Col, 30-31; instr anat, 32-35, assoc, 35-39, from asst prof to prof, 39-76, EMER PROF ANAT, DUKE UNIV, 76- *Concurrent Pos:* Vis prof, Univ Calif, Los Angeles, 52 & Univ Tenn, 54; sect ed, Biol Abstr, 56-71; assoc ed, Anat Rec, 57-63 & Biol of Reproduction, 68-69; vis lectr & consult, Lab Reproduction & Lactation, Mendoza, Arg, 82. *Honors & Awards:* Carl G Hartman Lect Award, Soc Study Reprod, 71; Fred Conrad Koch Medal, Endocrine Soc, 73; Sir Henry Dale Medal, Soc Endocrinol, 77; Henry Gray Award, Am Asn Anatomists, 85; Fel, Am Acad Arts & Sci. *Mem:* Am Asn Anatomists (pres, 77-78); Endocrine Soc; Am Physiol Soc; hon mem Int Soc Neuroendocrinol; fel NY Acad Sci; fel AAAS. *Res:* Physiology of reproduction; endocrinology of the ovary and hypophysis; hypothalamic control of hypophysis. *Mailing Add:* Box 2917 Duke Univ Med Ctr Durham NC 27710

EVERETT, K R, b Corning, NY, Jan 8, 34; m 56; c 2. GEOLOGY. *Educ:* Univ Buffalo, BA, 55; Univ Utah, MS, 58; Ohio State Univ, PhD(geol), 63. *Prof Exp:* Polar & mt geologist, US Army Natick Labs, 64-67; from asst prof to assoc prof, 67-78, PROF AGRON, COL AGR, OHIO STATE UNIV, 78-, RES SCIENTIST, BYRD POLAR RES CTR, 61 - *Mem:* Fel Arctic Inst NAm. *Res:* Geomorphology and pedology; genesis, classification and distribution of polar and mountain soils, primarily Alaska, Canadian Arctic, Greenland and Antarctica; slope morphology, mass wasting, permafrost and patterned ground development; active layer hydrology. *Mailing Add:* Dept Agron 2021 Coffey Rd 422C Ohio State Univ Columbus OH 43210

EVERETT, KENNETH GARY, b Vicksburg, Miss, Nov 25, 42. INORGANIC CHEMISTRY. *Educ:* Washington & Lee Univ, BS, 64; Stanford Univ, PhD(chem), 68. *Prof Exp:* Asst prof chem, Northeast La State Col, 68-69; asst prof, 69-77, PROF CHEM, STETSON UNIV, 77- *Concurrent Pos:* Consult, Columbian Carbon Co, La, 69- *Mem:* Am Chem Soc. *Res:* Chemical kinetics and mechanisms of inorganic reactions. *Mailing Add:* Dept Chem Stetson Univ De Land FL 32720-3799

EVERETT, LORNE GORDON, b Thunder Bay, Ont, Jan 1, 43; m 69; c 2. HYDROLOGY, WATER QUALITY. *Educ:* Lakehead Univ, BSc, 66, Hons, 68; Univ Ariz, MS, 69, PhD(hydrol), 72; Tucson Gen Hosp, ASMT, 70. *Prof Exp:* Chemist water qual, Great Lakes Paper Co, Can, 66-67; asst prof hydrol, Univ Ariz, 72-74, Off Water res grant, 72-73; prof staff hydrol, Ctr Advan Studies, Kaman Tempo, 74-76, mgr, Water Resources Prog, Gen Elec-Tempo, 76-78, mgr, Advan Energy Prog & Int Prog, 78-81, MGR, NAT RESOURCES PROG, KAMAN TEMPO, 81- *Concurrent Pos:* Collabr, Nat Park Serv, 71-74; consult, Bell Eng Co, 71-74, Col Eng, Utah State Univ, 72-73, Develop & Assistance Co, 73-74 & CODECU Int Inc, 73-74; proj mgr groundwater monitoring strategies, Environ Protection Agency, 73-76, prog mgr, 76-81; Invited dir, UNESCO Int Symp on water pollution control, Paris, 83. *Mem:* Am Water Resources Asn; Am Med Lab Asn; Am Asn Univ Professors; Am Soc Civil Eng; Int Water Resources Asn. *Res:* Water quality investigations utilizing monitoring, aquatic ecosystem modeling, eutrophication process studies, physical, biological and chemical reservoir models, remote sensing, saturated and unsaturated flow equations; published over 80 professional journals. *Mailing Add:* Kaman Tempo 1312 Portesuello St Santa Barbara CA 93105

EVERETT, MARK ALLEN, b Oklahoma City, Okla, May 30, 28; c 1. MEDICINE, DERMATOLOGY. *Educ:* Univ Okla, BA, 47, MD, 51; Tulane Univ, 52; Am Bd Dermat, dipl, 58. *Prof Exp:* Intern pediat, Univ Mich, 51-52, resident dermat, 54-56, instr, 56-57; from instr to assoc prof, Med Sch, Univ Okla, 57-68, dir res labs, 59, dir resident training & res, 63-64, regents prof dermat & path, 80, PROF DERMAT, MED SCH, UNIV OKLA, 68-, HEAD DEPT, 64- *Concurrent Pos:* Consult, Vet Admin Hosp, Oklahoma City & St Anthony Hosp, Oklahoma City; chmn fac bd, Univ Okla, 74-90; dir & trustee, Am Bd Dermat, 86-; prof & interin head path,Univ Hosp, 78-82, dermatologist-in-chief, 70- *Mem:* AMA; Am Acad Dermat; Am Dermat Asn; Soc Invest Dermat; Asn Prof Dermat (pres, 76-78); Sigma Xi; Am Soc Dermatopath (pres, 80-81). *Res:* Cutaneous photobiology; ultraviolet erythema; clinical dermatology; dermatopathology; medical education and organization; pigment biology; lymphoma; ontanios. *Mailing Add:* 619 NE 13th St Oklahoma City OK 73104

EVERETT, PAUL HARRISON, soil fertility, vegetable crops, for more information see previous edition

EVERETT, PAUL MARVIN, b Toledo, Ohio, Mar 15, 40; m 68; c 2. SOLID STATE PHYSICS. *Educ:* Case Inst Technol, BS, 62; Case Western Reserve Univ, PhD(solid state physics), 68. *Prof Exp:* Res assoc physics, La State Univ, Baton Rouge, 68-71, admin asst, 71-72; asst prof physics, Univ Ky, 72-79; mem tech staff, Tex Instruments, Inc, 79-83; unit mgr electronics, McDonnell Douglas, 83-85, sect mgr, 85-87, br mgr, 87-91; CHIEF SCIENTIST, MAGNAVOX NEW ENGLAND RES CTR, 91- *Mem:* Am Phys Soc; sr mem Inst Elec & Electronics Engrs. *Res:* Properties of electrons in metals; study of Fermi surfaces via the de Haas-van Alphen and galvanomagnetic effects; electron bombarded countercurrent distribution imagers, infrared component identification imagers. *Mailing Add:* 1608 Forestview Ridge Lane Manchester MO 63021

EVERETT, ROBERT LINE, mechanical engineering, for more information see previous edition

EVERETT, ROBERT R(IVERS), b Yonkers, NY, June 26, 21; m 44, 72, 82; c 6. ELECTRICAL ENGINEERING. *Educ:* Duke Univ, BS, 42; Mass Inst Technol, MS, 43. *Hon Degrees:* DEng, Northeastern Univ, 86. *Prof Exp:* Res & develop engr servomechanism lab, Mass Inst Tech, 42-51, assoc dir digital comput lab & assoc head digital comput div, Lincoln Lab, 51-56, head digital comput div, 56-58; tech dir command & control systs, Mitre Corp, 58-59, vpres tech opers, 59-69, exec vpres, 69, pres, 69-86; RETIRED. *Concurrent Pos:* Consult, Air Defense Panel, President's Sci Adv Comt, 59-60 & Air Force Systs Command Range Tech Adv Group, 62-68; mem, Air Traffic Control Adv Comt, US Dept Transp, Off Dir Defense Res & Eng, Systs Eng Mgt Panel & Defense Sci Bd Task Force Res & Develop Mgt, 68-69; mem sci adv bd, US Air Force, 69-; mem, Adv Coun Panel Major Systs Acquisition of the Comn Govt Procurement, 70-72 & NASA Tracking & Data Acquisition Adv Panel, 71-72; trustee, Mitre, 69-, mem bd trustees, Northern Energy corp, 76-80; sr scientist, US Air Force Sci Adv Bd; mem, Defense Commun Agency's Sci Adv Group, 80-86, Defense Sci Bd, 87-; mem bd dirs, Inst Educ Serv; mem McLean Hosp vis comt Arlington Sch, 75-76, MGH Corp, 80- *Mem:* Nat Acad Eng; fel Inst Elec & Electronics Engrs; Asn Comput Mach; Sigma Xi; AAAS; Defense Sci Bd; Cosmos Club. *Res:* Computer technology; military command control, surveillance and communications systems. *Mailing Add:* 80 Rollingwood Lane Concord MA 01742

EVERETT, ROBERT W, JR, b New Orleans, La, June 13, 21; m 52; c 1. MICROPALEONTOLOGY. *Educ:* Tulane Univ, BS, 42. *Prof Exp:* Seismic computor, Ark Fuel Oil, 42-43; micropaleontologist, Texaco, Inc, 46-53, res geologist, 53-58, micropaleontologist in charge of lab, 59-67, micropaleontologist, 67-70, sr paleontologist, 70-85; RETIRED. *Mem:* Am Asn Petrol Geol. *Res:* Foraminifera as used in economic work in oil industry; salt dome research; Gulf Coast geology and paleontology; nannofossil research in the Tertiary of Gulf Coast and especially South Louisiana; subsurface deltaic research. *Mailing Add:* 6511 General Diaz New Orleans LA 70124

EVERETT, WARREN S, b Wichita, Kans, Oct 19, 10; m 35; c 3. SPACE TRANSPORTATION, SENSOR TECHNOLOGY. *Educ:* Munic Univ Wichita, BA, 33; US Mil Acad, BS, 35; Cornell Univ, MS, 39. *Prof Exp:* Instr transp eng, US Army Eng Sch, 40-41, instr war gaming, US Army War Col, 56-59, dir US Army Construct Agency, France, 59-61, dist engr, US Army CEngr, Vicksburg, Miss, 61-64; chief engr, USOM, Vietnam, USAID, Nigeria & Wash, 64-68; eng consult, Off Emergency Preparedness, Exec Off Pres, 68-69; dir, Pac Architects & Engrs, Vietnam, 69-71; dep dir, US Property Disposal Agency, Vietnam, 71-74; off dir & div chief, Defense Property Disposal Serv, 74-85; NAT SECURITY COORDR, HIGH FRONTIER, 85- *Concurrent Pos:* Exec dir, Sci & Eng Adv Bd, Coalition Strategic Defense Initiative, 85- *Honors & Awards:* Legion of Merit, UN Command, Tokyo, 53 & US CEngr, Wash, 62. *Mem:* Fel Am Soc Civil Engrs; fel Soc Am Mil Engrs; Nat Soc Prof Engrs; Am Inst Aeronaut & Astronaut; Am Defense Preparedness Asn. *Res:* Space technology, in order to expedite exploitation of the future benefits with respect to elimination of the proliferating global ballistic missile threat, remote sensing, environmental protection, health, reduction in space transportation costs and recovery of rare materials in space. *Mailing Add:* Coalition for the Strategic Defense Initiative 2800 Shirlington Rd Suite 405 Arlington VA 22206

EVERETT, WILBUR WAYNE, b Benton, Ark, Mar 4, 32; m 54; c 2. BIOPHYSICAL CHEMISTRY. *Educ:* Ouachita Baptist Col, BS, 54; Purdue Univ, PhD(chem), 59. *Prof Exp:* Instr chem, Purdue Univ, 55-56; asst scientist, Geront Br, Nat Heart Inst, 59-61; PROF CHEM, OUACHITA BAPTIST UNIV, 61-, CHMN DEPT, 66- *Mem:* Fel Am Inst Chemists; Am Chem Soc. *Res:* Physical chemistry; structure of proteins and carbohydrate high polymers; application of thermodynamics and hydrodynamics to solutions of macromolecules. *Mailing Add:* Dept Chem Ouachita Baptist Univ Arkadelphia AR 71923-4099

EVERETT, WOODROW W, III, b Biloxi, Miss, Jan 11, 60. ELECTRICAL ENGINEERING. *Educ:* Univ Cent Fla, Orlando, BS, 81, Univ Kent, MS, 83. *Prof Exp:* SR STAFF MEM, PAR GOVT SYST CORP, 86- *Mem:* Inst Elec & Electronics Engrs. *Res:* Signal processing, especially filter designs in electro- optical systems. *Mailing Add:* Par Gov't Syst Corp 220 Seneca Hwy New Hartford NY 13413

EVERETT, WOODROW WILSON, JR, b Newton, Miss, Oct 11, 37; m 58; c 2. ELECTROMAGNETIC COMPATIBILITY, ORGANIZATIONAL MANAGEMENT. *Educ:* George Washington Univ, BEE, 59; Cornell Univ, MS, 65, PhD(indust rels, mgt & quantum anal), 68. *Prof Exp:* Commun & electronics officer, USAF, 59-62; microwave engr, Ithaca Res Labs, Atlantic Res Corp, 62-64; res engr & dir, Rome Air develop Ctr, 64-76; CHMN, SOUTHEASTERN CTR ELEC ENG EDUC & NORTHEAST CONSORTIUM ENG EDUC, 76- *Mem:* Fel Inst Elec & Electronics Engrs; Am Soc Eng Educ. *Res:* Author of five books on communications-electronics; technical management and education. *Mailing Add:* Southeastern Ctr Elec Eng Educ 1101 Massachusetts Ave St Cloud FL 34769

EVERHARD, MARTIN EDWARD, b Pittsburgh, Pa, Jan 28, 33; m 68; c 5. PHYSICAL CHEMISTRY, SURGERY. *Educ:* Col William & Mary, BS, 53; Univ Va, PhD(phys chem), 59; NY Univ, MD, 67; Am Bd Surg, dipl. *Prof Exp:* Sr res scientist phys chem, Squibb Inst Med Res, Olin Mathieson Chem Corp, 59-63; Squibb Inst grant, NY Univ, 63-65; mem surg house staff, St Vincent's Hosp, Bridgeport, Conn, 67-68; surg resident, 68-71, CHIEF RESIDENT SURG, MONTEFIORE HOSP & MED CTR, 71-; CHIEF GEN SURG, PHELPS MEM HOSP, TARRYTOWN, 84-, DIR SURG, 85- *Concurrent Pos:* Rubin scholar med, NY Univ, 63-64, univ merit scholar, 64-67; attend surgeon, NY Med Col, Valhalla, 72- & Phelps Mem Hosp, Tarrytown, 72-; chief surg, Phelps Mem Hosp; asst clin prof surg, NY Med Col; dir, Intensive Care Unit, Phelps Mem Hosp, 74-84. *Mem:* Am Chem NY Acad Sci; AMA; fel Am Col Surgeons; fel Int Col Surgeons; Laser Inst Am. *Res:* Aqueous solution theory; kinetics of color reactions; differential thermal analysis; medical applications of transport through lipid-like membranes; vitamin B-12 like compounds; clinical and vascular surgery; intestinal blood flow. *Mailing Add:* 308 Chappaqua Rd Briarcliff NY 10510

EVERHART, DONALD LEE, b Erie, Pa, Jan 27, 32; m 55. IMMUNOLOGY, BIOCHEMISTRY. *Educ:* Grove City Col, BS, 54; Boston Univ, AM, 58, PhD(immunochem, biochem), 61. *Prof Exp:* Res assoc, Univ Tenn Mem Res Ctr & Hosp, 61-63; res immunochemist, Res Inst, Ill Inst Technol, 63-66; asst prof microbiol, Med Col Va, 66-72; assoc prof, 72-78, PROF MICROBIOL, COL DENT, NY UNIV, 78-, CHMN DEPT, 72- *Mem:* AAAS; Int Asn Dent Res; Am Chem Soc; NY Acad Sci. *Res:* Immunoglobulin A and its productive effects in oral disease. *Mailing Add:* Dept Microbiol NY Univ Col Dent New York NY 10010

EVERHART, DONALD LOUGH, b Troy, Ohio, July 18, 17; m 42; c 4. ECONOMIC GEOLOGY. *Educ:* Denison Univ, AB, 39; Harvard Univ, AM, 42, PhD(geol), 53. *Hon Degrees:* DSc, Denison Univ, 86. *Prof Exp:* Asst geol, Denison Univ, 37-39; teaching fel mineral, Harvard Univ, 40-42; geologist, US Geol Surv, 42-48; geologist & chief, Geol Br, AEC, 49-54, Geol Adv Div Raw Mat, Atomic Energy Comm, 54-59; chief geologist, Int Minerals & Chem Corp, 59-70, div vpres, Mining & Explor Div, 70-73 & Geol & Explor Div, 73-77; proj mgr, Nat Uranium Resource Eval Prog, US Dept Energy, 77-79, mgr, Grand Junction Off, 79-81; CONSULT GEOLOGIST, 81- *Mem:* Fel Geol Soc Am; sr fel Soc Econ Geologists (pres, 89-). *Res:* Petrology and geology of batholithic igneous rocks; geology of the Franciscan group of California; geology of quicksilver and uranium ore deposits; genesis and economic geology of uranium deposits; economic geology of phosphate and potash deposits; economic geology of ferro alloy metal deposits; geologic site studies for nuclear waste depositories. *Mailing Add:* Unit 10D 2700 G Rd Grand Junction CO 81506

EVERHART, EDGAR, celestial mechanics, astronomy; deceased, see previous edition for last biography

EVERHART, JAMES G, b Pittsburgh, Penn, Aug 29, 15. MANAGEMENT, TRANSFORMERS. *Educ:* Pa State Univ, BS, 38. *Prof Exp:* Gen mgr, Transformer div, Line Mat Indust, McGraw Edison Co, 38-69; gen mgr, A B Chance Co, 70-75; pres pitman div, Emerson Elec Co, 75-80, exec vpres, A B Chance Co, Subsid Emerson Elec Co, 75-80; CONSULT, 80. *Mem:* Fel Inst Elec & Electronics Engrs; fel Am Mgt Asn (bd of trustee, 75). *Mailing Add:* 902 Eastmont Dr Centralia MO 65240

EVERHART, LEIGHTON PHREANER, JR, b Charleston, WVa, Sept 22, 42. BIOLOGY. *Educ:* Univ Del, BA, 63, MS, 66; Univ Calif, Berkeley, PhD(zool), 70. *Prof Exp:* Res assoc cell biol, Univ Colo, 70-73; biologist cent res, 73-76, regist specialist agr chem, 76-78, supvr prod regist, 78, asst mgr prod regist, 78-80, regional mkt mgr, 80-83, bus mgr biotechnol, 85-87, MGR REGULATORY AFFAIRS, E I DU PONT DE NEMOURS & CO, INC, 87- *Concurrent Pos:* NIH fel, 67-70 & 71-73. *Mem:* Sigma Xi; Am Soc Cell Biol; Am Chem Soc. *Res:* Control of cell division, cell cycle-cell surface interactions; human health and environmental safety aspects of pesticides. *Mailing Add:* 67 Ringtail Run Kennett Square PA 19348

EVERHART, THOMAS E(UGENE), b Kansas City, Mo, Feb 15, 32; m 53; c 4. ELECTRICAL ENGINEERING, APPLIED PHYSICS. *Educ:* Harvard Univ, AB, 53; Univ Calif, Los Angeles, MSc, 55; Cambridge Univ, PhD(eng), 58. *Hon Degrees:* LLD, Ill Wesleyan Univ, 90 & Pepperdine Univ, 90, DEng, Colo Sch of Mines, 90. *Prof Exp:* Mem tech staff, Hughes Res Labs, 53-55; from asst prof to prof elec eng, Univ Calif, Berkeley, 58- 78, Miller res prof, 69-70, chmn, Dept Elec Eng & Computer Sci, 72-77; dean, Col Eng, Cornell Univ, Ithaca, 79-84; chancellor, Univ Ill, Urbana-Champaign, 84-87; PROF ELEC ENG & APPL PHYS & PRES, CALIF INST OF TECHNOL, 87- *Concurrent Pos:* Fel scientist, Westinghouse Res Labs, 62-63; NSF sr fel & guest prof, Univ Tübingen, 66-67; consult, Hughes Res Labs; Guggenheim Mem Found fel, 74-75; sci & ed adv comt, Lawrence Berkeley Labs, 78-85, chmn, 80- 85; mem, Gen Motors Sci Adv Comt, 80-, chmn, 84-; mem, R R Donnelley Tech Adv Comt, 82- *Honors & Awards:* Centennial Medal, Inst Elec & Electronics Engrs, 84. *Mem:* Nat Acad Eng; AAAS; fel Inst Elec & Electronics Engrs; Electron Micros Soc Am (pres-elect, 76, pres, 77, past pres, 78). *Res:* Scanning electron microscopy; electron physics; electron beam recording; semiconductor electronics; microfabrication. *Mailing Add:* Calif Inst Technol Pasadena CA 91125

EVERING, FREDERICK CHRISTIAN, JR, b Baltimore, Md, Mar 20, 36; m 65. ELECTRICAL ENGINEERING. *Educ:* Johns Hopkins Univ, BES, 58, MSE, 60, PhD(elec eng), 65. *Prof Exp:* Electronic engr, US Dept Defense, 60-62; instr elec eng, Johns Hopkins Univ, 62-65; from asst prof to assoc prof, 65-77, PROF ELEC ENG, UNIV VT, 77- *Mem:* Inst Elec & Electronics Engrs; Am Soc Eng Educ. *Res:* Microwave diffraction; low noise systems; special purpose computers; bioengineering; psychological and neurological instrumentation. *Mailing Add:* Dept Elec Eng Univ Vt Votey Hall 381 Burlington VT 05405

EVERITT, C W FRANCIS, b Sevenoaks, Eng, Mar 8, 34. SPACE PHYSICS, LOW TEMPERATURE PHYSICS. *Educ:* Univ London, BSc, 55, PhD(physics), 59, ARCS, Royal Col Sci, 55, DIC, 58. *Prof Exp:* Vis res assoc, Phys-Tech Inst, Bundesanstalt, WGer, 55; res assoc, Imp Col, Univ London, 58-60; res assoc & instr, Univ Pa, 60-62; res assoc, 62-66, res physicist, 66-67, sr res physicist, 67-74, adj prof, 74-82, PROF RES, STANFORD UNIV, 82- *Concurrent Pos:* Instr physics, Univ Pa, 61-63; mem space relativity comt, Int Acad Astronaut, 65-; Guggenheim fel, 76-77; mem, MOWGSA, 76-79, Int Ctr Rel Astrophys, 85-, NASA Astrophys Coun, 85-; co-dir res, Int Ctr Rel astrophys, 85- *Honors & Awards:* Tyndall prize, Imperial Col, London, 55. *Mem:* Am Phys Soc; Sigma Xi; Am Asn Physics Teachers. *Res:* Electron optics; paleomagnetism; liquid helium; low temperature and space physics; history of physics. *Mailing Add:* H-E-P-L Stanford Univ Stanford CA 94305

EVERLY, CHARLES RAY, b Oklahoma City, Okla, Oct 13, 44; m 65; c 2. ORGANIC CHEMISTRY. *Educ:* Phillips Univ, BA, 66; Univ Ark, PhD(org chem), 70. *Prof Exp:* Asst prof, 69-73, assoc prof chem, Phillips Univ, 73-78; res chemist, 78-80, sr res chemist & supvr, 80-89, CHEM RES & DEVELOP DIR, ETHYL CORP, 89- *Concurrent Pos:* Vis Scholar, Louisiana State Univ, 76-77. *Mem:* Am Chem Soc; Sigma Xi. *Res:* Organic synthesis, organic reaction mechanics. *Mailing Add:* PO Box 14799 Baton Rouge LA 70898

EVERMANN, JAMES FREDERICK, b Van Nuys, Calif, Nov 18, 44; m 65; c 4. CLINICAL VIROLOGY, COMPARATIVE VIROLOGY. *Educ:* Univ Nev, Reno, BS, 69; Univ Wyo, Laramie, MS, 71; Purdue Univ, West Lafayette, PhD(virol), 74. *Prof Exp:* Persistent viral infections, Health Sci Ctr, Univ Ore, 74-76; asst prof, 76-81, assoc prof, 81-86, PROF CLIN VIROL, WASH STATE UNIV, 86- *Concurrent Pos:* Sr scientist, NIH-FCRF, 87-88. *Mem:* Am Asn Vet Lab Diagnosis; US Animal Health Asn; Am Soc Virol. *Res:* Viruses of domestic animals and wildlife species (bovine herpesviruses, bovine leukosis virus, bovine viral diarrhea, border disease viruses, canine encephalitogenic parainfluenza and canine parvovirus type-enteritis-myocarditis) including identification, development of diagnostic capabilities and public health concerns. *Mailing Add:* Dept Vet Clin Med & Surg Col Vet Med Wash State Univ Pullman WA 99164

EVERS, CARL GUSTAV, b Lake Benton, Minn, July 30, 34; m 60; c 3. MEDICINE, PATHOLOGY. *Educ:* Mankato State Col, BA, 55; Univ Minn, MD, 59; Am Bd Path, dipl, 65. *Prof Exp:* Intern, 59-60, resident, 60-64, from instr to assoc prof, 64-74, PROF PATH, MED CTR, UNIV MISS, 74-, ASSOC DEAN, SCH MED, 73- *Concurrent Pos:* USPHS trainee anat & exp path, 63-64, proj dir, Training Proj Cytotech grant, 66- *Mem:* AAAS; Int Acad Path; Sigma Xi. *Res:* Pathogenesis of human immune disorders and tumor immunology. *Mailing Add:* Dept Path Univ Miss Med Ctr Jackson MS 39216

EVERS, ROBERT C, b St Henry, Ohio, Nov 10, 39; m 62; c 5. POLYMER CHEMISTRY. *Educ:* Univ Dayton, BS, 61; Univ Notre Dame, PhD(org chem), 65. *Prof Exp:* Res chemist, 65-72, group leader, Air Force Mat Lab, 72-89, RES LEADER, POLYMER BR, WRIGHT-PATTERSON AFB, 89- *Mem:* AAAS; Am Chem Soc. *Res:* Synthesis of heterocyclic and fluorocarbon monomers and polymers for high temperature applications. *Mailing Add:* 839 Silver Leaf Dr Dayton OH 45431

EVERS, WILLIAM JOHN, b Long Branch, NJ, Sept 3, 32; m 71; c 2. ORGANIC CHEMISTRY. *Educ:* Monmouth Col, NJ, BS, 60; Univ Maine, MS, 62, PhD(org chem), 65. *Prof Exp:* Chemist, Chem Res Ctr, Edgewood Arsenal, Dept of Army, 65-66; SR PROJ LEADER NATURAL PROD & ORG SYNTHESIS, INT FLAVORS & FRAGRANCES, INC, 66. *Mem:* Sigma Xi; Am Chem Soc. *Res:* Heterocyclic and organosulfur chemistry; natural product chemistry of flavors and fragrances. *Mailing Add:* 1515 State Highway No 36 Union Beach NJ 07735

EVERS, WILLIAM L, b Pittsburgh, Pa, Aug 13, 06; m 64; c 1. RESEARCH ADMINISTRATION, POLYMERS. *Educ:* Univ Akron, BS, 28; Northwestern Univ, MS, 29; Pa State Univ, PhD(org chem), 32. *Prof Exp:* Res chemist, Socony Mobil Co, Inc, 31-35; res mgr, Rohm and Haas Co, 35-52; res mgr, Celanese Corp, 52-64, dir univ sci rels, 64-68; exec dir, 68-87, CONSULT, CAMILLE & HENRY DREYFUS FOUND, 87- *Concurrent Pos:* Mem vis comt phys sci, Chicago Univ; mem vis comt chem, Harvard Univ, 82. *Mem:* Am Chem Soc; Asn Res Dirs. *Mailing Add:* 104 Essex Rd Summit NJ 07901

EVERSE, JOHANNES, b Yerseke, Netherlands, Dec 2, 31; m 64; c 3. ENZYMOLOGY. *Educ:* Brandeis Univ, Mass, MA, 71; Univ Calif, San Diego, PhD(chem), 73. *Prof Exp:* Res technician biochem, Philips-Duphar Pharm Co, Holland, 52-60; res assoc, Brandeis Univ, Mass, 60-69; assoc specialist chem, Univ, Calif, San Diego, 69-73, asst res chemist biochem, 73-76; assoc prof, 76-80, PROF BIOCHEM, HEALTH SCI CTR, TEX TECH UNIV, 80- *Concurrent Pos:* NATO sr vis prof, Univ Milan, Italy, 80-81; vis scientist, Letterman Army Inst Res, San Francisco, Calif, 89-91. *Mem:* Am Soc Biol Chemists; Am Chem Soc; AAAS; Soc Exp Biol Med; Am Asn Cancer Res; Oxygen Soc. *Res:* Relationship between structure and function of enzymes and enzyme mechanisms; application of immobilized enzymes for chemotherapeutic and diagnostic purposes; role of the immune system in cancer development and regression. *Mailing Add:* Dept Biochem Tex Tech Univ Health Sci Ctr Lubbock TX 79430

EVERSMEYER, HAROLD EDWIN, b Randolph, Kans, July 7, 27; m 53; c 4. PLANT PATHOLOGY, BOTANY. *Educ:* Kans State Univ, BS, 51, PhD(plant path), 65. *Prof Exp:* County 4-H Club agent, Exten Serv, Kans State Univ, Olathe, 51-54, Emporia, 56-60; assoc prof, 64-70, PROF BIOL SCI, MURRAY STATE UNIV, 71- *Res:* Phytonematology-occurrence and damage by plant parasitic nematodes; aeromycology. *Mailing Add:* Dept Biol Sci Murray State Univ Murray KY 42071

EVERSON, ALAN RAY, b Ft Dodge, Iowa, Aug 11, 43; div; c 1. FOREST RECREATION. *Educ:* Iowa State Univ, BS, 65; Univ Mich, MFor, 67; Tex A&M Univ, PhD(recreation resources), 78. *Prof Exp:* Recreation resource specialist, Bur Outdoor Recreation, 66-68; dir field serv, Rocky Mountain Ctr Environ, 68; naturalist, Nat Park Serv, 69; state outdoor recreation planner, Colo Parks & Outdoor Recreation, 69-77; ASSOC PROF OUTDOOR RECREATION & LAND USE, UNIV MO, 77- *Mem:* Soc Am Foresters; Nat Recreation & Park Asn. *Res:* Human carrying capacities of forested lands. *Mailing Add:* 1-30 Agr Bldg Sch Forestry Univ Mo Columbia MO 65211

EVERSON, DALE O, b Geneva Lake, Wis, Feb 1, 30; m 54; c 2. STATISTICS, APPLIED STATISTICS. *Educ:* Univ Idaho, BS, 52, MS, 56; Iowa State Univ, PhD(animal breeding), 60. *Prof Exp:* Biometrician, Agr Res Serv, USDA, 60-62; assoc exp sta statistician, 62-66, PROF STATIST & EXP STA STATISTICIAN, UNIV IDAHO, 66- *Concurrent Pos:* Vis prof, Dept of Statist, Ore State Univ, 73-74. *Mem:* Am Dairy Sci Asn; Am Soc Animal Sci; Am Statist Asn; Sigma Xi. *Res:* Animal breeding and statistic methodology. *Mailing Add:* Dept Math & Appl Statist Univ Idaho Moscow ID 83844

EVERSON, EVERETT HENRY, b Whitehall, Wis, Oct 8, 23; m 47; c 2. GENETICS, PLANT BREEDING. *Educ:* Univ Wis, BS, 49; Univ Calif, PhD(genetics), 52. *Prof Exp:* Res agronomist, Pillsbury Mills, Inc, 49; asst, Univ Calif, 49-52; asst prof weed control & plant breeding, Univ Ariz, 52-54; res agronomist genetics & plant breeding, USDA, Wash State Univ, 54-56; assoc prof, 56-63, PROF CROP SCI, GENETICS & PLANT BREEDING, MICH STATE UNIV, 64- *Concurrent Pos:* Consult Int Agr, arid regions. *Mem:* Am Soc Agron; Am Genetic Asn; Crop Sci Soc Am. *Res:* Genetics and plant breeding; major organism; wheat; genus Triticum. *Mailing Add:* 1048 Wildwood Dr East Lansing MI 48823

EVERSON, HOWARD E, b Milan, Ohio, Feb 26, 18; m 47; c 3. INORGANIC CHEMISTRY, PHYSICAL CHEMISTRY. *Educ:* Western Reserve Univ, BA, 40, MS, 47, PhD(chem), 48. *Prof Exp:* Res chemist, Wyandotte Chems Corp, 40-42; asst prof chem, Univ Cincinnati, 48-51; res chemist & group leader, Diamond Alkali Co, 51-55; asst to res dir, Res Ctr, Diamond Shamrock Corp, 55-56, chief staff engr, 56-58, from asst dir to dir res, 58-67, tech dir, 67-73, dir safety & environ eng, Diamond Shamrock Chem Corp, Cleveland, 73-78; CONSULT, 80- *Mem:* Am Chem Soc. *Res:* Hydrotropic solvents; effect of salts on the aqueous solubility of non-electrolytes. *Mailing Add:* 6123 Campbell Dr North Madison OH 44057

EVERSON, RONALD WARD, b Dodgeville, Wis, Sept 14, 31. OPTOMETRY, PHYSIOLOGICAL OPTICS. *Educ:* Chicago Col Optom, BS, 53, OD, 54; Ind Univ, MS, 59. *Prof Exp:* Lectr optom, Ind Univ, 61-64; assoc prof optom & dir grad prog physiol optics, Pac Univ, 64-67; asst prof, 68-73, dir internal affairs, 73-78, ASSOC PROF OPTOM, IND UNIV, BLOOMINGTON, 73- *Concurrent Pos:* Mem ed coun, Am J Optom & Physiol Optics, 76-84. *Mem:* Am Optom Asn; fel Am Acad Optom. *Res:* Human color vision; illumination principles; visual acuity; visual system contrast sensitivity; aniseikonia. *Mailing Add:* Sch Optom Ind Univ Bloomington IN 47405

EVERSTINE, GORDON CARL, b Baltimore, Md, Mar 30, 43; m 68; c 2. COMPUTATIONAL MECHANICS. *Educ:* Lehigh Univ, Pa, BS, 64; Purdue Univ, MS, 66; Brown Univ, PhD(appl math), 71. *Prof Exp:* Mem tech staff, Bell Tel Labs, Inc, 64-66; mathematician conputational mech div, 68-90, SR SCIENTIST, DAVID TAYLOR RES CTR, 90- *Concurrent Pos:* Prof lectr, George Wash Univ, 79- *Honors & Awards:* Sci Achievement Award Math & Comput Sci, Acad Sci, 84. *Mem:* Int Asn Comput Mech. *Res:* Computational mechanics; finite element method; fluid-structure interaction; structural dynamics; computational structural acoustics. *Mailing Add:* Computational Mech Div 128 David Taylor Res Ctr Bethesda MD 20084-5000

EVERT, HENRY EARL, b Sherwood, Mich, Mar 2, 15; m 46. ORGANIC CHEMISTRY. *Educ:* Mich State Col, BS, 37, MS, 38; Univ Iowa, PhD(sanit chem), 41. *Prof Exp:* Asst org chem, Mich State Col, 37-38; asst inorg chem, Univ Iowa, 38-39, asst sanit chem, 39-40, storeroom asst, 40-41; res chemist, Masonite Corp, 41-43; USPHS consult chemist, Eng Dept, Johns Hopkins Univ, 46-47; asst prof chem, George Washington Univ, 47-48; prof physiol chem, Univ Scranton, 48-49; asst pharmacol, Univ Va, 49-52; asst prof biochem, Sch Med, State Univ NY, 52-62; PROF CHEM & CHMN DEPT, NASSAU COMMUNITY COL, 62- *Mem:* Fel AAAS; fel Am Inst Chemists; Am Chem Soc; Sigma Xi. *Res:* Carbohydrates; catalysis; wood chemistry; enzymology; chemical oceanography. *Mailing Add:* PO Box 326 Garden City NY 11530

EVERT, RAY FRANKLIN, b Mt Carmel, Pa, Feb 20, 31; m 60; c 2. PLANT ANATOMY. *Educ:* Pa State Univ, BS, 52, MS, 54; Univ Calif, Davis, PhD(bot), 58. *Prof Exp:* From instr to asst prof bot, Mont State Col, 58-60; from asst prof to prof bot, Univ Wis-Madison, 60-88, chmn dept, 73-74, & 74-79, prof bot & plant path, 77-88, KATHERINE ESAU PROF BOT & PLANT PATH, UNIV WIS-MADISON, 88- *Concurrent Pos:* NSF res grants, 59-93, mem cell biol fel rev panel, NIH, 64-68; Guggenheim Found fel, 65-66; vis prof, Univ Natal, 71 & Univ Gottingen, 71, 74-75 & 88; Alexander von Humboldt Award, 74-75; sci ed, Physiologia Plantarum, 83-; NSF adv comt, Biol Res Ctr Prog, 87-88. *Honors & Awards:* Merit Award, Bot Soc Am, 82; Bessey lectr, Iowa State Univ, 84; Benjamin Minge Duggar lectr, Auburn Univ, 85; Michael A Cichan Mem Lectr, 89. *Mem:* Fel AAAS; Bot Soc Am (pres, 86-87); Am Soc Plant Physiologists; Am Inst Biol Sci; Int Asn Wood Anatomists; fel Am Acad Arts & Sci. *Res:* Light and electron microscopic investigations of the ontogeny, structure and seasonal development of the phloem; leaf structure in relation to solute transport and phloem loading; phloem and leaf structure and function. *Mailing Add:* Dept Bot Birge Hall Univ Wis 430 Lincoln Dr Madison WI 53706-1381

EVERTS, CRAIG HAMILTON, b Appleton, Wis, Apr 18, 39; m 70; c 2. COASTAL ENGINEERING, MARINE GEOLOGY. *Educ:* Univ Southern Calif, BS, 66; Univ Wis, MS, 68, PhD(geol, geophys), 71. *Prof Exp:* Res oceanogr, 71-76, supvry geologist, 76-83, CONSULT, COASTAL ENG RES CTR, 83- *Mem:* Am Soc Civil Engrs; Am Geophys Union. *Res:* Coastal engineering research in harbor and navigation-channel sedimentation, beach erosion and the design of beach restoration projects and coastal structures. *Mailing Add:* PO Box 7707 Long Beach CA 90807

EVESLAGE, SYLVESTER LEE, b Ripley, Ohio, Apr 25, 23; m 55; c 4. ORGANIC CHEMISTRY. *Educ:* Univ Notre Dame, BS, 44, MS, 45, PhD(org chem), 53. *Prof Exp:* From instr to assoc prof, 48-66, PROF CHEM, UNIV DAYTON, 66- *Mem:* AAAS; Am Chem Soc. *Res:* Ion exchange chromatography; synthesis of chemotherapeutic compounds. *Mailing Add:* 3401 Lenox Dr Kettering OH 45429-1511

EVETT, ARTHUR A, b Pasco, Wash, Apr 12, 25; m 46; c 2. PHYSICS. *Educ:* Wash State Univ, BS, 48, PhD(physics), 51. *Prof Exp:* Adv sci warfare, Weapons Systs Eval Group, US Dept Defense, DC, 51-52; instr physics, Yale Univ, 52-55; asst prof, Wash State Univ, 56-58; from assoc prof to prof, Univ Ariz, 58-68; PROF PHYSICS, CALIF STATE UNIV, DOMINGUEZ HILLS, 68- *Mem:* Am Phys Soc. *Res:* Relativity theory; intermolecular forces; surface physics. *Mailing Add:* Dept Physics Calif State Univ Dominguez Hills CA 90747

EVETT, JACK B(URNIE), b York, Pa, May 26, 42; m 69; c 1. CIVIL ENGINEERING. *Educ:* Univ SC, BS, 64, MS, 65; Tex A&M Univ, PhD(civil eng), 68. *Prof Exp:* Asst prof civil eng, 67-77, assoc prof environ eng, 77-80, ASSOC PROF CIVIL ENG, UNIV NC, CHARLOTTE, 80-; AT ENG DEAN OFF, UNIV NC, CHARLOTTE. *Mem:* Am Soc Civil Engrs; Am Soc Eng Educ; Sigma Xi. *Res:* Waste dispersion patterns in an estuarine system; water resources planning and development. *Mailing Add:* Eng Dean Off Univ NC College Station Charlotte NC 28223

EVETT, JAY FREDRICK, b Lewiston, Idaho, Nov 5, 31; m 57; c 1. PHYSICS, BIOPHYSICS. *Educ:* Wash State Univ, BS, 53 & 57; Northwestern Univ, MS, 58; Ore State Univ, PhD(biophys), 68. *Prof Exp:* Nuclear engr, US AEC, 60-61; from asst prof to assoc prof, Moorhead State Col, 61-66; assoc prof, 68-74, PROF PHYSICS, WESTERN ORE STATE COL, 74- *Mem:* AAAS; Am Asn Physics Teachers. *Mailing Add:* Div Natural Sci Western Ore State Col Monmouth OR 97361-1394

EVILIA, RONALD FRANK, b Meriden, Conn, Dec 28, 43; m 67; c 3. ANALYTICAL CHEMISTRY. *Educ:* Lehigh Univ, BA, 65, PhD(chem), 69. *Prof Exp:* Res asst chem, Lehigh Univ, 65-69; res assoc chem, Univ NC, 69-72; asst prof, 72-77, ASSOC PROF CHEM, UNIV NEW ORLEANS, 77- *Mem:* Am Chem Soc; Sigma Xi. *Res:* Utilization of nuclear magnetic resonance techniques for structural and dynamic investigations of coordination compounds; liquid chromatographic separations of enantiomeric substances and new methods of liquid chromatographic detection. *Mailing Add:* Dept Chem Univ New Orleans Lake Front New Orleans LA 70148

EVITT, WILLIAM ROBERT, b Baltimore, Md, Dec 9, 23; m 50; c 3. PALYNOLOGY, FOSSIL DINOFLAGELLATES. *Educ:* Johns Hopkins Univ, AB, 42, PhD(geol), 50. *Prof Exp:* Asst, Johns Hopkins Univ, 46-48; from instr to assoc prof geol, Univ Rochester, 48-56; sr res geologist, Jersey Prod Res Co, 56-59, res assoc, 59-62; PROF GEOL, STANFORD UNIV, 62- *Concurrent Pos:* Ed, Jour, Paleont Soc, 53-56. *Mem:* Paleont Soc (vpres, 58); Geol Soc Am; Am Asn Stratig Palynologists; Int Asn Plant Taxon. *Res:* Palynology; dinoflagellate morphology; invertebrate paleontology. *Mailing Add:* Dept Geol Stanford Univ Stanford CA 94305

EVLETH, EARL MANSFIELD, b Evanston, Ill, Dec 7, 31; m 55; c 1. THEORETICAL CHEMISTRY, ORGANIC CHEMISTRY. *Educ:* Calif Inst Technol, BS, 54; Univ Southern Calif, PhD(chem), 63. *Prof Exp:* Chemist, Shell Oil Co, 54-55; res chemist, Am Potash & Chem Co, 55-57 & 60-61 & Int Bus Mach Corp, 62-65; asst prof natural sci, Univ Calif, Santa Cruz, 67-72, assoc prof chem, 72-77; DIR DE RECHERCHE, CENTRE NATIONAL DE LA RECHERCHE SCIENTIFIQUE, 74- *Mem:* Am Chem Soc. *Res:* Proton transfers mechanism; small molecule photochemistry and energy transfer mechanisms; spectra-structure correlations; ab initio molecular orbital calculations of Valence and Ryberg states. *Mailing Add:* Tour 22 Universite Paris VI Four Place Jussieu Paris 75230 France

EVOY, WILLIAM (HARRINGTON), b Philadelphia, Pa, July 1, 38; m 64; c 2. NEUROPHYSIOLOGY, NEUROETHOLOGY. *Educ:* Reed Col, BA, 60; Univ Ore, MA, 62, PhD(biol), 64. *Prof Exp:* Res assoc comp neurophysiol, Stanford Univ, 64-66; from asst prof to assoc prof, 66-76, dir, Lab Grant Biol, 72-82, PROF BIOL, UNIV MIAMI, 76- *Concurrent Pos:* Benjamin Meeker vis prof, Univ Bristol, Eng, 84. *Mem:* Fel AAAS; Am Soc Zool; Int Soc Neuroethology; Soc Exp Biol; Soc Neurosci; Animal Behavior Soc; Nat Sci Teachers Asn. *Res:* Comparative neurobiology of crustacean, amphibian and insect nervous and neuromuscular and sensory systems, in behavior; coordinating mechanisms in locomotion and orientation. *Mailing Add:* Biol Dept Univ Miami PO Box 249118 Coral Gables FL 33124-0421

EVTUHOV, VIKTOR, b Poland, May 24, 35; US citizen; m 57; c 2. QUANTUM ELECTRONICS, NONLINEAL OPTICS. *Educ:* Univ Calif, Los Angeles, BS, 56; Calif Inst Technol, MS, 57, PhD(elec eng/physics), 61. *Prof Exp:* Mem tech staff, 56 & 60-65, sr staff physicist, 65-70, head, Quantum Electronics Sect, Laser Dept, Hughes Res Lab, 70-72, asst mgr/mgr, Opto-Electronics Dept, 72-77, sr scientist, Hughes Res Lab, 77-78, mgr tech planning, 78-80, asst dir, 80-86 DIR INDEP, RES & DEVELOP, HUGHES AIRCRAFT CO, 86 - *Concurrent Pos:* Res fel & instr elec eng, Calif Inst Technol, 60-61, sr res fel, 70-76. *Mem:* AAAS; Am Phys Soc; Inst Elec & Electronics Engrs; Optical Soc Am. *Res:* Physical electronics; secondary emission; theory of semiconductors; band structure; quantum electronics; lasers; nonlinear optics; integrated optics; fiber optics. *Mailing Add:* 549 Arbramar Ave Pacific Palisades CA 90272

EWALD, ARNO WILFRED, b Fond du Lac, Wis, May 14, 18; m 43; c 4. SOLID STATE PHYSICS. *Educ:* Wis State Col, BS, 41; Univ Mich, MS, 42, PhD(physics), 48. *Prof Exp:* Fel physics, Univ Mich, 41-44, res assoc, 44-48; from instr to assoc prof, 48-61, PROF PHYSICS, NORTHWESTERN UNIV, ILL, 61- *Concurrent Pos:* Res assoc, Nat Defense Res Comt, 44-45. *Mem:* Fel Am Phys Soc. *Res:* Semiconductors; photoconductivity; energy band structure determinations; infrared phenomena; crystal physics. *Mailing Add:* Dept Physics Northwestern Univ 633 Clark St Evanston IL 60201

EWALD, FRED PETERSON, JR, b Saginaw, Mich, Mar 25, 32; wid; c 1. ANALYTICAL CHEMISTRY, INFRARED SPECTROSCOPY. *Educ:* Aquinas Col, BS, 54; Univ Kans, PhD(anal chem), 62. *Prof Exp:* Technician anal chem, Haviland Prod, 52-54; asst chem, Univ Kans, 54-59; res assoc, 82-87, SUPVR ANALYTICAL CHEM, PPG INDUSTS INC, 60-, SR RES ASSOC, 87- *Mem:* Am Chem Soc. *Res:* Nonaqueous electrochemistry; gas chromatography; mass spectrometry; air and water pollution analysis; infrared spectroscopy, emission and absorption spectroscopy; herbicide residue analysis; polarography; uv-vis spectroscopy. *Mailing Add:* PO Box 752 NeeNah WI 54957-0752

EWALD, SANDRA J, IMMUNOCHEMISTRY, CELLULAR IMMUNOLOGY. *Educ:* Univ Tex, Austin, PhD(immunogenetics), 76. *Prof Exp:* ASSOC PROF IMMUNOL & HEMAT, MONT STATE UNIV, 79- *Mailing Add:* Dept Microbiol Mont State Univ Bozeman MT 59717

EWALD, WILLIAM PHILIP, b Whitestone, NY, Mar 27, 22; m 44; c 4. OPTICAL ENGINEERING. *Educ:* Univ Rochester, BS, 53. *Prof Exp:* sr supv engr, Eastman Kodak Co, 44-80; RETIRED. *Concurrent Pos:* Lectr, Univ Rochester, 53-66. *Honors & Awards:* David Richard Medal, Optical Soc Am, 79. *Mem:* Optical Soc Am; Soc Motion Picture & TV Eng; Photog Soc Am. *Mailing Add:* 58 Twin Shores Blvd Longboat Key FL 34228

EWALL, RALPH XAVIER, polymer & textile chemistry, for more information see previous edition

EWAN, GEORGE T, b Edinburgh, Scotland, May 6, 27; m 52; c 2. NUCLEAR SCIENCE, EDUCATION. *Educ:* Univ Edinburgh, BSc, 48, PhD(physics), 52. *Prof Exp:* Asst lectr physics, Univ Edinburgh, 50-52; res assoc, McGill Univ, 52-55; from asst res officer to sr res officer, Atomic Energy Can, Ltd, 55-70; head dept, 74-77, PROF PHYSICS, QUEEN'S UNIV, 70- *Concurrent Pos:* Nat Res Coun fel, 54-55; Ford Found fel, Niels Bohr Inst, Copenhagen, 61-62; vis scientist, Lawrence Radiation Lab, 66; res assoc, Europ Orgn Nuclear Res, 77-78. *Honors & Awards:* Radiation Indust Award, Am Nuclear Soc, 67; Gold Medal, Can Asn Physics, 87 fel Japan Soc Promotion of Sci. *Mem:* Fel Royal Soc Can; Am Inst Physics; Roy Soc Edinburgh; fel Royal Soc Arts; Can Asn Physicists. *Res:* Nuclear physics; high resolution beta and gamma ray spectroscopy; semiconductor detectors; applications of nuclear techniques; neutrino physics. *Mailing Add:* Stirling Hall Physics Queen's Univ Kingston ON K7L 3N6 Can

EWAN, JOSEPH (ANDORFER), b Philadelphia, Pa, Oct 24, 09; m 35; c 3. BOTANY, HISTORY OF BIOLOGY. *Educ:* Univ Calif, AB, 34. *Hon Degrees:* ScD, Col William and Mary, 72, Tulane Univ, 80. *Prof Exp:* Asst phanerogamic bot, Univ Calif, 33-37; instr biol, Univ Colo, 37-44; botanist, For Econ Admin, Colombia, 44-45; asst cur div plants, Smithsonian Inst, 45-46; assoc botanist, Bur Plant Indust, USDA, Md, 46-47; from asst prof to prof, 47-72, Ida A Richardson prof, 72-77, EMER PROF BOT, TULANE UNIV LA, 77-; RES ASSOC, MO BOT GARDEN, 86- *Concurrent Pos:* Am Philos Soc grant, 49-52, 54, 58, 64 & 72; Guggenheim fel, 54; off del, Int Conf, Nat Sci Res Ctr, France, 56; panelist, Lilly Conf Res Opportunities Am Cultural Hist, Mo, 59; NSF fel, 59-61; ed, Classica Bot Am, 66-; vis prof, Univ Hawaii, 67 & 74, Univ Ore, 78 & 81 & Ohio State Univ, 82; Smithsonian Regents fel, 84. *Honors & Awards:* Eloise Payne Luquer Medal, 78; Founders Medal, Soc Hist Natural Hist London, 86; Cert Merit, Bot Soc Am, 89. *Mem:* Fel Linnean Soc London; Am Fern Soc (vpres, 41-47, pres, 58-59); Am Antiquarian Soc; Cooper Ornith Soc; His Sci Soc; Soc Hist Natural Hist London. *Res:* Taxonomy of delphinium, vismia and American gentianaceae; biography and bibliography of naturalists; expeditions and travels of naturalists in Americas. *Mailing Add:* Missouri Botanical Garden P O Box 299 St Louis MO 63166-0299

EWAN, RICHARD COLIN, b Cuba, Ill, Sept 10, 34; m 56; c 4. ANIMAL NUTRITION, BIOCHEMISTRY. *Educ:* Univ Ill, BS, 56, MS, 57; Univ Wis, PhD(animal sci & biochem), 66. *Prof Exp:* PROF ANIMAL SCI, IOWA STATE UNIV, 66- *Mem:* Am Soc Animal Sci; Am Inst Nutrit. *Res:* Vitamin and mineral interactions in nutrition; energy metabolism and utilization. *Mailing Add:* Dept Animal Sci Iowa State Univ Ames IA 50011

EWART, HUGH WALLACE, JR, b Decatur, Ill, Oct 18, 39; m 63; c 2. ORGANIC CHEMISTRY. *Educ:* Trinity Col, Conn, BS, 61; Yale Univ, MS, 63, PhD(org chem), 67. *Prof Exp:* Res chemist, Olympic Res Div, ITT Rayonier Inc, 68-81; dir tech serv, Tree Top Inc, 81-89; VPRES SCI AFFAIRS, NORTHWEST HORT COUN, 89- *Mem:* Am Chem Soc; Sigma Xi; Royal Soc Chem; Inst Food Technol; Am Soc Enologists. *Res:* Fermentation of wood sugars; pulping and bleaching of wood and wood pulp; synthetic chemistry of small ring systems via cycloaddition reactions; separation and identification of natural products and derivatives; photochemistry of organic compounds. *Mailing Add:* 750 Selah Heights Rd Selah WA 98942

EWART, MERVYN H, b Can, Dec 23, 20; m 57; c 2. BIOCHEMISTRY. *Educ:* Ont Agr Col, BSA, 44; McGill Univ, MSc, 46; Univ Minn, PhD(agr biochem), 51. *Prof Exp:* Res chemist, Can Dept Health & Welfare, 51-54; from asst prof to assoc prof, 54-65, PROF CHEM & CHMN DEPT, ALBANY COL PHARM, UNION UNIV, NY, 65- *Concurrent Pos:* Vis prof, State Univ Col Educ, Albany, 59; consult, Christian Hansen Labs, 56. *Mem:* Am Chem Soc. *Res:* Enzymes; metabolism; chemical analysis. *Mailing Add:* Box 378 Meadowdale Rd Rd 2 Altamont NY 12009-9534

EWART, R BRADLEY, b Mt Pleasant, Iowa, Mar 4, 32; m 64. SCIENCE WRITING, BOTANY. *Educ:* Univ Iowa, BA, 56; Wash Univ, MA, 62, PhD(bot), 69. *Prof Exp:* Instr high sch, Ill, 56-59, 60-63 & 64-66 & Guatemala, Cent Am, 63-64; teaching asst biol, Wash Univ, 66-68; asst prof biol & bot, Northwest Mo State Univ, 69-74; instr biol, Mo Western State Col, 75-77, FREELANCE SCI WRITER, 74- *Concurrent Pos:* Adj instr, Edison Community Col, 86- *Res:* Anatomical studies of the fossil genus Scolecopteris; developmental studies of cell wall synthesis in the zoospore of Vaucheria sessilis; photography of algae, fungi and lower vascular plants; history of machines that play chess. *Mailing Add:* 1926 SE 43rd Ave Apt 223 Cape Coral FL 33904

EWART, TERRY E, b Seattle, Wash, July 6, 34; m 54; c 4. PHYSICS. *Educ:* Univ Wash, BS, 59, PhD(physics), 65. *Prof Exp:* Sr physicist, 59-68, HEAD OCEAN PHYSICS DEPT, APPL PHYSICS LAB, UNIV WASH, 68-, SR RES ASSOC, DEPT OCEANOG, 70- *Concurrent Pos:* Consult, State Bur Fisheries, 62-65. *Mem:* AAAS; Marine Technol Soc; Am Phys Soc. *Res:* High energy particle physics; underwater acoustics. *Mailing Add:* Appl Physics Lab 1013 E 40th St Seattle WA 98195

EWBANK, WESLEY BRUCE, b Olivet, Kans, Sept 21, 32; m 57; c 2. NUCLEAR PHYSICS, ATOMIC PHYSICS. *Educ:* Univ Kans, BS, 54; Univ Calif, Berkeley, PhD(physics), 60. *Prof Exp:* Asst, Univ Calif, Berkeley, 54-59, res assoc nuclear physics, Lawrence Radiation Lab, 59-62; res assoc, Nuclear Data Proj, Nat Acad Sci-Nat Res Coun, 62-63; res staff mem nuclear physics, 64-68, asst dir, 68-75, dir, Nuclear Data Proj, 75-80, TECH ASST, INFO DIV, OAK RIDGE NAT LAB, 80- *Concurrent Pos:* Lectr, Univ Calif, Berkeley, 61-62. *Mem:* Am Phys Soc; AAAS; Am Soc Info Sci. *Res:* Measurement of nuclear moments by atomic beam spectroscopy; collection and synthesis of experimental results in nuclear-structure physics for publication of nuclear data sheets; use of computers to manage and communicate scientific technical information. *Mailing Add:* 104 Canterbury Rd Oakridge TN 37830

EWELL, JAMES JOHN, JR, b Baytown, Tex, Apr 3, 42; m 63; c 1. COMPUTER PROGRAM DESIGN. *Educ:* Univ Tex, Austin, BS, 65, MA, 69. *Prof Exp:* STAFF MGR, MCDONNELL DOUGLAS SPACE SYSTS CO, 74- *Res:* Statistical data processing applications to position determination; computer program design, development, test and evaluation. *Mailing Add:* McDonnell Douglas Corp PO Box 57568 Webster TX 77598

EWEN, ALWYN BRADLEY, b Saskatoon, Sask, Oct 24, 32; div; c 2. INSECT PATHOLOGY, INSECT PHYSIOLOGY. *Educ:* Univ Sask, BA, 55, MA, 57; Univ Alta, PhD(insect physiol), 61. *Prof Exp:* Res officer, 57-65, RES SCIENTIST, CAN DEPT AGR, 65- *Concurrent Pos:* Ed, The Can Entomologist, 85- *Mem:* Pan-Am Acridological Soc; Entom Soc Can; Soc Invert Path; fel Royal Entom Soc London. *Res:* Grasshopper diseases, especially the Microsporidia, pathology and physiology; insect cancers; physiology of insect reproduction; insect hormones; integrated pest management. *Mailing Add:* PO Box 509 Dalmeny SK S0K 1E0 Can

EWEN, HAROLD IRVING, b Chicopee, Mass, Mar 5, 22; m 56; c 8. MICROWAVE RADIOMETRY, RADIO ASTRONOMY. *Educ:* Amherst Col, BA, 43; Harvard Univ, MA, 48, PhD(physics), 51. *Prof Exp:* Instr math, Amherst Col, 42-43; instr astron, Harvard Univ, 52-57, assoc, 57-80; pres, Ewen Knight Corp, 52-85 & Ewen Dae Corp, 58-85; EXEC VPRES, MILLITECH CORP, 88- *Concurrent Pos:* US mem, Int Astron Union, 53- *Honors & Awards:* Morris E Leeds Award, Inst Elec & Electronics Engrs, 70. *Mem:* Fel Am Acad Arts & Sci; fel AAAS; fel Inst Elec & Electronics Engrs. *Res:* Millimeter spectrum of solar proton flares; microwave characteristics of terrain materials; millimeter wave characteristics of the atmosphere. *Mailing Add:* Hillcrest Dr South Deerfield MA 01373

EWERS, RALPH O, b Cincinnati, Ohio, Aug 14, 36; m; c 1. GROUNDWATER BASINS. *Educ:* Univ Cincinnati, BS, 69, MS, 72; McMaster Univ, Can PhD(geomorphol), 82. *Prof Exp:* Exed dir, Sci Mus Palm Beach, 75-80; asst prof, 80-84, ASSOC PROF, GEOL & HYDROGEOL, EASTERN KY UNIV, 84- *Concurrent Pos:* Adj prof, geol, Fla Atlantic Univ, 78-79, 79-80; adv coun mem, Ky Water Resources Res Inst, 83-; mem, Ky Water Well Drillers Cent Bd, 84- *Mem:* AAAS; Am Geophys Union; Geol Soc Am. *Res:* Published numerous articles in various journals. *Mailing Add:* Dept Geol Eastern Ky Univ Richmond KY 40475

EWERT, ADAM, b Mt Lake, Minn, Dec 1, 27; m 60; c 2. PARASITOLOGY. *Educ:* Tabor Col, BA, 51; Univ Tex, MA, 60; Tulane Univ, PhD(parasitol), 63. *Prof Exp:* Asst prof parasitol & lectr, Fac Med, Univ Singapore, 64-67; from asst prof to assoc prof, 67-79, PROF MICROBIOL, UNIV TEX MED BR, GALVESTON, 79- *Concurrent Pos:* Vis prof, Yangming Med Col, Taipei, 86; vis prof, Shandong Med Univ, China, 87. *Mem:* Am Soc Trop Med & Hyg; Am Soc Parasitol; Int Filariasis Asn. *Res:* Lymphatic filariae; host-parasite interactions involving microfilaria and mosquitoes. *Mailing Add:* Dept Microbiol Univ Tex Med Br Galveston TX 77550-2782

EWIG, CARL STEPHEN, b Elmira, NY, May 28, 45; m 78; c 3. COMPUTATIONAL CHEMISTRY. *Educ:* Univ Rochester, BS, 67; Univ Calif, Santa Barbara, PhD(chem), 73. *Prof Exp:* From res assoc to res asst prof, 73-82, RES ASSOC PROF CHEM, VANDERBILT UNIV, 82- *Mem:* Am Chem Soc; Am Phys Soc; Sigma Xi. *Res:* Use of theoretical techniques in quantum mechanics to predict molecular electronic structure and properties; related computer applications in chemistry and physics. *Mailing Add:* Dept Chem Box 1822 Vanderbilt Univ Nashville TN 37235

EWING, ANDREW GRAHAM, b Huntington, NY, Jan 19, 57; m 77; c 4. ELECTROCHEMISTRY, NEUROCHEMISTRY. *Educ:* St Lawrence Univ, BS, 79; Ind Univ, PhD(anal chem), 83. *Prof Exp:* Res assoc, Univ NC, Chapel Hill, 83-84; asst prof, 84-89, ASSOC PROF CHEM, PA STATE UNIV, 89-, ASST HEAD DEPT & DIR GRAD ADMIN CHEM, 90- *Concurrent Pos:* Res fel, Alfred P Sloan, 89. *Honors & Awards:* Presidential Young Investr

Award, 87. *Mem:* Am Chem Soc; AAAS; Soc Electroanal Chem; Sigma Xi; NY Acad Sci; Soc Neurosci. *Res:* Design and application of extremely small electrochemical and separation-based methods for chemical analysis in ultra-small environments; development of novel polymer-modified electrodes for chemical analysis. *Mailing Add:* 152 Davey Lab Pa State Univ University Park PA 16802

EWING, BEN B, b Donna, Tex, Apr 4, 24; m 47; c 3. CIVIL ENGINEERING, ENVIRONMENTAL ENGINEERING. *Educ:* Univ Tex, BSCE, 44, MS, 49; Univ Calif, PhD(sanit eng), 59; Am Acad Environ Engrs, dipl. *Prof Exp:* From instr to asst prof civil eng, Univ Tex, 47-52; civil engr, Hqs Fourth Army, Ft Sam Houston, 52-53; asst prof civil eng, Univ Tex, 53-55 & 58; assoc, Univ Calif, 55-56, asst res engr, 56-58; assoc prof sanit eng, 58-61, dir water resources ctr, 66-72, prof sanit eng & nuclear eng, 67-85, dir, Inst Environ Studies, 72-85, PROF CIVIL ENG, UNIV ILL, URBANA, 61-, EMER PROF SANIT ENG, 85-, EMER DIR, INST ENVIRON STUDIES, 85- *Concurrent Pos:* Pub mem, Water Resources Comn, State of Ill, 74-; chmn comt on lead in human environ, Nat Acad Sci-Nat Res Coun-Environ Studies Bd, 78-80. *Honors & Awards:* Harrison Prescott Eddy Award for Noteworthy Res, Water Pollution Control Fedn, 68. *Mem:* Water Pollution Control Fedn; fel Soc Civil Engrs; Am Soc Eng Educ; Am Water Works Asn; Asn Prof Environ Engrs. *Res:* Water quality and pollution; reactions between organic compounds and clay minerals; radioactive waste disposal; ground water pollution recharge of ground water supplies; water quality management; lead in environment. *Mailing Add:* 1810 Glenwood Oaks Ct B Urbana IL 61801

EWING, CHANNING LESTER, b Jefferson City, Mo, May 28, 27; m 56; c 2. BIOENGINEERING, AEROSPACE MEDICINE. *Educ:* Med Col Va, MD, 52; Johns Hopkins Univ, MPH, 63; Am Bd Prev Med, dipl, 64. *Prof Exp:* US Navy, 52-, intern, US Naval Hosp, Portsmouth, Va, 52-53, flight surgeon, VMA-211 & CVG-3, USS Wright & USS Ticonderoga, 53-58, resident prev med, Aerospace Crew Equip Lab, Naval Air Eng Ctr, Philadelphia, 58-60, sr med officer, USS Essex, 60-62, resident, Aerospace Crew Equip Lab, Naval Air Eng Ctr, 63-64, asst dir training, Naval Aerospace Med Inst, 64-67, chief bioeng sci div, Res Dept, Naval Aerospace & Med Res Lab, 67-69, dist med officer, 17th Naval Dist, Kodiak, Alaska, 69-70, chief bioeng sci div, Naval Aerospace Med Inst, 70-71, officer in chg, 71-76, sci dir, Naval Aerospace Med Res Lab Detachment, Naval Aerospace Med Res Lab, 76-82, chief scientist, Naval Biodynamics Lab, 82-84; PRES, EWING BIODYNAMICS CORP, 84- *Concurrent Pos:* Partic, NASA Gemini 5 Proj, 65, Gemini 9 Proj, 66; mem bd dir & dir res & develop, Snell Mem Found, 89- *Honors & Awards:* Liljenkrantz Award, Aerospace Med Asn, 77; Legion of Merit, 77. *Mem:* Fel Am Col Prev Med; fel Aerospace Med Asn; Am Nat Standards Inst. *Res:* Biomechanics; physiological basis of protective equipment design, especially dynamic response of living human head neck and torso to impact acceleration; mechanism of ejection vertebral fracture, and head and spine protection against crash injury. *Mailing Add:* Ewing Biodynamics Corp 1018 Napoleon Ave New Orleans LA 70115-2819

EWING, CLAIR EUGENE, b Blue Rapids, Kans, Sept 20, 15; m 42; c 5. GEODESY. *Educ:* Kans State Univ, BS, 41; Univ Colo, MS, 50; Ohio State Univ, PhD(geod), 55. *Prof Exp:* Dir range develop, US Air Force, Atlantic Missile Range, Patrick AFB, Fla, 55-58, comdr, Air Technol Intel Ctr, Tex, 59-60, dep comdr, Navy Pac Missile Range, 60-67, vcomdr, Air Force Western Test Range, 67-69; chief scientist, Fed Elec Corp, 69-72; CONSULT & LECTR 72- *Concurrent Pos:* Adj prof, Golden Gate Univ, 76- *Mem:* Sigma Xi; Am Geophys Union. *Res:* Missile and space range instrumentation; electronic surveying; geodetic computations. *Mailing Add:* 4344 Sirius Ave Lompoc CA 93436-1025

EWING, DAVID LEON, b Shreveport, La, Aug 20, 41; m 65; c 2. RADIATION BIOLOGY. *Educ:* Centenary Col, BS, 63; Univ Calif, Berkeley, MS, 65; Univ Tex, Austin, PhD(zool), 69. *Prof Exp:* Fel radiation biol, Inst Cancer Res, Surrey, Eng, 70-72; asst prof zool, Univ Tex, Austin, 72-76; assoc prof radiation ther, 76-83, PROF RADIATION ONCOL, HAHNEMANN UNIV, 83- *Concurrent Pos:* Radiation Res Soc travel award, 77 & 79. *Mem:* Radiation Res Soc; AAAS; Environ Mutagen Soc; NY Acad Sci; Soc Free Radical Res; Oxygen Soc. *Res:* Chemical mechanisms of radiation damage in cells; modification of radiation sensitivity by chemical processes. *Mailing Add:* Dept Radiation Oncol Hahnemann Univ MS 102 Philadelphia PA 19102

EWING, DEAN EDGAR, b Ft Wayne, Ind, Aug 15, 32; m 69. MEDICAL RESEARCH, AVIAN VETERINARY MEDICINE. *Educ:* Mich State Univ, BS, 54, DVM, 56; Univ Rochester, MS, 62. *Prof Exp:* Vet, Bur Poultry Inspection, Dept Agr, State of Calif, 56-58; Chief vet serv, 3605th Air Force Hosp, Ellington AFB, Tex, 58-59 & 821st Med Group, Ellsworth AFB, SDak, 59-61, res vet, 6570th Aerospace Med Res Lab, Wright-Patterson AFB, Ohio, 62-63, chief bioastronaut group, Air Force Weapons Lab, Kirtland AFB, NMex, 63-68, chief vet civic action, 606th Spec Opers Squadron, Nakhon Phanom Royal Thai AFB, Thailand, 68-69 & bioenviron br, Air Force Weapons Lab, 69, chief biomed br, 69-73, med res officer, Defense Civil Preparedness Agency, US Air Force, 73-77, chief, Vet Support Group, Naval Ocean Syst Ctr, San Diego, 77-79; pres, Birdlife Inc, 83-88; dir, The Bird Ctr, San Diego, Calif, 86-89. *Concurrent Pos:* Proprietor vet house call service avian med, 79- *Mem:* Am Vet Med Asn; Asn Avian Vet; fel Explorers Club. *Res:* Research management in areas ranging from basic biology to radiation effects in both natural and man-made environments; veterinary medicine; emergency and disaster medicine; conservation-ecology; avian medicine and nutrition. *Mailing Add:* 11262 Via Carroza San Diego CA 92124

EWING, DONALD J(AMES), JR, b Toledo, Ohio, Jan 7, 31; m 57; c 2. ELECTRICAL ENGINEERING. *Educ:* Univ Toledo, BS, 52; Mass Inst Technol, MS, 54; Univ Wis, PhD(elec eng), 71. *Prof Exp:* Asst elec eng, Mass Inst Technol, 52-54; from instr to assoc prof, 54-72, PROF ELEC ENG, UNIV TOLEDO, 72- *Concurrent Pos:* NSF fac fel, 62-63. *Mem:* Inst Elec & Electronics Engrs; Asn Comput Mach; Int Asn Math & Comput Simulation; Soc Comput Simulation; Sigma Xi. *Res:* Feed back control systems and computers. *Mailing Add:* 608 Pierce St Maumee OH 43537

EWING, ELMER ELLIS, b Normal, Ill, Sept 16, 31; m 55; c 5. VEGETABLE CROPS, PLANT PHYSIOLOGY. *Educ:* Univ Ill, BS, 53, MS, 54; Cornell Univ, PhD(veg crops), 59. *Prof Exp:* Sci aide, Agr Prog for Latin Am, Rockefeller Found, 56-57; from instr to assoc prof, 58-72, PROF VEG CROPS, CORNELL UNIV, 72-, DEPT CHMN, 82- *Concurrent Pos:* NSF fel hort, Purdue Univ, 65-66; vis prof, Dept Bot & Microbiol, Univ Col Wales, Aberystwyth, Wales, 74-75, Dept Veg Crops, Univ Calif, Davis, 80 & Dept Crop Sci, Ore State Univ, Corvallis, 88. *Mem:* Am Soc Hort Sci; Potato Asn Am (pres, 85-86); Am Soc Plant Physiol; Europ Asn Potato Res. *Res:* Physiological problems of potato. *Mailing Add:* Dept Veg Crops Cornell Univ Ithaca NY 14853-0327

EWING, GALEN WOOD, b Boston, Mass, Mar 14, 14; m 42; c 3. ANALYTICAL CHEMISTRY, PHYSICAL CHEMISTRY. *Educ:* Col William & Mary, BS, 36; Univ Chicago, PhD(phys chem), 39. *Prof Exp:* Prof chem & physics, Blackburn Col, 39-42; res phys chemist, Sterling Winthrop Res Inst, 42-46; from asst prof to assoc prof chem, Union Col, NY, 46-57; prof, NMex Highlands Univ, 57-64, chmn dept, 59-61; prof, 64-79, chmn dept, 69-72, EMER PROF CHEM, SETON HALL UNIV, 79- *Concurrent Pos:* Adj prof, NMex Highlands Univ, 79-; vis prof, Carleton Col, 83-84. *Mem:* AAAS; Soc Appl Spectros; Am Chem Soc; Sigma Xi. *Res:* Analytical instrumentation for educational laboratories; analytical chemistry. *Mailing Add:* 3009 Raybun St Las Vegas NM 87701-5119

EWING, GEORGE EDWARD, b Charlotte, NC, Nov 28, 33; c 6. PHYSICAL CHEMISTRY. *Educ:* Yale Univ, BS, 56; Univ Calif, Berkeley, PhD(chem), 60. *Prof Exp:* Sr scientist, Jet Propulsion Lab, 60-63; from instr to assoc prof, 63-71, PROF CHEM, IND UNIV, BLOOMINGTON, 71- *Concurrent Pos:* Mem tech staff, Bell Tel Labs, NJ, 69-70; dir res, Polytech Sch, Palaiseau, France, 76-77. *Mem:* Am Chem Soc; Am Phys Soc. *Res:* Molecular spectroscopy; low temperature chemistry; energy transfer mechanisms. *Mailing Add:* Dept Chem Ind Univ Bloomington IN 47405

EWING, GEORGE MCNAUGHT, b Lexington, Mo, Sept 30, 07; m 37; c 3. MATHEMATICS. *Educ:* Univ Mo, AB, 29, AM, 30, PhD(math), 35. *Prof Exp:* From instr to prof math, Univ Mo, 30-58; res assoc, Okla Res Inst, Combat Develop Dept, US Army Artillery & Missile Sch, 57-60; prof, 60-63, George L Cross res prof, 63-77, EMER GEORGE L CROSS RES PROF MATH, UNIV OKLA, 77- *Concurrent Pos:* Instr, Princeton Univ, 40-41; mem, Inst Advan Study, 40-41; mathematician, Naval Ord Lab, 44-45, Sandia Corp, 51-52 & Ramo-Wooldridge Corp, 54. *Mem:* Am Math Soc; Soc Indust & Appl Math; Math Asn Am. *Res:* Calculus of variations; optimal control theory; ordinary differential equations. *Mailing Add:* 816 Col Ave Norman OK 73069

EWING, GERALD DEAN, b Alliance, Nebr, Jan 6, 32; m 52; c 3. ELECTRONICS, INSTRUMENTATION. *Educ:* Univ Calif, Berkeley, BSEE, 57, MSEE, 59; Ore State Univ, EE, 62, PhD(elec eng), 64. *Prof Exp:* Elec engr, Lawrence Radiation Lab, Univ Calif, 56-58; electronics engr, Electronics Defense Lab, Sylvania Elec Prod, Inc, Gen Tel & Electronics Corp, 58-60; semiconductor appln engr, Rheem Semiconductor Corp, 60-61; supvr appln eng, Shockley Transistor, Clevelite Corp, 61; instr elec eng, Ore State Univ, 61-63; ASSOC PROF ELEC ENG, NAVAL POSTGRAD SCH, 63- *Concurrent Pos:* Lectr, Foothill Col, 58-61; consult, Lind Instrument Corp, Calif, 60-61 & Sylvania Elec Prod, Inc, Gen Tel & Electronics Corp, 63; lectr, Hartnel Col, 64-; mem staff, Behav Sci Inst, Calif, 66- *Mem:* Inst Elec & Electronics Engrs. *Res:* Solid state devices; oceanography. *Mailing Add:* Dept Elec Eng Naval Postgrad Sch Monterey CA 93940

EWING, GORDON J, b Smithfield, Utah, Nov 1, 31; m 68; c 12. PHYSICAL CHEMISTRY. *Educ:* Utah State Univ, BS, 54, MS, 57; Pa State Univ, PhD(thermodynamics), 60. *Prof Exp:* Asst prof, 62-68, ASSOC PROF CHEM, NMEX STATE UNIV, 68- *Mem:* AAAS; Am Chem Soc; Sigma Xi. *Res:* Interactions of gases with respiratory pigments; thermometric titrations; solution thermodynamics; alternate methods for determining chemical oxygen demand. *Mailing Add:* Chem Dept NMex State Univ Las Cruces NM 88003

EWING, J J, b Morristown, NJ, Dec 15, 42; m 67; c 1. CHEMICAL PHYSICS. *Educ:* Univ Calif, BA, 64; Univ Chicago, PhD(phys chem), 69. *Prof Exp:* Vis asst prof chem, Univ Ill, 69-71; asst prof, Univ Del, 71-72; prin res scientist atomic & molecular physics, Avco Everett Res Lab, 72-76; mem staff, Laser Fusion Prog, Lawrence Livermore Lab, 76-79; VPRES, LASER PROGS, MSNW, STI OPRONICS, 79- *Mem:* Am Phys Soc; AAAS; Sigma Xi. *Res:* Laser research and development; ultra-violet laser; laser radar; solid state lasers; electronic energy transfer; kinetics. *Mailing Add:* 5416 143rd Ave SE Bellevue WA 98006-4378

EWING, JOAN ROSE, electrophysics, for more information see previous edition

EWING, JOHN ALEXANDER, b Fife, Scotland, Mar 17, 23; US citizen; m 46; c 2. PSYCHIATRY. *Educ:* Univ Edinburgh, MB & ChB, 46, MD, 54; Univ London, dipl (psychol med), 50. *Prof Exp:* Res psychiat, Cherry Knowle Hosp, Eng, 47-51; sr physician, John Umstead Hosp, Butner, NC, 51-54; clin instr, 53-54, from instr to prof, 54-84, chmn dept, 65-70, dir, Ctr Alcohol Studies, 71-84, EMER PROF PSYCHIAT, SCH MED, UNIV NC, CHAPEL HILL, 84-; PSYCHIATRIST, PVT PRACT, WILMINGTON, NC, 87- *Concurrent Pos:* Asst physician, Psychiat Clin, Sunderland Royal Infirmary, Eng, 49-51 & S Shields Gen Hosp, 49-51; psychiatrist, NC Alcoholic Rehab Ctr, 51-54; dir, Psychiat In-Patient Serv, NC Mem Hosp, 57-64; consult psychiatrist, Watts Hosp, Durham, NC, 57-84. *Mem:* Fel Am Psychiat Asn; fel Royal Col Psychiatrists; Am Med Soc Alcoholism; Am Asn Social Psychiat; fel Am Col Psychiat. *Res:* Alcoholism and drug dependency; application of psychiatric principles in medical practice; psychoanalysis. *Mailing Add:* 2311 Canterwood Dr Wilmington NC 28401-2329

EWING, JOHN ARTHUR, b Euchee, Tenn, June 24, 12; m 39; c 3. AGRONOMY, RESEARCH ADMINISTRATION. *Educ:* Univ Tenn, BSA, 33, MS, 46; Harvard Univ, DPA, 56. *Prof Exp:* Teacher high sch, Tenn, 34-35; asst county agent, Demonstration Prog, Tenn Valley Authority, Agr Exten Serv, 35-44, from asst supt to supt, Middle Tenn Exp Sta, 44-49, asst dir, Agr Exp Sta, 49-55, sr vdean col agr, 55-57, dir, Agr Exp Sta, 57-68, dean, 68-75, EMER DEAN, AGR EXP STA, UNIV TENN, KNOXVILLE, 75- *Concurrent Pos:* Admin adv, Water Resource Res South, 53-, exp sta rep, Soybean Res, South, 54- prog leader, Comp Animal Res Lab, Energy Res & Develop Admin, Univ Tenn, 55-, exp sta rep, Seed & Plant Irradiation South, 56- & admin adv Grain Mkt South, 60-; chmn, South Agr Exp Sta Dirs, 61-62 & Exp Sta Sect, Land Grant Col Asn, 62-63; mem, Nat Tobacco Adv Comt, 62-64, Nat Cotton Seed Policy Comt, 65- & agr adv comt, South Regional Educ Bd; adminr adv, Southern Land Econ Res Comt; mem, Bd Univ Tenn Res Corp & State Tenn Air Pollution Bd; pres, Southern Asn Agr Scientist, 71; mem bd, Meigs County Farm Bur. *Mem:* Am Soc Agron; Sigma Xi. *Res:* Irrigation of pastures; effects of fluorine effluents on Cattle and crops; atomic energy in agricultural research. *Mailing Add:* PO Box 6 Ten Mile TN 37880

EWING, JOHN I, b Lockney, Tex, July 5, 24; m 48; c 3. GEOPHYSICS. *Educ:* Harvard Univ, BS, 50. *Prof Exp:* Res assoc geophysics, Lamont-Doherty Geol Observ, Columbia Univ, 50-58, sr res scientist, Woods Hole Oceanog Inst, 58-64, sr res assoc, 64-73, adj prof geol, 74-80, assoc dir res, 73-76; chmn, Dept Geol & Geophys, 76-81, sr scientist, 82-89, EMER SR SCIENTIST, WOODS HOLE OCEANOG INST, 90- *Honors & Awards:* Francis P Shepard Medal, Soc Econ Paleontologist & Mineralogists; Maurice Ewing Medal, Am Geophys Union & US Navy. *Mem:* Soc Explor Geophys; fel Am Geophys Union; fel AAAS. *Res:* Structure and constitution of the earth; underwater sound propagation. *Mailing Add:* Woods Hole Oceanog Inst Woods Hole MA 02543

EWING, JUNE SWIFT, b Fayetteville, Ark, July 19, 38; m 78; c 1. ELECTRON MICROSCOPY. *Educ:* Univ Wis-Madison, BS, 59; Univ Calif, Berkeley, MS, 61; Univ Colo, Boulder, MPA, 76. *Prof Exp:* Prog consult, Harvey Mudd Col, 60-63; electron microscopist, Ind Univ, Bloomington, 63-66, proj mgr, 67-68; instr chem, Rutgers Univ, 69-70; mgr lab cell biol, Univ Colo, Boulder, 71-74; mgt intern, NASA-Goddard Space Flight Ctr, 75-76; prog mgr, Univ Space Res Asn, 76-77, Sci Appln Inc, Va, 77-79; STAFF OFFICER, NAT ACAD SCI, WASHINGTON, DC, 79- *Concurrent Pos:* Vpres, Environ Sci Commun, Inc, 78-; secy, Appl Sci & Technol, Inc, 82- *Mem:* AAAS; Int Brain Res Orgn. *Res:* Electron microscopy of brain tissue. *Mailing Add:* Dept Health & Human Serv PHS ADA MHA 5600 Risher Rd Rm 13-103 Rockville MD 20857

EWING, LARRY LARUE, b Valley, Nebr, July 10, 36; m 54; c 4. PHYSIOLOGY, ENDOCRINOLOGY. *Educ:* Univ Nebr, BS, 58; Univ Ill, MS, 60, PhD(agr), 62. *Prof Exp:* From asst prof to prof physiol, Okla State Univ, 62-72; PROF POP DYNAMICS, SCH HYG, JOHNS HOPKINS UNIV, 72- *Concurrent Pos:* NIH trainee, Dept Biochem, Univ Utah, 64-65; NIH spec res fel, Dept Pharmacol, Johns Hopkins Univ, 68-69; NIH career develop award, 71; consult, Reprod Biol Study Sect, NIH, 73-77 & NSF, 73- *Mem:* Am Physiol Soc; Endocrine Soc; Soc Study Fertil; Soc Study Reprod. *Res:* Reproductive physiology and endocrinology; male contraception. *Mailing Add:* Dept Pop Dynamics Sch Hyg & Pub Health Johns Hopkins Univ Baltimore MD 21205

EWING, MARTIN SIPPLE, b Albany, NY, May 4, 45; m 66; c 3. RADIO ASTRONOMY. *Educ:* Swarthmore Col, BA, 66; Mass Inst Technol, PhD(physics), 71. *Prof Exp:* Res asst radio astron, Mass Inst Technol, 66-71; res fel radio astron, Calif Inst Technol, 71-73, sr res engr radio astron, 73-75, mem prof staff, 75-89; staff mem, Owens Valley Radio Observ, 75-89; DIR, SCI & ENG COMPUT FACIL, YALE UNIV, 89- *Mem:* Int Union Radio Sci; Int Astron Union; Inst Elec & Electronics Engrs; AAAS; Asn Comput Machinery. *Res:* Instrumentation for radio astronomy; very long baseline interferometry of extragalactic objects and pulsars; interstellar scintillation of pulsars; computer control systems. *Mailing Add:* Yale Univ PO Box 1968 New Haven CT 06520-1968

EWING, R(OBERT) A(RNO), b Washington, DC, June 20, 15; m 47; c 3. CHEMICAL ENGINEERING. *Educ:* Ohio State Univ, BSc, 36, MSc, 37. *Prof Exp:* Res engr, Monsanto Chem Co, 37-41; res engr, Eagle-Picher Co, 41-44, plant mgr, 44-47; res engr, 47-49, asst div chief, 49-60, res assoc, 61-65, fel, 65-71, sr res engr, Battelle Mem Inst, 71-80; RETIRED. *Mem:* Am Inst Chem Engrs. *Res:* Manufacture of elemental phosphorus and phosphates; manufacture of mineral wool insulation; recovery of uranium and thorium from ores; pollution control; environmental effects of toxic substances; environmental impact assessment. *Mailing Add:* 2435 Haverford Columbus OH 43220

EWING, RICHARD DWIGHT, b Lansing, Mich, Jan 17, 30; m 54; c 3. PHYSICS. *Educ:* Univ Chicago, BA, 50; Mich State Univ, MS, 54, PhD(physics), 58. *Prof Exp:* Res found grant, Univ Del, 59-60, from asst prof to assoc prof physics, 58-91; RETIRED. *Mem:* Am Phys Soc. *Res:* Electron spin and nuclear magnetic resonance. *Mailing Add:* 19 Helios Ct Treetop Newark DE 19711

EWING, RICHARD EDWARD, b Kingsville, Tex, Nov 24, 46; m 70; c 3. SCIENTIFIC COMPUTATION, MATHEMATICAL MODELING. *Educ:* Univ Tex, Austin, BA, 69, MA, 72, PhD(math), 74. *Prof Exp:* Asst prof math, Oakland Univ, 74-77; from asst prof to assoc prof math, Ohio State Univ, 77-80; sr res math, Mobil Res & Develop Corp, 80-83; PROF MATH, CHEM & PETROL ENG, UNIV WYO, 83-, DIR, ENHANCED OIL RECOV INST, 84-, J E WARREN DISTINGUISHED PROF ENERGY & ENVIRON, 84-, WOLD CENTENNIAL CHAIR ENERGY, 90- *Concurrent Pos:* Prin investr numerous orgns, 78-; ed, Soc Indust & Appl Math, 83-, Computer Methods Appl Mech & Eng, 85-, Numerical Methods Partial Differential Equations, 86-; dir, Inst Sci Computation, Univ Wyo, 86-, NSF Coop Res Ctr Math Modeling, 86-89, Wyo Ctr Energy Res, 89-; hon prof, Shandong Univ, People's Repub China, 87. *Mem:* Soc Indust & Appl Math; Int Asn Computational Mech; Am Math Soc; Math Asn Am; Am Geophys Union; Soc Petrol Engrs. *Res:* Mathematical modeling, numerical analysis, large scale scientific computation, environmental engineering, and enhanced oil recovery. *Mailing Add:* 1055 Granito Laramie WY 82070

EWING, RODNEY CHARLES, b Abilene, Tex, Sept 20, 46; div; c 2. MINERALOGY, GEOCHEMISTRY. *Educ:* Tex Christian Univ, BS, 68; Stanford Univ, MS, 72, PhD(geol), 74. *Prof Exp:* Cur mineral, Stanford Univ, 73-74; from asst prof to assoc prof, 74-84, chmn dept, 79-84, PROF MINERAL, UNIV NMEX, 84- *Concurrent Pos:* Prin investr, Sandia Corp, 74-81 & 84, Battelle Mem Labs, 77-80 & Argonne Nat Lab, 84-86, guest scientist, Hahn-Meitner Inst, Berlin, 79-88, Ctr D'Etudes Nucl-e01aires, Paris, 89, Japan Atomic Energy Res Inst, 90, Kernforschungszentrum Karlsruhe, Ger, 90. *Mem:* Fel Geol Soc Am; fel Mineral Soc Am; Mineral Asn Can; Sigma Xi; AAAS; Mat Res Soc; Nat Asn Geol Teachers; Geochemical Soc. *Res:* Rare-earth, metamict minerals; process of radiation damage in natural materials; relation of texture and fabric of fine grained sedimentary rocks to clay mineralogy and diagenetic history; nuclear waste disposal; corrosion of glasses; materials science. *Mailing Add:* Dept Geol Univ NMex Albuquerque NM 87131

EWING, RONALD IRA, b Dallas, Tex, July 13, 35; m 57; c 4. PHYSICS. *Educ:* Rice Inst, BA, 56, MA, 57, PhD(physics), 59. *Prof Exp:* SR MEM, TECH STAFF, SANDIA NAT LABS, 59- *Mem:* Am Phys Soc; Am Nuclear Soc. *Res:* Nuclear Physics. *Mailing Add:* 6715 Ruby NE Albuquerque NM 87109

EWING, SIDNEY ALTON, b Emory University, Ga, Dec 1, 34; m 63; c 3. VETERINARY PARASITOLOGY. *Educ:* Univ Ga, BSA & DVM, 58; Univ Wis, MS, 60; Okla State Univ, PhD(vet parasitol), 64. *Prof Exp:* Asst vet sci, Univ Wis, 58-60; instr vet parasitol, Okla State Univ, 60-61, asst prof, 61-64, assoc prof vet parasitol & pub health, 64-65; assoc prof vet path, parasitol & pub health, Kans State Univ, 65-67; prof vet sci & head dept, Miss State Univ, 67-68; prof vet parasitol & pub health & head dept, Okla State Univ, 68-72; prof & dean Col Vet Med, Univ Minn, St Paul, 72-78; dept head, 79-84, PROF VET PARASITOL, MICROBIOL & PUB HEALTH, OKLA STATE UNIV, 79- *Concurrent Pos:* Mem adv bd, Morris Animal Found, 67-69, consult, 69-78; mem comt animal health, Nat Res Coun-Nat Acad Sci, 71-75; mem adv, panel vet med, US Pharmacopeial Comt Rev, 80-95 & coun, Conf Res Workers Animal Dis, 80-85, vpres, 83-84, pres, 84-85; mem adv bd, Am Coun Sci & Health. *Mem:* Am Vet Med Asn; Am Soc Vet Parasitol; Am Soc Parasitol; Conf Res Workers Animal Dis (vpres, 83-84, pres, 84-85); Sigma Xi. *Res:* Swine lungworms; canine babesiosis; canine rickettsiosis; anaplasmosis; zoonotic potential of ehrlichiosis. *Mailing Add:* Col Vet Med Okla State Univ Dept PARA/MIC/PUB Stillwater OK 74078

EWING, SOLON ALEXANDER, b Headrick, Okla, July 21, 30; m 52; c 2. ANIMAL NUTRITION. *Educ:* Okla State Univ, BS, 52, MS, 56, PhD(animal nutrit), 58. *Prof Exp:* Instr animal sci, Okla State Univ, 56-58; from asst prof to assoc prof, Iowa State Univ, 58-64; prof, Okla State Univ, 64-68; asst dir, Iowa Agr & Home Econ Exp Sta, 68-73, HEAD DEPT ANIMAL SCI, IOWA STATE UNIV, 73- *Mem:* Am Soc Animal Sci. *Res:* Ruminant nutrition studies. *Mailing Add:* 101 Kildee Hall Iowa State Univ Ames IA 50010

EWING, THOMAS EDWARD, b Elgin, Ill, Aug 6, 54; m; c 1. TECTONICS, REGIONAL ANALYSIS. *Educ:* Colo Col, BA, 75; NMex Inst Mining & Technol, MS, 77; Univ BC, PhD(geol sci), 81. *Prof Exp:* Res assoc scientist, Tex Bur Econ Geol, 80-81, res assoc, 81-85; PRES, FRONTERA EXPLOR SERV, 85- *Concurrent Pos:* Adj prof, Univ Tex, San Antonio, 89- *Mem:* Geol Soc Am; Am Asn Petrol Geologists; Geol Asn Can; Asn Geoscientists Int Develop. *Res:* Regional analysis of geological structure, tectonics and structural development, presently focused on the Texas Gulf Coast and in the Pacific Northwest; volcanology and volcanic petrology; petroleum geology. *Mailing Add:* Frontera Explor 900 NE Loop 410 Suite D-303 San Antonio TX 78209

EWING, WILLIAM HOWELL, b Carnegie, Pa, Oct 4, 14; m 42; c 2. MEDICAL MICROBIOLOGY. *Educ:* Washington & Jefferson Col, AB, 37, MA, 39; Cornell Univ, PhD(bact), 48. *Prof Exp:* Instr biol, Washington & Jefferson Col, 37-39; asst bact, Univ Mich, 39-41; assoc prof, NY State Vet Col, Cornell Univ, 46-47, from instr to asst prof, 47-48; bacteriologist & asst in charge enteric bact unit, Ctr Dis Control, USPHS, 48-62, in charge Int Shigella Ctr, 50-74, chief enteric bact unit, 62-69, consult & res microbiologist, 69-74; CONSULT MICROBIOLOGIST, 74- *Concurrent Pos:* Mem, Int Subcomt Enterobacteriaceae, 50-, past secy; consult, WHO, 55, adv, 58, expert adv panel enteric dis, 64-; assoc prof, Sch Pub Health, Univ NC, 62-; Linton fel from Washington & Jefferson Col, Marine Biol Lab, Woods Hole. *Honors & Awards:* Kimble Methodology Res Award, 56; USPHS Meritorious Serv Award, 63; Wyeth Award Clin Microbiol, Am Soc Microbiol, 75; Bergey Award, 86. *Mem:* Am Soc Microbiol; Am Acad Microbiol; Can Soc Microbiol; Brit Soc Gen Microbiol; Infectious Dis Soc Am. *Res:* Antigenic analyses; antigens of enterobacteriaceae, especially Shigella, Escherichia, Salmonella; relationship of these bacteria to disease in man; biochemical characteristics and differentiation; classification and nomenclature. *Mailing Add:* 2364 Wineleas Rd Decatur GA 30033

EXARHOS, GREGORY JAMES, b Milwaukee, Wis, Oct 27, 48; m 79; c 3. PHYSICAL CHEMISTRY, SOLID STATE CHEMISTRY. *Educ:* Lawrence Univ, AB, 70; Brown Univ, PhD(chem), 74. *Prof Exp:* Res asst chem, Argonne Nat Lab, 69-70; asst prof chem, Harvard Univ, 74-80; sr res scientist, 80-86, STAFF SCIENTIST, PACIFIC NORTHWEST LAB, BATTELLE MEM INST, 86-, MGR MAT RES SECT, 91- *Concurrent Pos:* Assoc ed, Mat Lett; consult, NSDI, 80; prin investr, Dept Energy Progs. *Mem:* Am Chem Soc; Am Ceramic Soc; AAAS; Mat Res Soc; Microbeam Anal Soc. *Res:* Investigations of condensed phases, surfaces and interfaces by molecular spectroscopy; resonance Raman effects, infrared and magnetic resonance spectroscopy; phase transformation phenomena at high temperatures and pressures, materials preparation and solid state chemistry of glasses, polymers and dielectric films; non-linear optics measurements. *Mailing Add:* Battelle NW K2-44 Richland WA 99352

EXTON, JOHN HOWARD, b Auckland, NZ, Aug 29, 33; m 57; c 4. BIOCHEMISTRY. *Educ:* Univ NZ, BMedSc, 55; Univ Otago, NZ, MB, ChB, 58, PhD(biochem), 63; MD, 84. *Prof Exp:* Asst lectr biochem, Univ Otago, NZ, 61-63; from instr to assoc prof physiol, 63-70, PROF PHYSIOL, SCH MED, VANDERBILT UNIV, 70- *Concurrent Pos:* Investr, Howard Hughes Med Inst, 68-74 & 76- *Honors & Awards:* Lilly Award, Am Diabetes Asn, 72. *Mem:* Am Physiol Soc; Am Diabetes Asn; Am Soc Biol Chemists; Biochem Soc. *Res:* Control of metabolism; mechanisms of hormone action; gluconeogenesis, glycogen metabolism, and ketogenesis; effects of glucagon, catecholamines, glucocorticoids, and insulin on liver; signal transduction mechanisms; roles of phospholipids and calcium. *Mailing Add:* Dept Physiol & Investr Howard Hughes Med Inst Vanderbilt Univ 831 Light Hall Nashville TN 37232-0295

EXTON, REGINALD JOHN, b Folsom, NJ, July 17, 35; m 57; c 3. SPECTROSCOPY, OPTICS. *Educ:* Univ Richmond, BS, 58; WVa Univ, MS, 61, PhD(physics), 72. *Prof Exp:* PHYSICIST, LANGLEY RES CTR, NASA, 61- *Concurrent Pos:* Pres, Res Ventures Inc, 77- *Mem:* Am Phys Soc; AAAS. *Res:* Atomic and molecular spectroscopy; laser spectroscopy; environmental research. *Mailing Add:* NASA Langley Res Ctr MS 235A Hampton VA 23665

EYDE, RICHARD HUSTED, plant anatomy, paleobotany; deceased, see previous edition for last biography

EYDGAHI, ALI MOHAMMADZADEH, b Tehran, May, 1957; m 83; c 2. MULTIDIMENSIONAL SIGNAL PROCESSING, APPLICATIONS COMPUTER ALGEBRA & SYMBOLIC COMPUTATION. *Educ:* Detroit Inst Technol, BS, 79; Wayne State Univ, MS, 81, PhD(elec eng), 86. *Prof Exp:* Vis prof computer appln digital systs, Rensselear Polytech Inst, 85-86; PROF ROBOTICS & CONTROL SYSTS, STATE UNIV NY, 86- *Concurrent Pos:* Instr, Wayne State Univ, 80-85 & Engrs Training Prog, Ford Motor Co, 84-85; lectr, Wayne County Community Col, 84-85; prin investr, Very Large Scale Integration Rooting, IBM, 87-88; reviewer, Inst Elec & Electronics Engrs Trans, 87-; referee, Int J Computer Aided Very Large Scale Integration Design, 88-; chmn, Experts Rev Panel, US Dept Educ, 89, 90, 91; presiding chair, Third Nat Conf on Undergrad Res, Eureka, 89. *Honors & Awards:* Dow Outstanding Young Fac Award, Am Soc Eng Educ, 90. *Mem:* Sr mem Inst Elec & Electronics Engrs; sr mem Soc Mfg Engrs; Am Soc Eng Educ; Sigma Xi. *Res:* Development of computer simulation and animation for robotic and automated systems; professional computer-based software for control systems design and analysis; very large scale integration implementation of linear systems; distributed system program applications in control and robotic systems; multidimensional signal processing; computer algebra applications in control and systems. *Mailing Add:* Dept Elec Eng State Univ NY New Paltz NY 12561

EYE, JOHN DAVID, sanitary engineering, environmental health, for more information see previous edition

EYER, JAMES ARTHUR, b Rochester, NY, Dec 18, 29; m 60; c 1. OPTICS, PHOTOGRAPHY. *Educ:* Mass Inst Technol, BS, 51; Univ Rochester, PhD(optics, physics), 57. *Prof Exp:* Res assoc optics, Inst Optics, Univ Rochester, 57-61, asst prof, 61-63, asst dir, 63-65; private consult, NY, 65-67; assoc prof optical sci, Univ Ariz, 67-69, assoc dir, Optical Sci Ctr, 67-73, prof 69-75, spec asst to vpres res, 73-74; consult indust & govt, 74-79; RETIRED. *Concurrent Pos:* Chmn, Yotes County Civil Serv Comn, 84- *Mem:* Sigma Xi. *Res:* Optical image evaluation; photographic image structure; photographic theory; applications of information theory to optical systems; photointerpretation techniques. *Mailing Add:* PO Box 129 Dundee NY 14837-0129

EYER, JEROME ARLAN, b Orchard, Nebr, Aug 10, 34; m 57; c 2. EXPLORATION GEOLOGY, EXPLORATION GEOPHYSICS. *Educ:* Univ Mo, Columbia, BA, 60, MA, 61; Univ Colo, Boulder, PhD(geol, astrogeophys), 64. *Prof Exp:* Explor geologist, Humble Oil Co, 64-66; from res scientist to group supvr, Continental Oil Co, 66-69, dir geol res, 69-75; chmn, Dept Geol & Geophys, Univ Mo, Rolla, 75-77; explor mgr, Terra Resources, Inc, 77-78; chief geologist, Grace Petrol Corp, 78-79, vpres explor, 79-83; reg mgr, J A Eyer & Assoc, 83-90 & Forest Oil Corp, 87-90; ASSOC DIR & RES PROF, EARTH SCI & RESOURCE INST, UNIV SC, COLUMBIA, 90- *Concurrent Pos:* Mem speakers coun, Okla Petrol Coun, 71-80; consult, geothermal resources, State NMex, 73-74 & NASA, 73-75; mem bd dirs, Geosat Comt, Inc, 77-83; chmn, Sci Geophys Data Panel, Nat Acad Sci, 80-; vis prof, Univ Okla, 82-85. *Mem:* Am Geophys Union; Am Asn Petrol Geologists; Am Inst Prof Geologists; Soc Explor Geophysicists; Sigma Xi. *Res:* Petroleum source rock and paleotemperature documentation and determinations; remote detection of chemical variable; petroleum and mineral resource evaluations. *Mailing Add:* Earth Sci & Resources Inst Univ SC Columbia SC 29208

EYER, LESTER EMERY, b Ithaca, Mich, Apr 9, 12; m 40; c 3. ORNITHOLOGY. *Educ:* Alma Col, Mich, BS, 36; Univ Mich, MS, 42; Mich State Univ, PhD(zool), 54. *Prof Exp:* Teacher pub sch, Mich, 36-43; from instr to prof, 46-77, head dept, 51-71, EMER PROF BIOL, ALMA COL, MICH, 77- *Mem:* Am Ornith Union; Wilson Ornith Soc; Sigma Xi. *Res:* Ecology of birds. *Mailing Add:* 5355 Blue Heron Dr Alma MI 48801

EYERLY, GEORGE B(ROWN), b Canton, Ill, Mar 20, 17; m 47. CERAMICS ENGINEERING. *Educ:* Univ Ill, BS, 40; Univ Wash, MS, 41. *Prof Exp:* US Bur Mines fel, Northwestern Exp Sta, 40-41; chief ceramist, Refractories Corp, Calif, 41-42; chief metall sect, Manhattan Proj & US Atomic Energy Comn, Oak Ridge, 46-47; asst prof ceramic eng, Univ Wash, 47-48; assoc ceramist, Argonne Nat Lab, 48-52; ceramic engr, Allen Bradley Co, 52-55; secy & chief engr, Malvern Brick & Tile Co, 55-67; sr engr, D M Steward Mfg Co, 67-73; MGR ENERGY CONSERV DEPT, TEMTEK-ALLIED DIV, FERRO CORP, 73- *Concurrent Pos:* Consult, US Atomic Energy Comn, 59-62; mem, Ark State Geol Comn, 67 & Gov Tech Adv Comt, 67. *Mem:* AAAS; Am Ceramic Soc; Am Soc Metals; Am Chem Soc. *Res:* High temperature materials; materials for reactor applications; ferromagnetic and nonmetallic ferroelectric materials; structural clay products. *Mailing Add:* 634 Hurricane Creek Rd Chattanooga TN 37421

EYESTONE, WILLARD HALSEY, b Mulberry, Kans, Jan 7, 18; m 52; c 4. PATHOLOGY. *Educ:* Kans State Col, BS, 39, DVM, 41; Harvard Univ, MPH, 47; Univ Wis, PhD(path), 49. *Prof Exp:* Instr vet sci, Univ Wis, 41; res assoc vet path, Univ Ill, 42; res assoc, Univ Wis, 47-49; head comp path lab, Nat Cancer Inst, NIH, 49-55 & vet & chief lab aids br, 55-59, chief regional primate res ctrs br, Nat Heart Inst, 59-62, chief animal resources br, Div Res Facil & Resources, 62-71; prof & chmn dept, 72-83, interim dean, 81, EMER PROF VET PATH, COL VET MED, UNIV MO, 83-; CHIEF OPTOM, PHARM, PODIATRY & VET MED EDUC BR, DIV PHYSICIAN & HEALTH PROF EDUC, BUR HEALTH MANPOWER EDUC, NIH, 71- *Concurrent Pos:* Pathologist, Nat Zool Park, DC, 50-59; lab consult, Pan-Am Sanit Bur, Ecuador, 52; mem comt vet med res & educ, Nat Res Coun, 68-; mem, Nat Adv Coun Health Prof Educ, Dept Health, Educ & Welfare, 75- *Honors & Awards:* Charles A Griffin Award, Am Asn Lab Animal Sci, 70. *Mem:* Am Vet Med Asn; Am Col Vet Path (pres, 61-62); Am Asn Path & Bact; Am Asn Lab Animal Sci; Int Acad Path. *Res:* Cancer; pathogenesis of tumors; comparative pathology; research administration. *Mailing Add:* 912 Hulen Dr Columbia MO 65203

EYGES, LEONARD JAMES, b Chelsea, Mass, Oct 30, 20; m 43, 68; c 1. THEORETICAL PHYSICS. *Educ:* Univ Mich, BS, 42; Brown Univ, MS, 43; Cornell Univ, PhD, 48. *Prof Exp:* Staff mem, Radiation Lab, Mass Inst Technol, 43-46; AEC fel, 48-49; res physicist, Radiation Lab, Univ Calif, 49-52; sci attache, State Dept Paris, 52-53; instr physics, Mass Inst Technol, 53-55, staff mem, Lincoln Lab, 55-56 & 57-63; NSF sr fel, 56-57; sr physicist, Air Force Cambridge Res Labs, 63-82; NAVIG WRITER, 82- *Mem:* Fel Am Phys Soc. *Res:* Scattering theory; fluid flow. *Mailing Add:* 31 Denton Rd Wellesley MA 02181

EYKHOFF, PIETER, b The Hague, Neth, Apr 9, 29; m 55; c 2. SYSTEM IDENTIFICATION, PARAMETER ESTIMATION. *Educ:* Delft Univ, Neth, BSc, 55, MSc, 56; Univ Calif, Berkeley, PhD(elec eng), 61. *Hon Degrees:* Dr, Free Univ Brussels, 90. *Prof Exp:* Chief sci officer electronics, Delft Univ Technol, 56-64; vis res fel control, Nat Acad Sci, Univ Calif, Berkeley, 58-60; dean elec eng, 77-80, PROF CONTROL, EINDHOVEN UNIV TECHNOL, 64- *Concurrent Pos:* Consult, Orgn Econ Corp & Develop, Paris, 62-64; vis prof, Elec Eng Dept, Univ Waterloo, Can, 68; Fullbright fel, Univ Calif Berkeley, Southern Calif, Los Angeles & Wis, Madison, 75; res fel, Dept Appl Math & Physics, Japan Soc Prom Sci, Kyoto Univ, 74 & Dept Elec Eng, 85; hon prof, Xi'an Jiaotong Univ, Xi'an, China, 86. *Mem:* Fel Inst Elec & Electronics Engrs; Sigma Xi. *Res:* System identification and parameter estimation, ie deriving models for processes based on observations/measurements on such a process; diagnosis; monitoring; prediction; control. *Mailing Add:* Eindhoven Univ Technol PO Box 513 Eindhoven NL-5600 MB Netherlands

EYLER, JOHN ROBERT, b Wilmington, Del, May 29, 45; m 67; c 2. PHYSICAL CHEMISTRY. *Educ:* Calif Inst Technol, BS, 67; Stanford Univ, PhD(chem physics), 72. *Prof Exp:* Nat Res Coun-Nat Bur Stand assoc chem, Nat Bur Stand, 72-74; from asst prof to assoc prof, 74-87, PROF CHEM, UNIV FLA, 87- *Mem:* Am Chem Soc; Inter-Am Photochem Soc; Am Soc Mass Spectrometry. *Res:* Study of gaseous ions utilizing the techniques of ion cyclotron resonance mass spectrometry, with particular emphasis on the interaction of tunable laser radiation with gaseous ions; applications of Fourier transform ion cyclotron resonance mass spectrometry. *Mailing Add:* Dept Chem Univ Fla Gainesville FL 32611

EYMAN, DARRELL PAUL, b Mason Co, Ill, Dec 18, 37; m 59; c 3. SYNTHETIC INORGANIC CHEMISTRY, ORGANOMETALLIC CHEMISTRY. *Educ:* Eureka Col, BS, 59; Univ Ill, PhD(inorg chem), 64. *Prof Exp:* Asst prof, 64-69, ASSOC PROF CHEM, UNIV IOWA, 69- *Mem:* Am Chem Soc. *Res:* Structure and reactivity of organo-metallic compounds having two or more transition metal atoms. *Mailing Add:* Dept Chem Univ Iowa Iowa City IA 52242-1000

EYMAN, EARL DUANE, b Canton, Ill, Sept 24, 25; m 51; c 2. ENGINEERING, MATHEMATICS. *Educ:* Univ Ill, BS, 49, MS, 50; Univ Colo, PhD(elec eng), 66. *Prof Exp:* Scientist, Atomic Power Div, Westinghouse Elec Corp, 50-51; res engr, Caterpillar Tractor Co, 51-61, proj engr control, 61-66; assoc prof elec & mech eng, 66-69, prof elec eng & head dept, 69-75, PROF ELEC & COMPUT ENG, UNIV IOWA, 75- *Concurrent Pos:* Consult, Inst Telecommun Sci. *Mem:* Sr mem Inst Elec & Electronics Engrs. *Res:* Fluid mechanics; computers; electronics; classical and modern control systems; remote control; writer of textbooks in electrical engineering and computers. *Mailing Add:* PO Box 3282 Estes Park CO 80517

EYMAN, LYLE DEAN, b Petersburg, Ill, May 9, 41; m 61; c 2. LIMNOLOGY, AQUATIC ECOLOGY. *Educ:* Bradley Univ, BS, 64; Mich State Univ, MS, 69, PhD(limnol), 72. *Prof Exp:* Biologist aquatic ecol, Water Qual Sect, Ill State Water Surv, 64-67; res scientist limnol, 72-78, PROG MGR NUCLEAR WASTE, ENVIRON SCI DIV, OAK RIDGE NAT LAB, 78- *Mem:* Ecol Soc Am; Am Soc Limnol & Oceanog; Int Soc Theoret & Appl Limnol; Sigma Xi. *Res:* Environmental transport and fate of contaminants generated from advanced energy fuel cycles; disposal of solid wastes from these technologies in an environmentally and socially acceptable manner. *Mailing Add:* 2820 Wellesley Ct Blacksburg VA 24062-0004

EYRE, PETER, b Glossop, Eng, Oct 23, 36; m 63; c 2. PHARMACOLOGY, IMMUNOLOGY. *Educ:* Univ Edinburgh, BVM, 60, BSc, 62, PhD(pharmacol), 65; Royal Col Vet Surg, MRCVS, 60. *Prof Exp:* From asst lectr to lectr, Univ Edinburgh, 62-68; assoc prof, 68-73, PROF BIOMED SCI UNIV GUELPH, 74- *Mem:* Brit Pharmacol Soc; Can Physiol Soc; Am Soc Vet Physiol & Pharmacol. *Res:* Chemotherapy of parasitic diseases; pharmacologic actions of anti-parasitic drugs; pharmacologic mechanisms in allergy and anaphylaxis in domesticated herbivores. *Mailing Add:* Dept Vet Med VA Polytech Inst & State Univ Blacksburg VA 24061

EYRING, EDWARD J, b Oakland, Calif, Dec 25, 34; m 59; c 5. ORTHOPEDICS, PHYSIOLOGICAL CHEMISTRY. *Educ:* Princeton Univ, BA, 55; Harvard Univ, MD, 59; Univ Calif, San Francisco, PhD(biochem), 67. *Prof Exp:* Asst prof orthop & physiol chem, Ohio State Univ, 67-73; ASST PROF ORTHOP & CONSULT BIOENG, UNIV TENN, 74- *Mem:* Am Rheumatism Asn; Orthop Res Soc; Am Acad Orthop Surg; Am Asn Col Podiatric Med; Am Acad Pediat. *Res:* Anti-inflammatory drug metabolism; surgical implants; pediatric orthopedics; arthritis. *Mailing Add:* 2100 Clinch Ave E Tenn Child Hosp Prof Bldg Knoxville IN 37916

EYRING, EDWARD M, b Oakland, Calif, Jan 7, 31; m 54; c 4. CHEMICAL KINETICS, SURFACE SPECTROSCOPY. *Educ:* Univ Utah, BA, 55, MS, 56, PhD(chem), 60. *Prof Exp:* NSF fel phys chem, Univ Goettingen, 60-61; from asst prof to assoc prof, 61-68, PROF PHYS CHEM, UNIV UTAH, 68- *Concurrent Pos:* Dept chmn chem, Univ Utah, 73-76; Guggenheim fel, 82. *Mem:* Am Chem Soc; Am Phys Soc; AAAS; Soc Appl Spectros. *Res:* Photoacoustic spectroscopy, particularly of solid surfaces and infrared wavelengths; applications in heterogeneous catalysis, organic semiconductors, metal corrosion, and biocompatible polymers; rapid reactions of macrocyclic ligands in nonaqueous media studied by relaxation techniques and calorimetry. *Mailing Add:* Dept Chem Univ Utah Salt Lake City UT 84112-1102

EYRING, LEROY, b Pima, Ariz, Dec 26, 19; m 41; c 4. PHYSICAL CHEMISTRY, SOLID STATE CHEMISTRY. *Educ:* Univ Ariz, BS, 43; Univ Calif, PhD(chem), 49. *Prof Exp:* Asst, Univ Calif, Berkeley, 43-44, chemist, Radiation Lab, 46-49; from asst prof to assoc prof chem, Univ Iowa, 49-61; Prof, 61-88, REGENTS PROF CHEM, ARIZ STATE UNIV, 88- *Concurrent Pos:* Sr fel, NSF, 58-59; Guggenheim fel, 59-60; Fulbright award, 59-60. *Mem:* Am Chem Soc; AAAS. *Res:* Solid state chemistry, especially electron microscopic, thermodynamic, kinetic and high temperature studies of solid state chemical reactions. *Mailing Add:* 6995 E Jackrabbit Rd Scottsdale AZ 85253

EYSTER, EUGENE HENDERSON, b Wheaton, Minn, Mar 21, 14; m 42; c 5. PHYSICAL CHEMISTRY. *Educ:* Univ Minn, BChem, 35; Calif Inst Technol, PhD(phys chem), 38. *Prof Exp:* Nat Res fel molecular spectra, Dept Physics, Univ Mich, 39-40; Hale fel, Calif Inst Technol, 41, group leader explosives res lab, Nat Defense Res Comt & Off Sci Res & Develop, 42-45; subdiv chief explosives div, US Naval Ord Lab, 46-48; alt div leader, Los Alamos Nat Lab, 49-70, div leader, 70-80; RETIRED. *Mem:* AAAS; Am Chem Soc. *Res:* Molecular spectra and structure; physical chemistry of explosives and detonation. *Mailing Add:* 1437 41st St Los Alamos NM 87544

EYSTER, HENRY CLYDE, b Dornsife, Pa, July 10, 10; m 38; c 1. PLANT PHYSIOLOGY, PHYCOLOGY. *Educ:* Bucknell Univ, AB, 32; Univ Ill, AM, 34, PhD(bot), 36. *Prof Exp:* Asst and teaching fel, Univ Ill, 32-36; instr bot, NC State Col, 36-37; from asst prof to assoc prof bot & head dept, Univ SDak, 37-46; res plant physiologist, Charles F Kettering Found, 46-62; sr res biologist, Monsanto Res Corp, 62-66; chmn div natural sci, 70-76, prof biol, 66-80, distinquished lectr, 76-80, EMER PROF BIOL, MOBILE COL, 80- *Concurrent Pos:* From assoc prof to prof biol, Antioch Col, 46-62. *Mem:* Am Phycol Soc; Bot Soc Am; Am Soc Plant Physiol. *Res:* Photosynthesis; auxins; plant enzymes; hybrid vigor; plant genetics; Hill reaction; photophosphorylation; mineral and trace element nutrition of algae and duckweeds; nitrogen fixation; water pollution and eutrophication; aquaculture and mariculture. *Mailing Add:* 417 S Sage Ave Mobile AL 36606

EYSTER, MARSHALL BLACKWELL, b Toledo, Ohio, Sept 25, 23; m 47; c 3. ORNITHOLOGY. *Educ:* Univ Chicago, BS, 45; Univ Ill, MS, 50, PhD(zool), 52. *Prof Exp:* Asst zool, Univ Ill, 46-50; from asst prof to assoc prof, 50-66, PROF BIOL, UNIV SOUTHWESTERN LA, 66- *Mem:* Am Ornith Union; Wildlife Soc; Wilson Ornith Soc; Cooper Ornith Soc; Sigma Xi. *Res:* Daily rhythm and nocturnal unrest in birds; ecological distribution of mammals. *Mailing Add:* 226 Monteigne Dr Lafayette LA 70506

EYZAGUIRRE, CARLOS, b Santiago, Chile, Apr 28, 23; m 47; c 3. NEUROPHYSIOLOGY. *Educ:* Univ Chile, MD, 47. *Hon Degrees:* DSc, Cath Univ Chile, 72; DHon Causa(med), Univ Madrid, 75. *Prof Exp:* Fel med, Johns Hopkins Hosp, 47-50; from asst prof to assoc prof neurophysiol & pharmacol, Cath Univ Chile, 50-57; asst res prof, 57-62, PROF PHYSIOL, COL MED, UNIV UTAH, 62-, HEAD DEPT, 65- *Concurrent Pos:* Guggenheim fel, 53-55; vis praelector, Univ St Andrews, 65; consult, Training Comt B, Nat Inst Neurol & Communicative Dis, NIH, 66-69 & Neurol A Study Sect, 70-74, chmn sect, 73-74; Japan Soc Promotion Sci vis prof, 78; vis prof, A Rosenblueth-Grass Found, 79; mem, Dir Adv Comt, NIH, 80-82. *Honors & Awards:* Distinguished Res Award, Univ Utah, 74; Givaudan lectr, 80. *Mem:* Am Physiol Soc; Soc Neurosci; Sigma Xi. *Res:* Physiology of sensory receptors; physiology of chemoreceptors. *Mailing Add:* Physiol Dept Univ Utah-Research Park 410 Chipeta Way Salt Lake City UT 84108

EZEKOWITZ, MICHAEL DAVID, b Durban, Repub SAfrica, Jan 27, 46; m 71; c 2. CARDIOLOGY. *Educ:* Univ Cape Town, MB CHB, 70; Univ London, PhD, 76; MRCP, FRCP. *Prof Exp:* Intern & resident med, Univ Cape Town, 71-72; resident, Univ Natal, 72-73; res fel sci, Univ London, 73-76; fel med, Johns Hopkins Hosp, 76-78; asst prof med, Health Sci Ctr, Univ Okla, 78-82; assoc prof radiol, 82-90, PROF MED, YALE UNIV SCH MED, 90- *Res:* Arterial wall physiology; platelet and vascular disease; restenosis after angioplasty and imaging thrombosis. *Mailing Add:* Fitrin 3 Cardiol Sect Yale Univ Sch Med New Haven CT 06510

EZELL, RONNIE LEE, b Clarksville, Tenn, Nov 6, 44; m 73; c 1. NUCLEAR PHYSICS. *Educ:* Austin Peay State Univ, BA, 66; Univ Ga, PhD(physics), 73. *Prof Exp:* Asst prof, 73-77, ASSOC PROF PHYSICS, AUGUSTA COL, 77- *Mem:* Sigma Xi; Am Phys Soc. *Res:* Gamma-ray decay of nuclei; neutron spin-flip measurements. *Mailing Add:* Dept Chem & Physics Augusta Col Augusta GA 30910

EZELL, WAYLAND LEE, b Stockton, Calif, Dec 31, 37; m 61; c 3. SYSTEMATIC BOTANY, EVOLUTIONARY BIOLOGY. *Educ:* Univ of the Pac, BA, 59, MA, 63; Ore State Univ, PhD(syst bot), 70. *Prof Exp:* Instr biol & bot, Ventura Col, 62-66; res asst bot, Ore State Univ, 67-68; from asst prof to assoc prof, 70-79, actg assoc grad dean, 80-81, PROF & CHMN BIOL, ST CLOUD STATE UNIV, 79- *Concurrent Pos:* Deleg, Int Bot Cong, Seattle, 69; lectr, Sigma Xi Regional Lect Exchange Prog, 71. *Mem:* Bot Soc Am; Am Soc Plant Taxonomists; Int Asn Plant Taxonomists; AAAS; Sigma Xi. *Res:* Genetics, evolution and taxonomy of angiosperms, especially genus Mimulus of sections Eunanus and Eumimulus, using field herbarium, hybridization, cytological, biochemical and numerical methods. *Mailing Add:* 1720 Tenth Ave SE St Cloud MN 56304

EZRA, ARTHUR ABRAHAM, b Calcutta, India, July 9, 25; US citizen; m 56; c 5. CIVIL & MECHANICAL ENGINEERING. *Educ:* Univ Calcutta, BE, 46; Univ Mich, MSE, 48; Stanford Univ, PhD(eng mech), 58. *Prof Exp:* Engr, Int Eng Co, 48-49; struct & hydraul engr, US Corps Engrs, 49-51; sr struct engr, San Francisco Harbor, 51-56; staff engr, Martin Co, 57-60, chief technol develop, 60-62, mgr aeromech & mat res, 62-66; head mech div, Denver Res Inst & prof mech eng, Univ Denver, 66-72, chmn dept mech sci & environ eng, 68-72; head, Off Res & Develop Incentives, NSF, 72-74, prog dir water res, Urban & Environ Eng, 74-84, dir div Fundamental Res for Emerging & Critical Eng Systs, 84-87; DEAN, SCH ENG TECHNOL, STATE UNIV NY, 87- *Mem:* Am Soc Eng Educ; Am Soc Civil Engrs; Am Soc Mech Engrs; AAAS. *Res:* Technology transfer; explosive forming and welding of metals; energy absorbing devices and systems; ultra low cost housing; crashworthiness of motor vehicles; solar energy; water resources and environmental engineering. *Mailing Add:* State Univ NY Melville Rd Farmingdale NY 11735

EZRA, GREGORY SION, b London, Eng, Sept 16, 53; m 80; c 1. INTRAMOLECULAR DYNAMICS. *Educ:* Oxford Univ, BA, 76, DPhil(chem), 80. *Prof Exp:* NATO fel, Univ Chicago, 80-82; asst prof, 82-88, ASSOC PROF CHEM, CORNELL UNIV, 88- *Mem:* Am Phys Soc; Am Chem Soc; Royal Soc Chem; Sigma Xi. *Res:* Theoretical chemistry: intramolecular dynamics; highly-excited vibration-rotation states; semiclassical mechanics; nondiabatic interactions in photodisociation and collisions; group theory; few-body problems. *Mailing Add:* Dept Chem Cornell Univ Baker Lab Ithaca NY 14853

EZRIN, CALVIN, b Toronto, Ont, Oct 1, 26; m 46; c 8. ENDOCRINOLOGY, INTERNAL MEDICINE. *Educ:* Univ Toronto, MD, 49; FRCP(C), 54. *Prof Exp:* Res assoc path, Div Neuropath, Univ Toronto, 53-65, assoc med, dept med, 59-76, asst prof path, 65-68, asst prof path & med, 68-70, assoc prof, 70-76, prof med, 76-77; CLIN PROF, UNIV CALIF, LOS ANGELES, 77- *Concurrent Pos:* Stengel res fel, Am Col Physicians, 53-54; physician, Toronto Gen Hosp, 54-; consult, Dept Vet Affairs, Sunnybrook Hosp, Toronto, 54-; mem, Int Comt Nomenclature of Adenohypophysis, 63-; vis lectr path, Harvard Med Sch, 74-81. *Mem:* Endocrine Soc; Am Diabetes Asn; Am Thyroid Asn. *Res:* Anterior pituitary cytology; insulin resistance; metabolic effects of glucagon; obesity; serum binding and kinetics of thyroid hormones; biology of trans-sexualism. *Mailing Add:* 18372 Clark St Suite 226 Tarzana CA 91356

EZRIN, MYER, b Boston, Mass, June 23, 26; m 46; c 3. ANALYTICAL CHEMISTRY, PLASTICS CHEMISTRY. *Educ:* Tufts Col, BS, 48; Yale Univ, PhD(chem), 54. *Prof Exp:* Asst, Org Chem Lab, Yale Univ, 50-52; chemist, coated fabric appln of chlorosulfonated polyethylene, E I du Pont de Nemours & Co, Inc, 48-50 & Plastics Div, Monsanto Co, 53-65; anal group leader, DeBell & Richardson, Inc, 65-69, mgr, Anal Testing Div, 69-72 & Anal Dept, DeBell & Richardson Testing Inst, 72-77; mgr polymer characterization, SL Testing Inst & qual assurance dir, Springorn Inst Bioresearch, 78-80; DIR, ASSOCS PROG & PROG MGR, ELEC INSULATION RES CTR, INST MAT SCI, UNIV CONN, 80- *Mem:* Am Chem Soc; Soc Plastics Engrs; Sigma Xi. *Res:* Polymer analysis and characterization; molecular weight and molecular weight distribution; thermal analysis; polymers in patent infringement and product liability litigation; electrical insulation. *Mailing Add:* 173 Academy Dr Longmeadow MA 01106-2117

EZZAT, HAZEM AHMED, b Cairo, Egypt, July 12, 42; US citizen; c 2. MECHANICAL ENGINEERING, ENGINEERING MECHANICS. *Educ:* Cairo Univ, BSc, 63; Univ Wis, MS, 67, PhD(mech eng), 71. *Prof Exp:* Proj engr, Suez Canel Authority, Egypt, 63-65; instr theory mach, Cairo Univ, 65-66; assoc sr res engr, 70-73, sr res engr, 73-77, staff res engr, 77-81, asst head engr, mech dept, 81-84, HEAD, POWER SYSTS RES DEPT, GEN MOTORS RES LABS, 84- *Honors & Awards:* Henry Hess Award, Am Soc Mech Engrs, 73. *Mem:* Am Soc Mech Engrs; Soc Automotive Engrs; Am Acad Mech; Sigma Xi. *Res:* Tribology, particularly lubrication theory; dynamics of physical systems; optimization theory and its application to engineering design. *Mailing Add:* Power Systs Res Dept Gen Motors Res Labs Warren MI 48090

EZZELL, ROBERT MARVIN, CONTRACTILE PROTEINS, CELLULAR REGULATION. *Educ:* Univ Calif, Berkeley, PhD(zool), 84. *Prof Exp:* RES FEL, HEMAT-ONCOL UNIT, MASS GEN HOSP & DEPT MED, SCH MED, HARVARD UNIV, 84- *Res:* Cell motility during development. *Mailing Add:* Dept Surg & Anat & Cell Biol Mass Gen Hosp Harvard Med Sch Surg Res Unit L E Martin Labs 149 13th St Charlestown MA 02129

F

FAABORG, JOHN RAYNOR, b Hampton, Iowa, Jan 23, 49; m 69, 80; c 3. COMMUNITY ECOLOGY, BIOGEOGRAPHY. *Educ:* Iowa State Univ, BS, 71; Princeton Univ, PhD(ecol), 75. *Prof Exp:* From asst prof to assoc prof, 75-89, PROF BIOL, UNIV MO, COLUMBIA, 90- *Mem:* Fel Am Ornithologists Union; Wilson Ornith Soc; Cooper Ornith Soc. *Res:* Ecology of island bird communitites, focusing on the role of competition in structuring communities; applied biogeography and non-game bird management; behavioral ecology, particularly the evolution of cooperative polyandry; author of a textbook and separate laboratory manual of ornithology. *Mailing Add:* 110 Tucker Hall Univ Mo Columbia MO 65211

FAANES, RONALD, b Minneapolis, MN, Oct 4, 41. IMMUNOLOGY. *Educ:* Univ Minn, PhD(microbiol), 70. *Prof Exp:* SR PRIN SCIENTIST, BOEHRINGER INGELHEIM, INC, 81- *Mailing Add:* Boehringer Ingelheim Ltd 175 Briar Ridge Rd PO Box 368 Ridgefield CT 06877

FA'ARMAN, ALFRED, b New York, NY, Nov 29, 17; m 42; c 2. APPLIED PHYSICS, ELECTRONICS ENGINEERING. *Educ:* Brooklyn Col, BA, 39; NY Univ, PhD(physics), 55. *Prof Exp:* Phys sci aide, Eng Bd, US War Dept, 41-43; res adminr, Off Naval Res, 46-49; res assoc, Physics Dept, NY Univ, 49-55; mem tech staff, Hughes Res Labs, 55-65, consult, Hughes Res & Develop Labs, 66-71; physicist, Lawrence Radiation Lab, Univ Calif, 72-74; DIR, APPL PHYSICS CONSULTS, CASTRO VALLEY, CALIF, 75-; ELEC ENGR, CALIF PUB UTILITIES COMN, 81- *Concurrent Pos:* Lectr, Univ Calif, Los Angeles, 56-67 & NATO Advan Study Inst, 66, 67 & 70; res grant, Sci Affairs Div, NATO, 71; adv, French Govt, Nat Off Aerospace Studies & Res, 78-80. *Mem:* Am Phys Soc; NY Acad Sci; Sigma Xi; Am Asn Physics Teachers. *Res:* Applied physics/engineering particularly radiation effects, plasma physics, high voltage technology, liquid dielectrics, solid state physics, reliability, test and measurement procedures, electron and microwave tubes, cosmic ray and nuclear physics. *Mailing Add:* 2510 Delmer St Oakland CA 94602

FAAS, RICHARD WILLIAM, b Appleton, Wis, Nov 8, 31; m 55; c 3. GEOLOGY, PALEONTOLOGY. *Educ:* Lawrence Col, BA, 53; Iowa State Univ, MS, 62, PhD(geol, ecol), 64. *Prof Exp:* From asst prof to assoc prof, 64-74, PROF GEOL, LAFAYETTE COL, 74-, HEAD DEPT, 70- *Mem:* Geol Soc Am; Soc Econ Paleontologists & Mineralogists; Int Asn Sedimentologist. *Res:* Triassic stratigraphy; estuarine and near-shore marine sedimentation; organic-inorganic interrelationships; paleoecology. *Mailing Add:* Dept Geol Lafayette Col Easton PA 18042

FABBI, BRENT PETER, b Reno, Nev, Mar 1, 38; m 62, 84; c 4. MINING ENGINEERING. *Educ:* Univ Nev, Reno, BS, 63, MS, 65. *Prof Exp:* Sr res assoc, Nev Bur Mines, 63-67; proj chief, US Geol Surv, 64-74, asst br chief, 74-76, br chief, 76-82; dir X-ray prod, Bausch & Lomb, 82-83; VPRES, SHAH INC, 85- *Concurrent Pos:* Consult, 84. *Honors & Awards:* Safety Award, US Geol Surv, 81. *Mem:* Soc Appl Spectros; Am Chem Soc; Spectros Soc Can. *Res:* Fundamental and applied x-ray fluorescence and geochemical studies of matrix effects; mathematical algorithms; instrumental design and automation; sample communinution and fusion; quantitative analysis of major through trace elements; international consulting on instrumental methods and lab management. *Mailing Add:* 13802 Poplar Tree Rd Chantilly VA 22021

FABER, ALBERT JOHN, b North Battleford, Sask, Can, Oct 8, 45; div; c 3. ANALYTICAL CHEMISTRY, ENVIRONMENTAL SCIENCES. *Educ:* Univ Sask, BSc, 66 & 67; Univ Alta, PhD(biochem), 74. *Prof Exp:* Biochem technologist, McEachern Cancer Res Unit, 67-69, grad student, 69-74; postdoctorate fel biochem & cell biol, Swiss Cancer Inst, Lausanne, Switz, 74-77; asst prof biochem, Montreal Cancer Inst, 77-81; SUPVR, ANALYTICAL SERV LAB, HUSKY OIL OPERS LTD, 81- *Concurrent Pos:* Asst prof, Fac Med, Univ Montreal, 77-81; mem, Road Comt, Can Gen Standards Bd, 83- *Mem:* Can Biochem Soc; Can Tech Asphalt Asn; Am Chem Soc; Asn Asphalt Paving Technologists; Am Wood Preservers' Asn. *Res:* Organic and inorganic chemistry; applied petroleum chemistry to evaluate heavy crude oils and distillation asphalts; environmental test procedures; government regulations. *Mailing Add:* Anal Serv Lab Husky Oil Opers Ltd Lloydminster SK S9V 0Z8 Can

FABER, BETTY LANE, b Chicago, Ill, Feb 7, 44; m 69; c 1. ANIMAL BEHAVIOR, ENTOMOLOGY. *Educ:* Col William & Mary, BS, 66; Rutgers Univ, MS, 70, PhD(entom), 75. *Prof Exp:* Assoc lectr, Columbia Univ, 73-74, lectr biol sci, 77-81; res assoc animal behav, Am Mus Natural Hist, 75-85; ASST DIR, DOUGLAS PROJ, RUTGERS UNIV, 90- *Honors & Awards:* Explorers Club grant. *Mem:* Animal Behav Soc; Entom Soc Am; NY Acad Sci. *Res:* Ecology and behavior of individual animals in a population of wild American cockroaches; ecology of Trinidad cockroaches. *Mailing Add:* Douglas Proj Rutgers Women Math Sci & Engr Box 270 Douglas Col New Brunswick NJ 08903

FABER, DONALD S, b Buffalo, NY, Mar 3, 43; m 64; c 2. NEUROBIOLOGY. *Educ:* Mass Inst Technol, BS, 64; State Univ NY Buffalo, PhD(physiol), 68. *Prof Exp:* Nat Inst Neurol Dis & Stroke fel, Lab Neurobiol, State Univ NY Buffalo, 68-70; vis res scientist neurophysiol, Max Planck Inst Brain Res, Frankfurt, Ger, 70-72; vis scientist Univ Paris, 72; asst prof physiol, Med Ctr, Univ Cincinnati, 72-74; assoc res scientist neurophysiol, Res Inst Alcoholism, 74-78; res assoc prof, 75-78, assoc prof, 78-81, PROF PHYSIOL & HEAD, DIV NEUROBIOL, STATE UNIV NY BUFFALO, 78- *Concurrent Pos:* Grass Found fel, 69. *Mem:* AAAS; Soc Neurosci; Am Physiol Soc. *Res:* Neuronal excitability; synaptic transmission. *Mailing Add:* Dept Physiol SUNY at Buffalo 313 Cary Hall Neurobiol Lab Buffalo NY 14214

FABER, JAMES EDWARD, b Oct 23, 51; m 77; c 2. MICROCIRCULATION, NEURAL REGULATION. *Educ:* Univ Mo, PhD(physiol), 80. *Prof Exp:* asst prof 82-89, ASSOC PROF PHYSIOL, UNIV NC, 89- *Mem:* Am Phsiol Soc; Microcirculation Soc; Am Heart Asn. *Res:* Vascular smooth muscle and microvascular function with emphasis on microvascular neuroeffector junction, neurotransmitter and humoral receptors, local regulation, second messengers, receptor localization. *Mailing Add:* Dept Physiol Univ NC Med Res Bldg 7545 Chapel Hill NC 27599-7545

FABER, JAN JOB, b The Hague, Neth, June 16, 34; m 57; c 2. PHYSIOLOGY, BIOPHYSICS. *Educ:* Univ Amsterdam, Drs, 57, MD, 60; Univ Western Ont, PhD(biophys), 63. *Prof Exp:* Res asst biophys, Univ Western Ont, 60-62, res assoc, 62-63; from instr to asst prof phys med, Univ Wash, 63-66; from asst prof to assoc prof, 66-73, PROF PHYSIOL, MED SCH, ORE HEALTH SCI UNIV, PORTLAND, 73- *Concurrent Pos:* Estab investr, Am Heart Asn, 68-73. *Mem:* Am Physiol Soc; Biophys Soc; Am Heart Asn. *Res:* Cardiovascular physiology; prenatal physiology; membrane physiology. *Mailing Add:* Dept Physiol Oregon Health Sci Univ Sch Med 3181 SW Sam Jackson Park Rd Portland OR 97201

FABER, LEE EDWARD, b Detroit, Mich, Sept 21, 42; m 85. ENDOCRINOLOGY, BIOCHEMISTRY. *Educ:* Duke Univ, AB, 64; Bowling Green State Univ, MA, 67; Ind Univ, PhD(zool), 70. *Prof Exp:* Fel steroid metab, Syntex Res Div, 69-71; sr res fel steroid receptors, Inst Med Res, Toledo Hosp, 71-74, actg dir, 73-74; from asst to assoc prof, 74-78, PROF OBSTET & GYNEC, MED COL OHIO, 90-, DIR, ENDOCRINE RES UNIT, 74- *Mem:* Endocrine Soc; AAAS; Am Soc Zoologists. *Res:* Characterization of steroid receptors and their role in mechanism of steroid action. *Mailing Add:* Dept Obstet & Gynec Med Col Ohio PO Box 10008 Toledo OH 43699-0008

FABER, MARCEL D, microbial biochemistry; deceased, see previous edition for last biography

FABER, RICHARD LEON, b Winthrop, Mass, May 7, 40; m 64; c 1. MATHEMATICS. *Educ:* Mass Inst Technol, BS, 60; Brandeis Univ, MA, 62, PhD(category theory), 65. *Prof Exp:* Instr math, Regis Col, Mass, 64-65 & Univ Pa, 65-67; asst prof, Univ Calif, San Diego, 67-68; from asst prof to assoc prof, 68-85, PROF MATH, BOSTON COL, 85- *Mem:* Am Math Soc; Math Asn Am; Sigma Xi. *Res:* Computer science. *Mailing Add:* 48 Chinian Path Newton Centre MA 02159

FABER, ROGER JACK, b Grand Rapids, Mich, Oct 4, 31; div; c 3. CHEMICAL PHYSICS. *Educ:* Calvin Col, AB, 53; Mich State Univ, PhD(chem), 58. *Prof Exp:* Res instr chem, Mich State Univ, 57-58; instr physics, Calvin Col, 58-60, from asst prof to assoc prof, 60-64; NSF fac fel chem, Columbia Univ, 64-65; from asst prof to assoc prof, 65-72, PROF PHYSICS, LAKE FOREST COL, 72- *Concurrent Pos:* Res assoc, Philos Dept, Boston Univ, 73-74, 80-81. *Mem:* AAAS; Am Phys Soc; Am Asn Physics Teachers; Philos Sci Asn. *Res:* Electron spin resonance; atomic collision processes. *Mailing Add:* Dept Physics Lake Forest Col Lake Forest IL 60045

FABER, SANDRA MOORE, b Boston, Mass, Dec 28, 44; m 67; c 2. ASTRONOMY. *Educ:* Swarthmore Col, BA, 66; Harvard Univ, PhD(astron), 72. *Hon Degrees:* DSc, Swarthmore Col, 86. *Prof Exp:* From asst prof/asst astronomer to assoc prof/assoc astronomer, 72-79, PROF & ASTRONOMER, LICK OBSERV, UNIV CALIF, SANTA CRUZ, 79- *Concurrent Pos:* Alfred P Sloan Found fel, 77-; NSF Astron Adv Panel, 79-81; field comn, NAS Astron Surv Panel, 79-82, policy comt, Astron Surv Comt, 89-90; vis lectr, Univ Hawaii, 83; chmn, vis comt, Space Telescope Sci Inst, 83-84; sci adv comt, Nat New Technol Telescope, 83-84; vis prof, Arizona State Univ, 85; assoc ed, Astrophys J Letters, 82-86; bd trustees, Carnegie Inst Wash, 85-, Bd dirs, Ann Rev, 89-; chair, Keck Telescope Sci Steering Comt, 87-90; mem, Hubble Space Telescope Strategy Panel, 90, Users Comt, 90- & Wide Field Camera Team, 85- *Honors & Awards:* Bart J Bok Prize, Harvard Univ, 78; Heineman Prize, Am Astron Soc, 86; Phillips vis lectr, Haverford Col, 82; Fesbach Lectr, Mass Inst Technol, 90; Darwin Lectr, Royal Astron Soc, 91. *Mem:* Nat Acad Sci; Inst Astron Union; Am Astron Soc; Am Acad Arts & Sci. *Res:* Formation and evolution of normal galaxies; stellar populations in galaxies; galactic structure; stellar spectroscopy; clusters of galaxies; cosmology. *Mailing Add:* Lick Observ Univ Calif Santa Cruz CA 95064

FABER, SHEPARD MAZOR, b Brooklyn, NY, Aug 8, 28; m 53; c 4. SCIENCE EDUCATION. *Educ:* Emory Univ, BA, 49; Columbia Univ, MA, 50; Univ Fla, EdD(sci ed), 60. *Prof Exp:* Instr sci ed, Univ Fla, 55; assoc prof sci, ECarolina Col, 59-62; assoc prof, 62-70, PROF PHYS SCI, UNIV MIAMI, 70- *Concurrent Pos:* Consult, physics & phys sci teaching. *Mem:* Fel AAAS; Nat Asn Res Sci Teaching. *Res:* Science curriculum and instruction. *Mailing Add:* Dept Physics Univ Miami Univ Sta Coral Gables FL 33124

FABER, VANCE, b Buffalo, NY, Dec 1, 44; m 79; c 1. PARALLEL & IMAGE PROCESSING. *Educ:* Wash Univ, BA, 66, MA, 69, PhD(math), 71. *Prof Exp:* Assoc prof, Univ Colo, 70-79; STAFF MEM, LOS ALAMOS NAT LAB, 79- *Mem:* Am Math Soc; Math Asn Am. *Res:* Applied Math. *Mailing Add:* Comput Res & Appln Los Alamos Nat Lab MS BZ265 Los Alamos NM 87545

FABES, EUGENE BARRY, b Detroit, Mich, Feb 6, 37; m 59; c 3. MATHEMATICAL ANALYSIS. *Educ:* Harvard Univ, AB, 59; Univ Chicago, MS, 62, PhD(math), 65. *Prof Exp:* Asst prof math, Rice Univ, 65-66; from asst prof to assoc prof, 66-74, PROF MATH, INST TECHNOL, UNIV MINN, MINNEAPOLIS, 74- *Concurrent Pos:* Prin investr, NSF grant. *Res:* Singular integrals and partial differential equations. *Mailing Add:* Univ Minn Minneapolis MN 55455

FABIAN, LEONARD WILLIAM, b North Little Rock, Ark, Nov 12, 23; m 47; c 3. ANESTHESIOLOGY. *Educ:* Univ Ark, BS, 47, MD, 51; Am Bd Anesthesiol, dipl. *Prof Exp:* Intern, Univ Ark, 51-52, resident anesthesiol, 52-54; fel, Philadelphia Childrens Hosp, 54; instr, Sch Med, Univ Ark, 54-55; asst prof, Duke Univ, 55-58; prof, Univ Miss, 58-71; PROF ANESTHESIOL, SCH MED, WASHINGTON UNIV, 71- *Mem:* Am Soc Anesthesiol; fel Am Col Anesthesiol; AMA. *Res:* Chemistry and pharmacology of anesthetic drugs. *Mailing Add:* Dept Anesthesiol Wash Univ Sch Med 660 S Euclid St Louis MO 63110

FABIAN, MICHAEL WILLIAM, b Mercer, Pa, Sept 27, 31; m 52; c 2. VERTEBRATE ZOOLOGY, ECOLOGY. *Educ:* Grove City Col, BS, 52; Mich State Univ, MS, 54; Ohio State Univ, PhD(zool), 64. *Prof Exp:* Asst, Mich State Univ, 52-54; teacher pub schs, Ohio, 54-56; instr gen biol, Ariz State Univ, 56-57; asst prof biol, Geneva Col, 57-61; assoc prof physiol, Westminster Col, 61-64; assoc prof physiol & zool, 64-67, chmn dept biol, 64-89, PROF ECOL, GROVE CITY COL, 67- *Mem:* AAAS; Ecol Soc Am; Am Inst Biol Sci. *Res:* Predatory or carnivorous activity of Crustacea on vertebrates; physiology. *Mailing Add:* Dept Biol Grove City Col Grove City PA 16127

FABIAN, ROBERT JOHN, b Cleveland, Ohio, Mar 21, 39; m 62. INFORMATION SYSTEMS. *Educ:* Case Western Reserve Univ, BS, 61, MS, 63, PhD(math), 65. *Prof Exp:* Asst prof math, Smith Col, 64-70; ASSOC PROF COMPUT SCI, YORK UNIV, ONT, 70-; AT ROBERT J FABIAN ASSOC. *Concurrent Pos:* Vis asst prof, Dept Appl Anal & Comput Sci, Univ Waterloo, 68-69; planning consult, Bank of NS, 76; sr consult, Art Benjamin Assoc, 77; mgr, Gulf Can, 81-82; pvt consult, Parther, Cellman, Hayward & Partners Ltd, 87-90. *Mem:* Inst Elec & Electronics Engrs; Can Info Processing Soc; Asn Comput Mach. *Res:* Organizational use of information systems. *Mailing Add:* Robert Fabian Assocs 584 Church St Toronto ON M4Y 2E5 Can

FABIC, STANISLAV, b Tuzla, Yugoslavia, Nov 14, 25; US citizen; m 61; c 3. NUCLEAR ENGINEERING, MECHANICAL ENGINEERING. *Educ:* Univ Melbourne, Australia, BS, 54, MS, 58; Univ Calif, Berkeley, MS, 59, PhD(nuclear eng), 64. *Prof Exp:* Exp engr automotive res, Standard Motor Co, Port Melbourne, Australia, 54-58; assoc engr nuclear eng, Kaiser Eng, Oakland, Calif, 63-67; adv engr reactor safety, Westinghouse Elec Corp, Pittsburgh, 67-73; chief, Anal Models Br, Off Res, Nuclear Regulatory Comn, Washington, DC, 73-81; PRES, DYNATREK, INC, GAITHERSBURG, MD, 81- *Mem:* Am Nuclear Soc; Sigma Xi. *Mailing Add:* 19152 Roman Way Gaithersburg MD 20879

FABISH, THOMAS JOHN, b Youngstown, Ohio, Feb 27, 38; m 61; c 3. PHYSICAL CHEMISTRY, SURFACE PHYSICS. *Educ:* Ohio State Univ, BAeroEng, 60, MS, 66; Univ Rochester, PhD(mat sci), 75. *Prof Exp:* Res engr aerodynamics, NAm Aviation, Ohio, 61-63; res engr thermal properties of mat, Battelle Mem Inst, 66-69; scientist solid state physics, Xerox Corp, 69-80; sr res chemist, Ashland Chem Co, 80-83; prin scientist, Am Cyanamid, 83-85; TECH SPECIALIST, ALCOA, 85- *Mem:* Am Chem Soc; Adhesion Soc; Am Phys Soc. *Res:* Solid state and surface physics, with emphasis on the electronic and surface properties of polymers; chemistry physics of carbon; particulate-polymer interactions; optical properties; metal oxides; metal-polymer interfaces. *Mailing Add:* 4917 Simmons Dr Murrysville PA 15632

FABRE, LOUIS FERNAND, JR, b Akron, Ohio, Sept 13, 41; m 85; c 4. PSYCHIATRY, PSYCHOPHARMACOLOGY. *Educ:* Univ Akron, BS, 63; Western Reserve Univ, PhD(physiol), 66; Baylor Univ, MD, 69. *Prof Exp:* Res specialist, Tex Res Inst Ment Sci, 65-67, from actg chief to chief neuroendocrinol, 67-70, from assoc head to head, Div Ment Retardation, 70-73; ASST PROF MENT SCI, UNIV TEX GRAD SCH BIOMED SCI, 69- *Concurrent Pos:* NSF res grants, 67-71; intern, Methodist Hosp, Houston, Tex, 69-70; Kelsey Leary Found grant, 69-71; NIMH grant, 70-71; resident, Baylor Univ, 70-73, clin asst prof psychiat, 73-; clin assoc prof, Univ Tex Med Sch Houston; med dir, Tex Alcoholism Found, Inc, Fabre Clin & Res Testing, Inc. *Mem:* Am Physiol Soc; Am Psychiat Asn; Am Group Psychother Soc; Endocrine Soc; Aerospace Med Soc. *Res:* Endocrine function in alcoholism; clinical double-blind studies of new psychopharmacologic agents for anxiety, depression, insomnia, schizophrenia and alcoholism; phase one inpatient studies of new drugs. *Mailing Add:* Fabre Clin 5503 Crawford St Houston TX 77004

FABREY, JAMES DOUGLAS, b New York, NY, Oct 29, 43; m 67; c 2. COMPUTER SCIENCE, APPLIED MATHEMATICS. *Educ:* Cornell Univ, AB, 65; Mass Inst Technol, PhD(math), 69. *Prof Exp:* Asst prof math, Univ NC, Chapel Hill, 69-75; from asst prof to assoc prof, 75-83, prof math & comput sci, dir acad comput, 86-89, EXEC DIR ACAD COMPUT, WEST CHESTER UNIV, 89- *Concurrent Pos:* Lectr comput sci, Villanova Univ, 79-81; comput analyst, BenePac, Inc, 78-88 & RAI, Inc, 88- *Res:* Numerical analysis; software engineering; mathematical physics; applied mathematics. *Mailing Add:* Dir Acad Comput West Chester Univ West Chester PA 19383

FABRICAND, BURTON PAUL, b New York, NY, Nov 22, 23; m 52; c 2. PHYSICS. *Educ:* Columbia Univ, AB, 47, AM, 49, PhD(physics), 53. *Prof Exp:* Proj engr, Philco Corp, 52-54; lectr & res assoc physics, Univ Pa, 54-56; sr res scientist, Hudson Lab, Columbia Univ, 57-69; chmn dept, 69-71, PROF PHYSICS, PRATT INST, 71- *Concurrent Pos:* Consult, Moore Sch Elec Eng, Univ Pa, 54-60 & Indust Electronic Hardware Corp, 60-64. *Mem:* Am Phys Soc. *Res:* Nuclear magnetic resonance; atomic absorption spectroscopy; internal reflection spectroscopy; dosimetry; semiconductors; molecular beams; photonuclear reactions. *Mailing Add:* Dept Physics Pratt Inst 200 Willoughby Ave Brooklyn NY 11205

FABRICANT, BARBARA LOUISE, b Ithaca, NY, Mar 2, 50. SPECTROSCOPY & SPECTROMETRY. *Educ:* Wells Col, BA, 72; Univ Rochester, MS, 74, PhD(phys chem), 77. *Prof Exp:* Advan scientist glass res & develop, Owens-Corning Fiberglass Tech Ctr, 77-84, sr scientist, 84-90; RES ASSOC, ZEON CHEM, USA, INC, 91- *Honors & Awards:* Outstanding Serv Award, NAm Thermal Anal Soc, 89. *Mem:* Am Chem Soc; NAm Thermal Anal Soc; Am Soc Mass Spectrometry; Sigma Xi. *Res:* Analytical chemistry to analyze products to assist in product development and problem solving. *Mailing Add:* 507 Hudson Ave Newark OH 43055

FABRICANT, CATHERINE G, b Davoli, Italy, Sept 24, 19; US citizen; m 45; c 2. HERPESVIRUS, INDUCED ATHEROSCLEROSIS. *Educ:* Cornell Univ, BS, 42, MS, 48. *Prof Exp:* Chief technician, Cornell Univ Infirmary & Clin, 42-44; res bacteriologist, Univ Ill Med Sch, Chicago, 44-45; teaching asst, Div Bact, NY State Col Life Sci, Cornell Univ, 47-49 & 59-60, res assoc, 60-62; asst Dept Path & Bact, NY State Col Vet Med, Cornell Univ, 45-47, actg asst prof, 62-63, res assoc, Dept Microbiol, 65-73, sr res assoc, 73-85, VIS FEL, DEPT MICROBIOL, IMMUNOL & PARASITOL, NY STATE COL VET MED, CORNELL UNIV, 86- *Concurrent Pos:* Res microbiol, Inst Med Microbiol, Univ Aarhus, Denmark, 64-65, vis sr res assoc, 73; fel Morris Animal Found, 71-; prin investr herpesvirus induced feline urolithiasis, Ralston Purina, 75-82; prin investr pathogenesis herpesvirus induced atherosclerosis, Nat Heart, Lung & Blood Inst, NIH, 75-83, co-investr, 83-85, consult, 79, 81, 84-86. *Honors & Awards:* La Croix Award, Am Animal Hosp Asn, 76,. *Mem:* Am Soc Pathologists; Fedn Am Socs Exp Biol; Sigma Xi. *Res:* Hypothesized and demonstrated in repeated experiments that a herpesvirus infection causes atherosclerosis remarkably like that of humans in a pathogen-free normocholesterolemic animal model; involving a viral mechanism altering arterial smooth muscle cell metabolism causing cellular cholesterol and cholesteryl ester accumulations; herpesvirus induced Feline urolithiasis. *Mailing Add:* Dept Microbiol Immunol & Parasitol NY State Col Vet Med Cornell Univ Ithaca NY 14853

FABRICANT, JULIUS, b Philadelphia, Pa, Mar 30, 19; m 45; c 2. POULTRY PATHOLOGY. *Educ:* Univ Pa, VMD, 42; Pa State Col, BS, 45; Cornell Univ, MS, 47, PhD(poultry path), 49. *Prof Exp:* Asst animal path, Pa State Univ, 44-45; asst, 46-49, from asst prof to assoc prof, 49-60, prof poultry dis, 60-86, EMER PROF POULTRY DIS, NY STATE COL VET MED, CORNELL UNIV, 87- *Concurrent Pos:* NIH fel, Inst Gen Path, Aarhus Univ, 64-65. *Mem:* AAAS; Am Vet Med Asn; Poultry Sci Asn; Am Col Vet Microbiol; Am Asn Pathologists; Am Asn Avian Pathologists. *Res:* Poultry diseases, especially Newcastle, infectious bronchitis, chronic respiratory disease and infectious hepatitis of ducks; mycoplasma; arteriosclerosis; avian tumor viruses. *Mailing Add:* 802 Hanshaw Rd Ithaca NY 14850

FABRICIUS, DIETRICH M, b Bucholz-Aller, Ger, Nov 17, 50; US citizen; m 72; c 2. ORGANIC CHEMISTRY. *Educ:* Luther Col, BA, 73; Northwestern Univ, PhD(org chem), 78. *Prof Exp:* RES ASSOC ORG/DYE CHEM, IMAGING SYSTS, E I DU PONT DE NEMOURS & CO, INC, 78- *Mem:* Am Chem Soc; Soc Imaging Sci & Technol. *Res:* Dye chemistry; photographing chemicals. *Mailing Add:* 305 Silver Pine Dr Hendersonville NC 28739

FABRIKANT, ILYA I, b Riga, Latvia, Feb 15, 49; m 81; c 3. ELECTRONIC & ATOMIC COLLISIONS, INTERACTION OF RADIATION WITH ATOMS. *Educ:* Latvian State Univ, MS, 71; Inst Physics, Riga, PhD(physics), 74. *Prof Exp:* Jr researcher, Inst Physics, Latvian Acad Sci, 71-83, sr researcher, 83-88; vis scientist, Harvard-Smithsonian Ctr Astrophys, 89; ASSOC PROF PHYSICS, UNIV NEBR, 89- *Mem:* Am Phys Soc. *Res:* Scattering of low-energy electrons by atoms and molecules; excitation and dissociation of molecules by electron impact; photodetachment of negative ions. *Mailing Add:* Dept Physics & Astron Univ Nebr Lincoln NE 68588-0111

FABRIKANT, IRENE BERGER, b Jan 19, 33; m 56. PREVENTIVE MEDICINE, PUBLIC HEALTH & EPIDEMIOLOGY. *Educ:* McGill Univ, BSc, 54, MSc, 56; Univ Md, PhD(microbiol), 66. *Prof Exp:* USPHS fel microbiol, Sch Med, Univ Md, 58-61 & 65-67, instr, 67-70; asst prof med, Sch Med, Univ Conn, 70-75; asst prof microbiol & immunol, Fac Med, McGill Univ, 75-78, sci exec secy, Biohazards Comt, Fac Grad Studies, 77-78; USPHS vis fel, Ctr Dis Control, San Juan Labs, PR, 78-79; res assoc, Sch Pub Health, Univ Calif, Berkeley, 79-80; res assoc immunologist, Univ Calif Sch Med, San Francisco, 81-82; CONSULT, ENVIRON & HEALTH, BERKELEY, CALIF, 82-; SCI CONSULT, CROSBY, HEAFEY, ROACH & MAY, OAKLAND, CALIF, 85- *Concurrent Pos:* Consult typhus, WHO, 69, Pan-Am Health Orgn, 71; assoc mem & ad hoc comt mem typhus vaccines, Comn Rickettsial Dis, Armed Forces Epidemiol Bd, 66-73; hon res fel tumor immunol, Dept Zool, Univ Col London, 73-75; specialist pub health & med lab microbiol, Am Acad Microbiol, Nat Registry Microbiologist, 74-; mem & chmn, State of Calif, Dept Health Serv, Sexually Transmitted Dis Adv Coun, 85-; mem, State of Calif, Dept Consumer Affairs, Struct Pest Control Bd, 85-, vpres, 88-89, pres, 89-90 & mem, US Selective Serv Syst, 90- *Honors & Awards:* J Howard Brown Award, Soc Am Bacteriol, 60. *Mem:* Fel Royal Soc Trop Med & Hyg; Am Soc Microbiol; Am Soc Trop Med & Hyg; Brit Soc Immunol. *Res:* Rickettsiology and arbovirology; public health epidemiology preventive medicine of infectious disease; environmental health sciences. *Mailing Add:* 135 Alvarado Rd Berkeley CA 94705-1510

FABRIKANT, VALERY ISAAK, b Minsk, USSR, Jan 28, 40; Can citizen; m 81; c 2. PROGRAMMING. *Educ:* Ivanovo Power Inst, Bachelor, 62; Power Inst, Moscow, USSR, PhD(appl math & eng mech), 66. *Prof Exp:* Asst prof eng mech, Aviation Technol Inst, USSR, 67-69; prof eng mech, Polytech Inst, USSR, 70-73; sr res software systs, Res Inst, USSR, 73-78; res assoc eng mech, 80-84, RES PROF ENG, CONCORDIA UNIV, 84- *Concurrent Pos:* Prin investr, Power Inst, Ivanavo, USSR, 71-73, adj prof, 76-77; spec bursary, Ministry Educ, USSR, 57-62. *Mem:* Int Union Theoret & Appl Mech; Soc Eng Sci. *Res:* Elasticity theory; exact solutions of two-dimensional integral equations; contact problems for non-homogeneous bodies; numerical methods of solution of singular integral equations. *Mailing Add:* Dept Mech Eng Concordia Univ Montreal PQ H3G 1M8 Can

FABRIS, HUBERT, b Vienna, Austria, Sept 16, 26; US citizen; m 57; c 2. POLYMERIC MATERIALS. *Educ:* Univ Vienna, PhD(org chem), 56. *Prof Exp:* Sr res chemist, Gen Tire & Rubber Co, 58-65, group leader, 65-66, sect head, 67-80; mgr mat chem, 81-85, ASSOC DIR RES, GEN CORP, INC, 86- *Mem:* Am Chem Soc. *Res:* Synthesis, property and structure relationships of polymeric materials; polyurethanes, coatings, adhesives, foams; anionic polymerization, latex; polymer analysis and characterization. *Mailing Add:* 2260 Roundrock Dr Akron OH 44313

FABRIZIO, ANGELINA MARIA, b Italy; US citizen. MEDICAL MICROBIOLOGY, EXPERIMENTAL PATHOLOGY & TISSUE CULTURE. *Educ:* Villa Maria Col, BS, 44; Univ Ky, MS, 47; Univ Pa, PhD(med microbiol), 52; Hahnemann Med Col & Hosp, cert, 55. *Prof Exp:* Asst bact, Univ Ky, 45-46, instr Italian, 46-47; res bacteriologist antibiotics, Col Med, Univ Cincinnati & Cincinnati Gen Hosp, 47-48; res assoc exp cancer & tissue cult, Presbyterian Hosp, Philadelphia, 51-65; fac, Col Grad Studies, Thomas Jefferson Univ, 67-88; res assoc exp cancer & tissue cult, 65-67, asst prof path, 67-88, HON ASST PROF PATH & CELL BIOL, JEFFERSON MED COL, 88- *Concurrent Pos:* Instr, Sch Med, Univ Pa, 60; res consult, Vet Admin Hosp, Coatesville, Pa, 68-; vpres, RHO Chap, Philadelphia, pres-elect, 90-91, AAAS, 85- 90; Villa Maria Col career award. *Mem:* Fel AAAS; Am Soc Microbiol; Tissue Cult Asn; Am Soc Pathologists; Grad Women Sci (past pres); NY Acad Sci. *Res:* Acetylmethylcarbinol production by coliforms; sensitivities of bacteria to antibiotics; tuberculin sensitivity; tissue culture; experimental tumors; experimental heart research. *Mailing Add:* 2045 Spruce St Philadelphia PA 19103

FABRO, SERGIO, b Trieste, Italy, Sept 3, 31; m 58. PHARMACOLOGY, OBSTETRICS & GYNECOLOGY. *Educ:* Univ Milan, MD, 56; Univ Rome, PhD(biol chem), 66 & PhD(pharmacol), 68; Univ London, PhD(biochem), 67. *Prof Exp:* Intern med, Univ Milan, 56-57; asst prof path, Univ Modena, 58-59, asst prof biochem, 60-61; res asst, St Mary's Hosp Med Sch, Univ London, 63-67; RES PROF PHARMACOL, SCH MED, GEORGETOWN UNIV, 67-, PROF OBSTET & GYNEC & PROG DIR, COLUMBIA HOSP WOMEN, 74- *Honors & Awards:* Biochem Award, Nat Acad Lincei, 65. *Res:* Biochemistry of development; teratology; perinatal pharmacology. *Mailing Add:* Columbia Hosp Women Dept Obstet & Gynec Georgetown Univ 2425 L St NW Washington DC 20037

FABRY, ANDRAS, b Budapest, Hungary, Jan 10, 37; div; c 2. TOXICOLOGIC PATHOLOGY, PRECLINICAL TOXICOLOGY. *Educ:* Univ Liverpool, BVSc, 62; Colo State Univ, MS, 79; Am Col Vet Pathologists, dipl, 79. *Prof Exp:* Assoc vet, Gen Pract, 63-65; vet, Meat Inspection Div, Can Dept Agr, 65-67; asst toxicologist, Mason Res Inst, 67-68; supv toxicologist, Schering Corp, 68-76; resident, Dept Path, Col Vet Med & Biomed Sci, Colo State Univ, 76-79; sr res fel, Dept Safety Assessment, Merck Sharp & Dohme Res Labs, West Point, PA, 79-85; sr pathologist, Pharmaceuticals Div, Ciba-Geigy Corp, Summit, NJ, 85-86; SR RES PATHOLOGIST, SCHERING-PLOUGH CORP, LAFAYETTE, NJ, 86- *Mem:* Am Vet Med Asn; Brit Vet Asn; Am Col Vet Pathologists; Soc Toxicol Pathologists. *Res:* Clinical aspects of anesthetics; toxicological studies of cancer chemotherapeutic agents and new pharmaceuticals; pathological studies of new pharmaceuticals, agricultural chemicals and liotechnology products. *Mailing Add:* PO Box 262 Lafayette NJ 07848-0262

FABRY, MARY E RIEPE, b Iowa, May 17, 42. BIOPHYSICAL CHEMISTRY. *Educ:* Stanford Univ, BA, 64; Yale Univ, PhD(molecular physics), 67. *Prof Exp:* Lectr chem, Yale Univ, 67-68; res assoc biophysics, IBM Watson Lab, Columbia Univ, 68-69; assoc biol sci, 69-71, assoc med, 71-80, asst prof, 80-84, ASSOC PROF, ALBERT EINSTEIN COL MED, 84- *Concurrent Pos:* NY Heart Asn Sr Investr, 71-73; NIH spec fel, 74-75; mem, Biomed Sci Study Sect, NIH, 81-83. *Mem:* Am Chem Soc; Biophys Soc; Sigma Xi; Soc Magnetic Resonance in Med; Am Fedn Clin Res. *Res:* Heme proteins, membrane permeability to small molecules of biological membranes, particularly red blood cells; NMR relaxation; high resolution nuclear magnetic resonance of biological systems; magnetic resonance imaging; sickle cell anemia; enzyme mechanisms and metallo-proteins. *Mailing Add:* Dept Med Albert Einstein Col Med 1300 Morris Park Ave Bronx NY 10461

FABRY, THOMAS LESTER, b Budapest, Hungary, May 30, 37; m 67; c 3. GASTROENTEROLOGY, NUTRITION. *Educ:* St Andrews Univ, BSc, 61; Yale Univ, PhD(phys chem), 63; Albert Einstein Col Med, MD, 73; Am Bd Internal Med, Am Bd Gastroenterology, cert. *Prof Exp:* Fel Yale Univ, 63-64, lectr, 64-65; mem res staff biophys chem, Int Bus Mach Corp, 65-70; resident, Mt Sinai Hosp, 73-76, assoc, 76-79, ASSOC CLIN PROF MED, MT SINAI SCH MED, 79- *Concurrent Pos:* Adj asst prof, Rockefeller Univ, 70-72; res assoc & physician, Bronx Vet Admin Hosp, 76-79; DuPont fel. *Honors & Awards:* Irvine Medal, 61. *Mem:* Fel Am Col Gastroenterol. *Res:* Gastric acid secretion and gastric physiology; computer assisted instruction and artificial intelligence in medicine; trace metals and nutrition; oxygen and carbon dioxide transport by erythrocytes. *Mailing Add:* 853 Fifth Ave New York NY 10021

FABRYCKY, WOLTER J, b Queens, NY, Dec 6, 32; m 54; c 2. INDUSTRIAL ENGINEERING, SYSTEMS DESIGN & MANUFACTURING SYSTEMS. *Educ:* Wichita State Univ, BSIE, 57; Univ Ark, MSIE, 58; Okla State Univ, PhD(eng), 62. *Prof Exp:* Design engr, Cessna Aircraft Co, 54-57; instr indust eng, Univ Ark, 57-60; from asst prof to assoc prof indust eng, Okla State Univ, 62-65; assoc dean eng, 70-76, dean res, 76-81, PROF INDUST & SYSTEMS ENG, VA POLYTECH INST & STATE UNIV, 65- *Concurrent Pos:* Prin engr, Brown Eng Co, Ala, 62-65; ser ed, Prentice-Hall, Inc, 70- *Mem:* Am Inst Indust Engrs (exec vpres, 80-82); Am Soc Eng Educ (vpres, 75-76); Opers Res Soc Am; fel AAAS; fel Inst Indust Engrs. *Res:* Concurred engineering; manufacturing systems analysis and control; inquiry into the process of system/product life-cycle engineering with an orientation toward life-cycle engineering economics; engineering for total quality. *Mailing Add:* Dept Indust Syst Eng Va Polytech Inst & State Univ Blacksburg VA 24061

FABUNMI, JAMES AYINDE, b Ile-Ife, Nigeria, Sept 1, 50. AIRCRAFT PROPULSION, ROTORCRAFT DYNAMICS. *Educ:* Kiev Inst Civil Aviation Engrs, USSR, MechE & MSc, 74; Mass Inst Technol, PhD(aeronaut & astronaut), 78. *Prof Exp:* Res assoc fel, Mass Inst Technol, 78; res engr specialist, Kaman Aerospace Corp, 78-80; PRIN CONSULT ASSOC, FABUNMI & ASSOC, 80-; ASST PROF FLIGHT DYNAMICS & FLIGHT PROPULSION, AEROSPACE ENG DEPT, UNIV MD, 81- *Mem:* Am Helicopter Soc. *Res:* Aeroelastic stability of aircraft engines, such as high aspect ratio turbofans; theoretical and experimental methods for rotorcraft dynamics and vibration studies; unsteady aerodynamics of ducted fans; new propulsion concepts. *Mailing Add:* Aedar Corp 8401 Corporate Dr Suite 460 Landover MD 20785

FACKELMAN, GUSTAVE EDWARD, b Freeport, NY, May 19, 41; m 64; c 1. VETERINARY SURGERY. *Educ:* Cornell Univ, DVM, 64; Univ Zurich, Dr Med Vet, 71; Univ Pa, MA, 73; Am Col Vet Surgeons, dipl, 75. *Prof Exp:* Practr vet med, Crawford Animal Hosp, 64-65; asst prof surg clinician, Kans State Univ, 65-67; assoc prof surg, Univ Zurich, 67-73; ASSOC PROF ORTHOP SURG VET MED, UNIV PA, 73- *Concurrent Pos:* Res consult, Solco Basel AG, Basel, Switz & Straumann Inst, Waldenburg, Switz, 71- *Mem:* Am Vet Med Asn; Am Col Vet Surgeons; Asn Study Internal Fixation Vet; Vet Orthop Soc. *Res:* Fate of transplanted cancellous bone in the horse; joint mechanics and the biomechanics of equine articular fractures; spinal fusion in the horse; equine osteoarthritis. *Mailing Add:* Box 1269 Greenville ME 04441

FACKLER, MARTIN L, b York, Pa, Apr 8, 33; m 64. HISTORICAL RESEARCH, FORENSIC SCIENCE. *Educ:* Gettysburg Col, AB, 55; Yale Univ, MD, 59. *Prof Exp:* Intern, Univ Ore Med Sch Hosp, Portland, 60; residency surg, US Naval Hosp 61-65, plastic surg, Bethesda Naval Hosp, 65-67; chief surg, US Army Hosp, Landstuhl, Ger, 78-80, Ft Carson Army Hosp, Colo, 80-81; dir, Wound Ballistics Lab, Letterman Army Inst Res, 81-91; CONSULT, FORENSIC SCI & LAW ENFORCEMENT, 84- *Concurrent Pos:* Expert witness wound ballistics & surg, 83-; special consult wound ballistics, Int Defense Rev, 88-; ed, Wound Ballistics Rev, 90-; tech adv, Asn Firearm & Toolmark Examiners. *Mem:* Fel Am Col Surgeons; Asn Firearm & Toolmark Examiners; Int Wound Ballistics Asn (pres, 90-); Inst Res Small Arms Int Security. *Res:* All aspects of the effects of penetration projectiles on the human body including actual laboratory work, historical research, and collection of data from shootings; over 100 articles and book chapters published. *Mailing Add:* 1809 Wyman Ave San Francisco CA 94129

FACKLER, WALTER VALENTINE, JR, b Oak Park, Ill, Mar 19, 20; m 47; c 3. SURFACE CHEMISTRY. *Educ:* Iowa State Col, PhD(phys chem), 53. *Prof Exp:* Res chemist, Bauer & Black, 53-56; res supvr, Toni Co, 56-61; adv scientist, Continental Can Co, 61-66; chemist, Van Straaten Chem Co, 66-70; SR RES PROF, ARMOUR-DIAL, INC, 70- *Mem:* AAAS; Am Chem Soc. *Res:* Detergents; polymers; plastic films; adhesives. *Mailing Add:* 8348 E Via de Sereno Scottsdale AZ 85258

FACTOR, ARNOLD, b Boston, Mass, Apr 1, 36; m 61; c 2. PHYSICAL ORGANIC CHEMISTRY. *Educ:* Brandeis Univ, BA, 58; Harvard Univ, MA, 60, PhD(chem), 63. *Prof Exp:* NIH fel, Univ Calif, San Diego, 63-64; res chemist, Gen Elec Res & Develop Ctr, 64-73, mgr polymer flammability & stabilization proj, 73-80, mgr, polymer & surface stabilization unit, 80-86, res chemist, 86-91. *Mem:* Am Chem Soc. *Res:* Autoxidation; free radical chemistry; phenol oxidations; polymer stabilization; redox polymers; polymer flammability; polymer coatings; hertic chemistry; lubrication. *Mailing Add:* Gen Elec Res & Develop Ctr PO Box 8 Schenectady NY 12301

FACTOR, JAN ROBERT, b Brooklyn, NY, Mar 27, 50. HISTOLOGY, ELECTRON MICROSCOPY. *Educ:* Brooklyn Col, BS, 73; Cornell Univ, MS, 77, PhD(zool), 79. *Prof Exp:* Postdoctoral fel, Marine Sta, Link Port, Smithsonian Inst, 79-81; lectr invert zool, Sect Genetics & Develop, Cornell Univ, 81-82; asst prof, 82-89, ASSOC PROF BIOL, STATE UNIV NY, PURCHASE, 89-, CHAIR BIOL, 90- *Concurrent Pos:* Core fac, Shoals Marine Lab, Cornell Univ, 81-; dir, Lab Electron Micros, State Univ NY, Purchase, 82-; assoc ed, J Morphol, 91- *Mem:* Am Soc Zoologists; AAAS; Sigma Xi; Crustacean Soc. *Res:* Development, functional morphology, histology and ultrastructure of the feeding apparatus and digestive system of lobsters and other crustaceans; structure, function and systematic implications of complex basement membranes in arthropods. *Mailing Add:* Div Natural Sci State Univ NY Purchase NY 10577

FACTOR, STEPHEN M, b New York, NY, Oct 28, 42. PATHOLOGY & MEDICAL. *Educ:* City Univ NY, BA, 64; Albert Einstein Col Med, Bronx, NY, MD, 68. *Prof Exp:* Intern, Dept Surg, Univ Mich Hosps, Ann Arbor, 68-69, resident, 69-70; vis pathologist, Homer Cobb Mem Hosp, Phenix City, Ala, 72-73; resident anat & clin path, Bronx Munic Hosp Ctr & Weiler Hosp, Albert Einstein Col Med, Bronx, NY, 70-71, resident, 73-75, chief resident, 74-75, asst instr, Dept Path, Albert Einstein Col Med, 74-75, from asst prof to assoc prof, 75-85, assoc prof, Div Cardiol, Dept Med, 85-87, PROF, DEPT PATH, ALBERT EINSTEIN COL MED, BRONX, NY, 85-, PROF, DIV CARDIOL, DEPT MED, 87- *Concurrent Pos:* Abraham & Joseph Spector fel path, Albert Einstein Col Med, Bronx, NY, 78-; ad hoc mem, Spec Study Sect, NIH, 84, Cardiovasc A Study Sect, 90; consult & panel pathologist, Multictr Myocarditis Treatment Trial, 86-; ed-in-chief, Cardiovasc Path, 91-96. *Mem:* Fel Am Col Cardiol; fel Col Am Pathologists; Fedn Am Soc Exp Biol; Am Heart Asn; Int Acad Path; Int Soc Heart Res; Int Asn Cardiac Biol Implants; Soc Cardiovasc Path (vpres, 87-88, pres, 88-89). *Res:* Cardiomyopathy; diabetic and hypertensive heart disease; myocardial ischemia and infarction; microcirculation; myocardial connective tissue matrix; atherosclerosis; pulmonary circulation. *Mailing Add:* Dept Path & Med Albert Einstein Col Med 1300 Morris Park Ave Bronx NY 10461

FADDICK, ROBERT RAYMOND, b Sudbury, Ont, May 18, 38; m 65; c 6. FLUID MECHANICS, HYDRAULIC ENGINEERING. *Educ:* Queen's Univ, Ont, BApS, 61, MApS, 63; Mont State Univ, PhD(civil eng), 70. *Prof Exp:* Res asst hydraul eng, Queen's Univ, Ont, 61-63; hydraul engr, Alden Hydraul Lab, Worcester Polytech Inst, 63-66; res asst civil eng, Mont State Univ, 66-69; asst prof, 69-81, PROF CIVIL ENG, COLO SCH MINES, 81- *Concurrent Pos:* Pres, Slurry Pipeline Corp. *Mem:* Hon mem Soc Am Mil Engrs; Am Soc Civil Engrs; Soc Rheology; Slurry Transp Asn; hon mem Brit Hydromech Res Asn. *Res:* Transportation of solids in pipelines; rheology of mineral solids; compilation of a slurry pipeline computer data bank; pipeline and rheological aspects of mineral slurries; hydraulic aspects of thermal pollution; pneumotransport and capsule pipelining; slurry pump testing; slurry wear testing. *Mailing Add:* Slurry Pipeline Corp 2373 Coors Dr Golden CO 80401

FADELL, ALBERT GEORGE, b Niagara Falls, NY, Jan 5, 28; m 71. MATHEMATICS. *Educ:* Univ Buffalo, AB, 49, MA, 51; Ohio State Univ, PhD, 54. *Prof Exp:* From asst prof to assoc prof, 54-76, vchmn dept, 69-70, assoc provost fac natural sci math, PROF MATH, STATE UNIV NY BUFFALO, 76- *Mem:* Am Math Soc; Math Asn Am. *Res:* Real variables; measure theory. *Mailing Add:* Dept Math SUNY at Buffalo 106 Diefendorf Hall Buffalo NY 14214

FADELL, EDWARD RICHARD, b Niagara Falls, NY, March 8, 26; m 53; c 2. MATHEMATICS. *Educ:* Univ Buffalo, BA, 48; Ohio State Univ, MA, 50, PhD(math), 52. *Prof Exp:* Peirce instr math, Harvard Univ, 52-55; from instr to assoc prof, 55-62, PROF MATH, UNIV WIS- MADISON, 62- *Mem:* Am Math Soc; Math Asn Am. *Res:* Algebraic topology; fixed point theory; topological methods in critical point theory. *Mailing Add:* Dept Math Univ Wis 325 Van Vleck Madison WI 53706

FADEN, HOWARD, PEDIATRICS. *Prof Exp:* CO-DIR, DIV INFECTIOUS DIS, DEPT PEDIAT, STATE UNIV NY CHILDREN'S HOSP, 78- *Mailing Add:* Dept Pediat State Univ NY Children's Hosp 219 Bryant St Buffalo NY 14226

FADER, WALTER JOHN, theoretical physics; deceased, see previous edition for last biography

FADLEY, CHARLES SHERWOOD, b Norwalk, Ohio, Sept 4, 41; div. CHEMICAL PHYSICS. *Educ:* Mass Inst Technol, SB, 63; Univ Calif, MS, 65, PhD(chem), 70. *Prof Exp:* Res fel solid state physics, Dept Physics, Chalmers Inst Technol, Gothenburg, Sweden, 70-71; sr lectr physics, Univ Dar es Salaam, Tanzania, 71-72; asst prof, 72-74, assoc prof, 74-78, PROF CHEM, UNIV HAWAII, HONOLULU, 78- *Concurrent Pos:* Vchmn, Gordon Conf Electron Spectros, 74-76, chmn, 76-78; res fel, Alfred P Sloan Found, 75-77; vis prof, Univ Paris, 78-79 & Univ Utah, 79-80. *Mem:* Am Chem Soc; Am Phys Soc; Am Vacuum Soc; AAAS; Sigma Xi. *Res:* Experimental and theoretical studies involving the application of angular-dependent electron spectroscopy to problems in surface chemistry, surface physics, and solid state physics; experimental techniques include x-ray and ultraviolet photoelectron spectroscopy using laboratory sources and synchrotron radiation, low energy electron diffraction, and energy loss spectroscopy. *Mailing Add:* Dept Chem Univ Hawaii Honolulu HI 96822

FADLY, ALY MAHMOUD, b Cairo, Egypt, Nov 3, 41; US citizen; m 71; c 4. POULTRY DISEASES, AVIAN VIROLOGY. *Educ:* Cairo Univ, DVM, 64; Purdue Univ, MS, 73, PhD(vet microbiol), 75. *Prof Exp:* Res vet virol, Vet Res Inst, Cairo, 64-69; asst poultry dis, Sch Vet Med, Purdue Univ, 69-75, asst prof, 75-76; VET RES SCIENTIST POULTRY DIS, REGIONAL POULTRY RES LAB, USDA, 76- *Concurrent Pos:* Assoc ed, Poultry Sci Asn; consult, USAID & UN Food & Agr Orgn, UN Develop Programme, Int Agr Exchange Asn, Egypt & Brazil, 78- & Coop State Res serv Tech Adv Comt on Animal Health, USDA, 83 & 84; vis scientist, Houghton Poultry Res Sta, Huntingdon, Eng, 85-86. *Mem:* Am Asn Avian Pathologists; Am Vet Med Asn; Sci Soc NAm; Poultry Sci Asn; Int Asn Comp Res Leukemia & Related Dis; World Vet Poultry Asn. *Res:* Viral induced diseases of poultry; avian tumor virology; avian diseases in general. *Mailing Add:* USDA Regional Poultry Res Lab 3606 E Mount Hope Rd East Lansing MI 48823

FADNER, THOMAS ALAN, b Milwaukee, Wis, Aug 19, 29; div; c 2. POLYMER CHEMISTRY, PRINTING SCIENCES. *Educ:* Univ Wis, Oskkosh, BS, 51; Polytech Inst Brooklyn, PhD(polymer chem), 61. *Prof Exp:* From asst chemist to chemist, S C Johnson & Son, Inc, 51-57, sr chemist, 60-62; lab head paper specialties, Oxford Paper Co, 62-63, from asst dir to dir tech specialties dept, 63-70; mgr, Prod Develop Dept, Fiber Prod Res & Develop, Kendall Co, 70-72; pres, FDX Corp, Bethel, Maine, 72-75; mgr, Chem Div, Graphic Arts Technol Found, 75-79; res scientist, Tech Ctr, Am Can Co, 79-81; MGR MAT RES TO STAFF SCIENTIST, PRINTING PROCESSES/MAT, ROCKWELL GRAPHIC SYSTS, 81- *Concurrent Pos:* Bd dir, Tech Asn Graphic Arts, 86-88, fel Comt, 86; mem, res steering comt Graphic Arts Tech Found. *Honors & Awards:* Honors Award, Tech Asn Graphic Arts, 90. *Mem:* Tech Asn Pulp & Paper Indust; Am Chem Soc; Tech Asn Graphic Arts; Soc Mfg Engrs. *Res:* Application of physical properties of polymeric materials to commercial specialty products; paper-plastic, solid-state polymerization; colloidal phenomena; coating development; non-woven fabrics design; lithography and other printing processes. *Mailing Add:* 700 Oakmont Lane Westmont IL 60559-5546

FADUM, RALPH EIGIL, b Pittsburgh, Pa, July 19, 12; m 39; c 1. CIVIL ENGINEERING. *Educ:* Univ Ill, BSCE, 35; Harvard Univ, MS, 37, SD, 41. *Hon Degrees:* DEng, Purdue Univ, 63. *Prof Exp:* Asst civil eng, Harvard Univ, 35-37, instr, 37-41, fac instr, 41-43; from asst prof to prof soil mech, Purdue Univ, 43-49; head, Dept Eng, NC State Univ, 49-62, prof civil eng, 49-62, dean, Sch Eng, 62-78; CONSULT. *Concurrent Pos:* Consult, Dept Defense, US CEngrs; mem sci adv panel, Dept Army, 59-74; mem res adv comt, Fed Hwy Admin, 63-70; vchmn, Army Sci Adv Panel, Dept Army, 66-70, chmn adv group to comdr gen, Tank Automotive Command, 67-70; chmn, NC Water Control Adv Coun; mem bd dirs, Nat Driving Ctr, 73-; pres, Atlantic Coast Conf, 66-67, 71-72; chmn bd dirs, NC Water Resources Res Inst, Univ NC. *Mem:* Nat Acad Eng; fel Am Soc Civil Engrs; US Nat Coun Soil Mech & Found Eng; Nat Soc Prof Engrs; Am Soc Eng Educ. *Mailing Add:* Dept Civil Eng NC State Univ Raleigh NC 27650

FAETH, GERARD MICHAEL, b New York, NY, July 5, 36; m 59; c 3. MECHANICAL ENGINEERING, COMBUSTION. *Educ:* Union Col, BME, 58; Pa State Univ, MS, 61, PhD(mech eng), 64. *Prof Exp:* Res asst, Pa State Univ, 58-64, from asst prof to assoc prof, 64-75, consult, Ord Res Lab, 64-85, prof, 75-85, EMER PROF MECH ENG, PA STATE UNIV, PARK, 85-; ARTHUR B MODINE PROF AEROSPACE ENG & HEAD GAS DYNAMICS LAB, UNIV MICH, ANN ARBOR, 85- *Concurrent Pos:* Resident prof, Gen Motors Corp, 83; vis prof, Off Sci Res, USAF, Washington, DC, 83-84. *Honors & Awards:* Heat Transfer Mem Award, Am Soc Mech Engrs, 88. *Mem:* Nat Acad Eng; fel Am Soc Mech Engrs; fel Am Inst Aeronaut & Astronaut; Combustion Inst; Sigma Xi; fel AAAS. *Res:* Explosion hazards; physical aspects of fires; evaporation and combustion of sprays; combustion of metals; turbulent natural convection processes; two-phase flow and heat transfer. *Mailing Add:* 217 Aerospace Eng Bldg Univ Mich Ann Arbor MI 48109-2140

FAETH, STANLEY HERMAN, b Covington, Ky, June 1, 51; m 84; c 3. PLANT & INSECT RELATIONSHIPS, INSECT ECOLOGY. *Educ:* Univ Cincinnati, BS, 73 & MS, 77; Fla State Univ, PhD(biol), 80. *Prof Exp:* Biologist, US Environ Protection Agency, 75; entomologist, Cincinnati Mus Natural Hist, 77; asst prof, 80-86, ASSOC PROF ECOL, ARIZ STATE UNIV, 86- *Mem:* Assoc mem Ecol Soc Am; assoc mem AAAS; Sigma Xi; assoc mem Entom Soc Am. *Res:* Population dynamics and community ecology of phytophagous insects. *Mailing Add:* Dept Zool Ariz State Univ Tempe AZ 85287-1501

FAFLICK, CARL E(DWARD), b Cleveland, Ohio, Mar 10, 22; m 53; c 3. COMMUNICATIONS, SYSTEMS ENGINEERING. *Educ:* Oberlin Col, BA, 43; Harvard Univ, MA, 48, PhD(appl sci), 53. *Prof Exp:* Instr & res assoc, NMex State Univ, 47; Sylvania E Electronic Systs, Gen Tel & Electronics Inc, 55-69; vpres eng, Gen Tel & Electronics Int Systs Corp, 69-83; VPRES ENG, MEGAPULSE INC, 83- *Mem:* Inst Elec & Electronics Engrs. *Res:* Antenna; propagation; microwave; space communications; data communications. *Mailing Add:* 28 Moon Hill Rd Lexington MA 02173

FAGAN, J(OHN) R(OBERT), b Omaha, Nebr, Sept 17, 35; m 58; c 5. CHEMICAL & NUCLEAR ENGINEERING. *Educ:* Univ Nebr, BS, 57; Kans State Univ, MS, 62; Purdue Univ, PhD(nuclear eng), 68. *Prof Exp:* Jr chem engr, Argonne Nat Lab, Ill, 57-59; from instr to asst prof, Kans State Univ, 59-63; sr res engr, 63-66, res scientist, 66-67, sect chief, 67-68, mgr theoret res, 68-72, mgr turbine res & develop, 72-73, chief aerothermodyn res, 73-77, CHIEF PROJ ENGR ADVAN TURBINE ENGINE GAS GENERATOR PROG, DETROIT DIESEL ALLISON DIV, GEN MOTORS CORP, 77- *Mem:* Am Inst Aeronaut & Astronaut; Am Nuclear Soc; Sigma Xi. *Res:* Gas turbine engine design; internal aerodynamics of turbomachinery; nuclear reactor design. *Mailing Add:* 5833 N La Salle Indianapolis IN 46220

FAGAN, JOHN EDWARD, b Manchester, Conn, Oct 17, 43; m 69; c 2. SYSTEMS ENGINEERING. *Educ:* Univ Tex, Arlington, BS, 67, MS, 72, PhD(elec eng), 76. *Prof Exp:* Design engr antisubmarine warfare radar, Tex Instruments Inc, 67-69; design engr comput control test equip, USAF, 69-71; prof, Col Eng, Univ Okla, 75-78, assoc dean, 78-80; DIR, OKLA POWER SYSTS RES CTR, 80- *Honors & Awards:* Dow Award, Dow Chem Co, 78; C Holmes McDonald Award, 78; Jasper P Baldwin Award, 81. *Mem:* Inst Elec & Electronics Engrs; Sigma Xi; Conf Int des Grandes Reseaux Electriques; Am Soc Eng Educ. *Res:* Power system dynamics simulation and use of adaptive real-time control techniques to the control of power systems; transmission line wind loading; development of data aquisition systems. *Mailing Add:* 2329 Blue Creek Pkwy Norman OK 73071

FAGAN, JOHN J, geology, for more information see previous edition

FAGAN, PAUL V, b Newark, NJ, May 22, 27; m 53; c 4. BIOCHEMISTRY, ORGANIC CHEMISTRY. *Educ:* Seton Hall Univ, BS, 49; Fordham Univ, MS, 54. *Prof Exp:* Chemist, Sterone Corp, 51-53; jr scientist, Ethicon Inc, 53-55, asst scientist, 55-59, assoc scientist, Devro Div, 59-66, sr scientist, Johnson & Johnson, 66-83; RETIRED. *Honors & Awards:* Johnson Medal for Res & Develop, Johnson & Johnson, 73. *Mem:* Am Chem Soc; Inst Food Technologists. *Res:* Process development, synthesis of pharmaceuticals and fine chemicals; fiber forming polymer synthesis; chemical modification of natural fibers, edible film and casing; collagen sausage casing technology; tanning processes of proteins. *Mailing Add:* 997 Shadow Oak Lane Bridgewater NJ 08807

FAGAN, RAYMOND, b Brooklyn, NY, Dec 27, 14; m 36; c 3. EPIDEMIOLOGY, ENVIRONMENTAL SCIENCES. *Educ:* NY Univ, BA, 35; Cornell Univ, DVM, 39; Harvard Univ, MPH, 49. *Prof Exp:* Jr vet, Meat Inspection, USDA, 39-41; jr vet, Milk Sanitation, USPHS, 42, vet officer & epidemiologist, 46-54; assoc prof prev med & hyg & chmn dept, Sch Vet Med, Univ Pa, 54-56; sr investr virol, Wyeth Inst Med Res, 56-67; prin scientist biol, Philip Morris Res Ctr, 67-85; RETIRED. *Concurrent Pos:* Chmn pub health comn, Health & Welfare Coun, Chester County, Pa, 57-60; chmn adv coun, Philadelphia Dept Health, 63-67; consult, WHO, 63-70 & Nat Inst Environ Health Sci, 70-71; adj prof, Drexel Univ, 64-67; adj assoc prof, Med Col Va, 75-; mem, Air Pollution Control Bd, Richmond, 75-81. *Mem:* Am Pub Health Asn; Am Vet Med Asn; Sigma Xi. *Res:* Environmental sciences; host-parasite relationships; epidemiology; virology; education. *Mailing Add:* 8554 Old Spring Rd Richmond VA 23235

FAGAN, TIMOTHY CHARLES, b Phoenix, Ariz, Mar 22, 47; m 70; c 3. ANTI-HYPERTENSIVE DRUGS. *Educ:* Stanford Univ, AB, 69, Univ Calif, Los Angeles, MD, 73. *Prof Exp:* Resident internal med, Vet Admin Med Ctr, Los Angeles, 73-76; fel clin pharmacol, 76-78, asst prof pharmacol & med, Med Univ SC, Charleston, 78-81; asst prof, 81-87, ASSOC PROF INTERNAL MED & PHARMACOL, ARIZ HEALTH SCI CTR, 87- *Mem:* Fel Am Col Physicians; Am Soc Clin Pharmacol & Therapeut; Am Soc Pharmacol & Exp Therapeut; Am Soc Hypertension; Int Soc Cardiovasc Pharmacotherapy. *Res:* Studies of new anti-hypertension drugs particularly pharmacokinetics and mechanisms of action; studies of the hemodynamic and hormonal effects of food and pharmacokinetic and pharmacodynamic interactions with drugs. *Mailing Add:* Dept Internal Med Univ Ariz Health Sci Ctr 1501 N Campbell Ave Tucson AZ 85724

FAGEN, ROBERT, b 1945; US citizen. ANIMAL PLAY, DANCE. *Educ:* Mass Inst Technol, BS, 67; Univ Mich, MS, 68; Harvard Univ, PhD(math biol), 74. *Prof Exp:* Asst prof biol, Univ Ill, 74-78; asst prof anat, Univ Pa, 78-82; ASSOC PROF, SCH FISHERIES, UNIV ALASKA, FAIRBANKS, 82- *Mem:* Animal Behav Soc; Am Statist Asn; Sigma Xi; Am Fisheries Soc; Wildlife Soc. *Res:* Evolution and ethology of play in animals, and interdisciplinary synthetic approaches to human dance, Alaskan fisheries management, renewable natural resource applications of exploratory data analysis methods. *Mailing Add:* Univ Alaska Juneau AK 99801

FAGER, LEI YEN, biochemistry, for more information see previous edition

FAGERBERG, WAYNE ROBERT, b Denver, Colo, May 16, 44; m 71. MORPHOMETRICS, STEREOLOGY. *Educ:* Univ Wyo, BS, 67; Univ SFla, 72, PhD(algal develop cytol), 75. *Prof Exp:* Fel microecol, Univ Tex, Arlington, 75-77, vis prof biol, 78-79; dir, Electron Micros Lab, Southern Methodist Univ, 79-84; asst prof, 84-89, ASSOC PROF, UNIV NH, 89- *Mem:* Am Bot Soc; Int Soc Stereology; Sigma Xi; AAAS. *Res:* Quantitative morphometric analysis of plant cell structure with the purpose of determining the degree to which internal and external factors affect or change cell structure; correlation of cell structure and function. *Mailing Add:* Dept Plant Biol Univ NH Durham NH 03824

FAGERBURG, DAVID RICHARD, b Rockford, Ill, Aug 5, 42; m 65; c 6. CONDENSATION POLYMERS, POLYMER CHEMISTRY & PHYSICS PROBLEMS. *Educ:* Calif State Col, Long Beach, BS, 67; Univ Wash, Seattle, PhD(chem), 70. *Prof Exp:* Teaching asst chem, Univ Wash, 68-70; res chemist, 70-73, sr res chemist, 73-79, RES ASSOC, EASTMAN CHEM DIV, EASTMAN KODAK, 79- *Concurrent Pos:* Adj prof org chem, NE State Community Col, 89- *Honors & Awards:* Chem Medal, Am Inst Chem, 67. *Mem:* Am Chem Soc; affil Int Union Pure & Appl Chem; fel Am Inst Chemists. *Res:* Discovery and development of new condensation polymer systems; kinetics of polymer reactions; quantum mechanical approaches to understanding polymer reactions and properties. *Mailing Add:* 3812 Cimmaron Dr Kingsport TN 37664

FAGERSON, IRVING SEYMOUR, b Lawrence, Mass, June 7, 20; m 53; c 2. FOOD SCIENCE. *Educ:* Mass Inst Technol, SB, 42; Univ Mass, MS, 48, PhD(food technol), 50. *Prof Exp:* Asst food tech, Mass Inst Technol, 42; mkt specialist, USDA, 42-43; from asst prof to prof, 49-85, EMER PROF FOOD CHEM, UNIV MASS, AMHERST, 85- *Mem:* AAAS; Am Chem Soc; Inst Food Technol; fel Am Inst Chem; Am Soc Mass Spectrometry. *Res:* Nutritive value of foods; chemistry of flavor; analysis instrumentation. *Mailing Add:* 51 Jeffery Lane Amherst MA 01002

FAGERSTROM, JOHN ALFRED, b Ypsilanti, Mich, Jan 4, 30; m 53; c 3. INVERTEBRATE PALEONTOLOGY, MARINE ECOLOGY. *Educ:* Oberlin Col, AB, 52; Univ Tenn, MS, 53; Univ Mich, PhD(geol), 59. *Prof Exp:* From instr to prof geol, Univ Nebr-Lincoln, 58-88, chmn dept, 70-74; ADJ PROF GEOL, UNIV COLO-BOULDER, 88- *Mem:* Geol Soc Am; Soc Econ Paleontologists & Mineralogists; Int Coral Reef Soc. *Res:* Biostratigraphy and paleoecology of the Devonian rocks of the Great Lakes region; ecology of reefs. *Mailing Add:* Dept Geol Sci Univ Colo Boulder CO 80309-0077

FAGET, MAXIME A(LLAN), b Stann Creek, British Honduras, Aug 26, 21; US citizen; m 47; c 4. AERONAUTICAL ENGINEERING. *Educ:* La State Univ, BS, 43. *Hon Degrees:* DEng, Univ Pittsburgh, 66 & La State Univ, 72. *Prof Exp:* Aeronaut res scientist, Langley Res Ctr, Hampton, Va, 46-58; chief, Flight Systs Div, NASA, 58-61, dir eng & develop, Manned Spacecraft Ctr, Houston, Tex, 61-81; vpres new systs develop, Eagle Eng, Inc, 81-84; PRES & CHIEF EXEC OFFICER, SPACE INDUSTS, INC, 82- *Concurrent Pos:* Dir, Eagle Eng, Inc, Houston; trustee, Houston Mus Natural Sci; mem, Space Sci & Indust Comn. *Honors & Awards:* Arthur S Flemming Award, 59; Golden Plate Award, Acad Achievement, 62; Spacecraft Design Award, Am Inst Aeronaut & Astronaut, 70, Daniel & Florence Guggenheim Award, 73, Goddard Astronaut Award, 79; William Randolph Lovelace II Award, Am Astronaut Soc, 71, Lloyd V Berkner Award, 87; Gold Medal, Am Soc Mech Engrs, 75; Albert F Sperry Medal, Instrument, Soc Am, 76; Harry Diamond Award, Inst Elec & Electronics Engrs, 76; Jack Swigert Mem Award, 88. *Mem:* Nat Acad Eng; fel Am Inst Aeronaut & Astronaut; fel Am Astronaut Soc; Int Acad Astronaut. *Res:* Manned space flight; reentry aerodynamics; propulsion; space power systems; guidance and control; life support systems; engineering and space systems development; technical management; author of numerous technical publications. *Mailing Add:* Space Industs Inc 711 W Bay Area Blvd Suite 320 Webster TX 77598

FAGG, LAWRENCE WELLBURN, b NJ, Oct 10, 23; m 50, 58. NUCLEAR PHYSICS. *Educ:* US Mil Acad, BS, 45; Univ Md, MS, 47; Univ Ill, MA, 48; Johns Hopkins Univ, PhD(physics), 53; George Washington Univ, MA, 81. *Prof Exp:* Physicist coulomb excitation, Naval Res Lab, 53-58; physicist plasma physics, Atlantic Res Corp, 58-63; physicist electron nuclear scattering, US Naval Res Lab, Washington, DC, 63-76; RES PROF ELECTRON NUCLEAR SCATTERING, CATH UNIV AM, 77- *Concurrent Pos:* Vpres, Inst Relig Age Sci, 82-86. *Honors & Awards:* Meritorious Civilian Serv Award, Naval Res Lab, 75. *Mem:* Fel Am Phys Soc. *Res:* Assignment of nuclear energy levels and transition rates; plasma ion density studies; electron-nuclear scattering. *Mailing Add:* Dept Physics Cath Univ Am Washington DC 20064

FAGIN, CLAIRE M, b New York, NY, Nov 25, 26; m 52; c 2. PSYCHIATRIC NURSING, PEDIATRIC NURSING. *Educ:* Wagner Col, BS, 48; Columbia Univ, MA, 51; NY Univ, PhD(psychiat nursing), 64. *Hon Degrees:* DSc, Lycoming Col, 83, Cedar Crest Col, 87, Univ Rochester, 87, Med Col Pa, 89. *Prof Exp:* Staff nurse, Seaview Hosp, 47; clin instr nursing, Bellevue Hosp, 48-50; consult, Nat League Nursing, 51-52; asst chief psychiat, Clin Ctr, NIH, 53-54; res proj coordr, Children's Hosp, Washington, DC, 56-58; instr psychiat & mental health, NY Univ, 56-58, from asst prof to assoc prof nursing, 64-69, dir psychol & mental health, 65-69; chmn & prof nursing, Lehman Col, City Univ New York, 69-77, dir health prof, 75-77; DEAN & PROF NURSING, SCH NURSING, UNIV PA, 77- *Concurrent Pos:* Mem expert adv panel nursing, WHO, 74-; mem bd dirs, Provident Mutual Ins Co, 77-, audit comt, 78-, chmn audit comt, 82-; mem at large, Nat Bd Med Examiners, 80-83; mem bd ment health & behav med, Inst Med-Nat Acad Sci, 81-88, gov coun, 81-83, Nat Ment Health Adv Coun, 84-88; mem bd dirs, Daltex Med Sci, Inc, 84-, compensation comt, 84-; mem comn, Elderly People Living Alone Prog, Commonwealth Found, 86-; sr adv, Nurses Am, 89- *Honors & Awards:* Am Psychiat Nursing Asn Award, 88; Maes-MacInnes Award, 90. *Mem:* Inst Med-Nat Acad Sci; Am Acad Nursing. *Res:* Affects of maternal attendance during children's hospitalization. *Mailing Add:* Nursing Educ Bldg S2 Sch Nursing Univ Pa Philadelphia PA 19104

FAGIN, KARIN, b Brooklyn, NY, Jan 6, 49; m 79; c 2. MEDICINE. *Educ:* Univ Fla, BS, 77, MS, 80; Univ NMex, MD, 91. *Prof Exp:* HOUSE OFFICER I PATH, SHANDS HOSP, 91- *Mem:* AMA. *Res:* Clinical behavior induced by elileptogenic drugs including kainic acid, strychnine, pentylene tetrazole and cis-platin using the frog(Rana pipiens) as a model; histology effects of these drugs on the central nervous system. *Mailing Add:* 2600 SW Williston Rd Apt 203 Gainesville FL 32608

FAGIN, KATHERINE DIANE, b New York, NY. DRUG DELIVERY, PHARMACODYNAMICS OF THERAPEUTIC PROTEINS & PEPTIDES. *Educ:* Cornell Univ, AB, 76; Emory Univ, PhD(physiol), 80. *Prof Exp:* Res asst physiol & endocrinol, Emory Univ, 76-79; res asst endocrinol, Univ Ala, Birmingham, 79-80; postdoctoral fel endocrinol, Univ Calif, San Francisco, 80-83, res assoc endocrinol, 83-84; res scientist pharmacol, Amgen Corp, 84-89; RES SCIENTIST BIOL SCI, ALZA CORP, 89- *Concurrent Pos:* Ad hoc ed reviewer, Endocrinol, 81- *Mem:* AAAS; NY Acad Sci; Endocrine Soc; Am Diabetes Asn. *Res:* Development of delivery system strategies for protein and peptide therapeutics; discovery research and evaluation of physiology and pharmacodynamics of recombinant proteins. *Mailing Add:* Alza Corp 950 Page Mill Rd PO Box 10950 Palo Alto CA 94303

FAGIN, RONALD, b Oklahoma City, Okla, May 1, 45; m 85; c 2. LOGIC. *Educ:* Dartmouth Col, BA, 67; Univ Calif, Berkeley, PhD(math), 73. *Prof Exp:* Mem res staff, IBM Watson Res Ctr, Yorktown Heights, NY, 73-75, & IBM San Jose Res Lab, 75-79, MGR, FOUND COMPUT SCI, ALMADEN RES CTR, IBM, SAN JOSE, 79- *Concurrent Pos:* Asst ed, J Comput & Systs Sci, 84- *Honors & Awards:* Three Outstanding Innovation Awards, IBM, 81, 87. *Res:* Theory of knowledge and its applications to distributed computer systems; relational database theory; finite model theory. *Mailing Add:* IBM Res Div Almaden Res Ctr 650 Harry Rd San Jose CA 95120-6099

FAGLEY, THOMAS FISHER, b Mt Carmel, Pa, Sept 7, 13. PHYSICAL CHEMISTRY. *Educ:* Bucknell Univ, BS, 35, MS, 37; Univ Chicago, PhD(chem), 49. *Prof Exp:* Instr chem, Bucknell Univ, 38-40 & 42-46; instr, Univ Col, Univ Chicago, 47, teaching fel, 48-49; from asst prof to prof chem, Tulane Univ, 49-79; RETIRED. *Mem:* Fel AAAS; Am Chem Soc; Sigma Xi. *Res:* Microcalorimetry; heats of combustion; kinetics; thermodynamics of solutions. *Mailing Add:* 620 Dumaine St New Orleans LA 70116

FAGOT, WILFRED CLARK, b New Orleans, La, Dec 5, 24. APPLIED MATHEMATICS, MATHEMATICAL PHYSICS. *Educ:* Univ Tex, BS, 45; Tulane Univ, MS, 47; Univ Chicago, MS, 52. *Prof Exp:* Physicist, Enrico Fermi Inst Nuclear Studies, 52-53, physicist, Chicago Midway Labs, 53-54; sr systs physicist, Norden Div, United Aircraft Corp, 56-61; sr staff scientist, Kearfott Div, Gen Precision Aerospace, 61-63; prin scientist, Bedford Labs, Missile Systs Div, Raytheon Co, 63-65; asst prof math, Juniata Col, 65-66, chmn dept, 71-74, assoc prof, 66-83; SR STAFF ENGR, NBC-TV, 83- *Concurrent Pos:* NSF sci fac fel, Pa State Univ, 70-71. *Mem:* Soc Motion Pictures & TV Engrs; sr mem Inst Elec & Electronics Engrs. *Res:* Stochastic processes, electromagnetic theory, antennas, propagation, scattering, noise and radar clutter; perturbation theory; cosmic rays; radar and inertial systems analysis; advanced televison systems. *Mailing Add:* NBC-TV 30 Rockefeller Plaza Rm 1600W New York NY 10020

FAGUET, GUY B, b Turin, Italy, Mar 24, 39; m; c 2. INTERNAL MEDICINE. *Educ:* Xavier Univ, MD, 67. *Prof Exp:* PROF MED, MED COL GA, VET ADMIN MED CTR, 71- *Concurrent Pos:* Polycythemia Vera Task Force, 73-87; prin investr, Veterans Admin, 80-, clin investr; consult-lectr, Dept Army & Dept Human Resources of Ga; mem, Med Ctr, Savannah, Ga. *Mem:* Am Soc Immunol; Am Soc Hemat; Am Fed Clin Investr. *Res:* Cancer immunology, particularly in chronic lymphatic leukemia; have identified novel chronic lymphatic leukemia associated antigen and developed monochronal antibodies against the antigen. *Mailing Add:* Med Col Ga Vet Admin Med Ctr Rm 6B 135 Augusta GA 30912

FAHERTY, KEITH F, b Platteville, Wis, Dec 7, 31; m 51; c 5. CIVIL ENGINEERING, STRUCTURAL ENGINEERING. *Educ:* Wis State Univ, Platteville, BS, 54; Univ Ill, Urbana, MS, 62. *Prof Exp:* Trainee, US Gypsum Co, 54-55, authorities engr, 55, construct foreman, 55-57; from instr to assoc prof civil eng, 57-73, chmn dept, 66-80, PROF CIVIL ENG, UNIV WIS, PLATTEVILLE, 73-; AT DEPT CIVIL ENG, MARQUETTE UNIV. *Mem:* Am Soc Civil Engrs; Nat Soc Prof Engrs; Am Soc Eng Educ; Am Concrete Inst. *Res:* Making mathematical models of structures; wood and wood structures. *Mailing Add:* 17825 Wessex Dr Brookfield WI 53005

FAHEY, CHARLES J, b Baltimore, Md, Apr 13, 33. AGING STUDIES. *Educ:* St Bernard's Sem, BA, 55, MDiv, 82; Cath Univ Am, MSW, 63. *Hon Degrees:* LLD, St Thomas Univ, 82, D'Youville Col, 87; DDiv, St Bernard's Inst, 86. *Prof Exp:* Intern, Social Serv Dept, St Elizabeth's Hosp, Washington, DC, 61-62, Comn Aging, City Hall, Baltimore, Md, 62-63; asst pastor, St Vincent De Paul Church, Syracuse, NY, 59-61; asst dir, Cath Charities, Diocese Syracuse, 61-79; DIR, THIRD AGE CTR, FORDHAM UNIV, NEW YORK, NY, 79-, MARIE WARD DOTY PROF AGING STUDIES, GRAD SCH SOCIAL SERV, 80- *Concurrent Pos:* Partic, Nat Meetings Long Term Care, 65-; mem, Fed Coun Aging, 74-82, chmn, 80-81; mem & frequent spokesperson, Holy See's Deleg, '82 World Assembly Aging; mem, Comt Long Term Care, Cath Health Asn, 80-85, bd dirs, 83-89, mem, Task Force Long Term Care, 88, Task Force Nat Health Ins, 91-; mem, Comt Qual Assurance & Qual Improv Medicare Prog, 87-90; lectr, numerous US univs & cols. *Honors & Awards:* Nat Award of Honor, Am Asn Homes for Aging, 72; John McDowell Mem Lectr, Nat Interfaith Coalition Aging; Arthur Fleming Lectr, Nat Asn State Units Aging & Asn Area Wide Agencies Aging, 80; Ellen Winston Award, Nat Coun Aging, 86; Nat Award, Am Soc Aging, 89. *Mem:* Inst Med-Nat Acad Sci; assoc fel NY Acad Med; Nat Asn Social Work; fel Geront Soc Am; fel Am Col Health Care Adminr; Nat Coun Aging; Am Pub Health Asn. *Res:* Development of concepts, policies and programs in the area of long term care. *Mailing Add:* Third Age Ctr Fordham Univ 113 W 60th St New York NY 10023

FAHEY, DARRYL RICHARD, b Grand Forks, NDak, July 13, 42; m 66; c 1. ORGANOMETALLIC CHEMISTRY, POLYMER CHEMISTRY. *Educ:* Univ NDak, BS, 64, PhD(org chem), 69. *Prof Exp:* Res chemist, Phillips Petrol Co, 68-73, sr chemist, 73-83, res assoc, 83-89, SECT SUPVR, PHILLIPS PETROL CO, 88- *Mem:* Am Chem Soc. *Res:* Homogeneous and heterogeneous catalysis; polymer synthesis. *Mailing Add:* 6301 King Dr Bartlesville OK 74006

FAHEY, GEORGE CHRISTOPHER, JR, b Weston, WVa, Nov 12, 49. ANIMAL NUTRITION, NUTRITIONAL BIOCHEMISTRY. *Educ:* WVa Univ, BA, 71, MS, 74, PhD(animal nutrit), 76. *Prof Exp:* Res asst agr biochem, WVa Univ, 71-74, res asst animal nutrit, 74-76; asst prof, 76-80, ASSOC PROF ANIMAL SCI, UNIV ILL, URBANA, 80- *Mem:* Am Soc Animal Sci; Am Chem Soc; AAAS; Animal Nutrit Res Coun; Sigma Xi. *Res:* Fiber utilization by ruminant and non-ruminant animals. *Mailing Add:* Animal Sci Dept Univ Ill 124 Animal Sci Lab Urbana IL 61801

FAHEY, JAMES R, b Newark, NJ, July 25, 51. EXPERIMENTAL BIOLOGY. *Educ:* St Leo Col, BA, 73; Seton Hall Univ, MS, 78; Rutgers Univ, PhD(immune response), 82. *Prof Exp:* Res technician, Makari Can Res Lab, 73-74; res asst, Schering Corp, 77-78; postdoctoral fel & res assoc, Trudeau Inst, 82-87; sr res immunologist, Lederle Lab, 87-90; DEPT HEAD VIRAL VACCINE, INTERVET AM INC, 90- *Mem:* Am Asn Immunol; Am Soc Microbiol. *Mailing Add:* Virol Dept Intervet Am Inc 405 State St PO Box 318 Millsboro DE 19966

FAHEY, JOHN LEONARD, b Co Durham, UK, Mar 4, 44. CLINICAL RESEARCH, MEDICINAL & SYNTHETIC ORGANIC CHEMISTRY. *Educ:* Univ St Andrews, Scotland, BSc, 67; Stevens Inst Technol, PhD(chem), 72. *Prof Exp:* Lab asst inorg & anal chem, Imperial Chem Industs, UK, 60-67; res asst, Warner-Lambert Res Inst, 68-70; sr res chem, Merck, Sharp & Dohme Res Labs, 72-75; res scientist, Stevens Inst Technol, NJ, 75-78; assoc prof med chem, Fairleigh Dickinson Univ, 78-84; clin proj dir, Arzo Pharma, NJ, 85-89; DIR CLIN RES, COMMUNITY RES INITIATIVE, NY, 90- *Concurrent Pos:* Teaching asst, Stevens Inst Technol, 70-71, Delft fel, 71-72; mem adj fac, Kean Col, NJ, 74-75. *Mem:* Sigma Xi (vpres, 76-77, pres, 78); NY Acad Sci; AAAS; fel Chem Soc; Am Chem Soc. *Res:* Medicinal chemistry of beta-lactam antibiotics, antihypertensives and anti-inflammatories; synthetic organic chemistry of natural products, interpretive spectroscopy; phase II-III studies on antithrombotics; phase II-III studies in AIDS with antiretrovirals, immunomodulators, agents treating opportunistic infections. *Mailing Add:* 622 S Atlantic Ave Aberdeen Township NJ 07747

FAHEY, JOHN LESLIE, b Cleveland, Ohio, Sept 8, 24; m 54; c 3. IMMUNOLOGY. *Educ:* Wayne State Univ, MS, 49; Harvard Univ, MD, 51. *Prof Exp:* Intern med, Presby Hosp, NY, 51-52, asst res, 52-53; clin assoc, Nat Cancer Inst, NIH, 53-54, sr investr metab, 54-63, chief immunol br, 64-71; prof microbiol & immunol & chmn dept, Sch Med, 71-81, DIR, CTR INTERDISCIPLINARY RES IMMUNOL DIS, UNIV CALIF, LOS ANGELES, 78- *Mem:* Am Physiol Soc; Soc Exp Biol & Med; Am Asn Cancer Res; Am Fedn Clin Res; Am Soc Clin Invest. *Res:* Immunology; oncology. *Mailing Add:* Dept Microbiol & Immunol Sch Med Univ Calif Los Angeles CA 90024

FAHEY, JOHN VINCENT, b New York, NY, May 29, 48; m 76; c 2. PHYSIOLOGY, IMMUNOLOGY. *Educ:* Mt St Mary's Col, BS, 70; Univ Vt, MS, 74, PhD(cell biol), 77. *Prof Exp:* Teacher biol, Paramus Cath High Sch, 70-71; technologist viral oncol, Pub Health Res Inst, NY, 71-72; teaching fel med microbiol, Univ Vt, 72-74, res fel rheumatology, 74-77; mem fac biol, Clark Univ, 77-78; res assoc physiol, Dartmouth Med Sch, 78-81; mem fac sci, Bennington Col, 81-89; RES ASSOC, VERAX CORP, 89- *Concurrent Pos:* Andrew W Mellon teaching fel, Clark Univ, 77-78. *Mem:* AAAS; Sigma Xi. *Res:* Mediation and modulation of physiological and immune responses by prostaglandins and cyclic nucleotides. *Mailing Add:* Verax Corp Six Etna Rd Lebanon NH 03766

FAHEY, PAUL FARRELL, b Lock Haven, Pa, July 2, 42; m 65; c 4. PHYSICS. *Educ:* Univ Scranton, BS, 64; Univ Va, MS, 66, PhD(physics), 68. *Prof Exp:* From asst prof to assoc prof, 68-77, dept chmn, 82-88, PROF PHYSICS, UNIV SCRANTON, 77-, DEAN, COL ARTS & SCI, 89- *Concurrent Pos:* Fac sci fel, NSF & vis assoc prof, Cornell Univ, 75-76, vis prof, 77, 78, 80 & 81; vis scientist, Bell Labs, Murray Hill, 82; fac sci develop fel, NSF, 82. *Mem:* Am Phys Soc; Acoust Soc Am; AAAS; Am Asn Physics Teachers; Asn Res Otolaryngol. *Res:* Physical properties of biopolymers; pressure-volume-temperature relations of liquids; fluorescence correlation spectroscopy; lipid bilayer characterizations; cochlear mechanics and biophysics of middle and inner ear mechanical signals. *Mailing Add:* Col Arts & Sci Univ Scranton Scranton PA 18510-4672

FAHEY, ROBERT C, b Sacramento, Calif, Feb 8, 36; m 83; c 2. ORGANIC CHEMISTRY, BIOCHEMISTRY. *Educ:* Univ Calif, Berkeley, BS, 57; Univ Chicago, PhD(org chem), 63. *Prof Exp:* From asst prof to assoc prof, 63-82, PROF CHEM, UNIV CALIF, SAN DIEGO, 82- *Concurrent Pos:* Alfred P Sloan Found fel, 66-68; John Simon Guggenheim Found fel, 70-71; Am Cancer Soc scholar, 88-89. *Mem:* AAAS; Am Chem Soc; The Chem Soc; Am Soc Biol Chemists; Radiation Res Soc. *Res:* Biological chemistry of thiols and disulfides. *Mailing Add:* Dept Chem-0506 Univ Calif San Diego La Jolla CA 92093

FAHEY, WALTER JOHN, b Winnipeg, Man, Apr 10, 27; US citizen; m 49; c 4. ELECTRICAL ENGINEERING. *Educ:* Case Inst Technol, BS, 57, MS, 59, PhD(elec eng), 63. *Prof Exp:* Instr elec eng, Case Inst Technol, 59-62; from asst prof to prof, Ohio Univ, 63-69, chmn dept, 66-67, dean col eng & technol, 67-68; dean col eng, 69-77, PROF ELEC ENG, UNIV ARIZ, 69- *Concurrent Pos:* Instr adult div, Cleveland Pub Schs, 59-62; Am Coun Educ fel, 67-68; mem, Ohio Crime Comn, 67-68 & Ariz State Bd Tech Registr, 69-70; dir, Aviation Res & Educ Found, Ariz, 70- *Mem:* Inst Elec & Electronics Engrs; Am Soc Eng Educ; Sigma Xi. *Res:* Electrical properties of materials; theoretical electromagnets; plasma dynamics and gaseous electronics; methods of engineering education; biomedical instrumentation and measurements engineering; fiber optics; power electronics. *Mailing Add:* 6802 E Opatas Tucson AZ 85715

FAHIDY, THOMAS Z(OLTAN), b Budapest, Hungary, June 17, 34; Can citizen; m 62; c 1. CHEMICAL ENGINEERING, APPLIED MATHEMATICS. *Educ:* Queen's Univ, Ont, BSc, 59, MSc, 61; Univ Ill, Urbana, PhD(chem eng), 65. *Prof Exp:* From asst prof to assoc prof, 64-71, PROF CHEM ENG, UNIV WATERLOO, 71- *Concurrent Pos:* Hon res assoc & Shell vis fel, Univ Col, London, 68-69; partic, Can-France Sci Exchange Prog, 75-76, 77 & 83; assoc ed, Can J Chem Eng, 72-80, 90-; assoc prof, Univ Rheims, France, 83, IUT, St Nazaire, France, 90; invited prof, EPFL, Switz, 86. *Mem:* Am Inst Chem Engrs; Can Soc Chem Engrs; Electrochem Soc; NY Acad Sci; fel Chem Inst Can. *Res:* Electrochemical engineering; applied mathematics; process dynamics and control. *Mailing Add:* Dept Chem Eng Univ Waterloo Waterloo ON N2L 3G1 Can

FAHIEN, LEONARD A, b St Louis, Mo, July 26, 34; m 58; c 3. PHARMACOLOGY, BIOCHEMISTRY. *Educ:* Wash Univ, AB, 56, MD, 60. *Prof Exp:* Intern med, Univ Wis, 60-61; NIH fel biochem, Wash Univ, 61-62 & physiol chem, Univ Wis, 62-64; from asst prof to assoc prof, 66-74, assoc dean, Sch Med, 79-83, PROF PHARMACOL, UNIV WIS-MADISON, 74- *Concurrent Pos:* NIH res grant, 66-, career develop award, 68-73; univ prof, Auburn Univ, Ala, 85; vis prof, Inst Protein Res, Osaka, Japan, 91. *Mem:* AAAS. *Res:* Regulation of enzyme activity; studies of milti-enzyme complexes in tumor versus normal cells; effects of drugs on enzyme catalyzed reaction; regulation of insulin release. *Mailing Add:* Dept Pharmacol Univ Wis Madison WI 53706

FAHIEN, RAY W, b St Louis, Mo, Dec 26, 23. CHEMICAL ENGINEERING. *Educ:* Wash Univ, St Louis, BSChE, 47; Mo Sch Mines, MSChE, 50; Purdue Univ, PhD, 54. *Prof Exp:* Instr chem eng, Mo Sch Mines, 47-50; chem engr, Ethyl Corp, 53-54; from asst prof to prof chem eng, Iowa State Univ, 54-64; chmn dept, 64-69, PROF CHEM ENG, UNIV FLA, 64- *Concurrent Pos:* Sr engr, Ames Lab, US Atomic Energy Comn, 54-64; vis prof, Univ Wis, 59-60; Fulbright lectr, Brazil, 64; ed, Chem Eng Educ J, 67-; vis prof, Univ Minn, 77-78; consult, UNESCO, Univ de Oriente, Puerto La Cruz, Venezuela, 75. *Honors & Awards:* Distinguished Serv Award, Am Soc Eng Educ, 90- *Mem:* AAAS; Am Inst Chem Engrs; Am Chem Soc; fel Am Soc Eng Educ. *Res:* Transport of heat; mass and momentum; turbulent diffusion; applied mathematics; chemical reactor design. *Mailing Add:* Dept Chem Eng Univ Fla Gainesville FL 32603

FAHIM, MOSTAFA SAFWAT, b Cairo, Egypt, Oct 7, 31; m 60; c 1. REPRODUCTIVE BIOLOGY. *Educ:* Cairo Univ, BS, 53; Univ Mo-Columbia, MS, 58, PhD(reproductive biol), 61. *Prof Exp:* Asst dir animal sci dept, Ministry of Land Reform, Egypt, 61-63; dir, Animal Reproduction Dept, Off of the Pres, Algeria, 63-66; res assoc, 66-68, from asst prof to assoc prof, 68-74, PROF OBSTET & GYNEC, SCH MED UNIV MO-COLUMBIA, 74-, DIR, CTR REPRODUCTIVE SCI & TECHNOL, 87- *Concurrent Pos:* Prof, Univ Ain Shams, Cairo, 61-63; consult, Inst de Serotherapie de Toulouse, France & NAfrica Div, Russel Pharmaceut Co, 63-66; hon prof, Xian Med Univ, People's Repub China. *Mem:* Am Col Clin Pharmacol; Am Soc Pharmacol & Exp Therapeut; Soc Environ Geochem & Health; NY Acad Sci; Am Soc Androl. *Res:* Human reproduction biology; effects of drugs and environmental chemicals on the fetus and newborn. *Mailing Add:* Ctr Reproductive Sci & Technol 111 Allton Bldg Univ Mo Health Sci Ctr Sch Med Columbia MO 65212

FAHIMI, HOSSEIN DARIUSH, b Teheran, Iran, May 7, 33; m 64; c 2. PATHOLOGY, ELECTRON MICROSCOPY. *Educ:* Univ Heidelberg, MD, 58. *Prof Exp:* Assoc, Harvard Med Sch, 66-69, from asst prof to assoc prof path, 69-75; dean, preclin fac, 87-89, PROF ANAT & CHMN DEPT II DIV, MED SCH, UNIV HEIDELBERG, 75- *Concurrent Pos:* Assoc vis physician, Mallory Inst Path, 66-75; NIH career res develop award, 71-75; vis prof anat, Univ Heidelberg, 74-75. *Mem:* AAAS; Am Soc Cell Biol; Histochem Soc; Am Asn Path; NY Acad Sci; Ger Soc Cell Biol (pres, 87-90). *Res:* Experimental cell research; histochemistry; cytochemistry; immunoelectronmicroscopy; investigation of peroxisomes. *Mailing Add:* Dept Anat & Cell Biol II Div Univ Heildelberg Heidelberg 6900 Germany

FAHL, CHARLES BYRON, b Warsaw, NY, Dec 16, 39; m 66; c 2. CLIMATE CHANGE, CLIMATOLOGY. *Educ:* Antioch Col, BS, 63; Univ Alaska, Fairbanks, MS, 69, PhD(geophys & meteorol), 73. *Prof Exp:* Meteorologist, Dames & Moore Consult Engrs, Anchorage, 73-80; mgr air qual serv, Earth Technol Corp, Seattle, 80-81; vpres, Variance Corp, Anchorage, 81-82; asst prof math, 83-84, assoc prof math & natural sci, 84-89, PROF & CHAIR NATURAL SCI & MATH, ALASKA PAC UNIV, ANCHORAGE, 89- *Concurrent Pos:* Consult, N Slope Borough, 82, Shell Oil Co, 82-86, US Geol Surv, 85-, Tryck, Nyman & Hayes, 85, Anchorage Olympic Organizing Comt, 88, Montgomery Engrs, 89-, Hughes et al, attys, 90- *Mem:* Am Meteorol Soc; Math Asn Am; AAAS. *Res:* Detection of regional (Alaska) climatic change through frequency analysis of circulation patterns; climatology of Alaska, particularly precipitation; relationship between climatic change and glacier variability in Alaska. *Mailing Add:* Alaska Pac Univ 4101 Univ Dr Anchorage AK 99508

FAHL, ROY JACKSON, JR, b Richmond, Va, Oct 8, 25; m 53; c 2. INDUSTRIAL CHEMISTRY. *Educ:* Washington & Lee Univ, BS, 48; Univ NC, PhD(chem), 53. *Prof Exp:* Asst, Univ NC, 48-51; res chemist metallo-org compounds, 52-56, supvr pigments lab, 56-62, mgr, 63, tech mgr white pigments, 63-69, asst dir tech serv lab, 69-71, lab dir, Pigments Dept, 71-73, mgr titanium dioxide prods, Pigments Dept, 73-77, mkt mgr white pigments, 77-79, mkt mgr specialty chem prod, 79-83, res mgr, new opportunities, Chem & Pigments Dept, 83-85, consult, E I du Pont de Nemours & Co, Inc, 85-86; INDEPENDENT CONSULT, 86- *Mem:* Am Chem Soc; NY Acad Sci. *Res:* Pigments. *Mailing Add:* 1106 Barton Circle Wilmington DE 19807

FAHL, WILLIAM EDWIN, b Milwaukee, Wis, Apr 13, 50; m 74; c 3. CHEMICAL CARCINOGENESIS, CELL TRANSFORMATION. *Educ:* Univ Wis-Madison, BS, 72, PhD(physiol & oncol), 75. *Prof Exp:* Fel, Dept Pharmacol, Med Sch, Univ Wis-Madison, 75-79; from asst prof to assoc prof pharmacol, Cancer Ctr, Med Sch, Northwestern Univ, 79-85; asst prof, 85-87, ASSOC PROF ONCOL, MCARDLE LAB CANCER RES, UNIV WIS, 87- *Concurrent Pos:* Health Sci Rev Panel, Environ Protection Agency. *Honors & Awards:* Merit Award, NIH. *Mem:* Am Asn Cancer Res; AAAS; Am Soc Biochem Molecular Biol. *Res:* Mutagen-activated oncogenes in human cells; molecularly augmented carcinogen detoxification; cell transformation. *Mailing Add:* McArdle Lab Cancer Res Univ Wis Madison WI 53706

FAHLBERG, WILLSON JOEL, b Madison, Wis, July 20, 18; c 4. IMMUNOLOGY. *Educ:* Univ Wis, PhB, 48, MS, 49, PhD, 51; Am Bd Microbiol, dipl. *Prof Exp:* Asst microbiol, Univ Wis, 48-50; from instr to asst prof, 51-60, ASSOC PROF MICROBIOL, BAYLOR COL MED, 60-; DIR MED AFFAIRS, MEM BAPTIST HOSP SYST, 69- *Concurrent Pos:* Consult microbiologist, Methodist Hosp, 53-, Vet Admin Lab, 54-, Mem Hosp, Univ Tex M D Anderson Hosp & Tumor Inst, Houston State Psychiat Inst, Jefferson Davis Hosp Houston, St Tex Inst Rehab & Res. *Mem:* Fel Royal Soc Health; fel NY Acad Sci; fel Am Acad Microbiol; Sigma Xi. *Res:* Allergy; infectious diseases; hospital infection control. *Mailing Add:* 3746 Darcus Houston TX 77005

FAHLEN, THEODORE STAUFFER, b San Francisco, Calif, Sept 5, 41; m 63; c 3. LASERS. *Educ:* Stanford Univ, BS, 63; Univ NMex, MS, 64, PhD(physics), 67. *Prof Exp:* Res asst physics, Univ NMex, 64-67; sr engr, Aerojet-Gen Corp, 67-70; SR ENG SPECIALIST, GTE SYLVANIA INC, 70- *Mem:* Optical Soc Am. *Res:* Single particle scattering; laser applications; high power gas laser research and development, particularly carbon dioxide, copper vapor, xenon, nitrogen and excimons. *Mailing Add:* Res & Develop XMR 5403 Betsy Ross Santa Clara CA 95054

FAHLMAN, GREGORY GAYLORD, b Lethbridge, Alta, Oct 4, 44; m 69; c 2. ASTRONOMY. *Educ:* Univ BC, BSc, 66; Univ Toronto, MSc, 67, PhD(astron), 70. *Prof Exp:* Fel, Inst Theoret Astron, Cambridge Univ, 70-71; asst prof, 71-76, ASSOC PROF ASTRON, UNIV BC, 76- *Mem:* Am Astron Soc; Can Astron Soc. *Res:* Applications of signal processing theory to astronomical data; time resolved astronomical spectroscopy; observational and theoretical studies of high energy phenomena in astronomy. *Mailing Add:* Dept Astro-Geophys Univ BC 2075 Westbrook Pl Vancouver BC V6T 1W5 Can

FAHLQUIST, DAVIS A, b Providence, RI, July 16, 26; wid; c 5. GEOPHYSICS, OCEANOGRAPHY. *Educ:* Brown Univ, BS, 50; Mass Inst Technol, PhD(geophys), 63. *Prof Exp:* Engr, Owens Corning Fiberglas Corp, 51-53; from res asst to res assoc geophys, Woods Hole Oceanog Inst, 58-63; from asst prof to assoc prof geophys, Tex A&M Univ, 63-74, asst prof oceanog, 69-74, asst dean, Col Geosci, 79-82, PROF GEOPHYS & OCEANOG, TEX A&M UNIV, 74-, ASSOC DEAN, COL GEOSCI, 82- *Concurrent Pos:* Mem joint oceanog deep earth sampling prog site selection panel, NSF, 75-78. *Mem:* AAAS; Am Geophys Union; Oceanog Soc; Soc Explor Geophys; Sigma Xi. *Res:* Marine geophysics; seismic refraction studies in deep water areas; continuous seismic reflection profiling in ocean areas. *Mailing Add:* 835 Tanglewood Bryan TX 77802

FAHMY, ABDEL AZIZ, b Giza, Egypt, Apr 24, 25; m 53, 68; c 2. METALLURGY, NUCLEAR ENGINEERING. *Educ:* Cairo Univ, BEng, 47; Sheffield Univ, PhD(metall), 53. *Prof Exp:* Demonstr metall, Cairo Univ, 47-53, lectr, 53-60, assoc prof, 60-65, chair prof, 65-68; vis prof metall eng, 68-69, PROF MAT ENG, NC STATE UNIV, 69- *Concurrent Pos:* Partic int sch nuclear sci & eng, Atoms for Peace Prog, NC State Univ & Argonne Nat Lab, 56-57; spec lectr, NC State Univ, 57-59; resident res assoc, Argonne Nat Lab, 63-64; prof, Am Univ Cairo, 66-68; mem state awards comt eng sci, Egypt, 67-68; consult, Argonne Nat Lab, 68-70, IBM Corp, 68- & US Army & Batelle Mem Inst, 75- *Mem:* Sigma Xi. *Res:* Formation of austenite; structure of cement; effect of irradiation on properties of materials; interphase stresses in multiphase materials; thermal expansion of multiphase and composite materials; x-ray stress measurement; fiber reinforced composites; thermal conductivity of composite materials; wear resistance of composites. *Mailing Add:* 508 Dixie Trail Raleigh NC 27607-4151

FAHMY, ALY, b Cairo, Egypt; US citizen; c 4. SURGICAL PATHOLOGY, CLINICAL PATHOLOGY. *Educ:* Fouad Univ, Egypt, MD, 49; Univ London, PhD(med), 56. *Prof Exp:* From instr to asst prof path, Fac Med, Fouad Univ, Egypt, 56-61; Alexander von Humboldt Found res fel, Sch Med, Univ Dusseldorf, 61-62; asst prof, Sch Med, Emory Univ, 62-63; sr pathologist, Sch Med, Univ Dusseldorf, 63-64; from asst prof to assoc prof, Sch Med, Vanderbilt Univ, 65-70; prof, Sch Med Univ Sherbrooke, 70-72; PATHOLOGIST, PATH SERV, VET ADMIN HOSP, OKLAHOMA CITY, 72-; PROF PATH, COL MED, UNIV OKLA, 72- *Concurrent Pos:* Vis prof, Int Tech Coop Prog, Paris, 61; from asst prof to assoc prof, Meharry Med Col, 64-70; asst chief lab serv, Vet Admin Hosp, Nashville, Tenn, 65-70; consult, Bone & Joint Panel, Can Tumor Ref Ctr. *Mem:* Fel Col Am Path; fel Am Soc Clin Path; Int Acad Path; Electron Micros Soc Am; AMA. *Res:* Pathology of tumors; bone and joint pathology; application of electron microscopy to surgical pathology; bone growth problems; hormonal and genetic skeletal disturbances; cancer family syndromes; genodermatoses. *Mailing Add:* Dept Path Vet Admin Hosp 921 NE 13th St Oklahoma City OK 73104

FAHMY, MOHAMED HAMED, b Ismailia, Egypt, Dec 14, 40; m 67; c 3. GENETICS, ANIMAL HUSBANDRY. *Educ:* Ain Shams Univ, Cairo, BSc, 60, MSc, 64, PhD(animal breeding), 67. *Prof Exp:* Researcher animal breeding, Desert Inst, Egypt, 61-67; head meat animal sect, 83-88, RES SCIENTIST, CAN DEPT AGR, 68- *Concurrent Pos:* Vis scientist, ABRO, Scotland & INRA France, 74-75. *Mem:* Am Soc Animal Sci; Can Soc Animal Sci. *Res:* Swine, beef cattle, dairy cattle and sheep breeding. *Mailing Add:* Res Sta Can Dept Agr PO Box 90 Lennoxville PQ J1M 1Z3 Can

FAHMY, MOUSTAFA MAHMOUD, b Alexandria, Egypt, Mar 13, 29; Can citizen. DIGITAL SIGNAL PROCESSING, COMPUTER-AIDED FILTER DESIGN. *Educ:* Univ Alexandria, Egypt, BSc, 50; Univ Mich, MSc, 63; PhD(elec eng), 66. *Prof Exp:* Instr elec eng, Univ Alexandria, 50-60; res assoc control systs, Inst Sci & Tech, Univ Mich, 62-65; from asst prof to assoc prof, 66-76, PROF ELEC ENG, QUEENS UNIV, CAN, 76-, CHAIR, DIV III ENG DEPTS, SCH GRAD STUDIES & RES, 89- *Concurrent Pos:* Vis prof, Swiss Fed Inst Technol, 71-72, fac eng, Univ Alexandria, Egypt, 74-75, fac eng & petrol, Kuwait Univ, 80-81. *Mem:* Fel Inst Elec & Electronic Engrs; Sigma Xi; Can Asn Univ Teachers. *Res:* Digital signal processing; computer-aided techniques for the analysis and design of two-dimensional digital filters. *Mailing Add:* Dept Elec Eng Queens Univ Kingston ON K7L 3N6 Can

FAHN, STANLEY, b Sacramento, Calif, Nov 6, 33; m 58; c 2. NEUROLOGY, NEUROPHARMACOLOGY. *Educ:* Univ Calif, Berkeley, BA, 55; Univ Calif, San Francisco, MD, 58. *Prof Exp:* Intern, Philadelphia Gen Hosp, 58-59; resident neurol, Neurol Inst, Presby Hosp, 59-62; res assoc neurochem, NIH, 62-65; res assoc neurol, Columbia Univ, 65-67, asst prof, 67-68; from asst prof to assoc prof, Univ Pa, 68-73; prof, 73-78, H HOUSTON MERRITT PROF NEUROL, COLUMBIA UNIV, 78- *Concurrent Pos:* USPHS res grants, 70, 74, 80, 84, 87 & 88; mem sci coun, Comt Combat Huntington's Dis, 72-81, chmn, 74-81; attend neurologist, Neurol Inst, Presby Hosp, 73-; mem sci coun, Dystonia Med Res Found, 76-; Fogarty Sr Int fel, 75-76; mem adv coun, Nat Fedn Jewish Genetic Dis, 75-84; scientific dir & mem bd dirs, Parkinson's Dis Found, 75-; mem sci adv bd, hereditary Dis Fedn, 75-83; dir, Dystonia Res Ctr, 81-; assoc ed, Neurol, 76-86; chief ed, Movement Dis, 85-; chmn, educ comt, Am Acad Neurol, 87-; pres, Movement Dis Soc, 88-91. *Honors & Awards:* Levy Lectr, Wash Univ, 81; Greenberg Lectr, Univ Okla, 81; Druker Lectr, Harvard, 85; Wartenberg Lectr, Am Acad Neurol, 86; Soriano Lectr, Am Neurol Asn, 88; Hassel Lectr, Univ Pa, 90. *Mem:* AAAS; Am Neurol Asn; Am Acad Neurol; Soc Neurosci; Am Soc Neurochem; Movement Dis Soc. *Res:* Parkinsonism, Huntington's chorea, dystonia, myoclonus and other movement disorders; neurochemistry; neuropharmacology. *Mailing Add:* 155 Edgars Lane Hastings-on-Hudson NY 10706

FAHNESTOCK, GEORGE REEDER, forestry; deceased, see previous edition for last biography

FAHNESTOCK, STEPHEN RICHARD, BIOCHEMISTRY, MICROBULE GENETICS. *Educ:* Mass Inst Technol, PhD(biochem), 70. *Prof Exp:* Res dir, Genex Corp, 82-89; PRIN INVESTR, DUPONT, 89- *Mailing Add:* E I du Pont Exp Sta PO Box 80228 Wilmington DE 19880-0173

FAHNING, MELVYN LUVERNE, b St Peter, Minn, Apr 28, 36; m 56; c 4. REPRODUCTIVE PHYSIOLOGY, VETERINARY MEDICINE. *Educ:* Univ Minn, BS, 58, MS, 60, DVM & PhD, 64. *Prof Exp:* Asst prof dairy husb, Univ Minn, St Paul, 64-65, asst prof dairy husb & vet anat, 65-66, asst prof vet obstet & gynec, 66-70, assoc prof, 70-72; vpres & dir res, Int Cryobiol Serv, Inc, 72-76; vpres & dir, Tech Serv, Agro-K Corp, 76-77; pres, 77-85, clin assoc prof large animal clin sci, 78-80, prof, Div Theriogeneology, Ovatech, Inc, River Falls, Wis, 80-85; PRES, CRYOVATECH INT, INC, HUDSON, WIS, 84- *Mem:* Am Dairy Sci Asn; Am Soc Animal Sci; Am Vet Med Asn; Soc Study Reproduction; Brit Soc Study Fertil. *Res:* Ovum transplantation; superovulation; synchronization of estrus of cattle and swine; insemination of swine; collection and chemical analysis of female reproductive tract fluids; canine semen freezing. *Mailing Add:* Cryovatech Int Inc 592 Hwy 35 S Hudson WI 54016

FAHRENBACH, MARVIN JAY, b Buena Vista, Va, Apr 11, 18; m 49; c 3. CLINICAL PHARMACOLOGY. *Educ:* Yale Univ, BS, 39, PhD(org chem), 42. *Prof Exp:* ASST DIR MED RES, SANDOZ, INC, 74- *Concurrent Pos:* Mem, Arteriosclerosis Coun, Am Heart Asn. *Mem:* AAAS; Am Chem Soc;

Am Inst Chemists; NY Acad Sci; Am Heart Asn. *Res:* Pharmaceuticals; sulfonamides; vitamins; hormones; antiseptics of the quaternary ammonium salt type; anticoagulants; lipid metabolism; arteriosclerosis. *Mailing Add:* Buckberg Rd RR 2 Box 81 Tomkins Cove NY 10986-9707

FAHRENBACH, WOLF HENRICH, b Berlin, Ger, Apr 21, 32; US citizen; m 77; c 1. HISTOLOGY, NEUROCYTOLOGY. *Educ:* Univ Calif, Berkeley, BA, 54; Univ Wash, PhD(invert zool), 61. *Prof Exp:* Asst zool, Univ Wash, 57-60; NSF fel anat, Harvard Med Sch, 61-63; from assoc prof to prof exp biol, Med Sch, Univ Ore, 73-80; SCIENTIST ELECTRON MICROS & CHMN LAB, ORE REGIONAL PRIMATE RES CTR, 67-, RES PROF OPHTHAL, ORE HEALTH SCI UNIV, 87- *Mem:* AAAS; Am Asn Anat; Soc Neurosci; Asn Res Vision & Ophthal. *Res:* Vertebrate and invertebrate histology and cytology; electron microscopy of invertebrate visual systems and vasculature of mammalian eye. *Mailing Add:* Lab Electron Micros Ore Regional Primate Res Ctr 505 NW 185th Beaverton OR 97006

FAHRENHOLTZ, KENNETH EARL, b Peoria, Ill, Aug 9, 34; m 58. PHARMACEUTICAL & ORGANIC CHEMISTRY. *Educ:* Bradley Univ, BS, 56; Univ Rochester, PhD(org chem), 60. *Prof Exp:* Sr chemist, Strasenburgh Labs, 60-62; sr chemist, 62-71, RES FEL, RES DIV, HOFFMANN-LA ROCHE INC, 71- *Mem:* NY Acad Sci; Am Chem Soc; Sigma Xi. *Res:* Organic synthesis; heterocyclic compounds; medicinal chemistry. *Mailing Add:* 28 Winding Lane Bloomfield NJ 07003

FAHRENHOLTZ, SUSAN ROSENO, b Cologne, Ger; US citizen; m 58. POLYMER CHEMISTRY, ORGANIC CHEMISTRY. *Educ:* Cornell Univ, BA, 57; Univ Rochester, MS, 60. *Prof Exp:* Res chemist, Eastman Kodak Co, 60-62; res assoc chem, AT&T Bell Labs, 62-83, mem tech staff, 83-87; adj instr, 72-80, ADJ ASST PROF CHEM, FORDHAM UNIV, 80- *Mem:* Am Chem Soc. *Res:* Singlet oxygen reactions; chemically induced dynamic nuclear polarization; positive photoresists and electron beam resists; novolak polymers; materials science engineering; science education. *Mailing Add:* 28 Winding Lane Bloomfield NJ 07003

FAHRNEY, DAVID EMORY, b Stapleton, Nebr, Feb 1, 34; m 69. BIOCHEMISTRY. *Educ:* Reed Col, BA, 59; Columbia Univ, PhD(biochem), 63. *Prof Exp:* NIH fel biophys chem, Univ Calif, San Diego, 63-64; asst prof chem, Univ Calif, Los Angeles, 64-69; assoc prof, 69-80, PROF BIOCHEM, COLO STATE UNIV, 80- *Mem:* Am Chem Soc; Am Soc Biol Chemists. *Res:* Structure and function of enzymes; identification of functional groups in active sites of enzymes and proteins. *Mailing Add:* Dept Biochem Colo State Univ Ft Collins CO 80523

FAHSELT, DIANNE, b Cabri, Sask, May 12, 41; m 68. BOTANY. *Educ:* Univ Sask, BA, 63, Hons, 64; Wash State Univ, PhD(bot), 67. *Prof Exp:* From asst prof to assoc prof bot, 67-91, PROF PLANT SCI, UNIV WESTERN ONT, 91- *Concurrent Pos:* Grants, Nat Sci & Eng Res Coun Can, 67-74 & 76-, Dept Univ Affairs, 69, Ont Heritage Found, 85 & Northern Affairs, 89-; vis scholar, Duke Univ, 75-76; vis scientist, Waseda Univ, Japan, 89-90. *Honors & Awards:* Cooley Award, Am Soc Plant Taxonomists, 68; Dimond Award, Bot Soc Am, 75; Mary E Elliot Award, Can Bot Asn, 86. *Mem:* Can Bot Asn; Int Asn Lichenology; Int Asn Plant Taxon. *Res:* Lichen systematics, ecology; natural area conservation. *Mailing Add:* Dept Plant Sci Univ Western Ont London ON N6A 5B7 Can

FAICH, GERALD ALAN, b Milwaukee, Wis, Oct 7, 42. INFECTIOUS DISEASE. *Educ:* Univ Wis, BS, 64, MD, 68; Harvard Sch Pub Health, MPH, 76. *Prof Exp:* Intern med, Boston City Hosp, 68-70; with prev med, Ctr Dis Control, 72-74; with internal med, Bath Israel Hosp, Boston, 74-75; med epidemiologist, Ctr Dis Control, 70-76; asst prof community med, Brown Univ, 76-88; chief, Div Epidemiol, 76-78, ASSOC DIR, RI DEPT HEALTH, 78-; DIR, OFF BIOLOGISTIC, DRUG EVAL RES, 88- *Concurrent Pos:* Career develop award, USPHS, 74; mem, Nat Inst Health Study Sect, 81- *Mem:* Am Pub Health Asn; fel Am Col Physicians; fel Am Col Prev Med; fel Am Col Epidemiol; Am Coun Sci & Health. *Res:* Epidemiology of infectious and chronic diseases; disease control in Latin America and the United States. *Mailing Add:* 711 Marshall Ave Rockville MD 20851

FAIFERMAN, ISIDORE, CHEMISTRY. *Prof Exp:* CLIN INVESTR, SMITH KLINE BEECHAM PHARMACEUT, 87- *Mailing Add:* Smith Kline & French Lab Res & Develop L-217 PO Box 7929 Philadelphia PA 19101

FAIG, WOLFGANG, b Crailsheim, Ger, Apr 27, 39; Can citizen; m 66; c 3. ENGINEERING SURVEYING, PHOTOGRAMMETRY. *Educ:* Technion Univ Stuttgart, Dipl Ing, 62; Univ NB, MScE, 65; Stuttgart Univ, Dr Ing(photogram), 69. *Prof Exp:* Res assoc photogram, Dept Civil Eng, Univ NB, 65 & Inst Appl Geod, Stuttgart, 66-69; asst prof civil eng, Univ Ill, Champaign-Urbana, 70-71; from asst prof to assoc prof surv & photogram, 71-78, assoc dean eng, 81-90, PROF SURV ENG, UNIV NB, 78-, DEAN ENG, 90- *Concurrent Pos:* Prin investr, numerous res contracts, 71-; chmn, Working Group V-2, Int Soc Photogram & Remote Sensing, 72-76, nat reporter, 80-; vis prof, Sch Surv, Univ New South Wales, Sydney, Australia, 84-85 & Fac Eng Surv, Wuhan Tech Univ Surv & Mapping, 86; mem, Comt Int Rels, Nat Sci & Eng Res Coun Can, 88- *Mem:* Am Soc Photogram & Remote Sensing; Can Inst Surv & Mapping. *Res:* Self-calibration of amateur cameras and their use for precision photogrammetry; modeling of systematic errors, combined adjustment of vastly different observables; four-dimensional photogrammetry in deformation studies. *Mailing Add:* Fac Eng Univ NB PO Box 4400 Fredericton NB E3B 5A3 Can

FAIL, PATRICIA A, b Los Angeles, Calif, June 12, 42. REPRODUCTIVE ENDOCRINOLOGY & TOXICOLOGY, TESTICULAR FUNCTION. *Educ:* Mich State Univ, PhD(physiol), 75. *Prof Exp:* Asst prof physiol, NC State Univ, 80-85; sr reproductive biologist, 85-90, MGR, REPRODUCTIVE BIOL, RES TRIANGLE INST, 90- *Concurrent Pos:* Res assoc biochem, State Univ Utrecht, Neth, 76-78; asst prof, Lansing Community Col, 77-78; res assoc obstet & gynec, Mich State Univ, 78-80. *Mem:* Soc Andrology; Am Physiol Soc; Soc Study Reproduction; Asn Study Animal Sci. *Res:* Endocrine challenge test for toxicologic induced endocrinopathies; endocrine basis of genectically induced infertiligy; testing potential toxins for influence on reproduction and fertility using the continuous breeding protocol; exterioceptive control of gonadal function during pubertal development in male rodents. *Mailing Add:* Res Triangle Inst PO Box 12194 Bldg 3 Research Triangle Park NC 27709-2194

FAILL, RODGER TANNER, b Niagara Falls, NY, May 1, 36; m 63; c 4. STRUCTURAL GEOLOGY. *Educ:* Columbia Univ, BS, 61, MA, 64, PhD(geol), 66. *Prof Exp:* GEOLOGIST, PA GEOL SURV, 65- *Mem:* Geol Soc Am; Sigma Xi; Am Geophys Union. *Res:* Tectonic analysis of Appalachian mountains, including fold style, faulting and fossil deformation; experimental rock deformation. *Mailing Add:* Pa Geol Surv PO Box 2357 Harrisburg PA 17120

FAILLA, PATRICIA MCCLEMENT, b New York, NY, Dec 22, 25; wid. BIOPHYSICS. *Educ:* Barnard Col, Columbia Univ, AB, 46, Columbia Univ, PhD(biophys), 58; Univ Chicago, MBA, 76. *Prof Exp:* Asst physicist, Physics Lab, Dept Hosps, NY, 46-48; res scientist, Radiol Res Lab, Col Physicians & Surgeons, Columbia Univ, 50-60; assoc biophysicist, Radiol Physics Div, 60-71, asst dir, Radiol & Environ Res Div, 71-73, prog coordr res, Off Dir, 73-74, asst to lab dir, 74-78, asst lab dir, 78-80, prog coordr, Biomed & Environ Res, Off Dir, Argonne Nat Lab, 80-86; RETIRED. *Concurrent Pos:* Mem corp, Marine Biol Lab, Woods Hole; mem, Tech Electronic Prod Radiation Safety Standards Comt, Bur Radiol Health, 73-75; mem bd dirs, Sigma Xi, 75-78 & 80-83; Coun-at-large, Radiation Res Soc, 76-79; mem, Gen Res Support Rev Comt, NIH, 78-82; mem, Pres Adv Comt, Med Univ SC, 87- *Mem:* Radiation Res Soc; Sigma Xi; Health Physics Soc. *Res:* Dosimetry of ionizing radiations and the effects of radiation on biological systems. *Mailing Add:* 2149 Loblolly Lane Johns Island SC 29455

FAILLACE, LOUIS A, b Brooklyn, NY, June 7, 32; m 63; c 3. PSYCHIATRY. *Educ:* Marquette Univ, MD, 57; Am Bd Psychiat & Neurol, dipl, 70. *Prof Exp:* Intern med, Boston City Hosp, 57-58; asst resident psychiat, Bellevue Hosp, NY, 58-59; res fel neurochem, Mass Ment Health Ctr, 59-61; asst resident psychiat, Johns Hopkins Hosp, 61-63, resident psychiatrist, 63-64; clin assoc psychopharmacol, Clin Neuropharmacol Res Ctr, NIMH, St Elizabeth's Hosp, 64-66; assoc physician, Psychosom Clin, Johns Hopkins Univ, 66-67; chief psychiat, Baltimore City Hosps, 67-71; PROF PSYCHIAT & CHMN DEPT, UNIV TEX MED SCH HOUSTON, 71- *Concurrent Pos:* Res fels, Harvard Univ & NIH, 60-61; assoc prof, Johns Hopkins Univ, 68-71; consult psychiatrist, Good Samaritan Hosp, 68-71. *Mem:* Am Psychiat Asn; assoc Am Geriat Soc; Am Psychosom Soc; NY Acad Sci. *Res:* Clinical effects of psychoactive drugs; psychological, behavioral and metabolic factors in alcoholism. *Mailing Add:* 6400 W Cullen Houston TX 77025

FAILS, THOMAS GLENN, b Unity Twp, Ohio, Feb 28, 28; m 60; c 2. DIAPIRIC STRUCTURE TECTONICS. *Educ:* Columbia Univ, MA, 55. *Prof Exp:* Sr geologist, Explor Dept, Shell Oil Co, 55-66; dist geologist & vpres, Trend Explor, 67-75; PRES, RAVEN EXPLOR CORP, 77- *Concurrent Pos:* Dir, Petrol Explor Soc Gt Brit, 74. *Mem:* Am Inst Prof Geologists; Am Asn Petrol Geologists; fel Geol Soc London; Petrol Explor Soc Gt Brit. *Res:* Development or growth of diapiric structures in the Coastal Salt Basin, US Gulf Coast; special emphasis on how development, faulting and hydrocarbon distribution vary with respect to salt diapir type. *Mailing Add:* 1777 Larimer St Suite 1203 Denver CO 80202

FAIMAN, CHARLES, b Winnipeg, Man, Dec 6, 39; m 63; c 3. ENDOCRINOLOGY, PHYSIOLOGY. *Educ:* Univ Man, BSc & MD, 62, MSc, 66. *Prof Exp:* Res fel physiol, Univ Man, 64-65; res asst med, Univ Ill, 65-67; res assoc endocrinol, Mayo Found, 67-68; asst prof, 68-71, assoc prof physiol & med, 71-75, assoc prof med, 75-78, PROF PHYSIOL, UNIV MAN, 75-, PROF MED & HEAD, SECT ENDOCRINOL, HEALTH SCI CTR, 78- *Concurrent Pos:* Med Res Coun Can fel, Univ Man, 64-68, scholar, 68-73; dir clin invest unit, Winnipeg Gen Hosp, 71-74; head, Sect Endocrinol & Metab, Univ Man & Health Sci Ctr, 78- *Honors & Awards:* Prowse Prize, Univ Man, 66. *Mem:* Endocrine Soc; Can Soc Endocrinol Metab; Am Fedn Clin Res; Soc Exp Biol & Med; Can Soc Clin Invest. *Res:* Reproductive endocrinology, especially gonadotropin regulation in humans in health and in disease; human fetal pituitary gonadal interrelationships; the use of non-human primates as models for the study of reproductive physiology. *Mailing Add:* Sect Endocrinol & Metab Univ Man Winnipeg MB R3E 0Z3 Can

FAIMAN, MICHAEL, b London, Eng, Mar 27, 35; m 65; c 2. COMPUTER SCIENCE. *Educ:* Cambridge Univ, BA, 56; Univ Ill, Urbana-Champaign, MS, 64, PhD, 66. *Prof Exp:* Engr, Elliott Automation, Eng, 56-60; asst prof comput sci, 66-70, ASSOC PROF COMPUT SCI, UNIV ILL, URBANA-CHAMPAIGN, 70- *Res:* Computer logic and hardware; digital-analog systems; optical information processing. *Mailing Add:* Comput Sci Dept Univ Ill 1304 W Springfield Ave Urbana IL 61801

FAIMAN, MORRIS DAVID, b Winnipeg, Man, June 24, 32; m 62; c 2. PHARMACOLOGY, TOXICOLOGY. *Educ:* Univ Man, BS, 55; Univ Minn, MS, 61, PhD, 65. *Prof Exp:* From asst prof to assoc prof pharm, 65-73, PROF PHARMACOL & TOXICOL, UNIV KANS, 73- *Mem:* Fel AAAS; Am Soc Pharmacol & Exp Therapeut; Undersea Med Soc; Soc Toxicol; Aerospace Med Asn. *Res:* Drug metabolism; pharmacokinetics; oxygen toxicity; biogenic amines; alcoholism mechanisms. *Mailing Add:* Dept Pharmacol & Toxicol Sch Pharm Univ Kans 5012 Mol Univ Kans Lawrence KS 66045

FAIMAN, ROBERT N(EIL), b Excelsior, Minn, June 25, 23; m 44; c 2. ELECTRICAL ENGINEERING. *Educ:* NDak State Col, BSEE, 47; Univ Wash, MSEE, 48; Purdue Univ, PhD, 56. *Prof Exp:* Assoc elec eng, Univ Wash, 47-48; from asst prof to prof & chmn dept, NDak State Col, 48-58;

dean col technol, Univ NH, 59-67, vpres res, 67-74; dir acad affairs, 74-90, EMER DIR ACAD AFFAIRS, AIR FORCE INST TECHNOL, 90- *Concurrent Pos:* Engr, eng sci prog, NSF, 57-59; mem, NH Bd Registr Prof Engrs, 64-74. *Mem:* Am Soc Eng Educ; sr mem Inst Elec & Electronics Engrs; Soc Am Mil Engrs; Sigma Xi; AAAS; Nat Soc Prof Engrs; Am Soc Eng Educ. *Res:* Circuit analysis and synthesis; control systems; managerial systems. *Mailing Add:* Air Force Inst Technol CF-E Wright-Patterson AFB OH 45433-6583

FAIN, GORDON LEE, b Washington, DC, Nov 24, 46; m 68; c 2. PHOTORECEPTORS, VERTEBRATE RETINA. *Educ:* Stanford Univ, BA, 68; Johns Hopkins Univ, PhD(biophys), 73. *Prof Exp:* Predoctoral fel biophys, Johns Hopkins Univ, 68-73; postdoctoral fel biol, Harvard Univ, 73-74; grass fel biol, Marine Biol Lab, Woods Hole, 74; exchange fel biol, Ecole Normale Supericure, Paris, 74-75; from asst prof to assoc prof, 75-82, PROF OPHTHAL, SCH MED, UNIV CALIF, LOS ANGELES, 82-, ASSOC DIR, JULES STEIN EYE INST, 85- *Honors & Awards:* Adams Award, Res to Prevent Blindness, 77; Merit Award, NIH, 89. *Mem:* Soc Neurosci; Biophys Soc; Asn Res Vision & Ophthal; Am Physiol Soc; AAAS; Eng Physiol Soc. *Res:* Physiology of photoreceptors; mechanism of light adaptation; integration in vertebrate retina; physiology of synaptic transmission; ion channels; ion and water transport in ocular epithelia. *Mailing Add:* Dept Ophthal Jules Stein Eye Inst Univ Calif Los Angeles CA 90024

FAIN, JOHN NICHOLAS, b Jefferson City, Tenn, Aug 18, 34; m 58; c 3. BIOCHEMISTRY. *Educ:* Carson-Newman Col, BS, 56; Emory Univ, PhD(biochem), 60. *Prof Exp:* Res assoc biochem, Emory Univ, 60-61; NSF fel, 61-62; USPHS fel, 62-63; chemist, Nat Inst Arthritis & Metab Dis, 63-65; from asst prof to prof, med sci, Brown Univ, 65-85, chmn biochem, 74-84; VAN VLEET PROF & DEPT BIOCHEM, CTR HEALTH SCI, UNIV TENN, 85- *Concurrent Pos:* Macy fac scholar & vis fel, Clare Hall, Cambridge Univ, 77-78 & 84-85; Fogarty Int Fel NIH & prof biochem, Univ Nottingham. *Mem:* Am Soc Biol Chemists. *Res:* Transmembrane signalling via phosphoinositide hydrolysis; purification and regulation of membrane bound phospholipase C. *Mailing Add:* Dept Biochem Univ Tenn Ctr Health Sci 858 Madison Ave Suite G01 Memphis TN 38163

FAIN, ROBERT C, b Santa Rosa, Tex, Oct 19, 36; m 58; c 2. ORGANOMETALLIC CHEMISTRY. *Educ:* Southwest Tex State Univ, BS, 58, MA, 59; Univ Tex, PhD(chem), 65. *Prof Exp:* Instr chem, Southwest Tex State Univ, 59-61; res chemist, Celanese Corp, 65-66; assoc prof chem, 66-68, prof phys sci & head dept, 68-74, dean, Sch Arts & Sci, 70-81, VPRES ACAD AFFAIRS, TARLETON STATE UNIV, 79- *Mem:* Am Chem Soc. *Res:* Pi complexes of transition metals. *Mailing Add:* Dept Chem Tarleton State Univ Stephenville TX 76402

FAIN, SAMUEL CLARK, JR, b Jefferson City, Tenn, Aug 13, 42; m; c 2. SOLID STATE PHYSICS. *Educ:* Reed Col, BA, 65; Univ Ill, Urbana-Champaign, MS, 66, PhD(physics), 69. *Prof Exp:* NATO fel physics, Natuurkundig Lab, Univ Amsterdam, 69-70; from asst prof to assoc prof, 70-80, PROF PHYSICS, UNIV WASH, 80- *Concurrent Pos:* Alfred P Sloan res fel, 71-75. *Mem:* Fel Am Phys Soc; Am Vacuum Soc; Sigma Xi. *Res:* Surface physics; structures of physically absorbed monolayers on solid surfaces at low temperature. *Mailing Add:* Dept Physics Univ Wash FM 15 Seattle WA 98195

FAIN, WILLIAM WHARTON, b Augusta, Ga, Apr 23, 27; m 87; c 3. OPERATIONS RESEARCH, TECHNICAL MANAGEMENT. *Educ:* Univ Tex, BA, 50, MA, 51, PhD(physics), 55. *Prof Exp:* Res scientist, Electro-Mech Co, Tex, 53-56; eng specialist, Chance-Vought Aircraft, 56 & 57-59; mem tech staff, Pac Missile Range, Land-Air Inc, Point Mugu, 60-61; scientist, Supreme Hq, Allied Powers Europe Tech Ctr, Holland, 61-63; exec adv, Douglas Aircraft Co, Calif, 63-66; mem prof staff, Ctr Naval Anal, Va, 66-69; mgr opers res dept, Caci, Inc, Arlington, 69-71, exec vpres, 71-72, pres & chief exec officer, 72-81; PRES, OPTIONS ENTERPRISES, VIRGINIA CITY, NV, 81- *Concurrent Pos:* Inst Defense Anal fel opers anal, 61-63; consult, Steering Comt, Spring Joint Comput Conf, 71, Can Dept Nat Defence, United Aircraft Co, US Army War Col, Indust Col Armed Forces & Univ Southern Calif. *Mem:* Sigma Xi. *Res:* Simulation and gaming; military operations research; quantitative techniques in international studies; applications of game theory; computer applications. *Mailing Add:* PO Box 740 Virginia City NV 89440

FAINBERG, ANTHONY, b London, Eng, Jan 14, 44; US citizen; m 64, 86; c 2. NUCLEAR SAFEGUARDS, GENERAL PHYSICS. *Educ:* NY Univ, AB, 64; Univ Calif, Berkeley, PhD(high energy physics), 69. *Prof Exp:* Res assoc high energy physics, Lawrence Radiation Lab, 69 & Univ Turin, Italy, 70-72; res asst prof, Syracuse Univ, 73-77; physicist safeguards, Dept Nuclear Energy, Brookhaven Nat Lab, 77-84; SR ASSOC, OFF TECHNOL ASSESSMENT, US CONG, 84- *Concurrent Pos:* Fel NSF, 64-67; adj assoc prof, Syracuse Univ, 77-78; Cong sci fel, 83-84; panel public affairs, Am Phys Soc. *Mem:* AAAS; Inst Nuclear Mat Mgt. *Res:* Strategic defenses; nuclear safeguards; counter terrorism. *Mailing Add:* 326 Ninth St SE Washington DC 20003

FAINBERG, ARNOLD HAROLD, physical organic chemistry; deceased, see previous edition for last biography

FAINBERG, JOSEPH, b Passaic, NJ, Oct 18, 30; m 56; c 3. RADIO ASTRONOMY. *Educ:* Univ Chicago, AB, 50, BS, 51, MS, 53; Johns Hopkins Univ, PhD(elec eng), 65. *Prof Exp:* Res asst cosmic rays, Univ Chicago, 50-53, res asst meson physics, 53-57; res scientist, Johns Hopkins Univ, 57-66; PHYSICIST, GODDARD SPACE FLIGHT CTR, NASA, 66- *Concurrent Pos:* Instr, Roosevelt Univ, 53-54. *Mem:* AAAS; Int Astron Union; Am Phys Soc; Inst Elec & Electronics Engrs; Am Geophys Union. *Res:* Cosmic rays; meson physics; scattering and diffraction of electromagnetic waves; space physics. *Mailing Add:* 4000 Virgilia Chevy Chase MD 20815

FAINGOLD, CARL L, b Chicago, Ill, Feb 1, 43; m 64; c 3. NEUROPHARMACOLOGY, NEUROPHYSIOLOGY. *Educ:* Univ Ill, BS, 65; Northwestern Univ, PhD(pharmacol), 70. *Prof Exp:* Fel neuropharmacol, Inst Psychiat, Univ Mo, 70-72; asst prof, 72-76, actg chmn, dept pharmacol, 81-83, assoc prof, 76-87, PROF PHARMACOL, MED SCH, SOUTHERN ILL UNIV, 87- *Concurrent Pos:* NIH prin investr, Nat Inst Neurol & Commun Dis & Stroke, 79-; prin investr, Deafness Res Found & Am Heart Asn, 84-; vis scientist, Dept Neurol, Inst Psychiat, London, 84. *Mem:* Am Soc Pharmacol & Exp Therapeut; Soc Neurosci; AAAS; Am Epilepsy Soc; NY Acad Sci; Sigma Xi. *Res:* Examination of the specific mechanisms of anticonvulsant and convulsant drug action of brainstem neurons; elucidation of the changes in neuronal response in the brainstem reticular formation and inferior colliculus which subserve the development of seizures; investigations of brainstem role in epileptic seizures and the role of neurotransmitters in inferior colliculus in hearing and sound induced seizures. *Mailing Add:* Dept Pharmacol Sch Med Southern Ill Univ PO Box 19230 Springfield IL 62794-9230

FAINSTAT, THEODORE, b Montreal, Que, July 14, 29. OBSTETRICS & GYNECOLOGY. *Educ:* McGill Univ, BSc, 50, MSc, 51, MD, 55; Univ Cambridge, PhD, 70. *Prof Exp:* Instr genetics, McGill Univ, 50-51; intern, Univ Montreal, 55-56; asst obstet & gynec, Sch Med, Harvard Univ, 64-66; from assoc prof to prof, Med Sch, Northwestern Univ, Chicago, 67-74; prof obstet & gynec, Univ Kans Med Ctr, Kansas City, 74-77; PROF OBSTET & GYNEC, STANFORD UNIV, 77-; CHMN OBSTET & GYNEC, SANTA CLARA VALLEY MED CTR, 77- *Concurrent Pos:* Asst resident med, Boston City Hosp, 56-57, surgeon, 60; resident, Boston Lying-In Hosp, 64-66; sr attend, Chicago Wesley Mem Hosp, 69-; fel obstet & gynec, Sch Med, Harvard Univ, 56-63, Josiah Macy Jr Found fel, 57-63, res fel endocrinol, Biol Labs, 57-60; sr investr, NIH grant, 58-63; Am Cancer Soc scholar, Strangeways Res Labs, Eng, 64-66; Kellogg fac fel, Ctr Teaching Prof, Northwestern Univ, 70-73. *Honors & Awards:* Pres Award, Am Col Obstet & Gynec. *Mem:* Soc Study Reprod; Am Col Obstet & Gynec; Soc Gynec Invest; Endocrine Soc; Teratology Soc. *Res:* Biology of reproduction. *Mailing Add:* 751 S Bascom Ave San Jose CA 95128

FAIR, DARYL S, b Santa Ana, Calif, Oct 28, 47. BLOOD COAGULATION. *Educ:* Univ Wash, PhD(microbiol), 75. *Prof Exp:* Asst mem immunol, Res Inst Scripps Clin, 78-86; PROF BIOCHEM, HEALTH CTR, UNIV TEX, 86- *Mem:* Am Heart Asn; Am Soc Hemat; Am Soc Biochem & Molecular Biol; Protein Soc; Int Soc Thrombosis Haemostatis. *Res:* Biochemistry and immunochemistry of vitamin K-dependent blood coagulation proteins. *Mailing Add:* Dept Biochem Univ Tex Health Ctr PO Box 2003 Tyler TX 75710

FAIR, HARRY DAVID, JR, b Indiana, Pa, Dec 2, 36; m 64; c 1. SOLID STATE PHYSICS. *Educ:* Indiana Univ Pa, BS, 58; Univ Del, MS, 60, PhD(solid state physics), 67. *Prof Exp:* Teaching asst physics, Univ Del, 58-59; res physicist, Picatinny Arsenal, 60-62; vis scientist, Univ Del, 62-65; solid state physicist, Explosives Lab, 65-69, chief point defect & electron energy level sect, 69-71, chief solid state br, Picatinny Arsenal, 71-74; chief, Propulsion Technol Lab, Arradcom, 77-88; VIS PROF, ELEC & COMPUT ENG DEPT, UNIV TEX, 88- *Concurrent Pos:* Vis prof, Univ Paris, 74 & Royal Inst Gr Brit, 75. *Mem:* Am Phys Soc; Chem Soc; Sigma Xi. *Res:* Impurity levels in solids; electron spin resonance; optical and electronic of II-VI compound semiconductors and explosive solids; fast reactions in solids; combustion and ignition processes; high pressure physics; photophysics of energetic materials. *Mailing Add:* 1301 Ridgecrest Austin TX 78746

FAIR, JAMES R(UTHERFORD), b Charleston, Mo, Oct 14, 20; m 50; c 3. CHEMICAL ENGINEERING. *Educ:* Ga Inst Technol, BS, 42; Univ Mich, MS, 49; Univ Tex, PhD(chem eng), 54. *Hon Degrees:* DSc, Wash Univ, 77; DHL, Clemson Univ, 88. *Prof Exp:* Chemist & res engr, Monsanto Co, Marshall, Tex, 42-43, res & design engr, 43-45, develop assoc, Mo, 45-47, proj leader, engr, Texas City, 47-52; process design engr, Shell Develop Co, Calif, 54-56; res group leader & sect leader, Monsanto Co, 56-61, eng mgr, 61-69, eng dir, 69-79; MCKETTA CENTENNIAL ENERGY CHAIR, UNIV TEX, AUSTIN, 79- *Concurrent Pos:* Affil prof, Wash Univ, 64-79. *Honors & Awards:* Walker Award, Am Inst Chem Engrs, 73, Chem Eng Pract Award, 75 & Founders Award, 76 & Inst lect, 79. *Mem:* Am Chem Soc; Am Inst Chem Engrs; Nat Acad Eng; Nat Soc Prof Engrs. *Res:* Physical separation methods; heat transfer equipment; chemical reactor design; hydrocarbon pyrolysis operations. *Mailing Add:* Dept Chem Eng Univ Tex Austin TX 78712

FAIR, RICHARD BARTON, b Los Angeles, Calif, Sept 12, 42; m 64; c 4. SEMICONDUCTORS, ELECTRICAL ENGINEERING. *Educ:* Duke Univ, BS, 64, PhD(elec eng), 69; Pa State Univ, MS, 66. *Prof Exp:* mem staff, Bell Labs, 69-73, supvr semiconductors, 73-81; PROF ELEC ENG, DUKE UNIV, 81-; VPRES, MICROELECTRONICS CTR NC, 81- *Concurrent Pos:* Chmn, Solid State Device Subcomt, Int Electron Device Mgt, Inst Elec & Electronics Engrs, 77-78; symp chmn, Mat Res Soc, 84; mem, Electronic Mats Comt, Am Inst Metall Engrs, 85- & Electronics Exec Comt, Electrochem Soc, 85-; symp co-chmn, Fifth Int Symp on Silicon Mat Sci & Technol, 86. *Mem:* Sigma Xi; fel Inst Elec & Electronics Engrs; Electrochem Soc; Mat Res Soc. *Res:* Fundamental studies of the diffusion of Group III and V impurities in silicon; solubility effects and electrical activity of these impurities; theory and design of semiconductor devices such as transistors, diodes and integrated circuits. *Mailing Add:* 3414 Cambridge Rd Durham NC 27707

FAIRAND, BARRY PHILIP, b Watertown, NY, May 20, 34; m 59; c 5. LASERS. *Educ:* LeMoyne Col, BS, 55; Univ Detroit, MS, 57; Ohio State Univ, PhD(physics), 69. *Prof Exp:* SR SCIENTIST PHYSICS, BATTELLE MEM INST, 57- *Mem:* Am Phys Soc; Soc Photo-Optical Inst Engrs; Mat Res Soc; NY Acad Sci; Sigma Xi. *Res:* High-power laser applications, particularly in the areas of laser shock processing, thermal treatment of materials, and nondestructive evaluation. *Mailing Add:* 41473 Tiber Creek Terr Fremont CA 94539-4581

FAIRBAIRN, HAROLD WILLIAMS, b Ottawa, Ont, July 10, 06; nat US; m 39; c 4. PETROLOGY. *Educ:* Queen's Univ, Ont, BSc, 29; Harvard Univ, AM, 31, PhD(mineral & geol), 32. *Prof Exp:* Field asst, Geol Surv Can, 26-32; Royal Soc Can traveling fel, Innsbruck Univ, Univ Gottingen & Univ Berlin, 32-34; instr mineral, Queen's Univ, Ont, 34-37; from asst prof to prof, 37-72, EMER PROF PETROL, MASS INST TECHNOL, 72- *Concurrent Pos:* Surv chief, Ont Dept Mines, 35-39, Que Dept Mines, 40 & Geol Surv Can, 42; petrographer, Manhattan Dist Proj, 44. *Mem:* Fel Geol Soc Am; fel Mineral Soc Am; Am Acad Arts & Sci. *Res:* Structural petrology; optical crystallography; metamorphism; geochronology. *Mailing Add:* Dept Earth & Planetary Sci Mass Inst Technol Cambridge MA 02139

FAIRBAIRN, JOHN F, II, b Buffalo, NY, Nov 2, 22; m 60; c 2. MEDICINE. *Educ:* Univ Buffalo, MD, 45. *Prof Exp:* Pvt pract, Buffalo, NY, 52-53, 54; Nat Heart Inst trainee, Mayo Clin, 54-55; pvt pract, Buffalo, 55-56; from instr to asst prof med, 59-66, ASSOC PROF CLIN MED, MAYO GRAD SCH MED, UNIV MINN, 66-; CONSULT INTERNAL MED, MAYO CLIN, 56- *Concurrent Pos:* Attend physician & consult, St Mary's & Methodist Hosps, Rochester, 56-; head sect peripheral vascular dis, Mayo Clin, 60-74; mem coun arteriosclerosis, Am Heart Asn, 63, adv bd coun circulation, 64. *Mem:* Am Fedn Clin Res; Am Heart Asn; fel Am Col Physicians. *Res:* Atherosclerosis, hypertension and vascular diseases in general. *Mailing Add:* 200 First St SW Rochester MN 55901

FAIRBANK, HENRY ALAN, b Lewistown, Mont, Nov 9, 18; m 43; c 3. LOW TEMPERATURE PHYSICS. *Educ:* Whitman Col, AB, 40; Yale Univ, PhD(physics), 44; Oxford Univ, MA, 53. *Hon Degrees:* DSc, Whitman Col, 71. *Prof Exp:* Staff mem, Los Alamos Lab, NMex, 44-45; from instr to assoc prof physics, Yale Univ, 45-62; chmn dept, 62-73, prof, 62-88, EMER PROF PHYSICS, DUKE UNIV, 88- *Concurrent Pos:* Instr physics, Yale Univ, 42-44; Guggenheim fel, Oxford Univ, 53-54; consult, Los Alamos Sci Lab, 57-65. *Mem:* Fel Am Phys Soc; Am Asn Physics Teachers; fel AAAS. *Res:* Low temperature physics; properties of liquid and solid helium-3 and helium-4; heat transfer in solids and quantum fluids. *Mailing Add:* Dept Physics Duke Univ Durham NC 27706

FAIRBANK, WILLIAM MARTIN, cryogenics, gravitation; deceased, see previous edition for last biography

FAIRBANK, WILLIAM MARTIN, JR, b New Haven, Conn, Jan 7, 46; m 75; c 3. QUANTUM OPTICS. *Educ:* Pomona Col, BA, 68; Stanford Univ, MS, 69, PhD(physics), 74. *Prof Exp:* Res assoc physics, Optical Sci Ctr, Univ Ariz, 74-75; from asst prof to assoc prof, 75-83, PROF PHYSICS, COLO STATE UNIV, 83- *Mem:* Optical Soc Am; Am Phys Soc. *Res:* High resolution laser spectroscopy; new applications of tunable dye lasers; single atom detection. *Mailing Add:* Physics Dept Colo State Univ Ft Collins CO 80521

FAIRBANKS, GILBERT WAYNE, b Hartford, Conn, July 26, 37; m 69. PHYSIOLOGY. *Educ:* Trinity Col, Conn, BS, 59; Wesleyan Univ, MA, 61; Univ SC, PhD, 64. *Prof Exp:* ASSOC PROF BIOL, FURMAN UNIV, 64-, PREMED ADV, 75- *Mem:* AAAS. *Res:* Effect of hyperthermic conditions on lipids in animal tissues; membrane lipids of protozoans. *Mailing Add:* Dept Biol Furman Univ Greenville SC 29613

FAIRBANKS, GRANT, b Iowa City, Iowa, May 17, 40; m 62; c 2. MEMBRANE BIOCHEMISTRY, HEMATOLOGY. *Educ:* Grinnell Col, BA, 61; Mass Inst Technol, PhD(biophys), 69. *Prof Exp:* Res fel, biochem res, Mass Gen Hosp & dept biol chem, Sch Med, Harvard Univ, 69-70; res assoc, 70-71, staff scientist, 71-77, SR SCIENTIST, CELL BIOL GROUP, WORCESTER FOUND EXP BIOL, 77- *Concurrent Pos:* Prin investr, grants & contracts, NIH, Bethesda, Md, 72-, consult, 77- *Mem:* Am Soc Cell Biol; Am Soc Biochem & Molecular Biol; Sigma Xi; Fedn Am Scientists. *Res:* Organization and function of plasma membranes of mammalian cells; biochemical mechanism of shape determination in human erythrocytes; role of passive cation transport in volume regulation by normal and abnormal human erythrocytes; role of the epididymis in maturation of the ram sperm plasma membrane. *Mailing Add:* Worcester Found Exp Biol 222 Maple Ave Shewsbury MA 01545-2001

FAIRBANKS, HAROLD V, b Des Plaines, Ill, Dec 7, 15; m 51. METALLURGY. *Educ:* Mich State Col, BS, 37, MS, 39. *Prof Exp:* Instr gen chem, Mich State Col, 37-39; instr chem eng, Univ Louisville, 40-42; asst prof, Rose Polytech Inst, 42-47; from asst prof to assoc prof metall, 47-55, prof metall, 55-78, EMER PROF, WVA UNIV, 78- *Concurrent Pos:* Adv, Purdue Team Int Coop Admin, Chen Kung Univ, Tainan, Taiwan, 57-59; co-dir grad studies mat sci eng, WVa Univ, 63; secy-treas, J B H, Inc, 84-; consult ultrasonic indust applications, 78-; consult, Fairbanks Metall Lab, Cheng Kung Univ, Taiwan. *Mem:* Nat Asn Corrosion Engrs; Am Soc Metals; Int Solar Energy Soc; Am Inst Chem Engrs. *Res:* Application of high intensity ultrasonics to various processes. *Mailing Add:* Spring North Mesa 262 E Brown Rd Apt 232 Mesa AZ 85201

FAIRBANKS, LAURENCE DEE, b Wilson Co, Kans, May 23, 26; m 53; c 3. INVERTEBRATE ZOOLOGY. *Educ:* Univ Kans, AB, 49, MA, 56; Tulane Univ, PhD(zool), 59. *Prof Exp:* Instr zool, 58-66, asst prof med, 66-70, ASSOC PROF MED, SCH MED, TULANE UNIV, 70- *Mem:* AAAS; Am Soc Zoologists; Entom Soc Am; Am Inst Biol Sci; Sigma Xi. *Res:* Physiology and ecology of invertebrates, Mollusca and Insecta; general biology of aging; human and environmental biology. *Mailing Add:* 4539 General Meyers New Orleans LA 70114

FAIRBANKS, VIRGIL, b Ann Arbor, Mich, June 7, 30; m 55; c 3. INTERNAL MEDICINE, HEMATOLOGY. *Educ:* Univ Utah, BA, 51; Univ Mich, MD, 54. *Prof Exp:* Intern internal med, Bellevue Hosp, NY, 54-55; resident, Col Med, Univ Utah, 57-59; fel hemat, Scripps Clin, La Jolla, Calif, 59-60; asst physician, City of Hope Med Ctr, Duarte, 60-61, assoc physician, 61-63; asst prof internal med, Calif Col Med, 63-64; assoc prof, 72-78, PROF INTERNAL MED & LAB MED, MAYO MED SCH, 78-, MEM RES & TEACHING STAFF, MAYO CLIN, 65- *Concurrent Pos:* Sr attend physician, Los Angeles County Hosp & consult, Vet Admin Hosp, Long Beach, 63-64. *Mem:* AAAS; Am Soc Hemat; Am Fedn Clin Res; Am Col Physicians; Int Soc Hemat. *Res:* Pharmacology of veratrum alkaloids; glycolytic functions of human erythrocyte; hemoglobin structure and function; iron metabolism; human genetics. *Mailing Add:* Dept Lab Med Mayo Clin Rochester MN 55905

FAIRBRIDGE, RHODES WHITMORE, b Pinjarra, Australia, May 21, 14; m 43; c 1. GEOLOGY. *Educ:* Queen's Univ, BA, 36; Univ Oxford, BS, 40; Univ Western Australia, DSc, 44. *Hon Degrees:* Dr, Univ Gothenburg, Sweden, 84. *Prof Exp:* Field geologist, Iraq Petrol Co, 38-41; lectr geol, Univ Western Australia, 46-53; assoc prof, Univ Ill, 53-54; prof geol, Columbia Univ, 55-82; RETIRED. *Concurrent Pos:* Consult, Hydro-Elec Comn Tasmania, 47, Richfield Oil Co, 48, Australian Bur Mineral Resources, 50, Snowy Mountains Hydro-Elec Auth, 51, Pure Oil Co, 55-56, Life Mag, 56-, Nat Acad Sci, 58-59, Off Naval Res, 56-, Readers Digest Books, 72-, Fabbri Publ Co, Milano, 78- & Random House, 83-; leader, Nile Exped, Columbia Univ, 61; vis prof, Univ Sorbonne, 62; ed, Geol & Benchmark Series, Hutchinson & Ross Publ Co, 58- 86, Van Nostrand Reinhold Encycl Earth Sci, 63- *Honors & Awards:* Alexander von Humboldt Prize, 77. *Mem:* Fel Geol Soc Am; Soc Econ Paleontologists & Mineralogists; Am Asn Petrol Geologists; Am Meteorol Soc. *Res:* Gravitational processes in sedimentation and tectonics; littoral sedimentation; coral reefs; eustatic changes of sea-level; paleoclimatology; geomorphology; world geotectonics. *Mailing Add:* Dept Geol Columbia Univ New York NY 10027

FAIRBROTHERS, DAVID EARL, b Absecon, NJ, Sept 24, 25; m 49; c 2. BOTANY. *Educ:* Syracuse Univ, BS, 50; Cornell Univ, MS, 52, PhD(bot), 54. *Prof Exp:* Asst bot, Cornell Univ, 50-54, instr, 54; from instr to assoc prof, 54-65, chmn dept, 72-78, PROF BOT, RUTGERS UNIV, 65- *Concurrent Pos:* Rockefeller Res Found grant, 52; NSF grant, 57, 60, 63, 65, 67, 69, 71, 73 & 75. *Mem:* Int Soc Plant Taxonomists; Bot Soc Am; Am Soc Plant Taxonomists; Soc Study Evolution; Torrey Bot Club; Sigma Xi. *Res:* Chemosystematics; experimental taxonomy; rare and endangered plant species of New Jersey; scanning electron microscope. *Mailing Add:* Dept Biol Sci Rutgers Univ New Brunswick NJ 08903

FAIRCHILD, CLIFFORD EUGENE, b Philip, SDak, Sept 19, 34; m 60; c 2. OPTICS. *Educ:* Fresno State Col, BA, 56; Univ Wash, PhD(physics), 62. *Prof Exp:* From asst prof to assoc prof, 62-77, PROF PHYSICS, ORE STATE UNIV, 77- *Concurrent Pos:* Sr postdoctoral assoc, USAF Cambridge Res Labs, 68-69; vis fel, Joint Inst Lab Astrophys & Lab Atmospheric & Space Physics, 75-76. *Mem:* Am Phys Soc; Optical Soc Am. *Res:* Optics; lasers; fiber optics. *Mailing Add:* Dept Physics Ore State Univ Corvallis OR 97331

FAIRCHILD, DAVID GEORGE, b Albany, Calif, Jan 7, 39; m 61; c 4. VETERINARY PATHOLOGY, TOXICOLOGY. *Educ:* Univ Calif, BS, 60, DVM, 62; Tex A&M Univ, MS, 68. *Prof Exp:* Chief path br, Res & Nutrit Lab, US Army, Denver, 62-65; chief path div, Biomed Lab, Edgewood Arsenal, MD, 68-71; chief vet med dept, US Naval Res Lab, Cairo, Egypt, 71-72; chief vet path, Biomed Lab, Edgewood Arsenal, 72-74 & Letterman Inst Res, Presidio of San Francisco, 74-76; pathologist, Syntex Labs, 76-86, HEAD DEPT ANAT PATH, SYNTEX RES, 86- *Mem:* Am Vet Med Asn; Wildlife Dis Asn; Soc Pharmacol & Environ Pathologists; Am Col Vet Pathologists; Int Acad Path; Soc Toxicol Pathologists. *Res:* Industrial pharmaceutical toxicology; laboratory animal pathology; wildlife diseases; oncology; companion pet pathology. *Mailing Add:* Syntex Res Labs 3401 Hillview Palo Alto CA 94304

FAIRCHILD, EDWARD ELWOOD, JR, b Morgantown, WVa, Sept 21, 49; m 70; c 4. MATHEMATICAL PHYSICS. *Educ:* WVa Univ, BS, 71, MS, 72; Univ Tex, Austin, PhD(physics), 75. *Prof Exp:* ASST PROF PHYSICS, WASH UNIV, 75- *Concurrent Pos:* NATO fel, Dept Astrophys, Oxford Univ, 76-77. *Mem:* Am Phys Soc; Int Soc Gen Relativity & Gravitation. *Res:* Quantization of the gravitational field. *Mailing Add:* Dept Environ Servs Univ Tex Health Sci Ctr PO Box 20036 Houston TX 77225

FAIRCHILD, EDWARD H, b Wadsworth, Ohio, Oct 31, 43; m 67; c 2. NUCLEAR MAGNETIC RESONANCE, MASS SPECTROSCOPY. *Educ:* Wright State Univ, BS, 70; Ohio State Univ, MS, 72, PhD(natural prod chem), 75. *Prof Exp:* Mgr anal instrumentation, natural prod chem, Ohio State Univ, Columbus, 75-77; sr res assoc pharmacol, Case Western Reserve Univ, Cleveland, Ohio, 77-79; group leader anal chem, Sherex Chem Co, Dublin, Ohio, 79-87; mgr anal res, Henkel Corp, Cincinnati, Ohio, 87-89; DIR SUPPORT SCI, LONZA, INC, ANNANDALE, NJ, 89- *Mem:* Am Oil Chemists Soc; Am Chem Soc. *Res:* Characterization of oleochemicals; structure elucidation using nuclear magnetic resonance, MS, etc; modelling of EO/PO addition reactions; NMR of ongoing reactions; natural product structure elucidation. *Mailing Add:* Lonza Inc 79 Rte 22 E Annandale NJ 08801

FAIRCHILD, GRAHAM BELL, b Washington, DC, Aug 17, 06; m 38; c 2. ENTOMOLOGY. *Educ:* Harvard Univ, BS, 32, MS, 34, PhD, 42. *Prof Exp:* Asst entomologist, Exp Sta, Univ Fla, 34-35; entomologist, Yellow Fever Serv, Int Health Div, Rockefeller Found, 35-37; jr med entomologist & entomologist, Gorgas Mem Lab, 38-71; emer adj prof entom, Univ Fla, 71-80, RES ASSOC, FLA STATE COL OF ARTHROPODS, 80- *Concurrent Pos:* Asst prof, Univ Minn, 49-50 & Douglas Lake Biol Sta, Univ Mich, 56-57; assoc dir, Gorgas Mem Lab, 58-71; mem trop med & parasitol sect, Nat Inst Allergy & Infectious Dis, 61-65; res assoc, Mus Comp Zool, Harvard Univ, 65- & Fla State Collection Arthropods, 70- *Honors & Awards:* Founders Mem Award, Entom Soc Am, 68. *Mem:* AAAS; emer mem Entom Soc Am; Soc Syst Zool. *Res:* Taxonomy of arthropods of medical importance; insects affecting man and animals, especially Tabanidae, Simuliidae, Psychodidae and Ixodidae. *Mailing Add:* Dept Entom Div Plant Indust PO Box 1269 Gainesville FL 32602

FAIRCHILD, HOMER EATON, b Sioux Rapids, Iowa, Oct 1, 21; m 49; c 3. DEVELOPMENT OF PEST CONTROL TECHNOLOGIES, BIOCONTROL. *Educ:* SDak State Univ, BS, 44; Kans State Univ, PhD(entom), 53. *Prof Exp:* Res assoc, Iowa State Univ, 44-45; entomologist, Bur Entom & Plant Quarentine, USDA, 46-49; res asst, Sch Vet Med, Kans State Univ, 50-51; res entomologist, E I du Pont de Nemours & Co, Inc, 52-60; tech serv mgr, Geigy Chem Corp, NY, 60-64, staff specialist, 64-66; prod mgr, Union Carbide Corp, NY, 66-70, mkt mgr, 70-71; coordr, Spec Pesticides Review Group, Off Pesticides Prog, US Environ Protection Agency, 71-73, chief liaison officer indust affairs, 73-75, asst chief, Ecol Effects Br, 75-76; chief officer methods develop staff, 76-82, asst dir technol anal & develop staff, 82-87, CHIEF STAFF OFFICER, TECHNOL DEVELOP STAFF, PLANT PROTECTION & QUARANTINE ANIMAL & PLANT HEALTH INSPECTOR SERV, USDA, 87- *Concurrent Pos:* Mem & chmn imported fire ant working group, USDA, 77- *Mem:* AAAS; Entom Soc Am; Weed Sci Soc Am; NY Acad Sci; Am Mgt Asn. *Res:* Environmental monitoring (chemistry) and technology development (entomology, weed science, biocontrol and guarantine specialists); legislation, toxicology and residues of chemicals in relation to the environment. *Mailing Add:* 3314 Bermuda Village Advance NC 27006

FAIRCHILD, JACK, b Houston, Tex, Oct 25, 28; m 51; c 7. AERONAUTICAL ENGINEERING. *Educ:* Univ Tex, BS, 53; Univ Southern Calif, MS, 59; Univ Okla, PhD(eng sci), 64. *Prof Exp:* Aerodynamicist, Bell Helicopter Corp, 53-54; sr aerodynamics engr, Chance Vought Aircraft Corp, 54-56; lectr aerospace eng, Aviation Safety Div, Univ Southern Calif, 56-60; instr & res engr, Univ Okla, 60-62; from assoc prof to prof, 70-89, EMER PROF AEROSPACE ENG, UNIV TEX, ARLINGTON, 89-; CONSULT, 89- *Concurrent Pos:* Adv, US Army Bd Aircraft Accident Invest, 57-59; consult, Aircraft Div, Hughes Tool Co, 58-59, Vought Aeronaut Corp, 64-70, Bell Helicopter-Textron, 77-79 & Am Airlines Flight Acad, 66-80; vis prof, US Mil Acad, West Point, NY, 85-86; eng specialist, Gen Dynamics Corp, 83-84. *Mem:* Am Inst Aeronaut & Astronaut; Am Soc Eng Educ; Nat Soc Prof Engrs. *Res:* Flight mechanics; operational aerodynamics; automobile simulation and experimental testing; helicopter anti-torque systems; unsteady aerodynamics; gas turbine vibration analysis. *Mailing Add:* PO Box 435 Martindale TX 78655

FAIRCHILD, JOSEPH VIRGIL, JR, b New Orleans, La, Nov 26, 33; m 61; c 3. GEOLOGY, MANAGEMENT SCIENCE. *Educ:* La State Univ, BS, 56, MBA, 63, PhD(accounting), 75. *Prof Exp:* Geologist, United Core Inc, 56-57; assoc acct, Humble Oil & Refining Co, 63-64; sr acct, L A Champagne & Co CPA's, 64-68, partner, 68-69; from asst prof to assoc prof, 69-76, prof, 76-84, DISTINGUISHED PROF ACCT, NICHOLLS STATE UNIV, 84- *Res:* Business research, particularly asset revaluation and depreciation. *Mailing Add:* Dept Acct Nicholls State Univ Thibodaux LA 70310

FAIRCHILD, MAHLON LOWELL, b Spencer, Iowa, Oct 13, 30; m 54; c 3. ENTOMOLOGY. *Educ:* Iowa State Univ, BS, 52, MS, 53, PhD(entom), 59. *Prof Exp:* Asst entom, Iowa State Univ, 52-53 & 55-56; entomologist Europ corn borer res lab, Agr Res Serv, USDA, 57-59; from asst prof to assoc prof, 59-67, chmn dept, 69-80, PROF ENTOM, UNIV MO-COLUMBIA, 67-, COORDR INTEGRATED PEST MGT, 80- *Mem:* Am Entom Soc. *Res:* Biology and control of insects; pest management. *Mailing Add:* Pest Mgt Univ Mo 45 Agr Bldg Columbia MO 65201

FAIRCHILD, RALPH GRANDISON, radiological physics; deceased, see previous edition for last biography

FAIRCHILD, ROBERT WAYNE, b Williamsport, Pa; m 75; c 2. EXPERIMENTAL PHYSICS, COMPUTER INTERFACING. *Educ:* Rensselaer Polytech Inst, BS, 71; Cornell Univ, PhD(appl physics), 75. *Prof Exp:* Dir comput serv, 80-88, actg chair, Physics Dept, 90-91, ASST PROF PHYSICS, NEBR WESLEYAN UNIV, 75- *Concurrent Pos:* Res assoc, Univ Nebr, Lincoln, 76-78; prin investr, NSF, 80-83. *Mem:* Am Phys Soc; Am Asn Physics Teachers. *Res:* Computer interfacing and experiment control. *Mailing Add:* Dept Physics Nebr Wesleyan Univ Lincoln NE 68504-2796

FAIRCHILD, WILLIAM WARREN, b Rutland, Vt, July 30, 38. MATHEMATICS. *Educ:* Swarthmore Col, BA, 60; Univ Pa, MA, 63; Univ Ill, PhD(math), 67. *Prof Exp:* Physicist, Bartol Res Found, Franklin Inst, 60-62; asst prof math, Northwestern Univ, 67-70; ASSOC PROF MATH, UNION COL, NY, 70-, CHMN DEPT, 80- *Mem:* Am Math Soc. *Res:* Functional analysis, particularly convolution algebras. *Mailing Add:* Dept Math Union Col Bailey Hall 208C Schenectady NY 12308

FAIRCLOTH, WAYNE REYNOLDS, b Whigham, Ga, Jan 15, 32; m 66; c 3. SYSTEMATIC BOTANY. *Educ:* Valdosta State Col, BS, 55; Univ NC, MEd, 59; Univ Ga, PhD(bot), 71. *Prof Exp:* Teacher high sch, Ga, 51-61; from asst prof to prof, 61-84, cur herbarium, 71-84, PROF BIOL & CHMN DEPT, VALDOSTA STATE COL, 84- *Mem:* AAAS; Bot Soc Am; Am Fern Soc. *Res:* Ecology and systematics of vascular flora of the Atlantic and Gulf Coastal plains, particularly phytogeography; taxonomy of the genus Ophioglossum; Baptisia arachnifera. *Mailing Add:* Dept Biol Valdosta State Col Valdosta GA 31698

FAIRES, BARBARA TRADER, b Washington, DC, Apr 25, 43; m; c 1. PURE MATHEMATICS, GENERAL MATHEMATICS. *Educ:* ECarolina Univ, BS, 65; Univ SC, MS, 68; Kent State Univ, PhD(math), 74. *Prof Exp:* Instr math, Governor's Sch NC, 68; instr math, Westminster Col, 70-71; asst prof math, Carnegie-Mellon Univ, 74-76; from asst prof to asssoc prof, 76-85, PROF MATH, WESTMINISTER COL, 86-, CHAIR DEPT, 87- *Mem:* Am Math Soc; Math Asn Am; Asn Women Math; Sigma Xi; Soc Indust & Appl Math. *Res:* Theory of vector measures and applications of vector measures to Banach space theory and control theory. *Mailing Add:* VPres Acad Affairs Westminster Col New Wilmington PA 16172

FAIRES, JOHN DOUGLAS, b Sharon, Pa, Apr 27, 41; m 64, 69; c 1. MATHEMATICAL ANALYSIS. *Educ:* Youngstown State Univ, BS, 63; Univ SC, MS, 65, PhD(math), 70. *Prof Exp:* From asst prof to assoc prof, 69-79, PROF MATH, YOUNGSTOWN STATE UNIV, 79- *Mem:* Am Math Soc; Math Asn Am; Sigma Xi; Soc Indust & Appl Math. *Res:* Functional analysis; numerical analysis. *Mailing Add:* Dept Math Youngstown State Univ Youngstown OH 44503

FAIRES, WESLEY LEE, b El Dorado, Kans, Aug 28, 32; m 55; c 5. SPEECH PATHOLOGY. *Educ:* Wichita State Univ, BS, 58, MS, 62, PhD(speech path), 65. *Prof Exp:* Speech pathologist, Inst Logopedics, 58-62, preceptor, 62-65, dir clin serv, 65-70; asst prof, 65-75, ASSOC PROF LOGOPEDICS, WICHITA STATE UNIV, 75- *Mem:* Am Speech & Hearing Asn. *Res:* Auditory processing ability of cerebral palsied children. *Mailing Add:* Dept Commun Dis Wichita State Univ Wichita KS 67208

FAIRFAX, SALLY KIRK, b Bainbridge, Md, Feb 21, 44. FORESTRY. *Educ:* Hood Col, BA, 65; NY Univ, MA, 69; Duke Univ, MA(forestry), & PhD(polit sci), 74. *Prof Exp:* Asst prof natural resources, Univ Mich, Ann Arbor, 74-78; asst prof & asst economist, Agr Exp Sta, 78-80, ASSOC PROF NATURAL RESOURCE POLICY, COL NATURAL RESOURCES, UNIV CALIF, BERKELEY, 80- *Mem:* Soc Am Foresters; Am Polit Sci Asn. *Res:* Public involvement; natural resource law and environmental regulation; public land management; water law and policy range management policy. *Mailing Add:* Dept Landscape Design Univ Calif Berkeley CA 94720

FAIRHALL, ARTHUR WILLIAM, chemistry; deceased, see previous edition for last biography

FAIRHURST, C(HARLES), b Aug 5, 29; m 57; c 7. MINING ENGINEERING. *Educ:* Sheffield Univ, BEng, 52, PhD(mining), 55. *Prof Exp:* Mining engr, Northwest Div, Nat Coal Bd, Gt Brit, 55-56; res fel mineral & metall eng, 56-57, from asst prof to prof, 57-70, from assoc head dept to head dept, 65-70, head dept mining eng & rock mech, 72-87, PROF CIVIL & MINERAL ENG & E P PFLEIDER PROF MINING ENG & ROCK MECH, UNIV MINN, MINNEAPOLIS, 84- *Concurrent Pos:* Consult, tunneling & underground excavation. *Honors & Awards:* Outstanding Achievement Rock mech, Am Inst Mining, Metall & Petrol Engrs. *Mem:* Nat Acad Eng; Am Inst Mining, Metall & Petrol Engrs; Am Soc Civil Engrs; Am Underground-Space Asn (pres, 76-77, past pres); foreign mem Royal Swed Acad Eng Sci; Int Soc Rock Mech. *Res:* Rock mechanics. *Mailing Add:* Dept Civil & Mineral Eng Univ Minn 500 Pillsbury Dr Minneapolis MN 55455-0220

FAIRLEY, HENRY BARRIE FLEMING, b London, Eng, Apr 24, 27; US citizen; m 50; c 3. ANESTHESIOLOGY. *Educ:* Univ London, MB & BS, 49; FFARCS, 54; FRCP(C), 73. *Prof Exp:* Clin asst anaesthesia, Univ Toronto, 55-56, clin teacher, 56-61, assoc, 61-64, prof, 64-69; prof anesthesiol, 69-85, assoc dean, Univ Calif, San Francisco, 79-84; PROF & CHMN, ANESTHESIA, STANFORD UNIV, 85- *Mem:* Asn Univ Anesthesiologists; Can Anaesthetists Soc; Am Soc Anesthesiologists; hon mem Australian & NZ Socs Anaesthetists. *Res:* Respiratory physiology as applied to anesthesia and the management of respiratory failure. *Mailing Add:* Dept Anesthesiol Rm S278 Stanford Univ Sch Med Stanford CA 94305

FAIRLEY, JAMES LAFAYETTE, JR, b Orland, Calif, Oct 15, 20; m 48. BIOCHEMISTRY. *Educ:* San Jose State Col, AB, 42; Stanford Univ, PhD(chem), 50. *Prof Exp:* Instr meteorol, Univ Calif, Los Angeles, 43-44; instr physics & chem, San Jose State Col, 46-47; asst biochem, Stanford Univ, 47-49, res assoc, 49-51; res biochemist, Radiation Lab, Univ Calif, 51-52; from asst prof to assoc prof chem, 52-62, PROF BIOCHEM, MICH STATE UNIV, 62- *Mem:* AAAS; Am Chem Soc; Am Soc Biol Chemists; Sigma Xi. *Res:* Deoxyribonucleases; pyrimidine biosynthesis; glycoprotein structure and function. *Mailing Add:* 4363 Greenwood Dr Okemos MI 48864

FAIRLEY, WILLIAM MERLE, b Millinocket, Maine, Oct 13, 28; m 60; c 2. GEOLOGY. *Educ:* Colby Col, AB, 49; Univ Maine, MS, 51; Johns Hopkins Univ, PhD, 62. *Prof Exp:* Geologist, Ga Marble Co, 57-58; asst prof, 58-65, asst dean, Col Sci, 75-79, ASSOC PROF GEOL, UNIV NOTRE DAME, 65- *Mem:* Geol Soc Am; Sigma Xi. *Res:* Petrology and structure of metamorphic rocks, especially marble and soapstone. *Mailing Add:* Dept Earth Sci Univ Notre Dame Notre Dame IN 46556

FAIRMAN, FREDERICK WALKER, b Montreal, Que, May 29, 35; m 57; c 2. ELECTRICAL ENGINEERING, APPLIED MATHEMATICS. *Educ:* McGill Univ, BE, 59; Univ Pa, MSEE, 62, PhD(elec eng), 68. *Prof Exp:* Engr, Honeywell, Inc, 60-62; instr elec eng, Drexel Univ, 62-67; lectr, Univ Del, 67-69; assoc prof, 69-80, PROF, QUEEN'S UNIV, ONT, 80- *Mem:* Inst Elec & Electronics Engrs; Can Asn Univ Teachers. *Res:* On-line identification of dynamic system models; design of observers for estimating system states from input-output data. *Mailing Add:* Dept Elec Eng Queen's Univ Kingston ON K7L 3N6 Can

FAIRMAN, WILLIAM DUANE, b Paducah, Ky, June 7, 29; m 57; c 5. ANALYTICAL CHEMISTRY, RADIOLOGICAL HEALTH & INTERNAL DOSIMETRY. *Educ:* Marquette Univ, BS, 50, MS, 58. *Prof Exp:* From asst chemist to assoc chemist, 58-72, bioassay group leader, 74-87, CHEMIST, ARGONNE NAT LAB, 73-, INTERNAL DOSIMETRY GROUP LEADER, 87- *Concurrent Pos:* Abstractor, Chem Abstracts, 65-70; Dept Energy expert group, Internal Dosimetry, 86- *Mem:* Sigma Xi; Health Physics Soc. *Res:* Radiochemical and instrumental determination of radionuclides in humans; analytical tracer chemistry; radiation spectroscopy; computer applications in radiation measurement and in the laboratory; whole body and wound counting; internal dosimetry. *Mailing Add:* Argonne Nat Lab Bldg 200 9700 S Cass Ave Argonne IL 60439

FAIRWEATHER, GRAEME, b Dundee, Scotland, Apr 18, 42; m 65; c 4. NUMERICAL ANALYSIS. *Educ:* Univ St Andrews, BSc, 63, PhD(appl math), 65. *Prof Exp:* Lectr appl math, Univ St Andrews, 65-66; vis lectr math, Rice Univ, 66-67; lectr appl math, Univ St Andrews, 67-69; asst prof math, Rice Univ, 69-71; assoc prof, 71-81, PROF MATH, UNIV KY, 82-, ASSOC DIR, CTR COMPUTATIONAL SCI, 90- *Concurrent Pos:* Vacation assoc, UK Atomic Energy Auth, 68; vis sr res officer, Numerical Anal Div, Nat Res Inst Math Sci, SAfrica, 71, vis scientist, 76; vis prof, Univ Tulsa, 77-78; vis assoc prof, Univ Toronto, 80-81; actg dir, Ctr Computational Sci, Univ Ky, 86-87; vis prof, Sch Advan Studies, Indust & Appl Math, Valenzano Bari, Italy, 87; Fulbright scholar, Univ Valladolid, Valladolid, Spain, 88; vis prof, Univ New South Wales, Sydney, Australia, 89. *Mem:* Soc Indust & Appl Math; Can Appl Math Soc; SAfrican Math Soc; SAfrican Soc Numerical Math; Int Asn Computational Mech; fel Inst Math & Appln. *Res:* Numerical solution of partial differential equations. *Mailing Add:* Dept Math Univ Ky Lexington KY 40506

FAIRWEATHER, WILLIAM ROSS, US citizen. STATISTICS. *Educ:* Univ Calif, Berkeley, AB, 64; Cornell Univ, MS, 66; Univ Wash, PhD(statist), 73. *Prof Exp:* Syst analyst, MITRE Corp, 66-68; statistician, Nat Heart & Lung Inst, NIH, 68-70; statistician, 73-79, CHIEF, STATIST APPL & RES BR, US FOOD & DRUG ADMIN, 79- *Mem:* Sigma Xi; Am Statist Asn; Biomet Soc; Inst Math Statist. *Res:* Statistical methodology for longitudinal data; multivariate analysis; mathematical modeling; epidemiology. *Mailing Add:* Food & Drug Admin HFD-715 5600 Fishers Lane Rockville MD 20857

FAIRWEATHER-TAIT, SUSAN JANE, b King's Lynn, Eng, July 17, 49; m 86; c 1. MINERAL NUTRITION, INORGANIC STABLE ISOTOPES. *Educ:* Univ London, UK, BSc, 73, MSc, 74, PhD(nutrit), 78. *Prof Exp:* Res nutritionist, Beecham Prod Res & Develop, London, 78-79; sr sci officer, 79-86, UNIFIED GRADE & HEAD MINERAL NUTRIT, INST FOOD RES AGR & FOOD RES COUN, NORWICH, 86- *Concurrent Pos:* Mem, Com A Comt Recommended Daily Amounts Energy & Nutrients & vchmn, Minerals Panel, 87-91; mem, Task Force Calcium, Brit Nutrit Found, 87-89, Task Force Complex Carboyhdrates, 88-90; comt mem, Micronutrient Group, Brit Nutrit Soc, 88; hon lectr, Univ E Anglia, Norwich, UK, 90. *Mem:* Am Inst Nutrit; Soc Chem Indust. *Res:* Mineral nutrition, primarily calcium, iron, zinc and aluminum; bioavailability of minerals; development of methods to study mineral metabolism using stable isotopes; iron requirements in infants and children; dietary calcium and peak bone mass. *Mailing Add:* Agr & Food Res Coun Inst Food Res Colney Lane Norwich NR4 7UA England

FAISSLER, WILLIAM L, b Hammond, Ind, Nov 23, 38; m 66. COMPUTER INTERFACING, ELECTRONICS. *Educ:* Oberlin Col, BA, 61; Harvard Univ, MA, 62, PhD(physics), 67. *Prof Exp:* Res assoc, 67-70, asst prof, 70-74, assoc prof, 74-78, PROF PHYSICS, NORTHEASTERN UNIV, 78- *Concurrent Pos:* Vis prof, Max Planck Inst High Energy Physics, 77-78. *Mem:* Am Phys Soc; AAAS. *Mailing Add:* 47 Turkey Shore Rd Ipswich MA 01938

FAITH, CARL CLIFTON, b Covington, Ky, Apr 28, 27; m 51, 87; c 6. THEORY OF EQUATIONS, RING & MODULE THEORY. *Educ:* Univ Ky, BS, 51; Purdue Univ, MS, 53, PhD(math), 56. *Prof Exp:* Asst math, Purdue Univ, 51-55; from instr to asst prof, Mich State Univ, 55-57; from asst prof to assoc prof, Pa State Univ, 57-62; PROF MATH, RUTGERS UNIV, 62 - *Concurrent Pos:* NATO fel, Univ Heidelberg, 59-60; NSF fel, Inst Advan Study, 60-61, mem, 61-62; consult, Inst Defense Anal, 64; Rutgers fac fel, Univ Calif, Berkeley 65-66 & 69-70; mem screening comt int exchange of persons, Sr Fulbright Awards, 70-73; consult, US/AID, India, 68; vis prof, Tulane Univ, 70; vis mem, Inst Advan Study, 73-74 & 77-78, Israel Inst Technol, 76 & Autonoma Univ, Barcelona, 86 & 89-90. *Mem:* Am Math Soc; assoc mem Inst Advan Study; Can Math Soc; London Math Soc; Japanese Math Soc; Math Asn Am. *Res:* Galois theory; ring theory; commutativity theorems; structure of injective and projective modules; quotient rings; semiprime Noetherian and quasi-Frobenius rings; Dedekind prime rings; module theory; category theory; maximal von Neumann regular subrings of self-insective rings; various topics of commutative algebra. *Mailing Add:* 199 Longview Dr Princeton NJ 08540

FAJANS, EDGAR W, industrial chemistry; deceased, see previous edition for last biography

FAJANS, JACK, b USA, Nov 17, 22; m 44; c 2. PHYSICS. *Educ:* City Col, BChE, 44; Mass Inst Technol, PhD(physics), 50. *Prof Exp:* Sr engr, Sylvania Elec Co, 50-53; from assoc prof to prof physics, Stevens Inst Technol, 53-88, assoc dean grad studies, 74-77, dean, 77-84; CONSULT, HUYCK INDUST CONTROLS, 56- *Concurrent Pos:* Prof physics, Kabul Univ, Afghanistan, 63-69. *Mem:* Optical Soc Am; Am Asn Physics Teachers. *Res:* Low-temperature physics; superfluid helium; optical systems; instrumentation. *Mailing Add:* 1133 Magnolia Road Teaneck NJ 07666

FAJANS, STEFAN STANISLAUS, b Munich, Ger, Mar 15, 18; nat US; m 47; c 2. INTERNAL MEDICINE. *Educ:* Univ Mich, BS, 38, MD, 42; Am Bd Internal Med, dipl, 51. *Prof Exp:* Am Col Physicians res fel, 49-50; from asst prof to prof, 51-88, chief, Div Endocrin & Metab, 73-87, EMER PROF INTERNAL MED, SCH MED, UNIV MICH, 88- *Concurrent Pos:* Life Ins Med Res fel, 50-51; consult to Surgeon Gen, USPHS, 58-62 & 66-70; dir, Mich Diabetes Res & Training Ctr, 77-86; mem, Comt Planning & Orgn, Am Diabetes Asn, 77-79; guest prof, Univ Guadlajara, 79; mem, Int Sci Prog Adv Comt, XI Int Diabetes Fedn Cong, Kenya, 80-82, Regional Adv Comt, Can-USA, XII Cong, Spain, 84-85; mem, Vet Admin Med Res Serv Career Develop Comt, 87-91; mem, Nat Diabetes Adv Bd, NIH, 87-91; numerous lectrs, foreign & US. *Honors & Awards:* Banting Mem Award, Am Diabetes Asn, 72 & 78; Mary Jane Kugel Award, Juv Diabetes Found, 78; Henry Russel Lectr, Univ Mich, 83. *Mem:* Sr mem Inst Med-Nat Acad Sci; Endocrine Soc; Am Diabetes Asn (pres, 71-72); Asn Am Physicians; Am Fedn Clin Res; Am Soc Clin Invest; master Am Col Physicians; Cent Soc Clin Res; Sigma Xi; AAAS. *Res:* Endocrinology and metabolism; carbohydrate metabolism; diabetes; hypoglycemia; pituitary adrenal function; author or co-author of over 130 publications. *Mailing Add:* Univ Mich Hosp Ann Arbor MI 48109-0354

FAJER, ABRAM BENCJAN, b Piaski, Poland, Sept 12, 26; US citizen; m 56; c 3. PHYSIOLOGY, ENDOCRINOLOGY. *Educ:* Univ Sao Paulo, MD, 51. *Prof Exp:* Asst prof physiol, Sch Med, Univ Sao Paulo, 51-59, head lab exp endocrinol, 59-63; scientist, Worcester Found Exp Biol, 63-64; from asst prof to assoc prof, 64-74, PROF PHYSIOL & OBSTET-GYNEC, SCH MED, UNIV MD, BALTIMORE, 74- *Concurrent Pos:* Res fel, Inst Biol & Exp Med, Arg, 54, pharmacol lab, Univ Edinburgh, 56-58. *Mem:* Endocrinol Soc; Am Physiol Soc. *Res:* Ovarian and steroid physiology; cell toxicology. *Mailing Add:* Dept Physiol Univ Md Sch Med Baltimore MD 21201

FAJER, JACK, b Brussels, Belg, June 22, 36; US citizen; m 59; c 3. BIOPHYSICS. *Educ:* City Col New York, BS, 57; Brandeis Univ, PhD(phys chem), 63. *Prof Exp:* Res assoc, 62-64, from asst chemist to chemist, 64-76, GROUP LEADER, BROOKHAVEN NAT LAB, 76-, SR SCIENTIST, 80- *Concurrent Pos:* Adj prof, State Univ NY Stony Brook, 89; vis scholar, Stanford Univ, 79, Hebrew Univ, Jerusalem, 90. *Mem:* Am Chem Soc; Biophys Soc. *Res:* Electron spin resonance; electronic spectroscopy of metalloporphyrins and chlorophylls; photochemistry; theory and structure of porphyrins; mechanisms of photosynthetic and enzymatic reactions. *Mailing Add:* Brookhaven Nat Lab Upton NY 11973

FAKUNDING, JOHN LEONARD, b Fresno, Calif, Apr 12, 45; m 83. ENDOCRINOLOGY, CARDIOLOGY. *Educ:* State Univ Calif, Fresno, BA, 69; Univ Calif, Davis, PhD(biochem), 74. *Prof Exp:* Fel, Baylor Col Med, 73-77; staff fel, Nat Inst Child Health & Human Develop, NIH, 77-82, grants assoc, div res grants, 82-83, exec secy, div extramural affairs, Nat Heart, Lung & Blood Inst, 83-85, HEALTH SCIENTIST ADMINR, CARDIAC DIS BR, DHVD, NAT HEART, LUNG & BLOOD INST, NIH, 85- *Mem:* Endocrine Soc; Am Soc Cell Biol; AAAS; Am Heart Asn. *Res:* Molecular action of hormones; grant and contract administration; hormone receptors and second messengers. *Mailing Add:* NIH Nat Heart Lung & Blood Inst Div Heart & Vascular Dis NIH Fed Bldg Rm 4C04 Bethesda MD 20892

FAKUNDINY, ROBERT HARRY, b Manitowoc, Wis, Feb 11, 40; m. ENVIRONMENTAL GEOLOGY, REGIONAL GEOLOGY. *Educ:* Univ Calif, Riverside, BA, 62; Univ Tex, Austin, MA, 67, PhD(geol), 70. *Prof Exp:* Vol regional geol, US Peace Corps, Ghana, 63-65; sr scientist, 71-73, assoc scientist environ geol, 74-78, PRIN SCIENTIST & HEAD, ENERGY & ENVIRON GEOL SECT, NY STATE MUS-GEOL SURV & STATE GEOLOGIST & CHIEF, NY STATE GEOL SURV, 78- *Concurrent Pos:* Consult, Dow Chem Co, 67-, Los Alamos Nat Lab, 85; Hogg fel, Univ Tex, 70, assoc lectr, 67; adj asst prof environ geol, State Univ NY Albany, 75-87; ed, Northeastern Geol, 78-, Geol, 82-; chmn, NAm Comn Stratig Nomenclature, 88; dir, Northeastern Sci Found, 88; trustee, Int Basement Tectonics Asn, Inc, 91; mem, Multi Agency Group Neotectonics, Eastern Can. *Mem:* Geol Soc Am; Asn Am State Geologists; Am Asn Petrol Geologists; Am Geophys Union; Am Inst Prof Geologists; Sigma Xi; AAAS; Nat Asn Geol Teachers; NY Acad Sci; Asn Earth Sci Eds. *Res:* Geology, geohydrology, geomorphology, and radionuclide migration at radioactive waste burial ground, West Valley, NY; study of brittle and ductile structural features in western NY and Adirondack mountains; development of seismic hazard potential in the northeastern United States; sand and gravel economics. *Mailing Add:* NY State Geol Surv CEC 3136 ESP Albany NY 12230

FALB, PETER L, b New York, NY, July 26, 36; m 71; c 2. APPLIED MATHEMATICS. *Educ:* Harvard Univ, AB, 56, MA, 57, PhD(math), 61. *Prof Exp:* Mem staff, Lincoln Lab, Mass Inst Technol, 60-65; assoc prof info & control, Univ Mich, 65-67; assoc prof, 67-69, PROF APPL MATH, BROWN UNIV, 69- *Concurrent Pos:* Consult, Electronics Res Ctr, NASA, 65-70 & Bolt Beranek & Newman, Inc, 66-71; chmn & treas, Barberry Corp, 68-85; dir, Data Ledger, Inc, 70-72; vis prof, Lund Inst Technol, Sweden, 71, 72, 74, 76 & 78; prin, Dane, Falb, Stone & Co, 77- *Mem:* Am Math Soc; Inst Elec & Electronics Engrs; Soc Indust & Appl Math. *Res:* Control theory; control system design; algebraic geometry; human factors; convertible securities. *Mailing Add:* Dept Appl Math Box F Brown Univ Providence RI 02912

FALB, RICHARD D, biochemistry, biomedical engineering, for more information see previous edition

FALCHUK, KENNETH H, b New York, NY, June 17, 40. BIOPHYSICS. *Educ:* Harvard Univ, MD. *Prof Exp:* ASSOC PROF MED & DIR MED CLERKSHIP, HARVARD MED SCH. *Mailing Add:* Dept Med Harvard Med Sch Brigham & Women's Hosp 35 Huntington Ave Boston MA 02115

FALCI, KENNETH JOSEPH, b New York, NY; m 87; c 2. FOOD & COLOR ADDITIVE REGULATION, ADDITIVES GENERALLY RECOGNIZED AS SAFE. *Educ:* Marist Col, BA, 68; Fordham Univ, PhD(chem), 76. *Prof Exp:* First Lieutenant & field artillery comdr, US Army, 69-71; phosphate bus mgr & sr res chemist, Olin Corp, 76-77; SUPVR & CONSUMER SAFETY OFFICER, FOOD ADDITIVE REGULATION, FOOD & DRUG ADMIN, 77- *Concurrent Pos:* Adj prof, State Univ NY, Purchase, 74-75. *Mem:* Inst Food Technol; Am Chem Soc; Sigma Xi. *Res:* Write regulations which detail the toxicological, microbiological and chemical aspects of the additives used in the food supply. *Mailing Add:* 989 Paulsboro Dr Rockville MD 20850

FALCK, FRANK JAMES, b New York, NY, Oct 27, 25; m 50; c 1. AUDIOLOGY, SPEECH PATHOLOGY. *Educ:* Univ Ky, AB, 50, MA, 51; Pa State Univ, PhD(speech path), 55. *Prof Exp:* Asst prof & dir speech path, Sch Med, Vanderbilt Univ, 55-57; assoc prof & dir speech & hearing, Col Med, Univ Vt, 57-69; PROF SPEECH PATH & AUDIOL, UNIV HOUSTON, 69- *Concurrent Pos:* State consult, Vt Dept Health, 57- *Mem:* Am Speech & Hearing Asn; Am Psychol Asn. *Res:* Localization of auditory lesions; stuttering; audile-visile modability. *Mailing Add:* 2805 Bissonnet St Houston TX 77005

FALCO, CHARLES MAURICE, b Ft Dodge, Iowa, Aug 17, 48; m 73; c 2. PHYSICS. *Educ:* Univ Calif, Irvine, BA, 70, MA, 71, PhD(physics), 74. *Prof Exp:* asst physicist, Solid State Sci Div, Argonne Nat Lab, 74-77, physicist, 77-82; PROF, DEPT PHYSICS, OPTICAL SCI CTR & ARIZ RES LABS, UNIV ARIZ, 82-, DIR, LAB X-RAY OPTICS, 86- *Concurrent Pos:* Group leader, Tunneling & Transport Properties Group, 78-82, chmn, Surface Sci Div, 82- *Honors & Awards:* Indust Res 100 Award, 77; Technol 100 Award, 81; Alexander von Humboldt Sr Distinguished US Scientist Award, 89. *Mem:* Fel Am Phys Soc; sr mem Inst Elec & Electronics Engrs; Am Vacuum Soc; Mat Res Soc; Soc Photo-Optical Instrumentation Engrs. *Res:* Artificially layered metallic superlattices, x-ray optics; superconductivity, especially transport properties, Josephson effects, superconductive devices and high temperature superconductors; transport, electronic magnetic and phonon properties of artificially structured materials. *Mailing Add:* Dept Physics Univ Ariz Tucson AZ 85721

FALCO, JAMES WILLIAM, b Chicago, Ill, May 14, 42; m 75. CHEMICAL ENGINEERING, APPLIED MATHEMATICS. *Educ:* Univ Tenn, BS, 64; Univ Fla, MS, 69, PhD(chem eng), 71. *Prof Exp:* Exp test engr, Jet & Rocket Res, Pratt & Whitney Aircraft, 64-67; res engr, Environ Res, US Environ Protection Agency, 71-73 & US Army CEngr, 73-74; res engr, Off Environ Processes & Effects Res, US Environ Protection Agency, 74-81, dir exposure assessment, 81-85, actg dir, 85-89; MGR, EARTH ENVIRON SCI CTR, PATEL-PAC NW LAB, 89- *Mem:* Am Chem Soc; Am Inst Chem Engrs; Sigma Xi; AAAS. *Res:* Mathematical modeling of environmental systems; development protocols and models to assess toxic chemicals' impact on human health and the environment. *Mailing Add:* 1619 Butternut Ave Richland WA 99352

FALCON, CARROLL JAMES, b Rayne, La, Mar 15, 41; m 68; c 2. REPRODUCTIVE PHYSIOLOGY, ANIMAL SCIENCE. *Educ:* Univ Southwestern La, BS, 63; Univ Ky, MS, 65, PhD(genetics), 67. *Prof Exp:* Res asst animal sci, Univ Ky, 63-67; from asst prof to assoc prof animal sci, head dept agr, 71-77, PROF ANIMAL SCI, NICHOLLS STATE UNIV, 75-, DEAN COL LIFE SCI & TECHNOL, DEPT AGR, 77- *Mem:* AAAS; Am Inst Biol Sci; Am Soc Animal Sci; Am Soc Study Reproduction; Am Dairy Sci Asn. *Res:* Role of the uterus in hormone metabolism; sexual behavior of animals; growth stimulants; environmental influences on fertility; nutrition of ruminants. *Mailing Add:* Dean Technol Nicholls State Univ Thibodaux LA 70301

FALCON, LOUIS A, b Tarrytown, NY, Sept 6, 32; m 59; c 4. ENTOMOLOGY. *Educ:* Univ Calif, Berkeley, BS, 59, PhD(entom), 64. *Prof Exp:* Res asst, 59-63, asst insect pathologist, 63-70, lectr, 68-77, assoc insect pathologist, 70-77, PROF ENTOM, UNIV CALIF, BERKELEY, 77-, INSECT PATHOLOGIST, 77- *Concurrent Pos:* Consult, FAO, Nicaragua, 70-78, Univ Calif/USAID, 72-, Cent Am Res Inst Indust, 73-75 & Rome, 77-78. *Mem:* Entom Soc Am; Soc Invert Path; Am Soc Appl Sci. *Res:* Development and implementation of microbial control and integrated control, especially cotton and pome fruit; application of insect pathogens; applied ecology; systems analyses; developing countries; Latin America; prediction models. *Mailing Add:* Dept Entom Univ Calif Berkeley CA 94720

FALCONE, A B, b Bryn Mawr, Pa; c 2. ENDOCRINOLOGY, BIOCHEMISTRY. *Educ:* Temple Univ, AB, 44, MD, 47; Univ Minn, PhD(biochem), 54. *Prof Exp:* Intern, Philadelphia Gen Hosp, 47-48, resident physician, 48-49; teaching fel internal med, Univ Hosps, Univ Minn, 49-51, fel biochem, 51-54; asst prof, Inst Enzyme Res, Univ Wis-Madison, 63-66, vis prof, 66-67; CONSULT PRACTR ENDOCRINOL & METAB DIS, FRESNO, CALIF, 68- *Concurrent Pos:* Asst clin prof internal med, Univ Wis-Madison, 56-59, assoc clin prof, 59-63; mem hon staff, Valley Med Ctr, Fresno, 68-; mem staff, St Agnes Hosp, Fresno, 68-; mem staff, Fresno Community Hosp, 68-, chmn dept med, 73; sr corresp, Ettore Majorana Ctr Sci Cult, Erice, Italy, 77- *Mem:* Fel Am Col Physicians; Am Soc Biol Chem; Cent Soc Clin Res; Am Asn Study Liver Dis; Am Fedn Clin Res. *Res:* Mechanisms of adenosine triphosphate synthesis; oxidative phosphorylation; biological energy transformation mechanisms; mechanisms of drug action; membrane biochemistry; mechanism of enzyme action and use of radioactive and stable isotopes in biological systems. *Mailing Add:* Metab & Endocrine Dis 2240 E Illinois Ave Fresno CA 93701

FALCONE, DOMENICK JOSEPH, CELL BIOLOGY, PATHOLOGY. *Educ:* Cornell Univ, PhD(cell biol), 81. *Prof Exp:* ASST PROF MICROS ANAT & PATH, MED COL, CORNELL UNIV, 84- *Mailing Add:* Dept Path Cornell Univ Med Col 1300 York Ave New York NY 10021

FALCONE, JAMES SALVATORE, JR, b Bryn Mawr, Pa, Sept 17, 46; m 70; c 2. PHYSICAL & INORGANIC CHEMISTRY, TECHNOLOGY MANAGEMENT & PLANNING. *Educ:* Univ Pa, BS, 68; Univ Del, PhD(chem), 72. *Prof Exp:* Res assoc chem, Univ Fla, 72-73; res scientist, Union Camp Corp, 73-74; sr chemist, PQ Corp, 74-76, res supvr, 76-77, tech mgr, 77-87, corp planning mgr, 85-87, sr res fel & corp res prog dir, 88-89; INSTR, W CHESTER UNIV, 90- *Concurrent Pos:* Consult, 90- *Mem:* Am Chem Soc; Sigma Xi; fel Royal Soc Chem. *Res:* Properties of aqueous electrolyte solutions in general and soluble silicates in particular and the applications of these properties to industrial processes eg, detergency, waste and water treatment and oil recovery. *Mailing Add:* 222 Conestoga Rd Devon PA 19333

FALCONE, PATRICIA KUNTZ, b Mobile, Ala, Dec 28, 52; m 77; c 2. ENERGY SYSTEMS, THERMAL SCIENCES. *Educ:* Princeton Univ, BSE, 74: Stanford Univ, MS, 75, PhD(mech eng), 81. *Prof Exp:* MEM, TECH STAFF, SANDIA NAT LABS, LIVERMORE, CALIF, 80- *Concurrent Pos:* Distinguished mem tech staff, Sandia Nat Labs, 89. *Mem:* Am Soc Mech Engrs; AAAS; AIAA. *Res:* Energy conversion; high temperature gasdynamics. *Mailing Add:* Sandia Nat Labs Div 8435 PO Box 969 Livermore CA 94550

FALCONER, DAVID DUNCAN, b Moose Jaw, Saskatchewan, Can, Aug 15, 40; m 65; c 2. UNIVERSITY TEACHING, COMMUNICATIONS ENGINEERING. *Educ:* Univ Toronto, BASC, 62; Mass Instit Technol, PhDd(elec eng), 67. *Prof Exp:* Postdoctoral fel, elec eng, Roxal Instit Technol, Stockholm, 66-67; mem tec staff & supvr, Bell Labs, Holmdel, NJ, 67-80; PROF, SYSTS & COMPUT ENG, CARLETON UNIV, OTTAWA, 80- *Concurrent Pos:* Vis prof, elec eng, Linkoping Univ, Sweden, 76-77; consult, Bell-Northern Res, Ottawa, 86-87; dir, Ottawa-Carleton Ctr Comm Res, 87-90. *Mem:* Fel, Inst Elec & Electronics Engrs. *Res:* Communication theory and adaptive signal processing applied to digital communications. *Mailing Add:* Dept Systs & Comput Eng Carleton Univ Ottawa ON K1S 5B6 Can

FALCONER, ETTA ZUBER, b Tupelo, Miss, Nov 21. 33; m 55; c 3. QUASIGROUPS & LOOPS. *Educ:* Fisk Univ, BA, 53; Univ Wis, MS, 54; Emory Univ, PhD(math), 69. *Prof Exp:* Instr math, Okolona Jr Col, 54-63; teacher, Chattanooga Pub Schs, 63-64; assoc prof, Spelman Col, 65-71 & Norfolk State Col, 71-72; chmn dept math, 72-82, PROF, SPELMAN COL, 82- *Mem:* Math Asn Am; Nat Asn Mathematicians (secy, 70-72); Am Math Soc; Asn Women Math; AAAS. *Res:* Quasigroups and loops. *Mailing Add:* Dept Math Spelman Col 350 Spelman Lane Atlanta GA 30314-4399

FALCONER, JOHN LUCIEN, b Baltimore, Md, Aug 2, 46. CHEMICAL ENGINEERING. *Educ:* Johns Hopkins Univ, BES, 67; Stanford Univ, MS, 69, PhD(chem eng), 74. *Prof Exp:* Res chem engr catalysis, Exxon Res & Eng, 67; res assoc chem kinetics, Stanford Univ, 68-69; process res engr semiconductors, Fairchild Semiconductor Res & Develop, 69; fel catalysis, Stanford Res Inst, 74-75; from asst prof to assoc prof, 75-85, PROF CHEM ENG, UNIV COLO, 85- *Mem:* Am Chem Soc; Am Inst Chem Engrs; Am Vacuum Soc. *Res:* Mechanisms of catalytic reactions on supported metal catalysts; reactions of gases with well-defined surfaces using surface analysis techniques. *Mailing Add:* Dept Chem Eng Univ Colo Box 424 Boulder CO 80309-0424

FALCONER, THOMAS HUGH, b Erie, Pa, June 27, 35; m 57; c 3. MAGNET SYSTEM DESIGN, VIBRATORY EQUIPMENT DESIGN. *Educ:* Gannon Univ, BEE, 62; Pa State Univ, MSES, 74. *Prof Exp:* Elec engr, US Army Ord, 82-83 & NASA, 83-85; MGR RES & DEVELOP, ERIEZ MAGNETICS, 85- *Mem:* Inst Elec & Electronics Engrs; AAAS; Nat Soc Prof Engrs; Am Soc Mech Engrs. *Res:* Magnetic separation systems including superconducting for the beneficiation of minerals; vibratory conveying systems for the transport of bulk materials. *Mailing Add:* Res & Develop Eriez Mfg Co Asbury Rd at Airport Erie PA 16514

FALCONER, WARREN EDGAR, b Brandon, Man, Apr 13, 36; m 57; c 2. SYSTEMS ENGINEERING, COMMUNICATIONS NETWORK PLANNING. *Educ:* Univ Man, BSc, 57, MSc, 58; Univ Edinburgh, PhD, 61. *Prof Exp:* Res officer kinetics & catalysis, Nat Res Coun Can, 61-63; mem tech staff, Bell Tel Labs, Inc, 63-69, head phys chem res & develop dept, 69-73, asst chem dir, Bell Labs, 73, dir, phys chem res, 73-78, dir, Network Configuration Planning Ctr, 78-81, dir, Transmission Facil Planning, 81-86, exec dir, Transmission Systs Eng, 86-89; EXEC DIR NETWORK PLANNING, AT&T BELL LABS, 89- *Concurrent Pos:* NATO sci fel, Cath Univ Louvain, 61. *Mem:* Am Phys Soc; Inst Elec & Electronics Engrs; Cosmos Club. *Res:* Photochemistry; chemical kinetics; free and trapped radicals; combustion chemistry; noble gas chemistry; flouride chemistry; ion molecule reactions; telecommunications network planning. *Mailing Add:* At&T Bell Labs 2D-532C Crawfords Corner Rd Holmdel NJ 07733-1988

FALEK, ARTHUR, b New York, NY, Mar 23, 24; m 49; c 2. HUMAN GENETICS. *Educ:* Queen's Col, NY, BS, 48; NY Univ, MA, 49; Columbia Univ, PhD(human genetics), 57. *Prof Exp:* Res worker med genetics, NY State Psychiat Inst, 49-50, psychol asst, 50-51, res scientist, NY State Dept Ment Hyg, 51-57, sr res scientist, 58-65; asst prof, 65-67, assoc prof, 67-75, PROF PSYCHIAT, EMORY UNIV, 76- CHIEF DIV HUMAN GENETICS, GA MENT HEALTH INST, 65- *Concurrent Pos:* Asst, Columbia Univ, 50-53, assoc, 53-65; consult, Ct Dept Health, 61-; mem, Inst Study Human Variation; mem comt Huntington's chorea, WHO; genetic consult, Nat Inst Drug Abuse, 74, mem study sect, 47-; ed, Soc Biol, 77- *Mem:* AAAS; Am Soc Human Genetics; Am Eugenics Soc; fel Gerontol Soc; Soc Biol Psychiat; Sigma Xi. *Res:* Medical genetics; cytogenetics, psychogenetics and related fields. *Mailing Add:* Ga Ment Health Inst 1256 Briarcliffe Rd Atlanta GA 30306

FALER, KENNETH TURNER, b Rock Springs, Wyo, Mar 13, 31. NUCLEAR CHEMISTRY. *Educ:* Idaho State Univ, BS, 53; Univ Calif, Berkeley, PhD(chem), 59. *Prof Exp:* Jr chemist, Am Cyanamid Co & Phillips Petrol Co Chem Processing Plant, Idaho, 53-54, res chemist mat testing reactor, Atomic Energy Div, Phillips Petrol Co, 59-61, group leader nuclear chem group, 61-67; affil asst prof, 67-69, assoc dean, Col Lib Arts, 78-83, PROF CHEM, IDAHO STATE UNIV, 69- *Concurrent Pos:* Mem staff, Idaho Nuclear Corp, 67-69. *Mem:* Am Chem Soc. *Res:* Nuclear fission and decay schemes; fuel waste disposal; neutron cross sections; biological use of radioisotopes; biological fluid and electrolyte balance; geothermal site location using slow chemical kinetic systems. *Mailing Add:* 541 S Tenth Pocatello ID 83201

FALES, FRANK WECK, b Missoula, Mont, Sept 24, 14. BIOCHEMISTRY, CLINICAL CHEMISTRY. *Educ:* Ore State Univ, BS, 39; Stanford Univ, MA, 41, PhD(physiol), 51. *Prof Exp:* Res assoc physiol, Stanford Univ, 50-51; instr, 51-54, asst prof, 54-65, ASSOC PROF BIOCHEM, EMORY UNIV, 65- *Concurrent Pos:* Clin chemist, Hosp & Clin Res Ctr, Emory Univ, 51-70, NSF grant, 53-56. *Mem:* Fel AAAS; Am Soc Biol Chemists; Am Asn Clin Chem; fel Am Inst Chemists; Nat Acad Clin Biochem; Sigma Xi. *Res:* Physical and chemical characteristics of S-hemoglobin and sickle cell erythrocytes; clinical chemistry methods; cellular metabolism; relationship between structure and iodine staining of the amyloglucans. *Mailing Add:* Dept Biochem Emory Univ Atlanta GA 30322

FALES, HENRY MARSHALL, b New York, NY, Feb 12, 27; m 48; c 3. MASS SPECTROMETRY. *Educ:* Rutgers Univ, BSc, 48, PhD(org chem), 52. *Prof Exp:* Lectr org chem, Rutgers Univ, 52-53; CHIEF LAB BIOPHYS CHEM, NAT HEART, LUNG & BLOOD INST, 53- *Concurrent Pos:* Adv panel, NSF, 74-77; consult, NIH florence agreement comt, 64-69. *Mem:* Am Chem Soc; Am Soc Mass Spectrometry (secy, 77-81, pres-elect, 90-); Int Soc Chem Ecol. *Res:* Structure elucidation of natural products, especially using mass spectrometry and nuclear magnetic resonance. *Mailing Add:* NIH Bldg 10 Rm 7N318 Bethesda MD 20892

FALES, STEVEN LEWIS, b Providence, RI, Mar 14, 47; m 74. FORAGE CROP PHYSIOLOGY. *Educ:* Univ RI, BA, 70, MS, 77; Purdue Univ, PhD(agron), 80. *Prof Exp:* Res assoc, Univ RI, 77; asst prof, Dept Agron, Exp Sta, Univ Ga, 80-84; ASSOC PROF, DEPT AGRON, PA STATE UNIV, 84- *Honors & Awards:* Scarseth Award, Purdue Res Found, 79. *Mem:* Am Soc Agron; Crop Sci Soc Am; Int Grassland Cong; Sigma Xi. *Res:* Identifying environmental and physiological constraints to the production of high quality forage feedstuff for ruminant livestock; systems analysis of livestock forage systems. *Mailing Add:* Dept Agron Pa State Univ University Park PA 16802

FALES, WILLIAM HAROLD, b Redding, Calif, Dec 29, 40; m 63; c 2. VETERINARY MICROBIOLOGY. *Educ:* San Jose State Univ, BA, 64; Univ Idaho, MS, 71, PhD(bact), 74. *Prof Exp:* Microbiologist, Orange County Health Dept, Calif, 64-66; captain, US Army Med Serv Corp, 66-69; res assoc, 74-75, asst prof, 75-81, assoc prof, 81-86, PROF VET MICROBIOL & CLIN MICROBIOLOGIST VET MED, DIAG LAB, COL VET MED, UNIV MO, 86- *Honors & Awards:* Sigma Xi Res Award, Univ Idaho, 73. *Mem:* Am Soc Microbiol; Sigma Xi; AAAS; NY Acad Sci; US Animal Health Asn; Am Asn Vet Lab Diagnosticians; Am Acad Microbiol. *Res:* Clinical veterinary microbiology; anaerobic bacteriology; antimicrobial resistance. *Mailing Add:* Diag Lab Col Vet Med Univ Mo Columbia MO 65211

FALESCHINI, RICHARD JOHN, neuroscience, cytogenetics, for more information see previous edition

FALETTI, DUANE W, b Spring Valley, Ill, Apr 3, 34; m 59; c 2. CHEMICAL ENGINEERING. *Educ:* Univ Ill, BS, 56; Univ Wash, PhD(chem eng), 59. *Prof Exp:* Res engr, Boeing Airplane Co, Wash, 59-60; chem engr, Appl Physics Lab, Univ Wash, 60-62, sr chem engr, 62-74; sr res engr, Battelle Mem Inst, Wash, 74-88; PRIN ENGR, WESTINGHOUSE HAWFORD CO, 88- *Mem:* Am Inst Chem Engrs. *Res:* Power plant systems and economics; nuclear power plant operations and licensing; underwater ordnance and thermal propulsion; electrochemistry; two-phase critical flow; radioactive material transport. *Mailing Add:* 514 Doubletree Ct Richland WA 99352

FALICOV, LEOPOLDO MAXIMO, b Buenos Aires, Arg, June 24, 33; US citizen; m 59; c 2. THEORETICAL PHYSICS, SOLID STATE PHYSICS. *Educ:* Univ Buenos Aires, Lic en cie, 57; Balseiro Inst Physics, Arg, PhD(physics), 58; Cambridge Univ, PhD(physics), 60, ScD, 77. *Prof Exp:* Res assoc physics, Inst Study Metals, Univ Chicago, 60-61, from instr to prof, 61-69; Miller res prof, 79-80, chmn, 81-83, PROF PHYSICS, UNIV CALIF, BERKELEY, 69- *Concurrent Pos:* Vis staff mem, Bell Labs, 61; Sloan res fel, 64-68; vis mem, Cavendish Lab, Cambridge Univ, 66 & 76-77; vis fel, Fitzwilliam Col, 66 & Clare Hall, 76-77; Fulbright fel, Univ Antioquia, Colombia, 69 & 87; Nordita vis prof, Univ Copenhagen, 71-72 & Univ Antioquia, 87; coordr workshop, Valence Fluctuations in Solids, Inst Theoret Physics, Univ Calif, Santa Barbara, 80; lectr, NATO Sch, Mich State Univ, 81; mem adv comt, Div Mat Res, NSF, 78; mem rev comt, Solid State Sci Div, Argonne Nat Lab, 78-84, chmn, 80; mem rev panel, Solid State Physics & Chem, Exxon Labs, 79 & 81; vis prof, Univ Paris, Orsay, 84. *Honors & Awards:* Fulbright lectr, Spain, 72. *Mem:* Nat Acad Sci; Am Phys Soc; Third World Acad Sci; Royal Danish Acad Sci & Lett. *Res:* Electronic band structure of solids; superconductivity; many-body physics; theroretical chemistry. *Mailing Add:* Dept Physics Univ Calif Berkeley CA 94720

FALK, CATHERINE T, b Louisville, Ky, Aug 9, 39; m 63. POPULATION GENETICS. *Educ:* Pomona Col, BA, 61; Univ Pittsburgh, PhD(human genetics), 68. *Prof Exp:* Res fel, 68-70, res assoc, 70-81, assoc investr, 81-85, INVESTR GENETICS, NY BLOOD CTR, 86- *Concurrent Pos:* Vis investr, Rockefeller Univ, 69-71; assoc ed, Am Soc Human Genetics, 84-86. *Mem:* Genetics Soc Am; Biomet Soc; Am Soc Human Genetics. *Res:* Study of mathematical models of genetic populations; analysis and computer simulation; analysis of human genetics data to estimate linkage between known genetic markers; genetics of complex traits. *Mailing Add:* NY Blood Ctr 310 E 67th St New York NY 10021

FALK, CHARLES DAVID, b Chicago, Ill, July 18, 39; m 65, 81; c 1. PHYSICAL INORGANIC CHEMISTRY. *Educ:* Univ Chicago, BS, 61, PhD(chem), 66. *Prof Exp:* USPHS fel coord chem, Univ Sussex, 66-67; chemist, Explosives Dept, Exp Sta, E I du Pont de Nemours & Co, Inc, 67-70; mgr & adminr, Riverton Labs, 70-74; chemist-mgr, Engelhard Industs, Edison, 74-84; VPRES, RES & DEVELOP, LOPAT ENTERPRISES, WANAMASSA, NJ, 85- *Concurrent Pos:* Consult, 84- *Mem:* Am Chem Soc; Royal Soc Chem. *Res:* Kinetics and mechanisms of coordination complexes; ligand exchange reactions; hydride transfer; oxidation reactions with metal complexes; homogeneous and heterogeneous catalysis; auto exhaust catalysis; remediation, solidification and stabilization of hazardous waste. *Mailing Add:* Three Jacata Rd Marlboro NJ 07746

FALK, CHARLES EUGENE, b Hamm, Ger, Oct 20, 23; nat US; m 48; c 3. PHYSICS, SCIENCE POLICY. *Educ:* NY Univ, BA, 44, MS, 46; Carnegie Inst Technol, DSc(physics), 50. *Prof Exp:* Instr physics, Carnegie Inst Technol, 49-50; assoc physicist, Brookhaven Nat Lab, 50-53, admin scientist, 53-56, admin scientist, Div Res, US AEC, 56-58, asst to dir of lab, 58-61, from asst dir to assoc dir, 61-66; planning dir, NSF, 66-70, dir div sci resources studies, 70-85; CONSULT, 85- *Concurrent Pos:* Vis fel, Sci Policy Res Unit, Univ Sussex, 72-73; consult, Orgn Econ Develop & Coop, France, 72-, Natural Res Coun, 85- *Honors & Awards:* Distinguished Service Award, NSF, 80. *Mem:* AAAS; Am Phys Soc; Sigma Xi; NY Acad Sci. *Res:* High energy physics; particle accelerators; neutron scattering; deuteron stripping; research administration; science and technology manpower and policy; science and technology indicators. *Mailing Add:* 8116 Lilly Stone Dr Bethesda MD 20034

FALK, DARREL ROSS, b New Westminster, BC, Aug 25, 46; m 67; c 2. DEVELOPMENTAL GENETICS. *Educ:* Simon Fraser Univ, BSc, 69; Univ Alta, PhD(genetics), 73. *Prof Exp:* Fel genetics, Univ BC, 73-74; res scientist genetics, Univ Calif, Irvine, 74-76, asst prof, 76-81, ASSOC PROF BIOL, BIOL RES LAB, SYRACUSE UNIV, 81-; AT DEPT BIOL, MT VERNON NAZARENE COL. *Mem:* Genetics Soc Am. *Res:* Mutagenesis and gene organization in Drosophilo melanogaster. *Mailing Add:* Dept Biol Point Loma Nazarene Col 3900 Lomaland Dr San Diego CA 92106

FALK, EDWARD D, b Tonopah, Nev, Mar 13, 25; m 52; c 1. INSTRUMENTATION. *Educ:* Univ Calif, Berkeley, AB, 52; Ore State Col, MS, 56. *Prof Exp:* Physicist, Gen Elec Co, Wash, 52-54; sr scientist, Lockheed Aircraft Corp, Calif, 55-56; res engr, Atomics Int Div, NAm Aviation, Inc, Calif, 56-57, sr res engr, 57, eng supvr, 57-64, exec adv tech planning, 64-65, eng supvr irradiation testing, 65-68; from asst dir to exec dir, 68-86, DIR & SR SCIENTIST, INSTRUMENTATION SYSTS CTR, UNIV WIS-MADISON, 86- *Concurrent Pos:* Consult, var Far East & Mid East nations, 77- *Res:* Design and development of instruments and instrumentation systems used in interdisciplinary research. *Mailing Add:* 601 Morningstar Lane Madison WI 53704

FALK, GERTRUDE, b New York, NY, Aug 24, 25. BIOPHYSICS. *Educ:* Antioch Col, BS, 47; Univ Rochester, PhD(physiol), 52. *Prof Exp:* Fel physiol, Univ Chicago, 52, instr natural sci, 53-54; Porter fel psychiat, Med Sch, Univ Ill, 52-53; from instr to asst prof pharmacol, Univ Wash, 54-61; hon res asst, Univ Col, London, 61-72, lectr, 72-85, reader biophys, 85-89, PROF BIOPHYS, UNIV COL, LONDON, 89- *Concurrent Pos:* Nat Inst Neurol Dis & Blindness spec fel, 61-63; Guggenheim fel, 63-64; Brit Nat Comn Biophys, 87- *Mem:* Am Physiol Soc; Brit Biophys Soc; Brit Photobiol Soc; Brit Physiol Soc; Int Soc Exp Eye Res. *Res:* Excitation and contraction of muscle; visual excitation; synaptic transmission. *Mailing Add:* Dept Biophys Univ Col Univ London London WC1E 6BT England

FALK, HAROLD, b 1933. STATISTICAL MECHANICS. *Educ:* Iowa State Univ, BS, 56; Univ Ariz, MS, 57; Univ Wash, PhD(physics), 62. *Prof Exp:* Mem tech staff, Bell Tel Labs, Inc, 60; res asst theoret physics, Univ Wash, 60-62; res assoc, Univ Pittsburgh, 64-66; PROF THEORET PHYSICS, CITY COL NEW YORK, 66- *Mem:* Am Phys Soc; Math Asn Am. *Res:* Statistical mechanics of model systems. *Mailing Add:* Dept Physics City Col New York New York NY 10031

FALK, HAROLD CHARLES, b Mitchell, SDak, May 9, 34; m 56; c 2. ELECTRICAL ENGINEERING. *Educ:* SDak State Univ, BS, 56, MS, 58; Okla State Univ, PhD(elec eng), 66. *Prof Exp:* Instr elec eng, SDak State Univ, 58-59; electronics engr, Aeronaut Systs Div, Wright-Patterson AFB, 59-63; from instr to assoc prof, USAF Acad, 66-71; assoc prof, Pakistan AF Col Aeronaut Eng, 71-73; chief software br, 73-76, dep dir, Directorate Avionics Eng, Aero Systs Div, 76-80, dep elect tech, Rome Air Develop Ctr, 80-81, DIR ADVAN CONCEPTS & TECHNOL, ELEC SYSTS DIV, 81- *Mem:* Inst Elec & Electronics Engrs. *Res:* Integrated circuits systems design, systems analysis; engineering management. *Mailing Add:* Softech Inc 2940 Presidential Dr Suite 100 Fairburn OH 45324

FALK, HENRY, b New York, NY, Feb 7, 43; m 71; c 3. PEDIATRICS, PUBLIC HEALTH. *Educ:* Yeshiva Col, BA, 64; Albert Einstein Col Med, MD, 68; Harvard Sch Pub Health, MPH, 76. *Prof Exp:* Intern, Children's Hosp, Philadelphia, 68-69; resident, Bronx Munic Hosp Ctr, 69-72; MED EPIDEMIOLOGIST, CTR DIS CONTROL, 72-75 & 76- *Concurrent Pos:* Liaison mem, Comt Environ Health, Am Acad Pediat, 78- *Mem:* Am Acad Pediat; Soc Pediat Res; Am Col Epidemiol; Soc Epidemiol Res; Am Pub Health Asn. *Res:* Epidemiologic research on the etiology of cancer; environmental and occupational exposures; evaluation of vinyl chloride exposed individuals and development of hepatic tumors. *Mailing Add:* Ctr Dis Control NIH USPHS 1600 Clifton Rd NE Atlanta GA 30333

FALK, JOHN CARL, b Algonac, Mich, May 28, 38. ORGANIC CHEMISTRY. *Educ:* Kalamazoo Col, BA, 60; Univ Mich, MS, 62, PhD(org chem), 64. *Prof Exp:* Fel phys biochem, Northwestern Univ, 64-65; res chemist, Dow Chem Co, 65-67; sr res chemist, 67-70, group leader polymer synthesis, 70-74, SECT MGR, BORG WARNER RES CTR, 74- *Mem:* AAAS; Am Chem Soc; Sigma Xi. *Res:* Organic reaction mechanisms; physical biochemistry; polymer synthesis. *Mailing Add:* 3608 Russett Lane Northbrook IL 60062-4230

FALK, JOHN L, b Toronto, Ont, Dec 27, 27. PSYCHOLOGY. *Educ:* McGill Univ, BA, 50, MA, 52; Univ Ill, PhD(exp psychol), 53. *Prof Exp:* PROF PSYCHOL, DEPT PSYCHOL, RUTGERS UNIV, 69- *Concurrent Pos:* Res sci award, Nat Inst Drug Abuse, 90- *Mem:* Am Psychol Asn; Am Col Neuropsychopharmacol; Am Soc Pharmacol & Exp Therapeut; Am Phys Soc; Behav Pharmacol Soc. *Mailing Add:* Dept Psychol Rutgers Univ Psychol Bldg-Busch New Brunswick NJ 08903

FALK, LAWRENCE A, JR, b Houston, Tex, May 5, 38. VIROLOGY, IMMUNOLOGY. *Educ:* Centenary Col La, BA, 62; Univ Houston, MS, 66; Univ Ark, PhD(microbiol), 70. *Prof Exp:* Asst prof, 71-75, Rush-Presby-St Luke's Med Ctr, 71-75, assoc prof microbiol, 75-78; ASSOC PROF MICROBIOL & MOLECULAR GENETICS & CHMN, DIV MICROBIOL, NEW ENG REGIONAL PRIMATE RES CTR, HARVARD MED SCH, 78-, ASSOC PROF MICROBIOL, HARVARD SCH PUB HEALTH, 80-; FOUND/DIR HIV-AIDS RES INST, 89- *Concurrent Pos:* Sr virologist, Abbott Labs, 83-89. *Mem:* AAAS; Am Soc

Microbiol; Am Asn Immunol; Am Asn Cancer Res; Soc Exp Biol & Med. *Res:* Herpesviruses; lymphotropic viruses; oncogenic viruses of man and animals; Human Immunodeficiency Virus. *Mailing Add:* PO Box 148369 Chicago IL 60614-8369

FALK, LESLIE ALAN, b St Louis, Mo, Apr 19, 15; m 42; c 4. OCCUPATIONAL HEALTH, COMMUNITY HEALTH. *Educ:* Univ Ill, AB, 35; Oxford Univ, Eng, DPhil, 40; Johns Hopkins Univ, MD, 42. *Prof Exp:* Intern med, Johns Hopkins Hosp, 42-43; mem staff, subcomt health, US Senate, Med Corps, US Army, 43-46, captain, 43-45; med dir, UN Relief & Rehab Admin Mission to Byelorussia, 46-47; mem res staff group pract, USPHS, 46-47, med dir, Migratory Labor Health Asn, Southeast Region, Atlanta, 47-48; area med adminr, Welfare Fund, United Mine Workers Am, Pittsburgh, 48-67; prof occup & community health & chmn dept, 67-85, EMER PROF, MEHARRY MED COL, 85- *Concurrent Pos:* Fel med econ & admin, Med Admin Serv & Comt Res in Med Educ, 43-44; mem staff, Nat Study Med Group Pract, 46-48; from adj asst prof to adj prof health serv, Sch Pub Health, Univ Pittsburg, 48-82 & prev med, Vanderbilt Univ, 70-85; coordr primary medicare, health consult, Food Employees Health & Welfare Fund, Pittsburgh Pa, 54-70 & Off Econ Opportunity, 70-72; proj dir, Matthew Walker Ctr, Nashville, 67-68, co-dir, 68-70; vis prof, Univ WI, Jamaica, 77, 85 & 88. *Mem:* Am Col Prev Med; Am Asn Hist Med; Soc Occup & Environ Health; Am Pub Health Asn; Am Occup Med Asn; Asn Teachers Prev Med; Am Acad Family Physicians. *Res:* Penicillin lysozyme; antibiotics; medical care organization; international health; social history of medicine, especially black medical history. *Mailing Add:* Two Cedar St Montpelier VT 05602

FALK, LLOYD L(EOPOLD), b Ocean Grove, NJ, Nov 6, 19; m 45; c 3. ENVIRONMENTAL ENGINEERING. *Educ:* Rutgers Univ, BSc, 41, PhD(sanit), 49. *Prof Exp:* Res assoc, Rutgers Univ, 45-49; consult, Eng Dept, E I du Pont de Nemours & Co, 49-67, sr consult, 67-75, prin consult, 75-81; CONSULT, 82- *Mem:* Am Chem Soc; Water Pollution Control Fedn; Sigma Xi. *Res:* Air pollution; waste and sewage treatment; industrial climatology and meteorology; water pollution control. *Mailing Add:* 123 Bette Rd Wilmington DE 19803

FALK, MARSHALL ALLEN, b Chicago, Ill, May 23, 29; c 2. PSYCHIATRY. *Educ:* Bradley Univ, BS, 50; Univ Ill, Urbana, MS, 52; Chicago Med Sch, MD, 56; Am Bd Psychiat, dipl, 69. *Prof Exp:* Med dir, London Mem Hosp, 64-74; PROF PSYCHIAT & DEAN, EXEC UNIV HEALTH SCI, CHICAGO MED SCH, 74-, EXEC VPRES, 82- *Concurrent Pos:* Chmn coun ment health & addiction, Ill State Med Soc, 69-74; actg chmn dept psychiat, Chicago Med Sch, 73-75; Ill Hosp Licencing bd; pres, Ill Coun Deans, 81-83. *Honors & Awards:* Physician's Recognition Award Continuing Med Educ, AMA, 69 & 72. *Mem:* Fel Am Psychiat Asn; AMA; Am Asn Univ Professors; fel Am Col Psychiatrists. *Mailing Add:* Chicago Med Sch 3333 Green Bay Rd North Chicago IL 60064

FALK, MICHAEL, b Warsaw, Poland, Sept 22, 31; Can citizen; m 59; c 2. PHYSICAL CHEMISTRY. *Educ:* McGill Univ, BSc, 52; Laval Univ, DSc(chem), 58. *Prof Exp:* Res chemist, Can Copper Refiners, Montreal E, 52-54; fel spectros, Nat Res Coun Can, Ottawa, 58-60; res assoc, Mass Inst Technol, 60-62; assoc res officer, 62-68, SR RES OFFICER, ATLANTIC REGIONAL LAB, NAT RES COUN CAN, HALIFAX, 68- *Concurrent Pos:* Vis scholar phys chem, Univ Cambridge, 76-77. *Mem:* Chem Inst Can; Spectros Soc Can. *Res:* Molecular structure; infrared spectroscopy; hydration of biopolymers; hydrogen bonding; crystalline hydrates; water. *Mailing Add:* Nat Res Coun 1411 Oxford St Halifax NS B3H 3Z1 Can

FALK, RICHARD H, b Peru, Ill, Oct 12, 38; m 58; c 5. BOTANY, CYTOLOGY. *Educ:* Univ Ill, Urbana, BS, 64, MS, 65, PhD(bot), 68. *Prof Exp:* NIH fel, Harvard Univ, 68-69; from asst prof to assoc prof, 69-79, PROF BOT, UNIV CALIF, DAVIS, 79- *Mem:* AAAS; Bot Soc Am; Am Soc Plant Physiologists. *Res:* Biological ultrastructure; scanning electron microscopy; x-ray microanalysis. *Mailing Add:* Dept Bot Univ Calif Davis CA 95616

FALK, THEODORE J(OHN), b Meriden, Conn, Oct 9, 31; m 55; c 3. AERODYNAMICS, PHYSICS. *Educ:* Rensselaer Polytech Inst, BAero Eng, 53; Cornell Univ, MAeroEng, 56, PhD(aero eng), 63. *Prof Exp:* Res engr, Res Dept, United Aircraft Corp, 53-54; sr aerodyn engr, Gen Dynamics/Convair, 56-59; prin aerodynamicist, Calspan Corp, Buffalo, 63-81; SR MEM TECH STAFF, WILSON GREATBATCH LTD, CLARENCE, NY, 81- *Mem:* Am Inst Aeronaut & Astronaut; Am Soc Mech Engrs. *Res:* Shock tubes, chemical lasers and chemical transfer lasers; metallic evaporation and condensation; electromechanical devices; bioengineering. *Mailing Add:* 10880 Boyd Dr Clarence NY 14031

FALK, THOMAS, b Budapest, Hungary, Feb 19, 26; US citizen; m 60; c 2. INFORMATION THEORY, IMAGE PROCESSING. *Educ:* Graz Tech Univ, Austria, Dipl Ing, 51; Univ Buffalo, MSc, 61; Columbia Univ, DSc, 67. *Prof Exp:* Develop supvr analog electronics, Am Optical Co, Buffalo, NY, 58-60; staff consult infrared instrumentation, Barnes Eng Co, Stanford, Conn, 60-67; sr engr radar, Norden, Div United Technol, 67-78; SR CONSULT SCIENTIST, SCI APPL, INC, 78- *Mem:* Inst Elec & Electronics Engrs; Sigma Xi; Optical Soc Am. *Res:* High resolution radar imagery; data compression; transmission; change detection; electronic counter-counter measures aspects of synthetic aperture systems; radar applications in oceanography. *Mailing Add:* 4925 Calle Bendita Tucson AZ 85718

FALK, WILLIE ROBERT, b Dundurn, Sask, Mar 12, 37; Can citizen; m 60; c 3. NUCLEAR PHYSICS, INTEMEDIATE ENERGY NUCLEAR PHYSICS. *Educ:* Univ Sask, BSc, 59, MSc, 62; Univ BC, PhD(physics), 65. *Prof Exp:* Nat Res Coun Can overseas fel nuclear physics, Swiss Fed Inst Technol, 65-67; from asst prof to assoc prof, 67-78, PROF PHYSICS, UNIV MAN, 78- *Concurrent Pos:* Vis scientist, Inst Nuclear Physics, Julich, WGer, 73-74, Tri Univ Meson Facil, Univ BC, 80-81. *Mem:* Am Phys Soc; Can Asn Physicists; Sigma Xi. *Res:* Polarization phenomena and nuclear structure studies using direct reactions; pion production at intermediate energies. *Mailing Add:* Dept Physics Univ Man Winnipeg MB R3T 2N2 Can

FALKE, ERNEST VICTOR, b Brooklyn, NY, Mar 31, 42; m 72; c 2. GENETICS. *Educ:* Cornell Univ, BS, 64, MS, 65; Univ Va, PhD(genetics), 75. *Prof Exp:* Fel genetics, Albert Einstein Col Med, 75-78; res scientist, 78-80, TOXIC EFFECTS BR CHIEF, US ENVIRON PROTECTION AGENCY, 80- *Res:* Mutagenicity data; determination of genetic risk to humans due to exposure to chemicals; regulation of RNA and ribosome synthesis; governmental regulation of toxic chemicals. *Mailing Add:* US Environ Protection Agency (TS-796) 401 M St SW Washington DC 20460

FALKEHAG, S INGEMAR, b Falkenberg, Sweden, May 28, 30; m 55, 84; c 2. ORGANIC CHEMISTRY, SURFACE CHEMISTRY. *Educ:* Chalmers Univ Technol, Sweden, MS, 59, PhD(org chem), 62. *Prof Exp:* Res chemist, Res Dept, WVa Pulp & Paper Co, 62-64, group leader lignin & pulping, res ctr, Westvaco, Inc, 64-78; INDEPENDENT CONSULT, 78- *Mem:* AAAS; Tech Asn Pulp & Paper Indust; Soc Gen Systems Res; Am Chem Soc; World Acad Arts & Sci; World Future Studies Fedn; Electronic Networking Asn; Swed Soc Future Res. *Res:* Structure, reactions and technical utilization of wood chemicals; natural polymers; renewable resources; dietary fiber; nutrition; solution properties of macromolecules; research management and technology transfer; self-organizing and evolutionary processes; future studies; systems science; computer conferencing. *Mailing Add:* PO Box 953 Folly Beach SC 29439

FALKENBACH, GEORGE J(OSEPH), b Columbus, Ohio, July 24, 27; m 53; c 4. ELECTRICAL ENGINEERING. *Educ:* Univ Dayton, BEE, 48; Univ Ill, MSEE, 50. *Prof Exp:* Asst instr elec eng, Univ Ill, 48-50; electronics engr, Bell Aircraft Corp, NY, 50-55; res assoc, Lee Labs, Pa, 55; prin elec engr, Columbus Div, Battelle Mem Inst, 55-59, sr elec engr, 59-66, assoc chief, Electromagnetics Div, 66-74, prin elec engr, Phys Sci Sect, 74-76, res elec engr, Qual Assurance Sect, 76-82 & Nondestructive Testing Sect, 82-86, Electronics Technol Sect, 86-90 & Electronics Dept, Battelle Columbus Opers, 90; RETIRED. *Mem:* Inst Elec & Electronics Engrs. *Res:* Electronics; ferromagnetism; radar; microwave engineering; ultrahigh-frequency techniques; electromagnetic theory; dielectric and magnetic materials; magnetic device design. *Mailing Add:* 2671 Chester Rd N Columbus OH 43221-3307

FALKENSTEIN, GARY LEE, b Bottineau, NDak, Oct 8, 37; m 69, 81; c 2. CHEMICAL ENGINEERING, POLYMER PROCESSING. *Educ:* Mass Inst Technol, SB, 59, SM, 61, PhD(chem eng), 64. *Prof Exp:* Asst dir sch chem eng pract, Mass Inst Technol, 60-61; sr res engr, Res Dept, Rocketdyne Div, NAm Rockwell Corp, 63-66, prin scientist, 66-67, mgr advan progs, Canoga Park, 67-71; mgr, Sesame Div, Polaroid Corp, 71-77; managing dir, 77-80, vpres res & develop, 80-83, vpres mkt, 83-90, VPRES SALES & MKT, FLEXIBLE & GRAPHICS PACKAGING, AM NAT CAN CO, 90- *Mem:* Am Inst Chem Engrs. *Res:* Polymer processing; coatings; surface chemistry; adhesion; diffusion in plastics. *Mailing Add:* Am Nat Can Co 1275 King St Greenwich CT 06836-2600

FALKENSTEIN, KATHY FAY, b Frederick, Md, Feb 17, 50; m 79; c 1. PLANT PHYSIOLOGY. *Educ:* Gettysburg Col, BA, 72; WVa Univ, MS, 74; Pa State Univ, PhD(bot), 79. *Prof Exp:* Res fel & assoc lectr biol, Princeton Univ, 79-81; ASST PROF BIOL, HOOD COL, 81- *Mem:* Am Soc Plant Physiologists. *Res:* Phyto-hormone interaction and plant tissue culture; extraction of auxins from caulerpa, a coenocytic alga; crown gall induction; plasmid isolation and recombinant DNA research; genetics of phytophthora infestans. *Mailing Add:* Biol Dept Hood Col Frederick MD 21701

FALKIE, THOMAS VICTOR, b Mt Carmel, Pa, Sept 5, 34; m 57; c 5. MINING ENGINEERING, MANAGEMENT SCIENCE. *Educ:* Pa State Univ, BS, 56, MS, 58, PhD(mining eng), 61. *Prof Exp:* Res asst mining, Pa State Univ, 56-58; opers res consult, Int Minerals & Chem Corp, 61-62, oper res engr, 63-64, chief minerals planning, 64-65, asst mgr spec projs, 65-66, minerals planning & prod control mgr, 66-68, prod supt, 68-69; prof mineral eng & head dept & chmn mineral eng mgt prog, Col Earth & Mineral Sci, Pa State Univ, 69-74; dir, Bur Mines, Dept Interior, 74-77; PRES BERWIND NAT RESOURCES CO, 77- *Concurrent Pos:* Mem indust adv comt, Col Eng, Univ SFla, 65-69, adj prof, 66; consult, Fla State Bd Regents, 66-67; US del, Conf on Tunnelling, Orgn Econ Coop & Develop, 70; invited partic, Panel on Mineral Sci & Technol Educ Policy, Nat Acad Eng & US Bur Mines, 71; consult, UN, 71-73; neutral chmn, BCOA/UMWA Jt Comt Health & Safety, 73; mem, Nat Acad Sci-Nat Acad Eng Comt Mine Waste Disposal, 73-74; chmn, US Govt Interagency Task Force Coal, 74-75; mem, Bd Mineral & Energy Resources, Nat Res Coun, 81-88; dir, Foote Mineral Co, 84-88; bd dirs, Nat Coal Asn, 81-, Cyprus Mineral Co, 88-; bd gov, Am Asn Eng Socs, 87; chmn, Am Mining Cong Comt, Mining & Mineral Educ, 88-; mem, Int Dept Adv Comt mining & Mineral Res. *Honors & Awards:* Henry Krumb Lectr, Am Inst Mining, Metall & Petrol Engrs, 78; Distinguished mem, Soc Mining Engrs. *Mem:* Nat Acad Eng; Metall & Petrol Engrs (pres, 88); Soc Mining Engrs (pres, 85); Mining & Metall Soc Am. *Res:* Operations research; mine systems engineering; land reclamation and other phases of mine environmental control; industrial engineering; economic analysis; mineral resource management; surface and underground mining; national mineral and coal policy. *Mailing Add:* Berwind Corp 3000 Center Square W Philadelphia PA 19102

FALKIEWICZ, MICHAEL JOSEPH, b Brooklyn, NY, Oct 8, 42; m 66; c 2. COLLOID SCIENCE, TECHNICAL MANAGEMENT. *Educ:* City Col NY, BS, 65; Syracuse Univ, PhD(phys chem), 70. *Prof Exp:* Res chemist, Colgate-Palmolive Co, 70-75; proj mgr, Church & Dwight Co, Inc, 75-78; mgr phys chem, 78-88, RES ASSOC, FMC CORP, 88- *Concurrent Pos:* Consult surface chem, 78. *Mem:* Am Chem Soc; Royal Soc Chem; Soc Cosmetic Chemists; Am Oil Chemists Soc. *Res:* Absolute viscosities of molten metals and alloys by an oscillating closed cup method; surface chemistry of glass surface in contact with liquid phase; consumer product development; food, personal care products, rheology and colloid science, pharmaceuticals and coatings; kinetics of release; detergency; phosphate hydration. *Mailing Add:* 65 W Norton Dr Churchville PA 18966

FALKINHAM, JOSEPH OLIVER, III, b Oakland, Calif, May 3, 42; m 67; c 1. MICROBIAL GENETICS. *Educ:* Univ Calif, Berkeley, AB, 64, PhD(microbiol), 69. *Prof Exp:* Teaching asst bact, Univ Calif, Berkeley, 66-67; dir clin lab, David Grant Med Ctr, USAF, 69-71; dir lab serv, USAF Hosp, Castle AFB, Calif, 71-72; fel, Univ Ala Med Ctr, Birmingham, 72-74; asst prof, 74-80, ASSOC PROF MICROBIOL, VA POLYTECH INST & STATE UNIV, 80- *Mem:* Am Soc Microbiol; Genetics Soc Am. *Res:* Physiology, genetics and regulation of microbes, especially Esherichia coli; genetic linkage relationships in male strains of Escherichia coli; DNA replication in bacteria and its regulation; plasmid replication and inheritance; mechanism of gene transmission; structure and function of membranes; biosynthesis of alanine. *Mailing Add:* Dept Biol Va Polytech Inst Blacksburg VA 24061

FALKLER, WILLIAM ALEXANDER, JR, b York, Pa, Sept 9, 44; m 69; c 2. IMMUNOLOGY. *Educ:* Western Md Col, BA, 66; Univ Md, MS, 69, PhD(immunol), 71. *Prof Exp:* Fel & Clin instr immunol, dept trop med & med microbiol, Sch Med, Univ Hawaii, 71-73; from asst prof to assoc prof, 73-84, CHMN, DEPT MICROBIOL, DENT SCH, UNIV MD, 81-, PROF, 84- *Honors & Awards:* J Howard Brown Award, Am Soc Microbiol, 71. *Mem:* Am Soc Microbiol; Sigma Xi; Int Asn Dent Res; Am Asn Dent Schs. *Res:* Immunogenic relationships of oral anaerobic cocci; immunology and periodontal disease; micro-organisms involved in anaerobic infections; colonization mechanisms of oral micro-organisms. *Mailing Add:* Dept Microbiol Dent Sch Univ Md Baltimore MD 21201

FALKNER, FRANK TARDREW, b Hale, Eng, Oct 27, 18; nat US; m 47; c 2. PEDIATRICS. *Educ:* Cambridge Univ, BA, 45; Univ London, LRCP & MRCS, 45, MRCP, 62; FRCP, 72. *Prof Exp:* Chief res, London Hosp, 45-48; res physician, Childrens Hosp & Res Found, Cincinnati, Ohio, 48-49; res med officer, Inst Child Health, Univ London, 51-53, lectr child health, 53-56; from asst prof to prof pediat & chmn dept, Univ Louisville, 56-68; assoc dir, Nat Inst Child Health, 68-71; dir, Fels Res Inst, Yellow Springs & fels prof, Col Med, Univ Cincinnati, 71-79; prof maternal & child health, Univ MIch, Ann Arbor, 79-81; EMER PROF MATERNAL & CHILD HEALTH, UNIV CALIF, BERKELEY, 81- *Concurrent Pos:* Res scholar pediat, Univ Liverpool, 49-51; Markle scholar, 57-61; asst consult, Hosp Sick Children, France, 53; dir, Growth Study Sect, Int Ctr Children, Paris, 53, coord officer, 54; asst, Hosp Sick Children, London, 54-56; consult, Nat Inst Child Health, Nat Inst Neurol Dis & Stroke & Maternal & Child Health Servs & HEW; chmn comt, Comn Human Develop, Int Union Nutrit Sci; co-ed, Mod Probs Pediat; vis prof pediat, Col Med, Univ Cincinnati, 79-81; prof pediat, Univ Calif, San Francisco; chmn, dept admin health sci, Univ Cal, Berkeley, 83-87; ed-in-chief, Int Child Health, 90- *Mem:* Sr mem Inst Med Nat Acad Sci; Pediat Soc France; Brit Pediat Asn; Am Pediat Soc; fel Am Acad Pediat; Soc Pediat Res. *Res:* Prenatal biology; normal-abnormal growth. *Mailing Add:* Dept Mat & Child Health Univ Calif Warren Hall Berkeley CA 94720

FALKOW, STANLEY, b Albany, NY, Jan 24, 34; m 78; c 2. MICROBIOLOGY. *Educ:* Univ Maine, BS, 55; Brown Univ, MS, 59, PhD(biol), 60. *Hon Degrees:* ScD, Univ Maine, 79; MDGC, Univ Umea, Sweden, 89. *Prof Exp:* Chief, Dept Bact, Newport Hosp, Newport, RI, 56-58; dir, Med Technol Educ, Truesdale Hosp, Fall River, Mass, 56-58; res microbiologist, Dept Bact Immunol, Walter Reed Army Inst Res, Washington, DC, 61-62, asst chief, 63-66; prof lectr, Dept Microbiol, Georgetown Univ, Washington, DC, 65, from assoc prof to prof microbiol, Sch Med, 66-72; prof microbiol & med, Dept Microbiol & Immunol, Sch Med, Univ Wash, Seattle, 72-81; chmn, Dept Med Microbiol, 81-85, PROF MICROBIOL, IMMUNOL & MED, SCH MED, STANFORD UNIV, 81- *Concurrent Pos:* Soc Am Bacteriologists president's fel, 60-61; mem comn enteric infections, Armed Forces Epidemiol Bd; mem, Recombinant DNA Molecule Adv Comt, NIH, 75-76, Microbiol Immunol Adv Comt, 81- *Honors & Awards:* Paul Ehrlich-Ludwig Darmstaeder Prize, WGer, 81; Sommer mem lectr, Univ Ore Med Sch, 79; Weinstein lectr, Tuft Univ, 80; Stanhope Bayne-Jones mem lectr, John Hopkins Univ Sch Med, 82; Karl Meyer Lectr, Univ Calif, San Francisco Med Ctr, 85; Becton Dickinson Award Clin Microbiol, Am Soc Microbiol, 86; Squibb Award Infectious Dis Soc Am, 79. *Mem:* Nat Acad Sci; AAAS; fel Am Soc Microbiol; Genetics Soc Am; Am Inst Biol Sci; Infectious Dis Soc Am; Am Soc Biol Chemists; Am Acad Microbiol; Sigma Xi. *Res:* Molecular biology of bacterial plasmids; pathogenesis of enteric infections; antibiotic resistance of microorganisms; microbial genetics; molecular biology. *Mailing Add:* Dept Med Microbiol & Immunol Stanford Univ Stanford CA 94305-5402

FALKOWSKI, PAUL GORDON, b New York, NY, Jan 4, 51; m; c 1. BIOLOGY, OCEANOGRAPHY. *Educ:* City Col NY, BS, 72, MA, 73; Univ BC, PhD(biol), 75. *Prof Exp:* Res assoc oceanog, Univ RI, 75-76; SCIENTIST OCEANOG, BROOKHAVEN NAT LAB, 76- *Concurrent Pos:* Adj prof, State Univ NY, Stony Brook, 76- *Mem:* Am Soc Limnol & Oceanog; Am Soc Plant Physiol. *Res:* Marine phytoplankton physiology, biochemistry and ecology. *Mailing Add:* Three Cedar Hill Rd Stony Brook NY 11790

FALL, HARRY H, b Lucenec, Czech, Dec 8, 20; nat US; m 47; c 3. PHYSICAL ORGANIC CHEMISTRY. *Educ:* Pa State Col, BS, 42, PhD(chem), 50. *Prof Exp:* Res chemist, Sylvania Indust Corp, 42-44; sr res chemist, Upjohn Co, 51-56; sr res chemist, Mobay Chem Co, 56-57; sr res chemist, Gen Tire & Rubber Co, 57-64; sr res chemist, Goodyear Tire & Rubber Co, 64-80; consult, Omnitronics Res Corp, 81-83; dir res, 85-86, VPRES RES & DEV, HOLLAND LABS, INC, 87- *Mem:* Am Chem Soc; fel Am Inst Chemists. *Res:* Cellulose ethers; cellophane coatings; origin of petroleum; pterine chemistry; chemotherapy of experimental neoplastic diseases; kinetics of polyether glycol formation and polymerization; block and graft polymers; thermoplastic elastomers; pharmaceuticals. *Mailing Add:* 3959 Cardinal Rd Akron OH 44333

FALL, MICHAEL WILLIAM, b Port Clinton, Ohio, Dec 17, 42; m 66; c 1. PEST MANAGEMENT. *Educ:* Bowling Green State Univ, BS, 63, MA, 66; Pa State Univ, PhD(pest mgt), 78; Univ Denver, cert admin mgt, 85. *Prof Exp:* Wildlife biologist rodent control, Philipine Rodent Res Ctr, AID, 71-75; wildlife biologist animal damage control, 70-71, int progs vert pest control, 75-81, CHIEF, PREDATOR CONTROL RES, DENVER WILDLIFE RES CTR, US DEPT AGR, 81- *Concurrent Pos:* Grad res adv entom & appl zool, Univ Philippines, 72-75, vis asst prof zool, 75; affil fac, Fishery & wildlife Biol, Colo State Univ, 82-; adj prof zool, Univ Wyo, 86. *Mem:* Am Soc Mammalogists; AAAS; Wildlife Soc. *Res:* Ecology and behavior of pest vertebrates; development of vertebrate pest management methods and programs. *Mailing Add:* Denver Wildlife Res Ctr PO Box 25266 Lakewood CO 80225-0266

FALL, R RAY, b Los Angeles, Calif, Aug 15, 43. CHEMISTRY. *Educ:* Univ Calif, Los Angeles, BA, 66, PhD(biochem), 70. *Prof Exp:* NIH trainee & teaching asst, Univ Calif, Los Angeles, 67-70; NIH postdoctoral fel, Sch Med, Wash Univ, St Louis, Mo, 70-72, postdoctoral res fel, 72-73; from asst prof to assoc prof, 73-85, PROF, UNIV COLO, 85- *Concurrent Pos:* Postdoctoral fel, Nat Heart & Lung Inst, 70-72; fel, Coop Inst Res Environ Sci, 81-; vis res fel, Solar Energy Res Inst, 81-82; sr associateship, Nat Res Coun, 87-88. *Mem:* Sigma Xi; Am Soc Biol Chemists; Am Chem Soc; AAAS; Am Soc Microbiol. *Mailing Add:* Dept Chem & Biochem Univ Colo Campus Box 215 Boulder CO 80309

FALLDING, MARGARET HURLSTONE HARDY, b Sydney, Australia, July 18, 20; m 54; c 3. DEVELOPMENTAL BIOLOGY, HISTOLOGY. *Educ:* Univ Queensland, BSc, 42, MSc, 43; Cambridge Univ, PhD(zool), 49. *Prof Exp:* Res officer sheep biol, Commonwealth Sci & Indust Res Orgn, Australia, 45-50, sr res officer exp histol, 50-54, prin res officer, 54-55; res assoc anat, Col Physicians & Surgeons, Columbia Univ, 64-66; from asst prof to prof, 66-85, EMER PROF BIOMED SCI, UNIV GUELPH, 86- *Concurrent Pos:* Nat Sci Eng Res Coun grants, 67-; asst ed, In Vitro, 70-86; life mem, Clare Hall, Cambridge Univ, 71-; vis prof, Dept Zool, Univ Grenoble, France, 76-77; distinguished vis prof, dept biochem, Univ Adelaide, Australia, 85; mem, grants comt, Path & Morphol Med Res Coun, Can. *Mem:* Sigma Xi; Soc Develop Biol; Int Soc Develop Biol; emer mem Tissue Cult Asn; Can Asn Anatomists; Soc Investigative Dermat. *Res:* Histology of mammalian integument; organ culture of epithelial tissues, especially of skin and hair follicles; vitamin A and hormone effects on differentiation. *Mailing Add:* Dept Biomed Sci Univ Guelph Guelph ON N1G 2W2 Can

FALLER, ALAN JUDSON, b Boston, Mass, Mar 4, 29; m 51; c 4. METEOROLOGY, OCEANOGRAPHY. *Educ:* Mass Inst Technol, SB, 51, MS, 53, ScD(meteorol), 57. *Prof Exp:* Asst, Univ Chicago, 57-58 & Woods Hole Oceanog Inst, 58-63; res assoc prof, 63-66, RES PROF, INST PHYS SCI & TECHNOL, UNIV MD, COLLEGE PARK, 66- *Concurrent Pos:* Guggenheim fel, 60-61; consult, meteorologist. *Mem:* AAAS; fel Am Meteorol Soc; fel Am Phys Soc; Sigma Xi. *Res:* Hydrodynamic model experiments applied to the circulations of the oceans and the atmosphere. *Mailing Add:* 6705 Wells Pkwy Univ Park Hyattsville MD 20782

FALLER, JAMES E, b Mishawaka, Ind, Jan 17, 34; div; c 2. PHYSICS, ASTROPHYSICS. *Educ:* Ind Univ, AB, 55; Princeton Univ, MA, 57, PhD(physics), 63. *Hon Degrees:* MA, Wesleyan Univ, 72. *Prof Exp:* Instr physics, Princeton Univ, 59-62; Nat Res Coun fel, Joint Inst Lab Astrophys, Nat Bur Standards, 63-64, physicist, 64-66; from asst prof to prof physics, Wesleyan Univ, 66-73; FEL JOINT INST LAB ASTROPHYS & ADJOINT PROF PHYSICS & ASTROPHYS, UNIV COLO, BOULDER, 72- *Concurrent Pos:* Alfred P Sloan fel, 72-73. *Honors & Awards:* Arnold O Beckman Award, Instrument Soc Am, 70; Medal Except Sci Achievement, NASA, 73; Gold Medal, Dept Com, 90. *Mem:* AAAS; Am Phys Soc; Optical Soc Am; Sigma Xi; Am Geophys Union; Int Astron Union. *Res:* Precision experiments; optics; gravitation; fundamental constants and invariants; geophysics; experimental relativity; geodesy. *Mailing Add:* Joint Inst Lab Astrophys Univ Colo Box 440 Boulder CO 80309-0440

FALLER, JAMES GEORGE, b Wheeling, WVa, Sept 29, 34; m 59. MATERIALS ENGINEERING. *Educ:* WVa Univ, BChE, 59; Univ Del, MChE, 62, PhD, 66. *Prof Exp:* Phys metallurgist, Ballistics Lab, US Army Terminal, 64-66; sr mat engr, Vertol Div, Boeing Co, 66-70; pvt pract eng consult, 70-72; asst prof mech eng, US Naval Acad, 72-74; mat engr, David Taylor Naval Ship Res & Develop Ctr, 74-78; SR MECH ISR MAT ENGR, US ARMY COMBAT SYSTS TEST ACTIV, 78- *Concurrent Pos:* Resident indust scientist, State Pa, 69; adv, Armed Serv Comt, US House Rep, 75-78. *Mem:* Sigma Xi; Am Soc Metals; Am Defense Preparedness Asn; Int Test & Eval Asn; Am Ceramic Soc. *Res:* Nondestructive testing; failure of materials, ballistics and blast; composites, fatigue, thermal measurements and experimental stress analysis. *Mailing Add:* 317 Wilson Rd Newark DE 19711

FALLER, JOHN WILLIAM, b Louisville, Ky, Jan 7, 42; m. INORGANIC CHEMISTRY, ORGANOMETALLIC CHEMISTRY. *Educ:* Univ Louisville, BS, 63, MS, 64; Mass Inst Technol, PhD(inorg & phys chem), 67. *Prof Exp:* Asst prof, 66-71, assoc prof, 71-76, PROF CHEM, YALE UNIV, 76- *Concurrent Pos:* Petrol Res Fund grant, 67-; A P Sloan fel, 70-72; NSF grant, 72-; Guggenheim fel, 72-73. *Mem:* Am Chem Soc. *Res:* Synthesis and elucidation of structure and bonding of inorganic and organometallic compounds; stereospecific synthesis using transition metal complexes; mechanisms of catalysis; intramolecular rearrangement mechanisms of organometallics. *Mailing Add:* Dept Chem Yale Univ New Haven CT 06520

FALLETTA, CHARLES EDWARD, b Phoenix, Ariz, Feb 8, 44; m 66; c 3. PHYSICAL & INORGANIC CHEMISTRY. *Educ:* Johns Hopkins Univ, AB, 66; Univ Pittsburgh, PhD(chem), 72. *Prof Exp:* Jr res chemist catalysis, W R Grace & Co, 66-68; asst prof inorg chem, Denison Univ, 72-74; asst prof anal chem, Bates Col, 74-75; sr res chemist inorg chem, Foote Mineral Co, 75-77; sr res chemist catalysis, Johnson-Matthey, Inc, 77-; STAFF MEM, CERMET. *Res:* Precious metal catalysis. *Mailing Add:* Cermet Six Meco Wilmington DE 19804

FALLETTA, JOHN MATTHEW, b Arma, Kans, Sept 3, 40; m 63; c 2. ONCOLOGY, HEMATOLOGY. *Educ:* Univ Kans, Lawrence, AB, 62;xUni 62, MD, 66. *Prof Exp:* Intern mixed med, Kans Univ Med Ctr, 66-67; from asst instr to resident pediat, Baylor Col Med, 67-71, fel hemat oncol, 71-73, asst prof pediat, 73-76; assoc prof, 76-83, PROF PEDIAT, DUKE UNIV MED CTR, 84-, CHIEF, DIV PEDIAT HEMAT-ONCOL, 76- *Mem:* Int Soc Exp Hemat; Soc Pediat Res; Am Asn Cancer Res; Am Pediat Soc; Am Soc Clin Oncol. *Res:* Immunologic features of childhood cancer and its therapy; epidemiology of childhood cancer. *Mailing Add:* Duke Univ Med Ctr PO Box 2916 Durham NC 27710

FALLIEROS, STAVROS, b Athens, Greece, July 6, 27. THEORETICAL PHYSICS. *Educ:* Univ Md, BA, PhD(theoret physics), 59. *Prof Exp:* PROF PHYSICS, BROWN UNIV, 68- *Mem:* Fel Am Phys Soc. *Mailing Add:* Phys Dept Brown Univ Providence RI 02912

FALLIERS, CONSTANTINE J, b Athens, Greece, Dec 10, 24; nat US; m 53; c 4. ALLERGY. *Educ:* Nat Univ Athens, MD, 51; Am Bd Pediat, dipl, 58, cert pediat allergy, 65. *Prof Exp:* Intern, Evangelismos Hosp, Athens, Greece, 51; Fulbright fel basic med sci & clin pediat & resident, Med Ctr, Univ Colo, 51-53; resident pediat, Calif Babies & Children's Hosp, Los Angeles, 53-54; resident, Kaiser Found Hosp, Oakland, 54-55; mem Permanente Med Group, Walnut Creek, Calif, 55-57; dir clin serv, 59-63, med dir, children's Asthma Res Inst, 63-69; clin prof pediat, Univ Colo Med Ctr, Denver, 73-77; DIR CLIN RES, CHILDREN'S ASTHMA RES INST & HOSP, 69- *Concurrent Pos:* Fel pediat allergy, Jewish Nat Home Asthmatic Children & Children's Asthma Res Inst & Hosp, Denver, 57-59; co-investr, USPHS grants, 62-68, res support grant, 64; co-investr, Fleischman Found grant, 64-65; instr, Univ Colo Med Ctr, Denver, 61-64, asst clin prof pediat, 64-73; ed, J Asthma; Consult, Drug Eval, AMA. *Mem:* Am Acad Allergy; Am Col Allergists; Soc Biol Rhythm. *Res:* Longitudinal study of asthma; growth and development, especially in relation to allergy and immunology; metabolic and endocrine aspects of allergic disease and therapy; clinical pharmacology; cybernetics and information theory as applied to biology and medicine. *Mailing Add:* 360 S Garfield St No 670 Denver CO 80209-3136

FALLIS, ALEXANDER GRAHAM, b Toronto, Ont, Aug 20, 40; m 67; c 2. ORGANIC CHEMISTRY. *BSc:* Univ Toronto, BSc, 63, MA, 64, PhD(org chem), 67. *Prof Exp:* Nat Res Coun Can fel, Oxford Univ, 67-69; from asst prof to assoc prof, 69-78, prof org chem, Mem Univ Nfld, 78-88; PROF ORG CHEM, UNIV OTTAWA, 88- *Concurrent Pos:* Nat Res Coun Can res grants, 69- & IBM Can, 75-78; Prof Org Chem, Nfld Inst Cold Ocean Sci, 78-84. *Mem:* Am Chem Soc; Chem Inst Can; Royal Soc Chem; fel Chem Inst Can. *Res:* Structural and synthetic organic chemistry especially the total synthesis of terpenes; marine natural products and compounds of medical interest; synthetic methods particularly pericyclic free-radical and intramolecular reactions. *Mailing Add:* Dept Chem Univ Ottawa Ottawa ON K1N 6N5 Can

FALLON, ANN MARIE, b Westerly, RI, Nov 10, 49. ENTOMOLOGY, GENETICS. *Educ:* Univ Conn, BA, 72; Yale, MS, 74; Queen's Univ, Kingston, Ont, PhD(biol), 76. *Prof Exp:* Postdoctoral fel, Am Cancer Soc Fel, Univ Wis-Madison, 76-78; postdoctoral fel, NSF-Nat Needs Postdoctoral Fel, Tex A&M Univ, College Station, 78-79; postdoctoral fel, NIH Postdoctoral Fel, Univ Med & Dent, NJ, 79-81, from asst prof to assoc prof microbiol, 82-87; ASSOC PROF ENTOM, UNIV MINN, 87- *Concurrent Pos:* Chair, Invert Div, Tissue Cult Asn, 88-90; mem, Sci Meetings Comt, Tissue Cult Asn, 90- *Mem:* Tissue Cult Asn; Entom Soc Am; Am Soc Zoologists; Am Soc Microbiol; Am Mosquito Control Asn; Asn Women Sci. *Res:* Molecular genetics of mosquitoes; insect cell culture; gene transfer technology; genetic mechanisms of insecticide resistance. *Mailing Add:* Dept Entom Univ Minn 1980 Folwell Ave St Paul MN 55108

FALLON, FREDERICK WALTER, b Washington, DC. ASTRONOMY. *Educ:* Harvard Univ, AB, 61; Univ SFla, MA, 72; Univ Fla, PhD(astron), 75. *Prof Exp:* Astronr, GS 5-12, US Army Map Serv, Defense Mapping Agency, 61-69; asst astron, Univ Fla, 71-74; instr, Univ SFla, 74-77, asst prof astron, 78-80; MEM STAFF, LAB ASTRON, GODDARD SPACE FLIGHT CTR, NASA, 80- *Mem:* Am Astron Soc; Royal Astron Soc; Sigma Xi. *Res:* Astrometry and stellar kinematics; photographic and optical techniques. *Mailing Add:* 1700 Pumona Pl Bowie MD 20716

FALLON, HAROLD JOSEPH, b New York, NY, Aug 13, 31; m 55; c 4. PHARMACOLOGY. *Educ:* Yale Univ, BA, 53, MD, 57; Am Bd Internal Med, dipl, 65. *Prof Exp:* Instr internal med, Univ NC, Chapel Hill, 61-62, asst prof internal med, 64-69, prof med & pharmacol, 69-74; WILLIAM BRANCH PORTER PROF MED & CHMN, DEPT MED, MED COL VA, VA COMMONWEALTH UNIV, RICHMOND, 74- *Concurrent Pos:* Fel liver dis, Yale Univ, 62-63; NIH spec fel biochem, Duke Univ, 63-64; Sinsheimer Fund award, 65-70; res career develop award, NIH, 68; Burroughs Wellcome clin pharmacol award, 69-73; mem, Lipid Res Clins, Safety & Data Monitoring Bd, Nat Heart, Lung & Blood Inst, 72-82, chmn, 84-88, Lipid Adv Coun, 72-77, chmn, 76-77; mem, Comt Gastrointestinal Drugs, Food & Drug Admin, 75-78; comnr, Nat Digestive Dis Comn, 77-79; mem, Med Test Comt, Nat Bd Med Examiners, 78-81; mem adv coun, Nat Inst Arthritis, Diabetes, Digestive & Kidney Dis, 82-86; chmn elect, Am Bd Internal Med, 85-86, chmn, 86-87; Inglefinger prof, Boston Univ, Mass, 87; Strauss prof, Baylor Univ Hosp, Dallas, Tex, 88; vis prof, Univ Ill, Chicago, 89, Univ Ark, 89, Univ Brisbane, Australia, 90. *Honors & Awards:* Vilter Lectr, Univ Cincinnati, 87; Beeson Lectr, Yale Univ, 88. *Mem:* Inst Med-Nat Acad Sci; master Am Col Physicians; Asn Am Physicians (pres, 90-); Am Soc Clin Invest (vpres, 72-75); Am Asn Study Liver Dis (pres, 79); Am Clin & Climat Asn; Am Fed Clin Res; AAAS; Am Soc Pharmacol & Exp Therapeut; AMA. *Res:* Liver disease; Pyrimidine metabolism in man; regulation of metabolic pathways; lipid biosynthesis; serine metabolism; gastroenterology. *Mailing Add:* Dept Med Med Col Va Box 500 MCV Sta Richmond VA 23298-0500

FALLON, JAMES HARRY, b Poughkeepsie, NY, Oct 18, 47; m 69; c 3. NEUROSCIENCE, NEUROLOGY. *Educ:* St Michael's Col, Vt, BS, 69; Rennselaer Polytech Inst, NY, MS, 72; Univ Ill, PhD(neuroanat & neurophysiol), 75. *Prof Exp:* Teaching fel neurosci, San Diego, 78, PROF ANAT & NEUROBIOL, UNIV CALIF, IRVINE, 78- *Concurrent Pos:* Prin investr, NIH, 78- *Honors & Awards:* Sloan Scholar, Sloan Found, 77-78; Kaiser-Permanente Found Award, 79 & 86; NIH Career Develop Award, 80-86. *Mem:* Soc Neurosci; Europ Neurosci Asn; Am Asn Anatomists; NY Acad Sci; Int Basal Ganglia Soc; Raoul Garcia Y Vega Soc. *Res:* Neuroscience and physiology; neurotransmitters, limbic-basal ganglia brain systems, growth factors, endorphins, development, circadian rhythms, sensory-motor systems in mammalian and human brains. *Mailing Add:* Dept Anat Univ Calif Irving Col Med Irvine CA 92717

FALLON, JOHN FRANCIS, LIMB DEVELOPMENT, PATTERN FORMATION. *Educ:* Marquette Univ, PhD(biol), 66. *Prof Exp:* PROF ANAT, UNIV WIS-MADISON, 69-, ASST DEAN GRAD STUDIES, 89- *Mailing Add:* Dept Anat Univ Wis 351 Bardeen Med Lab Madison WI 53706

FALLON, JOHN T, b Providence, RI, Dec 27, 46. CARDIAC DISEASES. *Educ:* Albany Med Col, MD, 74. *Prof Exp:* Asst prof path, Sch Med, Harvard Univ, 76-82; ASSOC PROF PATH, MASS GEN HOSP, BOSTON, 83- *Mailing Add:* Dept Path Mass Gen Hosp Fruit St Boston MA 02114

FALLON, JOSEPH GREENLEAF, b Los Angeles, Calif, Oct 2, 11; m 34; c 3. PUBLIC HEALTH. *Educ:* Pac Union Col, BA, 38; Mass Inst Technol, MPH, 44. *Prof Exp:* Asst prof biol, 38-42, assoc prof biol & health, 44-45, assoc prof biol, nursing & health, 47-55, assoc prof biol, 55-57, assoc prof biol & health, 57-65, prof biol, 65-76, EMER PROF BIOL, PAC UNION COL, 76- *Concurrent Pos:* Fel, Int Union for Health Educ of Public, Int Comn UNESCO; instr, Sch Med, Loma Linda Univ; mem, UN Relief & Rehab Admin, China, 46; lay health officer, Pac Union Col, 47-55; vis prof, Pac Union Col Exten, Honolulu, 50. *Mem:* AAAS; Am Soc Microbiologists; Am Soc Parasitologists; fel Am Pub Health Asn; fel Royal Soc Health; fel Am Soc Pub Health Educrs; fel Am Sch Health Asn; fel Royal Trop Med. *Res:* Pollution studies on rivers; health surveys. *Mailing Add:* PO Box 203 Angwin CA 94508

FALLON, MICHAEL DAVID, METABOLIC BONE DISEASE, BONE TUMOR. *Educ:* St Louis Univ, MD, 77. *Prof Exp:* ASSOC PROF PATH, SCH MED, UNIV PA HOSP, 83- *Res:* Bone metabolism. *Mailing Add:* Dept Surg Path Thomas Jefferson Univ Hosp 1115 11th St Rm 7608 NHB Philadelphia PA 19107

FALLS, JAMES BRUCE, b Toronto, Ont, Dec 18, 23; m 52; c 3. ANIMAL BEHAVIOR, ECOLOGY. *Educ:* Univ Toronto, BA, 48, PhD(zool), 53. *Prof Exp:* Lectr zool, Univ Toronto, 52-53; Nat Res Coun Can Overseas fel, 53-54; mgr & adminr, Riverton Labs, 70-74; chemist-mgr, Engelhard Industs, Edison, 74-84; EMER PROF DEPT ZOOL, UNIV TORONTO, 84- *Concurrent Pos:* Royal Soc & Nuffield Found Commonwealth bursary, 64; vis fel, Wolfson Col, Oxford, 81. *Mem:* Am Ornith Union; Am Soc Mammal; Ecol Soc Am; Animal Behav Soc; Can Soc Zoologists. *Res:* Behavior contributing to population regulation, dispersion and resource use by animals; bioacoustics (structure and function of signals, individual recognition, repertoires) and territoriality in birds; populations, behavior and activity of mammals. *Mailing Add:* Dept Zool Univ Toronto George Campus Harbord St Toronto ON M5S 1A1 Can

FALLS, WILLIAM MCKENZIE, b Muncie, Ind, May 13, 48; m 89; c 1. NEUROANATOMY, NEUROCYTOLOGY. *Educ:* Hanover Col, BA, 70; Ohio State Univ, MS, 73, PhD(anat), 75. *Prof Exp:* Grad teaching assoc anat, Ohio State Univ, 71-75; staff fel neuroanat, Nat Inst Dent Res, 75-78, sr staff fel, 78-79; asst prof, 79-84, assoc prof anat, 84-90, PROF ANAT, MICH STATE UNIV, 90- *Mem:* Soc Neurosci; Am Asn Anatomists; AAAS. *Res:* Golgi technique, Horseradish Peroxidase technique and electron microscopic analysis of sensory trigeminal nuclei; elucidation of basic pain and tactile mechanisms and understanding of chronic pain states which affect the face and oral cavity. *Mailing Add:* Dept Anat Mich State Univ East Lansing MI 48824

FALLS, WILLIAM RANDOLPH, SR, b Ironton, Ohio, Sept 15, 29; m 56; c 3. SCIENCE EDUCATION. *Educ:* Rio Grande Col, BS, 53; Marshall Univ, MA, 59; Ind Univ, Bloomington, EdD(sci educ), 70. *Prof Exp:* Assoc prof, 61-75, PROF SCI, MOREHEAD STATE UNIV, 75-, CHAIR, DEPT PHYS SCI, 61-; Assoc prof, Morehead State Univ, 61-75, prof sci, 75-, chair, Dept Phys Sci, 61-; RETIRED. *Concurrent Pos:* AEC res grant & resident physicist, 65- *Mem:* AAAS; Nat Sci Teachers Asn; Am Asn Physics Teachers. *Res:* Trace elements present in post oak trees as revealed through activation analysis; intense gamma field's effects on metals. *Mailing Add:* 201 W Sycamore Ave Foley AL 36535

FALOON, WILLIAM WASSELL, b Pittsburgh, Pa, July 6, 20; m 48; c 3. GASTROENTEROLOGY. *Educ:* Allegheny Col, AB, 41; Harvard Univ, MD, 44; Am Bd Internal Med, dipl. *Prof Exp:* Intern, Pa Hosp, Philadelphia, 44-45; asst resident med, Albany Hosp, NY, 45-46, resident, 46-47; res fel, Thorndike Mem Lab, Harvard Med Sch & Boston City Hosp, 47-48; asst prof oncol & instr med, Albany Med Col, 48-50; from instr to prof med, State Univ NY Upstate Med Ctr, 50-68; physician-in-chief, Cottage & Gen Hosp, Santa Barbara, Calif, 68-69; chief med, 70-80, dir gastroenterol & nutrit, Highland Hosp, NY, 70-86; PROF SCH MED & DENT, UNIV ROCHESTER, 69- *Concurrent Pos:* Consult, Adv Comt, Surgeon Gen; vis prof, Cleveland Clin Found, 79, Univ Kans, 80 & Pensacola, Fla, 81; vis lectr, numerous med schs. *Mem:* Fel Am Col Physicians; Am Gastroenterol Asn; Am Inst Nutrit; Am Fedn Clin Res; Am Asn Study Liver Dis. *Res:* Nutrition; gastrointestinal and metabolic disease. *Mailing Add:* Dept Med Univ Rochester Sch Med Genesee Hosp 224 Alexander St Rochester NY 14607

FALSETTI, HERMAN LEO, b Niagara Falls, NY, Oct 8, 34; c 5. SPORTS MEDICINE. *Educ:* Univ Rochester, BA, 57, MD, 60. *Prof Exp:* Chief med serv, Edwards AFB Hosp, USAF, 64-66; res instr med, State Univ NY, Buffalo & Buffalo Gen Hosp, 66-68, asst res prof, 68-69; asst prof med, State Univ NY, Buffalo & E J Meyer Mem Hosp, 69-73; assoc prof, 73-75, PROF MED, UNIV IOWA HOSPS & CLINS, 75-; PRES, HEALTH CORP, 85- *Concurrent Pos:* Adj prof eng, 77- *Mem:* Am Heart Asn; Fedn Clin Res; Am Col Physicians; Am Col Cardiol; Am Physiol Soc; Am Col Sports Med. *Res:* Cardiac mechanics; blood velocity-flow relationship; coronary artery disease; computer applications in cardiology. *Mailing Add:* Health Corp Four Jenner Suite 110 Irvine CA 92718

FALTER, JOHN MAX, b West Bend, Wis, Dec 13, 30; m 59; c 3. ENTOMOLOGY. *Educ:* Univ Wis, BS, 52, MS, 59, PhD(entom), 64. *Prof Exp:* Instr entom, Univ Wis, 61-64; asst prof, 64-67, ASSOC PROF ENTOM, NC STATE UNIV, 67- *Mem:* Entom Soc Am. *Res:* Economic entomology; biology and control of economic pests. *Mailing Add:* State Rd 1010 Apex NC 27502

FALTINGS, GERD, b Gelsenkirchen-Buer, WGer, July 28, 54. ALGEBRAIC GEOMETRY, DIOPHANTINE PROBLEMS. *Educ:* Univ Munster, dipl & PhD(math), 78, Dr Hab, 81. *Prof Exp:* Asst math, Univ Munster, 79-82; prof, Univ Wuppertal, 82-84; PROF MATH, PRINCETON UNIV, 85- *Concurrent Pos:* Co-ed, Math Z, 84-; asst ed, J Am Math Soc, 88-; Guggenheim fel, 88-89. *Honors & Awards:* Danny Heineman Prize, Acad Sci, Gottingen, 83; Fields-Medal, Int Math Union, 86. *Mem:* German Math Union. *Res:* Commutative algebra; algebraic geometry; arithmetic. *Mailing Add:* Dept Math Princeton Univ Washington Rd Princeton NJ 08544

FALTYNEK, ROBERT ALLEN, b Chicago, Ill, Sept 1, 48; m 70. INORGANIC & ORGANOMETALLIC CHEMISTRY. *Educ:* Augustana Col, AB, 70; Univ Minn, Minneapolis, PhD(inorg chem), 76. *Prof Exp:* Instr chem, Coe Col, Iowa, 75-76; fel inorg chem, Mass Inst Technol, 76-78; staff chemist, Corp Res & Develop, Gen Elec, 78-84; AT NAT BUR STANDARDS, 85- *Concurrent Pos:* Assist prof, Am Univ, 84-85. *Mem:* Am Chem Soc; AAAS; NY Acad Sci. *Res:* Synthetic and structural inorganic and organometallic chemistry; transition metal photochemistry; silicon, silicone chemistry. *Mailing Add:* Philadelphia Col Pharm Sci 4300 Woodland Ave Philadelphia PA 19104

FAMA, DONALD FRANCIS, b Ilion, NY, Oct 12, 38; m 64; c 2. THREE DIMENSIONAL TRANSPARENCIES. *Educ:* Syracuse Univ, BA, 61, MS, 65; Univ Ill, MET, 69; State Univ NY, Brockport, MAdm, 74; Clarkson Univ, IFRICS, 85. *Prof Exp:* Instr math, Manlius Mil Inst, 61-62; chmn dept math-eng sci-drafting & design, 71-88, PROF MATH COMPUTER SCI, CAYUGA COMMUNITY COL, 65- *Concurrent Pos:* At Sperry Rand Univac, 61. *Mem:* Nat Educ Asn. *Res:* Matrix polynomials and matrix series; computer graphics; C language; local area networks. *Mailing Add:* 75 Steel St Auburn NY 13021

FAMBROUGH, DOUGLAS MCINTOSH, b Durham, NC, July 22, 41; c 2. NEUROBIOLOGY. *Educ:* Univ NC, AB, 63; Calif Inst Technol, PhD(biochem), 68. *Prof Exp:* Staff mem embryol, Carnegie Inst Wash, 68-85; from asst prof to assoc prof biol & biophys, 70-78, PROF BIOL & BIOPHYS, JOHNS HOPKINS UNIV, 78- *Concurrent Pos:* Sci Adv Bd, Muscular Dystrophy Asn, 82-88. *Honors & Awards:* Javits Neurosci Investr Award. *Mem:* Soc Develop Biol; Soc Neurosci; Soc Gen Physiologists (pres, 88-89); Am Soc Cell Biol. *Res:* Nerve-muscle interactions; physical and metabolic aspects of cell membranes. *Mailing Add:* Dept Biol Johns Hopkins Univ Charles & 34th St Baltimore MD 21218

FAMIGHETTI, LOUISE ORTO, b New York, NY, Sept 25, 30; m 65. LABORATORY MEDICINE. *Educ:* Col New Rochelle, BA, 52; Columbia Univ, MS, 63. *Prof Exp:* Res assoc human nutrit biochem, Nutrit Res Lab, St Luke's Convalescent Hosp, Greenwich, Conn, 52-60; asst dir res, Geriat Nutrit Lab, Osborn Mem Home, 52-74; asst dir res, Nutrit & Metab Res Div, 60-74, DIR CLIN LAB, BURKE REHAB CTR, 74- *Concurrent Pos:* Managing ed, Nutrit Reports Int, 69-74, assoc ed, 75-89; cert, Nat Registry Clin Chem & Nat Cert Agency for Med Lab Personnel. *Mem:* Fel Am Inst Chem; Am Inst Nutrit; Clin Lab Mgt Asn; Am Asn Clin Chem. *Res:* Proteins; amino acids; nutritional requirements. *Mailing Add:* Clin Lab Burke Rehab Ctr White Plains NY 10605

FAMIGLIETTI, EDWARD VIRGIL, JR, b Providence, RI, June 2, 43. NEUROBIOLOGY. *Educ:* Yale Univ, BA, 65; Boston Univ, MD, 72, PhD(neuroanat), 72. *Prof Exp:* Commissioned officer, USPHS, NIH, 72-74, fel, Nat Eye Inst, 74-76; fel, Keio Univ Sch Med, Tokyo, 77 & Wash Univ Sch Med, 77-79; ASST PROF, DEPT ANAT, WAYNE STATE UNIV SCH MED, 79- *Mem:* Asn Res Vision & Ophthal; Am Asn Anatomists; NY Acad Sci; Soc Neurosci. *Mailing Add:* Dept Anat Fac Med Univ Calgary Calgary AB T2N 4N1 Can

FAMILY, FEREYDOON, b Tehran, Iran, Sept 18, 45; m; c 2. CONDENSED MATTER PHYSICS, THEORETICAL SOLID STATE PHYSICS. *Educ:* Worcester Polytech Inst, BS, 68; Tufts Univ, MS, 70; Clark Univ, PhD(physics), 74. *Prof Exp:* Asst prof physics, Cent New Eng Col, 70-74; res assoc, Mass Inst Technol, 74-75; head, div solid state physics, Nuclear Res Ctr, Atomic Energy Orgn Iran, 75-79; res scientist, Boston Univ, 79-81; from asst prof to prof, 81-90, SAMUEL CANDLER DOBBS PROF CONDENSED MATTER PHYSICS, EMORY UNIV, 90- *Concurrent Pos:* Vis asst prof, Worcester Polytech Inst, 80-81; vis res physicist, Inst Theoret Physics, Univ Calif, 82-83; vis assoc prof chem, Mass Inst Technol, 85-86. *Honors & Awards:* Lawton-Plimpton Prize, 68. *Mem:* Fel Am Phys Soc; Am Asn Physics Teachers; Mat Res Soc; Sigma Xi. *Res:* Theory of condensed matter; random materials and disorderly growth phenomena, fractals, pattern formation. *Mailing Add:* Dept Physics Emory Univ Atlanta GA 30322

FAMULARO, KENDALL FERRIS, nuclear physics; deceased, see previous edition for last biography

FAN, CHANG-YUN, b Nantong, Jiangsu, China, Jan 7, 18; m 50; c 3. HIGH ENERGY ASTROPHYSICS, ATOMIC MOLECULAR PHYSICS. *Educ:* Nanking Univ, BA, 42; Univ Chicago, MS, 50, PhD(physics), 52. *Prof Exp:* Res assoc astrophysics, Univ Chicago, 52-57; asst prof physics, Univ Ark, 57-58; sr physicist space physics, Lab Appl Sci, Univ Chicago, 58-67; PROF PHYSICS, UNIV ARIZ, 67- *Mem:* Am Phys Soc; Am Geophys Union. *Res:* Measurements of the compositions and energies of charged particles in space for the understanding of plasma processes which energize these particles; determination of the molecular composition of interplanetary dust particles; searching for the origin of life. *Mailing Add:* 1760 N Potter Pl Tucson AZ 85719

FAN, CHIEN, b Kiang-Su, China, Apr 1 30; m 58; c 3. ENGINEERING SCIENCE, MECHANICAL ENGINEERING. *Educ:* Univ Taiwan, BS, 54; Univ Ill, MS, 58, PhD(mech eng), 64. *Prof Exp:* Asst prof eng sci, Fla State Univ, 61-65; res specialist, 65-78, staff engr, 78-87, GROUP ENGR, LOCKHEED MISSILES & SPACE CO, 87- *Concurrent Pos:* Assoc prof, Univ Ala, Huntsville, 67-69; vis scientist, Repub China, 70. *Mem:* NY Acad Sci; Am Inst Aeronaut & Astronaut; Am Soc Mech Engrs. *Res:* Fluid mechanics; heat transfer; reliability engineering; spacecraft thermodynamics. *Mailing Add:* Lockheed Missiles & Space Co PO Box 504 Sunnyvale CA 94086

FAN, DAH-NIEN, b Hubei, China, Sept 3, 37; c 1. FLUID MECHANICS, AEROSPACE ENGINEERING. *Educ:* Nat Taiwan Univ, BS, ME, 58; Cornell Univ, MAeroE, 63, PhD(aerospace eng), 66. *Prof Exp:* Res assoc, Cornell Univ, 66-68 & 68-69; from asst prof to assoc prof, 69-77, PROF MECH ENG, HOWARD UNIV, 77- *Concurrent Pos:* Eng consult, 75-; consult, Resource Recovery Servs Inc, 75-77 & QuesTech Inc, 90-; pres, WHF & Assocs, Inc, 77-85; expert consult, US Naval Res Lab, 85-90. *Mem:* AAAS. *Res:* Aerodynamics of sounding rockets; tether dynamics in aerospace systems; velocity bias in laser volocimetry; resource recovery from municipal solid waste; magneto-fluid dynamics. *Mailing Add:* Dept Mech Eng Howard Univ Washington DC 20059

FAN, DAVID P, b Hong Kong, Jan 18, 42; US citizen; m 69. BIOLOGY. *Educ:* Purdue Univ, BS, 61; Mass Inst Technol, PhD(biol), 65. *Prof Exp:* Fel, Med Res Coun Lab Molecular Biol, Eng, 65-67, fel, Univ Geneva, 67-69; asst prof, 69-73, assoc prof, 73-77, PROF GENETICS & CELL BIOL, UNIV MINN, ST PAUL, 77- *Mem:* Am Asn Pub Opinion Res. *Res:* Effect of the mass media on public opinion and behavior; mathematical modeling of AIDS; computer content analysis of text. *Mailing Add:* Dept Genetics & Cell Biol Univ Minn 1445 Gorther Ave St Paul MN 55108

FAN, HSIN YA, b Kiangsi, China. APPLIED MATHEMATICS. *Educ:* Ordnance Eng Col, China, BS, 56; Chenkung Univ, China, MS, 61; Univ Calif, Los Angeles, PhD(math), 70. *Prof Exp:* Instr mech eng, Ordnance Eng Col, 56-59; physicist, US Army Cold Regions Res & Eng Lab, 63; assoc prof, 69-77, PROF MATH, CALIF STATE POLYTECH UNIV, 77- *Honors & Awards:* Jeme Tien Yow Gold Medal, Chinese Inst Engrs, 61. *Mem:* Soc Indust & Appl Math. *Res:* Design and production of instructional models for mathematics. *Mailing Add:* Dept Math Calif State Polytech Univ 3501 W Temple Ave Pomona CA 91768

FAN, HSING YUN, b Changsha, China, Apr 27, 14; m 47; c 1. AGRICULTURAL BIOCHEMISTRY. *Educ:* Nat Tsing Hua Univ, BS, 35; Univ Minn, PhD(agr biochem), 45. *Prof Exp:* Asst, Nat Tsing Hua Univ, 35-40; China Found res fel insect physiol, Univ Minn, 45-46, res assoc, 48; res chemist, Julius Hyman & Co, 48-52; SR RES CHEMIST, SHELL DEVELOP CO, 53- *Mem:* Am Chem Soc. *Res:* Organic analysis; vitamins; pesticides; permeability of insect cuticle. *Mailing Add:* 830 Claribel Rd Modesto CA 95356-9613

FAN, HSU YUN, b Shanghai, China, July 15, 12; nat US; m 41; c 4. SOLID STATE PHYSICS. *Educ:* Mass Inst Technol, MS, 34, ScD, 37. *Hon Degrees:* ScD, Purdue Univ, 90. *Prof Exp:* From assoc prof to prof, Nat Tsing Hua Univ, China, 37-47; mem staff, Electron Res Lab, Mass Inst Technol, 48; vis prof, 48-49, from assoc prof to prof, 49-63, Duncan distinguished prof, 63-78, EMER PROF PHYSICS, PURDUE UNIV, 78- *Concurrent Pos:* Past mem var comts & panels, Nat Acad Sci-NSF; Past cor mem, Semiconductor Comn, Int Union Pure & Appl Physics; Past mem, Solid State Sci Panel, Nat Res Coun. *Mem:* Am Phys Soc. *Res:* Semiconductors. *Mailing Add:* Dept Physics Purdue Univ Lafayette IN 47907

FAN, JOHN C C, b Shanghai, China, Dec 5, 43; US citizen. ENERGY TECHNOLOGY. *Educ:* Univ Calif, Berkeley, BS, 66; Harvard Univ, MS, 67, PhD(applied physics), 72. *Prof Exp:* Tech staff, Mass Inst Technol, asst group leader, 80-82, assoc group leader electron mat, Lincoln Lab, 82-85; CHMN KOPIN CORP, 85- *Mem:* AAAS; Sigma Xi; Electrochem Soc; Mat Res Soc. *Res:* Thin films; electronic materials; solar selective coatings; solar cells; materials for integrated circuits. *Mailing Add:* Kopin Corp 695 Myles Standish Blvd Taunton MA 02780

FAN, JOYCE WANG, b China, Oct 2, 19; nat US; m 43; c 2. ORGANIC CHEMISTRY. *Educ:* Wheaton Col, BS, 42; Univ Iowa, MS, 44, PhD(chem), 46. *Prof Exp:* Fel, Northwestern Univ, 46-47; lectr chem, Univ Southern Calif, 47-48; from asst prof to assoc prof, Univ Houston, 49-63; prof chem & chmn Premed Adv Comt, Houston Baptist Univ, 63-83, head dept, 67-83; RETIRED. *Mem:* Am Chem Soc; fel Am Inst Chemists. *Res:* Polarography as applied to organic compounds. *Mailing Add:* 4323 Nenana Houston TX 77035

FAN, KY, b Hangchow, China, Sept 19, 14; nat US; m 36. MATHEMATICS. *Educ:* Univ Peking, BS, 36; Univ Paris, DSc(math), 41. *Prof Exp:* Fr Nat Sci fel, Univ Paris & Inst Henri Poincare, 42-45; mem, Inst Advan Study, 45-47; from asst prof to prof math, Univ Notre Dame, 47-60; prof, Wayne State Univ, 60-61; prof, Northwestern Univ, 61-65; chmn dept, 68-69, prof, 65-85 EMER PROF MATH, UNIV CALIF, SANTA BARBARA, 85- *Concurrent*

Pos: Assoc ed, J Math Anal & Appln, 60- & Linear Algebra & Its Appln, 68-; dir, Inst Math, Acad Sinica, Taiwan, 78-84; vis prof, Univ Tex, Austin, 65, Hamburg Univ, Ger, 72, Univ Paris IX, Dauphine, France, 81 & Univ Perugia, Italy, 85 & 87; hon prof, Peking Univ & Peking Normal Univ, China, 89- *Mem:* Am Math Soc; Math Asn Am; Acad Sinica. *Res:* Functional analysis. *Mailing Add:* Dept Math Univ Calif Santa Barbara CA 93106

FAN, LIANG-SHIH, b Fu-Wei, Taiwan, Dec 15, 47; m 78; c 2. REACTION ENGINEERING, FLUIDIZATION. *Educ:* Nat Taiwan Univ, BS, 70; WVa Univ, MS, 73, PhD(chem eng), 75; Kans State Univ, MS, 78. *Prof Exp:* Res asst chem eng, WVa Univ, 71-75, res assoc, 75; vis asst prof, Kans State Univ, 76-78, adv & judge, 76-77, coord grad res, 76-78; from asst prof to assoc prof, 78-85, PROF CHEM ENG, OHIO STATE UNIV, 85- *Concurrent Pos:* Res eng, Amoco Res Center, 79; Sigma Xi grant, 79; prin investr, Off Water Res & Technol, Dept Interior, 79-82, NSF, 79-82, Battelle Columbus Lab, 80-82; res assoc Argonne Nat Lab, 80; co-investr, Nat Sci Found, 81-88; consult, Argonne Nat Lab Chem Eng Div, 80-81, Battelle Mem Inst, 81-; Fulbright scholar, UK, 90; sr vis scientist award, Japan Soc Prom Sci, 90. *Honors & Awards:* Res Award, Battelle Mem Inst, 80; Chem Eng Innovation Award, Am Inst Chem Engrs, 85; Harrison Fac Award, Ohio State Univ, 86. *Mem:* Sigma Xi; Am Inst Chem Engrs. *Res:* Energy conversion, fluidization, fluid solid reaction and mathematical modeling; author and co-author of over 180 technical papers and two books. *Mailing Add:* 1286 Castleton Rd N Columbus OH 43220-1180

FAN, LIANG-TSENG, b Taiwan, Aug 7, 29; nat US; m 58; c 2. CHEMICAL ENGINEERING, MATHEMATICS. *Educ:* Nat Taiwan Univ, BS, 51; Kans State Univ, MS, 54; WVa Univ, PhD(chem eng), 57, MS(math), 58. *Prof Exp:* Jr chem engr, Taiwan Agr Chem Works, 51-52; asst chem eng, Kans State Univ, 52-54 & WVa Univ, 54-58; from instr to prof, 58-67, Kans Power & Light distinguished prof, 67-73, UNIV DISTINGUISHED PROF CHEM ENG, KANS STATE UNIV, 84-, DIR, INST SYSTS DESIGN & OPTIMIZATION, 67-, HEAD DEPT CHEM ENG, 68- *Concurrent Pos:* Phys chemist, US Bur Mines, 56-58, chem engr, 58-59; consult, Nat Air Pollution Control Admin, 69-75. *Honors & Awards:* Iinoya Award, Soc Powder Technol, Japan; Irvin E Youngberg Res Award, Univ Kans, 87. *Mem:* Fel AAAS; Am Chem Soc; fel Am Inst Chem Engrs; Soc Eng Sci; Japanese Soc Chem Engrs. *Res:* Mass and heat transfer; fluidization; chemical process design; applied mathematics; optimization; chemical process dynamics; mathematical optimization technique; air and water pollution control; desalination; energy resources conversion. *Mailing Add:* Dept Chem Eng Kans State Univ Manhattan KS 66504

FAN, NINGPING, b Shanghai, China, Dec 3, 54. MEDICAL IMAGING-IMAGE ANALYSIS & PATTERN RECOGNITION, PHYSIOLOGICAL SIGNAL PROCESSING. *Educ:* Xilan Jiaotong Univ, BS, 82; Univ Pittsburgh, MS, 86, PhD(elec eng), 91. *Prof Exp:* Teaching asst electronics & signal processing, Dept Elec Eng, Univ Pittsburgh, 83-86, res asst computer vision pattern recognition, 86-87, teaching fel Electronics & Signal Processing Lab, 87-89, RES FEL MED IMAGING DEPT CARDIOL, UNIV PITTSBURGH, 89- *Mem:* Inst Elec & Electronics Engrs Computer Soc; assoc mem Sigma Xi. *Res:* Machine recognition of human left ventricle from cineangiograms; image compression; animation; 3D reconstruction of left ventricle from two orthogonal projections; left ventricle wall motion; greyscale morphological operator for angiogram enhancement; design an integrated medical imaging workstation both hardware and software. *Mailing Add:* PO Box 7102 Pittsburgh PA 15213

FAN, POW-FOONG, b Palembang, Indonesia, Sept 4, 33; m 65. GEOLOGY. *Educ:* Wheaton Col, Ill, BS, 55; Univ Calif, Los Angeles, MA, 63, PhD(geol), 65. *Prof Exp:* Asst geophysicist, Univ Hawaii, 65-70, asst prof, 66-70, assoc prof & assoc geophysicist, 70-87, PROF, DEPT GEOL & GEOPHYS, UNIV HAWAII, 87- *Mem:* Geol Soc Am; Soc Econ Paleontologists & Mineralogists. *Res:* Mineralogy of sediments; marine geology; hydrothermal alteration of geothermal fields; geology and mineral resources of China and Asia. *Mailing Add:* Dept Geol & Geophys Univ Hawaii Honolulu HI 96822

FAN, STEPHEN S(HU-TU), b Shanghai, China, Jan 2, 34; m 59; c 4. CHEMICAL ENGINEERING. *Educ:* Stanford Univ, BS, 57, MS, 60, PhD(chem eng), 62. *Prof Exp:* From asst prof to assoc prof, 62-77, actg chmn dept, 70-71, PROF CHEM ENG, UNIV NH, 77-, CHMN DEPT, 71- *Mem:* Am Inst Chem Engrs; Am Soc Eng Educ; Am Asn Univ Profs. *Res:* Properties and heat transfer in chemically reacting systems; flow through porous media; applied kinetics; nuclear power. *Mailing Add:* Dept Chem Eng Univ NH Durham NH 03824

FANAROFF, AVROY A, b Bloemfontein, Repub S Africa, May 22, 38. PEDIATRIC, NEONATAL-PERINATAL MEDICINE. *Educ:* Univ Witwatersrand, MB & BCh, 60. *Prof Exp:* From asst prof to assoc prof, 71-81, PROF PEDIAT, NEONATAL & PERINATAL MED, CASE WESTERN RESERVE UNIV SCH MED, 81- *Concurrent Pos:* Assoc dir pediat, Univ Hosp, 73-74; co dir, Cleveland Regional Perinatal Network, 74-81; res assoc, Sch Eng, Case Western Reserve Univ, 75, assoc prof neonatal/perinatal med & reproductive biol, Sch Med, 75-81. *Honors & Awards:* Golden Stethoscope Award, 85. *Mem:* Am Pediat Soc; Brit Med Asn; SAfrican Med Asn; Soc Pediat Res; hon mem Perinatal Asn Peru. *Res:* Outcome of lowbirth weight infants; prevention of morbidity in low birthweight infants. *Mailing Add:* Rainbow Babies & Childrens Hosp 2101 Adelbert Rd Cleveland OH 44106

FANCHER, LLEWELLYN W, b Merced, Calif, Mar 30, 17; m 42; c 1. ORGANIC CHEMISTRY, AGRICULTURAL CMEMISTRY. *Educ:* Univ Calif, Berkeley, AB, 41. *Prof Exp:* Observer, US Steel Corp, 41-42, raw mat inspector, 42-44; res chemist, Ragooland-Broy Labs, 44-45; res chemist, Multiphase, Inc, 45-52; group leader, Stauffer Chem Co, 52-59, sect leader, 59-64, res assoc, 64-70, sr res assoc, 70-79; RETIRED. *Mem:* Am Chem Soc. *Res:* Industrial products and processes; new agricultural compounds such as pesticides, herbicides, fungicides and bactericides. *Mailing Add:* 233 Old Oak Rd New Castle CA 94658

FANCHER, OTIS EARL, b McCool, Miss, Jan 17, 16; m 46. CHEMISTRY. *Educ:* Miss State Col, BS, 38; Northwestern Univ, PhD(org chem), 42. *Prof Exp:* Res assoc, Nat Defense Res Comt Contract, Northwestern Univ, 42-45; res chemist, G D Searle & Co, 45-46; asst prof chem, Miss State Col, 46-47; head org sect, Miles-Ames Res Lab, 47-59, dir chem therapeut res lab, Miles Labs, Inc, 59-63, dir therapeut res labs, 63-66; vpres & sci dir, Indust Biotest Labs, Inc, 66-72, consult, 72-81; RETIRED. *Mem:* Am Chem Soc; Soc Toxicol. *Res:* Synthesis of pharmacologically active organic compounds; toxicological studies. *Mailing Add:* 16531 E Sullivan Dr Fountain Hills AZ 85268

FANCHER, PAUL S(TRIMPLE), b San Antonio, Tex, Jan 5, 32; m 54; c 4. INSTRUMENTATION, ENGINEERING. *Educ:* Univ Mich, BSE, 53, MSE, 59, InstmE, 64. *Prof Exp:* Asst, Univ Mich, 57-59, res assoc, 59-61, assoc res engr, 61-70, RES SCIENTIST, TRANSP RES INST, UNIV MICH, ANN ARBOR, 70-, ASSOC HEAD, ENG RES DIV, 89- *Mem:* Soc Comput Simulation; Am Soc Testing & Mat; Int Asn Vehicle Syst Dynamics; Soc Automotive Engrs. *Res:* Highway vehicle dynamics; analysis, control, and simulation of dynamical systems. *Mailing Add:* Univ Mich Transp Res Inst 2901 Baxter Rd Ann Arbor MI 48109-2150

FANCONI, BRUNO MARIO, b Merced, Calif, May 20, 39; m 65; c 3. POLYMER PHYSICS. *Educ:* Univ Calif, BS, 62; Univ Wash, PhD(chem), 68. *Prof Exp:* Teaching asst chem, Univ Wash, 63-64; from res asst to res assoc, Univ Ore, 64-71; res chemist polymers, 71-76, sect chief, 76-83, DEP CHIEF, POLYMERS DIV, NAT BUR STANDARDS, 83- *Mem:* AAAS; fel Am Phys Soc. *Res:* Vibrational spectroscopy of polymers. *Mailing Add:* Nat Inst Standards & Technol Bldg 224 Rm 209A Washington DC 20234

FAND, RICHARD MEYER, b Janow, Poland, Aug 13, 23; US citizen; m 66; c 3. HEAT TRANSFER, FLUID MECHANICS. *Educ:* Rensselaer Polytech Inst, BS, 45; Columbia Univ, MS, 49; Cornell Univ, PhD(mech eng), 59. *Prof Exp:* Mech engr, US Civil Serv, 46-48; res engr, Bendix Aviation, 51; engr dynamics, Canadair Ltd, 52; instr mech eng, Cornell Univ, 52-55; res engr, Mass Inst Technol, 55-61; sr eng scientist, Bolt Beranet & Newman Inc, 61-66; PROF MECH ENG, UNIV HAWAII, 66-, CHMN DEPT, 85- *Mem:* Am Soc Mech Engrs; Am Soc Eng Educ. *Res:* Influence of vibrations and intense sound, combined boiling and forced convection; combined forced and natural convection; influence of property variation on heat transfer; removal of scale by acoustically induced cavitation; fluid flow and heat transfer in porous media. *Mailing Add:* Dept Mech Eng Univ Hawaii Manoa Hol 304 2500 Campus Rd Hawaii HI 96822

FAND, THEODORE IRA, b Brooklyn, NY, Dec 1, 15; m 41; c 2. ORGANIC CHEMISTRY, PHARMACEUTICAL CHEMISTRY. *Educ:* Brooklyn Col, BS, 35; Univ Mich, MS, 36; Polytech Inst Brooklyn, PhD(chem), 54. *Prof Exp:* Jr chemist anal, Vet Admin, 36-38; asst chemist, Bd Transp, NY, 38-42; asst dir develop & eng, Nepera Chem Co, Inc, 42-57; from dir prod develop to dir pharmaceut res & develop, 57-74, vpres, new prod develop, Warner-Lambert Res Inst,74-83; RETIRED. *Mem:* AAAS; Am Chem Soc; Am Pharmaceut Asn; Soc Cosmetic Chem; NY Acad Sci. *Res:* Pharmaceutical research and development; new dosage forms; drug absorption; pharmaceutical technology; pyridine chemistry; heterocycles and vitamins. *Mailing Add:* 18 Braidburn Way Convent Station NJ 07961

FANELLI, GEORGE MARION, JR, b Pelham, NY, Oct 5, 26; c 4. PHARMACOLOGY. *Educ:* George Washington Univ, BS, 50; NY Univ, MS, 57, PhD(biol), 62. *Prof Exp:* Biologist, Hazleton Labs, Inc, 51-52; pharmacologist, Chas Pfizer & Co, Inc, 52-58; res scientist, Lederle Labs Div, Am Cyanamid Co, 58-63; SR INVESTR PHARMACOL, MERCK SHARP & DOHME RES LABS, 65- *Concurrent Pos:* USPHS res fel physiol, Harvard Med Sch, 63-65. *Mem:* Am Physiol Soc; Am Soc Pharmacol & Exp Therapeut; Am Soc Nephrology; Brit Pharmacol Soc; Brit Primate Soc. *Res:* Comparative renal physiology; effects of drugs on renal transport processes; organic acid transport especially uric acid in nonhuman primates; chimpanzee renal function; pharmacology of diuretics; renin-angiotensin-aldosterone system. *Mailing Add:* Merck Sharp & Dohme Res Labs Ten Sentry Pkwy Blue Bell PA 19422

FANESTIL, DARRELL DEAN, b Great Bend, Kans, Oct 31, 33; m 55; c 4. INTERNAL MEDICINE, NEPHROLOGY. *Educ:* Univ Kans, BA, 55, MD, 58. *Prof Exp:* Intern internal med, Los Angeles County Gen Hosp, 58-59; resident, Lahey Clin, 59-60; trainee cardiol, Scripps Clin & Res Found, 60-61; from asst prof to assoc prof internal med, Sch Med, Univ Kans, 66-70; assoc prof, 70-72, PROF INTERNAL MED, SCH MED, UNIV CALIF, SAN DIEGO, 72- *Concurrent Pos:* USPHS res fel biochem, Scripps Clin & Res Found, 61-62; Am Heart Asn adv res fel nephrology, Univ Calif Med Ctr, San Francisco, 64-66; Markle scholar acad med, 66-71; Am Heart Asn estab investr, 66-71. *Mem:* AAAS; Am Soc Clin Invest; Am Fedn Clin Res; Am Physiol Soc; Am Soc Nephrology. *Res:* Mechanism of action of hormones and drugs on transport of sodium and hydrogen ions and of water. *Mailing Add:* Dept Med Univ Calif San Diego M-023-B La Jolla CA 92093-0623

FANG, CHENG-SHEN, b Taipei, Taiwan, Mar 29, 36; US citizen; m 72. CHEMICAL ENGINEERING. *Educ:* Nat Taiwan Univ, BS, 58; Univ Houston, MS, 65, PhD(chem eng), 68. *Prof Exp:* Supvr chem eng, Taiwan Fertilizer Co, 60-62; ELOI GIRAD PROF CHEM ENG, UNIV SOUTHWESTERN LA, 69- *Concurrent Pos:* Fel chem eng dept, Univ Houston, 68-69; prin investr, Dept Conserv, State La, 75-76 & 79-80 & 88-89, NSF, 88-90; consult, Col Com, Univ Southwestern La, 77-78; tech consult, Chem Eng Dept, Lamar Univ, Tex, 77-78. *Mem:* Am Inst Chem Engrs; Sigma Xi; Soc Petrol Engrs. *Res:* Energy conservation and alternative energy sources; environmental engineering; process design. *Mailing Add:* Dept Chem Eng Univ SouthwesternLA Box 44130 Lafayette LA 70504

FANG, CHING SENG, b China, Nov 23, 38; US citizen; m 66; c 2. HYDROMECHANICS, ENVIRONMENTAL ENGINEERING. *Educ:* Nat Taiwan Univ, BS, 61; NC State Univ, MS, 64, PhD(hydromech), 68. *Prof Exp:* Water resources engr, Camp, Dresser & McKee, 68-69; assoc scientist marine sci, 69-70, SR SCIENTIST PHYS OCEANOG & HYDRAUL & HEAD DEPT, VA INST MARINE SCI, 70-; PROF MARINE SCI, WILLIAM & MARY COL, 79- *Concurrent Pos:* Gen mgr & treas, Coastal Environ Asn, Inc, 71-; assoc prof marine sci, Univ Va, 72-78. *Mem:* Am Soc Civil Engrs; Am Geophys Union. *Res:* Water resource systems analysis; mathematical models for thermal and water pollution management; flood routing, wave mechanics; ground water and estuarine circulation. *Mailing Add:* 325 Yorkville Rd Grafton VA 23692

FANG, FABIAN TIEN-HWA, b Nanking, China, Oct 14, 29; nat US; m 55; c 2. POLYMER CHEMISTRY. *Educ:* Nat Cent Univ, China, BS, 49; Univ Ill, MS, 52, PhD(chem), 54. *Prof Exp:* Res assoc chem, Univ Notre Dame, 54-55; res assoc, Iowa State Univ, 55-57; sr res chemist, Rohm & Haas Co, 57-64; asst prof chem, Univ Wis, Milwaukee, 64-69; vis assoc, Calif Inst Technol, 70; assoc prof, 70-72, PROF CHEM, CALIF STATE UNIV, BAKERSFIELD, 72-, CHMN DEPT, 70- *Concurrent Pos:* Sr vis prof chem, Tunghai Univ, Taiwan, 82-83. *Mem:* Am Chem Soc; Am Inst Chemists; Catalysis Soc. *Res:* Macromolecular chemistry; heterogeneous catalysis; physical organic chemistry; computational chemistry. *Mailing Add:* Dept Chem Calif State Univ Bakersfield CA 93311-1099

FANG, FRANK F, b Peiping, China, Sept 11, 30; m 57; c 4. SEMICONDUCTOR PHYSICS. *Educ:* Nat Univ Taiwan, BS, 51; Univ Notre Dame, MS, 54; Univ Ill, PhD(elec eng), 59. *Hon Degrees:* DSc, Nat Chiaotung Univ, 89. *Prof Exp:* Assoc phys electronics, Univ Ill, 57-59; res engr, Boeing Airplane Co, 59-60; RES STAFF MEM, IBM CORP, 60- *Honors & Awards:* John Price Wetherill Medal, Franklin Inst, 81; David Sarnoff Award, Inst Elec & Electronics Engrs, 87; Oliver E Buckley Prize, Am Phys Soc, 88; Alexander von Humboldt Award, 90. *Mem:* Nat Acad Eng; Sigma Xi; Am Phys Soc; Franklin Inst; Inst Elec & Electronics Engrs. *Res:* Solid state science; semiconductor physics and devices; solid state electronics. *Mailing Add:* Thomas J Watson Res Ctr IBM Corp PO Box 218 Yorktown Heights NY 10598

FANG, JEN-HO, b Tainan, Formosa, Oct 21, 29; m 61; c 3. MINERALOGY. *Educ:* Nat Taiwan Univ, BS, 53; Univ Minn, MS, 57; Pa State Univ, PhD(geochem), 61. *Prof Exp:* Res assoc chem, Boston Univ, 61-62; mem staff, Mass Inst Technol, 62-64; from asst prof to prof geol, Southern Ill Univ, 64-82; PROF GEOL, UNIV ALA, TUSCALOOSA, 82- *Mem:* Am Crystallog Asn; fel Mineral Soc Am; Am Geochem Soc. *Res:* X-ray crystallography; x-ray and neutron diffraction; physics of minerals; geostatistics. *Mailing Add:* Box 870338 Tuscaloosa AL 35487-0338

FANG, JOONG J, b Piongyang, Korea, Mar 30, 23; US citizen; m 56; c 2. MATHEMATICS, PHILOSOPHY. *Educ:* Yale Univ, MA, 50; Univ Mainz, Dr Phil, 57. *Prof Exp:* Asst prof math, Jinhae Col & Pusan Nat Univ, Korea, 45-48, Defiance Col, 57-58, Valparaiso Univ, 58-59 & St John's Univ, Minn, 59-62; assoc prof, Northern Ill Univ, 62-67; from assoc prof to prof philos & math, Memphis State Univ, 67-73; prof, 74-90, EMER PROF PHILOS, OLD DOMINION UNIV, 90- *Concurrent Pos:* Ed, Philosophia Mathematica, Asn Philos Math, 64-; vis prof math, Univ Munster, Ger, 71. *Mem:* Math Asn Am; Am Math Soc; Am Philos Asn. *Res:* Foundation, philosophy and sociology of mathematics; sociology, history, and philosophy of science; emphasis on interdisciplinary studies, moving from mathematics to physics to philosophy to history to sociology of mathematics in particular and science in general, then to philosophy of culture in general. *Mailing Add:* Woods Cross Rds VA 23190-0206

FANG, SHENG CHUNG, b Foochow, Fukien, China, June 27, 16; nat US; m 48. AGRICULTURAL CHEMISTRY. *Educ:* Fukien Christian Univ, BS, 37; Ore State Col, MS, 44, PhD(biochem), 48. *Prof Exp:* Instr, 48-53, from asst prof & asst chemist to assoc prof & assoc chemist, 53-71, prof agr chem, 71-82, EMER PROF AGR CHEM, STATE UNIV, 82- *Mem:* Weed Sci Soc Am; Am Soc Plant Physiologists; Am Chem Soc; Am Soc Biol Chemists. *Res:* Radioactive tracer studies in agricultural and biological chemistry; herbicides; plant growth regulators. *Mailing Add:* 2920 NW Ashwood Dr Corvallis OR 97330

FANG, TA-YUN, protein chemistry, enzymology, for more information see previous edition

FANG, TSUN CHUN, applied mechanics, applied mathematics, for more information see previous edition

FANG, VICTOR SHENGKUEN, b Chekiang, China, Oct 10, 29; US citizen; m 59; c 2. HORMONE ASSAYS, ADRENAL CELL CULTURE. *Educ:* Nat Defense Med Ctr, BS, 51; Mass Col Pharm, MS, 64, PhD(pharmacol), 67. *Prof Exp:* Postdoctoral fel endocrinol, Harvard Med Sch, 67-70; assoc endocrinol, Harvard Sch Pub Health, 70-71; from asst prof to assoc prof med, 71-86, assoc prof psychiat, 78-86, PROF MED, UNIV CHICAGO, 86-, PROF PSYCHIAT, 86- *Concurrent Pos:* Adj prof, Shanghai Second Med Univ, 80- & Zhejiang Med Univ, 85- *Mem:* AAAS; Soc Endocrinol Eng; Endocrine Soc; Am Soc Pharmacol & Exp Therapeut; Int Soc Psychoneuroendocrinol; Am Asn Clin Chem; Am Soc Clin Chem. *Res:* Biological aspects of psychiatric disorders; relationship between adrenal hyperfunction and depressive episodes reveal a yet unrecognized extra pituitary, non-adrenocorticotrophic hormone regulatory pathway for hypercortisolism. *Mailing Add:* 5841 S Maryland Ave Chicago IL 60637

FANGBONER, RAYMOND FRANKLIN, b Waukegan, Ill, July 10, 43. DEVELOPMENTAL BIOLOGY & NEUROBIOLOGY. *Educ:* Transylvania Col, BA, 65; Purdue Univ, MS, 68, PhD(biol), 72. *Prof Exp:* asst prof, 72-84, ASSOC PROF BIOL, TRENTON STATE COL, 84- *Mem:* Am Soc Zoologists; Am Inst Biol Sci; AAAS; NY Acad Sci; Soc Develop Biol; Asn Res Vision & Ophthal. *Res:* Studies in in vivo nerve growth patterns; resolution of competition between appropriate and inappropriate innervation for a given end-organ; extraocular muscle regeneration. *Mailing Add:* Dept Biol Trenton State Col Trenton NJ 08625

FANGER, BRADFORD OTTO, b Lafayette, Ind, May 14, 56. HORMONE RECEPTORS. *Educ:* Univ Colo, BA, 79; Univ Vt, PhD(biochem), 84. *Prof Exp:* Guest researcher, Lab Chemoprev, NIH, 84-86; res fel, Dept Biochem, Vanderbilt Univ, 86-87; SR RES SCIENTIST, MARION MERRELL DOW RES INST, 87- *Concurrent Pos:* Vol asst prof, Dept Anat & Cell Biol, Univ Cincinnati, 88- *Mem:* Am Soc Biochem & Molecular Biol; Am Chem Soc; AAAS; Union Concerned Scientists. *Res:* Characterizing the receptor for bombesin and developing receptor antagonist to treat small cell lung carcinoma and peptic ulcers; role of clustering in the activation of the epidermal growth factor receptor tyrosine kinase. *Mailing Add:* Dept Cancer Biol Marion Merrell Dow Res Inst 2110 E Galbraith Rd Cincinnati OH 45215-6300

FANGER, CARLETON G(EORGE), b Wolsey, SDak, Mar 22, 24; m 45; c 3. MECHANICAL ENGINEERING, ENGINEERING MECHANICS. *Educ:* Ore State Univ, BS, 47, MS, 48. *Prof Exp:* Asst prof eng, Vanport Exten Ctr, 48-52 & Portland State Exten Ctr, 52-55; from assoc prof to prof, 55-87, EMER PROF MECH ENG, PORTLAND STATE UNIV, 87- *Concurrent Pos:* Qual instr nuclear defense, Defense Civil Preparedness Agency, 66-87. *Mem:* Am Soc Mech Engrs; Nat Soc Prof Engrs; Am Soc Eng Educ. *Mailing Add:* 10234 SE Market Dr Portland OR 97216

FANGER, MICHAEL WALTER, b Ft Wayne, Ind, July 3, 40; m 62; c 2. BIOCHEMISTRY, IMMUNOLOGY. *Educ:* Wabash Col, BA, 62; Yale Univ, PhD(biochem), 67. *Prof Exp:* Asst prof microbiol, Case Western Reserve Univ, 70-76, assoc prof, 76-81; PROF MICROBIOL & MED, DARTMOUTH MED SCH, 81- *Concurrent Pos:* NIH fel, Nat Inst Med Res, London, 67-68; fel, Med Sch, Univ Ill, 68-69, NIH fel, 69-70. *Mem:* Am Asn Immunol; Fedn Am Socs Exp Biol. *Res:* Characterization of lymphocyte subpopulations; initiation of the immune response by the mechanism of transformation of the small lymphocyte. *Mailing Add:* Dept Microbiol Dartmouth Med Sch HB 7550 Hanover NH 03756

FANGMAN, WALTON L, b Louisville, Ky, Sept, 23, 39. GENETICS. *Educ:* Bellarmine Col, BA, 62; Purdue Univ, PhD(microbiol), 65. *Prof Exp:* Fel & res assoc, Inst Molecular Biol, Univ Ore, 65-67; from asst prof to assoc prof, 67-78, PROF, DEPT GENETICS, UNIV WASH, 78- *Concurrent Pos:* Chmn, Dept Genetics, Univ Wash, 85-90. *Mem:* AAAS; Am Soc Cell Biol; Am Soc Microbiol; Genetics Soc Am. *Res:* Temporal regulation of chromosome replication; mitochondrial DNA segregation and replication. *Mailing Add:* Dept Genetics SK-50 Univ Wash Seattle WA 98195

FANGMEIER, DELMAR DEAN, b Hubbell, Nebr, Oct 27, 32; m 69; c 2. AGRICULTURAL ENGINEERING, IRRIGATION. *Educ:* Univ Nebr, BSc, 54 & 60, MSc, 61; Univ Calif, Davis, PhD(eng), 67. *Prof Exp:* Agr engr, Agr Res Serv, USDA, 61; asst prof civil eng, Univ Wyo, 66-68; assoc prof agr eng, 68-72, PROF AGR ENG, UNIV ARIZ, 72- *Concurrent Pos:* Agr Water Conserv Panel, Calif Dept Water Res, 79; chmn, Soil & Water Div, Am Soc Agr Engrs, 78-80 & Guayale Rubber Soc, 87-88. *Mem:* Am Soc Agr Engrs; Am Soc Civil Engrs; Am Soc Eng Educ; US Comn Irrig & Drainage. *Res:* Hydraulics of surface irrigation; water and energy requirements for irrigation; irrigation management. *Mailing Add:* Agr & Biosysts Eng Dept Univ Ariz Tucson AZ 85721

FANGUY, ROY CHARLES, b New Orleans, La, Nov 23, 29; m 51; c 2. IMMUNOGENETICS. *Educ:* Miss State Univ, BS, 51; Auburn Univ, MS, 53; Tex A&M Univ, PhD(poultry breeding), 58. *Prof Exp:* From asst prof to assoc prof immunogenetics, 58-74, ASSOC PROF POULTRY SCI, TEX A&M UNIV, 74-, ASSOC PROF GENETICS, 80- *Concurrent Pos:* USPHS grant, 58- *Mem:* Poultry Sci Asn. *Res:* Poultry breeding; physiology. *Mailing Add:* Dept Poultry Sci Tex A&M Univ College Station TX 77843

FANKBONER, PETER VAUGHN, b Mare Island, Calif, Feb 9, 38; m 67; c 2. PEARL CULTURE AQUACULTURE. *Educ:* Calif Polytech State Univ, BS, 64; Univ Pac, MS, 70; Univ Victoria, PhD(invert biol), 72. *Prof Exp:* Marine tech, Scripps Inst Oceanog, 64-65; PROF BIOL, SIMON FRASER UNIV, BC, 72-; PRES, PAC PEARL CULT, LTD, 88- *Concurrent Pos:* Lectr, Univ Victoria, BC, 71; researcher, Cambridge Univ, 71-72; vis prof, Stazione Zoologica, Napoli, 72. *Mem:* Western Soc Naturalists. *Res:* Culture of pearls in the abalone Haliotis; cellular mechanisms of pearl production in mollusca; aquaculture of abalone. *Mailing Add:* Dept Biol Sci Simon Fraser Univ Burnaby BC V5A 1S6 Can

FANN, HUOO-LONG, b Formosa, China, Mar 29, 31; m 62. NUCLEAR PHYSICS. *Educ:* Taiwan Norm Univ, BS, 56; Univ Md, PhD(physics), 64. *Prof Exp:* Asst prof, 64-68, ASSOC PROF PHYSICS, MERRIMACK COL, 68- *Mem:* Am Phys Soc. *Res:* Experimental nuclear physics concerning reaction theories for light nuclei. *Mailing Add:* 20 Lisa Lane North Andover MA 01845

FANNIN, BOB M(EREDITH), b Midland, Tex, June 9, 22; m 47; c 4. ELECTRICAL ENGINEERING. *Educ:* Univ Tex, BS, 44, MS, 47, PhD(elec eng), 56. *Prof Exp:* Instr math, Arlington State Col, 47-48; res assoc, Sch Elec Eng, Cornell Univ, 48-51; res engr, Elec Eng Res Lab, Univ Tex, 51-56; assoc prof elec eng, Univ NMex, 56-58; from assoc prof to prof elec eng, Univ Tex Austin, 58-86, eng counr, 77-86; RETIRED. *Concurrent Pos:* Mem comn II, US Nat Comt, Int Sci Radio Union. *Res:* Tropospheric radio wave program. *Mailing Add:* 4709 Crestway Austin TX 78731

FANNING, DELVIN SEYMOUR, b Copenhagen, NY, July 13, 31; m 58; c 3. SOIL SCIENCE. *Educ:* Cornell Univ, BS, 54, MS, 59; Univ Wis, PhD(soil sci), 64. *Prof Exp:* Soil scientist, Soil Conserv Serv, USDA, 53-63; from asst prof to assoc prof, 64-77, PROF SOIL MINERAL & CLASSIFICATION,

UNIV MD, COLLEGE PARK, 77- *Concurrent Pos:* Guest prof, Soil Sci Inst, Munich Tech Univ, 71-72; res assoc, Tex A&M Univ, 79; sabbatical with Soil Conserv Serv, USDA, 86. *Mem:* AAAS; Am Soc Agron; fel Soil Sci Soc Am; Mineral Soc Am; Clay Minerals Soc; Sigma Xi. *Res:* Mineralogy of soils in relation to their genesis; clay mineral and x-ray spectroscopic analytical techniques; highly man-influenced soils; soil moisture regimes-hydric soils; acid sulfate soils, micas. *Mailing Add:* Dept Agron Univ Md College Park MD 20742

FANNING, GEORGE RICHARD, b Princeton, WVa, Sept 2, 36; m 62; c 2. MOLECULAR BIOLOGY, BIOCHEMISTRY. *Educ:* Concord Col, Athens, WVa, BS, 58; George Washington Univ, MS, 63. *Prof Exp:* Chemist, NIH, 60-63; BIOCHEMIST, WALTER REED ARMY INST RES, 63- *Concurrent Pos:* Instr chem, Am Univ, Washington, DC, 67-69. *Mem:* Am Soc Microbiol. *Res:* Identification and classification of bacteria, specifically those in the families Enterobacteriaceae and Vibrionaceae, using DNA hybridization methodology. *Mailing Add:* Dept Biochem Walter Reed Army Inst Res Washington DC 20307-5100

FANNING, JAMES COLLIER, b Atlanta, Ga, Nov 8, 31; m 57; c 3. SOLID STATE REACTIONS, IRON CHEMISTRY. *Educ:* The Citadel, BS, 53; Ga Inst Technol, MS, 56, PhD(chem), 60. *Prof Exp:* Instr chem, Ga Inst Technol, 57-59; fel, Tulane Univ, 60-61; from asst prof to assoc prof, 61-71, PROF CHEM, CLEMSON UNIV, 71- *Concurrent Pos:* Vis lectr, Univ Ill, 66-67; vis scientist, Nat Cancer Inst, 79-80. *Mem:* Am Chem Soc; Sigma Xi; Am Asn Univ Professors. *Res:* Chemistry of transition metals; focuses on chemistry of nitrate, nitrite and nitrosyl metal complexes, especially those of iron; the nature of iron in iron containing glass. *Mailing Add:* Dept Chem Clemson Univ 223 Hunter Lab Clemson SC 29631

FANNING, KENT ABRAM, b Lawton, Okla, May 22, 41; m 62; c 2. OCEANOGRAPHY, GEOCHEMISTRY. *Educ:* Colo Sch Mines, BS, 64; Univ RI, PhD(oceanog), 73. *Prof Exp:* Oceanog specialist marine chem, Univ RI, 64-66; from asst prof to assoc prof, 73-84, PROF MARINE SCI, UNIV SFLA, 84- *Concurrent Pos:* Prin investr, NSF, 74-, Off Naval Res, 76-78, US Dept Energy, 78-81; Nat Res Coun Marine Bd Comt, Seabed Utilization Exclusive Econ Zone. *Honors & Awards:* Sigma Xi. *Mem:* Am Soc Limnol & Oceanog; Geochem Soc; Am Geophys Union. *Res:* Chemical oceanography; interstitial chemistry of sediments; transport processes in sediments; anoxic basins; geochemistry of river plumes; marine geochemistry of radionuclides. *Mailing Add:* Dept Marine Sci Univ SFla St Petersburg FL 33701

FANO, ROBERT M(ARIO), b Torino, Italy, Nov 11, 17; nat US; m 49; c 3. EDUCATION, INFORMATION THEORY. *Educ:* Mass Inst Technol, SB, 41, ScD(elec eng), 47. *Prof Exp:* Asst elec eng, 41-43, instr, 43-44, mem staff, Radiation Lab, 44-46, res assoc, Res Lab Electronics, 46-47, from asst prof to prof elec commun, 47-62, dir, Proj MAC, 63-68, group leader, Lincoln Lab, 51-53, assoc head, Elec Eng Dept, head comput sci & eng, 71-74, Ford prof eng, 62-84, EMER PROF ENG, MASS INST TECHNOL, 84- *Concurrent Pos:* Report Rev Comt, Nat Acad Sci, 82-90; educ adv bd, Nat Acad Eng, 84-86; IBM Zurich Res Lab, 74-75. *Honors & Awards:* Shannon Lectr & Educ Medal, Inst Elec & Electronics Engrs. *Mem:* Nat Acad Sci; Nat Acad Eng; Am Acad Arts & Sci; Asn Comput Mach; fel Inst Elec & Electronics Engrs. *Res:* Microwave circuit components; network synthesis; transmission of information; computer sciences; theoretical limitations on the broad band matching of arbitrary impedances. *Mailing Add:* Dept Elec Eng Mass Inst Technol Cambridge MA 02139

FANO, UGO, b Torino, Italy, July 28, 12; nat US; m 39; c 2. THEORETICAL PHYSICS. *Educ:* Univ Turin, DSc(math), 34. *Hon Degrees:* Queen's Univ, Belfast, DSc, 78; Dr, Univ Pierre, Marie Curie, Paris, 79. *Prof Exp:* Ital Dept Educ Int fel, Univ Leipzig, 36-37; instr physics, Univ Rome, 38-39; res assoc, Wash Biophys Inst, 39-40; res fel genetics, Carnegie Inst, 40-41, res assoc, 42-43, physicist & mathematician, 43-45; consult & ballistician, Ballistics Res Lab, Aberdeen Proving Ground, 44-45; res assoc, Carnegie Inst, 46; physicist, Nat Bur Standards, 46-66; chmn dept, 72-74, prof, 66-82, EMER PROF PHYSICS, UNIV CHICAGO, 82- *Concurrent Pos:* Prof lectr, George Washington Univ, 46-47 & 57-58 & Univ Calif, 58 & 68; vis prof, Cath Univ Am, 63-64; assoc ed, Rev Modern Physics. *Honors & Awards:* Rockefeller Pub Serv Award, 56-57. *Mem:* Nat Acad Sci; Am Acad Arts & Sci; Am Phys Soc; Radiation Res Soc. *Res:* Atomic physics; molecular physics; quantum physics; radiological physics; statistical mechanics; genetics; radiobiology. *Mailing Add:* James Franck Inst Univ Chicago Chicago IL 60637

FANSHAWE, JOHN RICHARDSON, II, b Philadelphia, Pa, Oct 20, 06; m 37; c 3. PETROLEUM GEOLOGY. *Educ:* Princeton Univ, AB, 29, MA, 31, PhD(geol), 39; Univ Lille, Dr es Sc, 30. *Hon Degrees:* ScD, Rocky Mountain Col, 88. *Prof Exp:* Asst geol, Princeton Univ, 30-32; master physics & geol, Deerfield Acad, Mass, 33-35; instr geol, Williams Col, 35-39; geologist, Ohio Oil Co, 40-42; dist geologist & dist dir reserves, Dist IV, Petrol Admin War, 43-45; sr geologist, Gen Petrol Corp, 45-47; dist mgr & mgr explor, Seaboard Oil Co of Del, 47-51; consult geologist, 51-62; vpres & res mgr, Forest Cyprus Corp, 63-64; staff geologist, Mont Power Co, 64-71; CONSULT GEOLOGIST, 71- *Concurrent Pos:* Vis prof, Rocky Mt Col, 71-80. *Mem:* Am Asn Petrol Geologists; Am Inst Prof Geologists; Yellowstone-Bighorn Res Asn. *Res:* Structural theory and interpretation; regional stratigraphy; application of geophysics to petrol exploration; geology of fossil fuels, and geothermal resources. *Mailing Add:* 3116 E MacDonald Dr Billings MT 59102

FANSHAWE, WILLIAM JOSEPH, b Brooklyn, NY, May 7, 26; m 52; c 8. ORGANIC CHEMISTRY. *Educ:* St John's Univ, BS, 50. *Prof Exp:* SR RES CHEMIST, LEDERLE LABS, AM CYANAMID CO, 51- *Mem:* Am Chem Soc. *Res:* Chemical synthesis of central nervous system and hypoglycemic agents. *Mailing Add:* Lederle Labs Pearl River NY 10965

FANSLER, BRADFORD S, biochemistry, genetics; deceased, see previous edition for last biography

FANSLER, KEVIN SPAIN, b Thomas, Okla, Jan 13, 38; m 69; c 2. AEROSPACE ENGINEERING, PHYSICS. *Educ:* Okla State Univ, BS, 60; Univ Hawaii, MS, 64; Univ Del, PhD(aerospace eng), 74. *Prof Exp:* Physicist, Naval Ord Lab, Silver Spring, MD, 60-62; res & teaching asst, dept physics, Univ Hawaii, 62-64; physicist, US Army Electronics Lab, Ft Monmouth, NJ, 65 & US Army Chem Lab, Edgewood Arsenal, Md, 65-67; RES PHYSICIST, BALLISTICS RES, US ARMY BALLISTIC RES LAB, ABERDEEN PROVING GROUND, 67- *Honors & Awards:* Army Meritorious Civilian Serv Medal. *Mem:* Am Inst Astronaut & Aeronaut; Am Defense Preparedness Asn. *Res:* Pressure fields about weapons; effect of muzzle blast on trajectory of projectiles; motions of projectiles within guntubes; electronic populations in nuclear blast and aerosol dispersion. *Mailing Add:* 4044 Wilkinson Rd Havre de Grace MD 21078

FANSLOW, DON J, b Yankton, SDak, Apr 16, 36. ZOOLOGY, ENDOCRINOLOGY. *Educ:* Yankton Col, BA, 58; Univ SDak, MA, 60; Ind Univ, PhD(zool), 65. *Prof Exp:* Assoc prof, 65-77, prof biol, 77-, AT BRYN MAWR-ST LOUIS, NORTHEASTERN ILL UNIV. *Mem:* AAAS; Am Soc Zoologists; Sigma Xi; Am Asn Univ Professors; Nat Asn Adv Health Prof. *Mailing Add:* Northeastern Ill Univ 5500 N St Louis Chicago IL 60625

FANSLOW, GLENN E, b Minot, NDak, Sept 5, 27; m 60; c 2. ELECTRICAL ENGINEERING, PHYSICS. *Educ:* NDak Agr Col, BS, 53; Iowa State Univ, MS, 57, PhD(elec eng), 62. *Prof Exp:* Elec engr, Gen Elec Co, 53-55; from instr to assoc prof, 55-84, prof elec eng, 84-90, EMER PROF ELEC ENG, IOWA STATE UNIV, 90- *Concurrent Pos:* Vis res prof, Nat Taiwan Inst Tech, 89. *Mem:* Inst Elec & Electronics Engrs; Sigma Xi; Int Microwave Power Inst. *Res:* Applications of microwave power in the processing of materials; microwave generation, instrumentation and design. *Mailing Add:* Dept Elec Eng Iowa State Univ Coover Hall Ames IA 50011

FANTA, GEORGE FREDERICK, b Chicago, Ill, Aug 30, 34; m 57; c 3. ORGANIC POLYMER CHEMISTRY. *Educ:* Purdue Univ, BS, 56; Univ Ill, PhD(org chem), 60. *Prof Exp:* Res chemist, Ethyl Corp, 60-63; RES CHEMIST, NORTHERN REGIONAL RES CTR, US DEPT AGR, 63- *Honors & Awards:* IR-100 Award, Indust Res Mag, 75. *Mem:* Am Chem Soc. *Res:* Chemistry of starch and starch derivatives; synthesis and properties of starch graft copolymers. *Mailing Add:* 1815 N University St Peoria IL 61604

FANTA, PAUL EDWARD, b Chicago, Ill, July 24, 21; m 49; c 2. ORGANIC CHEMISTRY. *Educ:* Univ Ill, BS, 42, Univ Rochester, PhD(chem), 46. *Prof Exp:* Asst chem, Univ Rochester, 42-44, asst, Manhattan Proj, 44-46, fel, 46-47; instr, Harvard Univ, 47-48; from asst prof to prof, 48-84, EMER PROF CHEM, ILL INST TECHNOL, 84- *Concurrent Pos:* NSF fel, Imp Col, Univ London, 56-57; exchange scholar, Czech Acad Sci, 63-64 & Acad Sci USSR, 70-71. *Mem:* Am Chem Soc. *Res:* Nitrogen heterocycles; stereochemistry; chemical information. *Mailing Add:* Dept Chem Ill Inst Technol Chicago IL 60616

FANTE, RONALD LOUIS, b Philadelphia, Pa, Oct 27, 36; m 61; c 3. MICROWAVE PHYSICS. *Educ:* Univ Pa, BS, 58; Mass Inst Technol, MS, 60; Princeton Univ, PhD(elec eng), 63. *Prof Exp:* Sr physicist, Space Sci Inc, 63-64; staff scientist, Res & Adv Develop Div, Avco Corp, 64-70, sr consult scientist, Avco Systs Div, 70-71; sr scientist, USAF, Cambridge Res Lab, 71-80; asst vpres, Avco Systs Div, 80-88; FEL, MITRE CORP, 88- *Concurrent Pos:* Adj prof, Univ Mass, 73- *Honors & Awards:* Marcus O'Day Prize, USAF, 75. *Mem:* Inst Elec & Electronics Engrs; Optical Soc Am; Sigma Xi. *Res:* Propagation of microwaves and laser beams on turbulent media. *Mailing Add:* 26 Sherwood Rd Reading MA 01867

FANTINI, AMEDEO ALEXANDER, b New York, NY, Feb 11, 22; m 54; c 3. MICROBIAL GENETICS. *Educ:* NY Univ, BA, 52; Columbia Univ, MA, 59, PhD(genetics), 61. *Prof Exp:* Virol, Chas Pfizer & Co, 46-50; biologist cancer res, Lederle Labs Div, Am Cyanamid Co, 52-54, biologist mycol, 54-56; asst, Columbia Univ, 56-58; from res microbiologist to sr res microbiologist, 60-73, PRIN RES MICROBIOLOGIST, LEDERLE LABS DIV, AM CYANAMID CO, 73- *Mem:* Genetics Soc Am; Am Soc Microbiol; Soc Indust Microbiol; Sigma Xi. *Res:* Genetics and physiology of fungi and streptomyces in relation to increased yields of antibiotics; microbial fermentations; bioconversions. *Mailing Add:* No Z The Glen New City NY 10956

FANTZ, PAUL RICHARD, b St Louis, Mo, Mar 12, 41; m 65; c 3. PLANT SYSTEMATICS. *Educ:* Southern Ill Univ, BSEd, 64, MSEd, 69; Wash Univ, MA, 72; Univ Fla, PhD(bot), 77. *Prof Exp:* Sci teacher, Mehlville Sch Dist, St Louis, 64-72, sci coordr, 67-72; grad instr biol sci, Univ Fla, 74-76, adj asst prof bot, 77-78; res assoc taxonomy & hort, Fairchild Trop Garden, 78-79; from asst prof to assoc prof, 79-90, PROF HORT SCI, NC STATE UNIV, 90- *Concurrent Pos:* Collabr, Flora Panama, Nicaragua, Venezuelan, Guayana & Mesoamerica; Teaching fel, Nat Asn Cols & Teachers Agr, 89. *Mem:* Am Asn Plant Taxonomists; Bot Soc Am; Nat Asn Cols & Teachers Agr; Sigma Xi; Am Soc Hort Sci. *Res:* Systematic biology of the Glycineae; monographic studies on Clitoria, Centrosema, Rarbieria; systematics of horticultural plants-Liriope, Ophiopogon, Cercis. *Mailing Add:* Dept Hort Sci NC State Univ Box 7609 Raleigh NC 27695-7609

FANUCCI, JEROME B(EN), b Glen Lyon, Pa, Oct 7, 24; m 52; c 2. AERODYNAMICS. *Educ:* Pa State Univ, BS, 44, MS, 52, PhD(aeronaut eng), 56. *Prof Exp:* Aeronaut engr, Eastern Aircraft Corp, NJ, 44-45 & Repub Aviation Corp, NY, 47-50; instr aeronaut eng, Pa State Univ, 52-56, asst prof, 56-57; res eng gas dynamics, Missile & Space Vehicle Div, Gen Elec Co, Pa, 57-59; sr res scientist, plasma & space appl physics, Radio Corp Am, NJ, 59-64; prof aerospace eng & chmn dept, 64-81, PROF MECH & AEROSPACE ENG, WVA UNIV, 81- *Concurrent Pos:* Consult, RCA Corp, NJ, 64- *Mem:* Sigma Xi; assoc fel Am Inst Aeronaut & Astronaut. *Res:* Heat transfer; Laminar incompressible and compressible boundary layer theory; mass addition in boundary layer; ablation of reentry vehicles; blast wave theory of conducting fluids in magnetic fields; magnetohydrodynamic alternating current power generation; low speed aerodynamics. *Mailing Add:* Dept Mech Eng WVa Univ Box 6101 Morgantown WV 26506

FARADAY, BRUCE (JOHN), b New York, NY, Dec 9, 19; m 50; c 5. SOLID STATE PHYSICS. *Educ:* Fordham Univ, AB, 40, MS, 47; Cath Univ, PhD, 63. *Prof Exp:* Solid state supvry physicist, US Naval Res Lab, 48-72, head semiconductor sect, 65-72, consult radiation effects, 72-74, head, Radiation Effects Br, 74-80, prog magr low observables, 80-86; PRES, FARADAY ASSOCS INC, 86- *Concurrent Pos:* Lectr, Prince George's Community Col, 60-79, Northern Va Community Col, 70-; lectr, Univ Md, 67-70; sci prog adminr, Off Naval Res, 70-71; prog adminr electronics support, Naval Mat Command, 73. *Honors & Awards:* Super Civilian Serv Award, Dept Navy, 81. *Mem:* Fel Am Phys Soc; Sigma Xi; NY Acad Sci; AAAS; Asn Old Crows. *Res:* Radiation effects in materials and devices; modification of materials by radiation; energy conversion; radar absorbing materials; low observables technology. *Mailing Add:* Faraday Assocs Inc 8607 Sinon St Annandale VA 22003

FARAG, IHAB HANNA, b Cairo, Egypt, Aug 17, 46; US citizen; m 74; c 2. COMPUTER SIMULATION & GRAPHICS, REAL-TIME ACQUISITION & CONTROL. *Educ:* Cairo Univ, Egypt, BS, 67; Mass Inst Technol, MS, 70, ScD(chem eng), 76. *Prof Exp:* Instr chem eng, Mass Inst Technol, 73-75; asst prof, 76-82, ASSOC PROF CHEM ENG, UNIV NH, 82- *Concurrent Pos:* Vis scientist, chem eng dept, Mass Inst Technol, 82-83; consult, Owens-Ill, Inc, Gilette, Inc, FMC, Riley-Stoker & Mass Inst Technol; fac mem, Morgantown Energy Technol Ctr, WVa, 84. *Honors & Awards:* Goodwin Medal; Baker Award. *Mem:* Am Inst Chem Engrs. *Res:* Coal-water slurries and fluidized bed combustion; multistage fluidized bed calciner; furnace modeling, including Monte Carlo; finite element applications; heat transfer in glass forming and in solar energy; microcomputer for real-time data acquisition; staged cascaded fluidized bed combustor modeling; process simulation using ASPEN. *Mailing Add:* Chem Eng Dept Univ NH Durham NH 03824-3591

FARAGO, JOHN, b Budapest, Hungary, Sept 12, 17; nat US; m 45; c 2. POLYMER CHEMISTRY, APPLIED PSYCHOLOGY. *Educ:* Budapest Tech Univ, Dipl Chem Eng, 39, Dr Tech Sc, 47; Va Commonwealth Univ, MS, 68. *Prof Exp:* Asst to org chair, Univ Sci, Budapest, 39-40; res supvr, Egger Pharmaceut, 40-43; res engr, Grab Textile Factory, Hungary, 44; asst to org chair, Univ Sci, Budapest, 44-45; asst dir, Chem Inst City Budapest, 46-47; res assoc, George Washington Univ, 47-51, res prof, 52; from res chemist to sr res chemist, E I du Pont de Nemours & Co, Inc, 52-60, res assoc, 60, res supvr, 61-70, res fel, Textile Fibers Dept, 70-83; PRES, JFC INC, 88- *Concurrent Pos:* Lectr, George Washington Univ, 48-52. *Mem:* Am Psychol Asn; Biofeedback Soc Am. *Res:* Structure, application and analysis of polymers; learning; motivation. *Mailing Add:* 10424 Iron Mill Rd Richmond VA 23235

FARAH, ALFRED EMIL, b Nazareth, Palestine, July 10, 14; m 71. MEDICAL RESEARCH, PHARMACOLOGY. *Educ:* Am Univ Beirut, BA, 37, MD, 40. *Prof Exp:* From instr to asst prof pharmacol, Am Univ Beirut, 40-45; vis lectr, Harvard Med Sch, 45-47; asst prof, Med Sch, Univ Wash, 47-50; from assoc prof to prof, State Univ NY Upstate Med Ctr, 50-68, chmn dept, 53-68; dir biol div, Sterling-Winthrop Res Inst, 68-71, chmn, 72-78; vpres res, Sterling Drug Inc, 78-84; CONSULT, 84- *Concurrent Pos:* Res fel, Harvard Med Sch, 45-47; secy, State Univ NY, Albany Found, 75- *Mem:* Am Soc Pharmacol; Soc Exp Biol & Med; Cardiac Muscle Soc; hon mem Ger Pharmacol Soc; Sigma Xi; Am Heart Asn. *Res:* Cardiac and kidney pharmacology; cardiac glycosides; mercurial diuretics; secretory activity of kidney; pharmacology of enzyme inhibitors; drug development; inotropic agents. *Mailing Add:* 6001 Pelican Bay Blvd Apt 1406 Naples FL 33963-8168

FARAH, BADIE NAIEM, US citizen. EXPERT SYSTEMS, SOFTWARE SYSTEMS. *Educ:* Univ Damascus, BSc, 67, MA, 68; Wayne State Univ, MS, 73; Ohio State Univ, MSIE, 76, PhD(indust & systs eng), 77. *Prof Exp:* Teaching asst, Computer Sci Dept, Wayne State Univ, 71-73; res assoc, Instr & Res Computer Ctr, Ohio State Univ, 73-77; sr systs analyst, Gen Motors Corp, 77-78; asst prof, Col Bus, Oakland Univ, 78-82; from asst prof to assoc prof, 82-91, PROF, COL BUS, EASTERN MICH UNIV, 91- *Concurrent Pos:* Consult, 77-; vis gen mgr & consult, S&G Grocer Co, 78-82. *Mem:* Asn Comput Mach; Inst Indust Engrs; Opers Res Soc Am; Inst Mgt Sci. *Res:* Knowledge-based systems and their potential use in computer integrated manufacturing and flexible manufacturing systems; application of information systems technology to computer integrated manufacturing. *Mailing Add:* Dept Opers Res & Info Systs Eastern Mich Univ Ypsilanti MI 48197

FARAH, FUAD SALIM, b Apr 5, 29. DERMATOLOGY. *Educ:* Am Univ Beirut, MD, 54. *Prof Exp:* PROF MED & DERMAT, HEALTH SCI CTR, STATE UNIV NY, SYRACUSE, 76- *Res:* Host parasite relationships; cellular immunology. *Mailing Add:* Health Sci Ctr State Univ NY 750 E Adams St Syracuse NY 13210

FARAJ, BAHJAT ALFRED, ALCOHOLISM, REYES SYNDROME. *Educ:* Univ Kans, PhD(med chem), 71. *Prof Exp:* ASSOC PROF RADIOL & DIR IN-VITRO NUCLEAR MED, EMORY UNIV, 72- *Res:* Metabolic disorders. *Mailing Add:* Dept Radiol Emory Univ Sch Med 467 Woodruf Mem Bldg Atlanta GA 30322

FARAS, ANTHONY JAMES, b Chisholm, Minn, Dec 23, 42; m 66; c 2. VIROLOGY, MOLECULAR BIOLOGY. *Educ:* Univ Minn, BA, 65; Univ Colo, PhD(path), 70. *Prof Exp:* Fel microbiol & virol, Med Sch, Univ Calif, 70-73; asst prof, Med Sch, Univ Mich, 73-75; assoc prof, 75-78, PROF MICROBIOL & VIROL & DIR, INST HUMAN GENETICS, MED SCH, UNIV MINN, 78- *Concurrent Pos:* Consult med oncol, Med Sch, Univ Minn, 75-; mem, Exp Virol Study Sect, NIH, 79-83; mem adv bd, Battell Int Conf Genetic Eng; assoc ed, Virol; founder, Molecular Genetics Inc, 79. *Mem:* Am Soc Microbiol; AAAS; Sigma Xi; NY Acad Sci. *Res:* Molecular mechanisms by which viruses, particularly RNA tumor viruses and papilloma viruses, replicate and induce disease. *Mailing Add:* Dir Inst Human Genetics Univ Minn Box 206 Harvard St & E River Rd Minneapolis MN 55455

FARB, EDITH, b Philadelphia, Pa, Aug 7, 28; div; c 2. ORGANIC CHEMISTRY. *Educ:* Univ Pa, BA, 49, MS, 51; Bryn Mawr Col, PhD(phys & org chem), 58. *Prof Exp:* Asst chem, Rohm & Haas Co, 49-50; demonstr, Bryn Mawr Col, 51-52; asst prof, Long Island Univ, 58-67; lectr, Hunter Col, 67-70; lit chemist, Texaco Develop Corp, 70-73; analyst, New York Dept Health Environ Health Serv, 73-76, analyst, Bur Lead Poisoning Control, 76-78, sr res assoc & chief, Environ Unit, 78-83; forensic chemist, Lab, New York Police Dept, 83-88; CHEMIST, NEW YORK CITY DEPT ENVIRON PROTECTION, BUR WASTEWATER TREATMENT, 88- *Concurrent Pos:* Adj lectr, Barnard Col, 88. *Mem:* Am Chem Soc. *Res:* Aromatic substitution; physical properties of aromatic compounds; toxic materials; air contaminants; water quality. *Mailing Add:* 63-58 78th St Middle Village NY 11379

FARBER, ANDREW R, b New York, NY, 53. COMPUTER SCIENCE, ACCOUNTING. *Educ:* Columbia Col, BA, 76, MBA, 81. *Prof Exp:* Mgt consult, Bennett Kielson & Co, 87-89; MGT CONSULT, SMA MGT SYSTS, INC, 89- *Mem:* Asn Comput Mach. *Mailing Add:* 50 Main St White Plains NY 10606

FARBER, ELLIOT, b New York, NY, May 7, 32. PHYSICS. *Educ:* Brooklyn Col, BS, 54; Columbia Univ, AM, 56; Stevens Inst Technol, PhD(physics), 66. *Prof Exp:* From jr physicist to physicist, Naval Res Lab, 53-54; microwave engr, Sylvania Elec Prods, Inc, 56; from instr to asst prof physics, Pratt Inst, 57-67; ASSOC PROF PHYSICS, NJ INST TECHNOL, 67- *Concurrent Pos:* Res assoc, Stevens Inst Technol, 65-67. *Mem:* Am Asn Physics Teachers. *Res:* Plasma physics. *Mailing Add:* Dept Physics NJ Inst Technol Newark NJ 07102

FARBER, ELLIOTT, b NY, Mar 12, 28; m 55; c 4. POLYMER CHEMISTRY. *Educ:* NY Univ, BA, 50; Polytechn Inst Brooklyn, MS, 52, PhD(polymer chem), 59. *Prof Exp:* Chemist, FMC, 60-62; mgr polymer res & develop, Tenneco Inc, 62-70; dir res & develop, Lamaur Inc, 70-77; mgr cosmetic develop, Gen Nutrit Corp, 77-88; DIR TECH OPERS, SWEEN CORP, 88- *Concurrent Pos:* Lectr, Cosmetics for Pharmacists, N Dak State Univ, 84-85; lectr, Pharmaceut Cosmetics, Univ Minn, 91- *Mem:* Am Chem Soc; Soc Cosmetics Chemists. *Res:* Developed complete line of natural cosmetics for general nutrition stores; developing line of personal care items and over the counter drugs for hospitals and nursing homes; author of five papers; granted six patents. *Mailing Add:* Sween Corp 1940 Commerce Dr North Mankato MN 56003

FARBER, EMMANUEL, b Toronto, Ont, Oct 19, 18; US citizen; m 42; c 1. BIOCHEMISTRY, PATHOLOGY. *Educ:* Univ Toronto, MD, 42; Univ Calif, Berkeley, PhD(biochem), 49; FRCP(C), 75. *Hon Degrees:* Dr Med & Surg, Univ Torino, Italy, 85. *Prof Exp:* Instr path, Sch Med, Tulane Univ, 50-51, asst prof path & lectr biochem, 51-55, assoc prof path & biochem, 55-59, Am Cancer Soc res prof, 59-61; prof path & chmn dept, Sch Med, Univ Pittsburgh, 61-70; Am Cancer Soc res prof path & biochem & sr investr, Fels Res Inst, Sch Med, Temple Univ, 70-74, prof path & biochem & dir, 74-75; chmn dept path, 75-85, PROF, DEPT PATH, UNIV TORONTO, 75-, DEPT BIOCHEM, 85- *Concurrent Pos:* Am Cancer Soc fel cancer res, Cook County Hosp, Ill, 49-50; mem adv comt smoking & health, Surgeon Gen, 62; chmn path B study sect, NIH, 62-66; consult, Div Chronic Dis, Dept HEW; vis prof, Middlesex Hosp, Med Sch, Univ London, 68-69; vpres, Asn Cancer Res, 71-72, bd dirs, 70-73; mem panel D, Nat Cancer Inst, Can, 77-79; chmn panel 1, Comn Food Safety & Food Safety Policy, Nat Acad Sci, 78-79; mem, Chem Path Study Sect, NIH, 80-82; assoc dir, Can Ctr Toxicol, 80-; cancer coordr, Univ Toronto, 82-85. *Honors & Awards:* Second Annual Parke-Davis Award, 58; Fourth Annual Teplitz Mem Award, 61; Samuel R Noble Found Award, 76; Rous-Whipple Award, Am Asn Pathologists, 82; G H A Clowes Mem Award, Am Asn Cancer Res, 84. *Mem:* Am Gastroenterol Asn; Am Soc Biol Chem; Am Chem Soc; Am Soc Exp Pathologists (vpres, 72-73, pres, 73-74); Am Asn Path & Bact; fel Royal Soc Can; hon mem Soc Toxicol Pathologists; Am Asn Pathologists; Am Asn Cancer Res (pres, 72-73). *Res:* Biochemical pathology; carcinogenesis, cytochemistry and histochemistry. *Mailing Add:* Path-Med Sci Bldg One Kings College Circle Toronto ON M5S 1A8 Can

FARBER, ERICH A(LEXANDER), b Vienna, Austria, Sept 7, 21; nat US; m 49; c 2. MECHANICAL ENGINEERING. *Educ:* Univ Mo, BS, 43, MS, 46; Univ Iowa, PhD(mech eng), 49. *Prof Exp:* Mach operator, Erving Paper Mills, Mass, 40-41; drafting & blueprinting, City of Columbia, Mo & Univ Mo, 41-43; instr physics & math, Univ Mo, 43-46; instr mech eng, Univ Iowa, 46-49; asst prof, Univ Wis, 49-54, assoc prof, 54; prof mech eng & res prof, 54-80, DISTINGUISHED SERV PROF, UNIV FLA, 80-, DIR SOLAR ENERGY LAB, 65- *Concurrent Pos:* Consult to various indust & govt agencies, 46- *Honors & Awards:* Worcester Reed Warner Gold Medal, Am Soc Mech Engrs. *Mem:* Fel Am Soc Mech Engrs; Am Soc Eng Educ; Solar Energy Soc. *Res:* Heat transfer; solar energy; fluid flow; thermodynamics; energy conversion. *Mailing Add:* Dept Mech Eng Univ Fla Gainesville FL 32611

FARBER, EUGENE M, b Buffalo, NY, July 24, 17; m 44; c 4. DERMATOLOGY. *Educ:* Oberlin Col, AB, 39; Univ Buffalo, MD, 43; Univ Minn, MS, 46. *Prof Exp:* Intern, Buffalo Gen Hosp, NY, 43-44; fel dermat & syphilol, Mayo Clin, 44-48, asst, 47-48; instr dermat, Sch Med, Stanford Univ, 48, asst prof path, 49-50, assoc prof dermat, 50-54, clin prof, 54-59, prof dermat, 59-85, chmn dept, 50-85, dir, Psoriasis Day Care Ctr, 73-85, PRES, PSORIASIS RES INST, STANFORD UNIV, 85- *Concurrent Pos:* Consult, Surg Gen, USAF, 57-64; consult, Calif State Dept Pub Health, 63-; mem gen clin res ctr comt, NIH, 65; pres, Found Int Dermat Educ; pres, Orinoco Found; Howard Fox Mem lectr, NY Acad Med, 71; founder & ed, Int Psoriasis Bull, 73- *Honors & Awards:* Jose Maria Vargas Award, Cent Univ, Caracas, 72; Taub Int Mem Award Psoriasis Res, 74. *Mem:* Soc Exp Biol & Med; Am Soc Clin Invest; Soc Invest Dermat (vpres, 65); Am Acad Dermat; Am Asn Prof Dermat (secy, 67, pres, 68); Sigma Xi. *Res:* Peripheral vascular diseases; psoriasis; mycosis fungoides; tropical dermatology; epidemiology; cutaneous blood flow in various dermatoses and psoriasis. *Mailing Add:* Psoriasis Res Inst Stanford Univ 600 Town & Country Village Palo Alto CA 94301

FARBER, FLORENCE EILEEN, b New York, NY, Aug 11, 39. MOLECULAR BIOLOGY, VIROLOGY. *Educ:* Mt Holyoke Col, AB, 61; Columbia Univ, PhD(biochem), 66. *Prof Exp:* Lectr cell physiol, Univ Calif, Berkeley, 68-69; asst prof physiol & biophys, Col Med, Univ Vt, 69-71; asst prof genetics & microbiol, Mt Holyoke Col, 71-72; res assoc, Baylor Col Med, 72-73, asst prof virol, 74-78; asst prof genetics & microbiol, Carleton Col, 78-81; ASST PROF MICROBIOL & GENETICS, UNIV NH, 81- *Concurrent Pos:* NSF fel, Free Univ Brussels, 66-68; Am Heart Asn sr fel, Univ Calif, Berkeley, 68-69; Am Cancer Soc grant, Col Med, Univ Vt, 70-71; Edmond de Rothschild Found int observer, Weizmann Inst Sci, 68. *Mem:* Am Soc Microbiol. *Res:* Mechanism of herpes simplex virus-induced chromosomal alterations; mechanism of herpes simplex virus virion assembly. *Mailing Add:* 5605 Eldorado Ave Altoona PA 16601

FARBER, HERMAN, b New York, NY, Dec 3, 19; m 43; c 2. ELECTROPHYSICS. *Educ:* Brooklyn Col, BA, 41; Polytech Inst Brooklyn, MEE, 52. *Prof Exp:* Res engr, Bristol Co, 46-49; from asst to assoc prof elec eng, 54-60, assoc prof, 60-86, dir eve elec eng studies, 73-86, dir elec eng labs, 74-86, EMER PROF ELECTROPHYS, POLYTECH INST NY, BROOKLYN, 86. *Concurrent Pos:* Res chemist, Manhattan Proj, 43-46. *Mem:* NY Acad Sci; Am Asn Physics Teachers; sr mem Inst Elec & Electronics Engrs; AAAS; Sigma Xi. *Res:* Electromagnetic properties of materials, including the electric strength of solids and liquids at microwave frequencies; cryogenic engineering; plasma diagnostics; chemical synthesis using discharge and plasmas; relaxation phenomena at microwave frequencies. *Mailing Add:* 147C Maclura Plaza Cranbury NJ 08512

FARBER, HUGH ARTHUR, b Muskegon, Mich, Oct 6, 33; m 54; c 3. ORGANIC CHEMISTRY, TOXICOLOGY. *Educ:* Mich State Univ, BSCh, 56; Northwestern Univ, PhD(org chem), 60. *Prof Exp:* Res org chemist, Dow Chem, 59-66, proj leader, 66-68, group leader res & develop, 68-76, mgr environ affairs, 76-80, sr mgr environ affairs, 80-85; CONSULT ENVIRON & HEALTH, 86- *Mem:* Am Chem Soc; Sigma Xi; Am Asn Advan Sci. *Res:* Preparation of monomer, polymer modification and agricultural chemicals; study of reaction mechanisms; chlorinated solvents; ecology of chlorinated solvents; toxicology and environmental effects research; product stewardship; government regulation and legislative liaison/management. *Mailing Add:* 2807 Highbrook Dr Midland MI 48640

FARBER, JAY PAUL, b New York, NY, Jan 25, 42; m 68; c 1. RESPIRATORY PHYSIOLOGY, NEUROPHYSIOLOGY. *Educ:* City Col New York, BS, 63; State Univ NY, Buffalo, PhD(physiol), 69. *Prof Exp:* Fel respiratory physiol, Dartmouth Med Sch, 69-71; asst prof, Univ Iowa, 71-78; assoc prof, 78-84, PROF PHYSIOL, UNIV OKLA HEALTH SCI CTR, 84- *Concurrent Pos:* Res grants, NIH, 72-, res career develop award, 80-84. *Mem:* Am Physiol Soc; Am Col Sports Med; AAAS; Soc Neurosci. *Res:* The control of breathing and its development, emphasizing the functional naturation of respiratory neruons. *Mailing Add:* Dept Physiol & Biophysics Univ Okla Health Sci Ctr PO Box 26901 Oklahoma City OK 73190

FARBER, JOHN LEWIS, CELL INJURY, CARCINOGENESIS. *Educ:* Univ Calif, San Francisco, MD, 64. *Prof Exp:* Prof path, Sch Med, Hahnemann Univ, 82-86; PROF PATH, JEFFERSON MED COL, THOMAS JEFFERSON UNIV, 86- *Mailing Add:* Dept Path Thomas Jefferson Univ Rm 203A Main Bldg Philadelphia PA 19107

FARBER, JORGE, b Buenos Aires, Arg, Feb 29, 48; US citizen; m 77; c 1. NEUROPHYSIOLOGY, SLEEP. *Educ:* City Col NY, BA, 70, PhD(psychol), 75. *Prof Exp:* Adj lectr psychol, City Col NY, 72-74; NIMH fel, Albert Einstein Col Med, 74-76 & Montefiore Hosp & Med Ctr, 76-77; asst prof psychiat, Health Sci Ctr, Univ Tex, 77-84; RETIRED. *Concurrent Pos:* Res fel, Alfred P Sloan Found, 77-80. *Honors & Awards:* Gardner Murphy Award; Henry L Moses Award; Olto Peterson Award. *Mem:* NY Acad Sci; Soc Neurosci; AAAS; Asn Psychophysiol Study of Sleep. *Res:* Neurophysiological and neurochemical mechanisms involved in the mediation of the sleep and wake state; the role of rapid eye movement sleep in the maturation of the central nervous system in the newborn. *Mailing Add:* 415 Beach 146th St Nesconset NY 11694

FARBER, JOSEPH, b Newark, NJ, June 1, 24; m 51; c 2. ENERGY CONVERSION, FLUID PHYSICS. *Educ:* City Col New York, BS, 45; Univ Wis, PhD(phys chem), 51. *Prof Exp:* From res engr to sr thermodyn engr, Convair Div, Gen Dynamics Corp, Calif, 51-55; mgr real gas eng, Gen Elec Co, 55-56, mgr aerophys sect, Space Sci Lab, 56-64, mgr adv systs eng, 64-67; chief engr, Space & Reentry Systs Div, Philco-Ford Corp, 67-69, prog mgr, Mid-Course Surv Systs, 69-70; pres & gen mgr, KMS Technol Ctr, Div KMS Industs, Inc, 70-73; consult & lectr, Space Defense Systs, J F Assocs, 73-84; proj mgr, Solar Energy, Am Diversified, 84-86; SR SCIENTIST & ACTG DIR, GEN RES CORP, 86- *Concurrent Pos:* pres, Solar Res Systs, 75-84. *Mem:* Am Chem Soc; Am Phys Soc; Am Inst Aeronaut & Astronaut; Int Solar Energy Soc; Sigma Xi. *Res:* Systems engineering; fluid dynamics; space science; aerophysics; magnetohydrodynamics; solar thermal power generation; solar plastic heaters. *Mailing Add:* 1605 Sherrington Pl Suite Y212 Newport Beach CA 92663-6004

FARBER, MARILYN DIANE, b Los Angeles, Calif, Apr 21, 45. EPIDEMIOLOGY, PUBLIC HEALTH. *Educ:* Univ Calif, Los Angeles, BA, 67, MPH, 71, DrPH(epidemiol), 77. *Prof Exp:* Surv supvr hypertension res, Univ Calif, Los Angeles, Sch Pub Health, 73-77; instr, 77-78, ASST PROF EPIDEMIOL, MED CTR, SCH PUB HEALTH, UNIV ILL, CHICAGO, 78-, ASST PROF EPIDEMIOL & OPHTHAL, 83- *Mem:* Int Epidemiol Asn; Soc Epidemiol Res; Asn Researchers in Vision & Ophthal; Int Soc & Fedn Cardiol; Sigma Xi. *Res:* Ocular epidemiology; cancer, especially cancer of the breast; sickle cell anemia; sickle cell retinopathy; compliance studies in hypertension; ocular epidemiology. *Mailing Add:* Eye/Ear Infirmary Univ Ill 1855 W Taylor Chicago IL 60612

FARBER, MILTON, b Los Angeles, Calif, Oct 6, 16; m 42; c 3. PHYSICAL CHEMISTRY. *Educ:* Univ Calif, BS, 38; Univ Minn, MS, 39; Calif Western Univ, PhD, 76. *Prof Exp:* Chief chem engr, Colloidal Prod, 41-42; area supvr, Ky Ord Works, 42-43; sr res engr, Manhattan Proj, 43-46; sr res engr, Jet Propulsion Lab, Calif Inst Technol, 46-55; assoc dir res, Aerojet-Gen Corp Div, Gen Tire & Rubber Co, 55-57, head propulsion lab, Hughes Tool Co, 57-59; vpres, Maremont Corp, Rocket Power, Inc, 59-67; PRES, SPACE SCI, INC, 67- *Mem:* Am Chem Soc; Am Inst Physics; Am Chem Soc; The Chem Soc; Am Inst Aeronaut & Astronaut. *Res:* Thermodynamics; separation of isotopes; thermal diffusion; mass spectroscopy; kinetics; air pollution. *Mailing Add:* Space Sci Inc 135 W Maple Ave Monrovia CA 91016-3426

FARBER, PAUL ALAN, b Brooklyn, NY, Sept 13, 38; m 60; c 2. MICROBIOLOGY, DENTISTRY. *Educ:* Univ Mich, AB, 60, DDS, 62; Univ Rochester, PhD(microbiol), 67. *Prof Exp:* Asst prof microbiol, 67-72, assoc prof, 72-78, PROF PATH, SCH DENT, TEMPLE UNIV, 78-; CLIN PROF ORAL MED, DENT COL, NY UNIV, 80- *Concurrent Pos:* NIH spec fel, Nat Inst Dent Res, 70-71; guest worker, Albert Einstein Med Ctr, 75-76; distinguished vis scholar, Univ Adelaide, Australia, 83; guest worker, Nat Inst Arthritis, Digestive Diabetes & Kidney Dis, NIH, 85. *Mem:* AAAS; Am Soc Microbiol; Am Asn Pathologists; Int Asn Dent Res. *Res:* Periodontal disease; immunology of bacterial and virus infection. *Mailing Add:* 21 Glenn Circle Philadelphia PA 19118

FARBER, PHILLIP ANDREW, b Wilkes-Barre, Pa, Sept 19, 34; m; c 4. HUMAN GENETICS & CYTOGENETICS, HISTOLOGY. *Educ:* King's Col, BS, 56; Boston Col, MS, 58; Cath Univ Am, PhD(biol), 63. *Prof Exp:* Teaching asst biol, Boston Col, 56-57, St Louis Univ, 58-59 & Cath Univ Am, 59-63; asst instr, Georgetown Univ, 62-63; res biologist, Lab Perinatal Physiol, Nat Inst Neurol Dis & Blindness, NIH, 63-64; res instr, phys med & rehab, Med Ctr, NY Univ, 64-66; PROF BIOL & ALLIED HEALTH SCI, BLOOMSBURGH UNIV, PA, 66- *Concurrent Pos:* Summer res biologist, Comn Officer Student Training & Extern Prog, USPHS, NIH, 60; res fel, NSF, 62; USPHS res grant, 65; consult clin cytogenetics, Dept Lab Med & Path, Geisinger Med Ctr, 67-; Summer res fel, radiation biol, Space Radiation Effects Lab, NASA, Newport News, Va, 69. *Mem:* Nat Soc Histotechnol; Teratology Soc; Sigma Xi; Am Soc Human Genetics; AAAS; Asn Cytogenetic Technologists. *Res:* Cytogenetics of human chromosomal syndromes and leukemias; cytogenetics of Peromyscus leucopus (white footed mouse); cytogenetics of Maccaca mulatta (rhesus monkey); neurosecretion in annelid regeneration; effects of carbon tetrachloride on mitoses in Crocus zonatus. *Mailing Add:* Dept Biol & Allied Health Sci Bloomsburg Univ Bloomsburg PA 17815

FARBER, ROBERT JAMES, b Oak Ridge, Tenn, May 2, 46; m; c 2. VISIBILITY, TECHNICAL MANAGEMENT. *Educ:* Yale Univ, BS, 68; Univ Wash, MS, 71, PhD(aerosol chem), 75. *Prof Exp:* Res asst, Univ Wash, 68-72; mgr environ sci, STD Res Corp, 75-76; SR RES SCIENTIST, SOUTHERN CALIF EDISON CO, 76- *Mem:* Air Pollution Control Asn; Am Meteorol Soc; Sigma Xi. *Res:* Air pollution, meteorology, cloud physics and weather modification as these affect the environment; quantifying the measurement and monitoring of visibility impairment and sky coloration phenomena; chemical formations of major constituents in particulate smog in the Los Angeles area. *Mailing Add:* Southern Calif Edison Co 2244 Walnut Grove Ave Rosemead CA 91770

FARBER, ROSANN ALEXANDER, b Charlotte, NC, Nov 21, 44; m 73; c 2. HUMAN GENETICS, SOMATIC CELL GENETICS. *Educ:* Oberlin Col, AB, 66; Univ Wash, PhD(genetics), 73. *Prof Exp:* Fel genetics, Nat Inst Med Res, Mill Hill, London, 73-75 & Children's Hosp Med Ctr, Boston, 75-77; asst prof microbiol & genetics, Univ Chicago, 77-84, assoc prof molecular genetics & cell biol, 84-87, assoc prof obstet & gynec, 87-88; ASSOC PROF PATH, CURRIC IN GENETICS, UNIV NC, 88- *Concurrent Pos:* Fel, Jane Coffin Childs Mem Fund Med Res, 73-75; career develop res award, NIH, 81-86, mem, Mammalian Genetics Study Sect, 85-89. *Mem:* AAAS; Am Soc Human Genetics. *Res:* Human molecular genetics; somatic cell genetics; molecular genetics of therapy-related acute myelord leukemia. *Mailing Add:* Dept Path Univ NC CB No 7525 Brinkhous Bullitt Bldg Chapel Hill NC 27599-7525

FARBER, SAUL JOSEPH, b New York, NY, Feb 11, 18; m 49; c 2. MEDICINE. *Educ:* NY Univ, AB, 38, MD, 42; Am Bd Internal Med, dipl, 55. *Hon Degrees:* DPhil, Tel Aviv Univ, 83. *Prof Exp:* From instr to prof med, 50-66, actg dean, 63-66, 79-81 & 82-87, PROF INTERNAL MED & CHMN DEPT, SCH MED, NY UNIV, 66-, DIR MED, UNIV HOSP & BELLEVUE HOSP, 66- FREDERICK H KING PROF & DEAN, ACAD AFFAIRS, 78-, DEAN, 87- *Concurrent Pos:* Estab Investr Award, Am Heart Asn, 60-66; mem bd dirs, NY Heart Asn, 63-, pres, 73-75; mem, Nat Adv Res Resources Coun, 67-71; mem adv coun, NY Kidney Dis Inst, 68-; mem adv comt, Inter-Soc Comn Heart Dis Resources, 68-; mem, Am Bd Internal Med, 68-76, chmn, 73-76; mem med adv comt, Hosp Corp Task Force, NY, 69-71; mem bd dirs, Russell Sage Inst Path, 70-; mem med adv bd, Found Study Wilson's Dis, 71- & Hadassah; mem, Riverside Res Inst, 71-, trustee, 79-; mem, Irma T Hirschl Charitable Trust Sci Adv Comt, 72-; chmn comt resource requirements Vet Admin health care syst, Nat Res Coun, 74-77; ed, Am J Med Sci; mem adv comt, Spaciality & Geog Distribution Physicians, Inst Med, 74-; mem, Health Adv Coun, NY State, 75-80; mem bd trustees, Sackler Sch Med, Tel Aviv Univ, 77-, mem bd gov, 79-; mem adv comt long term care-chronic illness, Robert Johnson Found, 79-; mem, Comn on Nursing, 80-; mem, New York City Bd Health, 82-; spec adv med educ, Naval Med Command, Washington, DC, 85- *Mem:* Inst Med-Nat Acad Sci; Am Soc Clin Invest (secy-treas, 57-60); Harvey Soc (treas, 63-67, vpres, 67-68, pres, 68-69); master Am Col Physicians (pres, 84-85); Asn Prof Med (pres, 73-74); Asn Am Physicians; Am Clin & Climat Asn; Int Soc Technol Assessment Health Care. *Res:* Physiological and biochemical clinical investigation related to human disease; quality of health care and manpower on local, state and national levels; cardiovascular-renal diseases; renal prostaglandins. *Mailing Add:* Dept Internal Med Sch Med NY Univ 550 First Ave New York NY 10016

FARBER, SERGIO JULIO, b Arg, Jan 30, 38; m 60; c 3. ORGANIC CHEMISTRY, CLINICAL PATHOLOGY & BIOCHEMISTRY. *Educ:* Univ Buenos Aires, MS, 62, PhD(org chem), 65. *Prof Exp:* Teaching asst org chem, Univ Buenos Aires, 62-64, instr, 64-66; res chemist, Univ Calif, Santa Barbara, 67-68; sr res chemist, Calbiochem, Inc, 68-70; lab dir, Nuclear Dynamics, Inc, El Monte, 70-72; res assoc biochem procedures, 72-74, biochemist, 74-87, chief biochemist, 80-87, ASSOC DIR CLIN CHEM, CEDARS-SINAI MED CTR, LOS ANGELES, 87- *Mem:* NY Acad Sci; Am Asn Clin Chem; Nat Acad Clin Biochem. *Res:* Current methodology in clinical chemistry with emphasis in immunoassays and enzymology. *Mailing Add:* 475 Ladera St Monterey Park CA 91754

FARBER, SEYMOUR MORGAN, b Buffalo, NY, June 3, 12; m 40; c 3. THORACIC DISEASES & PULMONARY DISEASES. *Educ:* Univ Buffalo, BA, 31; Harvard Univ, MD, 39. *Prof Exp:* From instr to prof, 42-62, vchancellor pub prog & continuing educ, 73-78, CLIN PROF MED, SCH MED, UNIV CALIF, SAN FRANCISCO, 62- *Concurrent Pos:* In-chg tuberc & chest serv, San Francisco Gen Hosp, 45-62; lectr, Sch Pub Health, Univ Calif, Berkeley, 48-60, spec asst to pres, 64-; spec consult, Nat Cancer Inst, 58-60; nat consult to Surgeon Gen, USAF, 62-68; mem President's Comt Status of Women, 62-63. *Mem:* Am Col Chest Physicians (pres elect, 59, pres, 59-60); Am Col Cardiol; AMA; Am Fedn Clin Res; NY Acad Sci. *Res:* Cancer of lung; pulmonary cytology; chemotherapy of lung cancer; chemotherapy of tuberculosis pulmonary pathophysiology; continuing education in medicine and the health sciences. *Mailing Add:* 26303 Esperanza Dr Los Altos Hills CA 94022

FARBER, THEODORE MYLES, b New York, NY, July 20, 35; m 60; c 3. PHARMACOLOGY, TOXICOLOGY. *Educ:* Long Island Univ, BS, 57; Med Col Va, PhD(parmacol), 62. *Prof Exp:* Instr pharmacol, Med Col Va, 60-61; asst prof, George Washington Univ Med Sch, 61-65; pharmacologist, 65-73, sr res pharmacologist, 73-78, SUPVR PHARMACOLOGIST, FOOD & DRUG ADMIN, 78- *Concurrent Pos:* Lectr, Howard Univ Med Sch, 75-, US Dept Agr Grad Sch, 73- & NIH Eve Sch, 76- *Mem:* Sigma Xi; Soc Toxicol; Am Soc Pharmacol & Exp Therapeut; Soc Exp Biol & Med. *Res:* Drug metabolism; drug interactions; biochemical pharmacology and toxicology of drugs; food additives and veterinary drugs. *Mailing Add:* Sci Regulatory Serv Inst 1625 K St NW Washington DC 20036

FARBMAN, ALBERT IRVING, b Boston, Mass, Aug 25, 34; div; c 3. HISTOLOGY, CYTOLOGY. *Educ:* Harvard Univ, AB, 55, DMD, 59; NY Univ, MS, 61, PhD(basic med sci), 64. *Prof Exp:* Instr anat, Sch Med, NY Univ, 62-64; from asst prof to prof anat, 64-72, ASSOC DEAN GRAD SCH, NORTHWESTERN UNIV, 75-, PROF NEUROBIOL & PHYSIOL, 81- *Concurrent Pos:* Res career develop award, Nat Inst Dent Res, 66-71; vis scientist, Strangeways Res Lab, Cambridge, 68-69; mem bd dir, McGaw Med Ctr, 75-79. *Mem:* AAAS; Am Asn Anat; Am Soc Cell Biol; Soc Neurosi; Soc Develop Biol. *Res:* Cytodifferentiation of taste buds; differentiation of olfactory mucosa; neurosciences. *Mailing Add:* Dept Neurobiol & Physiol Northwestern Univ 2153 Sheridan Rd Evanston IL 60208

FARCASIU, DAN, b Carei, Romania, Aug 22, 37; US citizen; m 60; c 1. CATALYSIS. *Educ:* Polytech Inst, Bucharest, BS & MS, 59; Polytech Inst, Timisoara, Romania, PhD(org chem), 69. *Prof Exp:* Researcher, Inst Atomic Physics, Romania, 64-69; res assoc, City Univ New York, 69-71; res assoc, Princeton Univ, 71-72, instr, 72-73; mem staff org chem, Squibb Inst Med Res, 73-74; staff chemist, Corp Res Labs, Exxon Res & Eng Co, 74-86; prof, dept chem, Clarkson Univ, 86-89; RES PROF, DEPT CHEM ENG, UNIV PITTSBURGH, 90- *Mem:* Am Chem Soc. *Res:* Physical organic chemistry, especially organic reaction mechanisms; acid crystals; synthetic organic chemistry. *Mailing Add:* Dept Chem Eng Univ Pittsburgh 1249 Benedum Hall Pittsburgh PA 15261

FAREL, PAUL BERTRAND, b Camden, NJ, Sept 29, 44; m 67; c 2. NEUROSCIENCES. *Educ:* Univ Calif, Berkeley, AB, 66; Univ Calif, Los Angeles, MA, 67, PhD(psychol), 70. *Prof Exp:* from instr,72-73, from asst prof to assoc prof, 73-87, PROF PHYSIOL, SCH MED, UNIV NC, CHAPEL HILL, 87- *Concurrent Pos:* NIMH fel, Univ Calif, Irvine, 70-72; NSF res grant, Univ NC, Chapel Hill, 74-77 & 77-80; NIH res grant, 80- *Mem:* AAAS; Soc Neurosci. *Res:* Development and regeneration of specific neuronal connections in spinal cord. *Mailing Add:* Dept Physiol-CB7545 Univ NC Sch Med Chapel Hill NC 27599

FAREWELL, JOHN P, b Worcester, Mass, May 29, 42; m 68; c 2. PAPER CHEMISTRY, PHYSICAL CHEMISTRY. *Educ:* State Univ NY Col Plattsburgh, BSEd, 64; State Univ NY Buffalo, PhD(chem), 69. *Prof Exp:* Res scientist, 68-72, res engr, Bleached Papers, 72-76; sr res scientist, Am Cyanamid Corp, 76-88; APPLNS CHEMIST, A E STALEY MFG CO, 88- *Honors & Awards:* Hugh Camp Award Singular Achievement in Res, Union Camp Corp, 73. *Mem:* Am Chem Soc; Tech Asn Pulp & Paper Indust. *Res:* Physical chemistry of polymers; properties of paper; process modeling and optimization; physical chemistry of coating processes; emulsion stability. *Mailing Add:* Box 225E RR1 West Okaw Estates Lovington IL 61937-9801

FARHADIEH, ROUYENTAN, b Tehran, Iran, Nov 3, 44; US citizen; m 76; c 4. HEAT TRANSFER, FIBER OPTICS. *Educ:* Univ Ariz, BS, 68; Stanford Univ, MS, 70; Northwestern Univ, PhD(mech eng), 74. *Prof Exp:* mech engr, reactor anal & safety, Argonne Nat Lab, 75-84; engr consult, 84-86; sr scientist, Amphendl Corp, 86-88; vpres & dir, Forss Inc, 90; PRES, SINEX ENG CO, 88- *Concurrent Pos:* Fel, Argonne Nat Lab, 74-75. *Mem:* Am Soc Mech Eng; Am Soc Heating, Refrig & Airconditioning Engrs; Sigma Xi; Nat Soc Prof Engrs. *Res:* Natural laminar or turbulent convection; melting and freezing coupled with natural convection; high temperature studies; energy management; fiber optics. *Mailing Add:* 352 63rd St Clarendon Hills IL 60514

FARHATAZIZ, MR, b Amritsar, India, Dec 19, 32; nat US; m 70; c 3. PHYSICAL CHEMISTRY, RADIATION CHEMISTRY. *Educ:* Panjab Univ, Pakistan, BSc, 53, MSc, 54; Cambridge Univ, PhD(chem), 59; Univ Tex, MD, 88. *Prof Exp:* Lectr pharmaceut chem, Panjab Univ, 54-56, lectr chem, 59-60; sr sci officer, Pakistan AEC, 60-68, prin sci officer, 68-76; asst prof specialist, Univ Notre Dame, 74-76; asst prof, Tex Woman's Univ, 77-81, assoc prof, 81-84; res internal med, St Louis Hosp, Dallas, 88-90; RES RADIATION ONCOL, UNIV TEX HEALTH SCI CTR, SAN ANTONIO, 90- *Concurrent Pos:* Columbo Plan fel, UK Atomic Energy Authority, 61-62; fel Radiation Lab, Univ Notre Dame, 64-67, 70-74; prin investr, Robert A Welch Found, Tex Woman's Univ, 78- *Mem:* Am Chem Soc; AAAS; Radiation Res Soc. *Res:* Primary processes in radiation chemistry; fast chemical kinetics; chemical and physical effects at high pressures up to 7000 atmospheres and low temperatures down to 4 degrees K; management of cancer. *Mailing Add:* 2321 Georgetown Denton TX 76204

FARHI, DIANE C, b Cleveland, Ohio, Nov 27, 51; m; c 2. HEMATOPATHOLOGY, SURGICAL PATHOLOGY. *Educ:* Univ Mich, MD, 77. *Prof Exp:* Asst prof, Univ Colo Health Sci Ctr, 81-84; from asst prof to assoc prof path, Case Western Reserve Univ, 84-90; ASSOC PROF PATH, UNIV TENN SOUTHWESTERN MED CTR, 90- *Mem:* Int Acad Path; AAAS; Soc Hematopath. *Mailing Add:* Southwestern Med Ctr Univ Tex 5323 Harry Hines Blvd Dallas TX 75235-9072

FARHI, EDWARD, b New York, NY, June 26, 52. PARTICLES AND FIELDS. *Educ:* Brandeis Univ, BA, 73, MA, 73; Harvard Univ, PhD(physics), 78. *Prof Exp:* Res assoc, Stanford Linear Acceleration Ctr, 78-80; sci assoc, CERN, Geneva, Switz, 80-81; postdoctorate physics, 81-82, asst prof physics, 82-86, ASSOC PROF PHYSICS MASS INST TECHNOL, 86- *Mem:* Am Phys Soc. *Res:* Theory of elementary particles and cosmology. *Mailing Add:* Ctr Theoret Phys Mass Inst Technol 6-302 Cambridge MA 02139

FARHI, LEON ELIE, b Cairo, Egypt, Oct 9, 23; US citizen; m 49; c 2. CARDIOPULMONARY & ENVIRONMENTAL PHYSIOLOGY. *Educ:* Am Univ Beirut, BSc, 40; Univ St Joseph, Lebanon, MD, 47. *Prof Exp:* Resident, Hadassah Univ Hosp, Israel, 50-52; res fel, Trudeau Sanatorium, 53; res fel physiol, Univ Rochester, 53-54; res fel & asst physician, Sch Med, Johns Hopkins Univ, 54-55; instr physiol & pulmonary dis, Hebrew Univ, Israel, 56-58; from asst prof to prof physiol, 58-89, chmn, 82-91, DISTINGUISHED PROF PHYSIOL, STATE UNIV NY BUFFALO, 89- *Concurrent Pos:* Vis prof, Univ Fribourg & NSF sr fel, 65-66; consult, Erie County Health Dept & USPHS. *Mem:* Am Physiol Soc; Aerospace Med Asn; Undersea Med Soc; Biomed Eng Soc; Am Heart Asn. *Res:* Pulmonary physiology and physiopathology; environmental and cardio-respiratory physiology. *Mailing Add:* Dept Physiol 124 Sherman Hall State Univ NY 3435 Main St Buffalo NY 14214

FARID, NADIR R, b Dilling, Sudan, Apr 4, 44; Can citizen; c 2. IMMUNOGENETICS. *Educ:* Univ Khartoum, MS, 67; MRCP, 71; 73; AM FACP, 81. *Prof Exp:* Intern med, Univ Kartoum Teaching Hosp, 67-68; registrar internal med, Khartoum Civil Hosp, 68-69; med registrar med & endocrinol, Royal Victoria Infirmary Newcastle-upon-Tyne, UK, 69-71, res registrar metab 71-72; res fel immunol & endocrinol, Univ Toronto, 72-74; from asst prof to assoc prof med & endocrinol, Mem Univ Nfld, 74-84, chief Div Endocrinol & Metab, 81-91, metab prof med, 84-91; PROF & CHMN, DEPT MED, KING FAISAL SPECIALIST HOSP & RES CTR, RIYADH, SAUDI ARABIA, 91- *Concurrent Pos:* Dir, Thyroid Res Lab, Health Sci Ctr, 74-, med dir, Clin Invest Unit, 77-, consult, Div Nuclear med, 78-, Chief, Div Endocrinol & Metab, Gen Hosp, 81-, prof immunol, Div Basic Sci, 88- *Mem:* Am Soc Clin Invest; Am Thyroid Asn; Endocrine Soc; Am Diabetes Asn; Can Soc Immunol; Can Soc Clin Invest. *Res:* Immunology and immunogenetics of endocrine disorders; major histocompatibility complex function and association with human diseases. *Mailing Add:* King Faisal Specialist Hosp Dept Med-MBC 46 PO Box 3354 Riyadh 11211 Saudi Arabia

FARIES, DILLARD WAYNE, b Mooreland, Okla, Sept 28, 41; m 65; c 3. QUANTUM OPTICS. *Educ:* Rice Univ, BA, 63; Univ Calif, Berkeley, PhD(physics), 69. *Prof Exp:* Geophys trainee, Shell Oil Co, 63; res asst physics, Univ Calif, Berkeley, 65-69; from asst prof to assoc prof, 69-81, PROF PHYSICS & CHMN DEPT, WHEATON COL, 81- *Mem:* Am Asn Physics Teachers; Am Sci Affiliation. *Res:* Nonlinear interaction of electromagnetic fields with matter. *Mailing Add:* Dept Physics & Geol Wheaton Col Wheaton IL 60187

FARINA, JOSEPH PETER, b Queens, NY, May 11, 31; m 55; c 4. MICROBIAL PHYSIOLOGY, HEALTH SCIENCES. *Educ:* St John's Col, BS, 53; St John's Col, NY, MS, 58, MSE, 60, PhD(microbiol), 67. *Prof Exp:* Teacher, NY High Sch, 58-67; assoc prof biol, Univ Lowell, 67-71, dir, med technol maj, 70-75, prof biol, 71-90, spec asst to dean, Sch Allied Health & Natural Sci, 74-75, prof, clin lab sci, 74-90, chmn, dept clin lab sci, Col Health Professions, 75-85, EMER PROF CLIN LAB SCI, UNIV LOWELL, 90- *Concurrent Pos:* Sr lab technician, Mercy Hosp, NY, 58-63, instr, 62-67, instr radiol physics, 67-; instr radiol physics, Peninsular Gen Hosp, NY, 65-66; chmn prof educ comt, Am Cancer Soc, 73-75; admin internship allied health professions, State Univ NY, Albany, 74-75. *Mem:* AAAS; Am Soc Zoologists; NY Acad Sci; Soc Protozoologists; Am Inst Biol Sci. *Res:* Axenic cultivation and nutritional requirements of Blepharisma, a pink ciliate; hematology. *Mailing Add:* Seven Singlefoot Rd Chelmsford MA 01824

FARINA, PETER R, b New York, NY, Apr 30, 46; m 68; c 2. BIO-ORGANIC CHEMISTRY. *Educ:* Hofstra Univ, BS, 67; State Univ NY Buffalo, PhD(org chem), 72. *Prof Exp:* Fel, Pa State Univ, 71-73; res scientist, Corp Res Lab, Union Carbide Corp, 74-78, group leader med prod div, 78-79; sr prin biochemist, 80-84, group leader. 84-90, DIR BIOCHEM, BOEHRINGER INGELHEIM PHARMACEUT INC, 90- *Mem:* Am Chem Soc; AAAS; Am Soc Pharmacol & Exp Therapeut. *Res:* Discovery and elucidation of mechanism actions for anti-inflammatory and anti-viral agents; development of bioanalytical methods for drug discovery. *Mailing Add:* RFD 1 Sunset Dr North Salem NY 10560

FARINA, ROBERT DONALD, b Schenectady, NY, Sept 29, 34; m 66; c 3. BIOINORGANIC CHEMISTRY. *Educ:* Rensselaer Polytech Inst, BChE, 57; Union Col, MS, 63; State Univ NY, Buffalo, PhD(chem), 68. *Prof Exp:* Engr, Stauffer Chem Co, 57-60; test engr, Knolls Atomic Power Lab, Gen Elec Co, 60-63; fels chem, Univ Calif, 67-69 & Univ Utah, 69; PROF CHEM, W KY UNIV, 69- *Mem:* Am Chem Soc; Am Inst Chem Eng. *Res:* Kinetic studies of fast reactions in solution; coordination chemistry of transition metal complexes; kinetic studies of metalloenzymes in biological systems. *Mailing Add:* Dept Chem Western Ky Univ Bowling Green KY 42101

FARINA, THOMAS EDWARD, b Brooklyn, NY, Dec 9, 41; m 66; c 2. ORGANIC CHEMISTRY. *Educ:* St Bernadine of Siena Col, BS, 64; State Univ NY Buffalo, PhD(org chem), 70. *Prof Exp:* Res scientist, Union Camp Corp, 69-78; GROUP LEADER, GLYCO CHEM, INC, 78- *Mem:* Am Chem Soc; Am Oil Chemists Soc. *Res:* Fatty acid; rosin acid; hydantoin chemistry. *Mailing Add:* Lonza Inc R&D 79 Rte 22 E Annandale NJ 08801-9603

FARIS, JOHN JAY, b Grandview, Wash, Nov 7, 21; m 42; c 4. PHYSICS. *Educ:* Reed Col, BA, 43; Univ Wash, PhD(physics), 51. *Prof Exp:* Assoc prof physics, Pac Univ, 50-54; from asst prof to prof, Colo State Univ, 54-68; chmn dept, 68-80, PROF PHYSICS, UNIV WIS-STOUT, 68- *Mem:* Am Phys Soc; Am Asn Physics Teachers. *Res:* Secondary emission of electrons; electroretinagram; microwaves; magnetism. *Mailing Add:* c/o Johnann F McKee 4545 W 29th Ave Denver CO 80212

FARIS, SAM RUSSELL, b Moore, Okla, Nov 5, 17; m 42; c 5. PHYSICAL CHEMISTRY, ANALYTICAL CHEMISTRY. *Educ:* Univ Okla, BS, 42, MS, 46, PhD(chem), 49. *Prof Exp:* Sr res technologist, 49-54, RES ASSOC, FIELD RES LAB, MOBIL RES & DEVELOP CORP, 54- *Mem:* Fel AAAS; Am Chem Soc; NY Acad Sci. *Res:* Electrochemistry; electrokinetics. *Mailing Add:* 2541 Delmac Dr Dallas TX 75233

FARIS, WILLIAM GUIGNARD, b Montreal, Que, Nov 22, 39; US citizen; m 72; c 2. MATHEMATICAL PHYSICS. *Educ:* Univ Wash, BA, 60; Princeton Univ, PhD(math), 65. *Prof Exp:* Asst prof math, Cornell Univ, 64-70; mathematician, Battelle Inst, Geneva, Switz, 70-74; assoc prof, 74-80, PROF MATH, UNIV ARIZ, 80- *Mem:* Am Math Soc; Int Asn Math Physics. *Res:* Operator theory and quantum mechanics; probability and statistical mechanics. *Mailing Add:* Dept Math Univ Ariz Tucson AZ 85721

FARISH, DONALD JAMES, b Winnipeg, Man, Dec 7, 42; m 64, 83; c 3. ETHOLOGY, POPULATION GENETICS. *Educ:* Univ BC, BSc, 63; NC State Univ, MS, 65; Harvard Univ, 69; Univ Mo, JD, 76. 69. *Prof Exp:* Instr zool & entom, Univ Mo-Columbia, 68-69, asst prof, 70-74, chmn physiol & behav sect, 74-75, assoc prof biol sci & entom, 74-79; assoc dean arts & sci, Univ RI, 79-83; DEAN, SCH NATURAL SCI & PROF BIOL, SONOMA STATE UNIV, 83-, ACTG VPRES, ACAD AFFAIRS, 90- *Concurrent Pos:* Univ Mo Res Coun grant, 69, 70 & 71; NIMH grant, 72-73; Grant Found award, 75-77. *Mem:* AAAS. *Res:* Insect behavioral genetics; evolution of behavior; polymorphism; biology and law. *Mailing Add:* Office of Dean, Sch Natural Scis Sonoma State Univ Rohnert Park CA 94928

FARISON, JAMES BLAIR, b McClure, Ohio, May 26, 38; m 61; c 2. SYSTEMS ENGINEERING, BIOMEDICAL ENGINEERING. *Educ:* Univ Toledo, BS, 60; Stanford Univ, MS, 61, PhD(elec eng), 64. *Prof Exp:* From asst prof to assoc prof elec eng, 64-74, asst dean grad studies, 69-70, actg dean, 70-71, dean, 71-80, PROF ELEC ENG, UNIV TOLEDO, 74- *Concurrent Pos:* Consult, Med Col Ohio, 81; prin investr, NSF res grants, 66-67, Am Heart Asn, 83-84 & Edison Indust Systs Ctr, 88-90; adj prof, Med Col Ohio, 87-; collateral fac, Ohio Aerospace Inst, 90- *Honors & Awards:* Young Engr of the Year Award, Ohio Soc Prof Engrs, 73. *Mem:* Sr mem Inst Elec & Electronics Engrs; Am Soc Eng Educ; sr mem Instrument Soc Am; Nat Soc Prof Engrs; Machine Vision Asn Soc Mfg Engrs. *Res:* Biomedical modeling and medical imaging; image processing; stability robustness and control of discrete-time systems; machine vision and industrial tomography applications. *Mailing Add:* Col Eng Univ Toledo Toledo OH 43606-3390

FARISS, BRUCE LINDSAY, b Allisonia, Va, July 22, 34; c 8. ENDOCRINOLOGY, INTERNAL MEDICINE. *Educ:* Roanoke Col, BS, 57; Univ Va, MD, 61; Am Bd Internal Med, dipl. *Prof Exp:* Gen med officer, US Army Hosp, Ft Monroe, Va, 62-63; chief endocrine serv, Madigan Army Med Ctr, Tacoma, Wash, 68-76; consult endocrinol & internal med, Hq US Army Med Command, Europe, 76-79; Chief, Dept Clin Invest & dir, Endocrine Fel Prog, Madigan Army Med Ctr, Tacoma, 79-84; consult endocrinol, Surgeon Gen US Army, 79-84; ADJ PROF, DEPT BIOL, VA POLY-TECH INST, BLACKSBURG, VA, 84- *Concurrent Pos:* Intern, Univ Va Hosp, 61-62; resident internal med, Brooke Gen Hosp, Ft Sam Houston, Tex, 63-66; fel, Endocrinol Metab Res Unit, Univ Calif, San Francisco, 66-68; consult surg gen, US Army Endocrinol, 79-84; bd dirs Va affl, Am Diabetes Asn; bd supvr, Pulaski County, Va, 87-91. *Mem:* Am Fedn Clin Res; Endocrine Soc; Am Diabetes Asn; fel Am Col Physicians. *Res:* Endocrinology and metabolism; carbohydrate metabolism; adrenal gland; testicular function; pineal gland. *Mailing Add:* Rte HC02 Box 31 Allisonia VA 24347

FARISS, ROBERT HARDY, b St Louis, Mo, May 31, 28; m 50; c 5. CHEMICAL ENGINEERING. *Educ:* Wash Univ, BS, 50; Mass Inst Technol, MS, 51, DSc(chem eng), 54. *Prof Exp:* Chem engr, Mallinckrodt Chem Works, 54-59; chem engr, 59-67, TECHNOL DIR, MONSANTO CO, 67- *Mem:* Am Inst Chem Engrs; Am Chem Soc; Soc Plastics Engrs. *Res:* Thermodynamics of phase equilibria; kinetics of heterogeneous catalysis; applied mathematics; optimization; statistics. *Mailing Add:* Monsanto Chem Co 730 Worcester St Springfield MA 01151

FARKAS, DANIEL FREDERICK, food dehydration, food preservation; deceased, see previous edition for last biography

FARKAS, EUGENE, b Melvindale, Mich, Dec 11, 26; m 56; c 3. ORGANIC CHEMISTRY, MEDICINAL CHEMISTRY. *Educ:* Wayne State Univ, BS, 49, PhD(org chem), 52. *Prof Exp:* Res assoc, Mass Inst Technol, 52-53; res assoc, Wayne State Univ, 53-54; sr scientist, 54-70, sr scientist, Anal Develop Metab, 70-80, SR SCIENTIST PROCESS RES, LILLY RES LABS, 80- *Mem:* Am Chem Soc. *Res:* Steroids; alkaloids and natural products; organic synthesis. *Mailing Add:* Dept IC 742 Bldg 110 Eli Lilly & Co Indianapolis IN 46285

FARKAS, LESLIE GABRIEL, b Ruzomberok, Czech, Apr 18, 15; Can citizen; m 71; c 1. PLASTIC SURGERY. *Educ:* Univ Bratislava, MD, 41; Charles Univ, Prague, CSc, 59, DSc(plastic surg), 68; FRCS(C), 73. *Prof Exp:* Resident surg, Mil Hosp & Field Serv, Slovakia, 41-45; resident plastic surg, Charles Univ, Prague, 45-48, from asst prof to assoc prof, 48-68; asst prof, 70-78, assoc prof, Plastic Surg, 78-80, EMER ASSOC PROF PLASTIC SURG & SPEC LECTR, DEPT SURG, UNIV TORONTO, 80- *Concurrent Pos:* Dep dir, Plastic Surg Res Lab & dir, Div Congenital Anomalies, Acad Sci, Prague, 63-68; clin fel, Div Plastic Surg, Hosp Sick Children, Toronto, 68-69, res fel, Div Exp Surg, 69-70, asst scientist, Res Inst, 70-77, sr scientist, 77-, dir, Plastic Surg Res Lab, Res Inst, 70-81; consult, Cleft Palate Prog, Univ Iowa, 75-78, Res Inst, Univ Toronto, 80- & Behav Neurol Dept, Eunice Kennedy Sr Ctr, Waltham, Boston, 84- *Honors & Awards:* Award Excellence for Res in Attractive Face, Am Soc Aesthetic Plastic Surg, 85. *Mem:* Am Soc Plastic & Reconstruct Surgeons; Can Soc Plastic Surgeons; Plastic Surg Res Coun; Can Asn Anatomists; Biomat Soc Can; Can Craniofacial Soc. *Res:* Experimental plastic surgery; tendon repair; quantitative anatomical and functional assessment; anatomy and growth of experimentally reconstructed urethra; quantitative surface anatomy of growing, healthy, congenitally and traumatically damaged face; use of anthropometry in evaluation of morphological changes in patients with congenital anomalies of the cranio-orbito-facial complex; application of newly established rules about facial proportions in study of facial syndromes. *Mailing Add:* Dept Surg 555 Univ Toronto Toronto ON M5S 1A8 Can

FARKAS, WALTER ROBERT, b New York, NY, June 30, 33; m 56; c 2. BIOCHEMISTRY. *Educ:* City Col New York, BS, 55; Duke Univ, PhD(biochem), 60. *Prof Exp:* Res asst biochem, Duke Univ, 55-60; res assoc, Sch Med, NY Univ, 60-63; fel hemat, Col Physicians & Surgeons, Columbia Univ, 63-66; from asst prof to assoc prof, 66-75, actg head dept, 81-84, PROF MED BIOL, MEM RES CTR, UNIV TENN, 75-, PROF ENVIRON PRACT, 85- *Concurrent Pos:* Life Ins Med Res Fund grant, 67-70; NIH grant, 68-75; Am Cancer Soc res grant, 77-80; NIH res grants, 75-78, 79-82 & 81-90, USDA grant, 80-82. *Honors & Awards:* Beecham Pharmaceut Award, Res Excellence, 86. *Mem:* Am Chem Soc; Am Soc Biol Chemists; Am Col Toxicologists; AAAS; Soc Exp Hemat. *Res:* Saturnine gout; biosynthesis of lysine; control mechanisms in hemoglobin biosynthesis; biological effects of metals; transfer RNA of erythroid cells; discovery of guanylation of tRNA; discovery that lead depolymerizes RNA. *Mailing Add:* Dept Environ Pract Univ Tenn Col Vet Med Knoxville TN 37901-1071

FARKAS-HIMSLEY, HANNAH, b Moscow, Russia, May 3, 18; Can citizen; m 40; c 2. MEDICAL MICROBIOLOGY, ONCOLOGY. *Educ:* Hebrew Univ, Israel, MSc, 40, PhD(bact), 46. *Prof Exp:* Asst bact, Hadassah Med Sch, Hebrew Univ, 49-53; attache sci, Israel Embassy, Eng, 53-55; asst bact, Hadassah Med Sch, Hebrew Univ, 55-57; fel, 58-59, grant, 59, from asst prof to assoc prof, 59-83, EMER PROF MICROBIOL, FAC MED, UNIV TORONTO, 83-; ASSOC SCIENTIST, MT SINAI HOSP, 81- *Concurrent Pos:* Res grants, Nat Res Coun Can, 60-61, Med Res Coun Can, 61-70, Can Nat Health, 67-69, 73-75, WHO, 72-73 & Nat Cancer Inst, 75-78 & 81-84, Hosp Sick Children Found, 85-87, Univ Res Incentive Fund, 85-88, Atkinson Charitable Found, 89; vis prof, Lautenberg Ctr Gen & Tumor Immunol, Hebrew Univ Med Sch, Jerusalem, 78-79. *Mem:* NY Acad Sci; Can Soc Microbiol; Can Col Microbiologists; Royal Inst Gt Brit. *Res:* Streptomycin and penicillin resistance; Vibriocin production by Vibrio comma, mode of action; halogen resistant bacteria; microassay in-vitro for recognition of enterotoxins; bacterial proteinaceous products as cytotoxic agents of neoplasia; diagnosis of pseudomallei; survey of staphylococci on Easter Island; bacterial proteinaceous products in diagnosis and selective therapy of neoplasia and virus-infected cells, such as HIV infected cells. *Mailing Add:* Dept Microbiol Univ Toronto Fac Med Toronto ON M5S 1A8 Can

FARKASS, IMRE, b Budapest, Hungary, Sept 26, 19; nat US; m 51. APPLIED PHYSICS, TELECOMMUNICATIONS. *Educ:* Budapest Tech Univ, Dipl, 42. *Prof Exp:* Asst prof physics, Budapest Tech Univ, 42-46, lectr, Inst Physics, 47-49, assoc head, Vacuum Res Lab, 50-56; head dept math & physics, Agr Univ Budapest, 54-56; sr physicist, Nat Res Corp, 57-60; dir appl physics dept, Ilikon Corp, 61-65; MEM TECH STAFF, BELL LABS, 65- *Concurrent Pos:* Invited lectr, Mass Inst Technol, 64, Northeastern Univ, 65 & Johns Hopkins Univ, 69. *Mem:* Am Phys Soc. *Res:* Applied physics; vacuum physics and technology. *Mailing Add:* Four Stonehedge Lane Madison NJ 07940

FARLEE, RODNEY DALE, b Albany, Ore, Oct 9, 52. SPECTROSCOPY & NUCLEAR MAGNETIC RESONANCE, MOLECULAR STRUCTURE & SPECTROSCOPY DATABASES. *Educ:* Univ Idaho, BS, 74; Univ Ill, Urbana, PhD, 79. *Prof Exp:* CHEMIST, CENT RES & DEVELOP DEPT, E I DU PONT DE NEMOURS & CO, INC, 79-90. *Mem:* Am Chem Soc; Sigma Xi. *Res:* Spectroscopic, particularly nuclear magnetic resonance; studies of solid materials, including polymers, fossil fuels and catalysts. *Mailing Add:* 3316 Spring Garden St Bio-Rad Sadtler Div Philadelphia PA 19104

FARLEY, BELMONT GREENLEE, b Cape Girardeau, Mo, Dec 29, 20; m 53; c 3. BIOPHYSICS, INFORMATION SCIENCE. *Educ:* Univ Md, BS, 41; Yale Univ, MS, 46, PhD(physics), 48. *Prof Exp:* Asst math, Mass Inst Technol, 41-42, mem staff, Radiation Lab, 42-45; instr physics, Yale Univ, 47-48; mem tech staff, Bell Labs, 48-53; mem staff, Lincoln Labs, Mass Inst Technol, 53-64; assoc prof biophys, Johnson Found, Sch Med, Univ Pa,

64-70; from assoc prof to prof, 70-87, EMER PROF INFO SCI, TEMPLE UNIV, 87- *Mem:* Am Math Soc; Am Physiol Soc; Inst Elec & Electronics Engrs; Sigma Xi. *Res:* Theoretical and experimental neurophysiology. *Mailing Add:* 319 Bala Ave Bala Cynwyd PA 19004-2648

FARLEY, DONALD T, JR, b New York, NY, Oct 26, 33; m 56; c 3. IONOSPHERIC PHYSICS, PLASMA PHYSICS. *Educ:* Cornell Univ, BEngPhys, 56, PhD(ionospheric physics), 60. *Prof Exp:* NATO fel ionospheric physics, Cambridge Univ, 59-60; docent, Chalmers Univ Technol, Sweden, 60-61; physicist, Jicamarca Radar Observ, Lima, Peru, US Nat Bur Standards, 61-64, dir, 64-67; PROF ELEC ENG, CORNELL UNIV, 67- *Concurrent Pos:* Assoc ed, Rev Geophys & Space Physics, 63-69 & J Geophys Res, 74-77 & Radio Sci, 76-78 & 83-85; mem comn III & IV, Int Sci Radio Union, Exec Comt Comn IV, 66-69; Tage Erlander vis prof, Uppsala Ionospheric Observ, Sweden, 85-86. *Honors & Awards:* Nat Bur Standards Award, 63; Environ Sci Serv Admin Award, 64; US Dept Commerce Gold Medal, 67. *Mem:* Am Geophys Union; Inst Elec & Electronics Engrs; AAAS. *Res:* Scattering of radio waves from thermal fluctuations in a plasma; plasma instabilities in the ionosphere. *Mailing Add:* Sch Elec Eng Cornell Univ Ithaca NY 14853-3801

FARLEY, EUGENE SHEDDEN, JR, b Upland Borough, Pa, Feb 6, 27; m 55; c 4. MEDICINE, PUBLIC HEALTH. *Educ:* Swarthmore Col, BA, 50; Univ Rochester, MD, 54; Johns Hopkins Univ, MPH, 67. *Prof Exp:* Intern, Philadelphia Gen Hosp, 54-55; resident gen pract, Med Ctr, Univ Colo, 55-56; asst instr prev med, Med Col, Cornell Univ, 56-58; resident internal med, Univ Vt & De Goesbriand Hosp, 58-59; pvt pract, 59-66; resident fel pub health, Sch Hyg & Pub Health, Johns Hopkins Univ, 66-67; from assoc prof to prof family med, Sch Med, Univ Rochester, 67-78, dir family med prog, 67-78; prof & chmn dept family med, Univ Colo Med Ctr, 78-82; PROF & CHMN DEPT FAMILY MED & PRACT, UNIV WIS, 82- *Concurrent Pos:* Vis prof dept social & prev med, Univ WI, Kingston, Jamaica, 77-78. *Mem:* Am Acad Gen Practice; Asn Am Med Cols; Am Pub Health Asn; Asn Teachers Prev Med; Soc Teachers Family Med. *Res:* Development and implementation of systems of primary care; use of ancillaries who allow provision of more efficient and effective medical care to those needing it; built-in research potential to all primary care practices. *Mailing Add:* Dept Family Pract Univ Wis 777 S Mills St Madison WI 53715

FARLEY, JAMES D, b Olney, Ill, Sept 28, 38; m 65; c 2. PHYTOPATHOLOGY. *Educ:* Ill Wesleyan Univ, BS, 61; Mich State Univ, MS, 63, PhD(plant path), 68. *Prof Exp:* Asst res plant path, Univ Calif, Berkeley, 68-69; assoc prof plant path, Ohio State Univ, 69-82; PRES, DE RUITER SEEDS INC, 82- *Mem:* Am Phytopath Soc. *Res:* Soil borne diseases; tomato breeding. *Mailing Add:* 2954 Scioto Pl Columbus OH 43221

FARLEY, JERRY MICHAEL, REGULATION OF MUSCLE RECEPTORS, TRACHEAL EPITHELIUM. *Educ:* WVa Univ, PhD(physiol & biophys), 76. *Prof Exp:* ASSOC PROF MED PHARMACOL, MED CTR, UNIV MISS, 80- *Res:* Physiology of smooth muscle. *Mailing Add:* Dept Pharmacol Univ Miss Med Ctr 2500 N State St Jackson MS 39216

FARLEY, JOHN, b Eng, Apr 23, 36; Can citizen; m 60; c 4. HISTORY OF BIOLOGY, MEDICINE. *Educ:* Univ Sheffield, BSc, 59; Univ Western Ont, MSc, 61; Univ Man, PhD(zool), 64. *Prof Exp:* From asst prof to assoc prof, 64-78, PROF BIOL, DALHOUSIE UNIV, 78- *Concurrent Pos:* Res fel hist of sci, Harvard Univ, 70-71 & 77-78; Can Coun fel, 77-78; SSHRC fel, 85-86. *Honors & Awards:* Hannah Medal, Royal Soc Can. *Mem:* Hist Sci Soc; Am Soc Hist Med; Can Soc Hist & Philos Sci. *Res:* History of parasitology and tropical medicine; history of international health agencies; germ theory of disease. *Mailing Add:* Dept Biol Dalhousie Univ Halifax NS B3H 3J5 Can

FARLEY, JOHN WILLIAM, b Brooklyn, NY, Feb 7, 48; m; c 2. HIGH PRECISION MEASUREMENTS, LASER SPECTROSCOPY. *Educ:* Harvard Col, BA, 70; Columbia Univ, MA & MPh, 74, PhD(physics), 77. *Prof Exp:* Res assoc physics, Univ Ariz, 76-78, NSF teaching fel, 78-79, teaching fel physics & optical sci, dept physics & Optical Sci Ctr, 79-80, res asst prof, 80-81; asst prof physics, Dept Physics & Chem Physics Inst, Univ Ore, 81-87; ASSOC PROF PHYSICS, DEPT PHYSICS, UNIV NEV, LAS VEGAS, 87- *Mem:* Am Phys Soc; Inst Elec & Electronics Engrs; Laser & Electro-optic Soc; Optical Soc Am. *Res:* Laser physics; quantum optics; high precision measurements; unconventional spectroscopy; highly excited atoms; molecular ions; double resonance. *Mailing Add:* Dept Physics Univ Nev Las Vegas NV 89154

FARLEY, REUBEN WILLIAM, b Richmond, Va, Sept 21, 40; m 81; c 2. MATHEMATICS. *Educ:* Randolph-Macon Col, BS, 61; Univ Tenn, MA, 65, PhD(math), 68. *Prof Exp:* Instr math, Randolph-Macon Col, 61-63; asst prof math, Mary Washington Col, 66-67; assoc prof, 68-81, PROF MATH SCI, VA COMMONWEALTH UNIV, 81-, CHMN DEPT, 90- *Mem:* Am Math Soc; Math Asn Am. *Res:* Topological semigroups. *Mailing Add:* 1015 W Main St Richmond VA 23284-2014

FARLEY, ROGER DEAN, b Jefferson, Iowa. ANIMAL PHYSIOLOGY. *Educ:* Univ Northern Iowa, BA, 57; Univ Iowa, MS, 62; Univ Calif, Santa Barbara, PhD(neurophysiol), 66. *Prof Exp:* NIH fel, Tufts Univ, 66-67; asst prof, 67-74, ASSOC PROF ZOOL, UNIV CALIF, RIVERSIDE, 74- *Mem:* Am Soc Zoologists; Am Inst Biol Sci; AAAS; Soc Neuroethology; Soc Neurosci; Am Soc Arachnologists. *Res:* Behavior and physiology of desert scorpions. *Mailing Add:* Dept Biol Univ Calif Riverside CA 92521

FARLEY, THOMAS ALBERT, b Washington, DC, Feb 10, 33; m 58; c 2. TECHNOLOGY PLANNING. *Educ:* George Washington Univ, BS, 54; Mass Inst Technol, PhD(physics), 59; Univ Rochester, MBA, 82. *Prof Exp:* Mem tech staff, Space Technol Labs, Thompson-Ramo-Wooldridge, Inc, 59-61; res geophysicist, Inst Geophys & Planetary Physics, Univ Calif, Los Angeles, 61-73; mgr, Technol Ctr, Xerox Corp, 73-77, mgr advan develop, 76-81, prin scientist, 81-87, mgr technol planning, 87-91; RETIRED. *Mem:* Am Phys Soc. *Mailing Add:* 287 Brooksboro Dr Webster NY 14580

FARLOW, STANLEY JEROME, b Emmetsburg, Iowa, Mar 7, 37; m 67. APPLIED MATHEMATICS. *Educ:* Iowa State Univ, BS, 59; Univ Iowa, MS, 62; Ore State Univ, PhD(math), 67. *Prof Exp:* Mathematician, NIH, 62-68; ASST PROF MATH, UNIV MAINE, ORONO, 68- *Mem:* Am Math Soc; Math Asn Am. *Res:* Mathematical modeling of biological systems; partial differential equations; control theory; numerical analysis; computer systems. *Mailing Add:* 104 Forest Ave Orono ME 04473

FARM, RAYMOND JOHN, b Mankato, Minn, Sept 1, 39; m 61; c 3. ANALYSIS OF PATENT OR OTHER LITIGATION RELATED SAMPLES. *Educ:* Univ Minn, BChem, 61, MS, 63. *Prof Exp:* Analytical chemist, Dow Chem Co, 63-67; sr chemist, 67-72, from res specialist to sr res specialist, 72-83, STAFF SCIENTIST, 3M, 83- *Mem:* Am Chem Soc. *Res:* Development of analytical methods for polymers, commercial products and complex mixtures; thin layer chromatography, hydrolysis and derivatization techniques. *Mailing Add:* 3M Ctr Bldg 201-BS-07 St Paul MN 55144

FARMAN, ALLAN GEORGE, b Birmingham, Eng, July 26, 49; US citizen; m 72; c 2. ORAL & MAXILLOFACIAL RADIOLOGY. *Educ:* Univ Birmingham, BDS, 71; Royal Col Surgeons, LDSRCS, 72; Univ Stellenbosch, SAfrica, PhD(oral pathol), 77; Univ Louisville, EDS, 83, MBA, 87. *Prof Exp:* Pvt pract dent surg, Eng, 71-72, 77-78; asst prof,oral path, Univ Witwaterstrand, 72-74; assoc prof, Univ Stellenbosch, SAfrica, 74-77; head oral biol & path dept, Univ Riyadh, Saudi Arabia, 78-79; from asst prof to assoc prof, 80-82, PROF ORAL & MAXILLOFACIAL RADIO, UNIV LOUISVILLE, KY, 84- *Concurrent Pos:* Extern examnr oral path, Univ Western Cape, 76-77, Univ Pretoria, 74; consult radiol & sr specialist, Cape Prov-Tygerberg Hosp, 75-77; vis prof, Dept Dent Diag Sci, Univ Tex, San Antonio, 81, Dept Oral Surg, Univ Stellenbosch, 88 & Fac Dent, Bagota, Columbia, 88; Fulbright vis prof, US Educ Found India, 87; nat bd test consult, Joint Comn Dental Exam, 86- & Dent Assisting Nat Bd, 89-; ed, Oral Surg, Oral Med & Oral Path, Am Dent Asst Asn Dent Radiography Modular Ser, 89 & Kodak Dent Radiography Ser, 89; chairperson, Oral Radiol Sect, Am Asn Dent Schs; adj prof diag radiol & anat sci & neurobiol, Univ Louisville, 90-; med staff mem, Humana Hosp Univ, 89-, Jewish Hosp, Norton Hosp & Kosair Hosp, 90- *Honors & Awards:* Leith-Neumann Prize, 70; Philip-Jennens Prize, 71; Frank Stammers Prize, 71; Harry Crussley Award, 75 & 76. *Mem:* Am Asn Dent Schs; Am Acad Oral & Maxillofacial Radiol; Int Asn Dentomaxillofacial Radiol; Int Asn Dent Res; Am Dent Asn; Int Asn Oral Path. *Res:* Dental oncology; effects of restorative materials on therapeutic radiation treatment planning; effects of therapeutic radiations on composite materials; direct digital radiology; teleradiology. *Mailing Add:* Univ Louisville Sch Dent Louisville KY 40292

FARMANFARMAIAN, ALLAHVERDI, b Teheran, Iran, June 10, 29; m 58; c 2. PHYSIOLOGY, MARINE SCIENCES. *Educ:* Reed Col, BA, 52; Stanford Univ, MA, 55, PhD(physiol), 59. *Prof Exp:* Assoc prof physiol & chmn, Shiraz Univ, Iran, 61-66 & Univ Teheran, 66-67; assoc prof, Rutgers Univ, 67-72, chmn sect physiol, 74-75 & 84-86, chmn & grad dir, 79-81, PROF PHYSIOL, RUTGERS UNIV, 72- *Concurrent Pos:* Mem, Marine Biol Lab Corp, 63; US Agency Internat Develop fel, 63; prin investr, NSF grant, 66-69, co-prin investr, 68-69, Nat Inst Child Health & Human Develop, 69-70; sr investr, Marine Biol Lab, Woods Hole, Mass, 66-83; assoc ed, J Exp Zool, 74-78; NSF grant, 74-78; Sci Educ Admin grant, USDA, 79-81; NOAA-sea grants; NJSC grant; vis prof, Stanford Univ, Calif, 72; vis prof, Arc Inst Animal Physiol, Cambridge, Eng, 76; vis prof, Duke Univ Marine Lab, 77; vis prof, Princeton Univ, 82. *Mem:* Fel AAAS; Am Physiol Soc; Soc Gen Physiol; NY Acad Sci; World Aquacult Soc. *Res:* Comparative approach to the mechanisms of membrane transport; nutrition and ecological physiology of marine animals. *Mailing Add:* Dept Biol Nelson Bldg Rutgers Univ Busch Campus Piscataway NJ 08855-1059

FARMER, BARRY LOUIS, b Louisville, Ky, Feb 1, 47; m 69; c 2. POLYMER PHYSICS, POLYMER STRUCTURE. *Educ:* Case Inst Technol, BS, 69, MS, 72; Case Western Reserve Univ, PhD, 74. *Prof Exp:* Nat Res Coun res assoc, 73-76, CONSULT, POLYMERS DIV, NAT BUR STANDARDS, 76-; ASSOC PROF MAT SCI, WASH STATE UNIV, 76- *Concurrent Pos:* Asst prof mat sci, Wash State Univ; assoc ed, New Mat & Components Rev of Sci Instruments, 79-; vis scientist, IBM Almaden Res Ctr, 85-86. *Mem:* Am Phys Soc; Am Chem Soc. *Res:* Polymer structure and properties; crystalline structure of polymers by x-ray diffraction and computerized molecular modeling. *Mailing Add:* Dept Mat Sci Univ Vet Admin Thornton Hall Charlottesville VA 22901

FARMER, CHANTAL, b Montreal, Que, July 15, 59. ENDOCRINOLOGY, FETAL DEVELOPMENT. *Educ:* McGill Univ, BSc, 80; Univ Sask, MSc, 82; Pa State Univ, PhD(animal indust), 86. *Prof Exp:* Biologist 1, 81-86, RES SCIENTIST, AGR CAN, LENNOXVILLE RES STA, 86- *Mem:* Am Soc Animal Sci; Can Soc Animal Sci. *Res:* Pre-weaning piglet mortality and performance; endocrinology of the sow; piglet and fetus; fetal development and growth; peripartum behavior and housing. *Mailing Add:* PO Box 90 Lennoxville PQ J1M 1Z3 Can

FARMER, CROFTON BERNARD, b Rumney, Wales. INFRARED SPECTROSCOPY, RADIATION TRANSPORT. *Educ:* Univ London, BSc, 52, PhD(physics), 68. *Prof Exp:* Head dept infrared res, EMI Electronics Eng, 60-67; mem tech staff, Space Sci Div, Jet Propulsion Lab, Calif Inst Technol, 67-70, mgr planetary atmospheres sect, 70-73, sr mem tech staff, Space Sci Div, 73-899. *Concurrent Pos:* Mem, Adv Comt, NASA, 68-81; prin investr, Viking, 76, Mars Atmospheric Water Exp, Spacelab, 69-78 & 78-; vis prof, Div Geol & Planetary Sci, Calif Inst Technol, 78-81; mem Int Comn Planetary Atmospheres, 80- *Mem:* Am Astron Soc; AAAS. *Res:* Remote sensing of planetary atmospheres; radiative transfer; infrared spectroscopy; spectroscopy of earth's upper atmosphere; history and present distribution of water on Mars. *Mailing Add:* 2525 Hollister Terr Glendale CA 91206

FARMER, DONALD JACKSON, b Morenci, Ariz, Apr 7, 25; m 49; c 3. PHYSICS. *Educ:* Univ Wash, BS, 50, PhD(physics), 54. *Prof Exp:* Res assoc, Univ Wash, 54-55; mem tech staff, Ramo-Wooldridge Corp, 55-58; sr staff, Space Tech Labs, Inc, 58-60, assoc dept mgr, 59-60; mgr quantum electronics, Gen Tech Corp, 60-63, from vpres to pres, 63-69; from vpres to sr vpres, Tracor, Inc, 64-71; consult, 71-74; PRES, EXTEK MICROSYSTS, INC, 74- *Mem:* Am Phys Soc. *Res:* Atomic and nuclear physics; upper atmosphere physics; radio propagation; physical electronics. *Mailing Add:* 6955 Hayvenhurst Ave Van Nuys CA 91406

FARMER, JAMES BERNARD, b Liverpool, Eng, Dec 13, 28. PHYSICAL CHEMISTRY. *Educ:* Univ Liverpool, BSc, 50, PhD(chem), 53. *Prof Exp:* Fel chem, Nat Res Coun Can, 53-55 & Laval Univ, 55-56; fel, 56-57, from asst prof to assoc prof, 57-69, PROF CHEM, UNIV BC, 69- *Mem:* Am Phys Soc; The Chem Soc. *Res:* Electron spin resonance spectrometry. *Mailing Add:* Dept Chem Univ BC 2075 Westbrook Pl Vancouver BC V6T 1W5 Can

FARMER, JAMES LEE, b South Gate, Calif, Aug 8, 38; m 67; c 5. POPULATION CONTROL RESEARCH. *Educ:* Calif Inst Technol, BS, 60; Brown Univ, PhD(biol), 66. *Prof Exp:* Instr biophys, Med Ctr, Univ Colo, 66-68; from asst prof to assoc prof, 69-87, PROF ZOOL, BRIGHAM YOUNG UNIV, 87- *Mem:* AAAS; Genetics Soc Am. *Res:* Evolution of chromosome structure and molecular diversity of natural populations. *Mailing Add:* Dept Zool Brigham Young Univ Provo UT 84602

FARMER, JOHN JAMES, III, b Newnan, Ga, Aug 12, 43; m 68; c 2. MICROBIOLOGY. *Educ:* Ga Inst Technol, BS, 65; Univ Ga, PhD(microbiol), 68. *Prof Exp:* Sr asst scientist, USPHS, Environ Serv Br, NIH, 68-70; asst prof microbiol, Univ Ala, University, 70-72; Hosp Infections Prog, Ctr Infectious Dis, 72-85, DIR, WHO NAT LAB ENTERIC PHAGE TYPING & CHIEF, ENTERIC IDENTIFICATION LAB, ENTERIC BACT SECT, ENTERIC DIS BR, DIV BACT DIS, CTR INFECTIOUS DIS, CTRS DIS CONTROL, ATLANTA, GA, 85- *Concurrent Pos:* Attend microbiologist, Druid City Hosp, Tuscaloosa, Ala, 70-72; mem, working group standardization Pseudomonas aeruginosa serotyping, Int Subcomt Taxon Pseudomonas, 71-; adj assoc prof, Col Arts & Sci, Univ Ala, 71- & Sch Pub Health, Univ NC, 74-; scientist, dir & capt, Commissioned Corps, USPHS, 72-; mem safety comt, Centers Dis Control, 74-80; secy, Int Subcomt Taxon Vibrionaceae; mem, Int Working Group Standardization Vibri ocholearae Serotyping; mem, Proteus-Providencia-Morganella, Edwardsiella-Citrobacter & new designated genera Enterobacteriaceae working group, Int Subcomt Taxon Enterobacteriaceae; NSF fel; mem, Conf Pub Health Lab Dirs; reviewer, New Eng J Med, Infection & Immunity, J Infectious Dis, J Pediat, J Urol, Int J Systematic Bact, Europ J Clin Microbiol. *Honors & Awards:* Bausch & Lomb Sci Medal, 61; Commendation Medal, USPHS, 77; Chem Rubber Co Chem Achievement Award. *Mem:* AAAS; Am Soc Microbiol; Sigma Xi; Int Fedn Enteric Phage Typing; Conf Pub Health Lab Dirs. *Res:* Enterobacteriaceae and Vibrionaceae-their isolation, classification, identification, ecology, epidemiology, and role in human disease; concept of family, genus, species, subspecies, and clone in Enterobacteriaceae and Vibrionaceae; typhoid fever, salmonellosis, yersiniosis, cholera and other diseases due to Vibrio species, diarrhea, dysentery, and nosocomia infections; typing methods for Enterobacteriaceae and Vibrionaceae and their application in epidemiological analysis; new causes of diarrhea, intestinal infections, and extraintestinal infections; role of Aeromonas, Plesiomonas, Edwardsiella, Hafnia, Citrobacter and other Enterobacteriaceae and Vibrionaceae in diarrhea; laboratory methods of Vibrio Cholerae and other Vibrio species; formulation of new media for isolation and identification; new approaches in clinical microbiology, classification and identification; immunological response to Enterobacteriaceae and Vibrionaceae as an indication of infection; US patent. *Mailing Add:* Bldg 1 Rm B-307 Ctr Dis Control USPHS B-307 1600 Clifton Rd Atlanta GA 30333

FARMER, JOHN WILLIAM, b Springfield, Mo, May 2, 47; m 72; c 2. PHYSICS. *Educ:* Kans State Univ, PhD(physics), 74. *Prof Exp:* Fel, Argonne Nat Lab, 74-76; res physicist, Univ Dayton, 76-80; SR RES SCIENTIST GEN PHYSICS, RES REACTOR, UNIV MO, 80- *Concurrent Pos:* Prin investr, NSF Grant, 85- *Mem:* Am Phys Soc. *Res:* Defects in solids, in particular semiconductors; electron irradiation induced; neutron irradiation induced; intrinsic; defraction limited thermograph system studies of defects in semiconductors; uniaxial stress combined with defraction limited thermograph systems to obtain symmetry information. *Mailing Add:* Res Reactor Univ Mo Columbia MO 65211

FARMER, LARRY BERT, b Greenville, SC, Jan 5, 36; m 60; c 3. ORGANIC CHEMISTRY. *Educ:* Wofford Col, BS, 58; Univ Tenn, PhD(org chem), 63. *Prof Exp:* Res chemist, 64-65, sr res chemist, Chem Div, 65-70 & Res Div, 70-77, dept mgr, Res Serv Div, 77-80, proj mgr, 80-85, DEPT MGR, RES SERV DIV, MILLIKEN RES CORP, 85- *Mem:* Am Chem Soc; Am Asn Textile Chemists & Colorists; Am Soc Testing & Mat. *Res:* Organic syntheses; natural products; textile fibers. *Mailing Add:* Milliken Res Corp, M-425 PO Box 1927 Spartanburg SC 29304

FARMER, MARY WINIFRED, b Springfield, Mo, Aug 10, 39. ESTUARINE WATER-QUALITY MONITORING, LARGE-SCALE OCEAN CIRCULATION. *Educ:* Drury Col, Springfield, Mo, BS, 61; City Col City Univ NY, PhD(biol oceanog), 78. *Prof Exp:* Staff scientist oceanog, Sea Educ Asn, 77-86; WILLIAMS-MYSTIC PROG, MYSTIC SEAPORT MUS, 87- *Concurrent Pos:* Guest investr, Woods Hole Oceanog Inst, 85-86; watch coordr, Mystic River, 90-, educ software designer, 90- *Mem:* Am Soc Zoologists; Asn Women Sci; Crustacean Soc. *Res:* Water quality indicators; dispersal of the phyllosoma larvae of spiny lobsters-implications for fisheries, biogeography and evolution; larval dispersal mechanisms. *Mailing Add:* Williams-Mystic Prog Mystic Seaport Mus Mystic CT 06355

FARMER, PATRICK STEWART, b Saskatoon, Sask, Jan 27, 42; m 67. MEDICINAL CHEMISTRY. *Educ:* Univ Sask, BSP, 62, MSc, 64; Portsmouth Col Tech, Eng, PhD(pharm & chem), 68. *Prof Exp:* Asst prof, 68-75, ASSOC PROF PHARMACEUT CHEM, COL PHARM, DALHOUSIE UNIV, 75- *Concurrent Pos:* Res fel, ICI Pharmaceut Div, Alderly Park, 77-78. *Mem:* Acad Pharmacuet Sci; Asn Fac Pharm Can. *Res:* Synthesis of conformationally constrained butyrophenone analogs and benzothiazepines. *Mailing Add:* Col Pharm Dalhousie Univ Halifax NS B3H 3J5 Can

FARMER, RICHARD GILBERT, b Kokomo, Ind, Sept 29, 31; m 58; c 2. INTERNAL MEDICINE, GASTROENTEROLOGY. *Educ:* Univ Md, MD, 56; Univ Minn, MS, 60; Am Bd Internal Med, dipl, 63, gastroenterol, 68. *Prof Exp:* Chmn, dept gastroenterol, 72-82, MEM STAFF, CLEVELAND CLIN FOUND, 62-, CHMN, DIV MED, 75-; ASSOC CLIN PROF MED, SCH MED, CASE WESTERN RESERVE UNIV, 80- *Concurrent Pos:* Gov, Am Col Physicians, 84, regent, 85-91. *Mem:* Inst Med-Nat Acad Sci; Am Col Gastroenterol (pres, 78-79); Asn Prog Dis Internal Med (pres, 77-79); Am Gastroenterol Asn; fel Am Col Physicians; mem Int Orgn Study Inflammatory Bowel Dis. *Res:* Inflammatory bowel disease. *Mailing Add:* Cleveland Clin Found 9500 Euclid Ave Cleveland OH 44195-5123

FARMER, ROBERT E, JR, b Rehoboth Beach, Del, Dec 3, 30; m 60; c 2. FORESTRY, PLANT PHYSIOLOGY. *Educ:* Univ Mich, BSF, 53, MF, 57, PhD(forestry), 61. *Prof Exp:* Res forester, Southern Hardwoods Lab, US Forest Serv, 61-67; plant physiologist, Tenn Valley Authority, 67-81; PROF, SCH FORESTRY, LAKEHEAD UNIV, ONT, 81- *Mem:* Ecol Soc Am; Soc Am Foresters. *Res:* Genecology of plants in the North American forest. *Mailing Add:* Sch Forestry Lakehead Univ Thunder Bay ON P7B 5E1 Can

FARMER, THOMAS WOHLSEN, b Lancaster, Pa, Sept 18, 14; m 41; c 2. NEUROLOGY. *Educ:* Harvard Univ, AB, 35, MD, 41; Duke Univ, MA, 37. *Prof Exp:* From asst prof to prof neurol, Southwestern Med Sch, Tex, 48-52, prof med & actg chmn dept, 51-52; PROF NEUROL MED, SCH MED, UNIV NC, CHAPEL HILL, 52- *Concurrent Pos:* NIH spec fel, Inst Neurophysiol, Denmark, 57-58. *Mem:* AMA; Am Col Physicians; Am Neurol Asn; Am Acad Neurol (secy, 55-57). *Res:* Virus infections of the nervous system, including lymphocytic choriomeningitis and Coxsackie viruses; neurosyphilis; radioactive iodine in brain tumor localizations; electrophysiology of muscle. *Mailing Add:* Dept Neurol Sch Med Univ NC Chapel Hill NC 27599-7025

FARMER, WALTER ASHFORD, science education; deceased, see previous edition for last biography

FARMER, WALTER JOSEPH, b Anderson, Ind, Nov 1, 38; m 62; c 2. SOIL CHEMISTRY. *Educ:* Ind Univ, BS, 61; Purdue Univ, PhD(soil chem), 66. *Prof Exp:* Asst chemist, 66-70, from asst prof to assoc prof, 70-82, PROF SOIL SCI, UNIV CALIF, RIVERSIDE, 82- *Mem:* Am Soc Agron; Soil Sci Soc Am; Am Chem Soc; Sigma Xi; Int Soc Soil Sci. *Res:* Adsorption, movement and volatilization of organic pesticides and other synthetic organic compounds in soil-water systems and their reactions with soils, clays and other soil materials; fate of pesticides in soils. *Mailing Add:* Dept Soil & Environ Sci Univ Calif Riverside CA 92521

FARMER, WILLIAM MICHAEL, b Nashville, Tenn, May 10, 44; m 66; c 2. AEROSOLS, ELECTRO-OPTICAL INSTRUMENTATION. *Educ:* Univ Tenn, BS, 67, MS, 68, PhD(physics), 73. *Prof Exp:* Coop engr, Arnold Res Orgn, 62-66; res asst, Space Inst, Univ Tenn, 67-68; res engr, Arnold Res Orgn, 68-73; staff scientist, Sci Applications, Inc, 73-75; lab mgr & res scientist, Spectron Develop Labs, 75-77; staff scientist, 77-78, assoc prof physics, Optics & Classical Electrodynamics, Space Inst, Univ Tenn, 78-84; tech dir, Sci & Technol Corp, 84-89; PROG MGR & SR SCIENTIST, BIONETICS CORP, 89- *Concurrent Pos:* Consult, Solar-Div Int Harvestor, 77-79; mem, Plasma Dynamics Tech Comt, Am Inst Aeronaut & Astronaut, 79-81; tech adv, NAtlantic Treaty Orgn, 80-89, mem, indust adv group subpanels 27 & 34, 89-91. *Mem:* Am Optical Soc; Soc Photog Inst Engrs. *Res:* Development of laser instrumentation for aerosol measurement; aerosol research using wide spectrum of instrumentation for the measurement of smoke and obscurants; remote detection of chemical agents. *Mailing Add:* 4309 Mission Bell Ave Las Cruces NM 88001

FARMER, WILLIAM S(ILAS), JR, b Waterville, Maine, Apr 16, 22; m 47; c 4. ENGINEERING DESIGN, ENGINEERING MANAGEMENT. *Educ:* Tufts Univ, BS, 44; Univ Tenn, MS, 50. *Prof Exp:* Chem engr, Texas Co, 44; tech supvr, Fercleve Corp, 45; develop engr, Oak Ridge Nat Lab, 46-49, sr develop engr, 50-53; proj engr, Advan Nuclear Design Sect, Pratt & Whitney Aircraft Co, 54-55, proj engr, Aircraft Nuclear Propulsion, 56-57; sr nuclear engr & sect head, ACF Industs, Inc, 58; proj mgr, Elk River Nuclear Power Reactor, Allis-Chalmers Mfg Co, 59, mgr planning dept, 60-62, tech dir, 62-70; mem staff, AEC, 70-71, dept mgr, nuclear proj, Potomac Elec, 72-73, PROG MGR, US NUCLEAR REGULATORY COMN, 74- *Mem:* Am Chem Soc; Am Nuclear Soc; Am Inst Chem Engrs. *Res:* Nuclear engineering; heat transfer and fluid mechanics; chemical engineering. *Mailing Add:* 10115 Green Forest Dr Silver Spring MD 20903

FARN, CHARLES LUH-SUN, b Checkiang, China, Sept 19, 34; m 62; c 2. FLUID MECHANICS. *Educ:* Nat Taiwan Univ, BS, 58; NC State Col, MS, 62; Univ Mich, PhD(mech eng), 65. *Prof Exp:* Asst prof fluid mech, Carnegie-Mellon Univ, 65-68; sr engr, 68-76, MGR FLUID DYNAMICS RES, WESTINGHOUSE RES & DEVELOP CTR, 76- *Mem:* Am Soc Mech Engrs. *Res:* Boundary layers; magnetohydrodynamic generators; wave propagations; electrokinetics; turbulence; aerodynamics of turbomachines; steam turbine; gas turbine; pump; axial flow fan; gas dynamics; numerical analysis. *Mailing Add:* 999 Evergreen Dr Pittsburgh PA 15235

FARNELL, ALBERT BENNETT, b Shreveport, La, July 18, 17; m 41; c 3. MATHEMATICS. *Educ:* Centenary Col, AB, 38; La State Univ, MS, 40; Univ Calif, PhD(math), 44. *Prof Exp:* Asst prof math, Univ Colo, 46-48; lectr, Princeton Univ, 48-49; asst prof, Univ Colo, 49-51; sr res engr, NAm Aviation, 53-56; sr staff scientist, Convair Sci Res Lab Div, Gen Dynamics Corp, 56-63; prof math, Colo State Univ, 63-65; sr staff scientist, Convair Sci Res Lab Div, Gen Dynamics Corp, 65-66; prof, 66-77, EMER PROF MATH, COLO STATE UNIV, 77- *Mem:* Math Asn Am. *Res:* Matrix algebra; nonlinear differential equations; characteristic roots of matrix polynomials and infinite matrices; digital computation. *Mailing Add:* 3500 Carlton Ave M-71 Ft Collins CO 80525

FARNELL, DANIEL REESE, b Mobile, Ala, Feb 7, 32; m 54; c 4. TOXICOLOGIC PATHOLOGY, LABORATORY ANIMAL PATHOLOGY. *Educ:* Auburn Univ, DVM, 57, MS, 62; Mich State Univ, PhD(path), 69; Am Col Lab Animal Med, dipl, 66. *Prof Exp:* Res scientist, Southern Res Inst, 57-61; assoc prof animal dis res, Auburn Univ, 62-67; prof vet sci & head dept, Miss State Univ, 69-73; prof comp med, Col Med, Univ SAla, 73-75; dir animal resources, 73-78, PATHOLOGIST, SOUTHERN RES INST, 78- *Concurrent Pos:* Consult path, dept nutrit sci, Univ Ala, Birmingham, 85. *Mem:* Am Vet Med Asn; Am Asn Lab Animal Sci; Soc Toxicol Pathologists. *Res:* Pathology and oncology; nutrition; toxicology; endocrinology. *Mailing Add:* Southern Res Inst PO Box 55305 Birmingham AL 35255-5305

FARNELL, G(ERALD) W(ILLIAM), b Toronto, Ont, Aug 31, 25; m 48; c 2. ELECTRICAL ENGINEERING, PHYSICS. *Educ:* Univ Toronto, BASc, 48; Mass Inst Technol, SM, 50; McGill Univ, PhD, 57. *Prof Exp:* Asst, Res Lab Electronics, Mass Inst Technol, 48-50; lectr elec eng, McGill Univ, 50-54, from asst prof to assoc prof elec eng & physics, 54-61, chmn dept elec eng, 67-73, dean, Fac Eng, 73-84, PROF ENG PHYSICS, McGILL UNIV, 61- *Concurrent Pos:* Nuffield fel, Clarendon Lab, Oxford, 60-61. *Mem:* Fel Inst Elec & Electronics Engrs; Sigma Xi. *Res:* Solid state electronics; ultrasonics and elastic surface waves. *Mailing Add:* Elec Eng McGill Univ 3480 University St Montreal PQ H3A 2A7 Can

FARNER, DONALD SANKEY, physiology, zoology; deceased, see previous edition for last biography

FARNES, PATRICIA, hematology; deceased, see previous edition for last biography

FARNG, RICHARD KWANG, b Apr 1, 43; US citizen; m 72; c 3. PHARMACEUTICS. *Educ:* Nat Taiwan Univ, BS, 67; Univ Minn, PhD(pharm), 73. *Prof Exp:* Scientist pharm res & develop, 3M Co, 73-76; sr scientist res & develop, Johnson & Johnson, 76-; R W JOHNSON PHARMACEUT RES INST. *Mem:* Am Pharmaceut Asn; Acad Pharmaceut Sci. *Res:* Diffusion; skin penetration; preformulation; formulation; parenteral peptides. *Mailing Add:* R W Johnson Pharmaceut Rte 202 Box 300 Raritan NJ 08869-0602

FARNHAM, PAUL REX, b St Louis, Mo, Nov 2, 31; m 55. STRUCTURAL GEOLOGY, GEOPHYSICS. *Educ:* Western Md Col, BS, 53; Va Polytech Inst, MS, 60; Univ Minn, PhD(struct geol), 67. *Prof Exp:* Assoc engr, Martin Co, Md, 56-57; teaching asst phys geol & mineral, Va Polytech Inst, 58-59, instr, 59-61; teaching assoc phys geol, Univ Minn, Minneapolis, 61-66; res geophysicist, Seismic Data Lab, Earth Sci Co, Teledyne, Inc, 67-69; dir tech educ, Bison Instruments, 69-70; asst prof, 70-77, mem fac geol, 77-80, ASST PROF GEOL, COL ST THOMAS, 80- *Res:* Instrumentation and data processing in earthquake seismology and synthesis of geophysical and geological data to establish structural geologic relationships. *Mailing Add:* Dept Geol Col St Thomas 2115 Summit Ave St Paul MN 55105

FARNHAM, ROUSE SMITH, b Evergreen, Ala, Jan 29, 18; m 51; c 2. SOIL SCIENCE. *Educ:* Ala Polytech Inst, BS, 41; Ohio State Univ, PhD(agron), 51. *Prof Exp:* Soil surveyor, Agr Exp Sta, Ala Polytech Inst, 41-42 & 46; soil scientist, Soil Conserv Surv, 50-58, asst prof, 58-69, PROF SOIL SCI, UNIV MINN, MINNEAPOLIS, 69- *Res:* Soil genesis and classification; peat and organic soils, use for energy and biomass production for energy. *Mailing Add:* 1904 Eustis St Lauderdale MN 55108

FARNHAM, SHERMAN B, b New Haven, Conn, June 23, 12; m 36; c 2. DESIGN OF UTILITY SYSTEMS & LARGE HYDROELECTRIC PLANTS. *Educ:* Yale Univ, BS, 33, EE, 35. *Prof Exp:* Mgr eng div, New Eng Dist, Gen Elec Co, 35-66; chief elec engr, Hydroelec Div, Charles T Main, 66-78; RETIRED. *Honors & Awards:* Centennial Medal, Inst Elec & Electronics Engrs, 84. *Mem:* Fel Inst Elec & Electronics Engrs. *Mailing Add:* 15 Woodridge Rd Wellesley MA 02181

FARNSWORTH, ARCHIE VERDELL, JR, b Mesa, Ariz, July 12, 41; m 66; c 7. FLUID MECHANICS, HEAT TRANSFER. *Educ:* Ariz State Univ, BSE, 66, MSE, 68; Brown Univ, PhD(eng), 71. *Prof Exp:* MEM TECH STAFF ENG, SANDIA LABS, 70- *Honors & Awards:* Award of Excellence, Dept Energy, 88. *Res:* High temperature fluid flow and radiative transfer; laser, particle beam, or intense radiative interactions with matter and resulting flows; nuclear weapon effects, especially resulting ground shock; computational high temperature flows. *Mailing Add:* Sandia Labs Org 1542 Albuquerque NM 87185

FARNSWORTH, CARL LEON, b Lincoln, Nebr, Aug 5, 30; c 1. ORGANIC CHEMISTRY. *Educ:* Univ Tex, BS, 55, MA, 58, PhD(chem), 61. *Prof Exp:* Res chemist, Spruance Film Res & Develop Lab, Film Dept, E I du Pont de Nemours & Co, Inc, 60-66; chemist, W H Brady Co, 66-67; ASSOC PROF CHEM, UNIV WIS-STEVENS POINT, 67- *Mem:* Am Chem Soc. *Mailing Add:* Dept Chem Univ Wis Stevens Point WI 54481

FARNSWORTH, HARRISON E, b Greenlake, Wis, Mar 24, 96. SOLD STATE PHYSICS. *Educ:* Irbin Col, BA, 18; Univ Wis, MS, 21, PhD(physics), 22. *Prof Exp:* Prof physics, Brown Univ, 30-70; RETIRED. *Honors & Awards:* Welch Medal, Am Vacuum Soc. *Mem:* Am Vacuum Soc; Am Phys Soc; Sigma Xi. *Mailing Add:* 3940 E Ina Rd Tucson AZ 85718

FARNSWORTH, MARIE, b Holden, Mo, July 19, 96. CHEMISTRY. *Educ:* Univ Chicago, BS, 18, PhD(chem), 22. *Prof Exp:* Instr chem, Iowa State Col, 22-23; res chemist, US Bur Mines, 23-26; instr chem, NY Univ, 26-35; res, Munich & London, 35-36; res chemist, Fogg Art Mus, Harvard Univ, 36-37; chemist, Agora Excavations, Greece, 38-40; res supvr, Res & Develop Lab, M & T Corp, 40-61; Ford Found fel, 61-64; res assoc dept hist & archaeol, Columbia Univ, 64-79; RETIRED. *Concurrent Pos:* Vis prof, Univ Mo-Columbia, 70-74. *Honors & Awards:* First Pomerance Award, Archaeol Inst Am, 80. *Mem:* Am Chem Soc; Archaeol Inst Am. *Res:* Analytical chemistry; technical problems of archaeology. *Mailing Add:* Kingswood Manor 10000 Wornall Rd No 3319 Kansas City MO 64114-4368

FARNSWORTH, MARJORIE WHYTE, b Detroit, Mich, Nov 18, 21; m 45; c 2. GENETICS. *Educ:* Mt Holyoke Col, BA, 44; Cornell Univ, MS, 46; Univ Mo, PhD(zool), 51. *Prof Exp:* Asst entom, Cornell Univ, 44-46; instr zool, Univ Mo, 46-49, res fel, 49-50, from instr to asst prof, 50-52; cytologist, Roswell Park Mem Inst, 52-53; lectr & res assoc biol, State Univ NY Buffalo, 53-64, assoc prof, 65-79, adj assoc prof, 79-81; RETIRED. *Mem:* Fel AAAS; Genetics Soc Am; Am Soc Zoologists. *Res:* Biochemical genetics of Drosophila; author of one book on genetics. *Mailing Add:* Three Elm Creek Dr Elmhurst IL 60126

FARNSWORTH, NORMAN R, b Lynn, Mass, Mar 23, 30; m 53. PHARMACOGNOSY, PHYTOCHEMISTRY. *Educ:* Mass Col Pharm, BS, 53, MS, 55; Univ Pittsburgh, PhD(pharm), 59. *Hon Degrees:* Dr, Univ Paris, 78 & Uppsala Univ, 82 & Mass Col Pharm, 89. *Prof Exp:* Instr biol sci, Univ Pittsburgh, 55-59, from asst prof to prof pharmacog, 59-70, chmn dept, 64-70; prof pharmacog & head dept pharmacog & pharmacol, 70-82, DIR, PROG COLLAB RES PHARMACEUT SCI, COL PHARM, UNIV ILL, CHICAGO, 82- *Concurrent Pos:* Consult, Schering AG, Berlin, Amazon Natural Drug Co, Gillette, WHO, Nat Cancer Inst, A D Little, World Bank & Reader's Digest. *Mem:* Am Soc Pharmacog (vpres, 59-61, pres, 61-62); Soc Econ Bot; Am Pharmaceut Asn; fel Acad Pharm Sci. *Res:* Evaluation of medicinal folklore; computer analysis of chemical and biological data on natural products; isolation, identification and structure elucidation of biologically active plant constituents; investigation of plants for biologically active substances. *Mailing Add:* Col Pharm Univ Ill Chicago MC 877 Chicago IL 60612

FARNSWORTH, PATRICIA NORDSTROM, b Sioux City, Iowa, Aug 17, 30; m 52; c 4. PHYSIOLOGY, BIOCHEMISTRY. *Educ:* Morningside Col, BA, 51; Columbia Univ, MS, 52, PhD(physiol), 60. *Prof Exp:* Instr zool, Hofstra Col, 52-54; instr genetics & zool, Marymount Col, 56-57; asst physiol, Columbia Univ, 58-59, instr occup & phys ther, 59-60, fel med, Col Physicians & Surgeons, 62-63; asst prof genetics & physiol, Fairleigh Dickinson Univ, 63-67; asst prof physiol, Barnard Col, Columbia Univ, 67-69, asst prof biol, 69-73; assoc prof, 73-76, PROF PHYSIOL & OPHTHAL, COL MED & DENT NJ, NEWARK, 76- *Concurrent Pos:* Consult, Dept Health, Englewood, NJ, 66- & Arthur D Little, Inc, 71; vis lectr, Harvard Univ, 70-72; NIH sr res fel, 71; bd sci coun, Nat Eye Inst, 80-84; consult, Nat Adv Eye Coun, 80. *Honors & Awards:* Alcon Res Award. *Mem:* NY Acad Sci; Am Physiol Soc; Sigma Xi; Int Soc Eye Res; Asn Res Vision & Ophthal; Biophys Soc. *Res:* Animal physiology; molecular organization of cytoplasm. *Mailing Add:* Dept Physiol & Ophthal UMDNJ NJ Med Sch 185 S Orange Ave Newark NJ 07103

FARNSWORTH, PAUL BURTON, b Orem, Utah, Aug 3, 53. ATOMIC SPECTROSCOPY, INSTRUMENTATION. *Educ:* Brigham Young Univ, BS, 77; Univ Wis, Madison, PhD(chem), 81. *Prof Exp:* Res assoc, Ind Univ, 81-82; ASST PROF CHEM, BRIGHAM YOUNG UNIV, 83- *Mem:* Am Chem Soc; Soc Appl Spectroscopy. *Res:* Fundamental processes that control the behavior of atomic emission sources with the objective of developing new sources or enhancing the performance of existing ones. *Mailing Add:* Chem Dept Brigham Young Univ Provo UT 84602

FARNSWORTH, RICHARD KENT, b Salt Lake City, Utah, July 21, 34; m 59; c 4. HYDROLOGY, REMOTE SENSING. *Educ:* Univ Utah, BA, 61; Brigham Young Univ, MS, 66; Univ Mich, PhD(natural resources), 76. *Prof Exp:* Supvr physics, Tritium Lab, Water Resources Div, US Geol Surv, 65-67; RES HYDROLOGIST, HYDROL RES LAB, DEPUTY CHIEF HYDROLOGIC RES LAB, NAT WEATHER SERV, NAT OCEANIC & ATMOSPHERIC ADMIN, 67- *Mem:* Am Geophys Union. *Res:* Mathematical modeling of soil surface layers to compute the likelihood of signigicant changes in permeability due to impervious frost; satellite estimates of current rainfall; implications of tritium in natural waters; determination practical applications of remote sensing to flood forecast hydrology; precipitation assessment; quality control of radar rainfall using satellite IR data. *Mailing Add:* Nat Weather Serv W23 Silver Spring MD 20910

FARNSWORTH, ROY LOTHROP, b Shirley, Mass, Mar 4, 28; m 57; c 3. GEOLOGY. *Educ:* Boston Univ, AB, 49, AM, 56, PhD(geol), 61. *Prof Exp:* Pub sch teacher, Mass, 53-55; instr geol, Trinity Col, Conn, 57-59, pub sch teacher, Mass, 60-61; asst prof geol, 61-68, ASSOC PROF GEOL, BATES COL, 68- *Mem:* Soc Econ Paleontologists & Mineralogists; Am Asn Univ Prof; Geol Soc Am; Nat Asn Geol Teachers; Am Quaternary Asn. *Res:* Glacial geology; sedimentology; environmental geology; coastal geology; genesis analysis of land forms. *Mailing Add:* Dept Geol Bates Col Lewiston ME 04240

FARNSWORTH, WELLS EUGENE, b Hartford, Conn, July 10, 21; m 45; c 2. BIOCHEMISTRY. *Educ:* Trinity Col, Conn, BS, 46; Univ Mo, MA, 49, PhD(endocrine physiol, chem), 51. *Prof Exp:* Endocrinologist, Wm S Merrell Co, Ohio, 51-52; res biochemist, US Vet Admin Hosp, 52-61, dir biochem res, 61-80; assoc prof biochem, State Univ NY Buffalo, 67-80, assoc res prof urol, 75-80; PROF & CHMN BIOCHEM, CHICAGO COL OSTEOP MED, 80- *Concurrent Pos:* Consult, Edward J Meyer Hosp, 57-58; asst prof, State Univ NY Buffalo, 64-67. *Mem:* Endocrine Soc; Am Physiol Soc; Am Soc Biol Chem; Am Asn Cancer Res. *Res:* Steroid metabolism; influence of steroid on prostate metabolism and protein synthesis in vitro. *Mailing Add:* Dept Biochem Chicago Col Osteopathic Med 1122 E 53rd St Chicago IL 60615

FARNUM, BRUCE WAYNE, b Fargo, NDak, Apr 5, 35; m 59; c 1. ORGANIC CHEMISTRY, POLYMERS. *Educ:* NDak State Univ, BS, 57, MS, 59; Univ Del, PhD(org chem), 69. *Prof Exp:* Instr chem, Moorhead State Col, 59; from assoc prof to prof chem, Minot State Col, 64-76; res chemist, 76-80, CHIEF ORG ANALYSIS BR, GRAND FORKS ENERGY TECHNOL CTR, US DEPT ENERGY, 80-; SR RES ASSOC, ENERGY RES CTR, UNIV NDAK, 85- *Mem:* Am Chem Soc; Soc Appl Spectros. *Res:* Mechanism of coal liquefaction; characterization of gasifier tar, waste water and coal liquids; formulation of fire resistant coatings; nuclear magnetic spectroscopy. *Mailing Add:* 543 Quixote Ave N Lakeland MN 55043

FARNUM, DONALD G, b Oakland, Calif, Apr 3, 34; m 53; c 4. ORGANIC CHEMISTRY. *Educ:* Harvard Univ, AB, 56, PhD(org chem), 59. *Prof Exp:* Chemist, Arthur D Little, Inc, 53-59; from instr to asst prof chem, Cornell Univ, 59-66; assoc prof, 66-72, PROF CHEM, MICH STATE UNIV, 72- *Concurrent Pos:* Sloan Found fel, 62. *Mem:* Am Chem Soc. *Res:* Organic synthesis; stable carbonium ions; structures which test current concepts; bridged polycyclic olefins; pheromone synthesis and structure; reactive intermediates; photochemical synthesis. *Mailing Add:* Dept Chem Mich State Univ East Lansing MI 48824

FARNUM, EUGENE HOWARD, b Athol, Mass, July 7, 42; m 65; c 2. MATERIAL SCIENCE ENGINEERING. *Educ:* Clark Univ, AB, 64; Princeton Univ, MS, 66, PhD(solid state & mat sci), 68. *Prof Exp:* Mem staff mat sci, Sandia Labs, 68-73; mem staff laser fusion, 73-77, assoc group leader laser fusion, 77-79, dep group leader target fabrication, 79-83, mem staff weapons advan concepts, 83-85, MEM STAFF CONVENTIONAL WEAPONS DEVELOP, LOS ALAMOS NAT LAB, 88- *Concurrent Pos:* prog mgr armor & anti-armor, Defense Advan Res Proj Agency, 86-88. *Mem:* Am Phys Soc; Mat Res Soc; Am Vacuum Soc. *Res:* Radiation damage in materials; defects in crystalline solids; physical properties of ceramics; hydrogen and hydrogen isotope diffusion; permeation and embrittlement in metals; microscopic machining and assembly techniques; cleanliness control; technical management; inertially confined fusion target fabrication; armor and anti-armor weapons research, development and management. *Mailing Add:* 25 Karen Circle Los Alamos NM 87544

FARNUM, PETER, b Orange, NJ, Sept 24, 46; m 71; c 2. FORESTRY. *Educ:* Princeton Univ, AB, 68; Univ Washington, PhD(forestry), 77. *Prof Exp:* RES SCIENTIST, WEYERHAEUSER FORESTRY RES CENTRALIA, 77- *Mem:* Soc Am Foresters; Am Statist Asn. *Res:* Plant water relations, silviculture, and forest water management. *Mailing Add:* Weynerhaeuser Co Tacoma WA 98477

FARNUM, SYLVIA A, b St Paul, Minn, Dec 29, 36; m 59; c 1. POLYMER & SYNFUELS CHARACTERIZATIONS. *Educ:* NDak State Univ, BS, 58, MS, 59; Univ NDak, PhD(phys chem), 79. *Prof Exp:* Instr chem, Washington Col, Md, 60-61; chemist org synthesis, US Army Ballistics Res Lab, 61-62; res assoc, Univ Del, 62-64; instr chem & physics, Minot State Col, 64-76; res fel enzyme kinetics, Univ NDak, 76-78; res chemist coal liquefaction, Grand Forks Energy Technol Ctr, US Dept Energy, 78-83; res supvr, Univ NDak Energy Res Ctr, 83-86; sr chemist, 86-88, ANALYTICAL RES SPECIALIST, 3M, 88- *Mem:* Am Chem Soc; Int Soc Magnetic Resonance; Sigma Xi. *Res:* Nuclear magnetic resonance; mechanism and kinetics of coal conversion processes; high field nuclear magnetic resonance research on liquids and solids; fourth derivative ultra violet; separations of synfuel products; extraction and GC-MS of biomarkers in fossil fuels; adhesives, polymer, characterization, nuclear magnetic resonance spectroscopy, fine chemicals, specialty chemicals. *Mailing Add:* 3M Ctr 236-2B-11 St Paul MN 55144

FARNWORTH, EDWARD ROBERT, b Thorold, Ont, Aug 13, 47; m 70; c 2. NUTRITION, AGRICULTURE. *Educ:* Brock Univ, BSc, 70; McMaster Univ, MSc, 72; Guelph Univ, PhD(nutrit), 78. *Prof Exp:* RES SCIENTIST NUTRIT, AGR CAN, 78- *Concurrent Pos:* Tutor chem, Univ Papua, New Guinea, 72-74. *Mem:* Can Soc Nutrit Sci; Agr Inst Can; Can Soc Animal Sci. *Res:* Nutrition and biochemistry of lipids; fetal metabolism and development. *Mailing Add:* Animal Res Ctr K W Neatby Bldg Ottawa ON K1A 0C6 Can

FAROKHI, SAEED, b Tehran, Iran, Feb 26, 52; US citizen; m 71; c 3. DEVELOPMENT SMART COMPONENTS, PROPULSION SYSTEM ANALYSIS & DESIGN. *Educ:* Univ Ill, Urbana, BS, 75; Mass Inst Technol, MS, 76; PhD(aero & astro), 81. *Prof Exp:* Res & develop engr gas turbine, Brown, Boveri & Co, Switz, 81-84; asst prof, 84-87, ASSOC PROF AEROSPACE ENG, UNIV KANS, 87-, DIR, FLIGHT RES LAB, 90- *Concurrent Pos:* Prin investr, Lewis Res grants, NASA, 85-, Langley Res grants, 88-90 & GE-Aircraft Engines Res Contract, 86-90; lectr & organizer, Propulsion Short Courses for Int Audience, 86-; consult, GE-Aircraft Engines, Keddeg Corp, Aerotech Res, 88-; proprietor, Aerotech Eng & Res, 88-; tech comt mem, Aircraft Engines, Am Soc Mech Engrs, 90; pres, Sci Modeling & Res Corp, 91- *Mem:* Am Inst Aeronaut & Astronaut; Am Soc Mech Engrs; Soc Automotive Engrs; Am Soc Eng Educ; Nat Acad Mech; Int Gas Turbine Inst. *Res:* Aerodynamics of internal and external flows; active flow control in separated and free shear layers with applications to delta wings, subsonic diffusers and turbulent jets; unsteady aerodynamics and aeroacoustics of pusher propellers are investigated numerically; research involved flight testing a pusher propeller configuration; turbulence modeling with curvature effects which included CFD-code implementation and validation. *Mailing Add:* Flight Res Lab Raymond Nichols Hall Univ Kans 2291 Irving Hill Dr Lawrence KS 66045

FARONA, MICHAEL F, b Cleveland, Ohio, Jan 30, 35; m 60; c 2. INORGANIC CHEMISTRY, POLYMER CHEMISTRY. *Educ:* Western Reserve Univ, BS, 56; Ohio State Univ, MS, 62, PhD(chem), 64. *Prof Exp:* From asst prof to prof chem, Univ Akron, 64-90, head dept, 76-84; HEAD, DEPT CHEM, UNIV NC, GREENSBORO, 90- *Concurrent Pos:* Consult, Goodyear Tire & Rubber Co, 71-78 & Dow Chem, 88-90. *Mem:* Catalysis Soc; Am Chem Soc; Royal Soc Chem. *Res:* Transition metal chemistry, including homogeneous catalysis, metal carbonyls and organometallics; polymer synthesis. *Mailing Add:* Dept Chem Univ NC-Greensboro Greensboro NC 27412

FARONE, WILLIAM ANTHONY, b Cortland, NY, Feb 1, 40; m 83; c 3. TECHNOLOGY COMMERCIALIZATION. *Educ:* Clarkson Tech, BS, 61, MS, 63, PhD(chem), 65. *Prof Exp:* Res physicist, US Army Electronics Res & Develop Activity, White Sands Missile Range, NMex, 64-65; assoc prof phys chem, Va State Univ, 65-67; mgr sci res, Lever Bros Co, 67-72, dir, 72-75; vpres res & develop, Chem Specialties Div, PVO Int, Inc, 75-76; dir, appl res, Philip Morris, Inc, 76-84; sr vpres, Bio-Solar, Inc, 84-86; PRES, APPL POWER CONCEPTS, INC, 87- *Concurrent Pos:* Prin investr, Res Corp res grant, 65-67; co-prin investr, NSF res grant, 66-67. *Mem:* Am Optical Soc; Sigma Xi; Am Chem Soc; AAAS; Am Inst Physics. *Res:* Colloid science, particularly use of electromagnetic scattering to determine particle properties and interactions; renewable resource technology, particularly fermentation and liquid separation technology. *Mailing Add:* Appl Power Concepts Inc PO Box 18288 Irvine CA 92713-8288

FAROUK, BAKHTIER, b Dhaka, Bangladesh, Dec 5, 51; US citizen; m 82; c 1. HEAT TRANSFER & FLUID MECHANICS, COMPUTATIONAL METHODS. *Educ:* Bangladesh Univ Eng & Technol, BSME, 75; Univ Del, MS, 78, PhD(mech eng), 81. *Prof Exp:* Res asst, mech eng, Univ Houston, 76-77; teaching asst, Univ Del, 77-81; from asst prof to assoc prof, 81-89, PROF MECH ENG, DREXEL UNIV, 89- *Concurrent Pos:* Summer res fel, Inst Energy Conversion, Univ Del, 78; summer fac res assoc, Naval Res Lab, 88. *Honors & Awards:* Ralph Tector Award, Soc Automotive Eng, 86; Henry Marion Howe Medal, Am Soc Metals, 89. *Mem:* Am Soc Mech Engrs; Combustion Inst; Soc Automotive Eng; Am Inst Aeronaut & Astronaut. *Res:* Investigation of fundamental problems in heat transfer and fluid mechanics for improved design of engineering thermal systems; interests include convective heat transfer, turbulence modeling, combustion and fires and application of computational methods in heat transfer and fluid mechanics. *Mailing Add:* Dept Mech Eng Drexel Univ Philadelphia PA 19104

FARQUHAR, GALE BURTON, b Oakland, Calif, Jan 5, 27; m 51; c 3. CORROSION ENGINEERING. *Educ:* Occidental AB, 50, MA, 51. *Prof Exp:* Res & develop engr, Superior Oil Co, 51-60, from staff corrosion engr to sr corrosion engr, 60-72, gen corrosion engr, 72-78; res scientist, explor & prod res lab, Getty Oil Co, 78-84; res consult, EPT Div, Texaco, 84-88; ENG CONSULT, ARAMCO, SAUDI ARABIA, 88- *Honors & Awards:* Distinguished Serv Award, Nat Asn Corrosion Engrs, 86. *Mem:* Am Chem Soc; Nat Asn Corrosion Engineers. *Res:* Waterflood and bacterial corrosion; cathodic protection; inhibitor evaluation; chemical treatment of oil field production; corrosion control education. *Mailing Add:* Aramco PO Box 1791 Dhahran 31311 Saudi Arabia

FARQUHAR, JOHN WILLIAM, b Winnipeg, Man, June 13, 27; nat US. MEDICINE, PREVENTIVE MEDICINE. *Educ:* Univ Calif, AB, 49, MD, 52. *Prof Exp:* Fel med, Univ Minn, 54-56; fel, Univ Calif, 56-57, instr & chief resident, 57-58; from asst prof to assoc prof, 62-73, PROF MED, RES & POLICY, SCH MED, STANFORD UNIV, 73-, PROF HEALTH RES & POLICY, 88- & C F REHNBORG PROF DIS PREV, 89- *Concurrent Pos:* Res career develop award, USPHS, 63-71; prog dir, Stanford Gen Clin Res Ctr, Sch Med, Stanford Univ, 63-73, dir, Stanford Ctr Res & Dis Prev, 73- & Health Improvement Prog, 81-; mem, numerous comts & adv bds, 65-; visitor-sabbatical leave, Dept Epidemiol & Med Statist, London Sch Hyg & Trop Med, 68-69; consult, Ger Am Sci Treaty, 83-, US Task Force Prev Serv, USPHS, 84-90, Blue Cross Northern Calif, 84-87, Off Substance Abuse Prev & Nat Inst Alcohol Abuse & Alcoholism, 89-; adv, Citizens Pub Action on Blood Pressure & Cholesterol, 90-, Prev Res Ctr, Univ Ill, Chicago, 90-, Coun Health Prom, Ctr Corp Involvement, Am Coun Life Ins Health Ins Asn Am, 90- & Prog Human Biol Middle Grades Curric, 90- *Honors & Awards:* James D Bruce Mem Award for Distinguished Contributions in Prev Med, Am Col Physicians, 83; Helmut Schumann Lectr, Dartmouth Med Sch, 83; Rodale Lectr Prev Med, Med Sch, Boston Univ, 85; Myrdal Prize for Eval Pract, Am Eval Asn, 86; Charles A Dana Award for Pioneering Achievements in Health, 91. *Mem:* Inst Med-Nat Acad Sci; Am Soc Clin Invest; Inst Med; Am Heart Asn; Sigma Xi; Harvey Soc; Am Fedn Clin Res; Acad Behav Med Res; fel Soc Behav Med (pres-elect, 90); Soc Prev Cardiol. *Res:* Epidemiology of cardiovascular disease, human nutrition and atherosclerosis; behavior and communication in respect to human health. *Mailing Add:* Stanford Ctr Res & Dis Prev 1000 Welch Rd Palo Alto CA 94304-1885

FARQUHAR, MARILYN GIST, b Tulare, Calif, July 11, 28; m 51, 70; c 2. CELL BIOLOGY, EXPERIMENTAL PATHOLOGY. *Educ:* Univ Calif, MA, 53, PhD(exp path), 55. *Prof Exp:* Jr res pathologist, Univ Calif, 53-54, asst, 55-58; res assoc, Dept Cell Biol, Rockefeller Univ, 58-62, prof, 70-73; assoc res pathologist, Univ Calif, San Francisco, 62-64, assoc prof path, 64-68, prof in residence, 68-70; prof cell biol & path, Yale Univ Sch Med, 73-87, Sterling prof cell biol & path, 87-90; PROF PATH, DIV CELLULAR MOLECULAR MED, UNIV CALIF SAN DIEGO SCH MED, 90- *Concurrent Pos:* coun, Am Soc Cell Biol, 66-70, 80-83. *Honors & Awards:* E B Wilson Medal, Am Soc Cell Biol, 87; Distinguished Scientist Medal, Electron Microsci Soc Am, 87; Homer Smith Award, Am Soc Nephrology, 88. *Mem:* Nat Acad Sci; Am Soc Cell Biol (pres, 81-82); Am Asn Path; Am Asn Anat; Am Soc Nephrol; Endocrine Soc; Histochem Soc; Int Acad Path. *Res:* Intracellular membrane along the exocytic and endocytic; cellular and molecular bases of glomerular permeability and pathology; structure and function of Golgi complex and lysosomes; composition of glomerular basement membrane. *Mailing Add:* Div Cellular Molecular Med 0651 Univ Calif San Diego La Jolla CA 92093

FARQUHAR, OSWALD CORNELL, b Gt Brit, Sept 26, 21; m 53; c 3. ECONOMIC GEOLOGY. *Educ:* Univ Oxford, BA, 40, MA, 47; Aberdeen Univ, PhD(geol), 51. *Prof Exp:* Lectr geol & mineral, Aberdeen Univ, 48-53; asst prof geol, Univ Kans, 54-57; from assoc to prof geol, Univ Mass, Amherst, 61-88; RETIRED. *Concurrent Pos:* Sr res fel, Dept Sci & Indust Res, Wellington, NZ, 63-64. *Mem:* Soc Econ Geologists; Geol Soc Am; Am Asn Petrol Geologists; Brit Geol Soc; Sigma Xi. *Res:* Engineering geology; mineral deposits. *Mailing Add:* 288 Shays St Amherst MA 01002

FARQUHAR, PETER HENRY, b Boston, Mass, May 7, 47; m 69; c 3. MANAGEMENT SCIENCE, MARKETING. *Educ:* Tufts Univ, BS, 69; Cornell Univ, MS, 72, PhD(oper res), 74. *Prof Exp:* Assoc mathematician res & consult, Info Sci Dept, Rand Corp, 74-75; asst prof, Dept Indust Eng & Mgt Sci, Northwestern Univ, 75-78; asst prof, Grad Sch Bus Admin, Harvard Univ, 78-80; assoc prof, Grad Sch Admin & Dept Agr Econ, Univ Calif, Davis, 80-84; ASSOC PROF & DIR, PROD DEVELOP CTR, GRAD SCH INDUST ADMIN, CARNEGIE-MELLON UNIV, 84- *Concurrent Pos:* Consult, Rand Corp, 75-76; vis scholar, Mass Inst Technol, 82. *Mem:* Oper Res Soc Am; Inst Mgt Sci; Psychometric Soc; Am Mkt Asn. *Res:* Decision analysis; product management and consumer behavior; marketing strategy; multivariate statistical methods; risk analysis. *Mailing Add:* Grad Sch Industrial Admin Carnegie-Mellon Univ Pittsburgh PA 15213

FARQUHAR, RONALD MCCUNN, b Montreal, Que, Nov 25, 29; m 55; c 3. GEOCHRONOLOGY, ARCHAEOMETRY. *Educ:* Univ Toronto, BA, 51, MA, 52, PhD(physics), 54. *Prof Exp:* Fel physics, McMaster Univ, 54-55; lectr, 55-56, from asst prof to assoc prof, 56-71, assoc chmn dept, 74-78, PROF PHYSICS, UNIV TORONTO, 71- *Concurrent Pos:* Royal Soc Can traveling fel, 64; ed, Can Geophys Bull, 75-78. *Mem:* Fel Royal Soc Can; Can Geophys Union. *Res:* Geological age determinations; mass spectrometry; natural isotopic variations; archaeometry. *Mailing Add:* Dept Physics Univ Toronto Toronto ON M5S 1A7 Can

FARR, ANDREW GRANT, LYMPHOSYTE DIFFERENTIATION. *Educ:* Univ Chicago, PhD(anat), 75. *Prof Exp:* ASST PROF BIOL STRUCT, UNIV WASH, 82- *Mailing Add:* Dept Biol Struct Univ Wash SM 20 Seattle WA 98195

FARR, DAVID FREDERICK, b Salinas, Calif, Nov 15, 41; m 63. MYCOLOGY. *Educ:* Humboldt State Col, BA, 63; Univ Kans, MA, 65; Va Polytech Inst & State Univ, PhD(bot), 75. *Prof Exp:* RES BOTANIST MYCOL, SCI & EDUC ADMIN-AGR RES SERV, USDA, 74- *Mem:* Mycol Soc Am; Am Inst Biol Sci; Int Asn Plant Taxon; Sigma Xi. *Res:* Taxonomy, morphology and speciation in basidiomycetous fungi through study of cultural characters, genetics and fruit body production in conjunction with field and herbarium work. *Mailing Add:* Mycology Lab Bldg 011A-ARC Beltsville MD 20705

FARR, KENNETH E(DWARD), b Philadelphia, Pa, May 6, 17; m 68. ELECTRONICS ENGINEERING. *Educ:* Bucknell Univ, BSEE, 59; Drexel Inst, MSEE, 63. *Prof Exp:* Field serv engr, Philco Corp, 40-41; inspector radio mat, US Army Signal Corps, 42; sr engr licensee lab, adv develop, field serv engr & engr in-chg field serv sch, Hazeltine Corp, 42-48; eng sect mgr advan develop, Westinghouse Elec Co, 48-57; sr eng specialist underwater systs eng, Philco Corp, 60-61; mgr indust prod eng, Jerrold Electronics Corp, 61-63; adv engr control & instrumentation res, Westinghouse Elec Co, 63-84; RETIRED. *Mem:* Sr mem Inst Elec & Electronics Engrs. *Res:* Color, slow-scan and educational television; band-width compression; magnetic video recording; internal connection design; computer-aided design, computer-aided manufacturing application. *Mailing Add:* 732 Garden City Dr Monroeville PA 15146

FARR, LYNNE ADAMS, b June 10, 39; m. CHRONOBIOLOGY. *Educ:* Univ Tex Austin, BS, 68; Creighton Univ, PhD(physiol), 77. *Prof Exp:* ASSOC PROF, COL NURSING, MED CTR, UNIV NEBR, 79- *Mem:* Am Physiol Soc; Int Soc Chronobiol; Fedn Am Socs Exp Biol. *Res:* Chronobiology. *Mailing Add:* Col Nursing Univ Nebr Med Ctr 42nd Dewey Ave Omaha NE 68105

FARR, RICHARD STUDLEY, b Detroit, Mich, Oct 30, 22; m 44; c 2. MEDICINE, ALLERGY. *Educ:* Univ Chicago, BS, 45, MD, 46. *Prof Exp:* Intern, US Navy Hosp, Annapolis, 46-47, hematologist, Naval Med Res Inst, 47-48; res assoc anat, Univ Chicago, 48-50; hematologist, Naval Med Res Inst, 50-53; from instr to asst prof med, Univ Chicago, 54-56; assoc res prof anat, Univ Pittsburgh, 56-57, asst prof med & head sect clin immunol, 57-62, head div allergy, immunol & rheumatology, Scripps Clin & Res Found, 62-69, chmn dept clin biol, 66-69; chmn dept med, Nat Jewish Hosp & Res Ctr, Denver, 69-77, staff physician, 77-80, prof med, 80-; prof med, Univ Colo Sch Med, Denver, 69-; RES INST SCRIPPS CLIN. *Concurrent Pos:* Sr res fel chem & immunochem, Calif Inst Technol, 53-54. *Honors & Awards:* Borden Award, Chicago, 46. *Mem:* AAAS; Am Soc Clin Invest; Am Acad Allergy (pres, 70); Am Asn Immunol; hon fel Can Soc Allergy & Clin Immunol. *Res:* Morphogenesis of blood cells; irradiation illness; body temperature regulating mechanisms; immunochemistry; clinical immunology; platelet activity factor. *Mailing Add:* Dept Molecular Biol-Z Res Inst Scripps Clin 10666 N Torrey Pines Rd La Jolla CA 92037

FARR, WILLIAM MORRIS, b Kansas City, Mo, Oct 20, 38; m 60; c 2. PLASMA PHYSICS, REACTOR PHYSICS. *Educ:* Rice Univ, BA, 60; Univ Mich, MS, 62, PhD(nuclear sci), 66. *Prof Exp:* Res engr, Atomics Int, 62-63; physicist, Oak Ridge Nat Lab, 66-69; asst prof nuclear eng, 69-74, ASSOC PROF NUCLEAR ENG, UNIV ARIZ, 74- *Mem:* Am Phys Soc; Am Nuclear Soc. *Res:* Theoretical study of microinstabilities in plasma with emphasis on those instabilities of importance to controlled fusion. *Mailing Add:* Dept Nuclear Eng Univ Ariz Tucson AZ 85721

FARRAND, STEPHEN KENDALL, b Bremerton, Wash, Nov 28, 45; m 70. MICROBIOLOGY, MOLECULAR BIOLOGY. *Educ:* Whitman Col, AB, 67; Univ Rochester, PhD(microbiol), 73. *Prof Exp:* Nat Cancer Inst fel microbiol, Univ Wash, 72-73, sr res fel, 73-74, Nat Cancer Inst fel, 74-75; asst prof, 75-80, ASSOC PROF MICROBIOL, STRITCH SCH MED, LOYOLA UNIV CHICAGO, 80- *Concurrent Pos:* Distinguished vis scientist, Univ Adelaide, South Australia, 82-83. *Honors & Awards:* Lectr, Founds in Microbiol, Am Soc Microbiol, 85-86. *Mem:* Am Soc Microbiol. *Res:* Plasmids of Agrobacterium tumefaciens as they related to virulence; biological control and plant genetic engineering. *Mailing Add:* Dept Microbiol Univ Ill Urbana Campus 407 S Goodwin Ave Urbana IL 61801

FARRAND, WILLIAM RICHARD, b Columbus, Ohio, Apr 27, 31; m; c 2. QUATERNARY GEOLOGY, GEOARCHAEOLOGY. *Educ:* Ohio State Univ, BSc, 55, MSc, 56; Univ Mich, PhD(Pleistocene geol), 60. *Prof Exp:* Res assoc Pleistocene geol, Lamont Geol Observ, Columbia Univ, 60-61, asst prof, Dept Geol, 61-64; vis prof, Inst Geol, Univ Strasbourg, 64-65; from asst prof to assoc prof pleistocene geol, 65-74, PROF GEOL SCI & CUR MUS ANTHROP, UNIV MICH, 74- *Concurrent Pos:* Nat Acad Sci res fel, Univ Strasbourg, 63-64; vis prof, Hebrew Univ, 71-72, Univ Colo, 83, Ind Univ, 85, Univ Tex, Austin, 86. *Honors & Awards:* Archaeol Geol Div Award, Geol Soc Am, 86. *Mem:* AAAS; Geol Soc Am; Am Quaternary Asn. *Res:* Pleistocene geology; history of the Great Lakes and glacio-isostatic rebound; Quaternary paleoclimate; geoarchaeology and archaeological site geology. *Mailing Add:* 3909 Crestridge Dr Midland TX 79707-2727

FARRAR, DAVID TURNER, b Nashville, Tenn, Feb 4, 41; m 64; c 3. PHYSICAL CHEMISTRY. *Educ:* Vanderbilt Univ, BA, 63; Univ SC, PhD(phys inorg chem), 68. *Prof Exp:* From asst prof to assoc prof, 68-78, prof chem, Tenn Technol Univ, 78-; PROF CHEM, CUMBERLAND COL. *Mem:* Am Chem Soc; Sigma Xi. *Res:* Analysis of environmental plutonium. *Mailing Add:* Dept Chem Cumberland Col Williamsburg KY 40769

FARRAR, GEORGE ELBERT, JR, b Winter Park, Fla, Mar 12, 06; m 33; c 2. INTERNAL MEDICINE. *Educ:* Wesleyan Univ, BS, 27; Johns Hopkins Univ, MD, 31. *Prof Exp:* Asst pharmacol, Johns Hopkins Univ, 28-31; intern, Univ Mich Hosp, 31-32, asst resident med, 32-33, instr, Univ & T H Simpson Mem Inst, 33-35; assoc pharmacologist, Food & Drug Admin, USDA, 35-36; from asst prof to clin prof med, Sch Med, Temple Univ, 36-71; field rep, Joint Comn Accreditation Hosps, 71 & Dept Grad Educ, AMA, 72-74; dir med affairs, Excerpta Medica, 74-78, med consult, 78-88; RETIRED. *Concurrent Pos:* Consult pharmaceut advert, Wyeth Labs, Inc, 40-49, dir med serv, 49-71; assoc physician, Episcopal Hosp, Philadelphia, 44-46, chief med serv, 46-49; mem rev comt, US Pharmacopoeia, 50-60, vpres, 70-75; ed-in-chief, Clin Therapeut J, Princeton, NJ, 74-88. *Mem:* AAAS; Am Soc Pharmacol & Exp Therapeut; fel AMA; Am Rheumatism Asn; fel Am Col Physicians. *Res:* Metabolism of iron; nutritional anemia; osteoporosis in adults; hyaluronidase in rheumatic diseases. *Mailing Add:* Pennswood Village A No 106 Newtown PA 18940

FARRAR, GROVER LOUIS, b Lynchburg, Va, May 22, 36; m 63; c 2. ORGANIC CHEMISTRY, TECHNICAL MANAGEMENT. *Educ:* Randolph-Macon Col, BS, 56; Calif Inst Technol, PhD(org chem), 61. *Prof Exp:* Res chemist, Nitrogen Div, Allied Chem Corp, Va, 60-64 & Denver Res Ctr, Marathon Oil Co, 64-67; sr res chemist, Celanese Plastics Co, 67-70, group leader, 70-78, res & develop supt, 78-80; sr res assoc, Am Hoechst Corp, 80-87, SR RES ASSOC HOECHST CELANESE CORP, 87- *Mem:* Am Chem Soc. *Res:* Chemistry of organic nitrogen compounds; reagent additions to olefins; chlorination of aralkyl compounds; naphthalene dicarboxylates; polyester film; surface chemistry; adhesion, coatings and product safety. *Mailing Add:* Hoechst Celanese Corp Box 1400 Greer SC 29652

FARRAR, JAMES MARTIN, b Pittsburgh, Pa, June 15, 48; m 71; c 2. PHYSICAL CHEMISTRY. *Educ:* Wash Univ, AB, 70; Univ Chicago, MS, 72, PhD(chem), 74. *Prof Exp:* Res asst chem, Univ Chicago, 70-74; res assoc, Univ Calif, Berkeley, 74-76; from asst prof to assoc prof, 76-86, PROF CHEM, UNIV ROCHESTER, 86- *Concurrent Pos:* Alfred P Sloan res fel, 81-85; NSF fel, Univ Chicago, 70-73; vis fel, Joint Inst, Lab Astrophysics, 87-88. *Mem:* Fel Am Phys Soc; Am Chem Soc. *Res:* Molecular beam kinetics; ion molecule reaction dynamics; photodissociation of gas phase cations; state-to-state chemistry; size-selected cluster photochemistry. *Mailing Add:* Dept Chem Univ Rochester Rochester NY 14627

FARRAR, JOHN, b Manchester, Eng, Mar 30, 27; US citizen; m 55; c 3. ELECTRONIC ASSEMBLY PROCESSES. *Educ:* Univ Manchester, BSc, 48, PhD(chem), 51. *Prof Exp:* Chemist, E I du Pont de Nemours & Co, 52-53, Shell Develop Co, 53-56 & Technicolor Corp, 56-59; prin scientist, Rocketdyne Div, NAm Rockwell Corp, 59-67, MGR AUTONETICS DIV, ROCKWELL INT CORP, 67- *Mem:* Am Chem Soc; Electrochem Soc. *Res:* Electrochemistry and microelectronics; process development; desalination; batteries; photographic materials; crystal growth; microelectronics; solar energy conversion; materials and processes associated with electronics, computers, navigation and control systems used in ballistic missiles, avionics and electrooptical hardward- These processes include soldering, plating, printed wiring boards, factory automation, etc; failure analysis and chemical-metallurgical-electrical analysis of components and subsystems. *Mailing Add:* 1700 S Clementine St No 521 Anaheim CA 92802

FARRAR, JOHN J, BIOCHEMISTRY. *Prof Exp:* VPRES BIOL RES, STERLING RES GROUP, 87- *Mailing Add:* Dept Biol Sci Sterling Res Group Nine Great Valley Pkwy Great Valley PA 19355

FARRAR, JOHN KEITH, b Columbus, Ohio, Oct 9, 48; m 71; c 2. CEREBRAL BLOOD FLOW, CEREBRAL METABOLISM. *Educ:* Univ Western Ont, BSc, 71, PhD(biophysics), 74. *Prof Exp:* Res asst, Dept Physiol, Univ Glasgow & Dept Neurosurg, Southern Gen Hosp, 74-75; lectr, clin neurol sci/biophysics, 75-77, ASST PROF, UNIV WESTERN ONT, 77- *Mem:* Can Stroke Soc; fel Am Heart Asn; Am Stroke Soc; Int Soc Cerebral Blood Flow & Metabolism; sr fel Can Heart Found. *Res:* The study of cerebral blood flow and metabolism and physiological control and alterations during ischemia-stroke. *Mailing Add:* Dept Clin Neurol Sci Univ Hosp 339 Windermere Rd London ON N6A 5A5 Can

FARRAR, JOHN LAIRD, b Hamilton, Ont, Dec 31, 13; m 46. FORESTRY. *Educ:* Univ Toronto, BScF, 36; Yale Univ, MF, 39, PhD, 55. *Prof Exp:* Forester, Can Int Paper Co, 36-37; forest ecol, Can Forestry Br, 37-41 & 45-56; prof forestry, Univ Toronto, 56-78; RETIRED. *Concurrent Pos:* Ed, Can J Forest Res, 70-80. *Mem:* Sigma Xi; Can Inst Forestry; Can Bot Asn. *Res:* Forest ecology and tree physiology. *Mailing Add:* 255 Bamburgh Cr Unit 306 Scarborough ON M1W 3T6 Can

FARRAR, JOHN THRUSTON, b St Louis, Mo, June 26, 20; m 47; c 2. INTERNAL MEDICINE, GASTROENTEROLOGY. *Educ:* Princeton Univ, AB, 42; Univ Wash, MD, 45; Am Bd Internal Med, dipl, 54; Am Bd Gastroenterol, dipl, 62. *Prof Exp:* Asst res path, Boston City Hosp, 48-49; intern med, Mass Mem Hosp, Boston, 49-50, asst resident, 50-51, res assoc, 51-54; chief gastroenterol sect, Med Serv, Vet Admin Hosp, New York, 55-63, asst dir prof serv res, 56-63; assoc prof med, Med Col Va, 63-65, prof, 65-80, chief Gastroenterol Sect, 63-80; mem fac, Health Sci Div, Va Commonwealth Univ, 80-; AT VA MED CTR. *Concurrent Pos:* Asst, Sch Med, Boston Univ, 50-54, instr, 54-55; ed, Am J Digestive Dis, 68- *Mem:* AMA; Am Gastroenterol Asn; Am Fedn Clin Res; Inst Elec & Electronics Engrs. *Res:* Gastrointestinal physiology, particularly absorption and motility; pancreatic exocrine physiology; medical electronics. *Mailing Add:* Va Med Ctr Va Med Col 1201 Broad Rock Rd Richmond VA 23249

FARRAR, R(ICHARD) E(DWARD), b Lynchburg, Va, Mar 13, 17; m 46; c 1. CHEMICAL ENGINEERING. *Educ:* Va Polytech Inst, BS, 38; Johns Hopkins Univ, ME, 50, DrEng, 51. *Prof Exp:* Asst chemist, Control Div, Mead Corp, 38-40; chief chemist, Columbia Paper Co, 40-41; prod supvr, Prod Div, US Army Chem Ctr, 41-43, area engr, Develop Div, 45-48; mgr res & develop, Colgate-Palmolive Co, 51-60, res coordr, Household Prod Div, 60-62, dir res & develop, Europ Div, London, Colgate-Palmolive Int, Inc, 62-67; dir prod develop, R J Reynolds Tobacco, ND, 67-70; mem staff, Dairy Res, Inc, 70, exec vpres, 70-77; CONSULT PROD & PROCESS DEVELOP, RICHARD E FARRAR, CONSULT, 77- *Concurrent Pos:* Vpres prod develop, Lisher & Co Inc, New York, 77- *Mem:* Am Chem Soc; Am Inst Chem Engrs; Sigma Xi. *Res:* Administration of research and development. *Mailing Add:* Bermuda Run Box 717 Advance NC 27006

FARRAR, RALPH COLEMAN, b Cabool, Mo, Sept 22, 30; m 53; c 2. ORGANIC CHEMISTRY. *Educ:* Univ Wichita, BS, 53; Univ Ill, PhD(org chem), 56. *Prof Exp:* Chemist, Rubber Res Prog, Univ Ill, 54-56; res chemist, Phillips Petrol Co, 56-79, res assoc synthetic rubber, Res & Develop Br, 79-82, sect supvr rubber synthesis, 82, SECT SUPVR POLYOLEFINS BR RES & DEVELOP DIV, PHILLIPS PETROL CO, 83- *Mem:* Am Chem Soc; Am Inst Chemists; Sigma Xi. *Res:* New types of synthetic rubber and plastics. *Mailing Add:* 2352 London Lane Bartlesville OK 74006

FARRAR, ROBERT LYNN, JR, b Nashville, Tenn, Sept 3, 21; m 45; c 3. PHYSICAL CHEMISTRY. *Educ:* Vanderbilt Univ, AB, 43; Univ Wash, MS, 48; Univ Tenn, PhD(chem), 53. *Prof Exp:* Chemist, Carbide & Carbon Chem Corp, 47-54, res chemist, 54-58; head phys chem sect, Oak Ridge Gaseous Diffusion Plant, Union Carbide Nuclear Co, 58-61, spec assignment, Y12 Plant, 61-62, develop consult chem, 62-83; RETIRED. *Concurrent Pos:* Instr, Eve Sch, Univ Tenn, 57, vis assoc prof, Oak Ridge Grad Prog, 58. *Mem:* AAAS; Am Chem Soc; Sigma Xi; fel Am Inst Chemists. *Res:* Fluorine and uranium chemistry; kinetics of solid-gas reactions; adsorption; corrosion; physical properties of inorganic fluorine compounds. *Mailing Add:* PO Box 397 Walland TN 37886

FARRAR, THOMAS C, b Independence, Kans, Jan 14, 33; m 63; c 3. PHYSICAL CHEMISTRY. *Educ:* Univ Wichita, BS, 54; Univ Ill, PhD(phys chem), 59. *Prof Exp:* NSF fel, Cambridge Univ, 59-61; asst prof chem, Univ Ore, 61-63; chemist, Nat Bur Stand, 63-67, chemist, Inorg Chem Sect, 67-69, head magnetism group, 69-71; dir res & develop, Japan Electron Optics Lab Co, Inc, 71-76; dir, Chem Instrumentation Prog, NSF, 76-79; PROF CHEM, UNIV WIS-MADISON, 79- *Concurrent Pos:* Chmn, Enc Inc, 73-74; chmn, Nat Magnet Lab Adv Comt, 80- *Honors & Awards:* US Dept Com Silver Medal; Indust Res Mag Award, 75; NSF Silver Medal, 79. *Mem:* Fel Am Inst Chemists; Am Phys Soc; Am Chem Soc; NY Acad Sci. *Res:* Experimental and theoretical nuclear magnetic resonance spectroscopy; theoretical chemistry; research and development in analytical spectroscopy (ms, ir, nmr); development of new instrumentation. *Mailing Add:* Dept Chem Univ Wis Madison WI 53706

FARRAR, WILLIAM EDMUND, b Macon, Ga, May 28, 33; m 57; c 2. INFECTIOUS DISEASE, MICROBIOLOGY. *Educ:* Mercer Univ, BS, 55; Med Col Ga, MD, 58. *Prof Exp:* Intern med, Talmadge Mem Hosp, Augusta, 58-59, asst resident, 59-60; sr asst resident, Grady Mem Hosp, Atlanta, 60-61; assoc med, Emory Univ, 65-67, from asst prof to assoc prof prev med, 65-71, from asst prof to assoc prof med, 67-71, dir div infectious dis, Dept Med, 69-71; PROF MED & MICROBIOL & DIR, DIV INFECTIOUS DIS, DEPT MED, MED UNIV SC, 72- *Concurrent Pos:* USPHS fel, Sch Med, Emory Univ, 61-62; vis mem staff, Grady Mem Hosp, 65-; partic, Int Conf Molecular Biol of Gram-Negative Bacteria, NY Acad Sci, 65, Int Conf Biol Effects of Gram-Negative Bacteria, Nat Ctr Sci Res, France, 68 & Int Conf Recombinant DNA Molecules, Nat Res Coun, Nat Acad Sci, Calif, 75; consult, Emory Univ Clin, 65-71; consult malaria res prog, NIH, Atlanta Fed Prison, 67-; consult infectious dis, St George's Hosp, London, Eng, Atkinson Morley's Hosp & Royal Dent Hosp, 78-79. *Mem:* AAAS; Am Fedn Clin Res; Royal Soc Med; Am Soc Microbiol; fel Am Col Physicians; fel Infectious Dis Soc Am. *Res:* Resistance of bacteria to antibiotics; bacterial flora of gastrointestinal tract; bacterial endotoxins; infections due to gram-negative bacteria. *Mailing Add:* Dept Med Med Univ SC 171 Ashley Ave Charleston SC 29425

FARRAR, WILLIAM L, b Monterey, Calif, Nov 12, 50. EXPERIMENTAL BIOLOGY. *Educ:* Va Polytech Inst, PhD(biochem), 78. *Prof Exp:* HEAD, CYTOKINES MECH SECT, NAT CANCER INST, 83- *Mem:* Am Asn Immunologists. *Mailing Add:* Cytokines Mech Sect Nat Cancer Inst-FCRDC PO Box B Bldg 560 Frederick MD 21701

FARRAR, WILLIAM WESLEY, b Birmingham, Ala, July 29, 40; 67; c 2. BIOCHEMISTRY, ANATOMY. *Educ:* Samford Univ, BS, 64; Med Col Va, MS, 68; Va Polytech Inst & State Univ, PhD(biochem), 70. *Prof Exp:* NIH trainee biochem, Univ Utah, 70-71; NIH fel, Mich State Univ, 71-74; res assoc oncol, Med Col Va, 74-76; from asst prof to assoc prof, 76-85, PROF BIOL, EASTERN KY UNIV, 85- *Mem:* Sigma Xi; Biochem Soc; Int Soc Neurochem. *Res:* Physical, chemical and kinetic properties of glycolytic enzyme; protein folding. *Mailing Add:* Dept Biol Eastern Ky Univ Richmond KY 40475

FARRAUTO, ROBERT JOSEPH, b New York, NY, Nov 22, 41; m 65; c 2. AIR POLLUTION CONTROL USING CATALYSIS, CHEMICAL PRODUCTION USING CATALYSIS. *Educ:* Manhattan Col, Bronx, BS, 64; Rensselaer Polytech Inst, PhD(chem), 68. *Prof Exp:* Res mgr, 78-88, sr res assoc, 89-90, PRIN SCIENTIST, ENGELHARD CORP, 90- *Concurrent Pos:* Adj prof chem eng, Stevens Inst Technol, Hoboken, NJ, 89-, NJ Inst Technol, Newark, 89- *Mem:* Am Chem Soc; Am Inst Chem Engrs. *Res:* Industrial and basic catalysis; environmental catalysis; chemical catalysis. *Mailing Add:* Menlo Park Res & Develop Engelhard Corp Edison NJ 08818

FARRELL, DAVID E, b May 9, 39; US citizen. PHYSICS. *Educ:* Univ London, BSc, 60, PhD(physics), 64. *Prof Exp:* Res asst physics, Univ London, 64; fel, Western Reserve Univ, 64-66, instr, 66-67; asst prof, 67-72, assoc prof, 72-80, PROF PHYSICS, CASE WESTERN RESERVE UNIV, 80- *Mem:* Am Phys Soc; NY Acad Sci. *Res:* Low temperature physics; superconductivity; biomagnetism. *Mailing Add:* Dept Physics Case Western Reserve Univ Cleveland OH 44106

FARRELL, EDWARD JOSEPH, b San Francisco, Calif, Mar 28, 17; m 54; c 2. MATHEMATICS. *Educ:* Univ San Francisco, BSc, 39; Stanford Univ, MA, 42. *Prof Exp:* Dir math inst, 60-75, from assoc prof to prof, 41-82, EMER PROF MATH, UNIV SAN FRANCISCO, 82- *Concurrent Pos:* NSF dir inst progs, 60- & dir NSF Summer Conf Geom, 67-75. *Mem:* AAAS; Am Inst Phys; Math Asn Am. *Res:* Differential equations. *Mailing Add:* Dept Math Univ San Francisco San Francisco CA 94117

FARRELL, EUGENE PATRICK, b Wamego, Kans, July 3, 11; m 35; c 5. FOOD ENGINEERING. *Educ:* Kans State Univ, BS, 35, MS, 53. *Prof Exp:* Trainee, Gen Mills, Inc, Minn, 35-38, miller, NY, 38-42, plant supt, Ky, 42-43 & Iowa, 43-45, div supt, Minn, 45-47; prod mgr, Maney Milling Co, Nebr, 47-49; milling technologist, 49-53, assoc prof flour & feed milling indust, 53-67, prof, 67-81, EMER PROF GRAIN SCI & INDUST, KANS STATE UNIV, 81- *Concurrent Pos:* Consult, 51-85. *Honors & Awards:* Gold Medal Award, Asn Oper Millers, 76. *Mem:* Asn Oper Millers. *Res:* Flour milling; flow diagrams; equipment layout and operation; wheat quality evaluation and conditioning; feed grain grinding; corn and sorghum milling. *Mailing Add:* 805 Houston Manhattan KS 66502

FARRELL, F THOMAS, b 1929. MATHEMATICS. *Educ:* Univ Iowa, PhD(math), 59. *Prof Exp:* PROF MATH, CORNELL UNIV, 59- *Mem:* Am Math Asn. *Mailing Add:* Math Dept State Univ NY Binghamton NY 13901

FARRELL, HAROLD MARON, JR, b Pottsville, Pa, Sept 5, 40; m 63; c 2. PROTEIN CHEMISTRY, ENZYMOLOGY. *Educ:* Mt St Mary's Col, Md, BS, 62; Pa State Univ, MS, 65, PhD(biochem), 68. *Prof Exp:* Asst biochem, Pa State Univ, 63-66, res asst, 66-67; res assoc protein chem, 67-69, res chemist, 69-75, SUPVRY RES CHEMIST, EASTERN REGIONAL LAB, USDA, 75- *Honors & Awards:* Borden Award, Am Dairy Sci Asn, 85. *Mem:* Am Chem Soc; Am Dairy Sci Asn; Am Soc Biol Chem. *Res:* Protein chemistry, especially the relation of protein structure to biological function; milk proteins and enzymes of lactation. *Mailing Add:* Eastern Regional Lab USDA 600 E Mermaid Lane Philadelphia PA 19118

FARRELL, J(OSEPH) B(RENDAN), b New York, NY, May 3, 23; m 51; c 7. CHEMICAL ENGINEERING. *Educ:* Univ Notre Dame, BS, 43; Mass Inst Technol, MS, 47; Cornell Univ, PhD(chem eng), 54. *Prof Exp:* Chem engr, Kellex Corp, 44-45; instr chem eng, Univ Notre Dame, 47-49; res chem engr, M W Kellogg Co, 51-55; res assoc, Am Mach & Foundry Co, 56-61; assoc prof chem eng, Manhattan Col, 61-67; chem engr, RA, Taft Water Res Ctr, US Dept Interior, Fed Water Pollution Control Admin, 67-75; SECT CHIEF, US ENVIRON PROTECTION AGENCY, 75- *Concurrent Pos:* Indust consult; adj prof chem eng, Univ Cincinnati, 79- *Mem:* Am Chem Soc; Am Inst Chem Engrs; Water Pollution Control Fedn. *Res:* Flow of non-Newtonian liquids; film casting; drying; electrodialysis; wet oxidation of graphite; ultimate disposal of solid and liquid wastes from conventional and advanced methods for treatment of wastewater to eliminate pollution and pathogenic organisms. *Mailing Add:* 1117 Stormy Way Cincinnati OH 45230

FARRELL, JAY PHILLIP, PARASITOLOGY. *Educ:* Rutgers Univ, PhD(zool), 72. *Prof Exp:* ASSOC PROF PATHOBIOL, SCH VET MED, UNIV PA, 83-, HEAD LAB PARASITOL & CHMN GRAD GROUP, 84- *Mailing Add:* Dept Parasitol Univ Pa Sch Vet Med 388 Spruce St Philadelphia PA 19104

FARRELL, JOHN A, b Ft Worth, Tex, Dec 25, 35; m 58; c 1. NUCLEAR PHYSICS. *Educ:* Tex Christian Univ, BA, 59; Duke Univ, PhD(physics), 64. *Prof Exp:* Res assoc physics, Duke Univ, 64-66; STAFF MEM GROUP W-8, LOS ALAMOS SCI LAB, UNIV CALIF, 66- *Mem:* Am Phys Soc. *Res:* Neutron total cross sections in the kilovolt energy region. *Mailing Add:* Los Alamos Nat Lab PO Box 503 Los Alamos NM 87545

FARRELL, LARRY DON, b Woodward, Okla, Nov 5, 42; m 65; c 2. MICROBIAL GENETICS, BACTERIAL VIROLOGY. *Educ:* Univ Okla, BS, 64, MS, 66; Univ Calif, Los Angeles, PhD(bact), 70. *Prof Exp:* Fel & instr microbiol, Univ Ill Col Med, 70-72; asst prof, Idaho State Univ, 72-78, chmn dept microbiol & biochem, 77-84, asst chmn dept biol sci, 88-90, assoc prof, 78-89, PROF MICROBIOL, IDAHO STATE UNIV, 89- *Mem:* Am Soc Microbiol; AAAS; Sigma Xi; Int Soc AIDS Educ. *Res:* Bacterial virology; isolation and characterization of new bacteriophages (viruses) for Sphaerotilus natans; development of AIDS education programs for public schools. *Mailing Add:* Dept Biol Sci Idaho State Univ Pocatello ID 83209-0009

FARRELL, MARGARET ALICE, b Troy, NY, Mar 9, 32. MATHEMATICS EDUCATION. *Educ:* Col St Rose, Albany, AB, 53; Boston Col, MEd, 54; Ind Univ, PhD(math educ), 67. *Prof Exp:* Teacher math, Brasher Falls Cent Sch, NY, 54-55, Chatham Cent Sch, NY, 55-58 & Shaker High Sch, NY, 58-60; from asst prof to assoc prof, 60-73, prof math educ, State Univ NY Albany, 73-89. *Concurrent Pos:* Mem, Nat Comn Educ Teachers, 75- *Mem:* Sigma Xi. *Res:* Piagetian research in mathematics education; mathematics teacher education. *Mailing Add:* 32 Oak Rd Delmar NY 12054

FARRELL, RICHARD ALFRED, b Providence, RI, Apr 22, 39; m 61; c 2. PHYSICS, BIOPHYSICS. *Educ:* Providence Col, BS, 60; Univ Mass, MS, 62; Cath Univ Am, PhD(physics), 65. *Prof Exp:* NASA grant, Cath Univ Am, 64-65; sr staff physicist, 65-70, prin prof staff physicist, 70-77, SUPVR THEORET PROBS GROUP, APPL PHYSICS LAB, JOHNS HOPKINS UNIV, 77- *Concurrent Pos:* Prin investr, USPHS grant, Nat Eye Inst, NIH, 73-; prin investr, US Army Med Res & Develop Command Contract, 77-89 & Geosci Div, US Army Res Off Contract, 80-82. *Mem:* Am Phys Soc; NY Acad Sci; Asn Res Vision & Ophthal; Int Soc Eye Res; Optical Soc Am. *Res:* Application of theoretical physics to the fields of eye research; statistical mechanics and light scattering. *Mailing Add:* Appl Physics Lab Johns Hopkins Rd Laurel MD 20723-6099

FARRELL, ROBERT LAWRENCE, b Zanesville, Ohio, Oct 6, 25; m 45; c 4. VETERINARY PATHOLOGY. *Educ:* Ohio State Univ, DVM, 50, MSc, 51, PhD(vet path), 54. *Prof Exp:* From instr to prof vet path, Ohio State Univ, 51-73; PROF, DEPT PATH, UNIV GA, 73- *Concurrent Pos:* Consult, Procter & Gamble Co, 73- *Mem:* AAAS; Am Col Vet Path; Am Vet Med Asn; Conf Res Workers Animal Dis; Int Acad Path. *Res:* Effects of toxic aerosols on respiratory tract of animals; study of bovine pneumonia. *Mailing Add:* Dept Path Col Vet Med Univ Ga Athens GA 30602

FARRELL, ROBERT MICHAEL, b Honesdale, Pa, Nov 5, 47; m 72; c 2. COMPUTER SCIENCE. *Educ:* Villanova Univ, BEE, 69; Univ Pa, MSE, 71, PhD(sci eng), 79. *Prof Exp:* Engr, Gen Elec, 69-74, sr engr, 74-79, mgr advan technol, space div, 79-85, mgr technol ctr, 85-86; VPRES & CHIEF OPERATING OFFICER, MRI, INC, 86- *Concurrent Pos:* Lectr comput archit & orgn, Villanova Univ, 82- *Mem:* Inst Elec & Electronics Engrs; Asn Comput Mach. *Res:* Computer network mathematical models, performance estimation, and optimization; distributed database modeling, performance estimation and optimization; computer network architectures and designs to meet specific requirements. *Mailing Add:* 11611 Auburn Grove Ct Reston VA 22094

FARRELL, ROGER HAMLIN, b Greensboro, NC, July 23, 29; m 67. MATHEMATICAL STATISTICS. *Educ:* Univ Chicago, PhB, 47, MS, 51; Univ Ill, PhD(math), 59. *Prof Exp:* From instr to assoc prof, 59-67, PROF MATH, CORNELL UNIV, 67- *Mem:* Am Math Soc; Inst Math Statist; Am Statist Asn. *Res:* Measure theory; probability theory; mathematical statistics. *Mailing Add:* Dept Math Cornell Univ Ithaca NY 14853

FARRELLY, JAMES GERARD, b Jan 8, 39; div; c 3. METABOLISM OF NITROSAMINES, BIOCHEMISTRY. *Educ:* Manhattan Col, BS(chem), 61; Univ Tenn, PhD(biochem), 68. *Prof Exp:* SR RES SCIENTIST, DEPT CHEM CARCINOGENESIS, NAT CANCER INST, 76- *Mem:* Am Soc Biochem & Molecular Biol; Am Asn Cancer Res. *Res:* Study of the metabolism of nitrosamine and the interaction of metabolite with cellular macromolecules. *Mailing Add:* Dept ABL Basic Res Prog NCI-FCRDC PO Box B Frederick MD 21701

FARREN, ANN LOUISE, b Portage, Pa, Dec 5, 26. BIOCHEMISTRY, INFORMATION SCIENCE. *Educ:* Univ Pa, AB, 48. *Prof Exp:* Biochemist, Valley Forge US Army Hosp, Pa, 49-50 & Jefferson Med Col, 50-52; org chemist, Smith Kline & French Labs, 52-53; anal chemist, Rohm & Haas Co, 53-56; with info off pub rels, News Serv, Am Chem Soc, NY, 56-59; asst to dir, Biol Abstracts, Inc, 59-61, actg head, Lit Acquisition Dept, 61-62, prof rels off, 62-74, mgr educ bur, 74-78, mgr, User Commun, 78-80, SR EDUC SPECIALIST, BIOSIS, 80- *Concurrent Pos:* Mem, Nat Fedn Sci Abstracting & Indexing Serv. *Mem:* Fel AAAS; Am Chem Soc; Nat Asn Sci Writers; Am Inst Biol Sci. *Res:* Hepatic and metabolic diseases; organic syntheses; colchicine derivatives; ion exchange resins; abstracting, indexing, information retrieval; education. *Mailing Add:* Biosis 2100 Arch St Philadelphia PA 19103

FARRIER, MAURICE HUGH, b Washington Co, Iowa, Sept 18, 26; m 56; c 1. ACAROLOGY, FOREST ENTOMOLOGY. *Educ:* Iowa State Univ, BS, 48, MS, 50; NC State Col, PhD(entom), 55. *Prof Exp:* Asst to state entomologist, State of Iowa, 50-52; asst entom, 54-55, asst res prof, 55-60, assoc prof, 60-71, PROF ENTOM, NC STATE UNIV, 71- *Honors & Awards:* Southern Insect Work Conf Outstanding Contrib Award, 70. *Mem:* Entom Soc Am; Nat Agr Col Teach Asn. *Res:* Taxonomy of Veigaiidae and other gamasid mites in forest soils; forest entomology and bibliographic documentation in biology. *Mailing Add:* Box 7613 Raleigh NC 27695-7613

FARRIER, NOEL JOHN, b Pittsburgh, Pa, Dec 9, 37; m 73. ENVIRONMENTAL HEALTH, BIOINORGANIC CHEMISTRY. *Educ:* Carnegie Inst Technol, BS, 60, Univ Pittsburgh, MS, 64, MPH, 80; Ohio State Univ, PhD(inorg coord chem), 69. *Prof Exp:* Lab technician chem, Mellon Inst, 61; trainee metalloenzyme chem, Clin Study Ctr, Children's Hosp, Columbus, Ohio, 64; asst prof chem, Wittenberg Univ, 68-69; vis res assoc physiol chem, Col Med, Ohio State Univ, 69-70; asst prof chem, Raymond Walters Gen & Tech Col, Univ Cincinnati, 70-76, instr nutrit, 75-76; instr chem & phys sci, Edison State Community Col, Piqua, Ohio, 76-78; FEL, DEPT EPIDEMIOL, GRAD SCH PUB HEALTH, UNIV PITTSBURGH, 78- *Mem:* Am Chem Soc; Soc Epidemiol Res. *Res:* Coordination compounds of cobalt III, copper II and copper I containing unusual monodentate ligands; kinetics; infrared studies; blood lead studies in children; trace metals and cardiovascular diseases; trace metals in foods; carbon monoxide in blood of smokers, former smokers and never smokers. *Mailing Add:* 3089 Revlon Dr Kettering OH 45420-1244

FARRINGER, LELAND DWIGHT, b Lena, Ill, May 28, 27; m 50; c 3. PHYSICS. *Educ:* Manchester Col, BA, 49; Bethany Biblical Sem, BD, 52; Ohio State Univ, MA, 55, PhD(physics), 58. *Prof Exp:* Asst physics, Ohio State Univ, 52-58; from asst prof to assoc prof, 58-69, chmn, div sci, 80-84, PROF PHYSICS, MANCHESTER COL, 69-, head dept, 59-87. *Mem:* Am Phys Soc; Am Asn Physics Teachers. *Res:* Nuclear magnetic resonance; physical electronics; electronic circuitry; electrical measurements. *Mailing Add:* Dept Physics Manchester Col North Manchester IN 46962

FARRINGTON, GREGORY CHARLES, b Bronxville, NY, Aug 4, 46; m 70; c 1. PHYSICAL CHEMISTRY, ELECTROCHEMISTRY. *Educ:* Clarkson Col, BS, 68; Harvard Univ, AM, 70, PhD(chem), 72. *Hon Degrees:* PhD, Univ Uppsala, Sweden, 84. *Prof Exp:* Res chemist, Gen Elec Res & Develop Corp, 72-79; PROF, DEPT MAT SCI, UNIV PA, PHILADELPHIA, 79- *Mem:* AAAS; Electrochem Soc; Am Chem Soc; Mat Res Soc. *Res:* Conduction in solid electrolytes; interfacial electrochemistry of Na beta alumina and derivatives; solid state batteries; dielectric properties of solids; conductive polymers. *Mailing Add:* Dept Mat Sci Eng Univ Pa Philadelphia PA 19104

FARRINGTON, JOHN WILLIAM, b New Bedford, Mass, Sept 25, 44; m 66; c 2. MARINE GEOCHEMISTRY, GENERAL ENVIRONMENTAL SCIENCE & EDUCATIONAL ADMINISTRATION. *Educ:* Southeastern Mass Univ, BS, 66, MS, 68; Univ RI, PhD(oceanog), 72. *Prof Exp:* Investr, 71-72, from asst scientist to assoc scientist, 72-82, sr scientist chem, Woods Hole Oceonog Inst, 82-88, dir, Coastal Res Ctr, 81-87; Michael P Walsh prof, Univ Mass, Boston, 88-90; ASSOC DIR EDUC & DEAN GRAD STUDIES, WOODS HOLE OCEANOGR INST, 90- *Concurrent Pos:* Adj prof, Ctr Bioorganic Studies, Univ New Orleans, 78-81; environ studies bd, Nat Res Coun, 82-85; chair, US Nat Comt, Sci Comt Probs Environ, 83-86; consult, nat & int pub agencies & indust; adj prof, Environ Sci, Univ Mass, Boston, 90-; adv comm marine pollution, IOC/Unesco, 88- *Mem:* Sigma Xi; AAAS; Am Chem Soc; Am Geophys Union. *Res:* Sediment-water interface; organic geochemical processes in the marine environment; marine biochemistry; environmental quality; science education. *Mailing Add:* Educ Off Woods Hole Oceanogr Inst Woods Hole MA 02543

FARRINGTON, JOSEPH KIRBY, b Jacksonville, Fla, Oct 5, 48; m 76. PURIFIED WATER SYSTEMS, ASEPTIC MANUFACTURING. *Educ:* LaGrange Col, BS, 71; Clemson Univ, MS, 73; Auburn Univ, PhD(microbiol), 77. *Prof Exp:* Microbiologist, Kellogg Co, 76-77; ASSOC DIR, SCHERING-PLOUGH HEALTHCARE PROD, 77-, VPRES, PLOUGH LABS, 84- *Concurrent Pos:* Adj prof, Univ Miss Sch Pharm, 81-, Univ Tenn Sch Pharm, 84-; prof, Memphis State Univ, 82- *Mem:* Am Soc Microbiol; Cosmetic Toiletry & Fragrance Asn; Inst Food Technologists; Southern Asn Clin Microbiol; Soc Cosmetic Chemists. *Res:* Antimicrobial preservative systems; purified water systems; computer systems for use in pharmaceutical quality control. *Mailing Add:* Schering-Plough Corp PO Box 377 Memphis TN 38151

FARRINGTON, PAUL STEPHEN, b Indianapolis, Ind, May 9, 19; m 46, 77; c 5. ANALYTICAL CHEMISTRY. *Educ:* Calif Inst Technol, BS, 41, MS, 47, ChE, 48, PhD(chem), 50. *Prof Exp:* Control & res chemist, Kelco Co, 41-42; asst chem, Calif Inst Technol, 42-45 & 48-49; from instr to assoc prof, 50-62, assoc dean, Col Letters & Sci, 62-82, PROF CHEM, UNIV CALIF, LOS ANGELES, 62- *Concurrent Pos:* Guggenheim Found fel, Ger, 58-59. *Mem:* Am Chem Soc. *Res:* Coulometric and other instrumental methods of analysis, inorganic complexes. *Mailing Add:* 11809 Stanwood Dr Los Angeles CA 90066-1121

FARRINGTON, THOMAS ALLAN, b Newark, NJ, Nov 26, 41; m; c 1. WATER TREATMENT. *Educ:* Col William & Mary, BS, 63; Rutgers Univ, MS, 66, MBA, 74. *Prof Exp:* Chemist, Am Oil & Supply, 69-73; sr res chemist, 73-75, res assoc, 75-76, proj mgr, 77-85, mgr res, div Ashland Chem, 85-88, DIR RES, DREW CHEM, 88- *Mem:* Am Chem Soc; Water Pollution Control Fedn. *Res:* Development of coagulants; flocculants and application of same in solids/liquids separation. *Mailing Add:* Ashland Oil Inc Drew Chemical Lab One Drew Plaza Boonton NJ 07005

FARRINGTON, WILLIAM BENFORD, b New York, NY, Mar 10, 21; m 79; c 3. STRUCTURAL GEOLOGY. *Educ:* Cornell Univ, BCE, 47, MS, 49; Mass Inst Technol, PhD(geol), 53. *Prof Exp:* Radio engr radar design, Naval Res Labs, 42-43; plant engr, Hope's Windows, Inc, 50-51; instr geol, Univ Mass, 53-54; res geophysicist, Humble Oil & Ref Co, 54-56; investment analyst, Continental Res Corp, 56-61; vpres, Empire Resources Corp, Empire Trust Co, 61-64; sci dir, US Cong House Select Comt Govt Res, 64-65; PARTNER, FARRINGTON ASSOCS, 67- *Concurrent Pos:* Eve lectr, Univ Houston, 55-56; pres, Farrington Eng Corp, 58-67; eve lectr, Univ Calif, Los Angeles, 68- *Mem:* Geol Soc Am; Am Inst Aeronaut & Astronaut; Am Petrol Inst; fel AAAS; fel Financial Analyst Fedn. *Res:* Theory and measurement of stress distribution in rocks. *Mailing Add:* 1565 Skyline Dr Laguna Beach CA 92651

FARRIS, DAVID ALLEN, b Bloomington, Ind, Mar 26, 28; m 56; c 2. FISHERIES. *Educ:* Ind Univ, AB, 50; Stanford Univ, PhD(biol), 58. *Prof Exp:* Field asst fisheries, Ind Lake & Stream Surv, 43-50; asst biol, Stanford Univ, 53-54; res fisheries biologist, Bur Commercial Fisheries, 55-60; from asst prof to assoc prof biol, 60-66, PROF BIOL, SAN DIEGO STATE UNIV, 66- *Mem:* AAAS; Soc Conserv Biol; Am Inst Fishery Res Biol; Sigma Xi. *Res:* Population dynamics of egg and larval fish populations; the intrinsic and extrinsic factors which govern growth and death of fish larvae; factors which govern changes in the biochemistry of fishes; population dynamics of California spiny lobsters; fish community structure of small ponds. *Mailing Add:* Dept Biol San Diego State Univ San Diego CA 92182

FARRIS, RICHARD AUSTIN, b Baltimore, Md, July 24, 43; m 66; c 2. MARINE ECOLOGY. *Educ:* Calif Luthern Col, BA, 66; Calif State Univ, Humboldt, MA, 68; Univ NC, Chapel Hill, PhD(zool), 76. *Prof Exp:* Teacher biol, Newbury Park High Sch, 68-70; chmn deot, 83-86, ASSOC PROF BIOL, LINFIELD COL, 74- *Concurrent Pos:* Montgomery-Moore fel, Bermuda Biol Sta, 74; chmn, Sci State Adv Bd, 79-81. *Mem:* Asn Meiobenthologists; Sigma Xi; AAAS; Am Soc Zoologists. *Res:* Ecology and systematics of marine interstitial organisms in the Northwest and Bermuda, with emphasis on the phylum Gnathostomulida. *Mailing Add:* Dept of Biol Linfield Col McMinnville OR 97128

FARRISSEY, WILLIAM JOSEPH, JR, b Fall River, Mass, Nov 19, 31; m 57; c 4. ORGANIC POLYMER CHEMISTRY. *Educ:* Yale Univ, BS, 53, MS, 55, PhD, 57. *Prof Exp:* Fel, Rice Univ, 56-57; instr chem, Rutgers Univ, 57-58; res chemist, Humble Oil & Refining Co, 58-61, sr res chemist, 61-63; group leader, 63-64, head new prod, 64-66, mgr process res, 66-69, mgr polymer res, Carwin Res Labs, 69-78, group mgr polymer res & res admin, 78-84, dir, Polymer Res, D S Gilmore Labs, Upjohn Co, 84-85; NORTH HAVEN LABS, DOW CHEM, 85- *Mem:* Am Chem Soc. *Res:* High temperature polymers; polyimides; polyurethanes, isocyanate derived polymers. *Mailing Add:* Dow Chemical B-1610 Freeport TX 77541-3257

FARROW, JOHN H, b Chicago, Ill, Jan 23, 35; m 56; c 5. QUALITY MANAGEMENT, STATISTICAL PROCESS CONTROL. *Educ:* Marquette Univ, BEE, 56, MS, 80. *Prof Exp:* Engr motor design, Jack & Heintz, Inc, 60-61; supvr, Cutler-Hammer, Inc, 61-69; engr valuation, Am Appraisal, 69-70; mgr, aual control, RTE Corp, 70-81; dir indust eng, Marquette Univ, 81-88; DIR & CHMN, MECH ENG DEPT, MILWAUKEE SCH ENG, 88- *Concurrent Pos:* Mem, Prod Liability Loss Prev Comn, Am Soc Qual Control, 77- , Qual Audit Tech Comn, 78-; exec dir, Engrs & Scientists Milwaukee, 89-90. *Mem:* Am Soc Qual Control; Soc Reliability Engrs; Inst Indust Engrs; Soc Mfg Engrs; Am Soc Mech Engrs; Am Soc Eng Educ. *Res:* Quality auditing and its relationship to total quality programs; statistical process control; quality management; product liability loss prevention. *Mailing Add:* Dept Mech Eng Milwaukee Sch Eng PO Box 644 Milwaukee WI 53201

FARROW, LEONILDA ALTMAN, b Brooklyn, NY, May 11, 29; m 56. CHEMICAL PHYSICS. *Educ:* Cornell Univ, BEngPhysics), 51; Mass Inst Technol, PhD(physics), 56. *Prof Exp:* Res asst physics, Mass Inst Technol, 51-56; mem tech staff res chem, Bell Labs, 56-83, Bellcore, 83-91; PRES, NUMBREX, INC, 91- *Mem:* Am Phys Soc. *Res:* Chemical kinetics; reaction rate measurements; absorption spectroscopy; infared lasers; molecular spectroscopy; optoacoustics; atmospheric chemical reactions; chemistry of the contaminated troposphere; computer simulation; environmental physics and chemistry; plasma etching; raman spectroscopy of semi conductors and superconductors. *Mailing Add:* 33 Monmouth Hills Highlands NJ 07732

FARROW, WENDALL MOORE, b Winchester, Mass, June 20, 22; m 49; c 2. MICROBIOLOGY, ANIMAL GENETICS. *Educ:* Univ Iowa, BA, 49, MS, 51, PhD(bot, mycol), 53. *Prof Exp:* Asst, Univ Iowa, 49-53; microbiologist, Commercial Solvents Corp, 53-56, Hoffman-La Roche, Inc, 56-63, Marine Lab, Fla Bd Conserv, 63-65 & Germ-free Prod, Inc, 65-68; res assoc, Life Sci, Inc, 68-80, dir lab & safety off, 80-85, dir animal develop & prod, 82-90, CONSULT, LIFE SCI, INC, 90- *Mem:* Mycol Soc Am; Am Soc Microbiol. *Res:* Microbial fermentations; fungicides; carotenoids of microorganisms; soil fungi; marine bacteria and phytoplankton; animal breeding and development; gnotobiology. *Mailing Add:* Life Sci Inc 2900 72nd St N St Petersburg FL 33710

FARRUKH, USAMAH OMAR, b Beirut, Lebanon, Aug 24, 44; US citizen; m 80; c 3. TRANSIENT & SPATIAL TEMPERATURE VARIATIONS IN LASER RODS. *Educ:* Am Univ Beirut, BE, 67; Univ Southern Calif, PhD(elec eng), 74. *Prof Exp:* Lead analyst, Wolf Res & Develop, EG&G, 75-76; sci specialist, Phoenix Corp, McLean, Va, 76-77; staff scientist, Inst Atmospheric Optics & Remote Sensing, Hampton, Va, 78-85; ASSOC PROF ELEC ENG, HAMPTON UNIV, 85- *Concurrent Pos:* Prin investr, NASA, 86-90 & 90- & Battelle Army Res Off, 90-; mem fac, US Armament Res & Develop Ctr, 89. *Honors & Awards:* Cert of Recognition, NASA, 90. *Mem:* Optical Soc Am; Inst Elec & Electronic Engrs. *Res:* Analysis of temperature distribution and thermal lensing in pulsed solid state laser systems; diffraction and propagation of laser beams; laser propagation in scattering media; analysis of quantum well based detectors. *Mailing Add:* 236 Charlotte Dr Newport News VA 23601

FARTHING, BARTON ROBY, b Watauga Co, NC, Feb 20, 16; m 38; c 3. APPLIED STATISTICS. *Educ:* Wake Forest Col, BS, 38; NC State Col, BS 52, MS, 54, PhD(animal breeding), 58. *Prof Exp:* Res instr, NC State Col, 54-58, asst prof, 58-59; exp sta statistician, La State Univ, Baton Rouge, 59-64, head dept, 64-77, prof exp statist, 77-81; RETIRED. *Mem:* Biomet Soc; Am Soc Animal Sci; Am Dairy Sci Asn. *Res:* Design of experiments; analysis and interpretation of data. *Mailing Add:* Rte 2 Box 279 Vilas NC 28692

FARVOLDEN, ROBERT NORMAN, b Forestburg, Alta, May 22, 28; m 54; c 2. HYDROGEOLOGY. *Educ:* Univ Alta, BSc, 51, MSc, 58; Univ Ill, PhD, 63. *Prof Exp:* Head groundwater div, Res Coun Alta, 56-60; res assoc, Desert Res Inst, Univ Nev, 62-64; from asst prof to assoc prof geol, Univ Ill, Urbana, 64-67; assoc prof, Univ Western Ont, 67-70; chmn dept earth sci, 70-76, dean fac sci, 77-82, ASSOC PROF GEOL, UNIV WATERLOO, 70- *Mem:* Fel Geol Soc Am; Am Geophys Union; fel Geol Asn Can. *Res:* Water resources development; solid-waste disposal; scientific hydrology. *Mailing Add:* Dept of Earth Sci Univ of Waterloo Waterloo ON N2L 3G1 Can

FARWELL, GEORGE WELLS, b Oakland, Calif, Feb 15, 20; m 45, 86; c 4. ENVIRONMENTAL, EARTH & MARINE SCIENCES. *Educ:* Harvard Univ, SB, 41; Univ Calif, SB, 42; Univ Chicago, PhD(physics), 48. *Prof Exp:* Asst physics, Radiation Lab, Univ Calif, 42-43, physicist, Manhattan Proj, Los Alamos Sci Lab, NMex, 43-46; from asst prof to assoc prof, 48-59, assoc dean, Grad Sch, 59-65, asst vpres, 65-67, vpres res, 67-76, prof physics, 59-86, EMER PROF PHYSICS & VPRES RES, UNIV WASH, 87- *Concurrent Pos:* Sr fel, NSF, 60-61, Niels Bohr Inst, Copenhagen. *Mem:* Fel Am Phys Soc; AAAS. *Res:* Nuclear physics; ultrasensitive mass spectrometry with accelerators; radiocarbon and radioberyllium in the environment. *Mailing Add:* Dept Physics Univ Wash FM-15 Seattle WA 98195

FARWELL, ROBERT WILLIAM, b Providence, RI, May 11, 27; div; c 2. PHYSICS. *Educ:* Yale Univ, BS, 50; Pa State Univ, MS, 55, PhD(physics), 60. *Prof Exp:* Res asst physics, Ord Res Lab, Pa State Univ, 51, res assoc, 51-58, asst dept physics, 58-59, from asst prof to assoc prof eng res, Appl Res Lab, 60-83; RES PHYSICIST, NAVAL OCEANOG & ATMOSPHERIC RES LAB, 83- *Mem:* Acoust Soc Am; Sigma Xi. *Res:* Underwater sound transmission; reverberation; sound scattering; transducer calibrations; acoustic torpedoes; sonar; effect of sound on tissues; enzyme kinetics; ultraviolet and visible spectrophotometers; acoustical oceanography. *Mailing Add:* 500 Aries Dr Unit 4B Mandeville LA 70448

FARWELL, SHERRY OWEN, b Miles City, Mont, June 1, 44; m 63; c 2. ANALYTICAL CHEMISTRY, ATMOSPHERIC CHEMISTRY. *Educ:* SDak Sch Mines & Technol, BS, 66, MS, 69; Mont State Univ, PhD(chem), 73. *Prof Exp:* Assoc chemist, John Deere Co, 66-67; asst prof air chem, Wash State Univ, 73-77; from asst prof to assoc prof, 77-84, PROF ANAL CHEM, UNIV IDAHO, 85- *Concurrent Pos:* Fel, air pollution res sect, Wash State Univ, 73-74; consult, Hewlett-Packard, 78-, K-Prime Inc, 89-, Exxon Chem, 90-, Sievers Res, 90- & Potlatch Corp, 90- *Mem:* Am Chem Soc; Am Soc Testing Mat; Air Pollution Control Asn; Sigma Xi; Asn Petrol Geochem Explorationists; Asn Explor Geochemists; Asn Off Anal Chemists. *Res:* Development, optimization and validation of analytical methods for sulfur-containing and halogen-containing compounds; applications of modern analytical chemistry to atmospheric science, geochemical exploration, and food science; biogeochemistry. *Mailing Add:* 820 Courtney St Moscow ID 83843

FARY, RAYMOND W, JR, geology, for more information see previous edition

FASANG, PATRICK PAD, b Bangkok, Thailand; US citizen; c 3. VLSI DESIGN TESTABILITY, VLSI BUILT SELF TEST. *Educ:* Calif State Univ, Fresno, BSEE, 65; Calif State Univ, San Jose, MSEE, 68; Oregon State Univ, PhD(elec eng), 74. *Prof Exp:* Engr, Sprague Elec Co, 67-69; assoc prof, elec eng, Univ Portland, 72-78; admin, RCA Corp, 78-79; group leader, Siemens Corp, 80-83; mgr, Gen Elec Co, 84-86; mgr, Nat Semiconductor Corp, 87-90; DIR, NCHIP INC, 90- *Concurrent Pos:* Consult, US Dept Energy, Bonneville Power Admin, 74-78; assoc ed, IEEE Micro Mag, 80-83; assoc ed, IEEE Transactions on Industrial Electronics. *Honors & Awards:* Centennial Medal, IEEE, 84. *Mem:* Fel Inst Elec & Electronics Engrs; Inst Elec & Electronics Engrs Industr Electronics Soc(vpres 82-86); Inst Elec & Electronics Comput Soc. *Res:* Design for testibility of large scale integrated circuits made up of both digital and analog circuits, built-in self testing of VLSI circuits. *Mailing Add:* 19915 Wellington Ct Saratoga CA 95070

FASCHING, JAMES LE ROY, b Dickinson, NDak, Mar 15, 42; m 67; c 1. ANALYTICAL CHEMISTRY, CHEMOMETRICS. *Educ:* NDak State Univ, BS, 64; Mass Inst Technol, SM, 67, PhD(chem), 70. *Prof Exp:* Res asst chem, NDak State Univ, 62-63, instr, 63; asst, Mass Inst Technol, 64-70; from instr to prof, 69-82, CHMN DEPT CHEM, UNIV RI, 82- *Concurrent Pos:* Vis scientist, Univ Wash & Oak Ridge Nat Lab, 75-76. *Mem:* Sigma Xi; AAAS; Am Phys Soc; Am Chem Soc; Soc Appl Spectros. *Res:* Neutron activation analysis of biological samples; effects of trace element on cancer pattern recognition techniques; minicomputers; microprocessors for data acquisition and control and development of piezoelectric sorption devices; aerosols, global pollution transport; intelligent computer-aided instruction; expert systems. *Mailing Add:* Three Beach St Oceanside 23 Narragansett RI 02882

FASCO, MICHAEL JOHN, MECHANISM OF ACTION OF ANTICOAGULANT DRUGS. *Educ:* Rensselaer Polytech Inst, PhD(chem), 75. *Prof Exp:* RES SCIENTIST IV BIOCHEM & GENETIC TOXICOL, WADSWORTH CTR LABS & RES, 82- *Mailing Add:* Wadsworth Ctr Labs & Res Empire State Plaza Albany NY 12201

FASHENA, GLADYS JEANNETTE, b New York, NY, June 3, 10; m 38; c 2. PEDIATRICS, PEDIATRIC CARDIOLOGY. *Educ:* Hunter Col, BA, 29; Columbia Univ, MA, 30; Cornell Univ, MD, 34. *Prof Exp:* Instr pediat, Med Col, Cornell Univ, 37-38; instr, Baylor Col Med, 39-43; from instr to assoc prof, 43-49, PROF PEDIAT, UNIV TEX SOUTHWESTERN MED SCH, DALLAS, 49-, DIR REGIONAL CONGENITAL HEART DIS PROG, 52- *Honors & Awards:* Piper Prof, Minnie Stevans Piper Found, 74. *Mem:* Am Acad Pediat; Am Pediat Soc; Soc Pediat Res; Am Soc Clin Invest; Am Col Cardiol. *Res:* Development of methods for assay of various substances in blood; bilirubin metabolism in infancy; physiological aspects of congenital heart disease. *Mailing Add:* 11550 Wander Lane Dallas TX 75230

FASHING, NORMAN JAMES, b Walker, Minn, Aug 14, 43; m 69; c 3. ACAROLOGY. *Educ:* Calif State Univ, Chico, BA, 65, MA, 67; Univ Kans, PhD(entom), 73. *Prof Exp:* PROF BIOL, COL WILLIAM & MARY, 73- *Mem:* Acarological Soc Am (pres, 83, 90); Entom Soc Am; Europ Asn Acarologists. *Res:* Orientation behavior of insects and arachnids; taxonomy and biology of astigmarid. *Mailing Add:* Dept Biol Col William & Mary Williamsburg VA 23185

FASK, ALAN S, b New York, NY, Oct 26, 45. STATISTICS, OPERATIONS RESEARCH. *Educ:* City Col New York, BS, 68; NY Univ, MS, 72, PhD(statist), 73. *Prof Exp:* ASSOC PROF COMPUT & DECISION SYSTS, FAIRLEIGH DICKINSON UNIV, 73- *Mem:* Am Statist Asn; Opers Res Soc Am; Inst Mgt Sci; Economet Soc. *Res:* Applications of multivariate and time series statistical techniques to economics and marketing; optimal control of industrial processes. *Mailing Add:* Dept Info Sci Fairleigh Dickinson Madison Campus Madison NJ 07940

FASMAN, GERALD DAVID, b Drumheller, Alta, May 28, 25; nat US; m 53; c 3. BIOCHEMISTRY, BIOPHYSICS. *Educ:* Univ Alta, BSc, 48; Calif Inst Technol, PhD(chem), 52. *Prof Exp:* Royal Soc Can scholar, Cambridge Univ, 51-53; Merck fel natural sci, Eng Tech Inst, Zurich, 53-54; Weizmann fel, Weizmann Inst, 54-55; asst, Children's Res Found, Boston, 55-56, res assoc, 57-61, asst head biophys chem, 59-61; from asst prof to prof biochem, 61-71, ROSENFIELD PROF BIOCHEM, BRANDEIS UNIV, 71- *Concurrent Pos:* Asst, Med Sch, Harvard Univ, 57-58, res assoc, 58-61, tutor, univ, 60-61;

estab investr, Am Heart Asn, 61-66; ed, Biol Macromolecules & Critical Rev in Biochem, 72-; ed, Handbk Biochem & Molecular Biol, 73-; John Guggenheim fel, 74-75 & 88-89; res scholar, Japan Soc Prom Sci, 79; mem, NSF Adv Panel, 80-82; mem, adv bd, Am Cancer Soc, 80-83. *Mem:* Fel AAAS; fel Am Inst Chemists; Am Chem Soc; Am Soc Biochem & Molecular Biol Chem; NY Acad Sci; Royal Soc Chem; Biophys Soc; Protein Soc. *Res:* Enzymes, proteins, nucleic and polyamino acids; chromatin; conformational studies of biopolymers. *Mailing Add:* Grad Dept of Biochem Brandeis Univ Waltham MA 02154

FASOLA, ALFRED FRANCIS, b Pa, Mar 29, 19; c 2. CARDIOPULMONARY PHYSIOLOGY. *Educ:* Oberlin Col, AB, 48; Ohio State Univ, MS, 50, PhD(physiol), 53, MD, 57. *Prof Exp:* Res assoc, Ohio State Univ, 53-55, instr physiol, 55-57; intern, Mt Carmel Hosp, Columbus, Ohio, 57-58; resident internal med, Wishard Mem Hosp, Indianapolis, 59-60 & 60-61; ASST PROF MED, MED CTR, IND UNIV, INDIANAPOLIS, 68- *Concurrent Pos:* Res fel cardiol, Robert Moore Heart Clin, Indianapolis, 58-59; staff physician, Lilly Lab Clin Res, Wishard Mem Hosp, Indianapolis, 61-66, sr physician, 66-75, sr clin pharmacologist, 75-; mem coun arteriosclerosis, Am Heart Asn; mem, Med Adv Bd Coun High Blood Pressure. *Mem:* Am Heart Asn; Aerospace Med Asn; fel Am Col Cardiol; Am Col Physicians. *Res:* Space research, effects of deceleration and negative gravity; general cardiovascular research; renin-angiotensin system in hypertension. *Mailing Add:* 460 Braeside Dr N Indianapolis IN 46260

FASS, ARNOLD LIONEL, b New York, NY, Apr 2, 22. MATHEMATICS. *Educ:* City Col New York, BS, 42; Columbia Univ, MA, 47, PhD(math), 51. *Prof Exp:* Lectr math, Columbia Univ, 48-51; tutor, 51, from instr to prof, 51-87, EMER PROF MATH, QUEENS COL, NY, 87- *Concurrent Pos:* Lectr, NSF Inst Teacher Training, 61-62. *Mem:* AAAS; Am Math Soc; Math Asn Am. *Res:* Homology and cohomology of linear algebras; topological methods in algebra; general linear algebra. *Mailing Add:* Apt B324 8500 Royal Palm Blvd Coral Springs FL 33065

FASS, DAVID N, b New York, NY. BIOCHEMISTRY. *Educ:* Brooklyn Col, New York, NY, BS, 61; Fla State Univ, Tallahassee, PhD, 69. *Prof Exp:* Teaching asst gen biol & gen genetics, Fla State Univ, 62-64, instr, Dept Biol Sci, 67-69; postdoctoral trainee, Dept Genetics & Cell Biol, Univ Minn, 69-72; res assoc, Dept Hemat Res, Mayo Clin, Rochester, Minn, 72-75, asst prof med & biochem, Mayo Med Sch, 76-82, assoc prof cell biol, 81-86, assoc prof biochem & molecular biol, 86-87, PROF BIOCHEM & MOLECULAR BIOL, MAYO MED SCH, ROCHESTER, MINN, 87-, HEAD, SECT HEMAT RES, DIV HEMAT & INT MED, MAYO CLIN, 89- *Concurrent Pos:* Assoc consult, Dept Hemat Res, Mayo Clin, Rochester, Minn, 75-78, consult hemat res, 78-, cell biol, 80-85, biochem & molecular biol, 85-; mem, Great Plains Res Rev & Adv Comt, Am Heart Asn, 81-85, Peer Rev Comt, Am Heart Asn-Bugher Found Ctr, 91; mem, Med & Sci Adv Coun, Res Rev Panel, Nat Hemophillia Found, 82-83, Res Rev Comt B, Nat Heart, Lung & Blood Inst, 83-87, Sci & Standardization Comt, Int Soc Thrombosis & Harmostasis, 89-90. *Honors & Awards:* Medal, Hemophilia Asn France, 88; Dr Murray Thelin Award, Nat Hemophilia Found, 89. *Mem:* Am Soc Biochem & Molecular Biol; Am Asn Lab Animal Sci; Am Soc Hemat; Sigma Xi. *Res:* Hemostasis and the regulation of clotting factor levels and activities; genetics and gene expression in Von Willebradis disease. *Mailing Add:* Dept Hemat Mayo Clin & Found Rochester MN 55905

FASS, RICHARD A, b Brooklyn, NY, June 13, 43; m 64; c 1. PHYSICAL CHEMISTRY. *Educ:* Cooper Union, BE; Univ Wis, PhD(phys chem), 69. *Prof Exp:* Res assoc photochem kinetics & radiation chem, Univ Wis, 69; asst prof chem, 69-73 & 77-79, assoc dean, 73-76, DEAN STUDENTS, POMONA COL, 76-, VPRES, 79- *Res:* Reactions of hot and thermal hydrogen atoms produced by photolysis of hydrogen halides; reactions of hot free radicals in the gas phase; gas kinetics. *Mailing Add:* 156 E Seventh St Claremont CA 91711-6309

FASS, STEPHEN M, b New York, NY, Aug 22, 38; m 66; c 2. CHEMICAL ENGINEERING. *Educ:* Cooper Union, BChE, 60; Univ Colo, MS, 64, PhD(chem eng), 67. *Prof Exp:* Engr, FMC Corp, 60-62; sr res engr, US Steel Corp, 66-68; sr res engr, Gulf Res & Develop Co, 68-73; sr processing engr, Ralph M Parsons Co, 73-77, Crawford & Russell, 77-78; prin processing engr, 78-81, sect mgr, 81-88, PRIN PROCESS ENGR, FMC CORP, NJ, 88- *Mem:* Am Inst Chem Engrs. *Res:* Reaction kinetics; reactor design and process research in steel, petroleum and organic chemicals industries. *Mailing Add:* FMC Corp PO Box 8 Princeton NJ 08543

FASSEL, VELMER ARTHUR, b Frohna, Mo, Apr 26, 19; m 43. PHYSICAL CHEMISTRY, ANALYTICAL CHEMISTRY. *Educ:* Southeast Mo State Univ, BA, 41; Iowa State Univ, PhD(phys chem), 47. *Prof Exp:* Asst spectros, Nat Defense Res Comt contract, 42-43, jr chemist, Manhattan Proj, 43-46, assoc chemist, Inst Atomic Res, 46-47, from asst prof to prof chem, 47-87, dep dir Energy & Mineral Resources Res Inst & Ames Lab, 69-83, prin scientist, Ames Lab, Dept Energy, 83-87, EMER PROF CHEM, IOWA STATE UNIV, 87-, DISTINGUISHED PROF SCI & HUMANITIES, 76- *Concurrent Pos:* Lectr tour, Sci Coun Japan, 62; US ed, Spectrochimica Acta. *Honors & Awards:* Soc Appl Spectros Medal Award, 64; Pittsburgh Spectros Award, 69; Maurice F Hasler Award, 71; Anachem Award, 71; Fisher Award Anal Chem, Am Chem Soc, 78; Japan Soc Anal Chem Medal; Chem Instrumentation Award, Am Chem Soc, 83; Harvey W Wiley Award, Asn Off Anal Chem, 83; Lester Strock Award, 86; IR-100 Award, 86; Gold Medal, 65. *Mem:* Fel AAAS; Am Chem Soc; fel Optical Soc Am; hon mem Soc Appl Spectros; hon mem Japan Soc Anal Chem. *Res:* Analytical atomic spectroscopy; plasma analytical spectroscopy; fluorescence spectroscopy; analytical chemistry. *Mailing Add:* 17755 Rosedown Pl San Diego CA 92128

FASSETT, DAVID WALTER, b Broadalbin, NY, Nov 13, 08; m 34; c 3. TOXICOLOGY. *Educ:* Columbia Col, AB, 33; NY Univ, MD, 40. *Prof Exp:* Biochemist, Dept Exp Surg, NY Univ, 35-36; pharmacologist, Wellcome Res Lab, 36-38; intern, Bellevue Hosp, 40-41; asst therapeut, Col Med, NY Univ, 42-45; cardiologist, James M Jackson Mem Hosp, Fla, 45-48; dir health & safety lab, Eastman Kodak Co, 48-73; consult, Indust Toxicol, 73-85; RETIRED. *Concurrent Pos:* Wellcome fel therapeut, NY Univ, 41; actg chief div pharmacol, Food & Drug Admin, Fed Security Agency, DC, 42-44; clin assoc prof prev med, Univ Rochester, 65-73; vis lectr, Harvard Sch Pub Health, 65-83; consult toxicol, Nat Acad Sci, 73-76. *Honors & Awards:* Cummings Award, Am Indust Hyg Asn, 78. *Mem:* Fel AAAS; Am Soc Pharmacol & Exp Therapeut; Sigma Xi; Am Indust Hyg Asn (pres, 69-70); Soc Toxicol. *Res:* Pharmacology; industrial toxicology; industrial hygiene. *Mailing Add:* 13 Summer St Box 739 Kennebunk ME 04043

FASSETT, JAMES ERNEST, b Dearborn, Mich, May 1, 33. BASIN ANALYSIS STUDIES, EVENT STRATIGRAPHY. *Educ:* Wayne State Univ, BS, 59, MS, 64. *Prof Exp:* Geologist, 60-81, dep minerals mgr resource eval, Conserv Div, 81-83, chief, Br Tech Reports, 83-87, RES GEOLOGIST, US GEOL SURV, 87- *Concurrent Pos:* Gilbert fel, US Geol Surv, 87- *Mem:* Fel Geol Soc Am; Am Asn Petrol Geologists; Soc Explor Petrologists & Mineralogists. *Res:* Stratigraphy & sedimentology of upper cretaceous rocks in the western US, especially in New Mexico; environments & depositions of upper cretaceous rocks; events surrounding cretaceous tertiary, boundary & extinction of dinasours in New Mexico. *Mailing Add:* 29 S Joyce St Golden CO 80901

FASSETT, JOHN DAVID, b Ft Eustis, Va, Mar 21, 51; m 75; c 2. INORGANIC MASS SPECTROMETRY, ISOTOPE RATIO MEASUREMENT. *Educ:* Brown Univ, ScB, 73; Cornell Univ, PhD(anal chem), 78. *Prof Exp:* Res chemist, Nat Bur Standards, 78-85; GROUP LEADER, NAT INST STANDARDS & TECHNOL, 85- *Mem:* Am Chem Soc; Soc Appl Spectros; AAAS; Am Soc Mass Spectrometry. *Res:* Inorganic mass spectrometry and elemental isotopic ratio measurement; laser, secondary and thermal ionization mass spectrometry; ultratrace chemical analysis using isotope dilution mass spectrometry; standards certification. *Mailing Add:* Nat Inst Standards & Technol Bldg 221 Rm A21 Gaithersburg MD 20899

FASSNACHT, JOHN HARTWELL, b Wenonah, NJ, Oct 22, 33; m 60; c 2. ORGANIC CHEMISTRY. *Educ:* Middlebury Col, AB, 55; Mass Inst Technol, PhD(org chem), 59. *Prof Exp:* Res chemist, Org Chem Dept, Jackson Lab, 59-62, res chemist, Freon Prod Div, 62-67, proj leader, 67-68, tech asst, 68-69, tech assoc, 70-74, territory mgr, 75-78, sr territory mgr, Freon Prod Div, 78-80, AREA MGR, INTERMEDIATES PROD DIV, E I DU PONT DE NEMOURS & CO INC, 80- *Mem:* Am Chem Soc; Sigma Xi. *Mailing Add:* 451 Beverly Pl Lake Forest IL 60045

FAST, ARLO WADE, b Mt Clemens, Mich, Nov 22, 40; div. AQUACULTURE, LIMNOLOGY. *Educ:* Mich State Univ, BS, 62, PhD(limnol), 71; San Diego State Univ, MS, 68. *Prof Exp:* Fishery biologist, Calif Dept Fish & Game, 63-68; aquatic ecologist, Union Carbide Corp, 71-75; owner, Limnol Assocs, 75-78; AQUACULTURIST, UNIV HAWAII, 78- *Mem:* Am Soc Limnol & Oceanog; Am Fisheries Soc; Soc Int Limnol; World Maricult Soc; Am Inst Fishery Res Biologists. *Res:* Environmental impact assessments; assessments of freshwater habitats; lake restoration; lake aeration technology; aquaculture; marine shrimp culture. *Mailing Add:* Univ Hawaii PO Box 1346 Kaneohe HI 96744

FAST, C(LARENCE) R(OBERT), b Tulsa, Okla, Feb 12, 21; m 44; c 4. PETROLEUM ENGINEERING. *Educ:* Univ Tulsa, BS, 43. *Prof Exp:* Liaison engr, Douglas Aircraft Co, 43; res engr, Standard Oil Co, Ind, 43-56, res group supvr, Amoco Prod Co, 56-77; OWNER, FAST ENG, CONSULT PETROL ENGRS, 77- *Concurrent Pos:* Chmn nat subcomt perforating, Am Petrol Inst, exec comt prod pract; Soc Petrol Engrs distinguished lectr, 70-71; consult petrol eng probs. *Honors & Awards:* Uren Award, Soc Petrol Engrs; Distinguished Mem, Soc Petrol Engrs, 85. *Mem:* Soc Petrol Engrs; Am Petrol Inst. *Res:* Direct research on drilling, well completion, stimulation, well operation and production practice. *Mailing Add:* 4504 Gran Tara Afton OK 74331

FAST, DALE EUGENE, b Kansas City, Kans, Nov 2, 45; m 70; c 2. GENETICS, DEVELOPMENTAL BIOLOGY. *Educ:* Tabor Col, BA, 67; Univ Chicago, PhD(biol), 78. *Prof Exp:* Teacher & asst dir sci, Kikwit Sec Sch, Dem Rep Congo, 67-69; ASSOC PROF BIOL, ST XAVIER COL, 76- *Mem:* Sci for the People; AAAS. *Res:* Genetics and biochemistry of yeast sporulation. *Mailing Add:* Dept Sci St Xavier Col 3700 W 103rd St Chicago IL 60655

FAST, HENRYK, b Bochnia, Poland, Oct 4, 25; nat US; div; c 1. MATHEMATICAL ANALYSIS. *Educ:* Univ Wroclaw, PhM, 50; Polish Acad Sci, PhD(math), 58. *Prof Exp:* Asst math, Univ Wroclaw, 50-51, adj, 56-60; sr asst, Polish Acad Sci, 51-55; asst prof, Univ Notre Dame, 62-66; ASSOC PROF MATH, WAYNE STATE UNIV, 66- *Concurrent Pos:* Vis scholar, Stichting Math Centrum, Amsterdam, Holland, 73-74. *Mem:* Am Math Soc; Math Asn Am. *Res:* General measure and integration, especially geometrical measure theory; functions of real variables; convex sets and set theoretical geometry; integral geometry. *Mailing Add:* Dept Math Wayne State Univ 5950 Cass Ave Detroit MI 48202

FAST, PATRICIA E, b Santa Monica, Calif, July 21, 43; m; c 1. IMMUNOLOGY. *Educ:* Univ Calif, Los Angeles, AB, 65, PhD(microbiol), 69, MD, 84. *Prof Exp:* Teaching asst bact, Univ Calif, Los Angeles, 66-67; asst prof microbiol, Calif State Univ, Los Angeles, 69-70; asst res biologist, Univ Calif, Los Angeles, 70; staff scientist, Wellcome Res Labs, 70-72; asst prof pediat, microbiol & immunol, Harbor Gen Hosp, Univ Calif, Los Angeles, 72-75; res scientist, Upjohn Co, 75-81; med staff fel, Nat Cancer Inst, 87-89, MED OFFICER, VACCINE RES DEVELOP BR, DIV AIDS, NAT INST ALLERGY & INFECTIOUS DIS, NIH, 89- *Concurrent Pos:* Nat Res Coun Italy grant, 74. *Honors & Awards:* Glasgow Award. *Mem:* AAAS; fel Am Acad Pediat; Am Asn Immunol; Am Soc Microbiol; Brit Soc Immunol. *Mailing Add:* Nat Inst Allergy & Infectious Dis Div AIDS NIH Vaccine Br 6003 Executive Blvd Rm 203E Bethesda MD 20892

FAST, RONALD WALTER, b Toledo, Ohio, Apr 2, 34; m; c 2. CRYOGENIC ENGINEERING. *Educ:* Washington & Lee Univ, BS, 56; Univ Va, MS, 58, PhD(physics), 60. *Prof Exp:* Physicist, US Army Nuclear Defense Lab, 60-62 & Midwestern Univs Res Asn, 62-67; assoc scientist, Phys Sci Lab, Univ Wis, 67-69; APPL SCIENTIST, FERMI NAT ACCELERATOR LAB, 69- *Mem:* Cryogenic Eng Conf; Sigma Xi. *Res:* Cryogenic engineering and applied superconductivity. *Mailing Add:* Fermilab Mail Stop 219 PO Box 500 Batavia IL 60510

FAST, THOMAS NORMAND, b Selma, Calif, Sept 10, 22; m 52; c 2. MARINE BIOLOGY. *Educ:* Univ Santa Clara, BS, 49; Stanford Univ, PhD, 60. *Prof Exp:* Oceanog technician, Stanford Univ, 52-56; instr biol, 57-67, chmn dept, 73-77, ASSOC PROF BIOL, UNIV SANTA CLARA, 67-, CHMN DEPT, 80- *Mem:* AAAS; Am Soc Ichthyologists & Herpetologists; Sigma Xi. *Res:* Ecology of bathypelagic fishes; cardiovascular dynamics in stress. *Mailing Add:* 552 Bean Creek Rd 148 Scotts Valley CA 95066

FATELEY, WILLIAM GENE, b Franklin, Ind, May 17, 29; m 53; c 5. STRUCTURAL CHEMISTRY. *Educ:* Franklin Col, AB, 51; Kans State Univ, PhD(chem), 56. *Hon Degrees:* DSc, Franklin Col, 65. *Prof Exp:* Asst chem, Northwestern Univ, 51-53 & Kans State Univ, 53-55; res assoc, Univ Md, 56; res fel, Univ Minn, 56-57; res chemist & head, Spectros Lab, James River Div, Dow Chem Co, 57-60; fel chem, Carnegie-Mellon Univ, 60-62, head sci rels, 62-63, asst to pres, 63-67, sr fel, 65-67, from assoc prof to prof chem, 67-72, asst to vpres res, 67-72; head dept, 72-79, prof chem, 72-89, DISTINGUISHED PROF, KANS STATE UNIV, 89- *Concurrent Pos:* Treas, Fourier Transform Users Group, 70-; vis prof, Univ Tokyo, 72-73 & 81; ed-in-chief, J Appl Spectros, 74-; pres, DOM Assocs, Int, 79-; lectr, Boudoin Col, 81, Colo State, 85. *Honors & Awards:* Coblentz Award, 65; Gulf Award, 69; Spectros Soc Pittsburgh Award, 76; H H King Award, 79; Gold Medal, Eastern Anal Soc Appl Spectros, 87 & William F Meggers Award, 88; THE Award, Soc Automotive Engrs, 89. *Mem:* Am Chem Soc; fel Optical Soc; Soc Appl Spectros; hon mem Coblentz Soc; hon mem Soc Appl Spectros. *Res:* Infrared and Raman spectroscopy; structure of matter; Hadamard transfom spectroscopy. *Mailing Add:* Dept Chem Kans State Univ Manhattan KS 66506

FATEMAN, RICHARD J, b New York, NY, Nov 4, 46; m 68; c 2. COMPUTER SCIENCE, MATHEMATICS. *Educ:* Union Col, NY, BS, 66; Harvard Univ, PhD(appl math), 71. *Prof Exp:* Lectr math, Mass Inst Technol, 71-74; from asst prof to assoc prof, 74-85, PROF COMPUT SCI, UNIV CALIF, BERKELEY, 85-, CHMN DEPT, 88- *Concurrent Pos:* Mem staff comput sci, Proj MAC, Mass Inst Technol, 69-73. *Mem:* Asn Comput Mach; Soc Indust & Appl Math. *Res:* Algebraic manipulation by computer; programming languages; analysis of algorithms; scientific software. *Mailing Add:* Elec Eng & Comput Sci Univ of Calif Berkeley CA 94720

FATH, GEORGE R, b Cincinnati, Ohio, July 9, 38. ELECTRICAL ENGINEERING. *Educ:* Univ Miami, BSEE, 60; Univ Syracuse, MSEE, 66, PhD(elec eng), 69. *Prof Exp:* Mgr countermeasures develop eng, Gen Elec Co, Utica, NY, 72-75, mgr advan technol, Info Systs Prog, Arlington, Va, 75-77, consult engr, Utica, NY, 77, mgr avionics develop eng, 77-80, mgr systs concept & anal, Electronics Lab, Syracuse, NY, 80-83, mgr digital design & advan prog, Lynchburg, Va, 83-84, mgr advan eng, 85, MGR PUB SERV PROD ENG, GEN ELEC CO, LYNCHBURG, VA, 85- *Concurrent Pos:* Mem tech comt comput, Am Inst Aeronaut & Astronaut, 76-79. *Res:* Electrical engineering. *Mailing Add:* 102 Millview Terr Forest VA 24551

FATH, JOSEPH, b Frankfurt, Ger, Aug 31, 25; m 45; c 4. ORGANIC CHEMISTRY. *Educ:* Cornell Univ, BChem, 44. *Prof Exp:* Res chemist, Montrose Chem Co, 46-50; group leader, org sect, Nuodex Prods Co, 50-55; dir res, Thompson Chem Co, 55-65, vpres, 65-69; vpres bus develop & planning, Nuodex Div, Tenneco Chem, Inc, 69-71; gen mgr, Org & Polymers Div & vpres, Piscataway, 71-75, vpres, Planning & Develop, 76-80, SR VPRES PLANNING & DEVELOP, TENNECO CHEM CO, SADDLEBROOK, 80- *Mem:* Am Chem Soc. *Res:* Vinyl plasticizers and resins; organic intermediates; metalloorganics. *Mailing Add:* 90 Olden Lane Princeton NJ 08540-4940

FATHMAN, C GARRISON, b Aug 30, 42. IMMUNOLOGY. *Educ:* Univ Ky, Lexington, BA, 64; Wash Univ, St Louis, Mo, MD, 69. *Prof Exp:* Intern & resident med, Dartmouth Affil Hosp, Hanover, NH, 69-71; clin assoc, Immunol Br, Nat Cancer Inst, NIH, Bethesda, Md, 73-75; mem, Basil Inst Immunol, Switz, 75-77; assoc prof, Dept Immunol, Mayo Clin, Rochester, Minn, 77-81; postdoctoral fel immunol, 71-73, assoc prof med, 81-89, PROF MED IMMUNOL, SCH MED, STANFORD UNIV, CALIF, 89- *Concurrent Pos:* Co-chmn, Study Group Lymphocyte Cloning, Nat Inst Allergy & Infectious Dis, 80; assoc ed, Ann Rev Immunol, 81-; mem, Coun Midwinter Conf Immunologists, 81-86, Coun Am Soc Clin Invest, 84-87; dir, Biomed Res Unit, Stanford Multipurpose Arthritis Ctr, 83-, Prog Proj Grant Autoimmunity, 84-89 & bd dirs, Am Diabetes Asn, Calif Affil, 91-93; sect chief, Cellular Immunol, 86-90; mem, Immunol Sci Study Sect, 86-90, chmn, 89-90; chmn, Arthritis Found Task Force Postgrad Training, 86. *Mem:* Am Rheumatism Asn; Am Asn Immunol; Transplantation Soc; Am Fedn Clin Res; Am Soc Clin Invest; Am Diabetes Asn; Clin Immunol Soc; Am Asn Physicians. *Mailing Add:* Dept Immunol & Rheumatology Stanford Univ Sch Med Stanford CA 94305-5111

FATIADI, ALEXANDER JOHANN, b Kharkov, Ukraine, Oct 22, 23; US citizen; m 52; c 4. ORGANIC CHEMISTRY. *Educ:* Tech Husbandry Inst-Univ Mainz, Ger, DrNatSc, 50; George Wash Univ, BS, 57, MS, 59. *Hon Degrees:* DSc, World Univ, Tucson, Ariz, 85. *Prof Exp:* Res asst chemist, George Washington Univ, 56-59; org chemist, Nat Bur Stand, 59-67, res chemist & proj leader, 68-87; CONSULT, 87- *Honors & Awards:* Hillebrand Award, 81; Int Scholars Award, Int Sci Soc, Strasbourg, France, 73. *Mem:* Am Chem Soc; NY Acad Sci; Royal Soc Chem. *Res:* Periodic acid for oxidation of polycyclic aromatic hydrocarbons; clinical standards and chemistry; cyclic ketones; aromatization of cyclitols; polyhydroxy phenols; phenylhydrazine osazones and osotriazoles; oxidation of polycyclic aromatic hydrocarbons; stable free radicals; oxidation mechanisms; malononitrile and tetracyanoethylene chemistry; semiconducting oxocarbons, electrochemistry and mass spectrometry of oxocarbons and pseudo-oxocarbons; active manganese dioxide; biomedical technology; higher carbon sugars and their interaction with amino acids and polyamines; confirmational studies. *Mailing Add:* 7516 Carroll Ave Takoma Park MD 20912

FATT, IRVING, b Chicago, Ill, Sept 16, 20; m 42; c 1. PHYSIOLOGICAL OPTICS. *Educ:* Univ Calif, Los Angeles, BS, 47, MS, 48; Univ Southern Calif, PhD, 55. *Prof Exp:* Instr commun, Yale Univ, 44-45; sr res chemist, Calif Res Corp, 48-57; prof eng sci, Col Eng, 57-80, prof, 67-83, EMER PROF PHYSIOL OPTICS, SCH OPTOM, UNIV CALIF, BERKELEY, 83- *Honors & Awards:* Prentice Medal; Herschel Medal; Ruben Medal; Dallos Medal. *Mem:* Bioeng Soc; Am Acad Optom. *Res:* Fluid flow through porous media; structure of porous media; bioengineering; corneal physiology; contact lenses. *Mailing Add:* 360 Minor Hall Univ Calif Berkeley CA 94720

FATTIG, W DONALD, b DeKalb Co, Ga, Feb 22, 36; m 58; c 1. GENETICS. *Educ:* Emory Univ, AB, 59, MS, 60, PhD(genetics), 63. *Prof Exp:* Fel microbiol, Sch Med, Emory Univ, 63-64; asst prof biol, Southwestern Univ, Memphis, 64-68; assoc prof, 68-74, assoc dean, Sch Natural Sci & Math, 74-76, PROF BIOL, UNIV ALA, BIRMINGHAM, 74- *Concurrent Pos:* Danforth Assoc. *Mem:* AAAS; Am Genetic Asn; Am Inst Biol Sci; Am Soc Human Genetics; Genetics Soc Am. *Res:* Bacteriophage, human and drosophilia genetics; biological controls. *Mailing Add:* 1919 Seventh Ave S Univ Ala Birmingham AL 35233

FATTORINI, HECTOR OSVALDO, b Buenos Aires, Arg, Oct 28, 38; m 61; c 3. APPLIED MATHEMATICS, SYSTEMS DESIGN & SYSTEMS SCIENCE. *Educ:* Univ Buenos Aires, Lic en Mat, 60; NY Univ, PhD(math), 65. *Prof Exp:* Res assoc math, Nat Sci & Tech Res Coun, Arg & assoc prof, Sch Exact & Natural Sci, Univ Buenos Aires, 65-66; res assoc, Brown Univ, 67; from asst prof to assoc prof, 67-77, PROF MATH, ENG & APPL SCI, UNIV CALIF, LOS ANGELES, 77- *Concurrent Pos:* US Air Force Off Sci Res & Off Aerospace Res grant, 67; NASA grant, 67; Off Naval Res contract, 67-68; NSF grant, 69-; prof, Sch Exact & Natural Sci, Univ Buenos Aires, 70-75. *Mem:* Am Math Soc; Soc Indust & Appl Math; Arg Math Union. *Res:* Control theory; differential equations in linear topological spaces; partial differential equations; control systems in infinite dimensional spaces; system theory. *Mailing Add:* Dept Math Univ Calif 405 Hilgard Ave Los Angeles CA 90024

FATZINGER, CARL WARREN, b Albany, NY, June 9, 38. FOREST ENTOMOLOGY. *Educ:* Univ Mich, BS, 60, MS, 61; Nova Univ, MS, 90; NC State Univ, PhD(entom), 68. *Prof Exp:* Entomologist, NC, 62, Fla, 62-64, res entomologist, 64, Res Triangle Park, 64-68, Fla, 68-77, Athens, Ga, 77-79, PRIN INSECT ECOLOGIST, SOUTHEASTERN FOREST EXP STA, FLA, 79- *Mem:* Entom Soc Am; Soc Am Foresters. *Res:* Behavior of insects affecting pine cones; effects of semiochemicals, light and temperature on insect behavior; trapping systems for bark beetles; survey and sampling methods for forest insects. *Mailing Add:* Southeastern Forest Exp Sta PO Box 70 Olustee FL 32072

FAUBION, BILLY DON, b Breckenridge, Tex, May 24, 42; m 63; c 3. PHYSICAL CHEMISTRY. *Educ:* Tex A&M Univ, BS, 64, MS, 65, PhD(phys chem), 68. *Prof Exp:* Asst prof phys chem, Adams State Col, 68-70; SCIENTIST, MASON & HANGER-SILAS MASON CO, INC, 70- *Mem:* Am Chem Soc. *Res:* Molecular and electronic structure of molecules; thermal analysis and compatibility of high explosives. *Mailing Add:* 4619 Oregon Trail Amarillo TX 79109

FAUBL, HERMANN, b Hungary, Feb 8, 42; US citizen; m 66; c 3. ORGANIC CHEMISTRY, BIOCHEMISTRY. *Educ:* Loyola Univ, Ill, BS, 65; Northwestern Univ, PhD(chem), 69. *Prof Exp:* Sr res scientist, Pfizer, Inc, 69-77; sr res scientist, Clinton Corn Processing Co, 77-80; GROUP LEADER, ABBOTT LABS, 80- *Mem:* AAAS; Am Mgt Asn. Am Chem Soc; Sigma Xi. *Res:* Synthetic organic chemistry; natural product synthesis; synthesis and properties of strained olefins; pharmaceutical chemistry; chemotherapy; antibiotics; antivirals; carbohydrate chemistry; protein isolation. *Mailing Add:* Abbott Labs North Chicago IL 60064

FAUCETT, ROBERT E, b Dearborn, Mich, Nov 21, 26; m 46, 65; c 3. ELECTRICAL ENGINEERING. *Educ:* Case Inst Technol, BS, 47, MS, 51. *Prof Exp:* Eng asst, Cleveland Elec Illum Co, Ohio, 47-49, 50-51; mgr photom lab, Tech Develop & Eval Ctr, Civil Aeronaut Admin, Ind, 51-53; asst mgr photom dept, Elec Testing Labs, NY, 53-55; chief eng res & develop lab, US Corps Engrs, Va, 55-56; sr appln engr, Lighting Systs Dept, Gen Elec Co, 56-75; Pres, Independent Testing Lab Inc, 75-87, chief exec officer & pub rels officer, 80-87; RETIRED. *Mem:* Fel Illum Eng Soc. *Res:* Basic glare research; photometry; testing of lighting systems; design of outdoor floodlighting and roadway lighting systems; interior commerical and industrial lighting systems; computer application for solving complex lighting problems; photometric testing of lighting equipment; general lighting consulting and computer services. *Mailing Add:* 261 Window Rock Ct Grand Junction CO 81503

FAUCETT, T(HOMAS) R(ICHARD), b Hatton, Mo, Aug 22, 20; m 42; c 4. MECHANICAL ENGINEERING. *Educ:* Univ Mo, BS, 42; Purdue Univ, MS, 49, PhD(mech eng), 52. *Prof Exp:* Design analyst, Cleveland Diesel Engine Div, Gen Motors Corp, 42-46; instr mech eng, Purdue Univ, 46-52; assoc prof, Univ Rochester, 52-60; prof sch mines, Univ Mo-Rolla, 60-62; prof & head dept, Univ Iowa, 62-65; chmn dept mech & aerospace eng, 65-78, prof mech eng, 65-85, EMER PROF MECH & AERO ENG, UNIV MO-ROLLA, 85- *Mem:* Am Soc Mech Engrs; Am Soc Eng Educ; Sigma Xi. *Res:* Vibrations; dynamics; stress analysis; computer applications in engineering and manufacturing. *Mailing Add:* Dept Mech & Aerospace Eng Univ Mo Rolla MO 65401

FAUCETT, WILLIAM MUNROE, b Union, SC, Sept 16, 16; m 46; c 1. MATHEMATICS. *Educ:* Univ SC, BS, 42, MS, 50; Tulane Univ, PhD(math), 54. *Prof Exp:* Asst, Off Naval Res Contract, Tulane Univ, 52-54; asst prof math, Univ Ky, 54-55; sr aerophysics engr, Gen Dynamics, Ft Worth, 55-58, sr opers analyst, 58-61, proj opers analyst, 61-63, design specialist, 63-77, sr eng specialist, 77-78, staff consult, 78-81; RETIRED. *Concurrent Pos:* Consult, 87- *Mem:* Am Math Soc; Sigma Xi. *Res:* Operations research; management science. *Mailing Add:* 3500 Creston Ave Ft Worth TX 76133-1415

FAUCHALD, KRISTIAN, b Oslo, Norway, July 1, 35; nat US. SYSTEMATIC ZOOLOGY, MARINE BIOLOGY. *Educ:* Univ Bergen, Cand Mag, 59, Cand Real, 61; Univ Southern Calif, PhD(biol), 69. *Prof Exp:* Vitenskapelig asst biol, Univ Bergen, 59-64, amanuensis, 64-65; from res assoc to assoc prof biol, Univ Southern Calif, 75-79; CUR INVERT ZOOL, NAT MUS NATURAL HIST, SMITHSONIAN INST, 79- *Mem:* Am Soc Zool; Sigma Xi; fel AAAS; Syst Zool Soc. *Res:* Systematics and biology of polychaetous annelids from world wide areas; benthic ecology. *Mailing Add:* Dept Invert Zool Nat Mus Natural Hist Smithsonian Inst Washington DC 20560

FAUCI, ANTHONY S, b Brooklyn, NY, Dec 24,40; c 2. IMMUNOLOGY, INFECTIOUS DISEASE. *Educ:* Col Holy Cross, AB, 62; Cornell Univ Med Col, MD, 66. *Prof Exp:* Clin assoc, Lab Clin Invest, 68-70, sr staff fel, 70-71, sr investr, 72-74, head clin physiol sect, 74-80, dep clin dir, 77-84, CHIEF, LAB IMMUNOREGULATION, NAT INST ALLERGY & INFECTIOUS DIS, NIH, 80-, DIR, 84-; DIR, OFF AIDS RES & ASSOC DIR OF NIH AIDS RES, 88- *Concurrent Pos:* Chief resident, Dept Med, NY Hosp-Cornell Med Ctr, 71-72; instr med, Cornell Med Col, 71-72; clin prof, Dept Med, Div Rheumatology, Immunol & Allergy, Georgetown Univ Sch Med, 84-89, George Washington Univ Sch Med & Health Sci, 85-; assoc ed, J Immunol, 79; vis prof, Univ Mich Med Ctr, 82, Univ Colo, 82, Tufts-New Eng Med Ctr, 83 & 84, Baylor Univ Med Ctr, 84, Case Western Reserve Univ Med Ctr, 86. *Honors & Awards:* Paul B Beeson Lectr, Yale Univ Sch Med, 82; Squibb Award, Infectious Dis Soc Am, 83; John S Lawrence Lectr, Univ Calif, 84; Irwin H & Martha L Lepow Lectr, Univ Conn Sch Med, 84; Robert A Cooke MD Medal, Am Acad Allergy, 84; George Thorn Lectr, Brigham & Women's Hosp & Harvard Med Sch, 85; Leo H Criep MD Lectr, Montefiore Hosp-Univ Pittsburgh Sch Med, 86; E Ross Kyger Jr MD Lectr, Univ Pa Med Ctr, 87. *Mem:* Inst Med-Nat Acad Sci; Am Fedn Clin Res (pres, 80-81); Am Asn Immunologists; Am Acad Allergy & Immunol; AAAS; Am Soc Clin Invest; Am Asn Physicians; Am Rheumatism Asn; Infectious Dis Soc Am. *Res:* Study of the mechanisms of immunosuppression; the hyperosinophilic syndrome and other host defense defects; the pathogenesis and therapy of hypersensitivity vasculitic and granulomatous disease; the clinical, pathophysiologic, immunologic, molecular biologic, virologic, and therapeutic aspects of AIDS; delineation of the mechanisms of activation, proliferation and differentiation of human lymphoid cells. *Mailing Add:* Nat Inst Allergy & Infectious Dis NIH Bldg 31 Rm 7A32 Bethesda MD 20892

FAUDREE, RALPH JASPER, JR, b Durant, Okla, Aug 23, 39; m 62; c 2. GRAPH THEORY. *Educ:* Okla Baptist Univ, BS, 61; Purdue Univ, MS, 63, PhD(math), 64. *Prof Exp:* Instr math, Univ Calif, Berkeley, 64-66; asst prof, Univ Ill, Urbana, 66-70; assoc prof, 70-75, PROF MATH, MEMPHIS STATE UNIV, 75-, CHMN DEPT, 83- *Concurrent Pos:* Vis prof math, Univ Aberdeen, Scotland, 80; res mathematician, Hungarian Acad Sci, Budapest, 81. *Mem:* Am Math Soc. *Res:* Graph theory. *Mailing Add:* Memphis State Univ Memphis TN 38152

FAUGHT, JOHN BRIAN, b Toronto, Ont, Mar 18, 42; m 68. INORGANIC CHEMISTRY, X-RAY CRYSTALLOGRAPHY. *Educ:* Univ Windsor, BSc, 65; Univ Ill, Urbana, MS, 67, PhD(inorg chem), 69; St Mary's Univ, BEd, 77. *Prof Exp:* Res assoc, Univ Fla, 69-70; asst prof chem, Dalhousie Univ, 70-76; proj leader & chemist, Environ Can, Environ Protection Serv, 77-78; teacher chem & math, Dartmouth Acad, 78-79; HEAD SCI DEPT, HALIFAX GRAMMAR SCHOOL, 79- *Concurrent Pos:* Lectr chem, St Mary's Univ, 80- *Mem:* Am Chem Soc; Am Crystallog Asn. *Res:* Synthesis and structural studies of phosphorus-nitrogen and arsenic-nitrogen compounds, structural analysis utilizing the techniques of x-ray crystallography, infra-red and Raman spectroscopy. *Mailing Add:* 8 Overdale Lane Dartmouth NS B3A 3V3 Can

FAULCONER, ROBERT JAMIESON, b Sussex, Eng, July 11, 23; nat US; m 45; c 4. PATHOLOGY. *Educ:* Col of William & Mary, BS, 43; Johns Hopkins Univ, MD, 47. *Prof Exp:* Instr path, Univ Pa, 49-52; pathologist, DePaul Hosp, 54-78, dir labs, 66-78; PROF PATH, EASTERN VA MED SCH, 74-, CHMN DEPT, 78- *Concurrent Pos:* Fel gynecol path, Johns Hopkins Univ, 48-49; clin assoc, Med Col Va, 66-70, clin prof, 72-79; consult pathologist, US Naval Hosp, Portsmouth & Vet Admin Ctr, Hampton. *Honors & Awards:* J Shelton Horsley Award; Am Cancer Soc. *Mem:* AAAS; fel Am Soc Clin Path; Am Asn Anat; Am Asn Hist Med; Col Am Path; Am Asn Pathologists; Sigma Xi. *Res:* Pathologic anatomy; human embryology; immunology of cancer; pathology of endocrine organs. *Mailing Add:* Dept Path 700 Olney Rd Norfolk VA 23507-1980

FAULDERS, CHARLES R(AYMOND), b Spokane, Wash, May 21, 27; m 54; c 5. MECHANICAL ENGINEERING. *Educ:* Univ Calif, BS, 48; Mass Inst Technol, SM, 50, ME, 51. *Hon Degrees:* ScD, Mass Inst Technol,54. *Prof Exp:* Asst mech eng, Mass Inst Technol, 49-54; aerodynamicist Aerophysics Dept, NAm Aviation, Inc, 54-55, supvr propulsion aerodyn, Missile Div, 55-57, res specialist Eng Dept, 57-58, res specialist Aerospace Labs, 58-60; assoc prof, Univ Aix-Marseille, 60-62; mgr flight technol, Paraglider Prog, NAm Aviation, Inc, 63-65, mgr systs eng, 65-66, mgr flight sci, Space Div, NAm Rockwell Corp, 66-73, proj engr, Energy Systs Studies, Space Div, Rockwell Int, 73-76, proj mgr, Energy Conserv Systs, 76-77, mkt rep advan progs, 77-80, proj monitor, oil shale prog, Energy Systs Group, 80-83, PROJ ENGR, MONITORING GEOTHERMAL FOAM PROJ GUARANTEE PROJS, ROCKWELL INT, 84- *Mem:* Sigma Xi. *Res:* Fluid mechanics; aerodynamics of compressors and turbines; space flight mechanics; optimization of powered space flight trajectories; magnetohydrodynamics; atmospheric flight dynamics of lifting vehicles; thermodynamics; cost analysis; coal gasification and liquefaction; oil shale; geothermal power plant requirements. *Mailing Add:* 621 Trueno Ave Camarillo CA 93010

FAULK, DENNIS DERWIN, b Searcy, Ark, Nov 29, 36; m 58; c 2. ORGANIC CHEMISTRY. *Educ:* Ark State Teachers Col, BS, 58; Univ Ark, PhD(org chem), 66. *Prof Exp:* Chemist, Gulf Oil Corp, 58-61; asst chem, Univ Ark, 61-66; res chemist, Shell Oil Co, Deer Park, 66-67; asst chem, 67, asst prof, 67-71, assoc prof, 71-81, PROF CHEM, CENT MO STATE UNIV, 81- *Mem:* Am Chem Soc; Sigma Xi. *Res:* Heterogeneous catalysis; organic reaction mechanisms; acid-catalysed ketone rearrangements; spectroscopic identifications and correlations; structural correlations using nuclear magnetic resonance shift reagents; base catalyzed condensations of ketones and primary alcohols. *Mailing Add:* Dept Chem Cent Mo State Univ Warrensburg MO 64093

FAULKENBERRY, GERALD DAVID, b Wapanucka, Okla, May 28, 37; m 61; c 2. STATISTICS. *Educ:* Southeastern State Col, BS, 59; Okla State Univ, MS, 61, PhD(statist), 65. *Prof Exp:* Asst prof statist, Okla State Univ, 65; asst prof statist, Ore State Univ, 65-69; supvr statist group, Litton Sci Support Lab, Litton Industs, Calif, 69-71; vis assoc prof, Ore State Univ, 71-72, from assoc prof to prof statist & chmn dept, 81-91; DEPT HEAD, NMEX STATE UNIV, 91- *Mem:* Am Statist Asn. *Res:* Statistical theory and methodology, particularly estimation; predictive inferences: analysis of survey data; survey research. *Mailing Add:* Dept Exp Statist Univ Sta Ctr NMex State Univ Box 3003 Dept 3130 Las Cruces NM 88003-0003

FAULKIN, LESLIE J, JR, b Peoria, Ill, July 4, 30; m 50; c 2. ANATOMY. *Educ:* Univ Calif, Berkeley, BA, 55, MA, 57, PhD(zool), 64. *Prof Exp:* From asst prof to assoc prof, 64-85, chmn dept, 72-78, PROF ANAT, SCH VET MED, UNIV CALIF, DAVIS, 85- *Mem:* AAAS; Am Asn Cancer Res; Am Asn Anat. *Res:* Experimental oncology; growth and development; growth regulation; endocrinology. *Mailing Add:* Dept Anat Univ Calif Sch Vet Med Davis CA 95616

FAULKNER, D JOHN, b Bournemouth, Eng, June 10, 42; m 66. ORGANIC CHEMISTRY. *Educ:* Imp Col, London, BSc, 62; Univ London, PhD(org chem), 65. *Prof Exp:* Fel, Harvard Univ, 65-67; fel, Stanford Univ, 67-68; assoc prof marine chem, 68-80, PROF, SCRIPPS INST OCEANOG, UNIV CALIF, 80- *Mem:* Am Chem Soc; The Chem Soc. *Res:* Isolation and identification of natural products; chemical ecology. *Mailing Add:* Scripps Inst Oceanog San Diego CA 92093-0212

FAULKNER, FRANK DAVID, b Humansville, Mo, Apr 6, 15; m 41; c 6. APPLIED MATHEMATICS. *Educ:* Kans State Teachers Col, BS, 40; Kans State Col, MS, 42; Univ Mich, Ann Arbor, PhD(appl math), 69. *Prof Exp:* Jr physicist, Appl Physics Lab, Johns Hopkins Univ, 44-46; res mathematician, Eng Res Inst, Univ Mich, 46-50; from assoc prof to distinguished prof math, Naval Postgrad Sch, 50-81; RETIRED. *Mem:* Am Math Soc; Math Asn Am. *Res:* Numerical methods in optimization; mechanics; calculus of variations; applications in missile problems; numerical methods applied to partial differential equations. *Mailing Add:* Dept Math US Naval Postgrad Sch Monterey CA 93943

FAULKNER, GARY DOYLE, b Aberdeen, Miss, Sept 20, 44; m 72; c 2. MATHEMATICS. *Educ:* Ga State Univ, BS, 71; Univ SC, MS, 73; Ga Inst Technol, PhD(math), 76. *Prof Exp:* ASSOC PROF MATH, NC STATE UNIV, 76- *Mem:* Am Math Soc. *Res:* Analysis; functional analysis; topology and applied mathematics. *Mailing Add:* 1509 Chester Rd Raleigh NC 27608

FAULKNER, GAVIN JOHN, b Edinburgh, Scotland; Brit citizen. ROCK MECHANICS. *Educ:* Univ Strathclyde, Glasgow, BS, 77, PhD(mining eng), 80. *Prof Exp:* asst prof mining eng, Va Polytech Inst & State Univ, 80-88; PRES, ROWAN MOUNTAIN, INC, 88- *Mem:* Am Inst Mining Engrs; Inst Mining & Metall UK. *Res:* Mine support timber evaluation and improvement; longwall support characteristics; underground measuring techniques. *Mailing Add:* 2010 Broken Oak Dr Blacksburg VA 24060

FAULKNER, JAMES EARL, statistics; deceased, see previous edition for last biography

FAULKNER, JOHN, b Hayes, Middlesex, Eng, Apr 29, 37; m 66; c 2. ASTROPHYSICS, THEORETICAL PHYSICS. *Educ:* Cambridge Univ, BA, 59, MA, 63, PhD(appl math & theoret physics), 64. *Prof Exp:* Asst res astrophys, Cambridge Univ, 63-64; res fel physics, Calif Inst Technol, 64-66; staff mem astrophys, Inst Theoret Astron, Cambridge, 67-69; assoc prof, 69-73, prof astron & astrophys, 73-80, PROF STELLAR STRUCT & RELATIVITY, UNIV CALIF, SANTA CRUZ, 80- *Concurrent Pos:* Vis res assoc, Mass Inst Technol, 65-67; vis res scientist, Nat Radio Astron Observ, 77. *Honors & Awards:* William Stone Prize, Peterhouse, Cambridge, 65; Gravity Prize, Gravity Found, 72; Judy A Seydoux Mem Prize, Griffith Observ, 72-73. *Mem:* Fel Royal Astron Soc; fel Am Astron Soc; Int Astron Union. *Res:* Newtonian and general relativistic stellar structure and evolution; horizontal branch, helium content, dwarf novae and gravitational radiation; tides in general relativity, occasional general relativity and/or cosmology. *Mailing Add:* Dept Astrophys Univ Calif Santa Cruz CA 95064

FAULKNER, JOHN A, b Kingston, Ont, Dec 12, 23; m 55; c 2. PHYSIOLOGY. *Educ:* Queen's Univ, Ont, BA, 49, BPHE, 50; Ont Col Ed, Toronto, cert, 51; Univ Mich, MS, 56, PhD(educ), 62. *Prof Exp:* Teacher high sch, Ont, 51-52; teacher sci & phys educ, Glebe Collegiate Inst, Ont, 52-56; asst prof phys educ, Univ Western Ont, 56-60; asst prof educ, 60-66, assoc prof physiol, 66-71, PROF PHYSIOL, UNIV MICH, 71-; RES SCIENTIST, INST GERONT, 86-, ASSOC DIR BIOL RES, 90- *Concurrent Pos:* Mich Heart Asn grants, 63-67 & 69-81; NIH grants, 63-67, 71-; Muscular Dystrophy Asn of Am grant, 71-79; pres, Am Col Sports Med, 71-72; actg dir, Inst Geront, 88-89. *Honors & Awards:* Burke Aaron Hinsdake Scholar, Univ Mich, 62; Citation Award, Am Col Sports Med, 78. *Mem:* Fel AAAS; Am Physiol Soc; Biophys Soc; Geront Soc Am. *Res:* Physiological adaptation to exercise and hypoxia; skeletal muscle transplantation and regeneration; contractile and biochemical properties of skeletal muscles in aged rodents; injury and repair of muscle fibers following lengthening contractions. *Mailing Add:* Dept Physiol Univ Mich Med Sch Ann Arbor MI 48104-0622

FAULKNER, JOHN EDWARD, b Plattsburg, Ohio, Oct 5, 20; m 46; c 2. APPLIED PHYSICS. *Educ:* Oberlin Col, BA, 42; Univ Wis, PhD(physics), 50. *Prof Exp:* Mem staff radiation lab, Mass Inst Technol, 42-46; physicist, Hanford Labs, Gen Elec Co, Wash, 50-52, supvr exp nuclear physics, 52-57, mgr nuclear physics res, 57-63; consult exp physics, 63-64, mgr safeguards eng, 64-66, consult appl physics, Astronuclear Lab, 66-77, consult appl physics, Adv Energy Systs Div, Westinghouse Elec Co, 77-84; CONSULT APPL PHYSICS, 84- *Concurrent Pos:* Consult, Secy War, 45. *Mem:* Am Phys Soc. *Res:* Low energy neutron physics; reactor physics; x-rays. *Mailing Add:* 811 S Speed Santa Maria CA 93454

FAULKNER, JOHN SAMUEL, b Memphis, Tenn, Sept 30, 32; m 57, 88; c 2. PHYSICS. *Educ:* Auburn Univ, BS, 54, MS, 55; Ohio State Univ, PhD(physics), 59. *Prof Exp:* Asst prof physics, Univ Fla, 59-62; mem staff, Metals & Ceramics Div, Oak Ridge Nat Lab, 62-86; PROF PHYSICS, FLA ATLANTIC UNIV, 86- *Concurrent Pos:* Sr Fulbright res scholar, Univ Sheffield, UK, 68-69; vis prof, Univ Bristol, UK, 76-77; vis scientist, KFA Juelich, Ger, 85-86; co-dir, Alloy Res Ctr, 88- *Honors & Awards:* Award Best Sustained Res, US Dept Energy, 82. *Mem:* Fel AAAS; fel Am Phys Soc; Mat Res Soc. *Res:* Theoretical solid state physics; electronic states in ordered and disordered systems. *Mailing Add:* Dept Physics Fla Atlantic Univ Boca Raton FL 33431

FAULKNER, KENNETH KEITH, anatomy; deceased, see previous edition for last biography

FAULKNER, LARRY RAY, b Shreveport, La, Nov 26, 44; m 65; c 2. ELECTROCHEMISTRY, LUMINESCENCE SPECTROSCOPY. *Educ:* Southern Methodist Univ, BS, 66; Univ Tex, Austin, PhD(chem), 69. *Prof Exp:* Asst prof chem, Harvard Univ, 69-73; from asst prof to assoc prof, 73-79, dept head, Mat Res Lab, 84-89, PROF CHEM, UNIV ILL, URBANA-CHAMPAIGN, 79-, MEM MAT RES LAB, 84-; DEAN, COL LIB ARTS & SCI, 89- *Concurrent Pos:* Div ed, J Electrochem Soc, 74-80; US regional ed, J Electroanal Chem, 80-85; prof chem, Univ Texas, Austin, 83-84. *Honors & Awards:* Young Authors' Prize, Electrochem Soc, 76. *Mem:* Am Chem Soc; Electrochem Soc (vpres, 88-); Soc Electroanal Chem (treas, 84-88); AAAS. *Res:* Chemical reactions of excited states; electron transfer processes in systems of controlled architecture; fluorescence and phosphorescence phenomena and techniques; electrochemistry and electroanalytical chemistry. *Mailing Add:* Dept Chem Univ Ill 1209 W Calif St Urbana IL 61801

FAULKNER, LLOYD (CLARENCE), b Longmont, Colo, Oct 24, 26; m 54; c 5. REPRODUCTIVE PHYSIOLOGY. *Educ:* Colo State Univ, DVM, 52; Cornell Univ, PhD(animal physiol), 63; Am Col Theriogenologists, dipl, 71. *Prof Exp:* From asst prof to assoc prof vet clins & surg, Colo State Univ, 55-63, assoc prof vet clins, surg, physiol & endocrinol, 63 & 66, prof vet clins & surg, 66-70, prof physiol & biophys & chmn dept, 70-78; assoc dean vet res & asst dir, Mo Agr Exp Sta, Univ Mo, 79-81; DIR VET RES & GRAD STUDIES, OKLA STATE UNIV, 81-, DIR OKLA ANIMAL DIS DIAG LAB, 91-; CONSULT. *Concurrent Pos:* Consult, Quaker Oats, 71-75, Ft Dodge Labs, 73-75; Cong Sci & Eng fel, Fedn Am Socs Exp Biol; dir & officer, Intermountain Vet Med Asn; mem exec bd & res coun, Am Vet Med Asn; mem bd & exec comt, Am Asn Accredited Lab Animal Care. *Honors & Awards:* David E Bartlett Award, 88. *Mem:* Fel AAAS; Am Vet Med Asn; Am Soc Animal Sci; Soc Study Reproduction; Am Col Theriogenologists (secy, 71-74, pres, 74-75); Sigma Xi. *Res:* Testis-accessory sex gland relationships; hypothalamo-hypophyseal relationships in the rat; alterations in semen quality in bulls with lesions of the reproductive system; population control in companion animals. *Mailing Add:* Assoc Dean Vet Res & Asst Dir OK Agr Exp Sta Okla State Univ Stillwater OK 74078

FAULKNER, LYNN L, b Ft Wayne, Ind, June 24, 41; m 61; c 2. MECHANICAL ENGINEERING, ACOUSTICS. *Educ:* Purdue Univ, BS, 65, MS, 66, PhD(mech eng), 69. *Prof Exp:* Apprentice draftsman eng, Gen Elec Co, Ind, 59-63; res asst mech eng, Herrick Labs, Purdue Univ, 65-70; from asst prof to assoc prof mech eng, Ohio State Univ, 70-78; MGR, DYNAMICS SECT, BATTELLE MEM INST, 78- *Concurrent Pos:* Fel, Purdue Univ, 69-70. *Mem:* Am Soc Mech Engrs; Acoust Soc Am; Soc Exp Stress Anal; Am Soc Eng Educ. *Res:* Engineering acoustics; noise control; noise analysis of household appliances; vehicle noise; building acoustics; outdoor power equipment; machinery dynamics; machinery vibration; ultrasonics. *Mailing Add:* Battelle Mem Inst 505 King Ave Columbus OH 43201

FAULKNER, PETER, b Cardiff, Wales, July 8, 29; m 50; c 4. VIROLOGY, MOLECULAR BIOLOGY. *Educ:* Univ London, BSc, 50; McGill Univ, PhD(biochem), 54. *Prof Exp:* Res student, Montreal Gen Hosp, 50-54; agr res officer, Lab Insect Path, Can Dept Agr, 54-63; mem sci staff med res coun, Virus Res Unit, Carshalton Surv, Eng, 63-65; assoc prof, 65-80, CAREER INVESTR, MED RES COUN CAN, 65-; PROF MICROBIOL, QUEEN'S UNIV, ONT, 73- *Mem:* Brit Biochem Soc; Soc Gen Microbiol; Can Soc Microbiol; Am Soc Virol; Am Soc Microbiol; Soc Invert Path. *Res:* Insect virus molecular genetics; baculoviruses as microbiol pesticides; other agricultural and forest sciences. *Mailing Add:* Dept Microbiol Queen's Univ Kingston ON K7L 3N6 Can

FAULKNER, RUSSELL CONKLIN, JR, b Barbourville, Ky, Jan 31, 20; m 54; c 2. ZOOLOGY, EXPERIMENTAL MORPHOLOGY. *Educ:* Lincoln Mem Univ, BS, 48; Univ Okla, MS, 52, PhD(zool), 58. *Prof Exp:* Asst zool, Univ Okla, 48-55; asst prof biol, Okla Baptist Univ, 55-57; from asst prof to assoc prof, Tex Christian Univ, 57-67; prof biol, Stephen F Austin State Univ, 67-90; RETIRED. *Mem:* Sigma Xi. *Res:* Effects of radioisotopes on growth and development and use of histochemical methods in detecting effects; microtechnique; radioecological techniques. *Mailing Add:* 1816 Sheffield Dr Nacogdoches TX 75961

FAULKNER, THOMAS RICHARD, b Detroit, Mich, Dec 3, 47; m 74. SOFTWARE DEVELOPMENT, COMPUTER SYSTEMS. *Educ:* Oakland Univ, BA & MS, 70; Univ Minn, PhD(chem), 76. *Prof Exp:* Fel chem, Univ Minn, 76; instr chem, Univ Va, 76-79; software analyst, 79-80, ANAL MGR, CRAY RES INC, 81- *Mem:* Am Chem Soc; Am Phys Soc; AAAS; Asn Comput Mach. *Res:* Theoretical and experimental spectroscopy; molecular vibrations; quantum chemistry; design, development and installation of computer software. *Mailing Add:* 2615 Marsh Dr San Ramon CA 94583

FAULKNER, WILLARD RILEY, b Jerry, Wash, Jan 2, 15; m 50. BIOCHEMISTRY, CLINICAL CHEMISTRY. *Educ:* Univ Idaho, BS, 40; Univ Denver, MS, 50; Vanderbilt Univ, PhD, 56; Am Bd Clin Chem, dipl. *Prof Exp:* Asst biochem, 52-56, assoc prof, Vanderbilt Sch Med, 68-80; dir microchem lab, Cleveland Clin, 56-67; dir clin chem labs, 68-75, actg dir clins labs, 71-73; dir res & develop clin chem, Vanderbilt Univ Med Ctr, 75-77; EMER PROF, VANDERBILT SCH MED, 80- *Concurrent Pos:* Mem, med bd dirs, Nat Registry Clin Chem, 68-77; chmn, Llicensure Aadv Ccomt, Tenn Med Lab Act, 69-79; co-ed, CRC, Critical Rev Clin Lab Sci, 70-75; mem educ bd, Standard Math Clin Chem, 65-86, chmn, 83-85. *Honors & Awards:* Bronze Award, Am Soc Clin Pathologists & Col Am Pathologists, 58 & 65; Fisher Award, Am Asn Clin Chem, 80. *Mem:* AAAS; Am Microchem Soc; Am Chem Soc; Am Asn Clin Chemists; NY Acad Sci; Asn Clin Scientists. *Res:* Myoglobin; blood pH, urinary amino acids; blood ammonium; clinical microchemistry; normal clinical laboratory values; ionic calcium. *Mailing Add:* Emer Off Vanderbilt Univ Rm 607 Med Arts Bldg Nashville TN 37212

FAUPEL, JOSEPH H(ERMAN), b Waukegan, Ill, Oct 25, 16; m 40; c 9. ENGINEERING MECHANICS. *Educ:* Pa State Univ, BS, 39, PhD(eng mech), 48; Univ Pittsburgh, MS, 42. *Prof Exp:* Res metallurgist, Aluminum Co Am, 39-42; asst, Pa State Univ, 45-47; res engr, 48-52, proj engr, 52-55, res assoc, 55-62, prin design consult, E I du Pont de Nemours & Co, Inc, 62-80; PVT CONSULT, 80- *Concurrent Pos:* Du Pont fel eng, 46-48; spec lectr mech eng, Univ Del, 49-51; Am Soc Mech Engrs, Boiler & Pressure Vessel Comt, 68-80; mem, Pressure Vessel Res Comt, Boiler & Pressure Vessel Comt & indust & prof adv comt, Pa State Univ. *Mem:* Am Soc Metals; fel Am Soc Mech Engrs; Am Soc Testing & Mat; Sigma Xi. *Res:* Mechanics of materials; stress analysis; pressure vessels; plasticity; elasticity; viscoelasticity; limit design; high temperature and high pressure mechanics. *Mailing Add:* 400 Crest Rd Carrcroft Wilmington DE 19803

FAURE, GUNTER, b Tallinn, Estonia, May 11, 34; US citizen; m; c 4. GEOCHEMISTRY, ISOTOPE GEOLOGY. *Educ:* Western Ont Univ, BSc, 57; Mass Inst Technol, PhD(geol), 61. *Prof Exp:* Res assoc geochronology, Mass Inst Technol, 61-62; from asst prof to assoc prof, 62-68, PROF GEOL, OHIO STATE UNIV, 68- *Concurrent Pos:* NSF res grants, 64-85; ed-in-chief, Isotope Geosci, Elsevier, Amsterdam, 83-88; exec ed, Geochimica et Cosmochimica Acta, Pergamon, Oxford 88- *Mem:* Planetary Soc; Geol Soc Am; Geochem Soc. *Res:* Isotopic composition of strontium in volcanic rocks, in oceans and fresh water on continents, and its isotope geochemistry in the base metal deposits of the Red Sea; petrogenesis of basalt in Antarctica; sediment mixing in the Ross Sea and Black Sea; provenance of feldspar in glacial deposits of Ohio and Antarctica; meteorite ablation spherules; East Antarctic ice sheet. *Mailing Add:* Dept Geol Sci Ohio State Univ 125 S Oval Mall Columbus OH 43210

FAUSCH, HOMER DAVID, b Buffalo Center, Iowa, Apr 5, 19; m 43; c 3. ANIMAL GENETICS. *Educ:* Univ Minn, BS, 47, MS, 50, PhD(animal breeding), 53. *Prof Exp:* Assoc prof & animal husbandman, Northwest Exp Sta, Univ Minn, 47-56; dir res, 68-74, PROF ANIMAL SCI, CALIF STATE POLYTECH UNIV, POMONA, 56- *Concurrent Pos:* Lectr, Univ Alta, 64-65. *Mem:* AAAS; Am Soc Animal Sci; Sigma Xi. *Res:* Effect of inbreeding on variability of economic traits in the Minnesota number one and number two breeds of swine; lipid metabolism in swine. *Mailing Add:* 5082 Ebel Way Northfield MN 55057

FAUSETT, DONALD WRIGHT, b Sulphur Springs, Tex, Sept 20, 40; m 83; c 2. MATHEMATICAL MODELLING, NUMERICAL ANALYSIS. *Educ:* Arlington State Col, Tex, BS, 63; Univ Tex, Arlington, MA, 68; Univ Wyo, PhD(math), 74. *Prof Exp:* Math Statist asst, US Army, 63-66; res mathematician, Sun Oil Co, 66-70; mathematician, Laramie Energy Res Ctr, 71-72, fel math, US Bur Mines, 72-74; asst prof math, Colo Sch Mines, 74-78; res assoc, Laramie Energy Technol Ctr, 78-83; assoc prof, 83-86, actg head, 85-86, PROF MATH SCI, FLA INST TECHNOL, 86- *Concurrent Pos:* Fac partic, Energy Res & Develop Admin Fac Participation Prog, Assoc Western Univs, 76; adj assoc prof, dept chem eng, Univ Wyo; prin investr, oil shale retort modelling, ERDA, 76-78, hydroacoustic field phenomena, US Navy, 85- *Mem:* Soc Indust & Appl Math; Math Asn Am; Sigma Xi. *Res:* Mathematical modelling of energy recovery processes and hydroacoustic field phenomena. *Mailing Add:* 611 Xavier Ave Melbourne FL 32901

FAUSETT, LAURENE VAN CAMP, b Sacramento, Calif, Mar 2, 43. MATHEMATICAL MODELING, APPLIED MATHEMATICS. *Educ:* Univ Calif, Berkeley, BA, 64; Univ Wyo, MST, 75, Phd(math), 84. *Prof Exp:* Teacher math, Albany Pub Schs, 65-67; teaching asst math, Univ Wyo, 77-82; mathematician, Laramie Energy Technol Ctr, 75-76 & 81-83; Brevard Community Col, 83-84; ASST PROF MATH, 84-, CHMN UNDERGRAD PROG MATH, FLA INST TECHNOL, 87- *Mem:* Soc Indust & Appl Math; Math Asn Am. *Res:* Underground coal gasification; constrained layer damping technique for an infinite flat plate with fluid loading on either or both sides. *Mailing Add:* 611 Xavier Ave Melbourne FL 32901

FAUSEY, NORMAN RAY, b Fremont, Ohio, Oct 28, 38; m 59; c 7. SOIL PHYSICS. *Educ:* Ohio State Univ, BS, 62, MS, 66, PhD(agron), 75. *Prof Exp:* SOIL DRAINAGE SCIENTIST, USDA, 67- *Mem:* Am Soc Agron; Soil Sci Soc Am; Soil Conserv Soc Am; Am Soc Agr Engrs. *Res:* Agricultural water management for more efficient production and water quality protection. *Mailing Add:* 5383 Amy Lane Columbus OH 43220

FAUST, CHARLES HARRY, JR, b Allentown, Pa, Jan 10, 43. MOLECULAR BIOLOGY, IMMUNOLOGY. *Educ:* Franklin & Marshall Col, AB, 64; Colo State Univ, PhD(biochem), 70. *Prof Exp:* Asst prof, 74-80, assoc prof surg, biochem & microbiol, Med Sch, Univ Ore, 80-83; ASSOC PROF DEPT BIOCHEM, TEX TECH UNIV HEALTH SCI CTR, 84- *Concurrent Pos:* Max Planck Soc fel, Max Planck Inst Exp Med, Ger, 70-71; Am Cancer Soc fel, Dept Path, Univ Geneva, 71-73, Am-Swiss Coun Sci Exchange fel, 73-74. *Mem:* NY Acad Sci; Am Chem Soc; Sigma Xi; Am Asn for the Advan of Sci; Am Asn Immunologist; Am Soc Microbiol; Am Soc for Biochem & Molecular Biol. *Res:* Molecular biology of normal and malignant animal cells with emphasis on control of gene expression. *Mailing Add:* Dept Biochem Health Sci Ctr Tex Tech Univ Lubbock TX 79430

FAUST, CHARLES L(AWSON), b St Louis, Mo, Nov 8, 06; m 34; c 3. CHEMICAL ENGINEERING. *Educ:* Univ Washington, St Louis, BS, 30, MS, 31; Univ Minn, PhD(chem eng), 34. *Prof Exp:* Grad asst, Univ Minn, 31-34; res engr, Battelle Mem Inst, 34-41, asst supvr electrochem res, 41-44, supvr electrochem eng res, 44-53, chief electrochem eng div, 53-69, assoc mgr chem & chem eng dept, 69-71; consult, 71-86; RETIRED. *Concurrent Pos:* Mem Int Coun Electrodeposition, 64-67; consult, Mining & Mat Adv Bd. *Honors & Awards:* Atcheson Medal, Electrochem Soc, 62; Proctor award, Am Electroplaters & Surface Finishers Soc, 42, Gold Medal, 51, Heusner Award, 55, sci award, 61; Res Gold Medal, Soc Mfg Engrs, 66; Hothersall Medal, Brit Inst Metal Finishing, 67. *Mem:* Am Chem Soc; hon mem Electrochem Soc (vpres, 48-50, pres, 50-51); Am Electroplaters Soc; Soc Mfg Engrs; Am Soc Metals. *Res:* Electrodeposition of metals and alloys; electroforming; electropolishing; pickling; metal finishing; protection; alumina from clays; fuel cells; batteries; electrolysis; electrowinning; electrorefining; electrochemical machining. *Mailing Add:* 1802 Riverside Dr Apt 18 Upper Arlington OH 43212-1856

FAUST, CHARLES R, GEOLOGY. *Educ:* Pa State Univ, BS, 67, PhD(geol), 76. *Prof Exp:* Geologist, US Geol Survey, Reston, Va, 71-79; res asst geosci, Pa State Univ, 72-74; PRIN HYDROGEOLOGIST & EXEC VPRES, GEO TRANS INC, HERNDON, VA, 79- *Concurrent Pos:* Co-convenor, Geol Soc Am; co chmn, Am Geophys Union, chmn, Ground Water Session, 84. *Honors & Awards:* Wesley W Horner Award, Am Soc Civil Engrs, 85; P D Krynine Res Fund Award. *Mem:* Fel, Geol Soc Am; Soc Petrol Engrs; Am Geophys Union. *Res:* Author of several articles. *Mailing Add:* Rte One Box 228 Hamilton VA 22068

FAUST, GEORGE TOBIAS, b Philadelphia, Pa, Aug 27, 08; m 36; c 5. MINERALOGY, GEOLOGY. *Educ:* Pa State Univ, BS, 30; Univ Mich, MS, 31, PhD(mineral), 34. *Prof Exp:* Asst mineralogist, Univ Mich, 30-35; asst prof ceramic mineral, Rutgers Univ, 35-38; asst chemist-petrogr, US Bur Mines, Ala, 38-40; asst mineralogist-petrogr, Bur Plant Indust, USDA, Md, 40-42; from asst mineralogist to head mineral group, US Geol Surv, 42-53, staff assoc solid state group, 53-60, geologist, 60-63, res geologist, 63-77; RETIRED. *Concurrent Pos:* Instr, Gemological Inst, Am, 35-38; lectr grad sch, Am Univ, 57. *Mem:* Fel Mineral Soc (vpres, 64, pres, 65); fel Ceramic Soc; fel Geol Soc; Geochem Soc (treas, 55-61); Crystallog Asn. *Res:* Mineralogy, geochemistry and petrology; Watchung basalt flows of New Jersey. *Mailing Add:* 80 S Maple Ave Basking Ridge NJ 07920

FAUST, IRVING M, CELL BIOLOGY, NUTRITION. *Educ:* Cornell Univ, PhD(physiol psychol), 73. *Prof Exp:* ASSOC PROF, ROCKEFELLER UNIV, 83- *Mailing Add:* Lewis Rd Irvington NY 10533

FAUST, JOHN PHILIP, b New Orleans, La, Sept 26, 24; m 52; c 3. WATER CHEMISTRY. *Educ:* Loyola Univ, La, BS, 44; Univ Ill, MS, 48; Univ Notre Dame, PhD(chem), 52. *Prof Exp:* Chemist, Olin Mathieson Chem Corp, 52-53, res group leader, 53-54, chief inorg res sect, Olin Corp, 54-57, proj supvr, 58-62, mem staff, 63-71, sr res assoc, 71-78, consult scientist pool chem, Olin Corp, 78-85; RETIRED. *Concurrent Pos:* Chmn, New Haven Sect, Am Chem Soc; consult, 85- *Mem:* Am Chem Soc; Sigma Xi; fel Am Inst Chemists. *Res:* Infrared and ultraviolet studies of inorganic coordination compounds in the solid state; metal hydrides, boranes; catalysis high vacuum techniques; high energy fuels; high energy oxidizers; fluorine and pesticides; borazine polymers; hypochlorites; sanitizers; pollution control; water purification; swimming pool treatment; product liability; laboratory and plant safety; disposal of hazardous chemicals. *Mailing Add:* 100 Chestnut Lane Hamden CT 06518

FAUST, JOHN WILLIAM, JR, b Pittsburgh, Pa, July 25, 22; m 47; c 8. SURFACE PHYSICS & CHEMISTRY, CRYSTAL GROWTH. *Educ:* Purdue Univ, BS, 44; Univ Mo, MA, 49, PhD(phys chem), 51. *Prof Exp:* Res engr, Westinghouse Elec Corp, 51-59, supvry engr, 59-63, mgr, Mat Characterization Lab, 63-65, mgr semiconductor crystal growth, 65-67; prof solid state sci, Mat Res Labs, Pa State Univ, 67-69; PROF ENG, UNIV SC, 69-, GRAD DIR & ASSOC CHMN, ELEC & COMPUTER ENG DEPT, 90- *Concurrent Pos:* Consult, Silag Corp, 79-81 & res & develop; res physicist, Naval Res Labs, Washington, DC, 80-81, Morgan Semiconductor div, 82-88; chmn, Int Comt Silicon Carbide, 69-75. *Honors & Awards:* Gordon Conf Cert Recognition. *Mem:* Am Inst Mining, Metall & Petrol Engrs; Electrochem Soc; Sigma Xi; fel Am Inst Chemists; Am Chem Soc; Int Soc Hybrid Micro-electronics; Mat Res Soc. *Res:* Etching and surface phenomena of semiconductors, metals and ceramics; growth of metal and semiconductor crystals; submicron electronics processing; characterization of materials; solar cell materials and processing. *Mailing Add:* Swearingen Eng Bldg Univ SC Columbia SC 29208

FAUST, MARIA ANNA, b Budapest, Hungary; US citizen. MARINE MICROBIOLOGY, MARINE PHYCOLOGY. *Educ:* Agr Univ Budapest, BS, 51; Rutgers Univ, MS, 62; Univ Md, PhD(microbiol), 70. *Prof Exp:* Biologist, Ethicon, Inc, NJ, 59-61; res asst soil microbiol, Rutgers Univ, 61-62; res assoc microbiol, Cornell Univ, 62-66; teaching asst, Univ Md, 67-70; res assoc phycologist, Radiation Biol Lab, 71-72, microbiologist, Chesapeake Bay Ctr, Environ Studies, 73-88, DEPT BOT, MUS SUPPORT CTR, SMITHSONIAN INST, 88- *Mem:* Sigma Xi; Am Soc Microbiol; Phycol Soc Am; Am Soc Limnol & Oceanog; Estuarine Res Fed. *Res:* Structure and function of phytoplankton and bacterial communities in a watershed-estuarine ecosystem; effects of environmental stress on microorganisms; ultrastructure of flagellates and nannoplankton; photoadaptation of phytoplankton. *Mailing Add:* Dept Bot Mus Support Ctr Smithsonian Inst 4201 Silverhill Rd Sutland ND 20560

FAUST, MIKLOS, b Nagybereny, Hungary, Dec 12, 27; m 54; c 1. POMOLOGY. *Educ:* Agr Univ Budapest, BS, 52; Rutgers Univ, MS, 60; Cornell Univ, PhD(pomol), 65. *Prof Exp:* Mgr, Csaszartoltes State Farm, Hungary, 52-54, regional supvr, Ministry of State Farms, 55-57; res assoc, Rutgers Univ, 58-60; res horticulturist, United Fruit Co, NY, 60-62; res assoc, Cornell Univ, 63-65 & NY State Agr Exp Sta, 65-66; res assoc, 66-69, leader pome fruit invests, 69-72, CHIEF FRUIT LAB, BELTSVILLE AGR RES CTR, AGR RES SERV, USDA, 73- *Concurrent Pos:* Pres, Plant Nutrit Coun, 82-86. *Honors & Awards:* Gourley Award, Am Soc Hort Sci, 66, 68. *Mem:* Am Soc Hort Sci; Int Soc Hort Sci. *Res:* Postharvest physiology; biochemistry of fruits; metabolic changes in fruits exposed to different environmental conditions. *Mailing Add:* Agr Res Serv USDA Beltsville MD 20705

FAUST, RICHARD AHLVERS, b Terre Haute, Ind, Sept 6, 21; m 43; c 2. MICROBIOLOGY. *Educ:* Purdue Univ, BS, 48, MS, 52, PhD(bact), 58. *Prof Exp:* Instr bact, Purdue Univ, 51-55; from asst prof to prof, 58-86, EMER PROF MICROBIOL, UNIV MONT, 86- *Mem:* AAAS; Am Soc Microbiol. *Res:* Physiology of Bordetella pertussis; microbial ecology of alpine soils. *Mailing Add:* Dept Microbiol Univ Mont Missoula MT 59812

FAUST, RICHARD EDWARD, b Greenfield, Mass, Oct 26, 27; m 53; c 3. PHARMACEUTICAL CHEMISTRY. *Educ:* Mass Col Pharm, BS, 51; Purdue Univ, MS, 53, PhD(pharmaceut chem), 55; Columbia Univ, MBA, 68. *Hon Degrees:* Dr Pharm, Mass Col Pharm, 90. *Prof Exp:* Res assoc, Sterling-Winthrop Res Inst, 54-55; asst prof pharm, Ferris State Col, 55-57; dir res, Potter Drug & Chem Corp, 57-61; mgr new prod creation, Merck & Co, 61-63; asst dir res, Johnson & Johnson Res Ctr, 63-66, consult, 66-67; corp planning mgr, 68-69, dir res planning & develop, Hoffmann-La Roche, Inc, 69-85; PRES, AM FOUND PHARMACEUT EDUC, 86- *Concurrent Pos:* Prof lectr, Mass Col Pharm, 58-61; adj prof, Fairleigh Dickinson Univ, 76- *Mem:* AAAS; Am Pharmaceut Asn; Am Chem Soc; Am Asn Pharmaceut Scientists; Acad Pharmaceut Sci; Am Asn Pharmaceut Scientists; Am Asn Cols Pharm; Drug Info Asn. *Res:* Cosmetic and dermatologic preparations; topical therapeutics; toiletries; soaps; research administration and planning. *Mailing Add:* 80 Bayberry Lane Watchung NJ 07060

FAUST, ROBERT GILBERT, b Brooklyn, NY, Nov 9, 32; m 56; c 1. CELL PHYSIOLOGY, BIOPHYSICS. *Educ:* NY Univ, AB, 53; Univ Southern Calif, MS, 57; Princeton Univ, PhD(biol), 60. *Prof Exp:* Asst zool, Univ Southern Calif, 55-57; asst biol, Princeton Univ, 57-59; res assoc biophys, Harvard Med Sch, 62-63; asst prof physiol, 63-68, dir space sci prog, 68-72, assoc prof, 68-75,dir grad studies, 84-86, PROF PHYSIOL, SCH MED, UNIVNC, CHAPEL HILL, 75-,. *Concurrent Pos:* NIH fel, Oxford Univ, 60-62 & Harvard Med Sch, 62-63; NIH grants, 64-83; NASA grant, 68-72; mem physiol study sect, NIH, 70-74; NIH sr int fel, Max-Planck Inst Biophys, Frankfurt, Ger, 76-77; Erna & Jakob Michael vis prof, Weizman Inst Sci, Rehovot, Israel, 82. *Mem:* AAAS; Am Physiol Soc. *Res:* Permeability of cells and tissues; mechanisms of active transport of electrolytes and non-electrolytes; use of model and reconstituted systems to interpret solute penetration into cells and tissues. *Mailing Add:* Dept Physiol Univ NC Sch Med Chapel Hill NC 27514

FAUST, SAMUEL DENTON, b Shiloh, NJ, Aug 11, 29; m 60; c 1. ENVIRONMENTAL SCIENCES. *Educ:* Gettysburg Col, BS, 50, PhD(environ sci), 58. *Prof Exp:* Chemist, W A Taylor & Co, Md, 50-54; res fel, 54-58, from asst prof to assoc prof, 58-65, prof environ sci, 65-80, PROF ENVIRON SCI, RUTGERS UNIV, 80- *Mem:* Am Chem Soc; Am Soc Limnol & Oceanog; Am Water Works Asn; Am Geophys Union; Am Soc Testing & Mat. *Res:* Water chemistry; water quality management; water resources; acidic deposition. *Mailing Add:* PO Box 94 Changewater NJ 07831-0094

FAUST, SISTER CLAUDE MARIE, b San Antonio, Tex, Nov 18, 17. MATHEMATICS. *Educ:* Incarnate Word Col, BA, 39; Cath Univ Am, MA, 54; Marquette Univ, MS, 55; Univ Notre Dame, PhD(math), 61. *Prof Exp:* From instr to prof math, Incarnate Word Col, 46-88, Minnie Stevens Piper prof, 67, dir, NSF In-Serv Insts High Sch Teachers Math, 63-71; RETIRED. *Concurrent Pos:* Vis scientist, NSF Vis Scientist Prog, Tex Acad Sci, 64-65; dir, Tex Ctr for Minn Math & Sci Teaching Proj, NSF, 63-68. *Mem:* Math Asn Am; Am Math Soc. *Res:* Complex analysis; boundary behavior of holomorphic functions in the unit disc. *Mailing Add:* Incarnate World Motherhouse 4707 Broadway San Antonio TX 78209

FAUST, WALTER LUCK, b Benton, Ark, Feb 13, 34; m 57; c 3. LASERS, EXPERIMENTAL PHYSICS. *Educ:* Columbia Col, AB, 56; Columbia Univ, PhD(physics), 61. *Prof Exp:* Mem tech staff, Bell Labs, 61-67; from assoc prof to prof physics & elec eng, Univ Southern Calif, 67-72; head, Optical Physics Br, 72-75, SR SCIENTIST, OPTICAL SCI DIV, NAVAL RES LAB, 75- *Honors & Awards:* Alan Berman Res Award, 74, 89. *Mem:* Fel Am Phys Soc; Am Chem Soc. *Res:* Short pulse optical spectroscopy. *Mailing Add:* Naval Res Lab Code 6540 Washington DC 20375

FAUST, WILLIAM R, b Shawnee, Okla, Mar 9, 18; m 42; c 3. NUCLEAR PHYSICS, PLASMA PHYSICS. *Educ:* Okla State Univ, BS, 39; Ill Inst Technol, MAS, 41; Univ Md, PhD(physics), 49. *Prof Exp:* Physicist, US Naval Res Lab, 41-73; RETIRED. *Honors & Awards:* Hulbert Award. *Mem:* Am Phys Soc. *Res:* Electrodynamics of plasma; shockwaves in plasma; neutron cross-sections for nuclear physics; author of approx 50 papers in various journals and extensive naval research lab reports. *Mailing Add:* 5907 Walnut St Temple Hills MD 20748-4843

FAUSTO, NELSON, b Sao Paulo, Brazil, Dec 12, 36; m 66. BIOCHEMICAL PATHOLOGY. *Educ:* Univ Sao Paulo, Brazil, BS, 54, MD, 60. *Prof Exp:* Asst prof, Dept Histol & Embroyol, Med Sch, Univ Sao Paulo, 61; res assoc, Dept Path & Regional Primate Ctr, Med Sch, Univ Wis, 62-63, res assoc & fel, Damon Runyon Mem Fund Cancer Res, 64-65, instr med sci, 66; asst prof, 66-71, assoc prof, 71-75, PROF MED SCI, DEPT PATH, BROWN UNIV, 75-, CHMN, SECT PATH, 82- *Concurrent Pos:* Mem, Path B Study Sect, NIH, 75-79, chmn, Clin Sci Fel Review Comt, Div Cancer Biol & Diag, 80-81, consult, Bd Sci Counselors, Nat Cancer Inst, 81; vis prof, Dept Path, MD Anderson Hosp & Tumor Inst, Univ Tex, 79. *Mem:* Am Soc Cell Biol; Am Asn Pathologists; NY Acad Sci. *Res:* Gene expression in regenerating, neoplastic and fetal liver. *Mailing Add:* Dept Pathol & Lab Med Brown Univ Div Biol & Med Providence RI 02912

FAUSTO-STERLING, ANNE, b New York City, NY, July 30, 44; m 66. EMBRYOLOGY, GENDER. *Educ:* Univ Wis, BA, 65; Brown Univ, PhD(develop genetics), 70. *Prof Exp:* From asst prof to assoc prof, 71-86, PROF BIOL & MED, BROWN UNIV, 86- *Concurrent Pos:* Vis prof, Univ Amsterdam, 86. *Mem:* Soc Develop Biol; Cell Biol; Nat Women's Studies Asn; Int Soc Develop Biol; AAAS; Hist Sci Soc; Asn Women Sci. *Res:* Developmental biology; biological theories about women; gender, race and science. *Mailing Add:* Div Biol & Med Brown Univ Box G Providence RI 02912

FAUT, OWEN DONALD, b Allentown, Pa, July 8, 36; m 59; c 4. INORGANIC CHEMISTRY. *Educ:* Muhlenburg Col, BS, 58; Mass Inst Technol, PhD(chem), 62. *Prof Exp:* From asst prof to assoc prof chem, Hanover Col, 62-67; assoc prof, 67-77, PROF CHEM, WILKES UNIV, 77-, PROF ENG, 82-, CHMN, 90- *Concurrent Pos:* Goddard Space Flight Ctr, 79, 80, Lewis Res Ctr, 81-82, 83, 84 & 86-88; Sr Assoc, Nat Res Coun, 81-82 & 86-88. *Mem:* Am Chem Soc; Royal Soc; Sigma Xi. *Res:* Electronic and molecular structures of first row transition metal compounds; tribiology. *Mailing Add:* Dept Chem Wilkes Univ Wilkes-Barre PA 18766

FAUTH, DAVID JONATHAN, b Erie, Pa, Aug 9, 51; m 76; c 2. INORGANIC CHEMISTRY, ORGANOMETALLIC CHEMISTRY. *Educ:* Thiel Col, BA, 73; Univ SC, PhD(chem), 78. *Prof Exp:* Res chemist nuclear chem, Tech Qual Assurance & Qual Control, Savannah River Plant, E I du Pont de Nemours & Co, Inc, 80-90; MGR ANAL & ENVIRON LABS, WESTVALLEY NUCLEAR DIV, WESTINGHOUSE ELEC CORP, 90- *Mem:* Am Chem Soc. *Res:* Metal cluster chemistry, catalysis, metal carbonyl, thiocarbonyl and nitrosyl chemistry, phase transfer catalysis; solvent extraction. *Mailing Add:* Westvalley Nuclear Westinghouse Elec Corp PO Box 191 West Valley NY 14171-0191

FAUTH, MAE IRENE, b Wrightsville, Pa, June 12, 13. ANALYTICAL CHEMISTRY, RESOURCE MANAGEMENT. *Educ:* Lebanon Valley Col, BS, 33; Columbia Univ, AM, 46; Pa State Univ, PhD(chem), 55. *Prof Exp:* Attend nursing, Wernersville State Hosp, Pa, 35-43; asst engr elec eng, Western Elec Co, NJ, 43-45; head, dept sci, NY Pub Sch, 46-47; instr chem, Hazleton Ctr, Pa State Univ, 47-49, instr, Univ, 49-55; mgr, Anal Br, Naval Ord Sta, Indian Head, Md, 55-72; asst to pres environ res, Charles County Community Col, Md, 72-73; head pollution control group, Naval Surface Weapons Ctr, White Oak Lab, Indian Head, 73-90; RES CHEMIST, NAVAL ORD STA, INDIAN HEAD, MD, 90- *Concurrent Pos:* Prof lectr, Charles County Community Col, 59-83. *Mem:* Am Chem Soc; Am Defense Preparedness Asn. *Res:* Chemistry of rocket fuels; analysis and evaluation of solid propellants; propellant reclamation; pollution abatement; environmental effects of chemicals; environmental fate of energetic materials. *Mailing Add:* Box 217 Indian Head MD 20640-0217

FAUTIN, DAPHNE GAIL, b Urbana, Ill, May 25, 46; m 86. MARINE INVERTEBRATE ZOOLOGY. *Educ:* Beloit Col, BS, 66; Univ Calif, Berkeley, PhD(zool), 72. *Prof Exp:* NIH fel marine pharmacol, Univ Hawaii Sch Med, 72-73; res assoc zool, Univ Malaya, 73-75; res biologist, Calif Acad Sci, 75-80, asst cur invert zool, 80-84, assoc cur, 84-89, cur, 89-90, dir res, 88-90; ADJ PROF, UNIV KANS, 90- *Concurrent Pos:* NSF grant, Calif Acad Sci, 77-79; from actg asst prof to actg assoc prof, Beloit Col, 83-84; counr, Soc Syst Biol, 87-90. *Mem:* Soc Syst Biol; Sigma Xi; Am Soc Zoologists; Coun Biol Ed; Int Soc Reef Studies. *Res:* Marine invertebrate reproduction particularly brooding and biogeography; systematics of coelenterates, especially anthozoa; marine symbiology; tropical marine ecosystems. *Mailing Add:* Systematics & Ecol Snow Hall Univ Kans Lawrence KS 66045

FAUVER, VERNON A(RTHUR), b Hammond, Ind, Mar 25, 28; m 47; c 2. CHEMICAL ENGINEERING. *Educ:* Purdue Univ, BS, 52, MS, 53. *Prof Exp:* Chem engr, Eastman Kodak Co, NY, 53; chem engr, Styrene Polymerization Lab, 54-58, Process Eng Dept, 58-59 & C J Strosacker Res & Develop Lab, 59-61, res engr, Edgar C Britton Res Lab, 61-66, process specialist, Process Eng Dept, 66-72, sr process specialist, process eng dept, 72-83, SR PROCESS SPECIALIST, SPECIALTY PROD DEPT, DOW CHEM USA, 83- *Mem:* Am Inst Chem Engrs; Am Chem Soc; Nat Soc Prof Engrs. *Res:* Process development; reaction engineering; separations processes; plant start-up; fluid flow/pumps. *Mailing Add:* Dow Chem USA Mich Div Bldg 1261 Midland MI 48667

FAVALE, ANTHONY JOHN, b New York, NY, Feb 28, 35; m 72; c 3. NUCLEAR PHYSICS. *Educ:* Polytech Inst Brooklyn, BS, 56; NY Univ, MS, 63. *Prof Exp:* Res physicist, 56-62, head high energy physics & astrophysics, 62-74, HEAD, ENERGY RES OFF, GRUMMAN AEROSPACE CORP, GRUMMAN CORP, 74-, DEP DIR FUSION, 81- *Concurrent Pos:* Guest scientist, Reactor Div, Brookhaven Nat Lab, 61-64; lectr, C W Post Col, 64- *Mem:* Am Phys Soc; Inst Elec & Electronics Engrs; Am Nuclear Soc. *Res:* High energy physics experiments in space; gamma ray astronomy; solar energy research; fusion energy development. *Mailing Add:* 22 Major Trescott Lane Northport NY 11768

FAVAZZA, ARMANDO RICCARDO, b New York, NY, Apr 14, 41; m 71; c 2. PSYCHIATRY. *Educ:* Columbia Univ, BA, 62; Univ Va, MD, 66; Univ Mich, MPH, 71. *Prof Exp:* Resident psychiat, Univ Mich, 66-71; staff psychiatrist, US Naval Hosp, Oakland, Calif, 71-73; assoc prof, 73-80, PROF PSYCHIAT, UNIV MO-COLUMBIA, 80- *Concurrent Pos:* Psychiat consult, Vet Admin Hosp, Columbia, 73-; assoc ed, MD Mag, 73-; ed-in-chief, J Oper Psychiat, 73- *Honors & Awards:* George Kunkel Award Advances in Med Sci, Harrisburg Hosp. *Mem:* Fel Am Psychiat Asn; fel Am Col Psychiatrists; fel Am Asn Social Psychiat; Am Anthrop Asn; Sigma Xi. *Res:* Cultural psychiatry; deliberate self-harm. *Mailing Add:* Dept Psychiat Univ Mo Sch Med Columbia MO 65201

FAVORITE, FELIX, b Quincy, Mass, Mar 18, 25; m 51; c 4. OCEANOGRAPHY, ECOLOGY. *Educ:* Mass Maritime Acad, BS, 46; Univ Wash, BS, 55, MS, 65; Ore State Univ, PhD(oceanog, meteorol), 68. *Prof Exp:* Res asst oceanog, Univ Wash, 55-57; oceanogr phys oceanog, Seattle Biol Lab, Bur Com Fisheries, Northwest & Alaska Fisheries Ctr, Nat Marine Fisheries Serv, 57-59, chief oceanog invest, 59-70, prog dir oceanog, Seattle Biol Lab, Nat Marine Fisheries Serv, 70-75, coordr resource ecol, 75-80; CONSULT OCEANOGR, 80- *Concurrent Pos:* Expert oceanog, Int NPac Fish Comn, 57-76; partic, US-Japan Coop Sci Prog & US-USSR Oceanog Exchange Prog, 64 & US-Japan Bering Sea Prog & Nat Oceanic & Atmospheric Admin-Bur Land Mgt Outer Continental Shelf Environ Assessment Prog, 73-80; res consult, 81- *Honors & Awards:* Silver Medal, US Dept Com, 73. *Mem:* Sigma Xi; Oceanog Soc Japan; Am Inst Fishery Res Biologists. *Res:* Physical oceanography and resource ecology studies in the North Pacific Ocean and Bering Sea. *Mailing Add:* 16103 41st N E Seattle WA 98155

FAVORITE, JOHN R, b Muskegon, Mich, June 26, 16; m 42; c 2. CHEMICAL ENGINEERING. *Educ:* Purdue Univ, BS, 38. *Prof Exp:* Engr, Goodyear Tire & Rubber Co, 38-49 & Clopay Co, 49-50; supvr, Thermo-Fax prod develop, 50-55, tech dir, Duplicating Prod Div, 55-64, group mkt mgr duplicating & microfilm prod, Int Div, 64-67, PROJ MGR PHOTOG PROD DIV, MINN MINING & MFG CO, 67- *Mem:* Am Chem Soc; Am Inst Chem Engrs. *Res:* Product research and development. *Mailing Add:* 12062 Verano Ct San Diego CA 92128

FAVOUR, CUTTING BROAD, b Toreva, Ariz, July 19, 13; m 41, 74; c 3. MEDICINE. *Educ:* Hendrix Col, AB, 36; Johns Hopkins Univ, MD, 40; Am Bd Internal Med, dipl, 54. *Prof Exp:* Intern, Osler Wards, Hopkins Hosp, 40-41; asst resident, Peter Bent Brigham Hosp, Boston, 41-42, resident, 42-43; instr & asst, Harvard Med Sch, 43-47, assoc, 47-54; asst clin prof, Med Sch, Stanford Univ, 55-60; prof prev med & chmn dept, Sch Med, Georgetown Univ, 60-62; chief dept physiol, Nat Jewish Hosp, Denver, 62-64, chief dept exp epidemiol, 64-66; dir med educ, St Mary's Hosp, 66-70; med dir, Kaiser Industs Corp, 70-73; dir ambulatory serv, San Joaquin County Gen Hosp, 73; dir, Ambulatory Serv, Scenic Gen Hosp, 73-74; ASSOC CLIN PROF MED, SCH MED, UNIV CALIF, SAN FRANCISCO, 75- *Concurrent Pos:* Head dept immunol, Palo Alto Med Res Found, 54-60; asst vis physician, Stanford Serv, City & County Hosp, San Francisco, 55-60; mem staff, Palo Alto Hosp, 55-56; lectr & consult, US Navy Hosp, Oak Knoll, Calif, 56-60; chief, Georgetown Med Serv, Washington, DC, Gen Hosp, 60-62; mem attend staff, St Mary's Hosp, San Francisco, 66-74, consult staff, 75-, San Francisco Gen Hosp, 69-73, Highland Gen Hosp, Oakland, 70-73 & St Joseph's Hosp, 71-73; mem, Emergency Med Systs, Inc. 73-; active staff, Scenic Gen Hosp, Modesto, 73-; courtesy staff, San Francisco Gen Hosp & Mem Hosps, Modesto, 76-; consult staff, Oak Valley District Hosp, Oakdale, 74-75, active staff, 76-; med dir, Driftwood Convalescent Hosp, Modesto, 76- & Oakdale Convalescent Hosp, Oakdale, 79-; pvt pract, Oakdale 76-; consult tuberculosis, Pub Health Dept Stanislaus County, Modesto, 77- *Mem:* Soc Exp Biol & Med; fel Am Col Physicians; Am Thoracic Soc; affil Royal Soc Med; Am Rheumatism Asn. *Res:* Immunology and microbiology applied to clinical medicine. *Mailing Add:* PO Box 399 Oakdale CA 95361

FAVRE, HENRI ALBERT, b Payerne, Switz, Dec 4, 26; m 58; c 3. ORGANIC CHEMISTRY. *Educ:* Swiss Fed Inst Technol, Ing Chem Dipl, 48, DrSc, 51. *Prof Exp:* Brit Coun student, Sheffield Univ, 51-52; from asst prof to assoc prof org chem, 52-60, dir dept chem, 59-63, PROF ORG CHEM, UNIV MONTREAL, 60- *Concurrent Pos:* mem, Comn Nomenclature Org Chem, Int Union Pure & Appl Chem. *Mem:* Chem Inst Can. *Res:* Stereochemistry; nomenclature. *Mailing Add:* Univ Montreal Dept Chem PO Box 6128 Montreal PQ H3C 3T7 Can

FAVRET, A(NDREW) G(ILLIGAN), b Cincinnati, Ohio, May 9, 25; m 49; c 11. ELECTRONICS. *Educ:* US Mil Acad, BS, 45; Univ Pa, MS, 50; Cath Univ Am, DEng(elec eng), 64. *Prof Exp:* Staff mem, Lincoln Lab, Mass Inst Technol, 54-55; actg dir planning, Defense Prod Group, Am Mach & Foundry Co, 55-56, mgr syst anal dept, Alexandria Div, 56-59; sr sci adv to asst chief staff intel, US Dept Army, 59-63; assoc prof elec eng, Cath Univ Am, 63-67, dir, comput ctr, 68-73, dean, Sch Eng & Archit, 81, 84-88, PROF, 67-90, EMER PROF ELEC ENG, CATH UNIV AM, 90- *Concurrent Pos:* Sr analyst, Cent Intel Agency, 73-78; mem, Army Sci Bd. *Mem:* Sr mem Inst Elec & Electronic Engrs; Asn Comput Mach. *Res:* Digital computer applications; statistical decision theory; signal processing; digital computer systems; biomedical instrumentation; computer simulation. *Mailing Add:* 13290 Scott Rd Waynesboro PA 17268-9557

FAVRO, LAWRENCE DALE, b Pittsburgh, Pa, Apr 17, 32; m 57; c 2. THEORETICAL PHYSICS. *Educ:* Harvard Univ, AB, 54, AM, 55, PhD(physics), 59. *Prof Exp:* Instr physics, Columbia Univ, 59-62; from asst prof to assoc prof, 62-72, PROF PHYSICS, WAYNE STATE UNIV, 72- *Mem:* Am Phys Soc. *Res:* Stochastic processes; statistical theory of energy levels; coherent processes in particle beams; acoustics; optics; photoacoustics. *Mailing Add:* Dept Physics Wayne State Univ Detroit MI 48202

FAW, RICHARD E, b Adams Co, Ohio, June 22, 36; m 61; c 2. CHEMICAL & NUCLEAR ENGINEERING. *Educ:* Univ Cincinnati, BS, 59; Univ Minn, PhD(chem eng), 62. *Prof Exp:* From asst prof to assoc prof, 62-68, PROF NUCLEAR ENG, KANS STATE UNIV, 68- *Mem:* Am Nuclear Soc; Am Soc Eng Educ. *Res:* Radiation protection; nuclear reactor safety. *Mailing Add:* Nuclear Eng Dept Kansas State Univ Manhattan KS 66506

FAW, WADE FARRIS, b Eubank, Ky, Feb 23, 42; m 65; c 1. AGRONOMY, PLANT PHYSIOLOGY. *Educ:* Berea Col, BS, 65; WVa Univ, PhD(agron), 69. *Prof Exp:* Trainee farm planning, US Soil Conserv Serv, 63-65; asst prof agron, Rice Br Exp Sta, Univ Ark, 69-74; exten agronomist, Auburn Univ, 74-75; assoc prof & chmn, Plant Sci Dept, Tenn Technol Univ, 75-77; exten agronomist, Auburn Univ, 77-80; ASSOC SPECIALIST AGRON, LA STATE UNIV, 80- *Mem:* Am Soc Agron; Crop Sci Soc Am; Am Forage & Grassland Coun; Weed Sci Soc Am. *Res:* Response of plants to their environment; physiology of grain and forage crops. *Mailing Add:* 246 Rue de La Place Baton Rouge LA 70810

FAWAZ, GEORGE, b Deirminas, Lebanon, Nov 22, 13; nat US; m 46; c 2. PHARMACOLOGY, BIOCHEMISTRY. *Educ:* Am Univ, Beirut, AB, 33, MS, 35; Graz Univ, PhD (org chem), 36; Univ Heidelberg, MD, 55. *Prof Exp:* From instr to assc prof pharmacol, 39-53, PROF PHARMACOL, AM UNIV, BEIRUT, 53- *Concurrent Pos:* Rockefeller fel, Harvard Univ, 46-47. *Mem:* Corresp mem Ger Pharmacol Soc. *Res:* Organic phosphorus compounds of biological interest; cardiac and renal pharmacology and metabolism; synthetic antimalarials. *Mailing Add:* Dept Pharmacol Am Univ Beirut Beirut Lebanon

FAWCETT, COLVIN PETER, b Blyth, Eng, Feb 16, 35; m 61; c 2. ENDOCRINOLOGY, NEUROENDOCRINOLOGY. *Educ:* Univ Durham, BSc, 56; Univ Newcastle, PhD(org chem), 59. *Prof Exp:* Fel biochem, Brandeis Univ, 59-61; mem sci staff, Nat Inst Med Res, Eng, 61-66; vis asst prof neurochem, Western Reserve Univ, 66-67; asst prof physiol, 67-75, dir, grad prog physiol & biophysics, 75-83, ASSOC PROF PHYSIOL, UNIV TEX HEALTH SCI CTR, DALLAS, 75- *Concurrent Pos:* Assoc ed, Neuroendocrinol. *Mem:* Endocrine Soc; Am Physiol Soc; Int Soc Neuroendocrinol. *Res:* Neuropeptides and control of hormonal secretion from endocrine pancreas; isolation and characterization of peptides from the hypothalamus. *Mailing Add:* Dept Physiol Univ Tex Southwestern Med Ctr 5323 Harry Himes Blvd Dallas TX 75235

FAWCETT, DON WAYNE, b Springdale, Iowa, Mar 14, 17; m 42; c 4. ANATOMY, HISTOLOGY. *Educ:* Harvard Univ, AB, 38, MD, 42. *Hon Degrees:* DSc, Univ Siena, 74, NY Med Col, 75, Univ Chicago & Univ Cordoba, 78, Georgetown Univ, 85; DVetMed, Justus Liebig Univ, 77; MD, Univ Heidelberg, 77. *Prof Exp:* Surg intern, Mass Gen Hosp, 42-43; from instr to asst prof anat, Harvard Med Sch, 46-55; prof & head dept, Med Col, Cornell Univ, 55-59; cur, Warren Anat Mus, 61-70, sr assoc dean preclin affairs, Harvard Med Sch, 75-77, Hersey prof anat & head dept, 59-81, James Stillman prof comp anat, 62-81; sr scientist, Int Lab Res Animal Dis, Nairobi, Kenya, 80-85; RETIRED. *Concurrent Pos:* Res fel anat, Harvard Med Sch, 46; Markle scholar med sci, 49-54; Lederle Med Fac Award, 54-56; consult, NIH, 55-59, 64-; emer lectr vet, Univ Nairobi, Kenya, 81- *Honors & Awards:* Ferris Lectr, Yale Univ, 57, Phillips Lectr, Haverford Col, 58, Christiana Smith Lectr, Mt Holyoke Col, 66, Charnock Bradley Lectr, Royal (DIC) Sch Vet Studies, Edinburgh, Sigmund Pollitzer Lectr, NY Univ, Adam Miller Lectr, State Univ NY NY Downstate Med Ctr, 69, Robert Terry Lectr, Wash Univ, Daniel Kempner Lectr, Univ Tex Med Br, Galveston & Harold Chaffer Lectr, Univ Otago, NZ, 70; Henry Gray Award, Am Asn Anatomists, 83 & Centennial Medal, 87; Carl Hartman Award, Soc Study Reproduction, 85; Distinguished Scientist Award Life Sci, Electron Micros Soc Am, 89. *Mem:* Nat Acad Sci; Fedn Socs Electron Micros (pres, 75-78); Am Asn Anat (1st vpres, 59-60, pres, 65-66); Tissue Cult Asn (vpres, 54-55); Am Soc Cell Biol (pres, 61-62). *Res:* Electron microscopy; cytology; growth and differentiation; spermatogenesis; histophysiology of male reproductive tract; ultrastructure of cardiac muscle; ultrastructure of liver; host-parasite relations in Theileriosis. *Mailing Add:* 1224 Lincoln Rd Missoula MT 59802

FAWCETT, ERIC, b Blackburn, Eng, Aug 23, 27; m 54; c 3. EXPERIMENTAL MAGNETISM & METAL PHYSICS, PEACE RESEARCH. *Educ:* Cambridge Univ, MA, 52, PhD(physics), 54. *Prof Exp:* Fel, Div Low Temperature & Solid State Physics, Nat Res Coun Can, 54-56; sci officer physics, Royal Radar Estab, Eng, 56-61; mem tech staff, Bell Tel Labs, Inc, 61-70; PROF PHYSICS, UNIV TORONTO, 70- *Concurrent Pos:* Assoc prof, Univ de Paris, France, 67; vis prof, Univ Nagoya, Japan, 73 & 83, Technion, Haifa, Israel, 78, Univ Kyoto, Japan, 82 & H C Oersted Inst Denmark, 85; founding chmn, Canadian Comt Scientists & Scholars, 80-84, foreign secy, 84-; founding pres, Sci for Peace, 81-84, vpres, 90-; vis scientist, Inst Laue-Langevin, France, 83, Commonwealth Sci & Indust Res Orgn, Sydney, Australia, 84-85, Oersted Lab, Univ Copenhagen, 85, Rand Afrikaans Univ, Johannesburg, SAfrica, 89, Inst Phys Probs, Moscow, USSR, 90 & BARC, Bombay, India, 91. *Mem:* Fel Am Phys Soc; Can Asn Physicists; fel Brit Inst Physics & Phys Soc; Europ Physics Soc; fel AAAS; Australia Asn Advan Sci; NZ Asn Advan Sci. *Res:* Experimental study of electronic structure of metals, including neutron scattering, thermal expansion, magnetostriction and sound velocity, especially in transition and magnetically ordered metals. *Mailing Add:* Dept Physics Univ Toronto Toronto ON M5S 1A7 Can

FAWCETT, JAMES DAVIDSON, b New Plymouth, Taranaki, NZ. REPRODUCTIVE BIOLOGY OF REPTILES, ANATOMY OF REPTILES. *Educ:* Univ NZ, BSc, 60; Univ Auckland, MSc, 64; Univ Colo, PhD(biol), 75. *Prof Exp:* Head, Dept Biol, Kings Col, Auckland, 60-61; demonstr biol, Univ Auckland, NZ, 61-62, sr demonstr zool, 63-64; teaching asst zool, Univ Ill, 65-67, Univ Colo, 68-69; instr biol, 72-75, asst prof, 75-81, ASSOC PROF BIOL, UNIV NEBR, OMAHA, 81- *Mem:* AAAS; Am Soc Zoologists; Soc Study Amphibians & Reptiles; Am Soc Ichthyologists & Herpetologists; Sigma Xi. *Res:* Aspects of the reproductive biology of reptiles; compiling a synopsis of the New Zealand herpetofauna. *Mailing Add:* Dept Biol Univ Nebr Omaha NE 68182-0040

FAWCETT, JAMES JEFFREY, b Blyth, Eng, July 6, 36; m 60; c 2. GEOLOGY. *Educ:* Univ Manchester, BSc, 57, PhD(geol), 61. *Prof Exp:* Asst geol, Univ Manchester, 60-61; fel, Carnegie Inst Geophys Lab, 61-64; assoc chmn dept, 70-75, assoc dean, Sch Grad Studies, 77-80, PROF GEOL, UNIV TORONTO, 64-, assoc dean sci, 80-86, vice prin acad, Erindale Col, 85-86. *Mem:* Am Geophys Union; Mineral Soc Am; Geol Asn Can; Mineral Soc Gt Brit & Ireland; Mineral Asn Can. *Res:* Application of high temperature and pressure studies of rocks and minerals to problems of igneous and metamorphic petrogenesis. *Mailing Add:* Dept Geol Univ Toronto Toronto ON M5S 1A1 Can

FAWCETT, MARK STANLEY, b Jamestown, NDak, Oct 17, 32; m; c 2. ORGANIC CHEMISTRY. *Educ:* Northwestern Univ, BS, 54; Univ Minn, PhD(org chem), 58. *Prof Exp:* Res chemist, Elastomers Dept, E I du Pont de Nemours & Co, Inc, Wilmington, 58-64, mkt develop asst, 64-70, mem staff mkt res, 70-80, mkt res prog mgr, 80-85; CONTRACT MKT RES, FAWCETT ASSOCS, 85- *Res:* Structure of polyphenyl cyclopentadienes; elastomeric polymers. *Mailing Add:* Arthur Dr Wellington Hills Hockessin DE 19707

FAWCETT, NEWTON CREIG, b Fargo, NDak. NUCLEIC ACIDS, PIEZOELECTRIC DETECTION. *Educ:* Univ Denver, BS, 64; Univ NMex, MS, 72, PhD(chem), 73. *Prof Exp:* Staff asst, Sandia Corp, 65-68; staff mem, Los Alamos Sci Lab, 72-75; asst prof chem, Southwest Tex State Univ, 75-76; from asst prof to assoc prof, 76-85, PROF CHEM UNIV SOUTHERN MISS, 86- *Concurrent Pos:* Consult, Tex Res Inst, 76-; Robert A Welch fel, Southwest Tex State Univ, 76; consult, INDAL Aluminum Corp, 85- *Mem:* Am Chem Soc. *Res:* Nucleic acid hybridization and detection using piezoelectric crystals; corrosion studies of aluminum and steel. *Mailing Add:* Southern Sta Box 5043 Hattiesburg MS 39406

FAWCETT, RICHARD STEVEN, b Iowa City, Iowa, Apr 26, 48; m 72; c 2. WEED SCIENCE. *Educ:* Iowa State Univ, BS, 70; Univ Ill, PhD(agron), 74. *Prof Exp:* Asst prof res & exten weed control specialist, Dept Agron, Univ Wis-Madison, 74-766; prof & exten weed control specialist, Dept Plant Path Seed & Weed Sci, Iowa State Univ, Ames, 76-87, prof & exten weed scientist, Dept Agron, 87-89; AGR CONSULT & FARM J STAFF ENVIRON SPECIALIST, HUXLEY, IOWA, 89- *Concurrent Pos:* Fac exchange, Univ Costa Rica, San Jose, 80; vis prof, Shenyang Agr Col, Shenyang, Peoples Repub China, 83; bd dirs, Weed Sci Soc Am, 82-85; mem, Sci Adv Panel Groundwater, US Cong Off Technol Assessment, 88-89 & Sci Policy Bd, Am Coun Sci & Health, 89- *Honors & Awards:* Am Soybean Asn Award, 81; Outstanding Extension Award, Weed Sci Soc Am, 85. *Mem:* Weed Sci Soc Am; Am Soc Agron; Crop Sci Soc Am; Am Chem Soc. *Res:* Weed control systems for corn and soybeans in conservation tillage; herbicide interactions with soil microorganisms; selective applicators for herbicides; perennial weed control; ground water contamination by pesticides. *Mailing Add:* Fawcett Consult Rte 1 Box 44 Huxley IA 50124

FAWCETT, SHERWOOD LUTHER, b Youngstown, Ohio, Dec 25, 19; m 53; c 3. PHYSICS, RESEARCH ADMINISTRATION. *Educ:* Ohio State Univ, BS, 41; Case Inst Technol, MS, 48, PhD(physics), 50. *Hon Degrees:* DSc, Ohio State Univ, 71; DPS, Detroit Inst Technol. 74; DL Otterbein Col, 77; Whitman Col, 80; Gonzaga Univ, 82; Ohio Dominican Col, 84. *Prof Exp:* Instr physics, Case Inst Technol, 46-48; physicist, Battelle Mem Inst, 50-52, from asst chief to chief eng mech div, 52-57, from asst mgr to mgr physics dept, 57-62, mgr metall & physics dept, 62-64, dir, Pac Northwest Lab, 65-67, exec vpres, Inst, 67-68, pres, 68-81, chmn & chief exec officer, 81-84, chmn Inst, 84-87; RETIRED. *Honors & Awards:* Medal for Advan Res, Am Soc Metals, 77. *Mem:* Am Phys Soc; Am Nuclear Soc; Am Soc Metals; AAAS (vpres, 70); Metall Soc Am Inst Metall Engrs. *Res:* Reactor engineering; heat transfer; fluid flow; methods of electronic beam ejection from Betatrons; reactor irradiation experiments on reactor fuel elements; research and development management. *Mailing Add:* 2820 Margate Rd Columbus OH 43221

FAWCETT, TIMOTHY GOSS, b Boston, Mass, Nov 18, 53; m 77. STRUCTURAL CHEMISTRY, X-RAY DIFFRACTION ANALYSES. *Educ:* Univ Mass, Amherst, BS, 75; Rutgers Univ, PhD(chem), 79. *Prof Exp:* Teaching asst chem, Douglass Col, 75-78, res asst bioinorg chem, Rutgers Univ, 78-79; sr res chemist, Dow Chem Co, 79-83, proj leader, 83-85, group leader, 85-90, RES MGR, ADVAN CERAMICS LAB, 90- *Concurrent Pos:* Mem, Joint Comt Powder Diffraction Standards, Int Ctr Diffraction Data, 82-, bd dir, 86-88, chmn, Long Range Planning, 90- *Honors & Awards:* V A Stenger Award, Dow Chem Co, 85; IR-100, 87. *Mem:* Am Chem Soc; Sigma Xi. *Res:* Managing several groups developing advanced ceramics materials; development and application of high resolution, x-ray powder diffraction techniques used to analyze structural property relationships; co-inventor of the simultaneous DSC/XRD/MS instrument. *Mailing Add:* Cent Res Advan Ceramics Lab Dow Chem Co 1776 Bldg Midland MI 48674

FAWLEY, JOHN PHILIP, b Auburn, NY, July 23, 45; m 68; c 1. PHYSIOLOGY. *Educ:* Kent State Univ, BS, 67, MS, 70, PhD(physiol), 72. *Prof Exp:* Instr biol, Kent State Univ, 68-72; asst prof biol, 72-80, assoc prof, 80-84, PROF BIOL, WESTMINSTER COL, 84- *Mem:* Nat Speleol Soc; Nat Sci Teachers Asn; Soc Col Sci Teachers; NAm Biospeleol Soc; Am Col Sports Med. *Res:* Responses of rodents to exercise and environmental stress. *Mailing Add:* Dept Biol Westminster Col New Wilmington PA 16142

FAWWAZ, RASHID, b Sao Paulo, Brazil, May 19, 35; US citizen; m 66; c 2. NUCLEAR MEDICINE. *Educ:* Am Univ Beirut, MD, 60; Univ Calif, Berkeley, PhD(med physics), 69. *Prof Exp:* Res physician nuclear med, Donner Lab, Univ Calif, Berkeley, 66-76; from asst prof to assoc prof, 76-85, PROF RADIOL & NUCLEAR MED, COLUMBIA UNIV, 85-; RES COLLABR NUCLEAR MED, BROOKHAVEN NAT LAB, NY, 85- *Concurrent Pos:* Int Atomic Energy fel, Univ Calif, Berkeley, 63-64, Donner fel, 64-65; mem, Adverse Reactions Comt, Soc Nuclear Med, 83-; reviewer, J Nuclear Med, 85- *Mem:* Am Asn Cancer Res; Soc Nuclear Med; Transplantation Soc; AMA; AAAS. *Res:* Use of radionuclides in selective lymphoid ablation in control of transplant rejection; use of radiolabeled compounds for detection and treatment of cancer. *Mailing Add:* 622 W 168th St New York NY 10032

FAY, ALICE D AWTREY, b New York, NY, Nov 14, 26; m 54; c 5. PTERIDOLOGY, CHEMICAL TAXONOMY. *Educ:* Harvard Univ, AB, 47; Univ Calif, Berkeley, PhD(chem), 50. *Prof Exp:* Instr chem, Univ Calif, Berkeley, 50-51; res fel chem, Cornell Univ, 51-52; asst prof, Iowa State Univ, 52-55; instr, Mich Christian Jr Col, 64-65 & Cuttington Col, Liberia, 65-67; lectr, Univ WI, Trinidad, 68-73; assoc prof, Furman Univ, 73-74; head dept chem, Anderson Col, 74-88; ASSOC PROF, GA MIL COL, 88- *Concurrent Pos:* Fullbright prof, Njala Col, Univ Sierra Leone, 84-85. *Mem:* Am Chem Soc; Am Fern Soc. *Res:* Collection and identification of tropical ferns; identification of flavonoids by chromatography; kinetics of inorganic reactions. *Mailing Add:* PO Box 622 Milledgeville GA 31061

FAY, FRANCIS HOLLIS, b Melrose, Mass, Nov 18, 27; m 52; c 2. MARINE MAMMALOGY. *Educ:* Univ NH, BS, 50; Univ Mass, MS, 52; Univ BC, PhD(vert zool), 55. *Prof Exp:* Med biologist, Arctic Health Res Ctr, USPHS, 55-67, res biologist, 67-74; assoc prof, 74-83, PROF, INST MARINE SCI, UNIV ALASKA, FAIRBANKS, 83- *Concurrent Pos:* Mem comt sci adv, Marine Mammal Comn, 75-77, Comnr, 87- *Mem:* Fel AAAS; Am Soc Mammal; Ecol Soc Am; Wildlife Dis Asn; fel Arctic Inst NAm. *Res:* Biology of pinnipeds; vertebrate populations; animal ecology. *Mailing Add:* Inst Marine Sci Univ Alaska Fairbanks AK 99775-1080

FAY, FREDRIC S, PHYSIOLOGY. *Prof Exp:* PROF PHYSIOL & PHARMACOL & DIR BIOMED IMAGING GROUP, MED SCH, UNIV MASS, 70- *Mailing Add:* Dept Physiol Med Sch Univ Mass Worcester MA 01605

FAY, HOMER, b Brooklyn, NY, Aug 3, 28; m 55; c 2. GAS SEPARATION & PURIFICATION. *Educ:* Bowdoin Col, AB, 49; Mass Inst Technol, PhD(anal chem), 53. *Prof Exp:* Asst anal chem, Mass Inst Technol, 50-53; chemist, Union Carbide Corp, Tonawanda, NY, 53-70, sr scientist, Res Inst, Tarrytown, 70-79, sr res assoc, Linde Div, 79-85; PRES, FALITE INSTRUMENTS, BUFFALO, NY, 86- *Concurrent Pos:* Chemist, Ionics, Inc, 51. *Mem:* AAAS; Am Chem Soc; Am Phys Soc; Sigma Xi. *Res:* Materials and processes for gas separation and purification; adsorption and absorption; instrumentation for gas analysis; optical instrumentation for the detection of fluorescent minerals; solid-state materials synthesis and properties; dielectrics and ferroelectrics; electrical and optical properties of crystals; electrochemical behavior of semiconductors. *Mailing Add:* Falite Instruments 347 Brantwood Rd Buffalo NY 14226

FAY, JAMES A(LAN), b Southold, NY, Nov 1, 23; m 46; c 6. FLUID MECHANICS, HEAT TRANSFER. *Educ:* Webb Inst Naval Archit, BS, 44; Mass Inst Technol, MS, 47; Cornell Univ, PhD(mech eng), 51. *Prof Exp:* Res engr, Lima-Hamilton Corp, 47-49; asst prof eng mech, Cornell Univ, 51-55; prof, 55-89, EMER PROF MECH ENG, MASS INST TECHNOL, 89- *Concurrent Pos:* Consult, Avco-Everett Res Lab, 55-69, Exec Dept, State of Maine, 71-72, Natural Resources Coun Maine, 77-83 & Mass Energy Facil Siting Coun, 77-83; chmn, Boston Air Pollution Control Comn, 69-72 & Mass Port Authority, 72-77; mem, Environ Studies Bd, Nat Res Coun, 73-78 & 80-83, Comt Environ Decision Making, 76-78, Comt Urban Waterfront Lands & Comt Envrion Res & Develop, 78-79, comt on Radioactive Waste Mgt, 78-81, Panel Social & Econ Aspects Radioactive Waste Mgt, 80-84, Comt Risk Assessment & Commun, 87-88; dir, Union Concerned Scientists, 78-; dir, SCA Servs, Inc, 77-84, Conserv Law Found, 84- *Mem:* Am Soc Mech Engrs; fel Am Phys Soc; fel Am Inst Aeronaut & Astronaut; fel Am Acad Arts & Sci; fel AAAS. *Res:* Gaseous detonations; hypersonic heat transfer; magnetohydrodynamics; plasma physics; acid rain; air and oil pollution; liquified energy gas safety. *Mailing Add:* Rm 3-258 Mass Inst Technol Cambridge MA 02139

FAY, JOHN EDWARD, II, b Rochester, NY, July 28, 34. CHEMICAL ENGINEERING. *Educ:* Univ Mich, Ann Arbor, BSE, 56, MSE, 57; Mass Inst Technol, ScD, 71. *Prof Exp:* Process engr chem eng, capital budgets coordr, economist, process supvr & sr engr, Humble Oil & Refining Co, Bayonne Refinery, NJ, 57-68; res engr chem & metall, ASARCO Inc, 72-73, sect head, proj eval, 73-79, sect head, 80-81, supt, chem eng, cent res labs, 82-86, SR PROJ LEADER, PROD DEVELOP, TECH SERV CTR, ASARCO, INC, 87- *Concurrent Pos:* Chmn, Subcomt Energy Conserv Non-Ferrous Metals, Am Mining Cong. *Mem:* Am Inst Mining, Metall & Petrol Engrs; Am Soc Qual Control; Am Inst Chem Engrs; AAAS. *Res:* Economic and technical evaluation of research projects; energy conservation; process metallurgy; hydrometallury; statistical process control; process modeling & optimization. *Mailing Add:* ASARCO Inc Tech Serv Ctr 3422 South 700 W Salt Lake City UT 84119

FAY, MARCUS J, b Adair, Iowa, July 5, 21; m 44; c 2. PLANT TAXONOMY. *Educ:* Univ Iowa, PhD(bot), 53. *Prof Exp:* Asst, Univ Iowa, 49-53, instr, 53; from asst prof to assoc prof, 53-57, head dept, 53-55, PROF BIOL, UNIV WIS-EAU CLAIRE, 57-, HEAD DEPT, 57- *Mem:* Sigma Xi. *Res:* Floristics and plant distribution studies. *Mailing Add:* Dept Biol Univ Wis Eau Claire WI 54701

FAY, PHILIP S, b Ballard, Wash, Jan 24, 21; m 42; c 3. PETROLEUM. *Educ:* Cornell Col, AB, 41; Western Reserve Univ, PhD(chem), 49. *Prof Exp:* PROJ LEADER, CHEM & PHYS RES DIV, STANDARD OIL CO OHIO, 42-44 & 49- *Mem:* Am Chem Soc; Sigma Xi. *Res:* Reaction of phosphorous pentasulfide with olefins; gasoline additives; combustion chamber deposits; reaction of hydrocarbons; organoboron chemistry; petrochemicals; geochemical instrumentation and exploration. *Mailing Add:* 1163 Churchill Rd Lyndhurst OH 44124

FAY, RICHARD ROZZELL, b Holden, Mass, May 5, 44; m 68; c 2. PSYCHOPHYSIOLOGY, NEUROSCIENCES. *Educ:* Bowdoin Col, BA, 66; Conn Col, MA, 68; Princeton Univ, PhD(psychol), 70. *Prof Exp:* Res asst prof sensory sci, Sensory Sci Lab, Univ Hawaii, 72-74; asst prof otolaryngol, Bowman Gray Sch Med, 74-75; assoc prof, 75-80, PROF PSYCHOL, LOYOLA UNIV, CHICAGO, 80- *Concurrent Pos:* USPHS fel, Auditory Res Labs, Princeton Univ, 70-72; res career develop award, Nat Inst Neurol & Commun Dis & Stroke, 80- *Mem:* Fel Acoust Soc Am; Am Psychol Asn; Sigma Xi; Soc Neurosci; Asn Res Otolaryngol. *Res:* Processing by the brain of sensory information and the coding of information by the nervous system; neurophysiological correlates of sensory behavior. *Mailing Add:* Dept Psychol Loyola Univ 6525 N Sheridan Rd Chicago IL 60626

FAY, ROBERT CLINTON, b Kenosha, Wis, Mar 14, 36; m 60. INORGANIC CHEMISTRY. *Educ:* Oberlin Col, AB, 57; Univ Ill, MS, 60, PhD(inorg chem), 62. *Prof Exp:* From asst prof to assoc prof, 62-75, PROF CHEM, CORNELL UNIV, 75- *Concurrent Pos:* NSF fac fel, Univ E Anglia & Univ Sussex, 69-70; SERC vis fel, NATO/Heineman sr fel, Univ Oxford, 82-83; vis prof, Dept Chem, Harvard Univ, 90-91. *Mem:* Am Chem Soc; Royal Soc Chem; Am Crystallog Asn. *Res:* Stereochemistry of metal complexes; applications of nuclear magnetic resonance spectroscopy to inorganic chemistry; chemistry and structure of early transition metal complexes. *Mailing Add:* 318 Eastwood Ave Ithaca NY 14850

FAY, TEMPLE HAROLD, b 1960. IMAGE PROCESSING, REMOTE SENSING. *Educ:* Guilford Col, BS, 63; Wake Forrest Univ, MA, 64; Univ Fla, PhD(math), 71. *Prof Exp:* Asst prof math, Hendrix Col, 70-76; vis sr lectr, Univ Cape Town, 76-77; vis asst prof, NMex State Univ, 78-79; assoc prof, 79-82, PROF MATH, UNIV SOUTHERN MISS, 82- *Concurrent Pos:* Expert consult, Naval Oceanogr & Atmospheric Res Lab, 84-; consult, Inst Defense Anal, Supercomput Res Ctr, 88-89. *Mem:* Am Math Soc; Sigma Xi; SAfrican Math Soc. *Res:* Group theory and modules; pattern recognition; image processing. *Mailing Add:* Math Dept Univ Southern Miss Southern Sta Box 5045 Hattiesburg MS 39406

FAY, WARREN HENRY, b Scottsbluff, Nebr, Jan 3, 29; m 52; c 1. SPEECH PATHOLOGY, INFANTILE AUTISM. *Educ:* Colo State Col, BA, 51; Univ Ore, MEd, 59; Purdue Univ, PhD(speech path, audiol), 63. *Prof Exp:* High sch instr, Ore, 56-58; speech therapist, Pub Schs, 58-60; instr, Ore Health Sci Univ, 62-67, assoc prof, 67-78, prof speech path, 78-, head dept pediat; RETIRED. *Mem:* Am Speech & Hearing Asn. *Res:* Physiological aspects of language development and disorders, especially in the areas of echolalia and temporal coding of linguistic units; speech of childhood autism. *Mailing Add:* 3636 SW Beaverton Ave Portland OR 97201-1504

FAYER, MICHAEL DAVID, b Los Angeles, Calif, Sept 12, 47; m 68; c 2. CHEMICAL PHYSICS. *Educ:* Univ Calif, Berkeley, BS, 69, PhD(chem), 74. *Prof Exp:* From asst prof to assoc prof, 74-83, PROF CHEM, STANFORD UNIV, 83- *Mem:* Fel Am Phys Soc; Am Chem Soc. *Res:* Dynamics in solids and liquids, liquid helium to room temperatures, utilizing optical spectroscopic, picosecond, and feintosecond nonlinear laser techniques; development of nonlinear techniques; statistical mechanics theory; picosecond nonlinear studies of flames. *Mailing Add:* Dept Chem Stanford Univ Stanford CA 94305

FAYER, RONALD, b Philadelphia, Pa, Oct 12, 39. PARASITOLOGY, PROTOZOOLOGY. *Educ:* Univ Alaska, BS, 62; Utah State Univ, MS, 64, PhD(zool), 68. *Prof Exp:* Parasitologist, Beltsville Lab, USDA, 68-72, proj leader, 72-78, lab chief, 78-81 & 83, res prog leader, Agr Res Serv, USDA, 82, inst dir, Animal Parasitol Inst, 84-88, RES LEADER, ZOONOTIC DIS LAB, USDA, 88- *Concurrent Pos:* Adj prof, Sch Vet Med, Univ Pa, 78-; bd dir, Am Type Cult Col, 85-, exec comt, 88- *Honors & Awards:* USDA Cert of Merit, 74 & 78 & Super Serv Award, 78; H B Ward Medal, Am Soc Parasitologists, 78. *Mem:* Am Soc Parasitologists; Soc Protozoologists; Wildlife Dis Asn; Am Asn Vet Parasitologists; World Asn Advan Vet Parasitologists. *Res:* Veterinary and wildlife parasitology; pathology; immunology; in vitro cultivation. *Mailing Add:* Livestock & Poultry Sci Inst USDA 10300 Baltimore Blvd Beltsville MD 20705-2350

FAYLE, HARLAN DOWNING, b Hibbing, Minn, July 24, 07; m 36; c 1. BIOCHEMISTRY, PHARMACOLOGY. *Educ:* Hamline Univ, BA, 31; Univ Minn, MS, 36, PhD(biochem & pharmacol), 63. *Prof Exp:* Instr chem, Eveleth Jr Col, 33-36, Hibbing Jr Col, 36-46; & Duluth Jr Col, 46-48; from instr to asst prof, Univ Minn, Duluth, 48-56; asst, Univ Minn, Minneapolis, 54-55; from assoc prof to prof, 56-73, chmn sect, 56-73, EMER PROF CHEM, PURDUE UNIV, CALUMET CAMPUS, 73- *Concurrent Pos:* Consult, Butler Mining Co, 40-42, Elliott Packing Co, 48-50 & Mitchell Oil Co, 50-51; rep pres adv coun on retirement, Purdue Univ, Calumet Campus, 76-; adj prof chem, Ind Univ Northwest, 78. *Mem:* AAAS; Am Chem Soc; fel Am Inst Chemists; NY Acad Sci; Am Asn Univ Professors; Intercontinental Biol Asn Eng. *Res:* Distribution and determination of cadmium in biological material; toxicology of cadmium. *Mailing Add:* Purdue Univ Calument Campus 2233 171st St Hammond IN 46323

FAYMON, KARL A(LOIS), b St Louis, Mo, Sept 9, 27; m 60; c 2. SYSTEMS ANALYSIS. *Educ:* Univ Mich, BS, 51, MS, 52; Case Inst Technol, PhD(aerodyn), 57. *Prof Exp:* Instr, St Louis Univ, 52-53; instr & asst, Case Inst Technol, 53-57; theoret aerodynamicist, Convair Div, Gen Dynamics Corp, 57-60; systs analyst, Thompson-Ramo-Wooldridge, Inc, 60-63; chief systs anal off, Launch Vehicles Div, 63-74, asst chief, Systs Anal & Assessment Off, 74-80, DEP CHIEF, PLANNING, ANALYSIS, & SYSTS OFF, ENERGY PROGS DIRECTORATE, LEWIS RES CTR, NASA, CLEVELAND, 81- *Mem:* Sigma Xi. *Res:* Trajectory analysis; thermodynamics; applied mathematics; space launch vehicles systems analysis; structural and control dynamics; energy systems analysis; propulsion systems analysis; power and energy conversion systems analysis. *Mailing Add:* 2066 Marshfield Rd Cleveland OH 44124

FAYON, A(BRAM) M(IKO), b Sofia, Bulgaria, Apr 1, 20; US citizen; m 52, 58, 84; c 3. CHEMICAL ENGINEERING. *Educ:* Istanbul Univ, BS, 47; Johns Hopkins Univ, MSE, 52; NY Univ, EngScD(chem eng), 59. *Prof Exp:* Chemist, Baltimore Paint & Color Works, 48-51; chem engr res & develop, US Indust Chem, 51-52 & Chem Construct Corp, 52-54; asst chem eng, NY Univ, 56-57, from instr to asst prof, 57-60; res chem engr, Am Cyanamid Co, Conn, 60-62, sr res chem engr, 62-65; sr process engr, Sci Design Co, Inc, 65-66; assoc engr, Mobil Oil Corp, 66-67 & Mobil Res & Develop Corp, 67-70, opers res coordr, 70-75, sr planning assoc, New York, 75-80, sr eng assoc, Mobil Chem Co, Houston, Tex, 80-84; REMEDIAL PROJ MGR, US ENVIRON PROTECTION AGENCY, 87- *Concurrent Pos:* Consult, Nuclear Energy Prod Div, Am Car & Foundry, 56-57; assoc prof chem eng, Manhattan Col, 84-85, adj assoc prof & consult, 85-87. *Mem:* Am Inst Chem Engrs; Sigma Xi. *Res:* Thermodynamics; heat transfer; fluid flow; process simulation of petroleum and petrochemical plants; development of computer methods related to process and process design problems; financial and technical appraisal of petrochemical processes. *Mailing Add:* 510 Siwanoy Pl Pelham Manor NY 10803

FAYOS, JUAN VALLVEY, b Camaguey, Cuba, Nov 24, 29; US citizen; m 61; c 4. RADIOLOGY. *Educ:* Inst de Camaguey, BS, 48; Univ Havana, MD, 55. *Prof Exp:* From instr to prof radiol, Univ Mich, Ann Arbor, 61-86, dir radiation ther div, 72-86; SELF EMPLOYED, 86- *Concurrent Pos:* Mem staff, Wayne County Gen Hosp, 66; consult, Vet Admin Hosp, Ann Arbor. *Mem:* AMA; Radiol Soc NAm; Am Radium Soc; Am Soc Therapeut Radiol. *Res:* Clinical radiation therapy. *Mailing Add:* 6400 SW 144th St Miami FL 33183

FAYTER, RICHARD GEORGE, JR, b Woodbury, NJ, Nov 26, 37; m 61; c 2. ORGANIC CHEMISTRY. *Educ:* Temple Univ, AB, 69; Brown Univ, PhD(org chem), 73. *Prof Exp:* Sr res chemist resins group, 73-74, group leader ctr res, 74-79, mgr, 79-80, dir ctr res, 80-85, asst res dir, 85-88, SR RES DIR, QUANTUM CHEMICALS CO, EMERY DIV, OHIO, 88- *Mem:* Am Chem Soc; Sigma Xi. *Res:* Mechanistic organic chemistry; bridged polycyclic compounds; molecular rearrangements and ionic additions; synthetic organic chemistry; organic ultrasonic chemistry; agriculture chemistry; pesticide chemistry; polymer chemistry. *Mailing Add:* 2746 Saturn Dr Fairfield OH 45014

FAZEKAS, ARPAD GYULA, b Szeged, Hungary, Aug 30, 36; Can citizen; m 66; c 1. EXPERIMENTAL SURGERY, BIOCHEMISTRY. *Educ:* Med Univ Szeged, MD, 60; Hungarian Acad Sci, CMedSci, 65. *Prof Exp:* Asst lectr biochem, Med Univ Szeged, 60-62, lectr, 62-63; lectr path, Royal Infirmary, Univ Glasgow, 64-65; lectr biochem, Med Univ Szeged, 65-68; asst prof, 74-80, ASSOC PROF EXP SURG, MCGILL UNIV, 80-; ASST PROF MED, UNIV MONTREAL, 69- *Concurrent Pos:* Res assoc med, Columbia Univ, NY, 71-74. *Mem:* Endocrine Soc; Can Soc Endocrinol & Metab; Brit Biochem Soc. *Res:* Biosynthesis, metabolism and mechanism of action of steroid hormones; biosynthesis of flavin coenzymes, riboflavin metabolism. *Mailing Add:* 5025 Sherbrooke W Montreal PQ H4A 1S9 Can

FAZIO, GIOVANNI GENE, b San Antonio, Tex, May 26, 33; wid; c 2. INFRARED ASTRONOMY, HIGH ENERGY ASTROPHYSICS. *Educ:* St Mary's Univ, BS & BA, 54; Mass Inst Technol, PhD(physics), 59. *Prof Exp:* Res assoc, Univ Rochester, 59, from instr to asst prof physics, 59-62; physicist, 62-83, SR PHYSICIST, SMITHSONIAN CTR ASTROPHYS, 83-; LECTR ASTRON, HARVARD UNIV, 62- *Concurrent Pos:* NATO sr fel sci, 68 & 70; mem, NASA Airborne Astron Mgt Oper Working Group, 76-; prin investr, Spacelab 2 infrared telescope & balloon-borne telescope; mem sci adv comt, Max Planck Inst, 81-; mem, Balloon Working Group, NASA, 83-; mem, Astrophys Coun, NASA, 85-; mem, Adv Comt Large Optical Infrared Telescopes, NSF, 85- *Mem:* Am Astron Soc; Int Astron Union; fel Am Phys Soc; fel Royal Astron Soc; fel AAAS; Optical Soc Am. *Res:* Infrared astronomy; gamma-ray astronomy. *Mailing Add:* 244 Franklin St Newton MA 02158

FAZIO, PAUL (PALMERINO), b Italy, Apr 1, 39; Can citizen; m 66; c 3. SOLID MECHANICS. *Educ:* Univ Windsor, BASc, 63, MASc, 64, PhD(struct), 68. *Prof Exp:* Sessional instr, Univ Windsor, 64-67; from asst prof to assoc prof, 67-74, chmn dept civil eng, 73-77, PROF CIVIL ENG, CONCORDIA UNIV, 74-, DIR CTR FOR BLDG STUDIES, 77- *Concurrent Pos:* Jr struct engr, Can Bridge Co, 63; struct engr, Montreal Eng Co, 68; mem cent comt, Can Cong Appl Mech, 71-77; pres, Siricon; mem coun sci & technol, Que, 81-84; mem Construct Industs Develop Coun, 81-; mem grant selection comt civil, Natural Sci & Eng Res Coun Can, 82-85; mem bd, Ctr Int Grounds projs, 84-; mem & vchmn, Can Construct Res Bd, 86-; mem building indust strategy bd, Ont Ministry Housing, 86-89; vpres, Que Region, Can Soc Civil Eng, 87-89. *Honors & Awards:* Galbraith Prize, Eng Inst Can, 67. *Mem:* Am Soc Civil Engrs; Eng Inst Can; Am Concrete Inst; Am Soc Eng Educ; Am Soc Heating, Refriger & Air-Conditioning Engrs. *Res:* Building engineering, the building envelope; building science; energy conservation; management; structures; development of wall panels; panel connections. *Mailing Add:* Ctr Bldg Studies Concordia Univ Sir George Williams Campus 1455 de Maisonneuve Blvd W Montreal PQ H3G 1M8 Can

FAZIO, STEVE, b Phoenix, Ariz, Sept 2, 16; m 40; c 2. HORTICULTURE. *Educ:* Univ Ariz, BS, 40, MS, 51. *Prof Exp:* Head dept, 65-71, PROF HORT, UNIV ARIZ, 64- HORTICULTURIST & HORT SPECIALIST, AGR EXTEN SERV, 72- *Mem:* Am Soc Hort Sci; Int Plant Propagators Soc (vpres, 78-79, pres, 79). *Res:* Propagation of woody and herbaceous plants in arid desert regions. *Mailing Add:* 3554 E Calle Alarcon Tucson AZ 85706

FEAD, JOHN WILLIAM NORMAN, b Wetaskiwin, Alta, Oct 23, 23; m 47; c 3. CIVIL ENGINEERING. *Educ:* Univ Alta, BS, 45, MS, 49; Northwestern Univ, PhD, 57. *Prof Exp:* Instr civil eng, Univ Alta, 46-48 & Univ Sask, 48-49; from instr to assoc prof, SDak State Col, 49-57; lectr, Northwestern Univ, 51-53; assoc civil eng & asst res engr, Inst Transp & Traffic Eng, Univ Calif, 54-55; assoc prof, 57-61, head dept, 63-84, PROF CIVIL ENG, COLO STATE UNIV, 61- *Concurrent Pos:* Mem city coun, Ft Collins, Colo, 71-75, mayor, 74-75. *Mem:* Am Soc Civil Engrs; Am Soc Eng Educ; Am Concrete Inst. *Res:* Structures and structural mechanics; soil mechanics and foundations. *Mailing Add:* Dept Civil Eng Colo State Univ Ft Collins CO 80523

FEAGANS, WILLIAM MARION, b Fortescue, Mo, Feb 2, 27; m 50; c 3. ANATOMY. *Educ:* Univ Mo-Kansas City, DDS, 54; Med Col Va, PhD(anat), 60. *Prof Exp:* From instr to asst prof clin dent, Sch Dent, Univ Mo-Kansas City, 54-56; from instr to assoc prof anat, Med Col Va, 58-66, curric coordr med educ, 64-66; assoc prof anat, Schs Med & Dent Med, Tufts Univ, 66-70, from asst dean to assoc dean, Sch Dent Med, 66-70; PROF ANAT SCI, SCH MED & DEAN SCH DENT, STATE UNIV NY, BUFFALO, 70- *Concurrent Pos:* Consult, Surg Serv, US Naval Hosp, Va, 58-66, Dent Serv, 60-66. *Mem:* Histochem Soc; Am Asn Anat; NY Acad Sci. *Res:* Physiology of male reproduction; histochemistry and electron microscopy of oral tissues. *Mailing Add:* Dept Dent State Univ NY at Buffalo 134 Capen Hall Buffalo NY 14214

FEAGIN, FRANK J, b Kaufman, Tex, Mar 14, 14; m 36; c 3. GEOPHYSICS. *Educ:* Tex A&M Col, BSEE, 34. *Prof Exp:* Grad asst elec eng, Agr & Mech Col Tex, 34-35; attached helper, Seismic Party, Humble Oil & Refining Co, 35-37, seismic operator, 37-41; asst prof elec eng, Agr & Mech Col Tex, 41-42; electronic develop work, War Contracts, Humble Oil & Refining Co, 42-46, res specialist, 46-54, sr res specialist, Geophys Res Sect, 54-57, asst chief geophysics res, 57-60, chief geophys res & eng sect, 60-64; mgr appl geophys div, Esso Prod Res Co, 64-66, sr res assoc, 66-73, sr res adv, 73-79; RETIRED. *Mem:* Inst Elec & Electronics Engrs. *Res:* Geophysical instruments; geophysical displays; research management. *Mailing Add:* Seven Beavertail Pt Houston TX 77024

FEAGIN, FREDERICK F, b Pike Co, Ala, Nov 22, 31; c 3. DENTISTRY. *Educ:* Auburn Univ, BS, 58; Univ Ala, Birmingham, DMD, 64, PhD(physiol & pharmacol), 69. *Prof Exp:* Res asst, 60-64, res assoc & instr, 65-69, asst prof clin dent, Sch Dent, 69-77, investr, Inst Dent Res, 69-77, asst prof physiol & biophys, Grad Fac, 69-77, PROF PHYSIOL & BIOPHYS, SCH MED & SCH DENT & ASSOC PROF DENT, UNIV ALA, BIRMINGHAM, 77- *Concurrent Pos:* Staff dentist & prin investr, Vet Admin Hosp, Birmingham, Ala, 68-69, clin investr, 69-72. *Mem:* Int Asn Dent Res. *Res:* Mechanisms of biological calcifications. *Mailing Add:* Dept Dent Univ Ala 1919 S Seventh Ave Birmingham AL 35233

FEAGIN, ROY C(HESTER), b Andersonville, Ga, July 23, 14; m 38; c 2. CHEMICAL ENGINEERING. *Educ:* Ala Polytech Inst, BS, 36, MS, 37. *Prof Exp:* Res chemist, Gen Elec Co, NY, 37-39; res chemist, Austenal Microcast Div, Howmet Corp, 39-45, chief chemist, 45-56, mgr chem res, 56-67, assoc res dir, 67-68, mgr int opers, Superalloy Group, 68-73; consult, 73-75; tech dir, 75-77, vpres, 77-80, CONSULT, REMET CORP, 80- *Mem:* Am Chem Soc; Am Ceramic Soc; Int Asn Dent Res; Am Soc Testing & Mat; Am Inst Ceramic Engrs. *Res:* Development of compositions and applications of plastics, cements, refractories to dental and high temperature alloy casting field. *Mailing Add:* Edgewater Pointe Estates 23343 Blue Water Circle B232 Boca Raton FL 33433

FEAGIN, TERRY, b Houston, Tex, Mar 27, 45; m 68; c 2. COMPUTER SCIENCE, AEROSPACE ENGINEERING. *Educ:* Rice Univ, BA, 67; Univ Tex, Austin, MA, 69, PhD(aeorspace eng), 72. *Prof Exp:* Nat Acad Sci-Nat Res Coun res assoc astrodynamics, 72-73; asst prof comput sci & aerospace eng, 73-78, assoc prof comput sci & aerospace eng & assoc dir res & devel, Comput Ctr, 78-79, HEAD COMPUT SCI DEPT, UNIV TENN, KNOXVILLE, 80- *Mem:* Am Inst Aeronaut & Astronaut; Asn Comput Mach; Am Astron Soc; Am Phys Soc. *Res:* Astrodyn; optimal control theory; numerical analysis; small computer systems; parallel processing. *Mailing Add:* Univ Houston-Clear Lake 2700 Bay Area Blvd Houston TX 77058

FEAIRHELLER, STEPHEN HENRY, b Philadelphia, Pa, Nov 4, 33; m 59; c 3. NATURAL POLYMER CHEMISTRY. *Educ:* Pa State Univ, BS, 60; Mass Inst Technol, PhD(org chem), 64. *Prof Exp:* Res chemist, 64-73, RES LEADER, EASTERN REGIONAL RES CTR USDA, 73- *Concurrent Pos:* Asst Prof chem, Ogontz Campus, Pa State Univ, 69-70. *Honors & Awards:* Alsop Award, Am Leather Chemists Asn, 73. *Mem:* Am Chem Soc; Am Leather Chem Assoc (pres, 86-88). *Res:* Chemical modification of proteins; enzymatic modification of lipids. *Mailing Add:* 804 Preston Rd Erdenheim Philadelphia PA 19118-1329

FEAIRHELLER, WILLIAM RUSSELL, JR, b Camden, NJ, Nov 25, 31; m 56; c 4. ENVIRONMENTAL CHEMISTRY. *Educ:* Rutgers Univ, BA, 54; Univ Md, MS, 58. *Prof Exp:* Analytical chemist, Am Viscose Corp, 59-60; chemist, Avisun Corp, 60-63; sr res chemist, 63-68, res group leader, 68-74, sr res group leader, 74-77, RES SPECIALIST, MONSANTO RES CORP, 77- *Mem:* Air Pollution Control Asn. *Res:* Environmental monitoring and testing; air, stationary source and water emissions measurement; environmental source assessment; project and government contract management. *Mailing Add:* 4711 Gleanheath Dr Dayton OH 45440

FEAKES, FRANK, b York, Western Australia, Feb 2, 23; m 56; c 4. INORGANIC CHEMISTRY, RESEARCH ADMINISTRATION. *Educ:* Univ Western Australia, BSc, 44, BA, 48, MSc, 51; Mass Inst Technol, DSc(chem eng), 56. *Prof Exp:* Res chemist, Western Australia State Alunite Indust, 45-47, chief chemist, 47-49, plant supt, 49-50; res officer, Western Australia Dept Indust Develop, 50-51, sr res officer, 57-59; res assoc, dept chem eng, Mass Inst Technol, 56-57 & 59-62; sr res chemist, Nat Res Corp,

62-66, prog dir, 66-67; asst dir res, Norton Res Corp, 67-68; group leader, Cabot Corp, 70-72, mem tech staff, 72-74, dir corp res, 74-83, chief scientist, 83-88; RETIRED. *Mem:* Am Inst Chem Engrs; Am Vacuum Soc; NY Acad Sci. *Res:* Kinetics of decomposition of solids involving chemical reactions; high vacuum techniques; cryogenic techniques; vacuum evaporations; thin films; advanced composites; liquified natural gas; environmental safety; crystal nucleation and growth. *Mailing Add:* Six Juniper St Lexington MA 02173

FEAR, J(AMES) VAN DYCK, b Morgantown, WVa, Nov 7, 25; m 52, 82; c 3. CHEMICAL ENGINEERING. *Educ:* Univ Louisville, BChE, 45, MChE, 47. *Prof Exp:* Res engr, Sun Co, 48-60, sect chief process develop, 60-66, mgr process develop, 66-71, mgr res & develop spec projs, 71-, mgr raw mat supply & distrib, 71-75, vpres res & eng, Suntech, Inc, 75-77, vpres mkt, 77-80, vpres fuels, Sun Petrol Prod Co, 80-81, vpres, Fuels Div, Sun Refining & Mkt Co, 81-88; CONSULT, 88- *Mem:* Am Inst Chem Engrs; Am Petrol Inst. *Res:* Development of petroleum processes, especially in lubricating and waxes; hydrogenation and reforming; recovery of tar from Athabasca tar sands. *Mailing Add:* 1425 Carroll Brown Way West Chester PA 19382

FEARING, OLIN S, b Lawrence, Kans, Mar 30, 28; m 54; c 2. BOTANY. *Educ:* Univ Kans, AB, 50, MA, 51; Univ Tex, PhD, 59. *Prof Exp:* From asst prof to assoc prof, 59-75, PROF BIOL, TRINITY UNIV, TEX, 75- *Mem:* AAAS; Am Soc Plant Taxonomists; Bot Soc Am. *Res:* Cytotaxonomy of flowering plants; specifically the genus Cologania, Amphicarpaea and related genera; physiology and development of lichen symbionts. *Mailing Add:* Dept Biol Trinity Univ 715 Stadium Dr San Antonio TX 78284

FEARING, RALPH BURTON, b Oak Park, Ill, Aug 20, 18; m 46; c 2. ORGANIC CHEMISTRY, TEXTILE CHEMISTRY. *Educ:* Univ Chicago, BS, 40, MS, 43; Iowa State Col, PhD(org chem), 51. *Prof Exp:* Res assoc war gases, Univ Chicago, 43; res assoc anal chem, Los Alamos Sci Lab, 44-45; res assoc org synthesis, Monsanto Chem Co, 46-47; instr chem, Iowa State Col, 47-51; asst prof org chem, Utica Col, 51-56; res chemist, Victor Div, Eastern Res Labs, Stauffer Chem Co, Dobbs Ferry, 57-65, sr chemist, 65-82; RETIRED. *Res:* Organic synthesis; organic phosphorus chemistry; pesticide chemistry; textile finishing chemistry; synthesis of organophosphorus compounds and oligomers. *Mailing Add:* 527 Brittany Dr State College PA 16803-1421

FEARN, JAMES ERNEST, b Chattanooga, Tenn, Nov 21, 20; m 40; c 3. PHYSICAL ORGANIC CHEMISTRY, POLYMER CHEMISTRY. *Educ:* Howard Univ, BS, 49, MS, 50; Cath Univ Am, PhD(chem, physics), 54. *Prof Exp:* Org chemist, Nat Cancer Inst, 50-52; res chemist, Patuxent Res Refuge, US Dept Interior, 55-57, res chemist, Polymer Chem Sect, 57-73, RES CHEMIST, MAT & COMPOSITES SECT, NAT BUR STANDARDS, 73- *Mem:* AAAS; Am Chem Soc; Royal Soc Chem; fel Am Inst Chemists; Sigma Xi. *Res:* Synthesis of fluorocarbons and high thermostable polymers; kinetic studies; mechanism studies, free radical polymerization; high pressure reactions; abrasion of rubber; microstructure studies of building materials; microscopy; porosimetry. *Mailing Add:* 4446 Alabama Ave SE Washington DC 20019

FEARN, RICHARD L(EE), b Mobile, Ala, Mar 24, 37; m 69; c 2. AERODYNAMICS. *Educ:* Auburn Univ, BS & MS, 60; Univ Fla, PhD(physics), 65. *Prof Exp:* Asst prof, 65-75, assoc prof, 75-80, PROF ENG SCI, UNIV FLA, 80- *Mem:* Am Inst Aeronaut & Astronaut; Am Soc Eng Educ. *Res:* Jet in a cross flow; lifting surface theory. *Mailing Add:* Dept Eng Sci Univ Fla Gainesville FL 32611

FEARNLEY, LAWRENCE, b Bradford, Eng, July 18, 32; m 56. TOPOLOGY. *Educ:* Univ London, BSc, 53, PhD, 70; Univ Utah, PhD(math, topology), 59. *Prof Exp:* Asst math, Univ Utah, 54-57; from asst prof to assoc prof, 57-70, PROF MATH, BRIGHAM YOUNG UNIV, 70- *Mem:* Am Math Soc; Math Asn Am. *Res:* Topology of manifolds, limit spaces, continua. *Mailing Add:* Dept Math Brigham Young Univ TMCB 304 Provo UT 84602

FEARNOT, NEAL EDWARD, b Salem, Mass, June 7, 53; m 75; c 3. CARDIOVASCULAR INSTRUMENTATION. *Educ:* Purdue Univ, BS, 75, MS, 78, PhD(elec eng), 80. *Prof Exp:* CTS Found fel, 78-80, George A Cook Mem res fel, 80, ASSOC RES SCHOLAR, HILLENBRAND BIOMED ENG CTR, 81-; PRES, MED ENG & DEVELOP INST, INC, 83- *Concurrent Pos:* Vis prof elec eng & mech eng, 81-83, faculty, Purdue Univ, 86-; Consult engr, Cook Group Corp, 81-83. *Honors & Awards:* Harold Lamport Young Investr Award, Biomed Eng Soc, 82. *Mem:* Inst Elec & Electronics Engrs; Eng in Biol & Med Soc; Asn Advan Med Instrumentation; NAm Soc Pacing & Electrophysiol; fel Am Col Cardiol; Int Clin Hyperthermia Soc; Regulatory Affairs Prof Soc. *Res:* Biomedical studies in cardiology and cancer therapy necessary for the development of new medical devices. *Mailing Add:* Med Eng & Develop Inst PO Box 2402 West Lafayette IN 47906

FEARNSIDES, JOHN JOSEPH, b Philadelphia, Pa. AUTOMATIC CONTROL SYSTEMS. *Educ:* Drexel Univ, BSEE, 62, MSEE, 64; Univ Md, PhD(elec eng), 71. *Prof Exp:* Instr elec eng, Drexel Univ, 62-64; instr, Univ Md, 64-68; mem tech staff controls, Bellcomm, Inc, 68-71; mem tech staff decision theory, Bell Tel Labs, 71-72; mgr advan res prog transp systs anal, US Dept Transp, 72-75, chief res & develop policy div, 74-75, exec asst to dep secy, 75-77, dep under secy & chief scientist, 77-79; dir Economies & Mgt Sci, Analytic Sci Corp, 80, DIR PLANNING & POLICY ANAL, MITRE CORP, 80- *Concurrent Pos:* Assoc ed Trans, Inst Elec & Electronics Engrs, 76 & 77; NSF. *Mem:* Inst Elec & Electronics Engrs; Sigma Xi. *Res:* Control systems analysis, especially aerospace; investment decision analysis under uncertainty; application of research and development to transportation problems. *Mailing Add:* Civil Systs Div Mitre Corp 1820 Dolley Madison Blvd McLean VA 22102

FEARON, DOUGLAS T, b Brooklyn, NY, Oct 16, 42. RHEUMATOLOGY. *Educ:* Williams Col, Williamstown, Mass, BA, 64; Johns Hopkins Univ, Baltimore, MD, 68; Harvard Univ, MA, 84. *Prof Exp:* Intern med, Osler Med Serv, Johns Hopkins Hosp, 68-69, asst resident med, 69-70; res fel med, Harvard Med Sch & Robert B Brigham Hosp, Boston, Mass, 72-75; from instr to prof med, Harvard Med Sch, Boston, Mass, 75-87; PROF MED & JOINT APPT MOLECULAR BIOL & GENETICS, SCH MED, JOHNS HOPKINS UNIV, BALTIMORE, MD, 87-, DIR, GRAD PROG IMMUNOL, 88- *Concurrent Pos:* Helen Hay Whitney Found Postdoctoral Res Fel, 74-77; asst physician, Robert B Brigham Hosp, Boston, Mass, 75-79; mem, Bd Tutors Biochem Sci, Harvard Univ, Cambridge, Mass, 76-86; Res Career Develop Award, NIH, 77-82; assoc physician, Brigham & Women's Hosp, Boston, Mass, 79-82, rheumatologist & immunologist, 82-87, dep chairperson, Dept Rheumatology & Immunol, 84-87; mem, Prog Comt, Am Asn Immunologists, 79-82 & 84, Allergy & Immunol Study Sect, NIH, 81-84, Res Comt, Arthritis Found, 83-85 & 88-91 & Subcomt Arthritis Found Clin Res Ctrs, 83-86; sect ed, J Immunol, 83-87; chairperson, Subcomt Nomenclature Complement Receptors, World Health Orgn, 85; 05296729x Comt, Am Asn Immunologists, 86-89, Immunol & Microbiol Res Study Comt, Am Heart Asn, 88-91, Res Comt, Am Col Rheumatology, 89; dir, Div Molecular & Clin Rheumatology, Dept Med, Johns Hopkins Hosp, Baltimore, Md, 87; co-chmn, Immunol & Microbiol Res Study Comt, Am Heart Asn, 89-90; assoc ed, J Exp Med, 90. *Honors & Awards:* John Sheldon Mem Lectr, Am Acad Allergy, 85; First Manfred Mayer Mem Lectr, Johns Hopkins Univ, 86; Ecker Lectr, Case Western Reserve Med Sch, 89; Marion Hargrove Mem Lectr, La State Univ, 91. *Mem:* Fel AAAS; Am Asn Immunologists; Am Soc Clin Invest; Am Rheumatism Asn; Asn Am Physician. *Res:* Rheumatoid arthritis; author of various publications. *Mailing Add:* Sch Med Div Molecular & Clin Rheumatology Johns Hopkins Univ 725 N Wolfe St 617 Hunterian Bldg Baltimore MD 21205

FEARON, FREDERICK WILLIAM GORDON, b London, Eng, June 4, 38; m 63; c 2. CHEMISTRY, MATERIALS SCIENCE. *Educ:* Univ Leeds, BSc, 61; Univ Wales, PhD(chem), 65. *Prof Exp:* Chemist res, Laporte Chem, Luton, Eng, 61-63; demonstr chem, Univ Col Wales, 63-65; res assoc, Iowa State Univ, 65-67; chem polymer res, Dow Corning Corp, 68-70, from group leader to sr group leader fluids, 70-74, res & develop mgr new ventures, 74-78; dir res, Cent Chem Div, BEE Chem 78-80; technol mgr new ventures, Dow Corning Corp, 80-82, mgr elastomers res, 82-86, vpres & tech dir, Japan, 87-90, DIR CENT RES & DEVELOP, DOW CORNING CORP, 91- *Mem:* Am Chem Soc; Soc Plastic Engrs; Am Ceramic Soc; Soc Advan Mat & Process Eng. *Res:* Structure property relationships in organosilicon; ceramics. *Mailing Add:* Dow Corning Mail CO43MZ Midland MI 48686

FEASLEY, CHARLES FREDERICK, b Lexington, Ill, Jan 21, 15; m 40; c 4. ENVIRONMENTAL HEALTH. *Educ:* Hanover Col, AB, 37; Purdue Univ, MS, 40, PhD(org chem), 42. *Prof Exp:* Asst chem, Purdue Univ, 37-41; res chemist, Res Dept, Mobil Oil Corp, 41-44, sr chemist, 45-57, supv technologist, 57-62, asst to mgr toxicol & pollution, 63-68, toxicol adv, Res Dept, 68-70, toxicol adv, Med Dept, 70-76, sr toxicologist, Med Dept, 76-78, sr toxicologist, Environ Affairs & Toxicol, 76-80; RETIRED. *Mem:* AAAS; Am Indust Hyg Asn; Soc Toxicol; Am Chem Soc; Am Soc Testing & Mat. *Res:* Antiseptics; utilization of nitroalkanes; chemicals from petroleum; synthetic fuels from natural gas and coal; condensation of aryl diazonium salts and hydroxides with secondary nitroalkanes; development of laboratory tests for petroleum products; air and water pollution; toxicology and labeling of chemicals and petroleum products. *Mailing Add:* 37 N Columbia St Woodbury NJ 08096

FEASTER, CARL VANCE, b Monon, Ind, Aug 11, 21; m 78; c 3. PLANT BREEDING. *Educ:* Purdue Univ, BS, 44; Univ Mo, MA, 47, PhD(field crops), 50. *Prof Exp:* Argonomist, Soybean Res, Bur Plant Indust, Soils & Agr Eng, USDA, 44-50, Field Crops Res Br, 50-56, Agronomist Cotton Breeding, Sci & Educ Admin, Agr Res, 56-; prof plant sci, Ariz State Univ, 73-89; RETIRED. *Mem:* Sigma Xi. *Res:* Genetics; fiber plants. *Mailing Add:* 9715 E Michigan Ave Sun Lake AZ 85224

FEASTER, GENE R(ICHARD), b Winfield, Kans, Sept 19, 18; m 51. MEDICAL PHYSICS, RADIATION THERAPY. *Educ:* Univ Kans, BS, 40, PhD(physics), 53. *Prof Exp:* Physicist thermionic tubes, RCA Corp, NJ, 42-47; instr physics, Univ Kans, 47-52; adv engr thermionic tubes, Westinghouse Elec Corp, 53-66; adv engr image intensifier & thermionic tubes, 66-67; eng assoc, Corning Glass Works, 67-70; res assoc, Med Ctr, Univ Va, 76-77; asst prof, Med Ctr, Univ Kans, 77-88; RETIRED. *Mem:* Am Asn Physicists Med. *Res:* Thermionic emission; photon beam compensators in radiation therapy; hematoporphyrin and photoradiation therapy. *Mailing Add:* 7411 Long Ave Kansas City KS 66102

FEASTER, JOHN PIPKIN, b St Petersburg, Fla, Oct 1, 20; m 44; c 3. BIOLOGICAL CHEMISTRY. *Educ:* Col of William & Mary, BA, 43; Emory Univ, MS, 48; Univ NC, PhD(biochem), 51. *Prof Exp:* Assoc biochemist, Agr Exp Sta, 51-68, biochemist & prof, 68-82, EMER PROF, INST FOOD & AGR SCI, UNIV FLA, 82- *Mem:* Am Chem Soc; Am Soc Animal Sci; Am Inst Nutrit. *Res:* Mineral nutrition; placental transfer; lipid metabolism; pesticide toxicity. *Mailing Add:* 3021 S W 70th Lane Gainesville FL 32608

FEATHER, A(LAN) L(EE), food technology, for more information see previous edition

FEATHER, DAVID HOOVER, b Orange, NJ, Apr 27, 43. MATERIALS SCIENCE. *Educ:* Alfred Univ, BS, 68; Univ Calif, Berkeley, MS, 69, PhD(mat sci), 72. *Prof Exp:* Scientist phys chem, Aluminum Co Am; SCIENTIST HIGH TEMP CHEM, JOSEPH C WILSON CTR TECHNOL, XEROX CORP, 75- *Res:* Heterogeneous kinetics and high temperature chemistry; free surface and equilibrium vaporization processes. *Mailing Add:* 9899 Caminito Rogelto San Diego CA 92131

FEATHER, MILTON S, b Massillon, Ohio, Mar 14, 36; m 57; c 1. BIOCHEMISTRY. *Educ:* Heidelberg Col, BS, 58; Purdue Univ, MS, 61, PhD(biochem), 63. *Prof Exp:* Chemist, US Forest Serv, 63-67; from asst prof to assoc prof agr chem, 67-73, PROF BIOCHEM, UNIV MO-COLUMBIA, 73- *Concurrent Pos:* USDA trainee, Swedish Forest Prod Res Lab, Stockholm, 64-65; vis prof, Univ Calif-Davis, 87. *Mem:* Am Chem Soc; Fedn Am Soc Exp Biol. *Res:* Chemistry of nonenzymatic browning and dehydration reactions; structural studies on polysaccharides. *Mailing Add:* Dept Biochem Univ Mo 322 Chem Bldg Columbia MO 65211

FEATHERS, WILLIAM D, b Pittsburgh, Pa, Sept 14, 27; m 50; c 6. CHEMICAL ENGINEERING. *Educ:* Univ Pittsburgh, BS, 50. *Prof Exp:* Technician, Color Unlimited, Pa, 47-50; engr, Mellon Inst, 50; from engr to sr res engr, Photo Prod Dept, E I du Pont de Nemours & Co, Inc, Parlin, NJ, 51-69, Wilmington, 69-73 & Towanda, Pa, 73-74, res assoc, 74-83, sr res assoc, 83-90; RETIRED. *Concurrent Pos:* Consult, J F Beeman & Assocs, Wysox, Pa, 90. *Res:* Coating application methods and equipment development for photographic products. *Mailing Add:* RD One Box 76A Towanda PA 18848

FEATHERSTON, FRANK HUNTER, b Washington, DC, Mar 9, 29; m 51; c 2. PHYSICS, SCIENCE POLICY. *Educ:* US Naval Acad, BS, 50; US Naval Postgrad Sch, MS, 57, PhD(physics), 63; Sch Law, Univ Va, JD, 82. *Prof Exp:* Naval aviator, US Navy, 50-75, Patrol Squad Twenty-One, 52-54, asst nuclear physics, Radiation Lab, Univ Calif, Berkeley, 56-57, mem opers res staff, Off Comdr, Aircraft Early Warning Barrier, Pac, 57-58, SNAP reactor proj officer, Div Reactor Develop, US AEC, 58-60, res officer astronaut, US Naval Missile Ctr, Calif, 63-65, Phoenix Proj Mgr, Hq, Naval Mat Command, DC, 66-68, dep for avionics & armament, F-14/Phoenix Proj, Naval Air Systs Command, 68-70, commanding officer, Naval Training Device Ctr, 70-72, chief, Naval Training Support, Off Naval Res, 72-73, dep & asst chief, 74-75; consult, Veda, Inc, 75-85; lectr systs eng, Sch Eng & Appl Sci, 77-78, RES ASSOC, CTR LAW & NAT SECURITY, SCH LAW, UNIV VA, 84- *Mem:* Am Phys Soc; Inst Elec & Electronics Engrs. *Res:* K-meson research in nuclear emulsions; low temperature elastic constants of body-centered cubic transition metals; law and policy study of military major system acquisitions. *Mailing Add:* 15 Deer Path Charlottesville VA 22901

FEATHERSTONE, JOHN DOUGLAS BERNARD, b Stratford, NZ, Apr 26, 44; m 67; c 2. CHEMISTRY OF DENTAL DECAY, FLUORIDE CHEMISTRY. *Educ:* Victoria Univ Wellington, NZ, BSc, 65, PhD(chem), 77; Univ Manchester, Eng, MSc, 75. *Prof Exp:* Qual control chemist, Unilever Ltd, NZ, 64-66; tech mgr, Chem Indust Ltd, NZ, 66-72; prod mgr, Quinoderm Pharmaceut, Eng, 72-75; lectr pharmaceut chem, Cent Inst Technol, NZ, 77-79; Med Res Coun sr fel, Med Res Coun Dent Res Unit, NZ, 79-80; sr res assoc, 80-88, CHMN DEPT ORAL BIOL & SR SCIENTIST, EASTMAN DENT CTR, ROCHESTER, NY, 88- *Concurrent Pos:* Res fel, Sandoz Pharmaceut Sch, Univ Manchester, 74-75; Med Res Coun Training fel & lectr chem, Victoria Univ Wellington, 75-77; dir res proj, Med Res Coun, New Zealand, 77-80, sr fel, 79-80; mem, Flouridation Adv Comt, New Zealand Govt, 79-, Expert Groups Flouridated Dentifries, WHO, 81-; prin investr, NIH & Nat Inst Dent Res Grant, 80-; asst prof dent sci, Dept Dent Res, Univ Rochester, 80-83, assoc prof, 83- *Honors & Awards:* Colgate Res Prize, Int Asn Dent Res Australia, 76, Edward Hattone Award, 77; Hamilton Award, Royal Soc NZ, 79. *Mem:* Fel NZ Inst Chem; Int Asn Dent Res; Europ Orgn Caries Res; NY Acad Sci; AAAS. *Res:* Prevention of dental decay; fundamental mechanism of dental decay; fluoride and dental health; calcium phosphate chemistry; saliva chemistry. *Mailing Add:* Eastman Dent Ctr 625 Elmwood Ave Rochester NY 14620

FEAY, DARRELL CHARLES, b Larchwood, Iowa, Mar 13, 27; m 52. POLYMER CHEMISTRY. *Educ:* Univ Iowa, BS, 50; Univ Calif, PhD(chem), 54. *Prof Exp:* Asst chem, Univ Calif, 50-51, chemist, radiation lab, 51-54; res chemist, 54-65, proj leader, 64-70, res specialist, Dow Chem Co, 70-79, res leader, 79-82, RES ASSOC, DOW CHEM, USA, 82- *Mem:* Soc Plastic Engrs; Am Chem Soc. *Res:* Application and development of polymerization catalysts; stereoselective catalysts; inorganic and polymer solvent extraction; plastics extrusion and molding; spectroscopy; analytical chemistry; water purification; separation membranes; polymer characterization. *Mailing Add:* 67 La Cuesta Orinda CA 94563

FEAZEL, CHARLES ELMO, JR, b Crockett, Tex, Aug 10, 21; m 42; c 2. ORGANIC CHEMISTRY. *Educ:* Harvard Univ, AB, 41; Univ Md, MS, 50, PhD(chem), 53. *Prof Exp:* Chemist, B F Goodrich Co, 41-46; chemist, Appl Physics Lab, Johns Hopkins Univ, 46-53; res chemist & dir phys sci res, Southern Res Inst, 53-78; consult, 78-88; RETIRED. *Concurrent Pos:* Prof, Birmingham-Southern Col, 54-55 & 59-60. *Res:* Organic synthesis; cellulose; plastics; polymer synthesis; technical editing and writing. *Mailing Add:* 3929 Knollwood Trace Birmingham AL 35243-1140

FEAZEL, CHARLES TIBBALS, b Akron, Ohio, Apr 30, 45; m 68; c 2. PETROLEUM EXPLORATION, CABONATE. *Educ:* Ohio Wesleyan Univ, BA, 67; Johns Hopkins Univ, MA, 69, PhD(geol), 75. *Prof Exp:* From res geologist to sr geologist, Phillips Petrol Co, 75-81, res supvr, 81-85, chief geologist, 85-89, DIR DEVELOP GEOL, PHILLIPS PETROL CO, 89- *Concurrent Pos:* Adj prof, Univ Minn, 78-79. *Mem:* Soc Econ Paleontologists & Mineralogists; Am Asn Petrol Geologists. *Res:* Sedimentary basin analysis, including sedimentation, stratigraphy, diagenesis, geochemistry, geophysics and numerical modeling; primary interest in carbonate rocks; science writing and science education. *Mailing Add:* 1531 Hoveden Dr Katy TX 77450

FECHHEIMER, MARCUS, b Columbus, Ohio, Mar 13, 52. CELL MOTILITY, ACTIN-BINDING PROTEINS. *Educ:* Wash Univ, BA, 74; Johns Hopkins Univ, PhD(biol), 80. *Prof Exp:* Postdoctoral cell biol, Harvard Univ, 81-82, Carnegie-Mellon Univ, 83-84; asst prof zool, 84-90, ASSOC PROF ZOOL, UNIV GA, 90- *Concurrent Pos:* Adj app biochem, Univ Ga, 86- *Mem:* AAAS; Am Soc Cell Biol. *Res:* Cell biology; biochemical, cellular and genetic aspects of cytoplasmic structure and organization; molecular mechanisms of cell movements. *Mailing Add:* Dept Zool Univ Ga Athens GA 30602

FECHHEIMER, NATHAN S, b Cincinnati, Ohio, May 24, 25; m 46; c 2. ANIMAL GENETICS, CYTOGENETICS. *Educ:* Ohio State Univ, BS, 49, MSc, 50, PhD(dairy sci), 57. *Prof Exp:* From instr to assoc prof, 52-65, dir, Animal Reproduction Teaching & Res Ctr, 74, PROF DAIRY SCI, OHIO STATE UNIV, 65- *Concurrent Pos:* NATO fel, Univ Edinburgh, 59-60, sr res fels genetics, 70-71 & 77-78; mem animal health comt, Nat Res Coun, 68-72; vis prof, Fed Inst Technol, Zurich, Switz, 88. *Mem:* Fel AAAS; Genetics Soc Am; Am Soc Animal Sci; Soc Study Reproduction; Am Dairy Sci Asn. *Res:* Mammalian and avian genetics and cytogenetics; genetic influence on reproductive performance. *Mailing Add:* Dept Dairy Sci Ohio State Univ 2027 Coffey Rd Columbus OH 43210-1094

FECHNER, GILBERT HENRY, b Northbrook, Ill, Dec 20, 22; m 48; c 3. FOREST GENETICS. *Educ:* Colo State Univ, BS, 47, MS, 55; Univ Minn, PhD(forestry), 64. *Prof Exp:* Wood technologist, Hallack & Howard Lumber Co, 47-49; staff forester, Tenn Valley Authority, 49-53; from instr to assoc prof forest mgt, 54-67, prof forest genetics, 67-89, EMER PROF FOREST GENETICS, COLO STATE UNIV, 89- *Concurrent Pos:* Consult, US Peace Corps, Peru, 74, Rep Chile, 75, FAO, Rome, 85 & 88, Argonne Nat Lab, 86-90; NSF Sci Fac Fel, 60-61. *Mem:* Fel Soc Am Foresters; Bot Soc Am. *Res:* Ecotypic variation of coniferous and broadleaved species. *Mailing Add:* Dept Forest & Wood Sci Colo State Univ Ft Collins CO 80523

FECHTER, ALAN EDWARD, b New York, NY, Oct 19, 34; m 72; c 4. LABOR MARKETS FOR SCIENTISTS & ENGINEERS, CAREER CHOICE. *Educ:* Col City New York, BBA, 58; Univ Chicago, AM, 62. *Prof Exp:* Asst prof econ, Univ Pittsburgh, 62-64 & Inst Defense Anal, 64-72; sr res assoc econ, Urban Inst, 72-78; sect head, div sci resources studies, NSF, 78-83; EXEC DIR, NAT RES COUN, OFF SCI & ENG PERSONNEL, NAT ACAD SCI, 83- *Concurrent Pos:* Prof lectr, Am Univ, 72-73; econ consult, NSF, 74. *Mem:* Am Econ Asn; Soc Govt Economists (vpres, 73); Indust Relations Res Asn; Am Asn Higher Educ; AAAS. *Res:* Analysis of the socioeconomic characteristics and behavioral parameters of scientific and technical personnel; contribution of these personnel to the production of scientific knowledge and to technological change. *Mailing Add:* 2101 Constitution Ave NW Washington DC 20418

FECHTER, LAURENCE DAVID, b New York NY. HEARING RESEARCH, NEUROTOXICOLOGY. *Educ:* Clark Univ, BA, 67; Kent State Univ, MA, 69; Univ Rochester, PhD(biopsychol), 73. *Prof Exp:* Instr social sci, Mohawk Valley Community Col, 68-69; asst prof, Genesee Community Col, 69-72; fel pharmacol, Biomed Ctr, Uppsala Univ, 73-74; fel physiol, 74-76, res assoc toxicol, 76-77, asst prof neurotoxicol, 77-83, ASSOC PROF NEUROTOXICOL, JOHNS HOPKINS UNIV, 84-, ASSOC PROF OTOLARYNGOL, 86- *Concurrent Pos:* Consult, Environ Protection Agency, 79, 84- & Nat Comn Air Quality, 81; vis assoc prof, Univ Mich Med Sch, 84-85. *Mem:* Soc Neurosci; Soc Toxicol; Behav Toxicol Soc; Asn Res Otolaryngol; Acoust Soc Am; fel Acad Toxicol Sci. *Res:* Mechanisms of auditory toxicity; mechanisms of auditory system injury; effects of environmental pollutants and contaminants. *Mailing Add:* 615 N Wolfe St Sch Hyg Johns Hopkins Univ Baltimore MD 21205

FECHTER, ROBERT BERNARD, b Cleveland, Ohio, Oct 8, 40; m 79; c 1. PHENOLIC RESIN SYNTHESIS & APPLICATIONS. *Educ:* Western Reserve Univ, BA, 62; Ohio State Univ, PhD(org chem), 66. *Prof Exp:* Res assoc nitrogen fixation, Stanford Univ, 66-69; sr scientist org & polymer chem, Owens-Ill Inc, 69-80; SR CHEMIST FOUNDRY RES, ASHLAND CHEM INC, 80- *Concurrent Pos:* Instr, Univ Toledo, 75-79. *Mem:* Am Chem Soc; Sigma Xi. *Res:* Synthesis, characterization and evaluation of phenolic resins; factors controlling phenolic resin structure and performance, especially in foundry binders. *Mailing Add:* Ashland Chem Inc PO Box 2219 Columbus OH 43216

FEDAK, GEORGE, b Hudson Bay, Sask, Dec 28, 40; m 69; c 2. CYTOGENETICS. *Educ:* Univ Saskatchewan, BSA, 63, MSc, 65; Univ Man, PhD(cytogenetics), 69. *Prof Exp:* Fel cytogenetics, 69-70, res scientist, 70-79, SECT HEAD CYTOGENETICS, OTTAWA RES BR AGR CAN, 79- *Mem:* Genetics Soc Can. *Res:* Conducting cytogenetic research on intergeneric hybrids in cereal crop species. *Mailing Add:* Cytogenetics Sect Plant Res Ctr Agr Can Ottawa ON K1A 0C6 Can

FEDDE, MARION ROGER, b Ionia, Kans, Oct 1, 35; m 56; c 3. PULMONARY PHYSIOLOGY. *Educ:* Kans State Univ, BS, 57; Univ Minn, MS, 59, PhD(avian physiol), 63. *Prof Exp:* Asst prof vet anat, Univ Minn, 63-64; from asst prof to assoc prof, 64-73, PROF PHYSIOL, KANS STATE UNIV, 73- *Concurrent Pos:* Sr Int fel, Fogarty Int Ctr, 84. *Honors & Awards:* Res Award, Poultry Sci Asn, 64; Fac Res Award, Col Vet Med, Kans State Univ, 70 & 79; US Sr Scientist Award, Alexander von Humboldt Found, 73. *Mem:* AAAS; Poultry Sci Asn; Am Physiol Soc; Am Soc Vet Physiol & Pharmacol. *Res:* Avian physiology, respiration; respiratory physiology; exercise physiology. *Mailing Add:* Dept Anat & Physiol Kans State Univ Manhattan KS 66506

FEDDER, STEVEN LEE, b Denver, Colo, July 28, 50. GEOCHEMISTRY, ANALYTICAL CHEMISTRY. *Educ:* Colo Col, BA, 72; Ariz State Univ, PhD(chem), 78. *Prof Exp:* Teaching & res assoc chem, Ariz State Univ, 72-78, vis asst prof, 78-82; AT DEPT CHEM, SANTA CLARA. *Mem:* Am Chem Soc; Soc Appl Spectros; Sigma Xi. *Res:* Trace metal behavior in natural waters with special interest in adsorption processes; trace metal detection by flameless atomic absorption. *Mailing Add:* Dept Chem Ariz State Univ Tempe AZ 85287

FEDDERN, HENRY A, b Poughkeepsie, NY, May 22, 38; m 68; c 1. ICHTHYOLOGY. *Educ:* Univ Miami, BS, 60, MS, 63, PhD(ichthyol), 68. *Prof Exp:* Asst, Inst Marine Sci, Univ Miami, 66-68; aquatic biologist, Precision Valve Corp, 68-69, dir marine lab, 69-73; aquaculturist, Neptunian Maricult, 73; COLLECTOR-AQUACULTURIST, 73- *Concurrent Pos:* Mem adv panel, Coral Mgt Plan, Gulf of Mex Fishery Mgt Coun; vchmn adv panel, Trop Reef Fish Mgt Plan, Gulf of Mex Fishery Mgt Coun. *Mem:* Am Soc Ichthyologists & Herpetologists; Am Littoral Soc. *Res:* Tolerances of marine inshore fishes to insecticides; factors influencing the survival and breeding of marine coral fishes and the mass-culture of the same. *Mailing Add:* 156 Dove Ave Tavernier FL 33070

FEDDERS, PETER ALAN, b Minneapolis, Minn, Feb 8, 39; m 63; c 2. PHYSICS. *Educ:* Yale Univ, BS, 61; Harvard Univ, MA, 62, PhD(physics), 65. *Prof Exp:* Res assoc physics, Princeton Univ, 65-66, instr, 66-68; from asst prof to assoc prof, 68-74, PROF PHYSICS, WASHINGTON UNIV, 74- *Mem:* Am Phys Soc. *Res:* Solid state physics. *Mailing Add:* Dept of Physics Washington Univ St Louis MO 63130

FEDER, DAVID O, b New York, NY, Oct 6, 24; m 52; c 4. RECHARGEABLE BATTERIES, PRIMARY BATTERIES. *Educ:* Columbia Univ, BS, 46, MS, 48, PhD(chem eng), 59. *Prof Exp:* Process engr fused salt electrolysis, Am Electrometals Corp, NY, 48-49; mem tech staff, electron tube mat & process, Bell Tel Labs, 54-61, supvr battery design & appln, 61-68, head dept battery develop & appln, 68-83; mgr, dept battery appln, AT&T Technologies, 83-84; CONSULT BATTERY TECHNOL & PRES, ELECTROCHEM ENERGY STORAGE SYSTS, 84- *Concurrent Pos:* Mem mgt comt, Int Telecommunications Energy Conf, 78-82, adv bd, 81-83 & 85-, prog comt, 84, 86 & 88; mem, Nat Battery adv comt, 78-84, Battery Mat Task Force, Nat Mat adv bd, Nat Acad Sci, 79-82, US Dept Energy Proj rev comt, 82-83, Am Nat Standards Inst, Battery comt & Inst Elec & Electronics Engrs Battery Standards Working comt, 82-, Secy IEC/TC82WG4, US Rep IEC/TC21, 87- *Honors & Awards:* Silver Medal Award, Am Electroplates Soc, 60. *Mem:* Sigma Xi; Electrochem Soc; Am Chem Soc; Am Inst of Chemists; Sr mem, Inst Elec & Electronics Engrs. *Res:* Battery research, development, design and introduction into manufacture; customer applications, usage and problem solving; lead acid, nickel-cadmium and lithium batteries. *Mailing Add:* Electrochem Energy Storage Systs Inc 35 Ridgedale Ave Madison NJ 07940

FEDER, HARVEY HERMAN, b New York, NY, Mar 28, 40; m 61; c 3. ANATOMY, REPRODUCTIVE PHYSIOLOGY. *Educ:* City Col New York, BS, 61; Univ Ore, PhD(anat), 66. *Prof Exp:* Asst scientist, Ore Regional Primate Res Ctr, 63-70; assoc prof psychol, 70-74, PROF PSYCHOL, INST ANIMAL BEHAV, RUTGERS UNIV, 74- *Concurrent Pos:* USPHS res fel, 65-67, at Oxford Univ, 66-67; NIH res grant, 69-72; NIMH career develop award, 70- *Mem:* Endocrine Soc; Int Soc Psychoneuroendocrinol; AAAS; Brit Soc Endocrinol. *Res:* Role of gonadal steroids in differentiation of sexual behavior of females; estimations of circulation gonadal steroids. *Mailing Add:* Dept Biol Sci Rutgers Univ Newark Campus Newark NJ 07102

FEDER, HOWARD MITCHELL, b New York, NY, June 8, 22; m 85; c 4. MARINE BIOLOGY. *Educ:* Univ Calif, Los Angeles, AB, 48, MA, 51; Stanford Univ, PhD(marine biol), 57. *Prof Exp:* Asst gen zool, Univ Calif, Los Angeles, 48, asst protozool & human anat, 50-51; asst marine biol, Kerckhoff Marine Lab, Calif Inst Technol, 49; asst, Arctic Res Lab, Point Barrow, Alaska, 49-50; oceanographic technician, Hopkins Marine Sta, Stanford Univ, 51-52, asst marine invert zool, 52 & 54; instr biol, Hartnell Col, 55-65, prof, 65-70; assoc prof zool & marine sci, 70-76, PROF ZOOL & MARINE SCI, UNIV ALASKA, FAIRBANKS, 76- *Concurrent Pos:* Am Acad Arts & Sci res grants, 60-64; NSF res grants, 62-67 & 79-81; Sea Grant res grant, 70-84, Marine Biol Lab, Helsingor, Denmark, 64-66, & Dunstaffnage Marine Biol Lab, Oban, Scotland, 80. *Mem:* Marine Biol Asn UK; Sigma Xi; Nat Shellfisheries Asn; AAAS. *Res:* Benthic biology of the Gulf of Alaska, the Bering Sea, and the Beaufort Sea; Scottish fjords; intertidal biology; sea star and brittle star biology; fisheries biology clams; feeding biology of shrimps, crabs, demersal fishes. *Mailing Add:* Dept Biol Sci Univ Alaska Fairbanks AK 99701

FEDER, JOSEPH, b St Louis, Mo, Feb 20, 32; m 53; c 4. BIOCHEMISTRY. *Educ:* Roosevelt Univ, BS, 53; Ill Inst Technol, MS, 61, PhD(biochem), 64. *Prof Exp:* Res biochemist, 65-70, res group leader, 70-73, sci fel, 73-77, SR SCI FEL, 77-, DISTINGUISHED FEL, DIR BIOCHEM/CELL CULTURE, MONSANTO CO, 86-; from asst prof to assoc prof, 66-75, PROF BIOCHEM, UNIV MO, ST LOUIS, 75-; PRES & CHIEF EXEC OFFICER, INVITRON CORP, 89- *Concurrent Pos:* From asst prof to assoc prof biochem, Univ Mo, St Louis, 66-75, prof 75- *Mem:* Am Soc Biol Chemists; AAAS; Am Chem Soc; Sigma Xi; Tissue Cult Asn; Am Soc Cell Biol. *Res:* Cell culture: the development of large scale systems for growth and maintenance of mammalian cells; biochemistry of in vitro grown and maintained animal cells; cell growth factors; biochemistry of angiogenesis; endothelial cell growth factors; fibrinolysis; role of glycosylation on plasminogen activators; proteolytic enzymes, mechanisms, synthetic substitutes and collagenases. *Mailing Add:* Invitron Corp 311 N Lindbergh Blvd St Louis MO 63141

FEDER, RALPH, b Philadelphia, Pa, Jan 12, 22; m 50; c 3. MICROSCOPY. *Educ:* Ind Univ, BA, 50; Univ Pa, MS, 55. *Prof Exp:* Physicist, Pitman-Dunn Lab, Frankford Arsenal, Philadelphia, 49-61; physicist, Thomas J Watson Res Ctr, IBM Corp, 61-87; RETIRED. *Mem:* NY Acad Sci; fel Am Phys Soc; Am Vacuum Soc; Sigma Xi. *Res:* X-ray microscopy and x-ray lithography. *Mailing Add:* River Rd Hyde Park NY 12538

FEDER, RAYMOND L, b New York, NY, Apr 1, 20; m 43; c 2. CHEMICAL ENGINEERING. *Educ:* Polytech Inst Brooklyn, BChE, 43, MChE, 47, DChE, 49. *Prof Exp:* Sr engr penicillin develop, Schenley Lab, 43-46; plant engr indust enzymes, Takamine Lab, 48-51; supvr process develop, Plastics Div, Allied Chem Corp, 51-62, mgr develop lab, 62-63 & nylon res & develop, 63-65, mgr polyolefins res & develop, 65-67, tech mgr, Frankford Plant, 67-72, plant mgr, 72-76, tech dir, 77-79; CONSULT, ALLIED-SIGNAL, INC. *Concurrent Pos:* Instr, Eve Div, Drexel Inst Technol, 52-58, adj prof, 58-62. *Mem:* Am Chem Soc; Am Inst Chem Engrs. *Res:* Process development of organic chemicals; polyolefins and nylons; economic evaluation and process design; process improvement; quality control; process engineering for phenol from cumene, phthalic anhydride from naphthalene; natural tar acids from carbolic oil; environmental science relative to inorganic and organic chemicals. *Mailing Add:* Myer-4, Allied-Signal Inc PO Box 1139 Morristown NJ 07962

FEDER, WILLIAM ADOLPH, b New York, NY, Oct 15, 20; m 45; c 3. POLLUTION BIOLOGY. *Educ:* Johns Hopkins Univ, AB, 41; Univ Calif, PhD(plant path), 50. *Prof Exp:* Asst bot & plant physiol, Univ Calif, 48-49; storage dis pathologist, Indust Res Adv Coun, Exp Sta, Univ Hawaii, 50-51; from asst prof to assoc prof plant path, NY State Col Agr, Cornell Univ, 51-54; plant pathologist, Crops Res Div, Agr Res Serv, USDA, 54-66; PROF PLANT PATH & LEADER AIR POLLUTION RES PROJS, UNIV MASS, AMHERST, 66- *Concurrent Pos:* NSF sr fel, Eng, 58-59; Fulbright res scholar, Israel, 64- *Mem:* AAAS; Am Phytopath Soc; Bot Soc Am; Air Pollution Control Asn. *Res:* Environmental impact of saline aerosols from cooling towers; biological control of plant diseases; effects of air pollution on plant growth; environmental impact of waste to energy incineration. *Mailing Add:* Suburban Exp Sta Univ Mass 240 Beaver St Waltham MA 02254

FEDERBUSH, PAUL GERARD, b Newark, NJ, Mar 23, 34; m 56; c 2. THEORETICAL PHYSICS. *Educ:* Mass Inst Technol, BS, 55; Princeton Univ, PhD(physics), 58. *Prof Exp:* From instr to asst prof physics, Mass Inst Technol, 58-65; from lectr to assoc prof, 65-71, PROF MATH, UNIV MICH, ANN ARBOR, 71- *Mem:* Am Phys Soc; Am Math Soc. *Res:* Constructive field theory. *Mailing Add:* Dept Math Univ Mich Ann Arbor MI 48109-1092

FEDERER, C ANTHONY, b New York, NY, Jan 19, 39; m 60; c 2. FOREST SOILS, FOREST METEOROLOGY. *Educ:* Univ Mass, BS, 59; Univ Wis, MS, 62, PhD(soils), 64. *Prof Exp:* Assoc meteorologist, 64-70, prin meteorologist, 70-80, PRIN SOIL SCIENTIST, NORTHEASTERN FOREST EXP STA, US FOREST SERV, 80- *Concurrent Pos:* Adj assoc prof, Univ NH, 70- *Mem:* Am Meteorol Soc; Am Geophys Union; Am Soc Agron; Soil Sci Soc Am. *Res:* Weather and tree growth; nitrogen in forest soils; nutrient cycling in forests; water relations of trees; evapotranspiration from forests. *Mailing Add:* Northeastern Forest Exp Sta PO Box 640 Durham NH 03824

FEDERER, HERBERT, b Vienna, Austria, July 23, 20; nat US; m 49; c 3. MATHEMATICAL ANALYSIS, GEOMETRY. *Educ:* Univ Calif, AB, 42, PhD(math), 44. *Prof Exp:* From instr to prof, 45-66, Florence Pirce Grant Univ prof, 66-85, EMER FLORENCE PIRCE GRANT UNIV PROF MATH, BROWN UNIV, 85- *Concurrent Pos:* Sloan res fel, 57-60; NSF fel, 64-65; mem, Nat Res Coun, 66-69; Guggenheim fel, 75-76; colloquim lectr, Am Math Soc, 77. *Honors & Awards:* Steele Prize, Am Math Soc, 87. *Mem:* Nat Acad Sci; Am Math Soc (assoc secy, 67-68); Am Acad Arts & Sci. *Res:* Geometric measure theory. *Mailing Add:* Dept Math Brown Univ Providence RI 02912

FEDERER, WALTER THEODORE, b Cheyenne, Wyo, Aug 23, 15; m 82; c 1. BIOSTATISTICS. *Educ:* Colo State Col, BS, 39; Kans State Col, MS, 41; Iowa State Univ, PhD(math statist), 48. *Prof Exp:* Asst corn invests, USDA & Kans State Col, 39-41, Bur Agr Econ & Iowa State Col, 41-42, assoc geneticist, Spec Guayule Res Proj, Bur Plant Indust, USDA, 42-44, assoc agr statistician, Bur Agr Econ & Iowa State Col, 44-48; prof biostatist & in charge Biomet Unit, 48-77 & 81-86, LIBERTY HYDE BAILEY PROF BIOSTATIST, CORNELL UNIV, 77-, EMER PROF, 86- *Concurrent Pos:* Mem consult panel, 53-61; head dept exp statist, Hawaiian Sugar Planters' Asn & consult, Pineapple Res Inst, 54-55; reviewer, Math Rev & Math Tables & Other Aids Comput, 57-58; prof, Univ & US Army Math Res Ctr, Univ Wis, 62-63 & 69-70; chmn & exec secy, Comt Pres of Statist Socs, 64-71; bk rev ed, Biomet, 64-72; assoc ed, Biomet, 72-76; assoc ed, J Statist Planning & Inference, 77- & Int J Math & Statist, 79- *Mem:* Fel AAAS; Int Statist Inst; fel Am Statist Asn; fel Inst Math Statist; fel Royal Statist Soc. *Res:* Statistical design, statistical education and statistical analyses. *Mailing Add:* 337 Warren Hall Cornell Univ Ithaca NY 14850

FEDERICI, BRIAN ANTHONY, b Paterson, NJ, May 28, 43. INVERTEBRATE PATHOLOGY, VIROLOGY. *Educ:* Rutgers Univ, New Brunswick, BS, 66; Univ Fla, MS, 67, PhD(med entom), 70. *Prof Exp:* NIH fel, Boyce Thompson Inst, Yonkers, NY, 72-74; asst prof, 74-80, ASSOC PROF ENTOM, UNIV CALIF, RIVERSIDE, 80- *Mem:* Entom Soc Am; Soc Invert Path; AAAS; Sigma Xi. *Res:* Pathogens and pathology of invertebrates, particularly aquatic invertebrates; diseases of medically important human and animal disease vectors, especially mosquitoes. *Mailing Add:* 4995 Chicago Ave Riverside CA 92507

FEDERICO, OLGA MARIA, b New York, NY, Dec 12, 23. ANIMAL PHYSIOLOGY. *Educ:* Hunter Col, BA, 46; Long Island Univ, MS, 60; NY Univ, PhD(biol), 68. *Prof Exp:* Res asst rheumatologic dis, Hosp Spec Surg, 50-60; instr biol, Long Island Univ, 60-65; USPHS trainee hemat, NY Univ, 65-68; instr biol, Hunter Col, 66-67; instr biol, Queens Col, NY, 67-68; asst prof, 68-72, ASSOC PROF BIOL SCI & GEOL, QUEENSBOROUGH COMMUNITY COL, 72- *Res:* Hematology; medical laboratory science; plant ecology; environmental science. *Mailing Add:* Dept Biol Sci Queensborough Community Col 56th Ave Springfield Bayside NY 11364

FEDERIGHI, ENRICO THOMAS, b Norfolk, Va, Nov 1, 27; m 65; c 1. FIBONACCI SERIES, HIGH POWER RESIDUES. *Educ:* Antioch Col, AB, 50; Johns Hopkins Univ, MA, 54; Ind Univ, PhD, 57. *Prof Exp:* Jr instr math, Johns Hopkins Univ, 50-53; jr engr, Bendix Radio, 53-55, asst proj engr, 56-57; SR MATHEMATICIAN, APPL PHYSICS LAB, JOHNS HOPKINS UNIV, 58- *Mem:* Math Asn Am. *Res:* Algebra and number theory; statistics. *Mailing Add:* 5029 Round Tower Pl Columbia MD 21044-1322

FEDERIGHI, FRANCIS D, b Xenia, Ohio, Oct 19, 31; m 55; c 2. PHYSICS. *Educ:* Oberlin Col, BA, 53; Harvard Univ, MA, 55, PhD(physics), 61. *Prof Exp:* Theoret physicist, Knolls Atomic Power Lab, Gen Elec Co, 59-66; assoc prof comput sci, State Univ NY Albany, 66-81, PROF COMPUT SCI, UNION COL, SCHENECTADY, NY, 82- *Concurrent Pos:* Guest scientist, Swiss Fed Inst Reactor Res, 63-64. *Honors & Awards:* Mgt Award, Gen Elec Co, 60. *Mem:* Asn Comput Mach; Math Asn Am; Am Phys Soc; Inst Elec & Electronics Engrs Computer Soc. *Res:* Nuclear reactor physics; programming languages; numerical methods. *Mailing Add:* 2109 Baker Ave Schenectady NY 12309-2301

FEDERLE, THOMAS WALTER, b Cincinnati, Ohio, June 5, 52; m 76; c 2. ENVIRONMENTAL MICROBIOLOGY, MICROBIAL ECOLOGY. *Educ:* Univ Cincinnati, BS, 74, MS, 76, PhD(biol & microbiol), 81. *Prof Exp:* Instr microbial ecol, Univ Cincinnati, 81; res assoc, Fla State Univ, 81-82; asst prof, Univ Ala, Birmingham, 82-85; STAFF SCIENTIST, PROCTER & GAMBLE CO, 85- *Mem:* Sigma Xi; Am Soc Microbiol; Am Soc Limnol & Oceanog. *Res:* Ecology of microorganisms including the role of the microorganism in the degradation of natural and xenobiotic compounds; effect of pollutants on microbial processes. *Mailing Add:* Proctor & Gamble Environ Safety Dept Ivorydale Tech Ctr Cincinnati OH 45217

FEDERMAN, DANIEL D, b New York, NY, Apr 16, 28. MEDICAL ADMINISTRATION. *Educ:* Harvard Univ, BS, 49, MD, 53; Am Bd Internal Med, cert, 62. *Prof Exp:* Intern med, Mass Gen Hosp, Boston, 53-54, asst res med, 54-55, fel med, 58-60; sr asst surg endocrinol, USPHS, NIH, 55-57; prof & chmn dept med, Stanford Univ, 72-77; from instr to assoc prof, 60-72, dean, students & alumni, 77-89, PROF MED, MED SCH, HARVARD UNIV, 77-, DEAN MED EDUC, 89-; PHYSICIAN PETER BENT BRIGHAM HOSP, BOSTON, 77- *Concurrent Pos:* Asst med, Mass Gen Hosp, 60-63, asst physician, 64-66, chief, Endocrine Unit, 64-67, assoc physician, 66-70, asst chief med serv, 67-72, physician, 70-72 & 77-; vis prof med, Middlesex Hosp Med Sch, Univ London, 72-73; physician-in-chief, Stanford Univ Hosp, 73-77, Arthur L Bloomfield prof med, Sch Med, 75-77; mem, Endocrine Test Comt, Am Bd Internal Med, 74-76, chmn, 77-78; chmn, Federated Coun Internal Med, 77-78. *Mem:* Inst Med-Nat Acad Sci; master Am Col Physicians (pres, 82-83); Endocrine Soc; Am Soc Human Genetics; Asn Am Physicians; Am Clin & Climat Asn; Am Geriat Soc. *Mailing Add:* Off Dean Students Harvard Med Sch 25 Shattuck St Boston MA 02115

FEDERMAN, MICHELINE, b Paris, France, Jan 1, 39; US citizen. CELL BIOLOGY, ELECTRON MICROSCOPY. *Educ:* Long Island Univ, BS, 61; Rutgers Univ, PhD(cell biol), 66. *Prof Exp:* Res assoc cell biol & electron micros, Douglass Col, Rutgers Univ, 66-67 & Rutgers Med Sch, 67-68; RES ASSOC CELL BIOL & ELECTRON MICROS, CANCER RES INST, NEW ENG DEACONESS HOSP, 68-, SCI ASSOC, DEPT PATH, 75-; ASSOC PATH, HARVARD MED SCH, 75- *Mem:* AAAS; Electron Micros Soc Am; Am Soc Cell Biol. *Res:* Ultrastructural aspects and function of normal and pathological tissues. *Mailing Add:* Cancer Res Inst New Eng Deaconess Hosp Boston MA 02215

FEDEROWICZ, ALEXANDER JOHN, b Hartford, Conn, May 28, 35; m 59; c 2. APPLIED MATHEMATICS. *Educ:* Carnegie Inst Technol, BS, 57, MS, 58, PhD(math), 63. *Prof Exp:* Sr mathematician, 62-73, FEL MATHEMATICIAN, WESTINGHOUSE RES & DEVELOP CTR, 73- *Mem:* Sigma Xi; Math Prog Soc. *Res:* Mathematical programming techniques in electrical and nuclear fission and fusion design problems; geometric, integer and linear programming; electrical utility planning techniques. *Mailing Add:* 113 Country Club Dr Pittsburgh PA 15235

FEDERSPIEL, CHARLES FOSTER, b Flint, Mich, May 3, 29; m 57; c 1. OCCUPATIONAL HEALTH DISABILITY DATA. *Educ:* Univ Mich, AB, 50, AM, 52; NC State Col, PhD(statist), 59. *Prof Exp:* Statistician, Commun Dis Ctr, USPHS, 52-54; assoc prof biostatist, 59-76, PROF BIOSTATIST, SCH MED, VANDERBILT UNIV, 76- *Mem:* Am Statist Asn; Am Pub Health Asn; Am Col Epidemiol. *Res:* Worker's compensation statistics; disability data bases; health services research. *Mailing Add:* Div Biostatist Vanderbilt Med Ctr Nashville TN 37232-2637

FEDINEC, ALEXANDER, b Uzhorod, Czech, Jan 29, 26; nat US; m 52; c 2. ANATOMY. *Educ:* Univ Kans, MA, 57, PhD(anat), 58. *Prof Exp:* Instr anat, State Univ NY Downstate Med Ctr, 58-60; asst prof, Hahnemann Med Col, 60-62; from asst prof to assoc prof, 62-74, asst dean col med, 71-73, PROF ANAT, CTR HEALTH SCI, UNIV TENN, MEMPHIS, 74-, ASSOC DEAN STUDENT AFFAIRS, COL MED, 73- *Mem:* Am Soc Microbiol; Int Soc Toxicol; Soc Neurosci; Am Asn Anat; Am Soc Exp Path; Sigma Xi. *Res:* Mode of dispersal and action of neurotropic toxins; blood-brain barrier and placental permeability; teratology. *Mailing Add:* 5995 Quince Rd Memphis TN 38119

FEDOR, EDWARD JOHN, physiology; deceased, see previous edition for last biography

FEDOR, LEO RICHARD, b Boston, Mass, Jan 11, 34; m 62; c 2. ORGANIC CHEMISTRY. *Educ:* Mass Col Pharm, BS, 55, MS, 57; Ind Univ, PhD(org chem), 63. *Prof Exp:* Instr chem, ETex State Col, 58-59; fel, Cornell Univ, 63-64 & Univ Calif, Santa Barbara, 64-65; from asst prof to assoc prof, 65-84, PROF MED CHEM, STATE UNIV NY, BUFFALO, 84-, ASSOC DEAN STUDENT AFFAIRS, 90- *Mem:* Am Chem Soc. *Res:* Mechanisms of organic reactions related to enzymic reactions; design and synthesis of prodrugs. *Mailing Add:* Sch Pharm Cooke 457 State Univ NY Buffalo NY 14260

FEDOROFF, NINA V, b Cleveland, Ohio, Apr 9, 42; div; c 2. MOLECULAR BIOLOGY. *Educ:* Syracuse Univ, BS, 66; Rockefeller Univ, PhD(molecular biol), 72. *Prof Exp:* Asst prof biol, Univ Calif, Los Angeles, 72-74, Damon Runyan fel molecular biol, Sch Med, 74-75; NIH fel, 75-77, res assoc, 77-78, STAFF SCIENTIST, CARNEGIE INST WASH, 78- *Concurrent Pos:* Mem, Develop Biol Panel, NSF, 79-80; mem, Sci Adv Panel Appl Genetics, Office Technol Assessment, Cong US, 79-80; mem, NIH Recombinant DNA Adv Comt, 80-85; Phi Beta Kappa vis scholar, 84-85; mem, Coun Life Sci & Bd Basic Biol, Nat Acad Sci-Nat Res Coun, 85-90; mem, sci adv comt, Japanese Human Frontier Sci Prog, 88; bd dirs, Genetics Soc Am, 90- & Biosis, 90- *Honors & Awards:* Merit Award, NIH, 89; Howard Taylor Ricketts Award, 90. *Mem:* Nat Acad Sci; AAAS; Am Acad Arts & Sci; Sigma Xi. *Res:* Transposable elements in maize. *Mailing Add:* Dept Embryol 115 W University Pkwy Baltimore MD 21210

FEDOROFF, SERGEY, b Daugavpils, Latvia, Feb 20, 25; nat Can; m 54; c 4. EMBRYOLOGY, HISTOLOGY. *Educ:* Univ Sask, BA, 52, MA, 55, PhD(histol), 58. *Prof Exp:* Demonstr histol, 53-55, instr anat, 55-57, spec lectr, 57-58, from asst prof to assoc prof, 58-64, admin asst to dean med, 60-62, asst dean Col Med, 70-77, dir cell biol study Prog, 73-77, PROF ANAT & HEAD DEPT, UNIV SASK, 64- *Concurrent Pos:* Lederle med fac award, 57-60; mem steering comt study basic biol res Can, Sci Secretariat, 66; mem, Med Res Coun Assessment Group Anat Res Can, 67; mem studentship comt, Med Res Coun Can, 69-, mem, Coun, 73-; mem bd gov, W Alton Jones Cell Sci Ctr, NY, 70-72, vchmn, 80- *Mem:* Can Asn Anat (vpres, 65-66, pres, 66-67); Am Asn Anatomists; Pan-Am Asn Anat (pres, 72-75, hon pres, 75-); Can Soc Cell Biologists; Tissue Cult Asn (vpres, 64-68, pres, 68-72); Fedn Am Socs Exp Biol. *Res:* Cytogenetics; immunobiology; tissue culture; cell differentiation; author or coauthor of numerous scientific publications. *Mailing Add:* Dept Anat Univ Sask Saskatoon SK S7N 0W0 Can

FEDORS, ROBERT FRANCIS, b Bayonne, NJ, Jan 16, 34; m 61; c 3. POLYMER CHEMISTRY. *Educ:* Purdue Univ, BS, 55; Akron Univ, PhD(polymer chem), 62. *Prof Exp:* Chemist plastics, Gen Cable Corp, 55-56; chemist rubber & plastics, Edgewood Arsenal, 56-58; MEM TECH STAFF POLYMERS, JET PROPULSION LAB, CALIF INST TECHNOL, 62- *Mem:* Am Chem Soc. *Res:* Studies of the time dependent physical and mechanical properties of polymers, including extensive work on the time and temperature dependence of the ultimate properties of elastomers. *Mailing Add:* 2127 Pine Crest Dr Altadena CA 91001

FEDRICK, JAMES LOVE, b Lordsburg, NMex, Apr 17, 30; m 64; c 4. MEDICINAL CHEMISTRY, PHARMACEUTICS. *Educ:* Univ Ariz, BS, 53, MS, 55; Univ Ill, PhD(chem), 59. *Prof Exp:* Org chemist, Org Chem Res Sect, Lederle Lab Am Cyanamid Co, Stamford, Conn, 59-61, group leader, 61-63, tech dir fine chem dept, Pearl River, NY, 63-66, dir prod & process develop, Med Res, Lederle Labs, Cyanamid Int, 66-74, dir Pharmaceut & Mech Develop, 75-77, foreign plant coordr & agent, 77-90; CONSULT, 90- *Concurrent Pos:* Teacher, Bergen Col. *Mem:* Am Chem Soc; NY Acad Sci. *Res:* Heterocyclic synthesis; medicinal chemistry; management of international research and development; pharmaceutical product and process; foreign Patent Agent. *Mailing Add:* 51 Sparrow Bush Rd Mahwah NJ 07430

FEDUCCIA, JOHN ALAN, b Mobile, Ala, Apr 25, 43; m 76. ZOOLOGY, EVOLUTIONARY BIOLOGY. *Educ:* La State Univ, BS, 65; Univ Mich, MA, PhD(zool), 69. *Prof Exp:* Lectr zool, Univ Mich, 69; asst prof biol, Southern Methodist Univ, 69-71; from asst prof to assoc prof, 71-79, PROF BIOL, UNIV NC, CHAPEL HILL, 79-, ASSOC CHMN DEPT, 81- *Concurrent Pos:* Res assoc, Nat Mus Natural Hist, Smithsonian Inst, 78-87. *Mem:* Fel Am Ornith Union; Soc Vert Paleont; AAAS; Sigma Xi. *Res:* Avian evolution and systematics; avian paleontology. *Mailing Add:* Dept Biol CB No 3280 Coker Hall Univ NC Chapel Hill NC 27599-3280

FEE, JAMES ARTHUR, b Nokomis, Sask, Aug 30, 39; US citizen; m 60; c 2. PHYSICAL BIOCHEMISTRY. *Educ:* Pasadena Col, BA, 61; Univ Southern Calif, PhD(biochem), 67. *Prof Exp:* NSF fel biochem, Gothenburg Univ, 67-69; NIH trainee & res assoc biophys, Univ Mich, Ann Arbor, 69-70; asst prof chem, Rensselaer Polytech Inst, 70-74; assoc prof biol chem & assoc res biophysicist, 74-81, prof biol chem & res biophysicist, Univ Mich, Ann Arbor, 81-85; DIR, NIH RES STABLE ISOTOPE RESOURCE, LOS ALAMOS NAT LAB, 84- *Mem:* Am Chem Soc; Am Soc Biol Chemists. *Res:* Oxygen metabolism; role of metal ions in biological systems; mechanistic aspects of enzymatically catalyzed oxidation-reduction reactions. *Mailing Add:* Isotope & Nuclear Chem Los Alamos Nat Lab Los Alamos NM 87545

FEELEY, JOHN CORNELIUS, b Los Angeles, Calif, Mar 7, 33; m 57; c 3. IMMUNOLOGY. *Educ:* Univ Calif, Los Angeles, AB, 55, PhD, 58; Am Bd Microbiol, dipl, 64. *Prof Exp:* Asst bact, Univ Calif, Los Angeles, 55-58; from sr asst scientist to scientist, Div Biol Standards, NIH, 58-65, chief sect bact vaccines, 65-71, mem cholera adv comt, 68-72, chief bact immunol br, 71-81, asst dir, 81-84, DIR, DIV BACT DIS, CTR DIS CONTROL, USPHS, 84- *Concurrent Pos:* Mem cholera panel, US-Japan Coop Med Sci Prog, 65-73 & 78-84; sr scientist, USPHS, 66-73, scientist dir, 73-; mem, WHO Expert Panel Bact Dis, 67-; adj assoc prof parasitol & lab pract, Sch Pub Health, Univ NC, 68-; consult, US AID, 70-72 & Food & Drug Admin, 72-81; clin assoc prof path, Sch Med, Emory Univ, 77-; adj assoc prof biol, Ga State Univ, 82- *Mem:* AAAS; fel Am Acad Microbiol; Soc Exp Biol & Med; Am Soc Microbiol; fel Infectious Dis Soc Am; Sigma Xi; Am Asn Immunologists. *Res:* Hypersensitivity; hemagglutination; serum bactericidal action; standardization of biological and immunodiagnostic products; bacteriology and immunology of enteric diseases. *Mailing Add:* Div Bacterial Dis 1600 Clifton Rd NE Atlanta GA 30333

FEELY, FRANK JOSEPH, JR, b Chicago, Ill, Aug 26, 18; m 69; c 7. MECHANICAL ENGINEERING. *Educ:* Univ Mich, BS, 40. *Prof Exp:* Engr, Design Div, Standard Oil Develop Co, 40-48, group head, 48-51, asst supv engr, 51-55, asst dir, 55-59, assoc dir planning, Eng Div, Esso Res & Eng Co, 59-61, dir, Gen Eng Div, 61-64, asst gen mgr, Gen Mgr Off, 64-66, vpres & dir, 66-71; mgr opers coordr, Logistics Dept, Standard Oil Co, NJ, 71-77; vpres eng, Exxon Res & Eng Co, Florham Park, 77-81; RETIRED. *Concurrent Pos:* Mem, Am Petrol Inst, 53-; vpres & mem bd dir & exec comt, Am Nat Standards Inst, bd dir, 67-81. *Mem:* Nat Acad Eng; fel Am Soc Mech Engrs; Am Nat Standards Inst (pres, 78, 79). *Res:* Mechanical engineering developments in catalytic cracking including stress analysis of piping expansion joints; brittle fracture in steel. *Mailing Add:* PO Box 130 Center Harbor NH 03226

FEELY, HERBERT WILLIAM, b Brooklyn, NY, Apr 29, 28; m 67; c 4. GEOCHEMISTRY. *Educ:* City Col NY, BS, 50; Columbia Univ, MA, 52, PhD(geol), 56. *Prof Exp:* Instr, Upsala Col, 55-57; sr res geochemist, Isotopes, Inc, 57-67; assoc prof earth sci, Queens Col, NY, 67-76; RES CHEMIST, US DEPT ENERGY, NEW YORK, 76- *Concurrent Pos:* Res assoc, Columbia Univ, 56-57, sr res assoc, 67-76. *Mem:* Am Geophys Union; Am Meteorol Soc; Air & Waste Mgt Asn. *Res:* Atmospheric and marine geochemistry. *Mailing Add:* 31 Seneca Ave Emerson NJ 07630

FEELY, RICHARD ALAN, b Farmington, Minn, Feb 26, 47; m 71; c 2. CHEMICAL OCEANOGRAPHY, GEOCHEMISTRY. *Educ:* Col St Thomas, BA, 69; Tex A&M Univ, MS, 71, PhD(oceanog), 74. *Prof Exp:* Res asst chem oceanog, Tex A&M Univ, 69-73, res assoc, 73-74; RES ASSOC CHEM OCEANOG, UNIV WASH, 74-; OCEANOGR, PAC MARINE ENVIRON LAB, NAT OCEANOG & ATMOSPHERIC ADMIN, 74- *Mem:* AAAS; Am Geophy Union; Am Soc Limnol & Oceanog. *Res:* Factors influencing the major and trace element composition of marine particulate matter and sediments; reactions of trace metals at the freshwater-seawater interface; exchange of elements between sediments and seawater. *Mailing Add:* Pac Marine Environ Lab 7600 Sand Point Way NE Seattle WA 98115

FEELY, WAYNE E, b Brooklyn, NY, May 24, 31; m 55; c 5. PHOTOIMAGING & PHOTOCHEMISTRY, ORGANIC & POLYMER SYNTHESIS. *Educ:* Polytech Inst Brooklyn, BS, 53; Univ Rochester, PhD(chem), 57. *Prof Exp:* SR RES FEL CHEM, ROHM & HAAS CO, 57- *Mem:* Am Chem Soc; Soc Photog Scientists & Engrs; Int Soc Optical Engrs. *Res:* Novel photoresists for electronic imaging; general organic synthesis; polymer synthesis and applications. *Mailing Add:* 1172 Lindsay Lane Rydal PA 19046-1839

FEEMAN, GEORGE FRANKLIN, b Lebanon, Pa, Apr 16, 30; c 4. MATHEMATICS. *Educ:* Muhlenberg Col, BS, 51; Lehigh Univ, MS, 53, PhD(math), 58. *Prof Exp:* From instr to asst prof math, Muhlenberg Col, 54-59; instr, Mass Inst Technol, 59-61; from asst prof to assoc prof, Williams Col, 61-69; from actg chmn to chmn dept, 71-75, chmn dept, 78-81, assoc provost, 81-82, vprovost & grad dean, 82-84, PROF MATH, OAKLAND UNIV, 69- *Concurrent Pos:* NSF sci fac fel, 65-66; math coordr, African Math Proj, 65-66, Detroit Teacher Intern Proj, 69-70 & Urban Corps Proj, 69-70 & 70-71; math specialist, USAID team, Nepal, 75-77; math evaluator, Univ Qatar, 81; head math dept, Egyptian Air Acad Modernization Proj, 84-85, actg dean, 85, dean, 85- *Mem:* Am Math Soc; Math Asn Am. *Res:* Differential geometry, especially geometry of Riemannian manifolds; problems in mathematics education. *Mailing Add:* PO Box 258 Lake Orion MI 48035

FEEMAN, JAMES FREDERIC, b Lebanon, Pa, June 1, 22; m 47; c 4. INDUSTRIAL ORGANIC CHEMISTRY, RESEARCH ADMINISTRATION. *Educ:* Muhlenberg Col, BS, 45; Lehigh Univ, MS, 47, PhD(chem), 49. *Prof Exp:* Res Found fel, Ohio State Univ, 49-50; res chemist, Althouse Div, Crompton & Knowles Corp, Reading, Pa, 50-68, from asst dir to dir res, 68-80, vpres res & develop, Dyes & Chem Div, 80-86, sr scientist, 86-88; CONSULT SYNTHETIC DYES & CHEMICALS, 88- *Mem:* Fel AAAS; NY Acad Sci; fel Am Inst Chemists; Am Chem Soc; Am Asn Textile Chem & Colorists. *Res:* Synthetic dyes and specialty chemicals; textile chemistry; dye intermediates; chemical information management; synthetic organic chemistry. *Mailing Add:* Six Oriole Dr Wyomissing PA 19610

FEENEY, GLORIA COMULADA, b NJ, July 11, 25; div; c 2. PHARMACOLOGY, TOXICOLOGY. *Educ:* Univ Va, BS, 45; George Washington Univ, MS, 48, PhD(pharmacol), 53, MA, 84. *Prof Exp:* Instr pharmacol, Sch Med & Dent, Georgetown Univ, 54-57, asst prof, 57-59; toxicologist-pharmacologist, Bur Foods, Div Toxicol, Food & Drug Admin, 73-74; info scientist life sci, Smithsonian Sci Info Exchange, 74-75; consult info scientist, Technassociates, Inc, 78-79 & JRB Assocs, Inc, 79-80; zoo intern, Nat Zool Park, Washington, DC, 84; TECH ASST, ENVIRON PROTECTION AGENCY, 88- *Mem:* Soc Exp Biol & Med; Soc Petrol Engrs; Soc Cosmetic Chemists; Sigma Xi; Am Asn Mus. *Res:* Endocrinological influence of salicylates; cocaine and the autonomic nervous system; environmental, earth and marine sciences; ecology; technical management; profile of Asian small-clawed otter (Aonyx cinerea). *Mailing Add:* 618 Harvard Hall 1650 Harvard St NW Washington DC 20009-3740

FEENEY, ROBERT EARL, b Oak Park, Ill, Aug 30, 13; m 54; c 2. BIOCHEMISTRY, PROTEIN CHEMISTRY. *Educ:* Northwestern Univ, BS, 38; Univ Wis, MS, 39, PhD(biochem), 42. *Prof Exp:* Asst, Univ Wis, 40-42; res assoc, Harvard Med Sch, 42-43; chemist, Western Regional Res Lab, Bur Agr & Indust Chem, USDA, 46-53; prof chem & chmn dept biochem & nutrit, Univ Nebr, 53-60; prof, 60-84, RES BIOCHEMIST, AGR EXP STA & EMER PROF FOOD SCI & TECHNOL, UNIV CALIF, DAVIS, 84- *Concurrent Pos:* Chemist, Exp Sta, Univ Nebr, 53-60; prin investr, NIH & NSF, 56-; vis scholar, Univ Cambridge, Eng, 71; vis prof, Swiss Fed Inst Technol, Zurich, 72, Univ Bergen, 76, Univ Bielefeld, 79 & Hokkaido Univ, 81; distinguished vis prof, Mem Univ Nfld, 78-79. *Mem:* Am Soc Biochem & Molecular Biol; Am Chem Soc; Inst Food Technol. *Res:* Investigation of chemical, biochemical and physical techniques and their effects on proteins, including chemical improvement of food proteins for better function and nutritional properties; iron-binding sites of transferrins; water ice proteins interactions of antifreeze proteins; chemical modifications of proteins. *Mailing Add:* Dept Food Sci & Technol Univ Calif 109 Food Sci & Technol Bldg Davis CA 95616-5224

FEENEY, ROBERT K, b Albany, Ga, July 6, 38. MICROWAVE ELECTRICAL ENGINEERING. *Educ:* Ga Inst Technol, BAEE, 61, MSEE, 64, PhD(elec eng), 70. *Prof Exp:* PROF, SCH ELEC ENG, GA INST TECHNOL, 70- *Mem:* Inst Elec & Electronics Engrs; Am Phys Soc; Sigma Xi. *Mailing Add:* 762 Summer Dr Acworth GA 30101

FEENEY-BURNS, MARY LYNETTE, b Burnsville, WVa, Mar 5, 31; m 78. CELL BIOLOGY, OPHTHALMOLOGY. *Educ:* Col Mt St Joseph-on-the-Ohio, BA, 53; Univ Calif, San Francisco, MA, 64, PhD(endocrinol), 68. *Prof Exp:* From asst prof to assoc prof ophthal, Med Sch, Univ Ore, 70-79; assoc prof, 79-81, PROF OPHTHAL, SCH MED, UNIV MO-COLUMBIA, 81- *Concurrent Pos:* Nat Inst Neurol Dis & Blindness fel, Harvard Univ, 69-70; mem visual sci A study sect, NIH, 74-78; mem vision res rev comt, Nat Eye Inst, 85-89; res to prevent blindness, sr sci investr award. *Mem:* NY Acad Sci; AAAS; Am Soc Cell Biol; Asn Res Vision & Ophthal. *Res:* Correlated morphological and biochemical studies of normal and pathological ocular tissues. *Mailing Add:* Dept Ophthal Univ Mo Sch Med One Hospital Dr Columbia MO 65212

FEENSTRA, ERNEST STAR, b Grand Rapids, Mich, Oct 22, 17; m 44; c 3. VETERINARY PATHOLOGY. *Educ:* Mich State Univ, DVM, 42, MS, 44, PhD(animal path), 47; Am Col Vet Pathologists, dipl; Am Col Lab Animal Med, dipl. *Prof Exp:* Asst animal path, Mich State Univ, 42-47, asst prof, 47-48; sect head res div, Upjohn Co, 48-56, mgr path & toxicol res, 56-81; RETIRED. *Concurrent Pos:* Res prof path, Mich State Univ, 70- *Mem:* Int Acad Path; Am Sci Affiliation; Am Vet Med Asn; Am Asn Pathologists & Bacteriologists; Sigma Xi. *Res:* Experimental pathology and toxicology; diseases of laboratory animals. *Mailing Add:* 4374 Fox Farm Rd Manistee MI 49660

FEENY, PAUL PATRICK, b Birmingham, Eng, Feb 8, 40; US citizen; m 68; c 2. CHEMICAL ECOLOGY. *Educ:* Oxford Univ, BA, 60 & 63, BSc, 61, MA, 63, PhD(zool), 66. *Prof Exp:* Asst prof ecol, 67-72, fac trustee, 71-73, assoc prof, 72-78, PROF ENTOM, ECOL & SYSTS, CORNELL UNIV, 78-, CHAIR, 88- *Concurrent Pos:* Co-chmn, Gordon Conf, 80; John Simon Guggenheim fel, 83-84; vis fel, Cambridge Univ, Clare Hall, Eng, 83-84; mem, Panel Ecol Prog, NSF, 82-83, & prin investr, grant ecol prog, 72-; Cornell Biotech Prog, 85-88. *Mem:* AAAS; Ecol Soc Am; Soc Study Evolution. *Res:* Chemical ecology; evolution and ecological significance of secondary plant and animal chemical compounds, including attractants, repellents and toxins; roles of plant chemistry in insect evolution. *Mailing Add:* Sect Ecol & Systs Cornell Univ Corson Hall Ithaca NY 14853

FEERO, WILLIAM E, b Old Town, Maine, May 31, 38; m 66; c 3. ELECTRICAL ENGINEERING, TECHNICAL MANAGEMENT. *Educ:* Univ Maine, BSEE, 60; Univ Pittsburgh, MSEE, 69; Mass Inst Technol, EE, 72. *Prof Exp:* Staff engr, Westinghouse Elec Corp, 60-72, mgr, Transient Anal Sect, 72-76; chief overhead transmission br, Div Elec Energy Systs, Dept Energy, 76-78, prog mgr power supply integration prog, 78-79; vpres, Systs Eng Power, Inc, 79-80; PRES, ELEC RES & MGT, INC, 80- *Concurrent Pos:* Mem study comt, Nat Coun Radiation Protection, 85-; chmn, Working Group Biol Effects Power Frequency Elec & Magnetic Fields, Inst Elec & Electronics Engrs, mem Power Syst Relay Comt, chmn, Consumer Interface Protection Subcomt; mem, Int Conf Large High Voltage Elec Systs. *Mem:* Fel Inst Elec & Electronics Engrs. *Mailing Add:* 487 Nimitz Ave State College PA 16801

FEERST, IRWIN, b New York, NY, Nov 18, 27; m 50; c 2. ELECTRONICS ENGINEERING. *Educ:* City Col New York, BEE, 51; NY Univ, MEE, 55; Polytech Inst New York, MSEE, 74. *Prof Exp:* Pvt consult, 60-62; asst prof physics, Adelphi Univ, 62-69; PVT CONSULT, 69- *Concurrent Pos:* Consult, Am Mach & Foundry Co, 62-64 & Bucode Co, 69-75; NSF grant, 64-66; contrib ed, New Engr, 77-79; vis lectr, Nat Proj Philos & Eng Ethics, 78. *Honors & Awards:* Centennial Gadfly Award, Inst Elec & Electronics Engrs. *Mem:* Inst Elec & Electronics Engrs. *Res:* Servomechanisms; tape transport design and development; cathode ray tube deflection systems; spectral analysis of signals; pulse and digital circuits and systems; electronic warfare. *Mailing Add:* 368 Euclid Ave Massapequa Park NY 11762

FEESE, BENNIE TAYLOR, b Cane Valley, Ky, Dec 21, 37; m 60; c 2. MOLECULAR BIOLOGY. *Educ:* Centre Col, AB, 59; Wash Univ, PhD(molecular biol), 65. *Prof Exp:* Chmn life sci prog, 67-69, from asst prof to assoc prof, 64-76, PROF BIOL, CENTRE COL, 76-, CHMN MOLECULAR BIOL PROG COMT, 69-73 & 85- *Mem:* AAAS. *Res:* Erythropoietic aspects of amphibian metamorphosis. *Mailing Add:* Sci Div Centre Col Danville KY 40422

FEESER, LARRY JAMES, b Hanover, Pa, Feb 23, 37; m 61; c 2. CIVIL ENGINEERING, COMPUTERS. *Educ:* Lehigh Univ, BS, 58; Univ Colo, MS, 61; Carnegie Inst Technol, PhD(civil eng), 65. *Prof Exp:* From instr to prof civil eng, Univ Colo, Boulder, 58-74, res assoc, Comput Ctr, 70-73; prof civil eng & chmn dept, Sch Eng, 74-82, assoc dean, 82-85, VPROVOST, COMPUT & INFO TECHNOL, RENSSELAER POLYTECH INST, 85- *Concurrent Pos:* Mem, Nat Coop Hwy Res Prog; mem adv comt interactive comput graphics, NSF fac fel, Swiss Fed Inst Technol, 71-72. *Mem:* Am Soc Civil Engrs (nat dir, 79-82); Am Concrete Inst; Am Soc Eng Educ; Sigma Xi. *Res:* Dynamics and optimization of structures; computer applications; interactive computer graphics; highway computer graphics. *Mailing Add:* Dept Civil Eng Rensselaer Polytech Inst Troy NY 12180

FEFERMAN, SOLOMON, b New York, NY, Dec 13, 28; m 48; c 2. PROOF THEORY, CONSTRUCTIVE MATHEMATICS. *Educ:* Calif Inst Technol, BS, 48; Univ Calif, Berkeley, PhD(math), 57. *Prof Exp:* From instr asst prof to assoc prof, 56-68, PROF MATH, STANFORD UNIV, 68-, CHMN DEPT, 85- *Concurrent Pos:* Prin invest, Army Res Off grants, 56-67, NSF grants, 68-; NSF fel, Inst Advan Study, Princeton, 59-60, sr fel Univ Paris & Univ Amsterdam, 64-65; consult, Stanford Res Inst, 58-63; vis assoc prof, Mass Inst Technol, 67-68; Guggenheim fel, Univ Oxford & Univ Paris, 71-72, ETH Zurich & Univ Rome, 87-88; ed, Perspectives in Math Logic, 86-; fel, Stanford Humanities Ctr, 90-91; ed-in-chief, Collected Works Kurt Gödel, 82-; ed, Ergebrisse Mathematik, 86- *Mem:* Asn Symbolic Logic (pres 80-82); Am Math Soc; Math Asn Am; fel Am Acad Arts & Sci; Asn Comput Mach; Hist Sci Soc. *Res:* Foundations of explicit mathematics (recursive, constructive, predicative, hyper-arithmatic and so on); proof theory; type-free theories; history of modern logic; philosophy of mathematics; applications of logic to computer science. *Mailing Add:* Dept Math Stanford Univ Stanford CA 94305

FEFFERMAN, CHARLES LOUIS, b Washington, DC, Apr 18, 49; m 75; c 2. MATHEMATICAL ANALYSIS. *Educ:* Univ Md, BS, 66; Princeton Univ, PhD(math), 69. *Hon Degrees:* PhD, Univ Md, 79, Knox Col, 81, Bar-Ilan Univ, 85, Univ Madrid, 90. *Prof Exp:* Lectr math, Princeton Univ, 69-70; from asst prof to prof, Univ Chicago, 70-73; prof, 73-74, HERBERT JONES UNIV PROF, PRINCETON UNIV, 84- *Concurrent Pos:* Alfred P Sloan fel, 70-71; NATO fel, 71; vis prof, Calif Inst Technol, NY Univ, Univ Paris, Mittag-Leffler Inst, Sweden & Weitzmann Inst, Israel; Wilson Elkins vis prof, Univ Md; numerous invited lectr, US & foreign cols, insts & univs. *Honors & Awards:* Salem Prize, 71; First Recipient, Alan T Waterman Award, 76; Fields Medal, 78. *Mem:* Nat Acad Sci; Am Math Soc; Am Acad Arts & Sci; Am Philos Soc. *Res:* Fourier analysis; partial differential equations; several complex variables. *Mailing Add:* Dept Math Fine Hall Princeton Univ Washington Rd Princeton NJ 08544-1000

FEGER, CLAUDIUS, b Konstanz, Ger, June 3, 48; m 89; c 3. ELECTRONIC PACKAGING, OPTICAL WAVEGUIDES. *Educ:* Albert-Ludwigs Univ, Freiburg, Ger, dipl chem, 76; Inst Macromolecular Chem, Univ Freiburg, Ger, Dr rer nat, 81. *Prof Exp:* Postdoctoral, Polymer Sci Eng Dept, Univ Mass Amherst, 81-84; RES STAFF MEM, IBM T J WATSON RES CTR, 84- *Concurrent Pos:* Vis prof, Chem Dept, Univ Fed Rio Grande do Sal, Porto Alegie, Brazil, 84; adj prof, Elec Eng Dept, Univ Maine, Orono, 90-95; tech prog chair, Soc Plastics Eng, Plastics Anal Div, 90-91. *Mem:* AAAS; Am Chem Soc; Mat Res Soc. *Res:* Polyimides: cure, structure-properties, synthesis, waveguiding, characterization; electronic packaging: polymeric dielectrics. *Mailing Add:* IBM Corp T J Watson Res Ctr PO Box 218 Yorktown Heights NY 10598

FEGLEY, KENNETH A(LLEN), b Mont Clare, Pa, Feb 14, 23; m 51; c 3. ELECTRICAL ENGINEERING, SYSTEMS ENGINEERING. *Educ:* Univ Pa, BS, 47, MS, 50, PhD(elec eng), 55. *Prof Exp:* Instr, Univ Pa, 47-53, assoc, 53-55, from asst prof to assoc prof, 55-66, prof elec eng, Moore Sch Elec Eng, 66-86, PROF & CHAIR, DEPT SYSTS, UNIV PA, 86-, JOSEPH MOORE PROF SYSTS, 90- *Mem:* Am Soc Eng Educ; fel Inst Elec & Electronics Engrs; fel AAAS; Am Asn Univ Professors. *Res:* Navigation, control, optimization, modeling and simulation; applications of mathematical programming and other optimization techniques to engineering and medical problems; hazard analysis. *Mailing Add:* Sch Eng & Appl Sci Univ Pa Philadelphia PA 19104

FEHER, ELSA, b Buenos Aires, Arg, Dec 1, 32; US citizen; m 61; c 2. RESEARCH IN SCIENCE LEARNING, DEVELOPMENT OF INTERACTIVE MUSEUM EXHIBITS. *Educ:* Univ Buenos Aires, BA, 56; Columbia Univ, PhD(physics), 64. *Prof Exp:* Res asst physics, Columbia Univ, 59-61; res physicist, Univ Calif, San Diego, 61-67; Radcliffe scholar, Radcliffe Inst Independent Study, 67-68; lectr, 69-71, chmn, Natural Sci Dept, 80-83, PROF NATURAL SCI, SAN DIEGO STATE UNIV, 84- *Concurrent Pos:* Dir Sci Ctr, Reuben Fleet Space Theater & Sci Ctr, San Diego, 83- *Mem:* Am Asn Physics Teachers; Nat Sci Teacher's Asn; Asn Women Sci. *Res:* Science learning in both formal (classroom) and informal (hands-on museum or science center) environments. *Mailing Add:* Dept Natural Sci San Diego State Univ 5300 Campanile Dr San Diego CA 92182-0324

FEHER, FRANK J, b Waynesboro, Pa, Sept 30, 58. MOLECULAR ANALOGS FOR HETEROGENEOUS CATALYSTS. *Educ:* Rensselaer Polytech Inst, BS, 80; Univ Rochester, MS, 82, PhD(inorg chem), 84. *Prof Exp:* Postdoctoral researcher, Univ Bristol, Eng, 84-85; asst prof, 85-90, ASSOC PROF INORG CHEM, UNIV CALIF, IRVINE, 90- *Concurrent Pos:* NSF presidential young investr, 87; consult, Aerospace Corp, 87 & E I duPont de Nemours & Co, 90-; Alfred P Sloan Found fel, 90. *Mem:* Am Chem Soc; Royal Soc Chem. *Res:* Development of molecular analogs for heterogeneous catalysts; molecular approaches to surface chemistry; transition-metal catalyzed reactions of organic molecules. *Mailing Add:* Dept Chem Univ Calif Irvine CA 92717

FEHER, GEORGE, b Czech, May 29, 24; m 49; c 3. PHYSICS, BIOPHYSICS. *Educ:* Univ Calif, BS, 50, MS, 51, PhD(physics), 54. *Prof Exp:* Res physicist, Bell Tel Labs, 54-60; PROF SOLID STATE PHYSICS & BIOPHYS, UNIV CALIF, SAN DIEGO, 60- *Concurrent Pos:* Vis assoc prof, Columbia Univ, 59-60; vis prof, Mass Inst Technol, 67-68; mem bd gov, Israel Inst Technol, 68-, Weizman Inst, Rehovot, Israel. *Honors & Awards:* Prize, Am Phys Soc, 60, Buckley Prize, 75, Biol Physics Prize, 83; fel, AAAS, 86. *Mem:* Nat Acad Sci; fel Am Acad Arts & Sci; Am Phys Soc. *Res:* Biophysics; paramagnetic resonance; photosynthesis. *Mailing Add:* Dept Physics Univ Calif San Diego La Jolla CA 92037

FEHER, JOSEPH JOHN, b Derby, Conn, Apr 2, 49. CARDIOVASCULAR PHYSIOLOGY, CELL PHYSIOLOGY. *Educ:* Cornell Univ, BS, 71, MNS, 73, PhD(nutrit), 78. *Prof Exp:* From instr to asst prof, 78-85, ASSOC PROF RENAL PHYSIOL, DEPT PHYSIOL, MED COL VA, 85- *Mem:* Biophys Soc; AAAS; Am Physiol Soc. *Res:* Characterization of the mechanism of calcium uptake and release by sarcoplasmic reticulum membranes. *Mailing Add:* Dept Physiol & Biophysics Med Col Va VCU Box 551 MCV Sta Richmond VA 23298

FÉHETTE, H(OWELLS ACHILLE) VAN DERCK, b Ottawa, Ont, Jan 5, 16; nat US; m 40; c 5. CERAMIC SCIENCE, FRACTOGRAPHY. *Educ:* Alfred Univ, BS, 39; Univ Ill, MS, 40, PhD(ceramic eng), 42. *Hon Degrees:* DSc, Alfred Univ, 91. *Prof Exp:* Asst, Eng Exp Sta, Univ Ill, 38-40; res physicist, Corning Glass Works, NY, 42-44; asst instr petrog, 41-42, prof, 44-87, EMER PROF CERAMIC SCI, NY STATE COL CERAMICS, ALFRED UNIV, 87- *Concurrent Pos:* Chmn solid state studies, Gordon Res Conf, 55; Fulbright prof, Inst Phys Chem, Gottingen, Ger, 55-56; guest prof, Max Planck Inst Silicates, 65-66 & Univ Erlangen-Nurnberg, 73. *Honors & Awards:* Western Elec Award, Am Soc Eng Educ, 68; A V Bleininger Award, Am Ceramic Soc, 91. *Mem:* Nat Inst Ceramic Engrs; fel Am Ceramic Soc; Swed Royal Acad Sci; Ger Ceramic Soc; NY Acad Sci. *Res:* Petrography and mineralogy of ceramic materials; color of crystalline inorganic materials; solid state reactivity; refractories; tempering glass products; fractography. *Mailing Add:* NY State Col Ceramics Alfred Univ Alfred NY 14802

FEHLER, MICHAEL C, b Riverside, Calif, Nov 15, 51; m 91. GEOPHYSICS, SEISMOLOGY. *Educ:* Reed Col, BA, 74; Mass Inst Technol, PhD(geophysics), 79. *Prof Exp:* asst prof seismol, Ore State Univ, 79-84; STAFF MEM, LOS ALAMOS NAT LAB, 84- *Concurrent Pos:* Res fel, Sci & Technol Agency, Japan, 89-90. *Mem:* Am Geophy Union; Seismol Soc Am; Soc Explor Geophys. *Res:* Geothermal energy; volcanoes; earthquakes. *Mailing Add:* MS D443 Los Alamos Nat Lab Los Alamos NM 87545

FEHLNER, FRANCIS PAUL, b Dolgeville, NY, Aug 3, 34; div; c 3. PHYSICAL CHEMISTRY. *Educ:* Col Holy Cross, BS, 56; Rensselaer Polytech Inst, PhD(phys chem), 59. *Prof Exp:* MEM STAFF, RES, DEVELOP & ENG DIV, CORNING INC, 62- *Mem:* Am Chem Soc; Am Vacuum Soc; Electrochem Soc. *Res:* Glass chemistry; chemical kinetics; ultra-high vacuum; thin films; oxidation of metals. *Mailing Add:* 83 E Fourth St Corning NY 14830

FEHLNER, THOMAS PATRICK, b Dolgeville, NY, May 28, 37; m 62; c 2. PHYSICAL INORGANIC CHEMISTRY. *Educ:* Siena Col, BS, 59; Johns Hopkins Univ, MA, 61, PhD(phys chem), 63. *Prof Exp:* Res assoc inst coop res, Johns Hopkins Univ, 63-64; from asst prof to prof, 75-87, GRACE-RUPLEY, PROF CHEM, UNIV NOTRE DAME, 87- *Honors & Awards:* Fel, AAAS, 85; Guggenheim Fel, 88-89. *Mem:* Am Chem Soc; AAAS; Mat Res Soc. *Res:* Dynamics of unstable species; ultraviolet photoelectron spectroscopy of inorganic species; synthesis of metallaborane clusters; thin films from cluster precursors. *Mailing Add:* Dept Chem Univ Notre Dame Notre Dame IN 46556

FEHNEL, EDWARD ADAM, b Bethlehem, Pa, Apr 22, 22; m 44; c 2. ORGANIC CHEMISTRY. *Educ:* Lehigh Univ, BS, 43, MS, 44, PhD(org chem), 46. *Prof Exp:* Instr, Moravian Prep Sch, Pa, 43-44; res chemist, Cent Res Lab, Allied Chem & Dye Corp, NJ, 44-45; lectr chem & Am Chem Soc fel, Univ Pa, 46-48; from asst prof to prof, 48-72, Edmund Allen prof, 72-84, EMER PROF CHEM, SWARTHMORE COL, 84- *Concurrent Pos:* NSF sci fac fel, Cambridge Univ, 62; vis prof, Ind Univ, 68; lectr chem, Swathmore Col, 84-87. *Mem:* Am Chem Soc; Sigma Xi. *Res:* Synthetic organic chemistry; preparation and properties of organic sulfur compounds; ultraviolet absorption spectroscopy; quinoline analogs of podophyllotoxin; Friedlander reactions arene oxide rearrangements; photocycloaddition reactions of norbornadiene and quadricyclane. *Mailing Add:* 120 Paxon Hollow Rd Media PA 19063-1114

FEHON, JACK HAROLD, b Irvington, NJ, Dec 14, 26; m 50; c 3. ZOOLOGY. *Educ:* Univ Fla, BS, 50, MS, 52; Fla State Univ, PhD, 55. *Prof Exp:* Instr biol, Fla State Univ, 55-56; from assoc prof to prof, 56-62, DANA PROF BIOL, QUEENS COL, NC, 62-, HEAD DEPT, 59- *Concurrent Pos:* Chmn, Div Natural Sci & Math, 76- *Res:* Amino acid flux in marine invertebrates. *Mailing Add:* 2411 Vernon Dr Charlotte NC 28211

FEHR, ROBERT O, b Germany, Sept 12, 11; nat US; m 41; c 1. ACOUSTICS. *Educ:* Inst Technol, Berlin, Vordiplom, 33; Swiss Fed Inst Technol, Dipl Ing, 34; Dr Tech Sc, 39. *Prof Exp:* Test engr, Gen Elec Co, 37-38, develop engr, Gen Eng Lab, 39-46, sect engr, 46-53, consult engr, 53-56, mgr, Mech Eng Lab, 56-61; prof eng, Cornell Univ, 61-64; vpres Europ opers, Branson Europa NV, Branson Instruments, Inc, Neth, 64-66, vpres int opers, Conn, 66-68; pres, Fehr & Fiske Inc, 68-76; CONSULT ENGR, 76- *Concurrent Pos:* Ed, J Audio Engrs Soc. *Honors & Awards:* Audio Engrs Soc Award. *Mem:* Fel Acoust Soc Am; fel Audio Eng Soc; Inst Noise Control Engrs; Inst Elec & Electronics Engrs; Am Soc Mech Engrs. *Res:* Acoustics; vibration; mechanical shock. *Mailing Add:* 294 Round Hill Rd Greenwich CT 06830

FEHR, WALTER R, b East Grand Fork, Minn, Dec 4, 39. PLANT BREEDING. *Educ:* Univ Minn, BS, 61, MS, 62; Iowa State Univ, PhD(plant breeding), 67. *Prof Exp:* Res asst, Univ Minn, 61-62; agronomist, Congo Polytech Inst, Zaire, 62-64; res assoc, 64-67, from asst prof to assoc prof, 67-74, PROF AGRON, IOWA STATE UNIV, 74- *Concurrent Pos:* Mem, Nat Soybean Res Coord Comt; mem corn-soybean study team, Nat Acad Sci; mem, Nat Cert Soybean Variety Rev Bd; assoc ed, Crop Sci; mem, Germplasm Team, People's Repub China, USDA. *Res:* Development of superior soybeans. *Mailing Add:* Dept Agron 120 Agronomy Iowa State Univ Ames IA 50011

FEIBELMAN, PETER JULIAN, b New York, NY, Nov 12, 42; m 71; c 2. THEORETICAL SOLID STATE PHYSICS. *Educ:* Columbia Univ, BA, 63; Univ Calif, San Diego, PhD(physics), 67. *Prof Exp:* NSF fel physics, Saclay Nuclear Res Ctr, France, 68-69, Nat Ctr Sci Res researcher, 69; res asst prof, Univ Ill, Urbana-Champaign, 69-71; asst prof, State Univ NY Stony Brook, 71-74; MEM TECH STAFF SANDIA LABS, 74- *Honors & Awards:* Davisson-Germer Prize, Am Phys Soc, 89. *Mem:* Fel Am Phys Soc; Am Vacuum Soc. *Res:* Theoretical surface physics; dielectric properties of surfaces; photoemission spectroscopy; auger spectroscopy; electron stimulated desorption; electronic structure of surfaces; many-electron problems; diffusion on surfaces. *Mailing Add:* Div 1151 Sandia Lab Albuquerque NM 87185

FEIBELMAN, WALTER A, b Berlin, Ger, Oct 30, 25; US citizen. ASTRONOMY, ASTROPHYSICS. *Educ:* Carnegie Inst Technol, BS, 56. *Prof Exp:* Res specialist, 48-56, res engr, Westinghouse Res Labs, 56-63; asst res prof physics, Univ Pittsburgh, 64-69, observer, Allegheny Observ, 55-69; PHYSICIST, LAB OPTICAL ASTRON, GODDARD SPACE FLIGHT CTR, NASA, 69- *Concurrent Pos:* Consult, Westinghouse Elec Corp, 66- *Mem:* Am Astron Soc; fel Meteoritical Soc; Royal Astron Soc Can; Int Astron Union. *Res:* Microwave spectroscopy; thin films; planetary nebulae; infrared, visible and ultraviolet image converters; cold cathode electron emission; meteor spectroscopy; astrometry; interferometry; astronomical and upper-atmosphere instrumentation; discovered E-ring of Saturn. *Mailing Add:* Lab Astron & Solar Physics 684 Goddard Space Flight Ctr Greenbelt MD 20771

FEIBES, WALTER, b Aachen, Ger, Jan 26, 28; US citizen; m 50; c 3. STATISTICS, OPERATIONS RESEARCH. *Educ:* Union Col, NY, BS, 52; Western Reserve Univ, MS, 53; State Univ NY Buffalo, PhD(opers res), 68. *Prof Exp:* Doc analyst, Libr Cong, 53-55; head librn, Appliance Park, Gen Elec Co, 55-62; assoc prof math, 67-76, PROF MATH, WESTERN KY UNIV, 76-, PROF COMPUT SCI, 80-; PROF OPERS RES, BELLARMINE COL, LOUISVILLE, KY, 82- *Mem:* Math Asn Am; Am Statist Asn; Inst Mgt Sci; Opers Res Soc Am; Sigma Xi. *Res:* Applied statistics; time series. *Mailing Add:* 2376 Valley Vista Rd Louisville KY 40205

FEICHTNER, JOHN DAVID, b Erie, Pa, July 6, 30; m 57. QUANTUM ELECTRONICS. *Educ:* Stanford Univ, BS, 53; NMex State Univ, MS, 60; Univ Colo, PhD(physics), 64. *Prof Exp:* Physicist, Gen Elec Co, 55-57; res asst physics, Univ Colo, 60-65; sr res scientist, 65-71, fel scientist, 71-74, MGR OPTICAL PHYSICS, WESTINGHOUSE RES LABS, 74- *Mem:* Am Phys Soc; Inst Elec & Electronics Engrs; Optical Soc Am; Soc Photog Instrumentation Engrs; Am Asn Physics Teachers. *Res:* Atomic and molecular physics; nonlinear and acoustooptic materials; laser chemistry; laser isotope separation. *Mailing Add:* Lockheed Res & Develop Div 3251 Hanover St MS 0-9701 Bldg 201 Palo Alto CA 94304

FEIERTAG, THOMAS HAROLD, b Rockford, Ill, Sept 25, 35; m 68; c 2. ELECTRICAL ENGINEERING, NONDESTRUCTIVE TESTING. *Educ:* Monmouth Col, BA & Case Inst Technol, BS, 61; New York Univ, MS, 63. *Prof Exp:* Mem tech staff elec eng, Bell Tel Labs, 61-66; MEM STAFF ELEC ENG, LOS ALAMOS NAT LAB, 66- *Concurrent Pos:* Mem, Acoust Emission Working Group. *Mem:* Am Soc Nondestructive Testing. *Res:* Acoustic emission testing; stress emission from metals and the practical application of this phenomenon. *Mailing Add:* Los Alamos Nat Lab Box 1663 MS-914 Los Alamos NM 87545

FEIFEL, HERMAN, b New York, NY, Nov 4, 15. THANATOLOGY, PERSONALITY & MENTAL HEALTH. *Educ:* City Col New York, BA, 35; Columbia Univ, MA, 39, PhD(psychol), 48. *Hon Degrees:* Univ Judaism, DH lett, 84. *Prof Exp:* Aviation psychologist, US Army Air Force, 42-44, clin psychologist, US Army Adj Gen's Off, 44-46; res psychologist, Personnel res sect, Adj Gen Off, War Dept, 46-49; supvry clin psychologist, Winter Vet Admin Hosp, Topeka, 50-54, res & clin psychologist, Vet Admin Ment Hyg Clin, Los Angeles, 55-59, CHIEF PSYCHOLOGIST, VET ADMIN OUTPATIENT CLIN, LOS ANGELES, 61- *Concurrent Pos:* Vis sr res scientist, Res Ctr Ment Health, NY Univ, 59-60, prin investr, Nat Inst Ment Health, 59-62; emer clin prof psychiat & behav sci, Univ Southern Calif Sch Med, 65-, vis prof psychol, Univ Southern Calif, 66-67; consult, Peer Rev Comt, suicide prevention, NIMH, 67-70, adv consult ed, J Consult & Clin Psychiat, Hosp, Omega, Death Studies, An Int Quarterly; nat lectr, Sigma Xi, 75-77, chmn Res Comt Int Work Group Death, Dying & Bereavement, 79-84,; consult ed, McGraw-Hill Corp, 82- *Honors & Awards:* Distinguished Professorial Contrib to Knowledge Award, Am Psychol Asn, 88; Cert Recognition & Serv Award, Secy Vet Affairs, 90; Distinguished Sci Contrib to Clin Psychol Award, Div Clin Psychol, Am Psychol Asn, 90. *Mem:* Am Psychol Asn; AAAS; fel Gerontol Soc; fel Soc Sci Study Religion; fel Sigma Xi; Nat Acad Pract Psychol. *Res:* The psychological and philosophical meaning of death, dying and grief in contemporary society; impact of development and old age on personality and behavior; relationships between value-belief systems and religion and mental health. *Mailing Add:* Vet Admin Outpatient Clin-116B 425 S Hill St Los Angeles CA 90013

FEIG, GERALD, b Newark, NJ, July 29, 32; m 56; c 2. ORGANIC CHEMISTRY. *Educ:* Rutgers Univ, BA, 54, MS, 57, PhD(org chem), 59. *Prof Exp:* Sr res chemist, Nat Cash Register Co, 59-62; res chemist, Sun Chem Corp, 62-63, res group leader, 63-66, res sect head, 66-69, dir corp res lab, 69-75, resin prog mgr, 75-78; mgr, Specialty Prod Div, Polychrome Corp, 78-80; mgr tech oper, Gotham Ink & Color Co, 84-85; tech mgr, Flint Ink Corp, 86-88; TECH DIR, CONVERTERS INK CO, 88- *Concurrent Pos:* Consult, 80-84. *Mem:* Textile Res Inst; Tech Asn Graphic Arts; Am Chem Soc; Soc Photog Sci & Eng. *Res:* Mechanisms of organic peroxide decompositions; photopolymerization systems and photoinitiators; photochromic compounds; thermography; electrostatic printing; polymer synthesis; textile chemical finishes; graphic arts; printing ink. *Mailing Add:* Ten Eton Pl Springfield NJ 07081

FEIG, LARRY ALLEN, b New York, NY, Oct 17, 52; m 81; c 2. MOLECULAR ONCOGENESIS, GTP BINDING PROTEINS. *Educ:* Columbia Univ, BS, 74; Mass Inst Technol, MS, 76; Harvard Univ, PhD(physiol), 82. *Prof Exp:* Postdoctoral fel molecular biol, Dana Farber Cancer Inst, 82-87; ASST PROF BIOCHEM & MOLECULAR BIOL, DEPT BIOCHEM, TUFTS UNIV SCH MED, 87- *Mem:* AAAS. *Res:* Molecular basis of cancer with specific interest in a family of proteins that are regulated by binding guanine nucleotides; oncogenes. *Mailing Add:* Dept Biochem Tufts Univ Sch Med Boston MA 02111

FEIGAL, DAVID WILLIAM, JR, b Minneapolis, Minn, Aug 14, 49; m 80; c 1. INTERNAL MEDICINE, EPIDEMIOLOGY. *Educ:* Univ Minn, BA, 72; Stanford Univ, MD, 76; Univ Calif, Berkeley, MPH, 83. *Prof Exp:* Resident internal med, Univ Calif, Davis, 76-79, chief resident, 79-80, asst prof, 80-82, Mellon scholar epidemiol, 82-84, ADJ ASST PROF EPIDEMIOL & MED, UNIV CALIF, SAN FRANCISCO, 84- *Concurrent Pos:* Training prog coordr, dept med, Univ Calif, Davis, 80-82, clin epidemiologist, Systolic Hypertension Elderly Proj Pilot Study, Univ Calif, San Francisco, 82-, fac mem, Inst Health Policy, 84- & Clin Pharmacol Prog, 84-, co-prin investr, Mild Hypertension Proj, Dept Med, 85- & dir, Acquired Immune Deficiency Syndrome Clin Trials Res Unit, 86- *Mem:* Soc Epidemiol Res; Am Col Physicians; Am Heart Asn; Am Diabetes Asn; Soc Clin Trials; AAAS. *Res:* Epidemiology of iatrogenic events, particularly adverse effects of pharmacologic agents studied in clinical trials; natural history of chronic diseases studied by epidemiologic methods, particularly hypertension and rheumatologic diseases. *Mailing Add:* Univ Calif San Francisco Prev Sci Group 74 New Montgomery St Suite 600 San Francisco CA 94105

FEIGAL, ELLEN G, b Van Nuys, Calif, May 27, 54; m 80; c 2. MEDICINE. *Educ:* Univ Calif, Irvine, BS, 76, MS, 77; Univ Calif, Davis, MD, 81. *Prof Exp:* Resident internal med, Univ Calif, Davis, 81-83 & Stanford Univ, 83-84; fel hemat & oncol, Univ Calif, San Francisco, 85-87, instr in residence, Div AIDS & Oncol, 88-89; ASST ADJ PROF MED HEMAT & ONCOL, UNIV CALIF, SAN DIEGO, 89- *Concurrent Pos:* Physicians Res Training Grant, Am Cancer Soc, 87-89; AIDS Clin Res Ctr Grant, 88-89; training grant, Universitywide Task Force on AIDS, 88-90; prin investr, Navel Therapeut Approaches in Human Immunovirus Associated Lymphomas, NIH, 91- *Mem:* Am Col Physicians; Am Med Asns; Int AIDS Soc. *Res:* Pathogenesis of Bcell lynphoma in human immunovirus infected individuals and novel therapeutic approaches; pathogenesis of Kaposi's sarcoma and treatment strategies. *Mailing Add:* 225 Dickinson St H811K San Diego CA 92103

FEIGE, NORMAN G, b Evansville, Ind, Oct 16, 31. MATERIALS SCIENCE ENGINEERING. *Educ:* Univ Wis, BS, 54. *Prof Exp:* CONSULT, NORMAN G FEIGE ASSOCS, 77- *Mem:* Fel Am Soc Metals. *Mailing Add:* Norman G Feige Assoc Ridgefield Ave South Salem NY 10590

FEIGELSON, ERIC DENNIS, b Springfield, Ohio, April 23, 53; m 87. X-RAY ASTRONOMY, RADIO ASTRONOMY. *Educ:* Haverford Col, BA, 75; Harvard Univ, AM, 78, PhD(astron), 80. *Prof Exp:* Res asst astron, Smithsonian Astrophys Observ, 77-80; res scientist, Ctr Space Res, Mass Inst Technol, 80-82; asst prof, 82-85, ASSOC PROF ASTRON, PA STATE UNIV, 85- *Concurrent Pos:* Teaching asst astron, Harvard Univ, 76-77; prin investr, NASA, 83-; NSF Presidential young investr award, 84. *Mem:* Am Astron Soc; Int Astron Union; Fedn Am Scientists. *Res:* Non-thermal processes in galaxies and stars using space-based x-ray and ground-based radio telescopes; statistical methodology for astronomy; carbon chemistry in astronomy. *Mailing Add:* Dept Astron & Astrophys Pa State Univ University Park PA 16802

FEIGELSON, MURIEL, b New York, NY, July 15, 26; m 47; c 2. BIOCHEMISTRY. *Educ:* Queen's Col, NY, BS, 46; NY Univ, MS, 57, PhD(cell physiol), 61. *Prof Exp:* Trainee oncol biochem, 60-62, res assoc biochem, 63-71, asst prof biochem, 71-80, sr res assoc, 80-83, RES SCIENTIST, DEPT OBSTET & GYNEC, COL PHYSICIANS & SURGEONS, COLUMBIA UNIV, 83-; DIR, BASIC RES LAB, DEPT OBSTET & GYNEC, ST LUKES ROOSEVELT HOSP CTR, 73- *Concurrent Pos:* USPHS fel, Lab Exp Embryol, Col France, 62-63; dir res & clin labs, Dept Obstet & Gynec, Harlem Hosp, New York, 65-73; mem maternal & child health res comt, Nat Inst Child Health Develop, 80-83; bd gov, NY Acad Sci, 82-; mem, Res Ctrs in Minority Insts Comt, NIH,. *Mem:* Fel AAAS; Am Soc Biol Chem; Soc Study Reproduction; fel NY Acad Sci; Harvey Soc; Endocrine Soc; Am Soc Cell Biol; Soc Develop. *Res:* Developmental endocrine and reproductive biochemistry. *Mailing Add:* 265 Tenafly Rd Tenafly NJ 07670

FEIGELSON, PHILIP, b New York, NY, Apr 20, 25; m 47; c 2. BIOCHEMISTRY. *Educ:* Queens Col, NY, BS, 47; Syracuse Univ, MS, 48; Univ Wis, PhD(biochem), 51. *Prof Exp:* Asst prof biochem, Antioch Col, 51-54; res assoc, Fels Res Inst, 51-54; from asst prof to assoc prof, 54-70, PROF BIOCHEM, COL PHYSICIANS & SURGEONS, COLUMBIA UNIV, 70-, ASST VPRES & ASSOC DEAN GRAD AFFAIRS, 87- *Concurrent Pos:* Career investr, Health Res Coun, NY, 59-75. *Mem:* Fel NY Acad Sci (pres, 75); fel AAAS; Am Soc Biol Chem; Am Asn Cancer Res; Harvey Soc; fel World Acad Arts & Sci. *Res:* Mechanism of hormone action; control of gene expression in normal and neoplastic cells. *Mailing Add:* Dept Biochem Columbia Univ 630 W 168th St New York NY 10032

FEIGELSON, ROBERT SAUL, b New York, NY, Dec 3, 35; m 57; c 3. CRYSTAL GROWTH, CRYSTAL CHARACTERIZATION. *Educ:* Ga Inst Tech, BS, 57; Mass Inst Technol, MS, 61; Stanford Univ, PhD(mats sci), 74. *Prof Exp:* Mem tech staff res, Gen Dynamics Corp, 57-58, WaterTown Arsenal, 58-61, Sperry Rand Res Ctr, 61-63; DIR RES CRYSTAL SCI DEPT, CTR MATS RES, STANFORD UNIV, 63-, RES PROF, MAT SCI & ENG DEPT, 76- *Concurrent Pos:* Pres & founder, N Calif Crystal Growers, 74-77; assoc ed, J Crystal Growth, 74-; consult industrial orgn. *Honors & Awards:* Tech Achievement Award, NASA, 85 & 87. *Mem:* Am Asn Crystal Growth (pres, 80-83, treas, 74-80); Mat Res Soc. *Res:* Synthesis and crystal growth of wide variety of materials including semiconductors, superconductors, proteins, dielectrics, metals and ceramics in the form of fibers, thin films or bulk crystals; relationship between processing parameters and materials properties on properties and defect structure of crystals. *Mailing Add:* Ctr Mats Res Stanford Univ Stanford CA 94305

FEIGEN, LARRY PHILIP, b Everett, Mass, Mar 27, 42; m 65; c 2. PLATELET PHARMACOLOGY, HEMOSTASIS. *Educ:* Northeastern Univ, BA, 64, MS, 66; Chicago Med Sch, Univ Health Sci, PhD(physiol & biophys), 74. *Prof Exp:* Instr physiol, Chicago Med Sch, Univ Health Sci, 73-74; from asst prof to assoc prof physiol, Med Sch, Tulane Univ, 81-85; sr res scientist, 85-86, thrombosis team leader, 86-89, FEL, SEARLE RES & DEVELOP, 89- *Concurrent Pos:* Prin investr, NIH, 81-83, Am Heart Asn, 83-86. *Mem:* Am Heart Asn; AAAS; fel Am Physiol Soc; NY Acad Sci; Sigma Xi; Am Soc Pharmacol & Exp Therapeut. *Res:* Pharmacology of platelet function, pharmacology of integrin receptors; blood coagulation. *Mailing Add:* PHD Fel & Thrombosis Team Leader Cardiovasc Dis Searle Res & Develop 4901 Searle Pkwy Skokie IL 60077

FEIGENBAUM, ABRAHAM SAMUEL, b New York, NY, Mar 11, 29; m 52; c 3. RESEARCH ADMINISTRATION, NUTRITION. *Educ:* Rutgers Univ, BS, 51, MS, 59, PhD(nutrit), 62. *Prof Exp:* Inspection chemist, E R Squibb & Sons Div, Olin Mathieson Chem Corp, 54-56, assay methods anal chemist, 56-57; res asst nutrit, Poultry Dept, Rutgers Univ, 57-61; res scientist, NJ Bur Res Neurol & Psychiat, 61-73; assoc dir med servs, Warren-Teed Labs, Inc, 73-81; clin proj dir, Pharmaceut Res Inst, 81-83; DIR RES, ORGANON INC, 83- *Mem:* NY Acad Sci; Am Inst Nutrit; Am Chem Soc; Soc Exp Biol & Med; Am Oil Chem Soc. *Res:* Lipid metabolism in chickens, including fatty liver etiology and atherosclerosis; genetic variation in susceptibility to experimentally-induced atherosclerosis and spontaneous arteriosclerosis in rabbits; potassium metabolism; clinical use chemically defined diets. *Mailing Add:* 22 Berkshire Rd Maplewood NJ 07040

FEIGENBAUM, EDWARD A(LBERT), b Weehawken, NJ, Jan 20, 36; m 58; c 2. COMPUTER SCIENCE, PSYCHOLOGY. *Educ:* Carnegie Inst Technol, BS, 56, PhD(indust admin), 60. *Hon Degrees:* DSc, Aston Univ, Eng, 89. *Prof Exp:* Fulbright res scholar, Gt Brit, 59-60; asst prof bus admin, Univ Calif, Berkeley, 60-64; assoc prof comput sci, Stanford Univ, 65-69, dir comput ctr, 65-69, chmn dept, 76-81 PROF COMPUT SCI, STANFORD UNIV, 69- *Concurrent Pos:* Consult indust & govt, 57-; prin investr, Heuristic Prog Proj, Stanford Univ, 65-; mem, Computer & Biomath Sci Study Sect, NIH, 68-72, Comt Math Social Sci, Social Sci Res Coun, 77-78, Computer Sci Adv Comt, NSF, 77-80, Adv Comt Math Naval Res, Nat Res Coun, Off Naval Res & Adv Comt Info Sci & Technol, Defense Advan Res Projs Agency, 87-; res prof, Japan Inst Prom Sci, 79; Lee Kuan Yew distinguished prof, Nat Univ Singapore, 86. *Mem:* Nat Acad Eng; Asn Comput Mach; Am Psychol Asn; fel AAAS; fel Am Col Med Informatics. *Res:* Information processing models of cognitive processes; artificial intelligence; programming languages; verbal learning; models of human memory; author of various technical publications. *Mailing Add:* Knowledge Systs Lab 701-C Welch Rd Palo Alto CA 94304

FEIGENBAUM, HARVEY, b East Chicago, Ind, Nov 20, 33; m 57; c 3. MEDICINE. *Educ:* Ind Univ, AB, 55, MD, 58. *Prof Exp:* Sr res assoc, Krannert Inst Cardiol, 62; from instr to prof, 62-80, DISTINGUISHED PROF MED, MED CTR, IND UNIV, 80- *Concurrent Pos:* Assoc med, Wishard Mem Hosp, 65. *Mem:* Am Fedn Clin Res; fel Am Col Physicians; fel Am Col Cardiol; Am Soc Echocardiography; Am Soc Clin Invest. *Res:* Clinical cardiology; electrophysiology; hemodynamics; echocardiography. *Mailing Add:* 926 W Michigan St Indianapolis IN 46223

FEIGENBAUM, MITCHELL JAY, b Philadelphia, Pa, Dec 19, 44. DYNAMICAL SYSTEMS, CHAOS. *Educ:* City Col, NY, BEE, 64; Mass Inst Technol, PhD(theoret physics), 70. *Prof Exp:* Res assoc & instr physics, Cornell Univ, 70-72; res assoc physics, Va Polytech State Univ, 72-74; staff mem theoret physics, Los Alamos Nat Lab, 74-80, lab fel, 81-82; prof, physics, Cornell Univ, 82-86; PROF MATH & PHYSICS, ROCKEFELLER UNIV, NY, 86- *Concurrent Pos:* Vis mem, Inst Advan Studies, Princeton, 78 & Inst des Hautes Etudes Sci, France, 80-81; vis prof, Physics Dept, Cornell Univ, 81; ed, J Statist Physics, 81-; distinguished lectr, Univ Chicago, 83, Case Univ, 84, Yale Univ, 85, NY Univ, 87. *Honors & Awards:* Distinguished Performance Award, Los Alamos Nat Lab, 80; E O Lawrence Mem Award, US Dept Energy, 82; MacArthur Found Award, 84; Wolf Found Prize in Physics, 86; Dickson Prize, Carnegie-Mellon Univ, 87. *Mem:* Nat Acad Sci; NY Acad Sci; AAAS; Sigma Xi. *Res:* Discoverer of metrically universal behaviors in dynamical systems; subject matter represent the major thrust in the theory of the onset of chaos, leading towards an understanding of turbulence in flows and other contexts. *Mailing Add:* Dept Physics Rockefeller Univ 1230 York Ave Box 75 New York NY 10021-6399

FEIGENSON, GERALD WILLIAM, b Washington, DC, Aug 2, 46; m 85; c 1. BIOMEMBRANES. *Educ:* Rensselaer Polytech Inst, BS, 68; Calif Inst Technol, PhD(chem), 74. *Prof Exp:* From asst prof to assoc prof, 74-89, PROF BIOCHEM, CORNELL UNIV, 90- *Concurrent Pos:* Vis scientist, Scripps Inst Oceanog, 83-84; regular mem biophys panel, NSF, 84-89; vis prof, Dept Physics, Keio Univ, Japan, 89-90. *Honors & Awards:* Estab investr, Am Heart Asn, 80. *Mem:* Am Soc Biochem & Molecular Biol; Biophys Soc; AAAS. *Res:* Non-ideality of mixing of components of bilayer model membranes; measure high-affinity binding of calcium ions between bilayers and the consequent membrane rearrangements, including gel phase formation and protein redistribution. *Mailing Add:* 201 Biotech Bldg Cornell Univ Ithaca NY 14853

FEIGERLE, CHARLES STEPHEN, b Chicago, Ill, July 8, 50; m 76; c 2. ELECTRON SPECTROSCOPY, CHEMICAL DYNAMICS. *Educ:* Univ Ill, Chicago, BS, 77; Univ Colo, PhD(chem physics), 83. *Prof Exp:* Res physicist, Nat Bur Standards, 83-85; ASST PROF CHEM, UNIV TENN, KNOXVILLE, 85- *Honors & Awards:* IR-100 Award, Res & Develop Mag, 85. *Mem:* Am Chem Soc; Am Vacuum Soc. *Res:* Measurement ofkinetics parameters for single crystal surface catalyzed chemical reactions using thermal desorption mass spectrometry; mutiphoton ionization spectroscopy of excited molecules and free radicals. *Mailing Add:* Dept Chem Univ Tenn 322 Buehler Hall Knoxville TN 37996-1600

FEIGHAN, MARIA JOSITA, b Philadelphia, Pa, Aug 29, 32. PHYSICAL CHEMISTRY. *Educ:* Immaculata Col, Pa, AB, 64; Harvard Univ, MEd, 67; St Louis Univ, PhD(phys chem), 69. *Prof Exp:* Teacher high schs, Pa, 57-66; assoc prof chem, Immaculata Col, 69-80, asst acad dean, 74-76, acad dean, 83-90, PROF CHEM, IMMACULATA COL, 80- *Concurrent Pos:* Dir, evening div, Immaculata Col, 76-82. *Mem:* Am Chem Soc. *Res:* Educational programs on pollution; spectroscopic studies of nitro compounds. *Mailing Add:* Dept Chem Immaculata Col Immaculata PA 19345

FEIGHNER, SCOTT DENNIS, b Alma, Mich, Apr 1, 51; m 72; c 3. ANAEROBIC BACTERIOLOGY, INFECTIOUS DISEASES. *Educ:* Western Mich Univ, BS, 73, MA, 75; Va Commmonwealth Univ, PhD(microbiol), 79. *Prof Exp:* Res assoc, Mich State Univ, 79-80, Syracuse Res Corp, 80; sr res microbiol, 81-86, RES FEL, MERCK SHARP & DOHME RES LABS, MERCK & CO, INC, 86- *Mem:* Am Soc Microbiol; AAAS; NY Acad Sci. *Res:* Elucidation of the mechanisms of action of growth permittants in monogastric animals; interactions between gastrointestinal microflora and their hosts. *Mailing Add:* Merck Sharp & Dohme Res Labs PO Box 2000 Rahway NJ 07065

FEIGIN, IRWIN HARRIS, b New York, NY, May 13, 15; m 49; c 2. NEUROPATHOLOGY. *Educ:* Columbia Univ, BA, 34; NY Univ, MD, 38. *Prof Exp:* Res neuropathologist, Vet Admin Hosp, Bronx, 47-51; asst prof, Col Physicians & Surgeons, Columbia Univ, 52-56; assoc prof, 56-59, PROF NEUROPATH, COL MED, NY UNIV, 59- *Concurrent Pos:* Neuropathologist, Bellevue Hosp, 59-; pathologist, Sydenham Hosp, New York, 50-51; assoc pathologist, Mt Sinai Hosp, 51-56. *Mem:* Histochem Soc; Am Asn Neuropath. *Mailing Add:* Dept Path Bellevue Hosp 550 First Ave New York NY 10016

FEIGIN, RALPH DAVID, b New York, NY, Apr 3, 38; m 60; c 3. PEDIATRICS, INFECTIOUS DISEASES. *Educ:* Columbia Univ, AB, 58; Boston Univ, MD, 62. *Prof Exp:* From intern to resident, Boston City Hosp, 62-64; resident, Children's Serv, Mass Gen Hosp, 64-65, chief resident, 67-68; from instr to prof pediat, Washington Univ, St Louis, 68-77, dir, Div Infectious Dis, Dept Pediat, 73-77; J S ABERCROMBIE PROF PEDIAT & CHMN DEPT, BAYLOR COL MED, 77- , DISTINGUISHED SERV PROF, 91- *Concurrent Pos:* Fel pediat, Harvard Med Sch, 64-65 & 67-68; clin asst, Mass Gen Hosp, Boston, 67-68; asst pediatrician, St Louis Children's, St Louis Maternity & McMillan Hosps, 68; assoc pediatrician, Mo Crippled Children's Serv, 68; assoc dir, Clin Res Ctr, Wash Univ, 68; vis prof pediat, Univ Calif, San Francisco, 72, Duke Univ Sch Med, 76, Emory Med Sch, 76, Univ Wis Med Sch, 76, NY Univ Sch Med, 77, Univ Tex Health Sci Ctr, Galveston, 77 & San Antonio, 78, Univ Calif, Los Angeles-Harbor Gen Hosp, 79, Ohio State Univ & Children's Hosp, 81; J Pediat Educ Found prof, Loyola Univ, 73, vis prof, Columbia Univ Col Physicians & Surgeons, 80; A Ashley Weech vis prof, Univ Cincinnati Sch Med, Cincinnati Children's Hosp, 77; physician-in-chief, Tex Children's Hosp, 77- & Pediat Serv, Harris County Hosp Dist, Houston, 77-90, Pediat Serv, Ben Taub Gen Hosp, 90-; chief, Pediat Serv, Methodist Hosp, 79-; Abraham Finkelstein Mem vis prof, Univ Md Sch Med, 82; pres, Pediat Res Found, 82-; Charles C & Mary Elizabeth Lovely Verstandig distinguished vis prof, Univ Tenn Col Med, 84; Brooksaler vis prof pediat, Univ Tex Health Sci Ctr, Dallas, 85; vis prof, Univ Md, 87; Robert N Ganz prof & joint prog neonatology manual vis prof, Harvard Med Sch, 89. *Honors & Awards:* USPHS Res Career Develop Award, Nat Inst Allergy & Infectious Dis, 70; Amberg-Helmholz Lectr, Mayo Clin & Mayo Med Sch, 84. *Mem:* AAAS; Am Fedn Clin Res; Am Soc Microbiol; Soc Pediat Res (pres-elect, 81, pres, 82-83); Infectious Dis Soc Am; Am Pediat Soc; Am Acad Pediat; Asn Med Sch Pediat Dept Chmn (pres elect, 89-91, pres, 91-93). *Res:* Metabolic response of the host to infectious diseases, including the effect of time of infectious exposure upon the outcome of disease. *Mailing Add:* Dept Pediat Baylor Col Med One Baylor Plaza Houston TX 77030

FEIGL, DOROTHY M, b Evanston, Ill, Feb 25, 38. ORGANIC CHEMISTRY. *Educ:* Loyola Univ, Ill, BS, 61; Stanford Univ, PhD(org chem), 66. *Prof Exp:* Res assoc org chem, NC State Univ, 65-66; from asst prof to assoc prof, 66-75, chmn dept chem & physics, 77-85, actg vpres & dean fac, 85-87, PROF CHEM, ST MARY'S COL, IND, 75-, VPRES & DEAN FAC, 87- *Concurrent Pos:* Consult, Int Bakers Serv, Inc, 74-76; extramural assoc, NIH, 81-82; mem, Am Conf Acad Deans & Am Asn Higher Educ. *Mem:* Am Chem Soc; Royal Soc Chem; Sigma Xi; Int Union Pure & Appl Chem. *Res:* Structure of the Grignard reagent; mechanism of the Grignard reduction reaction; steroid chemistry; polymer chemistry; flavor chemistry. *Mailing Add:* Rm 130 LeMans Hall St Mary's Col Notre Dame IN 46556

FEIGL, ERIC O, b Iowa City, Iowa, June 5, 33; m 57; c 2. PHYSIOLOGY. *Educ:* Univ Minn, BA & BS, 54, MD, 58. *Prof Exp:* Intern, Philadelphia Gen Hosp, 58-59; instr physiol, Sch Med, Univ Pa, 59-61; vis scientist, Gothenburg Univ, 61-62; instr physiol, Sch Med, George Washington Univ, 62-64; asst prof, Sch Med, Univ Pa, 64-69; assoc prof, 69-72, PROF PHYSIOL, SCH MED, UNIV WASH, 72- *Concurrent Pos:* NIH fel, 59-62 & res career develop award, 64-69; res assoc, Nat Heart Inst, 62-64; mem, Basic Sci Coun, Circulation Coun & adv bd, High Blood Pressure Coun, Am Heart Asn. *Mem:* AAAS; Am Physiol Soc; Microcirculation Soc; Cardiac Muscle Soc. *Res:* Coronary circulation; neural control of the circulation; cardiovascular instrumentation. *Mailing Add:* Dept Physiol SJ-40 Univ Wash Med Sch Seattle WA 98195

FEIGL, FRANK JOSEPH, solid state physics; deceased, see previous edition for last biography

FEIGL, POLLY CATHERINE, b Minneapolis, Minn, July 30, 35; m 57; c 2. BIOSTATISTICS. *Educ:* Univ Chicago, BA & BS, 56; Univ Minn, MA, 57, PhD(biostatist), 61. *Prof Exp:* Statistician, Smith Kline & French Labs, 58-61; math statistician, Nat Cancer Inst, 62-64; Res Assoc Med Statist, Sch Med, Univ Pa, 64-67, asst prof, 67-69; from asst prof to assoc prof, 69-77, PROF BIOSTATIST, SCH PUB HEALTH & COMMUNITY MED, UNIV WASH, 77- *Mem:* Am Statist Asn; Biomet Soc; fel Am Statist Asn; mem Int Statist Inst; Sigma Xi. *Res:* Statistical design and analysis of biomedical experiments. *Mailing Add:* Dept Biostatist SC32 Univ Wash Seattle WA 98105

FEIGN, DAVID, b New York, NY, Apr 28, 23; m 54; c 4. COMPUTER SCIENCE, SOFTWARE ENGINEERING. *Educ:* City Col New York, BMechE, 44; Univ Buffalo, MS, 53; Univ Calif, Irvine, PhD(comput sci), 80. *Prof Exp:* Res aerodynamicist, Nat Adv Comt Aeronaut, 44-48; Cornell Aeronaut Labs, 48-59, head digital comput sect, 53-59; head, Software Develop Autonetics Div, NAm Aviation, Inc, 59-75; mem tech staff, Space Div, Rockwell Int, 75-79; CONSULT SOFTWARE & SYSTS, 67- *Concurrent Pos:* Instr, Canisius Col, 58, Univ Calif, Irvine, 66-69 & West Coast Univ, 75-85; teacher, Eldorado Sch Gifted Child, 61-66; assoc prof & dir, Comput Ctr, Champman Col, 78-79; res engr, Univ Calif, Irvine, 69-71; N Am Aviation Grad Study fel, 65-67; Forensic consult, comput matters, 82-91; chmn, Independent Comput Consult Asn. *Mem:* Asn Comput Mach; Inst Elec & Electronics Engrs; Independent Comput Consult Asn. *Res:* Digital computers; artificial intelligence; software engineering; programming languages; aircraft dynamic stability and control. *Mailing Add:* 1301 Landfair Circle Santa Ana CA 92705

FEIKER, GEORGE E(DWARD), JR, b Northampton, Mass, May 6, 18; m 40; c 3. ELECTRICAL ENGINEERING. *Educ:* Worcester Polytech Inst, BS, 39; Harvard Univ, MS, 40. *Prof Exp:* Asst elec eng, Calif Inst Technol, 40-41; engr, 41-48, mgr microwave eng, 48-57, mgr radio frequency & commun eng, 57-64, mgr electronic eng lab, 64-65, mgr electronic physics lab, 65-69, mgr advan systs studies, 69-78, CONSULT, GEN ELEC CO, 78- *Concurrent Pos:* Instr, Rensselaer Polytech Inst, 46-48. *Honors & Awards:* Coffin Award, Gen Elec Co. *Mem:* Inst Elec & Electronics Engrs. *Res:* Research and development in electrical power conversion and control; communications; microwave generation and radiation. *Mailing Add:* 883 Inman Rd Schenectady NY 12309

FEILER, WILLIAM A, JR, b Paducah, Ky, Oct 9, 40; m 62; c 3. INDUSTRIAL CHEMISTRY. *Educ:* Univ Ky, BS, 62; Univ Fla, PhD(chem), 65. *Prof Exp:* Res chemist, Monsanto Co, 65-70, res group leader, 70-79, mgr technol, 79-81; DIR PROCESS & ANALYTICAL RES, EDWIN COOPER DIV, ETHYL CORP, 81- *Mem:* Am Chem Soc; Am Inst Chem Engrs. *Res:* Process development, with emphasis on solvent extraction technology for inorganic materials; phosphate salts, household detergents and elemental phosphorus; oil additives. *Mailing Add:* 1612 LaCanada Brea CA 92621-1824

FEIN, BURTON IRA, b New York, NY, May 20, 40; m 68; c 1. ALGEBRA. *Educ:* Polytech Inst Brooklyn, BSc, 61; Univ Wis, MSc, 62; Univ Ore, PhD(math), 65. *Prof Exp:* Asst prof math, Univ Calif, Los Angeles, 65-70; assoc prof, 70-77, PROF MATH ORE STATE UNIV, 77- *Concurrent Pos:* Res fel, Alexander von Humboldt Found, 76-77, 83-84. *Mem:* Am Math Soc. *Res:* Brauer groups of fields, division algebras, algebraic number theory; Schur index questions in the representation theory of finite groups. *Mailing Add:* Dept Math Ore State Univ Cowallis OR 97330

FEIN, HARVEY L(ESTER), b Washington, DC, Mar 4, 36; m 73; c 2. CHEMICAL ENGINEERING. *Educ:* Cornell Univ, BChE, 59; Mass Inst Technol, SM, 61, ScD(chem eng), 63. *Prof Exp:* Res engr, Atlantic Res Corp, 63-66, head thermodyn sect, 66-67, staff scientist, Propulsion & Eng Res Dept, 67-75; sr chem engr & task mgr, Energy Eng Div, TRW, Inc, 75-80, mgr process eng, 80-81; PROJ LEADER, US SYNTHETIC FUELS CORP, 81- *Mem:* Am Chem Soc; Am Inst Aeronaut & Astronaut; Am Inst Chem Engrs. *Res:* Coal gasification; coal liquefaction; systems analysis; thermodynamic cycle analysis; energy conversion processes; theoretical and experimental combustion kinetics; gas dynamics; propulsion system analysis; solid propellants; chemical thermodynamics; polymer physics; diffusion. *Mailing Add:* 6444 Elmdale Rd Alexandria VA 22312

FEIN, JACK M, b New York, NY, Mar 10, 40; c 1. SURGERY. *Educ:* New York Univ, BA, 61, MD, 65. *Prof Exp:* Instr surg neurosurg & attend surg, Georgetown Univ Sch Med, 73-74; asst prof neurosurg, 74-77, ASST PROF VASCULAR SURG, A EINSTEIN COL MED, 77-, ASSOC PROF NEUROSURG, 78- *Concurrent Pos:* Vis prof microsurg, Theo Gildred Ctr, Univ Fla, 77, Downstate Med Ctr, NY & Karolinska Inst, Sweden, 78, Loyola Univ & NY Univ Med Ctr, 79, Neurol Inst, Columbia Univ & Jefferson Med Col, 80; res career develop award, NIH, 76-81; consult, The Pres Comn Ethical Issues in Med, 81-85. *Honors & Awards:* Irving Wright Award Res Cerebrovascular Dis, Am Heart Asn, 74. *Res:* Cerebral ischemia and the biochemical and physiological changes associated with its treatment. *Mailing Add:* Dept Neurol Surg Hosp Einstein Col Med Bronx NY 10461

FEIN, JAY SHELDON, b Brooklyn, NY, July 6, 37; m 66. ATMOSPHERIC SCIENCE. *Educ:* Rutgers Univ, BS, 62; Fla State Univ, MS, 66, PhD(meteorol), 72. *Prof Exp:* Weather forecaster, USAF, 62-65; asst prof meteorol, Univ Okla, 73-79; assoc prog dir, atmospheric Sci, NSF, 76-80, prog dir, 80-; PROG DIR, CLIMATE DYNAMICS PROG. *Mem:* Am Meteorol Soc. *Res:* Large-scale atmospheric circulations; geophysical fluid dynamics. *Mailing Add:* Climate Dynamics Prog 1800 G St NW Washington DC 20550

FEIN, MARVIN MICHAEL, organic chemistry; deceased, see previous edition for last biography

FEIN, R(ICHARD) S(AUL), b Milwaukee, Wis, Apr 25, 23; m 48; c 3. ENVIRONMENTAL ENERGY & SCIENCE. *Educ:* Univ Wis, BS, 47, PhD(chem eng), 49. *Prof Exp:* Asst, Naval Res Lab, Univ Wis, 46-49; chem engr, Texaco, Inc, 49-57, group leader, 57-64, res assoc, 64-68, fundamental res supvr, 68-77, sr res assoc, 77-82; RETIRED. *Concurrent Pos:* Adj prof mech eng, Columbia Univ, 82-90; consult tribology & environ, 82-; mem, Eng Found Bd, 84, chmn, 89-91; certified hazardous material mgr, 85. *Honors & Awards:* Hunt Award, Soc Tribologists & Lubrication Engrs, 66. *Mem:* AAAS; Am Chem Soc; Sigma Xi; fel Soc Tribologists & Lubrication Engrs; Am Soc Mech Engrs; NY Acad Sci; Soc Automotive Engrs. *Res:* Laminar flame speeds and temperatures; fuel applications, primarily in engines and vehicles; engine lubrication; wear and boundary lubrication; combustion chemistry; atmospheric chemistry and physics. *Mailing Add:* 35 Sheldon Dr Poughkeepsie NY 12603

FEIN, RASHI, b New York, NY, Feb 6, 26. MEDICAL ECONOMICS. *Educ:* Johns Hopkins, PhD, 48. *Prof Exp:* Lectr to assoc prof, Univ NC, 52-61; sr staff, Pres Kennedy's Coun Econ Advisors, 61-63; sr fel, Econ Studies Prog, Brookings Inst, 63-68; PROF ECON MED, DEPT SOCIAL MED, SCH MED, HARVARD UNIV, 68- *Concurrent Pos:* Traveling fel, WHO, 71; mem bd trustees, Hebrew Rehab Ctr Aged, Boston & mem, Clin Invest Comt; mem bd trustees, Beth Israel Hosp, Boston; mem spec med adv group to chief med dir, Vet Admin, Washington, DC. *Honors & Awards:* John M Russell Medal, Markle Scholars, 71; Martin E Rehfuss Lectr, Jefferson Med Sch, 73; Theobald Smith Lectr, Albany Med Col, 73; Heath Clark Lectr, London Sch Hyg & Trop Med, 80. *Mem:* Inst Med-Nat Acad Sci. *Res:* Author or co-author of 7 books. *Mailing Add:* Dept Social Med Harvard Med Sch 25 Shattuck St Boston MA 02115

FEINBERG, BARRY N, ENGINEERING ANALYSIS. *Educ:* Univ Mich, BSE, BSEE; Univ Louisville, MSEE; Case Western Reserve Univ, PhD(eng). *Prof Exp:* Instr math, Dept Eng Math, Sch Eng, Univ Louisville, 62-64; instr electronic circuit anal & design, Case Inst Technol, 64-68; vis prof elec eng, Univ Edinburgh, Scotland, 68-69; from asst prof to assoc prof elec eng, Cleveland Univ, 69-76, head, Col Eng Comput & Plotting Facil & dir, Clin Eng Prog, 72-76; assoc prof elec eng & dir, Clin & Med Radiation Eng Prog, Purdue Univ, 76-81; sr corp scientist, Barrington Res Ctr, Kendall Co, 81-86; PRIN CONSULT ENGR, CTR ENG ANALYSIS, 86- *Concurrent Pos:* Chmn, Int Cert Comn, Bd Examiners Clin Eng Cert, 80-83, Eng Med Biol Soc Conf, 85, Int Conf Med & Bio Eng & Int Conf Med Physics, World Cong Med Physics & Biol Eng, 88; mem, Health Care Eng Policy Comt, Inst Elec & Electronics Engrs, 83-88, Accreditation Bd Eng & Technol & Standards Comt Med Instrumentation & Elec Safety, Asn Advan Med Instrumentation; assoc ed, Inst Elec & Electronics Engrs, Eng Med & Biol Mag. *Mem:* Fel Inst Elec & Electronics Engrs; Inst Elec & Electronics Engrs Eng Med & Biol Soc (vpres, 89-90); Am Asn Advan Eng; Inst Elec & Electronics Engrs Power Eng Soc; Nat Soc Prof Engrs. *Res:* Engineering analysis and testing; applied clinical engineering; author of numerous technical articles and six books. *Mailing Add:* Ctr Eng Analysis 4007 Rutgers Lane Northbrook IL 60062

FEINBERG, BENJAMIN ALLEN, b St Louis, Mo, Mar 23, 44; m 81. ANALYTICAL CHEMISTRY, BIOCHEMISTRY. *Educ:* Wash Univ, AB, 66; Univ Kans, PhD(chem), 71. *Prof Exp:* Res & teaching fel, Pa State Univ, 71-72; res fel, Northwestern Univ, 72-75; ASSOC PROF CHEM, UNIV WIS-MILWAUKEE, 75- *Mem:* Am Chem Soc; Am Soc Biol Chemists. *Res:* Biochemical, biophysical and electroanalytical study of electron transfer mechanisms of redox proteins and enzymes; study of the influence of ionic strength upon redox protein reactions; redox protein structure-function relationships. *Mailing Add:* Dept Chem Univ Wis PO Box 413 Milwaukee WI 53201

FEINBERG, GERALD, b New York, NY, May 27, 33; m 68; c 2. THEORETICAL HIGH ENERGY PHYSICS. *Educ:* Columbia Univ, BA, 53, MA, 54, PhD(physics), 57. *Prof Exp:* Mem sch math, Inst Advan Study, Princeton Univ, 56-57; res assoc, Brookhaven Nat Lab, 57-59; from asst prof to assoc prof, 59-65, chmn, 80-83, PROF PHYSICS, COLUMBIA UNIV, 65-,. *Concurrent Pos:* Adj asst prof & consult physics, NY Univ, 59; consult, Brookhaven Nat Lab, 59-74; overseas fel, Sloan fel, 60-64 & Churchill Col, Cambridge, 63-64; vis prof, Rockefeller Univ, 73-74 & 86; Guggenheim fel, 73-74; div assoc ed, Phys Rev Lett, 83-86. *Mem:* Am Phys Soc. *Res:* Elementary particles; field theory. *Mailing Add:* Dept Physics Columbia Univ Broadway & W 116th New York NY 10027

FEINBERG, HAROLD, b Chicago, Ill, June 20, 22; div; c 2. PHARMACOLOGY. *Educ:* Univ Calif, Los Angeles, BA, 48, MA, 50; Univ Calif, Berkeley, PhD(physiol), 52. *Prof Exp:* Life Ins med res fel, Univ Calif, 52-53; chief biochemist, Children's Mem Hosp & instr biochem, Med Sch, Northwestern Univ, 53-55; res assoc, Cardiovasc Dept, Med Res Inst, Michael Reese Hosp, 55-61; res assoc biochem, Univ Birmingham, 61-63; assoc prof, 63-70, PROF PHARMACOL, UNIV ILL COL MED, 70- *Concurrent Pos:* Am Heart Asn res fel, 56-58 & adv res fel, 58-60, estab investr, 60-65. *Mem:* Am Chem Soc; Am Physiol Soc; Am Soc Pharmacol & Exp Therapeut; Am Asn Advan Animal Lab Care (pres, 79-80). *Res:* Cardiovascular physiology, platelets and hemostasis; metabolism of cardiac muscle; ionflux of human blood platelets in response to physiologic stimuli; metabolism and function of heart. *Mailing Add:* Dept Pharm Univ Ill Sch Med 835 S Wolcott Ave Chicago IL 60612

FEINBERG, MARTIN ROBERT, b New York, NY, Apr 2, 42; m 65. APPLIED MATHEMATICS. *Educ:* Cooper Union, BChE, 62; Purdue Univ, MS, 63; Princeton Univ, PhD(chem eng), 68. *Prof Exp:* from asst prof to assoc prof, 67-80, PROF CHEM ENG, UNIV ROCHESTER, 80- *Concurrent Pos:* Dreyfus teacher-scholar, Camille & Henry Dreyfus Found, 73. *Mem:* Am Inst Chem Engrs; Soc Natural Philos. *Res:* Applied mathematics; thermodynamics; mathematics of complex chemical reaction systems. *Mailing Add:* Dept Chem Eng Univ Rochester Rochester NY 14627

FEINBERG, ROBERT JACOB, b Chelsea, Mass, Apr 6, 31; m 64; c 2. HEALTH PHYSICS, RADIOLOGICAL ENGINEERING. *Educ:* Boston Col, BS, 53, MS, 54; Univ Rochester, MS, 55; Oak Ridge Sch Reactor Technol, dipl(nuclear eng), 56; Am Bd Health Physics, cert, 61, 80, 84 & 89. *Prof Exp:* Weapons physicist, Picatinny Arsenal, Divor, NJ, 51-52; physicist, Nat Bur Standards, 52-53; astrophysicist, Air Force Cambridge Res Ctr, 53-54; health physicist, Brookhaven Nat Lab, 54-55; nuclear engr, Oak Ridge Nat Lab, 55-56; physicist, 56-58, supv physicist radiol physics & eng, 58-62, proj engr radiol eng, 62-63, supvr nuclear & radiol safety, 63-66, MGR HEALTH PHYSICS & NUCLEAR SAFETY, GEN ELEC CO, KNOLLS ATOMIC POWER LAB, 66- *Concurrent Pos:* Mem, Atomic Indust Forum, 58- *Mem:* Int Health Physics Soc; Am Nuclear Soc. *Res:* Radiological dosimetry; environmental hazards analyses; nuclear criticality safety evaluation; health effects of low-level ionizing radiation; atmospheric diffusion of effluents; internal dosimetry. *Mailing Add:* 1223 Godfrey Lane Schenectady NY 12309

FEINBERG, ROBERT SAMUEL, b Baltimore, Md, June 10, 40; m 71; c 1. BIO-ORGANIC CHEMISTRY. *Educ:* Harvard Univ, BA, 61; Oxford Univ, PhD(org chem), 65. *Prof Exp:* Res assoc, McCollum Pratt Inst, Johns Hopkins Univ, 65-66 & org chem dept, 66-69; res assoc, Rockefeller Univ, 69-74, asst prof org chem, 74-76; secy-treas, 76-80, PRES, DURON, INC, 81- *Mem:* Am Chem Soc. *Res:* Hemoglobin model compounds; solid-phase synthesis of peptides and proteins. *Mailing Add:* Duron Inc 10406 Tucker St Beltsville MD 20705

FEINBERG, STEWART CARL, b Brooklyn, NY, July 7, 47; m 72; c 2. ORGANIC POLYMER CHEMISTRY. *Educ:* Brooklyn Col, BS, 69; Univ Akron, PhD(polymer sci), 76. *Prof Exp:* res chemist org polymers, Washington Res Ctr, W R Grace & Co, 75-79; RES CHEMIST, NEW

POLYMER PROD RES & DEVELOP, ARCO CHEM CO, PA, 79-; TECH ASSOC, E I du Pont. *Mem:* Am Chem Soc. *Res:* Applied research and product development in thermoplastic elastomers. *Mailing Add:* E I du Pont PO Box 80-269 Wilmington DE 19880-0269

FEINBLATT, JOEL DAVID, ENDOCRINOLOGY, CALCIUM METABOLISM. *Educ:* Univ Pa, PhD(physiol), 69. *Prof Exp:* ASSOC PROF PHYSIOL, HEALTH SCI SCHS, TUFTS UNIV, 81-, DEAN MED EDUC, SCH MED, 83- *Res:* Bone physiology. *Mailing Add:* Proj Coord Mgr Sandoz Res Inst Rte 10 East Hanover NJ 07936

FEINBLUM, DAVID ALAN, b New York, NY, May 20, 40; m 64; c 2. APPLIED PHYSICS, ENGINEERING. *Educ:* Cooper Union, BME, 60; Rensselaer Polytech Inst, PhD(physics), 66. *Prof Exp:* Asst prof physics, State Univ NY, Albany, 64-71; prin mathematician mil res & develop, Sperry/Univac, 71-75; TECH DIR, MIL RES & DEVELOP, XYBION CORP, 75- *Concurrent Pos:* Consult, Nuclear Fuel Serv, Inc, 65-66; vis fel, Princeton Univ, 67-68; fel, Weitzman Inst Sci, 69-70. *Mem:* Sigma Xi. *Res:* Engineering of software systems; signal processing, dynamics. *Mailing Add:* Xybion Corp 240 Cedar Knolls Rd Cedar Knolls NJ 07927

FEINDEL, WILLIAM HOWARD, b Bridgewater, NS, July 12, 18; m 45; c 6. NEUROSURGERY, NEUROHISTORY. *Educ:* Acadia Univ, BA, 39; Dalhousie Univ, MSc, 42; McGill Univ, MD & CM, 45; Oxford Univ, DPhil(neuroanat), 49; Am Bd Neurol Surg, dipl 55; FRCS(C), 55; FACS, 62. *Hon Degrees:* DSc, Acadia Univ, 63, McGill Univ, 84; LLD, Mt Allison Univ, 83, Univ Sask, 86. *Prof Exp:* Med researcher, Exp Head Injuries, Montreal Neurol Inst, 42-44; lectr neurosurg, McGill Univ, 52-55; prof, Univ Sask, 55-59; chmn dept neurol & neurosurg, McGill Univ, 72-77, Cone prof neurosurg, 59-88, dir Brain Imaging Ctr, 60-87, dir, Montreal Neurol Inst & Hosp, 72-84, dir PET Unit, 75-86, DIR NEUROHIST PROJ, MCGILL UNIV, 87- *Concurrent Pos:* Nat Res Coun Can fel med, Montreal Neurol Inst, 49-50, Reford fel, 53-55; Med Res Coun Can res grant, 56-; Can Cancer Found grant, 56-59; founder & dir, Cone Lab Neurosurgical Res, Montreal Neurol Inst, 59-84, neurosurgeon-in-chief, 63-72; cur, Osler Libr, McGill Univ, 63-, hon asst librn, 64-66; vis lectr, Columbia, 63, Yale Univ, 66, Univ Calif, San Francisco, 71, Sask, 86, Zagreb, 86, Belgrade, 86 & Taipei, 90; dir, Fourth Can Cong Neurol Sci, Montreal, 69; vis prof, Univ BC, 70-71; cur & executor, Wilder Penfield Papers, 76-; consult, neurosci, WHO, 74-; mem sci comt, Found Study Nerv Syst, Geneva, 82-; mem bd gov, Acadia Univ, 83-; rev, Proj Rev Neurol A Comt, NIH, 83-88; med consult, Nat Film Bd Can, 77-; pres, Brain, Int CBF Cong, Montreal, 87; prin investr, Brain Tumor Proj, NIH, 86-89; co-investr, Montreal Neurol Inst, 89-93. *Honors & Awards:* Elsberg lectr, NY Neurosurg Soc, 83; Penfield lectr, Can Neurosurg Soc, 84; First Penfield Lectr, McGill Univ, 84; Hugh Jackson Lectr, Montreal Neurol Inst, 88. *Mem:* Am Neurol Asn (vpres, 77); Am Acad Neurol Surg (pres, 76); Soc Neurol Surg (vpres, 78); Am Asn Neurol Surg; Can Neurosurg Soc (pres, 68); Am Osler Soc. *Res:* Clinical neurosurgery; cerebral edema; immersion foot; peripheral nervous system; pain; amygdala and temporal lobe; cerebral localization and circulation; medical history, Thomas Willis, William Osler, Wilder Penfield and history of Montreal Neurological Institute. *Mailing Add:* Montreal Neurol Inst 3801 University St Montreal PQ H3A 2B4 Can

FEINER, ALEXANDER, b Vienna, Austria, Apr 12, 28; m 69; c 3. SOFTWARE SYSTEMS. *Educ:* Techische Hochschule, BS, 50; Columbia Univ, MS, 52. *Prof Exp:* Mem tech staff, AT&T Bell Labs, 53-55, supvr, 55-59, dept head, 59-69, dir, 69-83, EXEC DIR, AT&T BELL LABS, 83- *Honors & Awards:* Eng Excellence, Inst Elec & Electronics Engrs, 91. *Mem:* Nat Acad Eng; fel Inst Elec & Electronics Engrs; Sigma Xi. *Res:* Development of stored-program controlled business communications systems; design of switching networks. *Mailing Add:* Four Sailers Way Rumson NJ 07760

FEINGOLD, ADOLPH, b Poltava, USSR, Mar 8, 20; m 64; c 2. MECHANICAL ENGINEERING, ENGINEERING ECONOMICS. *Educ:* Univ Genoa, Italy, DEng(naval archit & mech eng), 53. *Prof Exp:* From assoc prof to prof mech eng, Univ Mo-Rolla, 63-66; prof civil eng, 66-67, chmn, Dept Mech Eng, 67-71, gov, 76-79, PROF MECH ENG, UNIV OTTAWA, 67-; mgr, Design Eng Div, Syncrude Can Ltd, 80-86; RETIRED. *Mem:* Brit Inst Marine Eng; Israel Soc Naval Archit & Marine Eng (pres, 61); Am Soc Mech Engrs; fel Eng Inst Can; fel Can Soc Mech Eng (vpres, 77-79). *Res:* Internal combustion engines; application of computers in design; heat transfer; arctic technology; low temperature environmental problems; marine engineering; ocean bed technology; resource policy alternatives. *Mailing Add:* 511-10160 114th St Edmonton AB T5K 2L2 Can

FEINGOLD, ALEX JAY, b Baltimore, Md, Apr 1, 50; m 77; c 1. ALGEBRA. *Educ:* Johns Hopkins Univ, BA & MA, 71; Yale Univ, PhD(math), 77. *Prof Exp:* Jr instr math, Johns Hopkins Univ, 70-71; officer comput, USPHS, 71-73; teaching asst math, Yale Univ, 75-77; asst prof, Drexel Univ, 77-79; ASST PROF MATH, STATE UNIV NY, 79- *Concurrent Pos:* prin investr, NFS grant, Drexel Univ, 78-79 & State Univ NY, Binghamton, 80-82, 85-87. *Mem:* Math Asn Am; Am Math Soc; Sigma Xi. *Res:* Pure mathematics in an area of algebra known as Kac-Moody Lie algebras; application to elementary particle physics; superstring theory. *Mailing Add:* 45 Matthews St Binghamton NY 13905

FEINGOLD, ALFRED, b Sellersville, Pa, Jan 31, 41; m 66; c 2. ANESTHESIOLOGY. *Educ:* Dartmouth Col, BA, 62; Tufts Univ, MD, 66; Northwestern Univ, MS, 71. *Prof Exp:* Intern med, Univ Chicago Hosps & Clins, 66-67, resident anesthesia, 67-69; fel, Northwestern Univ Sch Med, 69-70; physician, Wright-Patterson AFB Hosp, 70-72; from asst prof to assoc prof anesthesia, Sch Med, Univ Miami, 72-77; dir anesthesia, 80-84, STAFF MEM, CEDARS MED CTR, 77- *Concurrent Pos:* Consult, Panel Eval Fentanyl, Am Med Asn, 70; fel biomed eng, Technol Inst, Northwestern Univ, 70; staff mem, Jackson Mem Hosp & Miami Vet Admin Hosp, 72-77; dir anesthesiol, Bascom Palmer Eye Inst-Ann Bates Leach Eye Hosp, Miami, 76; adj assoc prof biomed eng, Univ Miami, 77- *Mem:* Am Soc Anesthesiol; Am Soc Cardiovasc Anethesiologists. *Res:* Pharmacokinetics of anesthetics; biotransformation of anesthetics; computer applications to anesthesiology and medicine; biomedical engineering; human factors engineering. *Mailing Add:* Dept Anesthesiol Cedars Lebanon Health Care Ctr Miami FL 33136

FEINGOLD, ARNOLD MOSES, b Brooklyn, NY, Dec 30, 20; m 54; c 2. NUCLEAR PHYSICS. *Educ:* Brooklyn Col, AB, 41; Princeton Univ, MS, 48, PhD(physics), 52. *Prof Exp:* Physicist, Langley Mem Aeronaut Lab, 41-46; from instr to asst prof, Univ Pa, 50-55; asst prof, Univ Ill, 55-57; assoc prof, Univ Utah, 57-60; PROF PHYSICS, STATE UNIV NY STONY BROOK, 60- *Mem:* AAAS; fel Am Phys Soc; Am Asn Physics Teachers; Sigma Xi. *Res:* Theoretical nuclear physics; tensor force effects in nuclei. *Mailing Add:* 22 Brewster Hill Rd Setauket NY 11733

FEINGOLD, BEN F, allergy, immunology; deceased, see previous edition for last biography

FEINGOLD, DAVID SIDNEY, b Chelsea, Mass, Nov 15, 22; m 49; c 3. BIOCHEMISTRY. *Educ:* Mass Inst Technol, BS, 44; Hebrew Univ, PhD(biochem), 56. *Prof Exp:* Chemist, Lucidol Corp, 44; instr, Northeastern Univ, 46-47; chemist, Hadassah Hosp, Jerusalem, 50-51; asst, Med Sch, Hebrew Univ, 51-56; jr res biochemist, Univ Calif, Berkeley, 56-58, asst res biochemist, 58-60; from asst prof to prof biol, Fac Arts & Sci, 60-76, PROF MICROBIOL, SCH MED, UNIV PITTSBURGH, 66- *Concurrent Pos:* NIH res career develop award, 65-75. *Honors & Awards:* State of Israel Prize, 57. *Mem:* Am Chem Soc; Am Soc Biol Chem; Infectious Dis Soc Am; Protein Soc; Am Soc Microbiol; Int Endotoxin Soc. *Res:* Intermediary carbohydrate metabolism; enzymic transformations of carbohydrates; biosynthesis of poly- and oligosaccharides; transglycosylation; structure of polysaccharides; enzyme mechanism; dehydrogenases. *Mailing Add:* Dept Molecular Genetics & Biochem Univ Pittsburgh Sch Med Pittsburgh PA 15261-2072

FEINGOLD, EARL, b Philadelphia, Pa, Dec 4, 24; m 47; c 2. SOLID STATE PHYSICS. *Educ:* Temple Univ, AB, 50, AM, 51, PhD(physics), 59. *Prof Exp:* Asst instr physics, Temple Univ, 49-51; res physicist, Brown Instrument Div, Minneapolis Honeywell Regulator Co, 51-52 & res ctr, Burroughs Corp, 52-54; solid state res physicist, Atlantic Refinery Co, 54-59; SOLID STATE PHYSICIST & TECH STAFF SCIENTIST ASTRO SPACE DIV, GEN ELEC CO, 59- *Mem:* AAAS; Am Phys Soc; Sigma Xi; Int Metallog Soc. *Res:* Physical and chemical characterization of matter; electron and optical microscopy; x-ray diffraction and spectroscopy; refractory-high-strength-special purpose materials; materials sciences; surface properties; thin films; physical metallurgy. *Mailing Add:* 420 Anthony Rd King of Prussia PA 19406

FEININGER, TOMAS, b Stockholm, Sweden, Sept 21, 35; Can citizen; m 63; c 3. GEOLOGY. *Educ:* Middlebury Col, BA, 56; Brown Univ, MSc, 60, PhD(geol), 64. *Prof Exp:* Geologist, Eng Geol Br, US Geol Surv, 56-59, Br Regional Geol, New Eng, 60-64 & Off Int Geol, 64-69; vis res assoc, Smithsonian Inst, 69-70; chmn, Dept Geol, Mining & Petrol, Nat Polytech Sch, Quito, Ecuador, 70-78; researcher, Dept Geol, Laval Univ, Que, 78-83; AT GEOL SURV, CAN, 83- *Mem:* AAAS; Geol Soc Am; Ecuadorian Inst Natural Sci; Geol Asn Can; Mineral Soc Am; Sigma Xi. *Res:* Metamorphic petrology and regional geology of the northern Andes; regional geology of eastern Can. *Mailing Add:* Geol Surv Can CP 7500 Quebec PQ G1V 4C7 Can

FEINLAND, RAYMOND, b New York, NY, Oct 12, 28; m 59. COSMETIC CHEMISTRY. *Educ:* Brooklyn Col, BS, 49; Polytech Inst Brooklyn, MS, 52, PhD(chem), 57. *Prof Exp:* Sr res chemist, Am Cyanamid Co, Conn, 57-65; mgr anal res, 65-71, assoc dir, 71-80, dir, prod develop, 80-85, DIR, EXPLOR DEVELOP, CLAIROL INC, 85- *Mem:* Soc Cosmetic Chemists; Am Chem Soc. *Res:* Cosmetic products; ion exchange resins; hair bleaching; chromatography; spectrophotometry; general analytical techniques; hair color products. *Mailing Add:* Clairol Inc Two Blachley Rd Stamford CT 06902

FEINLEIB, JULIUS, b Brooklyn, NY, Sept 17, 36; m 61, 83; c 1. SOLID STATE PHYSICS. *Educ:* Cornell Univ, BEngPhys, 58; Harvard Univ, MA, 59, PhD(appl & solid state physics), 63. *Prof Exp:* Fel, Harvard Univ, 63-64; staff physicist, Lincoln Lab, Mass Inst Technol, 64-69; mem staff, Energy Conversion Devices, Inc, 69-71; MGR, PHYSICS LAB, ITEK CORP, 71- *Concurrent Pos:* Pres Adoptive Optics Assoc, Inc/United Technol Corp, 76- *Mem:* Am Phys Soc. *Res:* High pressure physics of semiconductors; nuclear resonance in magnetic materials; optical properties of magnetic materials; amorphous semiconductors and optical computer memories; adaptive optical systems. *Mailing Add:* Adaptive Optics Assoc Inc 54 Cambridge Park Dr Cambridge MA 02140

FEINLEIB, MANNING, b Brooklyn, NY, July 19, 35; m 57; c 3. BIOSTATISTICS, EPIDEMIOLOGY. *Educ:* Cornell Univ, AB, 56; State Univ NY Downstate Med Ctr, MD, 61; Harvard Univ, MPH, 63, DrPH, 66. *Prof Exp:* Instr statist, State Univ NY, 58-59; med officer epidemiol, Nat Heart Lung & Blood Inst, NIH, 66-68, chief, Field Epidemiol Res Sect, 68-83, chief, Epidemiol Br, 79-83; DIR, NAT CTR HEALTH STATIST, 83- *Concurrent Pos:* Fel epidemiol, Sch Pub Health, Harvard Univ, 63-66; asst med, Peter Bent Brigham Hosp, 62-66; assoc registr, Mass Tumor Registry, 64-66; asst prof, Sch Pub Health, Harvard Univ, 66-68. *Honors & Awards:* Mortimer Spiegelman Gold Medal Award, Statist Sect, Am Pub Health Asn, 72. *Mem:* AAAS; Am Pub Health Asn; Am Statist Asn; Biomet Soc; Am Epidemiol Soc. *Res:* Epidemiology of heart disease. *Mailing Add:* Nat Ctr Health Statist 6525 Belcrest Rd Hyattsville MD 20782

FEINLEIB, MARY ELLA (HARMAN), b Italy, May 21, 38; US citizen. PLANT PHYSIOLOGY. *Educ:* Cornell Univ, AB, 59; Radcliffe Col, AM, 61; Harvard Univ, PhD(biol), 66. *Prof Exp:* From instr to assoc prof, 65-82, chmn dept, 76-82, PROF BIOL & DEAN, COL LIB ARTS, TUFTS UNIV, 82- *Concurrent Pos:* Actg dean, Fac Arts, Sci & Technol, Tufts Univ, 90-91; fel & grants, Woodrow Wilson, NIH & NSF. *Mem:* AAAS; Am Soc Photobiol; Sigma Xi. *Res:* Phototaxis in algae, specifically in Chlamydomonas. *Mailing Add:* Ballou Hall Tufts Univ Medford MA 02155

FEINLEIB, MORRIS, b Berlin, Ger, July 16, 24; nat US; wid; c 2. CHEMICAL ENGINEERING, ELECTROCHEMISTRY. *Educ:* Columbia Univ, BS, 44, MS, 45, PhD(chem eng), 48. *Prof Exp:* Res electrochemist, metals res dept, Armour Res Found, Ill, 48-52; res electrochemist & group leader, reduction res sect, Kaiser Aluminum & Chem Corp, Wash & Calif, 52-58; res scientist, electrochem sect, missiles & space div, Lockheed Aircraft Corp, 58-59; sr res engr, tube res dept, 59-63, mgr tube process develop, 63-67, chem vapor deposition pilot facility, cent res, 67-69, mgr photoconductor prep, electrophotog unit, 69-75, mgr mat & processes, Graphics Div, 75-79, mgr mat res & develop, Palo Alto Microwave Tube Div, Varian Assocs, 79-85; RETIRED. *Concurrent Pos:* Consult scientist mat res & develop, Microwave Tube Div, Varian Assocs, 85- *Mem:* Electrochem Soc; Am Chem Soc. *Res:* High temperature materials and methods, including fused salts, chemical vapor deposition; batteries; electroplating; materials and chemical processes in electronics; special photoconductors and materials for electrophotography and electrostatic printing. *Mailing Add:* 1575 Murre Lane Sunnyvale CA 94087-4849

FEINMAN, J(EROME), b Brooklyn, NY, Mar 18, 28; m 54; c 3. FLUIDIZED BED TECHNOLOGY, PLASMA TECHNOLOGY. *Educ:* Polytech Inst Brooklyn, BChE, 49; Ill Inst Technol, MSChE, 56; Univ Pittsburgh, PhD(chem eng), 64. *Prof Exp:* From proj engr to sr proj engr, US Steel Corp, 52-60, sect head, ore reduction technol sect, 60-64, sr process engr process eng div, 64-65, assoc res consult, 65-, res consult, Raw Mat & Ore Reduction Div, Res Lab, 76-80; dir, tech develop, Occidental Oil Shale, Inc, 80-82; PRES & CHIEF CONSULT, J FEINMAN & ASSOC INC, 82- *Concurrent Pos:* Mem bd dir & tech adv comt, Particulate Solid Res, Inc, 77-79; chmn, publ comt, Iron & Steel Soc, Am Inst Mining & Metall & Petrol Engrs; ed, Plasma Technol Metall Processing, 87. *Mem:* Am Chem Soc; Am Inst Chem Engrs; Iron & Steel Soc; Am Inst Mining, Metall & Petrol Engrs. *Res:* Ferrous and nonferrous metallurgical process development and design; direct reduction of iron ore; fluidization; coal gasification and high temperature cleanup; process technical and economic evaluation and engineering; shale oil recovery and treatment; application of plasma technology to metallurgical processing. *Mailing Add:* 120 Kelly Ct Monroeville PA 15146-1352

FEINMAN, MAX L, b New York, NY, May 13, 05; m 33; c 2. SURGERY, ANATOMY. *Educ:* State Univ NY Downstate Med Ctr, MD, 28. *Prof Exp:* Assoc anat, Long Island Col Hosp, 50-54; attend surgeon, Wyckoff Heights Hosp, Brooklyn, NY, 58-64; DIR MED EDUC, DOCTORS HOSP, 64-; prof anat & surg, M J Lewi Col Podiatry, 65-81; RETIRED. *Concurrent Pos:* Fel anat, Long Island Col Hosp, 47-51; surgeon, Coney Island Hosp, 49-54; dir surg, Lefferts Gen Hosp, Brooklyn, NY, 58-63. *Res:* General surgery; traumatic surgery. *Mailing Add:* 25 Sutton Pl S New York NY 10022

FEINMAN, RICHARD DAVID, b Brooklyn, NY, July 19, 40. BIOCHEMISTRY. *Educ:* Univ Rochester, BA, 63; Univ Ore, PhD(chem), 69. *Prof Exp:* asst prof, 69-75, assoc prof, 75-80, PROF BIOCHEM, STATE UNIV NY DOWNSTATE MED CTR, 80- *Concurrent Pos:* Fac mem, New Sch Social Res, 73- *Res:* Enzyme mechanism; mechanism of inhibitors of proteolytic enzymes from blood plasma and stimulus response coupling in blood platelets and neurons; invertebrate behavior; chemical basis of learning and memory. *Mailing Add:* Biochem Dept State Univ NY Downstate Med Ctr Brooklyn NY 11203

FEINMAN, SUSAN (BIRNBAUM), b Atlanta, Ga, Sept 16, 30; c 3. MICROBIOLOGY. *Educ:* Wellesley Col, BA, 51; George Washington Univ, MS, 52, PhD(microbiol), 69. *Prof Exp:* Res asst pharmacol, Beth Israel Hosp, Harvard Med Sch, 52-53; teaching fel microbiol, Med Sch, George Washington Univ, 66-68; supvry microbiologist immunol, Lab Serol, Div Labs, Govt Washington, DC, 68-69; tech info specialist pharmacol, NIMH, HEW, 71-74; microbiologist drug resistance antibiotics, Bur Vet Med, Food & Drug Admin, 74-79; biologist supvr immunol/toxicol, health sci, US Consumer Prod Safety Comn, 79-86; biologist supvr, Ctr Food Safety & Nutrit, Food & Drug Admin, 87-89; HEALTH SCIENTIST ADMINR, NAT CANCER INST, NIH, 90- *Concurrent Pos:* Nat Cancer Inst fel, Lab Microbiol, Nat Inst Allergy & Infectious Dis, NIH, Bethesda, 69-71. *Honors & Awards:* Commendation, Food & Drug Admin, 77; Qual Increase Award, Bur Vet Med, 77. *Mem:* Am Soc Microbiol; AAAS; Sigma Xi; Am Col Toxicol; Soc Toxicol. *Res:* Effect of antibiotics in animal feeds on drug resistance; environmental impacts of antibiotics; arsenic-resistant E coli R-plasmids; hypersensitivity effects and toxicology of environmental chemicals; immunology of streptococcal L-forms; immunology of atypical mycobacteria. *Mailing Add:* Nat Cancer Inst 5333 Westbard Ave Rm 809B Bethesda MD 20892

FEINSINGER, PETER, b Madison, Wis, July 16, 48; m; c 2. COMMUNITY ECOLOGY, POLLINATION ECOLOGY. *Educ:* Colo Col, BA, 69; Cornell Univ, PhD(ecol, evolutionary biol), 74. *Prof Exp:* Vis asst prof zool, Ind Univ, 74; asst prof biol sci, Univ Denver, 75-76; from asst prof to assoc prof, 76-84, PROF ZOOL, UNIV FLA, 84- *Concurrent Pos:* NSF grants, 76, 80, 86 & 88. *Mem:* Ecol Soc Am; Soc Study Evolution; Asn Trop Biol; Am Ornithologists Union; Brit Ecol Soc; Am Soc Naturalists; Soc Conserv Biol. *Res:* Ecology of tropical communities: mutualistic interactions between plants and animals, relation to community dynamics; tropical conservation ecology. *Mailing Add:* Dept Zool Univ Fla 411 Bar Bldg Gainesville FL 32611

FEINSTEIN, ALEJANDRO, b La Plata, Arg, May 30, 29; m 59; c 3. ASTROPHYSICS. *Educ:* Univ La Plata, astronr, 56, DrAstron, 60. *Prof Exp:* Observer asst astron, 48-56, instr, 52-56, tech asst, 56-62, asst prof astrophys, 62-63, instr math, Fac Phys Sci & Math, 58-61, PROF ASTROPHYS, ASTRON OBSERV, UNIV LA PLATA, 63- *Concurrent Pos:* Nat Coun Sci & Tech Res, Arg fel, Lick Observ, Univ Calif, 61-62; Guggenheim Mem Found fel, Steward Observ, Univ Ariz, 69-70; supv investr, Photom & Structured Galactic Prog, Conicet, 83- *Mem:* Arg Astron Asn; Royal Astron Soc; Astron Soc Pac; Int Astron Union; NY Acad Sci. *Res:* Stellar photoelectric photometry, including infrared; open clusters; galactic structure; stellar formation. *Mailing Add:* Observ Astron Paseo Del Bosque La Plata 1900 Argentina

FEINSTEIN, ALLEN IRWIN, b Brooklyn, NY, Apr 9, 40; m 62; c 3. ORGANIC CHEMISTRY. *Educ:* Brooklyn Col, BS, 62; Iowa State Univ, PhD(alkaloid biosynthesis), 67. *Prof Exp:* Wax chemist, Austenal Co Div, Howe Sound Corp, 62; res chemist, 67-78, SR RES CHEMIST, AMOCO CHEM CORP, 78- *Concurrent Pos:* Instr, Aurora Col, 70- & chem, Elmhurst Col, 75- *Mem:* Am Chem Soc. *Res:* Vapor phase oxidations; heterogeneous catalysis. *Mailing Add:* 1903 Cheshire Lane Wheaton IL 60187-8550

FEINSTEIN, ALVAN RICHARD, b Philadelphia, Pa, Dec 4, 25. INTERNAL MEDICINE. *Educ:* Univ Chicago, BS, 47, MS, 48, MD, 52. *Prof Exp:* Asst med, Rockefeller Inst, 54-56; clin dir rheumatic heart dis, Irvington House, 56-60, med dir, 60-62; clin biostatist, Eastern Res Support Ctr, Vet Admin Hosp, West Haven, 62-67, chief, 67-74; assoc prof, 64-69, PROF MED & EPIDEMIOL, SCH MED, YALE UNIV, 69-, DIR JOHNSON CLIN SCHOLAR PROG, 74- *Concurrent Pos:* Asst prof Internal med, Col Med, NY Univ Med Ctr, 59-62. *Honors & Awards:* J Allyn Taylor Int Prize, Soc Gen Internal Med. *Mem:* Inst Med-Nat Acad Sci; Am Fedn Clin Res; Asn Am Physicians; Am Epidemiol Soc; Am Soc Clin Invest. *Res:* Rheumatic fever; obesity; prognosis of cancer; clinical pharmacology and biostatistics; clinical epidemiology; clinimetrics. *Mailing Add:* Dept Med Yale Univ Sch Med New Haven CT 06510

FEINSTEIN, CHARLES DAVID, b New York, NY, Oct 23, 46; m 80. OPTIMIZATION THEORY, MATHEMATICAL MODELING. *Educ:* Cooper Union, BSME, 67; Stanford Univ, MS, 68, MS, 78, PhD(eng & econ syst), 80. *Prof Exp:* Engr, Stanford Res Inst, 75-76, Palo Alto Res Ctr, Xerox Corp, 76-81; sr decision analyst, Appl Decision Anal, Inc, 81-82; ASSOC PROF, UNIV SANTA CLARA, 82- *Concurrent Pos:* Consult asst prof, Eng & Econ Syst, Stanford Univ, 80- *Res:* Optimal control theory; mathematical programming; dynamic systems; forecasting; stochastic choice theory. *Mailing Add:* 390 Bethany Dr Scotts Valley CA 95066

FEINSTEIN, HYMAN ISRAEL, b Brooklyn, NY, Apr 9, 11; m 43. ANALYTICAL CHEMISTRY. *Educ:* Univ Mich, AB, 30; Columbia Univ, MA, 32. *Prof Exp:* Asst instr, Long Island Univ, 30-36; chemist, Nat Bur Stand, 36-40, Customs Bur Labs, 40-41 & Nat Bur Stand, 41-54; anal chemist, US Geol Surv, 54-58; lectr & assoc prof chem, George Mason Univ, 58-79, RETIRED. *Concurrent Pos:* Instr, Howard Univ, 43-48 & grad sch, Nat Bur Stand, 50-53; lectr, Cath Univ Am, 54. *Mem:* Am Chem Soc; Nat Sci Teachers Asn; Am Inst Chem. *Res:* Rare elements; trace analysis; mineral and rock analysis; uranium; thorium; vanadium; coordination compounds of chromium; chemical literature and education; chemical microscopy; microchemistry. *Mailing Add:* 10411 Forest Ave Fairfax VA 22030-3614

FEINSTEIN, IRWIN K, b Chicago, Ill, Aug 29, 14; m 54; c 2. MATHEMATICAL ANALYSIS. *Educ:* Ill Inst Technol, BS, 36; Northwestern Univ, MA, 46, PhD(math educ), 52. *Prof Exp:* From instr to asst prof math, 46-60, from assoc prof to prof math educ, 60-66, PROF MATH, UNIV ILL, CHICAGO CIRCLE, 66- *Concurrent Pos:* Schwab Found lectr, Mus Sci & Indust, Ill, 65-67. *Mem:* AAAS; Am Math Asn; Nat Coun Teachers Math; Sch Sci & Math Asn; Am Asn Univ Professors. *Mailing Add:* 735 Lamon Ave Wilmette IL 60091

FEINSTEIN, JOSEPH, b New York, NY, July 8, 25; m 52; c 3. ELECTROMAGNETIC THEORY, PARTICLE BEAM PHYSICS. *Educ:* Cooper Union, BEE, 44; Columbia Univ, MA, 47; NY Univ, PhD(physics), 51. *Prof Exp:* Res physicist, Nat Bur Standards, 49-54; mem tech staff, Bell Tel Labs, 54-59; dir res, S-F-D Labs, 59-64; vpres res, Varian Assocs, 64-80; dir, electronics & phys sci, Off Under Secy Defense Res & Eng, Res & Advan Technol, Dept Defense, Washington, DC, 80-83; CONSULT PROF ELEC ENG, STANFORD UNIV, 83- *Concurrent Pos:* Consult, electronic indust, 83- *Mem:* Nat Acad Eng; fel Inst Elec & Electronics Engrs. *Res:* Electromagnetic theory; microwave electron tubes; power conversion; relativistic beams. *Mailing Add:* Dept Elec Eng Stanford Univ Stanford CA 94305

FEINSTEIN, LOUIS, b Philadelphia, Pa, Apr 20, 12; m 35; c 2. BIOCHEMISTRY. *Educ:* Univ Pa, AB, 33, BS, 34, MS, 39; Georgetown Univ, PhD(chem), 46. *Prof Exp:* Res chemist, Barrett Co, 34-35 & Sch Med, Univ Pa, 35-39; from jr chemist to chemist, Grain Br, Agr Res Serv, USDA, 39-47 & Bur Entom & Plant Quarantine, 47-53, sr res biochemist animal & poultry husb res, 53-56, supvry chemist & prin biochemist, Biol Sci Br, 56-60, asst chief, Field Crops & Animal Prods Res Br, Agr Res Serv, 60-67, chief, 67-73, chief, Seed Qual Lab, 73-74; CONSULT, 74- *Concurrent Pos:* Mem, Working Party of Experts, Pesticides, UN Food & Agr Orgn, 72-73. *Mem:* AAAS; Am Chem Soc; Entom Soc Am; Asn Off Anal Chem; fel Am Inst Chem; Sigma Xi. *Res:* Coal tar chemicals; liver and kidney metabolism; cereal chemistry; vitamins; plant alkaloids and insecticides; insect attractants and repellants; animal composition; quality evaluation of agricultural products. *Mailing Add:* 14757 Wild Flower Lane Delray Beach FL 33446

FEINSTEIN, MAURICE B, b New York, NY, Nov 28, 29; m 51; c 3. BIOCHEMISTRY. *Educ:* Columbia Univ, BS, 52, MS, 54; State Univ NY, PhD(pharmacol), 60. *Prof Exp:* Res assoc physiol, Inst Muscle Dis, 60-62; from instr to assoc prof pharmacol, State Univ NY Downstate Med Ctr, 62-69; assoc prof, 69-76, prof pharmacol, Schs Med & Dent Med, 76-90, HEAD, DEPT PHARMACOL, UNIV CONN, 90- *Mem:* Am Pharmacol Soc; Harvey Soc. *Res:* Protein phosphorylation; cytoskeleton; tyrosine phosphorylation; cyclic nucleotides; calcium phosphoinositide metabolism; IP3 receptor; blood platelets; monoclonal antibodies. *Mailing Add:* Dept Pharmacol Univ Conn Health Ctr 263 Farmington Ave Farmington CT 06030

FEINSTEIN, MYRON ELLIOT, b New York, NY, Jan 7, 43; m 64; c 2. SURFACE CHEMISTRY. *Educ:* City Col New York, BS, 63, MA, 65; City Univ New York, PhD(phys chem), 67; Chapman Col, MBA, 84. *Prof Exp:* Lectr chem, City Col New York, 64-67; from res chemist to sr res chemist, Allied Chem Corp, 67-68; estab scientist, Unilever Res Lab, Eng, 68-70, mgr process develop soaps & detergents, Unilever Scand, 70-74, area prod mgr, Lever-Gibbs, Italy, 74-77, mfg mgr, Hammond, Ind, 77-81, plant mgr, Lever Bros Co, Los Angeles, 81-87; DIR TECHNOL & FACIL PLANNING, LEVER BROS CO, NY, 87- *Concurrent Pos:* Res asst, E I du Pont de Nemour, 66. *Mem:* AAAS. *Res:* Production control; chemical process engineering; wetting and detergency; colloidal surfactants; emulsions and foams; insoluble monolayers; surface potentials; fluorinated surfactants. *Mailing Add:* 100 Hope St No 20 Stamford CT 06906

FEINSTEIN, ROBERT NORMAN, b Milwaukee, Wis, Aug 10, 15; m 40; c 2. ENZYMOLOGY, BIOCHEMISTRY. *Educ:* Univ Wis, BS, 37, MS, 38, PhD(physiol chem), 40. *Prof Exp:* Asst, McArdle Mem Inst, Univ Wis, 38-39; res assoc metab & endocrinol, Michael Reese Hosp, Chicago, 40-41 & May Inst Med Res, Cincinnati, 46; from instr to assoc prof biochem, Univ Chicago, 47-79, researcher, USAF Radiation Lab, 47-54, res assoc, 47-79; sr biochemist, Argonne Nat Lab, 59-79; RETIRED. *Concurrent Pos:* Assoc scientist, Argonne Nat Lab, 54-59; Guggenheim fel, Inst Radium, Paris, 59-60. *Mem:* AAAS; Am Chem Soc; Am Soc Biol Chem. *Res:* Radiation; enzymes; cancer; isozymes. *Mailing Add:* 4624 Highland Ave Downers Grove IL 60515

FEINSTEIN, SHELDON ISRAEL, b Brooklyn, NY, Sept 17, 50; m 83; c 4. MOLECULAR GENETICS, GENETIC ENGINEERING. *Educ:* Yeshiva Univ, BA, 71; Yale Univ, MPhil, 74, PhD(biol), 77. *Prof Exp:* Assoc bact genetics, Radiobiol Labs, Dept Therapeut Radiol, Sch Med, Yale Univ, 77-80; staff assoc DNA methylation, dept human genetics & develop, Col Physicians & Surgeons, Columbia Univ, 83-84; asst prof biochem genetics & metab, Rockefeller Univ, 84-87; ASST PROF HUMAN GENETICS, INST ENVIRON MED, UNIV PA, 88- *Concurrent Pos:* Vis scientist, Weizmann Inst Sci, Rehovot, Isreal, 80-82; mem, Arteriosclerosis Coun, Am Heart Asn. *Mem:* Sigma Xi; Am Heart Asn. *Res:* Suppression of nonsense mutations; mutations that increase genetic recombination; mutator genes; genetics of human interferon and apolipoprotein genes; isolation, mapping and expression; methylation of mammalian DNA. *Mailing Add:* 7661 Brookhaven Rd Overbrook PA 19151-2023

FEINSTONE, STEPHEN MARK, b Greenwich, Conn, July 24, 44. VIRAL HEPATITIS. *Prof Exp:* CHIEF, LAB HEPATITIS RES, DIV VIROL, NIH, 89- *Mailing Add:* Ctr Biol Eval & Res Div Virol Lab Hepatitis Res NIH Bldg 29A Rm 2D12 Bethesda MD 20892

FEIOCK, FRANK DONALD, b Murray, Ky, June 22, 36; m 54; c 2. PHYSICS, OPTICS. *Educ:* Murray State Col, AB, 58; Univ Iowa, PhD(physics), 64. *Prof Exp:* Res assoc physics, Tufts Univ, 64-65; assoc res scientist, Univ Notre Dame, 65-69; staff physicist, KMS Technol Ctr, KMS Industs, Inc, 69-72; sr scientist, McDonnell Douglas Astronaut, 72-77; assoc group leader, Lawrence Livermore Lab, Univ Calif, 77-78; prog mgr, Rocketdyne Lasers, 78-82; dept mgr, Maxwell Labs, 82-83, prog mgr, W J Schafer Assoc, 83-84, group leader, Los Alamos Nat Lab, 84-87; DIV MGR, SAIC, 87- *Concurrent Pos:* Instr mgt, Univ Southern Calif. *Mem:* Am Phys Soc; Optical Soc Am; Am Soc Mech Eng. *Res:* Lasers; optics; atomic physics; materials response; statistical mechanics; numerical analysis. *Mailing Add:* 4763 Crisp Way San Diego CA 92117

FEIR, DOROTHY JEAN, b St Louis, Mo, Jan 29, 29. INSECT PHYSIOLOGY. *Educ:* Univ Mich, BS, 50; Univ Wyo, MS, 56; Univ Wis, PhD(entom), 60. *Prof Exp:* Instr biol, Univ Buffalo, 60-61; from asst prof to assoc prof, 61-67, PROF BIOL, ST LOUIS UNIV, 67- *Concurrent Pos:* Ed, Environ Entom, 77-84. *Mem:* AAAS; NY Acad Sci; Am Physiol Soc; Entom Soc Am (pres-elect, 88, pres, 89); Sigma Xi. *Res:* Feeding behavior of insects; hematology and immunology of insects; action of hormones in insects; air pollutants and insects; lyme disease. *Mailing Add:* Dept Biol St Louis Univ St Louis MO 63103

FEISEL, LYLE DEAN, b Tama, Iowa, Oct 16, 35; m 57; c 3. ELECTRICAL ENGINEERING. *Educ:* Iowa State Univ, BS, 61, MS, 63, PhD(elec eng), 64. *Prof Exp:* From asst prof to prof elec eng, SDak Sch Mines & Technol, 64-83, head dept, 76-83; DEAN WATSON SCH, STATE UNIV NY, BINGHAMTON, 83- *Concurrent Pos:* Nat vis prof, Cheng Kung Univ, Taiwan, 69-70; Wachtmeister chair eng, Va Mil Inst, 82. *Honors & Awards:* Ben Dasher Award; Centennial Medal, Inst Elec & Electronics Engrs; Meritorious Serv Award, Inst Elec & Electronics Engrs Educ Soc; Ronald J Schmitz Award. *Mem:* Fel Inst Elec & Electronics Engrs; Am Vacuum Soc; fel Am Soc Eng Educ. *Res:* Thin film circuitry and components, especially phenomena in thin insulating films with emphasis on active components. *Mailing Add:* Watson Sch State Univ NY Binghamton NY 13902-6000

FEISS, PAUL GEOFFREY, b Cleveland, Ohio, Mar 7, 43; m 87; c 4. ECONOMIC GEOLOGY, GEOCHEMISTRY. *Educ:* Princeton Univ, AB, 65; Harvard Univ, MA, 67, PhD(geol), 70. *Prof Exp:* Asst prof geol, Albion Col, 70-75; asst prof, 75-78, PROF GEOL & CHMN DEPT, UNIV NC, CHAPEL HILL, 78- *Mem:* Sigma Xi; fel Geol Soc Am; AAAS; Soc Econ Geologists; Am Geophys Union. *Res:* Metallogeny; ore deposits of Southern Appalachians; igneous processes and ore deposition. *Mailing Add:* Dept Geol Univ NC CB No 3315 Chapel Hill NC 27599

FEIST, DALE DANIEL, b Cincinnati, Ohio, Feb 24, 38; m 62; c 2. ZOOLOGY, PHYSIOLOGY. *Educ:* Univ Cincinnati, AB, 60; Univ Calif, Berkeley, PhD(zool), 69. *Prof Exp:* NIH fel physiol, Karolinska Inst, Sweden, 70-71; from asst prof to assoc prof, 71-83, PROF ZOOPHYSIOL, INST ARCTIC BIOL, UNIV ALASKA, 83- *Mem:* AAAS; Am Soc Zool; Am Soc Mammal; Am Physiol Soc; Soc Neurosci. *Res:* Physiological mechanisms in cold acclimation and hibernation; neuroendocrine aspects of acclimatization and adaptation to cold; physiology of seasonal reproduction. *Mailing Add:* Inst Arctic Biol Univ Alaska Fairbanks AK 99775

FEIST, WILLIAM CHARLES, b St Paul, Minn, Nov 13, 34; wid; c 2. WOOD CHEMISTRY, POLYMER CHEMISTRY. *Educ:* Hamline Univ, BS, 56; Univ Colo, PhD(org chem), 61. *Prof Exp:* Res chemist, Esso Res & Eng Co, Stand Oil Co NJ, 60-64; res chemist, 64-83, SUPVRY RES CHEMIST, FOREST PROD LAB, US FOREST SERV, 83- *Mem:* Am Chem Soc; Soc Wood Sci & Technol; Forest Prod Res Soc; Sigma Xi; Int Res Group Wood Preserv; Inst Wood Sci. *Res:* Wood and water interactions; durability and dimensional stabilization of wood; weathering of wood; interaction of polymers with wood substance; wood finishing; surface chemistry of wood. *Mailing Add:* 6809 Forest Glade Ct Middleton WI 53562

FEIST, WOLFGANG MARTIN, b Oppeln, Ger, Mar 12, 27; m 57; c 3. SOLID STATE PHYSICS, SOLID STATE ELECTRONICS. *Educ:* Univ Frankfurt, dipl, 54; Univ Mainz, Dr rer nat, 57. *Prof Exp:* Electronic scientist, Diamond Ord Fuze Labs, DC, 57-60; prin scientist, 60-79, CONSULT SCIENTIST, RES DIV, RAYTHEON CO, 80- *Mem:* Inst Elec & Electronics Engrs. *Res:* Solid state devices; integrated circuits; thin films. *Mailing Add:* Raytheon Co Res Div 131 Spring St Lexington MA 02173

FEIT, CARL, b New York, NY, Oct 3, 45; m 68; c 2. IMMUNOLOGY. *Educ:* Yeshiva Univ, BA, 67; Rutgers Univ, PhD(microbiol), 73. *Prof Exp:* Res fel immunol, Waksman Inst Microbiol, Rutgers Univ, 73-75; RES ASSOC IMMUNODIAG, SLOAN-KETTERING INST, 75- *Concurrent Pos:* Vis asst prof biol, Yeshiva Univ, 75- *Mem:* Am Soc Microbiol. *Res:* Detection and analysis of tumor specific and tumor associated antigens; modulation of the immune response by tumors. *Mailing Add:* Dept Nat Sci Yeshiva Univ 500 W 185th St New York NY 10033

FEIT, DAVID, b New York, NY, Sept 17, 37; m 59; c 3. ACOUSTICS, MECHANICS. *Educ:* Columbia Univ, MS, 61, EngScD(eng mech), 64. *Prof Exp:* Res engr, Davidson Lab, Stevens Inst Technol, 59-60; res scientist, TRG, Inc, NY, 60-61; sr scientist, Cambridge Acoust Assocs, 64-73; supvry mech engr, David W Taylor Naval Ship Res & Develop Ctr, 73-83, res scientist, 83-87; LIASON SCIENTIST, OFF NAVAL RES, LONDON BR OFF, 88- *Mem:* Am Acad Mech; fel Acoust Soc Am; Am Soc Mech Engrs. *Res:* Vibrations; mathematics. *Mailing Add:* David W Taylor Naval Ship Res & Develop Ctr Bethesda MD 20084

FEIT, EUGENE DAVID, b Chicago, Ill, Sept 23, 35; m 64; c 3. ORGANIC POLYMER CHEMISTRY. *Educ:* Univ Chicago, MS, 64, PhD(phys org chem), 68. *Prof Exp:* Chemist, Am Meat Inst Found, 62-64; chemist, Bell Labs, 68-79; sr scientist, Harris Semiconductor, 79-83, dir, 83-86, sr scientist, VHSIC Fabrication, 87-88; DIR, UNIV & NAT LAB PROGS, SEMATECH, 88- *Honors & Awards:* IR100 Award, 77. *Mem:* Am Chem Soc; Electrochem Soc; Am Vacuum Soc; Inst Elec & Electronics Engrs. *Res:* Microlithography, plasma etching; radiation chemistry; photopolymerization; organic photochemistry; process developments for integrated circuits; complementary metal-oxide semiconductor technology; microelectronics. *Mailing Add:* 425 Rio Casa Indialantic FL 32903

FEIT, IRA (NATHAN), b Brooklyn, NY, Feb 28, 40; m 71; c 2. DEVELOPMENTAL BIOLOGY, MICROBIOLOGY. *Educ:* Brooklyn Col, BS, 60; Princeton Univ, MA, 64, PhD(biol), 69. *Prof Exp:* From instr to asst prof, 64-75, ASSOC PROF BIOL, FRANKLIN & MARSHALL COL, 75- *Concurrent Pos:* NSF trainee lab quant biol, Univ Miami, 69-70. *Mem:* AAAS; Sigma Xi. *Res:* Developmental control mechanisms in the cellular slime molds; small molecular signals in cellular slime mold development. *Mailing Add:* 1949 Crooked Oak Dr Lancaster PA 17601

FEIT, IRVING N, b Providence, RI, Oct 15, 42. ORGANIC CHEMISTRY. *Educ:* Univ RI, BS, 64; Univ Rochester, PhD(chem), 69, St John's Univ, JD, 79. *Prof Exp:* Fel, Univ Calif, Santa Cruz, 69-70; asst prof, 70-74, assoc prof chem, C W Post Col, Long Island Univ, 74-79; patent atty, Lever Bros, 80-82 & Ciba-Geigy, 82-88; PATENT ATTY, IMCLONE, 88- *Mem:* Am Chem Soc; Am Intellectual Property Law Asn. *Mailing Add:* 455 W 43rd St New York NY 10036

FEIT, JULIUS, b New York, NY, Nov 24, 19; m 53; c 2. SPACE PHYSICS, SOLAR PHYSICS. *Educ:* City Col New York, BS, 42; Columbia Univ, MA, 59; Adelphi Univ, MS, 63, PhD(physics), 67. *Prof Exp:* Lectr physics, Adelphi Univ, 65-66; assoc prof, 67-70, PROF PHYSICS, QUEENSBOROUGH COMMUNITY COL, CITY UNIV NEW YORK, 70- *Mem:* Am Geophys Union; Am Asn Physics Teachers. *Res:* Diffusion of solar flare cosmic rays through interplanetary space; effects of slow acceleration, external and internal boundaries and convection and energy loss in diffusion; modulation of galactic cosmic rays. *Mailing Add:* Hofstra Univ 1000 Fulton Ave Hempstead NY 11550

FEIT, MICHAEL DENNIS, b Easton, Pa, Nov 15, 42; m 67; c 2. FIBER & INTEGRATED OPTICS, LASERS. *Educ:* Lehigh Univ, BA, 64; Rensselaer Polytech Inst, PhD(physics), 70. *Prof Exp:* Res asst physics, Rensselaer Polytech Inst, 66-69; res assoc, Univ Ill, Urbana, 69-72; RES PHYSICIST, LAWRENCE LIVERMORE LAB, 72- *Concurrent Pos:* Lectr, Univ Calif, Davis/Livermore, 84- *Mem:* AAAS; fel Am Phys Soc; Sigma Xi; Optical Soc Am. *Res:* fiber and integrated optics; numerical methods; nonlinear propagation; self-focusing; quantum mechanics; chaos. *Mailing Add:* Lawrence Livermore Lab PO Box 808 L-296 Livermore CA 94550

FEIT, SIDNIE MARILYN, b Hackensack, NJ, Nov 29, 35; m 57; c 2. COMPUTER SCIENCE. *Educ:* Cornell Univ, BA, 57, MA, 63, PhD(math), 68. *Prof Exp:* Mathematician, Labs Appl Sci, Univ Chicago, 60-61; lectr math, Cornell Univ, 61-63; mathematician, Inst Naval Studies, 63-64; asst prof math, Quinnipiac Col, 68-70; asst prof math, Albertus Magnus Col, 70-77; res assoc, dept math, Yale Univ, 77-81; sr tech staff, ITT Inc, 81-85; prin consult, Ramesis Inc, 86-90; DIR, SYSTS INTEGRATION, OMNEE CORP. *Concurrent Pos:* Consult, Yale Univ, 75. *Mem:* Inst Elec & Electronics Engrs; Asn Comput Mach. *Res:* Topology; geometry; use of computers in teaching mathematics; data communications. *Mailing Add:* 36 Laurel Rd Hamden CT 06517

FEIT, WALTER, b Vienna, Austria, Oct 26, 30; nat US; m 57; c 2. GROUP THEORY. *Educ:* Univ Chicago, BA & MS, 51; Univ Mich, PhD(math), 54. *Prof Exp:* Instr math, Cornell Univ, 53-55, from asst prof to assoc prof, 56-64; PROF MATH, YALE UNIV, 64- *Concurrent Pos:* NSF fel, Inst Advan Study, 58-59. *Honors & Awards:* Cole Prize, Am Math Soc, 65. *Mem:* Nat Acad Sci; Am Math Soc; Math Asn Am; Am Adac Arts & Sci. *Res:* Group theory and algebra. *Mailing Add:* Dept Math Yale Univ New Haven CT 06520

FEITLER, DAVID, b Chicago, Ill, Oct 5, 52. ORGANIC CHEMISTRY, CATALYSIS. *Educ:* Univ Calif, BS, 73; Mass Inst Technol, PhD(org chem), 77. *Prof Exp:* Res chemist, 77-80, SR RES CHEMIST CATALYSIS, AIR PROD & CHEM, 80- *Mem:* Am Chem Soc; Sigma Xi. *Res:* Homogeneous catalysis; design and synthesis of catalytic materials; reactions of small molecules; synthetic inorganic and organometallic chemistry; zeolites. *Mailing Add:* 49 Clintonwood Dr New Windsor NY 12550

FEJER, STEPHEN OSCAR, b Budapest, Hungary, Dec 27, 16; Can citizen; m 50; c 1. GENETICS. *Educ:* Pazmany Peter Univ, Budapest, Dr jur, 39; Univ NZ, MAgrSci, 52; Swiss Fed Inst Technol, DrTechSc(genetics), 67. *Prof Exp:* Plant breeder grass genetics, Grasslands Div, Dept Sci & Indust Res, NZ, 50-61; geneticist, Res Sta, Can Dept Agr, 62-81, Univ Guelph, 82-84. *Concurrent Pos:* Ger Acad Exchange Serv vis scientist, Tech Univ Munich, 75 & 79. *Mem:* Genetics Soc Am; Genetics Soc Can; Can Soc Hort Sci. *Res:* Plant breeding for yield and its components; quantitative genetics and selection index; relations to environmental factors, adaptation and homeostasis; biochemical background of plant hormones; competition; frost resistance. *Mailing Add:* Nine LeRoy St Ottawa ON K1J 6X1 Can

FEKETE, ANTAL E, b Budapest, Hungary, Sept 8, 32. STATISTICS. *Educ:* Univ Waterloo, BA, 55, MA, 58. *Prof Exp:* PROF MATH, MEM UNIV, 58- *Mem:* Am Math Asn. *Mailing Add:* Mem Univ Nfld St Johns NF A1C 5S7 Can

FEKETY, F ROBERT, JR, b Pittsburgh, Pa, June 29, 29; m 54; c 2. INTERNAL MEDICINE, INFECTIOUS DISEASES. *Educ:* Wesleyan Univ, AB, 51; Yale Univ, MD, 55. *Prof Exp:* Intern & resident internal med, Yale Med Ctr, 55-56, 58-60; res fel, infectious dis, Sch Med, Johns Hopkins Univ, 57-58, from instr to asst prof med, 60-67; assoc prof internal med, Sch Med, 67-70, PROF INTERNAL MED & HEAD SECT INFECTIOUS DIS, MED CTR, UNIV MICH, ANN ARBOR, 70- *Mem:* AAAS; Infectious Dis Soc Am; Am Soc Microbiol. *Res:* Bacterial infections; epidemiology; clinical pharmacology of antibiotics; cellular immunology. *Mailing Add:* 3116 Taubman Infectious Dis Ctr Univ Mich Med Ctr E Medical Center Dr Ann Arbor MI 48109-0378

FELBECK, DAVID K(NISELEY), b Mt Vernon, NY, Apr 2, 26; m 75; c 5. METALLURGICAL FAILURE ANALYSIS. *Educ:* Cornell Univ, BME, 48; Mass Inst Technol, MS, 49, MechE, 51, ScD, 52. *Prof Exp:* Fulbright lectr, Delft Univ Technol, 52-53; asst prof mech eng, Mass Inst Technol, 53-55; exec dir, Nat Acad Sci-Nat Res Coun, 55-61; assoc prof mech eng, 61-65, PROF MECH ENG, UNIV MICH, ANN ARBOR, 65- *Concurrent Pos:* Consult & expert witness in over 500 cases involving metall failure, 61- *Honors & Awards:* Wilson Award, Am Soc Metals, 73. *Mem:* Fel Am Soc Mech Engrs; Am Soc Metals; Am Inst Mining, Metall & Petrol Engrs; Soc Automotive Engrs. *Res:* Fractography and failure analysis; high-performance composites. *Mailing Add:* 2060 Scottwood Ann Arbor MI 48104

FELBECK, GEORGE THEODORE, JR, b Buffalo, NY, Sept 18, 24; m 50; c 3. ORGANIC GEOCHEMISTRY. *Educ:* Mass Inst Technol, BS, 49; Pa State Univ, MS, 55, PhD(agron), 57. *Prof Exp:* Asst prof agron, WVa Univ, 56-57; lectr biol, Yonsei Univ, Korea, 57-58; asst prof agron, Univ Del, 58-64; assoc prof agr chem, 64-70, chmn dept, 72-83, PROF SOIL SCI, UNIV RI, 70-; TECH DIR, KINGSTON RES ASSOC, 80- *Mem:* Am Soc Agron; Soil Sci Soc Am; Geochem Soc. *Res:* Chemistry of soil organic matter; organic geochemistry. *Mailing Add:* Kingston Res Assoc PO Box 245 Kingston RI 02811-0245

FELBER, FRANKLIN STANTON, b Newark, NJ, June 11, 50; m 84; c 2. DIRECTED ENERGY, THEORETICAL PHYSICS. *Educ:* Princeton Univ, AB, 72; Univ Southern Calif, MA, 73, PhD(physics), 75; Univ Chicago, MS, 74. *Prof Exp:* Fel, Univ Southern Calif, 75-76; Nat Res Coun assoc, Naval Res Lab, 76-77; sr scientist, Gen Atomic Co, 77-79; sr staff scientist, Maxwell Labs, 79-81; theory group mgr, Western Res Corp, 81-83; DIRECTED ENERGY DIV MGR, JAYCOR, 84- *Concurrent Pos:* Nat Res Coun Assoc, 76; NATO fel, 77. *Honors & Awards:* Res Award, Sigma Xi, 75. *Mem:* Am Phys Soc; Sigma Xi; Inst Elec & Electronics Engrs; Optical Soc Am; Am Inst Aeronaut & Astronaut. *Res:* Plasma physics and instabilities; fusion; imploding plasmas; magneto-hydrodynamics; high-current accelerators; interaction of intense radiation with matter; nonlinear optics; space systems; survivability. *Mailing Add:* 16985 Manresa Court San Diego CA 92128

FELCH, WILLIAM C, INTERNAL MEDICINE. *Educ:* Columbia Univ, MD, 53. *Prof Exp:* Intern, St Luke's Hosp, NY, 45-51, asst res med, 45-46, chief res med, 48-50, assoc med, 50-51; Captain, Vet Admin Hosp Mass, Bedford, 46-48; att med & trustee, United Hosp Port Chester, NY; ED, J CONTINUING EDUC IN HEALTH PROFESSION, 90- *Mem:* Nat Acad Sci; fel Am Col Physicians; AMA; Am Heart Asn; Am Soc Intern Med. *Mailing Add:* 26337 Carmelo St Carmel CA 93923

FELCHER, GIAN PIERO, b Milano, Italy, June 14, 36; m 70; c 2. SOLID STATE PHYSICS. *Educ:* Univ Milan, Dr physics, 58. *Prof Exp:* Res assoc physics, Italian Ctr Exp Studies, 59-62; res assoc, Nat Ctr Nuclear Energy, 62-63; fel, Brookhaven Nat Lab, 63-66; asst physicist, 66-71, physicist, 71-86, SR PHYSICIST, ARGONNE NAT LAB, 87- *Concurrent Pos:* Int Atomic Energy Agency fel, 60-62; asst prof, Univ Rome, 62-63; vis prof, NATO, 75-76, res grant, 77-79. *Honors & Awards:* IR 100, 87. *Mem:* Fel Am Phys Soc; Sigma Xi. *Res:* Neutron scattering from solids; magnetism and magnetic structures; properties of surfaces and their interaction with gases. *Mailing Add:* Solid State Div Argonne Nat Lab Argonne IL 60439

FELD, BERNARD TAUB, b New York, NY, Dec 21, 19; m 47; c 2. ELEMENTARY PARTICLE PHYSICS, HIGH ENERGY PHYSICS. *Educ:* City Col New York, BS, 39; Columbia Univ, PhD(physics), 45. *Hon Degrees:* DSc, City Col New York, 75,. *Prof Exp:* Instr eve session, City Col New York, 40-41; res assoc, Columbia Univ, 41-42; physicist metall lab, Univ Chicago, 42-44 & Los Alamos Lab, Univ Calif, 44-46; from instr to assoc prof physics, 46-57, actg dir lab nuclear sci, 61-62, PROF PHYSICS, MASS INST TECHNOL, 57- *Concurrent Pos:* Consult, Brookhaven Nat Lab, 48-; asst ed, Ann of Physics; Guggenheim fel & vis prof, Univ Rome, 53-54; vis scientist & Ford fel, Europ Orgn Nuclear Res, 60-61; vis prof, Polytech Sch Paris, 66-67; vis prof theoret physics, Imp Col Sci & Technol, London, 73-75, Fulbright-Hays res fel, 74-75; secy-gen, Pugwash Conf on Sci & World Affairs, 72-77; ed-in-chief, Bull Atomic Scientists, 75-85. *Honors & Awards:* Leo Szilard Award, Am Phys Soc, 75. *Mem:* Fel Am Acad Arts & Sci; fel Am Phys Soc; Fedn Am Scientists. *Res:* Neutron physics; atomic and molecular hyperfine structure and nuclear moments; meson physics and elementary particles, theory and experiment. *Mailing Add:* Dept Physics Mass Inst Technol Cambridge MA 02139

FELD, MICHAEL S, b New York, NY, Nov 11, 40; m 80; c 3. LASER PHYSICS & SPECTROSCOPY, LASER-MEDICAL RESEARCH. *Educ:* Mass Inst Technol, SB & SM, 63, PhD(physics), 67. *Prof Exp:* Fel physics, 67-68, from asst prof to assoc prof, 68-79, PROF PHYSICS, MASS INST TECHNOL, 79- *Concurrent Pos:* Sloan fel, 73-; dir, Spectros Lab, 76-, Laser Res Ctr, 79- & Laser Biomed Res Ctr, Mass Inst Tech, 85- *Honors & Awards:* Gordon Y Billard Award, 82. *Mem:* AAAS; fel Optical Soc Am; fel Am Phys Soc; fel Sigma Xi. *Res:* Laser physics; quantum electronics; superradiance; lasers in medicine. *Mailing Add:* Spectros Lab Mass Inst Technol Cambridge MA 02139

FELD, WILLIAM ADAM, b Sandwich, Ill, Nov 20, 44; m 67; c 2. SYNTHESIS, MICROCOMPUTER APPLICATIONS. *Educ:* Loras Col, Dubuque, Iowa, BS, 66; Univ Iowa, Iowa City, PhD(chem), 71. *Prof Exp:* Asst prof org & analytical chem, Northern State Col, SDak, 71-72; adj asst prof, 72-77, from asst prof to assoc prof, 82-90, PROF ORG CHEM, WRIGHT STATE UNIV, 90- *Mem:* Am Chem Soc; Sigma Xi. *Res:* Synthesis of thermally stable polymer precursors; application of microcomputers in organic and polymer chemistry. *Mailing Add:* Dept Chem Wright State Univ Dayton OH 45435

FELDBERG, ROSS SHELDON, b Chicago, Ill, Sept 7, 43; m 66; c 1. ENZYMOLOGY, NUCLEIC ACIDS. *Educ:* Univ Ill, BS, 65; Univ Mich, Ann Arbor, PhD(biochem), 70. *Prof Exp:* Fel biochem, Univ Aberdeen, 70-72; fel, Brandeis Univ, 72-75; asst prof biochem, 75-81, ASSOC PROF BIOL, TUFTS UNIV, 81- *Concurrent Pos:* NIH fel, Brandeis Univ, 74-75. *Mem:* AAAS; Am Soc Microbiol. *Res:* Histidine metabolism in mast cells , alliin lyase enzyme of garlic; protein targetting to plant vacoule. *Mailing Add:* Dept Biol Tufts Univ Medford MA 02155

FELDBERG, STEPHEN WILLIAM, b New York, NY, July 22, 37. ELECTROCHEMISTRY. *Educ:* Princeton Univ, AB, 58, PhD(chem), 61. *Prof Exp:* Res assoc, Brookhaven Nat Lab, 61-63; vis asst prof, Univ Kans, 64; CHEMIST, BROOKHAVEN NAT LAB, 64-, SR SCIENTIST, 85- *Concurrent Pos:* Vis assoc chemist, Univ Kans, 68; vis prof, Colo State Univ, 72. *Mem:* Am Chem Soc; Electrochem Soc. *Res:* Studies of mechanism and kinetics of chemical reactions coupled with heterogeneous electron transfer; computer simulation techniques; studies of semi-conductor photophenomena; electronically conducting polymers; laser induced interfacial temperature jump studies of electrochemical phenomena. *Mailing Add:* Brookhaven Nat Lab 60 Rutherford Dr Upton NY 11973

FELDBUSH, THOMAS LEE, b Canton, Ohio, Apr 18, 39; m 63; c 3. IMMUNOBIOLOGY. *Educ:* Mt Union Col, BS, 61; Ohio State Univ, MSc, 64, PhD(microbiol), 66. *Prof Exp:* Sr res microbiologist, Merck Inst Therapeut Res, 66-70; from asst prof to assoc prof, 71-79, prof immunol-microbiol, Dept Microbiol, Col Med, Univ Iowa, 79-87; assoc chief staff, res & develop, Harry S Truman Mem Vet Hosp, Columbia, Mo, 88-90; asst dean res & acad affairs, Sch Med, Univ Mo Columbia, 88-90, prof, Dept Microbiol, 88-90; ASSOC DEAN RES, SCH MED, NORTHWESTERN UNIV CHICAGO, 90-, PROF, DEPT MICROBIOL & IMMUNOL, 90- *Concurrent Pos:* Vis lectr, Dept Microbiol, Rutgers Univ, 68-70; vis worker, Sir William Dunn Sch Path, Oxford, Eng, 69-70; Res Career Scientist Award, Vet Admin. *Mem:* Brit Soc Immunol; Am Asn Immunologists; Am Soc Microbiol; AAAS; Sigma Xi. *Res:* Influence of antigen on development, maintenance and characteristics of immunologic memory; regulation of B-cell mediated immunol response. *Mailing Add:* Ward 4-153 303 E Chicago Ave Chicago IL 60611-3008

FELDER, DARRYL LAMBERT, b Kingsville, Tex, Oct 21, 47; m 70. SYSTEMATIC ZOOLOGY, INVERTEBRATE PHYSIOLOGY. *Educ:* Tex A&I Univ, BS, 69, MS, 71; La State Univ, PhD(zool), 75. *Prof Exp:* Asst prof, 75-79, ASSOC PROF BIOL, UNIV SOUTHWESTERN LA, 79- *Concurrent Pos:* Taxon consult, US Bur Land Mgt Proj, Univ Tex Marine Sci Inst, Port Aransas, Tex, 76-77 & Southwest Res Inst, Houston, 78-80; dir planning off, La Maine Consortium, 79-80. *Mem:* Am Soc Zoologists; AAAS; Estuarine Res Fedn. *Res:* Systematics of decapod crustaceans; ecology of decapod crustaceans in the Gulf of Mexico; physioecology of estuarine invertebrates. *Mailing Add:* Dept Biol 42451 Univ Southwestern La Lafayette LA 70504

FELDER, RICHARD MARK, b New York, NY, July 21, 39; div; c 3. CHEMICAL ENGINEERING. *Educ:* City Col New York, BChE, 62; Princeton Univ, PhD(chem eng), 66. *Prof Exp:* NATO fel, theoret physics div, Atomic Energy Res Estab, Eng, 66-67; res scientist, nuclear eng dept, Brookhaven Nat Lab, 67-69; PROF CHEM ENG, NC STATE UNIV, 69- *Honors & Awards:* R J Reynolds Indust Award, 82; AT&T Found Award, Am Soc Engr Educ, 85, Corcoran Award, 85, Wickenden Award, 88 & 89, Dasher Award, 89; Nat Catalyst Award, Chem Mfrs Asn, 89. *Mem:* Am Inst Chem Engrs; Am Soc Engr Educ; Sigma Xi. *Res:* Computer-aided chemical manufacturing; process simulation and optimization. *Mailing Add:* Dept Chem Eng NC State Univ Box 7905 Raleigh NC 27695-7905

FELDER, WILLIAM, b Greensburg, Pa, Mar 20, 43; m 67; c 2. PHYSICAL CHEMISTRY. *Educ:* Univ Pa, AB, 64; Mass Inst Technol, SM, 66, PhD(phys chem), 69. *Prof Exp:* Fel physics & chem, Ctr Res Exp Space Sci, York Univ, Toronto, Can, 69-72; RES SCIENTIST, AEROCHEM RES LABS, INC, 72- *Mem:* Am Phys Soc; Am Chem Soc. *Res:* Chemical kinetics of high temperature gas phase reactions; excited species reactions; chemiluminescence; ion-molecule reactions; atmospheric chemistry; chemical lasers. *Mailing Add:* AeroChem Res Labs PO Box 12 Princeton NJ 08542

FELDHAMER, GEORGE ALAN, b Minneapolis, Minn, Feb 20, 47; m 74; c 2. WILDLIFE BIOLOGY. *Educ:* Univ Minn, BS, 69; Idaho State Univ, MS, 72; Ore State Univ, PhD(wildlife biol), 77. *Prof Exp:* Res assoc, Ctr Environ & Estuarine Studies, Univ Md, 77-80, asst prof wildlife ecol, 80-; AT DEPT ZOOL, SOUTHERN ILL UNIV, CARBONDALE. *Mem:* Am Soc Mammalogists; Wildlife Soc; Sigma Xi. *Res:* Ecology of sika deer; white-tailed deer. *Mailing Add:* Dept Zool Southern Ill Univ Carbondale IL 62901

FELDHAUS, RICHARD JOSEPH, b Omaha, Nebr, May 6, 29; m 53; c 5. SURGERY. *Educ:* Creighton Univ, BS, 53, MS, 55, MD, 59. *Prof Exp:* Intern, St Joseph Mem Hosp, Omaha, Nebr, 59-60; resident surg, Creighton Univ Affil Hosps, 60-62; resident, Vet Admin Hosp, 62-64, staff surgeon, 64-65; chief surg, Vet Admin Hosp, Phoenix, Ariz, 67-71; dir surg educ, Good Samaritan Hosp, 71-74; ASSOC PROF SURG, SCH MED, CREIGHTON UNIV, 74- *Concurrent Pos:* From instr to asst prof, Sch Med, Creighton Univ, 62-67. *Mem:* Fel Am Col Surgeons. *Res:* Peripheral vascular surgery; esophageal surgery. *Mailing Add:* 720 N 87th St Suite 201 Omaha NE 68114

FELDHERR, CARL M, b New York, NY, Jan 3, 34; m 59; c 1. ANATOMY. *Educ:* Hartwick Col, BA, 55; Univ Pa, PhD(physiol), 60. *Prof Exp:* Damon Runyon fel, Univ Pa, 60-62; asst prof physiol, Univ Alta, 62-65; asst prof anat, Sch Med, Univ Pa, 65-67; asst prof, 67-70, assoc prof, 70-77, PROF ANAT, COL MED, UNIV FLA, 77- *Mem:* AAAS; Am Soc Cell Biol. *Res:* Permeability characteristics of nuclear envelope. *Mailing Add:* Anat Dept J235 JHMHC Bldg Univ Fla Gainesville FL 32601

FELDLAUFER, MARK FRANCIS, b Bronx, NY, Nov 17, 48. ENTOMOLOGY, BIOCHEMISTRY. *Educ:* Rutgers Univ, BS, 70, MS, 74; Univ Calif, Davis, PhD(entom), 79. *Prof Exp:* Res fel, NY State Agr Exp Sta, Geneva, 79-82; RES ENTOMOLOGIST, INSECT PHYSIOL LAB, AGR RES SERV, USDA, 82- *Mem:* AAAS; Entom Soc Am; Am Chem Soc. *Res:* Insect biochemistry, physiology and endocrinology; comparative sterol and steroid metabolism in insects. *Mailing Add:* USDA Agr Res Serv Bldg 467 BARC-East Beltsville MD 20705

FELDMAN, ALAN SIDNEY, b New York, NY, Feb 19, 27; m 50; c 2. AUDIOLOGY, SPEECH PATHOLOGY. *Educ:* Syracuse Univ, AB, 49, MS, 51, PhD(audiol, speech path), 56. *Prof Exp:* Audiologist & speech pathologist, Mass Eye & Ear Infirmary, 52-58; from asst prof to prof, 58-82, dir, Commun Dis Unit, 64-82, EMER PROF, OTOLARYNGOL, STATE UNIV NY UPSTATE MED CTR, 82-; PRES, ALAN S FELDMAN, PHD, PC, 72- *Concurrent Pos:* Res fel, Lab Sensory Commun, Syracuse Univ, 58-; consult, Syracuse Vet Admin Hosp, 58- *Mem:* Am Speech & Hearing Asn. *Res:* Audition; problems in measurement and differential diagnosis of auditory disorders and speech communication; occupational hearing loss. *Mailing Add:* 404 Univ Ave Syracuse NY 13210

FELDMAN, ALBERT, b Jersey City, NJ, May 20, 36; m 59; c 3. THIN DIAMOND FILMS, OPTICAL MATERIALS RESEARCH. *Educ:* City Col New York, BS, 59; Univ Chicago, MS, 60, PhD(physics), 66. *Prof Exp:* Physicist, 66-91, GROUP LEADER OPTICAL MAT GROUP, NAT BUR STANDARDS, 85- *Concurrent Pos:* Conf chmn, topical conf basic properties of optical mat, 85; mem, Nat Mat Adv bd comt, Superhard Mat; conf ser chmn, Diamond Optics, Int Soc Optical Eng; conf chmn, Appl Diamond Conf, 91. *Honors & Awards:* Spec Achievement Award, US Dept Com, Nat Bur Standards, 72; Bronze Medal, US Dept Com, 80. *Mem:* Am Phys Soc; Optical Soc Am; Am Soc Testing & Mat; Mat Res Soc. *Res:* Optical properties of solid materials; diamond film deposition and characterization; optical properties of thin films. *Mailing Add:* Nat Bur Standards A329 Mat Bldg Gaithersburg MD 20899

FELDMAN, ALFRED PHILIP, b Hamburg, Ger, Aug 7, 23; US citizen; m 54; c 2. ORGANIC CHEMISTRY. *Educ:* Univ Chicago, MS, 56; Johns Hopkins Univ, MS, 78. *Prof Exp:* Asst ed, Chem Abstr Serv, 56-60; chief coding sect, 60-69, asst to dir, Div Biomet & Med Info Processing, 70-74, systs analyst, 74-79, HEAD, CHEM INFO SECT, NAT CANCER INST, NIH, 79- *Honors & Awards:* US Army Res & Develop Award. *Mem:* AAAS; Am Chem Soc; Asn Comput Mach. *Res:* Management of scientific information; man-machine interaction; chemical data processing. *Mailing Add:* 5133 Darting Bird Lane Columbia MD 21044-1503

FELDMAN, ARTHUR, b St Louis, Mo, Apr 6, 31; m 53; c 1. STRUCTURAL ENGINEERING, NUCLEAR WEAPONS EFFECTS. *Educ:* Wash Univ, St Louis, BS, 52; Univ Ill, MS, 54, PhD(civil eng), 60. *Prof Exp:* Asst civil eng, Univ Ill, 52-54, res assoc, 54-59, asst prof, 59-61; assoc prof, Univ Denver, 61-63; asst res scientist, 63-66, sr res scientist, 66-81, DEPT STAFF ENGR, MARTIN MARIETTA CORP, 81- *Concurrent Pos:* Instr, Off Civil Defense, 62- *Honors & Awards:* Martin Author Award, 68, 69 & 87. *Mem:* Soc Exp Mech; Am Soc Civil Engrs; Sigma Xi. *Res:* Composite pressure vessels; buckling and vibration of composite materials; reinforced concrete, blast and earthquake resistance; prestressed concrete; radiation protection. *Mailing Add:* 2045 S Fillmore Denver CO 80210-3516

FELDMAN, B ROBERT, b New York, NY, July 5, 34; m 60; c 2. MEDICINE. *Educ:* Col William & Mary, BS, 55; Chicago Med Sch, MD, 59. *Prof Exp:* Intern, Michael Reese Hosp & Med Ctr, 59-60, resident pediat, 60-62, fel allergy, 62-63; fel pediatric allergy, Babies Hosp, Columbia-Presby Med Ctr, 63-64; asst prof, 68-87, ASSOC PROF CLIN PEDIAT, COLUMBIA COL PHYSICIANS & SURGEONS, 87- *Concurrent Pos:* Am Acad Pediat fel, 68 & Am Acad Allergy, 75; consult allergist, St Mary's Hosp Children, 75-; assoc dir, Div Allergy, Babies Hosp, Columbia-Presby Med Ctr, 74- *Mem:* Fel Am Acad Allergy; fel Am Acad Pediat. *Res:* Pharmacotherapy of bronchial asthma; clinical research on newer therapeutic agents. *Mailing Add:* Babies Hosp 3959 Broadway New York NY 10032

FELDMAN, BARRY JOEL, b Providence, RI, Feb 25, 44; m 68; c 1. QUANTUM OPTICS, LASERS. *Educ:* Brown Univ, ScB, 65; Mass Inst Technol, PhD(physics), 71. *Prof Exp:* Staff physicist, Los Alamos Sci Lab, 71-83; HEAD, LASER PHYSICS BR, NAVAL RES LAB, 83- *Mem:* Am Phys Soc; Optics Soc Am. *Res:* Theoretical and experimental research in nonlinear interactions of photons and matter; laser isotope separation; laser fusion; new laser development; coherent multiphoton processes. *Mailing Add:* Naval Res Lab Code 6540 Washington DC 20375-5000

FELDMAN, BERNARD JOSEPH, b San Francisco, Calif, July 20, 46; m 70; c 3. SEMICONDUCTOR PHYSICS. *Educ:* Univ Calif, Berkeley, AB, 67; Harvard Univ, AM, 69, PhD(physics), 72. *Prof Exp:* Asst res physicist, Univ Calif, Berkeley, 72-74; from asst prof to assoc prof, 80-85, PROF PHYSICS, UNIV MO, ST LOUIS, 85-, CHAIRPERSON, 90- *Concurrent Pos:* Equip grant, Res Corp, 74-; NSF travel grant, 75. *Mem:* Am Phys Soc; Am Asn Physics Teachers. *Res:* Experimental studies of optical and electrical properties of thin amorphous films. *Mailing Add:* Dept Physics Univ Mo St Louis MO 63121-4499

FELDMAN, CHARLES, b Baltimore, Md, Mar 20, 24; m 46; c 2. SOLID STATE PHYSICS. *Educ:* Johns Hopkins Univ, AB, 44, AM, 49; Univ Paris, PhD(physics), 52. *Prof Exp:* Physicist, Aberdeen Proving Ground, 48; asst inst coop res, Johns Hopkins Univ, 49-50; sr physics master grammar sch, Eng, 50-51; asst, Nat Ctr Sci Res, Paris, France, 51-53; physicist, Crystal Br, Naval Res Lab, 53-60; sect head, Melpar, Inc, 60-63, lab mgr, 64-67; sr staff mem, John Hopkins Univ, 67-68, prin prof staff mem, 68-72, supvr solid state res group, Appl Physics Lab, 72-85; CONSULT, 88- *Concurrent Pos:* Adj prof, George Washington Univ, 69-76; vis prof, Nat Univ Singapore, 83; consult, 85- *Honors & Awards:* Award, Sci Res Soc Am, 58. *Mem:* Am Phys Soc; Sigma Xi; Optical Soc Am; Am Vac Soc. *Res:* Metal films; dielectric films; solid state physics, luminescence; thin film electronics; amorphous semiconductors; silicon thin films. *Mailing Add:* 2855 Davenport St NW Washington DC 20008

FELDMAN, CHARLES LAWRENCE, b Yonkers, NY, Dec 18, 35; m 56, 83; c 4. CARDIOVASCULAR INSTRUMENTATION, CARDIOLOGY. *Educ:* Mass Inst Technol, SB & SM, 58, MechE, 60, ScD(mech eng), 62. *Prof Exp:* Res asst, Brookhaven Nat Lab, 56; scientist commun theory, Edgerton, Germeshausen & Grier, Inc, 58-59; asst, Mass Inst Technol, 59-60; sr prof engr, Joseph Kaye & Co, Inc, 61-62, dir res, 62-65; from asst prof to prof mech eng, Worcester Polytech Inst, 65-78; sr vpres, Electronics for Med, Inc, 71-81; pres, Cardio Data Corp, 81-90; LECTR MED, HARVARD MED SCH, 90- *Concurrent Pos:* Instr, Lowell Inst, 58-59; consult, Autonetics Div, NAm Aviation, Inc, 59-61 & Joseph Kaye & Co, Inc, 60-61; lectr indust mgt, Northeastern Univ, 62-63, lectr eng, 63-; res assoc biomath, med sch, Harvard Univ, 66-69, lectr, 69-72; consult math, dept psychiat, Mass Gen Hosp, 67-72; adj prof, Univ Mass Med Sch, 76- & Worcester Polytech Inst, 78- *Mem:* Assoc fel Am Col Cardiol; Asn Advan Med Instrumentation. *Res:* Application of statistical communication theory to computers and medicine. *Mailing Add:* Div Cardiol Brigham & Women's Hosp 75 Francis St Boston MA 02115

FELDMAN, CHESTER, b South Bend, Ind, June 26, 20; m 54; c 1. MATHEMATICS. *Educ:* Univ Chicago, SB, 40, SM, 41, PhD(math), 50. *Prof Exp:* Asst prof, Antioch Col, 51-52; consult, Air-Craft-Marine Prod, Inc, 52-53; instr math, Purdue Univ, 53-55; asst prof, Univ NH, 55-57 & Univ Conn, 57-63; assoc prof, Kent State Univ, 63-68; assoc ed math rev, Univ Mich, Ann Arbor, 68-78; assoc prof, Ind Univ, Kokomo, 81-83; RETIRED. *Mem:* Fel AAAS; Am Math Soc; Sigma Xi. *Res:* Topological and Banach algebras. *Mailing Add:* 3157 Homestead Commons Dr Ann Arbor MI 48108-1791

FELDMAN, DANIEL S, b Philadelphia, Pa, Feb 26, 26; m 57; c 3. NEUROLOGY, NEUROPHYSIOLOGY. *Educ:* Univ Pa, AB, 45, MD, 49. *Prof Exp:* From asst prof to assoc prof neurol, State Univ NY Downstate Med Ctr, 61-72; PROF NEUROL & MED, MED COL GA, 72- *Concurrent Pos:* Fel, Nat Inst Neurol Dis & Blindness, 53-54; Abrahamson fel, Mt Sinai Hosp, NY, 56; dir neurol, Maimonides Hosp, Brooklyn, 60-66; mem med adv bd, Myasthenia Gravis Found, 62-, chmn, 80-81; guest investr, Dept Pharmacol, Univ Lund, 64-65; career scientist, Health Res Coun, City of New York, 66-72; consult, Maimonides-Coney Island Med Ctr, Brooklyn, 66-72; vis neurologist, Kings County Hosp, 66-72; attend neurologist, State Univ Hosp, 66-72 & E Talmadge Hosp, Augusta, Ga, 72-; consult neurol, Vet Admin Hosp, Augusta, US Army Hosp, Ft Gordon & Univ Hosp, Augusta, 72- *Mem:* Am Physiol Soc; Am Neurol Asn; Am Acad Neurol; Am Fedn Clin Res; AMA; Sigma Xi. *Res:* Electrophysiological studies in vitro of parameters affecting chemical transmitter release and post-synaptic response in vertebrate, including human muscle; application of physiological systems and methods to study of human neuromuscular function in health and disease. *Mailing Add:* Dept Neurol Med Col Ga Augusta GA 30912

FELDMAN, DAVID, b Brooklyn, NY, June 16, 21; m 46; c 2. THEORETICAL PHYSICS. *Educ:* City Col New York, BS, 40; NY Univ, MS, 46; Harvard Univ, PhD(physics), 49. *Prof Exp:* AEC fel, Inst Advan Study, 49-50; asst prof physics, Univ Rochester, 50-56; assoc prof, 56-59, PROF PHYSICS, BROWN UNIV, 59- *Concurrent Pos:* NSF sr fel, Univ Paris, 62-63; mem adv panel for physics, NSF, 68- *Mem:* Fel Am Phys Soc; Ital Phys Soc; Sigma Xi. *Res:* Quantum theory of fields; nuclear and high-energy physics. *Mailing Add:* Dept Physics Brown Univ Providence RI 02912

FELDMAN, DAVID, b New York, NY, Oct 16, 27; m 51; c 3. ELECTRICAL ENGINEERING. *Educ:* Newark Col Eng, BS, 47, MS, 49. *Prof Exp:* Asst prof elec eng, Cooper Union, 49-54; res engr non-linear magnetics, Polytech Res & Develop Co, 54-56; dir components lab, Bell Tel Labs, Inc, 56-88; RETIRED. *Concurrent Pos:* Consult, Aerospace Industs Asn, 62-66, Electronic Industs Asn, 66-71 & Fedn Mat Sci. *Honors & Awards:* Contrib Award, Inst Elec & Electronics Engrs. 77. *Mem:* Fel Inst Elec & Electronics Engrs. *Res:* Thick and thin film component development, electronic materials, feedback control. *Mailing Add:* 5222 Moya Laguna Hills CA 92653

FELDMAN, DONALD WILLIAM, b Memphis, Tenn, Oct 5, 31; m 55; c 2. PHYSICS. *Educ:* Southwestern at Memphis, BS, 52; Pa State Univ, MS, 54; Univ Calif, Berkeley, PhD(physics), 59. *Prof Exp:* PHYSICIST, SOLID STATE SCI DEPT, WESTINGHOUSE RES LABS, 59- *Mem:* Am Phys Soc. *Res:* Solid state physics; electron spin resonance; nuclear magnetic resonance; Raman spectroscopy. *Mailing Add:* MS J5 Los Alamos Nat Lab Los Alamos NM 87545

FELDMAN, DOREL, b Romania, Mar 16, 24; Can citizen; m 79; c 1. POLYMERS APPLICATION, MATERIAL PROPERTIES. *Educ:* Polytech Inst, Iasi, Romania, BA, 49, DrEng, 58, DrSci, 71. *Prof Exp:* Prof polymer chem & technol, Polytech Inst, Iasi, Romania, 50-78; PROF ENG MAT, MODERN MAT & POLYMER CHEM , ENG FAC, CONCORDIA UNIV, MONTREAL, 78-, GRAD PROG DIR, 80- *Concurrent Pos:* Researcher, Inst Macromolecular Chem, Iasi, Romania, 52-58, head Polymerization Dept, 58-78. *Mem:* Am Chem Soc; Plastics & Rubber Inst; Cancer Res Soc; Soc Plastics Engrs. *Res:* Polymer technology; polymerization; copolymerization; graft-copolymerization; photopolymerization; polymer blends; polymer compatibility; applications as plastics, synthetic fibers, adhesives, sealants; polymer characterization. *Mailing Add:* Concordia Univ 1455 De Maisonneuve Blvd W Montreal PQ H3G 1M8 Can

FELDMAN, EDGAR A, b New York, NY, May 11, 37; m 62; c 2. GEOMETRY. *Educ:* Mass Inst Technol, BS, 58; Columbia Univ, PhD(math), 63. *Prof Exp:* Instr math, Princeton Univ, 63-65; from asst prof to assoc prof, 65-74, PROF MATH, CITY UNIV NEW YORK, GRAD CTR, 75- *Mem:* Am Math Soc; Math Asn Am. *Res:* Relationship between the low eigen values of the Laplacian of a compact Riemannian manifold and its topology and differential geometry; heat kernel in Riemannian geometry. *Mailing Add:* Dept Math 33 W 42nd St New York NY 10036

FELDMAN, EDWIN B(ARRY), b Atlanta, Ga, Apr 30, 25; m 47; c 3. INDUSTRIAL ENGINEERING. *Educ:* Ga Inst Technol, BIE, 50. *Prof Exp:* Plant engr, Puritan Chem Co, 49-52, plant mgr, 52-58, vpres & dir eng, 58-60; PRES, SERV ENG ASSOCS, INC, 61- *Honors & Awards:* Outstanding Serv Award, Nat Soc Prof Engrs, 67; Distinguished achievement award, Educ Press Asn Am, 74. *Mem:* Nat Soc Prof Engrs; Am Inst Indust Engrs; Am Inst Plant Engrs; Environ Mgt Asn. *Res:* Application of engineering principles to industrial and institutional housekeeping and sanitation; facilities maintenance; author of over 200 publications. *Mailing Add:* 2765 Peachtree Rd Atlanta GA 30305

FELDMAN, ELAINE BOSSAK, b New York, NY, Dec 9, 26; m 57; c 3. MEDICINE. *Educ:* NY Univ, AB, 45, MS, 48, MD, 51. *Prof Exp:* From intern to resident path, Mt Sinai Hosp, NY, 51-52, asst resident med, 53, asst, 55-57; from instr to assoc prof, State Univ NY Downstate Med Ctr, 57-72; PROF MED, MED COL GA, 72-, CHIEF SECT NUTRIT, 77- *Concurrent Pos:* Fel, Mt Sinai Hosp, NY, 54-55, NY Heart Asn res fel, 55-57; USPHS spec fel physiol chem, Univ Lund, 64-65; from asst vis physician to assoc vis physician, Kings County Hosp, 57-72; attend physician, State Univ Hosp, 66-72 & Eugene Talmadge Mem Hosp; mem nutrit study sect, NIH, 76-80; dir, Ga Inst Human Nutrit, 78- & Clin Nutrit Res Unit, 80-86; mem, Geriat & Geront Rev Comt, NIA, 86-90; bd sci counselors, Div Cancer Prev & Control, Nat Cancer Inst, 90-94, Dept Agr, 90-92. *Honors & Awards:* Joseph Goldberger Award, Am Med Asn, 90; Nat Dairy Coun Award, Excellence Med/Dent Nutrit Educ, Am Soc Clin Nutrit, 91. *Mem:* Endocrine Soc; Am Heart Asn; Am Soc Clin Nutrit; Sigma Xi. *Res:* Nutrition and metabolism, especially the effect of diet and drugs on serum lipids; dietary control of cholesterol absorption and intestinal lipoprotein formation. *Mailing Add:* Dept Med Med Col Ga Augusta GA 30912

FELDMAN, FRED, b Baku, USSR, Dec 4, 42; US citizen; m 65; c 1. BIOCHEMISTRY, BIOTECHNOLOGY. *Educ:* Univ Chicago, BS, 66; Purdue Univ, PhD(biochem), 71. *Prof Exp:* Teaching asst biochem, Purdue Univ, 67; res assoc biochem, Ind Univ, 71-74; sr res scientist process develop, 74-77, prin scientist, 78-79, tech mgr, 79-83, DIR PROD DEVELOP, PHARMACEUT CO, 83- *Mem:* Am Chem Soc; AAAS; World Fedn Hemophilia; Int Soc Thrombosis & Hemostasis. *Res:* Human plasma protein purification and characterization; protein chemistry; cell biology; membrane-protein interactions; mitochondrial protein synthesis; immuno affinity purifications; industrial technology; protein scale up; coagulation research; biotechnology. *Mailing Add:* 870 St Andrews Way Frankfort IL 60423

FELDMAN, FREDRIC J, b Brooklyn, NY, Feb 9, 40; m 62; c 2. ANALYTICAL CHEMISTRY, INORGANIC CHEMISTRY. *Educ:* Brooklyn Col, BS, 60; Univ Md, MS, 64, PhD(anal chem), 67. *Prof Exp:* Prin investr, Walter Reed Army Inst Res, 64-68; pres, Chrisfeld Precision Instruments Corp, 64-70, dir anal lab, Instrumentation Lab, 68-70; prog mgr, Beckman Instruments Inc, 70-77, opers mgr & dir Europ Opers, 81-84; pres, Instrumentation Labs, 84-88; CHIEF EXEC OFF, MICROGENICS CORP, 88- *Mem:* AAAS; Am Asn Clin Chem. *Res:* Research and development of atomic absorption and emission methods for the analysis of substances of clinical, biological and industrial importance; physiological mechanisms of chromium; coulometric and polarographic studies of biologically important substances. *Mailing Add:* 1275 Pacific Ave Laguna Beach CA 92651

FELDMAN, GARY JAY, b Cheyenne, Wyo, Mar 22, 42; m 67; c 2. ELEMENTARY PARTICLE PHYSICS. *Educ:* Univ Chicago, BS, 64; Harvard Univ, AM, 65, PhD(physics), 71. *Prof Exp:* Res assoc physics, Stanford LInear Accelerator Ctr, Stanford Univ, 71-74, staff physicist, 74-79, assoc prof, 79-83, prof, 83-90; PROF, HARVARD UNIV, 90- *Concurrent Pos:* Sci assoc, Europ Orgn Nuclear Res, Geneva, Switz, 82-83. *Mem:* Am Phys Soc. *Res:* Electroproduction of hadrons, electron-positron annihilation, and hadron-hadron collisions. *Mailing Add:* Lyman Lab Harvard Univ Cambridge MA 20138

FELDMAN, GORDON, b Windsor, Ont, Dec 6, 28; c 3. THEORETICAL PHYSICS. *Educ:* Univ Toronto, BA, 50, MA, 51; Univ Birmingham, PhD(physics), 53. *Prof Exp:* Asst physics, Univ Birmingham, 53-55; mem, Inst Advan Study, Princeton Univ, 55-56; res assoc physics, Univ Wis, 56-57; asst prof, 57-64, PROF PHYSICS, JOHNS HOPKINS UNIV, 65- *Concurrent Pos:* Guggenheim fel, 62-63; vis prof physics, Imp Col, London, 68-69 & 74-75; Royal Soc guest res fel, Univ Cambridge, 84-85. *Res:* High energy physics; elementary particles and field theory. *Mailing Add:* Dept Physics Johns Hopkins Univ Baltimore MD 21218

FELDMAN, HENRY ROBERT, b New York, NY, June 28, 32; m 56; c 1. UNDERWATER ACOUSTICS. *Educ:* Harvard Univ, AB, 53; Columbia Univ, AM, 58, PhD(physics), 63. *Prof Exp:* Res asst prof physics, 63-65, sr physicist, Appl Physics Lab, Univ Wash, 66-84; CONSULT, 84- *Mem:* Sigma Xi; Acoust Soc Am; Am Phys Soc. *Res:* Acoustic lenses; underwater acoustic studies in the Arctic; marine bioacoustics. *Mailing Add:* 4823 NE 42nd St Seattle WA 98105

FELDMAN, ISAAC, b Washington, DC, Mar 6, 18; m 56. CHEMISTRY, BIOPHYSICS. *Educ:* George Washington Univ, BS, 41; Univ Ill, Urbana, PhD(phys chem), 47. *Prof Exp:* Instr electronics, US Army Air Force Tech Sch, Scott Field, Ill, 42-43; instr chem, George Washington Univ, 43-44; jr scientist toxicol, 47-48, from instr to assoc prof phys chem, 48-58, assoc prof radiation bio, 58-65, prof radiation biol & biophys 65-85, EMER PROF BIOPHYS, SCH MED & DENT, UNIV ROCHESTER, 85- *Concurrent Pos:* USPHS spec res fel, Univ Col London, 62-63. *Mem:* Biophys Soc; Asn Univ Col London Chemists; Sigma Xi. *Res:* Bioinorganic chemistry; application of magnetic resonance and fluorescence to biophysics; role of metal ions in normal and abnormal biochemistry. *Mailing Add:* Dept Biophys Sch Med & Dent Univ Rochester Rochester NY 14642

FELDMAN, JACK L, b Brooklyn, NY, Jan 6, 48. CONTROL MOVEMENT OF HOMEOSTASIS. *Educ:* Univ Chicago, PhD(physics), 73. *Prof Exp:* From assoc prof to prof physiol, Nortwestern Univ, 82-86; PROF NEUROSCI, UNIV CALIF, LOS ANGELES, 86- *Mailing Add:* Dept Kinesiology Univ Calif 405 Hilgard Ave Los Angeles CA 90024

FELDMAN, JACOB, b Philadelphia, Pa, Jan 10, 28; div; c 2. MATHEMATICS. *Educ:* Univ Pa, BA, 50; Univ Ill, MA, 51; Univ Chicago, PhD(math), 54. *Prof Exp:* NSF fel, Inst Advan Study, 54-56; vis asst prof, Columbia Univ, 56-57; asst prof, 57-64, PROF MATH, UNIV CALIF, BERKELEY, 64- *Concurrent Pos:* NSF fel, Inst Advan Study, 60-61; exchange visitor, USSR, 67. *Mem:* Am Math Soc; AAAS; Am Asn Univ Professors. *Res:* Stochastic processes; operator algebras; ergodic theory. *Mailing Add:* Dept Math Univ Calif 2120 Oxford St Berkeley CA 94704

FELDMAN, JACOB J, MEDICAL STATISTICAL ANALYSIS. *Prof Exp:* ASSOC DIR, ANALYTICAL & EPIDEMIOL PROG, NAT CTR HEALTH STATIST, US DEPT HEALTH & HUMAN SERV, 88- *Mem:* Inst Med-Nat Acad Sci. *Mailing Add:* Analytical & Epidemiol Div Nat Ctr Health Statist 3700 EW Hwy Hyattsville MD 20782

FELDMAN, JAMES MICHAEL, b Pittsburgh, Pa, Sept 29, 33; m 55; c 2. COMPUTER ARCHITECTURE, ELECTRICAL ENGINEERING. *Educ:* Carnegie Inst Technol, BS, 57, MS, 58, PhD(elec eng), 60. *Prof Exp:* Asst prof elec eng, Carnegie Inst Technol, 60-65; assoc prof, 65-71, dir, Power Systs Eng Prog, 74-82, PROF ELEC ENG, NORTHEASTERN UNIV, 71- *Concurrent Pos:* Res engr quantum electronics, Westinghouse Res Labs, 61-64; vis prof, Technion, Haifa, Israel, 71-72; vis prof, Telaviv Univ, 80-81; vis consult engr, Digital Equip Corp, Jerusalem, Israel, 85-86. *Mem:* Am Phys Soc; Inst Elec & Electronics Engrs; Sigma Xi. *Res:* Computer architecture and engineering; Semiconductor devices; optical properties of semiconductors. *Mailing Add:* Northeastern Univ Boston MA 02115

FELDMAN, JEROME A, b Pittsburgh, Pa, Dec 5, 38; m 61; c 2. COMPUTER SCIENCE, MATHEMATICS. *Educ:* Univ Rochester, BA, 60; Univ Pittsburgh, MA, 61; Carnegie Inst Technol, PhD(comput sci), 64. *Prof Exp:* Res scientist comput sci, Carnegie Inst Technol, 63-64; staff mem data systs, Lincoln Lab, Mass Inst Technol, 64-66; from asst prof to assoc prof comput sci, Stanford Univ, 66-75; chmn dept, Univ Rochester, 75-81, John H Dessauer prof, 82-88, PROF COMPUT SCI, UNIV ROCHESTER, 75- *Concurrent Pos:* Fulbright lectr & vis prof math & comput sci, Hebrew Univ, Jerusalem, 70-71. *Mem:* Asn Comput Mach. *Res:* Programming languages; aritifical intelligence. *Mailing Add:* Int Comput Sci Inst 1947 Center St Suite 600 Berkeley CA 94704-1105

FELDMAN, JERRY F, b Philadelphia, Pa, May 11, 42. GENETICS. *Educ:* Swarthmore Col, BA, 63; Princeton Univ, MA, 65, PhD(biol), 67. *Prof Exp:* USPHS res fel biol, Calif Inst Technol, 67-69; asst prof, State Univ NY Albany, 69-74; asst prof, 75-77, assoc prof, 77-81, PROF BIOL, UNIV CALIF, SANTA CRUZ, 81- *Mem:* Am Soc Plant Physiologists; Am Soc Cell Biol; Am Soc Microbiol; Genetics Soc Am; Biophys Soc. *Res:* Biological clocks - genetic and biochemical approaches in neurospora; sexual differentiation in neurospora. *Mailing Add:* Thimann Labs Univ Calif Santa Cruz CA 95064

FELDMAN, JOSE M, b Entre Rios, Arg, Sept 27, 27; m 54, 68; c 5. PLANT VIROLOGY. *Educ:* Cuyo Univ, IngAgron, 55. *Prof Exp:* From asst to assoc prof, 56-75, assoc pathologist, fac Agron, Nat Univ Cuyo, 63-75; pathologist, 61-75, PATHOLOGIST, NAT COUN SCI & TECHNOL RES, 84- *Concurrent Pos:* Nat Coun Sci & Technol Res grants, Arg, 59, 62 & 63, fel, 62-63, lectr, 64-65, prin researcher, 68; consult, Plant Path, Food & Agr Orgn, UN, 81. *Mem:* Arg Soc Plant Physiol; Latin Am Phytopath Asn; Int Soc Hort Sci. *Res:* Plant virus diseases, especially vegetable crops and grapes; plant virus serology; inhibition of plant viruses; physiology of virus-diseased plants. *Mailing Add:* Loria 5882 5505 Chacras de Coria Mendoza Argentina

FELDMAN, JOSEPH AARON, b Fall River, Mass, Feb 2, 25; m 48; c 3. PHARMACEUTICAL CHEMISTRY. *Educ:* RI Col Pharm, BS, 50; Univ Wis, MS, 52, PhD(pharm), 56. *Prof Exp:* Assoc prof, 55-64, PROF PHARMACEUT CHEM, DUQUESNE UNIV, 64- *Concurrent Pos:* Vis analytical consult, Ciba-Geigy Co, Basel, Switz, 71. *Mem:* Am Chem Soc; Am Pharmaceut Asn; Acad Pharmaceut Sci; Sigma Xi. *Res:* Nonaqueous titration; analysis using chelometric methods; fluorimetry; pharmaceutical analysis; effects of solvent systems on drug stability. *Mailing Add:* Sch Pharm Duquesne Univ Pittsburgh PA 15219

FELDMAN, JOSEPH DAVID, b Hartford, Conn, Dec 13, 16; m 49; c 3. PATHOLOGY. *Educ:* Yale Univ, BA, 37; LI Col Med, MD, 41. *Prof Exp:* From assoc prof to prof path, Sch Med, Univ Pittsburgh, 54-61; chmn, Dept Immunopath, 76-79, MEM STAFF, SCRIPPS CLIN & RES FOUND, LA JOLLA, 61- *Concurrent Pos:* Lectr, Hadassah Med Sch, Hebrew Univ Jerusalem, 50-54; consult, USPHS, 67-, chmn, Path B Study Sect, 67-70; adj prof path, Univ Calif, San Diego, 68-; ed-in-chief, J Immunol, 71-; consult, Nat Cancer Inst Virus Cancer Prog, Sci Rev Comt & Sci Adv Bd, Coun Tobacco Res, 74. *Mem:* Int Acad Path; Histochem Soc; Endocrine Soc; Electron Micros Soc Am; Am Asn Path. *Res:* Immunopathology; cytology. *Mailing Add:* 13030 Via Grimaldi Del Mar CA 92014

FELDMAN, JOSEPH GERALD, b New York, NY, Sept 5, 40; m 64; c 2. BIOSTATISTICS, EPIDEMIOLOGY. *Educ:* Lehman Col, BA, 62; City Univ New York, MBA, 67; Univ NC, DrPH(biostatist), 72. *Prof Exp:* Res assoc statist, Health & Hosp Coun South NY, 63-67; instr statist, Dept Social & Prev Med, State Univ NY Buffalo, 67-71; from asst prof to assoc prof, 72-85, PROF BIOSTATIST & EPIDEMIOL, DEPT ENVIRON MED & COMMUNITY HEALTH, DOWNSTATE MED CTR, STATE UNIV NY, 85- *Concurrent Pos:* Statist consult, Eval Training Prog, Dept Epidemiol, Sch Pub Health, Univ NC, 72- & Manhattan Vet Admin Hosp, 74-; consult, Breast Cancer Prog Eval, Div Cancer Control, Nat Cancer Inst, 75-, Kings County Med Soc, 80- & data mgt, Kings County Health Care Rev Orgn, 80- *Mem:* Am Statist Asn; Biomet Soc; Am Pub Health Asn; Int Epidemiol Asn. *Res:* Application of biostatistics to problems in epidemiology and public health with emphasis on the epidemiology of cancer; coronary heart disease; evaluation of health services. *Mailing Add:* Downstate Med Ctr Box 43 450 Clarkson Ave Brooklyn NY 11203

FELDMAN, JOSEPH LOUIS, b New York, NY, June 6, 38; m 63; c 2. SOLID STATE PHYSICS. *Educ:* Queens Col, BS, 60; Rutgers Univ, MS, 62, PhD(physics), 66. *Prof Exp:* Res fel physics, Rensselaer Polytech Inst, 65-68; RES PHYSICIST, US NAVAL RES LAB, 68- *Mem:* Sigma Xi; Am Phys Soc. *Res:* Lattice vibrations and thermal properties of solids; electronic properties of solids. *Mailing Add:* US Naval Res Lab Complex Systs Theory Br Code 4691 Washington DC 20375

FELDMAN, JULIAN, b Brooklyn, NY, May 24, 15; m 44; c 3. ORGANIC CHEMISTRY. *Educ:* City Col New York, BS, 35; Brooklyn Col, AM, 40; Univ Pittsburgh, PhD(chem), 50. *Prof Exp:* Chief chemist, Pro-Medico Labs, 35-40 & Gold-Leaf Pharmacal Co, 40-41; asst chemist, Bur Animal Indust, USDA, 41; asst plant mgr, Trubek Labs, 41-42; res assoc explosives res lab, Nat Defense Res Comn, 42-45; res chemist, Bur Mines, 45-53; group leader, Nat Distillers Prod & Chem Corp, 53-57, res supvr, 57-59, sr res assoc, 59-74, mgr explor chem res, US Indust Chem Co Div, Nat Distillers & Chem Corp, 74-81, res scientist, USI Chem Co, 81-82, RES ASSOC, UNIV CINCINNATI, 82- *Mem:* Am Chem Soc. *Res:* Separations and purifications; phase equilibria; distillation; kinetics; isomerization; catalysis; composition of fuels and carbonization products; molecular complexes; polynuclear compounds; catalytic organic processes; organometallic chemistry. *Mailing Add:* 7511 Sagamore Dr Cincinnati OH 45236

FELDMAN, KENNETH SCOTT, b Miami Beach, Fla, Dec 22, 56. CHEMISTRY. *Educ:* Harvey Mudd Col, BS, 78; Stanford Univ, PhD(chem), 83. *Prof Exp:* Res fel, E I du Pont de Nemours & Co, 83-84; ASST PROF CHEM, PA STATE UNIV, 84- *Mem:* Am Chem Soc. *Res:* Design and synthesis of structurally complex organic molecules. *Mailing Add:* Dept Chem 152 Davey Lab Pa State Univ University Park PA 16802-6302

FELDMAN, LARRY HOWARD, b Brooklyn, NY, Feb 25, 42; m 64; c 3. PHOTOGRAPHIC CHEMISTRY. *Educ:* Brooklyn Col, BS, 62; Mich State Univ, PhD(chem), 66. *Prof Exp:* Sr res chemist, Res Labs, 66-70, sr develop engr, 70-73, tech assoc, 73-76, supvr, 76-83, DIR, KODAK PARK DIV, EASTMAN KODAK CO, 83- *Mem:* Am Chem Soc; Soc Photog Sci & Eng; Am Soc Qual Control. *Res:* Photographic science and technology of silver halide systems; electron attachment reactions in anhydrous ethylenediamine; photographic film and paper quality control and product development. *Mailing Add:* 99 Hillhurst Lane Rochester NY 14617

FELDMAN, LAWRENCE, b Havana, Cuba, Apr 5, 22; US citizen; m 52. NUCLEAR PHYSICS, ACCELERATORS. *Educ:* Brooklyn Col, BA, 43; Univ NC, MS, 44; Columbia Univ, PhD(physics), 50. *Prof Exp:* Instr physics, Univ NC, 43-44; physicist, Naval Ord Lab, 44-46; asst physics, Columbia Univ, 46-48, res assoc, 50-63, Higgins fel, 51-52, dir 36" cyclotron, 52-63; sr res assoc, Univ Pa, 63-64; assoc prof, 64-69, PROF PHYSICS, ST JOHN'S UNIV, NY, 69- *Concurrent Pos:* Lectr, City Col New York, 52-64. *Mem:* Am Phys Soc. *Res:* Lambda beta decay; weak interactions; nuclear reactions; giant resonance phenomena; C-W cyclotron; high forbidden nuclear beta decay; calculations of atomic energy levels in helium. *Mailing Add:* Dept Physics St John's Univ Jamaica NY 11439

FELDMAN, LAWRENCE A, b Brooklyn, NY, Oct 11, 38; m 61; c 1. MICROBIOLOGY, VIROLOGY. *Educ:* Univ Wis, BS, 60; Pa State Univ, MA, 62, PhD(microbiol), 64. *Prof Exp:* USPHS training fel virol & epidemiol, Col Med, Baylor Univ, 64-66; from instr to asst prof microbiol, 66-71, asst dean student affairs, 71-73, assoc prof microbiol, 71-78, asst dean admin, 73-76, assoc dean planning & mgt, 76-77, PROF MICROBIOL, COL MED & DENT NJ, NEWARK, 78- *Mem:* AAAS; Am Soc Microbiol; Brit Soc Gen Microbiol. *Res:* Viral diseases of the central nervous system; viral oncogenesis; abortive viral infections. *Mailing Add:* 100 Bergen St NJ Med Sch Newark NJ 07103

FELDMAN, LEONARD, b Jamaica, NY, Oct 3, 23; m 48; c 2. MATHEMATICS EDUCATION, MATHEMATICAL LITERACY. *Educ:* Queen's Col, NY, BS, 48; Columbia Univ, MA, 50; Univ Calif, Berkeley, EdD(coun psychol), 57. *Prof Exp:* Tech writer electron tube group, NY Univ, 48-49; teacher pub schs, NY & Calif, 49-57; from asst prof to assoc prof, 57-70, PROF MATH, SAN JOSE STATE UNIV, 70- *Concurrent Pos:* Dir, NSF Acad Year Inst, 62-63, 66-68, NSF sci fac fel, 64-65; fel, Queen Mary Col, London, 63-64, Univ Calif, Berkeley, 64-65; vis assoc prof, Columbia Univ on assignment to Makerere Univ, Uganda, 68-70; consult, Calif Sch Dist, math publishers, Makerere Univ & Inst Teacher Educ, Uganda, 90. *Mem:* Asn Women in Math; Am Math Asn Two Yr Cols; Math Asn Am; Nat Coun Teachers Math; Res Coun Diag & Remedial Math. *Res:* Evaluation of mathematics education; in-service education for teachers; learning theory; remedial and developmental mathematics; mathematics for reluctant and anxious learners; equity in mathematics education; psychology and mathematics education. *Mailing Add:* 12 Summit Lane Berkeley CA 94708-2213

FELDMAN, LEONARD CECIL, b New York, NY, June 8, 39; m 64; c 2. SURFACE SCIENCE, SOLID STATE PHYSICS. *Educ:* Drew Univ, BA, 61; Rutgers Univ, MS, 63, PhD(atomic physics), 67. *Prof Exp:* Mem tech staff solid state physics, Bell Labs, 67-83, supvr, mat interface characterization group, 83-84, head, Mat Interfaces & Ceramics Res Dept, 84-88, HEAD, SILICON DEVICE RES DEPT, AT&T BELL LABS, 88- *Concurrent Pos:* Res assoc physics, Rutgers Univ, 69-; guest scientist solid state physics, Aarhus Univ, Denmark, 70-71; guest lectr physics, Drew Univ, 75 & 77; mem, Int Comn Ion Beam Anal Conf, 78- & Int Comn Atomic Collisions in Solids Conf, 75-85; chmn, Gordon Conf Particle-Solid Interactions, 76-78; res assoc, Univ Guelph, 77-79; vis prof mat sci, Cornell Univ, 81; ed, Appl Surface Sci; chmn, Surface Sci Div, Am Vacuum Soc, 86-88; counr, Mat Res Soc, 85-87; prin ed, J Mat Res, 88- *Honors & Awards:* Fel, Am Phys Soc. *Mem:* Am Phys Soc; Am Vacuum Soc; AAAS; Mat Res Soc; Am Ceramics Soc. *Res:* Fundamental studies of semiconductor thin films and interfaces and epitaxial growth with and emphasis on the use of energetic ion beams as a probe; physics of ion-solid interactions; surface science. *Mailing Add:* Rm 1D-145 AT&T Bell Labs Murray Hill NJ 07974

FELDMAN, LOUIS A, b Bay City, Mich, Nov 26, 41; m 88. TOPOLOGY, GEOMETRY. *Educ:* Univ Mich, BS, 63; Univ Calif, Berkeley, MA, 65, PhD(math), 69. *Prof Exp:* Asst math, Univ Calif, Berkeley, 64-68; from asst prof to assoc prof 68-76, PROF MATH, CALIF STATE UNIV, STANISLAUS, 76- *Mem:* Am Math Soc. *Res:* Algebraic topology; applications of mathematics. *Mailing Add:* Dept Math Calif State Univ Stanislaus Turlock CA 95380

FELDMAN, MARCUS WILLIAM, b Perth, Australia, Nov 14, 42; m 64; c 3. MATHEMATICAL BIOLOGY. *Educ:* Univ Western Australia, BSc, 64; Monash Univ, Australia, MSc, 66; Stanford Univ, PhD(math biol), 69. *Prof Exp:* Tutor math, Monash Univ, Australia, 64-65; res asst, 65-69, asst prof, 69-74, assoc prof, 74-77, PROF BIOL, STANFORD UNIV, 77-; PROF & DIR, MORRISON INST POP & RESOURCE STUDIES, 86- *Concurrent Pos:* Ed, Theoret Pop Biol & Am Naturalist; Guggenheim fel, 76-77; fel, Ctr Advan Study Behav Sci, 83-84. *Mem:* Am Soc Naturalists; Genetics Soc Am; Am Soc Human Genetics; Soc Study Evolution; fel AAAS; fel Am Acad Arts & Sci. *Res:* Mathematical models of genetical phenomena, primarily selection and recombination; population genetics and ecology of a theoretical nature; theory of cultural evolution. *Mailing Add:* Dept Biol Stanford Univ Stanford CA 94305

FELDMAN, MARTIN, b New York, NY, July 13, 35; m 61; c 3. PHYSICS. *Educ:* Rensselaer Polytech Inst, BS, 57; Cornell Univ, PhD(physics), 62. *Prof Exp:* Res assoc physics, Cornell Univ, 62-63; asst prof, Univ Pa, 63-68; MEM TECH STAFF, BELL LABS, 68- *Mem:* Am Phys Soc; Optical Soc Am; Inst Elec & Electronics Engrs. *Res:* Experimental high energy physics; electronics; optics. *Mailing Add:* 141 Murray Hill Blvd Murray Hill NJ 07974

FELDMAN, MARTIN LEONARD, b New York, NY, May 22, 37; c 2. NEUROANATOMY. *Educ:* Brown Univ, AB, 58; Boston Univ, MA, 64, PhD(psychol), 69. *Prof Exp:* Res assoc, 70-71, asst prof, 71-77, ASSOC PROF ANAT, BOSTON UNIV, 77- *Concurrent Pos:* NIH fels, New Eng Regional Primate Res Ctr, 68-69 & Boston Univ, 69-70. *Mem:* AAAS; Am Asn Anat; Int Brain Res Orgn; Gerontol Soc. *Res:* Neurocytology of the aging brain; neuroanatomy of the central auditory system and cerebral cortex. *Mailing Add:* Dept Anat Boston Univ Sch Med 80 E Concord St Boston MA 02118

FELDMAN, MARTIN LOUIS, b Brooklyn, NY, June 14, 41; m 62; c 2. ORGANIC CHEMISTRY. *Educ:* City Col New York, BS, 63; Univ Pittsburgh, PhD(org chem), 66. *Prof Exp:* Chemist, Am Cyanamid Corp, 66-72; sr chemist, Tenneco Chem, 72-77, lab mgr, org & polymers div, 77-82; MGR, COLORANTS RES & DEVELOP, HUODEX, 82- *Mem:* Soc Coatings Technol; Am Chem Soc; Soc Plastics Engrs. *Res:* Pigments; colorants for coatings; thermoset and thermoplastics; coating additives. *Mailing Add:* Nine Fieldcrest Dr East Brunswick NJ 08816-3511

FELDMAN, MARTIN ROBERT, b New York, NY, Apr 23, 38; m 59; c 2. ORGANIC CHEMISTRY, HISTORY SCIENCE. *Educ:* Columbia Univ, AB, 58; Univ Calif, Los Angeles, PhD(phys & org chem), 63. *Prof Exp:* Res chemist, Univ Calif, Berkeley, 62-63; from asst prof to assoc prof, 63-71, PROF CHEM, HOWARD UNIV, 71- *Concurrent Pos:* NSF sci fac fel, Univ Calif, Irvine, 69-70; vis scientist, King's Col, London, 77-78; fac fel, Smithsonian Inst, 87. *Mem:* Am Chem Soc; Hist Sci Soc; Soc Hist Technol. *Res:* Physical-organic chemistry. *Mailing Add:* Dept Chem Howard Univ 2400 Sixth St NW Washington DC 20059

FELDMAN, MILTON H, b New York, NY, Mar 17, 18; m 46; c 4. PHYSICAL CHEMISTRY. *Educ:* NY Univ, BS, 39, PhD(phys chem), 44. *Prof Exp:* Scientist, Oak Ridge Nat Lab, 46-49; res engr, NAm Aviation Co, 49-53; adv scientist, Bettis Atomic Power Lab, 53-59; dep tech dir, Defense Atomic Support Agency, 59-60; tech dir, Winchester Environ Lab, 60-61; chief radiation & radiochem sect, Walter Reed Army Inst Res, 61-66; res chemist, Environ Protection Agency, 66-81; CONSULT, COMPUT ELECTRONIC DATA PROCESSING, 81- *Concurrent Pos:* Consult indust req, Dept Defense, Environ Protection Agency & Occupational Safety & Health Admin. *Mem:* Fel AAAS; fel Am Inst Chem; fel Royal Soc Arts. *Res:* Radiation effects; nuclear technology; environmental and planetary sciences; biophysics; chemical and biological oceanography; ecological response of trace materials; electronic data processing usage; computer sciences. *Mailing Add:* 960 NW 178th St Beaverton OR 97006

FELDMAN, NATHANIEL E, b New London, Conn, Oct 7, 25; m 46; c 4. ELECTRONICS, ATMOSPHERIC SCIENCES. *Educ:* Univ Calif, Berkeley, BS, 48, MS, 50. *Prof Exp:* Asst elec eng, Univ Calif, Berkeley, 49-50, engr, Lawrence Radiation Lb, 51-54; instr fire control radar, Hughes Aircraft Co, 55; leader adv develop defense electronic prod div, Radio Corp Am, 56-60; proj leader, systs anal, Rand Corp, 60-78; chief scientist, Systs Res Oper, Sci Appln Inc, 78-81; systs dir, advan space commun, 81-83, SR ENG SPECIALIST, ELECTRONICS & OPTICS DIV, AEROSPACE CORP, 84- *Concurrent Pos:* Mem, Mil Commun Conf Bd, 86, Secy, 87-92. *Mem:* Sr mem Inst Elec & Electronics Engrs; assoc fel Am Inst Aeronaut & Astronaut; Sigma Xi; Armed Forces Commun & Electronics Asn. *Res:* Satellite communications; components for microwave systems for radar, electronic counter measures and aerospace communications; systems analysis and exploratory design; military hardware design and development, including technological options, economic tradeoffs and policy issues; applications of new technology in coding, modulation, interleaving and equalization; CATV systems designs for the Dayton metropolitan areas; millimeter wave communications. *Mailing Add:* Aerospace Corp M4/928 PO Box 92957 Los Angeles CA 90009

FELDMAN, NICHOLAS, b Mukacevo, Czech, Sept 25, 24; US citizen; m 58; c 2. PETROLEUM CHEMISTRY, ANALYTICAL CHEMISTRY. *Educ:* Prague Tech Univ, MS, 49. *Prof Exp:* Anal chemist, Fiber Chem Corp, NJ, 50-52, chief chemist, 52-57; from res chemist to sr res chemist, 57-68, res assoc, 68-75, SR RES ASSOC, PROD RES DIV, EXXON RES & ENG CO, 75- *Mem:* Am Chem Soc. *Res:* Effect of fuel and lube additives on product quality; interaction of additives; mechanism of wax crystal modification; leather and textile specialties; analytical methods; compositional analysis. *Mailing Add:* 48 Hunter Lane Woodbridge NJ 07095

FELDMAN, PAUL ARNOLD, b Everett, Mass, Apr 22, 40; c 2. RADIO & MM-WAVE, ASTRONOMY & ASTROPHYSICS. *Educ:* Mass Inst Technol, BS, 61; Stanford Univ, PhD(physics), 69. *Prof Exp:* Res fel astrophys, Inst Theoret Astron, Univ Cambridge, 68-70 & Dept Physics, Queen's Univ, Ont, 70-72; lectr astron & astrophys, York Univ, 72-74; res officer astron, Dominion Radio Astron Observ, Penticton, BC, 74-75; SR RES OFFICER ASTRON, HERZBERG INST ASTROPHYS, NAT RES COUN CAN, 75- *Concurrent Pos:* NATO fel sci, Univ Cambridge, 68-69. *Honors & Awards:* First Prize, Gravity Res Found, 70. *Mem:* Int Astron Union; Am Astron Soc; Can Astron Soc; Royal Astron Soc Can. *Res:* Molecular-line radio and mm-wave astronomy; radio astronomy of active stars; general radio astronomy and astrophysics. *Mailing Add:* Herzberg Inst Astrophys Nat Res Coun Can 9100 Ottawa ON K1A 0R6 Can

FELDMAN, PAUL DONALD, b Brooklyn, NY, Nov 4, 39; m 65; c 2. PLANETARY ATMOSPHERES, ULTRAVIOLET ASTRONOMY. *Educ:* Columbia Univ, AB, 60, MA, 62, PhD(atomic physics), 64. *Prof Exp:* Instr physics, Columbia Univ, 64-65, res assoc, Naval Res Lab, 65-67; from asst prof to assoc prof, 67-76, PROF PHYSICS, JOHNS HOPKINS UNIV, 76- *Concurrent Pos:* Sloan Found res fel, 69-73. *Mem:* Am Geophys Union; Am Astron Soc; Am Phys Soc; Int Astron Union. *Res:* Spectroscopy of planetary and earth atmospheres; comets; ultraviolet astronomy. *Mailing Add:* Dept Physics Johns Hopkins Univ 3400 N Charles St Baltimore MD 21218

FELDMAN, RICHARD MARTIN, b Wyandotte, Mich, July 19, 44; m 67; c 2. OPERATIONS RESEARCH, PROBABILITY. *Educ:* Hope Col, AB, 66; Mich State Univ, MS, 67; Ohio Univ, MS, 70; Northwestern Univ, PhD(indust eng), 75. *Prof Exp:* Mathematician, Goodyear Atomic Corp, 67-70; instr indust eng, Ohio Univ, 70-72; opers res analyst, Michael Reese Med Ctr, 74-75; from asst prof to assoc prof, 75-85, PROF INDUST ENG, TEX A&M UNIV, 85- *Mem:* Opers Res Soc Am; Am Inst Indust Engrs; Sigma Xi; Inst Mgt Sci. *Res:* Optimal replacement and maintenance analysis; applied probability modeling and simulation. *Mailing Add:* Dept Indust Eng Tex A&M Univ College Station TX 77843-3131

FELDMAN, ROBERT H L, b Brooklyn, NY, Feb 10, 43. HEALTH PSYCHOLOGY, OCCUPATIONAL HEALTH PSYCHOLOGY. *Educ:* Brooklyn Col, BA, 64; Pa State Univ, MA, 66; Syracuse Univ, MS, 72, PhD(psychol), 74. *Prof Exp:* Asst prof psychol, State Univ NY, Utica & Rome, 74-77; asst prof health educ & occup med, Johns Hopkins Univ, 78-79; asst prof, 79-84, assoc prof, 84-90, PROF, HEALTH EDUC, UNIV MD, 90- *Concurrent Pos:* Res assoc, Univ Nairobi, Kenya, 73; post-doctoral fel, Univ Conn Health Ctr, 77-78. *Mem:* Am Psychol Asn; Soc Behav Med; Am Pub Health Asn. *Res:* Effect of social support on smoking cessation, social psychological models of behavior and the effect of social support on health behavior; application of health psychology to workplace health programs. *Mailing Add:* Dept Health Educ Univ Md College Park MD 20742

FELDMAN, SAMUEL M, b Philadelphia, Pa, Sept 26, 33; div; c 2. ANIMAL PHYSIOLOGY, BEHAVIOR-ETHOLOGY. *Educ:* Univ Pa, AB, 54; Northwestern Univ, AM, 55; McGill Univ, PhD(physiol psychol), 59. *Prof Exp:* Fel, Univ Wash, 58-60; from instr to assoc prof physiol, Albert Einstein Col Med, 60-71; PROF PSYCHOL, NY UNIV, 71-, PROF NEURAL SCI, 87-, DIR GRAD STUDIES, CTR NEURAL SCI, 88- *Concurrent Pos:* Consult, NIMH, 68-72, 74-78 & 80-84; assoc ed, Brain Res, 70-74. *Mem:* Soc Neurosci; Am Physiol Soc; NY Acad Sci; Sigma Xi. *Res:* The role of catecholamines in the regulation of cerebral circulation and behavior; role of the locus coeruleus in behavior; aging brain; cellular activity of neurons in midbrain. *Mailing Add:* Ctr Neural Sci NY Univ New York NY 10003

FELDMAN, STUART, b Bronx, NY, Feb 14, 41; m 63; c 2. PHARMACEUTICS, BIOPHARMACEUTICS. *Educ:* Columbia Univ, BS, 62, MS, 66; State Univ NY Buffalo, PhD(pharmaceut), 69. *Prof Exp:* From asst prof to assoc prof biopharmaceut, Sch Pharm, Temple Univ, 69-74; assoc prof, 74-82, PROF PHARMACEUT, DEPT PHARMACEUT, COL PHARM, UNIV HOUSTON, 82-, CHMN DEPT, 76- *Mem:* Fel Acad Pharmaceut Sci; fel AAAS; Am Pharmaceut Asn; Am Col Clin Pharmacol; NY Acad Sci; Am Soc Clin Pharmacol & Therapeutics; fel Am Asn Pharmaceut Scientists; Am Soc Pharmacol & Exp Therapeut; fel Acad Pharmaceut Res & Sci. *Res:* Effect of surface active agents on drug absorption; study of the biopharmaceutical factors influencing drug absorption, distribution and elimination; pharmacokinetics; effect of disease on drug disposition, immunologic alterations of drug disposition; drug therapy in AIDS. *Mailing Add:* Dept Pharmaceut Univ Houston 1441 Moursund Houston TX 77030

FELDMAN, SUSAN C, b Brooklyn, NY, Oct 1, 43. NEUROANATOMY, DEVELOPMENTAL NEUROBIOLOGY. *Educ:* Hofstra Univ, BA, 63; Rutgers Univ, MS, 67; City Univ NY, PhD(biol), 76. *Prof Exp:* Fel neurol, Albert Einstein Col Med, Col Physicians & Surgeons, Columbia Univ, 75-77, fel anat, 77-79; asst prof, 79-86, ASSOC PROF ANAT, UNIV MED & DENT NJ, 86- *Mem:* Am Asn Anatomists; AAAS; Soc Neurosci; Sigma Xi; Am Soc Cell Biol. *Res:* Organization of the central nervous system by determining the biochemistry of neurons using immunohistochemistry and related techniques to identify transmitters and other molecules intrinsic to particular neuronal populations; development of neurons in vitro as a means of studying the organization of neurons and the regulation of gene expression. *Mailing Add:* Dept Anat Univ Med & Dent NJ NJ Med Sch 100 Bergen St Newark NJ 07103

FELDMAN, URI, b Tel-Aviv, Israel, 1935. ATOMIC SPECTROSCOPY, SOLAR PHYSICS. *Educ:* Hebrew Univ Jerusalem, MSc, 63, PhD(physics), 65. *Prof Exp:* Fel spectros, Goddard Space Flight Ctr, Greenbelt, Md, 65-67; lectr spectros, 67-69, sr lectr astrophys, 69-71, dir, Wise Astron Observ, 69-72, assoc prof astrophys, Tel-Aviv Univ, 71-74; ASTROPHYSICIST, NAVAL RES LAB, 74- *Mem:* Fel Optical Soc Am; Int Astron Union. *Res:* Investigation of ultraviolet and x-ray from the sun; physical conditions in laser produced plasmas, and tokamak plasmas, spectroscopy. *Mailing Add:* Code 4174 Naval Res Lab Washington DC 20375-5000

FELDMAN, WILLIAM, physics, research administration; deceased, see previous edition for last biography

FELDMAN, WILLIAM A, b Brookline, Mass, Mar 28, 45; m 76; c 3. FUNCTION SPACES. *Educ:* Tufts Univ, BS, 67; Northwestern Univ, MS, 68; Queen's Univ, PhD(math), 71. *Prof Exp:* Asst prof, 71-76, assoc prof, 76-81, PROF MATH, UNIV ARK, 81- *Mem:* Math Asn Am; Am Math Soc. *Res:* Functional analysis and topology on function spaces. *Mailing Add:* Univ Ark Philadelphia Fayetteville AR 72701

FELDMAN, WILLIAM CHARLES, b New York, NY, Apr 15, 40; m 65; c 2. SPACE PHYSICS. *Educ:* Mass Inst Technol, BS, 61; Stanford Univ, PhD(physics), 68. *Prof Exp:* Fel space physics, Univ Wis, 70-71; MEM STAFF, SPACE PLASMA PHYSICS, LOS ALAMOS SCI LAB, 71- *Mem:* Am Geophys Union. *Res:* Solar and interplanetary physics. *Mailing Add:* SST-8 Mail Stop D438 Los Alamos Nat Lab Los Alamos NM 87545

FELDMANN, EDWARD GEORGE, b Chicago, Ill, Oct 13, 30; m 52; c 4. PHARMACEUTICAL CHEMISTRY. *Educ:* Loyola Univ, Ill, BS, 52; Univ Wis, MS, 54, PhD(pharmaceut chem & biochem), 55. *Prof Exp:* Lab asst, Loyola Univ, Ill, 51-52; Alumni Res Found asst, Univ Wis, 52-53; sr chemist, Chem Div, Am Dent Asn, 55-58, dir, 58-59; chmn elect comt nat formulary & assoc dir rev, Am Pharmaceut Asn, 59-60, chmn comt & dir rev, 60-69, ed, J Pharmaceut Sci, 60-74, asst exec dir sci affairs, 69-70, assoc exec dir, 70-83, vpres sci affairs, 83-85; PRES, PHARMACEUT CONSULT SERV, 84- *Concurrent Pos:* Spec lectr, George Washington Univ, 60-65; deleg, US Pharmacopeia, 70-85; mem, Nat Res Coun, 71-; mem, Nat Coun on Drugs, 76-83; mem food codex panel, Nat Acad Sci-Nat Res Coun & expert panel pharmaceut, WHO; consult, 85-; dir, Am Asn Pharmaceut Scientists, 86- *Honors & Awards:* Spec Citation, Food & Drug Admin, 75. *Mem:* Am Chem Soc; Am Pharmaceut Asn; NY Acad Sci; assoc mem AMA; Int Pharmaceut Fedn; fel Acad Pharmaceut Sci (exec secy, 83-85); Am Asn Pharmaceut Scientists. *Res:* Analysis of pharmaceutical products; standards and specifications for drugs and dosage forms; chemistry of local anesthetics; chemical structure-therapeutic activity relationships; synthetic organic medicinal chemistry; federal drug law. *Mailing Add:* Pharmaceut Consult Serv 6306 Crosswoods Falls Church VA 22044-1302

FELDMANN, RODNEY MANSFIELD, b Steele, NDak, Nov 19, 39; m 64; c 1. INVERTEBRATE PALEONTOLOGY. *Educ:* Univ NDak, BS, 61, MS, 63, PhD(paleont), 67. *Prof Exp:* Teaching asst geol, Univ NDak, 62-65; from instr to assoc prof, 65-75, asst dean, Col Arts & Sci, 66-67, PROF GEOL, KENT STATE UNIV, 75-, ASST CHMN DEPT, 76- *Concurrent Pos:* Ed, Compass, Sigma Gamma Epsilon, 72-76; co-ed, J Paleont, Paleont Soc, 77-83. *Mem:* Fel Geol Soc Am; Paleont Soc; Am Asn Petrol Geol. *Res:* Cretaceous stratigraphy of the midcontinent; taxonomy, biogeography and paleoecology of decapod crustaceans; antarctic invertebrate paleontology. *Mailing Add:* Kent State Univ Main Campus Kent State Univ Kent OH 44242

FELDMEIER, JOSEPH ROBERT, b Niles, Ohio, Feb 17, 16; m 42; c 5. NUCLEAR PHYSICS. *Educ:* Carnegie Inst Technol, BS, 38; Univ Notre Dame, MS, 40, PhD(nuclear physics), 42. *Prof Exp:* Asst physics, Univ Notre Dame, 38-42; staff mem, Radiation Lab, Mass Inst Technol, 42-46; asst prof physics, Rutgers Univ, 46-48; prof & chmn dept, Col St Thomas, 48-52; physics mgr, Bettis Atomic Power Lab, Westinghouse Elec Corp, 52-60; assoc dir & dir sci lab, Philco Corp, 60-64; dir res labs, Franklin Inst, 64-73, vpres, Inst, 67-73; vpres, STV Engrs, 73-76; dean, Montgomery County Community Col, 76-82; CONSULT, 82- *Concurrent Pos:* Tech consult, Fed Telecommun Labs, NJ, 46-48; mem adv coun sci, Univ Notre Dame, 70-85. *Mem:* AAAS; Am Nuclear Soc; fel Am Phys Soc; sr mem Inst Elec & Electronics Engrs; Sigma Xi. *Res:* Nuclear physics using Van der Graaf generator; design of radar transmitters; low temperature research using Collins liquefier; excitation of nuclei by x-ray; reactor physics; electronics. *Mailing Add:* 631 Midway Lane Blue Bell PA 19422-2027

FELDMESSER, JULIUS, b New York, NY, Oct 23, 18; m 44; c 2. PLANT NEMATOLOGY, INVERTEBRATE ZOOLOGY. *Educ:* Brooklyn Col, AB, 40; NY Univ, MS, 51, PhD(invert zool, parasitol), 53. *Prof Exp:* Asst biol, NY Univ, 48-49; from jr nematologist to head nematologist, 47-78, tech adv, res leader, Nat Res Progs Agr Chem Technol, 75-86, RES NEMATOLOGIST, AGR RES SERV, USDA, 63- *Concurrent Pos:* Consult, Spencer Chem Co, 57 & 58 & Int Mineral & Chem Corp, 64; chmn & prin investr, tech eval groups providing assessments environ effects nematicides aid in US dept Agr policies, 78-86; consult, 86- *Mem:* Fel AAAS; Soc Nematol; Am Soc Zool; Sigma Xi; Controlled Release Soc; Am Phytopath Soc. *Res:* Parasitology; nematicide evaluation and development of evaluation techniques; host-parasite relationships; zoology and chemical control of plant-parasitic nematodes; controlled release nematicide formulations; forage grass nematode biology; new safer nematicides. *Mailing Add:* 13309 Clifton Rd Silver Spring MD 20904

FELDMETH, CARL ROBERT, b Los Angeles, Calif, Mar 16, 42; m 64. ZOOLOGY, ENVIRONMENTAL PHYSIOLOGY. *Educ:* Calif State Col, Los Angeles, BS, 64; Univ Toronto, MSc, 66, PhD(zool), 68. *Prof Exp:* Lectr zool, Univ Calif, Los Angeles, 68-71; from asst prof to assoc prof biol, 71-81, PROF BIOL, JOINT SCI DEPT, CLAREMONT COL, 81- *Mem:* Am Soc Limnol & Oceanog; Can Soc Zoologists. *Res:* Environmental physiology; marine crustacea; aquatic insects; respiratory and osmotic regulation; effect of thermal effluent on marine invertebrates and fish; comparative energy studies on swimming marine and freshwater fish; thermal tolerance of desert pupfish. *Mailing Add:* Dept Biol Claremont McKenna Col Bauer Ctr Claremont CA 91711

FELDSTEIN, ALAN, b Pittsburgh, Pa, Feb 28, 33; m 55; c 4. COMPUTER SCIENCE, NUMERICAL ANALYSIS. *Educ:* Ariz State Univ, BA, 54; Univ Calif, Los Angeles, PhD(math), 64. *Prof Exp:* Staff mem comput prog, Los Alamos Sci Lab, 56-60; res asst numerical anal, Univ Calif, Los Angeles, 60-64; asst prof math, Univ Calif, Los Angeles, 64-65; asst prof appl math, Brown Univ, 65-68; assoc prof comput sci, Univ Va, 68-70; assoc prof, 70-74, PROF MATH, ARIZ STATE UNIV, 74- *Concurrent Pos:* Consult, US Naval Res Lab, 70-74. *Mem:* Asn Comput Mach; Soc Indust & Appl Math. *Res:* Computer arithmetic; numerical solution of and theory of functional differential equations; iteration theory; parallel algorithms. *Mailing Add:* Dept Math Ariz State Univ Tempe AZ 85287

FELDSTEIN, NATHAN, b Haifa, Israel, Jan 4, 37; US citizen; m 61; c 3. PHYSICAL CHEMISTRY. *Educ:* City Col New York, BChE, 60; NY Univ, MS, 64, PhD(chem), 66. *Prof Exp:* Engr, Corning Glass Works, 60; lectr chem, Brooklyn Col, 64-66; res scientist, David Sarnoff Res Ctr, RCA Labs, 66-73; PRES, SURFACE TECHNOL, INC, 73- *Concurrent Pos:* Div ed, J Electrochem Soc. *Honors & Awards:* David Sarnoff Achievement Awards, 68, 72 & 73. *Mem:* Am Chem Soc; Electrochem Soc; fel Am Inst Chemists. *Res:* Adsorption effects in electrochemistry; electroless plating catalysis; composite electroless deposition. *Mailing Add:* Surface Technol Inc PO Box 8585 Trenton NJ 08650

FELDT, LEONARD SAMUEL, b Long Branch, NJ, Nov 2, 25; m 54; c 2. APPLIED STATISTICS. *Educ:* Rutgers Univ, BSc, 50, MEd, 51; Univ Iowa, PhD(educ), 54. *Prof Exp:* From asst prof to prof educ measurement, 55-81, chmn, Div Educ Psychol, 77-81, E F LINDQUIST DISTINGUISHED PROF EDUC MEASUREMENT & DIR, IOWA TESTING PROGS, UNIV IOWA, 81-, CHMN, DIV PSYCHOL & QUANT FOUNDS, 87- *Concurrent Pos:* Pres, Iowa Measurement Res Found, 78- *Mem:* Psychomet Soc; Am Educ Res Asn; Am Statist Asn; Inst Math Statist. *Res:* Experimental design; educational measurement; psychometrics. *Mailing Add:* Col Educ Univ Iowa Iowa City IA 52242

FELGER, MAURICE MONROE, b Ft Wayne, Ind, Feb 5, 08; m 33; c 3. POLYMER CHEMISTRY. *Educ:* Ind Univ, AB, 30, AM, 31, PhD(chem), 33. *Prof Exp:* Instr chem, Exten Div, Ind Univ, 33-39, asst prof, Ft Wayne Ctr, 39-41; assoc chemist, US Nitrate Plant, Wilson Dam, 41-45; develop chemist, Gen Elec Co, 45-73; consult polymer chem & plastics, 73-86; RETIRED. *Mem:* Am Chem Soc; Soc Plastics Eng. *Res:* Electrolytic deposition of metals; high pressure synthesis; synthetic resins as electrical insulation. *Mailing Add:* 7515 Winchester Rd No 305 Ft Wayne IN 46819

FELGNER, PHILIP LOUIS, b Frankenmuth, Mich, Feb 7, 50; m 82; c 2. GENE THERAPEUTICS. *Educ:* Mich State Univ, BS, 72, MS, 74, PhD(biochem), 78. *Prof Exp:* Postdoctoral fel biophys, Univ Va, Charlottesville, 78-81; staff scientist bioorg chem, Syntex Res, 81-88; dir pharmaceut develop, 88-91, SR DIR GEN THERAPEUT DIV, VICAL INC, 91- *Res:* Protein biochemistry and membrane reconstitution; membrane biophysics; pharmaceutical formulation development; liposomes; DNA delivery and RNA delivery; lipofection reagent. *Mailing Add:* Vical Inc 9373 Town Centre Dr Suite 100 San Diego CA 92121

FELICETTA, VINCENT FRANK, b Seattle, Wash, July 20, 19; m 42; c 2. ORGANIC POLYMER CHEMISTRY. *Educ:* Univ Wash, BS, 42; MS, 51. *Prof Exp:* Chemist, US Rubber Co, Mich, 42-45; res chemist, Pulp Mills Res Proj, 45-53, res assoc, 53-57, res asst prof, Univ Wash, 57-59; from res chemist to sr chemist, Bellingham Div, Ga Pac Corp 59-69, admin asst, 69-72, coordr chem res, 72-76, assoc dir prod develop, 76-83; RETIRED. *Mem:* Am Chem Soc. *Res:* Organic natural polymers, structure and ulitization of lignin, building materials, polymer fractionation and characterization; product development. *Mailing Add:* 2958 Plymouth Dr Bellingham WA 98225

FELICIOTTI, ENIO, b Southbridge, Mass, Oct 9, 26; m 50; c 3. FOOD CHEMISTRY. *Educ:* Univ Boston, AB, 49, AM, 52; Univ Mass, PhD(food technol), 56. *Prof Exp:* Instr food technol, Univ Mass, 53-55; mgr customer res, Hazel Atlas Glass Div, Continental Can Co, Inc, 55-60; from mgr prod develop to vpres res, Thomas J Lipton, Inc, 60-77, vpres, 77-81, sr vpres res, develop & qual assurance, 81-88; CONSULT, 88- *Concurrent Pos:* Tech consult, Agr Serv, USDA, 53-54. *Mem:* AAAS; Am Chem Soc; Inst Food Technol; Sigma Xi. *Res:* Food research; administration. *Mailing Add:* PO Box 226 Center Harbor NH 03226-0226

FELIG, PHILIP, b New York, NY, Dec 18, 36; m 58; c 3. INTERNAL MEDICINE, ENDOCRINOLOGY. *Educ:* Princeton Univ, AB, 57; Yale Univ, MD, 61. *Hon Degrees:* DMed, Karolinska Inst, Sweden, 78. *Prof Exp:* From intern to asst resident internal med, Yale-New Haven Hosp, Conn, 61-63, asst resident to chief resident, 65-67; from asst prof to assoc prof internal med, Sch Med, Yale Univ, 69-75, vchmn dept, 75-80, prof, 75-; PRES, SANDOZ RES INST, SANDOZ INC. *Concurrent Pos:* USPHS spec res fel metab-endocrinol, Joslin Res Lab, Harvard Med Sch & Peter Bent Brigham Hosp, Boston, Mass, 67-69; Am Col Physicians teaching & res scholar, 69-72; estab investr award, Am Diabetes Asn, 77; nat counr, Am Soc Clin Invest, 78-81. *Honors & Awards:* Alvarenga Prize, Swed Med Asn, 75; Lilly Award, Am Diabetes Asn, 76; John Claude Kellion lectr, Australian Diabetes Asn, 77. *Mem:* Am Fedn Clin Res; Am Diabetes Asn; Endocrine Soc; Am Physiol Soc. *Res:* Regulation of gluconeogenesis; amino acid metabolism in the regulation of insulin secretion and gluconeogenesis; clinical diabetes mellitus; development of an insulin infusion pump; metabolism of exercise. *Mailing Add:* Dept Internal Med NY Med Col 1056 Fifth Ave New York NY 10028

FELIX, ARTHUR M, b New York, NY, June 15, 38; m 67; c 2. BIO-ORGANIC CHEMISTRY. *Educ:* NY Univ, BA, 59; Polytech Inst NY, PhD(chem), 64. *Prof Exp:* NIH res fel, Harvard Univ, 64-66; sr res chemist, 66-76, res fel, 76-80, group chief, 80-85, DEPT HEAD, HOFFMANN-LA ROCHE, INC, 85- *Concurrent Pos:* Guest investr, Rockefeller Univ, 68-69; adj asst prof, Fairleigh Dickinson Univ, 68-78, Hunter Col, 74-75. *Mem:* Am Chem Soc; Royal Soc Chemists; NY Acad Sci; Am Inst Chemists; Sigma Xi; Am Peptide Soc. *Res:* Classical and solid phase peptide synthesis; synthesis and physical chemistry of polypeptides; development of pharmaceuticals. *Mailing Add:* Hoffmann-La Roche Inc Bldg 86 Nutley NJ 07110

FELIX, JEANETTE S, b Wausau, Wis, Feb 13, 44; m 68; c 2. HUMAN GENETICS. *Educ:* Univ Wis-Madison, PhD(genetics), 72. *Prof Exp:* ASST PROF PEDIAT, SCH MED, JOHNS HOPKINS UNIV, 78-; SCI DIR, RETINITIS PIGMENTOSA FOUND, 86- *Concurrent Pos:* Expert consult, Genetics & Teratology Br, Nat Inst Child Health & Human Develop, NIH, 83-86. *Mem:* Am Soc Human Genetics; AAAS; Asn Res Vision & Opthal. *Res:* Administrates research programs for RP Foundation Fighting Blindness; programs focus on multi-disciplinary research centers and molecular genetic projects to determine causes of human inherited retinal degenerations. *Mailing Add:* Retinitis Pigmentosa Found 1401 Mt Royal Ave Baltimore MD 21217-4245

FELIX, RAYMOND ANTHONY, b Los Angeles, Calif, Aug 25, 46. ORGANIC CHEMISTRY. *Educ:* Calif State Col, Los Angeles, BS, 69; Univ Southern Calif, PhD(chem), 72. *Prof Exp:* Res chemist, ICI Americus, 72-80, sr res chemist, 80-82, prin res chemist, 83-85, RES ASSOC, ICI AMERICUS, 86- *Mem:* Am Chem Soc. *Res:* Preparation of novel agrochemicals; phase transfer catalysis; homogeneous catalysis; organic synthesis. *Mailing Add:* Dept Chem ICI Americus 1200 S 47th St Richmond CA 94804

FELIX, ROBERT HANNA, b Downs, Kans, May 29, 04; m 33; c 1. PSYCHIATRY. *Educ:* Univ Colo, AB, 26, MD, 30; Johns Hopkins Univ, MPH, 42. *Hon Degrees:* ScD, Univ Colo, 53, Boston Univ, 53, Univ Rochester, 64; LLD, Univ Chattanooga, 57, Ripon Col, 59, St Louis Univ, 75. *Prof Exp:* Staff psychiatrist, Hosp for Fed Prisoners, Mo, 33-35, clin dir, 35-36; chief psychiat serv, USPHS Hosp, Ky, 36-38, dir res, 38-40, exec officer, 40-41; psychiatrist, US Coast Guard Acad, 42-43, sr med officer, 43-44; asst chief hosp div, USPHS, 44, chief ment hyg div, 44-49, dir, NIMH, 49-64; prof psychiat & dean sch med, 64-74, dir bi-state regional med prog, 74-76, EMER PROF PSYCHIAT & EMER DEAN SCH MED, ST LOUIS UNIV, 74- *Concurrent Pos:* Commonwealth Fund fel, Colo Psychopath Hosp, 31-33. *Mem:* Fel AMA; fel Am Psychiat Asn (treas, 58-59, pres, 60-61); Am Pub Health Asn; fel Am Col Physicians. *Res:* Medical education; drug addiction; mental public health; mental hygiene and socio-environmental factors; medical school administration. *Mailing Add:* St Louis Univ 10501 Indian Wells Dr Sun City AZ 85373

FELKER, JEAN HOWARD, b Centralia, Ill, Mar 14, 19; m 43; c 2. ELECTRICAL ENGINEERING. *Educ:* Wash Univ, BEE, 41. *Prof Exp:* Mem staff, Bell Tel Labs, 45-59; mem staff, Am Tel & Tel, 59-60, asst chief engr, 60-62; vpres opers & dir, NJ Bell Tel Co, 62-69; bus consult, 69-71; vpres, Bell Labs, 71-81; RETIRED. *Concurrent Pos:* Dir & mem exec comt, Colonial Life Ins Co Am & Shulman Transport Enterprises. *Mem:* Nat Acad Eng; fel Inst Elec & Electronics Engrs. *Mailing Add:* PO Box 86 Durham PA 18039

FELL, COLIN, physiology; deceased, see previous edition for last biography

FELL, GEORGE BRADY, b Elgin, Ill, Sept 27, 16; m 48. ECOLOGY. *Educ:* Univ Ill, BS, 38; Univ Mich, MS, 40. *Prof Exp:* Teacher pub sch, Ill, 40-41; technician, City Health Dept Lab, Rockford, Ill, 46-48; soil conservationist, US Soil Conserv Serv, 48-49; exec dir, Nature Conserv, 50-58; DIR, NATURAL LAND INST, 58- *Concurrent Pos:* Mem, Ill Nature Preserves Comn, 64-70, exec secy, 70-82. *Honors & Awards:* Am Motors Conserv Award, 58. *Mem:* Wildlife Soc; Ecol Soc Am; Nat Areas Asn (secy-treas, 78-). *Res:* Methods of preserving and maintaining natural areas; analysis and management of natural vegetation. *Mailing Add:* 320 S Third St Rockford IL 61104-2063

FELL, HOWARD BARRACLOUGH, b Lewes, Eng, June 6, 17; m 42; c 3. INVERTEBRATE ZOOLOGY. *Educ:* Univ NZ, BSc, 38, MSc, 39; Univ Edinburgh, PhD(zool), 41, DSc(zool), 55. *Hon Degrees:* AM, Harvard Univ, 65. *Prof Exp:* Demonstr zool, Univ Edinburgh, 39-41; sr lectr, Univ Victoria, NZ, 45-57, assoc prof, 57-64; cur invert zool, Mus Comp Zool, 64-65, prof, 65-77, cur, Mus Comp Zool, 74-77, EMER PROF INVERT ZOOL, MUS COMP ZOOL, HARVARD UNIV, 77- *Honors & Awards:* Hector Medal & Prize, Royal Soc NZ, 59, Hutton Medal, 62; Gold Medal, Inst Studies Am Cults, 90. *Mem:* Emer fel Am Acad Arts & Sci; fel Royal Soc NZ; NZ Asn Scientists (secy, 46, pres, 48); fel Explorer's Club; fel Sci Explor Soc; hon fel Portugese Soc Anthrop & Ethnol; Epigraphic Soc, (pres). *Res:* Evolution; general problems of marine biogeography; biology of deep-sea bottom faunas; systematics, morphology and paleontology of the Echinodermata; epigraphy. *Mailing Add:* 6625 Bamburgh Dr San Diego CA 92117-5105

FELL, JAMES MICHAEL GARDNER, b Vancouver, BC, Dec 4, 23; m 57; c 2. PURE MATHEMATICS. *Educ:* Univ BC, BA, 43; Univ Calif, Berkeley, MA, 45, PhD(math), 50. *Prof Exp:* Jr res physicist, Nat Res Coun, Can, 45-46; instr, Calif Inst Technol, 53-55; res assoc, Univ Chicago, 55-56; from asst prof to prof math, Univ Wash, 56-65; PROF MATH, UNIV PA, 65- *Mem:* Am Math Soc. *Res:* Functional analysis; group representations. *Mailing Add:* Dept Math Univ Pa Philadelphia PA 19174

FELL, PAUL ERVEN, b Richmond, Va, Oct 4, 37; m 71; c 2. INVERTEBRATE EMBRYOLOGY, MARINE BIOLOGY. *Educ:* Hope Col, BA, 60; Stanford Univ, PhD(biol), 68. *Prof Exp:* Instr biol, Stanford Univ, 64-65; res assoc, Univ Calif, San Diego, 66-68; from asst prof to assoc prof, 68-78, PROF ZOOL, CONN COL, 78- *Mem:* Am Soc Zool; Estuarine Res Fedn; Int Soc Invert Reproduction. *Res:* Reproduction and dormancy in marine sponges and tidal marsh ecology. *Mailing Add:* Dept Zool Box 1484 Conn Col New London CT 06320

FELL, RONALD DEAN, b Clinton, Iowa, May 4, 49; m 72; c 2. EXERCISE PHYSIOLOGY, MUSCLE METABOLISM. *Educ:* Iowa State Univ, BS, 72, MS, 74, PhD(zool), 77. *Prof Exp:* Muscular Dystrophy Asn fel, Sch Med, Wash Univ, 77-79; ASSOC PROF, UNIV LOUISVILLE, 88- *Mem:* Sigma Xi; Am Physiol Soc; Am Col Sports Med. *Res:* Physiological and biochemical adaptations which occur in response to exercise training; muscle utilization of energy providing substrates during and after chronic exercise training is being investigated. *Mailing Add:* Exercise Physiol Lab Crawford Gym Univ Louisville Belknap Campus Louisville KY 40292

FELLER, DANNIS R, b Monroe, Wis, Oct 27, 41; m 63; c 2. BIOCHEMICAL PHARMACOLOGY, LIPID PHARMACOLOGY. *Educ:* Univ Wis, BS, 63, MS, 66, PhD(pharmacol), 68. *Prof Exp:* Guest worker, Nat Heart Inst, NIH, 67-69; asst prof, 69-74, assoc prof, 74-80, PROF PHARMACOL, OHIO STATE UNIV, 80- *Mem:* Am Soc Pharmacol & Exp Therapeut; Am Pharmaceut Asn; Sigma Xi. *Res:* Evaluation of the mechanism of adrenoceptor, antilipemic and antiplatelet actions of pharmacologically active drugs; biochemical aspects of drug action and relationship to toxicity. *Mailing Add:* 2846 Wellesley Dr Columbus OH 43221

FELLER, RALPH PAUL, b Quincy, Mass, Aug 31, 34; m 59; c 3. PROSTHODONTICS. *Educ:* Tufts Univ, BS, 56, DMD, 64; Univ Tex, MS & cert prosthodontics, 75; Loma Linda Univ, MPH, 81. *Prof Exp:* Res asst biochem, Protein Found, 56-57; res asst dent, Sch Dent Med, Tufts Univ, 60-64, res assoc biochem, 64-66, clin instr oral diag, 66-69; coordr, Vet Admin Dent Res Trainee Prog & chief, Oral Biol Res Lab, Vet Admin Outpatient Clin, Boston, 69-71; asst prof community dent, Univ Tex, Houston, 71-74; chief, Dent Serv, Vet Admin Hosp, Lyons, NJ, 75-77; assoc prof prosthodontics, Sch Dent, Fairleigh Dickinson Univ, 75-77; CHIEF, DENT SERV, VET ADMIN HOSP, LOMA LINDA, CA, 77-,; PROF EDUC SERVS, SCH DENT, LOMA LINDA UNIV, 77-; PROG SPECIALIST RES DENT, VET ADMIN CENT OFF, WASHINGTON, DC, 87- *Concurrent Pos:* Sigma Xi res grant-in-aid, 63-64; gen pract, Hingham, Mass, 64-69; res dentist, Boston Vet Admin Hosp, 66-69; asst prof prosthetic dent, Sch Dent, Harvard Univ, 69-71; clin investr, Houston Vet Admin Hosp, 71-74. *Mem:* Am Dent Asn; Int Asn Dent Res; Am Col Prosthodontists; Paul Harris Fel, Rotary Int, 85; Am Asn Pub Health Dentists; Am Asn Geriat Dent; Europ Asn Caries Res. *Res:* Salivary gland physiology; olfaction and taste; preventive dentistry. *Mailing Add:* Dent Serv Pettis Mem Vet Hosp Loma Linda CA 92357

FELLER, ROBERT JARMAN, b Fayetteville, NC, Jan 31, 45; m 69; c 2. BIOLOGICAL OCEANOGRAPHY, BENTHOS ECOLOGY. *Educ:* Univ Va, Charlottesville, BA, 66; Univ Wash, Seattle, MS, 72, PhD(oceanog), 77. *Prof Exp:* Fel res assoc, Dept Oceanog, Univ Wash, 77-79; from res asst prof to asst prof, 79-87, ASSOC PROF BIOL & MARINE SCI PROG, UNIV SC, 87- *Mem:* AAAS; Am Soc Limnol & Oceanog; Sigma Xi; Ecol Soc Am. *Res:* Applications of immunological methods for investigating marine benthic food webs; secondary production processes; benthos ecology; biological oceanography. *Mailing Add:* Belle W Baruch Inst Univ SC Columbia SC 29208

FELLER, ROBERT LIVINGSTON, b Newark, NJ, Dec 27, 19; m 75. PHYSICAL ORGANIC CHEMISTRY. *Educ:* Dartmouth Col, AB, 41; Rutgers Univ, MS, 43, PhD(chem), 50. *Prof Exp:* Instr & lectr, Rutgers Univ, 46-49; Nat Gallery Art fel, 50-63, SR FEL, CARNEGIE-MELLON INST RES, 63- *Concurrent Pos:* Vis scientist, Inst Fine Arts, NY Univ, 61; pres comt conserv, Int Coun Mus, 69-75; pres, Nat Conserv Adv Coun, 75-79; dir, Res Ctr Mat Artist & Conservator, Mellon Inst, Carnegie-Mellon Univ, 76-88, emer dir, 88- *Honors & Awards:* Pittsburgh Award, Am Chem Soc, 83. *Mem:* Fel Illum Eng Soc; fel Int Inst Conserv; hon fel Am Inst Conserv; AAAS; Am Chem Soc; Fedn Socs Coatings Technol. *Res:* Properties of methacrylate polymers; scientific examination of materials in the fine arts, particularly spirit varnishes; artists' pigments; photochemical deterioration of museum objects. *Mailing Add:* Mellon Inst Carnegie-Mellon Univ 4400 Fifth Ave Pittsburgh PA 15213

FELLER, WILLIAM, b St Paul, Minn, Nov 2, 25; m 64; c 2. SURGERY, CANCER RESEARCH. *Educ:* Univ Minn, BA, 48, BS, 52, MD, 54, PhD(surg), 62. *Prof Exp:* Instr surg, Med Sch, Univ Minn, 61-62; asst prof, 64-69, ASSOC PROF SURG, SCH MED, GEORGETOWN UNIV, 69- *Concurrent Pos:* Vis scientist, Nat Cancer Inst, 62-64. *Mem:* AAAS; NY Acad Sci; Am Asn Cancer Res; Am Col Surg; AMA; Am Cancer Soc. *Res:* Cancer virology; human breast cancer; breast cancer tumor markers. *Mailing Add:* Dept Surg Georgetown Univ Hosp Washington DC 20007

FELLERS, DAVID ANTHONY, b Northampton, Mass, Jan 3, 35; m 61; c 2. FOOD SCIENCE. *Educ:* Univ Mass, BS, 57; Rutgers Univ, MS, 62, PhD(food sci), 64. *Prof Exp:* Res food technologist, 63-69, chief cereal lab, 69-73, RES FOOD TECHNOLOGIST, WESTERN REGIONAL RES CTR, USDA, 73- *Honors & Awards:* Distinguished Serv Award, USDA. *Mem:* Inst Food Technol; Am Asn Cereal Chem. *Res:* Discoloration of beef by myoglobin auto-oxidation; development of protein concentrates and high protein foods from wheat and rice; nutrient composition, enrichment, and stability of cereals; extrusion and other means to precook cereals; technology transfer and technical assistance in food technology to developing countries. *Mailing Add:* 8306 Chivalry Rd 01CD Annandale VA 22003

FELLERS, FRANCIS XAVIER, b Seattle, Wash, Feb 6, 22; m 74; c 2. NEPHROLOGY, GERIATRICS. *Educ:* Amherst Col, BA, 44; Cornell Univ, MD, 46. *Prof Exp:* Intern pediat, New York Hosp, 46-47; asst resident, Boston Floating Hosp, 53-56; from asst to assoc, Harvard Med Sch, 58-63, asst prof pediat, 63-69, assoc prof clin pediat, 69-75; med dir, Montello Manor Nursing Home, 74-85; MED DIR, RUSSELL PARK MANOR, 81- *Concurrent Pos:* Res fel, Harvard Med Sch, 56-58; asst, Med Col, Tufts Univ, 54-55; from asst to sr assoc physician, Children's Med Ctr, 56- *Mem:* AAAS. *Res:* Metabolic and Renal disease in children. *Mailing Add:* PO Box 107 New Gloucester ME 04260

FELLERS, RUFUS GUSTAVUS, b Columbia, SC, Sept 26, 20; m 67; c 3. ELECTRICAL ENGINEERING. *Educ:* Univ SC, BS, 41; Yale Univ, PhD(elec eng), 43. *Prof Exp:* Asst instr elec eng, Yale Univ, 42-43, instr, 43-44; electronic scientist & sect head, res & develop ultra-high frequencies, Naval Res Lab, 44-55; chmn dept elec eng, Univ SC, 55-60, dean col, 60-69, chmn elec & comput eng, 76-79 & 87-89, PROF ELEC ENG, UNIV SC, 55- *Concurrent Pos:* Lectr, Univ Md, 48-54; mem nat comt, Int Sci Radio Union, 68-71; mem tech staff, Bell Tel Labs, 70-71. *Honors & Awards:* Outstanding Engr Award, Inst Elec & Electronics Engrs, 72, Centennial Medal, 89. *Mem:* Fel Inst Elec & Electronics Engrs; Int Sci Radio Union; Am Soc Eng Educ; Sigma Xi. *Res:* Millimeter waves; electromagnetic theory; microwave techniques. *Mailing Add:* Col Eng Univ SC Columbia SC 29208

FELLEY, DONALD LOUIS, b Memphis, Tenn, Feb 7, 21; m 49; c 5. ORGANIC CHEMISTRY. *Educ:* Ark State Col, BS, 41; Univ Ill, PhD(org chem), 49. *Prof Exp:* Mem staff tech sales, Rohm & Haas Co, 49-57, mgr Societe Minoc Div, France, 57-64,,asst gen mgr foreign opers div, 64-68, vpres & prod mgr chem div, 68-71, bd dirs, 71, vpres & gen mgr int div, 71-76, vpres & dir, NAm region, 76-78, group vpres, 77, pres & chief operating officer, 78-86; RETIRED. *Mem:* Am Chem Soc; Soc Chem Indust; Soc des Amis de la Maison de Chimie; Sigma Xi. *Res:* Synthesis of heterocyclic nitrogen compounds. *Mailing Add:* 1152 Sewell Lane Rydal PA 19046

FELLING, WILLIAM E(DWARD), b St Louis, Mo, Nov 26, 24; m 48; c 8. RESEARCH & DEVELOPMENT MANAGEMENT. *Educ:* Iowa State Univ, BS, 45; St Louis Univ, MS, 49, PhD(math), 59. *Prof Exp:* Dept dir, Parks Col, St Louis Univ, 48-58; res sci dynamical astron, McDonnell Aircraft Corp, 58-61; corp dir res indust R&D, Raytheon Co, 61-65; prog officer sci & eng, Ford Found, 65-67, res & environ, 67-75; assoc dir to exec dir, Oak Ridge Assoc Univs, 75-88; CONSULT, 88- *Concurrent Pos:* Consult, Holcomb Res Inst & Int Inst Applied Syst Anal, 75-78, Sci Comt Problems Environ, 76-78; pres, Felling Assocs, Inc & Found Adv Serv, 75- *Mem:* Fel AAAS; Sigma Xi; Am Soc Eng Educ; Am Inst Aeronaut & Astronaut; Am Mgt Asn; Soc Res Admin. *Res:* Regional environmental resource management; energy policy analysis; social consequences of energy policies; organizing for multi-institutional research; technology transfer; university-industry research collaboration; university consortial arrangements. *Mailing Add:* Three Heyward Pl Hilton Head Island SC 29928

FELLINGER, L(OWELL) L(EE), b Norris City, Ill, Sept 7, 15; m 41; c 2. CHEMICAL ENGINEERING. *Educ:* Univ Ill, BS, 37; Mass Inst Technol, ScD(chem eng), 41. *Prof Exp:* Res chem engr, Monsanto Chem Co, 42-43, group leader interim prod, 43-47, asst dir res, 47-51, asst eng mgr, 51-57, proj sect mgr, 57-60, asst dir eng, 60-62, eng proj mgr, 62-65, dir process eng dept, Monsanto Co, 65-67, mgr proj sect, Cent Eng Dept, 67-70, mgr chem eng sect, 70-81; RETIRED. *Mem:* Am Inst Chem Engrs. *Res:* Unit operations; process development; engineering project management. *Mailing Add:* 1289 Weatherby Dr St Louis MO 63146

FELLINGER, ROBERT C(ECIL), b Burlington, Iowa, Aug 10, 22; m 91; c 2. MECHANICAL ENGINEERING. *Educ:* Univ Iowa, BS, 47; Iowa State Univ, MS, 48. *Prof Exp:* Mech engr, Manhattan Proj, Chicago, 44, res mech eng, Mass Inst Technol, 44-46; from instr to prof, 47-84, actg chmn, dept mech eng, 79-80, div leader thermodyn & energy utilization, 72-84, EMER PROF MECH ENG, IOWA STATE UNIV, 84- *Concurrent Pos:* Consult, prod liability, fires & explosions, alcoa prof. *Mem:* Fel Am Soc Mech Engrs; Sigma Xi. *Res:* Thermodynamics; fuels and combustion; compressible flow. *Mailing Add:* III Lynn No 408 Ames IA 50011

FELLMANN, ROBERT PAUL, b Chicago, Ill, July 22, 24; m 47. CHEMISTRY. *Educ:* Univ Pa, AB, 47. *Prof Exp:* Group leader, res depy, Rohn & Haas Co, 56-84; RETIRED. *Mem:* Am Chem Soc. *Res:* Polymerization of acrylic monomers; monomer synthesis; polymer chemistry. *Mailing Add:* Six Aspen Ct Newtown PA 18940-3217

FELLNER, SUSAN K, b Hartford, Conn, Nov 14, 36; div; c 2. NEPHROLOGY. *Educ:* Smith Col, AB, 58; Univ Fla, MD, 66. *Prof Exp:* Res asst pharmacol, Med Sch, Univ Fla, 58-61; med intern, Duval Med Ctr, 66-67; fel cardiol, Med Sch, Emory Univ, 67-68, res med, 68-69, fel nephrol, 69-71, from asst prof to assoc prof med, 71-83; PROF MED, UNIV CHICAGO, 83- *Mem:* Am Soc Nephrology; Int Soc Nephrology. *Res:* Cardiovascular physiology in uremia. *Mailing Add:* Med Dept BH S509 Box 28 Univ Chicago 5841 S Maryland Ave Chicago IL 60637

FELLNER, WILLIAM HENRY, b New York, NY, Sept 27, 42; m 80. QUALITY TECHNOLOGY. *Educ:* Brooklyn Col, BS, 63; Univ Calif, Berkeley, MA, 65, PhD(biostatist), 71. *Prof Exp:* Asst prof math, Univ Calif, Irvine, 69-73; asst prof biostatist, Med Col Va, Richmond, 73-76; statistician, Nat Inst Occup Safety & Health, Morgantown, WVa, 76-77; tech serv statistician, 77-80, sr statistician, 80-86, CONSULT, QUAL MGT & TECHNOL CTR, ENG DEPT, E I DU PONT DE NEMOURS & CO, 86- *Mem:* Am Statist Asn; Sigma Xi; Am Soc Qual Control. *Res:* Robust estimation of variance components; robust product design; process control. *Mailing Add:* Eng Dept E I du Pont de Nemours & Co Inc PO Box 6091 Newark DE 19714-6091

FELLOWS, JOHN A(LBERT), metallurgical engineering, failure analysis; deceased, see previous edition for last biography

FELLOWS, LARRY DEAN, b Magnolia, Iowa, May 29, 34; m 60; c 3. STRATIGRAPHY, ENVIRONMENTAL GEOLOGY. *Educ:* Iowa State Univ, BS, 55; Univ Mich, MA, 57; Univ Wis, PhD(geol), 63. *Prof Exp:* Geologist, Carter Oil Co, 57-59; asst prof geol, Southwest Mo State Col, 62-65; chief stratig, Mo Geol Surv & Water Resources, 66-71, asst state geologist, 71-79; asst dir & state geologist, Ariz Bur Geol & Mineral Technol, 79-88; DIR & STATE GEOLOGIST, ARIZ GEOL SURV, 88- *Mem:* Geol Soc Am; Soc Econ Paleontologists & Mineralogists; Am Inst Prof Geologists; Asn Am State Geologists. *Mailing Add:* Ariz Geol Surv 845 N Park Ave No 100 Tucson AZ 85719

FELLOWS, ROBERT ELLIS, JR, b Syracuse, NY, Aug 4, 33; c 2. ENDOCRINOLOGY, NEUROSCIENCE. *Educ:* Hamilton Col, AB, 55; McGill Univ, MD, 59; Duke Univ, PhD(biochem), 69. *Prof Exp:* Asst prof physiol & med, 66-69, assoc prof physiol & asst prof med, Sch Med, Duke Univ, 70-76; PROF PHYSIOL & BIOPHYS & CHMN DEPT, COL MED, UNIV IOWA, 76- *Concurrent Pos:* Dir, med scientist training prog, Univ Iowa, 77-, physician scientist prog, 84-90, neurosci grad prog, 85-89; mem, Population Res Comt, Nat Inst Child Health & Human Develop & NIH, 81-88, Vet Admin Career Develop Comt, 84-89. *Mem:* Soc Neurosci; Endocrine Soc; Am Soc Cell Biol; Am Physiol Soc; Am Soc Biol Chem; Biophys Soc. *Res:* Chemistry and structure-function relationship of protein and peptide hormones and growth factors; cellular and molecular neurobiology of central nervous system development. *Mailing Add:* Dept Physiol & Biophys Univ Iowa Col Med Iowa City IA 52242

FELMAN, YEHUDI M, b Bridgeport, Conn, July 11, 38; m 62; c 2. DERMATOLOGY. *Educ:* Yeshiva Univ, BA, 59; Albert Einstein Col Med, MD, 63; Columbia Univ, MA, 66, MPhil, 75; Am Bd Dermatol, cert, 68. *Prof Exp:* Asst instr dermat, Col Physicians & Surgeons, Columbia Univ, 65-67; instr, 67-70, asst prof, 70-72, asst clin prof, 72-77, clin assoc prof, 78-81, CLIN PROF DERMAT, STATE UNIV NY, DOWNSTATE MED CTR, BROOKLYN, 82-; PROF DERMAT, TOURO COL HEALTH SCI, NY, 73-; AT DEPT PUB HEALTH, COLUMBIA UNIV. *Concurrent Pos:* Dir, Bur Venereal Disease Control, New York City Health Dept, 76-; lectr health admin, Fac Med Sch Pub Health, Columbia Univ, NY, 81- *Mem:* Fel Am Col Physicians; fel Am Acad Dermat; fel Am Col Prev Med; fel Soc Trop Med & Hyg; fel Int Soc Trop Dermat. *Mailing Add:* 8100 Bay Pkwy Brooklyn NY 11214

FELMLEE, WILLIAM JOHN, b Bay City, Mich, Feb 14, 30; m 57; c 2. PHYSICS, CHEMISTRY. *Educ:* Alma Col, BS, 53; Univ Mich, Ann Arbor, MS, 57. *Prof Exp:* Res physicist, Dow Chem USA, 57-72; sales eng mgr, Audn Corp, 72-74; sr engr, 74-80, SR RES SCIENTIST, KMS FUSION, INC, KMS INDUST, 80- *Res:* Preparation of DT solid layer cryogenic laser fusion targets complete manufacture of targets for laser fusion energy research; DT gas permeation into hollow glass and polymeric shells. *Mailing Add:* 700 KMS Pl PO Box 1778 Ann Arbor MI 48106-1778

FELS, STEPHEN BROOK, meteorology, planetary atmospheres; deceased, see previous edition for last biography

FELSEN, LEOPOLD BENNO, b 1924; US citizen. ELECTRICAL ENGINEERING, ELECTROPHYSICS. *Educ:* Polytech Inst Brooklyn, DEE, 52. *Hon Degrees:* Dr tecnices, Tech Univ Denmark, 79. *Prof Exp:* Prof, 61-78, dean eng, 74-78, INST PROF, POLYTECH INST NY, 78- *Concurrent Pos:* Guggenheim Mem fel, 73-74; vis mem, Fac Math & Physics, Charles Univ, Prague, Czech, 84; vis Sackler fel, Tel Aviv Univ, 85; vis scholar, Acoust Inst, Academia Sinica, Beijing, 85; vis prof, Nat Defense Acad, Japan, 85; vis lectr, Soviet Acad Sci, 88. *Honors & Awards:* Distinguished Res Citation, Sigma Xi, 73; Van der Pol Gold Medal, Int Union Radio Sci, 75; Humboldt Sr Scientist Award, 80; Heinrich Hertz Medal, Inst Elec & Electronics Engrs, 91. *Mem:* Nat Acad Eng; fel Inst Elec & Electronics Engrs; Int Union Radio Sci; fel Optical Soc Am; Am Soc Eng Educ. *Res:* Electromagnetics; optics. *Mailing Add:* Polytech Univ Rte 110 Farmingdale NY 11735

FELSENFELD, GARY, b New York, NY, Nov 18, 29; m 56; c 3. BIOPHYSICS. *Educ:* Harvard Univ, AB, 51; Calif Inst Technol, PhD(phys chem), 55. *Prof Exp:* Asst scientist, USPHS, Lab Neurochem, NIMH, 55-57, sr asst scientist, 57-58; asst prof biophys, Univ Pittsburgh, 58-61; CHIEF, SECT PHYS CHEM, LAB MOLECULAR BIOL, NAT INST DIABETES & DIGESTIVE & KIDNEY DIS, NIH, 61- *Concurrent Pos:* Vis prof chem, Harvard Univ, 63, Cornell Univ, 67 & vis prof, Univ Fla, 77; co-chmn, Gordon Conf on Phys Chem of Biolymers, 70 & chmn, Gordon Conf on Nuclear Proteins, Chromatin Struct & Gene Regulation, 78; mem, Chem Adv Coun, Polytechnic Inst NY, 75-81; lectr, Fedn Europ Biochem Socs, 79; mem vis comt, Dept Biochem & Molecular Biol, Harvard Univ, 80-86; distinguished fac lectr, Tex Med Ctr, Univ Tex, 86; coun mem, Biophys Soc. *Honors & Awards:* Merck Distinguished Lectr, Rutgers Univ, 77; Meritorious Presidential Rank Award, US Govt, 85 & Distinguished Presidential Rank Award, 88; Merck Award, Am Soc Biol Chemists, 87. *Mem:* Nat Acad Sci; Am Soc Biol Chem & Molecular Biol; Am Chem Soc; Biophys Soc; fel AAAS; fel Am Acad Arts & Sci; Am Soc Biol Chemists. *Res:* Physical chemistry of nucleic acids and proteins; chromatin structure; nucleoprotein complexes; gene expression in eukaryotes. *Mailing Add:* Lab Molecular Biol Nat Inst Diabetes Digestive & Kidney Dis NIH Bethesda MD 20892

FELSENSTEIN, JOSEPH, b Philadelphia, Pa, May 9, 42. POPULATION GENETICS,EVOLUTION. *Educ:* Univ Wis-Madison, BS, 64; Univ Chicago, PhD(zool), 68. *Prof Exp:* NIH fel, Inst Animal Genetics, Scotland, 67-68; from asst prof to assoc prof, 67-78, PROF GENETICS, UNIV WASH, 78- *Concurrent Pos:* Assoc ed, Theoret Pop Biol, 75-86 & Evolution, 78-79, 83-86. *Mem:* Am Soc Naturalists; Genetics Soc Am; Soc Study Evolution (vpres, 86); Soc Syst Zool. *Res:* Statistical estimation of evolutionary trees; theoretical population genetics, applied to evolution. *Mailing Add:* Dept Genetics Univ Wash Seattle WA 98195

FELSENTHAL, GERALD, b New York, NY, Aug, 27, 41; m 64; c 3. ELECTRODIAGNOSTIC MEDICINE, GERIATRIC MEDICINE. *Educ:* NY Univ, BA, 63; Albany Med Col, MD, 67; Am Bd Phys Med & Rehab, cert, 75. *Prof Exp:* Phys med & rehab residency, Bronx Munic Hosp Ctr, Albert Einstein Col Med, 73; assoc physiatrists, dept rehab med, 73-76, assoc chief, 76-86, actg chief, 86-87, CHIEF PHYS MED & REHAB, DEPT REHAB MED, SINAI HOSP, BALTIMORE, 87- *Concurrent Pos:* Prog dir, Sinai Hosp-Johns Hopkins Phys Med & Rehab Residency Training Prog, 86-; head div rehab med, Levindale Hebrew Geriat Ctr & Hosp, 87-; assoc exam writer, Am Bd Phys Med & Rehab, 87-; bd dirs, Am Asn Electrodiag Med, 90-93; residency rev comt, Phys Med Rehab, 90-; clin assoc prof, Sch Med, Univ Md, 87-; assoc prof, Sch Med, Johns Hopkins Univ, 89- *Mem:* Am Acad Phys Med & Rehab; Am Asn Electromyography & Electrodiag; AMA; Am Geriat Soc; Asn Acad Physiatrists; Am Cong Rehab Med. *Res:* Electrodiagnostic evaluation of nerve entrapment syndromes and; peripheral neuropathy; geriatrics, elderly and rehabilitation. *Mailing Add:* Dept Rehab Med Sinai Hosp Belvedere & Greenspring Baltimore MD 21215

FELSHER, MURRAY, b New York, NY, Oct 8, 36; m 61; c 3. GEOLOGY, OCEANOGRAPHY. *Educ:* City Col New York, BS, 59; Univ Mass, Amherst, MS, 63; Univ Tex, Austin, PhD(geol & oceanog), 71. *Prof Exp:* Asst prof geol, Syracuse Univ, 67-69; assoc dir coun educ in geol sci, Am Geol Inst, 69-71; sr staff scientist, US Environ Protection Agency, 71-75; fed affairs officer, Off Appln, NASA, 75-77, chief, Geol & Energy Appln, 77-80; PRES, ASSOC TECH CONSULTS, 80- *Concurrent Pos:* Consult oceanog, Syracuse Univ Res Corp, 67-69; publ, Washington Fed Sci Newslett, 80- & Washington Remote Sensing Lett, 80-; postdoctoral fel, McMaster Univ, Hamilton, Ont, Can. *Mem:* AAAS; sr mem Am Astronaut Soc; fel Geol Soc Am; Am Geophys Union; Am Inst Aeronaut & Astronaut; Marine Technol Soc; Nat Defense Preparedness Asn; Am Soc Photogram & Remote Sensing; Brit Interplanetary Soc; Asn Old Crows; Armed Forces Commun & Electronics Asn; Nat Mil Intel Asn; Nat Press Club; Nat Space Club. *Res:* Coastal morphology; marine geology; remote sensing; educational administration; sedimentology and sedimentary petrology; cosmology and planetary geology; science policy. *Mailing Add:* Pres Assoc Tech Consults PO Box 20 Germantown MD 20875-0020

FELT, ROWLAND EARL, b Idaho Falls, Idaho, Aug 3, 36; m 66; c 5. CHEMICAL ENGINEERING. *Educ:* Univ Idaho, BS, 58, MS, 59; Iowa State Univ, PhD(chem eng), 64. *Prof Exp:* Instr chem eng, Univ Idaho, 59-60; res asst, Ames Lab, 60-64; staff engr, Gen Elec Co, Wash, 64-66 & Isochem Inc, 66-67; staff engr, Atlantic Richfield Hanford Co, 67-73, mgr plutonium process eng, 73-77; mgr eng develop, Rockwell Hanford Opers, 77-78; mgr chem engr, 78-81, mgr process eng, 81-87, MGR CHEM DEVELOP, EXXON NUCLEAR, 87-; PRES ADV ENG, WESTINGHOUSE IDAHO NUCLEAR. *Mem:* Am Inst Chem Engrs; Am Nuclear Soc; Sigma Xi. *Res:* Plutonium processing; plutonium scrap management; nuclear fuel cycle technology; nuclear fuel fabrication. *Mailing Add:* MS 3301 PO Box 4000 Idaho Falls ID 83403

FELTEN, DAVID L, b Sheboygan, Wis, Feb 28, 48; m 68. NEUROIMMUNE INTERACTIONS, NEURAL PLASTICITY IN AGING. *Educ:* Mass Inst Technol, BS, 69; Univ Pa, MD, 73 & PhD(anat), 74. *Prof Exp:* From asst prof to prof anat, Ind Univ, Sch Med, 74-83; PROF NEUROBIOL & ANAT, DEPT NEUROBIOL & ANAT, SCH MED, UNIV ROCHESTER, 83- *Concurrent Pos:* Alfred P Sloan Found fel, 79-80; mem Subcomt 1, Behav & Neurosci Study Sect, NIH, 84-88; Andrew W Mellon Found fel, 85-88; merit award, Nat Inst Aging, 86- & NIMH, 90-; assoc ed, Brain Behav & Immunity, 87-; mem, AIDS Study Sect, NIMH, 89- *Honors & Awards:* John D & Catherine T MacArthur Found Prize Fel, 83-88. *Mem:* AAAS; Am Asn Anat; Soc Neurosci. *Res:* Innervation of lymphoid organs and neural control of immune system; aging of catecholamine and indoleamine systems in the central nervous system related to models of neurodegenerative diseases. *Mailing Add:* Dept Neurobiol & Anat Sch Med Box 603 Univ Rochester 601 Elmwood Ave Rochester NY 14642

FELTEN, JAMES EDGAR, b Duluth, Minn, Sept 8, 34. ASTROPHYSICS. *Educ:* Univ Minn, Duluth, BA, 56; Cornell Univ, PhD(astrophys), 65. *Prof Exp:* Asst res physicist, Univ Calif, San Diego, 65-68; vis fel, Inst Theoret Astron, Cambridge Univ, 68-70; vis assoc prof, Univ Ariz, 70-72, assoc prof astron, 72-75, assoc astronr, 70-75; Nat Acad Sci/Nat Res Coun sr resident res assoc, Goddard Space Flight Ctr, 76-78; vis prof astron, Univ Md, 78-79, sr res assoc, 78-85; CONSULT, 85-; RES SCIENTIST, UNIV SPACE RES ASN, 89- *Concurrent Pos:* Int Astron Union vis fel, Tata Inst Fundamental Res, Bombay, 70; vis prof, Univs Padua & Bologna, 70; vis res physicist, Univ Calif, San Diego, 80; vis scientist, Ohio State Univ, 87. *Mem:* Am Astron Soc; Int Astron Union; fel Am Phys Soc. *Res:* Theoretical high-energy astrophysics; theories of celestial x-rays and gamma rays; cosmic rays; interstellar and intergalactic medium; galaxies and cosmology. *Mailing Add:* Code 685 Goddard Space Flight Ctr Greenbelt MD 20771

FELTEN, JOHN JAMES, b Louisville, Ky, July 19, 37; m 67; c 4. ORGANOMETALLIC CHEMISTRY. *Educ:* St Meinrad Col, AB, 59; Ind Univ, Bloomington, MS, 68; Univ Del, PhD(chem), 73. *Prof Exp:* Teaching asst chem, Ind Univ, Bloomington, 65-68; inorg chemist, Pigments Dept, E I du Pont de Nemours & Co, Del, 68-69; teaching asst chem, Univ Del, 69-73; res chemist, Photo Prod Dept, E I du Pont de Nemours & Co, Inc, Del, 73-74, res chemist, Niagara Falls, 74-76, sr res chemist, 76-80, RES ASSOC, PHOTO PROD DEPT, ELECTRONIC MAT DIV, E I DU PONT DE NEMOURS & CO, INC, WILMINGTON, DE, 82- *Mem:* Am Chem Soc; Int Soc Hybrid Microelectronics; AAAS; Sigma Xi. *Res:* Properties and composition of materials used in production of thick film microcircuits, both during processing and in functioning circuits. *Mailing Add:* 25 Quartz Mill Rd-Lamatan Newark DE 19711

FELTHAM, LEWELLYN ALLISTER WOODROW, b Nfld, Oct 23, 26; m 53; c 3. BIOCHEMISTRY. *Educ:* Dalhousie Univ, BSc, 47; Univ Toronto, MA, 52, PhD(path chem), 60. *Prof Exp:* Chemist, Pub Health Labs, St John's, Nfld, 49-50, biochemist, 52-55, chief biochemist, 58-67; head dept biochem, Mem Univ NFLD, 67-73, prof biochem, 67-87; RETIRED. *Concurrent Pos:* Consult biochemist, Grace Hosp, St John's, 58-66 & St Clare's Mercy Hosp, St John's, 58-67; vis lectr biol, Mem Univ Nfld, 59-62, part-time assoc prof, 62-67. *Mem:* Can Biochem Soc; Biochem Soc. *Res:* Properties of regulatory enzymes in marine animals; seasonal amino acid variation in marine fish; pheromones in marine fish; the effect of temperature on proteolytic enzymes in marine fish. *Mailing Add:* Six Wexford St St Johns NF A1B 1W5 Can

FELTHAM, ROBERT DEAN, b Roswell, NMex, Nov 18, 32; m 54; c 2. INORGANIC CHEMISTRY, PHYSICAL CHEMISTRY. *Educ:* Univ NMex, BSc, 54; Univ Calif, PhD(chem), 57. *Prof Exp:* Asst, Univ NMex, 54 & Univ Calif, 54-57; Fulbright study grant, Denmark, 57-58; res fel, Mellon Inst, 58-64; from asst prof to assoc prof, 64-70, PROF CHEM, UNIV ARIZ, 71- *Concurrent Pos:* Fulbright Award, 57; NATO fel, Univ Col, London, 63-64; vis prof, Univ Paul Sabatier, 81; consult, Lawrence Berkeley Labs. *Honors & Awards:* Welch Lectr, 83. *Mem:* Am Chem Soc; Am Asn Univ Professors; NY Acad Sci. *Res:* Synthesis and spectroscopic studies of transition metal derivatives of the group V elements; infrared, electron spin resonance and photoelectron spectroscopy; bioinorganic chemistry. *Mailing Add:* Dept Chem Univ Ariz Tucson AZ 85721

FELTHOUSE, TIMOTHY R, b Berkeley, Calif, Sept 25, 51. SOLID STATE CHEMISTRY & CATALYSIS. *Educ:* Univ Pac, BS, 73; Univ Ill, Urbana, PhD(chem), 78. *Prof Exp:* Grad teaching & res asst, Univ Ill, Urbana, 73-78; res assoc & Robert A Welch Found fel, Tex A&M Univ, College Station, 78-80; sr res chemist, Corp Res Labs, 80-83, res specialist, Cent Res Labs, 83-87, sr res specialist, 87-88, ASSOC FEL, MONSANTO CHEM CO, 88- *Mem:* Am Chem Soc; Sigma Xi; NY Acad Sci; Catalysis Soc; Am Inst Chemists. *Res:* Solid state chemistry, surface chemistry, and catalysis applied to supported metals, metal oxides, molecular sieves, and carbons; catalytic oxidations with molecular oxygen; structural, electronic, magnetic, and catalytic properties of the transition metals; applications of sol-gel technology. *Mailing Add:* Monsanto Chem Co 800 N Lindbergh Blvd Mail Zone Q4E St Louis MO 63167

FELTMAN, REUBEN, oral medicine; deceased, see previous edition for last biography

FELTNER, KURT C, b Rock Springs, Wyo, May 23, 31; m 51; c 3. AGRONOMY. *Educ:* Univ Wyo, BS, 56, MS, 59; Univ Ariz, PhD(crop physiol), 63. *Prof Exp:* Mgr seed cert serv, Univ Wyo, 57-60, asst prof crop physiol, 62-64; assoc prof plant sci, Kans State Univ, 65-70, prof, 70-71; prof agron & head dept plant & soil sci, Mont State Univ, 71-79; dean, col life sci & agr & dir, NH agr exp sta, Univ NH, 79-; AT KANS STATE UNIV. *Mem:* Weed Sci Soc Am; Am Soc Agron; Int Crop Improv Asn; Sigma Xi. *Res:* Physiology and ecology of higher economic plants, especially hardiness and competition. *Mailing Add:* 113 Waters Hall Kans State Univ Manhattan KS 66506

FELTON, JAMES STEVEN, b San Francisco, Calif, Jan 31, 45; m 68; c 2. BIOCHEMISTRY, TOXICOLOGY. *Educ:* Univ Calif, Berkeley, AB, 67; State Univ NY Buffalo, PhD(molecular biol), 73. *Prof Exp:* Staff fel develop pharmacol, Nat Inst Child Health & Human Develop, 73-76; sr biomed res scientist biochem & toxicol, 76-87, SECT LEADER & MOLECULAR BIOL, LAWRENCE LIVERMORE NAT LAB, UNIV CALIF, 87- *Concurrent Pos:* Pres, Gen Environ Toxicol Asn, 85-86; adj assoc prof toxicol, San Jose State Univ, 82-; mem sci adv bd, DCE, Nat Cancer Inst, 89- *Mem:* Environ Mutagen Soc; Am Asn Cancer Res; AAAS; Am Chem Soc; Inst Food Technologists. *Res:* Biochemical and genetic aspects of drug and carcinogen metabolism; purification, identification and risk analysis of mutagens-carcinogens in the diet; molecular mechanisms of mutagenesis. *Mailing Add:* L-452 Biomed Div PO Box 5507 Livermore CA 94550

FELTON, JEAN SPENCER, b Oakland, Calif, Apr 27, 11; m 37; c 3. OCCUPATIONAL MEDICINE. *Educ:* Stanford Univ, AB, 31, MD, 35. *Prof Exp:* Intern & resident surg, Mt Zion & Dante Hosps, San Francisco, 35-38; pract physician & surgeon, Calif, 36-40; med dir, Oak Ridge Nat Lab, 46-53; prof, Dept Med & Dept Prev Med & Pub Health & dir employees health serv, Sch Med, Univ Okla, 53-58; prof occup health, Dept Prev Med & Pub Health, Sch Med, Univ Calif, Los Angeles, 58-68; dir occup health serv, Dept Personnel, County of Los Angeles, 68-74; chief occup health serv, Naval Regional Med Ctr, Long Beach & med dir, Br Clin, Terminal Island, 74-78; clin prof, 68-83, emer prof commun med, Sch Med, Univ Southern Calif, 83-86; CLIN PROF COMMUNITY MED, COL MED, UNIV CALIF, IRVINE, 75- *Concurrent Pos:* Lectr sociol, Univ Tenn, 46-53; mem Nat Safety Coun Comt Indust Eye Protection, 47-51; mem Nat Publicity Coun Health & Welfare Servs, 47-49, dir, 49-54; chmn adv bd, Family Serv Bur, Oak Ridge, 47-48, mem, 48-49; consult, Atlanta area, US Vet Admin, 49-53, St Louis area, 53-58, Vet Admin Ctr, Los Angeles, 65-, Oak Ridge Hosp & Okla State Dept Health, 53-58; ed, Indust Med & Surg, 50-51; occup health adminr, Calif State Dept Health, 58- & NASA, 64-; mem, President's Comt on Employ of Handicapped. *Honors & Awards:* Knudsen Award, Indust Med Asn, 68; Physicians Award, IMA, 71, 78, 81 & 85; Katherine Boucot Sturgis Lectr, Am Col Prev Med, 81; Physician of the Year, Pres Comt Employ Handicapped, 78. *Mem:* Fel Am Pub Health Asn; Am Indust Hyg Asn; fel Am Occup Med Asn; fel Am Acad Occup Med; Nat Rehab Asn; Sigma Xi. *Res:* Occupational health methods and practices; health status of employee groups; mental health in industry; job performances of the physically impaired; communication in occupational health; history of occupational medicine; public speaking and medical writing. *Mailing Add:* S Occup Ctr Col Med Univ Calif Irvine CA 92717

FELTON, KENNETH E(UGENE), b Kenton, Ohio, Aug 18, 20; m 43; c 1. AGRICULTURAL ENGINEERING. *Educ:* Univ Md, BS(agr), 50, BS(civil eng), 51; Pa State Univ, MS, 62. *Prof Exp:* Engr, eng div, Assoc Factory Mutual Fire Ins Co, 51-52; sanit engr, Interstate Comn, Potomac River Basin, 52-54; agr engr irrig, Southern States Co-op, 54; asst prof, 54-63, assoc prof, 63-80, PROF AGR ENG, UNIV MD, 80- *Concurrent Pos:* Adv, Repub Korea on rice storage, 75-76, Repub Somalia, 78. *Mem:* Am Soc Agr Engrs. *Res:* Structural components of farm and light industrial buildings; environmental requirements of poultry for maximum performance; proper design and arrangement of farm structures for optimum operation; application of solar energy to poultry production. *Mailing Add:* 8775 20th St 256 Vero Beach FL 32960

FELTON, LEWIS P(ETER), b Brooklyn, NY, Dec 14, 38; m 60; c 2. STRUCTURAL MECHANICS, OPTIMUM STRUCTURAL DESIGN. *Educ:* Cooper Union, BCE, 59; Carnegie Inst Technol, MS, 61, PhD(civil eng), 64. *Prof Exp:* Mem tech staff, appl mech div, Aerospace Corp, 63-64; asst prof eng, 64-71, ASSOC PROF CIVIL ENG, UNIV CALIF, LOS ANGELES, 71- *Mem:* Am Soc Civil Engrs; Am Inst Aeronaut & Astronaut. *Res:* Structural mechanics and design; optimum structural design. *Mailing Add:* Dept Civil Eng Univ Calif 405 Hilgard Ave Los Angeles CA 90024

FELTON, RONALD H, b Washington, DC, Jan 12, 38. CHEMICAL PHYSICS. *Educ:* Mass Inst Technol, BS, 58; Harvard Univ, PhD(chem physics), 64. *Prof Exp:* Fel chem, Brandeis Univ, 65-67; NSF fel, Mass Inst Technol, 67-68; from asst to assoc prof, 68-77, PROF CHEM, GA INST TECHNOL, 77- *Concurrent Pos:* Res collabr, Brookhaven Nat Lab, 68- *Mem:* Am Inst Physics; AAAS. *Res:* EXAFS; nanoclusters. *Mailing Add:* Sch Chem Ga Inst Technol 225 North Ave NW Atlanta GA 30332

FELTON, SAMUEL PAGE, b Petersburg, Va, Sept 7, 19; m 55; c 1. AQUATIC CHEMISTRY, BIOLUMINESCENCE. *Educ:* Univ Wash, BS, 51. *Prof Exp:* Res assoc, Scripps Clin & Res Found, 62-64, asst mem, 64-66; asst biochemist, Childrens Orthop Hosp, Seattle, 66-68; res technician, 52-55, from res asst to res assoc, 55-62, res assoc, Dept Anesthesiol, 69-73, res assoc, Fisheries Res Inst, 73-78, DIR WATER QUAL LAB, UNIV WASH, 76-, SR RES ASSOC, FISHERIES RES INST, 78- *Concurrent Pos:* Vis scientist, Col William & Mary, 85. *Mem:* Am Chem Soc; NY Acad Sci; Am Inst Fishery Res Biologists; Am Inst Chemists. *Res:* Biochemistry (enzymology); fish, including environmental chemistry, the effects of nutrition in diseases, drugs and aquatic pollutant interactions, elucidation of vitamin C's role in detoxification and tumorgenesis, and fish as a model in biomedical research. *Mailing Add:* 8415 Talbot Rd Edmonds WA 98026

FELTON, STALEY LEE, b Whaleyville, Va, Oct 23, 20; m 51; c 4. AGRICULTURAL CHEMISTRY. *Educ:* Va Polytech Inst, BS, 50. *Prof Exp:* Chemist, Va Dept Agr, 50-51 & Tobacco Byprod & Chem Corp, 51-55; asst mgr res & develop dept, Diamond Black Leaf Co, 55-57; sr chemist, Va-Carolina Chem Corp, 57-64; prod mgr pesticides, Indust Chem Div, 64-81, coordr int sales, Mobil Chem Co, 81-; RETIRED. *Mem:* Am Chem Soc. *Res:* Formulation of agricultural chemicals; field development of new agricultural chemicals such as insecticides, nematocides, herbicides, repellents and plant growth regulators. *Mailing Add:* 1704 Brentwood Rd Richmond VA 23222

FELTS, JAMES MARTIN, physiology, biochemistry; deceased, see previous edition for last biography

FELTS, JOHN HARVEY, b Lumberton, NC, Apr 2, 24; m 55; c 2. INTERNAL MEDICINE. *Educ:* Wofford Col, BS, 49; Med Col SC, MD, 49; Am Bd Internal Med, dipl, 57. *Prof Exp:* Intern, Walter Reed Army Hosp, Washington, DC, 49-50; intern, NC Baptist Hosp, Winston-Salem, 50-51, from asst resident to resident med, 51-53; resident physician, Western NC Sanatorium Treat Tuberc, Black Mountain, 53; from instr to assoc prof, 55-70, PROF INTERNAL MED, BOWMAN GRAY SCH MED, 70-, ASSOC DEAN ADMIS, 78- *Concurrent Pos:* Attend physician cardiac & diabetic clin, Regional Off, Vet Admin Hosp, Winston-Salem, 52-53; consult physician, Vet Admin Hosp, Salisbury, 66-74; ed, NC Med J, 74-82. *Mem:* Am Fedn Clin Res; fel Am Col Physicians; Am Soc Artificial Internal Organs; Sigma Xi. *Res:* Renal disease and toxicology. *Mailing Add:* Bowman Gray Sch Med Wake Forest Univ Winston-Salem NC 27103

FELTS, WAYNE MOORE, b Oakland, Calif, Aug 5, 12; m 42; c 2. MINEROLOGY & PETROLOGY, GEOLOGY. *Educ:* Ore State Col, BS, 34, MS, 36; Univ Cincinnati, PhD(geol), 38. *Prof Exp:* Mem, Columbia River Ethnol Exped, Smithsonian Inst, 34-35; asst geol, Univ Cincinnati, 36-38, instr, Col Eng & Commerce, 38-41; assoc geologist, Ohio River Div, Cinnannati Testing Lab, US Eng Corps, 42-43; mem geol dept, Phillips Petrol Co, 43-47; geologist, Texas Co, 47-58; div geologist, Los Angeles Div, Texaco, Inc, 58-62, asst to div mgr, 62-66, mgr hard mineral explor, 66-73; owner, W M Felts Geol Serv, 79-88; RETIRED. *Concurrent Pos:* Vice-chmn, Alaska Comt, Western Oil & Gas Asn, 62-63. *Mem:* Fel Geol Soc Am; Am Asn Petrol Geologists; fel AAAS; Am Inst Prof Geologists. *Res:* Discovery of oil and gas accumulations; economic analysis of oil shales and bituminous rocks of NAm; petrography of pottery fragments from ancient Troy and elsewhere. *Mailing Add:* 1438 Sorrel Rd Boulder City NV 89005

FELTS, WILLIAM JOSEPH LAWRENCE, b Saginaw, Mich, Dec 29, 24; m 46; c 3. ANATOMY. *Educ:* Univ Mich, AB, 48, AM, 51, PhD(anat), 52. *Prof Exp:* Instr anat, Ind Univ, 51-52; instr, Tulane Univ, 52-55; from asst prof to prof, Univ Minn, Minneapolis, 55-68; chmn dept, 68-75, PROF ANAT SCI, HEALTH SCI CTR, UNIV OKLA, 75- *Concurrent Pos:* Res assoc, Eng Res Inst, Univ Mich, 52; vis prof, Univ Otago, NZ, 75, Univ Autonoma de Guadalajara, 85 & Univ Ponce, PR, 90. *Mem:* Am Asn Anat; Orthop Res Soc; Am Asn Phys Anthrop; Am Soc Zool. *Res:* Growth processes in bone and cartilage; mechanical organization of bone; skeletal aging; functional anatomy and adaptation in marine mammals. *Mailing Add:* Dept Anat Sci PO Box 26901 Oklahoma City OK 73190

FELTS, WILLIAM ROBERT, b Judsonia, Ark, Apr 24, 23; m 87; c 4. INTERNAL MEDICINE, RHEUMATOLOGY. *Educ:* Univ Ark, BS, 45, MD, 46. *Prof Exp:* Intern, Garfield Mem Hosp, Washington, DC, 46-47; from jr resident to resident med, Gallinger Munic Hosp, Washington, DC, 49-51; from resident to chief resident, George Washington Univ Hosp, 51-53; asst chief arthritis res unit, Vet Admin Hosp, Washington, DC, 53-54; trainee rehab med, rheumatol, Univ Hosp, 55-57, from instr to assoc prof, 58-79, dir div rheumatol, 70-79, PROF MED, SCH MED, GEORGE WASHINGTON UNIV, 80- *Concurrent Pos:* Consult-lectr, US Naval Hosp, Md, 57-70; chief arthritis res unit, Vet Admin Hosp, 58-62; mem, Nat Comn Arthritis & Related Musculoskeletal Dis, 75-77, prof adv bd, Control Data Corp, 76-83, adv bd, Nat Arthritis, 77-84 & Nat Comt Vital & Health Statist, 83-88; pres, Am Soc Int Med, 76-77, Nat Capital Med Found, 80-81; consult, Nat Ctr Health Serv, Res & Technol Assessment, 84-; chmn, CPT ed adv panel, AMA, 80-, Coun Legis, 85-90, chmn, 85-87. *Mem:* Inst Med-Nat Acad Sci; AMA; fel Am Col Rheumatology; Am Soc Internal Med (pres, 76-77); Am Fedn Clin Res; Nat Acad Pract Med; fel Am Col Med Info. *Res:* Arthritis and rheumatic diseases; computers in medicine. *Mailing Add:* Dept Med George Washington Univ Sch Med Washington DC 20037

FELTY, EVAN J, b Columbus, Ohio, Dec 22, 32; m 58; c 3. PHYSICAL CHEMISTRY, ELECTROPHOTOGRAPHY. *Educ:* Bowling Green State Univ, BA, 54; Ohio State Univ, PhD(phys chem), 63. *Prof Exp:* Sr chemist mat res, Xerox Corp, 63-64, scientist, 64-66, mgr photoconductor systs develop, 66-68, mgr mat sci lab, 68-71, mgr photoreceptor technol, 71-73, mgr explor photoreceptors, 73-80, mgr mat sci lab, 80-87; RETIRED. *Concurrent Pos:* Mem panel tellurium, Comt Tech Aspects Critical & Strategic Mat, Nat Mat Adv Bd, Nat Res Coun-Nat Acad Sci, 69 & Comt Fundamentals Amorphous Mat, 70-71. *Mem:* AAAS; Sigma Xi; fel Am Inst Chem; Am Chem Soc. *Res:* Electrical, optical and structural properties of photoconductor materials and devices; structure of inorganic materials by x-ray diffraction techniques. *Mailing Add:* 169 Shirewood Dr Rochester NY 14625

FELTY, WAYNE LEE, b Harrisburg, Pa, Aug 27, 43; m 67; c 2. LECTURE DEMONSTRATIONS, COMPUTER APPLICATIONS. *Educ:* Lebanon Valley Col, BS, 65; Ohio State Univ, MSc, 68, PhD(analytical chem), 71. *Prof Exp:* Teaching asst analytical chem, Ohio State Univ, 65-67 & 69-70, asst instr, 67 & 69, from res asst to res assoc, 67-71; postdoctoral res asst & instr, Pa State Univ, University Park, 71-72; asst prof gen & anal chem, Mansfield State Col, Pa, 72-73; ASST PROF GEN & ANALYTICAL CHEM, PA STATE UNIV, WILKES-BARRE, 73- *Concurrent Pos:* Referee & tester, J Chem Educ, Tested Demonstrator's Feature, 75-88. *Mem:* Am Chem Soc. *Res:* Chemical education, equilibria and kinetics in solution involving transition metal complexation and protonation; textbook reviewer; computer applications in analytical chemistry and education. *Mailing Add:* Pa State Univ Wilkes-Barre Campus Box PSU Lehman PA 18627-0217

FELTZ, DONALD EVERETT, b Sherman, Tex, Aug 23, 33; m 55; c 2. NUCLEAR & MECHANICAL ENGINEERING. *Educ:* Tex A&M Univ, BS, 59, MS, 63. *Prof Exp:* Reactor supvr, 61-63, chief facil opers, 63-65, asst dir, 65-78, ASSOC DIR, NUCLEAR SCI CTR, TEX A&M UNIV, 78- *Mem:* Am Soc Mech Engrs; Am Nuclear Soc; Sigma Xi. *Res:* Investigations of neutron flux perturbations and the development on non-perturbing foils; reactor design, installation and operations; mechanical heat transfer systems; isotope production and applications; systems engineering. *Mailing Add:* 206 Fireside Circle College Station TX 77840

FELTZIN, JOSEPH, b New York, NY, Mar 2, 21; m 47; c 1. ORGANIC CHEMISTRY. *Educ:* Brooklyn Col, BA, 43, MA, 50; Polytech Inst Brooklyn, PhD(chem), 54. *Prof Exp:* Asst pharmaceut chem, E R Squibb & Sons, 47-51; res chemist, Norda Chem Co, 53-58; res chemist, Aero-Jet Gen Corp, Gen Tire & Rubber Co, 58-64, group leader mat res & develop, Struct Mat Div, 64-65; supvr new prod develop, Atlas Chem Indust, 65-75; MGR NEW PROD DEVELOP, ICI US, 75- *Mem:* Am Chem Soc; NY Acad Sci. *Res:* Epoxies; phenolics and related polymers; reinforced plastics applications; polyesters; urethane foams; elastomers; chemical additives to resins and plastics; thermoset and thermoplastic materials; polymer chemistry. *Mailing Add:* 102 E Sutton Pl Wilmington DE 19810

FEMAN, STEPHEN SOSIN, b Perth Amboy, NJ, Aug 3, 40; c 2. OPHTHALMOLOGY, RETINAL SURGERY. *Educ:* Franklin & Marshall Col, BA, 62; Univ of Pa, MD, 66. *Prof Exp:* Asst prof ophthal, Albany Med Col Union Univ, 74-78; assoc prof, 78-87, PROF OPHTHAL & DIR RETINAL SERV, SCH MED, VANDERBILT UNIV, 88- *Concurrent Pos:* Consult retinal dis, Vet Admin Hosp, Nashville, Metro Nashville Gen Hosp, 78-, NIH fel, 68-69, Heed Found fel, Jules Stein Eye Inst, 72-73, Seeing Eye Found fel, Wilmer Ophthal Inst, Johns Hopkins Univ Hosp, 73-74. *Honors & Awards:* Hon Award, Am Acad Ophthal, 84. *Mem:* AMA; Asn Res Vision & Ophthal; Am Acad Ophthal; fel Am Col Surg; Macula Soc. *Res:* Diabetic retinal disease; rhegmatogenous retinal disease; vitreo-retinal disease. *Mailing Add:* Sch Med Vanderbilt Univ Nashville TN 37232-2540

FEMENIA, JOSE, b New York, NY, May 24, 42; m 72; c 3. MARINE ENGINEERING. *Educ:* State Univ NY, BE, 64; City Col New York, MS, 67. *Prof Exp:* Asst instr, Maritime Col, State Univ NY, 64-67, instr, 67-68, from asst prof to assoc prof, 68-79, PROF ENG, MARITIME COL, STATE UNIV NY, 79-, CHMN ENG, 74- *Concurrent Pos:* Adj asst prof & res assoc, Webb Inst Naval Archit, 70-74; mem, Comt Alt Fuels Maritime Use, Maritime Transp Res Bd (NAS), 79-80, Comt Strategies to Improve Res & Develop, 85-87; vis prof, World Maritime Univ-Malno, Sweden, 83; mem, State Bd Eng & Land Surv, NY, 91- *Honors & Awards:* Bliss Award, Soc Am Mil Engrs, 90. *Mem:* Soc Naval Architects & Marine Engrs (vpres, 86-89); Am Soc Eng Educ; Soc Marine Port Engrs. *Res:* Power plant economics; power plant thermodynamics; maritime education; engineering education. *Mailing Add:* Maritime Col State Univ NY Bronx NY 10465

FENBURR, HERBERT L(ESTER), chemical engineering; deceased, see previous edition for last biography

FENCL, VLADIMIR, PULMONARY DISEASE, ACID-BASE BALANCE. *Educ:* Charles Univ, MD, 49. *Prof Exp:* ASSOC PROF MED, MED SCH, HARVARD UNIV, 62- *Mailing Add:* Dept Biol Med Sch Harvard Univ Boston MA 02115

FENDALL, ROGER K, b Newberg, Ore, Aug 20, 35; m 57; c 4. AGRONOMY. *Educ:* Ore State Univ, BS, 60; NDak State Univ, PhD(agron), 64. *Prof Exp:* Asst prof agron, Wash State Univ, 64-68; from asst prof to assoc prof, 68-77, PROF AGRON, ORE STATE UNIV, 77-, ASST DEAN SCH AGR & HEAD ADV, 70- *Mem:* Am Soc Agron. *Res:* Physiology of seed germination; mechanisms of seed dormancy, inhibition and stimulation; physiology of flowering. *Mailing Add:* Dept Crop Sci Ore State Univ Corvallis OR 97331

FENDEL, DANIEL, b New York, NY, April 30, 46. MATHEMATICS. *Educ:* Harvard Univ, BA, 66; Yale Univ, PhD(math), 70. *Prof Exp:* PROF MATH, SAN FRANCISCO STATE UNIV, 73- *Mem:* Math Asn Am. *Mailing Add:* Math Dept San Francisco State Univ 1600 Holloway Ave San Francisco CA 94132

FENDER, DEREK HENRY, b Hethe, Eng, Dec 4, 18; m 44; c 2. BIOLOGY, APPLIED SCIENCE. *Educ:* Univ Reading, BSc, 39 & 47, PhD(physics), 56. *Prof Exp:* Sr lectr physics, Royal Mil Col Sci, Eng, 46-53; lectr, Univ Reading, 53-61; assoc prof biol & elec eng, 61-66, prof biol & Appl sci, 66-, EMER PROF APPL SCI, CALIF INST TECHNOL. *Concurrent Pos:* NIH grant, Univ Reading, 56-61 & Calif Inst Technol, 61-; mem comt, Photobiol Group, 58-61; mem res bd, Dept Sci & Indust Res, Gt Brit, 59-61; consult, Electronic Color Assocs, 63- *Mem:* Human Factors Soc; Optical Soc Am; Brit Biol Eng Soc. *Res:* Interaction between the scanning motions of the human eye and the pattern recognition processes of which it is capable. *Mailing Add:* Div Appl Sci 286-80 Calif Inst Technol 1201 California Blvd Pasadena CA 91125

FENDERSON, BRUCE ANDREW, b Minneapolis, Minn, June 22, 52. GLYCOCONJUGATE RESEARCH, MONOCLONAL ANTIBODIES. *Educ:* Mich State Univ, BS, 74; Johns Hopkins Univ, PhD(develop biol), 80. *Prof Exp:* Sr fel, Univ Wash, 80-83; staff scientist, Fred Hutchinson Cancer Res Ctr, 83-87; head cell biol, Biomembrane Inst, 87-90; ASSOC PROF, THOMAS JEFFERSON UNIV, 90- *Mem:* Am Soc Cell Biol. *Res:* Mechanisms of mammalian embryonic development; role of cell surface glycolipids, glycoproteins, and proteoglycans in cell physiology and cell-cell interactions. *Mailing Add:* Dept Path & Cell Biol Thomas Jefferson Univ 1020 Locust St Rm 275 Philadelphia PA 19107

FENDLER, ELEANOR JOHNSON, b Danville, Pa, June 27, 39; div; c 2. PHYSICAL ORGANIC CHEMISTRY, BIO-ORGANIC CHEMISTRY. *Educ:* Bucknell Univ, BA, 61; Univ Calif, Santa Barbara, PhD(phys org chem), 66. *Prof Exp:* Res assoc statist, Upper Susquehanna Valley Prog Coop Res, Bucknell Univ, 61-62; teacher pub schs, Pa, 62-63; part-time instr chem, Univ Calif, Santa Barbara, 63-64, res assoc, 65-66; NASA fel, Univ Pittsburgh, 66-68, res asst prof, 68-70; vis assoc prof chem, Tex A&M Univ, 70-74, assoc prof, 74-80; mem staff, Kimberly Clark Pioneering Res, 80-; AT SOHIO RES CTR. *Concurrent Pos:* Health Res Serv Found grant, 68-70; Soc Sigma Xi grant-in-aid, 70-71; NIH career develop award, 71-76. *Mem:* Am Chem Soc; The Chem Soc; Am Asn Univ Prof; NY Acad Sci; Sigma Xi. *Res:* Bio-organic and physical organic reaction mechanisms; biomedical chemistry; interactions and catalyses in bile salt systems; drug interactions and transport;

colon cancer; micellar interactions and catalysis; nucleophilic aromatic substitution; H-1 and C-13 nuclear magnetic resonance of reactive bioorganic substrates and micellar systems. *Mailing Add:* Sohio Res Ctr 4440 Warrensville Ctr-Rd Cleveland OH 44128

FENDLER, JANOS HUGO, b Budapest, Hungary, Aug 12, 37; m 75. ORGANIC BIOCHEMISTRY. *Educ:* Leicester Univ, BSc, 60; Leicester Col Technol, Eng, dipl radiochem, 62; Univ London, PhD(phys-org chem), 64; Univ London, DSc(membrane mimetic chemistry), 78. *Prof Exp:* NSF fel phys-org chem, Univ Calif, Santa Barbara, 64-66; fel radiation chem, Radiation Res Labs, Mellon Inst Sci, Carnegie-Mellon Univ, 66-70; assoc prof, Tex A&M Univ, 70-75, prof chem, 75-81; prof chem, Clarkson Col Technol, 81-85; DISTINGUISHED PROF CHEM & DIR, CTR MEMBRANE ENG & SCI, SYRACUSE UNIV, 85- *Concurrent Pos:* Abstractor, Chem Abstr, 65-70; assoc prof, Univ Montreal. *Honors & Awards:* Nat Award, Collon or Surface Sci, Am Chem Soc, 83. *Mem:* AAAS; Royal Soc Chem; Am Chem Soc; Mat Res Soc. *Res:* Membrane mimetic chemistry; characterization and utilization for synthesis, energy conversion and drug delivery; stereochemistry in the excited state; circularly polarized laser induced excited state processes; photophysical investigations of chiral discrimination and enantiomeric recognition; picosecond spectroscopy; fluorescence detected circular dichroism and circularly polarized luminescence. *Mailing Add:* Dept Chem Syracuse Univ Syracuse NY 13210

FENDLEY, TED WYATT, b Eatonton, Ga, Oct 28, 39; m 64; c 1. CLINICAL CHEMISTRY, BIOCHEMISTRY. *Educ:* Mercer Univ, BA, 61; Auburn Univ, MS, 68, PhD(biochem), 71; Am Bd Clin Chem, cert. *Prof Exp:* Fel clin chem, Med Lab Assoc, 71-73, asst dir, 73-75, assoc dir, 75-76; regional tech dir clin lab, Med Diag Serv, 76-78; TECH DIR CLIN LABS, PATH & CYTOL LABS, 78- *Concurrent Pos:* Clin instr, Univ Ala, Birmingham, 73-76; mem adv coun, Auburn Univ Arts & Sci Health Prof, 73-76, vchmn, 76; instr, Jefferson State Jr Col, 72-76. *Mem:* Am Asn Clin Chem; Am Chem Soc; Am Asn Bioanal. *Res:* Analytical methods in endocrinology; clinical toxicology. *Mailing Add:* Roche Biomed Labs 1801 First Ave S Birmingham AL 35233-1901

FENECH, HENRI J, b Alexandria, Egypt, Mar 14, 25; US citizen; m 52; c 2. NUCLEAR ENGINEERING, ENGINEERING SCIENCES. *Educ:* Nat d'Ingenieurs des Arts et Metiers, France, Dipl Ecole, 46; Mass Inst Technol, SM, 57. *Hon Degrees:* DSc, Mass Inst Technol,59. *Prof Exp:* Engr, Foster Wheeler, France, 52-55; staff mem, Gen Atomic Div, Gen Dynamics Corp, 59-60; from asst prof to assoc prof nuclear eng, Mass Inst Technol, 60-69; PROF NUCLEAR ENG, UNIV CALIF, SANTA BARBARA, 69-, VCHMN DEPT, 70- *Concurrent Pos:* AEC res grant, 57-62; NATO sr sci fel, 69; lectr, USSR, Acad Sci, Moscow-Leningrad, 71. *Mem:* Fel Am Nuclear Soc; Am Soc Mech Engrs. *Res:* Thermal contact resistance between surfaces; analysis of nuclear reactors; methods of optimization of power systems; nuclear power safety; thermal and hydraulics in energy systems; nuclear fuel management. *Mailing Add:* Dept Chem & Nuclear Eng Univ Calif Santa Barbara CA 93106

FENG, ALBERT SHIH-HUNG, b Indonesia; US citizen; m; c 2. NEUROBIOLOGY, BIOENGINEERING. *Educ:* Univ Miami, BS, 68, MS, 70; Cornell Univ, PhD(neurobiol), 75. *Prof Exp:* Asst res neuroscientist, Univ Calif, San Diego, 74-76; trainee neurobiol, Wash Univ, 76-77; from asst prof to assoc prof physiol, 77-89, chmn, Neural & Behav Biol Prog, 87-90, PROF PHYSIOL, UNIV ILL, URBANA-CHAMPAIGN, 89-, ASSOC DIR, BECKMAN INST, 90- *Concurrent Pos:* NIH fel, 75; Alexander von Humboldt fel. *Mem:* AAAS; Soc Neurosci; Sigma Xi; Acoust Soc Am; Asn Res Otolaryngol. *Res:* Neural basis of acoustic communication; development of the nervous system. *Mailing Add:* Beckman Inst Univ Ill 405 N Mathews Urbana IL 61801

FENG, CHUAN C(HUNG) D(AVID), b Shanghai, China, Sept 15, 22; c 3. STRUCTURAL & CIVIL ENGINEERING. *Educ:* Chiao Tung Univ, BS, 45; Univ Mo, MS, 55, PhD(civil eng), 59. *Prof Exp:* Assoc prof, 63-67, PROF CIVIL ENG, UNIV COLO, BOULDER, 67- *Concurrent Pos:* Mem staff, Calif Inst Technol & Stanford Univ. *Mem:* Am Soc Civil Engrs; Am Concrete Inst; Am Soc Eng Educ. *Res:* Relaxation method for structural problems; optimization; flow graph analysis; systems engineering analysis. *Mailing Add:* Civil Engr Box 428 Univ Colo Boulder Boulder CO 80309-0428

FENG, DA-FEI, b Shanghai, China; US citizen. RADIATION CHEMISTRY. *Educ:* DePauw Univ, BA, 67; Wayne State Univ, PhD(radiation chem), 73. *Prof Exp:* Fel radiation chem, Wayne State Univ, 73-74; fel radiochem, Univ Calif, Davis, 74-76; FEL RADIOCHEM, SAN DIEGO MESA COL, 76- *Res:* Theoretical studies of high energy reactions. *Mailing Add:* Dept Phys Sci San Diego Mesa Col 7250 Mesa Col Dr San Diego CA 92111

FENG, DAHSUAN, b New Delhi, India; US citizen; m 72; c 2. THEORETICAL NUCLEAR PHYSICS, THEORETICAL QUANTUM OPTICS. *Educ:* Drew Univ, BA, 68; Univ Minn, Minneapolis, PhD(physics), 72. *Prof Exp:* UK Sci Res Coun res fel theoret physics, Univ Manchester, 72-74; res assoc, Ctr Nuclear Studies, Univ Tex, Austin, 74-76; prog dir, Theoret Physics, NSF, 83-85; PROF PHYSICS & ATMOSPHERIC SCI, DREXEL UNIV, 86- *Concurrent Pos:* Chmn Planning Comt, Franklin Inst. *Mem:* Southeast Asia Theoret Physics Asn; Sigma Xi; Am Phys Soc. *Res:* Theoretical studies of nuclear reaction theories; nuclear structure; statistical mechanics and thermodynamics; foundations of Quantum mechanics; nonlinear dynamics. *Mailing Add:* Dept Physics & Atmospheric Sci Drexel Univ Philadelphia PA 19104-9984

FENG, JOSEPH SHAO-YING, b Peiping, China; US citizen. DATA STORAGE, MICROMAGNETICS. *Educ:* Calif Inst Technol, BS, 69, Northwestern Univ, MS, 70; Calif Inst Technol, MS, 71, PhD(elec eng), 75. *Prof Exp:* Res staff mem, T J Watson Res Lab, Yorktown Heights, NY, 74-77, SCIENTIST/ENGR, GEN PROD DIV & STORAGE SYSTS PROD DIV, IBM, 77- *Mem:* Am Phys Soc; Inst Elec & Electronics Engrs. *Res:* Magnetic recording; recording heads; analog and digital electronics. *Mailing Add:* 4251 Norwalk Dr Apt 302 San Jose CA 95193

FENG, PAUL YEN-HSIUNG, b Peking, China, Aug 29, 26; nat US; m 47; c 3. NUCLEAR CHEMISTRY, RADIATION CHEMISTRY. *Educ:* Cath Univ, China, BS, 47; Wash Univ, PhD(chem), 54. *Prof Exp:* Asst chem, Wash Univ, 50-51, res chemist geochem, 51-54; chemist, Manu Mine Res & Develop Co, 54, chief chemist, 54-55, tech dir, 55; assoc physicist, IIT Res Inst, 55-56, res physicist, 56-57, group leader, 57-58, asst supvr nuclear physics, 58-59, supvr chem physics, 59-62, sci adv, 62-67; from assoc prof to prof chem, Marquette Univ, 67-88; RETIRED. *Concurrent Pos:* Asst, AEC Res Prog, Wash Univ, 53-54; lectr, Ill Inst Technol, 56-58, adj assoc prof, 66-67; vis prof, Tsinghus Univ Inst Nuclear Sci, Formosa, 58; tech adv US deleg, Int Conf Peaceful Uses Atomic Energy, Geneva, Switz, 58; Fulbright lectr, Taiwan Nat Univ, 65. *Mem:* AAAS; The Chem Soc; Am Asn Physics Teachers; Am Chem Soc; Radiation Res Soc. *Res:* Radiation effects; electric discharge and electron impact phenomena; high polymer physics; mass spectrometry; geochemistry and silicate chemistry; reactions at extreme temperatures. *Mailing Add:* 88 Robsart Rd Kenilworth IL 60043

FENG, RONG, b Shouguang Shandong Prov, China, Feb 19, 62. MASS SPECTROMETRY, BIOCHEMICALS ANALYSIS. *Educ:* Lanzhou Univ, BS, 82; Cornell Univ, MS, 84, PhD (chem), 88. *Prof Exp:* Res fel environ carcinogenesis, Am Health Found, Naylor Dana Inst Dis Prev, 88-89; ASSOC RES OFFICER, PROTEIN ENG, NAT RES COUN CAN, BIOTECH RES INST, 89- *Concurrent Pos:* Vis scientist, Sciex, Toronto, 89. *Mem:* Am Chem Soc; Am Soc Mass Spectrometry; AAAS. *Res:* Develop and apply mass spectrometric methods for chemical analysis with emphasis on solving problems of biological and environmental significance; gas phase chemistry of ions and unconventional neutrals. *Mailing Add:* Nat Res Council Can Biotech Res Inst 6100 Royalmont Ave Montreal PQ H4P 2R2 Can

FENG, SUNG YEN, b Shanghai, China, Oct 1, 29; US citizen; m 63; c 1. PHYSIOLOGICAL ECOLOGY. *Educ:* Univ Taiwan, BS, 54; Col William & Mary, MA, 58; Rutgers Univ, PhD(parasitol), 62. *Prof Exp:* Res asst, Dept Zool & NJ Oyster Res Lab, Rutgers Univ, 60-62, res assoc, NJ Oyster Res Lab, 62-66; asst prof systs & environ biol, Marine Sci Inst & Biol Sci Group, Univ Conn, 66-68, assoc prof biol, 68-74, asst dir, Marine Sci Inst, 72-75, dir, Marine Sci Inst, 75-85, prof, Marine Sci Head, Marine Sci Dept, 77-85, PROF BIOL SCI, MARINE SCI INST, 74- *Mem:* Am Soc Parasitol; Am Soc Protozool; Nat Shellfisheries Asn; Soc Invert Path; Am Soc Zool; Sigma Xi. *Res:* Invertebrate pathobiology; diseases, pathology and defense mechanisms of marine molluscs; physiological ecology of marine molluscs; pathobiology of invertebrates. *Mailing Add:* Marine Sci Univ Conn Main Campus Storrs CT 06268

FENG, TSE-YUN, b Hangzhou, China, Feb 26, 28; US citizen; m 65; c 4. INTERCONNECTION NETWORKS, SEARCH ALGORITHMS. *Educ:* Nat Taiwan Univ, BS, 50; Okla State Univ, MS, 57; Univ Mich, PhD (computer eng), 67. *Prof Exp:* From asst prof to assoc prof computer eng, Syracuse Univ, 67-75; prof, Wayne State Univ, 75-79; prof computer sci, Ohio State Univ, 80-84; BINDER PROF COMPUTER ENG, PA STATE UNIV, 84- *Concurrent Pos:* Ed in chief, Inst Elec & Electronics Engrs Trans Computers, 82-86 & Trans Parallel & Distrib Systs, 90-; mem, Tech Act Bd, Inst Elec & Electronics Engrs, 79-80, Publ Bd, 83-85, chmn, Distinguished Visitors Prog, Computer Soc, 87-; mem, bd dirs, Am Fedn Info Processing Socs, 79-87. *Honors & Awards:* Centennial Medal, Inst Elec & Electronics Engrs, 84; R E Merwin Distinguished Serv Award, Inst Elec & Electronics Engrs, Computer Soc, 85, Meritorious Serv Award, 87. *Mem:* Fel Inst Elec & Electronics Engrs; Inst Elec & Electronics Engrs Computer Soc (vpres, 78, pres, 79-80); Am Fedn Info Processing Socs. *Res:* Parallel processors and processing; parallel processor architecture; parallel algorithms; sorting and sorting networks; interconnection networks; search and retrieval algorithms; associative memories and processors; distributed logic systems; switching theory and logic design. *Mailing Add:* 319 Christopher Lane State College PA 16803

FENICHEL, GERALD M, b New York, NY, May 11, 35; m 58; c 3. NEUROLOGY. *Educ:* Johns Hopkins Univ, AB, 55; Yale Univ, MD, 59. *Prof Exp:* Fel neurol, Sch Med, Yale Univ, 63-64. Prof Exp: From instr to asst prof neurol, Sch Med, George Washington Univ, 64-69; PROF NEUROL, SCH MED, VANDERBILT UNIV, 69- *Mem:* Am Acad Neurol; Am Neurol Asn; AMA; Am Acad Cerebral Palsy. *Res:* Muscle development; neuromuscular diseases of infancy and childhood. *Mailing Add:* Dept Neurol Vanderbilt Hosp 2100 Pierce Ave Nashville TN 37212

FENICHEL, HENRY, b The Hague, Neth, Apr 13, 38; US citizen; m 61; c 2. LOW TEMPERATURE PHYSICS, HOLOGRAPHY. *Educ:* Brooklyn Col, BS, 60; Rutgers Univ, MS, 62, PhD(physics), 65. *Prof Exp:* From asst prof to assoc prof, 65-76, PROF PHYSICS, UNIV CINCINNATI, 76- *Mem:* Optical Soc Am; Am Phys Soc; Am Asn Physics Teachers. *Res:* Lattice dynamics; measurements of specific heats of inert gas solids; thermal properties of solids at low temperatures; holographic interferometry with application to diffusion in liquids. *Mailing Add:* Dept Physics Univ Cincinnati Cincinnati OH 45221-0011

FENICHEL, RICHARD LEE, b New York, NY, July 23, 25; m 51; c 2. BIOCHEMISTRY, PHYSIOLOGY. *Educ:* NY Univ, AB, 47; Polytech Inst Brooklyn, MS, 51; Wayne State Univ, PhD(physiol, biochem), 56. *Prof Exp:* Biochemist, Med Dept, Chrysler Corp, 51-54; asst, Wayne State Univ, 54-56, res assoc, 56-57; investr, Aviation Med Accelerator Lab, 57-59; sr res scientist, Ortho Res Found, 59-63; SR RES FEL, WYETH-AYERST LABS, 63- *Honors & Awards:* Angus McClean Award, Wayne Univ, 56. *Mem:* AAAS; Am Chem Soc; Am Soc Biol Chemists; NY Acad Sci. *Res:* Protein isolation and characterization; blood coagulation; enzymatic studies; mechanisms of protein interactions; related animal models. *Mailing Add:* Wyeth-Ayerst Res CN 8000 Princeton NJ 08543-8000

FENIMORE, DAVID CLARKE, b Evansville, Ind, Jan 27, 30; m 65. ANALYTICAL CHEMISTRY, BIOCHEMISTRY. *Educ:* DePauw Univ, BA, 52; Univ Houston, PhD(chem), 66. *Prof Exp:* Chemist, Thomas & Skinner, Inc, 54-57; sect leader instrumental anal, Baroid Div, Nat Lead Co, 57-64; fel chem, Univ Houston, 66-67; sect head anal chem, Tex Res Instrumental Sci, 67-69, div head instrumental anal, 69-81; PRES, CLARK ANALYTICAL SYSTS, 81- *Concurrent Pos:* Instr, Grad Sch Biomed Sci, Univ Tex, 67-69, asst prof, 69-; clin asst prof, Dept Biophys, Univ Houston, 70-; mem biomed res review comt, Nat Inst Drug Abuse. *Mem:* AAAS; Am Chem Soc; fel Am Inst Chemists; NY Acad Sci; Am Soc Mass Spectrometry; Sigma Xi. *Res:* Gas chromatographic instrumentation with emphasis on electron capture detector design and studies of electron capture processes; application of electron capture detection to ultramicro biochemical analysis; development of high performance thin-layer chromatographic instrumentation. *Mailing Add:* PO Box 744 Sierra Madre CA 91024

FENN, H(OWARD) N(ATHAN), b Milford, Conn, Nov 11, 07; m 48; c 1. CHEMICAL ENGINEERING. *Educ:* Yale Univ, BS, 29. *Prof Exp:* Chem engr org chem, Dow Chem Co, 29-37, develop group engr cellulose prods, 37-43, supt silicone prod & mgr silicone mfg, Dow Corning Corp, 44-62, dir mfg, 62-63, vpres mfg, 63-65, vpres eng & mfg, 65-71, vpres & asst to pres, 71-72; MGT CONSULT, 73- *Concurrent Pos:* Consult, Tri-City Plastics Inc, & Plastic Prod Div, Kardon Industs, Inc. *Mem:* AAAS; Am Inst Chem Engrs; Am Chem Soc; NY Acad Sci. *Res:* Manufacture of silicones; development of silicone products; industrial management; industrial safety. *Mailing Add:* 2502 South Point Dr Midland MI 48640

FENN, JOHN BENNETT, b New York, NY, June 15, 17; m 39; c 3. CHEMISTRY. *Educ:* Berea Col, AB, 37; Yale Univ, PhD(phys chem), 40. *Prof Exp:* Res chemist, Monsanto Chem Co, Ala, 40-42 & Sharples Chem, Inc, Mich, 43-45; vpres & res supvr, Exp, Inc, Va, 45-52; dir proj Squid, Forrestal Res Ctr, Princeton Univ, 52-62, lectr aerospace & mech sci, 59-60, prof, 60-67; PROF APPL SCI & CHEM, YALE UNIV, 67- *Honors & Awards:* Sr Scientist Award, Alexander von Humboldt Found, 84. *Mem:* AAAS; Am Chem Soc; Combustion Inst; Am Inst Chem Engrs. *Res:* Combustion thermodynamics; propulsion; kinetics; molecular beams; rarefied gas dynamics. *Mailing Add:* Dept Chem Eng PO Box 2159 Yale Sta New Haven CT 06520-2159

FENN, ROBERT WILLIAM, III, b Philadelphia, Pa, Feb 1, 41; m 64; c 3. CHEMICAL ENGINEERING. *Educ:* Villanova Univ, BE, 62; Carnegie-Mellon Univ, MS, 65; Univ Rochester, PhD(chem eng), 68. *Prof Exp:* Sr res engr, 67-76, res supvr electrochem res & develop, 76-77, group leader, 77-78, mgr electrochem res & develop, 78-81, SPEC PROJS LEADER, DIAMOND SHAMROCK CORP, 81- *Mem:* Am Inst Chem Engrs; Electrochem Soc. *Res:* Application of electrochemistry and electrochemical engineering principles to water and air pollution control and energy storage and conservation as well as research and development on industrial electrochemical processes. *Mailing Add:* 6335 Coleridge Rd Painesville OH 44077

FENNA, ROGER EDWARD, b Stafford, Eng, Jan 6, 47; m 72. MOLECULAR BIOLOGY, X-RAY CRYSTALLOGRAPHY. *Educ:* Univ Leeds, BSc, 69; Univ Oxford, DPhil(molecular biophys), 73. *Prof Exp:* Res assoc molecular biol, Univ Ore, 73-76; res assoc molecular biol, Univ Calif, Los Angeles, 76-80; MEM FAC, MED SCH, UNIV MIAMI, 80- *Mem:* Fedn Am Socs Exp Biol. *Mailing Add:* 7055 SW 84th Ave Miami FL 33143

FENNEL, WILLIAM EDWARD, b Moberly, Mo, Mar 4, 23. INVERTEBRATE ZOOLOGY. *Educ:* Univ Mo, AB, 46, MA, 49; Univ Mich, PhD(zool), 59. *Prof Exp:* Asst zool, Univ Mo, 46-49; instr, Eastern Ill State Col, 48, Univ Mo, 49-50 & biol, Flint Jr Col, Mich, 50-53; from instr to asst prof, Brooklyn Col, 58-67; from assoc prof to prof, Pace Col, 67-70; PROF BIOL, EASTERN MICH UNIV, 70- *Mem:* Am Micros Soc; NAm Benthological Soc; Ecol Soc Am; Am Soc Zool; Sigma Xi. *Res:* Aquatic invertebrate ecology. *Mailing Add:* Dept Biol Eastern Mich Univ Ypsilanti MI 48197

FENNELL, ROBERT E, b Peoria, Ill, Apr 21, 42; m 69; c 3. ANALYSIS & FUNCTIONAL ANALYSIS. *Educ:* Bradley Univ, BA, 64; Univ Iowa, MS, 66, PhD(math), 69. *Prof Exp:* Instr math, Grinnell Col, 68-69; from asst prof to assoc prof math, 69-86, PROF MATH SCI, CLEMSON UNIV, 86- *Concurrent Pos:* Sabbatical leave, Langley Res Ctr, NASA, 79-80; vis scientist, Inst Comput Appln Sci & Eng, 86-87. *Mem:* Am Math Soc; Soc Indust & Appl Math; Math Asn Am; Sigma Xi; Inst Elec & Electronics Engrs; Am Inst Aeronaut & Astronaut. *Res:* Differential equations; control theory. *Mailing Add:* 413 Skyview Dr Clemson SC 29631

FENNELLY, PAUL FRANCIS, b Frackville, Pa, Aug 1, 45; m 70; c 4. POLLUTION CONTROL. *Educ:* Villanova Univ, BS, 67; Brandeis Univ, MA, 68, PhD(chem), 72. *Prof Exp:* Teaching fel phys chem, Brandeis Univ, 67-72, Ctr Res Exp Space Sci, York Univ, Toronto, 72-73; phys chemist environ sci, Aero Chem Res Lab, Princeton, NJ, 73-74; prog mgr & prin scientist energy & environ sci, GCA Corp, 74-85; gen mgr, Air Toxics Sci & Eng, Environ Res & Technol, Inc, 85-88; VPRES, ENSR CORP, 88- *Concurrent Pos:* Nat Res Coun Can fel, 72-73; vis lectr air pollution sci, Univ Lowell, 77-78. *Mem:* Air Pollution Control Asn; Am Chem Soc; Am Inst Chem Engrs. *Res:* Fluidized bed combustion; coal combustion; air pollution control; incineration of chemical waste; atmospheric chemistry; air toxics releases. *Mailing Add:* 97 Gray Arlington MA 02174

FENNEMA, OWEN RICHARD, b Hinsdale, Ill, Jan 23, 29; m 48; c 3. FOOD SCIENCE. *Educ:* Kans State Univ, BS, 50; Univ Wis, MS, 51, PhD(food sci), 60. *Prof Exp:* Proj leader food process res, Pillsbury Co, Minn, 54-57; from asst prof to assoc prof, 60-69, chmn dept, 77-81, PROF FOOD SCI, UNIV WIS, MADISON, 69- *Honors & Awards:* Cruess Award; Carl Fellers Award; Nicholas Appert Award. *Mem:* Fel Inst Food Technol (pres, 82-83); fel Am Chem Soc; Am Dairy Sci Asn; Soc Cryobiol; Am Inst Nutrit. *Res:* Low temperature preservation; protein alteration during processing; edible packaging films, fiber-lipid interactions. *Mailing Add:* Dept Food Sci Babcock Hall Univ Wis 1605 Linden Dr Madison WI 53706

FENNER, HEINRICH, animal nutrition; deceased, see previous edition for last biography

FENNER, PETER, b Zurich, Switz, Oct 2, 37; US citizen; c 3. GEOLOGY, SCIENCE EDUCATION. *Educ:* City Col New York, BS, 59; Univ Ill, MS, 61, PhD(sedimentology & clay mineral), 63. *Prof Exp:* From instr to asst prof geol, Univ Pa, 63-67; from assoc dir to exec dir, Coun Educ Geol Sci, Am Geol Inst, 67-70; from asst dean to dean, Col Environ & Appl Sci, Governors State Univ, 70-79, prof geol, 70-81, spec asst to provost, 79-81; VCHANCELLOR ACAD AFFAIRS & PROF GEOL SCI, PURDUE UNIV, CALUMET, 81- *Concurrent Pos:* Chmn instructional mat panel, Coun Educ Geol Sci; adv geol ed, Appleton-Century-Crofts; scientist, Smithsonain Inst-Coast Guard Oceanog Res Cruises; mem bd adv, Nat Study Math Req of Scientists & Engrs. *Mem:* AAAS; Geol Soc Am; Soc Econ Paleont & Mineral; World Future Soc; Sigma Xi. *Res:* Application of quantitative methods to geology, particularly in clay mineral and trace element studies; environmental and earth-science education; oceanography; higher education planning, especially implementing basic skills; developmental studies programs. *Mailing Add:* 682 Windings Lane Cincinnati OH 45220-1083

FENNER, WAYNE ROBERT, b Butte, Mont, Aug 6, 39; m 61; c 3. INTEGRATED OPTICS, OPTICAL COMMUNICATIONS. *Educ:* Univ Calif, Berkeley, BS, 62; Univ Ill, MS, 64, PhD(physics), 69. *Prof Exp:* Res assoc physics, Univ Southern Calif, 69-71; mem tech staff elec eng, 72-80, mgr signal processing, 80-81, HEAD, DEPT LASERS & OPTICS, ELECTRONICS RES LAB, AEROSPACE CORP, 81- *Concurrent Pos:* Lectr, Univ Calif, Los Angeles, 77; instr, El Camino Col, 78-79. *Mem:* Am Phys Soc; Inst Elec & Electronics Engrs. *Res:* Laser raman spectroscopy; ultrasonic imaging; digital signal processing; adaptive antenna arrays; integrated optics; optical communications. *Mailing Add:* Electronics Res Lab/MZ-246 Aerospace Corp 2350 E El Segundo Blvd El Segundo CA 90245

FENNER-CRISP, PENELOPE ANN, b Milwaukee, Wis, Apr 18, 39; m 65; c 2. NEUROTOXICOLOGY, REPRODUCTIVE TOXICOLOGY. *Educ:* Univ Wis, Milwaukee, BS, 62; Univ Tex Med Br Galveston, MS, 64, PhD(pharmacol), 68. *Prof Exp:* Fel pharmacol-morphol, Sch Med & Dent, Georgetown Univ, 71-73, adj instr, 73-74, res assoc pharmacol, 76-78; pharmacologist, 78-80, SR TOXICOLOGIST REGULATORY AFFAIRS, US ENVIRON PROTECTION AGENCY, 80- *Concurrent Pos:* Consult, USV Pharmaceut Co, Tuckahoe, NY, 70; vis scientist, Dept Physiol, Univ Birmingham, 74-75. *Mem:* Sigma Xi; AAAS. *Res:* Regulations and non-regulatory health effects guidance concerning chemical contamination of drinking water. *Mailing Add:* 5920 35th St N Arlington VA 22207

FENNESSEY, PAUL V, b Oklahoma City, Okla, Oct 3, 42; m 62; c 3. MASS SPECTROMETRY. *Educ:* Univ Okla, BS, 64; Mass Inst Technol, PhD(chem), 68. *Prof Exp:* Fel, Mass Inst Technol, 64-69; sr res scientist, Monsanto Corp, 68-69; asst prog scientist, Martin Marietta Corp, 69-72, prog scientist, 72-74; from asst prof to assoc prof, 75-91, PROF PEDIAT PHARMACOL, SCH MED, UNIV COLO, 91- *Mem:* Am Soc Mass Spectros; AAAS; Soc Inherited Metab Dis; Am Chem Soc; Am Soc Pharmacol & Exp Therapeut. *Res:* Application of mass spectrometry to clinical medicine; inborn errors of metabolism; steroid imbalances; trace elements in human nutrition. *Mailing Add:* Mass Spectra Res Resource Med Sch Univ Colo 4200 Ninth Ave Denver CO 80262

FENNESSY, JOHN JAMES, b Clonmel, Ireland, Mar 8, 33; m 60; c 7. RADIOLOGY. *Educ:* Nat Univ Ireland, MB, BCH & BAO, 57. *Prof Exp:* From instr to assoc prof radiol, 63-74, chief chest & gastrointestinal radiol, 73-74, actg chief diag radiol, 74, PROF RADIOL & CHMN DEPT, UNIV CHICAGO HOSPS & CLINS, 74- *Honors & Awards:* McClintock Award; H Cline Fixott Sr Mem lectr, Am Acad Dent Radiol. *Mem:* Am Asn Univ Radiologists; Am Gastroenterol Soc; Am Med Asn; Am Col Cardiol; Am Col Radiol; Sigma Xi; Am Heart Asn; Radiol Soc NAm; Soc Thoracic Radiol; hon fel Royal Col Surgeons Ireland. *Mailing Add:* Dept Radiol Univ Chicago Hosp & Clin Chicago IL 60637

FENNEWALD, MICHAEL ANDREW, b Jefferson City, Mo, Sept 8, 51; m 74. MICROBIOLOGY. *Educ:* Carleton Col, BA, 73; Univ Chicago, PhD(microbiol), 79. *Prof Exp:* Fel biochem, Univ Chicago, 79-81; asst prof microbiol, Univ Notre Dame, 81-88; ASSOC PROF, UNIV HEALTH SCI/CHICAGO MED SCH, 88- *Mem:* Am Soc Microbiol; Genetics Soc Am; AAAS. *Res:* Transposons and insertion elements; nucleic acid enzymes; site-specific recombination. *Mailing Add:* Dept Microbiol & Immunol Univ Health Sci Chicago Med Sch 3333 Green Bay Rd North Chicago IL 60064

FENNEY, NICHOLAS WILLIAM, b New Haven, Conn, July 18, 06; m 30; c 2. PHARMACY. *Educ:* Columbia Univ, PhG, 25; Univ Conn, PhC, 30, BS, 42; Yale Univ, MPH, 46. *Prof Exp:* From instr to prof, 25-68, EMER PROF PHARM, UNIV CONN, 68- *Concurrent Pos:* Vis lectr, Dept Pharmacol, Sch Med, Yale Univ, 35-42, Dept Pub Health, 48, Cancer Control Sect, 48-49; mem, Conn Adv Comt Foods & Drugs, 50-; estab Nicholas W Fenney scholar, Conn Pharm Asn, 65; mem, Conn Regional Med Prog, 66-70 & Conn Comprehensive Health Planning Coun, 68-72; pharmaceut consult, Conn Blue Cross/Blue Shield, 68-, mem, 71- *Honors & Awards:* Sydney Rome Achievement Award, 64; Bowl of Hygeia Award, 69; Nard-Lederle Nat Interprof Serv Award, 69. *Mem:* Am Pharmaceut Asn; Am Pub Health Asn; fel Am Col Apothecaries. *Res:* Detoxication of toxic chemicals with vitamin C; bacteriology; sanitation; public health. *Mailing Add:* 62 Broadfield Rd Hamden CT 06517

FENOGLIO, CECILIA M, b New York, NY, Nov 28, 43; m; c 4. PATHOLOGY, IMMUNOLOGY. *Educ:* Col Saint Elizabeth, BS, 65; Sch Med, Georgetown Univ, MD, 69. *Prof Exp:* Instr path, Col Physicians & Surgeons, Columbia Univ, 73-74, asst prof, 74-77; co-dir, path, Col Physicians & Surgeons, Columbia Univ, 78- 83, prof, 82-83; chief, Lab Serv, Vet Admin Med Ctr, Albuquerque, 83-90; DIR, ELECTRON MICRO LAB, INT INST

HUMAN REPRODUCTION, 78-; CHMN, DEPT PATH & LAB MED, UNIV CINCINNATI, 90- *Concurrent Pos:* Fel immunol, Mem Sloan Kettering Cancer Ctr, 73; asst attend pathologist, 74-77, assoc attend pathologist, Presby Hosp, 77-; mem, path comt, Nat Cancer Inst, Div Cancer Control & Rehab, 75-, Nat Ileitis & Colitis Found, 78-, NIH study sect, cancer control intervention, 80- & path B; mem, Nat Adv Comt Colorectal Cancer, Am Cancer Soc, Career Develop Award Grant Comt, Sci Adv Comt Clin Invests, Grants Rev Comt, Prev Diag & Ther. *Honors & Awards:* Timely Topics lectr, US Can Asn Pathologists. *Mem:* Am Asn Pathologists; Int Acad Path (pres, 89-90); NY Acad Med; NY Acad Sci; Sigma Xi; AMA. *Res:* Problems in cancer research and tumor markers; gastrointestinal cancer. *Mailing Add:* Dept Path & Lab Med Univ Cincinnati Cincinnati OH 45221

FENOGLIO, RICHARD ANDREW, b Joliet, Ill, Jan 4, 41; m 65; c 2. ORGANIC CHEMISTRY. *Educ:* Univ Ill, BS, 62; Yale Univ, PhD(org chem), 67. *Prof Exp:* RES CHEMIST, E I DU PONT DE NEMOURS & CO, INC, 67- *Mem:* Am Chem Soc. *Res:* Study of solvolyses and thermochemistry of small ring compounds; study of dye chemistry; study of titanium dioxide pigments. *Mailing Add:* 1024 Radley Dr W Chester PA 19382-8089

FENRICH, RICHARD KARL, b Painesville, Ohio, Nov 15, 62; m 85; c 1. CHARACTER RECOGNITION, PROTOTYPE SYSTEM DESIGN. *Educ:* Allegheny Col, BS, 85; Ohio State Univ, MS, 87; State Univ NY, Buffalo, MS, 89. *Prof Exp:* Software qual engr, Bell Aerospace, Textron, 88-89; SR RES SCIENTIST, RES FOUND STATE UNIV NY, BUFFALO, 89- *Concurrent Pos:* Co-prin investr three res projs US Postal Serv, Res Found, State Univ NY, Buffalo, 89- *Mem:* Inst Elec & Electronics Engrs; Asn Comput Mach. *Res:* Development of state-of-the-art alphanumeric character recognition algorithms; automatic text interpretation tools; real time systems; to enable automatic processing of handwritten and machine printed mail. *Mailing Add:* Dept Computer Sci 226 Bell Hall State Univ NY Buffalo Buffalo NY 14260

FENRICK, HAROLD WILLIAM, b Janesville, Wis, Mar 31, 35; m 63; c 3. PHYSICAL CHEMISTRY. *Educ:* Beloit Col, BS, 57; Univ Wis-Madison, PhD(chem), 65. *Prof Exp:* Instr chem, Carleton Col, 65-66; asst prof, Monmouth Col, Ill, 66-68; assoc prof, 68-80, PROF CHEM, UNIV WIS-PLATTEVILLE, 80- *Mem:* Am Chem Soc. *Res:* Radiation chemistry of solids; electron spin resonance studies of free radicals. *Mailing Add:* Dept Chem Wis State Univ Platteville WI 53818

FENRICK, MAUREEN HELEN, b Toronto, Ont, Feb 8, 46; US citizen. ALGEBRA. *Educ:* Edgewood Col, BS, 67; Northern Ill Univ, MS, 69; Univ Fla, PhD(math), 73. *Prof Exp:* Instr, Wichita State Univ, 73-74, asst prof math, 74-87; ASSOC PROF MATH, MANKATO STATE UNIV, 87- *Concurrent Pos:* Vis asst prof math, Colo Col, 81-82. *Mem:* Am Math Soc. *Res:* Noncommutative ring theory; author and co-author of several books. *Mailing Add:* Dept Math Mankato State Univ Mankato MN 56001

FENSELAU, ALLAN HERMAN, b Philadelphia, Pa, May 17, 37; div; c 2. ORGANIC CHEMISTRY. *Educ:* Yale Univ, BS, 58; Stanford Univ, PhD(org chem), 64. *Prof Exp:* Teacher, Univ Sch, 58-60; fel org chem, Inst Molecular Biol, Syntex Res Ctr, Calif, 64-65; fel biochem, Univ Calif, Berkeley, 65-67; from asst prof to assoc prof physiol chem, Johns Hopkins Univ, 73-81; sr scientist, Papanicolaou Cancer Res Inst, 81-; assoc prof anat & cell biol, Sch Med, Univ Miami, 81-; RES BIOCHEMIST, EMERSON FARMS, BALTIMORE, MD. *Concurrent Pos:* USPHS res career develop award, 72-77. *Honors & Awards:* Myers Honor Award Ophthal. *Mem:* AAAS; Am Chem Soc; Royal Soc Chem; Am Soc Biol Chemists. *Res:* Protein chemistry; cancer; biochemistry of angiogenesis. *Mailing Add:* Emerson Farms-Unit 2 800 Greenspring Valley Rd Timonium MD 21093

FENSELAU, CATHERINE CLARKE, b York, Nebr, Apr 15, 39; m 62; c 2. BIOORGANIC CHEMISTRY, MASS SPECTROMETRY. *Educ:* Bryn Mawr Col, AB, 61; Stanford Univ, PhD(org chem), 65. *Prof Exp:* Res chemist, NASA, Univ Calif, Berkeley, 66-67; from instr to prof pharmacol, Sch Med, Johns Hopkins Univ, 67-87; PROF & CHMN, DEPT CHEM, UNIV MD, BALTIMORE COUNTY, 87- *Concurrent Pos:* Am Asn Univ Women fel, Univ Calif, Berkeley, 65-66; USPHS res career develop award, 72-77; ed-in-chief, Biomed Mass Spectrometry, 73-89; consult, Med Chem Study Sect, NIH, 75-79; vis prof, Univ Warwick, 80, Kansai Univ, Osaka, 86. *Honors & Awards:* Garvan Medal, Am Chem Soc, 85; Outstanding Medal Chem, Am Chem Soc, 89. *Mem:* AAAS; Am Chem Soc; Am Soc Mass Spectrometry; Am Soc Pharmacol & Exp Therapeut. *Res:* Biomedical applications of mass spectrometry; drug metabolism; chemistry of gaseous ions; post translational modifications of proteins. *Mailing Add:* Dept Chem Univ Md Baltimore MD 21228

FENSKE, GEORGE R, b Green Bay, Wis. TRIBOLOGICAL COATINGS, X-RAY OPTICAL COATINGS. *Educ:* Univ Ill, BS, 72, MS, 75, PhD(nuclear eng), 80. *Prof Exp:* Nuclear engr, 79-86, metallurgist, 86-89, MAT SCIENTIST, ARGONNE NAT LAB, 89- *Mem:* Am Soc Metals; Soc Tribologists & Lubrication Engrs; Am Phys Soc. *Res:* Uses of advanced coating processes for friction, wear, and x-ray optical applications. *Mailing Add:* MCT-212 Argonne Nat Lab Argonne IL 60439

FENSKE, PAUL RODERICK, b Ellensberg, Wash, May 15, 25; m 52; c 4. HYDROGEOLOGY. *Educ:* SDak Sch Mines & Technol, BS, 50; Univ Mich, MS, 51; Univ Colo, PhD(geol), 63. *Prof Exp:* Geologist, Magnolia Petrol Co, 51-56 & Delfern Oil Co, 56-59; asst prof geol, Idaho State Univ, 63-65; mgr earth sci & eng, Teledyne Isotopes Palo Alto Labs, 65-71; res assoc, 71-73, asst exec dir, 79-80, dep exec dir, 80-82, actg exec dir, 82-83 RES PROF, WATER RES CTR, DESERT RES INT, UNIV NEV, RENO, 73-, EXEC DIR, 83- *Concurrent Pos:* Consult hydrogeol, mining hydrol & waste mgt. *Mem:* Am Inst Mining Engrs; Am Geophys Union; Am Water Resources Asn; Sigma Xi. *Res:* Geochemistry of sedimentary rocks; hydrogeochemistry; hydrologic systems analysis; groundwater transport of contaminants; well hydraulics; porous media flow. *Mailing Add:* Water Res Ctr Desert Res Inst Univ Nev Syst Reno NV 89507

FENSOM, DAVID STRATHERN, b Toronto, Ont, Apr 10, 16; m 44; c 2. PLANT BIOPHYSICS & PHYSIOLOGY, EDUCATION PHILOSOPHY. *Educ:* Univ Toronto, BASc, 38. *Hon Degrees:* FRSChem(UK), 53; LLD, Dalhousie Univ, 88. *Prof Exp:* Master sci & head dept, Ridley Col, Ont, 46-63; from assoc prof to prof biol, 63-88, head dept, 69-79, EMER PROF BIOL, MT ALLISON UNIV, 88- *Concurrent Pos:* Nuffied Travel Sch, 60-61; Sr Killim res fel, 73-74. *Mem:* Fel Royal Inst Chem; fel Royal Soc Arts; Can Soc Plant Physiologists (secy-treas, 69-71, pres, 80); Brit Soc Exp Biol. *Res:* Electroosmosis, electrophysiology; long distance transport in plants; membrane phenomena; Adam architecture. *Mailing Add:* Dept Biol Mt Allison Univ Sackville NB E0A 3C0 Can

FENSTER, AARON, b Munich, WGer, Sept 12, 47; Can citizen; c 1. MEDICAL PHYSICS, RADIOLOGY. *Educ:* Univ Toronto, BSc, 71, MSc, 73, PhD(med biophys), 76. *Prof Exp:* DIR, IMAGING RES LABS, JOHN P ROBARTS RES INST, 87-; PROF, DEPT MED BIOPHYS & DEPT DIAG RADIOL & NUCLEAR MED, UNIV WESTERN ONT, 88- *Concurrent Pos:* Career scientist award, Ont Ministry Health, 83-87; Adj prof, Dept Physics, Univ Western Ont, 88-, hon lectr, Dept Oncol & chmn, Biomed Eng Res & Grad Studies Group, Fac Eng, 90-; hon lectr, Can Col Physicists Med, 89- *Honors & Awards:* Sylvia Fedoruk Prize in Med Physics, 88. *Mem:* Can Orgn Physicists Med; Am Asn Physicists Med; Soc Photographic Scientists & Engrs; Soc Photo-Optical Instrumentation Engrs; Can Col Physicists Med; Inst Elec & Electronics Engrs. *Res:* New systems and techniques to be used for vascular imaging; laboratory volume ct scanner for in-vitro imaging of intact vessels and three dimensional colour Doppler ultrasound. *Mailing Add:* Imaging Res Labs PO Box 5015 100 Perth Dr London ON N6A 5K8 Can

FENSTER, SAUL K, b New York, NY, Mar 22, 33; m 59; c 3. MECHANICAL ENGINEERING. *Educ:* City Col New York, BME, 53; Columbia Univ, MS, 55; Univ Mich, PhD(heat transfer), 59. *Prof Exp:* Tool designer, Tech Facilities, Inc, 52; lectr mech eng, City Col New York, 53-56; mech engr, res inst, Univ Mich, 58; res engr aerospace, Sperry Gyroscope Div, Sperry Rand Corp, 59-62; prof physics, Fairleigh Dickinson Univ, 62-63, chmn dept physics, 62-63, chmn dept mech eng, 63-70, prof, 63-72, grad admin asst to dean col sci & eng, 65-70, assoc dean col sci & eng, 70-71, exec asst to pres, 71-72, provost, Rutherford Campus, 72-78; PRES, NJ INST TECHNOL, 78- *Concurrent Pos:* Indust consult, 62-; mem bd dirs, NJ Asn Cols & Univs; chmn, Gov Citizens Task Force on Water Mgt Emergency, Am Asn State Cols & Univs; mem, NJ Comn Sci & Technol; mem, Res & Develop Coun, Pub Technol, Inc, 88-; mem, Coun Competitiveness, 88-; bd dirs, Nat Action Coun Minorities in Eng, 88-; bd gov, Union County Col, 89- *Honors & Awards:* Presidential Recognition Award Community Serv, 84. *Mem:* Am Soc Mech Engrs; Am Soc Eng Educ; AAAS; Sigma Xi; Soc Mfg Engrs; fel Am Soc Mech Engrs. *Res:* Heat transfer; machine dynamics and design; structural analysis; cryogenic. *Mailing Add:* 524 Bernita Dr River Vale NJ 07675

FENSTERMACHER, CHARLES ALVIN, b Scranton, Pa, Mar 31, 28. EXPERIMENTAL PHYSICS. *Educ:* Philadelphia Col Pharm, BS, 50; Swarthmore Col, BA, 53; Yale Univ, MS, 55, PhD(physics), 57. *Prof Exp:* Staff mem weapons test div, Los Alamos Nat Lab, 57-59, group leader test div, Rover Prog, 5969, group leader res & develop electrically excited gas lasers, 69-75, assoc div leader, Laser Div, 75-79, prog mgr Advan Lasers, 79-83, PROG DIR, NEW SOURCES, LOS ALAMOS NAT LAB, 83- *Mem:* Am Phys Soc. *Res:* Gamma ray spectroscopy of neutron capture gamma rays from rare earths; weapons test measurements; Rover nuclear rocket reactor systems design and test; high energy lasers. *Mailing Add:* 3215 Arizona Los Alamos NM 87544

FENSTERMACHER, JOSEPH DON, b Rochester, Minn, Aug 30, 34; m 57; c 4. CEREBRAL BLOOD FLOW, BLOOD-BRAIN BARRIER. *Educ:* Augustana Col, BA, 56; Mich State Univ, MS, 59; Univ Minn, PhD(physiol), 64. *Prof Exp:* Res fel, Univ Minn, 64-65, Univ Bern Switz, 65-66, Nat Cancer Inst, NIH, 72-83; staff fel, 68-72, head, membrane transp sect, Nat Cancer Inst, NIH, 72-83; PROF, DEPT NEUROSURG, STONY BROOK UNIV, 83-, PROF, DEPT PHYSIOL & BIOPHYS, 84- *Concurrent Pos:* Trustee, 70-76, mem, Mt Desert Island Biol Lab, Salsbury Cove, Maine, 71-74; co-ed, Fluid Environ Brain, 75, co-ed, Ocular & cerebrospinal fluids, Acad Press, 77; prin invest NIH Grants NS 21157, NS 26004, HL 35791. *Honors & Awards:* Senator Jacob J Javits Neurosc Investr Award, 84; Heinz Karger Award (co-winner), 86; K A C Elliott lectr, 89. *Mem:* Am Physiol Soc; Microcirculatory Soc; Soc Neurosci; Am Soc Neurochem; Biophys Soc; Sigma Xi. *Res:* Physiology and pathophysiology of cerebal microcirculation, cerebrospinal fluid and brain fluids and focuses on the distribution of nutrients, drugs and other materials among blood brain, and cerebrospinal fluids in normal animals and animals with cerebral lesions such as brain tumors. *Mailing Add:* Dept Neurol Surg Health Sci Ctr State Univ NY Stony Brook NY 11794-8122

FENSTERMACHER, ROBERT LANE, b Scranton, Pa, May 30, 41; m 83; c 2. SOLID STATE PHYSICS, RADIO ASTRONOMY. *Educ:* Drew Univ, BA, 63; Pa State Univ, PhD(physics), 68. *Prof Exp:* From asst prof to assoc prof, 68-80, PROF PHYSICS, DREW UNIV, 80-, CHMN DEPT, 75- *Concurrent Pos:* NASA fel, Jet Propulsion Lab, 79; NSF Fac fel, Bell Tele Labs, 80-81; dir, NJ Gov Sch Sci, 82-87. *Mem:* AAAS; Am Asn Univ Professors; Am Asn Physics Teachers; Am Phys Soc. *Res:* Particle-solid interactions; solar radio astronomy. *Mailing Add:* Dept Physics Drew Univ Madison NJ 07940

FENSTERMAKER, ROGER WILLIAM, b Akron, Ohio, Aug 17, 42; m 64; c 1. CHEMISTRY. *Educ:* Case Inst Technol, BS, 64; Univ Wis, PhD(chem), 70. *Prof Exp:* Res assoc, Wayne State Univ, 70-71; fel, Rice Univ, 71-72; asst prof chem, Northern Mich Univ, 72-73; RES CHEMIST, PHILLIPS PETROL CO, 73- *Mem:* Soc Automotive Engrs; Am Chem Soc; Sigma Xi. *Res:* Fundamental and applied studies of combustion. *Mailing Add:* 2929 Ridge Ct Bartlesville OK 74006

FENTER, FELIX WEST, b Paris, Tex, Sept 16, 26; m 51; c 1. AERONAUTICAL ENGINEERING. *Educ:* Univ Tex, BS, 53, MS, 54, PhD(aeAonaut & space eng), 60. *Prof Exp:* Asst aeromech div, Defense Res Lab, Tex, 52-53, res engr, 53-55, syst develop specialist, 55-58; eng specialist, aerodyn sect, Chance Vought Aircraft, Inc, 58-60, sr scientist, Vought Res Ctr, 60-61, supvr, aerophysics group, Ling-Temco-Vought Res Ctr, 61-62, asst dir, 62-66, assoc dir, LTV Res Ctr, Ling-Temco-Vought, Inc, 66-71, vpres Advan Technol Ctr, Inc, 71-76; CHMN BD & PRES, ADVAN TECHNOL CTR, INC, 73-; VPRES RES & ADVAN TECHNOL, VOUGHT CORP, 76- *Mem:* AAAS; Am Ord Asn; Am Inst Aeronaut & Astronaut. *Res:* Supersonic and hypersonic aerodynamics; mechanics of viscous fluids. *Mailing Add:* 4114 Wingren St Irving TX 75062

FENTERS, JAMES DEAN, b Attica, Ind, Sept 23, 36. VIROLOGY, MICROBIOLOGY. *Educ:* Purdue Univ, BS, 58; Univ Iowa, MS, 61, PhD(bact), 62. *Prof Exp:* Res virologist, Abbott Labs, 62-67; res virologist, 67-74, head microbiol & immunol res, 74-84, HEAD, TOXICOL & ENVIRON HEALTH, IIT RES INST, 84, DIR, LIFE SCI RES, 89- *Mem:* AAAS; Am Soc Microbiol; fel Am Acad Microbiol; Sigma Xi; Soc Toxicol. *Res:* Environmental pollutant effects on infectious processes; water microbiology, viral vaccines, immune responses and serology, antiviral chemotherapy, organ and tissue culture, in vitro bioassay systems. *Mailing Add:* IIT Res Inst Ten W 35th St Chicago IL 60616-3799

FENTIMAN, ALLISON FOULDS, JR, b Pittsburgh, Pa, Apr 21, 37; m 84. SYNTHESIS & SPECTROMETRIC ANALYSIS. *Educ:* Muskingum Col, BS, 61; Ohio State Univ, MS, 66, PhD(org chem), 69. *Prof Exp:* SR RES CHEMIST ORG CHEM, COLUMBUS DIV, BATTELLE MEM INST, 63- *Concurrent Pos:* Lectr, Denison Univ, 82, Ohio State Univ, 88. *Mem:* AAAS; Am Chem Soc; Sigma Xi. *Res:* Organic synthesis, mechanism, and spectrometric analysis, including synthesis of deuterium-labeled drugs; mechanistic study of the competition between aryl substituents and neighboring groups; identification of insect pheromones and marihuana constituents and metabolites; isolation, identification and synthesis of drug and pesticide metabolites. *Mailing Add:* Mead Imaging 3385 Newmark Dr Miamisburg OH 45342-5497

FENTON, DAVID GEORGE, b London, Eng, June 8, 32; m; m; c 2. PHYSICS. *Educ:* Univ London, Eng, BSc, 55; Univ Conn, PhD, 64. *Prof Exp:* Asst, Purdue Univ, 55-57; prod engr, Taylor Instrument Co, 57-58; from instr to assoc prof, 58-73, PROF PHYSICS, CONN COL, 73- *Mem:* Hist Sci Soc; Am Asn Physics Teachers. *Res:* Quantum mechanics of the electronic structure of molecules and of atomic collision processes; history of 20th century physics. *Mailing Add:* Dept Physics Box 5441 Conn Col New London CT 06320

FENTON, DONALD MASON, b Los Angeles, Calif, May 23, 29; m 53; c 2. PETROLEUM CHEMISTRY. *Educ:* Univ Calif, Los Angeles, BS, 52, PhD, 58. *Prof Exp:* Res chemist, Rohm and Haas Co, 58-61; res chemist, 62-82, mgr planning develop, 82-84, MGR NEW TECHNOL DEVELOP, UNION OIL CO, 84- *Concurrent Pos:* Adv bd, Petrol Res Fund, Am Chem Soc; cofounder, Petrol Environ Res Forum. *Mem:* Am Chem Soc; Am Petrol Inst; fel Am Inst Chemists. *Res:* Organometallic chemistry; environmental chemistry. *Mailing Add:* Unocal Corp PO Box 76 Brea CA 92621

FENTON, EDWARD WARREN, b Lucky Lake, Sask, Mar 5, 37; m 66; c 2. SOLID STATE PHYSICS. *Educ:* Univ Alta, BSc, 59, MSc, 62, PhD(physics), 65. *Prof Exp:* Nat Res Coun fel physics, Simon Fraser Univ, Vancouver, 65-67; res officer physics, Noranda Res Ctr, Montreal, 67; from asst res officer to sr res officer, 67-90, PRIN RES OFFICER PHYSICS, NAT RES COUN CAN, 90-, HEAD, CONDENSED MATTER SECT, 84- *Mem:* Am Phys Soc. *Res:* Theoretical solid state physics, including properties of semiconductors, metals, and itinerant-electron condensates. *Mailing Add:* Nat Res Coun Can Ottawa ON K1A 0R6 Can

FENTON, JOHN WILLIAM, II, b Salamanaca, NY, Feb 4, 39; m 65; c 2. BIOCHEMISTRY, IMMUNOLOGY. *Educ:* Cornell Univ, BS, 61; Univ Wis-Madison, MS, 64; Univ Calif, San Diego, PhD(biochem), 68. *Prof Exp:* SR RES SCIENTIST, DIV LABS & RES, NY STATE DEPT HEALTH, 68- *Concurrent Pos:* USPHS grants, 69-; Brown-Hazen Fund grant, 74; adj asst prof, Dept Microbiol, Albany Med Col, 70-74, adj prof, Dept Physiol, 85-, Dept Biochem, 90- *Mem:* Am Heart Asn; AAAS; Am Soc Biol Chemists; NY Acad Sci; Am Chem Soc; Am Soc Clin Chemists; Int Soc Thrombosis & Hemostasis; Nat Acad Clin Biochem; Biochem Soc. *Res:* Structures, specificity and functionality of human thrombins and other blood components; coagulation; purification of antibodies and diagnostic application; antibody and enzyme active sites, thrombosis, hemostasis and wound healing. *Mailing Add:* Div Labs & Res NY State Dept Health Empire State Plaza Albany NY 12201

FENTON, M BROCK, b Guyana, Oct 20, 43; Can citizen; m 69. MAMMALOGY. *Educ:* Queen's Univ, Ont, BSc, 65; Univ Toronto, MSc, 66, PhD(zool), 69. *Prof Exp:* prof biol, Carleton Univ, 69-86; PROF & CHMN BIOL, YORK UNIV, 86- *Concurrent Pos:* Res assoc mammal, Royal Ont Mus. *Honors & Awards:* A B Howell Award, Am Soc Mammal, 69. *Mem:* AAAS; Am Soc Mammal; Can Soc Zool; Animal Behav Soc; Asn Trop Biol. *Res:* Chiroptology; ecology of bats with special references to their behavior; echolocation. *Mailing Add:* Dept Biol York Univ North York ON M3J 1P3 Can

FENTON, MARILYN RUTH, b Salem, Ohio, May 17, 42; m 76. ZINC & MEMBRANE STRUCTURE & FUNCTION, ZINC & IMMUNE RESPONSE. *Educ:* Capitol Univ, BS, 64; Ohio State Univ, MSc, 66, PhD(physiol chem), 69. *Prof Exp:* Instr, Dept Biochem, Thomas Jefferson Univ, 69-71 & Dept Microbiol & Immunol, Sch Med, Temple Univ, 71-73; from asst prof to assoc prof biochem, 73-81, PROF BIOCHEM, DEPT PHYSIOL SCI, PA COL PODIATRIC MED, 81-, CHMN, 85- *Concurrent Pos:* Adj assoc prof, Dept Microbiol & Immunol, Sch Med, Temple Univ, 77-; prin investr, NIH BRDG grant, 80-83 & Am Podiatric Med Asn grant, 89-90; vis prof, Dept Med, Res Lab, Grad Hosp, Philadelphia, Pa, 89. *Mem:* Am Asn Immunologists; NY Acad Sci; Am Inst Nutrit. *Res:* Zinc in cell structure and function; biomembranes, alterations in trace mineral metabolism in disease states such as tumor growth and diabetes; nutrition in prevention and therapy of disease. *Mailing Add:* Podiatric Med Pa Col Eighth at Race St Philadelphia PA 19107

FENTON, MATTHEW JOHN, b Hartford, Conn, Dec 26, 54; m 84. MOLECULAR IMMUNOLOGY. *Educ:* Boston Univ, PhD(biochem), 85. *Prof Exp:* Postdoctoral fel, Mass Inst Technol, 84-86, postdoctoral assoc, 86-88; ASST PROF MED, BOSTON UNIV SCH MED, 88- *Concurrent Pos:* Asst prof biochem, Boston Univ, 89-; indust consult; prin investr NIH grants. *Honors & Awards:* Whitaker Health Sci Fund Award, 88; Aid for Cancer Res Young Investr Award, 89. *Mem:* Am Asn Immunologists; Am Soc Microbiol; Sigma Xi; Soc Leukocyte Biol. *Res:* Molecular basis for immune cell activation; role of transcriptional regulatory factors and selective mRNA degradation in the expression of proinflammatory cytokines. *Mailing Add:* Immunol Unit E-337 Boston Univ Med Ctr 88 E Newton St Boston MA 02118

FENTON, ROBERT E, b Brooklyn, NY, Sept 30, 33; m 55; c 2. ELECTRICAL ENGINEERING. *Educ:* Ohio State Univ, BEE, 57, MSc, 60, PhD(elec eng), 65. *Prof Exp:* Res engr, NAm Aviation, Inc, 57; from instr to assoc prof elec eng, 60-73, asst supvr, Commun & Controls Syst Labs, 65-67, actg dir, 67-70, dir, transp control lab, 70-82, PROF ELEC ENG, OHIO STATE UNIV, 73- *Concurrent Pos:* Mem, Hwy Res Bd Comt Hwy Commun, 66-82; consult, Transp Systs Div, Gen Motors Corp, 73-74 & 76-80. *Mem:* Fel Inst Elec & Electronics Engrs; Vehicular Technol Soc (treas, 81-83, vpres, 83-85, pres, 85-87); fel Radio Club Am. *Res:* Automatic control and computer systems; automated ground transportation systems. *Mailing Add:* Dept Elec Eng 2015 Neil Ave Columbus OH 43210

FENTON, ROBERT GEORGE, b Budapest, Hungary, Apr 3, 31; m 57; c 2. MECHANICAL ENGINEERING. *Educ:* Univ Budapest, Dipl Ing, 53; Univ NSW, PhD(mech eng), 68. *Prof Exp:* Sr lectr mech eng, Univ NSW, 62-68; assoc prof, 68-81, PROF MECH ENG, UNIV TORONTO, 81- *Honors & Awards:* Water Arbit Prize, Inst Mech Engrs, 69; Handerson Mem Prize, In1t Mech Engrs, 70. *Mem:* Advan Process Eng Order; Soc Mech Engrs. *Res:* Experimental and analytical investigation of metal flow during machining and forming; yield criteria, thermal effects; machine tool vibration; wear analysis; tool life and machining economy; computerized design; reliability; simulation models; computer aided design; robotics. *Mailing Add:* Dept Mech Eng Univ Toronto Toronto ON M5S 1A4 Can

FENTON, STUART WILLIAM, b London, Ont, Apr 29, 22; m 62. ORGANIC CHEMISTRY. *Educ:* Queen's Univ, Ont, BSc, 45, MSc, 46; Mass Inst Technol, PhD(chem), 50. *Prof Exp:* Asst prof org chem, 50-55, assoc prof chem & assoc chmn dept, 55-61, chmn dept, 61-68, prof chem, 61-86, EMER PROF CHEM, UNIV MINN, MINNEAPOLIS, 86- *Mem:* AAAS; Am Chem Soc; The Chem Soc; Sigma Xi. *Res:* Organic synthesis and peroxides. *Mailing Add:* 139 Chem Univ Minn Minneapolis MN 55455

FENTON, THOMAS E, b Cabery, Ill, July 19, 33; m 59; c 4. SOIL GENESIS, SOIL CLASSIFICATION. *Educ:* Univ Ill, Urbana, BS, 59, MS, 60; Iowa State Univ, PhD(soil genesis & classification), 66. *Prof Exp:* Instr, US Army Europe Qm Sch, 54-55; res soil scientist, USDA, summers 61-64; res assoc agron, 64-66, from asst prof to assoc prof soil genesis & classification, 66-74, asst prof, Res Found, 68, PROF SOIL GENESIS & CLASSIFICATION, IOWA STATE UNIV, 74- *Concurrent Pos:* Consult, Attorney Gen Off, State of Iowa, 67-69. *Mem:* AAAS; Clay Mineral Soc; Am Soc Agron; fel Soil Sci Soc Am; Soil Conserv Soc Am. *Res:* Soil genesis and classification combined with geomorphology. *Mailing Add:* Dept Agron Iowa State Univ Ames IA 50011

FENTON, WAYNE ALEXANDER, b New Castle, Del, Nov 3, 45. MOLECULAR GENETICS, CELL BIOLOGY. *Educ:* Univ Del, BS, 67; Brandeis Univ, PhD(biochem), 74. *Prof Exp:* Fel, 73-76, res assoc, 76-84, RES SCIENTIST HUMAN GENETICS, SCH MED, YALE UNIV, 84- *Mem:* Am Soc Human Genetics. *Res:* Human inherited metabolic disease (methylmalonic acidemia and hyperammonemia), including diagnosis and treatment; definition of the pathways of cobalamin metabolism in mammals; delineation of the biogenesis of mitochondrial enzymes. *Mailing Add:* Dept Human Genetics Sch Med Yale Univ 333 Cedar St New Haven CT 06510

FENTRESS, JOHN CARROLL, b East Chicago, Ind, Feb 4, 39. ETHOLOGY, NEUROBIOLOGY. *Educ:* Amherst Col, BA, 61; Cambridge Univ, PhD(zool), 65. *Prof Exp:* From asst prof to assoc prof biol-psychol, Univ Ore, 67-75; PROF PSYCHOL & CHMN DEPT, DALHOUSIE UNIV, 75- *Concurrent Pos:* USPHS fel, Ctr Brain Res, Univ Rochester, 65-67; mem bd dirs, Wild Canid Survival & Res Ctr, 71- & Ore Zool Res Ctr, 72-; assoc ed, Behav Biol, 74-; res award, Nat Res Coun, 75- *Mem:* AAAS; Animal Behav Soc; Am Soc Zool; Soc Neurosci; Sigma Xi. *Res:* Integration and development of species-characteristic behaviors; brain mechanisms and behavior. *Mailing Add:* Psychol Dept Dalhousie Univ Halifax NS B3H 4H6 Can

FENVES, STEVEN J(OSEPH), b Subotica, Yugoslavia, June 6, 31; US citizen; m 55; c 4. CIVIL & SOFTWARE ENGINEERING. *Educ:* Univ Ill, BS, 57, MS, 58, PhD(eng), 61. *Prof Exp:* Draftsman, Erik Floor & Assoc, Ill, 50-52; from asst prof to prof civil eng, Univ Ill, 57-71; prof & head dept, 72-75, SUN CO, UNIV PROF CIVIL ENG, CARNEGIE-MELLON UNIV, 75- *Concurrent Pos:* Indust & govt consult, 57-; vis prof, Mass Inst Technol, 62-63, Nat Univ Mex, 65 & Cornell Univ, 70-71. *Mem:* Nat Acad Eng; Am Soc Civil Engrs. *Res:* Computer applications in civil engineering; expert systems and AI in civil engineering; design theory. *Mailing Add:* Dept Civil Eng Carnegie-Mellon Univ Pittsburgh PA 15213

FENWICK, HARRY, b Filer, Idaho, Sept 24, 22; m 46; c 3. PLANT PATHOLOGY. *Educ:* Mont State Univ, BS, 49, MS, 52; Ore State Col, PhD(plant path), 56. *Prof Exp:* Assoc prof plant path, 68-72, PROF PLANT SCI, UNIV IDAHO, 72-, EXTEN PLANT PATHOLOGIST, 56- *Mem:* Am Phytopath Soc; Sigma Xi. *Res:* Seedling diseases of sugarbeets; control of cereal rusts by chemotherapy; studies of dwarf bunt in cereals and grasses. *Mailing Add:* Dept Plant Sci Univ Idaho Moscow ID 83843

FENWICK, JAMES CLARKE, b New West Minster, BC, Jan 5, 40; m 67; c 2. COMPARATIVE ENDOCRINOLOGY. *Educ:* Univ Man, BS, 62, MS, 65; Univ BC, PhD(zool), 69. *Prof Exp:* Lectr biol, St Johns Col, Man, 64-65; fel, Nat Res Coun Can, 69-70; asst prof, 70-75, assoc prof, 75-81, PROF BIOL, UNIV OTTAWA, 81- *Concurrent Pos:* Vis prof, Univ Nijmegen, The Netherlands, 81. *Mem:* Can Soc Zoologists; Am Zool Soc; Can Physiol Soc; Europ Soc Comp Biochem Physiol. *Res:* Calcium metabolism in lower vertebrates with special emphasis on transport processes in fish. *Mailing Add:* Dept Biol Sci Univ Ottawa Ottawa ON K1N 6N5 Can

FENWICK, ROBERT B, b Indianapolis, Ind, Apr 13, 36; m 60; c 3. ELECTRICAL ENGINEERING. *Educ:* Purdue Univ, BS, 58; Stanford Univ, MS, 59, PhD(elec eng), 63. *Prof Exp:* Res assoc, Radiosci Lab, Stanford Univ, 60-68; vpres, 66-76, PRES, BR COMMUN, 76-85, CHMN, 85- *Mem:* Am Geophys Union; Inst Elec & Electronics Engrs. *Res:* High frequency radio wave propagation via the ionosphere; techniques of ionospheric measurement, especially ionospheric sounders; frequency management systems including methods of measuring spectrum occupancy; data modem research. *Mailing Add:* 28011 Elena Rd Los Altos Hills CA 94022

FENYES, JOSEPH GABRIEL EGON, b Paszto, Hungary, Mar 19, 25; US citizen; m 57; c 3. AGRICULTURAL CHEMISTRY. *Educ:* Univ Szeged, BSc, 48; McGill Univ, PhD(org chem), 55. *Prof Exp:* Lectr org chem, Royal Mil Col, Ont, 55-56; res chemist, Shawinigan Chem Ltd, Que, 56-58; res group leader, Hyman Labs, Inc, Calif, Berkeley, 58-62; res chemist, Ortho Div, Calif Chem Co, Richmond, 62 & Chevron Chem Co, Calif, 62-69; CHIEF ORG RES CHEMIST, BUCKMAN LABS, INC, MEMPHIS, TENN, 69- *Mem:* Am Chem Soc; Sigma Xi. *Res:* Synthesis of biologically active and agriculturally useful organic compounds; sulfenyl halides; halogen and sulfur containing fungicides; plant growth regulators; organophosphorous insecticides; B-substituted naphthalenes; heterocyclic chemicals; plastics additives; ultraviolet light absorbers; fire retardants; agricultural and industrial microbicides and herbicides. *Mailing Add:* 1827 Oakhill Cove Germantown TN 38138

FENYVES, ERVIN J, b Budapest, Hungary, Aug 29, 24; m 51; c 2. HIGH ENERGY PHYSICS, COSMIC RAY PHYSICS. *Educ:* Eotvos Lorand Univ, Budapest, MS, 46, PhD(physics), 50; Hungarian Acad Sci, Cand Phys Sci, 55, Dr Phys Sci, 60. *Prof Exp:* Asst prof physics, Eotvos Lorand Univ, 46-51; res fel, Cent Res Inst Physics, Hungarian Acad Sci, 51-59, head lab cosmic rays, 59-65, dep sci dir, 65-69; res fel physics, Univ Pa, 69-70; actg dir, Ctr Environ Studies, 77-80, PROF PHYSICS, UNIV TEX, DALLAS, 70- *Concurrent Pos:* Assoc prof, Eotvos Lorand Univ, 60-64, prof, 64-69; cor mem high energy nuclear physics comn, Int Union Pure & Appl Physics, 63-69, mem cosmic ray comn, 66-70; vdir, Joint Inst Nuclear Res, Dubna, 64-66; head physics sect, Int Atomic Energy Agency, Vienna, 68-69. *Honors & Awards:* Brodi-Schmidt Prize, Hungarian Phys Soc, 52; Nat Kossuth Prize, Hungarian Govt, 65; Prize for Books, Hungarian Acad Sci, 67. *Mem:* Fel Am Phys Soc; NY Acad Sci. *Res:* Cosmic rays; high energy physics. *Mailing Add:* Univ Tex-Dallas PO Box 830688 Richardson TX 75083

FEOLA, JOSE MARIA, b Buenos Aires, Arg, May 30, 26; US citizen; m 50; c 3. RADIATION BIOLOGY, PHYSICS. *Educ:* Univ Rochester, MS, 61; Univ La Plata, Arg, licenciate physics & math, 63; Univ Minn, Minneapolis, PhD(environ health), 74. *Prof Exp:* Mem res staff radiobiol, Arg AEC, 56-65, Donner Lab, Univ Calif, Berkeley, 65-69; instr radiobiol, Univ Minn, Minneapolis, 70-73, instr sci methods, Exten Div, 74-75; ASSOC PROF CLIN MED, DEPT RADIATION MED, UNIV KY, 75- *Concurrent Pos:* Asst prof, Fac Eng, Univ Buenos Aires, 56-64; Nat Cancer Inst grant, 77-80; Univ Ky Res Found grant, 78, NIH grant, 78-79 & 84-85 & 87-89; Kircher fund grant, 80-81, Electro-Biol grant, 80-85, Am Cancer Soc grant, 81-82 & 81-84 & Int Atomic Energy Agency grant, 84-85. *Mem:* Radiation Res Soc; Sigma Xi. *Res:* Cancer radiation and chemotherapy in experimental systems; biological effects of low doses of ionizing radiation of high and low linear energy transfer; biological effects of magnetic fields. *Mailing Add:* Dept Radiation Med Univ Ky Med Ctr 800 Rose St Lexington KY 40536-0084

FEOLA, MARIO, SURGERY. *Prof Exp:* PROF SURG, HEALTH SCI CTR, TEX TECH UNIV, 81- *Mailing Add:* Dept Surg Health Sci Ctr Tex Tech Univ Fourth & Indiana Lubbock TX 79430

FERADAY, MELVILLE ALBERT, b Toronto, Ont, Jan 13, 29; m 58; c 3. MECHANICAL ENGINEERING, MATERIALS SCIENCE. *Educ:* Queen's Univ, Ont, BSc, 54; Univ Waterloo, MSc, 70. *Prof Exp:* Engr, Chalk River Nuclear Labs, Atomic Energy Can, Ltd, 56-58, commissioning engr, Colombo Plan, India, 59-60, fuel develop engr, 60-80, mgr nuclear waste, 80-89; RETIRED. *Res:* Design, development, metallurgy and irradiation of uranium based fuel elements, particularly metallic fuels and powder packed uranium dioxide fuels; design of a remotely operated plant to fabricate gamma active nuclear fuels; long term environmental planning of low level radioactive waste management sites. *Mailing Add:* 2045 Lakeside Blvd Apt 207 Toronto ON M8V 2Z6 Can

FERAMISCO, JAMES ROBERT, b Fresno, Calif, June 3, 52; m 70; c 2. CANCER BIOLOGY. *Educ:* Univ Calif, BS & BA, 74, PhD(biochem), 79. *Prof Exp:* Fel biochem, Cold Spring Harbor Lab, 79-80, staff investr biochem, 80-81, sr staff investr cell biol & biochem, 81-84, sr scientist & head, Cell Biol Group, 84-88; CONSULT, NIH, 84-; PROF PHARMACOL & MED, CANCER CTR, UNIV CALIF, SAN DIEGO, 88- *Mem:* Am Chem Soc. *Res:* Determining growth control mechanisms in human cells; involvement of oncogene proteins (from human tumors) in cancer. *Mailing Add:* Univ Calif-San Diego 9500 Gilman Dr La Jolla CA 92093-0636

FERBEL, THOMAS, b Radom, Poland, Dec 12, 37; US citizen; m 63; c 2. PARTICLE PHYSICS. *Educ:* Queens Col, NY, BS, 59; Yale Univ, MS, 60, PhD(physics), 63. *Prof Exp:* Res staff physicist, Yale Univ, 63-65; from asst prof to assoc prof, 65-73, PROF PHYSICS, UNIV ROCHESTER, 73-, ASSOC DEAN GRAD STUDIES, COL ARTS & SCI, 89- *Concurrent Pos:* Sloan Found res fel, 70; Guggenheim Found fel, 71; mem, High Energy Adv Comt, Stanford Linear Accel Ctr, 74-76 & Brookhaven Lab, 81-84; Mgt Dept contract, Dept Energy, 77-80; chmn, Fermilab Users Orgn, 86-87; ed, Phys Rev, 78-80, Zeitschriff fur Physik, 81-85; sci dir, Advan Study Inst Tech & Concepts High Energy Physics, St Croix, 80, 84, 86, 88, 90 & Adirondacks, 82; sci assoc, Europ Orgn Nuclear Res Lab, Switz, 80-81; vis scientist, Cent Design Group, 88-89; secy-treas, Particles & Fields Div, Am Phys Soc, 83-86. *Mem:* Fel Am Phys Soc. *Res:* Experimental elementary particle physics; phenomenology of strong interactions. *Mailing Add:* Dept Physics & Astron Univ Rochester Rochester NY 14627

FERBER, KELVIN HALKET, b Brooklyn, NY, Oct 18, 10; m 37; c 2. INDUSTRIAL HYGIENE, OCCUPATIONAL HEALTH. *Educ:* Cornell Univ, BChem, 32; State Univ NY Buffalo, MBA, 71. *Prof Exp:* Supvr anal lab, Allied Chem Corp, 38-42, supt qual control, 42-44, asst to plant mgr, 45-49, supt tests & inspections, 49-64, tech mgr, 64-71, tech asst to mgr, 71-72, mgr occup health, 72-75; CONSULT OCCUP HEALTH, 76- *Mem:* Am Chem Soc; fel Am Soc Testing & Mat; AAAS; Am Indust Hyg Asn; Air Pollution Control Asn. *Res:* Industrial toxicology; environmental control; occupational hazards and industrial hygiene control of carcinogens. *Mailing Add:* 24 Linden Ave Buffalo NY 14214-1502

FERBER, ROBERT R, b Monongahela, Pa, June 11, 35; m 64; c 2. ENGINEERING PHYSICS. *Educ:* Univ Pittsburgh, BS, 58; Carnegie-Mellon Univ, MS, 66, PhD(elec eng), 67. *Prof Exp:* Audio engr, Radio Sta WWSW, 52-56; head dept eng, WRS Motion Picture Labs, 54-58; res engr radiation detection, Westinghouse Res Labs, 58-66, adv engr, Westinghouse Astronuclear Labs, 67-71, power systs planning, Westinghouse Elec Corp, 71-77; mem tech staff, Photovoltaic Lead Ctr, 77-80, MGR PHOTOVOLTAIC COLLECTOR TECHNOL DEVELOP, JET PROPULSION LAB, 80- *Concurrent Pos:* Lectr, West Coast Univ, 78-81; adj asst prof, Inst Safety & Systs Mgt, Univ Southern Calif, 81-, Univ Denver, 88- *Mem:* Inst Elec & Electronics Engrs; Int Solar Energy Soc. *Res:* Advanced concepts in electric power generation; nuclear radiation effects; nuclear power; solar power; semiconductor physics. *Mailing Add:* Jet Propulsion Lab 264-358 4800 Oak Grove Dr Pasadena CA 91109

FERCHAU, HUGO ALFRED, b Mineola, NY, July 22, 29; m 52; c 4. BOTANY. *Educ:* Col William & Mary, BS, 51; Duke Univ, PhD(bot), 59. *Prof Exp:* Actg jr botanist, Div Indust Res, Wash State Univ, 53-54; assoc prof biol, Wofford Col, 58-62; from asst prof to assoc prof, 62-69, PROF BIOL, WESTERN STATE COL COLO, 69- *Concurrent Pos:* Dir, Bact Testing Serv, Western State Col, 62-; mem fac, Rocky Mountain Biol Lab, 63-65; mem fac, Inst Mountain Ecol, 72-73, fac affil agron, Colo State Univ, Ft Collins, 75-; vis prof environ sci, Colo Sch Mines, Golden, 81-82. *Mem:* Am Inst Biol Sci; Ecol Soc Am; Bot Soc Am; Soc Range Mgt; Sigma Xi; Soc Southwestern Naturalists. *Res:* Ecology of mycorrhizae; vegetation analysis; environmental analysis; reclamation mines; ecology of hot springs. *Mailing Add:* Div Natural Sci Western State Col Colo Gunnison CO 81230

FERCHAUD, JOHN B(ARTHOLOMEW), b New Orleans, La, June 14, 12; m 47; c 3. CHEMICAL ENGINEERING. *Educ:* La State Univ, BS, 35. *Prof Exp:* Process engr, Standard Oil Co, La, 35-36; plant chemist, Ark Fuel Oil Co, 36-37; state chem eng, Conserv Dept, La, 37-42; supt construct & oper, Chem Construct Corp, NY, 42-51; tech asst to mfg mgr, Lion Oil Co Div, Monsanto Co, 51-56; eng mgr, 56-61, eng mgr agr div, 61-65 & cent eng dept, 65-77; RETIRED. *Mem:* Am Inst Chem Engrs. *Res:* Engineering and construction of manufacturing units. *Mailing Add:* 182 Meadow Lark Dr St Louis MO 63146

FERDINANDI, ECKHARDT STEVAN, DRUG METABOLISM, ANALYTICAL CHEMISTRY. *Educ:* McGill Univ, PHD(org chem), 69. *Prof Exp:* RES ASSOC, AYERST LABS RES, INC, 70- *Res:* Biotransformation. *Mailing Add:* Metab Dept Bio-Res Labs 87 Senneville Rd Senneville PQ H9X 3R3 Can

FERENCE, MICHAEL, JR, b Whiting, Ind, Nov 6, 11; m 37; c 5. PHYSICS. *Educ:* Univ Chicago, BS, 33, MA, 34, PhD(physics), 37. *Hon Degrees:* DSc, Kenyon Col, 69. *Prof Exp:* From instr to assoc prof physics, Univ Chicago, 37-46; chief meteorol br, Signal Corps Eng Labs, Evans Signal Lab, 46-48, chief scientist, 48-51, tech dir, 51-53; chief scientist, Ford Motor Co, 53-54, from assoc dir to exec dir, 54-62, vpres res, sci lab, 62-80; RETIRED. *Concurrent Pos:* Trustee, Rand Corp; mem panel, Res & Develop Bd, tech panel earth satellite prog & comt atmospheric sci, Nat Acad Sci & res panel, Signal Corps Res & Develop Adv Coun; mem adv group on weather modification, NSF, spec adv comt to Dept of Comt, Rocket & Satellite Res Panel, adv comt, US Weather Bur & President's Sci Adv Comt; mem bd of trustees, Carnegie Inst, Dirs Indust Res & adv comt, PR Nuclear Ctr, Univ PR. *Mem:* Nat Acad Eng; Am Phys Soc; Am Geophys Soc; Am Meteorol Soc; Soc Automotive Eng; fel Inst Elec & Electronics Engrs. *Res:* Physics of the upper atmosphere; experimental hydrodynamics; designs of radiosondes; radar; electronics; microwave propagation; x-ray spectroscopy. *Mailing Add:* Anglers Cove 12-203 1456 NE Ocean Blvd Stuart FL 34996

FERENCZ, CHARLOTTE, b Budapest, Hungary, Oct 28, 21; US citizen. PEDIATRIC CARDIOLOGY, EPIDEMIOLOGY. *Educ:* McGill Univ, BSc, 44, MD, CM, 45; Johns Hopkins Univ, MPH, 70. *Prof Exp:* Demonstr pediat, McGill Univ, 52-54; asst prof pediat, Johns Hopkins Univ, 54-59; asst prof, Med Col, Univ Cincinnati, 59-60; from asst prof to assoc prof, Sch Med, State Univ NY Buffalo, 60-73, clin asst prof social & prev med, 71-73; assoc prof prev med, 73-74, PROF PEDIAT, SCH MED, UNIV MD, 73-, PROF PREV MED, 74- *Concurrent Pos:* Fogarty Int Ctr,Individual Health Sci Exchange Prog, US-Hungary, 88. *Honors & Awards:* Merit Award, Nat Heart

Lung & Blood Inst, 87. *Mem:* Am Acad Pediat; Am Col Cardiol; Am Pub Health Asn. *Res:* Changes in the pulmonary vascular bed associated with congenital heart disease; normal growth of pulmonary vessels; epidemiology of congenital heart disease; environmental teratology. *Mailing Add:* Dept Prev Med Univ Md Sch Med 655 W Baltimore St Baltimore MD 21201

FERENCZ, NICHOLAS, b Cleveland, Ohio, Apr 22, 37; div; c 2. DENTISTRY. *Educ:* Hiram Col, AB, 59; Cath Univ Am, MS, 62, PhD(cell biol), 67; Case Western Reserve Univ, DDS, 74. *Prof Exp:* Instr radiol, Sch Dent, Case Western Reserve Univ, 67-69, asst prof radiation biol, 69-71, clin instr, 74-77, asst clin prof, 77-82, assoc clin prof prosthodont, 82-89; PVT PRACT, 89- *Honors & Awards:* Callahan Award, 74. *Res:* Tissue culture; radiology; prosthodontics; dental materials. *Mailing Add:* 24300 Chagrin Blvd Suite 107 Beachwood OH 44122

FERENTZ, MELVIN, b New York, NY, Oct 14, 28; m 48, 64; c 3. COMPUTER SCIENCE. *Educ:* Brooklyn Col, BS, 49; Univ Pa, PhD(physics), 53. *Prof Exp:* Asst instr physics, Univ Pa, 48-50; assoc physicist, Argonne Nat Lab, 52-57; sr mathematician analyst, Int Bus Mach Corp, 57-59; assoc prof physics, St John's Univ, NY, 59-65, prof & chmn dept, 65-66; dept chmn grad studies, Dept Physics, Brooklyn Col, 67-74, prof physics, 66-79; DIR COMPUT SERV, ROCKEFELLER UNIV, 78- *Concurrent Pos:* Res assoc, Columbia Univ, 64-66; vis prof, NY Univ, 66; assoc dir, City & State Univs NY Joint Inst Learning & Instr, 67-70; dir, Nysernet. *Mem:* Asn Comput Mach; Sigma Xi. *Res:* Networking; digital computers. *Mailing Add:* 535 East 86 New York NY 10028

FERER, KENNETH MICHAEL, b Pittsburgh, Pa, Dec 12, 37; m 62; c 2. TECHNICAL MANAGEMENT, RESEARCH ADMINISTRATION. *Educ:* George Washington Univ, BS, 72; Univ Southern Miss, MS, 82. *Prof Exp:* Proj mgr, Naval Res Lab, 72-75, Naval Oceanog Res & Develop Activ, 75-78; PROG MGR, ANTI-SUBMARINE WARFARE, OCEANOG CHIEF NAVAL OPERS, 78- *Concurrent Pos:* Mem, buoy comt, Marine Technol Soc, 70-79, cable comt, 75-79, educ comt, 82- *Mem:* Am Geol Union; Marine Technol Soc. *Res:* Development of sensors, instrument and systems to support the US Navy's interest in deep ocean search capability and non-acoustic submarine warfare; oceanographic engineering. *Mailing Add:* Dir ASW Oceanog Prog Off Naval Oceanog & Atmospheric Res Lab NSTL Station MS 39529-5004

FERGIN, RICHARD KENNETH, b Tacoma, Wash, Dec 9, 33; m 67; c 1. ENGINEERING, PHYSICAL CHEMISTRY. *Educ:* Wash State Col, BS, 55; NMex State Univ, MS, 61, ScD(mech eng), 64. *Prof Exp:* Tool engr, Boeing Airplane Co, 55-57; mech engr, Ft Belvoir, Va, 58-59; res assoc, Rocket Sect, Phys Sci Lab, NMex State Univ, 59-60, asst mech eng, 60-61, instr, 61-63; assoc prof eng, San Diego State Col, 64-68; assoc prof math, US Int Univ, 68; staff scientist, Geosci Ltd, 69-72; dir res & develop, NRG Technol, 73-74; sr thermodyn engr, Teledyne Ryan, 75-76; chief mech engr, HVAC, Carter Engrs, 76-79; LEAD ENGR & MGR ENERGY CONSERV, NAVY PUB WORKS CTR, 80- *Mem:* Am Soc Heat, Refrig & Air-Conditioning Engrs; Am Soc Eng Educ; Am Soc Mech Engrs; Sigma Xi. *Res:* Binary liquid jet refrigeration using immiscible fluids; heat, mass and momentum transfer and enhancement with applications in energy recovery; micrometeorology; air and thermal pollution; solar energy; environmental control; energy optimization. *Mailing Add:* 1779 Ocean Front St San Diego CA 92107

FERGUSON, ALASTAIR VICTOR, b Walton upon Thames, Eng, Oct 29, 55; Can citizen; m. NEUROENDOCRINOLOGY, NEUROSCIENCES. *Educ:* Birmingham Univ, Eng, BSc Hons, 77; Univ Calgary, PhD(neurosci), 82. *Prof Exp:* Res asst neurosci, Univ Calgary, Alta, 77-80, Heritage student, 80-82; res fel neurol, Montreal Gen Hosp & McGill Univ, 82-84; asst prof, 84-88, ASSOC PROF PHYSIOL, QUEEN'S UNIV, 88- *Mem:* Can Asn Neurosci; Am Physiol Soc; Can Physiol Soc; Int Brain Res Orgn; Soc Neurosci. *Res:* Central nervous system control of the autonomic nervous system. *Mailing Add:* Dept Physiol Queen's Univ Kingston ON K7L 3N6 Can

FERGUSON, ALBERT BARNETT, b New York, NY, June 10, 19; m 43; c 3. ORTHOPEDIC SURGERY. *Educ:* Dartmouth Col, BA, 41; Harvard Univ, MD, 43. *Prof Exp:* Asst orthop surg, Harvard Univ, 51-52; from assoc prof to prof, 53-58, SILVER PROF ORTHOP SURG & CHMN DEPT, SCH MED, UNIV PITTSBURGH, 58- *Mem:* Am Acad Orthop Surg; Am Acad Pediat; Am Acad Neurol; Am Bd Orthop Surg (past pres); Am Orthop Asn (past pres). *Res:* Physiology of muscles; growth. *Mailing Add:* Dept Orthop Surg Univ Pittsburgh Health Ctr Hosps 300 Fox Chapel Rd Pittsburgh PA 15238

FERGUSON, ALBERT HAYDEN, b Big Timber, Mont, Sept 12, 28; m 51; c 2. SOIL PHYSICS. *Educ:* Mont State Col, BS, 50; Wash State Univ, MS, 56, PhD(soils), 59. *Prof Exp:* Soil scientist, US Bur Reclamation, 50-51; asst soils, Wash State Univ, 53-58; PROF SOILS, MONT STATE UNIV, 58- *Mem:* Fel Am Soc Agron; Soil Sci Soc Am. *Res:* Water movement in soils; soil-plant relationships. *Mailing Add:* Agron/Soils Dept Mont State Univ Bozeman MT 59715

FERGUSON, ALEXANDER CUNNINGHAM, b Scotland; m 68; c 3. ALLERGY, IMMUNOLOGY. *Educ:* Univ Glasgow, MB, ChB, 67; Royal Col Physicians Glasgow, DCH, 69; FRCP(C), 72; Am Bd Allergy & Immunol, dipl, 75. *Prof Exp:* Intern/resident med, surg, pediat & infectious dis, Univ Glasgow, 67-69; resident pediat, Univ Western Ont, 69-70 & Hosp for Sick Children, Toronto, 70-72; res fel pediat immunol/allergy, Univ Calif, Los Angeles, 72-74; lectr pediat, Queen's Univ, 74-76; asst prof pediat, 76-82, ASSOC PROF PEDIAT, UNIV BC, 82- *Concurrent Pos:* Can Pediat Soc fel, 72-74; grants, Med Res Coun Can, 75-, BC Health Res Found, BC Med Serv Found & BC Lung Asn, 77- *Honors & Awards:* Ross Award, Can Pediat Soc, 74. *Mem:* Am Acad Allergy; Can Pediat Soc; Can Soc Allergy & Clin Immunol; Royal Col Physicians Can; Can Med Asn. *Res:* Developmental immunology; immunologic function and nutrition; pathophysiology and treatment of asthma; growth and development in allergic children. *Mailing Add:* Dept Pediat Univ BC 4480 Oak St Vancouver BC V6H 3V4 Can

FERGUSON, COLIN C, b Winnipeg, Man, Oct 3, 21; m 49; c 4. SURGERY. *Educ:* Univ Man, MD, 45; McGill Univ, dipl, 52; FRCS(C), 53. *Prof Exp:* Demonstr path, Univ Man, 44-45, head dept, 53-69, prof surg, 53-85; RETIRED. *Concurrent Pos:* Harrison fel surg, Univ Pa, 48-49; res fel pediat surg, Boston Children's Hosp, 51-52; teaching fel surg, Harvard Univ, 52-53; surgeon-in-chief, Children's Hosp, 54-73; guest lectr, Univ Edinburgh, 56; mem coun, Royal Col Physicians & Surgeons Can, 62-70, chmn comt gen surg, 68-72; mem, Can Coun Hosp Accreditation, 70-74; head pediat surg, Children's Ctr, 73-81. *Mem:* Am Col Surg; Am Surg Asn. *Res:* Cardiac and pediatric surgery. *Mailing Add:* Health Sci Ctr 9-525 Wellington Crescent Winnipeg MB R3M 0A1 Can

FERGUSON, DALE CURTIS, b Dayton, Ohio, Aug 21, 48; m 71, 81; c 2. RADIO ASTRONOMY, AERONAUTICAL & ASTRONAUTICAL ENGINEERING. *Educ:* Case Western Reserve Univ, BS, 70; Univ Ariz, PhD(astrophys), 74. *Prof Exp:* Vis asst prof astron, La State Univ, 74-75; res fel, Max Planck Inst Radio Astron, 75-77; asst prof physics, NY Univ, 77-78; resident pulsar-observer, Nat Astron & Ionosphere Ct, Cornell Univ, Arecibo, 78-81; asst prof physics, Southeast Mo State Univ, 81-82; sr res assoc, Case Western Reserve Univ, 82-83; PHYSICIST, NASA LEWIS RES CTR, 83-; ASTRON INSTR, BALDWIN-WALLACE COL, 85- *Mem:* AAAS; Am Astron Soc; Royal Astron Soc. *Res:* Pulsars; special relativity; polarization of light; atomic oxygen degradation of materials; arcing on space materials. *Mailing Add:* 168 Edgewood Dr Berea OH 44017-1412

FERGUSON, DALE VERNON, b Tulsa, Okla, Nov 29, 43; m 63; c 2. MICROBIOLOGY. *Educ:* Okla State Univ, BS, 66, MS, 67, PhD(microbiol), 70. *Prof Exp:* Res scientist virol, Armour-Baldwin Labs, 67-68; assoc prof biol, 70-80, PROF BIOL, UNIV ARK, LITTLE ROCK, 81- *Concurrent Pos:* Vis scientist, Nat Toxicol Ctr. *Mem:* Am Soc Microbiol. *Res:* antimicrobial action of antibiotics and chemotherapeutic agents; immunotoxicology. *Mailing Add:* Dept Biol Univ Ark 33rd & University Ave Little Rock AR 72204

FERGUSON, DAVID B, b Conrad, Mont, May 19, 26; m 50; c 2. PLANT BREEDING. *Educ:* Mont State Univ, BS, 50; Univ Minn, PhD(plant genetics), 62. *Prof Exp:* Asst agron, Mont State Univ, 54-57; asst plant genetics, Univ Minn, 57-62; supt plant breeding, Plains Br Sta, NMex State Univ, 62-66; res mgr plant breeding, Northrup, King & Co, Lubbock, Tex Br, 66-70 & Woodland, Calif Br, 70-74; plant breeder, David & Sons, Inc, Fresno, Calif, 74-80 & Seeds Group, Stauffer Chem Co, 80-83; RETIRED. *Mem:* AAAS; Am Soc Agron; Crop Sci Soc Am; Am Genetic Asn. *Res:* Corn breeding; wheat, sorghum and sudan breeding; breeding cucurbita for seed; breeding confectionary sunflowers; breeding oil sunflowers. *Mailing Add:* 6225 N Callisch Ave Fresno CA 93710

FERGUSON, DAVID JOHN, b Sandwich, Ill, May 24, 39; m 60; c 2. MATHEMATICS. *Educ:* Univ Idaho, BS, 64, PhD(math), 71. *Prof Exp:* Asst prof, 70-74, ASSOC PROF MATH, BOISE STATE UNIV, 74- *Mem:* Am Math Soc; Math Asn Am. *Res:* Structure of nonassociative nilalgebras. *Mailing Add:* Dept Math Boise State Univ Boise ID 83701

FERGUSON, DAVID LAWRENCE, b St Louis, Mo, Aug 19, 49. INTELLIGENT SYSTEMS. *Educ:* Southeast Mo State Univ, BS, 71; Univ Calif, Los Angeles, MA, 75; Univ Calif, Berkeley, PhD(math educ), 80. *Prof Exp:* Consult math & sci, Harper & Row, Scott Foresman & Co, 74-76; res assoc group math sci educ, Univ Calif, Berkeley, 77-79, teaching assoc math, 79-81; asst prof, 81-87, ASSOC PROF, TECHNOL SOC, STATE UNIV NY, STONY BROOK, 87- *Concurrent Pos:* Dir acad math & sci, Partnership Prog, Univ Calif, Berkeley, 79-81. *Mem:* Math Asn Am; Fedn Am Scientists; Am Statist Asn; AAAS; Asn Comput Mach; Inst Elec & Electronics Engrs Comput Soc. *Res:* Quantitative methods; applications of cognitive science and artificial intelligence to problem solving in mathematics, science, and engineering; mathematics, science, and engineering education; evaluation of computer-based learning environments. *Mailing Add:* Dept Technol Soc Col Eng & Appl Sci State Univ NY Stony Brook NY 11794-2250

FERGUSON, DONALD JOHN, b Minneapolis, Minn, Nov 19, 16; m 43; c 3. SURGERY. *Educ:* Yale Univ, BS, 39; Univ Minn, MD, 43, MS & PhD(surg), 51. *Prof Exp:* From asst prof to prof surg, Univ Minn, 52-60; prof, 60-87, EMER PROF SURG, UNIV CHICAGO, 87- *Mem:* Am Col Surg; Am Surg Asn; Soc Univ Surg. *Res:* Surgical research. *Mailing Add:* 5629 S Blackstone Chicago IL 60637

FERGUSON, DONALD LEON, veterinary parasitology; deceased, see previous edition for last biography

FERGUSON, EARL J, b Dallas, Tex, June 30, 25; m 46; c 3. INDUSTRIAL ENGINEERING. *Educ:* Tex A&M Univ, BS, 49; Okla State Univ, MS, 59, PhD(indust eng), 64. *Prof Exp:* Time study engr, Montgomery Ward & Co, 49-51; mfg engr, Gen Dynamics Corp, 51-56; assoc prof indust eng, Okla State Univ, 56-69, prof indust eng & mgt, 69-86; RETIRED. *Mem:* Am Inst Indust Engrs; Nat Soc Prof Engrs. *Res:* Management and management science; safety engineering; statistical quality control. *Mailing Add:* 3020 N Lincoln Stillwater OK 74075

FERGUSON, EARL WILSON, b Lebanon, Pa, Aug 29, 43; m 65; c 3. INTERNAL MEDICINE, CARDIOVASCULAR DISEASES. *Educ:* Baylor Univ, BA, 65; Univ Tex Med Br, Galveston, MD, 70, PhD(physiol), 70. *Prof Exp:* Intern, Univ Tex Med Br, Galveston, 70-71; resident, Duke Univ Med Ctr, NC, 71-73, fel cardiol, 73-75; cardiologist, Wilford Hall USAF Med Ctr, Lackland AFB, 75-76; asst prof biochem & med, 76-80, ASSOC PROF, PHYSIOL & MED, UNIFORMED SERV HEALTH SCI UNIV, BETHESDA, MD, 80- *Concurrent Pos:* Consult, Surgeon Gen USAF Internal Med, Physiol & Cardiol, 80-; rep, Interagency Tech Comt Heart, Blood Vessel, Lung, Blood Dis & Blood Resources, 81-; nat fac, Advan Cardiac Life Support, Am Heart Asn, 81- *Mem:* Am Physiol Soc; Am Fedn Clin Res; fel Am Col Cardiol; fel Am Col Physicians; Aerospace Med Asn. *Res:* Fibrinogen structure and fibrin crosslinking; evaluating effects of exercise and physical conditioning on blood clotting, fibrinolysis, platelets, red blood cells, selected endocrine studies and thermal adaptations. *Mailing Add:* 7100 CSW Med Ctr Wiesbaden APO New York NY 09220-5300

FERGUSON, EDWARD C, III, b Beaumont, Tex, Mar 11, 26. OPHTHALMOLOGY. *Educ:* Northwestern Univ, BS, 46, BM, 49, MD, 50; Am Bd Ophthal, dipl, 57. *Prof Exp:* Asst prof ophthal, Col Med, Univ Iowa, 56-57; assoc prof, 64-69, PROF OPHTHAL, UNIV TEX MED BR GALVESTON, 69-, CHMN DEPT, 64- *Concurrent Pos:* Fel, Howe Lab Ophthal, Mass Eye & Ear Infirmary, 56-57; Head fel ophthal, 57. *Mem:* AMA; Asn Res Vision & Ophthal; Am Acad Ophthal & Otolaryngol; Am Col Surgeons; Sigma Xi. *Mailing Add:* Dept Ophthal Univ Tex Med Br Galveston TX 77550

FERGUSON, FREDERICK PALMER, physiology, health science administration; deceased, see previous edition for last biography

FERGUSON, GARY GENE, b East St Louis, Ill, Jan 2, 40; m 63; c 1. PHARMACOLOGY. *Educ:* Univ Houston, BS, 63; Baylor Univ, MS, 65; Univ Colo, Boulder, PhD(pharmacol), 69. *Prof Exp:* Instr pharmacol, Univ NMex, 65-69; asst prof, 69-73, ASSOC PROF PHARMACOL, NORTHEAST LA UNIV, 73- *Concurrent Pos:* Consult pharmacol, Enviro-Med Labs, Ruston, La, 80- *Res:* Cardiovascular pharmacology; psychopharmacology; drug screening. *Mailing Add:* Dept Pharm Northeast La Univ Monroe LA 71209

FERGUSON, GARY GILBERT, b London, Ont, Aug 30, 41; m 64; c 3. NEUROSURGERY, BIOPHYSICS. *Educ:* Univ Western Ont, BA, 61, MD, 65, PhD(biophysics), 70; FRCS(C), 73, FACS, 77. *Prof Exp:* asst prof, 73-81, assoc prof neurosurg & biophys, 81-84, ASST PROF SURG, UNIV WESTERN ONT, 74-, PROF NEUROSURG & BIOPHYS, 84- *Honors & Awards:* Ann Award, Am Acad Neurol Surg, 70. *Mem:* Can Med Asn; Can Neurosurg Soc; Cong Neurol Surgeons; Royal Col Physicians & Surgeons Can; Am Asn Neurol Surgeons. *Res:* Cerebral blood flow; application to cerebrovascular disease. *Mailing Add:* Univ Hosp 339 Windermere Rd London ON N6A 5A5 Can

FERGUSON, GEORGE ALONZO, b Washington, DC, May 25, 23; m 66; c 5. NUCLEAR ENGINEERING, SOLID STATE PHYSICS. *Educ:* Howard Univ, BS, 47, MS, 48; Cath Univ Am, PhD(physics), 65. *Prof Exp:* Res asst physics, Univ Pa, 48-50; chmn dept physics, Clark Col, Ga, 50-53; res scientist, Naval Res Lab, Washington, DC, 54-67; prof physics, 67-80, PROF ENG, HOWARD UNIV, 80- *Concurrent Pos:* AEC/NASA/Pepco res grants, 67-; consult, Nuclear Regulatory Comn, 73- *Mem:* Am Phys Soc; AAAS; Am Nuclear Soc; Am Asn Physics Teachers; Sigma Xi. *Res:* Properties of materials and instrumentation in nuclear engineering; solid state physics, especially structure determination. *Mailing Add:* Sch Eng Howard Univ Washington DC 20059

FERGUSON, GEORGE E(RNEST), b Stillwater, Minn, Apr 2, 06; m 29; c 1. HYDRAULIC ENGINEERING. *Educ:* Univ Minn, BCE, 28. *Prof Exp:* Hydraul engr, Water Resources Div, Tex, 28-31, Hawaii, 31-37 & Washington, DC, 37-40, dist engr, Fla, 40-46, staff officer, 46-48, chief prog control br, 48-55, regional hydrologist, Atlantic Coast, 55-72, HISTORIAN-WRITER, US GEOL SURV, 74- *Mem:* Am Soc Civil Engrs; hon mem Am Water Works Asn; Am Geophys Union. *Res:* Hydrologic investigations. *Mailing Add:* US Geol Surv Nat Ctr Mail Stop 425A Reston VA 22092

FERGUSON, GEORGE RAY, b Bolivar, La, Jan 8, 15; m 56; c 5. ENTOMOLOGY. *Educ:* Ore State Col, BS, 36, MS, 39; Ohio State Univ, PhD(entom), 41. *Prof Exp:* Asst, Ore State Col, 37-39, asst entomologist, 41-43; res assoc, Crop Protection Inst, 43-45; chief entomologist, Geigy Chem Corp, 45-47, tech dir, 48-53, pres agr chem div, 53-69, exec vpres corp, 69-70, vpres, Ciba-Geigy Corp, 70-72; PROF ENTOM, ORE STATE UNIV, 73- *Mem:* Fel AAAS; Entom Soc Am; Weed Sci Soc Am; Am Chem Soc; Sigma Xi. *Res:* Insecticides; agriculture pest control. *Mailing Add:* 3343 NW Walnut Blvd Corvallis OR 97330

FERGUSON, HARRY, b Dayton, Ohio, May 1, 14; m 41. MATHEMATICS, MECHANICS. *Educ:* Boston Univ, BS, 39; Harvard Univ, AM, 49; Univ Pittsburgh, PhD(math), 58. *Prof Exp:* Instr math, Northeastern Univ, 39-43, Bowdoin Col, 43-44, Tufts Univ, 44-47, Northeastern Univ, 47-48 & Ohio Univ, 48-50; mathematician, Wright-Patterson AFB, Ohio, 50-56, aeronaut res engr fluid mech, 56-59; assoc prof math, 59-66, PROF ENG SCI, UNIV CINCINNATI, 66- *Concurrent Pos:* Mem, Appl Math Br, Wright Air Develop Ctr, 50-59. *Mem:* Am Math Soc; Math Asn Am; Am Soc Eng Educ. *Res:* Boundary value problems; turbulence; complex variables; Laplace transform; rigid body mechanics. *Mailing Add:* 5105 Weston Circle Dayton OH 45429

FERGUSON, HARRY IAN SYMONS, physics; deceased, see previous edition for last biography

FERGUSON, HELAMAN ROLFE PRATT, b Salt Lake City, Utah, Aug 11, 40; m 63; c 7. COMPUTER SCIENCES. *Educ:* Hamilton Col, AB, 62; Brigham Young Univ, MS, 66; Univ Wash, MS & PhD(math), 71. *Prof Exp:* Asst & instr, Univ Wash, 66-71; prof, Princeton Univ, 83-84; asst math, Brigham Young Univ, 65-66, from asst prof to prof math, 71-91; RES SCIENTIST, SUPERCOMPUTING RES CTR, 88- *Concurrent Pos:* Consult, Impact Resources, Inc, 81-82, Knowledge Eng, Inc, 84-86, Computer Systs Architects, 85-87, Nat Security Agency, 87-88. *Mem:* Am Math Soc; Math Asn Am; Soc Indust & Appl Math; Int Sculpture Ctr. *Res:* Harmonic analysis; group representations; computational number theory; plasma physics; computer algorithms; mathematical sculpture. *Mailing Add:* Supercomputing Res Ctr 17100 Science Dr Bowie MD 20715-4300

FERGUSON, HERMAN WHITE, b Chapel Hill, Tenn, Dec 28, 16; m 43; c 2. ECONOMIC GEOLOGY. *Educ:* Vanderbilt Univ, BA, 39, MS, 40. *Prof Exp:* Geol aide to asst geologist, Div Geol, Tenn Dept Conserv, 40-46, asst state geologist, 46-51, state geologist, 51-52; geologist, Tenn Coal & Iron Div, 52-57, sr geologist, Mich Limestone Div, 57-64, mgr geol invests stone & coal, 64-65, Sr geologist, Int & Resource Develop Dept, US Steel Corp, 65-; RETIRED. *Mem:* Inst Mining Engrs; fel Geol Soc Am; Soc Econ Geologists. *Res:* Structural and economic geology. *Mailing Add:* 600 Chad Dr Rocky Mt NC 27803

FERGUSON, HUGH CARSON, b Detroit, Mich, July 13, 21; m 55; c 2. PHARMACOLOGY. *Educ:* Wayne State Univ, BS, 48; Purdue Univ, MS, 50, PhD, 52. *Prof Exp:* Asst prof pharmacol, Univ NMex, 52-58; assoc prof, Col Pharm, Ohio Northern Univ, 58-61; head, Pharmacol Lab, Distillation Prod Indust Div, Eastman Kodak Co, 61-66; group leader, 66-69, sect leader, 69-70, prin investr, 70-74, SR PRIN INVESTR PHARMACOL, RES CTR, MEAD JOHNSON & CO, 74- *Mem:* Am Pharmaceut Asn; Am Chem Soc; Am Soc Pharmacol & Exp Therapeut. *Res:* Pharmacological investigation of plant products; cause of hypertension with relation to kidney function; relationship of stress to disease; new methods to evaluate drug action. *Mailing Add:* 3030 W Wardcliffe Dr Peoria IL 61604-2160

FERGUSON, JAMES HOMER, b San Antonio, Tex, July 26, 36; m 59; c 4. ENVIRONMENTAL PHYSIOLOGY. *Educ:* Sul Ross State Univ, BS, 58; Univ Ariz, PhD(zool), 64. *Prof Exp:* Asst zool, Univ Ariz, 61-64; from asst prof to assoc prof zool, Univ Idaho, 64-73; prof, Univ Tex, San Antonio, 73-74; prof zool, Univ Idaho, 75-83; CONSULT, 83- *Concurrent Pos:* Vis asst prof, Univ Iowa, 69; NIH spec fel, 69. *Mem:* AAAS; Am Soc Ichthyol & Herpet; Am Soc Zool; Am Physiol Soc. *Res:* Evolution; physiology of temperature adaptations in vertebrates; mammalian and environmental physiology. *Mailing Add:* MGK-3F Orofino ID 83544

FERGUSON, JAMES JOSEPH, JR, b Glen Cove, NY, Feb 1, 26; m 52; c 4. BIOCHEMISTRY, MEDICINE. *Educ:* Univ Rochester, BA, 46, MD, 50. *Prof Exp:* From intern to asst resident, Mass Gen Hosp, 50-52, resident, 55; assoc, Univ Pa Sch Med, 59-63, from asst prof to prof biochem & med, 63-87, assoc dean instrnl res & grad educ, 75-87; AT NAT LIBR MED, BETHESDA, MD, 86- *Concurrent Pos:* Res fel biochem, Western Reserve Univ, 56-58; Markle fel, 60-64. *Mem:* Am Soc Biol Chem; Endocrine Soc. *Res:* Metabolic regulation; hormone action; information management; database design. *Mailing Add:* Spec Info Serv Nat Libr Med 5600 Wisconsin Ave Apt 1607 Chevy Chase MD 20815

FERGUSON, JAMES KENNETH WALLACE, b Tamsui, Formosa, Japan, Mar 19, 07; m 33; c 4. PHARMACOLOGY. *Educ:* Univ Toronto, BA, 28, MA, 29, MD, 32. *Prof Exp:* From instr to asst prof physiol, Sch Med, Univ Western Ont, 34-36; asst prof, Ohio State Univ, 36-38; asst prof pharmacol, Univ Toronto, 38-41, prof & head dept, 45-55; dir, Connaught Med Res Labs, 55-72; RETIRED. *Concurrent Pos:* Nat Res Coun fel, 33-34. *Mem:* Am Soc Pharmacol & Exp Therapeut; Am Physiol Soc; Can Physiol Soc; fel Royal Soc Can. *Res:* Anoxia and oxygen equipment; antithyroid drugs; carbon dioxide in tissues; uterine motility; anti-alcoholic drugs. *Mailing Add:* 56 Clarkehaven St Thornhill ON L4J 2B4 Can

FERGUSON, JAMES L, b Thomas, Okla, July 9, 47; m 78; c 2. CARDIOVASCULAR PHYSIOLOGY, CIRCULATORY SHOCK. *Educ:* Southwestern State Col, Okla, BS, 69; NTex State Univ, MS, 72; Purdue Univ, PhD(physiol), 75. *Prof Exp:* Fel, La State Univ Med Ctr, 75-78; ASST PROF PHYSIOL, UNIV ILL MED CTR, 78- *Concurrent Pos:* Lectr, Am Physiol Soc; prin investr, Chicago Heart Asn, 81-83. *Mem:* Am Physiol Soc; Shock Soc; Sigma Xi. *Res:* Study of alterations of regional distribution of blood flow during states of shock and trauma; mechanisms of control of skeletal muscle, splanchnic and cerebral blood flow; pharmacological approaches to altering blood flow to the above vascular beds. *Mailing Add:* Dept Physiol & Biophys Univ Ill Health Sci Ctr PO Box 6998 Chicago IL 60680

FERGUSON, JAMES MALCOLM, b Chicago, Ill, June 6, 31; m 70. NUCLEAR PHYSICS. *Educ:* Antioch Col, BS, 53; Mass Inst Technol, PhD, 57. *Prof Exp:* Physicist, US Naval Radiol Defense Lab, 57-69; PHYSICIST, LAWRENCE LIVERMORE LAB, 69- *Mem:* Fel Am Phys Soc; Am Nuclear Soc. *Res:* Neutron transport; nuclear reactions; theory; nuclear fission. *Mailing Add:* 1856 Grand View Dr Oakland CA 94618

FERGUSON, JAMES MECHAM, b Washington, DC, Apr 16, 41; m 68; c 2. PSYCHIATRY, BEHAVIORAL BIOLOGY. *Educ:* Stanford Univ, BA, 64, MD, 71. *Prof Exp:* Res asst & assoc psychiat, Med Sch, Stanford Univ, 64-69; intern med & surg, Med Sch, Univ Utah, 71-72; resident, Med Sch, Stanford Univ, 72-75; ASST PROF PSYCHIAT, MED SCH, UNIV CALIF SAN DIEGO, 75-; CHIEF AMBULATORY PSYCHIAT & MENT HYG CLIN, VET ADMIN HOSP, SAN DIEGO, 75- *Concurrent Pos:* Attend physician, Univ Hosp, San Diego, 75- *Mem:* Am Psychiat Asn; Asn Advan Behav Ther; Asn Psychophysiol Study Sleep; Biofeedback Res Soc. *Res:* Behavior therapy; behavioral medicine; development of behavioral medicine and behavior change programs for use by non-physicians. *Mailing Add:* Dept Psychiat Univ Calif San Diego Sch Med 9834 Genesee Ave No 427 La Jolla CA 92037

FERGUSON, JOHN ALLEN, b Cincinnati, Ohio, Dec 20, 45; m 68; c 5. INDUSTRIAL CHEMISTRY. *Educ:* Univ Cincinnati, BS, 67; Univ NC, PhD(chem), 71. *Prof Exp:* Fel chem, Univ NC, Chapel Hill, 71; sr res chemist, Drackett Co, 73-75; group leader prod develop, Clairol Res Labs, 75-76, sect head prod develop, 76-78; sect mgr prod develop, 78-83, RES DIR, DRACKETT CO, 83- *Mem:* Soc Cosmetic Chemists; Am Chem Soc. *Res:* Hazard evaluation; analytical techniques; qfd in product development-goal/qpc; emulsion theory; sensory evaluation. *Mailing Add:* Drackett Res Labs 5020 Spring Grove Ave Cincinnati OH 45232

FERGUSON, JOHN BARCLAY, b Baltimore, Md, July 5, 47; m 70; c 2. BIOCHEMISTRY. *Educ:* Brown Univ, ScB, 69; Yale Univ, M Phil, 71, PhD(biol), 73. *Prof Exp:* Res fel chem, Harvard Univ, 73-77; asst prof, 77-83, ASSOC PROF BIOL, BARD COL, 83- *Concurrent Pos:* NIH fel, 74-76. *Mem:* AAAS; Am Soc Microbiol; NY Acad Sci. *Res:* Enzymology; purification and characterization of enzymes; substrate specificity; active-site-directed inhibition. *Mailing Add:* Dept Biol Bard Col Annandale-on-Hudson NY 12504

FERGUSON, JOHN CARRUTHERS, b Tuscaloosa, Ala, Mar 2, 37; m 61; c 3. INVERTEBRATE ZOOLOGY, PHYSIOLOGY. *Educ:* Duke Univ, BA, 58; Cornell Univ, MA, 61, PhD(invert zool), 63. *Prof Exp:* From asst prof to assoc prof, 63-72, PROF BIOL, ECKERD COL, 72-, COORDR BIOL PROG, 85- *Concurrent Pos:* NSF grants, 64-; vis investr, Marine Biol Lab, Woods Hole, 66; proj leader, Marine Biol Prog, Jamaica, 68 & 69; vis investr, Friday Harbor Labs, Wash, 70 & 88; fac dir, Eckerd Col London Study Centre, 90. *Mem:* AAAS; Am Soc Zoologists; Am Micros Soc. *Res:* Physiology and ecology of starfish; nutrient translocation; utilization of dissolved nutrients by marine invertebrates; water balance. *Mailing Add:* Dept Biol Eckerd Col St Petersburg FL 33733

FERGUSON, JOHN HOWARD, b Edinburgh, Scotland, Mar 1, 02; nat US; m 27, 55; c 6. PHYSIOLOGY. *Educ:* Univ Cape Town, BA, 21, DrSc, 57; Oxford Univ, BA, 25, MA, 31; Harvard Univ, MD, 28. *Prof Exp:* Lectr pharmacol, Univ Cape Town, 23, asst prof bact, 28-31; asst path, Harvard Univ, 26-28; instr physiol, Sch Med, Yale Univ, 31-34; asst prof physiol & pharmacol, Sch Med, Univ Ala, 34-35, assoc prof, 35-37; asst prof pharmacol, Univ Mich, 37-43; from actg head dept to head dept, Med Sch, Univ NC, Chapel Hill, 43-67, prof physiol, 43-70, emer prof physiol, 70-; RETIRED. *Mem:* Emer mem AAAS; emer fel Am Col Physicians; emer mem Am Physiol Soc; emer mem Int Soc Hemat. *Res:* Blood coagulation and related fields. *Mailing Add:* 226 Glandon Dr Chapel Hill NC 27514

FERGUSON, JOSEPH GANTT, b Charleston, WVa, May 26, 21; m 48, 79; c 1. RESEARCH ADMINISTRATION. *Educ:* Clemson Univ, BS, 42. *Prof Exp:* Lab analyst mfg tech, Southern Kraft Div, 47-51, develop engr, 51-54, second asst chief chemist, 54-55, res proj leader pulp & paper res & develop, 55-57, first asst chief chemist mfg tech, 57, group leader, res assoc, sr res assoc, chief paper res & first asst dir res, 57-66, dir res res admin, 66-75, mgr mfg res, Mfg & Eng Serv, 75-77, MGR TECH & ADMIN SERV, ERLING RIIS RES LAB, INT PAPER CO, 77- *Mem:* Tech Asn Pulp & Paper Indust. *Res:* Management of applied research for the pulp, paper and wood products business; chemical and mechanical engineering; wood and paper chemistry; polymer and environmental sciences; management of creativity. *Mailing Add:* Corp Eng PO Box 160707 Mobile AL 36616

FERGUSON, JOSEPH LUTHER, JR, b Utica, Miss, May 22, 41. PLASMA PHYSICS, OPTICS. *Educ:* Miss State Univ, BS, 63; Vanderbilt Univ, PhD(plasma physics), 69. *Prof Exp:* From asst prof to assoc prof, 68-80, PROF PHYSICS, MISS STATE UNIV, 80- *Mem:* Am Asn Physics Teachers. *Res:* Electromagnetic shock tube studies and laser-produced plasma studies. *Mailing Add:* Dept Physics Miss State Univ Box 5167 Miss State MS 39762

FERGUSON, KAREN ANNE, b Nebo, Ill, Oct 25, 42. BIOCHEMISTRY, LIPID CHEMISTRY. *Educ:* Western Ill Univ, BS, 63; Bryn Mawr Col, PhD(biochem), 71. *Prof Exp:* Fel biochem, Wash Univ, 72-74; asst prof chem, Eastern Ill Univ, 74-78; ASST PROF BIOCHEM, STATE UNIV NY, BUFFALO, 78- *Mem:* AAAS; NY Acad Sci; Am Chem Soc. *Res:* Membrane-bound enzymes; fatty acid desaturation and elongation in Tetrahymena; metabolic control by sterols. *Mailing Add:* 116 Park Ave State Univ of NY Hightstown NJ 08520-4122

FERGUSON, LAING, b Dunfermline, Scotland, Apr 25, 35; m 60; c 3. GEOLOGY, PALEONTOLOGY. *Educ:* Univ Edinburgh, BSc, 57, PhD(paleont), 60. *Prof Exp:* Nat Res Coun Can fel, Univ Alta, 60-62; from asst prof to assoc prof, 62-78, head dept, 73-88, PROF GEOL, MT ALLISON UNIV, 78-; SIR JAMES DUNN CHAIR, 82- *Concurrent Pos:* Fel, Univ Edinburgh, 69-70. *Mem:* fel Geol Asn Can (chmn paleont div, 86-87); Soc Econ Paleont & Mineral; Asn Petrol Geol; fel Geol Soc London; fel Geol Soc Am; fel Linnean Soc London. *Res:* Upper Paleozoic paleoecology, especially brachiopods; effects of environmental factors on criteria used in brachiopod taxonomy; Permian faunas from the high Arctic; distortion of fossils by compaction; trace fossils; Carboniferous ostracods; fossil trees and sedimentation rates. *Mailing Add:* Dept Geol Mt Allison Univ Sackville NB E0A 3C0 Can

FERGUSON, LE BARON O, b Oak Bluffs, Mass, Apr 20, 39; div; c 3. MATHEMATICS. *Educ:* Mass Inst Technol, SB, 61; Univ Wash, MA, 63, PhD(math), 65. *Prof Exp:* Assoc res engr, Boeing Airplane Co, 61-62; asst prof, 65-70, ASSOC PROF MATH, UNIV CALIF, RIVERSIDE, 70- *Concurrent Pos:* Vis assoc prof, Rensselaer Polytech Inst, 70-71 & Univ Nancy, 71-72; USAF Off Sci Res grant, 71- *Mem:* Am Math Soc; Math Asn Am; Sigma Xi. *Res:* Approximation theory. *Mailing Add:* Dept Math Univ Calif Riverside CA 92521

FERGUSON, LLOYD NOEL, b Oakland, Calif, Feb 9, 18; m 44; c 3. CHEMISTRY. *Educ:* Univ Calif, BS, 40, PhD(chem), 43. *Hon Degrees:* DSc, Howard Univ, 70, Coe Col, 79. *Prof Exp:* Asst chem, Nat Defense Res Comt Proj, Univ Calif, 41-44; asst prof, Agr & Tech Col, NC, 44-45; from asst prof to prof, Howard Univ, 45-65, head dept, 58-65; chmn dept, 68-71, prof chem, 65-86, EMER PROF CHEM, CALIF STATE UNIV, LOS ANGELES, 86- *Concurrent Pos:* Guggenheim fel, Carlsberg Lab, Copenhagen, 53-54; NSF fac fel, Swiss Fed Inst Technol, 61-62; vis prof, Univ Nairobi, 71-72; consult, Col Chem Consult Serv; vis scientist, Div Chem Educ, Am Chem Soc; mem chemother adv comt, Nat Cancer Inst, 72-75; mem US nat comt, Int Union Pure & Appl Chem, 73-76; mem, US Nat Sea Grant Rev Panel, 78-81; mem bd sci counselors, Nat Inst Environ Health Sci, 79-83; chmn, Div Chem Educ, Am Chem Soc, 80; United Negro Col Fund Scholar-at-Large, Bennett Col, Greensboro, NC, 84-85. *Honors & Awards:* Am Chem Soc Award in Chem Educ, 78; Distinguished Am Medallion, Am Found Negro Affairs, 76; Outstanding Contrib to Chem Educ Award, Coun Black Prof Engrs, 83. *Mem:* Fel AAAS; Am Chem Soc; fel Royal Soc Chem. *Res:* Taste and molecular properties; homoconjugation; chemistry of alicycles; cancer chemotherapy. *Mailing Add:* Dept Chem Calif State Univ 5151 State University Dr Los Angeles CA 90032

FERGUSON, MALCOLM STUART, b St Thomas, Ont, Apr 1, 08; nat US; m 38. MEDICAL PARASITOLOGY, COMMUNICATION SCIENCE. *Educ:* Univ Western Ont, BA, 32, MA, 34; Univ Ill, PhD(invert zool, parasitol), 37. *Prof Exp:* Demonstr zool, Univ Western Ont, 32-34; asst, Univ Ill, 34-36; res investr, Inst Parasitol, Macdonald Col, McGill Univ, 38; Royal Soc Can fel, Rockefeller Inst, 38-39, fel, 39-40, asst, 40-47; scientist dir, Commun Dis Ctr, USPHS, 47-61, chief med arts & photog br, NIH, 61-67; multi-media specialist, Nat Libr Med, 67-72; admin dir, Am Sci Film Asn, 72-74; expert prev med, Nat Naval Med Ctr, 74-80. *Mem:* Fel AAAS; Am Soc Parasitol; Am Soc Trop Med & Hyg. *Res:* Life cycles of trematodes; sterile culture of helminths; migration and localization of parasites within the host; application of the motion picture camera as a tool in study of living organisms; production of technical motion pictures on medical and public health subjects; development of learning resource centers in libraries; local and national environmental concerns. *Mailing Add:* 23053 Westchester Apt No R114 Port Charlotte FL 33980

FERGUSON, MARION LEE, b Washington, Iowa, Nov 27, 23; m 46; c 2. PHSIOLOGY. *Educ:* Univ Iowa, BA, 49; Iowa State Univ, MS, 55, PhD(physiol), 59. *Prof Exp:* Instr zool, Iowa State Univ, 53-55 & 57-60; head physiol lab, Life Sci Res Dept, Goodyear Aerospace Corp, 60-65; from assoc prof to prof, 69-85, dir conserv lab, 67-84, EMER PROF BIOL, KENT STATE UNIV, 85- *Concurrent Pos:* Vis lectr, Univ Colo, 61; dir joint flight res prog, USAF-Goodyear Aerospace Corp, 62-63; res investr, NASA Biosatellite Prog, Ames Res Ctr, 63-65; partic, BS/MD Prog, Col Med, Northeastern Ohio Univ, 75-84. *Res:* General and protozoan physiology; environmental biology. *Mailing Add:* 235 Seneca Trail Hartville OH 44632

FERGUSON, MARY HOBSON, b Canton, Miss, Apr 3, 27; m 46, 69; c 1. PHARMACY, COMMUNICATION SCIENCE. *Educ:* Univ Miss, BS, 61, PhD(pharm), 64. *Prof Exp:* Sr scientist, Alcon Labs, 64-67; from asst ed to ed, J Pharmaceut Sci, Am Pharmaceut Asn, 67-82, lit scientist, sci div, 67-82; RETIRED. *Res:* Pharmaceutical product development; scientific technical writing and editing; scientific information computerized systems. *Mailing Add:* 825 N Kathy Circle Canton MS 39046

FERGUSON, MICHAEL JOHN, b Toronto, Ont, May 7, 41; m 69; c 1. COMMUNICATION NETWORKS. *Educ:* Univ Toronto, BASc, 62; Calif Inst Technol, MSc, 63; Stanford Univ, PhD(electronic eng), 66. *Prof Exp:* Eng specialist, Ford Aerospace, Palo Alto, 66-68; assoc prof commun systs, McGill Univ, 68-76; res assoc, Aloha Proj, Univ Hawaii, 74-77; res scientist, Inst Appl Systs Anal, Austria, 76-78; mgr systs anal, Bell-Northern Res, 78-82; PROF TELECOMMUN, INRS-TELECOMMUN, 82- *Mem:* Fel Inst Elec & Electronics Engrs; Asn Comput Mach; Sigma Xi. *Res:* Computer communication performance analysis; overseeing models of protocols. *Mailing Add:* INRS Telecommun Three Place du Commerce Verdun PQ H3E 1H6 Can

FERGUSON, MICHAEL WILLIAM, b Chelsea, Mass, Sept 29, 47; c 2. PHYTOPATHOLOGY, BACTERIOLOGY. *Educ:* Calif State Univ, Long Beach, BA, 74; Calif State Polytech Univ, MS, 76; Kans State Univ, PhD(plant path), 81. *Prof Exp:* ASST PROF PLANT PATH, SDAK STATE UNIV, 81- *Mem:* Am Plytopath Soc. *Res:* Host-parasite interaction; plant disease physiology. *Mailing Add:* 1431 Seventh St Brookings SD 57006-1619

FERGUSON, NOEL MOORE, pharmacognosy, botany; deceased, see previous edition for last biography

FERGUSON, PAMELA A, b Berwyn, Ill, May 5, 43; m 65; c 2. FINITE GROUP THEORY, COMBINATORIES. *Educ:* Wellesley Col, BA, 65; Univ Chicago, MS, 66, PhD(math), 69. *Prof Exp:* Asst prof math, Northwestern Univ, 69-70; vis asst prof, 70-72, from asst prof to assoc prof, 72-81, PROF MATH, UNIV MIAMI, 81-, ASSOC PROVOST & DEAN GRAD SCH, 87- *Concurrent Pos:* Dir honors & privileged studies, Univ Miami, 85-87; chmn NSF liaison comt, Asn Women Math; NSF fel reviewer, 85-86, prin investr, NSF. *Mem:* Am Math Soc; Asn Women Math. *Res:* Finite group theory; combinatories; education and improved teaching. *Mailing Add:* 210 Ferre Bldg Grad Sch Univ Miami Coral Gables FL 33124-2220

FERGUSON, RAYMOND CRAIG, b Ft Morgan, Colo, May 28, 22; m 55, 68; c 2. MOLECULAR SPECTROSCOPY, NUCLEAR MAGNETIC RESONANCE. *Educ:* Iowa State Univ, BS, 48, MS, 50; Harvard Univ, PhD(phys chem), 53. *Prof Exp:* Jr chemist, Ames Lab, AEC, 48-50; teaching fel, Harvard Univ, 50-53; chemist, E I du Pont de Nemours & Co, Inc, 53-68, supvr, 68-74, res assoc, 74-85; CONSULT, CONDUX INC, 85-; ORAL HISTORIAN, CTR HIS CHEM, 86- *Mem:* AAAS; Am Chem Soc; Am Phys Soc; Soc Appl Spectros; Am Inst Chemists. *Res:* Nuclear magnetic resonance; infared and microwave spectroscopy; polymer structure analysis; computer analysis of spectra. *Mailing Add:* 300 Whitby Dr Wilmington DE 19803

FERGUSON, ROBERT BURY, b Cambridge, Ont, Feb 5, 20; m 48; c 3. MINERALOGY, CRYSTALLOGRAPHY. *Educ:* Univ Toronto, BA, 42, MA, 43, PhD(mineral), 48. *Prof Exp:* Asst, Royal Ont Mus Mineral, 42-43; meteorologist, Can Dept Transp, 43-45; demonstr geol sci, Univ Toronto, 45-47; from asst prof to prof, 47-85, EMER PROF MINERAL, UNIV MAN, 86- *Concurrent Pos:* Nat Res Coun Can fel, Crystallog Lab, Cambridge Univ, 50-51; vis scientist, Dept Geol & Mineral, Oxford Univ, 72-73 & Dept Geol & Mineral, Adelaide Univ, Australia, 79-80. *Honors & Awards:* Hawley Award, Mineral Asn Can, 80. *Mem:* Am Crystallog Asn; fel Mineral Soc Am; Mineral Asn Can (pres, 77); fel Royal Soc Can. *Res:* Crystal structures and crystal chemistry of rock-forming silicate minerals. *Mailing Add:* Dept Geol Sci Univ Man Winnipeg MB R3T 2N2 Can

FERGUSON, ROBERT LYNN, b Mayfield, Ky, Apr 1, 32; m 54, 86; c 2. RADIOBIOASSAY, DATA ACQUISITION. *Educ:* Murray State Col, BA, 54; Wash Univ, PhD(chem), 59. *Prof Exp:* Chemist, 59-66, RES STAFF MEM, OAK RIDGE NAT LAB, 66- *Concurrent Pos:* Vis scientist, Tech Univ Munich, 70-71, Europ Ctr Nuclear Res, Geneva, 85-86. *Mem:* Am

Chem Soc; Am Phys Soc. *Res:* Fission; heavy-ion induced nuclear reactions; ultrarelativistic nucleus-nucleus reactions; data management and process automation for radiobioassay. *Mailing Add:* Oak Ridge Nat Lab PO Box 2008 Oak Ridge TN 37831-6105

FERGUSON, ROBERT NICHOLAS, b Washington, DC, Apr 17, 44; m 71; c 1. BIO-ORGANIC CHEMISTRY, ANALYTICAL CHEMISTRY. *Educ:* Cath Univ Am, BA, 66; Johns Hopkins Univ, PhD(org chem), 71. *Prof Exp:* Staff fel biochem, Nat Inst Arthritis, Metab & Digestive Dis, 71-74; res scientist chem, Philip Morris, USA, 74-79, assoc sr scientist res & develop, 79-83, sr scientist, 83-85, mgr chem res, 85-90, MGR ANALYTICAL RES, PHILIP MORRIS, USA, 90- *Mem:* AAAS; Am Chem Soc; Sigma Xi. *Res:* Intramolecular isotope effects; protein structure and function; structure biological activity relationships; isolation and identification of natural products; pyrolysis reactions. *Mailing Add:* Philip Morris Res Ctr PO Box 26583 Richmond VA 23261

FERGUSON, ROGER K, b Oklahoma City, Okla, May 21, 37; c 2. CLINICAL PHARMACOLOGY, TOXICOLOGY. *Educ:* Univ Utah, MD, 65; Univ Iowa, MS, 70. *Prof Exp:* Dir clin pharmacol, Sch Med, Thomas Jefferson Univ, 77-85; chmn dept internal med, Univ Nev, Reno, 85-86. *Mem:* Am Soc Clin Pharmacol; Am Soc Pharmacol & Exp Therapeut; Int Soc Hypertension; fel Am Col Physicians; fel Am Col Chest Physicians. *Res:* Internal medicine. *Mailing Add:* Dept Med Savitt Med Sci Bldg Univ Nev Sch Med Reno NV 89557

FERGUSON, RONALD JAMES, b Toronto, Ont, Sept 18, 39. PREVENTIVE MEDICINE. *Educ:* Univ Western Ont, BA, 63; Univ Mich, MSc, 65, PhD(phys educ), 67. *Prof Exp:* Asst prof, 67-70, ASSOC PROF PHYS EDUC, UNIV MONTREAL, 70-, assoc prof prev med, 74-84. *Concurrent Pos:* Res assoc, Montreal Heart Inst, 70-; chmn res comt, Can Asn Sport Sci, 71-73. *Mem:* Fel Am Col Sports Med; Can Asn Sport Sci. *Res:* Exercise physiology; cardiac rehabilitation. *Mailing Add:* Dept Phys Educ Univ Montreal 2100 Edward Mont Peit Montreal PQ H3T 1J3 Can

FERGUSON, RONALD M, SURGERY. *Prof Exp:* PROF SURG & DIR, DIV TRANSPLANTS, OHIO STATE UNIV HOSP, 82- *Mailing Add:* Dept Surg Div Transplantation Ohio State Univ Hosp 259 Means Hall 1654 Upham Dr Columbus OH 43210

FERGUSON, SAMUEL A, pharmacology, physiology, for more information see previous edition

FERGUSON, SHIRLEY MARTHA, b Syracuse, NY, Mar 9, 23; m 51; c 3. PSYCHIATRY, NEUROPSYCHIATRY. *Educ:* Syracuse Univ, BA, 45, MD, 47; McGill Univ, dipl psychiat, 55. *Prof Exp:* Res fel psychiat, McGill Univ, 54-55; res fel neurol, Montreal Neurol Inst, 57-58; res assoc, Columbia Univ, 58-60; res assoc psychiat, Albert Einstein Col Med, 60-65, asst clin prof, 65-68; PROF PSYCHIAT & NEUROL SURG & DIR, NEUROPSYCHIAT DIV, MED COL OHIO, 68- *Concurrent Pos:* Spec fel, Interdept Inst Training & Res Behav & Neurol Sci, 63-65; adj prof, Mt Zion Hosp & Med Ctr, 69; dir med col units, Toledo Ment Health Ctr, 69- *Mem:* Soc Biol Psychiat; Am Psychiat Asn; Am Epilepsy Soc; Neurobehav Soc; Am Neuropsychiat Soc. *Res:* Epilepsy. *Mailing Add:* Med Col Ohio CS 10008 Toledo OH 43699-0008

FERGUSON, STEPHEN MASON, b Bozeman, Mont, Oct 29, 39; m 63; c 2. MATERIAL ANALYSIS. *Educ:* Mont State Univ, BS, 62; Univ Wash, MS, 64, PhD(nuclear physics), 69. *Prof Exp:* Assoc atomic physics, Kans State Univ, 69-71; res fel nuclear physics, Australian Nat Univ, 71-74; ACCELERATOR PHYSICIST, WESTERN MICH UNIV, 74- *Concurrent Pos:* Vis scholar, Atomic Energy Res Estab, Harwell Eng, 80-81. *Mem:* Am Phys Soc; Am Asn Physics Teachers. *Res:* Practical applications of particle accelerators; trace element analysis with particle induced x-ray emission. *Mailing Add:* Dept Physics Western Mich Univ Kalamazoo MI 49008-5151

FERGUSON, THOMAS, b Union, SC, June 30, 21. ZOOLOGY. *Educ:* Fisk Univ, BA, 43; Univ Iowa, MS, 48, PhD(zool), 55. *Prof Exp:* From instr to prof, J C Smith Univ, 48-62; PROF BIOL, DEL STATE COL, 62-, CHMN DEPT, 74- *Mem:* Am Soc Zoologists; Nat Inst Sci. *Res:* Neuroembryology. *Mailing Add:* Dept Biol Del State Col Dover DE 19901

FERGUSON, THOMAS LEE, b Dallas, Tex, Nov 8, 42; m 68; c 3. CHEMICAL ENGINEERING. *Educ:* Tex A&M Univ, BS, 64; Univ Mo, Columbia, MS, 82. *Prof Exp:* Chem engr petrochem prod, Monsanto Co, 64-65; chem engr viral & rickettsial agent prod, US Army Chem Corps, 65-67; chem engr petrochem prod, Monsanto Co, 67-68; chem engr, 68-76, prin chem engr, 76-79, head, Process Assessment Sect, 80-86, MGR, HAZARDOUS WASTE PROG, 86-; SR PROG MGR, RADIAN CORP. *Concurrent Pos:* Consult expert, US Environ Protection Agency, 75. *Mem:* Am Inst Chem Engrs. *Res:* Systems studies of the operation and environmental control of chemical and biological production facilities, especially pesticides, toxic chemicals, and hazardous wastes. *Mailing Add:* Radian Corp 2455 Horsepen Rd Suite 250 Herndon VA 22071

FERGUSON, THOMAS S, b Oakland, Calif, Dec 14, 29; m 58; c 2. MATHEMATICAL STATISTICS. *Educ:* Univ Calif, PhD(statist), 56. *Prof Exp:* PROF MATH, UNIV CALIF, LOS ANGELES, 56- *Res:* Game theory; decision theory. *Mailing Add:* Dept Math Univ Calif Los Angeles CA 90024

FERGUSON, WILLIAM ALLEN, b Roanoke, Mo, Mar 6, 17; m 42; c 4. MATHEMATICS. *Educ:* Mo Valley Col, BA, 37; Univ Ill, MA, 38, PhD(math), 46. *Prof Exp:* Asst instr, 38-41, 45-46, from instr to asst prof, 46-55, assoc prof math, Univ Ill, Urbana, 55-, Exec Secy Dept, 68-; RETIRED. *Mem:* Math Asn Am. *Mailing Add:* 1007 W Hill Champaign IL 61821

FERGUSON, WILLIAM E, b Oakland, Calif, July 11, 21; m 47; c 2. ENTOMOLOGY, ZOOLOGY. *Educ:* Univ Calif, Berkeley, BS, 46, MS, 56, PhD(entom), 61. *Prof Exp:* Teacher high sch, Calif, 47-56, counsr, 48-56; sr lab technician entom, Univ Calif, Berkeley, 56-60, NIH fel, 60-62; from asst prof to assoc prof, 62-70, NSF res grant, 64-66, dir ctr res & advan studies, 66-68, prof entom, 70-81, EMER PROF, SAN JOSE STATE UNIV, 81- *Mem:* Entom Soc Am; Soc Syst Zool; Asn Trop Biol; Sigma Xi. *Res:* Systematic entomology; biology and systematics of the hymenopterous family Mutillidae; insect parasite-predator-pathogen complexes; insect postembryonic development, especially differentiation of environmental from genetic influences; insect photography; protective coloration, form and behavior. *Mailing Add:* 245 Vista de Sierra Los Gatos CA 95030

FERGUSON, WILLIAM SIDNEY, b Chatfield, Minn, Nov 15, 27; m 48; c 2. AUTOMATED SYSTEMS INTEGRATION, ANALYTICAL CHEMISTRY. *Educ:* Ore State Univ, BS, 49, MS, 53; Univ Ill, PhD(chem), 56. *Prof Exp:* Chemist, Gen Elec Co, 49-53; res geochemist, Marathon Oil Co, 56-70; dir, Anal Chem Facil, Colo State Univ, 70-79; res supvr, 80-86, RES ASSOC, AMOCO PROD CO, RES, 87- *Mem:* AAAS; Am Chem Soc; Soc Mfg Engrs. *Res:* Design and custom integration of automated systems and robotics; environmental chemistry; trace metal analysis; microanalytical techniques; oceanography; organic geochemistry; fused salt electrochemistry. *Mailing Add:* AMOCO Prod Res PO Box 3385 Tulsa OK 74102

FERGUSON-MILLER, SHELAGH MARY, b Toronto, Ont, Mar 12, 42; m 73; c 2. BIOENERGETICS, MEMBRANE PROTEINS. *Educ:* Univ Toronto, BSc, 64, MA, 66; Univ Wis, PhD(biochem), 71. *Prof Exp:* Fel, Dept Biochem, Univ Oxford, Eng, 71-72; res assoc, Dept Chem, Northwestern Univ, Evanston, Ill, 72-77; from asst to assoc prof, 78-87, PROF, DEPT BIOCHEM, MICH STATE UNIV, EAST LANSING, 87-, ASSOC CHAIR DEPT, 90- *Mem:* Am Soc Biochem & Molecular Biol; NY Acad Sci; Sigma Xi; Am Chem Soc; Biophys Soc. *Res:* Structural and kinetic properties of electron transfer proteins involved in energy conservation in mitochondria; purified membrane proteins and their interaction with detergents and membranes. *Mailing Add:* Biochem Dept Biochem Bldg Mich State Univ East Lansing MI 48824

FERGUSSON, WILLIAM BLAKE, b Boston, Mass, Apr 24, 24; m 48; c 6. GEOLOGY. *Educ:* Boston Univ, BA, 52, MA, 53; Univ Ariz, PhD(geol), 65. *Prof Exp:* Instr geol, Boston Univ, 52-53; explor geologist, USAEC, 53-54; explor geologist & consult, Grand Junction, Colo, 54-55; instr geol, Univ Ariz, 55-59; geologist, Pa RR Co, 59-67; ASSOC PROF GEOL, VILLANOVA UNIV, 67- *Concurrent Pos:* Consult geologist; adj prof, Drexel Univ, 77-78; Fulbright sr scholar, 82-83. *Mem:* Fel Geol Soc Am; Asn Eng Geologists; Int Asn Eng Geol. *Res:* Geological processes relative to engineering and environmental geology, particularly slope stability, Karst topography, groundwater, weathering and erosion. *Mailing Add:* Dept Civil Eng Villanova Univ Villanova PA 19085

FERIN, JURAJ, b Topolcany, Czech; m 53; c 2. ENVIRONMENTAL MEDICINE, INHALATION TOXICOLOGY. *Educ:* Slovak Univ, Bratislava, MD, 50; Charles Univ, Prague, PhD(indust med), 55. *Prof Exp:* Physician, Med Sch, Slovak Univ, Bratislava, 50-51; res scientist, Inst Indust Med & Occup Dis, Bratislava, 54-59; sr scientist environ health, Inst Exp Hyg, Slovak Acad Sci, 59-68; prof environ med, dept radiation biol & biophys, 68-80, PROF TOXICOL, DEPT BIOPHYS & ENVIRON HEALTH SCI CTR, SCH MED & DENT UNIV ROCHESTER, 80- *Concurrent Pos:* Environ Res & Develop Agency & Nat Inst Environ Health Sci res grant, Sch Med & Dent, Univ Rochester, 69-74, US Dept Energy & Environ Protection Agency grants, 74-81; dep dir, Inst Exp Med, Slovak Acad Sci, 61-65; assoc prof, Sch Med, Slovak Univ, 65-68. *Mem:* AAAS; Int Soc for Aerosols Med; Am Thoracic Soc. *Res:* Air pollutants and the response of the respiratory system; particle clearance from the lung; occupational and environmental health; experimental pathology of the lung. *Mailing Add:* Dept Biophys Univ Rochester Sch Med & Dent Box EHSC Rochester NY 14642

FERINGA, EARL ROBERT, b Grand Rapids, Mich, May 30, 32. NEUROLOGY. *Educ:* Calvin Col, BS, 53; Northwestern Univ, MD, 57. *Prof Exp:* Intern, Philadelphia Gen Hosp, Pa, 57-58; clin pharmacologist cancer chemother, Nat Serv Ctr, NIH, 58, neuroanatomist, Lab Neuroanat Sci, 59-62; neuroanatomist, Col Physicians & Surgeons, Columbia Univ, 59; resident & assoc neurol, Colo Med Ctr, 62-64; from instr to prof neurol, Med Ctr, Univ Mich, Ann Arbor, 64-79, prof neurol, Dept Path, 75-79, chief neurol serv, Ann Arbor Vet Admin Hosp, 64-79; chief neurol serv, San Diego Vet Admin Hosp, 79-84; prof neurol, Univ Calif San Diego, 79-84; chief neuro serv, Vet Admin Med Ctr, Augusta, 84-90; PROF NEUROL, MED COL GA, AUGUSTA, 84- *Concurrent Pos:* Consult, Wayne County Gen Hosp, 66- *Mem:* Fel Am Acad Neurol; fel Am Col Physicians; Soc Clin Neurol; Soc Neurosci. *Res:* Neurology and neuroanatomy; regeneration of central nervous system, particularly the spinal cord. *Mailing Add:* Dept Neurol BIW-338 Med Col Ga Augusta GA 30912

FERINGTON, THOMAS EDWIN, b Lockport, NY, Oct 19, 26; m 49; c 2. POLYMER CHEMISTRY. *Educ:* Univ Buffalo, BA, 49; Calif Inst Technol, MS, 52; Princeton Univ, PhD(chem), 57. *Prof Exp:* Res chemist, Chemstrand Corp, 51-54; from instr to asst prof chem, Col Wooster, 57-61; res chemist, 61-65, res supvr, 65-69, res mgr, 69-73, SR DEVELOP ASSOC, W R GRACE & CO, 73- *Mem:* Am Chem Soc. *Res:* Physical chemistry of high polymers; dilute solution properties; kinetics; chemistry of organic sulfur compounds; rheology; chemistry of metallorganic compounds; materials science; processing and properties of polymers; photopolymerization; cross-linked systems. *Mailing Add:* 4011 Randolph Rd Wheaton MD 20902

FERKO, ANDREW PAUL, b Trenton, NJ, Aug 19, 42; m 71; c 3. DRUG ABUSE, TOXICOLOGY. *Educ:* Philadelphia Col Pharm & Sci, BS, 65; Hahnemann Med Col, PhD(pharmacol), 69. *Prof Exp:* From jr instr to asst prof, 67-81, ASSOC PROF PHARMACOL, HAHNEMANN UNIV, 81- *Concurrent Pos:* Lindback Found Teaching Award, 81. *Mem:* Sigma Xi; Am

Soc Pharmacol & Exp Therapeut; Soc Toxicol; Int Soc Study Xenobiotics. *Res:* Narcotic analgesics; dopamine; levodopa; lead; abused drugs; ethanol and physical dependence on ethanol. *Mailing Add:* Hahnemann Univ Sch Med Broad & Vine Philadelphia PA 19102

FERL, ROBERT JOSEPH, b Conneaut, Ohio, Jan 19, 54; m 80; c 1. RECOMBINANT DNA. *Educ:* Hiram Col, BA, 76; Ind Univ, MA, 78, PhD(genetics), 80. *Prof Exp:* Asst prof genetics, Dept Bot, Univ Fla, 80-85, assoc prof bot, 85-87, actg chmn bot, 86-87, assoc prof veg crops, 87-89, PROF VEG CROPS, UNIV FLA, 90- *Concurrent Pos:* Vis scientist, Div Plant Indust, Commonwealth Sci & Indust Res Orgn, Canberra, Australia, 81. *Mem:* Genetics Soc Am; Bot Soc Am; Am Asn Plant Physiologists. *Res:* Gene expression, structure and function in plants. *Mailing Add:* Dept Vegetable Crops Univ Fla Gainesville FL 32611

FERLAND, GARY JOSEPH, b Washington, DC, Oct 5, 51; m 82; c 1. INTERSTELLAR ASTRONOMY, SPECTROSCOPY. *Educ:* Univ Tex, BS, 73, PhD(astron), 78. *Prof Exp:* Staff mem, Inst Astron, Cambridge Univ, 78-80; from asst prof to assoc prof astron & physics, Univ Ky, 80-85; ASSOC PROF ASTRON, OHIO STATE UNIV, 85- *Concurrent Pos:* Consult, NASA, 81-83; prin investr, NSF, 80-82; sr vis fel, Cambridge Univ, 87. *Mem:* Am Astron Soc; Royal Astron Soc; Int Astron Union. *Res:* Numerical simulations of processes in ionized gasses, with application to interpreting the spectra of active galactic nuclei. *Mailing Add:* Astron 5040 Smit Lab Ohio State Univ Main Campus Columbus OH 43210

FERM, JOHN CHARLES, b East Liverpool, Ohio, Mar 21, 25; m 49; c 3. GEOLOGY OF COAL. *Educ:* Pa State Univ, BS, 46, MS, 48, PhD(mineral), 57. *Prof Exp:* Instr mineral, Pa State Univ, 51-52; geologist, US Geol Surv, 52-57; from asst prof to prof geol, La State Univ, 57-69; dir grad studies geol, Univ SC, 69-74, prof geol, 69-80, dir, Caro Coal Group, 77-80; dir grad studies, 81-84, PROF GEOL, UNIV KY, 80- *Concurrent Pos:* Consult var industs & pub agencies. *Honors & Awards:* Esso Distinguished Lectr, 81. *Mem:* AAAS; Geol Soc Am; Sigma Xi. *Res:* Coal geology; fluvial deltaic sedimentation; carboniferous stratigraphy; underground mine planning; resource estimates; computer applications. *Mailing Add:* Dept Geol Sci Univ Ky Lexington KY 40506

FERM, RICHARD L, b Kansas City, Mo, June 18, 24; m 53; c 3. CHEMICAL ENGINEERING. *Educ:* Univ Kans, BS, 44, MS, 45, PhD(chem), 48. *Prof Exp:* From instr to asst prof chem eng, Univ NMex, 48-55; res engr, Calif Res Corp, 55-61, sr res chemist, 61-69, sr res assoc, Chevron Res Co, 69-83,; DIR, US PARA PLATE CORP, 80-; PRES, INT FOUND EARTH CONSTRUCT, 83-, DIR, BLIP CORP, 87- *Concurrent Pos:* Consult, var cos. *Mem:* Am Chem Soc. *Res:* Petroleum products; emulsions and surface chemistry; ecological uses of asphalt; sulfur products; building materials; fuel chemistry and additives. *Mailing Add:* 3282 Theresa Lane Lafayette CA 94549

FERM, VERGIL HARKNESS, b West Haven, Conn, Sept 13, 24; m 48; c 4. BIOLOGY, EMBRYOLOGY. *Educ:* Col Wooster, BA, 46; Western Reserve Univ, MD, 48; Univ Wis, MS, 50, PhD, 55; Dartmouth Col, MA, 66. *Prof Exp:* Intern, St Luke's Hosp, Cleveland, Ohio, 48-49; asst zool, Univ Wis, 49-51 & 54-55; asst prof anat, Ind Univ, 55-57; assoc prof, Univ Fla, 57-61; assoc prof path, 61-66, PROF ANAT-EMBRYOL & CHMN DEPT, DARTMOUTH MED SCH, 66- *Concurrent Pos:* USPHS sr res fel, Univ Fla, 58-63. *Mem:* Am Asn Anat; Am Soc Zool; Am Soc Human Genetics; Am Soc Exp Path; Teratology Soc. *Res:* Placental physiology; teratology; experimental embryology. *Mailing Add:* Dept Anat-Cytol Dartmouth Med Sch Hanover NH 03756

FERMI, GIULIO, b Rome, Italy, Feb 16, 36; US citizen; m 60; c 2. MOLECULAR BIOLOGY, SYSTEMS ANALYSIS. *Educ:* Princeton Univ, MA, 57; Univ Calif, Berkeley, PhD(biophys), 62. *Prof Exp:* NSF fel, Max Planck Inst Biol, 61-62, Nat Inst Neurol Dis & Blindness fel, 62-63; staff mem, Inst Defense Anal, 64-69; dir systs eval group, Ctr Naval Anal, 69; consult systs anal, Inst Defense Anal, 70-71; TECH OFFICER, LAB MOLECULAR BIOL, MED RES COUN, 71- *Res:* Mutation and replication mechanisms in bacteriophages; optomotor reaction of musca domestica; system evaluation techniques for military communications and antisubmarine warfare; crystal structure of haemoglobin and derivatives. *Mailing Add:* Med Res Coun Lab Molecular Biol Hills Rd Cambridge England

FERMIN, CESAR DANILO, NEURO SCIENCES, DEVELOPMENTAL NEUROANATOMY. *Educ:* Fla Inst Technol, PhD(cell biol), 81. *Prof Exp:* ASST PROF COMMUN SCI, BAYLOR COL MED, 81- *Res:* Neuronal plasticity. *Mailing Add:* Dept Path Tulane Univ Med Sch 1430 Tulane Ave New Orleans LA 70112-2699

FERNALD, ARTHUR THOMAS, b Nottingham, NH, Dec 13, 17; m 55; c 1. GEOLOGY, APPLIED, ECONOMIC & ENGINEERING. *Educ:* Univ NH, BS, 41; Harvard Univ, AM, 51, PhD(geomorphol), 56. *Prof Exp:* Instr geol, Colby Col, 46-47; geologist, US Geol Surv, 51-83; RETIRED. *Mem:* Geol Soc Am. *Res:* Geologic, geomorphic and engineering studies of polar and desert regions, particularly Alaska, Greenland, Nevada, New Mexico. *Mailing Add:* 95 S Carr Lakewood CO 80226

FERNALD, RUSSELL DAWSON, b Chuquicamata, Chile, Nov 20, 41; US citizen; m 69; c 2. NEUROBIOLOGY. *Educ:* Swarthmore Col, BSEE, 63; Univ Pa, PhD(biophys), 68. *Prof Exp:* Fel neurophysiol of visual cortex, Max Planck Inst Psychiat, 68-70; staff scientist neurobiol, Mac Planck Inst Behav Physiol, 70-76; ASSOC PROF BIOL, UNIV ORE, 76- *Concurrent Pos:* Proj leader, Cybernetic Study Group 50, 72-76; Fogarty sr res fel, 84-85; dir, Inst Neurosci, 86-; Career Develop Award, NIH. *Mem:* Animal Behav Soc; NY Acad Sci; Neurosci Soc; AAAS; Europ Brain & Behav Soc; Int Soc Neuroethnol. *Res:* Quantitative analysis of visual communication signals in fish and the central nervous system processing of those signals. *Mailing Add:* Dept Biol Univ Ore Eugene OR 97403

FERNANDES, DANIEL JAMES, b Fall River, Mass, June 22, 48; m 82; c 1. DRUG DEVELOPMENT. *Educ:* Providence Col, BS, 70; George Washington Univ, PhD(pharmacol), 78. *Hon Degrees:* DSc, Univ Mass, Dartmouth, 90. *Prof Exp:* Res assoc pharmacol, Yale Univ Sch Med, 77-80; asst prof, 80-86, ASSOC PROF BIOCHEM, BOWMAN GRAY SCH MED, 86-; PRIN, PHARMACOL CONSULTS, 90- *Concurrent Pos:* Distinguished lectr, Lederle Labs, Am Cyanimid, 85-; prin investr, Nat Cancer Inst & Am Cancer Soc, 86-; Leukemia Soc Am Scholar, 86; consult, Device Labs Inc, 87-89, Nat Cancer Inst, 88-, Lipogen, Inc & Triton Biosci, 89, Berlox Biosci, 90- *Mem:* Am Asn Cancer Res; Am Soc Biochem & Molecular Biol. *Res:* Design and discovery of anticancer drugs; biochemical pharmacology and mechanisms of action of anticancer drugs; mechanisms of DNA replication; development of medical devices. *Mailing Add:* Bowman Gray Sch Med 300 S Hawthorne Rd Winston-Salem NC 27103

FERNANDES, GABRIEL, b Berkur, India, Mar 25, 36. AGING, AUTOIMMUNE DISEASES. *Educ:* Univ Minn, BS, 73, MS, 75; Univ Bombay, PhD(philos), 78. *Prof Exp:* Instr, Dept Lab Med & Path, Univ Minn, 75-77; assoc, Sloan Kettering Inst, 78-81; assoc prof med-immunol, 81-88, PROF MED, MICROBIOL & PHYSIOL, UNIV TEX, 88- *Concurrent Pos:* Vis investr, Sloan Kettering Inst, 77-88; asst prof biol, Cornell Univ, 79-81; res grant, NIH, 80-; mem study sect, Nat Inst Aging, 86-89. *Mem:* Nat Inst Aging; Am Asn Immunologists; Am Asn Gerontologists; Am Chem Soc. *Mailing Add:* Dept Med & Physiol Univ Tex Health Sci Ctr 7703 Floyd Curl Dr San Antonio TX 78284

FERNANDES, JOHN HENRY, b Tiverton, RI, Aug 21, 24; m 47; c 6. ENERGY ENGINEERING. *Educ:* RI State Col, BS, 49; Lehigh Univ, MS, 53; Calvin Coolidge Col, ScD, 60. *Prof Exp:* Serv engr, Combustion Eng, Inc, 49-50; from instr to assoc prof mech eng, Lafayette Col, 50-60; prof & head dept mech eng, Manhattan Col, 60-63; chief proj engr, New Prod Div, Combustion Eng, Inc, 63-68, admin asst to vpres, 68-70, dir tech activities industrial group, 70-72, corp coordr, Environ Control Systs, 72-76, corp dir, Corp Tech Transfer, 76-81, dir corp tech resources, 81-84; vpres energy div, Maguire Group, Inc, 84-90; CONSULT MECH ENG, 90- *Concurrent Pos:* Consult, McGinley Mills, NJ, 54-56; dir eve col, Lafayette Col, 57-60; gen mgr, C E Raymond, 69. *Honors & Awards:* Centennial Medal, Am Soc Mech Engrs, Performance Test Code Medal & Codes & Standards Medal. *Mem:* Fel Am Soc Mech Engrs; Nat Soc Prof Engrs; Air Pollution Control Asn; Am Acad Environ Engrs. *Res:* Broad areas of thermodynamics and compressible flow; particularly steam power plants and gas turbines; broad fields of waste and fuel combustion; power generation and related equipment. *Mailing Add:* 294 Riverside Dr Tiverton RI 02878-4206

FERNANDES, PRABHAVATHI BHAT, b Mangalore, India, Apr 11, 49; m 71; c 2. PHARMACEUTICAL PRODUCTS. *Educ:* Bangalore Univ, India, BSc, 68; Madras Univ, MSc, 71; Thomas Jefferson Univ, PhD(microbiol), 75. *Prof Exp:* Res asst immunochem, Univ Ghent, Belg, 71-72; fel microbiol, Thomas Jefferson Univ, 75-76, Inst Cancer Res, Philadelphia, 76-78, clin microbiol, Temple Univ, 78-80; RES INVESTR MICROBIOL, SQUIBB INST MED RES, 80-; DIR MICROBIOL, MOLECULAR BIOL & NATURAL PROD RES, PHARMACEUT RES INST, BRISTOL-MYERS SQUIBB, 89- *Concurrent Pos:* Res instr, Temple Univ, 78-80; sr proj leader, anti-infective res, Abbott Labs, Ill, 80-88. *Mem:* Am Soc Microbiol; Soc Gen Microbiol; NY Acad Sci; Am Chem Soc. *Res:* Virulence factors of gram negative bacteria such as exotoxins; role of outer membrane proteins; lipopolysaccharide and capsule gram; negative bacteria in host parasite interactions; antibiotics; quinolones; macrolides; natural products. *Mailing Add:* Pharmaceut Res Inst Bristol-Myers Squibb Princeton NJ 08543-4000

FERNANDEZ, ALBERTO ANTONIO, b Buenos Aires, Arg, July 12, 25; m 49; c 1. BIOCHEMISTRY, CLINICAL CHEMISTRY. *Educ:* Univ Buenos Aires, MS, 54, PhD(chem), 60. *Prof Exp:* Chemist prod control, Behring Inst, Arg, 52-55, chief control, 55-58; chief chem res, Quimica Estrella, 58-61; res chemist, Bio-Sci Labs, 61-63, chief chem res, 63-66; head res & develop, Pfizer Int Subsidiaries, Arg, 66-67; dir res, Biochem Procedures, Inc, North Hollywood, 67-77; CONSULT, RIA, 77-; CANCER RESEARCHER, UNIV SOUTHERN CALIF, 79- *Mem:* Am Asn Clin Chem. *Res:* Chemical methods of clinical analysis; industrial organic synthesis; basic biochemical research; radioimmunoassays; cancer research. *Mailing Add:* 6322 Ellenview Ave Canoga Park CA 91307

FERNANDEZ, HECTOR R C, b Jaruco, Cuba, Feb 18, 37; US citizen. VISUAL PHYSIOLOGY, COMPARATIVE PHYSIOLOGY. *Educ:* Univ Miami, BS, 60, PhD(zool), 65. *Prof Exp:* Postdoctoral fel biol, Yale Univ, 65-67, res assoc biol, 67-69; asst prof biol, Univ Southern Calif, 69-76; assoc prof biol, 76-91, UNDERGRAD OFFICER, BIOL DEPT, WAYNE STATE UNIV, 90- *Concurrent Pos:* Fel, NIH, 65-67 & Japan Soc Prom Sci, 81; dir, Arctic Proj, Univ Southern Calif, 72-76; co-dir, Bermuda Prog, Wayne State Univ, 78-, chmn, Div Biophys, 78-81, grad officer, Biol Dept, 86-89, dir, Summer Teacher Insts, 85-; counr, Am Soc Photobiologists, 78-80; vis prof, Physics Dept, Konstanz Univ, Ger, 80 & Biol Dept, Keio Univ, Japan. *Mem:* Am Soc Zoologists; fel AAAS; Am Soc Photobiologists; Sigma Xi. *Res:* Animal physiology. *Mailing Add:* Dept Biol Wayne State Univ Detroit MI 48202

FERNANDEZ, HUGO LATHROP, b Rengo, Chile, Jan 16, 40; m 61; c 2. NEUROSCIENCE, NEUROBIOLOGY. *Educ:* Cath Univ Valparaiso, BS, 65; Mass Inst Technol, MS, 69; Univ Kans, PhD(physiol & cell biol), 71. *Prof Exp:* Res assoc neurosci, Dept Physiol, Univ Kans, 71; asst prof, Dept Neurobiol, Cath Univ, 72-73, assoc prof neurophysiol, 74-76; asst prof neurol, 77-79, ASSOC PROF PHYSIOL, UNIV KANS MED CTR, 79-; ASST PROF PHYSIOL, NEUROSCI RES LAB, VET ADMIN MED CTR, 80- *Concurrent Pos:* Prin investr, UNESCO-PNUD, Cath Univ grant, 74-76; assoc investr, Muscular Dystrophy Asn Am grant, 77-79; prin investr, Vet Admin Merit Rev grant, 79-, NIH grant, 79-81; co-dir, Neurosci Res Lab, Vet Admin Med Ctr, 79-80. *Honors & Awards:* P Newmark Award, Univ Kans, 71. *Mem:* AAAS; Soc Neurosci; Int Brain Res Org; Sigma Xi; NY Acad Sci. *Mailing Add:* Neurosci Res Lab Vet Admin Med Ctr 4801 Linwood Blvd Kansas City MO 64128

FERNANDEZ, JACK EUGENE, b Tampa, Fla, May 18, 30; m 51; c 3. ORGANIC CHEMISTRY. *Educ:* Univ Fla, BSCh, 51, MS, 52, PhD(chem), 54. *Prof Exp:* Chemist, Naval Stores Sta, USDA, 54; instr & res assoc, Duke Univ, 56-57; chemist, Tenn Eastman Co, 57-60; from asst prof to assoc prof, 60-72, PROF CHEM, UNIV SFLA, 72- *Mem:* Am Chem Soc. *Res:* Kinetics and mechanisms; infrared spectroscopy; polymers; history of chemistry; chemistry for the non-scientist. *Mailing Add:* 5113 Rolling Hill Ct Tampa FL 33617-1024

FERNANDEZ, JOSE MARTIN, b La Coruna, Spain; US citizen. ENVIRONMENTAL HEALTH, ORGANIC CHEMISTRY. *Educ:* City Col New York, BS, 63; Yale Univ, PhD(org chem), 68. *Prof Exp:* Sr res chemist, 68-73, res assoc org chem, 73-78, MEM STAFF, HEALTH & SAFETY LAB, EASTMAN KODAK CO RES LABS, 78- *Mem:* Am Chem Soc; Sigma Xi. *Res:* Couplers and developers for the photographic system; dyes; environmental regulations; toxicology; environmental sciences. *Mailing Add:* 525 Fox Meadow Rd Rochester NY 14626

FERNANDEZ, LOUIS AGNELO, b Karachi, Pakistan, May 30, 44; Can citizen; m 73; c 2. HEMATOLOGY. *Educ:* Univ Karachi, MB & BS, 66; FRCP(C), FACP, cert med, 73, hemat, 74. *Prof Exp:* Lectr, 74-76, from asst prof to assoc prof, 76-89, PROF MED, DALHOUSIE UNIV, 89-, DIV HEAD HEMAT & MED ONCOL, 86- *Concurrent Pos:* Consult hemat, Camp Hill Hosp, Dalhousie Univ, 74-, Victoria Gen Hosp, 77-, Halifax Infirmary, 84-, Archie Macullum Hosp, 86- *Mem:* Can Soc Hemat; fel Am Col Physicians; Am Soc Hemat; Can Soc Clin Invest. *Res:* Characterization of lymphocyte subpopulations and functions in chronic lymphocytic leukemia; normal young and aging individuals. *Mailing Add:* Bldg 7 Camp Hill Hosp 1763 Robie St Halifax NS B3H 3G2 Can

FERNANDEZ, LOUIS ANTHONY, b New York, NY, Oct 5, 39; m 65; c 1. GEOLOGY, PETROLOGY. *Educ:* City Col New York, BS, 62; Univ Tulsa, MS, 64; Syracuse Univ, PhD(geol), 69. *Prof Exp:* Res geologist, Yale Univ, 68-71; asst prof, 71-74, actg asst dean, Col Sci, 76, assoc prof, 74-77, PROF GEOL, UNIV NEW ORLEANS, 77-, CHMN DEPT, 81- *Mem:* Geol Soc Am; Mineral Soc Am; Nat Asn Geol Teachers; Sigma Xi. *Res:* Igneous petrology; volcanology. *Mailing Add:* Dept Earth Sci Univ New Orleans New Orleans LA 70122

FERNANDEZ-BACA, JAIME ALBERTO, b Lima, Peru, May 23, 54; US citizen; m 83. MAGNETISM, NEUTRON SCATTERING. *Educ:* Nat Univ Eng, Lima, Peru, BS, 77; Univ Md, MS, 83, PhD(physics), 86. *Prof Exp:* Fel, Int Atomic Energy Agency, 79-81; res asst, Univ Md, 81-86; RES STAFF, OAK RIDGE NAT LAB, 86- *Mem:* Am Phys Soc. *Res:* Magnetic systems with competition of ferromagnetic and antiferromagnetic interactions; amorphous magnets; reentrant spin glasses; spin dynamics of rare earth metals. *Mailing Add:* Oak Ridge Nat Lab MS-6393 PO Box 2008 Oak Ridge TN 37831-6393

FERNÂNDEZ-CRUZ, EDUARDO P, b Santiago de Compostela La Coruña, July 16, 46; m 73; c 3. IMMUNOLOGY. *Educ:* Loyola Col, BSc, 63; Univ Madrid, MA, 70; Complutense Univ Madrid, doctoral, 73, specialist, 82. *Prof Exp:* Res fel immunol, Middlesex Hosp Med Sch, 73-74; asst prof immunopath, Clin Puerto Hierro, Madrid, 76-78; res assoc immunopath, Scripps Clin & Res Found, La Jolla, Calif, 79-82, sci assoc immunol, 82-84; clin asst immunol & allergy, Univ Calif, San Diego, 82-84; CHIEF PROF, DIV IMMUNOL, HOSP GEN GREGORIO MARAÑÓN, 84-; ASSOC PROF MED, COMPLUTENSE UNIV MADRID, 86- *Concurrent Pos:* Mem, Comt Tumor Immunol, Am Col Allergists, 82-84, sci fel, 84; sci consult, Med Biol Inst, 82; vis prof, Nat Cancer Inst (FCRF), 87; mem, Comn Sci & Technol, 88-90 & Comn Basic Res, Clin Trials Community Madrid, 91-; vis prof, M D Anderson Cancer Ctr, 90; prin investr, Multicentric Europ Study, ZIDON, 90- *Mem:* Span Soc Immunol; Brit Soc Immunol; Am Asn Immunologists; Span Soc Cancer Res; Span AIDS Soc; Soc Biol Theory. *Res:* Immunological (basic and clinical) studies on AIDS and human immunovirus infection; biological markers predictive of progression to AIDS; clinical trials; alterations in production of prostaglandin E2 and monokines by monocytes infected by human immunovirus; experimental and clinical models of immunotherapy of cancer. *Mailing Add:* Hosp Gen Gregorio Marañón Dr Esquerdo 46 Madrid 28007 Spain

FERNANDEZ-MADRID, FELIX, b Buenos Aires, Arg, Nov 28, 27; US citizen; m 54; c 4. RHEUMATOLOGY, BIOCHEMISTRY. *Educ:* Univ Buenos Aires, MD, 53; Univ Miami, Fla, PhD(physiol, cell & molecular biol), 66. *Prof Exp:* Instr med, Hosp de Clinicas, Buenos Aires, 54-57; Howard Hughes fel, Howard Hughes Med Inst, Coral Gables, Fla, 61-66; asst prof med & physiol, Univ Miami, 66-68; assoc prof & chief rheumatology, 68-72, PROF MED, WAYNE STATE UNIV SCH MED, 72- *Mem:* Fel Am Col Rheumatology; Arthritis Found; fel Am Col Physicians; Am Fedn Clin Res; AAAS. *Res:* Collagen biosynthesis & fibrillogenesis; macromolecules of the connective tissue; pathogenesis of the connective tissue diseases; history of medicine; medical quackery. *Mailing Add:* 4201 Stantoine Detroit MI 48201

FERNANDEZ-POL, JOSE ALBERTO, b Buenos Aires, Arg, Mar 17, 43; m 71; c 2. GROWTH FACTORS, ONCOGENES. *Educ:* Col Nat de Vincente Lopez, Buenos Aires, BA, 63; Univ de Buenos Aires, MD, 69. *Hon Degrees:* Diploma de Honor, Univ Buenos Aires, 69. *Prof Exp:* Physician, Hosp Escuela Jose de San Martin, Univ Buenos Aires, 69-70, res assoc, Endocrine Res Lab, Cent de Med Nuclear, Univ Buenos Aires, 70-71; resident physician, State Univ NY at Buffalo, 71-72, res fel, Nuclear Med Lab, 72-75; asst prof med, St Louis Univ, 77-80; assoc prof med, 80-85, assoc prof radiol, 81-87, PROF MED, ST LOUIS UNIV, 85-, PROF RADIOL & NUCLEAR MED, 88- *Concurrent Pos:* Fel, Dept Internal Med, Inst Med Sci, Can, 72; dir, Immunoassay Lab & prin investr, Vet Admin Med Ctr, St Louis, Mo, 77-; chief, Lab Molecular Oncol, 87-; fel, Lab Molecular Biol, Nat Cancer Inst, NIH, 75-77. *Mem:* Am Asn Cancer Res; Am Soc Biol Chem & Mol Biol; Soc Nuclear Med. *Res:* Growth control mechanisms in normal and cancerous cells; oncogenes; immunology. *Mailing Add:* Vet Admin Med Ctr Lab Molecular Oncol St Louis Univ 915 N Grand Blvd St Louis MO 63106

FERNÁNDEZ-REPOLLET, EMMA D, b Ciales, PR, Apr 3, 51; m 90. RENAL PHYSIOLOGY, RENAL PHARMACOLOGY. *Educ:* Univ PR, Rfo Piedras, BS, 72; Univ PR, MS, 77, PhD(physiol), 79. *Prof Exp:* Sci teacher chem, Fernando Collazo HS, 72-74; postdoctoral renal morphol, Duke Univ, NC, 79-80; postdoctoral renal micropuncture, Univ NC, 80-82; teaching asst renal physiol, 78-79, asst prof, 82-86, ASSOC PROF PHARMACOL, SCH MED, UNIV PR, 86- *Concurrent Pos:* Chief, Micropuncture Lab, Vet Admin Ctr, San Juan, PR, 82-88; sci adv, Animal Res Facil, Vet Admin Med, Ctr, 83-88; prin investr, NIH-NIRA grant, 86-89, Vet Admin Merit Rev Award, 86-89, NIH-Minority Biomed Res Support Prog grant, 86- & BAXTER grant, 91-; assoc dir, Res Ctrs Minority Inst, Sch Med, Univ PR, 86- *Mem:* Am Soc Nephrology; Int Soc Nephrology; Soc Anal Cytol; Sigma Xi; Am Heart Asn; Am Diabetic Asn. *Res:* Glomerular function and structure in a variety of physiological and pathophysiological conditions; determine the influence of vasoactive hormones on the regulation of glomerular filtration rate in experimental models such as cortical necrosis, acute renal failure and diabetes mellitus; effects of aging, protein intake and systemic hypertension on the changes in glomerular function associated with these conditions and performing morphological studies to correlate the alterations in glomerular function with structural changes. *Mailing Add:* Dept Pharmacol Sch Med Univ PR PO Box 365067 San Juan PR 00936-5067

FERNANDEZ Y COSSIO, HECTOR RAFAEL, b Jaruco, Cuba, Feb 18, 37. PHYSIOLOGY. *Educ:* Univ Miami, Fla, BSc, 60, PhD(zool), 65. *Prof Exp:* Teaching asst zool, Univ Miami, Fla, 60-65; NIH fel, 65-67; staff res biologist, Yale Univ, 67-69; asst prof biol, Univ Southern Calif, 69-76, dir, Arctic Res Proj, 71-77; assoc prof biol & chmn div regulatory biol & biophys, Wayne State Univ, 77-81. *Mem:* Fel AAAS; Am Soc Zool; Am Soc Photobiol; Soc Res Vision & Ophthal; Soc Neurosci; Sigma Xi; fel Japan Soc for Prom of Sci; Japan Zool Soc. *Res:* Biochemical and electrophysiological properties of photoreceptors; physiological adaptations of arctic and deep sea organisms; visual physiology, extraction and identification of crustacean visual pigments; electrophysiological studies of arthropod photoreceptors; plant physiology, study of electrical properties of plant cell membranes; role of calcium in photoreceptor function. *Mailing Add:* Dept Biol Wayne State Univ Detroit MI 48202

FERNANDO, CONSTANTINE HERBERT, b Colombo, Sri Lanka, Apr 4, 29; m 57; c 3. LIMNOLOGY, RESERVOIR FISHERIES. *Educ:* Univ Ceylon, BSc, 52; Oxford Univ, DPhil(entom), 56. *Prof Exp:* Asst lectr zool, Univ Ceylon, 56-59; lectr, Univ Singapore, 60-64; sr res officer, Dept Fisheries, Ceylon, 64-65; assoc prof, 65-67, PROF BIOL, UNIV WATERLOO, 67- *Concurrent Pos:* Consult, Filariasis Res Unit, WHO, Rangoon, Burma, 63, Freshwater Fisheries, 79; dir, Can Int Develop Agency proj, ADDIS, ABABA, Ethiopia, 82-87. *Mem:* Int Soc Limnol; Brit Freshwater Biol Asn; Can Soc Zool; Am Fisheries Soc. *Res:* Reservoir ecology; tropical lake and reservoir fisheries; introduced fishes; lacustrine fishes; freshwater zooplankton systematics and distribution. *Mailing Add:* Dept Biol Univ Waterloo Waterloo ON N2L 3G1 Can

FERNANDO, HARINDRA JOSEPH, b Sri Lanka, Oct 19, 55; m 85; c 1. GEOPHYSICAL FLUID DYNAMICS, STRATIFIED & ROTATING FLOWS. *Educ:* Univ Sri Lanka, BSc, 79; Johns Hopkins Univ, MA, 82, PhD(fluid mech), 84. *Prof Exp:* Asst lectr mech eng, Univ Sri Lanka, 79-80; res asst fluid mech, Johns Hopkins Univ, 80-83; postdoctoral fel, Calif Inst Technol, 83-84; asst prof, 84-87, ASSOC PROF MECH ENG, ARIZ STATE UNIV, 88- *Concurrent Pos:* Prin investr, NSF, 85- & US Off Naval Res, 87-; consult, Solar Energy Res Inst, 87-90; sr vis, Univ Cambridge, UK, 90; tech ed, Appl Mech Reviews. *Mem:* Am Soc Mech Engrs; Am Geophys Union; Am Phys Soc; Europ Geophys Union. *Res:* Turbulent mixing in stratified and rotating flows; geophysical fluid dynamics. *Mailing Add:* Dept Mech & Aerospace Eng Ariz State Univ Tempe AZ 85287-6106

FERNANDO, QUINTUS, b Colombo, Ceylon, Nov 23, 26; m 50; c 2. ANALYTICAL CHEMISTRY. *Educ:* Univ Ceylon, BS, 49; Univ Louisville, MS, 51, PhD(chem), 53. *Hon Degrees:* Dr, Univ Autonoma, Guadalajara, Mex. *Prof Exp:* Lectr chem, Univ Ceylon, 49; asst analyst, Govt Ceylon, 53; lectr chem, Univ Ceylon, 54-57; res assoc, Univ Pittsburgh, 57-58, asst prof, 58-61; assoc prof, 61-64, PROF CHEM, UNIV ARIZ, 64-, PROF TOXICOL & FORENSIC SCI, 80- *Mem:* Fel Royal Soc Chem; Am Chem Soc; Am Crystallog Asn. *Res:* Development of techniques for trace element analysis; multielement analysis of environmental and forensic samples by proton induced x-ray emission/x-ray crystallographic structure determination and photoelectron spectroscopy of metal complexes of importance in analytical chemistry; molecular emission spectroscopy; kinetics and thermodynamics of metal complex reactions; separations with supercritical fluids. *Mailing Add:* Dept Chem Univ Ariz Tucson AZ 85721

FERNANDO, ROHAN LUIGI, b Colombo, Sri Lanka, Jan 19, 52. STATISTICAL GENETICS, MIXED MODEL METHODOLOGY. *Educ:* Calif State Univ, Fresno, BS, 78; Univ Ill, Urbana-Champaign, MS, 81, PhD(animal sci), 84. *Prof Exp:* ASST PROF STATIST, DEPT ANIMAL SCI, UNIV ILL, 84- *Concurrent Pos:* Statist consult. *Mem:* Am Stat Asn; Biometric Soc. *Res:* Development of statistical methods for genetic improvement of livestock. *Mailing Add:* Dept Animal Sci Univ Ill Urbana Campus 1301 W Gregory Dr Urbana IL 61801

FERNBACH, DONALD JOSEPH, b Brooklyn, NY, Apr 10, 25; m 54; c 4. PEDIATRICS, HEMATOLOGY. *Educ:* Tusculum Col, AB, 48; George Washington Univ, MD, 52. *Prof Exp:* Instr zool, Tusculum Col, 47-48; Jesse Jones fel pediat hemat, Children's Med Ctr, Boston, Mass, 56-57; from instr to assoc prof, 57-71, PROF PEDIAT, BAYLOR COL MED, 71- *Concurrent Pos:* Consult, Wilford Hall Army Hosp, Lackland AFB, 60-75. *Mem:* Am Acad Pediat; Am Pediat Soc; AMA; Am Asn Cancer Res; Am Soc Hemat; Am Soc Pediat Hemat. *Res:* Cancer chemotherapy in children; possible viral etiology; clinical and epidemiological investigations of pediatric neoplasia, especially childhood leukemia. *Mailing Add:* Dept Pediat Baylor Col Med Tex Childrens Hosp 6621 Fannin St Houston TX 77030

FERNBACH, SIDNEY, b Philadelphia, Pa, Aug 4, 17; m 55; c 3. THEORETICAL PHYSICS, COMPUTER SCIENCE. *Educ:* Temple Univ, AM, 40; Univ Calif, PhD(physics), 52. *Prof Exp:* Physicist, Frankford Arsenal, 40-43; asst instr, Univ Pa, 43-44; asst, Univ Calif, 46-48, physicist, Lawrence Radiation Lab, 48-51; mem staff, Stanford Univ, 51-52; head, Theoret Div, Lawrence Livermore Nat Lab, 58-68, head, Comput Dept, 62-77, dep assoc dir sci support, 77-79; CONSULT, INFO SYSTS, 79- *Concurrent Pos:* Mem bd dirs, Advan Memory Systs, Sunnyvale, Calif, 69-76; consult, US Dept of Defense, 71-; consult, Comput Systs Tech Adv Comt, Domestic & Int Bus Admin Bur East West Trade, US Dept Com, 72-80; mem steering comt, Comput Sci & Eng Res Study, NSF, 74-80; mem bd dir, Solar Energy Sales, 78-80. *Mem:* Fel Am Phys Soc; Asn Comput Mach; fel AAAS; Math Asn Am; Inst Elec & Electronics Engrs; Sigma Xi. *Res:* Neutron physics; cosmic ray shower theory; computation. *Mailing Add:* Four Holiday Dr Alamo CA 94507

FERNELIUS, NILS CONARD, b Columbus, Ohio, Nov 10, 34. SURFACE PHYSICS, EXPERIMENTAL SOLID STATE PHYSICS. *Educ:* Harvard Univ, AB, 56; Univ Ill, Urbana, MS, 59, PhD(physics), 66. *Prof Exp:* Res assoc, dept physics, Univ Ill, Urbana, 66-67; asst physicist, Mat Sci Div, Argonne Nat Lab, 68-71; vpres, Res Consults, Inc, 71-72; Nat Res Coun sr res assoc, Aerospace Res Lab, Wright-Patterson AFB, 73-75, Mat Lab, 82-85, vis scientist, Air Force Mat Lab, 75-77 & 87-88; res physicist, Univ Dayton Res Inst, 77-82; physicist, Stolle Corp, 85-86; PHYSICIST MAT LAB, WRIGHT PATTERSON AFB, OH, 88- *Concurrent Pos:* Adj assoc prof, phys dept, Wright State Univ, 79, 87. *Mem:* Am Phys Soc; Optical Soc Am; Am Asn Physics Teachers; Sigma Xi; Am Vacuum Soc; Inst Elec & Electronics Engrs; Mat Res Soc; Soc Optic Engr; Am Soc Appl Spectros; Int Soc Optical Soc. *Res:* Studies of point defects using nuclear magnetic resonance and nuclear quadrupole resonance; laser applications including laser speckle, holography, laser annealing and laser windows; x-ray photoelectron, auger electron, appearance potential and photoacoustic spectroscopies; magnetic recording; superconductivity. *Mailing Add:* 1528 Sussex Rd Troy OH 45373-2446

FERNER, JOHN WILLIAM, US citizen. ECOLOGY. *Educ:* Col Wooster, BA, 67; Univ Colo, PhD(biol), 72. *Prof Exp:* Asst prof, Wash & Jefferson Col, 72-73 & Franklin & Marshall Col, 73-77; assoc prof, 77-88, PROF BIOL & CHMN DEPT, THOMAS MOORE COL, 88- *Concurrent Pos:* Vis lectr, Univ Colo, 72; vis prof, Njala Univ Col, Sierra Leone, 88. *Mem:* Soc for Study Amphibians & Reptiles; Am Soc Ichthyologists & Herpetologists. *Mailing Add:* Dept Biol Thomas Moore Col Crestview Hills KY 41017

FERNIE, BRUCE FRANK, b Manchester, NH, Sept 28, 48; m 70; c 3. VIROLOGY. *Educ:* Carnegie-Mellon Univ, BS, 70; Univ Pittsburgh, ScD(hyg), 76. *Prof Exp:* Res fel, Med Ctr, Johns Hopkins Univ, 76-78; from res asst prof to res assoc prof, 78-89, ASST DIR RES, MED SCH, GEORGETOWN UNIV, 87-, RES ASSOC PROF, 89- *Concurrent Pos:* Consult, Bone Physiol Lab, Orthop Hosp, Univ Southern Calif, 81-, Ortho Diag Systs, Inc, 85-86. *Mem:* Am Soc Microbiol; Soc Gen Microbiol; Am Asn Immunologists. *Res:* Synthesis, maturation and structure of respiratory syncytial virus protein S; virion proteins of human T-cell leukemia III-lymphoma virus; confirmatory testing for AIDS. *Mailing Add:* Div Molecular Virol & Immunol Sch Med Georgetown Univ 5640 Fishers Lane Rockville MD 20852-1770

FERNIE, JOHN DONALD, b Pretoria, SAfrica, Nov 13, 33; Can citizen; m 55; c 2. ASTRONOMY. *Educ:* Univ Cape Town, BSc, 53, Hons, 54, MSc, 55; Ind Univ, PhD(astron), 58. *Prof Exp:* Lectr astron, Univ Cape Town, 58-61; asst prof, 61-64, assoc prof, 64-67, dir observ & chmn Astron Dept, 78-88, PROF ASTRON, DAVID DUNLAP OBSERV, UNIV TORONTO, 67- *Honors & Awards:* Fel Royal Soc Can. *Mem:* Am Astron Soc; fel Royal Astron Soc; Int Astron Union; Royal Astron Soc Can (pres, 74-76); Can Astron Soc. *Res:* History of modern astronomy; variable stars; galactic structure; photoelectric photometry. *Mailing Add:* David Dunlap Observ Univ Toronto Box 360 Richmond Hill ON L4C 4Y6 Can

FERNOW, RICHARD CLINTON, b Newark, Ohio, Feb 5, 47; m 70; c 2. LASERS, SUPERCONDUCTING MAGNETS. *Educ:* Ohio Univ, BS, 69; Syracuse Univ, MS, 71, PhD(physics), 73. *Prof Exp:* Res assoc physics, Univ Mich, 73-78; assoc physicist, 78-81, PHYSICIST, BROOKHAVEN NAT LAB, 81- *Mem:* Am Phys Soc. *Res:* Experimental particle physics; laser acceleration of particles; super conducting; polarized target development. *Mailing Add:* Physics Dept Bldg 510 Brookhaven Nat Lab Upton NY 11973

FERNSTROM, JOHN DICKSON, b New York, NY, July 9, 47; m 78; c 2. NEUROPHARMACOLOGY, NEUROENDOCRINOLOGY. *Educ:* Mass Inst Technol, SB, 69, PhD(nutrit biochem & metab), 72. *Prof Exp:* Fel neuroendocrinol, Roche Inst Molecular Biol, Hoffmann-LaRoche Inc, NJ, 72-73; asst prof neuroendocrinol, Mass Inst Technol, 73-77, assoc prof, 77-82; assoc prof, 82-87, PROF PSYCHIAT, SCH MED, UNIV PITTSBURGH, 87- *Concurrent Pos:* Fel neurochem, Alfred P Sloan Found, 74-76; mem, Neurol Dis Spec Projs A Study Sect, Nat Inst Neurol & Commun Dis & Stroke, 78-82, chmn, 81-82; NIMH res scientist develop award, 79-88, res scientist award, 89-; mem, Life Sci Adv Comt, NASA, 80-86, Pharmacol Rev Panel, US Army Med Res & Develop Command, 85-, Nat Adv Coun, Monell Chem Senses Ctr, Philadelphia, 87-, Adv Panel, Nutrit Res Planning, Nat Inst Child Health & Human Develop, 89-90; chmn, Nervous Syst Sect, Am Inst Nutrit, 91- *Honors & Awards:* Mead-Johnson Award, Am Inst Nutrit, 80. *Mem:* Am Physiol Soc; Endocrine Soc; Am Soc Pharmacol & Exp Therapeut; Soc Neurosci; Am Inst Nutrit; Am Soc Neurochem. *Res:* Effects of diet and drugs on brain neurotransmitter synthesis and utilization; control of pituitary hormone secretion by the brain; malnutrition and brain function; amino acid utilization in the body; nutrition; neuropeptide synthesis and function in the brain. *Mailing Add:* Western Psychiat Inst & Clin 3811 O'Hara St Pittsburgh PA 15213

FEROE, JOHN ALBERT, b Grand Forks, NDak, Dec 6, 46. MATHEMATICS. *Educ:* St Olaf Col, BS, 68; Univ Calif, San Diego, MA, 70, PhD(math), 74. *Prof Exp:* ASST PROF MATH, VASSAR COL, 74- *Mem:* Am Math Soc; Math Asn Am; Soc Indust & Appl Math. *Res:* Differential equations; reaction-diffusion equations; mathematical modeling. *Mailing Add:* Dept Math Box 335 Vassar Col Poughkeepsie NY 12601

FERONE, ROBERT, b Mt Vernon, NY, Nov 8, 36; m 60; c 4. MICROBIOLOGY, BIOCHEMISTRY. *Educ:* NY Univ, BA, 58, MS, 63. *Prof Exp:* Res microbiologist, 58-68, sr res microbiologist, 68-79, group leader, Wellcome Res Labs, 79-90, ASST DIV DIR, DIV MOLECULAR GENETICS & MICROBIOL, BURROUGHS WELLCOME CO, 90- *Mem:* Am Soc Microbiol; Am Soc Biol Chem; Am Asn Cancer Res. *Res:* Chemotherapy; antifolate inhibitors; folate metabolism in microorganisms and tumors; antimalarial drugs and antitumor drugs. *Mailing Add:* Burroughs Wellcome & Co Research Triangle Park NC 27709

FERRANS, VICTOR JOAQUIN, b Barranquilla, Colombia, June 7, 37; US citizen; m 60; c 4. ELECTRON MICROSCOPY. *Educ:* Tulane Univ, MS & MD, 60, PhD(anat), 63. *Prof Exp:* From instr to asst prof med, Sch Med, Tulane Univ, 61-69; vis scientist, 69-74, sr scientist, 74-78, CHIEF ULTRASTRUCT SECT, NAT HEART, LUNG & BLOOD INST, NIH, 78- *Mem:* Int Soc Heart Res; Am Soc Cell Biol; Am Col Cardiol. *Res:* Electron microscopy of cardiovascular, pulmonary and metabolic diseases. *Mailing Add:* NIH 9000 Rockville Pike NIH Bldg 10 Rm 7N- 236 Bethesda MD 20892

FERRANTE, MICHAEL JOHN, b New York, NY, Feb 1, 30; m 53; c 3. THERMODYNAMICS. *Educ:* Eastern Wash Col, BA(chem) & BA(educ), 52. *Prof Exp:* Res chemist metall, 56-69, res chemist thermodyn, 69-86, SPECTROS, US BUR MINES, ALBANY METALL RES CTR, 86- *Mem:* Sigma Xi; Am Inst Chem Engrs. *Res:* Experimental and theoretical studies of thermodynamic properties of substances by conducting research in high-temperature calorimetry. *Mailing Add:* Albany Metall Res Ctr 1450 Queen Ave SW Albany OR 97321-2198

FERRANTE, W(ILLIAM) R(OBERT), b Providence, RI, Mar 9, 28; m 68; c 2. ENGINEERING MECHANICS. *Educ:* Univ RI, ScB, 49; Brown Univ, ScM, 55; Va Polytech Inst, PhD(eng mech), 62. *Prof Exp:* Instr math, Univ RI, 49-50; asst prof mech, Lafayette Col, 52-56; assoc prof mech eng, 56-68, assoc dean, 67-71, vpres acad affairs, 71-73, pres, 73-74, PROF MECH ENG & APPL MECH, UNIV RI, 68-, VPRES ACAD AFFAIRS, 74- *Concurrent Pos:* Fulbright fel, Al-Hikma Univ, Baghdad, 63-64. *Mem:* Asn Higher Educ; Am Soc Mech Engrs; Am Soc Eng Educ; Sigma Xi. *Res:* Applied mechanics; elasticity; shell theory; applied mathematics. *Mailing Add:* Dept Mech Eng Univ RI Kingston RI 02881

FERRAR, JOSEPH C, b Lansing, Mich, Dec 10, 39; m 61; c 4. MATHEMATICS, GEOMETRY. *Educ:* Mich State Univ, BS, 60; Yale Univ, PhD(math), 66. *Prof Exp:* assoc prof, 65-80, PROF MATH, OHIO STATE UNIV, 80- *Concurrent Pos:* Vis prof, State Univ Utrecht, 67-68, 76-77 & 82-83. *Mem:* Am Math Soc. *Res:* Classification of Lie algebras and the related algebraic structures. *Mailing Add:* Dept Math Ohio State Univ Columbus OH 43210

FERRARA, LOUIS W, b Chicago, Ill, Aug 27, 23; m 53; c 6. ANALYTICAL CHEMISTRY, BIOCHEMISTRY. *Educ:* Univ Ill, BS, 52. *Prof Exp:* Sr res chemist, Int Minerals & Chem Corp, 52-65, supvr anal chem, 65-67, anal serv specialist, 67-68, mgr anal serv, 68-78, mgr anal chem tech develop, 78-87; RETIRED. *Mem:* Am Chem Soc; Am Soc Testing & Mat; Asn Off Agr Chem; Inst Food Technol; Soc Appl Spectros. *Res:* Organic and inorganic chemistry; microbiology; plants and soils. *Mailing Add:* 905 Rose Lane Wheeling IL 60090-5521

FERRARA, THOMAS CIRO, b Sacramento, Calif, Mar 27, 47; m 70; c 2. TRANSPORTATION ENGINEERING. *Educ:* Univ Calif, Davis, BS, 69, MS, 70, PhD(civil eng), 75. *Prof Exp:* Instr eng, Ft Lewis Col, 70-71; from asst prof to assoc prof, 71-82, PROF CIVIL ENG, CALIF STATE UNIV, CHICO, 82-, CHMN, 84- *Concurrent Pos:* Consult traffic engr, 71- *Mem:* Am Soc Civil Engrs; Inst Transp Engrs; Am Soc Eng Educ; Transp Res Bd. *Res:* Bicycle traffic. *Mailing Add:* Calif State Univ Calif State Univ Chico CA 95929-0930

FERRARI, DOMENICO, b Gragnano Tr, Piacenza, Italy, Aug 31, 40; m 66; c 2. PERFORMANCE EVALUATION, COMPUTER NETWORKS. *Educ:* Milan Polytech, Italy, 63; Italy Govt, Rome, Libera docenza(comput sci), 69. *Prof Exp:* Res fel comput sci, Milan Polytech, Italy, 64-67, asst prof, 67-70; from asst prof to assoc prof, 70-79, vchmn, Dept Elec Eng & Comput Sci, 77-79, chmn, Dept Comput Sci, dir comput systs res group, 83-87, PROF COMPUT SCI, UNIV CALIF, BERKELEY, 79- *Concurrent Pos:* Vis prof, Univ Pisa, 73, Milan Polytech, Italy, 76-77 & Univ Pavia, 79-80; prin investr, NSF, 74-; co-prin investr, Exp Comput Sci grant, Defense Advan Res Prog Agency, 83- *Honors & Awards:* Ottavio Bonazzi Award, Ital Asn Elec, 70; A A Michelson Award, Comput Mgt Group, 87. *Mem:* Asn Computing Mach; fel Comput Soc Inst Elec & Electronics Engrs. *Res:* Evaluation of computer systems performance (and in particular that of distributed systems); design of high-speed computer networks. *Mailing Add:* 839 Indian Rock Ave Berkeley CA 94707

FERRARI, HARRY M, b Detroit, Mich, May 20, 32; m 60; c 2. METALLURGICAL ENGINEERING. *Educ:* Wayne State Univ, BS, 54; Univ Mich, MS, 55, PhD(metall eng), 58. *Prof Exp:* Metallurgist, Pontiac Div, Gen Motors Corp, 53-54; assoc metallurgist, eng res inst, Univ Mich, 54-58; sr engr thermoelec mat, 58-60, tech adv nuclear mat, European Atomic Energy Comn, Brussels, Belg, 60-63, supvry engr, 63-64, mgr nuclear fuel tech, 64-69 & fuel assembly develop, 69-72, CONSULT ENGR, ATOMIC POWER DIV, WESTINGHOUSE ELEC CO, 72 - *Concurrent Pos:* Fel, Univ Mich, 58; lectr, Milan Polytech Inst, 61-62; consult, US Dept Energy,

79 - *Mem:* Fel Am Nuclear Soc; Am Soc Metals; Am Inst Mining, Metall & Petrol Engrs. *Res:* Studies on nuclear fuel materials, advanced fuel assembly designs and effects of irradiation on material properties. *Mailing Add:* 121 Evergreen Pittsburgh PA 15238

FERRARI, LAWRENCE A, b Hackensack, NJ, Nov 30, 37; m 64; c 4. PHYSICS, ELECTROMAGNETISM. *Educ:* Stevens Inst Technol, ME, 58, MS, 60, PhD(physics), 65. *Prof Exp:* Vis res assoc physics, Plasma Physics Lab, Princeton Univ, 64-65; asst prof, 65-69, chmn dept, 70-77, PROF PHYSICS, QUEENS COL, NY, 69- *Mem:* Am Phys Soc; Sigma Xi; NY Acad Sci. *Res:* Experimental plasma physics; wave propagation in random media. *Mailing Add:* Dept Physics Queens Col Flushing NY 11367

FERRARI, LEONARD, b New York, NY, April 8, 43. ELECTRICAL ENGINEERING. *Educ:* Northeastern Univ, BA, 67; Univ Calif, PhD(elec eng), 80. *Prof Exp:* Res prof elec eng & radio, 80-85, PROF ELEC ENG, UNIV CALIF, IRVINE, 85- *Mailing Add:* Elect Eng Dept Elect Eng Dept IERF 208A Irvine CA 92717

FERRARI, RICHARD ALAN, b Minneapolis, Minn, June 13, 32; m 58; c 3. BIOCHEMICAL PHARMACOLOGY. *Educ:* Cornell Univ, BA, 54, MFS, 55; Pa State Univ, PhD(biochem), 59. *Prof Exp:* From assoc res biologist to sr res biologist biochem, Sterling Winthrop Res Inst, 59-76, group leader pharmacol, 76-90; SR RES INVESTR, STERLING RES GROUP, 90- *Mem:* AAAS; Am Chem Soc; Sigma Xi. *Res:* purification and kinetics of enzymes; central nervous system biochemistry; radiobiology; biochemical pharmacology. *Mailing Add:* Sterling Res Group Rensselaer NY 12144

FERRARIO, CARLOS MARIA, US citizen. CARDIOVASCULAR PHYSIOLOGY, CARDIOVASCULAR DISEASES & CARDIOLOGY. *Educ:* Mariano Moreno Col, Agr, BS, 56; Univ Buenos Aires, MD, 63. *Prof Exp:* Asst physiol, Med Sch, Univ Buenos Aires, 63-64; assoc mem res staff, 70-71, actg chmn, dept cardiovascular res, 81-83, MEM RES STAFF, CLEVELAND CLIN RES DIV, 71-, CHMN, DEPT BRAIN & VASCULAR RES, 83- *Concurrent Pos:* Nat Sci Coun Agr fel, Univ Buenos Aires, 63-64; Swedish Int Agency fel, Gothenberg Univ, 64-66; Nat Lung & Heart Insts grant, Cleveland Clin Found, 67-; estab investr cardiovasc res, Am Heart Asn, 72-77, mem, Med Adv Bd, Coun High Blood Pressure Res. *Honors & Awards:* H Goldblatt Award, Am Heart Asn, 79. *Mem:* Am Heart Asn; Am Col Cardiol; Physiol Soc; InterAm Soc Hypertension (pres); NY Acad Sci; Fedn Am Socs Exp Biol; Soc Neurosci. *Res:* Investigation into the cause of arterial hypertension, especially in reference to its hemodynamic, humoral and nervous system participation. *Mailing Add:* Cleveland Clin Res Div 9500 Euclid Ave Cleveland OH 44195-5070

FERRARIS, JOHN PATRICK, b Stamford, Conn, Apr 6, 47. ORGANIC CHEMISTRY, SOLID STATE CHEMISTRY. *Educ:* St Michael's Col, Vt, BA, 69; Johns Hopkins Univ, MA, 71, PhD(chem), 73. *Prof Exp:* Nat Res Coun res assoc, Nat Bur Standards, 73-75; asst prof chem, 75-80, ASSOC PROF CHEM & PHYSICS, UNIV TEX, DALLAS, 80- *Mem:* Am Chem Soc; Royal Soc Chem; AAAS; Sigma Xi; Am Phys Soc. *Res:* Design, synthesis and characterization of organic charge transfer materials; investigation of transport, optical and magnetic properties of organic metals and semiconductors; study of piezo- and pyroelectricity in polar polymers. *Mailing Add:* Univ Tex Dallas PO Box 830688 Richardson TX 75083-9500

FERRARO, CHARLES FRANK, b New York, NY, Sept 14, 24; m 59; c 3. ANALYTICAL CHEMISTRY. *Educ:* Polytech Inst Brooklyn, BS, 44; Columbia Univ, MA, 46, PhD(chem), 50. *Prof Exp:* From instr to asst prof chem, Fordham Univ, 49-56; mgr anal serv, Ctr Tech Dept, FMC Corp, 56-90; RETIRED. *Mailing Add:* 12 Millbrook Lane Lawrenceville NJ 08648

FERRARO, DOUGLAS P, PSYCHOLOGY, HEALTH PSYCHOLOGY. *Educ:* Columbia Univ, PhD(psychol), 65. *Prof Exp:* Prof psychol & chmn dept, Univ NMex, 65-90; DEAN, COL ARTS & SCI, WESTERN MICH UNIV, 90- *Res:* Psychophysiology. *Mailing Add:* Col Arts & Sci Western Mich Univ 2020 Friedmann Hall Kalamazoo MI 49008-5010

FERRARO, JOHN ANTHONY, b Pueblo, Colo, Dec 21, 46; m 73; c 2. NEUROPHYSIOLOGY. *Educ:* Southern Colo State Col, BS, 68; Univ Denver, MS, 70, PhD(speech & hearing sci), 72. *Prof Exp:* NIH fel, Northwestern Univ, 72-74; asst prof speech & hearing sci, 74-78, ASSOC PROF, OHIO STATE UNIV, 78-; PROF & CHMN, DEPT HEARING & SPEECH & ASSOC DEAN, SCH ALLIED HEALTH, UNIV KANS MED CTR. *Concurrent Pos:* NIH res asst, Nat Inst Neurol Dis & Stroke, 72-73, prin investr, 73-74; consult, Speech & Hearing Sci Sect, Dept Otolaryngol, Ohio State Univ, 74-81, asst dir, 80-81; clin neurophysiologist, Swed Med Ctr, 81- *Mem:* Acoust Soc Am; Am Speech & Hearing Asn; Asn Res Otolaryngol; Sigma Xi; Int Elec Response Audiometry Study Group. *Res:* Clinical applications of auditory evoked potentials. *Mailing Add:* Dept Hearing & Speech Univ Kans Med Ctr 39th & Rainbow Blvd Kansas City KS 66103

FERRARO, JOHN J, b New York, NY, Apr 20, 31; m 54; c 5. ORGANIC CHEMISTRY, BIOCHEMISTRY. *Educ:* Fordham Univ, BS, 52; Polytech Inst Brooklyn, PhD(org chem), 61. *Prof Exp:* Res assoc peptide chem, Med Col, Cornell Univ, 62-63, instr biochem, 63-64; asst prof biochem & org chem, St John's Univ, NY, 64-69; assoc prof, 69-74, PROF CHEM, LONG ISLAND UNIV, 69- *Mem:* AAAS; Am Chem Soc; Royal Soc Chem. *Mailing Add:* 57-25 228th St Bayside NY 11364-2468

FERRARO, JOHN RALPH, b Chicago, Ill, Jan 27, 18; m 47; c 3. MOLECULAR SPECTROSCOPY. *Educ:* Ill Inst Technol, BS, 41, PhD(chem), 54; Northwestern Univ, MS, 48. *Prof Exp:* Supv chemist, Tetryl Labs, Kankakee Arsenal, Ill, 41-43; lab asst, Northwestern Univ, 46-48; assoc chemist, Argonnne Nat Lab, 48-68, sr scientist, 68-81; Searle prof, 81-86, EMER PROF CHEM, LOYOLA UNIV, CHICAGO, 86- *Concurrent Pos:* Vis prof, Univ Rome, Italy, 66-67; ed jour, Soc Appl Spectros, 68-74; mem spec adv bd, Chem Rubber Co, 71-; vis prof, Univ Ariz, 73-74, adj prof planetary sci, 74-87; vis prof, Univ Florence, 78, Univ Rome, 84, 88, Aachem Univ, Ger, 87. *Honors & Awards:* Spectros Award, Soc Appl Spectros, 70, Meggers Award & Prof Achievement Award Spectros, 75; Distinguished Scientist Award, Argonne Univs Asn, 73; NATO recipient, Sr Scientist Award, 78, 84, 87. *Mem:* Am Chem Soc; AAAS; NY Acad Sci; hon mem Soc Appl Spectros (pres, 65); hon mem Coblentz Soc. *Res:* Infrared spectroscopy of inorganic complexes, coordination compounds and electrical conductors; Raman spectra; vibrational spectroscopy at high pressures; fourier transform infrared spectroscopy. *Mailing Add:* 568 Saylor Ave Elmhurst IL 60126

FERRAUDI, GUILLERMO JORGE, b Buenos Aires, Argentina, Dec 9, 42; m 72; c 2. INORGANIC CHEMISTRY, PHYSICAL CHEMISTRY. *Educ:* Univ Buenos Aires, MS, 70; Univ Chile, PhD(inorg chem), 74. *Prof Exp:* Res assoc, Wayne State Univ, 74-76; asst prof, Fac Sci, Univ Chile, 76-78; SCIENTIST, RADIATION LAB, UNIV NOTRE DAME, 78- *Mem:* Am Chem Soc. *Res:* Physical inorganic chemistry; photochemistry of coordination complexes; reactions of radicals with coordination complexes; chemistry of the complexes in unusual oxidation states; magnetic field effects on the thermal and photochemical reactivity of coordination complexes; biphotonic photochemistry. *Mailing Add:* Radiation Lab Univ Notre Dame Notre Dame IN 46556-0768

FERREE, CAROLYN RUTH, b Liberty, NC, Jan 29, 44; m 68. RADIATION ONCOLOGY. *Educ:* Univ NC, Greensboro, BA, 66; Bowman Gray Sch Med, MD, 70. *Prof Exp:* Intern med, NC Baptist Hosp, Winston-Salem, 70-71, resident radiation ther, 71-74; instr, 74-75, asst prof, 75-81, assoc prof, 81-87, PROF, RADIATION THER, BOWMAN GRAY SCH MED, 87- *Concurrent Pos:* Am Cancer Soc jr fac clin fel, 76-79. *Mem:* Am Col Radiol; Am Soc Therapeut Radiologists. *Res:* Psychology of terminal patients; Hodgkin's disease; conservative management of breast cancer. *Mailing Add:* Bowman Gray Sch of Med 300 Hawthorne Rd Winston-Salem NC 27103

FERREE, DAVID C, b Lock Haven, Pa, Feb 9, 43; m 68; c 3. HORTICULTURE, POMOLOGY. *Educ:* Pa State Univ, BS, 65; Univ Md, College Park, MS, 68, PhD, 69. *Prof Exp:* Assoc prof, 71-80, PROF HORT, OHIO AGR RES & DEVELOP CTR, 80- *Honors & Awards:* Joseph Gurley Award, Amer Soc Hort Sci, 81, Stark Award, 83; Distinguished Res, Int Dwarf Fruit Tree Asn, 90. *Mem:* Fel Am Soc Hort Sci, 82. *Res:* Integrating present knowledge and techniques of apple production management into efficient systems for high density orchards, with emphasis on light relations as affected by pruning, training systems and root stocks. *Mailing Add:* Dept Hort Ohio Agr Res & Develop Ctr Wooster OH 44691

FERREIRA, LAURENCE E, b Livermore, Calif, Mar 23, 28; m 54; c 3. CERAMICS ENGINEERING. *Educ:* Univ Calif, Berkeley, BS, 52, MS, 55. *Prof Exp:* Ceramic engr, Western Gold & Platinum Co, Calif, 53-56; dir res, Coors Porcelain Co, 56-71, consult ceramist, 71-72; dir, Tech Ctr, Interpace Corp, 72-77; vpres & gen mgr, Cyprus Mines Res Co, 77-81, gen consult, 81-84, dir prod develop, San Jose Delta, 84-86; GEN CONSULT & MGR SALES MKT, CURTIS TECH, 86- *Honors & Awards:* PACE Award, Nat Inst Ceramics Engrs. *Mem:* Fel Am Ceramic Soc; Am Soc Testing & Mat; Sigma Xi. *Res:* Research management development of oxide ceramics for industrial use; ceramic processing method development. *Mailing Add:* 695 S Nardo Ave G6 Solano Beach CA 92075-2307

FERRELL, BLAINE RICHARD, b Upper Darby, Pa, Feb 2, 51; m 73; c 1. PHYSIOLOGICAL ECOLOGY, CIRCADIAN RHYTHMS. *Educ:* Univ Pa, BA, 73; Western Ky Univ, MS, 75; La State Univ, PhD(vert zool), 79. *Prof Exp:* ASST PROF ZOOL BEHAV, WESTERN KY UNIV, 78- *Mem:* Am Soc Zoologists; Am Ornith Union; Coopers Ornith Soc; Wilson Ornith Soc; Sigma Xi. *Res:* Mechanisms by which photoperiod length and temperature time the occurrence of seasonal physiological and behavioral condition in vertebrates, the approach used is based on the premise that such mechanisms have a circadian basis. *Mailing Add:* Dept Biol Western Ky Univ Bowling Green KY 42101

FERRELL, CALVIN L, b Okla, Oct, 25, 49; c 4. FETAL & MATERNAL RELATIONSHIPS. *Educ:* Univ Calif, Davis, PhD(nutrit), 75. *Prof Exp:* RES ANIMAL SCIENTIST, ROMAN L HRUSKA US MEAT ANIMAL RES CTR, 75- *Honors & Awards:* Outstanding Young Res Award, Am Soc Animal Sci, midwest sect, 86; Nutrit Res Award, Am Soc Animal Sci, 89. *Mem:* Am Soc Animal Sci; Am Inst Nutrit. *Res:* Energy and protein requirements of ruminant animals. *Mailing Add:* Roman L Hruska US Meat Animal Res Ctr PO Box 166 Clay Center NE 68933

FERRELL, D THOMAS, JR, b Durham, NC, Sept 28, 22; m 65; c 1. ELECTROCHEMISTRY. *Educ:* Eastern Ky State Col, BS, 43; Duke Univ, AM, 48, PhD(chem), 50. *Prof Exp:* Chemist, Naval Ord Lab, Md, 50-56; mgr res & develop, Battery Lab, Am Mach & Foundry Co, 57-58, asst lab mgr, 58-59; asst gen mgr, Missile Battery Div, Elec Storage Battery Co, 59-60, from assoc dir to asst dir eng, Exide Indust Div, 60-65, tech coordr, ESB Inc, 65-69, mgr lead acid eng, Exide Power Systs Div, 69-74, assoc dir technol, 74-78, vpres, ESB Technol Co, 78- 82; CONSULT, 82- *Mem:* Am Chem Soc; Electrochem Soc. *Res:* Electrode reactions; batteries. *Mailing Add:* 3871 Dempsey Lane Huntingdon Valley PA 19006-1901

FERRELL, EDWARD F(RANCIS), b Louisa, Ky, Apr 19, 14; m 45; c 3. PHYSICAL METALLURGY. *Educ:* Western Ky State Col, BS, 39; Vanderbilt Univ, MS, 40, PhD, 50. *Prof Exp:* Analyst, Electro Metall Co, WVa, 41-42; mem staff, Battelle Mem Inst, 47-55; instr chem, Bowling Green State Univ, 55-57; asst prof, Colo Sch Mines, 57-64; phys metallurgist, US Bur Mines, 64-86; RETIRED. *Mem:* Am Chem Soc; Am Soc Eng Educ; Metall Soc; Am Ceramic Soc. *Res:* Properties of metallic carbides and silicides; cermets; refractory metals; powder metallurgy; vapor phase deposition; oxidation rare earth metals; metal-ceramic bonding; refractory metals; ceramics. *Mailing Add:* 2775 Mayberry Reno NV 89509

FERRELL, HOWARD H, b Shreveport, La, Apr 11, 29; m 54; c 3. PETROLEUM ENGINEERING. *Educ:* Okla State Univ, BS, 51, MS, 57; Tex A&M Univ, PhD(petrol eng), 61. *Prof Exp:* Eng trainee, Stanolind Oil & Gas Co, 51-52; res engr,Continental Oil Co, 59-61, res group supvr, 61-65, res group leader, 65-77, Sr Staff Engr, 77-85; sr consult engr, K & A Energy Consult, 85-91; RETIRED. *Mem:* Soc Petrol Engrs. *Res:* Water-oil displacement in porous underground rock; application of the displacement mechanism for development of oil recovery processes; techniques to evaluate the characteristics of underground oil bearing strata; design of surface equipment to control underground extraction of oil. *Mailing Add:* 5813 S Lakewood Tulsa OK 74135-7661

FERRELL, JAMES K(IO), b Maryville, Mo, Jan 18, 23; m 43; c 2. CHEMICAL ENGINEERING. *Educ:* Univ Mo, BS, 48, MS, 49; NC State Univ, PhD(chem eng), 54. *Prof Exp:* Asst prof chem eng, NC State Univ, 53-56; design specialist, Martin Marietta Corp, 56-58; sect chief design, Babcock & Wilcox, Co, 58-61; PROF CHEM ENG, NC STATE UNIV, 61-, DEPT HEAD, 66-, ASSOC DEAN GRAD PROG & RES, NC STATE UNIV. *Concurrent Pos:* Pres, Triangle Univ Comput Ctr, 65-66; dept head, NC State Univ, 66- *Honors & Awards:* Alcoa Res Award, 82. *Mem:* Am Inst Chem Engrs; Am Chem Soc; Sigma Xi. *Res:* Heat transfer applied to the chemical industry; coal gasification and environmental control for the coal processing industry, process control. *Mailing Add:* NC State Univ PO Box 7905 Raleigh NC 27695-7905

FERRELL, RAY EDWARD, JR, b New Orleans, La, Jan 29, 41; m 62; c 4. MINERALOGY, GEOCHEMISTRY. *Educ:* Univ Southwestern La, BS, 62; Univ Ill, MS, 65, PhD(geol), 66. *Prof Exp:* From asst prof to assoc prof, 66-77, chmn dept, 78-82, PROF GEOL, LA STATE UNIV, BATON ROUGE, 78- *Concurrent Pos:* Vis prof, Sheffield, Eng, Oslo, Norway & Bern, Switz; Fulbright sr res award. *Mem:* Clay Minerals Soc; Am Asn Petrol Geologists; Geol Soc Am. *Res:* Origin, geologic importance, and environmental applications of clay minerals. *Mailing Add:* Dept Geol & Geophysics La State Univ Baton Rouge LA 70803

FERRELL, RICHARD ALLAN, b Santa Ana, Calif, Apr 28, 26; m 52; c 2. SUPERCONDUCTIVITY & THE JOSEPHSON EFFECT, CRITICAL DYNAMICS OF TRANSPORT COEFFICIENTS NEAR SECOND ORDER PHASE TRANSITIONS. *Educ:* Calif Inst Technol, BS, 48, MS, 49; Princeton Univ, PhD(physics), 52. *Prof Exp:* PROF PHYSICS, UNIV MD, 53- *Concurrent Pos:* Vis prof physics, Univ Va, 66-67 & 74, Univ Paris, 68-68, Tech Univ, Munich, Ger, 76-77, Univ Calif, Berkeley, 64, Univ Tokyo, 81, Tech Univ, Karlsruhe, 69, Max Planck Inst Solid State Res, 84-85, Univ Calif, Santa Barbara, 71; Inst Theoret Physics, Univ Calif, Santa Barbara, 79-80; prin investr, Nat Sci Found Grant Basic Res, 72-; consult, NASA, 87-; Humboldt Found award, 75. *Honors & Awards:* Award Sci Achievement, Sigma Xi, 72. *Mem:* Fel Am Phys Soc; Am Asn Physics Teachers; Am Asn Univ Prof. *Res:* Studing the critical phenomena exhibited by materials undergoing phase transitions, such as magnets, superfluids and superconductors; critical dynamics of transport coefficients near second order phase transitions. *Mailing Add:* Dept Physics & Astron Univ Md College Park MD 20742

FERRELL, ROBERT EDWARD, b Meridian, Miss, July 15, 43; m 70. BIOCHEMICAL GENETICS. *Educ:* Miss Col, Clinton, BS, 66; Univ Tex, Austin, PhD(biochem), 71. *Prof Exp:* Fel human genetics, Dept Human Genetics, Univ Mich, Ann Arbor, 70-72, res assoc, 72-75; asst prof pop genetics & head genetic markers lab, 75-79, ASSOC PROF POP GENETICS, CTR POP & DEMOG GENETICS, HEALTH SCI CTR, UNIV TEX, HOUSTON, 79- *Mem:* AAAS; Am Chem Soc; Sigma Xi; Am Asn Phys Anthrop. *Res:* Nature and meaning of genetically determined biochemical variability in natural populations with particular interest in human populations; biochemical evolution of human and other populations. *Mailing Add:* Dept Biostatist Grad Sch Pub Health Univ Pittsburgh Pittsburgh PA 15261

FERRELL, WILLIAM JAMES, b Wheeling, WVa, Apr 7, 40; m 60; c 3. BIOCHEMISTRY, ORGANIC CHEMISTRY. *Educ:* West Libery State Col, BS, 61; WVa Univ, MS, 63; Univ Pittsburgh, PhD(biochem), 67. *Prof Exp:* Asst prof biochem, Univ Detroit, 67-74; asst prof biol chem, Univ Mich, Ann Arbor, 74-81; assoc prof path, Wayne State Univ, 81-86; DIR CLIN CHEM, CHILDREN'S HOSP, DETROIT, 81- *Concurrent Pos:* Grant-in-aid, Mich Heart Asn, 68-69 & 71-72 & Licensed Beverage Industs, 68-69 & 70-71; NIH grant, 70-73. *Mem:* Am Asn Clin Chemists. *Res:* Isolation, characterization and biosynthesis of aldehydogenic lipids and glycerol thioethers; biosynthesis and metabolism of free fatty aldehydes; changes in plasma and heart adenosine triphosphate, adenosine diphosphate and creatine phosphate during shock; enzymatic interconversion of fatty acids, alcohols and aldehydes; applications of high performance liquid chromatography to therapeutic drug monitoring. *Mailing Add:* 22131 Jerome St Oak Park MI 48237

FERRELL, WILLIAM KREITER, b Barberton, Ohio, Oct 18, 19; m 49; c 3. FORESTRY. *Educ:* Univ Mich, BS, 41; Duke Univ, MF, 46, PhD(forest soils), 49. *Prof Exp:* Asst forest soils specialist, Univ Idaho, 48-53, asst prof forestry, 53-56; from asst prof to prof, 56-80, EMER PROF FOREST MGR, ORE STATE UNIV, 80- *Concurrent Pos:* NSF sci fac fel, Denmark, 64-65; vis prof ecol syst, Cornell Univ, 71-72; ed forest sci, Soc Am Foresters, 75-79; vis prof, Northern Ariz Univ, 82-83; pres, Northwest Sci Asn, 59. *Mem:* Fel AAAS; Soc Am Foresters; Am Soc Plant Physiol; Ecol Soc Am. *Res:* Photosynthesis, respiration and drought resistance studies of tree seedlings; ecotypic variation; forest carbon budgets. *Mailing Add:* PO Box 207 Philomath OR 97370

FERRELL, WILLIAM RUSSELL, b Cleveland, Ohio, June 19, 32; m 54; c 2. SYSTEMS & HUMAN FACTORS ENGINEERING. *Educ:* Swarthmore Col, BA, 54; Mass Inst Technol, SB & SM, 61, ME, 63, PhD(mech eng), 64. *Prof Exp:* Res asst mech eng, Mass Inst Technol, 61-62, from instr to assoc prof, 62-69; PROF SYSTS & INDUST ENG, UNIV ARIZ, 69- *Concurrent Pos:* Ford fel, Mass Inst Technol, 64-66; ed, Trans Man-Mach Systs, 69-70, co-ed, Trans Systs, Man & Cybernetics, 70-71; dept ed, J Forecasting, 89- *Mem:* Fel Human Factors Soc; Inst Elec & Electronics Engrs; Asn Comput Mach. *Res:* Human performance in engineering systems; remote manipulation; decision making; subjective probability and utility; mathematical models of performance; work methods and measurement. *Mailing Add:* Dept Systs & Indust Eng Univ Ariz Tucson AZ 85721

FERREN, LARRY GENE, b Emporia, Kans, Oct 13, 48; m 69; c 2. BIOCHEMISTRY. *Educ:* Univ Mo, BS, 70, PhD(biochem), 74. *Prof Exp:* Fel biochem, Univ Iowa, 74-75; instr chem radiation controls, Westinghouse, 80-82; from asst prof to assoc prof, 75-80, PROF CHEM, OLIVET NAZARENE COL, 82- *Concurrent Pos:* Instr chem, Idaho State Univ, 80-81; part-time instr chem, St Xavier Col, 83- *Mem:* Am Chem Soc; Nat Assn Health Professions Adv. *Res:* Metal ion catalyzed hydrolysis of dipeptides; metalloenzymes; proteolytic enzymes requiring metal ions for activity. *Mailing Add:* Dept Chem Olivet Nazarene Col Kankakee IL 60901

FERREN, RICHARD ANTHONY, b New Brunswick, NJ, Jan 19, 31; m 56; c 5. PHYSICAL ORGANIC CHEMISTRY. *Educ:* Villanova Univ, 52; Univ Pa, MS, PhD(chem), 56. *Prof Exp:* RES CHEMIST, PENNWALT CHEM CO, 56- *Mem:* Am Chem Soc; Soc Rheol. *Res:* Polymer chemistry. *Mailing Add:* 817 Warren Rd Ambler PA 19002-2206

FERRENDELLI, JAMES ANTHONY, b Trinidad, Colo, Dec 5, 36. NEUROLOGY, NEUROPHARMACOLOGY. *Educ:* Univ Colo, BA, 58, MD, 62. *Prof Exp:* Intern, Med Ctr, Univ Ky, 62-63; med officer, US Army, 63-65; resident neurol & neuropath, Cleveland Metrop Gen Hosp, 65-68; USPHS fel pharmacol, 68-71, from asst prof to assoc prof neurol & pharmacol, 70-77, PROF NEUROL & PHARMACOL & SEAY PROF CLIN NEUROPHARMACOL, SCH MED, WASHINGTON UNIV, 77- *Concurrent Pos:* Asst neurologist, Barnes Hosp, 72-75, assoc neurologist, 75-78, neurologist, 78-; fel adv comn, Pharmacol-Morphol, Pharmaceut Mgrs Asn Found Inc, 79-; mem comt, Epilepsy Adv Comn, NINCDS. *Honors & Awards:* Epilepsy Award, Am Soc Pharmacol & Exp Therapeut, 81. *Mem:* Am Neurol Asn; Am Acad Neurol; Am Soc Pharmacol & Exp Therapeut; Soc Neurosci; Am Soc Neurochem; Am Epilepsy Soc. *Res:* Neurochemistry and neuropharmacology of epilepsy; role of cyclic nucleotides in nervous tissue; mechanisms of action of neuropharmacological agents; drug development; pathophysiological mechanisms of seizure disorders. *Mailing Add:* Dept Neurol & Neurol Surg Sch Med Wash Univ 660 Euclid Ave Box 8111 St Louis MO 63110

FERRER, JORGE F, MEDICAL DISEASES. *Prof Exp:* HEAD, COMPARATIVE LEUKEMIA & RETROVIRUS UNIT, UNIV PA, 69- *Mailing Add:* Comparative Leukemia & Retrovirus Unit Univ Pa New Bolton Ctr 382 W St Rd Kennett Square PA 19348

FERRETTI, ALDO, b Rome, Italy, Jan 22, 29; US citizen; m 56; c 2. ORGANIC CHEMISTRY. *Educ:* Univ Rome, PhD(chem), 53. *Prof Exp:* Org chemist, Nat Hydrocarbon Corp, Italy, 53-57; Fulbright fel, Univ Ill, 57-59; res chemist, Farmitalia Div, Montecatini SA, 59-62; res assoc, Vanderbilt Univ, 62-63; res chemist, US Naval Propellant Plant, 63-64; sr scientist, Tenco Div, Coca-Cola Co, NJ, 64-68; RES CHEMIST, USDA, 68- *Mem:* Am Chem Soc; NY Acad Sci; Am Oil Chemists' Soc; The Chem Soc; Am Inst Chemists. *Res:* Synthetic organic chemistry; chemistry and biochemistry of natural products; lipid chemistry. *Mailing Add:* 8516 Howell Rd Bethesda MD 20817-6804

FERRETTI, JAMES ALFRED, b Sacramento, Calif, Aug 1, 39; m 70; c 2. PHYSICAL & ANALYTICAL CHEMISTRY. *Educ:* San Jose State Col, BS, 61; Univ Calif, Berkeley, PhD(chem), 65. *Prof Exp:* Bio-org chemist, Stanford Res Inst, 61-62; res asst chem, Lawrence Radiation Lab, 62-65; instr, Univ Naples, Italy, 65-66; chemist, Food & Drug Admin, 66-67; RES CHEMIST, NIH, 67- *Mem:* Am Chem Soc; Am Phys Soc; Biophys Soc; Sigma Xi; NY Acad Sci; Soc Magnetic Resonance in Med. *Res:* Nuclear magnetic resonance spectroscopy applied to biological molecules, the conformation of intermediate molecular weight peptides, and in vivo systems. *Mailing Add:* NIH Bldg 10 Rm 7N315 Bethesda MD 20892

FERRETTI, JOSEPH JEROME, b Chicago, Ill, Dec 23, 37; m 65; c 2. BIOCHEMISTRY, MICROBIOLOGY. *Educ:* Loyola Univ Chicago, BS, 60; Univ Minn, Minneapolis, MS, 65, PhD(biochem), 67. *Prof Exp:* Res asst biochem, Med Sch, Northwestern Univ, 60-62; assoc prof microbiol, 69-76, PROF MICROBIOL, HEALTH SCI CTR, UNIV OKLA, 76-, CHMN, 83- *Concurrent Pos:* USPHS fel, Johns Hopkins Univ, 67-69; NSF grant, 73; NIH grants, 73-; Nat Acad Sci vis scientist, German Dem Repub, 80; mem, NSF US-France Coop Sci Prog, Pasteur Inst, 82, NIH Study Sect; George Lynn Cross Res Prof. *Honors & Awards:* Golden Key Nat Hon Soc Res Award. *Mem:* AAAS; Am Soc Microbiol. *Res:* Plasmids in Streptococci; immunologic cross-reactions with Streptococcal antigens; genetic studies on group A Streptococci and their bacteriophage; molecular cloning of virulence determinants in streptococci and their bacteriophage; streptocci and dental caries; streptokinase. *Mailing Add:* Dept Microbiol PO Box 26901 Oklahoma City OK 73190

FERRIANS, OSCAR JOHN, JR, b Touchet, Wash, Mar 9, 28; m 53. GEOLOGY. *Educ:* State Col Wash, BS, 52, MS, 58. *Prof Exp:* GEOLOGIST, US GEOL SURV, 53- *Mem:* AAAS; Geol Soc Am; Arctic Inst NAm; Am Quaternary Asn. *Res:* Areal, surficial, glacial, economic, engineering and environmental geology; geomorphology; permafrost; earthquake effects; gold placer deposits; remote sensing. *Mailing Add:* US Geol Surv 4200 University Dr Anchorage AK 99508

FERRIER, BARBARA MAY, b Edinburgh, Scotland, Aug 7, 32; m 63; c 2. BIOCHEMISTRY. *Educ:* Univ Edinburgh, BSc, 54, PhD(chem), 58. *Prof Exp:* Asst chem, Univ Edinburgh, 54-58; res fel, Hickrill Chem Res Found, NY, 58-59; asst lectr, Bedford Col, Univ London, 59-61; res fel, Yale Univ, 61-62; res assoc biochem, Med Col, Cornell Univ, 62-63, instr, 63-66; res assoc obstet physiol, Clin Hosp, Montevideo, Uruguay, 66-68; assoc prof biochem, 69-76, PROF BIOCHEM, MCMASTER UNIV, 76- *Mem:* Can Biochem Soc; Can Soc Res Med Educ. *Res:* Neuropeptides and memory, career choice. *Mailing Add:* Dept Biochem Fac Health Ctr McMaster Univ 1200 Main St W Hamilton ON L8N 3Z5 Can

FERRIER, GREGORY R, b Winnipeg, Man, July 17, 43; m 69; c 4. CARDIAC ELECTROPHYSIOLOGY, CARDIAC PHARMACOLOGY & CONTRACTION. *Educ:* Univ Man, BSc, 66, PhD(pharmacol), 70. *Prof Exp:* Fel cardiol, Masonic Med Res Lab, Utica, NY, 70-72, Res Sci, 72-81; PROF PHARMACOL, DALHOUSIE UNIV, HALIFAX, NS, 81- *Concurrent Pos:* Consult, Nat Heart, Lung & Blood Inst, 75-, Can Heart Found, 75- & Med Res Coun Can, 81-; mem, res comt, NY affil, Am Heart Asn, 79-81; Van Horne Investr, Cent NY Chap, Am Heart Asn, 79; mem, sci rev comt, Can Heart Found, 81-84; med adv comt, Nova Scotia Heart Found, 84-90. *Honors & Awards:* Van Horne Investr, Am Heart Asn, Cent NY Chap, 79. *Mem:* Am Physiol Soc; Cardiac Electrophysiol Soc; Int Soc for Heart Res, Am Sect; Can Cardiac Electrophysiology Soc; Biophys Soc. *Res:* Investigation of cellular electrophysiological mechanisms underlying disruptions of normal cardiac rate and rhythm (arrhythmias); investigation of mechanisms of action of drugs and physiological mediatiors on cardiac arrhythmias and heart failure; mechanisms of cardiac excitation-contraction coupling. *Mailing Add:* Dept Pharmacol Charles Tupper Med Bldg Dalhousie Univ Halifax NS B3H 4H7

FERRIER, JACK MORELAND, b Cleveland, Ohio, Aug 8, 43; Can citizen; m 66; c 1. ELECTROPHYSIOLOGY, THEORETICAL BIOLOGY. *Educ:* Ohio State Univ, BSc, 66, MSc, 68, PhD(physics), 73. *Prof Exp:* Lectr physics, Dept Physics, Ohio State Univ, 73-75; res assoc, Bot Dept, 75-79, asst prof, 79-84, ASSOC PROF PERIODONT PHYSIOL, UNIV TORONTO, 84- *Mem:* Am Phys Soc. *Res:* Ion and water transport in biological cells and tissues: electrophysiology of mammalian cells and algal cells; elasticity and water transport in plant and animal tissue; galvanotaxis of osteoclasts and osteoblasts. *Mailing Add:* 4384 Med Sci Bldg Univ Toronto One Kings College Circle Toronto ON M5S 1A8 Can

FERRIER, LESLIE KENNETH, b Welland, Ont, Apr 13, 41; m 74; c 3. FOOD SCIENCE, PROTEIN CHEMISTRY. *Educ:* Univ Guelph, BSA, 63, MSc, 65; Univ Wis, PhD(food sci), 72. *Prof Exp:* Res chemist brewing res, Can Breweries Ltd, 65-67; fel enzym, Dept Food Sci, Univ Wis, 71-72; asst prof food sci, Dept Food Sci & Int Soybean Prog, Univ Ill, 72-77; proj leader, Gen Foods Tech Ctr, Tarrytown, NY, 77-81, group leader, 81-85; ASSOC PROF FOOD SCI, UNIV GUELPH, 85- *Concurrent Pos:* Adj prof, Dept Chem, Pace Univ, 80-85 & Dept Nutrit, NY Med Col, 84-85. *Mem:* Inst Food Sci; Am Chem Soc; AAAS; Can Inst Food Sci & Technol; Am Asn Cereal Chemists. *Res:* Food irradiation; frozen doughs; processing soybeans for human food; enzymes for food processing; egg protein chemisty and technology; effects of processing on protein functionality and nutrition. *Mailing Add:* Dept Poultry Sci Univ Guelph Guelph ON N1G 2W1 Can

FERRIGNO, PETER D, b Washington, DC, Aug 26, 27; m 52; c 4. DENTISTRY. *Educ:* Georgetown Univ, BS, 50, DDS, 55; Cath Univ Am, MS, 52; Ohio State Univ, MS, 57; Am Bd Endodont, dipl, 65. *Prof Exp:* From instr to prof periodont, Sch Dent, Georgetown Univ, 59-90, chmn dept, 68-90; PVT PRACT, 90- *Concurrent Pos:* Consult, Vet Admin, 63-, Montgomery Jr Col, 61 & US Navy. *Mem:* Am Dent Asn; Am Asn Endodont; Am Acad Periodont; Int Asn Dent Res. *Res:* Fibrogenesis on periodontal ligament in rats by radioautolosis. *Mailing Add:* 4400 Jennifer St W Suite 335 Washington DC 20015

FERRIGNO, THOMAS HOWARD, b Newark, NJ, Dec 3, 25; m 47; c 2. ORGANIC CHEMISTRY. *Educ:* Seton Hall Univ, BS, 51. *Prof Exp:* Plant chemist, Pabco Prod, Inc, 50-55; res supvr minerals & chem, Philipp Corp, 55-63; asst tech dir, United Clay Mines Corp, 63-65; mgr appl res, United Sierra Div, Cyprus Mines Corp, 65-67; group leader, Tenneco Plastics Div, 67-70; MINERAL INDUST CONSULT, 71- *Mem:* Soc Plastics Engrs; fel Am Inst Chemists; Am Chem Soc; Soc Plastics Indust. *Res:* Development of fillers and application in paints, plastics, rubber and allied fields; surface treatment of minerals; product development and promotion. *Mailing Add:* Improde 29 Clover Hill Circle Trenton NJ 08638

FERRIS, ANN M, b Stoneham, Mass, Sept 18, 45; m; c 2. HUMAN LACTATION. *Educ:* Univ Mass, PhD(food sci & nutrit), 78. *Prof Exp:* Asst prof, 78-84, ASSOC PROF NUTRIT SCI, UNIV CONN, 84- *Mem:* Am Inst Nutrit; Am Diatetic Asn. *Mailing Add:* Dept Nutrit Sci Univ Conn 3624 Horsebarn Rd Exten Storrs CT 06269-4017

FERRIS, BENJAMIN GREELEY, JR, b Watertown, Mass, Jan 24, 19; m 42; c 5. ENVIRONMENTAL HEALTH, PULMONARY DISEASES. *Educ:* Harvard Col, AB, 40, Harvard Med Sch, MD, 43. *Hon Degrees:* Dr, Bordeaux II, 83. *Prof Exp:* From intern to asst resident pediat, Children's Hosp, Boston, 43-48; from asst prof to assoc prof, 50-71, PROF ENVIRON HEALTH, SCH PUB HEALTH, HARVARD UNIV, 71- *Concurrent Pos:* Res fel physiol, Sch Pub Health, Harvard Univ, 48-50; asst physician med, Phillips Acad, Andover, Ma, 49-50; dir res & med care, Mary MacArthur Respirator Unit, 50-58; indust res physician, Ludlow Jute Co, India, 51; consult, Mass Gen Hosp, Lemuel Shattuck Hosp & Children's Med Ctr, 56-; dir environ health & safety, Harvard Univ Health Servs, 58-; lectr med, Med Sch, Tufts Univ, 65-; vis prof, Univ BC, 72-78. *Mem:* AAAS; Am Physiol Soc; Am Pub Health Asn; Am Epidemiol Soc; Int Epidemiol Asn. *Res:* Effects of air-borne pollutants on human health; low levels of air pollution and exposures at work. *Mailing Add:* Dept Physiol Harvard Univ Sch Pub Health Boston MA 02115

FERRIS, BERNARD JOE, b Denver, Colo, Nov 16, 22; m; m 44; c 2. PETROLEUM GEOLOGY, ORGANIC GEOCHEMISTRY. *Educ:* Colo Sch Mines, GeolE, 47, MGeolE, 48. *Prof Exp:* Geologist, Shell Oil Co, 48-53, dist geologist, 53-54, div explor mgr, 54-56, sr geologist, 56-60, mgr geol dept, Shell Develop Co, 60-63, chief geologist, Midland Area, Shell Oil Co, 63-66, div explor mgr, 66-69, sr res assoc, Shell Develop Co, 69-70, mgr geol dept, 70-71, mgr explor training dept, Shell Oil Co, 71-73, chief geologist, Western Region, 73-76, chief geologist, Int Region, 76-79; independent consult, 79-82; vpres & sr vpres, McAdams, Roux & Assocs, 82-85; INDEPENDENT GEOL CONSULT, 85- *Mem:* Am Asn Petrol Geologists; Geol Soc Am. *Res:* Use of fossil calcareous algae in biostratigraphy; geological and chemical investigations of the origin and migration of petroleum hydrocarbons in the subsurface. *Mailing Add:* 6586 S Garfield Way Littleton CO 80121

FERRIS, CLIFFORD D, b Philadelphia, Pa, Nov 19, 35. BIOENGINEERING, ELECTRICAL ENGINEERING. *Educ:* Univ Pa, BS, 57, MS, 58; George Washington Univ, DSc(math, eng, physics), 62. *Prof Exp:* Engr, electronic instrument div, Burroughs Corp, Pa, 53, basic physics div, res ctr, 55-56; res asst, Univ Pa, 56-57, res assoc electromed res lab & consult, dept pharmacol & dept therapeut res, Univ Hosp, 57-59, asst instr elec eng, 58-59; from instr to assoc prof, George Washington Univ, 59-63; assoc prof, Drexel Inst Technol, 63-64 & Univ Md, 64-68; assoc prof, Univ Wyo, 68-73, actg dean, Col Eng, 73-74, actg head Elec Eng, 87-89, PROF ELEC ENG, UNIV WYO, 73-, DIR BIOENG PROG, 74- *Concurrent Pos:* Systs analyst, res lab, Melpar, Inc, Va, 59; consult, instrument eng & develop br, NIH, 60-62; vis scientist, Armed Forces Inst Path, Walter Reed Army Med Ctr, DC, 61-68. *Mem:* AAAS; Inst Elec & Electronics Engrs; NY Acad Sci. *Res:* Interaction of electromagnetic fields with biological systems; bioinstrumentation; bioelectrodes. *Mailing Add:* Dept Elec Eng Univ Wyo Laramie WY 82071

FERRIS, CLINTON S, JR, b Chicago, Ill, Aug 13, 33; m 59; c 3. EXPLORATION GEOLOGY. *Educ:* Colo Col, BS, 55; Univ Sask, MSc, 61; Univ Wyo, PhD(geol), 65. *Prof Exp:* From geologist to sr geologist, Kerr-McGee Corp, 64-69, explor adv, 70-71; proj mgr, Tex Gulf Inc, 72-73; chief geologist, Urania Explor Inc, 73-77; CHIEF GEOLOGIST, RESERVE INDUST, 77- *Mem:* Soc Econ Geol; Am Inst Prof Geologists. *Res:* Sandstone uranium genesis and ore controls; porphyry copper genesis and relation to rock alteration; structural geology; occurrences and formation of the uranium mineral brannerite; Athabasca region uranium geology. *Mailing Add:* 11427 W 59th Ave Arvada CO 80004

FERRIS, DEAM HUNTER, b Mankato, Minn, July 8, 12; m 35; c 4. VETERINARY MICROBIOLOGY & TECHNICAL ENGLISH. *Educ:* Drake Univ, BA, 34, MA, 38; Univ Wis, PhD(vet sci, zool), 53. *Prof Exp:* Teacher high schs, Iowa & Kans, 35-42; instr AV educ, Univ Wis, 46-48; prof biol & natural hist, Graceland Col, 47-57; microbiologist, Plum Island Animal Dis Ctr, USDA, 73-85; from assoc prof to prof, 57-74, EMER PROF VET PATH & HYG, COL VET MED, UNIV ILL, URBANA, 74- *Concurrent Pos:* Virologist, Near East Animal Health Inst, Food & Agr Orgn, UN, 64-66; vis prof, Dept Microbiol, Soochow Univ Taiwan. *Mem:* Fel AAAS; Am Soc Microbiol; Am Soc Parasitol; Wildlife Dis Asn; fel Royal Soc Health; Am Vet Med Asn; Am Micros Soc. *Res:* Host range and pathobiology of African malignant catarrhal fever and African swine fever; foot-and-mouth disease and rinderpest in deer; cytauxzoon-like protozoon of domestic cat; development of microfiche; handbooks and manuals on emergency diseases as well as research on computer-based and self-instructional autotutorial systems; development of a new immunofluorescence and peroxidase for trypanosomarypanosoma vivax and the adaptation of Colombian bovine T vivax to mice; development of simplified immunoelectroosmophoresis methods for diagnosis of African Swine Fever; lyophilization of reference antisera and African mink cell focus-inducing virus; world domestic animal diseases; Cytauxzoon felis, leptospirosis, Coxiella burneti, Plasmodium gallinacium, Micronema deletrix in the horse; abortive equine rabies; epizootiology of transmissible gastroenteritis in swine. *Mailing Add:* 11 Oak Hill Cluster Independence MO 64050

FERRIS, FREDERICK L, b Princeton, NJ, Apr 24, 46. EYE RESEARCH. *Educ:* Princeton Univ, AB, 68; Johns Hopkins Univ, MD, 72; Am Bd Ophthal, cert, 79. *Prof Exp:* Fel, Off Biometry & Epidemiol, Nat Eye Inst, NIH, 73-75; resident, 75-78, fel, 78-82, ASST PROF RETINAL VASCULAR SERV DEPT OPHTHAL, WILMER INST, JOHNS HOPKINS HOSP, 83-; CHIEF, CLIN TRIALS BR, BIOMETRY & EPIDEMIOL PROG, NAT EYE INST, NIH, 85- *Concurrent Pos:* Fel, Dept Med, Johns Hopkins Hosp, 72-73; mem, Vitrectomy Adv Group, Nat Eye Inst, 74, Data & Safety Qual Rev Group, Diabetes Control & Complications Trial, 82-, Data Monitoring Comt, Collab Ocular Melanoma Study, 86-; co-chmn, Early Treat Diabetic Retinopathy Study, 84-, Krypton Argon Regression Neovascularization Study, 84- *Honors & Awards:* Honor Award, Am Acad Ophthal, 84. *Mem:* Fel Am Acad Ophthalmol; fel Col Epidemiol; Sigma Xi; AMA; Soc Epidemiol Res. *Res:* DNA-cellulose chromatography; isolation of DNA repair enzyme of T4 bacteriophage; electron microscopy. *Mailing Add:* NIH Nat Eye Inst Clin Trials Br Bldg 31 Rm 6A24 Bethesda MD 20892

FERRIS, HORACE GARFIELD, b Los Angeles, Calif, Aug 3, 13; m 45; c 2. PHYSICS. *Educ:* Pomona Col, BA, 36; Univ Calif, Los Angeles, MA, 39, PhD(physics), 49. *Prof Exp:* Asst physics, Univ Calif, Los Angeles, 39-40; instr, Long Beach Jr Col, 41-42; civilian physicist, US Navy, 42-43, instr mil training prog, Pomona Col, 43-44; res assoc, Calif Inst Technol, 44-45; physicist, US Naval Ord Test Sta, 45-46; lectr physics, Univ Southern Calif, 46-49; instr, San Diego State Col, 49-51; asst res physicist, Scripps Inst, Univ Calif, 51-55; assoc prof physics, Chapman Col, 55-58; from asst prof to prof, 58-78, EMER PROF PHYSICS, CALIF STATE POLYTECH UNIV, POMONA, 78- *Concurrent Pos:* Consult, Robert Shaw Fulton Controls Co, 56-57, US Naval Radiol Defense Lab, 56-59, Civil Eng Lab, 59-60 & Hughes Aircraft Co, 57-69; assoc instr, Univ Calif, Los Angeles, 64-66. *Mem:* AAAS; Acoust Soc Am; Am Asn Physics Teachers. *Res:* Geothermal gradient in ocean floor by probe methods; hydrodynamics of stratified fluids including the ocean; partial differential equations of physics; methods of mathematical physics; measurement of thermal conductivities at high temperatures by transient methods; underwater acoustics, including theory of sound transmission in the ocean. *Mailing Add:* 4934 Gentry Ave North Hollywood CA 91607

FERRIS, JAMES PETER, b Nyack, NY, July 25, 32; div; c 2. BIOCHEMISTRY, ORGANIC CHEMISTRY. *Educ:* Univ Pa, BS, 54; Ind Univ, PhD, 58. *Prof Exp:* Lectr chem, Ind Univ, 58; res assoc, Mass Inst Technol, 58-59; instr org chem, Fla State Univ, 59-60, asst prof, 61-64; res assoc, Salk Inst Biol Studies, 64-67; assoc prof, 67-74, chmn dept, 80-83, PROF CHEM, RENSSELAER POLYTECH INST, 74- *Concurrent Pos:* USPHS career award, 69-74; sr res assoc, NASA Ames Res Ctr, Moffett Field, Calif, 76; ed bd, Biosystems, 77-; ed, Origins Life & Evol Biosphere, 83-; mem, Space Sci Bd, Nat Res Coun Task Force Life Sci, 84-86 & Life Sci Adv Comt, NASA, 87-; vis prof, ETH Lab Org Chem, Zurich, 85-86. *Mem:* Inter-Am Photochem Soc; Am Chem Soc; Planetary Soc; Int Soc Study Origins Life (treas, 83-); AAAS; Clay Mineral Soc. *Res:* Chemistry of the origins of life; atmospheric photochemistry; chemistry of hydrocyanic acid and nitriles. *Mailing Add:* Dept Chem Rensselaer Polytech Inst Troy NY 12180

FERRIS, JOHN MASON, US citizen; m 53; c 2. ECOLOGY, NEMATOLOGY. *Educ:* Cornell Univ, BS, 51, PhD(plant path), 56. *Prof Exp:* Asst plant path, Cornell Univ, 51-56; asst plant pathologist, State Natural Hist Surv, Ill, 57-58; from asst prof to assoc prof, 58-75, PROF NEMATOL, PURDUE UNIV, 75- *Concurrent Pos:* Ed, Nematol News Lett, Soc Nematol, 63-65; mem, Inter Soc Plant Protection, 80-82, Chmn, 82. *Honors & Awards:* Indiana Acad Sci Fel, 73. *Mem:* AAAS; Am Phytopath Soc; Soc Nematol (treas, 74-77, secy, 80, vpres, 79-80, pres, 80-81); Soc Europ Nematol; Sigma Xi. *Res:* Ecology of soil and freshwater nematodes; integrated pest management. *Mailing Add:* Dept Entom Purdue Univ West Lafayette IN 47907-1158

FERRIS, PHILIP, b New York, NY, Mar 17, 30; m 53; c 2. ONCOLOGY. *Educ:* City Col New York, BS, 52, MS, 57; NY Univ, PhD(biol), 70. *Prof Exp:* Pub sch teacher, NY, 53-59; res asst, 60-69, RES ASSOC ONCOL, WALDEMAR MED RES FOUND, 69- *Concurrent Pos:* NSF res grant, 65-71; adj assoc prof cell biol, State Univ NY Col Old Westbury, 78. *Mem:* NY Acad Sci; Sigma Xi. *Res:* Cancer virology; immunology and cytokinetics of chloroleukemia; biomedical research. *Mailing Add:* 20 Daley Pl LynbrooK NY 11563-2217

FERRIS, ROBERT MONSOUR, b Omaha, Nebr, Sept 1, 38; m 61; c 6. BIOCHEMISTRY, NEUROCHEMISTRY. *Educ:* Creighton Univ, BS, 60; Univ Nebr, MS, 63, PhD(biochem), 67. *Prof Exp:* NIH neurosci res fel, Sch Med, Duke Univ, 67-70; ASSOC HEAD PHARMACOL, BURROUGHS WELLCOME & CO USA, INC, 70- *Mem:* Neurochem Soc; Am Soc Pharmaceut Exp. *Res:* Biochemical and pharmacological properties governing the uptake, storage, synthesis and release of neurotransmitters in peripheral and central nervous systems; receptor pharmacology; pharmacology of mental diseases; neurosciences. *Mailing Add:* Dept Pharmacol Wellcome Res Labs 3030 Cornwellis Rd Research Triangle Park NC 27709

FERRIS, STEVEN HOWARD, b New York, NY, June 27, 43; c 2. GERONTOLOGY, ALZHEIMERS DISEASE. *Educ:* Rensselaer Polytech Inst, BA, 65; Queens Col, MA, 67; City Univ NY, PhD(exp psychol), 70. *Prof Exp:* NRC res associateship, Submarine Med Res Lab, Naval Submarine Med Ctr, 70-72; from asst prof to assoc prof psychiat, 79-89, EXEC DIR, AGING & DEMENTIA RES CTR, MED CTR NY UNIV, 73-, PROF PSYCHIAT, 90- *Concurrent Pos:* Vis scientist, Brookhaven Nat Lab, 79-; consult, Psychiat Serv, Manhattan Vet Admin Med Ctr, 80-; res scientist, Nathan Kline Inst, 89- *Mem:* Am Psychol Asn; Am Geront Soc; Neurosci Soc; Int Neuropsychol Soc; Am Col Neuropsychopharmacology. *Res:* Gerontology; neuropsychology, neurobiology and psychopharmacology of memory; cognitive deficits in aging and senile dementia; Alzheimers disease. *Mailing Add:* Dept Psychiat NY Univ Med Ctr 550 First Ave New York NY 10016

FERRIS, THOMAS FRANCIS, b Boston, Mass, Dec 27, 30; m 57; c 4. INTERNAL MEDICINE. *Educ:* Georgetown Univ, AB, 52; Yale Univ, MD, 56. *Prof Exp:* Intern, Osler Serv, Johns Hopkins Hosp, Baltimore, Md, 56-57; resident, New Haven Hosp, Conn, 60-62; from instr to asst prof med, Sch Med, Yale Univ, 63-67; from assoc prof to prof med, Col Med, Ohio State Univ, 67-78, dir renal dis, Univ Hosp, 67-78; PROF MED & CHMN DEPT, UNIV MINN, MINNEAPOLIS, 78- *Concurrent Pos:* USPHS clin fel renal dis, New Haven Hosp, Conn, 59-60, USPHS res fel, 62-63; John & Mary Markle scholar acad med, 64-69; vis investr, Regius Dept Med, Oxford Univ, 66-67. *Mem:* Am Fedn Clin Res; Am Soc Clin Invest; Am Soc Nephrology (pres); Asn Am Physicians; Asn Prof Med (pres). *Res:* Renal physiology; hypertension; diseases of the kidney. *Mailing Add:* Univ Hosp Harvard St at East River Rd Minneapolis MN 55455

FERRIS, VIRGINIA ROGERS, b Abilene, Kans; m 53; c 2. NEMATOLOGY, MOLECULAR BIOLOGY. *Educ:* Wellesley Col, BA, 49; Cornell Univ, MS, 52, PhD(plant path), 54. *Prof Exp:* Asst plant path, Cornell Univ, 49-52, asst prof, 54-56; consult plant path, 56-65; asst dean, Grad Sch, 71-75, asst provost, 76-79, from asst prof to assoc prof, 65-74, PROF ENTOM, PURDUE UNIV, WEST LAFAYETTE, 74- *Concurrent Pos:* Consult, NSF, 79-83, 85 & 88; bd dirs, Asn Systematics Collections, 75 & 78-81. *Mem:* Soc Nematol (secy, 65-68, vpres, 68-69, pres, 69-70); Am Phytopath Soc; Ecol Soc Am; Soc Syst Zool (coun, 79-81); Asn Systematics Collections; Soc Study Evolution; fel Soc Nematologists. *Res:* Evolutionary biology of nematodes using macromolecular data; plant diseases caused by nematodes; ecology and historical biogeography of soil and freshwater nematodes. *Mailing Add:* 2237 Delaware Dr West Lafayette IN 47906-1917

FERRIS, WAYNE ROBERT, b Lockman, Iowa. CYTOLOGY. *Educ:* Univ Chicago, PhD(zool), 59. *Prof Exp:* Asst zool, Univ Chicago, 50-58; from instr to assoc prof, 58-66, PROF ZOOL, UNIV ARIZ, 66- *Mem:* Am Soc Cell Biol. *Res:* Myogenesis of vertebrates striated muscle; cytochemistry and ultrastructural aspects of induction mechanisms; electron microscopy. *Mailing Add:* Dept Cell Biol Univ Ariz Col Med 1501 N Campbell Tucson AZ 85724

FERRIS-PRABHU, ALBERT VICTOR MICHAEL, b Meerut, Uttar Pradesh, India; m 67; c 1. SOLID STATE PHYSICS, MEMORY TECHNOLOGY. *Educ:* Univ Dayton, BME, 57; Princeton Univ, MSE & MA, 60; Cath Univ Am, PhD(solid state physics), 63. *Prof Exp:* Res assoc mat theory group, Mass Inst Technol, 63-64; physicist, Goddard Space Flight Ctr, NASA, 64-66; assoc prof appl sci, George Washington Univ, 66-68; ADV PHYSICIST, COMPONENTS DIV, IBM CORP, 68- *Concurrent Pos:* Consult, Wolf Res & Develop Corp, 67-68 & Adaptronics Inc, 68; adj prof, Univ Vt, 78- *Mem:* Am Phys Soc; sr mem Inst Elec & Electronics Engrs; fel AAAS. *Res:* Diffusion kinetics; semiconductors; nonvolatile memories; magnetism; error correction; systems performance evaluation; bipolar products yield management; over 100 published papers. *Mailing Add:* Gen Tech Div NO2/861-2 IBM Corp Essex Junction VT 05452

FERRISS, DONALD P, b Rutherford, NJ, Apr 17, 24; m 52; c 3. PHYSICAL METALLURGY, CERAMICS. *Educ:* Stevens Inst Technol, ME, 50, MS, 54; Mass Inst Technol, ScD(metall), 61. *Prof Exp:* Res engr powder metall, Stevens Inst Technol, 50-57; instr metall, Mass Inst Technol, 57-61; res metallurgist, 61-66, res assoc, 66-77, SR RES ASSOC, ENG TECHNOL LAB, E I DU PONT DE NEMOURS & CO INC, 77- *Mem:* Am Soc Metals; Am Inst Mining, Metall & Petrol Engrs; Sigma Xi; Am Powder Metall Inst. *Res:* Special processing of metal powders; deformation and fracture of metals; mechanical properties of ceramics; wear of materials. *Mailing Add:* 411 Baynard Blvd Wilmington DE 19803-4303

FERRISS, GREGORY STARK, b Summit, NJ, Aug 2, 24; m 57; c 2. NEUROLOGY, SLEEP DISORDERS MEDICINE. *Educ:* Harvard Univ, AB, 45; Tulane Univ, MD, 51. *Prof Exp:* From instr to prof neurol, Sch Med, La State Univ, 55-82; CLIN PROF NEUROL, TULANE UNIV SCH MED, 82-, DIR, SLEEP DIS CTR, 83- *Concurrent Pos:* Vis physician, Charity Hosp La, New Orleans, 55- & South Baptist Hosp, 63-; consult, DePaul Hosp, Coliseum House, Childrens' Hosp. *Mem:* Am Acad Neurol; Am Electroencephalog Soc; Am Epilepsy Soc; Asn Res Nerv & Ment Dis; Clin Sleep Soc. *Res:* Clinical electroencephalography, epilepsy and sleep. *Mailing Add:* 2820 Napolean Ave Suite 420 New Orleans LA 70115

FERRO, DAVID NEWTON, b San Bernardino, Calif, Sept 26, 46; div; c 2. ENTOMOLOGY. *Educ:* San Jose State Univ, BA, 69; Wash State Univ, MS, 73, PhD(entom), 75. *Prof Exp:* Technician entom, Univ Calif, Riverside, 68-70; res asst, Wash State Univ, 70-74; lectr, Lincoln Col, Univ Canterbury, 74-77; from asst prof to assoc prof, 78-88, PROF ENTOM, UNIV MASS, 88- *Honors & Awards:* Distinguished Achievement Award, Entom Soc Am. *Mem:* Entom Soc Am; Sigma Xi; NY Acad Sci. *Res:* Insect pest management; applied ecology; foraging behavior of natural enemies. *Mailing Add:* Dept Entom Univ Mass Amherst MA 01003

FERRON, JEAN H, b Montreal, Que, Mar 5, 48; m 77; c 2. ETHOLOGY, WILDLIFE MANAGEMENT. *Educ:* Univ Montreal, BSc, 69, MSc, 71, PhD(biol & ethol), 74. *Prof Exp:* Res fel, Dept Zool, Univ Alta, 75; PROF ETHOL & ECOL, DEPT BIOL, UNIV QUE, RIMOUSKI, 75- *Concurrent Pos:* Actg dean, res & grad studies, Univ Que, 87 & 89; invited prof, Lab Ethol & Sociobiol, Univ Paris XIII, 91. *Mem:* Animal Behav Soc; Am Soc Mammal; Wildlife Soc. *Res:* An evolutionary and ecological approach of the ethology of Sciurid rodents; habitat selection and use of space by mammals; snowshoe hare social behavior; behavioral ecology applied to wildlife management. *Mailing Add:* Dept Biol Univ Que 300 av des Ursulines Rimouski PQ G5L 3A1 Can

FERRON, JOHN R(OYAL), b Anoka, Minn, Oct 29, 26; m 51; c 4. CHEMICAL ENGINEERING. *Educ:* Univ Minn, BChE, 48, MS, 50; Univ Wis, PhD(chem eng), 58. *Prof Exp:* Instr chem, Macalester Col, 48-49; asst chem eng, Univ Minn, 49-50; engr, E I du Pont de Nemours & Co, 50-54; instr chem eng, Univ Wis, 54-57; from asst prof to prof, Univ Del, 58-69; PROF CHEM ENG, UNIV ROCHESTER, 69- *Concurrent Pos:* NSF sci fac fel, Univ Naples, 66-67; vis prof, Calif Inst Technol & Univ Calif, Berkeley, 78-79. *Mem:* Am Chem Soc; Am Inst Chem Engrs; NY Acad Sci. *Res:* High temperature reactions and transport phenomena; fluidization; optimization; applied mathematics. *Mailing Add:* Dept Chem Eng Univ Rochester Rochester NY 14627

FERRONE, FRANK ANTHONY, b New York, NY, Aug 13, 47; m; c 2. BIOLOGICAL SELF ASSEMBLY, MACROMOLECULAR DYNAMICS. *Educ:* Manhattan Col, BS, 69; Princeton Univ, MA, 71, PhD(physics), 74. *Prof Exp:* Instr physics, Princeton Univ, 74-75, vis fel, 75-76; staff fel, Lab Chem Physics, NIH, 76-80; from asst prof to assoc prof, 80-90, PROF PHYSICS, DREXEL UNIV, 90- *Mem:* Am Phys Soc; Biophys Soc; Sigma Xi. *Res:* Conformational kinetics of normal hemoglobin & kinetic mechanism of the self assembly of sickle-cell hemoglobin using optical techniques. *Mailing Add:* 501 S Taney St Philadelphia PA 19146

FERRONE, SOLDANO, b Bella, Italy, April 4, 40; US citizen. MEDICINE. *Educ:* Univ Milan, MD, 64, PhD, 71. *Prof Exp:* Asst prof, Scripps Found, La Jolla, Calif, 72-76, assoc mem, 76-81; prof, dept path & surg, Columbia Univ, 81-83; PROF & CHMN, DEPT MICROBIOL & IMMUNOL, NY MED COL, 83- *Concurrent Pos:* Adj assoc prof, Univ Calif, San Diego, 80-81; vis prof, Inst Giessen, WGer, 85-86; estab investr award, Am Heart Asn, 76-81; sr fel, Am Cancer Soc, Calif Div, 74-76. *Honors & Awards:* Res Award, Italian Res Coun, 70; Res Award, Inst Social Med, 71; Cabrini Gold Medal Award. *Mem:* Am Asn Immunologists; Reticuloendothelial Soc; Am Soc Hematologists. *Res:* Immunochemical analysis with monoclonal antibodies of the antigenic profile of tumor cells; immunodiagnosis & immunotherapy of melanoma with monoclonal anitbodies; immunotherapy of autoimmune diseases with antidiotypic monoclonal antibodies. *Mailing Add:* Dept Microbiol & Immunol NY Med Col Basic Sci Bldg Valhalla NY 10595

FERRY, ANDREW P, b New York, NY, June 15, 29; m 64. OPHTHALMOLOGY, PATHOLOGY. *Educ:* Manhattan Col, BS, 50; Georgetown Univ, MD, 54. *Prof Exp:* Intern med, Duke Univ Hosp, 54-55; asst resident, Hosp, Univ Mich, Ann Arbor, 57-58; resident ophthal, NY Hosp-Cornell Med Ctr, 58-61; dir Jordan Eye Bank, St John Ophthalmic Hosp, Jerusalem, Jordan, 64-65; assoc prof ophthal, Mt Sinai Sch Med, 65-76, prof, 76-78, asst prof path, 65-78; PROF & CHMN, DEPT OPHTHAL, MED COL, VA COMMONWEALTH UNIV, 78-, PROF PATHOL, 80- *Concurrent Pos:* Instr, Col Med, Cornell Univ, 60-61; inst Neurol Dis & Blindness spec fel, Armed Forces Inst Path, 61-64; consult, Manhattan Eye, Ear & Throat Hosp, NY, 66-78, Beth Israel Med Ctr, 67-78, Am Acad Ophthal & Otolaryngol, 69-82, City Hosp Ctr, Elmhurst, 74-78 & Richmond Vet Admin Hosp, 78-; vis prof, var univs, 72-87; guest of hon, Europ Ophthalmic PathSoc, 78 & Pac Coast Oto- Ophthal Soc, 81; pres, Verhoeff Ophthalmic Path Soc, 78; cur, Pharmaceut Div, Mus Found Am Acad Ophthal, 83-; mem bd dirs, Am Asn Ophthalmic Pathologists, 83- & Am Registry Path, 85- *Honors & Awards:* Kober Medal & Gold Medal Prev Med, Sch Med, Georgetown Univ, 54; Billings Bronze Medal, AMA, 65; appointed by Queen Elizabeth II to position of Officer in Grand Priory in Brit Realm of Most Venerable Order of Hosp of St John of Jerusalem, 68 and promoted to Comdr, 74; W Yerby Jones lectr, State Univ NY Buffalo Sch Med, 70; John M McLean Mem lectr, NY Acad Med, 73, Charles May Mem lectr, 89; Knighted by Queen Elizabeth II, 81; Sr Hon Award, Am Acad Ophthal, 85; Lois Young mem lectr, Howard Univ Sch Med, 90. *Mem:* Fel Am Col Surg; Asn Res Vision & Ophthal; AMA; Am Ophthal Soc; Am Asn Ophthal Pathologists (pres, 85-87); fel Am Acad Ophthal. *Res:* Ophthalmic pathology, especially pathology of ocular tumors. *Mailing Add:* Dept Ophthal Med Col Va Richmond VA 23298

FERRY, DAVID K, b San Antonio, Tex, Oct 25, 40; m 62; c 2. ELECTRICAL ENGINEERING, SOLID STATE PHYSICS. *Educ:* Tex Tech Col, BS, 62, MS, 63; Univ Tex, Austin, PhD(elec eng), 66. *Prof Exp:* NSF fel physics, Univ Vienna, 66-67; from asst prof to assoc prof elec eng, Tex Tech Univ, 66-73; with res off, Off Naval Res, 73-77; prof elec eng & head dept, Colo State Univ, 77-83; dir, Ctr Solid State Electronics, 83-89, REGENTS PROF & CHMN ELEC ENG, ARIZ STATE UNIV, 89- *Mem:* Fel Am Phys Soc; Inst Elec & Electronics Engrs. *Res:* High field transport in semiconductors, physics and modeling of sub-micron semiconductor devices; optical interactions in semiconductors. *Mailing Add:* Elec Engr Dept Ariz State Univ Tempe AZ 85287

FERRY, JAMES A, b Mazomanie, Wis, Sept 9, 37; m 64; c 2. NUCLEAR PHYSICS. *Educ:* Univ Wis, BS, 59, MS, 62, PhD(physics), 65. *Prof Exp:* Res assoc physics, Univ Wis, 65-66; mgr accelerator div & vpres, 66-70, vpres prod, 70-78, EXEC VPRES & CHIEF OPERATING OFFICER, NAT ELECTROSTATICS CORP, 70- *Mem:* Am Phys Soc. *Res:* Low energy nuclear physics particle accelerator design and construction; ultra high vacuum technology. *Mailing Add:* Nat Electrostatics Corp Box 310 Middleton WI 53562-0310

FERRY, JOHN DOUGLASS, b Dawson, YT, May 4, 12; US citizen; m 44; c 2. POLYMER CHEMISTRY. *Educ:* Stanford Univ, AB, 32, PhD(chem), 35. *Prof Exp:* Attached worker, Nat Inst Med Res, London, 32-34; pvt asst, Hopkins Marine Sta, Stanford, 35-36; instr biochem sci, Harvard Univ, 36-38, jr fel, Soc Fels, 38-41; assoc chemist, Oceanog Inst, Woods Hole, 41-45; from asst prof to assoc prof, 46-73, chmn dept, 59-67, Farrington Daniels res prof, 73-82, EMER PROF CHEM, UNIV WIS-MADISON, 82- *Concurrent Pos:* NSF fel, Brussels, 59, Strasbourg & Kyoto, 68; vis lectr, Kyoto, 68; chmn, Int Comt Rheology, 63-68; vis lectr, Univ Grenoble, 73. *Honors & Awards:* Lilly Award, Am Chem Soc, 46, Kendall Award, 60 & Witco Award, 74; Bingham Medal, Soc Rheol, 53; High Polymer Physics Prize, Am Phys Soc, 66; Colwyn Medal, Brit Inst of the Rubber Indust, 72; Tech Award, Int Inst Synthetic Rubber Producers, 77; Charles Goodyear Medal, Rubber Div, Am Chem Soc, 81, Polymer Div Chem Award, 84. *Mem:* Nat Acad Sci; fel Am Acad Arts & Sci; Am Chem Soc; Am Soc Biol Chemists; hon mem Fr Group Rheol Soc; Rheol Soc Japan; fel Am Phys Soc. *Res:* Ultrafiltration; polymers of high molecular weight; proteins; mechanical properties of viscoelastic materials. *Mailing Add:* Dept Chem Univ Wis Madison WI 53706

FERRY, JOHN MOTT, b Madison, Wis, Mar 21, 49; m 77; c 2. PETROLOGY, GEOCHEMISTRY. *Educ:* Stanford Univ, BS & MS, 71; Harvard Univ, PhD(geol sci), 75. *Prof Exp:* Fel petrol, Geophys Lab, Carnegie Inst Washington, 75-77; assoc prof, 84-86, PROF DEPT EARTH & PLANETARY SCI, JOHNS HOPKINS UNIV, 86- *Honors & Awards:* Mineral Soc of Am Award, 85. *Mem:* Mineral Soc Am; Am Geophys Union; AAAS; Sigma Xi; Geochem Soc. *Res:* Determination of temperature, pressure, fluid compositions, mass transfer, heat transfer and mineral reactions in the earth's crust during metamorphism. *Mailing Add:* Dept Earth & Planetary Sci Johns Hopkins Univ Baltimore MD 21218

FERSTANDIG, LOUIS LLOYD, b Brooklyn, NY, Apr 26, 24; m 46; c 3. ORGANIC CHEMISTRY. *Educ:* Univ Ill, BS, 44; Cornell Univ, PhD(chem), 49. *Prof Exp:* Res assoc, Univ Minn, 49-50; res chemist, Calif Res Corp, 50-56, group supvr, 56-60, res assoc, 60-64; res dir, 64-70, TECH DIR, 70-, VPRES, HALOCARBON PROD CORP, 77- *Mem:* Am Chem Soc; Am Soc Anesthesiologists. *Res:* Chemicals for use in plastics, fiber and surface coating; stereospecific polymers; isocyanates; molecular complexes; allophanates; hydrogen bonding; organic halogen compounds; toxicity of trace concentrations of anesthetics; metabolism of anesthetics. *Mailing Add:* Halocarbon Prod Corp 82 Burlews Ct Hackensack NJ 07601-4829

FERTIG, STANFORD NEWTON, b Marlinton, WVa, July 10, 19; m 49; c 1. WEED SCIENCE. *Educ:* Univ WVa, BS, 46, MS, 47; Cornell Univ, PhD(weed sci), 50. *Prof Exp:* From asst prof to prof agron, Cornell Univ, 50-66; dir, Res & Develop, Am Chem, Inc, 66-78; chief, Pesticide Impact Assessment Staff, USDA, 76-85; RETIRED. *Concurrent Pos:* Vis assoc prof bot, Univ Philippines, Los Banos, 54-56; mem weeds sub-comt, Nat Acad Sci, 64-66, mem pesticide study panel, Agr Res Inst, 72-, secy, 73-75, chmn, 76 & 80, mem gov bd, 81; mem sub-comt fertilizers & pesticides, Nat Indust Pollution Control Coun, Dept of Com, 71-72; mem res dir comn, NACA, 73-77; mem, Environ Protection Agency comt study pesticides, 72-73; US rep, Codex Comt Pesticides, 79-85; dir, USDA Tech Off Int Trade, 84-91; res leader, Pesticide Assessment Lab, 85- *Mem:* Weed Sci Soc Am; Sigma Xi; Philippines Asn Adv Res; fel Weed Sci Soc Am. *Res:* Weed control; research in herbicides and growth regulators for agriculture; benefits of pesticides in agriculture. *Mailing Add:* Agr Res Ctr-USDA Bldg 1070 BARC-E Beltsville MD 20705

FERTIS, DEMETER G(EORGE), b Athens, Greece, July 25, 26; nat US; m 53; c 2. STRUCTURAL DYNAMICS. *Educ:* Mich State Univ, BS, 52, MS, 55; Nat Tech Univ Athens, DE, 64. *Prof Exp:* Bridge design engr, Mich State Hwy Dept, 52-55, phys res engr, 56-57; asst prof eng mech, Wayne State Univ, 57-63; assoc prof civil eng, Univ Iowa, 64-66; PROF CIVIL ENG, UNIV AKRON, 66- *Concurrent Pos:* Consult, Atomic Power Develop Assocs, 57-, Power Reactor Develop Co, 57-, Ford Motor Res Ctr, 61-, Gen Motors Proving Grounds, 62-, NASA, 63-, Boeing Aircraft Co, 65-, Lockheed Co, 66- & Dept Navy, 77- *Mem:* Am Soc Civil Engrs; Am Concrete Inst; Indust Math Soc; NY Acad Sci; Am Soc Eng Educ. *Res:* Structures; vibrations; urban planning; theoretical mechanics; acoustic stochastic method and approach to determine concrete material properties; developer of the method of the equivalent systems and the concept of the dynamic hinge; developed new aerodynamic wing designs for airplanes and other applications. *Mailing Add:* Dept Civil Eng Univ Akron Akron OH 44325

FERTL, WALTER HANS, b Vienna, Austria, Mar 16, 40; m 65; c 2. GEOPHYSICS. *Educ:* Mining & Metall Col, Austria, Dipl Ing, 63, Dr mont, 71; Univ Tex, Austin, MS, 66, PhD(petrol eng), 68. *Hon Degrees:* Dr, Pepperdine Univ, Calif, 82 & Bedford Univ, Ariz, 83. *Prof Exp:* Asst mgr, Well Serv & Workover Dept, Austrian State Oil Co, Vienna, 63-65; res scientist, Prod Res Div, Continental Oil Co, 68-72, sr res scientist, 72-76; dir, Interpretation & Field Develop, 76-81, vpres, petrol eng serv, 81-83, vpres, 83-87, EXEC VPRES, DRESSER ATLAS, 87- *Concurrent Pos:* Lectr, Univ Zulia, Venezuela, 67 & 73; guest lectr, Mining & Metall Col, Austria, 71 & Tech Univ Istanbul, 71; ed, Log Analyst, Soc Prof Well Log Analysts. *Mem:* Soc Prof Well Log Analysts (pres, 78-81); Soc Petrol Eng; Can Well Logging Soc; Soc Explor Geophys; Am Asn Petrol Geologists; Soc Am Asn Petrol Geologists. *Res:* Geophysical well logging research and tool responses; development and improvement of interpretation methods; laboratory and field investigations of physical and chemical rock properties with special emphasis on abnormally pressured formations. *Mailing Add:* Piroska OES 1627 Scenic Ridge Houston TX 77043

FERTZIGER, ALLEN PHILIP, b New York, NY, June 27, 41; m 70; c 2. NEUROSCIENCES, THANATOLOGY. *Educ:* City Col New York, BS, 63; Univ Mich, PhD(physiol), 68. *Prof Exp:* Fel anat, Albert Einstein Col Med, 68-70; asst prof physiol, Med Sch, Univ Md, Baltimore, 70-78; fel, Johns Hopkins Sch Pub Health & Hyg, 78-79; asst prof health educ, Col Perh, Univ Md, 79-85, dir develop, 87-90; INDEPENDENT CONSULT, 90- *Mem:* Am Physiol Soc; Electroencephalog Soc; Soc Neurosci; Am Inst Biol Sci; Found Thanatology. *Res:* Electrophysiology of epilepsy; mechanism of action of anticonvulsant drugs; developmental neurobiology; health promotion and wellness; health education; death & dying. *Mailing Add:* 3609 Taylor St Chevy Chase MD 20815

FERY, RICHARD LEE, b Salem, Ore, Dec 4, 43; div; c 1. GENETICS, PLANT BREEDING. *Educ:* Ore State Univ, BS, 66; Purdue Univ, PhD(plant genetics & breeding), 70. *Prof Exp:* Res horticulturist, 70-72, RES GENETICIST, US VEG LAB, AGR RES SERV, USDA, 72- *Concurrent Pos:* Assoc ed, J Am Soc Hort Sci, 81-85, Hortsci, 81-85; adj prof hort, Clemson Univ, 85- *Honors & Awards:* Marian W Meadows Award, Am Soc Hort Sci, 71, Asgrow Award, 76. *Mem:* AAAS; fel Am Soc Hort Sci; Crops Sci Soc Am; Am Soc Agron; Int Soc Hort Sci. *Res:* Genetics of plant resistance to diseases and insects; breeding vegetable crop plants for disease and insect resistance. *Mailing Add:* 2875 Savannah Hwy Charleston SC 29414-5334

FERZIGER, JOEL H(ENRY), b Brooklyn, NY, Mar 24, 37; div; c 3. MECHANICAL ENGINEERING. *Educ:* Cooper Union, BChE, 57; Univ Mich, MSE, 59, PhD(nuclear eng), 62. *Prof Exp:* From asst prof to assoc prof nuclear eng, 61-72, PROF MECH ENG, STANFORD UNIV, 72- *Concurrent Pos:* Consult, Gen Elec Co & Encyclopedia Britannica Films, 61-67 & Nielsen Eng & Res, 75-; Fulbright res fel, Neth, 67-68; vis prof, Queen Mary Col, London, 79, Ecole Centrale de Lyon, 86; Alexander von Humboldt fel, Erlangen, 87. *Mem:* Fel Am Phys Soc; Soc Indust & Appl Math; Am Inst Aeronaut & Astronaut; Am Soc Mech Engr. *Res:* Turbulent flow; heat transfer; numerical fluid mechanics. *Mailing Add:* 271 Santa Rita Ave Palo Alto CA 94301

FESHBACH, HERMAN, b New York, NY, Feb 2, 17; m 40; c 3. NUCLEAR PHYSICS. *Educ:* City Col New York, BS, 37; Mass Inst Technol, PhD(physics), 42. *Hon Degrees:* ScD, Lowell Technol Univ. *Prof Exp:* Tutor physics, City Col New York, 37-38; from asst prof to prof, Mass Inst Technol, 45-76, dir, Ctr Theoret Physics, 67-73, head physics dept, 73-87, Cecil & Ida Green prof, 76-83, inst prof physics, 83-87; RETIRED. *Concurrent Pos:* Guggenheim fel, 54-55; Ford Found fel, 62-63; mem bd trustees, Assoc Univ Inc, 74-87 & 90-; mem nuclear sci adv comt, Dept Energy/NSF, chmn, 79-83; chmn, nuclear physics div, Int Union Pure & Appl Physics, 84-90. *Honors & Awards:* Bonner Prize, Am Phys Soc, 73; Townsend-Harris Medal, City Col New York, 77; Nat Medal Sci, Presented by President, USA, Ronald Reagan,

86. *Mem:* Nat Acad Sci; Am Acad Arts & Sci (vpres, 73-76, pres, 82-86); fel Am Phys Soc (vpres, 79, pres, 80); AAAS; Am Acad Arts & Sci, (pres, 82-86). *Res:* Theoretical nuclear physics. *Mailing Add:* Dept Physics 6-307 Mass Inst Technol Cambridge MA 02139

FESJIAN, SEZAR, b Dec 7, 43. STATISTICAL MECHANICS. *Educ:* Ohio Univ, BS, 66; Yeshiva Univ, PhD(physics), 73. *Prof Exp:* Res assoc statist mech, Res Found State of NY, State Univ NY Albany, 72-78; res assoc, Dept Chem, Univ Minn, Minneapolis, 78-81; ASST PROF, DEPT PHYSICS, MANHATTAN COL, BRONX, NY, 81- *Res:* Equilibrium and nonequilibrium fluids; polymer physics. *Mailing Add:* Dept Physics Manhattan Col Manhattan College Pkwy Bronx NY 10471

FESSENDEN, PETER, b Newton, Mass, Sept 5, 37; m 59; c 2. RADIOLOGIC PHYSICS, PHYSICS. *Educ:* Williams Col, AB, 59; Brown Univ, ScM, 63, PhD(physics), 65. *Prof Exp:* Fel nuclear physics, Los Alamos Sci Lab, 65-66; assoc prof physics, Ore State Univ, 67-72; postdoctoral appointment, Dept Radiol, 72-74, asst prof radiol, 74-81, PROF & DIR RADIOL PHYSICS, SCH MED, STANFORD UNIV, 82- *Concurrent Pos:* Vis scholar, Swiss Inst for Nuclear Res, 79-80. *Mem:* Am Asn Physicists Med; Am Phys Soc; Radiation Res Soc; NAm Hyperthermia Group. *Res:* Particle radiotherapy research; hyperthermia. *Mailing Add:* Dept Radiation Oncol Stanford Univ Sch of Med Stanford CA 94305

FESSENDEN, RALPH JAMES, b Chicago, Ill, Oct 25, 32; m 55; c 2. CHEMISTRY. *Educ:* Univ Ill, BS, 55; Univ Calif, PhD(chem), 58. *Prof Exp:* From asst prof to assoc prof chem, San Jose State Col, 58-67; chmn dept, 67-73, PROF CHEM, UNIV MONT, 67- *Concurrent Pos:* Alfred P Sloan fel, 65-67. *Mem:* Am Chem Soc. *Res:* Organo-metallic chemistry; organo-silicon chemistry; author of numerous textbooks. *Mailing Add:* Dept Chem Univ Mont Missoula MT 59801

FESSENDEN, RICHARD WARREN, b Northampton, Mass, Jan 22, 34; m 57; c 2. PHYSICAL CHEMISTRY. *Educ:* Univ Mass, BS, 55; Mass Inst Technol, PhD(phys chem), 58. *Prof Exp:* NSF fel, Calif Inst Technol, 58-59; fel, Mellon Inst, 59-62, sr fel, Radiation Res Labs, 62-78, prof chem, Carnegie-Mellon Univ, 67-78; MEM STAFF, RADIATION LAB, UNIV NOTRE DAME, 78- *Mem:* AAAS; Am Chem Soc; Am Phys Soc; Sigma Xi. *Res:* Electron spin resonance; radiation chemistry. *Mailing Add:* Radiation Lab Univ Notre Dame Notre Dame IN 46556

FESSENDEN-MACDONALD, JUNE MARION, b Whitinsville, Mass, Sept 2, 37; m 88; c 1. BIOCHEMISTRY, TOXICOLOGY. *Educ:* Brown Univ, AB, 59; Tufts Univ, PhD(biochem), 63. *Prof Exp:* NSF fel biochem, Pub Health Res Inst of City of New York, Inc, 63-65 & Am Cancer Soc fel, 65-66; consult, Am Pub Health Serv, 66; asst prof, 66-74, NIH career develop award, 70-74, assoc dir acad affairs, Div Biol Sci, 74-75, vprovost undergrad educ, 75-78, ASSOC PROF BIOCHEM, BIOL & SOC, CORNELL UNIV, 74- *Concurrent Pos:* Mem, NIH Biochem Study Sect, 75-77, Middle States, 78-; vis res scholar humanities, Dartmouth Col, 81-83; NSF interdisciplinary incentive award, 82-83; chair, Cornell Recombinant DNA Res Comt, biol & soc prog, 86-89; Dep Dir, Nat Agr Biotechnol Coun, 89- *Mem:* AAAS; Am Soc Biol Chemists; Asn Women Sci; Sigma Xi; Soc Risk Anal. *Res:* Environmental and risk commuication; setting risk priorities; role of women; biotechnology and risk communication; biology and society. *Mailing Add:* 159 Biotechnol Bldg Cornell Univ Ithaca NY 14853

FESSLER, JOHN HANS, b Vienna, Austria, June 15, 28; m 58; c 2. MOLECULAR BIOLOGY OF CONNECTIVE TISSUE DURING DEVELOPMENT. *Educ:* Oxford Univ, BA, 49, BSc, 51, BA & MA, 52, PhD(biochem), 56. *Prof Exp:* Res fel biochem & med, Mass Gen Hosp & Harvard Med Sch, 56-58; sci officer biochem, Unit Body Temperature Res, Med Res Coun Eng, 58-61; sr res fel biophys & molecular biol, Calif Inst Technol, 61-66; assoc prof, 66-70, PROF MOLECULAR BIOL & BIOL, UNIV CALIF, LOS ANGELES, 70- *Concurrent Pos:* Arthritis & Rheumatism Found fel & Fulbright grant, 56-58; Wellcome fel, Royal Soc Med, 59-60; Sen Fogarty Int fel, 85-86. *Mem:* Soc Biol Chem; Brit Biochem Soc; Soc Develop Biol. *Res:* Macromolecules of connective tissue; physical chemistry of proteins; developmental biology. *Mailing Add:* Molecular Biol Inst Dept Biol Univ Calif Los Angeles CA 90024-1570

FESSLER, LISELOTTE I, MOLECULAR BIOLOGY. *Educ:* Univ Wis, PhD(oncol), 55. *Prof Exp:* RES MOLECULAR BIOLOGIST, DEPT BIOL, UNIV CALIF, LOS ANGELES, 70- *Mem:* Soc Biol Chem. *Res:* Role of extra cellular matrix in development. *Mailing Add:* Biol Dept Univ Calif 405 Hilgard Ave Los Angeles CA 90024-1570

FESSLER, RAYMOND R, PHYSICAL METALLURGICAL ENGINEERING. *Educ:* Carnegie Inst Technol, BS; Mass Inst Technol, PhD(metall). *Prof Exp:* Staff mem, Batelle Columbus Div, 65-68, assoc mgr, Ferrous Metall Sect, 68-77, mgr, Phys Metall Sect, 77-82, assoc dir progs, Corp Tech Develop, 82-83, mgr, Transp & Struct Dept, 83-85, mgr, Advan Mat Dept, 85; DIR BASIC INDUST RES LAB, NORTHWESTERN UNIV, 85- *Mem:* Fel Am Soc Metals Int. *Res:* Physical metallurgy of steels, high temperature alloys and nonferrous metals; fracture toughness and metal physics on optical and electron metallography; advanced ceramics; process and physical metallurgy; polymers; corrosion; electrochemistry; mechanics. *Mailing Add:* Basic Indust Res Lab Northwestern Univ 1801 Maple Ave Evanston IL 60201

FESTA, ROGER REGINALD, b Norwalk, Conn, Sept 6, 50. SCIENTIFIC PUBLISHING, PROFESSIONAL DEVELOPMENT. *Educ:* St Michael's Col, BA, 72; Univ Vt, MA, 79; Fairfield Univ, CAS, 81; Univ Conn, PhD(chem educ), 82. *Prof Exp:* Instr chem, Cent Cath High Sch, 77-79 & Brien McMahon High Sch, 79-83; ASSOC PROF CHEM, NORTHEAST MO STATE UNIV, 83- *Concurrent Pos:* Assoc ed, J Chem Educ, 80-89, Hexagon, 84-; educ ed, Chemist, 80-; contrib ed, Educ Chem, 84-90; consult, Can Chem News, 84-90. *Mem:* Fel Am Inst Chemists Found (secy, 91-); Am Chem Soc; Royal Soc Chem; Chem Inst Can. *Res:* Reconstruction of the national archives of the American Institute of Chemists to study the socio-economic climate leading to the formation of the American Institute of Chemists as distinct from the American Chemical Society; professional development of undergraduate chemistry majors (ethics). *Mailing Add:* Dept Chem Northeast Mo State Univ Kirksville MO 63501-0828

FESTER, DALE A(RTHUR), b Sterling, Colo, Nov 7, 32; m 53; c 4. AEROSPACE ENGINEERING. *Educ:* Univ Denver, BS, 53, MS, 61; Colo State Univ, Exec Develop Inst, 83. *Prof Exp:* Engr, Phillips Petrol Co, 53; proj assoc, GOG Lof, Consult Chem Eng, 56-61; sr engr, 61-62, design specialist, 62-67, staff engr, 67-70, sr res scientist, 70-72, sr group engr, 72-77, prog mgr, 77-79, MGR, MARTIN MARIETTA DENVER AEROSPACE, 79- *Concurrent Pos:* Proj asst solar energy, Univ Wis, 56-61; assoc fel, Am Inst Aeronaut & Astronaut; mem, Space Transp Comt, Inst Astronaut Fedn. *Honors & Awards:* First Shuttle Flight Achievement Award, NASA. *Mem:* Am Inst Aeronaut & Astronaut; Air Force Asn; Cryogenic Soc Am. *Res:* Low gravity fluid behavior, high energy liquid propellants; orbital fluid management; cryogenic systems; pressurization systems; fluid mechanics; heat transfer; mass transfer; pressurized gas absorption; material compatibility; solar energy utilization. *Mailing Add:* 2916 S Fenton Denver CO 80227

FESTER, KEITH EDWARD, b Glendale, Calif, Nov 16, 42; m 67; c 3. ELECTROCHEMISTRY. *Educ:* Loyola Univ, Calif, BS, 64; Creighton Univ, MS, 66; Univ of the Pac, PhD(phys chem), 68. *Prof Exp:* ELECTROCHEMIST, MEDTRONIC, INC, 69- *Mem:* Am Chem Soc; Electrochem Soc. *Res:* Complex ion polarography; applied research in the design and use of chemical cells as implantable power sources. *Mailing Add:* 5141 Brighton Lane New Brighton MN 55112-4845

FETCHER, E(DWIN) S(TANTON), b Winnetka, Ill, Aug 9, 09; m 53; c 4. HEALTH SCIENCES. *Educ:* Harvard Univ, BS, 31; Univ Chicago, PhD(phys chem), 34. *Prof Exp:* Res chemist, Universal Oil Prod Co, 35-36 & Rockefeller Inst Med Res, 37; instr physiol, Univ Chicago, 37-40 & Univ Minn, 40-43; res physiologist, environ physiol & equip develop, Air Mat Command, USAF, 43-49; rancher & stockman, Fetcher Ranch, 49-62; res assoc, lab physiol hyg sch pub health, Univ Minn, Minneapolis, 62-70; cur, environ sci, Sci Mus Minn, St Paul, 70-72; prog exec, Mt Sinai Hosp, Minneapolis & Univ Minn, 72-75; consult, 75-79; RETIRED. *Mem:* AAAS. *Res:* Membrane chemistry and physiology; water balance of marine mammals; emergency ocean survival personal equipment; special protective flight clothing; livestock production; agribusiness; epidemiology of cardiovascular diseases. *Mailing Add:* 12 Crocus Hill St Paul MN 55102-2809

FETCHO, JOSEPH ROBERT, b Bethlehem, Pa, June 28, 57; m 83; c 1. MOTOR SYSTEMS, NEURAL CIRCUITS. *Educ:* Lehigh Univ, Bethlehem, BS, 79; Univ Mich, Ann Arbor, PhD(biol sci), 85. *Prof Exp:* Postdoctoral fel neurobiol, State Univ NY, Buffalo, 85-90; ASST PROF NEUROBIOL, STATE UNIV NY, STONY BROOK, 90- *Mem:* AAAS; Soc Neurosci; Am Soc Zoologists. *Res:* Neural circuits in spinal cord that are responsible for generating movements. *Mailing Add:* Dept Neurobiol & Behav Life Sci Bldg State Univ NY Stony Brook NY 11794-5230

FETH, GEORGE C(LARENCE), b Pittsburgh, Pa, Aug 17, 31; m 57; c 3. PACKAGING OF INTEGRATED CIRCUITS & DIGITAL INTEGRATED CIRCUITS DEVICES. *Educ:* Carnegie Inst Technol, BS, 53, MS, 54, PhD(elec eng), 56. *Prof Exp:* Engr, Gen Elec Co, 56-61; mgr magnetic film devices res, IBM Corp, 61-62, exploratory memory res, 62-64, integrated circuits & systs res, 64-65, res tech planning staff, 65-67, res staff mem memory & storage res, 67-69, mgr eng & syst anal, 69-71, res staff mem memory & storage res, 71-76, controlled transformer, 76-77, self-aligned bipolar transistors & circuits, VLSI, 77-87, RES STAFF MEM PACKAGING LOW-END DIGITAL ELECTRONICS, SILICON MULTI CHIP CARRIERS, THOMAS J WATSON RES CTR, IBM CORP, 87- *Mem:* Sr mem Inst Elec & Electronics Engrs. *Res:* Packaging for low end digital electronics-multi chip modules; silicon carriers. *Mailing Add:* Thomas J Watson Res Ctr IBM Corp PO Box 218 Yorktown Heights NY 10598

FETKOVICH, JOHN GABRIEL, b Aliquippa, Pa, June 9, 31; m 58; c 2. SURFACE PHYSICS, ENERGY CONVERSION. *Educ:* Carnegie Inst Technol, BS, 53, MS, 55, PhD(physics), 59. *Prof Exp:* Res physicist, 59-61, from asst prof to assoc prof, 61-68, PROF PHYSICS, CARNEGIE-MELLON UNIV, 68- *Concurrent Pos:* Consult, Argonne Nat Lab, 60-70, Argonne Univs Asn appointee, 70-71; consult, Rutherford High Energy Lab, Eng, 71-72. *Mem:* AAAS; Am Phys Soc; Am Asn Physics Teachers; Sigma Xi. *Res:* Surface physics; physics of elementary particles; ocean thermal energy conversion. *Mailing Add:* Dept Physics Carnegie Inst Tech Pittsburgh PA 15213

FETNER, ROBERT HENRY, b Savannah, Ga, Feb 22, 22; m 44; c 1. BIOLOGY. *Educ:* Univ Miami, BS, 50, MS, 52; Emory Univ, PhD(biol), 55. *Prof Exp:* Res asst prof appl biol, Eng Exp Sta, Ga Inst Technol, 55-58, res prof, 64-77, dir, Sch Biol, 65-77, prof biol, Nuclear Res Ctr, 77-81; RETIRED. *Res:* Radiation biology; cellular physiology; cytogenetics. *Mailing Add:* 2219 Walker Dr Lawrenceville GA 30245

FETSCHER, CHARLES ARTHUR, b New York, NY, Dec 7, 12; m 42; c 3. ORGANIC POLYMER CHEMISTRY. *Educ:* Col of the Holy Cross, BS, 34, MS, 35; Columbia Univ, PhD(org chem), 38. *Prof Exp:* Res chemist, Cuban Mining Co, Cuba, 38-40 & Shawinigan Resins Corp, Mass, 41-45; asst dir res, Cluett Peabody & Co, NY, 45-55; dir cent res labs, Nopco Chem Co, 55-65; group mgr res, Hysol Div, Dexter Corp, 65-76; RETIRED. *Mem:* Fel AAAS; Fiber Soc; Am Chem Soc. *Res:* Resins, plastics and additives. *Mailing Add:* Ashmere Manor Hinsdale MA 01235

FETSKO, JACQUELINE MARIE, b Allentown, Pa, Jan 14, 26. PHYSICAL CHEMISTRY. *Educ:* Univ Pa, BA, 46; Lehigh Univ, MS, 53. *Prof Exp:* Tech asst, 49-55, res supvr, 55-61, asst res dir, 61-66, asst to dir, Ctr for Surface & Coating Res, 66-69, ed, 69-74, ASST RES DIR, NAT PRINTING INK RES INST, LEHIGH UNIV, 74-, ADMIN ASST, CTR SURFACE & COATINGS RES, 76- *Concurrent Pos:* Asst ed, Encycl Org Coatings, 56-58; consult, Scott Paper Co, 56-58 & Handy & Harman, 59-60. *Res:* Tech Asn Pulp & Paper Indust. Res: Printing ink-paper relationships; transfer; surface strength; optical properties. *Mailing Add:* 811 Delaware Ave Bethlehem PA 18015

FETT, JOHN D, b New York, NY, Mar 2, 33; m 56; c 3. GEOLOGY, GEOPHYSICS. *Educ:* Redlands Univ, BS, 54; Univ Calif, Riverside, MS, 67. *Prof Exp:* Chemist, Naval Ord Test Sta, Calif, 54-55; gravity observer, Lamont Geol Observ, Columbia Univ, 56-57; res asst, Inst Geophys, Univ Calif, Los Angeles, 57-59; geologist, Beylik Drilling Co, 59-60; consult, John D Fett & Assocs, 60-69; prin consult, Earth Sci Assocs, 69-73; PRES, EARTH SCI & ENG, 73- *Concurrent Pos:* Dir & vpres, Eastern Munic Water Dist. *Mem:* Fel Geol Soc Am; Soc Explor Geophys; Seismol Soc Am; Am Geophys Union. *Res:* New techniques for the application of geophysics to problems in engineering geology and ground water studies. *Mailing Add:* La Costa & Romberg Bldg 2 4807 Spicewood Spring Rd Austin TX 78759-8404

FETT, WILLIAM FREDERICK, b Hinsdale, Ill, Oct 30, 52; m 77; c 2. PHYTOBACTERIOLOGY, PHYSIOLOGICAL PLANT PATHOLOGY. *Educ:* Univ Ill, Champaign-Urbana, BS, 74; Univ Wis-Madison, MS, 77, PhD(phytopath), 79. *Prof Exp:* Nat Res Coun res fel, 79-80, RES PLANT PATHOLOGIST, EASTERN REGIONAL RES CTR, USDA, 80- *Mem:* Am Phytopath Soc; Am Soc Microbiol; Am Soc Plant Physiologists. *Res:* Bacterial pathogen-plant interactions; biochemical, physiological and ultrastructural features leading to resistance or susceptibility; bacterial extracellular polysaccharides. *Mailing Add:* Eastern Regional Res Ctr Agr Res Serv USDA 600 E Mermaid Lane Philadelphia PA 19118

FETTER, ALEXANDER LEES, b Philadelphia, Pa, May 16, 37; m 62; c 2. THEORETICAL CONDENSED MATTER PHYSICS. *Educ:* Williams Col, AB, 58; Rhodes Scholar, Oxford Univ, BA, 60, MA, 64; Harvard Univ, PhD(physics), 63. *Prof Exp:* Miller res fel physics, Univ Calif, Berkeley, 63-65; from asst prof to assoc prof, 65-74, assoc dean undergrad studies, 76-79, chmn dept, 85-90, PROF PHYSICS, STANFORD UNIV, 74-, ASSOC DEAN HUMANITIES & SCI, 90- *Concurrent Pos:* Sloan res fel, 68-70; trustee, Williams Col, 74-79; Nordita quest prof, Helsinki Tech Univ, 76; chmn fac Senate, Stanford Univ, 82-83; vice-chair, Div Condensed Matter Physics, 90-91. *Mem:* Fel Am Phys Soc; fel AAAS; Sigma Xi. *Res:* Low-temperature behavior of quantum fluids, especially superfluid helium and type-II superconductors; general theory of many-particle systems; quantum hydrodynamics. *Mailing Add:* Dept Physics Stanford Univ Stanford CA 94305-2196

FETTER, BERNARD FRANK, b Baltimore, Md, Jan 21, 21; m 45; c 4. PATHOLOGY. *Educ:* Johns Hopkins Univ, AB, 41; Duke Univ, MD, 44. *Prof Exp:* From instr to assoc prof, 51-67, PROF PATH, MED CTR, DUKE UNIV, 67- *Mem:* AMA; Am Asn Path; Col Am Path; Am Soc Clin Path; Am Soc Dermatopathology. *Res:* Surgical pathology. *Mailing Add:* 3836 Somerset Dr Durham NC 27707

FETTER, CHARLES WILLARD, JR, b Dayton, Ohio, Apr 15, 42; m 65; c 3. HYDROGEOLOGY, HYDROGEOCHEMISTRY. *Educ:* DePauw Univ, AB, 64; Ind Univ, MA, 66, PhD(hydrol), 71. *Prof Exp:* Staff hydrogeologist, Holzmacher, McLendon & Murrell, 66-70; from asst prof to assoc prof, 71-78, dept chmn, 78-81, PROF GEOL, UNIV WIS-OSHKOSH, 78-, DEPT CHMN, 84- *Concurrent Pos:* Prog mgr, Law Eng Testing Co, 81-82. *Mem:* Am Inst Prof Geologists; Am Geophys Union; Am Water Resources Asn; Am Water Works Asn; Nat Water Well Asn; Am Inst Hydrol. *Res:* Contaminant transport in groundwater flow systems; monitoring of groundwater quality; regional groundwater flow systems; groundwater flow models. *Mailing Add:* Dept Geol Univ Wis Oshkosh WI 54901

FETTER, WILLIAM ALLAN, b Independence, Mo, Mar 14, 28; m 63; c 2. COMPUTER GRAPHICS, HUMAN FIGURE SIMULATION. *Educ:* Univ Ill, Urbana, BFA, 52. *Prof Exp:* Designer, Univ Ill Press Art, 50-54; art dir, Family Weekly Mag, 54-58 & John Higgs Studios, 58-59; supvr, Boeing Co, 59-69; vpres, Graphcomp Sci Corp, 69-70; chair, Design Dept, 70-77, PRES, SOUTHERN ILL RES INST, 77- *Concurrent Pos:* Adj prof computer sci, Univ Utah, Salt Lake City, 72-73 & Dept Archit, Univ Wash, 80-81; mem, US Comn Unesco Technol Proj, 73-83 & Master Resources Coun Int, 80-90. *Mem:* Assoc fel Am Inst Aeronaut & Astronaut; Indust Design Soc Am; Environ Design Res Asn; Soc Info Display. *Res:* Computer graphics; cockpit visibility simulations; human figure simulations; hemispheric projections. *Mailing Add:* 2165 156th Ave SE Bellevue WA 98007

FETTERMAN, HAROLD RALPH, b Jamaica, NY, Jan 17, 41; m 65; c 2. LASERS, EXPERIMENTAL SOLID STATE PHYSICS. *Educ:* Brandeis Univ, BA, 62; Cornell Univ, PhD(physics), 67. *Prof Exp:* Asst prof physics, Univ Calif, Los Angeles, 67-69; staff physicist, Lincoln Lab, Mass Inst Technol, 69-82; assoc dean, 86-89, PROF ELEC ENG, SCH ENG, UNIV CALIF, LOS ANGELES, 82- *Mem:* Am Phys Soc; fel Optical Soc Am; fel Inst Elec & Electronics Engrs; Sigma Xi; AAAS; NY Acad Sci. *Res:* Development and application of millimeter sources and solid state devices; field-effect transistors; picosecond electronics; high temperature superconducting circuits. *Mailing Add:* Elec Eng Dept Univ Calif Eng IV Rm 66-1475 Los Angeles CA 90024

FETTEROLF, CARLOS DE LA MESA, JR, b Glenridge, NJ, Dec 28, 26; m 52; c 4. FISHERIES, POLLUTION BIOLOGY. *Educ:* Univ Conn, BS, 50; Mich State Univ, MS, 52. *Prof Exp:* From fishery biologist reservoir res to dist fishery biologist, Tenn Game & Fish Comn, 52-57; from chief biologist to supvr water qual appraisal, Bur Water Mgt, Mich Dept Natural Resources, 57-71; sci coordr water qual criteria develop, Nat Acad Sci & Eng, 71-72; chief environ scientist, Bur Water Mgt, Mich Dept Natural Resources, 72-75; EXEC SECY, GREAT LAKES FISHERY COMN, 75- *Concurrent Pos:* Pres, Southern Div, Am Fisheries Soc, 56, vpres, Water Qual Sect, 83, pres, 84-85; mem, US Govt Interstate Water Pollution Tech Comts, 66-71; mem sci adv bd, Int Joint Comn, 72-84, chmn water qual objectives comt, 72-75; mem comt stand methods, Am Pub Health Asn, 72-73; lectr, Univ Mich, 83- *Honors & Awards:* Anderson-Everett Award, Int Asn Great Lakes Res, 89. *Mem:* NAm Benthol Soc (vpres, 64, pres, 66); Int Asn Great Lakes Res (vpres, 75, pres, 76); Am Fisheries Soc; fel Am Inst Fishery Res Biologists. *Res:* Interdisciplinary, interagency, international research and management coordination on the Great Lakes. *Mailing Add:* 8200 Pinecross Ann Arbor MI 48103

FETTERS, KARL L(EROY), metallurgical engineering; deceased, see previous edition for last biography

FETTERS, LEWIS, b Toledo, Ohio, Mar 29, 36. PHYSICAL CHEMISTRY, POLYMER CHEMISTRY. *Educ:* Col Wooster, AB, 58; Univ Akron, PhD(chem), 62. *Prof Exp:* NSF fel polymer chem, Univ Akron, 62-63; Nat Acad Sci-Nat Res Coun fel, Nat Bur Standards, 63-65; prof polymer chem, Univ Akron, 67-77, prof chem & polymer sci, Inst Polymer Sci, 77-83; SR RES ASSOC, EXXON RES & ENGR CO, 83- *Res:* Ionic polymerization kinetics; polymer rheology and synthesis; solution properties of polymers; radiation polymerization. *Mailing Add:* Exxon Res & Engr Co Clinton Township Rte 22E No LC-134 Annandale NJ 08801

FETTES, EDWARD MACKAY, b Brooklyn, NY, Jan 10, 18; m 41, 53, 80, 83; c 13. POLYMER CHEMISTRY. *Educ:* Mass Inst Technol, SB, 40; Polytech Inst Brooklyn, PhD(chem), 57. *Prof Exp:* Develop chemist, US Rubber Co, RI, 40-41; res chemist, Kendall Co, Mass, 41-42; res chemist, 42-44, group leader, 44-45, asst develop mgr, 45-46, mgr develop dept, 46-50 & res develop dept, 51-58, dir res & develop, Thiokol Chem Corp, NJ, 58-60; mgr plastics res, Koppers Co, Inc, 60-70; tech dir, Northern Petrochem Co, 70-81; RETIRED. *Concurrent Pos:* Consult, 81- *Mem:* Am Chem Soc; Soc Plastics Engrs; Sigma Xi. *Res:* Elastomers; plastics; resins. *Mailing Add:* 33 Gunwale Way Yarmouthport MA 02675

FETTWEIS, ALFRED LEO MARIA, b Eupen, Belg, Nov 27, 26; Ger citizen; m 57; c 5. CIRCUIT THEORY & DESIGN, DIGITAL SIGNAL PROCESSING. *Educ:* Cath Univ Louvain, Belg, ingénieur civil électricien, 51, Docteur en sci appliquées, 63. *Hon Degrees:* Dr, Linköping Univ, Sweden, 86, Polytech Inst Mons, Belg & Cath Univ Leuven, Belg, 88. *Prof Exp:* Develop engr, Int Tel & Tel Corp, 51-63; prof theoret elec, Eindhoven Univ Technol, Neth, 63-67; PROF TELECOMMUN, RUHR UNIV BOCHUM, GER, 67- *Concurrent Pos:* Consult, Int Tel & Tel Corp, Belg, 63-66 & Siemens, Ger, 68-; mem tech staff, Bell Labs, N Andover, Mass, 74; co-ed, Commun & Control Eng, 82- *Honors & Awards:* Centennial Medal, Inst Elec & Electronics Engrs, 84; Tech Achievement Award, Inst Elec & Electronics Engrs Circuits & Systs Soc, 88. *Mem:* Fel Inst Elec & Electronics Engrs; Inst Elec & Electronics Engrs Circuits & Systs Soc (vpres, 87); Europ Asn Signal Processing. *Res:* Circuits and systems; communications; digital signal processing; author of one book and numerous publications; holder of 30 patents. *Mailing Add:* Ruhr Univ Bochum Lehrst f Nachrichtentechnik PO Box 102148 Bochum D-4630 Germany

FETZER, HOMER D, b San Antonio, Tex, Oct 19, 32; m 54; c 5. ATOMIC PHYSICS. *Educ:* St Mary's Univ, Tex, BS, 54; Univ Tex, MA, 59, PhD(electron scattering), 65. *Prof Exp:* From instr to assoc prof, 59-69, chmn dept, 65-69 & 73-75, PROF PHYSICS, ST MARY'S UNIV, 70-, CHMN DEPT, 77- *Concurrent Pos:* NSF sci fac fel, 63-65; Minnie Stevens Piper prof award, 83. *Mem:* Am Asn Physics Teachers. *Res:* Ion mass and energy analysis. *Mailing Add:* Dept Physics St Mary's Univ San Antonio TX 78284

FETZER, JOHN CHARLES, b Leesville, La, Apr 19, 53; m 84. POLYCYCLIC-POLYNUCLEAR AROMATICS. *Educ:* Univ Ark, BSc, 76; Univ Ga, PhD(chem), 80. *Prof Exp:* RES CHEMIST, CHEVRON CORP, 80- *Concurrent Pos:* Lectr, Coutra Costa Community Col, 90- *Mem:* Am Chem Soc; Soc Appl Spectros. *Res:* Synthesis, separation, and identification of polycyclic aromatic hydrocarbons; environmental and petroleum processing; author of over 50 publications; awarded one patent. *Mailing Add:* Chevron Res & Technol Co PO Box 1627 Richmond CA 94802-0627

FEUCHT, DONALD LEE, b Akron, Ohio, Aug 25, 33; m 58; c 2. ELECTRICAL ENGINEERING. *Educ:* Valparaiso Univ, BS, 55; Carnegie Inst Technol, MS, 56, PhD(elec eng), 61. *Prof Exp:* From instr to prof elec eng, Carnegie-Mellon Univ, 58-77, assoc head dept, 69-73, assoc dean eng, 73-77; br chief photovoltaic res & develop, Solar Technol Div, US Dept Energy, 77-78; mgr, Photovoltaic Prog Off, 78-80, MGR, PHOTOVOLTAICS DIV, 80-, ASSOC DIR, RES & DEVELOP, 81-, DEP DIR, SOLAR ENERGY RES INST, 81- *Concurrent Pos:* Consult, Technograph Printed Circuits, Inc, 61, Union Carbide Corp, 63, Power Components, 65-67, PPG Industs, 67-69, Aluminum Asn Am, 67-69 & Essex Int, 67-74. *Mem:* fel Inst Elec & Electronics Engrs; Am Solar Energy Soc. *Res:* Theoretical and experimental properties of semiconductor heterojunctions; fabrication and electrical and optical properties of semiconductor devices; integrated circuits; solar energy; photovoltaic devices. *Mailing Add:* Dirs Off Solar Energy Res Inst 1617 Cole Blvd Golden CO 80401

FEUCHT, JAMES ROGER, b Denver, Colo, June 6, 33; m 56; c 3. HORTICULTURE, BOTANY. *Educ:* Colo State Univ, BS, 56; Mich State Univ, MS, 57, PhD(hort), 60. *Prof Exp:* Teaching asst hort, Mich State Univ, 56-60; asst prof, Western Ill Univ, 60 & Rutgers Univ, 61-66; EXTEN PROF HORT & EXTEN HORTICULTURIST, COLO STATE UNIV, 66- *Concurrent Pos:* Consult landscape mgt. *Res:* Landscape horticulture; air-layering of pine and spruce using several potential rooting hormones; effects of gibberellin A-3 on cell growth in Phaseolus vulgaris L, an anatomical study; various applied research in insect control on landscape plants. *Mailing Add:* 15200 W Sixth Ave Golden CO 80401

FEUCHTWANG, THOMAS EMANUEL, b Budapest, Hungary, May 21, 30; m 53; c 4. SOLID STATE PHYSICS, SURFACE PHYSICS. *Educ:* Ga Inst Technol, BEE, 53; Calif Inst Technol, MS, 54; Stanford Univ, PhD(microwave theory), 60. *Prof Exp:* Res asst, Microwave Lab, Stanford Univ, 54-59; res assoc physics, Univ Ill, 60-62; asst prof physics, Univ Minn, Minneapolis, 62-65; assoc prof, 65-70, PROF PHYSICS, PA STATE UNIV, UNIVERSITY PARK, 70- *Concurrent Pos:* Vis prof, Dept Physics, Tel-Aviv Univ, Israel, 71-72; Lady Davis vis prof, Israel Inst Technol, 78-79; vis prof, Dept Physics, Univ Hawaii, 80-81; Lady Davis vis prof, Israel Inst Technol, 86-87; vis prof, The Cavendish Lab, Univ Cambridge, UK, 88. *Mem:* Fel Am Phys Soc; fel Sigma Xi. *Res:* Lattice dynamics; electronic structure of point defects in ionic crystals; theory of low energy electron scattering from single crystals; many and single particle tunneling phenomena across interfaces; field and photoemission; theory and application of scanning tunnel microscope; electronic properties of suburicron systems. *Mailing Add:* Dept Physics Pa State Univ 104 Davey Lab University Park PA 16802

FEUER, GEORGE, b Szeged, Hungary, Mar 12, 21; m 49; c 3. BIOCHEMICAL PHARMACOLOGY, CLINICAL BIOCHEMISTRY & TOXICOLOGY. *Educ:* Univ Szeged, BS, 43, PhD(phys chem), 44, CMedSc(biochem), 52. *Prof Exp:* Asst lectr org chem, Univ Szeged, 45-46, sr lectr biochem, 46-48; sr lectr, Univ Budapest, 48-50; sr res assoc, Hungarian Acad Sci, 50-53, head muscular & neurochemistry, 53-56; head biochem, Cancer Res Inst, Budapest, 56; guest worker, Pasteur Inst, Paris, 57; sr res assoc neurochem, Inst Psychiat, Univ London, 57-62; sr res assoc neuropsychiat, Med Res Coun, Carshalton, Eng, 62-63; head biochem, Brit Indust Biol Res Asn, Carshalton, 63-68; from asst prof to assoc prof path chem, 68-74, PROF CLIN BIOCHEM, UNIV TORONTO, 74-, PROF PHARMACOL & TOXICOL, 79- *Concurrent Pos:* Prof biochem, Eotvos Lorand Univ, 50-53; chmn session, Fourth Int Goitre Conf, London, Eng, 60; vis prof, Warner-Lambert Res Inst, Can, 69-73 & Food & Drug Admin, 75-77; mem, Med Nobel Prize Comt, 76. *Mem:* Fel Royal Inst Chem; NY Acad Sci; Am Chem Soc; Soc Toxicol; Can Biochem Soc; Can Toxicol Soc. *Res:* Mechanism of muscular contraction; neurochemistry; metabolism of thyroid and other hormones in connection with emotional behavior; biochemical organization of the liver function in relation to its response to drugs and toxic compounds; iatrogenic diseases; carcinogenesis; malignant melanoma. *Mailing Add:* Dept Clin Biochem Univ Toronto Rm 521 Banting Inst Toronto ON M5G 1L5 Can

FEUER, HENRY, b Stanislau, Austria, Apr 4, 12; nat US; m 46. ORGANIC CHEMISTRY. *Educ:* Univ Vienna, MS, 34, PhD(org chem), 37. *Prof Exp:* Fel, Sorbonne, 39 & Purdue Univ, 43-46; pharmacist, Toledo Hosp, Ohio, 41-43; instr chem, Univ Exten, Ind Univ, 46; from asst prof to prof, 46-78, EMER PROF CHEM, PURDUE UNIV, WEST LAFAYETTE, 78- *Concurrent Pos:* Vis prof chem, Hebrew Univ, Israel, 67 & 72, Univ Tokyo, 71, Inst Technol, Kanpur, India, 71 & Beijing Inst Technol, China, 79; managing ed, Org Nitro Chem, VCH Publishers Inc, 82. *Mem:* Fel AAAS; Am Chem Soc; Sigma Xi; fel Chem Soc London. *Res:* Organic nitrogen compounds; synthesis and reactions of nitro compounds and heterocyclic systems. *Mailing Add:* Dept Chem Purdue Univ West Lafayette IN 47907

FEUER, MARK DAVID, b Philadelphia, Pa, 1953. III-V TRANSISTORS, MILLIMETER-WAVE PROBING. *Educ:* Harvard Univ, BA, 74; Yale Univ, PhD(solid-state physics), 80. *Prof Exp:* Mem tech staff, Murray Hill, 80-84, MEM TECH STAFF, AT&T BELL LABS, HOLMDEL, 85- *Concurrent Pos:* Assoc ed, Inst Elec & Electronics Engrs Trans on Electron Devices, 90- *Mem:* Inst Elec & Electronics Engrs; Am Phys Soc. *Res:* Fabrication and characterization of high-speed electron devices, especially those based on heterojunctions; InP-based transistors and millimeter-wave probing. *Mailing Add:* AT&T Bell Labs M/S 4F-423 Holmdel NJ 07733-1988

FEUER, PAULA BERGER, b New York, NY, Feb 11, 22; m 46. PHYSICS. *Educ:* Hunter Col, BA, 41; Purdue Univ, MS, 46, PhD(physics), 51. *Prof Exp:* Instr physics, 46-55, from asst prof to prof eng sci, 55-87, EMER PROF, PURDUE UNIV, WEST LAFAYETTE, 87- *Concurrent Pos:* Vis prof, Hebrew Univ, Israel, 64. *Mem:* Am Phys Soc; Soc Eng Sci (treas, 64-69); Sigma Xi. *Res:* Solid state physics; electronic properties solids; gas-surface interactions. *Mailing Add:* 726 Princess Dr West Lafayette IN 47907

FEUER, RICHARD DENNIS, b New York, NY, Sept 2, 40. ALGEBRA. *Educ:* Cornell Univ, BA, 62; Courant Inst, NY Univ, MS, 65, PhD(math), 70. *Prof Exp:* Adj Lectr math, New York City Community Col, 70-74; instr, Queens Col, City Univ New York, 71-77; adj asst prof, Medgar Evers Col & York Col, City Univ New York, 72-74; ASST PROF MATH, NY INST TECHNOL, 74- *Mem:* Math Asn Am; Am Math Soc. *Res:* Combinational group theory. *Mailing Add:* 230 West End Ave New York NY 10023

FEUER, ROBERT CHARLES, b New York, NY, Feb 23, 36; m 69. ENVIRONMENTAL BIOLOGY, HERPETOLOGY. *Educ:* Cornell Univ, BS, 56; Tulane Univ, MS, 58. Univ Utah, PhD, 66. *Prof Exp:* Asst, Tulane Univ, 56-58; asst prof biol, Hartwick Col, 60-61; asst, Univ Utah, 61-63; instr biol sci, Purdue Univ, Calumet Campus, 63-64; from instr to asst prof, Philadelphia Col Pharm, 64-74. *Concurrent Pos:* Ed, Bull Philadelphia Herpet Soc, 69-84; consult, McCormick, Taylor Assocs, Inc, 74-76, Berger Group, 77-81, Biomet Systs, 83-, Schmid & Co, 84- & Newbery Eng, 85-; writer, var periodicals. *Mem:* Sigma Xi; Soc Study Amphibians & Reptiles; Am Soc Ichthyol & Herpet. *Res:* Taxonomic herpetology, especially Chelydridae; ecology; author of various papers regarding ecology, fishing and horticulture. *Mailing Add:* 102 S New Ardmore Ave Broomall PA 19008

FEUERSTEIN, IRWIN, b New York, NY, Sept 18, 39. CHEMICAL ENGINEERING, FLUID MECHANICS. *Educ:* City Col New York, BChE, 62; Newark Col Eng, MSChE, 65; Univ Mass, Amherst, PhD(chem eng), 69. *Prof Exp:* Process engr, Esso Res & Eng Co, 62-64; fel exp med, McGill Univ, 69-70; asst prof, 70-77, assoc prof, 77-82, PROF CHEM ENG, MCMASTER UNIV, 82- *Concurrent Pos:* Can Heart Found sr res fel, 78. *Mem:* Int Soc Thrombosis & Haemostasis; Am Inst Chem Engrs; Can Soc Chem Eng. *Res:* Biological fluid mechanics, epifluorescent video microscopy; blood platelet transport and adhesion to artificial surfaces; white blood cell damage. *Mailing Add:* Dept Chem Eng McMaster Univ Hamilton ON L8S 4L7 Can

FEUERSTEIN, SEYMOUR, b New York, NY, Dec 12, 31; m 57; c 3. SPACECRAFT MATERIALS, SPACECRAFT TECHNOLOGY. *Educ:* Univ Ariz, BS, 53; Univ Calif, Berkeley, MS, 58, PhD(metall), 62. *Prof Exp:* Metallurgist, Gen Elec Co, NY, 53 & Goodyear Aircraft Co, Ariz, 56; res engr, Univ Calif, 56-61, res engr, Lawrence Radiation Lab, Univ Calif, Berkeley, 60-61; mem tech staff, Mat Sci Lab, Aerospace Corp, Calif, 61-66, head, Surface & Lubrication Phenomena Sect, Chem & Physics Lab, 67-75, head Interfacial Sci Dept, 76-80, dir, Chem & Physics Lab, 81-89, DIR, MAT SCI LAB, AEROSPACE CORP, CALIF, 90- *Concurrent Pos:* Lectr eng exten, Univ Calif, Los Angeles, 64-68. *Mem:* Am Vacuum Soc. *Res:* Mechanical properties of materials; crystal and surface physics; crystal growth; vacuum and radiation effects on materials; lubrication phenomena; adhesion, electronic and infrared sensor materials, battery electrochemistry; composite materials. *Mailing Add:* c/o Aerospace Corp M2/247 Box 92957 Los Angeles CA 90009

FEUSTEL, EDWARD ALVIN, b Fort Wayne, Ind, June 18, 40; m 71; c 1. COMPUTER SCIENCE, ELECTRICAL ENGINEERING. *Educ:* Mass Inst Technol, BEE & MEE, 64; Princeton Univ, MA, 66, PhD(elec eng), 67. *Prof Exp:* Res fel, Calif Inst Technol, 67; syst programmer, Commun Res Div, Inst Defense Anal, NJ, 68; asst prof comput sci, Rice Univ, 68-73, assoc prof elec eng & comput sci, 73-78; mem tech staff, Commun Res Div, Inst Defense Anal, 79-80; SR RES CONSULT, PRIME COMPUT, 80- *Concurrent Pos:* Lectr elec eng, Princeton Univ, 68; consult, Inst Defense Anal, 68-80, Los Alamos Nat Labs, 73-76, NSF, 75-76, 86-87 & Dept Pub Welfare, State Tex, 76; syst programmer, Lawrence Livermore Labs, 72. *Mem:* Asn Comput Mach; Inst Elec & Electronics Engrs; Soc Indust & Appl Math; Brit Comput Soc; Australian Comput Soc; Am Asn Artificial Intel; IFIP. *Res:* Computer architecture; protection in operating systems; high-level-language computer architecture; programming language design; artificial intelligence. *Mailing Add:* 26 Dopping Brook Rd Sherborn MA 01770-1047

FEVOLD, HARRY RICHARD, b Madison, Wis, Jan 28, 35; m 58; c 3. BIOCHEMISTRY, MOLECULAR BIOLOGY. *Educ:* Univ Mont, BS, 56; Univ Utah, PhD(biochem), 61. *Prof Exp:* Asst biochem, Univ Utah, 61; NIH fel, Biochem Inst, Uppsala Univ, 61-63; from asst prof to assoc prof, Univ Mont, 63-70; proj assoc, McArdle Labs, Univ Wis-Madison, 70-71; prof chem, 71-88, PROF BIOCHEM, UNIV MONT, 88- *Concurrent Pos:* NIH res grant, 64-78, career develop award, 65-70; NSF res grant, 64-70; vis prof biochem, Univ Tex Southwestern Med Sch, Dallas, 87-88. *Mem:* AAAS; Am Soc Biochem & Molecular Biol; Endocrine Soc; Am Chem Soc. *Res:* Steroidogenic enzymes and their control; avian endocrinology; metabolic regulation during food and water deprivation. *Mailing Add:* Div Biol Sci Univ Mont Missoula MT 59812

FEW, ARTHUR ALLEN, b Jasper, Tex, Nov 30, 39; m 62; c 3. THUNDERSTORMS, LIGHTNING. *Educ:* Southwestern Univ, Tex, BS, 62; Univ Colo, MBS, 65; Rice Univ, PhD(space sci), 69. *Prof Exp:* Syst analyst gas pipeline simulation, Tenneco, Inc, 62-63; res assoc thunderstorm res, Rice Univ, 68-70, from adj asst prof to adj assoc prof, 70-78, assoc prof space physics, 78-84, PROF SPACE PHYSICS, RICE UNIV, 84- *Concurrent Pos:* Prin investr, thunderstorm elec res, NSF & Off Naval Res, 70-, NASA, 78-81. *Honors & Awards:* Mitchell Prize, Third Biennial Woodlands Conf Growth Policy, 79. *Mem:* Am Geophys Union; Am Meterol Soc; Sigma Xi; AAAS; Am Forestry Asn. *Res:* Experimental and theoretical research in the areas of atmospheric electricity and atmospheric acoustics, especially lightning and thunderstorm research. *Mailing Add:* Dept Space Physics & Astron Rice Univ PO Box 1892 Houston TX 77251

FEWER, DARRELL R(AYMOND), b Perth, NB, Mar 12, 23; m 49; c 1. PHYSICS, ELECTRONICS. *Educ:* Univ NB, BSc, 50; Univ Western Ont, MSc, 52. *Prof Exp:* Mem tech staff, Bell Tel Labs, NJ, 52-57; sr proj engr, semiconductor components div, Tex Instruments Inc, 57-58, br mgr semiconductor surface studies, res & eng dept, 58-60; vpres res & eng & mem bd dirs, Tex Res & Electronic Corp, 60-63; br mgr reliability sci, semiconductor res & develop labs, Tex Instruments Inc, 63-69, br mgr optoelectronics dept, 69-74; pres, Ensa Corp, 74-84; sr proj mgr, Nat Res Coun, Can, 84-90; CONSULT, DRAY ASSOC, 90- *Honors & Awards:* W R G Baker Award, Inst Elec & Electronics Engrs, 57. *Mem:* Sr mem Inst Elec & Electronics Engrs. *Res:* Semiconductor device characterization and applications; semiconductor surface physics; electrochemical devices. *Mailing Add:* 1617 Sunview Dr Gloucester ON K1C 5B9 Can

FEWKES, ROBERT CHARLES JOSEPH, b Colonial Manor, NJ, Mar 25, 35; m 70; c 3. PROCESS ENGINEERING. *Educ:* Univ Del, BchE, 62; Mass Inst Technol, MS, 72, PhD(biochem eng), 77. *Prof Exp:* Proj engr chem eng, Gulf Res & Develop Co, 62-70; sr res engr biochem eng, Lederle Labs, 76-80; res assoc biochem eng, Eastman Kodak Res Labs, 80-86, MGR, BIOPROCESSING TECHNOL, EASTMAN KODAK BIO-PROD DIV, 80- *Mem:* Am Chem Soc; Am Inst Chem Engrs; Inst Food Technol. *Res:* Enzyme and bioconversion processes; process development; process scaleup. *Mailing Add:* Genencor Int Cambridge Pl 1850 S Winton Rd Rochester NY 14650-0832

FEX, JORGEN, b Stockholm, Sweden, Mar 24, 24; US citizen; m 57; c 3. HEARING, NEUROLOGY. *Educ:* Univ Stockholm, BA, 45; Univ Lund, MD, 52; Karolinska Inst, PhD(neurophysiol), 62. *Prof Exp:* Docent neurophysiol, Nobel Inst Neurophysiol, Karolinska Inst, Stockholm, 62-64; sr res fel, Dept Physiol, John Curtin Sch Med Res, Australian Nat Univ, 64-66; vis scientist, Lab Neurobiol, Dept Health, Educ & Welfare, NIMH, 66-69; prof neural sci, Ind Univ, 69-73; CHIEF, LAB NEURO-OTOLARYNGOL, PUB HEALTH SERV, DEPT HEALTH & HUMAN SERV, NAT INST NEUROL & COMMUN DIS & STROKE, NIH, 73- *Mem:* NY Acad Sci; fel Acoust Soc Am; Am Soc Neurochem; Soc Neurosci. *Res:* Function and mechanisms of the mammalian organ of hearing from a multidisciplinary point of view; using experimental techniques in the fields of anatomy, biochemistry, genetics, immunology, molecular biology, pharmacology and physiology. *Mailing Add:* Bldg 36 Rm 5-D-32 Div Intramural Res LNO NIH Nat Inst Neurol Disorders & Stroke Bethesda MD 20892

FEY, CURT F, b Berlin, Germany, May 19, 32; nat US. OPERATIONS RESEARCH, ELECTRONICS. *Educ:* Haverford Col, BS Hons, 54; Univ Pa, PhD(opers res), 60; Univ Rochester, MBA, 76, CFA, 84. *Prof Exp:* Prin scientist mach intel, Gen Dynamics Corp, 60-61; adv scientist opers res, IBM Corp, 61-65; mem corp staff, Tex Instruments, Inc, 65-70; MGR TECHNOL STRATEGY, XEROX CORP, 70- *Concurrent Pos:* Prof & lectr math, Am Univ, 63-65; lectr, Indust Col Armed Forces, 64; lectr opers res, NTex State Univ, 68-69. *Mem:* Inst Elec & Electronics Engrs; Inst Mgt Sci. *Res:* Corporate planning; securities analysis; operations res; economics of microelectronics; management of technology portfolios. *Mailing Add:* PO Box 22853 Rochester NY 14692

FEY, GEORGE, b Rietberg, West Germany, June 1, 44. MOLECULAR GENETICS. *Educ:* Univ Erlangen, PhD(biochem), 73. *Prof Exp:* ASST MEM, DEPT IMMUNOL, SCRIPPS CLIN & RES FOUND, 83- *Mem:* Am Soc Biol Chemists; German Soc Biol Chemists; Swiss Soc Cellular Molecular Biologists. *Mailing Add:* Dept Immunol IMM14 Scripps Clin & Res Found 10666 N Torrey Pines Rd La Jolla CA 92037

FEY, GEORGE TING-KUO, b Shanghai, China, July 4, 40; Can & Chinese citizen; c 1. CHEMISTRY, CHEMICAL ENGINEERING. *Educ:* Nat Taiwan Univ, BS, 65; Univ Mass, MS, 70, PhD(chem), 73. *Prof Exp:* Fel coordr chem, Univ Guelph, 73-74; lectr & fel electrochem, Univ Western Ont, 74-77; res assoc organometallic chem, Univ Guelph, 77-78; res assoc catalysis chem eng, Univ Waterloo, 78-80; anal chemist gas chromatography, Labstat Inc, Kitchener, Ont, 80-81; RES CHEMIST, LITHIUM SEC BATTERY, BALLARD RES INC, NORTH VANCOUVER, 81- *Mem:* Am Chem Soc. *Res:* Vibrational spectroscopy; pentacoordinated phosphorus compounds; phosphine transition metal complexes; platinum metals chemistry; organometallic electrochemistry; homogeneous catalysis of transition metal complexes. *Mailing Add:* Nat Taiwan Univ One Roosevelt Rd IV Taipei Taiwan

FEYERHERM, ARLIN MARTIN, b West Point, Nebr, May 21, 25; m 51; c 3. EXPERIMENTAL STATISTICS. *Educ:* Univ Minn, BS, 46; State Univ Iowa, MS, 48; Iowa State Univ, PhD(math), 52. *Prof Exp:* Instr math, Northwest Mo State Col, 48-49; asst prof, Iowa State Univ, 52-53; from asst prof to prof, 53-64, PROF STATIST, KANS STATE UNIV, 53- *Concurrent Pos:* Consult, USAF, 51-70. *Mem:* Am Statist Asn; Inst Math Statist; Am Soc Agron. *Res:* Operations research, statistical climatology and crop yield modeling. *Mailing Add:* Dept Statistics Dickens Hall Kans State Univ Manhattan KS 66506

FEYNMAN, JOAN, b New York, NY, Mar 30, 27; div; c 3. SOLAR PHYSICS, SPACE SCIENCE. *Educ:* Oberlin Univ, BA, 48; Syracuse Univ, PhD(theoret physics), 58. *Prof Exp:* Instr physics, Syracuse Univ, 58-59; geophysicist, Lamont Observ, Columbia Univ, 61-64; geophysicist, Ames Res Ctr, NASA, 64-72; res scientist, High Altitude Observ, Nat Ctr Atmospheric Res, 72-76; phys sci adminr, NSF, 76-78, sr res physicist, Boston Col, 79-84; JET PROPULSION LAB, 84- *Concurrent Pos:* Res assoc appl mech, Stanford Univ, 69-71; sr resident res assoc, Nat Res Coun-Nat Acad Sci, 71-, lectr, Colo Univ, 76-78. *Mem:* Am Geophys Union; Am Astron Soc. *Res:* Geomagnetics and space research; solar and interplanetary and magnetospheric physics. *Mailing Add:* Jet Propulsion Lab 4800 Oak Grove Dr Mail Stop 169-506 Pasadena CA 91109

FEYNS, LIVIU VALENTIN, b Bucharest, Romania, May 5, 38; m 64; c 2. ANALYTICAL CHEMISTRY, PHARMACEUTICAL CHEMISTRY. *Educ:* Polytech Inst, Bucharest, MS, 69; Timishoara, Polytech Inst, PhD(org chem), 73. *Prof Exp:* Chemist drugs mfg, Biofarm, Bucharest, 59-64; chemist design & synthesis anticancer agents, Oncological Inst, Bucharest, 64-71, head, Anal Lab Cancer Res, 71-76; vis fel, Drug Design & Chem Sect, Nat Cancer Inst, NIH, 76-78; assoc scientist drugs anal, 78-80, SUPVR METHOD DEVELOP, US PHARMACOPEIA, 80- *Mem:* Am Chem Soc. *Res:* Drug design; synthesis and analysis; analytical methods development; drug purity profiles. *Mailing Add:* 498 W Deer Park Rd Gaithersburg MD 20877-1631

FEYOCK, STEFAN, b Austria, June 24, 42; m 68; c 3. ARTIFICIAL INTELLIGENCE. *Educ:* Colo Col, BA, 64; Univ Kans, MA, 66; Univ Wis, PhD(comput sci), 71. *Prof Exp:* Asst prof comput sci, Univ Okla, 71-76, ASSOC PROF MATH & COMPUT SCI, COL WILLIAM & MARY, 76- *Concurrent Pos:* Consult, ICASE, 83-; pres, Va Artificial Intel Res, Inc, 84-; vis sr scientist, Sci & Technol Corp, 85- *Mem:* Asn Comput Mach; Am Asn Artificial Intelligence. *Res:* Artificial intelligence; integration of logic programming and expert systems techniques with existing data base systems; intelligent databases; automated fault diagnosis by means of model-based reasoning. *Mailing Add:* Dept Comput Sci Col William and Mary Williamsburg VA 23185

FEZER, KARL DIETRICH, b Englewood, NJ, Mar 2, 30; m 52; c 4. SCIENCE EDUCATION, PHILOSOPHY OF SCIENCE. *Educ:* Cornell Univ, BS, 51, PhD(plant path), 57; Haverford Col, MA, 53. *Prof Exp:* Asst plant path, Cornell Univ, 52-55; from instr to asst prof, Univ Minn, St Paul, 57-62; div sci & math, Univ Minn, Morris, 62-63; tutor, St John's Col, 63-66; assoc prof, 66-67, prof & chmn, 67-85, PROF DEPT BIOL, CONCORD COL, 85- *Concurrent Pos:* vis scholar, Harvard Divinity Sch, 80-81; NSF-NEH Ethics & Values in Sci & Technol, Interdisciplinary Incentive Award, 80-81. *Mem:* AAAS; Sigma Xi; Nat Ctr Sci Educ. *Res:* Philosophy and history of science in relation to science education; interaction between science and religion. *Mailing Add:* Dept Biol Concord Col Athens WV 24712

FIALA, ALAN DALE, b Beatrice, Nebr, Nov 9, 42. ASTRONOMY. *Educ:* Carleton Col, BA, 63; Yale Univ, MS, 64, PhD(astron), 68. *Prof Exp:* ASTRONR, US NAVAL OBSERV, 63-67 & 68- *Concurrent Pos:* Instr, US Dept Agr Grad Sch, 69- *Mem:* AAAS; Am Astron Soc; Am Inst Navig; Am Inst Aeronaut & Astronaut; NY Acad Sci; Sigma Xi. *Res:* Celestial mechanics; general and special perturbations; numerical analysis; ephemerides; astronomical constants; eclipses and transits; computerized data handling; natural satellite dynamics. *Mailing Add:* US Naval Observ Washington DC 20392-5100

FIALA, EMERICH SILVIO, b Prague, Czechoslovakia, Aug 24, 38; US citizen; m 63; c 3. CANCER RESEARCH, PHARMACOLOGY. *Educ:* Columbia Col, BS, 59; Rutgers Univ New Brunswick, PhD(biochem), 64. *Prof Exp:* Res assoc, Fels Res Inst, 64-67; res chemist, Veterans Admin Hosp, 67-73; res assoc, 73-76, head sec biochem & pharmacol, 76-80, assoc chief, Div Molecular Biol & Pharmacol, Naylor Dana Inst, 80-84, CHIEF, DIV BIOCHEM PHARMACOL, AM HEALTH FOUND, 84- *Concurrent Pos:* adj assoc prof Dept Pharmacol, NY Med Col, 80-; Bd Dirs, NY Chromatography Club, 81-; Metab Path Study Sect, NIH, 85- *Mem:* Am Assoc Cancer Res; Am Chem Soc; Am Soc Biochem & Molecular Biol. *Res:* Mechanics of action of chemical carcinogens; pharmacology; analytical chemistry; toxicology; biochemistry; metabolism. *Mailing Add:* Am Health Found One Dana Rd Valhalla NY 10595

FIALER, PHILIP A, b San Francisco, Calif, Nov 6, 38; m 67, 85; c 2. RADIO SCIENCE. *Educ:* Stanford Univ, BS, 60, MS, 64, PhD(elec eng), 70. *Prof Exp:* Sr res engr electronics, Lockheed Missiles & Space Co, Calif, 61-67; res assoc, radiosci lab, Stanford Univ, 67-70; asst dir, SRI Int, 70-80, dep dir remote measurements lab, 80-84, staff scientist, 84-88; PRES, MIRAGE SYSTEMS, 84- *Concurrent Pos:* Consult, 68- *Mem:* Inst Elec & Electronics Engrs. *Res:* Ionospheric structure and physics; radar cross-section analysis; radio propagation in irregular media; computer methods in signal and image processing. *Mailing Add:* Mirage Systems 537 Lakeside Ave Sunnyvale CA 94086

FIALKOW, AARON DAVID, b New York, NY, Aug 9, 11; m 40; c 3. MATHEMATICS. *Educ:* City Col New York, BS & MS, 31; Columbia Univ, PhD(math), 36. *Prof Exp:* Teacher high sch, NY, 32-33 & 35; Nat Res fel, Inst Advan Study & Princeton Univ, 36-37; instr, Brooklyn Col, 37-42; lectr, Columbia Univ, 42-45; math res engr, Fed Telecommun Labs, 45-46; head math sect, Control Instrument Co, 46-54; adj prof, 46-47, from assoc prof to prof, 47-76, EMER PROF MATH, POLYTECH UNIV, 76- *Mem:* Am Math Soc. *Res:* Differential geometry; electric network theory. *Mailing Add:* 3125 Tibbett Ave Bronx NY 10463

FIALKOW, PHILIP JACK, b New York, NY, Aug 20, 34; m 60; c 2. INTERNAL MEDICINE, MEDICAL GENETICS. *Educ:* Univ Pa, AB, 56; Tufts Univ, MD, 60. *Prof Exp:* From intern to resident med, Sch Med, Univ Calif, San Francisco, 60-62; resident, Univ Wash Hosp, Seattle, 62-63, fel med genetics, Sch Med, 63-65, from instr to assoc prof med & genetics, 65-72, vchmn, Dept Med, 74-80, chmn, Dept Med & physician-in-chief, 80-90, PROF MED, SCH MED, UNIV WASH, 72-, DEAN, SCH MED, 90- *Concurrent Pos:* Chief med serv, Seattle Vet Admin Hosp, 74-80. *Mem:* Inst Med-Nat Acad Sci; Am Soc Human Genetics; Am Soc Clin Invest; Asn Am Physicians; fel Am Col Physicians; Am Fed Clin Res; Am Soc Hemat; AAAS. *Res:* Human genetics; etiology of chromosomal abnormalities; origin and development of tumors. *Mailing Add:* Off Dean SC-64 Univ Wash Seattle WA 98195

FICENEC, JOHN ROBERT, b Rochester, Minn, Oct 29, 38; m 59; c 4. EXPERIMENTAL HIGH ENERGY PHYSICS. *Educ:* St John's Univ, Minn, BS, 60; Univ Ill, Urbana, MS, 61, PhD(physics), 66. *Prof Exp:* From asst prof to assoc prof, 68-85, dept head, 78-79, PROF PHYSICS, VA POLYTECH INST & STATE UNIV, 85- *Concurrent Pos:* Guest asst physicist, Brookhaven Nat Lab, 68-71; Res Corp grants, 69-70; Petrol Res Fund grants, 69-72; assoc investr, NSF grants, 72-74, prin investr, 79-86. *Mem:* AAAS; Am Phys Soc; Sigma Xi; Am Asn Univ Professors. *Res:* Bubble chamber physics; electron scattering from nickel, tin and zirconium isotopes at 300 MeV/c; counter-wire chamber experiments to study high momentum transfer, high multiplicity p-p interactions; magnetic monopole search; search for high mass resonance states; proton-antiproton collisions at two trillion electron volts; electron-proton collisions; electromagnetic inhimetions in nuclei. *Mailing Add:* Dept Physics Va Polytech Inst & State Univ Blacksburg VA 24061

FICH, SYLVAN, b New York, NY, Aug 8, 10; wid. ELECTRICAL ENGINEERING. *Educ:* Cooper Union, BSc, 31; Rutgers Univ, MSc, 32. *Prof Exp:* Instr math & eng, Union Jr Col, 33-39, dean eve session, 40-42; instr elec eng, Rutgers Univ, 42-44; mem staff, wave propagation res, Columbia Univ, 44-45; from asst prof to assoc prof, Rutgers Univ, 45-80, dir grad progs elec eng, 68-80, emer prof elec eng, 80-; RETIRED. *Concurrent Pos:* Consult, Essex Electronics Co, 49-52 & Plastron Co, 69-70. *Honors & Awards:* Columbia Res Medal, 46. *Mem:* Inst Elec & Electronics Engrs; Soc Math Biol; Int Soc Ventricular Dynamics; NY Acad Sci. *Res:* Electromagnetic waves and radiation; biomedical engineering; hemodynamics and models of the circulatory system; wave phenomena in biomedical systems. *Mailing Add:* 37 Hamlin Rd Edison NJ 08817

FICHER, MIGUEL, b Buenos Aires, Arg, July 24, 22; US citizen; m 50; c 3. ENDOCRINOLOGY, CHEMISTRY. *Educ:* Univ Buenos Aires, Lic in Chem, 48, PhD, 50. *Prof Exp:* Instr biol anal, Sch Chem, Univ Buenos Aires, 50; sub-dir clin lab, Moron County Hosp, Arg, 56-57; dir clin lab, Pvt Policlin, Arg, 57-61; dir endocrine diag lab, Jewish Hosp, St Louis, Mo, 61-65; res mem endocrinol, Univ Buenos Aires, 65-66; res mem steroid chem, Dept Endocrinol, Albert Einstein Med Ctr, 66-71; assoc prof pediat, Temple Univ, 71-74; med res scientist, Div Psychoendocrinol, Eastern Pa Psychiat Inst, 74-80; res assoc prof, 75-80, RES PROF PSYCHIAT & HUMAN BEHAV, JEFFERSON MED COL, THOMAS JEFFERSON UNIV, 80- *Concurrent Pos:* Consult chem invest, St Louis State Sch & Hosp, 64-65; consult psychoendocrinol, Vet Admin Hosp, Coatesville, Pa, 76- *Mem:* Am Asn Clin Chem; Endocrine Soc; Soc Study Reproduction; Am Soc Andrology; AAAS. *Res:* Biosynthesis of steroid hormones by gonads in humans and animals; study of pituitary-gonad axis; reproduction and spermatogenesis; psychoendocrinology of sexual functioning, of sleep, and of aging. *Mailing Add:* 502 Lombard St Philadelphia PA 19147

FICHTEL, CARL EDWIN, b St Louis, Mo, July 13, 33; m 88. ASTROPHYSICS. *Educ:* Wash Univ, BS, 55, PhD(physics), 60. *Prof Exp:* Asst physics, Wash Univ, 56-59; scientist, 59-60, head nuclear emulsion sect, 60-66, head gamma ray & nuclear emulsion sect, 66-70, HEAD GAMMA RAY & NUCLEAR EMULSION BR, GODDARD SPACE FLIGHT CTR, NASA, 70- CHIEF SCIENTIST, LAB HIGH ENERGY ASTROPHYS, 83-, GODDARD SR FEL, 88- *Concurrent Pos:* Vis lectr, Univ Md, 63-; chmn, High Energy Astrophys Div, Am Astron Soc, 80-81; chmn, Astrophys Div, Am Phys Soc, 82-83; chief scientist, High Energy Astrophys Div, 83- *Honors & Awards:* John C Lindsay Mem Award, Goddard Space Flight Ctr, 68; Exceptional Sci Achievement Medal, NASA, 71; Spec Achievement Award, NASA Goddard Space Flight Ctr, 78. *Mem:* Int Astron Union; Am Phys Soc; Am Astron Soc; Sigma Xi. *Res:* High energy astrophysics; galactic cosmic rays; solar cosmic rays, especially solar particle composition and its relation to solar abundances; gamma ray astronomy; cosmic ray propagation and confinement; cosmic ray and gamma ray instrumentation. *Mailing Add:* Goddard Space Flight Ctr NASA Code 660 Greenbelt MD 20771

FICK, GARY WARREN, b O'Neill, Nebr, July 10, 43; m 69; c 3. AGRONOMY, CROP ECOLOGY. *Educ:* Univ Nebr, Lincoln, BS, 65; Massey Univ, NZ, Dipl Agr Sci, 68; Univ Calif, Davis, PhD(plant physiol), 71. *Prof Exp:* From asst prof to assoc prof, 71-84, PROF AGRON, CORNELL UNIV, 84- *Concurrent Pos:* Vis scientist, Lincoln Col, 77-78; mem, Coun Agr Sci & Technol. *Honors & Awards:* Merit Cert, Am Forage & Grassland Coun, 89. *Mem:* Am Soc Agron; Crop Sci Soc Am; Am Forage & Grassland Coun. *Res:* Forage crop ecology; computer modelling of crop growth; forage nutritional value. *Mailing Add:* Dept Soil Crop & Atmospheric Sci Cornell Univ Ithaca NY 14853

FICK, HERBERT JOHN, b Minn, Jan 4, 37; m; c 3. PLASTICS CHEMISTRY, POLYMER CHEMISTRY. *Educ:* St Olaf Col, BA, 59; Univ Chicago, MS, 60. *Prof Exp:* Supvr corp chem lab, GT Schjeldahl Co, 63-65, mgr mat eng, Elec Prod Div, 65-66, eng mgr, UK Div, 66-69, chemist, 69, mgr display devices mfg, 69-71, prog mgr new process develop, 71-72, mgr tape & laminations prod line, 72-74, sr staff tech specialist, Sheldahl Co, 75-81; consult, Thermically Conductive Dielectrics, 81-84; DEVELOP MGR, BERGQUIST CO, MINNEAPOLIS, 84- *Concurrent Pos:* Consult, elec insulation prod for rotating equipment; DuPont fel, Univ Chicago, 59. *Mem:* Am Chem Soc; Am Soc Testing & Mat; Am Soc Metals Int. *Res:* Flexible printed wiring; adhesives; electroluminescent display; electrical insulating materials; materials for flexible, inflatable structures; adhesives; adhesives; silicone rubber compounding and products; thermal management and thermal management (conductive) dielectrics. *Mailing Add:* 519 E Eighth St Northfield MN 55057

FICKEISEN, DUANE H, aquatic ecology, for more information see previous edition

FICKEN, MILLICENT SIGLER, b Washington, DC, July 27, 33; div; c 2. BEHAVIORAL ECOLOGY. *Educ:* Cornell Univ, BS, 55, PhD(zool), 60. *Prof Exp:* Res assoc zool, Univ Md, 63-68; from asst prof to assoc prof, 67-75, PROF BIOL SCI, UNIV WIS-MILWAUKEE, 75-, DIR FIELD STA, 67- *Mem:* Fel Am Ornith Union; Soc Study Evolution; fel Animal Behav Soc; AAAS. *Res:* Ornithology; animal communication; sociobiology. *Mailing Add:* Dept Biol Sci Univ Wis Milwaukee WI 53201

FICKEN, ROBERT W, b Brooklyn, NY, Feb 26, 32; m 55; c 2. ETHOLOGY. *Educ:* Cornell Univ, BS, 53, PhD(vert zool), 60. *Prof Exp:* Res assoc animal behav, Cornell Univ, 60-62; asst prof zool, Univ Md, 62-68; res assoc zool, Univ Wis-Milwaukee, 68-87; RETIRED. *Mem:* AAAS; Am Ornith Union. *Res:* Comparative avian ethology; ecology; interspecific communication; evolution of signalling systems; analysis of vocalizations. *Mailing Add:* 1129 N Jackson St Apt 1403 Milwaukee WI 53202

FICKESS, DOUGLAS RICARDO, b Piedmont, Okla, Aug 25, 31; m 53; c 2. ZOOLOGY, PHYSIOLOGY. *Educ:* Univ Okla, BS, 54, MS, 56; Univ Mo, PhD(zool), 63. *Prof Exp:* From asst prof to assoc prof, 62-69, chmn dept, 66-74, PROF BIOL, WESTMINSTER COL, MO, 69- *Mem:* Am Soc Mammalogists; Am Soc Zoologists; Sigma Xi. *Res:* Effects of population density dependent factors on the histology and physiology of adrenal glands in mammals and reptiles. *Mailing Add:* Dept Biol Westminster Col Fulton MO 65251

FICKETT, FREDERICK ROLAND, b Portland, Maine, Sept 30, 37; m 61; c 2. PHYSICS. *Educ:* Univ NH, BS, 60; Univ Ariz, MS, 62; Ore State Univ, PhD(physics), 67. *Prof Exp:* Nat Bur Standards-Nat Res Coun res assoc, 67-69; mem staff, Cryogenics Div, 69-80, MEM STAFF, ELECTROMAGNETIC TECHNOL DIV, NAT BUR STANDARDS, 80- *Honors & Awards:* Silver Medal, Dept Com, 85. *Mem:* Fel Am Phys Soc; Inst Elec & Electronics Engrs. *Res:* Properties of metals and alloys at cryogenic temperatures; materials science; solid state physics; magnetics. *Mailing Add:* 3660 Cloverleaf Dr Boulder CO 80304

FICKETT, WILDON, b Tucson, Ariz, Mar 25, 27; m 45; c 5. FLUID PHYSICS. *Educ:* Univ Ariz, BS, 48; Calif Inst Technol, PhD(chem), 51. *Prof Exp:* MEM STAFF, LOS ALAMOS NAT LAB, UNIV CALIF, 51- *Mailing Add:* Los Alamos Nat Lab Mail Stop P952 PO Box 1663 Los Alamos NM 87545

FICKIES, ROBERT H, b Brooklyn, NY, Mar 23, 44; m 69; c 2. GEOLOGY, ENGINEERING. *Educ:* Brooklyn Col, BS, 66, MS, 69. *Prof Exp:* Lectr geol, City Univ NY, Brooklyn Col, 66-69; eng geologist, Gerard Eng Corp, 69-70; geologist, Dunn Geosci Corp, 70-74, mgr geotech group, 74-78; ASSOC SCIENTIST, NY STATE GEOL SURV, 78-, ASST CHIEF, 88- *Concurrent Pos:* Eng geologist, Tippetts, Abbett, McCarthy & Stratton, 70; proj dir, US Environ Protection Agency grant, 78-81, US Nuclear Regulatory Comn grant, 78-81; prin investr, US Geol Surv grant, 82-83 & 82-84; mem Govs Select Sci Adv Comn Dynamics Long Island Barrier Syst, 87-; newsletter ed, Geol Soc Am, Eng Geol Div, 79-85. *Honors & Awards:* Meritorious Serv Award, Geol Soc Am, 85. *Mem:* Am Inst Prof Geologists; Asn Eng Geologists; Fel Geol Soc Am; Int Asn Eng Geol; Am Soc Civil Eng. *Res:* Engineering geology of the soil and bedrock of NY with emphasis on solving problems in slope stability, radioactive and hazardous waste disposal. *Mailing Add:* RD 3 Box 52 Averill Park NY 12018-9507

FICKINGER, WILLIAM JOSEPH, b New York, NY, July 18, 34. HIGH ENERGY PHYSICS. *Educ:* Manhattan Col, BS, 55; Yale Univ, PhD(physics), 61. *Prof Exp:* Asst prof physics, Univ Ky, 61-62; asst physicist, Brookhaven Nat Lab, 62-63 & 64-65; assoc physicist, Saclay Nuclear Res Ctr, France, 63-64 & 65-66; asst prof physics, Vanderbilt Univ, 66-67; assoc prof, 67-76, PROF PHYSICS, CASE WESTERN RESERVE UNIV, 76- *Mem:* Am Phys Soc; Am Asn Univ Professors. *Res:* Experimental particle physics; intermediate energy scattering processes and particle production. *Mailing Add:* Dept Physics Case Western Reserve Univ Cleveland OH 44106

FICSOR, GYULA, b Kiskunhalas, Hungary, Apr 11, 36; US citizen; m 65; c 2. GENETIC TOXICOLOGY. *Educ:* Colo State Univ, 60; Univ Mo, Columbia, PhD(genetics), 65. *Prof Exp:* NIH fel, 65-66; res assoc, Molecular Biol Lab, Univ Wis, 67; asst prof, 67-71, assoc prof biol, 72-80, PROF BIOMED SCI, WESTERN MICH UNIV, 80- *Concurrent Pos:* Nat Acad Sci exchange scientist, Hungary, 73; consult, Nat Inst Environ Health Sci, 76-78. *Mem:* Environ Mutagen Soc. *Res:* Methods development for detecting mutagens in mammalian sperm. *Mailing Add:* Dept Biomed Sci Western Mich Univ Kalamazoo MI 49008

FIDDLER, WALTER, b Vienna, Austria, Feb 5, 36; US citizen; c 4. ANALYTICAL CHEMISTRY. *Educ:* Temple Univ, AB, 59, PhD(org chem), 65. *Prof Exp:* RES CHEMIST, MEAT LAB, FOOD SAFETY LAB, USDA, 65- *Mem:* Am Chem Soc; Inst Food Technol; Am Meat Sci Asn; Assoc Off Anal Chem. *Res:* Pyrimidine containing sulfonamides; chemical information retrieval; smoke flavor investigations; studies on carcinogens in cured meat products. *Mailing Add:* Eastern Regional Res Ctr USDA 600 E Mermaid Lane Philadelphia PA 19118

FIDELLE, THOMAS PATRICK, b Boston, Mass, July 14, 39; m 65; c 2. FLAME RETARDANTS, WATER TREATMENT CHEMICALS. *Educ:* Tufts Univ, BS, 61; Univ Mass, Amherst, MS, 67, PhD(chem eng), 69. *Prof Exp:* Res mgr polymers, Celanese Corp, 72-82; DIR RES, GREAT LAKES CHEM CORP, 83- *Concurrent Pos:* Adj prof environ eng, Univ NC, 70-72. *Mem:* Am Inst Chem Engrs; Soc Petrol Engrs; Soc Plastics Engrs. *Res:* Polymer additives, materials science and water treatment chemicals. *Mailing Add:* Box 2628 West Lafayette IN 47906

FIDLAR, MARION M, b Vincennes, Ind, June 7, 09; m 37. GEOLOGY. *Educ:* Indiana Univ, BA, 34, MA, 36, PhD(physiog), 42. *Prof Exp:* Asst state geologist, Ind, 36-38; dist geologist, Ohio Oil Co (Marathon), 38-43; sr geologist, Mountain Fuel Supply Co, 43-58, exec vpres, 58-62, pres, 62-72, chmn bd & chief exec officer, 72-74; sr dir, Questar Corp, 86-91; dir, Questar Pipeline Co, 88-91; RETIRED. *Concurrent Pos:* Dir, First Security Corp, Salt Lake, 64-76; mem, bd dirs, US Chamber Com, 69-77. *Honors & Awards:* Richard Owen Award, Ind Univ, 86. *Mem:* Sigma Xi; sr fel, Geol Soc Am; Am Asn Petrol Geologists; Am Inst Mining, Metall & Petrol Engrs; Am Inst Prof Geologists. *Mailing Add:* 908 E S Temple Number 6-E Salt Lake City UT 84102

FIDLER, ISAIAH J, b Jerusalem, Israel, Dec 4, 36; m 75; c 3. CANCER BIOLOGY, IMMUNOLOGY. *Educ:* Okla State Univ, BS, 61, DVM, 63; Univ Pa, PhD(path), 70. *Prof Exp:* Instr vet surg, Sch Med, Univ Pa, 66-68, from asst prof to assoc prof path, 70-75, adj prof path, 81-84; head lab, Frederick Cancer Res Ctr, Nat Cancer Inst, 75-80, dir lab, Cancer Metastasis & Treatment Prog, 80-83; adj prof path, Sch Med, Univ Md, Baltimore, 81-85; PROF & CHMN, DEPT CELL BIOL, DIR PROG INTERSERON RES & R E BOB SMITH CHAIR, CELL BIOL, M D ANDERSON CANCER CTR, UNIV TEX, HOUSTON, 83- *Concurrent Pos:* Mem, Study Sect Path B, NIH, 77-81; chief ed, Cancer Metastasis Review, 81- *Honors & Awards:* Bertner Prize, 83; Joseph Steiner Prize, 89. *Mem:* Am Vet Med Asn; Am Asn Cancer Res (pres, 85); Am Asn Path; AAAS. *Res:* Mechanisms of cancer metastasis; investigations into the nature of the metastatic cancer cell and its relationship with host factors; the manipulation of host defense toward eradication of metastases. *Mailing Add:* Dept Cell Biol Inst M D Anderson Cancer Ctr Univ Tex Houston TX 77030

FIDLER, JOHN M, b Baltimore, Md, Aug 16, 47. IMMUNOREGULATION & IMMUNOPHARMACOLOGY, ANIMAL EFFICACY MODELS. *Educ:* Purdue Univ, PhD, 82. *Prof Exp:* Asst mem, dept immunopath, Scripps Clin & Res Found, La Jolla, Calif, 76-82; group leader, cell & molecular immunregulator res unit, inflammation & immunol res dept, Lederle Labs, Cyanamid Co, 82-88; SR SCIENTIST & GROUP LEADER, DEPT PHARMACOL, CETUS CORP, EMERYVILLE, CALIF, 88- *Mem:* Am Asn Immunologists; Soc Anal Cytometry; Int Soc Exp Hemat; Inflammation Res Asn; AAAS; NY Acad Sci; Int Soc Immunopharmacol. *Res:* Xerobiotic, recombinant, cytokine and biological therapeutics in treatment of human disease using animal models for preclinical research, including graft in host disease in bone marrow transplantation, infectious disease and septic. *Mailing Add:* Dept Pharmacol c/o Cetus Corp 1400 53rd St Emeryville CA 94608

FIDONE, SALVATORE JOSEPH, b New York, NY, June 10, 39; m 62; c 2. PHYSIOLOGY, NEUROCHEMISTRY. *Educ:* Georgetown Univ, BS, 62; State Univ NY, PhD(physiol), 67. *Prof Exp:* From instr to assoc prof, 69-81, PROF PHYSIOL, SCH MED, UNIV UTAH, 81- *Concurrent Pos:* NIH fel neurophysiol, Col Med, Univ Utah, 67-69, prin investr, NIH res grants, 75-93; co-investr, NIH prog proj grants, 68-84 & 84-; mem, Neurol A Study Sect, NIH, 81-85. *Honors & Awards:* Javits Neurosci Investr Award, 86-93. *Mem:* Soc Neurosci; Am Physiol Soc. *Res:* Neurochemistry and neurophysiology of arterial chemoreception. *Mailing Add:* Dept Physiol Univ Utah Sch Med Salt Lake City UT 84112

FIEDEL, BARRY ALLEN, INFLAMMATION, WOUND HEALING. *Educ:* Syracuse Univ, PhD(immunol), 73. *Prof Exp:* Assoc prof immunol & biol, Rush Med Col, Chicago, 80-85; ASSOC MEM, IMMUNOL DIV, JAMES N GAMBLE INST MED RES, CINCINNATI, 85- *Mailing Add:* Toltzis Commun 137 S Easton Rd Glenside PA 19038

FIEDELMAN, HOWARD W(ILLIAM), b Sheboygan, Wis, Apr 23, 16; m 47; c 2. CHEMICAL ENGINEERING. *Educ:* Univ Wis, BS, 38. *Prof Exp:* From chem engr to res supvr, cent res lab, Morton Salt Co, 39-67, dir salt res, Morton Salt Div, Morton-Norwich Prod, Inc, 67-77; EXEC DIR, SOLUTION MINING RES INST, 78- *Res:* Salt technology; purification of salt from brines; theoretical aspects of caking. *Mailing Add:* 812 Muriel St Woodstock IL 60098

FIEDLER, GEORGE J, electrical engineering; deceased, see previous edition for last biography

FIEDLER, H(OWARD) C(HARLES), b Chicago, Ill, June 24, 24; m 49; c 5. PHYSICAL METALLURGY. *Educ:* Purdue Univ, BS, 49; Mass Inst Technol, MS, 50, ScD(metall), 53. *Prof Exp:* Res assoc, metal dept res lab, Gen Elec Co, 53-59, metallurgist, 59-66, metallurgist, Res & Develop Ctr, 66-89; RETIRED. *Mem:* Am Soc Metals; Am Inst Mining, Metall & Petrol Engrs. *Res:* Soft magnetic materials; recrystallization; surface alloying; amorphous metals. *Mailing Add:* 2280 Berkeley Ave Schenectady NY 12309

FIEDLER, HAROLD JOSEPH, b Detroit, Mich, Apr 29, 24; m 49; c 3. POWER SYSTEM COMMUNICATIONS, POWER SYSTEM AUTOMATION. *Educ:* Univ Detroit, BEE, 51. *Prof Exp:* Appln engr, Power Gen Syst Protection & Control, Gen Elec Co, 51-64, Syst Automation Commun & Control, 64-67, sr appln engr, 67-70, mgr, Syst Automation Oper, 70-73, sr appln engr, Power Line Carrier Systs Mgt, Metering High Voltage Direct Current Commun, 73-86 & Fiber Optics, 80-86; RETIRED. *Concurrent Pos:* Chmn, Gen Elec Task Force Automated Distrib Systs, 80-81; expert adv, Int Conf Large High Voltage Elec Systs, 73-86. *Mem:* Power Eng Soc, Inst Elec & Electronics Engrs; Nat Soc Prof Engrs. *Mailing Add:* 22 Beechwood Ave Rte 1 Ballston Lake NY 12019

FIEDLER, PAUL CHARLES, b Schenectady, NY, Sept 15, 53; m 82; c 2. FISHERY OCEANOGRAPHY, REMOTE SENSING. *Educ:* Dartmouth Col, BA, 75; Univ Calif, San Diego, PhD(oceanog), 82. *Prof Exp:* Res asst, Scripps Inst Oceanog, 78-81; OCEANOGRAPHER, NAT OCEANIC & ATMOSPHERIC ADMIN, NAT MARINE FISHERIES SERV, SOUTHWEST FISHERIES CTR, 82-86, 87- *Concurrent Pos:* Vis lectr zooplankton ecol & sampling tech, Ctr Sci Invest & Grad Res, Ensenada, 81-82; fel, Naval Ocean Systs Ctr, 86-87. *Mem:* Am Geophys Union; Am Soc Limnol & Oceanog; AAAS. *Res:* Biological oceanography of the eastern tropical Pacific and effects on marine mammals; fisheries applications of oceanographic satellite data; effects of mesoscale events and seasonal patterns on the distribution, spawning, survival and availability of fish and marine mammals. *Mailing Add:* Nat Oceanic & Atmospheric Admin/Nat Marine Fisheries Serv SW Fisheries Ctr PO Box 271 La Jolla CA 92038

FIEDLER, VIRGINIA CAROL, b Salt Lake City, Utah. IMMUNOLOGY. *Educ:* Northwestern Univ, BS, 70, MD, 73. *Prof Exp:* Assoc prof,Col Med Univ of Ill, 77-88; CLIN RES MGR, UPJOHN CO, 88- *Mem:* Am Acad Dermat; Soc Invest Dermat. *Res:* Alopecia. *Mailing Add:* Upjohn Co-Derm Kalamazoo MI 49001

FIEDLER-NAGY, CHRISTA, b Marienbad, Czech, July 8, 43; Ger citizen; m 69; c 1. INFLAMMATION DERMATOLOGY. *Educ:* Fairleigh Dickinson Univ, BS, 67, MS, 74; Rutgers Univ, PhD(biochem), 81. *Prof Exp:* Sr scientist pharmacol, 81-88, ASSOC RES INVESTR, DERMAT, HOFFMANN-LA ROCHE, INC, 89- *Mem:* AAAS; Am Soc Bio Chem; NY Acad Sci; Soc Investigative Dermat. *Res:* Mediators of inflammation and hypersensitivity reactions, arachidonic acid metabolites, cytokines, etc; cutaneous inflammation-antiinflammatory agents. *Mailing Add:* Hoffmann-La Roche Inc Nutley NJ 07110

FIELD, A J, b Long Beach, Calif, Jan 6, 24. OIL & GAS OPERATIONS, MARINE ENGINEERING. *Educ:* Calif Inst Technol, BS, 44; Stanford Univ, MS, 47. *Prof Exp:* Pres, Global Marine, Inc, 60-77; petrol consult, 77-81; vpres Offshore Develop, Santa Fe Drilling Co, 81-86; CONSULT ENGR, 86- *Concurrent Pos:* Chmn engr adv comt, Harvey Mudd Col, 80. *Mem:* Nat Acad Eng; Soc Petrol Engrs; Am Petrol Inst; Am Inst Mining, Metall & Petrol Engrs; Am Asn Petrol Geologists. *Mailing Add:* 487 Lincoln Dr Ventura CA 93001

FIELD, ARTHUR KIRK, b North Adams, Mass, Jan 6, 38; m 60; c 2. IMMUNOBIOLOGY, VIROLOGY. *Educ:* Cornell Univ, BS, 60, MS, 61; Univ Calif, Berkeley, PhD(virol, biochem), 65. *Prof Exp:* Res fel, Merck Inst Therapeut Res, 65-72, sr res fel, 72-80, sr investr, 80-86, dir, 86-87, EXEC DIR BRISTOL-MYERS SQUIBB PHARM RES INST, 88- *Concurrent Pos:* Instr, Gwynedd Mercy Col, 71- *Mem:* Soc Exp Biol & Med; Am Asn Immunol; Am Soc Microbiol; NY Acad Sci; Sigma Xi; Am Soc Virol. *Res:* Mode of action of interferon and means of interferon induction; cellular immune response to viral antigens; antiviral chemotherapy; herpesvirus replication. *Mailing Add:* Dept Virol Bristol-Myers Squibb Pharmaceut Res Inst Box 4000 Princeton NJ 08543-4000

FIELD, BYRON DUSTIN, b Charlotte, Mich, June 2, 18; m 41; c 2. SPECTROSCOPY. *Educ:* Mich State Univ, BS, 39, MS, 42. *Prof Exp:* Asst spectroscopist, Wyandotte Chem Corp, Mich, 41-43; spectroscopist, Mallinckrodt Inc, 43-56, head phys methods lab, 56-82; CONSULT, 82- *Mem:* Am Chem Soc. *Res:* Spectrochemical analysis of chemical products; x-ray diffraction and fluorescence spectroscopy; general instrumental and analytical chemistry; atomic absorption spectroscopy; computer applications in analytical chemistry. *Mailing Add:* 38 Harneywold Dr St Louis MO 63136-2402

FIELD, CHRISTOPHER BOWER, b Dinuba, Calif, Mar 12, 53. PHYSIOLOGICAL PLANT ECOLOGY. *Educ:* Harvard Univ, AB, 75; Stanford Univ, PhD(biol), 81. *Prof Exp:* Asst prof plant ecol, dept biol, Univ Utah, 81-; AT SOUTHWEST MO STATE UNIV. *Mem:* AAAS; Am Soc Plant Physiologists; Sigma Xi. *Res:* Ways that physiological processes determine the distribution and abundance of plant species, specifically photosynthesis, water relations and nutrient uptake. *Mailing Add:* Biomed Sci SW MO State Univ Springfield MO 65804

FIELD, CYRUS WEST, b Duluth, Minn, May 5, 33; m 58; c 4. ECONOMIC GEOLOGY, GEOCHEMISTRY. *Educ:* Dartmouth Col, BA, 56; Yale Univ, MS, 57, PhD(geol), 61. *Prof Exp:* Res asst geol, Yale Univ, 58-60; geologist, Bear Creek Mining Co, Kennecott Copper Corp, 60-63; from asst prof to assoc prof, 63-76, PROF GEOL, ORE STATE UNIV, 76- *Mem:* Soc Econ Geol (vpres, 81); Geochem Soc; Geol Soc Am; Am Inst Mining, Metall & Petrol Engrs. *Res:* Sulfur isotope abundances in minerals; geology, geochemistry and mineralogy of metallic ore deposits. *Mailing Add:* Dept Geosci Ore State Univ Corvallis OR 97331-5506

FIELD, DAVID ANTHONY, US citizen. COMPUTATIONAL GEOMETRY, COMPUTER AIDED DESIGN. *Educ:* Bowdoin Col, AB, 65; Oakland Univ, MS, 66; Univ Colo, PhD(math), 71. *Prof Exp:* Asst prof math, Col Holy Cross, 71-78; STAFF RES SCIENTIST, GEN MOTORS RES LABS, 78- *Mem:* Math Asn Am; Soc Indust & Appl Math; Sigma Xi. *Res:* Computer software which assists automotive design and manufacturing processes. *Mailing Add:* 1732 Norfolk Birmingham MI 48009

FIELD, FRANK HENRY, b Keansburg, NJ, Feb 27, 22; m 44, 59, 77; c 2. PHYSICAL CHEMISTRY. *Educ:* Duke Univ, BS, 43, MA, 44, PhD(chem), 48. *Prof Exp:* Instr & asst prof chem, Univ Tex, 47-52; res chemist, Humble Oil & Refining Co, 52-53, sr res chemist, 53-60, res specialist, 60-62, res assoc, 62-65 & Esso Res & Eng Co, 65-68, sr res assoc, 68-70; prof, 70-88, Camille & Henry Dreyfus prof, 88-89, EMER PROF CHEM, ROCKEFELLER UNIV, 89- *Concurrent Pos:* Guggenheim fel, 63-64. *Honors & Awards:* Frank H Field & Joe L Franklin Award, Am Chem Soc. *Mem:* Am Chem Soc; Am Soc Mass Spectrometry (vpres, 70-72, pres, 72-74, past pres, 74-76). *Res:* Mass spectrometry, particle bombardment, electron impact studies, energies and reactions of gaseous ions; radiation chemistry. *Mailing Add:* 13 Moore Lane Oak Ridge TN 37830

FIELD, GEORGE BROOKS, b Providence, RI, Oct 25, 29; m 56; c 2. THEORETICAL ASTROPHYSICS. *Educ:* Mass Inst Technol, BS, 51; Princeton Univ, PhD(astron), 55. *Prof Exp:* Physicist, Naval Ord Lab, 51-52; res asst, Princeton Univ, 52-54; jr fel astron, Harvard Soc Fels, 55-57; from asst prof to assoc prof astron, Princeton Univ, 57-65; prof, Univ Calif, Berkeley, 65-72, chmn dept, 70-71; prof astron, 72-73, dir, Ctr Astrophys, Harvard Col Observ & Smithsonian Astrophys Observ & Paine prof practical astron, 73-82, ROBERT WHEELER WILLSON PROF APPL ASTRON & SR PHYSICIST, SMITHSONIAN ASTROPHYS OBSERV, HARVARD UNIV, 82- *Concurrent Pos:* Vis prof astron, Calif Inst Technol, 64; mem planetology subcomt, Space Sci Steering Comt, NASA, 64-66, mem astron missions bd, 68-70, mem space telescope working group, 73-77, mem shuttle astron working group, 74-76 & space prog adv coun, 75-77, chmn phys sci comt, 75-77; Phillips visitor, Haverford Col, 65 & 71; mem panel on astron adv to Off Naval Res, 65-66, mem physics surv comt, 69-72, mem astron surv comt, 69-72, mem panel on radio astron, Astron Surv Comt, 69-72, chmn panel astrophys & relativity, Physics & Astron Survs, 69-72; NSF grants astrophys, 65-73, mem astron panel, NSF, 66-67, chmn, 67-69; mem space sci panel, President's Sci Adv Comt, 66-67; mem vis comt, Nat Radio Astron Observ, 67-69; co-ed, Gordon & Breach Ser Astrophys & Space Sci, 68-75; corresp, Comments on Astrophys & Space Sci, 68-71; vis prof, Cambridge Univ, 69; trustee-at-large, Assoc Univs, Inc, 69-72; trustee, mem corp & exec comt, Aspen Ctr Physics, 72-74; mem adv bd, Nat Astron & Ionosphere Ctr, 71-74; mem vis comt div physics, math & astron, Calif Inst Technol, 71-; vpres comn 34, Int Astron Union, 73-76, pres, 76-79; lectr, Sch Theoret Physics, Les Houches, France, 74; chmn, Boyden Observ Coun, 74-76; mem vis comt, Lowell Observ, 75; mem comt on study of uses for balloons for sci purposes, Nat Acad Sci, 75-76; mem adv panel on orientation & roles, Univ Space Res Asn, 75; mem corp vis comt, Dept of Physics, Mass Inst Technol, 76-79; liaison rep NASA phys sci comt, Space Sci Bd, Nat Res Coun, 76-77; mem vis comt, Dept Physics & Astrophys, Univ Colo, Boulder, 77, subcomt, Nat Optical Observ, 88; mem exec comt, Space Adv Coun, NASA, 78-; chmn, Nat Res Coun astron surv, Nat Acad Sci, 78-82, chmn Briefing Panel on Astron & Astrophys, 82-83; Adv Comt Space Telescope Sci Inst, 83- (chmn, 84); President's Nat Comn on Space, 85-86. *Honors & Awards:* NASA Pub Serv Medal, 77; Joseph Henry Medal of Smithsonian Inst, 82. *Mem:* Nat Acad Sci; fel Am Phys Soc; Am Astron Soc; Int Astron Union; Sigma Xi; assoc Royal Astron Soc; Am Acad Arts & Sci; AAAS. *Res:* Dynamics of interstellar matter, including galaxy and star formation; instabilities in dilute gases; cosmology, including background radiation and intergalactic matter. *Mailing Add:* Smithsonian Ctr Astrophys Harvard Univ 60 Garden St Cambridge MA 02138

FIELD, GEORGE FRANCIS, b Stockton, Calif, Dec 26, 34; m 62; c 2. ORGANIC CHEMISTRY. *Educ:* Pomona Col, AB, 56; Harvard Univ, PhD(org chem), 62. *Prof Exp:* Sr chemist, 62-70, RES FEL, HOFFMAN- LA ROCHE, INC, 70- *Mem:* Am Chem Soc. *Res:* Medicinal chemistry; heterocyclic synthesis. *Mailing Add:* Lawrence Livermore Lab PO Box 808 L-463 Livermore CA 94550-0622

FIELD, GEORGE ROBERT, b Vienna, Austria, Oct 16, 19; US citizen; m 46; c 3. ELECTRICAL ENGINEERING. *Educ:* Washington Univ, St Louis, BSEE, 48; Drexel Univ, MSEE, 60, MSEngMgt, 71, MBA, 78. *Prof Exp:* Design & develop engr, RCA Corp, 48-55, design & develop leader, 55-55-56, mgr radar data conversion eng, 56-61, design support & integration, 61-62, eng opers, 62-63 & design & develop eng, 63-64, mgr tech assurance, 64-69, mgr tech opers, Missile & Surface Radar Div, 69-84; CONSULT, 84- *Mem:* Sr mem Inst Elec & Electronics Engrs. *Res:* Design and development of electronic equipment for missile range instrumentation; strategic and defensive systems and space exploration. *Mailing Add:* 305 Parry Dr Moorestown NJ 08057

FIELD, HERBERT CYRE, b Cleveland, Ohio, Nov 2, 30; m 53; c 4. SAFETY MANAGEMENT. *Educ:* Case Univ, BS, 53; Purdue Univ, MS, 57. *Prof Exp:* Physicist exp physics div, Lawrence Livermore Nat Lab, Calif, 53-54 & exp shock hydrodyn div, 54-55; physicist exp physics group, Atomic Int Div, NAm Aviation, Inc, 57-59, sr physicist reactor physics group, 59-63; reactor safety specialist, Div Oper Safety, US Atomic Energy Comn, US Dept Energy, 63-77, sr exec, Off Environ & Safety, 77-82, sr tech adv, Nuclear Safety Off, 82-86, safety appraisal team leader, 86-88, gen engr, Off Energy Res, 89-90, SAFETY MGT CONSULT, US DEPT ENERGY, 91- *Concurrent Pos:* Consult, US Navy, 70-71. *Mem:* AAAS; Sigma Xi; Am Nuclear Soc; Antarctican Soc; NY Acad Sci. *Res:* Administration of safety programs at basic research laboratories; experimental neutron physics; nuclear reactors, especially their safety evaluation. *Mailing Add:* 8185 Inverness Ridge Rd Potomac MD 20854

FIELD, HERMANN HAVILAND, b Zurich, Switz, Apr 13, 10; US citizen; m 40; c 3. ENVIRONMENTAL PLANNING, HEALTH FACILITIES PLANNING. *Educ:* Harvard Univ, BA, 33; Swiss Fed Politech Inst, dipl archit, 36. *Prof Exp:* Dir planning, Cleveland Col Western Reserve Univ, 47-49; dir planning off, Urban & Inst Planning, Tufts New Eng Med Ctr, 61-72; dir grant, New Urban Sch Concept, US Off Educ & Boston Sch Dept, 64-71; dir, Prog Urban & Environ Policy, 72-78, prof polit sci, 72-78, EMER PROF ENVIRON PLANNING, TUFTS UNIV, 78- *Concurrent Pos:* Prin investr, Proji New Concepts of Pediat Hosp Design, NIH, 63-67; proj dir, Grant US Off Transp, 65-67; vpres, Cambridge Interfaith Housing Corp, 68-70; mem, Mass Governor's Task Force on Transp, 70; coordr & initiator, Spec Course on World Conserv Strategy, Tufts Univ, 80-81; comn mem, Comn Environ Planning of Int Union for Conserv of Nature, 81- *Mem:* Am Inst Certified Planners; fel Am Inst Architects. *Res:* Developing institutional, policy, political strategies for global sustainable use of natural resources from macro to micro level at the local and regional level; autobiographical relating to issues and insights arising out of incident; global environmental issues. *Mailing Add:* 110 Center Rd Shirley MA 01464

FIELD, HOWARD LAWRENCE, b Buffalo, NY, May 1, 28; m 60; c 3. PSYCHIATRY. *Educ:* Univ NC, AB, 49; Jefferson Med Col, MD, 54. *Prof Exp:* PROF PSYCHIAT, JEFFERSON MED COL, 60- *Mem:* AAAS; Am Psychiat Asn; Am Med Asn. *Res:* Psychiatric treatment of patients in a general hospital. *Mailing Add:* Dept Psychiat Jefferson Med Col Philadelphia PA 19107

FIELD, JACK EVERETT, b Gary, Ind, June 3, 27; m 58; c 3. INORGANIC CHEMISTRY, TRACE METAL ANALYSIS. *Educ:* Harvard Univ, AB, 52; Tulane Univ, MS, 55, PhD(inorg chem), 57. *Prof Exp:* Chemist, E I du Pont de Nemours & Co, 57-63; res chemist, Corning Glass Works, 64-66; prof phys sci, 66-71, head dept, 71-77, PROF CHEM, NICHOLLS STATE UNIV, 71- *Mem:* AAAS; Am Chem Soc; Am Soc Testing & Mat. *Res:* Inorganic chemistry of the solid states and surfaces, including crystal growth; water pollution. *Mailing Add:* 110 Garden Circle Thibodaux LA 70301-3728

FIELD, JAMES BERNARD, b Ft Wayne, Ind, May 28, 26; m 54; c 4. MEDICINE. *Educ:* Harvard Univ, MD, 51. *Prof Exp:* From intern to resident, Mass Gen Hosp, 51-54; sr investr clin endocrinol br, NIH, 54-62; from assoc prof to prof med, Sch Med, Univ Pittsburgh, 61-78, dir clin res unit, 62-78; Rutherford Prof Med, Baylor Col Med, 78-89; RETIRED. *Concurrent Pos:* USPHS training grant, 70-74; consult, diabetes & arthritis prog, Div Chronic Dis, USPHS, 62-68, mem, endocrinol study sect, 65-69 & chmn, 68-69, mem, diabetes & endocrinol training grant comt, 70-74; ed, Metab, 70-; res collabr, Brookhaven Nat Lab, 72-85; mem clin sci panel, Utilization Human Resources Comn, Nat Res Coun, 76-79; mem, gen clin res ctr review comt, USPHS, 76-80; mem, nat diabetes adv bd, USPHS, 77-; assoc ed, Clin Thyroidology; instr med, Harvard Med Sch, 91- *Honors & Awards:* Eli Lilly Award, Am Diabetes Asn, 58; Van Meter Prize, Am Goiter Asn, 61. *Mem:* Am Soc Clin Invest; Endocrine Soc; Am Physiol Soc; Am Diabetes Asn; Asn Am Physicians; Am Clin & Climat Asn. *Res:* Diabetes mellitus and endocrinology. *Mailing Add:* 241 Perkins St Unit I-101 Boston MA 02130

FIELD, JAY ERNEST, b Roanoke, Va, Nov 4, 47; m 70; c 3. PHYSICAL CHEMISTRY. *Educ:* Va Polytech Inst, BS, 70; Univ Fla, PhD(phys chem), 75. *Prof Exp:* Res chemist hermetic systs, Major Appliances Lab, 76-84, MGR, AIR CONDITIONING PLASTIC APPLNS DEVELOP, GEN ELEC CO, 84- *Mem:* Am Soc Testing & Mat. *Res:* Chemical aspects of hermetic refrigeration systems and electrical properties evaluations of electrical insulating materials; applications of plastics to residential air conditioning systems. *Mailing Add:* Technol Develop & Appln Labs Gen Elec Co Appliance Park 35-1004 Louisville KY 40225

FIELD, KATHARINE G, b Pittsfield, Mass, Dec 20, 51; m 82; c 2. MOLECULAR EVOLUTION. *Educ:* Yale Univ, BA, 75; Boston Univ, MA, 78; Univ Ore, PhD(biol), 85. *Prof Exp:* Vis asst prof, Earlham Col, 84-85; fel biol, Ind Univ, 85-88; ASST PROF MICROBIOL & ZOOL, ORE STATE UNIV, 88- *Concurrent Pos:* Lectr marine evolution, Marine Biol Labs, 88 & 92, lectr marine ecol, 89. *Mem:* Soc Study Evolution; Am Soc Zoologists. *Res:* Molecular phylogeny of the animal kingdom; molecular approaches to the study of evolution and ecology. *Mailing Add:* Depts Microbiol & Zool Ore State Univ Corvallis OR 97331

FIELD, KURT WILLIAM, b Akron, Ohio, Mar 29, 44; m 67; c 2. CHEMISTRY. *Educ:* Hiram Col, BS, 66; Case Western Reserve Univ, PhD(org chem), 70. *Prof Exp:* Fel org chem, Johns Hopkins Univ, 70-72; asst prof, Univ Wis-Whitewater, 72-76; from asst prof to assoc prof, 76-90, PROF ORG CHEM, BRADLEY UNIV, 90- *Concurrent Pos:* Vis scientist, Univ Ill, 84-85; vis prof, Univ Ill, 86-88 & 90, Univ Hawaii, 88-89. *Mem:* Am Chem Soc. *Res:* The formation and cleavage of carbon-nitrogen bonds; the chlorination of alkenes with nitrogen trichloride; rotational isomerism in substituted acylamides; synthesis of small ring heterocycles; organic reaction mechanisms; organic synthesis via organometallics and carbenes. *Mailing Add:* Dept Chem Bradley Univ Peoria IL 61625

FIELD, LAMAR, b Montgomery, Ala, July 19, 22; m 48; c 2. MEDICINAL CHEMISTRY, ORGANOSULFUR CHEMISTRY. *Educ:* Mass Inst Technol, SB, 44, PhD(chem), 49. *Prof Exp:* Asst chemist res lab, Merck & Co, Inc, 44-46; Socony-Vacuum fel, Mass Inst Technol, 47-48; from instr to assoc prof, 49-59, chmn, 61-67, prof, 59-89, EMER PROF CHEM, VANDERBILT UNIV, 89- *Concurrent Pos:* Consult chem indust, 55-; vis assoc, Am Chem Soc, 63-82, vis scientist, 66-73; consult, NIH, 65-69, NY State Educ Dept, 74 & Southern Asn Cols & Schs, 76 & 83; Coulter lectr, Univ Miss, 68; vis fel, John Curtin Sch Med Res, Australian Nat Univ & hon fel, Res Sch Chem, 74. *Honors & Awards:* William Barton Rogers Award, 44; L C Glenn Award, 52. *Mem:* Am Chem Soc; Sigma Xi. *Res:* Synthetic, structural, medicinal and biological aspects of organic chemistry, especially with reference to organic sulfur compounds; medicinal chemistry. *Mailing Add:* Box 1507 Sta B Vanderbilt Univ Nashville TN 37235-0001

FIELD, LESTER, b Ill, Feb 9, 18. AERONAUTICS, ACOUSTICS. *Educ:* Purdue Univ, BSEE, 39; Stanford Univ, PhD(physics & math), 44. *Prof Exp:* Asst prof, Stanford Univ, 40-44; mem tech staff, Phys Res Dept, Bell Tel Labs, 44-46; from assoc prof to prof, Stanford Univ, 46-52; prof, Calif Inst Technol, 52-60; dir, Microwave Tube Div, Hughes Aircraft Co, 55-62, assoc dir, Com Prod Group, 60-62, vpres, 60-74, dir diversification, 62-64, co-dir res labs, 64-72, chief scientist, 72-74; consult, 76-86, SCIENTIST, M & K CORP, 86- *Mem:* Nat Acad Eng; Inst Elec & Electronics Engrs; Am Phys Soc; Audio Eng Soc. *Res:* Consulting in acoustics, electromagnetic radiation, electronics and television display tube technology. *Mailing Add:* 8660 Burton Way Apt 306 Los Angeles CA 90048

FIELD, MARVIN FREDERICK, b Manistee, Mich, Oct 3, 26; m 51; c 3. MICROBIOLOGY. *Educ:* Cent Mich Univ, AB, 48; Mich State Univ, MS, 50; Univ Minn, PhD(microbiol), 57. *Prof Exp:* Instr microbiol, Univ Rochester, 50-51; res microbiologist, Chas Pfizer & Co, NY, 51-53, microbiologist, Ind, 57-62; sr virologist, Jensen Salsbery Labs Div, Richardson-Merrell, Inc, 62-63; asst prof dent, Univ Mo, Kansas City, 63-70, assoc prof, 70-73; res mgr, Haver Lockhart Labs, 73-76; res dir, Douglas Industs, 76-80; RES DIR, CEVA LABS, 80- *Concurrent Pos:* Asst prof microbiol, Med Ctr, Univ Kans, 63-69. *Mem:* AAAS; Am Chem Soc; Am Soc Microbiol. *Res:* Electron microscopy; virology. *Mailing Add:* Sanofi Animal Health 12300 Santa Fe Dr Lenexa KS 66215

FIELD, MICHAEL, b New York, NY, Feb 21, 14; m 42; c 2. MANUFACTURING ENGINEERING. *Educ:* City Col New York, BME, 37; Columbia Univ, MS, 38; Univ Cincinnati, PhD(physics), 48. *Prof Exp:* Res engr, Cincinnati Milling Mach Co, 38-45, res physicist, 46-48; partner, Metcut Res Assocs, Inc, 48-58, pres, 59-78, chmn bd & chief exec officer, 78-82; CONSULT, 82- *Mem:* Nat Acad Eng; Am Soc Mech Eng; Am Soc Metals; Soc Mfg Eng. *Res:* Physics and mechanics of metal cutting; machinability; high temperature mechanical testing; machine tools; metallurgy. *Mailing Add:* 9060 Spooky Ridge Lane Cincinnati OH 45242

FIELD, MICHAEL, b London, UK, July 20, 33; US citizen. EPITHELIAL MEMBRANE TRANSPORT, GASTROENTEROLOGY. *Educ:* Univ Chicago, AB, 53; Boston Univ, MD, 59. *Prof Exp:* Postdoctoral fel membrane biophys, Biophys Lab, Harvard Med Sch, 64-66, from asst prof to assoc prof med, 69-77; postdoctoral fel gastroenterol, Beth Israel Hosp & Harvard Med Sch, 66-68; prof med & prof pharmacol & physiol sci, Div Biol Sci, Univ Chicago, 77-84; PROF PHYSIOL & CELLULAR BIOPHYS, COL PHYSICIANS & SURGEONS, COLUMBIA UNIV, 84-, PROF MED, 84- *Honors & Awards:* Distinguished Achievement Award, Am Gastroenterol Asn, 84. *Mem:* Emer mem Am Soc Clin Invest; Asn Am Physicians; Am Physiol Soc. *Res:* Cell physiology of epithelial ion transport; pathophysiology of diarrheal diseases; pathophysiology of cystic fibrosis; studies of intestinal absorption/secretion. *Mailing Add:* 630 W 168th St PS 10-508 New York NY 10032

FIELD, MICHAEL EHRENHART, b Baltimore, Md, June 27, 45; m 67; c 2. MARINE GEOLOGY. *Educ:* Univ Del, BS, 67; Duke Univ, MA, 69; George Washington Univ, PhD(geol), 76. *Prof Exp:* Res geologist, US Army Coastal Eng Res Ctr, 69-75; MARINE GEOLOGIST, US GEOL SURV, 75- *Mem:* Geol Soc Am; Soc Econ Paleontologists & Mineralogists; Sigma Xi. *Res:* Sedimentary processes and patterns on continental margins; quaternary evolution of coastal areas and continental shelves; modern tectonic and sedimentary processes on the continental margin as they relate to basin evolution, hydrocarbon accumulation, placer deposits of minerals, and geologic hazards. *Mailing Add:* 1561 Samedra 345 Middlefield Rd Sunnyvale CA 94087

FIELD, NATHAN DAVID, b New York, NY, Aug 21, 25; m 49; c 3. POLYMER SCIENCE, ORGANIC CHEMISTRY. *Educ:* City Col New York, BS, 45; Columbia Univ, AM, 49; Polytech Inst Brooklyn, PhD(polymer chem), 56. *Prof Exp:* Res chemist polymer chem & textile fibers, E I du Pont de Nemours, Inc, 56-62; supvr polymer chem, Atlantic Refining Co, 62-64; mgr, GAF Corp, 64-68, mgr, Cent Res Lab, 68-70; res dir chem & consumer prod, Int Playtex Co, Inc, 70-80; prof, Dept Chem Eng, City Univ New York, 81-85; vpres res & develop, Dartco Mfg Inc, 85-88; ADJ PROF, DEPT CHEM, TEMPLE UNIV, 89- *Concurrent Pos:* Mem adv bd, Jour Polymer Sci, 64-, Sch Theoret & Appl Sci, Ramapo Col, 77- *Mem:* Am Chem Soc; Soc Plastics Engrs; AAAS; Sigma Xi. *Res:* Polymer science; liquid crystalline polymers, polymer blends, high temperature polymers, water soluble and swellable polymers; materials; surface active agents; absorbency; organic chemistry; consumer products. *Mailing Add:* 373 Linden Dr Elkins Park PA 19117

FIELD, NORMAN J, b New York, NY, Dec 5, 22; m 46; c 4. PHYSICS, RESEARCH ADMINISTRATION. *Educ:* City Col New York, BS, 42; Polytech Inst Brooklyn, MS, 59. *Hon Degrees:* LHD, Monmouth Col, 79. *Prof Exp:* Electronic engr, Signal Corps Radar Lab, 42-44, chief optical micros, Eng Labs, 46-53, external res, 53-54, asst to dir res, 54-58; asst dir inst explor res, US Army Signal Res & Develop Lab, 58-62, dep dir res, US Army

Electronics Labs, 62-65, chief sci & technol, US Army Electronics Command, 65-70, dir prog mgt, 70-74, spec asst logistics, 74-76, DIR INT LOGISTICS, US ARMY ELECTRONICS COMMAND, 76- *Concurrent Pos:* Adj prof, Monmouth Col, 56- *Mem:* Am Phys Soc; Optical Soc Am; Am Chem Soc; Am Asn Physics Teachers; NY Acad Sci. *Res:* Crystal physics and chemistry; optical properties of solids; administration of research. *Mailing Add:* 726 Sycamore Ave Shrewsbury NJ 07702

FIELD, PAUL EUGENE, b Easton, Pa, June 16, 34; m 60; c 3. PHYSICAL CHEMISTRY. *Educ:* Moravian Col, BS, 57; Pa State Univ, MS, 61, PhD(chem), 63. *Prof Exp:* Asst prof, 63-68, ASSOC PROF CHEM, VA POLYTECH INST & STATE UNIV, 68- *Mem:* AAAS; Am Chem Soc; Sigma Xi. *Res:* Thermodynamic properties of solutions; high temperature calorimetry; microcomputer interfacing for measurement and control. *Mailing Add:* Dept Chem Va Polytech Inst & State Univ Blacksburg VA 24061

FIELD, RAY A, b Ogden, Utah, Dec 15, 33; m 58; c 5. MEAT SCIENCE. *Educ:* Brigham Young Univ, BS, 58; Univ Ky, MS, 60, PhD(animal sci), 63. *Prof Exp:* Meat specialist, Nat Livestock & Meat Bd, 58; res asst, Univ Ky, 59-62; from asst prof to prof meat sci, 62-90, HEAD DEPT ANIMAL SCI, UNIV WYO, 91- *Concurrent Pos:* Consult meat deboning, Beehive Equip Co, 74- *Honors & Awards:* Meat Sci Award, Am Soc Animal Sci, 75. *Mem:* Am Meat Sci Asn; Inst Food Technol; Am Soc Animal Sci. *Res:* Live animal and carcass evaluation; inheritance of quality in meat; consumer acceptance, and palatability characteristics of meat; postmortem biochemistry of collagen as related to tenderness; mechanically deboned meat. *Mailing Add:* Dept Meats Univ Wyo Univ Sta Box 3684 Laramie WY 82071

FIELD, RICHARD D, b Pasadena, Calif, Apr 13, 44; m 66; c 3. ELEMENTARY PARTICLE PHYSICS. *Educ:* Univ Calif, Berkeley, BS, 66, PhD(theoret physics), 71. *Prof Exp:* Fel physics res, Brookhaven Nat Lab, 71-73; res fel physics, Calif Inst Technol, 73-78, res assoc prof, 78-80; PROF PHYSICS, UNIV FLA, 80- *Mem:* fel Am Phys Soc. *Res:* Theory and phenomenology of the strong, weak, and electromagnetic forces in elementary particle physics. *Mailing Add:* Dept Physics 215 WM Univ Fla Gainesville FL 32611

FIELD, RICHARD JEFFREY, b Attleboro, Mass, Oct 26, 41; m 66; c 2. CHEMICAL KINETICS. *Educ:* Univ Mass, BS, 63; Col Holy Cross, MS, 64; Univ RI, PhD(phys chem), 68. *Prof Exp:* Res assoc phys chem, Univ Ore, 68-74 & vis asst prof, 70-73; sr res chemist, Radiation Res Lab, Carnegie-Mellon Univ, 74-75; from asst prof to assoc prof, 75-83, PROF CHEM, UNIV MONT, 84-, CHMN, 90- *Concurrent Pos:* Vis prof, Univ Würzburg, 85-86; ed, Oscillations & Traveling Waves In Chem Systs, 85. *Mem:* Am Chem Soc; Am Phys Soc. *Res:* Experimental gas and solution phase kinetics of complex chemical reactions, especially those exhibiting oscillatory behavior related to an unstable or excitable steady state; thermo-chemical kinetics, diffusive properties of free radicals and numerical methods in physical chemistry. *Mailing Add:* Dept Chem Univ Mont Missoula MT 59812-1006

FIELD, ROBERT WARREN, b Wilmington, Del, June 13, 44; m 84. CHEMICAL PHYSICS, LASER SPECTROSCOPY. *Educ:* Amherst Col, BA, 65; Harvard Univ, MA & PhD(phys chem), 72. *Prof Exp:* Fel spectros, Quantum Inst, Univ Calif, Santa Barbara, 71-74; from asst prof to assoc prof, 74-82, PROF CHEM, MASS INST TECHNOL, 82- *Concurrent Pos:* Sloan Found res fel, 75. *Honors & Awards:* H P Broida Atomic & Molecular Spectros Chem Physics, Am Phys Soc, 80, E K Plyler Prize Molecular Spectros, 88; E Lippincott Award, Optical Soc Am, 90. *Mem:* Am Chem Soc; fel Am Phys Soc. *Res:* Tunable laser spectroscopy, particularly optical-optical double resonance on small, gas phase molecules; vibrationally highly excited polyatomic molecules; collision induced transitions; quantum ergodicity and intramolecular vibrational redistribution; analysis and consequences of intramolecular perturbations. *Mailing Add:* Dept Chem 6-219 Mass Inst Technol Cambridge MA 02139

FIELD, RONALD JAMES, b Grand Rapids, Mich, Jan 3, 46; m 69; c 1. WILDLIFE MANAGEMENT, ENVIRONMENTAL SCIENCES. *Educ:* Mich State Univ, BS, 68, MS, 70, PhD(zool), 71. *Prof Exp:* Asst prof zool, Howard Univ, 71-73; chmn, Dept Earth & Life Sci, Univ DC, 73-78; head, Dept Agr Sci, Tuskegee Inst, 78-80; prog mgr, Wildlife Res Develop, Tenn Valley Auth, 80-88; DEAN COL LIFE SCI, UNIV WASHINGTON DC. *Concurrent Pos:* Consult rodent control, Synecology, Inc, 70-72; res grants, US Dept Interior, 72, NIH, 72-73; bd dirs, Friends Nat Zoo, 72-78; wildlife res, US Fish & Wildlife Serv, 72; assoc dean, Agr & Nat Res, Washington Tech Inst, 75-76; vis prof wildlife mgt, Tenn Technol Univ, 76; asst pres & asst vpres, Univ DC, 76-77; consult furbearers, Tenn Valley Auth, 76. *Mem:* Wildlife Soc; Am Soc Mammalogists; Soc Am Foresters; Sigma Xi; AAAS. *Res:* Population structures and ecology of several species of wildlife, especially semiaquatic furbearers; population demography and control through chemical sterilization in wild rats. *Mailing Add:* Col Life Sci Univ Wash DC 4200 Connecticut Ave Washington DC 20008

FIELD, RUTH BISEN, b Brooklyn, NY, Nov 6, 27; m 52; c 3. LIPASES, LIPID BIOCHEMISTRY. *Educ:* Brooklyn Col, BS, 48; Georgetown Univ, MS, 53; Univ Md, PhD(biochem), 77. *Prof Exp:* Chemist biochem, Nat Cancer Inst, NIH, 49-53, res assoc microbiol, Cancer Chemother Dept, 67-72, postdoctoral fel, Nat Inst Arthritis, Diabetes & Kidney Dis, 79-83, staff fel, Nat Inst Dent Res, 83-85, sr staff fel, 85-87; teaching & res asst, Dept Chem, Biochem Lab, Univ Md, 72-77; staff fel, Bur Foods, Food & Drug Admin, 78-79; res assoc, 87-89, RES ASST PROF, DEPT PEDIAT, GEORGETOWN UNIV MED CTR, 89-; RES SCIENTIST, ORAL PATH RES LAB, DEPT VET AFFAIRS MED CTR, 88- *Concurrent Pos:* Consult, Oral Path Res Lab, Dept Vet Affairs, 87. *Mem:* Am Soc Biochem & Molecular Biol; Int Asn Dent Res; Am Asn Dent Res; Am Chem Soc; Int Asn Women Bioscientists; AAAS. *Res:* Regulation of secretion of the digestive enzymes, lingual lipase and amylase, from von Ebner's gland of rat tongue; phosphatidyl inositol-protein kinase C pathway using this gland as a model system; lipases, activity, purification and assays. *Mailing Add:* Oral Path Res Lab 151-I Dept Vet Affairs Med Ctr 50 Irving St NW Washington DC 20422

FIELDER, D(ANIEL) C(URTIS), b North Kingstown, RI, Oct 9, 17; m 44. ELECTRICAL ENGINEERING. *Educ:* Univ RI, BS, 40, EE, 50; Ga Inst Technol, MS, 48, PhD(elec eng), 57. *Prof Exp:* Jr design engr transformers & switchgear, Westinghouse Elec Corp, 40-41; design engr degaussing & compass compensation, Bur Ships, Washington, DC, 41-46; asst elec eng, Ga Inst Technol, 46-47; instr, Syracuse Univ, 47-48; from asst prof to assoc prof, 48-63, PROF ELEC ENG, GA INST TECHNOL, 63- *Mem:* Sr mem Inst Elec & Electronics Engrs; Math Asn Am; Sigma Xi. *Res:* Graph theory; combinatorics; electric circuit theory; digital computer theory. *Mailing Add:* 1304 Kittredge Ct NE Atlanta GA 30329

FIELDER, DOUGLAS STRATTON, b Washington, DC, July 22, 40; m 68; c 1. PHYSICS. *Educ:* Va Mil Inst, BS, 62; Univ Va, MS, 64, PhD(physics), 67. *Prof Exp:* Asst prof, 69-70, assoc prof, 70-86, Chmn Dept, 70-90, PROF PHYSICS, STATE UNIV NY COL ONEONTA, 86- *Mem:* AAAS; Am Phys Soc; Am Asn Physics Teachers. *Res:* Determination of absolute cross sections for several photoneutron reactions; photonuclear physics; low energy nuclear reactions. *Mailing Add:* Dept Physics & Astron State Univ NY Oneonta NY 13820

FIELDHOUSE, DONALD JOHN, b Dodgeville, Wis, Nov 18, 25; m 49; c 4. SEED-BORNE BACTERIAL DISEASES, VEGETABLE PHYSIOLOGY. *Educ:* Univ Wis, PhD(hort), 54. *Prof Exp:* From asst prof to prof Hort, Univ Del, 54-85; RETIRED. *Mem:* Am Soc Hort Sci; Am Phytopath Soc. *Res:* Irrigation of vegetable crops; vegetable physiology; seed technology; soil, plant, and water relationships; plant growth regulators; crop physiology; plant pathology. *Mailing Add:* Plant Sci Dept Univ Del Newark DE 19711

FIELDHOUSE, JOHN W, b Rensselaer, Ind, Oct 17, 41; m 63; c 1. SYNTHETIC ORGANIC CHEMISTRY. *Educ:* Hope Col, BA, 63; Purdue Univ, PhD(sulfenes), 68. *Prof Exp:* Res scientist, 68-73, sr res scientist, 73-78, ASSOC SCIENTIST, FIRESTONE TIRE & RUBBER CO, 78- *Mem:* Am Chem Soc. *Res:* Olefin and rubber intermediates synthesis; synthesis of polyfluoroalkoxyphosphazenes; emulsion polymerization of olefins. *Mailing Add:* Polysar Adhesives & Sealants 1020 Evans Ave Akron OH 44305-1021

FIELDING, CHRISTOPHER J, b Cheadle, Eng, Jan 26, 42; m 68. METABOLISM, BIOCHEMISTRY. *Educ:* Univ London, BSc, 62, PhD(genetics), 65. *Hon Degrees:* MA, Oxford Univ, 65. *Prof Exp:* Am Heart Asn estab investr, Dept Med, Univ Chicago, 70-71; Am Heart Asn estab investr biochem, Sch Med, 71-75, asst prof physiol & assoc staff, 75-77, ASSOC PROF PHYSIOL & SR MEM STAFF, CARDIOVASC RES INST, UNIV CALIF, SAN FRANCISCO, 78- *Concurrent Pos:* Res fel biochem, Oxford Univ, 65-67, fel, New Col, 65-69; Med Res Coun grant, 67-69; Am Heart Asn grant, 71-75; USPHS grant, Nat Heart & Lung Inst, 71- *Mem:* Am Soc Biol Chemists; fel Am Heart Asn. *Res:* Lipid metabolism; metabolism and structure of plasma lipoproteins; enzymology of lipases and acyl transferases; cellular cholesterol metabolism. *Mailing Add:* Cardiovasc Res Inst 1315-M Univ Calif Med Ctr San Francisco CA 94143

FIELDING, STUART, b Bronx, NY, Oct 31, 39; m 62; c 2. PSYCHOPHARMACOLOGY, NEUROPHARMACOLOGY. *Educ:* Monmouth Col, BA, 62; Howard Univ, MS, 64; Univ Del, PhD(psychol), 68. *Prof Exp:* Teaching asst psychol, Howard Univ, 62-64; res psychologist, Commun Sci Res Ctr, 64; mgr psychopharmacol, Ciba-Geigy, 67-75; asst dir pharmacol, 75-76, assoc dir biol sci, 77-84, DIR PHARMACOL, HOECHST-ROUSSEL PHARMACEUT, INC, 84-, DIR BIOL RES. *Concurrent Pos:* Adj asst prof, Fairleigh Dickinson Univ, 72-74, adj assoc prof, 74-; adj prof drug develop res, Univ RI, 80- *Mem:* Fel Am Psychol Asn; Soc Neurosci; Behav Pharmacologist Soc; Am Soc Pharmacol & Exp Therapeut; NY Acad Sci; Am Chem Soc. *Res:* Psychopharmacology, physiological basis of behavior, aggression, analgesia and operant behavior; aging. *Mailing Add:* Interneuron Pharmaceut 99 Hayden Ave Suite 340 Lexington MA 02173-7966

FIELDING-RUSSELL, GEORGE SAMUEL, b Calcutta, India, Dec 10, 39; Brit citizen; m 66; c 2. POLYMER SCIENCE. *Educ:* Loughborough Univ Technol, BSc Hons & dipl, 63, PhD(polymer sci), 67; Univ Leicester, MSc, 65. *Prof Exp:* Indust chemist, Brit Celanese Ltd, 61-62; sr res chemist polymer sci, 68-79, sect head mat physics, 79-83 , MGR, RUBBER MAT SCI, RES DIV, GOODYEAR TIRE & RUBBER CO, 83- *Mem:* Am Chem Soc; Tire Soc; Soc Rheology. *Res:* Physics of high polymers; mechanical, dielectric, and viscoelastic properties of solid polymers; polymer crystallization and morphology; polymer characterization via thermal methods; polymer processing rheology; polymer tribology; tire performance. *Mailing Add:* Goodyear Tire & Rubber Co Dept 410F 1144 E Market St Akron OH 44316-0001

FIELDS, ALFRED E, b Rochester, NY, July 2, 37. POLYMER PHYSICS. *Educ:* St John Fisher Col, NY, BS, 59; Univ Akron, PhD(polymer sci), 67. *Prof Exp:* Mgr, Voplex Corp, NY, 67-69 & Polymer Prod Corp, 69-72; sr res chemist, Photomat Lab, 72-73, sect coordr, Environ Chem Sect, 74, group leader, 75, SUPVR ENVIRON SERV LAB, EASTMAN KODAK CO, 76- *Res:* Environmental studies and physical chemistry of polymers in the photographic industry. *Mailing Add:* Eastman Kodak Co Kodak Park Div Bldg 320 First Fl Rochester NY 14652-3615

FIELDS, BERNARD N, b Brooklyn, NY, Mar 24, 38; c 5. VIROLOGY, GENETICS. *Educ:* Brandeis Univ, AB, 58; NY Univ, MD, 62. *Hon Degrees:* AM, Harvard Univ, 76. *Prof Exp:* From asst to assoc prof med & cell biol, Albert Einstein Col Med, 67-75; PROF MICROBIOL, HARVARD MED SCH, 75-, CHMN, DEPT MICROBIOL & MOLECULAR GENETICS, 82- *Concurrent Pos:* Mem, NSF Genetic Biol Adv Panel, 73-76, Multiple Sclerosis Soc Adv Panel Fundamental Res, 76- & NIH Task Force Virol, 76-77; assoc ed, J Infectious Dis, 76-; vis prof, Wash Univ, 77; NIH grants; ed, Infection & Immunity, 79-, J Virol, 83-; mem, Bd Scientific Counr, Nat Inst Allergy & Infectious Dis, NIH, 81, Nat Adv, 87- *Honors & Awards:* Wellcome Lectr, Am Soc Microbiol, 82; Lippard Lectr, Columbia Univ, 83;

Thayer Lectr, Johns Hopkins Univ Sch Med, 84; Dyer Lectr, NIH, 86. *Mem:* Nat Acad Sci; Am Soc Clin Invest; Infectious Dis Soc Am; Harvey Soc; Am Asn Immunologists; AAAS; Asn Am Physicians; Am Soc Virol (pres, 90-91); Am Acad Arts & Sci. *Res:* Viral genetics; virus-host interaction; infectious disease. *Mailing Add:* Harvard Med Sch Boston MA 02115

FIELDS, CLARK LEROY, b Pipestone, Minn, Oct 25, 37; m 63; c 1. INORGANIC CHEMISTRY. *Educ:* Pasadena Col, BA, 59; Univ Iowa, MS, 62, PhD(inorg chem), 64. *Prof Exp:* Assoc prof, 64-74, PROF CHEM, UNIV NORTHERN COLO, 74- *Mem:* AAAS; Am Chem Soc. *Res:* Chemistry of boron and transition metal hydrides. *Mailing Add:* Dept Chem Univ Northern Colo Greeley CO 80639

FIELDS, DAVID EDWARD, b Hickman, Ky, Oct 25, 44; div; c 1. APPLIED PHYSICS. *Educ:* Murray State Univ, BS, 66; Univ Wis, Madison, MS, 68, PhD(solid state physics), 72. *Prof Exp:* Teaching asst physics, Univ Wis, 66-67, res asst, 67-72; APPL PHYSICIST, HEALTH & SAFETY RES DIV, OAK RIDGE NAT LAB, 72- *Concurrent Pos:* Vis lectr electronics & comput sci, Murray State Univ, Murray, Ky, 76-77; fel & past-pres, Tenn Acad Sci. *Mem:* Am Inst Physics; Am Geophys Union; Am Phys Soc. *Res:* Environmental modeling; numerical analysis; hydrologic sediment transport; windblown seed dispersal; solid state physics; pattern recognition; solar energy usage. *Mailing Add:* Box 2008 Oak Ridge Nat Lab Oak Ridge TN 37831-6383

FIELDS, DAVIS S(TUART), JR, b Lexington, Ky, Apr 17, 29; m 55; c 3. PHYSICAL METALLURGY. *Educ:* Univ Ky, BS, 50; Mass Inst Technol, MS, 54, ScD(metall), 57. *Prof Exp:* Phys metallurgist, Watertown Arsenal Labs, 51-52; res metallurgist, Res Labs, Aluminum Co Am, 57-60; from asst prof to assoc prof metall, Univ Ky, 60-62; staff metallurgist, IBM Corp, 62-64, adv metallurgist, 64-67, sr engr, 67-85; PROF MECH ENG, COLO STATE UNIV, FT COLLINS, 85- *Concurrent Pos:* Lectr, Univ Ky, 62-67; adj prof, Univ Ariz, 80. *Mem:* Am Soc Metals; Sigma Xi. *Res:* Mechanical metallurgy; strain hardening behavior of metals and alloys; relationship to working and forming operations; relationship to microstructural characteristics; practical applications of superplastic alloys; reliability modeling and maintenance philosophy of computer systems; technology advancement in digital magnetic recording systems; physics of non-impact printing; manufacturing systems engineering and materials in design. *Mailing Add:* 4328 Picadilly Dr Ft Collins CO 80526

FIELDS, DONALD LEE, b Louisville, Ky, May 16, 32; m 53; c 4. ORGANIC CHEMISTRY. *Educ:* East Ky State Col, BS, 54; Ohio State Univ, PhD(chem), 58. *Prof Exp:* Res chemist, 58-60, sr res chemist, 60-64, res assoc chem, 64-74, lab head, 74-75, SR LAB HEAD, CORP RES LABS, EASTMAN KODAK CO, 75- *Mem:* Am Chem Soc. *Res:* Exploratory research in molecular design. *Mailing Add:* Five Brackley Circle Fairport NY 14450-4127

FIELDS, ELLIS KIRBY, b Chicago, Ill, May 10, 17; m 39; c 3. ORGANIC CHEMISTRY. *Educ:* Univ Chicago, BS, 36, PhD(org chem), 38. *Prof Exp:* Eli Lilly res fel, Univ Chicago, 38-41; dir res, Develop Lab, Univ Chicago, 41-50; res assoc, Amoco Chem Corp Div, Standard Oil Co, Ind, 50-62, sr res assoc, 62-75; RES CONSULT, AMOCO CHEM CO, AMOCO CORP, 75- *Concurrent Pos:* Mem staff, Kings Col, London, 62-; mem petrol res fund adv bd, Am Petrol Inst; assoc ed, Chem Revs, Petrol Preprints, Am Chem Soc; Almquist lectr, Univ Idaho, 79. *Honors & Awards:* Petrol Chem Award, Am Chem Soc, 78. *Mem:* AAAS; Am Chem Soc; Faraday Soc; Royal Soc Chem; fel Acad Arts & Sci. *Res:* Petrochemicals; oxidation processes; lube oils and additives; catalysis; photochemistry. *Mailing Add:* Davenport House 559 Ashland Ave River Forest IL 60305

FIELDS, GERALD S, b 1929; m 51; c 2. FLUID DYNAMICS. *Educ:* Wayne Univ, BS, 56; Univ Colo, MS, 65. *Prof Exp:* VPRES ENG & ORIGINAL EQUIPMENT SALES, CHAMPION LABS, INC, 78- *Mem:* Nat Fluid Power Soc. *Res:* Fluid filtration. *Mailing Add:* Champion Labs Inc PO Box 307 West Salem IL 62476

FIELDS, HOWARD LINCOLN, b Chicago, Ill, Dec 12, 39; m 66; c 2. NEUROPHYSIOLOGY. *Educ:* Univ Chicago, BS, 60; Stanford Univ, MD, 65, PhD(neurosci), 66. *Prof Exp:* Fel neurol, Harvard Med Sch, 70-72; from instr to assoc prof, 72-82, PROF NEUROL & PHYSIOL, UNIV CALIF, SAN FRANCISCO, 82- *Concurrent Pos:* Macy Found fel, 79; fel, Clare Hall Col, Univ Cambridge. *Honors & Awards:* Nat Migraine Found Award, 84; J J Bonica Lectr, 87; Bristol-Myers Award, 88; J Elliot Royer Award, 89; Gunn-Loke lectr, 89. *Mem:* Soc Neurosci; Int Asn Study Pain; Am Acad Neurol; Am Neurol Asn; Am Soc Clin Invest; Am Pain Soc. *Res:* Neurophysiological mechanisms of pain transmission; modulation of sensory transmission by central nervous system; mechanism of action of analgesic drugs. *Mailing Add:* Dept Neurol Box 0114 Univ Calif Med Sch San Francisco CA 94143

FIELDS, JAMES PERRY, b Sherman, Tex, July 30, 32; m 58; c 2. DERMATOLOGY, DERMATOPATHOLOGY. *Educ:* Univ Tex, Austin, BS, 53, MS, 57; Univ Tex Med Br, Galveston, MD, 58. *Prof Exp:* Asst dermat to assoc clin prof dermat & path, Columbia Univ Col Physicians & Surgeons, New York, 63-80; dir dermat, USPHS Hosp, 64-79; assoc prof med & path, 79-88, ASSOC CLIN PROF MED, VANDERBILT UNIV MED SCH, 88- *Concurrent Pos:* J D Lane Clin Res Award, USPHS, 78. *Mem:* Fel Am Col Physicians; fel Am Acad Dermat; fel Am Soc Dermatopath; fel Am Acad Allergy; fel Am Col Allergists. *Res:* Pharmacology of cobalt; immunopathology of leprosy; cancer of the skin (leiomyosarcoma). *Mailing Add:* 4301 Hillsboro Rd Suite 200 Nashville TN 37215

FIELDS, JERRY L, mathematics; deceased, see previous edition for last biography

FIELDS, KAY LOUISE, b Palo Alto, Calif, July 22, 41; c 1. MOLECULAR BIOLOGY, IMMUNOLOGY. *Educ:* Radcliffe Col, AB, 63; Mass Inst Technol, PhD(biol), 68. *Prof Exp:* Am Cancer Soc fel molecular biol, Univ Geneva, Switz, 68-70; res assoc, Inst Molecular Biol, Univ Geneva, 70-71; hon res fel neuroimmunol, Univ Col, London, 71-77; from asst prof to prof neurol & neurosci, Albert Einstein Col Med, 86-90; CHIEF, MOLECULAR & CELLULAR NEUROSCI RES BR, EXTRAMURAL RES, DIV BASIC BRAIN & BEHAV SCI, NIMH, 90- *Mem:* Am Soc Neurochem; Int Soc Neurochem; Am Soc Cell Biol; Soc Neurosci. *Res:* Definition and characterization of cell surface antigens and receptors on normal and tumor cells of the nervous system using biochemical and immunological approaches. *Mailing Add:* Molecular & Cellular Neurosci Res Br NIMH 5600 Fishers Lane 11-105 Rockville MD 20857

FIELDS, MARION LEE, b Plainfield, Ind, Dec 15, 26; m 46; c 4. FOOD SCIENCE, MICROBIOLOGY. *Educ:* Univ Ind, AB, 50; Purdue Univ, MS, 56, PhD(food tech), 59. *Prof Exp:* Sanitarian, Ind State Bd Health, 51-54; asst hort, Purdue Univ, 54-56, instr, 56-59, asst prof, 59-60; from asst prof to assoc prof hort, 60-67, assoc prof food sci & nutrit, 67-69, PROF FOOD SCI & NUTRIT, UNIV MO, COLUMBIA, 69- *Mem:* Inst Food Technol; Am Soc Microbiol. *Res:* Microbial ecology; heat resistance of bacterial spores; protein quality evaluation; microbial protein for food for man. *Mailing Add:* Dept Food Sci & Nutrit Univ Mo 219 Eckles Hall Columbia MO 65211

FIELDS, PAUL ROBERT, b Chicago, Ill, Feb 4, 19; m 43; c 3. HEAVY ELEMENT CHEMISTRY, NUCLEAR CHEMISTRY. *Educ:* Univ Chicago, BS, 41. *Prof Exp:* Chemist, Tenn Valley Authority, 41-43, Metall Lab, 43-45, Monsanto Chem Co, 45; Standard Oil Ind, 45-46; chemist, 46-48, assoc chemist, 48-58, sr chemist, 58-71, dir, Chem Div, 71-80, assoc lab dir, Physical Res, 80-81, dir, Sci Support Div, 82-83, SR CHEMIST, ARGONNE NAT LAB, 83- *Concurrent Pos:* Mem, Adv Comt Transplutonium Elements, 58-76; mem, Nat Acad Sci Ad Hoc Comt Heavy Ion Facil & Res, 70-71, Ad Hoc Comt Trans Plutonium Elements, 83; consult, Simon & Schuster, 83- *Honors & Awards:* Nuclear Appln Award, Am Chem Soc, 70. *Mem:* Am Chem Soc; Am Phys Soc; fel Am Nuclear Soc; fel AAAS. *Res:* Heavy element chemistry; nuclear chemistry; chemistry of rare gas compounds; lunar soil analysis. *Mailing Add:* 7308 N California Ave Chicago IL 60645

FIELDS, R(ANCE) WAYNE, b Tulsa, Okla, Feb 1, 41; m 67; c 1. BUSINESS DEVELOPMENT, STARTUP & BUSINESS FUNDING. *Educ:* Ore State Univ, BS, 63; Ore Health Sci Univ, PhD(neurophysiol), 69. *Prof Exp:* Staff scientist, NASA, 69-70; dir, Biophys Lab, Ore Health Sci Univ, 70-78; mgr advan develop, B-D Drake Willock, 78-84; vpres, Premium Equity Corp, 84-87; PRES & CO-OWNER, VENTURE SOLUTIONS, LTD, 87- *Concurrent Pos:* Consult bus develop, 84-; vpres res & develop, Electronic Med Instruments, Inc, 88-; dir, Enhanced Facil Mgt, Inc, 88-, Optec Wall, Inc, 90-, Tactron, Inc, 90- & Diesel Systs, Ltd, 91-; dir & treas, Univend, Inc, 90- *Res:* Bringing new products to market, design, production, marketing, strategic management, venture funding; author and co-author of ten federal grants; writer of acclaimed business plans. *Mailing Add:* 6490 Chessington Lane Gladstone OR 97027-1011

FIELDS, REUBEN ELBERT, b Society Hill, SC, Sept 30, 16; m 45; c 3. NUCLEAR SCIENCE, NUCLEAR STRUCTURE. *Educ:* Ga Inst Technol, BS, 40; Univ Wis, MS, 51, PhD(physics), 54. *Prof Exp:* Assoc engr, Rural Electrification Admin, 40-42; physicist nuclear physics, Argonne Nat Lab, 44-46 & 47-49; STAFF SCIENTIST NUCLEAR PHYSICS, FT WORTH DIV, GEN DYNAMICS CORP, 53- *Concurrent Pos:* Lectr, Southern Methodist Univ; adj prof, Tex Christian Univ. *Mem:* Am Phys Soc; Am Nuclear Soc; Am Asn Physics Teachers; Sigma Xi. *Res:* Photoneutron sources; neutron cross sections; nuclear forces; nuclear reactor hazards; analysis of survivability/vulnerability of military aircraft subjected to nuclear weapon threat fields, including blast, thermal, neutron, gamma and x-radiation, and electromagnetic pulse. *Mailing Add:* 4132 Clayton Rd W Ft Worth TX 76116

FIELDS, RICHARD JOEL, b Philadelphia, Pa, Nov 21, 47; m 75; c 4. METALLURGY, CHEMISTRY. *Educ:* Univ Pa, BA & Bsc, 71; Harvard Univ, MSc, 73; Cambridge Univ, PhD(eng), 77. *Prof Exp:* Res assoc metall, 77-79, METALLURGIST, NAT BUR STANDARDS, 79- *Concurrent Pos:* Consult, Automotive Training Ctr, 76- *Honors & Awards:* Bronze Medal Nat Bur of Standard. *Mem:* Am Soc Metals; Am Soc Test & Mat; Am Ceramic Soc. *Res:* High temperature fracture and effects of inclusions on strength of metals; dynamic fracture and crack arrest in pressure vessel steel; precipitation kinetics in high-strength low-alloy steel. *Mailing Add:* 15812 Amelung Lane Rockville MD 20855

FIELDS, ROBERT WILLIAM, b San Leandro, Calif, Sept 17, 20; m 45; c 2. PALEONTOLOGY, SEDIMENTOLOGY. *Educ:* Univ Calif, BA, 49, PhD(vert paleont), 52. *Prof Exp:* Asst teacher vert paleont, Univ Calif, 49-51; geologist, Shell Oil Co, 52-55; from asst prof to prof, Univ Mont, 55-82, chmn dept, 64-70 & 77-79, emer prof geol, 82-; RETIRED. *Concurrent Pos:* Field leader exped, Colombia, SAm, Univ Calif; geol consult, var petrol & mineral co, 70- *Mem:* Geol Soc Am. *Res:* Evolution, sedimentology and paleoecology of Tertiary intermontane basins of the Northern Rocky Mountain province; mammalian vertebrate paleontology; evolution of South American hystricomorph rodents; comparative anatomy and evolution of Tertiary mammals. *Mailing Add:* 5005 Doon Way Anacortes WA 98221

FIELDS, THEODORE, b Chicago, Ill, Jan 23, 22; m 45, 85; c 3. MEDICAL PHYSICS. *Educ:* Univ Chicago, BS, 42; DePaul Univ, MS, 53; Am Bd Radiol, dipl, 50; Am Bd Health Physics, dipl, 60, Am Bd Med Physics, dipl, 90. *Prof Exp:* Physicist, Manhattan Proj, Univ Chicago, 43-45; chief engr, Precision Radiation Instrument Co, 48-49; instr radiation physics, Med Sch, Northwestern Univ, 49-61; asst prof, 61-74, clin assoc prof radiol, Stritch Sch Med, Loyola Univ, 74-78; lectr radiol, Univ Chicago, 80-87; LECTR RADIOL, RUSH MED SCH, 87- *Concurrent Pos:* Chief physics sect, Vet Admin Hines Hosp, 49-65; physicist, Radiation Ctr, Cook County Hosp,

Chicago, 58-85; pres, Health Physics Assocs, 60-; pres, Isotope Measurement Labs, Highland Park, 69-87; pres, Mobile Imaging Labs, Highland Park, 77-87. *Mem:* Am Phys Soc; Am Pub Health Asn; Inst Elec & Electronics Engrs; fel Am Col Radiol; Sigma Xi; Am Asn Physicists Med; Am Col Med Physicists. *Res:* Radiation safety in medicine, industry, and teaching; quality control evaluations of x-ray diagnostic procedures and nuclear medicine. *Mailing Add:* 1141 Hohlfelder Rd Glencoe IL 60022-1018

FIELDS, THOMAS HENRY, b Kearny, NJ, Oct 23, 30; m 58; c 4. EXPERIMENTAL PHYSICS, COMPUTERS. *Educ:* Carnegie-Mellon Univ, BS, 51, PhD(physics), 55. *Prof Exp:* Assoc prof physics, Carnegie-Mellon Univ, 58-60; prof physics, Northwestern Univ, 60-69; dir, High Energy Physics Div, Argonne Nat Lab, 65-74, assoc lab dir, 74-77, div dir, 85-89, SR PHYSICIST, ARGONNE NAT LAB, 77-85, 89- *Concurrent Pos:* Vis scientist, Univ Birmingham, 70-71 & Europ Orgn Nuclear Res, Switzerland, 77-78. *Mem:* Am Phys Soc. *Res:* Experimental study of antiproton annihilation, large transverse momentum hadron reactions and proton decay. *Mailing Add:* 5749 Dearborn Downers Grove IL 60516

FIELDS, THOMAS LYNN, b Baltimore, Md, Oct 12, 25; m 47; c 2. ORGANIC CHEMISTRY. *Educ:* Ala Polytech Inst, BS, 52, MS, 54. *Prof Exp:* RES CHEMIST, LEDERLE LABS, AM CYANAMID CO, 54- *Mem:* Am Chem Soc. *Res:* Chemistry of natural products; total synthesis and chemical modifications of tetracyclines; medicinal chemistry. *Mailing Add:* Lederle Labs Pearl River NY 10965-1299

FIELDS, WILLIAM GORDON, invertebrate zoology; deceased, see previous edition for last biography

FIELDS, WILLIAM STRAUS, b Baltimore, Md, Aug 18, 13; m 41, 74; c 3. NEUROLOGY. *Educ:* Harvard Univ, AB, 34, MD, 38; Am Bd Psychiat & Neurol, dipl, 50. *Prof Exp:* From assoc prof to prof, Baylor Col Med, 49-67, chmn dept, 59-65; prof, Univ Tex Med Sch Dallas, 67-69; prof, Univ Tex Grad Sch Biomed Sci Houston, 69-70; prof neurol, Med Sch, Univ Tex, Houston, 70-82, chmn dept, 72-82, chmn dept, M D Anderson Cancer Ctr, 82-88, PROF, UNIV TEX, M D ANDERSON CANCER CTR, 82- *Concurrent Pos:* Rockefeller fel, Sch Med, Washington Univ, St Louis, 46-49; consult neurologist, St Luke's Episcopal Hosp, Tex Children's Hosp; chief neurol serv, Hermann Hosp, St Anthony Ctr, Houston, 70-79; mem exec comt & study group chmn, Joint Comt Stroke Res, Washington, DC, 68-76; mem adv comt, Psychiat, Neurol & Psychol Serv, Vet Admin, 66-74; consult, Hearings & Appeals Bds, Social Security Admin, US Dept Health, Educ & Welfare, 65-78; mem exec comt, Stroke Coun Am Heart Asn, 85-88. *Honors & Awards:* Order of Brit Empire, 46. *Mem:* Am Acad Neurol; Am Asn Neurol Surg; Am Epilepsy Soc; AMA; Brazilian Acad Neurol; Can Neurol Soc; Korean Med Asn; Neurol Soc Colombia; Psychiat, Neurol & Neurosurg Soc Peru; World Fed Neurol Res Group Cerebral Circulation. *Res:* Epilepsy; cerebrovascular disease; anti-thrombotic drugs; neuro-oncology. *Mailing Add:* M D Anderson Cancer Ctr Univ Tex Houston TX 77030

FIENBERG, STEPHEN ELLIOTT, b Toronto, Ont, Nov 27, 42; m 65; c 2. STATISTICS. *Educ:* Univ Toronto, BSc, 64; Harvard Univ, AM, 65, PhD(statist), 68. *Prof Exp:* Instr psychol, Wellesley Col, 68; asst prof statist & theoret biol, Univ Chicago, 68-72; assoc prof, Univ Minn, 72-76, chmn dept, 72-78, prof appl statist, 78-80; head, dept statist, 81-84, prof, 80-85, MAURICE FALK PROF STATIST & SOCIAL SCI, CARNEGIE MELLON UNIV, 85-, DEAN, COL HUMANITIES & SOCIAL SCI, 87- *Concurrent Pos:* Mem adv & planning comt social indicators, Social Sci Res Coun, 72-77 & bd dirs, 80-86; vis assoc dir, Ctr Health Practs, Sch Pub Health, Harvard Univ & vis lectr, Dept Statist, 75-76; coord & appl ed, J Am Statist Asn, 77-79, co-ed, Chance, 87-; mem bd dirs, Social Sci Res Coun, 80-86 & Nat Inst Statist Sci, 90-; chmn, Comt Nat Statist, Nat Res Coun, Nat Acad Sci, 81-87, chair Sect U, AAAS, 81; fel, Ctr for Advan Study in the Behav Sci, 84-85; Guggenheim fel, 84-85; Berman vis prof, Hebrew Univ, Jerusalem, 89-90. *Mem:* Fel AAAS; fel Am Statist Asn (vpres, 86-88); Biomet Soc; fel Inst Math Statist; Psychomet Soc; Statist Soc Can; Int Statist Inst; fel Inst Statisticians; fel Royal Statist Soc. *Res:* Analysis of cross classified data; Bayesian inference; data analysis; stochastic modelling; federal Statistics; statistics and the law; cognitive aspects of survey design. *Mailing Add:* Dept Statist Carnegie Mellon Univ Pittsburgh PA 15213-3890

FIENUP, JAMES R, b St Louis, Mo, Apr 17, 48; m 70; c 4. INFORMATION PROCESSING, HOLOGRAPHY. *Educ:* Holy Cross Col, BA, 70; Stanford Univ, MS, 72, PhD(appl physics), 75. *Prof Exp:* Res asst, Stanford Electronics Lab, Stanford Univ, 72-75; RES PHYSICIST, ENVIRON RES INST MICH, 75- *Concurrent Pos:* NSF grad fel, 70-72; assoc ed, Optics Lett, 83-85. *Honors & Awards:* Rudolf Kingslake Medal & Prize, Soc Photo-Optical Instrumentation Engrs, 80; Int Prize Optics, Int Comn Optics, 83. *Mem:* Fel Optical Soc Am; fel Soc Photo-Optical Instrumentation Engrs; Sigma Xi; Inst Elec & Electronics Engrs. *Res:* Digital and coherent optical image processing; phase retrieval image reconstruction; holographic optical elements; computer-generated holograms. *Mailing Add:* Environ Res Inst Mich PO Box 8618 Ann Arbor MI 48107-8618

FIERER, JOSHUA A, b New York, NY, Nov 25, 37; m 59; c 4. PATHOLOGY, IMMUNOPATHOLOGY. *Educ:* Alfred Univ, BA, 59; State Univ NY Downstate Med Ctr, MD, 63. *Prof Exp:* Asst surg, Sch Med, Univ Rochester, 64-65; resident path, Columbia-Presby Med Ctr, 67-68, vis fel, 68-70; asst prof path, Columbia Univ Col Physicians & Surgeons, 70-76; prof path, Sch Med, Creighton Univ, 76-78; PROF & CHMN DEPT PATH, COL MED, UNIV ILL, PEORIA, 78- *Concurrent Pos:* Vis lectr path, Sch Dent, Fairleigh Dickinson Univ, 68-72; guest investr immunol, Rockefeller Univ, 69-70; asst med examr, Off Chief Med Examr, City of New York, 69-76; dir immunopath, Columbia-Presby Med Ctr, 72-76; consult renal path, Holy Name Hosp, 72-76; consult immunopath, Kidney Hypertension Unit, Lenox Hill Hosp, 73-75; assoc dir labs, Francis Delafield Div, Columbia-Presby Med Ctr, 74-75; dir, Div Anat Path, Creighton-Omaha Med Ctr, 76-78; dir lab, Col Med, Univ Ill, 78- *Mem:* Am Asn Pathologists & Bacteriologists; Am Asn Immunologists; fel Col Am Pathologists; Int Acad Path; Sigma Xi. *Res:* Ultrastructure of lung connective tissue; tumor associated antigens; proteolytic enzymes and enzyme inhibitors; cellular aspects of aging. *Mailing Add:* PO Box 6365 Peoria IL 61601

FIERING, MYRON B, b New York, NY, Apr 5, 34; m 57; c 2. CIVIL ENGINEERING, APPLIED MATHEMATICS. *Educ:* Harvard Univ, AB, 55, SM, 58, PhD(eng), 60. *Prof Exp:* Soils engr, Tippetts, Abbott, McCarthy, Stratton, 55-57; asst prof eng, Univ Calif, Los Angeles, 60-61; res fel environ eng, 61-62, lectr appl math, 62-63, asst prof eng & appl math, 63-70, GORDON McKAY PROF ENG & APPL MATH, HARVARD UNIV, 70- *Concurrent Pos:* Consult, USPHS, 63-; Fulbright lectr & vis assoc prof, Univ New South Wales, Australia, 64. *Honors & Awards:* Gold Medal, Am Asn Sanitary Engrs, 62. *Mem:* Am Soc Civil Engrs; Am Geophys Union; Sigma Xi; Int Asn Sci Hydrol. *Res:* Water resources; statistics. *Mailing Add:* Appl Sci Pierce Hall Harvard Univ Cambridge MA 02138

FIERO, GEORGE WILLIAM, JR, b Buffalo, NY, Jan 16, 36; m 59; c 3. GEOLOGY, ENVIRONMENTAL SCIENCES. *Educ:* Dartmouth Col, BA, 57; Univ Wyo, MS, 59; Univ Wis, PhD(geol), 68. *Prof Exp:* Explor geologist, Texaco, Inc, 59-61, exploitation geologist, 61-63; teacher geol, Mt Hermon Sch, 63-66; PROF GEOL, UNIV NEV, LAS VEGAS, 68- *Honors & Awards:* Stanley A Tyler Award, Univ Wis, 68. *Mem:* AAAS; Am Asn Petrol Geol; Am Geophys Union; Am Water Resources Asn; Geol Soc Am. *Res:* Ground water effects on geothermal gradients; remote sensing in water resource studies; hydrogeological aspects of water pollution; regional geology of the Great Basin; geology of the national parks. *Mailing Add:* 527 Hotel Plaza Boulder City NV 89005

FIERSTINE, HARRY LEE, b Long Beach, Calif, Aug 14, 32; m 58; c 3. COMPARATIVE ANATOMY, ICHTHYOLOGY. *Educ:* Long Beach State Col, BS, 57; Univ Calif, Los Angeles, MA, 61, PhD(zool), 65. *Prof Exp:* From instr to asst prof zool, Calif State Col Long Beach, 64-66; from asst prof to assoc prof, 66-74, PROF ZOOL, CALIF POLYTECH STATE UNIV, SAN LUIS OBISPO, 74-, AT SCH SCI & MATH. *Mem:* Am Soc Zool; Am Soc Ichthyol & Herpet; Soc Vert Paleont; Sigma Xi. *Res:* Functional fish anatomy; tertiary fish paleontology. *Mailing Add:* Sch Sci & Math Calif Polytech State Univ San Luis Obispo CA 93407

FIESINGER, DONALD WILLIAM, b Corning, NY, Aug 4, 43; m 67; c 2. IGNEOUS PETROLOGY. *Educ:* State Univ NY Col, Potsdam, BA, 66; Wayne State Univ, MS, 69; Univ Calgary, PhD(geol), 75. *Prof Exp:* Asst prof geol, State Univ NY Col, New Paltz, 74-76; asst prof, 76-82, ASSOC PROF GEOL & HEAD DEPT, UTAH STATE UNIV, 82- *Concurrent Pos:* Investr fac res fel & grant-in-aid, Res Found State Univ NY, 76-77; co-investr fac res grant, Utah State Univ, 77-79 & 88-89; co-investr, US Geol Surv res grant, 78-79; Indo-Am fel, 84. *Mem:* Geol Soc Am; Am Geophys Union; Geol Asn Can; Mineral Asn Can; Geol Soc India. *Res:* Chemical evolution of magma systems; study of Tertiary volcanism in the Basin and Range Province; computer modeling of igneous processes. *Mailing Add:* Dept Geol Utah State Univ Logan UT 84322-4505

FIESLER, EMILE, b Amsterdam, Neth. NEURAL COMPUTING, GRAPH THEORY. *Educ:* Univ Amsterdam, BS, 80, BS & MS, 86, PhD(computer sci), 91. *Prof Exp:* Internship, User Interface Mgt Systs, In Coop with Philips, Intern, Eindhoven, 85-86; RES FEL GRA NEURAL COMPUT, UNIV ALA HUNTSVILLE, 86-91. *Concurrent Pos:* Prog comt mem, Int Asn Sci & Technol Develop, 89-; Comt Neural Network Standardization, Inst Elec & Electronics Engrs, 91- *Mem:* Inst Elec & Electronics Engrs; Soc Photo-optical Instrumentation Engrs. *Res:* Topological, architectural, graph theoretical, computational, formalization and implementation aspects of neural networks; neural computing; connectionism; mathematical optimization and modeling; massive parallelism; networking; user interface management systems. *Mailing Add:* Computer Sci Dept Univ Ala Huntsville AL 35899

FIESS, HAROLD ALVIN, b Ringoes, NJ, Apr 28, 17; m 43; c 4. CHEMISTRY. *Educ:* Wheaton Col, Ill, BS, 39; Univ Ill, MS, 42, PhD(chem), 44. *Prof Exp:* Chemist, Am Cyanamid Co, NJ, 39-40; asst chem, Univ Ill, 40-43; from instr to prof, 44-85, chmn dept, 75-80, EMER PROF CHEM, WHEATON COL, 85- *Concurrent Pos:* Res assoc, Northwestern Univ, 49-58. *Mem:* Am Chem Soc. *Res:* Analytical chemistry. *Mailing Add:* Dept Chem Wheaton Col Wheaton IL 60187

FIETZ, WILLIAM ADOLF, b Corinna, Maine, Sept 20, 31; m 60; c 2. LOW TEMPERATURE PHYSICS. *Educ:* Cornell Univ, BEE, 57, MS, 63, PhD(appl physics), 67. *Prof Exp:* Field engr, Fla Power & Light Co, Miami, 57-60; res physicist, Linde Div Lab, Union Carbide Corp, NY, 66-71; staff scientist, Intermagnetics Gen Corp, 71-74; group leader, Oak Ridge Nat Lab, 74-90; DEP DEPT HEAD CRYOGENICS, SUPERCONDUCTING SUPERCOLLIDER LAB, 90- *Mem:* AAAS; Am Phys Soc; Inst Elec & Electronics Eng. *Res:* Superconductivity of high field-high current materials; properties of plasma-plated metal powders. *Mailing Add:* 1404 Horton Dr Cedar Hill TX 75104

FIEVE, RONALD ROBERT, b Stevens Point, Wis, Mar 5, 30; m 63; c 2. PSYCHIATRY. *Educ:* Univ Wis, BMS, 51; Harvard Med Sch, MD, 55; Am Bd Psychiat & Neurol, dipl, 62. *Prof Exp:* Intern, Columbia Med Div, Bellevue Hosp, 55-56; asst resident internal med, NY Hosp, Cornell Univ, 56-57; resident psychiat, NY State Psychiat Inst, 57-60; instr, 60-62 & assoc, 64-67, from asst prof to assoc prof, 67-73, PROF CLIN PSYCHIAT, COL PHYSICIANS & SURGEONS, COLUMBIA UNIV, 73-; CHIEF PSYCHIAT RES, NY STATE PSYCHIAT INST, 76- *Concurrent Pos:* NIMH res grant biogenetics of depression, 81-; dir personnel med clin & acute psychiat serv, NY State Psychiat Inst, 60-; prin res scientist & chief psychiat res, Lithium Clin, 62-, dir metab res serv, 63-; mem, WHO, 60-; asst exam, Am Bd Psychiat & Neurol, 66- *Honors & Awards:* Richard H Hutchings Award. *Mem:* AMA; Am Psychiat Asn; Acad Psychoanal; Am Col

Neuropsychopharmacol; Am Psychopath Asn. *Res:* Behavioral and biological psychiatry; lithium; manic depressive illness and psychopharmacology; depressive disorders: evaluation and treatment; evaluation of new antidepressive and antianxiety compounds; alcohol, cocaine, substance abuse; complicating depressive disorders. *Mailing Add:* Found Depression 161 Ft Washington Ave New York NY 10032

FIFE, PAUL CHASE, b Cedar City, Utah, Feb 14, 30; m 59; c 3. MATHEMATICS, CONTINUUM MECHANICS. *Educ:* Univ Chicago, BA, 51; Univ Calif, Berkeley, BA, 53; NY Univ, PhD(math), 59. *Prof Exp:* From instr to asst prof math, Stanford Univ, 59-63; from asst prof to assoc prof, Univ Minn, 63-68; prof math, Univ Ariz, 69-88; PROF MATH, UNIV UTAH, 88- *Concurrent Pos:* Res grants, Off Naval Res, Air Force Off Sci Res & NSF, 59-; US Dept State Fulbright res grant, Ger, 66-67; teaching grant, Peru, 71; sr res fel, UK Sci Res Coun, 74-75, 82-83; comnr, Sch Comn with People's Repub China, 83. *Mem:* Am Math Soc; Soc Natural Philos; Soc Indust & Appl Math. *Res:* Various aspects of partial differential equations of elliptic and parabolic types; singular perturbations of partial differential equations; reacting and diffusing systems; phase changes. *Mailing Add:* Dept Math Univ Utah Salt Lake City UT 84112

FIFE, THOMAS HARLEY, b Oak Park, Ill, Feb 6, 31; m 58; c 2. PHYSICAL ORGANIC. *Educ:* Univ Ill, BS, 55; Univ Minn, PhD(chem), 59. *Prof Exp:* Fel, Johns Hopkins Univ, 59-60, Cornell Univ, 60-62; from asst prof to assoc prof, 62-70, PROF BIOCHEM, UNIV SOUTHERN CALIF, 70- *Mem:* Am Chem Soc; Am Soc Biol Chemists. *Res:* Kinetics and mechanisms of chemical and enzymatic reactions. *Mailing Add:* Dept Biochem Univ Southern Calif Mudd-612 2025 Zonal Ave Los Angeles CA 90033

FIFE, WILLIAM PAUL, b Plymouth, Ind, Nov 23, 17; m 47; c 2. PHYSIOLOGY, ANATOMY. *Educ:* Univ Ore, BS, 56; George Washington Univ, 56-58; Ohio State Univ, PhD(physiol), 62. *Prof Exp:* Asst chief aerospace med res div, USAF Sch Aerospace Med, 62-67; asst dir, Inst Life Sci, 67-77, PROF BIOL, TEX A&M UNIV, 70-, ASSOC DEAN, COL SCI & DIR, HYPERBARIC LAB, 77- *Concurrent Pos:* Actg head dept biol, Tex A&M Univ, 67-70 & 73-74; interim head dept, 80-81. *Mem:* Am Physiol Soc; Aerospace Med Asn; Undersea Med Soc. *Res:* Cardiovascular and diving physiology, including environmental physiology; study of deep diving and use of hydrogen-oxygen mixtures for diving. *Mailing Add:* 2350 W Briargate Bryan TX 77801

FIFE, WILMER KRAFFT, b Wellsville, Ohio, Oct 19, 33; m 59; c 3. ORGANIC SYNTHESIS, POLYMER CATALYSTS. *Educ:* Case Inst Technol, BSc, 55; Ohio State Univ, PhD(org chem), 60. *Prof Exp:* From asst prof to prof chem, Muskingum Col, 60-71 & chmn dept, 66-71; chmn dept, 71-80, PROF CHEM, IND UNIV-PURDUE UNIV, INDIANAPOLIS, 71- *Concurrent Pos:* Res grants, Res Corp, 60-61, 83-85, NSF, 62-66, 71 & 75, Off Naval Res 88-; NIH spec fel, Harvard Univ, 65-66 & Chandler Lab, Columbia Univ, 68-69; prin res fel, Am Chem Soc, 86-88. *Mem:* AAAS; Am Chem Soc. *Res:* New synthetic methodology with emphasis on multiple-phase media; new routes to pyridine derivatives; transacylation reactions; polymer-mediated organic synthesis. *Mailing Add:* Dept Chem Ind Univ-Purdue Univ 1125 E 38th St Indianapolis IN 46205-2810

FIFER, ROBERT ALAN, b Abington, Pa, Nov 25, 43; m 65, 80; c 3. PHYSICAL & ANALYTICAL CHEMISTRY & SPECTROSCOPY. *Educ:* Gordon Col, BS, 65; Temple Univ, PhD(phys chem), 69. *Prof Exp:* Res fel chem, Cornell Univ, 69-71; sr res assoc, Boston Col, 71-72; res chemist, US Army Frankford Arsenal, 72-76, RES CHEMIST, US ARMY BALLISTIC RES LAB, 77- *Concurrent Pos:* NSF presidential internship, 72. *Mem:* Am Chem Soc; Am Phys Soc; NAm Thermal Anal Soc; Soc Appl Spectros. *Res:* High temperature chemical kinetics behind shock waves; kinetics and modelins of high energy materials and air polluting systems; analytical chemistry techniques for combustion and pyrolysis diagnostics; vibrational spectroscopy and fourier transform infrared techniques; surface spectroscopy; thermodynamics. *Mailing Add:* Ballistic Res Lab Int Ballistics Div Bldg 390 Aberdeen Proving Ground MD 21005

FIFKOVA, EVA, b Prague, Czech, May 21, 32; US citizen. NEUROCYTOLOGY, NEURONAL PLASTICITY. *Educ:* Sch Med, Charles Univ, Prague, Czech, MD, 57; Inst Physiol, Czech Acad Sci, PhD(physiol), 63. *Prof Exp:* Lectr anat, Sch Med, Prague, Czech, 57-60; res fel physiol, Czech Acad Sci, 60-63, staff mem, 63-68; sr res fel biol, Calif Inst Technol, 68-72, res assoc, 72-74; from asst prof to assoc prof, 74-78, PROF PSYCHOL, UNIV COLO, BOULDER, 78- *Concurrent Pos:* Mem, Cajal Club, 80-, neurobiol adv Panel, NSF, 82-85, Alcohol Biomed Res Rev Comt, NIAAA, 88-89, Neurol A Study Sect, NIH, 90-94. *Mem:* AAAS; Am Physiol Soc; Soc Neurosci; Am Asn Anatomists; Electron Micros Soc Am; Int Brain Res Orgn. *Res:* Plastic changes in the fine structure of the nervous tissue with respect to functional requirements; contractile proteins like actin, myosin and cytosolic calcium are involved in the phenomena of neuronal plasticity; actin and calcium are being studied in the hippocampus following electrical stimulation, alcohol treatment and during aging. *Mailing Add:* Dept Psychol Univ Colo Campus Box 345 Boulder CO 80309

FIGDOR, SANFORD KERMIT, b New York, NY, Mar 3, 26; m 48; c 2. ORGANIC CHEMISTRY, DRUG METABOLISM. *Educ:* NY Univ, BA, 49; Univ New Brunswick, MS, 51, PhD(chem), 53. *Prof Exp:* Am Heart Asn fel, Wayne State Univ, 53-54; res chemist, 54-68, head radiochem, Cent Res, 68-86, ADMINR COMPLIANCE & EDUC, PFIZER INC, 86- *Honors & Awards:* Outstanding Achievement Award, Int Soc Study Xenobiotics, 89. *Mem:* Am Chem Soc; Int Soc Study Xenobiotics; Int Isotope Soc. *Res:* Natural products; alkaloids; steroids; organic synthesis medicinals; radiochemistry. *Mailing Add:* Cent Res Pfizer Inc Groton CT 06340

FIGGE, DAVID C, b Twin Falls, Idaho, May 20, 25; m 51; c 2. OBSTETRICS & GYNECOLOGY. *Educ:* Northwestern Univ, BS & MD, 50. *Prof Exp:* Am Cancer Soc fel gynec, 54-56; res fel obstet & gynec, 55-56, from instr to assoc prof, 56-69, PROF OBSTET & GYNEC, SCH MED, UNIV WASH, 70-, DIR GYNEC, UNIV HOSP, 69-, DIR GYNEC ONCOL, 74- *Concurrent Pos:* Obstetrician-gynecologist-in-chief, King's County Hosp, 63-69. *Res:* Investigation and management of gynecologic cancer; tissue culture investigations in uterine tissues. *Mailing Add:* 1959 Pacific NE Univ Hosp Seattle WA 98105

FIGLEY, MELVIN MORGAN, b Toledo, Ohio, Dec 5, 20; m 46; c 3. RADIOLOGY. *Educ:* Harvard Univ, MD, 44. *Prof Exp:* From asst prof to assoc prof radiol, Univ Mich, 48-58; prof radiol & chmn dept, 58-78, PROF RADIOL & MED, UNIV WASH, 79- *Concurrent Pos:* Markle Found scholar, 53-58; mem bd trustees, Am Bd Radiol, 67-73; mem bd dirs, James Picker Found, 70-79; ed, Am J Roentgenology, 76-85; mem, Radiation Study Sect, NIH; chmn comt radiol, Nat Acad Sci-Nat Res Coun, 64-70. *Honors & Awards:* Gold Medal, Radiol Soc NAm, Asn Univ Radiologists, Am Roentgen Ray Soc & Am Col Radiol. *Mem:* Radiol Soc NAm; Fleischner Soc; Am Roentgen Ray Soc; hon mem Royal Soc Med; Asn Univ Radiologists; fel Royal Col Radiol; fel Royal Australasian Col Radiol Sydney. *Res:* Thoracic and cardiovascular radiology. *Mailing Add:* PO Box 10100 Bainbridge Island WA 98110

FIGUEIRA, JOSEPH FRANKLIN, b Paris, Ill, Mar 7, 43; m 65; c 2. LASERS PHYSICS. *Educ:* Univ Ill, BS, 65; Cornell Univ, MS & PhD(physics), 71. *Prof Exp:* Staff mem, 71-75, alt group leader, 75-76, group leader, 76-88, PROG MGR, LOS ALAMOS NAT LAB, 88- *Mem:* Am Phys Soc; Optical Soc Am. *Res:* Ultraviolet, visible and infrared lasers; laser systems and laser effects in direct support of Los Alamos programs in inertial confinement fusion; advanced isotope separation; directed energy weapons. *Mailing Add:* Rte 5 Box 322-Q Santa Fe NM 87501

FIGUERAS, JOHN, b Rochester, NY, Oct 28, 24; m 70; c 4. STRUCTURE ELUCIDATION. *Educ:* Univ Rochester, BS, 49; Univ Ill, MS, 50, PhD(chem), 52. *Prof Exp:* Instr chem, Ill Inst Technol, 52-53; sr res assoc, Eastman Kodak Co, 53-83; ASST PROF, STATE UNIV NY, GENESEO, 83- *Mem:* AAAS; Am Chem Soc. *Res:* Computerized chemical information systems for personal computers. *Mailing Add:* 65 Steele Rd Victor NY 14564

FIGUERAS, PATRICIA ANN MCVEIGH, organic chemistry, mathematics; deceased, see previous edition for last biography

FIGUEROA, WILLIAM GUTIERREZ, b El Paso, Tex, May 25, 21; m 46; c 4. INTERNAL MEDICINE. *Educ:* Tex Western Col, AB, 42; St Louis Univ, MD, 46; Am Bd Internal Med, dipl. *Prof Exp:* Intern, Robert B Green Hosp, San Antonio, Tex, 46 & Los Angeles County Gen Hosp, 46-47; resident internal med, Vet Admin Hosp, Los Angeles, 50-53; jr res metabolist, 52-53, from instr to asst prof, 53-61, assoc prof, 61-75, PROF MED, UNIV CALIF, LOS ANGELES, 75- *Concurrent Pos:* Attend physician, Vet Admin Hosp, Los Angeles, 53- *Mem:* AMA; Sigma Xi. *Res:* Diseases of metabolism; iron metabolism in various diseases states using radioactive iron; the nature and mechanism of anemia in various blood dyscrasias. *Mailing Add:* 453 Loring Ave Los Angeles CA 90024

FIGURSKI, DAVID HENRY, b Erie, Pa, Apr 12, 47; m 69; c 2. MOLECULAR BIOLOGY. *Educ:* Univ Pittsburgh, BS, 69; Univ Rochester, PhD(microbiol), 74. *Prof Exp:* USPHS fel molecular biol, Univ Calif, San Diego, 74-78; asst prof, 78-85, ASSOC PROF MICROBIOL, COL PHYSICIANS & SURGEONS , COLUMBIA UNIV, 85- *Concurrent Pos:* Ed bd, J Bacteriol, 85-; adv comt Microbiol & Virol, Am Cancer Soc, 87-90; fac res award, Am Cancer Soc, 85-90; ed, Plasmid, 90- *Honors & Awards:* Lambert Award for Basic Research, 84. *Mem:* Am Soc Microbiol; AAAS. *Res:* Replication of extrachromosomal elements in bacteria; broad host range R-plasmids. *Mailing Add:* Dept Microbiol Columbia Univ New York NY 10032

FIGWER, J(OZEF) JACEK, b Mielec, Poland, Mar 16, 28; US citizen; m 55; c 2. ACOUSTICS, ELECTROACOUSTICS. *Educ:* Silesian Polytech Inst, Gliwice, Poland, MS, 51; NIKFI, Moscow, USSR, DS, 58. *Prof Exp:* Supervry consult acoustics, Bolt Beranek Newman, Inc, 62-78; PRIN CONSULT ACOUSTICS, JACEK FIGWER ASSOC, INC, 78- *Concurrent Pos:* Consult, Inst Elec & Electronics Engrs, Inc, 72-75. *Mem:* fel Acoust Soc Am; fel Audio Eng Soc. *Res:* Architectural acoustics. *Mailing Add:* Jacek Figwer Assoc Inc 85 The Valley Rd Concord MA 01742

FIKE, HAROLD LESTER, b Toledo, Ohio, Jan 30, 26; m 56; c 4. ORGANIC CHEMISTRY. *Educ:* Univ Toledo, BS, 50; Northwestern Univ, MBA, 63. *Prof Exp:* Res chemist, Int Mineral & Chem corp, 52-56, sr res chemist, 56-59, asst dir res, 60-63, economist, 63-68; dir res, 68-80, vpres, 80-84, PRES, SULPHUR INST, 84- *Mem:* Am Chem Soc; Chem Mkt Res Asn. *Res:* Development of processes for the recovery of amino acids from natural sources and resolution of racemic mixtures of amino acids; mechanisms of sulfur reactions and development of new uses for sulfur. *Mailing Add:* 1894 Milboro Rockville MD 20854

FIKE, WILLIAM THOMAS, JR, b McKeesport, Pa, Nov 22, 28; m 52; c 1. AGRONOMY. *Educ:* Pa State Univ, BS, 52, MS, 56; Univ Minn, PhD(agron), 62. *Prof Exp:* From asst prof to assoc prof, 59-76, PROF CROP SCI, NC STATE UNIV, 76- *Mem:* Am Soc Agron. *Res:* Introduction, evaluation and improvement of new and established crops; oil, fiber, forage and feed for industrial and agricultural uses; maximizing net return per acre for corn, sorghum and small grains. *Mailing Add:* Box 7620 Crop Sci NC State Univ Raleigh Main Campus Raleigh NC 27695-7620

FIKE, WINSTON, b Pittsfield, Mass, Feb 15, 21; m; c 4. ANALYTICAL CHEMISTRY. *Educ:* NY State Col, Albany, AB, 41; Rensselaer Polytech Inst, PhD(chem), 50. *Prof Exp:* Analyst pharmaceut control, Sterling-Winthrop Corp, 41-45; anal chemist, Tenn Eastman Corp, 45-46; asst prof org chem, Thiel Col, 50-52; res chemist, Ansco Div, Gen Aniline & Film Corp, 52-55; asst prof anal chem, Utica Col, 55-63; res assoc chromatog, County Coroner's Off, 63-65; chemist Anal Methods Develop, Merrell Dow Pharm, 65-86; RETIRED. *Concurrent Pos:* NIH grant, 63-65; consult, Sch Med, Case Western Reserve Univ, 63-65. *Mem:* Sigma Xi. *Res:* Thin layer, gas, and liquid chromatography. *Mailing Add:* 7507 Loannes Dr Cincinnati OH 45243

FILACHIONE, EDWARD MARIO, chemistry; deceased, see previous edition for last biography

FILADELFI-KESZI, MARY ANN STEPHANIE, b Montreal, Que, Sept 2, 52; m 85. SENSORY EVALUATION, FLAVOR CHEMISTRY & FOOD TOXICOLOGY. *Educ:* Concordia Univ, BSc, 74; McGill Univ, MSc, 77; Univ Guelph, PhD(hort sci), 80. *Prof Exp:* Asst prof res, McGill Univ, 80-83, univ lectr, 83-85; asst prof, Dept Food Sci & Human Nutrit, Mich State Univ, 85-88; LECTR, FOOD SCI DEPT, UNIV OTAGO, 88- *Concurrent Pos:* Regional contact, Inst Food Technologists. *Mem:* Inst Food Technologists; Am Soc Hort Sci; Int Soc Hort Sci; Can Inst Food Sci & Technol. *Res:* Sensory evaluation of food and chemistry of food flavor as affected by post-harvest physiology processing and storage; analysis of naturally occuring food toxicants in plant products; potato glycoalkaloids. *Mailing Add:* Dept Food Sci Univ Otago PO Box 56 Dunedin New Zealand

FILANDRO, ANTHONY SALVATORE, b New York, NY, Oct 10, 30; m 58; c 2. ORGANIC CHEMISTRY. *Educ:* Fordham Univ, BS, 50. *Prof Exp:* Chemist, Florasynth labs, 50-51; chemist, 51-56, chief chemist, 56-64, tech dir, 64-77, V PRES, VIRGINIA DARE EXTRACT CO, 77- *Mem:* Inst Food Technologists; Flavor & Extract Mfrs Asn US (vpres); NY Acad Sci; Soc Soft Drink Technologists; Nat Asn Fruits Flavors & Syrups. *Res:* Flavoring extracts; natural and synthetic flavoring compounds; analytical methods for flavors. *Mailing Add:* Virginia Dare Extract Co 882 Third Ave Brooklyn NY 11232

FILANO, ALBERT E, b Penfield, Pa, Aug 17, 25; m 50; c 4. MATHEMATICS. *Educ:* Univ Pa, BS, 48, MS, 49; Pa State Univ, PhD(math ed), 54. *Prof Exp:* Instr math, Hobart Col, 49-50, State Univ NY Oswego, 50-52 & Pa State Univ, 52-56; chmn dept math, 58-69, dir div sci & math, 67-69, interim dean fac & acad affairs, 69-70, vpres acad affairs, 70-74, PROF MATH, WEST CHESTER STATE COL, 56- *Concurrent Pos:* Dir & prof, Math Insts, NSF, 62-71. *Mem:* Math Asn Am; Nat Coun Teachers Math. *Res:* Mathematics education. *Mailing Add:* Dept Math Univ Pa West Chester PA 19383

FILAS, ROBERT WILLIAM, b Jersey City, NJ, Oct 10, 47. LIQUID CRYSTALS. *Educ:* Wesleyan Univ, BA, 69; Princeton Univ, MA, 71, PhD(chem), 79. *Prof Exp:* Res assoc chem, Temple Univ, 78-80; MEM TECH STAFF, AT&T BELL LABS, 80- *Mem:* Am Chem Soc; Am Phys Soc; Sigma Xi. *Res:* Silicone polymers for fiber optic applications and nonhermetic packaging; magnetohydrodynamics of liquid-crystalline polymers; polymer-surface interactions; liquid crystal displays. *Mailing Add:* AT&T Bell Labs Rm 7F-322 600 Mountain Ave Murray Hill NJ 07974-2070

FILBERT, AUGUSTUS MYERS, b Hazleton, Pa, Sept 28, 33; m 63; c 2. PHYSICAL CHEMISTRY. *Educ:* Lehigh Univ, BS, 55; Univ Pa, PhD(phys chem), 62. *Prof Exp:* Lab asst, Univ Pa, 55-61; from sr chemist to sr res chemist, 61-73, res supvr, 73-74, mgr biomat res & develop, 74-77, mgr, Tech Staffs Div, 78, dir, Res & Develop Anal Serv, 78-87, DIR, RES & DEVELOP ENG SERV, CORNING GLASS WORKS, 87- *Mem:* AAAS; Am Chem Soc; Am Ceramic Soc; Am Ord Asn. *Res:* Metal-ammonia solutions; solution chemistry; glass composition; chemistry and physics of glass; conductivity studies; surface chemistry of glass; immobilized enzymes; electrophoresis; biochemistry; immunochemistry; analytical chemistry. *Mailing Add:* 210 W Fourth St Corning NY 14830

FILBEY, ALLEN HOWARD, organic chemistry; deceased, see previous edition for last biography

FILBY, ROYSTON HERBERT, b London, Eng, Feb 16, 34; m 78; c 2. NUCLEAR CHEMISTRY, GEOCHEMISTRY. *Educ:* Univ London, BSc, 55; McMaster Univ, MSc, 57; Wash State Univ, PhD(chem), 71. *Prof Exp:* Res fel geochem, Univ Oslo, 61-64; head dept chem, Univ El Salvador, 64-67; chemist, 67-70, asst prof chem, 70-71, assoc prof chem & asst dir Nuclear Radiation Ctr, 70-74, prof chem & assoc dir nuclear radiation ctr, 74-76, prof chem & dir Nuclear Radiation Ctr, 76-88, PROF & CHMN, DEPT CHEM, WASH STATE UNIV, 88- *Concurrent Pos:* Org Europ Econ Coop sr vis fel, Europ Atomic Energy Comn, Mol, Belg, 62; guest worker, Nat Bur Stand, Washington, DC, 75-76; vis prof, Univ Port Harcourt, Nigeria, 81; chmn, Div Geochem, Am Chem Soc, 87; chmn, Div Isotopes & Radiation, Am Nuclear Soc 88. *Mem:* Fel, AAAS; Geochem Soc; Am Chem Soc; Am Asn Petrol Geologists; Am Nuclear Soc. *Res:* Neutron activation analysis; petroleum geochemistry; geochemistry of metal species in petroleum generation. *Mailing Add:* Dept Chem Wash State Univ Pullman WA 99164-4630

FILE, JOSEPH, b Lecce, Italy, May 6, 23; US citizen; m 44; c 3. MECHANICAL ENGINEERING, SUPERCONDUCTIVITY. *Educ:* Cornell Univ, BME, 44; Columbia Univ, MS, 55, PhD(mech eng), 68. *Hon Degrees:* Dr Physics, Univ Lecce, 78. *Prof Exp:* Design engr mech eng, Petro-Chem Develop Co, 48-56; RES ASSOC & SR TECH STAFF MEM FUSION ENERGY RES, PLASMA PHYSICS LAB, PRINCETON UNIV, 56- *Concurrent Pos:* Sr consult, Radio Corp Am, 67-70 & Univ Calif, Los Angeles, 78-; Fulbright-Hays teaching grant, 78. *Mem:* Sigma Xi. *Res:* Application of superconducting magnets to fusion reactors; research in fusion technology; design of fusion energy research devices. *Mailing Add:* Plasma Physics Lab Princeton Univ Princeton NJ 08543

FILER, CRIST NICHOLAS, b Hartford, Conn, May 12, 49; m 75; c 1. SYNTHETIC ORGANIC CHEMISTRY, MEDICINAL CHEMISTRY. *Educ:* Trinity Col, Conn, BS, 71; Mass Inst Technol, Phd(org chem), 75. *Prof Exp:* Sr chemist org synthesis, 77-78, supvr ligands, 78-85, SUPVR CUSTOM SYNTHESIS, E I DU PONT, NEW PROD, BOSTON, 85- *Concurrent Pos:* Res assoc med chem, Dept Med Chem, Northeastern Univ, 75-77. *Mem:* Am Chem Soc. *Res:* Central nervous system radiolabeled (carbon 14, tritium) neurotransmitter synthesis. *Mailing Add:* E I du Pont, NEW Prod 549 Albany St Boston MA 02118

FILER, LLOYD JACKSON, JR, b Grove City, Pa, Sept 30, 19; m 42; c 3. PEDIATRICS, NUTRITION. *Educ:* Univ Pittsburgh, BS, 41, PhD(biochem), 44; Univ Rochester, MD, 52. *Prof Exp:* Intern, Strong Mem Hosp, 52-53; med dir, Ross Labs, 53-65; prof, 65-86, EMER PROF PEDIAT, COL MED, UNIV IOWA, 86- *Concurrent Pos:* Sr res fel, Univ Pittsburgh, 44-45; res fel, Univ Rochester, 45-52, instr physiol, 45-47; asst clin prof pediat, Sch Med, Ohio State Univ, 53-65; exec dir, Int Life Sci Nutrit Found, 86-90. *Honors & Awards:* Goldberger Award, Am Med Asn, 78; Award, Am Col Nutrit, 79; hon mem, Am Dietetic Asn, 87; Nutrit Award, Am Acad of Pediat. *Mem:* Am Pediat Soc; Soc Pediat Res; Am Soc Clin Nutrit; AMA; Am Inst Nutrit; Inst Food Technologists. *Res:* Autooxidation of fats and oils; x-ray diffraction studies in glycerides; vitamin E in animal and human nutrition; nutritional studies on infants; studies on body composition; food additives; biochemistry. *Mailing Add:* Univ Hosp Univ Iowa Iowa City IA 52242

FILER, THEODORE H, JR, b Galveston, Tex, Aug 15, 28; m 50; c 2. PLANT PATHOLOGY, SOILS. *Educ:* Tex A&M Univ, BS, 50, MS, 58; Wash State Univ, PhD(plant path), 64. *Prof Exp:* Instr forest & gen path, Wash State Univ, 62-63; plant pathologist, 63-74, PRIN PLANT PATHOLOGIST & PROJ LEADER, US FOREST SERV, 74- *Mem:* Am Phytopath Soc; Mycol Soc Am; Am Forestry Asn; Int Soc Plant Path. *Res:* Factors concerned with development and control of root rot, decay and cankers of hardwood trees; fungi and significance of hardwood mycorrhizae; regeneration diseases in hardwood nurseries and plantations. *Mailing Add:* US Forest Serv Box 227 Stoneville MS 38776

FILGO, HOLLAND CLEVELAND, JR, b Van Alstyne, Tex, Mar 28, 26. MATHEMATICAL ANALYSIS. *Educ:* Baylor Univ, BS, 48; Rice Univ, MA, 51, PhD(math), 53. *Prof Exp:* Asst math, Rice Univ, 51-53; from asst prof to assoc prof, Univ Ala, 53-60; asst prof, Univ Ga, 60-61; assoc prof, 61-70, PROF MATH, NORTHEASTERN UNIV, 70- *Mem:* Am Math Soc; Math Asn Am. *Res:* Complex analysis. *Mailing Add:* Dept Math Northeastern Univ Boston MA 02115

FILICE, FRANCIS P, b Hollister, Calif, Aug 19, 22; m 47; c 6. ZOOLOGY. *Educ:* Univ San Francisco, BS, 43; Univ Calif, MA, 45, PhD(zool), 49. *Prof Exp:* From instr to prof, 47-78, EMER PROF BIOL, UNIV SAN FRANCISCO, 78- *Res:* Protozoan cytology; invertebrate ecology. *Mailing Add:* San Francisco State Univ 1600 Holloway Ave San Francisco CA 94117-1080

FILIPESCU, NICOLAE, b Predeal, Romania, July 30, 35; US citizen; m 63; c 3. ORGANIC CHEMISTRY, MEDICINE. *Educ:* Bucharest Polytech Inst, MS, 57, George Washington Univ, PhD(phys & org chem), 64, MD, 75. *Prof Exp:* Chem engr, Res Labs, Anticorrosion Plant, Romania, 57-59; sr chemist, Res Inst Construct Mat, 59; sr scientist, Melpar, Inc, 60-63; from asst prof to assoc prof, 63-71, PROF CHEM, GEORGE WASHINGTON UNIV, 71- *Concurrent Pos:* Res grant & consult, Goddard Space Flight Ctr, NASA, 64-; consult, Lockheed Elec Co, 66-70 & TAAG, Inc, 70-71; res grant, AEC, 69-75 & Energy Res & Develop Admin, 75-79. *Honors & Awards:* Hillebrand Prize, Chem Soc Wash, 71. *Mem:* Am Chem Soc; NY Acad Sci; Am Med Asn. *Res:* Photochemistry; intra- and intermolecular energy transfer; molecular spectroscopy; reaction mechanisms; endocrinology; obstetrics and gynecology. *Mailing Add:* Dept Chem George Washington Univ Washington DC 20052

FILIPOVICH, ALEXANDRA H, IMMUNODEFICIENCY, BONE MARROW TRANSPLANTATION. *Educ:* Univ Minn, MD, 74. *Prof Exp:* ASSOC PROF PEDIAT, HOSP & CLIN, UNIV MINN, 85- *Mailing Add:* Dept Pediat Div Immunol Univ Minn Box 261 Mayo Mem Bldg 420 Delaware St SE Minneapolis MN 55455

FILIPPENKO, ALEXEI VLADIMIR, b Berkeley, Calif, July 25, 58; m 89. ACTIVE GALACTIC NUCLEI, SUPERNOVAE. *Educ:* Univ Calif, Santa Barbara, BA, 79; Calif Inst Technol, PhD(astron), 84. *Prof Exp:* Hertz fel, Fannie & John Hertz Found, 79-84; Miller res fel, 84-86, asst prof, 86-88, ASSOC PROF ASTRON, UNIV CALIF, BERKELEY, 88- *Concurrent Pos:* Res fel astron, Calif Inst Technol, 84. *Mem:* Am Astron Soc. *Res:* Physical processes which occur in the nuclei of nearby galaxies; possible evolutionary relationship of nearby galaxies to distant quasars; optical spectra of supernovae. *Mailing Add:* Dept Astron Univ Calif Berkeley CA 94720

FILIPPENKO, VLADIMIR I, b Belgrade, Yugoslavia, Aug 23, 30; US citizen; m 52; c 2. APPLIED MATHEMATICS. *Educ:* Univ Calif, Berkeley, BA, 53, PhD(appl math), 64. *Prof Exp:* Res mathematician, Inst Eng Res, Univ Calif, Berkeley, 57-59, assoc math, 59-62, res asst, 62-64; sr res mathematician high velocity physics, Gen Motors Corp, 64-68; asst prof, 68-71, ASSOC PROF MATH, CALIF STATE UNIV, NORTHRIDGE, 71- *Concurrent Pos:* Off Naval Res fel, 64. *Mem:* Am Math Soc. *Res:* Partial and ordinary differential equations; fluid mechanics. *Mailing Add:* 5903 Via Lemora Goleta CA 93117

FILIPPOU, FILIP C, b Thessaloniki, Greece, July 14, 55; US citizen; m 84; c 2. STRUCTURAL ENGINEERING, EARTHQUAKE ENGINEERING. *Educ:* Tech Univ Munich, Ger, Dipl Ing, 78; Univ Calif, Berkeley, PhD(civil eng), 83. *Prof Exp:* Asst prof, 83-89, ASSOC PROF STRUCT, UNIV CALIF, BERKELEY, 89- *Concurrent Pos:* Eng, Tylin Int, 83-84; NSF presidential young investr, 87; vis prof, Univ Rome, Italy, 88. *Honors & Awards:* Alfred

Noble Prize, Am Soc Civil Engrs, 88. *Mem:* Am Soc Civil Engrs; Am Concrete Inst; Prestressed Concrete Inst; Earthquake Eng Res Inst. *Res:* Analysis and behavior of reinforced and prestressed concrete structures under normal and extreme loadings; development of models and effective simulation strategies; design guidelines for structures, particularly, under earthquake loads. *Mailing Add:* 731 Davis Hall Dept Civil Eng Univ Calif Berkeley CA 94720

FILISKO, FRANK EDWARD, b Lorain, Ohio, Jan 29, 42; m 70; c 3. POLYMER PHYSICAL CHEMISTRY & PROCESSING, ELECTRORHEOLOGICAL MATERIALS. *Educ:* Colgate Univ, BA, 64; Purdue Univ, MS, 66; Case Western Reserve Univ, PhD(physics), 69. *Prof Exp:* From asst prof to assoc prof, 70-84, PROF MAT SCI & ENG, UNIV MICH, 84-, ACTG DIR, MACROMOLECULAR SCI & ENG, 88- *Mem:* Am Chem Soc; Am Phys Soc; Soc Rheology. *Res:* Physical chemical characterization, physical aging and processing of glassy polymers and polymer blends; compatability of polymer blends and solutions; electrorheologically active materials; protein adsorption on biomedical materials. *Mailing Add:* Mat Sci & Eng Univ Mich Ann Arbor MI 48109-2136

FILKINS, JAMES P, b Milwaukee, Wis, Apr 10, 36; m 59; c 6. PHYSIOLOGY. *Educ:* Marquette Univ, BS, 57, MS, 59, PhD(physiol), 64. *Prof Exp:* Instr physiol, Sch Med, Marquette Univ, 64-65; from asst prof to assoc prof physiol & biophys, Univ Tenn, 65-71; assoc prof, 71-75, PROF PHYSIOL & CHMN DEPT, STRITCH SCH MED, LOYOLA UNIV CHICAGO, 75- *Concurrent Pos:* USPHS fel, 64-65. *Mem:* AAAS; Reticuloendothelial Soc; Am Physiol Soc. *Res:* Physiopathology of the reticuloendothelial system; mechanism of hepatic phagocytosis; physiology of shock; insulin and glucose regulation. *Mailing Add:* Loyola Univ Stritch Sch Med 2160 S First Ave Maywood IL 60153

FILKKE, ARNOLD M(AURICE), b Viroqua, Wis, July 8, 19; m 42; c 3. AGRICULTURAL ENGINEERING. *Educ:* Univ Wis, BS, 41; Univ Minn, MS, 43; Auburn Univ, PhD(agr eng), 72. *Prof Exp:* From instr to prof agr eng & rural electrification, Univ Minn, 46-84, dept head, 72-84; RETIRED. *Mem:* Fel Am Soc Agr Engrs; Am Soc Eng Educ; Inst Elec & Electronics Engrs. *Res:* Application of electricity to agriculture; alternate energy from agricultural biomass; fermentation and combustion; power and machinery design; soil, machine dynamics. *Mailing Add:* 3409 Downers Dr NE Minneapolis MN 55418

FILLER, LEWIS, b New York, NY, Feb 15, 28. APPLIED MECHANICS. *Educ:* NY Univ, BAE, 51, MAE, 53, DEngSci, 58. *Prof Exp:* Instr aeronaut eng, NY Univ, 52-57; res specialist, Boeing Airplane Co, 57-59; vis asst prof aeronaut eng, Univ Wash, 59; assoc res scientist, Denver Div, Martin Co, 59-62; assoc prof, 26-68, PROF MECH ENG, SEATTLE UNIV, 68-, CHMN DEPT, 83- *Concurrent Pos:* Lectr, Univ Colo, 60-62. *Mem:* Am Phys Soc; Am Soc Mech Eng; NY Acad Sci. *Res:* Fluid mechanics; applied mathematics. *Mailing Add:* Dept Mech Eng Seattle Univ 12th & E Columbia Seattle WA 98122

FILLER, ROBERT, b Brooklyn, NY, Feb 2, 23; m 45, 59; c 5. ORGANIC CHEMISTRY. *Educ:* City Col New York, BS, 43; Univ Iowa, MS, 47, PhD(chem), 49. *Prof Exp:* Asst chemist, Off Sci Res & Develop, Columbia Univ, 43-44; asst dept pharmacol, Univ Iowa Col Med, 47-49; asst prof chem, Albany Col Pharm, 49-50; res fel, Purdue Univ, 50-51; res chemist, Wright Air Develop Ctr, USAF, 51-53; asst prof chem, Ohio Wesleyan Univ, 53-55; from asst prof to assoc prof, 55-66, chmn dept, 68-76, PROF CHEM, ILL INST TECHNOL, 66- *Concurrent Pos:* Actg chmn dept chem, Ill Inst Technol, 66-68, 90-91; NIH spec fel, Cambridge Univ, 62-63; vis scientist, Weizmann Inst Sci, Israel, 74; chmn div fluorine chem, Am Chem Soc, 76; dean, Lewis Col Sci & Letters, 76-86; guest prof, Ruhr Univ, WGer, 87. *Mem:* AAAS; Am Chem Soc; The Chem Soc; NY Acad Sci; Sigma Xi. *Res:* Organic fluorine chemistry; heterocyclic chemistry; synthetic and medicinal chemistry. *Mailing Add:* Dept Chem Ill Inst Technol Chicago IL 60616-3793

FILLEY, GILES FRANKLIN, physiology; deceased, see previous edition for last biography

FILLIBEN, JAMES JOHN, b Philadelphia, Pa, Dec 14, 43; m 67; c 4. COMPUTER LANGUAGE. *Educ:* LaSalle Col, BA, 65; Princeton Univ, PhD(statist), 69. *Prof Exp:* Programmer, Radio Corp Am, 65; mathematician, Bell Telephone Res Lab, 66, Rand Corp, 67; STATIST CONSULT, NAT BUR STANDARDS, 69- *Mem:* Am Statist Soc; Am Comput Mach. *Res:* Design and development of dataplot-A widely used comprehensive, high-level (English-syntax), portable computer language with extensive capabilities in graphics, non-linear fitting, data analysis and mathematics. *Mailing Add:* Admin Bldg Rm A337 Statist Eng Lab Nat Inst Technol Standards Washington DC 20034

FILLIOS, LOUIS CHARLES, b Boston, Mass, July 1, 23; m 47; c 3. BIOCHEMISTRY, NUTRITION. *Educ:* Harvard Univ, AB, 48, MS, 53, ScD, 56. *Prof Exp:* From asst to assoc nutrit, Harvard Univ, 53-60; res assoc biochem, Mass Inst Technol, 60-61, asst prof physiol chem, 61-64, assoc prof, 64-66; assoc res prof biochem & path, Sch Med, 66-68, assoc prof biochem, Sch Med & assoc prof nutrit, Sch Grad Dent, 68-69, dir Div Basic Sci, 69-74, co-chmn div med & dent sci, Grad Sch, 70-71, PROF BIOCHEM, SCH MED & SCH GRAD DENT, BOSTON UNIV, 69-, CHMN DEPT NUTRIT SCI, 74- *Concurrent Pos:* Res fel, Harvard Univ, 56-57; Am Heart Asn estab investr, 61-66. *Mem:* Fel AAAS; Am Inst Nutrit; fel Am Heart Asn; Sigma Xi. *Res:* Endocrinology; experimental atherosclerosis; lipid, cholesterol, thyroid and nucleic acid metabolism; protein and lipoprotein synthesis; growth and development; periodontal res; glyprotein met; bioflavonoid met. *Mailing Add:* Boston Univ Med Ctr Boston MA 02118

FILLIPI, GORDON MICHAEL, b Warren, Minn, Oct 17, 40; m 60; c 5. MICROBIOLOGY, MEDICAL TECHNOLOGY. *Educ:* Univ NDak, BS, 62, MS, 70, PhD(microbiol), 73. *Prof Exp:* Chief med technologist, Clin Tuberc Lab, Nopeming Sanatorium, 62-68; teaching asst microbiol, Univ NDak, 68-69; instr, 73-74, ASST PROF PATH & MICROBIOL, MED SCH, UNIV NDAK, 74-; DIR CLIN MICROBIOL, UNITED HOSP, 73- *Concurrent Pos:* NIH trainee, 73. *Mem:* Am Soc Microbiol; affil Am Soc Clin Path. *Mailing Add:* United Hosp Microbiol Lab 1200 S Columbia Rd Grand Forks ND 58201

FILLIPPONE, WALTER R, b Newark, NJ, Mar 17, 21; m 43; c 2. GEOLOGY, GEOPHYSICS. *Educ:* Marietta Col, BA, 42; Calif Inst Technol, MSc, 44. *Prof Exp:* Seismologist, United Geophys Co, 43-45, party chief, 46-51, area mgr, Alaska, 52 & Rocky Mts, 53-55; sr geophysicist, Union Oil Co Calif, 55-60, dir geologist, 61-65, sr res assoc, 65-84; GEOPHSY CONSULT, 84- *Mem:* Soc Explor Geophys. *Res:* Applied geophysics; exploration seismic interpretation and techniques; data processing and interpretation of remote sensing from space. *Mailing Add:* 421 Larry Lane Placentia CA 92670-3426

FILLIUS, WALKER, b Washington, DC, Mar 22, 37; m 84; c 1. MAGNETOSPHERIC PHYSICS, SCIENTIFIC INSTRUMENTATION. *Educ:* Cornell Univ, BEP, 60; State Univ Iowa, MS, 63, PhD(physics), 65. *Prof Exp:* From asst res physicist to assoc res physicist, 65-80, RES PHYSICIST, UNIV CALIF, SAN DIEGO, 80- *Honors & Awards:* Exceptional Sci Achievement Medal, NASA, 74. *Mem:* Am Geophys Union; Sigma Xi; Am Phys Soc. *Res:* Measuring the properties of energetic charged particles trapped in the magnetospheres of earth, Jupiter, Saturn, and the sun; building instrumentation to make these measurements aboard satellites and space probes. *Mailing Add:* 6737 Welmer St San Diego CA 92122

FILLMORE, PETER ARTHUR, b Moncton, NB, Oct 28, 36; m 60; c 3. ALGEBRA. *Educ:* Dalhousie Univ, BSc, 57; Univ Minn, Minneapolis, MA, 60, PhD(math), 62. *Prof Exp:* Instr math, Univ Chicago, 62-64; from asst prof to prof math, Ind Univ, 64-72; sr fel, 72-73, Killam Res prof, 73-76, PROF MATH, DALHOUSIE UNIV, 76-, DEPT CHMN, 87- *Concurrent Pos:* Vis assoc prof, Univ Toronto, 70-71; sr vis fel, Univ Edinburgh, 77; vis mem, Math & Sci Res Inst, Berkeley, 84-85; vis prof, Univ Copenhagen, 90. *Mem:* Am Math Soc; Can Math Soc (vpres, 73-75); fel Royal Soc Can. *Res:* Functional analysis; operator theory. *Mailing Add:* Dept Math & Statist Comput Sci Dalhousie Univ Halifax NS B3H 3J5 Can

FILMER, DAVID LEE, b Youngstown, Ohio, June 4, 32; m; c 3. BIOCHEMISTRY, BIOPHYSICS. *Educ:* Youngstown Univ, AB, 54; Univ Wis, MS, 58, PhD(biochem), 61. *Prof Exp:* Res assoc enzyme mechanisms, Brookhaven Nat Lab, 61-63, asst biochemist, 63-65; asst prof, 65-68, ASSOC PROF BIOPHYS, PURDUE UNIV, 68- *Concurrent Pos:* Mem analog comput educ users group. *Mem:* AAAS; Asn Comput Mach; Soc Comput Simulation; Inst Elec & Electronics Engrs. *Res:* Physical biochemistry; biomedical applications of computers; fast reaction kinetics; enzyme mechanisms; bio-instrumentation; computer applications to life science and medicine. *Mailing Add:* Dept Biol Sci Purdue Univ Lafayette IN 47907

FILNER, BARBARA, b Philadelphia, Pa, Nov 15, 41; m 84. HEALTH SCIENCES POLICY. *Educ:* Queen's Col, BS, 62; Brandeis Univ, PhD(biol), 67. *Prof Exp:* Res assoc, Atomic Energy Comn Plant Res Lab, Mich State Univ, 67-69; Pub Health Serv fel, Inst Cancer Res, 69-71; asst prof biol, Columbia Univ, 71-76; asst prof cell & molecular biol, Kalamazoo Col, 77-78; staff officer, Div Health Promotion & Dis Prevention, 78-79, sr staff officer, 80-81, assoc dir, 81-84, DIR, DIV HEALTH SCI POLICY, INST MED, NAT ACAD SCI, 84- *Concurrent Pos:* Prin investr, Columbia Univ, 72-75; consult, Robert Wood Johnson Found, 84. *Mem:* AAAS; NY Acad Sci; Am Pub Health Asn; Asn Women Sci (pres-elect, 84-85). *Res:* Health sciences research policy; science based health policy; aging; child health; medical education; alcohol abuse; gene therapy; drug development. *Mailing Add:* Prog Dir Howard Hughes Med Inst 6701 Rockledge Dr Bethesda MD 20817

FILNER, PHILIP, b Philadelphia, Pa, July 12, 39; m 67; c 3. BIOCHEMISTRY. *Educ:* Johns Hopkins Univ, BA, 60; Calif Inst Technol, PhD(biochem), 65. *Prof Exp:* From asst prof to prof biochem, Mich State Univ, AEC Plant Res Lab, Mich State Univ, 73-82, actg dir, Dept Energy Plant Res Lab, 79-80; ASSOC DIR, ARCO PLANT CELL RES INST, 81- *Concurrent Pos:* NIH spec fel, Univ Copenhagen, 72-73. *Mem:* Am Soc Biol Chemists; Am Soc Plant Physiologists. *Res:* Plant biochemistry; mechanisms of enzyme regulation and role of enzymes in development; biochemistry of cultured plant cells; enzymes of nitrate and sulfate assimilation; enzyme amplification mechanisms. *Mailing Add:* Dept Molecular Biol Correlation Genetics Corp 2050 Concourse Dr San Jose CA 95131

FILSETH, STEPHEN V, b Portland, Ore, Nov 7, 36; m 58; c 3. PHYSICAL CHEMISTRY. *Educ:* Stanford Univ, BS, 58; Univ Wis, PhD(phys chem), 62. *Prof Exp:* From asst prof to assoc prof chem, Harvey Mudd Col, 62-71; assoc prof, 71-81, PROF CHEM, YORK UNIV, ONT, 81- *Mem:* Am Phys Soc. *Res:* Kinetics of elementary, homogeneous gas phase reactions; laser chemistry; gas phase photochemistry and energy transfer. *Mailing Add:* Dept Chem York Univ Toronto ON M3J 1P3 Can

FILSON, DON P, b Chicago, Ill, Feb 6, 31; div; c 5. PHYSICAL CHEMISTRY. *Educ:* Park Col, AB, 52; Univ Ill, MS, 62, PhD, 67. *Prof Exp:* Assoc prof, 60-67, prof chem, 67-77, STRAWN PROF CHEM, ILL COL, 77- *Mem:* Am Chem Soc; Asn Comput Mach. *Res:* Precipitation problems; computer-based instruction in chemistry; electrochemical methods of analysis. *Mailing Add:* Ill Col 1101 W College Jacksonville IL 62650-2299

FILSON, MALCOLM HAROLD, b Chattanooga, Tenn, Oct 19, 07; m 32, 60; c 3. CHEMISTRY. *Educ:* Univ Ky, BS, 29, MS, 31; Univ Mich, PhD(anal chem), 36. *Prof Exp:* Asst anal chem, Univ Ky, 29-31, asst res chemist, Exp Sta, 29; prof inorg chem, Ohio Northern Univ, 34-35; prof chem & head dept chem & physics, Miss Woman's Col, 35-36; asst prof, 35-42, prof chem, 42-72, chmn dept chem & physics, 55-65, chmn dept chem, 65-71, EMER PROF CHEM, CENT MICH UNIV, 72- *Concurrent Pos:* Consult, chem eng; assoc mem, Winter Haven Libr Bd & Winter Haven Human Rels Comt & Community Develop Comn. *Mem:* Fel AAAS; Am Chem Soc; Sigma Xi; Nat Sci Teachers Asn. *Res:* Analytical chemistry; chromyl chloride; chromium plating baths; chromyl fluoride; periodates; perchloric acid in analytical chemistry; solubilities and fiscosity as applied to silver, lead and mercury compounds. *Mailing Add:* 442 Ave A NE Winter Haven FL 33881

FILTEAU, GABRIEL, b Quebec, Que, Oct 16, 18; m 46; c 2. MARINE ECOLOGY. *Educ:* Laval Univ, BA, 39, BScA, 44, PhD(biol), 51. *Prof Exp:* Asst biologist, Biol Sta, St Laurent, Que, 44-50; lectr zool, 46-50, asst prof, 50-55, asst dean fac sci, 70- 77, prof zool, Laval Univ, 55-86, prof biol, 81-86, Head Dept Biol, 61-86; RETIRED. *Concurrent Pos:* Fel, Nuffield Found, Eng, 56; dir res, Fisheries & Oceans, Can, 77-; dir gen, Que Region Fisheries & Oceans, Can, 79-81; pres, Interuniv Res Group, 81. *Mem:* French-Can Asn Advan Sci; Can Soc Zool; Royal Soc Can. *Res:* Freshwater and marine biology, especially plankton; invertebrate zoology. *Mailing Add:* 1247 Des Gouverneurs Sillery PQ G1T 2G2 Can

FIMIAN, WALTER JOSEPH, JR, b New York, NY, May 22, 26; m 50; c 5. MORPHOLOGY, RADIOBIOLOGY. *Educ:* Univ Vt, AB, 50; Notre Dame Univ, MS, 52, PhD, 55. *Prof Exp:* Instr & res investr, Marquette Univ, 54-55; asst prof, 55-60, ASSOC PROF ZOOL, BOSTON COL, 60- *Concurrent Pos:* Premed & predent adv, Boston Col, 74- *Mem:* AAAS; Radiation Res Soc; Am Soc Zool. *Res:* Tissue culture studies; amphibian regeneration; melanin synthesis; experimental morphogenesis; radiation biology. *Mailing Add:* Dept Biol Boston Col Chestnut Hill MA 02167

FINA, LOUIS R, b Cleveland, Ohio, Dec 13, 18; m 46; c 3. MICROBIAL PHYSIOLOGY. *Educ:* Univ Ill, AB, 42, MS, 48, PhD(bact, chem), 50. *Prof Exp:* Asst prof bact, Univ Wyo, 50-52 & Univ Ark, 52-54; PROF BACT & MICROBIOLOGIST, KANS STATE UNIV, 54- *Concurrent Pos:* Vis prof, Rowett Res Inst, Scotland, 62. *Mem:* Am Soc Microbiol; Int Water Resources Asn; Sigma Xi. *Res:* Methane fermentation; industrial problems; Rumen microbiology; disinfectants. *Mailing Add:* 744 Crestline Dr Manhattan KS 66502

FINANDO, STEVEN J, b New York, NY, Mar 13, 48; m 70; c 1. ORIENTAL MEDICINE, HOLISTIC HEALTH. *Educ:* Queens Col, BA, 68, MA, 70; Fla State Univ PhD(commun res), 73. *Prof Exp:* Assoc dir, Holistic Health Ctr, 77-88; PRES, NEW CTR HOLISTIC HEALTH EDUC & RES, 82- *Concurrent Pos:* Dir res, NY Chiropractic Col, 84-85; chmn, Nat Comn for the Cert Acupuncturists, 83-87. *Mem:* Biofeedback Soc Am; Am Asn Acupucture & Oriental Med; Int Kirlian Res Asn. *Res:* Oriental medicine; innovative and iminimally invasive diagnostics and treatment. *Mailing Add:* 11 Hill Lane Roslyn Heights NY 11577

FINBERG, LAURENCE, b Chicago, Ill, May 20, 23; m 45; c 3. PEDIATRICS, PHYSIOLOGY. *Educ:* Univ Chicago, BS, 44, MD, 46. *Prof Exp:* From instr to asst prof pediat, Sch Med, Johns Hopkins Univ, 51-63, pediatrician, Hosp, 51-63; prof pediat, Albert Einstein Col Med, 63-; chmn pediat dept, Montefiore Hosp & Med Ctr, 63-; CHMN DEPT PEDIAT, BROOKLIN MED CTR, STATE UNIV NY, 82-; DEAN, COL MED, STATE UNIV NY, BROOKLYN, 88-3. *Concurrent Pos:* From asst chief pediatrician to assoc chief pediatrician, Baltimore City Hosp, 51-63; chmn, Am Bd Pediat, 78 & 87; mem, Nat Cholesterol Educ Panels. *Mem:* Am Pediat Soc; Soc Pediat Res; Am Fedn Clin Res; Am Acad Pediat; Am Inst Nutrit. *Res:* Electrolyte physiology, especially relating to disturbances of sodium and water osmotic equilibrium; metabolic and infectious diseases. *Mailing Add:* Dept Pediat Box 49 State Univ NY Brooklyn Med Ctr 450 Clarkson Ave Brooklyn NY 11203

FINBERG, ROBERT WILLIAM, b Baltimore, Md, 1950; m; c 3. INFECTIOUS DISEASES. *Educ:* Albert Einstein Col Med, MD, 74. *Prof Exp:* CHIEF, LAB INFECTIOUS DIS, SCH MED, HARVARD UNIV, 80- *Concurrent Pos:* Assoc prof med, Harvard Med Sch. *Mem:* Am Asn Immunologists; Am Soc Clinical Invest. *Res:* Immunology of infectious diseases; regulation of t-cell responses; control of responses to polysaccharides. *Mailing Add:* Dana-Farber Cancer Inst 44 Binney St Boston MA 02115

FINCH, C(HARLES) R(ICHARD), b Memphis, Tenn, Nov 30, 28; m 56; c 4. CHEMICAL ENGINEERING. *Educ:* Univ Md, BS, 50, PhD(chem eng), 55. *Prof Exp:* Asst chem eng, Univ Md, 51-53, instr, 53-54; proj leader, Dow Chem Co, 56-60, group leader, 60-86; CONSULT, PLASTICS TECHNOL, 86- *Mem:* Soc Plastics Engrs; Am Chem Soc. *Res:* Plastics development. *Mailing Add:* 1280 E Chippewa River Midland MI 48640

FINCH, CALEB ELLICOTT, b July 4, 39; US citizen; m. ENDOCRINOLOGY, GERONTOLOGY. *Educ:* Yale Univ, BS, 61; Rockefeller Univ, PhD(cell biol), 69. *Prof Exp:* Asst prof anat, Med Col, Cornell Univ, 70-72; from asst prof to assoc prof, 72-78, PROF BIOL, UNIV SOUTHERN CALIF, LOS ANGELES, 78-, ARCO & WILLIAM KIESCHNICK PROF NEUROBIOL AGING, 85- *Concurrent Pos:* NIH fel, Rockefeller Univ, 69-70; Pvt Investr, Alzheimer Dis Res Ctr Southern Calif, Nat Adv Coun on Aging, 87- *Honors & Awards:* Kleemeier Award, Geront Soc, 84; Brookdale Award for Contrib to Geront, 85; Allied-Signal Award, 88. *Mem:* Geront Soc; Endocrine Soc; Sigma Xi; Neurosci Soc; Neuroendocrine Soc. *Res:* Neuroendocrine control of aging. *Mailing Add:* Andrews Geront Ctr Univ Southern Calif Los Angeles CA 90089-0191

FINCH, CLEMENT ALFRED, b Broadalbin, NY, July 4, 15. HEMATOLOGY. *Educ:* Union Col, BA, 36; Univ Rochester, MD, 41. *Prof Exp:* From intern to asst resident med, Peter Bent Brigham Hosp, Boston, 41-43, resident, 44-46; instr, Harvard Med Sch, 46-48, assoc, 48-49; assoc prof med, 49-55, prof hemat & head div, 55-81, prof med, 81-87, EMER PROF MED, SCH MED, UNIV WASH, 87- *Concurrent Pos:* Res fel hemat, Evans Mem Hosp, 43-44; dir hemat res, Providence Hosp, 81-88. *Mem:* Nat Acad Sci; Am Soc Clin Invest; Asn Am Physicians; Am Soc Hemat; Int Soc Hemat. *Mailing Add:* HC 60 Box 7736 Cle Elum WA 98922

FINCH, GAYLORD KIRKWOOD, b Owosso, Mich, Nov 16, 23; m 45; c 4. ORGANIC CHEMISTRY. *Educ:* Univ Mich, BSChE, 45, MS, 48, PhD(chem), 54. *Prof Exp:* Ord engr, US Govt, 46; sr chemist, 50-60, chief chemist, 60-64, supvr acid develop & control dept, 64-66, asst div supt, Acid Div, 66-70, asst div supt, Polymers Div, 70-71, asst div supt, Org Chem Div, 71-72, asst gen mgr chem, Europ Region, Kodak Int Photog Div, 72-75, staff asst, 75-76, ASST DIR RES LABS, EASTMAN CHEMS DIV, KINGSPORT, TENN, 76- *Mem:* Am Chem Soc; Am Inst Chem Engrs; Sigma Xi. *Res:* Synthesis, development and process improvement of aliphatic and aromatic oxygenated compounds; statistical design and interpretation of experiments; polymer chemistry; manufacture of polyesters; synthesis of dyes and fine organic chemicals; manufacture of photographic chemicals. *Mailing Add:* 472 Lakeside Dr Kingsport TN 37664

FINCH, HARRY C, b Ringsted, Iowa, Jan 15, 17; m 45; c 2. PLANT PATHOLOGY. *Educ:* Iowa State Univ, BA, 46, MS, 47, PhD(plant path), 50. *Prof Exp:* Asst prof plant path, NC State Univ, 50-54; assoc prof, Pa State Univ, 54-59; proj leader fungicides-nematicides, Monsanto Chem Co, 59-62; from asst prof to assoc prof, 62-74, PROF BIOL SCI, CALIF POLYTECH STATE UNIV, SAN LUIS OBISPO, 74- *Concurrent Pos:* US AID technologist, Guatemala, 70-71; vis prof, Honduras, 73. *Mem:* AAAS; Am Phytopath Soc; Mycol Soc Am. *Res:* Diseases of fruit trees; virus diseases of plants; fungicide-nematicides. *Mailing Add:* PO Box 3202 Shell Beach CA 93448

FINCH, JACK NORMAN, b Esbon, Kans, Sept 26, 23; m 47; c 2. PHYSICAL CHEMISTRY. *Educ:* Ft Hays State Univ, BS, 48, MS, 49; Kans State Univ, PhD, 57. *Prof Exp:* Instr chem, Northern Okla Jr Col, 49-53; asst prof, Northeastern State Univ, 53; instr, Kans State Univ, 55-56; sr res chemist, Phillips Petrol Co, 56-85; RETIRED. *Mem:* Am Chem Soc. *Res:* Molecular structure, spectroscopy, catalysis and synthetic fuels. *Mailing Add:* 4762 Dartmouth Dr Bartlesville OK 74006

FINCH, JOHN VERNOR, mathematics; deceased, see previous edition for last biography

FINCH, LEIKO HATTA, applied mathematics, for more information see previous edition

FINCH, ROBERT ALLEN, b Cleveland, Ohio, Mar 15, 41; m 68; c 2. ANATOMY, CELL BIOLOGY. *Educ:* Oberlin Col, AB, 63; Case Western Reserve Univ, PhD(anat), 68; Am Bd Toxicol, dipl, 85. *Prof Exp:* NIH trainee biol, Brandeis Univ, 68-70; asst prof anat, Bowman Gray Sch Med, 70-78; Nat Inst Environ Health Sci fel toxicol, Univ Rochester, 78-79; head genetic toxicol, Raltech Sci Serv, Madison, Wis, 79-82; pres, Finch, Biesemeier & Assocs, 82-83; RES TOXICOLOGIST & PRIN INVESTR, US ARMY BIOMED RES & DEVELOP LAB, HEALTH EFFECTS RES DIV, FT DETRICK, FREDERICK, MD, 83- *Mem:* Tissue Culture Asn; Am Col Toxicol; Environ Mutagen Soc; Genetic Toxicol Asn; Am Soc Test & Mat. *Res:* Cell and tissue culture; in vitro and in vivo mutagenesis and carcinogenesis; alternative toxicity test methods. *Mailing Add:* US Army Biomed Res & Develop Lab Health Effect Res Div, Bldg 568, Ft Detrick Frederick MD 21702-5010

FINCH, ROBERT D, b Westcliff, Eng, Aug 18, 38; US citizen; m 63; c 2. ENGINEERNIG ACOUSTICS, PHYSICAL ACOUSTICS. *Educ:* Imperial Col, Univ London, BS, 59, PhD(physics), 63; Chelsea Col, Univ London, MSc, 60. *Prof Exp:* Fel physics, Univ Calif, Los Angeles, 63-65; PROF MECH ENG, UNIV HOUSTON, 65- *Concurrent Pos:* Asst Exec Dir, Gov Energy Adv Coun, Tex, 73-75; chmn dept mech eng, Univ Houston, 76-79. *Honors & Awards:* Biennial Award, Acoust Soc Am, 72. *Mem:* Fel Acoust Soc Am; Am Soc Mech Engrs; Am Phys Soc. *Res:* Engineering and physical acoustics; transducer design; signal processing and ultrasonic testing and processing; acoustic signature monitoring. *Mailing Add:* Dept Mech Eng Univ Houston Houston TX 77204-4792

FINCH, ROGERS B(URTON), b Broadalbin, NY, Apr 16, 20; m 42; c 5. TEXTILE TECHNOLOGY, ASSOCIATION MANAGEMENT. *Educ:* Mass Inst Technol, SB, 41, SM, 47, ScD, 50. *Prof Exp:* Textile technologist, Broadalbin Knitting Co, Inc, NY, 41; dir, Heavy Textile Res & Develop, Eng Div, Jeffersonville, Ind Quartermaster Depot, 43-46; res assoc mech eng, textile div, Mass Inst Technol, 46-47, asst prof textile tech, 47-53; dir, US Opers Mission, Burma, 53-54; asst dir res div, Rensselaer Polytech Inst, 54-58, assoc dean, sch sci, 58-60, dir res div, 60-61; dir univ rels, Peace Corps, DC, 61-63; assoc dean, Hartford Grad Ctr, Rensselaer Polytech Inst, 63-66, dir acad planning, 66-70, vpres planning, 70-72; exec dir, Am Soc Mech Engrs, 72-81; exec vpres, Illuminating Eng Soc, 82-87; PRES, FINCH CONSULTS, 87- *Honors & Awards:* Key Award, Am Soc Asn Execs. *Mem:* Fel AAAS; fel Am Soc Mech Engrs; Am Soc Eng Educ; Coun Eng & Sci Soc Execs (pres, 79-80); Sigma Xi. *Res:* Association management; long-range planning. *Mailing Add:* 12 Sherwood Rd Little Silver NJ 07739-1309

FINCH, STEPHEN JOSEPH, b St Louis, Mo, Mar 20, 45. STATISTICS. *Educ:* St Louis Univ, BS, 67; Princeton Univ, MA, 69, PhD(statist), 74. *Prof Exp:* Asst prof, 74-80, ASSOC PROF STATISTIC, STATE UNIV NY, STONY BROOK, 80- *Concurrent Pos:* Consult, Brookhaven Nat Labs, 75- *Mem:* Am Statist Asn; Am Pub Health Asn; AAAS. *Res:* Robust tests of structural properties of random variables, such as symmetry; applications of statistics in policy studies. *Mailing Add:* Dept Math & Statist State Univ NY Stony Brook Main Stony Brook NY 11794

FINCH, STUART CECIL, b Broadalbin, NY, Aug 6, 21; m 46; c 4. INTERNAL MEDICINE. *Educ:* Univ Rochester, MD, 44. *Prof Exp:* Intern surg, Baltimore City Hosp, Md, 44-45, asst resident path, 45-46; res fel med, Harvard Med Sch, 48-49; asst, Dept Hemat, Peter Bent Brigham Hosp, 48-49, asst resident, 49-50; asst, Sch Med, Boston Univ, 50-52, instr, 52-53; from asst prof to assoc prof med, Sch Med, Yale Univ, 53-67, prof, 67-; AT DEPT MED, COOPER MED CTR, CAMDEN, NJ. *Concurrent Pos:* Res assoc, Evans Mem-Mass Mem Hosps, 50-52; asst mem, 52-53; assoc physician, Grace-New Haven Community Hosp, 53-; consult, West Haven Vet Admin Hosp, Conn, 53-, Laurel Heights Hosp, Derby, 58- & Meriden Hosps, 58-; clin prof, Yale Univ, 78-79. *Mem:* Am Soc Hemat; Am Soc Clin Invest; Asn Am Physicians; Am Fedn Clin Res; Int Soc Hemat; Sigma Xi. *Res:* Iron metabolism; leukemia; leukocyte kinetics and immunology. *Mailing Add:* Cooper Hosp Univ Med Ctr One Cooper Plaza Camden NJ 08103

FINCH, STUART MCINTYRE, b Salt Lake City, Utah, Aug 16, 19; m 41; c 2. PSYCHIATRY. *Educ:* Univ Colo, MD, 43. *Prof Exp:* Intern, Alameda County Hosp, Oakland, Calif, 43-44; resident psychiat, Sch Med, Temple Univ, 46-49, from instr to assoc prof, 49-50; from assoc prof to prof, Med Sch, Univ Mich, Ann Arbor, 56-73; lectr med, Col Med, Univ Ariz, 73-85; prof psychiat, Med Col Ga, Augusta, 85-90; RETIRED. *Concurrent Pos:* Dir child psychiat, Sch Med, Temple Univ, 49-56; attend psychiatrist, St Christopher's Hosp for Children, 53-56; consult, Vet Admin Hosp, Battle Creek, 56-58, Battle Creek Child Guid Clin, 56-58, Kalamazoo State Hosp, 56-58 & fresh air camp, Univ Mich, 57-; mem comt cert child psychiat, Am Bd Psychiat & Neurol; mem, Gov Ment Health Adv Coun. *Mem:* Am Psychosom Soc; AMA; Am Psychiat Asn; Am Psychoanal Asn; Am Orthopsychiat Asn. *Res:* Psychophysiologic illness. *Mailing Add:* 745 W Paseo Del Canto Green Valley AZ 85614-1721

FINCH, THOMAS LASSFOLK, b Madison, Wis, Nov 26, 26. PHYSICS. *Educ:* Univ Wis, BA, 47, MA, 49, PhD(physics), 57. *Prof Exp:* Asst prof physics, Union Univ, NY, 55-57; asst prof, 57-60, ASSOC PROF PHYSICS, ST LAWRENCE UNIV, 60- *Concurrent Pos:* Vis lectr, Univ Col NWales, 64-65. *Mem:* AAAS; Am Phys Soc; Am Asn Physics Teachers; Acoust Soc Am; Sigma Xi. *Res:* Low-temperature physics; intermediate state of super conductors; musical acoustics. *Mailing Add:* 11 College St Canton NY 13617

FINCH, THOMAS WESLEY, b Alhambra, Calif, Dec 17, 46; m 79; c 5. CATHODIC PROTECTION ENGINEERING, CORROSION CONTROL-MITIGATION. *Educ:* Colo Sch Mines, BS, 68. *Prof Exp:* Engr, US Army CEngr, 68-72; area mgr, Corrintec, 80-83; dist mgr, Gen Cathodic Protection Serv, 83-86; engr, 73-80, DIST MGR, CATHODIC PROTECTION SERV CO, 86- *Concurrent Pos:* Corrosion technologist, Nat Asn Corrosion Engrs, 74-; mem, Nat Adv Bd, Am Security Coun, 78- *Mem:* Soc Am Mil Engrs. *Res:* Cathodic protection; external corrosion mitigation for pipe lines, well casings and underground storage tanks. *Mailing Add:* 1710 E 22nd St Farmington NM 87401

FINCH, WARREN IRVIN, b Union Co, SDak, Oct 27, 24; m 51; c 3. GEOLOGY. *Educ:* SDak Sch Mines & Technol, BS, 48; Univ Calif, MS, 54. *Prof Exp:* Geologist, US Geol Surv, 48-72, chief supvry geologist, 72-77, supvry res geologist, 83-88, URANIUM COMMODITY GEOLOGIST, BR SEDIMENTARY PROCESSES, US GEOL SURV, 77-, RES GEOLOGIST, 89- *Concurrent Pos:* Chmn, Int Atomic Energy Agency Working Group Sandstone Uranium Deposits, 79-90; chmn consult groups, Int Atomic Energy Agency Mission, Indonesia, 88, China, 90. *Honors & Awards:* Meritorious Serv Award, Dept Interior, 81. *Mem:* Fel Geol Soc Am; fel Soc Econ Geol; Int Asn Genesis Ore Deposits; AAAS. *Res:* Exploration for uranium deposits of the Colorado Plateau and beryllium; geology of uranium deposits in sandstone formations of the United States; regional geology of Jackson Purchase region, Kentucky; resource assessment methodology. *Mailing Add:* US Geol Surv Box 25046 MS 939 Denver Fed Ctr Denver CO 80225-0046

FINCHER, BOBBY LEE, b Clarksville, Ark, Aug 28, 34; m 54; c 2. ALGEBRA. *Educ:* Univ Ark, BS, 56; Okla State Univ, MS, 62; Ind Univ, PhD(algebra), 72. *Prof Exp:* Radiation engr, McCullough Tool Co, 56-58; teacher geol & math, Ft Smith High Sch, Ark, 58-60; from instr to assoc prof, 63-88, PROF MATH, TEX WOMAN'S UNIV, 88- *Mem:* Math Asn Am; Sigma Xi. *Res:* Relationship between mathematics and music. *Mailing Add:* Box 23434 Tex Woman's Univ Denton TX 76204

FINCHER, EDWARD LESTER, b Atlanta, Ga, Jan 16, 21; m 48; c 2. MICROBIOLOGY. *Educ:* Mercer Univ, BA, 48; Emory Univ, MSc, 49; Univ Ga, PhD(bact), 62. *Prof Exp:* Res biologist, Eng Exp Sta, Ga Inst Technol, 50-58, asst res scientist, 58-60, asst res prof biol, 60-63; res microbiologist, Biophys Sect, Tech Br, Commun Dis Ctr, USPHS, 64-65; assoc prof environ sci & eng, Univ NC, 65-68; actg dir dept, 70-72, assoc prof, 68-80, prof biol, 80-84, EMER PROF APPL BIOL, GA INST TECHNOL, 84- *Concurrent Pos:* USPHS trainee, 58-59, res fel, 59-60. *Mem:* AAAS; Am Soc Microbiol. *Res:* Aerobiology; bacterial growth and cytology; infection, host-parasite relationships; microbiology of the hospital environment; bacterial degradation of synthetic molecules; effect of chemical and physical factors on bacterial cells; electron microscopy. *Mailing Add:* 6380 Bridewood Valley Rd NW Atlanta GA 30328

FINCHER, GEORGE TRUMAN, b Arlington, Ga, Sept 28, 39; m 63; c 3. VETERINARY ENTOMOLOGY, PARASITOLOGY. *Educ:* Univ Ga, BSA, 61, MS, 66, PhD(entom), 68. *Prof Exp:* Zoologist parasitol, 68-78, RES ENTOMOLOGIST BIO-CONTROL, USDA, 78- *Mem:* Entom Soc Am; Am Soc Parasitologists; Sigma Xi. *Res:* Use of dung-burying beetles as biological control agents for control of livestock inst pests that breed in cattle dung. *Mailing Add:* USDA-ARS-FAPRL RR 5 Box 810 College Station TX 77845-9594

FINCHER, JOHN ALBERT, b Union, SC, Sept 8, 11; m 39; c 3. ZOOLOGY. *Educ:* Univ SC, BS, 33, MS, 35; Univ NC, PhD(zool), 39. *Prof Exp:* Prin pub sch, SC, 33-34; instr biol, Univ SC, 34-35; lab asst zool, Univ NC, 35-39; instr biol, Cumberland Col, 39-40; from asst prof to assoc prof, Millsaps Col, 40-46; prof & head biol dept, Samford Univ, 45-57, asst to the pres, 55-57, dean, 57-68, prof, Sch Anesthesia, 78-82; pres, Carson-Newman Col, 68-77, emer pres, 77-; RETIRED. *Concurrent Pos:* Fel, Highlands Biol Lab, 39; trustee, Gorgas Scholar Found, State Sci Talent Search, 52-68 & East End Mem Hosp, Birmingham, Ala; educ consult & prof biol, Baptist Med Ctr, Sch Anesthesia, Samford Univ, 78-82; acad vpres, Baptist Col, Charleston, SC, 83, actg pres, 83-84, exec vpres, 84-87; interim pres, N Grenville Col, 87-88; acad vpres, Limestone Col, 88-89. *Mem:* Am Suffolk Sheep Soc; Am Soc Zool. *Res:* Invertebrate zoology; cell behavior and gametogenesis in sponges; growth of sponges from gemmules. *Mailing Add:* 457 Hawthorne Rd Spartanburg SC 29303-2662

FINCHER, JULIAN H, b Cross Keys, SC, July 22, 35; m 66; c 3. PHYSICAL PHARMACY. *Educ:* Univ SC, BS, 58; Univ Ga, MS, 62; Univ Conn, PhD(pharm), 64. *Prof Exp:* Instr pharm, Univ SC, 58-59 & Univ Ga, 59-61; from asst prof to assoc prof pharmaceut, Univ Miss, 64-72 & chmn dept, 70-72; PROF PHARMACEUT, UNIV SC, 72-, DEAN & PROF, COL PHARM, 72- *Concurrent Pos:* Chmn, Coun Deans, Am Asn Col Pharm, 81-82. *Mem:* Am Pharmaceut Asn; Am Asn Cols Pharm; Am Assoc Pharmaceut Scientists; Am Soc Hosp Pharmacists Soc; Am Soc Pharmacognosy; Am Inst Hist Pharmacy. *Res:* Biopharmaceutics; physical and chemical properties affecting drug release and absorption from dosage forms; history of pharmacy and drugs; dictionary pharmacy compilation. *Mailing Add:* Col Pharm Univ of SC Columbia SC 29208

FINCKE, MARGARET LOUISE, chemistry, nutrition; deceased, see previous edition for last biography

FINCO, ARTHUR A, b Ely, Minn, Mar 1, 32; m 60; c 2. MATHEMATICS. *Educ:* St Cloud State Univ, BS, 53; Univ Northern Iowa, MA, 59; Purdue Univ, PhD(math educ), 66. *Prof Exp:* Instr high sch, Minn, 56-58; instr math, Mankato State Univ, 59-61; from asst prof to assoc prof, 65-74, PROF MATH & MATH EDUC, PURDUE UNIV, FT WAYNE, 74-, ASSOC CHMN, DEPT MATH SCI, 89- *Mem:* Math Asn Am; Nat Coun Teachers Math. *Res:* Mathematics education. *Mailing Add:* 2101 Coliseum Blvd E Ft Wayne IN 46805

FINCO, DELMAR R, b Roundup, Mont, Nov 5, 36; div; c 4. ANIMAL PHYSIOLOGY. *Educ:* Univ Minn, St Paul, BS, 57, DVM, 59, PhD(canine leptospirosis), 66. *Prof Exp:* Fel, Univ Minn, St Paul, 61-66, asst prof clin vet med, 66-70; prof clin vet med, 70-75, PROF PHYSIOL, COL VET MED, UNIV GA, 75- *Concurrent Pos:* Vis res prof, Wash Univ, St Louis, 80; dipl, Am Col Vet Internal Med. *Honors & Awards:* Gaines Award, 83. *Mem:* Am Vet Med Asn; Am Soc Nephrology; Int Soc Nephrology. *Res:* Canine leptospirosis and studies of leptospira canicola; urogenital diseases of dog and cat; renal physiology. *Mailing Add:* Dept Physiol Univ Ga Col Vet Med Athens GA 30602

FINDLAY, JOHN A, b Manchester, Eng, May 29, 36; Can citizen; m 63; c 3. ORGANIC CHEMISTRY. *Educ:* Univ NB, BSc, 58, PhD(org chem), 62. *Prof Exp:* NATO-Dept Indust & Sci Res UK fel org chem, Cambridge Univ, Eng, 62-63; lectr, 63-65, from asst prof to assoc prof, 65-74, PROF CHEM, UNIV NB, 74- *Concurrent Pos:* Asst dean sci, Univ NB, 73-75, actg chmn chem dept, 75, chmn dept, 82-85. *Mem:* Fel Chem Inst Can; Am Soc Pharmacognosy; Am Chem Soc. *Res:* Structural elucidation and synthesis of natural products of biological interest, including those of marine origin and from fungi. *Mailing Add:* Dept Chem Univ NB Fredericton NB E3B 6E2

FINDLAY, JOHN W A, b Mar 27, 45. BIOCHEMISTRY. *Educ:* Univ Aberdeen, BSc Hons, 66, PhD(org chem), 69. *Prof Exp:* Postdoctoral res assoc med chem, Univ Va, 69-72; res biochemist, Radioimmunoassay Sect, Dept Path, Norfolk Gen Hosp, 72-73; res scientist, Pharmaceut Dept, Radiochem Ctr, Amersham, Eng, 73-75; res scientist III, Dept Med Biochem, Wellcome Res Labs, Burroughs Wellcome Co, 75-77, res scientist IV, 77-81, group leader, 81-86, ASST HEAD, DIV PHARMACOKINETICS & DRUG METAB, WELLCOME RES LABS, BURROUGHS WELLCOME CO, 86- *Concurrent Pos:* Consult, Coun Tobacco Res, 78; mem, Site Visit Team, NIH, 84. *Mem:* Am Soc Pharmacol & Exp Therapeut; AAAS. *Res:* Development of drug analytical methodologies and their application to preclinical and clinical drug disposition studies; drug disposition and metabolism and their relationship to pharmacology or toxicology; pharmacokinetics; drug excretion in breast milk; narcotics. *Mailing Add:* Div Pharmacokinetics & Drug Metab Burroughs Wellcome Co 3030 Cornwallis Rd Research Triangle Park NC 27709

FINDLAY, JOHN WILSON, b Kineton, Eng, Oct 22, 15; nat US; m 53; c 2. PHYSICS. *Educ:* Cambridge Univ, BA, 37, PhD(physics), 50. *Prof Exp:* Researcher, Cambridge Univ, 37-39, univ demonstr & fel & lectr physics, Queens' Col, 45-53; mem staff, Brit Ministry Supply, 54-56; mem staff, Nat Radio Astron Observ, 57-65; dir, Arecibo Ionospheric Observ, 65-66; sr scientist, Nat Radio Astron Observ, 66-86; RETIRED. *Concurrent Pos:* Mem, Order of the Brit Empire. *Mem:* Fel Inst Elec & Electronics Engrs. *Res:* Physics of the ionosphere; radio astronomy. *Mailing Add:* Millbank Box 317 Greenwood VA 22943

FINDLAY, RAYMOND D(AVID), b Toronto, Ont, Aug 10, 38; m 61; c 4. ELECTRICAL ENGINEERING. *Educ:* Univ Toronto, BASc, 63, MASc, 65, PhD(elec eng), 68. *Prof Exp:* Lectr elec eng, Univ NB, 67-68, asst prof elec eng & comput sci, 68-70, from asst prof to assoc prof elec eng, 70-78, dir grad studies, elec eng dept, 75-78, prof, 78-81; asst dean, 84-87, PROF, DEPT ELEC ENG, MCMASTER UNIV, 83- *Concurrent Pos:* Nat Res Coun Can operating grant, 68-; Nat Res Coun sr indust fel & proj dir, Can Gen Elec Co, 72-73; comt chair, Inst Elec & Electronics Engrs Student Activ, 90- 91. *Honors & Awards:* Am Soc Eng Educ/DOW Award, 72. *Mem:* Am Soc Eng

Educ; Inst Elec & Electronics Engrs; fel Univ Southampton, Eng, 79-80; res fel CSIRO, Australia, 88. *Res:* Electrical power and apparatus; pole face losses in synchronous machines; current and loss distributions in aluminum conductor steel-reinforced cable. *Mailing Add:* Dept Elec Eng McMaster Univ 1280 Main St W Hamilton ON L8S 4K1 Can

FINDLER, NICHOLAS VICTOR, b Budapest, Hungary, Nov 24, 30; US citizen; m 55; c 2. COMPUTER SCIENCE, ARTIFICIAL INTELLIGENCE. *Educ:* Budapest Tech Univ, BEE, 53, PhD(math, physics), 56. *Prof Exp:* Sr lectr, Theoret Physics, Budapest Tech Univ, 56; vis lectr, Univ Vienna, 56-57; res fel theoret physics & comput sci, Univ Sydney, 57-59; staff appl mathematician, CSR Co, Ltd, Australia, 59-63; res assoc comput sci, Univ Pittsburgh, 64; assoc prof comput sci & math, Univ Ky, 64-66; prof, State Univ NY, Buffalo, 66-82; PROF COMPUT SCI & MATH & DIR, ARTIFICIAL INTEL LAB, ARIZ STATE UNIV, TEMPE, 82- *Concurrent Pos:* Adj asst prof math, Univ New South Wales, 62-63; res scientist & Adv Res Proj Agency fel, Carnegie Inst Technol, 63-64; invited lectr, NATO Adv Study Insts, 65-90, dir, 70; consult, var indust firms, 64-; sr Fulbright scholar vis prof, Tech Univ Vienna, 72-73, Univ Amsterdam & Free Univ Amsterdam, 79-80, Hungarian Acad Sci, 88-89; vis prof, Univ Zurich, 88-89. *Mem:* AAAS; Asn Comput Mach; fel Brit Comput Soc; Asn Comput Linguistics; Am Asn Artificial Intel; fel Inst Elec & Electronics Engrs. *Res:* Heuristic programming; special purpose computer languages; simulation of human cognitive behavior; theory of competition and strategies; man-machine relations; computer graphics; self-adaptive systems; computational linguistics; information retrieval; mathematical physics; mathematical psychology; mathematical biophysics. *Mailing Add:* Dept Computer Sci & Eng Ariz State Univ Tempe AZ 85287-5406

FINDLEY, JAMES SMITH, b Cleveland, Ohio, Dec 28, 26; m 49; c 4. VERTEBRATE ECOLOGY, MAMMALOGY. *Educ:* Western Reserve Univ, AB, 50; Univ Kans, PhD, 55. *Prof Exp:* Asst instr, Univ Kans, 50-53, asst, Mus Natural Hist, 53-54; instr zool, Univ SDak, 54-55; from asst prof to assoc prof, 55-70, chmn dept, 78-82, PROF BIOL, UNIV NMEX, 70- *Honors & Awards:* C Hart Merriam Award, Am Soc Mammalogists, 78; Leopold Conservation Award, Nature Conservancy, 87. *Mem:* AAAS; Am Soc Mammalogists (pres, 80-82); Ecol Soc Am; Soc Syst Zool; Am Soc Naturalists; Soc Study Evolution; Am Soc Mammal. *Res:* Ecology and zoogeography; mammalogy. *Mailing Add:* Dept Biol Univ NMex Albuquerque NM 87131

FINDLEY, MARSHALL E(WING), b Arkansas City, Kans, Oct 13, 27; m 55; c 2. CHEMICAL ENGINEERING. *Educ:* Agr & Mech Col Tex, BS, 49; Inst Paper Chem, MS, 51; Univ Fla, PhD(chem eng), 55. *Prof Exp:* Chem engr, Celotex Corp, La, 51-52; asst, Univ Fla, 52-55; res engr, E I du Pont de Nemours & Co, Tenn, 55-58; assoc res prof chem eng, Auburn Univ, 58-65; from asst prof to assoc prof, Univ Mo, Rolla, 65-69, prof chem eng, 69-90; RETIRED. *Concurrent Pos:* Fulbright lectr, Univ Alexandria, 62-63; eng educ adv, USAID, SVietnam, 69-71; prof, Inst Algerien du Petrole, 78-79; vis prof, Univ PR, Mayaguez, 83-84. *Mem:* Am Chem Soc; Am Soc Eng Educ. *Res:* Process control; water desalination; adsorption. *Mailing Add:* 1004 Lynwood Rolla MO 65401

FINDLEY, WILLIAM N(ICHOLS), b Mankato, Minn, Feb 12, 14; m 39; c 1. APPLIED MECHANICS, MECHANICS OF MATERIALS. *Educ:* Ill Col, AB, 36; Univ Mich, BSE(mech eng) & BSE(math), 37; Cornell Univ, MS, 39. *Hon Degrees:* DSc, Ill Col, 70. *Prof Exp:* Asst eng mech, Univ Mich, 36-37; instr civil eng, George Washington Univ, 38-39; from instr to asst prof theoret & appl mech, Univ Ill, 39-47, res assoc prof, 47-54; prof, 54-84, dir, Ctr Facility Mech Testing,66-69, EMER PROF ENG, BROWN UNIV, 84- *Concurrent Pos:* mem sci adv coun, Picatinny Arsenal, 51-63; consult, Lawrence Radiation Lab, 62-78; mem orgn comt, Joint Int Conf Creep, NY & London, 63; res scholar struct mech, ONR/Am Inst Aeronaut & Astronaut, 79. *Honors & Awards:* Prize paper, Soc Plastics Engrs, 49, 50; Dudley Medal, Soc Testing & Mat, 45, Templin Award, 53 & 64. *Mem:* Am Soc Testing & Mat; Soc Exp Stress Anal; Am Soc Eng Educ; Soc Rheol; fel Am Soc Mech Engrs; Am Acad Mech. *Res:* Creep, fatigue and other strength properties of plastics at various temperatures; fatigue and creep of metals in combined stress; photoelasticity; mechanics of creep; theories of fatigue; viscoelasticity. *Mailing Add:* Div Eng Box D Brown Univ Providence RI 02912

FINDLEY, WILLIAM RAY, JR, b Manhattan, Kans, June 26, 20; m 43; c 2. PLANT BREEDING. *Educ:* Kans State Univ, BS, 49, MS, 50; Univ Md, PhD(plant breeding), 60. *Prof Exp:* From asst agronomist to assoc agronomist, Div Cereal Crops & Dis, 50-56, RES AGRONOMIST, CEREAL CROPS RES BR, USDA, 56- *Mem:* Am Soc Agron. *Res:* Corn breeding and genetics, with emphasis on virus disease of corn. *Mailing Add:* Dept Agron Ohio Agron Res & Dev Ctr Wooster OH 44691

FINDLEY, WILLIAM ROBERT, b Detroit, Mich, Apr 26, 35. APPLIED CHEMISTRY. *Educ:* Denison Univ, BS, 57; Ohio Univ, MSc, 61, PhD(chem), 63. *Prof Exp:* Chemist, dyestuffs & Chem Div, 63-64, prod mgr chelates, 64-66, tech develop mgr whiteners, 66-74, tech develop mgr, 74-88, DIR, TECH DEVELOP & SERV, SPEC CHEMS, CIBA-GEIGY CORP, 88- *Mem:* Am Oil Chemists Soc. *Res:* Chelates. *Mailing Add:* Ciba-Geigy Corp 410 Swing Rd Greensboro NC 27419

FINE, ADRIAN, b Dublin, Ireland, June 28, 45; m 68; c 2. MEDICINE. *Educ:* Trinity Col, Dublin, MB, BCh & BAO Hons, 68, MD, 72. *Prof Exp:* Med Res Coun Ireland fel, 69-70; sr registrar nephrology med, Royal Infirmary, Glasgow, 74-77; consult, Greater Glasgow Health Bd, 77-78; asst prof med, Mem Univ Nfld, 78-; AT DEPT MED & PHARMACOL, UNIV MAN. *Concurrent Pos:* Med Res Coun fel, Univ West Indies, 75-76. *Mem:* Int Soc Nephrology; Renal Asn Brit. *Res:* Intermediary metabolism of the kidney; hormone metabolism by the kidney; drug metabolism in uraemia. *Mailing Add:* Dept Med Nephrol Fac Med Univ Man 753 McDermot Ave Winnepeg MB R3E 0W3 Can

FINE, ALBERT SAMUEL, b Philadelphia, Pa, Oct 24, 23; m 59; c 1. BIOCHEMISTRY. *Educ:* Brooklyn Col, BA, 50, MA, 53; NY Univ, PhD, 70. *Prof Exp:* Asst biochem & enzymol, Col Med, NY Univ, 50-52; biochemist, Col Physicians & Surgeons, Columbia Univ, 52-56; biochemist, Spec Dent Res Prog, 56-66 & Spec Res Lab Oral Tissue Metab, 66-70, CHIEF DENT RES LAB, VET ADMIN HOSP, 70- *Concurrent Pos:* Assoc prof periodont, histol & cell biol, Dent Ctr, NY Univ. *Mem:* AAAS; Am Chem Soc; NY Acad Sci; Am Inst Biol Sci; Int Asn Dent Res; Sigma Xi. *Res:* Intermediary metabolism of the oral tissues; biochemistry of oxidative and electron transport enzymes of oral mucosa; collagen synthesis during wound healing of epithelium; cyclic nucleotide and phosphatidyl inositol regulation of cellular proliferation during regeneration of oral tissues. *Mailing Add:* 2928 W Fifth St Brooklyn NY 11224

FINE, BEN SION, b Peterborough, Ont, Sept 29, 28. OPHTHALMOLOGY. *Educ:* Univ Toronto, MD, 53; Am Bd Ophthal, dipl, 59. *Prof Exp:* Assoc prof, 64-70, ASSOC RES PROF OPHTHAL, GEORGE WASHINGTON UNIV, 70-, RES ASSOC, ARMED FORCES INST PATH, 58- *Concurrent Pos:* Clin prof pathol, Uniformed Serv Univ Health Sci, Bethesda, Md, 77- *Honors & Awards:* Cert Award, Am Acad Ophthal, 83. *Mem:* Electron Micros Soc Am; Am Acad Ophthal. *Res:* Ophthalmic pathology; investigations into the structure of the eye, both normal and abnormal. *Mailing Add:* George Washington Univ Wash Hosp Ctr 1133 20th St NW No B150 Washington DC 20036

FINE, BENJAMIN, b New York, NY, Oct 12, 48; m 70; c 2. GROUP THEORY. *Educ:* Brooklyn Col, BS, 69; NY Univ, MS, 71, PhD(math), 73. *Prof Exp:* Instr math, Fashion Inst Technol, 73-74; from asst prof to assoc prof, 74-81, chmn dept, 78-81, PROF MATH, FAIRFIELD UNIV, 81- *Concurrent Pos:* Lilly Found res grant, Yale Univ, 77, assoc fel, 77-78; consult statist, Ctr Creative Living, 77-; prof, Univ Calif, Santa Barbara, 84-85; vis prof, Courant Inst, 87. *Mem:* Am Math Soc; Math Asn Am; Am Soc Qual Control. *Res:* Combinatorial group theory; structure of discrete groups of matrices; applications of discrete group theory to ring theory and number theory; quality control and industrial statistics. *Mailing Add:* 24 West Bank Lane Stamford CT 06902

FINE, DAVID H, b Johannesburg, SAfrica, Sept 17, 42; US citizen; m 64; c 2. CHEMISTRY, CHEMICAL ENGINEERING. *Educ:* Univ Witwatersrand, BSc, 64; Univ Leeds, PhD(chem), 67. *Prof Exp:* Fel chem, Univ Leeds, 67-68; fel, Univ Man, 68-69; res assoc chem eng, Mass Inst Technol, 69-72; sr scientist chem & head, Cancer Res Div, Thermo Electron Corp, 72-79; dir res, New Eng Inst Life Sci, 79-; AT THERMEDICS INC. *Mem:* Am Chem Soc; Soc Occup & Environ Health; Am Inst Chem Engrs; The Chem Soc. *Res:* Nitrosamines; nitroso compounds; worker and environmental exposure; analytical chemistry; combustion; explosions; air pollution chemistry; instrumentation; author of over 50 research papers. *Mailing Add:* Thermedics Inc 470 Wildwood St Woburn MA 01801-1799

FINE, DONALD LEE, b Nanticoke, Pa, Jan 14, 43; m 65; c 3. MICROBIOLOGY. *Educ:* Wilkes Col, AB, 64; Pa State Univ, MS, 66, PhD(microbiol), 68. *Prof Exp:* Res asst limnol, Wilkes Col, 62-64; trainee microbiol, Pa State Univ, 64-68, res assoc, 68-69; proj leader, Virus & Rickettsia Div, US Army Biol Ctr, 69-71; chief oncol, virol & cell biol, Bionetics Res Labs, Inc, 71-72; sect head, RNA Virus Lab, 72-79, MGR, NIH INTRAMURAL RES PROG, FREDERICK CANCER RES CTR, NAT CANCER INST, 79-, DIR RES SUPPORT PROG, 82- *Concurrent Pos:* Dep dir, sci mgt. *Mem:* AAAS; Sigma Xi; NY Acad Sci; Am Soc Microbiol; Am Asn Cancer Res. *Res:* Limnology; ecology; cell physiology; virus biochemistry; virology and immunology with particular reference to arboviruses; viral oncology, with special interest in primates; immunology, biochemistry and virology of mammary tumor viruses; in vitro anti-cancer and anti-acids drug screening. *Mailing Add:* Frederick Cancer Res & Develop Ctr PO Box B Frederick MD 21702

FINE, DOUGLAS P, b Cleburne, Tex, Mar 15, 43; m; c 2. MEDICAL. *Educ:* Univ Tex, BA, 66, MD, 68; Am Bd Internal Med, dipl, 72 & 78. *Prof Exp:* Intern med, Med Br, Univ Tex, Galveston, 68-69, from asst prof to assoc prof, Dept Med, 75-82; resident med, Vanderbilt Univ, Nashville, Tenn, 69-71, instr internal med, 71-72; fel infectious dis, Vet Admin Med Ctr & Vanderbilt Univ, Nashville, Tenn, 71-72; prin investr, Med Res Inst Infectious Dis, US Army, Frederick, Md, 72-75; CHIEF, INFECTIOUS DIS SECT, DEPT VET AFFAIRS MED CTR, OKLAHOMA CITY, OKLA, 82-; PROF, DEPT MED & CHIEF, INFECTIOUS DIS SECT, HEALTH SCI CTR, UNIV OKLA, OKLAHOMA CITY, 82- *Concurrent Pos:* Adj prof, Dept Microbiol & Immunol, Health Sci Ctr, Univ Okla, Oklahoma City, 84- *Mem:* AMA; Am Fedn Clin Res; fel Am Col Physicians; Am Soc Microbiol; Am Asn Immunologists; fel Infectious Dis Soc Am; Vet Affairs Soc Physicians Infectious Dis. *Mailing Add:* Infectious Dis Sect Univ Okla Health Sci Ctr 921 NE 13th St Oklahoma City OK 73104

FINE, DWIGHT ALBERT, b Los Angeles, Calif, Sept 17, 33. INORGANIC CHEMISTRY, ANALYTICAL CHEMISTRY. *Educ:* Univ Calif, Los Angeles, BS, 56; Univ Calif, Berkeley, PhD(inorg chem), 60. *Prof Exp:* Asst chem, Univ Calif, Berkeley, 56-57 & Lawrence Radiation Lab, 57-60; CHEMIST, NAVAL WEAPONS CTR, 60- *Concurrent Pos:* Vis res & Teaching assoc, Univ Calif, Irvine, 73-74. *Mem:* Am Chem Soc; AAAS. *Res:* Coordination chemistry; solution chemistry of metals of Group VIII; electroanalytical chemistry; mass spectroscopy. *Mailing Add:* 541 Rio Bravo St Ridgecrest CA 93555-3330

FINE, JO-DAVID, b Louisville, Ky, Apr 9, 50; m 72; c 3. IMMUNODERMATOLOGY, INTERNAL MEDICINE. *Educ:* Yale Col, BS, 72; Univ Ky, Lexington, MD, 76. *Prof Exp:* Intern, Duke Univ Med Ctr, Durham, NC, 76-77, jr asst resident internal med, 77-78; resident dermat, Harvard Med Sch/Mass Gen Hosp, 78-80, sr resident, Lahey Clin Found, 80-81; med staff fel dermat, Nat Cancer Inst, NIH, Bethesda, Md, 81-83; asst prof, 83-85, ASSOC PROF DERMAT, MED SCH, UNIV ALA

BIRMINGHAM, 85-, DIR DERMAT RES, 83-; CHIEF DERMAT SECT, MED SERV, BIRMINGHAM VET ADMIN MED CTR, 84- *Mem:* Soc Invest Dermat; fel Am Col Physicians; fel Am Acad Dermat; Am Fedn Clin Res; Dermat Found. *Res:* Blistering diseases of the skin; epidermolysis bullosa; immunology, biochemistry, ultrastructure and function of basement membrane in both normal and diseased skin; production of monoclonal antibodies to new basement membrane and extracellular matrix components. *Mailing Add:* Univ NC Rm 137 NC Mem Hosp Chapel Hill NC 27514

FINE, LAWRENCE OLIVER, b Sheyenne, NDak, May 14, 17; m 41; c 3. SOIL CHEMISTRY, WATER CHEMISTRY. *Educ:* NDak Agr Col, BS, 38; Univ Wis, PhD(soil chem), 41. *Prof Exp:* Soil surveyor, USDA, 41-42; resident instr agron, Univ Ark, 42 & asst agronomist, 45-46; from asst prof to assoc prof, 46-53, prof agron, SDak State Univ, 53-; RETIRED. *Concurrent Pos:* Assoc agronomist, SDak State Univ, 48-50, agronomist, Exp Sta, 53-69, head dept agron, 58-69; collabr soil & water div, Agr Res Serv, USDA, 53-84. *Mem:* Fel Am Soc Agron; AAAS; Soil Sci Soc Am; Am Inst Chemists. *Res:* Salinity, phosphate supply and nutrition; general soil fertility; irrigation. *Mailing Add:* Plant Sci Dept SDak State Univ Box 2207-A Brookings SD 57007

FINE, LEONARD W, b Bridgeport, Conn, Apr 19, 35; m 58; c 3. ORGANIC CHEMISTRY, POLYMER CHEMISTRY. *Educ:* Marietta Col, BS, 58; Univ Md, PhD(org chem), 62. *Prof Exp:* Res chemist, Harris Res Lab, Gillette Co, 62-64, Ethyl Corp, Mich, 64-66 & Am Cyanamid Co, Bridgeport, 66-70; from assoc prof to prof, Housatanic Community Col, 76-82; DIR, UNDERGRAD STUDIES CHEM, COLUMBIA UNIV, 82-, PROF, 88- *Concurrent Pos:* Lectr, Columbia Univ, 75-81, vis scholar, 78-79, sr lectr, 82-88. *Mem:* AAAS; Am Chem Soc; NY Acad Sci; Hist Sci Soc; Soc Hist Technol. *Res:* Synthesis and reactions of pyrimidine heterocycles; peroxide oxidations of organic compounds; synthesis of alpha-amino acids; natural auxins and synthetic plant growth regulators; organometallic synthesis; catalysis; engineering plastics and polymer synthesis chemical education; history of chemistry. *Mailing Add:* 15 Grey Hollow Rd Norwalk CT 06850

FINE, MICHAEL LAWRENCE, b Brooklyn, NY, Feb 10, 46; m 77; c 1. NEUROETHOLOGY, BIOCOMMUNICATIONS. *Educ:* Univ Md, BS, 67; Col William & Mary, MA, 70; Univ RI, PhD(oceanog), 76. *Prof Exp:* Assoc, Cornell Univ, 76-79; asst prof, 79-85, ASSOC PROF, VA COMMONWEALTH UNIV, 85- *Concurrent Pos:* Oceanographer, US Naval Oceanog Off, 76-77; res physiologist, Tunison Lab Fish Nutrit, Fish & Wildlife Serv, 79. *Mem:* Soc Neurosci; Animal Behav Soc; Am Soc Ichthyologists & Herpetologists. *Res:* In the oyster toadfish-described seasonal and geographical variation of mating call, evoked sound production by electrical brain stimulation, recorded extracellularly from single units of VIIIth nerve, localized testosterone and estrogen targets neurons in brains, and described development of sexually dimorphic sonic motor nucleus and sonic muscles. *Mailing Add:* Dept Biol Va Commonwealth Univ Richmond VA 23284-2012

FINE, MORRIS EUGENE, b Jamestown, NDak, Apr 12, 18; m 50; c 2. METALLURGY & MATERIALS SCIENCE. *Educ:* Univ Minn, BMetE, 40, MS, 42, PhD(phys metall), 43. *Prof Exp:* Instr phys metall, Univ Minn, 42-44; assoc metallurgist, Manhattan Proj, Chicago, 44-45; assoc metallurgist, Univ Calif, Los Alamos, NMex, 45; mem tech staff, Bell Tel Labs, 46-54; prof metall, Tech Inst, 54-64, chmn, Dept Mat Sci, 54-60, chmn res ctr, 60-64, assoc dean, Technol Inst Res & Grad Studies, 74-85, prof, 85-90, EMER PROF TECH, INST, 90-, WALTER P MURPHY PROF MAT SCI & ENG, NORTHWESTERN UNIV, EVANSTON, 64- *Concurrent Pos:* Mem mat adv bd, Nat Acad Sci, 65-69; vis prof, Stanford Univ, 67-68, Japan Soc for Prom Sci, 79, Univ Tex, Austin, 84-88; chmn mem comt, Nat Acad Eng, 79; chmn, Steel Resource Ctr, Northwestern Univ, 86- *Honors & Awards:* Campbell lectr, Am Soc Metals, 79; Mathewson Gold Medal, Am Inst Mining, Metall & Petrol Engrs Metall Soc, 81; Douglas Gold Medal, Am Inst Mech Engrs, 82; Gold Medal, Am Soc Metals, 86. *Mem:* Nat Acad Eng; fel Am Soc Metals; fel Am Phys Soc; fel Metall Soc; fel Am Ceramic Soc; hon mem Japan Inst Metals,87. *Res:* Phase transformations in solids; precipitation hardening; theory of the strength of metals and alloys; elasticity and internal friction of solids; magnetic properties of metals and ceramics; materials science; fatigue of metals; tribology of ceramic and composites. *Mailing Add:* Dept Mat Sci & Eng Northwestern Univ Evanston IL 60201

FINE, MORRIS M(ILTON), b St Louis, Mo, Nov 15, 14; m 37; c 1. EXTRACTIVE METALLURGY. *Educ:* Wash Univ, BS, 35. *Prof Exp:* Chemist, Mo, US Bur Mines, 35-40, metallurgist, 41-55, sect chief mineral dressing, 56-59, proj coordr metall, Minn, 59-70, res dir, Rolla Metall Res Ctr, 71-79; RETIRED. *Mem:* Am Inst Mining, Metall & Petrol Eng; Sigma Xi (pres, 59). *Res:* Agglomeration and reduction of iron ore raw materials; extractive metallurgy of lead and zinc; iron and steel process metallurgy. *Mailing Add:* 1405 Liberty Dr Rolla MO 65401

FINE, RANA ARNOLD, b New York, NY, Apr 17, 44; m 83. PHYSICAL OCEANOGRAPHY, CHEMICAL OCEANOGRAPHY. *Educ:* New York Univ, BA, 65; Univ Miami, MA, 73, PhD(marine sci), 75. *Prof Exp:* Post doc chem oceanog, Rosenstiel Sch Marine & Atmospheric Sci, Univ Miami, 76-77, res asst prof, 77-80, res assoc prof, 80-81; assoc prog dir phys oceanog, NSF, 81-83; assoc prof, 83-90, PROF, RSMAS, UNIV MIAMI, 90-, CHAIRPERSON, MARINE-ATM CHEM, 90- *Mem:* Am Geophys Union; AAAS; Oceanog Soc. *Res:* Use of anthropogenic tracers found in the oceans, such as chlorofluoromethanes, tritium and radiocarbon, to study circulation and mixing. *Mailing Add:* RSMAS Mac Univ Miami 4600 Rickenbacker Causeway Miami FL 33149

FINE, RICHARD ELIOT, b Pueblo, Colo, May 27, 42; m 69; c 2. CELL BIOLOGY, BIOCHEMISTRY. *Educ:* Univ Calif, Berkeley, AB, 64; Brandeis Univ, PhD(biochem), 69. *Prof Exp:* Am Cancer Soc fel, Med Res Coun Lab Molecular Biol, Cambridge, Eng, 69-71; from asst prof to assoc prof physiol, 71-84, prof biochem, 78-88, PROF BIOCHEM & NEUROL, SCH MED, BOSTON UNIV, 88- *Mem:* AAAS; Fedn Am Socs Exp Biol. *Mailing Add:* Dept Biochem & Neurol Boston Univ Med Sch 80 E Concord St Bldg K Rm 407 Boston MA 02118

FINE, SAMUEL, b Baranowiczach, Poland, Jan 21, 25; US citizen. BIOMEDICAL ENGINEERING, BIOPHYSICS. *Educ:* Univ Toronto, BASc, 46, MD, 57; Mass Inst Technol, SM, 53. *Prof Exp:* Jr engr, Can Gen Elec Co, 47-48; staff mem, Res Lab Electronics, Mass Inst Technol, 51-53; civil serv app, NIH, 58-59; res assoc med, Brookhaven Nat Lab, 59-61; assoc prof elec eng, 61-64, PROF BIOMED ENG, NORTHEASTERN UNIV, 64-, CHMN DEPT, 66- *Concurrent Pos:* Mem adv comt optical masers, Surgeon Gen, US Army, 64-70; founding dir, Laser Inst Asn, 67-71; consult, Mass Dept Pub Health, 66, Inst for Serv to Educ, Wash, 70-75; dir biomet, 68-72; dir adv technol pub, 68-81; clin & res fel, Mass Gen Hosp, Boston, 69-70; Z136 comt, Am Nat Standards Inst, NY, 72- *Honors & Awards:* Klein Lectr, 69; Laser Med & Surg Award, 86. *Mem:* Inst Elec & Electronics Engrs; Soc Nuclear Med; Sigma Xi; Am Soc Exp Path. *Res:* Electrical engineering; biomedical engineering; nuclear medicine; biological effects of laser radiation. *Mailing Add:* 360 Huntington Ave Boston MA 02115

FINE, TERRENCE LEON, b New York, NY, Mar 9, 39; div; c 2. ELECTRICAL ENGINEERING. *Educ:* City Col New York, BEE, 58; Harvard Univ, SM, 59, PhD(appl physics), 63. *Prof Exp:* Res fel, Harvard Univ, 63-64; Miller Inst fel, Univ Calif, Berkeley, 64-66; from asst prof to assoc prof, 66-79, PROF ELEC ENG, CORNELL UNIV, 79- *Concurrent Pos:* Consult, TASC, Inc, IBM Watson Res Ctr, Signatron, Inc & Lawrence Livermore Lab; vis prof, Stanford Univ, 79-80; assoc ed, Inst Elec & Electronics Engrs Trans Info Theory; pres, Inst Elect & Electronics Engrs Info Theory Group. *Mem:* Fel Inst Elec & Electronics Engrs; Inst Math Statist. *Res:* Neural Networks; foundations of probability and statistics. *Mailing Add:* Sch Elec Eng Cornell Univ Col Eng Ithaca NY 14853

FINEBERG, CHARLES, b Philadelphia, Pa, Jan 1, 21; m 46; c 3. THORACIC SURGERY. *Educ:* Wake Forest Col, BS, 40; Hahnemann Med Col, MD, 50; Am Bd Surg, dipl, 56; Am Bd Thoracic Surg, dipl, 57. *Prof Exp:* Intern med, Mt Sinai Hosp, 50-51; resident surg, 51-55, asst, 55-56, from instr to assoc prof, 55-72, PROF SURG, THOMAS JEFFERSON UNIV, 72-; CHMN DEPT SURG, DAROFF DIV, ALBERT EINSTEIN MED CTR, 68- *Concurrent Pos:* Res fel, Thomas Jefferson Univ, 51-52; USPHS fel, Nat Cancer Inst, 52-54; dir, Clin Cancer Training, Tumor Adv Group, 66-68; dir, oper rms, Thomas Jefferson Univ Hosp. *Mem:* Fel Am Col Surgeons; fel Am Asn Thoracic Surgeons; AMA; Sigma Xi; Royal Col Med. *Res:* Surgical correction of coronary artery disease and malabsorption syndromes; surgical uses of hyperbaric oxygenation; carcinoma of the breast. *Mailing Add:* 902 Locust St Philadelphia PA 19107

FINEBERG, HARVEY V, b Pittsburgh, Pa, Sept 15, 45. PUBLIC HEALTH. *Educ:* Harvard Univ, BA, 67, MA & MD, 72, PhD, 80. *Hon Degrees:* DHC, Univ Bordeaux II, 87. *Prof Exp:* From asst prof to assoc prof, 73-81, dir grad prog, 75-78, PROF HEALTH POLICY & MGT, SCH PUB HEALTH, HARVARD UNIV, 82-, DEAN, 84- *Honors & Awards:* Joseph Mountin Prize, 88; Wade Hampton Frost Prize, 88. *Mem:* Inst Med-Nat Acad Sci; Soc Med Decision-Making. *Mailing Add:* Sch Pub Health Harvard Univ 677 Huntington Ave Boston MA 02115

FINEBERG, HERBERT, b Portland, Maine, Jan 16, 15; m 41; c 2. ORGANIC CHEMISTRY, PHYSICS. *Educ:* Trinity Col, BS, 35; Univ Ill, PhD(org chem), 41. *Prof Exp:* Res chemist org, Eastman Kodak Co, 35-38; res mgr, Conn Hard Rubber Co, 41-45; pres, Geral Chem Co, 45-48; vpres res, Glyco Chem Co, 48-62; res mgr fat derivatives, Archer-Daniels-Midland Co, 62-67; res mgr fat derivatives, 67-75, RES MGR, TECH INFO CTR & PLANNING, ASHLAND CHEM CO, 75- *Concurrent Pos:* Prin, Proj Seed, Am Chem Soc, 91- *Mem:* Am Oil Chemists Asn; Am Chem Soc; Lic Exec Soc; Chem Mgt & Resources Asn. *Res:* Technical information storage and retrieval and technical and economic evaluations of projects in fatty derivatives and petrochemicals. *Mailing Add:* 2848 Maryland Ave Columbus OH 43209

FINEBERG, RICHARD ARNOLD, biochemistry; deceased, see previous edition for last biography

FINEG, JERRY, b Buffalo, NY, Jan 7, 28; m 55; c 4. LABORATORY ANIMAL SCIENCE. *Educ:* Tex A&M Univ, BS, 49, DVM, 53; Univ Southern Calif, MS, 64. *Prof Exp:* Chief vet sci, 6571st Aeromed Res Lab, USAF, 58-62, chief biodynamics, 64-67, chief, Vet Sci Div, Sch Aerospace Med, Brooks AFB, Tex, 69-73; DIR, ANIMAL RESOURCES CTR & PROF PHARM & PHYSIOL, COL PHARM, UNIV TEX, AUSTIN, 73- *Concurrent Pos:* Consult lab animal facil, Architects-Barnes, Landes, Goodman, Youngblood, White, Budd VanNess Partnership, 76- *Mem:* Am Vet Med Asn; Am Asn Lab Animal Sci; Int Primatol Soc; Am Soc Lab Animal Practitioners. *Res:* Drug effects and distribution in disease states of laboratory animals. *Mailing Add:* Animal Resources Ctr Univ Tex Austin TX 78712

FINEGOLD, HAROLD, molecular spectrometry, for more information see previous edition

FINEGOLD, LEONARD X, b London, Eng, Feb 15, 35; m 65; c 2. MOLECULAR BIOPHYSICS, POLYMER PHYSICS. *Educ:* Univ London, BSc, 56, PhD(physics), 59. *Prof Exp:* Res fel, Div Eng & Appl Physics, Harvard Univ, 59-62; res assoc, Lawrence Radiation Lab, Univ Calif, Berkeley, 62-65; from asst prof to assoc prof physics, Univ Colo, Boulder, 65-74; PROF PHYSICS, DREXEL UNIV, 74- *Concurrent Pos:* Am Cancer Soc res scholar biol, Univ Calif, San Diego, 71-72 & NIH spec fel biol, 72-73; secy, Div Biol Physics, Am Phys Soc, 75-80; Watkins prof, 78; Queen's Quest scholar, Queen's Univ, Kingston, Ont, 84; vis fel, Clare Hall, Univ Cambridge, 87-88. *Mem:* Am Phys Soc; Biophys Soc; Calorimetry Conf; AAAS; Fedn Am Scientists; Soc Basic Irreproducible Res. *Res:* Liposomes as models for cell membranes; metastable phase properties; cholesterol in liposomes; membranes of organisms from Antarctic rocks; internal dynamics of proteins; synchrocyclotron x-ray diffraction on liposomes. *Mailing Add:* Dept Physics Drexel Univ 31st & Chestnut St Philadelphia PA 19104

FINEGOLD, MILTON J, b New York, NY, Jan 13, 35; m 82; c 3. DEVELOPMENTAL PATHOLOGY, HEPATOLOGY. *Educ:* Columbia Univ, AB, 55; Univ Rochester, MD, 60. *Prof Exp:* From asst prof to prof path, NY Univ Med Sch, 67-79; dir, dept path, Bellevue Hosp, 76-79; PROF PATH & PEDIAT, BAYLOR COL MED, 79-; HEAD, DEPT PATH, TEX CHILDREN'S HOSP, 79- *Concurrent Pos:* Consult, St Clare's Hosp, 72-79, Manhattan Vet Admin Hosp, 75-79. *Mem:* Soc Pediat Path; Am Asn Pathologists; Soc Pediat Res; Int Acad Path; fel NY Acad Sci; Am Soc Clin Pathologists. *Res:* Somatic cell gene expression in normal and neoplastic liver, using cell culture; molecular probes; immunochemistry and ultrastructure to study regulation and control mechanisms. *Mailing Add:* Dept Path Baylor Col Med Houston TX 70030

FINEGOLD, SYDNEY MARTIN, b New York, NY, Aug 12, 21; m 47; c 3. INFECTIOUS DISEASES, MICROBIOLOGY. *Educ:* Univ Calif, Los Angeles, AB, 43; Univ Tex, MD, 49; Am Bd Internal Med, dipl, 57; Am Bd Med Microbiol, dipl, 78. *Prof Exp:* Intern, US Marine Hosp, 49-50; resident, Wadsworth Vet Hosp, Los Angeles, 53-54; from instr to assoc prof, Univ Calif, Los Angeles, 55-68; sect chief infectious dis, 57-86, ASSOC CHIEF OF STAFF RES & DEVELOP, WADSWORTH VET HOSP, 86-; PROF MED, SCH MED, UNIV CALIF, LOS ANGELES, 68- *Concurrent Pos:* Fel med, Med Sch, Univ Minn, 50-52; attend physician, Minn Gen Hosp, 51-52; mem comt infectious dis res prog, Vet Admin, 61-65, mem, Merit Rev Bd, 72-74 & Infectious Dis Adv Comt, 74-87; mem adv panel, US Pharmacopeia, 70-75; mem, Nat Res Coun-Nat Acad Sci Drug Efficacy Study Group, 66-69; mem subcomt, Int Comn Nomenclature Bacteria, 66-, chmn, 72-78, mem Sci Adv Comt, Am Found AIDS Res, 85-88; chmn working group, Anaerobic Suscept Test Methods, Nat Comt Clin Lab Standards, 87-; chair, Univ Calif, Los Angeles Acad Senate, 87-88; ed, Rev Infectious Dis, 90- *Honors & Awards:* V A William S Middleton Award Biomed Res, 84; Bristol Award, Infect Dis Soc Am, 87; Mayo Soley Award, 88. *Mem:* Fel AAAS; fel Am Acad Microbiol; master Am Col Physicians; Infectious Dis Soc Am (pres, 81); Soc Intestinal Microbiol Ecol & Dis (pres); Asn Am Physicians. *Res:* Infectious diseases; intestinal flora; anaerobic bacteria and anaerobic infections. *Mailing Add:* Wadsworth Vet Hosp W151 Los Angeles CA 90073

FINELLI, ANTHONY FRANCIS, b Newton, Mass, June 18, 22; m 56; c 3. ORGANIC CHEMISTRY, POLYMER CHEMISTRY. *Educ:* Boston Col, BS, 43; Univ Pa, MS, 47, PhD(chem), 50. *Prof Exp:* Org res chemist, Goodyear Tire & Rubber Co, 50-64, head urethane elastomers res, 64-74, mgr, rubber applns res, 74-77, mgr urethanes, 77-78, mgr urethane prod & polymer characterization, 78-82, mgr urethane & polyester prod, 82-85, mgr external polyesters, 85-86, mgr new tire prod, 86-87; RETIRED. *Mem:* Am Chem Soc; Am Inst Chemists; Am Soc Artificial Internal Organs; Soc Automotive Engrs. *Res:* Artificial heart valves, ventricles and arterial grafts; morphine chemistry; vinyl chloride polymers; natural and synthetic latices, natural and synthetic rubbers, urethane elastomers; foams and coatings; polyester characterization; polyesters for solution adhesives, hot melt adhesives, powder coatings; thermoplastic elastomers; stereospecific polymers for tires and industrial products. *Mailing Add:* 575 Castle Blvd Akron OH 44313

FINEMAN, MANUEL NATHAN, b Montreal, Que, Sept 9, 20; US citizen; m 45; c 3. SURFACTANTS, SOAPS & DETERGENTS. *Educ:* McGill Univ, BSc, 41, PhD(phys chem), 44. *Prof Exp:* Chemist colloids, Nat Res Coun, Ottawa, 44-46; fel phys chem, Stanford Univ, 46-47; res chem soaps & detergents, Rohm & Haas Co, Philadelphia, 47-51, group leader, 51-56; tech dir, Puritan Chem Co, Atlanta, 56-68, vpres, 68-70; dir res & develop, Avmor Ltd, Montreal, 70-73; vpres res & develop, ZEP Mfg Co, Atlanta, 73-90; CONSULT & EXPERT WITNESS, CHEM PROD LIABILITY CASES, 91- *Mem:* Am Chem Soc; Chem Inst Can; Am Inst Chemists; Sigma Xi; Chem Specialties Mfrs Asn; Southern Aerosol Tech Asn. *Res:* Physical chemistry of surfactants including mechanisms of foam formation and of the detergency process. *Mailing Add:* 55 Forrest Lake Dr NW Atlanta GA 30327-3310

FINEMAN, MORTON A, b Kearny, NJ, Aug 9, 19; m 49; c 2. ATOMIC & MOLECULAR PHYSICS, SOLID STATE PHYSICS. *Educ:* Ind Univ, BA, 41; Univ Pittsburgh, PhD(chem), 48. *Prof Exp:* Asst, Univ Pittsburgh, 41-44; mem res staff, Sprague Elec Co, 48-50; res fel, Univ Minn, 50-52; from asst prof to prof chem, Providence Col, 52-61; sr res staff mem, Gen Atomic Div, Gen Dynamics Corp, 61-66; chmn dept physics, 66-75, prof physics, Lycoming Col, 66-84; adj prof, San Diego State Univ, 84-89; CONSULT, SEPARATION SYSTS TECH, 90- *Concurrent Pos:* Vis prof physics, Univ Calif, San Diego, 74-75. *Mem:* Am Phys Soc; Am Chem Soc. *Res:* Thermodynamics of solid solutions; chemical kinetics; electron impact studies of gases; negative ions; chemical reactive cross sections via crossed beam techniques; angular scattering of electrons from atoms; nuclear magnetic resonance; reverse osmosis. *Mailing Add:* 4085-262 Rosenda Ct San Diego CA 92122

FINERMAN, A(ARON), b New York, NY, Apr 1, 25; m 68; c 2. COMPUTER SCIENCE. *Educ:* City Col New York, BCE, 48; Mass Inst Technol, SM, 51, ScD(civil eng), 56. *Prof Exp:* Instr, City Col New York, 48-49, lectr, 51-54; engr, Thomas Worcester & Co, 49; proj engr struct design, Voorhees, Walker, Foley & Smith, 51-54; res engr, Mass Inst Technol, 54-55, instr struct eng, 55-56; mgr digital comput & data process, Repub Aviation Corp, 56-61; prof eng & dir comput ctr, State Univ NY, Stony Brook, 61-69, prof comput sci, 69-71; mgr off comput & info systs, Jet Propulsion Lab, 71-73; prof comput sci, State Univ NY, 73-78, chmn dept, 75-77; dir comput ctr & prof comput & commun sci, 78-86, PROF ELEC ENG & COMPUT SCI, UNIV MICH, 86-, SPEC ASSOC INFO TECHNOL, 86- *Concurrent Pos:* Pres, SHARE, 61-62; sr res assoc, Jet Propulsion Lab, 68-69; mem, Nat Coun Asn Comput Mach, 57-83. *Mem:* Fel AAAS; Asn Comput Mach (treas, 73-); Am Fed Info Processing Soc; Asn Am Univ Profs. *Res:* Management of computing ctr; education in computing science and data processing; computing applications. *Mailing Add:* Univ Mich 2220 Bonisteel Blvd 5113 IST Univ Mich 2220 Bonisteel Blvd Ann Arbor MI 48109-2099

FINERTY, JOHN CHARLES, b Chicago, Ill, Oct 20, 14; m 40; c 2. ANATOMY. *Educ:* Kalamazoo Col, AB, 37; Kans State Col, MS, 39; Univ Wis, PhD(zool), 42. *Prof Exp:* Asst zool, Kans State Univ, 37-39; asst, Univ Wis, 39-42, asst physiol, 42; instr, Univ Mich, 43, instr anat, 43-46; from asst prof to assoc prof, Sch Med, Wash Univ, 46-49; from assoc prof to prof, Univ Tex Med Sch, 49-56, asst dean, 54-56; prof, Sch Med, Univ Miami, 56-66, chmn dept, 56-66, from asst dean to assoc dean, 56-66; dean, Sch Med, La State Univ Med Ctr, New Orleans, 66-71, prof anat, 66-84, vchancellor acad affairs, 71-84, dean, Sch Grad Studies, 74-84, emer prof, 84-; RETIRED. *Concurrent Pos:* Rackham Found fel, Univ Mich, 42-43. *Mem:* AAAS; Am Asn Anat (pres, 75-76); Am Physiol Soc; Radiation Res Soc; Endocrine Soc. *Res:* Experimental endocrinology; pituitary cytophysiology; correlation of microscopic structure and function; neurohumoral control of respiration; gross human anatomy; parabiosis; protection from x-ray irradiation; effects of fat deficiency. *Mailing Add:* 1110 Bethlehem St Houston TX 77018

FINESTONE, ALBERT JUSTIN, b Philadelphia, Pa, May 12, 21; m 51; c 3. MEDICINE. *Educ:* Temple Univ, AB, 42, MD, 45, MSc, 51. *Prof Exp:* Asst prof, 58-64, CLIN PROF MED, SCH MED, TEMPLE UNIV, 64-, DEAN CONTINUING MED EDUC, 73- *Concurrent Pos:* Dir continuing educ, Col Physicians Philadelphia, 74-79. *Mem:* AMA; Am Diabetes Asn; Am Col Physicians; Am Psychosom Soc; Am Fedn Clin Res. *Res:* Metabolic diseases. *Mailing Add:* Temple Univ Sch Med Temple Univ Hosp 3401 N Broad St Philadelphia PA 19140

FINGAR, WALTER WIGGS, b Nashville, Tenn, Jan 14, 34; m 55; c 6. DENTISTRY. *Educ:* Univ Tenn, DDS, 56; Univ Iowa, MS, 65. *Prof Exp:* From instr to asst prof dent technol, Col Dent, Univ Iowa, 64-67; asst prof oper dent, Col Dent, Univ Tenn, 67-68; assoc prof, 68-73, chmn dept, 71-77, assoc dean, 76-88, PROF OPER DENT, MED UNIV SC, 73-, DEAN, 88- *Concurrent Pos:* Dir maternity & infant care proj, Dent Clin, Med Univ SC, 68-70; mem nat rev comt, Proj ACORDE. *Mem:* Int Col Dentists. *Res:* Dental materials especially composite resins and pit and fissure sealants; pulpal response to dental materials and operations. *Mailing Add:* Off Dean Univ SC 171 Ashley Ave Charleston SC 29425

FINGER, IRVING, b Peekskill, NY, Sept 23, 24. GENETICS. *Educ:* Swarthmore Col, BA, 50; Univ Pa, PhD(zool), 55. *Prof Exp:* Am Cancer Soc fel, Columbia Univ, 56-58; from asst prof to assoc prof, 57-69, PROF BIOL, HAVERFORD COL, 69- *Concurrent Pos:* USPHS spec fel, Univ Calif, San Diego, 63-64. *Mem:* Genetics Soc Am; Soc Protozool; Am Soc Zool. *Res:* Immunology; microbial genetics; protozoology; cellular differentiation. *Mailing Add:* Dept Biol Haverford Col Haverford PA 19041

FINGER, KENNETH F, b Antigo, Wis, Jan 2, 29; m 51; c 1. BIOCHEMICAL PHARMACOLOGY. *Educ:* Univ Wis, BS, 51, MS, 53, PhD(pharm, pharmacol), 55. *Prof Exp:* Sr investr pharmaceut chem, Chas Pfizer & Co, 55-57, res supvr, 59-61, res mgr, 61-63; guest worker, Nat Heart Inst, 57-59; from assoc prof to prof pharmacol, Sch Pharm, Univ Wis, Madison, 63-68; dean Col Pharm, 68-74, ASSOC VPRES HEALTH AFFAIRS, UNIV FLA, 74- *Mem:* Am Soc Pharmacol & Exp Therapeut; Am Pharmaceut Asn. *Res:* Biochemistry of function; drug metabolism; drug-receptor interactions; catecholamines; anti-diabetic drugs and drugs affecting behavior. *Mailing Add:* Box J-14 JHMHC Univ Fla Gainesville FL 32610

FINGER, LARRY W, b Terril, Iowa, May 22, 40; m 62; c 2. MINERALOGY, CRYSTALLOGRAPHY. *Educ:* Univ Minn, BPhys, 62, PhD(crystal struct), 67. *Prof Exp:* Fel, Geophys Lab, Carnegie Inst, 67-69, mem staff, 69-; AT GEOPHYS LAB, WASHINGTON, DC. *Mem:* AAAS; Mineral Soc Am; Am Crystallog Asn; Am Geophys Union; Sigma Xi. *Res:* Crystallography and crystal structure of minerals at elevated pressure and temperature and the computations required in these studies. *Mailing Add:* Geophys Lab 5251 Broad Branch Rd NW Washington DC 20015-1305

FINGER, TERRY RICHARD, b Saugerties, NY, July 21, 48; m 79; c 2. FISH ECOLOGY, STREAM ECOLOGY. *Educ:* Northeastern Univ, BS, 71; State Univ NY, Syracuse, MS, 75; Ore State Univ, PhD(fisheries), 79. *Prof Exp:* Res assoc, Dept Fisheries & Wildlife, Ore State Univ, 79-80; asst prof Pop Dynamics & Fishery Biol, Sch Forestry, Fisheries & Wildlife, Univ Mo, 80-87; ASST PROF BIOL, WESTMINSTER COL, 89- *Mem:* Am Soc Ichthyologists & Herpetologists; Am Fisheries Soc; Sigma Xi. *Res:* Investigation of ecology of stream fishes, especially species interactions among benthic species; relations between ecology and systematics of stream fishes; ecology of larval fishes. *Mailing Add:* Dept Biol Westminster Col 501 Westminster Ave Fulton MO 65251-1299

FINGER, THOMAS EMANUEL, b Orange, NJ, Aug 30, 49; m 74; c 3. NEUROANATOMY, IMMUNOCYTOCHEMISTRY. *Educ:* Mass Inst Technol, SB, 72, SM, 73, PhD(psychol & brain sci), 75. *Prof Exp:* Res assoc psychiat, Res Found State Univ NY, Stony Brook, 75; NIH fel vision res, Dept Anat, Sch Med, Wash Univ, 75-78; from asst prof to assoc prof, 78-89, PROF, DEPT CELLULAR & STRUCT BIOL, MED SCH, UNIV COLO, 90- *Concurrent Pos:* Grass Found fel neurobiol, 76; investr, Marine Biol Lab, Woods Hole, Mass, 81-82; ed, J Comp Neurol, 82-; NIH career develop award, 83-87; mem, NIH study sect, 85-88; sr fel, Karolinska Inst, Swedish Med Res Coun, 87; dir, Rocky Mountain Taste & Smell Ctr, 88- *Mem:* Soc Neurosci; Asn Chemoreception Sci (treas, 88-); AAAS. *Res:* Comparative neuroanatomy especially of teleost fish; organization of taste buds; localization of neuropeptides and neurotransmitters; chemosensory systems; lateral line system; cerebellum; evolution of the nervous system. *Mailing Add:* Dept Cellular & Struct Biol Med Sch Univ Colo Denver CO 80262

FINGERMAN, MILTON, b Boston, Mass, May 21, 28; m 58; c 2. COMPARATIVE PHYSIOLOGY. *Educ:* Boston Col, BS, 48; Northwestern Univ, MS, 49, PhD(biol), 52. *Prof Exp:* Asst, Northwestern Univ, 49-51; from instr to assoc prof, 54-63, chmn dept, 66-69 & 80-85, PROF BIOL, TULANE UNIV, 63-, CHMN DEPT, 90- *Concurrent Pos:* Mem adv panel regulatory biol, NSF, 66-69; mem supply dept comt, Marine Biol Lab, Woods Hole, 70-

73 & comt chmn, 71-73; chmn nominating comt, Div Comp Endocrinol, Am Soc Zoologists, 72, prog officer, 77-78; Comt Animal Models Biomed Res Invertebrate, Inst Lab Animal Resources, Nat Res Coun, 72-73, comt marine invertebrate, 76-81; Environ Sci Prog Planning Coun, Gulf Univ Res Consortium, 77-78; managing ed, Am Zoologist, 81; mem exec comt, Am Soc Zoologists, 81; petrie chair vis prof, Technion, Haifa, Israel, 86; assoc ed, Pigment Cell Res, 86- *Mem:* AAAS; Am Inst Biol Sci; Sigma Xi; Am Soc Zool; Crustacean Soc. *Res:* Comparative endocrinology; biological chronometry; animal color changes; crustacean physiology and endocrinology; chromatophores. *Mailing Add:* Dept Ecol Evolution & Organismal Biol Tulane Univ New Orleans LA 70118

FINGERMAN, SUE WHITSELL, b Earlington, Ky, May 4, 32; m 58; c 2. ENVIRONMENTAL TOXICOLOGY. *Educ:* Transylvania Col, BA, 55; Tulane Univ, MS, 59, PhD(biol), 75. *Prof Exp:* Instr biol, Ursuline Acad, 67-68, Dominican Col, 68-69 & Xavier Univ, 69-71; res assoc biol, Tulane Univ, 72-80; res fac, Okla State Univ, 80-81; BIOL CONSULT, 80- *Concurrent Pos:* Vis prof, Tulane Univ, 84. *Mem:* Sigma Xi; Soc Environ Toxicol & Chem; Am Soc Zoologists. *Res:* Study of effects of insecticides and other environmental pollutants on the physiology and behavior of fishes and crabs. *Mailing Add:* 1730 Broadway New Orleans LA 70118

FINGL, EDWARD (GEORGE), b Oak Park, Ill, Oct 24, 23; m 56. PHARMACOLOGY. *Educ:* Purdue Univ, BS, 43, MS, 49; Univ Utah, PhD(pharmacol), 52. *Prof Exp:* Asst pharmaceut chem, Purdue Univ, 46-49; asst pharmacol, 49-51, USPHS fel, Col Med, 52-53, asst res prof, 53-54, from asst prof to assoc prof, 54-72, PROF PHARMACOL, COL MED, UNIV UTAH, 72- *Mem:* AAAS; Am Soc Pharmacol & Exp Therapeut. *Res:* Pharmacodynamics; pharmacokinetics; biostatistics; drug tolerance. *Mailing Add:* Em Pharmacol 2567 Beacon Dr Salt Lake City UT 84108

FINHOLT, ALBERT EDWARD, b Chicago, Ill, Jan 28, 18; m 41; c 4. CHEMISTRY. *Educ:* Knox Col, AB, 38; Univ Chicago, PhD(chem), 46. *Hon Degrees:* DSc, Knox Col, 87. *Prof Exp:* Res chemist, Gen Printing Ink Co, Ill, 39-42; res assoc chem dept staff & assoc dir hydride proj, US Navy, 45-47; chief chemist, Metal Hydrides, Inc, 47-49; assoc prof chem, 49-54, prof, 54-64, chmn dept, 57-64, dean, 64-71, vpres, 66-71, prof, St Olaf Col, 71-85; RETIRED. *Mem:* AAAS; Am Chem Soc. *Res:* Inorganic and metal hydrides; preparation of aluminum hydrides and their use in inorganic and organic chemistry; organometallics. *Mailing Add:* Dept Chem St Olaf Col Northfield MN 55057

FINHOLT, JAMES E, b Oak Park, Ill, Oct 28, 33; m 56; c 1. PHYSICAL CHEMISTRY, INORGANIC CHEMISTRY. *Educ:* St Olaf Col, BA, 55; Univ Calif, Berkeley, PhD(chem), 60. *Prof Exp:* Asst prof chem, Albion Col, 59-60; from asst prof to assoc prof, 60-72, chmn dept, 75-78 & 86-90, PROF CHEM, CARLETON COL, 73- *Concurrent Pos:* Petrol Res Found grant theoret inorg chem, Columbia Univ, 65-66; AEC res assoc, Argonne Nat Lab, 72-73; vis scientist, Oak Ridge Nat Lab, 80; vis prof, Newcastle Univ, Newcastle-upon-Tyne, England, 81; vis prof, Calif Inst Technol, 91. *Mem:* Am Chem Soc; Sigma Xi; AAAS. *Res:* Physical properties of coordination compounds in aqueous solution; chromium species; x-ray diffraction. *Mailing Add:* 411 Prairie St Northfield MN 55057

FININ, TIMOTHY WILKING, b Decatur, Ill, Aug 4, 49; m 72; c 3. KNOWLEDGE-BASED SYSTEMS, COMPUTATIONAL LINGUISTICS. *Educ:* Mass Inst Technol, SB, 71; Univ Ill, Urbana-Champaign, 77, PhD(computer sci), 80. *Prof Exp:* Res assoc, Artificial Intel Lab, Mass Inst Technol, 71-74; res asst, Coord Sci Lab, Univ Ill, 74-80; res staff, IBM Res Lab, San Jose, 77; asst prof computer sci, Computer & Info Sci, Univ Pa, 80-87; tech dir, Ctr Advan Info Technol, Unisys, 87-91; PROF & CHAIR COMPUTER SCI, UNIV MD, BALTIMORE COUNTY, 91- *Concurrent Pos:* Adj assoc prof, Computer & Info Sci, Univ Pa, 87-; assoc ed, Artificial Intel Rev, 89-; lectr decision sci, Wharton Sch Bus, 90- *Mem:* Am Asn Artificial Intel; Inst Elec & Electronics Engrs, Computer Soc; Asn Comput Mach. *Res:* Development of tools and techniques for building Knowledge-based systems, especially for representing and reasoning about knowledge and the integration with data base management systems; intelligent interfaces, including natural language processing. *Mailing Add:* 4524 Locust St Philadelphia PA 19139

FINITZO-HIEBER, TERESE, b Chicago, Ill, Apr 8, 47; m 69; c 1. AUDIOLOGY & SPEECH PATHOLOGY. *Educ:* Northwestern Univ, BS, 69, MA, 71, PhD(auditory res), 75. *Prof Exp:* Clin audiologist, Otol Prof Asn, 71-73; from asst prof to assoc prof audiol, 75-86, RES SCIENTIST, NEUROSCI RES CTR, UNIV TEX, DALLAS, 86-; RES CONSULT PEDIAT, AUDIOL, SOUTHWESTERN MED UNIV, 74- *Concurrent Pos:* Consult, Suburban Low Incidence Develop Exemplary Serv, 70-71, Children's Med Ctr, 76-; grant, Univ Tex, Dallas, 76-77, 77-78, Schering Drug Co, 78-79. *Mem:* Am Speech & Hearing Asn; Am Acad Audiol. *Res:* Audiological assessment and habilitation of infants using auditory evoked potentials; quantitative electrophysiology. *Mailing Add:* Neuro Sci Res Ctr 9705 Harry Hines Blvd Suite 203 Dallas TX 75220

FINIZIO, MICHAEL, b Naples Italy, Nov 27, 38; m 66; c 2. ORGANIC CHEMISTRY, MEDICINAL CHEMISTRY. *Educ:* Univ Naples, PhD(org chem), 62. *Prof Exp:* Res chemist, Endo Labs, Inc, 63-76, sr res chemist med chem, 76-81, proj mgr, 81-86, SR PROJ MGR, E I DUPONT DE NEMOURS & CO, 87- *Mem:* Am Chem Soc. *Res:* Synthesis of biological active compounds. *Mailing Add:* DuPont Co Med Prod Barley Mill Plaza P27-2112 Wilmington DE 19880

FINK, ANTHONY LAWRENCE, b Hertford, Eng, Jan 25, 43; m 66; c 1. BIOCHEMISTRY. *Educ:* Queen's Univ (Ont), BSc, 64, PhD(org chem), 68. *Prof Exp:* Nat Res Coun Can fel, Northwestern Univ, 67-69; from asst prof to assoc prof, 69-82, PROF CHEM, UNIV CALIF, SANTA CRUZ, 82-, CHMN DEPT, 88- *Concurrent Pos:* Staff, Lab Molecular Biophysics, Univ Oxford, England, 78-79; vis fel, All Souls Col, Oxford, 80-81. *Mem:* AAAS; Am Chem Soc; Biochem Soc; Am Soc Biochem & Molecular Biol; Biophys Soc; Am Soc Microbiol. *Res:* Mechanisms of enzyme reactions; cryobiochemistry; protein folding and stability; cryoenzymology; development of anti-penicillinase compounds; heat shock proteins. *Mailing Add:* Dept Chem Univ Calif Santa Cruz CA 95064

FINK, ARLINGTON M, b Armour, SDak, Dec 15, 32; m 66; c 2. MATHEMATICS. *Educ:* Wartburg Col, BA, 56; Iowa State Univ, MS, 58, PhD(math), 60. *Prof Exp:* Res assoc math, Princeton, NJ, 60; instr, Univ Va, 60-62; asst prof, Univ Nebr, Lincoln, 62-67; assoc prof, 67-71, PROF MATH, IOWA STATE UNIV, 71- *Mem:* Am Math Soc; Math Asn Am. *Res:* Almost periodic functions; ordinary differential equations; inequalities. *Mailing Add:* Dept Math Iowa State Univ Ames IA 50011

FINK, AUSTIN IRA, b New York, NY, Nov 18, 20; m 56; c 3. OPHTHALMOLOGY. *Educ:* Univ Mich, AB, 42; Long Island Col Med, MD, 44. *Prof Exp:* Instr, Med Col, Cornell Univ, 50-57; from asst prof to assoc prof, 55-72, PROF OPHTHAL, COL MED, STATE UNIV NY DOWNSTATE MED CTR, 72- *Res:* Blood vessels of conjunctiva and retina in relation to disease; surgery of congenital cataracts; electron microscopy of Schlemms canal and surrounding structures; lasers and glaucoma. *Mailing Add:* Dept Ophthal State Univ NY Downstate Med Ctr Brooklyn NY 11203

FINK, BERNARD RAYMOND, b London, Eng, May 25, 14; nat US; m 44; c 2. ANESTHESIOLOGY. *Educ:* Univ London, BSc, 35, MB & BS, 38. *Prof Exp:* Med supt, Methodist Mission Hosp, SAfrica, 47-49; from instr to assoc prof anesthesiol, Columbia Univ, 52-64; prof, 64-84, EMER PROF ANESTHESIOL, SCH MED, UNIV WASH, 84- *Concurrent Pos:* Commonwealth Fund fel, Monaco, 63; assoc prof anesthesiologist, Presby Hosp, NY, 55-64; fel fac anaesthetists, Royal Col Surgeons, Eng, 80. *Honors & Awards:* Fulbright lectr, Turku, 59; Excellence Res, Am Soc Anesthesiologists, 86. *Mem:* AAAS; Am Soc Pharmacol; Am Physiol Soc; Soc Neurosci; Am Soc Neurochem; hon mem Int Asn Study Pain. *Res:* Cellular biology of anesthesia; rapid axoplasmic transport; neuropharmacology of local anesthetics; mechanics of the human larynx. *Mailing Add:* Dept Anesthesiol RN-10 Univ Wash Sch Med Seattle WA 98195

FINK, CHARLES LLOYD, b Pittsburgh, Pa, May 29, 44; m 67. ACCELERATOR DESIGN, NUCLEAR INSTRUMENTATION. *Educ:* Univ Pittsburgh, BS, 66, PhD(nuclear physics), 71. *Prof Exp:* Fel nuclear physics, Argonne Nat Lab, 71-73, Los Alamos Nat Lab, 73-74; physicist reactor res, 74-85, ACCELERATOR PHYSICIST, ARGONNE NAT LAB, 85- *Mem:* Am Phys Soc; Am Nuclear Soc; Inst Elec & Electronics Engrs. *Res:* Development and testing of high current, high brightness linear particle accelerators and associated beam transport systems; development of new diagnostic devices for beam transport systems; computer data analysis. *Mailing Add:* Eng Dept Res Bldg 207 Argonne Nat Lab Argonne IL 60439

FINK, CHESTER WALTER, b New York, NY, May 6, 28; m 55; c 3. PEDIATRICS, RHEUMATOLOGY. *Educ:* Duke Univ, BA, 47, MD, 51. *Prof Exp:* From instr to assoc prof, 57-71, PROF PEDIAT, UNIV TEX HEALTH SCI CTR DALLAS, 71- *Concurrent Pos:* Fel pediat, Sch Med, Western Reserve Univ, 56-57. *Mem:* Am Pediat Soc; Soc Pediat Res; Am Col of Rheumatology. *Res:* Rheumatoid arthritis and allied diseases in children. *Mailing Add:* Dept Pediat Univ Tex Southwestern Med Ctr Dallas TX 75235

FINK, COLIN ETHELBERT, b Columbia, Pa, Aug 21, 10; m 48; c 1. INORGANIC CHEMISTRY. *Educ:* Pa State Col, BS, 32; Mass Inst Technol, MS, 33; Columbia Univ, PhD(chem eng), 44. *Prof Exp:* Prod engr, Cracking Dept, Atlantic Refining Co, 33-36; prod engr res lubricating oils, Pa State Univ, 36-40; asst chem eng, Columbia Univ, 40-44; asst pilot plant & develop, Am Cyanamid Co, 44-45; asst res & develop floor coverings, Armstrong Cork Co, 45-46; asst develop & prod cathode ray tubes, Radio Corp Am, 46-47; assoc prof chem eng, Drexel Inst, 47-51, prof, 51-57; from assoc prof to prof chem, 57-76, EMER PROF CHEM, FRANKLIN & MARSHALL COL, 76- *Mem:* AAAS; Am Chem Soc; Am Soc Eng Educ; Am Inst Chem Eng. *Res:* Viscosity-density-pressure characteristics of lubrication oils; binders in activated carbons; pulsating flow of fluids; heart-lung machine for cardiac surgery. *Mailing Add:* Maple Farm Akron PA 17501

FINK, DANIEL J, b Jersey City, NJ, Dec 13, 26; m 51; c 3. AERONAUTICAL & ASTRONAUTICAL ENGINEERING. *Educ:* Mass Inst Technol, BS, 48 & MS, 49. *Prof Exp:* Group leader, Aircraft Dynamics, Bell Aircraft Corp, 49-52; vpres, Allied Res Assocs Inc, 52-63; dep dir, Defense Res & Eng Strategic & Space Systs & asst dir, Defensive Systs, Dept Defense, Washington, DC, 63-67; vpres & gen mgr, Space Div, Gen Elec Co, 68-77; vpres & group exec, Aerospace Group, Gen Elec, Fairfield, 77-79, sr vpres, Corp Planning & Develop, 79-82; PRES, D J FINK ASSOCS INC, 82- *Concurrent Pos:* Mem, Defense Sci Bd, 69-72 & sr consult, 77-; mem, NASA Adv Coun, 78-81 & chmn, 82-88; dir, Titan Corp, Orbital Sci Corp; chmn, Bd Telecommun & Comput Appln, Nat Res Coun, 84-87, Dept Adv Bd, RPI, 81-83; Army Sci Adv Panel, 71-76; vis comt, dept aeronaut & astronaut, Mass Inst Technol, 72-83 & Sloan Sch Mgt, 82-85; consult, pres sci adv comt; mem, Comn on Future of US Space Prog, 90. *Honors & Awards:* Collier Trophy, Nat Aeronaut Asn, 76; von Karman lectr, Am Inst Astronauts & Aeronauts, 80. *Mem:* Nat Acad Eng; hon fel Am Inst Aeronaut & Astronaut; fel AAAS (pres, 74); corresp mem Int Acad Astronaut; Nat Soc Prof Engrs. *Res:* Author of 30 publications. *Mailing Add:* D J Fink Assoc Inc 8016 Matterhorn Ct Potomac MD 20854

FINK, DAVID JORDAN, b Columbus Ohio, Aug 5, 43; m 67. BIOCHEMICAL ENGINEERING. *Educ:* Univ Cincinnati, BSChE, 66; Univ Mich, MSE, 68, PhD(chem eng), 73. *Prof Exp:* Staff scientist, Battelle Mem Inst, 74-77, prin scientist, 77-81, projs mgr biochem eng, Columbus Labs, 81-86; PRES, BIO-INTEGRATION INC, 86- *Concurrent Pos:* Res assoc, Dept Biochem, Purdue Univ, 73-74; pres, Collatek Inc, 88. *Mem:* Am Inst Chem Engrs; Soc Biomat; Mat Res Soc; Controlled Release Soc. *Res:*

Enzyme technology in industrial processing, medical and analytical applications; immunoassay systems; controlled-release technology; microencapsulation; biocompatible materials. *Mailing Add:* Bio-Integration Inc 1445 Summit St Columbus OH 43210

FINK, DAVID WARREN, b Brooklyn, NY, Mar 30, 44; m 67; c 2. ANALYTICAL CHEMISTRY. *Educ:* Brooklyn Col, BS, 64; Lehigh Univ, PhD(anal chem), 69. *Prof Exp:* Sr res chemist, Res Ctr, Lever Bros Co, Edgewater, 69-71; sr res chemist, 71-72, sect leader, 72-75, asst dir, 75-84, ASSOC DIR, MERCK, SHARP & DOHME RES LABS, 84- *Concurrent Pos:* Fel, Asn Off Anal Chem, Collaborative Study Award. *Mem:* Am Chem Soc; Am Pharmaceut Asn. *Res:* Luminescence of metal chelates; properties of excited states; analytical applications of fluorescence spectroscopy. *Mailing Add:* 31 Ethan Allen Rd Freehold NJ 07728

FINK, DON ROGER, b Reading, Pa, Apr 27, 31; m 52; c 6. GEOPHYSICS, ELECTRONICS. *Educ:* Harvard Univ, AB, 52; Wash Univ, AM, 54. *Prof Exp:* Asst geophys anal group, Mass Inst Technol, 54-56; asst geophysicist, Woods Hole Oceanog Inst, 56-59; res geophysicist, Humble Oil & Refining Co, 59-63; res scientist, Melpar Inc, 63-64, sr scientist, 64; sr eng specialist, Philco Corp, 64-65, proj mgr, 65-67; sr eng specialist, Gen Atronics Corp, 67-70, chief geophysicist, Magnavox, 70-72; mgr oceanog & environ serv, Raytheon Co, 72-78; exec dir, New Eng Conf Pub Utilities Comnrs Inc, 78-84; SR PROG MGR, MARTIN MARIETTA CORP, 84- *Mem:* Soc Explor Geophys; Inst Elec & Electronics Engrs. *Res:* Elastic wave propagation; geomagnetism; oceanography and underwater systems; information theory; computer applications; communication systems; geophysical exploration. *Mailing Add:* Martin Marietta Corp 6711 Baymeadow Dr Glen Burnie MD 21061

FINK, DONALD G(LEN), b Englewood, NJ, Nov 8, 11; m 48; c 3. ELECTRICAL ENGINEERING, ELECTRONICS. *Educ:* Mass Inst Technol, BSc, 33; Columbia Univ, MSc, 42. *Prof Exp:* Asst, Mass Inst Technol, 33-34, staff mem, Radiation Lab, 41-43; asst ed, McGraw-Hill Publ Co, 34-37, managing ed electronics, 37-41, exec ed, 45-46, ed-in-chief, 46-52; expert consult, Off Secy War, 43-45; consult, Bur Ships US Navy, 46; dir res, Philco Corp, 52-61, vpres, 61; gen mgr, 62-63, gen mgr & exec dir, 63-75, exec consult, 75-76, EMER DIR, INST ELEC & ELECTRONICS ENGRS, 74- *Concurrent Pos:* Consult, TV standards, Belg Govt, 52; ed, Standard Handbk for Elec Engrs, 68-; mem & consult, Army Sci Adv Panel, 57-78; ed, Proc Inst Radio Engrs, 56-57; mem, Bd Int Orgns & Progs, Nat Acad Sci, 74-81; mem, US Nat Comn, UNESCO, 76-81. *Mem:* Nat Acad Eng; fel Inst Elec & Electronics Engrs (pres, 58); Soc Motion Picture & TV Engrs. *Res:* Radio and radar navigation systems; television, color systems and standards; stereophonic sound systems. *Mailing Add:* 103-B Heritage Hills Somers NY 10589

FINK, DWAYNE HAROLD, b Albert Lea, Minn, June 15, 32; m 60; c 3. AGRONOMY. *Educ:* Univ Minn, BS, 60; Va Polytech Inst, MS, 63, PhD(agron), 65. *Prof Exp:* Fel, Univ Minn, 65-66; soil scientist, US Water Conserv Lab, Agr Res Serv, 66-88, WESTERN COTTON RES TECH, USDA, 88- *Mem:* Clay Minerals Soc; Int Asn Study Clays; Am Soc Agron. *Res:* Surface chemistry of clay minerals, especially new methods for determination of surface area and adsorption of pollutants; reduction of surface energy of natural soils to induce precipitation runoff; establishment of new crops in arid lands. *Mailing Add:* Western Cotton Res Lab 4135 E Broadway Rd Phoenix AZ 85040

FINK, GERALD RALPH, b Brooklyn, NY, July 1, 40; m 61; c 2. GENETICS. *Educ:* Amherst Col, BA, 62; Yale Univ, MS, 64, PhD(genetics), 65. *Hon Degrees:* DSc, Amherst Col, 82. *Prof Exp:* Fel biochem genetics, NIH, 65-67; from asst prof to prof genetics, Cornell Univ, 67-69, prof biochem, 79-82; Am Cancer Soc prof genetics, 82-90, DIR, WHITEHEAD INST BIOMED RES, 90- *Concurrent Pos:* Mem, Adv Panel Genetic Biol & Genetics Study Sect, NSF, 70-73; assoc ed, J Genetics Soc Am, 70-73, Genetics 70-74 & J Bact, 73-78; Guggenheim fel, 74-75; chmn, Int Conf Molecular Biol of Yeast, 75, Adv Group Virol & Cell Biol, Am Cancer Soc, 76-77 & Corp Vis Comt Biol Sci, Yale Univ, 84-; co-chmn, Gordon Conf Biol Regulation Mechanisms, 77; ed, Gene, 78-89, Molecular & Gen Genetics, 79-90; comt mem, Panel Genetic Stock Ctrs, Genetics Soc Am, 78-, vis comt, Carnegie Inst, 83-, sci adv panel, Searle Scholars Prog, 84- & bd sci adv, Jane Coffin Childs Mem Fund Med Res, 84-; mem, Genetics Study Sect, NIH, 79-84 & vis comt, Bd Overseers, Dept Cell & Develop Biol, Harvard Univ, 88-; lectr, Europ Molecular Biol Orgn, 80. *Honors & Awards:* Ciba-Geigy Lectr Microbial Biochem, 80; US Steel Prize Molecular Biol, Nat Acad Sci, 81; Genetics Soc Am Medal, 82; Sci & Eng Award, Yale Univ, 84; Emil Christian Hansen Found Award, 86; Am Cyanamid Lectr, Princeton Univ, 91. *Mem:* Nat Acad Sci; Genetics Soc Am (secy, 77-80, vpres, 86-87, pres, 88-89); Am Acad Arts & Sci. *Res:* Regulation of gene activity in eucaryotes; control of histidine biosynthesis in yeast. *Mailing Add:* Whitehead Inst Biomed Res Nine Cambridge Ctr Cambridge MA 02142

FINK, GREGORY BURNELL, b Outlook, Mont, Aug 3, 28; m 57; c 3. PHARMACOLOGY. *Educ:* Mont State Univ, BS, 50; Univ Utah, PhD(pharmacol), 60. *Prof Exp:* Asst prof, Wash State Univ, 60-63; asst prof, Univ Kans, 64; head dept, 64-70, PROF PHARMACOL, ORE STATE UNIV, 64- *Concurrent Pos:* Consult, Ore State Drug Adv Coun. *Mem:* Am Pharmaceut Asn; Am Soc Pharmacol & Exp Therapeut. *Res:* Neuropharmacology; psychopharmacology. *Mailing Add:* Ore State Univ Col Pharm Corvallis OR 97331

FINK, HERMAN JOSEPH, b Neutitschein, Czech, Aug 16, 30; c 3. SOLID STATE PHYSICS. *Educ:* Univ BC, BASc, 55, MASc, 56, PhD(physics), 59. *Prof Exp:* Nat Res Coun Can fel, Oxford Univ, 59-61; mem tech staff parametric amplifiers & magnetic properties solids, Bell Tel Labs, NJ, 61-63; res specialist electronic structure solids, Atomics Int Div, NAm Aviation, Inc, 63-69; actg prof, 69-70, PROF ELEC ENG & COMP SCI, UNIV CALIF, DAVIS, 70- *Mem:* Fel Am Phys Soc. *Res:* Properties of superconductors, semiconductors at low temperatures and microwave frequencies; resonance properties of magnetic materials and magnetic levitation. *Mailing Add:* Dept Elec Eng & Comput Sci Univ Calif Davis CA 95616

FINK, JAMES BREWSTER, b Los Angeles, Calif, Jan 12, 43; m 70; c 1. IMPEDANCE SPECTROSCOPY. *Educ:* Univ Ariz, BSc, 69, PhD(geol eng & hydrol), 89; Univ Witwatersrand, SAfrica, MSc, 80. *Prof Exp:* Geophysicist, Geo-Comp Explor Inc, 69-70 & Ifex-Geotechnica, SAm, 70-71; chief geophysicist, Mining Geophys Surv, 71-72; consult, Mining & Groundwater Geophys, 72-76; sr mining geophysicist, Esso Eastern Inc, 76-79; sr res geophysicist, Exxon Prod Res Co, 79-80; pres, Geophynque Int, 81-91; PRES, HYDROGEOPHYS CO, 91- *Concurrent Pos:* Consult, Dept Elec Eng, Univ Ariz, 84-85, adj lectr, Dept Mining & Geol Eng, 85-87, assoc instr, Dept Geosci, 86-87 & consult, Dept Hydrol, 87; consult, Ensco, 86-88 & Los Alamos Nat Lab, 87-91. *Mem:* Soc Explor Geophysicists; Europ Asn Explor Geophysicists; Am Geophys Union; Minerals & Geotech Logging Soc; Asn Petrol Geochem Explorationists. *Res:* Application of electrical, magnetic, electromagnetic, gravity, and refraction seismologic geophysical methods to geotechnical, environmental, and hydrological problems; development of microcomputer software and hardware. *Mailing Add:* HydroGeophys Co 5865 S Old Spanish Trail Tucson AZ 85747

FINK, JAMES PAUL, b Calumet, Mich, Nov 10, 40; m 66; c 1. APPLIED MATHEMATICS. *Educ:* Drexel Inst Technol, BS, 63; Stanford Univ, MS, 65, PhD(math), 67. *Prof Exp:* Eng technician, Clifton Precision Prod Co, Inc, 59-62; instr physics, Drexel Univ, 62-63; instr math, Stanford Univ, 66-67; from asst prof to assoc prof math, Univ Pittsburgh, 73-86; PROF MATH & DEPT HEAD, BUTLER UNIV, 86- *Mem:* Am Math Soc; Soc Indust & Appl Math; Math Asn Am; hon mem Sigma Xi. *Res:* Differential equations; functional analysis; nonlinear differential equations; mathematical modeling; analysis, bifurcation theory, continuation methods and numerical analysis. *Mailing Add:* Dept Math Sci Butler Univ Indianapolis IN 46208

FINK, JOANNE KRUPEY, b Greensburg, Pa, Feb 19, 45; m 67. NUCLEAR ENGINEERING, PHYSICS. *Educ:* Univ Pittsburgh, BS, 66, PhD(physics), 72. *Prof Exp:* Appointee physics, Argonne Nat Lab, 72-73; prof classical mech, Univ NMex, Los Alamos Exten, 74; appointee physics, Los Alamos Sci Lab, 74; res assoc, 74-75, asst physicist chem eng, 75-79, chem engr, 79-86, PHYSICIST, ARGONNE NAT LAB, 86- *Mem:* Am Phys Soc; Am Asn Aero Res Soc. *Res:* Experimental and theoretical determination of physical properties used in reactor safety analysis; material interaction experiments; analysis of light water reactor fission product; source term data; experimental aerosol research related to reactor safety; aerosol data analysis. *Mailing Add:* Reactor Eng Div 9700 S Cass Ave Argonne IL 60439-4840

FINK, JONATHAN HARRY, b New York, NY, May 2, 51; m 86. VOLCANOLOGY, PLANETARY GEOLOGY. *Educ:* Colby Col, BA, 73; Stanford Univ, PhD(geol), 79. *Prof Exp:* Postdoctoral res assoc appl math, Weizmann Inst, Rehovot, Israel, 79; postdoctoral res assoc, Ariz State Univ, 80-82, vis assoc prof, 82-83, from asst prof to assoc prof, 83-91, actg chair geol, 90-91, PROF GEOL ARIZ STATE UNIV, 91- *Concurrent Pos:* Prin investr geochem prog, NSF, 82-, planetary geol & geophysics prog, NASA, 83-, Dept Energy, 85-87; field geologist, US Geol Surv, Menlo Park, Calif, 82; adj prof, Dept Chem Eng, Univ Colo, Boulder, 85-87; vis res fel, Australian Nat Univ, Canberra, 88-90; assoc ed, J Geophys Res, 90-93. *Mem:* AAAS; Geol Soc Am; Am Geophys Union; Sigma Xi. *Res:* Volcanic hazards; lava flow emplacement and rheology; growth of lava domes; laboratory simulation and theoretical modeling of volcanic processes; planetary volcanism; tectonics of icy satellites; mechanisms of igneous intrusion. *Mailing Add:* Dept Geol Ariz State Univ Tempe AZ 85287

FINK, JORDAN NORMAN, b Milwaukee, Wis, Oct 13, 34; m 56; c 3. ALLERGY, IMMUNOLOGY. *Educ:* Univ Wis, BS, 56, MD, 59. *Prof Exp:* Asst instr, Sch Med, Marquette Univ, 60-63; NIH fel allergy & immunol & assoc, Med Sch, Northwestern Univ, 63-65; from instr to assoc prof med, 65-73, PROF MED, MED COL WIS, 73- *Concurrent Pos:* Clin investr, Vet Admin, 65-68. *Mem:* AAAS; Am Asn Immunologists; Am Asn Clin Invest; fel Am Acad Allergy; Assoc Am Physicians; Am Thoracic Soc. *Res:* Basic mechanisms in human hypersensitivity, and their diagnosis and treatment; internal medicine. *Mailing Add:* 8700 W Wisconsin Milwaukee WI 53226

FINK, KATHRYN FERGUSON, biochemistry; deceased, see previous edition for last biography

FINK, KENNETH HOWARD, b Omaha, Nebr, Dec 11, 14; m 40; c 2. FOOD CHEMISTRY. *Educ:* Univ Denver, BS, 36. *Prof Exp:* Analyst, Great Western Sugar Co, 38-39 & Wilson & Co, 39-41; analyst, Joliet Ord Works, Army Ord, 41-44, resident inspector, 44-45; res engr, Podbielniak, Inc, 45-49; res chemist, Food Res Div, Armour & Co, 49-54, chief chemist, Refinery, 54-56, asst sect head fats & oils, Food Res Div, 56-67, res analyst, 67-79; RETIRED. *Mem:* Am Chem Soc; Am Oil Chemists Soc. *Res:* Technology of edible fats and oils; food analysis. *Mailing Add:* 1461 Selleck St Crete IL 60417

FINK, LESTER HAROLD, b Philadelphia, Pa, May 3, 25; m 55; c 2. POWER SYSTEM OPERATION & CONTROL, POWER NETWORKS DYNAMIC PHENOMENA. *Educ:* Univ Pa, BSEE, 50, MSEE, 61. *Prof Exp:* Supv engr, Philadelphia Elec Co, 50-74; asst dir, Elec Energy Syst Div, US Dept Energy, 74-79; pres, Systs Eng Power, Inc, 79-83; chmn, Carlsen & Fink Assocs, Inc, 83-89; EXEC VPRES, ECC, INC, 89- *Concurrent Pos:* Adj prof, Drexel Univ, 61-74 & Univ Md, 79-80. *Mem:* Fel Inst Elec & Electronics Engrs; fel Instrument Soc Am; NY Acad Sci. *Res:* Electric energy systems engineering, analysis, and control, including underground high voltage cable; power plant, power system, and interconnected power systems modeling, dynamics, operation, security, reliability and control. *Mailing Add:* 11304 Full Cry Ct Oakton VA 22124

FINK, LINDA SUSAN, b Great Neck, NY, June 3, 58. EVOLUTIONARY ECOLOGY, FIELD NATURAL HISTORY. *Educ:* Amherst Col, BA, 80; Univ Fla, MS, 84, PhD(zool), 89. *Prof Exp:* Vis asst prof biol, Middlebury Col, 88-90; ASST PROF BIOL, SWEET BRIAR COL, 90- *Concurrent Pos:* Pratt postdoctoral fel, Mountain Lake Biol Sta, Univ Va, 90. *Mem:* Am Inst Biol

Sci; Am Soc Zoologists; Animal Behav Soc; Ecol Soc Am; Lepidopterists' Soc; AAAS. *Res:* Ecology and evolution of arthropod defenses against predators and parasites; polymorphism and polyphenism in herbivorous insects, especially hawkmoth caterpillars. *Mailing Add:* Dept Biol Sweet Briar Col Sweet Briar VA 24595

FINK, LOUIS MAIER, b Brooklyn, NY, Mar 28, 42; m 62; c 3. PATHOLOGY. *Educ:* Boston Univ, BA, 61; Albany Med Col, MD, 65. *Prof Exp:* Asst prof path, Col Physicians & Surgeons, Columbia Univ, 70-72; from asst prof to prof, Dept Path, Med Sch, Univ Colo, 72-77; from assoc prof to prof path, Med Sch, Vanderbilt Univ, 77-81; PROF, SCH MED, UNIV ARK, 81- *Mem:* Am Asn Cancer Res; Am Asn Pathologists; Harvey Soc; Int Acad Pathologists; Col Am Pathologists. *Res:* Plasma membrane protein interactions and neoplastic processes. *Mailing Add:* Dept Path Little Rock VAMC 4300 W Seventh Ave Little Rock AR 72205

FINK, LOYD KENNETH, JR, b Canton, Ill, Dec 26, 36; m 63; c 1. MARINE GEOLOGY. *Educ:* Univ Ill, BS, 61; Univ Miami, PhD(marine geol, geophys), 68. *Prof Exp:* Fel geophys, Inst Marine Sci, Univ Miami, 68-69; asst prof, 69-77, ASSOC PROF OCEANOG, UNIV MAINE, ORONO, 77-, COOP ASST PROF GEOL SCI, IRA C DARLING RES CTR, WALPOLE, 74- *Mem:* Geol Soc Am; Am Geophys Union. *Res:* Tectonics of the seafloor; origin of island arcs. *Mailing Add:* Dept Oceanog Univ Maine Six Coburn Hall Orono ME 04469

FINK, LYMAN R, b Elk Point, SDak, Nov 14, 12; m 37; c 3. ELECTRICAL ENGINEERING. *Educ:* Univ Calif, Berkeley, BS, 33, MS, 34, EE, 35, PhD(elec eng), 37. *Prof Exp:* Mgr, Electronics Lab, Gen Elec Co, 47-49, chief engr, Radio & TV Dept, 49-55, mgr, Res Appln Dept, Res Lab, 55-57, gen mgr, X-ray Dept & pres, Gen Elec X-Ray, Can, 57-59 & Atomic Prod Div, 59-63; vpres, Otis Elevator Co, 63-66; vpres & chief tech officer, Singer Co, 66-68, group vpres, Off Equip Group, 68-70; exec vpres, Church's Inc, 70-73; pres & owner, Diversitek Co, 73-87; ADV COUN, UNIV TEX, SAN ANTONIO, 78- *Concurrent Pos:* Chmn, ad hoc comt digital comput, Res & Develop Bd, 49-50; mem, comt atomic energy, US Chamber Com, 59-63; dir, Atomic Indust Forum, 60-63, vpres, 63; trustee, Southwest Res Inst, 76-87. *Mem:* Fel Inst Elec & Electronics Engrs. *Res:* Administration of engineering and research; television receivers; x-ray and medical electronics; commercial applications of atomic energy; atomic power plants; atomic fuel cycle; computer music. *Mailing Add:* 4309 Muirfield San Antonio TX 78229

FINK, MANFRED, b Berlin, Ger, Aug 16, 37; m 64; c 2. ATOMIC PHYSICS. *Educ:* Univ Karlsruhe, vordiplom, 58, diplom, 61, PhD, 64. *Prof Exp:* Fel physics, Ind Univ, 65-66, res assoc, 66-67; fac assoc, 67-69, asst prof, 69-73, assoc prof, 73-80, PROF PHYSICS, UNIV TEX, AUSTIN, 80- *Mem:* Am Phys Soc; Ger Phys Soc. *Res:* Electron scattering from gases and numerical evaluation of the scattering theories; amorphous materials. *Mailing Add:* Dept Physics Univ Tex Austin TX 78712

FINK, MARTIN RONALD, b New York, NY, Apr 27, 31; m 52; c 3. AERODYNAMICS, ACOUSTICS. *Educ:* Mass Inst Technol, BS & MS, 53. *Prof Exp:* Res asst, Transonic Aircraft Control Proj, Mass Inst Technol, 52-53; res engr, United Aircraft Labs, 53-58, supvr, Missile Aerodyn, 59-63, supvr, Aerodyn, 64-67; sr consult, United Technologies Res Ctr, 67-80; chief aerodyn, Norden Systs, 80-84, advan systs engr, 84-89; CONSULT, 89- *Mem:* Am Inst Aeronaut & Astronaut. *Res:* Methods used for predicting airframe noise and powered-lift noise, including an understanding of processes that cause such noise; methods for predicting air-conditioner fan noise. *Mailing Add:* 183 Woody Lane Fairfield CT 06432

FINK, MARY ALEXANDER, b Camden, Tenn, Oct 18, 19; m 50. CANCER. *Educ:* Okla Agr & Mech Col, BS, 39; Univ Mich, MS, 45; George Washington Univ, PhD(bact), 49. *Prof Exp:* Immunologist, Camp Detrick, 46-49; res assoc, R B Jackson Mem Lab, 49-51 & Dept Microbiol, Univ Colo Sch Med, 51-58; immunologist, Nat Cancer Inst, 59-66, head immunol sect, Viral Leukemia & Lymphoma Br, 66-70, prog dir immunol, Extramural Area, 70-74, assoc dir res progs, 74-77, spec asst for Spec Projs, Div Extramural Activ, Nat Cancer Inst, 77-84; RETIRED. *Mem:* Emer mem Am Asn Immunologists; emer mem Soc Exp Biol & Med; emer mem Am Asn Cancer Res; emer mem Am Acad Microbiol; emer mem Brit Soc Immunol; emer mem Sigma Xi. *Res:* Tularemia in humans; immunology and hypersensitivity, particularly in mice; immune response to tumors; oncogenic viruses, especially leukemia. *Mailing Add:* PO Box 386 Ogunquit ME 03907

FINK, MAX, b Vienna, Austria, Jan 16, 23; nat US; m 49; c 3. NEUROPSYCHIATRY. *Educ:* NY Univ, BA, 42, MD, 45. *Prof Exp:* Supvry psychiatrist, Dept Exp Psychiat, Hillside Hosp, 53-54, dir, 54-62; res prof psychiat, Sch Med, Wash Univ, 62-66; prof psychiat, New York Med Col, 66-72, dir div biol psychiat, 67-72; PROF PSYCHIAT, HEALTH SCI CTR, STATE UNIV NY STONY BROOK, 72- *Concurrent Pos:* Dir, Mo Inst Psychiat, 62-65; mem comt clin drug eval, NIMH, 62-65; prof, Univ Mo, 65-66; exec dir, Int Asn Psychiat Res, 67- & Div Clin Sci, Long Island Res Inst, 76-83; ed-in-chief, Convulsive Ther, 85- *Honors & Awards:* Award, Electroshock Res Asn, 56; Bennett Award, Soc Biol Psychiat, 58; S Hamilton Award, Am Psychopath Asn, 74; Anna Monika Prize, 79; Taylor Manor Hosp Psychiat Award, 83; Meduna Prize, 87. *Mem:* Soc Biol Psychiat; fel Am Psychiat Asn; Am Psychopath Asn; Am Col Neuropsychopharmacol. *Res:* Experimental alteration of human behavior by neurophysiologic agents; evaluation of psychiatric therapies; opiate dependence; cannabis research; convulsive therapy; electroencephalography. *Mailing Add:* State Univ NY Stony Brook Univ Hosp 36 Spring Hollow Rd PO Box 457 St James NY 11780

FINK, RACHEL DEBORAH, b New York, NY, Sept 29, 56; m 89. CELL REARRANGEMENTS DURING GASTRULATION, VIDEO MICROSCOPY. *Educ:* Cornell Univ, BA, 78; Duke Univ, PhD(zool), 84. *Prof Exp:* Postdoctoral fel, Yale Univ, 84-86; ASST PROF, DEPT BIOL SCI, MT HOLYOKE COL, 86- *Concurrent Pos:* Investr, Marine Biol Lab, Woods Hole, 84-; NSF presidential young investr award, 89-94. *Mem:* Soc Develop Biol; Am Soc Zoologists; Am Soc Cell Biol; AAAS; Am Inst Biol Sci; Sigma Xi. *Res:* Embryonic cell rearrangements during fish gastrulation; time-lapse video microscopy and image processing of embryonic cells; membrane dynamics studied with fluorescent probes. *Mailing Add:* Dept Biol Sci Mt Holyoke Col South Hadley MA 01075

FINK, RICHARD DAVID, b New York, NY, July 14, 36; m 61; c 2. PHYSICAL CHEMISTRY. *Educ:* Harvard Univ, AB, 58; Mass Inst Technol, PhD(nuclear chem), 62. *Hon Degrees:* MA, Amherst Col, 71, LHD, Doshisha Univ, Kyoto. *Prof Exp:* NSF fel chem, Yale Univ, 62-63, NIH fel, 63-64; from asst prof to prof chem, Amherst Col, 64-71, chmn dept, 70-73 & 80-82, mellon prof chem, 77-80, dean fac, 83-88, COREY PROF CHEM, AMHERST COL, 88- *Concurrent Pos:* Lectr, Yale Univ, 62-64; NSF sci fac fel, King's Col, Univ London, 68-69 & Amherst Col, 76-77; Sloan res fel, 70-74; Dreyfus Teacher-Scholar Prize, 71-75; vis prof, King's Col, Univ London, 72-73 & 80-81; vis lectr sci & humanities, Univ Kans, 81; vis scholar, Sci Technol & Soc, Mass Inst Technol, 88-90. *Mem:* Am Phys Soc; Am Chem Soc. *Res:* Chemistry of molecular beams; photochemistry of atomic and molecular interactions. *Mailing Add:* Chem Dept Amherst Col Amherst MA 01002

FINK, RICHARD WALTER, nuclear chemistry, physics; deceased, see previous edition for last biography

FINK, ROBERT DAVID, b Brooklyn, NY, Oct 1, 42; c 4. BIOCHEMISTRY. *Educ:* St Lawrence Univ, BS, 64; Univ Tenn, MD, 67. *Prof Exp:* Intern, Baptist Mem Hosp, 67-68; resident psychiat, Univ Tenn, 68-71; dir psychiat serv, alcohol & drug unit, 71-74, clin dir, 74-76, asst supt adult serv, 76-77, dep supt, 77-78, SUPT PSYCHIAT, MEMPHIS MENT HEALTH INST, 78- *Concurrent Pos:* Am Psychiat Asn Falk fel, 70-71. *Mem:* Am Med Soc Alcoholism. *Res:* Alcohol and drug abuse; psychopharmacology. *Mailing Add:* 539 Yates Rd S Memphis TN 38119

FINK, ROBERT M, b Greenville, Ill, Sept 27, 15. BIOLOGICAL CHEMISTRY. *Educ:* Univ Ill, AB, 37; Univ Rochester, PhD(biochem), 42. *Prof Exp:* prof, 47-78, EMER PROF BIOL CHEM, UNIV CALIF, LOS ANGELES, 78- *Concurrent Pos:* Mem, Subcomt Int Dose, Nat Coun Radiation Protection & Measurements, NBS, 47-49; res biochemist, Vet Admin, 47-54. *Mem:* Am Soc Biol Chemists. *Mailing Add:* 17774 Tramonto Dr Pacific Palisades CA 90272

FINK, RODNEY JAMES, b Oregon, Mo, Apr 10, 34; m 58; c 4. AGRONOMY, WEED SCIENCE. *Educ:* Univ Mo, BS, 56, MS, 61, PhD(field crops), 66. *Prof Exp:* Exten agronomist, Coop Exten Serv, Iowa State Univ, 61-64; instr field crops, Univ Mo, 64-66; assoc prof agr, Murray State Univ, 66-68, Univ Found grant, 66-67; from assoc prof to prof agr & chmn dept, Western Ill Univ, 68-74, dean appl sci, 74-90; agr adv to NWFP, Agr Univ, Peshawar, Pakistan, 90- *Concurrent Pos:* Partic conf comt educ in agr & natural resources, Nat Res Coun, 67; from chmn resident educ sect to vpres, North Cent Weed Control Conf, 74-76, pres, 76-; Western Ill Univ contract, Agency Int Develop Progs, USDA, 74-; contact, Univ Title XII & Inst Soc Econ Change. *Mem:* Weed Sci Soc Am; Soil Conserv Soc Am; Sigma Xi. *Res:* Effectuating new, better and safer methods of controlling weeds in crop and non-crop areas; effects of herbicides on crop species; effects of fertilizer sources on nitrate pollution in soil; factors affecting soil persistence of herbicides; administration of academic and international programs in agriculture and applied sciences. *Mailing Add:* Col Appl Sci Western Ill Univ Macomb IL 61455

FINK, THOMAS ROBERT, b Indianapolis, Ind, June 5, 43; m 67; c 2. BIOPHYSICAL CHEMISTRY, SURFACE CHEMISTRY. *Educ:* Ind Univ, AB, 65; Yale Univ, PhD(biophys chem), 70. *Prof Exp:* Fel biophys chem, Wash State Univ, 70-72; asst prof biochem, Univ Tulsa, 72-74; sr res chemist, Res & Develop Dept, Atlantic Richfield Corp, 74-78,; mgr, environ conserv, 78-88, MGR, ENVIRON SCI, AREO ALASKA INC, 88- *Concurrent Pos:* NIH & NSF fel, 65-70; NIH fel, 71-72. *Mem:* Am Chem Soc. *Res:* Physical polymer chemistry of polynucleotides, helix-coil transitions; surface chemistry of enhanced recovery of petroleum, chemical flooding; chemical dispersion of oil spills; Arctic environmental research. *Mailing Add:* 6359 Colgate Anchorage AK 99504-3305

FINK, WILLIAM HENRY, b Marshfield, Wis, Nov 4, 41; m 68. CHEMICAL PHYSICS, THEORETICAL CHEMISTRY. *Educ:* Univ Wis, BS, 63; Princeton Univ, MS, 65, PhD(phys chem), 66. *Prof Exp:* Res assoc chem, Princeton Univ, 66-67; NATO res fel appl math, Queen's Univ Belfast, 67-68; from asst prof to assoc prof, 68-80, PROF CHEM, UNIV CALIF, DAVIS, 80- *Mem:* Am Chem Soc; Am Phys Soc; Sigma Xi. *Res:* Application of quantum mechanical methods to problems of chemical interest; especially ab initio molecular structure calculations. *Mailing Add:* Dept Chem Univ Calif Davis CA 95616

FINK, WILLIAM LEE, b Coleman, Tex, July 22, 46; m 72; c 1. ICHTHYOLOGY, SYSTEMATIC BIOLOGY. *Educ:* Univ Miami, BS, 67; Univ Southern Miss, MS, 69; George Wash Univ, PhD(biol), 76. *Prof Exp:* Asst cur fishes & asst prof, Mus Comp Zool, Harvard Univ, 76-80, assoc prof biol, 80-; ASSOC PROF & ASSOC CURATOR FISHES, MUS ZOOL, UNIV MICH, ANN ARBOR, 81- *Concurrent Pos:* Prin & co-prin investr, NSF grants, 78-91. *Mem:* Am Soc Ichthyol & Herpetol; Soc Syst Zool (pres, 90); Soc Study Evolution; Am Soc Zool; Sigma Xi; fel AAAS. *Res:* Systematics of neotropical freshwater fishes; systematics of mesopelagic fishes; systematic theory. *Mailing Add:* Mus Zool Univ Mich Ann Arbor MI 48109

FINKBEINER, HERMAN LAWRENCE, b Syracuse, NY, July 20, 31; m 54. PHYSICAL CHEMISTRY, ORGANIC CHEMISTRY. *Educ:* Park Col, AB, 52; Univ Mich, PhD(chem), 60. *Prof Exp:* Staff asst org chem, Spencer Chem Co, 53-56; res scientist, Corp Res & Develop, 59-70, mgr chem synthesis & processing opers, 70-72, mgr synthesis & characterization br, 72-76, mgr employee rels oper, CRD, 76-78, mgr planning & resources, mat sci & eng, 78-82, MGR BIOL SCI, GEN ELEC CO, 82- *Mem:* Am Chem Soc. *Res:* Action of metal ions on organic reactions, especially biological reactions. *Mailing Add:* Gen Elec Res & Dev Ctr Box 8 Bldg K-1 Rm 3B35 Schenectady NY 12301-0008

FINKBEINER, JOHN A, b Freeport, Pa, Sept 3, 17; wid; c 2. MEDICINE. *Educ:* Univ Pittsburgh, BS, 39; Western Reserve Univ, MD, 42. *Prof Exp:* Asst med, Harrisburg Polyclin Hosp, 46-51; ASST PROF CLIN MED, COL MED, CORNELL UNIV, 56- *Concurrent Pos:* Dir tumor clin, Harrisburg Polyclin Hosp, 49-51, consult, 53-; clin asst, Mem & James Ewing Hosps, 53-55, asst attend physician, 55-74, assoc attend physician, 74-; assoc, Sloan-Kettering Inst Cancer Res, 53-62, asst clinician, 62-65; adj physician, Lenox Hill Hosp, 64-66, assoc physician, 66-78, physician, 78-, chief med neoplasia serv, 65-86. *Honors & Awards:* Meritorious Service Award, Am Cancer Soc. *Mem:* NY Acad Sci; AMA; Soc Surg Oncol; fel Am Col Physicians; Am Soc Cytol; Am Soc Clin Oncol; Am Soc Int Med; Am Radium Soc. *Res:* Medical oncology; Hodgkin's disease and other lymphomas; therapy of advanced cancer. *Mailing Add:* 1015 Madison Ave, Suite 302 New York NY 10021

FINKE, GUENTHER BRUNO, b Minden, Ger, May 31, 30; m 57; c 3. SOLID STATE PHYSICS, METALLURGY. *Educ:* Brunswick Tech Univ, dipl, 55, Dr rer Nat(semiconductors), 57. *Prof Exp:* Asst tech physics, Brunswick Tech Univ, 55-57; asst magnetic mat, Widia-Factory-Krupp-Essen, Ger, 57-58, asst dept head, 58-60; VPRES RES & DEVELOP, MAGNETIC METALS CO, 60- *Mem:* Inst Elec & Electronics Engrs; Ger Phys Soc. *Res:* Fuel cells; semiconductors; magnetic alloys and devices. *Mailing Add:* Magnetic Metals Co 21st & Hayes Ave Camden NJ 08101

FINKE, JAMES HAROLD, b Joplin, Mo, July 20, 44; m 67; c 1. IMMUNOLOGY. *Educ:* Kans State Col Pittsburg, BS, 67; Univ Mo, Kansas City, MS, 70; Univ Mo, Columbia, PhD(immunol), 73. *Prof Exp:* Fel immunol, Albert Einstein Col Med, 73-74; fel, 74-75, ASSOC STAFF IMMUNOL, CLEVELAND CLIN FOUND, 75- *Mem:* Sigma Xi; Fedn Am Socs Exp Biol. *Res:* Determining the nature of the cell-cell interaction between lymphocytes and target cells necessary for the generation of cytotoxic lymphocytes. *Mailing Add:* Immunol Res Dev Cleveland Clin Found Cleveland OH 44106

FINKE, REINALD GUY, b San Francisco, Calif, June 10, 28; m 84; c 3. SYSTEMS MODELING, TRAJECTORY ANALYSIS. *Educ:* Univ Calif, Berkeley, AB, 49, MA, 51, PhD(nuclear physics), 54. *Prof Exp:* Teaching asst physics, Univ Calif, Berkeley, physicist, Radiation Lab, 50-54, sr physicist, Radiation Lab, Livermore, 54-62; MEM SR TECH STAFF, INST DEFENSE ANALYSIS, 82- *Mem:* Am Astronaut Soc; Sigma Xi. *Res:* Study of operations and configurations of military aerospace systems, particularly space launch vehicles, spacecraft, and supporting ground facilities, with emphasis on performance, and technological and economic feasibility. *Mailing Add:* 5040 Fernleaf Ct Woodbridge VA 22192

FINKE, RICHARD GERALD, b Scottsbluff, Nebr, July 1, 50; m 73; c 1. BIOINORGANIC & ORGANOTRANSITION METAL CHEMISTRY. *Educ:* Univ Colo, BA, 72; Stanford Univ, PhD(chem), 76. *Prof Exp:* NSF fel chem, Stanford Univ, 76-77; asst prof, 77-82, ASSOC PROF CHEM, UNIV ORE, 82- *Concurrent Pos:* NSF postdoctoral fel, 77, Alfred P Sloan fel, 82-84, Dreyfus Teacher-Scholar fel, 82-87, Guggenheim fel, 85-86. *Mem:* Am Chem Soc. *Mailing Add:* Dept Chem Univ Ore Eugene OR 97403-1210

FINKEL, ASHER JOSEPH, medicine, zoology; deceased, see previous edition for last biography

FINKEL, HERMAN J(ACOB), b Chicago, Ill, Mar 9, 18; m 41; c 2. SOIL & WATER, RELOCATING HEAVY STRUCTURES. *Educ:* Univ Ill, BS, 40; Hebrew Univ, PhD, 57. *Prof Exp:* Dairy farm inspector, Health Dept, St Louis, Mo, 40-41; civil engr, Great Lakes Div, Corps Eng, Ill, 41-43; agr engr & farm planner, Soil Conserv Serv, NY, 43-46; consult structural engr, Ill, 46-49; chief engr, Soil Conserv Serv, Israel Ministry Agr, 49-52; lectr agr eng, Israel Inst Technol, 52-57; expert irrig adv to Govt Peru, Food & Agr Orgn, UN, 57-59; assoc prof agr eng, Israel Inst Technol, 59-67, prof & consult engr, 67-70, vpres acad affairs, 70-73, dean fac agr eng, 73-77, EMER PROF AGR ENG, ISRAEL INST TECHNOL, 77- *Concurrent Pos:* Consult, Water Dept, Israel Ministry Agr, 56-57 & UN Spec Fund Peru, 60; mem agr mission Haute Volta, WAfrica, 61; mem UNESCO teams ed planning, Chile, Peru, Trinidad, Jamaica & Brit Honduras, 64; mem agr mission & consult, Brit WIndies, Thailand & Panama, 62-64; consult engr, Cyprus, Nigeria, Iran, Uganda, Ivory Coast, Thailand & Brazil, 70-; head, Finkel & Finkel, Consult Engrs, Haifa; consult long distance relocation heavy struct. *Res:* Irrigation methods and efficiency; soil conservation, especially sand-dune control and economic water utilization; development of underground dams; modernization of the traditional Arab village; relocating heavy structures. *Mailing Add:* Yoqneam 20600 Israel

FINKEL, LEONARD, b Brooklyn, NY, Jan 7, 31; m 49; c 2. ELECTRONICS. *Educ:* Rutgers Univ, BS, 52; Harvard Univ, MS, 53. *Prof Exp:* Test engr, Chelsea Fan & Blower Co, NJ, 51-52; develop engr, Raymond Rosen Eng Prod, Pa, 53-56, group supvr, 56-58; sect mgr, Teledynamics, Inc, 58-60, asst to chief engr, Tele Dynamics Div, Am Bosch Arma Corp, Pa, 60-62, chief engr, 62-66, dir tech planning, 66-71; prog mgr advan systs develop, RCA Corp, 71-75, mgr advan progs, Missile & Surface Radar Div, 75-81, mgr systs concepts, Gov Systs Div, 81-87; MGR ADV SYST ENG, GOV ELECTRONICS SYST DIV, GEN ELEC CORP, 87- *Mem:* Inst Elec & Electronics Engrs. *Res:* Radar; weapon systems; instrumentation; air traffic control; communications; space systems; telemetry. *Mailing Add:* Combat Systems Eng RCA Corp Marne Hwy 108-104 Moorestown NJ 08057

FINKEL, MADELON LUBIN, b Mt Vernon, NY, Oct 11, 49; m 73; c 1. DEMOGRAPHY. *Educ:* NY Univ, BA, 71; Grad Sch Pub Admin, MPA, 73, Grad Sch Arts & Sci, PhD(health serv res), 80. *Prof Exp:* Res asst pub health & demog), Grad Sch Pub Health, NY Univ, 72-74; staff assoc pub health, Sch Pub Health, Columbia Univ, 74-75; res consult demog, Dept Health, New York City, 75-77; res assoc, 77-80, asst prof, 80-86, CLIN ASSOC PROF, PUB HEALTH & EPIDEMIOL, CORNELL UNIV, MED COL, 86-; PRES, SECOND OPINION CONSULTS, INC, 84- *Concurrent Pos:* Organizer & co-sponsor, Conf Second Surg Opinion Progs, 81; consult, Amalgamated Meat Cutters, 80-81, Mobil Oil Corp, 80-81, Am Pub Health Asn, 80-; consult, Cath Med Ctr, 85-87. *Mem:* Am Public Health Asn; Am Fedn Clin Res; Pop Asn Am; Int Epidemol Asn; Asn Health Ser & Res; Am Col Epidemiol. *Res:* Medical care (health services); demographic research; occupational epidemiological studies. *Mailing Add:* Dept Pub Health Med Col Cornell Univ 1300 York Ave New York NY 10021

FINKEL, RAPHAEL ARI, b Chicago, Ill, June 13, 51; m 84; c 2. COMPUTER SCIENCE. *Educ:* Univ Chicago, BA & MAT, 72; Stanford Univ, PhD(comput sci), 76. *Prof Exp:* Assoc prof comput sci, Univ Wis-Madison, 76-86; PROF, UNIV KY, LEXINGTON, 87- *Mem:* Sigma Xi; Asn Comput Mach; AAAS; Am Asn Prof Yiddish; Inst Elec & Electronics Engrs. *Res:* Operating systems; distributed computing; programming languages. *Mailing Add:* Dept Computer Sci Univ Kentucky 959 POT Lexington KY 40506-0027

FINKELMAN, FRED DOUGLASS, b New York, NY, Mar 4, 47. IMMUNOLOGY, RHEUMATOLOGY. *Educ:* Yale Univ, MD, 71. *Prof Exp:* Assoc prof, 80-86, PROF MED, UNIFORMED SERV UNIV HEALTH SCI, 86- *Mailing Add:* Dept Med Uniformed Serv Univ Health Sci 4301 Jones Bridge Rd Bethesda MD 20814

FINKELMAN, ROBERT BARRY, b New York, NY, Feb 17, 43; m 65; c 2. COAL SCIENCE, MICROMINERALOGY. *Educ:* City Col New York, BS, 65; George Washington Univ, MS, 70; Univ Md, PhD(chem), 80. *Prof Exp:* Cartographer, US Coast & Geodetic Surv, 65; geologist, US Geol Surv, 65-80; sr res specialist, 80-86, environ consult, 86-87, RES CHEMIST, US GEOL SURV, EXXON PROD RES CO, 87- *Concurrent Pos:* Prin investr, Nat Aeronaut & Space Admin, 74-75; lectr geol, Northern Va Community Col, Loudoun Campus, 74-80. *Honors & Awards:* Nininger Meteorite Award, 70. *Mem:* Mineral Soc Am; Geochem Soc; Geol Soc Am. *Res:* Geochemistry of coal, peat and burning coal-waste banks; mineralogy and petrography of lunar soil and cosmic dust; mineralogy of geodes; micromineralogical techniques; acid mine damage prediction. *Mailing Add:* 11907 Blue Spruce Rd Reston VA 22091

FINKELSTEIN, ABRAHAM BERNARD, b New York, NY, Feb 11, 23; m 47; c 3. APPLIED MATHEMATICS. *Educ:* City Col New York, BS, 43; NY Univ, MS, 47, PhD(math), 53. *Prof Exp:* Asst prof math, Long Island Univ, 48-53; sr res assoc aero eng & appl mech, Polytech Inst Brooklyn, 53-57; prof eng sci, 57-64, PROF MATH, PRATT INST, 64-, CHMN DEPT, 66- *Concurrent Pos:* Consult, Eastern Res Group, 55- *Mem:* Am Math Soc; Math Asn Am. *Res:* Water waves; underwater propulsion systems; transpiration cooling; differential equations. *Mailing Add:* 992 E 21st St Brooklyn NY 11210

FINKELSTEIN, DAVID, b New York, NY, July 19, 29; m 47, 81; c 4. ELEMENTARY PARTICLE PHYSICS. *Educ:* City Col New York, BS, 49; Mass Inst Technol, PhD(physics), 53. *Prof Exp:* Asst physics, Mass Inst Technol, 50-53; from instr to assoc prof, Stevens Inst Technol, 60-79; from assoc prof to prof physics, Yeshiva Univ, 60-78, dean, 78-79; dir, Sch Physics, 79-80, PROF PHYSICS, GA INST TECHNOL, 79- *Concurrent Pos:* Res assoc, NY Univ, 54-56; consult, Brookhaven Nat Lab, 57-; Ford fel, Europ Orgn Nuclear Res, Geneva, 58-59; ed, Int J Theoret Physics, 77-; vis prof, Oxford Univ, UK, 89-90; coun, Int Quantum Logics Asn, 91- *Mem:* Am Phys Soc. *Res:* Quantum mechanics; general relativity; quantum logics; quantum topology; discovered physical meaning of Schwarzschild singularity, topological spin-statistics theorem. *Mailing Add:* Sch Physics Ga Inst Technol Atlanta GA 30332-0430

FINKELSTEIN, DAVID, b Philadelphia, Pa, Apr 27, 11; m 39; c 1. CARDIOLOGY. *Educ:* Temple Univ, BS, 32, MD, 35; Am Bd Internal Med, dipl. *Prof Exp:* From instr to asst prof, Sch Med, Univ Pa, 50-67, assoc prof cardiol, 67-87; RETIRED. *Concurrent Pos:* Asst cardiologist, Philadelphia Gen Hosp, 37-75; chief cardiac clin, St Luke's & Children's Hosps, Philadelphia, 47-50; chief cardiac clin, Grad Hosp Univ Pa, 64- *Mem:* AMA; Am Heart Asn; Am Col Physicians; Am Col Cardiol. *Res:* Clinical research in relationship to cardiovascular parameters. *Mailing Add:* 1120 Hillcrest Rd Narberth PA 19072

FINKELSTEIN, DAVID B, b New York, NY, Dec 12, 45; m 65; c 3. MOLECULAR BIOLOGY. *Educ:* Univ of Wis, BS, 66; Albert Einstein Col Med, PhD(biochem), 72. *Prof Exp:* instr, Dept Biochem, Univ Tex Health Sci Ctr, Dallas, 75-77, asst prof, 77-84; dir fungal res, Pan Lab Res, 84-86; dir, Molecular Biol Serv Div, Panlabs, Inc, 86-88, sci dir, US Labs, 88-89, corp mgr, sci affairs, 89-90, PRIN SCIENTIST, PANLABS, INC, 90- *Mem:* Am Soc Biochem & Molecular Biol; Am Soc Microbiol; Soc Indust Microbiol. *Res:* Application of molecular biology to industrially relevant microorganisms. *Mailing Add:* Panlabs Inc 11804 N Creek Pkwy S Bothell WA 98011-8805

FINKELSTEIN, JACOB, b New York, NY, Oct 27, 10; m 45; c 2. ORGANIC CHEMISTRY, MEDICINAL CHEMISTRY. *Educ:* City Col New York, BS, 33; Columbia Univ, AM, 34, PhD(chem), 39. *Prof Exp:* Res chemist, Res Corp, NY, 35 & Merck & Co, NJ, 35-43; res scientist, Hoffmann-La Roche, Inc, 43-75; prof org chem, St Peter's Col, Jersey City, 77-79; CHEM CONSULT, 77- *Concurrent Pos:* Mem fac, Fairleigh Dickinson Univ. *Mem:* AAAS; fel Am Inst Chemists; Am Chem Soc; Sigma Xi. *Res:* Vitamins; sulfanilamides, antimalarials; contrast media; renal clearance; synthetic organic chemicals; antihistamines; antispasmodics; analgesics; alkaloidal salts of organic acids and quaternary hydroxides in the resolution of optically active substances; glucose lowering and antiviral agents; tetracyclines; stimulants; anticholesterol and antihypertensive drugs; penicillins; cephalosporins; antibiotics; patent evaluation; Food and Drug Administration regluatory matters. *Mailing Add:* 648 Sunderland Rd Teaneck NJ 07666

FINKELSTEIN, JACOB NOAH, b New York, NY, Mar 18, 49; m 71; c 2. CELL BIOLOGY, INHALATION TOXICOLOGY. *Educ:* Carnegie Mellon Univ, BS, 71; Northwestern Univ Med Sch, PhD(biochem), 76. *Prof Exp:* Fel radiat biol & biophys, Sch Med, Univ Rochester, 76-78, asst prof pediat & radiation biol & biophys, 78-85, scientist, 85-86, ASSOC PROF PEDIAT & TOXICOL, UNIV ROCHESTER, 86- *Concurrent Pos:* Grad Res Award, Sigma Xi, 74. *Mem:* Am Soc Biochem & Molecular Biol; AAAS; Am Chem Soc; NY Acad Sci; Am Soc Cell Biol; Soc Toxicol. *Res:* Understanding the mechanisms through which eukaryotic cells respond to external signals, which can take the form of normal physiologic stimuli or related to toxicologic stresses; responses of the lung to inhaled or systemic agents. *Mailing Add:* 74 Westerloe Ave Rochester NY 14620

FINKELSTEIN, JAMES DAVID, b New York, NY, Oct 16, 33; m 59; c 2. GASTROENTEROLOGY, BIOCHEMISTRY. *Educ:* Harvard Univ, AB, 54; Columbia Univ, MD, 58; Am Bd Internal Med, dipl, 67. *Prof Exp:* From intern to asst resident med, Presby Hosp, NY, 58-61; clin assoc, Nat Inst Arthritis & Metab Dis, 63-65; from asst prof to assoc prof, 66-74, PROF MED, GEORGE WASHINGTON UNIV, 74-; CHIEF, BIOCHEM RES LAB, VET ADMIN HOSP, 68-, CHIEF MED SERV, 79- *Concurrent Pos:* USPHS trainee gastroenterol, Col Physicians & Surgeons, Columbia Univ, 61-63; clin investr, Vet Admin, 65-68, med investr, 70-75; consult, Children Hosp, Washington, DC, 67-80 & Nat Inst Arthritis, Metab & Digestive Dis, 73-77; prof med, Howard Univ, 83-; clin prof med, Georgetown Univ, 82-; chief, gastroenterol & hepatology, Vet Admin Hosp, 70-79, assoc chief staff res, 75-80. *Honors & Awards:* Arthur S Flemming Award, 71. *Mem:* AAAS; Am Fedn Clin Res; Am Gastroenterol Asn; Am Soc Clin Invest; Am Soc Clin Nutrit; Asn Am Physicians. *Res:* Sulfur amino acid metabolism; active transport across cell membranes; intestinal malabsorption. *Mailing Add:* Vet Admin Hosp 50 Irving St NW Washington DC 20422

FINKELSTEIN, JAY LAURENCE, b Brooklyn, NY, Mar 26, 38; m 80; c 4. COMPUTER SCIENCES, TECHNICAL MANAGEMENT. *Educ:* William Marsh Rice Inst, BA, 60; Rice Univ, BS, 61; Calif Inst Technol, MS, 65. *Prof Exp:* Aerospace engr, Naval Missile Ctr, Ft Mugu, Calif 61-65; proj engr, Navy Space Systs Activ, Los Angeles, 65-74, Washington rep, 75-78; sr analyst, Naval Space Proj Off, 79-88, ASST PROG MGR, SPACE & SENSORS DIR, SPACE & NAVAL WARFARE SYST COMN, WASHINGTON, DC, 88- *Concurrent Pos:* Officer & dir, Syst Synthesis Inc, Calif, 66-68; comput consult, 80- *Mem:* Fel AAAS. *Res:* Cost and effectiveness of satellite surveillance systems with respect to the accomplishment of military missions. *Mailing Add:* 12202 Nutmeg Lane Reston VA 22091-1207

FINKELSTEIN, JEROME, b Long Branch, NJ, Aug 22, 41; m 66; c 2. HIGH-ENERGY THEORY, MULTIPARTICLE PRODUCTION. *Educ:* Columbia Univ, BA, 63; Univ Calif, Berkeley, PhD(physics), 67. *Prof Exp:* Asst prof physics, Stanford Univ, 68-72 & Columbia Univ, 72-79; from asst prof to assoc prof, 80-85, PROF PHYSICS, SAN JOSE STATE UNIV, 85- *Concurrent Pos:* Vis Scientist physics, Europ Organ Nuclear Res-Cern, 67-68, 70-71; guest prof, Univ Helsinki, 74; partic guest, Lawrence Berkeley Lab, 80- *Res:* Theory of elementary particles, with particular emphasis on events at the highest available energies. *Mailing Add:* Dept Physics San Jose State Univ San Jose CA 95192

FINKELSTEIN, MANUEL, b Scranton, Pa, Oct 18, 28; m 58; c 2. ORGANIC ELECTROCHEMISTRY. *Educ:* Univ Scranton, BS, 50; Williams Col, MA, 52; Yale Univ, PhD(chem), 56. *Prof Exp:* Asst chem, Williams Col, 50-52; sr engr org chem res, 56-62, sr res scientist, Sprague Elec Co, 62-87; consult, 87-88; CONSULT, MRA, 90- *Concurrent Pos:* Vis instr, Williams Col, 57-61 & 68-75, lectr, 75- *Mem:* Am Chem Soc. *Res:* Organic electrochemistry; organic synthesis; reaction mechanisms. *Mailing Add:* Four Birchwood Terrace Box 170 North Adams MA 01247-0170

FINKELSTEIN, PAUL, b New York, NY, Nov 20, 22; m 55; c 2. BIOCHEMISTRY. *Educ:* Brooklyn Col, BA, 43; Polytech Inst Brooklyn, PhD(chem), 49. *Prof Exp:* Res assoc, Ohio State Univ, 50-51; res biochemist, US Naval Med Res, 51-52; assoc dir res, Toni Co Div, Gillette Co, 52-64, lab dir, Gillette Med Res Inst, 64-68; dir clin dermat, Carter Prod Res Div, Carter-Wallace, Inc, 68-74; asst dir med res, Health Care Div, 74-77, ASSOC DIR MED RES, PROF PROD RES & DEVELOPMENT, JOHNSON & JOHNSON, INC, 77- *Concurrent Pos:* Fel, Univ Chicago, 49-50. *Mem:* Am Chem Soc; Soc Cosmetic Chem; Soc Invest Dermat; Am Soc Clin Pharmacol & Therapeut; Am Asn Dent Res. *Res:* Photochemistry of proteins; biochemical effects of radiation; protein isolation and characterization; structure and functions of skin; clinical dermatological evaluations; pharmacology; toxicology; oral hygeine research and product development. *Mailing Add:* Ten Springwood Dr Princeton Junction NJ 08550-1196

FINKELSTEIN, RICHARD ALAN, b New York, NY, Mar 5, 30; m 52, 76; c 4. MICROBIOLOGY. *Educ:* Univ Okla, BS, 50; Univ Tex, MA, 52, PhD(bact), 55; Am Bd Med Microbiol, dipl. *Prof Exp:* Res scientist, Univ Tex, 52-55, fel microbiol, Med Sch, Dallas, 55-58, instr, 58; chief bio-assay sect, Div Commun Dis & Immunol, Walter Reed Army Inst Res, 58-67; assoc prof microbiol, Univ Tex Health Sci Ctr Dallas, 67-74, prof microbiol, 74-79; PROF & CHMN, DEPT MOLECULAR MICROBIOL & IMMUNOL, SCH MED, UNIV MO, COLUMBIA, 79-, MILLSAP DISTINGUISHED PROF, 85-, CUR PROF, 90- *Concurrent Pos:* Dep chief & chief dept bact & immunol, Med Res Lab, SEATO, Thailand, 64-67; vis assoc prof, Fac Grad Studies, Univ Med Sci, Bangkok, 64-67; mem, Nat Comt Cholera Res, Thailand, 64-67; consult, WHO; mem, NIH Cholera Adv Comt, 71-; consult to comndg gen, US Army Med Res & Develop Command, 75-; vis scientist, Japanese Sci Coun, 76; vis prof, Med Sch, Univ Chicago, 77; chmn, Prog Comt, 79-82; consult, Inst Med, Nat Acad Sci, 84-85. *Honors & Awards:* Robert Koch Prize Sci & Med, Ger, 76; Res Award, Sigma Xi, 86. *Mem:* Am Soc Microbiol; Soc Gen Microbiol; Am Asn Immunol; Path Sci Gt Brit & Ireland; fel Am Acad Microbiol. *Res:* Pathogenesis and immunology of cholera, enteric diseases and gonorrhea; role of iron in microbial-host interactions; antimicrobial activity of human milk. *Mailing Add:* Dept Molecular Microl & Immunol M652 Sch Med Univ Mo, Columbia Columbia MO 65212

FINKELSTEIN, ROBERT, b New York, NY, Oct 13, 42; m 64; c 3. OPERATIONS RESEARCH. *Educ:* Temple Univ, BA, 64; Lowell Technol Inst, MS, 66; George Washington Univ, MS, 74, Ap Sci, 77. *Prof Exp:* Systs analyst mil syst, Missile Intel Agency, 66-68; physicist, Mass Inst Technol Draper Lab, 68-70; task leader space physics, Comput Sci Corp, 70-72; proj mgr oper res, Atlantic Res Corp, 72-75 & Ketron, Inc, 75-76; proj mgr opers res, Mantech NJ Corp, 76-77; systs scientist, Mitre Corp, McLean, Va, 77-85; PRES, ROBOTIC TECHNOL INC, POTOMAC, MD, 85- *Concurrent Pos:* Instr physics, Lowell Technol Inst & Univ Ala, 64-68; prof lectr mgt sci, Southeastern Univ, 75-76; lectr oper res, Cent Mich Univ, 75-78; lectr mgt, Prince Georges Community Col, 75-80; adj lectr mgt, George Washington Univ, 83-85. *Mem:* Oper Res Soc Am; Inst Mgt Sci; Am Phys Soc; AAAS; Am Asn Artificial Intel; Asn Unmanned Vehicle Systs. *Res:* Unmanned weapons systems; robotics artificial intelligence and cybernetics; countermeasures for sensor-dependent weapons; combat modeling; command, control, communications and intelligence. *Mailing Add:* 10001 Crestleigh Lane Potomac MD 20854

FINKELSTEIN, ROBERT JAY, b Pittsfield, Mass, Mar 26, 16. PHYSICS. *Educ:* Dartmouth Col, BA, 37; Harvard Univ, PhD(physics), 41. *Prof Exp:* Theoret physicist, Bur Ord, US Navy Dept, 41-46 & Argonne Nat Lab, 46-47; fel, Inst Advan Study, 47-48 & Calif Inst Technol, 48-49; from asst prof to assoc prof physics, 49-57, PROF PHYSICS, UNIV CALIF, LOS ANGELES, 57- *Mem:* Sigma Xi. *Res:* Quantum theory; elementary particle theory. *Mailing Add:* Dept Physics Univ Calif Los Angeles CA 90024

FINKELSTEIN, STANLEY MICHAEL, b Brooklyn, NY, June 16, 41; m 67; c 2. BIOENGINEERING, BIOMEDICAL COMPUTING. *Educ:* Polytech Inst Brooklyn, BS, 62, MS, 64, PhD(elec eng, syst sci), 69. *Prof Exp:* Instr mech eng, Polytech Inst NY, 65-67, instr elec eng, 67-68, from asst prof to assoc prof bioeng & syst eng, 68-77; assoc prof, 77-89, PROF LAB MED, UNIV MINN, 90- *Concurrent Pos:* Mem fac NIH Bioeng Training Prog, Mt Sinai Sch Med, 66-72, lectr, 74-77; prin investr NSF res grant, 69-70, fel, 75-77; co-investr, NIH res grant, 71-73, NIH training prog, 77-, NIH prog proj, 80-, prin investr, NIH grant, 81- *Mem:* Inst Elec & Electronics Engrs; Eng Med & Biol Soc; Biomed Eng Soc; AAAS; NY Acad Sci. *Res:* Biomedical engineering; signal processing; computer applications in health science; simulation of physiological systems; patient monitoring systems; computer assisted instruction; electrocardiographic analysis. *Mailing Add:* Div Health Comput Sci 420 Delaware St SE Minneapolis MN 55455

FINKL, CHARLES WILLIAM, II, b Chicago, Ill, Sept 19, 41; m 65; c 2. COASTAL GEOMORPHOLOGY, COASTAL ZONE MANAGEMENT. *Educ:* Ore State Univ, BSc, 64, MSc, 66; Univ Western Australia, PhD(soil sci), 71. *Prof Exp:* Instr natural resources, Ore State Univ, 66-67; demonstr phys geog, Univ Western Australia, 67-68; staff geochemist, explor geochem, Int Nickel Australia, 70-74; dir & prof, Marine Sci Inst Coastal Studies, Nova Univ, 79-83; EXEC DIR & VPRES, COASTAL EDUC & RES FOUND INC, CHARLOTTESVILLE, VA, 83- *Concurrent Pos:* Ed, Encyclopedia Earth Sci Series, 74-87; courtesy prof, Fla Int Univ, 76-; prof, dept geol, Fla Atlantic Univ, 83-; ed-in-chief, J Coastal Res, Int Forum Littoral Sci, Coastal Educ & Res Found, Charlottesville, Va, 83-; managing ed, Bull Paleomalacology, 87-88. *Mem:* Am Geophys Union; Geol Soc Am; Soil Sci Soc Am; Soc Econ Paleontologists & Mineralogists; Soc Mining Engrs. *Res:* Soil-geomorphic relationships in tropical cratonic regions, modes of landscape sculpture in deeply weathered terranes, soil micromorphology and soil stratigraphy; coastal and submarine soils are of current interest. *Mailing Add:* Coastal Educ & Res Found PO Box 8068 Charlottesville VA 22906

FINKLE, BERNARD JOSEPH, b Chicago, Ill, Mar 17, 21; m 44; c 2. CRYOBIOLOGY, TISSUE CULTURE. *Educ:* Univ Chicago, BS, 42; Univ Calif, Los Angeles, PhD(plant biochem), 50. *Prof Exp:* Chemist, Manhattan Proj, Univ Chicago & Oak Ridge Nat Lab, 43-46; AEC fel, Molteno Inst, Eng, 50; biochemist, Univ Calif, 51-53; lab dir, Atomic Res Lab, 53-54; biochemist, Univ Utah, 54-57; chemist, fruit lab, USDA, 57-83; assoc plant biochemist & pomol, Univ Calif, Davis, 83-89; CONSULT, 83- *Concurrent Pos:* Fel, US-Japan Coop Sci Prog, Dept Med Chem, Kyoto Nat Univ, 66-67; consult cryobiol, tissue cult & food biochem, 83- *Honors & Awards:* Japan Soc Prom Sci Award, 74. *Mem:* Am Soc Biol Chem; Phytochem Soc NAm (vpres, 64-65, pres, 65-66); Am Soc Plant Physiologists; Soc Cryobiol. *Res:* Factors in tissue freezing viability; biosynthesis and metabolism of phenolic acids, ascorbate and oxalate; thiol groups of papain; chlorophyll biogenesis; culture of green algae; food science and technology; plant physiology. *Mailing Add:* 21 Kingston Rd Berkeley CA 94707

FINKLEA, HARRY OSBORN, b Columbia, SC, Feb 26, 49; m 76; c 2. ELECTROCHEMISTRY, SURFACE ANALYSIS. *Educ:* Duke Univ, BS, 70; Calif Inst Technol, PhD(chem), 76. *Prof Exp:* NSF res fel, Royal Inst Gt Brit, London, 75-76; res assoc, Univ NC, Chapel Hill, 76-78 & Fla Atlantic Univ, 78-79; ASST PROF ANALYTICAL CHEM, VA POLYTECH INST & STATE UNIV, 79- *Mem:* Am Chem Soc; Electrochem Soc. *Res:* Semiconductor, electrochemistry and photoelectrochemistry; photoelectrochemical solar cells; chemically modified electrodes. *Mailing Add:* Chemistry Dept West Virginia Univ PO Box 6045 Morgantown WV 26506-6045

FINKLEA, JOHN F, b Florence, SC, Aug 27, 33; m 58; c 2. PREVENTIVE MEDICINE, RESEARCH ADMINISTRATION. *Educ:* Davidson Col, BS, 54; Med Univ SC, MD, 58; Univ Mich, DPH, 66. *Prof Exp:* Assoc med, Sch Med, Northwestern Univ, 66-67; from asst prof to assoc prof prev med, Med Col SC, 66-69; chief ecol res br environ health, Nat Air Pollution Control Admin, 69-71; from dir div health effects res to dir environ health, Nat Environ Res Ctr, Environ Protection Agency, 71-75; dir health res, Nat Inst Occup Safety & Health, 75-78; PROF PUB HEALTH, UNIV ALA, BIRMINGHAM, 78- *Concurrent Pos:* Teaching asst epidemiol, Univ Mich Sch Pub Health, 66; consult epidemiologist, Northwestern Univ, 67-71; mem, Nat Ctr Health Statist Adv Bd, 71-; chmn, Nat Air Qual Adv Comt, 72-73; mem, Adv Comt Atomic Bomb Casualty Comn, Nat Acad Sci, 72-74 & Comt on Hearing, Bioacoust & Biomech, 72-; consult & sci adv, Environ Health Prog, 73-74; mem adv comt conserv of energy, Nat Power Surv, Fed Power Comn, 74; examr, Am Bd Prev Med, 75. *Mem:* Soc Occup & Environ Health; Am Occup Med Asn; Am Pub Health Asn; Am Col Prev Med; Am Acad Pediat. *Res:* Occupational health research with emphasis on epidemiology of chronic diseases; health effects of air pollutants and pesticides. *Mailing Add:* 1717 11th Ave S 717-MTB Birmingham AL 35294

FINKLER, PAUL, b Brooklyn, NY, Nov 29, 36; m 65; c 1. PARTICLE PHYSICS, NUCLEAR PHYSICS. *Educ:* Brooklyn Col, BS, 58; Purdue Univ, PhD(physics), 63. *Prof Exp:* Res appointee theoret physics, Lawrence Radiation Lab, Univ Calif, 63-65; asst prof, 65-74, ASSOC PROF PHYSICS, UNIV NEBR, LINCOLN, 74- *Mem:* Am Phys Soc. *Res:* High energy physics; dispersion relations; Regge pole theory; high energy phenomenology; electromagnetic interactions of particles. *Mailing Add:* Dept Physics & Astron Univ Nebr Lincoln NE 68588-0111

FINKNER, MORRIS DALE, b Akron, Colo, Feb 11, 21; m 49; c 2. EXPERIMENTAL STATISTICS. *Educ:* Colo State Univ, BS, 43; Kans State Univ, MS, 47; NC State Univ, PhD(agron), 52. *Prof Exp:* Asst prof forage crops, Miss State Col, 47-49; asst cotton breeding, NC State Col, 49-52; asst prof exp statist, NMex Agr & Mech Col, 52-55; biometrician, Agr Res Serv, USDA, Md, 56-58; from assoc prof to prof exp statist, NMex State Univ, 64-84, head, dept exp statist & dir, statist ctr, 70-84; RETIRED. *Mem:* Am Soc Agron; Biomet Soc; Crop Sci Soc Am; Am Statist Asn. *Res:* Design and analysis of experiments; data processing; climatology; plant breeding. *Mailing Add:* 1802 Half Moon Dr Las Cruces NM 88005-3311

FINKNER, RALPH EUGENE, b Akron, Colo, Mar 24, 25; m 50; c 3. PLANT BREEDING, AGRONOMY. *Educ:* Colo State Univ, BS, 50; Iowa State Col, MS, 52, PhD(plant breeding, path), 53. *Prof Exp:* Asst agron, Iowa State Col, 50-53; plant breeder, Am Crystal Sugar Co, 53-55, chief plant breeder, 55-56, res sta mgr, 56-66; SUPT & PROF AGRON, AGR SCI CTR, NMEX STATE UNIV, 66- *Concurrent Pos:* US AID consult, Paraguay, 70, Turkey, 71 & Egypt, 78-79 & 82. *Mem:* Am Soc Sugar Beet Technol; Crop Sci Soc Am; Am Soc Agron; Sigma Xi; Am Peanut Res & Educ Soc. *Res:* Development of improved beet varieties; development of high yielding, high protein grain sorghum hybrids. *Mailing Add:* 2604 E 21st St Rte 4 Box 211 Clovis NM 88101

FINKS, ROBERT MELVIN, b Portland, Maine, May 12, 27. INVERTEBRATE PALEONTOLOGY, PALEOECOLOGY. *Educ:* Queens Col, NY, BS, 47; Columbia Univ, MA, 54, PhD(geol), 59. *Prof Exp:* Asst zool, Columbia Univ, 47-48, cur asst paleontol, 48-49; lectr geol, Hofstra Col, 49-50; asst instr, Rutgers Univ, 50-54; lectr, Brooklyn Col, 55-58, instr, 59-61; lectr, 61-62, from asst prof to assoc prof, 62-70, PROF GEOL, QUEENS COL, NY, 71- *Concurrent Pos:* Lectr, Hunter Col, 52-54; geologist, US Geol Surv, 52-54, 63-; mus aide, Smithsonian Inst, 56-57, res assoc, 68-; res assoc, Am Mus Natural Hist, 61-77; vis prof, Syracuse Univ, 62 & Univ Wyo, 65; chmn, Northeastern Sect, Paleont Soc, 78-79; mem doctoral fac, City Univ NY, 83- *Mem:* Fel AAAS; fel Geol Soc Am; Int Paleont Asn; Paleont Soc; Brit Paleont Asn; Sigma Xi. *Res:* Fossil sponges; paleoecology; longevity and growth-rates in fossil populations; natural selection in fossil populations; biostratigraphy; evolution of reef ecosystems; Paleozoic corals. *Mailing Add:* Dept Geol Queens Col Flushing NY 11367-0904

FINLAY, BARTON BRETT, b Edmonton, Alta, Apr 4, 59; m 84; c 2. INFECTIOUS DISEASES, MICROBIAL PATHOGENESIS. *Educ:* Univ Alta, BSc Hons, 81, PhD(biochem), 86. *Prof Exp:* Postdoctoral fel, Dept Microbiol & Immunol, Stanford Univ, 86-89; ASST PROF BIOCHEM, MICROBIOL & BIOTECHNOL, UNIV BC, 89- *Concurrent Pos:* Howard Hughes int res scholar, 91. *Honors & Awards:* Fisher Sci Award, Can Soc Microbiologists, 91. *Mem:* Am Soc Microbiol; Am Soc Cell Biol. *Res:* Host-pathogen interactions in microbial pathogens and infectious diseases; invasion, intracellular survival, and intracellular replication of intracellular parasites. *Mailing Add:* Biotechnol Lab Univ BC Westbrook Bldg Vancouver BC V6T 1Z3 Can

FINLAY, JOSEPH BRYAN, b Meriden, Eng, Feb 28, 43; m 68; c 3. ORTHOPAEDIC BIOMECHANICS, CLINICAL ENGINEERING. *Educ:* Lanchester Polytech, BSc, 66; Univ Strathclyde, PhD(bioeng), 70. *Prof Exp:* Control systs engr, Gyroscope Div, Sperry Rand, 66-67; lectr, Univ Strathclyde, 67-72; mgr biomed eng, Univ Hosp, London, Ont, 72-82; assoc prof, 76-82, PROF SURG, MED BIOPHYS, & MECH ENG, UNIV WESTERN ONT, 82-, GWYNNETH D RORABECK SCIENTIST ORTHOP, 90-; DIR ORTHOP RES, UNIV HOSP, LONDON, ONT, 82- *Concurrent Pos:* Mem, Health Devices J, Emergency Res Care Inst, 74-82; co-prin investr, Med Res Coun Can grants, 74- & Ont Heart Found grant, 77-78; assoc prof, dept biophys, Univ Western Ont, 76-82; chmn, technol adv comt, Fanshawe Community Col, London, Ont, 80-81; mem, biomed eng grant comt, Med Res Coun Can, 81-84, 87, prog grants comt, 88. *Honors & Awards:* John Charnley Award, Hip Soc, 85. *Mem:* Can Med & Biol Eng Soc; Inst Mech Engrs; Inst Elec Engrs; Am Soc Mech Engrs; Inst Elec & Electronics Engrs; Can Biomaterials Soc; Orthop Res Soc; Soc Exp Mech. *Res:* Distribution of the principal strains in the human bones as they are affected by surgical procedures; the implantation of prosthetic components, optimum design of prostheses for the hip and knee; development of externally-applied fracture-fixation frames. *Mailing Add:* 61 Hampton Crescent London ON N6H 2P1 Can

FINLAY, JOSEPH BURTON, b Collins, Ohio, Sept 15, 21; m 42; c 2. ORGANIC CHEMISTRY. *Educ:* Bowling Green State Univ, BA, 43; Univ NC, PhD(org chem), 52. *Prof Exp:* Instr chem, Bowling Green State Univ, 46-48; chemist, E I Du Pont de Nemours & Co, Inc, 51-79, res fel, 79-83; RETIRED. *Mem:* AAAS; Am Chem Soc; Sigma Xi. *Res:* Elastomeric materials. *Mailing Add:* 1229 Lakewood Dr Wilmington DE 19803

FINLAY, MARY FLEMING, b Columbia, SC, Oct 3, 44; m 66; c 2. DEVELOPMENTAL GENETICS. *Educ:* Sweet Briar Col, BA, 66; Univ SC, PhD(zool), 70. *Prof Exp:* Res trainee genetics, Population Coun, Rockefeller Univ, 66; instr biol, Allen Univ, 69; from asst prof to assoc prof, 70-76, PROF BIOL, BENEDICT COL, 76- *Concurrent Pos:* Prin investr, Nat Inst Child Health & Human Develop, NIH, 70-72, Div Res Resources, 74-, Nat Inst Gen Med Sci, 77-; prog dir, Minority Biomed Support Prog, 74-; prog dir, Minority Access Res Careers, 77-85, fac fel, 79-80; chair, NIH DRR Subcomt, 88-89. *Mem:* Sigma Xi; Am Soc Zoologists; Am Genetics Asn. *Res:* Developmental changes in isozymes in rodents, particularly as markers of stress or dysfunction; circadian patterns of activity, and hybridization in Peromyscus. *Mailing Add:* Dept Biol Benedict Col Columbia SC 29204

FINLAY, PETER STEVENSON, b Montclair, NJ, Oct 12, 24; m 56; c 3. ZOOLOGY. *Educ:* Williams Col, Mass, BA, 49; Univ Vt, MS, 53; Syracuse Univ, PhD(zool), 57. *Prof Exp:* Assoc prof, 56-70, PROF BIOL, ALFRED UNIV, 70- *Mem:* Soc Protozool; Am Soc Parasitol; Am Soc Zool; Sigma Xi. *Res:* Red cell parasites of amphibia. *Mailing Add:* 37 Sayles St Alfred NY 14802

FINLAY, ROGER W, b Pittsburgh, Pa, Oct 22, 35; div; c 2. NUCLEAR PHYSICS. *Educ:* Johns Hopkins Univ, AB, 57, PhD(physics), 62. *Prof Exp:* Asst, Johns Hopkins Univ, 60-62; from asst prof to assoc prof, 62-69, PROF PHYSICS, OHIO UNIV, 69- *Concurrent Pos:* Vis scientist, Max Planck Inst Nuclear Physics, Heidelberg, 68-69 & Nuclear Res Ctr, Karlsruhe, 85-86; consult, Lawrence Livermore Nat Lab, 86- *Mem:* Fel Am Phys Soc. *Res:* Nuclear structure and nuclear reaction mechanisms; nuclear radiation detectors. *Mailing Add:* Dept Physics Ohio Univ Athens OH 45701

FINLAY, THOMAS HIRAM, b Brooklyn, NY, Nov 28, 38; m 61; c 2. MOLECULAR ENDOCRINOLOGY. *Educ:* Univ Md, BS, 64; PhD(biochem), 69. *Prof Exp:* Fel biochem, Brandeis Univ, 69-71; asst prof, Rockefeller Univ, 71-74; from asst to assoc prof, 75-89, PROF OBSTET & GYNECOL, MED CTR, NY UNIV, 89- *Concurrent Pos:* Res career develop award, Med Ctr, Nat Heart, Lung & Blood Inst, 77-81. *Mem:* Am Soc Biol Chemists; Am Chem Soc; NY Acad Sci; Endocrine Soc; Soc Gynec Invest. *Res:* Role of proteolytic enzymes and inhibitors of proteolytic enzymes in the regulation of cellular activity. *Mailing Add:* Dept Obstet & Gynecol NY Univ Med Ctr 550 First Ave New York NY 10016

FINLAY, WALTER L(EONARD), b Brooklyn, NY, Mar 20, 23; m 47; c 2. PHYSICAL METALLURGY, MINING ENGINEERING. *Educ:* Lehigh Univ, BS, 46; Yale Univ, MSc, 47, DEng, 48. *Prof Exp:* Supvr chem & metall res, Remington Arms Co, 49-51; res mgr, Rem-Cru Titanium, Inc, 51-55, vpres res, 54-58; dir res, Crucible Steel Co Am, 58-62, asst vpres adv technol, 62-65; asst vpres res, Copper Range Co, 65-67, dir res, 67-70, vpres res & develop, 70-77, dir res, 77-; RETIRED. *Concurrent Pos:* Chmn nat adv bd, Nat Acad Sci, 67-68; vpres mining res, White Pine Copper Co, 70-78; pres, Contemp Metals Res, Inc, 70-78; consult, 78-90. *Mem:* Am Soc Metals; Electrochem Soc; Am Inst Mining, Metall & Petrol Engrs; Brit Inst Metals; Brit Iron & Steel Inst. *Res:* Small arms ballistics; titanium; steels; refractory metals; several dozen patents and publications in the field of ammunition, titatium and steel. *Mailing Add:* Queensborough 101 W Olympic Pl Apt 706 Seattle WA 98119

FINLAYSON, BIRDWELL, b Pocatello, Idaho, Oct 28, 32; m 55; c 2. UROLOGY, BIOPHYSICS. *Educ:* Univ Chicago, MD, 57, PhD(biophys), 67. *Prof Exp:* From intern to resident, Univ Chicago, 57-63; from asst prof to prof, 67-86, GRA RES PROF, UNIV FLA, 86- *Concurrent Pos:* Mem, Nat Kidney Found; bd mem, Uro-Tech Mgt Corp. *Mem:* Am Urol Asn; Am Chem Soc. *Res:* Urolithiasis; 173 manuscripts. *Mailing Add:* Div Urol Univ Fla Sch Med Box J-247 Gainesville FL 32610

FINLAYSON, BRUCE ALAN, b Waterloo, Iowa, July 18, 39; m 61; c 3. CHEMICAL ENGINEERING, APPLIED MATH. *Educ:* Rice Univ, BA, 61, MS, 63; Univ Minn, PhD(chem eng), 65. *Prof Exp:* Proj officer appl physics, Off Naval Res, 65-67; from asst prof to assoc prof chem, Univ Wash, 66-77, prof chem eng & appl math, 77-85, prof chem eng, 85-89, CHMN CHEM ENG, UNIV WASH, 89- *Concurrent Pos:* Co-holder, Rehnberg Chair, Univ Wash, 82- *Honors & Awards:* William H Walker Award, Am Inst Chem Engrs, 83. *Mem:* Soc Rheol; Am Inst Chem Engrs; Am Chem Soc; Soc Indust & Appl Math. *Res:* Fluid mechanics; rheology; approximate and variational methods of analysis; finite element methods; flow through porous media. *Mailing Add:* Dept Chem Eng BF-10 Univ Wash Seattle WA 98195

FINLAYSON, HENRY C, b Vulcan, Alta, July 5, 30; m 57; c 3. MATHEMATICAL ANALYSIS. *Educ:* Univ Alta, BSc, 52, MSc, 54; Univ Minn, PhD(math), 64. *Prof Exp:* Lectr math, Univ Alta, 54-55; lectr, 56-62, asst prof, 62-69, ASSOC PROF MATH, UNIV MAN, 69- *Mem:* Math Asn Am; Am Math Soc. *Res:* Integration in function space. *Mailing Add:* Fac Math Univ Man Winnipeg MB R3T 0K1 Can

FINLAYSON, JAMES BRUCE, b Montrose, Colo, July 3, 37; m 70; c 2. ANALYTICAL CHEMISTRY, GEOCHEMISTRY. *Educ:* Univ Ore, BA, 59; La State Univ, Baton Rouge, MS, 62; Univ Hawaii, PhD(anal chem), 67. *Prof Exp:* Scientist, Chem Div, Dept Sci & Indust Res, NZ Govt, 67-70; from asst prof to assoc prof, chem, 70-75, Hilo Col, Univ Hawaii, 70-75; SCIENTIST, CHEM DIV, DEPT SCI & INDUST RES, NZ, 75- *Mem:* Am Chem Soc; NZ Geochem Group. *Res:* Volcano chemistry; volcanic gases and volatiles; gas analysis by gas chromatography; geothermal chemistry; trace methods of analysis; water analysis, especially trace metals in natural waters; inorganic environmental water analysis. *Mailing Add:* Dept Sci & Indust Res, Chem Geothermal Res Ctr Private Bag 2000 Taupo New Zealand

FINLAYSON, JOHN SYLVESTER, b Philadelphia, Pa, Sept 19, 33; m 57; c 2. BIOCHEMISTRY. *Educ:* Marietta Col, BA, 53; Univ Wis, MS, 55, PhD(biochem), 57. *Prof Exp:* Asst, Wis Alumni Res Found, 53-55; Nat Cancer Inst fel, Univ Wis, 55-57 & Inst Radiophysics, Stockholm, Sweden, 57-58; biochemist, Lab Blood & Blood Prod, Div Biol Standards, NIH, 58-72; res chemist, 72-75, chief plasma proteins lab, 75-85, ASSOC DIR SCI, DIV HEMATOL, CTR BIOLOGICS & EVAL RES, FOOD & DRUG ADMIN, 85- *Concurrent Pos:* Instr, Found Advan Educ Sci, 61-75, 86-; vis scientist & prof, Inst Protein Res, Osaka Univ, Japan, 76. *Mem:* Soc Exp Biol & Med; Int Soc Thrombosis & Hemostasis; Coun Thrombosis Am Heart Asn. *Res:* Plasma proteins; blood coagulation. *Mailing Add:* Ctr Biologics Eval & Res Food & Drug Admin Bethesda MD 20892

FINLEY, ARLINGTON LEVART, b Newburn, NC, Mar 17, 48; m 70; c 2. SOLID STATE CHEMISTRY, INORGANIC CHEMISTRY. *Educ:* Tougaloo Col, BS, 70; Brown Univ, PhD(inorg chem), 75. *Prof Exp:* Res asst chem, Brown Univ, 70-74; proj leader, 74-84, RES SCIENTIST, DOW CHEM CO, 84- *Mem:* AAAS; Am Chem Soc. *Res:* Development of solid state technology in Dow processes; use of solid state technology for cigarette design to filter technology at Philip Morris. *Mailing Add:* Philip Morris USA Res Ctr Box 26583 Richmond VA 23261-6583

FINLEY, ARTHUR MARION, b La Monte, Mo, Apr 15, 18; m 43; c 6. PLANT PATHOLOGY. *Educ:* Univ Mo, BS, 41, MA, 48, PhD(plant path), 50. *Prof Exp:* Asst plant pathologist, 50-54, assoc prof plant path & assoc plant pathologist, 54-55, head dept plant path, 55-63, head dept plant sci, 63-71, PROF PLANT PATH & PLANT PATHOLOGIST, UNIV IDAHO, 55- *Mem:* Am Phytopath Soc; Sigma Xi. *Res:* Diseases of vegetable crops; soil borne pathogenic organisms. *Mailing Add:* 819 Harold Moscow ID 83843

FINLEY, DAVID EMANUEL, b Springfield, Ill, July 23, 35; m 60; c 2. BIOLOGY, BOTANY. *Educ:* Western Ill Univ, BS, 62, MS, 64; Univ Ill, PhD(bot), 67. *Prof Exp:* Teaching asst bot, Univ Ill, Urbana, 63-67; assoc prof biol, 67-73, PROF BIOL, LINCOLN UNIV, MO, 73- *Mem:* Mycol Soc Am; Bot Soc Am. *Res:* Taxonomy of the Stilbellaceae; fungal cytology. *Mailing Add:* Dept Biol Lincoln Univ Jefferson City MO 65101

FINLEY, JAMES DANIEL, III, b Louisville, Ky, Aug 2, 41; div; c 2. THEORETICAL PHYSICS, RELATIVITY. *Educ:* Univ Tex, Austin, BS & BA, 63; Univ Calif, Berkeley, PhD(physics), 68. *Prof Exp:* From asst prof to assoc prof, 68-78, PROF PHYSICS, UNIV NMEX, 78-, CHAIRPERSON DEPT, 85- *Concurrent Pos:* Vis prof, Centro de IEA del Inst Politecnico Nacional, 75, 82. *Mem:* Am Phys Soc; Soc Gen Relativity & Gravitation. *Res:* General relativity; methods of complex-valued differential geometry to investigate existence and structure of exact solutions to Einstein's field equations and perturbations to them, especially those describing gravitational waves; symmetics of non-linear equations investigated using techniques from differential geometry and Lie algebras. *Mailing Add:* Dept Physics & Astron Univ NMex Albuquerque NM 87131

FINLEY, JOANNE ELIZABETH, b Brockport, NY, Dec 28, 22; m 50; c 4. PUBLIC HEALTH, PREVENTIVE MEDICINE. *Educ:* Antioch Col, BA, 44; Yale Univ, MPH, 51; Case-Western Reserve Univ, MD, 62; Am Bd Prev Med, Dipl, 72. *Prof Exp:* Econ Res Anal, US Off Alien Property Custodian, 44-45; admin asst, US House of Reps, 45-48; field dir, Nat Inst Soc Rels, 48-49; Pub Affairs Anal, Nat Comt Effective Cong, 49-50; dir health educ, Montgomery County TB & Heart Asn, 52-55; exec dir, Montgomery County Planned Parenthood League, 55-56; founder & dir, Parent & Child, Inc, 56-57; res dir, Cleveland Health Goals Proj, 63-66; dep comnr & actg comnr, Cleveland Div Pub Health, 66-68; dir health planning, Dept Pub Health, Philadelphia, 68-72; vpres med affairs, Blue Cross, Philadelphia, 72-73; dir, Pub Health New Haven Dept Health, 73-74; COMNR NJ STATE DEPT OF HEALTH, 74- *Concurrent Pos:* Asst clin prof, Med Col Pa, 69- vis lectr, NJ Col Med & Dent, 74- comnr, Kellogg Found Comn Educ Health Admin, 72-74; mem, HEW Task Force Future Nat Immunization Pol, 77; mem adv comn immunization practices, Ctr Dis Control, USPHS, 81. *Mem:* Asn State & Territorial Health Off; Am Pub Health Asn; Am Soc Pub Admin. *Mailing Add:* 4201 Underwood Rd Baltimore MD 21218

FINLEY, JOHN P, b Lunenburg, NS, Aug 2, 45; m 75; c 2. PEDIATRICS, CARDIOLOGY. *Educ:* Dalhousie Univ, BSc, 67, MSc, 68; McGill Univ, MD, CM, 73; FRCP(C), 77. *Prof Exp:* PEDIAT CARDIOLOGIST, DALHOUSIE UNIV & IWK HOSP FOR CHILDREN, HALIFAX, NS, 78-, HEAD CARDIOL, IWK HOSP, 86- *Concurrent Pos:* Mem, sci rev comt, Can Heart Found, 79-82. *Mem:* Can Cardiovasc Soc; Can Pediat Cardiol Soc; Am Physiol Soc. *Res:* Autonomic control of heart in children; breathing patterns in infants. *Mailing Add:* Dalhousie Univ Cardiologist Izoak Walton Killam Hosp Children 5850 University Ave Halifax NS B3J 3G9 Can

FINLEY, JOHN WESTCOTT, b Auburn, NY, Sept 21, 42; m 67; c 2. FOOD SCIENCE. *Educ:* Le Moyne Col, BS, 64; Cornell Univ, PhD(food sci), 68. *Prof Exp:* Fel food sci, Mich State Univ, 68-69; res chemist, Sci & Educ Admin-Agr Res, USDA, 69-81; SCIENTIST, RALSTON PURINA CO, ST LOUIS, 81- *Mem:* Am Chem Soc; Inst Food Technologists; Am Asn Cereal Chemists; NY Acad Sci; AAAS. *Res:* Effects of food processing on the lipids and proteins in foods, especially amino acids and their degradation products with regard to nutrition and food safety. *Mailing Add:* Three Old Stone Lane Whippany NJ 07981-1849

FINLEY, JOSEPH HOWARD, b Newark, NJ, Jan 15, 31; m 54; c 6. ORGANIC CHEMISTRY. *Educ:* Seton Hall Univ, BS, 53, MS, 61; Rutgers Univ, PhD(org chem), 68. *Prof Exp:* Chemist control, Am Cyanamid Co, 55; chemist anal, Air Reduction Co, Inc, 55-60; RES ASSOC PROCESS & PROD DEVELOP, FMC CORP, 80- *Mem:* Am Chem Soc. *Res:* Active chlorine and peroxygen bleaches; development of new detergent products; development of new industrial processes. *Mailing Add:* 65 Sharon Ct Metuchen NJ 08840-1763

FINLEY, KAY THOMAS, b Elmira, NY, Aug 29, 34; m 78; c 3. PHYSICAL ORGANIC CHEMISTRY. *Educ:* Rochester Inst Technol, BS, 59; Univ Rochester, PhD(chem), 63. *Prof Exp:* From asst prof to assoc prof chem, Rochester Inst Technol, 62-66; sr res chemist, Eastman Kodak Co, 66-70; PROF CHEM, STATE UNIV NY COL BROCKPORT, 70- *Res:* Nucleophilic addition to quinonoid systems; reviews of organic chemical literature; biography chemists and women scientists. *Mailing Add:* Dept Chem State Univ NY Col Brockport Brockport NY 14420

FINLEY, ROBERT BYRON, JR, b Pittsfield, Mass, Aug 22, 17; m 46, 74; c 3. VERTEBRATE ZOOLOGY. *Educ:* Univ Calif, AB, 41; Univ Kans, PhD(zool), 56. *Prof Exp:* Instr zool & asst cur mus natural hist, Univ Kans, 50-51; sci intel analyst, Off Naval Intel, 55-59; mem staff, Denver Wildlife Res Ctr, US Fish & Wildlife Serv, 59-65, chief, Sect Wildlife Ecol Pub Lands, 65-73, zoologist, Nat Fish & Wildlife Lab, 73-82; RETIRED. *Mem:* Soc Syst Zool; Am Soc Mammologists; Wildlife Soc; Sigma Xi; Am Soc Ichthyol & Herpet. *Res:* Biogeography; evolution of mammals and reptiles; paleobiology of Pleistocene; ecology; wildlife management; effects of chemicals on wildlife. *Mailing Add:* 1050 Arapahoe Apt 405 Boulder CO 80302

FINLEY, ROBERT JAMES, b New York, NY, Apr 14, 47; m 68; c 2. GEOLOGICAL REMOTE SENSING, PETROLEUM GEOLOGY. *Educ:* City Univ NY, BS, 67; Syracuse Univ, MS, 69; Univ SC, PhD(geol), 75. *Prof Exp:* Geologist, Chevron Oil Co, 69-71; assoc res scientist, 75-78, prog dir, 84-88, RES SCIENTIST, BUR ECON GEOL, 78-, ASSOC DIR, UNIV TEX, 88- *Concurrent Pos:* Mem, Austin City Environ Bd, 77-81, co-chmn, 78-79; mem, Remote Sensing & Cartog Comt, Tex Natural Resources Info Syst, 77-, instr, 79-81; consult, Computech Energy & Explor, 80-82; mem, Comt Undiscovered Oil & Gas Resources, Nat Acad Sci, 88-91. *Mem:* Am Asn Petrol Geologists; Soc Econ Paleontologists & Mineralogists; Int Asn Sedimentologists; Soc Petrol Engr; Sigma Xi. *Res:* Interpretation of clastic depositional systems and their relationship to hydrocarbon and geothermal resources; interpretation of reservoir heterogeneity; applications of remote sensing to resource exploration through geomorphic, structural and surface analysis. *Mailing Add:* Bur Econ Geol Univ Tex PO Box X Austin TX 78713

FINLEY, SARA CREWS, b Lineville, Ala, Feb 26, 30; m 52; c 2. MEDICAL GENETICS, PEDIATRICS. *Educ:* Univ Ala, BS, 51; Med Col Ala, MD, 55. *Prof Exp:* From instr to assoc prof, 60-75, PROF PEDIAT, SCH MED, UNIV ALA, BIRMINGHAM, 75-, CO-DIR, LAB MED GENETICS, 66- *Concurrent Pos:* NIH fel pediat, Sch Med, Univ Ala, Birmingham, 56-60 & NIH trainee, Inst Med Genetics, Univ Uppsala, 61-62; mem, White House Conf Health, 65; mem res manpower rev comt, Nat Cancer Inst, 77-81; mem, NIH Adv Comt Sickle Cell Dis, 83-87; Wayne H & Sara C Finley Chair Med Genetics, 86- *Mem:* Am Soc Human Genetics; Am Fedn Clin Res; NY Acad Sci. *Res:* Cytogenetics; congenital malformations; human growth and development; genetic counseling. *Mailing Add:* 3412 Brookwood Rd Birmingham AL 35223

FINLEY, WAYNE HOUSE, b Goodwater, Ala, Apr 7, 27; m 52; c 2. MEDICAL GENETICS, BIOCHEMISTRY. *Educ:* Jacksonville State Univ, BS,47; Univ Ala, MA, 50, MS, 55, PhD(biochem), 58, MD, 60. *Prof Exp:* Sr high sch teacher, Ala, 49-51; from asst prof to assoc prof pediat, 62-66, asst prof biochem, 65-75, asst prof physiol & biophys, 67-75, assoc prof biochem, 75-77, PROF PEDIAT, SCH MED, UNIV ALA, BIRMINGHAM, 70-, ASSOC PROF PHYSIOL & BIOPHYS, 75-, PROF PUB HEALTH & EPIDEMIOL, 75-, DIR LAB MED GENETICS, 66-, PROF BIOCHEM, 77-, ADJ PROF BIOL, 82- *Concurrent Pos:* Fel, Inst Med Genetics, Univ Uppsala, 61-62; prin investr, USPHS res grant, 62-68; co-prin investr, US Dept Health & Human Serv, 64-; sr scientist, Comprehensive Cancer Ctr; mem, Nat Adv Res Resources Coun, NIH, 76-80; sr scientist, Cystic Fibrosis Res Ctr; dir, Med Genetics Grad Prog, Sch Med, Univ Ala, Birmingham, Med Genetics Prog, 78- *Mem:* AAAS; Am Soc Human Genetics; Am Fedn Clin Res; Am Chem Soc; Am Inst Chem; Sigma Xi; NY Acad Sci. *Res:* Medical genetics; congenital malformations; cell culture techniques and human cytogenetics. *Mailing Add:* Lab Med Genetics Univ Ala Birmingham AL 35294

FINLON, FRANCIS P(AUL), b Carbondale, Pa, Sept 2, 24; m 47; c 2. ELECTRICAL ENGINEERING. *Educ:* Pa State Univ, BS, 47, MS, 48. *Prof Exp:* Assoc prof, Pa State Univ, 47-64, prof eng res, 64-, head dept acoust, 76-84; consult, US Navy; RETIRED. *Mem:* Inst Elec & Electronics Engr. *Res:* Underwater ordnance. *Mailing Add:* 3216 S Lakeview Circle Ft Pierce FL 34949

FINN, ARTHUR LEONARD, b Boston, Mass, Mar 24, 34; m 56; c 3. PHYSIOLOGY, BIOPHYSICS. *Educ:* Harvard Univ, AB, 54; Boston Univ, MD, 58. *Prof Exp:* Intern med, Mass Mem Hosp, Boston, 58-59; resident, Duke Hosp, Durham, 59-60 & 62-63; asst prof physiol & med, Sch Med, Yale Univ, 65-70; assoc prof, 70-74, PROF MED & PHYSIOL, SCH MED, UNIV NC, CHAPEL HILL, 74- *Concurrent Pos:* Fel metab, Univ NC, Chapel Hill, 60-62; adv res fel membrane transport, Nat Heart Inst, 63-65. *Mem:* AAAS; Biophys Soc; Am Physiol Soc; Soc Gen Physiol; Am Soc Clin Invest. *Res:* Membrane biophysics; transport of ions and water in the isolated amphibian urinary and gall bladder. *Mailing Add:* Univ NC Sch Med 3034 Old Clin Bldg CB No 7155 Chapel Hill NC 27599

FINN, D(AVID) L(ESTER), b Memphis, Tenn, Mar 24, 24; m 48; c 3. ELECTRICAL ENGINEERING. *Educ:* Purdue Univ, BSEE, 48, MS, 50, PhD(elec eng), 52. *Prof Exp:* Prof elec eng, Ga Inst Technol, 52-; AT LOCKHEED GA CO. *Mem:* Inst Elec & Electronics Eng. *Res:* Applied mathematics; random processes; sampling theory. *Mailing Add:* 646 Longwood Dr NW Atlanta GA 30305

FINN, EDWARD J, b Ridgefield Park, NJ, July 24, 30; m 54; c 4. NEURAL NETWORK THEORY. *Educ:* Col Holy Cross, BS, 51; Catholic Univ, MS, 55; Georgetown Univ, PhD(physics), 62. *Prof Exp:* Instr physics, Georgetown Univ, 52-54 & St Vincent Col, 54-55; physicist, Naval Res Lab, 55-56; from asst prof to assoc prof, 56-77, PROF PHYSICS, GEORGETOWN UNIV, 77-, CHAIR PHYSICS DEPT, 90- *Concurrent Pos:* Physicist, Naval Res Lab, 56-80; lectr, Nat Univ Mex, 71, Univ Rio Grande del Sul, Brazil, 82; prog mgr, NSF, 80-82; physicist, NIH, 82-87; consult, Mass Inst Technol/Lincoln Labs, 87-88. *Mem:* Am Asn Physics Teachers; Am Phys Soc; Sigma Xi; Int Neural Network Soc. *Res:* Computer managed instruction; direction of scientific symposia; vibrational energies of diatomic molecules; theoretical analysis of positron emission scans; science curriculum development; management of bilateral United States-Latin American scientific projects; investigation of metallic clips embedded in humans; neurocomputation-massively parallel processing. *Mailing Add:* Dept Physics Georgetown Univ Washington DC 20057

FINN, FRANCES M, b Pittsburgh, Pa, May 6, 37; m 65. BIOCHEMISTRY. *Educ:* Univ Pittsburgh, BS, 59, MS, 61, PhD(biochem), 64. *Prof Exp:* USPHS fel, Harvard Univ, 64-65; res assoc, Univ Pittsburgh, 65-69, asst prof, 69-73, res assoc prof biochem, 73-79, assoc prof med & biochem, 79-88, PROF MED & BIOCHEM, UNIV PITTSBURGH, 88- *Mem:* Am Chem Soc; Endocrine Soc; Am Soc Biochem & Molecular Biol. *Res:* Structure-function studies with peptide hormone receptors; mechanism of action of peptide hormones. *Mailing Add:* Protein Res Lab 1276-A Scaife Hall Univ Pittsburgh Pittsburgh PA 15261

FINN, HAROLD M, b New York, NY, June 3, 22. RADAR SYSTEM DETECTION TECHNIQUES. *Educ:* City Col NY, BEE, 50; Univ Md, MS, 55; Univ Pa, PhD(elec eng), 65. *Prof Exp:* Sr engr, Eng & Res Corp, 50-53; Medpar Co, 53-56; prin engr, Emerson Res Lab, 56-59; sr staff engr, RCA Missile & Surface Radar Div, 58-68; mgr, Advan Prog Div, Technol Serv Corp, 68-80; PRES, ADAPTIVE SENSORS INC, 80- *Mem:* Fel Inst Elec & Electronics Engrs; AAAS. *Res:* Statistical decision processes as applied to radar; electronic-counter-counter-measures; constant false alarm rate detection systems. *Mailing Add:* Adaptive Sensors Inc 216 Pico Blvd Santa Monica CA 90405

FINN, JAMES CRAMPTON, JR, b Detroit, Mich, Oct 18, 24; m 55; c 2. PLANT PHYSIOLOGY. *Educ:* Mich State Univ, BS, 52, MS, 53; Univ Calif, Los Angeles, PhD(bot), 58. *Prof Exp:* Plant physiologist cotton physiol, Delta Exp Sta, USDA, Miss, 58; asst prof hort, Agr & Mech Col, Tex, 58-59; prin scientist bio, Aero-space Labs, Missile Div, NAm Aviation, Inc, 59-62, asst dir life sci dept, Space & Info Systs Div, 62-63, tech dir res planning aerospace sci, 63-65, staff scientist, Life Sci Systs Opers, 65-68, sci adv, Autonetics Div, NAm Rockwell Corp, 68; asst mgr microbics mkt & admin, Microbics Opers, 68-75, asst mgr mkt planning, Biol & Fine Chems Div, 75-81, prog mgr microbics, 81-85, MGR MFG STRATEGIC PLANNING, BECKMAN INSTRUMENTS, INC, 85- *Mem:* AAAS. *Res:* Photoperiodism; plant growth regulators; biological rhythms; closed ecological systems; research administration. *Mailing Add:* 1540 Harbor Dr N Box 131 Oceanside CA 92054

FINN, JAMES WALTER, b St Paul, Minn, Oct 3, 46; m 80; c 2. CHEMISTRY. *Educ:* Univ Minn, BChem, 70; Kans State Univ, PhD(chem), 81. *Prof Exp:* Res prof chemist, Pillsbury Co, 71-77; res chemist, 81-84, GROUP LEADER, ICI AMERICAS, 84- *Mem:* Am Chem Soc; Am Asn Cereal Chemists; Asn Off Anal Chemists; Sigma Xi. *Res:* High pressure liquid chromatography of polymeric materials; fluorine combustion; elemental analysis by gas chromatography; surfactants, biocides, carbohydrates analyses. *Mailing Add:* 2011 Foulk Rd Wilmington DE 19810

FINN, JOHN MCMASTER, b Washington, DC, Jan 8, 47; m 77; c 4. NONLINEAR PLASMA PHYSICS. *Educ:* Ga Inst Technol, BS, 69; Univ Houston, MS, 71; Univ Md, PhD(physics), 74. *Prof Exp:* Res assoc plasma physics, Princeton Univ, 74-76 & Cornell Univ, 76-79; physicist, Sci Applns Inc, 79-81; physicist, Naval Res Lab, 81-82; assoc res scientist, 82-88, SR RES SCIENTIST, UNIV MD, 88- *Mem:* Am Phys Soc; fel Am Phys Soc. *Res:* Ideal and resistive magneto hydrodynamic stability of tokamaks, FRC's, spheromaks and RFP's (types of magnetic fusion devices); the effects of ergodic particle orbits on such instabilities; the destruction of magnetic surfaces due to such instabilities; turbulent relaxation and reconnection in plasmas; application of nonlinear dynamics to plasma physics; solar magnetohydrodynamics; dynamo theory; three dimensional magnetic reconnection. *Mailing Add:* Lab for Plasma Res Univ Md College Park MD 20742

FINN, JOHN MARTIN, b Philadelphia, Pa, Nov 16, 19; m 54; c 3. APPLIED CHEMISTRY, RESEARCH ADMINISTRATION. *Educ:* Harvard Univ, AB, 48; Univ Pa, MS, 49, PhD(chem), 53. *Prof Exp:* Chemist, E I du Pont de Nemours & Co, 40-42, shift supvr, 42-43, area engr, 43-44; asst instr, Univ Pa, 48-51, asst, 51-52; proj supvr chem, Horizons, Inc, 52-55; res chemist, Union Carbide Corp, 55-70, sr res scientist, 70-72; asst prof eng, Kent State Univ, 72-73; dir inorg res & develop, Harshaw Chem Co, 73-80, dir corp res, 80-83; sr consult, Reid, Whitehurst & Co, 83-88; ADJ INSTR, CUYAMOGA COMMUNITY COL, 84- *Mem:* Am Chem Soc. *Res:* Liquid ammonia chemistry; phosphide chemistry; complex ions; boron chemistry; anode processes; fused salts and glasses; high temperature chemistry; pigments; fire retardant chemistry; polymer and composite technology. *Mailing Add:* 7663 Alan Pkwy Cleveland OH 44130

FINN, JOHN THOMAS, b Washington, DC, Oct 23, 48; m 71; c 1. SYSTEMS ECOLOGY, DIGITAL REMOTE SENSING. *Educ:* Georgetown Univ, BS, 70; Univ Ga, PhD(ecol), 77. *Prof Exp:* Res asst membrane transp, Renal Lab, Beth Israel Hosp & Harvard Med Sch, 71-73; assoc systs ecol Ecosystems Ctr, Marine Biol Lab, 77-78; from asst prof to assoc prof, 78-90, PROF SYSTS ECOL, DEPT FORESTRY & WILDLIFE MGT, UNIV MASS, 90- *Concurrent Pos:* Res consult, Ecol Simulations, Inc, 76-; vis assoc prof, Ctr Wetlands, La State Univ, 85. *Mem:* AAAS; Am Inst Biol Sci; Ecol Soc Am; Int Soc Trop Ecol; Soc Comput Simulation; Wildlife Soc. *Res:* Application of general systems theory, modeling, simulation and systems analysis to ecological problems; particular interest in nutrient cycling, watershed management, remote sensing and methods of qualitative analysis including flow analysis, loop analysis and path analysis. *Mailing Add:* Dept Forestry & Wildlife Mgt Univ Mass Amherst MA 01003

FINN, PATRICIA ANN, b Oak Park, Ill. THERMODYNAMICS & MATERIAL PROPERTIES, CHEMICAL ENGINEERING. *Educ:* Mundelein Col, BS, 67; Univ Calif, Berkeley, PhD(phys inorg chem), 71; Univ Chicago, MBA, 86. *Prof Exp:* Fel organometallic chem, Chem Dept, Iowa State Univ, 72-73; fel, Chem Div, 73-75, CHEMIST, CHEM ENG DIV, ARGONNE NAT LAB, 75- *Mem:* Am Nuclear Soc; Sigma Xi; Am Ceramic Soc; Am Chem Soc. *Res:* System design for nuclear fusion reactors; lithium systems, physical properties, corrosion, for use in fusion reactors as breeding medium; tritium systems in fusion reactors; surface oxidation. *Mailing Add:* 6925 Fairmount Downers Grove IL 60516

FINN, R(OBERT) K(AUL), b Waukesha, Wis, May 3, 20; m 49; c 4. CHEMICAL ENGINEERING. *Educ:* Cornell Univ, BCh, 41, ChE, 42; Univ Minn, PhD(chem eng), 49. *Prof Exp:* Res chem engr, Merck & Co, Inc, 42-46; asst prof chem eng, Univ Ill, 49-55; assoc prof, 55-65, PROF CHEM ENG, CORNELL UNIV, 65- *Concurrent Pos:* Consult; Fulbright fel, 62-63; vis prof, Univ Calif, Berkeley, 69; Guggenheim fel, 75-76. *Honors & Awards:* James Van Lanen Award, Am Chem Soc, 82; Food, Pharmaceut & Bioengineering Award, Am Inst Chem Engrs, 85. *Mem:* AAAS; Am Chem Soc; Am Soc Microbiol; Am Inst Chem Engrs. *Res:* Applied microbiology; fermentation engineering; biodegradation of hazardous wastes. *Mailing Add:* 172 Olin Hall Cornell Univ Ithaca NY 14850-2488

FINN, ROBERT, b Buffalo, NY. MATHEMATICAL PROBLEMS. *Educ:* Syracuse Univ, PhD(math), 51. *Prof Exp:* Vis mem, Inst Advan Study, 51-53; res assoc, Inst Fluid Dynamics, Univ Md, 53-54; asst prof math, Univ Southern Calif, 54-56; assoc prof, Calif Inst Technol, 56-59; PROF MATH, STANFORD UNIV, 59- *Concurrent Pos:* Fel, Guggenheim Found, 58-59 & 65-66, fel, Fulbright Comn, 87; vis prof, various univs, Europe, Asia & USA, 58-87; exchange fel, USSR Nat Acad Sci, 78 & EGer Nat Acad Sci, 87; ed, Pac J Math, Int J Math & Math Sci, 79-, Zeit Anal Anwendungen, 90- *Res:* Mathematical problems arising in fluid dynamics; differential geometry and capillarity. *Mailing Add:* Math Dept Stanford Univ Stanford CA 94305-2125

FINN, RONALD DENNET, b Weymouth, Mass, Aug 15, 44; m 69. RADIOCHEMISTRY. *Educ:* Worcester Polytech Inst, BS, 66; Va Polytech Inst, PhD(nuclear & radiochem), 71. *Prof Exp:* Res assoc chem, Brookhaven Nat Lab, 71-72, res assoc med, 72-73, assoc chemist solid state physics, 73-74; radiochemist, Mt Sinai Med Ctr, Fla, 74-75, tech dir radiochem & radiopharmacol, 75-79, dir cyclotron facil, 79-85; dir radiopharmaceut chem, Dept Nuclear Med, NIH, 85-89; assoc prof, Med Sch, Univ Miami, 84; ATTEND RADIOCHEMIST; ASSOC PROF, CORNELL UNIV. *Concurrent Pos:* Collabr dept chem, Brookhaven Nat Lab, 74-; asst res prof, Med Sch, Univ Miami, 74-77, asst prof, 78-84. *Mem:* AAAS; Sigma Xi; NY Acad Sci; Soc Nuclear Med. *Res:* Reactions of hot atoms produced by nuclear processes; synthesis of labelled compounds incorporating short half-life nuclides for potential clinical utility. *Mailing Add:* Mem Sloan-Kettering 1275 York Ave New York NY 10021

FINN, WILLIAM DANIEL LIAM, b Cork, Ireland, Aug 25, 33. SOIL MECHANICS. *Educ:* Nat Univ Ireland, BE, 54; Univ Wash, MSc, 57, PhD, 60. *Prof Exp:* Instr, Univ Wash, 56-60; asst prof civil eng, 61, head dept, 64-70, PROF CIVIL ENG, UNIV BC, 64- *Concurrent Pos:* Partner, Pan-Am Eng & Comput Serv, Ltd. *Mem:* Am Soc Civil Engrs; Am Soc Eng Educ. *Res:* Creep of soils; plasticity theory in soil mechanics; soil structure interaction during earthquakes; seismic response of earth dams; ocean engineering. *Mailing Add:* Dept Civil Eng Univ BC 2075 Westbrook Pl Vancouver BC V6T 1W5 Can

FINNEGAN, CYRIL VINCENT, b Dover, NH, July 17, 22; m 47; c 9. EXPERIMENTAL EMBRYOLOGY, DEVELOPMENTAL BIOLOGY. *Educ:* Bates Col, BS, 46; Notre Dame Univ, MS, 48, PhD(zool), 51. *Prof Exp:* Instr gen biol, Wabash Col, 49-50; res fel, Stanford Univ, 51-52; from instr to asst prof embryol & anat, St Louis Univ, 52-56; asst prof, Notre Dame Univ, 56-58; from asst prof to prof zool, Univ BC, 58-64, chmn biol prog, 69-78, from assoc dean to dean fac sci, 72-85, assoc vpres academic, 86-88, prof, 64-88, EMER PROF DEVELOP BIOL & MORPHOGENESIS & EMER DEAN, UNIV BC, 88- *Mem:* Am Soc Zoologists; Soc Develop Biol; Int Soc Develop Biol; Tissue Cult Asn; Can Soc Cell Biol. *Res:* Tissue interactions in induction in Amphibia; history and philosophy of biology. *Mailing Add:* Dept Zool 2354 Univ BC 6270 University Blvd Vancouver BC V6T 2A9 Can

FINNEGAN, MICHAEL, b Appleton, Wis, Jan 17, 41; m 71; c 2. FORENSIC ANTHROPOLOGY. *Educ:* Univ Colo, BA, 67, MA, 70, PhD(anthrop), 72; Am Bd Forensic Anthrop, dipl. *Prof Exp:* Fel, Smithsonian Res Found, 72-73; from asst prof to assoc prof, 73-82, PROF ANTHROP, KANS STATE UNIV, 82- *Concurrent Pos:* Osteologist, archaeologist, Kans Hist Soc, 73-; consult, osteological, Off Kans State Archaeologist, 73-, forensic osteologist, Kans Bur Invest, 74-, Van Doren Archaeol Surv, 75-, Prairie Land II, 77- & var co in Kans & Pa; field consult, Van Doren-Hazard-Stallings, 75; consult, Dept Army, 90-; bd dirs, Am Bd Forensic Anthrop, 87 & Kans Div, Int Asn Identification, 89- *Mem:* Am Asn Phys Anthropologists; Am Acad Forensic Sci; Int Asn Identification; Soc Am Archaeol; Paleopath Asn; Am Bd Forensic Anthrop (pres, 89-); Sigma Xi. *Res:* Non-metric skeletal variation; forensic osteology; paleopathology. *Mailing Add:* Anthropology Dept Kans State Univ Manhattan KS 66506

FINNEGAN, RICHARD ALLEN, organic chemistry; deceased, see previous edition for last biography

FINNEGAN, WALTER DANIEL, b Anaconda, Mont, Mar 31, 23; m 48; c 2. METALLURGY. *Educ:* Mont Col Mineral Sci & Tech, BS, 48, MS, 49. *Prof Exp:* Engr metall, US Bur Mines, 49-52; res metallurgist, 52-55, from asst supvr to supvr joining br, Dept Metall Res, Wash, 55-70, SR RES ASSOC, CTR TECHNOL, KAISER ALUMINUM & CHEM CORP, CALIF, 70- *Mem:* Am Welding Soc; Am Soc Metals. *Res:* Metallurgy and joining of aluminum alloys. *Mailing Add:* 3227 Inverness Dr Walnut Creek CA 94598

FINNEMORE, DOUGLAS K, b Cuba, NY, Sept 9, 34; m 56; c 3. SOLID STATE PHYSICS. *Educ:* Pa State Univ, BS, 56; Univ Ill, MS, 58, PhD(superconductivity), 62. *Prof Exp:* Res assoc physics, Univ Ill, 62 & Ames Lab, 62-63; from asst prof to assoc prof, 63-68, PROF PHYSICS, IOWA STATE UNIV, 68- *Concurrent Pos:* Prog dir, Quantum Solids & Liquids, NSF, 76-77; prog dir, Solid State Physics, Ames Lab, Iowa State Univ, 77-83, assoc dir, Sci & Technol, 83- *Mem:* Fel Am Phys Soc. *Res:* Superconductivity and magnetism. *Mailing Add:* Dept Physics Iowa State Univ Ames IA 50012

FINNERTY, FRANK AMBROSE, JR, b Montclair, NJ, Nov 3, 23; m 46, 75; c 6. CARDIOLOGY. *Educ:* Georgetown Univ, AB, 43, MD, 47; Am Bd Internal Med, cert, 55. *Prof Exp:* Vis physician, Georgetown Med Div, Gen Hosp, Washington, DC, 52, chief cardiovascular res, 52-76; asst prof med & pharmacol, Med Sch, Georgetown Univ, 55-65, prof med & obstet & gynec, 65-72, prof obstet & gynec, 72-76; chief med, Columbia Hosp Women, 63-73; CLIN PROF MED, GEORGE WASHINGTON UNIV MED CTR, 77-; DIR, HYPERTENSION CTR, WASHINGTON, DC, 77- *Concurrent Pos:* Vis physician, Med Ctr, Georgetown Univ, 52-76; Am Heart Asn res fel, 55, estab investr, 57; adj prof med, Grad Sch Nursing, Cath Univ, 75-; mem, Coun High Blood Pressure Res & Task Force Hypertension, Am Heart Asn; charter mem, Task Force Nat Hypertension Info & Educ Prog & mem Task Force Drug Protocol, Nat High Blood Pressure Info Prog, NIH; assoc ed, Dialogues in Hypertension. *Mem:* AMA; fel Am Col Cardiol; fel Am Col Physicians; fel Am Col Angiol. *Res:* Cardiovascular research hypertension toxemias of pregnancy; postural hypotension; patient compliance. *Mailing Add:* 4900 Mass Ave NW Washington DC 20016

FINNERTY, JAMES LAWRENCE, b Sioux Falls, SDak, Mar 9, 27; m 75; c 2. BIOCHEMISTRY, INFORMATION SCIENCE. *Educ:* Marquette Univ, BS, 48; Univ Ill, MS, 50; Loyola Univ, Ill, PhD(chem), 60. *Prof Exp:* Prof, Sogang Univ, Korea, 66-75; prof, Basic Sci & librn, Health Sci, Peoria Sch Med, Univ Ill, 75-80; DIR MED LIBR, COLUMBUS HOSP, CHICAGO, ILL, 80- *Concurrent Pos:* Fulbright sci educ prof, Korea, 66-68. *Mem:* Am Chem Soc; Med Libr Asn; Spec Libr Asn; Sigma Xi. *Res:* Information retrieval in the health sciences. *Mailing Add:* Columbus Hosp Med Libr 2520 N Lakeview Chicago IL 61614

FINNERTY, WILLIAM ROBERT, b Keokuk, Iowa, May 2, 29; m 53; c 6. MICROBIAL PHYSIOLOGY, BIOCHEMISTRY. *Educ:* Univ Iowa, BA, 55, PhD(microbiol), 61. *Prof Exp:* Teaching fel microbiol, Univ Iowa, 57-58, res asst, 58-60, instr, 60, res assoc, 60-61; USPHS fel biochem & enzym, Oak Ridge Nat Labs, 61-62; from asst prof to assoc prof microbiol, Ind Univ Sch Med, 62-68; assoc prof, 68-75, head dept, 77-88, PROF MICROBIOL, UNIV GA, 75-; DIR, RES & DEV, SURAX INC. *Concurrent Pos:* Vis prof, Univ Gottingen, 73. *Honors & Awards:* P R Edwards Award, Am Soc Microbiol, 80. *Mem:* AAAS; Am Soc Microbiol; Am Soc Biol Chemists; Am Chem Soc; fel Am Inst Chemists. *Res:* Mechanism of oxidation of aliphatic hydrocarbons by microorganisms; molecular genetics; lipids; petroleum microbiology; membranes. *Mailing Add:* 126 Chinquapin Pl Athens GA 30605

FINNEY, ESSEX EUGENE, JR, b Powhatan, Va, May 16, 37; m 59; c 2. AGRICULTURAL ENGINEERING. *Educ:* Va Polytech Inst, BS, 59; Pa State Univ, MS, 61; Mich State Univ, PhD(eng), 63. *Prof Exp:* Res agr engr, Sci & Educ Admin-Agr Res, 65-77, from asst dir to assoc dir, 77-89, DIR, BELTSVILLE AGR RES CTR, USDA, 89- *Concurrent Pos:* Sr policy anal, Presidential Sci Adv Off, Washington, DC, 80-81. *Mem:* Am Soc Agr Engrs; Inst Food Technol. *Res:* Instrumentation and techniques for measuring physical properties and characteristics associated with quality within agricultural and food products. *Mailing Add:* Agr Res Ctr W Beltsville MD 20705

FINNEY, JOSEPH J, b New York, NY, Mar 11, 27; m 61; c 3. GEOLOGY, MINERALOGY. *Educ:* US Merchant Marine Acad, BS, 50; Univ NMex, MS, 59; Univ Wis, PhD(struct mineral), 62. *Prof Exp:* prof geol, Colo Sch Mines, 62-86; ASSOC DIR, POTENTIAL GAS AGENCY, 85- *Concurrent Pos:* Res grants, Colo Sch Mines Found, Inc, 62- & Res Corp, 63; Eng Accreditation Comn, 86- *Mem:* Mineral Soc Am; Mineral Asn Can; Mineral Soc Gt Brit & Ireland; Soc Mining Engrs. *Res:* Structural mineralogy; investigations in structure of minerals; crystal chemistry. *Mailing Add:* Dept Civil Agr Civil Eng NMex State Univ Box 30001 Dept CE Las Cruces NM 88003

FINNEY, KARL FREDERICK, b Salina, Kans, July 25, 11; m 35; c 3. CEREAL CHEMISTRY. *Educ:* Kans Wesleyan Univ, AB, 35; Kans State Univ, BS, 36, MS, 37. *Prof Exp:* Assoc chemist, Hard Winter Wheat Qual Lab, USDA, Kans State Univ, 38-43; from assoc chemist to chemist, Soft Wheat Lab, Ohio Exp Sta, 43-46; prof grain sci, Kans State Univ, 46-83, res chemist in chg hard winter wheat qual lab, 46-72, res chemist & res leader, grain qual & end use properties unit, 72-83, EMER PROF GRAIN SCI, KANS STATE UNIV, 83- *Concurrent Pos:* Collabr & consult, US Grain Mkt Res Ctr, USDA, 83- *Honors & Awards:* Wheat Award & Steve Vesecky Award, Hard Winter Wheat Adv Coun; Thomas Burr Osborne Medal, Am Asn Cereal Chemists; Owen Wimberly Award, Nat Asn Wheat Growers. *Mem:* Fel AAAS; Am Chem Soc; Am Soc Agron; Am Asn Cereal Chemists; Sigma Xi. *Res:* Role of wheat flour components in breadmaking by fractionating and reconstituting techniques; optimized physical and breadmaking methods for evaluating functional properties of wheat; effects of environmental factors and varying stages of wheat maturity on quality. *Mailing Add:* 2315 Bailey Dr Manhattan KS 66502

FINNEY, ROSS LEE, III, b Springfield, Mass, May 31, 33; m 63; c 2. TOPOLOGY. *Educ:* Univ Mich, BA, 54, MA, 55, PhD(math), 62. *Prof Exp:* Instr math, Mass Inst Technol, 61-63 & Princeton Univ, 63-66; from asst prof to assoc prof math, Univ Ill, Urbana-Champaign, 68-80; sr lectr math, Mass Inst Technol, 80-90; US NAVAL POSTGRAD SCH, 90- *Concurrent Pos:* Fulbright scholar, Poincare Inst, 56-57, lectr, Bo, Sierra Leone, 68 & 69 & Addis Ababa, Ethiopia, 71; chmn curriculum comt, African Math Prog, USAID, 62, secondary C writing group, 65-67; codir, Ghana Teaching Intern Prog, 71-74; sr mathematician, Proj CALC, 76; dir NSF Undergrad Math Appln Proj (UMAP), 77-84; founding ed, The UMAP J, 80-; chmn, Math Adv Bd of Educ Develop Ctr, Tech Assistance Ctr, Ford Found Urban Math Collaborative, 86-90; Sigma Xi nat lectr, 85-88; deleg, People-to-People Univ Math Educ Deleg to the People's Repub China, 83. *Mem:* Math Asn Am; Nat Coun Teachers Math; Sigma Xi; NY Acad Sci; Am Math Asn of Two-Yr Cols. *Res:* Author of numerous high school and university mathematics textbooks. *Mailing Add:* Four Goose Pond Rd Lincoln MA 01773

FINNEY, ROY PELHAM, b Gaffney, SC, Dec 7, 24; m 63; c 5. SURGERY, BIOENGINEERING. *Educ:* Univ SC, MD, 52. *Prof Exp:* Inst urol, Johns Hopkins Hosp, 56-57; dir, div urol, 72-82, PROF SURG, UNIV S FLA, 72- *Mem:* Am Urol Asn; Soc Univ Urologists; Int Soc Urol; Urodynamic Soc; Int Continence Soc; Am Col Surgeons. *Res:* Design of urologic devices as Double J Ureteral Stent, Flexirod and Flexiflate Penile Prostheses for impotence. *Mailing Add:* 4382 Aztec Ct Bayport FL 34607

FINNEY, STANLEY CHARLES, b Chula Vista, Calif, Oct 14, 47; m 77; c 2. GEOLOGY, PALEONTOLOGY. *Educ:* Univ Calif, Riverside, BS, 69, MS, 71; Ohio State Univ, PhD(geol), 77. *Prof Exp:* Res assoc geol, Field Mus Natural Hist, 78-79; instr, Ohio State Univ, 79-80; asst prof, Northern Ariz Univ, 80-81; assoc prof geol, Okla State Univ, 81-86; PROF GEOL, 86-, CHMN DEPT, CALIF STATE UNIV, LONG BEACH, 88- *Concurrent Pos:* Fels, Mem Univ Nfld, 77-78 & Field Mus Natural Hist, 78-79. *Mem:* Geol Soc Am; Am Asn Petrol Geologists; Paleont Soc; Paleaont Asn; Int Paleont Asn. *Res:* Invertebrate paleontology-stratigraphy; graptolites- morphology, taxonomy, biostratigraph; ordovician biostratigraphy and biogeography. *Mailing Add:* Dept Geol Calif State Univ Long Beach CA 90840

FINNIE, I(AIN), b Hong Kong, July 18, 28; nat US; m 69; c 1. MECHANICAL ENGINEERING. *Educ:* Glasgow Univ, BSc, 49; Mass Inst Technol, SM, 50, ME, 51, ScD(mech eng), 53. *Hon Degrees:* DSc, Glasgow Univ, 75. *Prof Exp:* Instr mech eng, Mass Inst Technol, 52-53; engr, Shell Develop Co, 53-61; assoc prof mech eng, 61-63, PROF MECH ENG, UNIV CALIF, BERKELEY, 63- *Mem:* Hon mem Am Soc Mech Engrs; Nat Acad Eng. *Res:* Mechanical behavior of engineering materials, especially creep, wear and fracture. *Mailing Add:* Dept Mech Eng Univ Calif Berkeley CA 94720

FINNIGAN, J(EROME) W(OODRUFF), b Oak Park, Ill, Feb 9, 24; m 49; c 4. CHEMICAL ENGINEERING. *Educ:* Northwestern Univ, BS, 50; Univ Idaho, MS, 53; Ore State Univ, PhD(chem eng), 58. *Prof Exp:* Reactor engr, Hanford Atomic Prod Oper, Gen Elec Co, 50-56, supvr reactor tech develop oper, 58-62; proj mgr & res & develop mgr, TRW Systs, 62-70; mgr Fuels & Mat Dept, Battelle-Northwest Labs, 70-81; DEAN & RESIDENT DIR, JOINT CTR GRAD STUDY, 81- *Mem:* Am Nuclear Soc; Am Inst Chem Engrs; Am Ord Asn. *Res:* Nuclear reactor and systems engineering; materials development and technology applied to nuclear and non-nuclear projects for government and industry. *Mailing Add:* 2322 Enterprise Dr Richland WA 99352

FINS, LAUREN, b New York, NY, April 14, 45; m 74; c 1. TREE BREEDING, TREE IMPROVEMENT. *Educ:* NY Univ, BA, 65; Colo State Univ, MS, 73; Univ Calif, Berkeley, PhD(genetics), 79. *Prof Exp:* Asst prof, 79-84, ASSOC PROF FOREST GENETICS, UNIV IDAHO, 84-, DIR, INLAND EMPIRE TREE IMPROVEMENT COOP, 79. *Concurrent Pos:* Instr forestry, Colo State Univ, 72; teaching asst genetics, Univ Calif, Berkeley, 73-79; assoc ed, Western J Appl Forestry, Soc Am Forestry, 86-, mem educ Policies Nat Comt, 86-, chmn-elect, Tree Improvement Working Group, 88-90, chair Soc Am Forestry, 91- *Mem:* Soc Am Foresters. *Res:* Investigations of population architecture of forest tree species; applied research in forest tree improvement including techniques for vegetative propagation; changes in growth patterns over time. *Mailing Add:* Dept Forest Resources Col Forestry Wildlife & Range Sci Univ Idaho Moscow ID 83843

FINSETH, DENNIS HENRY, b Bremerton, Wash, July 24, 45; m 69; c 3. SPECTROCHEMISTRY. *Educ:* Western Wash State Col, BA, 67; Univ Pittsburgh, MS, 69, PhD(chem), 73. *Prof Exp:* Instr chem, Univ Pittsburgh, 73-76; RES CHEMIST DEPT ENERGY, PITTSBURGH ENERGY TECHNOL CTR, 76- *Mem:* Am Chem Soc; Sigma Xi. *Res:* Vibrational spectroscopy of theoretically interesting small molecules; analytical applications of spectroscopy; chemistry of coal conversion. *Mailing Add:* 1425 Browning Rd Pittsburgh PA 15206

FINSTEIN, MELVIN S, b Cambridge, Mass, June 25, 31; m 51; c 3. MICROBIOLOGY. *Educ:* Cornell Univ, BS, 59, MS, 61; Univ Calif, Berkeley, PhD(soil microbiol), 64. *Prof Exp:* Asst prof, 64-70, assoc prof, 70-73, PROF ENVIRON SCI, RUTGERS UNIV, NEW BRUNSWICK, 73- *Mem:* Am Soc Microbiol; Am Soc Limnol & Oceanog. *Res:* Physiology of autotrophic bacteria; microbiology of polluted waters; waste water treatment processes; ecology of nitrifying waste treatment and bacteria in polluted water; composting for the treatment of sludge and solid waste; environmental regulation and management. *Mailing Add:* Dept Environ Sci Rutgers Univ New Brunswick NJ 08903

FINSTER, MIECZYSLAW, b Lwow, Poland, Aug 1, 24; US citizen; m 51; c 2. PHYSIOLOGY, PHARMACOLOGY. *Educ:* Univ Geneva, MD, 57. *Prof Exp:* Resident anesthesiol, Columbia Univ-Presby Med Ctr, 58-60; from instr to assoc prof, 61-75, prof anesthesia, 75-78, PROF ANESTHESIA, OBSTET & GYNEC, COL PHYSICIANS & SURGEONS, COLUMBIA UNIV, 78-

Concurrent Pos: NIH res fel anat, Col Physicians & Surgeons, Columbia Univ, 60-61. *Mem:* Am Soc Anesthesiologists; Asn Univ Anesthetists; Am Soc Pharmacol & Exp Therapeut; Soc Obstet Anesthesiologists & Perinatologists; assoc fel Am Col Obstetricians & Gynecologists. *Res:* Transmission of drugs across the placenta; neonatal pharmacology. *Mailing Add:* 622 W 168th St New York NY 10032

FINSTON, HARMON LEO, b Chicago, Ill, Feb 16, 22; m 50; c 4. NUCLEAR CHEMISTRY. *Educ:* Ill Inst Technol, BSAS, 43; Ohio State Univ, PhD(chem), 50. *Prof Exp:* Cur chem, Lewis Inst Br, Ill Inst Technol, 42-43; jr chemist, Manhattan Dist, Metall Lab, Univ Chicago, 43-45; asst cyclotron chem, Ohio State Univ, 45-50; assoc chemist, Dept Chem, Brookhaven Nat Lab, 50-51, assoc chemist & supvr, Radiochem Anal Sect, Dept Nuclear Eng, 51-58, chemist, 58-60, leader, Radiochem Anal Group, 60-63; PROF CHEM, BROOKLYN COL, CITY UNIV NEW YORK, 63- *Concurrent Pos:* Mem subcmt radiochem, Nat Res Coun, 57-60 & subcmt use radioactivity standards, 62; fel Israel AEC, 65; consult, US Army Nuclear Defense Lab; I M Kolthoff fel & vis prof inorg & anal chem, Hebrew Univ, 73; fel, Japan Soc Prom Sci, Kyoto Univ, 72. *Mem:* Am Chem Soc; Am Nuclear Soc; fel Japan Soc Prom Sci. *Res:* Nuclear, radio and analytical chemistry; mechanisms of synergic solvent extraction; homogeneous liquid-liquid extraction and extraction at near neutral pH; acid-base theory. *Mailing Add:* 18 Springfield Ave No 3F Cranford NJ 07016-2151

FINSTON, ROLAND A, b Chicago, Ill, Jan 27, 37; m 60; c 2. HEALTH PHYSICS, RADIOLOGICAL PHYSICS. *Educ:* Univ Chicago, AB & SB, 57; Vanderbilt Univ, MS, 59; Cornell Univ, PhD(biophys), 65. *Prof Exp:* Assoc prof radiol physics, Ore State Univ, 65-66; sr health physicist, 66-77, LECTR RADIOL, STANFORD UNIV, 70-, DIR HEALTH PHYSICS, 77- *Mem:* Health Physics Soc; Am Asn Physicists in Med. *Res:* Radiation dosimetry; radiation biology; radiological health; radiological physics; dosimetry in nuclear medicine. *Mailing Add:* Health Physics Off Environ Safety Fac Stanford Univ Oak Road St Stanford CA 94305-8006

FINZEL, RODNEY BRIAN, b Monroe, Mich, Dec 23, 51; m 80; c 2. PHYSICAL ORGANIC CHEMISTRY, COMPUTER APPLICATIONS IN CHEMISTRY. *Educ:* Eastern Mich Univ, BS, 73, MS, 77; Northwestern Univ, PhD(chem), 81. *Prof Exp:* Instr org chem, Northwestern Univ, 80; asst prof, 81-90, ASSOC PROF ORG CHEM, HOFSTRA UNIV, 90- *Mem:* Am Chem So. *Res:* Organic reaction mechanism; computer applications in chemistry; molecular orbital theory. *Mailing Add:* Dept Chem Hofstra Univ Hempstead NY 11550

FIORE, ANTHONY WILLIAM, b Youngstown, Ohio, Oct 26, 20; m 44. AERONAUTICAL & ASTRONAUTICAL ENGINEERING. *Educ:* Univ Cincinnati, BS, 48, MS, 49; Ohio State Univ, PhD(aeronaut & astronaut eng), 66. *Prof Exp:* Instr aeronaut, Col Eng, Univ Cincinnati, 48-49; aeronaut develop engr, Aircraft Lab, Air Mat Command, 49-52, aeronaut res engr, Aircraft Lab, Wright Air Develop Ctr, 52-54, aeronaut res engr, Aeronaut Res Lab, 54-65, asst for exp aerodyn res, Hypersonic Res Lab, Aerospace Res Lab, Air Force Syst Command, 65-75, tech mgr, 75-82, SR AERODYNAMIC RES SCIENTIST, AIR FORCE FLIGHT DYNAMICS LAB, USAF, WRIGHT PATTERSON AFB, 82- *Mem:* Assoc fel Am Inst Aeronaut & Astronaut; Sigma Xi. *Res:* Theoretical and experimental basic research in aerodynamics; subsonic, transonic, supersonic and hypersonic aerodynamic research. *Mailing Add:* 5463 Grantland Dr Dayton OH 45429

FIORE, CARL, b New Haven, Conn, Sept 9, 28; m 56. INSECT PHYSIOLOGY, INSECT DEVELOPMENT. *Educ:* Yale Univ, BA, 50; Fordham Univ, MS, 56, PhD(physiol), 59. *Prof Exp:* Instr physiol, Fordham Univ Col Pharm, 56-58, res assoc biol, 58-61; asst prof, City Univ New York, 61-63; fel, Develop Biol Ctr, Western Reserve Univ, 63-65; researcher, Am Mus Natural Hist, 65-66; assoc prof, 66-69, PROF BIOL, CENT CONN STATE UNIV, 69- *Mem:* AAAS; Am Soc Zool; Entom Soc Am; Sigma Xi. *Res:* Physiology and biochemistry of aging; environmental effects on development; cell biology. *Mailing Add:* Dept Biol Sci Cent Conn State Univ New Britain CT 06050

FIORE, JOSEPH VINCENT, b NY, Oct 9, 20; m 50; c 3. BIOCHEMISTRY, SEPARATION PROCESSES. *Educ:* Fordham Univ, BS, 43, MS, 47, PhD(chem), 50. *Prof Exp:* Chief biochem & clin biochem, Rochester Gen Hosp, 50-53; group leader cereal chem, Fleischmann Labs Div, Standard Brands, Inc, 53-60; mgr tobacco res group, AMF Inc, 60-64, area mgr chem develop lab, 64-66, mgr food & tobacco lab, Res Div, 66-70, mgr tobacco & anal lab, 70-78, dir appl chem, 78-; RETIRED. *Mem:* Am Chem Soc; Inst Food Technol; NY Acad Sci; Water Pollution Control Fedn; Tech Asn Pulp & Paper Indust. *Res:* Enzymology; fermentation; clinical biochemistry; ecological sciences; analytical chemistry. *Mailing Add:* 205 Steiner St Fairfield CT 06430

FIORE, NICHOLAS F, b Pittsburgh, Pa, Sept 24, 39; m 60; c 4. MATERIALS ENGINEERING, PHYSICAL METALLURGY. *Educ:* Carnegie Inst Technol, BS, 60, MS, 63, PhD(metall), 64. *Hon Degrees:* DEng, Univ Waterloo, Waterloo, Ont, Can, 85. *Prof Exp:* From asst prof to prof metall, Univ Notre Dame, 66-82, chmn, Dept Metall Eng & Mat Sci, 69-82; vpres, Cabot Corp, Boston, Mass, 82-89; managing dir mat & appl physics, Arthur D Little Co, Cambridge, Mass, 89-90; VPRES RES & DEVELOP, CARPENTER TECHNOL CORP, READING, PA, 90- *Concurrent Pos:* Pres, Electron Devices Soc, Inst Elec & Electronics Engrs, 80-81; mem, Nat Sci & Eng Res Coun Can, (Commun & Comput), 81-84; mem, Queens Univ Adv Coun on Eng, 80-86, chmn, Exec Comt, 85; Allegheny Ludlum Corp fel, 63. *Mem:* Fel Am Soc Metals; Am Welding Soc; Am Inst Mining, Metall & Petrol Engrs. *Res:* Internal friction, non-destructive testing, wear, corrosion, environment-sensitive degradation of materials; optoelectronics, fiber optics and telecommunications systems applications; management and administration of advanced technology research laboratory & telecommunications systems research programs. *Mailing Add:* Carpenter Technol Corp PO Box 14662 Reading PA 19612-4662

FIORI, BART J, b Passaic, NJ, Dec 7, 30; m 56; c 4. ENTOMOLOGY. *Educ:* Univ Ga, BSA, 54; Cornell Univ, PhD(entom), 63. *Prof Exp:* Field res specialist, Ortho Div, Calif Chem Co, 63-64; RES ENTOMOLOGIST & RES LEADER, SCI & EDUC ADMIN AGR RES, USDA, 64- *Concurrent Pos:* Assoc prof, Cornell Univ, 68- *Mem:* Entom Soc Am; Sigma Xi. *Res:* Petroleum oils as ovicides; evaluation of insecticides; physical, chemical and biological control of the European chafer Amphimallon majalis; plant resistance against insects. *Mailing Add:* 2565 Melvin Hill Rd Geneva NY 14456

FIORINDO, ROBERT PHILIP, b Pottstown, Pa, May 9, 35; m 62; c 3. NEUROENDOCRINOLOGY, ENDOCRINOLOGY. *Educ:* Albright Col, BS, 58; Univ Md, MS, 65; Univ Calif, Berkeley, PhD(physiol), 70. *Prof Exp:* Vis prof, Mills Col, Oakland, Calif, 68-69; res fel, Inst Pharmacol, Univ Milan, Italy, 69-70, neuroendocrinol, Physiol Labs Univ Cambridge, UK, 70-71 & dept physiol, Univ Md, 71; asst prof physiol, Univ Notre Dame, 71-74; asst prof physiol, Ohio State Univ, Columbus, 74-81; ASSOC PROF PHYSIOL, COL OSTEOPATH MED PAC, POMONA, CALIF, 81- *Concurrent Pos:* Fel, Pop Coun, NY, Univ Milan, Italy & Univ Cambridge, UK, 69-71; fel, Nat Inst Child Health & Human Develop, Sch Med, Univ Md, 71; prin investr, NSF grant, Univ Notre Dame, 72-74, NIH grant, Univ Notre Dame, 73-74 & Ohio State Univ, 75-78 & Inst grant, Ohio State Univ, 74-76. *Mem:* Am Physiol Soc; Endocrine Soc. *Res:* Neural control of anterior pituitary secretion of prolactin, gonadotropins and growth hormone; comparative aspects of the hypothalamic control of prolactin secretion; comparative endocrinology. *Mailing Add:* Dept Physiol Col Osteop Med Pac 309 Pomona Mall E Pomona CA 91766-1889

FIORITO, RALPH BRUNO, b New York, NY, Oct 24, 41. PHYSICS. *Educ:* Col Holy Cross, BS, 63; Cath Univ Am, MSE, 66, PhD(physics), 71. *Prof Exp:* Res asst space sci physics, Cath Univ Am, 65-71; physicist acoust, 73-78, RES PHYSICIST ELECTRON BEAM PROPAGATION, WHITE OAK LAB, NAVAL SURFACE WEAPONS CTR, 76- *Concurrent Pos:* Lectr, Cath Univ, 73-74; vis res physicist electron beam propagation, Naval Res Lab, 78-80 & Lawrence Livermore Nat Lab, 81- *Mem:* Am Phys Soc; Acoust Soc Am; Sigma Xi. *Res:* Intensive relativistic electron beam diagnostics and propagation; acoustic wave propagation in materials. *Mailing Add:* Three Lauer Terr Silver Spring MD 20901

FIRBY, JAMES R, b Detroit, Mich, Nov 28, 33; m 56; c 1. INVERTEBRATE PALEONTOLOGY, STRATIGRAPHY. *Educ:* San Francisco State Col, BA, 60; Univ Calif, Berkeley, MA, 63, PhD(paleont), 69. *Prof Exp:* Asst prof, 66-74, asst dean, 74-80, ASSOC PROF GEOL, MACKAY SCH MINES, UNIV NEV, RENO, 74- *Mem:* Am Asn Petrol Geologists. *Res:* Cenozoic non-marine Mollusca, especially of Western North America. *Mailing Add:* Mackay Sch Mines Univ Nev-Geol Reno NV 89507

FIRDMAN, HENRY ERIC, b Leningrad, USSR, Oct 24, 36; US citizen; m 75; c 5. HUMAN ASPECTS OF INFORMATION SYSTEMS, TECHNOLOGY MANAGEMENT. *Educ:* Leningrad Polytech Inst, MSEE, 69; Ctr Microelectronics, Moscow, PhD(microelectronics), 66; Inst Electronic Design, Moscow, PhD(computer sci), 69. *Prof Exp:* Mem staff, Leningrad Design Bur, USSR, 59-64, lab dir, 64-68, dept mgr, 68-71, vpres, 71-73; founding dir, AI Lab, Far East Res Ctr, USSR, 74-77; mem tech staff, Hewlett Packard Labs, Palo Alto, 81-83; dir, Access Technologies Inc, 83-84; pres, Henry Firdman & Assocs, 84-91; DIV MGR, PAC BELL, SAN RAMON, 91- *Concurrent Pos:* Prof computer sci, Vladivostok State Univ, USSR, 75-77; instr, Indust Rels Ctr, Calif Inst Technol, 85- *Mem:* Am Asn Artificial Intel; Asn Comput Mach; Inst Elec & Electronics Engrs Computer Soc. *Res:* Information technology and computer science; artificial intelligence; strategic information systems; systems analysis and design methodology; information infrastructure design. *Mailing Add:* Pac Bell 2600 Camino Ramon Rm 1E954 San Ramon CA 94583

FIRE, PHILIP, b Paterson, NJ, Dec 18, 25; m 51; c 2. ELECTRICAL ENGINEERING. *Educ:* Mass Inst Technol, BS & MS, 52; Stanford Univ. PhD(elec eng), 64. *Prof Exp:* Staff engr, Lincoln Lab, Mass Inst Technol, 52-54; sr scientist, 55-80, PRIN SCIENTIST, GTE-GOVT SYSTS, WESTERN DIV, GTE CORP, 82- *Concurrent Pos:* Liason scientist, Off Naval Res, London, 80-82; ed, Trans on Commun, Inst Elec & Electronics Engrs, 78-79. *Mem:* Fel Inst Elec & Electronics Engrs. *Res:* Error-correcting codes for communication systems; analysis and synthesis of electronic systems. *Mailing Add:* GTE-Govt Systs PO Box 7188 Mountain View CA 94039

FIREBAUGH, MORRIS W, b Freeport, Ill, July 5, 37; m 60; c 2. COMPUTER SCIENCES. *Educ:* Manchester Col, AB, 59; Univ Ill, MS, 60, PhD(physics), 66. *Prof Exp:* Res assoc high-energy physics, Univ Ill, 66-67; instr physics, Univ Wis, 67-69, prof physics, 69-82, PROF OF COMPUT SCI, UNIV WIS-PARKSIDE, 82- *Concurrent Pos:* Vis scientist, Inst Energy Anal, Oak Ridge, Tenn, 79-80. *Mem:* Am Phys Soc; Am Asn Physics Teachers; Sigma Xi; Am Asn Artificial Intel; Asn Comput Mach. *Res:* Artificial intelligence; neural networks; expert systems; computer graphics. *Mailing Add:* Dept Comput Sci Box 2000 Univ of Wis-Parkside Kenosha WI 53141

FIREHAMMER, BURTON DEFOREST, veterinary microbiology, veterinary immunology, for more information see previous edition

FIREMAN, EDWARD LEONARD, physics; deceased, see previous edition for last biography

FIREMAN, PHILIP, b Pittsburgh, Pa, Feb 28, 32; m 57; c 5. PEDIATRICS, IMMUNOLOGY. *Educ:* Univ Pittsburgh, BS, 53; Univ Chicago, MD, 57. *Prof Exp:* Clin assoc, NIH, 60-62; from instr to assoc prof, 63-74, PROF PEDIAT, SCH MED, UNIV PITTSBURGH, 74- DIR ALLERGY-IMMUNOL, CHILDREN'S HOSP, 68- *Concurrent Pos:* Mead-Johnson fel pediat, 58-60; USPHS res fel, Harvard Med Sch, 62-63; Interstate Postgrad Med Asn res award, 64; USPHS res career develop award, 65; res collabr,

Brookhaven Nat Lab, 64-69 & Univ Lausanne & Swiss Inst Cancer Res, 72-73. *Res:* Immediate and delayed hypersensitivity; clinical immunology; allergy. *Mailing Add:* Dept Pediat Univ Pittsburgh Sch Med Childrens Hosp 3705 Fifth Ave at DeSoto St Pittsburgh PA 15213

FIRESTONE, ALEXANDER, b New York, NY, July 22, 40. HIGH ENERGY PHYSICS. *Educ:* Columbia Univ, BS, 62; Yale Univ, MA, 64, PhD(physics), 66. *Prof Exp:* Res physicist, Lawrence Berkeley Lab, 66-71; asst prof high energy physics, Calif Inst Technol, 71-75; assoc prof, 75-78, PROG DIR, AMES LAB, IOWA STATE UNIV, 78-, PROF HIGH ENERGY PHYSICS, 79- *Mem:* Fel Am Phys Soc. *Res:* Electron-positron annihilations at CERN LEP collider; problems in strong interactions; particle dynamics, resonance production, diffractive effects, correlations, high transverse momentum effects and jet structure. *Mailing Add:* Dept Physics Iowa State Univ Ames IA 50011

FIRESTONE, RAYMOND A, b New York, NY, Jan 20, 31; m 87; c 3. PHARMACEUTICAL CHEMISTRY. *Educ:* Cornell Univ, AB, 51; Columbia Univ, PhD(org chem), 54. *Prof Exp:* Sr chemist, Merck & Co, Rahway, 56-73, res fel, 73-77, sr res fel, 77-80, sr investr, 80-87; DISTINGUISHED RES FEL, BRISTOL-MYERS, 87- *Concurrent Pos:* Chmn, Gordon Heterocyclic Conf, 78; vis prof, Univ Wis-Madison, 83. *Honors & Awards:* Directors Award, Merck Sharp & Dohme, 68. *Mem:* The Chem Soc. *Res:* Synthesis and mechanism in organic chemistry; enzyme inhibitors; cancer chemotherapy. *Mailing Add:* Pharmaceut Res & Develop Dept Bristol-Myers Co PO Box 5100 Wallingford CT 06492-7660

FIRESTONE, RICHARD FRANCIS, b Canton, Ohio, June 18, 26; m 54; c 3. PHYSICAL CHEMISTRY. *Educ:* Oberlin Col, AB, 50; Univ Wis, PhD, 54. *Prof Exp:* Asst chem, Univ Wis, 50-51, asst radiochem, 51-54; resident res assoc chem div, Argonne Nat Lab, 54-55, instr, Int Sch Nuclear Sci & Eng, 55-56; asst prof chem, Western Reserve Univ, 56-60; assoc prof, 61-67, PROF CHEM, OHIO STATE UNIV, 67- *Mem:* Fel AAAS; Am Phys Soc; Am Chem Soc; Sigma Xi. *Res:* Radiation chemistry; kinetics of ionizing-radiation induced reactions; fast reactions of excited atoms in the gaseous phase; rare gas excimer formation and decay. *Mailing Add:* Dept Chem Ohio State Univ 120 W 18th Ave Columbus OH 43210

FIRESTONE, ROSS FRANCIS, b Freeport, Ill, Oct 10, 32; m 57; c 4. CERAMIC SCIENCE, GLASS TECHNOLOGY. *Educ:* Univ Chicago, BA, 53, BS, 56, MBA, 65; Case Inst Technol, PhD (mat sci), 74. *Prof Exp:* Geol field asst, US Geol Surv, 56-57; assoc scientist, Armour Res Found, 57-65; sr res chemist, Standard Oil Co, 65-68; grad res fel ceramic sci, Case Inst Technol, 68-71; asst prof ceramics, Univ Pittsburgh, 71-77; mgr, IIT Res Inst, 77-83 & L J Broutman & Assoc, 83-86; CONSULT SCIENTIST MAT SCI, PVT PRACT, 86- *Concurrent Pos:* Eng ed, J Irreproducible Results, 55- *Mem:* Am Ceramic Soc; Soc Glass Technol; Royal Micros Soc; Am Soc Metals; Sigma Xi; Brit Ceramic Soc; Int Asn Fire Safety Sci. *Res:* Materials at high temperatures; fire science. *Mailing Add:* 188 Mary St Winnetka IL 60093-1520

FIRESTONE, WILLIAM L(OUIS), b Chicago, Ill, June 20, 21; m 53; c 3. ELECTRICAL ENGINEERING. *Educ:* Univ Colo, BS, 46; Ill Inst Technol, MS, 49; Northwestern Univ, PhD(elec eng), 52. *Prof Exp:* Elec engr, Manhattan Proj, Univ Calif, 46; microwave engr, Motorola Radio, 46-49; grad asst, Northwestern Univ, 49-50; chief engr, Res Dept, Motorola Inc, Ill, 51-60, dir eng, Commun Div, 60-62, asst gen mgr, Chicago Mil Electronics Ctr, 62-64; vpres, Hallecrafte Corp, 65-66; vpres, Whittaker Corp, Calif, 66-70, gen mgr, Tech Prod Div, 66-69, group exec, Corp, 69-70; vpres & gen mgr, F W Sickles Div, Gen Instruments Corp, 70-71, vpres & group exec entertainment prod, Chicopee, 71-75; div v pres & gen mgr, Avionics Systs, RCA Inc, 75-; PRES, TELOC, INC. *Concurrent Pos:* Instr, US Radio Mat Sch, 42 & Ill Inst Technol 48-49 & 55-56; lectr, Northwestern Univ, 57. *Mem:* Fel Inst Elec & Electronics Engrs. *Res:* Communication engineering; communication theory and single sideband; engineering managements. *Mailing Add:* Texscan Inc 10841 Pellicano St El Paso TX 79935

FIREY, JOSEPH CARL, b Roundup, Mont, Oct 22, 18; m 51; c 3. MECHANICAL ENGINEERING. *Educ:* Univ Wash, BS, 40; Univ Wis, MS, 41. *Prof Exp:* Instr mech eng, Univ Wis, 41-42; res engr, Calif Res Corp, 42-43, 46-54; from asst prof to prof mech eng, 54-88, EMER PROF MECH ENG, UNIV WASH, 80- *Mem:* Am Soc Mech Engrs; Soc Automotive Engrs. *Res:* Combustion and lubrication problems in power generating equipment. *Mailing Add:* Dept Mech Eng FU-10 Univ Wash Seattle WA 98195

FIREY, WILLIAM JAMES, b Roundup, Mont, Jan 23, 23; m 46; c 2. MATHEMATICS. *Educ:* Univ Wash, BS, 49; Univ Toronto, MA, 50; Stanford Univ, PhD(math), 54. *Prof Exp:* From instr to asst prof math, Wash State Univ, 53-61; assoc prof, 61-64, PROF MATH, ORE STATE UNIV, 64- *Concurrent Pos:* Mem staff, Fulbright Res Sch, Univ Otago, NZ, 67; vis prof, Mich State Univ, 69-70, Univ Freiburg, WGer, 75-76. *Mem:* Am Math Soc; Math Asn Am; Can Math Cong. *Res:* Theory of convex sets and integral geometry. *Mailing Add:* Dept Math Ore State Univ Corvallis OR 97331

FIRK, FRANK WILLIAM KENNETH, b London, Eng, Nov 2, 30; m 52; c 3. NUCLEAR PHYSICS. *Educ:* Univ London, BSc, 56, MSc, 65, PhD(physics), 67. *Prof Exp:* Asst exp officer physics, Atomic Energy Res Estab, Eng, 52-56, exp officer, 56-59, sr sci officer, 59-62, prin sci officer, 62-65; sr res assoc, 65-68, assoc prof, 68-77, PROF PHYSICS, YALE UNIV, 77-, DIR, ELECTRON ACCELERATOR LAB, 76-, CHMN DEPT, 80- *Concurrent Pos:* Vis scientist, Oak Ridge Nat Lab, 60-61. *Mem:* Am Inst Physics; Brit Inst Physics & Phys Soc. *Res:* Low energy neutron spectroscopy; nuclear photo-disintegration. *Mailing Add:* Dept Physics Yale Univ New Haven CT 06520

FIRKINS, JOHN FREDERICK, b Olympia Wash, Jan 31, 35; m 63; c 3. MATHEMATICS EDUCATION, PROBLEM SOLVING & TACTILE LEARNING. *Educ:* St Martin Col, BS, 57; Univ Miami, MS, 59; Gonzaga Univ, MEd, 70. *Hon Degrees:* LLD, Gonzaga Univ, 87. *Prof Exp:* From instr to assoc prof math, 61-77, PROF MATH, GONZAGA UNIV, 77-, ARNOLD PROF HUMANITIES, 86- *Concurrent Pos:* Pres, Wash State Math Couns, 87-; mem, Proj Nat Teachers Math; ed, Nat Coun of Teachers of Math Addenda Mat, 89-92; western dir, Nat Coun Supvr of Math, 90-93; co dir, Wash State Math Coalition, 90-91; found, Wash State Math Leadership Conf. *Mem:* Nat Coun Teachers Math; Math Asn Am; Nat Coun Supvr Math. *Res:* How children learn math; creating problem solving activities to maximize learning. *Mailing Add:* Dept Math Gonzaga Univ Spokane WA 99258

FIRKINS, JOHN LIONEL, b Victoria, BC, Feb 13, 42; m 66; c 2. PHYSICAL ORGANIC CHEMISTRY. *Educ:* Univ Victoria, BC, BSc, 65; Calif Inst Technol, PhD(chem, bus econ), 70. *Prof Exp:* Sr res & develop chemist, Celanese Fibers Co, 69-75; mem staff fiber chem, Weyerhaeuser Co, 75-78, group leader paper, 79-80, corp metric coordr, 80-; at Thilmay Pulp & Paper Co 87; AT INT PAPER CO, 87- *Mem:* AAAS; Am Chem Soc; Chem Inst Can; Tech Asn Pulp & Paper Indust. *Res:* Free radical reaction mechanisms; wood and cellulose chemistry; wood pulps for disposable products; chemical derivatives of renewable resources; cellulose acetate and other textile fibers. *Mailing Add:* A E Staley Mfg Co 2200 E Eldorado St Decatur IL 62521-1578

FIRLE, TOMAS E(RASMUS), b Berlin, Ger, July 4, 26; nat US; div; c 3. ENVIRONMENTAL SCIENCE, COMMUNICATIONS. *Educ:* Univ Calif, Los Angeles, BS, 52, MS, 55. *Prof Exp:* Res physicist & head develop group, Semiconductor Lab, Hughes Aircraft Co, 52-58; res physicist, Gulf Energy & Environ Systs Inc, San Diego, 58-72; spec asst to supt, San Dieguito High Sch, 72-73; head environ anal, City San Diego, Environ Qual Div, 73-74; DEPT HEAD ENVIRON MGT, PORT OF SAN DIEGO, 74- *Concurrent Pos:* Lectr, Univ Calif, Los Angeles, 57-58 & Univ Calif, San Diego, 59-61; discussion leader, Univ Calif, San Diego & univ exten, 62-; trustee, Bd Educ, Del Mar Union Sch Dist, 67; mem ombudsman comt, Dept Educ, 69; bd dirs, Indust-Educ Coun, 69; pres, Inter-Focus, Calif, 70-; mem bd dirs & regional dir, Calif Asn Environ Profs, 74-77; lectr environ mgt, Sch Bus Admin, San Diego State Univ, 75-; environ comt chmn, Airport Opers Coun Int, 75-77; vchmn environ comt, Pac Coast Airport Authorities, 78-; mem environ comt, Calif Asn Port Authorities. *Mem:* AAAS; Am Phys Soc; Sigma Xi; Marine Technol Soc. *Res:* Environmental effects; interpersonal human relationships; group and individual behaviorism; social and political psychology; human communications; effectiveness training; management consulting; establishment of integrated eco-systems; environmental management; salt marsh and shoreline rehabilitation. *Mailing Add:* PO Box 9782 San Diego CA 92169-0782

FIRLIT, CASIMIR FRANCIS, b Chicago, Ill, Dec 7, 39; m 65; c 2. PEDIATRICS, UROLOGY. *Educ:* Loyola Univ, Chicago, MS & MD, 65, PhD(pharmacol), 71. *Prof Exp:* Intern med, Mercy Hosp Med Ctr, Chicago, 65-66; resident gen surg, Vet Admin Hosp, Hines, 68-70; from instr to asst prof pharmacol, Stritch Sch Med, Loyola Univ, Chicago, 70-73; asst prof urol & physiol, 73-80, PROF UROL, NORTHWESTERN UNIV, CHICAGO, 80-; HEAD PEDIAT RENAL TRANSPLANTATION, CHILDREN'S MEM HOSP, 73-, UROL FEL, 77-, CHMN PEDIAT UROL, 76- *Concurrent Pos:* Resident urol & res & educ assoc, Vet Admin Hosp, Hines, 70-73; attend pediat urologist, Children's Mem Hosp, Chicago, 73-; attend urologist, Northwestern Mem Hosp, 73- & Vet Admin Lakeside Hosp, 73-; head pediat urol, Childrens Mem Hosp; lectr urol & pharmacol, Stritch Sch Med, Loyola Univ, 73-; consult urol, Vet Admin Hosp, Hines, 73-; attend urologist, Loyola McGaw Med Ctr, Maywood, 73- *Honors & Awards:* Resident Prize, Chicago Urol Soc, 72. *Mem:* fel Am Acad Pediat; fel Am Bd Urol; Am Soc Transplant Surgeons; Am Fedn Clin Res. *Res:* Testicular metabolism and male fertility; neurogenic bladder research; pediatric renal transplantation; pediatric embryology. *Mailing Add:* 8501 Oak Knoll Dr Hinsdale IL 60521

FIRMAGE, D(AVID) ALLAN, b Nephi, Utah, Feb 15, 18; m 40; c 6. CIVIL ENGINEERING. *Educ:* Univ Utah, BS, 40; Mass Inst Technol, MS, 41. *Prof Exp:* Stress analyst, E G Budd Mfg Co, Pa, 41-42; res struct eng, Res & Develop Labs, Ft Belvoir, Va, 42-47; assoc prof civil eng, Univ Fla, 47-52; res asst, Mass Inst Technol, 52-53; asst chief struct engr, Patchen & Zimmerman, Engrs, Ga, 53-55; assoc prof civil eng, Brigham Young Univ, 55-57; chief bridge design engr, Capitol Engrs, Saigon, Vietnam, 57-59; from assoc prof to prof, 59-83, chmn dept, 69-72, EMER PROF CIVIL ENG, BRIGHAM YOUNG UNIV, 83- *Concurrent Pos:* Adv, Guindy Col Eng, India, 63-64; nat dir, Am Soc Civil Engrs, 78-81; partner, LFDB Consult Engrs, Provo, Utah & Grand Junction, Colo, 84-87; consult bridge engr, Provo, Utah, 87- *Mem:* Am Soc Eng Educ; Am Soc Civil Engrs; Int Asn Bridge & Struct Eng. *Res:* Structural design; design of highway bridges; textbooks; design and evaluation of bridge structures; structural engineering. *Mailing Add:* 1079 Ash Provo UT 84604

FIRMENT, LAWRENCE EDWARD, b Elyria, Ohio, Mar 24, 50; m. MATERIAL SURFACE SCIENCE, POLYMER INTERFACE CHARACTERIZATION. *Educ:* Mass Inst Technol, SB, 72; Univ Calif, Berkeley, PhD(phys chem), 77. *Prof Exp:* CHEMIST, E I DU PONT DE NEMOURS & CO, INC, 77- *Mem:* Am Chem Soc; Am Phys Soc; Am Vacuum Soc. *Res:* Surface chemistry and surface analysis of materials, especially polymers. *Mailing Add:* Cent Res & Develop Dept E I du Pont de Nemours & Co Inc Wilmington DE 19880-0228

FIRMINGER, HARLAN IRWIN, b Minneapolis, Minn, Dec 31, 18; m 42; c 3. PATHOLOGY. *Educ:* Wash Univ, AB, 39, MD, 43; Am Bd Path, dipl, 49. *Prof Exp:* Asst path, Sch Med, Wash Univ, 43-44; sr resident, Mass Gen Hosp, 46-47; pathologist, Nat Cancer Inst, 48-51; from asst prof to prof path & oncol, Med Sch, Univ Kans, 51-57; head dept path, Sch Med, Univ Md,

Baltimore, 57-67, prof path, 57-75; dir anat path, Gen Rose Mem Hosp, 75-76; prof, 75-89, EMER PROF PATH, UNIV COLO SCH MED, 89- *Concurrent Pos:* Consult, Ft Howard Vet Hosp; mem sci adv bd consults, Armed Forces Inst Path, 65-70, ed, Atlas Tumor Path, 66-75; mem comt path, Nat Acad Sci-Nat Res Coun, 65-71. *Mem:* Am Asn Pathologists; Am Asn Cancer Res; Int Acad Path; Sigma Xi. *Res:* Testicular tumors; chemical carcinogenesis; induced tumors of the liver; endocrine pathology. *Mailing Add:* 4200 E Ninth Ave Denver CO 80262

FIRNKAS, SEPP, b Rinnberg, Ger, Nov 16, 25; US citizen. CIVIL ENGINEERING. *Educ:* Munich Tech Univ, Dipl Ing, 53; Columbia Univ, CE, 78. *Prof Exp:* Assoc struct eng, CETBA, Algeria, 54-56; proj engr, Raymond Int Inc, NY, & Venezuela, 56-59, chief engr, Northeast Concrete Prod, 59-61; Pres, Sepp Firnkas Eng Inc, 61; assoc prof civil eng, Northeastern Univ, 62-75; RETIRED. *Honors & Awards:* Progressive Archit Struct Design Award, 64; Prestressed Concrete Inst Struct Design Award, 66. *Mem:* Am Concrete Inst; Int Asn Shell Struct; Prestressed Concrete Asn. *Res:* Design and applications of reinforced and prestressed concrete; precast structural systems; analysis and design for seismic forces. *Mailing Add:* Three Bellingham Pl Boston MA 02114

FIROR, JOHN WILLIAM, b Athens, Ga, Oct 18, 27; m 50, 83; c 4. SOLAR PHYSICS, CLIMATE CHANGE. *Educ:* Ga Inst Technol, BS, 49; Univ Chicago, PhD(physics), 54. *Prof Exp:* Staff mem, Dept Terrestrial magnetism, Carnegie Inst, 53-61; from assoc dir to dir, 61-74, dir high altitude observ, 61-68, exec dir, 74-80, DIR ADVAN STUDY PROG, NAT CTR ATMOSPHERIC RES, 80- *Concurrent Pos:* Mem US nat comt, on Global Atmospheric Res Prog, 68-72; Am Astron Soc vis prof, adj prof astrophys, Univ Colo, 62-68; mem, Dept of Com Weather Modification Adv Bd, 77-78; chmn, NASA Space & Terrestrial Appln Adv Comt, 78-81; mem, NASA Adv Coun, 78-81; trustee, Environ Defense Fund, & mem, Exec Comt, 74-, chmn, 75-80; sr fel, Hubert Humphrey Inst Pub Affairs, Univ Minn, 81-82; trustee, World Resources Inst, 82- *Mem:* Am Astron Soc; Am Geophys Union; Am Meteorol Soc; Int Astron Union. *Res:* Physical conditions in solar atmosphere; solar-terrestrial relations; physics of earth's atmosphere; impact of climate change. *Mailing Add:* Nat Ctr for Atmospheric Res PO Box 3000 Boulder CO 80307

FIRRIOLO, DOMENIC, b Brooklyn, NY, Sept 4, 33; m 59; c 2. PHYSIOLOGY, ANATOMY. *Educ:* St Francis Col, NY, BS, 54; St John's Univ, NY, MS, 56, PhD(physiol), 64. *Prof Exp:* From instr to assoc prof, 56-66, asst dean, Col Lib Arts & Sci, 66-70, chmn dept, 76, 82-85, PROF BIOL, LONG ISLAND UNIV, 70-, DIR DIV SCI. *Mem:* AAAS. *Res:* Comparative hematology; vertebrate erythropoiesis. *Mailing Add:* Dept Biol Long Island Univ Brooklyn Ctr Brooklyn NY 11201

FIRSCHEIN, HILLIARD E, b Brooklyn, NY, Apr 7, 27. BIOCHEMISTRY, CLINICAL RESEARCH. *Educ:* Ohio State Univ, BS, 48; Univ Wis, MS, 50; Univ Rochester, PhD(biochem), 58. *Prof Exp:* Asst, Univ Wis, 48-50; biochemist, Army Med Res Lab, Ft Knox, Ky, 51-55; res assoc, Atomic Energy Proj, Univ Rochester, 55-58, asst scientist, 58-60, instr biochem, Sch Med & Dent, 58-60, instr radiation biol, 59-60; from instr to asst prof biochem, Sch Med, NY Univ, 60-64; asst prof biochem orthop surg, Med Col, Cornell Univ, 64-69; assoc res biochemist, Univ Calif, Los Angeles, 69-71; asst dir, 72-80, ASSOC DIR, CIBA-GEIGY CORP, 72- *Concurrent Pos:* Sr scientist, Hosp for Spec Surg, 64-69. *Mem:* Am Chem Soc; Am Physiol Soc. *Res:* Endocrinology; vitamins; metabolism; clinical research. *Mailing Add:* 141 Laauwe Ave Wayne NJ 07470

FIRSCHING, FERDINAND HENRY, b Utica, NY, June 22, 23; m 55; c 6. ANALYTICAL CHEMISTRY. *Educ:* Syracuse Univ, MS, 51, PhD(chem), 55. *Prof Exp:* Res chemist, Skenandoa Rayon Corp, NY, 51-52; anal chemist, Cowles Chem Co, 52-53; sr chemist, Diamond Alkali Co, Ohio, 55-58; asst prof anal chem, Univ Ga, 58-63; assoc prof, 63-69, PROF ANALYTICAL CHEM, SCH SCI, SOUTHERN ILL UNIV, EDWARDSVILLE, 69- *Mem:* Am Chem Soc. *Res:* Precipitation from homogeneous solution; development of analytical methods; use of radioactive tracers. *Mailing Add:* Sch Sci Southern Ill Univ Edwardsville IL 62026-1001

FIRSHEIN, WILLIAM, b New York, NY, Aug 28, 30; m 55, 71. MICROBIOLOGY. *Educ:* Brooklyn Col, BS, 52; Rutgers Univ, MS, 53, PhD, 58. *Prof Exp:* Microbiol physiologist, Camp Detrick, Md, 54-55; asst prof, 58-65, assoc prof, 65-70, chmn dept, 70-74, PROF BIOL, WESLEYAN UNIV, 70- *Concurrent Pos:* USPHS career develop award, 65-70. *Mem:* AAAS; Am Soc Microbiol; Am Soc Biochemists. *Res:* Microbial genetics; biochemistry; DNA synthesis. *Mailing Add:* Dept Molecular Biol & Biochem Wesleyan Univ Middletown CT 06457

FIRST, MELVIN WILLIAM, b Boston, Mass, Dec 23, 14; m 38; c 2. PUBLIC HEALTH. *Educ:* Mass Inst Technol, BS, 36; Harvard Univ, MS, 47, ScD(indust hyg eng), 50; Environ Eng Intersoc Bd, dipl; Am Bd Indust Hyg, dipl. *Prof Exp:* Indust hyg engr, Dept of Health, Detroit, Mich, 36-39 & Mich State Health Dept, 39-41; res assoc, Sch Pub Health, Harvard Univ, 50-53; CONSULT ENGR, 53-; PROF ENVIRON HEALTH ENG, SCH PUB HEALTH, HARVARD UNIV, 71- *Concurrent Pos:* Assoc prof appl indust hyg, Harvard Univ, 62-71. *Mem:* Am Chem Soc; Nat Soc Prof Engrs; Am Indust Hyg Asn; Air Pollution Control Asn; Sigma Xi. *Res:* Air and gas purification equipment and techniques for control of industrial atmospheres and prevention of air pollution. *Mailing Add:* 295 Upland Ave Newton Highlands Boston MA 02161-2002

FIRST, NEAL L, b Ionia, Mich, Oct 8, 30; m 51; c 4. REPRODUCTIVE PHYSIOLOGY. *Educ:* Mich State Univ, BS, 52, MS, 57, PhD(animal physiol), 59. *Prof Exp:* Instr animal husb, Mich State Univ, 59-60; from asst prof to prof meat & animal sci, 60-89, LE CASIDA PROF REPRODUCTIVE BIOL & BIOTECHNOL, UNIV WIS-MADISON, 89- *Concurrent Pos:* Von Humboldt award agr, 88; res scientist award, Soc Study Reproduction, 91. *Honors & Awards:* Physiol & Endocrinol Award, Am Soc Animal Sci, 77; Nat Asn Animal Breeders Award, 86. *Mem:* Nat Acad Sci; Am Genetic Asn; Am Soc Study Reproduction; Soc Study Fertil; Am Soc Animal Sci. *Res:* Reproductive physiology, especially of male livestock; artificial insemination. *Mailing Add:* Dept Animal Sci Univ Wis 752A Sci Bldg Madison WI 53706

FIRSTBROOK, JOHN BRADSHAW, physiology, internal medicine, for more information see previous edition

FIRSTENBERGER, B(URNETT) G(EORGE), b Seneca, Kans, July 18, 17; m 43; c 3. ENGINEERING, CHEMISTRY. *Educ:* Univ Kans, BS, 39; Iowa State Col, PhD(chem eng), 42. *Prof Exp:* Chemist res & develop, Nat Aniline Div, Allied Chem Corp, Morristown, 42-44, engr & group leader eng res, 45-48, asst operating supvr, Detergents Div, 48-53, chief chemist, Moundsville Plant, WVa, 53-57, supt quality control, 57-60, tech asst to plant mgr, 60-69, environ engr, Corp Eng Dept, 69-80, prin engr, environ allied chem eng dept, 80-82; RETIRED. *Mem:* Am Inst Chem Engrs. *Res:* Utilization of agricultural wastes; synthetic organics; synthetic detergents; design, construction and operation of pilot plants; process design for pollution abatement. *Mailing Add:* RR 1 Sloan IA 51055

FIRSTMAN, SIDNEY I(RVING), b Chicago, Ill, Dec 28, 31; m 55; c 7. SYSTEMS DESIGN & SYSTEMS SCIENCE. *Educ:* Univ Calif, Los Angeles, BS, 54, MS, 58; Stanford Univ, PhD(eng opers res), 64. *Prof Exp:* Mem tech staff, Rand Corp, 56-63, head opers group, 63-65, sr staff, 67-68; mgr, man-mach systs, Planning Res Corp, 65-67, corp dir, Urban Progs, 68-72; dir, urban technol, Aerospace Corp, 72-76; mgr, Energy Appl Div, Sci Appln, Inc, 76-81; dir, Technol Appl Lab, Ga Tech Res Inst, 81-85; dir new bus develop, 85-88, VPRES CORP BUS DEVELOP, IIT RES INST, 88- *Concurrent Pos:* Consult bldg mats & interior air qual. *Mem:* Am Defense Preparedness Asn; Sigma Xi. *Res:* Automation and robotics; renewable energy; man-machine systems; operations research-system analysis; manufacturing and urban technology. *Mailing Add:* IIT Res Inst Ten W 35th St Chicago IL 60616

FIRTH, WILLIAM CHARLES, JR, b Buffalo, NY, Apr 9, 34; m 61; c 3. CHEMISTRY. *Educ:* Rensselaer Polytech Inst, BS, 56; Univ Colo, PhD(org chem), 60. *Prof Exp:* Asst gen chem, Univ Colo, 56-57; res chemist, Am Cyanamid Co, 60-64, sr res chemist, 64-76; res scientist, Union Camp Corp, 76-86; RETIRED. *Mem:* Am Chem Soc. *Res:* Stereochemistry of reactions in a bridged polycyclic system; brominative decarboxylation reactions; fluorine chemistry; polymer synthesis; product and process development; bleaching technology; monomer synthesis; applications of humic substances. *Mailing Add:* 40 Galston Dr RD 4 Robbinsville NJ 08691

FISANICK, GEORGIA JEANNE, b New York, NY, Dec 29, 50. LASER-INDUCED CHEMISTRY. *Educ:* Polytech Inst Brooklyn, BS, 70, MS, 70; Princeton Univ, MA, 72, PhD(chem), 75. *Prof Exp:* MEM TECH STAFF, BELL LABS, 74- *Mem:* Am Chem Soc; Am Phys Soc; Sigma Xi. *Res:* Multiphoton ionization mass spectroscopy; chemical dynamics; non-linear laser absorption effects in molecules. *Mailing Add:* Five Sunrise Dr Warren NJ 07060

FISCH, CHARLES, b Poland, May 11, 21; nat US; m 43; c 3. MEDICINE. *Educ:* Ind Univ, AB, 42, MD, 44; Am Bd Internal Med, dipl, 53, cert cardiovasc med, 60. *Prof Exp:* Resident internal med, Vet Admin Hosp, Indianapolis, 48-50; fel gastroenterol, Marion County Gen Hosp, 50-51, fel cardiol, 51-53; prof med & dir, Cardiovasc Div, 63-80, DISTINGUISHED PROF MED, SCH MED, IND UNIV, 80-; DIR, KRANNERT HEART RES INST, 53- *Concurrent Pos:* Consult, La Rue Carter & St Vincent's Hosps; fel coun cardiol, Am Heart Asn. *Mem:* Am Fedn Clin Res; fel Am Col Physicians; Am Col Cardiol (pres 75-77); Am Physiol Soc; Sigma Xi. *Res:* Electrolytes and drugs in cardiovascular disease. *Mailing Add:* Med Dept-UH 485 Ind Univ Sch Med 926 W Michigan St Indianapolis IN 46223

FISCH, FOREST NORLAND, b Cope, Colo, July 6, 18; m 43; c 2. MATHEMATICS. *Educ:* Univ Northern Colo, AB, 40, MA, 47. *Prof Exp:* From asst prof to assoc prof math, Univ Northern Colo, 47-69, Chmn Dept, 66-69, prof math, 69-80; RETIRED. *Mem:* Math Asn Am. *Res:* General mathematics. *Mailing Add:* 18 Levis Rd Greeley CO 80631

FISCH, HERBERT A(LBERT), b Cleveland, Ohio, Oct 6, 23; m 51; c 3. CHEMICAL ENGINEERING. *Educ:* Case Inst Technol, BSc, 44; Ohio State Univ, MSc, 48, PhD(chem eng), 51. *Prof Exp:* Asst chem eng, Ohio State Univ, 47-48; chem engr, Lubrizol Corp, Ohio, 48-49; chem engr, Gaseous Diffusion Plant, Union Carbide Nuclear Co, Tenn, 51-53; phys chemist, Oak Ridge Nat Lab, 53-56; chem engr, US Atomic Energy Comn, NY, 56-58; phys chemist, Knolls Atomic Power Lab, Gen Elec Co, 58-62, res metallurgist, Lamp Metals & Components Div, 62-67; prin engr, TRW, Inc, 67-; RETIRED. *Res:* Reactions between metals and gaseous or liquid environments; effects of nuclear radiation on corrosion; vacuum technology; high temperature oxidation/sulfidation resistant coatings; electrocoating. *Mailing Add:* 3590 Beacon Dr Beachwood OH 44122

FISCH, OSCAR, b Buenos Aires, Arg, Oct 22, 34; US citizen; m 64; c 2. ENGINEERING ECONOMICS, TRANSPORTATION ECONOMICS. *Educ:* Univ Buenos Aires, MS; Harvard Univ, MS, 68; Univ Calif, Berkeley, PhD, 72. *Prof Exp:* From asst prof to assoc prof, Ohio State Univ, 72-87; PROF ECON, LEHMAN COL & GRAD CTR, CITY UNIV NY, 87-, DEAN NATURAL & SOC SCI, LEHMAN COL, 87- *Concurrent Pos:* Distinguished prof, Univ New Orleans, 86- *Mailing Add:* Herbert H Lehman Col City Univ NY 250 Bedford Park Blvd W Bronx NY 10468

FISCH, ROBERT O, b Budapest, Hungary, June 12, 25; div; c 1. PEDIATRICS. *Educ:* Med Sch Budapest, MD, 51. *Prof Exp:* Resident, 59-60, res fel, 61, from instr to assoc prof, 61-78, PROF PEDIAT, UNIV MINN, MINNEAPOLIS, 78- *Concurrent Pos:* Pediat coord, Child Develop Study, 61-63, dir, 63-65; dir, Phenylketonuric Clin, 61-, Child Care Clin, 72- &

Speech, Language & Hearing Proj, 75-77. *Mem:* Am Acad Pediat; Asn Ambulatory Pediat Serv; Am Med Asn; Am Pediat Soc. *Res:* Phenylketonuria; child development. *Mailing Add:* Mayo Mem Hosp Univ Minn Box 384 Minneapolis MN 55455

FISCH, RONALD, b Forest Hills, NY, Jan 24, 51. THEORETICAL CONDENSED MATTER PHYSICS. *Educ:* Cornell Univ, AB, 72; Univ Pa, PhD(physics), 77. *Prof Exp:* Res assoc, Princeton Univ, 77-80; asst prof physics, Wash Univ, 80-84; CONSULT, 84- *Concurrent Pos:* Consult, Bell Tel Labs, 78-80. *Mem:* Am Phys Soc. *Res:* Theoretical condensed matter physics; phase transitions and cooperative phenomena; non-crystalline solids, glass, spin-glass and random magnetism. *Mailing Add:* Dept Physics Washington Univ St Louis MO 63130

FISCHBACH, DAVID BIBB, b Beckley, WVa, Oct 28, 26; m 52; c 2. MATERIALS SCIENCE ENGINEERING. *Educ:* Denison Univ, BA, 50; Yale Univ, MS, 51, PhD(physics), 55. *Prof Exp:* From res engr to res specialist & mem tech staff, Mat Sect, Jet Propulsion Lab, Calif Inst Technol, 55-68; res assoc prof, 69-77, RES PROF CERAMIC ENG, UNIV WASH, 77- *Concurrent Pos:* Vis res fel, Dept Metall, Univ Col, Swansea, Wales, 67-68; assoc ed, Carbon J, 79-; assoc prof, Paul Pascal Res Ctr, Univ Bordeaux I, France, 84. *Honors & Awards:* George D Graffin Lectureship Carbon Sci & Eng, Am Carbon Soc, 81. *Mem:* Am Phys Soc; Metall Soc; Am Inst Mining, Metall & Petrol Engrs; Am Ceramic Soc; Am Carbon Soc; Sigma Xi. *Res:* Properties and structure of carbons and graphite and other ceramic materials; composite materials; internal friction; ferromagnetic materials; defects in solids. *Mailing Add:* 335 Roberts Hall FB-10 Univ Wash Seattle WA 98195

FISCHBACH, EPHRAIM, b Brooklyn, NY, Mar 29, 42; m 71; c 3. PHYSICS. *Educ:* Columbia Univ, AB, 63; Univ Pa, MS, 64, PhD(physics) 67. *Prof Exp:* Res assoc physics, Inst Theoret Physics, State Univ NY, Stony Brook, 67-69 & Niels Bohr Inst, Copenhagen, 69-70; asst prof, Purdue Univ, West Lafayette, 70-74, assoc prof, 74-78, prof physics, 79-; AT DEPT PHYSICS, UNIV WASH, SEATTLE. *Concurrent Pos:* Vis assoc prof physics, Inst Theoret Physics, State Univ NY, Stony Brook, 78-79. *Mem:* Am Phys Soc. *Res:* Elementary particle theory. *Mailing Add:* Dept Physics Purdue Univ W Lafayette IN 47907

FISCHBACH, FRITZ ALBERT, b Kenosha, Wis, June 16, 37; m 63; c 2. BIOPHYSICS, BIOCHEMISTRY. *Educ:* Univ Wis, BS, 59, MS, 61, PhD(biophys), 65. *Prof Exp:* NIH fels biophys, Univ Sheffield, 65-66, 67-68 & Purdue Univ, 66-67; asst prof physics, 68-70, assoc prof, 71-73, assoc prof ecosysts anal, 73-77, ASSOC PROF SCI & ENVIRON CHANGE, UNIV WIS-GREEN BAY, 77- *Mem:* AAAS; Biophys Soc. *Res:* Structure of large biological molecules, viruses and iron storage molecules; mineral structure of biological iron crystals; natural atmospheric minerals and aerosols. *Mailing Add:* Dept Physics Univ Wis Green Bay Green Bay WI 54301

FISCHBACH, GERALD DAVID, b New Rochelle, NY, Nov 15, 38. NEUROBIOLOGY. *Educ:* Colgate Univ, AB, 60; Cornell Univ, MD, 65. *Hon Degrees:* MA, Harvard Univ, 78. *Prof Exp:* Intern, Univ Wash Hosp, 65-66; sr surgeon, Lab Neurophysiol, Pub Health Serv, Nat Inst Neurol Dis & Stroke, NIH, 66-69; staff fel, Behav Biol Br, Nat Inst Child Health, 69-73; assoc prof pharmacol, Med Sch, Harvard Univ, 73-78, prof, 78-81; Edison prof neurobiol & chmn, Dept Anat & Neurobiol, Sch Med, Wash Univ, 81-90; NATHAN MARSH PUSEY PROF NEUROBIOL & CHAIR, DEPT NEUROBIOL, MED SCH & MASS GEN HOSP, HARVARD UNIV, 90- *Concurrent Pos:* Assoc ed, Develop Biol, 74-78, J Neurophysiol, 75-81; vis prof, Univ Calif, San Francisco, 78 & Univ Ariz, 84; mem, Neurol B Study Sect, NIH, 78-80, Sci Adv Bd, Nat Spinal Cord Injury Asn, 78- & Fidia Res Found, 86-, Neurosci Rev Bd, Alfred P Sloan Found, 84-89; chmn, Gordon Conf Molecular Pharmacol, 83; dir, Ctr Cellular & Molecular Neurobiol, Wash Univ, 83-90, Jacob Javits Ctr Excellence Neurosci, 85-90 & Ctr Higher Brain Function, 88-90; dir, Neurosci Ctr, Mass Gen Hosp, 90- *Honors & Awards:* Mathilde Solowey Award in Neurosci, 75; Alden Spencer Lectr, Columbia Univ, 81; Nelson Distinguished Lectr, Univ Mo, Columbia, 84; Otto Loewi Mem Lectr, NY Univ, 86; Stephen M Schuetze Mem Lectr, Columbia Univ, 88; Stephen W Kuffler Lectr, Harvard Univ, 90. *Mem:* Nat Acad Sci; Inst Med-Nat Acad Sci; Soc Neurosci; Am Acad Arts & Sci; Am Asn Anatomists; Am Physiol Soc. *Res:* Neuroscience; neuropharmacology; author of numerous technical publications. *Mailing Add:* Dept Neurobiol Med Sch Harvard Univ 220 Longwood Ave Boston MA 02115

FISCHBACH, HENRY, b New York, NY, May 2, 14; m 45; c 3. ANALYTICAL CHEMISTRY. *Educ:* Ind Univ, AB, 35, AM, 36, PhD(inorg chem), 38. *Prof Exp:* Asst chem, Ind Univ, 35-38; dir educ prog, Joseph E Seagram & Sons, Inc, 39; food & drug inspector, Food & Drug Admin, Dept Health, Educ & Welfare, 39-41, res chemist, Washington, DC, 41-45, in chg chem res antibiotics, Med Div, 45-53, Antibiotics & Alkaloids Div Pharmaceut Chem, 53-56, asst to dir, Bur Biol & Phys Sci, 56-59, dir, Div Food, 59-69, Div Pesticides, Bur Sci, 69-71, Dir Off Sci, Bur Foods, 71-72, asst dir phys sci, 72-73, assoc dir, 74; RETIRED. *Concurrent Pos:* Secy, Tech Comt Lab Equipment & Supplies, Fed Specifications Bd, 48-54; mem panel, Bd US Civil Serv Exam, 57-66; mem, Adv Panel & chmn, Chem Comt, Food Chems Codex, Food Protection Comt, Nat Acad Sci-Nat Res Coun, 61-73, chmn, 65-73; chmn, Trace Substances Comn, Int Union Pure & Appl Chem, 64-73; mem, Subcomt Ref Mat, Div Anal Chem, Nat Acad Sci-Nat Res Coun, 65-69; chmn, Joint Asn Off Anal Chem-Am Oil Chemists Soc-Am Asn Cereal Chemists, Mycotoxin Comt, 65-74; mem, Gov's Northern Va Potomac River Basin Comt, 85- *Mem:* AAAS; Am Chem Soc; fel Asn Off Anal Chem; Inst Food Technologists; Sigma Xi. *Res:* Antibiotics; food chemistry; chlorophyllins; continuous ascending chromatography; aflatoxins; pesticides; analytical research. *Mailing Add:* 5627 Bradley Blvd Alexandria VA 22311

FISCHBACH, JOSEPH W(INSTON), general engineering; deceased, see previous edition for last biography

FISCHBACK, BRYANT C, b Alhambra, Calif, Nov 29, 26; m 50; c 3. HAZARDOUS WASTE & MATERIALS MANAGEMENT. *Educ:* Univ Calif, Los Angeles, BS, 49. *Prof Exp:* Chemist, res dept, 49-56, proj leader, 56-68, tech res & develop mgr, prod depts-west, 68-72, res mgr, 72-79, mgr environ serv, 79-88, ASSOC ENVIRON CONSULT, WESTERN DIV, DOW CHEM USA, 88- *Concurrent Pos:* Appointee, Hazardous Waste Mgt Coun, Calif, 82-84; consult, Calif Senate Select Comt Long Range Planning, 85-86, & Calif Inst Pub Affairs, 85-; mem, Univ Calif Toxic Substances Res & Teaching Prog Adv Comt, 88-; comnr, Contra Costa County Hazardous Mats Comn, 87- *Mem:* Am Chem Soc; Royal Soc Chem. *Res:* Organic synthesis of new insecticides; herbicides; general agricultural chemicals; coccidiostats; animal health products; pharmaceuticals; mechanism of nitration reactions; synthesis of ore flotation agents; research and development of secondary oil recovery, mining and environmental control systems; development of hazardous waste management technologies and systems; organic chemistry. *Mailing Add:* Dow Chem USA PO Box 1398 Pittsburg CA 94565

FISCHBARG, JORGE, b Buenos Aires, Arg, Aug 14, 35; nat US; m 64; c 2. PHYSIOLOGY & BIOPHYSICS, OPHTHALMOLOGY. *Educ:* Univ Buenos Aires, BS, 53, MD, 62; Univ Chicago, PhD(physiol), 71. *Prof Exp:* Asst biophys, Sch Med, Univ Buenos Aires, 62-64; trainee ophthal, Eye Res Lab, Univ Louisville, 64-65; trainee math biol & physiol, Univ Chicago, 65-70; asst prof ophthal, 70-73, asst prof physiol, 73-78, assoc prof physiol & ophthal, 78-84, PROF PHYSIOL & OPHTHAL, COLUMBIA UNIV, 84- *Concurrent Pos:* Vis Scientist, Dept Biol, Centre D'Energie Nucleaire, Saclay, France, 73 & 78; consult, Nat Adv Eye Coun's Vision Res Planning Subcomt, NIH, USPHS, 74 & 90; USPHS Res Career Develop Award, Nat Eye Inst, 75-80; fel commoner, Churchill Col, Cambridge, Eng, 76-77; mem, prog comt, Int Soc Eye res, 80; mem, Prog Comt Physiol & Pharmacol, 80-83; mem, Visual Sci A Study Sect, NIH, 80-84; secy treas, Membrane Subgroup, Biophys Soc, 85-88; Alcon res award, 86; exec ed, Exp Eye Res, 88- *Mem:* Am Physiol Soc; NY Acad Sci; Biophys Soc; Asn Res Vision & Ophthal; Am Physiol Soc; Int Soc Eye Res; Biophys Soc; AAAS. *Res:* Molecular biophysics of cell membrane proteins; water permeation, water channels; transport of fluid and electrolytes across epithelia; theoretical modeling and simulation of channels. *Mailing Add:* Dept Physiol Col P&S Columbia Univ 630 W 168th St New York NY 10032

FISCHBECK, HELMUT J, b Tubingen, Ger, Oct 19, 28; m 55; c 2. PHYSICS. *Educ:* Univ Heidelberg, MA, 55; Ind Univ, PhD(physics), 60. *Prof Exp:* Instr physics, Univ Mich, 60-61, asst prof, 62-66; assoc prof, 66-71, PROF PHYSICS, UNIV OKLA, 71-, CHMN, ENG PHYSICS, 80- *Concurrent Pos:* Vis scientist, Argonne Nat Lab, 75. *Mem:* Am Phys Soc; Am Soc Eng Educ; Sigma Xi. *Res:* Nuclear spectroscopy; atomic physics; solid state physics; accelerator based material characterization. *Mailing Add:* Dept Physics & Astron Univ Okla Norman OK 73019

FISCHEL, DAVID, b Du Bois, Pa, Sept 12, 36; m 60; c 3. IMAGE PROCESSING, REMOTE SENSING. *Educ:* Brown Univ, ScB, 58; Ind Univ, MA, 61, PhD(astrophys), 63. *Prof Exp:* Res scientist astrophys, Space Sci Div, Ames Res Ctr, 63-65, head anal sect, 70-78, res scientist astrophys lab astron & solar physics, Data Anal & Observ Br, 65-78, sr phys scientist & landsat-D assessment syst sci mgr, 78-84, head res systs br, Info Extinction Div, 83-84, Goddard Space Flight Ctr, sr comput scientist, Systs & Appl Sci Corp, 84-85; sr systs engr, 85-89, CHIEF SCIENTIST, EARTH OBSERV SATELLITE CO, 89- *Mem:* Soc Photo-Optical Instrument Engrs; Int Astron Union; Sigma Xi; Am Soc Photogram & Remote Sensing; Inst Elec & Electronics Engrs. *Res:* Theoretical model stellar atmospheres and interiors; image processing; numerical methods for electronic computers; plasma thermodynamics. *Mailing Add:* 6072 Warm Stone Columbia MD 21045

FISCHEL, EDWARD ELLIOT, b New York, NY, July 29, 20; m 43; c 2. INTERNAL MEDICINE. *Educ:* Columbia Univ, BA, 41, MD, 44, ScD(med), 48. *Prof Exp:* Asst physician, Presby Hosp, 47-54; assoc clin prof, Albert Einstein Col Med, 57-69; assoc clin prof med, Columbia Univ, 69-72; prof med, Albert Einstein Col Med, 72-81; dir, Dept Med, Bronx-Lebanon Hosp Ctr, 54-80; prof med, Sch Med, Univ Conn, 80-; chief, dept med, Mt Sinai Hosp, Hartford, 80-83; chief staff, Vet Admin Med Ctr, Northport, 83-91; RETIRED. *Concurrent Pos:* Assoc, Dept Med, Columbia Univ, 50-55; chmn med admin coun, Arthritis Found, 68-69; mem med & sci adv comt, Arthritis & Rheumatism Found; mem coun rheumatic fever & congenital heart dis, Am Heart Asn. *Mem:* AAAS; Am Soc Clin Invest; Am Soc Exp Biol & Med; AMA; Am Rheumatism Asn (pres, 68-69). *Res:* Immunochemistry; hypersensitivity reactions; rheumatic diseases; nephritis; inflammation; serum complement activity; immunosuppression by cortisone. *Mailing Add:* 220 Little Neck Rd Centerport NY 11721-1145

FISCHELL, DAVID R, b Washington, DC, Dec 4, 53; c 2. MEDICAL DEVICE DEVELOPMENT. *Educ:* Cornell Univ, BS, 75, MS, 78, PhD(appl physics), 80. *Prof Exp:* Mem tech staff, AT&T Bell Labs, 79-82, supvr, 83-86 & 87-91, venture leader, AT&T Audiographic Commun Systs, 86-87; EXEC VPRES, MEDINTEC, INC, 91- *Mem:* Inst Elec & Electronics Engrs. *Res:* Laser induced fluorescence spectroscopy; exploratory development of products for audiographic teleconferencing; performance evaluation of new network service technologies; new product and service development; design and development of devices for atherectomy and other interventions. *Mailing Add:* 71 Riverlawn Dr Fair Haven NJ 07704

FISCHELL, ROBERT E, b New York, NY, Feb 10, 29; m 51; c 3. SPACE PHYSICS, BIOMEDICAL ENGINEERING. *Educ:* Duke Univ, BSME, 51; Univ Md, MS, 53. *Prof Exp:* Physicist, US Naval Ord Lab, Md, 51-56; prin staff physicist, Emerson Res Labs, 56-59; sr staff physicist, Johns Hopkins Univ, 59-60, proj supvr, 60-64, chief engr, Space Dept, 73-81, asst head dept, 78-88, chief technol transfer, 81-88, PRIN STAFF PHYSICIST, APPL PHYSICS LAB, JOHNS HOPKINS UNIV, 63-; PRES, MEDINTECH, INC, 90- *Concurrent Pos:* Consult, USAF, 59-61; mem ad hoc comt, NASA Dept Defense, 63; consult, French Space Agency, 65; proj scientist, Navy

Navig Satellites, 69-79; consult, NASA Hq, 71; consult, US Cong Tech Assessment, 84; dir, Pacesetter Syst, Inc, 70-86 & Patlex Corp, 81-87; med affil, Sci Res Comt, Am Heart Asn, 84-86; res assoc med, Johns Hopkins Sch Med, 83- & Yale Univ Sch Med, 89-; mem, Exec Panel, US Chief Naval Opers, 85-89; pres, Med Innovations, Inc, 86-90. *Honors & Awards:* Award, Am Soc Mech Engrs, 63; three IR-100 Awards, Indust Res Mag, 67-70; NASA Space Act Award & Engr Excel Medal, 84; Aerospace Sci Medal, NY Acad Sci, 87; Inducted Space Technol Hall Fame, 88. *Mem:* Nat Acad Eng; Int Soc Artificial Organs; Am Heart Asn. *Res:* Attitude control systems for earth satellites; space electric power systems; cardiac pacemakers; implantable biomedical instrumentation; artificial pancreas; atherectomy catheters. *Mailing Add:* Appl Physics Lab Johns Hopkins Univ Laurel MD 20723

FISCHELL, TIM ALEXANDER, b Washington, DC, Feb 10, 56; m 84. INTERNAL MEDICINE, CARDIOLOGY. *Educ:* Cornell Univ, AB, 78; Cornell Univ Med Col, MD, 81. *Prof Exp:* Resident, Mass Gen Hosp, 81-84; fel cardiol, 84-87, ASST PROF MED, STANFORD UNIV, 87- *Mem:* Am Col Cardiol; Am Fedn Clin Res; Am Col Physician; AAAS; Am Heart Asn Coun Circulation. *Res:* Laboratory and clinical investigation of the vasomotor consequences of arterial injury; effects of contrast agents on renal function and the design and development of a novel atherectomy catheter. *Mailing Add:* Falk Cardiovasc Res Ctr Cardiol Stanford Univ Med Ctr Stanford CA 94305

FISCHER, A(LBERT)ALAN, b Indianapolis, Ind, June 30, 28; c 4. FAMILY MEDICINE. *Educ:* Ind Univ, MD, 52; Am Bd Family Pract, dipl. *Prof Exp:* Intern, St Vincent Hosp, Indianapolis, 52-53; pvt pract, 53-70; dir family pract residency prog, St Vincent Hosp, 69-75; prof family med & chmn dept, Ind Univ, Indianapolis, 74-90; MED DIR, LAKEVIEW MANOR, 70- *Concurrent Pos:* Pvt pract family med, 53-; mem, Nat Joint Pract Comn, Nat Acad Sci; cert mem, Am Bd Family Pract, Am Acad Family Physicians. *Mem:* Inst Med-Nat Acad Sci; Am Acad Family Physicians (vpres, 71-72); AMA. *Mailing Add:* 3400 Lafayette Rd Indianapolis IN 46222

FISCHER, ALBERT G, b Ilmenau, Ger, July 5, 28; US citizen; m 56; c 3. SOLID STATE ELECTRONICS. *Educ:* Univ Giessen, Diplom-Physiker, 55, Dr phil nat, 57. *Prof Exp:* Scientist, Lamp Div, Gen Elec Co, 58-59; mem tech staff, RCA Labs, 59-71; adv scientist, Res & Develop Ctr, Westinghouse Elec Corp, 71-72; PROF ELEC ENG, UNIV DORTMUND, WGER, 73- *Mem:* Inst Elec & Electronics Engrs. *Res:* Electro luminescence; materials science; single crystal growth; films; liquid crystals; thin-film transistors; flat panel tv. *Mailing Add:* Dept Elec Eng Univ Dortmund Dortmund-Hombruch 46 Germany

FISCHER, ALBERT KARL, b Newark, NJ, Oct 15, 31; m 59; c 3. PHYSICAL INORGANIC CHEMISTRY, SURFACE CHEMISTRY. *Educ:* NY Univ, BA, 53; Harvard Univ, MA, 55, PhD(chem), 58. *Prof Exp:* Res chemist, Union Carbide Metals Co Div, Union Carbide Corp, 57-60 & Union Carbide Chem Co Div, 60-62; assoc chemist, 62-73, CHEMIST, ARGONNE NAT LAB, 73- *Mem:* Am Chem Soc; Sigma Xi. *Res:* Metal carbonyls; organometallics; fused salt and liquid metal chemistry; surface chemistry of liquid metals; morphology of electrodeposition; physical chemistry of tritium-ceramic systems. *Mailing Add:* Chem Technol Div Argonne Nat Lab Argonne IL 60439

FISCHER, ALFRED GEORGE, b Rothenburg, Ger, Dec 10, 20; US citizen; m 39; c 3. GEOLOGY. *Educ:* Univ Wis, BA, 39, MA, 40; Columbia Univ, PhD, 50. *Prof Exp:* Instr geol, Va Polytech Inst, 41-43; geologist, Stanolind Oil & Gas Co, Kans, 43-44, Fla, 44-46; instr geol, Univ Rochester, 47-48; from instr to asst prof, Univ Kans, 48-51; sr geologist, Int Petrol Co, Peru, 51-56; from asst prof to assoc prof geol, Princeton Univ, 56-63, prof, 63-84; stratigrapher & paleontologist, 84-91, EMER STRATIGRAPHER & PALEONTOLOGIST, UNIV SOUTHERN CALIF, 91- *Concurrent Pos:* Guggenheim fel, 69-70. *Honors & Awards:* Von Buch Medal; Verill Medal; Twenhofel Medal. *Mem:* Geol Soc Am; Soc Econ Paleont & Mineral; Paleont Soc; Am Asn Petrol Geol; AAAS. *Res:* Invertebrate paleontology; paleoecology; historical geology; carbonate sediments; sedimentation; history of outer earth; evolution of life, especially of pelagic community; paleoclimatic-paleoceanographic change as related to internal (plate tectonic) and external (orbital) forcing; cyclic patterns in stratigraphy. *Mailing Add:* Univ Southern Calif University Park Los Angeles CA 90089-0740

FISCHER, C RUTHERFORD, b New York, NY, June 21, 34; wid; c 1. MOLECULAR PHYSICS, SOLID STATE PHYSICS. *Educ:* City Col New York, BS, 54; Yale Univ, MS, 55, PhD(physics), 60. *Prof Exp:* Asst prof physics, NMex State Univ, 59-61 & Adelphi Univ, 61-64; assoc prof, 64-70, PROF PHYSICS, QUEENS COL, NY, 70- *Concurrent Pos:* NSF grant, Adelphi Univ, 62-64; consult, US Army Res Off, Durham, 70-71 & Sandia Labs, 75-82. *Mem:* Am Phys Soc. *Res:* Molecular structure and spectra; point defects in ionic crystal; ionic interactions; optical properties of solids; fast reactions in solids. *Mailing Add:* Dept Physics Queens Col Flushing NY 11367

FISCHER, CHARLOTTE FROESE, b Ukraine, Russia, Sept 21, 29; nat US; m 67; c 1. THEORETICAL ATOMIC PHYSICS. *Educ:* Univ BC, BA, 52, MA, 54; Cambridge Univ, PhD(appl math), 57. *Prof Exp:* From asst prof to prof math, Univ BC, 57-67; prof appl anal & comput sci, Univ Waterloo, 67-68; vis prof, Pa State Univ, 68-69, prof comput sci, 69-72, prof appl math, 72-75, prof comput sci, 74-79; PROF COMPUT SCI, VANDERBILT UNIV, 80- *Concurrent Pos:* Programmer, numerical analyst & consult, Pac Oceanog Group, BC, 57-59; res fel, Harvard, 63; Sloan fel, 64; grant, Nat Res Coun Can, 64-76, US Dept of Energy, 77- *Mem:* Asn Comput Mach; Am Phys Soc. *Res:* Atomic structure calculations; numerical analysis. *Mailing Add:* Dept Comput Sci Vanderbilt Univ Box 6035-B Nashville TN 37235

FISCHER, DAVID JOHN, b Jefferson City, Mo, Apr 30, 28; m 54; c 2. MATERIALS & PROCESS RESEARCH, QUALITY ASSURANCE. *Educ:* Univ Mo, BS, 50, MS, 52, PhD(phys chem), 54. *Prof Exp:* Proj group leader, Polymer Res Lab, Dow Corning Corp, 56-59, dept supvr, Hyper-Pure Silicon Div, 59-62; sr chemist, Midwest Res Inst, 62-65; mgr microcircuits res dept, Corning Glass Works, 65-68, mgr spec projs, Advan Mkt Develop, 68-69, res assoc, 69-70, mgr, Bio-Organic Develop, 70-73, phys scientist, Corning Mus Glass, 73-75; dir res & develop, Ophthalmic Group, Milton Roy Co, 75-79; pres, Universal Res & Develop Corp, 79-81; sr res advr, Gulf South Res Inst, 81-82; MGR ADVAN QUAL TECHNOL & LABS, MARTIN MARIETTA MANNED SPACE SYSTS, 82- *Mem:* Am Chem Soc; Electrochem Soc; Am Inst Conserv Hist & Artistic Works; Asn Advan Med Instrumentation; Am Inst Aeronaut & Astronaut. *Res:* Quality engineering, aerospace materials and processes; renewable energy sources, anaerobic digestion for producing methone; lasers for spectrophotometric measurements and machining plastics. *Mailing Add:* Martin Marietta Manned Space Systs Mail 3770 PO Box 29304 New Orleans LA 70189

FISCHER, DAVID LLOYD, nuclear physics, for more information see previous edition

FISCHER, DIANA BRADBURY, b Mt Vernon, NY, May 5, 34. STATISTICS, EPIDEMIOLOGY. *Educ:* Mt Holyoke Col, BA, 56; Yale Univ, MPH, 66, PhD(biomet), 74. *Prof Exp:* Res asst endocrinol, Med Sch, Harvard Univ, 56-63, comput programmer biostatist, Sch Pub Health, 63-64; from sr programmer statistician to mgr appln prog, Comput Ctr, Yale Univ, 66-71; res assoc fac statistician, Dept Epidemiol Pub Health & Yale Comprehensive Cancer Ctr, 74-75, DIR EPIDEMIOL STATIST UNIT, YALE COMPREHENSIVE CANCER CTR & ASST PROF PUB HEALTH, SCH MED, YALE UNIV, 76- *Mem:* Am Statist Asn; Biomet Soc; Sigma Xi. *Res:* Statistical methods for time-dose relationships in radiotherapy; life table and survivorship analysis with covariates; clinical trials methodology; computer methodology. *Mailing Add:* Dept Pub Health Sch Med Yale Univ 20 York St New Haven CT 06519

FISCHER, EDMOND H, b Shanghai, China, Apr 6, 20; m 48; c 2. BIOCHEMISTRY. *Educ:* State Col Geneva, Mat Fed, 39; Univ Geneva, Lic es Sc, 43, dipl, 44, PhD(chem), 47. *Hon Degrees:* Dr, Univ Montpellier, 85. *Prof Exp:* Asst org chem labs, Univ Geneva, 46-48, Swiss Found res fel chem, 48, pvt-docent, 50, Rockefeller Found res fel, 50-53; from asst prof to prof 53-90, EMER PROF BIOCHEM, UNIV WASH, 90- *Concurrent Pos:* Lederle med fac award, 56-59. *Honors & Awards:* Warner Medal, Swiss Chem Soc, 52; Jaubert Prize, 68; Steven C Beering Award, 91. *Mem:* Nat Acad Sci; Am Acad Arts & Sci; Am Chem Soc; Am Soc Biol Chemists; Swiss Chem Soc. *Res:* Regulation of cellular processes; hormone regulation; growth factors; hormone receptors; signal transduction; metabolic regulation. *Mailing Add:* Dept Biochem SJ-70 Univ Wash Seattle WA 98105

FISCHER, EDWARD G(EORGE), b New York, NY, Mar 31, 16; m 44; c 4. MECHANICAL ENGINEERING, SEISMIC ENGINEERING. *Educ:* Cooper Union, BS, 36, MS, 39, PhD(math eng), 46. *Prof Exp:* Consult engr, Res Lab, Westinghouse Elec Corp, 36-82; PVT CONSULT ENGR, 82- *Concurrent Pos:* Instr, Grad Sch, Univ Pittsburgh, 46-59. *Honors & Awards:* Cert Appreciation, US Corps Engrs. *Mem:* Am Soc Mech Engrs; Soc Exp Stress Anal; Inst Environ Sci; Seismol Soc Am; Inst Elec & Electronic Engrs. *Res:* Vibration; shock; noise; seismic design. *Mailing Add:* 5525 Third St Verona PA 15147

FISCHER, EUGENE CHARLES, b New York, NY, Apr 7, 40; m 64; c 4. RESEARCH ADMINISTRATION, MARINE BIOLOGY. *Educ:* Iona Col, BS, 61; St John's Col, NY, MS, 63, PhD(marine microbiol), 66. *Prof Exp:* Staff scientist microbiol, US Naval Appl Sci Lab, 64-65, res oceanogr, 65-70, HEAD OCEAN ENVIRON BR, US NAVAL SHIP RES & DEVELOP LAB, ANNAPOLIS, 70- *Concurrent Pos:* Cmndg officer & dir, US Naval Appl Sci Lab, 66. *Mem:* Am Soc Microbiol; Sigma Xi. *Res:* Deep ocean-high pressure microbial physiology; marine fouling; pollution abatement; polymer chemistry; antifouling materials. *Mailing Add:* 611 Breton Pl Arnold MD 21012-1537

FISCHER, FERDINAND JOSEPH, b Kansas City, Mo, June 13, 40; m 62; c 1. ENGINEERING MECHANICS, APPLIED MATHEMATICS. *Educ:* Univ Kans, BS, 62; Rice Univ, MS, 64; Harvard Univ, MA, 65, PhD(appl math), 68. *Prof Exp:* Res asst eng mech, Harvard Univ, 66-67; res engr, Shell Develop Co, 67-70, sr res engr, 70-73, staff res engr, 73-75, staff engr, 75-78, staff supvr, 79-81, sr staff supvr, 81-85, sr staff res engr, 85-90, RES ADV, SHELL DEVELOP CO, 90- *Concurrent Pos:* Lectr, Univ Houston, 68-71. *Mem:* Fel Am Soc Mech Engrs; Soc Eng Sci. *Res:* Elastic shell theory; continuum mechanics; numerical analysis; ocean engineering. *Mailing Add:* 13130 Rummel Creek Houston TX 77079

FISCHER, GEORGE A, b Cleveland, Ohio, Mar 27, 39; m 62; c 3. BIOCHEMISTRY, CLINICAL CHEMISTRY. *Educ:* Univ Detroit, BS, 61, MS, 64, PhD(biochem), 67; Am Bd Clin Chem, dipl, 72. *Prof Exp:* Dir chem, Grace Hosp, 75-82,; clin chemist, 66-68, dir, CPS Lab, 84-88, TECH DIR, DAMON CLIN LABS, HARPER HOSP, 88- *Mem:* AAAS: Am Chem Soc; Nat Acad Clin Biochem; Am Asn Clin Chemists; Clin Radioassay Soc. *Res:* Immunochemistry; mechanisms and chemistry of hypertension; radioligand binding analyses; laboratory computerization. *Mailing Add:* 19601 Cherry Hill Lane 3990 John R St Southfield MI 48076-5317

FISCHER, GEORGE J, b Bronx, NY, Mar 30, 25; m 48; c 2. METALLURGICAL ENGINEERING. *Educ:* Polytech Inst Brooklyn, BMetE, 49, MMetE, 53. *Prof Exp:* Instr metall, Polytech Inst Brooklyn, 48-50; plant metallurgist, Western Elec Co, 50-53; dept head, Sam Tour & Co, 53-55; from asst prof to assoc prof, 55-65, admin officer, Div Metall Eng, 61-71, head dept, 71-76, dean student serv, 76-80, prof eng metall, 65-, EMER PROF, POLYTECH INST BROOKLYN. *Concurrent Pos:* Res grants, Int Nickel Co, 60-61 & Curtiss-Wright Corp, 62-65; NSF res grant, 63-65. *Mem:* Am Soc Metals; Am Inst Mining, Metall & Petrol Engrs; Am Soc Testing & Mat. *Res:* Physical metallurgy; weldability criteria of metallic materials. *Mailing Add:* Polytech Univ 333 Jay St Brooklyn NY 11201

FISCHER, GEORGE K, b Uniondale, NY, May 20, 23; m 55; c 4. BEARING-ROTOR DYNAMICS, AERODYNAMIC MEASUREMENT. *Educ:* Polytech Inst NY, BME, 50; Fla Inst Technol, MSSM, 75. *Prof Exp:* Proj engr turbo/compressors, Stratus Div, Fairchild Engine, 50-59; vpres bearing design, Eastern Bearings & Mfg, Inc, 60; lectr & sr scientist bearing develop, Univ Va, 61; mech engr bearing design, Rocketdyne Div, NAm, 62; mat engr bearing res, Lewis Res Ctr, 63-66, aerospace engr transmissions, 79-83; syst engr space systs, 67-78; sr mech engr struct, United Space Boosters Inc, BPC, United Technol Inc, 83-85; CONSULT, 85- *Mem:* Am Soc Mech Engrs; Am Soc Lubrication Engrs; Soc Automotive Engrs. *Res:* High speed bearings, turbines and compressors; gas lubrication, transmission and electric vehicles; fluid film bearing design and rotor dynamics. *Mailing Add:* Ohlone Col PO Box 3909 38258 Ashford Way Fremont CA 94536

FISCHER, GERHARD EMIL, b Berlin, Ger, Mar 1, 28; nat US; m 51; c 2. HIGH ENERGY PHYSICS. *Educ:* Univ Toronto, BASc, 49; Univ Calif, PhD(physics), 54. *Prof Exp:* From instr to asst prof physics, Columbia Univ, 54-59; res fel, Harvard Univ, 59-63, sr res assoc, 63-65; STAFF MEM STANFORD LINEAR ACCELERATOR CTR, 65- *Concurrent Pos:* Vis scientist, DESY, Hamburg, Ger, 74, 87 & CERN, Geneva, Switz, 75, 88. *Mem:* Am Phys Soc. *Res:* Storage ring physics; linear colliders. *Mailing Add:* Stanford Linear Accelerator Ctr Box 4349 Stanford CA 94309

FISCHER, GLENN ALBERT, b Pritchett, Colo, Nov 18, 22; m 46; c 2. GENETICS, PHARMACOLOGY. *Educ:* Univ Colo, BS, 49; Calif Inst Technol, MS, 51, PhD(genetics), 54. *Prof Exp:* Res assoc, Univ Mich, 54; asst res prof, Sch Med, George Washington Univ, 55; asst prof pharmacol, Univ PR, 55; from asst prof to assoc prof, Sch Med, Yale Univ, 58-69; prof biochem pharmacol, 69-, EMER PROF, BROWN UNIV. *Mailing Add:* 14 Echo Dr Barrington RI 02806

FISCHER, GRACE MAE, b Weatherly, Pa, Nov 25, 27. PHYSIOLOGY, CARDIOVASCULAR DISEASE. *Educ:* Bucknell Univ, BS, 49; Temple Univ, MD, 53; Drexel Univ, MS, 64. *Prof Exp:* Res assoc, Sch Med, Univ Pa, 66-67, asst prof, 69-72, assoc prof physiol, 72-; RETIRED. *Concurrent Pos:* Nat Heart Inst fel, 63-64, spec fel, 65-66; res fel cardiovasc res, Bockus Res Inst & Dept Physiol, Sch Med, Univ Pa, 64-66. *Mem:* Am Heart Asn; Am Physiol Soc. *Res:* Arterial connective tissue, especially chemical and endocrine effects on arterial wall properties and collagen and elastin metabolism in arterial wall; regulation of cardiovascular processes. *Mailing Add:* 218 E Meade St Philadelphia PA 19118

FISCHER, HARRY WILLIAM, b St Louis, Mo, June 4, 21; m 43; c 5. RADIOLOGY. *Educ:* Univ Chicago, BS, 43, MD, 45. *Prof Exp:* Intern surg, Barnes Hosp, St Louis, 45-46; resident, Wash Univ, 48-51; resident radiol, St Louis City Hosp, 52-54, mem teaching staff, 56; pvt pract, 54-56; from asst prof to prof radiol, Col Med, Univ Iowa, 56-66; prof, Med Sch, Univ Mich, 66-71; prof radiol & chmn dept, Sch Med, Univ Rochester, 71-85; radiologist in chief, Strong Mem Hosp, 71-85; CONSULT. *Concurrent Pos:* Fel, Wash Univ, 48-51; fel pediat surg, Children's Mem Hosp, Chicago, 51-52; dir dept radiol, Wayne County Gen Hosp, 66-71. *Honors & Awards:* Gold Medal, Asn Univ Radiologists. *Mem:* Am Roentgen Ray Soc; Am Col Radiol; Radiol Soc NAm; Asn Univ Radiol; Uroradiological Soc; Int Soc Planning Radiol Depts. *Res:* Contrast visualization of liver, spleen and lymph nodes; toxicity of contrast media; excretion of contrast media by liver and kidney; adherence of contrast media to mucosal surfaces. *Mailing Add:* 405 Paseo Del Canto Green Valley AZ 85614

FISCHER, IMRE A, b Budapest, Hungary, Apr 12, 35; US citizen; m 61. CLINICAL BIOCHEMISTRY, QUALITY CONTROL. *Educ:* Univ Louvain, MS, 62, PhD(microbial biochem), 65. *Prof Exp:* Res asst microbial biochem, Univ Louvain, 62-65; res microbiologist, Univ Calif, Davis, 65-67; res biochemist, Sch Med, Univ Calif, Los Angeles, 67-69; biochemist, Xerox/Med Diag Opers, 69-70; res assoc, Harbor Gen Hosp, Torrance, Calif & Sch Med, Univ Calif, Los Angeles, 70; tech dir, Bio-Technics Labs, Inc, Los Angeles, 70-72; clin biochemist, Long Beach Gen Hosp, Calif, 72-81; asst prof path in residence, Sch Med, Univ Calif, Los Angeles, 78-81, head, Emergency Lab Serv Coun Harbor, Med Ctr, 78-81; DIR, QMT LAB SERV, LONG BEACH, CALIF, 84-, PRES, QUALIMEDTECH, INC, 82- *Concurrent Pos:* Asst prof, Calif State Univ, Dominguez Hills, 74-79; lectr, Calif State Univ, Fullerton, 82-88. *Honors & Awards:* Clinical Chemists Recognition Award, Am Asn Clin Chem, 81 & 86; Testimonial Award, Am Soc Qual Control, 78. *Mem:* Fel Am Soc Qual Control; Am Asn Clin Chem; fel Asn Clin Scientists; fel Nat Acad Clin Biochem; fel Inst Advan Eng; AAAS. *Res:* Enzyme isolation and kinetics; red blood cell aging; medical diagnostics; clinical chemistry; laboratory medicine; biomedical quality control and assurance; laboratory management. *Mailing Add:* OMT Lab Serv 1703 Termino Ave Suite 104 Long Beach CA 90804-2126

FISCHER, IRENE KAMINKA, b Vienna, Austria, July 27, 07; US citizen; m 30; c 2. GEODESY. *Educ:* Vienna Inst Technol, MA, 31. *Hon Degrees:* Hon Doctorate, Univ Karlsruhe, 75. *Prof Exp:* Mathematician, Geod Br, Defense Mapping Agency Topog Ctr, 52-58, geodesist, 58-62, supvry geodesist, 62-65, supvry res geodesist, Geod Br, 65-77, br chief, 62-77; RETIRED. *Concurrent Pos:* Mem & officer, Int Union Geod & Geophys, 54- & comt SAm datum, Pan-Am Inst Geog & Hist; mem spec study group hist geod, Int Asn Geod. *Mem:* Nat Acad Eng; fel Am Geophys Union; Int Asn Geod. *Res:* Figure of the earth; shape of the geoid; parallax and distance of the moon; geodetic world datum; Fischer ellipsoid; mercury datum for space flights; Mercury, Gemini and Apollo projects; South American datum of 1969; deflections at sea; mean sea level slopes; history of geodesy; over 100 technical publications. *Mailing Add:* 301 Philadelphia Ave Takoma Park MD 20912

FISCHER, IRWIN, b New York, NY, Nov 23, 27; m 54; c 2. MATHEMATICS. *Educ:* City Col New York, BS, 48; Harvard Univ, AM, 49, PhD(math), 53. *Prof Exp:* Mathematician, Air Force Cambridge Res Ctr, 52-54; instr math, Univ Minn, 54-55 & Dartmouth Col, 55-57; from asst prof to assoc prof, 57-69, PROF MATH, UNIV COLO, BOULDER, 69- *Mem:* Am Math Soc. *Res:* Algebraic geometry. *Mailing Add:* Dept Math Box 426 Univ Colo Boulder Boulder CO 80309-0426

FISCHER, JAMES JOSEPH, b Hazleton, Pa, Aug 17, 36. RADIOBIOLOGY, RADIOTHERAPY. *Educ:* Yale Univ, BS, 57; Harvard Med Sch, MD, 61; Harvard Univ, PhD(pharmacol), 64. *Hon Degrees:* MA, Yale Univ, 73. *Prof Exp:* PROF THERAPEUT RADIOL, SCH MED, YALE UNIV, 72- *Concurrent Pos:* Nat Cancer Inst clin & res fel radiother, Yale Univ, 65-68; consult, West Haven Vet Admin Hosp, 68-, Hosp St Raphael & Waterbury Hosp. *Mem:* AAAS; Am Asn Cancer Res; Am Soc Therapeut Radiol; Radiation Res Soc; Asn Univ Radiol. *Res:* Experimental therapeutic radiology; mammalian cell radiobiology; radioprotective and sensitizing agents; cell kinetics. *Mailing Add:* Dept Therapeut Radiol Yale Univ Sch Med New Haven CT 06510

FISCHER, JAMES ROLAND, b Washington, Mo, Feb 23, 45; m 69; c 3. AGRICULTURE ADMINISTRATION, AGRICULTURAL ENGINEERING. *Educ:* Univ Mo, BS, 67, MS, 69, PhD(agr eng), 72. *Prof Exp:* Assoc prof agr eng, Univ Mo, 74-85; assoc dir, Agr Exp Sta, Mich State Univ, 85-87, interim dir, Pesticide Res Ctr, 86-87; DIR, SC AGR EXP STA, CLEMSON UNIV, 87- *Concurrent Pos:* Chmn, Southern Regional Asn Agr Exp Sta Dirs, 91; mem gov bd, Agr Res Inst. *Mem:* AAAS; Am Soc Agr Engrs; Agr Res Inst; Soil & Water Conserv Soc Am. *Res:* Agricultural research administration; agriculture, natural resources and the rural environment. *Mailing Add:* 104 Barre Hall Clemson Univ Clemson SC 29634-0351

FISCHER, JANET JORDAN, b Pittsburgh, Pa, Apr 28, 23; m 51; c 5. INFECTIOUS DISEASES, PULMONARY MEDICINE. *Educ:* Vassar Col, AB, 44; Johns Hopkins Univ, MD, 48; Am Bd Internal Med, dipl, 56. *Prof Exp:* Asst physician, Outpatient Dept, Johns Hopkins Hosp, 50-52; from instr to assoc prof, 54-70, prof, 70-80, SARAH GRAHAM KENAN PROF MED, UNIV NC, CHAPEL HILL, 80-, PROF BACTERIOL & IMMUNOL, 88- *Concurrent Pos:* Fel infectious dis, Sch Med, Johns Hopkins Univ, 50-51; Nat Found Poliomyelitis Found fel, 51-52; fel infectious dis, Sch Med, Univ NC, 52-53; consult physician, Watts Hosp, Durham, NC, 54-; attend physician, NC Mem Hosp, Chapel Hill, 54- & Gravely Sanatorium, 54-74. *Mem:* AAAS; AMA; Am Fedn Clin Res; Am Thoracic Soc; Am Soc Microbiol. *Res:* Pyelonephritis; urinary tract infections; bacterial endocarditis; laboratory medicine in microbiology; antibiotic sensitivity testing; microbiological factors in Byssinosis (Brown lung disease of cotton worker); endotoxin as a factor in Byssinosis and in clinical disease; prevalence of endotoxin in the environment. *Mailing Add:* Dept Med Mem Hosp Univ NC Sch Med Chapel Hill NC 27514-6097

FISCHER, JOHN EDWARD, b Albany, NY, June 8, 39; m 63; c 3. SOLID STATE PHYSICS. *Educ:* Rensselaer Polytech Inst, BME, 61, PhD(nuclear sci & eng), 66; Calif Inst Technol, MS, 62. *Prof Exp:* Res assoc, Univ Paris, 66-67; res physicist, Michelson Lab, Naval Weapons Ctr, 67-70, br head, 70-73; assoc prof, 73-77, PROF ELEC ENG & SCI, UNIV PA, 77- *Concurrent Pos:* Vis assoc prof, Univ Pa, 71-72; consult, Naval Air Develop Ctr, 75- *Mem:* Am Phys Soc. *Res:* Optical properties of solids; characterization and electronic properties of synthetic metals, such as graphite intercalation compounds; point defects, compositional and topological disorder in semiconductors; radiation effects; thin films; energy band structure. *Mailing Add:* MSE/LRAM KI Univ Pa Philadelphia PA 19104

FISCHER, LAWRENCE J, b Chicago, Ill, Sept 2, 37; m 63; c 3. BIOCHEMICAL PHARMACOLOGY, TOXICOLOGY. *Educ:* Univ Ill, BS, 59, MS, 61; Univ Calif, PhD(pharmaceut chem), 65. *Prof Exp:* Sr res pharmacologist, Merck Inst Therapeut Res, 66-68; from asst prof to assoc prof, 68-76, PROF PHARMACOL, COL MED, UNIV IOWA, 76- *Concurrent Pos:* NIH fel biochem, St Mary's Hosp Med Sch, London, Eng, 65-66; vis prof, Univ Geneva, 76-77. *Mem:* AAAS; Am Soc Pharmacol & Exp Therapeut; Am Pharmaceut Asn; Soc Toxicol. *Res:* Absorption; distribution; metabolism and excretion of drugs and chemicals; toxicity of chemicals to the endocrine pancreas. *Mailing Add:* Dept Pharmacol & Toxicol C-231 Holden Hall Mich State Univ Ctr Environ Toxicol East Lansing MI 48824

FISCHER, LEEWELLYN C, b Litchfield, Minn, May 23, 37; div; c 2. PHOTOGRAPHIC CHEMISTRY. *Educ:* Univ Minn, BS, 59; Carnegie Inst Tech, MS, 61, PhD(phys chem), 64. *Prof Exp:* RES CHEMIST, PHOTO PRODS DEPT, E I DU PONT DE NEMOURS & CO, INC, 63- *Mem:* Soc Photog Scientists & Engrs. *Res:* Kinetics of fast reactions, specifically photolysis; photochemistry of liquids and thin films; free radical reactions in gas phase; photographic imaging and emulsion research. *Mailing Add:* E I du Pont de Nemours & Co Inc 666 Driving Park Ave Rochester NY 14613

FISCHER, MARK BENJAMIN, b San Francisco, Calif, Mar 23, 49. CATALYSIS, PHOTOGRAPHIC CHEMICALS. *Educ:* Univ Calif, Berkeley, BS, 71, Univ Wis-Madison, PhD(inorg chem), 78. *Prof Exp:* Fel dept chem, Univ Toronto, 78-79, Tulane Univ, 79-80; res chemist, MW Kellog Res & Develop Ctr, 80-82; RES CHEMIST, GEN PHOTO PROD CO, 82- *Mem:* Am Chem Soc; Sigma Xi; Am Inst Chemists. *Res:* Performance of experimental programs leading to process innovation in synfuels production and upgrading; development and evaluation of novel catalysts for process applications in synfuels production; production technical chemistry for photographic chemicals. *Mailing Add:* 2461 Bexford Pl Bexley OH 43209

FISCHER, MICHAEL, b Wisconsin, Oct16,47. MATHEMATICS, SCIENCE. *Educ:* Univ Wis, BA, 69; Univ Ore, PhD(math), 74. *Prof Exp:* PROF MATH, YALE UNIV, 74- *Mem:* Am Math Asn. *Mailing Add:* Dept Comp Sci Yale Univ Box 2158 Yale Sta New Haven CT 06520

FISCHER, NIKOLAUS HARTMUT, b Kunzendorf, Ger, Aug 8, 36; m. NATURAL PRODUCTS CHEMISTRY, SYNTHETIC ORGANIC CHEMISTRY. *Educ:* Univ Tuebingen, Ger, BS, 60, MS, 63, Dr Natural Sci(org chem), 65. *Prof Exp:* R A Welch fel natural prod chem, Univ Tex, Austin, 65-67; asst prof chem, La State Univ, 67-73; asst biosynthesis betalains, Univ Zurich, 68-70; assoc prof, 73-78, PROF NATURAL PROD CHEM, LA STATE UNIV, BATON ROUGE, 78- *Concurrent Pos:* Distinguished fac fel, La State Univ, 79. *Honors & Awards:* Charles E Coates Award, 87. *Mem:* Am Chem Soc; Am Soc Pharmacog; Phytochem Soc NAm; Int Soc Chem Ecol; Sigma Xi; Phytochem Soc Europe. *Res:* Isolation, structure elucidation and chemistry of natural products; biochemical systematics in the compositae; allelopathy biotechnology of root cultures; plant-insect and insect-insect interaction. *Mailing Add:* Dept Chem La State Univ Baton Rouge LA 70803-1804

FISCHER, PATRICK CARL, b St Louis, Mo, Dec 3, 35; m 67; c 2. COMPUTER SCIENCE. *Educ:* Univ Mich, BS, 57, MBA, 58; Mass Inst Technol, PhD(math), 62. *Prof Exp:* Asst prof appl math, Harvard Univ, 62-65; assoc prof comput sci, Cornell Univ, 65-68; vis prof, Univ Waterloo, 68-69, prof, 69-74, chmn dept appl anal & comput sci, 72-74; head, Dept Comput Sci, Pa State Univ, 74-78, prof, 74-79; PROF & CHMN, DEPT COMPUT SCI, VANDERBILT UNIV, 80- *Concurrent Pos:* Grants, NSF, 64-66, 66-68 & 79-84, Nat Res Coun Can, 68-76; vis assoc prof, Univ BC, 67-68; mem, Grant Selection Comt, Nat Res Coun Can, 73-76; ed-in-chief, Asn Comput Mach Spec Publ, 74-88; ed, J Comput Syst Sci & J Comput Lang; vis prof, Univ Calif Berkeley, Georgia Inst Tech, 86-87. *Mem:* Asn Comput Mach; Inst Elec & Electronics Engrs; Sigma Xi. *Res:* Data-base theory; user-oriented data-base systems; computational complexity. *Mailing Add:* Box 6026 Sta B Vanderbilt Univ Nashville TN 37235

FISCHER, PAUL HAMILTON, CANCER CHEMOTHERAPY, BIOCHEMISTRY. *Educ:* Univ Calif, San Francisco, PhD(pharmacol), 76. *Prof Exp:* ASSOC PROF HUMAN ONOCOL, UNIV WIS, 85- *Res:* Molecular pharmacology. *Mailing Add:* Dept Human Onocol Wis Clin Cancer Ctr 600 Highland Ave Madison WI 53792

FISCHER, R(OLAND) B(ARTON), b Denver, Colo, Feb 28, 20; m 42; c 2. PHYSICAL METALLURGY. *Educ:* Colo Sch Mines, MetE, 42. *Prof Exp:* Asst metallurgist, Am Smelting & Refining Co, El Paso, Tex, 46; res engr, Battelle Mem Inst, 46-49, asst supvr, 49-53, div chief, 53-63; sr develop specialist, Dow Chem Co, 63-66, sr res metallurgist, 66-75; sr res specialist, Rocky Flats Plant, Rockwell Int, 75-83; RESEARCHER, 83- *Res:* Non-ferrous metals, hard metals technology; tungsten carbide, powder metallurgy and others; high pressure research; nuclear materials; diamond technology; gems; minerology. *Mailing Add:* 12035 Applewood Knolls Dr Lakewood CO 80215

FISCHER, RICHARD BERNARD, b Boston, Mass, Jan 19, 19; m 53; c 3. BIOLOGY. *Educ:* Queens Col, NY, BS, 42; Columbia Univ, MA, 43; Cornell Univ, PhD(zool), 53. *Prof Exp:* Asst press serv, Cornell Univ, 51-52, asst biol, 52-53, prof environ educ, 53-85; RETIRED. *Honors & Awards:* Meritorious Serv Award, Asn Interpretive Naturalists. *Mem:* Sigma Xi; NAm Bluebird Soc; Am Nature Study Soc; Am Environ Defense Fund; Sierra Club Legal Defense Fund. *Res:* Chimney swift; environmental quality; ecology; natural history writing and photography; natural history. *Mailing Add:* 135 Pine Tree Rd Ithaca NY 14850

FISCHER, RICHARD MARTIN, JR, b Vancouver, Wash, Apr 23, 47; m 69; c 3. POLYMER & PHOTOCHEMISTRY. *Educ:* NDak State Univ, BS, 69, PhD(polymer chem), 74. *Prof Exp:* Process develop chemist, heterogeneous catalysis, Shell Chem Co, 69-70; sr res chemist, 74-80, res specialist, 80-84, SR RES SPECIALIST, POLYMER PHOTOCHEM, 3M CO, 84- *Concurrent Pos:* NSF grant, 70-74. *Honors & Awards:* Roon Award, Fedn Soc Paint Technol, 74. *Mem:* Am Soc Testing & Mat; Soc Automotive Engrs. *Res:* Ultraviolet curing chemistry; photodegradation of polymer systems; accelerated weathering and durability test development. *Mailing Add:* 803 Eighth St Hudson WI 54016

FISCHER, ROBERT B, b Easton, Pa, May 27, 17. METALLURGICAL ENGINEERING. *Educ:* Lafayette Col, BS, 39. *Prof Exp:* Mgr metall, Cast Div, Ingersoll-Rand Corp, 40-75, consult, 75-83; RETIRED. *Concurrent Pos:* Chmn, Brass & Bronze Div, Am Foundrymen's Soc, 60-61, dir, 75. *Honors & Awards:* Sci Merit Award, Am Foundrymen's Soc, 71, Seaman Gold Medal, 83. *Mem:* Am Foundrymen's Soc; fel Am Soc Metals. *Res:* Casting of bronze, iron and steel; sand casting. *Mailing Add:* 4573 Calvert Pl Center Valley PA 18034

FISCHER, ROBERT BLANCHARD, b Hartford, Conn, Oct 24, 20; m 46; c 5. ANALYTICAL CHEMISTRY. *Educ:* Wheaton Col, Ill, BS, 42; Univ Ill, PhD(anal chem), 46. *Prof Exp:* Radio broadcast engr, Chicago 41; asst anal chem, Univ Ill, 42-44, instr, 46-48; res chemist, Metall Lab, Univ Chicago, 44-46; from asst prof to prof chem, Ind Univ, 48-63; prof chem, 63-79, dean, Sch Natural Sci & Math, 63-79, EMER PROF CHEM, CALIF STATE UNIV, DOMINGUEZ HILLS, CARSON, 79-; PROVOST & SR VPRES, BIOLA UNIV, 79- *Concurrent Pos:* Vis assoc prof, Calif Inst Technol, 59-60. *Mem:* AAAS; Am Chem Soc; Am Sci Affil. *Res:* Science and society, quantitative chemical analysis; instrumentation. *Mailing Add:* 30238 Via Victoria Rancho Palos Verdes CA 90274-4440

FISCHER, ROBERT GEORGE, b St Paul, Minn, Oct 17, 20; m 47; c 3. MICROBIOLOGY. *Educ:* Univ Minn, BA, 42, MS, 47, PhD, 48; Am Bd Microbiol, dipl. *Prof Exp:* Asst bact, Univ Minn, 45-48; from asst prof to assoc prof, Univ NDak, 48-62, prof microbiol & chmn dept, 62-; RETIRED. *Mem:* Am Soc Microbiol; Am Acad Microbiol. *Res:* Viruses; experimental leukemia transmission; poliomyelitis; virus tumors. *Mailing Add:* 447 Campbell Dr Grand Forks ND 58201

FISCHER, ROBERT LEIGH, b Chicago, Ill, July 29, 26; m 54; c 3. CLINICAL CHEMISTRY. *Educ:* Northern Ill State Teachers Col, BS, 50; Univ Ill, MS, 51, PhD(biochem), 54. *Prof Exp:* Asst, Univ Ill, 51-54; res chemist, E I du Pont de Nemours & Co, 54-55; asst prof chem, Med Sch, Univ Tenn, 55-61, res assoc pediat, 55-61; sr res chemist, Campbell Soup Co, 61-64, div head proteins, 64-67; chief clin chemist, Philadelphia Gen Hosp, 67-78; dir, Philadelphia Police Lab, 78-80; RES SCIENTIST, NJ DEP, 80- *Mem:* Am Chem Soc. *Res:* Clinical chemistry methods. *Mailing Add:* 08077-3343 Cinnaminson NJ 08077

FISCHER, ROLAND LEE, b Detroit, Mich, Sept 9, 24; m 46; c 3. ENTOMOLOGY. *Educ:* Univ Mich, BS, 46; Mich State Col, MS, 48; Kans State Col, PhD(entom), 52. *Prof Exp:* Teacher pub schs, Mich, 46-47; asst entom, Mich State Col, 47-48; asst, Kans State Col, 48-50, instr, 51; res assoc, Univ Kans, 52; res fel, Univ Minn, 52-53; from asst prof to assoc prof entom, 53-64, cur entom, Mus, 59-64, PROF ENTOM, MICH STATE UNIV, 64- *Mem:* Entom Soc Am; Soc Syst Zool; Entom Soc Can; Int Union Study Social Insects; fel Royal Entom Soc London. *Res:* T Taxonomy of the Aculeate Hymenoptera, particularly the Apoidea; biological and phenological investigations in the Aculeate Hymenoptera; insect morphology; pollination of legume seed crops. *Mailing Add:* Dept Entom Mich State Univ East Lansing MI 48824

FISCHER, RONALD HOWARD, b New York, NY, May 1, 42. FUELS ENGINEERING, CHEMICAL ENGINEERING. *Educ:* City Col New York, BS, 63; Princeton Univ, MA, 66, PhD(chem), 67. *Prof Exp:* Sr res chemist, Mobil Res & Develop Corp, 67-75; proj mgr coal technol, 75-78, dir coal liquefaction, US Dept Energy, Washington, DC, 78-80; MGR PRODUCING & TECHNOL, MOBIL OIL CORP, 80- *Mem:* Am Inst Chem Engrs; AAAS; Am Chem Soc; Sigma Xi. *Res:* Upgrading of petroleum residue; synthetic fuels processing, refining of synthetic fuels, synthesis gas conversion reactions; conversion of coal to fuel oil and gasoline. *Mailing Add:* Mobil Res & Develop Corp Paulsboro NJ 08066

FISCHER, RUSSELL JON, b Peoria, Ill, Feb 19, 60. COMPOUND SEMICONDUCTOR EPITAXY, SEMICONDUCTOR LASERS. *Educ:* Univ Ill, BS, 82, MS, 83, PhD(elec eng), 86. *Prof Exp:* MEM TECH STAFF, AT&T BELL LABS, MURRAY HILL, NJ, 86- *Concurrent Pos:* Young scientist award, 89. *Mem:* Inst Elec & Electronics Engrs; Am Phys Soc. *Res:* Epitaxial growth of compound semiconductor lasers; vertical cavity surface emitting and shorter wavelength visible designs. *Mailing Add:* AT&T Bell Labs Rm 7C-229 600 Mountain Ave Murray Hill NJ 07974-2070

FISCHER, SUSAN MARIE, b Washington, DC, May 19, 47. TUMOR PROMOTION, EICOSANOIDS. *Educ:* High Point Col, BS, 69; Univ Wyo, MS, 72, PhD(physiol), 74. *Prof Exp:* Fel, Univ Tex M D Anderson Cancer Ctr, 74-76; staff scientist, Oak Ridge Nat Lab, 76-82; asst prof biochem, 82-87, ASSOC PROF BIOCHEM, UNIV TEX M D ANDERSON CANCER CTR, 87- *Concurrent Pos:* Vis scientist, Ger Cancer Ctr, Heidelberg, 80; mem, Chem Path Study Sect, NIH, 84- *Mem:* Am Asn Cancer Res; Sigma Xi; Am Oil Chemists Soc; AAAS. *Res:* Mechanism of action of tumor promoters in the multistage model of chemical carcinogenesis in mouse skin; role of inflammation and its mediators, eicosanoids and reactive oxygens. *Mailing Add:* Sci Park Res Div Univ Tex Syst Cancer Ctr PO Box 389 Smithville TX 78957

FISCHER, THEODORE VERNON, b Brillion, Wis, July 25, 39; m 69; c 2. HUMAN ANATOMY. *Educ:* NCent Col, Ill, BA, 61; Univ Wis, PhD(anat), 66. *Prof Exp:* Fel reproductive physiol, Univ Wis, 66-67; asst prof anat, 67-76, ASSOC PROF ANAT, UNIV MICH, 76- *Mem:* Am Asn Anat. *Res:* Mammalian placentation; computer assisted instructional materials. *Mailing Add:* Dept Anat Univ Mich Ann Arbor MI 48109

FISCHER, TRAUGOTT ERWIN, b Aarau, Switz, Jan 21, 32; m 58; c 3. SURFACE PHYSICS. *Educ:* Swiss Fed Inst Technol, dipl, 56, PhD(solid state physics), 63. *Prof Exp:* Asst solid state physics, Swiss Fed Inst Technol, 57-63; mem tech staff surface physics, Bell Labs, NJ, 63-66; assoc prof eng & appl sci, Yale Univ, 66-72; sr res assoc, Exxon Res & Eng Co, 72-86; PROF, STEVENS INST TECHNOL, 86- *Concurrent Pos:* Prog chmn, 29th Phys Electronics Conf, 69; organizing comt, NATO Conf, Hohegeiss, Ger, 75; planning panel, Catalysis, Nat Res Coun, 75; vis adv comt, Am Phys Soc, 72-75; prog chmn, Surface Sci Div, Am Vacuum Soc, 77, chmn, 78; adv comt, Dept Metall, Ohio State Univ; organizing comt, Atomistics Fracture, NATO Res Inst, Corsica, 81, Chem & Physics Fracture, Bad Reichenhall, Ger, 86; mem comt & referee panels, NSF, 86-87; mem comt, Lubrication Ceramics, Nat Res Coun, 86-87; prog comt, Soc Tribologists & Lubrication Engrs, 86-90, secy, Wear Comt, 86-88, vchmn, 88-90, chmn, 90; mem prog coun, NJ Advan Technol Ctr Surface Engineered Mat, 87-; res comt tribology, Am Soc Mech Engrs, 87, chmn subcomt, Nat Tribology Prog, 90. *Honors & Awards:* Kern Prize, Swiss Fed Inst Technol, 62. *Mem:* Am Vacuum Soc; fel Am Phys Soc; Swiss Phys Soc; Soc Tribologists & Lubrication Engrs; Am Ceramic Soc. *Res:* Physical properties of surfaces; optical properties and energy-band structure of solids; surface-physical aspects of catalysis and metallurgy; tribology. *Mailing Add:* Dept Mat Stevens Inst Technol Hoboken NJ 07030

FISCHER, WILLIAM HENRY, b Schenectady, NY, Dec 17, 21; m 47; c 2. ENERGY-ENVIRONMENT INTERFACES. *Educ:* Univ Colo, BS, 51; Univ Wash, MS, 52; Rensselaer Polytech Inst, PhD(phys chem), 61. *Prof Exp:* Phys chemist, Gen Elec Co, 52-62; sr scientist, Nat Ctr Atmospheric Res, 62-68; staff scientist, Bell Bros Res Corp, 69-70; sr res assoc, State Univ NY, Albany, 70-71; dir, Air Pollution Labs, WVa Univ, 71-72; sr scientist, Mech Technol Inc, 73-75; consult scientist, Gibert-Commonwealth Inc, 75-86; RETIRED. *Mem:* Am Chem Soc; fel AAAS; fel Am Inst Chemists. *Res:* Application of basic physical chemical principles to evaluation of fuel and flue gas processes relating to combustion; gasification; liquefaction; advanced power cycles; coal, refuse, biomass and thermal energy conversion processes. *Mailing Add:* PO Box 940 Basalt CO 81621

FISCHHOFF, DAVID ALLEN, b Detroit, Mich, Dec 12, 52. SCIENCE. *Educ:* Mass Instit Technol, BS, 74; Rockefeller Univ, PhD(genetics), 80. *Prof Exp:* Res assoc, Wash Univ Sch Med, 80-83; sr res biologist, 83-86, res specialist, 86-89, RES GROUP LEADER, MONSANTO CO, 89- *Res:* Genetic engineering of agronomically useful traits into transgenic crop plants. *Mailing Add:* Monsanto Co 700 Chesterfield Village Pkwy St Louis MO 63198

FISCHINGER, PETER JOHN, b Ljubljana, Yugoslavia, Sept 29, 37; US citizen; m 63. CANCER, ANIMAL VIROLOGY. *Educ:* Univ Ill, Chicago, MD, 63, PhD(microbiol), 66. *Prof Exp:* Fel, USPHS, 63-65, spec fel, 65-66, staff fel cancer, 66-68, med officer, Nat Cancer Inst, 68-70, head, Tumor Virus Sect, 70-75, ASSOC CHIEF, VIRAL CARCINOGENESIS LAB, NIH, 75- *Concurrent Pos:* Vis scientist, Max Planck Inst Virus Res, Tübingen, 70-71. *Mem:* AAAS; Am Soc Microbiol. *Res:* Interactions of cells with endogenous or exogenous oncornaviruses; molecular biology of viral transforming genes; nature and mechanisms of cellular reversion. *Mailing Add:* Off Director Nat Cancer Inst 9000 Rockville Pike Bldg 31 NIH Bethesda MD 20892

FISCHLER, DRAKE ANTHONY, b Sisak, Croatia, Apr 19, 23; m 61; c 1. COMPUTER TECHNOLOGY. *Educ:* Ind Univ, AB, 60, MBA, 62. *Prof Exp:* Res asst genetic res, dept zool, Ind Univ, 58-63; res assoc mkt res, Eli Lilly Int Corp, London, Eng, 63-66, staff assoc field tech serv, 66-70, res staff assoc, Int Field Res, Lilly Res Labs, 70-74, mgr, Int Agr Chem Regulatory Serv, 74-78, mgr, Doc Control Ctr, Agr Regulatory Serv, Lilly Res Labs, 78-88; RETIRED. *Mailing Add:* 4255 Cooper Rd Indianapolis IN 46208

FISCHLER, MARTIN A(LVIN), b New York, NY, Feb 15, 32; m 60; c 2. COMPUTER SCIENCE, ARTIFICIAL INTELLIGENCE. *Educ:* City Col New York, BEE, 54; Stanford Univ, MS, 58, PhD(elec eng), 62. *Prof Exp:* Electronic engr, Nat Bur Standards, 56; scientist, Lockheed Missiles & Space Co, 58-61, sr scientist, 61-62, res scientist, 62-71, staff scientist, 71-77; SR COMPUTER SCIENTIST, SRI INT, 77- *Mem:* Pattern Recognition Soc; Asn Comput Mach; Math Asn Am; Inst Elec & Electronics Engrs. *Res:* Artificial intelligence; switching theory; information theory; computer organization; information retrieval; operations research. *Mailing Add:* 966 Bonneville Way Sunnyvale CA 94087

FISCHLSCHWEIGER, WERNER, b Bremen, Ger, May 4, 32; m 60; c 2. ELECTRON MICROSCOPY. *Educ:* Graz Univ, PhD(zool), 57. *Prof Exp:* Dir tissue cult, Austrian Cancer Res Inst, 58-62; asst prof histol, Med Sch, Graz Univ, 62-63; asst prof, Med Sch, St Louis Univ, 63-65; assoc prof, Dent Sch, Univ Md, Baltimore, 65-69; assoc prof, 69-73, PROF ORAL BIOL, COL DENT, UNIV FLA, 73-, DIR, E M LAB, 74- *Mem:* Int Asn Dent Res. *Res:* Dental research. *Mailing Add:* 804 NW 36th Ter Gainesville FL 32605

FISCHMAN, DONALD A, b New York, NY, Apr 27, 36; m 60; c 3. MEDICINE. *Educ:* Kenyon Col, AB, 57; Cornell Univ, MD, 61. *Prof Exp:* Res fel anat, Med Col, Cornell Univ, 61-63; fel embryol, Strangeways Lab, Cambridge, Eng, 63-64; instr anat, Med Col, Cornell Univ, 64-65; asst prof zool, Univ Chicago, 65-66, asst prof biol & anat, 68-71, assoc prof biol & anat, 71-77, assoc dean curric, div biol sci, 70-77; mem fac, Downstate Med Ctr, State Univ NY, 77-80, prof & chmn, dept anat & cell biol, 80-; AT DEPT CELL BIOL & ANAT, MED COL, CORNELL UNIV. *Concurrent Pos:* NY Heart Asn res fel, 61-63; USPHS fel, 63-64 & res grant, 64-66 & 71-; res grants, NSF, 68- & Chicago Heart Asn, 69-71. *Mem:* Am Asn Anatomists; Am Soc Cell Biologists; Soc Develop Biol; Biophys Soc; Soc Gen Physiol. *Res:* Anatomy; developmental biology; electron microscopy; development, growth and physiology of muscle; ultrastructure of bacterial cell walls and membranes. *Mailing Add:* Anat Dept Cornell Univ Med Col 1300 York Ave New York NY 10021

FISCHMAN, HARLOW KENNETH, b New York, NY, Dec 19, 32; div; c 2. MEDICAL GENETICS, CYTOGENETICS. *Educ:* Brooklyn Col, BA, 55; Univ Ark, MS, 60; NY Univ, PhD(genetics), 68. *Prof Exp:* Sr res assoc cancer, Sloan-Kettering Inst, 60; instr biol, Long Island Univ, 60-61 & Nassau Community Col, 61-63; asst prof biol, Queensboro Community Col, 65-69; res assoc genetics, Col Physicians & Surgeons, Columbia Univ, 69-72; asst prof genetics, Fairleigh Dickinson Univ, 72-73; ASSOC RES SCIENTIST MED GENETICS, NY STATE PSYCHIAT INST, 73- *Concurrent Pos:* Fac res fel, State Univ NY, 68; res assoc, Col Physicians & Surgeons, Columbia Univ, 73-; NIH grant, 75-; travel grant, Environ Mutagenesis Symp, 81. *Mem:* Environ Mutagenesis Soc; Am Soc Human Genetics; Am Soc Cell Biologists; Somatic Cell Genetics Soc; Genetic Toxicol Asn. *Res:* Human chromosome structure and function; behavioral genetics; genetic toxicology; karyotype analyses; chromosome localization; meiosis; gene mapping; somatic cell hybridization. *Mailing Add:* NY State Psychiat Inst 722 W 168th St New York NY 10032

FISCHMAN, MARIAN WEINBAUM, b New York, NY, Oct 13, 39; c 3. BEHAVIORAL PHARMACOLOGY, PSYCHOLOGY. *Educ:* Barnard Col, BA, 60; Columbia Univ, MA, 62; Univ Chicago, PhD(psychol), 72. *Prof Exp:* Res assoc anat, Cornell Med Sch, 61-63; res asst psychol, Univ Chicago, 65-66; res assoc, Inst Behav Res, 66-68; res asst psychol, Univ Chicago, 68-69, res assoc & asst prof psychiat, 72-84; assoc prof, 84-88, PROF, DEPT PSYCHIAT & BEHAV SCI, JOHNS HOPKINS MED SCH, 90- *Mem:* Fel Am Psychol Asn; Behav Pharmacol Soc; Am Soc Pharmacol & Exp Therapeut; fel Am Col Neuropsychopharmacol; fel Am Psychol Soc; Soc Behav Med. *Res:* Behavioral pharmacology; drug abuse; psychology; behavioral medicine. *Mailing Add:* Dept Psychiat & Behav Sci Johns Hopkins Med Sch Houck Bldg (E-2) 600 N Wolfe St Baltimore MD 21205

FISCHMAN, STUART L, b Buffalo, NY, Nov 29, 35; m 60; c 1. ORAL PATHOLOGY. *Educ:* Harvard Univ, DMD, 60; Am Bd Oral Path, dipl, 69; Am Bd Oral Med, dipl, 86. *Prof Exp:* Intern dent, Boston Vet Admin Hosp, Mass, 60-61; instr clin dent, 61-64, from asst prof to assoc prof oral path, 64-72, asst dean acad develop & facil planning, 70-73, PROF ORAL MED, SCH DENT, STATE UNIV NY BUFFALO, 72- *Concurrent Pos:* Res assoc, Vet Admin Hosp, Buffalo, NY, 63-64; consult, USPHS, 65-70; vis prof, Nat Univ Asuncion, Paraguay, 69-; univ assoc dent, Buffalo Gen Hosp, 69-; consult, WHO, 70-; consult oral med, Univ PR, 74; dir dent, Meyer Mem Hosp, Buffalo, 74-78 & Erie County Med Ctr, Buffalo, 78-; consult, Coun Hosp Dent & Coun Dent Educ, Am Dent Asn, 76-90; Lady Davis fel, Hebrew Univ, Jerusalem, 81 & 89; consult, Lever Bros Co, 76- *Mem:* fel Am Col Dent; Fel Am Acad Oral Path; Am Dent Asn; Int Asn Dent Res; fel Int Col Dent. *Res:* Experimental oral pathology and periodontal diseases; clinical testing of therapeutic dentifrice; oral diseases; tropical oral pathology. *Mailing Add:* Dept Oral Med State Univ NY Sch Dent Buffalo NY 14214

FISCUS, ALVIN G, b Newell, SDak, July 6, 30; m 56; c 3. VIROLOGY, MICROBIOLOGY. *Educ:* SDak State Univ, BS, 56, MS, 57; Univ Ariz, PhD(microbiol), 66. *Prof Exp:* Dir lab clin microbiol, Tucson Med Ctr, Ariz, 66-67; assoc prof, 67-77, PROF MICROBIOL, MONT STATE UNIV, 77- *Concurrent Pos:* Consult, Mont Regional Med Prog, 67- *Mem:* Am Soc Microbiol. *Res:* Clinical microbiology; viral and chemical oncology. *Mailing Add:* Dept Microbiol Mont State Univ Bozeman MT 59717

FISCUS, EDWIN LAWSON, b Ellwood City, Pa, Jan 20, 42; m 69. PLANT PHYSIOLOGY. *Educ:* Slippery Rock State Col, BSEd, 64; Univ Ariz, MS, 66; Duke Univ, PhD(bot), 69. *Prof Exp:* Res assoc plant physiol, Auburn Univ, 69-70; res assoc plant physiol, Duke Univ, 70-76; PLANT PHYSIOLOGIST, AGR RES, SCI & EDUC ADMIN, USDA, 76- *Mem:* Sigma Xi; Am Soc Plant Physiologists; Am Inst Biol Sci; Am Soc Agron. *Res:* Plant and soil water relations; root-oxygen relations; factors influencing salt and water uptake by roots and water movement through the plant; stress physiology. *Mailing Add:* 100 Olympic Dr Cary NC 27513

FISER, PAUL S(TANLEY), b Roznava, Czech, Apr 30, 38; Can citizen; m 62; c 1. REPRODUCTIVE PHYSIOLOGY, GENETICS. *Educ:* Univ Agr, Brno, Ing Zoot, 63; Univ Guelph, PhD(genetics & reproductive physiol), 72. *Prof Exp:* Jr scientist biol, Vet Col, Brno, 63-69; res assoc reproductive physiol, Univ Guelph, 72-78; res scientist reproductive physiol, 78-90, TEAM LEADER, GAMETE-EMBRYO TECHNOL TEAM, ANIMAL RES CTR, OTTAWA, 90- *Concurrent Pos:* Consult, Hybrid Turkeys LTD, Kitchener, Ont, 76-78; assoc fac mem, Dept of Animal & Poultry Sci, Univ Guelph, 76- *Mem:* Soc Study Reproduction; Can Soc Animal Sci; World's Poultry Sci Asn; Soc Cryobiol. *Res:* Low temperature preservation of mammalian gametes; experimental embryology (mammalia and aves); lethal factors in avian species. *Mailing Add:* Animal Res Ctr Agr Can Ottawa ON K1A 0C6 Can

FISET, PAUL, b Quebec, Que, Nov 7, 22; m 53; c 3. MICROBIOLOGY. *Educ:* Laval Univ, BA, 44, MD, 49; Cambridge Univ, PhD(microbiol), 56. *Prof Exp:* Asst bact, Laval Univ, 55-57; asst prof, Sch Med & Dent, Univ Rochester, 58-64, asst med, 58-64; assoc prof, 64-75, PROF MICROBIOL, SCH MED, UNIV MD, BALTIMORE CITY, 75- *Concurrent Pos:* Actg chmn, Sch Med, Univ Md, 86- *Mem:* Am Soc Microbiol; Am Asn Immunol; Am Soc Trop Med & Hyg. *Mailing Add:* Dept Microbiol Univ Md Sch Med 655 W Baltimore St Baltimore MD 21201

FISH, BARBARA, b New York, NY, July 31, 20; m 53; c 2. PSYCHIATRY, CHILD PSYCHIATRY. *Educ:* Columbia Univ, BA, 42; NY Univ, MD, 45; Am Bd Psychiat & Neurol, dipl, 55, cert child psychiat, 60; W A White Inst Psychoanal, cert, 56. *Prof Exp:* Intern & asst resident med, Bellevue Hosp, 45-47; resident pediat, New York Hosp, 47-48 & NY Univ-Bellevue Med Ctr, 48-49, resident psychiat, Bellevue Hosp, 49-52, clin asst psychiat, 51-55; from instr to asst prof pediat & psychiat, Med Col, Cornell Univ, 55-60; assoc prof psychiat, Sch Med, NY Univ, 60-70, prof child psychiat, 70-72, adj prof child psychiat, 72-; prof psychiat, 72-89, DELLA MARTIN PROF PSYCHIAT, UNIV CALIF, LOS ANGELES, 89- *Concurrent Pos:* Dir child psychiat, NY Univ Med Ctr, 60-72; mem comt cert child psychiat, Am Bd Psychiat & Neurol, 69-77; prin investr, Schizophrenic Offspring From Birth To Adulthood, 52- & mem clin prog projs res rev comt, Nat Inst Mental Health, 76-79; mem, Risk Res Consortium Schizophrenia, 76-; prin investr, Childrens Psychopharmacol Res Unit, NY Univ Sch Med, 61-72; consult, WHO, 83- *Honors & Awards:* McGavin Award, Am Psychiat Asn, 87. *Mem:* Soc Res Child Develop; fel Am Psychiat Asn; Am Acad Child Psychiat; Asn Res Nerv & Ment Dis; fel Am Col Neuropsychopharmacol. *Res:* Child psychiatry, especially childhood schizophrenia and pharmacotherapy; infant development and the neurological antecedents of schizophrenia and other neuropsychiatric disorders in infancy. *Mailing Add:* Dept Psychiat Univ Calif Sch Med Los Angeles CA 90024-1759

FISH, DONALD C, b New York, NY, Apr 19, 37; m 58; c 3. MICROBIOLOGY, IMMUNOLOGY. *Educ:* Cornell Univ, BS, 58; Univ Mich, MS, 61, PhD(microbiol), 64. *Prof Exp:* Asst bact, Cornell Univ, 57-58; lab instr, Univ Mich, 58-61, asst, 61-64; microbiologist anthrax res, US Army Biol Labs, 64-71; dir microbiol div, Woodard Res Corp, 71-72; MGR DIS CONTROL LAB, FREDERICK CANCER RES CTR, 72- *Concurrent Pos:* Instr microbiol, Sci Prog, Interlochen Arts Acad, Univ Mich, 63. *Mem:* Tissue Cult Asn; Am Soc Microbiol; Am Chem Soc; NY Acad Sci; Sigma Xi. *Res:* Immunological, biochemical and genetic studies on the control over expression of endogenous virus and its relationship to spontaneous cancer; development of potential vaccines; monitor immune reactivity during cancer progression. *Mailing Add:* Box B Col Estates Post Off Frederick MD 21701

FISH, ELEANOR N, b London, UK, Sept 21, 52; m 76; c 3. IMMUNOLOGY, MICROBIOLOGY. *Educ:* Univ Manchester, UK, BSc, 74; Univ London, UK, MPhil, 77; Univ Toronto, Can, PhD(cell biol), 86. *Prof Exp:* Res assoc, Res Inst, Hosp Sick Children, Can, 81-87; lectr, Dept Pediat, 87-88, ASST PROF MICROBIOL, UNIV TORONTO, 88- *Mem:* Europ Sch Oncol; Int Soc Interferon Res. *Res:* Structure and function studies with a panel of recombinant DNA-derived biological response modifiers; molecular analyses of interferon and growth factor regulation of cell proliferation; characterization of cytokine involvement in the pathogenesis of rheumatoid arthritis; therapeutic application for cytokines in viral infections. *Mailing Add:* 20 Loganberry Crescent Willowdale ON M2H 3H1 Can

FISH, F(LOYD) H(AMILTON), JR, b Bryan, Tex, May 8, 23; m 53; c 5. MECHANICS. *Educ:* Va Polytech Inst, BS, 43, MS, 50. *Prof Exp:* Asst prof fluid mech, Va Polytech Inst, 48-50; sr res engr, Textile Fibers Dept, 50-60, RES SUPVR, TECH DIV FIBERS DEPT, E I DU PONT DE NEMOURS & CO, INC, 60- *Mem:* Am Soc Mech Engrs; Soc Automotive Engrs; Sigma Xi. *Res:* Mechanics of paper, fabrics and non-woven fabrics and cushioning structures, including manufacturing processes as well as products. *Mailing Add:* 115 Quintynnes Wilmington DE 19807

FISH, FEROL F, JR, b East Chicago, Ind, Jan 15, 30; m 56; c 5. THERMODYNAMICS & MATERIAL PROPERTIES. *Educ:* Ind Univ, BS, 55, AM, 57; Pa State Univ, PhD(geophys), 61. *Prof Exp:* Geophys investr, NJ Zinc Co, 60-63; sr res scientist, Gen Dynamics/Ft Worth, 63-64; sr scientist, Douglas Aircraft Co, 64-68; mgr appl physics, Roy C Ingersoll Res Ctr, Borg-Warner Corp, Des Plaines, 68-79; MGR PHYSICS, GAS RES INST, CHICAGO, 79- *Concurrent Pos:* Nat Sci Found fel, 58-60. *Mem:* Am Geophys Union; Sigma Xi; Soc Petrol Engrs. *Res:* Research administration of geosciences, materials, and thermophysical properties of fluids. *Mailing Add:* 804 N Kaspar Ave Arlington Heights IL 60004

FISH, FRANK ELIOT, b New York, NY, June 8, 53; m 84; c 2. FUNCTIONAL MORPHOLOGY, ECOLOGICAL PHYSIOLOGY. *Educ:* State Univ Col, Oswego, BA, 75; Mich State Univ, MS, 77, PhD(zool), 80. *Prof Exp:* PROF ZOOL, WEST CHESTER UNIV, 80- *Concurrent Pos:* Consult, Dames & Moore Environ Consult Co, 75; vis prof, Wallops Island Marine Sci Ctr, 82. *Honors & Awards:* Anne M Jackson Award, Am Soc Mammalogists. *Mem:* Am Soc Mammalogists; Am Soc Zoologists; Sigma Xi; Soc Marine Mammal. *Res:* Energetics, biomechanics and hydrodynamics of swimming by animals; evolution and adaptations of aquatic mammals. *Mailing Add:* Dept Biol West Chester Univ West Chester PA 19383

FISH, GORDON E, b Dayton, Tenn, Jan 4, 51; m 81. SOLID STATE PHYSICS, MAGNETIC MATERIALS. *Educ:* Wheaton Col, Ill, BS, 72; Univ Ill, Urbana, MS, 73, PhD(physics), 77. *Prof Exp:* Teaching asst & fel, Univ Ill, Urbana, 72-73, res asst physics, 73-77; physicist, US Nat Bur Standards, 77-79; physicist, 79-83, sr res physicist, 83-87, RES ASSOC, ALLIED-SIGNAL, INC, 87- *Concurrent Pos:* Nat Merit scholar, Wheaton Col, 68-72; Nat Bur Standards-Nat Res Coun fel, Nat Bur Standards, 77-79. *Mem:* Am Phys Soc; Am Sci Affil; Inst Elec & Electronics Engrs; Sigma Xi. *Res:* Magnetic materials: properties and applications; neutron scattering; structure of amorphous materials. *Mailing Add:* Metals & Ceramics Lab Corp Res Allied-Signal, Inc Morristown NJ 07960

FISH, IRVING, b Montreal, Can, May 1, 38; m 67; c 3. PEDIATRIC NEUROLOGY. *Educ:* McGill Univ, BSc, 59; Dalhousie Univ, MD, 64. *Prof Exp:* Instr Neurol, Cornell Univ, 67-69; asst prof, 69-76, ASSOC PROF NEUROL, MED CTR, NEW YORK UNIV, 76-, DIR PEDIAT NEUROL, 77- *Concurrent Pos:* Consult neurol, Vet Admin Hosp, 69- *Mem:* Am Acad Neurol. *Res:* Amino acid transport; effects of nutrition on development; spina bifida, hydrocephalus, and epilepsy. *Mailing Add:* 462 First Ave NY Univ Sch Med 550 First Ave New York NY 10016

FISH, JAMES E, b Ann Arbor, Mich, Nov 29, 45; m. PULMONARY. *Educ:* Univ Notre Dame, Ind, BS, 67; Northwestern Univ, Chicago, Ill, MD, 71. *Prof Exp:* Intern internal med, Duke Univ Med Ctr, Durham, NC, 71-72, resident, 72-73; clin assoc, NIH, Nat Inst Child Health & Human Develop, USPHS, Geront Res Ctr, Baltimore, Md, 73-75; res assoc pulmonary dis, Johns Hopkins Med Insts, Baltimore, Md, 73-75, asst prof med, Sch Med, Johns Hopkins Univ, 79-85, asst prof environ health sci, Sch Hyg & Pub Health, 80-85; fel pulmonary, Med Sch, Northwestern Univ, Chicago, Ill, 75-76, assoc med, 76-77, asst prof med, 77-79; PROF MED & DIR, PULMONARY MED & CRIT CARE, JEFFERSON MED COL, 85-, ACTG CHMN, DEPT MED, 90- *Concurrent Pos:* Fel, Am Lung Asn, 75-76, Owen L Coon Found, 75-76 & Edward L Trudeau, 76-80; pulmonary young investr award, NIH, 76-78; trustee, Asn Pulmonary Prog Dirs, 89-92. *Mem:* Fel Am Col Chest Physicians; fel Am Col Physicians; Am Thoracic Soc; Am Fedn Clin Res; AAAS; Am Acad Allergy & Clin Immunol; Am Physiol Soc; AMA; NY Acad Sci; Am Sleep Dis Asn. *Mailing Add:* Jefferson Med Col 1025 Walnut St Rm 821 Philadelphia PA 19107

FISH, JEFFERSON, b New York, NY, Nov 27, 42; m 70; c 1. CLINICAL PSYCHOLOGY, PSYCHOTHERAPY. *Educ:* City Col New York, BA, 63; Columbia Univ, MS, 66, PhD(clin psychol), 69; State Univ NY, Stony Brook, postdoctoral dipl, 69. *Prof Exp:* Asst prof psychol, State Univ NY, Stony Brook, 69-70; dir psychol, Suffolk Psychiat Hosp, 70-71; asst prof psychol, Hunter Col, 71-74; vis prof psychol, Pontifical Cath Univ Campinas, Brazil, 74-76; dir educ & training, Manhattan Psychiat Ctr, 77-78; dir clin psychol, 78-81, chairperson, 81-84, PROF PSYCHOL, ST JOHN'S UNIV, 78- *Concurrent Pos:* Mem, Blue Ribbon Panel, AASPB & PES, 81; chair, Psychol Sect, NY Acad Sci, 87; Fulbright fel, Coun Int Exchange Scholars, Fulbright Found, 87; mem, bd dirs, Am Bd Family Psychol, 87- *Mem:* NY Acad Sci; fel Am Psychol Soc; Int Coun Psychologists (treas, 87-90); Am Anthropol Asn; Sigma Xi. *Res:* Psychotherapy as a social influence process; hypnosis; placebo; behavior modification; cognitive therapy; cross-cultural psychology; strategic and systemic therapies. *Mailing Add:* Dept Psychol St John's Univ Jamaica NY 11439

FISH, JOHN G, b Chicago, Ill, Mar 30, 38; m 61; c 4. POLYMER CHEMISTRY. *Educ:* Western Mich Univ, BS, 62; Univ Cincinnati, PhD, 68. *Prof Exp:* Mem tech staff, Mat Group, 66-68, mem tech staff, Cent Anal & Charactization Lab, 68-79, MGR, UNIV PHD RECRUITING & UNIV RELS, TEX INSTRUMENTS, INC, 79- *Mem:* Am Chem Soc; Creation Res Soc. *Res:* Analysis and characterization of polymers and plastics used in electronic processing, equipment and devices. *Mailing Add:* 1815 Plymouth Rock Richardson TX 75081

FISH, JOSEPH LEROY, b Portland, Ore, Feb 10, 43; m 64; c 3. VERTEBRATE ECOLOGY, POPULATION ECOLOGY. *Educ:* Walla Walla Col, BS, 65, MA, 67; Wash State Univ, PhD(zool biomet), 72. *Prof Exp:* Instr Biol, Walla Walla Col, 67-68; asst prof biol, Oakwood Col, 72-77; assoc prof biol, Southwestern Adventist Col, 77-82; RETIRED. *Concurrent Pos:* Consult, Teledyne Brown Eng, 74-75; fel, Lawrence Livermore Lab, 77. *Mem:* Wildlife Soc; Nat Wildlife Fedn. *Res:* Population biology of coyotes (canis latrans) and heavy metal toxicity to freshwater and marine vertebrates. *Mailing Add:* 93903 Swamp Creek Rd Blachly OR 97412

FISH, RICHARD WAYNE, b Gowrie, Iowa, Aug 27, 34; m 64; c 2. ORGANIC CHEMISTRY. *Educ:* Iowa State Univ, BS, 56; Mich State Univ, PhD(org chem), 60. *Prof Exp:* Res chemist, Calif Res Corp, Standard Oil Co, Calif, 60-61; res assoc phys org chem, Brandeis Univ, 61-63; NSF fel, Univ Calif, Berkeley, 63-64; from asst prof to assoc prof, 64-70, dir, NSF Undergrad Res Partic Prog, 65-71, PROF CHEM, CALIF STATE UNIV, SACRAMENTO, 70- *Concurrent Pos:* NSF postdoctoral fel, 63-64; Petrol Res Fund grant, 64-66; sr res fel, Brandeis Univ, 71-72. *Mem:* Am Chem Soc. *Res:* Physical organic chemistry; synthesis and reactions of aromatic, nonclassical aromatic and organometallic compounds; electrophilic reactions of metallocenes; stabilized carbocation intermediates; benzyne intermediates and charge transfer studies. *Mailing Add:* Dept Chem Calif State Univ Sacramento CA 95819-6057

FISH, STEWART ALLISON, b Benton, Ill, Nov 4, 25; m 57; c 4. OBSTETRICS & GYNECOLOGY. *Educ:* Univ Pa, MD, 49. *Prof Exp:* Asst prof obstet & gynec, Univ Tex Southwestern Med Sch, 54-58, clin asst prof, 58-62; pvt pract, 56-62; asst prof, Univ Ark, 62-66; prof obstet & gynec & chmn dept, Univ Tenn, Memphis, 66-75; CHIEF DEPT GYNEC, NACOGDOCHES MED CTR HOSP, 75- *Mem:* Am Col Surg; AMA; Am Col Obstet & Gynec; Am Asn Obstet & Gynec; Sigma Xi. *Res:* Mammalian teratology; infectious diseases in pregnancy. *Mailing Add:* Free Womens Hosp Brookline 53 4800 NE Stallings Dr Suite 115 Nacogdoches TX 75961

FISH, WAYNE WILLIAM, b Helena, Okla, Apr 21, 41; m 62; c 3. BIOCHEMISTRY. *Educ:* Okla State Univ, BS, 63, PhD(biochem), 67. *Prof Exp:* Res assoc, Med Ctr, Duke Univ, 67-68; from asst prof to prof biochem, 78-86, assoc dean res, 83-86, ADJ PROF, MED UNIV SC, 86-; SECT LEADER BIOCHEM, PHILLIPS PETROL CO, 86- *Concurrent Pos:* NIH fel, Duke Univ, 68-70; Swedish Med Res Coun vis scientist fel, Univ Uppsala, 78. *Honors & Awards:* Outstanding chemist, Am Chem Soc, 86. *Mem:* Am Soc Biol Chem; Am Chem Soc. *Res:* Intrinsic physical and chemical properties of biological macromolecules, principally proteins; biodegradable plastics; biomediation; production of food & flavor ingredients from yeast. *Mailing Add:* 255 CPL Phillips Res & Develop Ctr Bartlesville OK 74004

FISH, WILLIAM, b Dania, Fla, Apr 26, 56. AQUATIC CHEMISTRY, ENVIRONMENTAL HYDROGEOCHEMISTRY. *Educ:* Univ Fla, BSE, 79; Mass Inst Technol, PhD(aquatic chem), 84. *Prof Exp:* Postdoctoral fel, 84-86, ASST PROF, ENVIRON CHEM, ORE GRAD INST, 86- *Mem:* AAAS; Am Soc Limnol & Oceanog; Oceanog Soc; Int Humic Substances Soc; Am Geophys Union; Am Chem Soc. *Res:* Transport and chemical cycling of trace and toxic metals in environmental systems; absorption to soils and mineral surfaces; interactions of toxic substances with natural organic matter. *Mailing Add:* Oregon Graduate Inst 19600 NW Von Neumann Dr Beaverton OR 97006-1999

FISH, WILLIAM ARTHUR, b Boston, Mass, Aug 2, 21; m 41; c 7. EMBRYOLOGY. *Educ:* Notre Dame Univ, BSc, 42; Ohio State Univ, MSc, 46, PhD, 48. *Prof Exp:* Asst zool, Ohio State Univ, 46, instr, 46-48; instr comp anat, 48-49, from asst prof to prof comp anat & embryol, 49- , EMER PROF COMP ANAT & EMBRYOL, PROVIDENCE COL. *Mem:* AAAS; NY Acad Sci; Sigma Xi. *Res:* Insect embryology; metabolism of thyroid in rats and of cholesterol in normal, neoplastic and embryonic tissue. *Mailing Add:* 168 Garden City Dr Cranston RI 02920

FISHBACK, WILLIAM THOMPSON, b Milwaukee, Wis, Jan 28, 22; m 60; c 1. MATHEMATICS. *Educ:* Oberlin Col, AB, 43; Harvard Univ, AM, 47, PhD(math), 52. *Prof Exp:* Mem staff, Radiation Lab, Mass Inst Technol, 43-46; from instr to asst prof math, Univ Vt, 50-53; from asst prof to prof, Ohio Univ, 53-66; prof math, 66-87, EMER PROF MATH, EARLHAM COL, 87- *Concurrent Pos:* Vis prof, New Paltz State Teachers Col, 54; vis lectr, Clark Univ, 59, Harvard Univ, 60 & Denison Univ, 64. *Mem:* Am Math Soc; Math Asn Am. *Res:* Mathematical education; geometry. *Mailing Add:* Dept 06184765x Earlham Col Richmond IN 47374-4095

FISHBEIN, EILEEN GREIF, b Baltimore, Md, April 17, 36; m 56; c 3. NURSING, EDUCATION. *Educ:* Univ Md, BS, 58, MS, 59; Cath Univ Am, DNSc, 82. *Prof Exp:* Mem staff, Obstet Nursing, Univ Md Hosp, 58-59 & in-serv educ instr, 61-62; instr nursing, Sch Nursing, Cath Univ Am, 71-73, asst prof, 73-78; asst prof nursing, Sch Nursing, Univ Md, 81-85; ASST PROF GEORGETOWN UNIV SCH NURSING, WASHINGTON, DC, 85- *Concurrent Pos:* Nat res serv award, Dept Educ & Welfare, 78-81; consult, Asn Am Med Cols, 77; guest worker, NIH, 78; consult, Vet Admin, 90. *Mem:* Nat League Nursing; Nurses Asn Am Col Obstet & Gynec; Am Asn Univ Prof. *Res:* Paternal parenting behavior by expectant couples and its relationship to attitudes toward womens roles and anxiety in the prospective father; exercise and pregnancy; women's health. *Mailing Add:* Georgetown Univ Sch Nursing 3700 Resevoir Rd NW Washington DC 20007

FISHBEIN, LAWRENCE, organic chemistry, biochemistry, for more information see previous edition

FISHBEIN, WILLIAM, b New York, NY. RADAR. *Educ:* City Col NY, BEE, 48; Rutgers Univ, MS, 55. *Prof Exp:* Supvr electronics eng, EW/RETA Ctr, US Army Commun & Electronics command, 64-74, dir, Radar Div, 74-81, assoc dir, Combat Surveillance & Target Acquistion Lab, US Army Electronics Res &Develop Command, 81-85, chief scientist, 85-87; PRIN RES ENG, GA TECH RES INST, 87- *Mem:* Sr mem Inst Elec & Electronics Engrs. *Res:* Development of radar and indentification equipments analysis of radar target signatures; design of signal processing systems for moving target detection radars; design of microwave antennas. *Mailing Add:* Csta Lab US Aerodcom Ft Monmouth NJ 07703

FISHBEIN, WILLIAM NICHOLS, b Baltimore, Md, July 21, 33; m 56; c 3. METABOLIC MYOPATHY, ENZYME DEFICIENCIES. *Educ:* Johns Hopkins Univ, BA, 53, MD, 57; Univ Md, Baltimore, PhD(biochem), 66. *Prof Exp:* From intern to resident med, Univ Hosp, Baltimore, Md, 57-60; clin assoc, Nat Cancer Inst, 62-64; instr biochem, Med Sch, Univ Md, Baltimore, 64-65; CHIEF BIOCHEM DIV, ARMED FORCES INST PATH, 65- *Concurrent Pos:* NIH trainee neurochem, Pediat Res Lab, Univ Hosp, Baltimore, Md, 60-62, NIH spec fel neurochem, 64-65; res grants cryobiol & hepatic coma, US Army Med Res & Develop Command, Washington, DC, 66-72; res grant urease, Nat Inst Arthritis, Metab & Digestive Dis, 67-81; Am Cancer Soc res grant cryobiol, 74-80; scientist assoc, Univs Assoc Res & Educ in Path, Md, 67-81; res grant adenylate deaminase, Muscular Dystrophy Asn, 79-82. *Honors & Awards:* A Cressy Morrison Award, NY Acad Sci, 68. *Mem:* Am Soc Biol Chem; Am Asn Path; Am Fedn Clin Res. *Res:* Enzyme structure and function; urease; cryobiology; hepatic coma; adenylate deaminase deficiency; hydroxamic acids; pathology. *Mailing Add:* Biochem Div Armed Forces Inst Path 14th NW & Alaska Ave Rm 3001 Washington DC 20306-6000

FISHBONE, LESLIE GARY, b Elizabeth, NJ, Feb 20, 46; m 77; c 2. NUCLEAR SAFEGUARDS, ENERGY ANALYSIS. *Educ:* Calif Inst Technol, BS, 68; Univ Md, PhD(physics), 72. *Prof Exp:* Fel physics, Univ Md, 72-73; exch fel physics, Landau Inst Theoret Physics, Acad Sci USSR, 73; res assoc & assoc instr physics, Univ Utah, 74-76; res assoc, Energy & Environ, Utah State Adv Coun Sci & Technol, 76-77; asst scientist energy & environ, 77-79, assoc scientist, 79-83, SCIENTIST, NUCLEAR ENERGY, BROOKHAVEN NAT LAB, 83 - *Mem:* Am Phys Soc; AAAS; Sigma Xi; Inst Nuclear Mat Mgt. *Res:* Research and development in support of domestic and international nuclear materials safeguards, computer modeling studies of national energy supply and demand systems; also black hole astrophysics. *Mailing Add:* Bldg 197C Brookhaven Nat Lab Upton Upton NY 11973

FISHBURNE, EDWARD STOKES, III, b Charleston, SC, Feb 8, 36; m 56; c 2. MOLECULAR PHYSICS, AERONAUTICAL ENGINEERING. *Educ:* The Citadel, BS, 57; Ohio State Univ, PhD(aeronaut & astronaut eng), 63; Hofstra Univ, MBA, 72. *Prof Exp:* Scientist, Crosley Div, Avco Corp, Ohio, 57 & Booz-Allen Appl Res Inc, 57-59; res assoc gas dynamics, Rocket Res Lab, Ohio State Univ, 59-63, asst supvr rocket propulsion, 63-67; asst prof aeronaut & astronaut eng, Univ, 63-67; res scientist, Grumman Aerospace Corp, 67-68, head aerophys sect, Res Dept, 68-73; vpres, Aeronaut Res Assocs Princeton, Inc, 73-79; PRES, SCI TECHNOL, INC, 79- *Mem:* AAAS; Am Phys Soc; Combustion Inst. *Res:* High temperature chemistry; energy exchange between molecules; spectroscopy; chemical physics; fluid dynamics; combustion. *Mailing Add:* 811 Ocean Ave Spring Lake NJ 07762

FISHEL, DERRY LEE, b Findlay, Ohio, July 15, 29; m 67; c 2. ORGANIC CHEMISTRY. *Educ:* Bowling Green State Univ, BA, 52; Ohio State Univ, PhD(org chem), 59. *Prof Exp:* Asst prof chem, SDak Univ, 58-60; from asst prof to assoc prof, 60-73, PROF CHEM, KENT STATE UNIV, 73- *Concurrent Pos:* Consult, D & S Consult, 75- *Mem:* Sigma Xi; Am Soc Mass Spectrometry; Am Chem Soc. *Res:* Syntheses and properties of oligomeric liquid crystals; particle induced desorption mass spectrometry; syntheses and properties of thermotropic mesogens; chemistry and mechanism of C-N rearrangement reactions. *Mailing Add:* 257 Woodridge Dr Tallmadge OH 44278

FISHEL, JOHN B, b Hagerstown, Md, Sept 24, 14; wid; c 2. ORGANIC CHEMISTRY. *Educ:* Lehigh Univ, BS, 36; Ohio State Univ, MS, 48. *Prof Exp:* Chemist, Neville Co, 38-39 & Gulf Res & Develop Co, 39-42; prin chemist, Battelle Mem Inst, 48-56; org res chemist, Res Lab, Gen Cigar & Tobacco Co 56-64, res mgr, 64-69, mgr res & develop projs, 69-79, consult, 80-84; RETIRED. *Mem:* Am Chem Soc; Sigma Xi. *Res:* Chemistry of tobacco and tobacco smoke. *Mailing Add:* 334 Valleybrook Dr Lancaster PA 17601-4633

FISHER, ARTHUR DOUGLAS, b Philadelphia, Pa, June 18, 52. OPTICAL COMPUTING, ARTIFICIAL NEURAL NETWORKS. *Educ:* Carnegie-Mellon Univ, BS(elec eng) & BS(physics), 74; Mass Inst Technol, SM, 76, PhD(elec eng & computer sci), 81. *Prof Exp:* Res asst, Res Lab Electronics, Mass Inst Technol, 74-77 & Ctr Mat Sci & Eng, 77-81; RES PHYSICIST, NAVAL RES LAB, 81- *Concurrent Pos:* Panelist, Neural Network Study, Defense Advan Res Proj Agency, 87-88; lectr, Int Comn Optics, 88; mem, Tech Prog Comt Topical Mkt Optical Comput, Optical Soc Am, 88-89 & Tech Prog Comt Topical Mkt Spatial Light Modulators, 89-90, chmn, 91-92; co-ed, Spec Issue Appl Optics, Optical Soc Am, 89-91. *Mem:* Sr mem Inst Elec & Electronics Engrs; Optical Soc Am; Sigma Xi. *Res:* Algorithms, architectures, and devices for optical information processing and optical computing; spatial light modulators including magnetostatic wave optical Bragg modulators, photo-emitter membrane light modulator, microchannel spatial light modulator; optical neural networks; high-resolution adaptive optical phase estimation and compensation; electromagnetic intensity propagation in inhomogeneous media. *Mailing Add:* 500 S Wayne St Arlington VA 22204

FISHER, BEN, b Chicago, Ill, Jan 8, 24. HEMATOLOGY. *Educ:* Univ Ill, BS, 46, Col Med, MD, 48. *Prof Exp:* Instr med, Sch Med, Tulane Univ, 49-50; instr path, Col Med, Univ Ill, 50-53; assoc med to assoc prof med, Sch Med, State Univ NY, Buffalo, 59-77; assoc dean & prof med & path, Chicago Med Sch, Univ Health Sci, 77-83; chief staff, Vet Admin Med Ctr, N Chicago, 77-83, STAFF PHYSICIAN, VET ADMIN WESTSIDE MED CTR, CHICAGO, 83-; PROF PATH, COL MED, UNIV ILL, 83- *Concurrent Pos:* Resident clin path, Marine Hosp, US Pub Health Serv, New Orleans, 49-50; fel path, Michael Reese Hosp, Chicago, 50-51 & asst dir hemat, 51-53; clin pathologist, Wayne Co Gen Hosp, 53-54; res hematologist, Roswell Park Mem Inst, 54-56; dir clin labs, Deaconess Hosp, Buffalo, NY, 59-68; chief hematol, Vet Admin Hosp, Buffalo, NY, 68-77. *Honors & Awards:* Borden Found Award, 48. *Mem:* Sigma Xi; Am Col Physicians; Col Am Pathologists; Int Soc Hemat; Am Soc Hemat. *Res:* Hereditary hemoglobinopathies. *Mailing Add:* 1112 Garfield Ave Libertyville IL 60048

FISHER, BERNARD, b Pittsburgh, Pa, Aug 23, 18; m 48; c 3. SURGERY. *Educ:* Univ Pittsburgh, BS, 40, MD, 43; Am Bd Surg, dipl, 52. *Hon Degrees:* DSc, Mt Sinai Med Sch Med, City Univ NY. *Prof Exp:* From asst instr to instr surg, Univ Pittsburgh, 47-53, asst prof surg & assoc dir Gibson Lab, 53-55, assoc prof, 56-59, dir lab surg res, 56-74, dir oncol, 74-78, dir Breast Cancer Detection Ctr, 75-85, prof surg, 59-86, DISTINGUISHED SERV PROF, UNIV PITTSBURGH, 86-, SR SCI ADV CLIN AFFAIRS-PC, 91- *Concurrent Pos:* Fel, Univ Pa, 50-52; Markle scholar, 53-58; chmn, Nat Surg Adj Breast & Bowel Proj, 67-; consult, Nat Cancer Inst; Pres Cancer Panel, 79-82; chmn, Coop Group Chmn, 85-89; Nat Cancer Adv Bd, 86-92; ICI Professorship Surg, Univ Pittsburgh. *Honors & Awards:* Karnofsky Award, Am Soc Clin Oncol, 80; Lucy Wortham James Clin Res Award, Soc Surg Oncol, 81; Albert Lasker Res Award, 85; Medal of Honor, Am Cancer Soc, 86; Hammer Cancer Prize, 88; Susan Komen Award Sci Distinction, 88; Cancer Res Award, Milken Family Med Found, 89; Bernard Fisher Lectr, Univ Pittsburgh. *Mem:* Nat Acad Sci; Soc Univ Surg; Asn Community Cancer Ctrs; AMA; fel Am Col Surg; fel AAAS; Am Soc Clin Oncol; Am Surg Asn; Soc Surg Oncol; Am Asn Cancer Res; Nat Asn Black County Officials. *Res:* Clinical trials; 456 publications. *Mailing Add:* Univ Pittsburgh 3500 Terrace St Rm 914 Pittsburgh PA 15261

FISHER, C(HARLES) PAGE, JR, b Richmond, Va, Sept 24, 21; m 72; c 2. CIVIL ENGINEERING. *Educ:* Univ Va, BSCE, 49; Harvard Univ, SM, 50; NC State Univ, PhD(civil eng, physics), 62. *Prof Exp:* Rodman, Va Hwy Dept, 39-41; land surv, W W La Prade & Bros, 45-46; engr, Metcalf & Eddy, 51; found engr, Robertson & Assocs, 52-53 & H S Porter, 53-55; from instr to assoc prof civil eng, NC State Univ, 55-69; CONSULT ENGR, 78- *Concurrent Pos:* Secy corp, Troxler Electronics Labs, Inc, 62-; chmn bd & pres, Geotech Eng Co, 63-78; prin, Gardner-Kline Assoc, 67-71. *Mem:* Fel Am Soc Civil Engrs; Am Soc Testing & Mat; fel Am Consult Eng Coun; Sigma Xi. *Res:* Measurement of the properties of earth materials in situ and in the laboratory; application of modern physics to civil engineering problems; quality control of engineering materials. *Mailing Add:* One Stoneridge Circle Durham NC 27705

FISHER, CHARLES HAROLD, b Hiawatha, WVa, Nov 20, 06; m 33, 68. ORGANIC CHEMISTRY. *Educ:* Roanoke Col, BS, 28; Univ Ill, MS, 29, PhD(org chem), 32. *Hon Degrees:* DSc, Tulane Univ, 53 & Roanoke Col, 63. *Prof Exp:* Instr org chem, Harvard Univ, 32-35; res group leader, US Bur Mines, 35-40; res group leader, Eastern Regional Ctr, Philadelphia, USDA, 40-50, Southern Utilization Res Div, New Orleans, 50-72; ADJ RES PROF, ROANOKE COL, SALEM, VA, 72- *Concurrent Pos:* Mem comt chem utilization of coal, Nat Res Coun, 39-41; chmn bd, Am Inst Chemists, 63 & 73-75; consult, textile res, Repub SAfrica, 67, food technol, Pan Am Union, 68 & paper technol, Libr Cong, 73-76; mem bd, Am Chem Soc, 69-71. *Honors & Awards:* Herty Award, 59; Chem Pioneer Award, Am Inst Chemists, 66. *Mem:* AAAS; Am Chem Soc; Am Asn Textile Chemists & Colorists; hon mem Am Inst Chemists (pres, 62-63); Royal Soc Chem; Am Inst Chem Engrs. *Res:* Cotton textiles; cellulose; chemistry and utilization of farm crops; polymers; organic synthesis. *Mailing Add:* Dept Chem Roanoke Col Salem VA 24153-3794

FISHER, CLARK ALAN, b Torrance, Calif, Sept 5, 33; m 61; c 1. MICROBIOLOGY, IMMUNOLOGY. *Educ:* Ore State Col, BS(pharm) & BS(bact), 60; Univ Calif, Berkeley, MA, 64, PhD(immunol), 69. *Prof Exp:* Asst res scientist microbiol, NY Univ Sch Med, 70-72; instr, 72-75, ASST PROF MICROBIOL, NEW YORK MED COL, 75- *Mem:* Am Chem Soc; Am Soc Microbiol; Int Leprosy Asn; NY Acad Sci; Sigma Xi. *Res:* Immunochemistry and biological properties of envelope antigens of anaerobic corynebacteria; pathogenesis of persistent bacterial infections. *Mailing Add:* PO Box 897 Philmont NY 12565-0897

FISHER, CLETUS G, b Canton, Ohio, Sept 22, 22; m 46; c 2. SPEECH PATHOLOGY, AUDIOLOGY. *Educ:* Kent State Univ, BS, 49; Univ Iowa, MA, 50; Ohio State Univ, PhD(audiol), 63. *Prof Exp:* Instr speech, Kans State Col, 50-53; exec dir, Hearing & Speech Ctr, Dayton, Ohio, 53-60; instr speech, Ohio State Univ, 62-63; asst prof speech path & audiol, Univ Iowa, 63-68; dir speech & hearing div, Dept Speech, Northern Ill Univ, 68-69, assoc prof to prof speech, 69-87, chmn dept commun dis, 74-87; RETIRED. *Mem:* Acad Rehab Audiol; Am Speech & Hearing Asn; Sigma Xi. *Res:* Visual and aural perception of oral signals, including those with normal or with pathological hearing. *Mailing Add:* 107 Cynthia Pl De Kalb IL 60115

FISHER, DALE JOHN, b Omro, Wis, June 4, 25; m 57; c 1. INSTRUMENTATION & MEDICAL DIAGNOSTIC DEVICES. *Educ:* Univ Wis-Oshkosh, BS, 47; Ind Univ, PhD(chem), 51. *Prof Exp:* Chemist, Ionic Anal Group, Oak Ridge Nat Lab, 51-52, group leader anal instrumentation, Anal Chem Div, 52-72, dirs staff, 72-73; physicist, Vet Admin Hosp, 73-74, tech dir nuclear med, 74-76; physicist, Bur Med Devices, Silver Spring, 76-83, PHYSICIST, DIV LIFE SCIS, OFF SCI & TECH, CTR FOR DEVICES & RADIOL HEALTH, FDA, ROCKVILLE, 83- *Concurrent Pos:* All univ fel, Ind Univ, 50-51. *Honors & Awards:* Am Chem Soc Award, 69. *Mem:* Am Chem Soc; Sigma Xi; Am Soc Testing & Mat. *Res:* Design and new applications of instruments and methods for chemical analysis and research; research with computer-based nuclear medicine instrumentation to obtain better diagnostic information for patient care; concerns with clinical laboratory instrument systems including hazards, problems, standards and performance requirements; research to evaluate medical device safety and effectiveness; toxicology; simulation models for device performance. *Mailing Add:* 6319 Golden Hook Columbia MD 21044

FISHER, DARRELL R, b Salt Lake City, Utah, Feb 12, 51; M; c 6. MEDICAL & HEALTH PHYSICS, RADIATION BIOLOGY. *Educ:* Univ Utah, BA, 75; Univ Fla, MS, 76, PhD(nuclear eng sci), 78. *Prof Exp:* SR RES SCIENTIST, DEPT HEALTH PHYSICS, BATTELLE MEM INST, PAC NORTHWEST LABS, RICHLAND, WASH, 78-; CONSULT, DIV NUCLEAR MED, UNIV HOSP, SEATTLE, 85- & FRED HUTCHINSON CANCER RES CTR, 86- *Concurrent Pos:* Assoc newslett ed, Health Physics

Soc, 81-86, ed, J Health Physics, 85-; affil asst prof radiol sci, Univ Wash Tri-Cities Univ Ctr, Richland, 85-; adj asst prof nuclear eng, Tex A&M Univ, Col Sta, 89-; mem bd dirs, Health Physics Soc. *Honors & Awards:* Elda Anderson Award, Health Physics Soc, 86. *Mem:* Radiation Res Soc; Health Physics Soc. *Res:* Dosimetry and health effects of internally deposited radionuclides; biological effects of radiation; internal microdosimetry; standards for radiation protection; medical internal dosimetry. *Mailing Add:* Battelle Mem Inst Pac Northwest Labs Battelle Blvd PO Box 999 Richland WA 99352

FISHER, DAVID E, b Philadelphia, Pa, June 22, 32; m 54; c 3. GEOCHRONOLOGY, METEORITICS. *Educ:* Trinity Col, BS, 54; Univ Fla, PhD(chem physics), 58. *Prof Exp:* Oak Ridge Inst Nuclear Studies fel, Oak Ridge Nat Lab, 57-58; res assoc, Brookhaven Nat Lab, 58-60; asst prof eng physics, Cornell Univ, 60-66; assoc prof, 66-70, PROF MARINE & GEOPHYS, INST MARINE SCI, UNIV MIAMI, 70- *Mem:* AAAS; Meteoretical Soc; Am Phys Soc; Am Chem Soc; Am Geophys Union. *Res:* Nuclear cosmochronology; cosmic-ray nuclear reactions in meteorites; activation analysis; isotopic abundances in meteorites and tektites; cosmic chemistry; marine geology and geochemistry; geochronology. *Mailing Add:* Inst Marine Sci Univ Miami Miami FL 33149

FISHER, DELBERT A, b Placerville, Calif, Aug 12, 28; m 51; c 3. MEDICINE, PEDIATRICS. *Educ:* Univ Calif, AB, 50, MD, 53; Am Bd Pediat, dipl 59, & cert, pediat endocrinol,, 78. *Prof Exp:* Instr & res assoc pediat, endocrinol & metab, Sch Med, Univ Ore, 57-60; from asst prof to prof pediat, Sch Med, Univ Ark, 60-68, asst dir clin study ctr, 62-68; prof pediat, Univ Calif, Los Angeles, 68-73, prof pediat & med, 73-91, chmn, Dept Pediat, Univ Calif Los Angeles Harbor Med Ctr, 85-89, EMER PROF PEDIAT & MED, UNIV CALIF, LOS ANGELES, 91-; SR SCIENTIST, WALTER P MARTIN RES CTR, 91- *Concurrent Pos:* USPHS career develop award, 64-68; examr, Am Bd Pediat, 70-; ed, J Clin Endocrinol Metab, 78-83, Pediat Res, 84-; res prof develop biol, Harbor Univ Calif, Los Angeles Med Ctr, 75-85; dir, Walter P Martin Res Ctr, 86-91; pres, Nichols Inst Reference Labs, San Juan Capistrano, Calif, 91- *Honors & Awards:* Nutrit Res Award, Am Acad Pediat, 81; Park Davis Award, Am Thyroid Asn. *Mem:* Inst Med-Nat Acad Sci; Soc Pediat Res (vpres, 73-74); Am Soc Clin Invest; Lawson Wilkins Pediat Endocrine Soc (pres, 82-83); Am Acad Pediat; Am Pediat Soc (pres-elect, 91-92); Sigma Xi; Endocrine Soc (pres, 83-84); Am Thyroid Asn (pres, 88); Asn Am Physicians. *Res:* Pediatric endocrinology and metabolism; thyroid disease; maternal-fetal metabolism and endocrinology. *Mailing Add:* Four Pear Tree Lane Rolling Hills Estates CA 90274

FISHER, DON LOWELL, b Salt Lake City, Utah, June 14, 41; m 67; c 4. EMBRYOLOGY, TERATOLOGY. *Educ:* Brigham Young Univ, BS, 66, MS, 68; Univ Minn, PhD(anat), 71. *Prof Exp:* Instr histol, Univ Mich, 71-72; ASST PROF GROSS ANAT, UNIV MINN, ANN ARBOR, 72-, PROF EMBRYOL, 78- *Concurrent Pos:* Consult embryol, Eastern Mich Univ, 81 & Oakland Univ, 81 & 82. *Mem:* Teratology Soc; Am Asn Anatomists; Soc Study Reproduction; Sigma Xi. *Res:* Early experimental embryology and teratology. *Mailing Add:* Dept Anat 4728 Med Sci II Univ Mich Ann Arbor MI 48104

FISHER, DONALD B, b Philadelphia, Pa, Oct 14, 35; m 63. BOTANY. *Educ:* Univ Wash, BS, 57; Univ Wis, MS, 61; Iowa State Univ, PhD(biochem), 65. *Prof Exp:* Asst bot, Univ Wis, 59-61 & biochem, Iowa State Univ, 61-63; NIH fel bot, Univ Calif, Berkeley, 65-67, res botanist, 67-68; from asst prof to assoc prof, Univ Ga, 68-78; assoc prof, 78-80, PROF BOT, WASH STATE UNIV, 80- *Mem:* AAAS; Am Soc Plant Physiol; Crop Sci Soc Am. *Res:* Transport of materials in plants. *Mailing Add:* Dept Bot Wash State Univ Pullman WA 99164-4238

FISHER, DONALD D, b Spokane, Wash, Dec 20, 29; m 51; c 3. INFORMATION PROCESSING, COMPUTER SCIENCE EDUCATION. *Educ:* Wash State Univ, BA, 51, MA, 53; Stanford Univ, PhD(math), 62. *Prof Exp:* Asst math, State Col Wash, 51-53; mathematician, Douglas Aircraft Co, Inc, 53-54; asst math, Stanford Univ, 54-57, actg instr, 57-58; appl sci rep, Int Bus Mach Corp, 58-60, mathematician, 60-62; res assoc comput sci, Stanford Univ, 62-65; dir res, Comput Ctr & assoc prof prev med & math, Ind Univ, 65-69; prof & head comput & info sci dept, 69-73, dir, Sch Math Sci, 73-81, head, comput & info sci dept, 81-85, regents serv prof, 85-88, EMER REGENTS SERV PROF, OKLA STATE UNIV, 88- *Concurrent Pos:* Vis mem tech staff, Bell Labs, 83; Fulbright scholar, Univ Ioannina, Greece, 84; vis fac mem, IBM Sci Ctr, Palo Alto, Calif, 84; gen chair workshop on appl comput, 86; WAC, 89. *Mem:* Asn Comput Mach; Soc Indust & Appl Math; Sigma Xi; Inst Elec & Electronics Engrs. *Res:* Computer architecture; data flow architecture; data structures; numerical solution of partial differential equations - particulary free boundary problems; computer-based scheduling and simulation; data structures, data bases, communications network, CAD, distributed systems, computer architecture, super computing. *Mailing Add:* Dept Comput & Info Sci 219 Math Sci Bldg Okla State Univ Stillwater OK 74078

FISHER, DONALD WILLIAM, b Schenectady, NY, Sept 8, 22; m 55; c 2. PALEONTOLOGY. *Educ:* Univ Buffalo, AB, 44, AM, 48; Univ Rochester, PhD(geol), 52. *Prof Exp:* Instr geol, Union Col, 49-51, asst prof, 52-53; sr paleontologist & stratigrapher, 53-55, STATE PALEONTOLOGIST, NY STATE MUS, STATE SCI SERV NY, 55- *Concurrent Pos:* Field geologist, State Geol Surv, NY, 47-53. *Mem:* AAAS; fel Geol Soc Am; Int Paleont Union; Paleont Soc; Geol Asn Can. *Res:* Cambrian, Ordovician and Silurian stratigraphy and paleontology of New York; Taconic geology; tentaculitids; New York Pleistocene mammals; history of New York geology and paleontology. *Mailing Add:* One Lindenwald Ct Kinderhook NY 12106

FISHER, E(ARL) EUGENE, b Monongahela, Pa, Dec 18, 22; m 44; c 4. ORGANIC CHEMISTRY, RESEARCH ADMINISTRATION. *Educ:* Washington & Jefferson Col, BS, 43; Ohio State Univ, MS, 44; Carnegie Inst Technol, PhD(chem), 48. *Prof Exp:* Asst, Ohio State Univ, 43-44; asst, Carnegie Inst Technol, 46-47, res assoc, 47-48; res chemist org chem, E I du Pont de Nemours & Co, Inc, 48-52, res chemist fabrics & finishes dept, 52-58; head explor sect, Chem Dept, A E Staley Mfg Co, 58-61, dir chem res, 61-69, dir food prod res & develop, 69-70; dir, Res & Develop, William Wrigley Jr Co, 70-73, vpres, 73-88; RETIRED. *Mem:* AAAS; Am Chem Soc; Sigma Xi; Inst Food Technologists; Am Soc Qual Control. *Res:* Organic syntheses; carbohydrates; polymers; analytical methods. *Mailing Add:* 19 Harbormist Circle Berlin MD 21811

FISHER, EDWARD, b Boston, Mass, Sept 3, 13; m 57. MATHEMATICAL PHYSICS. *Educ:* Mass Inst Technol, BS, 33; Cornell Univ, PhD(theoret physics), 45. *Prof Exp:* Instr physics, Univ Md, 42-43; instr, Cornell Univ, 43-45; asst prof, Univ Wyo, 46-48; assoc prof, Mo Sch Mines, 48-52; physicist, Gen Precision Lab, 52-53, consult, 53-54; physicist, Atomic Power Develop Assocs, 54-56; staff scientist, Lockheed Aircraft Corp, 57-59; prin scientist, United Tech Ctr, 59-63; PROF PHYSICS, STATE UNIV NY COL OSWEGO, 63- *Mem:* Am Phys Soc. *Res:* Concepts in wave mechanics; electrical rocket propulsion; nuclear reactor shielding; noise theory; calculation of the fine-structure constant. *Mailing Add:* 59 Roycroft Blvd Snyder NY 14226

FISHER, EDWARD ALLEN, b New York, NY, Nov 5, 50; m 85; c 3. PEDIATRIC LIPID METABOLISM. *Educ:* State Univ NY, BA, 71; NY Univ, MD, 75; Univ NC, MPH, 78; Mass Inst Technol, PhD(biochem & nutrit), 82. *Prof Exp:* Resident pediat, Duke Univ, 75-77; fel nutrit & metab, Boston Children's Hosp, 78-81; fel molecular biol, NIH, 81-84; asst prof pediat, Univ Pa, 84-87; asst prof, 87-91, ASSOC PROF BIOCHEM, MED COL PA, 91- *Concurrent Pos:* Assoc prof pediat, Med Col Pa, 91-; co-dir, Pediat Lipid Dis Clin, 91-05307752x; mem, Coun Arteriosclerosis, Am Heart Asn. *Mem:* AAAS; fel Am Heart Asn; Am Inst Nutrit; Am Soc Microbiol. *Res:* Genetic factors affecting lipoprotein metabolism; identification of causes of atherosclerosis. *Mailing Add:* Dept Biochem Med Col Pa 3300 Henry Ave Philadelphia PA 19129

FISHER, EDWARD RICHARD, b Detroit, Mich, Mar 24, 38; m 73; c 5. CHEMICAL PHYSICS, ENGINEERING. *Educ:* Univ Calif, Berkeley, BSc, 61; Johns Hopkins Univ, PhD(chem eng sci), 65. *Prof Exp:* Res chemist, Lawrence Radiation Lab, 61; asst prof chem, Univ Copenhagen, 65-66; phys chemist, Space Sci Lab, Gen Elec Co, Pa, 66-68; assoc prof chem eng, Wayne State Univ, 68-74, prof, 74-; dir, Res Inst Eng Sci, 77-; PROF & HEAD CHEM & CHEM ENG, MICH TECH UNIV, HOUGHTON. *Concurrent Pos:* Lectr, Univ Copenhagen, 66 & Gen Elec Grad Prog, Rensselaer Polytech Inst, 67. *Mem:* Am Inst Physics; Am Inst Chem Engrs; Sigma Xi. *Res:* Experimental and theoretical study of the details of energy transfer mechanisms in molecular collisions; plasma chemistry and molecular laser studies. *Mailing Add:* Prof & Head Chem & Chem Eng Mich Tech Univ Houghton MI 49931

FISHER, EDWARD S, b Minneapolis, Minn, Apr 23, 21; m 53; c 3. PHYSICAL METALLURGY. *Educ:* Univ Minn, BS, 48; Ill Inst Technol, MS, 54. *Prof Exp:* Asst metallurgist, Argonne Nat Lab, 48-55, assoc metallurgist, 55-80, res metallurgist, 80-81; OWNER, MAT CONSULT SERV, 81- *Mem:* Am Inst Mining, Metall & Petrol Engrs; Metall Soc. *Res:* Elastic constants of single crystals, particularly relation between elastic constant changes and phase changes; applications of refractory metals and precious metals in glass industry; forensic failure analysis. *Mailing Add:* Mat Consult Serv 2923 Cedar Knoll Ctr Minnetonka MN 55343

FISHER, EDWIN RALPH, b Pittsburgh, Pa, Sept 2, 23; m 53. PATHOLOGY. *Educ:* Univ Pittsburgh, BS, 45, MD, 47. *Prof Exp:* Assoc pathologist, Cleveland Clin, 52, 54 & USPHS, 52-54; dir labs, Shadyside Hosp, 70-; assoc prof, 54-58, PROF PATH, UNIV PITTSBURGH, 58- *Concurrent Pos:* Chief lab serv, Vet Admin Hosp, 54-70. *Honors & Awards:* Parke-Davis Award, 63. *Mem:* Am Soc Clin Pathologists; Am Soc Exp Pathologists; Am Asn Pathologists & Bacteriologists; Col Am Pathologists; Int Acad Path. *Res:* Histochemistry; pathology of gastrointestinal tract and thyroid. *Mailing Add:* 5812 Solway St Pittsburgh PA 15217

FISHER, ELLSWORTH HENRY, b Ottawa, Kans, Dec 30, 11; m 37; c 2. ENTOMOLOGY. *Educ:* Okla ECent State Teachers Col, BS, 34; Okla Agr & Mech Col, MS, 39; Univ Wis, Madison, PhD(entom), 48. *Prof Exp:* Entomologist & crop prod specialist, Stokely Van Camp, Indianapolis, Ind, 41-45; from asst prof to prof entom, Univ Wis-Madison, 45-78, coordr pesticide use educ, 65-78; RETIRED. *Mem:* Am Entom Soc; Entom Soc Am. *Res:* Insects of field crops, man, livestock and household; rats and mice; coordination of educational programs in the safe and economical uses of pesticides. *Mailing Add:* 5212 Manitowoc Pkwy Madison WI 53705

FISHER, ELWOOD, b New Martinsville, WVa, Apr 12, 26; m 56; c 1. PARASITOLOGY, ECOLOGY. *Educ:* Fairmont State Col, BSc, 53; Miami Univ, MSc, 60; Va Polytech Inst, PhD(biol), 67. *Prof Exp:* Instr high sch, Ohio, 53-59; asst prof, Miami Univ, 59-60; from asst prof to assoc prof, 60-77, PROF BIOL, MADISON UNIV, 77- *Mem:* Am Soc Parasitol; Am Pomol Soc. *Res:* Parasites of wildlife; genetics of fruit trees. *Mailing Add:* Dept Biol James Madison Univ Harrisonburg VA 22807

FISHER, FARLEY, b Cleveland, Ohio, Apr 30, 38. CHEMISTRY, CHEMICAL ENGINEERING. *Educ:* Mass Inst Technol, SB, 60; Univ Ill, PhD(chem), 65. *Prof Exp:* Asst prof chem, Tex A&M Univ, 65-69; vis lectr, Bucknell Univ, 69-70; res assoc, Ctr Sci Pub Interest, 71-72; chemist, Environ Protection Agency, 72-73, dir hazard assessment, 73-77; prog mgr chem threats, NSF, 77-79, prog dir kinetics, catalysis & reaction eng, 79-85, PROG DIR KINETICS & CATALYSIS, NSF, 85- *Concurrent Pos:* Res fel, Caltech, 65-66; vis prof chem eng, Univ Minn, 84-85. *Mem:* Am Chem Soc; Am Inst Chem Eng; Mat Res Soc; AAAS; Electrochem Soc. *Res:* Materials processing; catalysis; photochemistry; research management. *Mailing Add:* Div Chem & Therm Sys NSF Washington DC 20550

FISHER, FRANCIS JOHN FULTON, b Roxburgh, NZ, Oct 31, 26; Can citizen; m 49, 80, 82; c 6. EVOLUTIONARY BIOLOGY, PLANT PHYSIOLOGY. *Educ:* Univ Canterbury, BSc, 47, MSc, 49; Univ NZ, PhD(exp taxon), 54. *Prof Exp:* Lectr bot, Univ Melbourne, 54-57; sr lectr, Univ Tasmania, 57-58; prin sci off, Bot Div, NZ Dept Sci & Indust Res, 58-65; assoc prof, 65-67, PROF BIOL SCI, SIMON FRASER UNIV, 67- *Concurrent Pos:* Nuffield Found biol res grant, Australian Nat Univ, 54-55; Brit Coun travel grant, 55-56; Carnegie Corp travel grant, 56; Carnegie Inst Wash res fel, 56-57; dir, Tasmanian Bot Gardens, 57-58; USPHS int fel, 63-64, int follow-up grant, 66-67; biol consult, Sch Social Welfare, Univ Calgary, 67- & Can Environ Sci Ltd, 68-; consult, Environ Res Consults, Ltd, 71- *Mem:* AAAS; Am Inst Biol Sci; Can Bot Asn; Royal Soc NZ; fel Linnaean Soc London; fel Biogeographic Soc Paris. *Res:* Ecophysiology of New Zealand alpine screes; genecology of mountain ranunculus of New Zealand and North and South America; evolution of reproductive isolation in plants; morphogenesis of leaves; adaptive change processes in living systems including human; biological basis of value systems; physiology and biophysics of light induced movements of leaves. *Mailing Add:* Dept Biosci Simon Fraser Univ Burnaby BC V5A 1S6 Can

FISHER, FRANK M, JR, b Louisville, Ky, Oct 16, 31; m 56; c 2. PARASITOLOGY, INVERTEBRATE PHYSIOLOGY. *Educ:* Hanover Col, BA, 53; Purdue Univ, MS, 56, PhD(invert physiol), 61. *Prof Exp:* NIH fel, 61-63, asst prof, 63-69, assoc prof, 69-73, PROF BIOL, RICE UNIV, 73- *Concurrent Pos:* Instr invert zool, Marine Biol Lab, Woods Hole, Mass, 64-70; mem ARB study sect, NIH, 75-, chairperson, 76- *Res:* Chemical basis parasitism; distribution and flux of organic materials in wetland ecosystems; toxic materials in wetland environments. *Mailing Add:* Dept Biol Rice Univ PO Box 1892 Houston TX 77251

FISHER, FRANKLIN A, b Albuquerque, NMex, July 29, 29. HIGH VOLTAGE POWER FIELD. *Educ:* NMex State Univ, BS, 51. *Prof Exp:* Sr engr, High Voltage Lab, Gen Elec Co, 53-78, Elect Utilities Systs Eng Dept, 78-86; SR RES ENGR, LIGHTNING TECHNOLOGIES INC, 86- *Mem:* Fel Inst Elec & Electronics Engrs. *Res:* Lightning effects on air craft; high voltage effects on electrical apparatus. *Mailing Add:* Lightning Technologies Inc Ten Downing Parkway Pittsfield MA 01201

FISHER, FRANKLIN E(UGENE), b Robinson, Ill, Mar 8, 33; m 59; c 2. MECHANICAL ENGINEERING, ENGINEERING MECHANICS. *Educ:* Rose Polytech Inst, BS, 60; Univ Md, College Park, MS, 65, PhD(mech eng), 69. *Prof Exp:* Aerospace engr, Aeronaut Systs Div, Wright-Patterson AFB, Ohio, 60-63; instr mech eng, Univ Md, College Park, 64-69; assoc prof, 69-80, PROF MECH ENG & CHMN, LOYOLA MARYMOUNT UNIV, 80- *Concurrent Pos:* Res mech eng, Naval Res Lab, Washington, DC, 65-66; consult, Ralph M Parson Co, 73-75; mem tech staff, Hughes Aircraft Co, 77- *Mem:* Am Soc Mech Engrs; Am Soc Eng Educ; Soc Automotive Engrs. *Res:* Shock loading and vibration of structures as applied to mechanical engineering design. *Mailing Add:* Mathmedics Corp 1630 Borealis SE RioRancho NM 87124

FISHER, FREDERICK HENDRICK, b Aberdeen, Wash, Dec 30, 26; m 55; c 4. PHYSICS. *Educ:* Univ Wash, BS, 49, PhD(physics), 57. *Prof Exp:* Asst physics, Univ Calif, Los Angeles, 54-55; from grad res physicist to assoc res physicist, 55-68, assoc dir, 75-87, RES OCEANOGRAPHER & LECTR, MARINE PHYS LAB, SCRIPPS INST OCEANOG, UNIV CALIF, SAN DIEGO, 68-, DEPT DIR, 87- *Concurrent Pos:* Res fel acoust, Harvard Univ, 57-58; co-designer & proj scientist res platform FLIP, 60-62; dir res, reverse osmosis, Havens Indust, 63-64; prof & chmn dept physics, Univ RI, 70-71; assoc ed, J Acoust Soc Am, 69-76; mem exec coun, Acoust Soc Am, 76-79; mem gov bd, Am Inst Physics, 85-; ed, Inst Elec & Electronics Engrs J Oceanic Eng, 88- *Mem:* Fel Acoust Soc Am (vpres, 80-81, pres, 83-84); sr mem Inst Elec & Electronics Engrs; Am Phys Soc; Am Chem Soc; Am Geophys Union; Am Meterol Soc. *Res:* Acoustics; physical chemistry; underwater sound; oceanography; how acoustic propagation in the ocean is affected by oceanographic environmental fluctuations and chemical ion-speciation, especially at elevated pressures. *Mailing Add:* Marine Phys Lab Scripps Inst Oceanog Univ Calif LaJolla CA 92093-0701

FISHER, FREDERICK STEPHEN, b Highland Park, Mich, Aug 5, 37; m 61; c 2. GEOLOGY, GEOCHEMISTRY. *Educ:* Wayne State Univ, BS, 61, MS, 62; Univ Wyo, PhD(geol), 66. *Prof Exp:* Asst prof geol, Rocky Mountain Col, 75-78; GEOLOGIST, US GEOL SURV, 66- *Mem:* Geol Soc Am; Soc Econ Geologists; Am Inst Prof Geologists. *Res:* Geology and geochemistry of hydrothermal deposits. *Mailing Add:* 9920 E War Bonnet Lane Tucson AZ 85749

FISHER, GAIL FEIMSTER, b Washington, DC, Sept 18, 28; c 2. APPLIED STATISTICS, DEMOGRAPHY. *Educ:* Univ Md, BA, 49, MA, 51; Univ NC, PhD, 77. *Prof Exp:* Analyst health, Air Pollution Med Prog, Dept Health, Educ & Welfare, 56-58, analyst, Accident Prev Prog, 58-60, chief statistician, Diabetes & Arthritis Prog, 60-66, eval officer, Bur Health Servs, 66-68, dir off prog planning & eval, 68-73, assoc dir coop health statist prog, 73-81, ASSOC DIR PROG PLANNING, EVAL & COORD, NAT CTR HEALTH STATIST, HEW, 81- *Concurrent Pos:* Mem staff, President's Comn Health Care Facil, 67-68. *Res:* Health research; health statistics. *Mailing Add:* 2696 Cardinal Ridge Rd Charlottsville VA 22901

FISHER, GALEN BRUCE, b Bethesda, Md, Jan 14, 45; m 76; c 2. CATALYSIS, SURFACE SCIENCE. *Educ:* Pomona Col, BA, 66; Stanford Univ, MS, 67, PhD(solid state physics), 74. *Prof Exp:* Res assoc, Div Eng, Brown Univ, 72-75; physicist, Surface Sci Div, Nat Bur Standards, 75-78; sr res scientist, Gen Motors Res Labs, 78-82, staff res scientist, 82-85, sect mgr surface chem, phys chem dept, 83-90, SR STAFF RES SCIENTIST, GEN MOTORS RES LABS, 85-, SECT MGR, CATALYIC CHEM, PHYS CHEM DEPT, 90- *Concurrent Pos:* Lectr, NATO Advan Study Inst, 74; vis asst prof, Physics Dept, Brown Univ, 75; dir, Am Vacuum Soc, 86-87. *Mem:* Am Vacuum Soc; Am Phys Soc; Am Chem Soc; Catalysis Soc; Mat Res Soc. *Res:* Heterogeneous catalysis, surface science and chemisorption; the bonding, electronic structure, and vibrations of adsorbates on metal surfaces. *Mailing Add:* Phys Chem Dept Gen Motors Res Labs Warren MI 48090-9055

FISHER, GENE JORDAN, b Quitman, Miss, March 26, 31; m 54; c 2. ORGANIC CHEMISTRY. *Educ:* Univ Tex, BS, 52. *Prof Exp:* From res chemist to sr res chemist, Celanese Chem Co, 52-59, group leader, 59-67, res mgr, 67-77, dir res, 77-83, tech dir, 83-85; RETIRED. *Mem:* Am Chem Soc; fel Am Inst Chemists. *Res:* Reactions of ketene; reactions of formaldehyde; synthesis of monomers; heterogeneous catalysis; research management. *Mailing Add:* 417 Ashland Dr Corpus Christi TX 78412-2327

FISHER, GEORGE A, JR, b North Girard, Pa, Apr 9, 12. MATERIALS SCIENCE ENGINEERING. *Educ:* Purdue Univ, BS, 35. *Prof Exp:* Dist mgr, Inco, Inc, 45-74; pres & consult, Mapem Co, 74-82; RETIRED. *Mem:* Fel Am Soc Metals; Nat Asn Corrosion Engrs; Am Inst Mining Metall & Petrol Engrs. *Mailing Add:* 1987 W Alex-bell Rd Dayton OH 45459

FISHER, GEORGE HAROLD, JR, b San Antonio, Tex, July 19, 43. ORGANIC BIOCHEMISTRY. *Educ:* Rollins Col, Fla, BS, 65; Univ Fla, MS, 68; Univ Miami, PhD(org chem), 73. *Prof Exp:* Res asst org chem, Univ Fla, 68; res fel biomed chem, Inst Biomed Res, Univ Tex, Austin, 73-75; res scientist peptide chem, Papanicolaou Cancer Res Inst, 75-77; from res asst prof to res assoc prof chem & med, Univ Miami, 76-89; ASSOC PROF CHEM, BARRY UNIV, 88- *Concurrent Pos:* Vis assoc prof chem, Fla Int Univ, 87-88; adj assoc prof chem & med, Univ Miami, 89- *Honors & Awards:* Thomas R Baker Mem Prize in Chem, 64. *Mem:* Am Heart Asn; Am Chem Soc; Sigma Xi; Am Asn Advan Sci. *Res:* Bio-organic, neurochemical studies of amino acid alterations and protein changes in brain with age and neurological dysfunctions (Alzheimer's Disease); solid-phase peptide syntheses; studies of immunological fate and structure-activity of biologically active peptides with amino acid alterations. *Mailing Add:* Dept Chem Barry Univ 11300 NE Second Ave Miami Shores FL 33161

FISHER, GEORGE PHILLIP, b Quincy, Mass, May 10, 38; m 79. APPLIED PHYSICS. *Educ:* Mass Inst Technol, BS, 59; Univ Ill, MS, 61, PhD(physics), 65. *Prof Exp:* Res assoc physics, Univ Colo, Boulder, 64-65, asst prof, 65-71; assoc physicist, Physics Dept, Brookhaven Nat Lab, 71-73; PHYSICIST, R&D ASSOCS, 73- *Concurrent Pos:* NSF fel, Europ Ctr Nuclear Res, 67-68; US AEC nuclear educ & training appointment, Lawrence Radiation Lab, Berkeley, 69, Brookhaven Nat Lab, 70; consult, Dow Chem Co, 71. *Mem:* AAAS; Am Phys Soc; Am Asn Physics Teachers. *Res:* Pulsed power systems; x-ray sources; nuclear effects; broad range of problems in energy, fusion and national security. *Mailing Add:* 8512 Tuscany Ave No 405 Playa Del Rey CA 90293

FISHER, GEORGE ROBERT, dairy husbandry; deceased, see previous edition for last biography

FISHER, GEORGE WESCOTT, b New Haven, Conn, May 16, 37; m 59; c 3. METAMORPHIC PETROLOGY. *Educ:* Dartmouth Col, BA, 59; Johns Hopkins Univ, MA, 62, PhD(geol), 63. *Prof Exp:* Fel, Carnegie Inst Geophys Lab, 64-66; from lectr to assoc prof, Johns Hopkins Univ, 65-74, chmn dept Earth & planetary sci, 78-83, dean sch arts & sci, 83-87; PROF GEOL, JOHNS HOPKINS UNIV, 74- *Mem:* Mineral Soc Am; Geol Soc Am. *Res:* Field and laboratory studies of the chemistry, structure and origin of metamorphic and igneous rocks; kinetics of petrologic processes. *Mailing Add:* Dept Earth & Planetary Sci Johns Hopkins Univ Baltimore MD 21218

FISHER, GERALD LIONEL, b Malden, Mass, Aug 28, 46. ENVIRONMENTAL CHEMISTRY, METABOLISM. *Educ:* Northeastern Univ, BA, 68; Univ Calif, Santa Barbara, MA, 71; Univ Calif, Davis, Phd(chem ecol), 74. *Prof Exp:* ASSOC RES ENVIRON CHEMIST POLLUTION TOXICOL, RADIOBIOL LAB, UNIV CALIF, DAVIS, 75-; MGR, TOXICOL & HEALTH SCI SECT, BATTELLE MEM INST, COLUMBUS, 80- & ADJ PROF, PHARMACOL, OHIO STATE UNIV, 80- *Concurrent Pos:* Adj assoc prof, Div Environ Studies, Univ Calif, Davis, 75-80, fac mem, Biomed Eng Grad Group, 75-80 & Ecol Grad Group & Nutrit Grad Group, 76-80. *Mem:* NY Acad Sci; Am Chem Soc; Int Asn Bioinorg Scientists; Soc Environ Geochem & Health; AAAS. *Res:* Trace mineral metabolism and homeostasis; evaluation of health hazard and mechanisms of damage from pollutant exposure. *Mailing Add:* Toxicol/Pharmacol Sect Battelle Columbus Labs Columbus OH 43201

FISHER, GORDON MCCREA, b St Paul, Minn, Oct 5, 25; m 56; c 2. HISTORY OF MATHEMATICS, HISTORY OF SCIENCE. *Educ:* Univ Miami, BA, 51; La State Univ, PhD(math), 59. *Prof Exp:* Instr math, Miami Univ, 53 & 56-57; instr, La State Univ, 57-59; instr, Princeton Univ, 59-62; lectr math & hon lectr hist & philos sci, Univ Otago, NZ, 62-64, sr lectr hist & philos sci & hon sr lectr math, 64-66; sr lectr math & sr lectr hist philos sci, Univ Waikato, 66-67; prof math, 67-91, EMER PROF MATH & COMP SCI, JAMES MADISON UNIV, 91- *Mem:* Am Math Soc; Math Asn Am. *Res:* Analysis; history of mathematics; differential equations. *Mailing Add:* Dept Math & Computer Sci James Madison Univ Harrisonburg VA 22807

FISHER, GORDON P(AGE), b Baltimore, Md, July 26, 22; m 44; c 4. CIVIL ENGINEERING. *Educ:* Johns Hopkins Univ, BE, 42, DEng, 48. *Prof Exp:* Engr, Struct Res Div, NASA, 42-44; from asst prof to assoc prof civil eng, 48-59, assoc dean col eng, 60-66, dir water resources ctr, 62-64, head dept environ systs eng, 66-71, PROF CIVIL ENG, CORNELL UNIV, 59- *Concurrent Pos:* Sr engr, Pittsburgh Des Moines Steel Co, Pa, 54-55, consult, 57; consult, Power Reactor Develop Co, Mich, 57-64; vis prof, Chalmers Univ, 62-63; mem, Transp Res Forum & Hwy Res Bd, Nat Acad Sci-Nat Res Coun; expert mem, Europ Comt Concrete; sr researcher, Inst Eng, Nat Univ Mex, 76-77; fel, Japanese Soc Prom Sci, 83. *Honors & Awards:* Norman Medal, Am Soc Civil Engrs, 62. *Mem:* Am Soc Civil Engrs; Am Concrete Inst; Opers Res Soc Am. *Res:* Buckling; foundation design; matrix formulation of structural theory; industrialization of building construction; water resources planning and management; analysis and design of transportation systems; traffic flow theory. *Mailing Add:* 218 Hollister Hall Civil Engr Cornell Univ Main Campus Ithaca NY 14853

FISHER, HANS, b Ger, Mar 4, 28; nat US; m 50; c 3. NUTRITION. *Educ:* Rutgers Univ, BS, 50; Univ Conn, MS, 52; Univ Ill, PhD(nutrit), 54. *Prof Exp:* From asst prof to prof, 54-62, chmn nutrit coun, 61-63, chmn dept, 66-88, DISTINGUISHED PROF NUTRIT, RUTGERS UNIV, 72- *Concurrent Pos:* assoc ed, J Nutrit, 71-75; contrib ed, Nutrit Rev, 88- *Honors & Awards:* Am Feed Mfrs Award, 59; Poultry Sci Award, 57. *Mem:* Fel AAAS; Poultry Sci Asn; Am Inst Nutrit; fel NY Acad Sci; Am Bd Nutrit; Soc Exp Biol & Med; Am Chem Soc; Brit Nutrit Soc. *Res:* Amino acid requirements; cholesterol metabolism; histamine stress; carnosine metabolism; gastric acid and calcium interrelationship; tryptophan serotonin and tardive dyskinesia. *Mailing Add:* Dept Nutrit Sci Rutgers Univ New Brunswick NJ 08903

FISHER, HAROLD M, b Fayetteville, NC, Feb 19, 40; m 63; c 2. FIBER FINISHES, POLYMER TECHNOLOGY. *Educ:* Davidson Col, BS, 62; Univ Fla, MS, 64, PhD(inorg chem), 66. *Prof Exp:* Asst gen chem, Univ Fla, 62-66; res chemist, Atlantic-Richfield Co, 66; fiber engr, 69-77, POLYMER GROUP LEADER, FIBER INDUSTS, INC, 77- *Res:* Inorganic chemistry, transition metal complexes, magnetically anomalous cobalt II spectra; propellant chemistry, burning rate modifiers, polybutadine, ferrocene derivatives; homogeneous catalysis; fiber engineering, polyester, nylon; polyester polymer engineering; fiber finishes. *Mailing Add:* Dept Educ Univ Lethbridge 4401 Univ Dr Lethbridge AB T1K 3M4 Can

FISHER, HAROLD WALLACE, b Rutland, Vt, Oct 27, 04; m 30; c 1. CHEMICAL ENGINEERING. *Educ:* Mass Inst Technol, BS, 27. *Hon Degrees:* DSc, Clarkson Col Technol, 60. *Prof Exp:* Mem staff res lab, Standard Oil Co La, 27-29; mem staff, Standard Oil Develop Co, NJ, 29-30, mem staff hydro-eng & chem, 30-32, mem staff, Esso Labs, 32-33, asst dir, 35-36, mgr com dept, NY, 36-41, asst mgr sales eng, Standard Oil Co, NJ, 33-35, mgr chem prod dept, 41, dir, 45-47, mgr chem prod dept & vpres, Stanco Distributors, Inc, 41-44, pres, Standard Alcohol Co, 44-47, pres, Enjay Co, Inc, 47-48, dir mfg opers ECoast & mgr chem prod dept, Esso Standard Oil Co, 48-49, dept refining coordr, Standard Oil Co, NJ, 49-50, refining coordr, 50-54, rep in UK, 54-57, managing dir, Iraq Petrol Co, London, 57-59, dir contact for MidE refining & transp activities, Standard Oil Co, NJ, 59-62, vpres & dir chem res & refining, 62-69; RETIRED. *Concurrent Pos:* Trustee, Sloan-Kettering Inst Cancer Res, 64-, emer bd overseers, 64-, chmn bd, 70-74; vchmn & chmn exec comt, Community Blood Coun Greater New York, 69-71. *Honors & Awards:* Chem Indust Medal, 68. *Mem:* Nat Acad Eng; fel AAAS; Am Chem Soc; Am Inst Chem Engrs. *Mailing Add:* PO Box 1792 Duxburg MA 02331

FISHER, HAROLD WILBUR, b Galt, Ont, Nov 29, 28; US citizen; m 49; c 2. BIOPHYSICS, MOLECULAR BIOLOGY. *Educ:* Univ Mich, BS, 51, MS, 53; Univ Colo, PhD(biophys), 59. *Prof Exp:* Res assoc biophys, Univ Mich, 50-53; chemist, Eli Lilly Res Labs, 53-55; instr biophys, Univ Colo, 56-59, asst prof, 60-63; assoc prof, 63-68, PROF BIOPHYS, BIOCHEM & MICROBIOL, UNIV RI, 68- *Concurrent Pos:* Am Cancer Soc fel, 60-62. *Mem:* Biophys Soc; Am Soc Cell Biol. *Res:* Electron microscopy; bacteriophage; mammalian cell culture and genetics; cell fractionation and membranes; tumor viruses. *Mailing Add:* Dept Biochem & Biophys Univ RI 226 Morrill Kingston RI 02881

FISHER, HARVEY FRANKLIN, b Cleveland, Ohio, June 12, 23; m 64; c 1. BIOCHEMISTRY. *Educ:* Western Reserve Univ, BS, 47; Univ Chicago, PhD(biochem), 52. *Prof Exp:* Biochemist, Col Physicians & Surgeons, Columbia Univ, 52-54; res assoc chem, Univ Wis, 54-56; instr, Univ Mass, 56-57; sr assoc biochem, Edsel B Ford Inst, 57-63; assoc prof, 63-65, PROF BIOCHEM, SCH MED, UNIV KANS MED CTR, KANSAS CITY, 65-; DIR MOLECULAR BIOCHEM LAB, VET ADMIN HOSP, 63- *Concurrent Pos:* Mem, Nat Sci Found Molecular Biol adv panel, 77; mem, Vet Admin basic sci adv panel, 65-68, 81-84. *Mem:* Am Chem Soc; Am Soc Biol Chem. *Res:* Mechanisms and energetics of enzymatic catalyses. *Mailing Add:* Res Serv-151 Vet Admin Med Ctr 4801 Linwood Blvd Kansas City MO 64128

FISHER, IRVING SANBORN, b Augusta, Maine, May 21, 20; m 45; c 3. GEOLOGY. *Educ:* Bates Col, AB, 41; Harvard Univ, MA, 48, PhD(geol), 52. *Prof Exp:* Asst instr geol, Dartmouth Col, 41-42; field asst, US Geol Surv, 42; from asst prof to assoc prof geol, Univ Ky, 49-85; RETIRED. *Mem:* Fel AAAS; fel Meteoritical Soc; Mineral Soc Am; fel Geol Soc Am; Soc Econ Paleont & Mineral. *Res:* Petrology; mineralogy. *Mailing Add:* Nine Ricker Park Portland ME 04101

FISHER, JACK BERNARD, b New York, NY, July 13, 43; m 73; c 2. TROPICAL BOTANY, PLANT ANATOMY. *Educ:* Cornell Univ, BS, 65, MS, 66; Univ Calif, Davis, PhD(bot), 69. *Prof Exp:* Cabot Found res fel, Harvard & Fairchild Trop Garden, Miami, 69-71; PLANT MORPHOLOGIST, FAIRCHILD TROP GARDEN, MIAMI, 72- *Concurrent Pos:* Asst prof, Ohio Univ, 71-72; vis prof, Univ Calif, Berkeley, 76, Harvard Univ summer sch, 77, 79, 82 & Univ Guelph, 87; sr teaching fel, Nat Univ Singapore, 85-86. *Mem:* Bot Soc Am; Asn Trop Biol; Int Soc Plant Morphologists; Linnean Soc London; Int Asn Wood Anatomists. *Res:* Developmental anatomy and morphology, especially of monocotyledons and tropical trees; tropical botany and agriculture; functional anatomy especially of lianas. *Mailing Add:* Fairchild Trop Garden 11935 Old Cutler Rd Miami FL 33156

FISHER, JAMES DELBERT, b Lebanon, Ohio, June 23, 42; m 64; c 2. PESTICIDE CHEMISTRY. *Educ:* Ohio State Univ, BS, 64; Univ Ill, MS, 67, PhD(plant physiol), 70. *Prof Exp:* Group leader XT-11, 69-74, GROUP LEADER XT-10 PESTICIDE CHEM, ROHM AND HAAS CO, 74- *Mem:* Am Chem Soc. *Res:* To develop procedures for laboratory and field studies which delineate the environmental fate of pesticides; to supervise scientists in actually carrying out the above environmental studies. *Mailing Add:* Rohm and Haas Co Norristown & McKean Rds Spring House PA 19477

FISHER, JAMES HAROLD, b Mayfield, Ky, Nov 8, 19; m 43; c 3. PETROLEUM GEOLOGY. *Educ:* Univ Ill, AB, 43, BS, 47, MS, 49, PhD(geol), 53. *Prof Exp:* Asst, Univ Ill, 47-48; geologist, McCurtain Limestone Co & Pure Oil Co, 48-51; asst prof geol, Univ Ill, 52-55; asst prof, Univ Nebr, 55-57; from asst prof to prof, 57-84, EMER PROF GEOL, MICH STATE UNIV, 84- *Mem:* Geol Soc Am; An Asn Petrol Geologists; Am Inst Prof Geologists; Ont Petrol Inst. *Res:* Stratigraphy and structure of the Rocky Mountains and the Great Plains province; petroleum geology of the Michigan Basin; fuel energy resources of the US. *Mailing Add:* Dept Geol Mich State Univ East Lansing MI 48824

FISHER, JAMES RUSSELL, b New Castle, Pa, Sept 24, 40. GEOCHEMISTRY, PHYSICAL INORGANIC CHEMISTRY. *Educ:* Harvard Univ, AB, 62; Pa State Univ, PhD(geochem), 69. *Prof Exp:* Nat Acad Sci res assoc geochem, 69-71, GEOLOGIST, US GEOL SURV, 71- *Mem:* AAAS; Mineral Soc Am. *Res:* Chemistry of aqueous solutions at elevated temperature and pressure; hydrothermal synthesis of crystalline materials; pressure-volume-temperature properties of gases and aqueous solutions. *Mailing Add:* US Geol Surv Nat Ctr Stop 809 Reston VA 22092

FISHER, JAMES THOMAS, b Bryan, Tex, June 28, 46; m 65; c 3. FOREST TREE PHYSIOLOGY. *Educ:* WTex State Univ, BS, 69; Colo State Univ, MS, 72, PhD(forest tree physiol), 75. *Prof Exp:* Res fel forest tree physiol, Colo State Univ, 69-70; chmn biol, Le Tourneau Col, 71-72; res fel forest tree physiol, Colo State Univ, 72-75; PROF SILVICULT TREE PHYSIOL, NMEX STATE UNIV, 75-; MGR SILVICULT TREE PHYSIOL, MORA RES CTR, 75- *Concurrent Pos:* Adj asst prof forestry, NMex Highlands Univ, 76-78; mem, NMex Forest Planning Comt, 76-, Int Parasitic Seed Plant Res Group, 78-, NMex State Reforestation Comt, 82-, Res Adv Comt, NMex Agr Exp Sta, 82-; delegate, Eisenhower Consortium, 80-86; chmn, GP-13 Regional Tree Improvement Comt, 82-84; fac ath rep, NCAA/B16 West. *Mem:* Soc Am Foresters; Sigma Xi. *Res:* Forest tree physiology; physiological genetics and tree improvement as related to forestation and intensive silvicultural management of conifer plantations. *Mailing Add:* Box 3530 Dept Hort & Agron Home Econ NMex State Univ Las Cruces NM 88003

FISHER, JAMES W, b Startex, SC, May 22, 25; m 47; c 6. PHARMACOLOGY. *Educ:* Univ SC, BS, 47; Univ Louisville, PhD, 58. *Prof Exp:* Chemist org synthesis, Abbott Pharmaceut Labs, Ill, 48; develop chemist, Armour Pharmaceut Labs, 50-54; sr pharmacologist, Res Div, Lloyd Bros, Inc, Ohio, 54-56; from instr to prof pharmacol, Med Units, Univ Tenn, Memphis, 58-68; REGENTS PROF PHARMACOL & CHMN DEPT, SCH MED, TULANE UNIV, 68- *Concurrent Pos:* USPHS career develop award, Univ Tenn, 60-65; guest lectr & investr, Christie Hosp, Holt Radium Inst & Sch Med, Victoria Univ Manchester, 63-64; mem comt erythropoietin, Nat Heart & Lung Inst, 71-74; mem res comt, Cooley's Anemia Found, 74-; mem career develop award comt, Vet Admin; mem, Nat Heart & Lung Inst training grants adv comt, 75-; dir Pan Am Health Orgn training prog physiol sci, Tulane Univ-Nat Univ North-East, Arg, 72-; consult, Schering Corp, NJ, 76- & Upjohn Co, Mich, 77-; McDonnell Douglas, Life Sci Div, St Louis, Mo, 85-87; vis prof, Univ Zambia Sch Med, Lusaka, 87, Keio Med Sch, Tokyo, 87. *Honors & Awards:* Purkinje Medal, Czech Med Soc, 75. *Mem:* AAAS; Am Soc Hemat; Am Soc Pharmacol & Exp Therapeut; Int Soc Nephrology; Soc Exp Biol & Med; Am Soc Nephrol. *Res:* Hematopharmacology; erythropoietin, hormones and erythropoiesis; erythropoietic function of the kidney; adrenocortical steroids; anemia and kidney disease; kidney hormones. *Mailing Add:* Dept Pharmacol Sch Med Tulane Univ 1430 Tulane Ave New Orleans LA 70112

FISHER, JOHN BERTON, b McKees Rocks, Pa, Sept 28, 51; m 73. ENVIRONMENTAL SCIENCE. *Educ:* Yale Univ, BS, 73; Case Western Reserve Univ, MS, 76, PhD(geol), 79. *Prof Exp:* Res asst geol, Yale Univ, 71-73; res assoc environ sci, Normandeau Asn, Inc, 73-74; res asst, Case Western Reserve Univ, 75-78, res assoc geol, 78-80; res asst, Univ Calif, 80-81; STAFF RES SCIENTIST, AMOCO PROD CO, 81-; INSTR GEOL, TULSA JR COL, 82- *Concurrent Pos:* Asst, Case Western Reserve Univ, 75-78, fel, 78- *Honors & Awards:* Belknap Prize, Dept Geol & Geophys, Yale Univ, 73. *Mem:* Soc Petrol Engs; Am Geophys Union; Am Geol Inst. *Res:* Sedimentary geochemistry; early sediment diagenesis, animal-sediment interaction; water-rock interaction; environmental geochemistry; biotoxicity. *Mailing Add:* 5205 S Joplin Tulsa OK 74135

FISHER, JOHN CROCKER, b Ithaca, NY, Dec 19, 19; m 43; c 3. PHYSICS, NATURAL SCIENCE. *Educ:* Ohio State Univ, AB, 41; Mass Inst Technol, ScD, 47. *Prof Exp:* Res engr, Battelle Mem Inst, 41-42; asst & later instr mech, Mass Inst Technol, 42-47; res assoc, Gen Elec Co, 47-51, mgr phys metall sect, 51-57, physicist, 57-63, mgr liaison & transition, Res Lab, 63-64, mgr phys sci & info disciplines, Tech Mil Planning Oper, Calif, 64-68, consult scientist, Re-Entry & Environ Systs Prod Div, Pa, 69-72, mgr energy technol anal, Power Generation Bus Group, 72-78, consult res & develop, Corp Res & Develop, 78-81; consult & entrepreneur, 81-86; PHYSICIST, 86- *Concurrent Pos:* Chief scientist, USAF, Washington, DC, 68-69; Campbell mem lectr; Gillette mem lectr; mem, Bus & Indust Comt Energy & Raw Mat, 75-77 & Comn Sociotech Systs of Nat Res Coun, 76-78. *Honors & Awards:* Alfred Noble Prize. *Mem:* Nat Acad Eng; AAAS; Am Inst Physics; Am Phys Soc; Am Soc Metals. *Res:* Physical sciences. *Mailing Add:* 600 Arbol Verde Carpinteria CA 93013

FISHER, JOHN F, b East Liverpool, Ohio, July 23, 37; m 59; c 3. ANALYTICAL CHEMISTRY, PHYSICAL CHEMISTRY. *Educ:* Univ WVa, BS, 59, MS, 61, PhD(chem), 63. *Prof Exp:* Analytical chemist, Res & Develop Dept, 62-70, prin engr, Eng Dept, 70-87, MEM, AUTOMATED ANALYTICAL SYSTS, UNION CARBIDE CORP, 88- *Res:* Gas chromatography; non-aqueous titrations; spectrophotometry; infrared spectroscopy; analytical instrumentation; microprocessors; process instrumentation; process control computers; systems software; laboratory robotics. *Mailing Add:* 313 Parkview Dr St Albans WV 25177

FISHER, JOHN HERBERT, b Tipton, Mich, Dec 19, 21; m 47; c 4. SURGERY. *Educ:* Harvard Med Sch, MD, 46; Tufts Univ, MS, 52. *Prof Exp:* Surgeon, 56-59, pediat surgeon in chief, Boston Floating Hosp, 60-73; asst prof surg, Sch Med, Tufts Univ, 59-72; ASSOC CLIN PROF SURG, HARVARD UNIV, 77-, SR ASSOC SURG, CHILDREN'S HOSP MED CTR, 76- *Res:* Pediatric surgery. *Mailing Add:* Harvard Childrens Hosp Childrens Hosp Med Ctr 300 Longwood Ave Boston MA 02115

FISHER, JOHN WILLIAM, b Ancell, Mo, Feb 15, 31; m 52; c 4. CIVIL ENGINEERING. *Educ:* Washington Univ, BSCE, 56; Lehigh Univ, MS, 58, PhD(struct joints), 64. *Hon Degrees:* DSc, Swiss Fed Inst Technol, 88. *Prof Exp:* Struct res asst, Lehigh Univ, 56-58; asst bridge res engr, Hwy Res Bd, Nat Acad Sci-Nat Res Coun, 58-61; struct res instr, 61-64, from asst prof to assoc prof, 64-69, PROF CIVIL ENG, LEHIGH UNIV, 69-; DIR, ATLSS ENG RES CTR, 86- *Concurrent Pos:* Mem, Res Coun Riveted & Bolted Struct Joints; mem transp res bd, Nat Acad Sci-Nat Res Coun; consult, Conn Dept Transp, Triborough Bridge & Tunnel Auth, Can Nat RR, Bethlehem Steel, Allied Struct Steel & Ammann & Whitney, NY DOT, US Steel, La Dept Transp, Bechtel Power Corp, NJ Transit, Modjeski & Masters & DeLeuw-Cather, 77-81; vis prof, Swiss Fed Inst Technol, Lausanne, Switz, 82; Nat Sci Found Adv Comt Critical & Emerging Systs; chmn, Nat Acad Eng, Nat Res Coun Comt Int Construction Study. *Honors & Awards:* Walter L Huber Res Prize, Am Soc Civil Engrs, 69; A W S Adams Memorial Mem Award, Am Welding Soc, 74; Ernest E Howard Award, Am Soc Civil Engrs, 79; Raymond C Reese Res Prize, Am Soc Civil Engrs, 81. *Mem:* Nat Acad Eng; Am Soc Eng Educ; Nat Soc Prof Engrs; Am Welding Soc; Int Asn Bridge & Struct Engrs; fel Am Soc Civil Engrs; Am Railway Eng Asn; hon mem, Am Soc Civil Engrs. *Res:* Behavior of welded connections; fatigue studies of bridge components; behavior of bridge structures; high strength bolts; bolted joints and composite beams. *Mailing Add:* 117 Atlss Dr Lehigh Univ Bethlehem PA 18015

FISHER, JUNE MARION, b Elizabeth, NJ, June 10, 33. OCCUPATIONAL HEALTH. *Educ:* Downsate Med Ctr, State Univ NY, MD, 61. *Prof Exp:* DIR CTR MUNIC OCCUP SAFETY & HEALTH, SAN FRANCISCO GEN HOSP, 78-; ASSOC PROF CLIN MED, UNIV CALIF, SAN FRANCISCO, 85- *Mailing Add:* San Francisco Gen Hosp Third Floor Bldg Nine San Francisco CA 94110

FISHER, KATHLEEN MARY FLYNN, b Long Branch, NJ, Aug 4, 38; div; c 2. SCIENCE EDUCATION, BIOLOGY. *Educ:* Rutgers Univ, BS, 60; Univ Calif, PhD(genetics), 69. *Prof Exp:* Lab technician malaria res, Christ Hosp Inst Med Res, Cincinnati, 60-63; res specialist, Nat Ctr Primate Biol, Univ Calif, Davis, 63, lab technician, Dept Genetics, 64-68, res cytogeneticist, 69, AEC fel, 69-71, asst prof genetics, 71-75, dir, Teaching Res Ctr, 74-79, assoc prof biol sci & educ, 75-88; PROF BIOL SCI, SAN DIEGO STATE UNIV, 88- *Concurrent Pos:* Mem, Sesame Grad Group in Sci & Math Educ, Univ Calif, Berkeley, 75- & Sch Educ, 86-; prog assoc, Res Sci Educ, NSF, 80-81; dir, Sem Net Res Group, 83-; mem, Ctr Res Math & Sci Educ, San Diego State Univ, 88- *Honors & Awards:* Fulbright Sr Lectr, 80. *Mem:* AAAS; Am Educ Res Asn; Nat Asn Res Sci Teaching; Sigma Xi. *Res:* Development and testing of semantic networking software for representing knowledge; study of cognitive structure, conceptual change and alternate conceptions in biology learning; use of electronic media in science education molecular genetics and malaria research. *Mailing Add:* Dept Natural Sci San Diego State Univ San Diego CA 92182-0324

FISHER, KENNETH D, b Lowell, Mass, Mar 3, 32; m 56; c 3. RESEARCH ANALYSIS, SCIENCE COMMUNICATION. *Educ:* Oberlin Col, Univ Vt, BS, 53, MS, 55; NC State Univ, PhD(plant path), 60. *Prof Exp:* Asst plant path, NC State Univ, 57-60; asst prof, SDak State Univ, 60-63; asst bot, Univ Vt, 53-55, asst prof bot, 63-66, plant pathologist, Vt Agr Exp Sta & Exten Serv, 63-68; res assoc, 68-75, assoc dir, 75-77, DIR LIFE SCI RES OFF, FEDN AM SOCS EXP BIOL, 77- *Concurrent Pos:* Adj prof biol, Montgomery Col, 69-85. *Mem:* AAAS; Am Inst Nutrit; Soc Nematol; Am Soc Micros. *Res:* Research analysis and evaluation in the life sciences; research analysis and evaluation in the life sciences; phytonematology. *Mailing Add:* Life Sci Res Off Fedn Am Socs Exp Biol 9650 Rockville Pike Bethesda MD 20814

FISHER, KENNETH ROBERT STANLEY, b Toronto, Ont, Dec 8, 42; m; c 2. DEVELOPMENTAL BIOLOGY. *Educ:* Univ Toronto, BS, 63, MS, 67, PhD(zool), 73. *Prof Exp:* Researcher embryol, Animal Biol Lab, Nat Ctr Sci Res, Univ Paris, 64-65; demonstr zool, Univ Toronto, 65-71; instr anat, Univ Sask, 72-73, Med Res Coun Can fel, 73-76, asst prof, 76-86, ASSOC PROF MICROS ANAT, DEPT BIOMED SCI, ONT VET COL, UNIV GUELPH, 86- *Mem:* Can Asn Anatomists; Can Soc Cell Biol; Soc Develop Biol; Tissue Cult Asn; Am Genetics Asn. *Res:* Embryological development of biological control mechanisms, sexual organogenesis and developmental neurobiology. *Mailing Add:* Dept Biomed Sci Ont Vet Col Univ Guelph Guelph ON N1G 2W1 Can

FISHER, KENNETH WALTER, b Heston, Eng, Dec 30, 31; m 65; c 5. MOLECULAR GENETICS. *Educ:* Univ London, BSc, 53, MSc, 54, PhD(bact genetics), 57. *Prof Exp:* Res worker med res coun, Microbial Genetics Res Unit, Hammersmith Hosp, London, 57-66; assoc prof biol, Kans State Univ, 66-70; actg chmn dept biol, 72-73, chmn dept, 73-78, dir, Univ Grad Prog Microbiol, 75-78, PROF BIOL, FAC ARTS & SCI, RUTGERS UNIV, NEW BRUNSWICK, 70- *Concurrent Pos:* Vis worker, Physiol Microbial Serv, Pasteur Inst Paris, 57-58; Rockefeller traveling fel med, Group Biochem Studies, Princeton Univ, 62-63; temp assoc prof physics, Kans State Univ, 64; NSF res grant, 66-68; Eli Lilly grant-in-aid, 67-70; mem cause grant review panel, NSF, 76-78; mem fel review sect, Microbiol Chem Study Sect, NIH. *Mem:* Genetics Soc Am; Brit Soc Gen Microbiol; Brit Biochem Soc; Brit Genetical Soc; Am Soc Microbiol. *Res:* Bacterial genetics and chemistry; mechanism of conjugation in the bacterium Escherichia coli; isolation of auxotrophic mutants in Solanum tuberosum tissue cultures. *Mailing Add:* Dept Biol Rutgers Univ Douglass Campus New Brunswick NJ 08903

FISHER, KNUTE ADRIAN, b Woodland, Calif, Dec 2, 41; m 64; c 1. CELL BIOLOGY, BIOCHEMISTRY. *Educ:* Univ Calif, Berkeley, PhD(bot), 74. *Prof Exp:* ADJ ASSOC PROF BIOCHEM, DEPT BIOCHEM & PHYSICS, UNIV CALIF, SAN FRANCISCO, 74- *Concurrent Pos:* Estab investr, Am Heart Asn, 77. *Mem:* AAAS; Am Soc Cell Biol; Biophys Soc; Electron Micros Soc Am. *Res:* Cell membranes; transmembrane signaling. *Mailing Add:* Dept Biochem Univ Calif San Francisco CA 94143

FISHER, LEON HAROLD, b Montreal, Que, July 11, 18; nat US; m 41; c 4. ELECTRON PHYSICS. *Educ:* Univ Calif, BS, 38, MS, 40, PhD(physics), 43. *Prof Exp:* Instr physics, Univ Calif, 43-44, vis prof physics, summers, 49, 51, 55, 83; instr physics, Univ NMex, 44; assoc scientist, Los Alamos Sci Lab, Univ Calif, 44-46; from asst prof to prof physics, NY Univ, 46-61; mgr plasma physics, Lockheed Missiles & Space Co, 61-62; head plasma physics, Gen Tel & Electronics Labs, 62-63; sr mem & sr consult scientist, Lockheed Palo Alto Res Lab, 63-70, asst mgr, Electronic Sci Lab, 67-68; prof elec eng & head dept info eng, Univ Ill, Chicago Circle, 71; prof physics, 71-83, dean sch sci, 71-79, EMER PROF PHYSICS, CALIF STATE UNIV, HAYWARD, 83- *Concurrent Pos:* Chmn, Gaseous Electronics Conf, 48, 67 & 68; assoc ed Phys Review, 55-58; consult, EGG, 54-55, Harry Diamond Labs, 58-61, Xerox Corp, 58-61, Army Res Off Durham, 58-63, Rome Air Develop Ctr, 59-63, Re-entry Physics Panel, Nat Acad Sci, 65-66 & Monsanto Envirochem, 71; sr liaison scientist, Off Naval Res, Tokyo, 79-82; vis prof physics, Univ Wash, 84, Stanford Univ, 85-87; lectr, San Jose State Univ, 85-87. *Mem:* fel Am Phys Soc; Am Asn Physics Teachers. *Res:* Low temperature heat capacities; ionization coefficients in gases; mechanism of electrical breakdown in gases; formation of negative ions; corona discharges; plasma physics; contact potentials. *Mailing Add:* 102 Encinal Ave Atherton CA 94027

FISHER, LEONARD V, b Elizabeth, NJ, May 22, 29; m 63; c 2. MEDICINE. *Educ:* Rutgers Univ, BSc, 48; Yale Univ, MS, 50; Univ Chicago, MD, 54; Am Bd Internal Med, cert internal med, 64, cert endocrinol & metab, 72. *Prof Exp:* From intern to resident med, Montefiore Hosp, NY, 54-57; asst prof, Seton Hall Col Med & Dent, 60-61, clin asst prof, 61-64; asst prof, 64-70, ASSOC PROF MED, NY MED COL, 70-, ASST PROF RADIOL & NUCLEAR MED, 70-, CHIEF SECT ENDOCRINOL & METAB, B S COLER HOSP, 64-, ASSOC DIR MED, 70- *Concurrent Pos:* Vis res fel med, Col Physicians & Surgeons, Columbia Univ, 57-60; Nat Found fel, 57-59; spec fel, Nat Inst Arthritis & Metab Dis, 59-60; dir endocrinol & biochem, St Michael Hosp, Newark, NJ, 60-64; assoc vis physician, Metrop Hosp, 64- *Mem:* Soc Nuclear Med; Am Diabetes Asn; Endocrine Soc; Am Geriatrics Soc; Asn Aging Res. *Res:* Geriatric endocrinology. *Mailing Add:* Univ Chicago Pritzker Sch Med Oak Forest Hosp Geriat Med 15900 S Cicero Ave Oak Forest IL 60452

FISHER, LESLIE JOHN, b New York, NY, Feb 26, 40; m 62. NEUROBIOLOGY. *Educ:* Rensselaer Polytech Inst, BEE, 61; Tufts Univ, MS, 66, PhD(biol), 69. *Prof Exp:* Vis asst prof physiol, Sch Med, Univ Md, Baltimore, 71-72; asst prof anat, Med Sch, Univ Mich, Ann Arbor, 72-80; DIR RES, DEPT OPHTHAL, HENRY FORD HOSP, DETROIT, 80- *Concurrent Pos:* NIH fel, Johns Hopkins Univ Sch Med, 69-71; NIH & NSF res grant; NATO partic, Neurobiological Study Inst, 71; adj assoc prof anat, dept ophthal, Univ Mich Sch Med, 81- *Mem:* Soc Neuroscience; Asn Res Vision & Ophthal; Soc Anal Cytol. *Res:* Electron microscopic studies of the development of neurons and synaptic arrays in the retina; studies of the retinal pigment epithelium and eye diseases. *Mailing Add:* 1501 Morton Ann Arbor MI 48104

FISHER, LINDA E, b Kansas City, Kans, Dec 3, 47; m 70; c 1. VIROLOGY. *Educ:* Univ Kans, BA, 69, PhD(microbiol), 74. *Prof Exp:* Fel, Roche Inst Molecular Biol, 74-76 & Pa State Univ Col Med, 76-78; asst prof, 78-84, chmn dept biol sci, 86-89, ASSOC PROF MICROBIOL, UNIV MICH, DEARBORN, 84-, CHMN MICROBIOL PROG, 89- *Concurrent Pos:* Adj asst prof microbiol, Lebanon Valley Col, 77; prin investr res grant, Nat Multiple Sclerosis Soc, 80-82; vis scientist, St Jude Childrens Res Hosp, Memphis, 85-86; prin investr, Nat Inst Allergy & Infectious Dis res grant, 87-89; ad hoc reviewer, Virol & Exp Virol Study Sects, NIH, 88- *Mem:* Am Soc Microbiol; Sigma Xi; Am Soc Virol; AAAS. *Res:* Virus persistence; role of matrix protein in virus assembly and subacute infections; virus receptors. *Mailing Add:* Dept Natural Sci Univ Mich Dearborn MI 48128-1491

FISHER, LLOYD D, b Baltimore, Md, June 8, 39; m 61; c 2. BIOSTATISTICS. *Educ:* Mass Inst Technol, SB, 61; Dartmouth Col, MA, 65, PhD(math), 66. *Prof Exp:* Sci programmer, Lockheed Missiles & Space Co, 61-63; asst prof math, 66-70, assoc prof, 70-74, dir, Coord Ctr Collab Studies Coronary Artery Surg, 73-84, PROF BIOSTATIST, UNIV WASH, 75- *Concurrent Pos:* Consult, Aerospace Corp, 66-67, Minneapolis-Honeywell Regulator Co, 68, Vet Admin, 74-75, Fed Drug Admin, 75- & Am Arthritis Asn, 78, Mayo Clin, 84-85; mem, FDA Cardiorenal Adv Comt, 83- *Mem:* Fel Am Col Cardiol; Biomet Soc; fel Am Statist Asn; Royal Statist Soc; fel Inst Math Statist; fel Am Col Epidemiol. *Res:* Randomized clinical trials especially in cardiology; new drug evaluation; observational clinical data analysis; biostatistical methodology. *Mailing Add:* Dept Biostatist Univ Wash Seattle WA 98195

FISHER, LYMAN MCARTHUR, b Appin, Ont, Mar 21, 23; m 53; c 4. CLINICAL PATHOLOGY, HEMATOLOGY. *Educ:* Univ Western Ont, BA, 51; Univ Sask, MA, 54, PhD(physiol), 57, MD, 60; Am Bd Clin Path, dipl, 65. *Prof Exp:* Lectr physiol, Univ Sask, 57-58; from asst prof to assoc prof clin path, 60-70, PROF CLIN PATH, MED COL VA, 70- *Honors & Awards:* Fulbright Lectr, Norway, 72-73. *Mem:* Am Asn Blood Banks; Am Soc Exp Path; Am Soc Clin Path; Nat Hemophilia Found; Int Soc Thrombosis & Hemostasis. *Res:* Blood coagulation and atherosclerosis. *Mailing Add:* Med Col Va Box 597 Richmond VA 23298

FISHER, MARTIN JOSEPH, b Pittsburgh, Pa, Mar 18, 44; m 68; c 2. MEDICAL PHYSIOLOGY. *Educ:* Wheeling Col, BS, 66; WVa Univ, PhD(physiol & biophys), 71; World Univ, MD, 84. *Prof Exp:* Res asst biochem, WVa Univ, 66-67, res asst physiol & biophys, 67-71; instr anesthesiol physiol, Univ Ala, 71-72; NIH fel respiratory physiol, Univ Fla, 72-74, asst prof physiol, 74-81; assoc prof & chmn, Dept Physiol, Sch Med, World Univ, Santo Domingo, Dominican Repub, 81-83; house physician, Baptist Hosp, Miami, Fla, 84-85; resident pediat, Charleston Area Med Ctr, WVa, 86-89, chief resident, 88-89; RESIDENT ANESTHESIOL, RUBY MEM HOSP, MORGANTOWN, WVA, 89- *Concurrent Pos:* NIH fel, 71 & 74; fel respiratory physiol, Univ Fla, 74. *Res:* Respiratory physiology; pulmonary mechanics; enzyme inhibitors and pulmonary emphysema; models of lung disease and stereology. *Mailing Add:* Ruby Mem Hosp Morgantown WV 26505

FISHER, MICHAEL ELLIS, b Fyzabad, Trinidad, Sept 3, 31; m 54; c 4. THEORETICAL PHYSICS & CHEMISTRY. *Educ:* Univ London, BSc, 51, PhD(physics), 57. *Hon Degrees:* DSc, Yale Univ, 87. *Prof Exp:* Lectr math, Royal Air Force Tech Col, 52-53; tutor physics, King's Col, Univ London, 53-57, lectr theoret physics, 57-62, reader physics, 62-64, prof, 65-66; prof chem & math, Cornell Univ, 66-73, Horace White prof chem, physics & math, 73-87, chmn, Dept Chem, 75-78; WILSON H ELKINS DISTINGUISHED PROF, INST PHYS SCI & TECHNOL, UNIV MD, COLLEGE PARK, 87- *Concurrent Pos:* Dept Sci & Indust Res sr res fel, 56-58; guest investr, Rockefeller Inst, 63-64; John Simon Guggenheim mem fel & vis prof, Stanford Univ, 70-71; Guggenheim fel, 78-79; Fairchild scholar, Calif Inst Technol, 84; vis prof, Weizmann Inst & Univ Oxford, 85; Wilson H Elkins Prof, Univ Md, 87-88; mem numerous comts, Int Union Pure & Appl Chem, Yale Univ, Carnegie Mellon Univ, Cornell Univ, Am Phys Soc, NSF & Nat Res Coun. *Honors & Awards:* Irving Langmuir Prize, Am Phys Soc, 71; Phys & Math Sci Award, NY Acad Sci, 78; Bakerian lectr, Royal Soc, 79; Loeb lectr, Harvard Univ, 79; Guthrie Medal & Prize, Inst Physics, UK, 80; Wolf Prize in Physics, 80; James Murray Luck Award, Nat Acad Sci, 83; Boltzmann Medal, Int Union Pure & Appl Physics, 83. *Mem:* Foreign mem Nat Acad Sci; fel Royal Soc; Inst Physics; fel Am Phys Soc; Math Asn Am; Am Acad Arts & Sci; fel AAAS; Soc Indust & Appl Math; Am Chem Soc; NY Acad Sci. *Res:* Theory and practice of electronic analog computing; statistical mechanics of phase transitions and critical phenomena; magnetism; polymer configurations; combinatorial mathematics; mathematical foundations of statistical mechanics. *Mailing Add:* Inst Phys Sci & Technol Univ Md College Park MD 20742

FISHER, NEWMAN, b San Francisco, Calif, Mar 21, 28. MATHEMATICS. *Educ:* Univ Calif, AB, 50, MA, 51; Univ Idaho, PhD(math), 62. *Prof Exp:* Res aeronaut scientist, Ames Aeronaut Lab, NASA, 52-55; instr math, San Francisco State Col, 57-59; instr, Univ Idaho, 59-61; from asst prof to assoc prof, 61-71, chmn dept, 68-75, assoc dean sci, 75-81, PROF MATH, SAN FRANCISCO STATE UNIV, 71-, chmn dept, 83- *Mem:* Am Math Soc; Soc Indust & Appl Math; Math Asn Am; Nat Coun Math Teachers. *Res:* Transonic flow theory; stability theory of ordinary differential equations. *Mailing Add:* Dept Math San Francisco State Univ San Francisco CA 94132

FISHER, NICHOLAS SETH, b New York, NY, Apr 29, 49; m 74; c 2. PHYTOPLANKTON ECOLOGY, MARINE POLLUTION. *Educ:* Brandeis Univ, BA, 70; State Univ NY Stony Brook, PhD(marine biol), 74. *Prof Exp:* Postdoctoral investr, Woods Hole Oceanog Inst, 74-77; res scientist, Marine Sci Labs, Ministry Conserv, Australia, 77-80, Int Lab Marine Radioactivity, Int Atomic Energy Agency, Monaco, 80-85, oceanog sci div, Brookhaven Nat Lab, Upton NY, 85-87; ASSOC PROF, MARINE SCI RES CTR, STATE UNIV NY, STONY BROOK, 87- *Concurrent Pos:* Mem working group air-sea exchange processes, Group Experts Sci Aspects Marine Pollution, 83-85; mem staff & jury, Ecol Inst, Oldendorf, Ger, 85-; Comn Ecol, Int Union Conserv Nature Natural Res, 86-; Marine Radioactivity Comt, Int Comn Sci Explor Mediterranean Sea, 82- *Mem:* Phycological Soc Am; Am Soc Limnol & Oceanog. *Res:* Physiology and ecology of marine phytoplankton; marine pollution research; effects of metals on marine biota; radionuclides in the sea; community ecology; biogeochemistry. *Mailing Add:* Marine Sci Res Ctr State Univ NY Stony Brook NY 11794-5000

FISHER, PAUL ANDREW, b Schenectady, NY, Dec 3, 52; m 82; c 2. MOLECULAR CELL BIOLOGY, BIOCHEMISTRY. *Educ:* State Univ NY, Binghamton, BS, 74; Stanford Univ, MD, 80, PhD(biophys), 80. *Prof Exp:* Fel biochem, Sch Med, Stanford Univ, 79-80; fel cell biol, Rockefeller Univ, 80-81, asst prof, 81-83; ASST PROF PHARMACOL, STATE UNIV NY, STONY BROOK, 83- *Mem:* Am Chem Soc; Am Soc Cell Biol; assoc mem Am Soc Biol Chemists. *Res:* Structure and function of the karyoskeleton; DNA replication in eukaryotic cells. *Mailing Add:* Dept Pharmacol Sci Health Sci Ctr State Univ NY Stony Brook NY 11794

FISHER, PAUL DOUGLAS, b Vancouver, BC, June 2, 54; m 81; c 2. MEDICAL IMAGING, MICROWAVE BIOEFFECTS. *Educ:* Univ Victoria, BSc, 75; Univ Alta, MSc, 78, PhD(physiol), 82. *Prof Exp:* Res assoc, Sierra Pac Sci Corp, 83-84; res asst physics, Univ Victoria, 84-85; RES ASSOC MED IMAGING, VICTORIA GEN HOSP, 85-; ASST PROF, HEALTH INFO & SCI, UNIV VICTORIA. *Concurrent Pos:* Assoc ed, J Microwave Power, 84-; consult, OZY Holdings Ltd, 85- *Mem:* Inst Elec & Electronics Engrs. *Res:* Balance of system components for photovoltaic systems; high energy-plasma jet ignition systems for compressed natural gas powered vehicles; evaluation of digital imaging systems for medicine; low level microwave bioeffects. *Mailing Add:* 3850 Pitcombe Pl Victoria BC V8N 4B9 Can

FISHER, PEARL DAVIDOWITZ, b New York, NY, May 22, 20; m 41; c 2. BIOSTATISTICS. *Educ:* Brooklyn Col, BA, 41; Columbia Univ, MS, 51; Univ Okla, PhD(prev med, pub health), 58. *Prof Exp:* Asst parasitol & epidemiol, Sch Med, Univ Okla, 53-58, res assoc prev med & publ health, 58-60, asst prof, 60-64; life scientist advan studies, Marshall Space Flight Ctr, NASA, 64-66; math statistician & sr analyst, Off Biomet, Nat Inst Neurol Dis & Blindness, 67-71; chief, br serv delivery res, Nat Ctr Family Planning Serv, Health Serv & Ment Health Admin, Rockville, 71-75; chief appl statist Training Inst, Nat Ctr Health Statist, 75-82; RETIRED. *Concurrent Pos:* Instr statist, Grad Sch, NIH, 68-84. *Res:* Survival of man in space; radiation hazards; sterilization of spacecraft; duration of life studies on parasitic nematodes; survival evaluation of cancer patients; etiological factors of leukemia in children. *Mailing Add:* 13324 Sherwood Forest Dr Silver Spring MD 20904

FISHER, PERRY WRIGHT, b New York, NY; m 68. AIR POLLUTION, METEOROLOGY. *Educ:* Cornell Univ, BEngrPhysics, 62; Univ Mich, Ann Arbor, MS, 67, Am 68, PhD(meteorol), 70. *Prof Exp:* Res asst meteorologist, Univ Mich, Ann Arbor, 66-70; asst prof geophys & eng, Case Western Reserve Univ, 71-73; proj meteorologist, 73-80, assoc, 80-82, PARTNER, DAMES & MOORE, 82- *Concurrent Pos:* Cert consult meteorologist, 73- *Mem:* Am Meteorol Soc; Air Pollution Control Asn. *Mailing Add:* Dames & Moore 1550 Northwest Hwy Park Ridge IL 60068

FISHER, PHILIP CHAPIN, b Rochester, NY, Aug 3, 26; m 48; c 1. PHYSICS, ASTRONOMY. *Educ:* Univ Rochester, BS, 47; Univ Ill, MS, 48, PhD(physics), 53. *Prof Exp:* Staff mem, Los Alamos Sci Lab, Univ Calif, 53-59; consult scientist, Phys Sci Lab, Lockheed Missiles & Space Co, 59-74; physicist, Ruffner Assocs, 75-77; Sr physicist, Rasor Assocs, Inc, 76-77; Sensor eng mgr, Gas Tech Inc, 77-92; PHYSICIST, RUFFNER ASSOCS, 82- *Mem:* Am Phys Soc; Am Astron Soc; Int Astron Union; Inst Elec & Electronics Engrs; Soc Photo-Optical Inst Eng. *Res:* Bremsstrahlung; delayed gammas from fission; x-ray astronomy and instrumentation; space particle measurements; optics; astrophysics; gas-detection senors. *Mailing Add:* Ruffner Assocs PO Box 7070 Menlo Park CA 94026-7070

FISHER, RAY W, b Anamosa, Iowa, Nov 27, 21; m 45; c 4. COAL RESEARCH, ETHANOL PRODUCTION. *Educ:* Iowa State Univ, BS, 48. *Prof Exp:* Asst atomic res & phys chem, Iowa State Univ, 43-48, admin aide, 49-61, from asst prof to assoc prof mech eng, 61-90, plant mgr, Ames Lab, 69-77, dir, Fossil Energy Prog, 75-90, Mining & Mineral Resources Res Inst, 81-90; RETIRED. *Concurrent Pos:* Res assoc, Ames Lab, Iowa State Univ, 49-57, assoc engr, 57-62, engr, 62-64, assoc prof nuclear eng, 63-71, head bldg & eng serv, 64-69; prin lectr coal preparation res, The Royal Swedish Acad Eng Sci, Stockholm, 79. *Honors & Awards:* Clean Air Award, Am Lung Asn. *Mem:* Soc Mining Engrs. *Res:* High temperature corrosion studies of components used in molten metal reactor concepts; design of research facilities; production of pure chemicals; coal mining, preparation, and conversion; land restoration; ethanol separation by absorption. *Mailing Add:* 3612 Mary Circle Ames IA 50010

FISHER, REED EDWARD, b Bloomsburg, Pa, March 10, 32; m 61; c 2. ELECTRICAL ENGINEERING. *Educ:* Pa State Univ, BSEE, 58; NY Univ, MSEE, 62. *Prof Exp:* Mem tech staff, 58-69, SUPVR ELEC ENG, BELL LABS, 69- *Mem:* Fel Radio Club Am; fel Inst Elec & Electronics Engrs. *Res:* Design of cellular system architecture; call processing algorithms; base station radio control and test hardware; factory prototype of mobile telephone unit; cellular mobile telephone system development. *Mailing Add:* AT&T Bell Labs 3f-226 Whippany NJ 07981

FISHER, RICHARD FORREST, b Urbana, Ill, May 15, 41; m 59; c 3. TREE NUTRITION, ALLELOPATHY. *Educ:* Univ Ill, BS, 64; Cornell Univ, PhD(soil sci), 68. *Prof Exp:* Res scientist, Can Forestry Serv, 68-69; asst prof forest soils, Univ Ill, 69-72; assoc prof, Univ Toronto, 72-77; prof, Univ Fla, 77-82; PROF & HEAD DEPT FOREST RESOURCES, UTAH STATE UNIV, 82- *Concurrent Pos:* Instr & consult, tropical forest soils, Africa, SE Asia & Latin Am. *Mem:* Fel Soil Sci Soc Am; fel Soc Am Foresters; Ecol Soc Am; Asn Trop Biol; Int Soc Trop Foresters. *Res:* Soil plant relationships; plant-plant interactions; application of science to planning; value and use of experimentally derived knowledge to managing forest productivity; soil chemistry and biochemistry; nitrogen fixation; agroforestry. *Mailing Add:* Dept Forestry Utah State Univ Logan UT 84322-5215

FISHER, RICHARD GARY, b Philadelphia, Pa, Nov 24, 52. PROTEIN STRUCTURE, INTERACTIVE GRAPHICS. *Educ:* Northwestern Univ, BA & MS, 74; Univ Calif, Los Angeles, PhD(molecular biol), 80. *Prof Exp:* Fel, Biol Lab, Harvard Univ, 80; res assoc, Inst Molecular Biol, Univ Ore, 80-86; Smith Kline French Labs, Philadelphia, 86-88; SR SCIENTIST, KRAFT TECH CTR, GLENVIEW, ILL, 88- *Concurrent Pos:* Damon Runyon-Walter Winchell Cancer Fund, fel, 80; sci asv, Evans & Sutherland, Salt Lake City. *Mem:* AAAS; Sigma Xi; Am Crystallog Asn. *Res:* Development of computer methods in protein structure, in particular real-time interactive graphics; development of computer methods for simulating heating materials within a microwave field. *Mailing Add:* Kraft Technol Ctr 801 Waukegan Rd Glenview IL 60025

FISHER, RICHARD PAUL, b Alameda, Calif, Feb 10, 48; m 73. BORATE CHEMISTRY, BORANE CHEMISTRY. *Educ:* Univ Calif, Berkeley, BS, 70; Univ Calif, Davis, PhD(chem), 74. *Prof Exp:* Res chemist, 74-79, SR RES CHEMIST, US BORAX RES CORP, 74- *Mem:* Am Chem Soc. *Res:* Development of borate products and processes; application of analytical instrumentation to such processes; res analytical instrumentation. *Mailing Add:* 8850 Hickory St Hesperia CA 92345-3845

FISHER, RICHARD R, b Wichita, Kans, June 2, 41; m 62; c 1. ASTROPHYSICS, SOLAR ASTRONOMY. *Educ:* Grinnell Col, BA, 61; Univ Colo, PhD(astrophys), 65. *Prof Exp:* Res asst solar astrophys res, High Altitude Observ, Univ Colo, 62-65; from asst astrophysicist to assoc astrophysicist, Univ Hawaii, 69-71, resident astronr, Mees Solar Observ & asst fiscal off, Univ, 70-71; astrophysicist, Sacramento Peak Observ, 71-76; PROJ SCIENTIST, CORONAL DYNAMICS PROJ, HIGH ALTITUDE OBSERV, BOULDER, 76- *Concurrent Pos:* Lectr, Dept Gen Sci, Maui Community Col, 70; telescope scientist, Solar Optical Telescope Proj, NASA,

81-; sci leader, Joint US-USSR Siberian Eclipse Expedition, 81. *Mem:* Am Astron Soc; Optical Soc Am; Int Astron Union. *Res:* Solar research, including solar activity, coronal structure and chromospheric structure; instrument design and geometrical optics. *Mailing Add:* High Altitude Observ 1850 Table Mesa Dr Boulder CO 80303

FISHER, RICHARD VIRGIL, b Whittier, Calif, Aug 8, 28; m 47; c 4. GEOLOGY, STRATIGRAPHY-SEDIMENTATION. *Educ:* Occidental Col, BA, 52; Univ Wash, PhD, 57. *Prof Exp:* Actg instr geol, Univ Calif, Santa Barbara, 55-57, from instr to prof & chmn dept, 57-73, chmn dept, 79-80, 83-84, PROF GEOL SCI, UNIV CALIF, SANTA BARBARA, 73- *Concurrent Pos:* NSF grants, 61-67, 76-81, 82-91; vis assoc prof, Univ Hawaii, 65-66; res affil, Hawaii Inst Geophys, 65-80; NASA grant, 67-73; vis lectr, Am Geol Inst, 70; prin investr, NASA Apollo 12 lunar samples, consult, synthesis proposals rev panel, NASA, 73-77; mem, Working Group Explosive Volcanism, Int Asn Volcanol & Chem Earth's Interior, 83-; chmn, Volcanic Adv Group, Lithostratigraphic Subcomt, NAm Comn Stratigraphic Nomenclature, 79-80; mem comn syst igneous rocks, Int Union Geol Sci, 80-84; US sr scientist award, Alexander Von Humboldt Found, WGer, 80-81 & 88; invited vis prof, Nanjing Univ, China, 85; mem, Mt St Helens Adv Bd, 89; ed, J Volcanology & Geothermal Res, 90- *Honors & Awards:* N L Bower Award, Am Geophys Union, 85. *Mem:* Fel Geol Soc Am; Soc Econ Paleont & Mineral; Am Geophys Union. *Res:* Process oriented interdisciplinary field, volcanological sedimentology and stratigraphy (volcaniclastic rocks); pyroclastic flow and pyroclastic surge deposits, fallout tephra, turbidites and volcanic debris flows; physical volcanology. *Mailing Add:* Dept Geol Sci Univ Calif Santa Barbara CA 93106

FISHER, ROBERT ALAN, b Berkeley, Calif, Apr 19, 43; m 67; c 2. LASERS, OPTICS. *Educ:* Univ Calif, Berkeley, AB, 65, MA, 67, PhD(physics), 71. *Prof Exp:* Scientist laser physics, Lawrence Livermore Lab, 71-74; STAFF MEM GROUP CHM-6, LOS ALAMOS NAT LAB, 74- *Concurrent Pos:* Lectr solid state physics & nonlinear optics, Dept Appl Sci, Univ Calif, Davis, 72-73; consult pvt indust. *Mem:* Am Phys Soc; fel Optical Soc Am; sr mem Inst Elec & Electronics Engrs; Soc Photo-Optical Instrumentation Engrs. *Res:* Nonlinear optical effects; linear optics; spectroscopy; x-ray optics; optical phase conjugation. *Mailing Add:* 1841 Yosemite Rd Berkeley CA 94707

FISHER, ROBERT AMOS, JR, b Honey Grove, Pa, Mar 25, 34; m 58; c 1. LOW TEMPERATURE PHYSICS, THERMODYNAMICS. *Educ:* Juniata Col, BS, 56; Pa State Univ, PhD(phys chem), 61. *Prof Exp:* RES CHEMIST, UNIV CALIF, BERKELEY, 60- *Mem:* AAAS; Am Phys Soc; Sigma Xi. *Res:* low temperature, chemical and magneto thermodynamics; cryogenics; magnetism; calorimetry; single crystal growth. *Mailing Add:* Low Temperature Lab Univ Calif Berkeley CA 94720

FISHER, ROBERT CHARLES, mathematics, for more information see previous edition

FISHER, ROBERT EARL, b Salt Lake City, Utah, Apr 22, 39; m 60; c 8. CHEMICAL ENGINEERING. *Educ:* Univ Utah, BS, 61; Mass Inst Technol, ScD(chem eng), 66. *Prof Exp:* Sr process engr, Air Prod & Chem, Inc, 66-67, staff engr, Pa, 67-70; res engr, 70-71, res supvr, 71-76, RES DIV DIR, AMOCO CHEM CORP, 76- *Concurrent Pos:* Lectr, Lehigh Univ, 66-68. *Mem:* Am Inst Chem Engrs; Sigma Xi. *Res:* Chemical process engineering and development. *Mailing Add:* 25 W 124th Summit Naperville IL 60540

FISHER, ROBERT GEORGE, b Bound Brook, NJ, Jan 6, 17; m 42; c 3. NEUROSURGERY. *Educ:* Rutgers Univ, BS, 38; Univ Pa, MD, 42; Univ Minn, PhD, 51; Am Bd Neurol Surg, dipl, 68. *Prof Exp:* Instr neurosurg, Johns Hopkins Hosp, 49-51; from instr to prof, Dartmouth Med Sch, 51-67; prof surg & chmn dept neurosurg, Med Sch, Univ Okla, 67-74; CHIEF SURG, MUHLENBERG HOSP, PLAINFIELD, 74- *Concurrent Pos:* Chmn dept neurosurg, Hitchcock Clin, Dartmouth Med Sch; consult, Vet Hosp, White River Junction, Vt; prof clin surg & neurosurg, Rutgers Med Sch, Col Med & Dent, NJ, 74-; mem liaison comt grad med educ, AMA, 75- *Mem:* Am Asn Neurol Surgeons; AMA; Asn Res Nerv & Ment Disease; Am Acad Neurol; Am Acad Neurol Surg. *Res:* Cerebrovascular hemodynamics; brain circulatory system research; brain healing. *Mailing Add:* CMDNJ Rutgers Muhlenberg Hosp Rutgers Med Sch Piscataway NJ 08854

FISHER, ROBERT L, b San Jose, Calif; div; c 3. ANIMAL ECOLOGY, MAMMALOGY. *Educ:* San Jose State Col, AB, 58; Cornell Univ, PhD(vert zool), 68. *Prof Exp:* Assoc prof, 63-77, PROF BIOL, JUNIATA COL, 77- *Mem:* Am Soc Mammal; Sigma Xi; Orgn Inland Biol Field Stas; Japanese Soc Pop Ecol. *Res:* Ecology of microtine rodents. *Mailing Add:* Biol Dept Juniata Col 1700 Moore St Huntingdon PA 16652

FISHER, ROBERT LLOYD, b Alhambra, Calif, Aug 19, 25; m 86; c 1. MARINE GEOLOGY, GEOPHYSICS. *Educ:* Calif Inst Technol, BS, 49; Univ Calif, MS, 52, PhD(oceanog), 57. *Prof Exp:* Jr geologist, US Geol Surv, 49; assoc dir, 74-80, RES GEOLOGIST, SCRIPPS INST OCEANOG, UNIV CALIF, SAN DIEGO, 50- *Concurrent Pos:* Mem, Int Hydrographic Orgn-Intergovt Oceanog Comn, guiding comt, Gen Bathymetric Chart of Oceans, 78-; adv comt, Int Geol Correlation Proj, USSR, 82, Greece, 83, Turkey, 85. *Mem:* Geol Soc Am; Am Geophys Union; Explorers Club; Sigma Xi. *Res:* Shipborne geophysical explorations throughout Pacific, Indian and South Atlantic Oceans; deep-sea topography; crustal structure and igneous composition in oceanic regions; comparisons with ophiolites; oceanic trenches and fracture zones; tectonic evolution and plate motions. *Mailing Add:* Scripps Inst Oceanog Univ Calif San Diego La Jolla CA 92093-0215

FISHER, ROBERT STEPHEN, b Brooklyn, NY, July 28, 39; m 64; c 3. GASTROENTEROLOGY. *Educ:* Princeton Univ, BSE, 60; Univ Pa, MD, 64. *Prof Exp:* Intern, Chicago Wesley Mem Hosp, 64-65; resident, internal med, Temple Univ Hosp, 67-70; fel gastroenterol, Hosp Univ Pa, 70-72; from asst prof to assoc prof, 72-80, PROF MED, TEMPLE UNIV HOSP, 80-, DIR, FUNCTIONAL GASTROINTESTINAL DIS CTR, 84- *Concurrent Pos:* Co-chmn gastrointestinal sect, Temple Univ Hosp, 81-85, chmn, 85-; prin investr, NIH, 86- *Mem:* Am Gastroenterol Asn; Am Soc Gastrointestinal Endoscopy; Am Fedn Clin Res; Res Soc Alcoholism. *Res:* Quantitating the movement of the gastrointestinal contents transit including esophageal, stomach emptying, gall bladder evacuation, small intestinal transit, colonic transit; gastroesophageal reflux and enterogastric reflux; correlating symptoms with manometric and scintigraphic events; gastrointestinal motor disorders; functional gastrointestinal disorders. *Mailing Add:* Temple Univ Hosp 3401 N Broad St Philadelphia PA 19140

FISHER, ROBIN SCOTT, b Peebles, Ohio, Apr 22, 52; m 90. CELL BIOLOGY, DEVELOPMENTAL NEUROBIOLOGY. *Educ:* Ohio Univ, BGS, 74, MS, 76, PhD(psychol), 79. *Prof Exp:* Postdoctoral fel neuroanat, dept biol, Purdue Univ, 78-79; postdoctoral trainee neurophysiol, Brain Res Inst, 79-81; asst res neuroanat, Ment Retardation Res Ctr, 81-85; asst prof anat, Dept Anat, Psychiat & Biobehav Sci, 85-89, ASSOC PROF ANAT & PSYCHIAT, DEPT ANAT & PSYCHIAT, UNIV CALIF, LOS ANGELES, 89- *Mem:* Am Asn Anatomists; Soc Neurosci; Soc Exp Biol & Med. *Res:* Combinations of modern morphological methods to determine the developmental organization of synaptic connections in mammalian brain; the reorganization of circuits after brain damage and the neuropathology of human mental retardation. *Mailing Add:* NPI 58-258 760 Westwood Pl Los Angeles CA 90024

FISHER, RONALD RICHARD, biochemistry; deceased, see previous edition for last biography

FISHER, SALLIE ANN, b Green Bay, Wis, Sept 10, 23. WATER CHEMISTRY, ION EXCHANGE. *Educ:* Univ Wis, BS, 45, MS, 46, PhD(anal chem, inorg chem), 49. *Prof Exp:* Instr quant anal, Mt Holyoke Col, 49-50; asst prof, Univ Minn, Duluth, 50-51; res chemist, Rohm and Haas Co, 51-60; assoc dir res, Robinette Res Labs, Inc, 60-72; vpres, 72-76, PRES, PURICONS, INC, 76- *Honors & Awards:* Max Hecht Award, Am Soc Testing & Mat, 75. *Mem:* Am Chem Soc; fel Am Soc Testing & Mat; Nat Asn Corrosion Engrs; Am Water Works Asn; Soc Chem Indust UK. *Res:* Ion exchange applications; industrial water production; metal recovery. *Mailing Add:* Puricons Inc 101 Quaker Ave Malvern PA 19355

FISHER, SAMUEL STURM, b Pinconning, Mich, Feb 25, 38; m 59; c 3. FLUID MECHANICS. *Educ:* Univ Cincinnati, ME, 61; Univ Calif, Los Angeles, MS, 63, PhD(eng), 67. *Prof Exp:* From asst prof to assoc prof, 67-81, PROF AEROSPACE ENG, SCH ENG & APPL SCI, UNIV VA, 81- *Concurrent Pos:* Consult, Nuclear Div, Union Carbide Corp, 71-73; vis scientist, Los Alamos Sci Lab, 75-77. *Mem:* Am Phys Soc; Am Inst Aeronaut & Astronaut; Sigma Xi. *Res:* Low-density fluid mechanics; molecule-surface interactions; homogeneous condensation processes; flow-diagnostic techniques; isotope separation methods; electromagnetic suspension systems. *Mailing Add:* 2409 Kerry Lane Charlottesville VA 22901

FISHER, SAUL HARRISON, b Brooklyn, NY, Feb 20, 13; m 45; c 1. PSYCHIATRY. *Educ:* City Col New York, BS, 32; NY Univ, MD, 36. *Prof Exp:* Asst physiol, 36-37, instr med, 43-45, clin asst psychiat, 48-50, clin instr, 50-54, asst clin prof, 54-57, from asst prof to assoc prof clin psychiat, 57-72, CLIN PROF PSYCHIAT, SCH MED, NY UNIV, 72- *Mem:* AMA; Am Psychiat Asn; Am Orthopsychiat Asn; Acad Psychoanal. *Res:* Relationship of brain damage to personality; effect of severe stress on personality; biological studies in acute psychosis. *Mailing Add:* NY Univ Sch Med Univ Hosp 132 E 35th St New York NY 10016

FISHER, SEYMOUR, b Baltimore, Md, May 13, 22; m 47; c 2. PSYCHIATRY. *Educ:* Univ Chicago, PhD(psychol), 48. *Prof Exp:* Chief psychologist, Elgin State Hosp, Ill, 49-51; USPHS career res investr & assoc prof, Baylor Col Med, Houston, 57-61; PROF, DEPT PSYCHIAT, MED CTR, STATE UNIV NY, SYRACUSE, 61- *Concurrent Pos:* Adj prof, Syracuse Univ, NY, 72-; consult ed, J Nonverbal Behav, 76- *Mem:* Am Psychol Asn. *Res:* How people organize and integrate their body perceptions; nature of changes in body perception during stress, psychosis, and illness; personality factors involved in sexual responsiveness; determinants of psychosomatic symptom choice. *Mailing Add:* Dept Psychiat State Univ Hosp 750 E Adams St Syracuse NY 13210

FISHER, SEYMOUR, b New York, NY, Nov 4, 25; m 59; c 2. PSYCHOPHARMACOLOGY, PHARMACOEPIDEMIOLOGY. *Educ:* NY Univ, Bronx, BA, 48; Univ NC, Chapel Hill, PhD, 52. *Prof Exp:* Supv clin psychologist, Walter Reed Army Inst Res, 52-58; res psychologist, Psychopharmacol Serv Ctr, NIMH, 58-60, chief, spec studies, Psychopharmacol Res Br, 60-63; prof psychol, Boston Univ Sch Med, 63-78; PROF, DEPT PSYCHIAT & BEHAV SCI, UNIV TEX MED BR, GALVESTON, 78-, PROF & DIR, CTR MEDICATION MONITORING, 87- *Concurrent Pos:* Mem, adv bd Psychopharmacol, 63-76, Res Rev Comt, NIMH, 73-77 & 79-83, Hogg Found Comn Commun Care Ment Ill, 87-90; consult, NIMH, 64-66; pres, Boston Mental Health Found Inc, 70-72; assoc ed, Psychopharmacol Bull, 84-; vis prof, Harvard Univ & Boston Univ, 88. *Mem:* Am Col Neuropsychopharmacol (pres, 83-84); Int Soc Pharmacoepidemiol; Am Psychol Asn; Int Col Psychosom Med; Am Psychopath Asn; Col Int Neuropsychopharmacol. *Res:* Nationwide study of side effects and new benefits of prescription drugs used in elderly patients. *Mailing Add:* Dept Psychiat & Behav Sci D-41 Univ Tex Med Br Galveston TX 77550-2777

FISHER, STANLEY PARKINS, b Suffern, NY, Dec 9, 19; m 43; c 2. PETROLEUM GEOLOGY. *Educ:* Univ Va, BA, 42; Univ Okla, MS, 48; Cornell Univ, PhD(geol), 52. *Prof Exp:* Instr geol, Rutgers Univ, 47-49; asst prof geol, Univ NDak & asst state geologist, NDak Geol Surv, 52-53; geologist oil explor, Nat Petrol Coun, Brazil, 53-54; regional explor geologist, Mene Grande Oil Co Div, Gulf Oil Corp, 54-60; from asst prof to assoc prof,

60-68, chmn dept, 70-75, prof, 68-80, assoc dean arts & sci, 78-80, EMER PROF GEOL, OHIO UNIV, 80- *Mem:* Geol Soc Am; Am Asn Petrol Geologists. *Res:* Sedimentology & petroleum geology of northern Appalachian basin; structure and stratigraphy of Andes Mountains and Caribbean area; landslides of central Ohio River Valley; strip mine land reclamation. *Mailing Add:* 26011 Misty Way Dr Ft Mills SC 29715

FISHER, STEPHEN D, b Chicago, Ill, July 23, 41; m 65; c 3. MATHEMATICS. *Educ:* Mass Inst Technol, BS, 63; Univ Wis, Madison, PhD(math), 67. *Prof Exp:* Instr math, Mass Inst Technol, 67-69; from asst prof to assoc prof, 69-79, PROF MATH, NORTHWESTERN UNIV, 79- *Mem:* Am Math Soc. *Res:* Complex analysis; variational problems. *Mailing Add:* Dept Math Northwestern Univ Evanston IL 60208

FISHER, STEVEN KAY, b Rochester, Ind, July 18, 42; m 71; c 3. NEUROBIOLOGY. *Educ:* Purdue Univ, BS, 64, MS, 66, PhD(neurobiol), 69. *Prof Exp:* USPHS fel biophys, Johns Hopkins Univ, 69-71; PROF NEUROBIOL & DIR, NEUROSCI RES INST, UNIV CALIF, SANTA BARBARA, 71- *Concurrent Pos:* Prin investr, USPHS res grant, Nat Eye Inst, 72; res career develop award, NIH, 79; mem, Visual Sci A-2 Study Sect, NIH, 89, chmn, 90. *Honors & Awards:* Merit Award, NIH, 90. *Mem:* AAAS; Asn Res Vision & Ophthal; Electron Micros Soc Am; Am Soc Cell Biol. *Res:* Studies of the developing and adult vertebrate visual system using light and electron microscopy, autoradiography and immunocytochemistry; cell ultrastucture, protein synthesis, membrane renewal, retinal degeneration and regeneration, retinal detachment. *Mailing Add:* Dept Biol Sci Univ Calif Santa Barbara CA 93106

FISHER, STUART GORDON, b Elmhurst, Ill, Mar 1, 43; m 82; c 3. ECOLOGY, LIMNOLOGY. *Educ:* Wake Forest Col, BS, 65, MA, 67; Dartmouth Col, PhD(biol), 71. *Prof Exp:* Asst prof biol, Amherst Col, 70-76; assoc prof, 78-84, PROF ZOOL, ARIZ STATE UNIV, 84- *Concurrent Pos:* Adv panel, long term ecol res, Nat Sci Found Ecosyst Studies, 81-88; assoc ed, J NAm Benthological Soc, 85-86; ed, Ecol, 86- *Mem:* Ecol Soc Am; Am Soc Limnol & Oceanog; Int Asn Theoret & Appl Limnol. *Res:* Ecosystem biology; Aquatic ecology; energy flow and nutrient cycling in stream ecosystems; ecology of desert streams; metabolism of running water ecosystems; stream ecology. *Mailing Add:* Dept Zool Ariz State Univ Tempe AZ 85287

FISHER, T RICHARD, b Brownstown, Ill, Dec 23, 21; m 44; c 4. BOTANY, HORTICULTURE. *Educ:* Eastern Ill Univ, BS, 47; Ind Univ, PhD(bot), 54. *Prof Exp:* Asst prof, Appalachian State Teachers Col, 54-56; prof bot, Ohio State Univ, 56-68; chmn dept biol, Bowling Green State Univ, 68-74, prof bot, 68-88; RETIRED. *Concurrent Pos:* Consult, Aid India Prog, 66-; lectr, Urbana Col, Ohio, 82-; dir, Schedel Found: Arboretum & Gardens. *Mem:* Sigma Xi; Bot Soc Am; Am Soc Plant Taxon; Int Asn Plant Taxon; Am Asn Bot Gardens & Arboreta. *Res:* Biosystematic investigations of Silphium and other Compositae; taxonomy and cytotaxonomy of compositae. *Mailing Add:* PO Box 81 Elmore OH 43416

FISHER, THEODORE WILLIAM, b San Francisco, Calif, May 26, 21; m 46; c 1. ENTOMOLOGY. *Educ:* San Jose State Col, AB, 43; Univ Calif, Berkeley, PhD(entom), 52. *Prof Exp:* Prin lab technician, Citrus Exp Sta, Univ Calif, Riverside, 48-57, asst entomologist, 57-64, assoc specialist entom, 64-66, specialist entom biol control, 66-91; RETIRED. *Mem:* Am Entom Soc; Registry Prof Entomologists. *Res:* Biological control of pest mollusks; consulting of citrus pest management; taxonomy of Sciomyzidae (Diptera); consulting for escargot industry in US. *Mailing Add:* Dept Entom Univ Calif Riverside CA 92521-0317

FISHER, THOMAS HENRY, b Fulton, Mo, Aug 22, 38; m 56; c 3. ORGANIC CHEMISTRY, PHYSICAL CHEMISTRY. *Educ:* Westminster Col, Mo, BA, 60; Univ Ill, Urbana, MS, 62, PhD(org chem), 64. *Prof Exp:* Asst prof org chem, 66-70, assoc prof, 70-88, PROF CHEM, MISS STATE UNIV, 88- *Mem:* Am Chem Soc; Sigma Xi. *Res:* Organic reaction mechanisms; free radical reactions; factors affecting free radical stability. *Mailing Add:* Box CH Dept Chem Miss State Univ Miss State MS 39762

FISHER, THORNTON ROBERTS, b Santa Monica, Calif, Feb 16, 37; m; c 4. NUCLEAR PHYSICS, PLASMA PHYSICS. *Educ:* Wesleyan Univ, BA, 58; Calif Inst Technol, PhD(physics), 63. *Prof Exp:* Res assoc nuclear struct physics, Calif Inst Technol, 63; res assoc, Stanford Univ, 63-69; res scientist, Lockheed Palo Alto Res Lab, 69-75, staff scientist, 75-82, sr staff scientist, 82-84, consult scientist, 85-89, MGR, SPACE PAYLOADS, LOCKHEED PALO ALTO RES LAB, 89- *Mem:* Am Phys Soc; Int Soc Optical Eng; Int Soc Hybrid Microelectronics; Soc Indust & Appl Math. *Res:* Nuclear structure physics; charged particle, gamma ray, and neutron producing reactions; cryogenics; laser-driven fusion; space physics. *Mailing Add:* Lockheed Palo Alto Res Lab 3251 Hanover St Palo Alto CA 94304

FISHER, TOM LYONS, b Cincinnati, Ohio, Aug 13, 42; m 72. BIOCHEMISTRY. *Educ:* Old Dominion Col, BS, 64; Iowa State Univ, PhD(biochem), 71. *Prof Exp:* Res assoc, Va Polytech Inst, 70-72; asst prof, St Mary's Col Md, 72-76; from asst prof to assoc prof, 76-88, PROF, DEPT CHEM JUNIATA COL, 88- *Concurrent Pos:* Adj prof, Va Polytech, Reston, 73; grantee, Res Corp, 78-79; NSF grant, 80. *Mem:* Am Chem Soc. *Res:* Bioorganic chemistry; synthesis and characterization of analogs of vitamins, especially nicotinamide adenine dinucleotide; design of inexpensive laboratory instrumentation. *Mailing Add:* Dept Chem Juniata Col Huntingdon PA 16652

FISHER, VINCENT J, b New York, NY, Mar 31, 30. PHYSIOLOGY. *Educ:* Fordham Univ, BS, 51; State Univ NY, MD, 58. *Prof Exp:* Res trainee, NY Heart Asn, Cardiopulmonary Lab, NY Vet Admin Med Ctr, 62-64, clin investr, 70-72; Am Heart Asn advan res fel, Dept Physiol, Downstate Med Ctr, State Univ NY, Brooklyn, 64-66, asst instr physiol, 64-66, asst prof physiol; 66-72, asst prof physiol, Sch Grad Studies, 70-72, assoc prof physiol, Health Sci Ctr, 72-73, assoc prof physiol, Sch Grad Studies, Health Sci Ctr, 72-73; assoc prof, 73-79, PROF PHYSIOL & BIOPHYS, SCH MED, NY UNIV, 79- *Concurrent Pos:* Attend physician, Cardiopulmonary Lab, NY Vet Admin Med Ctr, 64-66, asst chief, Cardiol Sect, 66-71, asst chief, Cardiopulmonary Lab, 71-73, co-chief, Cardiol Sect, 71-89, assoc chief staff res & develop, 73-, chief, Cardiol Sect, 89-; sr lectr physiol, Downstate Med Ctr, State Univ NY, Brooklyn, 73-; course dir, Mammalian Physiol Course, Sch Med, NY Univ, 77-; mem, Res Coun, NY Heart Asn, 77-80, Vet Admin Rehab Eng Ctr Task Force, 82-83 & Inst Bd Res Assocs, Sch Med, NY Univ, 74- *Mem:* Am Physiol Soc; Cardiac Muscle Soc; Int Soc Heart Res; Asn Vet Admin Cardiologists. *Res:* Author of various publications. *Mailing Add:* Dept Physiol & Biophys NY Univ Sch Med 550 First Ave New York NY 10016

FISHER, WALDO REYNOLDS, b Philadelphia, Pa, Sept 10, 30; div; c 3. METABOLISM, ATHEROSCLEROSIS. *Educ:* Wesleyan Univ, BA, 52; Univ Pa, MD, 56, PhD(biochem), 64; Am Bd Internal Med, dipl, 66. *Prof Exp:* From intern to asst resident med, Presby Hosp, Philadelphia, 56-58; instr biochem, Sch Med, Univ Pa, 63-64; sr resident, Peter Bent Brigham Hosp, Boston, 64-65; asst prof med, 65-71, asst prof biochem, 68-71, actg chief div endocrinol, 74, assoc prof med & biochem, 71-78, PROF MED & BIOCHEM, COL MED, UNIV FLA, 78- *Concurrent Pos:* NIH res grants, 66-90, res career develop award, 67-72; Am Heart Asn res grant, 67-73; fac develop award, Univ Fla, 70; vis scientist, Lab Chem Biol, NIH, 71-72, Lab Theoret Biol, 86; mem coun atherosclerosis, Am Heart Asn; mem, NIH Metab Study Sect, 81-84. *Mem:* AAAS; Am Soc Biol Chem; Am Heart Asn. *Res:* Biochemistry and physiology of metabolic disease; structure and metabolism of plasma lipoproteins and pathophysiology of hyperlipemic diseases; apolipoprotein kinetics; tracer kinetics and comparimental modeling. *Mailing Add:* Dept Med Box J-226 Univ Fla Col Med Gainesville FL 32610

FISHER, WARNER DOUGLASS, b Sharon, Tenn, Aug 17, 23; m 48; c 4. AGRONOMY. *Educ:* Purdue Univ, BS, 47; Utah State Agr Col, MS, 49; Agr & Mech Col Tex, PhD(plant breeding), 54. *Prof Exp:* Instr agron, Univ Tenn, Martin, 48-51; asst, Agr & Mech Col Tex, 51-53; assoc prof, Ala Polytech Inst, 53-54; PLANT BREEDER, COTTON RES CTR, UNIV ARIZ, 54- *Mem:* Am Soc Agron. *Res:* Cotton production and breeding. *Mailing Add:* 1757 E Garnet Mesa AZ 85204

FISHER, WILLIAM FRANCIS, b Mobile, Ala, Nov 3, 41; m 62; c 5. IMMUNOLOGY, PARASITOLOGY. *Educ:* Col St Joseph, BS, 63; NMex Highlands Univ, MS, 65. *Prof Exp:* Res lab technician histol, Vet Admin Hosp, 65; lab instr biol, Univ Albuquerque, 65-66; biol lab technician parasitol, Agr Res Serv, Parasite Res Lab, 66-67, res asst, 67-77, RES SCIENTIST IMMUNOL & ACAROLOGY, US LIVESTOCK INSECTS LAB, AGR RES SERV, SCI & EDUC, USDA, 77- *Mem:* Am Soc Parasitologists; Acarological Soc Am; Entom Soc Am. *Res:* Immunological and physiological responses of animals to parasitic mites; host-parasite relationships; biology and control of parasitic mites. *Mailing Add:* PO Box 232 Kerrville TX 78029

FISHER, WILLIAM GARY, b Cleveland, Ohio, Aug 15, 47. MEDICAL PHYSICS. *Educ:* Ariz State Univ, BS, 69, MS, 71. *Prof Exp:* Res asst, Orange County Med Ctr, Calif, 72-73; dept supvr nuclear med, Mesa Gen Hosp, Mesa, Ariz, 74-78; PRES, FISHER MED PHYSICS, TEMPE, ARIZ, 79- *Concurrent Pos:* Lectr, Saint Joseph's Hosp & Med Ctr, Phoenix, Ariz, 76- *Mem:* Am Asn Physicists Med; Am Phys Soc; Health Physics Soc; Radiol Soc NAm; Soc Nuclear Med. *Mailing Add:* Fisher Med Phys 2082 E Balboa Dr Tempe AZ 85282

FISHER, WILLIAM LAWRENCE, b Marion, Ill, Sept 16, 32; m 54; c 3. GEOLOGY, PALEONTOLOGY. *Educ:* Southern Ill Univ, BS, 54; Univ Kans, MS, 58, PhD(geol), 61. *Hon Degrees:* DSc, Southern Ill Univ, 86. *Prof Exp:* Geologist, Aluminum Co Am, 57; asst, Kans Geol Surv, 57-58; from dep asst secy to asst secy energy & minerals, Dept Interior, 75-77; res assoc, Univ Tex, Austin, 60-64, res scientist, 64-68, from assoc dir to dir, 68-75, spec lectr geol & assoc mem grad fac, 64-69, prof pub affairs, 81-84, PROF GEOL SCI & MEM GRAD FAC, BUR ECON GEOL, UNIV TEX, AUSTIN, 69-, DIR, 77-, LEONIDAS BORROW CHAIR MINERAL TO RESOURCES, 86- *Concurrent Pos:* Outer Continental Shelf Policy Adv Bd, US Dept Interior, 79- & chmn, 82-84; adv comt, Coastal Zone Mgt, US Dept Commerce, 73-75; mem, Interstate Oil comp act Comn, 77-, Interstate Mining Compact Comn, 80-, US Nat Comt Geol, 81-90, bd mineral & energy resources, Nat Res Coun, 82-89, adv coun, Ga Res Inst, 86- & Nat Petrol Coun, 88-; chmn, Dept Geol Sci, Univ Texas, Austin, 84-90; distinguished lectr, Am Asn Petrol Geologists, 70-71 & 78-79; Morgan Davis centennial prof petrol geol, 82-86; White House Sci Coun, 89-90, secy, Energy Adv Bd, 89-; bd, Earth Sci & Resources, Nat Res Coun, co-chmn, 89-90, chmn, 90- *Honors & Awards:* Distinguished Serv Award, Am Asn Petrol Geologists; Pub Serv Award, Am Inst Prof Geologists. *Mem:* Geol Soc Am; hon mem Am Asn Petrol Geologists (pres, 85-86); Soc Econ Geol; Asn Am State Geol (pres, 81-82); Soc Mining Eng; Sigma Xi; Am Inst Prof Geologist. *Res:* Depositional systems; seismic stratigraphy; oil, gas and mineral exploration and development; energy and minerals policy. *Mailing Add:* Bur Economic Geol Univ Tex Austin TX 78712

FISHKIN, ARTHUR FREDERIC, b New York, NY, May 27, 30; m 56; c 4. BIOCHEMISTRY. *Educ:* Ind Univ, AB, 51, AM, 53; Univ Iowa, PhD(biochem), 57. *Prof Exp:* From instr to asst prof biochem, Sch Med, La State Univ, 58-64; asst prof chem, NMex State Univ, 64-68; ASSOC PROF BIOCHEM, SCH MED, CREIGHTON UNIV, 68-, CHMN, ANIMAL RES COMT, 83- *Mem:* Soc Complex Carbohydrates; NY Acad Sci; Am Chem Soc. *Res:* Glycoproteins of cardiovascular connective tissue; connective tissue biochemistry. *Mailing Add:* Creighton Univ Sch Med Omaha NE 68178

FISHMAN, ALFRED PAUL, b New York, NY, Sept 24, 18; wid; c 2. PHYSIOLOGY. *Educ:* Univ Mich, AB, 38, MS, 39; Univ Louisville, MD, 43; Am Bd Internal Med, cert, 53. *Prof Exp:* Intern med, Jewish Hosp, Brooklyn, NY, 43-44; from asst resident to resident med, Mt Sinai Hosp, 47-48; Am Heart Asn estab investr, Columbia Univ, 51-55, from asst prof to assoc prof med, Col Physicians & Surgeons, 55-67; prof, Univ Chicago, 67-69; prof med, 69-72, assoc dean sch med, 69-76, dir, Cardiovasc Pulmonary Div, 69-89, WILLIAM MAUL MEASEY PROF MED, SCH MED, UNIV PA, 72-, CHMN, REHAB MED, 89- *Concurrent Pos:* Dazian Found fel path, Mt Sinai Hosp, New York, 46-47; fel cardiovasc physiol, Michael Reese Hosp, Chicago, Ill, 48-49; Am Heart Asn fels physiol, Bellevue Hosp, New York, 49-50 & Harvard Univ, 50-51; Commonwealth fel physiol, Nuffield Inst Med Res, Eng, 64-65; dir cardio-respiratory lab, Columbia-Presby Med Ctr, 55-66; trustee, Mt Desert Island Biol Lab, 55-60; consult, Vet Admin Hosp, Bronx, NY, 61-67 & Off Sci Technol, 62-; hon consult med, St Mary's Hosp Med Sch, London, 64; Sir Ernest Finch prof, Univ Sheffield, 65; dir cardiovasc inst, Michael Reese Med Ctr, Chicago, 67-69; mem, Nat Adv Heart Coun, 68-72; James Howard Means vis prof, Harvard Med Sch, 70; Arthur E Strauss vis prof, Sch Med, Washington Univ, 73; mem adv panel cardiovasc dis, WHO, 73; consult to chancellor, Univ Mo-Kansas City, 73-78; mem adv comt respiratory dis coal miners, Gov Pa, 74; vis scientist, Johns Hopkins Univ & Univ Md, 74; Zyskind hon vis prof, Ben Gurion Univ, Israel, 75; chmn bd external reviewers, Cleveland Found, Case Western Reserve Med Sch, 78-; deleg, Nat Heart Lung Blood Inst, People's Repub China, 80, USSR, 85; chmn, Am Physiol Soc deleg to Nat Acad Sci, People's Repub China, 83; consult, Cleveland Found, Ohio, 85-, Health Comnr, NY State, 88-; chmn, Int Conf Pulmonary Hypertension & Nat Registry, Nat Heart, Lung & Blood Inst & Burroughs-Wellcome, 87- *Honors & Awards:* V D Mattle Mem Lectr, Rutgers Med Sch, 77; Ross Mem Distinguished Lectr, Univ Mo, 77; Dickinson W Richards Lectr, 79; Distinguished Achievement Award, Am Heart Asn Sci Councils, 80; Hans Hecht Mem Lectr, 80; Distinguished Serv Award, Am Heart Asn, 89. *Mem:* Inst Med-Nat Acad Sci; Am Soc Clin Invest; Am Physiol Soc (pres-elect, 82-83, pres, 83-84); Am Fedn Clin Res; fel Am Col Physicians; fel AAAS; Asn Am Physicians; hon fel Am Col Cardiol; hon fel Am Col Chest Physicians; Am Heart Asn. *Res:* Physiology of respiration and circulation; comparative physiology; internal medicine; interactions of the heart and lungs for purposes of gas exchange; regulation of the pulmonary circulation, heart failure, control of breathing in health and disease; comparative biology. *Mailing Add:* Rehab Dept Maloney Bldg Univ Pa 3600 Spruce St Philadelphia PA 19104

FISHMAN, DAVID H, b Brooklyn, NY, Aug 21, 39; m 62; c 3. POLYMER SCIENCE. *Educ:* Columbia Col, AB, 60; Pa State Univ, PhD(org chem), 64; Farleigh Dickinson Univ, MBA, 85. *Prof Exp:* Res chemist org chem dept, E I du Pont de Nemours & Co, Inc, 64-66; Celanese Plastics Co, 66-69, group leader plastic molding resins, 69-74, tech mgr, 74-76; mgr tech acquisitions, BASF Inmont Corp, 76-80, dir tech planning & develop, 80-85, dir cent res lab, 85-90; TECH DIR PACKAGING INKS, SUN CHEM CO, 91- *Mem:* Am Chem Soc; Soc Plastics Engrs; Licensing Exec Soc; Soc Chem Indust France. *Res:* Coatings; printing inks; sealants; adhesives; cellular rubber; colorants; concentrates; plastics; composites. *Mailing Add:* 155 Kline Blvd Berkley Hgts NJ 07922

FISHMAN, ELLIOT KEITH, b Brooklyn, NY, Feb 13, 53; m 82; c 2. COMPUTED TOMOGRAPHY, THREE-DIMENSIONAL IMAGING. *Educ:* Univ Md, BS, 73, MD, 77. *Prof Exp:* Fel comput tomography, 80-81, asst prof radiol, 81-85, ASSOC PROF RADIOL, JOHNS HOPKINS MED INST, 85- *Concurrent Pos:* Consult, Siemans Med Systs, 81-, Winthrop Labs, 83- & Pixar, 85- *Mem:* Radiol Soc NAm; Soc Comput Body Tomography; Nat Comput Graphics Asn; Am Roentgen Ray Soc. *Res:* Development of an advanced imaging system for three-dimensional display of medical images. *Mailing Add:* Dept Radiol John Hopkins Hosp Baltimore MD 21205

FISHMAN, FRANK J, JR, b Chicago, Ill, Jan 18, 31; m 54; c 5. THEORETICAL PHYSICS. *Educ:* Univ Ill, BS, 52; Harvard Univ, AM, 53, PhD(physics), 57. *Prof Exp:* Prin res scientist, Avco-Everett Res Lab, Avco Corp, 57-68; assoc prof physics, Adrian Col, 68-76; INSTR PHYS SCI, GRAND RAPIDS JR COL, 76- *Mem:* Am Phys Soc; Am Asn Physics Teachers; Am Sci Affiliation. *Res:* Magnetohydrodynamics; generators, shock wave structure, kinetic models of gases and plasmas; physics of music. *Mailing Add:* Grand Rapids Jr Col 143 Bostwick NE Grand Rapids MI 49502

FISHMAN, GEORGE SAMUEL, b Everett, Mass, July 3, 37; m 69; c 2. OPERATIONS RESEARCH, APPLIED STATISTICS. *Educ:* Mass Inst Technol, BS, 60; Stanford Univ, MA, 63; Univ Calif, Los Angeles, PhD(biostatist), 70. *Prof Exp:* Economist, Rand Corp, 62-70; assoc prof opers res and assoc dir health serv res training prog, Yale Univ, 70-74; chmn, Opers Res & Systs Anal, 80-90, PROF OPERS RES, ANALYSIS & SYSTS, UNIV NC, CHAPEL HILL, 74- *Concurrent Pos:* Consult, New York City Rand Inst, 70-71, Bur Manpower Educ, Health Serv & Ment Health Admin, 72-73 & Housing Auth, City of New Haven & Ford Found, 72-74, Xerox, 80, CACI, 83-88. *Mem:* Am Statist Asn; Opers Res Soc Am; Inst Mgt Sci. *Res:* Statistical analysis of discrete event digital simulation; experimental design and techniques aimed at statistical evaluation of simulation results. *Mailing Add:* Dept Oper Res Univ NC CB No 3180 Smith Bldg 128A Chapel Hill NC 27599

FISHMAN, GERALD JAY, b St Louis, Mo, Feb 10, 43; m 67; c 2. ASTROPHYSICS, GAMMA-RAY ASTRONOMY. *Educ:* Univ Mo-Columbia, BS, 65; Rice Univ, MS, 68, PhD(space sci), 69. *Prof Exp:* Res assoc space sci, Rice Univ, 69; sr physicist, Teledyne Brown Eng Co, 69-74; res physicist, 74-77, staff scientist, 77-78, RES PHYSICIST, MARSHALL SPACE FLIGHT CTR, NASA, 78- *Concurrent Pos:* Prin investr, Gamma Ray Observ, 78- *Honors & Awards:* NASA Medal, 83. *Mem:* AAAS; Am Astron Soc; Am Phys Soc; Int Astron Union; Sigma Xi. *Res:* Nuclear instrumentation; gamma-ray and x-ray astronomy; high energy and nuclear astrophysics; radiation in space. *Mailing Add:* 8024 Tea Garden Rd Huntsville AL 35802

FISHMAN, HARVEY MORTON, b Lynn, Mass, Oct 27, 37; m 62; c 3. BIOPHYSICS, PHYSIOLOGY. *Educ:* Mass Inst Technol, BS, 54; Univ Calif, Berkeley, MS, 64, PhD(biophys), 68. *Prof Exp:* Design engr, Hewlett-Packard Co, 59-63; staff fel, Lab Biophys, Nat Inst Neurol & Commun Dis & Stroke, 68-70; asst prof biol sci, State Univ NY Albany, 70-73; assoc prof, 73-77, PROF PHYSIOL & BIOPHYS, UNIV TEX MED BR GALVESTON, 77- *Concurrent Pos:* Develop engr, Space Sci Lab, Univ Calif, Berkeley, 64-66, vis prof, Dept Biophysics, 79; investr, Marine Biol Lab, Woods Hole, Mass, 68-, mem corp, 71-; assoc ed, J Neurosci, Sect Cellular Neurosci, 81-83; chmn, NIH Small Bus Res Grant Study Sect, 86; consult, FDA, Cardio-Renal Drug Prod Div, 86-91; guest investr, Centre Nat de la Recherche Scientifique, France, 89; track chair, Inst Elec & Electronics Engrs-EMBS, 87-88. *Mem:* Biophys Soc; Inst Elec & Electronics Engrs; Soc Gen Physiol; Sigma Xi; Am Phys Soc; NY Acad Sci; AAAS; Soc Neurosci. *Res:* Ion conduction in excitable membranes; membrane biophysics; investigation of ion-channel processes and dielectric phenomena in membranes and evalution of model mechanics; fluctuation and linear (complex admittance) analysis of membrane ion conduction; nerve membrane repair and degenerative processes after injury and in disease. *Mailing Add:* Dept Physiol & Biophys Univ Tex Med Br Galveston TX 77550-2779

FISHMAN, IRVING YALE, b Ardmore, Okla, Sept 12, 20; m 42; c 2. SENSORY PHYSIOLOGY. *Educ:* Univ Okla, BS, 42, MS, 48; Fla State Univ, PhD(physiol, biochem), 55. *Prof Exp:* Spec lectr physiol, Fla State Univ, 51-54; from instr to prof biol, Grinnell Col, 54-91, Stone prof biol, 70-91; RETIRED. *Concurrent Pos:* NSF res grants, 54-; prof physiol, Rush Med Col, 73-82. *Mem:* Am Physiol Soc. *Res:* Electrophysiological studies of chemoreception in mammals. *Mailing Add:* Dept Biol Grinnell Col Grinnell IA 50112

FISHMAN, JACK, b Poland, Sept 27, 30; nat US; m 64; c 3. ORGANIC CHEMISTRY, BIOCHEMISTRY. *Educ:* Yeshiva Univ, BA, 50; Columbia Univ, MA, 52; Wayne State Univ, PhD, 55. *Prof Exp:* Fel, Oxford Univ, 55-56; res assoc, Sloan-Kettering Inst Cancer Res, 56-58, asst, 58-62; assoc, Inst Steroid Res, Montefiore Hosp & Med Ctr, 62-74, dir, 74-78; PROF BIOCHEM, ROCKEFELLER UNIV, 78- *Concurrent Pos:* From assoc prof to prof biochem, Albert Einstein Col Med, Yeshiva Univ, 67-78. *Mem:* Endocrine Soc; Am Chem Soc; Am Soc Biol Chemists. *Res:* Natural products, particularly steroids and alkaloids; metabolism of steroids. *Mailing Add:* 8800 NW 36th St Miami FL 33178

FISHMAN, JERRY HASKEL, b June 21, 23. CHEMISTRY. *Educ:* Sir George Williams Univ, BSc, 52; Brooklyn Col, MS, 58; Stevens Inst Technol, PhD(chem), 60. *Prof Exp:* Res mgr, Leesona Moos Lab, Div Leesona Corp, 60-71; prin investr, Med Ctr, Montefiore Hosp, 71-73; res chemist, Inst Steroid Res, 74-77; sr res assoc, Rockefeller Univ, 77-89; MGR, HASKEL TECHNOL, 90- *Mem:* Electrochem Soc; Sigma Xi; Am Inst Chemists; NAm Membrane Soc. *Res:* Electrochemical reactions at interfaces; steroid hormones; hormone receptors and cancer; permeable membranes. *Mailing Add:* 227 Central Park W New York NY 10024

FISHMAN, MARSHALL LEWIS, b Philadelphia, Pa, July 2, 37; m 66; c 2. AGRICULTURAL & FOOD CHEMISTRY, POLYMER CHEMISTRY. *Educ:* Temple Univ, BA, 59; Villanova Univ, MS, 61; Polytech Inst Brooklyn, PhD, 69. *Prof Exp:* Res chemist, Agr Res Serv, East Reg Res Ctr, 69-71, res chemist, R B Russell Ctr, 71-80, RES CHEMIST, EAST REG RES CTR, USDA, 80- *Concurrent Pos:* Gen chmn, Fed Analyt Chem & Spectroscopy Soc, 85-86; pres, Del Valley Chrom Forum, 85-86. *Mem:* Am Chem Soc; Sigma Xi. *Res:* Isolation, purification, fractionation, and characterization of polysaccharides; study chemistry and function of pectins as plant cell wall polysaccharide and food fibers; study solution properties of polysaccharides and proteins by high performance size exclusion chromotography, membrane osometry, on line viscometry and light scattering; study interactions and aggregation properties of pectins. *Mailing Add:* 139 Woodland Dr Lansdale PA 19446

FISHMAN, MARVIN, b New York, NY, Mar 17, 28; m 57; c 2. IMMUNOLOGY. *Educ:* City Col New York, BS, 49; Univ Ky, MS, 51; Wash Univ, PhD(microbiol), 54. *Prof Exp:* Assoc mem, 62-66, MEM, PUB HEALTH RES INST, CITY OF NEW YORK, INC, 66-; ADJ PROF PATH, POSTGRAD MED SCH, NY UNIV, 77-; AT ST JUDE CHILDREN'S RES HOSP. *Concurrent Pos:* Res prof path, Postgrad Med Sch, NY Univ, 68-76. *Honors & Awards:* Selman A Waksman Award, 65. *Mem:* Am Asn Immunol; Fedn Am Socs Exp Biol. *Res:* Investigation in the mechanism of antibody formation and antibody action. *Mailing Add:* Div Immunol St Jude Children's Res Hosp 332 N Lauderdale PO Box 318 Memphis TN 38101

FISHMAN, MARVIN JOSEPH, b Denver, Colo, Apr 15, 32; m 57; c 3. WATER CHEMISTRY. *Educ:* Univ Colo, BA, 54, MS, 56. *Prof Exp:* Chemist, Water Resources Div, US Geol Surv, 56-90; RETIRED. *Honors & Awards:* Super Serv Award, US Dept Interior, 81; Award of Merit, Am Soc Testing & Mat, 89. *Mem:* Soc Appl Spectros; Am Soc Testing & Mat. *Res:* Water analysis; development of new or improved methods for determining inorganic constituents found in water; automation of analytical methods; development of atomic absorption methods for determining metals in water; determining anions in water by ion chromatography. *Mailing Add:* 3353 S Niagara Way Denver CO 80224

FISHMAN, MORRIS, b Montreal, Que, Mar 2, 39; m 67; c 2. ORGANIC CHEMISTRY. *Educ:* McGill Univ, BSc, 60; Univ NB, PhD(chem), 66. *Prof Exp:* Lab asst chem, Ayerst, McKenna & Harrison, Que, 60-62; sr res chemist, FMC Corp, 66-83; lab suprv, Angus Chem Co, 83-86; ASSOC PROF, NY UNIV, 86- *Mem:* Am Chem Soc; Chem Inst Can. *Res:* Synthesizing organic compounds to be evaluated in medicinal or agricultural biological programs; process development work. *Mailing Add:* 37 Maplestream Rd E Windsor NJ 08520

FISHMAN, MYER M, b Boston, Mass, Apr 27, 18; m 48; c 2. BIOCHEMISTRY. *Educ:* City Col New York, BS, 38; Univ Minn, MS, 40, PhD, 42. *Prof Exp:* Assoc chemist, Sig Corps, US Army, 42-43; chemist, Stein, Hall & Co, 43-44; from instr to assoc prof, 46-67, assoc dean, Col Lib Arts & Sci, 69-76, PROF CHEM, CITY COL NY, 67-, DIR, BIOMED RES PROGS, 79- *Mem:* Am Chem Soc; NY Acad Sci; Am Asn Cancer Res; Am Soc Biol Chem. *Res:* Physical biochemistry. *Mailing Add:* Dept Chem City Col New York New York NY 10031

FISHMAN, NORMAN, b Petaluma, Calif, July 18, 24; m 53; c 2. POLYMER TECHNOLOGY, CHEMICAL ENGINEERING. *Educ:* Univ Calif, BS, 48. *Prof Exp:* Chem engr, Western Regional Utilization Lab, USDA, 49-53; proj engr, Food Mach & Chem Corp, 53-54; sr chem engr, SRI Int, 54-61, mgr propellant eval, 61-67, mgr polymer technol, 67-76, sr indust economist, 76-79, dir, Polymers Dept, 79-82, sr consult, 82-88; INDEPENDENT CONSULT, 88- *Honors & Awards:* USDA Award, 54. *Mem:* Soc Plastics Engrs; Sigma Xi; Am Technion Soc; Am Chem Soc; Soc Advan Mat & Process Eng; ASM Int. *Res:* Applications of polymeric materials; fire retardance in polymers; market research; acquisition and diversification studies; technoeconomics. *Mailing Add:* 2316 Blueridge Ave Menlo Park CA 94025

FISHMAN, PETER HARVEY, b Boston, Mass, Dec 8, 39. BIOCHEMISTRY. *Educ:* Mass Inst Technol, BS, 61; George Washington Univ, MS, 67, PhD(biochem), 70. *Prof Exp:* From res asst to teaching fel biochem, George Washington Univ, 65-69; staff fel, 70-75, res assoc biochem, 75-80, CHIEF MEMBRANE BIOCHEM SECT, NAT INST NEUROL & COMMUN DIS & STROKE, 80- *Concurrent Pos:* Lectr biochem, George Washington Univ, 70-78; prof lectr, 78- *Mem:* Am Soc Biol Chemists; Sigma Xi. *Res:* Biosynthesis of complex carbohydrates; abnormal biosynthesis of gangliosides in neoplasia and genetic diseases; role of surface glycoconjugates in transmission of information across cell membranes; role of gangliosides as receptors and in cellular differentiation; regulation and desensitization of adenylate cyclase; mechanism of action of bacterial toxins. *Mailing Add:* 10644 Muirfield Dr Potomac MD 20854

FISHMAN, PHILIP M, b Brooklyn, NY, June 22, 43; m 67; c 3. STATISTICAL COMMUNICATION THEORY, COMMUNICATIONS SYSTEM ENGINEERING. *Educ:* Brooklyn Col, BS, 64; NY Univ, MS, 68; Wash Univ, DSc, 75. *Prof Exp:* Asst prof genetics psychiat, Med Sch, Wash Univ, 75-78; mem tech staff commun eng, Lincoln Lab, Mass Inst Technol, 78-87; LEAD ENGR COMMUN ENG, MITRE CORP, 87- *Mem:* Inst Elec & Electronics Engrs; Am Inst Aeronaut & Astronaut. *Res:* Satellite and terrestrial communications systems; spread spectrum communications; statistical detection theory; stochastic processes; genetic statistics-particularly as applied to human diseases. *Mailing Add:* 225 Mill St Newton MA 02160

FISHMAN, ROBERT ALLEN, b New York, NY, May 30, 24; m 56, 83; c 3. NEUROLOGY. *Educ:* Columbia Univ, AB, 44; Univ Pa, MD, 47. *Prof Exp:* Intern & asst resident med, New Haven Hosp, 47-49; resident neurol, Mass Gen Hosp, 49-50; asst resident neurol, Neurol Inst, 50-51; from instr to assoc prof neurol, Col Physicians & Surgeons, Columbia Univ, 54-66; PROF NEUROL & CHMN DEPT, SCH MED, UNIV CALIF, SAN FRANCISCO, 66- *Concurrent Pos:* Asst med, Yale Univ, 47-49; teaching fel, Harvard Med Sch, 49-50; neurophysiologist, Army Med Serv Grad Sch, DC, 51-53; chief res, Neurol Inst, 53-54, asst attend neurologist, 55-62, assoc attend neurologist, 62-66; Markle scholar med sci, 60-65; co-dir, Neurol Clin Res Ctr, 61-66; consult neurologist, San Francisco Gen Hosp, San Francisco Vet Admin Hosp & Letterman Gen Hosp, San Francisco; dir, Am Bd Psychiat & Neurol, 81-88, pres, 87; mem, Nat Adv Coun, Nat Inst Neurol & Commun Dis & Stroke, 84-87; dir, Brain Edema Clin Res Ctr, Univ Calif, San Francisco. *Honors & Awards:* Moses Res Prize, Columbia Univ, 70, Royer Award, 73; Jacoby Award, Am Neurol Asn, 86. *Mem:* Am Neurol Asn(pres, 84-85); Am Acad Neurol (vpres, 71-73, pres, 75-77). *Res:* Metabolic disorders of the nervous system; cerebrospinal fluid; brain edema. *Mailing Add:* Moffitt Long Hosp-Neurol M-794 Univ Calif San Francisco Med Ctr San Francisco CA 94143

FISHMAN, ROBERT SUMNER, b Boston, Mass, May 17, 32; m 57; c 4. MATHEMATICS. *Educ:* Northeastern Univ, BS, 54; Univ Vt, MA, 56; Boston Univ, PhD(math), 61. *Prof Exp:* From assoc prof to prof math, Emmanuel Col, Mass, 61-65; prof, Mass State Col Salem, 65-68; sr mathematician, Missle Systs Div, Raytheon Co, Bedford, 68-74; consult, QEI Consult Firm, 74-79; sr eng scientist, RCA, 79-83; sr eng scientist, TRW, 83-87; PRIN SCIENTIST/ANALYST, VANGUARD RES INC, 87- *Concurrent Pos:* Lectr, Mass Col Pharm, 61-65. *Mem:* Soc Indust & Appl Math; Am Math Soc; Math Asn Am; Asn Symbolic Logic. *Res:* The average of a function over a group of translations in a subinvariant measure space; developed new formulas for use in navigation and military maps; developed three-dimensional parametric equations of a conic section, the longitude being the parameter. *Mailing Add:* 20 Ruby Ave Marblehead MA 01945

FISHMAN, SHEROLD, b Winnipeg, Man, Jan 24, 25; m 54; c 2. NUCLEAR MEDICINE, INTERNAL MEDICINE. *Educ:* Univ Sask, BA, 45; McGill Univ, PhD(biochem), 53; Univ BC, MD, 59; FRCP(C), 65; Am Bd Nuclear Med, cert, 73. *Prof Exp:* Biochemist, Montreal Gen Hosp, 51-54; clin chemist, Univ BC, 60-62, partic residency prog internal med, 62-63; CONSULT NUCLEAR MED, SHAUGHNESSY HOSP, 65-; REGIONAL LAB COORDR, ROYAL COLUMBIAN HOSP, 72-, DIR NUCLEAR MED, 73- *Concurrent Pos:* Res fel, Univ Wash, 63-64; consult internal med & endocrinol, Vancouver Gen Hosp, 64-72. *Mem:* Sigma Xi. *Res:* Endocrinology. *Mailing Add:* 7170 Hudson St Vancouver BC V6P 4K8 Can

FISHMAN, STEVEN GERALD, b Liberty, NY, Jan 15, 42; m 66; c 2. METAL MATRIX COMPOSITES, CERAMIC COMPOSITES. *Educ:* Cornell Univ, BChE, 64; Univ Ill, MS, 66, PhD(metall eng), 69. *Prof Exp:* Res asst metall, Univ Ill, 64-69; res scientist, Naval Weapons Lab, 69-75; head metall mat, Naval Surface Weapons Ctr, 75-82; prog mgr, Defense Advan Res Projs Agency, 82-83; SCI OFFICER COMPOSITES, OFF NAVAL RES, 83- *Concurrent Pos:* Tech rep, Metal Matrix Composites Info Anal Ctr, Dept Defense, 79-82; vchmn, composites comt, Am Inst Mining, Metall & Petrol Engrs, 85- *Honors & Awards:* Am Ceramic Soc, Int Mat Reviews Comt. *Mem:* Am Inst Mining, Metall & Petrol Engrs; Am Soc Metals; Nat Asn Corrosion Engrs; Sigma Xi. *Res:* Metal and ceramic composite materials; material processing science; interfacial phenomena; micromechanics of materials; thermodynamics and kinetics of metallurgical processes; flaw initiation; growth in brittle materials and composites. *Mailing Add:* 9233 Sealed Message Rd Columbia MD 21045

FISHMAN, WILLIAM HAROLD, b Winnipeg, Man, Mar 2, 14; nat US; m 39; c 3. BIOCHEMISTRY, ONCOLOGY. *Educ:* Univ Sask, BS, 35; Univ Toronto, PhD(biochem), 39. *Hon Degrees:* Dr, Univ of Umea, Umea, Sweden. *Prof Exp:* From instr to asst prof biochem, Bowman Gray Sch Med, 41-45; res assoc biochem & asst prof surg, Univ Chicago, 45-48; res prof biochem & nutrit, Sch Med, Tufts Univ, 48-59, res prof oncol, 59-70, prof path & dir cancer res ctr, 70-77; pres, La Jolla Cancer Res Found, 76-89, Dir, Cancer Res Ctr, 81-89; RETIRED. *Concurrent Pos:* Fedn fel, Int Physiol Cong, Oxford, Eng, 47; assoc dir cancer res & chief biochemist, New Eng Ctr Hosp, 48-59, dir cancer res, 59-; vis prof, NSF travel award, Japan, 59; mem tissue & cell biol study sect, Vet Admin; mem, Int Study Group Carcinoembryonic Proteins; ed-in-chief, J Oncol Develop Biol & Med, 79- *Honors & Awards:* Gov Gen Medal, 31; Hon Mem, Int Soc Clin Enzym, 86. *Mem:* Am Soc Exp Path; Am Soc Biol Chemists; Am Chem Soc; Histochem Soc; Soc Exp Biol & Med; Histochemical Soc (pres, 83); fel Am Inst Chemists. *Res:* Enzymes; beta-glucuronidase; prostatic acid phosphatase; alkaline phosphatase isoenzymes; steroids; renal beta-glucuronidase response to androgens; glucuronic acid metabolism; biochemical diagnostic tests; enzyme histochemistry; enzymorphology; experimental pathology; oncodevelopmental gene expression. *Mailing Add:* La Jolla Cancer Res Found 10901 N Torrey Pines Rd La Jolla CA 92037

FISK, FRANK WILBUR, b Logan, Utah, Apr 15, 14; m 44; c 2. INSECT PHYSIOLOGY, INSECT TAXONOMY. *Educ:* Univ Ill, BS, 36; Univ Minn, MS, 39, PhD(entom), 49. *Prof Exp:* Jr entomologist, USPHS, 38-40, asst entomologist, 41-43, sanitarian, 43-45, assoc entomologist, 45-46; from asst prof to prof zool & entom, 49-76, emer prof entom, Ohio State Univ, 76; RETIRED. *Concurrent Pos:* Vis res assoc, Stetson Univ, Deland, FL, 88- *Mem:* Entom Soc Am. *Res:* Blattaria (cockroach) taxonomy. *Mailing Add:* 1175 W Minnesota Ave No 14 Deland FL 32720

FISK, GEORGE AYRS, b Baton Rouge, La, Jan 27, 41; m 63; c 2. CHEMICAL DYNAMICS, CHEMICAL KINETICS. *Educ:* Rice Univ, BA, 62; Univ Calif, Berkeley, PhD(phys chem), 66. *Prof Exp:* Fel phys chem, Harvard Univ, 66-68; asst prof chem, Cornell Univ, 68-74; staff scientist, 74-84, TECH SUPVR, SANDIA NAT LABS, 84- *Mem:* Am Phys Soc; Am Chem Soc. *Res:* Chemical kinetics and dynamics; intramolecular redistribution of vibrational energy on short time scales and studies of the chemical reactions that lead to soot in hydrocarbon flames. *Mailing Add:* Sandia Nat Labs Div 8353 Livermore CA 94551-0969

FISK, LANNY HERBERT, b Edmore, Mich, Feb 24, 44; m 67. PALYNOLOGY, PETROLEUM GEOLOGY. *Educ:* Andrews Univ, BA, 71; Loma Linda Univ, PhD(biol), 76. *Prof Exp:* Asst prof paleobiol, Walla Walla Col, 74-78, assoc prof, 78-79; assoc prof geol & biol, Loma Linda Univ, 79-90; PRES, NORWESTCO INC, 90- *Concurrent Pos:* Consult, Amoco Oil Co, 79-81 & Davis Oil Co, 81-82; pres, F & F GeoResource Assocs, Inc, 82- *Mem:* Am Asn Stratig Palynologists; Am Asn Petrol Geologists; Sigma Xi; Paleont Soc; Soc Econ Paleontologists & Mineralogists. *Res:* Paleoecological interpretation of the fossil plant record; Tertiary stratigraphic palynology; petroleum source rock analysis; petroleum potential of West Coast Cretaceous-Tertiary basins. *Mailing Add:* F&F GeoResource Assocs Inc River's Edge Bldg 2955 N Hwy 97 Suite B Bend OR 97701

FISK, ROBERT SPENCER, b Chicago, Ill, Dec 21, 39; c 1. APPLIED MATHEMATICS. *Educ:* Univ Wyo, BS, 60, MS, 62, PhD(math), 77. *Prof Exp:* Dean of Inst, Sheridan Col, 66-67; from instr to assoc prof, Pac Lutheran Univ, 68-78; ASST PROF MATH, COLO SCH MINES, 78- *Concurrent Pos:* Vis asst prof, Colo Sch Mines, 76-77; vis lectr, Univ Calif, Davis, 81-82; instrnl comput consult III, Comput & Telecommun Serv. *Mem:* Math Asn Am; Soc Indust & Appl Math; Sigma Xi. *Res:* Partial differential equations. *Mailing Add:* Comput & Telecommun Serv Calif State Univ-Stanislaus 801 W Monte Vista Turlock CA 95380

FISKE, MILAN DERBYSHIRE, b Sharon, Wis, Nov 15, 14; m 36; c 3. SOLID STATE PHYSICS. *Educ:* Beloit Col, BS, 37; Univ Wis, PhD(physics), 41. *Prof Exp:* Asst physics, Univ Wis, 37-40; res assoc, Res Lab, Gen Elec Co, 41-57, mgr personnel & sci rels, 57-59, res personnel, 59-62, physicist, 62-66, mgr personnel & admin, Gen Phys Lab, 66-68, progs & admin, Phys Sci & Eng, 68-69, mgr, Phys Sci Br, 69-77; CONSULT, 78- *Concurrent Pos:* Mem, US Nat Comt, Int Union Pure & Appl Physics, 68-74. *Mem:* Fel Am Phys Soc; AAAS. *Res:* Thermionics; gas discharges; T-R switches for radar; cryogenics; high-pressure effects at low temperature; radiation damage; superconductivity; management of multidisciplinary group engaged in exploratory research. *Mailing Add:* 215 Lake Hill Rd Burnt Hills NY 12027

FISKE, RICHARD SEWELL, b Baltimore, Md, Sept 5, 32; m 59; c 2. GEOLOGY. *Educ:* Princeton Univ, BSE, 54, MSE, 55; Johns Hopkins Univ, PhD(geol), 60. *Prof Exp:* Geologist, Union Oil Co, Calif, 55-56; Am Chem Soc Petrol Res Fund fel, Univ Tokyo, 60-61; res assoc, Johns Hopkins Univ, 61-63; geologist, US Geol Surv, 64-76; geologist, 76-80, dir, Nat Mus Nat Hist, 80-85, GEOLOGIST, SMITHSONIAN INST, 85- *Mem:* Geol Soc Am; Am Geophys Union. *Res:* Petrology; Cenozoic volcanic geology of the Pacific Northwest and Japan; volcanic sedimentation; applied geophysics; monitoring active Hawaiian and Caribbean volcanoes; Mesozoic volcanism of the Sierra Nevada. *Mailing Add:* Smithsonian Inst NHB-119 Washington DC 20560

FISKELL, JOHN GARTH AUSTIN, soil chemistry; deceased, see previous edition for last biography

FISLER, GEORGE FREDERICK, b Saginaw, Mich, Nov 29, 31. BIOLOGY, ZOOLOGY. *Educ:* Mich State Univ, BS, 54, MS, 56; Univ Calif, Berkeley, PhD(zool), 61. *Prof Exp:* Res zoologist, Hastings Natural Hist Reservation, Univ Calif, Carmel Valley, 60-61, jr res zoologist, 61-62; asst prof biol, Portland State Col, 62-64; from asst prof to assoc prof, 64-70, PROF BIOL, CALIF STATE UNIV, NORTHRIDGE, 70- *Mem:* Am Soc Mammal; Ecol Soc Am; Am Ornith Union; Cooper Ornith Soc; Animal Behav Soc. *Res:* Behavior, ecology and systematics of mammals; behavior of birds. *Mailing Add:* Dept Biol Calif State Univ Northridge 18111 Nordhoff St Northridge CA 91330

FISSER, HERBERT GEORGE, b Ames, Iowa, Mar 9, 26; m 52; c 3. RANGE MANAGEMENT, ECOLOGY. *Educ:* Mont State Col, BS, 58, MS, 61; Univ Wyo, PhD(range mgt), 62. *Prof Exp:* From instr to assoc prof, 59-70, PROF RANGE MGT, UNIV WYO, 70- *Concurrent Pos:* Consult, Rangeland Inventory; sabbatical leave, 83-84. *Mem:* Ecol Soc Am; Soc Range Mgt; Wildlife Soc; Sigma Xi; Int Asn Ecol. *Res:* Analysis of 21 year multiple site data sets; testing validation of application to hydrology base yield forecast model on vegetation and environment, at semiarid locations of western Wyoming; game range improvement; plant pattern and distribution; grazing systems and wildlife interrelationships; plant phenology and seed production; range site productivity; author of one publication. *Mailing Add:* Dept Range Mgt Univ Wyo Laramie WY 82071

FITCH, ALANAH, b Tucson, Ariz, June 28, 53; m; c 2. ELECTROCHEMISTRY, CLAY CHEMISTRY. *Educ:* Antioch Col, BA, 75; Univ Ariz, MS, 77; Univ Ill-Champaign, PhD(soil chem), 81. *Prof Exp:* Postdoctoral soil chem, Soil Sci Dept, Univ Wis-Madison, 81-82, electrochem, Chem Dept, 82-84 & Northwestern Univ, 84-85; asst prof, 85-91, ASSOC PROF, LOYOLA UNIV, CHICAGO, 91- *Mem:* Soil Sci Soc Am; Am Chem Soc; Electrochem Soc; Am Soc Agron; Asn Women Sci; Soc Electroanal Chem. *Res:* Modifying layers at electrodes, particularly clays and antibodies; dynamic shrink swell properties of clays and their relationship to diffusional transport of pollutants through clay beds. *Mailing Add:* Chem Dept Loyola Univ 6525 N Sheridan Chicago IL 60626

FITCH, COY DEAN, b Marthaville, La, Oct 5, 34; m 56; c 2. MEDICINE, BIOCHEMISTRY. *Educ:* Univ Ark, BS, 56, MS & MD, 58; Am Bd Internal Med, dipl, 65, dipl endocrinol, 72. *Prof Exp:* Instr biochem, Sch Med, Univ Ark, Little Rock, 59-62, from asst prof to assoc prof med & biochem, 62-67; assoc prof internal med & biochem, Sch Med, St Louis Univ, 67-73, dir div endocrinol, 77-85, acting chmn internal med, 85-88, PROF INTERNAL MED, SCH MED, ST LOUIS UNIV, 73-, PROF BIOCHEM, 76-, CHMN INTERNAL MED, 88- *Concurrent Pos:* Russell M Wilder-Nat Vitamin Found fel, 59-62; mem nutrit study sect, NIH, 67-71; dep dir, Div Biochem, Walter Reed Army Inst Res, 69. *Mem:* Endocrine Soc; Am Soc Clin Invest; Am Inst Nutrit; Am Soc Biol Chemists. *Res:* Metabolism and nutrition; membrane transport processes; role of vitamin E in hematopoiesis and muscle function; diseases of skeletal muscle; drug resistance in malaria. *Mailing Add:* Dept Internal Med 1402 S Grand Blvd St Louis MO 63104

FITCH, ERNEST CHESTER, JR, b Wichita, Kans, Nov 30, 24; m 48; c 3. MECHANICAL ENGINEERING, FLUID POWER ENGINEERING. *Educ:* Okla State Univ, BS, 50, MS, 51; Univ Okla, PhD(eng sci), 64. *Prof Exp:* Res engr, Jersey Prod Res Corp, Okla, 51-53; prof mech eng, Okla State Univ, 63-, dir res, Fluid Power Res Ctr, 65-, fluid power indust prof, 76-; RETIRED. *Concurrent Pos:* Deleg, US Fluid Power Indust, Int Orgn Standard, 71-80; chmn, US Dept Com, Int Trade Mission, Korea & Taiwan, 77; chmn, 7th Int Conf Fluid Mech & Fluid Power, Baroda, India, 77; mem bd adv, NSF, 77-78; consult. *Honors & Awards:* Arch T Colwell Award, Soc Automotive Engrs, 70; Ann Achievement Award, Nat Fluid Power Asn, 70. *Mem:* Nat Soc Prof Engrs; Am Soc Testing & Mat. *Res:* Fluid power component and system design; diagnostics and condition monitoring; system reliability and component service life; filtration and fluid contamination control. *Mailing Add:* 813 W Eskridge Stillwater OK 74075

FITCH, FRANK WESLEY, b Bushnell, Ill, May 30, 29; m 51; c 2. PATHOLOGY. *Educ:* Univ Chicago, MD, 53, MS, 57, PhD(path), 60. *Hon Degrees:* MD, Univ Lavsanne, Switz, 90. *Prof Exp:* Intern, Hosp, Univ Mich, 53-54; from instr to prof, 57-76, assoc dean educ affairs, 76-86, ALBERT D LASKER PROF, UNIV CHICAGO, 76-, DIR, BEN MAY INST, 86- *Concurrent Pos:* USPHS fel path, Univ Chicago, 54-57; Lederle med fac award, 58-61; Markle scholar acad med, 61-66; vis scientist, Inst Biochem, Univ Lausanne, 65-66; Guggenheim fel, 74-75. *Mem:* AAAS; Am Asn Pathologists; Am Asn Immunologists; Am Asn Cancer Res. *Res:* Experimental pathology; immunology. *Mailing Add:* Dept Path Univ Chicago Chicago IL 60637

FITCH, HENRY SHELDON, b Utica, NY, Dec 25, 09; m 46; c 3. ECOLOGY. *Educ:* Univ Ore, BA, 30; Univ Calif, MA, 33, PhD(zool), 37. *Prof Exp:* Wildlife technician, Hastings Wildlife Reserve, Univ Calif, 37-38; biologist, US Fish & Wildlife Serv, Wash, 38-47; from instr to assoc prof, 48-59, prof zool, 59-70, prof systs & ecol, 70-80, EMER PROF SYSTS & ECOL, UNIV KANS, 80-, SUPT NATURAL HIST RESERVATION, 48- *Mem:* Am Soc Ichthyol & Herpet; Ornith Soc; Cooper Ornith Soc; Soc for Study Amphibians & Reptiles; Sigma Xi. *Res:* Systematics and natural history of North American and neotropical reptiles; economics and ecology of rodent populations; predator ecology; reproductive cycles in reptiles. *Mailing Add:* Rte 3 Box 142 Lawrence KS 66044-9153

FITCH, JOHN HENRY, b Medford, Ore, Mar 1, 44; m 79. BEHAVIORAL ECOLOGY, ENVIRONMENTAL SCIENCE POLICY. *Educ:* Univ Kans, BA, 66; Mich State Univ, MA, 72, PhD(zool), 75. *Prof Exp:* Asst prof biol, Mich State Univ, 75-78; staff scientist ecol, Exec Off President, Coun Enivron Qual, 78-79; asst prof biol, Univ NDak, 79-80; DIR, ENIVRON SCI DEPT, MASS AUDUBON SOC, 80- *Concurrent Pos:* Bd mem, Detroit Zoo Med Adv Coun, 75-78; adj prof, Tufts Univ, 81- *Mem:* AAAS; Am Soc Mammalogists; Animal Behav Soc; Sigma Xi; Ecol Soc Am. *Res:* Studies of pelagic birds, bat homing behavior and population dynamics; behavioral ecology of rodents; sociobiology of ungulates. *Mailing Add:* Dir/Sr Scientist Environ Sci Dept Mass Audubon Soc Lincoln MA 01773

FITCH, JOHN WILLIAM, III, b San Antonio, Tex, July 28, 38. ORGANOMETALLIC CHEMISTRY. *Educ:* Univ Tex, Austin, BS, 60, PhD(chem), 65. *Prof Exp:* Res chemist, E I du Pont de Nemours & Co, Inc, 65-67; from asst prof to assoc prof chem, 67-77, PROF CHEM, SOUTHWEST TEX STATE UNIV, 77- *Concurrent Pos:* Robert A Welch Found grant, 69-81. *Mem:* Am Chem Soc. *Res:* Organometallic synthesis; unusual olefin-metal complexes; organosilicon chemistry. *Mailing Add:* Dept Chem Southwest Tex State Univ San Marcos TX 78666

FITCH, KENNETH LEONARD, b Genoa, Nebr, Mar 8, 29; m 50; c 3. ANIMAL PHYSIOLOGY, GENERAL MEDICAL SCIENCES. *Educ:* Univ Nebr, BS, 51; Univ Kans, MA, 52; Univ Mich, PhD(zool), 56. *Prof Exp:* Asst zool, Univ Kans, 51-52; instr, Ohio Univ, 55-56; from instr to asst prof anat, Univ Mo, 56-59; asst prof, Col Med, Univ Nebr, 59-63; assoc prof anat, Ill State Univ, 63-87; HEALTH SCIENTIST ADMIN, NAT HEART LUNG BLOOD INST, 87- *Concurrent Pos:* Fulbright lectr, Izmir, Turkey, 68-69. *Mem:* AAAS; Am Asn Anat; Am Soc Zool. *Res:* Biomechanics of the human foot; aging in the mouse and rat cornea. *Mailing Add:* NIH Div Res Grants Referral & Review Br Bldg WB Rm 2407B 5333 Westbard Ave Bethesda MD 20892

FITCH, ROBERT MCLELLAN, b Shanghai, China, Apr 30, 28; US citizen; m 55; c 3. POLYMER CHEMISTRY, COLLOID CHEMISTRY. *Educ:* Dartmouth Col, AB, 49; Univ Mich, PhD(chem), 54. *Prof Exp:* Res chemist, E I du Pont de Nemours & Co, Inc, 54-62; from asst prof to assoc prof chem, NDak State Univ, 62-67; from assoc prof to prof, Univ Conn, 67-83, asst head dept, 67-73; sr vpres, Res & Develop, S C Johnson Wax Co, 83-89; CHMN, NAT INDUST COUN SCI EDUC, 90-; PRES, FITCH & ASSOCS, CONSULT, 90- *Concurrent Pos:* Chmn polymer colloids, Gordon Res Conf, 81. *Mem:* Fel AAAS; Am Chem Soc; Indust Res Inst. *Res:* Mechanism of particle formation in polymer colloids; chemical and mechanical behavior of polymers at interfaces; dielectric spectroscopy of model colloids; chemical reactions at the polymer colloidal interface. *Mailing Add:* 5020 Wind Point Rd Racine WI 53402

FITCH, STEVEN JOSEPH, b Chicago, Ill, Feb 25, 30; m 52, 67; c 4. INDUSTRIAL CHEMISTRY. *Educ:* Univ Ill, BS, 52; Cornell Univ, PhD(inorg chem), 58. *Prof Exp:* Sr res chemist phosphorus chem, Monsanto Chem Co, 57-64; dir corp res dept, Glidden Co, 64-65, dir chem res dept, 65-66, mgr pigments res dept, Inorg Res Ctr, 66-68, mgr res dept, Glidden Pigments Group, SCM Corp, 68-87, MGR RES & DEVELOP, SCM CHEMICALS, 87- *Mem:* Am Chem Soc; Am Inst Chemists; Fine Particle Soc. *Res:* Titanium dioxide chemistry, technology and processes; pigments; inorganic and organic phosphorus chemistry; pollution control processes, research, development and engineering; waste acid neutralization; gypsum; iron precipitation; fine particle silica. *Mailing Add:* 3936 Cloverhill Rd Baltimore MD 21218

FITCH, VAL LOGSDON, b Merriman, Nebr, Mar 10, 23; m 49; c 2. PHYSICS. *Educ:* McGill Univ, BE, 48; Columbia Univ, PhD(physics), 54. *Prof Exp:* Mem staff, Los Alamos Sci Lab, 46-47; instr physics, Columbia Univ, 53-54; from instr to prof, 54-60, Cyrus Fogg Brackett prof, 76-83, JAMES S MCDONNELL DISTINGUISHED UNIV PROF PHYSICS, PRINCETON UNIV, 83- *Concurrent Pos:* Sloan fel, 60-64; trustee, Assoc Univs, Inc, 61-67; mem, President's Sci Adv Comt, 70-73; pres, Am Phys Soc, 87-88. *Honors & Awards:* Nobel Prize Physics, 80; E O Lawrence Award, AEC, 67; Res Corp Award, 68; Witherell Medal, 78. *Mem:* Nat Acad Sci; Am Acad Arts & Sci; fel Am Phys Soc. *Res:* Particle physics. *Mailing Add:* Jadwin Hall Princeton Univ PO Box 708 Princeton NJ 08540

FITCH, W(ILLIAM) CHESTER, b Billings, Mont, Nov 12, 16; m 46; c 1. INDUSTRIAL ENGINEERING. *Educ:* Mont State Col, BS, 38; Iowa State Col, MS, 39, PhD(eng valuation), 50. *Prof Exp:* Instr eng drawing, Iowa State Col, 39-41, instr mech eng, 41-45; asst prof indust eng, Mont State Col, 45-46; lectr mech eng, Univ Calif, 46-47; assoc prof, Iowa State Col, 47-52; asst dir, Valuation Div, Gannett, Fleming, Corddry & Carpenter, Inc, 52-58 & 59-64; prof mech eng & head dept, Utah State Univ, 58-59; prof mech eng & head dept, Mich Technol Univ, 64-68; prof eng & technol & chmn dept, Western Mich Univ, 68-72, dean, Col Eng & Appl Sci, 73-82; PRES, DEPRECIATION PROGS INC, 78- *Concurrent Pos:* Consult engr, 52- *Mem:* Am Soc Eng Educ; Am Soc Mech Engrs; Am Inst Indust Engrs; Nat Soc Prof Engrs. *Res:* Statistical analyses of rives of industrial property; depreciation and engineering economy. *Mailing Add:* 10237 N 106th Ave Sun City AZ 85351-4420

FITCH, WALTER M, b San Diego, Calif, May 21, 29; m 51; c 3. BIOCHEMISTRY, GENETICS. *Educ:* Univ Calif, Berkeley, AB, 53, PhD(comparative biochem), 58. *Prof Exp:* Lab technician, Calif Res Corp, 51-52 & Pac Guano Co, Calif, 53; res physiologist, Univ Calif, Berkeley, 57-58; NIH fels physiol, Univ Calif, Berkeley, 58-59 & pharmacol, Sch Med, Stanford Univ, 59-61; Fulbright fel & lectr biochem, Univ Col, London, 61-62; from asst prof to prof physiol chem, Sch Med, Univ Wis-Madison, 62-86; prof biol sci, Univ Southern Calif, 86-89; PROF ECOL & EVOLUTIONARY BIOL, UNIV CALIF, IRVINE, 89- *Concurrent Pos:* Fulbright scholar, 61-63, Nat Inst Neurol Dis & Blindness res grant, 63-69; NSF res grant, 65-; examr physiol chem, Wis State Bd Examrs, 66-75; mem adv bd, Biochem Genetics, 66-; mem subcomt cytochrome nomenclature, Int Univs Bur, 72-76; NIH spec fel & vis prof, Univ Hawaii & mem bd reviewers, Fed Proc, 73-74; NIH grant, 74-78; ed, Classification Lit Automated Res Serv, 75-; assoc ed, J Molecular Evolution, 70-82, assoc ed Syst Zool, 76-80, ed-in-chief, Molecular Biol & Evolution, 80-; vis prof, Univ Calif Los Angles, 81-82, Macy Scholar, 81-82, vis prof, Univ Southern Calif, 85. *Mem:* Nat Acad Sci; Am Chem Soc; classification Soc; Am Soc Biol Chemists; Soc Syst Zool; Sigma Xi; Genetics Soc Am; Soc Study Evolution; Soc Math Biol; AAAS; Am Soc Naturalists. *Res:* Enzymology; kinetics; molecular genetics; evolution; molecular biology; the utilization of protein and nucleic acid sequences to infer the course, rate and basis of evolutionary events. *Mailing Add:* Dept Ecol & Evolutionary Biol Univ Calif Irvine Irvine CA 92717

FITCH, WALTER STEWART, b Oak Park, Ill, Mar 6, 26; m 47; c 3. ASTRONOMY. *Educ:* Univ Chicago, AB, 48, PhD(astron), 55. *Prof Exp:* From instr to prof astron, Univ Ariz, 63-86; RETIRED. *Mem:* Am Astron Soc; Int Astron Union. *Res:* Astronomical photoelectric photometry; intrinsic variable stars. *Mailing Add:* PO Box 100 Oracle AZ 85623-0100

FITCHEN, DOUGLAS BEACH, b New York, NY, June 8, 36; m 58; c 3. SOLID STATE PHYSICS. *Educ:* Harvard Col, BA, 57; Univ Ill, PhD(physics), 62. *Prof Exp:* From asst prof to assoc prof physic, 62-71, PROF PHYSICS CORNELL UNIV, 71- *Concurrent Pos:* Vis prof, Clarendon Lab, Oxford Univ, 68-69, Univ Paris, Sud, 75-76; chmn, dept Physics, Cornell Univ, 77-82 & 86- *Res:* Laser studies of solids, especially raman scattering and time-resolved measurements of ultrafast processes. *Mailing Add:* Physics Dept Clark Hall Cornell Univ Ithaca NY 14853

FITCHEN, FRANKLIN CHARLES, b New Rochelle, NY, June 15, 28; m 50; c 3. ELECTRICAL ENGINEERING. *Educ:* Univ RI, BS, 50; Northeastern Univ, MS, 57; Yale Univ, DEng, 64. *Prof Exp:* Design engr, Meter & Instrument Dept, Gen Elec Co, 50-56; assoc prof elec eng, Univ RI, 56-65; prof & head dept, SDak State Univ, 65-72; dean eng, 72-80, PROF ELEC ENG, UNIV BRIDGEPORT, 81- *Concurrent Pos:* Consult tech educ, 72- *Mem:* Sr mem Inst Elec & Electronics Engrs. *Res:* Transistor and integrated circuits. *Mailing Add:* Col Eng Univ Bridgeport Bridgeport CT 06601

FITCHETT, GILMER TROWER, b Cape Charles, Va, June 16, 20; m 53; c 2. ORGANIC CHEMISTRY. *Educ:* Col William & Mary, BS, 42; Univ Va, PhD(chem), 51. *Prof Exp:* Anal chemist, Norfolk Naval Shipyard, 42-43; develop chemist, Pharmaceut Dept, Am Cyanamid Co, NJ, 51-66; develop chemist, Gane's Chem Works, Inc, NJ, 66-68; sr chemist, Ciba-Geigy Ltd, 68-73, sr staff scientist, 73-87; RETIRED. *Concurrent Pos:* Anal chemist, Hoechst Celanese, 87- *Res:* Phenylcyclopropyl amine derivatives and 5-, 8-dimethoxyquinoline derivatives. *Mailing Add:* 13 Briar Circle Green Brook NJ 08812

FITE, LLOYD EMERY, b Litchfield, Ill, Dec 12, 30; m 55. ELECTRICAL ENGINEERING. *Educ:* Agr & Mech Col Tex, BS, 60, MS, 61; Tex A&M Univ, PhD(elec eng), 68. *Prof Exp:* Staff, 58-60, chief engr & proj engr, 60-62, ASSOC HEAD, ACTIVATION ANAL RES LAB, TEX A&M UNIV, 62-, ASSOC PROF ELEC ENG, 69-, TRAINING SPECIALIST, INST ELECTRONIC SCI, 77- *Mem:* Inst Elec & Electronics Engrs; Sigma Xi. *Res:* Instrumentation and electronic data reduction equipment required to support and improve the technique of neutron activation analysis. *Mailing Add:* 14734 C Perthshire Houston TX 77079

FITE, WADE LANFORD, b Apperson, Okla, Oct 4, 25; m 47; c 4. PHYSICS. *Educ:* Univ Kans, AB, 47; Harvard Univ, MA, 49, PhD(physics), 51. *Prof Exp:* Physicist, Res Labs, Philco Corp, 51-52; instr physics, Univ Pa, 52-54; NSF hon res asst, Univ Col, Univ London, 54-55; mem staff, Gen Atomic Div, Gen Dynamics Corp, 56-63; PROF PHYSICS, UNIV PITTSBURGH, 63-; PRES, EXTRANUCLEAR LABS, INC, 72- *Concurrent Pos:* Chmn, Int Conf Physics Electronic & Atomic Collisions, 63-67, Extranuclear Labs, Inc, 67- & Nat Acad Sci, Nuclear Res Coun Comt, Atomic & Molecular Sci, 77-80. *Mem:* AAAS; fel Am Phys Soc; Am Soc Mass Spectrometry. *Res:* Plasma, atomic and atomic collision physics; collision properties of atomic hydrogen; laboratory experimentation on upper atmosphere and astrophysics; chemical reactions; gas-surface phenomena; atomic beams; aerosols; mass spectrometry. *Mailing Add:* Extrel Corp PO Box 11512 Pittsburgh PA 15238

FITT, PETER STANLEY, b La Punta, Peru, Oct 12, 31; Brit & Can citizen; m 61; c 2. NUCLEIC ACIDS, DNA REPAIR. *Educ:* Univ London, BSc & ARCS, 53, Imp Col, PhD(org chem) & DIC, 56. *Prof Exp:* Sci officer, Royal Aircraft Estab, Eng, 56-57; res assoc biochem, Med Col, Cornell Univ, 57-58; res chem, Imp Chem Industs, Ltd, 58-59; vis res fel, Sloan-Kettering Inst Cancer Res, 59-62; vis scientist, Inst Physico-Chem Biol, Paris, France, 62-64; from asst prof to assoc prof, 64-73, PROF BIOCHEM, UNIV OTTAWA, 73- *Concurrent Pos:* Vis assoc prof biochem virol, Baylor Col Med & Med Res Coun Can vis scientist, 70-71. *Mem:* AAAS; Can Biochem Soc; Brit Biochem Soc; The Chem Soc; Am Soc Biol Chemists. *Res:* Nucleic acid enzymology; molecular biology. *Mailing Add:* Dept Biochem Univ Ottawa 451 Smyth Rd Ottawa ON K1H 8M5 Can

FITT, WILLIAM K, b Pasadena, Calif. LARVAE, SYMBIOSES. *Educ:* Univ Calif, Santa Barbara, PhD(biol sci), 82. *Prof Exp:* ASST PROF ZOOL, UNIV GA, 87- *Mem:* Am Soc Zoologists; Nat Shellfish Asn. *Res:* Chemical induction of settlement and metamorphosis of marine invertebrate larvae; zooxantellae symbioses. *Mailing Add:* Dept Zool Univ Ga Athens GA 30602

FITTERER, G(EORGE) R(AYMOND), b Newark, Ohio, Apr 10, 01; m 25; c 1. METALLURGY. *Educ:* Rose Polytech Inst, BS, 24; Carnegie Inst Technol, MS, 27; Univ Pittsburgh, PhD(metall), 30. *Hon Degrees:* DSc, Rose Polytech Inst, 61; DEng, Universidad Technica Federico Santa Maria, Chile, 65; Dr, Cath Univ Cordoba, 67. *Prof Exp:* Metallurgist, Am Chain Co, Ind, 24-25; metallographer, Stanley Works, Conn, 25-26; from asst metallurgist to asst dir res, US Bur Mines, 27-31, head dept metall, 31-33; pres, Fitterer Pyrometer Co, 34-38; chmn dept metall eng, 38-51, dean schs eng & mines & dir eng res div, 51-63, first distinguished prof metall eng & dir ctr study thermodyn properties mat, 63-77, EMER PROF METALL ENG, SCH ENG, UNIV PITTSBURGH, 77- *Concurrent Pos:* Supvr navy res contracts, Nat Defense Res Comt, 42-; dir res, Acid Open Hearth Res Asn, 42-; eng ed contract, Int Co-op Admin, Chile; US deleg, UN, 1st Pan-Am Metall Conf, Bogota, Colombia & 2nd conf, Sao Paulo, Brazil; chmn, Fitterer Eng Asn, Inc, 69-79; pres, Sci Applns, Inc, 79-; pres, Thermosensors, Inc, 80- *Honors & Awards:* Campbell Mem Lecturer, Am Soc Metals. *Mem:* Am Soc Metals; Am Inst Mining, Metall & Petrol Engrs; Brit Iron & Steel Inst. *Res:* Metallurgical thermodynamics; physical chemistry of steel making; high temperature measurement in liquid metals; liquid steel process metallurgy. *Mailing Add:* 416 Isabelle St No 202 Oakmont PA 15139

FITTING-GIFFORD, MARJORIE ANN PREMO, b Detroit, Mich, Nov 29, 33; m 88; c 3. MATHEMATICS, COMPUTER SCIENCE EDUCATION. *Educ:* Mich State Univ, BS, 54, PhD(math educ), 68; Wayne State Univ, MEd, 58; Univ Mich, Ann Arbor, AM, 66. *Prof Exp:* Teacher var high schs, 54-60; instr math, Lawrence Inst Technol, 61-65; asst, Mich State Univ, 66-68; from asst prof to prof, 68-84, dir, Comput Sci Inst, 79-80 & 84-85, Emer prof, 84-, PROF, SAN JOSE, 86- *Concurrent Pos:* LC Plant Scholar, 54; NSF sci fac fel, 65-66; consult, NSF comprehensive grant, 69-73; dir, Metra Instruments, Inc, 71-81; chief exec officer, Métier Software, 83-; Fulbright Scholar, 85-86. *Mem:* Math Asn Am; Am Math Asn Two Yr Col; Nat Coun Teachers Math; Am Math Soc. *Res:* Curriculum study; encouraging undergraduate research on number theory and group theory; manipulatives in teaching mathematics; using microcomputers in teaching mathematics; computer science. *Mailing Add:* Dept Math & Comput Sci San Jose State Univ San Jose CA 95192

FITTS, CHARLES THOMAS, b Jackson, Tenn, July 4, 32; m 54; c 8. SURGERY. *Educ:* Princeton Univ, BA, 53; Univ Pa, MD, 57. *Prof Exp:* Chief trauma study br, US Army Surg Res Unit, Brooke Army Med Ctr, Ft Sam Houston, Tex, 63-65; from asst prof to assoc prof surg, 65-74, PROF SURG, MED UNIV SC, 74-, CHIEF TRANSPLANTATION SERV, 70- *Concurrent Pos:* Attend surgeon, Med Univ SC Hosp & Vet Admin Hosp, Charleston, 65-; consult surg, US Naval Hosp, Charleston, 65-; med dir, SC Org Proc Agency. *Honors & Awards:* Gold Medal, Southeastern Surg Cong, 61. *Mem:* Am Asn Surg Trauma; Am Col Surg; Soc Univ Surgeons; Am Surg Asn; Southern Surg Asn. *Res:* General surgery; transplantation immunology; burn therapy, fluid replacement therapy in hemorrhagic shock. *Mailing Add:* Dept Surg Med Univ SC 171 Ashley Ave Charleston SC 29425

FITTS, DONALD DENNIS, b Concord, NH, Sept 3, 32; m 64; c 2. PHYSICAL CHEMISTRY, CHEMICAL PHYSICS. *Educ:* Harvard Univ, AB, 54; Yale Univ, PhD(chem), 57. *Hon Degrees:* MA, Univ Pa, 71. *Prof Exp:* NSF fel, Univ Amsterdam, 57-58; res fel chem, Yale Univ, 58-59; from asst prof to assoc prof, 59-69, actg vdean, Col Lib Arts & Sci, 63-64, asst chmn dept, 65-72, assoc dean, 78-82, actg dean, 82-83, PROF CHEM, UNIV PA, 69-, ASSOC DEAN, SCH ARTS & SCI, 83- *Concurrent Pos:* NATO sr sci fel, Imp Col, Univ London, 71; acad vis, Oxford Univ (UK), 78. *Mem:* Am Phys Soc; Royal Soc Chem. *Res:* Quantum-statistical mechanics; thermodynamics and statistical theory of irreversible processes; theory of liquids; theoretical chemistry. *Mailing Add:* Dept Chem Univ Pa Philadelphia PA 19104-6323

FITTS, JAMES WALTER, b Ft Riley, Kans, July 17, 13; m 35; c 3. SOIL FERTILITY. *Educ:* Nebr State Teachers Col, BSc, 35; Univ Nebr, MS, 37; Iowa State Col, PhD, 52. *Prof Exp:* Instr agron, Univ Nebr, 37-42, asst prof agron & asst exten agronomist, 42-48; res asst prof agron, exten asst prof & in chg soil testing lab, Iowa State Col, 48-52; in chg soil testing div, Dept Agr, 52-56, prof soils & head dept, 56-64, dir int soil testing proj, 64-69, dir int soil fertility eval proj, 70-75, emer prof soils, NC State Univ, 75-87; pres, 73-, dir, res proj, phospho-gypsum for agr purposes, Agr Serv Int, 85-87; RETIRED. *Concurrent Pos:* Chmn, Nat Soil Test Work Group, 52-55; pres, Agr Serv Int Inc, 73- & Fla Imp Growers. *Mem:* Fel Soil Sci Soc Am (vpres, 59, pres, 60); fel Am Soc Agron; Int Soil Sci Soc. *Res:* Alkali soil; movement of water in soil during irrigation; commercial fertilizers; procedures and factors affecting soil testing; nitrogen availability. *Mailing Add:* Seven Flagler Lane Apt 102 Holly Hill FL 32017

FITTS, RICHARD EARL, b Montpelier, Vt, Nov 12, 31; m 53; c 7. ELECTRICAL ENGINEERING. *Educ:* Yale Univ, BE, 53; Mass Inst Technol, SM, 55, PhD(elec eng), 66. *Prof Exp:* Proj officer, Rome Air Develop Ctr, Griffiss AFB, NY, 55-57 & 61-63, co-pilot, Pease AFB, NH, 58-61, forward air controller, Da Nang Air Base, Vietnam, 66-67, from instr to assoc prof, USAF Acad, 67-73; staff analyst, Prog Anal & Eval, Off of Asst Secy Defense, 73-77; SR ANALYST, GEN RES CORP, 77- *Mem:* Inst Elec & Electronics Engrs; Sigma Xi. *Res:* Nonlinear feedback control theory; electronic warfare; military force balance analysis. *Mailing Add:* 4116 Duncan Dr Annandale VA 22003

FITZ, HAROLD CARLTON, JR, b Charleston, SC, Aug 30, 26; m 49; c 3. SPACE PHYSICS, ATMOSPHERIC PHYSICS. *Educ:* US Mil Acad, BS, 49; Univ Ala, MS, 55; Univ Va, PhD(solid state physics), 62. *Prof Exp:* Instr physics, Spec Weapons Proj, US Army, 55-57, syst analyst, 57-59, syst analyst, Satellite Commun Agency, 61-64; from asst prof to assoc prof physics, US Mil Acad, 64-67; test plans officer, Atmospheric Effects Div, Defense Atomic Support Agency, Washington, DC, 67-70; staff scientist, Gen Res Corp, Va, 70; Chief Atmospheric Effects Div, 70-77 & 79-81, Chief Electronics Vulnerability div Defense Nuclear Agency, Washington, DC, 77-79; VPRES, PHYS RES INC, 85- *Mem:* Am Phys Soc; Am Inst Aeronaut & Astronaut; Am Geophys Union; Sigma Xi. *Res:* Design and effects of nuclear weapons; atmospheric and space environment; optics; electromagnetism. *Mailing Add:* PO Box 503 Shady Side MD 20764-0503

FITZER, JACK, b Joplin, Mo, Oct 5, 26; m 50; c 3. ELECTRICAL ENGINEERING. *Educ:* Univ Mo, BS, 51; Wash Univ, MS, 60, DSc(elec eng), 62. *Prof Exp:* Design engr, Chance Vought Aircraft Co, 51-54; design engr, Emerson Elec Mfg Co, 54-59, consult, 59-62, sr res engr, 62-63; mem res & develop staff, LTV Electrosysts, 63-67; from asst prof to assoc prof elec eng, 67-77, PROF ELEC ENG, UNIV TEX, ARLINGTON, 77- *Mem:* Inst Elec & Electronics Engrs; Sigma Xi; Soc Eng Educ; Asn Unmanned Vehicle Systs. *Res:* Automatic control, guidance; network theory; photovoltaic power systems; robotics. *Mailing Add:* Dept Elec Eng Univ Tex Arlington TX 76010

FITZGERALD, CARL, MATHEMATICS. *Educ:* Stanford Univ, BA, 63, MA, 64, PhD(math), 67. *Prof Exp:* PROF MATH, UNIV CALIF, SAN DIEGO, 67- *Mem:* Am Asn Math. *Mailing Add:* Math Dept Univ Calif San Diego San Diego CA 92093

FITZGERALD, DAVID J P, b Dublin, Ireland, Mar 18, 56; US citizen; m. MICROBIOLOGY. *Educ:* Trinity Col, Univ Dublin, Ireland, BA, 77; Univ Cincinnati, PhD(microbiol), 82. *Prof Exp:* Staff fel, 82-84, sr staff fel, 85-87, MICROBIOLOGIST, LAB MOLECULAR BIOL, NAT CANCER INST, NIH, 87- *Concurrent Pos:* Mem, Study Sect Trop Med & Parasitol, 86, sect rev toxin-based grant proposals, 88; lab asst, Dept Microbiol, Univ Cincinnati; co-investr, FDA-PE-ANTI-TAC & OVB3-PE. *Mem:* Am Soc Microbiol; AAAS; Am Soc Biochem & Molecular Biol. *Res:* Structure-function relationships of Bacterial Protein Toxins; mechanisms of cell-killing by Pseudomonas exotoxin and diphtheria toxin; use of toxins includes the design, construction and testing of toxin-based conjugates as novel drugs to treat human and animal diseases; three patents. *Mailing Add:* Lab Molecular Biol Nat Cancer Inst NIH Bldg 37 Rm 4B27 Bethesda MD 20892

FITZGERALD, DONALD RAY, b Pomeroy, Wash, June 18, 23; m 55; c 2. ATMOSPHERIC PHYSICS. *Educ:* Univ Chicago, BS, 47, MS, 49, PhD(meteorol), 56. *Prof Exp:* Asst, Univ Chicago, 51-54, res assoc, 54-61; physicist, USAF Geophys Lab, 61-87; CONSULT, 87- *Mem:* Am Meteorol Soc; Am Geophys Union; Sigma Xi. *Res:* Electrical structure of thunderstorms; general atmospheric electrostatic phenomena; radar meteorology; instrumentation for cloud physics; research aircraft; pulsed carbon dioxide laser wind systems. *Mailing Add:* 71 Jennie Dugan Rd Concord MA 01742

FITZGERALD, DUANE GLENN, b Jackson, Mich, Oct 6, 31; m 55; c 4. NUCLEAR & ELECTRICAL ENGINEERING. *Educ:* Univ Mich, BS, 57, MS, 59; Univ Mo, PhD(elec eng), 66. *Prof Exp:* Proj engr, Bendix Corp, 59-62; reactor supvr, Univ Mo, 62-67; mgr systs anal, Eng Div, Nus Corp, 67-74, mgr int bus develop, 74-77, regional gen mgr, 77-79, vpres, Int Opers, 79-88; GEN ACCT OFF, US GOVT. *Mem:* Am Nuclear Soc. *Res:* Reliability analysis; protection system design and analysis; system kinetics analysis; direct digital control. *Mailing Add:* 5706 English Ct Bethesda MD 20817

FITZGERALD, EDWARD ALOYSIUS, b Washington, DC, Feb 11, 42; m 67; c 5. VIROLOGY, MICROBIOLOGY. *Educ:* Georgetown Univ, BS, 63; Cath Univ Am, PhD(microbiol), 70. *Prof Exp:* Contract officer res contracts, Div Biol Standards, NIH, 67-69, res microbiologist vaccine control, Lab Control Activ, 69-74; dep div dir vaccine control, Bur Biol, 74-85, actg dir, Div Prod Qual Control, 85-90, DIR DIV PROD QUAL CONTROL, CENT BIOLOGICS, FDA, 90- *Mem:* AAAS; Am Soc Microbiol; Sigma Xi; Int Asn Biol Standardization. *Res:* Control testing of biological products, specifically vaccines; development and testing of human rabies vaccine and rabies immune globulin; development and testing of reference preparations for use in biological testing. *Mailing Add:* Bldg 29A Rm 1A19 8800 Rockville Pike Bethesda MD 20892

FITZGERALD, EDWIN ROGER, b Oshkosh, Wis, July 14, 23; m 46; c 8. PHYSICS. *Educ:* Univ Wis, BS, 44, MS, 50, PhD(physics), 52. *Prof Exp:* Mem tech staff, B F Goodrich Co, 44-46; asst physics & math, Univ Wis, 46-48, asst chem, 48-51, proj assoc, 51-53; from asst prof to prof physics, Pa State Univ, 53-61; PROF MECHANICS, JOHNS HOPKINS UNIV, 61- *Concurrent Pos:* Vis prof chem, Univ Wis, Madison, 81; consult, indust & govt. *Mem:* Fel Am Phys Soc; Acoust Soc Am; Mats Res Soc. *Res:* Solid state physics; dynamic mechanical properties of solids; polymers, metals and long chain hydrocarbon compounds; dielectric properties of liquids and solids; wave mechanical explanations of deformation and other mechanical behavior. *Mailing Add:* Johns Hopkins Univ 20 Latrobe Hall Baltimore MD 21218

FITZGERALD, GEORGE PATRICK, b Milwaukee, Wis, Sept 22, 22; m 43; c 2. FRESH WATER ECOLOGY. *Educ:* Univ Wis, BS, 48, MS, 49, PhD(bot), 50. *Prof Exp:* Asst bot, Univ Wis-Madison, 48-50, res assoc, 50-68, sr scientist, 68-75; Indust & Govt Consult, 75-86; RETIRED. *Mem:* Phycol Soc Am. *Res:* Mineral nutrition of algae; control of algae. *Mailing Add:* 3644 Rivercrest Rd McFarland WI 53558

FITZGERALD, GERALD ROLAND, b Milwaukee, Wis, Aug 5, 30; m 86; c 14. VIRAL IMMUNOLOGY, MYCOPLASMOLOGY. *Educ:* Univ Notre Dame, BS, 52, MS, 61, PhD(microbiol), 67. *Prof Exp:* Staff fel, Nat Cancer Inst, NIH, Bethesda, Md, 66-68; res assoc, St Jude Childrens Res Hosp, 68-70; head microbiol sect, Path Dept, Mercy Hosp, Des Moines, Iowa, 70-75; SCI DIR, AMBICO, INC, 76- *Concurrent Pos:* Adj prof, Drake Univ, 73-75 & Univ Tenn, 68-70. *Mem:* Am Soc Microbiol; NY Acad Sci. *Res:* Host defense systems, especially against viral attack; interferon and its role in host defense; local immune responses and their role in host defense; development of viral vaccines that involve local immune responses for their efficacy. *Mailing Add:* Ambico Inc Rte 2 Box 522 Dallas Center IA 50063

FITZGERALD, GERARD JOHN, b Nfld, Sept 9, 49. ANIMAL BEHAVIOR, BEHAVIORAL ECOLOGY. *Educ:* Mem Univ, BSc, 70; McGill Univ, BSc, 72; Univ Western Ont, PhD(zool), 76. *Prof Exp:* PROF BIOL, LAVAL UNIV, 76- *Mem:* Animal Behav Soc; Can Soc Zoologists. *Res:* Reproduction and parental behavior in fish; fish ecology in unstable environments. *Mailing Add:* Dept Biol Laval Univ Quebec PQ G1K 7P4 Can

FITZGERALD, GLENNA GIBBS (CADY), b Westfield, Mass. TOXICOLOGY, NEUROPHARMACOLOGY. *Educ:* Univ Mass, Amherst, BS & MS; Yale Univ, PhD(pharmacol), 68. *Prof Exp:* Res assoc pharmacol, Yale Univ, 61-62; instr, George Washington Univ, 68-71; staff fel, Lab Cerebral Metab, Sect Develop Neurochem, NIMH, 71-77; TOXICOLOGIST & PHARMACOLOGIST, DIV NEUROPHARMACOL, OFF DRUG RES & REV, CTR DRUGS & BIOLOGICS, FDA, 77- *Mem:* Asn Govt Toxicologists; Am Soc Clin Pharmacol & Therapeut. *Res:* Neurochemistry; development and differentiation of the central nervous system; mechanism of action of centrally acting drugs. *Mailing Add:* FDA 5600 Fishers Lane Rm 10B32 Rockville MD 20857

FITZGERALD, J(OHN) EDMUND, b Chelsea, Mass, Sept 29, 23; m 45; c 4. CIVIL ENGINEERING, MATERIALS SCIENCE. *Educ:* Harvard Univ, MS, 47; Univ Col, Cork, MSc, 70; Nat Univ Ireland, DSc(math & physics), 72. *Prof Exp:* Mgr appl mech, Am Mach & Foundry Res Labs, 53-57 & Borg-Warner Res Labs, 57-60; dir res eng, Lockheed Propulsion Co, 60-66; assoc dean, Col Eng, Univ Utah, 66-75; PROF & ASSOC DEAN ENG, GA INST TECHNOL, 75- *Concurrent Pos:* Consult, Lockheed Propulsion Co, 66-70, Math Sci Corp, 66-75, US Army Missile Command, 67-77, Battelle Mem Inst, 68- 78, United Technol Corp, 70-75 & Aerojet Gen Corp, 76-78, chief, Naval Res, 66-72; US sr scientist, Alexander von Humboldt Inst, 73. *Honors & Awards:* Solid Rocket Tech Achievement Award, Inst Aeronaut & Astronaut. *Mem:* Fel Inst Physics; fel Am Soc Civil Engrs; assoc fel Am Inst Aeronaut & Astronaut; Am Phys Soc; Soc Rheology; Sigma Xi. *Res:* Characterization of irreversible effects in polymers especially damage; applications to solid propellants, asphalt concrete and fabrics. *Mailing Add:* Col Eng Ga Inst Technol Atlanta GA 30332-0360

FITZGERALD, JAMES ALLEN, b Albany, NY, Apr 7, 41; m 64; c 2. FIBER SCIENCE. *Educ:* Univ Rochester, BS, 62; Carnegie Inst Technol, MS, 64, PhD(org chem), 66. *Prof Exp:* Res chemist, 66-73, res supvr, 74-78, sr res supvr textile fibers, 78-79, RES MGR HIGH PERFORMANCE FIBERS, TEXTILE FIBERS DEPT, E I DU PONT DE NEMOURS & CO, INC, 79- *Mem:* Am Chem Soc; AAAS. *Res:* Research and supervision in organic and polymer chemistry, polymer processing, fiber science and process development; specialty areas include high performance polymers and fibers including Kevlar and Nomex. *Mailing Add:* 106 Marlbrooke Way Kennett Square PA 19348-1720

FITZGERALD, JAMES EDWARD, b Paris, Ill, Feb 14, 31; m 59; c 3. VETERINARY PATHOLOGY, TOXICOLOGY. *Educ:* Univ Ill, Urbana, BS, 53; DVM, 55, MS, 62, PhD(vet med), 64. *Prof Exp:* Instr vet clin med, Univ Ill, Urbana, 55-56; vet med dir, Agr Res Ctr, Pfizer, Inc, 58-60; NIH trainee vet path & hyg, Col Vet Med, Univ Ill, Urbana, 61-64; pathologist, Dept Toxicol, Parke Davis & Co, 64-72, asst dir 72-76, dir, Toxicol Lab, 77-84, dir, Comp Path, 85-87; RETIRED. *Mem:* Am Vet Med Asn; Am Col Vet Path; Soc Toxicol; Soc Toxicol Pathologists. *Res:* Experimental pathology and toxicology; animal neoplasia. *Mailing Add:* 1940 Ridge Rd Ann Arbor MI 48104

FITZGERALD, JAMES W, b New York, NY, May 1, 42; m 69; c 1. ATMOSPHERIC PHYSICS. *Educ:* City Col New York, BS, 63; Univ Chicago, MS, 64, PhD(atmospheric sci), 72. *Prof Exp:* RES PHYSICIST, US NAVAL RES LAB, 72- *Mem:* Am Meteorol Soc; Sigma Xi. *Res:* Cloud, fog and aerosol physics; physical meteorology; boundary layer meteorology; electro-optical meteorology. *Mailing Add:* US Naval Res Lab Code 4212 Washington DC 20375-5000

FITZGERALD, JERRY MACK, b Alliance, Nebr, Jan 20, 37; div; c 2. ANALYTICAL CHEMISTRY. *Educ:* Univ Colo, BA, 59; Princeton Univ, MA, 61, PhD(anal chem), 63. *Prof Exp:* Res assoc chem, Purdue Univ, 63-64; asst prof, Seton Hall Univ, 64-68; assoc prof, Univ Houston, 68-76; qual control dir, Nat Health Labs, Inc, Vienna, Va, 76-79; head, anal chem, Litton Bionetics, Inc, Kensington, Md, 79- 85; TECH STAFF, MITRE CORP, MCLEAN, VA, 85- *Mem:* Sigma Xi; Soc Appl Spectros; Am Chem Soc. *Res:* Photochemical reactions for analytical determinations; electron spin resonance and fluorescence spectroscopy. *Mailing Add:* 5805 W 39th Ave Denver CO 80212-7201

FITZGERALD, JOHN DESMOND, b Aberkenfig, Wales, UK, Oct 31, 28. ADRENOCEPTOR PHARMACOLOGY, CLINICAL PHARMACOLOGY. *Educ:* Nat Univ Ireland, BSc, 49, MBBCh, 54. *Hon Degrees:* FRCP, Edinburgh, 71; FFPM, 90. *Prof Exp:* Sect mgr, C V Res, Pharmaceut Div, ICI, 69-73; prof med pharmacol, McMaster Univ, Hamilton, 73-76; gen mgr res drug discovery, Pharmaceut Div, Imp Chem Industs Ltd, 76-83, int med dir drug eval, 83-86; CONSULT PHARMACEUT MED, MATERIA MEDICA, 86- *Concurrent Pos:* Vis prof, Dept Epidemiol & Biostatist, McGill Univ, Montreal, 88- & Dept Physiol & Pharmacol, Univ Strathclyde, 89-; coun mem, Brit Heart Found, 90-; mem, Sci Policy Support Group Task Force on Drug Innovation, 90-91. *Mem:* Am Soc Pharmacol & Exp Therapeut; fel Royal Soc Med. *Res:* Cardiovascular pharmacology; beta adrenoceptor ligands; management of drug research and drug evaluation. *Mailing Add:* Mere Croft Chester Rd Mere near Knutsford Cheshire WA16 6LG England

FITZGERALD, JOSEPH ARTHUR, b Cleveland, Ohio, Apr 22, 25; m 55; c 8. MEDICINE, PSYCHIATRY. *Educ:* Case Western Reserve Univ, BS, 46; Loyola Univ, Chicago, MD, 51; Am Bd Psychiat & Neurol, dipl, 59. *Prof Exp:* Intern gen med, Milwaukee County Gen Hosp, 51-52; resident psychiat, Cleveland Psychiat Inst & Hosp, 52-54; sr resident, 57-58, from instr to prof, 58-82, CLIN PROF PSYCHIAT, MED CTR, SCH MED, IND UNIV, INDIANOPOLIS, 82- *Concurrent Pos:* Actg dir, Riley Child Guid Clin, 58; psychiatrist, Larue D Carter Hosp, 58-82, dir outpatient clin, 62- 82; lectr, postgrad progs gen practitioners & nonpsychiat specialists, 60-; residency training comt psychiat, Med Sch, Ind Univ, 60-, lectr personality theory, psychopath & psychopharmacol, Sch Nursing, 67-; dir diag ctr, Marydale Sch for Girls, 61-68; mem attend staff, Tenth St Vet Admin Hosp, 62-70; consult, St Mary's Child Ctr, 65-; dir psychiat serv, Cath Social Serv, 73-; consult psychiatrist, Adult & Child Ctr, 81-82, staff psychiatrist, 82-89, med dir, 90. *Mem:* Fel Am Psychiat Asn. *Res:* Anxiety; group process; primary prevention of mental disability through identifying high risk groups from perinatal and other preschool studies. *Mailing Add:* 471 W 63rd St Indianapolis IN 46260

FITZGERALD, LAURENCE ROCKWELL, b Boston, Mass, Sept 15, 16; m 42; c 3. PHYSIOLOGY. *Educ:* Tufts Col, BS, 39; Univ Iowa, MS, 41, PhD(zool), 49. *Prof Exp:* Res assoc zool, Univ Iowa, 48-49; from instr to asst prof anat, Ctr Health Sci, Univ Tenn, Memphis, 49-82; RETIRED. *Concurrent Pos:* Chmn sect anat sci, Am Asn Dent Schs, 71-72. *Mem:* AAAS; Soc Develop Biol; Am Soc Zoologists; Am Asn Anatomists; Int Asn Dent Res; Sigma Xi. *Res:* Physiology of neonatal period. *Mailing Add:* 1463 Rosemary Lane Memphis TN 38104

FITZGERALD, LAWRENCE TERRELL, b Tampa, Fla, Sept 19, 38; m 87. MEDICAL RADIATION PHYSICS. *Educ:* Univ Tampa, BS, 60; Univ Fla, MS, 62, PhD(med radiation physics), 74. *Prof Exp:* Instr med radiation physics, 64-69, res assoc, 70-74, asst prof, 74-80, ASSOC PROF MED RADIATION PHYSICS, DEPT RADIOL, UNIV FLA, 80- *Mem:* Am Asn Physicists Med; Asn Comput Mach; Inst Elec & Electronics Engrs; Health Physics Soc; Am Col Radiol. *Res:* Medical imaging; medical radiation physics; radiation dosimetry. *Mailing Add:* Univ Fla-Shands Teaching Hosp Box J-385 Gainesville FL 32610

FITZGERALD, MARIE ANTON, b Boston, Mass, Apr 10, 22; m 45; c 3. ANATOMY AND PHYSIOLOGY. *Educ:* Columbia Univ, BS, 50; Mass Inst Technol, PhD(physiol), 54. *Prof Exp:* Asst biol, Mass Inst Technol, 52-54, res assoc, 54-55; fel biophys, New Eng Inst Med Res, 58-59; assoc, Inst Biophys Eng, 60-68; prof biol, Holyoke Community Col, 68-88; ADJ PROF BIOL, UNIV MASS, AMHERST, 89- *Concurrent Pos:* Comput assisted instr biol students, 86 -; consult, 86- *Mem:* Sigma Xi. *Res:* Primarily a writer. *Mailing Add:* 525 Hadley St South Hadley MA 01075

FITZGERALD, MAURICE E, b Holyoke, Mass, Dec 16, 32; m 60; c 2. RHEOLOGY. *Educ:* St Anselm's Col, BA, 54; St John's Univ NY, MS, 56. *Prof Exp:* Lab asst quant anal, St John's Univ NY, 54-56; develop chemist, Borden Chem Co, Pa, 56-57; jr chemist, 57-64, res chemist & group leader, Mass Spectros & Comput Lab, 64-67, sr chemist & supvr, Mass Spectros & Comput Lab, 67-78, sr chemist & supvr elemental anal, Res & Eng, PRIN RES SCIENTIST, POLYMER APPLNS ARCO CHEM CO, NEWTOWN SQUARE, PA, 82- *Concurrent Pos:* Ed, Annual Rev, Am Soc Testing & Mat, 65-67. *Mem:* Am Chem Soc; Soc Rheology. *Res:* Rheological characterizations of polymer systems; development and characterization of new composite formulations for conductive composites and enhanced mechanical properties; analysis of fundamental structural properties and effect of processing variables on ultimate physical properties. *Mailing Add:* 517 Upland Rd Havertown PA 19083

FITZGERALD, MAURICE PIM VALTER, b Manchester, Eng, Aug 14, 39; Can citizen; m 64; c 2. ASTROPHYSICS. *Educ:* Univ Toronto, BSc, 62, MA, 63; Case Western Reserve Univ, PhD(astron), 67. *Prof Exp:* Assoc prof, 67-80, PROF PHYSICS, UNIV WATERLOO, 80- *Mem:* AAAS; Am Astron Soc; Royal Astron Soc; Royal Astron Soc Can; Int Astron Union. *Res:* Galactic structure; interstellar dust; spectroscopic binary stars; photometry; space density studies; spectroscopy; radial velocities; open clusters; statistical astronomy. *Mailing Add:* Dept Physics Univ Waterloo Rm 252 Waterloo ON N2L 3G1 Can

FITZGERALD, PATRICK HENRY, b Utica, NY, Sept 5, 43; m 67; c 3. PHYSICAL ORGANIC CHEMISTRY. *Educ:* Univ Toronto, BS, 66, MS, 68, PhD(phys org chem), 73. *Prof Exp:* Res assoc, Dept Chem, Ill Inst Technol, 73; res chemist, Res & Develop, Pigments Dept, 73-77, sr res chemist, 77-81, res assoc, Res & Develop, C&P Dept, 81-88, SR RES ASSOC, E I DU PONT DE NEMOURS & CO, INC, 88- *Mem:* Am Chem Soc. *Res:* Organic reaction mechanisms; fluorochemicals synthesis and processing; organic analysis and instrumentation. *Mailing Add:* Jackson Lab Chambers Works E I du Pont C&P Deepwater NJ 08023

FITZGERALD, PATRICK JAMES, b Haverhill, Mass, Aug 9, 13; m 49; c 1. ONCOLOGY. *Educ:* Univ Mass, BS, 36; Tufts Univ, MD, 40. *Prof Exp:* Chmn path, State Univ NY, Downstate Med Ctr, Brooklyn, 53-72, Mem Hosp, New York, NY, 72-80; prof path, Dept Path & Oncol, Univ Kans Med Ctr, 80-83; prof, 83-91, EMER PROF PATH, CORNELL UNIV MED COL, 91- *Concurrent Pos:* Vis scientist biochem, Biochem Dept, Oxford, Eng, 59-60 & 68, Inst Biol Chem, Marseilles, France, 68-; mem oncol, Sloan-Kettering Inst, New York, NY, 72-80; Allan Christiansen fel, St Catherine's Col, Oxford Univ, Eng, 90. *Honors & Awards:* Gold Headed Cane Award, Am Asn Path, 68. *Mem:* AAAS; Am Asn Path (pres, 68); Int Asn Pathologists. *Res:* Oncology; use of autoradiography in nucleic acid metabolism; regeneration; cancer of the thyroid and pancreas. *Mailing Add:* 35 Wood Ridge Lane Sea Cliff NY 11579

FITZGERALD, PAUL JACKSON, b Nashville, Tenn, July 20, 24; m 50; c 3. PLANT BREEDING, PLANT PATHOLOGY. *Educ:* Univ Tenn, BS, 50; Purdue Univ, MS, 52, PhD(plant path, breeding & genetics), 54. *Prof Exp:* Breeder hard red winter wheat, Intermountain Area & res agronomist, Plant Sci Res Div, USDA, 54-60; forage improve leader, Rockefeller Found, Santiago, Chile, 60-62; mem staff, Northern Grain Insects Res Lab, SDak, 62-68; from asst chief to chief cereal crops res br, Plant Indust Sta, 68-72, area dir, Lake States Area, 72-74, assoc dep adminr, 74-80, REGIONAL ADMINR, NORTH CENT REGION, AGR RES SERV, USDA, 80-, AGARA SCI ADV PLANT GERM PLASMA, PURDUE UNIV, 87- *Concurrent Pos:* Agr Sci Adv Panel Germplasm, Agr Res Serv, USDA, 85. *Mem:* Fel Am Soc Agron; fel Crop Sci Soc Am. *Res:* Inheritance of resistance to insects and insect-transmitted disease in plants. *Mailing Add:* Dept Agron Purdue Univ West Lafayette IN 47907

FITZGERALD, PAUL RAY, b Elsinore, Utah, May 2, 20; m 41; c 4. ZOOLOGY, PARASITOLOGY. *Educ:* Utah State Univ, BS, 49, MS, 50; Univ Ill, PhD, 61. *Prof Exp:* Instr zool & biol, Utah State Univ, 49-53; coop agt, Animal Dis & Parasite Res Div, USDA, 50-53, res parasitologist, 53-66; prof vet parasitol, Col Vet Med, Univ Ill, Urbana-Champaign, 66-86; RETIRED. *Concurrent Pos:* Proj dir grants, NIH, State of Ill & commercial; Fulbright fel, Arg; La State Univ fel, 71. *Mem:* Am Soc Parasitol; Soc Protozool; Am Soc Zool; Am Inst Biol Sci; Sigma Xi. *Res:* Parasitic protozoa, helminths; arthropods; domestic livestock; wild ruminants; parasites of fish; physiology of parasites and effects upon hosts; studies in environmental biology; public health aspects of sewage sludge utilization; parasites, heavy metals in animal tissues. *Mailing Add:* 625 S Canyon Dr Springville UT 84633

FITZGERALD, PAULA MARIE DEAN, b Los Angeles, Calif, Nov 28, 49. MACROMOLECULAR CRYSTALLOGRAPHY. *Educ:* Stanford Univ, BS & BA, 72; Johns Hopkins Univ, PhD(biophys), 77. *Prof Exp:* Fel, dept biol chem, Milton S Hershey Med Ctr, Pa State Univ, 77-79; res fel, molecular biophys dept, Med Found Buffalo, 79-81, res scientist I, 80-84; prof assist, Dept Biochem, Univ Alta, 84-87; RES FEL, MERCK SHARP & DOHME RES LABS, 88- *Mem:* Am Crystallog Asn; Am Soc Biol Chemists. *Res:* X-ray crystallographic structure determination of macromolecules; molecular replacement methods of structure determination. *Mailing Add:* Merck Sharp & Dohme Res Lab PO Box 2000 R80M203 Rahway NJ 07065

FITZGERALD, ROBERT HANNON, JR, b Denver, Colo, Aug 25, 42; m 85; c 7. ORTHOPAEDIC SURGERY. *Educ:* Univ Notre Dame, BS, 63; Univ Kans, MD, 67; Univ Minn, MS(orthop surg), 74. *Prof Exp:* Intern, Univ Tenn, 67-68; resident, Mayo Grad Sch Med, 70-74; from instr to prof orthop, Mayo Med Sch, 74-89; PROF & CHMN, DEPT ORTHOP SURG, WAYNE STATE UNIV SCH MED, 89-; ORTHOP SURG-IN-CHIEF, DETROIT MED CTR, 89-; CHIEF ORTHOP SURG, HUNTZEL HOSP & DETROIT RECEIVING HOSP, 89- *Concurrent Pos:* Consult orthop surgeon, Mayo Clin, 74-89; consult, Ctr Dis Control, 81-; educ comt, Am Acad Orthop Surgeons, 83; chief, adult reconstructive surg, Dept Orthop, Mayo Clin, 86-; bd trustees, J Bone & Joint S, 87-; consult, NIH-NIAMSD, 87- *Honors & Awards:* Kappa Delta Award for Res, Am Acad Orthop Surgeons-Am Orthop Res Soc, 83; F Stinchfield Award for Res, Hip Soc, 84, Sir John Charnley Award for Res, 86. *Mem:* Am Acad Orthop Surgeons; Hip Soc, Bd Dir, 83-87; Am Orthop Asn; Inter-Urban Orthop Asn; Surg Infection Soc; Orthop Res Soc. *Res:* Pathophysiology of musculo-skeletal infections; alterations of cell function which creates an environment susceptible to infection. *Mailing Add:* Dept Orthop Surg 4707 St Antoine Blvd Detroit MI 48201

FITZGERALD, ROBERT JAMES, b New York, NY, Nov 3, 18; m 45. MICROBIOLOGY, ORAL BIOLOGY. *Educ:* Fordham Univ, BS, 39; Va Polytech Inst, MS, 41; Duke Univ, PhD(pharmacol), 48. *Prof Exp:* Microbiologist, Chemother Div, Am Cyanamid Co, 41-45; jr asst sanitarian, Tuberc Control, USPHS, 45-46, scientist to scientist dir dent res, NIH, 48-69; res prof oral biol, Sch Med, Univ Miami, 69-77, prof, 77-87; RES MICROBIOLOGIST ORAL BIOL, VET ADMIN HOSP, MIAMI, 69-; EMER PROF MICROBIOL, SCH MED, UNIV MIAMI, 88- *Concurrent Pos:* Clin prof endodont, Col Dent, Univ Fla; consult, Nat Inst Dent Res, 69-80, Merck & Co, 70-75 & Naval Dent Res Inst, 72-; Vet Admin career res scientist, 79-; mem Nat Adv Dent Res Coun, 83-86. *Honors & Awards:* Res Award, Chicago Dent Soc, 60; Albert Joachim Res Prize, Int Fedn Dent, 62; Caries Res Award, Int Asn Dent Res, 77. *Mem:* Fel AAAS; Int Asn Dent Res; Am Soc Microbiol; Asn Gnotobiotics; Acad Microbiol. *Res:* Etiology and control of oral diseases; chemotherapy; oral microbiology. *Mailing Add:* Vet Admin Hosp Dent Res Unit 1201 NW 16th St Miami FL 33125

FITZGERALD, ROBERT SCHAEFER, b Detroit, Mich, July 12, 31; m 87; c 1. ENVIRONMENTAL HEALTH, PUBLIC HEALTH & EPIDEMIOLOGY. *Educ:* Xavier Univ, Ohio, LittB, 54; Spring Hill Col, MA, 57; Univ Chicago, PhD(physiol), 63; Woodstock Col, STB, 65, STM, 67. *Prof Exp:* Fel physiol, 63-67, from asst prof to assoc prof, 67-77, assoc chmn, 80-88, PROF ENVIRON HEALTH SCI, JOHNS HOPKINS UNIV, 77-, ACTG CHMN, 90- *Concurrent Pos:* Vis scientist, Univ Wash, 66, Univ Nancy, 67 & Univ Paris, 68; USPHS res career develop award, 68-73; vis assoc prof physiol, Univ Calif, San Francisco, 73-74. *Mem:* AAAS; Am Physiol Soc; Am Thoracic Soc; Sigma Xi; fel Royal Col Med, Ireland. *Res:* Examination of the elements and reflexes involved in the control of respiration and circulation, especially in response to different gaseous environments. *Mailing Add:* Dept Environ Health Sci Johns Hopkins Univ 615 N Wolfe St Baltimore MD 21205

FITZGERALD, ROBERT WILLIAM, b Canton, Ohio, May 30, 31; m 54; c 2. FIRE PROTECTION ENGINEERING, STRUCTURAL ENGINEERING. *Educ:* Worcester Polytech Inst, BS, 53, MS, 60; Univ Conn, PhD(struct eng), 69. *Prof Exp:* Struct engr, Harvey & Tracy Consult Engrs, Mass, 55-58; instr mech, Worcester Jr Col, 58-62; instr struct eng & mech, Univ Conn, 62-63; from asst prof to assoc prof, 63-78, PROF CIVIL ENG & FIRE PROTECTION ENG, WORCESTER POLYTECH INST, 78- *Mem:* Am Soc Civil Engrs; Soc Fire Protection Engrs; Nat Fire Protection Asn. *Res:* Building design for fire safety; building codes; building design and construction; risk management and assessment. *Mailing Add:* Dept Civil Eng Worcester Polytech Inst 100 Institute Rd Worcester MA 01609

FITZGERALD, THOMAS JAMES, b Troy, NY, May 7, 38; m 74; c 4. PHARMACOLOGY, MEDICINAL CHEMISTRY. *Educ:* Union Univ, BS, 60; Ohio State Univ, MS, 62, PhD(med chem), 65. *Prof Exp:* NIH fel, Inst Cancer Res, Univ Heidelberg, Ger, 65-66, Alexander von Humboldt fel, 66-67; NIH trainee fel, Sch Pharm, Univ Minn, 68-69; from instr to asst prof pharmacol, Univ Kans Med Ctr, 69-73; from asst prof to assoc prof, 73-85, PROF PHARMACOL, COL PHARM, FLA A&M UNIV, 85- *Mem:* Am Chem Soc; Sigma Xi; Am Soc Pharmacol & Exp Therapeut; AAAS. *Res:* Molecular pharmacology; mitotic inhibitors; cancer chemotherapy. *Mailing Add:* Sch Pharm Fla A&M Univ Tallahassee FL 32307

FITZGERALD, THOMAS JAMES, b Chicago, Ill, Dec 12, 43; m 71. MICROBIOLOGY. *Educ:* St Mary's Col, Minn, BA, 66; Loyola Univ Ill, MS, 68, PhD(microbiol), 71. *Prof Exp:* Fel microbiol, Harvard Univ, 71-72; fel, Univ Calif, Los Angeles, Med Sch, 72-73, lectr, 73-74; res assoc, 74-80, ASST PROF MICROBIOL, MED SCH, UNIV MINN, 80- *Mem:* Sigma Xi; Am Soc Microbiol; AAAS; NY Acad Sci. *Res:* In vitro cultivation of treponema pallidum in tissue culture; pathogenic mechanisms of treponema pallidum with emphasis on cell mediated immunity. *Mailing Add:* 2432 E First Duluth MN 55812

FITZGERALD, THOMAS MICHAEL, b Boston, Mass, Oct 20, 36; m 58; c 3. ACOUSTICS. *Educ:* Boston Col, ScB, 58; Brown Univ, ScM, 61, PhD(physics), 63. *Prof Exp:* Res assoc solid state physics, Brown Univ, 63-65; physicist, NASA Electronics Res Ctr, 65-70; physicist, Naval Underwater Syst Ctr, 70-76 & Rand Corp, Calif, 76-77; physicist, Naval Underwater Ctr, 77-80, sci adv, Comdr-in-Chief, Atlantic Fleet, 81-83; PHYSICIST, NAVAL OCEAN SYST CTR, 83- *Mem:* Sigma Xi; Am Phys Soc. *Res:* Phonon-phonon, electron-phonon interactions; elastic constants; effects of irradiation on ultrasonic properties of materials; radiation damage; underwater acoustics; acousto-optics; systems analysis; microwave acoustics; infra-red spectroscopy. *Mailing Add:* 3310 Avenida Hacienda Escondido CA 92025

FITZGERALD, WILLIAM FRANCIS, b Boston, Mass; m 72; c 2. CHEMICAL OCEANOGRAPHY, MARINE & ENVIRONMENTAL GEOCHEMISTRY. *Educ:* Boston Col, BS, 60; Col of the Holy Cross, MS, 61; Mass Inst Technol, PhD(chem oceanog), 70; Woods Hole Oceanog Inst, PhD(chem oceanog), 70. *Prof Exp:* Res asst chem oceanog, Woods Hole Oceanog Inst, 61-66, investr, 69-70; PROF CHEM OCEANOG, MARINE SCI DEPT, UNIV CONN, 70- *Mem:* Sigma Xi; Am Chem Soc; AAAS; Am Geophys Union; Oceanog Soc. *Res:* Global cycling of mercury; chemical routes and transport of mercury to the marine environment; trace metal speciation in sea water; sedimentary geochemistry and the biological interactions of heavy metals in the coastal zones; marine pollution and air-sea exchange. *Mailing Add:* Marine Sci Dept Univ Conn Groton CT 06340

FITZGIBBON, WILLIAM EDWARD, III, b Cambridge, Mass, July 21, 45; div. MATHEMATICS. *Educ:* Vanderbilt Univ, BA, 68, PhD(math), 72. *Prof Exp:* From asst prof to assoc prof, 72-80, PROF MATH, UNIV HOUSTON, 81- *Concurrent Pos:* Vis assoc prof math, Univ Calif, San Diego, 80-81; vis assoc prof, Appl Math Div, Argonne Nat Lab, 80. *Mem:* Am Math Soc; Soc Indust & Appl Math; Sigma Xi. *Res:* Pure and applied mathematics; ordinary and partial differential equations; operator theory; nonlinear analysis approximation theory; integral equations. *Mailing Add:* Dept Math Univ Houston Houston TX 77004

FITZGIBBONS, J(OHN) D(AVID), b Ithaca, NY, Apr 18, 22; m 46; c 1. NUCLEAR ENGINEERING. *Educ:* NC State Univ, BS, 53; Oak Ridge Sch Reactor Technol, cert, 55. *Prof Exp:* Indust radiographer, Los Alamos Sci Lab, 46-47; nuclear engr, Atomic Energy Div, Babcock & Wilcox Co, 53-58, proj engr res reactors, 58-59, proj mgr & engr res reactors, 59-64, planner, Div Long Range Planning, 64-68, supvr systs data mgt, Naval Nuclear Fuel Div, 68-69, sr systs analyst, 69-72, proj adminr, 72-79, eng supvr, 79-82, classification adminr, Security, 82-85; RETIRED. *Res:* Research reactor design and development; systems analysis. *Mailing Add:* 3639 E Woodside Ave Lynchburg VA 24505

FITZHARRIS, TIMOTHY PATRICK, b New York, NY, Oct 31, 44; m 67; c 2. CELL BIOLOGY, DEVELOPMENTAL BIOLOGY. *Educ:* State Univ NY Col Potsdam, AB, 66; State Univ NY Albany, MS, 69, PhD(biol sci), 71. *Prof Exp:* NSF fel, Dept Molecular, Cellular & Develop Biol, Univ Colo, 71-73; assoc anat, Nat Inst Neurol Dis & Stroke grant, 74-76, from asst prof to prof anat, Med Univ SC, 74-84; DIR RES & DEVELOP, ORGANON TEKNIKA CORP, 84- *Concurrent Pos:* Affil marine scientist, SC Marine Resources Ctr, 73-; SC Heart Grant-in Aid, 77-78; estab investr, Am Heart Asn, 80-84; Am Heart-grant-in-aid, 79-82; exchange scientist, Nat Acad Sci, Charles Univ, Prague, 80. *Mem:* Fel AAAS; Am Soc Cell Biologists; Soc Develop Biol; NY Acad Sci; Am Asm Anatomists; Sigma Xi. *Res:* Extracellular matrix and morphogenesis; early heart development; cell motility and intracellular transport; congenital heart defects; diagnostic methods in AIDS research. *Mailing Add:* Organon Teknika Corp 100 Akzo Ave Durham NC 27704

FITZHUGH, HENRY ALLEN, b San Antonio, Tex, July, 2, 39; m 84; c 1. AGRICULTURAL SYSTEMS, ANIMAL BREEDING. *Educ:* Tex A&M Univ, BS, 61, MS, 63, PhD(animal breeding), 65. *Prof Exp:* Assoc prof animal sci statist, Tex A&M Univ, 65-73; exec vpres, Agri-Link Corp, 73-75; dir, Africa & Middle East Winrock Intl Agr Dev, 75-; AT ILCA, ETHIOPIA. *Concurrent Pos:* NATO fel, Agr Res Serv, Animal Breeding Res Orgn, 65-66; partner, Genetics Appl Prod Inc, 66-73; consult, Southwest Res Found, 71-72, US Feed Grains Coun, 76, Enterprise Brasil Invest Agr, Brasil, 79 & Inst Nat Invest Pecuarias, Mex, 80; res geneticist, US Meat Animal Res Ctr, Agr Res Serv, USDA, 75; US AID, 80- *Mem:* Am Soc Animal Sci; AAAS. *Res:* Development and testing of biological and socioeconomic interventions to improve efficiency of agricultual production systems. *Mailing Add:* Dept Dir Gen ILCA PO Box 5689 Addis Ababa Ethiopia

FITZHUGH, OSCAR GARTH, b Hood, Va, Aug 26, 01; m 29; c 4. PHARMACOLOGY. *Educ:* Univ Va, BS, 27, MS, 33, PhD(physiol), 36. *Prof Exp:* Instr physiol & pharmacol, Univ Vt, 34-36, asst prof, 36-37; res assoc pharmacol, Sch Med, Vanderbilt Univ, 37-39; pharmacologist, Food & Drug Admin, 39-47, chief chronic toxicity sect, Div Pharmacol, 47-53, chief toxicity br, 53-64, dep dir div toxicol eval, 64-68, toxicol dir, Bur Sci, 68-70, assoc dir toxicol res, Off Pesticides, Bur Foods & Pesticides, 70-71; toxicol adv, Off Pesticide Progs, Environ Protection Agency, 71-72; consult toxicol, 72-81; RETIRED. *Mem:* AAAS; Soc Toxicol; Inst Food Technol; Am Soc Pharmacol & Exp Therapeut; Soc Exp Biol & Med. *Res:* Adrenal cortex; anesthetics; chronic toxicology; coal tar colors; pesticides; food additives. *Mailing Add:* 3816 Old Gun Rd Midlothian VA 23113

FITZHUGH, RICHARD, b Concord, Mass, Mar 30, 22; m 63; c 2. BIOPHYSICS, BIOMATHEMATICS. *Educ:* Univ Colo, BA, 48; Johns Hopkins Univ, PhD(biophys), 53. *Prof Exp:* Instr physiol optics, Med Sch, Johns Hopkins Univ, 53-55; biophysicist, NIH, USPHS, 56-85; RETIRED. *Mem:* AAAS; Biophys Soc. *Res:* Physiology of the nerve membrane; mathematical models of nerve cells, statistical detection of single channel signals, and of muscular control. *Mailing Add:* 3806 Everett St Kensington MD 20895

FITZHUGH-BELL, KATHLEEN, b Kansas City, Mo. NEUROPSYCHOLOGY. *Educ:* Univ Mo, Kansas City, BA, 51, MA, 54; Purdue Univ, PhD(clin psychol), 58; Am Bd Prof Neuropsychol, dipl. *Prof Exp:* Clin & res psychologist, New Castle State Hosp, Ind, 58-66; vis assoc prof phys med, Sch Med, Tufts Univ, 69-70; from asst prof to assoc prof neurol & Clin psychol, dept neurol, 66-78,; CONSULT. *Concurrent Pos:* Consult, Marion County Asn Retarded Persons, Ind, 71-80, Larue Carter Mem Hosp, 74-80 & Cent State Hosp, 84- *Mem:* Am Psychol Asn; Am Acad Retardation; Am Asn Mental Deficiency; AAAS; Int Neuropsychol Soc; Nat Acad Neuropsychologists. *Res:* Brain behavior relationships; learning disabilities. *Mailing Add:* Raleigh Hosp for Children 702 Barnhill Dr Rm 375 Indianapolis IN 46202-5200

FITZ-JAMES, PHILIP CHESTER, b Vancouver, BC, Nov 25, 20; m 48, 83; c 6. MICROBIAL BIOCHEMISTRY. *Educ:* Univ BC, BSA, 43; Univ Toronto, MSA, 45; Univ Western Ont, MD, 49, PhD(bact, biochem), 53. *Prof Exp:* Asst penicillin prod, Banting Inst, Toronto, 43-44; assoc prof, 53-67, EMER PROF BACT, IMMUNOL, SCH MED, UNIV WESTERN ONT, 67- *Concurrent Pos:* Res assoc, Nat Res Coun Can, 56- *Honors & Awards:* Harrison Award, Royal Soc Can, 63. *Res:* Structure, composition and activities of bacteria, particularly the process of spore formation; role of the membrane in cell wall assembly; insect toxicity of the parasporal process; role of spore coat in germination. *Mailing Add:* Dept Microbiol & Immunol Health Sci Ctr London ON N6A 5B8 Can

FITZLOFF, JOHN FREDERICK, b Jan 7, 40; US citizen; m 69; c 3. MEDICINAL CHEMISTRY, TOXICOLOGY. *Educ:* San Jose State Col, BS, 61; Univ Calif, San Francisco, PhD(pharmaceut chem), 72. *Prof Exp:* Chemist, Stanford Res Inst, 61-65; teaching asst chem, Univ Calif, San Francisco, 66-68; asst prof, 72-77, ASSOC PROF MED CHEM, UNIV ILL MED CTR, 77- *Concurrent Pos:* Vis scientist, Inst Toxicol & Pharmacol, Phillips Univ, Fed Repub of Ger, 79-80; fel, Am Found Pharmaceut Educ. *Mem:* Sigma Xi; Am Asn Pharmaceut Scientists; Am Chem Soc; Int Soc Study Xenobiotics; Am Asn Col Pharm. *Res:* Metabolism of drugs and other foreign compounds; methods of analyses of foreign compounds in biological media; relationship of chemical structure and metabolizing enzymes. *Mailing Add:* Univ Ill Dept Med Chem/Pharmacog (P'cog) PO Box 6998 Rm 544 Pharm (M/C781) Chicago IL 60680-6998

FITZMAURICE, NESSAN, b Ireland, May 23, 59. DYNAMICAL SYSTEMS THEORY, SCIENTIFIC COMPUTATION. *Educ:* Nat Univ Ireland, Dublin, BSc, 80, MSc, 81; Cornell Univ, PhD(eng), 86. *Prof Exp:* Res asst, vortex flows, Cornell Univ, 81-85; staff scientist, Inst Computer Appl in Sci & Eng, NASA, Langley, 85-88; PROF MATH, CASE WESTERN RESERVE UNIV, 88- *Concurrent Pos:* Consult, Ohio Aerospace Inst, 85-; scientist, Inst Comput Mech & Propulsion, 88-90. *Mem:* Soc Indust Appl Math; Asn Comput Mach. *Res:* Expertise in applied mathematics; high performance scientific computation; nonlinear dynamical systems and chaos; hydrodynamic stability; turbulence modeling; large scale direct turbulance simulations and oceanography. *Mailing Add:* Math Dept Case Western Reserve Univ Cleveland OH 44106

FITZNER, RICHARD EARL, b Belding, Mich, June 25, 46; m 70. VERTEBRATE ECOLOGY, ZOOLOGY. *Educ:* Mich State Univ, BS, 69, MS, 70; Wash State Univ, PhD(zool), 78. *Prof Exp:* RES SCIENTIST TERRESTRIAL ECOL, PAC NORTHWEST LABS, BATTELLE MEM INST, 71- *Concurrent Pos:* Mem, Wash State, Natural Preserves Adv Comt, 77-80. *Mem:* Cooper Ornith Soc; Am Ornithologists Union; Wildlife Soc; Sigma Xi. *Res:* Vertebrates and their role in ecosystem functioning; environmental perturbations (land disturbances, and chemical contaminants) and their impact on the environment; natural history and behavior of vertebrates, particularly birds. *Mailing Add:* Battelle Northwest 3190 George Washington Way Richland WA 99352

FITZPATRICK, BEN, JR, b Miami, Fla, Sept 28, 32; m 53; c 3. MATHEMATICS. *Educ:* Ala Polytech Inst, BS, 52; Univ Tex, MA & PhD, 58. *Prof Exp:* Asst math, Univ Tex, 55-56, spec instr, 56-58, asst prof, 58-59; from asst prof to assoc prof, 59-66, PROF MATH, AUBURN UNIV, 66- & HEAD DEPT MATH, 78- *Mem:* Am Math Soc; Math Asn Am. *Res:* Continua and point set theory; abstract spaces; theory of integration. *Mailing Add:* Dept Math Auburn Univ Auburn AL 36849-3501

FITZPATRICK, FRANCIS ANTHONY, b Wilmington, Del, July 5, 47. ANALYTICAL BIOCHEMISTRY. *Educ:* Villanova Univ, BS, 69; Univ Mass, PhD(anal chem), 72. *Prof Exp:* Sr anal chemist, Vick Chem Co, Div Richardson-Merrill, 72-74; SCIENTIST & ANALYTICAL BIOCHEMIST, UPJOHN CO, 74- *Concurrent Pos:* Vis scientist, Karolinska Inst, Stockholm, Sweden, 82-83. *Honors & Awards:* UpJohn Award, 84. *Mem:* AAAS; Am Chem Soc; Am Soc Biol Chem. *Res:* Cell biology, pharmacology, biochemistry and analytical chemistry of lipids including steroids, prostaglandins, leukotrienes, thromboxanes, and glyceryl ether phospholipids; elucidation of mechanism of action of drugs, identification of modulators of lipid biosynthesis and pharmacology. *Mailing Add:* Univ Colo Health Sci Ctr 4200 E Ninth Ave Box C236 Denver CO 80262

FITZPATRICK, GARY OWEN, b Petaluma, Calif, Dec 20, 43; c 2. PHYSICS, ENGINEERING. *Educ:* Calif Inst Technol, BSc, 66. *Prof Exp:* Sr engr thermionics, Gulf Gen Atomic, 66-73; sr engr, EMC, Intelcom Rad Technol, 73-74; prog mgr, Direct Energy Conversion, 74-79, VPRES, RASOR ASSOC, INC, 79- *Mem:* AAAS; Inst Elec & Electronics Engrs. *Res:* Thermionics; high temperature materials; energy converters; systems behavior; electromagnetic pulse effects; astronomical instrumentation. *Mailing Add:* Rasor Associates Inc 253 Humboldt Ct Sunnyvale CA 94089

FITZPATRICK, GEORGE, b Trenton, NJ, Oct 7, 46; div; c 2. HORTICULTURE, ENVIRONMENTAL BIOLOGY. *Educ:* Trenton State Col, BA, 68, MA, 72; Rutgers Univ, PhD(entom & econ zool), 75. *Prof Exp:* Res assoc entom, Miss State Univ, 75-76; asst res scientist, 76-79, asst prof,

79-84, ASSOC PROF, INST FOOD & AGR SCI, UNIV FLA, 84- *Concurrent Pos:* Sci adv & consult, Agripost, Inc, 84-90 & Taino Farms, Freeport, Bahamas, 87- *Mem:* AAAS; Am Soc Hort Sci; Sigma Xi; Int Soc Hort Sci. *Res:* Horticultural substrates; waste product utilization in horticultural production; quantitative and qualitative water relations in plants. *Mailing Add:* Univ Fla Ft Lauderdale Res & Educ Ctr 3205 Col Ave Ft Lauderdale FL 33314

FITZPATRICK, HUGH MICHAEL, b Pittsburgh, Pa, Apr 22, 20; m 44; c 5. PHYSICS. *Educ:* George Washington Univ, BS, 44. *Prof Exp:* Naval architect, David Taylor Model Basin, 42-48, physicist, 48-59; physicist, Syst Eng Div, Cleveland Pneumatic Indust Inc, 59-63; physicist, Off Naval Res, 63-79; consult; RETIRED. *Mem:* Acoust Soc Am. *Res:* Fluid mechanics; acoustics; hydromechanics; cavitation; hydrodynamic noise; propulsion; radio; astronomy. *Mailing Add:* 4709 Merivale Rd Chevy Chase MD 20815

FITZPATRICK, J(OSEPH) F(ERRIS), JR, b New Orleans, La, Mar 8, 32; m 61; c 5. INVERTEBRATE ZOOLOGY, FRESHWATER CRUSTACEA. *Educ:* Tulane Univ, BS, 59, MS, 61; Univ Va, PhD(biol), 64. *Prof Exp:* Asst zool, Tulane Univ, 60, asst bot & zool, 61; asst biol, Univ Va, 62-64; instr, dept zool, Univ Ky, 64; asst prof zool, Miss State Univ, 64-69; assoc prof biol, Randolph-Macon Woman's Col, 69-73; assoc prof, 73-78, PROF, UNIV SALA, 78- *Concurrent Pos:* Assoc ed, Northeast Gulf Sci, 80-85. *Mem:* Fel AAAS; Sigma Xi; Soc Syst Zool; Crustacean Soc; Int Asn Astacol. *Res:* Systematics of North American crawfishes and other freshwater crustacea. *Mailing Add:* Dept Biol Sci Univ SAla Mobile AL 36688

FITZPATRICK, JIMMIE DOILE, b Jonesboro, La, Aug 6, 38; m 59; c 2. ORGANIC CHEMISTRY, ANALYTICAL CHEMISTRY. *Educ:* La Polytech Inst, BS, 60; Iowa State Univ, MS, 63; Univ Tex, Austin, PhD(org chem), 66. *Prof Exp:* Res chemist, Phillips Petrol Co, 65-66; res assoc, Purdue Univ, 66-67; res chemist, GAF Corp, 67-68; asst prof org chem, 68-76, ASSOC PROF CHEM, UNIV SOUTHWESTERN LA, 77- *Mem:* Am Chem Soc. *Res:* Organic synthesis; organometallic and polymer chemistry; homogenous and heterogenous catalysis; flavor chemistry. *Mailing Add:* Chem Dept Univ Southwestern La PO Box 44370 Lafayette LA 70504

FITZPATRICK, JOHN WEAVER, b St Paul, Minn, Sept 17, 51. ORNITHOLOGY, ECOLOGY. *Educ:* Harvard Univ, BA, 74; Princeton Univ, PhD(biol), 78. *Prof Exp:* Chmn, Dept Zool, 85-88, CUR BIRDS DIV, FIELD MUS NATURAL HIST, 78-; EXEC DIR, ARCHBOLD BIO STA, LAKE PLACID, FLA, 88- *Honors & Awards:* Brewster Award, Am Ornithologists Union, 85. *Mem:* Soc Syst Zool; Am Ornithologists Union; Wilson Ornith Soc; Cooper Ornith Soc; Sigma Xi; Soc Animal Behav; AAAS. *Res:* Systematics, ecology, and zoogeography of neotropical birds, especially Tyrannidae; adaptive radiation in morphology and behavior among birds; evolution of cooperative breeding social systems in birds; tropical bird community structure; ecological distribution neo-tropical birds; Florida Scrubjay conservation biology. *Mailing Add:* Archbold Biol Sta Old SR 8 County Rd Lake Placid FL 33852

FITZPATRICK, JOSEPH A, b Albany, NY, Oct 20, 44; m 67; c 6. BIOENVIRONMENTAL ENGINEERING, DISPERSE SYSTEMS ENGINEERING. *Educ:* Univ Notre Dame, BS, 66; Harvard Univ, MS, 67, PhD(eng), 72. *Prof Exp:* Res sanitary engr, Metcalf & Eddy Engrs, Boston, 66; asst prof civil engr, Clarkson Col, 71-72; asst prof, 72-76, ASSOC PROF CIVIL ENG, NORTHWESTERN UNIV, 77- *Concurrent Pos:* Sci adv, KOR, Inc, Cambridge, 70-; consult, Commonwealth Edison Co, 73-75, Environ Equip Div, FMC Corp, 74-, Tech Prod Div, Brunswick Corp, 75-77, Argonne Nat Lab, 77-78 & Hydro Clear Corp; prin investr, Environ Protection Agency, 74-80, co-prin investr, Am Iron & Steel Inst, 77-78, US Army, 74-76, Hydro Clear Corp, 78- 80, Ill Dept Energy, 84-88, Waste Mgt Corp, 85-86; vis res prof, Univ Ill, 84. *Mem:* Am Chem Soc; Am Inst Chem Engrs; Fine Particles Soc; Am Water Works Asn; Am Filtration Soc. *Res:* Chemistry, mechanics and removal of particulates in water and wastewater; solid-liquid separation process design; filtration theory and practice; ground water quality modeling and management. *Mailing Add:* 2633 Broadway Evanston IL 60201-1555

FITZPATRICK, LLOYD CHARLES, b Erie, Pa, May 12, 37; div; c 3. ECOLOGY. *Educ:* Mt Union Col, BS, 60; Kent State Univ, MA, 66; PhD(ecol), 70. *Prof Exp:* High sch instr, Fla, 60-61; jr high instr, 61-66; instr biol sci, Kent State Univ, 68-70; asst prof, 70-75, assoc dir res, Inst Environ Studies, 73-76, chmn ecol div, Dept Biol Sci, 76-80, assoc dir, Anal Water Qual Lab, 75-78, coordr grad studies, 78-79, assoc prof, 75-81, PROF BIOL, NTEX STATE UNIV, 81- *Mem:* Am Soc Naturalists; Ecol Soc Am; Am Soc Zool; Sigma Xi; NAm Benthological Soc. *Res:* Ecological, geographical and evolutionary effects on intra- and interspecific energetics, physiologies and reproductive strategies in animals; population dynamics and trophic dynamics of aquatic and terrestrial ecosystems; assessment of environmental impacts. *Mailing Add:* Dept Biol Sci Univ NTex Denton TX 76203

FITZPATRICK, PATRICK MICHAEL, b Youghal, Ireland, Mar 14, 46; US citizen; m 70; c 1. PURE MATHEMATICS. *Educ:* Rutgers Univ, BS, 66, PhD(math), 71. *Prof Exp:* Vis mem, Courant Inst, NY Univ, 71-72; lectr, Rutgers Univ, 72-73; L E Dickson instr, Univ Chicago, 73-75; ASST PROF MATH, UNIV MD, COLLEGE PARK, 75- *Concurrent Pos:* NSF res awards, 70-76. *Mem:* Am Math Soc; Am Math Asn. *Res:* Application of functional analysis and topological methods to the solution of abstract nonlinear equations and to integral and partial differential equations. *Mailing Add:* Dept Math Univ Md College Park MD 20742

FITZPATRICK, PHILIP MATTHEW, b New York, NY, Sept 17, 15; m 42; c 2. APPLIED MATHEMATICS. *Educ:* Univ Okla, BS, 50, MS, 51, PhD(physics), 55. *Prof Exp:* Asst prof physics, Univ Okla, 53-55; physicist, US Navy Mine Defense Lab, Fla, 55-59; physicist, Air Proving Ground Ctr, Eglin AFB, 59-62; from assoc prof to prof, 62-82, EMER PROF MATH, AUBURN UNIV, 82- *Mem:* Am Phys Soc; Math Asn Am. *Res:* Astrodynamics. *Mailing Add:* 107 Ryan St Auburn AL 36830

FITZPATRICK, ROBERT CHARLES, b Port Huron, Mich, Jan 18, 26; m; c 2. GEOPHYSICS. *Educ:* Univ Mich, BS, 48, MS, 50. *Prof Exp:* Packaging-mat handling engr, Kaiser-Frazer Corp, 50-51; geophysicist, Off Naval Res, 51-53; geophysicist, Air Force Cambridge Res Ctr, 53-54, chief geodesy sect, 54-56; res assoc, Seismol & Acoustics, Willow Run Labs, Univ Mich, 56-60, mem staff, Res Admin, 60-65, from asst dir to assoc dir, 65-69; mgr mkt res, Datamax Corp, 69-70; asst vpres res, 70-80, assoc vpres res, 80-87, ASSOC DEAN SCH DENTAL MED, STATE UNIV NY BUFFALO, 87-; ASSOC DEAN SCH DENTAL MED, STATE UNIV NY BUFFALO, 87- *Concurrent Pos:* Mem, Nat Coun Univ Res Adminrs. *Mem:* Geol Soc Am; Soc Explor Geophys; Int Asn Gt Lakes Res; Am Geophys Union. *Res:* Physical oceanography; gravimetry; seismology. *Mailing Add:* 99 Parkledge Dr Snyder NY 14226

FITZPATRICK, THOMAS BERNARD, b Madison, Wis, Dec 19, 19; m 44; c 5. MEDICINE. *Educ:* Univ Wis, AB, 41; Harvard Univ, MD, 45; Univ Minn, PhD, 52; Am Bd Dermat, dipl, 52. *Hon Degrees:* DSc, Univ Mass Med Sch. *Prof Exp:* Asst prof, Univ Mich, 51-52; prof & head div, Med Sch, Univ Ore, 52-58; Edward Wigglesworth prof dermat, Harvard Sch Med, 59-87; chief dermat serv, Mass Gen Hosp, 59-87; CONSULT, 87- *Concurrent Pos:* Chief dept dermat, Multnomah & Children's Hosps, Portland, Ore, 52-58; guest lectr, Univs Tokyo, Tohoku & Kyoto, Japan, 56; consult, Nat Inst Arthritis, Metab & Digestive Dis, 60-; mem dermat training grants comt, USPHS, 60-65; Pollitzer lect, NY Univ, 62; consult, Peter Bent Brigham Hosp, 62-; Prosser White Oration, Royal Soc Med, 64; mem, Climatic Impact Comt, Nat Acad Sci, 71-74 & Comt on Impacts of Stratospheric Change, 74-; pres, Dermat Found, 71-73; pres, Int Pigment Cell Soc, 78-; pres-elect, Asn Prof Dermat, 80; master dermat, Am Acad Dermat. *Honors & Awards:* Mayo Found Award, 51; Myron Gordon Award, Int Union Against Cancer, 65; Stephen Rothman Gold Medal Award Distinguished Achievement, Soc Invest Dermat, 70; Dome lectr, Am Acad Dermat, 79; Taub Psoriasis Award, 80. *Mem:* Soc Invest Dermat (pres, 59-60); Am Acad Arts & Sci; Am Dermat Asn; Am Soc Photobiology; Asn Am Physicians. *Res:* Melanin biosynthesis; normal and abnormal reactions of man to light; molecular biology of melanin; origin of racial color; dermatology. *Mailing Add:* Dept Dermat Mass Gen Hosp Boston MA 02114

FITZROY, NANCY DELOYE, b Pittsfield, Mass, Oct 5, 27; m 51. ENGINEERING. *Educ:* Rensselaer Polytech Inst, BChE, 49. *Hon Degrees:* DSc, NJ Inst Technol, Newark, 87; DEng, Rensselaer Polytech Inst, Troy, NY, 90. *Prof Exp:* Asst engr, Knolls Atomic Power Lab, 50-52; develop engr, Hermes Missile Proj, 52-53, develop engr, Gen Eng Lab, 53-63, heat transfer engr, Advan Technol Labs, 63-65, consult heat transfer, Res & Develop Ctr, 65-71, mgr heat transfer consult, 71-74, strategy planner, 74-76, advan concepts planner & proposal mgr, 76-79, prog develop mgr, Gas Turbine Div, 79-82, mgr energy & environ progs, Turbine Mkt & Projs Div, 82-87; CONSULT, 87- *Concurrent Pos:* Lectr, Advan Eng Course, Gen Elec Co, 62-67; res comt adv, NSF, 72-75; bd mem, Bd Eng Manpower & Educ Policy, Nat Acad Eng, 74-76. *Honors & Awards:* Achievement Award, Soc Women Engrs, 72, Fedn Prof Women, 84; Demers Medal, Rensselaer Polytech Inst, 75; Centennial Medallion, Am Soc Mech Engrs, 80. *Mem:* Fel Am Soc Mech Engrs (pres, 85-87); Am Inst Chem Engrs; Nat Soc Prof Engrs; Soc Women Engrs. *Res:* Heat transfer; thermal engineering; thermal properties of materials; high temperature radiation from nuclear source; cooling of integrated circuits; heat transfer in regenerator matrices; patentee in field of cooling of integrated circuits. *Mailing Add:* 2125 Rosendale Rd Schenectady NY 12309

FITZSIMMONS, DELBERT WAYNE, b Bazine, Kans, Jan 20, 32; m 52; c 4. AGRICULTURAL & IRRIGATION ENGINEERING. *Educ:* Univ Idaho, BS, 59, MS, 62; Wash State Univ, PhD(eng sci), 70. *Prof Exp:* From instr to assoc prof, 59-71, actg chmn, 70-72, PROF AGR ENG, UNIV IDAHO, 71-, CHMN DEPT, 72- *Mem:* Am Soc Agr Engrs; Am Soc Eng Educ; Nat Soc Prof Engrs; Sigma Xi. *Res:* Unsteady flow into and through porous media; hydraulic characteristics of porous media; mathematical modeling of flow through soils; effects of irrigated agriculture on water quality. *Mailing Add:* 2107 Robinson Park Rd Moscow ID 83843

FITZWATER, DONALD (ROBERT), b Kansas City, Mo, Oct 5, 30; m 51; c 2. COMPUTER SCIENCE. *Educ:* William Jewell Col, BA, 50; Iowa State Col, MS, 52, PhD(chem), 58. *Prof Exp:* Asst chem, Ames Lab, Atomic Energy Comn, 50-52, jr chemist, 51-52, 55-58, assoc chemist, 58-64, chemist, 64-67, from asst prof to assoc prof chem, Iowa State Univ, 59-67; ASSOC PROF COMPUT SCI, UNIV WIS-MADISON, 67- *Mem:* Asn Comput Mach. *Res:* Real time control; programming systems; information processing; system & language structures. *Mailing Add:* Dept Comput Sci Univ Wis Madison 1210 W Dayton St Madison WI 53706

FITZWATER, ROBERT N, b Elkins, WVa, Apr 8, 24; m 50; c 1. CHEMISTRY. *Educ:* Rollins Col, BS, 49; Univ Fla, PhD, 58. *Prof Exp:* Asst, Bur Entom & Plant Quarantine, USDA, 51-53; asst, Univ Fla, 53-55; sr chemist, Res Labs, Gen Motors Corp, 58-61 & Martin Col, 62; asst prof chem, Rollins Col, 62-65; asst prof pharmacol, Sch Med, Univ Miami, 65-69; assoc prof chem, 69-74, ASSOC PROF CHEM, GA SOUTHERN COL, 74- *Mailing Add:* PO Box 2202 Statesboro GA 30458-2202

FIUMARA, NICHOLAS J, b Boston, Mass, Oct 31, 12; m 44; c 1. PUBLIC HEALTH, VENEREAL DISEASES. *Educ:* Boston Col, AB, 34; Boston Univ, MD, 39; Harvard Univ, MPH, 47. *Prof Exp:* Assoc clin prof, 51-69, CLIN PROF DERMAT, SCH MED, BOSTON UNIV, 69-; INSTR EPIDEMIOL, SCH PUB HEALTH, HARVARD UNIV, 57- *Concurrent Pos:* Lectr, Sch Med, Tufts Univ, 52-; instr dermat, Harvard Med Sch, 63-; vis physician, Mass Mem Hosps, 57-; asst clin dermatologist, Mass Gen Hosp, 53-; assoc vis physician, Boston City Hosp, 58-; physician, Dept Dermat & Syphil, Boston Dispensary, 59-; dir div commun & venereal dis, Boston. *Mem:* Fel AMA; fel Am Venereal Dis Asn (pres, 73-74); Conf State & Territorial Epidemiol (pres, 71-73); fel Am Col Prev Med; assoc Am Acad Dermat. *Res:* Common health; clinical and public health aspects of communicable and venereal diseases and dermatology. *Mailing Add:* Boston Univ Sch Med Northeast Med Ctr Six Gale Rd Belmont MA 02178

FIVEL, DANIEL I, b Baltimore, Md, Oct 12, 32; m 57; c 1. THEORETICAL PHYSICS. *Educ:* Johns Hopkins Univ, PhD(physics), 59. *Prof Exp:* Res assoc physics, Univ Pa, 59-61; NSF fels, Synchrotron Lab, Frascatti, Rome, Italy, 61-62 & Mass Inst Technol, 62-63; asst prof, 63-68, ASSOC PROF PHYSICS, UNIV MD, COLLEGE PARK, 68- *Concurrent Pos:* Mem exec comt, Aspen Ctr Physics, Colo. *Mem:* Am Phys Soc. *Res:* Weak interactions; quantum field theory; dispersion relations; elementary particles; symmetry properties; potential theory. *Mailing Add:* Dept Physics & Astron Univ Md College Park MD 20742-4111

FIVES-TAYLOR, PAULA MARIE, b Brooklyn, NY, Oct 9, 33; m 71. MEDICAL MICROBIOLOGY. *Educ:* St Thomas Aquinas Col, BS, 58; Villanova Univ, MS, 65; Univ Vt, PhD(med microbiol), 73. *Prof Exp:* Sci teacher, Our Lady Perpetual Help Sch, 55-60, St Helena's High Sch, 60-68; instr biol, St Thomas Aquinas Col, 67; teaching asst, Univ Vt, 68-71; instr biol, Trinity Col, 72-74; asst prof med microbiol, 74-78, assoc prof, 78-84, PROF MED MICROBIOL, COL MED UNIV VT, 84- *Concurrent Pos:* Prin investr, NIH grant; vpres, Genetics Resources, Inc. *Mem:* Am Soc Microbiol; Sigma Xi; Int Asn Dent Res; Am Asn Sci. *Res:* Mechanisms involved in the assembly of surface structures on the bacterial cell and the role these structures play in adherence to mucous membranes and biological implants and cellulose. *Mailing Add:* Microbiol & Molecular Genetics Dept Given Bldg Univ Vt Burlington VT 05405

FIVIZZANI, ALBERT JOHN, JR, b Chicago, Ill, Sept 11, 46; m 71. ENDOCRINOLOGY, BEHAVIOR. *Educ:* DePaul Univ, BS, 68, MS, 71; La State Univ, PhD(physiol), 77. *Prof Exp:* Instr physiol, La State Univ, 76-78; asst prof Biol, 78-83, ASSOC PROF BIOL, UNIV NDAK, 83- *Mem:* AAAS; Am Fisheries Soc; Am Soc Zoologists; Soc Neurosci; Sigma Xi. *Res:* Hormones and behavior; animal migration and orientation; regulation of seasonal conditions of vertebrates; biological rhythms. *Mailing Add:* Biol Dept Univ NDak Grand Forks ND 58202-8238

FIVOZINSKY, SHERMAN PAUL, b Hartford, Conn, Aug 2, 38; m 61; c 2. NUCLEAR PHYSICS. *Educ:* Univ Conn, BA, 61, MS, 63, PhD(nuclear physics), 71. *Prof Exp:* Nuclear physicist electron scattering, Ctr Radiation Res, Nat Bur Standards, 66-76, asst to chief, Off Standard Ref Data, 76-84, prog mgr, 84-86; PROG MGR, DIV NUCLEAR PHYSICS, US DEPT ENERGY, 86- *Concurrent Pos:* Adj prof, Montgomery Col, Rockville, Md, 78-83; sci asst to dir, Nat Measurement Lab, Nat Bur Standards, 83. *Mem:* AAAS; Am Phys Soc; Am Asn Physicists Med. *Res:* Electron scattering, nuclear structure; medical physics; critical data evaluation. *Mailing Add:* 5503 Manorfield Rd Rockville MD 20853

FIX, DELBERT DALE, b Pierce, Nebr, Dec 10, 26; m 46; c 4. PHOTOGRAPHIC CHEMISTRY. *Educ:* Univ Nebr, BS, 48, MS, 50; Univ Colo, PhD(chem), 52. *Prof Exp:* Asst, Univ Nebr, 48-49; asst, Univ Colo, 49-52; res chemist, 52-65, RES ASSOC, EMULSION RES DIV, RES LAB, EASTMAN KODAK CO, 65- *Mem:* AAAS; Am Chem Soc; Soc Photographic Scientists & Engrs. *Res:* Organic synthesis; nitrogen heterocyclic chemistry; reaction mechanisms; photographic emulsions; black and white and color photographic systems. *Mailing Add:* 123 Parkmere Rd Rochester NY 14617

FIX, GEORGE JOSEPH, b Dallas, Tex, May 10, 39; m; c 2. FINITE ELEMENT METHODS, PARTIAL DIFFERENTIAL EQUATIONS. *Educ:* Tex A&M Univ, BS, 63; Rice Univ, MS, 65; Harvard Univ, PhD(math), 68. *Prof Exp:* Engr, Tex Instruments, 63-64; asst prof math, Harvard Univ, 68-72; assoc prof, Dept Comput Sci, Univ Md, College Park, 72-73; assoc prof, Dept Math, Univ Mich, Ann Arbor, 73-75; PROF & HEAD MATH, CARNEGIE-MELLON UNIV, 75- *Concurrent Pos:* Consult, Westinghouse Res & Develop Ctr, 75-; prin investr, NASA Langley Res Ctr, 72-; vis prof, Inst Angewandte Math, Univ Bonn, Ger, 76-77, 77-78 & 79-80. *Mem:* Am Math Soc; Soc Ind & Appl Math; AAAS. *Res:* Scientific computing with a special emphasis on acoustics, fluid dynamics, and free boundary problems; numerical analysis; author or coauthor of over 49 technical publications. *Mailing Add:* Dept Math Univ Tex Arlington TX 76019

FIX, JAMES D, b Atlantic City, NJ, Jan 24, 31; m 54; c 4. ANATOMY, PHYSICAL ANTHROPOLOGY. *Educ:* Univ Del, BA, 58; Univ Tuebingen, Dr rer nat(anat & phys anthrop), 67. *Prof Exp:* From instr to asst prof anat, Sch Med, Univ Louisville, 67-70, asst prof ophthal, 70-71; assoc prof path & anat, Sch Med, Ind Univ, 71-79; PROF & CHMN ANAT, SCH MED, MARSHALL UNIV, 80- *Concurrent Pos:* Assoc prof anat, Sch Med, ECarolina Univ, 69-80; NIH spec res fel opthal, 70-71. *Mem:* Am Asn Anatamists; Am Asn Neuropathologists; fel Royal Micros Soc. *Res:* Cytoarchitecture of the central nervous system; neuropathology; quantitative neuropathology of the autonomic nervous system in diabetes. *Mailing Add:* Dept Anat Marshall Univ Sch Med Huntington WV 25701

FIX, JOHN DEKLE, b Melrose Park, Ill, Dec 23, 41; m 67; c 3. ASTROPHYSICS. *Educ:* Purdue Univ, BS, 63; Ind Univ, Bloomington, MA, 67, PhD(astrophys), 69. *Prof Exp:* from asst prof to assoc prof, 69-80, PROF ASTRON, UNIV IOWA, 80- *Mem:* Am Astron Soc; Int Astron Union. *Res:* Particulate matter in stellar atmospheres and interstellar space; stellar masers. *Mailing Add:* Dept Physics & Astron Univ Iowa Iowa City IA 52242-1410

FIX, KATHLEEN A, b Sept 24, 60. CORROSION STUDIES, ELECTROANALYTICAL WORK. *Educ:* State Univ NY, Oswego, BS, 82; Mich State Univ, PhD(anal chem), 88. *Prof Exp:* Res asst, Chem Dept, Mich State Univ, 82-87; sr electrochemist, Eltron Res, Inc, 88-89; SR RES CHEMIST RES & DEVELOP, DREW DIV, ASHLAND OIL, 90- *Mem:* Am Chem Soc; Soc Electroanal Chem; Electrochem Soc. *Res:* Existing treatments and the development of new treatments for the inhibition of corrosion in boiler water and cooling water systems. *Mailing Add:* 500 Bensel Dr F-46 Landing NJ 07850

FIX, RICHARD CONRAD, b Milwaukee, Wis, Dec 26, 30; m 58; c 3. ENVIRONMENTAL CHEMISTRY. *Educ:* Univ Wis, BS, 52; Mass Inst Technol, PhD(chem), 56. *Prof Exp:* Sr scientist, Tracerlab, Inc, 56-57; asst tech dir, Controls for Radiation, Inc, 57-63, vpres & dir res & develop, 63-65; sr staff scientist, Tracerlab Div, Lab for Electronics, Inc, 65-69; mgr, Tech Serv Dept, ICN & Tracelab, 69-72; mgr environ sci, Interex Corp, Natick, 72-81 & Chem Waste Mgt, Inc, 81-85; tech mgr, Clean Harbors Natick, Inc, 85-87; dir, Tech Serv, Clean Harbors, Inc, 87-90; CONSULT, 90- *Mem:* Sigma Xi. *Res:* Radiation physics; nuclear and medical instrumentation; dosimetry; environmental radioactivity; site surveys. *Mailing Add:* 484 Hosmer St Marlboro MA 01752

FIX, RONALD EDWARD, b Dallas, Tex, Dec 27, 41. GEOGRAPHIC INFORMATION SYSTEMS, WATER RESOURCES. *Educ:* Tex A&M Univ, BS, 63; Univ Tex, Tyler, MS, 90. *Prof Exp:* Engr training, Tex Water Comn, 63-65; process engr, Eimco Corp, Salt Lake City, Utah, 65-68; proj engr, Albert H Halff & Assoc, Dallas, 68-69 & Wisenbaker, Fix & Assocs, 69-79; OWNER, RONALD E FIX & ASSOCS, 79- *Concurrent Pos:* Res asst water resource, environ health eng, Univ Tex, Austin, 63-65; chmn, Task Force Outreach Water Resources Tech Group, Am Soc Civil Engrs, 90-91, mem, Task Comt Digital Mapping Water Resources, 90-91; adj instr computer sci, Jacksonville Col, Tex, 90; teacher in training, Whitehouse High Sch, 91. *Mem:* Am Soc Civil Engrs; Math Asn Am; Asn Comput Mach; Inst Elec & Electronics Engrs; Am Water Works Asn. *Res:* Computer applications in engineering, especially with the application of geographic information systems technology to design of water resources and environment control. *Mailing Add:* PO Box 595 Whitehouse TX 75791-0595

FIXMAN, MARSHALL, b St Louis, Mo, Sept 21, 30; m 59, 74; c 3. THEORETICAL CHEMISTRY. *Educ:* Washington Univ, AB, 50; Mass Inst Technol, PhD(chem), 54. *Prof Exp:* Instr chem, Harvard Univ, 56-59; sr fel, Mellon Inst, 59-61; prof, Univ Ore, 61-65 & Yale Univ, 65-79; PROF CHEM, COLO STATE UNIV, FT COLLINS, 79- *Concurrent Pos:* Sloan vis prof chem, Harvard Univ, 65. *Honors & Awards:* Pure Chem Award, Am Chem Soc, 64, Polymer Chem Award, 91; Polymer Physics Award, Am Phys Soc, 80; Sr US Scientist Award, Humboldt Found, 88. *Mem:* Nat Acad Sci; Am Chem Soc; Am Phys Soc. *Res:* Statistical mechanics; polymer theory. *Mailing Add:* Dept Chem Colo State Univ Ft Collins CO 80523

FJARLIE, EARL J, b Nanaimo, BC, Apr 10, 32; m 59; c 3. APPLIED PHYSICS. *Educ:* Univ BC, BASc, 55, MASc, 58; Univ Sask, PhD(physics), 65. *Prof Exp:* Defense serv tech officer, Defense Res Bd-Defence Res Estab Valcartier, 55, defense serv sci officer, 57-59; mem II res & develop, RCA Ltd, 65-76; PROF MECH ENG, ROYAL MIL COL CAN, 76- *Concurrent Pos:* Part-time lectr physics, Sir George Williams Univ, 66-73; mem comt reflection spectros, Can Ctr Remote Sensing, 75-77; mem comt optical surveillance, Defense Res Bd, 75-76; mem comt phys sci Valcartier, Defense Res Estab Valcartier, 76-77; dir, Tekerg, Inc, PQ, 76-79; exchange prof, Royal Mil Col Sci, Shrivenham, Eng, 83-84; chmn, Div Optical Physics, Can Asn Physicists, 84-85; mem comt electro optics Ottawa, Nat Res Coun, 86-; sabbatical, Univ Louis Pasteur ENSPS, Strasbourg, France, 89-90. *Mem:* Optical Soc Am; Am Asn Physics Teachers; Can Asn Physicists; Soc Photo-Optical & Instrumentation Engrs. *Res:* Optoelectronic systems; infrared spectroscopy; radiometry; laser doppler anemometry; atmospheric and space physics; interferometry; remote sensing; fluid flow; liquid crystals; laser systems; holography; fiber-optic tranducers; diffraction. *Mailing Add:* 4 Copperfield Dr Kingston ON K7M 1M4 Can

FJELD, ROBERT ALAN, b Greensboro, NC, Dec 28, 47; m 68; c 2. NUCLEAR ENVIRONMENTAL SCIENCE ENGINEERING. *Educ:* NC State Univ, BS, 70; Pa State Univ, MS, 73, PhD(nuclear eng), 76. *Prof Exp:* Asst prof, Nuclear Eng, Tex A&M Univ, 76-80; ASSOC PROF ENVIRON SYSTS ENG, CLEMSON UNIV, SC, 81- *Concurrent Pos:* Consult to indust. *Mem:* Sigma Xi; Am Nuclear Soc; Health Phys Soc; Am Asn Aerosol Res; AAAS. *Res:* Nuclear environmental engineering; occupational and environmental health physics radiation waste management, risk assessment; theoretical and experimental aerosol physics. *Mailing Add:* Environ Systs Eng Clemson Univ Clemson SC 29634-0919

FLACCUS, EDWARD, b Lansdowne, Pa, Feb 3, 21; m 47; c 3. PLANT ECOLOGY. *Educ:* Haverford Col, BS, 42; Univ NH, MS, 52; Duke Univ, PhD(bot), 59. *Prof Exp:* Teacher pvt sch, 48-50 & 51-55; asst zool, Univ NH, 50-51; asst bot, Duke Univ, 55-57; asst prof, Univ Minn, Duluth, 58-62, assoc prof, 63-68; vis prof, State Univ NY, Stony Brook, 68-69; prof biol sci, Bennington Col, 69-86; RETIRED. *Concurrent Pos:* Vis scientist, Brookhaven Nat Lab, 68-69. *Mem:* AAAS; Ecol Soc Am; Bot Soc Am; Am Inst Biol Sci; Sigma Xi; Ny Acad Sci. *Res:* Plant successions; forest ecology. *Mailing Add:* 110 Putnam St Bennington VT 05201

FLACH, FREDERIC FRANCIS, b New York, NY, Jan 25, 27; m 51, 71; c 5. PSYCHIATRY. *Educ:* St Peters Col, BA, 47; Cornell Univ, MD, 51. *Prof Exp:* Intern, Bellevue Hosp, 51-52; from asst resident psychiatrist to resident psychiatrist, Payne Whitney Clin, 53-58; asst prof, 58-62, ADJ ASSOC PROF CLIN PSYCHIAT, MED COL, CORNELL UNIV, 62- *Concurrent Pos:* From asst attend psychiatrist to attend psychiatrist, Payne Whitney Clin, 58-; attend psychiatrist, St Vincent's Hosp, 58. *Mem:* Fel Am Psychiat Asn; AMA. *Res:* Electrolyte and endocrine metabolism in psychiatric disorders; integration of chemotherapy and analytic psychotherapy in clinical practice; development of programs in preventive psychiatry; visual perceptual and spatial control systems as related to stress, occupational performance, and psychiatric illness; creativity and stress management. *Mailing Add:* 420 E 51st St New York NY 10022

FLACH, KLAUS WERNER, b Kolbermoor, Ger, Mar 24, 27; nat US; m 59; c 2. SOIL SCIENCE. *Educ:* Munich Tech Univ, Dipl, 50; Cornell Univ, MS, 54, PhD(soils), 60. *Prof Exp:* Asst soils, Cornell Univ, 53-58; soil scientist, Soil Surv Lab, 58-72, dir soil survey invest, 72-76, asst adminr in chg of soil survey, 76-80, assoc dep chief natural resource assesments, 80-84, SPEC

ASST SCI & TECHNOL, SOIL CONSERV SERV, USDA, 84- *Mem:* Soil Sci Soc Am; fel Soil Sci Soc Am. *Res:* Soil and soil mineralogy; soil conservation; soil information systems; soil classification and mapping. *Mailing Add:* Dept Agron Colo State Univ Ft Collins CO 80523

FLACHSBART, PETER GEORGE, b St Louis, Mo, Apr 15, 44; m 76, 83. ENVIRONMENTAL PLANNING, LAND USE PLANNING. *Educ:* Wash Univ, BS, 66; Northwestern Univ, MS, 68, PhD(civil eng), 71. *Prof Exp:* Sr res assoc, Univ Southern Calif, 71-72; lectr urban & environ mgt, Calif State Univ, Dominguez Hills, 72-73, asst prof pub admin, 73-76; asst prof civil eng, Stanford Univ, 76-80; asst prof urban & regional planning, 80-83, ASSOC PROF URBAN & REGIONAL PLANNING, UNIV HAWAII, MANOA, 83- *Concurrent Pos:* Prin investr, Stanford Univ, 77-80 & Univ Hawaii, Manoa, 80-85; consult, US Environ Protection Agency, 79-81, 87. *Mem:* Sigma Xi; Sci Soc. *Res:* Development of methodologies and models to estimate human exposure to air pollution using personal monitoring instruments; public policy analysis for energy conservation in urban passenger travel. *Mailing Add:* Dept Urban Studies Univ Hawaii Manoa Honolulu HI 96822

FLACHSKAM, ROBERT LOUIS, JR, b Joliet, Ill, Oct 11, 46; m 70. ORGANIC CHEMISTRY. *Educ:* Lewis Univ, BA, 68; Ohio State Univ, PhD(chem), 73. *Prof Exp:* asst prof, 75-80, ASSOC PROF CHEM, KY WESLEYAN COL, 75- *Mem:* Am Chem Soc; Asn Astron Educr. *Mailing Add:* Dept Chem Ky Wesleyan Col Owensboro KY 42301-6055

FLACK, J(OHN) E(RNEST), b Ft Collins, Colo, Jan 28, 29; m 51; c 3. WATER RESOURCES ENGINEERING. *Educ:* Colo State Univ, BSc, 50; Univ Iowa, MSc, 54; Stanford Univ, PhD(civil eng), 65. *Prof Exp:* From asst prof to assoc prof civil eng, Univ Colo, Boulder, 57-68; vis prof eng & econ planning, Water Resources Ctr, Ga Inst Technol, 68-69; PROF CIVIL ENG, UNIV COLO, BOULDER, 69- *Concurrent Pos:* NSF fac fel, 62-63; trustee, Rocky Mt Hydraul Lab; deleg, Univs Coun Water Resources, exec bd, 72-75 & 77-80; Eng Found fel, 78-79; vis prof, New South Wales Inst Technol, Australia, 80. *Honors & Awards:* Amsbery Prize, Am Water Works Asn, 65. *Mem:* AAAS; fel Am Soc Civil Engrs; Am Geophys Union; Am Water Resources Asn; Sigma Xi. *Res:* Hydrology; fluid mechanics; water resources. *Mailing Add:* Dept Civil Environ & Archit Eng Campus Box 428 Univ Colo Boulder CO 80309-0428

FLACK, RONALD DUMONT, JR, b South Bend, Ind, Dec 24, 47; m 69; c 2. EXPERIMENTAL FLUID MECHANICS, LUBRICATION. *Educ:* Purdue Univ, BS, 70, MS, 73, PhD(mech eng), 75. *Prof Exp:* Anal design engr, Pratt & Whitney Aircraft, 70-71; teaching & res asst fluid mech, Purdue Univ, 71-75; from asst prof to assoc prof, 76-87, PROF MECH ENGR, UNIV VA, 87- *Concurrent Pos:* Prin investr res grants, NASA, NIH, Dept Energy, NSF, GM, 76-; consult several industs, 76-; Fulbright Scholar, 88-89. *Mem:* Am Soc Mech Engrs; Am Soc Eng Educ; Am Soc Lubrication Eng; Optical Soc Am. *Res:* Experimental fluid mechanics, gas dynamics, lubrication, rotor dynamics, convective heat transfer and optical instrumentation; application in rotating machinery. *Mailing Add:* Dept Mech & Aerospace Eng Thornton Hall McCormick Rd Univ Va Charlottesville VA 22901

FLACKE, JOAN WAREHAM, b Dec 16, 31; US citizen; m 57; c 3. CARDIAC ANESTHESIOLOGY. *Educ:* Scripps Col, BA, 53; Med Sch, Harvard Univ, MD, 59. *Prof Exp:* Res fel anesthesiol, Med Sch, Harvard Univ, 64-67, res assoc, 67-69, instr, 69-70; from asst prof to assoc prof, Univ Ark Med Sci, 72-77; from adj assoc prof to adj prof, 77-89, PROF-IN-RESIDENCE, ANESTHESIOL, SCH MED, UNIV CALIF, LOS ANGELES, 89- *Concurrent Pos:* Trainee, res anesthesia, NIH, Mass Gen Hosp, 53-55; assoc examr, Am Bd Anesthesiol, 74-76; vis prof, Univ Wis, 79; investr, clin study sufentanil for anesthesia; mem adv comt, Anesthetic Life Support Drugs, Food & Drug Admin, 85-89; dir, Cardiac Anesthesia, 90- *Mem:* Am Soc Anesthesiologists; Asn Univ Anesthetists; Soc Cardiovasc Anesthesiologists. *Res:* Physiology and pharmacology of the circulation and autonomic nervous systems; one lung anesthesia; narcotics in anesthesia; temperature regulation and hypothermia; alpha2 adrenergic agonists. *Mailing Add:* Dept Anesthesiol Sch Med Univ Calif BH-961 CHS Los Angeles CA 90024

FLACKE, WERNER ERNST, b Recke, Germany, July 14, 24; nat US; m 57; c 3. PHARMACOLOGY. *Educ:* Univ D sseldorf, MD, 50. *Prof Exp:* Asst pharmacol, Univ D sseldorf, 51-52; asst path, Koblenz, Rhein, 52-53; resident internal med, St Elisabeth Hosp, Essen, Ruhr, 53-54; from instr to assoc prof pharmacol, Harvard Med Sch, 55-70; prof pharmacol & chmn dept, Univ Ark, Little Rock, 70-77; PROF & VCHAIR ANESTHESIOL & PHARMACOL, UNIV CALIF, LOS ANGELES, 77- *Concurrent Pos:* Fel pharmacol, Harvard Med Sch, 54. *Mem:* Pavlovian Soc; Am Soc Pharmacol & Exp Therapeut; sr fel Max Planck Soc; Biophys Soc; Ger Pharmacol Soc. *Res:* Pharmacology of excitable membranes; circulation; sensory input to central nervous system; respiration; skeletal muscle. *Mailing Add:* Dept Anesthesiol Univ Calif Sch Med Ctr Health Sci BH-961 Los Angeles CA 90024

FLAGAN, RICHARD CHARLES, b Spokane, Wash, June 12, 47; m 79; c 3. COMBUSTION, AEROSOL PHYSICS. *Educ:* Univ Mich, BSME, 69; Mass Inst Technol, SM, 71, PhD(mech eng), 73. *Prof Exp:* Res assoc mech eng, Mass Inst Technol, 73-75, lectr, 75; from asst prof to prof environ sci & mech eng, 75-90, PROF CHEM ENG, CALIF INST TECHNOL, 90- *Honors & Awards:* Marian Smoluchowsky Award Aerosol Res, 90. *Mem:* Combustion Inst; Am Inst Chem Engrs; Am Asn Aerosol Res. *Res:* Formation, control and atmospheric transformations of gaseous and particulate pollutants; combustion processes; fundamental aerosol physics; aerosol processes for material synthesis. *Mailing Add:* Calif Inst Technol 210-41 Pasadena CA 91125

FLAGG, JOHN FERARD, b Wellsville, NY, Dec 30, 14; m 40; c 2. ANALYTICAL CHEMISTRY, RESEARCH ADMINISTRATION. *Educ:* Univ Rochester, BS, 36; Princeton Univ, AM, 37, PhD(chem), 39. *Prof Exp:* Instr chem, Univ Rochester, 39-43, asst prof, 43-46; res assoc, Gen Elec Co, 46-52, mgr chem & chem eng, Knolls Atomic Power Lab, 52-56, proj analyst, GE Res Lab, 56-59, mgr mat eng lab, 59-61; dir res, Cent Res Div, Am Cyanamid Co, 61-64; dir & gen mgr, 64-72; VPRES & DIR RES, UOP, INC, 72- *Mem:* Am Chem Soc. *Res:* Radioactivity in inorganic and analytical chemistry; organic analytical reagents; processing of nuclear fuels; planning and evaluation of research; materials development; fuels; petrochemicals; environmental control. *Mailing Add:* Box 92 East Sandwich MA 02537-0092

FLAGG, RAYMOND OSBOURN, b Martinsburg, WVa, Jan 31, 33; m 56; c 3. BOTANY, SCIENCE ADMINISTRATION. *Educ:* Shepherd Col, BA, 57; Univ Va, PhD(biol), 61. *Prof Exp:* Teacher, High Sch, Md, 57; res assoc biol, Univ Va, 61-62; head bot, 62-79, VPRES, CAROLINA BIOL SUPPLY CO, 80-; VPRES, WOLFE SALES CORP, 85- *Concurrent Pos:* Vpres, Found Educ Develop, 83-85. *Mem:* AAAS; Bot Soc Am; Am Inst Biol Scis. *Res:* Plant cytogenetics and bisystematics; nutrition of Drosophila. *Mailing Add:* Carolina Biol Supply Co Burlington NC 27215

FLAGLE, CHARLES D(ENHARD), b Scottdale, Pa, Apr 26, 19; m 46, 65; c 4. OPERATIONS RESEARCH, HEALTH SERVICES. *Educ:* Johns Hopkins Univ, BE, 40, MSc, 54, DrEng, 55. *Prof Exp:* Design engr mech eng, Westinghouse Elec Corp, 40-46; consult, Henry C Robinson Co, Invest Bankers, Conn, 46-47, John & Sjostrom Co, Inc, Philadelphia & Wayne Iron Works, Wayne Pa, 47-50; res assoc fluid mech, Inst Co-op Res, 50-53, mem staff, Opers Res Off, 53-55, consult, 55-62, dir, opers res, Johns Hopkins Hosp, 56-63, prof opers res & indust eng, Sch Eng Sci, 61-72, prof pub health admin, 63-83, prof math sci, 72-85, prof health policy & mgt, 83-84, EMER PROF HEALTH POLICY & MGT, JOHNS HOPKINS UNIV, 84- *Concurrent Pos:* Consult, USPHS, 58-, NY Health Dept, 62-65, Res Anal Corp, 62-65, Calif Dept Pub Health, 63-66, Vet Admin, 65-, Res & Health Technol Assessment, Nat Ctr Health Serv, 85-; mem, Nat Adv Comt Epidemiol & Biometry, 63-66; spec asst Surgeon Gen, USPHS, 67-68; mem adv panel occup health, WHO, 79-; sr res scientist, Vet Admin, 85-; sr res assoc,Health Servs Res & Develop Ctr, 85- *Honors & Awards:* Kinball Medal, Opers Res Soc Am, 84. *Mem:* Inst Med-Nat Acad Sci; Opers Res Soc Am; Asn Health Serv Res; Soc Advan Med Systs (pres, 74); emer mem Pavlovian Soc (pres, 81). *Res:* Operations research applied to health services, particularly applications of decision theory and stochastic processes; published numerous articles in various journals. *Mailing Add:* Dept Health Policy & Mgt Johns Hopkins Univ Baltimore MD 21205

FLAHERTY, FRANCIS JOSEPH, b Chicago, Ill, July 26, 35; m 62; c 2. GEOMETRY. *Educ:* Univ Wis-Madison, BA, 56; Univ Notre Dame, MS, 59; Univ Calif, Berkeley, PhD(math), 65. *Prof Exp:* Mathematician, Rand Corp, 56-57; asst prof math, San Francisco State Col, 59-65; asst prof, Univ Southern Calif, 65-67; from asst prof to assoc prof, 67-77, PROF MATH, ORE STATE UNIV, 77- *Concurrent Pos:* Vis prof, Math Inst, Univ Bonn, 71-72 & 75; vis scholar, Inst Astron, Cambridge, Eng, 75. *Mem:* Am Math Soc. *Res:* Differential and integral geometry; differential topology; general relativity. *Mailing Add:* Dept Math Kidder Hall 368 Ore State Univ Corvallis OR 97331

FLAHERTY, FRANKLIN TRIMBY, JR, b Philadelphia, Pa, Aug 7, 34; m 56; c 2. MECHANICAL ENGINEERING, APPLIED MATHEMATICS. *Educ:* Mass Inst Technol, BS, 56, MS, 58; NY Univ, EngScD, 63. *Prof Exp:* Res asst, Dynamic Anal & Control Labs, Mass Inst Technol, 56-58; mem tech staff, 58-65, supvr eng mech group, 65-69, HEAD POWER SYSTS PHYS DESIGN DEPT, BELL TEL LABS, 69- *Res:* Dynamic behavior of discrete and continuous mechanical systems; communications equipment design. *Mailing Add:* Woodland Rd Brookside NJ 07926

FLAHERTY, JOHN THOMAS, b Albany, NY, Sept 25, 41; m 63; c 3. MEDICAL RESEARCH, CARDIOVASCULAR DISEASE. *Educ:* Mass Inst Tech, BS, 63; Duke Univ, MD, 67. *Prof Exp:* Intern med, Johns Hopkins Hosp, 67-68; res assoc biophysics, NIH, 68-71; resident intern med, 71-72, fel cardiol, 72-74, asst prof, 74-80, ASSOC PROF MED, SCH MED, JOHNS HOPKINS UNIV, 80- *Concurrent Pos:* Res career develop, Nat Heart & Lung Inst, 75-80; fel, Am Col Cardiol, 77; fel, Coun on Circulat Am Heart Asn, 81; comt mem, Topics in Intern Med Steering Comt, 84-86; adv comt mem, Med Sch Admin Comt, 84-87; res comt mem, Am Heart Asn, Md Affiliate, 85-; study sect comt mem, Nat Heart Lung & Blood Inst, 88- *Mem:* Am Heart Asn; Am Fedn Clin Res; Am Physiol Soc; Int Soc Heart Res. *Res:* Myocardial reperfusion injury studies in animal models; 31-phosphorus nuclear magnetic resonance spectroscopy to assess metabolic changes during and following ischemia and electron; paramagnetic resonance techniques to measure oxygen free radicals; clinical trials of free radical scavengers and reperfusion therapies in critical care unit; clinical trials of vasodilators in critical care unit patients. *Mailing Add:* Halsted 500 Johns Hopkins Hosp 600 N Wolfe St Baltimore MD 21205

FLAHERTY, JOSEPH E, b Brooklyn, NY, Sept 16, 43; m 67; c 2. NUMERICAL ANALYSIS. *Educ:* Polytech Inst Brooklyn, BS, 64, MS, 66, PhD(appl math), 69. *Prof Exp:* Instr appl mech, NY Univ 68-70, asst prof math, 70-72; from asst prof to assoc prof, 72-83, PROF MATH, RENSSELAER POLYTECH INST, 83-, CHAIRPERSON, COMPUTER SCI, 84- *Concurrent Pos:* Lectr, Soc Indust Appl Math, 75-77; vis prof, Univ Ariz, 80-81 & Standard Univ, 81-82; consult, Benet Labs & Inst Comput Appl Sci & Eng, 80-; mem adv bd, Nat Res Coun, 86-88. *Mem:* Soc Indust Appl Math; Asn Comput Mach; Inst Elec & Electronics Engrs Computer Soc. *Res:* Numerical analysis, especially the solution of differential equations; adaptive and intelligent software for partial differential equations; numerical methods for singularly perturbed differential systems; parallel procedures for partial differential equations. *Mailing Add:* Rensselaer Polytech Inst Troy NY 12181

FLAHIBE, MARY ELIZABETH, b Brooklyn, NY, Feb 24, 48. NUMBER THEORY. *Educ:* St Joseph Col, NY, BA, 69; Ohio State Univ, MS, 71, PhD(math), 76. *Prof Exp:* Lectr math, Ohio State Univ, 76-77; asst prof math, Tex A&M Univ, 77-; ASSOC PROF MATH, ORE STATE UNIV. *Mem:* Am Math Soc; Math Asn Am; London Math Soc. *Res:* Diophantine approximation; elementary number theory. *Mailing Add:* Ore State Univ Kidder Hall Rm 368 Corvallis OR 97331-4605

FLAIM, FRANCIS RICHARD, b Sublet, Wyo, Nov 2, 13; m 44; c 2. ZOOLOGY. *Educ:* Univ Utah, BA, 36, MA, 38; Stanford Univ, PhD, 56. *Prof Exp:* Asst zool, Univ Utah, 34-35; instr, 38-44; food technologist, Food Mach Corp, Calif, 44-46; from asst prof to assoc prof, 46-56, chmn dept biol, 60-66, PROF ZOOL & BIOL, UNIV SANTA CLARA, 56- *Mem:* AAAS; Am Soc Mammal; Sigma Xi; Am Asn Univ Prof. *Res:* Anatomy of muskrat; problems in comparative anatomy. *Mailing Add:* Dept Biol Univ Santa Clara Santa Clara CA 95053

FLAIM, KATHRYN ERSKINE, b Oakland, Calif, Nov 5, 49; m 75; c 2. PHYSIOLOGY. *Educ:* Univ Calif, Davis, BS, 71, PhD(physiol), 75. *Prof Exp:* Res fel physiol, Milton S Hershey Med Ctr, Pa State Univ, 75-78, res assoc, 78-81, asst prof physiol, 81-; SR INVESTR, DEPT MOLECULAR PHARMACOL, SMITHKLINE & FRENCH LABS. *Res:* Roles of hormones in controlling protein metabolism in liver and skeletal muscle. *Mailing Add:* Dept Clin Invest SmithKline & French Labs Box 1539 King of Prussia PA 19406

FLAIM, STEPHEN FREDERICK, b San Jose, Calif, May 28, 48; m 75. PHYSIOLOGY. *Educ:* Univ Santa Clara, BS, 70, sec credential, 71; Univ Calif, Davis, PhD(physiol), 75. *Prof Exp:* Res assoc, Univ Calif, Davis, 71-72 & 73-75, teaching asst, 72-73; fel physiol, Sch Med, Pa State Univ, 75-78, asst prof med & physiol, 78-82; prin scientist, Dept Biol Res, McNeil Pharmaceut, 82-85, res fel, 85-87; sect head, Dept Pharmacol, Squibb Inst Med Res, 87-88, assoc dir, 88-90; DIV DIR, PHARMACOL & TOXICOL, ALLIANCE PHARMACEUT CORP, 90- *Concurrent Pos:* Adj instr, Sch Med, Pa State Univ, 75-78. *Mem:* Am Physiol Soc; fel Am Col Clin Pharmacol; AAAS; Am Asn Univ Profs; Am Heart Asn. *Res:* Regulation of the peripheral circulation in health and disease. *Mailing Add:* 4455 Foxhollow Ct San Diego CA 92130

FLAIM, THOMAS ALFRED, b Paris, Tex, July 11, 46; m 70; c 2. MATERIALS SCIENCE, SURFACE ANALYSIS. *Educ:* Univ Mo-Rolla, BS, 68, MS, 70, PhD(mat sci), 71; Univ Detroit, MBA, 74. *Prof Exp:* Staff res scientist, Metall Dept, Gen Motors Res Lab, 71-83, X-car eng invest team, current prod eng, 83-85, STAFF PROJ ENGR, BRAKE TECHNOL GROUP, BRAKE & BEARING SYSTS CTR, GEN MOTORS PROVING GROUNDS, 85- *Concurrent Pos:* Prof mgt sci, Interpersonal & Pub Commun Dept, Cent Mich Univ, 73-83; chmn, Troy Bldg Code Adv Comt, 75-76. *Mem:* Am Ceramic Soc; Am Vacuum Soc; Sigma Xi; Am Inst Physics. *Res:* Aluminum alloy processing and joining; surface studies of high temperature oxidation of metals and alloys and studies of brake friction and rotor materials; physics of the solid-solid interface; author of over 25 articles and books. *Mailing Add:* 5072 Blair St Windmill Point N Troy MI 48098

FLAKE, JOHN C, dairy science, for more information see previous edition

FLAKS, JOEL GEORGE, b New York, NY, Oct 20, 27; m 61; c 2. BIOCHEMISTRY. *Educ:* Brooklyn Col, BA, 50; Univ Pa, PhD(biochem), 57. *Prof Exp:* Asst instr biochem, Univ Pa, 51-53; instr, Mass Inst Technol, 55-57; Damon Runyon Mem Fund fel, 57-58; from instr to assoc prof, 58-73, PROF BIOCHEM, SCH MED, UNIV PA, 73- *Concurrent Pos:* USPHS career develop award, 61-70; consult microbial chem study sect, NIH, 69-73; ed, Antimicrobial Agents & Chemother, 71-; Fogarty Sr Int fel, Imp Cancer Res Fund, 77-78. *Mem:* AAAS; Am Chem Soc; Am Soc Biol Chemists; Am Soc Microbiol; Genetics Soc Am; Sigma Xi. *Res:* Purine and pyrimidine biosynthesis and metabolism; microbial metabolism and alterations induced by bacteriophage infection; antimetabolite action; ribosome structure, function and genetics. *Mailing Add:* Dept Biochem Univ Pa Sch Med Philadelphia PA 19104

FLAM, ERIC, b Vienna, Austria, Feb 14, 35; US citizen; m 63; c 2. BIOMEDICAL & CHEMICAL ENGINEERING. *Educ:* City Col New York, BChE, 56; Southern Methodist Univ, MS, 61; NY Univ, PhD(biomed eng), 68. *Prof Exp:* Engr aerospace/electronics, 56-63; sr res scientist biomed eng, Johnson & Johnson, 68-74, proj leader, 74-76, sr group leader, 76-80; qual assurance mgr, C R Bard, Inc, 81-83, tech mgr, 80-84; PRES, NTL ASSOCS, 84- *Concurrent Pos:* Adj lectr, Rutgers Univ, 70-71 & Polytech Inst NY, 71; res dir, Breast Exam Bras, Inc, 74-; adj asst prof, Rutgers Med Sch, Col Med & Dent NJ, 77- *Mem:* Am Acad Dermatol; Am Thermographic Soc; Biomed Eng Soc; Am Soc Testing & Mat; Asn Advan Med Instrumentation. *Res:* Biomaterials and connective tissues; biomechanics and support systems; absorbents; menstrual protection and fluid management; diagnostics and temperature measurement, hemostasis, ostomy management, incontinence control, wound management; implants. *Mailing Add:* 29 Ainsworth Ave East Brunswick NJ 08816

FLAMM, DANIEL LAWRENCE, b San Francisco, Calif, Sept 14, 43; m 65. ELECTRONIC MATERIALS PROCESSING, PLASMA CHEMISTRY. *Educ:* Mass Inst Technol, BS, 64, MS, 66, ScD(chem eng), 70. *Prof Exp:* Asst prof chem eng, Northeastern Univ, 69-70 & Tex A&M Univ, 72-77; sr design programmer, Foxboro Co, 70-72; distinguished mem tech staff, AT&T Bell Labs, 78-89; MCKAY LECTR ELEC ENG, UNIV CALIF, BERKELEY, 88- *Concurrent Pos:* Consult, Tex Air Pollution Control Serv, 72; fac fel, NASA Ames Res Ctr, Stanford Univ, 74 & 75; mem, Subcomt Plasma Chem, Int Union Pure & Appl Chem, 80-87 & adv bd, NSF, 81-83; co-dir, Adv Study Inst Plasma-Mat Interactions, NATO, 88; vpres, Crystalline Mat, 89-90; contrib ed, Solid State Technol, 90-; vpres technol, Mattson Technol, Sunnyvale, Calif, 91- *Mem:* Am Inst Chem Engrs; Am Vacuum Soc; Mat Res Soc. *Res:* Plasma processing and chemical vapor deposition for semiconductor applications; advanced plasma sources, diamond growth, and plasma-enhanced silicon oxide and nitride deposition. *Mailing Add:* Univ Calif 187M Cory Hall Berkeley CA 94720

FLAMMER, GORDON H(ANS), b St Johns, Ariz, June 9, 26; m 49; c 6. CIVIL ENGINEERING. *Educ:* Utah State Univ, BS, 52, MS, 53; Univ Minn, PhD(civil eng), 58. *Prof Exp:* Asst, Utah State Univ, 52-53; asst, Univ Minn, 53-55, instr civil eng, 55-58; from asst prof to prof civil eng, 58-89, EMER PROF CIVIL ENG, UTAH STATE UNIV, 89- *Concurrent Pos:* Prof, Asian Inst Technol, 66-67, acad dean, 67-68; vis prof, Stanford Univ, 70-71. *Mem:* Am Soc Civil Engrs; Am Soc Eng Educ. *Res:* Fluid mechanics; hydrodynamics; hydrology; hydraulic models; education research and methodology. *Mailing Add:* 559 E Center St Midvale UT 84047

FLANAGAN, CARROLL EDWARD, b Price Co, Wis, Dec 18, 11; m 39; c 2. MATHEMATICS. *Educ:* Univ Wis-Oshkosh, EdB, 33; Univ Wis, PhM, 43, PhD(math educ, statist), 60. *Prof Exp:* Prin, Waukau State Graded Sch, Wis, 33-36, teacher high sch, 36-42; instr math, Univ Wis-Superior, 42-44; from instr to prof, 46-82, dir lib arts & sec educ, 50-63, coordr sec educ, 63-79, EMER PROF MATH, UNIV WIS-WHITEWATER, 82- *Concurrent Pos:* Math curric consult, 38- & modern math elem schs, 62-; tech consult math, Northern Nigeria Teacher Educ Proj, 65-68. *Mem:* Am Math Soc; Math Asn Am; Nat Coun Teachers Math. *Res:* Mathematics curriculum; algebra. *Mailing Add:* 281 N Park St Whitewater WI 53190

FLANAGAN, JAMES L(OTON), b Greenwood, Miss, Aug 26, 25. ELECTRICAL ENGINEERING. *Educ:* Miss State Univ, BS, 48; Mass Inst Technol, SM, 50, ScD(elec eng), 55. *Prof Exp:* Res engr, Acoust Lab, Mass Inst Technol, 48-50, 53-54; asst prof elec eng, Miss State Univ, 50-52; electronic scientist, USAF Cambridge Res Ctr, 54-57; mem tech staff, Bell Tel Labs, 57-61, head speech & auditory res dept, 61-67, head acoust res dept, 67-85, DIR, INFO PRIN RES LABS, AT&T BELL LABS, 85- *Honors & Awards:* L M Ericsson Int Prize, 85; Edison Medal, Inst Elec & Electronics Engrs, 86; Gold Medal, Acoust Soc Am, 86. *Mem:* Nat Acad Sci; Nat Acad Eng; fel Inst Elec & Electronics Engrs; fel Acoust Soc Am (pres, 78-79). *Res:* Digital communications; optimal coding of speech signals; acoustic theory of speech production; psychoacoustics of speech perception; digital filtering; computer simulation; approximately 140 papers and 40 patents. *Mailing Add:* Info Prin Res Labs Rm 2D38 AT&T Bell Labs 600 Mountain Ave Murray Hill NJ 07974

FLANAGAN, PATRICK WILLIAM, mycology, administration research, for more information see previous edition

FLANAGAN, PETER RUTLEDGE, b Sligo, Ireland, Jan 16, 45; Can citizen; m 69; c 4. INTESTINAL ABSORPTION, IRON & ZINC NUTRITION. *Educ:* Univ Col, Dublin, BSc Hons, 67; Univ BC, MSc, 70, PhD(biochem), 74. *Prof Exp:* Fel, gastroenterol, Hosp for Sick Children, Toronto, 74-76; lectr, 76-78, asst prof, 78-82, ASSOC PROF MED, UNIV WESTERN ONT, 82-, HON LECTR BIOCHEM, 76- *Mem:* Am Inst Nutrit; Am Gastroenterol Asn; Biochem Soc; Can Biochem Soc; Can Soc Clin Invest. *Res:* Biochemical nutrition of trace elements, especially iron and zinc; intestinal absorption of essential and toxic minerals. *Mailing Add:* Dept Med Univ Hosp Univ Western Ont London ON N6A 5A5 Can

FLANAGAN, ROBERT JOSEPH, b Alexandria, Minn, Aug 10, 24; m 47, 77; c 7. SYSTEM ANALYSIS. *Educ:* Univ NMex, BS, 49. *Prof Exp:* Math analyst, Sandia Corp, 48-49, supvr data reduction, 49-52, staff mem systs anal, 52-57; res engr, Dikewood Industs, Inc, 57-59, sr res engr, 59-64, dir, 64-69, vpres, 69-72 & sr vpres, 72-80; sr vpres, Hancock/Dikewood Serv, Inc, 80-85; RETIRED. *Mem:* Opers Res Soc Am; Soc Advan Med Systs; AAAS. *Res:* Administration of systems analyses of military and civil systems; current emphasis on health systems. *Mailing Add:* 51 Monticello Dr NE Albuquerque NM 87123

FLANAGAN, STEVEN DOUGLAS, b Columbus, Ga, Sept 9, 48; m 74. NEUROSCIENCES, MEMBRANE STRUCTURE & FUNCTION. *Educ:* Calif Inst Technol, BSc, 70; Univ Calif, San Diego, PhD(biol), 75. *Prof Exp:* HEAD, MEMBRANE NEUROCHEM SECT, CITY OF HOPE RES INST, 77- *Mem:* Soc Neurosci. *Res:* The biochemical structure of synapses, with the long term aim of determining the supramolecular basis for synaptic activity and modulation. *Mailing Add:* Neurosci Div Beckman Res Inst City of Hope 1450 E Duarte Rd Duarte CA 91010

FLANAGAN, TED BENJAMIN, b Oakland, Calif, July 11, 29; m 55; c 3. PHYSICAL CHEMISTRY. *Educ:* Univ Calif, Berkeley, BS, 51; Univ Wash, PhD(phys chem), 55. *Prof Exp:* Res chemist, Picatinny Arsenal, NJ, 55-57; fel phys chem, Queen's Univ, Belfast, 57-59; assoc physicist, Brookhaven Nat Lab, 59-61; from asst prof to assoc prof phys chem, 61-68, PROF PHYS CHEM, UNIV VT, 68- *Concurrent Pos:* Fulbright res scholar, Univ Munster, 67-68; Petrol Res Fund int fac award, 67-68; Fulbright res scholar, Univ Utrecht, 77-78. *Mem:* Am Chem Soc; Am Phys Soc; Faraday Soc. *Res:* Heterogeneous catalysis; hydrogen storage in solids; hydrogen in Palladium and its alloys; kinetics of the thermal decomposition of solids; diffusion in solids. *Mailing Add:* Dept Chem Univ Vt Burlington VT 05405-0001

FLANAGAN, THEODORE ROSS, b New York, NY, May 9, 20; m 43; c 1. AGRONOMY. *Educ:* Rutgers Univ, BS, 48; Pa State Col, MS, 50, PhD, 51. *Prof Exp:* Plant mgr & field agronomist, Grange League Fedn, 51-52; asst prof plant & soil sci & asst agronomist, Univ Vt, 53-72, assoc prof plant & soil sci, 72-76, specialist, garden, crop & weeds, 76-82; RETIRED. *Mem:* AAAS; Weed Sci Soc Am; Am Soc Agron. *Res:* Cold resistance of Ladino clover; development and adaptation of birdsfoot trefoil; herbicides; weed control-extension; environmental pollution; ecology; natural resources. *Mailing Add:* Rte 2 Box 221 Ethridge TN 38456

FLANAGAN, THOMAS DONALD, b Providence, RI, Jan 6, 35; m 55; c 5. VIROLOGY, IMMUNOLOGY. *Educ:* Univ RI, AB, 57, MS, 62; State Univ NY Buffalo, PhD(virol), 65; Am Bd Med Microbiol, dipl, 78. *Prof Exp:* Chemist, US Rubber Co, 56-60; res asst, Dept Animal Path, Univ RI, 60-62; from instr to assoc prof, 66-76, PROF MICROBIOL, STATE UNIV NY BUFFALO, 76-, CHMN, DEPT MICROBIOL, 87- *Concurrent Pos:* Fel,

Dept Bacteriol & Immunol, State Univ NY Buffalo, 65-66; vis res worker, Dept Tumor Biol, Karolinska Inst, Sweden, 73-74; vis prof microbiol, Univ Linkoping, Sweden, 74; dir, Erie County Virol Lab, 74-80; vis res worker, Nat Inst Neurol & Commun Dis & Stroke, 82. *Mem:* Am Soc Microbiol; fel Am Acad Microbiol; Am Asn Immunol; fel Infectious Dis Soc Am; Am Soc Virol. *Res:* Biology of paramyxoviruses; cellular response to virus infection. *Mailing Add:* Dept Microbiol State Univ NY 203 Sherman Hall Buffalo NY 14214

FLANAGAN, THOMAS LEO, b Philadelphia, Pa, Sept 25, 16; m 47; c 9. BIOCHEMISTRY. *Educ:* Drexel Univ, dipl, 43. *Prof Exp:* Tech anal chem, Rohm & Haas Co, 35-38, jr chemist, 38-41, chemist anal & textile chem, 41-47; jr scientist anal chem, 48, biochemist, 48-50, sr scientist, 50-56, group leader biochem res, 56-62, asst sect head biochem serv, 62-67, from asst sect head to sect head, Drug Metab Sect, 63-67, assoc dir biochem, 67-75, ASSOC DIR BIOL SCI, SMITH KLINE & FRENCH LABS, 75- *Mem:* Am Chem Soc; NY Acad Sci. *Res:* Drug metabolism, biochemical methods development; isotope tracer studies; instrumental methodology. *Mailing Add:* 1330 E Montogomery Ave Philadelphia PA 19125

FLANAGAN, THOMAS RAYMOND, b Gardner, Mass, July 24, 51; m 84. TRANSPLANTATION, ENCAPSULATION. *Educ:* Southeastern Mass Univ, BS, 73; Univ Mass, Amherst, MS, 76; Wesleyan Univ, PhD(biol), 82. *Prof Exp:* Fel neurobiol, Cold Spring Harbor Labs, 82-86; res assoc neuroendocrinol, Univ NC, Chapel Hill, 86-87; res scientist neurosci, Burroughs Wellcome Labs, 87-90; GROUP LEADER NEUROCHEM, CELLULAR TRANSPLANTS, INC, 91- *Concurrent Pos:* Adj asst prof, Dept Psychiat, Brown Univ, 90- *Mem:* Sigma Xi; Soc Neurosci; AAAS. *Res:* Isolation and domination of neuromal and endocrine secretory products; development of hollow-fiber encapsulated cells for transplant therapies. *Mailing Add:* Cellular Transplants Inc Four Richmond Square Providence RI 02906

FLANAGAN, WILLIAM F(RANCIS), b Cambridge, Mass, Apr 27, 27; m 58; c 5. PHYSICAL METALLURGY. *Educ:* Mass Inst Technol, SB, 51, SM, 53, ScD(metall), 59. *Prof Exp:* Asst, Mass Inst Technol, 51-53, 55-59, instr phys metall, 53-55; asst prof metall eng, Univ Wash, 59-66; sr res physicist, Gen Motors Res Lab, 66-68; assoc prof mat sci & eng, 68-72, PROF MAT SCI & ENG, VANDERBILT UNIV, 72- *Mem:* Am Soc Metals; Am Inst Mining, Metall & Petrol Eng; Sigma Xi. *Res:* Materials engineering embracing solid state physics; deformation mechanisms; physics of solids; materials resources policy. *Mailing Add:* Vanderbilt Univ Box 17 Sta B Nashville TN 37235

FLANDERS, CLIFFORD AUTEN, b Rutherford, NJ, May 3, 11; m 38; c 4. ANALYTICAL CHEMISTRY. *Educ:* Wagner Col, BS, 33; Columbia Univ, MA, 34; WVa Univ, PhD(agr biochem), 51. *Prof Exp:* Asst instr chem, Wagner Col, 35-36; instr, NC State Col, 36-37 & 38-40; instr, Univ Louisville, 40-45; asst biochemist, Exp Sta, WVa Univ, 45-51; dir chem prod, Fisher Sci Co, 51-66; assoc prof chem, Pace Univ, 66-69; assoc prof chem, William Paterson Col NJ, 69-81; instr, Morris County Col, Dover, 81-83; RETIRED. *Mem:* Sigma Xi; Am Chem Soc; Am Inst Chemists. *Res:* Analytical chemistry. *Mailing Add:* 81 Beechwood Dr Wayne NJ 07470-5703

FLANDERS, HARLEY, b Chicago, Ill, Sept 13, 25; div; c 2. PURE MATHEMATICS, APPLIED MATHEMATICS. *Educ:* Univ Chicago, BS, 46, MS, 47, PhD(math), 49. *Prof Exp:* Bateman fel math, Calif Inst Technol, 49-51; from instr to assoc prof math, Univ Calif, 51-60; prof math, Purdue Univ, 60-70; prof math, Tel-Aviv Univ, 70-77; vis prof math, Fla Atlantic Univ, 78-85; adj prof, 87, VIS SCHOLAR, UNIV MICH, 85- *Concurrent Pos:* NSF fel, Cambridge Univ, 57-58; ed-in-chief, Am Math Monthly, 68-73; chair, Israel Math Union, 72-74. *Honors & Awards:* Lester Ford Award, 69. *Mem:* Am Math Soc; Math Asn Am (vpres, 59-61); London Math Soc; AAAS; Inst Elec & Electronics Engrs; Comput Soc; Asn Comput Mach. *Res:* Algebra; differential geometry; electric circuit theory; computer science. *Mailing Add:* Dept Math Univ Mich Ann Arbor MI 48109

FLANDERS, KATHLEEN COREY, b North Adams, Mass, Oct 31, 56; m. CHEMISTRY. *Educ:* Rensselaer Polytech Inst, BS, 79; Ind Univ, PhD(biol chem), 83. *Prof Exp:* Assoc instr, Ind Univ, 79-81, res asst, 80-83; res chemist, Vet Admin Med Ctr, Cleveland, 84; fel, Sch Med, Case Western Reserve Univ, Cleveland, 84-85; guest researcher, 85-87, SR STAFF FEL, LAB CHEMOPREV, NAT CANCER INST, DCE, NIH, 87- *Concurrent Pos:* NIH postdoctoral nat individual res serv award, 84-87; reviewer, Endocrinol, Growth Factors, Cell Regulation, Swiss Nat Sci Found & Develop Biol. *Honors & Awards:* John & Mary Cloke Prize, 79. *Mem:* Am Soc Cell Biol. *Res:* Biochemistry; numerous publications. *Mailing Add:* Lab Chemoprev Nat Cancer Inst NIH Bldg 41 Rm B1103 Bethesda MD 20892

FLANDERS, ROBERT VERN, b Orange, Calif, Oct 28, 47; m 69; c 2. BIOLOGICAL CONTROL. *Educ:* Calif State Polytech Univ, Pomona, BS, 74; Univ Calif, Riverside, MS, 76, PhD(biol control), 79. *Prof Exp:* Res assoc, Entom Dept, Univ Calif, Riverside, 74, res asst, Div Biol Control, 74-79; ASST PROF, DEPT ENTOM, PURDUE UNIV, 79- *Mem:* Entom Soc Am; Int Orgn Biol Control (vpres, 85-86); Entom Soc Can; Am Inst Biol Sci; AAAS; Sigma Xi. *Res:* Biological and ecological interactions between arthropod pests and their parasitoids, predators and/or pathogens in agricultural ecosystems. *Mailing Add:* Dept Entom Purdue Univ West Lafayette IN 47907

FLANDRO, GARY A, b Salt Lake City, Utah, Mar 30, 34; m 61; c 1. MECHANICAL & AERONAUTICAL ENGINEERING. *Educ:* Univ Utah, BS, 57; Calif Inst Technol, MS, 60, PhD(aeronaut), 67. *Prof Exp:* Sci teacher, Liahona Col, Tonga, SPac, 59; res engr, Jet Propulsion Lab, Calif Inst Technol, 60; sr proj engr, Sperry Utah Co, 60-61; instr mech eng, Univ Utah, 61-63; teaching asst aeronaut, Calif Inst Technol, 63-64, sr res engr, Jet Propulsion Lab, 64-66, mem tech staff, 66-67; assoc prof mech eng, 67-76, PROF MECH & INDUST ENG, UNIV UTAH, 76- *Mem:* Am Inst Aeronaut & Astronaut. *Res:* Acoustic combustion instability and internal ballistics of rocket motors; astrodynamics; low thrust trajectory and mission analysis for interplanetary space missions. *Mailing Add:* Dept Aerospace Studies Ga Inst Technol Atlanta GA 30332

FLANIGAN, EVERETT, organic chemistry, biochemistry, for more information see previous edition

FLANIGAN, NORBERT JAMES, b Green Bay, Wis, Aug 12, 18; m 46; c 6. ANATOMY. *Educ:* St Norbert Col, AB, 40; Univ Iowa, MS, 53, PhD, 58. *Prof Exp:* Instr biol, St Francis Col Col, Pa, 46-47 & Creighton Univ, 48-52; instr zool, Univ Iowa, 52-55; instr anat, Sch Med, Univ Southern Calif, 55-58, asst prof biol, 58-63; chmn dept, St Norbert Col, 65-70, from asst prof to prof biol, 63-; RETIRED. *Mem:* AAAS; Am Asn Anat; Am Soc Zool; Int Asn Aquatic Animal Med. *Res:* Comparative and functional neuroanatomy, especially Cetacea; Cetacean behavior. *Mailing Add:* 1527 Quinnette Lane De Pere WI 54115

FLANIGAN, STEVENSON, b Lebanon, Tenn, July 2, 26; m 48; c 3. MEDICINE. *Educ:* Wash Univ, AB, 49, MD, 53. *Prof Exp:* From asst prof to assoc prof, Sch Med, Yale Univ, 61-67; PROF NEUROSURG, MED CTR, UNIV ARK, LITTLE ROCK, 67-, HEAD DEPT, 72- *Concurrent Pos:* Consult, Vet Admin Hosp, West Haven, Conn, 61-; mem, Forum Univ Neurosurgeons, 64- *Mem:* Am Asn Neurol Surg; Cong Neurol Surg; Am Col Surg; Asn Res Nerv & Ment Dis. *Res:* Biology of glial tumor cells; electrophysiology of basal nuclear groups and the urinary bladder; clinical neurosurgery. *Mailing Add:* NS UAMS Mail Slot 507 Univ Ark Hosp 4301 W Markham St Little Rock AR 72205

FLANIGAN, VIRGIL JAMES, b Higginsville, Mo, Dec 31, 38; m 64; c 3. MECHANICAL ENGINEERING, SYSTEMS ENGINEERING. *Educ:* Mo Sch Mines, BSME, 60, MSME, 62; Univ Mo, Rolla, PhD(mech eng), 68. *Prof Exp:* Prod engr, Western Elec Co, 60-61; instr, Mech Eng Dept, Mo Sch Mines, 62-64; res engr, Boeing Co, 64-65; from instr to assoc prof, 65-77, PROF MECH ENG, UNIV MO, ROLLA, 77- *Concurrent Pos:* Consult, var law firms, 72-, Naval Weapons Ctr, 73, Noranda Aluminum & Sverdrup & Parcel & Assocs, 72-, Commonwealth Edison, US Dept Energy & USAID. *Mem:* Am Soc Mech Engrs; Forest Prod Res Soc; Sigma Xi. *Res:* Energy research with primary emphasis on coal, biomass and waste materials utilization including direct combustion, gasification and pyrolysis. *Mailing Add:* 221 Eng Res Lab Univ Mo Rolla MO 65401

FLANIGAN, WILLIAM FRANCIS, JR, b Montreal, Que, Nov 25, 39; US citizen; m 68; c 1. COMMAND CONTROL & COMMUNICATIONS, ANTENNA DESIGN. *Educ:* Clarkson Col Technol, BS, 61; Cornell Univ, MBA, 63; George Washington Univ, MA, 68; Univ Chicago, PhD(biol psychol), 73. *Prof Exp:* Financial analyst, Securities & Exchange Comn, 63-66; res psychologist, Nat Naval Med Ctr, 68; eng res psychologist, 73-78, SR ELECTRONIC ENGR, NAVAL OCEAN SYSTS, 79- *Mem:* AAAS; Armed Forces Com & Electronics Asn; Inst Elec & Electronics Engrs; NY Acad Sci. *Res:* Vulnerability reduction of communication, navigation, and electronic warfare antennas and electronic systems; nonconventional design, modification and reconfiguration. *Mailing Add:* 5313 Vickie Dr San Diego CA 92109

FLANIGAN, WILLIAM J, b Hot Springs, Ark, June 2, 30; c 5. MEDICINE, PHYSIOLOGY. *Educ:* Univ Ark, BS, 53, MD, 55. *Prof Exp:* Med house officer, Peter Bent Brigham Hosp, 55-56; asst resident med, Sch Med, Univ Ark, 56; asst, Peter Bent Brigham Hosp, 59-63; from asst prof to assoc prof, 63-72, PROF MED, SCH MED, UNIV ARK, LITTLE ROCK, 72-, DIR DIV DIALYSIS & TRANSPLANTATION, 74- *Concurrent Pos:* Res fels, Harvard Med Sch, 59 & 60-61; res fel, Peter Bent Brigham Hosp, 61-63 & NSF, 60-62; Am Heart Asn adv res fel, 62-63; chmn coun dialysis & transplantation, Nat Kidney Found, 73-74. *Mem:* Sigma Xi. *Res:* Renal physiology and tissue transplantation. *Mailing Add:* Univ Ark Med Ctr 4301 W Markham St Little Rock AR 72205

FLANIGEN, EDITH MARIE, b Buffalo, NY, Jan 28, 29. INORGANIC CHEMISTRY, STRUCTURAL CHEMISTRY. *Educ:* D'Youville Col, BA, 50; Syracuse Univ, MS, 52. *Hon Degrees:* Dsc, D'Youville Col, 83. *Prof Exp:* Res chemist, Union Carbide Corp, 52-60, sr res chemist, 60-62, res assoc, 62-67, sr res assoc, 67-69, sr res scientist, 69-73, corp res fel, 73-82, corp sr res fel, 82-88; SR RES FEL, UOP, INC, 88- *Honors & Awards:* Donald W Breck Award, Int Zealite Asn, 83; Chem Pioneer Award, Am Inst Chemists, 91. *Mem:* Nat Acad Eng; Mineral Soc Am; Am Chem Soc; Int Molecular Sieve Conf (pres, 70-73); Int Zeolite Asn (secy, 77-); AAAS. *Res:* Inorganic and physical chemistry research in crystalline molecular sieves, zeolites, adsorbents, and catalysts; hydrothermal synthesis of mineral phases, particularly silicates; crystal growth. *Mailing Add:* 502 Woodland Hills Rd White Plains NY 10603-3136

FLANK, WILLIAM H, b Akron, Ohio, Jan 7, 32; m 56; c 2. MOLECULAR SIEVES, CATALYSIS. *Educ:* Temple Univ, AB, 58; Univ Del, PhD(chem), 65. *Prof Exp:* Res chemist, Houdry Labs, Air Prod & Chem Inc, 58-71; res chemist, Union Carbide Corp, 71-72, sr staff chemist, 72-77, technol consult, Molecular Sieve Dept, 77-89; RETIRED. *Concurrent Pos:* Lectr, Temple Univ, 67-70; chmn, Indust & Eng Chem Div, Am Chem Soc, 85; subcomt chmn, Comt D-32 Catalysts, Am Soc Testing & Mat, 85-; adv bd, Chemtech, 86-; adj prof chem, Pace Univ, 88- *Mem:* Clay Minerals Soc; Am Chem Soc; Catalysis Soc; Int Zeolite Asn; Int Confedn Thermal Anal. *Res:* Zeolite chemistry; heterogeneous catalysis; surface science, clays; preparation, characterization and application of zeolites and catalytic materials; adsorption and catalytic mechanisms; thermoanalytical and other physicochemical characterization techniques; fluid cracking catalysts; statistics. *Mailing Add:* 1021 Hard Scrabble Rd Chappaqua NY 10514-2423

FLANNAGAN, JOHN FULLAN, b Glasgow, Scotland, Jan 10, 40; Can citizen; m 61; c 4. FRESHWATER BIOLOGY, ECOLOGY. *Educ:* Paisley Col Sci & Technol, Scotland, HNC, 66; Inst Biol, London, MIBiol, 74; Manitoba, MSc, 82. *Prof Exp:* Sr technician, Dept Zool, Glasgow Univ, 61-66, Can Land Inventory Proj, Govt Man, 66-67 & Freshwater Inst, Fisheries Res Bd Can, 67-72; RES SCIENTIST & PROJ LEADER, FRESHWATER INST, CAN DEPT FISH & OCEANS, 72-; ADJ PROF ENTOM, UNIV

MANITOBA, 82- *Concurrent Pos:* Mem, Alta Blackfly Control Res Comt, 74-; chmn, 3rd Int Mayfly Symp Orgn Comt, 75-79. *Mem:* Sigma Xi; Entom Soc Can; Can Coun Freshwater Fisheries Res; NAm Benthological Soc; Int Limnol Asn; NY Acad Sci. *Res:* Investigation and improvement of sampling methodology for macrobenthic research; investigation of effect of insecticides on freshwater ecosystems; development of environmental protocols to assess environmental impact of toxic substances; ecology of aquatic invertebrates; stream rehabilitation-walleye. *Mailing Add:* Freshwater Inst 501 University Crescent Winnipeg MB R3T 2N6 Can

FLANNELLY, WILLIAM G, b Scranton, Pa, May 15, 31; m 55; c 2. STRUCTURAL DYNAMICS. *Educ:* Rensselaer Polytech Inst, BS, 54. *Prof Exp:* Mech engr, Consol Molded Prod Corp, Pa, 54-55 & Sylvania Elec Prod Co, Pa, 55-56; design engr, Hamilton Standard Div, United Aircraft Corp, 56-57; stress engr, 57-58, dynamicist, 58-59, 60-63, test engr, 59-60, asst chief vibrations res, 63-69, staff res engr, 69-73, prin res engr, 73-80, PRES & CHIEF RES ENGR, KAMAN, 80- *Mem:* Sigma Xi. *Res:* Structural dynamics system identification; structural vibrations analyses; antiresonant theory; vibration control instruments; applied matrix algebra. *Mailing Add:* 108 Hilton Dr South Windsor CT 06074

FLANNERY, BRIAN PAUL, b Utica, NY, July 30, 48; m 70; c 2. APPLIED MATHEMATICS. *Educ:* Princeton Univ, BA, 70; Univ Calif, Santa Cruz, PhD(astrophysics), 74. *Prof Exp:* Res assoc, Inst Advan Study, Princeton, 74-76; asst prof astron, Harvard Univ, 76-80, assoc prof, 80; sr staff physicist, 80-82, RES ASSOC, 83-, SECT HEAD, EXXON RES & ENG CO, 88- *Mem:* Am Astron Soc; Int Astron Union; Am Phys Soc; AAAS. *Res:* radiative transfer fluids; climate; synchrotron imaging and microtomography; process models; applied math. *Mailing Add:* Exxon Res & Eng Co Rte 22 E Annandale NJ 08801

FLANNERY, JOHN B, JR, b Providence, RI, Mar 15, 41; m 68; c 3. RESEARCH ADMINISTRATION. *Educ:* St Vincent Col, BS, 62; Rensselaer Polytech Inst, PhD(phys chem), 65. *Prof Exp:* From assoc scientist to prin scientist, Res Labs, 65-71, mgr optical mat & devices res, 71-74, mgr, Mat Sci Lab, 74-78, mgr tech strategy, 78-82, mgr eng automation, 82-87, MGR PROD ARCHIT, XEROX CORP, ROCHESTER, 87- *Concurrent Pos:* Invited lectr, Eng & Sci Soc. *Mem:* AAAS; Am Chem Soc; Soc Info Display. *Res:* Engineering/manufacturing processes and automation; document processing systems; technical and business planning methodologies. *Mailing Add:* 1258 Wildflower Dr Webster NY 14580

FLANNERY, MARTIN RAYMOND, b Co Derry, Northern Ireland, Jan 8, 41; m 67; c 4. ATOMIC & MOLECULAR PHYSICS, THEOR ETICAL PHYSICS. *Educ:* Queens Univ, Belfast, BSc, 61, PhD(atomic & molecular physics), 64. *Prof Exp:* Lectr theoret physics, Queens Univ, 64-66; res assoc astrophys, Joint Inst Lab Astrophys, 66-67; physicist & lectr, Harvard Col Observ & Smithsonian Astrophys Observ, 68-71; from asst prof to assoc prof, 67-74, PROF PHYSICS, GA INST TECHNOL, 74- *Concurrent Pos:* Mem, exec comt gaseous electronics conf, 81-83, 86-88. *Honors & Awards:* Cert Higher Achievement, Smithsonian Inst, 72; Monie A Ferst Award, Sigma Xi, 74 & 75. *Mem:* Fel Inst Physics London; fel Am Phys Soc; Sigma Xi; fel Inst Physics. *Res:* Scattering theory; theoretical studies of basic atomic and molecular collision processes with application to laboratory and astrophysical plasmas; non-equilibrium recombination processes; ion-molecule reactions; collisions with Rydberg atoms; energy-transfer mechanisms; termolecular processes. *Mailing Add:* Sch Physics Ga Inst Technol Atlanta GA 30332

FLASAR, F MICHAEL, b East St Louis, Ill, Feb 20, 46; m 69; c 1. PLANETARY PHYSICS, ASTROPHYSICS. *Educ:* Mass Inst Technol, BS, 67, PhD(physics), 72. *Prof Exp:* Res fel planetary physics, Harvard Univ, 71-75; RES SCIENTIST PLANETARY PHYSICS & ASTROPHYS, GODDARD SPACE FLIGHT CTR, NASA, 75- *Concurrent Pos:* Res assoc, Nat Res Coun, 75-77. *Mem:* Am Astron Soc; Am Geophys Union. *Res:* Planetary interiors and atmospheres; geophysical and astrophysical fluid dynamics; planetary meteorology. *Mailing Add:* Goddard Space Flight Ctr Greenbelt Rd Mail Code 639 2 Bldg 2 Rm S109 Greenbelt MD 20771

FLASCHKA, HERMENEGILD ARVED, b Cilli, Austria, June 10, 15; US citizen; m 45; c 2. ANALYTICAL CHEMISTRY. *Educ:* Graz Univ, DrPhil, 38. *Prof Exp:* Demonstr, Graz Univ, 37-38; sci co-worker, Kaiser Wilhelm Inst, Ger, 45-46; chemist, Fa Schultz, 46-47; asst, Graz Univ, 48-49; res dir, Fa A Zankl-Sohne, Austria, 49-53; privatdozent, Graz Univ, 53; lectr, Graz Tech Univ, 53-55; head anal dept, Nat Res Ctr, Egypt, 55-57; guest prof, Univ NC, 57-58; from assoc prof to prof chem, Ga Inst Technol, 58-65, Regent's prof chem, 65-81; RETIRED. *Concurrent Pos:* Consult, Grazer Glass Works, Austria, 50-55 & J T Baker Chem Co, NJ, 58-76; co-ed, Microchem J, 63-79. *Honors & Awards:* Fritz Feigl Award, Austrian Chem Soc, 53; L Gordon Mem Award, 68; Benedetti-Pichler Award, Am Microchem Soc, 81. *Mem:* Am Chem Soc; Am Microchem Soc; Austrian Chem Soc. *Res:* Ethylenediamine-tetraacetic acid and related titrations; organic reagents; indicator theories; complexes in analytical chemistry; complementary tristimulus colorimetry; long-path photometry; photometric titrations. *Mailing Add:* 2318 Hunting Valley Dr Decatur GA 30033

FLASHMAN, STUART MILTON, b Boston, Mass, Feb 14, 48; m 85. SOMATIC CELL GENETICS, PLANT GENETIC ENGINEERING. *Educ:* Brown Univ, AB & ScM, 69; Harvard Univ, PhD(biochem), 76. *Prof Exp:* Teaching fel biol, Harvard Univ, 70-73, res asst biochem, 73-75; res assoc, ERDA Plant Res Lab, Mich State Univ, 75-76; asst prof genetics, Sch Agr, NC State Univ, 77-80; res biologist, 81-84, SR RES BIOLOGIST, STAUFFER CHEM CO, 84- *Mem:* Am Soc Plant Physiologists; Int Asn Plant Cell & Tissue Cult; Int Soc Plant Molecular Biol. *Res:* Application of plant somatic cell genetics to problems of crop improvements; genetic regulation in higher plants; identification, isolation and modification of agronomically important plant genes. *Mailing Add:* 1063 48th St Emeryville CA 94608

FLASHNER, MICHAEL, b Brooklyn, NY, Aug 16, 42. BIOCHEMISTRY. *Educ:* Brooklyn Col, BS, 65; Univ Mich, MA, 70, PhD(biochem), 71. *Prof Exp:* Fel microbiol, Univ Mich, 72-73; asst prof chem, 73-80, ASSOC PROF CHEM, COL ENVIRON SCI & FORESTRY, STATE UNIV NY, 80- *Mem:* Am Soc Biol Chem; Am Chem Soc; Am Soc Microbiol; Sigma Xi. *Res:* Basic enzymology; mechanisms of enzyme reactions; bacterial physiology; cell and molecular biology. *Mailing Add:* 4497 Juneberry Ct Concord CA 94521-4507

FLASPOHIER, DAVID, b Bethany, Ohio, Apr 12, 39. MATHEMATICS. *Educ:* Xavier Univ, BS, 61; John Carroll Univ, PhD(math), 71. *Prof Exp:* PROF MATH, XAVIER UNIV, 62- *Mailing Add:* Registrar Math Dept Xavier Univ 3800 Victory Pkwy Cincinnati OH 45207

FLATH, ROBERT ARTHUR, b St Louis, Mo, Mar 14, 33; m 55; c 2. NATURAL PRODUCTS CHEMISTRY. *Educ:* Concordia Teachers Col, Ill, BS, 55; Long Beach State Col, BS, 60; Univ Calif, Berkeley, PhD(org chem), 64. *Prof Exp:* RES CHEMIST FRUIT COMPOSITION INVEST, WESTERN UTILIZATION RES & DEVELOP DIV, USDA, 64- *Mem:* AAAS; Am Chem Soc; Royal Soc Chem; Am Soc Enologists; Am Phytochem Soc; Sigma Xi. *Res:* Terpenoid and food flavor chemistry; instrumental analysis. *Mailing Add:* 207 Colgate Ave Kensington CA 94708

FLATO, JUD B, b Brooklyn, NY, Feb 21, 40; m 63; c 2. PHYSICAL CHEMISTRY. *Educ:* Polytech Inst Brooklyn, BS, 61; NY Univ, MS, 63, PhD(phys chem), 68. *Prof Exp:* Sr res chemist, Technicon, Inc, 66-67; sr res chemist, EG&G & Princeton Appl Res Corp, 67-71, mgr chem instrument group, 71-72, mkt mgr, 72-73, vpres, 74-81, sr vpres, 81-83; pres & chief exec officer, Spex Group, Inc, 83-88; MANAGING PARTNER, J B FLATO & ASSOCS & INT MARKETING VENTURES MERGER & ACQUISITION OPERS, UPTON, FLATO & CO, INC, 88-, PRES, 90- *Mem:* Am Chem Soc; Instrument Soc Am; Soc Appl Spectros. *Res:* Paramagnetic resonance and absorption spectroscopy in molten salts; design and development of scientific instrumentation in electrochemistry and spectroscopy; applications of electronics to chemistry. *Mailing Add:* 244 Glenn Ave Lawrenceville NJ 08648

FLATT, ADRIAN EDE, b Frinton, Eng, Aug 26, 21; nat US; m 55; c 1. SURGERY. *Educ:* Cambridge Univ, BA, 42, MB, BCh & MA, 45, MD, 51; FRCS, 53. *Prof Exp:* Instr anat, Cambridge Univ, 50-51 & Royal Col Surgeons Eng, 51; from asst to prof surg, London Hosp, London Univ, 51-53, asst orthop, 53-56; from asst prof to assoc prof, 57-66, PROF ORTHOP SURG, UNIV IOWA, 66- *Concurrent Pos:* Fulbright grant, 54-55; civilian consult hand surg, USAF Aerospace Med Hosp, Wilford Hall, Tex. *Mem:* Am Soc Surg of the Hand; Am Soc Plastic & Reconstruct Surg; AMA; fel Am Col Surg; Sigma Xi. *Res:* Reconstructive surgery of the congenitally deformed, diseased or injured hand, especially the biomechanical influence of disease or injury on hand function. *Mailing Add:* Baylor Univ Med Ctr 3500 Gaston Ave Dallas TX 75246

FLATT, JEAN-PIERRE, b Basel, Switz, Mar 3, 33; m 63; c 3. ENERGY METABOLISM, OBESITY. *Educ:* Bymnase Cantonal, Switz, Baccalaureat, 52; Univ Lausanne, dipl, 56, DSc, 59. *Prof Exp:* Postdoctoral res fel biol chem, Harvard Med Sch, Boston, Mass, 60-63; chief travaux biochem, Inst Biochem, Univ Lausanne, Switz, 65-67; res assoc & asst prof physiol chem, Dept Nutrit & Food Sci, Mass Inst Technol, Cambridge, 67-70, assoc prof, 70-73; PROF BIOCHEM, UNIV MASS MED SCH, WORCESTER, 74- *Concurrent Pos:* Lectr biochem, Fac Sci, Univ Lausanne, Switz, 65-67, vis prof physiol, 77, 83 & 90; actg chair, Dept Biochem, Univ Mass Med Sch, Worcester, 80-81; sci adv, Nestle Found Study Probs Nutrit, 90- *Mem:* Am Soc Biochem & Molecular Biol; Am Inst Nutrit; Am Soc Clin Nutrit; Am Diabetes Asn. *Res:* Interactions between energy and protein metabolism in man; roles of dietary fat, carbohydrate balance and exercise on body weight maintenance; diet and obesity; costs of ATP synthesis. *Mailing Add:* Dept Biochem Univ Mass Med Ctr 55 Lake Ave N Worcester MA 01655

FLATT, WILLIAM PERRY, b Newbern, Tenn, June 17, 31; m 49; c 2. ANIMAL NUTRITION. *Educ:* Univ Tenn, BS, 52; Cornell Univ, PhD(animal nutrit), 55. *Prof Exp:* Res resident, Dept Animal Husb, Cornell Univ, 55; dairy cattle nutritionist & head, Energy Metab Lab, Nutrit & Physiol Sect, Dairy Cattle Res Br, Agr Res Serv, USDA, 56-68, asst dir animal husbandry res div, 68-69; prof animal sci & head dept, 69-70, dir, Agr Exp Sta, 70-81, DEAN & COORDR, COL AGR, UNIV GA, 81- *Concurrent Pos:* Chmn, Southern Asn Agr Exp Sta Dirs, 74-75 & Legis Subcomt Exp Stas Comt Orgn & Policy, 76-78; mem, Bd Agr & Renewable Resources, Nat Res Coun-Nat Acad Sci, 75-78; chmn coun admin heads agr, Southern Region, 87-88; chmn, Nat Coun Admin Heads Agr, 88-89, chmn div agr, 89-90. *Honors & Awards:* Hoblitzelle Nat Award Agr Res, 68. *Mem:* AAAS; Am Soc Animal Sci; Am Dairy Sci Asn; Am Inst Nutrit; Sigma Xi. *Res:* Biochemistry; physiology; energy metabolism of dairy cattle, especially the nutritive evaluation of forages and the energy requirements of cattle performing various functions; factors affecting rumen development. *Mailing Add:* 101 Conner Hall Col Agr Univ Ga Athens GA 30602

FLATTE, STANLEY MARTIN, b Los Angeles, Calif, Dec 2, 40; m 66; c 2. SEISMOLOGY, WAVES IN RANDOM MEDIA. *Educ:* Calif Inst Technol, BSc, 62; Univ Calif, Berkeley, PhD(physics), 66. *Prof Exp:* Res physicist, Lawrence Berkeley Lab, 66-71; from asst prof to assoc prof, 71-78, chmn dept, 86-89, PROF PHYSICS, UNIV CALIF, SANTA CRUZ, 78-, ASSOC DIR, INST TECTONICS, 88- *Concurrent Pos:* Guggenheim fel, 75; vis scientist, Orgn Europ Res Nuclear, 75-76 & La Jolla Inst, 80-81; sr vis, Univ Cambridge, 81; dir, Ctr Studies Nonlinear Dynamics, La Jolla, 82- *Mem:* Am Physical Soc; Sigma Xi; fel Acoust Soc Am; Am Geophys Union; AAAS. *Res:* Study of wave propagation through random media, including sound in the ocean, radio waves in plasmas, light and sound in the atmosphere and seismic waves for the purpose of understanding the physical processes in the media. *Mailing Add:* Natural Sci II Univ Calif Santa Cruz CA 95064

FLATTERY, DAVID KEVIN, b S Ruislip, Eng, July 10, 60; m 85. KINETICS IN NON ISOTHERMAL ENVIRONMENT, HEATING IN STRONG ELECTRICAL FIELD. *Educ:* Harvard Univ, BS, 80. *Prof Exp:* Res engr, 80-83, dir res, 85-88, VPRES RES, RADIANT TECHNOL CORP, 88- *Honors & Awards:* Tech Achievement Award, Int Soc Hybrid Microelec, 86. *Mem:* Int Soc Hybrid Mycroelec; Semiconductor Equip & Mat Inst; Inst Elec & Electronics Engrs; Int Electronic Pakaging Soc; Surface Mount Technol Asn. *Res:* Process of firing, joining, sintering of metal, ceramics, polymers as applied to microelectronics; emphasis on the thermal, radiative and electromagnetic environment and its effect on kinetics. *Mailing Add:* 910 Old Public Rd Hockessin DE 19707

FLATTO, LEOPOLD, b Antwerp, Belg, Aug 20, 29; US citizen; m 66; c 2. MATHEMATICS. *Educ:* City Col New York, BS, 50; Johns Hopkins Univ, MA, 51; Mass Inst Technol, PhD, 55. *Prof Exp:* Mathematician, Reeves Instrument Corp, 55-57; prof math, Brooklyn Polytech Inst, 57-60; mathematician, IBM Res Ctr Yorktown Heights, 60-61; prof math, Belfer Grad Sch Sci, Yeshiva Univ, 61-79; MATHEMATICIAN, BELL LABS, 79- *Concurrent Pos:* NSF grant, 63-68; vis prof, Hebrew Univ, 68-69. *Mem:* Am Math Soc. *Res:* Mean value problems related to harmonic functions; finite reflection groups and their invariant theory; applied probability; queuing theory; ergodic theory. *Mailing Add:* AT&T Bell Labs 600 Mountain Ave Murray Hill NJ 07974

FLAUGH, MICHAEL EDWARD, b Findlay, Ohio, Nov 14, 41; m 66. SYNTHETIC ORGANIC CHEMISTRY. *Educ:* Univ Dayton, BS, 63; Univ Calif, Berkeley, PhD(chem), 68. *Prof Exp:* SR CHEMIST, ELI LILLY RES LABS, 67- *Mem:* Am Chem Soc. *Res:* Synthesis of organic structures, particularly heterocyclic compounds. *Mailing Add:* 9224 Kinlock Dr Indianapolis IN 46256-2242

FLAUMENHAFT, EUGENE, CYTOCHEMISTRY, NUCLEAR BIOLOGY. *Educ:* Univ Chicago, PhD(cytochem), 58. *Prof Exp:* ASSOC PROF BIOL, UNIV AKRON. *Mailing Add:* Dept Biol Univ Akron Akron OH 44325-0001

FLAUTT, THOMAS JOSEPH, JR, b Glendora, Miss, Feb 8, 32. CITRUS TECHNOLOGY. *Educ:* Univ St Louis, BS, 53; Univ Calif, PhD(chem), 57. *Prof Exp:* Res chemist, 57-67, sect head, 67-72, HEAD PROD RES SECT, PROCTER & GAMBLE CO, 72- *Mem:* Int Microwave Power Inst; Am Chem Soc. *Res:* Nuclear magnetic resonance, mesmorphic phases; citrus and paper product development; microwave applications. *Mailing Add:* 2562 Observatory Ave Cincinnati OH 45208

FLAVELL, RICHARD ANTHONY, b Chelmsford, UK, Aug 23, 45; m; c 3. IMMUNOLOGY, BIOCHEMISTRY. *Educ:* Univ Hull, BSc, 67, PhD(biochem), 70. *Prof Exp:* Fel biochem, Royal Society Gt Brit Europ Prog, Univ Amsterdam, 70-72; fel, Inst Molecular Biol, Univ Zurich, Switz, 72-73; lectr, Univ Amsterdam, 73-79; lab head, Lab Gene Struct & Expression, Nat Inst Med Res, London, Eng, 79-82; pres, Biogen Res Corp, Cambridge, Mass, 82-88, chief sci officer, 84-88; CHMN, IMMUNOBIOL DEPT, YALE UNIV SCH MED, 88- *Concurrent Pos:* Ed, J Molecular & Appl Genetics, 83; ed, Nucleic Acids Res, 79-82. *Honors & Awards:* Colworth Medal, 80. *Mem:* Fel Royal Soc. *Res:* Analysis of the immune response using molecular genetics; role of MHC molecules in antigen presentation and T cell development; molecular genetics of T cell development. *Mailing Add:* Sect Immunobiol Yale Univ Sch Med 310 Cedar St New Haven CT 06510

FLAVIN, MARTIN, b Chicago, Ill, Mar 18, 20. BIOCHEMISTRY. *Educ:* Stanford Univ, BA, 44; Univ Calif, MD, 47; Columbia Univ, PhD(biochem), 51. *Prof Exp:* Sr asst surgeon, Nat Heart Inst, 52-54; mem dept biochem, NY Univ, 54-56; Am Heart Asn estab investr, Dept Agr Biochem, Univ Calif, 56-57; mem enzyme sect, 57-61, med dir, 61-77, HEAD SCIENTIST, ORGANELLE BIOCHEM, NAT HEART & LUNG INST, NIH, 78- *Mem:* Am Soc Biol Chem. *Res:* Enzymatic reaction mechanisms; chemical pathways of metabolism. *Mailing Add:* Lab Cell Biol Nat Heart Lung & Blood Inst NIH Bethesda MD 20892

FLAVIN, MICHAEL AUSTIN, b New York, NY, Nov 27, 32; m 55; c 4. ELECTRICAL ENGINEERING. *Educ:* Univ Conn, BSE, 54; Columbia Univ, MSEE, 61. *Prof Exp:* Mem tech staff commun, 54-65, dept head tel, 65-71, dir ocean technol, 71-74, dir semiconductors, 74-79, exec dir, bus commun, 79-87, EXEC DIR, BUS SYSTEM PERFORMANCE, AT&T INFO SYST, 88- *Mem:* Inst Elec & Electronics Engrs. *Res:* Semiconductor memory; metal oxide semiconductor integrated circuits; telephone systems. *Mailing Add:* AT&T Info Systs 11900 N Pecos St Denver CO 80234

FLAWN, PETER TYRRELL, b Miami, Fla, Feb 17, 26; m 46; c 2. GEOLOGY, ACADEMIC ADMINISTRATION. *Educ:* Oberlin Col, BA, 47; Yale Univ, MS, 48, PhD(geol), 51. *Prof Exp:* Jr geologist, US Geol Surv, 48-49; res scientist & geologist, Bur Econ Geol, Univ Tex, Austin, 49-60, dir bur & prof geol, 60-70, dir div natural resources & environ & vpres acad affairs, 70-72, prof geol sci & pub affairs, 70-85, exec vpres, 72-73, pres, Univ Tex, San Antonio, 73-77, pres, Univ Tex, Austin, 79-85, L T Barrow chair mineral resources, 79-85, EMER PROF GEOL SCI & EMER PRES, UNIV TEX, AUSTIN, 85- *Concurrent Pos:* Mem mineral resources bd & space appln bd, Nat Acad Sci-Nat Res Coun, 75-80; mem, Nat Sci Bd, 80-86. *Honors & Awards:* Ben H Parker Medal, Am Geol Inst, 89. *Mem:* Nat Acad Eng; hon mem Asn Am State Geologists; Geol Soc Am; Am Geol Inst (dir, 67-70); Soc Econ Geologists; hon mem Am Asn Petrol Geologists. *Res:* Economic and environmental geology; geology of Texas and Mexico. *Mailing Add:* 3718 Bridle Path Austin TX 78703

FLAX, ALEXANDER H(ENRY), b New York, NY, Jan 18, 21; m 51; c 1. AERONAUTICAL ENGINEERING. *Educ:* NY Univ, BAeE, 40; Univ Buffalo, PhD(physics), 58. *Prof Exp:* Stress analyst, Curtiss-Wright Corp, NY, 40-44; chief aerodyn & struct, Piasecki Helicopter Corp, Philadelphia, 44-46, asst head aeromech dept, Cornell Aeronaut Lab, 46-49, head aerodyn res dept, 49-55, asst dir, 55-59; chief scientist, US Dept Air Force, 59-61; vpres-tech dir, Cornell Aeronaut Labs, 61-63; asst secy res & develop, USAF, 64-69; vpres res, Inst Defense Anal, 69, pres, 69-83; HOME SECY, NAT ACAD ENG, 84- *Concurrent Pos:* Instr eng sci & mgt war training, Cornell Univ, 43, 48; instr exten course, Pa State Col, 45-46; mem comt aerodyn, Nat Adv Comt Aeronaut, 52-53, mem subcomt high speed aerodyn, 54-58; mem res adv comt aircraft aerodyn, NASA, 59-63; mem adv comt, Defense Sci Bd, Dept Defense, 70-; mem, Air Force Sci Adv Bd, 70-; adv, Dept of Transp, 70-74 & Off Sci Technol, 70-73. *Honors & Awards:* Sperry Award, Inst Aeronaut Sci, 49; Air Force Except Civilian Serv Award, 61 & 69; NASA Distinguished Serv Medal, 69; von Karman Medal, NATO Adv Group for Aerospace Res & Develop, 78; Clifford C Furnas Award, State Univ NY, Buffalo, 86. *Mem:* Nat Acad Engrs; fel Am Inst Aeronaut & Astronaut; Int Acad Astronaut. *Res:* Aircraft flutter and vibration; aero-elastic effects on stability and control; helicopter dynamics and aerodynamics; supersonic aerodynamics; missile dynamics; lifting surface theory; high temperature gas dynamics. *Mailing Add:* Off Home Secy, Nat Acad Eng 2101 Constitution Ave NW Washington DC 20418

FLAX, LAWRENCE, b Brooklyn, NY, May 17, 34; m 56; c 3. ACOUSTICS. *Educ:* Brooklyn Col, BS, 59; John Carroll Univ, MS, 66; Colo State Univ, PhD(physics), 69. *Prof Exp:* Theoret physicist, NASA, 60-72; physicist, Naval Res Lab, Washington, DC, 72-81; div head, Sensor & Countermeasure Naval Coastal Systs, Panama City, Fla, 81-90; INSTR, ST JOHNS RIVER COMMUNITY COL, 90- *Concurrent Pos:* Prof, Cuyahoga Community Col, 68-72; instr, Am Univ, 75-81. *Mem:* Am Phys Soc; Sigma Xi; fel Acoust Soc Am; Cong Int du Froid. *Res:* Acoustic reflection and scattering from finite bodies; acoustic field interaction with nonabsorptive and absorptive materials; fourier optical analysis of acoustic schlieren systems; theory of parametric sound transmission. *Mailing Add:* 337 Foxridge Rd Orange Park FL 32065-5737

FLAX, MARTIN HOWARD, b New York, NY, Jan 19, 28; m 55; c 3. PATHOLOGY. *Educ:* Cornell Univ, AB, 46; Columbia Univ, AM, 48, PhD, 53; Univ Chicago, MD, 55. *Prof Exp:* Asst zool, Columbia Univ, 46-49; intern, Mt Sinai Hosp, New York, 55-56; asst path, Med Sch, Univ Chicago, 56-57; chief biophys br, Armed Forces Inst Path, 57-59; res & clin fel path, Mass Gen Hosp, 59-61; from instr to assoc prof path, Harvard Med Sch, 61-70; PROF PATH & CHMN DEPT, SCH MED, TUFTS UNIV, 70- *Concurrent Pos:* Asst path, Mass Gen Hosp, 61-70, consult path B study sect, 70-74; NIH career develop award, 66- *Mem:* Am Soc Exp Path; Am Asn Path & Bact; Int Acad Path; Am Asn Immunologists; Fedn Am Socs Exp Biol. *Res:* Immunopathology; electron microscopy. *Mailing Add:* Dept Path Tufts Univ Sch Med & Vet Med 136 Harrison Ave Boston MA 02111

FLAX, STEPHEN WAYNE, b Denver, Colo, Apr 10, 46; m 71. ULTRASONICS, MEDICAL IMAGING. *Educ:* Univ Colo, BS, 68; Univ Wis, MS & PhD(elec eng), 75. *Prof Exp:* Asst prof eng, Portland State Univ, 75-78; SR PHYSICIST MED IMAGING, MED SYSTS GROUP, GEN ELEC, 78- *Mem:* Sigma Xi; Inst Elec & Electronics Engrs. *Res:* Medical imaging systems; basic ultrasound research and the computer modeling of ultrasound instrumentation as applied to medical applications. *Mailing Add:* Med Systs Div W-875 PO Box 414 Milwaukee WI 53201

FLAY, BRIAN RICHARD, b Hamilton, NZ, Feb 1, 47; m 78. HEALTH BEHAVIOR, HEALTH PROMOTION & DISEASE PREVENTION. *Educ:* Waikato Univ, BSoc Sc, 72, MSocSc, 73, DPhil, 76. *Prof Exp:* Asst prof health studies, Univ Waterloo, Ont, Can, 78-80; asst prof health behav, Univ South Calif, 80-84, assoc prof prev med & dep dir, Inst Prev Res, 84-87; ASSOC PROF PUB HEALTH & DIR, PREV RES CTR, UNIV ILL, CHICAGO, 87- *Mem:* Am Pub Health Asn; Am Psychol Asn; Int Union Health Educ; Soc Behav Med. *Res:* Health promotion and disease prevention; smoking and drug abuse prevention; AIDS prevention; use of mass media for health promotion; methodological aspects of prevention research. *Mailing Add:* Prev Res Ctr 850 W Jackson Suite 400 Chicago IL 60607

FLEAGLE, JOHN G, b US, April 30, 48; c 1. PHYSICAL ANTHROPOLOGY, BIOLOGY. *Educ:* Yale Univ, BS, 71; Harvard Univ, MS, 73, PhD(anthrop), 76. *Prof Exp:* Res consult, Dept Nutrit, Sch Health, Harvard Univ, 73-74; from lectr to assoc prof, 75-84, PROF, DEPT ANAT SCI, STATE UNIV NY, STONY BROOK, 85- *Concurrent Pos:* LSB Lenkey Fedn Sci & Grants exec comt, 85-; Guggenhein fel, 83; MacArthur fel, 88. *Mem:* Am Asn Phys Anthropologists; Soc Vert Paleont; Am Inst Biol Sci; Am Soc Mammalogists. *Res:* Evolution and radiation of higher primates; growth and development, and evolutionary biology of primates. *Mailing Add:* Dept Anat Sci Health Sci Ctr State Univ NY Stony Brook Stony Brook NY 11794

FLEAGLE, ROBERT GUTHRIE, b Baltimore, Md, Aug 16, 18; m 42; c 2. METEOROLOGY. *Educ:* Johns Hopkins Univ, AB, 40; NY Univ, MS, 44, PhD(physics), 49. *Prof Exp:* High sch teacher, Md, 40-41; asst, NY Univ, 46-48; from asst prof to prof, 48-87, chmn dept, 67-77, EMER PROF, ATMOSPHERIC SCI, UNIV WASH, 88- *Concurrent Pos:* NSF sr fel, Imp Col, Univ London, 58-59; tech asst, Off Sci & Technol, 63-64; chmn, comt atmospheric sci, Nat Acad Sci, 69-73; chmn, Atmospheric & Hydrol Sect, AAAS, 77-78. *Honors & Awards:* Meisinger Award, Am Meteorol Soc, 59, Cleveland Abbe Award, 72 & Charles Brooks Award, 85. *Mem:* Am Meteorol Soc (pres, 81-82); Am Geophys Union (pres, Metrol Sect, 68-70); Royal Meteorol Soc; AAAS. *Res:* Description and theory of large scale atmospheric motions; physics of air near the earth's surface; air sea interaction; atmospheric science policy problems. *Mailing Add:* Dept Atmospheric Sci Univ Wash Seattle WA 98195

FLECHSIG, ALFRED J, JR, b Tacoma, Wash. ELECTRICAL ENGINEERING. *Educ:* Wash State Univ, BS, 57, MS, 59; La State Univ, PhD(elec eng), 70. *Prof Exp:* assoc prof, 60-80, PROF ELEC ENG, WASH STATE UNIV, 80- *Mem:* Inst Elec & Electronics Engrs. *Res:* Power system protection; digital relaying. *Mailing Add:* Dept Elec & Comp Eng Wash State Univ Pullman WA 99164

FLECHTNER, THOMAS WELCH, b Fostoria, Ohio, July 7, 43; m 67; c 2. ORGANIC CHEMISTRY, PHOTOCHEMISTRY. *Educ:* Dartmouth Col, AB, 65; Univ Wis, PhD(chem), 70. *Prof Exp:* Technician, Upjohn Co, 65; res assoc chem, Columbia Univ, 70-72; asst prof, 72-78, ASSOC PROF CHEM, CLEVELAND STATE UNIV, 78-, ASSOC CHAIR, 84- *Concurrent Pos:* NIH res fel, 70-72. *Mem:* Am Chem Soc; Sigma Xi. *Res:* Exploratory organic photochemistry; development of new synthetic reactions; physical organic chemistry. *Mailing Add:* 1375 Inglewood Dr Cleveland OH 44121

FLECK, ARTHUR C, b Chicago, Ill, Oct 29, 36; m 57; c 3. DECLARATIVE PROGRAMMING, DATA ABSTRACTION. *Educ:* Univ Western Mich, BS, 59; Mich State Univ, MA, 60, PhD(math), 64. *Prof Exp:* Dir prog comput, Mich State Univ, 64-65; asst prof math & comput, Univ Iowa, 65-67, assoc prof, 67-71, chmn, 84-90, PROF COMPUT SCI, UNIV IOWA, 71- *Concurrent Pos:* Vis prof, Dept Appl Math & Comput Sci, Univ Va, 73-74. *Mem:* Asn Comput Mach; Inst Elec & Electronics Engrs Comput Soc; Europ Asn Theoret Comput Sci; Asn Logic Prog. *Res:* Automata theory; programming languages, theory data structures. *Mailing Add:* Comput Sci Dept Univ Iowa Iowa City IA 52242

FLECK, GEORGE MORRISON, b Warren, Ind, May 13, 34; m 59; c 2. PHYSICAL CHEMISTRY, HISTORY OF SCIENCE. *Educ:* Yale Univ, BS, 56; Univ Wis, PhD(phys chem), 61. *Prof Exp:* From asst prof to assoc prof, 61-76, PROF CHEM, SMITH COL, 76- *Mem:* Sigma Xi; Am Chem Soc. *Res:* Systems chemistry; kinetics and mechanisms of chemical reactions in solution; multiple equilibria involving metal complexes in solution; history of physical chemistry. *Mailing Add:* Clark Sci Ctr Northampton MA 01063

FLECK, JOSEPH AMADEUS, JR, b Kansas City, Mo, Mar 10, 28; m 61, 85; c 2. LASERS. *Educ:* Harvard Univ, AB, 48; Rice Inst, MA, 50, PhD(physics), 52. *Prof Exp:* Assoc physicist, Brookhaven Nat Lab, 52-57; physicist, 57-69, group leader theoret div, 69-74, ASSOC DIV LEADER, THEORET PHYSICS DIV, LAWRENCE LIVERMORE LAB, UNIV CALIF, 74-, PROJ LEADER, TRANSPORT & PROPAGATION PHYSICS, HIGH TEMPERATURE PHYSICS DIV, 85- *Concurrent Pos:* Fulbright adv res fel, Norway, 57-58; consult, AEC Halden Proj, Norway, 58-59 & Atomic Energy Res Inst, Japan, 61; lectr, Univ Calif, Davis, 67- *Mem:* Fel Am Phys Soc; Optical Soc Am. *Res:* Reactor and nuclear weapons physics; Monte Carlo methods; radiative transfer; laser physics and quantum optics; numerical methods; laser effects; atmospheric propagation of laser beams; optical wave guide theory; hydrodynamics and x-ray holography. *Mailing Add:* High Temperature Physics Div Lawrence Livermore Nat Lab Livermore CA 94550

FLECK, STEPHEN, b Frankfurt-am-Main, Ger, Sept 18, 12; nat US; m 45; c 3. PSYCHIATRY. *Educ:* Harvard Univ, MD, 40. *Prof Exp:* Intern med, Beth Israel Hosp, Boston, 40-42; intern, Johns Hopkins Hosp, 46, resident psychiat, 46-48, asst, 48, instr psychiat & asst med, 48-49; from instr to asst prof psychiat, Med Sch, Univ Wash, 49-53; assoc prof psychiat, 53-54, from assoc prof to prof, 54-83, psychiatrist-in-chief, Yale Psychiat Inst, 63-83 & Conn Ment Health Ctr, 69-83, EMER PROF PSYCHIAT & PUB HEALTH, SCH MED, YALE UNIV, 83- *Concurrent Pos:* Attend staff, King County Hosp & Vet Admin Hosp, Seattle, 51-53; consult, Vet Admin Hosp, West Haven, Conn, 53-, State Hosp, Middletown, Conn, 54-; med dir, Yale Psychiat Inst, 54-63. *Honors & Awards:* Am Family Ther Asn Award, AFTA, 85-; Silvano Arieti Award, Am Acad Psychoanal. *Mem:* Am Psychosom Soc; Am Psychiat Asn; Am Pub Health Asn. *Res:* Schizophrenia, emotional aspects of motherhood outside marriage; conditioning and integration of the central nervous system; social structure of mental hospitals; family psychiatry and medicine; population control. *Mailing Add:* Dept Psychiat Yale Univ Sch Med 25 Park St New Haven CT 06519

FLECK, WILLIAM G(EORGE), b Havertown, Pa, Nov 17, 40; m 67; c 3. CIVIL & STRUCTURAL ENGINEERING. *Educ:* Villanova Univ, BCE, 62; Carnegie-Mellon Univ, MS, 64, PhD(civil eng), 68. *Prof Exp:* ASSOC PROF CIVIL ENG, CLEVELAND STATE UNIV, 70- *Concurrent Pos:* Interim dean, Fenn Col Eng, 76-77. *Mem:* Am Soc Eng Educ. *Res:* Mechanics of materials; reinforced concrete. *Mailing Add:* Dept Civil Eng Cleveland State Univ Euclid Ave at E 24th Cleveland OH 44115

FLECKENSTEIN, ALBRECHT, b Aschaffenburg, Ger, May 3, 17. PHYSIOLOGY, PHARMACOLOGY. *Educ:* Univ Würzburg, MD, 42. *Hon Degrees:* Dr, Ludwig-Maximilian Univ, Munich, 84, Ruprecht-Karl Univ, Ger, 86, Rijksuniversiteit, Neth, 86, Nat Univ Plata, Arg, 87, Univ Basel, Switz, 90. *Prof Exp:* Res asst, Pharmacol Inst, Univ Würzburg, 43-45; lectr, Pharmacol Inst, Univ Heidelberg, 47-53, assoc prof pharmacol, 53-56; PROF PHYSIOL, UNIV FREIBURG, 56- *Concurrent Pos:* Brit Coun exchange lectr, Dept Pharmacol, Univ Oxford & Dept Biochem, Univ Sheffield, 51-52; chmn, Ger Physiol Soc, 59-62. *Honors & Awards:* Paul Morawitz Prize, Ger Cardiovasc Soc, 84, Carl Ludwig Medal, 89; Franz Gross Award, Ger League for Res on Antihypertensive Ther, 84; Carl Mannich Medal, Ger Pharmaceut Soc, 85; Hartmann Müller Prize for Med Res, Univ Zürich, Switz, 86; Ernst Jung Prize for Med Res, 86; Schmiedeberg medal, Ger Pharmacol Soc, 87; Award for Outstanding Cardiol Res, Europ Sect, Int Soc Heart Res, 87; Award for Outstanding Basic Pharmacol Invest, Am Soc Pharmacol & Exp Therapeut, 87; Spec Achievement Award, Am Soc Hypertension, 88; Peter Harris Distinguished Scientist Award, Int Soc Heart Res, 89; Karl Heinz Beckurts Prize, Ger Ministry Sci & Technol, 89; Distinguished Investr Award, Am Col Clin Pharmacol, 89. *Mem:* Hon mem Ger Physiol Soc; hon mem Egyptian Cardiol Soc. *Res:* Calcium anagonists. *Mailing Add:* Study Group Calcium Antagonism Physiol Inst Univ Freiburg Hermann-Herder-Str 7 Freiburg D-7800 Germany

FLEDDERMANN, RICHARD G(RAYSON), b Havana, Cuba, June 4, 22; US citizen; m 51; c 3. AERONAUTICAL ENGINEERING, PHYSICS. *Educ:* Loyola Univ, La, BS, 41; Univ Mich, BSE, 43, MSE, 47, PhD(aeronaut eng), 50. *Prof Exp:* Aerodynamicist, El Segundo Div, Douglas Aircraft Co, 43 & Higgins Industs Inc, La, 43-44; res assoc, Eng Res Inst, Univ Mich, 45-60; assoc prof aeronaut eng, Ga Inst Technol, 50-54; engr aeronaut res & develop, Arnold Eng Develop Ctr, USAF, 54-56; sr staff scientist & chmn theoret aerodyn group, Avco Mfg Co, 56-59; sr eng scientist, Radio Corp Am, 59-61; prin staff engr, staff of vpres eng, Martin Co, 61-65, staff of dir res & eng, Denver Div, 65-66, mgr space physics, 66-68; PROF AERONAUT SYSTS & CHMN DEPT, OMEGA COL, UNIV WFLA, 68-, PROF PHYSICS, 76- *Concurrent Pos:* Consult, Arnold Eng Develop Ctr, USAF, 51, 53-54. *Mem:* Am Inst Aeronaut & Astronaut. *Res:* Hypersonic aerodynamics and heat transfer; ablation materials; properties of turbulent flow; fuel sprays; history of thermodynamics. *Mailing Add:* Dept Physics Univ WFla Pensacola FL 32514

FLEEGER, JOHN WAYNE, b Ann Arbor, Mich, May 21, 49; m 69; c 1. ESTUARINE BIOLOGY, MARINE ECOLOGY. *Educ:* Slippery Rock Univ, BA, 71; Ohio Univ, MS, 74; Univ SC, PhD(biol), 77. *Prof Exp:* Res asst marine sci, Univ SC, 75-76; from asst prof to assoc prof, 77-88, PROF ZOOL & PHYSIOL, LA STATE UNIV, 88- *Concurrent Pos:* Chairperson, Int Asn Meiobenthologists, 87-89; adj prof, Inst Marine Sci, Univ Alaska, Fairbanks, 86- *Mem:* Ecol Soc Am; Am Soc Zoologists; Sigma Xi; Int Asn Meiobenthologists; Am Soc Limnol & Oceanog. *Res:* Population dynamics, energetics and community structure of estuarine meiofauna; fish feeding. *Mailing Add:* Dept Zool & Physiol La State Univ Baton Rouge LA 70803-1725

FLEEKER, JAMES R, b Emporia, Kans, Aug 11, 37; m 49; c 4. PESTICIDE BIOCHEMISTRY. *Educ:* Kans State Teachers Col, BS, 60; Mich State Univ, PhD(biochem), 65. *Prof Exp:* Res assoc, Mich State Univ, 65-66; asst prof, NDak State Univ, 66-71, assoc prof biochem, 71-78, PROF BIOCHEM, NDAK STATE UNIV, 78- *Mem:* AAAS. *Res:* Metabolism of pesticides; immunoassay of drugs and pesticides. *Mailing Add:* Dept Biochem NDak State Univ Main Campus Fargo ND 58105

FLEENOR, MARVIN BENSION, b Bristol, Tenn, May 15, 39; m 65; c 4. BIDEGRADATION, WASTE WATER TREATMENT. *Educ:* East Tenn State Univ, BS, 69; La State Univ, PhD(microbiol), 73. *Prof Exp:* Res assoc, Div Eng Res, La State Univ, 74-76; sr proj engr, Ralston Purina, St Louis, 76-80; TECH DIR, AQUA TERRA BIOCHEM (NCH), 80- *Mem:* Am Soc Microbiol; Soc Indust Microbiol; Appl Biotreat Asn. *Res:* Fermentation optimization, equipment design, process development solids liquid recovery and processing; bidegradation lab studies, product research and development; waste water treatment optimization and trouble shooting. *Mailing Add:* Aqua Terra Biochem PO Box 890 Lancaster TX 75146

FLEESON, WILLIAM, b Sterling, Kans, May 21, 15; m 43; c 5. MEDICINE, PSYCHIATRY. *Educ:* Univ Kans, AB, 37; Yale Univ, MD, 42; Am Bd Psychiat & Neurol, dipl, 58. *Prof Exp:* Intern, Univ Minn Hosps, 42-43; assoc psychiatrist & dir child guid div, Minn Psychiat Inst, 46-55; from asst to assoc prof psychiat, Col Med Sci, Univ Minn, 56-63, asst dean col, 60-63; assoc dean, Sch Med, 63-74, prof psychiat, 63-82, EMER PROF PSYCHIAT, SCH MED, UNIV CONN,FARMINGTON, 82-; ASSOC PSYCHIAT, VA MED CTR, NEWINGTON, CONN, 82- *Concurrent Pos:* USPHS fel, Judge Baker Guid Ctr, Boston, Mass, 50-51; NIH spec fel psychiat & rehab, Univ Minn, 55; mem adv mgt prog, Harvard Univ, 64; lectr psychiat, Sch Med, Yale Univ, 65-72; actg head dept psychiat & chief psychiat serv, Univ Conn-McCook Hosp, 67-68; bd dirs, Comt Int Med Exchange, 74-79; vis prof social & behav sci, Univ Lille, France, 74-79. *Mem:* Fel Am Psychiat Asn; Sigma Xi. *Res:* Medical education and administration. *Mailing Add:* 44 Outlook Ave West Hartford CT 06119

FLEETWOOD, CHARLES WESLEY, chemistry; deceased, see previous edition for last biography

FLEETWOOD, MILDRED KAISER, b Keyser, WVa, Jan 10, 45; m 67. IMMUNOPATHOLOGY. *Educ:* Univ Richmond, BS, 66; Med Col Va, MS, 68, PhD(clin path), 72. *Prof Exp:* ASSOC PATH, GEISINGER MED CTR, 72- *Mem:* Am Soc Microbiol; Sigma Xi; NY Acad Sci; Am Soc Clin Chemists. *Res:* Cellular immunity; development of immunology tests for the clinical laboratory. *Mailing Add:* 152 Colonial Ave Bloomsburg PA 17815

FLEGAL, ARTHUR RUSSELL, JR, b Oakland, Calif, Aug 30, 46; m 71; c 2. MARINE GEOCHEMISTRY, BIOLOGICAL OCEANOGRAPHY. *Educ:* Univ Calif, Santa Barbara, BA, 68; Calif State Univ, Hayward, MA, 76; Ore State Univ, PhD(oceanog), 79. *Prof Exp:* Res assoc oceanog, Moss Landing Marine Labs, Calif State Univs, 78-85; assoc res geochemist, 85-88, RES GEOCHEMIST, UNIV CALIF, SANTA CRUZ, 88- *Concurrent Pos:* Res fel geochem, Calif Inst Tech, 80, vis assoc geochem, 80-; vis assoc geochem, Swiss Fed Inst Technol, 88; comt mem, Nat Res Coun, 89-; mem group experts, Intergovt Oceanog Comn, 89- *Mem:* Am Geophys Union; Am Chem Soc; Soc Environ Toxicol & Chem. *Res:* Biogeochemical cycles of trace elements; radioecology; analytical and nuclear chemistry; environmental toxicology and health; earth science. *Mailing Add:* Inst Marine Sci Univ Calif Santa Cruz CA 95064

FLEGAL, CAL J, b Kalamazoo, Mich, Feb 25, 36; m 58; c 3. POULTRY NUTRITION. *Educ:* Mich State Univ, BS, 58, MS, 62, PhD(poultry nutrit), 65. *Prof Exp:* Asst, 60-63, nutrit technician, 63-65, EXTEN SPECIALIST, DEPT POULTRY SCI, COOP EXTEN SERV, MICH STATE UNIV, 65- *Mem:* Poultry Sci Asn; World Poultry Sci Asn. *Res:* Vitamin A-Beta-Carotene relationships and activity; utilization of raw soybeans in poultry rations, the nutritional aspects of the fatty-liver syndrome in chickens and recycling animal wastes, gomefied nutrition. *Mailing Add:* Dept Animal Sci Mich State Univ East Lansing MI 48824-1225

FLEGAL, ROBERT MELVIN, b Salt Lake City, Utah, Dec 22, 41; m 66; c 1. COMPUTER SCIENCE, OBJECT-ORIENTED DATABASES HYPERMEDIA. *Educ:* Univ Utah, BA, 66, MA, 68. *Prof Exp:* Teacher math, US Peace Corps, 68-69; mem staff comput sci, Univ Utah, 69-70; res scientist comput sci, 70-86, MGR, INFO SPACES AREA, XEROX PALO ALTO RES CTR, 86- *Res:* Computer graphics, mathematical representations of images, object oriented databases and hypermedia. *Mailing Add:* Xerox Palo Alto Res Ctr 3333 Coyote Hill Rd Palo Alto CA 94304

FLEGENHEIMER, HAROLD H(ANSLEO), b Heilbronn, Ger, Feb 18, 24; nat US; m 47; c 1. CHEMICAL ENGINEERING. *Educ:* Cooper Union, BChE, 44; Polytech Inst Brooklyn, MChE, 47. *Prof Exp:* Formulator org coatings, Devoe & Raynolds Co, Inc, 47-54, tech dir, Newark Plant, 54-66; Eastern tech mgr, Celanese Coatings Co, 66-70, mgr prod develop planning, 70-74, MGR ENVIRON, HEALTH & PROD SAFETY AFFAIRS, CELANESE PLASTICS & SPECIALTIES CO, 74- *Mem:* Am Chem Soc. *Res:* Development of protective and decorative coatings; related materials, especially polymers, resins, solvents and pigments. *Mailing Add:* 2556 Cherosen Rd Louisville KY 40205-1708

FLEHARTY, EUGENE, b Beaver Falls, Pa, Oct 16, 34; m 55; c 2. MAMMALIAN ECOLOGY. *Educ:* Hastings Col, BA, 56; Univ NMex, MS, 58, PhD(biol), 63. *Prof Exp:* Asst prof biol, Nebr Wesleyan Univ, 60-62; from asst prof to assoc prof, 62-70, chmn dept, 79-90, PROF ZOOL, FT HAYS STATE UNIV, 70- *Mem:* Am Soc Mammal; Ecol Soc Am; Southwestern Asn Naturalists. *Res:* Mammalogy and herpetology distribution, taxonomy and ecology. *Mailing Add:* Dept Zool Ft Hays State Univ Hays KS 67601-4099

FLEIG, ALBERT J, JR, b Rochester, NY, May 17, 37; m 65; c 2. SCIENCE MANAGEMENT, ATMOSPHERIC SCIENCE. *Educ:* Purdue Univ, BS, 58; Cath Univ, PhD, 68. *Prof Exp:* Design engr, Convair Astronaut, 58-60; systs engr, Vitro Corp Am, 60-62; systs analyst, 62-68, spec asst to dir admin & mgt, 68-70, head technol applns, 70-72, dep proj mgr, 72-76, MGR OZONE PROCESSING TEAM, GLOBAL ATMOS RES PROJ, GODDARD SPACE FLIGHT CTR, NASA, 76-, NIMBUS PROJ SCIENTIST, 79- *Concurrent Pos:* Lectr aerospace eng, Univ Md, 69- *Res:* Management of large scale scientific and/or technical research interests; development of algorithms and concepts for interpreting experimental data; atmospheric research in stratosphere and climate. *Mailing Add:* Code 910 Goddard Space Flight Ctr NASA Greenbelt Rd Greenbelt MD 20771

FLEISCH, JEROME HERBERT, b Bronx, NY, June 6, 41; c 2. PHARMACOLOGY. *Educ:* Columbia Univ, BS, 63; Georgetown Univ, PhD(pharmacol), 67. *Prof Exp:* Res fel pharmacol, Harvard Med Sch, 67-68; res assoc, Nat Heart & Lung Inst, 68-70, sr staff fel, 70-74; sr pharmacologist, 74-77, from res scientist to sr res scientist, 77-86, RES ADV, LILLY RES LABS, 86- *Concurrent Pos:* Lectr pharmacol & physiol sci, Univ Chicago, 75-78. *Mem:* AAAS; Am Soc Pharm & Exp Therapeut; Soc Exp Biol Med; Col Int Allergologicum; Am Acad Allergy & Immunol; Am Thoracic Soc. *Res:* Smooth muscle, respiratory and mediator release lung. *Mailing Add:* Lilly Res Labs MC 771 Indianapolis IN 46285

FLEISCHAUER, PAUL DELL, b Buffalo, NY, Sept 23, 42; m 78; c 2. SURFACE SPECTROSCOPY, LUBRICATION CHEMISTRY. *Educ:* Wesleyan Univ, BA, 64; Univ Southern Calif, PhD(phys chem), 68. *Prof Exp:* NSF exchange fel, Rome, 68-69; mem tech staff, 69-75, HEAD TRIBOLOGY & SURFACE CHEM SECT, AEROSPACE CORP, 75- *Concurrent Pos:* Vis asst prof, Univ Southern Calif, 75-76; mem, SDIO Technol Insertion Working Group for Tribological Mat, 86-; assoc ed, Tribiology Transactions, 87-89. *Honors & Awards:* Al Sonntag Award, Am Soc Lubrication Engrs, 85; Bunshah Award, Int Conf Metall Coatings, Am Vacuum Soc, 87. *Mem:* Soc Tribologists & Lubrication Engrs; Sigma Xi; Am Chem Soc; Am Vacuum Soc. *Res:* Characterization of chemical and physical processes, relating to lubrication of solid surfaces; identification of the chemical composition of interface states; infrared and electron spectroscopy of solids. *Mailing Add:* Aerospace Corp M2/271 PO Box 92957 Los Angeles CA 90009

FLEISCHER, ALLAN A, b Hartford, Conn, Feb 6, 31; m 56; c 2. NUCLEAR MEDICINE, NUCLEAR SCIENCE. *Educ:* Yale Univ, BS, 52, MS, 56, PhD(physics), 59. *Prof Exp:* Mem tech staff, Ramo-Wooldridge Corp, 59-60; dept mgr, Edgerton, Germeshavsen & Grier, 60-63, tech adv to mgr, 63; dir res, W M Brobeck & Assocs, 63-65; vpres res & develop, Cyclotron Corp, Calif, 65-70; pres & chief exec officer, Medi-Physics, Inc, 70-79. *Concurrent Pos:* Consult, 79- *Mem:* AAAS; Am Phys Soc; sr mem Inst Elec & Electronics Engrs; Am Nuclear Soc; Soc Nuclear Med. *Res:* Radiopharmaceuticals; radioisotopes; nucleonics; accelerator development. *Mailing Add:* 2365 Green St San Francisco CA 94123

FLEISCHER, ARTHUR C, b Miami, Fla, May 15, 52; m 75; c 3. FERTILITY AND REPRODUCTION RESEARCH. *Educ:* Emory Univ, BS, 73; Med Col Ga, MD, 76. *Prof Exp:* Pres, radiol, 76-80, from asst prof to assoc prof, 80-87, PROF RADIOL, VANDERBILT UNIV MED CTR, 87- *Concurrent Pos:* Clin fel, Diagnostic Ultrasound, Vanderbilt Univ Med Ctr, 80; bd gov, Am Inst Ultrasand Med, 87- *Honors & Awards:* Presidents Award, Am Roentgen Ray Soc, 77. *Mem:* Am Inst Ultrasound Med; Am Col Radiol; Soc Radiologists Ultrasound; Diagnostic Ultrasound; Clinical Application Operations. *Mailing Add:* Dept Radiol Vanderbilt Univ Med Ctr Nashville TN 37232-2675

FLEISCHER, BECCA CATHERINE, b Brooklyn, NY, Feb 12, 30; m 62. CELL BIOLOGY. *Educ:* Brooklyn Col, BS, 52; Ind Univ, MA, 55, PhD(biochem), 58. *Prof Exp:* Proj assoc, Dept Genetics, Univ Wis, 58-62, trainee, Enzyme Inst, 62-64; res assoc molecular biol, 64-72, res asst prof, 72-74; res assoc prof, 74-89, RES PROF MOLECULAR BIOL, VANDERBILT UNIV, 89- *Mem:* NY Acad Sci; Am Soc Cell Biol; AAAS; Am Chem Soc; Am Soc Biol Chemists. *Res:* Comparative biochemistry of membranes; isolation and characterization of subcellular organelles of liver and kidney, particularly Golgi apparatus; mechanism of glycosylation in Golgi; bile acid transport in rat liver. *Mailing Add:* Dept Molecular Biol Vanderbilt Univ Box 1702 Sta B Nashville TN 37235

FLEISCHER, EVERLY B, b Salt Lake City, Utah, June 5, 36; m 59; c 2. BIOINORGANIC CHEMISTRY. *Educ:* Yale Univ, BS, 58, MS, 59, PhD(chem), 61. *Prof Exp:* From asst prof to assoc prof chem, Univ Chicago, 61-71; chmn dept chem, Univ Calif, Irvine, 72-74, prof chem, 71-80, dean, sch phys sci, 74-80; dean arts & sci, Univ Colo, Boulder, 80-88; EXEC VCHANCELLOR, UNIV CALIF, RIVERSIDE, 88- *Concurrent Pos:* Sloan fel, Sloan Found, 67-68. *Mem:* Am Chem Soc; The Chem Soc; Sigma Xi. *Res:* Role of metal ions in biology; porphyrin and metalloporphyrin chemistry; coordination chemistry. *Mailing Add:* Univ Calif 4108 Admin Bldg Riverside CA 92521

FLEISCHER, GERALD A, b St Louis, Mo, Jan 7, 33; m 60; c 2. INDUSTRIAL ENGINEERING, SYSTEMS ANALYSIS. *Educ:* St Louis Univ, BS, 54; Univ Calif, Berkeley, MS, 59; Stanford Univ, PhD(indust eng, eng-econ planning), 62. *Prof Exp:* Instr eng ed, Heald Eng Col, 58-59; opers analyst, Consol Freightways, Inc, 59-61; asst prof indust eng, Univ Mich, 63-64; assoc prof, 64-71, dir, Traffic & Safety Ctr, 76-79, PROF INDUST & SYSTEMS ENG, UNIV SOUTHERN CALIF, 71- *Concurrent Pos:* Statist analyst, Maritime Cargo Transp Conf Div, Nat Acad Sci-Nat Res Coun, 58; res mgr, Hawaiian Marine Freightways, 60; actg asst prof, Stanford Univ, 60-62; consult, USAID, 65-66, US Navy, 66, Rand Corp, 66- & Nat Coop Hwy Res Prog, Nat Acad Sci, 70-; Fulbright sr lectr, Ecuador, 74; Dir, NATO Advan Study Inst, 79; vis prof, The Chinese Univ Hong Kong, 87. *Mem:* Fel Am Inst Indust Engrs; Am Soc Eng Educ; fel Inst Advan Eng; Inst Mgt Sci. *Res:* Engineering-economic planning and systems analysis, especially in public sector. *Mailing Add:* Dept Indust & Systs Eng Univ Southern Calif Los Angeles CA 90089-0193

FLEISCHER, HENRY, b Paris, France, Nov 10, 23; US citizen; m 49; c 2. PRECISION FINISHING OF METALS, FLUID DYNAMICS. *Educ:* City Col NY, BME, 48; Columbia Univ, MSc, 52. *Prof Exp:* Proj engr, Eversharp Res Lab, 48-49 & Trubenizing Process Corp, 49-51; asst chief engr, Swingline Prod Inc, 51-53; chief engr, Finkel Outdoor Prod Inc, 53-63; vpres eng, Flair Mfg Corp, 63-68; DIR ENG, NUMATICS INC, 68- *Concurrent Pos:* Mem, Standards Comt, Nat Fluid Power Asn, 68-75; lectr pneumatics numerous univs. *Mem:* Nat Soc Prof Engrs; Soc Mfg Engrs. *Res:* Precision metal removal such as grinding and honing; author of one book and 30 publications; techniques for optimizing pneumatic components to increase productivity, conserve energy and optimal utilization of compressed air. *Mailing Add:* 473 Ashley Dr Grand Blanc MI 48439

FLEISCHER, HERBERT OSWALD, b Lake Geneva, Wis, June 22, 13; m 40, 67; c 2. FORESTRY, GENERAL BIOLOGY. *Educ:* Northwestern Col, BA, 35; Univ Mich, BS & MF, 38; Yale Univ, PhD, 52. *Prof Exp:* Field asst, Lake States Forest & Range Exp Sta, US Forest Serv, 38, jr forester, Nicolet Nat Forest, Wis, 39; forester, Consol Water Power & Paper Co, 39-41; forest prod technologist, Forest Prod Lab, US Forest Serv, USDA, 42-57, chief div timber processing, 57-64; dir forest prod utilization & eng res, 64-67, dir, Forest Prod Lab, 67-75; RETIRED. *Concurrent Pos:* Tech consult, Foreign Econ Admin, Eng & Germany, 45, forestry & forest prod, UN Develop Prog, India, 73, 76 & 78 & Brazil, 78; lectr, dept forestry, Univ Wis-Madison, 50-75; mem comt res eval, US Dept Agr, 56; chmn wood-based panel prod comt, UN, 65-72; mem subcomt fed labs, Comt Sci & Technol, US Govt, 68-74; mem adv panel res mgt, AID, Nat Acad Sci, 73; mem, Bd Regents, Wis Lutheran Col, Milwaukee, Wis, 84- *Honors & Awards:* Wood Award, Forest Prod Res Soc, 67; Distinguished Serv Award, Soc Wood Sci & Technol, 83. *Mem:* Fel Soc Am Foresters; Forest Prod Res Soc (pres, 64-65); Soc Wood Sci & Technol; fel Int Acad Wood Sci; Int Union Forestry Res Orgns; Sigma Xi; Am Soc Testing & Mat. *Res:* Forest products utilization and engineering research, and research administration; history and philosophy of science. *Mailing Add:* 2508 Santa Maria Ct Middleton WI 53562

FLEISCHER, ISIDORE, b Leipzig, Ger, June 4, 27; US citizen; c 1. MATHEMATICS. *Educ:* Brooklyn Col, BSc, 48; Univ Chicago, MSc, 49, PhD(math), 52. *Prof Exp:* Teaching and res asst, Univ Chicago, 50-52; Nat Res Coun fel, Paris, France, 52-53; lectr math, Northwestern Univ, 55-56; mem tech staff, Bell Tel Labs, 56-57; mathematician, Eastern Res Group, 57-58 & Appl Res Lab, Sylvania Elec Prod, Inc, 59-60; assoc prof math, Purdue Univ, 61 & Univ Clermont-Ferrand, France, 61-62; prof agr & fel, Univ Montreal, 63-65; fel, Queen's Univ, Ont, 67-68; vis scientist, Dept Math, Univ Montreal, 68-73, vis scientist, Res Ctr, 74-85; ADJ PROF, DEPT MATH, UNIV WINDSOR, 85- *Mem:* Am Math Soc. *Res:* Ordered, universal, topological and logical algebra. *Mailing Add:* Dept Math Univ Windsor Windsor ON N9B 3P4 Can

FLEISCHER, MICHAEL, b Bridgeport, Conn, Feb 27, 08; m 34; c 2. MINERALOGY, GEOCHEMISTRY. *Educ:* Yale Univ, BS, 30, PhD(chem), 33. *Prof Exp:* Fel, Yale Univ, 33-34; asst to ed, Dana's Syst Mineral, 35-36; asst phys chemist, Geophys Lab, Carnegie Inst, Washington, DC, 36-39; geochemist, US Geol Surv, 39-78; RES ASSOC, DEPT MINERAL SCI, SMITHSONIAN INST, 78- *Concurrent Pos:* Asst ed, Chem Abstr, 40-; vpres comn geochem, Int Union Chem, 51-53, pres, 53-57; prof lectr, George Washington Univ, 57-65; chmn comn new minerals, Int Mineral Asn, 59-74; mem, US Nat Comt Geochem, 69-74. *Honors & Awards:* Roebling Medal, Mineral Soc Am, 75; Becke Medal, Mineral Soc Austria, 77. *Mem:* Am Chem Soc; fel Mineral Soc Am (vpres, 51, pres, 52); fel Geol Soc Am (vpres, 53); fel Soc Econ Geol; Geochem Soc (vpres, 63, pres, 64); hon fel Mineral Soc Gt Brit; hon fel Mineral Soc France. *Res:* Geochemical abundance and distribution of elements; chemical mineralogy. *Mailing Add:* Dept Mineral Sci Smithsonian Inst Washington DC 20560

FLEISCHER, PETER, b Coburg, Ger, Sept 10, 41; US citizen; m 72. GEOLOGICAL OCEANOGRAPHY, HIGH FREQUENCY ACOUSTICS. *Educ:* Univ Minn, BA, 63; Univ Southern Calif, PhD(geol), 70. *Prof Exp:* Lectr geol, Loyola Univ Los Angeles, 69 & Univ Calif, Los Angeles, 70; NSF fel biol oceanog, Marine Lab, Duke Univ, 70-71; asst prof geol oceanog, Old Dominion Univ, 71-78; GEOLOGIST, NAVAL OCEAN RES & DEVELOP ACTIV, 78- *Mem:* Geol Soc Am; Soc Econ Paleontologists & Mineralogists; Sigma Xi; Am Asn Petrol Geologists; AAAS. *Res:* Beach processes and beach erosion; continental margin sedimentation and structure; side-scan sonar and acoustic seafloor classification. *Mailing Add:* Seafloor Geosci Div Naval Oceanog & Atmospheric Res Lab Pass Christian MS 39529

FLEISCHER, ROBERT LOUIS, b Columbus, Ohio, July 8, 30; m 54; c 2. PHYSICS, GEOPHYSICS. *Educ:* Harvard Univ, AB, 52, AM, 53, PhD(appl physics), 56. *Prof Exp:* Asst prof metall, Mass Inst Technol, 56-60; RES LAB, GEN ELEC CO, 60- *Concurrent Pos:* Sr res fel physics, Calif Inst Technol, 65-66; adj prof, Rensselaer Polytech Inst, 67-68; consult, US Geol Surv, 67-70; assoc ed, Geochem & Cosmochem Acta, Lunar Sci Conf Proc, 70-73; vis scientist, Nat Ctr Atmospheric Res, Nat Oceanic & Atmospheric Admin & Atmospheric Physics & Chem Lab, 73-74; consult, Calif Inst Technol, 80; adj prof geol, State Univ NY, Albany, 81-87; adj prof mech eng & appl physics, Yale Univ, 84; mem, Int Comt, Nuclear Track Soc. *Honors & Awards:* Ernest O Lawrence Award, 71; Gen Elec Co Inventor's Awards, 67 & 71; Golden Plate Award, Am Acad Achievement, 72; Gen Elec Coolidge Medal, 72; Except Sci Achievement Medal, NASA, 73. *Mem:* Fel Am Geophys Union; fel Am Phys Soc; Am Soc Metals; Health Physics Soc; Am Inst Mining, Metall & Petrol Engrs; fel AAAS; fel Am Acad Arts & Sci. *Res:* Solid state physics; charged particle tracks in solids; superconductivity; crystal plasticity; geochronology and application of geochronology to anthropology; applications of charged particle tracks in space sciences; nuclear physics and engineering; cosmic ray physics; radiobiology of alpha emitters; uranium exploration; indoor radon and health, earthquake prediction; high temperature mechanical properties. *Mailing Add:* Res Lab Gen Elec Co Schenectady NY 12301

FLEISCHER, SIDNEY, b New York, NY, May 10, 30; m 62. STRUCTURE & FUNCTION OF MEMBRANES, MUSCLE PHYSIOLOGY. *Educ:* City Col New York, BS, 52; Ind Univ, PhD(biochem), 57. *Prof Exp:* Res biochemist, Antibiotics Res Div, Heyden Chem Corp, 52-53; asst, Ind Univ, 53-56, res asst with Prof F Haurowitz, 57-58; res fel, Inst Enzyme Res, Univ Wis, Madison, 58-60, asst prof, 60-64; assoc prof, 64-68, PROF MOLECULAR BIOL, VANDERBILT UNIV, NASHVILLE, 68- *Concurrent Pos:* Mem, Physiol Chem Study Sect, Am Heart Asn, 68-72; mem, Sci Adv Comt, Am Cancer Soc, 73-77, Biochem Study Sect, NIH, 74-78, Molecular Cytol Study Sect, NIH, 85-89; coun mem, Biophys Soc, 77-80; chmn, Gordon Conf Energy Coupling Mechanisms, 79; mem, Int Cell Res Orgn, UNESCO, 77-, Comt Int Sci Coop, 80- & Expert Comt Biomat, 83-; chmn Bioenergetics Group, Int Union Biochem & Int Union Pure & Appl Biophys, 80-83; mem, Prog Planning Comt, Am Soc Biol Chem, 82-83; mem, Spec Adv Coun, NIH/NIADDK, 84; coun mem, Biophys Soc, 85-87; mem, Molecular Cytology Study Sect, NIH, 85-89. *Honors & Awards:* Earl Sutherland Prize, 81. *Mem:* Am Soc Biol Chemists; Am Chem Soc; Am Soc Cell Biol; NY Acad Sci; Biophys Soc (pres-elect, 88-89); Int Cell Res Orgn, UNESCO. *Res:* Structure and function of membranes; functional role of lipids; organelles and cell function; transport across biological membranes; muscle physiology; excitation-contraction coupling; Ca2. *Mailing Add:* Dept Molecular Biol Vanderbilt Univ Nashville TN 37235

FLEISCHER, THOMAS B, b Oslo, Norway, Mar 27, 29; m 58; c 3. PAPER CHEMISTRY. *Educ:* Univ Oslo, BS, 50, 52 & 53, PhD(phys chem), 56. *Prof Exp:* Res chemist, WVa Pulp & Paper Co, Md, 57-59; res assoc papermaking, Scott Paper Co, Pa, 59-61; res group leader pulp-paper prod, A/B Borregaard, Sarpsborg, Norway, 61-66; mgr res & develop phys & chem res, Huyck Formex, 66-68, mgr felt & fabrics, 68-71, dir res, Huyck Res Ctr, 71-79, vpres technol, 80- 84; MGR PROD DEVELOP, HUYCH FELT, 84- *Mem:* Am Tech Asn Pulp & Paper Industs. *Res:* Structure of polymers; fillers and additives in paper; paper coating forming fabric development; dryer fabric development. *Mailing Add:* 240 Haywick Pl Wake Forest NC 27587

FLEISCHMAJER, RAUL, b Buenos Aires, Arg, Dec 17, 24; US citizen; m 57; c 1. DERMATOLOGY, BIOCHEMISTRY. *Educ:* NY Univ, dipl, 62. *Prof Exp:* Instr dermat, NY Univ, 60-62, asst prof dermat & res assoc biochem, 62-63; assoc prof dermat, Hahnemann Med Col, 63-69, prof med & dir dermat, 69-79; PROF DERMAT & CHMN DEPT, MT SINAI MED CTR, NY, 79- *Concurrent Pos:* Asst vis dermatologist, Bellevue Hosp, New York, 61-63; fel, Arthritis & Rheumatism Found, 62-64; asst attend physician, Philadelphia Gen Hosp, 64-67; consult, Vet Admin Hosps, Philadelphia, 64-67 & Wilkes-Barre, 65-67. *Honors & Awards:* Henry Silver Award, 63. *Mem:* AAAS; AMA; Soc Invest Dermat; Am Acad Dermat. *Res:* Disturbances of lipid metabolism; chemical structure of human connective tissue; role of connective tissue in experimental carcinogenesis; chemical structure of collagen in scleroderma; immunology; connective tissue disease. *Mailing Add:* Dept Dermat Mt Sinai Sch Med One Gustave Levy Plaza New York NY 10029

FLEISCHMAN, ALAN R, b NY, Mar 8, 46; m 68; c 1. PEDIATRICS, NEONATOLOGY. *Educ:* City Col New York, BS, 66; Albert Einstein Col Med, MD, 70. *Prof Exp:* Clin assoc pregnancy res br, Nat Inst Child Health & Human Develop, 72-74; assoc dir new born serv, Morrisania City Hosp, 75-76; assoc dir newborn serv, 76-78, dir, 78-81, DIR NEONATOL, MONTEFIORE HOSP & MED CTR, 81-; asst prof, 75-80, ASSOC PROF PEDIAT, ALBERT EINSTEIN COL MED, 80- *Concurrent Pos:* Royal Soc Med Found grant, 74. *Mem:* Am Acad Pediat; AAAS; NY Acad Sci; Am Fedn Clin Res. *Res:* Perinatal physiology; clinical neonatology; metabolic bone diseases; calcium and vitamin D metabolism. *Mailing Add:* Dept Pediat Div Neonatal Albert Einstin Col Med Montefiore Med Ctr 1825 Eastchester Rd Bronx NY 10461

FLEISCHMAN, DARRELL EUGENE, b Gooding, Idaho, Apr 7, 34; m 84. VISION, PHOTOSYNTHESIS. *Educ:* Calif Inst Technol, BS, 58; Univ Ariz, PhD(chem), 65. *Prof Exp:* Fel, 64-65, staff scientist, 65-70, investr, Charles F Kettering Res Lab, 70-82, prin res scientist, Battelle Kettering Res Lab, 83-85; adj assoc prof, 85-87, RES ASSOC PROF, WRIGHT STATE UNIV, 87- *Mem:* Biophys Soc; Soc Res Vision & Ophthal. *Res:* Photosynthesis; vision; nitrogen fixation; study of the properties of guanylate cyclase in vertebrate retinal rods; study of bacterial photosynthesis and of symbiotic association between nitrogen-fixing bacteria and plants. *Mailing Add:* Dept Biochem Wright State Univ Dayton OH 45435

FLEISCHMAN, JULIAN B, b Philadelphia, Pa, Dec 6, 33. MICROBIOLOGY. *Educ:* Yale Univ, BS, 55; Harvard Univ, PhD(biochem), 60. *Prof Exp:* NSF fel, Stanford Univ, 59-61; Am Cancer Soc fel, St Mary's Hosp, London & Pasteur Inst, Paris, 61-63; Weizmann Inst, Israel, 63-64; asst prof, 64-77, ASSOC PROF MICROBIOL & IMMUNOL, SCH MED, WASHINGTON UNIV, ST LOUIS, 78- *Mem:* AAAS; Am Asn Immunol. *Res:* Structure and biosynthesis of immunoglobulins and antibodies. *Mailing Add:* Dept Molecular Microbiol Wash Univ Sch Med St Louis MO 63110

FLEISCHMAN, MARVIN, b New York, NY, May 19, 37; c 3. OCCUPATIONAL HEALTH & SAFETY, WASTE MANAGEMENT. *Educ:* City Col New York, BChE, 59; Univ Cincinnati, MS, 63, PhD(chem eng), 68. *Prof Exp:* Res chemist, Monsanto Res Corp, 59-61; sr asst engr, USPHS, 61-63; chem engr, Exxon, 68-70; res engr, Amoco Chem Corp, 77-78; dir eng prof develop, 79-80, chmn, dept chem & environ eng, 80-85, PROF CHEM & ENVIRON ENG, UNIV LOUISVILLE, 70- *Concurrent Pos:* Eng dir, USPHS, NIOSH, 85-; prof engr, Commonwealth of Ky, 74-; proj dir, AID grant, 75-77; fac res assoc, US Army, 86, 87; dir, Ky Waste Redn Ctr, 88- *Mem:* Am Inst Chem Engrs; Am Soc Eng Ed; Air & Waste Mgt Asn. *Res:* Membrane separations; waste treatment, minimization utilization; engineering career development; health and safety. *Mailing Add:* Dept Chem Eng Univ Louisville Louisville KY 40292

FLEISCHMAN, ROBERT WERDER, b New York, NY, Feb 13, 37; m 63; c 2. VETERINARY PATHOLOGY. *Educ:* NY State Col Vet Med, Cornell Univ, DVM, 61. *Prof Exp:* Sr asst vet, Lab Perinatal Physiol, Nat Inst Neurol Dis & Blindness, 62-63; assoc bacteriologist, Nat Ctr Primate Biol, 64-65; fel comp path, Sch Med, Johns Hopkins Univ, 66-69; pathologist, EG&G Mason Res Inst, 69-86; NORTHBORO VET CLIN, 86-; VET PATH CONSULT, 86-; DIR, AAHA HOSP, 86- *Concurrent Pos:* Consult, Worcester Found Exp Biol, Mass, 70-83; dir, Northboro Vet Clinic, 70- *Mem:* Am Vet Med Asn; Am Col Vet Path; Am Animal Hosp Asn; Am Asn Feline Practr. *Res:* Primate care and pathology; laboratory animal medicine and pathology; drug induced toxicology and pathology. *Mailing Add:* Northboro Vet Clinic Inc 286 W Main St Northborough MA 01532-2195

FLEISCHMANN, CHARLES WERNER, b New York, NY, Jan 22, 36; m 59; c 3. PHYSICAL CHEMISTRY, ELECTROCHEMISTRY. *Educ:* Queens Col, NY, BS, 57; Polytech Inst NY, MS, 65, PhD(chem), 70. *Prof Exp:* Jr chemist bio org, Jacques Loewe Res Lab, 57-58; chemist fuel-cell batteries, Leesona Moos Labs, 58-59 & 62-66; lieutenant ammunition, USAF, 59-62; NASA trainee chem, Polytech Inst NY, 66-69; res engr, Mallory Battery Co, 69-70; res scientist, NL Industs, 70-74; mgr res, C & D Batteries Div, Eltra Corp, 74-84; mgr eng, Honeywell Power Serv Ctr, 84-88; dir res & develop, Exide Corp Tech Ctr, 88-90; CONSULT, 90- *Mem:* Am Chem Soc; Electrochem Soc; Sigma Xi; Am Soc Metals. *Res:* Electrochemical power sources, lead acid batteries; lithium cells/batteries alkaline zinc cells; fuel cells; magnesium cells; solid state cells; electrochemistry in molten salts; electrolytic production of magnesium and tungsten bronzes; magnetochemical measurement. *Mailing Add:* 715 Maple Hill Dr Blue Bell PA 19422

FLEISCHMANN, HANS HERMANN, b Munich, Ger, June 2, 33. NUCLEAR FUSION, PLASMA PHYSICS. *Educ:* Tech Univ, Munich, dipl, 59, Dr rer nat(physics), 62. *Prof Exp:* Res assoc physics, Tech Univ, Munich, 62-63; staff mem fusion, Gen Atomic Div, Gen Dyn Corp, 63-67; assoc prof, 67-81, PROF APPL PHYSICS, CORNELL UNIV, 81- *Concurrent Pos:* Consult, Rohde & Scharz, Munich, 60-63, Naval Res Lab, 69-75, McDonnel Douglas Corp, 81-82, Los Alamos Nat Lab, 82; pres, HHF Assocs. *Mem:* Am Nuclear Soc; fel Am Phys Soc; fel Inst Elec & Electronics Engrs. *Res:* Thermonuclear fusion, in particular application of intense relativistic electron and ion beams, field-reversed configurations, atomic collision physics, magnetic accelerators. *Mailing Add:* Appl Eng Dept Cornell Univ Clark Hall Ithaca NY 14853

FLEISCHMANN, WILLIAM ROBERT, JR, b Baltimore, Md, Oct 16, 44; m 66; c 1. VIROLOGY, CELL BIOLOGY. *Educ:* Capital Univ, BS, 66; Purdue Univ, PhD(viral genetics), 72. *Prof Exp:* NIH fel genetics, Ind Univ, 71-72,; asst prof microbiol, Idaho State Univ, 73-76; McLaughlin fel, 76-77, from asst prof to assoc prof, 77-86, PROF MICROBIOL, UNIV TEX MED BR, GALVESTON, 86- *Concurrent Pos:* Vis asst prof genetics & Virol, Ind Univ, 72-73. *Mem:* Am Soc Microbiol; NY Acad Sci; AAAS; Sigma Xi; Am Soc Virol; Int Soc Interferon Res; Soc Exp Biol & Med. *Res:* Interferon production and action, including studies of interferon inhibitors; cooperative effects of interferons and other lymphokines and cytokines (potentiation); interactions of interferons and hyperthermia. *Mailing Add:* Dept Microbiol Univ Tex Med Br Galveston TX 77550

FLEISHER, HAROLD, b Kharkov, Russia, Oct 12, 21; nat US; m 45; c 3. PHYSICS, MATHEMATICS. *Educ:* Univ Rochester, BA, 42, MS, 43; Case Inst Technol, PhD, 51. *Prof Exp:* Res assoc physics, Univ Rochester, 42-43; mem staff radar circuitry, Radiation Lab, Mass Inst Technol, 43-45; sr engr, Video Circuitry, Rauland Corp, 45-46; instr physics, Case Inst Technol, 46-50; mem staff physics & comput, Res Lab, Int Bus Mach Corp, 50-61, sr physicist, Systs Develop Div, IBM, 61-65, mgr adv tech develop, Poughkeepsie Systs Develop Div, 65-68, prog mgr, Systs Develop Div Lab, 68-74, fel, Data Systs Div, 74-87; PROF PHYSICS, STATE UNIV NY, COL AT NEW PALTZ, 87-; PRES, HUDSON VALLEY TECH DEV CTR, 87- *Concurrent Pos:* Vis prof, Vassar Col. *Mem:* Am Phys Soc; fel Inst Elec & Electronic Engrs; Sigma Xi; AAAS. *Res:* Semiconductors and cryogenic physics and devices; optics and quantum electronics; computer organization theory and design. *Mailing Add:* 30 Wilmot Terr Poughkeepsie NY 12603

FLEISHER, LYNN DALE, human biochemical genetics, for more information see previous edition

FLEISHER, MARTIN, b Liberty, NY, May 15, 35; m 58; c 3. CLINICAL CHEMISTRY, IMMUNOCHEMISTRY. *Educ:* Harpur Col, State Univ NY, BA, 58; NY Univ, MS, 61, PhD(biochem), 66. *Prof Exp:* Instr radiol, Col Med, NY Univ, 66-67; ASST PROF, SLOAN-KETTERING DIV, GRAD SCH MED SCI, CORNELL UNIV, 69-; assoc, 69-82, MEM, MEMORIAL SLOAN-KETTERING CANCER CTR, 90- *Concurrent Pos:* Assoc attend biochemist, Dept Clin Chem, Mem Hosp Cancer & Allied Dis, 72-81, attend clin chemist, 81-; adj prof clin chem, Richmond Col, 73-75; adj prof pharmaceut sci, Columbia Univ, 74-75. *Honors & Awards:* Cert Hon, Am Asn Clin Chem, 82; Am Asn Clin Chem Award, 87. *Mem:* Am Asn Clin Chemists; Am Chem Soc; NY Acad Sci; AAAS; Am Inst Chemists; fel Nat Acad Clin Biochemist. *Res:* Development of immunologic and enzymatic diagnostic laboratory procedure; characterization of tumor associated antigens; clinical chemical methodology. *Mailing Add:* Mem Hosp Dept Clin Chem 1275 York Ave New York NY 10021

FLEISHER, PENROD JAY, b Liberty, NY, Aug 2, 37; m 62; c 2. GEOMORPHOLOGY, GLACIOLOGY. *Educ:* St Lawrence Univ, BS, 61; Univ NC, Chapel Hill, MS, 63; Wash State Univ, PhD(glacial geol), 67. *Prof Exp:* Geophysicist, Pan Am Petrol Corp, 63-64; DISTINGUISHED TEACHING PROF GEOL, STATE UNIV NY COL ONEONTA, 67- *Concurrent Pos:* Juneau Ice Field Res Proj. *Mem:* Geol Soc Am; Nat Asn Geol Teachers; Sigma Xi. *Res:* Glacial geology. *Mailing Add:* Dept Earth Sci State Univ NY Col Oneonta NY 13820

FLEISHER, THOMAS A, b Rochester, Minn, Oct 3, 46; m 69; c 3. IMMUNOLOGY. *Educ:* Univ Minn, BS, 69, MD, 71. *Prof Exp:* CHIEF IMMUNOL SERV, NIH, BETHESDA, MD, 83- *Concurrent Pos:* Assoc prof pediat, Uniformed Serv Univ Health Sci, Bethesda, Md, 81- *Mem:* Am Asn Immunologists; Clin Immunol Soc; Am Fedn Clin Res; Soc Pediat Res; Am Acad Allergy & Immunol; Soc Anal Cytol. *Res:* Status of peripheral blood lymphocyte subpopulations in disease states. *Mailing Add:* NIH Bldg Ten Rm 2C410 Bethesda MD 20892

FLEISHMAN, BERNARD ABRAHAM, b New York, NY, June 16, 25; m 50; c 3. APPLIED MATHEMATICS, SCIENCE EDUCATION. *Educ:* City Col New York, BA, 44; NY Univ, MS, 48, PhD(math), 52. *Prof Exp:* Spec instr ling, US Army Specialized Training Prog, Univ Pa, 43-44; asst physics, NY Univ, 46-48, res asst & instr math, 48-52; sr staff mathematician, Appl Physics Lab, Johns Hopkins Univ, 52-55; from asst prof to assoc prof, 55-61, PROF MATH, RENSSELAER POLYTECH INST, 61- *Concurrent Pos:* Vis prof, US Army Math Res Ctr, Univ Wis, 61-62; Fulbright-Hays sr res scholar, Delft Tech Univ, 73; fac res partic & consult, Davidson Lab, Stevens Inst Technol, 74; consult, NY State Educ Dept, 83-; Rensselaer distinguished teaching fel, 88. *Mem:* Am Math Soc; Soc Indust & Appl Math; AAAS. *Res:* Nonlinear vibrations; diffusion and heat conduction. *Mailing Add:* Colehamer Ave Troy NY 12180

FLEISIG, ROSS, b Montreal, Can, Oct 12, 21; nat US; m 43; c 2. AERONAUTICAL ENGINEERING. *Educ:* Polytech Inst Brooklyn, BAeroEng, 42, MS, 55. *Prof Exp:* From aerodynamicist to aerodyn design engr, Chance Vought Aircraft, 42-50; flight control systs engr, Sperry Gyroscope Co, 50-54, eng sect head, Missile Systs, 54-58 & Astronaut Syst, 58-60; group head, Guidance Dynamics, Grumman Aerospace Corp, 60-62, head lunar excursion module dynamics & performance anal, 62-65, head lunar excursion module guid, Navig & Control Anal & Integration, 65-67, LM-5 space craft team mgr, 67-69, space systs proj mgr, 69-80, systs proj mgr, 80-84; PRES, AEROSPACE SYSTS CONSULTANCY, THERUS DYNAMICS, INC, 84- *Concurrent Pos:* Mem space rescue studies comt, Int Acad Astronaut; dir, Am Astronaut Soc, 59-67. *Honors & Awards:* Apollo 11 Manned Flight Awareness Award, 69; NASA Medallion, Lunar Module Prog, NASA, 73; Grumman Citation Excellence, Lunar Module Prog Proj Apollo, 73; Thomas Sanial Mem Award, Am Inst Aeronaut & Astronaut. *Mem:* Fel AAAS; fel Am Astronaut Soc (pres, 57-58); assoc fel Am Inst Aeronaut & Astronaut; fel Brit Interplanetary Soc; Int Acad Astronaut; Nat Soc Prof Engrs. *Res:* Aerodynamics; flight dynamics; flight control and guidance systems analyses and development; astronautics; aerospace systems integration. *Mailing Add:* 58 Kilburn Rd Garden City NY 11530

FLEISS, JOSEPH L, b Brooklyn, NY, Nov 13, 37; m; c 3. BIOSTATISTICS. *Educ:* Columbia Col, AB, 59; Columbia Univ, MS, 61, PhD(math statist), 67. *Prof Exp:* Asst, Div Biostatist, Sch Pub Health, Columbia Univ, 60-65; sr biostatistician, NY State Psychiat Inst, 61-65, assoc biostatistician, 65-66; instr, Columbia Univ, 65-68; prin biostatistician, NY State Psychiat Inst, 67, assoc res scientist, 67-76; from adj asst prof to adj assoc prof, 72-75, PROF BIOSTATIST & HEAD DIV, SCH PUB HEALTH, COLUMBIA UNIV, 75- *Concurrent Pos:* Assoc ed, Biometrics, 75-; mem data monitoring comt, Ctr Prev Premature Arteriosclerosis, 75-; biostatistician, Presby Hosp, New York, 76-; mem bd dirs, Nat Comn Confidentiality of Health Records, 79- *Honors & Awards:* Spiegelman Gold Medal, Am Pub Health Asn, 73. *Mem:* Am Col Neuropsychopharmacol; Am Psychopath Asn; Am Pub Health Asn; fel Am Statist Asn; Biometric Soc; Sigma Xi; Soc Clin Trials; Soc Epidemiol Res. *Mailing Add:* Div Biostatist Columbia Univ Sch Pub Health 600 W 168 St New York NY 10032

FLEIT, HOWARD B, b 1952. RECEPTORS, CELLULAR DIFFERENTIATION. *Educ:* NY Univ, PhD(cell biol), 80. *Prof Exp:* asst prof, 83-89, ASSOC PROF PATH, STATE UNIV NY, STONY BROOK, 89- *Concurrent Pos:* Postdoctoral fel, Rockefeller Univ, 80-83; Catacosinus Young Investr Award, 86; Sinsheimer Found Scholar Award, 86. *Mem:* Am Soc Cell Biol; Am Asn Immunologists; Soc Leukocyte Biol; Harvey Soc; Sigma Xi; NY Acad Sci; AAAS. *Res:* Leukocyte Fc receptors. *Mailing Add:* Path Dept State Univ NY Stony Brook NY 11794

FLEMAL, RONALD CHARLES, b Two Rivers, Wis, Feb 17, 42; m 64; c 4. GEOLOGY. *Educ:* Northwestern Univ, BA, 63; Princeton Univ, AM, 65, PhD(geol), 67. *Prof Exp:* Asst prof, 67-74, ASSOC PROF GEOL, NORTHERN ILL UNIV, 74- *Concurrent Pos:* Mem state board, Ill Pollution Control Bd, 85- *Mem:* AAAS; Geol Soc Am; Am Quaternary Asn. *Res:* Water quality; pollution control; application of quantitative techniques in geology and geomorphology; quaternary geology. *Mailing Add:* Dept Geol Northern Ill Univ Box 505 DeKalb IL 60115

FLEMING, ALAN WAYNE, b Kansas City, Mo, Dec 2, 39; m 61; c 2. CONTROL SYSTEMS, DYNAMICS. *Educ:* Univ Kans, BS, 61; Stanford Univ, MS, 62, PhD(aeronaut & astronaut sci), 66. *Prof Exp:* Asst engr, Boeing Co, 61; assoc eng, Lockheed Missiles & Space Co, 62-63; res engr, Stanford Univ, 66-72; staff engr, 72-77, mgr, control systs eng dept, 77-82, proj mgr, 82-85, MGR, CONTROL & SENSOR SYSTS LAB, TRW SPACE & TECHNOL GROUP, 85- *Mem:* Am Inst Aeronaut & Astronaut; Inst Elec & Electronics Engrs; Am Astronaut Soc. *Res:* Spacecraft and missile control system engineering; design; analysis; modelling; simulation instrumentation; testing; flexible structure control; attitude determination; navigation; microprocessor control implementation. *Mailing Add:* Lab Mgr TRW Space & Technol Group Eng 82/2076 One Space Park Redondo Beach CA 90278

FLEMING, ATTIE ANDERSON, b Russellville, Ala, Dec 3, 21. PLANT GENETICS. *Educ:* Auburn Univ, BS, 43, MS, 49; Univ Minn, PhD(plant genetics), 51. *Prof Exp:* Asst county agr agent, Calhoun County, Ala, 43-45; asst agron, Ala Polytech Inst, 47-49; asst agron & plant genetics, Univ Minn, 49-51; from asst prof to prof plant genetics, Univ Ga, 51-88; RETIRED. *Mem:* Am Soc Agron; Am Genetic Asn; Genetics Soc Can; Crop Sci Soc Am. *Res:* Corn improvement; genetics of corn; disease and insect resistance of corn; field-plot technique. *Mailing Add:* 140 Hope Ave Athens GA 30606

FLEMING, BRUCE INGRAM, b Bradford, Eng, Jan 30, 41; Can citizen; m 66; c 2. REDOX REACTIONS, CHEMICAL SENSORS. *Educ:* Oxford Univ, BA, Hons, 63, MA, 68; McGill Univ, PhD(org chem), 74. *Prof Exp:* Chemist, Shell Can, Ltd, 64-66; scientist, 74-84, SR SCIENTIST, PULP & PAPER RES INST CAN, 84-, DIR CHEM PULPING & BLEACHING DIV, 85-, DIV DIR CHEMICAL PULPING. *Honors & Awards:* Dave Welterhorn Award, 85; Douglas Jones Award, 86. *Mem:* Chem Inst Can; Can Pulp & Paper Asn; Tech Asn Pulp & Paper Indust. *Res:* Chemistry and technology of alkaline pulping and bleaching, including the chemistry of lignin, and the development of new technical processes to produce superior types of pulp; development of new sensors. *Mailing Add:* Pulp & Paper Res Inst Can 570 St John's Blvd Pointe Claire PQ H9R 3J9 Can

FLEMING, DONOVAN ERNEST, b Ogden, Utah, Aug 16, 32; m 55; c 3. PSYCHOPHYSIOLOGY. *Educ:* Brigham Young Univ, BS, 56, MS, 57; Wash State Univ, PhD(physiol psychol), 62. *Prof Exp:* Trainee, Vet Admin Hosp, Salt Lake City, Utah, 60-61; res psychologist brain res, 61-64, dir res unit, Phoenix, Ariz, 64-71; chmn dept,psychol, 77-87, PROF PSYCHOL, BRIGHAM YOUNG UNIV, 71- *Concurrent Pos:* Spec instr, Brigham Young Univ, 60-64; res instr, Div Neurol, Col Med & lectr psychol, Univ Utah, 62-64; res asst prof psychol, Ariz State Univ, 65-67, vis assoc prof, 67-71. *Mem:* Am Psychol Asn; Am Physiol Soc; Soc Neurosci; Sigma Xi. *Res:* Behavioral neurophysiology; physiological and psychological study of central nervous system functions. *Mailing Add:* Dean Soc Sci Brigham Young Univ Provo UT 84602

FLEMING, EDWARD HOMER, JR, b University City, Mo, Feb 27, 25. NUCLEAR CHEMISTRY, SCIENCE ADMINISTRATION. *Educ:* Wabash Col, AB, 49; Univ Calif, PhD(chem), 52. *Prof Exp:* Chemist, Lawrence Radiation Lab, Univ Calif, Berkeley, 49-52; res chemist, Calif Res Corp, 52-55; asst head dept chem, Lawrence Livermore Lab, Univ Calif, 55-7S; dir, Div Appl Technol, US AEC, 73-75; spec asst to dir, Lawrence Livermore Lab, Univ Calif, 75-85, sr scientist, 85-87; RETIRED. *Concurrent Pos:* Tech dir, Proj Sulky & chmn, Radioactivity Working Group, Atlantic-Pac Interoceanic Canal Studies Comn, 67-70. *Mem:* AAAS; Sigma Xi. *Res:* Half-lives of uranium isotopes; gamma-ray scattering; oil soluble tracers; fallout from nuclear explosions; fission product decay chains; radiation doses from radionuclides; international treaties on nuclear explosives. *Mailing Add:* 40 Golf Rd Pleasanton CA 94566

FLEMING, GORDON N, b Pittsburgh, Pa, Apr 14, 36; m 58, 84; c 2. THEORETICAL PHYSICS. *Educ:* Univ Pittsburgh, BS, 58; Univ Pa, PhD(physics), 64. *Prof Exp:* Res asst prof theoret physics, Univ Wash, 63-65; from asst prof to assoc prof, 65-74, PROF PHYSICS, PA STATE UNIV, UNIVERSITY PARK, 74- *Mem:* Am Phys Soc. *Res:* Relativistic quantum theory; theory of elementary particles; foundations of quantum mechanics. *Mailing Add:* Dept Physics Davey Lab Pa State Univ University Park PA 16802

FLEMING, GRAHAM RICHARD, b Barrow Cumbria, Eng, Dec 3, 49; m 78; c 1. DYNAMICS OF LIQUIDS & SOLUTIONS, PHOTOSYNTHESIS. *Educ:* Univ Bristol, BSc, 71; Univ London, PhD, 74. *Prof Exp:* Res fel, Calif Inst Technol, 74-75, Univ Melbourne, Australia, 75-76; Leverhulme fel, Royal Inst, 77-79; from asst prof to prof chem, 79-, chmn, 88-90, ARTHUR HOLLY COMPTON DISTINGUISHED SERV PROF, UNIV CHICAGO, 87- *Concurrent Pos:* Co-chmn, Ultrafast Phenomena V Meeting, 86; adv comt, Nat Sci Found, 90-92. *Honors & Awards:* Marlow Medal, Royal Soc Chem, 81; Coblentz Award, Coblentz Soc, 85. *Mem:* Optical Soc Am; Royal Soc Chem. *Res:* Ultrafast spectroscopy studies of reaction dynamics in solution; primary processes in photosynthesis and internal motions of protein and peptides. *Mailing Add:* Dept Chem Univ Chicago 5735 S Ellis Ave Chicago IL 60637

FLEMING, HENRY PRIDGEN, b New Haven, Conn, Aug 9, 32; m 54; c 3. FOOD SCIENCE, MICROBIOLOGY. *Educ:* NC State Univ, BS, 54, MS, 58; Univ Ill, PhD(food sci), 63. *Prof Exp:* Microbiologist, Merck & Co, Inc, 63-64; instr animal sci, 58-60, FOOD SCIENTIST, DEPT FOOD SCI, NC STATE UNIV, 64-, PROF, 77- *Concurrent Pos:* Sci adv, Pickle Packers Int, Inc, Nat Kraut Packers Asn; food scientist, Agr Res Serv, USDA, 64-, res leader, Food Fermentation Lab, 77- *Mem:* Inst Food Technologists; Am Soc Microbiol; Sigma Xi. *Res:* Microbiology and chemistry of vegetable fermentations; lactic acid bacteria; spore forming bacteria; yeasts; fungi; pickles; sauerkraut. *Mailing Add:* Dept Food Sci USDA-Agr Res Serv NC State Univ Box 7624 Raleigh NC 27695-7624

FLEMING, JAMES CHARLES, b Lancaster, Ohio, Aug 25, 38; m 68; c 1. PHOTOGRAPHIC CHEMISTRY, OPTICAL RECORDING. *Educ:* Ohio Univ, BS, 60; Ohio State Univ, PhD(diazooxides), 64. *Prof Exp:* Org res chemist, 65-70, RES ASSOC, EASTMAN KODAK CO, 70- *Mem:* Am Chem Soc; Soc Photographic Scientists & Engrs. *Res:* Photographic and synthetic organic chemistry. *Mailing Add:* 1241 Holley Rd Webster NY 14580-9554

FLEMING, JAMES EMMITT, b Richland, Wash, May 28, 47; m 67; c 1. GERONTOLOGY. *Educ:* West Valley Col, AS, 76, San Jose State Univ, BA, 78. *Prof Exp:* Res asst physiol, Ames Res Ctr, NASA, 76-78; sr tech specialist physiol, State Univ NY-NASA, 78-80; res scientist aging & develop, 80-85, DIR, AGING RES LAB, LINUS PAULING INST, 85-; PRES BIOTECHNOL, ELECTRO BIOTRANSFER, INC, 88- *Concurrent Pos:* Consult, Ames Res Ctr, NASA, 86; assoc ed, Age, Jour of Am Aging Asn, 86-; bd dirs, Electro Biotransfer, Inc, 88- *Mem:* Am Aging Asn; Am Soc Microbiol; AAAS. *Res:* Biochemical and molecular aspects of aging and development in Drosophilia; author of over 40 publications and 50 abstracts; development of methods and instrumentation for the separation and isolation of proteins and nucleic acids; two US patents. *Mailing Add:* 440 Page Mill Rd Palo Alto CA 94306

FLEMING, JAMES JOSEPH, b Chicago, Ill, Feb 26, 17; m 50; c 4. COMPUTER SCIENCE. *Educ:* Northwestern Univ, BS, 38. *Prof Exp:* Physicist, Naval Res Lab, 41-62, head, Opers Res Br, 45-62; assoc supt applns res div, 56-62; chief, Data Systs Div, NASA, Goddard Space Flight Ctr, 62-67, dept asst dir, 67-70, asst dir, Ctr Automatic Data Processing, 70-81; RETIRED. *Mem:* Am Phys Soc; Inst Elec & Electronics Engrs; Am Inst Aeronaut & Astronaut; Sigma Xi. *Res:* Systems engineering; electronics; radar; digital computers. *Mailing Add:* 7216 Windsor Lane Hyattsville MD 20782

FLEMING, JAMES STUART, b Buffalo, NY, Sept 1, 36; m 60; c 2. THROMBOSIS RESEARCH, PROJECT MANAGEMENT. *Educ:* Northwestern Univ, BA, 58; Univ Buffalo, MA, 62; Ohio State Univ, PhD(physiol), 65; Syracuse Univ, MBA, 83. *Prof Exp:* Res asst physiol, Ohio State Univ, 62-65; res scientist, 65-74, sr res scientist, 74-82, mgr, 82-85, proj coordr, 83-90, assoc dir, pharmacol dept, 85-90, ASSOC DIR, PHARMACEUT DEVELOP, BRISTOL-MYERS CO, 90- *Concurrent Pos:* Mem, Coun Thrombosis, Am Heart Asn. *Mem:* Am Soc Pharmacol Exp Therapeut; Int Soc Oxygen Transport to Tissue. *Res:* Indentification and development of inhibitors of blood platelet aggregations; orderly development of new drugs entering clinical trials; antithrombotic agents. *Mailing Add:* 27 Edgewood Lane Glastonbury CT 06033

FLEMING, JOHN F, b Indiana, Pa, Dec 15, 34; m 57; c 2. CIVIL ENGINEERING. *Educ:* Carnegie Inst Technol, BS, 57, MS, 58, PhD(civil eng), 60. *Prof Exp:* Assoc prof civil eng, Northwestern Univ, 60-65; proj engr, Gen Analytics, Inc, Pa, 65-69; ASSOC PROF CIVIL ENG, UNIV PITTSBURGH, 69- *Concurrent Pos:* Consult, Westinghouse Nuclear Systs. *Mem:* Am Soc Civil Engrs; Soc Exp Stress Anal. *Res:* Structural dynamics; computer applications in structural design; structural model analysis. *Mailing Add:* Dept Civil Eng Univ Pittsburgh Main Campus 949 Benedum Pittsburgh PA 15260

FLEMING, JUANITA W, NURSING. *Educ:* Hampton Inst, BS, 57; Univ Chacago, MA, 59; Cath Univ Am, PhD(educ except children), 69. *Prof Exp:* Staff nurse to head nurse, Med Surg Pediat Unit, Children's Hosp, Washington, DC, 57-58; pub health nurse, Bur Pub Health Nursing, 59-60; instr nursing children, Sch Nursing, Freedmen's Hosp, Washington, DC, 62-65; pub health nursing consult, Dept Pediat, Child Develop Clin, Howard Univ, 65-66; from asst prof to assoc prof, 69-73, PROF, COL NURSING, UNIV KY, 73-, SPEC ASST TO PRES ACAD AFFAIRS, CENT ADMIN, 91- *Concurrent Pos:* Prin investr, Am Nurses Found, 70-71; mem, Grad Fac, Col Nursing, Univ Ky, 71-, asst dean grad educ, 75-81, assoc dean & dir grad educ, 82-86, prof, Col Educ, Dept Educ Policy Studies & Eval, 79-, assoc vchancellor acad affairs, Med Ctr, 84-91; co-proj dir, Nursing Care High Risk Infants, State Maternal & Child Health Div, 72; proj dir, Advan Nurse Training Grant, Div Nursing, Dept Health Educ & Welfare, 77-80; vis prof, Case Western Reserve Univ, Cleveland, Ohio, 84; Martin Luther King/Rosa Parks/Cesar Chavez vis prof, Univ Mich, Ann Arbor, 89. *Mem:* Inst Med-Nat Acad Sci; Am Acad Nursing; AAAS. *Res:* Pediatrics in nursing. *Mailing Add:* Univ Ky 205 Admin Bldg Lexington KY 40506-0032

FLEMING, LAWRENCE THOMAS, b Tacoma, Wash, Sept 26, 13; m 37; c 1. APPLIED PHYSICS, RESEARCH ADMINISTRATION. *Educ:* Calif Inst Technol, BS, 37. *Prof Exp:* Examr, US Patent Off, DC, 37-41; engr, US Naval Ord Lab, 41-46, sect head, 46-50; head mech group, Electron Tube Lab, Nat Bur Stand, 50-53, group leader, 55-56; chief instrumentation sect, Diamond Ord Fuze Labs, US Army, 53-55; prin res engr, Res Ctr, Bell & Howell Co, 59-67; PRES, INNES INSTRUMENT CO, 67- *Mem:* Fel Acoust Soc Am. *Res:* Instrumentation; mechanical measurements; vibration and shock; history, sociology and legal aspects of invention. *Mailing Add:* 1830 Reiter Dr Pasadena CA 91106

FLEMING, MICHAEL PAUL, b Hingham, Mass, Feb 18, 48. CHEMISTRY. *Educ:* Mankato State Col, BS, 70; Ore State Univ, MS, 73; Univ Calif, Santa Cruz, PhD(chem), 76. *Prof Exp:* NIH fel, Colo State Univ, 77-79; res chemist, Arapahoe Chem, Inc, 79-; AT SYNTEX CHEM, INC. *Mem:* Am Chem Soc. *Res:* Synthetic organic research and development of pharmaceuticals and fine chemicals. *Mailing Add:* Syntex Chem Inc 2075 N 55th St Boulder CO 80301-2803

FLEMING, PATRICK JOHN, b Springfield, Mass, June 6, 47. BIOCHEMISTRY. *Educ:* Kalamazoo Col, BA, 69; Univ Mich, Ann Arbor, PhD(biochem,), 75. *Prof Exp:* Fel, Health Ctr, Univ Conn, 75-77, instr biochem, 78-79; asst prof, 79-84, ASSOC PROF BIOCHEM, MED SCH, GEORGETOWN UNIV, 84- *Concurrent Pos:* Estab investr, Am Heart Asn, 80-85. *Mem:* AAAS; NY Acad Sci; Am Soc Biol Chem. *Res:* Structure, function, and biosynthesis of membrane-bound proteins; mechanics of biology electron transfer. *Mailing Add:* Dept Biochem Georgetown Univ 353 Basic Sci Washington DC 20057

FLEMING, PAUL DANIEL, III, b Tampa, Fla, June 1, 43; m 67; c 3. COMPUTER SCIENCE, CHEMICAL PHYSICS. *Educ:* Ohio State Univ, BSc, 64; Harvard Univ, AM, 66, PhD(chem physics), 70. *Prof Exp:* Res assoc chem, Columbia Univ, 70-71 & Brown Univ, 71-74; res physicist enhanced oil recovery, Phillips Petrol Co, 74-79, res assoc, Prod Res Br, 79-86; GROUP LEADER, APPL MATH & COMPUTER APPLN, GENCORP RES, 86- *Mem:* Soc Petrol Engrs; Am Phys Soc. *Res:* Statistical mechanics; many body theory; theory of condensed matter; interfacial phenomena; phase transitions and critical phenomena; rate processes and fluid mechanics with applications to enhanced oil recovery; numerical methods. *Mailing Add:* Gencorp Res 2990 Gilchrist Rd Akron OH 44305

FLEMING, PETER B, b Trenton, NJ, Apr 9, 41; m 63; c 4. INORGANIC CHEMISTRY. *Educ:* Union BS, 63; Iowa State Univ, PhD(inorg chem), 68. *Prof Exp:* Sr chemist, Cent Res Lab, 68-73, res specialist, 73-74, res & develop supvr, 74-75, res & develop mgr, 75-80, MGR TECH CERAMICS LAB, 3M CO, 80- *Mem:* Am Chem Soc. *Res:* Metal-metal bonded complexes; niobium and tantalum clusters and complexes; copper chemistry; optical spectra and photochemistry of transition metal complexes; fire protection systems; intumescent compounds; ceramic materials. *Mailing Add:* 6517 Crackleberry Trail St Paul MN 55125

FLEMING, PHYLLIS JANE, b Shelbyville, Ind, Oct 9, 24. EXPERIMENTAL SOLID STATE PHYSICS. *Educ:* Hanover Col, BA, 46; Univ Wis, MS, 48, PhD, 54. *Prof Exp:* Instr physics, Mt Holyoke Col, 48-50; from instr to assoc prof, 53-67, dean col, 68-72, prof, 67-80, SARAH FRANCES WHITING PROF PHYSICS, WELLESLEY COL, 80-, DEPT CHMN, 80- *Mem:* Am Phys Soc; Sigma Xi. *Res:* Film flow of liquid helium II; photoconductivity of lead sulfide films. *Mailing Add:* Dept Physics Wellesley Col Wellesley MA 02181

FLEMING, RICHARD ALLAN, b Chicago, Ill, Sept 16, 29; m 53; c 3. PLASTICS CHEMISTRY. *Educ:* Knox Col, AB, 51; Iowa State Univ, MS, 55, PhD(phys chem), 57. *Prof Exp:* Res chemist, E I du Pont de Nemours & Co, Inc, 57-62, res assoc, 62-64, staff scientist, 64, res supvr, 64-71, sr res supvr, 71-82, TECH MGR, E I DU PONT DE NEMOURS & CO, INC, 82- *Mem:* Am Chem Soc; Soc Plastics Engrs. *Res:* Responsible for all technical aspects of Du Pont's programs on recycle of post-consumer plastic waste. *Mailing Add:* 19 Quail Crossing Wilmington DE 19807

FLEMING, RICHARD CORNWELL, b Blue Island, Ill, Mar 10, 32; m 58; c 5. ZOOLOGY, ENTOMOLOGY. *Educ:* Kalamazoo Col, AB, 54; Univ Kans, MA, 56; Mich State Univ, PhD, 68. *Prof Exp:* Asst instr zool, Univ Kans, 54-56; teacher, high sch, Mich, 58-61; from asst prof to assoc prof biol, 61-71, PROF BIOL, OLIVET COL, 71- *Concurrent Pos:* Instr, Western Mich Univ, 59-60. *Mem:* Lepidop Soc; Nat Audubon Soc; Sigma Xi. *Res:* Entomology, especially distribution, behavior and morphology of Lepidoptera. *Mailing Add:* PO Box 411 Olivet MI 49076

FLEMING, RICHARD J, b Dexter, Iowa, July 26, 38. MATHEMATICS. *Educ:* Northwest Mo State Col, BS, 60; Fla State Univ, MS, 62, PhD, 65. *Prof Exp:* Instr, Fla State Univ, 62-65; asst prof, Univ Mo, 65-71; from assoc prof to prof math sci, Col Arts & Sci, Memphis State Univ, 71-82; PROF & CHAIRPERSON, MATH DEPT, CENT MICH UNIV, 82- *Concurrent Pos:* Lectr, Univ Mo Columbia, 80, LeMoyne-Owen Col, 80, Memphis State Univ, 85, Calvin Col, 86, Univ Louisville, 87. *Mem:* Am Math Soc; Math Asn Am. *Mailing Add:* Dept Math Cent Mich Univ Mt Pleasant MI 48859

FLEMING, RICHARD SEAMAN, b La Jolla, Calif, Apr 30, 39; m 75; c 2. ECOLOGY, ENVIRONMENTAL SCIENCES. *Educ:* Univ Wash, BA, 61, PhD(zool), 73; Univ Alaska, MS, 68. *Prof Exp:* Inst Environ Studies, Univ Wash, 73-76, res asst prof, 76-77; prin scientist, R W Beck & Assoc, 78-82; dir Environ & Licensing, Alaska Lower Auth, 82-85; sr ecologist, Hosey & Assoc, 86-90; SR SCIENTIST, HARZA NORTHWEST, 91- *Concurrent Pos:* Dep Proj Mgr, Sustina Hydroelectric Proj, 82-85; licensing mgr, Elwha-Ghnas Canyon Hydroelectric Proj, 86-90. *Mem:* Ecol Soc Am; Wildlife Soc. *Res:* Assessment of water resource projects; use of environmental sciences in land use planning; environmental analysis of hydroelectric development; wildlife mitigation planning. *Mailing Add:* 5622 12th Ave NE Seattle WA 98105

FLEMING, ROBERT MCLEMORE, b Nashville, Tenn, June 1, 46; m 68; c 2. X-RAY SCATTERING. *Educ:* Vanderbilt Univ, BA, 68; Univ Va, PhD(physics), 76. *Prof Exp:* MEM TECH STAFF, AT&T BELL LABS, 77- *Mem:* Fel Am Phys Soc; Mat Res Soc. *Res:* Structure; physical and electrical properties novel inorganic materials including superconductors; Buckminster fullerenes; materials with charge-density waves and artificially modulated multilayers. *Mailing Add:* AT&T Bell Labs Rm 1E-334 600 Mountain Ave Murray Hill NJ 07974

FLEMING, ROBERT WILLERTON, b Bon Aire, Pa, May 28, 19; m 43; c 4. MEDICINAL CHEMISTRY. *Educ:* Univ Pittsburgh, BS, 40; Univ Cincinnati, MS, 41, PhD(org chem), 43. *Prof Exp:* Chemist asst, Calgon Inc, Pa, 37-42; chemist, Gen Dyestuff Corp, NY, 43-44; asst, Manhattan Dist, Rochester, 44-46; res chemist, Parke, Davis & Co, 46-63; head dept org res, William S Merrell Co Div, Richardson-Merrell, Inc, 63-72; res scientist, Warner-Lambert/Parke Davis, 72-77, sr res assoc, pharmaceut res div, 77-82; RETIRED. *Mem:* AAAS; Am Chem Soc. *Res:* Synthetic organic medicinals. *Mailing Add:* 6173 Dogwood Ridge Milford OH 45150

FLEMING, ROBERT WILLIAM, b Macomb, Ill, Oct 12, 36; m 58; c 3. GEOLOGY, ENGINEERING GEOLOGY. *Educ:* Okla Univ, BS, 62; Brown Univ, MS, 64; Stanford Univ, PhD(geol), 72. *Prof Exp:* Geologist, US Army Corps Engrs, Sacramento, 63-68, US Army Eng Nuclear Cratering Group, 68-69; asst prof geol, Univ Cincinnati, 72-74, actg head dept, 74-75; GEOLOGIST, US GEOL SURV, 75- *Concurrent Pos:* Mgr, Nat Prog Landslide Hazard Reduction, US Geol Surv, 75-79. *Mem:* Geol Soc Am; Int Asn Eng Geol. *Res:* Mass wasting processes on hillsides. *Mailing Add:* Geol Risk Br US Geol Surv MS 966 Box 25046 Denver Fed Ctr Denver CO 80225

FLEMING, SHARON ELAYNE, b Tisdale, Sask, Aug 20, 49. NUTRITION. *Educ:* Univ Saskatchewan, BSc, 71, PhD(food sci & nutrit), 75. *Prof Exp:* Food serv supvr, Univ Saskatchewan, 69-70; org chem lab asst, Saskatchewan Res Coun, 70-71; food scientist, Dept Crop Sci, Govt Can Res Grant, 73-74; res assoc, Prairie Regional Lab, Nat Res Coun Can, 75-79; ASST FOOD SCIENTIST, AGR EXP STA, UNIV CALIF, BERKELEY, 79- *Mem:* Can Inst Food Sci & Technol; Am Asn Cereal Chemists; Am Inst Nutrit; Inst Food Technol; Asn Anal Chemists; Sigma Xi. *Res:* Influence of food and their indigestible residues on gastrointestinal function; fermentation and the influence of the resulting compounds on health and nutrition. *Mailing Add:* Dept Nutrit Sci Univ Calif 119 Morgan Hall Berkeley CA 94720

FLEMING, SUZANNE M, b Detroit, Mich, Feb 4, 27. INORGANIC CHEMISTRY. *Educ:* Marygrove Col, BS, 57; Univ Mich, MS, 59, PhD(inorg chem), 63. *Prof Exp:* Teacher, St Mary Convent, Mich, 48-58; asst chem, Univ Mich, 61; from instr to assoc prof, 62-70, chmn natural sci div, 70-75, vpres & dean, 75-78, acad vpres, 78-80, PROF CHEM, MARYGROVE COL, 70-; ASST ACAD VPRES, EASTERN MICH UNIV, 80-, ADJ PROF CHEM, 80- *Concurrent Pos:* Res grants, Gulf Equip, 64, Sigma Xi, 64-65, NIH, 64-67 & Am Chem Soc Petrol Res Fund, 67-75. *Mem:* Am Chem Soc; Sigma Xi; fel Am Inst Chemists. *Res:* Synthesis and study of bonding characteristics of transition metal complexes and Lewis acid-Lewis base complexes; boron hydride chemistry; spectroscopic studies of bonding. *Mailing Add:* VChancellor Academic Affairs Univ Wisc Eau Claire 206 Schofield Eau Claire WI 54702

FLEMING, SYDNEY WINN, b Thomasville, Ga, July 12, 24; m 49; c 4. PROCESS ANALYTICAL CHEMISTRY, FTIR. *Educ:* Emory Univ, BA, 47, MS, 48; Univ Pa, PhD(chem), 54. *Prof Exp:* Instr chem, Emory Univ, 48-49; from res phys chemist to sr res phys chemist, 53-61, res supvr, 61-68, from res assoc to sr res assoc, 68-82, FEL RES, ENG PHYSICS LAB, EXP STA, E I DU PONT DE NEMOURS & CO, INC, 82- *Mem:* Am Chem Soc; Optical Soc Am; Coblentz Soc; Inst Soc Am; Soc Appl Spec. *Res:* Polymer physical chemistry; optics; color. *Mailing Add:* E I du Pont de Nemours & Co Inc Bldg E357-253 Wilmington DE 19880-0357

FLEMING, THEODORE HARRIS, b Detroit, Mich, Mar 27, 42; m 65; c 2. TROPICAL ECOLOGY, PLANT-ANIMAL INTERACTIONS. *Educ:* Albion Col, BA, 64; Univ Mich, MS, 68, PhD(zool), 69. *Prof Exp:* Syst zoologist, US Nat Mus, 66-67; asst prof biol, Univ Mo, St Louis, 69-74, assoc prof, 74-78; assoc prof, 78-80, PROF BIOL, UNIV MIAMI, 80- *Concurrent Pos:* Vis scientist, Oxford Univ, 77-78, Duke Univ, 84-85; sr Fulbright fel, 87-88. *Mem:* AAAS; Am Soc Mammal; Ecol Soc Am; Soc Study Evolution; Asn Trop Biol; Bot Soc Am; Brit Ecol Soc. *Res:* Foraging behavior of tropical bats; plant-animal coevolution; pollination biology; evolution of plant breeding systems. *Mailing Add:* Dept Biol Univ Miami Coral Gables FL 33124

FLEMING, WALTER, b Langham, Sask, May 20, 19; nat US; m 45; c 1. MATHEMATICS. *Educ:* Univ Sask, BA, 42; Univ Minn, MA, 44, PhD(math), 49. *Prof Exp:* Asst prof math, Univ NB, 45-46; lectr, Univ Man, 46-48; asst prof, Ft Hays Kans State Col, 48-49; assoc prof & chmn dept, Mankato State Col, 49-57; chmn dept, 57-73, PROF MATH, HAMLINE UNIV, 57- *Concurrent Pos:* Consult, St Paul high schs, 59- *Mem:* Math Asn Am; Nat Coun Teachers Am. *Res:* Integration in Wiener space. *Mailing Add:* 2788 Aglan Ave Roseville MN 55113

FLEMING, WARREN R, physiology; deceased, see previous edition for last biography

FLEMING, WENDELL HELMS, b Guthrie, Okla, Mar 7, 28; m 48; c 3. MATHEMATICS. *Educ:* Univ Wis, PhD(math), 51. *Prof Exp:* Mathematician, Rand Corp, 51-53; res proj assoc, Univ Wis, 53-54; mathematician, Rand Corp, 54-55; asst prof math, Purdue Univ, 55-58; from asst prof to assoc prof, 58-63, chmn dept, 65-68, 82-85, PROF MATH & APPL MATH, BROWN UNIV, 63- *Concurrent Pos:* Mem staff, Math Res Ctr, Univ Wis, 62-63; NSF fel, 68-69; Guggenheim fel, 76-77; vis prof, Mass Inst Technol, 80 & Univ Minn, 85. *Honors & Awards:* Steele Prize, Am Math Soc, 87. *Mem:* Soc Indust & Appl Math; Am Math Soc; Math Asn Am. *Res:* Stochastic control; calculus of variations; stochastic differential equations; population genetics theory. *Mailing Add:* Dept Math Brown Univ Providence RI 02912

FLEMING, WILLIAM HERBERT, b Galt, Ont, Sept 3, 25; m 45; c 1. PHYSICS. *Educ:* McMaster Univ, BSc, 50, MSc, 51, PhD(physics), 54. *Prof Exp:* Fel, 54-58, reactor supt, 58-69, ASSOC PROF APPL MATH, MCMASTER UNIV, 66-, MGR COMPUT SYSTS & PROG, 69- *Mem:* Am Nuclear Soc; Can Asn Physicists. *Res:* Nuclear reactor operation; digital computer programming systems. *Mailing Add:* 2121 Junction No 305 Burlington ON L7R 1C9 Can

FLEMING, WILLIAM LEROY, b Morgantown, WVa, Aug 29, 05; m 34; c 4. PREVENTIVE MEDICINE. *Educ:* Vanderbilt Univ, BA, 25, MS, 27, MD, 32. *Prof Exp:* Asst bact, Sch Med, Vanderbilt Univ, 25-28; intern & house physician, Bellevue Hosp, New York, 33-34; asst resident & resident physician med serv, Vanderbilt Univ Hosp, 34-37; Milbank fel syphilis clin, Johns Hopkins Hosp, 37-39, instr med, Sch Med, Johns Hopkins Univ, 37-39; mem staff internal health div, Rockefeller Found, 39; res prof syphil, Sch Pub Health, Univ NC, 39-45; assoc prof med, Sch Med, Boston Univ, 46-48, prof prev med & chmn dept, 48-52; chmn dept, 52-70, asst dean, 57-70, prof prev med, 52-78, EMER PROF PREV MED, SCH MED, UNIV NC, CHAPEL HILL, 78- *Concurrent Pos:* Mem staff, Evans Mem Hosp, Boston, 46-52; vis physician, Mass Mem Hosp, 46-52, chief genito infectious dis clin, 46-52; dir gen clin, NC Mem Hosp, 52-64; vis prof & consult prev med, Paulista Sch Med, Sao Paulo, 62; serv with Proj Hope, Ceylon, 68-69, Jamaica, 71, Natal, Brazil, 72, Maceio, Brazil, 73, Egypt & Tunisia, 77. *Mem:* Am Soc Clin Invest; Am Pub Health Asn; Am Venereal Dis Asn (pres, 54-55); Asn Teachers Prev Med (pres, 59-60); Am Col Prev Med. *Res:* Research bacteriology; venereal disease control; internal medicine; clinical experimental syphilology. *Mailing Add:* Dept Fam Med Univ NC Sch Med Chapel Hill NC 27514

FLEMING, WILLIAM WRIGHT, b Washington, DC, Jan 30, 32; m 52; c 3. PHARMACOLOGY. *Educ:* Harvard Univ, AB, 54; Princeton Univ, PhD(biol), 57. *Prof Exp:* NIH res fel pharmacol, Harvard Med Sch, 57-60; from asst prof to assoc prof, 60-66, prof pharmacol & toxicol & head dept, 66-86, PROF & MYLAN CHMN, MED SCH, WVA UNIV, 86- *Concurrent Pos:* Vis prof, Univ Melbourne, 69 & St George's Hosp Med Sch, Univ London, 78; specific field ed, Autonomic Pharmacol, 69-73; mem Pharmacol-Toxicol Prog Comt, Nat Inst Gen Med Sci, 73-77, chmn, 75-77; Fogarty Sr Int Fel, 78, NIH postdoctoral fel; rev comt, Drug Abuse Biomed Res, Nat Inst Drug Abuse, 86-90; vis prof, Univ Adelaide, 87, Flinders Univ, 87, Georgetown Univ, 88; mem, pharmacol study sect, NIH, 90-94. *Honors & Awards:* Otto Krayer Res Award, Am Soc Pharmacol & Exp Therapeut, 86. *Mem:* AAAS; Am Heart Asn; Am Soc Pharmacol & Exp Therapeut (counr, 75-78, pres-elect, 80-81, pres, 81-82); Fedn Am Soc Exp Biol; Asn Med Sch Pharmacol (exec comt, 77-79, treas, 78, pres, 86-88). *Res:* Autonomic and cardiovascular pharmacology; electrophysiology; cellular control of sensitivity. *Mailing Add:* Dept Pharmacol & Toxicol WVa Univ Med Ctr Morgantown WV 26506

FLEMINGER, ABRAHAM, invertebrate zoology; deceased, see previous edition for last biography

FLEMINGS, MERTON CORSON, b Syracuse, NY, Sept 20, 29; m; c 4. METALLURGY, MATERIALS ENGINEERING. *Educ:* Mass Inst Technol, SB, 51, SM, 52, ScD(metall), 54. *Prof Exp:* Metallurgist, Am Brake Shoe Res Lab, 54-56; from asst prof to assoc prof metall, Mass Inst Technol 56-69, Abex prof, 70-75, assoc dir, Ctr Mat Sci & Eng, 73-77, Ford Prof, 75-81, dir, Mat Processing Ctr, 79-82, PROF METALL, MASS INST TECHNOL, 69-, TOYOTA PROF MAT PROCESSING, 81- *Concurrent Pos:* Consult, var govt labs & industs, 70-; overseas fel, Churchill Col, Eng, 70-71; vis prof, Univ Cambridge, 70-71 & Univ Tokyo, 89; head, Dept Math Sci & Eng, 82-; mem & chmn var comts, govt agencies & nat socs; hon res prof, Academia Sinica, China, 88. *Honors & Awards:* Simpson Gold Medal, Am Foundrymen's Soc, 61; Mathewson Gold Medal, Am Inst Mining, Metall & Petrol Engrs, 69, James Douglas Gold Medal, 85; Henry Marion Howe Medal, Am Soc Metals Int, 73 & 90, Albert Sauveur Achievement Award, 78, Edward DeMille Campbell Mem Lectr, 90. *Mem:* Nat Acad Eng; Am Foundrymen's Soc; fel Am Soc Metals; Inst Metals London; Am Inst Metall Engrs; AAAS. *Res:* Materials processing; solidification processing; foundry and ingot-making; technical education; innovation in industrial research; primary materials production; author of 244 technical publications; awarded 24 patents. *Mailing Add:* Bldg Eight Rm 309 Mass Inst Technol Cambridge MA 02139

FLEMISTER, LAUNCELOT JOHNSON, b Atlanta, Ga, Dec 11, 13; m 41. PHYSIOLOGY. *Educ:* Duke Univ, AB, 35, MA, 39, PhD(physiol), 41. *Prof Exp:* Asst physiol, Duke Univ, 38-41; instr pharmacol, Sch Med, George Washington Univ, 41-43; res assoc, Sharp & Dohme, Philadelphia, 46-47; from asst prof to assoc prof, 47-66, PROF ZOOL, SWARTHMORE COL, 66- *Concurrent Pos:* Fulbright fel, Peru, 59-60; consult, NSF Facil Prog, 63-64; served to Lieut USNR physiol, 42-46. *Mem:* Fel AAAS; Am Physiol Soc; Am Soc Zool; Ecol Soc Am; Sigma Xi. *Res:* Comparative aspects of water balance and metabolism; environmental adaptation. *Mailing Add:* PO Box F Swarthmore PA 19081-0368

FLEMISTER, SARAH C, b Nashville, Tenn, Aug 31, 13; m 41. ANIMAL PHYSIOLOGY. *Educ:* Duke Univ, AB, 34, MA, 36, PhD(physiol), 38. *Prof Exp:* Asst zool, Duke Univ, 34-38, instr anat & physiol, 38-42; res asst physiol, Swarthmore Col, 46-48, instr biol, 48-52, lectr, 52-59; Fulbright lectr physiol, Peru, 59-60; lectr biol, Bryn Mawr Col, 60-64; RETIRED. *Concurrent Pos:* Researcher, Air Force Contract, Swarthmore Col, 46-47 & US Navy Contract, 47-48. *Mem:* Fel AAAS; Am Soc Zoologists; NY Acad Sci; Sigma Xi. *Res:* Anaerobics of fresh water mussel; micro blood gas methods; respirometry of human subjects; water and ion balance in marine crustacea; invertebrate physiology. *Mailing Add:* PO Box F Swarthmore PA 19081-0368

FLESCH, DAVID C, b Lancaster, Wis, Aug 13, 44; m 68; c 3. CELL BIOLOGY. *Educ:* Iowa State Univ, PhD(molecular & cellular & develop biol), 77. *Prof Exp:* PROF BIOL, MANSFIELD UNIV, 77-, DEPT CHAIR, 90- *Concurrent Pos:* Pres, Commonwealth Pa Univ Biologists, 89-90. *Mem:* Am Soc Cell Biol; Am Inst Biol Sci; AAAS; Nat Sci Teachers Asn. *Mailing Add:* Dept Biol Grant Sci Ctr Mansfield Univ Mansfield PA 16933

FLESHER, JAMES WENDELL, b Chicago, Ill, June 24, 25; m 52; c 3. PHARMACOLOGY. *Educ:* Northwestern Univ, BS, 49; Loyola Univ Chicago, PhD(pharmacol), 58. *Prof Exp:* Control chemist edible fats & oils, Lever Bros Co, 51-55; res assoc, Ben May Lab Cancer Res, Chicago, 58-62; asst prof, 62-67, assoc prof, 67-80, PROF PHARMACOL, UNIV KY, 80- *Concurrent Pos:* Vis prof, Inst Cancer Res, Columbia Univ, 70-71, Chester Beatty Res Inst, Royal Cancer Hosp, London, 83. *Mem:* Soc Toxicol; Am Asn Cancer Res; Am Chem Soc; Am Soc Pharmacol & Exp Therapeut. *Res:* Chemical carcinogenesis; steroid hormones; mechanism of drug action; bio-transformation of unsubstituted aromatic hydrocarbon preprocarcinogens to alkyl substituted procarcinogens especially meso-hydroxyalkyl metabolite formation and subsequent possible activation via electrophilic esters; development of drugs to inactivate and eliminate procarcinogens and the active metabolites. *Mailing Add:* Dept Pharmacol Univ KY Col Med Lexington KY 40536

FLESHLER, BERTRAM, b New York, NY, May 1, 28; m 56; c 2. GASTROENTEROLOGY. *Educ:* Univ Wis, AB, 48; Boston Univ, MD, 51. *Prof Exp:* Intern med, Mass Mem Hosp, 51-52; asst resident, Georgetown Univ Hosp, 52-53; asst resident, Mt Alto Vet Admin Hosp, 55-56; fel gastroenterol, Mass Mem Hosp, 56-58; from sr instr to assoc prof, 58-72, ASSOC CLIN PROF MED, SCH MED, CASE WESTERN RESERVE UNIV, 72-; SR STAFF GASTROENTEROL & DIR TRAINING PROG GASTROENTEROL, CLEVELAND CLIN FOUND, 76- *Concurrent Pos:* Assoc vis physician, Cleveland Metrop Gen Hosp, 58-59, dir gastroenterol, 58-72, vis physician, 59-; consult, Vet Admin Hosp, 58- & Lutheran Hosp, 63-; NIH res career develop award, 66-71; hon sr lectr & consult, Kings Col Hosp Med Sch, London, Eng, 66-69; dir dept med, St Vincent Charity Hosp, Cleveland, 72-75. *Mem:* AAAS; Soc Exp Biol in Med; Am Fedn Clin Res; Am Gastroenterol Asn; fel Am Col Physicians. *Res:* Esophageal motility studies; amino acid absorption. *Mailing Add:* Dept Gastroenterol Case Western Res Univ Cleveland Clin Cleveland OH 44106

FLESHMAN, JAMES WILSON, JR, SYNAPTIC CIRCUITS, SENSORIMOTOR. *Educ:* Univ Fla, PhD(med sci), 80. *Prof Exp:* ASST RES SCIENTIST, SYSTS RES CTR, UNIV MD, 86- *Res:* Recording technology. *Mailing Add:* NIH Lab Neural Control Bldg 36 Rm 4A29 Bethesda MD 20205

FLESSA, KARL WALTER, b Nuremberg, Ger, Aug 3, 46. PALEONTOLOGY. *Educ:* Lafayette Col, AB, 68; Brown Univ, PhD(geol sci), 73. *Prof Exp:* Asst prof earth & space sci, State Univ NY Stony Brook, 72-77; from asst prof to assoc prof, 77-87, PROF GEOSCI, UNIV ARIZ, 87-; PROG DIR, NSF, 88-90. *Concurrent Pos:* Humboldt Fel, Univ Tuebingen, Fed Repub Ger, 83-84. *Mem:* Paleont Asn; Geol Soc Am; Soc Econ Paleontologists & Mineralogists; Paleont Soc. *Res:* Quantitative analysis of taxonomic diversity, evolutionary rates and morphology; biogeography; taphonomy. *Mailing Add:* Dept Geosci Univ Ariz Tucson AZ 85721

FLETCHALL, OSCAR HALE, b Grant City, Mo, May 4, 20; m 50. WEED SCIENCE. *Educ:* Univ Mo, BS, 42, PhD(field crops), 54. *Prof Exp:* Asst field crops, USDA & Univ Mo, 38-42, asst instr & sci aide field crops, 47-51, instr field crops & agronomist, 52-54; from asst prof to assoc prof, 54-61, prof agron, 61-84, EMER PROF AGRON, UNIV MO, COLUMBIA, 84- *Concurrent Pos:* Consult on agr, lawyers. *Mem:* Am Soc Agron; fel Weed Sci Soc Am. *Res:* Development of more effective, more efficient, safer methods of controlling weeds in agronomic crops and on non-crop land; farmer. *Mailing Add:* Rte 13 4751 Hwy WW Columbia MO 65201

FLETCHER, AARON NATHANIEL, b Los Angeles, Calif, Dec 24, 25; m 51, 90; c 5. PHYSICAL CHEMISTRY, ANALYTICAL CHEMISTRY. *Educ:* Calif Inst Technol, BS, 49; Univ Calif, Los Angeles, PhD(anal chem), 61. *Prof Exp:* Lab technician, Southern Pac Co, 49-50, chemist, 50-54; res chemist, Naval Weapons Ctr, China Lake, 54-75, head gen res br, 75-78, head energy chem br, 78-86; RETIRED. *Mem:* Am Chem Soc. *Res:* Dye laser systems; photoluminescence; self-association of alcohols; chemiluminescence; photogalvanic and thermal batteries; quantitative spectrophotometry; electroanalytical chemistry. *Mailing Add:* PO Box 2010 Sparks NV 89432

FLETCHER, ALAN G(ORDON), b Gibson's Landing, BC, Jan 2, 25; m 49; c 4. FLUID MECHANICS, WATER RESOURCES. *Educ:* Univ BC, BASc, 48; Calif Inst Technol, MSc, 52; Northwestern Univ, PhD(civil eng), 65. *Prof Exp:* Engr-in-training, BC Elec Co, Ltd, 48-51, hydraul design engr, 52-56, supvr hydro-planning, 56-59; asst prof civil eng, Univ Idaho, 59-60, assoc prof, 60-62; assoc prof, Univ Utah, 64-69; dean, Sch Eng & Mines, Univ NDak, 68-89; RETIRED. *Concurrent Pos:* Danforth Assoc, 65. *Mem:* Am Soc Civil Engrs; Am Soc Eng Educ; Nat Soc Prof Engrs; Sigma Xi. *Res:* Hydraulic engineering; flood waves in natural channels. *Mailing Add:* Three Connie Lane Bella Vista AR 72714

FLETCHER, ANTHONY PHILLIPS, b Maidenhead, Eng, Feb 25, 19; nat US; m 61. INTERNAL MEDICINE. *Educ:* Univ London, MB, BS, 43, MD, 49. *Prof Exp:* Lectr human physiol, St Mary's Hosp Med Sch, London, 47-48; sr med registr, St Mary's Hosp & Wright-Fleming Inst Microbiol, 50-53; res assoc med, NY Univ, 54-56; asst prof, 56-62, ASSOC PROF MED, SCH MED, WASH UNIV, 62- *Concurrent Pos:* Merck Inst fel microbiol, Col Med, NY Univ, 53-54; asst attend physician, Barnes Hosp Group, 56- *Mem:* Am Physiol Soc; Am Soc Clin Invest; Am Fedn Clin Res. *Res:* Physiological fibrinolysis and the development of enzymatic methods for the treatment of thrombo-embolic vascular disease; blood coagulation and plasma proteins. *Mailing Add:* Clin Labs Vet Admin Med Ctr Jefferson Barracks St Louis MO 63125

FLETCHER, BARRY DAVIS, b London, Can, June 25, 35; m 78; c 3. DIAGNOSTIC RADIOLOGY, PEDIATRIC RADIOLOGY. *Educ:* Univ Western Ont, BA, 57; McGill Univ, Can, MD & CM, 61. *Prof Exp:* Radiologist & instr radiol, Johns Hopkins Hosp, 66-67, asst prof, 67; asst prof, McMaster Univ, 67-69; asst prof diag radiol, McGill Univ, Can, 69-74, assoc prof, 74-76; PROF RADIOL & PEDIAT, CASE WESTERN RESERVE UNIV, 76-; CHMN DIAGNOSTIC IMAGING, ST JUDE CHILDREN'S RES HOSP, MEMPHIS, PROF RADIOL, UNIV TENN. *Concurrent Pos:* Radiologist, Montreal Childrens Hosp, 69-76; dir, Div Pediat Radiol, Univ Hosp Cleveland, 76-87. *Mem:* Am Col Radiol; Radiol Soc NAm; Soc Pediat Radiol; Can Asn Radiologists; Asn Univ Radiologists. *Res:* Pulmonary maturation and development in neonates and infants using radiologic techniques with clinical research in pediatric radiology; applications of new technologic methods (digital radiography and nuclear magnetic resonance) to pediatric diagnosis; congenital heart disease and oncology. *Mailing Add:* Univ Hosp Cleveland OH 44106

FLETCHER, DEAN CHARLES, biochemistry, nutrition; deceased, see previous edition for last biography

FLETCHER, DONALD JAMES, b Atlanta, Ga, Mar 21, 51; m 74; c 1. ENDOCRINOLOGY, CELL BIOLOGY. *Educ:* Ga Inst Technol, BS, 73; Emory Univ, PhD(biol), 77. *Prof Exp:* Res fel cell biol, Dept Anat, Sch Med, Emory Univ, 77-79; res fel physiol, dept med, Med Col Va, 80-; AT DEPT ANAT, ECAROLINA UNIV MED SCH. *Concurrent Pos:* Nat Arthritis, Metab & Digestive Dis trainee grant, 77-79. *Mem:* Am Soc Zoologists; Am Soc Cell Biol; Am Diabetes Asn. *Res:* Separation of dispersed islet cells and characterization of secretion and attachment of those cells; parcrine aspects of hormone secretion in situ and in vitro. *Mailing Add:* Dept Anat & Cell Biology ECarolina Univ Med Sch Greenville NC 27858

FLETCHER, DONALD WARREN, b Phoenix, Ariz, June 8, 29; m 63; c 2. MICROBIAL ECOLOGY. *Educ:* Ore State Univ, BS, 51, MS, 53; Wash State Univ, PhD(bact), 56. *Prof Exp:* Instr bact, Wash State Univ, 56-59; assoc prof biol, San Francisco State Univ, 59-67, exec dir, Ctr Advan Med Technol & assoc dean, Sch Natural Sci, 70-74, prof biol, 67-75; dep dean extended educ, Off Chancellor, Calif State Univ, Long Beach, 75-; RETIRED. *Concurrent Pos:* Lectr & consult, Sch Med, Univ Calif, San Francisco, 62-64; Fulbright lectr, Univ Belgrade, 67-68; dean col arts & sci, Univ Bridgeport, 69-70; asst dir health manpower educ proj, Calif State Univ, 75- *Mem:* Fel AAAS; Am Soc Microbiol; Brit Soc Gen Microbiol; Sigma Xi. *Res:* Microbial physiology; bacterial ecology; rumen and soil microbiology; bacterial nutrition; microbiology of the gut of herbivorous fish; allied health nursing manpower; adult education and learning. *Mailing Add:* PO Box 636 Greenville CA 95947

FLETCHER, EDWARD A(BRAHAM), b Detroit, Mich, July 30, 24; m 48; c 3. SOLAR ENERGY, THERMOCHEMISTRY. *Educ:* Wayne State Univ, BS, 48; Purdue Univ, PhD(inorg chem), 52. *Prof Exp:* Aeronaut res scientist, NASA, 52-56, head flame mech sect, 56-57, head propellant chem sect, 57-59; assoc prof, 59-60, dir grad studies, 71-86, PROF MECH ENG, UNIV MINN, MINNEAPOLIS, 60- *Concurrent Pos:* Vis exchange scientist, Byelorussian Acad Sci, 64; vis prof, Univ Poitiers, 68; mem comt waste heat mgt, Dept Com, 74-75; co-chmn safety hydraul comn fluids, Nat Res Coun, Nat Acad Sci, 77-78; Fulbright sr res scientist, Weizmann Inst Sci, 89-90, vis prof, Rehovot, 91- *Honors & Awards:* Fulbright Lectr, Univ Ankara, 89. *Mem:* Am Chem Soc; Combustion Inst (secy, 68-69, vchmn, 78-, chmn, 80-82); AAAS; Am Solar Energy Soc; Int Solar Energy Soc. *Res:* Combustion; ignition; chemical kinetics; thermochemical solar energy storage; electrochemistry; fluidized bed combustion. *Mailing Add:* Univ Minn 111 Church St SE Minneapolis MN 55455

FLETCHER, EDWARD ROYCE, b Hays, Kans, May 20, 37. BIODYNAMICS. *Educ:* Univ NMex, BS, 58, MS, 60, PhD(physics), 64. *Prof Exp:* Electronic technician, Lovelace Biomed & Environ Res Inst, 56, math analyst, 57-60, physicist, 61-63, head theoret anal sect, Dept Physics, 64-70, head, Dept Physics, 71-86; CONSULT BIODYNAMICS, 86- *Mem:* Inst Elec & Electronics Engrs. *Res:* Analysis of biological and physical systems in terms of the physical process and the development of mathematical models to simulate these systems. *Mailing Add:* 3209 Madria Dr NE Albuquerque NM 87110

FLETCHER, FRANK WILLIAM, b Camden, NJ, Oct 7, 37; m 60; c 3. GEOLOGY. *Educ:* Lafayette Col, BA, 59, Univ Rochester, PhD(geol), 64. *Prof Exp:* Eng, Aide Soils Div, NJ State Hwy Dept, 58-59; asst geol, Univ Rochester, 59-62; geologist, NY State Mus & Sci Serv, 59-64; from instr to prof geol, Susquehanna Univ, 62-89, dir, Inst Environ Studies, 70-79, asst dean fac, 81-83, dean arts & sci, 83-85, C B DEGENSTEIN DISTINGUISHED PROF ENVIRON SCI, SUSQUEHANNA UNIV, 90- *Concurrent Pos:* Cooperating geologist, Pa Geol Surv, 66- *Mem:* Asn Groundwater Scientists & Engrs; Sigma Xi; Hazardous Mat Res Inst. *Res:* Groundwater hydrology; hazardous wastes; geological hazards. *Mailing Add:* Geol Dept Susquehanna Univ Selingsgrove PA 17870

FLETCHER, FREDERICK BENNETT, metallurgy, for more information see previous edition

FLETCHER, GARTH L, b Glasgow, Scotland, Apr 15, 36; Can citizen; m 64; c 2. ANIMAL PHYSIOLOGY. *Educ:* Univ BC, BSc, 63; Univ Calif, Santa Barbara, PhD(biol), 67. *Prof Exp:* Res scientist, Halifax Lab, Fisheries Res Bd, Can, 67-70, Marine Ecol Lab, Bedford Inst, 70-71; res scientist, Marine Sci Res Lab, 71-87, PROF, OCEAN SCI CENTRE, MEM UNIV NFLD, 88- *Mem:* Sigma Xi; Am Soc Zoologists; Can Soc Zoologists. *Res:* Water and electrolyte regulation in birds; steroid hormones in fish; toxicology of elemental phosphorus; mechanisms controlling heavy metal levels in fish; red blood cell production and senescence in fish; regulation of protein anti-freeze levels in fish blood; gene transfer in fish. *Mailing Add:* Marine Sci Res Lab Mem Univ St John's NF A1C 5S7 Can

FLETCHER, HARRY HUNTINGTON, organic polymer chemistry; deceased, see previous edition for last biography

FLETCHER, HARVEY JUNIOR, b New York, NY, Apr 9, 23; m 53; c 6. APPLIED MATHEMATICS. *Educ:* Mass Inst Technol, BS, 44; Calif Inst Technol, MS, 48; Univ Utah, PhD(math), 54. *Prof Exp:* Instr physics, Univ Utah, 53; from instr to prof math, Brigham Young Univ, 54-63, chmn dept, 58-61 & 62-63; mem tech staff, Bellcom, 63-64; prof math, Brigham Young Univ, 64-74; prof, Staten Island Community Col, 74-75; mem staff, Eyring Res Inst, 75-80; mem staff, Brigham Young Univ, 80-90; RETIRED. *Concurrent Pos:* Mem tech staff, Bell Labs, 61-62 & 73-74; sr tech specialist, Hercules, Inc, 67-68. *Mem:* Math Asn Am. *Res:* Bending of plates; Fourier series; attitude of satellite; Apollo trajectories; diffusion theory; minuteman simulation; intercept trajectories; delta modulation coding; beta battery; EM diffraction. *Mailing Add:* Dept Math Brigham Young Univ Provo UT 84602

FLETCHER, JAMES CHIPMAN, b Millburn, NJ, June 5, 19; m 46; c 4. SPACE & GUIDED MISSILES. *Educ:* Columbia Univ, AB, 40; Calif Inst Technol, PhD(physics), 48. *Hon Degrees:* Various from Univ Utah, Brigham Young Univ & Lehigh Univ. *Prof Exp:* Res physicist, Bur Ord US Navy, 40-41; spec res assoc, Cruft Lab, Harvard Univ, 41-42; instr, Princeton Univ, 42-45; teaching fel, Calif Inst Technol, 45-48; instr, Univ Calif, Los Angeles, 48-50; dir theory & anal lab, Hughes Aircraft Co, 48-54; from assoc dir guided missile lab to dir electronics, guided missile res div, Space Tech Labs, Ramo-Wooldridge Corp, 54-58; pres, Space Electronics Corp, 58-60; pres, Space-Gen Corp, 60-62, chmn bd, 60-64, vpres systs, Aerojet-Gen Corp, 62-64; pres, Univ Utah, 64-71; adminr, NASA, 71-77; dir & mem gov bd, Amoco, 77-86; adminr, NASA, 86-89; Whiteford Prof Technol & Energy Resources, Univ Pittsburgh, 77-; RETIRED. *Concurrent Pos:* Mem subcomt stability & control, Nat Adv Comt Aeronaut, 50-54; consult & mem, President's Sci Adv Comt, 58-70, mem strategic weapons panel, 59-61, command, control & intel panel, 62-63 & mil aircraft panel, 64-67; chmn ad hoc comt rev, Minuteman Command & Control Syst, 61; mem command, control & intel comt, US Dept Defense, 61-62; mem, Woods Hole Summer Study Group Arms Control, 62; consult, Arms Control & Disarmament Agency, 62-64; mem, Air Force sci adv bd, 63, Pan Am adv bd, 77-, technol adv assessment comn, US Cong Off Tech Assessment, 78-, Defense Sci Bd, 81- & nat adv comt, Gas Res Inst, 84-; mem, gov bd, Astrotech Int, 77-, Comarco, 77-, Fairchild Industs, 77-, Nat Res Coun, 78-84, exec comt, 81-85 & Argonne Nat Lab, 84-; chmn, energy res adv bd, Dept Energy, 78-83 & TMI 2 safety adv bd, 81-; mem, gov coun, Nat Acad Eng, 78-85; trustee, Rockefeller Found, 78-85 & Univ Corp Atmospheric Res, 82-; bd dirs, Raytheon Corp, 89-, Fairchild Indust, 89-, Evans & Sutherland Inc, 89- *Honors & Awards:* Arthur Beuche Award, Nat Acad Eng, 90. *Mem:* Nat Acad Eng; fel Inst Elec & Electronics Engrs; Am Astron Soc; hon fel Am Inst Aeronaut & Astronaut; fel Am Acad Arts & Sci. *Res:* Underwater acoustics; aerodynamics; shock waves; cosmic rays; magnetic survey of naval vessels; servomechanisms; radar; guided missiles and space electronics; guidance; instrumentation; adminstration of large scale research and development programs; communications, systems engineering, space technology and energy resources. *Mailing Add:* 7721 Falstaff Rd McLean VA 22102

FLETCHER, JAMES ERVING, b Logan, Utah, Nov 21, 35; m 63; c 3. PSYCHOPHYSIOLOGY OF COMMUNICATION, IMPACT OF MEDIA UPON AUDIENCES. *Educ:* Univ Ariz, BA, 57; Univ Utah, PhD (speech telecommun), 71. *Prof Exp:* Vis asst prof speech, Univ Utah, 70-71; asst prof telecommun, Univ Ky, 71-73; from asst prof to assoc prof, 74-84, PROF JOUR, UNIV GA, 84- *Concurrent Pos:* Chair, Mass Commun Div, Speech Commun Asn, 84-85; mem, Radio Audience Measurement Task Force, Nat Asn Broadcasters, 85-87; assoc ed, J Broadcasting & Electronic Media, 85-; commun adv, Int Orgn Psychophysiol 86-; res consult, Church Jesus Christ Latter-Day Saints, 87-; tech dir, Inner Response, Inc, 87-; legis coun, Speech Commun Asn, chair, pub rel comt, 79-83. *Mem:* Soc Psychophysiol Res; Int Organ Psychophysiol; Speech Commun Asn; Int Commun Asn (co-secy, 73-75). *Res:* Psychophysiological effects of being exposed to mass communication, including effects of popular music, commercials and entertainment programs; audience measurement of audiences of radio, television, cable television and motion pictures. *Mailing Add:* Col Jour & Mass Commun Univ Ga Athens GA 30602

FLETCHER, JAMES W, b Belleville, Ill, Oct 6, 43; m 67; c 3. NUCLEAR MEDICINE, INTERNAL MEDICINE. *Educ:* St Louis Univ, MD, 68. *Prof Exp:* Intern med, Sch Med, St Louis Univ, 68-69, asst resident, 69-70, resident fel nuclear med, 70-71; resident fel nuclear med, Harvard Med Sch, 71-72; from asst prof to assoc prof med, Sch Med, St Louis Univ, 72-83, assoc prof radiol, 78-84; asst chief, 76-79, CHIEF NUCLEAR MED, ST LOUIS VET ADMIN HOSP, 79-, DIR OPERS MAGNETIC RESONANCE, 83-; PROF MED, SCH MED, ST LOUIS UNIV, 83-, PROF RADIOL, 84-,. *Concurrent Pos:* Staff physician, St Louis Univ Hosp, 72-, actg dir, div nuclear med, 85-88, med dir, Nuclear Med Dept, 89- *Mem:* Am Fedn Clin Res; Soc Nuclear Med; AAAS; AMA; Sigma Xi; Radiol Soc NAm; Am Col Nuclear Physicians; Soc Magnetic Resonance Med. *Res:* Clinical investigation in nuclear medicine; medical applications; nuclear magnetic resonance; health care technology, assessment, diffusion and adoption; evaluation of diagnostic health care services, cost-effectiveness, efficacy; biomedical applications of radionuclides and nuclear magnetic resonance, radiologic sciences. *Mailing Add:* St Louis Med Ctr 3635 Vista Ave at Grand Blvd PO Box 15250 St Louis MO 63110-0250

FLETCHER, JOEL EUGENE, soils, for more information see previous edition

FLETCHER, JOHN EDWARD, b Banner Elk, NC, June 12, 37; m 64; c 2. APPLIED MATHEMATICS. *Educ:* NC State Univ, BS, 59, MS, 61; Univ Md, PhD(math), 72. *Prof Exp:* Res asst differential equations, NC State Univ, 59-61; sr analyst opers res, Adv Concepts Sect, Adv Studies Div, Lockheed-Ga Co, 64-66; res mathematician, 66-69, chief, appl math sect, 69-82, CHIEF, LAB APPL STUDIES, DIV COMPUT RES & TECHNOL, NIH, 82- *Concurrent Pos:* Chmn math & comp sci fac, Found Advan Educ in Sci, 75- *Mem:* Soc Indust & Appl Math; Int Soc Oxygen Transport to Tissue; Found Advan Educ Sci. *Res:* Application of deterministic mathematical models to problems of biological sciences; mathematical simulation models in biology and medicine; numerical methods; data fitting and analysis methods. *Mailing Add:* 12A 2041 Div Comput Res & Technol 9000 Rockville Pike Bethesda MD 20892

FLETCHER, JOHN GEORGE, b Aberdeen, SDak, Oct 28, 34; wid; c 3. COMPUTING NETWORKS, DISTRIBUTED COMPUTING. *Educ:* George Wash Univ, BS, 55; Princeton Univ, AM, 57, PhD(physics), 59. *Prof Exp:* Mem tech staff, Bell Tel Labs, 59; instr, Princeton Univ, 59-60; PHYSICIST, LAWRENCE LIVERMORE LAB, UNIV CALIF, 61- *Concurrent Pos:* Consult, Grumman Aircraft Eng Corp, 59-63 Ed Serv, Inc, 61-63; fel, Miller Inst, 60-62; lectr, Univ Calif, Davis & Univ Calif, Berkeley, 62-; mem comt, Am Nat Standards Inst, 70-73. *Mem:* Fel Am Phys Soc; Asn Comput Mach. *Res:* Computer networks and operating systems; symbol manipulation; relativity; automata. *Mailing Add:* L-60 Lawrence Livermore Lab Univ Calif Box 808 Livermore CA 94550-0662

FLETCHER, JOHN LYNN, b Springdale, Ark, Apr 18, 25; m 49; c 2. FISH & WILDLIFE SCIENCES, BEHAVIORAL ETHOLOGY. *Educ:* Univ Ark, Fayetteville, BA, 50, MA, 51; Univ Ky, Lexington, PhD(psychol), 55. *Prof Exp:* From lieutenant to lieutenant colonel, US Army Med Res Lab, 55-70; prof psychol, Memphis State Univ, 70-75 & Dept Otolaryngol,Ctr Health Servs, Univ Tenn, 75-80; PROF PSYCHOL, DEPT PSYCHOL, UNIV MO-ROLLA, 81-; LECTR PSYCHOL, SOUTHWESTERN TEX STATE UNIV. *Concurrent Pos:* Mem Comt Hearing & Bioacoust, Nat Res Coun-Nat Acad Sci, 55- & Int Comn Biol Effects Noise, 68-; consult, US Environ Protection Agency, 70-74, US Dept Labor, 71-; chair, Am Nat Standards Inst Comt High Frequency Audiom, 84-, Int Standards Orgn Study Group High Frequency Audiom, 85- *Mem:* Fel Acoust Soc Am; fel Am Speech, Lang & Hearing Asn; NY Acad Sci; Psychonomic Soc; Human Factors Soc. *Res:* Effects of noise on human hearing and performance. *Mailing Add:* Dept Psychol Southwestern Tex State Univ San Marcos TX 78666

FLETCHER, JOHN SAMUEL, b Columbus, Nebr, Jan 7, 38; m 60; c 4. CELL BIOLOGY. *Educ:* Ohio State Univ, RSE, 60; Ariz State Univ, MNS, 65; Purdue Univ, PhD(plant physiol), 69. *Prof Exp:* From asst prof, 69-82, PROF BOT & MICROBIOL, UNIV OKLA, 82- *Mem:* Am Soc Plant Physiologists; Am Inst Biol Sci. *Res:* Metabolism of environmental waste compounds by plants; growth and development of plant tissue cultures; fate of xenobiotic compounds in the environment; plant metabolism. *Mailing Add:* Dept Bot Univ Okla Main Campus Norman OK 73019

FLETCHER, KENNETH STEELE, III, b Springfield, Mass, May 4, 41; m 63; c 2. ANALYTICAL CHEMISTRY. *Educ:* Trinity Col, Conn, BS, 63; Univ Mass, Amherst, PhD(chem), 68. *Prof Exp:* RES CHEMIST, FOXBORO CO, 68- *Mem:* Am Chem Soc. *Res:* Solid state ionic conductivity; ion selective electrode materials; fiber optics sensors; voltammetric hydrodynamics; process analytical chemistry. *Mailing Add:* Foxboro Co 600 N Bedford St East Bridgewater MA 02333

FLETCHER, LEROY S(TEVENSON), b San Antonio, Tex, Oct 10, 36; m 66; c 2. MECHANICAL & AEROSPACE ENGINEERING. *Educ:* Tex A&M Univ, BS, 58; Stanford Univ, MS, 63, Engr, 64; Ariz State Univ, PhD(mech eng), 68. *Prof Exp:* Aeronaut engr, Ames Aeronaut Lab, Nat Adv Comt Aeronaut, 58-60, aerospace engr, Ames Res Ctr, NASA, 61-63; asst heat transfer, dept mech eng, Stanford Univ, 62-63, asst thermodynamics, 63-64; asst mech eng, Ariz State Univ, 64-65, instr, 65-68; from asst prof to prof aerospace eng, Rutgers Univ, 68-75, actg assoc dean, 74-75; prof & chmn mech eng dept, Univ Va, 75-77, prof mech & aero eng & chmn dept, 77-80; assoc dean, Col Eng, 80-88, assoc dir, Tex Eng Exp Sta, 85-88, THOMAS A DIETZ PROF, MECH ENG DEPT, TEX A&M UNIV, 88- *Concurrent Pos:* Mem President's adv comt energy conservation, Univ Va, 75-80; mem governor's comt conservation energy resources, Commonwealth Va, 76-80; bd dirs, Am Soc Eng Educ, 74-77 & 78-80, Am Inst Aeronaut & Astronaut, bd govs, Am Soc Mech Engrs, 83-87; series ed, Mechanics & Mech Eng, 81-, assoc tech ed, J Heat Transfer, 89-; hon prof, Ruhr Univ, Bochum, Ger, 88- *Honors & Awards:* Ralph R Teeter Award, Soc Automotive Engrs, 70; Centennial Medallion, Am Soc Mech Engrs, 80, Charles Russ Richards Mem Award, 82, Dedicated Serv Award, 88; George Westinghouse Award, Am Soc Eng Educ, 81, Ralph Coats Roe Award, 83, Donald E Marlowe Award, 86; Energy Systs Award, Am Inst Aeronaut & Astronaut, 84; Lee Atwood Award, Int Acad Astronaut, 82. *Mem:* Fel Am Soc Mech Engrs (pres, 85-86); fel Am Inst Aeronaut & Astronaut; fel Am Soc Eng Educ (vpres, 79-80); fel AAAS; fel Am Astronaut Soc; fel Inst Mech Engrs UK. *Res:* Heat transfer; conduction, convection, radiation; aerothermodynamics; energy conservation, energy systems, aerodynamics, fluid mechanics. *Mailing Add:* Dietz Prof Tex A&M Univ Mech Eng College Station TX 77843-3123

FLETCHER, LOWELL W, b Princeton, WVa, Aug 18, 20; m 49; c 4. ENTOMOLOGY. *Educ:* Concord Col, BS, 47; WVa Univ, MS, 58; Rutgers Univ, PhD(entom), 61. *Prof Exp:* Res med entomologist, USDA, 61-69, res entomologist, Tobacco Insect Invest, 69-70, INSECT ECOLOGIST, TOBACCO INSECT INVEST, USDA, 70- *Mem:* Entom Soc Am; AAAS. *Res:* Basic insect biology; physical and biological methods of insect control. *Mailing Add:* 11132 Robious Rd Richmond VA 23235

FLETCHER, MARTIN J, b New York, NY, Aug 24, 32; m 60; c 1. BIOCHEMISTRY. *Educ:* Columbia Col, AB, 53; Purdue Univ, MS, 58, PhD(biochem), 59. *Prof Exp:* Biochemist, NIH, 59-62; chemist natural prods, Wallace Labs, Carter Prods, Inc, 62-68; dir biochem, Affil Med Enterprises, Inc, 68-, vpres, 74; COOP DIR TOXICOL, FMC CORP. *Mem:* AAAS; Am Chem Soc; NY Acad Sci. *Res:* Biochemistry of energy metabolism; isolation, purification and identification of natural products. *Mailing Add:* FMC Corp US Hwy 1 Princeton NJ 08540

FLETCHER, MARY ANN, b Little Rock, Ark, July 23, 37; c 1. IMMUNOLOGY. *Educ:* Tex Technol Col, BS, 59; Univ Tex, MA, 61; Baylor Univ, PhD(microbiol, biochem), 66. *Prof Exp:* Res assoc immunochem, Evanston Hosp, 66-69; asst dir, Div Hemat, Michael Reese Hosp, Chicago, 69-72; from asst prof to assoc prof microbiol, 72-80, PROF MED, MICROBIOL, IMMUNOL & ONCOL, SCH MED, UNIV MIAMI, 80-, DIR, E M PAPPER CLIN IMMUNOL LAB, 83- *Concurrent Pos:* Asst prof, dept microbiol, Northwestern Univ, 67-69; asst prof, dept biol, Ill Inst Technol, 70-72. *Mem:* Am Asn Immunol; Am Soc Microbiol; Soc Complex Carbohydrates; AAAS; Am Asn Women Sci; Soc Exp Med Biol; Sigma Xi. *Res:* Immunochemistry of membrane antigens; clinical immunology; immune dysfunction in immunodeficiency syndromes; effects of stress on immune functions. *Mailing Add:* Dept Med Microbiol & Immunol R-42 Univ Miami Sch Med Clin Immunol Lab PO Box 016960 Miami FL 33101

FLETCHER, NEIL RUSSEL, b Morenci, Mich, Oct 17, 33; m 58; c 3. NUCLEAR PHYSICS. *Educ:* Mich State Univ, BS, 55; Duke Univ, PhD(physics), 61. *Prof Exp:* Res assoc physics, Duke Univ, 60-61; res assoc, 61-63, from asst prof to assoc prof, 63-74, PROF PHYSICS, FLA STATE UNIV, 74-, MEM GRAD FAC, 65- *Concurrent Pos:* Sr res fel, Oxford, 83. *Mem:* Am Phys Soc. *Res:* Low energy nuclear physics; direct reactions and reaction mechanisms; angular correlations; structure of light nuclei. *Mailing Add:* Dept Physics Fla State Univ Tallahassee FL 32306

FLETCHER, OSCAR JASPER, JR, b Bennettsville, SC, Oct 18, 38; m 63; c 2. PATHOLOGY. *Educ:* Wofford Col, BS, 60; Univ Ga, DVM, 64, MS, 65; Univ Wis, PhD(vet sci), 68; Am Col Vet Path, dipl. *Prof Exp:* Res asst path, Univ Ga, 64-65; trainee, Univ Wis, 65-68; asst prof, 68-72, assoc prof path, 75-78, assoc dean, Col Vet Med, 76-82, PROF AVIAN MED & PATH, UNIV GA, 78-, HEAD AVIAN MED, 82- *Mem:* Am Vet Med Asn; Am Asn Avian Pathologists; World Vet Poulty Asn. *Res:* Rheumatoid; neoplastic diseases of domestic animals especially serum protein alterations; immunopathology; avian histopathy. *Mailing Add:* 2236 Ironwood Ct Ames IA 50010

FLETCHER, PAUL CHIPMAN, b New York, NY, Jan 10, 26; m 55; c 4. SOLID STATE PHYSICS. *Educ:* Mass Inst Technol, BS, 47; Columbia Univ, PhD, 57. *Prof Exp:* Res physicist, Res Labs, Hughes Aircraft Co, 57-61; mgr quantum physics lab, Electro-Optical Systs, Inc, 61-66; chief space optics lab, NASA Electronic Res Ctr, 66, chief optics lab, 66-70; head Electromagnetic Systs Dept, Naval Electronics Lab Ctr, 70-80; HEAD DEPT ENG SCI, US NAVAL OCEAN SYSTS CTR, 80- *Mem:* AAAS; Am Phys Soc; sr mem Inst Elec & Electronics Engrs; Sigma Xi. *Res:* Gaseous microwave spectroscopy; magnetism; lasers; electro-optics. *Mailing Add:* 11359 Fuerte Dr El Cajon CA 92020

FLETCHER, PAUL LITTON, JR, b Raleigh, NC, Aug 19, 41; m 73; c 1. PROTEIN CHEMISTRY, CELL BIOLOGY. *Educ:* Va Polytech Inst, BS, 64; Univ NC, Greensboro, MA, 68; Vanderbilt Univ, PhD(microbiol), 71. *Prof Exp:* Res assoc biochem, Rockefeller Univ, 71-74; res assoc, Med Sch, Yale Univ, 74-75, asst prof cell biol, 75-79; ASSOC PROF MICROBIOL, SCH MED, EAST CAROLINA UNIV, GREENVILLE, NC, 79- *Concurrent Pos:* dir bioanal div, Shared Res Resource Labs, 85- *Mem:* Sigma Xi; NY Acad Sci; Am Chem Soc; Soc Anal Cytol; Am Soc Cell Biol; AAAS. *Res:* Isolation and characterization and sequence analysis of proteins from cell membranes; mechanisms of exocytosis from pancreatic acinar cells. *Mailing Add:* Bioanal Div Shared Res Resource Labs E Carolina Univ Sch Med Greenville NC 27858

FLETCHER, PAUL WAYNE, b Woodbury, NJ, Feb 20, 51; m 73. BIOCHEMICAL ENDOCRINOLOGY, GONADOTROPIC RECEPTORS. *Educ:* Univ Del, BS, 74; Clemson Univ, MS, 75; Univ Wyo, PhD(reproduction phys), 78. *Prof Exp:* Fel, Colo State Univ, 78-80; ASST PROF MED BIOCHEM, ALBANY MED COL, UNION UNIV, 80- *Mem:* Soc Study Reproduction. *Res:* Mechanism of gonadotropin action in the testis; regulation of hormone-receptor interaction; activation of adenylate cyclase. *Mailing Add:* Dept Biochem (A-10) Albany Med Col 47 New Scotland Ave Albany NY 12208

FLETCHER, PETER, b New York, NY, July 6, 39; m 62; c 2. TOPOLOGY. *Educ:* Washington & Lee Univ, BS, 62; Univ NC, Chapel Hill, MA, 64, PhD(math), 66. *Prof Exp:* assoc prof, 67-80, PROF MATH, VA POLYTECH INST & STATE UNIV, 80- *Mem:* Am Math Soc; Math Asn Am. *Res:* Quasi-uniform spaces. *Mailing Add:* Dept Math Va Polytech Inst Blacksburg VA 24061

FLETCHER, PETER C, b Shrewsbury, Eng, Nov 16, 35; m 67. PHYSICAL CHEMISTRY, CERAMICS. *Educ:* Univ Liverpool, BSc, 56, PhD(surface chem), 59. *Prof Exp:* Gulf Oil fel, State Univ NY, 59-60; sr scientist physics & surface chem, Owens-Ill Co, Ohio, 60-70; MAT ENGR ELECTRONIC MAT DIV, BELL & HOWELL CO, PASADENA, 70- *Mem:* Am Chem Soc; Am Ceramic Soc; sr mem Am Vacuum Soc. *Res:* Surface physics and chemistry; heterogeneous catalysis; glass. *Mailing Add:* 890 Ridgeside Dr Monrovia CA 91016-1724

FLETCHER, ROBERT CHIPMAN, b New York, NY, May 27, 21; m 45; c 8. PHYSICS. *Educ:* Mass Inst Technol, BS, 43, PhD(physics), 49. *Prof Exp:* Mem staff, Radiation Lab, Mass Inst Technol, 43-45, asst, Insulation Lab, 47-49; mem tech staff, Bell Tel Labs, 49-58, dir solid state device develop, 58-64; vpres res, Sandia Corp, 64-67; exec dir, Mil Systs Res Div & Ocean Systs div, Bell Labs, 67-71, exec dir, Integrated Circuit Design Div, 71-86; vpres res develop, Ceramics Process Systs, 86-89; RETIRED. *Mem:* Fel Am Phys Soc; fel Inst Elec & Electronics Engrs. *Res:* Electron dynamics; magnetrons; traveling wave tubes; gas discharge; impulse breakdown of air; semiconductors; magnetic and ultrasonic devices; masers; solid state devices; optical devices; integrated circuits. *Mailing Add:* 44 Hastings Rd Belmont MA 01278

FLETCHER, ROBERT HILLMAN, b Abington, Pa, Mar 26, 40; m 63; c 2. INTERNAL MEDICINE, EPIDEMIOLOGY. *Educ:* Wesleyan Univ, BA, 62; Harvard Univ, MD, 66; Johns Hopkins Univ, MSc, 73. *Prof Exp:* Asst prof med & epidemiol, McGill Univ, 73-78; co-chief, Div Gen Med, 78-84 & res assoc, Health Serv Res Ctr, 78- 84; assoc prof, 78-82, PROF MED & EPIDEMIOL, UNIV NC, CHAPEL HILL, 82-, DIR, ROBERT WOOD JOHNSON CLIN SCHOLARS PROG, 78- *Concurrent Pos:* Dir med poly clin, Royal Victoria Hosp, 73-78; co-ed, J Gen Internal Med; co-dir, Univ NC Training Ctr, Int Clin Epid Network. *Mem:* Fel Am Col Physicians; Am Fedn Clin Res; Sigma Xi; Asn Health Rec; Soc Res & Educ Primary Care Med. *Res:* Clinical epidemiology and health care research. *Mailing Add:* NC Mem Hosp Univ NC 5034 Old Clin Bldg 226H CB 7105 Chapel Hill NC 27599

FLETCHER, RONALD AUSTIN, b Cape Comorin, India, July 1, 31; Can citizen; m 62; c 2. PLANT PHYSIOLOGY. *Educ:* Univ Delhi, BSc, 54; Univ BC, MSc, 61; Univ Alta, PhD(physiol), 64. *Prof Exp:* Nat Res Coun Can-NATO overseas res fel sci, Eng, 64-65; from asst prof to assoc prof bot, 65-72, PROF ENVIRON BIOL, UNIV GUELPH, 72- *Concurrent Pos:* Nat Res Coun Can exchange fel, France, 73; consult, United Nat Develop Prog, Foreign Agr Orgn, 83; vis prof, Nat Res Coun, India, 86 & 90. *Mem:* Can Soc Plant Physiologists (secy, 67-68); Am Soc Plant Physiologists; Scand Soc Plant Physiologists. *Res:* Hormonal and chemical regulation of plant growth and development; stress physiology. *Mailing Add:* Dept Environ Biol Univ Guelph Guelph ON N1G 2W1 Can

FLETCHER, RONALD D, b Foxboro, Mass, Jan 18, 33; m 54; c 3. ENVIRONMENTAL BIOREMEDIATION. *Educ:* Univ Conn, BS, 54, MS, 59, PhD(bact), 63. *Prof Exp:* Nat Inst Allergy & Infectious Dis fel microbiol, Vet Bact Inst, Univ Zurich, 63-64; res virologist, Antimicrobial Ther Dept, Lederle Labs, Am Cyanamid Co, 64-67; prof microbiol & assoc head dept, Univ Pittsburgh, Sch Dent Med, 67-86; sr intel analyst, Dept Defense, Armed Forces Med Intel Ctr, 86-89; PROF MICROBIOL, DEPT MED TECHNOL, SCH HEALTH RELATED PROFESSIONS, UNIV PITTSBURGH, 89-; EXEC VPRES RES & DEVELOP, ATI BIOREMEDIATION INC, PITTSBURGH, 89- *Concurrent Pos:* Gen res support grant, Univ Pittsburgh, 67-68; Am Cancer Soc instnl res grant, 68-69; Dept Army life sci div grant, 69-72; Nat Inst Dent Res grant, 75-79, res develop fund, 80-81; dept state lectr, Med Sch, Turkey, 82; mem, Biotech Steering Comt, Dept Defense, 87-89; consult, Battelle Mem Inst, 89-90. *Honors & Awards:* Certificate of Achievement in Microbiol, Surgeon Gen, US Army, 73. *Mem:* Fel AAAS; fel Am Acad Microbiol; Am Soc Microbiol; Tissue Cult Asn; Int Asn Dent Res (pres, 79-80); Am Soc Cell Biol; Asn Mil Surgeons US; Soc Indust Microbiol; Nat Mil Intel Asn; Int Asn Chiefs Police. *Res:* Environmental bioremediation of petroleum hydrocarbons; microbial isolations from chemical contaminated soils; relationship between mycoplasma species and selected respiratory viruses; adhesion of human cells with inanimate materials; adhesion in biological systems; ascorbic acid and its effect on the metabolism of enteric bacteria; the effect of acyclovir and monoglycerides on herpes simplex virus. *Mailing Add:* Dept Med Technol 209 Pa Hall Univ Pittsburgh Sch Health Related Professions Pittsburgh PA 15261

FLETCHER, ROY JACKSON, b Red Deer, Alta, Feb 1, 35; c 3. CLIMATOLOGY. *Educ:* Univ Alta, BA, 57, Univ Minn, MA, 59; Clark Univ, PhD(geog), 68. *Prof Exp:* Lectr geog, State Univ NY Buffalo, 61-68; assoc prof, 68-76, PROF GEOG, UNIV LETHBRIDGE, 76- *Mem:* Asn Am Geog; Am Soc Photogram; Arctic Inst NAm; Can Asn Geog. *Res:* Synoptic and dynamic climatology; biometeorology; arctic, especially physical geography; transportation and human ecology. *Mailing Add:* Dept Geog Univ Lethbridge 4401 University Dr Lethbridge AB T1K 3M4 Can

FLETCHER, STEWART G(AILEY), b Wilkinsburg, Pa, Jan 20, 18; m 42; c 4. METALLURGY. *Educ:* Carnegie Inst Technol, BS, 38; Mass Inst Technol, ScD(phys metall), 43. *Prof Exp:* Lab instr metall, Carnegie Inst Technol, 37-39; asst, Mass Inst Technol, 39-42, res assoc, 42-45; chief res metallurgist, Latrobe Steel Co, 45-47, chief metallurgist, 47-57, vpres & tech dir, 57-73; sr vpres, Am Iron & Steel Inst, 73-80; Consult & Pres, Ferrotechnol Inc, 80-85; RETIRED. *Concurrent Pos:* Lab instr, Lowell Inst Sch, 39-41, instr, 42-44. *Honors & Awards:* Howe Medal, Am Soc Metals, 46 & 49. *Mem:* Fel AAAS; fel Am Soc Metals (secy, 62-64, vpres, 64-65, pres, 65-66); Am Iron & Steel Inst; Am Inst Mining, Metall & Petrol Engrs. *Res:* Heat treatment of tool and die steels; alloying tool steels; electric steel making; vacuum melting of steels; superalloys; electroslag melting; technical and engineering administration. *Mailing Add:* 4407 Tournay Rd Bethesda MD 20816

FLETCHER, SUZANNE W, b Jacksonville, Fla, Nov, 14, 40; m; c 2. MEDICAL ADMINISTRATION. *Educ:* Swarthmore Col, BA, 62; Harvard Med Sch, MD, 66; Am Bd Internal Med, dipl, 73. *Prof Exp:* Physician, Dept Army, Ger, 69-70; asst prof, epidemiol & health, 74-77, assoc prof, epidemiology, McGill Univ Fac Med, 77-78; assoc prof med, Univ NC, 78-83, prof med & clin prof epidemiol, 83-90, prog leader, Cancer Prev, Lineberger Cancer Res Ctr, 85-90; CO-ED, ANNALS INTERNAL MED, 90- *Concurrent Pos:* Asst prof med, Fac Med, McGill Univ, 73-78; dir, Med Clin, Dept Med, NC Mem Hosp, 78-82; res assoc, Univ NC, 78-90, co-chief, Div Gen Med, 78-86, vchmn, Serv Dept, 81-84; co-dir, Rockefeller Found Int Clin Epidemiol Network Prog, 86-90; vis prof, Med Col Va, 86, Univ Ala, 87, McMaster Univ, 87; chairperson, Comt Eval Spec Candidates, Inst Med, 85-87, Comt Core Internal Med Questions, 86-89; mem clin pract subcomt, Am Col Physicians, 88-89, publ subcomt, 88-89; mem, Panel Nat Health Care Surv, Nat Res Coun, 90-; mem, Comn Med Educ, Robert Wood Johnson Found, 90-; adj prof med, Univ Pa, 90- *Mem:* Inst Med-Nat Acad Sci; fel Am Col Physicians; fel Am Col Epidemiol; Am Geriat Soc; Am Fedn Clin Res; Am Pub Health Asn; Soc Gen Internal Med (pres-elect, 82-83, pres, 83-84). *Mailing Add:* Independence Mall W Sixth St at Race Philadelphia PA 19106-1572

FLETCHER, THOMAS FRANCIS, b New York, NY, Mar 26, 37; m 60; c 5. VETERINARY ANATOMY. *Educ:* Cornell Univ, DVM, 61; Univ Minn, PhD(vet anat), 65. *Prof Exp:* From asst prof to assoc prof, 65-72, assoc dean, 85-89, PROF VET ANAT, COL VET MED, UNIV MINN, ST PAUL, 72- *Concurrent Pos:* USPHS grants, 65-66 & 67-81. *Mem:* Am Asn Anatomists; Am Asn Vet Anat; World Asn Vet Anatomists; Am Vet Med Asn. *Res:* Neuroanatomy and neurophysiology of the spinal cord; anatomicophysiological basis of nervous disorders in domestic animals; morphometry. *Mailing Add:* Col Vet Med Univ Minn St Paul MN 55108

FLETCHER, THOMAS HARVEY, b Provo, Utah, July 26, 55; m 81; c 3. COAL COMBUSTION, COMPUTATIONAL FLUID MECHANICS. *Educ:* Brigham Young Univ, BS, 79, MS, 80, PhD(chem eng), 83. *Prof Exp:* Res assoc chem eng, Combustion Lab, Brigham Young Univ, 83-84; STAFF MEM, COMBUSTION RES DIV, SANDIA NAT LABS, 84- *Mem:* Combustion Inst. *Res:* Comprehensive modeling of coal combustors, including turbulence, fluid mechanics, chemistry, and radiation; detailed modeling of single particle combustion; experimental determination of coal devolatilization rates at high temperatures and high heating rates using optically measured particle sites and temperatures. *Mailing Add:* Sandia Nat Labs 8361 PO Box 969 Livermore CA 94551-0969

FLETCHER, THOMAS LLOYD, b Boydton, Va, Jan 4, 17; m 41; c 4. ORGANIC CHEMISTRY. *Educ:* Clark Univ, AB, 37, MA, 38; Univ Wis, PhD(biochem, org chem), 49. *Prof Exp:* Chemist, Lever Bros Co, Mass, 39-42; chemist, Colonial-Beacon Oil Co, 42; teacher chem, Adm Billard Acad, 42-43; chemist, Forest Prod Lab, US Forest Serv, 43-48; chemist, Pulp

Mills Res, 48-51, res chemist, Dept Surg, Sch Med, 51-55, from res assoc prof to res prof surg, 55-67, PROF SURG, CHEM RES LAB, SCH MED, UNIV WASH, 67-; MEM, FRED HUTCHINSON CANCER RES CTR, 72- *Concurrent Pos:* Nat Cancer Inst res career develop award, 61-71. *Mem:* Fel AAAS; Am Chem Soc; The Chem Soc; Soc Exp Biol & Med; Am Cancer Soc; Sigma Xi. *Res:* Chemistry of fluorene and other aryl polycyclics; carcinogenicity; cancer chemotherapy. *Mailing Add:* 2212 E Miller St Seattle WA 98112

FLETCHER, WILLIAM ELLIS, b Colfax, La, Nov 11, 36; m; c 3. HORTICULTURE. *Educ:* Univ Southwestern La, BS, 58; Iowa State Univ, MS, 61, PhD, 64. *Prof Exp:* Asst, Iowa State Univ, 58-63, instr, 63-64; asst prof ornamental hort, Univ Fla, 64-67; asst prof, 67-69, assoc prof, 69-76, PROF HORT, UNIV SOUTHWEST LA, 76- *Res:* Woody ornamentals; general nursery stock and foliage plants. *Mailing Add:* Agr Sci Univ Southwest La PO Box 44433 Lafayette LA 70504

FLETCHER, WILLIAM H, ANATOMY. *Educ:* Calif State Univ, Fresno, AB, 67; Univ Calif, Berkeley, PhD(anat), 72. *Prof Exp:* NIH postdoctoral fel, Med Ctr, Duke Univ, 72-73, res assoc obstet-gynec, 73-74, asst prof anat & Comprehensive Cancer Ctr, 76-77; Nat Inst Child Health & Human Develop Int fel, Duke Univ & Libre Univ Bruxelles, Belg, 74-75; asst prof biomed sci, Univ Calif, Riverside, 77-85; DIR, LAB MOLECULAR CYTOL, JERRY L PETTIS MEM VET HOSP, LOMA LINDA, CALIF, 85-; PROF ANAT, SCH MED, LOMA LINDA UNIV, 85-, ASSOC PROF PHYSIOL-PHARMACOL, SCH MED, 87- *Concurrent Pos:* Prin investr, Cancer Res Coord Comt, 78-79, NIH, 79-81, 82-86, 86-90, Vet Admin Merit Rev, 86-89 & 89-92, Seed Grant, Loma Linda Univ, 87-88 & 88-89; res career scientist awardee, Vet Admin, 89-96. *Mem:* Am Soc Cell Biol; Endocrine Soc; Am Soc Biol Chemists. *Res:* Role of intercellular signal transfer and protein kinases in regulating cell differentiation and function. *Mailing Add:* Dept Anat Res Serv 151 Sch Med Loma Linda Univ Jerry L Pettis Mem Vet Admin Hosps 11201 Benton St Loma Linda CA 92357

FLETCHER, WILLIAM HENRY, b Eureka, Kans, Apr 25, 16; m 49; c 4. PHYSICAL CHEMISTRY. *Educ:* Col Idaho, BS, 39; Univ Minn, PhD(chem), 49. *Prof Exp:* Res chemist, Norwich Pharmacal Co, 40-42; res chemist, Lubrizol Corp, 42-46; from asst prof to assoc prof, 49-59, PROF CHEM, UNIV TENN, KNOXVILLE, 59- *Concurrent Pos:* Consult, Union Carbide Nuclear Co, 57. *Mem:* Optical Soc Am; Coblentz Soc; Soc Appl Spectros; Am Chem Soc; Am Phys Soc. *Res:* Molecular spectroscopy; molecular force fields; high resolution Raman spectroscopy. *Mailing Add:* Dept Chem Univ Tenn Knoxville TN 37916

FLETCHER, WILLIAM L, US citizen. ENGINEERING. *Educ:* New England Col, BS, 57; Northeastern Univ, MS, 62. *Prof Exp:* Proj engr, Camp, Dresser & McKee, Inc, 57-67; prin sanit engr, NH Water Supply & Pollution Control Comn, 67-69; mem staff, Environ Engrs, Inc, 69-76; vpres, Weston, 76-86; mem staff, NH Dept Transp, 86-90; RETIRED. *Concurrent Pos:* Adv, Gov Coun Energy, 74-78; dipl, Am Acad Environ Engrs. *Mem:* Am Soc Civil Engrs. *Res:* Engineering management; project management; environmental engineering; civil engineering. *Mailing Add:* PO Box 123 Warner NH 03278

FLETCHER, WILLIAM SIGOURNEY, b Arlington, Mass, May 7, 27; div; c 3. SURGICAL ONCOLOGY. *Educ:* Dartmouth Col, AB, 52; Harvard Med Sch, MD, 55; Am Bd Surg, dipl, 51. *Prof Exp:* Intern surg, Harvard Surg Serv, Boston City Hosp, Mass, 55-56, jr asst resident surgeon, 56-57, sr asst resident surgeon, 57-58; chief resident surgeon, 59-60, from instr to assoc prof, 60-70, PROF SURG, HEALTH SCI CTR, UNIV ORE, 70-, HEAD DIV SURG ONCOL, 75-, CHIEF SURG, UNIV HOSP, 76- *Concurrent Pos:* Clin fel, Am Cancer Soc, Middlesex Hosp, London, 58-59 & advan clin fel, 60-63; Markle scholar med sci, 62-67. *Mem:* Am Col Surgeons; Soc Surg Oncol; Am Soc Clin Oncol; Am Fedn Clin Res; AMA; Am Asn Cancer Educ; Am Asn Cancer Res; Am Surg Asn; Am Trauma Soc; Soc Surg Alimentary Tract. *Res:* Medical education; basic experimental and clinical cancer research. *Mailing Add:* Dept Surg Ore Health Sci Univ Hosp N 3181 SW San Jackson Park Rd Portland OR 97201

FLETCHER, WILLIAM THOMAS, b Durham, NC, Aug 19, 34. MATHEMATICS. *Educ:* NC Col Durham, BS, 56, MS, 58; Univ Idaho, PhD(math), 66. *Prof Exp:* From assoc prof to prof math, LeMoyne-Owen Col, 57-72; PROF & CHMN DEPT MATH, NC CENT UNIV, 72- *Mem:* Math Asn Am; Am Math Soc. *Res:* Lie algebras; abstract algebra; associative and non associative. *Mailing Add:* Dept Math & Comp Sci NC Cent Univ Durham NC 27707

FLETTERICK, ROBERT JOHN, b Cleveland, Ohio, July 22, 43. MOLECULAR BIOPHYSICS. *Educ:* Marietta Col, BS, 65; Cornell Univ, PhD(chem), 70. *Prof Exp:* Fel biochem, Yale Univ, 70-74; asst prof biochem, Univ Alta, 74-79; assoc prof, 79-82, PROF, BIOCHEM, UNIV CALIF, SAN FRANCISCO, 83- *Mem:* Am Crystallog Asn; Brit Biophys Soc; Can Biochem Soc. *Res:* Structure and function of proteins; investigations using x-ray diffraction; regulation of glycogen metabolism. *Mailing Add:* Dept Biochem & Biophys Univ Calif San Francisco S-964 San Francisco CA 94143-0448

FLEURY, PAUL A, b Baltimore, Md, July 20, 39; m 64; c 3. SOLID STATE PHYSICS, SPECTROSCOPY. *Educ:* John Carroll Univ, BS, 60, MS, 62; Mass Inst Technol, PhD(physics), 65. *Prof Exp:* Mem tech staff, 65-70, head dept physics, 70-79, dir mat res, 79-84, DIR PHYSICS RES, AT&T BELL LABS, 84- *Honors & Awards:* Michelson Morley Award, 85. *Mem:* Fel Am Phys Soc; AAAS. *Res:* Nonlinear optical and inelastic laser light scattering studies of solids and simple fluids including semiconductors, ferroelectrics, magnets and their phase transitions; low temperature physics, biophysics. *Mailing Add:* Bell Labs Rm 1D-269 Murray Hill NJ 07974

FLEXMAN, EDMUND A, JR, b Harrisburg, Liberia, Aug 13, 40; US citizen; m 64; c 2. POLYMER CHEMISTRY. *Educ:* Bradley Univ, BA, 62; Ind Univ, Bloomington, PhD(org chem), 67. *Prof Exp:* Phys sci aide paper chem, Northern Regional Lab, USDA, 61-64; RES CHEMIST, EXP STA, E I DU PONT DE NEMOURS & CO, INC, 65- *Mem:* Am Chem Soc; Sigma Xi; Soc Plastic Eng. *Res:* Polymer chemistry and physics. *Mailing Add:* Ten Chestfield Rd Chestfield Wilmington DE 19803

FLEXNER, JOHN M, b Louisville, Ky, Mar 29, 26; m 54; c 4. INTERNAL MEDICINE, HEMATOLOGY. *Educ:* Yale Univ, BA, 50; Johns Hopkins Univ, MD, 54. *Prof Exp:* Intern, Vanderbilt Univ Hosp, 54-55, asst resident internal med, 55-56; Grace New Haven Hosp, Conn, 56-57; from instr to asst prof, 59-71, ASSOC PROF MED, SCH MED, VANDERBILT UNIV, 71- *Concurrent Pos:* USPHS fel hemat, Sch Med, Vanderbilt Univ, 57-59; asst & vis hematologist, Vanderbilt Univ Hosp, 57-59, vis hematologist & dir, Hemat Labs & Blood Bank, 59-; consult, Thayer Vet Admin Hosp, Nashville, Tenn, 59- & Regional Blood Ctr Labs, 65- *Mem:* AAAS; Am Soc Hemat. *Res:* Effect of oral fat intake on blood coagulation; megaloblastic anemia due to anti-convulsants; paroxysmal nocturnal hemoglobinuria; myleran induced pulmonary pneumonitis and fibrosis. *Mailing Add:* Dept Med-Hemat Vanderbilt Univ Sch Med 21st Ave S & Garland Nashville TN 37240

FLEXNER, LOUIS BARKHOUSE, b Louisville, Ky, Jan 7, 02; m 37. ANATOMY. *Educ:* Univ Chicago, BS, 22; Johns Hopkins Univ, MD, 27. *Hon Degrees:* LLD, Univ Pa, 74. *Prof Exp:* Loeb fel, Johns Hopkins Univ, 27-28; resident house officer, Clins, Chicago, 28-29; instr & assoc anat, Sch Med, Johns Hopkins Univ, 30-40; mem staff, Dept Embryol, Carnegie Inst, 40-51; chmn dept anat, 51-67, dir inst neurol sci, 53-65, PROF ANAT, SCH MED, UNIV PA, 51- *Concurrent Pos:* Mem tech aide comt aviation med, Nat Res Coun, 42-45; mem study sects, USPHS, 51-70; res assoc, Carnegie Inst, 51-; mem adv bd, United Cerebral Palsy Asn, 56 & Nat Found, 59-64. *Honors & Awards:* Weinstein Award, 57. *Mem:* Nat Acad Sci; Am Physiol Soc; Am Soc Biol Chemists; Am Acad Arts & Sci; Am Asn Anatomists (secy-treas, 56-64); Am Phys Soc. *Res:* Meninges and cerebrospinal fluid; learning and memory; fetal physiology. *Mailing Add:* Dept Anat Univ Pa Sch Med Philadelphia PA 19104

FLEXSER, LEO AARON, b New York, NY, June 20, 1910; m 39; c 1. CHEMISTRY. *Educ:* Columbia Univ, AB, 31, AM, 32, PhD(chem), 35. *Prof Exp:* Asst chem, Columbia Univ, 31-34; res chemist, Montrose Chem Co, NJ, 35-36, plant supt, 36-38; instr chem, Brooklyn Col, 38; res chemist, NY Quinine & Chem Co, 38-41; sr chemist, Hoffmann-La Roche, Inc, 41-45, group chief mfg & develop dept, 45-59, dir chem prod, 59-63, vpres, 63-75, consult, 75-77; RETIRED. *Concurrent Pos:* Mem bd trustees, Jersey City State Col, NJ, 68-82; pres, R P, Inc, Puerto Rico, 74-75, consult, 75-77. *Mem:* AAAS; Am Chem Soc; fel Am Inst Chemists. *Res:* Process research and development in synthesis and manufacture of vitamins and pharmaceutical chemicals; administration of fine chemical manufacturing and development. *Mailing Add:* 3 The Fairway Upper Montclair NJ 07043

FLICK, CARL, b Vienna, Austria, June 22, 26; US citizen; m 54; c 3. DESIGN CONCEPT FORMULATION. *Educ:* Polytech Inst Brooklyn, BEE, 51, MEE, 53. *Prof Exp:* Sr engr, generator eng, Westinghouse Elec Corp, 52-70, fel engr, 70-81, adv engr, 81-89; CONSULT, TECHNO-LEXIC, 89- *Concurrent Pos:* Lectr, Univ Pittsburgh, 56-57; instr, Polytech Inst Brooklyn, 51-52, res fel elec eng, Microwave Res Inst, 51-52; distinguished lectr, Inst Elec & Electronics Engrs, Power Eng Soc, 87- *Honors & Awards:* Centennial Medal, Inst Elec & Electronics Engrs, 84. *Mem:* Fel Inst Elec & Electronics Engrs; AAAS; Inst Elec Engrs Japan. *Res:* All phases of electric machine technology, with emphasis on large generators and motors, including a variety of special machine technologies, such as pulse generators and generators with superconducting field windings; international technology transfer and monitoring in these areas; design assurance and verification; design process analysis and standardization; technical translation. *Mailing Add:* 200 St Andrews Blvd 1902 Winter Park FL 32792

FLICK, CATHY, b Bryn Mawr, Pa, June 19, 49. PHYSICS, CHEMICAL PHYSICS. *Educ:* Marywood Col, BS, 71; Kent State Univ, MA, 74, PhD(chem physics), 76. *Prof Exp:* Asst prof physics, Earlham Col, 76-79; SCI TRANSLR & CONSULT, 78- *Mem:* Am Phys Soc; Am Chem Soc; Am Trans Assoc. *Res:* Liquid crystals, biomembrane model systems, magnetic resonance spectroscopy. *Mailing Add:* 302 S W Fifth St Richmond IN 47374

FLICK, CHRISTOPHER E, organelle genetics, for more information see previous edition

FLICK, GEORGE JOSEPH, JR, b New Orleans, La, May 30, 40; m 67; c 3. FOOD SCIENCE. *Educ:* La State Univ, Baton Rouge, BA, 63, MS, 66, PhD(food sci), 69. *Prof Exp:* From asst prof to assoc prof, 69-78, PROF FOOD SCI & TECHNOL, VA POLYTECH INST & STATE UNIV, 78- *Concurrent Pos:* Vis prof, Southern Regional Res Ctr, US Dept Agr, 73-; mem bd dirs, Nat Fisheries Educ & Res Found, 81-; vis prof, Univ Tokyo, 85. *Mem:* AAAS; Am Chem Soc; Inst Food Technol; Marine Technol Soc; Am Sch Food Serv Asn; Sigma Xi. *Res:* Separation and isolation of compounds in marine organisms; chemical and microbiological contaminants in processed seafood products; utilization and biochemical properties of marine products and sub-tropical fruits and vegetables. *Mailing Add:* Food Sci & Technol Va Polytech Inst & State Univ Blacksburg VA 24061-0418

FLICK, JOHN A, b Camden, NJ, May 10, 17; m 43, 67; c 5. IMMUNOLOGY. *Educ:* Haverford Col, BS, 39; Harvard Univ, MD, 43. *Prof Exp:* Instr, 44-53, chmn dept, Div Grad Med, 55-72, ASSOC PROF MED MICROBIOL, SCH MED, UNIV PA, 53- *Mem:* AAAS; Am Asn Immunol; Reticuloendothelial Soc. *Res:* Allergy; immunology. *Mailing Add:* 115 Glenwood Rd Merion PA 19066

FLICK, PARKE KINSEY, b French Lick, Ind, Oct 3, 46; m 77; c 2. RESEARCH ADMINISTRATION. *Educ:* Purdue Univ, BSChE, 68, MS, 70; Harvard Univ, PhD(chem), 75. *Prof Exp:* Fel biochem, Sch Med, Wash Univ, 75 & Merck Sharp & Dohme Res Labs, 75-77; asst prof biochem, Sch Med, Wright State Univ, 77-84; ASSOC SCI DIR, US BIOCHEM CORP, CLEVELAND, OH, 84- *Concurrent Pos:* NIH grant, 78-84. *Mem:* Am Chem Soc; AAAS; Am Soc Biochem & Molecular Biol. *Res:* Regulation of gene expression; molecular biology. *Mailing Add:* 33385 Rockford Dr Solon OH 44139-1934

FLICKER, HERBERT, b Brooklyn, NY, Jan 28, 30; m 53; c 2. LASERS, MOLECULAR SPECTROSCOPY. *Educ:* Cornell Univ, AB, 51; Univ Pa, MS, 53, PhD(physics), 59. *Prof Exp:* Mem tech staff, RCA Labs, 59-64; vis assoc prof eng, Brown Univ, 64-65; mem staff, TRW Systs Group, 65-66; mgr semiconductor dept, Electrooptical Systs, Xerox Corp, 66-68, chief scientist advan systs & requirements, 68-72; mem staff, Los Alamos Sci Lab, 72-86; MEM STAFF, NICHOLS RES CORP, 86- *Mem:* Am Phys Soc; Inst Elec & Electronics Engrs. *Res:* Semiconductor physics; radiation damage in semiconductors; cryogenics; optical and electrical properties of semiconductors; infrared photodetectors, systems analysis; molecular spectroscopy using tunable diode lasers, and fourier transform spectrometer. *Mailing Add:* 535 Totavi Los Alamos NM 87544-2641

FLICKER, YUVAL ZVI, b Kfar Sava, Israel, Jan 3, 55; m; c 3. AUTOMORPHIC FORMS. *Educ:* Tel Aviv Univ, BA, 73; Hebrew Univ, MA, 74; Cambridge Univ, PhD(math), 78. *Prof Exp:* Mem, Inst Advan Study, Princeton, 78-79; asst prof, Columbia Univ, 79-81; vis, Univ Paris, 80-81; asst prof, Princeton Univ, 81-85; asst prof, Harvard Univ, 85-87; ASSOC PROF, OHIO STATE UNIV, 87- *Res:* Automorphic representations, trace formula; elliptic modules. *Mailing Add:* Dept Math OSU 231 W 18th Ave Columbus OH 43210

FLICKINGER, CHARLES JOHN, b Bethlehem, Pa, July 13, 38; m 63; c 2. CELL BIOLOGY, REPRODUCTIVE BIOLOGY. *Educ:* Dartmouth Col, AB, 60; Harvard Univ, MD, 64. *Prof Exp:* Res fel, dept anat, Med Ctr, Univ Colo, 64-65, res assoc, Inst Develop Biol, 66-67, asst prof, 67-70; res fel, Med Sch, Harvard Univ, Boston, 65-66; assoc prof, 71-75, PROF DEPT ANAT, UNIV VA, 75-, DIR, CENT ELECTRON MICROS FACIL, SCH MED, 79-, CHMN, DEPT ANAT & CELL BIOL, 82- *Concurrent Pos:* Assoc ed, Anat Rec, 73- & J Andrology, 89-92; adv ed, Int Rev Cytol, 75-; mem, Reprod Biol Study Sect, NIH, 79-83, & Nat Bd Med Examr, Anat Test Com, 81-84. *Mem:* Am Soc Cell Biol; Am Asn Anatomists (vpres, 88-89); Soc Study Reprod; Am Soc Androl. *Res:* Male reproductive cell biology, including testicular and epididymal structure and function; effects of vasectomy and anti-fertility agents; cell ultrastructure; role of cell organelles in the secretory process. *Mailing Add:* Dept Anat & Cell Biol Univ Va Health Sci Ctr Box 439 Charlottesville VA 22908

FLICKINGER, GEORGE LATIMORE, JR, b Hanover, Pa, May 21, 33; m 58; c 3. COMPARATIVE PATHOLOGY, REPRODUCTIVE PHYSIOLOGY. *Educ:* Pa State Univ, BS, 54; Univ Pa, VMD, 58, PhD(path), 63. *Prof Exp:* Assoc dir path, Penrose Res Lab, 62-63; from instr to assoc path, 63-66, from assoc to asst prof obstet & gynec, 64-73, ASSOC PROF OBSTET & GYNEC, UNIV PA, 73- *Mem:* AAAS; Endocrine Soc. *Res:* Hormone secretion by gonadal tissues; relationship of social behavior to reproductive performance. *Mailing Add:* Dept Obstet & Gynec Univ Pa Sch Med 36th & Hamilton Wk Philadelphia PA 19104

FLICKINGER, REED ADAMS, b Council Bluffs, Iowa, Apr 5, 24; m 49; c 2. DEVELOPMENTAL BIOLOGY. *Educ:* Stanford Univ, BA, 46, PhD, 49. *Prof Exp:* Instr embryol, Univ Pa, 49-52; asst prof, Univ Calif, Los Angeles, 52-58; assoc prof, Univ Iowa, 58-61 & Univ Calif, Davis, 61-64; prof, 64-90, EMER PROF EMBRYOL, STATE UNIV NY, BUFFALO, 90- *Concurrent Pos:* Rockefeller res fel, Brussels, Belg, 50-51, Stockholm, Sweden, 51-52; USPHS spec fels, Paris, 61 & Honolulu, 72. *Mem:* Soc Develop Biol; Am Soc Cell Biologists. *Mailing Add:* PO Box 741 Captain Cook HI 96704

FLICKINGER, STEPHEN ALBERT, b Savanna, Ill, Feb 12, 42; m 66; c 1. FISH BIOLOGY. *Educ:* Southern Ill Univ, Carbondale, BA, 64, MA, 66; Colo State Univ, PhD(fishery biol), 69. *Prof Exp:* Asst prof, 70-77, ASSOC PROF FISHERY BIOL, COLO STATE UNIV, 77- *Mem:* Am Fisheries Soc. *Res:* Culture of food, bait, and sport fishes. *Mailing Add:* Dept Wildlife Mgt Colo State Univ Ft Collins CO 80523

FLIEDNER, LEONARD JOHN, JR, b Flushing, NY, Mar 16, 37. MEDICINAL CHEMISTRY, CLINICAL RESEARCH. *Educ:* Princeton Univ, AB, 58; Fordham Univ, MS, 60; Univ Mass, PhD(org chem), 65. *Prof Exp:* SR RES CHEMIST, ENDO LABS, INC, GARDEN CITY, 65- *Concurrent Pos:* Clin res monitor, Ayerst Labs, New York, NY, 81- *Mem:* Am Chem Soc; Am Soc Clin Pharmacol & Therapeut. *Res:* Synthetic organic chemistry; synthesis of fibrinolytics, analgetics and antidepressants; clinical research on analgesics. *Mailing Add:* 33-67 157 St Flushing NY 11354-3333

FLIEGEL, FREDERICK MARTIN, b Bellefonte, Centre Co, Pa, Sept 14, 56; m 81. ACOUSTICS. *Educ:* Univ Ill, Urbana-Champaign, BSEE, 79, MSEE, 82, PhD(elec eng), 87. *Prof Exp:* Res asst, Coord Sci Lab, Univ Ill, Urbana, 79-82, 84-87; mem tech staff, Tex Instruments, Dallas, 82-84; saw res & develop scientist, R F Monolithics, Inc, Dallas, Tex, 88-89; electronic staff engr, 89-91, PATENT ENGR, MOTOROLA GOVT ELECTRONICS GROUP, PHOENIX, ARIZ, 91- *Concurrent Pos:* Proj leader, Electronic Decisions, Inc, Urbana, Ill, 84-87; consult, Gould Res Corp, Rolling Meadows, Ill, 87. *Mem:* Sr mem Inst Elec & Electronics Engrs; Asn Old Crows; Am Vacuum Soc; Sigma Xi. *Res:* Acoustic charge transport; author of various publications. *Mailing Add:* 3917 S Mitchell Tempe AZ 85282-4762

FLIEGEL, HENRY FREDERICK, b Ridley Park, Pa, Apr 25, 36. ASTRONOMY, GEODESY. *Educ:* Univ Pa, BA, 58, PhD(astron), 63; Ohio State Univ, MA, 59. *Prof Exp:* Staff astronr, US Naval Observ, 63-65; asst prof astron, Georgetown Univ, 65-68; prof astron & physics, Del State Col, 68-69; mem tech staff earth physics appln, Jet Propulsion Lab, Calif Inst Technol, 69-81; PROJ ENGR, AEROSPACE CORP, 81- *Concurrent Pos:* Mem, Com l'Heure, Int Astron Union, 75- *Mem:* Am Astron Soc; Am Geophys Union. *Res:* Applications of astronomical data, especially optical observations of time and polar motion; radio interferometry to spacecraft navigation and studies of earth crustal motion. *Mailing Add:* PO Box 8682 La Crescenta CA 91214

FLINDERS, JERRAN T, b Salt Lake City, Utah, July 2, 37; m 56; c 5. WILDLIFE ECOLOGY, RANGE ECOLOGY. *Educ:* Univ Utah, BS, 67, MS, 68; Colo State Univ, PhD(animal ecol), 71. *Prof Exp:* Asst prof, Dept Range & Wildlife Mgt, Tex Tech Univ, 71-74; assoc prof wildlife resources & affil fac range resources, Col Forestry Wildlife & Range Resources, Univ Idaho, 74-76; assoc prof, 76-80, PROF & DIR WILDLIFE & RANGE RESOURCES GRAD PROG, BRIGHAM YOUNG UNIV, 76-, CHMN DEPT BOT & RANGE SCI, 80- *Mem:* Wildlife Soc; Soc Range Mgt; Am Soc Mammalogists; Nat Wildlife Fedn. *Res:* Mule deer habitat selection and deer fawn mortality; baseline wildlife habitat studies prior to strip mining; ecological studies of bobcats leading to management purposes; dietary studies of pygmy rabbits; habitat selection of sage grass in relation to fire and grazing practices. *Mailing Add:* Dept Zool Brigham Young Univ Provo UT 84602

FLINK, EDMUND BERNEY, b Isanti, Minn, Jan 27, 14; m 40; c 4. MEDICINE. *Educ:* Univ Minn, BS & MB, 37, MD, 38, PhD(internal med), 45. *Prof Exp:* From instr to asst prof internal med, Sch Med, Univ Minn, 42-45, from asst prof to prof med, 45-60; chmn dept, 60-76, PROF MED, MED CTR, W VA UNIV, 60- *Concurrent Pos:* Commonwealth Fund fel, Harvard Univ, 48-49; chief med serv, Vet Admin Hosp, Minn, 52-60. *Mem:* Endocrine Soc; Am Soc Clin Invest; fel Am Col Physicians; Asn Am Physicians. *Res:* Hemoglobin metabolism; clinical endocrinology; mineral metabolism. *Mailing Add:* Dept Med WVa Univ Med Ctr Morgantown WV 26506

FLINN, EDWARD AMBROSE, b Oklahoma City, Okla, Aug 27, 31; m 62; c 1. GEOPHYSICS. *Educ:* Mass Inst Technol, SB, 53; Calif Inst Technol, PhD(geophys), 60. *Prof Exp:* Seismologist, United Electrodynamics, Inc, 60-68, assoc dir, Alexandria Labs, Teledyne Geotech, Va, 68-74; dir lunar progs, 75, dep dir & chief scientist, Planetary Progs, 75-77, chief scientist, 77-86, CHIEF, GEODYNAMICS PROG, NASA HQ, 87- *Concurrent Pos:* Vis assoc prof geophys, Brown Univ, 69; vis res assoc, Calif Inst Technol, 78; secy comt math geophys, chmn comn planetary sci & ed, J Geophys Res, Int Union Geod & Geophys; secy-gen, Inter-Union Comn Lithosphere, Int Coun Sci Unions, 80-86. *Mem:* AAAS; Am Geophys Union; Seismol Soc Am; Royal Astron Soc. *Res:* Seismology; general geophysics; coastal movements. *Mailing Add:* Code 910 Goddard Space Flight Ctr NASA Greenbelt Rd Greenbelt MD 20771

FLINN, JAMES EDWIN, b Cincinnati, Ohio, Sept 3, 34; m 84; c 5. CHEMICAL ENGINEERING. *Educ:* Purdue Univ, BS, 56; Univ Cincinnati, PhD(chem eng), 65. *Prof Exp:* Design engr, Dow Corning Corp, Mich, 56-57; develop engr, Nat Lead Co, Ohio, 60-62; sr res engr, Battelle-Columbus Labs, 65-66, assoc div chief, 66-72, mgr,waste control & process technol sect, energy & environ systs sect & chem process develop sect, 73-80, assoc dir corp tech develop, Battelle Mem Inst, 80-82, sr prog mgr, Biotechol proj off, 82-84, res coun dir, 84-85; EXEC VPRES, BERNARD WOLNACK & ASSOC, CHICAGO, ILL, 85- *Mem:* AAAS; Am Inst Chem Engrs; Sigma Xi. *Res:* Biotechnology; separations technology; microencapsulation; environmental control technology; toxic and hazardous waste; oil shale and coal processing; fluidized bed technology; laboratory, bench and pilot scale process development. *Mailing Add:* 1210 Edward Rd Naperville IL 60540

FLINN, JANE MARGARET, b Warrington Lancs, UK, Feb 12, 38; m 62; c 1. PHYSIOLOGICAL PSYCHOLOGY, BRAIN SCIENCES. *Educ:* Oxford Univ, BA, 60; Univ Calif, Los Angeles, MSc, 62; Cath Univ Am, PhD(physics), 69; George Washington Univ, PhD(psychol), 74. *Prof Exp:* Lectr physics, 69-74, asst prof, 74-80, ASSOC PROF PHYSICS & PSYCHOL, GEORGE MASON UNIV, 80-, CHMN DEPT PSYCHOL, 83- *Mem:* Am Psychol Asn. *Res:* Brain function; evoked potentials; effect of malnutrition on development. *Mailing Add:* 3605 Tupelo Pl Alexandria VA 22304

FLINN, PAUL ANTHONY, b New York, NY, Mar 25, 26; m 49; c 5. PHYSICS, METALLURGY. *Educ:* Columbia Col, Ill, AB, 48, AM, 49; Mass Inst Technol, ScD, 52. *Prof Exp:* Asst prof physics, Wayne State Univ, Detroit, Mich, 52-54; res staff, Westinghouse Res Lab, Pittsburgh, Pa, 54-63; from assoc prof to prof physics & metall, Carnegie Inst Technol, 63-78; SR STAFF SCIENTIST, INTEL CORP, 78-; CONSULT PROF, DEPT MAT SCI, STANFORD UNIV, STANFORD, CALIF. *Concurrent Pos:* Vis prof, Univ Nancy, 67-68; distinquished appointment, Argonne Univs Asn, 71; mem adv bd, Mossbauer Effect, Int Comn Applications, 75-; vis scientist, Argonne Nat Lab, 77-78. *Mem:* AAAS; fel Am Phys Soc; Am Inst Metall, Mining & Petrol Eng; Mat Res Soc. *Res:* Structure of solids; Mossbauer effect; dynamics of highly viscous liquids; Raman spectroscopy; diffusion; mechanical properties of thin films; electromigration. *Mailing Add:* 2162 Coronet Dr San Jose CA 95124

FLINN, RICHARD A(LOYSIUS), b New York, NY, May 31, 16; m 44; c 5. METALLURGY. *Educ:* City Col New York, BS, 36; Mass Inst Technol, MS, 37, ScD(phys metall), 41. *Prof Exp:* Res metallurgist, Int Nickel Co, NJ, 37-39; asst phys metall, Mass Inst Technol, 39-41; from metall asst to asst chief metallurgist, Am Brake Shoe Co, 41-52; prof metall eng, 52-88, in-chg, Cast Metals Lab, 52-88, EMER PROF METALL ENG, UNIV MICH, 88- *Concurrent Pos:* With Atomic Energy Comn & Off Sci Res & Develop, 44. *Honors & Awards:* Howe Medal, Am Soc Metals, 44 & 63; Simpson Medal, Am Foundrymen's Soc, 47, Hoyt Mem lectr, 81. *Mem:* Am Foundrymen's

Soc; fel Am Soc Metals. *Res:* Failure analysis; physical and production metallurgy of cast metals; casting design and stress analysis; basic cupola melting; shell molding; basic oxygen steel making; metal casting; recipient of over 50 patents and product liability cases; commercial component development. *Mailing Add:* Dept Mat Eng Dow Bldg Univ Mich Ann Arbor MI 48109

FLINNER, JACK L, b Canton, Ohio, June 6, 31; m 55; c 3. NUCLEAR PHYSICS. *Educ:* Wittenberg Univ, AB, 53; Univ Ill, MS, 55; Ohio State Univ, PhD(physics), 65. *Prof Exp:* Prof physics, Wittenberg Univ, 57-69; PROF PHYSICS, MANKATO STATE UNIV, 69-, CHMN DEPT PHYSICS & ELECTRONIC ENG TECHNOL, 75- *Mem:* Sigma Xi; Am Asn Physics Teachers. *Res:* Low energy nuclear physics; gas scattering experiments; angular correlation. *Mailing Add:* 121 Cliff Ct North Mankato MN 56001-5905

FLINT, DELOS EDWARD, b Pasadena, Calif, Dec 5, 18; m 50; c 5. ECONOMIC GEOLOGY. *Educ:* Calif Inst Technol, BS, 39; Northwestern Univ, MS, 41. *Prof Exp:* Asst, Northwestern Univ, 40-41; jr geologist, US Geol Surv, 41-42, geologist, 42-57; geologist, Freeport Sulphur Co, 57-64, chief geologist, 64-70; chief geologist, Freeport Minerals Co, 70-83; RETIRED. *Concurrent Pos:* Consult, 85- *Mem:* Am Inst Mining, Metall & Petrol Engrs; fel Geol Soc Am; Soc Econ Geologists. *Res:* Areal, economic and military geology. *Mailing Add:* 1120 Mt Rose St Reno NV 89509

FLINT, ELIZABETH PARKER, b Washington, DC, July 1, 51; m 79. PLANT ECOLOGY, WEED SCIENCE. *Educ:* Mt Holyoke Col, AB, 72; Duke Univ, PhD(bot & plant ecol), 80. *Prof Exp:* RES ASSOC, DEPT BOT, DUKE UNIV, 80-, RES ASSOC, DEPT HIST, 85- *Concurrent Pos:* Res tech, Southern Weed Sci Lab, USDA, 80-81. *Mem:* Ecol Soc Am; Weed Sci Soc Am; Sigma Xi. *Res:* Ecological consequences of historical changes in land use and vegetation cover; carbon content of terrestrial vegetation; competition of weeds and crops in controlled environments. *Mailing Add:* Land Use Proj Duke Univ 104-A W Duke Bldg Durham NC 27708

FLINT, FRANKLIN FORD, b Va, Aug 4, 25; m 48; c 3. BIOLOGICAL STRUCTURE. *Educ:* Lynchburg Col, BS, 49; Univ Va, MS, 50, PhD(biol), 55. *Prof Exp:* From instr to prof biol, Randolph-Macon Woman's Col, 51-68, chmn, Biol Dept, 60-84; staff biologist, Comn Undergrad Educ Biol Sci, NSF, Washington, DC, 68-69; CHARLES A DANA PROF BIOL, RANDOLPH-MACON WOMAN'S COL, 75- *Concurrent Pos:* Res grants, Am Philos Soc, 57 & Am Acad Arts & Sci, 58; Fulbright res scholar, Portugal, 64-65; dir, Lee Soil & Water Conserv Dist, Va, 73-; bd dirs, Va Chap Nature Conservancy, 75-; extramural assoc, NIH, Bethesda, 81; pres, Va Asn Conserv Dists, 81-83 & coun, Am Inst Biol Sci, 82-85; summer fac res grant, Med Univ SC, Charleston, 85. *Honors & Awards:* Fel AAAS. *Mem:* Bot Soc Am. *Res:* Cell research; gametogenesis in the angiosperms. *Mailing Add:* Dept Biol Randolph-Macon Woman's Col Lynchburg VA 24503

FLINT, HARRISON LEIGH, b Barre, Vt, Nov 5, 29; m 55, 86; c 4. LANDSCAPE HORTICULTURE. *Educ:* Cornell Univ, BS, 51, PhD(floricult), 58; Mich State Univ, MS, 52. *Prof Exp:* Asst, Mich State Univ, 51-52 & Cornell Univ, 53 & 56-58; asst prof hort, Univ RI, 58-62; assoc, Univ Vt, 62-66; assoc horticulturist, Arnold Arboretum, Harvard Univ, 66-68; PROF HORT, PURDUE UNIV, WEST LAFAYETTE, 68- *Mem:* Am Soc Hort Sci; Am Asn Bot Gardens & Arboretums; Ecol Soc Am; Int Plant Propagators Soc; Am Soc Landscape Architects. *Res:* Woody landscape plants; adaptation and comparative physiology; cold resistance in plants. *Mailing Add:* Dept Hort Purdue Univ West Lafayette IN 47907

FLINT, HOLLIS MITCHELL, b Miami, Fla, May 28, 38; m 60; c 4. ENTOMOLOGY, RADIATION BIOLOGY. *Educ:* Stetson Univ, BS, 60; Univ Fla, PhD(entom), 64. *Prof Exp:* Res entomologist, metab & radiation res lab, Entom Res Div, Agr Res Serv, USDA, NDak, 64-71, RES ENTOMOLOGIST, ENTOM RES DIV, AGR RES SERV, USDA, WESTERN COTTON INSECTS LAB, 71- *Mem:* Entom Soc Am. *Res:* Effects of radiation on insect sterility, longevity, oviposition and mating behavior; use of sex pheromones for control of pest Lepidoptera; host plant resistance. *Mailing Add:* Western Cotton Res Lab 4135 E Broadway Phoenix AZ 85040

FLINT, JEAN-JACQUES, b Le Vesinet, France, Aug 14, 43. GEOMORPHOLOGY, HYDROLOGY. *Educ:* Colby Col, BA, 66; State Univ NY Binghamton, MSc, 68, PhD(geol), 72. *Prof Exp:* Res asst geol, State Univ NY Binghamton, 68-70; asst prof, 70-77, ASSOC PROF GEOL, BROCK UNIV, 77- *Mem:* Geol Soc Am; Am Geophys Union; Int Asn Great Lakes Res; Ont Asn Remote Sensing. *Res:* Lake Ontario water levels; ancestral draninage of the Niagara River. *Mailing Add:* Dept Geol Sci Brock Univ Merrittville Hwy St Catharines ON L2S 3A1 Can

FLINT, NORMAN KEITH, geology; deceased, see previous edition for last biography

FLINT, OLIVER SIMEON, JR, b Amherst, Mass, Oct 10, 31; m 71; c 3. SYSTEMATIC ENTOMOLOGY. *Educ:* Univ Mass, BS, 53, MS, 55; Cornell Univ, PhD(entom), 60. *Prof Exp:* Assoc cur, 60-65, CUR ENTOM, NAT MUS NATURAL HIST, SMITHSONIAN INST, 65-, IN CHG, DIV NEUROPTERIODS, 63- *Mem:* Am Entom Soc; Benthological Soc; Asn Trop Biol. *Res:* Taxonomy and biology of caddis flies and dobson flies, especially those of the New World. *Mailing Add:* Nat Mus Natural Hist Stop 105 Smithsonian Inst Washington DC 20560

FLINT, SARAH JANE, b Farnborough-Hampshire, Eng; c 2. MOLECULAR BIOLOGY, BIOCHEMISTRY. *Educ:* Univ Col London, BSc, 70, PhD(molecular biol), 73. *Prof Exp:* Fel, Cold Spring Harbor Lab, 73-74; fel, Ctr Cancer Res, Mass Inst Technol, 74-76; from asst prof to assoc prof, dept biochem sci, 77-84, assoc prof, dept molecular biol, 85-88, PROF DEPT MOLECULAR BIOL, PRINCETON UNIV, 88- *Concurrent Pos:* Assoc chmn dept biochem sci, Princeton Univ, 82-83, dir prog molecular biol, 83-84; mem molecular biol study sect, NIH, 87-91, res career develop award, 83-88; NATO fel, 73-75. *Mem:* AAAS; Am Soc Microbiol; Am Soc Virol. *Res:* Mechanisms of regulation of gene expression in mammalian cells, particularly the molecular roles of adenoviral oncogene products in transcriptional and post-transcriptional regulation. *Mailing Add:* Dept Molecular Biol Princeton Univ Princeton NJ 08544

FLIPO, JEAN, veterinary medicine, for more information see previous edition

FLIPPEN, RICHARD BERNARD, b Williamson, WVa, Aug 23, 30; m 61; c 2. SPECTROSCOPY. *Educ:* WVa Univ, AB, 53, MS, 54; Carnegie Inst Technol, PhD(physics), 60. *Prof Exp:* PHYSICIST, CENT RES DEPT, E I DU PONT DE NEMOURS & CO, INC, 60- *Mem:* Sigma Xi; Am Phys Soc; AAAS. *Res:* Magnetic properties of materials; magnetic ordering transitions; superconductivity; low temperature thermal properties of materials; electrical transport properties of materials; laser light scattering. *Mailing Add:* Exp Sta 228-332 E I du Pont de Nemours & Co Wilmington DE 19880-0228

FLIPPEN-ANDERSON, JUDITH LEE, b Winthrop, Mass, Apr 21, 41; m 78; c 1. X-RAY CRYSTALLOGRAPHY. *Educ:* Northeastern Univ, Boston, BA, 63; Ariz State Univ, MS, 66. *Prof Exp:* Lab technician, Cancer Res Inst, 58-63; CHEMIST X-RAY CRYSTALLOG, NAVAL RES LAB, 66- *Honors & Awards:* Super Serv Award, US Dept Agr, 84. *Mem:* Am Chem Soc; Am Crystallog Asn (vpres, 90, pres, 91); Sigma Xi. *Res:* X-ray diffraction analysis of small to medium sized organic and biochemical materials; adaptation and optimization of programs for a large vector processor. *Mailing Add:* Code 6030 Naval Res Lab Washington DC 20375-5000

FLIPSE, JOHN E, b Montville, NJ, Feb 4, 21; m; c 4. MARINE ENGINEERING, NAVAL ARCHITECTURE. *Educ:* Mass Inst Technol, SB, 42; NY Univ, MME, 48. *Prof Exp:* Res engr & dir res asst to pres, Dry Dock Co, Newport, Va, 57-68; chmn, chief exec officer & pres, Deep Sea Ventures Inc, 68-77; prof civil & ocean eng, Tex A&M Univ, 79-82, assoc dean eng, 84-88, assoc dep chancellor, 84-89, DISTINGUISHED PROF CIVIL & OCEAN ENG, TEX A&M UNIV, 82-, WOFFORD CAIN PROF ENG & DIR, OFFSHORE TECHNOL RES CTR, 88- *Concurrent Pos:* Consult torsional vibration, ship design & marine ins surv; vis prof civil & ocean eng, Tex A&M Univ, 78-79; vchmn, Marine Bd, Nat Res Coun, 81-82, chmn, 82-84; chmn, Tech & Res Steering Comt, Soc Naval Architects & Marine Engrs & Nat Adv Comt Oceans & Atmosphere, 85-87; pres & chief exec officer, Tex A&M Res Found, 83-84; dep dir, Tex Eng Exp Sta, 85-88. *Honors & Awards:* Schwab Mem Lectr, Am Iron & Steel Inst, 72; Doherty Lectr, 85; Blakely Smith Award, Soc Naval Architects & Marine Engrs, 89. *Mem:* Nat Acad Eng; fel Marine Technol Soc (pres, 85-87); fel Soc Naval Arch & Marine Engrs. *Res:* Technology frontier associated with mining deep ocean minerals; author of 40 technical publications; awarded 9 patents. *Mailing Add:* Eng Prog Off Tex A&M Univ College Station TX 77843-3126

FLIPSE, MARTIN EUGENE, b Monteville, NJ, Apr 27, 19; div; c 6. PREVENTIVE MEDICINE. *Educ:* Hope Col, AB, 40; Harvard Med Sch, MD, 43; Am Bd Internal Med, dipl. *Prof Exp:* Intern med, Boston City Hosp, Mass, 44, resident med & contagious dis, 44-45; asst res path, Jackson Mem Hosp, Miami, Fla, 48-49; fel med, Mayo Found & Clin, Univ Minn, 49-51; resident pulmonary dis, Nopening Sanatorium, 51; staff asst, Mayo Found & Clin, Univ Minn, 52; from asst prof to assoc prof, 56-59, PROF MED, SCH MED, UNIV MIAMI, 59-, DIR UNIV HEALTH SERV, 57- *Mem:* AMA; Am Col Health Asn; fel Am Col Physicians. *Res:* Pulmonary disease; public health. *Mailing Add:* Univ Miami Health Ctr 5513 Merrick Dr Coral Gables FL 33146

FLIPSE, ROBERT JOSEPH, b Topeka, Kans, July 18, 23; m 44; c 5. DAIRY SCIENCE. *Educ:* Kans State Col, BS, 47; Mich State Col, MS, 48, PhD(animal nutrit), 50. *Prof Exp:* Asst dairy husb, Mich State Col, 47-49; from asst prof to prof dairy, Pa State Univ, 50-88; asst dir, Agr Exp Sta, 68-88; RETIRED. *Concurrent Pos:* Mem, Sci Group Chem & Physiol Gametes, WHO, 65. *Honors & Awards:* Borden Award Res Dairy Sci, 69. *Mem:* Fel AAAS; Am Inst Biol Sci; Soc Study Reproduction; Am Soc Animal Sci; Am Dairy Sci Asn. *Res:* Cellular metabolism; nutrition and reproductive physiology; rumen physiology. *Mailing Add:* 832 Hemlock St Boalsburg PA 16827

FLISZAR, SANDOR, b Lugano, Switz, May 11, 27; Can citizen; m 61; c 3. PHYSICAL ORGANIC CHEMISTRY, QUANTUM CHEMISTRY. *Educ:* Univ Geneva, PhD(chem), 62. *Prof Exp:* Res chemist, Cyanamid Europ Res Inst, Geneva, Switz, 62-64; from asst prof to assoc prof, 64-71, PROF PHYS CHEM, UNIV MONTREAL, 71- *Res:* Ozone chemistry; theoretical. *Mailing Add:* Dept Chem Univ Montreal 6128 Succursala Montreal PQ H3C 3J7 Can

FLITCRAFT, R(ICHARD) K(IRBY), II, b Woodstown, NJ, Sept 5, 20; m 42; c 4. CHEMICAL ENGINEERING. *Educ:* Rutgers Univ, BS, 42; Washington Univ, St Louis, MS, 48. *Prof Exp:* Mem tech serv, Org Chem Div, Monsanto Co, 42-46, group leader, Tech Serv, 46-48, supvr prod dept, 48-50, supt tech serv, 50-52, asst dir res, Meramec Div, 52-55; asst dir res, Inorg Chem Div, 55-60, prod bd dir, 60, dir res, 60-65, dir mgt info & systs dept, 65-67, asst to pres, 67-68, group mgr, Electronics Enterprise, 68-69, gen mgr, Electronic Prod Div, 69-71; dir, Mound Labs & vpres, Monsanto Res Corp, 71-76, pres, 76-82; RETIRED. *Mem:* AAAS; Am Chem Soc; Am Inst Chem Engrs; Sigma Xi; NY Acad Sci. *Res:* Surface active agents and detergents; phosphates; production and process development; management information systems; electronic data processing; electronic materials; light emitting devices; environmental analyses; government contracts. *Mailing Add:* 6051 Kimway Dr Dayton OH 45459

FLITTER, DAVID, b Philadelphia, Pa, Aug 1, 13; m 42; c 3. ORGANIC CHEMISTRY. *Educ:* Univ Pa, BS, 34; Pa State Col, MS, 50, PhD(chem), 52. *Prof Exp:* Chemist, R T French Co, 34-38; asst, Sun Oil Co, 38-41; jr chemist, US Govt, 41-42; chemist, Celanese Corp Am, 42-45, supvr, 45-48; chemist, E I du Pont de Nemours & Co, 52-61; sr chemist, Wyeth Labs, Inc, 61-78; RETIRED. *Mem:* Am Chem Soc. *Res:* Organic coatings; development of new drugs. *Mailing Add:* 270 Ellis Rd Havertown PA 19083-1123

FLITTNER, GLENN ARDEN, b Los Angeles, Calif, Sept 10, 28; m 57; c 2. MARINE ECOLOGY, METEOROLOGY. *Educ:* Univ Calif, Berkeley, AB, 51, MA, 53; Univ Mich, PhD(fisheries), 64. *Prof Exp:* Fishery res biologist, Biol Lab, Bur Com Fisheries, US Dept Interior, Mich, 57-61 & Tuna Resources Lab, Calif, 61-70; dir bur marine sci, 70-74, prof biol, San Diego State Univ, 70-74; chief ocean serv div, Nat Weather Serv, 74-82, actg dir, Ocean Prod Ctr, Camp Springs, Md, 83-84, SPEC ASST TO DIR, OFF OCEAN SERV, NAT OCEANIC & ATMOSPHERIC ADMIN-NAT OCEAN SERV, WASHINGTON DC, 85- *Honors & Awards:* Commendation, Marine Technol Soc, 68. *Mem:* Am Inst Fishery Res Biol; Am Fisheries Soc. *Res:* Life history, ecology of marine fishery stocks; effects of physical characteristics of the environment on abundance, availability and distribution of stocks, special interest in forecasting availability and distribution of tuna stocks in relation to predicted changes in physical oceanographic parameters such as upper mixed layer temperature; oceanography. *Mailing Add:* 6105 Wayside Dr Rockville MD 20852

FLOCK, DONALD LOUIS, b Calgary, Alta, Feb 19, 30; m 52; c 4. PETROLEUM ENGINEERING. *Educ:* Univ Okla, BSc, 52, MSc, 53; Agr & Mech Col, Tex, PhD, 56. *Prof Exp:* Res engr, Cities Serv Res & Develop Co, Ltd, 56-57; from assoc prof to prof, 57-67, assoc dean fac eng, 69-73, prof petrol eng, 73-79, CHMN DEPT MINERAL ENG, UNIV ALTA, 79- *Mem:* Am Inst Mining, Metall & Petrol Engrs; Sigma Xi. *Res:* Petroleum production and reservoir engineering. *Mailing Add:* Dept Metall & Mat Univ Alta Edmonton AB T6G 2M7 Can

FLOCK, JOHN WILLIAM, b Denver Colo, Feb 4, 47; m 69; c 2. CHEMICAL ENGINEERING. *Educ:* Univ Colo, Boulder, BS, 69; Univ Calif, Berkeley, MS, 72, PhD(chem eng), 76. *Prof Exp:* Mem tech staff chem eng, TRW Systs Group, 70-71; res asst, Dept Chem Eng, Univ Calif, Berkeley, 71-76; staff chem engr, 76-80, GEN MGR, PROCESS TECHNOL, GEN ELEC CO, MT VERNON, IND, 80- *Mem:* Am Inst Chem Engrs; AAAS; Sigma Xi. *Res:* Thermodynamic analysis and development of closed loop chemical systems for transport and storage of energy; thermodynamic analysis of the irreversible processes associated with heat pumps and chemical processes. *Mailing Add:* Gen Elec Co One Lexan Lane Mt Vernon IN 47620

FLOCK, WARREN L(INCOLN), b Kellogg, Idaho, Oct 26, 20; m 57. ELECTRICAL ENGINEERING, GEOPHYSICS. *Educ:* Univ Wash, BS, 42; Univ Calif, Berkeley, MS, 48; Univ Calif, Los Angeles, PhD(eng), 60. *Prof Exp:* Staff mem radiation lab, Mass Inst Technol, 42-45; partner, Radar Engrs, Wash, 45-46; lectr & asst engr, Col Eng, Univ Calif, Los Angeles, 50-56, lectr assoc engr, 56-60; from assoc prof to prof geophys, Geophys Inst, Univ Alaska, 60-64; PROF ELEC ENG, UNIV COLO, BOULDER, 64- *Concurrent Pos:* Mem tech staff, Jet Propulsion Lab, Pasadena, CA, 80-81. *Mem:* AAAS; Inst Elec & Electronics Engrs; Am Geophys Union; Int Solar Energy Soc; Arctic Inst NAm. *Res:* Propagation effects on electromagnetic waves in satellite and deep-space communications; radio science; remote sensing of the environment; telecommunications; radar ornithology; solar energy. *Mailing Add:* Univ Colo Campus Box 425 Boulder CO 80309

FLOCKEN, JOHN W, b Sioux Falls, SDak, Sept 2, 39; m 64, 76; c 3. THEORETICAL SOLID STATE PHYSICS. *Educ:* Augustana Col, SDak, BA, 61; Univ Nebr, Lincoln, MS, 64, PhD(physics), 69. *Prof Exp:* Scientist, Raven Industs, SDak, 64-65; physicist, Bendix Labs, Mich, 65-66; from asst prof to assoc prof physics, Univ Nebr, Omaha, 69-75, asst dean arts & sci, 75, chmn dept, 80-83, PROF PHYSICS, UNIV NEBR, OMAHA, 75- *Mem:* Am Phys Soc. *Res:* Lattice statics calculations; determination of atomic displacements about point defects in metals; formation and activation energies; interactions between pairs of defects in crystals; structural phase transitions; ferroelectricity; high technetium superconductivity. *Mailing Add:* Dept Physics Univ Nebr Omaha NE 68182

FLODIN, N(ESTOR) W(INSTON), b Chicago, Ill, Jan 30, 15; m 41; c 2. BIOCHEMISTRY, NUTRITION. *Educ:* Univ Chicago, BS, 35, PhD(chem), 38. *Prof Exp:* Instr chem, Cent State Teachers Col, Stevens Point, 38-40; chemist electrochem, E I Du Pont de Nemours & Co, Inc, 40-61, chemist indust & biochem, 61-66, mkt res pharmaceut, 66-73; PROF BIOCHEM, UNIV SALA COL MED, 74- *Concurrent Pos:* Assoc Ed, J Am Col Nutrit; editor, Am Col Nutrit Newsletter. *Mem:* NY Acad Sci; AAAS; Am Col Nutrit; Am Inst Nutrit; Sigma Xi. *Res:* Organic boron compounds; vinyl monomers and polymers; amino acids in nutrition; pharmaceuticals marketing and research; kinetics of nutritional responses; vitamins; trace elements. *Mailing Add:* Univ SAla Col Med Dept Biochem Mobile AL 36688

FLOKSTRA, JOHN HILBERT, b Chicago, Ill, Apr 17, 25; m 48; c 3. CLINICAL CHEMISTRY. *Educ:* Calvin Col, AB, 47; Mich State Col, MS, 50, PhD(biochem), 52. *Prof Exp:* Asst biochem, Mich State Col, 47-51; res assoc dept biochem, Upjohn Co, 52-59, head clin res lab, 60-83; RETIRED. *Mem:* Am Chem Soc; Am Asn Clin Chem; Am Soc Clin Path; Can Soc Clin Chem. *Mailing Add:* Box 732 Seven Lakes West End NC 27376

FLOM, DONALD GORDON, b Kenyon, Minn, May 8, 24; m 57; c 3. TRIBOLOGY, MANUFACTURING ENGINEERING. *Educ:* St Olaf Col, BA, 44; Purdue Univ, MS, 49; Pa State Col, PhD(chem), 52. *Prof Exp:* Instr chem, St Olaf Col, 46-47; asst, Purdue Univ, 47-49; res assoc, Res Lab, Gen Elec Co, 51-59, phys chemist, 59-61, mgr chem res, Space Sci Lab, Valley Forge Space Tech Ctr, 61-71, mgr chem eng br, Res & Develop Ctr, 71-74, mgr mat removal & lubrication prog, 74-80, mgr machining technol appln, Res & Develop Ctr, 80-87; CONSULT, 87- *Honors & Awards:* Nat Award, Soc Tribologists & Lubrication Engrs; Mayo D Hersey Award, Am Soc Mech Engrs. *Mem:* Am Chem Soc; Soc Tribologists & Lubrication Engrs; Soc Mfg Engrs. *Res:* Vapor liquid equilibria; chemical use of high frequency oscillators; surface chemistry; electrical contacts; friction and wear; polymer chemistry; reinforced composites; ablation; machining; superhard materials. *Mailing Add:* 107 Oakwood Dr Scotia NY 12302-4711

FLOM, MERTON CLYDE, b Pittsburgh, Pa, Aug 19, 26; m 68; c 4. OPTOMETRY. *Educ:* Univ Calif, Berkeley, BS, 50, MOpt, 51, PhD(physiol optics), 57. *Prof Exp:* Clin instr optom, Sch Optom, Univ Calif, Berkeley, 51-57, instr, 57-58, from asst prof to assoc prof physiol optics & optom, 58-66, vchmn sch optom, 62-63, asst dean, 69-72, prof physiol optics & optom, 67-; AT COL OPTOM, UNIV HOUSTON CENT COL. *Concurrent Pos:* Pvt pract optom, 60-70; USPHS res grants, Sch Optom, Univ Calif, Berkeley, 62-67 & 77-80; NASA grant, Smith-Kettlewell Inst Visual Sci, 72, US Army Basic contract grants, 72-; assoc res mem, Inst Med Sci, Pac Med Ctr, San Francisco, 68-; consult, Optical Sci Group, San Rafael, 69-; ed, J Optom & Physiol Optics, 77- *Mem:* AAAS; Am Acad Optom (pres elect, 78); Am Asn Univ Prof; Am Optom Asn; Optical Soc Am. *Res:* Physiological optics; binocular vision and space perception. *Mailing Add:* Col Optom Univ Houston Cent Col 4800 Calhoun St Houston TX 77204-6052

FLOOD, BRIAN ROBERT, b Phelps, NY, Feb 21, 48; m 70; c 3. ENTOMOLOGY, WEED SCIENCE. *Educ:* Purdue Univ, Lafayette, BS, 70; Univ Wis-Madison, MS, 72, PhD(entom), 75. *Prof Exp:* Consult, Agr Chem Serv Corp, Sodus, NY, 74-75; SR ENTOMOLOGIST APPL RES, DEL MONTE, USA, ROCHELLE, ILL, 75-, AGR RES MGR, 80-, MGR PEST MGT-VEGS, 85- *Concurrent Pos:* Portable Containment Inc. *Mem:* Entom Soc Am; Am Registry Prof Entomologists; Coun Agr Sci & Technol; NCent Weed Sci Soc. *Res:* Develop, maintain and oversee effective economical weed and insect control programs and cultural practices for sweet corn, snap beans, lima beans, peas and cabbage; development of alternates for processing waste and vegetable processing methods; develop portable containment for chemicals. *Mailing Add:* Del Monte Foods PO Box 89 Rochelle IL 61068

FLOOD, CLIFFORD ARRINGTON, JR, b Jacksonville, Fla, Oct 17, 39; m 68; c 2. AGRICULTURAL ENGINEERING. *Educ:* Univ Fla, BAGE, 62; Univ Ky, MSAE, 66; Purdue Univ, PhD(agr eng), 71. *Prof Exp:* asst prof, 71-79, ASSOC PROF AGR ENG, AUBURN UNIV, 79- *Concurrent Pos:* Vis assoc prof agr eng, Univ Ky, 81-82. *Mem:* Am Soc Agr Engrs; Coun Agr Sci & Technol. *Res:* Heat and mass transfer in agricultural products; feed and grain drying and storage; environmental control in livestock housing; agricultural application of solar energy for heating and cooling. *Mailing Add:* 856 Janet Dr Auburn AL 36830

FLOOD, DOROTHY GARNETT, b Sayre, Pa, Oct 7, 51; m 83. NEUROSCIENCES, GERONTOLOGY. *Educ:* Lawrence Univ, BA, 73; Univ Ill, Champaign-Urbana, BA, 73; Univ Rochester, MS & PhD(anat), 80. *Prof Exp:* Sr instr anat, 80-83, sr instr, 84, asst prof, 84-90, ASSOC PROF NEUROL, UNIV ROCHESTER, 90- *Mem:* Soc Neurosci; Am Asn Anatomists; AAAS; Europ Neurosci Asn. *Res:* Plasticity of the nervous system, particularly in development and aging. *Mailing Add:* Dept Neurol Box 673 601 Elmwood Ave Rochester NY 14642

FLOOD, PETER FREDERICK, b Manchester, Eng, Apr 15, 40; Can citizen; m 61, 77; c 3. VETERINARY ANATOMY, REPRODUCTION IN UNGULATES. *Educ:* Univ Liverpool, BVSc, 63; Univ Wales, MSc, 67; Univ Bristol, PhD(anat), 71. *Prof Exp:* Lectr anat, Univ Bristol, 67-77; assoc prof, 77-80, PROF ANAT, 80- HEAD DEPT, WESTERN COL VET MED, UNIV SASK, 81. *Mem:* Soc Study Fertility, UK; Can Asn Anatomists; Am Asn Vet Anatomists; Wildlife Dis Asn. *Res:* Ecophysiology of the muskox, Ovibos moschatus, especially reproductive adaptations, role of body composition, reproductive and metabolic hormones and photoperiod; endocrinology of pregnancy in the muskox; placental morphology of ungulates. *Mailing Add:* Dept Vet Anat Univ Sask Saskatoon SK S7N 0W0 Can

FLOOD, THOMAS CHARLES, b Wilmington, Del, June 5, 45; m 67; c 2. ORGANIC & INORGANIC CHEMISTRY, ORGANOMETALLIC CHEMISTRY. *Educ:* Trinity Col, BS, 67; Mass Inst Technol, PhD(org chem), 72. *Prof Exp:* From asst prof to assoc prof, 72-88, PROF CHEM, UNIV SOUTHERN CALIF, 88- *Concurrent Pos:* Sloan fel, 77; Humboldt fel, 82. *Mem:* Am Chem Soc; The Chem Soc. *Res:* Stereochemistry and mechanism of organometallic reactions, homogeneous catalysis, models for metallo-enzymes; metal catalyzed oxidations. *Mailing Add:* Dept Chem Univ Southern Calif Los Angeles CA 90089-0744

FLOOD, WALTER A(LOYSIUS), b New York, NY, Apr 27, 27; m 54; c 3. ELECTRICAL ENGINEERING. *Educ:* Cornell Univ, BEE, 50, MEE, 52, PhD, 54. *Prof Exp:* Asst ionosphere lab, Cornell Univ, 50-54, res engr aeronaut lab, 54-59, head radio physics sect, 59-64, staff scientist, 64-67; prof elec eng, NC State Univ, 67-81; DIR, GEOSCI DIV, ARMY RES OFF, 81- *Concurrent Pos:* Consult, Res Triangle Inst, Stanford Res Inst & Gen Elec Tempo Ctr Advan Study, Calif, Appl Sci Assoc, NC. *Mem:* Fel Inst Elec & Electronic Engrs; Am Geophys Union; Sigma Xi; Am Optical Soc. *Res:* Upper atmospheric physics; ionospheric and tropospheric propagation; D-region backscatter; ionospheric drifts; rough surface backscatter; radio wave propagation; ionospheric scintillation and multipath phenomena. *Mailing Add:* PO Box 12211 Research Triangle Park NC 27709

FLOOK, WILLIAM MOWAT, JR, b Briarcliff, NY, July 7, 21; m 45; c 5. PHYSICS. *Educ:* Harvard Univ, AB, 43; Brown Univ, MS, 49, PhD(physics), 52. *Prof Exp:* Asst physics, Woods Hole Oceanog Inst, Mass, 46-47; res supvr, E I du Pont de Nemours & Co, 51-61, res assoc, 61-69, res assoc, Peltron Lab, 69-81; RETIRED. *Mem:* AAAS; Optical Soc Am; Sigma Xi; Am Inst Aeronaut & Astronaut. *Res:* Instrumentation; process control; optical and electronic measurements. *Mailing Add:* Rte 3 Box 278A Chestertown MD 21620

FLORA, EDWARD B(ENJAMIN), b Phillipsburg, Ohio, June 23, 29; m 52; c 3. STRUCTURAL MECHANICS, MECHANICAL ENGINEERING. *Educ:* Carnegie Inst Technol, BS, 51, MS, 53. *Prof Exp:* Engr, Nuclear Res Ctr, Carnegie Inst Technol, 51-52; proj engr, Nevis Cyclotron Lab, Columbia Univ, 53-58; sr engr, Dalmo Victor Co, 58-63; vpres, 63-82, mgr, 72-85, PRES, AEROSPACE STRUCTURES INFO & ANALYTICAL CTR, ANAMET LABS, 82- *Mem:* Am Soc Mech Engrs; Soc Exp Mech; Soc Automotive Engrs; Am Inst Aeronaut & Astronaut; Am Soc Testing & Mat. *Res:* Analytical and experimental stress analysis; vibration; failure analysis; fracture mechanics. *Mailing Add:* Anamet Laboratories 3400 Investment Blvd Hayward CA 94545-3811

FLORA, GEORGE CLAUDE, b Clark, SDak, Apr 8, 23; m 51; c 3. NEUROLOGY. *Educ:* Univ SDak, BS, 48; Temple Univ, MD, 50. *Prof Exp:* Prof neurol, Univ Minn, Minneapolis, 63-74; PROF NEUROL & HEAD, SCH MED, UNIV SOUTH DAK, 74- *Concurrent Pos:* Consult neurol, Fargo Vet Admin Hosp, 63-74; consult neuropath, Ramsey-Ancker County Hosp, St Paul, 63-68 & State of Minn, Anoka, 63-68; acad consult, Sioux Valley Hosp, McKennen Hosp, Vet Admin Hosp, Sioux Falls, Sacred Heart Hosp, Yankton, Human Serv Ctr, Yankton & St Annes Hosp, Watertown, 74- *Mem:* AMA; Acad Neurol; Am Soc Internal Med; Asn Univ Prof Neurol. *Res:* Cerebrovascular disease, epidemiology and ultrastruc- ultrastructure of the arthrosclerotic plaque; chemophilia; role of medical profession in the handling of alcoholism and chemical dependency. *Mailing Add:* Dept Neurol Univ SDak Sch Med 2501 W 22nd St Sioux Falls SD 57117

FLORA, JAIRUS DALE, JR, b Northfield, Minn, Mar 27, 44; m 67; c 1. BIOSTATISTICS, ENVIRONMENTAL STATISTICS. *Educ:* Midland Lutheran Col, BS, 65; Fla State Univ, MS, 68, PhD(statist), 71. *Prof Exp:* From asst prof to prof biostatist, Sch Pub Health, Univ Mich, Ann Arbor, 71-84, asst res scientist, 73-76, assoc res scientist, Hwy Safety Res Inst, 76-81, res scientist, Transportation Res Inst, 82-84; prin satisfaction, 84-90, SR ADV STATIST, MIDWEST RES INST, KANSAS CITY, MO, 91-; CLIN PROF BIOSTATIST, UNIV MO, KANSAS CITY, 84- *Concurrent Pos:* Dir statist anal & data collection, Inst Burn Med, 72-77; Bicycle Mfrs Asn grant bicycle accidents & injuries, Hwy Safety Res Inst, Univ Mich, 74-78; proj dir & prin investr res contracts, Nat Hwy Traffic Safety Admin, 74-84; dir data mgt & statist, Underground Storage Tank Surv, Environ Protection Agency, 85. *Honors & Awards:* Enterprise Award, Coun Prin Scientists, 85. *Mem:* Int Biomet Soc; Am Statist Asn; Inst Math Statist; Sigma Xi. *Res:* Statistical methodology; analysis of accident statistics; cooperative medical trials, particularly of burn treatment, cardiology and cancer treatments; analysis of accident statistics, evaluation of standards effectiveness; design and implementation and evaluation of prevention programs; applied statistics; chemometrics. *Mailing Add:* Midwest Res Inst 425 Volker Blvd Kansas City MO 64110

FLORA, LEWIS FRANKLIN, b Frederick, Md, June 9, 47; m 69; c 3. FOOD SAFETY & PROCESSING. *Educ:* Univ Md, BS, 69, MS, 71, PhD(food sci), 73. *Prof Exp:* Food technologist, US Food & Drug Admin, 74; asst prof, Dept Food Sci, Exp Sta, Univ Ga, 74-80; proj leader, Am Home Foods, 80-82; res scientist, Coca-Cola Co, 82-87; res & develop mgr, McCormick & Co, 87-89; FOOD SCIENTIST & PROG MGR, COOP ST RES SERV, USDA, 89- *Mem:* Inst Food Technologists. *Res:* Administration and coordination of federally supported research programs in food science at land-grant universities and state agricultural experiment stations; food and beverage research and development. *Mailing Add:* Coop State Res Serv USDA 901 D St SW Washington DC 20250-2200

FLORA, ROBERT HENRY, b Johnson City, NY, July 12, 44; m 68; c 2. QUENCH PROTECTION OF LARGE DISTRIBUTED SUPERCONDUCTING SYSTEMS. *Educ:* Clarkson Col, BS, 67, MS, 68. *Prof Exp:* Instr physics, De Vry Inst Technol, 68-70; opers engr, Brookhaven Nat Lab, 70-71; ENG PHYSICIST, FERMI NAT ACCELERATOR LAB, 71- *Res:* Application of superconductivity to high energy physics particle accelerators; mechanical loading during magnet excitation; quench development; quench protection of large magnet strings. *Mailing Add:* 815 Greenwood Ct Batavia IL 60510-3228

FLORA, ROBERT MONTGOMERY, b Richmond, Va, Oct 1, 38; m 65; c 2. BIOCHEMISTRY, ENZYMOLOGY. *Educ:* Bridgewater Col, BA, 60; Va Polytech Inst & State Univ, PhD(biochem), 65. *Prof Exp:* NIH fel microbiol, Vanderbilt Univ, 64-66; asst prof chem, Am Univ, 66-68; sr res biochemist, 68-72, dir res prod develop, 72-77, head biotechnol, 79-83, SR CONSULT SCIENTIST, WORTHINGTON BIOCHEM CORP, DIV MILLIPORE CORP, 83-; VPRES RES & DEVELOP, PHARMACIA LKB BIOTECHNOL INC, 84- *Mem:* AAAS; Am Chem Soc; NY Acad Sci; Sigma Xi. *Res:* Products, processes and applications, enzymes and biochemicals; separations and purifications products and applications. *Mailing Add:* 28 Stacy Dr Belle Mead NJ 08502

FLORA, ROGER E, b Roanoke, Va, Feb 5, 39; m 63; c 2. BIOSTATISTICS. *Educ:* Univ Va, BA, 60; Va Polytech Inst, MS, 64, PhD(statist), 66. *Prof Exp:* Asst prof statist, WVa Univ, 65-68; from asst prof to assoc prof biostatist, Med Col Va, Va Commonwealth Univ, 68-77; SR RES STATISTICIAN, A H ROBINS CO, 77- *Mem:* Am Statist Asn; Biostatist Soc. *Res:* Multivariate analysis; statistical inference; analysis of clinical trials. *Mailing Add:* 4505 Shirley Rd Richmond VA 23225-1056

FLORAN, ROBERT JOHN, b Bronx, NY, June 4, 47; m 71; c 2. BASIN MODELING. *Educ:* City Col New York, BS, 69; Columbia Univ, MA, 71; State Univ NY, Stony Brook, PhD(petrol), 75. *Prof Exp:* Res assoc petrol, Johnson Space Ctr, 75-76 & Am Mus Natural Hist, 76-77; res assoc geol, Oak Ridge Nat Lab, 78-80; RES ASSOC, UNOCAL, 81- *Concurrent Pos:* Nat Res Coun fel, 75-76; NASA res grants co-investr, 76-77. *Mem:* Am Asn Petrol Geologists. *Res:* Basin modeling; meteoritics; impact cratering processes; mineralogy and petrology of iron formations; radioactive waste management. *Mailing Add:* Unocal 376 S Valencia Ave Brea CA 92621

FLORANCE, EDWIN R, b Lewiston, Idaho, Mar 2, 42; m; c 1. CELL BIOLOGY, BOTANY. *Educ:* Eastern Wash State Col, BA, 65; Ore State Univ, MS, 71, PhD(bot), 74. *Prof Exp:* Asst instr biol, Eastern Wash State Col, 65-68; instr biol, Columbia Basin Community Col, 74-77; from asst prof to assoc prof, 77-86, PROF BIOL, LEWIS & CLARK COL, 86- *Concurrent Pos:* Res assoc, Wash State Univ Tree Fruit Res Ctr, 77-79; coop aid, USDA Forest Serv, 78-82; Grad res asst Ore State Univ, 71-73. *Mem:* Am Inst Biol Sci; Electron Micros Soc Am; Am Phytopath Soc. *Res:* General interests: developmental cytology and three-dimensional cell structure; specifically, developmental ultrastructural cytology of viruses, mycoplasma, bacteria, fungi and algae, cytopathology; the cytology of host parasite interactions. *Mailing Add:* Dept Biol Lewis & Clark Col 0615 SW Palatine Hill Rd Portland OR 97219

FLORANT, GREGORY LESTER, b New York, NY, Aug 31, 51; m; c 1. ENDOCRINOLOGY, NEUROPHYSIOLOGY. *Educ:* Cornell Univ, BS, 73; Stanford Univ, PhD(biol), 78. *Prof Exp:* Fel neurol res, Montefiore Hosp, NY, NIH, 78-80; asst prof biol, Swarthmore Col, 80-89; ASST PROF BIOL, TEMPLE UNIV, 90- *Mem:* Am Physiol Soc; AAAS; Am Soc Zoologists; Sigma Xi. *Res:* Hibernation physiology, specifically how hormones regulate metabolic processes during euthermia and hibernation. *Mailing Add:* Dept Biol Temple Univ Philadelphia PA 19122

FLOREA, HAROLD R(OBERT), b New York, NY, July 24, 14; m 47. ENGINEERING. *Educ:* Stevens Inst Technol, ME, 37. *Prof Exp:* Dist head, US Navy Field Off, Ind, 41-45, Vt, 45-46, & NY, 47-51; prof eng, Off Naval Res, 51-57, head weapon syst, Trainers Flight Br, 57-60, adv tech dept, 60-72; ENG CONSULT, 72- *Res:* Development and applications in training devices. *Mailing Add:* 1128 W Winged Foot Circle Winter Springs FL 32708

FLOREEN, STEPHEN, b Chicago, Ill, July 19, 32; m 61; c 1. PHYSICAL METALLURGY. *Educ:* Mass Inst Technol, BS, 54; Univ Mich, MS, 55, PhD(metall), 60. *Prof Exp:* Res asst, Univ Mich, 57-60; metallurgist steels & nickel alloys, 60-62, sect head, 62-64, res assoc, 64-73, RES FEL, INT NICKEL CO, 73-; PRIN ENGR, GEN ELEC CO, 84- *Mem:* Am Soc Metals; Am Inst Mining, Metall & Petrol Engrs; Sigma Xi. *Res:* High temperature metallurgy; surface effects; nickel alloys; high strength steels; fracture behavior. *Mailing Add:* 2501 Whamer Lane Schenectady NY 12309

FLORENTIN, RUDOLFO ARANDA, SURGICAL PATHOLOGY. *Educ:* Univ Philippines, MD(surg path), 53. *Prof Exp:* PROF & DIR SURG PATH, DEPT PATH, ALBANY MED COL, 55- *Mailing Add:* Dept Path Albany Med Col 47 New Scotland Ave Albany NY 12208

FLORES, IVAN, b New York, NY, Jan 3, 23; m 54; c 2. COMPUTER SCIENCE, ELECTRICAL ENGINEERING. *Educ:* Brooklyn Col, BA, 48; Columbia Univ, MA, 49; NY Univ, PhD(ed), 55. *Prof Exp:* Sr engr, Mergenthaler Linotype Co, 50-53; proj engr, Balco Labs, 53-55 & Nuclear Develop Corp, 55-57; proj supvr comput design, Remington Rand, 57-59; comput consult, Dunlap & Assocs, 59-60; assoc prof elec eng, Polytech Inst Brooklyn, 61-62; adj prof, NY Univ, 62-63; assoc prof, Stevens Inst Technol, 65-67; INDEPENDENT CONSULT COMPUT DESIGN, 60-; PROF STATIST, CITY UNIV NEW YORK, 67- *Concurrent Pos:* Ed, J Asn Comput Mach, 63-67; consult, US Army Sci Adv Panel, 69-70 & UN Develop Prog Surv, India & Philippines. *Mem:* AAAS; Inst Elec & Electronics Engrs; Asn Comput Mach; fel Brit Comput Soc. *Res:* Theory and design of general and special digital computers; programming computers; optical, magnetic character recognition; mathematical analysis, modeling; programming and software development; author or coauthor of seventy paper and twenty textbooks. *Mailing Add:* 108 Eighth Ave Brooklyn NY 11215

FLORES, ROMEO M, b San Fernando, Philippines, Apr 28, 39; c 1. SEDIMENTOLOGY, GEOLOGY. *Educ:* Univ Philippines, BS, 59; Univ Tulsa, MS, 62; La State Univ, PhD(geol), 66. *Prof Exp:* Jr geologist & micropaleontologist, Island Oil Co, Inc, Philippines, 59-60; from asst prof to prof geol, Sul Ross State Univ, 66-75, chmn dept, 67-75; RES GEOLOGIST, BR COAL RESOURCES, US GEOL SURVEY, 75- *Concurrent Pos:* Adj prof, Colo State Univ & NC State Univ, 78-; lectr, Am Asn Petrol Geologists, 81- *Mem:* Int Asn Math Geol; Am Asn Petrol Geol; Soc Econ Paleontologists & Mineralogists; Geol Soc Am. *Res:* Sedimentology and stratigraphy of the Paleozoic sedimentary rocks, Marathon Basin, Texas; fluvial and lake sedimentation of Lake Erie, New York; coal resources and coal depositional environments of Tertiary and Cretaceous basins in Alaska, Wyoming, Montana, North Dakota, Colorado, New Mexico, Utah, Kentucky, West Virginia and New Zealand; Fluvio-deltaic-barrier sediments of Rio Grande, Mustang Island and Gulf Coast continental shelf sediments; lectures in Brazil, China, New Zealand, Australia, India, South Africa and Belgium. *Mailing Add:* Br Coal Resources US Geol Survey MS 972 Fed Ctr Denver CO 80225

FLORES, SAMSON SOL, b Luzon, Philippines, Jan 6, 22; US citizen; m 51; c 2. PROSTHODONTICS. *Educ:* Centro Escolar Univ, Manila, DMD, 44; Univ Ill, DDS, 58. *Prof Exp:* From asst to assoc prof prosthodontics, Col Dent, Univ Ill Med Ctr, 48-68, consult-chancellor adv coun, 69-71, from clin dir to actg head dept, 68-75, prof prosthodontics, 68-, dir advan prosthodontics, 75-; RETIRED. *Concurrent Pos:* Consult, Mouth Guard Proj, USPHS, 63, Spec Patient Geriat Proj, Ill State Pub Health, 63 & VA Hosp, Chicago, 75; dir dent div, Pac Garden Mission Clins, Chicago; exec dir, Philippine Dent Soc Midwest, 72-77; consult, Ill State Dent Bd Examrs, 72- *Mem:* Am Col Dent; Am Dent Asn; Am Prosthodont Soc; Acad Gen Dent; Am Asn Univ Prof. *Res:* Oral surgical prosthesis after cancer patients are released; mouth protectors for use in general anesthesia. *Mailing Add:* 1415 Oakton Ave Evanston IL 60202

FLOREY, KLAUS, b Dresden, Germany, July 4, 19; nat US; m 56; c 2. PHARMACEUTICAL CHEMISTRY. *Educ:* Univ Heidelberg, dipl, 47; Univ Pa, PhD(biochem), 54. *Prof Exp:* Asst org chem, I G Farben, Germany, 44-45; res chemist, Merck & Co, 49-50; res asst, E R Squibb & Sons, 54-59, dir anal res & phys chem, 59-84, CONSULT, 84. *Concurrent Pos:* Mem, USP

Comt on Revision, 70-; ed, Anal Profiles of Drug Substances, 72-; mem expert advisory panel int pharmacopeia, WHO, 76-; chmn, Coun Sci Soc Pres, 83. *Honors & Awards:* Justin Powers Award in Pharmaceut Analysis, 87; Distinguished Serv Award, Am Asn Pharmaceut Scientists. *Mem:* fel AAAS; Am Chem Soc; fel Acad Pharmaceut Sci (pres, 80-81); Am Asn Pharmaceut Scientists. *Res:* Isolation, chemistry and development of natural products, steroids and medicinals; analytical research; chromatography; radiopharmaceuticals. *Mailing Add:* 151 Loomis Ct Princeton NJ 08540

FLORIDIS, THEMISTOCHLES PHILOMILOS, metallurgy, for more information see previous edition

FLORIG, HENRY KEITH, b Pittsburgh, Pa, July 15, 54; m 80; c 1. RISK MANAGEMENT, NONIONIZING RADIATION EXPOSURE ASSESSMENT. *Educ:* Carnegie Mellon Univ, BS, 76, MS, 79 & 81, PhD(eng & pub policy), 86. *Prof Exp:* Intern, Issue Anal & Mgt Dept, Westinghouse Power Systs Co, 79-81; assoc scientist, Inst Res, Carnegie Mellon Univ, 81-82, res asst, Dept Eng & Pub Policy, 83-86, res fel, 86-89, res scientist, 89-90; phys scientist, Int Nuclear Affairs Div, US Arms Control & Disarmament Agency, 90; FEL, CTR RISK MGT, RESOURCES FOR THE FUTURE, 90- *Concurrent Pos:* Consult, Consumers Union, Duquesne Light Co, Northern Elec Co, Westinghouse Elec Co, US Food & Drug Admin, World Bank & var other insts, 80-; vis scholar, Int Inst Appl Systs Anal, Laxenburg, Austria, 83; prin investr, Elec Power Res Inst, 88-90; mem, Comt on Health Effects of Ground-Wave Emergency Network, Nat Acad Sci, 91-; adj asst prof, Dept Eng & Pub Policy, Carnegie Mellon Univ, 91- *Mem:* AAAS; Bioelectromagnetics Soc; Health Physics Soc; Inst Elec & Electronics Engrs. *Res:* Technical aspects of environmental policy issues involving ionizing and non-ionizing radiation, including exposure assessment and risk management for radon and 60 Hertz electromagnetic fields. *Mailing Add:* Ctr Risk Management Resources for the Future 1616 P St NW Washington DC 20036

FLORIN, ROLAND ERIC, b Chicago, Ill, Jan 18, 15; m 49; c 2. POLYMER CHEMISTRY, ORGANIC CHEMISTRY. *Educ:* Univ Ill, BS, 36, PhD(phys chem), 48. *Prof Exp:* Abstractor, Standard Oil Co, Ill, 37-38, technologist, Ill, 38-41; asst, Univ Ill, 46-47; instr phys chem, Univ Nebr, 48-51; chemist, 51-80, GUEST WORKER, NAT BUR STANDARDS, 80- *Mem:* Am Chem Soc; Am Phys Sco. *Res:* Mechanism of polymerization; polymer degradation; deuterium compounds; fluorine compounds; copolymer composition; free radicals. *Mailing Add:* Polymer Stab & React Sect Gaithersburg MD 20899

FLORINI, JAMES RALPH, b Gillespie, Ill, Sept 22, 31; m 55; c 2. BIOCHEMISTRY. *Educ:* Blackburn Col, BA, 53; Univ Ill, PhD(biochem), 56. *Prof Exp:* Asst chem, Univ Ill, 53-56; res chemist, Lederle Labs, Am Cyanamid Co, 56-60, group leader & sr res chemist, 60-66; assoc prof, 66-70, PROF BIOCHEM, SYRACUSE UNIV, 70- *Concurrent Pos:* Chmn, Gordon Res Conf on Hormone Action, 69 & Gordon Res Conf on Biol of Aging, 80. *Mem:* Am Soc Cell Biol; Am Soc Biochem & Molecular Biol; Geront Soc; Endocrine Soc. *Res:* Control of muscle cell growth and differentiation in culture; hormone action; muscular dystrophy; aging; growth factors. *Mailing Add:* Dept Biol Syracuse Univ 130 College Pl Syracuse NY 13244

FLORIO, PASQUALE J, JR, b Palmyria, Pa, May 24, 36; m 74; c 2. APPLIED MATHEMATICS, NUMERICAL MODELING. *Educ:* Newark Col Eng, BSME, 59; NY Univ, MSME, 60, PhD(mech eng), 67. *Prof Exp:* Engr, NASA, 61-62; ASSOC PROF MECH ENG, NJ INST TECHNOL, 66- *Mem:* Am Soc Mech Engrs. *Res:* Heat transfer and fluid flow analysis; turbulent two phase flows. *Mailing Add:* 22 Swayze St West Orange NJ 07052

FLORMAN, ALFRED LEONARD, b Jersey City, NJ, Oct 11, 12; m 44; c 3. PEDIATRICS, INFECTIOUS DISEASES. *Educ:* Princeton Univ, AB, 34; Johns Hopkins Univ, MD, 38. *Prof Exp:* Intern, Johns Hopkins Hosp, 38-40; Dazian fel, Harvard Med Sch, 40-41; resident, Mt Sinai Hosp, 42; Welt fel, Rockefeller Inst Hosp & Mt Sinai Hosp, 46-47; adj pediatrician, Mt Sinai Hosp, 47-52, asst attend pediat, 52-56, assoc attend, 56-68; dir pediat, NShore Hosp, 52-68; prof pediat, Sch Med, NY Univ, 68-81; PROF PEDIAT, UNIV NMEX SCH MED, 82- *Mem:* Am Pediat Soc; Soc Pediat Res; fel Am Acad Pediat; NY Acad Med; fel Infectious Dis Soc Am. *Res:* Nosocomial infections; hospital epidemiology; clinical virology and bacteriology; neonatal and intrauterine infections. *Mailing Add:* Dept Pediat Univ NMex Sch Med Albuquerque NM 87131

FLORMAN, EDWIN F(RANK), b Venice, Ill, Feb 16, 04; m 47; c 3. ELECTRONICS. *Educ:* Washington Univ, MS, 34. *Prof Exp:* Res physicist, Western Cartridge Co, Ill, 34-41; radio engr, Nat Bur Standards, 41-44, Philco Radio Co, 44-46 & Nat Bur Standards, DC & Colo, 46-65; radio telecommun consult, Defense Commun Agency, 65-74; RETIRED. *Concurrent Pos:* Mem, Int Radio Consult Comt. *Honors & Awards:* Silver Medal, Dept Com, 56. *Mem:* AAAS; Sigma Xi; Inst Elec & Electronics Engrs. *Res:* Ballistic research; meteorological measurements; low frequency radio wave propagation studies; precise measurement of the velocity of propagation of radio waves; characteristics of atmospheric lightning discharges; tropospheric radio wave propagation; research and engineering of broadband telecommunication systems. *Mailing Add:* 10239 Brigade Dr Fairfax VA 22030

FLORMAN, MONTE, b New York, NY, Nov 19, 26; m; c 3. ENGINEERING ADMINISTRATION, ELECTRICAL ENGINEERING. *Educ:* NY Univ, BEE, 46. *Prof Exp:* Test engr bur standards, R H Macy & Co, 46-48; proj engr, Consumers Union US, Inc, 48-52; engr, Am Bosch Arma Corp, 52-53; proj engr, Consumers Union US, Inc, 53-57, head appliance div, 57-64, assoc tech dir, 64-70, tech dir, 71-83; RETIRED. *Concurrent Pos:* Auth, Consumer-directed books, 83-91. *Mem:* Inst Elec & Electronics Engrs. *Res:* Evaluation of electrical, automotive recreational, electronic electro-mechanical and medical products intended for consumer use. *Mailing Add:* 31 Vernon Pkwy Mt Vernon NY 10552

FLORMAN, SAMUEL C, b New York, NY, Jan 19, 25; m 51; c 2. CIVIL ENGINEERING. *Educ:* Dartmouth Col, BS, 44, CE, 73; Columbia Univ, MA, 47. *Hon Degrees:* DSc, Manhattan Col, 83, Clarkson Univ, 86. *Prof Exp:* VPRES, KREISLER BORG FLORMAN CONSTRUCT CO, 56- *Concurrent Pos:* Columnist, Technol Rev, Mass Inst Technol. *Honors & Awards:* Ralph Coats Roe Medal, Am Soc Mech Engrs, 82; Distinguished Lit Contrib, Inst Elec & Electronic Engrs, 90. *Mem:* Am Soc Civil Engrs; Am Soc Eng Educ; Am Soc Mech Engrs; Inst Elec & Electronics Engrs; Nat Soc Prof Engrs; Soc Am Mil Engrs. *Res:* Relationship of engineering and technology to the general culture. *Mailing Add:* 55 Central Park W New York NY 10023

FLORSCHUETZ, LEON W(ALTER), b Sublette, Ill, Aug 11, 35; m 57; c 3. MECHANICAL ENGINEERING. *Educ:* Univ Ill, BS, 58, MS, 59, PhD(eng), 64. *Prof Exp:* Asst mech eng, Univ Ill, 62-63; from asst prof to assoc prof, 64-77, PROF ENG, ARIZ STATE UNIV, 77- *Concurrent Pos:* NSF res grant, 65-67. *Mem:* Am Soc Eng Educ; Am Soc Mech Engrs. *Res:* Heat transfer; vapor bubble dynamics and boiling in single and binary component systems. *Mailing Add:* Dept Mech Eng Ariz State Univ Tempe AZ 85287

FLORSHEIM, WARNER HANNS, b Hamburg, Ger, Dec 11, 22; nat US; m 52; c 2. ENDOCRINOLOGY. *Educ:* Univ Calif, Los Angeles, BA, 43, MA, 44, PhD(chem), 48. *Prof Exp:* Asst chem, Univ Calif, Los Angeles, 43-46, res assoc zool, 48-51, anat, Med Sch, 51-53, asst clin prof biol chem, 55-71, assoc clin prof physiol, Irvine, 71-81; BIOCHEMIST, US VET ADMIN HOSP, LONG BEACH, CALIF, 53- *Concurrent Pos:* USPHS fel anat, Oxford Univ, 63-64. *Mem:* Endocrine Soc; Soc Exp Biol & Med; Am Thyroid Asn; Am Fedn Clin Res; Am Physiol Soc. *Res:* Neuroendocrinology; thyroid function. *Mailing Add:* Vet Admin Hosp 5901 E Seventh St Long Beach CA 90822

FLORY, LESLIE E(ARL), b Sawyer, Kans, Mar 17, 07; m 31; c 2. ELECTRICAL ENGINEERING. *Educ:* Univ Kans, BS, 30. *Prof Exp:* Engr, RCA Corp, Camden, 30-42; res engr, RCA Labs, 42-64, fel, 61, leader spec systs res, Astroelectronics Appl Res Lab, 64-67, chief scientist, RCA Med Electronics, 67-68; dir advan develop, Roche Med Electronics Div, Hoffmann-La Roche, Inc, 68-71; RETIRED. *Concurrent Pos:* Secy-gen, Int Fedn Med Electronics & Biol Eng. *Honors & Awards:* RCA Awards, 48, 50 & 57. *Mem:* Fel Inst Elec & Electronics Engrs; Sigma Xi; NY Acad Sci. *Res:* Television and television tubes; electronic computers; special electronic tubes and circuits; industrial television; transistor applications; medical electronics; astronomical and space television. *Mailing Add:* 153 Philip Dr Princeton NJ 08540-5422

FLORY, THOMAS REHERD, b Roanoke, Va, Apr 17, 46. NUCLEAR EFFECTS SURVIVABILITY. *Educ:* Wake Forest, BS, 67; Univ Va, MS, 71. *Prof Exp:* Res assoc physics, Univ Va, 67-68; electronics engr, US Army Missile Command, 68-69; mem tech staff, TRW Systs, 69-70; RES PHYSICIST, HARRY DIAMOND LABS, 71- *Concurrent Pos:* Consult, US Army Aviation Command & Commun Electron Command, 77-, US Army Armament Command & Missile Command, 79-; fel, InterAm Coord Comt, Argentina, 83; mem soc bd dirs, Sigma Xi, 88-89. *Mem:* Sigma Xi; Am Phys Soc; Inst Elec & Electronics Engrs. *Res:* Nuclear hardening problems; electromagnetic pulse; transient radiation effects of neutrons and gamma rays on electronics; thermal and shock phenomena. *Mailing Add:* 12515 Colby Dr Woodbridge VA 22192-2106

FLOSS, HEINZ G, b Berlin, Germany, Aug 28, 34; m 56; c 4. BIOCHEMISTRY. *Educ:* Tech Univ, Berlin, BS, 56, MS, 59; Munich Tech, PhD(org chem), 61. *Hon Degrees:* DSc, Purdue Univ, 86. *Prof Exp:* Sci asst, Munich Tech, 61-64 & 64-65, dozent, 66; from assoc prof to prof chem, Purdue Univ, West Lafayette, 66-77, head dept, 68-79, Lilly distinguished prof med chem, 77-82; prof chem, Ohio State Univ, 82-87, chmn dept, 82-86; PROF CHEM, UNIV WASH, 88- *Concurrent Pos:* Fel biochem, Univ Calif, Davis, 64-65; mem med chem study sect, NIH, 74-78, bio-organic & natural products chem study sect, 89- *Honors & Awards:* Res Achievement Award, Acad Pharmaceut Sci, 79; Volwiler Award, 79; Humboldt Sr Scientist Award, 80; Herbert Newby-McCoy Award, Am Soc Pharmacog, 88. *Mem:* Phytochem Soc NAm (pres, 74-75); Am Chem Soc; Am Soc Biol Chemists; fel Acad Pharmaceut Sci; Am Soc Pharmacog (pres, 77-78); fel AAAS. *Res:* Biosynthesis of secondary plant and mold metabolites; regulation of secondary metabolism; stereochemistry and mechanism of biological reactions; stereo chemistry and mechanism of enzyme reactions; stereochemical and isotope effect probes of the mechanism of alkylation of nucleic acids by carcinogenic nitrosamines; biosynthesis of natural products; studies on the physiology and regulation of secondary metabolism, and on the production of new natural products by genetic engineering; applications of radioactive and stable isotopes; applications of Fourier-transform NMR spectroscopy. *Mailing Add:* Dept Chem BG-10 Univ Wash Seattle WA 98195-0001

FLOURET, GEORGE R, b Rosario, Arg, Jan 5, 35; US citizen; m 63; c 3. BIOCHEMISTRY, ENDOCRINOLOGY. *Educ:* Columbia Univ, BS, 57, MS, 59; Univ Wis, PhD(org med chem), 63. *Prof Exp:* Instr org chem, Univ Wis, 61, asst prof org med chem, 62-63; res assoc biochem, Med Sch, Cornell Univ, 63-65, instr, 64-65; sr chemist, Abbott Labs, 65-71; assoc prof, 72-81, PROF PHYSIOL, NORTHWESTERN UNIV, CHICAGO, 81- *Res:* Chemistry, biochemistry and physiology of hypothalamic and neurohypophyseal peptide hormones. *Mailing Add:* Dept Physiol Northwestern Univ Med Sch 303 E Chicago Ave Chicago IL 60611-3064

FLOURNOY, DAYL JEAN, b San Antonio, Tex, Dec 17, 44; m 67, 90; c 3. CLINICAL MICROBIOLOGY. *Educ:* Southwest Tex State Univ, San Marcos, BSc, 65; Incarnate Word Col, MA, 68; Univ Houston, PhD(biol), 73. *Prof Exp:* Med technologist automated chem, Santa Rosa Med Ctr, 66-69; teaching fel microbiol, Univ Houston, 69-72; med tech, St Luke's Episcopal Hosp, 72-74, teaching fel, 74-75; DIR CLIN MICROBIOL, VET ADMIN MED CTR, OKLA, 75-; adj asst prof path & microbiol, Univ Okla Health Sci Ctr, 75-79, from asst prof path to assoc prof, 79-87, PROF PATH, UNIV

OKLA HEALTH SCI CTR, 87- *Concurrent Pos:* Contrib ed, Method Finding Exp Clin Pharmacol, 79-; adj prof microbiol & immunol, Univ Okla Health Sci Ctr, 88, adj prof clin lab sci, 88, adj prof biostatist & epidemiol, 88; clin lab dir, Am Bd Bioanal, 85; sci staff, Okla Med Res Found, 87- *Mem:* Am Soc Microbiol; fel Am Acad Microbiol; Sigma Xi. *Res:* Clinical microbiology-pathology, antimicrobials, infectious diseases, medical education and immunology-serology; septicemia, antimicrobial susceptibility testing, innate human serum bactericidal activity and organism survival. *Mailing Add:* Microbiol Sect 113 Vet Admin Med Ctr 921 NE 13th St Oklahoma City OK 73104

FLOURNOY, ROBERT WILSON, b Tulsa, Okla, Dec 12, 36; c 6. ENVIRONMENTAL TOXICOLOGY, HAZARDOUS WASTE MANAGEMENT. *Educ:* Tex A&M Univ, BS, 59, MS, 61, PhD(physiol), 66. *Prof Exp:* Instr biol, Tex A&M Univ, 61-65; from asst prof to assoc prof zool, La Tech Univ, 66-74, prof, 75-77; pres & chief exec officer, Found Testing Labs, Inc, 76-; PRES & CHIEF EXEC OFFICER, ENVIRO MED LABS INC, 74- & ENVIRON MGN INC, 86- *Concurrent Pos:* Teaching fel, Baylor Col Med, Dallas, 67, 68; consult toxicologist, Hazardous Waste Mgt Prog, La Dept Environ Qual, Baton Rouge, 78-81; consult environ scientist, La Dept Transp & Develop, Baton Rouge, 80-82; pres, Environ Mgt Inc, 85. *Mem:* Am Water Works Asn; Am Water Resources Asn; NY Acad Sci; Am Physiol Soc; AAAS; Soc Toxicol; Am Asbestos Coun. *Res:* Effects of sympathomimetic amines and related compounds on the cardiovascular system of mammals; action of saponins on uterine motility in rodents; effects of environmental products on mammalian systems; development of a chemical spill detector for industry; biogradation of hazardous waste. *Mailing Add:* Enviro-Med Labs Inc 414 W California Ruston LA 71270

FLOUTZ, WILLIAM VAUGHN, b Defiance, Ohio, Nov 12, 35; m 59; c 2. ANALYTICAL CHEMISTRY. *Educ:* Kent State Univ, BS, 57; Purdue Univ, MS, 59. *Prof Exp:* Chemist, 59-61, res chemist, 61-67, sr res chemist, 67-76, res assoc, 76-85, SR RES ASSOC, BASF WYANDOTTE CORP, 85- *Mem:* Am Chem Soc; Coblentz Soc; Sigma Xi; Fedn Anal Chem & Spectros Socs. *Res:* The use of infrared spectroscopy and nuclear magnetic resonance spectroscopy for the elucidation of chemical structure. *Mailing Add:* BASF Corp 1419 Biddle Ave Wyandotte MI 48192

FLOWER, PHILLIP JOHN, b Toledo, Ohio, Feb 4, 48; m 70; c 1. ASTROPHYSICS. *Educ:* Univ Toledo, BS, 70; Univ Wash, PhD(astron), 76. *Prof Exp:* Res assoc astrophys, Joint Inst Lab Astrophys, 76-78; ASST PROF PHYSICS & ASTRON, CLEMSON UNIV, 78- *Mem:* Am Astron Soc; Am Phys Soc. *Res:* Stellar evolution; stellar interiors; Magellanic clouds; dwarf galaxies; distance scale; evolution of galaxies. *Mailing Add:* Dept Physics & Astron Clemson Univ Main Campus Clemson SC 29634

FLOWER, ROBERT WALTER, b Baltimore, Md, July 6, 41; m 68; c 2. MEDICAL PHYSICS. *Educ:* Johns Hopkins Univ, BA, 66. *Prof Exp:* Assoc staff, Appl Physics Lab, 66-71, sr staff, 71-76, instr ophthal, Sch Med, 72-73, asst prof, 73-81, spec adv to dir biomed prog, 74-81, PRIN STAFF, APPL PHYSICS LAB, JOHNS HOPKINS UNIV, 76-, PROF, SCH MED, 81-, DIR BIOMED PROG, 81-, ASSOC PROF OPHTHAL, 81-, ASSOC PROF BIOMED ENG, 82- *Concurrent Pos:* Instr ophthal, Johns Hopkins Univ, 72-73, asst prof, 73-, spec adv to dir biomed eng prog, Appl Physics Lab, 74- *Honors & Awards:* Award for Achievement, Soc Tech Commun, 72; Meyers Honor Award for Basic Sci Res, 77. *Mem:* Asn Res Vision & Ophthal; Int Soc Oxygen Transport to Tissue; Wilmer Res Group; Sigma Xi; NY Acad Sci. *Res:* Relationship of ocular blood flow dynamics to distribution and maintenance of metabolites, particularly oxygen, within tissues of the eye during perinatal development; development of techniques for routine clinical monitoring of retinal and choroidal blood flow. *Mailing Add:* Appl Physics Lab Johns Hopkins Univ Laurel MD 20707

FLOWER, ROUSSEAU HAYNER, b Center Brunswick, NY, Mar 21, 13; m 50; c 2. GEOLOGY, PALEONTOLOGY. *Educ:* Cornell Univ, AB, 34, AM, 35; Univ Cincinnati, PhD(paleont), 39. *Prof Exp:* Temp paleontologist, NY State Mus, 38; cur univ mus, Univ Cincinnati, 40-43, 44; lectr, Bryn Mawr Col, 43-44; temp expert, NY State Mus, 44-45, asst state paleontologist, NY, 45-51; stratigraphic geologist, NMex Inst Mining & Technol, 51-66, sr paleontologist, 66-78, EMER SR PALEONTOLOGIST, NMEX BUR OF MINES & MINERAL RESOURCES, N MEX INST MINING & TECHNOL, 78- *Concurrent Pos:* NSF res grant, 67-71 & 78-83; paleontologist, US Geol Survey, 75- *Mem:* Soc Syst Zool; Paleont Soc; fel Geol Soc Am; Soc Study Evolution; Soc Geol France. *Res:* Paleozoic cephalopods, primarily Nautiloidea-evolution, taxonomy, morphology, stratigraphy, faunal realms with implicatios relating to plate tectonics and continental drift; Ordovician colonial corals; cyathaspids, graptolites, general stratigraphy and faunas Cambrian-Devonian. *Mailing Add:* 205 Pena Pl Socorro NM 87801

FLOWER, TERRENCE FREDERICK, b Chicago, Ill, Aug 24, 41; m 65; c 2. INFRARED ASTROPHYSICS, STELLAR STRUCTURE-EVOLUTION. *Educ:* USAF Acad, BS, 74; Univ Wyo, MS, 74, PhD(sci educ), 77. *Prof Exp:* Teacher physics/math, Wheatland High Sch, 76-79; PROF PHYSICS & ASTRON & CHAIR, DEPT PHYSICS, COL ST CATHERINE, 79- *Concurrent Pos:* Vis prof, Univ Minn, 87-88 & USAF Acad, 91-92; consult, rocketing & energy conserv, alternative energy systs. *Mem:* Am Asn Physics Teachers; Am Inst Physics; Nat Sci Teachers Asn. *Res:* Study of comets, supernovae and variable stars; study of symbiotic stars in the infrared. *Mailing Add:* 13875 Mississippi Trail Hastings MN 55033

FLOWERDAY, ALBERT DALE, b Nebr, June 14, 27; m 51; c 3. CROP PHYSIOLOGY. *Educ:* Univ Nebr, BS, 50, MS, 51, PhD(agron), 58. *Prof Exp:* From supt Northeastern Nebr Exp Sta to asst prof agron, Univ Nebr, Lincoln, 57-66, assoc prof, 66-73, mem staff, Mission to Nat Univ Colombia, 67-69, dir exten, 67-70, prof agron, 73-86; AGRON MGR, PIONEER HI-BRED INT, 86- *Mem:* Am Soc Agron. *Res:* Crop production system with special emphasis on conservation of energy, soil and water. *Mailing Add:* Pioneer Hi-Bred Int PO Box 5307 Lincoln NE 68505

FLOWERS, ALLAN DALE, b New Castle, Ind, Mar 21, 46; m 68; c 1. OPERATIONS RESEARCH. *Educ:* Ind Univ, BS, 67, MBA, 70, DBA(opers mgt), 72. *Prof Exp:* Systs coordr comput, Univ Div, Ind Univ, 70-72; from asst prof to assoc prof opers mgt, Tex Tech Univ, 72-78; ASSOC PROF OPERS MGT, CASE WESTERN RESERVE UNIV, 78- *Concurrent Pos:* Consult, Tex Instruments, Inc, 77-78, Devro, Inc, 77-78, Gen Elec, 79-81, TRW, 82-83, Sherwin-Williams, 83-84, Keithley Instruments, 83-86, IBM, 85-90 & Gen Motors, 85. *Honors & Awards:* Develop & Appln Award, Inst Indust Engrs, 81. *Mem:* Inst Indust Engrs; Opers Res Soc Am; Inst Mgt Sci; Prod & Opers Mgt Soc. *Res:* Production planning and inventory control; forecasting; operations scheduling; quality control; assembly line balancing; materials requirements planning; micro-computer applications. *Mailing Add:* Dept Opers Res Case Western Reserve Univ Cleveland OH 44106

FLOWERS, CHARLES E, JR, b Zebulon, NC, July 20, 20; m 72; c 2. OBSTETRICS & GYNECOLOGY. *Educ:* The Citadel, BS, 41; Johns Hopkins Univ, MD, 44. *Prof Exp:* Asst obstet & gynec, Johns Hopkins Hosp, 48-50; from instr to asst prof, Col Med, State Univ NY, 51-53; from assoc prof to prof, Sch Med, Univ NC, 53-66; prof obstet & gynec & chmn dept, Col Med, Baylor Univ, 66-69; obstetrician & gynecologist-in-chief, Ben Taub & Jefferson Davis Hosps, Houston, 66-69; PROF OBSTET & GYNEC & CHMN DEPT, MED CTR, UNIV ALA, BIRMINGHAM, 69- *Mem:* Fel Am Col Surgeons; fel Am Col Obstet & Gynec; fel Am Gynec Soc; fel Am Asn Obstet & Gynec. *Res:* Obstetrical anesthesia and analgesia; metabolism in toxemia of pregnancy; studies of the endometrium and the menstrual cycle. *Mailing Add:* Dept Obstet & Gynec Univ Ala Univ Hosp 3757 Rockhill Rd Birmingham AL 35233

FLOWERS, DANIEL F(ORT), b New York, NY, Jan 21, 20; m 58; c 3. MECHANICAL ENGINEERING. *Educ:* Va Mil Inst, BS, 40; Mass Inst Technol, SM, 42, ScD(mech eng), 49. *Prof Exp:* Engr, 46-47 & 49-56, vpres, 56-75, CHMN BD, DIFFERENTIAL CORP, 75- *Mem:* Am Soc Mech Engrs; Am Inst Mining, Metall & Petrol Engrs. *Res:* Aircraft vibration and flutter; gas turbine control; railway vehicles; oil and gas production. *Mailing Add:* Differential Corp 2001 Kirby Dr No 513 Houston TX 77019

FLOWERS, HAROLD L(EE), b Hickory, NC, June 25, 17; m 41; c 2. ELECTRONICS. *Educ:* Duke Univ, BS, 38; Univ Cincinnati, MS, 48. *Prof Exp:* Asst, Proctor & Swartz, Inc, Pa, 38-41; asst radio engr radio & radar, Off Chief Sig Off, US War Dept, Washington, DC, 41-42; br head missile command & report links, Naval Res Lab, 42-50; dir weapon syst, Goodyear Aircraft Corp, 50-61; gen eng mgr, Avco Electronics, Ohio, 61-63; eng mgr, McDonnell Aircraft Corp, Mo, 63-66, chief engr tactical missiles, Fla Div, McDonnell Douglas Corp, 66-69, dept prog mgr, 69-74, CHIEF PROG MGR, MCDONNELL DOUGLAS ASTRONAUT CO, MO, 74- *Concurrent Pos:* Mem guided missile comt, Res & Develop Bd, Dept Defense, 49-53. *Mem:* Assoc fel Am Inst Aeronaut & Astronaut; fel Inst Elec & Electronics Engrs. *Res:* Electronic guidance and missile systems; radar; weapon systems. *Mailing Add:* 1767 Golden Lake Ct Chesterfield MO 63017

FLOWERS, JOE D, b Sayre, Iowa, April 15, 43. MATHEMATICS. *Educ:* Iowa State Univ, BA, 66, PhD(math), 69. *Prof Exp:* PROF MATH, UNIV IOWA, 69- *Mem:* Math Asn Am. *Mailing Add:* Math & Comp Sci Div Northeast Mo St Univ Kirksville MO 63501

FLOWERS, JOHN WILSON, b Memphis, Tenn, Aug 20, 10; m 41; c 1. PHYSICS. *Educ:* Southwestern (Tenn), BS, 31; Univ Va, MS, 33, PhD(physics), 35. *Prof Exp:* Mem staff, Gen Elec Co, 37-47; from assoc prof to prof, 47-82, EMER PROF PHYSICS, UNIV FLA, 82- *Concurrent Pos:* Sr physicist, Aerojet Gen Corp, 61; consult, Union Carbide Co, 52- & Gen Dynamics/Ft Worth, 59, 60, Sperry Rand, 61 & Radiation Res Corp, 64- *Mem:* AAAS; Am Phys Soc. *Res:* Plasma, thermonuclear and space physics. *Mailing Add:* 2815 SW Eighth Dr Gainesville FL 32601

FLOWERS, NANCY CAROLYN, b McComb, Miss, Sept 28, 28; m 66; c 3. CARDIOVASCULAR DISEASES. *Educ:* Miss State Col Women, BS, 50; Univ Tenn, Memphis, MD, 58; Am Bd Internal Med, dipl, 66; Am Bd Cardiovasc Dis, dipl, 70. *Prof Exp:* Preceptorship under Dr Ralph R Braund, 58; intern med, Roanoke Mem Hosp, 58-59; preceptorship internal med, Beckley Mem Hosp, WVa, 60, resident, 60-62; instr med, Col Med, Univ Tenn, Memphis, 63-65, NIH fel & trainee, 62-65, asst prof, 65-67, asst prof physiol, 66-67; from assoc prof to prof med, Med Col Ga, 67-73; PROF MED & CHIEF DIV CARDIOL, SCH MED, UNIV LOUISVILLE, 73- *Concurrent Pos:* Res physician cardiol, Kennedy Vet Admin Hosp, Memphis, Tenn, 63-67; dir heart sta, John Gaston Hosp, 64-65; consult, William F Bowld Hosp, 65-66 & WTenn Tuberc Hosp, 66-67; prin investr, Am Heart Asn grant-in-aid, 69-72; sect chief cardiol, Forest Hills Div, Vet Admin Hosp, Augusta, Ga, 67-73, dir training prog, 70; co-investr, NIH grant-in-aid, 67-74; fel coun clin cardiol, Am Heart Asn, 70. *Mem:* AMA; Am Fedn Clin Res; fel Am Col Physicians; fel Am Col Cardiol; fel Am Col Chest Physicians. *Res:* Distribution of electrocardiographic potential on the body surface with respect to the limits of the contained information, equivalent cardiac generator representation, clinical significance and A-V conduction system of the heart. *Mailing Add:* Cardiol Med Col Ga Augusta GA 30912

FLOWERS, RALPH GRANT, physical chemistry, for more information see previous edition

FLOWERS, RALPH WILLS, b Pittsfield, Mass, Oct 2, 48; m 78; c 4. ENTOMOLOGY, MALACOLOGY. *Educ:* Cornell Univ, BS, 70; NC State Univ, MS, 72; Univ Wis, PhD(entom), 75. *Prof Exp:* Res asst entom, NC State Univ, 70-71 & Univ Wis, 72-75; from asst prof to assoc prof, 75-88, PROF ENTOM, FLA A&M UNIV, 88- *Mem:* Am Orchid Soc; Lepidopterists Soc; Asn Trop Biol; Coleopterists' Soc; NAm Benthological Soc. *Res:* Taxonomy and biology of Ephemeroptera; taxonomy of chrysomelidae. *Mailing Add:* Fla A&M Univ Tallahassee FL 32307

FLOWERS, RUSSELL SHERWOOD, JR, b Raleigh, NC, July 16, 51; div; c 2. FOOD SAFETY & QUALITY. *Educ:* NC State Univ, BA, 73; Univ Ill, Champaign, MS, 75, PhD(food sci & microbiol), 78. *Prof Exp:* Res asst food sci, Dept Food Sci, NC State Univ, 73-75; res teaching asst, Dept Food Sci, Univ Ill, 75-78; asst prof microbiol, Dept Microbiol & Dept Nutrit & Food Sci, Univ Ariz, 78-79; lab dir, Silliker Labs, Inc, 79-82, vpres & lab dir, 82-87, exec vpres & lab dir, 87-89, exec vpres & tech dir, 89-90, PRES, SILLIKER LABS GROUP, INC, 90- *Concurrent Pos:* Assoc referee, Comt Microbiol, Asn Off Anal Chemists. *Mem:* Am Asn Cereal Chemists; Am Soc Microbiol; Asn Off Anal Chemists; Inst Food Technologists; Int Asn Milk Food & Environ Sanitarians; Soc Indust Microbiol. *Res:* Food microbiology; food safety; new methods for detection of foodborne pathogens; author of many scientific publications. *Mailing Add:* Silliker Labs Group Inc 1304 Halsted St Chicago Heights IL 60411

FLOYD, ALTON DAVID, b Henderson, Ky, July 17, 41; m 62; c 2. ANATOMY, EMBRYOLOGY. *Educ:* Univ Ky, BS, 63; Univ Louisville, PhD(anat), 68. *Prof Exp:* From instr to asst prof anat, Univ Mich, Ann Arbor, 67-72; from asst prof to assoc prof anat, Med Sci Prog, Ind Univ, Bloomington, 72-83, assoc prof path, 83-84; sect head, Cell & Molecular Biol, Miles Sci Lab, 84-85, sr staff scientist, 86-89, CONSULT, CELL & TISSUE TECHNOL, MILE, INC, 89- *Mem:* Am Asn Anatomists; Tissue Cult Asn; Teratol Soc; Histochem Soc; Soc Anat Cytometry; Biol Stain Comn. *Res:* Cell nuclear differentiation and specialization; quantitative cytology; micro spetrophotometry; flow cytometry; histology. *Mailing Add:* 23126 S Shore Dr Edwardsburg MI 49112-9550

FLOYD, CAREY E, JR, b Nashville, Tenn; m 75. IMAGE RECONSTRUCTION ALGORITHMS, SCATTER COMPENSATION IN MEDICAL WAGING. *Educ:* Eckerd Col, BS, 76; Duke Univ, PhD(physics), 81. *Prof Exp:* Res assoc nuclear physics, Triangle Univ Nuclear Lab, Duke Univ, 82-87; res assoc, 83-85, ASST PROF MED IMAGING, DEPT RADIOL, DUKE UNIV MED CTR, 85- & DEPT BIOMED ENG, DUKE UNIV, 88- *Mem:* Inst Elec & Electronics Engrs; Soc Nuclear Med. *Res:* Three dimensional medical imaging including digital radiography and emission computed tomography. *Mailing Add:* Duke Univ Med Ctr Box 3949 Durham NC 27710

FLOYD, DON EDGAR, organic polymer chemistry, for more information see previous edition

FLOYD, EDWIN EARL, b Eufaula, Ala, May 8, 24; m 45; c 3. TOPOLOGY. *Educ:* Univ Ala, BA, 43; Univ Va, PhD(math), 48. *Prof Exp:* Instr math, Princeton Univ, 48-49; from asst prof to assoc prof, 53-56, chmn dept, 66-69, prof, 56-80, dean fac, Arts & Sci, 74-81, provost, 81-86, ROBERT C TAYLOR PROF MATH, UNIV VA, 80- *Concurrent Pos:* Mem, Inst Advan Study, 58-59 & 63-64; Sloan res fel, 60-64. *Mem:* Am Math Soc; Math Asn Am. *Res:* Differential topology; cobordism; periodic maps; transformation groups. *Mailing Add:* Dept Math Univ Va Math-Astr Bldg Charlottesville VA 22903-3199

FLOYD, GARY LEON, b Moline, Ill, Dec 23, 40; m 63. PHYCOLOGY, ALGAL ULTRASTRUCTURE & ALGAL EVOLUTION. *Educ:* Univ Northern Iowa, BA, 62; Univ Okla, MS, 66; Miami Univ, PhD(bot), 71. *Prof Exp:* Teacher biol, Grinnell Jr High, Iowa, 62-65; instr bot, Miami Univ, Ohio, 66-68; asst prof, Rutgers Univ, New Brunswick, 71-75; from asst prof to assoc prof, Ohio State Univ, Columbus, 75-83, assoc dean, Col Biol Sci, 86-90, PROF BOT, OHIO STATE UNIV, COLUMBUS, 83-, DEAN, COL BIOL SCI, 90- *Concurrent Pos:* Prin investr, NSF res grants, 76-; dir, Transmission Electron Microscope Lab, Ohio State Univ, 78-; assoc ed, J Phycol, 83-84; Nat lectr, Phycol Soc Am, 84-85; cur, teaching slide collection, Phycol Soc Am, 84- *Mem:* Am Soc Cell Biol; Phycol Soc Am; Int Phycol Soc; Bot Soc Am; Sigma Xi; AAAS; Am Inst Biol Sci. *Res:* Green algal ultrastructure with emphasis on systematics and evolution; fine structure of motile cells as related to function and evolution; developmental and molecular studies on green algae. *Mailing Add:* Dept Plant Biol Ohio State Univ 1735 Neil Ave Columbus OH 43210

FLOYD, J F R(ABARDY), b Pagosa Springs, Colo, May 22, 15; m 44; c 3. AERONAUTICAL ENGINEERING. *Educ:* Carnegie Inst Technol, BS, 37. *Prof Exp:* Aerodynamicist, Glenn L Martin Co, 37-43, chief aerodyn, 43-47; sr engr & mem prin prof staff, Appl Physics Lab, Johns Hopkins Univ, 47-85, missile syst proj engr, 67-85; RETIRED. *Mem:* Am Inst Aeronaut & Astronaut. *Res:* Design and development of aircraft and guided missiles, particularly missile system design and launching techniques. *Mailing Add:* 9217 Crownwood Rd Ellicott City MD 21043

FLOYD, JOHN CLAIBORNE, JR, b Olla, La, July 3, 27; m 52; c 4. DIABETOLOGY, ENDOCRINOLOGY. *Educ:* La State Univ, BS, 49, MD, 54. *Prof Exp:* Instr internal med, Med Ctr, Univ Mich, 58-59; instr internal med, Sch Med, La State Univ, 59-60; from instr to assoc prof, 60-70, PROF INTERNAL MED, MED CTR, UNIV MICH, 70- *Concurrent Pos:* Mem, bd dirs, Am Diabetes Asn, 74-80; vis prof, Univ Aarhus, 78-79. *Mem:* Am Diabetes Asn; Am Fedn Clin Res; Endocrine Soc; Cent Soc Clin Res; Sigma Xi; AAAS. *Res:* Research activities in physiological and pathophysiological regulation of endocrine pancreatic (islet) function in man (healthy, diabetes, obesity, starvation); evaluation of treatment of diabetes mellitus. *Mailing Add:* 3920 Taubman Ctr 1500 E Med Ctr Dr Box 0354 Ann Arbor MI 48109-0354

FLOYD, JOSEPH CALVIN, b La Grange, Ga, June 6, 41; m 65; c 2. ORGANIC CHEMISTRY. *Educ:* Ga Inst Technol, BChE, 65, PhD(org chem), 69. *Prof Exp:* Res chemist, Exxon Res & Eng Co, 68-74; staff chemist, 74-77, SR STAFF CHEMIST, EXXON CHEM CO, 77- *Mem:* Am Chem Soc; Am Inst Chem Engrs. *Res:* Organic synthesis; oxidative stabilization of polymers; polymer process development. *Mailing Add:* 5106 Arrowhead Baytown TX 77521-2904

FLOYD, ROBERT A, b Yosemite, Ky, Oct 7, 40; m 65; c 2. BIOCHEMISTRY. *Educ:* Univ Ky, BS, 63, MS, 65; Purdue Univ, PhD(agron), 69. *Prof Exp:* Fel agron, Univ Calif, Davis, 68-69; fel, Johnson Res Found, Univ Pa, 69-71; res assoc, Ctr Biol Natural Systs, Wash Univ, 71-74; asst prof, 74-78, PROF BIOCHEM MOLEC BIOL, HEALTH SCI CTR, UNIV OKLA, 78-; MEM, OKLA MED RES FOUND, 83- *Mem:* AAAS; Am Soc Biol Chemists; Am Chem Soc; Biophys Soc; Am Soc Photochem & Photobiol; Soc Toxicol; Oxygen Soc. *Res:* Carcinogenesis; carcinogen free radicals; spin trapping in biological systems; bioenergetics; ozone damage to plants; photosynthesis; electron spin resonance in biological systems; oxygen free radicals; aging brain membranes. *Mailing Add:* Dept Molecular Toxicol Okla Med Res Found 825 NE 13th St Oklahoma City OK 73104

FLOYD, ROBERT W, b New York, NY, June 8, 36; div; c 3. COMPUTER SCIENCE. *Educ:* Univ Chicago, BA, 55, BS, 58. *Prof Exp:* Elec engr, Westinghouse Elec, 55-56; comput oper programmer & analyst, Armour Res Found, 56-62; sr proj scientist, Comput Assoc, 62-65; assoc prof comput sci, Carnegie-Mellon Univ, 65-68; assoc prof, 68-70, chmn, 73-76, PROF COMPUT SCI, STANFORD UNIV, 70- *Concurrent Pos:* Assoc ed, J Asn Comput Mach, 67-69; fel, Guggenheim Found, 78. *Honors & Awards:* Turing Award, Asn Comput Mach, 78. *Mem:* Fel Am Acad Arts & Sci; Asn Comput Mach; AAAS. *Res:* Optimal methods for computation; computer programming languages. *Mailing Add:* 895 Allardice Way Stanford CA 94305

FLOYD, WILLIAM BECKWITH, b Atlanta, Ga, Dec 27, 30; m 53; c 3. COMPUTER SCIENCE. *Educ:* Harvard Univ, AB, 52; Emory Univ, MS, 53. *Prof Exp:* Sr res engr, Appl Sci Div, Melpar, Inc, 56-59, head systs res lab, 59-60; sr staff engr, Info Sci Lab, Litton Systs, Inc, 60-66, tech mgr, Signal Processing Dept, Data Systs Div, 66-67; assoc tech dir, B-D Spear Med Systs Div, Becton, Dickinson & Co, 67-73; pres, Ruskin Data Systs, Ltd, 73-90. *Mem:* Inst Elec & Electronics Engrs; Asn Comput Mach. *Res:* Clinical information system, design and development. *Mailing Add:* 59 Ash St Weston MA 02193

FLOYD, WILLIS WALDO, b Whitesboro, Tex, Aug 2, 03; m 32; c 4. NUCLEAR CHEMISTRY, AGRICULTURAL & FOOD CHEMISTRY. *Educ:* NTex State Univ, AB, 25; Univ Tex, Austin, AM, 28; Univ Iowa, PhD(chem), 31. *Prof Exp:* Instr gen sci, Sherman High Sch, Tex, 25-26; lab instr gen chem, Univ Tex, 26-29, res asst org chem, 27-28; lab instr gen & anal chem, Univ Iowa, 29-30, fel phys chem, 30-31; prof chem & head dept, Ottawa Univ, Kans, 31-36; prof physics & head dept, Sam Houston State Univ, 36-39, prof chem & head dept, 36-49; prof gen & anal chem, 49-73, EMER PROF CHEM, SAM HOUSTON STATE UNIV, 73- *Concurrent Pos:* Res prof phys chem & mech eng, Purdue Univ, 47-48; res partic nuclear chem, Oak Ridge Nat Lab, 53. *Mem:* AAAS; Am Chem Soc. *Res:* Petroleum nitrogen; activity coefficients; castor fiber separation; reactor thorium; copper recovery; oxinate solubilities; group eight congruency; 3-uridylic acid yields; ascorbic acid in raw and cooked vegetables; urethane yields; chemical education. *Mailing Add:* 1924 Ave N Huntsville TX 77340-5014

FLUCK, EUGENE RICHARDS, b Hazleton, Pa, Dec 10, 34; m 57; c 2. BIOCHEMISTRY. *Educ:* Pa State Univ, BS, 56, MS, 60, PhD(biochem), 62. *Prof Exp:* Metabolic chemist, R J Reynolds Tobacco Co, 62-69; METABOLIC CHEMIST, WYETH LABS, INC, 69- *Mem:* Am Chem Soc; NY Acad Sci. *Res:* Drug metabolism and drug safety evaluations. *Mailing Add:* Wyeth Labs PO Box 8299 Philadelphia PA 19101-8299

FLUCK, MICHELE MARGUERITE, b Geneva, Switzerland, Aug 5, 40. VIRAL ONCOLOGY. *Educ:* Univ Geneva, MS(physics), 64, MS(molecular biol), 66, PhD(molecular biol), 72. *Prof Exp:* Instr viral oncol, Harvard Med Sch, 72-78, asst prof, 78-79; from assoc prof to prof, 86-90, UNIV DISTINGUISHED PROF VIRAL ONCL, MICH STATE UNIV, 90- *Mem:* Am Women Sci; Am Soc Virologists. *Res:* Neoplastic transformation by polyoma virus: how a virus transforms a normal cell into a cancer cell. *Mailing Add:* Microbiol Dept Mich State Univ East Lansing MI 48824

FLUCK, RICHARD ALLEN, b Litchfield, Minn, June 21, 45; m 70; c 3. CELL PHYSIOLOGY. *Educ:* Iowa State Univ, BS, 66; Univ Calif, Berkeley, PhD(zool), 71. *Prof Exp:* Res assoc bot, Ohio Univ, 71-74; asst prof, 74-81, ASSOC PROF BIOL, FRANKLIN & MARSHALL COL, 81- *Concurrent Pos:* NSF fel, Ohio Univ, 71-72. *Mem:* Am Soc Cell Biologists; Soc Gen Physiologists; AAAS. *Res:* Ooplasmic segregation; cytokinesis. *Mailing Add:* Dept Biol Franklin & Marshall Col PO Box 3003 Lancaster PA 17604

FLUCK, RICHARD CONARD, b Clemmons, NC, May 22, 38; m 60; c 3. AGRICULTURAL ENGINEERING. *Educ:* NC State Univ, BS, 60, MS, 63, PhD(agr eng), 66. *Prof Exp:* Asst prof agr eng & asst agr engr, 65-69, assoc prof agr eng & assoc agr engr, 69-77, PROF & AGR ENGR, UNIV FLA, 77- *Mem:* Am Soc Agr Engrs; Sigma Xi; Asn Energy Engrs. *Res:* Agricultural energy analysis and management; agricultural machinery management and engineering economy; resource management; technology assessment. *Mailing Add:* 1400 NW 35 Ter Gainesville FL 32605-4830

FLUECK, JOHN A, b Apr 13, 33; US citizen; div; c 3. WEATHER STATISTICS, ENVIRONMENTAL STATISTICS. *Educ:* Beloit Col, BS, 55; Univ Chicago, MBA, 58, PhD (economet & statist), 67. *Prof Exp:* Lectr statist, 64-65, res assoc geophys, Univ Chicago, 65-69; from assoc prof to prof, Temple Univ, 68-85; sr res fel, Coop Inst Res Environ Sci, Univ Colo, 84-88; chief statist, Environ Sci Group, NOAA-ERL, 85-89; DIR ENVIRON STATIST & MODELING DIV, ENVIRON RES CTR, UNIV NEV, LAS VEGAS, 89- *Concurrent Pos:* Ford Found fel, 58-59; vis lectr, Univ Ill, Chicago, 67-68; consult, Panel Weather & Climate Modification, Nat Acad Sci, Nat Res Coun, 68-70; consult, Bur Budget, Washington, DC, 70, vis appointment, Off Mgt & Budget, Exec Off President, 70-71; dir, Data Anal Lab, Temple Univ, 72-82; consult, Campbell Inst Food Res, 73-81; chmn, Coun Am Statist Asn, 75; mem, Adv Panel Weather Modification, NSF,

75-78; sr res fel, Nat Acad Sci-Nat Res Coun, 82-84. *Mem:* Fel Am Statist Asn; Am Soc Qual Control; Biomet Soc; Am Meteorol Soc; Sigma Xi; fel AAAS. *Res:* Total quality improvement; environmental research; data analysis and estimation; cloud physics; forecast verification. *Mailing Add:* Environ Res Ctr 4505 S Maryland Pkwy Las Vegas NV 89154

FLUHARTY, ARVAN LAWRENCE, b Haines, Ore, June 10, 34; m 61; c 3. BIOCHEMISTRY, MENTAL RETARDATION. *Educ:* Univ Wash, BS, 56; Univ Calif, Berkeley, PhD(biochem), 59. *Prof Exp:* From asst prof to assoc prof biochem, Univ Southern Calif, 62-68; res specialist, Pac State Hosp-Calif State Dept Ment Hyg, 68-73, assoc res biochemist, 73-75, adj prof psychiat, Neuropsychiat Inst, Pac State Hosp Res Group, 75-79, PROF IN RESIDENCE, UCLA MENTAL RETARDATION RES CTR RES GROUP, LANTERMAN DEVELOP CTR, UNIV CALIF, LOS ANGELES, 79-, RES GROUP COORDR, 80- *Concurrent Pos:* Adj assoc prof biochem, Sch Med, Univ Southern Calif, 69-72, adj prof, 72-75. *Mem:* AAAS; Am Chem Soc; Am Soc Biol Chemists; Am Soc Neurochem; Biochem Soc. *Res:* Biochemistry of metabolic diseases; biochemistry of mental retardation; neurobiochemistry; metabolism of four carbon sugars; enzymatic dithiols; cellular energy transformations. *Mailing Add:* UCLA Ment Retardation Res Ctr Res Group Lanterman Develop Ctr Box 100-R Pomona CA 91769

FLUHARTY, DAVID LINCOLN, b Seattle, Wash, Jan 30, 46; m 85. OFFSHORE OIL, FISHERIES. *Educ:* Univ Wash, BA, 68, MA, 72; Univ Mich, Ann Arbor, PhD(natural resource conserv), 77. *Prof Exp:* Fel, Inst Marine Studies, 76-78, res assoc, 78-89, SR RES ASSOC, UNIV WASH, 90- *Concurrent Pos:* Fel, US Int Commun Agency, 77; lectr, Dept Geog, Univ Wash, 78-79; lectr, Inst Eviron Studies, Univ Wash, 85; vchmn, Puget Sound Water Qual Authority, 83-85. *Mem:* AAAS; Int Inst Fisheries Econ & Trade; Am Fisheries Soc; Int Studies Asn. *Res:* Marine natural resource management at the international level; integration of management policies for natural resources. *Mailing Add:* Sch Marine Affairs HF-05 Univ Wash Seattle WA 98195

FLUKE, DONALD JOHN, b Nankin, Ohio, Feb 17, 23; m 54; c 2. RADIATION BIOPHYSICS. *Educ:* Wooster Col, BA, 47; Yale Univ, MS, 48, PhD(physics), 50. *Prof Exp:* Instr physics, Yale Univ, 50-52; biophysicist, Brookhaven Nat Lab, 52-57; assoc prof zool, 58-65, chmn acad coun, 69-71, chmn dept, 69-78, PROF ZOOL, DUKE UNIV, 65- *Concurrent Pos:* Lectr, Univ Calif, 56-57, vis assoc prof, Donner Lab, 58; vis lectr, Am Inst Biol Scientists, 61-63; vis prof, Inst Molecular Biophy, Fla State Univ, 64-65; tech rep biophys, Div Biol & Med, US Atomic Energy Comn, 68-69; vis prof, Univ Utrecht, 73-74. *Mem:* AAAS; Radiation Res Soc; Biophys Soc; Am Asn Physics Teachers; Am Soc Photobiol; Sigma Xi. *Res:* Ultraviolet action spectroscopy; biophysical application of accelerated ions on virus and enzyme; radiation biology; temperature dependence, direct effect; ultraviolet photobiology. *Mailing Add:* Dept Zool Duke Univ Durham NC 27706

FLUM, ROBERT S(AMUEL), SR, b Indianapolis, Ind, July 3, 25; m 47; c 7. SYSTEMS ANALYSIS. *Educ:* Univ Ind, BS, 49, MS, 56. *Prof Exp:* Asst physics, Univ Ind, 49-50; asst discharge in gases, Univ Md, 51-52; proj engr underwater acoust, US Naval Ord Lab, 52, physicist aero boundary layer, 52-54, physicist ord eng, 54-55, staff consult aerodyn, 55-58 & aerodevelop eng, 58-61; mem staff, Univ Chicago, 61-63; physicist, US Naval Ord Lab, 63-65, sr analyst, Systs Anal Off, Anti-submarine Warfare, Spec Proj Off, 65-70, sr analyst, US Naval Ord Lab, 70-75, sr analyst, Navy Dept, 75-79, physical scientist, Anti-Submarine Warfare Systs Proj Off, 79-85; SENIOR PHYSICIST, VITRO CORP, 85- *Mem:* Opers Res Soc Am; Am Phys Soc; Mil Opers Res Soc. *Res:* Application of logic and common sense to the basic problems of optimization of the Naval antisubmarine warfare posture. *Mailing Add:* 11603 Georgetowne Ct Potomac MD 20854

FLUMERFELT, RAYMOND W, b Hobbs, NMex, Nov 18, 39; m 59; c 2. CHEMICAL ENGINEERING. *Educ:* Lamar State Col, BS, 61; Northwestern Univ, MS, 63, PhD, 65. *Prof Exp:* Asst prof eng sci, Univ Notre Dame, 65-67; fel, Univ Wis, 67-68; from asst prof to prof chem eng, Univ Houston, 68-85, assoc chmn dept, 77-79; PROF & CHMN DEPT CHEM ENG, UNIV TULSA, 85- *Mem:* Am Inst Chem Engrs; Soc Rheol; Am Chem Soc; Soc Petrol Engrs. *Res:* Enhanced oil recovery; fluid mechanics; interfacial phenomena. *Mailing Add:* 2706 Wingate Cir Col Sta TX 77840-3837

FLURRY, ROBERT LUTHER, JR, b Hattiesburg, Miss, Nov 15, 33; m 57; c 4. THEORETICAL CHEMISTRY. *Educ:* Emory Univ, AB, 58, MS, 59, PhD(org chem), 61. *Prof Exp:* NIH fel, 61-62; from asst prof to assoc prof, 62-70, assoc dean res, 83-86, assoc vice chancellor res, 86-89, PROF CHEM, UNIV NEW ORLEANS, 70- *Concurrent Pos:* Vis prof, Math Inst, Eng, 68 & Dartmouth Col, 80-81; pres, La Chap, Am Inst Chemists, 73-74. *Mem:* Am Chem Soc; NY Acad Sci; fel Am Inst Chemists. *Res:* Applications of quantum chemistry and group theory to chemical and biological problems. *Mailing Add:* Dept Chem Univ New Orleans New Orleans LA 70148

FLURY, ALVIN GODFREY, b Austin, Tex, Nov 1, 20; m 44; c 4. HERPETOLOGY. *Educ:* Univ Tex, BA, 48, MA, 51; Tex Tech Univ, PhD(zool), 72. *Prof Exp:* Aquatic biologist, Freshwater Fisheries, Tex Game & Fish Comn, 51-62; info-ed officer, Tex Parks & Wildlife Dept, 62-65; instr biol, Angelo State Col, from asst prof to assoc prof biol, 69-80; RETIRED. *Mem:* Am Soc Ichthyol & Herpet; Soc Study Evolution; Soc Study Amphibians & Reptiles; Asn Study Animal Behav; Ecol Soc Am. *Res:* Reptiles; amphibians; fish; fish and wildlife conservation. *Mailing Add:* 3523 Judith Lane San Angelo TX 76904

FLUSSER, PETER R, b Vienna, Austria, July 3, 30; US citizen; m 58; c 4. MATHEMATICS. *Educ:* Ottawa Univ, BA, 58; Univ Kans, MA, 60; Okla State Univ, EdD(higher educ), 71. *Prof Exp:* From asst prof to prof math, Ottawa Univ, 60-78; asst prof math, Fort Hays State Univ, 78-82; PROF MATH, KANS WESLEYAN UNIV, 86- *Mem:* Am Math Soc; Math Asn Am. *Res:* Probability theory; characterization theorems in probability, especially characterization theorems for random variables with values in topological groups. *Mailing Add:* Dept Math Kans Wesleyan Univ 100 E Claflin St Salina KS 67401

FLY, CLAUDE LEE, b Fulbright, Tex, June 23, 05; m 27; c 2. SOIL CHEMISTRY, PLANT PHYSIOLOGY. *Educ:* Okla Agr & Mech Col, BS, 27, MS, 28; Iowa State Col, PhD(soil chem), 31. *Prof Exp:* Prof chem & head dept sci, Panhandle Agr & Mech Col, 31-35; asst soil survr, Soil Conserv Serv, USDA, 35, assoc soil scientist, 35-39, soil scientist, Tex, 39-42, Nebr, 42-47, state soil scientist, Kans, 47-52; head land develop dept, Int Eng Co & Morrison-Knudsen-Afghanistan Co, 52-58; area dir, Northern Great Plains, Soil & Water Conserv, Agr Res Serv, USDA, 58-59, res proj leader, Western Br, 59-63; PRES, CLAUDE L FLY & ASSOCS, 63- *Concurrent Pos:* Soil scientist to Greece, Italy & Sicily, UN Relief & Rehab Admin, 46-47, Kaiser Engrs, Ghana, WAfrica, Ivory Coast, 63-64, Int Eng Co-Peru, 64-67; partner & dir, Agriconsult, 65-67; consult, Eng Consult, Inc, Jamaica & Turkey, 66-67, Food & Agr Orgn, UN, Jordan, Yugoslavia, Nigeria, Uruguay, Panama, 67-69; consult, Develop, Planning & Res Assocs, Brazil, 72, Great Western Sugar Co, 73, Develop Planning, Inc, Kans, 72-, Am Agr Industs, Ill, 75- & Platte River Power Authority, Colo, 75; lectr & writer; mem, Subcomt Southern Great Plains, President's Nat Resources Bd, 28. *Mem:* Hon mem Am Soc Agron; Soil Sci Soc Am; fel Soil Conserv Soc Am; Am Soc Agr Consult (pres, 75); Int Soc Soil Sci; Sigma Xi; Am Soc Agr Engrs; fel Am Inst Chemists. *Res:* Land and water resource development and conservation; research, program planning and administration; demonstration on the use of industrial wastes for soil improvement. *Mailing Add:* 415 S Howes St Apt 1107 Ft Collins CO 80521

FLYGER, VAGN FOLKMANN, b Aalborg, Denmark, Jan 4, 22; nat US; m 46; c 1. WILDLIFE ECOLOGY. *Educ:* Cornell Univ, BS, 48; Pa State Univ, MS, 52; Johns Hopkins Univ, ScD, 56. *Prof Exp:* Game biologist, Md Dept Res & Educ, 48-51 & Wildlife Res, Md Game & Inland Fish Dept, 54-55; sr biologist, Inland Res Div, Md Dept Res & Educ, 55-61; res assoc prof, 61-67, actg dir, 64-65, chmn dept forestry, Fish & Wildlife, 71-74, PROF, NATURAL RESOURCES INST, UNIV MD, 67-, PROF ANIMAL SCI, 79- *Mem:* Am Soc Mammal; Wildlife Soc; Am Inst Biol Sci. *Res:* Mammal behavior; factors influencing animal populations and especially biology of tree squirrels. *Mailing Add:* Animal Sci Dept Univ Md College Park MD 20742

FLYNN, COLIN PETER, b Stockton-on-Tees, Eng, Aug 18, 35; m 61, 71; c 1. SOLID STATE PHYSICS. *Educ:* Univ Leeds, BSc, 57, PhD(physics), 60; Cambridge Univ, MA, 66. *Prof Exp:* Res assoc, 60-62, res asst prof, 62-64, from asst prof to assoc prof, 64-68, PROF PHYSICS, UNIV ILL, URBANA, 68-, DIR, MAT RES LAB, 78- *Concurrent Pos:* Fel, Christ's Col, Cambridge Univ, 66-67; NSF int prog fel, Sao Carlos, Brazil, 77-78. *Mem:* Am Soc Metals; fel Am Phys Soc. *Res:* Impurities and thermal defect structure in crystals; impurity magnetism and nuclear magnetic resonance in metals; diffusive hopping of ions and electrons in crystals; kinetics of defect equilibration. *Mailing Add:* 206 Mat Res Lab Univ Ill 104 S Goodwin Ave Urbana IL 61801

FLYNN, EDWARD JOSEPH, b Waltham, Mass, Apr 4, 42; m 65; c 2. PHARMACOLOGY. *Educ:* Northeastern Univ, BS, 64, MS, 66; NY Univ, PhD(pharmacol), 71. *Prof Exp:* Pharmacologist, US Army Res Inst Environ Med, Natick, Mass, 66; asst prof, 71-80, ASSOC PROF PHARMACOL, NJ MED SCH, UNIV MED & DENT NJ, 80- *Concurrent Pos:* Fel, Roche Inst Molecular Biol, Nutley, NJ, 70-71; Pharmaceut Mfrs Asn Found starter grant, 74; NIH grant, 74; consult curric div & instr, NJ State Dept Educ, 71- *Mem:* AAAS; NY Acad Sci. *Res:* Drug metabolism; barbiturate pharmacology; immunopharmacology. *Mailing Add:* Dept Pharmacol & Toxicol UMDNJ NJ Med Sch 185 S Orange Ave Newark NJ 07103

FLYNN, EDWARD ROBERT, b Joliet, Ill, July 7, 34; m 54; c 6. PHYSICS. *Educ:* Univ Ill, BS, 56; Univ NMex, MS, 64, PhD(physics), 66. *Prof Exp:* Accelerator engr, Univ Ill, 56-58; mem staff, 58-63, FEL, LOS ALAMOS NAT LAB, 83- *Concurrent Pos:* NATO fel, Niels Bohr Inst, Copenhagen, 70-71; vis prof, univ colo, 77; Von Humboldt prize, Hahn-Meitner Inst, Berlin, Ger, 81-82; Wellcome Trust award, Strathclyde Univ, 90. *Honors & Awards:* Von Humboldt Prize, 81. *Mem:* Fel Am Phys Soc; AAAS; Am Asn Physicists Med; Neurosci Soc. *Res:* Neurophysics(MEG). *Mailing Add:* Los Alamos Sci Lab PO Box 1663 Los Alamos NM 87545

FLYNN, GARY ALAN, b Jacksonville, Ill, Nov 2, 50; m 72. ORGANIC CHEMISTRY. *Educ:* Northern Ill Univ, BS, 72, MS, 73; Northwestern Univ, PhD(chem), 77. *Prof Exp:* RES ASSOC, MARION MERRELL-DOW RES CTR, MARION MERRELL-DOW PHARMACEUT, INC, 77- *Mem:* Am Chem Soc; Sigma Xi. *Res:* Synthesis and biological evaluation of organic compounds related to or derived from naturally occurring substances; design and synthesis of enzyme inhibitors and inactivators. *Mailing Add:* Merrell Res Ctr 2110 E Galbraith Rd Cincinnati OH 45215

FLYNN, GEORGE P(ATRICK), b Fall River, Mass, Aug 12, 36. PHYSICAL CHEMISTRY. *Educ:* Providence Col, BS, 57; Brown Univ, PhD(phys chem), 62. *Prof Exp:* Res asst chem, Yale Univ, 61-63; res assoc, Brown Univ, 64-67; res assoc chem, Mass Inst Technol, 67-70 & 72-80. *Mem:* Am Chem Soc; Sigma Xi. *Res:* Textbook writing; viscosity of gases. *Mailing Add:* PO Box 1069 Kendall Sq Sta Cambridge MA 02142

FLYNN, GEORGE WILLIAM, b Hartford, Conn, July 11, 38; m 70. CHEMICAL PHYSICS. *Educ:* Yale Univ, BS, 60; Harvard Univ, AM & PhD(chem), 64. *Prof Exp:* Fel physics, Mass Inst Technol, 64-66; from asst prof to assoc prof, 67-76, PROF CHEM, 76-, THOMAS ALVA EDISON PROF, COLUMBIA UNIV, 86- *Concurrent Pos:* NSF fel, 64-65; Alfred P Sloan fel, 68-71; res collabr, Brookhaven Nat Lab, 69-; John Simon Guggenheim Found fel, 74-75; vis scientist, Mass Inst of Technol, 75; dir, Columbia Radiation Lab, 79-84 & co-dir, 84-; adv ed, Chem Physics, 78-, J Phys Chem, 80-84, J Chem Phys, 83-86, Chem Physics Lett, 81-84, Ann Rev

Phys Chem, 80-85; Distinguished Summer Lectr, Northwestern Univ, 86. *Honors & Awards:* A Cressy Morrison Award-Natural Sci, NY Acad Sci, 83; Reilly lectr, Univ Notre Dame, 83; William Pyle Phillips lectr, Haverford Col, 83,. *Mem:* Fel Am Phys Soc; Am Chem Soc; NY Acad Sci. *Res:* Relaxation phenomena in molecular systems; development and uses of lasers; photo fragmentation dynamics; matrix isolation studies; diode laser probes of collision dynamics. *Mailing Add:* Dept Chem Columbia Univ 315 Havemeyer Hall New York NY 10027

FLYNN, JAMES PATRICK, b Wilkes Barre, Pa, Aug 1, 24; m 54; c 3. PHYSICAL CHEMISTRY, INORGANIC CHEMISTRY. *Educ:* Bucknell Univ, BS, 48; Iowa State Univ, PhD(chem), 53. *Prof Exp:* Res engr, Battelle Mem Inst, 48-49; asst, Ames Lab, AEC, 49-53; res & develop engr, Magnesium Dept, 53-58, res chemist, Sci Proj Lab, 58-67, proj leader, Prod Dept Labs, 67-70, RES ASSOC, DOW CHEM CO, 70- *Mem:* Nat Acad Sci; Sigma Xi; Am Soc Testing & Mat; Am Chem Soc. *Res:* Magnesium alloy and recovery of metals; propellant testing and compatibility; evaluation of chemical hazards; hazardous waste disposal; health and environmental regulations. *Mailing Add:* 4513 Bond Ct Midland MI 48674

FLYNN, JOHN JOSEPH, JR, b Salida, Colo, Sept 16, 31. ORGANIC CHEMISTRY. *Educ:* Western State Col Colo, BA, 53; Okla State Univ, MS, 55; Purdue Univ, PhD(chem), 61. *Prof Exp:* ASSOC PROF CHEM, PURDUE UNIV FT WAYNE, 58- *Mem:* Am Chem Soc. *Res:* Diels-Alder reactions. *Mailing Add:* Dept Chem Ind Univ Purdue Ft Wayne 2101 Coliseum Blvd E Ft Wayne IN 46805

FLYNN, JOHN JOSEPH, b Wilkes-Barre, Pa, Aug 10, 55; m 82; c 1. MAMMALIAN PALEONTOLOGY, PALEOMAGNETISM. *Educ:* Yale Univ, BS, 77; Columbia Univ, MA, 80, PhD(geol sci), 83. *Prof Exp:* Asst prof geol, Rutgers Univ, 82-87; ASSOC CUR, FOSSIL MAMMALS FIELD MUSEUM, 88- *Concurrent Pos:* Vis lectr geol & geophys, Yale Univ, 82; res assoc, Am Mus Natural Hist, 85-; assoc ed, Jour Vert Paleont, 87-90; lectr, Comn Evol Biol & Biol Sci Col Div, Univ Chicago, 89- *Honors & Awards:* Alfred Sherwood Romer Prize, Soc Vert Paleont, 82. *Mem:* Soc Vert Paleont; Geol Soc Am; Am Geophys Union; Sigma Xi; AAAS; Paleont Soc. *Res:* Applications of paleomagnetism and mammalian biostratigraphy to studies of geochronology, tectonics, and stratigraphic correlation; Cenozoic time scale; Cenozoic mammalian evolution; South American geology and mammalian evolution. *Mailing Add:* Dept Geol Field Museum Nat Hist Chicago IL 60605-2496

FLYNN, JOHN M(ATHEW), b Cleveland, Ohio, Dec 9, 29; div; c 7. CHEMICAL ENGINEERING. *Educ:* Case Inst Technol, BSChE, 51, MSChE, 53, PhD, 56. *Prof Exp:* Instr, Case Inst Technol, 51-52, asst, 52, res assoc, 53-56; proj leader, High Pressure Lab, Dow Chem Co, 56-60, lab dir, 60-63, prod supt pelaspan, 63-66, sales mgr chlorine based polymers, 66-67, bus mgr, 67, mgr res & develop, Plastics Dept, 67-72, dir prod res, Res & Develop, 72-73, gen mgr, Styrene Plastics Dept, 73-77, PROD DIR, AGR CHEM, METALS & SPECIALTY PROD, EXEC DEPT, DOW CHEM CO, 77- *Mem:* Am Chem Soc; Sigma Xi; Soc Plastics Indust. *Res:* Plastics technology; polyolefins and polystyrene. *Mailing Add:* 738 St Joseph Ave No 36 Sutton's Bay MI 49682

FLYNN, JOHN THOMAS, b Chester, Pa, Mar 14, 48; m 70; c 2. CARDIOVASCULAR PHYSIOLOGY. *Educ:* Widener Univ, BS, 70; Hahnemann Med Col, PhD(physiol), 74. *Prof Exp:* Fel, Thomas Jefferson Univ, 74-75, NIH fel, 75-76, from asst prof to assoc prof, 76-87, PROF PHYSIOL, JEFFERSON MED COL, THOMAS JEFFERSON UNIV, 87- *Concurrent Pos:* Prin investr, NIH grants. *Mem:* Am Physiol Soc; Shock Soc; Am Fedn Clin Res; NY Acad Sci. *Res:* Role of prostglandin-like materials in circulatory shock and cellular injury; endothelial cell biology; arachidonic acid cascade; regulation of eicosanoid synthesis and metabolism. *Mailing Add:* Dept Physiol Thomas Jefferson Univ 1020 Locust St Philadelphia PA 19107

FLYNN, JOSEPH HENRY, b Washington, DC, Oct 28, 22; m 52; c 9. POLYMER SCIENCE, THERMAL SCIENCES. *Educ:* Georgetown Univ, BS, 43; Cath Univ, PhD(phys chem), 54. *Prof Exp:* Asst, Cath Univ, 46-50; res chemist, Nat Bur Standards, 52-84; prin res scientist, Johns Hopkins Univ, 84-85; sr res scientist, Georgetown Univ, 85-86; RETIRED. *Concurrent Pos:* Guest worker, Nat Bur Standards, 84-; counr, Int Confedn Thermal Anal, IUPAC. *Honors & Awards:* Mettler Award, 80. *Mem:* AAAS; Am Chem Soc; Am Soc Testing & Mat; fel NAm Thermal Anal Soc; Int Confedn Thermal Anal. *Res:* Thermal analysis; polymer degradation; chemical kinetics; photochemistry and radiation chemistry of polymers; diffusion in polymers. *Mailing Add:* 5309 Iroquois Rd Bethesda MD 20816

FLYNN, KEVIN FRANCIS, b Chicago, Ill, Oct 28, 27; m 50; c 4. NUCLEAR WASTE DISPOSAL, NUCLEAR FISSION. *Educ:* Ill Inst Technol, BS, 50, MS, 53, MS, 55. *Prof Exp:* NUCLEAR CHEMIST, ARGONNE NAT LAB, 51- *Concurrent Pos:* Vis scientist, Swiss Fed Inst Reactor Res, 67-68. *Mem:* Am Chem Soc; Am Nuclear Soc; Health Physics Soc; Sigma Xi. *Res:* Nuclear chemistry and physics; nuclear fission; nuclear decay schemes; natural radio activities; environmental effects of nuclear radiation; nuclear measurement techniques. *Mailing Add:* 10057 S Longwood Dr Chicago IL 60643

FLYNN, MARGARET A, b Hurley, Wis, Nov 22, 15; m 38; c 2. NUTRITION, CLINICAL. *Educ:* Col St Catherine, BS, 37; Univ Iowa, MS, 38; Univ Mo, PhD(nutrit), 66; Am Bd Nutrit, cert, 71, dipl. *Prof Exp:* Teaching dietitian, Levi Hosp, Hot Springs, Ark, 42-46 & Holy Name Hosp, 48-54; res assoc, 61-63, from asst prof to assoc prof nutrit, 66-75, PROF NUTRIT, UNIV MO-COLUMBIA, 75- *Mem:* Am Inst Nutrit; Sigma Xi; Am Col Nutrit. *Res:* Body composition changes with age and disease; blood lipids. *Mailing Add:* M228 MDE SCI Univ Mo Columbia MO 65212

FLYNN, MICHAEL J, b New York, NY, May 20, 34; m 57; c 4. COMPUTER SCIENCE, ELECTRICAL ENGINEERING. *Educ:* Manhattan Col, BS, 55; Syracuse Univ, MS, 60; Purdue Univ, PhD(elec eng), 61. *Prof Exp:* Engr & mgr, IBM Corp, 55-65; assoc prof systs eng, Univ Ill, Chicago Circle, 65-66; assoc prof indust & elec eng, Northwestern Univ, 66-70; prof computer sci, Johns Hopkins Univ, 70-74; PROF ELEC ENG, STANFORD UNIV, 75- *Mem:* Asn Comput Mach; fel Inst Elec & Electronics Engrs; Brit Comput Soc; fel Inst Engrs Ireland. *Res:* Organization of computer systems; design of high speed computing processors and high speed arithmetic algorithms. *Mailing Add:* Dept Elec Eng Stanford Univ Stanford CA 94305-4055

FLYNN, PATRICK, b Ireland, 1929; Can citizen; m 58; c 1. PSYCHOPHARMACOLOGY. *Educ:* Univ Col Dublin, MPSIPhC, 51, MDBChBAO, 57; FRCP(C), 62; Am Col Psychiat, FACP, 74. *Prof Exp:* Clin dir psychiat, Waterford Hosp, St John's, Nfld, 62-64; dir psychiat res, May & Baker, Ltd, Dogenham, Essex, 64-65; from asst prof to assoc prof, 65-81, PROF PSYCHIAT, MED SCH, DALHOUSE UNIV, 82- *Concurrent Pos:* Consult psychiatrist, Victoria Gen Hosp, Halifax Infirmary, NS Hosp & Camp Hill Hosp. *Mem:* Can Med Asn; Can Psychiat Asn. *Res:* Long acting neuroleptic agents; lithium toxicity. *Mailing Add:* Victoria Gen Hosp Centennial 9A Halifax NS B3H 2Y9 Can

FLYNN, PAUL D(AVID), b Baltimore, Md, Oct 23, 26. EXPERIMENTAL MECHANICS, PHOTOELASTICITY. *Educ:* Johns Hopkins Univ, BE, 48, MSE, 50; Ill Inst Technol, PhD(mech), 54. *Prof Exp:* Inst mech eng, Johns Hopkins Univ, 48-51; engr, Gen Elec Co, 54-59; assoc prof mech, Ill Inst Technol, 59-62; consult, Frankford Arsenal, 60-62, res physicist, 62-77; RETIRED. *Honors & Awards:* Karl Fairbanks Mem Award, Soc Photo-Optical Instrumentation Engrs, 64. *Res:* Engineering mechanics; theoretical and experimental stress analysis; photoelasticity; strain gages; ballistics; high-speed photography; high-speed radiography; instrumentation; residual stresses. *Mailing Add:* 215 Tuscany Rd Baltimore MD 21210-3009

FLYNN, ROBERT JAMES, b Chicago, Ill, Jan 8, 23; m 42; c 6. LABORATORY ANIMAL MEDICINE, VETERINARY PUBLIC HEALTH. *Educ:* Mich State Univ, DVM, 44; Am Col Lab Animal Med, dipl, 57. *Prof Exp:* Supvr animal facil, Argonne Nat Lab, 48-55, assoc vet, 48-66, asst dir, Div Biol & Med Res, 62-76, sr vet, 66-81; consult vet, La Rabida, Chicago, 81-83; RETIRED. *Concurrent Pos:* Vet inspector, State of Ill, 44-57; consult, Pan-Am Health Orgn, 56-62; secy-treas, Am Col Lab Animal Sci, 56-62, pres, 63; mem adv coun, Inst Lab Animal Resources, Nat Acad Sci-Nat Res Coun, 57-64; county vet, Lake County, Ill, 57-83, rabies inspector, 70-73, animal control adminr, 74-83; adv bd vet specialties, Am Vet Med Asn, 60-66; consult, NIH, 60-76 & Biomed Res Found, AMA, 64-66; mem coun accreditation, Am Asn Accreditation Lab Animal Care, 64-66, consult, 64-76; consult, Vet Admin, 64-76; mem, Nat Res Coun, 67-70 & Comt Vet Med Res & Educ, 68-71; mem, NIH Adult Develop & Aging Res & Training Comt, 70-74; ed, Lab Animal Sci, 76-78. *Honors & Awards:* Griffin Award, 68 & Robert J Flynn Award, 69, Am Asn Lab Animal Sci. *Mem:* AAAS; Am Asn Lab Animal Sci (secy-treas, 53-62, pres, 64); Am Vet Med Asn; Am Soc Lab Animal Practitioners; Soc Exp Biol & Med. *Res:* environmental impact studies; virology, molecular biology, aging; diseases of laboratory animals; husbandry of laboratory animals. *Mailing Add:* 421 E Westleigh Rd Lake Forest IL 60045

FLYNN, ROBERT W, b Brooklyn, NY, July 26, 34. PLASMA PHYSICS. *Educ:* US Naval Acad, BS, 58; Mass Inst Technol, SM, 65, ScD(nuclear eng), 68. *Prof Exp:* From asst prof to assoc prof, 68-76, PROF PHYSICS, UNIV SFLA, 76- *Concurrent Pos:* Scientist-aquanaut, Scientist-in-the-Sea Prog, 73- *Mem:* Am Phys Soc; Am Asn Physics Teachers; Sigma Xi. *Res:* Large amplitude plasma waves; plasma mode coupling; plasma turbulence. *Mailing Add:* Dept Physics Univ SFla Tampa FL 33620

FLYNN, RONALD THOMAS, b Cleveland, Ohio, Dec 6, 47; m 69; c 2. GEOCHEMISTRY, GLASS SCIENCE. *Educ:* Temple Univ, BA, 69, MA, 72; Penn State Univ, PhD(geochem), 77. *Prof Exp:* Teaching asst geol, Temple Univ, 70-72; res asst geochem, Penn State Univ, 72-77; adv scientist glass sci, 77-78, supvr glass structure, 78-80, supvr fundamental studies, 80-83, mgr prod develop, 83-86, LAB DIR, OWENS-CORNING FIBERGLAS CORP, 86- *Mem:* AAAS; Fiber Soc; Am Chem Soc. *Res:* Redox equilibria; experimental geochemistry; glass structure; fiberglass product development. *Mailing Add:* Owens-Corning Fiberglas Corp PO Box 415 Granville OH 43023

FLYNN, T(HOMAS) F(RANCIS), b New Haven, Conn, Feb 27, 27; m 50; c 6. ELECTRICAL ENGINEERING. *Educ:* Yale Univ, BE, 50, ME, 51. *Prof Exp:* Engr, Perkin-Elmer Corp. 51-56, group leader, 56-60, chief engr, 60-69, dir eng, Instrument Div, 70-80, gen mgr, Spectros Div, 80-; RETIRED. *Concurrent Pos:* Lectr, Univ Conn, 55-57. *Mem:* Optical Soc Am; Inst Elec & Electronics Engrs. *Res:* Scientific instrument development. *Mailing Add:* 38 Big Oak Circle Stamford CT 06903

FLYNN, THOMAS GEOFFREY, b Ystradgynlais, Wales, Feb 20, 37; m 61; c 3. BIOCHEMISTRY. *Educ:* Univ Wales, BSc, 60, MSc, 62, PhD(biochem), 66. *Prof Exp:* Res asst clin chem, Med Unit, Royal Infirmary, Cardiff, Wales, 60-62; fel biochem, Univ Col, Cardiff, 66-67; fel & lectr, Queen's Univ, Ont, 67-69; from asst prof to assoc prof, 69-77, PROF BIOCHEM, QUEEN'S UNIV, ONT, 77- *Concurrent Pos:* Del, NATO Conf Protein Struct & Function, Venice, Italy, 70; vis assoc prof, Dept Biochem, Univ Calif, Berkeley, 75-76; mem, special rev comt, Nat Inst Alcohol Abuse & Alcoholism, 83-85 & mem study sect, 85-89; chmn study sect, 87-89. *Honors & Awards:* Gairdner Int Award, 86. *Mem:* Am Chem Soc; Brit Biochem Soc; Can Biochem Soc; Am Soc Biol Chem; AAAS; Am Diabetes Asn. *Res:* Structure of enzymes in relation to their function, especially monomeric oxidoreductases aldehyde reductases; aldose reductase and the complications of diabetes; atrial natriuretic factor and its receptor. *Mailing Add:* Biochem Dept Queen's Univ Kingston ON K7L 3N6 Can

FLYNN, THOMAS M(URRAY), b Huntsville, Tex, July 19, 33; m 58; c 4. CHEMICAL ENGINEERING, CRYOGENIC ENGINEERING. *Educ:* Wm M Rice Univ, BA, 54, BS, 55; Univ Colo, MS, 56, PhD(chem eng), 58. *Prof Exp:* Asst explor res, Magnolia Petrol Co, 51-52; process engr catalytic cracking, Shell Oil Co, 53-54; instr chem eng, Univ Colo, 55-58; proj leader, Cryogenic Eng Labs, Nat Bur Standards, 56-61; mgr cryogenic res & develop, Bendix Corp, 61-63; chief cryogenic instrument sect, US Govt, Nat Bur Standards, 63-66, sr scientist, Inst Mat Res, 66-68, chief prog coord off, Inst Basic Standards, 68-70, dirs staff, 70-72, sr chem engr, 72-83; STAFF CONSULT CRYOGENICS, BALL AEROSPACE, 87- *Concurrent Pos:* NSF lectr, 64; lectr, Univ Colo, 64, Univ Calif, Los Angeles, 72-, Am Inst Chem Engrs, 80-; adj prof, Univ Colo, 81-, Colo Sch Mines, 85-, Nanjing Univ, People's Repub China, 85-; consult engr, 83- *Mem:* Am Inst Chem Engrs; Sigma Xi. *Res:* Cryogenic engineering; unit operations; cryogenic instrumentation; separation phenomena; membranes; thermo-dynamics. *Mailing Add:* 511 N Adams Aveve Louisville CO 80027-2441

FOA, J(OSEPH) V(ICTOR), b Turin, Italy, July 10, 09; nat US; m 42; c 4. AERONAUTICAL ENGINEERING, ENGINEERING PHYSICS. *Educ:* Univ Turin, PhD(mech eng), 31; Univ Rome, PhD(aeronaut eng), 33. *Prof Exp:* From res engr to proj engr, Piaggio Aircraft Co, Italy, 33-35 & 37-39; chief engr, Studi Caproni, 35-37; proj engr, Bellanca Aircraft Corp, Del, 39-40; chief engr, Am Aero-Marine Indust, Inc, Mass, 42-43; head design res, Curtiss-Wright Corp, NY, 42-45; head propulsion br, Cornell Aeronaut Lab, 45-52; prof aeronaut eng, Rensselaer Polytech Inst, 52-58, head dept aeronaut eng & astronaut, 58-67; PROF ENG & APPL SCI, GEORGE WASHINGTON UNIV, 70- *Honors & Awards:* Arch J Colwell Merit Award, Soc Automotive Engrs, 70. *Res:* Fluid mechanics; propulsion; transportation; pressure exchange; cryptosteady flow and interactions; heating and air conditioning; author and coauthor of professional books and journal articles. *Mailing Add:* Sch Eng & Appl Sci George Washington Univ Washington DC 20052

FOA, PIERO PIO, b Torino, Italy, Apr 13, 11; nat US; m 41; c 2. PHYSIOLOGY. *Educ:* Univ Milan, MD, 34, ScD(chem), 38. *Prof Exp:* Instr biochem, Univ Milan, 34-36; asst prof physiol, Univ Pavia, 36-38; Mendelson fel surg, Univ Mich, 39-42; from asst prof to prof physiol & pharmacol, Chicago Med Sch, 42-61; prof physiol, 61-81, interim chmn, 80-81, EMER PROF, SCH MED, WAYNE STATE UNIV, 81- *Concurrent Pos:* Fel med, Univ Mich, 42-43; consult in res; chmn dept res, Sinai Hosp Detroit, 61-66, mem attend staff, 61-80. *Mem:* Am Physiol Soc; Soc Exp Biol & Med; Endocrine Soc; Am Diabetes Asn; Am Fedn Clin Res; Sigma Xi; Am Chem Soc. *Res:* Arterial hypertension; metabolism of thiamine; choline deficiency; functional innervation of the bone marrow; metabolism of lactic and pyruvic acids; glucagon; insulin; growth hormone; oral antidiabetic drugs. *Mailing Add:* 2104 Rhine Rd West Bloomfield MI 48323

FOARD, DONALD EDWARD, b Alexandria, Va, Dec 17, 29; m 55; c 3. BOTANY. *Educ:* Univ Va, BA, 52, MA, 54; NC State Col, PhD, 60. *Prof Exp:* Asst bot, Longwood Col, 54-56; asst prof, Univ Tenn, 59-60; asst prof, Univ Calif, Los Angeles, 60-63; BIOLOGIST, OAK RIDGE NAT LAB, 63-; biologist comp animal res, Univ Tenn, 77-80; vis prof, Dept Bot Plant & Path, Perdue, 80-84; VIS SCIENTIST, ISTITUTO AGRONOMICO PER L'OL TREMAR, FLORENCE ITALY, 84- *Mem:* Am Soc Plant Physiol. *Res:* Morphogenesis; morphology; anatomy. *Mailing Add:* Bot & Plant Path Purdue Univ Lilly Hall West Lafayette IN 47907

FOBARE, WILLIAM FLOYD, b Harper Woods, Mich, Nov 26, 54; c 1. MEDICINAL CHEMISTRY. *Educ:* Hope Col, AB, 77; Univ SC, PhD(org chem), 84. *Prof Exp:* Postdoctoral org chem, Ind Univ, 84-86; sr chemist, 86-88, res scientist, 88-90, PRIN SCIENTIST, WYETH-AYERST RES, 90- *Mem:* Am Chem Soc; AAAS. *Res:* Use of heterocyclic chemistry for designing pharmacologically active substances; altherosclerosis and lipid altering drugs. *Mailing Add:* 24 Cheverny Ct Hamilton NJ 08619

FOBES, MELCHER PRINCE, b Portland, Maine, Sept 18, 11; m 42. MATHEMATICAL ANALYSIS. *Educ:* Bowdoin Col, AB, 32; Harvard Univ, AM, 33, PhD(math), 47. *Prof Exp:* Instr math, Harvard Univ, 34-37 & Bryn Mawr Col, 38-39; from instr to prof, 40-81, EMER PROF MATH, COL WOOSTER, 81- *Mem:* Math Asn Am. *Res:* Topology; a conjectured inequality related to the product of Lipschitz Skeleton Cochains. *Mailing Add:* 1560 Hawthorne Dr Wooster OH 44691

FOCELLA, ANTONINO, b Baucina-Palermo, Italy, Dec 11, 24; US citizen; m 55; c 4. SUGAR ANALYSIS. *Educ:* Univ Palermo, Italy, Dr(org chem & biochem), 55. *Prof Exp:* Chemist food chem, Pepsi-Cola, Co, 56-59; RES INVESTR, HOFFMAN-LA ROCHE, INC, 59- *Mem:* Am Chem Soc; Sigma Xi; Sci Res Soc. *Res:* Synthetic organic chemistry; food chemistry; oil emulsion for beverages. *Mailing Add:* Hoffmann La Roche Nutley NJ 07110

FOCHT, DENNIS DOUGLASS, b West Reading, Pa, Aug 30, 41; m 66. SOIL MICROBIOLOGY. *Educ:* Rutgers Univ, BS, 63; Iowa State Univ, MS, 65, PhD(bact), 68. *Prof Exp:* Fel microbiol, Cornell Univ, 68-70; asst prof soil microbiol, 70-74, asst microbiologist, Citrus Res Exp Sta, 70-74, assoc prof, 74-80, PROF SOIL MICROBIOL, UNIV CALIF, RIVERSIDE, 80- *Concurrent Pos:* Proj leader, Western Regional Res Comt on Nitrogen in Environ, 70-75; consult, Stanford Res Inst, 73-74. *Mem:* Am Soc Microbiol; Am Soc Agron; Can Soc Microbiologists. *Res:* Biodegradation; nitrogen transformations; microbial ecology; methylation of metals. *Mailing Add:* Dept Soil Environ Sci Univ Calif Riverside CA 92521-0118

FOCHT, JOHN ARNOLD, JR, b Rockwall, Tex, Aug 31, 23; m 50; c 2. GEOTECHNICAL ENGINEERING. *Educ:* Univ Tex, Austin, BS, 44; Harvard Univ, MS, 47. *Prof Exp:* Soils engr, US Army Waterways Exp Sta, 47-53; sr soils engr, McClelland Engrs Inc, 53-55, vpres, 55-71, exec vpres, 71-87, chmn bd, 87-90; CONSULT, 91- *Honors & Awards:* Middlebrooks Award, 57 & 76, Laurie Prize, 59 & State of Art Award, Am Soc Civil Engrs, 71-79. *Mem:* Nat Acad Eng; Am Soc Civil Engrs; Am Consult Engrs Coun; Nat Soc Prof Engrs. *Res:* Performance of pile foundations to axial and lateral loads for design of offshore drilling platforms. *Mailing Add:* PO Box 740010 Houston TX 77274

FOCKE, ALFRED BOSWORTH, acoustics; deceased, see previous edition for last biography

FOCKE, ARTHUR E(LDRIDGE), b Cleveland, Ohio, June 17, 04; m 29, 68. METALLURGY. *Educ:* Ohio State Univ, BMetE, 25, MS, 26, PhD(metall), 28. *Prof Exp:* Metallurgist, Cleveland Wire Div, Gen Elec Co, 27-29; chief engr, P R Mallory & Co, 29-30; res metallurgist, Diamond Chain & Mfg Co, 30-45, chief metallurgist, 45-51; mgr mat develop, Aircraft Nuclear Power Dept, Gen Elec Co, 51-61; from assoc prof to prof, 62-74, EMER PROF METALL ENG, UNIV CINCINNATI, 74- *Concurrent Pos:* Bur Standards fel; pres, A E Focke Corp, 63-89. *Mem:* Fel Am Soc Metals (pres, 50); Am Soc Testing & Mat; Am Inst Mining, Metall & Petrol Engrs; Am Nuclear Soc. *Res:* Mechanical metallurgy; wear and fatigue; factor influencing the quality of tungsten incandescent lamp filaments; selection and development of materials for aircraft nuclear power plants. *Mailing Add:* 7799 E Galbraith Rd Cincinnati OH 45243

FODDEN, JOHN HENRY, b Halifax, Eng, June 16, 18; m 42; c 6. PATHOLOGY. *Educ:* Univ Leeds, MB, ChB, 41, MD, 46; FRCP(C), 56. *Prof Exp:* Asst path, Univ Leeds, 41-44; asst clin pathologist, Royal Hosp, Sheffield, 45-46; asst prof path, Univ Liverpool, 46-48; assoc prof, Dalhousie Univ, 48-52; assoc prof, Med Sch, Univ SDak, 52-57; assoc dir labs, Mt Sinai Hosp, 57-60; dir labs, Manitowoc County Hosp, 60-65; dir labs, Manitowoc Mem, Two Rivers Community & Algoma Mem Hosps, 65-85; RETIRED. *Concurrent Pos:* Registr, Cancer Control Orgn, Radium Inst, 46-48; Markle Found scholar, Univ SDak, 51-56, Lederle Med Fac award scholar, 56-57; forensic pathologist, Manitowoc County, 70-85; consult, medico-legal path. *Mem:* Fel Col Am Path; Am Soc Exp Path; Am Soc Clin Path; Am Asn Path & Bact; Am Acad Forensic Sci. *Res:* Etiologic and pathogenetic factors in peptic ulceration; hormonal factors in abnormal carbohydrate metabolism; clinical and experimental pathology. *Mailing Add:* Box 148 A/Rte 1 Glenflora Rd Kiel WI 53042

FODERARO, ANTHONY HAROLDE, b Scranton, Pa, Apr 3, 26; m 53; c 3. PHYSICS, REACTOR ENGINEERING. *Educ:* Univ Scranton, BS, 50; Univ Pittsburgh, PhD(physics), 55. *Prof Exp:* Consult nuclear physics, Westinghouse Atomic Power Div, 52-53, supvry scientist radiation shielding, 54-65; sr nuclear physicist reactor physics, Gen Motors Res Labs, 56-60; PROF NUCLEAR ENG, PA STATE UNIV, 60- *Concurrent Pos:* Consult, Gilbert Assoc, Inc, 73-77, Westinghouse Elec Corp, 60-77 & Stone & Webster Eng Corp, 76-; tech expert, Int Atomic Energy Agency, 76-77. *Mem:* Fel Am Nuclear Soc; Am Phys Soc; Am Asn Physics Teachers. *Res:* Reactor physics, reactor safety, radiation transport and shielding. *Mailing Add:* 301 S Gill St State College PA 16801

FODOR, GABOR, b Budapest, Hungary, Dec 5, 15; US citizen; m 39, 64, 78, 88; c 5. ORGANIC CHEMISTRY. *Educ:* Graz Tech Univ, state exam, 34; Univ Szeged, PhD(org chem), 37, Veniam Legendi, 45; Hungarian Acad Sci, DSc(org chem), 52. *Hon Degrees:* Golden Doctoral Diploma, U Szeged, 88. *Prof Exp:* Univ demonstr org chem, Univ Szeged, 35-38, from assoc prof to prof, 45-57; res assoc, Chinoin Pharmaceut Ltd, Hungary, 38-45; head lab stereochem, Hungarian Acad Sci, 58-65; prof org chem, Laval Univ, 65-69; centennial prof 69-86, EMER PROF CHEM, WVA UNIV, 86- *Concurrent Pos:* Overseas fel, Churchill Col, 61; vis scientist, Nat Res Coun Can, 64-65; vis prof, Stevens Inst Technol, 68 & Polytech Inst, Darmstadt, Germany, 75-76; emer prof, Jozsef Attila Univ Szeged, 90. *Honors & Awards:* Kossuth Award, Hungary, 50 & 54. *Mem:* Am Chem Soc; Can Inst Chem; Hungarian Acad Sci; Swiss Chem Soc; Royal Soc Chem London; NY Acad Sci. *Res:* Constitutional and synthetic work in isoquinolines, ephedrines, adrenaline and its derivatives; elucidation of configuration, chloromycetine, the tropines, scopolamine, cocaines, sedridine and sphingosine; total synthesis of valeroidine scopolamine and hyoscine; selective quaternization of tertiary nitrogen; new reactions of ascorbic acid with highly electrophilic aldehydes leading to new, immunopotentiating gulonolactones. *Mailing Add:* Dept Chem WVa Univ Morgantown WV 26506-6045

FODOR, GEORGE E, b Mako, Hungary, Feb 13, 32; US citizen; m; c 2. ORGANIC CHEMISTRY. *Educ:* Univ Szeged, dipl, 55; Rice Univ, PhD(chem), 65. *Prof Exp:* Chemist, Hungarian Oil Refining Co, 55; mem sci staff, Hungarian Oil & Gas Res Inst, 56; chemist, Pontiac Refining Co, Tex, 57-60; asst org chem, Rice Univ, 60-62; res chemist, Photo Prod Dept, E I du Pont de Nemours & Co, Inc, NJ, 65-66; sr res scientist, 66-81, STAFF SCIENTIST, SOUTHWEST RES INST, 81- *Mem:* Am Chem Soc; Soc Tribologists & Lubrication Engrs. *Res:* Synthetic organic and petroleum chemistry. *Mailing Add:* 6220 Culebra Rd Box 28510 Southwest Res Inst San Antonio TX 78284

FODOR, LAWRENCE MARTIN, b Cleveland, Ohio, Dec 1, 37; m 62; c 3. INORGANIC CHEMISTRY, POLYMER CHEMISTRY. *Educ:* Western Reserve Univ, AB, 59; Cornell Univ, PhD(inorg chem), 63. *Prof Exp:* Res chemist, Phillips Petrol Co, 63-76, supvr chem res, 76-84, supvr stereoregular resins, 84-90, COORDR PLASTICS RECYCLING RES, PHILLIPS PETROL CO, 90- *Mem:* Am Chem Soc. *Res:* Olefin polymerization; organic chemicals processes; fertilizer research; impact polystyrene; organometallic compounds; polyolefin plastics technology. *Mailing Add:* Phillips Petrol Co 83E PRC Bartlesville OK 74004

FOECKE, HAROLD ANTHONY, b Crofton, Nebr, Mar 7, 26; m 51; c 7. SCIENCE EDUCATION. *Educ:* Iowa State Univ, BS, 45, MS, 48; Univ Notre Dame, PhD(educ), 62. *Prof Exp:* Jr res fel, Iowa State Univ, 46, instr elec eng, 46-48; sr res fel, Polytech Inst Brooklyn, 48-49; from instr to asst prof elec eng, 49-52; asst prof, Univ Notre Dame, 54-58, 61-62; proj dir comt develop, Eng Fac, Am Soc Eng Educ, 58-61; specialist eng educ, US Off Educ, 62-65; dean eng, Gonzaga Univ, 65-68; dir div sci teaching, 68-76, DEP ASST DIR-GEN, EDUC, UNESCO, PARIS, FRANCE, 76- *Honors & Awards:* Arthur L Williston Award, Am Soc Eng Educ, 68. *Mem:* AAAS; Am Soc Eng Educ. *Res:* Engineering education; higher education. *Mailing Add:* 78 Rue de la Federation Paris 75015 France

FOEGE, WILLIAM HERBERT, b Decorah, Iowa, Mar 12, 36; m 58; c 3. MEDICINE, EPIDEMIOLOGY. *Educ:* Pac Lutheran Univ, BA, 57; Univ Wash, MD, 61; Harvard Univ, MPH, 65. *Hon Degrees:* DSc, Calif Lutheran Col, 77, Wartburg Col, 79, Tulane Univ, 80, Augustana Col, 84, WVa Univ, 84, Emory Univ, 86. *Prof Exp:* Intern, USPHS Hosp, Staten Island, NY, 61-62; epidemiologist, Epidemic Intel Serv, Nat Commun Dis Ctr, 62-64; med officer, Emmanuel Med Ctr, Lutheran Church-Mo Synod, Nigeria Mission, 65-66; epidemiologist, Ctr Dis Control, 66-70, dir smallpox eradication prog, 70-73; med epidemiologist, Southeast Asia Smallpox Eradication Prog, WHO, New Delhi, India, 73-75, asst dir, Ctrs Dis Control, Atlanta, 75-77, dir, 77-83, ASST SURGEON GEN & SPEC ASST POLICY DEVELOP, CTRS DIS CONTROL, 83-; EXEC DIR, CARTER CTR, EMORY UNIV, ATLANTA, GA, 86- *Concurrent Pos:* Consult, Smallpox Eradication Prog, Nigeria, 66-67; exec dir, Task Force Child Survival, 84-; mem var bds & comts, pub health & int health probs. *Honors & Awards:* Dept Health, Educ & Welfare Superior Serv Award; Joseph C Wilson Award, 78. *Mem:* Inst Med Nat Acad Sci; Royal Soc Trop Med & Hyg; AAAS; Am Epidemiol Soc; Am Col Prev Med; Am Pub Health Asn; AMA; Am Col Epidemiol; Physicians Social Responsibility; Royal Soc Trop Med & Hyg. *Res:* Epidemiology and control of communicable diseases in the tropics; author over 75 publications in area of public health and international health. *Mailing Add:* One Copenhill Atlanta GA 30307

FOEGH, MARIE LADEFOGED, b Denmark, Nov 3, 42. IMMUNOLOGY & CYTOLOGY. *Educ:* Univ Copenhagen, Denmark, MD, 69, DSc, 83. *Prof Exp:* ASSOC DIR TRANSPLANTATION, GEORGETOWN UNIV HOSP, 83-, ASSOC PROF, DIV NEPHROLOGY, DEPT MED & SURG & DIR RES, DIV TRANSPLANTATION, 85- *Mem:* Endocrine Soc; Int Soc Study Hypertesion in Pregnant Women; AMA. *Mailing Add:* Dept Surg Georgetown Univ Sch Med 3900 Reservoir NW Washington DC 20007

FOEHR, EDWARD GOTTHARD, b Philadelphia, Pa, Sept 22, 17; m 41; c 1. CHEMISTRY. *Educ:* Pa State Col, BS, 38, MS, 41, PhD(org chem), 44. *Prof Exp:* Asst petrol refinery lab, Pa State Col, 38-44; RES CHEMIST & SUPVR RES, CHEVRON RES CO, 44- *Concurrent Pos:* With Off Sci Res & Develop, 44. *Mem:* Am Chem Soc; Am Soc Lubrication Engrs. *Res:* Type analysis of lubricating oils; industrial lubricants; hydraulic transmission fluids; analysis of lubricating oil additives. *Mailing Add:* 12 Culloden Park Rd San Rafael CA 94901-1958

FOELSCHE, HORST WILHELM JULIUS, b Darmstadt, Ger, Oct 28, 37; m 61; c 2. HIGH ENERGY PHYSICS. *Educ:* Univ NMex, BS, 59; Yale Univ, MS, 60, PhD(physics), 63. *Prof Exp:* Res assoc physics, Yale Univ, 63-65; from asst physicist to physicist, Accelerator Dept, 65-83, SR PHYSICIST, AGS DEPT, BROOKHAVEN NAT LAB, 83- *Mem:* Am Phys Soc. *Res:* Accelerator physics. *Mailing Add:* Bldg 911 Brookhaven Nat Lab Upton NY 11973

FOERNZLER, ERNEST CARL, b Indianapolis, Ind, Apr 12, 35; m 64; c 2. PHARMACEUTICAL CHEMISTRY. *Educ:* Purdue Univ, BS, 57, MS, 59, PhD(phys chem), 62. *Prof Exp:* Instr radiochem, Purdue Univ, 57-59; mem opers res staff, Arthur D Little, Inc, 62-65; MGR COMPUT SYSTS, HOFFMANN-LA ROCHE, INC, NUTLEY, 65- *Mem:* Am Pharmaceut Asn; Asn Comput Mach; Opers Res Soc Am; Am Chem Soc. *Res:* Biomedical computer techniques; operations research; quantum biochemistry; computer systems and programming. *Mailing Add:* Glaxo Inc Five Moore Dr Research Triangle Park NC 27709

FOERSTER, EDWARD L(EROY), SR, b Chicago, Ill, Sept 17, 19; m 48; c 3. CHEMICAL ENGINEERING. *Educ:* Univ Ill, BS, 41; Univ Va, ChemE, 42. *Prof Exp:* Asst chem engr, Merck & Co, Inc, 42-48, chem engr, 48-52, sect leader, 52-58, mgr develop & control, Electronic Chem Div, 58-59; consult, 59-60; INDUST & COMMERCIAL CONSULT, 60- *Mem:* Am Inst Chem Engrs; Am Chen Soc; Nat Soc Prof Engrs. *Res:* Waste disposal; by-product recovery; food processing; fermentation; ion exchange; organic chemical manufacturing; rendering. *Mailing Add:* Box 779 Harrisonburg VA 22801-0779

FOERSTER, GEORGE STEPHEN, b New Orleans, La, Sept 29, 29; div; c 6. METALLURGY. *Educ:* Tulane Univ, BS, 50. *Prof Exp:* Res & develop engr, Metall Lab, 50-56, group leader, 56-61, assoc scientist, 61-72; prin res scientist, Metall Lab, NL Industs, Inc, 72-79, sr process engr permanent mold castings, 79-82; SR RES & DEVELOP ENGR, FARLEY INDUSTS, 83- *Mem:* Fel Am Soc Metals. *Res:* Metal alloy design, magnesium, aluminum, lead other non-ferrous metals; metal processing, battery grids; lead shot and body solder; die casting; alumuminum permanent mold casting. *Mailing Add:* 29 Glen Mawr Dr Trenton NJ 08618-2006

FOFONOFF, NICHOLAS PAUL, b Queenstown, Alta, Aug 18, 29; m 51; c 4. MECHANICS. *Educ:* Univ BC, BA & MA, 51; Brown Univ, PhD(appl math), 55. *Prof Exp:* From asst scientist to sr scientist, Pac Oceanog Group, Fisheries Res Bd Can, 54-62; SR SCIENTIST, WOODS HOLE OCEANOG INST, 62- *Concurrent Pos:* Nat Res Coun Can overseas fel, 55; Gordon McKay prof pract of phys oceanog, Harvard Univ, 69-86. *Honors & Awards:* Ocean Sci Award, Am Geophys Union, 89. *Mem:* AAAS; Am Geophys Union. *Res:* Fluid mechanics; dynamics of ocean circulation; thermodynamics of sea water; measurement of ocean currents. *Mailing Add:* Dept Phys Oceanog Woods Hole Oceanog Inst Woods Hole MA 02543

FOFT, JOHN WILLIAM, b Los Angeles, Calif, May 13, 28; m 57; c 2. PATHOLOGY. *Educ:* Univ Nebr, BS, 51, MD, 54. *Prof Exp:* Pathologist, Sch Aerospace Med, US Air Force, 61-63, chief dept path, 63-64; asst prof path, Univ Chicago, 64-68; assoc prof, Univ Hosps, 68-70, prof clin path & chmn dept, Sch Med, Univ Ala, Birmingham, 70-77; CHMN DEPT PATH, CARRAWAY METHODIST MED CTR & NORWOOD CLIN, 77- *Mem:* Am Asn Pathologists; Am Soc Clin Path; Sigma Xi. *Res:* Computer applications in diagnosis and monitoring of disease; clinical laboratory equipment design for developing countries. *Mailing Add:* 3529 Spring Valley Ct Birmingham AL 35223

FOGARTY, JOHN CHARDE, b Leamington, UK, Dec 16, 34; US citizen. ALGEBRA. *Educ:* Harvard Univ, AB, 61, PhD(math), 66. *Prof Exp:* Asst prof math, Univ Pa, 66-71; assoc prof, 71-80, PROF MATH, UNIV MASS, 80- *Res:* Invariant theory. *Mailing Add:* Dept Math Univ Mass Amherst Campus Amherst MA 01003

FOGARTY, WILLIAM JOSEPH, b Toronto, Ont, Nov 18, 32; US citizen; m 58; c 3. CIVIL ENGINEERING. *Educ:* Univ Miami, BSCE, 58; Purdue Univ, MSCE, 61; Ga Inst Technol, PhD(transp), 68. *Prof Exp:* Engr trainee, Fla State Rd Dept, 58-59; instr civil eng, Purdue Univ, 59-61; from asst prof to assoc prof, 61-73, PROF CIVIL ENG, UNIV MIAMI, 73- *Concurrent Pos:* Pvt consult, 61- *Res:* Multidisciplinary accident analysis, vehicular crash incidents, human, vehicle and environmental factors; environmental wastewater operations; water pollution and purification. *Mailing Add:* Dept Civil Eng Univ Miami Coral Gables FL 33124

FOGEL, BERNARD J, b New York, NY, Nov 30, 36; m 58; c 3. IMMUNOLOGY, PEDIATRICS. *Educ:* Univ Miami, MD, 61. *Prof Exp:* Intern pediat, Jackson Mem Hosp, Miami, Fla, 61-62, resident, 62-63; fel, Sch Med, Johns Hopkins Univ, 63-64; asst chief pediat, Walter Reed Army Med Ctr & researcher immunol, Walter Reed Army Inst Res, 64-66; ASSOC PROF PEDIAT, SCH MED, UNIV MIAMI, 66-, ASSOC DEAN EDUC, 67-, ASST VPRES MED AFFAIRS, 74-, PROF PEDIAT, 74-, VPRES & DEAN, SCH MED, 81- *Concurrent Pos:* Chief resident, Sinai Hosp, Baltimore, Md, 63-64; Am Cancer Soc fel, Jackson Mem Hosp, Miami, Fla, 66-68. *Mem:* Am Fedn Clin Res. *Res:* Complement system in human and animal diseases; osmotic fragility of erythrocytes; malaria; the human neonate. *Mailing Add:* Dean Med Univ Miami Sch Med 1600 NW Tenth Ave Miami FL 33101

FOGEL, CHARLES M(ORTON), b Syracuse, NY, June 21, 13; m 47; c 3. CIVIL ENGINEERING. *Educ:* Univ Buffalo, BA, 35, MA, 38. *Prof Exp:* Teacher sci pub schs, Buffalo, 37-41; instr physics, Univ Buffalo, 41-44; res engr, Nat Union Radio Corp, 44-46; from asst prof to prof eng, State Univ NY, Buffalo, 46-84, asst dean, 47-52, dir, Div Gen & Tech Studies, 52-57, indust liaison officer, 52-58, asst exec vpres, 67-77, actg grad dean, 77-78, actg exec vpres, 78-84; RETIRED. *Mem:* Am Soc Eng Educ. *Res:* Tapered columns; general engineering. *Mailing Add:* Civil Engr-229 Froniczak Hall State Univ NY N Campus Buffalo NY 14260

FOGEL, NORMAN, b Chicago, Ill, May 20, 24; m 60; c 2. PHYSICAL INORGANIC CHEMISTRY. *Educ:* Univ Ill, BS, 50; Univ Wis, MS, 51, PhD(chem), 56. *Prof Exp:* Instr chem, Exten Div, Univ Wis, 52-53, asst, 53-54, Univ Wis Alumni Res Found asst, 54-56; from asst prof to prof chem, 56-88, EMER PROF CHEM, UNIV OKLA, 88- *Mem:* Chem Soc; Sigma Xi; Am Chem Soc; AAAS. *Res:* Inorganic complex ions in solution and in solid state; spectra and magnetism of transition metal; complexes and electronic configuration of transition metal complexes. *Mailing Add:* Dept Chem & Biochem Rm 208 620 Parrington Oval Univ Okla Norman OK 73019-0370

FOGEL, ROBERT DALE, b Spokane, Wash, Jan 16, 47; m 68; c 2. HYPOGEOUS FUNGI, FUNGAL & ECOSYSTEMS ECOLOGY. *Educ:* Ore State Univ, BS, 69, PhD(mycol), 75; Univ Colo, MA, 70. *Prof Exp:* Res asst, Sch Forestry, Ore State Univ, 72-75, res assoc, 75-78; asst prof & asst cur, 78-84, ASSOC PROF BOT & FUNGAL ECOL, DIV BIOL SCI & ASSOC CUR, HERBARIUM, UNIV MICH, 84-, ASSOC PROF, SCH NATURAL RESOURCES, 90- *Concurrent Pos:* Counr, ecol-plant path, Mycol Soc Am. *Honors & Awards:* Alexopoulos Prize, Mycol Soc Am, 84. *Mem:* Mycol Soc Am; British Mycol Soc; US Soil Biol Soc. *Res:* Ecology and systematics of hypogeous fungi-truffles; role of mycorrhizae in nutrient cycling in forest ecosystems; molecular biology. *Mailing Add:* Herbarium N Univ Bldg Univ Mich Ann Arbor MI 48109

FOGEL, SEYMOUR, b New York, NY, Sept 27, 19; m 62; c 2. GENETICS. *Educ:* Queen's Col, BS, 41; Univ Mo, PhD(genetics), 46. *Hon Degrees:* DSc, City Univ NY, 87 & Univ Mo, 90. *Prof Exp:* Asst biol, Univ Mo, 42-43, instr zool, 46; instr biol, Queen's Col, 46-50; from assoc prof to prof, Brooklyn Col, 50-69, exec officer PhD prog, 65-69; prof & chmn dept, 69-80, EMER PROF GENETICS, UNIV CALIF, BERKELEY, 90- *Concurrent Pos:* Sigma Xi grant in aid, Queen's Col, NY, 47; USPH training fel, 59, res grant, 60; vis assoc prof, Stanford Univ; Guggenheim fel, Univ Calif, 67-68; vis prof, Univ Paris-Sud, Orsay, 80; vis res fel, Stanford Univ, 80. *Mem:* Genetics Soc Am; Am Soc Microbiol; AAAS. *Res:* Molecular mechanisms of recombination; gene conversion and recombinant DNA in yeast; fungal genetics; genetics of wine yeasts; fermentations; nondisjunction; heavy metal resistance. *Mailing Add:* Dept Genetics 2120 Oxford St Berkeley CA 94720

FOGELSON, DAVID EUGENE, b St Paul, Minn, Sept 13, 26; m 56; c 1. GEOPHYSICS. *Educ:* Univ Minn, BA, 52, MS, 56. *Prof Exp:* Asst seismol oil explor, Shell Oil Co, 50-52; geologist, Minn Dept Hwys, 52-56; geophysicist, Appl Physics Lab, US Bur Mines, 57-61, res geophysicist, 61-68, head explosive fragmentation lab, 61-70, sci res mgr, 70-71, res supvr, Advan Fragmentation Tech, 71-74, res supvr, environ effects mining, 74-77; RES ASSOC, NUCLEAR REPOSITORY DESIGN STUDIES, UNIV MINN, 79- *Concurrent Pos:* Prog mgr, NASA Contract on Lunar Mining Studies, 70-73; consulting geophysicist, 77- *Mem:* Seismol Soc Am; Int Soc Rock Mech. *Res:* Engineering seismology; generation and propagation of explosive waves; damage to structure from blasting; rock mechanics; lunar mining studies. *Mailing Add:* 1800 Valley Curve St Paul MN 55118

FOGG, DONALD ERNEST, b Camden, NJ, Jan 26, 22; m 43; c 3. VETERINARY PATHOLOGY. *Educ:* Univ Pa, VMD, 45. *Prof Exp:* Pvt pract, 45-49; with Bur Animal Indust, USDA, 49; with State of Del, 50; mgr clin res, Merck & Co, Inc, 51-60; tech serv dir, Del Poultry Labs, Inc, 60-66; dir vet serv, Eshams Farms Corp, 66-71; vet med officer, Consumer & Mkt Serv, USDA, 71-73, supvry vet med officer, Animal & Plant Health Inspection Serv, 73-84; RETIRED. *Mem:* Am Asn Avian Path. *Res:* Diseases of poultry. *Mailing Add:* 128 Hunter Forge Rd Newark DE 19713

FOGG, EDWARD T(HOMPSON), b Salem, NJ, Mar 24, 27; m 50. CHEMICAL ENGINEERING. *Educ:* Univ Va, BChE, 49; Univ Pa, MS, 51, PhD(chem eng), 53. *Prof Exp:* Asst instr chem eng, Univ Pa, 49-50, instr, 50-51; res engr, Jackson Lab, E I du Pont de Nemours & Co, Inc, Chambers Works, 53-56, res supvr, 56-60, div head, 60-66, mgr lab, 66-68, design supt, 68-70, works engr, 70-73, gen supt, Environ Serv Dept, 73-78; RETIRED. *Mem:* Am Chem Soc; Am Inst Chem Engrs. *Mailing Add:* Greenwich St Box 530 Alloway NJ 08001-0530

FOGG, PETER JOHN, b Bristol, Eng, June 14, 31; m 58; c 3. WOOD SCIENCE, WOOD TECHNOLOGY. *Educ:* Univ Wales, BSc, 52; La State Univ, MF, 61, PhD(forestry), 68. *Prof Exp:* Asst conservator, Forest Dept, Govt of Brit Honduras, 55-58; from instr to asst prof, 60-72, assoc prof, 72-79, PROF FORESTRY, LA STATE UNIV, BATON ROUGE, 79- *Mem:* Forest Prod Res Soc; Soc Wood Sci & Technol; Sigma Xi; Int Soc Trop Foresters. *Res:* Wood quality in relation to genetic and environmental factors; permeability of wood to liquids and gases; relation of wood properties to anatomical structure. *Mailing Add:* Sch Forestry, Wildlife & Fisheries La State Univ 108 Forestry Bldg Baton Rouge LA 70803

FOGIEL, ADOLF W, POLYMER CHEMISTRY, RUBBER CHEMISTRY. *Educ:* Karlsruhe Tech Univ, dipl, 49; Wayne State Univ, PhD(phys chem), 62. *Prof Exp:* Chemist, By-Prod & Chem, Australia, 50-52; res chemist, Am Agr Chem Co, Mich, 54-60, res supvr, 60-62; res chemist, 62-79, RES ASSOC, E I DU PONT DE NEMOURS & CO, INC, 79- *Mem:* Am Chem Soc. *Res:* Physical chemistry of polymers, especially elastomers. *Mailing Add:* 3309 N Rockfield Dr Deunshire DE 19810

FOGLE, HAROLD WARMAN, b Morgantown, WVa, Apr 23, 18; m 47; c 2. PLANT BREEDING & GENETICS. *Educ:* WVa Univ, BS, 40, MS, 41; Univ Minn, PhD(hort, plant genetics), 49. *Prof Exp:* Asst WVa Univ, 38-41; asst county supvr, CharlesTown & Martinsburg, WVa, 41-42; county supvr, Farm Security Admin, USDA, West Union, Morgantown & Buckhannon, WVa, 42-46; asst, Univ Minn, 46-49; horticulturist, Irrigation Exp Sta, USDA & Wash State Univ, 49-63; stone fruit invest leader, Plant Sci Res Div, Plant Indust Sta, USDA, Beltsville, Md, 63-72, res horticulturist, Fruit Lab, Agr Res Ctr, 72-80; RETIRED. *Concurrent Pos:* Radio operator, US Army & USAF, Anchorage & Shemya, Ariz, 43-46. *Honors & Awards:* Wilder Medal, Am Pomol Soc, 78. *Mem:* Am Soc Hort Sci; Am Pomol Soc (vpres, pres, 73-75); Int Soc Hort Sci. *Res:* Stone fruit breeding and varietal investigations; winter hardiness of tree fruits; ultrastructure of fruit and pollen surfaces; inheritance studies with tomatoes and stone fruits; vegetative propagation of stone fruits. *Mailing Add:* 2014 Forest Dale Dr Silver Spring MD 20903-1529

FOGLEMAN, RALPH WILLIAM, b McDonald, Kans, Mar 18, 26; m; c 3. RESEARCH ADMINISTRATION, SCIENCE POLICY. *Educ:* Kans State Univ, DVM, 47. *Prof Exp:* Asst vet, R A Self Animal Hosp, Tex, 47-48; vet, R W Fogleman Small Animal Hosp, Nebr, 48-50; head agr chem dept, Hazleton Labs, Inc, Va, 53-57, mgr western div, 57-60; vpres, Hazleton-Nuclear Sci Corp, Calif, 60-61, sr res pharmacologist, Am Cyanamid Co, 61-62; vpres, AME Assocs, 62-65; pres, Biographics, Inc, 65; toxicologist & registrations supvr, CIBA Agrochem Co, 65-69; dir toxicol, 69-71, vpres, Affil Med Enterprises, Inc, 71-75; DIR & OWNER, R W FOGLEMAN & ASSOCS, 75- *Honors & Awards:* Fel, AAAS, 86. *Mem:* Am Vet Med Asn; Am Col Toxicol; Soc Toxicol; Coun Agr Sci & Technol; Am Soc Agr Consult. *Res:* Research administration; applied biological sciences; toxicology of pesticides, drugs and xenobiotics in man and animals; metabolism of pesticides in plants and animals; industrial hygiene. *Mailing Add:* River Rd Upper Black Eddy PA 18972

FOGLEMAN, WAVELL WAINWRIGHT, b Winston-Salem, NC, July 17, 42. ENVIRONMENTAL CHEMISTRY, INORGANIC CHEMISTRY. *Educ:* Univ NC, Chapel Hill, BS, 64, MSPH, 76; Tulane Univ, PhD(chem), 68. *Prof Exp:* Asst prof chem, Old Dominion Univ, 68-74; asst prof natural sci, 77-82, chmn dept, 84-86, assoc dean acad affairs, 86-88, ASSOC PROF CHEM, PLYMOUTH STATE COL, 82- *Mem:* Am Chem Soc; Water Pollution Control Fedn. *Res:* Analysis and impact assessment of metals, gases and nutrients in water. *Mailing Add:* Dept Natural Sci Plymouth State Col Plymouth NH 03264

FOGLER, HUGH SCOTT, b Normal, Ill, Oct 28, 39; m 62; c 3. CHEMICAL ENGINEERING. *Educ:* Univ Ill, BS, 62; Univ Colo, MS, 63, PhD(chem eng), 65. *Prof Exp:* Asst prof, Univ Mich, 65; mem res staff, Jet Propulsion Labs, US Army, Pasadena, 66-68; from asst prof to assoc prof, 68-75, PROF CHEM ENG, UNIV MICH, ANN ARBOR, 75-, DIR, COLLOID STABILITY LAB. *Concurrent Pos:* Consult, Packaging Corp Am, 65-66, Chevron Oil Field Res, 67- & Mich Consol Gas, 76-; Fulbright scholar, 74. *Honors & Awards:* Dow Outstanding Young Fac Award, Am Soc Eng Educ, 71. *Mem:* Am Inst Chem Engrs; Am Chem Soc; Am Soc Eng Educ. *Res:* Colloid stability; flow and reaction in porous media, dissolution kinetics; processing of microelectronic materials. *Mailing Add:* 534 Heritage Univ Mich 3074 Dow Bldg Ann Arbor MI 48105-2517

FOGLESONG, MARK ALLEN, b Keokuk, Iowa, June 20, 49; m 71; c 2. MICROBIOLOGY, FERMENTATION TECHNOLOGY. *Educ:* Univ Iowa, BS, 71, MS, 72, PhD(microbiol), 74. *Prof Exp:* sr microbiologist res & develop, 74-80, SR MICROBIOLOGIST FERMENTATION TECHNOL, ELI LILLY & CO, 80- *Mem:* Am Soc Microbiol. *Res:* Microbiological assay development and therapeutic drug monitoring; control and regulation of macrolide antibiotic synthesis. *Mailing Add:* 3827 Billingsley Rd Greenwood IN 46143

FOGLIA, THOMAS ANTHONY, b Philadelphia, Pa, Apr 11, 40; m 63; c 3. ORGANIC CHEMISTRY, AGRICULTURAL & FOOD CHEMISTRY. *Educ:* Drexel Univ, BS, 64; Temple Univ, PhD(org chem), 68. *Prof Exp:* RES LEADER, EASTERN REGIONAL RES CTR, NORTHEASTERN REGION, AGR RES SERV, USDA, 68-; PROF DEPT CHEM, UNIV COL ST JOSEPH UNIV. *Concurrent Pos:* Asst prof, Dept Chem, Eve Col, Drexel Univ, 68-; lectr, St Joseph's Univ. *Mem:* Am Chem Soc; Am Oil Chemist Soc; Sigma Xi; AAAS. *Res:* Chemistry of fats and oils; lipid chemistry; food safety and quality; interaction of food components; lipid-protein interactions; membrane structure and function. *Mailing Add:* Dept Chem Univ Col St Joseph Univ 5600 City Ave Philadelphia PA 19131

FOGLIO, MARIO EUSEBIO, b Buenos Aires, Arg, Apr 25, 31; m 57; c 3. INTERMEDIATE VALENCE, STRONGLY CORRELATED SYSTEMS. *Educ:* Univ Buenos Aires, Lic en Quimica, 54, Dr en Quimica, 58; Bristol Univ, PhD(physics), 62. *Prof Exp:* Researcher phys chem, Arg Atomic Energy Comn, 54-56, researcher physics, 56-61; res assoc, Bristol Univ, 61-62; researcher, Arg Atomic Energy Comn, 62-68; res assoc, Harvard Univ, 68-69; assoc prof, Southern Ill Univ, 69-74; MEM STAFF, INST FISICA, 74- *Concurrent Pos:* S J Guggenheim Mem Found fel, 68-69; sr assoc, Int Centre Theoret Physics, Trieste, Italy, 85-90 & 90-96. *Mem:* Am Phys Soc; NY Acad Sci; Phys Soc Brasil. *Res:* Spin lattice relaxation; electron spin resonance; magnetic impurities in dielectrics; ferrimagnetic relaxation in europium iron garnet; statistical mechanics of impurities; neutron scattering in intermediate valence compounds; cumulant expansions for Anderson lattice; Kadanoff-Baym expansion for Anderson and su(n) lattice and for Hubbard model. *Mailing Add:* Inst Fisica Unicamp-CP 6165 13081 Campinas Sao Paulo Brazil

FOGLIO, SUSANA, b Bahia Blanca, Arg, June 6, 24; m 57; c 3. MATHEMATICS, STATISTICS. *Educ:* Univ Buenos Aires, MS, 49, PhD(math), 59. *Prof Exp:* Asst prof math, Nat Univ of the South, Arg, 49-52 & Univ Buenos Aires, 52-54; assoc prof, Nat Univ of the South, Arg, 54-56 & Bariloche Atomic Ctr, Arg, 56-70; PROF MATH, FED UNIV SAO CARLOS, 77- *Mem:* Arg Math Union; Am Math Soc. *Res:* Theory of integration; stochastic integrals. *Mailing Add:* Dept de Estatistica Univ Fed de Sao Carlos Sao Carlos Brazil

FOGWELL, JOSEPH WRAY, mechanical engineering; deceased, see previous edition for last biography

FOHL, TIMOTHY, b Pittsburgh, Pa, Apr 21, 34; m 61; c 3. FLUID DYNAMICS. *Educ:* Dartmouth Col, AB, 56; Mass Inst Technol, MS, 59, PhD(geophys), 63. *Prof Exp:* Res scientist physics, Itek Corp, 62-63; res scientist & dir, Mt Auburn Res Assoc, Inc, 63-72; mgr, New Prod Res, 72-85, dir eng develop, GTE Prod Corp, 85-88, PRIN RES SCIENTIST, GTE LABS, INC, 88- *Concurrent Pos:* Res affil, Mass Inst Technol, 63-65. *Honors & Awards:* Leslie H Warner Award. *Mem:* Am Phys Soc. *Res:* Vortex motion and turbulent flows with application to atmospheric motions and flows in arcs; operations analysis. *Mailing Add:* 681 South St Carlisle MA 01741

FOHLEN, GEORGE MARCEL, b San Francisco, Calif, Jan 3, 19; m 46. ORGANIC CHEMISTRY, POLYMER CHEMISTRY. *Educ:* Univ Calif, BS, 40, PhD(pharmaceut chem), 44. *Prof Exp:* Asst pharm, Univ Calif, 40-43; res chemist, Oronite Chem Co, Calif, 44-45; sr res chemist, Sterling-Winthrop Inst, NY, 45-47, Res & Develop Dept, Barrett Div, Allied Chem & Dye Corp, 47-55 & Reichhold Chem, Inc, Calif, 55-60; mgr anal servs, Cutter Labs, Berkeley, 60-62; sr res chemist process develop, Kaiser Chem Corp, Calif, 62-67 & Appl Space Prod, Inc, 67-70; res scientist chem res projs, Ames Res Ctr, NASA, 70-84; RETIRED. *Concurrent Pos:* Civilian with Off Sci Res & Develop, 44; consult, 84- *Honors & Awards:* NASA Award for develop achievement, 69. *Mem:* Am Chem Soc; Sigma Xi. *Res:* Analytical methods; spectroscopy; organic synthesis; dipole moments and structures of urea, thiourea and some sex hormones; industrial organic chemistry; plastics; high temperature polymers; intumescent coatings; transparent fire-resistant polymers; phthalocyanine polymers; cyclotriphosphazene polymers. *Mailing Add:* 1307 Vista Grande Millbrae CA 94030-2137

FOHLMEISTER, JURGEN FRITZ, b Znaim, Ger, Nov 18, 41; US citizen. NEUROPHYSIOLOGY, BIOPHYSICS. *Educ:* Univ Minn, BS, 64, MS, 65, PhD(physics), 71. *Prof Exp:* Fel, 71-73, lectr, 73-78, ASST PROF NEUROPHYSIOL, UNIV MINN, 78- *Mem:* Soc Neurosci; Biophys Soc. *Res:* Basic mechanisms of neural excitation; membrane biophysics. *Mailing Add:* Dept Physiol 6-255 Millard Hall Univ Minn Minneapolis MN 55455

FOIL, ROBERT RODNEY, b Bogalusa, La, Aug 12, 34; m 59, 84; c 2. FORESTRY, RESEARCH ADMINISTRATION. *Educ:* La State Univ, Baton Rouge, BS, 56, MFor, 60; Duke Univ, DFor(forest soils), 65. *Prof Exp:* Forester, Union Bag Corp, 56; instr forestry, La State Univ, 59-62, asst prof, 62-67; assoc specialist, La Exten Serv, 67-68, specialist, 68-69; prof & head dept, Miss State Univ, 69-73, dean, Sch Forest Resources, 74-78; dir, Miss Agr & Forestry Exp Sta, 78-86; VPRES, AGR FORESTRY, VET MED, 86- *Mem:* Fel Soc Am Foresters; Am Forestry Asn. *Res:* Forest resource management and use, particularly efficient harvesting of timber crops without damaging soils or desirable vegetation; science policy; agricultural and forest policy. *Mailing Add:* Div Agr Forestry Vet Med Box 5386 Mississippi State MS 39762

FOILES, CARL LUTHER, b Hardin, Ill, Oct 1, 35; m 63; c 2. PHYSICS. *Educ:* Univ Ariz, BS, 57, MS, 60, PhD(physics), 64. *Prof Exp:* Res assoc physics, Mich State Univ, 64-66; NSF fel Imp Col London, 66-67; asst prof, 67-70, assoc prof, 70-76, PROF PHYSICS, MICH STATE UNIV, 76- *Concurrent Pos:* Vis assoc prof, Univ Wash, 73-74. *Mem:* Am Phys Soc. *Res:* Low temperature properties of metals and alloys; transport properties and magnetic properties. *Mailing Add:* Dept Physics Mich State Univ East Lansing MI 48824

FOILES, STEPHEN MARTIN, b Tonasket, Washington, July 20, 56; m 86; c 1. SOLID STATE PHYSICS. *Educ:* Stanford Univ, BS, 78; Cornell Univ, MS, 81, PhD(physics), 83. *Prof Exp:* SR MEM TECH STAFF, SANDIA NAT LAB, 83- *Honors & Awards:* Math Sci Award, US Dept Energy, 87. *Mem:* Am Phys Soc; Mat Res Soc. *Res:* Calculation of the structural and thermodynamic properties of defects and interfaces in metals and alloys. *Mailing Add:* Div 8341 PO Box 969 Livermore CA 94551-0969

FOISTER, ROBERT THOMAS, b High Point, NC, Aug 20, 49; m 74; c 2. COLLOID & INTERFACE SCIENCE, ADHESION SCIENCE. *Educ:* Guilford Col, Greensboro, BS, 74; Univ NC, Chapel Hill, PhD(phys chem), 77. *Prof Exp:* Postdoctoral fel, Dept Chem & Pulp & Paper, Res Inst Can, McGill Univ, Montreal, 77-79; asst prof chem, Dept Chem, Guilford Col, 79-80; RES SCIENTIST, POLYMERS DEPT, GEN MOTORS RES, 80- *Concurrent Pos:* John M Campbell res award, Gen Motors, 91. *Mem:* Am Chem Soc; Adhesion Soc; Sigma Xi. *Res:* Physical chemistry of adhesion; surface physics of wetting and spreading; adsorption; flow properties of colloidal dispersions; dielectric properties of colloidal dispersions; electrorheological fluids. *Mailing Add:* Polymers Dept Gen Motors Res Labs Warren MI 48090-9055

FOISY, HECTOR B, b Ft Covington, NY, Feb 8, 36; m 58; c 6. MATHEMATICS. *Educ:* St Michael's Col, Vt, BA, 58; Univ Ill, Urbana, MA, 62; George Peabody Col, PhD(math), 71. *Prof Exp:* Teacher & chmn dept, High Sch, NY, 58-61; assoc prof, 62-80, PROF MATH, STATE UNIV NY COL POTSDAM, 80- *Mem:* Math Asn Am; Nat Coun Teachers Math. *Res:* Qualifications, duties and training of elementary school mathematics specialists. *Mailing Add:* Dept Math State Univ NY Col Potsdam Potsdam NY 13676

FOIT, FRANKLIN FREDERICK, JR, b Buffalo, NY, Jan 15, 42; div; c 2. GEOLOGY, MINERALOGY. *Educ:* Univ Mich, BS, 64, MS, 65, PhD(mineral), 68. *Prof Exp:* from asst prof to assoc prof geol, 71-85, PROF & CHMN, WASH STATE UNIV, 85- *Concurrent Pos:* Res assoc, Va Polytech Inst & State Univ, 70-71; res assoc, Southern Ill Univ, 80-81. *Mem:* Mineral Asn Can; Mineral Soc Am; Sigma Xi. *Res:* Crystal chemistry of silicates and borosilicates mineralogy and chemistry of coal. *Mailing Add:* Dept Geol Wash State Univ Pullman WA 99164

FOK, AGNES KWAN, b Hong Kong, Brit Crown Colony, Dec 11, 40; m 65; c 2. CELL BIOLOGY. *Educ:* Col Great Falls, BA, 65; Utah State Univ, MS, 66; Univ Tex, Austin, PhD(biochem), 71. *Prof Exp:* from asst res prof to assoc res prof, 73-88, RES PROF, UNIV HAWAII, 88- *Concurrent Pos:* Fel, Ford Found, 75. *Mem:* Am Soc Cell Biologists; Soc Protozoologists; Sigma Xi. *Res:* Intracellular digestion using paramecium as a model cell system to understand this process using a multidisciplinary approach. *Mailing Add:* Snyder 306 2538 The Mall Dept Microbiol Univ Hawaii Honolulu HI 96822

FOK, SAMUEL S(HIU) M(ING), b Macao, China, Feb 15, 26; US citizen; m 52; c 3. CHEMICAL ENGINEERING. *Educ:* Ohio State Univ, BChe, 49; Case Inst Technol, MS, 51, PhD(chem eng), 55. *Prof Exp:* Res asst, Case Inst Technol, 50-54; res engr, Indust Rayon Corp, Ohio, 54-56; sr staff engr, Shockley Transistor Corp, Calif, 56-60; mem tech staff, Res & Develop Lab, Fairchild Semiconductor Corp, 60-70; microphotog consult, 70-71; group leader, Siliconix, Inc, 71-73, engr mgr, 73-74; consult, 74-77; mgr processing & mach acceptance, ETEC Corp, 77-79; mgr process technol, 79-80, sr staff engr, 80-84, PRIN ENGR, EBT DIV, PERKIN-ELMER ETEC, INC, 84- *Concurrent Pos:* Tech adv, Ledel Semiconductor, Hong Kong, 73; dep managing dir, F C T Bros Co; exec dir, Int Frozen Foods Co; dir, Inter-Asian Tobacco Exporters Co, Bangkok, Thailand, 75-81. *Mem:* Am Chem Soc; Electrochem Soc; Soc Photog Sci & Eng; Am Soc Test Mat; Semiconductor Equip Mat Inst; Sigma Xi. *Res:* Large scale integration, metal-oxide-silicon or LSI/MOS mask making; masking; photoresists; photofabrication; microphotography; semiconductor process development; plastic packaging; high polymers; chrome mask making; transparent mask; one to one projection system; electron beam mask making and processing; multi-level resist. *Mailing Add:* 820 Talisman Dr Palo Alto CA 94303

FOK, SIU YUEN, b Macau, China, Nov 23, 37; m; c 2. POLYMER PHYSICS, CHEMICAL ENGINEERING. *Educ:* Nat Taiwan Univ, BS, 60; Univ Cincinnati, MS, 62; Va Polytech Inst, PhD(chem eng), 65. *Prof Exp:* Fel chem, Clarkson Col Technol, 65-66; sr res chemist, Fibers & Laminates Div, Enjay Chem Co, 66-69; prin chemist, Dart Indust, 69-73; plant mgr, Beau Plastic Co, 73-81; GEN MGR, KEMCOAT CO, 81- *Mem:* Am Chem Soc; Am Inst Chem Engrs; Soc Plastics Engrs. *Res:* Polymer physics and characterization; fiber spinning; coated fabrics; vinyl technology. *Mailing Add:* Kemcoat Co 43 Cloudview Rd Rm B-14 Kowloon Hong Kong

FOK, THOMAS DSO YUN, b Canton, China, July 1, 21; nat US; m 49. CIVIL ENGINEERING. *Educ:* Nat Tung-Chi Univ, China, BEng, 45; Univ Ill, MS, 48; NY Univ, MBA, 50; Carnegie Inst, PhD(civil eng), 56. *Prof Exp:* Design engr, Richardson, Gordon & Assocs, Pa, 56-58; assoc prof civil eng, Youngstown Univ, 58-67, dir comput ctr, 63-67; partner, Mosure & Fok, consult engrs, 67-76; CHMN, THOMAS FOK & ASSOCS, LTD, 77- *Concurrent Pos:* Consult to indust & pub agencies Ohio & Pa, 58- *Mem:* Am Soc Eng Educ; Am Soc Civil Engrs; Am Concrete Inst; Int Asn Bridge & Struct Engrs. *Res:* Structural engineering; applied mechanics and programming for digital computers. *Mailing Add:* 325 S Canfield-Niles Rd Youngstown OH 44515

FOK, YU-SI, b China, Jan 15, 32; c 2. WATER RESOURCES, IRRIGATION. *Educ:* Nat Taiwan Univ, BS, 55; Utah State Univ, MS, 59, PhD(civil eng), 64. *Prof Exp:* Res asst civil & irrig eng, Utah State Univ, 57-63, asst prof agr & irrig eng & asst res engr, Utah Water Res Lab, 63-66; hydrologist, Ill State Water Surv, 66-68; head basin hydrol sect, Tex Water Rights Comn, Tex, 68-70; assoc prof, 70-77, PROF CIVIL ENG & RESEARCHER, WATER RESOURCES RES CTR, UNIV HAWAII, 77- *Concurrent Pos:* Res assoc, Univ Ill, Urbana, 66-68; pres, Int Rainwater Catchment Systs Asn, 89- *Mem:* Am Geophys Union; fel Int Water Resources Asn; fel Am Soc Civil Engrs; Am Soc Agr Eng; Sigma Xi; Am Water Works Asn. *Res:* Soil physics; stream system morphology-mechanics of stream morphological systems; hydrology-urban hydrology, flood hydrology, water conservation and availability and streamflow forecasting; water resources systems analysis-optimization of water resources; irrigation and drainage science; infiltration; surface irrigation systems design; stream systems hydraulic geometry; explicit algebraic infiltration equations in one, two and three dimensions; basin water availability assessments; watershed modeling; rain water cistern systems and flexible impermeable membrane for water separation and conservation in an embayment; water resources planning and management; low cost desalination of brackish and seawaters. *Mailing Add:* Dept Civil Eng Univ Hawaii 2540 Dole St Honolulu HI 96822

FOLAN, LORCAN MICHAEL, b Dublin, Ireland, Dec 4, 60; m 86. CHEMICAL PHYSICS, AEROSOL SCIENCE. *Educ:* Trinity Col, Dublin, Ireland, BSc, 81; Polytech Univ, Brooklyn, PhD(physics), 87. *Prof Exp:* ASST PROF PHYSICS, POLYTECH UNIV, 87- *Mem:* Am Phys Soc; Optical Soc Am; Sigma Xi. *Res:* Optical processes in micron sized particles; physical and chemical characterization of single aerosol particles. *Mailing Add:* Physics Dept Polytech Univ 333 Jay St Brooklyn NY 11201

FOLAND, KENNETH AUSTIN, b Frederick, Md, May 25, 45; m 68. ISOTOPE GEOCHEMISTRY. *Educ:* Bucknell Univ, BS, 67; Brown Univ, MSc, 69, PhD(geochem), 72. *Prof Exp:* Fel geol, Univ Penn, 72-73, from asst prof to assoc prof, 73-80; assoc prof, 80-87, PROF GEOL, OHIO STATE UNIV, 87- *Concurrent Pos:* Consult, Lawrence Livermore Nat Lab, 82-86, Nuclear Regulatory Comn, 90- *Mem:* Am Geophys Union; Geochem Soc; Geol Soc Am; Sigma Xi. *Res:* Use of isotope geochemistry and geochronology to understand Earth's processes; genesis of igneous rocks; structure and composition of the Earth's mantle; diffusional processes in rocks and minerals; development of dating techniques. *Mailing Add:* Dept Geol Sci Ohio State Univ Columbus OH 43210-1398

FOLAND, NEAL EUGENE, b Parnell, Mo, Dec 9, 29; m 56; c 5. MATHEMATICS. *Educ:* Northeast Mo State Teachers Col, BS, 54; Univ Mo, MA, 58, PhD(math), 61. *Prof Exp:* From asst prof to assoc prof math, Kans State Univ, 61-65; assoc prof, 65-70, PROF MATH, SOUTHERN ILL UNIV, 70-, CHMN DEPT, 71- *Mem:* Am Math Soc; Math Asn Am. *Res:* Topological dynamics. *Mailing Add:* Dept Math Southern Ill Univ Carbondale IL 62901-4408

FOLAND, WILLIAM DOUGLAS, b Knoxville, Tenn, Jan 15, 26; m 55. PHYSICS. *Educ:* Univ Tenn, AB, 51, MS, 55, PhD(physics), 58. *Prof Exp:* Asst physics, Univ Tenn, 51-55, instr, 55-58; from asst prof to assoc prof, Univ Mass, Amherst, 58-68; assoc prof, 68-76, PROF PHYSICS, WASHINGTON & JEFFERSON COL, 76-, DEPT CHMN, 75- *Res:* Atomic collisions; theory. *Mailing Add:* Dept Physics Washington & Jefferson Col Washington PA 15301

FOLDES, FRANCIS FERENC, b Budapest, Hungary, June 13, 10; nat US; m 38; c 3. ANESTHESIOLOGY. *Educ:* Univ Budapest, MD, 34. *Hon Degrees:* Dr, Sch Med, Univ Szeged, Hungary, 79; FRCS, 82. *Prof Exp:* Res physician, Svabhegyi Sanatorium, Budapest, 35-39; pvt pract, 39-41; res fel anesthesia, Mass Gen Hosp, Boston, 41-42, resident, 42-43, asst anesthetist, 44-47; dir dept anesthesiol, Mercy Hosp, Pittsburgh, Pa, 47-62; chmn dept anesthesiol, 62-75, CONSULT ANESTHESIOL, MONTEFIORE HOSP & MED CTR, 75-, EMER PROF ANESTHESIOL, ALBERT EINSTEIN COL MED, 75- *Concurrent Pos:* Asst, Harvard Med Sch, 43-47; assoc prof, Med Sch, Univ Pittsburgh, 48-57, clin prof, 57-62; mem med adv bd, Myasthenia Gravis Found, 59-, from vchmn to chmn, 62-68; vpres, World Fedn Socs Anesthesiologists, 60-64, mem sci adv & membership comts, 65-68, pres fedn, 68-72; mem subcomt anesthesiol, Comt Revision US Pharmacopeia, 60-75; clin prof, Col Physicians & Surgeons, Columbia Univ, 62-64; res assoc, Mt Sinai Hosp, NY, 62-75; prof anesthesiol, Albert Einstein Col Med, 64-75; vis prof, Cath Univ Nijmegen, Holland, 75; prof anesthesiol, Sch Med, Miami Univ, 76- *Honors & Awards:* Semmelweis Award, Am Hungarian Med Asn, 66, George Washington Award, 70; Am Soc Anesthesiol Distinguished Serv Award, 72; Span Soc Anesthesiologists Gold Medal, 74; Heidbrink Award, Am Dent Soc Anesthesiol, 75; Ralph M Waters Award, Int Anesthesia Award Comn, 76; Issekutz, Hungarian Pharmacologist Soc, 79; Harold R Griffith Medal, McGill Univ, 86; Res Award, Am Soc Anesthesiologists, 88. *Mem:* Am Soc Anesthesiol; Am Soc Pharmacol & Exp Therapeut; NY Acad Med; NY Acad Sci; Int Anesthesia Res Soc. *Res:* Neuropharmacology; enzymes; co-enzymes. *Mailing Add:* Dept Anesthesiol Montefiore Med Ctr 111 E 210th St Bronx NY 10467

FOLDI, ANDREW PETER, b Budapest, Hungary, Feb 24, 31; US citizen; m 66. ORGANIC CHEMISTRY, POLYMER CHEMISTRY. *Educ:* Budapest Tech Univ, dipl Chem Eng, 53; Univ Del, PhD(org chem), 63. *Prof Exp:* Design & prod chem engr, Chem Complex, Hungary, 53-56; chem engr, 57-60, sr res chemist, E I du Pont de Nemours & Co, 62-85; INDEPENDENT CONSULT, C & C CONSULTS, 85- *Mem:* Am Chem Soc. *Res:* Industrial fibers; rubber technology; thermoplastics; polyurethanes; pharmaceutical and synthetic organic chemistry; theoretical organic and industrial chemistry; textile technology. *Mailing Add:* 2833 W Oakland Dr Wilmington DE 19808-2422

FOLDS, JAMES DONALD, b Augusta, Ga, Sept 26, 40; m 62; c 2. MICROBIOLOGY, IMMUNOLOGY. *Educ:* Univ Ga, BS, 62; Med Col Ga, PhD(microbiol), 67. *Prof Exp:* USPHS fel immunol, Sch Med, Case Western Reserve Univ, 67-69; from instr to assoc prof Bact & Immunol, 69-82, PROF MICROBIOL & IMMUNOL, SCH MED, UNIV NC, CHAPEL HILL, 82- *Concurrent Pos:* Dir clin microbiol & immunol labs, NC Mem Hosp, 76-83. *Mem:* AAAS; Am Asn Immunol; Am Soc Microbiol. *Res:* host responses to hemophilus ducreyi; rapid methods of immunodiagnosis; immunity to treponema pallium. *Mailing Add:* Dept Microbiol & Immunol Univ NC Sch Med Chapel Hill NC 27514

FOLDVARY, ELMER, b Youngstown, Ohio, Mar 9, 35; m 65; c 2. ORGANIC CHEMISTRY. *Educ:* Youngstown State Univ, BS, 58; Tex A&M Univ, MS, 61, PhD(chem), 64. *Prof Exp:* Asst prof, 63-68, assoc prof, 68-77, PROF CHEM, YOUNGSTOWN STATE UNIV, 77- *Mem:* Am Chem Soc; Sigma Xi. *Res:* Linear free energy relationships; synthetics; polyester plastics; kinetics. *Mailing Add:* Dept Chem Youngstown State Univ Youngstown OH 44555

FOLDY, LESLIE LAWRANCE, b Sabinov, Czech, Oct 26, 19; US citizen; m 44; c 2. THEORETICAL PHYSICS. *Educ:* Case Inst Technol, BS, 41; Univ Wis, PhM, 42; Univ Calif, PhD(physics), 48. *Prof Exp:* Res physicist, Div War Res, Columbia Univ, 42-45; res physicist, Radiation Lab, Univ Calif, 45-46; from asst prof to prof, 48-66, EMER INST PROF PHYSICS, CASE WESTERN RESERVE UNIV, 66- *Concurrent Pos:* Fulbright & Guggenheim fel, Inst Theoret Physics, Copenhagen, 53-54; NSF sr fel, Europ Orgn Nuclear Res, Geneva, 63-64. *Honors & Awards:* William A Fouler Award, Am Phys Soc, 89. *Mem:* Fel Am Phys Soc; fel Acoust Soc Am; Fedn Am Sci; AAAS. *Res:* Acoustical theory; theoretical nuclear physics; quantum theory; quantum field theories; theory of high energy accelerators; elementary particle physics. *Mailing Add:* 2232 Elandon Dr Cleveland Heights OH 44106

FOLEN, VINCENT JAMES, b Scranton, Pa, Jan 17, 24; m 54. MAGNETISM. *Educ:* La Salle Col, BA, 49; Univ Pa, MA, 54; Am Univ, PhD(physics), 72. *Prof Exp:* Res asst, Univ Pa, 50-54; physicist, 54-59, HEAD FERROMAGNETISM SECT, US NAVAL RES LAB, 59- *Honors & Awards:* Res Publ Awards, Naval Res Lab, 68, 70 & 79; Pure Sci Award, Sci Res Soc Am, 71. *Mem:* Fel Am Phys Soc; Sigma Xi. *Res:* Electron spin and ferromagnetic resonance, magnetoelectric effect, magnetocrystalline anisotropy, quantum electronics. *Mailing Add:* 9342 Old Mt Vernon Rd Alexandria VA 22309

FOLEY, DEAN CARROLL, b Pomeroy, Wash, Nov 25, 25; m 55. PHYTOPATHOLOGY. *Educ:* Univ Idaho, BS, 49; WVa Univ, MS, 51; Pa State Univ, PhD(plant path), 55. *Prof Exp:* Asst plant path, WVa Univ, 49-51 & Pa State Univ, 51-55; from asst prof to assoc prof plant path, Iowa State Univ, 55-; RETIRED. *Mem:* AAAS; Am Phytopath Soc; Am Inst Biol Sci. *Res:* Diseases resistance; diseases of cereals. *Mailing Add:* W 2803 Longfellow Iowa State Univ Spokane WA 99205

FOLEY, DENNIS JOSEPH, b Brooklyn, NY, Dec 28, 45; m 68; c 3. PHARMACOLOGY. *Educ:* Fordham Univ, BS, 68; WVa Univ, PhD(pharm), 72. *Prof Exp:* Res pharmacologist, Walter Reed Army Inst Res, 72-75; group leader Clin Study Anal, 75-81, DEPT HEAD, CLIN RES INFO, MED RES DIV, LEDERLE LABS, AM CYANAMID CO, 81- *Mailing Add:* 9 Midway Rd Spring Valley NY 10977

FOLEY, DUANE H, CARDIOVASCULAR PHYSIOLOGY, BLOOD FLOW REGULATION. *Educ:* Univ Calif, Davis, BS, 69, PhD(physiol), 75. *Prof Exp:* ASSOC PROF PHYSIOL, COL OSTEOP MED PAC, 84- *Mailing Add:* Dept Physiol Col Osteo Med Pac 309 E College Plaza Pomona CA 91766-1889

FOLEY, EDWARD LEO, b Butte, Mont, Apr 15, 30; m 55; c 8. PHYSICS. *Educ:* Mont State Col, BS, 54; Lehigh Univ, MS, 57, PhD(physics), 62. *Prof Exp:* Instr physics, Lehigh Univ, 60-62; asst prof, Univ Vt, 62-66; from asst prof to assoc prof, 66-71, chmn dept, 67-77, PROF PHYSICS, ST MICHAEL'S COL, 71- *Mem:* Am Asn Physics Teachers; Int Solar Energy Soc. *Res:* Thermal diffusivities of metals; surface physics; solar energy, insolation on tilt. *Mailing Add:* Dept Physics St Michael's Col Winooski VT 05404

FOLEY, GEORGE EDWARD, medical microbiology; deceased, see previous edition for last biography

FOLEY, H THOMAS, b Pittsburgh, Pa, Mar 27, 33; m 54; c 2. INTERNAL MEDICINE. *Educ:* Univ Pittsburgh, BS, 54; NY Univ, MD, 60; Georgetown Univ, JD, 77. *Prof Exp:* Intern & resident, III & IV Med Div, Bellevue Hosp, New York, 60-63; liason officer to eastern solid tumor group, Nat Cancer Inst, 63-65, sr investr radiation & med br, 65-69; sr investr, Baltimore Cancer Res Ctr, 69-70; med officer, Georgetown Med Serv, DC Gen Hosp, 70-81; med officer, Armed Forces Inst Path, 81-91, MED OFFICER, OFF SURGEON GEN, DEPT ARMY, 91- *Concurrent Pos:* Clin asst prof, Georgetown Univ, 81- *Mem:* Fel Am Col Physicians; fel Am Col Legal Med. *Mailing Add:* CCRB Army Prof Health Support Agency Walter Reed Army Med Ctr Washington DC 20307

FOLEY, HENRY CHARLES, b Providence, RI, Jan 24, 56; m 79; c 2. INDUSTRIAL & ENGINEERING CHEMISTRY, CATALYTIC & SEPARATIVE MATERIALS DESIGN. *Educ:* Providence Col, BS, 77; Purdue Univ, MS, 79; Pa State Univ, PhD(phys & inorg chem), 82. *Prof Exp:* Res chemist, Chem Res Dept, Am Cyanamid Co, 82-84, sr res chemist, 84-85, group leader, 85-86; asst prof, 86-91, ASSOC PROF CHEM ENG & DIR, CTR CATALYTIC SCI & TECHNOL, UNIV DEL, 91- *Concurrent Pos:* Indust consult, 86-; NSF presidential young investr, 87; assoc dir, Ctr Catalytic Sci & Technol, Univ Del, 87-91, asst prof, 89- *Mem:* Am Inst Chem Engrs; Am Chem Soc; Mat Res Soc; Math Asn Am. *Res:* Design and synthesis of novel materials for catalysis and separation; inorganic and physical chemistry; chemical reaction engineering; materials science and engineering. *Mailing Add:* Dept Chem Eng Univ Del Newark DE 19716

FOLEY, HENRY MICHAEL, physics; deceased, see previous edition for last biography

FOLEY, JAMES DAVID, b Palmerton, Penn, July 20, 42; m 64; c 2. SOFTWARE SYSTEMS, INTELLIGENT SYSTEMS. *Educ:* Lehigh Univ, BS, 64; Univ Mich, MS, 65, PhD(comput info & control eng), 69. *Prof Exp:* Asst prof comput sci, Univ NC, 70-76; from assoc prof to prof elec eng & comput sci, George Wash Univ, 77-90, chmn, 88-90; PROF, COL COMPUT, GA INST TECHNOL, 90- *Concurrent Pos:* assoc ed, Transactions Graphics, 81-; pres, Comput Graphics Consult, Inc, 79- *Mem:* Asn Comput Mach; Fel Inst Elec & Electronics Engrs; Comput Soc; Nat Comput Graphics Asn; Human Factor Soc. *Res:* Interactive computer graphics software, hardware, systems, and applications; human factors of user-computer interfaces; cognitive psychology. *Mailing Add:* Col Comput Ga Inst Technol Atlanta GA 30332-0280

FOLEY, JAMES E, b Newburyport, Mass, Jan 4, 50; m 72; c 2. DIABETES, OBESITY. *Educ:* Merrimack Col, BA, 72; Dartmouth Med Sch, PhD(physiol), 76. *Prof Exp:* Res assoc, Dartmouth Med Sch, 76-77, fel, 77-79; guest scientist, Panum Inst, Univ Copenhagen & Univ Arhus, 79-80; res asst prof, Univ Tex Health Sci Ctr, Dallas, 80-81; sr staff fel, Phoenix Clin Res, NIH/Nat Inst Arthritis, Diabetes & Digestive & Kidney Dis, 81-85, sr scientist clin diabetes & nutrit, 85-86; diabetes group leader, 86, DIR, DIABETES DEPT, SANDOZ RES INST, 86- *Mem:* Am Diabetes Asn; Am Fedn Clin Res; NY Acad Sci; AAAS; Am Physiol Soc; Europ Asn Study Diabetes. *Res:* Mechanisms relevant to etiology of non-insulin-dependent diabetes mellitus and obesity and how these mechanisms can be modified pharmacologically in order to treat and possibly prevent these diseases. *Mailing Add:* Diabetes Dept Sandoz Res Inst East Hanover NJ 07936

FOLEY, JOHN F, b Buffalo, NY, Feb 1, 31; m 55; c 5. INTERNAL MEDICINE. *Educ:* Univ Buffalo, MD, 55; Univ Minn, PhD(internal med), 62. *Prof Exp:* Res specialist, Univ Minn, 60-62, res fel internal med, 62-63; asst prof med, 63-68, PROF MED, UNIV NEBR MED CTR, 68- *Mem:* Am Asn Cancer Res; Tissue Cult Asn; Am Soc Clin Oncol; Am Asn Cancer Educ; fel Am Col Physicians. *Res:* Tissue culture; cell interactions; chemical chemotherapy; preventive medicine; hormonal action on tissue culture cells. *Mailing Add:* Dept Internal Med Univ Nebr Med Ctr Omaha NE 68105

FOLEY, JOSEPH MICHAEL, b Dorchester, Mass, Mar 9, 16; m 44. NEUROLOGY. *Educ:* Col of the Holy Cross, AB, 37; Harvard Univ, MD, 41. *Hon Degrees:* ScD, Col of the Holy Cross, 62. *Prof Exp:* Intern, Bellevue Hosp, NY, 41-43; asst neurol, Harvard Med Sch, 46-48, from instr to asst prof, 48-59; prof, Seton Hall Col Med & Dent, 59-61; PROF NEUROL, SCH MED, CASE WESTERN RESERVE UNIV, 61-; EMER PROF NEUROL, UNIV HOSPS, CLEVELAND, 86- *Concurrent Pos:* Rockefeller asst, Boston City Hosp, 46-48; asst prof, Med Sch, Univ Boston, 48-51; dir dept neurol, Jersey City Med Ctr, 59-61; consult, US Air Force, 59-63; chmn med adv bd & res rev panel, Nat Multiple Sclerosis Soc; coordr continuing med educ, Case Western Reserve Univ, 67-76; mem bd trustees, Col Holy Cross; trustee, Cleveland Med Libr Asn, 72-84, pres, 82-83. *Honors & Awards:* Bronze Hope Chest Award, Nat Multiple Sclerosis Soc, 75; Arnold Heller Award, Menorah Park, 83; Salzman Award, Mt Sinai Hosp, 90. *Mem:* Am Neurol Asn (pres, 74-75); Am Asn Neuropath; Asn Res Nerv & Ment Dis; Am Acad Neurol (pres, 63-65); Am Fedn Clin Res. *Res:* Neurological medicine and pathology. *Mailing Add:* Dept Neurol Case Western Reserve Univ 2040 Adelbert Rd Cleveland OH 44106

FOLEY, KATHLEEN M, b Flushing, NY, Jan 28, 44; m 68; c 2. NEUROLOGY. *Educ:* St John's Univ, BA, 65; Cornell Univ, MD, 69. *Prof Exp:* Fel genetics, New York Hosp, 70-71, resident neurol, New York Hosp-Cornell Med Ctr, 71-74, chief resident neurol, 74; ASST PROF NEUROL, COL MED, CORNELL UNIV, 75-; ASST ATTEND PHYSICIAN, SLOAN KETTERING INST CANCER RES, 75- *Mem:* Am Soc Neurol; Int Asn Study Pain; Am Women's Med Asn; NY Acad Sci; Am Fedn Clin Res. *Res:* Cancer pain; neurological complications of cancer. *Mailing Add:* Dept Neurol Cornell Univ Med Col 1275 York Ave New York NY 10021

FOLEY, KENNETH JOHN, b Newport, Wales, UK, Mar 12, 36; US citizen; m 60; c 2. EXPERIMENTAL PARTICLE PHYSICS, HIGH ENERGY HEAVY ION EXPERIMENTS. *Educ:* Oxford Univ, BA, 58, DPhil(physics), 61. *Prof Exp:* Res assoc, 61-63, from asst physicist, to physicist, 63-83, SR PHYSICIST, BROOKHAVEN NAT LAB, 83- *Concurrent Pos:* Leader, Particle Spectrometer Group, Brookhaven Nat Lab & supvr, Multiparticle Spectrometer Facil. *Honors & Awards:* Fel, Am Phys Soc. *Mem:* Am Phys Soc; AAAS. *Res:* Study of the nature of the strong interaction force using high energy pion, kaon, proton, antiproton and nuclear beams. *Mailing Add:* Brookhaven Nat Lab Bldg 510 Upton NY 11973

FOLEY, MICHAEL GLEN, b Independence, Mo, July 23, 45; m 68. GEOMORPHOLOGY, GEOLOGY. *Educ:* Calif Inst Technol, BS, 67, MS, 68, PhD(geol), 76. *Prof Exp:* Res engr aeronaut, Boeing Co, 68-69; analyst, Jet Propulsion Lab, 69-70; asst prof geol, Univ Mo-Columbia, 76-80; sr geologist, 80-85, STAFF SCIENTIST, BATTELLE PAC NORTHWEST LABS, 85- *Mem:* Am Geophys Union; Am Quaternary Asn; Brit Geomorphol Res Group; Geol Soc Am. *Res:* Computer-aided remote geologic analysis; laboratory and deterministic stochastic mathematical modeling of sediment transport mechanics and fluvial system behavior; field study of modern and late Cenozoic channels and drainage basins; computer modeling of geologic/hydrologic systems for nuclear-waste containment studies. *Mailing Add:* 2305 Towhee Lane West Richland WA 99352

FOLEY, ROBERT THOMAS, b Turners Falls, Mass, Dec 21, 18; m 45; c 2. ELECTROCHEMISTRY. *Educ:* Univ Mass, BS, 40; Lafayette Col, MS, 41; Univ Tex, PhD(phys chem), 48. *Prof Exp:* Res chemist, Am Cyanamid Co, NJ, 41-45; instr phys chem, Univ Tex, 45-48; res chemist, Gen Elec Co, 48-52, supvr gen metall, 52-54, specialist surface chem, 54-61; supvr electrochem, Melpar, Inc, Va, 61-64; prof chem, Am Univ, 64-84; RETIRED. *Concurrent Pos:* Consult, Union Carbide Corp, US Army, Silver Inst, Washington, DC, 72-, W R Grace Co, 76- *Honors & Awards:* William Blum Award, 78. *Mem:* Am Chem Soc; Electrochem Soc; Int Electrochem Soc; Am Inst Chem; Nat Asn Corrosion Eng. *Res:* Energy conversion; fuel cells; batteries; corrosion mechanisms; fuel cell electrolytes. *Mailing Add:* 7509 Clifton Rd Clifton VA 22024

FOLEY, THOMAS PRESTON, JR, b Indianapolis, Ind, July 31, 37; m 85; c 1. PEDIATRIC THYROIDOLOGY, AUXOLOGY. *Educ:* Washington & Lee Univ, BS, 59; Univ Va, MD, 63. *Prof Exp:* Intern med & pediat, Med Ctr, Univ Kans, 63-64; resident pediat, Children's Hosp, Cincinnati, Ohio, 64-66; capt med corp pediat, Orlando AFB Hosp, 66-68; teaching fel pediat endocrinol, Johns Hopkins Univ, 68-71; from asst prof to assoc prof, 71-81, PROF PEDIAT, UNIV PITTSBURGH, 81-, DIR, CLIN RES CTR, 76- *Concurrent Pos:* Chmn, adv comt hypothyroid screening prog, Pa Dept Health, 78-; fel, Nat Cancer Inst, 79; pediatrician & pediat endocrinologist consult, Poland & China, Proj Hope, 81-; med adv bd, Proj Hope, 90- *Mem:* Am Pediat Soc; Lawson Wilkins Pediat Endocrine Soc; Endocrine Soc; Am Fedn Clin Res; Am Thyroid Asn; Pittsburgh Pediat Soc. *Res:* Initiate primary thyrotropic hormone screening for congenital hypothyroidism; etiology and pathogenesis of sporatic congenital hypothyroidism; pathogenesis and management of juvenile graves diseases and autoimmune thyroid disorders; auxology. *Mailing Add:* Children's Hosp 3705 Fifth Ave Pittsburgh PA 15213-2583

FOLEY, WILLIAM ARTHUR, ANATOMIC CLINICAL PATHOLOGY. *Educ:* Univ Minn, MD, 56. *Prof Exp:* Clin instr path, Univ Minn, 52-67; PATHOLOGIST, NORTHWESTERN HOSP, 67- *Mailing Add:* Dept Path Box 198 Mayo Univ Minn Med Sch 420 Delaware St SE Minneapolis MN 55455

FOLEY, WILLIAM M, b Hoquiam, Wash, Mar 12, 29; m 53; c 4. FLUID MECHANICS. *Educ:* Univ Minn, BS, 51; Stanford Univ, MS, 56, PhD(aeronaut & astronaut), 62. *Prof Exp:* Supvr gaseous physics group, Res Labs, United Aircraft Corp, 58-65, chief aerophysics sect, 65-66, mgr fluid dynamics lab, 67-76, DEP DIR RES, UNITED TECHNOL CORP, 76- *Concurrent Pos:* Mem res adv comt aircraft aerodyn, NASA, 66-71. *Mem:* Am Phys Soc; Am Inst Aeronaut & Astronaut. *Res:* Rarefied gas dynamics; fluid mechanics of gaseous core nuclear reactors; physics of gas-surface interactions. *Mailing Add:* 125 Farmstead Lane Glastonbury CT 06033

FOLEY, WILLIAM THOMAS, b New York, NY, Oct 30, 11; c 5. MEDICINE. *Educ:* Columbia Univ, AB, 33; Cornell Univ, MD, 37; Am Bd Internal Med, dipl. *Prof Exp:* from asst prof to assoc prof, 51-77, emer assoc prof, 77-84, EMER PROF CLIN MED, CORNELL UNIV, 84- *Concurrent Pos:* Attend physician, NY Hosp, 85; consult, Hosps. *Mem:* Sr mem Am Soc Clin Invest; Harvey Soc; Am Heart Asn; fel Am Col Physicians; Am Col Cardiol. *Res:* Clinical research in thromboembolism and vascular diseases; lymphedema; cerebrovascular disease. *Mailing Add:* 441 E 68th St New York NY 10021

FOLGER, DAVID W, b Woburn, Mass, Nov 21, 31; m 82; c 3. GEOLOGICAL OCEANOGRAPHY. *Educ:* Dartmouth Col, BA, 53; Columbia Univ, MA, 58, PhD(submarine geol), 68. *Prof Exp:* Petrol geologist, Chevron Oil Co, 58-63; res asst submarine geol & geophys, Lamont Geol Observ, Columbia Univ, 64-68; investr submarine geol, Woods Hole Oceanog Inst, 68-69; asst prof geol, Middlebury Col, 69-75, actg chmn, dept geol & geog, 70; coordr environ assessment, 75-77, chief, Atlantic & Gulf Br Marine Geol, 77-82, coordr wetak studies, 82-86, coordr DMA gravity studies, 86-88, COORDR GREAT LAKES STUDIES, US GEOL SURV, WOODS HOLE, 88- *Mem:* AAAS; Soc Econ Paleont & Mineral; fel Geol Soc Am; Am Asn Petrol Geologists. *Res:* Submarine geology; processes of estuarine, continental shelf and deep sea sedimentation; composition of air and waterborne particulate matter; environmental effects of continental shelf exploitation; coastal erosion. *Mailing Add:* Off Marine Geol US Geol Surv Woods Hole MA 02543

FOLINAS, HELEN, b Montpelier, Vt, June 5, 27. CELL PHYSIOLOGY. *Educ:* Trinity Col, Vt, BS, 52; St Michael's Col, Vt, MS, 59; Fordham Univ, PhD(physiol), 69. *Prof Exp:* From instr to asst prof, 57-71, assoc prof, 71-80, PROF BIOL, TRINITY COL, VT, 80-, CHMN DEPT, 65- *Mem:* AAAS; Am Soc Cell Biol; Am Soc Zool; Am Inst Biol Sci. *Res:* Cell physiology with emphasis on the physiological and cellular aspects of development and inheritance. *Mailing Add:* Trinity Col Colchester Ave Burlington VT 05401

FOLINSBEE, ROBERT EDWARD, b Edmonton, Alta, Apr 16, 17; m 42; c 4. GEOLOGY. *Educ:* Univ Alta, BSc, 38; Univ Minn, MS, 40, PhD(petrol), 42. *Hon Degrees:* LLD, Univ Windsor, 72; DSc, Alberta, 89. *Prof Exp:* Asst geologist, Geol Surv Can, 41-43; from asst prof to prof. 46-78, EMER PROF GEOL, UNIV ALBERTA, 78- *Concurrent Pos:* Pres, 24th Int Geol Cong, Montreal, 72. *Honors & Awards:* The Order of Can, Gov Gen of Can, 73. *Mem:* Geol Soc Am (pres, 76); Soc Econ Geol; Royal Soc Can (pres, 77); Can Inst Mining & Metall; Geol Asn Can. *Res:* Petrology; economic geology; geochemistry. *Mailing Add:* Dept Geol Univ Alta Edmonton AB T6G 2E3 Can

FOLK, EARL DONALD, b Corpus Christi, Tex, Mar 16, 39; m 60; c 2. AVIATION SAFETY. *Educ:* Okla State, BS, 61; Kans State, MS, 62; Univ Wis, MS, 64; Univ Okla, PhD(biostatist), 70. *Prof Exp:* Proj asst, Dept Neurol, Univ Wis, 62-66, statist asst, Dept Biostatist, 66-68; STATISTICIAN, FED AVIATION AUTHORITY, DEPT TRANSP, CIVIL AEROMED INST, US GOVT, 68- *Concurrent Pos:* Adj asst prof, Univ Okla, 74- *Mem:* Am Statist Asn. *Res:* Human factors relating to aviation and in particular to aviation safety. *Mailing Add:* 2825 NW 117 Oklahoma City OK 73120

FOLK, GEORGE EDGAR, JR, b Natick, Mass, Nov 12, 14; m; m 56; c 2. PHYSIOLOGY. *Educ:* Harvard Univ, AB, 37, MA, 40, PhD(biol), 47. *Prof Exp:* Instr biol, St Mark's Sch, 40-43; res assoc, Fatigue Lab, Harvard Univ, 43-47; asst prof biol, Bowdoin Col, 47-52, dir, Bowdoin Sci Sta; from assoc prof to prof, 52-85, EMER PROF PHYSIOL, UNIV IOWA, 85- *Concurrent Pos:* Physiologist, Climatic Res Lab, 47; Univ Iowa Col Med res fel, Kings Col, Univ London, 57-58; vis res prof, Arctic Aeromed Lab, Ft Wainwright, Alaska, 64-65. *Honors & Awards:* Fulbright Award, 74. *Mem:* AAAS; Am Physiol Soc; Am Soc Zool. *Res:* Environmental physiology; factors influencing dissipation of heat by the human body and lower animals; mammalian physiology; biological rhythms; global science. *Mailing Add:* Lab Environ Physiol Dept Physiol Univ Iowa Iowa City IA 52242

FOLK, JOHN EDWARD, b Washington, DC, Oct 29, 25; m 52; c 1. BIOCHEMISTRY. *Educ:* Georgetown Univ, BS, 48, MS, 50, PhD(biochem), 52. *Prof Exp:* Am Dent Asn fel, NIH, 52-59, biochemist, 59-67, CHIEF, SECT ENZYME CHEM, NAT INST DENT RES, NIH, 67- *Mem:* Am Soc Biol Chemists. *Res:* Enzyme mechanisms; relationships of enzyme structure and function. *Mailing Add:* Lab Cell Develop Oncol Nat Inst Dent Res NIH Bldg 30 Rm 211 Bethesda MD 20892

FOLK, ROBERT LOUIS, b Cleveland, Ohio, Sept 30, 25; m 46; c 3. SEDIMENTARY PETROLOGY. *Educ:* Pa State Col, BS, 46, MS, 50, PhD(mineral), 52. *Prof Exp:* Asst & instr mineral, Pa State Col, 46-50; res geologist, Gulf Res & Develop Co, 51-52; from asst prof to prof geol, Univ Tex, Austin, 52-77, Nalle Gregory prof sedimentary geol, 77-84, D Carleton prof geol, 84-88, EMER D CARLETON PROF GEOL, UNIV TEX, AUSTIN, 88- *Honors & Awards:* W H Twenhofel Medal, Soc Econ Paleontologists & Mineralogists, 79; Neil Miner Award, Nat Asn Geol Teachers, 89; H C Sorby Medal, Int Asn Sedimentologists, 90. *Res:* Petrography and genesis of limestones, dolomites, cherts and sandstones; particle size distribution of recent sediments; geomorphology; sedimentary petrology; electron microscopy; texture of carbonate sands; natural history; archaeological geology. *Mailing Add:* Dept Geol Univ Tex Austin TX 78712

FOLK, ROBERT THOMAS, b Reading, Pa, Oct 17, 27; m 58; c 3. NUCLEAR PHYSICS. *Educ:* Lehigh Univ, BS, 53 & 54, MS, 55, PhD(physics), 58. *Prof Exp:* Instr physics, Princeton Univ, 57-59, 60-61; NSF fel, Ger, 59-60; from asst prof to assoc prof, 61-66, PROF PHYSICS, LEHIGH UNIV, 66- *Mem:* Am Phys Soc. *Res:* Theoretical nuclear physics. *Mailing Add:* Dept Physics Bldg 16 Lehigh Univ Bethlehem PA 18015

FOLK, STEWART H, b Urbanette, Arkansas, Nov 30, 15; m 40; c 4. SULPHUR DEPOSITS, PETROLEUM EXPLORATION. *Educ:* Baylor Univ, BA, 36; Univ Iowa, MS, 38. *Prof Exp:* Field geologist, Magnolia Petrol Co, 38-42; ground water & petrol geologist, Ill State Geol Surv, 42-43; geologist, US Navy, Naval Petrol Reserve, 43- 45, geotech asst to officer in charge, 45-48; consult geologist & Mex rep, DeGolyer & McNaughton Co, 48-51; geol supvr, Tex Gulf Sulfur Co, 51-58; vpres, Tulm Corp, 58-62; frontier geologist, Union Tex Petrol Co, 62-65; consult geologist, 65-68; explor mgr, Jefferson Lake Sulfur Co, 68-70; CONSULT GEOLOGIST, STEWART FOLK & CO, 70- *Mem:* Fel Geol Soc Am; Am Asn Petrol Geologists; Am Inst Mining & Metall Engrs. *Res:* Elemental sulphur deposits; exploration, evaluation and appraisal of sulphur deposits; petroleum exploration and evaluation. *Mailing Add:* 11907 Longleaf Lane Houston TX 77024

FOLK, THEODORE LAMSON, b Aurora, Ill, June 19, 40. ANALYTICAL CHEMISTRY. *Educ:* Knox Col, Ill, BA, 62; Univ Ariz, MS, 65; Univ Hawaii, PhD(org chem), 68. *Prof Exp:* Asst chem, Univ Ariz, 62-64 & Univ Hawaii, 65-68; SR RES CHEMIST, GOODYEAR TIRE & RUBBER CO, 68- *Mem:* Am Chem Soc; Am Soc Mass Spectrometry; Sigma Xi. *Res:* Addition reactions of organolithium compounds on acetylenic systems; synthetic studies of polyhydroxynaphthoquinones; mass spectrometry of organic compounds. *Mailing Add:* Res Div Goodyear Tire & Rubber 142 Goodyear Blvd Akron OH 44305

FOLK, WILLIAM ROBERT, b Little Rock, Ark, July 4, 44; m 66; c 2. BIOCHEMICAL GENETICS. *Educ:* Rice Univ, BA, 66; Stanford Univ, PhD(biochem), 71. *Prof Exp:* Helen Hay Whitney fel, Imp Cancer Res Fund, London, 71-73; prof, Univ Mich, Ann Arbor, 73-83 & Univ Tex, Austin, 84-88; DEPT BIOCHEM, UNIV MO, COLUMBIA, 88- *Concurrent Pos:* Eleanor Roosevelt fel; Fac Res Award, Am Cancer Soc. *Mem:* AAAS; Am Soc Microbiol; Genetics Soc Am; Am Soc Biol Chemists; Am Cancer Soc fel; Int Soc Plant Molecular Biol. *Res:* Structure and function of nucleic acids and of macromolecules which regulate gene expression. *Mailing Add:* Dept Biochem Univ Mo Columbia MO 65211

FOLKERS, KARL AUGUST, b Decatur, Ill, Sept 1, 06; m 32; c 2. CHEMISTRY. *Educ:* Univ Ill, BS, 28; Univ Wis, PhD(org chem), 31. *Hon Degrees:* DSc, Philadelphia Col Pharm & Sci, 62, Univ Uppsala, Sweden, 69; Univ Wis, 69, Univ Ill,73; Hon Degree Med & Surg, Univ Bologna, Italy, 89. *Prof Exp:* Squibb & Lilly res fel, Yale Univ, 31-34; lab pure res, Merck & Co, Inc, 34-38, asst dir res, 38-45, dir org & biol chem res div, 53-56, exec dir fundamental res, 56-62, vpres explor res, 62-63; pres, Stanford Res Inst, 63-68; prof, 68-73, DIR, INST BIOMED RES, UNIV TEX, AUSTIN, 68-, ASHBEL SMITH PROF CHEM, 73- *Concurrent Pos:* Mem div 9, Nat Defense Res Comt, 43-46; Harrison-Howe lectr, 49; Baker nonresident lectr, Cornell Univ, 53; lectr med fac, Lund, Stockholm, Uppsala Gothenborg Univs, Sweden, 54; Sturmer lectr, 57; chmn adv coun, Dept Chem, Princeton Univ, 58-64; guest lectr, Am-Swiss Found Sci Exchange, 61; Robert A Welch Found lectr, 63; courtesy prof, Stanford Univ, 63-; courtesy lectr, Univ Calif, Berkeley, 63-; Marchon vis lectr, Univ Newcastle, 64; F F Nord lectr, Fordham Univ, 71; mem rev comt, US Pharmacopoeia; chmn, Nat Acad Sci, 75; Alexander von Humboldt-Stiftung award, Germany, 77. *Honors & Awards:* Co-recipient, Mead Johnson & Co Award, 40 & 49; Award for Meritorious Work in Pure Chem, Am Chem Soc, 41; Presidential Cert of Merit, 48; Merck & Co, Inc Award, 51; Spencer Award, 59; Perkin Medal, Soc Chem Indust, 60; co-recipient, Van Meter Prize, Am Thyroid Asn, 69; Robert A Welch Int Award & Medal, 72; Acad Pharmaceut Sci Res Achievemnt Award In Pharmaceut & Med Chem, APhA Found, 74; Nat Med of Sci, 90. *Mem:* Nat Acad Sci; AAAS; Am Soc Biol Chem; Am Inst Chemists; fel Am Inst Nutrit; Am Chem Soc (pres, 62); foreign mem Royal Swed Acad Eng Sci. *Res:* Organic chemistry; catalytic hydrogenation; pyrimidine; alkaloids; vitamins; synthetic medicinals; antibiotics; hormones; coenzymes. *Mailing Add:* Int Biomed Res Univ Tex Welch Hall Austin TX 78712

FOLKERT, JAY ERNEST, b Holland, Mich, Dec 16, 16; m 46; c 3. MATHEMATICS. *Educ:* Hope Col, AB, 39; Univ Mich, MA, 40; Mich State Univ, PhD(math), 55. *Prof Exp:* Teacher pub schs, 40-42, 46; chmn, Dept Math, 57-71, prof math, Hope Col, 46-81; emer prof math, Hope Col, 82-; RETIRED. *Mem:* Math Asn Am; Am Math Soc; Am Sci Affil; Nat Coun Teachers Math. *Res:* General mathematics; education. *Mailing Add:* Math Dept Hope Col 37 Graves Pl Holland MI 49423

FOLKERTS, GEORGE WILLIAM, b Beardstown, Ill, Nov 26, 38; m 65; c 3. HERPETOLOGY, AQUATIC COLEOPTERA. *Educ:* Southern Ill Univ, BA, 61, MA, 63; Auburn Univ, PhD(zool), 68. *Prof Exp:* Instr zool, Auburn Univ, 66-68; asst prof, Clemson Univ, 68-69; from asst prof to assoc prof, 69-77, ALUMNI PROF ZOOL, AUBURN UNIV, 77- *Mem:* Natural Areas Asn; Soc Wetland Scientists; Coleopterists Soc. *Res:* Systematics and ecology of reptiles and amphibians; systematics and ecology of aquatic coleoptera; ecology of southeastern bogs. *Mailing Add:* Dept Zool & Wildlife Sci Auburn Univ Auburn AL 36849-5414

FOLKERTS, THOMAS MASON, b Peoria, Ill, Sept 5, 26; m 48; c 7. VETERINARY MICROBIOLOGY. *Educ:* Univ Ill, BS, 50 & 52, DVM(vet med), 54. *Prof Exp:* Practr vet med, Wes-Lyn Animal Hosp, Ill, 54-61; teaching asst animal health & feeding, Southern Ill Univ, 61-62; field vet tech serv, 63-65, mgr, 65-66, head field res, 66-68, head vet res, 68-79, HEAD INT ANIMAL SCI FIELD RES, ELI LILLY & CO, 79- *Mem:* Am Soc Microbiol; AAAS; Am Vet Med Asn; Am Asn Bovine Practr; Am Col Vet Toxicologists. *Res:* Compound screening and testing for useful antibacterial and antiviral activity; target species toxicology of drugs being developed for animal agriculture; resistance studies on antibiotic compounds used and developed for animal agriculture; field evaluation of animal health products. *Mailing Add:* 8100 Allisonville Rd Indianapolis IN 46250

FOLKMAN, MOSES JUDAH, b Cleveland, Ohio, Feb 24, 33; m 60; c 1. SURGERY. *Educ:* Ohio State Univ, BA, 53; Harvard Med Sch, MD, 57; Am Bd Surg, dipl, 66; Am Bd Thoracic Surg, dipl, 68. *Prof Exp:* Intern surg, Mass Gen Hosp, Boston, 57-58, asst resident, 59-60, sr asst resident, 62-64, chief resident, 64-65; instr surg, 65-66, assoc & prof, 67, JULIA DYCKMAN ANDRUS PROF PEDIAT SURG, HARVARD MED SCH, BOSTON, 68-, PROF ANAT & CELL BIOL, 80-; SR ASSOC SURG & DIR, SURG RES LAB, CHILDREN'S HOSP, BOSTON, 81- *Concurrent Pos:* Nat Cancer Inst res career develop award, 65-70; assoc dir, Sears Surg Lab, Boston City Hosp, 66-67; surgeon-in-chief & chmn, Dept Surg, Children's Hosp, Boston, 67-81; chief resident pediat surg, Philadelphia Children's Hosp, Pa, 69; prof pediat surg, Div Health Sci & Technol, Mass Inst Technol, Harvard Univ, 79; mem, Path A Study Sect, Nat Cancer Inst, 90- *Honors & Awards:* Boylston Med Prize, 57; Soma Weiss Award, 57; Lila Gruber Award, Am Acad Dermat, 74; Simon M Shubitz Cancer Prize & Lect for Excellence in Cancer Res, Med Ctr, Univ Chicago, 82; Smith Kline & French Distinguished Lectr, 83; Peter Kiewit Mem Distinguished Lectr in Med, Eisenhower Med Ctr, 84; G H A Clowes Mem Award, Am Asn Cancer Res, 85; Sheen Award, 89. *Mem:* Nat Acad Sci; Inst Med-Nat Acad Sci; NY Acad Sci; fel Am Acad Arts & Sci; Am Col Chest Physicians; Am Surg Asn; Asn Acad Surg; Soc Univ Surgeons; fel Am Col Surgeons; Am Acad Pediat. *Res:* Tissue culture; diffusion of drugs and administration of anesthetics through silicone rubber; cancer research. *Mailing Add:* Children's Hosp 300 Longwood Ave Boston MA 02115-6737

FOLKS, HOMER CLIFTON, b Hydro, Okla, Aug 6, 23; m 50; c 3. SOILS. *Educ:* Okla State Univ, BS, 50; Iowa State Univ, PhD(soils), 54. *Prof Exp:* Instr soils, Iowa State Univ, 53-54, exten area agronomist, 54-55; asst prof soils, NC State Col, 55-58, assoc prof, 58-59, asst dir instr, 59-63; assoc dean, Col Agr, 63-80, PROF AGRON, UNIV MO-COLUMBIA, 63- *Mem:* Am Soc Agron. *Res:* Soil genesis and morphology; mechanics of soil profile development. *Mailing Add:* 103 W Burnam Rd Columbia MO 65203

FOLKS, JOHN LEROY, b Hydro, Okla, Oct 12, 29; m 56; c 4. ANALYTICAL STATISTICS, APPLIED STATISTICS. *Educ:* Okla State Univ, BA, 53, MS, 55; Iowa State Univ, PhD(statist), 58. *Prof Exp:* Scientist opers res, Tex Instruments, Inc, 58-60, mgr opers res, 60-61; assoc prof math & statist, Okla State Univ, 61-67, chmn statist unit, 69-73, PROF MATH & STATIST, OKLA STATE UNIV, 67-, CHMN STATIST DEPT, 73- *Concurrent Pos:* NSF sci fac fel, Univ Calif, Berkeley, 66-67. *Mem:* Fel Am Statist Asn; Inst Math Statist; Biomet Soc; Int Statist Inst; Royal Statist Soc. *Res:* Significance testing. *Mailing Add:* Statist Dept Okla State Univ Stillwater OK 74075

FOLLAND, GERALD BUDGE, b Salt Lake City, Utah, June 4, 47. MATHEMATICS. *Educ:* Harvard Univ, BA, 68; Princeton Univ, MA, 70, PhD(math), 71. *Prof Exp:* Instr math, Courant Inst, NY Univ, 71-73; from asst prof to assoc prof, 73-80, PROF MATH, UNIV WASH, 80- *Concurrent Pos:* Vis asst prof math, Princeton Univ, 74; vis mem, Inst Advan Study, 79; vis prof, Tata Inst Fundamental Res, 81 & Indian Statist Inst, 87. *Mem:* Am Math Soc; Math Asn Am. *Res:* Partial differential equations and Fourier analysis. *Mailing Add:* Dept Math Univ Wash Seattle WA 98195

FOLLAND, NATHAN ORLANDO, b Greenbush, Minn, Jan 12, 37; m 60, 80; c 3. PHYSICS. *Educ:* Concordia Col, BA, 59; Iowa State Univ, PhD(physics), 65. *Prof Exp:* NATO fel, Univ Messina, 65-66; from asst prof to assoc prof, 66-82, PROF PHYSICS, KANS STATE UNIV, 82-, ASSOC DEPT HEAD, 89- *Mem:* Am Phys Soc. *Res:* Theoretical solid state physics; band theory; transport; optical properties. *Mailing Add:* Dept Physics Cardwell Hall Kans State Univ Manhattan KS 66506

FOLLETT, ROY HUNTER, b Cowdrey, Colo, Feb 27, 35; m 59; c 2. AGRONOMY, SOIL SCIENCE. *Educ:* Colo State Univ, BS, 57, MS, 63, PhD(soil sci), 69. *Prof Exp:* Soil scientist, US Geol Surv, Norton, Kans, 57-58, Soil Conserv Serv, Ft Collins, Colo, 58-60, 63-64; jr agronomist, Colo State Univ, 60-63, exten agronomist soils, 64-70; asst prof, Ohio State Univ, 70-74; from assoc prof to prof soils, Kans State Univ, 74-81; PROF SOILS, COLO STATE UNIV, 81- *Concurrent Pos:* Assoc ed, J of Agron, ed, 78-84; fel, Soil & Water Conserv Soc, 79; fel, Am Soc Agron, 86; fel, Soil Sci Soc Am, 87. *Mem:* Am Soc Agron; Soil Sci Soc Am; fel Soil & Water Conserv Soc Am; Sigma Xi. *Res:* Soil fertility, fertilizer materials, rates and time of application; conservation tillage and reduced tillage system. *Mailing Add:* Dept Agron Colo State Univ Ft Collins KS 80523

FOLLETT, WILBUR IRVING, b Newark, NJ, Mar 10, 01; m 29. ICHTHYOLOGY. *Educ:* Univ Calif, Berkeley, AB, 24,. *Hon Degrees:* JD,Univ Calif, Berkeley, 26. *Prof Exp:* Asst ichthyol, Calif State Fish & Game Comn, 26; from actg cur to cur, 47-69, EMER CUR ICHTHYOL, CALIF ACAD SCI, 70- *Concurrent Pos:* Atty at law, 27-; deleg, Int Cong Zool & Colloquium Zool Nomenclature Denmark, 53. *Mem:* Soc Syst Zool (pres, 60); Am Soc Ichthyologists & Herpetologists (pres, 69); Sigma Xi. *Res:* Nomenclature, taxonomy, osteology and distribution of California fishes; identification of fish remains from archaeological sites. *Mailing Add:* Calif Acad Sci Golden Gate Park San Francisco CA 94118-4599

FOLLINGSTAD, HENRY GEORGE, b Wanamingo, Minn, Jan 6, 22; m 45; c 5. RELATIVITY & GRAVITATION. *Educ:* Univ Minn, BEE, 47, MS, 71. *Prof Exp:* Mem tech staff, BELL LABS, AT&T, Inc, 48-62; sci res consult, Honeywell, Inc, 64-81; emer prof math, Augsburg Col, 62-87; RETIRED. *Concurrent Pos:* Electronic res consult, N Star Res & Develop Inst, 65-66. *Mem:* Inst Elec & Electronics Engrs; Math Asn Am; Am Biog Inst Res Asn; Int Platform Asn. *Res:* Mathematical-logical-physical absurdities, experimental interpretation errors and inadequacies in Einstein special and general relativity with space-age alternatives; abuses of mathematics in scientific, academic, social, philosophical and theological areas with space-age alternatives; math-model analyses of complex physical systems. *Mailing Add:* 3506 Garfield Ave S Minneapolis MN 55408

FOLLOWS, ALAN GREAVES, b Normanton, Eng, Dec 20, 21; nat US; m 50; c 3. CHEMISTRY. *Educ:* Queen's Univ, Can, BSc, 44, MSc, 45; Univ Toronto, PhD(elec), 48. *Prof Exp:* Chem res, Nat Res Coun Can, Queen's Univ, 44-45; lectr phys chem, Univ Toronto, 47-48; res chemist, Solvay Process Div, 48-59, asst dir res, 59-69, mgr technol, 69-72; mgr tech serv, Indust Chem Div, Allied Chem Corp, 72-79; PRES, NEOS CHEM CORP, 79- *Mem:* Am Chem Soc. *Res:* Electrochemistry; industrial research and management; titanium; alkalies. *Mailing Add:* NEOS Chem Corp Box 201 Camillus NY 13031

FOLLSTAD, MERLE NORMAN, b Milwaukee, Wis, May 14, 31; m 63; c 2. BOTANY. *Educ:* Univ Wis, BS, 55; Univ Minn, MS, 61, PhD(plant path), 64. *Prof Exp:* Res asst, Inst Paper Chem, Lawrence Univ, 56-58; res asst plant path, Univ Minn, 58-64; res plant pathologist, Mkt Qual Res Div, US Dept Agr, 64-66; res assoc, NC State Univ, 66-67; microbiologist, Northern Regional Res Lab, Ill, 67-68; ASST PROF BIOL, UNIV WIS, WHITEWATER, 68- *Mem:* AAAS; Botanical Soc Am; Am Inst Biol Sci; Am Forestry Asn. *Res:* Developmental anatomy of seed plants; plant histochemistry. *Mailing Add:* Dept Biol Univ Wis Whitewater WI 53190

FOLLSTAEDT, DAVID MARTIN, b Haskell, Tex, Aug 18, 47; m 71; c 2. ION IMPLANTATION, RAPID QUENCHING. *Educ:* Tex Technol Univ, BS, 69; Univ Ill, MS, 70, PhD(physics), 75. *Prof Exp:* Staff mem, nuclear magnetic resonance, Sandia Nat Labs, 74-76, staff mem, Ion Implantation, 76-85. *Concurrent Pos:* NSF fel, 69-70. *Mem:* Am Phys Soc; Electron Micros Soc Am; Mat Res Soc. *Res:* Nuclear magnetic resonance; ion implantation in metals; electron microscopy; laser and electron beam pulsed melting of metals; rapid quenching; metastable phase formation; quasicrystal research. *Mailing Add:* Div 1112 Sandia Nat Labs PO Box 5800 Albuquerque NM 87185

FOLLWEILER, DOUGLAS MACARTHUR, b Allentown, Pa, June 28, 42; m 64; c 2. ANALYTICAL CHEMISTRY. *Educ:* Muhlenberg Col, BS, 64; Univ Pa, PhD(org chem), 68. *Prof Exp:* Fel org chem, Princeton Univ, 68-69; ENGR ANALYTICAL CHEM, BETHLEHEM STEEL CORP, 69- *Mem:* Sigma Xi. *Res:* Gas chromatography; liquid chromatography; minicomputers and analysis of industrial pollutants. *Mailing Add:* Box 428 Springtown PA 18081

FOLLWEILER, JOANNE SCHAAF, b Philadelphia, Pa, Aug 8, 42; m 64; c 2. ORGANIC CHEMISTRY. *Educ:* Muhlenberg Col, BS, 64; Univ Pa, MS, 68, PhD(chem), 77. *Prof Exp:* Libr chemist, E I du Pont de Nemours & Co, Inc, 64-66; lab asst chem, Radiocarbon Lab, Univ Pa, 66-67; consult, FMC Corp, 68; instr, Mercer County Community Col, 68-69; ASST PROF CHEM, LAFAYETTE COL, 70- *Concurrent Pos:* Consult, chem info and library sci & lit searches. *Mem:* Sigma Xi; Am Chem Soc. *Res:* Organo-sulfur compounds; organometallic compounds. *Mailing Add:* Dept Chem Lafayette Col Easton PA 18042

FOLSE, DEAN SYDNEY, b Kansas City, Mo, Dec 19, 21; m 47. VETERINARY PATHOLOGY, VETERINARY PARASITOLOGY. *Educ:* Agr & Mech Col, Univ Tex, BS & DVM, 45; Kans State Col, MS, 46; Univ Tex, PhD, 70. *Prof Exp:* Asst prof path, Ala Polytech Inst, 48-50, assoc prof, 50-52; assoc prof, Kans State Univ, 52-66; US AEC histopathologist, Inst Biol & Agr, Seibersdorf Reactor Ctr, Austria, 66-68; res instr path, 69-71, res asst prof, 71-82, asst prof, 83-85, assoc prof, 85-88, PROF PATH, UNIV TEX MED BR GALVESTON, 88- *Mem:* AAAS; Am Soc Parasitol; Am Vet Med Asn; Am Assoc Pathologists; Sigma Xi; Int Acad Path; Am Asn Lab Animal Sci; Int Soc Lymphology. *Mailing Add:* Dept Path Univ Tex Med Br Galveston TX 77550-2779

FOLSE, RAYMOND FRANCIS, JR, b New Orleans, La, Sept 4, 40; m 74; c 1. FLUID DYNAMICS. *Educ:* Loyola Univ South, BS, 63; La State Univ, PhD(physics), 68. *Prof Exp:* Vis asst prof physics, La State Univ, 68; asst prof, 68-70, ASSOC PROF PHYSICS, UNIV SOUTHERN MISS, 70- *Concurrent Pos:* NSF teacher proj dir, NSF & Univ Southern Miss, 74-75, NSF res grant proj dir, 87-89; vis assoc prof physics, La State Univ, 87. *Mem:* Am Phys Soc; AAAS; Sigma Xi. *Res:* The effects of unsteady flows in density stratified fluids studied experimentally in order to determine the importance of buoyant forces to various geophysical phenomena. *Mailing Add:* Box 5046 Southern Sta Hattiesburg MS 39406

FOLSOM, RICHARD G(ILMAN), b Los Angeles, Calif, Feb 3, 07; m 29; c 3. MECHANICAL ENGINEERING. *Educ:* Calif Inst Technol, BS, 28, MS, 29, PhD(mech eng), 32. *Hon Degrees:* DSc, Northwestern Univ, 62, Union Col, 64, Albany Med Col, 71 & Lehigh Univ, 72; DEng, Rose Polytech Inst, 71; LLD, Rensselaer Polytech Inst, 73. *Prof Exp:* Engr, Pasadena, Calif, 32-33; instr mech eng, Dept Eng, Univ Calif, Berkeley, 33-37, from asst prof to prof, 37-53, chmn div mech eng, 48-53; prof & dir eng res inst, Univ Mich, 53-58; pres, 58-71, EMER PRES, RENSSELAER POLYTECH INST, 71- *Concurrent Pos:* Staff engr, Jet Propulsion Lab, Calif Inst Technol, 43-44; pvt pract, 71-; counr, Nat Acad Eng, 75-78. *Honors & Awards:* Lamme Award, Am Soc Educ, 71. *Mem:* Nat Acad Eng; hon mem Am Soc Mech Engrs (pres, 72-73); hon mem & fel, Am Soc Eng Educ; assoc fel Inst Aeronaut & Astronaut; fel Am Inst Chem Engrs. *Res:* Pumps; fluid metering; fluid flow and heat transfer at low pressures; fundamental fluid mechanics; mechanical equipment; aeronautical engineering. *Mailing Add:* 585 Oakville Crossroad Napa CA 94558-9744

FOLSOM, THEODORE ROBERT, physics, oceanography; deceased, see previous edition for last biography

FOLSOME, CLAIR EDWIN, microbial genetics, exobiology; deceased, see previous edition for last biography

FOLTS, DWIGHT DAVID, b Fillmore, NY, Mar 24, 47; m 71; c 2. ENTOMOLOGY. *Educ:* State Univ NY Col Environ Sci & Forestry, BS, 69, MS, 73. *Prof Exp:* sr biologist entomol, 72-77, SYSTS ANALYST/ PROGRAMMER, AGR CHEM GROUP, FMC CORP, 77- *Mem:* Entomol Soc Am; Ecol Soc Am; Acarological Soc Am; Data Retrieval Syst Users Soc (pres, 80-82). *Res:* Screening of chemicals for potential use as insecticides and miticides; studies on the effects of organic pesticides on soil microarthropods. *Mailing Add:* Nine Oak St Geneseo NY 14454

FOLTS, JOHN D, b LaCrosse, Wis, Dec 11, 38; m 90. THROMBOSIS & THROMBOLYSIS, BIOMEDICAL ENGINEERING. *Educ:* Univ Wis, BS, 64, MS, 68, PhD(cardiovasc physiol & path), 72. *Prof Exp:* PROF MED CARDIOL, MED SCH, UNIV WIS, 83- *Concurrent Pos:* Prin investr, NIH grants, 73-92, reviewer & site visitor, 79-; vis prof, nine major Univ Med Ctrs, 80-91. *Mem:* Am Fedn Clin Res; fel Am Col Cardiol; fel Am Heart Asn; Am Physiol Soc; Int Soc Thrombosis & Hemostasis; Int Soc Heart Res. *Res:* Developed Folts model of experimental coronary artery stenosis and thrombosis; mechanisms that cause coronary artery; thrombosis and ways to prevent it; first to recommend that patients with coronary disease take one aspirin tablet daily to prevent coronary thrombosis; published 110 papers, 12 chapters in books and 140 scientific presentations; one US patent. *Mailing Add:* Clin Sci Ctr H6 Rm 379 600 Highland Ave Madison WI 53792

FOLTZ, CRAIG BILLIG, b Shamokin, Pa, June 28, 52; m 80; c 2. ASTRONOMICAL INSTRUMENTATION. *Educ:* Dartmouth Col, AB, 74; Ohio State Univ, PhD(astron), 79. *Prof Exp:* Fel, Dept Astron, Ohio State Univ, 79-80, instr, Dept Physics, 80, res assoc, Dept Astron, 81-82; fel, Steward Observ, 81-83, Staff Astronr, Mult Mirror Telescope Observ, 84-90, DEP DIR, MULT MIRROR TELESCOPE OBSERV, UNIV ARIZ, 90- *Concurrent Pos:* Asst prof, Dept Astron, Univ Ill, 83-84. *Mem:* Am Astron Soc; Int Astron Union; Astron Soc Pac. *Res:* Problems in quasar absorption-line spectra; analysis of emission-line spectra of active galactic nuclei and quasars; development of astronomical spectrophotometers; development of software for instrument control and data handling. *Mailing Add:* Mult Mirror Telescope Observ Univ Ariz Tucson AZ 85721-0465

FOLTZ, DONALD RICHARD, b Pittsburgh, Pa, Aug 4, 26; m 54; c 4. COMPRESSED GAS FILTRATION, FLUID MIXING TECHNOLOGY. *Educ:* Carnegie Mellon Univ, BS, 51. *Prof Exp:* Engr, Brown Fintube Co, 51-54; asst chief engr, Power Piping Div, Blaw-Knox Corp, 54-59; chief engr, Grafe-Weeks Corp, 59-61; vpres eng, Par-Dan Corp, 61-62; pres, Nat Banner Corp, 62-65 & Hankison Corp, 65-87; consult, 87-90; EXEC VPRES, ACT LABS, INC, 90- *Res:* Compressed gas purification including filtration and catalysis; fluid mixing technology (solids and liquids); heat transfer and fluid flow dynamics; awarded three Unites States patents; author of 14 publications. *Mailing Add:* 175 Reed Dr Pittsburgh PA 15205

FOLTZ, ELDON LEROY, b Ft Collins, Colo, Mar 28, 19; m 43; c 4. NEUROSURGERY. *Educ:* Mich State Col, BS, 41; Univ Mich, MD, 43. *Prof Exp:* Asst resident surg, Univ Mich, 46-47; neurosurg resident, Med Sch, Dartmouth Col, 47-49; clin asst, Univ Louisville, 49-50; NIMH fel & res assoc surg, Univ Wash, 50-51, from instr to prof neurosurg, 51-69; chmn div, 69-77, chief neurol surg, 69-80, PROF NEUROL SURG, UNIV CALIF, IRVINE-CALIF COL MED, 69- *Concurrent Pos:* NIH teaching fel, 51; Markle scholar, 54-59. *Mem:* Am Asn Neurol Surg; Am EEG Soc; Am Col Surgeons; Am Acad Neurol; Am Acad Neurol Surgeons. *Res:* Cerebral concussion; consciousness; central factors of psychosomatic diseases; hydrocephalus. *Mailing Add:* Surg-UCIMC 101 City Dr S Orange CA 92668

FOLTZ, FLOYD MATHEW, anatomy, for more information see previous edition

FOLTZ, GEORGE EDWARD, b Pittsburgh, Pa, May 20, 24; m 45, 55; c 2. SYNTHETIC ORGANIC CHEMISTRY. *Educ:* Carnegie Inst Technol, BS, 47, MS, 51, PhD(chem), 52. *Prof Exp:* Control chemist resins & solvents, Neville Co, 47-48; sr res chemist, Explor Org, Columbia-Southern Chem Corp, 52- 54; asst chemist, 54-80, MGR PROD DEVELOP & ANALYSIS LABS, NEVILLE CHEM CO, 80- *Res:* Polymer chemistry; antioxidants; ultraviolet stabilizers. *Mailing Add:* 339 Goldsmith Rd Pittsburgh PA 15237-3660

FOLTZ, NEVIN D, b Akron, Pa, Feb 12, 40; m 65; c 2. MOLECULAR PHYSICS. *Educ:* Pa State Univ, BSc, 62, MSc, 64, PhD(physics), 68. *Prof Exp:* Instr physics, Commonwealth Campus, Pa State Univ, 64-66; asst prof, 68-74, ASSOC PROF PHYSICS, MEM UNIV, NFLD, 74- *Mem:* Optical Soc Am; Am Inst Physics; Can Asn Physicists; Am Phys Soc. *Res:* Stimulated light scattering from liquids and gases; dye lasers; nonlinear optics techniques for Raman spectroscopy. *Mailing Add:* Dept Physics Mem Univ Nfld St John's NF A1B 3X7 Can

FOLTZ, RODGER L, b Milwaukee, Wis, Feb 10, 34; m 56; c 2. ANALYTICAL CHEMISTRY, MASS SPECTROMETRY. *Educ:* Mass Inst Technol, 56; Univ Wis, PhD(org chem), 61. *Prof Exp:* sr res leader chem, Battelle-Columbus Labs, 61-79; res assoc prof, Col Pharmacol, 80-85, ASSOC DIR, CTR HUMAN TOXICOL, UNIV UTAH, 79-, RES PROF, COL PHARMACOL, 85-; VPRES, NORTHWEST TOXICOL, INC, 87- *Concurrent Pos:* Adj prof, Col Pharm, Ohio State Univ, 72-76, adj assoc prof, Col Med, 76- *Mem:* Am Chem Soc; Am Soc Mass Spectrometry; Am Acad Forensic Sci. *Res:* Application of mass spectrometry to analysis of organic and biochemical materials. *Mailing Add:* Ctr Human Toxicol Univ Utah 417 Wakara Way Rm 290 Salt Lake City UT 84108

FOLTZ, THOMAS ROBERTS, JR, b Philadelphia, Pa, July 8, 20; m 45; c 2. ENVIRONMENTAL SCIENCES. *Educ:* Philadelphia Col Pharm, BS, 42; Temple Univ, AM, 52. *Prof Exp:* Mem staff, Gulf Refining Co, 41; res chemist, Philadelphia Quartz Co, 42-46; instr textile chem, Philadelphia Textile Inst, 46-49, asst prof, 49-54; chief chemist & dir chem res, Lockport Felt Co, 54-63; prod mgr, Armour Food Co, Collagen & Protein Specialties, 63-75, opers specialist, fresh meats div, 75-80; CHIEF CHEMIST, WASTE WATER TREAT DIV, CITY PHOENIX, 81- *Concurrent Pos:* Registered rep, Nat Asn State Develop, 82- *Mem:* Am Chem Soc; fel Am Inst Chemists; Am Leather Chem Asn; Inst Food Technologists; AAAS. *Res:* Analytical methodology of textile chemistry; wool protein; denaturation; wool fabric finishing; leather chemistry, by-product utilization, meat or poultry; microwave, collagen and protein technology; rendering technology; hazardous wastes technology and testing, permitting, and waste disposal. *Mailing Add:* 7726 E Oak St Scottsdale AZ 85257

FOLWEILER, ROBERT COOPER, b Tallahassee, Fla, Aug 2, 33; m 61; c 3. MATERIALS SCIENCE. *Educ:* Rice Univ, BA, 55, BScME, 56, MS, 58. *Prof Exp:* Res asst metall, Rice Univ, 56-58; metallurgist ceramics, Gen Elec Res Lab, 59-61; sr scientist, Res & Adv Develop Div, Avco Corp, 61; vpres & lab dir, Lexington Labs, Inc, 61-68; mgr mat technol, Defensive Systs Div, Sanders Assocs, Inc, 68-79; sr mat scientist, Microwave Assocs, Inc, 79-80; PRIN INVESTR, GTE LABS INC, 80- *Concurrent Pos:* Vis scientist, Mass Inst Technol, 69-74. *Mem:* Am Asn Crystal Growth; Am Ceramic Soc; AAAS; Sigma Xi. *Res:* Preparation of ultra pure fluoride compounds; preparation of heavy metal fluoride glass optical fiber; investigation of new optical materials for telecommunications; materials synthesis. *Mailing Add:* 45 Wildwood Dr Bedford MA 01730-1139

FOLZ, SYLVESTER D, b Marshfield, Wis, Feb 26, 41; m 62; c 3. VETERINARY PARASITOLOGY, MEDICAL PARASITOLOGY. *Educ:* Univ Wis, BS, 64, MS, 66, PhD(parasitol), 68. *Prof Exp:* Res asst parasitol, Univ Wis, 64-66, asst, 66-68; res assoc, 68-69, scientist, 69-72, res scientist, 72-75, res head, 75-84, SR RES SCIENTIST, UPJOHN CO, 84- *Concurrent Pos:* Ed, Vet Parasitol, 75-90, chief ed, 91-; secy & treas, Am Asn Vet Parasitol, 86- *Mem:* Am Soc Parasitol; Am Soc Protozool; Am Heartworm Soc; Wildlife Dis Asn; Australian Soc Parasitologists. *Res:* Helminthology; protozoology; coccidiosis; malaria; helminth studies in man and animals; veterinary and medical entomology. *Mailing Add:* The Upjohn Co Kalamazoo MI 99001

FOMON, SAMUEL JOSEPH, b Chicago, Ill, Mar 9, 23; m 48; c 5. PEDIATRICS, NUTRITION. *Educ:* Harvard Univ, AB, 45; Univ Pa, MD, 47. *Hon Degrees:* Dr, Cath Univ Cordoba, Arg, 74. *Prof Exp:* Resident pediat, Philadelphia Children's Hosp, 48-50; Res Found fel biochem, Cincinnati Children's Hosp, 50-52; from asst prof to assoc prof pediat, 54-61, PROF PEDIAT, UNIV IOWA, 61- *Concurrent Pos:* Chmn, Comt on Nutrit, Am Acad Pediat, 60-63; USPHS career develop award, 62-67; consult nutrit, Bur Community Health Serv, Health Serv Admin, 65-81; chmn sect II, panel 2, White House Conf on Food, Nutrit & Health, 69; vpres, 12th Int Cong Nutrit, 81; dir, Program Human Nutrit, 80-88. *Honors & Awards:* Borden Award, Am Acad Pediat, 66; Rosen von Rosenstein Medal, Swed Pediat Asn, 74; McCollum Award, Am Soc Clin Nutrit, 79; Award F Cuenca Villoro Found, Zaragoza, Spain, 81; Comnr Spec Citation, Food & Drug Admin, 84; Founder's Award, Midwest Soc Pediat Res, 86; Conrad A Elvehjem Award, 90; Nutricia Res Found Award, Netherlands, 91. *Mem:* Am Acad Pediat; Am Soc Clin Nutrit (pres, 81-82); Soc Pediat Res; Am Pediat Soc; fel Am Inst Nutrit (pres, 89-90); hon mem Am Dietetic Asn; fel AAAS. *Res:* Infant nutrition, growth, iron absorption. *Mailing Add:* Dept Pediat Univ Iowa Hosps Iowa City IA 52242

FONAROFF, LEONARD SCHUYLER, b Philadelphia, Pa, Nov 30, 28; m 51. BIOGEOGRAPHY. *Educ:* Univ Ariz, BA, 55; Johns Hopkins Univ, PhD(geog), 61. *Prof Exp:* From instr to asst prof anthrop & int health, Johns Hopkins Univ, 59-63; from asst prof to assoc prof geog, Calif State Univ, Northridge, 63-67; assoc prof, 67-69, PROF GEOG, UNIV MD, COLLEGE PARK, 69- *Concurrent Pos:* Clin assoc prof community med & int health, Med Sch, Georgetown Univ, 68-76; mem subcomt geochem environ & health & dis, Nat Comt Geochem, Nat Acad Sci, 70-72. *Mem:* Asn Am Geogr; Am Anthrop Asn; Brit Ornith Union; Int Soc Tropical Ecol; Am Geog Soc. *Res:* Disease ecology; zoogeography; animal ecology; man-environment relationships, especially adaption and faunal change. *Mailing Add:* Dept Geog Univ Md College Park MD 20742-8225

FONASH, STEPHAN J(OSEPH), b Philadelphia, Pa, Oct 28, 41; m 68; c 2. SOLID STATE ELECTRONICS, ENERGY CONVERSION. *Educ:* Pa State Univ, BS, 63; Univ Pa, PhD(eng), 68. *Prof Exp:* Fel surface physics, Univ Pa, 68; from asst prof to prof, 68-85, ALUMNI PROF ENG SCI, 85-, DIR, CTR ELECTRONIC MAT & DEVICES, PA STATE UNIV, 87- *Concurrent Pos:* NASA fac fel, Jet Propulsion Lab, 70; vis mem tech staff, Bell Tel Labs, Murray Hill, 72; vis fac mem, Dept Physics, Univ Lyon, France, 73;

assoc ed, IEEE Trans on Electron Devices, 84-; ed, J Electrochem Soc (Solid State), 86- *Mem:* Am Phys Soc; fel Inst Elec & Electronics Engrs; Am Vacuum Soc. *Res:* Solid state and thin film microelectronics devices; photovoltaics; interface and bulk phenomena exhibited by materials; solid state sensors. *Mailing Add:* Dept Eng Sci Pa State Univ University Park PA 16802

FONCK, EUGENE J, b Taipei, Taiwan, Feb 27, 54; m 90. DIGITAL & ANALOG CIRCUIT DESIGN. *Educ:* Tankang Univ, BS, 77; Ind Univ, MS, 83; Univ NMex, MS, 88. *Prof Exp:* Assoc instr physics, Ind Univ, 81-83; res asst microelectronics, Univ NMex, 87-88; ELEC ENGR, UTHE TECHNOL, INC, 88- *Mem:* Inst Elec & Electronics Engrs; Am Inst Physics. *Res:* Digital and analog circuit design; electrochemical transducer properties and its application in wire bonding; semiconductor device design and processing. *Mailing Add:* PO Box 611884 San Jose CA 95161

FONCK, RAYMOND JOHN, b Joliet, Ill, Nov 1, 51; m 77. ATOMIC PHYSICS, PLASMA PHYSICS. *Educ:* Univ Wis-Madison, BA, 73, PhD(physics), 78. *Prof Exp:* Res asst plasma physics, Dept Astrophys Sci, Princeton Univ, 73-74; NSF fel & res asst physics, Univ Wis-Madison, 74-78; res assoc, Plasma Physics Lab, Princeton Univ, 78-80, mem res staff, 80-; ASSOC PROF, DEPT NUCLEAR ENG & ENG PHYSICS, UNIV WIS. *Mem:* Am Phys Soc; Optical Soc Am. *Res:* Optical and ultraviolet atomic spectroscopy; atomic diamagnetism; spectroscopic instrumentation; plasma diagnostics. *Mailing Add:* Dept Nuclear Eng & Eng Physics Univ Wisconsin 1500 Johnson Dr Madison WI 53706

FONDA, MARGARET LEE, b Cleveland, Ohio, July 13, 42; m 76. ENZYMES, COENZYMES. *Educ:* Univ Del, BS, 64; Univ Tenn, Knoxville, PhD(biochem), 68. *Prof Exp:* Res assoc biochem, Iowa State Univ, 68-70; from, asst prof to assoc prof, 70-87, PROF BIOCHEM, SCH MED, UNIV LOUISVILLE, 87- *Concurrent Pos:* NIH res assoc, Iowa State Univ, 69-70. *Mem:* AAAS; Am Soc Biol Chemists; Am Chem Soc; Sigma Xi. *Res:* Mechanism of enzyme action; interaction of coenzymes and substrates with enzymes; control of enzyme action; metabolism of vitamins and coenzymes. *Mailing Add:* Dept Biochem Univ Louisville Sch Med Louisville KY 40292

FONDA, RICHARD WESTON, b Chicago, Ill, June 14, 40; m 63; c 2. PLANT ECOLOGY. *Educ:* Duke Univ, BA, 62; Univ Ill, MS, 65, PhD(bot), 67. *Prof Exp:* Asst prof biol, Western Ill Univ, 67-68; from asst prof to assoc prof biol, 68-79, PROF BIOL, WESTERN WASH UNIV, 79-, BIOL GRAD PROG DIR, 69- *Concurrent Pos:* Grants, Western Ill Univ Res Coun, 68, US Forest Serv, 78 & US Nat Park Serv, 81. *Mem:* Ecol Soc Am; Torrey Bot Club. *Res:* Vegetation ecology in western Washington especially forest, timberline and alpine; fire ecology. *Mailing Add:* Dept Biol Western Wash Univ 516 High St Bellingham WA 98225

FONDACARO, JOSEPH D, b Gloversville, NY, Nov 29, 43; m 68; c 3. ELECTROLYTE TRANSPORT, SUB-CELLULAR REGULATION. *Educ:* Univ Cincinnati, PhD(physiol), 72. *Prof Exp:* ASSOC FEL, SMITH, KLINE & FRENCH, 85-, INTERIM ASST DIR, DEPT PHARMACOL, 86- *Concurrent Pos:* Adj assoc prof physiol, Univ Cincinnati, Col Med, 84-, Univ Penn, Sch Med, 86- *Mem:* Am Physiol Soc; Soc Exp Biol & Med; NY Acad Sci; Am Gastroenterol Asn; Mid-Atlantic Pharmacol Soc. *Res:* Physiology, pharmacology and biochemistry of intestinal electrolyte transport and its regulation; role of alpha adrenoceptors, calcium and cyclic nucteotides in intestinal membrance transport; mediators of intestinal inflammation. *Mailing Add:* Dept Spons Ecternal Res Marion Labs Inc Park B PO Box 9627 Kansas City MO 64134

FONDAHL, JOHN W(ALKER), b Washington, DC, Nov 4, 24; m 46; c 4. CIVIL ENGINEERING. *Educ:* Dartmouth Col, BS, 47, MS, 48. *Prof Exp:* Struct detailer, Bridge Div, Am Bridge Co, 48; from instr to asst prof civil eng, Univ Hawaii, 48-51; engr & estimator heavy construct, Winston Bros Co, 51-52, off engr, 52; proj engr, Nimbus Dam & Powerhouse contract, Winston Johnson Construct Co, 53-55; from asst prof to prof, 55-77, Charles H Leavell prof civil eng & chair, 77-90, EMER PROF, STANFORD UNIV, 90- *Concurrent Pos:* Dir, Scott Co, Calif, 63-81; pres, Proj Mgt Inst, 74 & 75, chmn bd, 76 & 77; dir, Caterpillar Inc, 76-; fel award, Proj Mgt Inst, 82; Peurifoy Construct res award,Am Soc Civil Engrs, 90. *Honors & Awards:* Golden Beaver Award, Beavers Heavy Construct Soc, 76. *Mem:* Am Soc Civil Engrs; Proj Mgt Inst. *Res:* Construction engineering; administration; planning and scheduling. *Mailing Add:* 12810 Viscaino Rd Los Altos Hills CA 94022

FONDY, THOMAS PAUL, b Pittsburgh, Pa, Dec 24, 37; div; c 3. BIOCHEMISTRY, CHEMOTHERAPY. *Educ:* Duquesne Univ, BS, 59, PhD(biochem), 61. *Prof Exp:* Fel biochem, Duquesne Univ, 61-62 & Brandeis Univ, 62-65; from asst prof to assoc prof, 65-75, PROF BIOCHEM, SYRACUSE UNIV, 75- *Concurrent Pos:* Res trainee, Nat Inst Neurol Dis & Blindness, 62-65; Nat Cancer Inst career develop award, 72-77; vis prof pharmacol, Yale Univ Sch Med, 75-76. *Mem:* Am Soc Biol Chem; Am Asn Cancer Res. *Res:* carbohydrate analogs in cell membrane directed chemotherapy and chemoimmunotherapy of cancer; host response modifiers in host-tumor interaction; effects of Cytochalsins in vivo on host-tumor interactions; micro-encapulated agents in cancer chemotherapy. *Mailing Add:* Dept Biol Syracuse Univ Syracuse NY 13244

FONER, SAMUEL NEWTON, b New York, NY, Mar 21, 20. CHEMICAL PHYSICS, MASS SPECTROMETRY. *Educ:* Carnegie-Mellon Univ, BS, 40, MS, 41, DSc(physics), 45. *Prof Exp:* Asst physics, Carnegie-Mellon Univ, 40-43, instr, 43-44, res assoc, 44-45; physicist, Appl Physics Lab, Johns Hopkins Univ, 45-47, supvr, Mass Spectrometry Group, 47-52, supvr, Electronic Physics Group, Appl Physics Lab, 53-83, vchmn, Milton S Eisenhower Res Ctr, 75-83, sr scientist, Strategic Systs Dept, 84-90; RETIRED. *Concurrent Pos:* Mem comt, Nat Acad Sci-Nat Res Coun; adv, Army Res Off, 61-64. *Honors & Awards:* Phys Sci Award, Wash Acad Sci, 54. *Mem:* Fel AAAS; fel Am Phys Soc; Am Vacuum Soc; Combustion Inst. *Res:* Mass spectrometry of free radicals and excited state molecules; reaction kinetics; surface science; electron spin; free radicals; molecular beams; electronic physics; scanning electron microscopy; acoustics; solid propellants. *Mailing Add:* 11500 Summit West Blvd No 15B Temple Terrace FL 33617-2317

FONER, SIMON, b Pittsburgh, Pa, Aug 13, 25; m 55; c 2. SOLID STATE PHYSICS. *Educ:* Carnegie Inst Technol, BS, 47, MS, 48, DSc(physics), 52. *Prof Exp:* Asst physics, Carnegie Inst Technol, 47-52, res physicist, 52-53; physicist, Lincoln Lab, Mass Inst Technol, 53-61, group leader, Nat Magnet Lab, 61-63, leader transport & resonance group, 63-71, PROJ LEADER MAGNETISM & SUPERCONDUCTIVITY, NAT MAGNET LAB, MASS INST TECHNOL, 71-, CHIEF SCIENTIST & HEAD RES DIV, 77-, SR RES SCIENTIST, DEPT PHYSICS, 82-, ASSOC DIR, 89- *Concurrent Pos:* Mem adv comt to Army Res Off, Nat Res Coun, 67-73; dir, NATO Advan Study Inst, 70, 73, 76 & 80; consult ed, Rev Sci Instruments, 79-, chmn Div Condensed Matter Physics, 78-82, counr div, 82-86. *Mem:* Fel Am Phys Soc; sr mem Inst Elec & Electronics Engrs; Mat Res Soc. *Res:* Magnetism; high field superconductors; magnetic resonance; high magnetic fields; magnetometry. *Mailing Add:* Nat Magnet Lab Mass Inst Technol Cambridge MA 02139

FONG, CHING-YAO, b Soochow, China, May 23, 35. SOLID STATE PHYSICS. *Educ:* Nat Taiwan Univ, BS, 58; Univ Calif, Berkeley, MS, 62, PhD(physics), 68. *Prof Exp:* Res physicist, Lawrence Radiation Lab, Univ Calif, Berkeley, 68; from asst prof to assoc prof, 69-78, PROF PHYSICS, UNIV CALIF, DAVIS, 78-; CONSULT, LAWRENCE LIVERMORE NAT LAB, 82- *Concurrent Pos:* Vis physicist, Bell Lab, 75, Lawrence Livermore Nat Lab, 81 & Naval Res Lab, 84; vis scientist, Argonne Nat Lab, 80. *Mem:* Am Phys Soc. *Res:* Optical properties of solids; semiconductor alloys; amorphous semiconductor; semiconductor superlattices. *Mailing Add:* Dept Physics Univ Calif Davis CA 95616

FONG, FRANCIS K, b Shanghai, China, Mar 21, 38; m 63; c 3. CHEMICAL PHYSICS, BIOPHYSICS. *Educ:* Princeton Univ, AB, 59, PhD(chem), 62. *Prof Exp:* Res assoc, Princeton Univ, 62-63; mem tech staff, RCA Labs, 63-64 & NAm Aviation Sci Ctr, Calif, 64-68; assoc prof, 68-71, PROF CHEM, PURDUE UNIV, 71- *Mem:* Am Chem Soc; Am Phys Soc. *Res:* Photochemistry; primary light reaction in photosynthesis; radiationless processes in large molecules; chlorophyll chemistry; theory of molecular relaxation; solid state chemical physics. *Mailing Add:* Dept Chem Purdue Univ Main Campus West Lafayette IN 47907

FONG, FRANKLIN, b San Francisco, Calif, Apr 16, 47. PLANT PHYSIOLOGY. *Educ:* Univ Calif, Davis, BS, 69; Univ Calif, Riverside, PhD(biol), 75. *Prof Exp:* NSF fel, Inst Photobiol Cells & Organelles, Brandeis Univ, 75-78; asst prof, Dept Plant Sci, Tex A&M Univ, 78-85, assoc prof, Dept Soil & Crop Sci, 85-90; VIS ASSOC PROF, UNIV PORTLAND, 90- *Honors & Awards:* Menzel Award, Genetics Soc Am. *Mem:* Am Soc Plant Physiologists; AAAS; Bot Soc Am. *Res:* Air pollution physiology; carotenoid and abscisic acid metabolism; stress physiology; seed development. *Mailing Add:* Dept Phys & Life Sci Univ Portland Portland OR 97215

FONG, GODWIN WING-KIN, b Kwang-Tung, China, Oct 11, 50; US citizen. ANALYTICAL CHEMISTRY, PHARMACEUTICAL ANALYSIS. *Educ:* State Univ NY, Stony Brook, BS, 74; State Univ NY, Buffalo, PhD(chem), 80. *Prof Exp:* Res asst chem, State Univ NY, Buffalo; group leader anal res & develop, Ayerst Labs, Inc, Am Home Prod Corp, 78-84; asst dir, Smith Kline Consumer Prod, Res & Develop Ctr, 84-86, ASST DIR RES & DEVELOP, SMITH KLINE & FRENCH LAB, SMITH KLINE BECKMAN CORP, 86- *Concurrent Pos:* Mem HPLC comt, Qual Control Sect, Pharmaceut Mfg Asn, 79-84. *Mem:* Am Chem Soc; AAAS; Am Asn Pharmaceut Scientist; Asn Off Anal Chemists; NY Acad Sci. *Res:* Analytical chemistry, especially chromatography, both theoretical and experimental; bonded phase chemistry as applied in chromatography; pharmaceutical analysis; development and validation of analytical procedures; dissolution testing for solid dosages, especially controlled-released dosage forms; laboratory automation. *Mailing Add:* 42 Brookfield Rd Ft Salonga Northport NY 11768

FONG, HARRY H S, b Kwangtung, China, June 30, 35; US citizen; m 64; c 3. PHARMACOGNOSY. *Educ:* Univ Pittsburgh, BS, 59, MS, 62; Ohio State Univ, PhD(pharm), 65. *Prof Exp:* Res asst prof pharmacog, Sch Pharm, Univ Pittsburgh, 65-68, assoc prof, 68-70; assoc prof, 70-72, PROF PHARMACOG, COL PHARM, UNIV ILL, 72-, DIR PHARMACOG FIELD STA, 81- *Concurrent Pos:* Consult, WHO, 77; hon vis res fel, Inst Sci & Technol, Chinese Univ, Hong Kong, 77; temp adv, WHO, 79, 80, 82-84; roster of experts, UN Indust Develop Bd, 81-; coun mem, Soc Econ Bot, 82-83; consult, World Bank, 83. *Mem:* Am Soc Pharmacog (vpres, 77-78, pres, 78-79, past-pres, 79-80); Soc Econ Bot (treas, 75-77, vpres, 80-81, pres, 81-82, past pres, 82-83). *Res:* Phytochemistry, isolation, characterization and/or identification and biological evaluation of pharmacologically active principles from plants; biological and phytochemical evaluation of fertility regulating agents from plants. *Mailing Add:* Col Pharm Univ Ill Chicago 833 S Wood St M-C781 Chicago IL 60612

FONG, JACK SUN-CHIK, b Hong Kong, Jan 6, 41; Can citizen; m 68; c 3. PEDIATRICS, IMMUNOLOGY. *Educ:* McGill Univ, BS, 64, MS, MD & CM, 68. *Prof Exp:* Intern med, St Josephs Hosp, Ont, 68-69; res fel immunol, Univ Minn, Minneapolis, 69-72, med fel pediat, 72-74; from asst prof to assoc prof pediat, McGill Univ, 74-85; CHMN, DEPT PEDIAT, DANBURY HOSP, CONN, 85-; CLIN ASSOC PROF PEDIAT, YALE SCH MED, 86- *Mem:* Am Asn Immunologists; Am Fedn Clin Res; Soc Exp Biol & Med; Can Pediat Soc; Am Acad Pediat; Clin Immunol Soc. *Res:* Mechanism of immunologic injury, immune mediated coagulopathy, the complement system and immunodeficiency state. *Mailing Add:* Danbury Hosp Danbury CT 06810

FONG, JAMES T(SE-MING), b Shanghai, China, Aug 23, 27; nat US; m 64; c 3. ENGINEERING, MATERIALS SCIENCE. *Educ:* Mass Inst Technol, SB, 48; Columbia Univ, MS, 51. *Prof Exp:* Develop engr, Griscom-Russell Co, 48-51; mech engr, Burns & Roe, Inc, 51-54; proj engr, Res Dept, Foster Wheeler Corp, 54-57; fel engr, AIW Proj, 57-61; tech staff gen mgr, 61-66, supvr irradiated components eng, 66-68, mgr adv cores, 68-72, mgr adv submarine proj, 72-74, mgr mat technol, 74-79, CONSULT, BETTIS ATOMIC POWER LAB, WESTINGHOUSE ELEC CORP, 79- *Mem:* Am Soc Mech Engrs. *Res:* Nuclear reactor engineering; heat transfer; power plant design; project management. *Mailing Add:* Bettis Atomic Power Lab Westinghouse Elec Corp PO Box 79 West Mifflin PA 15122-0079

FONG, JEFFERY TSE-WEI, b Shanghai, China, Nov 24, 34; US citizen; c 2. ENGINEERING EXPERT SYSTEMS, FATIGUE OF ENGINEERING MATERIALS. *Educ:* Univ Hong Kong, BSc(Eng), 55; Columbia Univ, MS, 61; Stanford Univ, PhD(appl mech & math), 66. *Prof Exp:* Design engr powerplants, Ebasco Serv Inc, New York, NY, 55-63; res assoc appl mech, Stanford Univ, 64-66; res assoc appl math, US Nat Bur Standards, 66-68, PHYSICIST, US NAT INST STANDARDS & TECHNOL, 68- *Concurrent Pos:* Sr policy analyst, US Nuclear Regulatory Comn, 75-76; chair subcomt E9-01, Am Soc Testing & Mat, 78-80; chmn, PVP Div, Am Soc Mech Engrs, 86-87, distinguished lectr, 88-92. *Honors & Awards:* Silver Medal, US Dept Com, 79. *Mem:* Fel Am Soc Mech Engrs; fel Am Soc Testing & Mat; Am Phys Soc; Soc Rheology; Soc Indust & Appl Math; Am Statist Asn. *Res:* Computational mathematics, applied mechanics and statistical methods for modeling thermo-mechanical behavior of structural materials in both elastic and inelastic states as well as fatigue and fracture to improve quality and reliability. *Mailing Add:* US Nat Inst Standards & Technol MC A238/101 Gaithersburg MD 20899

FONG, JONES W, b Sacramento, Calif, Sept 7, 36; m 64; c 2. MICROENCAPSULATION, DRUG DELIVERY SYSTEM. *Educ:* Univ Calif, BS, 58; Rutgers Univ, PhD, 62. *Prof Exp:* Res chemist, E I du Pont de Nemours, 62-65; sr res chemist, US Chem Co, Tex, 65-72; res assoc, Albert Einstein Col of Med, 73-75; res scientist, NY State Basic Res Mental Retardation, 75-76; sr scientist A, 76-81, sr scientist B, 81-87, ASSOC FEL, SANDOZ PHARMACEUT CORP, 87- *Mem:* Am Chem Soc. *Res:* Microencapsulation; iontophoretic transdermal delivery. *Mailing Add:* 41 Normandy Dr Parsippany NJ 07054

FONG, KEI-LAI LAU, b Hong Kong, May 15, 50; US citizen; m 75. MEDICINAL CHEMISTRY, PHARMACOLOGY. *Educ:* Nat Taiwan Univ, BS, 72; Univ Minn, PhD(med chem), 78. *Prof Exp:* Proj investr develop therapeut, M D Anderson Hosp & Tumor Inst, 78-81; ASSOC SR INVESTR DRUG METAB, SMITH KLINE & FRENCH LABS, 81- *Mem:* Am Chem Soc; Am Asn Cancer Res; AAAS; Int Soc Study Xenobiotics. *Res:* Biological evaluation of novel thromobylosis agents; drug metabolism; pharmokinetics protein; peptide metabolism and disposition. *Mailing Add:* Smith Kline & French Drug Metas 709 Swedeland Rd Swedeland PA 19479

FONG, PETER, b Tungshang, China, Sept 3, 24; m 59; c 3. THEORETICAL PHYSICS, MOLECULAR BIOLOGY. *Educ:* Chekiang Univ, BS, 45; Univ Chicago, MS, 50, PhD(physics), 53. *Prof Exp:* Physicist, Inst Nuclear Studies, Univ Chicago, 54; from asst prof to prof physics, Utica Col, Syracuse Univ, 54-66; PROF PHYSICS, EMORY UNIV, 66- *Concurrent Pos:* NSF res grants, 55-66; res assoc, Mass Inst Technol, 56 & 57; res fel, Kellogg Radiation Lab, Calif Inst Technol, 57; vis prof, Cornell Univ, 63-64 & Univ Calif, Berkeley, 65-66. *Mem:* Fel Am Phys Soc; fel Biophys Soc; Am Geophys Union. *Res:* Nuclear physics; astrophysics; quantum mechanics; thermodynamics; DNA functions; brain memory mechanism; origin of ice ages; carbon dioxide greenhouse effect; philosophy and history. *Mailing Add:* Dept Physics Emory Univ Atlanta GA 30322

FONG, SHAO-LING, b China; US citizen. EYE RESEARCH, PROTEIN CHEMISTRY. *Educ:* Chung-Hsing Univ, BS, 66; Southern Ill Univ, MS, 73, PhD(biochem), 78. *Prof Exp:* Teaching asst, chem lab, Fu-Jen Univ, 67-69, Southern Ill Univ, 70-71; teaching asst & res asst, chem lab, Iowa State Univ, 71-74 , Southern Ill Univ, 75-77; res assoc, Baylor Col Med, 77-79, res instr, 79-80, res asst prof, 80-87; assoc res scientist, Purdue Univ, 87-89; ASST PROF OPTHAL, IND UNIV, 89- *Mem:* Asn Res Vision & Ophthal; Am Soc Biol Chemists; Int Soc Eye Res. *Res:* Metabolism and utilization of vitamin A in vision. *Mailing Add:* Dept Ophthal Ind Univ 702 Rotary Cir Indianapolis IN 46202

FONGER, WILLIAM HAMILTON, b Chicago, Ill, Sept 19, 25; wid; c 3. PHYSICS. *Educ:* Univ Chicago, SB, 48, SM, 50, PhD(physics), 53. *Prof Exp:* mem tech staff, Labs, RCA Corp, 53-87; RETIRED. *Mem:* Am Phys Soc. *Res:* Solid state physics; experimental and theoretical studies of phosphors; deposition of phosphors in color-tv picture tubes. *Mailing Add:* 174 Guyot Ave Princeton NJ 08504-3433

FONKALSRUD, ERIC W, b Baltimore, Md, Aug 31, 32; m 59; c 4. SURGERY. *Educ:* Univ Wash, BA, 53; Johns Hopkins Univ, MD, 57; Am Bd Surg, dipl, 64; Am Bd Thoracic Surg, dipl, 66, Am Bd Pediat Surg, 75 & 84. *Prof Exp:* Intern surg, Johns Hopkins Hosp, Baltimore, Md, 57-58, asst resident, 58-59; from asst resident to chief resident, 59-63, from asst prof to assoc prof, 63-71, PROF SURG, SCH MED, UNIV CALIF, LOS ANGELES, 71-, CHIEF PEDIAT SURG, 63- *Concurrent Pos:* Instr, Ohio State Univ & resident, Children's Hosp, Columbus, Ohio, 63-65; Mead Johnson grad training award surg, Univ Calif, Los Angeles, Markle scholar, 63-68; USPHS res grants, 65-75; Calif Inst Cancer Res & Los Angeles County Heart Asn grants; James IV surg traveller, Gt Brit, 71; mem surg study sect, NIH, 71-75; mem bd dirs, Martin Mem Found, Am Col Surgeons. *Honors & Awards:* Alpha Omega Alpha Golden Scalpel Award Surg. *Mem:* Fel Am Col Surg; Am Acad Pediat (prs, 87); Soc Univ Surg (pres, 77); Asn Acad Surg (pres, 72); Am Pediat Surg Asn(pres, 89); Am Asn Thoracic Surg; Sigma Xi. *Res:* Organ transplantation, experimental and clinical; pulmonary, hepatic and cardiac physiology; studies of fetal and neonatal physiology; development of new surgical techniques; inflammatory bowel disease; use of computer assisted instruction in medical education; author of one book, various journal articles and book chapters. *Mailing Add:* Dept Surg Univ Calif Sch Med Los Angeles CA 90024

FONKEN, DAVID W(ALTER), b Denver, Colo, Oct 9, 31; m 53; c 3. ENGINEERING HYDROLOGY, IRRIGATION ENGINEERING. *Educ:* Colo State Univ, BS, 53, MCE, 56; Univ Ariz, PhD(civil eng), 74. *Prof Exp:* Hydraul engr, US Bur Reclamation, 56-58; asst prof & asst agr engr, Dept Agr Eng, Univ Ariz, 58-61; hydrometeorologist, Harza Eng Co, 61-68; grad res asst, Dept Agr Eng, Univ Ariz, 68-71; asst prof & dist exten irrig engr, Univ Nebr, 71-75; systs analyst, 75-81, IRRIGATION ENGR & HYDROLOGIST, HARZA ENG CO, 81- *Concurrent Pos:* Mem US Comt on Irrigation & Drainage. *Mem:* Am Soc Civil Engrs; Am Soc Agr Engrs. *Res:* Hydrological studies in arid regions; irrigation; hydraulics of irrigation; irrigation scheduling; soil-water-plant relationships. *Mailing Add:* Harza Eng Co 233 S Wacker Dr Chicago IL 60606

FONKEN, GERHARD JOSEPH, b Ger, Aug 31, 28; nat US; m 52; c 5. ORGANIC CHEMISTRY. *Educ:* Univ Calif, BS, 54, PhD(chem), 57. *Prof Exp:* Res chemist, Procter & Gamble Co, 57-58 & Stanford Res Inst, 58-59; from instr to assoc prof, 59-74, exec asst to pres, 76-79, PROF CHEM & ASSOC PROVOST, UNIV TEX, AUSTIN, 74-, VPRES ACAD AFFAIRS & RES, 79- *Mem:* Am Chem Soc. *Res:* Natural products; structural determinations of organic compounds; organic photochemical transformations. *Mailing Add:* Dept Chem Univ Tex Main Bldg 201 Austin TX 78712-1111

FONKEN, GUNTHER SIEGFRIED, b Krefeld, Ger, Jan 29, 26; nat US; m 51, 84; c 3. RESEARCH ADMINISTRATION. *Educ:* Mass Inst Technol, BS, 46; Univ Wis, PhD(chem), 51. *Prof Exp:* Jr chemist, Res Dept, Merck & Co, Inc, 46-47; res chemist, Chem Dept, Upjohn Co, 51-60, sect head, Biochem Dept, 60-68, res mgr, Cancer Res Dept, 68-73, res mgr, Exp Biol Res Dept, 73-79, dir, Drug Safety, Planning & Admin Serv, 79-81, group dir, 81-82, vpres, Pharmaceut Res & Develop Admin & Support Opers, Upjohn Co, 82-86; RETIRED. *Mem:* Fel AAAS. *Res:* Pharmaceutical substances. *Mailing Add:* PO Box 2355 Kalamazoo MI 49003-2355

FONNESBECK, PAUL VANCE, b Weston, Idaho, Aug 4, 31; m 53; c 4. ANIMAL NUTRITION. *Educ:* Brigham Young Univ, BS, 53; Utah State Univ, MS, 59, PhD(animal nutrit), 62. *Prof Exp:* Asst prof nutrit, Rutgers Univ, 63-70, asst res prof, 70; res assoc animal sci, Utah State Univ, 70-77, res asst prof animal dairy & vet sci, 77-; RETIRED. *Mem:* Am Soc Animal Sci. *Res:* Nutrient content and utilization of foods and feeds; improving chemical methods for food analysis; alfalfa hay quality determination and production; feed nutrient information tables. *Mailing Add:* Ten E 7800 N Smithfield UT 84335

FONO, ANDREW, b Budapest, Hungary, Apr 24, 23; nat US; m 58; c 2. ORGANIC CHEMISTRY, ELECTRO CHEMISTRY. *Educ:* Pazmany Peter Univ, Hungary, PhD(chem), 45. *Prof Exp:* Asst, Inst Org Chem Res, Stockholm, 46-47; res fel chem, Univ Chicago, 47-51; head, Res Dept, Otto B May Inc, 51-56; indust res assoc, Inst Org Chem, Univ Chicago, 57; sr res scientist, Firestone Tire & Rubber Co, Ohio, 58-60; res dir, Otto B May, Inc, Newark, NJ, 60-75; TECH DIR, ROYCE ASSOCS, 75-, EXEC VPRES, PASSAIC COLOR, 78- *Mem:* Am Chem Soc; AAAS; Tech Asn Pulp & Paper Indust; Am Asn Textile Chemists & Colorists; NY Acad Sci. *Res:* Organic reaction mechanism; effect of metal ions and metallo-organic compounds on reactions in solution; polymer and dyestuff chemistry; colloid chemistry; textile chemistry, paper pulping and alkaline battery. *Mailing Add:* 207-215 Ave L Newark NJ 07105

FONONG, TEKUM, b Cameroon, Mbemi, Dec 14, 52; m 80; c 1. ENZYME IMMUNOASSAY, MICROANALYSIS OF ENVIRONMENTAL POLLUTANTS. *Educ:* Sioux Falls Col, BS, 77; Marquette Univ, MS, 79; Univ Iowa, PhD(chem), 83. *Prof Exp:* Res fel chem, Univ Del, 83-84, vis asst prof, 84-85; ASST PROF CHEM, ILL STATE UNIV, 85- *Concurrent Pos:* Lectr, Del Tech & Community Col, 84; consult, Lincoln Univ, Pa, 85. *Mem:* AAAS; Am Chem Soc; Int Asn African Scientists. *Res:* Use of the highly specific and selective properties of enzymes to design electrodes for the measurement of metabolites in physiological fluids. *Mailing Add:* Dept Chem Hampton Univ Hampton VA 23668

FONSECA, ANTHONY GUTIERRE, b Chattanooga, Tenn, Mar 31, 40; m 65; c 2. MINERAL PROCESSING. *Educ:* Univ Chattanooga, AB, 62; Univ Ga, MS, 66, PhD(inorg chem), 68. *Prof Exp:* Res scientist, Cent Res Div, 68-74, sr res scientist, Continental Oil Co, 74-75, res group leader, Mining Res Div, 76-81, dir spec projs, Conoco Inc, 81-82; DIR COALUTILIZATION, CONSOLIDATION COAL CO, 82- *Mem:* Am Chem Soc; Am Inst Mining, Metall & Petrol Engrs. *Res:* Investigations of mineral processing systems, plant testing, instrumentation and control of coal processing systems; combustion of coal; slagging, fouling, effects of coal quality on combustion; coal and lignite cleaning, desulfurization, processing and preparation; mineral processing and beneficiation, within the general area of mining engineering. *Mailing Add:* 2170 Meadowmont Dr Pittsburgh PA 15241

FONT, ROBERT G, b Havana, Cuba, May 5, 40. GEOPHYSICS. *Educ:* Baylor Univ, BS, 67, MS, 69; Tex A&M, PhD(eng biol), 73. *Prof Exp:* Exp petrol geologist, Conoco, 69-70; assoc prof geol, Baylor Univ, 73-80; sr staff geologist, 80-82, proj supvr, 82-85, area geologist, 85-87, EXEC VPRES, STRATEGIC PETROLEUM, CONOCO, INC, 87- *Mem:* Fel Geol Soc Am; Am Geophys Union; Soc Petrol Engrs; Asn Eng Geologists; Int Soc Rock Mech. *Res:* Exploration for oil & gas reserves. *Mailing Add:* 16316 Lauder Lane Dallas TX 75248-2349

FONT, WILLIAM FRANCIS, b New Orleans, La, Aug 11, 44; m 68; c 1. PARASITOLOGY. *Educ:* Tulane Univ, BS, 66; La State Univ, MS, 72, PhD(zool), 75. *Prof Exp:* From asst prof to assoc prof biol, Univ Wis, Eau Claire, 75-85; ASSOC PROF BIOL, SOUTHEASTERN LA UNIV, 85- *Honors & Awards:* Elon E Byrd Mem Award, Southeastern Soc Parasitologists, 75. *Mem:* Am Soc Parasitologists; Sigma Xi; Am Micros Soc. *Res:* Taxonomy and life history of digenetic trematodes. *Mailing Add:* Dept Biol Box 814 Southeastern La Univ Harmond LA 70402

FONTAINE, ARTHUR ROBERT, b Lawrence, Mass, Aug 1, 29; Can citizen; m 52; c 2. MARINE ZOOLOGY. *Educ:* McGill Univ, BSc, 52; Oxford Univ, DPhil(zool), 61. *Prof Exp:* Researcher marine biol, Inst Jamaica, Kingston, 52-53; instr biol, Victoria Col, 56-58, asst prof, 58-63; assoc prof, 63-68, assoc dean, 69-70, dean grad studies, 70-72, chmn dept, 77-79, PROF BIOL, UNIV VICTORIA, BC, 68- *Concurrent Pos:* Vis prof, Friday Harbor Labs, Univ Wash, 64, 65 & 69; vis zoologist, Dove Marine Lab, Newcastle-on-Tyne, 66-67; CIDA prof marine biol, Univ of Spac, Suva, Fiji, 75-77. *Mem:* AAAS; Soc Syst Zool; Marine Biol Asn UK; sci fel Zool Soc London. *Res:* Functional morphology and experimental biology of echinoderms. *Mailing Add:* Dept Biol Sci Univ Victoria Box 1700 Victoria BC V8W 2Y2 Can

FONTAINE, GILLES JOSEPH, b Levis, Que, Aug 13, 48; m 69; c 2. ASTROPHYSICS. *Educ:* Univ Laval, BS, 69; Univ Rochester, PhD(astrophys), 73. *Prof Exp:* Nat Res Coun Can fel, 73-75, res assoc astrophys, 75-77, from asst prof to assoc prof, 77-85, PROF ASTROPHYSICS, DEPT PHYSICS, UNIV MONTREAL, 85- *Concurrent Pos:* EWR Steacie Mem Fel, 86-88. *Honors & Awards:* Steacie Prize Winner, 88. *Mem:* Am Astron Soc; Can Asn Physicists; Can Astron Soc; Int Astron Union. *Res:* Stellar structure and stellar evolution; physics of dense matter; interiors and atmospheres of white dwarf stars; pulsating white dwarfs. *Mailing Add:* Dept Physics Univ Montreal PO Box 6128 Montreal PQ H3C 3J7 Can

FONTAINE, JULIA CLARE, b Chattanooga, Tenn, Dec 11, 20. ANATOMY, BIOLOGY. *Educ:* Nazareth Col, BS, 48; Cath Univ, MS, 56; Univ Louisville, PhD(anat), 69. *Prof Exp:* Asst prof biol, 59-66, PROF BIOL & CHMN DEPT, SPALDING COL, 69- *Mem:* AAAS; NY Acad Sci. *Res:* Fluctuations in amounts of DNA in mammalian peripheral leukocytes; DNA changes in non-replicating cells; circadian rhythms in DNA content of specialized cells. *Mailing Add:* 2002 Newburg Rd Apt 304 Louisville KY 40205

FONTAINE, MARC F(RANCIS), b Mexico City, Mex, Mar 27, 26; nat US; m 49; c 3. CHEMICAL ENGINEERING & PETROLEUM ENGINEERING. *Educ:* La State Univ, BS, 50, MS, 51; Okla State Univ, PhD(chem eng), 54. *Prof Exp:* Asst chem eng, Okla State Univ, 51-53, instr, 53, instr math, 53-54; res chem engr, Res Ctr, Texaco Inc, NY, 54-60, sr engr-sci liaison, Texaco UK Ltd, Eng, 60-62, res chem engr, Bellaire Res Labs, Tex, 62-65, group leader, 65-69, supvr, 69-79, sr supvr, 79-84, mgr, thermal miscilbe oil recovery process, Houston Res Ctr, 84-89, mgr, EOR-E&P Technol Div, 89-90, PROG MGR, EPTD, TEXACO INC, 90- *Mem:* AAAS; Am Chem Soc; Am Inst Chem Engrs; Am Math Soc; Am Inst Mining, Metall & Petrol Engrs. *Res:* Fuel technology; experimental design; production research and enhanced oil recovery, petroleum. *Mailing Add:* Texaco Inc PO Box 770070 Houston TX 77215-0070

FONTAINE, THOMAS DAVIS, b Utica, Miss, Apr 12, 16; m 41; c 2. BIOCHEMISTRY. *Educ:* Miss Col, AB, 37; Univ Pittsburgh, PhD(biochem), 42. *Prof Exp:* Asst chem, Univ Pittsburgh, 37-38; asst, Cotton Res Found, Mellon Inst, 38-41; asst chemist, Oil, Fat & Protein Div, Southern Regional Res Lab, Bur Agr & Indust Chem, USDA, 41-44, assoc chemist, 44-45, assoc chemist biol, Active Compounds Div, 45-46, chemist, 46-47, sr biochemist, 47-48, prin biochemist & head div, 48-52, prin chemist & head biol active chem compounds div, Eastern Utilization Res Br, 52-55; admin asst to US Sen John C Stennis, 55-57; head fel sect, Sci Personnel & Ed Div, NSF, 57-65, dir, Div Grad Educ Sci, 65-66, assoc dir sci educ, 66-69, dep asst dir, 69-71; asst dir, Div Sponsored Res, Univ Fla, 71-73, assoc dir, 73-75, dir, 75-76; RETIRED. *Mem:* Am Chem Soc; Am Soc Biol Chem. *Res:* Proteins; amino acids; enzymes; plant diseases; antibiotics from plants; plant growth modifiers; plant alkaloids; government. *Mailing Add:* Plymouth Harbor 700 John Ringling Blvd Sarasota FL 34236

FONTANA, MARIO H, b West Springfield, Mass, Mar 30, 33; m 58; c 2. NUCLEAR SAFETY, HEAT TRANSFER. *Educ:* Univ Mass, Amherst, BS, 55; Mass Inst Technol, SM, 57; Purdue Univ, PhD(mech eng), 68. *Prof Exp:* Assoc engr, Oak Ridge Nat Lab, 57-61; sr scientist, High Temp Mat, Inc, Mass, 61-62; mem sr staff nuclear safety, Oak Ridge Nat Lab, 62-63; sr scientist plasma chem, Avco Res & Advan Develop, Mass, 63-64; instr thermodyn, Purdue Univ, 64-65; asst dir, Nuclear Safety Progs, Oak Ridge Nat Lab, 65-71, mgr, Breeder Reactor Safety Progs, 71-80, head, Advan Reactor Systs Sect, 74-81; dir, Nat Indust Nuclear Reactor Degraded Core Rulemaking Prog, Technol Energy Corp, 81-84; vpres eng, Energex, 84-85; dir, Nuclear Safety, Risk Control Serv, Int Technol Inc, 85-88; dir Nuclear Safety Div, Technol Tenora, 88-90, ASST TO DIR, ENG TECHNOL, OAK RIDGE NAT LAB, 90- *Concurrent Pos:* Consult, Adv Comt Reactor Safeguards, 68-, Nuclear Regulatory Comn, 76-77 & Dept Energy, 85-88. *Mem:* Soc Risk Animals; Am Soc Mech Engrs; fel Am Nuclear Soc. *Res:* Safety of nuclear reactors; evaluation of potential reactor accidents and preventive safeguards; director of safety research programs. *Mailing Add:* 106 Caldwell Dr Oak Ridge TN 37830

FONTANA, PETER R, b Berne, Switz, Apr 20, 35; m; c 2. THEORETICAL PHYSICS. *Educ:* Univ Miami, Ohio, MS, 58; Yale Univ, PhD(physics), 60. *Prof Exp:* Res assoc physics, Univ Chicago, 60-62; asst prof, Univ Mich, Ann Arbor, 62-67; assoc prof, 67-74, PROF PHYSICS, ORE STATE UNIV, 74- *Concurrent Pos:* Prof, Swiss Inst Technol, Lausanne, Switz, 74-75, 76 & 78; Univ Tubingen, WGermany, 82. *Mem:* Fel Am Phys Soc. *Res:* Optical information processing; acoustic interferometry; atomic radiative decay processes. *Mailing Add:* Dept Physics Ore State Univ Corvallis OR 97330

FONTANA, ROBERT E, b Brooklyn, NY, Nov 26, 15; m 45; c 3. SOLID STATE ELECTRONICS, CONTROL SYSTEMS. *Educ:* NY Univ, BEE, 39; Univ Ill, MS, 47, PhD(elec eng), 49. *Prof Exp:* Res scientist, Sandia Corp, 49-54, asst for nuclear develop, Hqs, Washington, DC, 54-58, chief nuclear applns, Hq, Air Res & Develop Command, 58-61, dir, Aerospace Res Labs, 61-66; head dept, 66-84, PROF EMER, DEPT ELEC COMP ENG, AIR FORCE INST TECHNOL, 84- *Concurrent Pos:* Ed joint newslett, Inst Elec & Electronics Engrs-Am Soc Eng Educ, 70-80. *Honors & Awards:* Legion Merit Medals for Contribution to Air Force Res & Develop, 66 & 69. *Mem:* Am Soc Eng Educ; fel Inst Elec & Electronics Engrs. *Res:* Solid state electronics; control systems; computer aided designs of flight control systems. *Mailing Add:* Dept Elec & Comp Eng Air Force Inst Tech Aflt/Eng Wright Patterson AFB OH 45433-6583

FONTANA, VINCENT J, b New York, NY, Nov 19, 23. IMMUNOLOGY, PEDIATRICS. *Educ:* Long Island Col Med, MD, 47. *Hon Degrees:* LHD, St John's Univ, 80. *Prof Exp:* Fel, NY Univ Hosp, 51-52, assoc prof, 57-69, PROF CLIN PEDIAT, NY UNIV COL MED, 69-; MED DIR & PEDIATRICIAN IN CHIEF, NY FOUNDLING HOSP CTR PARENT & CHILD DEVELOP, 62- *Concurrent Pos:* Dir pediat, St Vincent's Hosp & Med Ctr, 62-74, emer dir pediat, 74-; med ed, Missing/Abused. *Honors & Awards:* Dr Frank L Babbott Mem Award; Brandt F Steele Award. *Mem:* Fel Am Acad Pediat; Fel Am Acad Allergy; Am Fedn Clin Res; Fel NY Acad Med; Harvey Soc. *Res:* Pediatric allergy and immunology; child abuse and neglect; author of over 150 scientific publications. *Mailing Add:* Box 336 55 Main St Stony Brook NY 11790

FONTANELLA, JOHN JOSEPH, b Rochester, Pa, Dec 20, 45; m 78; c 2. SOLID STATE PHYSICS. *Educ:* Westminster Col, Pa, BS, 67; Case Inst Technol, MS, 69, PhD(physics), 71. *Prof Exp:* From asst prof to assoc prof, 74-84, PROF PHYSICS, US NAVAL ACAD, 84- *Mem:* Am Phys Soc; Inst Elec & Electronics Engrs. *Res:* Real and imaginary parts of the dielectric constant for solids as functions of temperature, pressure and audio frequency. *Mailing Add:* Dept Physics US Naval Acad Annapolis MD 21402-5026

FONTENELLE, LYDIA JULIA, b New Orleans, La, May 28, 38. BIOCHEMISTRY. *Educ:* La State Univ, Baton Rouge, BS, 60; Tulane Univ, PhD(biochem), 67. *Prof Exp:* Teacher high schs, La, 60-61; lab technician biochem, Tulane Univ, 61-63; asst prof, 69-73, ASSOC PROF BIOCHEM & PHARMACOL, COL PHARM, IDAHO STATE UNIV, 73- *Concurrent Pos:* Nat Inst Can fel, McEachern Lab, Univ Alta, 67-69; Idaho State Univ fac grant, 70-71. *Mem:* Am Chem Soc; NY Acad Sci. *Res:* Metabolism of drugs and the effects of drugs on control mechanisms of purine biosynthesis de novo. *Mailing Add:* Col Pharm Idaho State Univ Pocatello ID 83201

FONTENOT, JOSEPH PAUL, b Mamou, La, May 11, 27; m 46; c 6. ANIMAL NUTRITION. *Educ:* Univ Southwestern La, BS, 51; Okla State Univ, MS, 53, PhD(animal nutrit), 54. *Prof Exp:* Instr physiol & pharmacol, Okla State Univ, 54-55; asst prof animal husb, Miss State Univ, 55-56; from assoc prof to prof, 56-86, JOHN W HANCOCK JR PROF ANIMAL SCI, VA POLYTECH INST & STATE UNIV, 86- *Concurrent Pos:* Comt animal nutrit, Nat Acad Sci, 76-82, chmn, 79-82 & mem bd on agr, 84-89. *Honors & Awards:* Am Feed Mfrs Asn Nutrit Res Award; Morrison Award, Am Soc Animal Sci; Am Forage & Grassland Coun Medallion. *Mem:* AAAS; Am Soc Animal Sci (pres, 85); Am Inst Nutrit; Animal Nutrit Res Coun; Equine Nutrit & Physiol Soc (pres, 83-85). *Res:* Ruminant nutrition; nitrogen requirement and metabolism; metabolic disturbances; administration of hormones and other drugs; forage utilization; cellulose digestion; recycling of animal wastes by feeding. *Mailing Add:* Dept Animal Sci Va Polytech Inst & State Univ Blacksburg VA 24061

FONTENOT, MARTIN MAYANCE, JR, b Mamou, La, May 30, 44; m 66; c 5. ENVIRONMENTAL ENGINEERING, ENVIRONMENTAL CHEMISTRY. *Educ:* Southern Univ, BS, 66, MS, 74; La State Univ, ME, 78. *Prof Exp:* Chemist, Uniroyal, Inc, 66-69; process chemist, Monsanto Co, 69-71; staff environ engr, Kaiser Aluminum & Chem Co, 80-83; qual control mgr, Rollins Environ Serv, 83-84; chemist & environ engr, 71-80, SR STAFF ENVIRON ENGR, CIBA-GEIGY CORP, 84- *Concurrent Pos:* Res assoc, Southern Univ, 76-84, adj prof environ chem, 79-; mem, Nat Alliance Bus, 78-79 & Chem Eng Prod Res Panel, 79; proj engr, W F Snyder & Assocs, 84-; deleg, USSR Citizen Ambassador Prog People to People Int, 90. *Mem:* Air & Waste Mgt Asn; Water Pollution Control Asn; Am Inst Chemists. *Res:* Environmental analytical chemistry; environmental chemistry and engineering solid waste management; incineration; recycling; waste minimization; ground water and soil assessments. *Mailing Add:* 9024 Staring Ct Baton Rouge LA 70810

FONTEYN, PAUL JOHN, b Williamstown, Mass, Aug 9, 46; m 72; c 2. PLANT ECOLOGY. *Educ:* Univ San Francisco, BS, 68; Univ Calif, Santa Barbara, MA, 74, PhD(biol), 78. *Prof Exp:* ASSOC PROF PLANT ECOL, SOUTHWEST TEX STATE UNIV, 78- *Mem:* Ecol Soc Am; Sigma Xi. *Res:* Study of individual species distributions, the physiological and abiotic factors influencing these distributions and the dynamic interrelationships such as competition among species. *Mailing Add:* Dept Biol SW Tex State Univ San Marcos TX 78666

FONTHEIM, ERNEST GUNTER, b Berlin, Ger, Oct 23, 22; US citizen; m 50; c 2. SPACE PLASMA PHYSICS. *Educ:* Southwest Mo State Col, AB & BS, 50; Lehigh Univ, MS, 52, PhD(physics), 60. *Prof Exp:* Assoc res physicist, Radiation Lab, Univ Mich, Ann Arbor, 60-62, RES PHYSICIST, SPACE PHYSICS RES LAB, 62- *Concurrent Pos:* Sr res assoc, Goddard Space Flight Ctr, Nat Acad Sci, 72-73 & 83-84; staff scientist, Planetary Atmospheres, NASA, 85. *Mem:* Am Phys Soc; Am Geophys Union; Int Union Radio Sci; Int Asn Geomag & Aeronomy. *Res:* Physics of the ionosphere, magnetosphere; solar wind interaction with planets; kinetic theory. *Mailing Add:* Space Phys Res Lab Univ Mich 2455 Hayward St Ann Arbor MI 48109-2143

FONTIJN, ARTHUR, b Amsterdam, Neth, Apr 3, 28; nat US; div; c 1. PHYSICAL CHEMISTRY, CHEMICAL KINETICS. *Educ:* Univ Amsterdam, BSc, 49. *Hon Degrees:* DSC, Univ Amsterdam, 57. *Prof Exp:* Nat Res Coun Can fel, Univ Sask, 55-57; res assoc upper atmosphere chem group, McGill Univ, 57-60; phys chemist, Aerochem Res Labs, Inc, 60-68, head reaction kinetics group, 68-79; sr vis fel, Queen Mary Col, London Univ, 79-81; actg chmn, 84-86, PROF, DEPT CHEM ENG & ENVIRON ENG,

RENSSELAER POLYTECH INST, 81- *Concurrent Pos:* Mem, Int Union Pure & Appl Chem, subcomt plasma chem, 75-83 & exec comt, Phys Chem Div, Am Chem Soc, 79-82, publ subcomt, Int Combustion Symp, 73-; vis prof, Univ Bordeaux, 88; sabbatical vis, Oxford Univ, 89. *Honors & Awards:* Silver Medal, Combustion Inst, 74; Am Chem Soc Award, 85. *Mem:* Royal Soc Chem; Am Chem Soc; Am Phys Soc; Royal Dutch Soc Chem; Am Inst Chem Engrs; Sigma Xi. *Res:* Gas kinetics high-temperature reactions; chemi-ionization and chemiluminescence; atmospheric chemistry; combustion. *Mailing Add:* Dept Chem Eng Rensselaer Polytech Inst Troy NY 12180-3590

FOODEN, JACK, b Chicago, Ill, May 21, 27; wid; c 2. ZOOLOGY. *Educ:* Chicago Teachers Col, MEd, 56; Univ Chicago, MA, 51, PhD(zool), 60. *Prof Exp:* Teacher, Chicago Pub Schs, 53-56; NIH res fel primate taxon, Univ Chicago & Chicago Natural Hist Mus, 60-62; emer prof zool, Chicago State, 62-84; RES ASSOC, DIV MAMMALS, FIELD MUS NATURAL HIST, 64 - *Concurrent Pos:* Indo-Am Fel, 79-80; grantee, Comt on Scholarly Commun with People's Repub China, 85-86. *Mem:* AAAS; Am Soc Mammal; Int Primatol Soc. *Res:* Primatology; mammalogy. *Mailing Add:* Div Mammals Field Mus Natural Hist Chicago IL 60605

FOOR, W EUGENE, b Wood, Pa, Feb 7, 36; m 58; c 2. ZOOLOGY, PARASITOLOGY. *Educ:* Shippensburg State Col, BSE, 59; Univ Mass, Amherst, PHD(zool), 66. *Prof Exp:* Teacher high schs, Pa, 59-62; fel parasitol, Med Sch, Tulane Univ, 66-67, instr, Sch Pub Health & Trop Med, 67-68, asst prof, 68-70; assoc prof, 70-75, prof biol, Wayne State Univ, 75-78; CHMN, DIV NATURAL SCI, UNIV PITTSBURG, JOHNSTOWN, 88- *Mem:* Am Soc Cell Biol; Am Soc Parasitol. *Res:* Ultrastructural studies of reproductive cells in parasitic nematodes; oogenesis, vitellogenesis and shell formation about newly fertilized eggs as well as the mechanism of oocyte penetration employed by spermatozoa. *Mailing Add:* Div Natural Sci Univ Pittsburg at Johnstown Johnstown PA 15904

FOOS, KENNETH MICHAEL, b Bellefontaine, Ohio, Jan 16, 43; m 66; c 1. BOTANY, MYCOLOGY. *Educ:* Ohio State Univ, BS, 65, MS, 70, PhD(bot), 72. *Prof Exp:* Sci teacher biol & chem, Bowling City Schs, 66-67 & bot, Jefferson Local Schs, 67-70; vis asst prof bot, Ohio State Univ, 72-73; from asst prof to assoc prof biol, Lake Erie Col 73-83; PROF & CHMN DEPT BIOL, IND UNIV EAST, 83- *Concurrent Pos:* Grant proj dir & Lake Erie Col, 76-78. *Mem:* Nat Asn Biol Teachers; Brit Mycol Soc; Am Soc Microbiologists; Mycol Soc Am; Sigma Xi; Soc Col Sci Teachers. *Res:* Effects of ecological factors on physiology and morphology of zygomycetous fungi; geographic distribution of fungi; epizootiologic effects of fungi. *Mailing Add:* Dept Biol Ind Univ 2325 Chester Blvd Richmond IN 47374

FOOS, RAYMOND ANTHONY, b Bowling Green, Ohio, Sept 30, 28; m 53; c 11. PHYSICAL CHEMISTRY. *Educ:* Xavier Univ, Ohio, BS, 50, MS, 53; Iowa State Univ, PhD(chem), 54. *Prof Exp:* Res chemist, AEC, 51-54; group leader, Union Carbide Metal Co, 54-57; supvr metals & inorg chem, US Indust Chem Co, 57-59, supvr polymer chem, 59-60; mgr extraction & oxide, Brush Beryllium Co, 60-66, mgr, Metal Oxide Dept, 66-69, dir corp res & develop, 69-70, vpres res & develop, 70-74, sr vpres, Friction & Crystal Prod, 74-76, sr vpres, 76-80, exec vpres, 80-86, PRES, BERYLLIUM PROD, BRUSH WELLMAN INC, 87- *Mem:* Am Chem Soc; Am Inst Metall & Petrol Engrs; Am Ceramic Soc; Soc Automotive Engrs; Int Soc Hybrid Microelectronics. *Res:* Extractive metallurgy of transition elements; sodium chemistry; electro chemistry; patents; liquid extraction; polymers; beryllium chemistry; manufacturing; ceramics; beryllium product metallurgy; powder technology; administration. *Mailing Add:* 30972 Pinehurst Dr Westlake OH 44145-1770

FOOS, ROBERT YOUNG, b Philadelphia, Pa, Nov 20, 22; m 45; c 4. OPHTHALMOLOGY, PATHOLOGY. *Educ:* Univ Calif, Davis, BS, 51, DVM, 53; Univ Calif, Los Angeles, MD, 63. *Prof Exp:* Nat Inst Neurol Dis & Stroke spec fel, 65-66; asst prof, 66-70, assoc prof, 70-76, PROF PATH, SCH MED, UNIV CALIF, LOS ANGELES, 76- DIR OPHTHALMIC PATH LABS, JULES STEIN EYE INST, 67- *Mem:* AAAS. *Res:* Diseases of the retina; experimental ophthalmic pathology. *Mailing Add:* Dept Path Jules Stein Eye Inst Univ Calif Los Angeles CA 90024

FOOSE, RICHARD MARTIN, b Lancaster, Pa, Oct 9, 15; m 43; c 4. GEOLOGY. *Educ:* Franklin & Marshall Col, BS, 37; Northwestern Univ MS, 39; Johns Hopkins Univ, PhD(struct & econ geol), 42, Amherst Col, MS, 64. *Prof Exp:* Asst instr geol, Northwestern Univ, 37-39; asst geologist, Pa State Topog & Geol Surv, 39-42, assoc geologist, 42-43, sr geologist, 43-46; prof geol & head dept, Franklin & Marshall Col, 46-57; chmn dept earth sci, Stanford Res Inst, 57-63; prof geol & chmn dept, Amherst Col, 63-81, Hitchcock prof geol, 81-; RETIRED. *Concurrent Pos:* Coop geologist, Pa State Turnpike Comn, 41; consult geologist, 42-, chmn, Indust Mining Div, Soc Mining Engrs, 50, Mineral Econ Div, 63; deleg, Int Geol Cong, 52, 56, 60, 68, 80, 84; Ford fel, Stanford Univ, 55-56; mem tech work group 2, Nuclear Test Ban Discussions, Geneva, 59; NSF sr fel, Swiss Fed Inst Technol, 62-63; vis prof, Univ Vienna, 68 & Am Univ Beirut, 69; Nat Acad Sci exchange fel, USSR, 69 & 76, Bulgaria, 72; mem, Nat Res Coun, 71-75; mem adv comt, Nat Acad Sci, USSR & Eastern Europe, 72-75. *Mem:* Am Inst Prof Geologists (secy, 68); fel AAAS; fel Geol Soc Am; Soc Econ Geol; fel Explorers Club; Soc Mining Engrs; Asn Eng Geologists. *Res:* Origin and occurrence of hialumina clays; manganese minerals of Pennsylvania; groundwater geology; tectonics of middle Rocky Mountains; catastrophic sinkhole development; tectonic evolution of the Mediterranean Basin. *Mailing Add:* No 7 Hunter Hill Circle Amherst Col Amherst MA 01002

FOOSE, THOMAS JOHN, b Waynesboro, Pa, Mar 7, 45; m 77; c 4. CONSERVATION BIOLOGY, POPULATION BIOLOGY. *Educ:* Princeton Univ, BA, 69; Univ Chicago, PhD(biol), 82. *Prof Exp:* Res consult, Philadelphia Zool Soc, 75-76; demog consult, Inventory Species Inventory Syst, 76; res asst, Cornell Univ, 77; zool curator, Okla City Zoo, 77-80; CONSERV COORDR, AM ASN ZOOL PARKS & AQUARIUMS, 81- *Concurrent Pos:* Vis researcher, Philadelphia Zool Soc, 75-76 & Toronto Zool Soc, 77; field researcher, Univ Chicago, 76-77; resident researcher, Okla City Zoo, 77-80. *Res:* Population biology, both analysis and management, of endangered species; development of demographic and genetic strategies and programs to propagate and preserve endangered species in captivity; trophic biology of ungulates. *Mailing Add:* 2813 Crater Ct Burnsville MN 55337

FOOTE, BENJAMIN ARCHER, b Delaware, Ohio, Oct 25, 28; m 54; c 4. ENTOMOLOGY. *Educ:* Ohio Wesleyan Univ, BA, 50; Ohio State Univ, MS, 52; Cornell Univ, PhD, 61. *Prof Exp:* Asst entom, Ohio State Univ, 50-52; asst limnol, Cornell Univ, 54-58; asst entomologist, Univ Idaho, 58-60; from asst prof to assoc prof, 61-72, PROF ENTOM, KENT STATE UNIV, 72- *Concurrent Pos:* NSF res grant, 69-; Nat Geog Soc grant, 71, 84. *Mem:* Entom Soc Am; Ecol Soc Am; NAm Benthological Soc. *Res:* Ecology of acalyptrate Diptera; biological control of ragweeds; biology of root nodule flies, acalyptrate Diptera of coastal habitats. *Mailing Add:* Dept Biol Sci Kent State Univ Main Campus Kent OH 44242

FOOTE, BOBBIE LEON, b Spavinaw, Okla, Jan 24, 40; m 68; c 3. INDUSTRIAL ENGINEERING, OPERATIONS RESEARCH. *Educ:* Univ Okla, BS, 61, MA, 63, PhD(eng), 67. *Prof Exp:* Instr math, NE Okla A&M Jr Col, 63-65; prof eng, Sch Indust Eng, Univ Okla, 67-78, dir, 77-81, prof indust eng, 82-87, dir, 87-89, ASSOC DIR, OKLA CTR INTEGRATED DESIGN & MFG, SCH INDUST ENG, UNIV OKLA, 90- *Concurrent Pos:* Engr, US Post Off Dept, 68; dir modeling studies, Univ Tex, 74-75; asst planning officer, Asst Secy Off, US Navy, Philadelphia, 81-82, 89. *Honors & Awards:* Merrick Found Award, 84. *Mem:* Opers Res Soc Am; Inst Mgt Sci; Int Inst Indust Engrs (secy, 73). *Res:* Plant and production planning; quality control; energy distribution; energy systems; math programming. *Mailing Add:* Sch Indust Eng 202 W Boyd Norman OK 73019

FOOTE, CARLTON DAN, b State Center, Iowa, Jan 16, 35; m 59; c 3. BIOCHEMISTRY, ORGANIC CHEMISTRY. *Educ:* Cent Col, BA, 57; Univ Ill, PhD(biochem), 63. *Prof Exp:* NIH fel, 63-64; asst prof chem, Union Col, Ky, 64-65; asst prof, 65-70, ASSOC PROF CHEM, EASTERN ILL UNIV, 70- *Mem:* Am Chem Soc; Sigma Xi. *Res:* Steroid biosynthesis; mechanisms of enzymes; cellular differentiation. *Mailing Add:* 2150 S 11th St Charleston IL 61920

FOOTE, CHRISTOPHER S, b Hartford, Conn, June 5, 35; m 60; c 2. ORGANIC CHEMISTRY. *Educ:* Yale Univ, BS, 57; Harvard Univ, AM & PhD(org chem), 62. *Prof Exp:* From instr to assoc prof, 61-69, chmn dept, 78-81, PROF CHEM, UNIV CALIF, LOS ANGELES, 69- *Concurrent Pos:* Alfred P Sloan Found fel, 65-67; J S Guggenheim Found fel, 67-68; consult, Clorox. *Honors & Awards:* Leo Hendrick Baekeland Medal, Am Chem Soc, 75. *Mem:* Fel AAAS; Am Chem Soc; Royal Soc Chem; Ger Chem Soc; Am Soc Photobiol. *Res:* Organic photochemistry; chemical generation of molecules in excited states; reactions of singlet oxygen; photodynamic effect. *Mailing Add:* Dept Chem Univ Calif Los Angeles CA 90024

FOOTE, FLORENCE MARTINDALE, b Montague, Mass, June 28, 11; wid. ANATOMY, EMBRYOLOGY. *Educ:* Mt Holyoke Col, AB, 32; AM, 34; Univ Iowa, PhD(embryol, endocrinol), 40. *Prof Exp:* Asst zool, Mt Holyoke Col, 32-34, instr, 35-38; instr, Univ Del, 40-41; instr biol, Wagner Col, 42-45, asst prof, 46-47; asst prof zool, Southern Ill Univ, 47-50, lectr physiol, 50-62, assoc prof, 63-68, actg chmn dept, 71-72, 75-76, prof, Sch Med, 71-76, prof physiol, 68-76, EMER PROF PHYSIOL, SOUTHERN ILL UNIV, 76- *Concurrent Pos:* Guest investr lab exp embryol, Col France, 60-61. *Mem:* Emer mem Am Asn Anat; emer mem Soc Develop Biol; emer mem Int Inst Embryol. *Res:* Vertebrate embryology. *Mailing Add:* 2012 Woodriver Dr Apt 1 Carbondale IL 62901

FOOTE, FREEMAN, b Orange, NJ, Nov 8, 08; m 39; c 1. GEOLOGY. *Educ:* Princeton Univ, AB, 31. *Prof Exp:* Asst geol, Columbia Univ, 34-37; from instr to prof geol, 37-68, chmn dept, 64-67, Edward Brust prof geol & mineral, 68-74, EMER PROF GEOL & MINERAL, WILLIAMS COL, 74- *Mem:* AAAS; fel Geol Soc Am; Nat Asn Geol Teachers (secy, 57-60); Sigma Xi. *Res:* Petrology; structural geology. *Mailing Add:* 1550 Cold Spring Rd Williamstown MA 01267

FOOTE, GARVIN BRANT, b Murray, Utah, Dec 30, 42; m 67; c 2. SEVERE STORMS, CLOUD PHYSICS. *Educ:* Univ Ariz, BS, 64, MS, 67, PhD(atmospheric sci), 71. *Prof Exp:* Res assoc cloud physics, Univ Ariz, 67-70; fel, Nat Ctr Atmospheric Res, 70-71, scientist, 71-74, proj leader cloud physics, 74-86, & 84-89, mgr, Field Observ Facil, 86-87, GROUP HEAD, RES APPLN PROG, NAT CTR ATMOSPHERIC RES, 90- *Concurrent Pos:* Mem, Comte Atmospheric Electricity, Am Meteorol Soc, 72-73; affil prof, Colo State Univ, 78; sci adv comn, Alberta Res Coun, 74-79; Weather Modification Adv Comn, State Colo, 77-85; comt on Planned & Inadvertant Weather Modification, Am Meteorol Soc, 88-86; ed, J Atmospheric Sci, 85-; lectr, Lanzhou Univ, PRC, 85; consult, World Meteorol Orgn, Nat Acad Sci, Int Comn Cloud Physics, 80-88. *Mem:* Fel Am Meteorol Soc. *Res:* Severe local storms; hail; weather modification; radar meteorology; short-range forecasting. *Mailing Add:* Nat Ctr Atmospheric Res PO Box 3000 Boulder CO 80307

FOOTE, JAMES HERBERT, b Tacoma, Wash, Dec 13, 29; m 54; c 2. PLASMA PHYSICS, ELEMENTARY PARTICLE PHYSICS. *Educ:* Univ Calif, Berkeley, AB, 53, PhD(physics), 61. *Prof Exp:* PHYSICIST, LAWRENCE LIVERMORE NAT LAB, 60- *Mem:* Am Phys Soc. *Res:* Experimental research and related computer applications; controlled magnetic fusion and high temperature plasma physics, specializing in plasma diagnostics. *Mailing Add:* Lawrence Livermore Nat Lab L0637 PO Box 5511 Livermore CA 94550

FOOTE, JOE REEDER, b Amarillo, Tex, Aug 17, 19; m 49; c 4. APPLIED MATHEMATICS. *Educ:* Tex Tech Col, BS, 40; Mass Inst Technol, PhD(math), 49. *Prof Exp:* Instr math, Univ Tex, 40-41, Univ Okla, 45-46 & Mass Inst Technol, 46-49; asst prof, Iowa State Col, 49-51; mathematician, Wright-Patterson AFB, 51-53; asst prof math, Univ Okla, 53-57; assoc prof,

Div Eng Sci, Purdue Univ, 57-58; prof math & dir, Holloman Grad Ctr, NMex, 58-66; prof math, Univ Mo-Rolla, 66-70, chmn dept, 67-68; chmn dept, 70-77, PROF MATH, UNIV NEW ORLEANS, 70- *Concurrent Pos:* Mem, Inst Math Sci, NY Univ, 56-57; consult, US Air Force Missile Develop Ctr, 56-65. *Mem:* Math Asn Am; Soc Indust & Appl Math. *Res:* Flight dynamics; differential equations; optimization; calculus of variations; operations research for ICBM defense. *Mailing Add:* Dept Math Univ New Orleans PO Box 593 New Orleans LA 70148

FOOTE, JOEL LINDSLEY, b Cleveland, Ohio, Jan 11, 28; m 51; c 2. BIOCHEMISTRY. *Educ:* Miami Univ, BS, 52; Case Inst Technol, PhD(org chem), 60. *Prof Exp:* Teacher pub schs, Ohio, 52-56; res assoc biochem, Univ Mich, 60-62, part-time instr, 62-65, asst res biochemist, Ment Health Res Inst, 63-65; asst assoc prof, dept chem, Western Mich Univ, 65-89; RETIRED. *Concurrent Pos:* NSF fel, 60-62; USPHS trainee, 62-64. *Mem:* Am Chem Soc; Am Soc Biochem & Molecular Biol; AAAS. *Res:* Lipid biochemistry; effect of lead on neurolipids. *Mailing Add:* Dept Chem Western Mich Univ Kalamazoo MI 49008

FOOTE, KENNETH GERALD, b Cleveland, Ohio, Mar 24, 35; m 59; c 3. BIOLOGY EDUCATION, BRYOPHYTES & LICHENS. *Educ:* Hiram Col, AB, 58; Univ Wis, MS, 62, PhD(bot), 63. *Prof Exp:* Asst prof biol, Sheboygan Ctr, Univ Wis, 63-66; assoc prof, 66-78, PROF BIOL, UNIV WIS-EAU CLAIRE, 78- *Mem:* Am Bryol & Lichenological Soc; Brit Lichenological Soc; Am Inst Biol Sci; Nat Asn Biol Teachers; Sigma Xi. *Res:* Ecology of saxicolous lichen and bryophyte communities; Bryophytes and lichens of Wisconsin; methods and investigations in biology teaching. *Mailing Add:* Univ Wis-Eau Claire Eau Claire WI 54701

FOOTE, MURRAY WILBUR, b Charlotte, Vt, Mar 22, 16; m 40; c 2. BIOCHEMISTRY. *Educ:* Univ Vt, BS, 38, MS, 50; Univ Conn, PhD, 54. *Prof Exp:* Asst chemist, 40-51, from asst prof biochem to assoc prof biochem, 53-69; assoc biochemist, 53-69, ASSOC PROF MICROBIOL & BIOCHEM, UNIV VT, 69-, ASSOC BIOCHEMIST, AGR EXP STA, 77- *Mem:* Am Soc Plant Physiol. *Res:* Minor elements in plant metabolism; plant proteins and nitrogen metabolism in plants. *Mailing Add:* Microbiol Biochem Col Agr Univ Vt Charlotte VT 05445

FOOTE, RICHARD HERBERT, b Bozeman, Mont, May 2, 18; m 43, 64; c 5. ENTOMOLOGY. *Educ:* Mont State Col, BS, 42; Johns Hopkins Univ, DSc(parasitol), 52. *Prof Exp:* Asst entom, Univ Minn, 42-43; entomologist, USPHS, 47-49; entomologist, Johns Hopkins Univ, 49-52; entomologist, Entom Res Div, Sci & Educ Admin-Agr, USDA, 52-67, asst chief insect identification & parasite introd res br, 67-72, chief syst entom lab, Insect Identification & Beneficial Insect Introd Inst, 72-76, res entomologist, 76-83; RETIRED. *Concurrent Pos:* Mem coun, Biol Sci Info; mem ed bd, Abstracts Entom. Honors &. *Mem:* Entom Soc Am; Sigma Xi. *Res:* Taxonomy of Tephritidae; information storage and retrieval for biology. *Mailing Add:* Box 166 Lake of the Woods Locust Grove VA 22508

FOOTE, RICHARD MARTIN, b Chester, Eng, June 22, 50; Can citizen; m 81. MATHEMATICS. *Educ:* Univ Toronto, BSc, 72; Univ Cambridge, Eng, PhD(math), 76. *Prof Exp:* Res fel math, Trinity Col, Cambridge, 75-76; instr, Calif Inst Technol, 76-78; res fel, Trinity Col, Cambridge, 78-79; asst prof, Rutgers Univ, 79-80; asst prof math, Univ Minn, 80-83; asst prof, 81-83, ASSOC PROF MATH, UNIV VT, 83- *Mem:* Asn Women in Math. *Res:* Finite simple groups; number theory. *Mailing Add:* Dept Math Univ Vt 16 Colchester Ave Burlington VT 05405

FOOTE, ROBERT HUTCHINSON, b Gilead, Conn, Aug 20, 22; m 46; c 2. ANIMAL PHYSIOLOGY. *Educ:* Univ Conn, BS, 43; Cornell Univ, MS, 47, PhD(animal breeding), 50. *Prof Exp:* Asst, 46-50, from asst prof to assoc prof, 50-63, PROF ANIMAL PHYSIOL, CORNELL UNIV, 63-, JACOB GOULD SCHURMAN PROF, 80- *Concurrent Pos:* Fulbright scholar, Denmark, 58-59; vis prof, Finland, 74 & Univ Calif, 78; vis scholar, Japan, 88. *Honors & Awards:* NY Farmers' Award, 69; Award, Endocrinol & Physiol, 70; Award, Am Dairy Sci Asn, 70; Upjohn Physiol Award, 85; Outstanding Andrologist Award, Am Sci Andrology, 85; Distinguished Serv Award, USDA, 88. *Mem:* Fel AAAS; Am Fertil Soc; Am Soc Animal Sci; Am Dairy Sci Asn; Soc Study Reproduction (pres, 86-87); Am Andrology Soc; Cryobiology Soc; Soc Study on Fertil. *Res:* Superovulation; cellular preservation and cryogenic effects on sperm cells and embryos; embryo culture; fertility, embryonic mortality and congenital defects; aging and reproductive failure; biotechnology of gametes. *Mailing Add:* Dept Animal Sci Cornell Univ Ithaca NY 14853-4801

FOOTE, ROBERT S, b Decatur, Ill, June 6, 22; m 53; c 4. PHYSICS. *Educ:* Univ Ill, BS, 48, MS, 49. *Prof Exp:* Mem tech staff, Nat Bur Standards, 49-55; sect head, Cent Res Lab, Tex Instruments, Inc, 57-60, chief geonuclear opers, Sci Serv Div, 61-68; pres, Geosensors, Inc, 69-73 & Geodata Int Inc, 73-84; PRES, GEOSCI & TECHNOL, INC, 84- *Mem:* Am Phys Soc; sr mem Inst Elec & Electronics Engrs. *Res:* Gamma ray spectroscopy; advanced electronic technology; instrumentation. *Mailing Add:* 1328 Etain Rd Irving TX 75060

FOOTE, VERNON STUART, JR, b East Grand Rapids, Mich, Dec 13, 24; m 55; c 3. PLASMA PHYSICS, ENGINEERING. *Educ:* George Washington Univ, BME, 51. *Prof Exp:* Armament engr ASW Ord, US Navy Bur Aeronaut, 51-52; sr designer ord, Nav Ord Lab, 52, 53-54; jr mech engr flight simulation, Melpar, Inc, 53; assoc engr, Washington Technol Assocs, Inc, 54-58; sr mech engr, Nat Co, Inc, Nat Radio Co, 58-59; mem prof tech staff, Plasma Physics Lab, Princeton Univ, 60-80, sr mem prof tech staff, 80-86; RETIRED. *Res:* Mechanisms; control theory; optics; electrooptics; hydraulics. *Mailing Add:* Castle Hill Rd RR 1, Box 34 Wilmington VT 05363

FOOTE, WARREN CHRISTOPHER, b Orderville, Utah, Oct 6, 27; m 49; c 6. ANIMAL PHYSIOLOGY. *Educ:* Utah State Univ, BS, 54; Univ Wis, MS, 55, PhD, 58. *Prof Exp:* From asst prof to assoc prof, 58-69, PROF ANIMAL PHYSIOL, UTAH STATE UNIV, 69- *Concurrent Pos:* Dir, Int Sheep & Goat Inst, 72-86; consult, US AID, Bolivia, San Marcos Univ Rockefeller Found, Portugal Min Agr & Off Int Coop & Develop. *Mem:* Am Soc Animal Sci. *Res:* Physiology and endocrinology of reproduction and genetics; mechanisms involved; influencing factors, their modification and control; systems of production. *Mailing Add:* Dept Animal Sci Utah State Univ Logan UT 84322-4815

FOOTE, WILFORD DARRELL, b Kanab, Utah, Jan 9, 31; m 53; c 7. ANIMAL PHYSIOLOGY. *Educ:* Utah State Univ, BS, 53; Univ Wis, MS, 56, PhD, 59. *Prof Exp:* Asst genetics & animal husb, Univ Wis, 55-59; PROF ANIMAL SCI & PHYSIOLOGIST, UNIV NEV, RENO, 59- *Mem:* AAAS; Am Soc Animal Sci; Soc Study Reproduction; Int Embryo Transfer Soc; Am Dairy Sci Asn. *Res:* Physiology of reproduction. *Mailing Add:* Dept Animal Sci Univ Nev Reno NV 89557

FOOTE, WILSON HOOVER, b Nephi, Utah, Jan 30, 20; m 52; c 2. AGRONOMY. *Educ:* Utah State Univ, BS, 42; Univ Minn, MS, 46, PhD, 48. *Prof Exp:* Field foreman, Nephi Dryland Exp Sta, Utah State Univ, 38-42; asst agronomist, Spec Guayule Res Proj, Bur Plant Indust, Soils & Agr Eng, USDA, 43-44; from asst prof to prof agron & agronomist, 48-85, assoc dir sta, 70-85, EMER PROF CROP SCI, AGR EXP STA, ORE STATE UNIV, 85- *Concurrent Pos:* Chmn, Nat Plant Germplasm comt, 79-84, mem Nat Plant Genetic Resources Bd, 79-84, bd trustees, Consortium Inst Develop, mem USDA, State Agr Exp Sta Comt Nine, 76-78, pres Western Soc Crop Sci, 62, secy, Agr Res Found, 58-, Escop Seed Policy Comt, 78-84. *Mem:* Fel Am Soc Agron. *Res:* Plant breeding; cereal crop production; research administration. *Mailing Add:* 830 NW Witham Dr Corvallis OR 97330

FOPEANO, JOHN VINCENT, JR, b Ann Arbor, Mich, Jan 29, 28; m 50; c 5. BIOCHEMISTRY. *Educ:* Yale Univ, BA, 50; Univ Mich, MS, 52, PhD(biochem), 55. *Prof Exp:* Instr biochem, State Univ NY, Buffalo, 54-57, from asst to assoc prof, 57-70, consult, Dept Surg, 54-57; prof med technol & chmn dept, Clin Ctr, Sch Health & Related Prof, State Univ NY Buffalo, 70-85; RETIRED. *Concurrent Pos:* Consult, Mt St Mary's Hosp, Lewiston, NY, 61-78; dir sub-bd I student health serv clin lab, State Univ NY Buffalo, 73-85; dir, Health Care Plan Clin Lab, West Seneca, NY, 78- *Mem:* Am Chem Soc; Am Asn Clin Chemists; Am Soc Med Technol. *Mailing Add:* Kraus Rd Clarence NY 14031

FORAL, RALPH FRANCIS, engineering mechanics, materials science; deceased, see previous edition for last biography

FORAN, JAMES MICHAEL, b Milwaukee, Wis, Oct 20, 42; m 71; c 1. MATHEMATICS. *Educ:* Marquette Univ, BA, 65; Univ Wis-Milwaukee, MS, 69, PhD(math), 73. *Prof Exp:* Teaching asst math, Univ Wis-Milwaukee, 67-73, lectr, 73-74; asst prof, 74-80, ASSOC PROF MATH, UNIV MO-KANSAS CITY, 80- *Concurrent Pos:* Consult, Archit & Design Graphics Corp, 73-78; assoc ed, Real Anal Exchange, 76-78. *Mem:* Am Math Soc. *Res:* Functions of a real variable, integration and differentiation theory, measure theory, set theory particularly subsets of the line. *Mailing Add:* Dept Math Univ Mo Kansas City MO 64110

FORBES, ALBERT RONALD, b Victoria, BC, Oct 16, 31. ENTOMOLOGY. *Educ:* Univ BC, BA, 52, Ore State Univ, MS, 55; Univ Calif, PhD(entom, zool), 63. *Prof Exp:* Asst entomologist, Field Crop Insect Lab, Victoria, BC, 52-55; assoc entomologist, Can Dept Agr, 56-67, res scientist, 67-73, sr res scientist, 73-91, asst dir, Res Sta, 84-91; RETIRED. *Mem:* Entom Soc Am; fel Entom Soc Can,. *Res:* Economic entomology; systematics of aphididae; ultrastructure of insects. *Mailing Add:* 3275 Capilano Rd North Vancouver BC V7R 4H4 Can

FORBES, ALLAN LOUIS, b Richmond, Va, July 28, 28; m 54; c 3. INTERNAL MEDICINE. *Educ:* McGill Univ, BSc, 49; Med Col Va, MD, 53, MS, 64; Nat War Col, dipl, 68. *Prof Exp:* Intern, Montreal Gen Hosp, 53-54; from jr asst resident to sr asst resident internal med, Med Col Va, 54-56; assoc med & attend physician, Med Ctr, Univ Colo, 56-58; clin investr, Vet Admin Hosp, Richmond, Va, 58-61; asst dir med prog, Interdept Comt Nutrit Nat Defense, NIH, 61-63; med officer, Dept Army, 63-67, chief sci anal br, Life Sci Div, 68-70; dir nutrit bur, Health Protection Br, Ottawa, Ont, 73-74; dept dir, div nutrit, Bur Foods, 70-73, assoc dir, Nutrit & Consumer Sci, 74-79, DIR, OFF NUTRIT & FOOD SCI, CTR FOOD SAFETY & APPL NUTRIT, FOOD & DRUG ADMIN, 79- *Concurrent Pos:* Intern, King Edward VII Mem Hosp, Bermuda, 53-54; chief clin physiol br, Physiol Div, US Army Med Res & Nutrit Lab, Denver, 56-58; lectr, Med Col Va, 58-73; assoc ed, Am J Clin Nutrit, 79-81; chmn, Food Standards Comn, Int Union Nutrit Sci, 78- *Honors & Awards:* William Branch Porter Award Internal Med, 53. *Mem:* Nutrit Soc Can; Am Inst Nutrit; Am Soc Clin Nutrit; Am Fedn Clin Res. *Res:* Clinical nutrition and metabolic disease; international and environmental medicine; food science and technology, food microbiology, food behavior and food marketing. *Mailing Add:* 11312 Farmland Dr Rockville MD 20852

FORBES, CHARLES EDWARD, b Baltimore, Md, Sept 24, 44; m 71. NUCLEAR MAGNETIC RESONANCE, ELECTRON SPIN RESONANCE. *Educ:* Franklin & Marshall Col, AB, 66; Mass Inst Technol, PhD(inorg chem,), 71. *Prof Exp:* Res fel phys chem, Leicester Univ, Eng, 72-73; teaching intern phys chem, Syracuse Univ, 73-74; sr anal chemist, W R Grace & Co, 74-75; res chemist, Allied Corp, 75-79, sr res chemist, 79-86; sr res chemist, 86-87, STAFF SCIENTIST, CELANESE RES CO, 87- *Mem:* Sigma Xi. *Res:* Electron spin resonance and nuclear magnetic resonance; applications of magnetic resonance to inorganic and polymeric materials. *Mailing Add:* 38 Park St 6A Florham Park NJ 07932

FORBES, DONNA JEAN, b Moline, Ill, June 12, 42; m 67; c 2. NEUROSCIENCE. *Educ:* Monmouth Col, BA, 64; Univ Wis-Madison, MS, 67, PhD(physiol), 71. *Prof Exp:* Med res assoc neurosci, Galesburg State Res Hosp, Ill, 64-65; NIH fel, Dept Anat, Univ Wis-Madison, 70-73, instr, 73; asst prof, 73-81, ASSOC PROF ANAT, UNIV MINN, DULUTH, 81- *Concurrent Pos:* Consult & lectr, Col St Scholastica. *Mem:* Sigma Xi; Soc Neurosci; Cajal Club; Am Asn Anatomists; Asn Res Vision & Ophthal. *Res:* Herpes Simplex infections of eye studied in tissue culture model; development of the rat somatosensory system. *Mailing Add:* Dept Anat Univ Minn Sch Med Duluth MN 55812

FORBES, GILBERT BURNETT, b Rochester, NY, Nov 9, 15; m 39; c 2. ENDOCRINOLOGY, NUTRITION. *Educ:* Univ Rochester, AB, 36, MD, 40. *Prof Exp:* Intern pediat, Strong Mem Hosp, Rochester, 40-41; asst pediatrician, St Louis Children's Hosp, 40-43, resident physician, 41-43; from instr to assoc prof pediat, Washington Univ, 43-50; prof & chmn dept, Univ Tex Southwestern Med Sch, 50-53; assoc prof, 53-57, PROF PEDIAT, SCH MED & DENT, UNIV ROCHESTER, 57-, PROF RADIATION BIOL & BIOPHYS, 70- *Concurrent Pos:* Physician, St Louis Children's Hosp, 43-46, 47-50; chief pediatrician, Los Alamos Hosp, 46-47; chief staff pediat, Parkland City-County Hosp & med dir, Children's Med Ctr, Dallas, 50-53; assoc pediatrician, Strong Mem Hosp, 53-57, pediatrician, 57-; consult hosps; mem comt infant nutrit, Food & Nutrit Bd, Nat Res Coun; mem sci adv comt, Nutrit Found; res fel, Oxford Univ, 70-71; Nat Inst Child Health & Human Develop res career award; chief ed, Am J Dis Child, 73-83. *Honors & Awards:* Borden Award, Am Acad Pediat, 64; Res Career Award, Nat Inst Health, 62. *Mem:* AAAS; Am Acad Pediat; Soc Pediat Res (pres, 60-61); Am Pediat Soc (vpres, 75-76); Human Biol Coun. *Res:* Metabolism in infancy and childhood; diagnosis and therapy of clinical pediatrics; pediatric endocrinology; chemical growth; body fluid physiology; infant nutrition; human body composition, particularly lean body mass and body fat as estimated by potassium. *Mailing Add:* Sch Med & Dent Univ Rochester Rochester NY 14642

FORBES, IAN, b Pittsburgh, Pa, Jan 16, 20; m 44; c 3. AGRONOMY, PLANT BREEDING. *Educ:* Univ Md, BSc, 41, MSc, 49, PhD(plant morphol), 54. *Prof Exp:* Mem res staff, Plant Indust Sta, USDA, Beltsville, 41-42, 45-47, res agronomist, 47-51, res agronomist, Coastal Plain Exp Sta, Univ Ga, 51-63, sr res agronomist, 63-78; RETIRED. *Honors & Awards:* Sears Roebuck Award, 60. *Mem:* Am Soc Agron; Am Genetic Asn; Crop Sci Soc Am. *Res:* Cytogenetics and breeding of forage legumes. *Mailing Add:* 200 Fulwood Blvd Tifton GA 31794

FORBES, JACK EDWIN, b Bloomington, Ill, Dec 11, 28; m 59. MATHEMATICS. *Educ:* Ill Wesleyan Univ, BS, 49; Bradley Univ, MS, 52; Purdue Univ, PhD(math), 57. *Hon Degrees:* DSc, Purdue Univ, 87. *Prof Exp:* Instr math, Bradley Univ, 51-52; from instr to asst prof, Purdue Univ, 55-60; dir sec math, Educ Res Coun, Cleveland, Ohio, 60-61; dir res math, Britannica Ctr Studies Learning, Calif, 61-63; math consult, Encycl Britannica Press, 63-64; prof, 64-86, EMER PROF MATH, PURDUE UNIV, CALUMET CAMPUS, 86- *Concurrent Pos:* Vis prof, Ball State Teachers Col, 58-60. *Mem:* Math Asn Am; Sigma Xi. *Res:* Foundations; mathematics education; programmed instruction; teacher education. *Mailing Add:* 6 Constitution Dr Michigan City IN 46360-3300

FORBES, JAMES, b Yonkers, NY, July 9, 10; m 37; c 1. ENTOMOLOGY. *Educ:* Fordham Univ, BS, 32, MS, 34, PhD(entom), 36. *Prof Exp:* From instr to prof, 36-79, EMER PROF, FORDHAM UNIV, 79- *Concurrent Pos:* Vpres & pres, NY Entomol Soc, 49-51, trustee, 52-60, hon mem, 84-; assoc ed & ed, J NY Entomol Soc, 58-70; treas, Coun Biol Editors, 65-71. *Mem:* Entom Soc Am; Sigma Xi. *Res:* Anatomy and histology of the ants; comparisons of the anatomies of male ants, the most conservative and least specialized of the nest forms. *Mailing Add:* Dept Biol Fordham Univ Bronx NY 10458

FORBES, JAMES FRANKLIN, b Berwyn, Ill, Feb 9, 41; m 78; c 4. DENTAL MATERIALS. *Educ:* Wheaton Col, BS, 62; Southern Ill Univ, PhD(inorg chem), 68. *Prof Exp:* Instr chem, Drew Univ, 66-68; proj chemist, Ametek Tech Prod, 68-73; res chemist, Lee Pharmaceut, 73-74; proj leader, Am Hosp Supply Corp, 74-77, res chemist, 77-78, proj scientist, Group Technol Ctr, 78-81; MGR, RES & DEVELOP, UNITEK CORP/3M, 81- *Mem:* Am Chem Soc; Int Asn Dent Res. *Res:* Formulation of orthodontic adhesives; development of polymeric and mineral systems for preventative and restorative dentistry; formulation of dental impression materials. *Mailing Add:* 2724 S Peck Rd Monrovia CA 91016-5005

FORBES, JERRY WAYNE, b Oquawka, Ill, July 12, 41; m 61; c 2. HIGH PRESSURE PHYSICS, SHOCKWAVE PHYSICS & DETONATION SCIENCE. *Educ:* Western Ill Univ, BS, 63; Univ Md, MS, 67; Wash State Univ, PhD(physics), 76. *Prof Exp:* RES PHYSICIST, US NAVAL SURFACE WARFARE CTR, WHITE OAK LAB, 63- *Concurrent Pos:* Secy-Treas of Am Phys Soc Topical Group on Shock Compression of Condensed Matter, 85- *Mem:* Am Phys Soc. *Res:* High pressure physics of solids and liquids, specifically phase transformations under shock wave loading; also, materials response due to interaction with particle beams; dynamic properties of energetic materials. *Mailing Add:* Naval Surface Warfare Ctr Bldg 319 Code R13 Silver Spring MD 20903-5000

FORBES, JUDITH L, b Fullerton, Calif, Sept 27, 42; c 3. INERTIAL GUIDANCE, SYSTEMS ENGINEERING. *Educ:* Calif State Univ, Fullerton, BA, 74, MS, 79; Univ Southern Calif, MBA, 83. *Prof Exp:* Mem tech staff, TRW, 80-81; engr, Electromech Div, Gen Res Corp, Northrop, 75-80, proj engr mgr systs eng, 81-87, proj engr, 87, proj mgr, El Segundo, 87-88; PROG MGR, TRW TECHNAR, 89- *Mem:* Soc Women Engrs (vpres); Am Inst Aeronaut & Astronaut; Soc Automotive Engrs. *Res:* Atmospheric radioactivity-use of naturally occuring isotopes as tracers for investigation of atmospheric phenomena; inertial guidance-classified studied of error mechanisms affecting accuracy of guidance systems. *Mailing Add:* 23557 Casa Loma Dr Diamond Bar CA 91765

FORBES, MALCOLM HOLLOWAY, b New Haven, Conn, Aug 20, 33; m 63; c 3. ORGANIC CHEMISTRY. *Educ:* Yale Univ, BS, 54; Trinity Col, Conn, MS, 58; Cambridge Univ, PhD(org chem), 60. *Prof Exp:* Res assoc antibiotics, Mass Inst Technol, 60-61, fel, 62-63; consult chem, Ed Serv, Inc, Mass, 63-65; acad dean, Cazenovia Col, 65-70; dean, Col Arts & Sci, Millikin Univ, 70-78; V PRES ACAD AFFAIRS, UNIV EVANSVILLE, 78- *Mem:* Am Asn Higher Educ; AAAS; Am Chem Soc; Sigma Xi. *Res:* College or university administration and improvement of academic programs; curriculum development and evaluation in the sciences. *Mailing Add:* Chief Academic Dean Roger Williams C Main Campus Bristol RI 02809

FORBES, MARTIN, b Brussels, Belgium, July 25, 20; nat US; m 49; c 2. MICROBIOLOGY. *Educ:* Moravian Col, BS, 47; Univ Pa, MS, 49, PhD(microbiol), 51. *Prof Exp:* US Army res fel microbiol, Univ Hosp, Pa, 51-53, dir animal res proj & res assoc dept microbiol, Sch Med, 53-58; asst prof microbiol, Sch Med, Temple Univ, 58-59; group leader, Dept Chemother, Am Cyanamid Co, 59-65, head, Dept Antimicrobial Ther, 65-69, dir, Infectious Dis Ther Res Sect, Lederle Labs, 69-81; mgr cancer chemother res, Glenolden Labs, E I du Pont de Nemours & Co, Inc, 82-89; RETIRED. *Concurrent Pos:* Res assoc, Walter Reed Army Inst Res, 53-58. *Mem:* Am Inst Nutrit; Am Soc Microbiol; NY Acad Sci; Sigma Xi. *Res:* Antimicrobial agents and cancer chemotherapy. *Mailing Add:* 607 Painters Crossing Chadds Ford PA 19317

FORBES, MICHAEL SHEPARD, b Washington, DC, Jan 21, 45; m 70; c 2. BIOLOGICAL STRUCTURE, CELL BIOLOGY. *Educ:* Univ Va, Charlottesville, BA, 66, PhD(biol), 71. *Prof Exp:* Fel physiol, Univ Va, 71-73, res assoc physiol & biol, 73-74; instr, Univ Md, 74-75, asst prof, 76-78; res asst prof, 79-85, RES ASSOC PROF PHYSIOL, UNIV VA, 85- *Concurrent Pos:* Res career develop award, NIH, 79-84. *Mem:* Am Heart Asn; Electron Micros Soc Am; Sigma Xi. *Res:* Fine structure and electron cytochemistry of cardiac muscle, skeletal muscle, smooth muscle and blood vessels; filaments and microtubules; sarcoplasmic reticulum; transverse tubules; cell-to-cell attachments. *Mailing Add:* Dept Physiol Univ Va Med Sch Charlottesville VA 22908

FORBES, MILTON L, b Aruba, Netherlands Antilles, Dec 14, 30; m 51; c 4. BIOLOGY. *Educ:* Iowa State Teachers Col, BA, 52, MA, 53; Fla State Univ, PhD(marine biol), 62. *Prof Exp:* Instr biol, Burlington High Sch & Jr Col, Iowa, 53-56 & Ill State Norm Univ, 56-57; assoc prof, Francis T Nicholls State Univ, 60-63; assoc prof biol & marine sci, Lamar Univ, 63-70; PROF SCI, UNIV VI, 70- *Concurrent Pos:* NSF grant investr, 76-77. *Mem:* AAAS; Ecol Soc Am; Sigma Xi. *Res:* Ecology of oysters; science education. *Mailing Add:* Univ VI Kingshill P O St Croix VI 00850

FORBES, OLIVER CLIFFORD, b Eureka, Calif, May 25, 27. GENETICS. *Educ:* Humboldt State Col, AB, 50; Univ Calif, MA, 52, PhD(zool), 58. *Prof Exp:* Instr biol sci, San Francisco State Col, 56-57; from instr to asst prof, 57-68, ASSOC PROF ZOOL, UNIV IDAHO, 68- *Res:* Drosophila genetics; chromosome segregation; mutation. *Mailing Add:* Dept Zool Univ Idaho Moscow ID 83844

FORBES, PAUL DONALD, b Binghamton, NY, Mar 3, 36; m 60; c 4. REGULATORY COMPLIANCE. *Educ:* Wheaton Col, Ill, BS, 57; Brown Univ, PhD, 61. *Prof Exp:* Instr anat, Sch Med, Temple Univ, 61-64; assoc prof biol, Barrington Col, 64-68; PROF DERMAT, SCH MED, TEMPLE UNIV, 87-, DIR CHEM HYG & SAFETY, 89- *Concurrent Pos:* USPHS traineeship cancer, Sch Med, Temple Univ, 61-64; radiobiologist, Roger Williams Gen Hosp, Providence, RI, 66-68. *Mem:* AAAS; Radiation Res Soc; Am Asn Cancer Res; Soc Invest Dermat; Am Soc Photobiol; Am Col Toxicol. *Res:* Chemical and physical carcinogenesis; photobiology of skin; laboratory safety; regulatory compliance. *Mailing Add:* 3307 N Broad St Temple Univ Philadelphia PA 19140

FORBES, RICHARD BRAINARD, b Ellington, NY, Aug 29, 21; m 50; c 2. SOIL FERTILITY, VEGETABLE CROPS. *Educ:* Rollins Col, BS, 43; Univ Fla, MS, 48; Pa State Univ, PhD(agron, soils), 56. *Prof Exp:* Asst soils chemist, Exp Sta, Univ Fla, 48-49, asst prof soils, Univ, 49-53; chemist & agent, US Regional Pasture Res Lab, Pa State Univ, 53-56; asst soils chemist, Cent Fla Exp Sta, Univ Fla, 56-67, assoc soils chemist, Inst Food & Agr Sci, 67-87; RETIRED. *Mem:* Am Soc Agron; Soil Sci Soc Am; Am Inst Chemists. *Res:* soil fertility; plant nutrition; vegetables; soybeans. *Mailing Add:* Agr Res & Educ Ctr Box 909 Sanford FL 32771

FORBES, RICHARD BRYAN, b Correctionville, Iowa, July 29, 36; m 60; c 2. VERTEBRATE ZOOLOGY, ECOLOGY. *Educ:* Univ SDak, AB, 58; Univ NMex, MS, 61; Univ Minn, PhD(ecol), 64. *Prof Exp:* From asst prof to assoc prof, 64-76, PROF BIOL, PORTLAND STATE UNIV, 76- *Mem:* Am Soc Naturalists; AAAS; Am Soc Mammal; Ecol Soc Am; Sigma Xi. *Res:* Ecology and life histories of terrestrial vertebrates, particularly mammals. *Mailing Add:* Dept Biol Portland State Univ PO Box 751 Portland OR 97207

FORBES, RICHARD MATHER, b Wooster, Ohio, Jan 8, 16; m 44; c 3. NUTRITION. *Educ:* Pa State Col, BS, 38, MS, 39; Cornell Univ, PhD(nutrit), 42. *Prof Exp:* Instr biochem, Wayne State Univ, 42; res fel, Cornell Univ, 42-43; asst prof animal husb, Univ Ky, 46-49; from assoc prof to prof, 49-85, EMER PROF NUTRIT BIOCHEM, UNIV ILL, URBANA-CHAMPAIGN, 85- *Concurrent Pos:* Mem comn animal nutrit, Nat Res Coun, 56-72; mem nutrit study sect, NIH, 74-78; competitive grants review comn, USDA, 79-82. *Honors & Awards:* Gustav Bohstedt Award, Am Soc Animal Sci, 68; H H Mitchell Award, Univ Ill, 81; Borden Award, Am Inst Nutr, 84. *Mem:* fel Am Inst Nutrit. *Res:* Mineral metabolism of animals. *Mailing Add:* 2005 S Vine St Urbana IL 61801

FORBES, ROBERT SHIRLEY, forest entomology, for more information see previous edition

FORBES, TERRY GENE, b Terre Haute, Ind, Sept 3, 46. SPACE PHYSICS, SOLAR PHYSICS. *Educ:* Purdue Univ, BS, 68; Univ Colo, MS, 70, PhD(astrogeophys), 78. *Prof Exp:* Res engr I, amphenol div, Bunker-Ramo Corp, 68-69; res fel, Los Alamos Nat Lab, 78-80 & dept appl math, Univ St Andrews, Scotland, 80-84; res scientist II, Space Sci Ctr, 84-86, assoc res prof, 86-89, RES PROF, UNIV NH, 89- *Concurrent Pos:* Prin investr grant, NSF, 85-; vis scientist, Observ de Paris-Meudon, 86. *Mem:* Am Astron Soc; Royal Astron Soc; Am Geophys Union; Int Astron Union. *Res:* Process of magnetic field annihilation in the solar atmosphere and the terrestrial magnetosphere. *Mailing Add:* Space Sci Ctr Univ NH SER Bldg Durham NH 03824

FORBES, THOMAS ROGERS, anatomy; deceased, see previous edition for last biography

FORBES, WARREN C, mineralogy; deceased, see previous edition for last biography

FORBES, WILLIAM FREDERICK, b Can, 1924; m 72; c 3. BIOMETRICS. *Educ:* London Univ, BSc, 46, BSc hons, 49, PhD(org chem), 52, DSc(phys org chem), 64. *Prof Exp:* Lectr chem, Univ Nottingham, Eng, 52-53; from assoc prof to prof, Univ Nfld, 53-59; prin res officer, Commonwealth Sci & Indust Res Orgn, Australia, 59-62; prof chem, 62-69, dean fac math, 72-80, PROF STATIST, UNIV WATERLOO, 66-, DIR, PROG IN GERONT, 83- *Concurrent Pos:* Vis prof biochem, Univ Rochester, 64-71; spec consult, Nat Inst Child Health, Develop & Aging, 66-70; mem, Comt Presidents of Univs of Ont, 70-71; chmn bd, Off Comput Coord, Coun Ont Univs, 71-73; hon assoc, Royal Holloway Col, London Univ, 71-; head, Red Assessment Smoking Habits, WHO Collaborating Ctr, 78-90; pres, Can Coun on Smoking & Health, 83-85. *Honors & Awards:* Award Outstanding Contrib Geront, Can Asn Geront, 81. *Mem:* Can Asn Geront (pres, 71-73); Geront Soc (vpres, 71-72). *Res:* Epidemiological studies on aging, in particular the effects of cigarette smoking and trace metals; use of mathematical modelling in the evaluation of risk factors; drug use in the elderly; evolution of assistive devices. *Mailing Add:* Geront Univ Waterloo Waterloo ON N2L 3G1 Can

FORBIS, ORIE LESTER, JR, b Encinal, Tex, Dec 25, 22; m 46; c 2. PEDIATRICS, PSYCHIATRY. *Educ:* Univ Tex, BA, 51, MD, 53. *Prof Exp:* Res pediat, John Sealy Hosp, Tex, 54-57, res pediat psychiat & neurol, 57-59; child psychiat, Hawthorn Ctr, Mich, 59-61; asst prof psychiat & pediat & dir child guid clin, Med Ctr, Univ Ark, 62-66; psychiatrist dir, Genesee County Ment Health Serv Bd, Mich, 66-67; dir children's serv, Ment Health-Ment Retardation Ctr Austin-Travis County, Austin, 67-69; chief child psychiatrist, Northwest Ment Health Ctr, 69-73; San Marcos Treat Ctr, 86-89; outreach psychiatrist, 73-86, PSYCHIATRIST, SAN ANTONIO STATE HOSP, 89- *Concurrent Pos:* Consult, Northville State Hosp & pub schs, Mich; dir ment health planning, Ark State Health Dept, 63-65; assoc prof pediat & psychiat, Med Sch, Univ Tex, San Antonio, 69-73 & clin assoc prof psychiat, 73- *Mem:* AMA; Am Psychiat Asn. *Res:* Child psychiatry; longitudinal studies of emotional development of children with special reference to socioeconomically-culturally deprived. *Mailing Add:* 314 E Lincoln St New Braunfels TX 78130

FORBRICH, CARL A, JR, b San Antonio, Tex, Nov 5, 39; m 62; c 2. AERONAUTICAL ENGINEERING. *Educ:* Univ Tex, Austin, BS, 61; Univ Okla, MS, 63; Stanford Univ, PhD(aeronaut eng), 67; Univ Colo, Boulder, MBA, 71. *Prof Exp:* Proj engr, Boeing Aircraft Co, Wash, 61; US Air Force, 61-, aerodynamicist, Tinker AFB, Okla, 61-64, from instr to assoc prof aeronaut, US Air Force Acad, 67-71, res assoc, F J Seiler Res Lab, 71-73, proj officer, Chem Laser Br, 73-74, sect chief, 74-75, br chief, Chem Laser Br, Air Force Weapons Lab, 75-79, chief, Munitions Div, Air Force Armament Lab, 79-82, DEP SYST PROG DIR, DEFENSE SATELLITE COMMUN SYST, SPACE DIV, LOS ANGELES, 82- *Mem:* Assoc fel Am Inst Aeronaut & Astronaut; Am Phys Soc; Sigma Xi. *Res:* Experimental measurement of atomic oscillator strengths in high temperature gases; spectral line broadening; high energy laser research and development. *Mailing Add:* 29516 Bernice Dr San Pedro CA 90732

FORBUSH, BLISS, III, b Pittsburgh, Pa, July 1, 45; m 67; c 2. MEMBRANE TRANSPORT. *Educ:* Harvard Col, AB, 67; Johns Hopkins Univ, PhD(physiol), 75. *Prof Exp:* Fel physiol, 75-79, asst prof, 79-, PROF PHYSIOL, YALE UNIV SCH MED. *Res:* Molecular mechanism of solute transport across cell membranes; mechanism of the sodium, potassium-pump, and characterization of a sodium, potassium, chlorine co-transport system. *Mailing Add:* Solar & Molecular Physiol Yale Univ Med Sch 333 Cedar St B157 New Haven CT 06510

FORCE, CARLTON GREGORY, b Gouverneur, NY, Aug 5, 26; m 53; c 4. COLLOID CHEMISTRY. *Educ:* Clarkson Col Technol, BS, 52, PhD(phys chem), 65; Univ Ill, MS, 57. *Prof Exp:* Chemist, Merck & Co, Inc, 52-56; res chemist, Esso Res & Eng Co, 57-59; res chemist, Latex Fiber Industs, Inc, 59-61, sr res chemist, 63-67; res chemist, 67-70, sr res chemist, 70-78, RES ASSOC, WESTVACO CORP, 78- *Mem:* Am Chem Soc. *Res:* Colloid chemistry particularly rubber latex systems and rosin and fatty acid emulsifiers; paper making industry. *Mailing Add:* 299 Hobcaw Dr Mt Pleasant SC 29464

FORCE, DON CLEMENT, b Clear Lake, SDak, July 5, 28; m 53; c 3. ENTOMOLOGY. *Educ:* Fresno State Col, BA, 54; Univ Calif, Davis, MS, 58; Univ Calif, Berkeley, PhD(entom), 63. *Prof Exp:* Entomologist, Stauffer Chem Co, 56-58; lab technologist biol control, Univ Calif, Berkeley, 58-62; entomologist, USDA, 62-65; from asst prof to assoc prof, 65-73, PROF BIOL SCI, CALIF STATE POLYTECH UNIV, 73- *Mem:* AAAS; Am Soc Naturalists; Ecol Soc Am; Sigma Xi. *Res:* Insect ecology and biological control. *Mailing Add:* Dept Biol Sci Calif State Polytech Univ Pomona CA 91768

FORCHHEIMER, OTTO LOUIS, b Nurnberg, Ger, Sept 18, 26; US citizen; m 57; c 2. PHYSICAL CHEMISTRY, INORGANIC CHEMISTRY. *Educ:* McGill Univ, BSc, 47; Brown Univ, PhD(chem), 51. *Prof Exp:* Res chemist, Univ Chicago, 51-53; sr res chemist, Gen Abrasive Co, 53-58, asst dir res, 58-59; mgr chem div, Trionics Corp, 59-62; tech dir, Dolomite Brick Corp Am, 62-67,; dir res, 62-80, VPRES MKT & TECH J E BAKER CO, 80-, DIR, 84-; VPRES & DIR, DOLOMITE BRICK CORP AM, 67- *Mem:* Fel Am Ceramic Soc; fel Am Soc Testing & Mat; Am Soc Testing & Mat; Refractories Inst. *Res:* Rates and mechanisms of inorganic reactions; chemistry of abrasives; refractories. *Mailing Add:* J E Baker Co PO Box 1189 York PA 17405

FORCHIELLI, AMERICO LEWIS, b Alpha, NJ, May 28, 22; m 42; c 1. CHEMISTRY. *Educ:* Lafayette Col, BA, 47. *Prof Exp:* Chemist resins, Hercules Powder Co, 47-48; group leader explosives, Picatinny Arsenal, 48-52; res dir waxes, Sure-Seal Corp, 52-53; res chemist, Cent Res Lab, Gen Aniline & Film Corp, 53-54; res chemist, Masury-Young Co, 54-56, dir res & mgr lab, 56-66, res dir, Masury-Columbia Co Div, Alberto-Culver Co, 66-69; prod mgr, Wayland Chem Div, Philip A Hunt Chem Co, Lincoln, 69-74; gen mgr opers, Butcher Polish Co, 74-78, exec tech dir, 78-80, vpres res & develop, 80-86; RETIRED. *Mem:* Am Chem Soc; fel Am Inst Chem; AAAS. *Res:* Organic and polymer research; detergents; textile chemicals; lubricants; waxes and specialty products; chemical specialties field. *Mailing Add:* Three Sherwood Dr Sterling MA 01564-2453

FORCIER, GEORGE ARTHUR, b Mapleville, RI, Nov 29, 38; m 64; c 2. ANALYTICAL CHEMISTRY. *Educ:* Providence Col, BS, 60; Univ Mass, MS, 64, PhD(anal chem), 66. *Prof Exp:* Res chemist, Pfizer, Inc, 66-70, proj leader, 70-79, mgr, 79-81, asst dir, 81-86, DIR, ANALYTICAL RES & DEVELOP, PFIZER, INC, 86- *Res:* Electroanalytical chemistry in acetonitrile; electroanalytical and other instrumental methods of analysis. *Mailing Add:* Pfizer Central Res Dept Eastern Pt Rd Groton CT 06340

FORCIER, LAWRENCE KENNETH, b Gardner, Mass, Aug 17, 43; m 65; c 3. NATURAL RESOURCE EDUCATION. *Educ:* Dartmouth Col, BA, 66; Yale Univ, MForSc, 68, MPhil, 70, PhD(forest ecol), 73. *Prof Exp:* From asst prof to assoc prof forest ecol, Univ Mont, Missoula, 70-76, actg dean forestry, 77; assoc prof ecol-silviculture, Univ Wis-Stevens Point, 75-76; from asst dir to assoc dir, 77-83, actg dir, 83-84, DEAN NATURAL RESOURCES, UNIV VT, 84- *Mem:* AAAS; Soc Am Foresters; Wildlife Soc. *Res:* Forest tree population dynamics and the response of forest ecosystems to disturbance. *Mailing Add:* Dean Natural Resources Univ Vt Burlington VT 05405

FORD, ALBERT LEWIS, JR, b Ft Worth, Tex, May 12, 46; m 68; c 3. COLLISION PHYSICS. *Educ:* Rice Univ, BA, 68; Univ Tex, Austin, PhD(chem physics), 72. *Prof Exp:* Res fel atomic & molecular physics, Harvard Col Observ, Harvard Univ, 72-73; asst prof, 73-78, ASSOC PROF, 79- 85, PROF PHYSICS, TEX A&M UNIV, 86- *Mem:* Am Phys Soc. *Res:* Ionization and charge transfer in atom-ion collisions. *Mailing Add:* Dept Physics Tex A&M Univ College Station TX 77843

FORD, ARTHUR B, b Seattle, Wash, Sept 4, 32; m 55; c 2. GEOLOGY, PETROLOGY. *Educ:* Univ Wash, BS, 54, MS, 57, PhD(geol), 59. *Prof Exp:* Asst prof geol, San Diego State Col, 58-60; GEOLOGIST, US GEOL SURV, 60- *Mem:* Geol Soc Am; Am Geophys Union. *Res:* Geology of Antarctica and Alaska; metamorphic and igneous petrology; petrology of Dufek intrusion, Antarctica; geology and petrology of Glacier Peak Wilderness Area, North Cascades, Washington, Juneau and Sitka regions, southeastern Alaska. *Mailing Add:* Alaskan Geol Br US Geol Surv 345 Middlefield Rd Menlo Park CA 94025

FORD, C(LARENCE) QUENTIN, b Glenwood, NMex, Aug 6, 23; m 50; c 2. MECHANICAL ENGINEERING. *Educ:* Merchant Marine Acad, BS, 44; NMex State Univ, BSME, 49; Univ Mo, MS, 50; Mich State Univ, PhD, 59. *Prof Exp:* Instr mech eng, Univ Mo, 49-50; from instr to asst prof, Wash State Univ, 50-56; instr, Mich State Univ, 56-59; chmn dept, 60-70, PROF MECH ENG, NMEX STATE UNIV, 59- ASSOC DEAN ENG, 74- *Mem:* AAAS; Am Soc Mech Engrs; Nat Soc Prof Engrs; Am Soc Eng Educ; Sigma Xi; Nat Coun Eng Examiners. *Res:* Combustion kinetics; rate of scale formation in boiling processes; heat transfer; thermodynamics. *Mailing Add:* NMex State Univ PO Box 3449 Las Cruces NM 88003

FORD, CLARK FUGIER, b Glendale, Calif, Aug 29, 53; m 76. GENETICS, MOLECULAR BIOLOGY. *Educ:* Univ Calif, Los Angeles, BA, 75; Univ Iowa, MS, 77, PhD(genetics), 81. *Prof Exp:* Res assoc, Dept Bot, Univ Iowa, 82 & Dept Microbiol, Univ Va, 82-85; ASST PROF, DEPT GENETICS, IOWA STATE UNIV, 85- *Mem:* Sigma Xi; AAAS; Am Soc Microbiol. *Res:* Genetics and molecular biology of the single-celled alga Chlamydomonas reinhardtii and the yeast Saccharomyces cerevisiae; genetic and biochemical analysis of pigment mutations in Chlamydomonas reinhardtii; isolation of histone mutants in Saccharomyces cerevisiae. *Mailing Add:* Genetics Dept Iowa State Univ Ames IA 50011

FORD, CLINITA ARNSBY, b Okmulgee, Okla, Sept 23, 28; m 51; c 3. NUTRITION. *Educ:* Lincoln Univ, Mo, BS; Columbia Univ, MS; Kans State Univ, PhD(foods & nutrit). *Prof Exp:* Asst prof home econ, Fla A&M Univ, 49-56; res asst, Kans State Univ, 56-59; prof home econ & chmn dept, 59-77, PROF CONSUMER SCI & TECHNOL & COORDR TITLE III PROG, FLA A&M UNIV, 77- *Concurrent Pos:* Dir & admin consult, Proj Upward Bound, Fla A&M Univ, 65-70; Fla del, White House Conf Aging, 71; mem, Nat Coun Admin Women in Educ. *Mem:* Am Dietetic Asn; Am Home Econ Asn; Am Voc Asn; Inst Food Technol; Am Voc Educ Res Asn; Sigma Xi. *Res:* Nutritional status of children as determined by dietary and biochemical analysis. *Mailing Add:* 209 Jackson Davis Hall Fla A&M Univ Tallahassee FL 32307

FORD, CLINTON BANKER, b Ann Arbor, Mich, Mar 1, 13; m 40, 61. ASTRONOMY. *Educ:* Univ Mich, AB, 35, MS, 36. *Hon Degrees:* ScD, Ithaca Col, 83. *Prof Exp:* Asst astron, Brown Univ, 39-41; instr, Smith Col, 41-42; group leader pure sci, Ordwes Labs, Wesleyan Univ, 46-50, asst dir, 50-53; vpres & res dir, Nikor Prod Co, 53-61, tech consult, 61-75; RETIRED. *Concurrent Pos:* Trustee, Ithaca Col, 67- *Mem:* AAAS; fel Am Astron Soc; Optical Soc Am; Am Inst Aeronaut & Astronaut; Am Asn Variable Star Observers (secy, 48-, pres, 61-62); Astron Soc Pac. *Res:* Photometry of variable stars; celestial navigation; highspeed photography; spectroscopy; photo developing equipment. *Mailing Add:* Ten Canterbury Lane Wilton CT 06897

FORD, DANIEL MORGAN, b Medford, Ore, Apr 5, 42; m 65; c 3. SKIN BIOLOGY, TOXICOLOGY. *Educ:* Ore State Univ, BA, 64; Ore Health Sci Univ, PhD(anat), 71. *Prof Exp:* Instr biol, Ore State Univ, 64-66; NIH trainee, Ore Regional Primate Res Ctr, 66-70; scientist, Syntex Res, 70-77; dept head, Int Flavors & Fragrances, Inc, 77-82; PRES, CUBICON, 82-; ADJ ASSOC PROF ANAT, UNIV MED & DENT, NJ, ROBERT WOOD JOHNSON MED SCH, 85- *Concurrent Pos:* Consult, Res Inst Fragrance Mat, 78-82, Fragrance Mat Asn, 80-81; Monmouth County Hazardous Waste Adv Comt, County Bd Health, 84-; teacher, Ctr Prof Advan. *Mem:* Soc Invest Dermat; Am Soc Photobiol; Am Col Toxicol; AAAS; NY Acad Sci. *Res:* Skin as a model for the study of cellular control mechanisms and immunology, as well as practical application in safety evaluation and claims subtantiation for pharmaceuticals and consumer products; clinical manifestations of skin allergy, photobiology, and percutaneous absorption; neurobehavioral toxicology. *Mailing Add:* 207 Deepdale Dr Middletown NJ 07748

FORD, DAVID A, b Pasadena, Calif, Oct 25, 35; m 64; c 2. MATHEMATICS. *Educ:* Occidental Col, AB, 56; Univ Utah, MS, 58, PhD(math), 62. *Prof Exp:* Asst prof, 65-71, ASSOC PROF MATH, EMORY UNIV, 71- *Mem:* Am Math Soc; Math Asn Am. *Res:* Functional analysis and ordinary differential equations; applications in the mathematical theory of optimal control. *Mailing Add:* Dept Math Emory Univ Atlanta GA 30322

FORD, DENYS KENSINGTON, b Newcastle, Eng, Aug 8, 23; Can citizen; m 55; c 3. MEDICINE, RHEUMATOLOGY. *Educ:* Cambridge Univ, BA, 44, MB, 47, MD, 53; FRCPS(C). *Prof Exp:* Registr med, London Hosp Med Col, Eng, 51-53; assoc prof, 60-73, PROF MED, FAC MED, UNIV BC, 73- *Concurrent Pos:* Fel, Fac Med, Univ BC, 54-60. *Mem:* Am Rheumatism Asn; Can Med Asn; Can Soc Clin Invest. *Res:* Arthritis; tissue culture; virology; mycoplasmology. *Mailing Add:* The Arthritis Ctr 895 W Tenth Ave Vancouver BC V5Z 1L7 Can

FORD, DONALD HOSKINS, b Upland, Calif, July 28, 30; m 53; c 3. PLANT PATHOLOGY. *Educ:* Pomona Col, BA, 52; Claremont Col, MA, 54; Univ Calif, PhD(plant path), 58. *Prof Exp:* Asst bot, Rancho Santa Ana Bot Garden, 53-54; res plant pathologist, Forest Serv, US Dept Agr, 54-55; asst plant path, Univ Calif, 55-58; sr plant pathologist, Chem Plant Dis Control, 58-64, plant sci rep, Calif, 64-66, regional coordr, Calif Res Sta, 66-71, res scientist, 71-73, res assoc, 73-80, RES ADV, CALIF RES STA, ELI LILLY & CO, 80- *Mem:* AAAS; Am Phytopath Soc; Am Inst Biol Sci; Sigma Xi. *Res:* Fungicides, herbicides and growth regulators; soil fungi in relation to plant roots; mycological interest in the Phycomycetes. *Mailing Add:* 655 W Morris Ave Fresno CA 93704

FORD, DWAIN, CHEMISTRY. *Educ:* Andrews Univ, BA, 49; Clark Univ, PhD(chem), 62. *Prof Exp:* Instr sci & math, Wis Acad, 49-58; from asst prof to assoc prof, 62-67, chmn dept, 63-71, acad dean, 71-74, dean, Col Arts & Sci, 74-81, PROF CHEM, ANDREWS UNIV, 67-; DEAN, UNIV OMBUDSMAN, 81- *Concurrent Pos:* Univ ombudsman, Andrews Univ, 81- *Mem:* Sigma Xi; Am Chem Soc. *Res:* Chemical carcinogenesis; peptide synthesis; organic reaction mechanisms. *Mailing Add:* 7041 Deans Hill Rd Berrien Center MI 49102

FORD, EMORY A, b South New Berlin, NY, Oct 17, 40; m 63; c 2. ORGANIC POLYMER CHEMISTRY. *Educ:* Hartwick Col, BA, 62; Syracuse Univ, PhD(org chem), 67. *Prof Exp:* Res chemist, 67-72, res group leader polymers, 72-78, technol mgr, 78-81, mgr res northern petrochem, Morris, Ill, 81-84; dir, Basic Res Chem, Rolling Meadows, Ill, 84-86; DIR, BASIC RES QUANTUM CHIMICA, USI DIV, 86- *Mem:* Am Chem Soc; AAAS; Soc Plastics Engrs. *Res:* Synthesis and evaluation of polymeric composites; organic reactions of high polymers. *Mailing Add:* Quantum 1275 Section Rd Cincinnati OH 45222

FORD, FLOYD MALLORY, zoology; deceased, see previous edition for last biography

FORD, GEORGE DUDLEY, b Morgantown, WVa, Aug 18, 40; m 65; c 2. PHYSIOLOGY, BIOPHYSICS. *Educ:* WVa Univ, BS, 61, PhD(pharmacol), 67; Univ Iowa, MS, 64. *Prof Exp:* Instr pharmacol, WVa Univ, 67; AEC fel biophys, Sch Med & Dent, Univ Rochester, 67-70; asst prof, 70-78, ASSOC PROF PHYSIOL, HEALTH SCI DIV, MED COL VA, VA COMMONWEALTH UNIV, 78- *Concurrent Pos:* Phys testing & des engr, Allegheny Ballistics Lab, Cumberland, Md, 60-61; Fulbright res scholar, Univ Leuven, Belg, 82-83. *Mem:* Am Physiol Soc; AAAS; Biophys Soc. *Res:* Physiology of vascular smooth muscle; mechanisms of action of vasoactive agonists; mathematical models of membrane transport and adrenergically mediated responses; bioinstrumentation. *Mailing Add:* Dept Physiol Med Col Va Va Commonwealth Univ Richmond VA 23298-0551

FORD, GEORGE PETER, b Ilford, Essex, England, Apr 2, 49. MOLECULAR ORBITAL THEORY, CHEMICAL CARCINOGENESIS. *Educ:* Univ London, England, BSc, 70; Univ East Anglia, England, PhD(chem), 74. *Prof Exp:* Fel, Univ Tex, Austin, 74-79; staff scientist, Pac Northwest Res Found & Hutchinson Cancer Res Ctr, 79-84; asst prof, 84-90, ASSOC PROF, DEPT CHEM, SOUTHERN METHODIST UNIV, 90- *Mem:* Am Chem Soc; Royal Soc Chem; Sigma Xi. *Res:* Quantum mechanical procedures to study the mechanisms of chemical and biochemical processes; mechanism of bioactivation of precursors of carcinogenic metabolites and the reactions of the latter with nucleic acid components. *Mailing Add:* Dept Chem Southern Methodist Univ Dallas TX 75275

FORD, GEORGE PRATT, b Leesburg, Ga, Apr 25, 19; m 44; c 2. CHEMISTRY. *Educ:* Ga Inst Technol, BS, 40; Columbia Univ, MA, 48, PhD(chem), 49. *Prof Exp:* Radio chemist, Los Alamos Sci Lab, Univ Calif, 49-78, consult, 78-90; RETIRED. *Mem:* Am Chem Soc; Am Math Soc; Am Phys Soc. *Res:* Aerosols; radio chemistry; fission; statistical treatment of data. *Mailing Add:* 1292 45th St Los Alamos NM 87544

FORD, GEORGE WILLARD, b Detroit, Mich, Jan 25, 27; m 53; c 4. THEORETICAL PHYSICS. *Educ:* Univ Mich, AB, 49, MS, 50, PhD, 55. *Prof Exp:* Asst prof physics, Univ Notre Dame, 54-58; from asst prof to assoc prof, 58-68, PROF PHYSICS, UNIV MICH, ANN ARBOR, 68- *Concurrent Pos:* Mem, Inst Advan Study, 55-56. *Mem:* Am Phys Soc. *Res:* Statistical mechanics; nuclear physics; graph theory. *Mailing Add:* Dept Physics Univ Mich Ann Arbor MI 48109-1120

FORD, GILBERT CLAYTON, b Hill City, Kans, Mar 31, 23; m 46; c 3. PHYSICS. *Educ:* Univ Colo, AB, 43; Harvard Univ, MA, 48, PhD(physics), 51. *Prof Exp:* Jr physicist, Manhattan Dist Proj, Univ Calif & Nat Bur Standards, 44-46; fac mem physics, 50-70, vpres acad affairs, 70-85, FAC MEM, ACAD AFFAIRS, NORTHWEST NAZARENE COL, 85- *Concurrent Pos:* Assoc res physicist, Univ Calif, Berkeley, 58; NSF & Res Corp basic res grants mass spectroscopy, 58-70. *Mem:* Am Phys Soc. *Res:* Mass spectroscopy. *Mailing Add:* Northwest Nazarene Col Nampa ID 83651

FORD, HOLLAND COLE, b Norman, Okla, Apr 12, 40. GALAXIES, PLANETARY NEBULAE. *Educ:* Univ Okla, BS, 62; Univ Wis, PhD(astron), 70. *Prof Exp:* Physicist, US Naval Ord Lab, Corona, 64-65; mem tech staff astrodynamics, TRW, Inc, 65-66; fel astron, Univ Ore, 70-71; asst prof, 71-76, ASSOC PROF ASTRON, UNIV CALIF, LOS ANGELES, 76- *Mem:* Int Astron Union; Am Astron Soc. *Res:* Gaseous nebulae and extragalactic astronomy. *Mailing Add:* Dept Physics Johns Hopkins Univ 34th & Charles St Baltimore MD 21218

FORD, JAMES, b Ryderwood, Wash, Sept 30, 27; m 51; c 4. BIOLOGY, ZOOLOGY. *Educ:* Western Wash State Col, BA(educ) & BA(biol), 51; Ore State Univ, MS, 53, PhD(zool), 62. *Prof Exp:* Instr biol, Wash High Sch, 51-54; instr, Skagit Valley Col, 54-56; instr, Ore Col Educ, 56-57; chmn dept, 57-66, chmn div natural sci, 64-65, dean instr, 65-77; PRES, SKAGIT VALLEY COL, 77- *Mem:* AAAS; Sigma Xi. *Res:* Chromosome study of pulmonate snails. *Mailing Add:* 2405 E College Way Mt Vernon WA 98273

FORD, JAMES ARTHUR, b Chicago, Ill, Feb 3, 34. MATERIALS SCIENCE, METALLURGY. *Educ:* Univ Mich, BS(chem eng) & BS(metall eng), 56, MS, 57, PhD(metall eng), 62. *Prof Exp:* Instr metall eng, Univ Mich, 59-61; sr res scientist, Mat Sci Sect, Res Labs, United Aircraft Corp, 61-64; supvr alloy systs group, 65-67, chief chem metall sect, 67-72, assoc dir, Olin Metals Res Labs, Olin Corp, 73-74; vpres res & develop, Conalco, Inc, 74-75; DIR RES & DEVELOP, COMPOSITE CAN DIV, BOISE CASCADE CORP, 75- *Concurrent Pos:* Adj assoc prof metall, Rensselear Polytech Inst, 61-71. *Mem:* Am Soc Testing Mat; Nat Asn Corrosion Engrs; Electrochem Soc; Am Soc Mech Engrs. *Res:* Research and development of composite cans; development of new and improved alloys; corrosion of alloys. *Mailing Add:* 703 Judith Dr Johnson City TN 37604

FORD, JAMES L C, JR, physics; deceased, see previous edition for last biography

FORD, JOHN ALBERT, JR, b Phoenixville, Pa, Jan 28, 31; m 55; c 3. ORGANIC CHEMISTRY. *Educ:* Hobart Col, BS, 53; Univ Del, MS, 56, PhD, 58. *Prof Exp:* RES CHEMIST, EASTMAN KODAK CO, 58- *Concurrent Pos:* Chmn, Gordon Res Conf Org Reactions & Processes, 75. *Mem:* Sigma Xi. *Res:* Organic synthesis of organophosphorus compounds, dyes and photographic chemicals; synthesis of organic compounds for color photographic products and processes. *Mailing Add:* 370 Pine Grove Ave Rochester NY 14617

FORD, JOHN PHILIP, b London, Eng, Mar 2, 30; m 58, 76. STRATIGRAPHY, REMOTE SENSING. *Educ:* Univ London, BSc, 59; Ohio State Univ, PhD(geol), 65. *Prof Exp:* engr, A V Roe, Can, 51-56; assoc dir extramural studies, Antioch Col, 60-64; asst prof geol, DePauw Univ, 65-66; from asst prof geol to assoc prof geog & geol, Eastern Ill Univ, 66-77; sr res assoc, 77-79, MEM TECH STAFF, PLANETOLOGY & OCEANOGRAPHY SECT, JET PROPULSION LAB, CALIF INST TECHNOL, 79- *Mem:* Fel Geol Soc Am; Am Asn Petrol Geol; Am Soc Photogram. *Res:* Ordovician stratigraphic geology, paleontology and sedimentation; continental interior of North America; glacial field mapping and survey work, Ill & Ohio; laboratory analysis of field data; geologic interpretation of side-looking airborne radar imagery; landsat multispectral imagery aerial photography, Western US. *Mailing Add:* Jet Propulsion Lab 4800 Oak Grove Dr Pasadena CA 91103

FORD, JOHNY JOE, b Orient, Iowa, Sept 3, 44; div; c 2. ENDOCRINOLOGY, ANIMAL REPRODUCTION. *Educ:* Iowa State Univ, BS, 66, PhD(reproductive physiol), 72. *Prof Exp:* Fel reproductive physiol, Harvard Med Sch, 72-74; res physiologist, Meat Animal Res Ctr, 74-78, res leader, 78-84, RES PHYSIOLOGIST, REPRODUCTIVE PHYSIOL, US DEPT AGR, 84- *Mem:* Soc Study Reproduction; Endocrine Soc; Am Soc Animal Sci; Soc Study Fertil; Soc Exp Biol Med. *Res:* Reproductive endocrinology in swine with emphasis on endocrine changes associated with sexual differentiation and development. *Mailing Add:* Reprod Res Unit US Meat Animal Res Ctr Clay Center NE 68933

FORD, JOSEPH, b Asheville, NC, Dec 18, 27; m 51, 76; c 4. THEORETICAL PHYSICS. *Educ:* Ga Inst Technol, BS, 52; Johns Hopkins Univ, PhD(physics), 56. *Prof Exp:* Res physicist, Electro Metall Res Labs, 56-58; asst prof physics, Univ Miami, 58-60; vis prof, Johns Hopkins Univ, 60-61; from assoc prof to prof, 61-78, REGENT'S PROF PHYSICS, GA INST TECHNOL, 78- *Concurrent Pos:* Consult, Solid State Div, Oak Ridge Nat Lab, 64-66. *Mem:* Fel AAAS; Am Phys Soc. *Res:* Statistical mechanics; solid state physics; nonlinear dynamics; experimental mathematics. *Mailing Add:* Sch Physics Ga Inst Technol Atlanta GA 30332

FORD, KENNETH WILLIAM, b West Palm Beach, Fla, May 1, 26; m 53, 62; c 7. THEORETICAL PHYSICS. *Educ:* Harvard Univ, BA, 48; Princeton Univ, PhD(physics), 53. *Prof Exp:* Asst, Los Alamos Sci Lab, 50-51; res assoc, Ind Univ, 53-54, from asst prof to assoc prof physics, 54-57; consult, Los Alamos Sci Lab, 57-58; from assoc prof to prof, Brandeis Univ, 58-64; prof, Univ Calif, Irvine, 64-70; prof, Univ Mass, Boston, 70-75; pres, NMex Inst Mining & Technol, 75-82; exec vpres, Univ Md, 82-83; pres, Molecular Biophysics Technol, 83-87; EXEC DIR, AM INST PHYSICS, 88- *Concurrent Pos:* Fulbright fel, Max Planck Inst, Goettingen, 55-56; NSF fel, Mass Inst Technol & Imp Col, London, 61-62; comnr, Col Physics, 68-71; chmn, Forum on Physics & Soc, Am Phys Soc, 82. *Honors & Awards:* Distinguished Serv Citation, Am Asn Physics Teachers, 76. *Mem:* Fel Am Phys Soc; Am Asn Physics Teachers (pres, 72). *Res:* Nuclear theory; elementary particle theory; physics education. *Mailing Add:* Am Inst Physics 500 Sunnyside Blvd Woodbury NY 11797

FORD, LAWRENCE HOWARD, b Logan, Utah, Feb 14, 48. RELATIVITY & COSMOLOGY, QUANTUM FIELD THEORY. *Educ:* Mich State Univ, BS, 70; Princeton Univ, MA, 71, PhD(physics), 74. *Prof Exp:* Fel, Univ Wis-Milwaukee, 74-77 & King's Col, London, 77-79; vis asst prof, dept physics & astron, Univ NC, Chapel Hill, 79-80; asst prof, 80-85, ASSOC PROF PHYSICS, TUFTS UNIV, 85- *Mem:* Am Phys Soc; Int Soc Gen Relativity & Gravitation; NY Acad Sci; Am Asn Univ Prof. *Res:* General relativity theory; quantum field theory; problems related to the very early universe and quantum gravity. *Mailing Add:* Dept Physics & Astron Tufts Univ Medford MA 02155

FORD, LESLIE, ONCOLOGY. *Prof Exp:* BR CHIEF COMMUNITY ONCOL & REHAB, NAT CANCER INST, 87- *Mailing Add:* NIH Nat Cancer Inst Community Oncol & Rehab Br Exec Plaza N Rm 300D Bethesda MD 20892

FORD, LESTER RANDOLPH, JR, b Houston, Tex, Sept 23, 27; m 50, 68; c 9. MATHEMATICS. *Educ:* Univ Chicago, PhB, 49, SM, 50; Univ Ill, PhD(math), 53. *Prof Exp:* Asst, Univ Ill, 50-53; res instr, Duke Univ, 53-54; mathematician, Rand Corp, 54-57; dir opers res, Gen Anal Corp, 57-60; proj mgr, CEIR, Inc, 60-63; head comput sci dept, 63-73, SR SCIENTIST, GEN RES CORP, 73- *Mem:* Soc Indust & Appl Math; Math Asn Am; Opers Res Soc Am. *Res:* Point set topology; operations research; network flow theory; system simulation. *Mailing Add:* Gen Res Corp 5383 Hollister Ave Santa Barbara CA 93111

FORD, LINCOLN EDMOND, b Boston, Mass, May 14, 38; m 73; c 4. MEDICAL PHYSIOLOGY. *Educ:* Harvard Univ, AB, 60; Univ Rochester, MD, 65. *Prof Exp:* Intern med & surg, Bassett Hosp, Cooperstown, NY, 65-66; staff assoc physiol, NIH, 66-68; NIH spec fel, Peter Bent Brigham Hosp, Boston, 68-70, res fel cardiovasc sci, 68-69, res assoc, 69-70, asst resident med, 70-71; hon res fel physiol, Univ Col, Univ London, 71-74, NIH spec fel, 71-73; asst prof, 74-80, ASSOC PROF MED & CARDIOL, UNIV CHICAGO, 81- *Concurrent Pos:* adj prof biomed eng, Northwestern Univ, 78-88; estab investr, Am Heart Asn, 75-80. *Mem:* Am Physiol Asn; Soc Gen Physiologist; Biophys Soc. *Res:* Muscle physiology; cardiovascular sciences. *Mailing Add:* Cardiol Sect Dept Med Box 249 Univ Chicago Hosps 5841 S Maryland Ave Chicago IL 60637

FORD, LORETTA C, b New York, NY, Dec 28, 20; m; c 1. PUBLIC HEALTH. *Educ:* Univ Colo, BS, 49, MS, 51, EdD, 61. *Prof Exp:* Staff nurse, Pub Health Nursing Serv, Boulder County, Colo, 48-50; supvry nursing dir, Boulder City County Health Dept, 51-55, 56-58; asst prof, Sch Nursing, Univ Colo, 55-56, 60-61, from assoc prof to prof, 62-72; prof & dir, Nursing, 72-86, actg dean, Grad Sch Educ & Human Develop, 88, EMER PROF NURSING, UNIV ROCHESTER, 86-, EMER DEAN, 88- *Concurrent Pos:* Vis prof, Univ Fla, 68 & Univ Wash, 75. *Honors & Awards:* Linda Richards Award, Nat League Nursing, 80; Gustave O Leinhard Award, Inst Med Nat Acad Sci, 90; Ruth B Freeman Award, Am Pub Health Asn. *Mem:* Inst Med-Nat Acad Sci; fel Acad Nursing; Am Nurses Asn; Am Pub Health Asn; Am Sch Health Asn. *Mailing Add:* 80 Dambury Circle S Rochester NY 14618

FORD, MICHAEL EDWARD, b Sunbury, Pa, Dec 27, 48. CHEMISTRY. *Educ:* Lehigh Univ, BA, 70, BA, 70; Univ EAnglia, PhD(chem), 73. *Prof Exp:* NIH res assoc chem, Colo State Univ, 73-75; RES CHEMIST, AIR PROD & CHEM, PA, 75- *Concurrent Pos:* Adj prof chem, Cedar Crest Col, 81-82. *Mem:* Am Chem Soc; Chem Soc London; Org Reactions Catalysis Soc; Catalysis Club. *Res:* Application of organic and organometallic principles to the discovery and optimization of novel routes to commercially important chemicals. *Mailing Add:* Air Prod Chem 7201 Hamilton Blvd Allentown PA 18195-1501

FORD, MILLER CLELL, JR, b Lake Village, Ark, Mar 24, 29; m 61; c 2. CIVIL ENGINEERING. *Educ:* Univ Ark, BSCE, 52, MSCE, 60; Okla State Univ, PhD, 73. *Prof Exp:* Engr, Chance Vought Aircraft Co, 52-55; from asst prof to assoc prof, 59-77, PROF CIVIL ENG, UNIV ARK, FAYETTEVILLE, 77- *Mem:* Am Soc Civil Engrs; Asn Asphalt Paving Technol; Am Soc Testing Mat; Am Cong Surv & Mapping. *Res:* Flexible pavement research on relationship of pavement and soil physical properties with pavement performance; asphalt paving durability and skid resistance investigation; evaluation of asphalt emulsion surface treatment characteristics and performance. *Mailing Add:* Dept Civil Eng Bel Eng Ctr Rm 4149 Univ Ark Fayetteville AR 72701

FORD, NEVILLE FINCH, b Greenock, Scotland, Nov 30, 34; US citizen; m 78. ORGANIC CHEMISTRY, MEDICINE. *Educ:* Univ Bristol, BSc, 55, PhD(org chem), 58; Wash Univ, MD, 85. *Hon Degrees:* DSc, Univ Bristol, 75. *Prof Exp:* Fel, Wayne State Univ, 58-59 & Stanford Univ, 59-60; sr res chemist, Ciba Pharmaceut Co, 60-68, mgr chem res, Ciba-Geigy Corp, 68-69, dir, 69-71, exec dir, Chem Res Pharm Div, 71-80; house physician, Jewish Hosp, St Louis, 85-88; assoc dir clin pharmacol, Squibb Inst Med Res, 88-90, DIR CLIN PHARMACOL, BRISTOL-MYERS SQUIBB PHARMACEUT RES INST, PRINCETON, NJ, 90- *Mem:* Am Chem Soc; Am Med Asn; Am Col Physicians; Am Soc Clin Pharmacol & Therapeut. *Res:* Clinical pharmacology and medicinal chemistry. *Mailing Add:* 47 Laurel Wood Dr Lawrenceville NJ 08648-1043

FORD, NORMAN CORNELL, JR, b Springfield, Mass, Feb 9, 32; m 55; c 3. BIOPHYSICS. *Educ:* Mass Inst Technol, BS, 53; Syracuse Univ, MS, 60; Univ Calif, Berkeley, PhD(physics), 64. *Prof Exp:* Mem staff physics, Mass Inst Technol, 64-65; from asst prof to assoc prof, Univ Mass, Amherst, 64-74, prof physics, 74-,; AMHERST PROCESS INSTRUMENTS. *Concurrent Pos:* Guggenheim fel, 71-72; pres, Langley-Ford Instruments, Inc, 77- *Mem:* Am Phys Soc. *Res:* Studies of conformational changes in biological macromolecules. *Mailing Add:* Amherst Process Instruments Hadley MA 01035

FORD, NORMAN LEE, b Osage City, Kans, Dec 18, 34; m 54; c 3. ORNITHOLOGY, ECOLOGY. *Educ:* Univ Kans, BA, 57; Univ Mich, MS, 62, PhD(zool), 67. *Prof Exp:* Tech asst ornith, Univ Mich Mus Zool, 57-67; from asst prof to assoc prof, 67-78, chmn dept, 72-79, PROF BIOL, ST JOHN'S UNIV, MINN, 78- *Mem:* Am Ornith Union; Wilson Ornith Soc; Cooper Ornith Soc; Evolution Soc Am; Animal Behav Soc. *Res:* Anatomy and systematics of owls; reproductive behavior and ecology of passerine birds. *Mailing Add:* Dept Biol St John's Univ Collegeville MN 56321

FORD, PATRICK LANG, b Lake Charles, La, Sept 23, 27; m 50; c 6. MATHEMATICS. *Educ:* La State Univ, BS, 48, MS, 49; Univ Mo, PhD, 61. *Prof Exp:* Instr math, Kemper Mil Sch, 49-50; asst prof, McNeese State Col, 50-55; instr, Univ Mo, 55-56; assoc prof, McNeese State Univ, 56-65, head dept, 62-80; RETIRED. *Concurrent Pos:* Dir, NSF Inserv Inst. *Mem:* Math Asn Am. *Res:* Algebra; statistics. *Mailing Add:* 401 Dolby St Lake Charles LA 70605

FORD, PETER CAMPBELL, b Salinas, Calif, July 10, 41; m 63; c 2. INORGANIC CHEMISTRY. *Educ:* Calif Inst Technol, BS, 62; Yale Univ, MS, 63, PhD(chem), 66. *Prof Exp:* NSF fel, 66-67; from asst prof to assoc prof, 67-77, PROF CHEM, UNIV CALIF, SANTA BARBARA, 77- *Concurrent Pos:* Camille & Henry Dreyfus Found teacher-scholar grant, 71; vis fel, Res Sch Chem, Australian Nat Univ, 74; guest prof, H C Oasted Inst, Univ Copenhagen, Denmark, 81. *Mem:* Am Chem Soc; Royal Soc Chem. *Res:* Transition metal chemistry; reactions of coordinated ligands; homogeneous catalysis; oxidation mechanisms; photochemistry of metal complexes. *Mailing Add:* Dept Chem Univ Calif Santa Barbara CA 93106

FORD, RICHARD EARL, b Des Moines, Iowa, May 25, 33; m 54; c 4. VIROLOGY, PLANT PATHOLOGY. *Educ:* Iowa State Univ, BS, 56; Cornell Univ, MS, 59, PhD(plant path), 61. *Prof Exp:* Res technician plant path, Iowa State Univ, 53-55, res technician plant breeding, 55, asst bot, 55-56, instr, 56; asst plant path, Cornell Univ, 56-59, asst plant virol, 59-61; asst prof plant virol, Ore State Univ & res plant pathologist, US Dept Agr, 61-65; from assoc prof to prof plant virol, Iowa State Univ, 65-72; PROF PLANT PATH & HEAD DEPT, UNIV ILL, URBANA, 72- *Concurrent Pos:* Mem, Int Working Group Legume Viruses; int consult plant path & gen agr, India, Pakistan, Sri Lanka, Thailand, Taiwan, South Korea, Peru, Iran, Yugoslavia, Peoples Repub China. *Honors & Awards:* Distinguished Serv Award, Am Phytopath Soc. *Mem:* Am Phytopath Soc (secy, 71-74, vpres, 80-81, pres elect, 81-82, pres, 82-83); fel AAAS; Am Inst Biol Sci; fel Nat Acad Sci, India. *Res:* Plant virus; legume and corn viruses; serology; electron microscopy; virus effects on host physiology; interaction of viruses and other pathogens. *Mailing Add:* Dept Plant Path Univ Ill 1102 S Goodwin Ave Urbana IL 61801

FORD, RICHARD FISKE, b Los Angeles, Calif, Mar 7, 34; m 57; c 2. MARINE ECOLOGY, WATER POLLUTION. *Educ:* Pomona Col, BA, 56; Stanford Univ, MA, 59; Scripps Inst Oceanog, Univ Calif, PhD(oceanog), 65. *Prof Exp:* From asst prof to assoc prof ecol, 64-71, dir, Ctr Marine Studies, 74-86, PROF ECOL, SAN DIEGO STATE UNIV, 71- *Concurrent Pos:* Prin investr, Marine Res Comt Grant, State of Calif, 65-67, NSF grants, 68-70, 80-81 & 86-87, Nat Oceanic & Atmospheric Agency grants, 70-; prin investr & consult, San Diego Gas & Elec Co Res Contracts, 68-; consult, Environ Eng Lab, Inc, 68-72, Bissett-Berman, Inc, 70-71, Ocean Sci & Eng, Inc, 70-, David D Smith & Assocs, 73-, Intersea Res Corp, 73-, State of Calif Water Qual Control Bd, 75-76, Calif State Coastal Conservancy, 88-90 & Michael Brandman Assocs, 87-; res contracts, Res & Develop Prog, Southern Calif Edison Co, 73-78, Calif Dept Fish & Game, 85-, USN, 87-, Calif Regional Water Qual Control Bd, 87-90. *Mem:* AAAS; Am Soc Limnol & Oceanog; Ecol Soc Am; Am Asn Univ Professors. *Res:* Population ecology, feeding relationships and related behavior of benthic marine animals; ecological effects and beneficial uses of thermal effluent; ecological effects of marine pollution. *Mailing Add:* Dept Biol San Diego State Univ San Diego CA 92182-0420

FORD, RICHARD WESTAWAY, b London, Ont, Dec 14, 30; m 52; c 2. INDUSTRIAL CHEMISTRY, POLYMER CHEMISTRY. *Educ:* Univ Western Ont, BSc, 52; Queen's Univ, Ont, MA, 54; McMaster Univ, PhD, 57. *Prof Exp:* Chemist, Dow Chem Can, Inc, 57-65, group leader Polymer Chem Sect, Res Dept, 65-72, mgr inorg chem prod distrib & planning, 72-74, develop mgr chlor-alkali prod, 74-79, mgr, plastics eng, 79-81, mgr prod flow, 81-85, community affairs mgr, 85-88, EXEC DIR, INST CHEM SCI & TECHNOL, DOW CHEM CAN, INC, 89- *Mem:* Fel Chem Inst Can. *Res:* Polymer chemistry. *Mailing Add:* 1371 Indian Rd N Sarnia ON N7V 4C7 Can

FORD, ROBERT SEDGWICK, b Pascagoula, Miss, Aug 8, 16; m 37; c 4. FOOD SCIENCE. *Educ:* Ga Sch Technol, BS, 38. *Prof Exp:* Jr marine engr, Navy Civil Serv, 38-41; chief mech engr, Ingalls Shipbuilding Corp, 41-48; chief marine engr, chief eng submarines & chief res engr, Ingalls-Litton Ship Building Corp, 52-66; OWNER, ROBERT FORD ASSOC, 48-; PRES, MAGNOLIA LAB, 61- *Concurrent Pos:* Chief mech engr, Ingalls Shipbuilding Corp, 52-60; chief submarine engr, Ingalls-Litton Shipbuilding Corp, 60-64; chief res engr, Ingalls-Litton Shipbuilding Corp, 64-66. *Mem:* AAAS. *Res:* Nutritional pathology; relationship of nutrition to degenerative diseases; engineering research and development; arteriosclerosis; development of internal combustion engines and geophysical power sources. *Mailing Add:* 701 Beach Blvd Pascagoula MS 39567

FORD, STEPHEN PAUL, b Palo Alto, Calif, Oct 11, 48; m 70; c 3. REPRODUCTIVE PHYSIOLOGY, BIOLOGY. *Educ:* Ore State Univ, BS, 71, PhD(reprod physiol), 77; WVa Univ, MS, 73. *Prof Exp:* Res asst, Dept Animal & Vet Sci, WVa Univ, 71-73 & Dept Animal Sci, Ore State Univ, 74-77; res physiologist, US Meat Animal Res Ctr, Sci & Educ Admin, USDA, 77-79; PROF, DEPT ANIMAL SCI, IOWA STATE UNIV, 79- *Concurrent Pos:* Younger animal sci res award, Midwest Sect, Am Soc Animal Sci, 88. *Mem:* Am Soc Animal Sci; Soc Study Reprod; Sigma Xi. *Res:* Role of the uterus and ovaries in potentiating conceptus survival in domestic animals, specifically, control of uterine and ovarian blood flow by the conceptus. *Mailing Add:* Dept Animal Sci Iowa State Univ Ames IA 50011

FORD, SUSAN HEIM, b New Castle, Ind, May 3, 43; m 68; c 2. BIOCHEMISTRY. *Educ:* Univ Mich, Ann Arbor, BS, 65; Univ Chicago, PhD(biochem), 69. *Prof Exp:* Grant, Dept Biochem, Univ Chicago, 69-71; instr biol, Kennedy-King Col, 73; res assoc, Dept Biochem, Univ Chicago, 73-77; ASST PROF CHEM, DEPT PHYS SCI, CHICAGO STATE UNIV, 78- *Concurrent Pos:* USPHS grant, Univ Chicago, 70-71; instr gen sci, Loyola Univ, Chicago, 75-77; res grant, Chicago State Univ, 83-87. *Mem:* Sigma Xi; Am Chem Soc. *Res:* Biosynthesis of porphyrins, corrins and chlorins; biosynthesis of precursors to these substances and related enzymology. *Mailing Add:* Dept Phys Sci Chicago State Univ 95th Atking Dr Chicago IL 60626

FORD, THOMAS AVEN, b Washington, DC, Aug 29, 17; m 44; c 4. CHEMISTRY. *Educ:* Univ Wyo, AB, 37; Yale Univ, PhD(chem), 40. *Prof Exp:* Asst chem, Yale Univ, 37-39; res chemist, Exp Sta, E I du Pont de Nemours & Co Inc, 40-64, supvr pressure res labs, 64-66, mgr, Tech Facil Div, 66-81; RETIRED. *Mem:* AAAS; Am Chem Soc. *Res:* Organic chemistry; fluorine chemistry; polymers. *Mailing Add:* 617 Horseshoe Hill Rd Hockessin DE 19707-9570

FORD, THOMAS MATTHEWS, b Cambridge, Ohio, Sept 11, 31; m 59; c 2. VETERINARY MEDICINE. *Educ:* Mich State Univ, BS, 53, DVM, 57, MS, 59; Am Col Lab Animal Med, dipl. *Prof Exp:* Asst, Mich State Univ, 57-58; asst, Res Dept, Kellogg Co, 58-60; instr vet hyg, Iowa State Univ, 60-62; fel, Bowman Gray Sch Med, 62-63; fel, Med Sch, Univ Mich, 63-64, dir, animal diag lab, 64-67; VET, ABBOTT LABS, 67- *Mem:* Am Vet Med Asn; Am Asn Lab Animal Sci. *Res:* Nutrition, pathology and diseases of fur bearing animals; diseases of laboratory animals. *Mailing Add:* Abbott Labs One Abbott Park Rd Abbott Park IL 60064

FORD, WARREN THOMAS, b Kalamazoo, Mich, Mar 22, 42; m 67; c 2. POLYMER CHEMISTRY, ORGANIC CHEMISTRY. *Educ:* Wabash Col, AB, 63; Univ Calif, Los Angeles, PhD(org chem), 67. *Prof Exp:* NSF res fel org chem, Harvard Univ, 67-68; asst prof chem, Univ Ill, Urbana, 68-75; res chemist, Rohm & Haas Co, 75-78; from asst prof to assoc prof, 78-83, PROF CHEM, OKLA STATE UNIV, 83- *Concurrent Pos:* Vis prof, Max Planck Inst Polymer Res, Mainz, Ger, 85-86. *Honors & Awards:* Fulbright Fel. *Mem:* Am Chem Soc; fel AAAS; Mat Res Soc. *Res:* Polymer and organic chemistry; polymeric reagents and catalysts; colloids, liquid crystals, NMR spectroscopy; current major projects are catalysis by polymer colloids and polymeric liquid crystals. *Mailing Add:* Chem Dept Okla State Univ Stillwater OK 74078

FORD, WAYNE, b Iowa City, Iowa, Feb 9, 31. MATHEMATICS. *Educ:* Iowa City Univ, BA, 52; Univ Iowa, BS, 53; Rice Univ, PhD(math), 64. *Prof Exp:* PROF MATH, TEX TECH UNIV, 67- *Mem:* Math Asn Am. *Mailing Add:* 3613 68th St Lubbock TX 79413

FORD, WAYNE KEITH, b Buffalo, NY, Dec 21, 53; m 76; c 3. SOLID STATE PHYSICS, ELECTRONIC STRUCTURE OF POLYMERS. *Educ:* Rochester Inst Technol, BS, 76; Brown Univ, ScM, 78 & PhD(physics), 81. *Prof Exp:* Fel solid state physics, Xerox Webster Res Ctr, 80-83; res investr solid state physics, Univ Pa, 83-84; SR MAT ENGR, INTEL CORP, 90- *Mem:* Am Phys Soc; Am Vacuum Soc. *Res:* Theoretical description of both the electronic properties of organic polymers and the atomic structure of semiconductor surfaces; theory of few-body systems; low energy electron diffraction theory and experiment. *Mailing Add:* AL3-15 Intel Corp 5200 NE Elam Young Hillsboro OR 97124

FORD, WILLIAM ELLSWORTH, b Richmond, Va, Nov 28, 50. PHOTOCHEMISTRY, ELECTRON TRANSFER. *Educ:* Cornell Univ, BS, 73; Univ Calif, Berkeley, PhD(phys chem), 82. *Prof Exp:* Postdoctoral fel, Dept Chem & Biochem, Univ Ariz, 80-85; res assoc, Radiation Lab, Univ Notre Dame, 85-87 & Ctr Fast Kinetics Res, Univ Tex, Austin, 87; SR RES ASSOC, CTR PHOTOCHEM SCI, BOWLING GREEN STATE UNIV, 88- *Mem:* Am Chem Soc; Am Soc Photobiol. *Res:* Design of new molecular assemblies for investigating mechanisms of light-initiated energy and electron transport through lipid bilayers using laser flash photolysis techniques. *Mailing Add:* Dept Chem Bowling Green State Univ Bowling Green OH 43403

FORD, WILLIAM FRANK, b Rockford, Ill, Jan 27, 34. THEORETICAL PHYSICS, APPLIED MATHEMATICS. *Educ:* John Carroll Univ, BS, 55, MS, 56; Case Inst Technol, PhD(physics), 61. *Prof Exp:* Instr physics, John Carroll Univ, 55-56; engr acoustics, Clevite Corp, 56-57; instr physics, Case Inst Technol, 59-61; engr reactor physics, 61-63, physicist nuclear physics, 63-73, mathematician, 73-75, HEAD MATH ANALYSIS SECT, LEWIS RES CTR, NASA, 75- *Mem:* Am Phys Soc. *Res:* Quantum theory of scattering; approximation theory; fluid mechanics. *Mailing Add:* Dept Math & Comp Sci Bucks Co Community Col Newton PA 18940

FORD, WILLIAM KENT, JR, b Clifton Forge, Va, Apr 8, 31; m 61; c 3. ASTRONOMY. *Educ:* Washington & Lee Univ, BA, 53; Univ Va, MS, 55, PhD(physics), 57. *Prof Exp:* STAFF MEM ASTRON, DEPT TERRESTRIAL MAGNETISM, CARNEGIE INST WASHINGTON, DC, 57- *Concurrent Pos:* Vis resident scientist, Kitt Peak Nat Observ, 73-74. *Mem:* Am Astron Soc; Am Phys Soc. *Res:* Astronomical instrumentation; electronic image intensification; galaxies; rotation of galaxies; redshifts of galaxies. *Mailing Add:* Dept Terrestrial Magnetism Carnegie Inst Washington 5241 Broad Branch Rd NW Washington DC 20015

FORDEMWALT, JAMES NEWTON, b Parsons, Kans, Oct 18, 32; m 63; c 2. INTEGRATED CIRCUITS PROCESSING. *Educ:* Univ Ariz, BS, 55, MS, 56; Univ Iowa, PhD(phys chem), 60. *Prof Exp:* Sr engr, Gen Elec Corp, Evandale, Ohio, 59-60 & US Semcor, Phoenix, Ariz, 60-61; sect mgr, Motorola Semiconductor Div, Phoenix, Ariz, 61-66; dept mgr, Philco-Ford Microelectronics, Santa Clara, Calif, 66-68; assoc dir res, Am Microsysts Inc, Santa Clara, Calif, 68-71; adj res prof, Inst Biomed Eng, Univ Utah, Salt Lake City, 71-76; dir microelectronic lab, elec & comput dept, Univ Ariz, Tucson, 76-87; ASSOC PROF ELEC & COMPUT TECHNOL & LAB DIR, ARIZ STATE UNIV, TEMPE, 87- *Concurrent Pos:* Instr, Ariz State Univ, Tempe, 62-64; consult, Integrated Circuits Eng Co, Scottsdale, Ariz, 76-, United Technologies Corp, E Hartford, Conn, 77-, Eastman Kodak Co, Rochester, NY, 79-80 & Hughes Semiconductor Div, Anaheim, Calif, 85-; UN develop prog expert. *Mem:* Sigma Xi; Electrochem Soc; Inst Elec & Electronics Engrs; Int Soc Hybrid Microelectronics. *Res:* Microelectronic devices, circuits and processing methods; integrated circuit device design. *Mailing Add:* 613 W Summit Pl Chandler AZ 85224

FORDHAM, CHRISTOPHER COLUMBUS, III, b Greensboro, NC, Nov 28, 26; m 47; c 3. INTERNAL MEDICINE. *Educ:* Harvard Med Sch, MD, 51. *Prof Exp:* From instr to prof med, Sch Med, Univ NC, Chapel Hill, 58-69, asst dean, 65-68, assoc dean clin sci, 68-69; prof med, vpres & dean sch med, Med Col Ga, 69-71; prof & dean med, 71-79, vchancellor health affairs, 79-80, chancellor, 80-88, EMER CHANCELLOR & PROF MED, UNIV NC, CHAPEL HILL, 88- *Concurrent Pos:* Actg asst to Secy Health, HEW, 77; mem, gov coun, IOM, NAS, 85-90; mem, Comn on Educ Quality, S Regional Educ Bd, 86-87. *Mem:* Inst Med-Nat Acad Sci; fel AAAS; fel Am Col Physicians; Am Fedn Clin Res. *Res:* Metabolism; renal physiology and disease; disease and disorders of water and electrolyte homeostasis. *Mailing Add:* Sch Med Univ NC 413 MacNider Bldg Campus Box 7000 Chapel Hill NC 27599-7000

FORDHAM, JAMES LYNN, b Rodney, Can, Mar 27, 24; US citizen; m 46; c 4. BIOELECTRIC DEVICES, TECHNOLOGY DEVELOPMENT. *Educ:* Univ Western Ont, BSc, 46. *Prof Exp:* Chemist, Uniroyal Can, Polymer Corp & Monsanto, 46-55; sr res chemist, Diamond Shamrock Corp, 55-56, group leader, 57-59, mgr polymer res & develop, 59-65, from asst dir to dir res, 65-69, vpres res & new bus, 69-81; pres, FTI, 81-83; PRES, BIO-ELECTRIC INC, 84- *Concurrent Pos:* Consult, 81- *Honors & Awards:* Eng Mat Achievement Award, Am Soc Metals, 76. *Mem:* Am Chem Soc. *Res:* Synthesis, structure, properties and uses of biochemicals, electrochemicals, polymers, specialty chemicals and related products; processes; induced electric fields and electrotherapy. *Mailing Add:* 9701 Franklin Hill Blvd Knoxville TN 37922-3332

FORDHAM, JOSEPH RAYMOND, b Hornell, NY, Sept 1, 37; m 62; c 4. NUTRITIONAL BIOCHEMISTRY, FOOD & BIOPESTICIDES. *Educ:* Col Holy Cross, BS, 59; Purdue Univ, MS, 63, PhD(biochem), 66. *Prof Exp:* Asst biochem, Purdue Univ, 60-65; res scientist, Joseph E Seagram & Sons, 65-69; asst prof nutrit & food sci, Univ Ky, 69-76; assoc prof food & nutrit, Ga Southern Col, 76-77; mgr tech regulatory compliance, Clinton Corn Processing Co, 77-82; mgr regulatory affairs, 82-88, DIR, REGULATORY AFFAIRS, NOVO NORDISK BIOINDUSTRIALS, INC, 88- *Mem:* Food & Drug Law Inst; Inst Food Technologists; Am Dietetics Asn. *Res:* Carbohydrate metabolism; rumen microbiology; short-chain fatty acid metabolism; diastatic and proteolytic enzymes; regulatory affairs relating to enzymes, biopesticides and other biotechnologyy products; novel plant protein products; nutrition status; regulatory affairs. *Mailing Add:* 370 Cross Hill Rd Monroe CT 06468

FORDHAM, WILLIAM DAVID, b Marietta, Ohio, Apr 24, 39; m 64. BIOPHYSICAL CHEMISTRY, BIOCHEMISTRY. *Educ:* Marietta Col, BS, 61; Yale Univ, MS, 62, PhD(biophys chem), 67. *Prof Exp:* Res chemist, Am Cyanamid Co, Bound Brook, 66-71; NIH spec res fel, Princeton Univ, 71-73; from asst prof to assoc prof, 73-82, PROF CHEM, FAIRLEIGH DICKINSON UNIV, 82- *Concurrent Pos:* Vis res prof, Univ Cambridge, Eng, 82, Columbia Univ, 87. *Res:* Protein structure and dynamics; molecular graphics; biological reaction mechanisms; chemical kinetics; structure of bacterial cell walls; antibiotics. *Mailing Add:* 216 Voorhis Ave Fairleigh Dickinson Univ Teaneck Campus River Edge NJ 07661

FORD-HUTCHINSON, ANTHONY W, b Eng, Feb 5, 47; Can & Brit citizen; m 70; c 3. LEUKOTRIENES, PULMONARY PHARMACOLOGY. *Educ:* Univ Birmingham, BSc, 68; Univ Warwick, MSc, 69; Univ London, PhD(biochem), 72. *Prof Exp:* Res asst chem path, St George's Hosp Med Sch, 69-72; res fel biochem pharmacol, King's Col Hosp Med Sch, 72-73, hon lectr, 75-77, lectr chem path, 77-81; dir, 81-84, SR DIR PHARMACOL, MERCK FROSST CAN INC, 84-; ADJ PROF, DEPT PHARMACOL &

THERAPEUT, MCGILL UNIV, 88- *Mem:* Brit Pharmacol Soc; Am Soc Pharmacol & Exp Therapeut; Am Asn Immunol; Pharmacol Soc Can; Skin Pharmacol Soc. *Res:* Arachidonic acid metabolism to leukotrienes; prostaglandins and thromboxanes; role of arachidonic acid metabolites in pulmonary, allergic, cardiovascular and inflammatory diseases; mechanisms involved in and models for bronchial asthma. *Mailing Add:* Dept Pharmacol Merck Frosst Can Inc PO Box 1005 Point Claire-Dorval PQ H9R 4P8 Can

FORDON, WILFRED AARON, b New York, NY, Sept 6, 27; m 50; c 2. PATTERN RECOGNITION, ELECTROMAGNETIC THEORY. *Educ:* City Col New York, BEE, 49; Hofstra Univ, MA, 54; Purdue Univ, PhD(elec eng), 76. *Prof Exp:* Jr engr, Link Radio Corp, 50; engr, Airborne Instruments Lab, 51-54, Teletronics Lab, 54-56; mem staff, Lincoln Lab, Mass Inst Technol, 56-58; supvr, Sylvania Electronic Systs, 58-61; mem tech staff, Mitre Corp, 61-67; prin engr, Raytheon Co, 67-69; teaching asst, Purdue Univ, 73-76; vis asst prof, 76-77, ASSOC PROF ELEC ENG, MICH TECHNOL UNIV, 77- *Mem:* Sr mem Inst Elec & Electronics Engrs; AAAS; NY Acad Sci; Sigma Xi. *Res:* Pattern recognition as applied to biomedical engineering; electromagnetic theory as applied to radiowave propagation. *Mailing Add:* RR 3 Box 291 Sag Harbor NY 11963-8864

FORDYCE, DAVID BUCHANAN, b Los Angeles, Calif, Oct 15, 24; m 47; c 3. PHYSICAL CHEMISTRY, ANALYTICAL CHEMISTRY. *Educ:* Univ BC, BASc, 46; Univ Wis, PhD(chem), 50. *Prof Exp:* Asst chem, Univ Wis, 47-48; res chemist, Rohm & Haas Co, 50-56, head lab, 56-63, res dept mgr, 63-85; RETIRED. *Mem:* Am Chem Soc. *Res:* Application properties of ion exchange resins, water soluble polymers and surfactants. *Mailing Add:* 813 12th Pl N Edmonds WA 98020

FORDYCE, JAMES STUART, b London, Eng, Dec 10, 31; nat US; m 54; c 2. PHYSICAL CHEMISTRY, ELECTROCHEMISTRY. *Educ:* Dartmouth Col, AB, 53; Mass Inst Technol, PhD(phys chem), 59. *Prof Exp:* Asst, Mass Inst Technol, 53-59; res chemist, Nat Carbon Co Div, Union Carbide Corp, 58-59, res chemist, Union Carbide Consumer Prod Co Div, 59-63, res chemist develop dept, 63-66; res scientist, NASA, 66-68, sect head, Direct Energy Conversion Div, 68-74, chief environ res off, 71-74, mgr environ monitoring systs off, energy conversion & environ systs div, Lewis Res Ctr, 74-76, chief electrochem br, 76-80, dep chief, 80-81, actg chief, solar & electrochem div, 81-84, chief, Space Power Technol Div, 84-85, DIR, AEROSPACE TECHNOL, NASA, 85- *Concurrent Pos:* Mem environ rev team jetport study, Lake Erie Regional Transp Auth, Cleveland, Ohio, 74-78; lectr, Electrochem Soc, 85; lectr, Int Space Univ, 88-; adj fac, Ohio Aerospace Inst, 90- *Honors & Awards:* Indust Res-100 Award, 77; Except Serv Medal, NASA, 84; Presidential Rank Meritorious Exec Award, 87. *Mem:* AAAS; Am Chem Soc; Electrochem Soc; Fedn Am Scientists; Am Inst Aeronaut & Astronaut. *Res:* Photochemistry; fast reaction kinetics; spectroscopy of molten salts and electrolyte solutions; solid electrolytes; urban air pollution; aircraft pollution; environmental monitoring technology; energy conversion; batteries, fuel cells, bulk energy storage systems; photovoltaic technology and applications; space power systems; technology transfer and national policy. *Mailing Add:* Lewis Res Ctr NASA 21000 Brookpark Rd Cleveland OH 44135

FORDYCE, WAYNE EDGAR, ACCLIMATIZATION TO CHRONIC HYPOXIA. *Educ:* Univ Southern Calif, BS, 70, MS, 71, PhD(biomed eng), 76. *Prof Exp:* ASST PROF PHYSIOL, HEALTH SCI CTR, STATE UNIV NY, SYRACUSE, 81- *Mailing Add:* Dept Pediat State Univ Ny Upstate Med Ctr 788 Irving Ave Syracuse NY 13210

FOREE, EDWARD G(OLDEN), b Sulphur, Ky, Feb 24, 41; m 62; c 2. ENVIRONMENTAL & CIVIL ENGINEERING. *Educ:* Univ Ky, BS, 64; Stanford Univ, MS, 65, PhD(civil eng), 68. *Prof Exp:* Asst prof, 68-74, assoc prof civil eng, Univ Ky, 74-80; PRES, COMMONWEALTH TECHNOL INC, 80- *Mem:* Am Soc Civil Engrs; Water Pollution Control Fedn; Nat Soc Prof Engrs; Am Acad Environ Engrs; Am Water Works Asn. *Res:* Growth and decomposition of algae and related effects on water quality; plant nutrients and eutrophication; acid mine drainage; biological waste treatment. *Mailing Add:* Commonwealth Technol Inc 2520 Regency Rd Lexington KY 40503-2961

FOREGGER, THOMAS H, b Madison, Wis, June 26, 46; m 71; c 1. DATA COMMUNICATIONS NETWORK DESIGN. *Educ:* Univ Wis-Madison, BA, 68, PhD(math), 73. *Prof Exp:* Lectr math, Univ Wis, 73-74; MEM TECH STAFF, AT&T BELL LABS, 74- *Concurrent Pos:* Prin investr, Univ Wis-Madison, 73-74. *Mem:* Math Asn Am; Soc Indust & Appl Math. *Res:* Nearly decomposable matrices and permanents of doubly stochastic matrices; Investigations of non-negative matrices. *Mailing Add:* 14 Dorset Rd Berkeley Heights NJ 07922

FORELLI, FRANK JOHN, b San Diego, Calif, Apr 8, 32; m 59; c 2. MATHEMATICS. *Educ:* Univ Calif, Berkeley, AB, 54, MA, 59, PhD(math), 61. *Prof Exp:* Instr, 61-63, from asst prof to assoc prof, 63-67, PROF MATH, UNIV WIS-MADISON, 67- *Mem:* Am Math Soc; Math Asn Am. *Res:* Mathematical analysis. *Mailing Add:* Dept Math Univ Wis 213 Van Vleeck Hall Madison WI 53706

FOREMAN, BRUCE MILBURN, JR, b Arkadelphia, Ark, Nov 10, 32; m 64; c 1. PHYSICS. *Educ:* Univ Calif, Berkeley, BS, 54, PhD(nuclear chem), 58. *Prof Exp:* Res assoc, Brookhaven Nat Lab, 58-60; res scientist chem, Columbia Univ, 60-64; asst prof physics, Univ Tex, Austin, 64-68; sr scientist, 68-75, sr systs analyst, 75-80, mgr Publ Data Processing, 80-82, MGR PUBL PROD, DIV 1, AM INST PHYSICS, 82- *Mem:* Am Phys Soc; Sigma Xi. *Res:* Mechanisms of nuclear reactions. *Mailing Add:* 72 E End Ave New York NY 10028

FOREMAN, CALVIN, mathematics; deceased, see previous edition for last biography

FOREMAN, CHARLES FREDERICK, b Blue Rapids, Kans, Nov 9, 20; m 46; c 3. DAIRY SCIENCE. *Educ:* Kans State Col, BS, 48, MS, 49; Univ Mo, PhD, 53. *Prof Exp:* Asst prof dairy exten, Kans State Col, 49-51; asst instr dairy husb, Univ Mo, 51-53; asst prof, Univ Minn, 53-54; from asst prof to prof dairy sci, Iowa State Univ, 55-68, prof animal sci, 68-; RETIRED. *Mem:* Am Soc Animal Sci; Am Dairy Sci Asn. *Res:* Dairy cattle nutrition and management. *Mailing Add:* 14025 N Cameo Dr Sun City AZ 85351

FOREMAN, CHARLES WILLIAM, b Norman Park, Ga, Nov 2, 23; m 51; c 3. GENETICS. *Educ:* Univ NC, BA, 49; Duke Univ, MA, 51, PhD(physiol), 54. *Prof Exp:* Asst prof biol, Wofford Col, 53-55; asst prof zool, Univ Md, 55-56; from assoc prof to prof biol, Pfeiffer Col, 56-63; from assoc prof to prof, 63-81, chmn dept, 76-83, WILLIAM HENDERSON PROF BIOL, UNIV OF THE SOUTH, 81- *Concurrent Pos:* Res partic, Oak Ridge Nat Lab, 60; trustee, Highlands Biol Sta, Univ NC, 61-; bd sci advisors, Highlands Biol Sta, 78- *Mem:* Fel AAAS; Am Soc Mammalogists; Genetics Soc Am; NY Acad Sci; Am Physiol Soc. *Res:* Taxonomic significance and genetic basis of hemoglobin structure. *Mailing Add:* Dept Biol Univ of the South Sewanee TN 37375

FOREMAN, DARHL LOIS, b Idaho Falls, Idaho, May 3, 24. ZOOLOGY, ENDOCRINOLOGY. *Educ:* George Washington Univ, BS, 46; Mt Holyoke Col, MA, 48; Univ Chicago, PhD(zool), 55. *Prof Exp:* From instr to asst prof, 54-70, ASSOC PROF BIOL, CASE WESTERN RESERVE UNIV, 70- *Mem:* Fel AAAS; Am Soc Zoologists; Endocrine Soc; Soc Study Reproduction. *Res:* Ovarian physiology; physiology of annual breeders; animal behavior and endocrine effects; endocrinology of sex and reproduction. *Mailing Add:* Dept Biol Case Western Reserve Univ Cleveland OH 44106

FOREMAN, DENNIS WALDEN, JR, b Akron, Ohio, Dec 27, 29; m 53; c 2. MINERALOGY, PHYSICAL CHEMISTRY. *Educ:* Mt Union Col, BS, 52; Ohio State Univ, MS, 60, PhD(mineral), 66. *Prof Exp:* Res assoc, 66-68, asst prof, 68-69, assoc prof, 69-76, PROF DENT, OHIO STATE UNIV, 76- *Mem:* AAAS; Am Crystallog Asn; Electron Micros Soc Am; Int Asn Dent Res; Sigma Xi. *Res:* Crystal structures of hydrogrossular, tin apatite; mineralization in biological environments; bone and tooth structure; crystalline water in human enamel. *Mailing Add:* Col Dent Box 186 Ohio State Univ 305 W 12th Ave Columbus OH 43210

FOREMAN, HARRY, b Winnipeg, Man, Mar 5, 15; US citizen; m 55; c 2. RADIOLOGICAL HEALTH, POPULATION STUDIES. *Educ:* Antioch Col, BS, 38; Ohio State Univ, PhD(biochem), 42; Univ Calif, San Francisco, MD, 47. *Prof Exp:* Asst prof biochem, Oglethorpe Univ, 42-43; res chemist, Eldorado Oil Co, 43-44; AEC fel, Univ Calif, Berkeley, 49-51; mem staff, Los Alamos Sci Lab, 51-62; assoc prof pub health, 62-73, assoc dean int progs, 66-69, assoc prof obstet & gynec, 69-73, prof, 73-85, DIR CTR POP STUDIES, UNIV MINN, MINNEAPOLIS, 69-, EMER PROF OBSTET & GYNEC & PUB HEALTH, 85- *Concurrent Pos:* Mem atomic safety & licensing bd panel, AEC, 71- *Honors & Awards:* McNight Award, Univ Minn Press, 72. *Mem:* Soc Exp Biol & Med; Health Physics Soc; Am Soc Pharmacol & Exp Therapeut; Am Physiol Soc. *Res:* Development of contraceptives; clinical trials, family planning acceptance and administration. *Mailing Add:* 1564 Burton Ave St Paul MN 55108

FOREMAN, JOHN E(DWARD) K(ENDALL), b Hamilton, Ont, Feb 14, 22; m 47; c 4. MECHANICAL ENGINEERING. *Educ:* Univ Toronto, BASc, 45; Cornell Univ, MME, 52. *Prof Exp:* Spec lectr, Univ Toronto, 46-49; instr & res asst, Cornell Univ, 49-52; res engr, George Kent, Ltd, Eng, 52-54, head res & develop, Can, 54-56; from asst prof to assoc prof eng sci, Univ Western Ont, 56-63, head mech group, 60-72, prof mech eng, 63-87, head, Sound & Vibration Lab, 72-87; RETIRED. *Concurrent Pos:* Proj coordr, Can Elec Asn, 76- *Mem:* Can Acoust Asn (pres, 72-74); Eng Inst Can. *Res:* Biomechanics; mechanical design; noise abatement. *Mailing Add:* RR 3 Denfield ON N0M 1P0 Can

FOREMAN, KENNETH M, b New York, NY, July 30, 25; m 54; c 2. AERONAUTICAL ENGINEERING, FLUID DYNAMICS. *Educ:* NY Univ, BA, 50, MA, 53; Polytech Univ, NY, MS, 81. *Prof Exp:* Res engr, Bendix Aviation Corp, 51-52; proj engr, Wright Aeronaut Div, Curtiss-Wright Corp, NJ, 52-56; res engr, Engine Div, Fairchild Engine & Airplane Corp, NY, 56-59; specialist & sci res engr, Repub Aviation Corp, 59-65; chief proj engr, Edo Corp, 65-66; staff scientist, 66-80, head, Advan Fluid Concepts Lab, 80-90, SR STAFF SCIENTIST, CORP RES CTR, GRUMMAN CORP, BETHPAGE, 90- *Concurrent Pos:* Abstractor, Fire Res Abstr & Revs, Nat Acad Sci. *Mem:* Am Soc Mech Engrs; assoc fel Am Inst Aeronaut & Astronaut; AAAS; Am Hist Soc. *Res:* Aerospace propulsion; gas dynamics; chemical reactions resulting in heat addition; flow instrumentation; planetary atmospheres; space sciences; systems engineering; wind energy conversion engineering. *Mailing Add:* 32 Stratford Ct North Bellmore NY 11710

FOREMAN, MATTHEW D, b Los Alamos, NMex, Mar 21, 57; m 84. LARGE CARDINALS. *Educ:* Univ Colo, BA, 75; Univ Calif, Berkeley, PhD(math), 80. *Prof Exp:* Hedrick asst prof math, Univ Calif, Los Angeles, 80-82; res fel, NSF, 82-83; fel, Hebrew Univ, Jerusalem, 83-84; Bantrell fel, Calif Inst Technol, 84-86; ASSOC PROF MATH, OHIO STATE UNIV, 86-, ASSOC PROF PHILOS 90- *Res:* Foundations of mathematics; primary set theory and applications to combinatorics, algebra and analysis. *Mailing Add:* Dept Math Ohio State Univ Columbus OH 43210

FOREMAN, ROBERT DALE, b Orange City, Iowa, Apr 27, 46; m 69; c 2. NEUROPHYSIOLOGY, CARDIOVASCULAR PHYSIOLOGY. *Educ:* Cent Col, Pella, BA, 69; Loyola Univ, Chicago, PhD(physiol), 73. *Prof Exp:* Mem staff neurophysiol, Marine Biomed Inst, Galveston, 75-77; asst prof physiol, Univ Tex Med Br, Galveston, 75-77; from asst prof to assoc prof neurophysiol, 77-86, PROF & CHMN, UNIV OKLA HEALTH SCI CTR, 86- *Concurrent Pos:* Fel, Marine Biomed Inst, Univ Tex Med Br Galveston, 73-75; asst prof, NIH, Nat Heart, Lung & Blood Inst, 75-90, Am Heart Asn,

75-77 & NIH, 76-77; mem Clin Sci Study Sec, NIH, 82-86; mem prog adv comt, Am Physiol Soc, 83-87; mem & chmn, Am Heart Asn, Southern Region comt, 85-88; affil mem, Okla Med Res Found, 84-; adj mem grad fac, Okla State Univ, 86; ed bd, J Appl Physiol, 84; ed bd circulation res, 86-; mem sect adv comt, Am Physiol Soc, 81-86; mem asn, chmn dept physiol, 85-; subcomt regional Nat Res Am Heart Asn, 86-88; Am State Univ Asn Lectr, 87-88; res career develop award, NIH; dept physiol, Am Phys Soc, 87-90; mem chap comt, Soc Neurosci; George Lynn Cross prof. *Honors & Awards:* Provost Res Award, Univ Okla Health Sci Ctr, 84. *Mem:* Am Physiol Soc; Soc Neurosci; Sigma Xi; Am Pain Soc; AAAS; Int Asn Study Pain; Am Col Sport Med. *Res:* Neural mechanisms underlying the cardiac pain associated with angina pectoris; neural control of heart and blood vessels; neural mechanisms underlying hypertension and arrhythmias. *Mailing Add:* Dept Physiol & Biophys Col Med Oklahoma City OK 73190

FOREMAN, ROBERT WALTER, b Flint, Mich, Jan 4, 23; m 45; c 2. METALLURGICAL CHEMISTRY. *Educ:* Univ Mich, BS, 44, MS, 46; Western Reserve Univ, PhD, 55. *Prof Exp:* Asst chem, Manhattan Proj, 44; res assoc eng res, Univ Mich, 44-45; res chemist, Standard Oil Co, 46-66; CONSULT, PARK CHEM CO, 66- *Mem:* Am Chem Soc; fel Am Soc Metals Int. *Res:* Metallurgy; organic chemistry; petroleum chemistry; molten salt chemistry; analytical chemistry; absorption spectroscopy. *Mailing Add:* 4353 Covered Bridge Rd Bloomfield Hills MI 48013-3399

FOREMAN, RONALD EUGENE, b Osceola, Iowa, May 28, 38; div; c 1. MARINE ECOLOGY. *Educ:* Univ Colo, BA, 62; Univ Calif, Berkeley, PhD(bot), 70. *Prof Exp:* Asst proj supvr alpine ecol, Inst Arctic & Alpine Res, 58-64; radiation biologist, US Naval Radiol Defense Lab, 65 & 66; teaching assoc bot, Univ Calif, Berkeley, 68-70; asst prof, 70-79, ASSOC PROF BOT & DIR, BAMFIELD MARINE STA, UNIV BC, 79- *Mem:* Ecol Soc Am; Brit Ecol Soc; Phycol Soc Am. *Res:* Investigation of spatial-temporal variations in nearshore marine benthic ecosystem structure and function; population dynamics of the kelp Nereocystis leutkeana; phycocolloid production by red algae. *Mailing Add:* Dept Bot Univ BC 2075 Westbrook Pl Vancouver BC V6T 1W5 Can

FOREMAN, RONALD LOUIS, b Chicago, Ill, Oct 26, 37; div; c 2. PHARMACOLOGY, TOXICOLOGY. *Educ:* Univ Ill, Chicago, BS, 60, MS, 64, PhD(pharm), 69. *Prof Exp:* NIH trainee pharmacol, 69-71, ASST PROF PHARMACOL, UNIV ILL MED CTR, 71-; TOXICOLOGIST, CHICAGO BD HEALTH LABS, 71- *Concurrent Pos:* Regist Pharmacist. *Mem:* AAAS; Am Chem Soc; Am Pharmaceut Asn. *Res:* Therapeutic and toxicological aspects of drug metabolism and distribution; epoxides as obligatory intermediates in the metabolism of olefins to glycols; public health aspects of drug use and toxicity. *Mailing Add:* City Chicago Dept Health Daley Ctr Rm LL-151 Chicago IL 60602

FOREMAN, WILLIAM EDWIN, b Columbus, Ohio, Sept 6, 29; m 57; c 3. MINING ENGINEERING, MINERAL PROCESSING. *Educ:* Ohio State Univ, BEM, 53, Va Polytech Inst, MS, 61; Pa State Univ, PhD(mineral prep), 65. *Prof Exp:* Res engr, Basic Inc, Cleveland, 53-57; ASSOC PROF MINING ENG, VA POLYTECH INST & STATE UNIV, 57- *Concurrent Pos:* Chmn, Inst Coal Mining, Health, Safety & Res, 75-78; Carborundum fel. *Mem:* Am Inst Mining, Petrol & Metall Engrs; Sigma Xi. *Res:* Health and safety in mining and engineering; mineral beneficiation; coal preparation. *Mailing Add:* 907 Mason Dr Blacksburg VA 24060

FORENZA, SALVATORE, b Taranto, Italy, June 22, 46; m 73; c 2. MEDICINAL CHEMISTRY. *Educ:* Univ Naples, Italy, Dr org chem, 70. *Prof Exp:* Fel natural prod, Coun Nat Res, 70-71; teach asst org chem, Yale Univ, 71-72; fel natural prod, Univ Wyo, 72-74; fel biosynthetics, Tulane Univ, 74; vis asst prof org chem, Col VI, 75-76; sr res scientist natural prod, Farmitalia-C-Erba, Italy, 76-80; sr res scientist natural prod, 80-83, ASSOC DIR, ANTITUMOR CHEM & FERMENT, BRISTOL-MYERS CO, 83- *Mem:* Am Chem Soc; AAAS; Soc Indust Biol. *Res:* Isolation and structure elucidation of naturally occuring antitumoral compounds; semisynthesis of analogs to elucidate structure-activity relationships; fermentation and biosynthesis of antitumor agents; development of screening methodologies. *Mailing Add:* PO Box 5100 Wallingford CT 06492

FORER, ARTHUR H, b Trenton, NJ, Dec 17, 35; m 64; c 2. CELL BIOLOGY. *Educ:* Mass Inst Technol, BS, 57; Dartmouth Med Sch, PhD(molecular biol), 64. *Prof Exp:* Am Cancer Soc fel, Carlsberg Found, Copenhagen, Denmark, 64-66; Helen Hay Whitney Found fel, Cambridge, Eng & Duke Univ, 67-70; assoc prof molecular biol, Odense Univ, Denmark, 70-72; assoc prof biol, 72-75, PROF BIOL, YORK UNIV, 75- *Concurrent Pos:* Hargitt res fel, Duke Univ, 69-70; mem, Cell Biol & Genetics Grant Selection Panel, Natural Sci & Eng Res Coun Can, 75-78; assoc ed, Biochem & Cell Biol, 80-; assoc ed, J Cell Sci, 70-82; assoc ed, Cell Biol Int Reports, 85- *Mem:* AAAS; Am Soc Cell Biol; Sci Peace Can; fel Royal Soc Can. *Res:* Chromosome movement during cell-division; roles of actin filaments and microtubules in cell motility; ultraviolet microbeam irradiations of living cells; effects of insecticides on laboratory animals; cell biology. *Mailing Add:* Dept Biol York Univ Downsview ON M3J 1P3 Can

FORERO, ENRIQUE, b Bogata, Colombia, Dec 7, 42; m 66; c 1. RESEARCH ADMINISTRATION, BOTANY-PHYROPATHOLOGY. *Educ:* Nat Univ Bogota, BS, 65; City Univ NY, PhD(biol), 72. *Prof Exp:* Asst prof bot, Inst Natoral Sci, Nat Univ, Bogota, 65-66, head, Nat Herbalium, 72-73, chmn,Bot Sect, 73-77,assoc prof, 77-82, dir grad prog, 81-82, prof, 82-86; DIR RES, MO BOT GARDEN, 86- *Concurrent Pos:* Sci adv, Flora, Reserv Fontes Ipiranga, Sao Paulo, Brazil, 80-; mem plant adv group, Int Union Conserv Nature & Worldwide Fund Nature, 84-; chmn, S Am Plant Specialist Group, Species Survival Comn, Int Union, 84- *Mem:* Am Soc Plant Taxonomists; Asn Trop Biol; Bot Soc Am; Orgn Flora Neotrop (pres, 81-). *Res:* Floristic projects on two plant families, the Connaraceae and Leguminosae; floristic inventory of Choco region of Colombia. *Mailing Add:* Mo Bot Garden PO Box 299 St Louis MO 63166-0299

FORESMAN, KERRY RYAN, b Minneapolis, Minn, May 25, 48; m 78. REPRODUCTIVE BIOLOGY, ENDOCRINOLOGY. *Educ:* Univ Mont, BA, 71; Univ Idaho, MS, 73, PhD(zool), 77. *Prof Exp:* Teaching fel, Univ Tenn-Knoxville, 77-78, vis asst prof physiol, 78-79; asst prof physiol, Univ RI, 79-83; asst prof physiol, 84-86, ASSOC PROF ZOOL, UNIV MONT, 86- *Mem:* Soc Study Reproduction; Am Soc Mammalogists. *Res:* Hormonal regulation of embryonic development; ovarian regulation of uterine protein synthesis during early stages of pregnancy in mammals; evolution of reproductive strategies in insectivores (soricids). *Mailing Add:* Div Biol Sci Univ Mont Missoula MT 59812

FOREST, CHARLENE LYNN, b Brooklyn, NY, Feb 27, 47; m 69. CELL BIOLOGY, INTERCELLULAR INTERACTIONS. *Educ:* Cornell Univ, BS, 68; Adelphi Univ, MS, 72; Indiana Univ, PhD(genetics), 76. *Prof Exp:* Fel, Harvard Univ, 76-79; asst prof, 79-83 & 84-85, prin investr & res assoc, Res Found, 83-84, ASSOC PROF BIOL, BROOKLYN COL, CITY UNIV NEW YORK, 86- *Concurrent Pos:* Prin investr NIH, 80-83 & NSF, 83-91. *Mem:* Genetics Soc Am; Am Soc Cell Biol; AAAS; NY Acad Sci; Sigma Xi; Asn Women Sci. *Res:* Adhesion and fusion of plasma membrane-located mating structures; intercellular signaling; isolation and analysis of temperature sensitive mutants with defects in gametogenesis and mating in Chlamydomonas. *Mailing Add:* Dept Biol Brooklyn Col City Univ New York Brooklyn NY 11210

FOREST, EDWARD, b New York, NY, Mar 6, 33; m 61; c 2. MATERIALS SCIENCE ENGINEERING. *Educ:* Brooklyn Col, BS, 55; Princeton Univ, MA, 62, PhD(phys chem), 63. *Prof Exp:* Instr chem, Brooklyn Col, 58-59; asst, Princeton Univ, 59-63; scientist, Xerox Corp, 63-69, mgr res div & lab mgr, 69-83; DIR GRAD PROG & RES, SOUTHERN METHODIST UNIV, 83- *Mem:* Am Chem Soc; Soc Photog Scientists & Engrs; Sigma Xi. *Res:* Microwave absorption and relaxation; liquid crystalline materials; charge transport in organic materials; photo responsive systems; non impact printing. *Mailing Add:* Sch Eng & Appl Sci Southern Methodist Univ Dallas TX 75275

FOREST, HERMAN SILVA, b Chattanooga, Tenn, Feb 18, 21; m 63; c 2. ENVIRONMENTAL BIOLOGY. *Educ:* Univ Tenn, BA, 42; Mich State Univ, MS, 48, PhD(bot), 51. *Prof Exp:* Asst prof biol, Col William & Mary, 53-54; instr bot, Univ Tenn, 54-55; instr bot & biol, Univ Okla, 55-58; dir students mus, Knoxville, 58-60; res asst, Med Ctr, Univ Okla, 60-61; res assoc biol, Univ Rochester, 61-65; PROF BIOL, STATE UNIV NY COL GENESEO, 65- *Concurrent Pos:* NIH fel, 61-62; AEC grant, 62-64; exchange scholar, Univ Moscow, 79, Bot Inst USSR, Leningrad, 64, Czechoslovakia, 80; NSF grant, 65; vis biologist, Am Inst Biol Sci, 69-71; prin scientist, Environ Resource Ctr, 71-81; mem adv panel, Ford Found Prog Assistance Local Conserv Comns, 72-74; fac vis scholar, State Univ NY, 74-; nat lectr, fel, Scientist Inst Pub Int, 75-; dir, Int Great Lakes Res Conf, 79; fel, Rochester Acad Sci, 81, ed, Proceedings, 82-85; consult assoc, IMS Engrs, 83-; sci adv, Fedn NY Lake Assoc & City of Rochester, 85-87. *Mem:* AAAS; Am Inst Biol Sci; Ecol Soc Am; Bot Soc Am; Phycol Soc Am; NAm Lake Mgt Asn. *Res:* Lake ecology; population ecology of sugar maples and American chestnut; rooted aquatic plants and non-marine algae; ecosystem perspective of lake ecology and management, theory and application. *Mailing Add:* Dept Biol State Univ NY Col Geneseo NY 14454

FORESTER, DONALD CHARLES, b Detroit, Mich, Feb 21, 43; m 66; c 2. BEHAVIORAL BIOLOGY, HERPETOLOGY. *Educ:* Tex Tech Univ, BA, 66, MS, 69; NC State Univ, PhD(zool), 74. *Prof Exp:* Instr biol, Hill Jr Col, 68-69, McLennan Community Col, 69-73; vis instr zool, NC State Univ, 73-74; from asst to assoc prof, 74-86,PROF BIOL TOWSON STATE UNIV, 86- *Mem:* Herpetologists League; Am Soc Ichthyologists & Herpetologists; Soc Study Amphibians & Reptiles; Sigma Xi; Am Soc Zoologists. *Res:* Sexual selection in anurans; parental care in salamanders. *Mailing Add:* Dept Biol Sci Towson State Univ Baltimore MD 21204

FORESTER, DONALD WAYNE, b Knoxville, Tenn, Apr 7, 37; m 59; c 2. MAGNETISM, MILLIMETER-WAVES. *Educ:* Berea Col, BA, 59; Univ Tenn, MS, 61, PhD(physics), 64. *Prof Exp:* Asst prof physics, Univ Nebr, Lincoln, 64-66; Res Corp grant & asst prof physics, Ga Inst Technol, 66-69; res physicist, solid state div, 69-78, SUPVRY RES PHYSICIST, MAGNETISM BR, ELECTRONICS TECHNOL DIV, NAVAL RES LAB, 78- *Mem:* Am Phys Sco; Sigma Xi. *Res:* Mossbauer experiments; low temperature physics; optical properties of solids; electrical discharges; magnetism in solids; magnetic silencing applications; radar absorbing materials; amorphous materials; thin-film technology. *Mailing Add:* Naval Res Lab Code 1504 Washington DC 20375-5000

FORESTER, RALPH H, b Chicago, Ill, Apr 11, 28; m 53; c 2. POLYMER CHEMISTRY. *Educ:* Carleton Col, BA, 50. *Prof Exp:* Chemist, G H Tennant Co, 55-66; sr chemist, North Star Div, Midwest Res Inst, 66-78; sr scientist, Filmtec Corp, 78-89; RETIRED. *Res:* Biophysics, low-level bioelectric phenomena; epoxy and urethane coatings; metal fibers, reinforced plastics, abrasives and filters; ultrathin polymer membrane technology; reverse osmosis & ultrafiltration. *Mailing Add:* 4829 Dupont Ave S Minneapolis MN 55409

FORESTER, RICHARD MONROE, b Hancock, NY, Dec 23, 47; m 72. MICROPALEONTOLOGY, PALEOECOLOGY. *Educ:* Syracuse Univ, BS, 69; Univ Ill, MS, 72, PhD(paleoecol), 75. *Prof Exp:* NAT RES COUN ASSOC FEL MICROPALEONT, US GEOL SURV, 75- *Mem:* Paleont Soc; Paleont Asn; Int Paleont Union; AAAS; Geol Soc Am. *Res:* Biostratigraphy, systematics and paleoecology of non-marine ostracodes and charaphytes. *Mailing Add:* Mail Stop 919 BPS US Geol Surv Denver Fed Ctr Denver CO 80225

FORESTER, ROBERT DONALD, geophysics, for more information see previous edition

FORESTI, ROY J(OSEPH), JR, b Baltimore, Md, Mar 25, 25; m 53; c 2. CHEMICAL ENGINEERING, THERMODYNAMICS & HEAT TRANSFER. *Educ:* Johns Hopkins Univ, BE, 47; Carnegie Inst Technol, MS, 48; Pa State Univ, PhD(fuel technol), 51. *Prof Exp:* Asst, Pa State Univ, 48-50; res chem engr, US Bur Mines, 51-54; res chem engr, Monsanto Chem Co, 54-55, group leader, 55-59; lectr chem eng, Univ Dayton & sr res scientist, Res Inst, 59-61; assoc prof chem eng, Univ Conn, 61-63; assoc prof chem eng & chmn dept, Cath Univ Am, 63-80; sr eng, Vitro Corp, 80-87; ENGR CONSULT, 87-; ADJ PROF, HOWARD UNIV, 90- *Mem:* Am Chem Soc; fel Am Inst Chem Engrs; Sigma Xi; Am Soc Heat Refrig Air Engrs. *Res:* Airborne solid dispersions; gasification of solids; oxidation of gases; nonideal liquids. *Mailing Add:* 301 Willington Dr Silver Spring MD 20904

FORESTIERI, AMERICO F, b Cleveland, Ohio, May 4, 29; m 56; c 3. ENERGY CONVERSION, RESEARCH ADMINISTRATION. *Educ:* Case Inst Technol, BS, 51, MS, 54. *Prof Exp:* Res staff solid state physics, Nat Adv Comt Areonaut, 51-65, HEAD PHOTO-VOLTAICS TECH, NASA, 65- *Concurrent Pos:* Chmn, Photovoltaic Specialists Conf, 75-76, publ chmn, 76-; chmn solar working group, Interagency Advan Power Group, 78-; asst chmn, Intersoc Energy Conversion Eng Conf, 81-82. *Mem:* Inst Elec & Electronics Engrs. *Res:* Develop technology programs to reduce cost and weight and increase efficiency and lifetime of solar cells for space applications. *Mailing Add:* 74 Barberry Dr Berea OH 44017

FORET, JAMES A, b Lutcher, La, Sept 3, 21; m 46; c 10. ORNAMENTAL HORTICULTURE, WEED SCIENCE. *Educ:* Univ Southwestern La, BS, 43; Iowa State Univ, MS, 47, PhD(plant physiol), 50. *Prof Exp:* Instr & res asst prof, Iowa State Univ, 46-50; assoc prof, Univ Southwestern La, 50-53, head dept gen agr, 53-64, head dept plant indust & gen agr, 69-74, prof hort, 53-80, dean col agr, 76-80; RETIRED. *Mem:* Am Soc Hort Sci. *Res:* Disposal of rice hulls; control of Elodea canadensis in lakes of Louisiana; control of alligatorweed in rice irrigation canals; herbicide residue studies in rice irrigation canals; aquatic weed control. *Mailing Add:* 256 Edgewood Dr Lafayette LA 70503

FORET, JOHN EMIL, b New York, NY, Nov 19, 37. DEVELOPMENTAL BIOLOGY. *Educ:* Univ NH, BA, 62, MS, 63; Princeton Univ, AM, 65, PhD(biol), 66. *Prof Exp:* NIH fel, 66-67; asst prof, 67-73, ASSOC PROF ZOOL, UNIV NH, 73-, CHMN DEPT, 78- *Mem:* Am Soc Zoologists; Am Soc Cell Biol; Soc Develop Biol; AAAS. *Res:* Cellular origins and control mechanisms in limb regeneration. *Mailing Add:* Dept Zool Univ NH Durham NH 03824

FORGAC, JOHN MICHAEL, b Lakewood, Ohio, May 30, 49; m 71. CHEMICAL ENGINEERING. *Educ:* Case Western Reserve Univ, BS, 71, MS, 75, PhD(chem eng), 78. *Prof Exp:* res engr coal liquefaction, 77-80, res engr oil shale processing, 80-84, process design & econ, 84-86, CATALYTIC CRACKING, AMOCO OIL CO, STANDARD OIL, IND, 86- *Mem:* Am Inst Chem Engrs; Am Chem Soc; AAAS; Sigma Xi; Catalysis Soc. *Res:* Oil shale and coal conversion; heterogeneous chemical reactions; reaction kinetics; crystal growth; petroleum processing; catalytic cracking. *Mailing Add:* 13 Oak Tree Ct Elmhurst IL 60126

FORGACS, GABOR, b Budapest, Hungary, June 13, 49; m; c 2. DISORDERED SYSTEMS, PHASE TRANSITIONS. *Educ:* Eötvös Roland Univ, BS & MS, 73, PhD(physics), 78. *Prof Exp:* Postdoctoral fel physics, Univ Ill, Urbana, 79-81; staff researcher, Cent Res Inst Physics, Budapest, 81-84; sr staff scientist, French Atomic Energy Agency, Saclay, 84-86; ASSOC PROF PHYSICS, DEPT PHYSICS, CLARKSON UNIV, 86- *Concurrent Pos:* Consult, French Atomic Energy Agency, Saclay, 86-; vis prof, Brookhaven Nat Lab, Upton, 86, Univ Paris, Orsay, 87; collabr, Los Alamos Nat Lab, 91- *Mem:* Am Phys Soc. *Res:* Disordered systems: phase transitions interplay between generic, physical, and genetic mechanisms in biological evolution, morphogenesis and pattern formation. *Mailing Add:* Physics Dept Clarkson Univ Potsdam NY 13676

FORGACS, JOSEPH, b Nokomis, Ill, Mar 20, 17; m 46; c 5. MEDICAL MICROBIOLOGY. *Educ:* Univ Ill, BS, 40, MS, 42, PhD(bact, hort), 44. *Prof Exp:* Asst hort, Univ Ill, 40-43, asst plant path, 43-44; bacteriologist & chief antibiotics & chemother, Camp Detrick, 46-54; mem staff res div, Am Cyanamid Co, 54-57; dir lab, Spring Valley Gen Hosp, New York, 58-60; mem staff res div, Am Cyanamid Co, 60; staff microbiologist & consult mycotoxocologist, Good Samaritan Hosp, Suffern, NY, 60-69; dir microbiol, Automated Biochem Labs Inc, 69-79; DIR MICROBIOL, DEPT MENT HEALTH, LETCHWORTH DEVELOP CTR, 79-, CONSULT, GEN FOODS CORP, 68- *Concurrent Pos:* Consult microbiologist, Tuxedo Mem Hosp, New York, 60-64; consult mycotoxicologist, Agr Res Serv, USDA, 65-; consult & attend microbiologist, Dept Mental Hyg, State of NY, 73- *Mem:* Am Soc Microbiologists; NY Acad Sci; Int Soc Trop Dermat. *Res:* Food bacteriology and chemistry; antibiotics and chemotherapy; mycology; mycotoxicoses; determine role of fungi in toxicoses of unknown etiology in animals and human beings. *Mailing Add:* 302 N Highland Ave Pearl River NY 10965

FORGACS, OTTO LIONEL, b Berlin, Ger, Jan 4, 31; Can citizen; m 60; c 3. PHYSICAL CHEMISTRY. *Educ:* Manchester Univ, BSc Tech, 55; McGill Univ, PhD(phys chem), 59. *Prof Exp:* Proj supvr, Pulp & Paper Res Inst Can, 58-63; head pulp & paper & allied bldg prod sect, Res Centre, Domtar Ltd, 63-70, mgr prod develop, 70-72, assoc res dir & dir tech planning, 72-73; res dir, 73-77, vpres, Res & Develop, 77-79, SR VPRES, RES & DEVELOP, MACMILLAN BLOEDEL LTD, 79- *Mem:* Fel Tech Asn Pulp & Paper Indust; fel Chem Inst Can; Can Pulp & Paper Asn; Can Forestry Adv Coun. *Res:* Pulp and paper technology. *Mailing Add:* 1843 Acadia Rd Vancouver BC V6T 1R2 Can

FORGASH, ANDREW JOHN, b Dunellen, NJ, Nov 21, 23; m 44; c 4. ENTOMOLOGY. *Educ:* Rutgers Univ, BS, 49, MS, 50, PhD(entom), 52. *Prof Exp:* Asst res specialist chem, Rutgers Univ, 52-54, asst res specialist entom, 54-58, assoc res specialist, 58-61, res specialist, 61-71, res prof entom & econ zool, 71-; RETIRED. *Mem:* Entom Soc Am. *Res:* Physiology; biochemistry; toxicology. *Mailing Add:* 2607 Park Ave South Plainfield NJ 07080

FORGENG, W(ILLIAM) D(ANIEL), astronomy; deceased, see previous edition for last biography

FORGET, BERNARD G, b Fall River, Mass, Mar 23, 39; m 65; c 3. BIOCHEMISTRY, MOLECULAR BIOLOGY. *Educ:* Univ Montreal, BA, 59; McGill Univ, MD, 63; Yale Univ, MA, 78. *Prof Exp:* Clin assoc biochem, Nat Cancer Inst, NIH, 65-67; asst prof, Harvard Med Sch, 72-75, assoc prof pediat, 75-76; assoc prof, 76-77, PROF MED & HUMAN GENETICS, SCH MED, YALE UNIV, 78- *Concurrent Pos:* Spec fel, NIH, 68-69; res fel, Harvard Med Sch, 69-71; fel Med Found, Inc, Boston, 70-72; NIH res career develop award, 72-76. *Mem:* Am Soc Biol Chemists; Am Soc Clin Invest; Am Soc Hematol; Am Fedn Clin Res; Asn Am Physicians. *Res:* Molecular genetics of human hemoglobin synthesis; nucleic acid sequencing; globin gene cloning and structural analysis; molecular basis of thalassemia. *Mailing Add:* Dept Med Yale Univ Sch Med 333 Cedar St PO Box 3333 New Haven CT 06510-8056

FORGHANI-ABKENARI, BAGHER, b Bandar Enzali, Iran, March 10, 36; m 69; c 2. VIROLOGY, IMMUNOLOGY. *Educ:* Justus Liebig Univ, WGer, MS, 61, PhD(virol), 65. *Prof Exp:* Fel virol, Utah State Univ, 65-67; asst prof biol, Nat Univ Iran, 67-69; res assoc virol, Utah State Univ, Logan, 69-70; fel, 70-72, res specialist, 72-83, RES SCIENTIST, DIV VIRUS RICKETTSIAL DIS LAB, CALIF STATE DEPT HEALTH SERV, 83- *Mem:* Am Soc Microbiol; Am Acad Microbiol. *Res:* Viruses including, Rubella, Hepatitis, Herpes, Measles, Cytomegalo Varicella-Zoster; degenerative disease; multiple sclerosis: using radio-or enzyme-immunoassay methodology; production and characterization of monoclonal antibody to varicella-zoster virus. *Mailing Add:* Calif State Dept Health Serv 2151 Berkeley Way Berkeley CA 94704

FORGOTSON, JAMES MORRIS, JR, b Albuquerque, NMex, Mar 17, 30; m 58; c 2. PETROLEUM GEOLOGY, PETROLEUM ENGINEERING. *Educ:* Wash Univ, AB, 51; Univ Tex, BS, 52; Northwestern Univ, MS, 54, PhD(geol), 56. *Prof Exp:* Explor geologist, 53-55; sr res engr, Pan Am Petrol Corp, 56-61, tech group supvr, 61-62, res group supvr, 62-68; vpres, Petrol Info Corp, 68-71; staff geologist, Geophys Serv Inc, 72-74; geologist petrol explor & prod, 74-80, mgr, Burk, Bakwin & Henry, 80-83; independent geologist & producer, 83-86, dir, Sch Geol & Geophysics, 86-89, PROF, UNIV OKLA, 89- *Concurrent Pos:* Adj prof, Univ Tex, Arlington, 74- & Southern Methodist Univ, 75-; pres, Bluebonnet Drilling Inc, 80-82, Hydrocarbon Transport Inc, 80-83. *Honors & Awards:* President's Award, Am Asn Petrol Geol, 64; Distinguished Serv Award, Am Asn Petrol Geol, 88. *Mem:* Am Asn Petrol Geol (secy-treas, 68-70, vpres, 86-87); Soc Econ Paleont & Mineral; Soc Petrol Eng; Soc Explor Geophys; Geol Soc Am. *Res:* Regional stratigraphic studies applied to petroleum exploration; new methods of presenting quantitative geological data; use of electronic computers to process and statistically analyze geological data; well log analysis; interpretation of gravity and magnetic data; data base systems for exploration; petroleum exploration and production; management of geological, land and engineering operations, reservoir characterization and basin analysis. *Mailing Add:* Eight Rustic Hills St Norman OK 73072-7411

FORGY, CHARLES LANNY, b Graham, Tex, Dec 29, 49; m 77. ARTIFICIAL INTELLIGENCE SYSTEMS. *Educ:* Univ Tex, Arlington, BS, 72; Carnegie-Mellon Univ, PhD(comput sci), 79. *Prof Exp:* Res assoc, 78-84, RES COMPUT SCIENTIST, CARNEGIE-MELLON UNIV, 84- *Mem:* Asn Comput Mach; Am Asn Artificial Intel. *Res:* Programming languages and computer architectures for artificial intelligence; applications of artificial intelligence. *Mailing Add:* 642 Gettysberg St Pittsburgh PA 15206

FORIST, ARLINGTON ARDEANE, b Lansing, Mich, Oct 14, 22; m 52; c 5. DRUG DEVELOPMENT. *Educ:* Mich State Col, BS, 48, PhD(biochem), 52. *Prof Exp:* Asst, Qm Food & Container Inst, Mich State Col, 48-51; res assoc, Upjohn Co, 51-60, res sect head, 60-68, res mgr phys & anal chem, 68-79, group mgr phys & anal chem & drug metab res, 79-84, asst to vpres, Pharmaceut Res & Develop, 84-85; RETIRED. *Mem:* Fel AAAS; Am Chem Soc; NY Acad Sci; Sigma Xi; fel Am Asn Pharmaceut Scientists. *Res:* Multidisciplinary research administration in analytical and physical chemistry; spectroscopy and organic structure elucidation; drug metabolism; biopharmaceutics; pharmacokinetics. *Mailing Add:* 3406 Croyden Ave Kalamazoo MI 49007

FORK, DAVID CHARLES, b Detroit, Mich, Mar 4, 29; m 58; c 3. PLANT PHYSIOLOGY. *Educ:* Univ Calif, Berkeley, AB, 51, PhD(bot), 61. *Prof Exp:* BIOLOGIST, DEPT PLANT BIOL, CARNEGIE INST WASHINGTON, 60- *Concurrent Pos:* Courtesy prof, Stanford Univ, 68- *Mem:* AAAS; Am Soc Plant Physiologists; Am Soc Photobiol; Sigma Xi. *Res:* Study of the basic mechanisms of photosynthesis. *Mailing Add:* Dept Plant Biol Carnegie Inst Washington Stanford CA 94305

FORK, RICHARD LYNN, b Dearborn, Mich, Sept 1, 35; m 57, 71; c 4. SOLID STATE PHYSICS, SPECTROSCOPY. *Educ:* Principia Col, BS, 57; Mass Inst Technol, PhD(physics), 62. *Prof Exp:* MEM TECH STAFF, BELL LABS, 62- *Mem:* Fel Am Phys Soc; Optical Soc Am. *Res:* Lasers in magnetic fields; semiconductor luminescence; solid state spectroscopy; femtosecond spectroscopy; semiconductor microstructures. *Mailing Add:* Bell Labs 4D-417 Holmdel NJ 07733

FORKER, E LEE, b Pittsburgh, Pa, Aug 28, 30; m 55; c 4. PHYSIOLOGY, INTERNAL MEDICINE. *Educ:* Haverford Col, BS, 53; Univ Pittsburgh, MD, 57. *Prof Exp:* Intern, Presby Hosp, Denver, Colo, 57-58; resident internal med, Hosp, Col Med, Univ Iowa, 60-63, Iowa Heart Asn fel physiol & biophys, 63-65, asst prof med, 65-69, asst prof physiol & biophys, 67-69, assoc prof internal med, physiol & biophys, 69-73, prof internal med, 73-79; prof med & dir, Div Gastroenterol & Liver Dis, 80-84, PROF MED & PHYSIOL & ASSOC DEAN RES, SCH MED, UNIV MO-COLUMBIA, 84- *Concurrent Pos:* Res assoc, Vet Admin Hosp, Iowa City, 65-67; Markle scholar acad med, 67; NIH career develop award, 68. *Mem:* Am Soc Clin Invest; Am Asn Study Liver Dis (pres, 83); Soc Math Biol; Am Physiol Soc; Am Gastroenterol Asn; Int Asn Study Liver Dis. *Res:* Hepatic physiology; bile formation; membrane transport. *Mailing Add:* Dept Physiol & Med Univ Mo Sch Med M463 Med Sci Columbia MO 65212

FORKEY, DAVID MEDRICK, b Minneapolis, Minn, Apr 9, 40; m 66. ORGANIC CHEMISTRY. *Educ:* St Olaf Col, BA, 62; Univ Wash, PhD(org chem), 67. *Prof Exp:* Nat Res Coun res assoc org synthesis & mass spectrometry, Naval Weapons Ctr, Calif, 67-69; from asst prof to assoc prof, 69-81, PROF CHEM, CALIF STATE UNIV, SACRAMENTO, 81- *Mem:* Am Chem Soc. *Res:* Syntheses and reactions of heterocyclic compounds; applications of nuclear magnetic resonance and mass spectrometry to organic structural problems; applications of gas chromatography, gas chromatography-mass spectrometry and high-performance liquid chromatography to chemical separations. *Mailing Add:* 4805 Thor Way Carmichael CA 95608

FORLAND, MARVIN, b Newark, NJ, Mar 29, 33; m 65; c 2. INTERNAL MEDICINE, NEPHROLOGY. *Educ:* Colgate Univ, AB, 54; Columbia Univ, MD, 58. *Prof Exp:* Intern & resident internal med, Univ Chicago Hosps, 59-62, asst prof med, Sch Med, Univ Chicago, 64-68; assoc prof & chief Renal & Electrolyte Sect, 68-72,dep chmn dept med, 75-86, PROF MED, UNIV TEX MED SCH, SAN ANTONIO, 72-, ASSOC DEAN CLIN AFFAIRS, 83- *Mem:* Cent Soc Clin Res; fel Am Col Physicians; Am Soc Nephrol; Int Soc Nephrol. *Res:* Clinical nephrology and metabolism; urinary tract infection. *Mailing Add:* Univ Tex Med Sch San Antonio TX 78284

FORMAL, SAMUEL BERNARD, b Providence, RI, Aug 28, 23; m 51; c 3. MICROBIOLOGY. *Educ:* Brown Univ, AB, 45, ScM, 48; Boston Univ, PhD(microbiol), 52; Am Bd Microbiol, dipl. *Prof Exp:* Bacteriologist, Food & Drug Admin, 48-49; bacteriologist, 52-56, chief dept appl immunol, 56-76, chief dept bact dis, 76-85, CHIEF DEPT ENTERIC INFECTIONS, WALTER REED ARMY INST RES, 85- *Concurrent Pos:* Mem comn enteric infections, Armed Forces Epidemiol; prof lectr, Sch Med, Georgetown Univ, 73-; mem ed bd, J Reticular Endothelial Soc, 74-84, J Infection & Immunity, 78- *Mem:* Am Soc Microbiologists; Am Asn Immunologists; Am Acad Microbiol; Infectious Dis Soc Am. *Res:* Pathogenesis and immunity in enteric infections. *Mailing Add:* 11405 Woodson Ave Kensington MD 20895

FORMAN, DAVID S, b Detroit, Mich, Dec 10, 42. CELL BIOLOGY, NEUROBIOLOGY. *Educ:* Harvard Col, BA, 64; Rockefeller Univ, PhD(life sci), 71. *Prof Exp:* Fel, Lab Neuropharmacol, NIMH, 71-75; res physiologist, Naval Med Res Inst, 75-81; assoc prof anat, Uniformed Serv Univ Health Sci, 81-88; CONSULT, 88- *Concurrent Pos:* Vis asst prof, Dept Anat, Sch Med, Howard Univ, 75-78; vis asst prof & lectr, Dept Physiol, Sch Med, George Washington Univ, 77; mem, Merit Rev Bd, neurobiol, Vet Admin, 78-81; mem adv panel cell biol, NSF, 84- *Mem:* Soc Neurosci; Am Soc Cell Biol; Am Soc Neurochem; Int Brain Res Orgn; AAAS. *Res:* Axonal transport, the movement of materials inside neurons; nerve regeneration and non-neural cell motility. *Mailing Add:* Finnegan Henderson Farnbow Garrett & Dunner 1300 Eye St NW Washington DC 20005-3315

FORMAN, DEBRA L, b Philadelphia, Pa, May 9, 57. TOXIOLOGY. *Educ:* Loyola Univ, BS, 79; Temple Univ, MS, 83, PhD(pulmonary physiol), 85. *Prof Exp:* Postdoctoral, Med Col Pa, 85-87, Smith Kline Beckman, 87-89; POSTDOCTORAL, HAZARDOUS WASTE MGT DEPT, US ENVIRON PROTECTION AGENCY, 89-, REG TOXICOLOGIST RISK ASSESSOR & TOXICOL, 89- *Concurrent Pos:* Fel, NIH, 85-87. *Mem:* Soc Toxicol; Am Physiol Soc. *Mailing Add:* Hazardous Waste Mgt Dept US EPA Region III 841 Chestnut St 3HW15 Philadelphia PA 19107

FORMAN, DONALD T, b New York, NY, Feb 27, 32; m 53; c 3. BIOCHEMISTRY, ANALYTICAL CHEMISTRY. *Educ:* Brooklyn Col, BS, 53; Wayne State Univ, MS, 57, PhD(biochem), 59. *Prof Exp:* Chief biochemist, Hazleton Lab, 59-60; from instr to asst prof, 61-69, ASSOC PROF BIOCHEM, MED SCH, NORTHWESTERN UNIV, EVANSTON, 69-; PROF PATH & BIOCHEM & DIR CLIN CHEM, DIV LAB MED, UNIV NC, CHAPEL HILL, 78- *Concurrent Pos:* NIH & Am Heart Asn res grants, 63-68; dir biochem, Mercy Hosp, 60-63; dir clin biochem, Evanston Hosp, 63-78. *Mem:* AAAS; Am Chem Soc; Am Asn Clin Chem; Sigma Xi. *Res:* Clinical biochemistry methods, regulation of alcohol metabolism; atherosclerosis. *Mailing Add:* 2559 Owens Ct Chapel Hill NC 27514

FORMAN, EARL JULIAN, b Hartford, Conn, June 22, 29; m 53; c 3. ANALYTICAL CHEMISTRY. *Educ:* Wesleyan Univ, BA, 53; Mass Inst Technol, PhD(anal chem), 57. *Prof Exp:* Res chemist, Hercules Powder Co, 57-70; res group leader, 70-77, mgr, Film Div Anal Serv, 78-81, MGR, CORP CHEM INSPECTION, POLAROID CORP, 81- *Mem:* Am Chem Soc. *Mailing Add:* 115 Loring Rd Weston MA 02193-2453

FORMAN, G LAWRENCE, b Hartford, Conn, Nov 13, 42; m 81; c 3. MAMMALOGY, COMPARATIVE ANATOMY. *Educ:* Univ Kans, BA, 64, MA, 67, PhD(mammal), 69. *Prof Exp:* Chmn dept biol, 71-90, chmn div sci & math, 76-79, PROF BIOL, ROCKFORD COL, 69- *Concurrent Pos:* Vis prof biol, Hofstra Univ, Hempstead, NY, 76; Cent Asn Adv to the Health Professions (secy, 88-91); mem, Comt Anat & Physiol, Am Soc Mammologists. *Mem:* AAAS; Am Soc Mammalogists; Sigma Xi; Am Inst Biol Sci; Soc Syst Zool; Nat Assoc Adv Health Prof. *Res:* Mammalian systematics; microanatomy of mammals; gastrointestinal morphology in relation to feeding habits; cellular morphology in mammals in relation to systematics; anatomy of desert mammals in relation to water conservation; anatomy of lymphoid organs. *Mailing Add:* Dept Biol Rockford Col Rockford IL 61108

FORMAN, GEORGE W, b Salt Lake City, Utah, Dec 9, 19; m 41; c 3. MECHANICAL ENGINEERING, ENGINEERING MECHANICS. *Educ:* Univ Ill, BS, 41; Univ Kans, MS, 58. *Prof Exp:* Design engr & supvr eng training, Hamilton Stand Div, United Aircraft Corp, 41-46; mgr mech eng, Marley Co, 46-53; res mgr, Butler Mfg Co, 53-55; prof mech eng, 55-88, EMER PROF MECH ENG, UNIV KANS, 88- *Concurrent Pos:* Indust consult, 55- *Mem:* Fel Am Soc Mech Engrs. *Res:* Machine design; composite materials research; structural response to impact loads. *Mailing Add:* 7007 Vintage Ct Lawrence KS 66047

FORMAN, GUY, b Dundee, Ky, Oct 23, 06; m 26. PHYSICS. *Educ:* Western Ky State Col, BS, 29; Ind Univ, MA, 31; Univ Ky, PhD, 50. *Prof Exp:* Assoc prof physics, Western Ky State Col, 29-43; assoc prof, Vanderbilt Univ, 43-62; chmn dept, 62-69, prof, 62-72, EMER PROF PHYSICS, UNIV SFLA, 72- *Mem:* Am Phys Asn; Optical Soc Am; Am Asn Physics Teachers. *Res:* Action of light on selenium cells as a surface effect; x-ray excited phosphorescence of natural crystals; teaching elementary and intermediate physics. *Mailing Add:* 12401 N 22nd Apt A-405 Tampa FL 33612

FORMAN, J(OSEPH) CHARLES, b Chicago, Ill, Dec 22, 31; m 53; c 3. CHEMICAL & BIOCHEMICAL ENGINEERING. *Educ:* Mass Inst Technol, SB, 53; Northwestern Univ, MS, 57, PhD(chem eng), 60. *Prof Exp:* Mem tech proj training prog, Dow Chem Co, Mich, 53-54; chem engr, Develop Div, Abbott Labs, 56-59, sr chem engr, Eng Develop Dept, 59-63, group leader, Extreme Conditions Lab, 63-67, group leader fermentation eng, Biol Develop Dept, 67-68, sect mgr & proj mgr, 68-74, opers mgr corp procurement, 74-75, dir mfg opers, Agr & Vet Prod Div, 75-77; assoc exec dir, 77-78, EXEC DIR & SECY, AM INST CHEM ENGRS, NEW YORK, 78- *Concurrent Pos:* Pres, Coun Eng & Sci Soc Execs, 85-86. *Mem:* Fel Am Inst Chem Engrs; Nat Soc Prof Engrs; Am Chem Soc; Am Soc Assoc Execs; Sigma Xi. *Res:* Association management; technical management; direction and evaluation of technical projects; biochemical engineering; industrial fermentations. *Mailing Add:* Forman Assoc Consult & Technol Servs 77 Stanton Rd Darien CT 06820

FORMAN, JAMES, b Watertown, Mass, Dec 23, 39; m 65; c 3. IMMUNOLOGY. *Educ:* Tufts Univ, BS, 61, DMD, 65, PhD(physiol), 71. *Prof Exp:* Fel immunol, Karolinska Inst, Sweden, 71-73; fel, 73-74, asst prof, 74-76, assoc prof, 76-80, PROF IMMUNOL, DEPT MICROBIOL, UNIV TEX HEALTH SCI CTR, DALLAS, 80- *Concurrent Pos:* NIH fel, 71-72. *Mem:* Am Asn Immunologists. *Res:* Cellular immunology; immunogenetics. *Mailing Add:* Dept Microbiol Univ Tex Southwestern Med Ctr 5323 Harry Hines Blvd Dallas TX 75235-9048

FORMAN, MCLAIN J, b Mt Kisco, NY, July 8, 28; c 3. SUBSURFACE EXPLORATION. *Educ:* Tulane Univ, BS, 52; Harvard Univ, MA, 53, PhD(geol), 57. *Prof Exp:* Geologist, Atwater, Cowan & Assoc, Consult Geologists, 55-60, McLain J Forman & Assoc, 60-72; PRES, FORMAN PETROL CORP, 72- *Mem:* Am Asn Petrol Geologists; fel Geol Soc Am; Sigma Xi; Soc Econ Paleontologists & Mineralogists; Am Inst Mining, Metall & Petrol Engrs. *Res:* Structural and stratigraphic studies of the subsurface geology of the Gulf Coastal Region, including salt dome structure, regional structure and regional stratigraphy, with emphasis on oil and gas exploration and exploitation. *Mailing Add:* 1010 Common St Suite 500 New Orleans LA 70112

FORMAN, MICHELE R, b NY, Aug 23, 50; m; c 1. EXPERIMENTAL BIOLOGY. *Educ:* Rutgers Univ, BA, 72; Univ NC, Chapel Hill, MSPH, 74, MA, 75, PhD(epidemiol), 77. *Prof Exp:* Teaching asst, Epidemiol Dept, Univ NC, 73, lectr med anthrop, 74-75, grad res assoc, Social Sci Res Inst, 75-76; instr appl epidemiol, Sch Med & Health Serv, George Washington Univ, 77; epidemiologist, Epidemiol & Biomet Res Prog, Nat Inst Child Health & Human Develop, 76-83; sr epidemiologist, Epidemiol Br, Div Nutrit, Ctr Health Prom & Educ, Ctr Dis Control, 83-86; assoc prof, Div Human Nutrit, Dept Int Health & Epidemiol, Mem Inst Int Prog Sch Hyg & Pub Health, Johns Hopkins Univ, 86-89; NUTRIT EPIDEMIOLOGIST, CANCER PREV STUDIES BR, DIV CANCER PREV & CONTROL, NAT CANCER INST, 89- *Concurrent Pos:* NIH environ epidemiol trainee, Sch Pub Health, Univ NC, 72-75, Pub Health Serv grant trainee, Inst Social Sci Res, 75-77; mem, Clin Res Rev Comt, Nat Inst Child Health & Human Develop, 78-83, Comt Rev Infant Feeding Res, Dept Health & Human Serv, 79-80, Agency Int Develop Rev Panel Four Country Infant Feeding Studies, 80-87; mem, Task Force Infant Feeding, Dept Health & Human Serv, 82-84, Task Force Women Sci, Pub Health Serv, 85-86. *Mem:* Am Inst Nutrit; Am Pub Health Asn; Am Soc Clin Nutrit; Int Epidemiol Asn; Soc Epidemiol Res. *Mailing Add:* Div Cancer Prev & Control Cancer Prev Studies Br Nat Cancer Inst NIH 61130 Executive Plaza Blvd Rockville MD 20892

FORMAN, MIRIAM AUSMAN, b San Francisco, Calif, Apr 12, 39; m 64; c 2. SPACE PHYSICS, ASTRONOMY. *Educ:* Univ Chicago, BS, 60, MS, 61; State Univ NY, Stony Brook, PhD(physics), 72. *Prof Exp:* Res scientist, SAfrican Coun Sci & Indust Res, 65-67; res assoc, Lab Space Res, Mass Inst Technol, 67-68; res assoc, 72-73, ADJ PROF SPACE SCI, STATE UNIV NY, STONY BROOK, 73- *Concurrent Pos:* NASA fel, 73-; fel, Max Planck Soc, 78-79; dep exec secy, Am Phys Soc, 85-; NRC comt solar-terrestrial relations, 86- *Mem:* Am Phys Soc; Am Astron Soc; Am Geophys Union; Asn Women Sci; Planetary Soc. *Res:* Solar effects on cosmic rays in interplanetary space; carbon-14 variations and solar activity; cosmic physics and high-energy astrophysics; solar terrestrial relations; acceleration of cosmic rays. *Mailing Add:* Dept Earth & Space Sci SUNY at Stony Brook Stony Brook NY 11794

FORMAN, PAUL, b Philadelphia, Pa, Mar 22, 37; m 88. HISTORY OF MODERN PHYSICS. *Educ:* Reed Col, BA, 59; Univ Calif, Berkeley, MA, 62, PhD(hist), 67. *Prof Exp:* Actg instr hist sci, Univ Calif, Berkeley, 65-66; asst prof, Univ Rochester, 67-72; assoc cur, 72-75, CUR, HIST MODERN PHYSICS, SMITHSONIAN INST, 75- *Concurrent Pos:* Fel, Inst Int Studies, Univ Calif, Berkeley, 72-73; assoc ed, Hist Studies Phys Sci, 81-; vis prof, Dept Physics, Univ Rome, 86 & Prog Hist Sci, Princeton Univ, 87; vis scholar, Dept Physics, New York Univ, 88-; adv ed, Isis: Off J Hist Sci Soc, 88- *Mem:* Fel Am Phys Soc. *Res:* History of modern physics, especially in relation to society and culture; physics in Germany after World War I and in the US after World War II; beginnings of quantum mechanics and quantum electronics. *Mailing Add:* Smithsonian Inst NMAH-5025 Washington DC 20560

FORMAN, RALPH, b New York, NY, Oct 4, 21; m 43; c 3. PHYSICS. *Educ:* Brooklyn Col, BA, 42; Univ Md, MS, 51, PhD(physics), 54. *Prof Exp:* Physicist, Nat Bur Standards, 42-55 & Union Carbide Corp, 55-64; physicist, Lewis Res Ctr, NASA, 64-83; CONSULT, ANALEX CORP, 83- *Mem:* Am Phys Soc. *Res:* irradiated gas plasma; High temperature heat pipes; high temperature thermal conductivity measurements; auger and electron spectroscopy for chemical analysis surface studies of impregnated tungsten thermionic cathodes; studies on the surface reactions involving barium-oxygen-tungsten; secondary- emission measurements on low secondary yield surfaces. *Mailing Add:* 21516 Hilliard Blvd Cleveland OH 44116

FORMAN, RICHARD ALLAN, b Brooklyn, NY, Mar 5, 39; m 60; c 1. COMPUTER SCIENCE. *Educ:* Polytech Univ, Brooklyn, NY, BS, 59; Univ Md, College Park, PhD(physics), 65. *Prof Exp:* Physicist solid state, Nat Bur Standards, 59-89; PRES, OPTIMUM TECH, INC, POTOMAC, MD, 88-; TECH STAFF SYSTS, MITRE CORP, 90- *Concurrent Pos:* Vis scientist, Brown Univ, 66-67; com sci & technol fel, US Dept Com, 76; com sci fel, Dept Defense, Off Under Secy Defense Res & Eng, 76-77; vis distinguished scholar, Ariz State Univ, 89-90. *Mem:* Am Phys Soc; Japan Soc Appl Physics; AAAS. *Res:* Architecture and design of imaging systems involving mass storage; image analysis; computer hardware; solid state physics; author of various publications. *Mailing Add:* 7525 Colshire Dr McLean VA 22102-3481

FORMAN, RICHARD T T, b Richmond, Va, Nov 10, 35; m 63; c 3. LANDSCAPE & FOREST ECOLOGY. *Educ:* Haverford Col, BS, 57; Univ Pa, PhD(bot), 61. *Hon Degrees:* MA, Harvard Univ, 85; LHD, Miami Univ, 87. *Prof Exp:* Asst instr bot, Univ Pa, 58-59, 60-61 & Duke Univ, 59-60; mem Am Friends Serv Comt, agr exten & community develop, Inst Agr Nat, Guatemala, 61-62; asst prof biol, Escuela Agr Pan Am, Honduras, 62-63; asst prof bot & zool, Univ Wis, 63-66; from asst prof to prof, Rutgers Univ, 66-84, dir, Hutcheson Mem Forest Ctr, 72-84, dir grad prog bot & plant physiol, 79-83; AT GRAD SCH DESIGN, HARVARD UNIV, 84- *Concurrent Pos:* Vis scientist, Orgn Trop Studies, Costa Rica, 70; Fulbright scholar, Bogota, Colombia, 70-71; vis scientist, WI Lab, VI, 73, Ft Burgwin Res Ctr, NMex, 76; ed, Ecol & Ecol Monographs, 73-77, Biosci, 78-84; CNRS chercheur, Centre Emberger CEPE, Montpellier, France, 77-78; fel, Clare Hall, Univ Cambridge, 85-; mem comt examrs grad record exam, Educ Testing Serv, 78-80. *Honors & Awards:* Lindback Found Award, 84. *Mem:* Fel AAAS; Ecol Soc Am (vpres, 82-83); Torrey Bot Club (pres, 80-81); Int Asn Landscape Ecol (vpres, 82-88). *Res:* Principles of landscape structure, function and change; forest communities and ecosystems. *Mailing Add:* Landscape Ecol Grad Sch Design Harvard Univ Cambridge MA 02138

FORMAN, WILLIAM, b New York, NY, Oct 16, 14; m 38; c 2. MATHEMATICS. *Educ:* Brooklyn Col, BS, 36, MA, 37; NY Univ, PhD, 53. *Prof Exp:* Lectr, 36-53, from instr to assoc prof, 53-71, prof math, 71-85, EMER PROF, BROOKLYN COL, 85- *Concurrent Pos:* Asst, Inst Math & Mech, NY Univ, 51-52; consult, Systs Res Group, 59-62. *Mem:* Am Math Soc; Math Asn Am; NY Acad Sci. *Res:* Theory of numbers; algebra. *Mailing Add:* Brooklyn Col Bedford Ave & Ave H Brooklyn NY 11210

FORMANEK, EDWARD WILLIAM, b Chicago, Ill, May 6, 42; m 86. ALGEBRA. *Educ:* Univ Chicago, BS, 63; DePaul Univ, MS, 65; Rice Univ, PhD(math), 70. *Prof Exp:* Asst prof math, Univ Mo-St Louis, 70-71; fel, Carleton Univ, 71-73; vis mem, Inst Advan Study, 73-74; res assoc, Univ Pisa, 74-75; res assoc math, Univ Chicago, 75-77; res assoc, Hebrew Univ Jerusalem, 77-78; PROF MATH, PA STATE UNIV, 78 - *Concurrent Pos:* Vis prof, Univ Calif, Santa Barbara, 80-81; res assoc, Bedford Col, Univ London, 81-82; Guggenheim fel, 83; vis, Math Sci Res Int, Berkeley, 88-89. *Mem:* Am Math Soc. *Res:* Group theory and ring theory. *Mailing Add:* McAllister Bldg Pa State Univ University Park PA 16802

FORMANEK, R(OBERT) J(OSEPH), b Schenectady, NY, Dec 25, 22; m 46; c 2. CHEMICAL ENGINEERING. *Educ:* Purdue Univ, BSChE, 43. *Prof Exp:* Trainee, B F Goodrich Co, 44-47, compounder, 47-51, textile engr, 51-54; mgr textiles, Dunlop Tire & Rubber Corp, 54-57, mgr compounding & textiles, 57-61, tech mgr, 61-64, asst vpres technol, 64-69, V PRES TECHNOL, DUNLOP TIRE & RUBBER CORP, 69- *Concurrent Pos:* Dir, Am Synthetic Rubber Corp. *Mem:* Rubber Mfg Asn. *Res:* Technical development of tires. *Mailing Add:* 10 N Colvin Pl Buffalo NY 14223

FORMANOWICZ, DANIEL ROBERT, JR, b Dunkirk, NY, Apr 22, 54. PREDATOR-PREY INTERACTIONS. *Educ:* State Univ Col NY, BS, 76; Adelphi Univ, MS, 78; State Univ NY, Albany, PhD(biol), 82. *Prof Exp:* Instr, biol, 84-85, vis asst prof, 85-86, ASST PROF, ECOL, UNIV TEX, ARLINGTON, 86- *Concurrent Pos:* Assoc ed, J Herpetologica, 89- *Mem:* AAAS; Ecol Soc Am; Animal Behav Soc; Entom Soc Am; Am Archeol Soc; Asn Trop Biol. *Res:* Behavioral ecology of invertebrate predator-prey interactions; study of the foraging behavior of insects and centipedes and the antipredator behavior of potential prey organisms. *Mailing Add:* Dept Biol Univ Tex Arlington TX 76019

FORMEISTER, RICHARD B, b Chicago, Ill, Jan 19, 46; m 83; c 2. LOCAL AREA NETWORKS, RF COMMUNICATION SYSTEMS. *Educ:* Univ Ill, BS, 68; Ariz State Univ, MS, 78. *Prof Exp:* Mem staff, Sperry Flight Systs, 68-79, Eng Sect head, 79-84; mem tech staff, 84-89, VPRES, FAIRCHILD DATA CORP, 89- *Concurrent Pos:* Fulbright/Hayes grant, Ariz State Univ, 78; lectr, Sperry Corp, 78; owner consult, Circuits & Software Resources Co, 84- *Mem:* Inst Elec & Electronics Engrs. *Res:* Communication systems and circuits with particular emphasis on local area networks for computer communication. *Mailing Add:* 6407 W Yucca St Glendale AZ 85304

FORMICA, JOSEPH VICTOR, b New York, NY, July 3, 29; m 56; c 2. BIOCHEMISTRY. *Educ:* Syracuse Univ, BS, 53, MS, 54; Georgetown Univ, PhD(microbiol), 67. *Prof Exp:* Supvr & bacteriologist, Ft Detrick, Md, 54-58; assoc neurochem, NIH, 58-63; assoc biochem, Georgetown Univ, 63-67, asst prof pediat virol, 67-69; from asst dean to assoc dean, Sch Basic Sci, 77-81, ASSOC PROF MICROBIOL, MED COL VA, 69- *Mem:* Am Soc Microbiol; fel Am Acad Microbiol; Soc Wine Educators. *Res:* Biosynthesis of antibiotics; biosynthesis of fuel and beverage alcohol. *Mailing Add:* Dept Microbiol & Immunol Box 678 Va Commonwealth Univ Richmond VA 23298

FORNACE, ALBERT J, JR, b Philadelphia, Pa, Apr 5, 49; m 80; c 2. DNA REPAIR, RADIOBIOLOGY. *Educ:* Pa State Univ, BS, 70; Jefferson Med Col, MD, 72. *Prof Exp:* SR INVESTR LAB, MOLECULAR PHARMACOL, NAT CANCER INST, NIH, BETHESDA, MD, 83- *Mem:* Am Asn Cancer Res; Radiation Res Soc; Am Soc Microbiol. *Res:* Mammalian cell DNA repair; molecular biology of the cellular response to injury. *Mailing Add:* Lab Molecular Pharmacol NIH 37-5C09 Bethesda MD 20892

FORNAESS, JOHN ERIK, b Hamar, Norway, Oct 14, 46; m 73, 85; c 4. MATHEMATICAL ANALYSIS. *Educ:* Univ Oslo, Cand Real, 70; Univ Wash, PhD(math), 74. *Prof Exp:* from instr to assoc prof, 74-82, PROF MATH, PRINCETON UNIV, 82- *Res:* Several complex variables. *Mailing Add:* Dept Math Princeton Univ Fine Hall Princeton NJ 08544

FORNBERG, BENGT, b Halmstad, Sweden, June 8, 46; m 74; c 3. NUMERICAL ANALYSIS, COMPUTATIONAL FLUID DYNAMICS. *Educ:* Univ Uppsala, Sweden, BS & MS, 69; PhD(numerical anal), 72. *Prof Exp:* Fel math, Europ Ctr Nuclear Res, Geneva, Switz, 72-74; res instr appl math, Calif Inst Technol, 74-75, from asst prof to assoc prof, 75-84; RES ASSOC APPL MATH, EXXON RES & ENG CO, 84- *Concurrent Pos:* John Simon Guggenheim Mem Found fel, 81. *Res:* Numerical analysis; computational fluid dynamics; parallel computing. *Mailing Add:* Exxon Res & Eng Co Rte 22 E Clinton Twp Annandale NJ 08801

FORNEFELD, EUGENE JOSEPH, b Sandusky, Ohio, May 4, 20; m 44; c 5. ORGANIC CHEMISTRY. *Educ:* Xavier Univ, Ohio, BS, 41; Univ Detroit, MS, 43; Univ Mich, PhD(org chem), 50. *Prof Exp:* Res assoc pharmacol, Univ Mich, 50-51; RES CHEMIST, ELI LILLY & CO, 51- *Mem:* Am Chem Soc. *Res:* Synthesis of steroids and natural products; structure-activity studies in the acetylcholine and analgesic areas; determination of morphine and physiologically active bases in biological media. *Mailing Add:* 7838 Forest Lane Indianapolis IN 46240

FORNES, RAYMOND EARL, b Greenville, NC, Jan 16, 43; m 66; c 3. POLYMER PHYSICS. *Educ:* ECarolina Univ, AB, 65; NC State Univ, PhD(physics), 70. *Prof Exp:* From asst prof to assoc prof textile technol & physics, NC State Univ, 70-79, prof textile mat, Mgt Physics, 79-84, assoc dean grad sch, 83-87, actg dean grad sch, 85, PROF PHYSICS, NC STATE UNIV, 84-, ASSOC DEAN RES, COL PHYSICS & MATH SCI, 89- *Concurrent Pos:* Mem, Nat Sci Comt on Byssinosis, 80-81; gen chair, Ann Meeting, Sigma Xi, 90. *Honors & Awards:* Distinguished Achievement Award, Fiber Soc. *Mem:* Am Phys Soc; The Fiber Soc; AAAS; Sigma Xi; Am Chem Soc. *Res:* Structure and physical properties of polymers; composites; x-ray diffraction; nuclear magnetic resonance. *Mailing Add:* Dept Physics NC State Univ Raleigh NC 27695-8202

FORNEY, ALBERT J, b Vanport, Pa, July 8, 15; m 42; c 3. CHEMICAL ENGINEERING. *Educ:* Geneva Col, BS, 47; Carnegie Inst Technol, BS, 57. *Prof Exp:* Chemist, US Bur Mines, 47-50, chem engr, 50-59, supvry chem engr, 59-76; RETIRED. *Concurrent Pos:* Consult, 76-83. *Mem:* Am Chem Soc. *Res:* Synthetic liquid and gaseous fuels and coal gasification; gas purification, fluidization, steam-iron process; sewage treatment; coal pretreatment; chemicals from coal. *Mailing Add:* 410 Marney Dr Coraopolis PA 15108

FORNEY, BILL E, b Beggs, Okla, June 23, 21; c 4. ENGINEERING. *Educ:* Univ Okla, BS(mech eng), 48 & BS(civil eng), 52. *Prof Exp:* PVT CONSULT, 59- *Concurrent Pos:* Div mgr, Nat Tank Co; mgr eng, Maloney-Crawford Tank Corp; pres, Southern Supply & Valve Corp; owner, BEF Eng Co; adv bd, Mech Eng Col, Tulsa Univ; chmn, Mid Cont Sect, Am Soc Mech Engrs; founding pres, Tulsa Eng Found. *Mem:* Am Soc Mech Engrs; Am Soc Civil Engrs; Am Inst Steel Construct; Am Welding Soc; Am Concrete Inst; Am Consult Engrs Coun; Nat Soc Prof Engrs; Am Acad Forensic Sci; Nat Acad Forensic Engrs. *Res:* Welded and bolted steel pressure vessels, tanks, piping, valves, instruments, machinery and plant process equipment; heating, ventilating, refrigeration and air conditioning units; concrete foundation design; metallurgical analysis; designed the world's largest bolted steel plate and sheet structures and bolted aluminum sheet structure; designed world's first large bolted steel grain storage tanks installed throughout US and other countries; author of various publications; holder of two US patents; foreign business. *Mailing Add:* 6011 E 57th St Tulsa OK 74135

FORNEY, G(EORGE) DAVID, JR, b New York, NY, Mar 6, 40; m; c 3. ELECTRONICS, COMMUNICATIONS. *Educ:* Princeton Univ, BSE, 61; Mass Inst Technol, MS, 63, ScD, 65. *Prof Exp:* Vpres res, Codex Corp, 70-75 & 79-82, vpres res & develop, 75-79; vpres & dir technol & planning, Motorola Inc, 82-86; VPRES, MOTOROLA INC, 86- *Concurrent Pos:* Vis scientist, Stanford Univ, 71-72. *Honors & Awards:* Centennial Medal, Inst Elec & Electronics Engrs, 84. *Mem:* Nat Acad Eng; fel Inst Elec & Electronics Engrs. *Mailing Add:* Motorola Codex Corp Mail Stop C-1-200 Mansfield MA 02048-1193

FORNEY, LARRY J, b Waterloo, Iowa, Nov 1, 44; m 74; c 1. ENGINEERING SCIENCE. *Educ:* Case Inst Technol, BS, 66; Mass Inst Technol, MS, 68, ME, 69; Harvard Univ, PhD(eng sci), 74. *Prof Exp:* Res engr, Gen Elec Lighting Res Lab, 66 & Nat Res Corp, 68; res asst, Mass Inst Technol, 66-68; asst prof, Univ Ill, Urbana, 74-79; ASSOC PROF CHEM ENG, GA INST TECHNOL, ATLANTA, 79- *Concurrent Pos:* Consult, Walden Res, Abcor, Inc, 72-74; NSF initiation grant, 75-77, Sverdrup Technol Inc, 82-86, Lockheed-Georgia Co, 82-83 & Georgia Pacific Co, 85; SCEEE fel, US Air Force, 82; ASEE/NASA fel, 88. *Mem:* Am Inst Chem Engrs; Am Asn Aersol Res. *Res:* Applied fluid mechanics; turbulent jets and plumes; aerosol mechanics; aerosol instrumentation; pipeline mixing. *Mailing Add:* 4029 Menlo Way Doraville GA 30340

FORNEY, R(OBERT) C(LYDE), b Chicago, Ill, Mar 13, 27; m 48; c 4. CHEMICAL ENGINEERING. *Educ:* Purdue Univ, BS, 47, MS, 48, PhD(chem eng), 50. *Hon Degrees:* Dr Eng, Purdue Univ, 81. *Prof Exp:* Res engr, Textile Fibers Dept, E I du Pont de Nemours & Co, Inc, 50-52, group supvr, 52, plant res supvr, 52-56, process supt, 56-59, asst plant mgr, 59-63, tech mgr, 63-64, prod mgr, 64-66, dir prod mkt div, 66-69, asst gen dir mkt div, 69-70, asst gen mgr, 70-75, vpres & gen mgr, 75-77, vpres, Plastics Prod & Resins Dept, 77-78, sr vpres admin dept, 79-81, exec vpres, 81-89; RETIRED. *Mem:* Am Chem Soc; Am Inst Chem Engrs; Soc of Chem Indust; Am Asn Advan Chem; Sigma Xi; Honor Sci Res Soc; Nat Acad Engr. *Res:* Industrial relations; reaction kinetics. *Mailing Add:* Hilltop View Rd Unionville PA 15375

FORNEY, ROBERT BURNS, b Ashley, Ind, July 9, 16; m 41; c 2. TOXICOLOGY. *Educ:* Ind Univ, PhD, 48. *Hon Degrees:* LLD, Ind Cent Col, 64. *Prof Exp:* From asst prof to assoc prof toxicol, 48-62, prof pharmacol & toxicol, Sch Med, Ind Univ-Purdue Univ, Indianapolis, 62-77; DISTINGUISHED PROF TOXICOL & DIR STATE TOXICOL LAB, SCH MED, IND UNIV, INDIANAPOLIS, 77- *Concurrent Pos:* Mem traffic safety comt & comt alcohol & drugs, Nat Safety Coun. *Mem:* Am Soc Pharmacol & Exp Therapeut; Soc Toxicol; fel Am Acad Forensic Sci; Am Soc Clin Pharmacol & Therapeut; Sigma Xi. *Res:* Effects of alcohol, marihuana and other drugs on performance. *Mailing Add:* Dept Pharm MS 410A Ind Univ Sch Med 1001 Walnut St Indianapolis IN 46223

FOROULIS, Z ANDREW, b Volos, Greece, Dec 6, 26; US citizen; m 62. CHEMICAL & METALLURGICAL ENGINEERING. *Educ:* Nat Univ Athens, Dipl, 54; Mass Inst Technol, MS & ChE, 56, MetE, 60, DSc(metall eng), 61. *Prof Exp:* Chem engr, Dewey & Almy Chem Div, W R Grace & Co, 56-57; engr, 61-62, sr engr, 62-63, eng assoc, 63-68, SR ENG ASSOC, EXXON RES & ENG CO, 68- *Concurrent Pos:* Adj prof, NY Univ, 68- *Mem:* Am Chem Soc; Am Inst Chem Engrs; Am Inst Mining, Metall & Petrol Engrs; Am Soc Mech Engrs. *Mailing Add:* Exxon Res & Eng Co PO Box 101 Florham Park NJ 07932-1093

FORQUER, SANDRA LYNNE, b New Haven, Conn, Sept 7, 43; m 65; c 2. QUALITY ASSURANCE, RISK FACTOR RESEARCH. *Educ:* Chatham Col, BA, 65; Univ Pittsburgh, MEd, 72, PhD(guid & coun), 77. *Prof Exp:* Proj dir, St Francis Community Ment Health Ctr, 77-79; med res scientist, Systs Res Unit, 79-82, asst res prof, 82 -86, ASSOC PROF, MED COL PA, 87-, DIR CONTINUING MENTAL HEALTH EDUC, 86- *Concurrent Pos:* Psychiat social worker, St Francis Community Hosp, 73-76; prin investr, St Francis Community Ment Health Ctr, 77-79; preceptor, Grad Sch Pub Health, Univ Pittsburgh, 78-79; adj prof, Univ Pittsburgh, 78-79; consult, Dept Human Resources, Ga, 79-81 & Region III Tech Asst Ctr, 80-81; prog consult, Div Continuing Ment Health, Med Col Pa, 81-82, assoc dir, 82-86. *Mem:* Am Psychol Asn. *Res:* Development of a mental health primary prevention screening instrument for use in pediatric clinics; development and research on the implementation of quality assurance systems. *Mailing Add:* NY Ment Health Qual Assurance Div 44 Holland Ave Eighth Floor Albany NY 12229

FORRAY, MARVIN JULIAN, b New York, NY, Apr 18, 22; m 45; c 4. MATHEMATICS. *Educ:* NY Univ, BA, 43, MS, 44; Columbia Univ, PhD(appl mech), 55. *Prof Exp:* Instr math, Polytech Inst Brooklyn, 44-53; sr engr struct, Repub Aviation Corp, NY, 53-55, prin engr, develop & res, 55-58, design specialist, 58-61, develop engr, 61-66; PROF MATH, C W POST COL, 66- *Concurrent Pos:* Lectr & prof, Adelphi Col, 55-65. *Mem:* Math Asn Am; Am Soc Civil Eng. *Res:* Applied mechanics; elasticity; thermoelasticity; plates and shells; instability of structures; vibrations; heat conduction; calculus of variations; differential equations; functions of a real and complex variable; transform theory. *Mailing Add:* 21 Edward St Lynbrook NY 11563

FORRER, MAX P(AUL), b St Gallen, Switz, Oct 15, 25; nat US; m 52; c 2. ELECTRICAL ENGINEERING. *Educ:* Swiss Fed Inst Technol, dipl ing, 50; Stanford Univ, PhD(elec eng), 59. *Prof Exp:* Develop engr commun, Standard Tel & Radio Corp, Switz, 51-52 & Western Elec Co, NJ, 52-55; proj engr microwave electronics, Microwave Lab, Gen Elec Co, 55-61; proj engr & mgr microwave component, Kane Eng Lab, 61-63; head circuits develop sect, 63-68, vdir, 65-68, dir, Centre Electronique Horloger, 68-84, DIR GEN & CHIEF EXEC OFFICER, SWISS CTR ELECTRONICS & MICROTECHNICS. *Concurrent Pos:* Asst, Hansen Labs, Stanford Univ, 57-58. *Honors & Awards:* Victor Kullberg Medal, Urmakare Ambete, Stockholm, 70. *Mem:* Sr mem Inst Elec & Electronics Engrs. *Res:* Microwave components; semiconductor devices; research and development management. *Mailing Add:* Ctr Suisse Delectronique Et De Microtechnique Sa/Csem 71 Rue Maladiere Neuchatel 2007 Switzerland

FORREST, BRUCE JAMES, b Clinton, Ont, Can, July 31, 50. NUCLEAR MAGNETIC RESONANCE, ELECTRON SPIN RESONANCE. *Educ:* Univ Western Ont, BSc, 71; Bishop's Univ, MSc, 73; Simon Fraser Univ, PhD(chem), 78. *Prof Exp:* ASST RES PROF CHEM, DALHOUSIE UNIV, 81- *Mem:* Am Chem Soc; Can Inst Chem; Spectros Soc Can. *Res:* Application of magnetic resonance to chemical and biochemical problems; studies of model and biological membranes; mode of action of pharmaceutical agents; liquid crystals; micelles; detergents. *Mailing Add:* Dept Chem Brandon Univ Brandon MB R7A 6A9 Can

FORREST, HUGH SOMMERVILLE, b Glasgow, Scotland, Apr 28, 24; m 54; c 3. MOLECULAR BIOLOGY. *Educ:* Glasgow Univ, BSc, 44; Univ London, PhD(chem), 48, DSc, 70; Cambridge Univ, PhD(chem), 51. *Prof Exp:* Mem sci staff, Nat Inst Med Res, Med Res Coun, Gt Brit, 45-48; res fel biol, Calif Inst Technol, 51-54, sr res fel, 55-56; res scientist, 56-57, assoc prof, 57-63, chmn dept, 74-78, PROF ZOOL, UNIV TEX, AUSTIN, 63- *Concurrent Pos:* Co-ed, Biochem Genetics, 70-78, ed, 79- *Mem:* Am Chem Soc; Royal Soc Chem; Am Soc Biol Chemists. *Res:* Chemistry and biology of pteridines and of a new coenzyme, methoxatin; chemistry of insect lipoproteins. *Mailing Add:* Dept Zool Univ Tex Austin TX 78712-1064

FORREST, IRENE STEPHANIE, b Charlottenburg, Ger, Aug 20, 08; nat US; m 34; c 1. BIOCHEMISTRY. *Educ:* Univ Berlin, PhD(chem), 32. *Prof Exp:* Researcher, microbiol & biochem, Pasteur Inst Paris & Munic Hosp (Hotel Dieu), Paris, France, 32-34; asst chemist, Istanbul, 35-36; researcher biochem, NY Univ, 41-43; researcher, Moraine Prod Div, Gen Motors Corp, Dayton, Ohio, 44; translator, US War Dept, 45-47; nucleic acids researcher, NY Univ, 48; endocrinol res, St Clare's Hosp, NY, 49-50, consult chem, 51-52; biochem res, Polytech Inst Brooklyn & NY Med Col, 53-56; chief biochem res lab, Vet Admin Hosp, Brockton, Mass, 57-61; chief, Biochem Res Lab, Vet Admin Hosp, Palo Alto, 61-80; RETIRED. *Concurrent Pos:* Res assoc psychiat, Sch Med, Stanford Univ, 61-73; sr res scientist, Inst Chem Biol, Univ San Francisco, 73- *Mem:* Am Soc Pharmacol & Exp Therapeut. *Res:* General biochemistry; psychopharmacology; drug metabolism; phenothiazine drugs. *Mailing Add:* 540 Ringwood Ave Menlo Park CA 94025

FORREST, JAMES BENJAMIN, b Newcastle, NB, Oct 8, 35; m 60, 70, 85; c 3. GEOTECHNOLOGY, STRUCTURES. *Educ:* Univ NB, BSc, 59; Univ BC, MASc, 63; Northwestern Univ, PhD(civil eng), 66. *Prof Exp:* Engr, Dept Nat Health & Welfare, Govt Can, 59-60; design engr, Montreal Eng Ltd, 60-61; asst prof eng, Carleton Univ, 65-68; res assoc engr, Univ NMex, 68-70; SR PROJ ENGR, NAVAL CIVIL ENG LAB, 70- *Concurrent Pos:* Vis lectr, Univ Calif, Los Angeles, 80; exchange scientist to USSR. *Mem:* Am Soc Civil Engrs; Am Soc Testing & Mat; Soc Am Mil Engrs. *Res:* Yielding of clays subjected to steady state vibratory, quasi-static and impulsive loadings; the constitutive relationships for pavement component materials; soil dynamics, applied mechanics; earthquake induced soil liquefaction; waterfront structures. *Mailing Add:* Naval Civil Eng Lab Code L42 Port Hueneme CA 93043-5003

FORREST, JOHN CHARLES, b Larned, Kans, July 30, 36; m 64; c 2. MEAT SCIENCE. *Educ:* Kans State Univ, BS, 60, MS, 62; Univ Wis, PhD(muscle physiol), 66. *Prof Exp:* Proj asst muscle chem & physiol, Univ Wis, 65-66; asst prof animal sci, Univ Minn, 66-67; assoc prof, 67-77, PROF ANIMAL SCI, PURDUE UNIV, 77- *Concurrent Pos:* Mem bd dirs, Am Meat Sci Asn, 84-86 & chmn, Reciprocal Meat Conf, 85. *Honors & Awards:* Exten Indust Serv Award, Am Meat Sci Asn, 83. *Mem:* Am Soc Animal Sci; Am Meat Sci Asn; Inst Food Technologists. *Res:* Influence of ante-mortem physiology and stress on the post-mortem metabolism of skeletal and cardiac muscle; effects of temperature on pre-rigor muscle systems; developing price discovery systems based upon carcass composition and yield. *Mailing Add:* Dept Animal Sci Purdue Univ West Lafayette IN 47907

FORREST, ROBERT NEAGLE, b Pendleton, Ore, Nov 25, 25; m 51; c 2. OPERATIONS RESEARCH, OPERATIONS ANALYSIS. *Educ:* Univ Ore, BS, 50, MS, 52 & 54, PhD(physics), 59. *Prof Exp:* Nuclear physicist, Lawrence Radiation Lab, Univ Calif, 58-59; mem staff geoastrophysics, Sci Lab, Boeing Airplane Co, 59; asst prof surface physics, Ore State Univ, 59-63; asst prof physics, Southern Ore Col, 63-64; asst prof, 64-80, PROF OPERS ANALYSIS, US NAVAL POSTGRAD SCH, 80- *Mem:* AAAS; Am Asn Physics Teachers; Am Vacuum Soc; Opers Res Soc Am; Sigma Xi. *Mailing Add:* Dept Opers Res US Naval Postgrad Sch Monterey CA 93940

FORREST, THOMAS DOUGLAS, b Murray, Ky, Mar 13, 33; m 64; c 2. MATHEMATICS. *Educ:* Murray State Univ, BS, 54; Univ Tenn, MMath, 63; George Peabody Col, PhD(math), 71. *Prof Exp:* Instr math, Marshall County Schs, Ky, 56-61, Murray High Sch, 63-64 & Murray State Univ, 64-67; assoc prof math, Mid Tenn State Univ, 68-89; RETIRED. *Concurrent Pos:* NSF panelist, 77. *Mem:* Nat Coun Math Teachers; Math Asn Am. *Mailing Add:* 2129 Harding Pl Murfreesboro TN 37129

FORRESTER, ALVIN THEODORE, plasma physics; deceased, see previous edition for last biography

FORRESTER, DONALD JASON, b Attleboro, Mass, Jan 31, 37; m 61; c 3. WILDLIFE PARASITOLOGY, WILDLIFE DISEASES. *Educ:* Univ Mass, Amherst, BS, 58; Univ Mont, MS, 60; Univ Calif, Davis, PhD(zool), 67. *Prof Exp:* Res assoc zool, Univ Mont, 61-63; asst prof, Clemson Univ, 67-69; asst prof, Div Biol Sci, 69-70, asst prof, Dept Vet Sci, 70-73, assoc prof, 73-79, PROF PARASITOL, COL VET MED, UNIV FLA, 79- *Concurrent Pos:* Ed, J Wildlife Dis, 81-86. *Honors & Awards:* Beecham Award for Res Excellence, 88. *Mem:* Am Soc Parasitologists; Wildlife Disease Asn (vpres, 77-79, pres, 79-81); Wildlife Soc. *Res:* Parasites and diseases of wild animals; epizootiology and ecology of transmission of protozoans and helminths of wild animals. *Mailing Add:* Dept Infectious Dis Col Vet Med Univ Fla Gainesville FL 32610-0137

FORRESTER, JAY W, b Climax, Nebr, July 14, 18; c 3. SYSTEM DYNAMICS. *Educ:* Univ Nebr, BSc, 39; Mass Inst Technol, SM, 45. *Hon Degrees:* DEng, Univ Nebr, 54, Newark Col Eng, 71, Univ Notre Dame, 74; DSc, Boston Univ, 69, Union Col, 73, Univ Mannheim, Ger, 79. *Prof Exp:* Mem staff, Servomechanisms Lab, 40-46, Digital Comput Lab, 46-51, head, Digital Comput Div, Lincoln Lab, 52-56, prof mgt, 56-72, GERMESHAUSEN PROF MGT, MASS INST TECHNOL, 77- *Concurrent Pos:* Hon prof, Shanghai Inst Technol, 87. *Honors & Awards:* Valdemar Poulsen Gold Medal, Danish Acad Tech Sci, 69; Medal of Honor & Systs, Man & Cybernetics Soc Award, Inst Elec & Electronics Engrs, 72;

Howard N Potts Award, The Franklin Inst, 74; Harry Goode Mem Award, Am Fedn Info Processing Socs, 77; Computer Pioneer Award, Comput Soc, Inst Elec & Electronics Engrs, 82; Info Storage Award, Inst Elec & Electronics Engrs Magnetics Soc, 88; Lord Found Award, 88; US Nat Medal Technol, 89; Pioneer Award, Inst Elec & Electronics Engrs Aerospace & Electronic Systs Soc, 90. *Mem:* Nat Acad Eng; fel Inst Elec & Electronic Engrs; fel Acad Mgt; fel Am Acad Arts & Sci; hon mem Soc Mfg Engrs. *Res:* Random-access; coincident-current magnetic storage used as standard memory device for digital computers; computer modeling to analyze social systems and further develop the field of system dynamics. *Mailing Add:* Syst Dynamics Group Mass Inst Technol Bldg E40 294 Cambridge MA 02139

FORRESTER, SHERRI RHODA, b New York, NY, Aug 8, 36. ORGANIC CHEMISTRY. *Educ:* Duke Univ, BS, 58; Northwestern Univ, PhD(org chem), 62. *Prof Exp:* Asst prof, 62-78, ASSOC PROF CHEM, UNIV NC, GREENSBORO, 78- *Mem:* Am Chem Soc. *Mailing Add:* Dept Chem Univ NC Greensboro Greensboro NC 27412-0002

FORRESTER, WARREN DAVID, b Hamilton, Ont, Mar 4, 25. OCEANOGRAPHY. *Educ:* Univ Toronto, BA, 47; Univ BC, MS, 61; Johns Hopkins Univ, PhD(oceanog), 67. *Prof Exp:* Engr, Geod Surv Can, 47-57 & Can Hydrographic Serv, 57-59; researcher, Bedford Inst Oceanog, 63-75; chief, Tides Currents & Water Levels, Can Hydrographic Serv, 75-81; PVT CONSULT PHYS OCEANOG, 81- *Concurrent Pos:* Assoc prof, McGill Univ, 69-74 & Dalhousie Univ, 71-75; consult, Phys Oceanog, 81- *Res:* Internal tides in Gulf of St Lawrence; currents and tides in coastal regions; application of geostrophic approximation to transport calculations; circulation in Gulf of St Lawrence; tidal analysis. *Mailing Add:* 500 Laurier W Apt 904 Ottawa ON K1R 5E1 Can

FORRETTE, JOHN ELMER, b Chicago, Ill, Sept 14, 22; m 48; c 1. ANALYTICAL CHEMISTRY. *Educ:* Loyola Univ, Ill, BS, 48; DePaul Univ, MS, 55; Pac Western Univ, PhD, 80. *Prof Exp:* Chemist, Div Hwys, State of Ill, 48-56; from res chemist to sr res chemist, Borg-Warner Corp, 57-63, group leader anal chem, 63-67; group leader, 67-70, MGR ANAL RES, VELSICOL CHEM CO, 70- *Mem:* Sigma Xi; Am Soc Appl Spectros; Am Chem Soc; fel Am Inst Chemists. *Res:* Chemical spectroscopy; gas chromatography; general analytical chemistry; analytical chemistry of agricultural chemicals. *Mailing Add:* 1590 Heights Blvd Winona MN 55987-2520

FORREY, ARDEN W, b Nampa, Idaho, Dec 19, 32; m 61; c 1. BIOCHEMISTRY. *Educ:* Univ Wash, AB, 55, PhD(biochem), 63. *Prof Exp:* Trainee biochem, Univ Wash, 63-64; Nat Inst Arthritis & Metab Dis fel, 64-65; univ fel, Univ Wash, 65-66, lectr biochem, 66-67, res assoc, 67-74, res asst prof, Dept Med, Clin Res Ctr, Harborview Hosp & Univ Wash, 74-77; clin comput specialist, Vet Admin Hosp, Seattle, 78-80; asst prof res, Dept Pediat, Georgetown Univ, 80-82, trauma regist dir, 83-87; MED CONSULT, 88- *Mem:* AAAS; Am Chem Soc; Brit Biochem Soc; Am Asn Clin Chem; Sigma Xi; Nat Acad Clin Biochem. *Res:* Structure and active sites of proteins, particularly muscle phosphorylase and actin; clinical chemistry; pharmacokinetics; clinical computing. *Mailing Add:* 4916 Purdue Ave NE Seattle WA 98105

FORRO, FREDERICK, JR, b Woonsocket, RI, May 17, 24; m 44; c 4. DNA REPLICATION, RADIATION BIOLOGY. *Educ:* Yale Univ, MD, 49. *Prof Exp:* USPHS fel, Yale Univ, 49-51; assoc biophysicist, Brookhaven Nat Lab, 51-55; from asst prof to assoc prof biophys, Yale Univ, 55-68; chmn dept, 68-77, prof, Dept Genetics & Cell Biol, Univ Minn, St Paul, 68-83, dir minority affairs, Col Biol Sci, 77-83; RETIRED. *Concurrent Pos:* Agent Europ Atomic Energy Comn, Dept Biol, Univ Brussels, 66-67. *Mem:* AAAS; Genetics Soc Am; Biophys Soc; Sigma Xi. *Res:* Molecular biology, particularly self-duplication; radiation biology; origin of life. *Mailing Add:* 43 E Pleasant Lake Rd St Paul MN 55127

FORS, ELTON W, b Drake, NDak, Jan 29, 34; m 61. MATHEMATICS. *Educ:* Univ NDak, BS, 56, MS, 59; Univ Okla, PhD(math educ), 69. *Prof Exp:* Assoc prof, 60-70, PROF MATH, NORTHERN STATE COL, 70- *Mem:* Math Asn Am; Nat Coun Math Teachers. *Mailing Add:* Dept Math Northern State Univ Aberdeen SD 57401

FORSBERG, CECIL WALLACE, b Kinistino, Sask, Mar 24, 42; m 70; c 2. MICROBIAL PHYSIOLOGY. *Educ:* Univ Sask, BSA, 64, MSc, 66; McGill Univ, PhD(microbial physiol), 69. *Prof Exp:* Fel microbial physiol, Nat Inst Med Res, Med Res Coun Eng, 69-71, scientist, 71-73; PROF MICROBIOL ECOL, UNIV GUELPH, 73-, INSTR, MICROBIOL ECOL & MICROBIOL TECHNOL, 79- *Concurrent Pos:* Sect ed, Can J Microbiol. *Mem:* Can Soc Microbiol; Am Soc Microbiol. *Res:* Physiology of anaerobic rumen bacteria, with emphasis on the physiological and genetic characterization of the cellulase(s) and hemicellulase(s) of the rumen bacterium Bacteroides succinogenes. *Mailing Add:* Dept Microbiol Univ Guelph Guelph ON N1G 2W1 Can

FORSBERG, JOHN HERBERT, b Duluth, Minn, Apr 24, 42; m 66. INORGANIC CHEMISTRY. *Educ:* Univ Minn, Duluth, BA, 64; Univ Ill, MS, 66, PhD(chem), 68. *Prof Exp:* Assoc prof, 68-74, PROF CHEM, ST LOUIS UNIV, 74-, CHMN DEPT, 76- *Concurrent Pos:* Res Corp grant, 68- *Mem:* Am Chem Soc. *Res:* Coordination chemistry in nonaqueous solvent; lanthanide chemistry. *Mailing Add:* Dept Chem St Louis Univ St Louis MO 63103-3096

FORSBERG, KEVIN, b Oakland, Calif, July 20, 34; m 81; c 3. ENGINEERING MANAGEMENT, SOLID MECHANICS. *Educ:* Mass Inst Technol, SB, 56; Stanford Univ, MS, 58, PhD(eng mech), 61. *Prof Exp:* Scientist, Lockheed Missiles & Space Co, 56-61, res specialist shell dynamics, 62-64, mgr solid mech lab, 64-71, asst dir mat & struct directorate, 71-73, dep prog mgr thermal protection systs, 73-75, prog mgr, 75-79, prog mgr new bus, 79-84; vpres, Consult Resources Inc, 84-89; COFOUNDER, CTR SYSTS MGT, 89- *Mem:* Fel Am Soc Mech Engrs; Am Inst Aeronautics; NY Acad Sci; Am Inst Aeronaut & Astronaut. *Res:* Static and dynamic behavior of shell structures; application of computer technology and active computer graphics to engineering problems; development and production of thermal protection system for space shuttle orbiter. *Mailing Add:* 1225 Vienna Dr Sunnyvale CA 94089

FORSBERG, ROBERT ARNOLD, b Julesburg, Colo, Nov 6, 30; m 58; c 4. AGRONOMY, PLANT GENETICS. *Educ:* Univ Wis, BS, 52, MS, 57, PhD(agron, plant breeding), 61. *Prof Exp:* Fel statist genetics, NC State Univ, 61-63; from asst prof to assoc prof agron, 63-69, chmn dept, 79-89, PROF AGRON, UNIV WIS-MADISON, 73- *Mem:* Fel Am Soc Agron; fel Crop Sci Soc Am. *Res:* Interspecific hybridization in Avena; biometrical procedures as applied to quantitative inheritance and plant breeding; breeding and genetics of wheat, oats, barley, rye and triticale. *Mailing Add:* Dept Agron Univ Wis Madison 1575 Linden Dr 373 Moore Hall Madison WI 53706

FORSBLAD, INGEMAR BJORN, b Gothenburg, Sweden, June 18, 27; m 57; c 2. ORGANIC CHEMISTRY. *Educ:* Univ Uppsala, MS, 55, PhD(org chem), 58. *Prof Exp:* Technician chem, Svenska Oljeslageri AB, Sweden, 45-46 & Skanska Attiksfabrik AB, 48; engr org chem, AB Bofors Nobelkrut, 50-51 & Osterreichische Stickstoffweke, Austria, 55; fel, Fla State Univ, 58-59; res assoc, Upjohn Co, 59-66, head fine chem, 66-71, mgr specialty chem, 71-79, prod mgr fine chem, 79-85, group mgr chem prod, 85-90; RETIRED. *Honors & Awards:* Upjohn Award, 75. *Mem:* Am Inst Chem Engrs; Am Chem Soc; Swedish Chem Soc; Sigma Xi; Int Union Pure & Appl Chem. *Res:* Mechanism of alkylation of beta-ketoesters; process development and production of steroids and aromatic compounds. *Mailing Add:* 9414 E Chore Dr Kalamazoo MI 49002

FORSCHER, BERNARD KRONMAN, b New York, NY, Nov 15, 27; m 48; c 5. BIOCHEMISTRY. *Educ:* NY Univ, BA, 46; Northwestern Univ, PhD(chem), 52. *Prof Exp:* Res chemist, Chas Pfizer & Co, 46-48, chemist, Eng Res & Develop Lab, 47; asst instr, Dent Sch, Northwestern Univ, 48-52; coordr res, Sch Dent, Georgetown Univ, 52-54; chemist, Nat Inst Dent Res, NIH, 54-57; prof biochem & chmn dept, Sch Dent, Univ Kansas City, 57-62; ed, Scott & White Mem Hosp, 75-76; managing ed, Proc Nat Acad Sci, 76-83; consult, Sect Publ, Mayo Clin, 62- 75, head, 83-90; RETIRED. *Mem:* AAAS. *Res:* Chemistry of inflammation; chemistry of bone. *Mailing Add:* Five Conchas Ct Mayo Clinic Santa Fe NM 87505-8803

FORSCHER, FREDERICK, b Vienna, Austria, May 20, 18; nat US; m 44; c 3. ENGINEERING. *Educ:* Princeton Univ, BS & MS, 47; Columbia Univ, PhD(appl mech), 53. *Prof Exp:* Instr civil eng, Columbia Univ, 47-52; supvr mech metall, Atomic Power Div, Westinghouse Elec Corp, 52-57; vpres, Nuclear Mat & Equip Corp, 57-67; mgr adv fuels, Westinghouse Elec Corp, 67-71; energy mgr consult, 71-79; guality assurance specialist, Nat Res Coun, 79-86; MGT CONSULT, 86- *Concurrent Pos:* Staff mem, Kemeny Comn on Nuclear Three Mile Island-2 accident, 79. *Mem:* Am Soc Metals; Am Soc Mech Engrs; Am Nuclear Soc; hon mem Am Soc Testing & Mat; Sigma Xi. *Res:* Social metabolism; nucleon cluster model. *Mailing Add:* 144 N Dithridge St Pittsburgh PA 15213

FORSDYKE, DONALD ROY, b London, Eng, Oct 9, 38; m 64; c 4. MOLECULAR BIOLOGY, BIOCHEMISTRY. *Educ:* Univ London, MB, BS, 61; Cambridge Univ, BA, 65, PhD(biochem), 67. *Prof Exp:* House physician, St Mary's Hosp, London, 61; house surgeon, Addenbrooke's Hosp, Cambridge, 63; sci officer exp path, Inst Animal Physiol, Babraham, Cambridge, 67-68; ASSOC PROF BIOCHEM, QUEEN'S UNIV, ONT, 68- *Mem:* Can Biochem Soc; Brit Biochem Soc; Can Soc Immunol; Am Asn Biol Chem. *Res:* Cell control mechanisms, especially lymphoid tissue and nucleic acids; cytokines and growth factors; regulation of gene expression; theoretical immunology. *Mailing Add:* Dept Biochem Queen's Univ Fac Med Kingston ON K7L 3N6 Can

FORSEN, HAROLD K(AY), b St Joseph, Mo, Sept 19, 32; m 52; c 3. PLASMA PHYSICS, NUCLEAR ENGINEERING. *Educ:* Calif Inst Technol, BS, 58, MS, 59; Univ Calif, Berkeley, PhD(elec eng), 65. *Prof Exp:* Staff assoc exp physics, Gen Atomic Div, Gen Dynamics Corp, 59-62; res assoc elec eng, Univ Calif, Berkeley, 62-65; from assoc prof to prof nuclear eng, Univ Wis-Madison, 65-73, dir phys sci lab, 70-72; mgr eng, Exxon Nuclear Co, Bellevue, 73-75, vpres, 75-80, mem bd dir, 78-80; dir, Jersey Nuclear-Avco Isotopes Inc, 73-81, exec vpres, 74-80, pres, 81; mgr eng & mat, res & eng, Bechtel Corp, 81-83, dep mgr, 83, mgr advan systs, 84-85, vpres, 84-86, MGR RES & DEVELOP, BECHTEL CORP, 85-, SR VPRES, 86- *Concurrent Pos:* Mem tech staff, Hughes Aircraft Co, 58-59; consult, Gen Atomic Div, Gen Dynamics Corp, 62-64, Lawrence Radiation Lab, 64-65, Oak Ridge Nat Lab, 69-72; Argonne Nat Lab, 70-72, Jersey Nuclear Co, 70-73 & Battelle Mem Inst, 71-72; fusion adv to dir, Oak Ridge Nat Lab, 77-84, mem, Argonne Univ Asn Spec Comt on Fusion, 80-90; mem, Dept Energy Fusion Sr Rev Comt, 77; mem bd trustees, Pac Sci Ctr Found, 77, vpres, 77, pres, 78-80, chmn, 81; mem vis comt, Col Eng, Univ Wash, 81-; mem adv comt magnetic fusion, Dept Energy, 83-89, fusion policy, 90; Res Coun Comt on Magnetic Fusion in Energy Policy, 87-88; mem bd dirs, Bay Area Sci Fair, 89-; technol policy options comt, Nat Acad Eng, 90-91. *Honors & Awards:* Arthur Holly Compton Award, Am Nuclear Soc, 73. *Mem:* Nat Acad Eng; Am Nuclear Soc; sr mem Inst Elec & Electronics Engrs; fel Am Phys Soc. *Res:* Plasma physics; ion beams; plasma sources; thermonuclear fusion; reactor concepts; laser isotope separation. *Mailing Add:* 255 Tim Ct Danville CA 94526

FORSHAM, PETER HUGH, b New Orleans, La, Nov 15, 15; m 47; c 3. INTERNAL MEDICINE, ENDOCRINOLOGY. *Educ:* Cambridge Univ, BS, 37, MA, 41; Harvard Med Sch, MD, 43; Am Bd Internal Med, dipl, 51. *Prof Exp:* Res assoc physiol chem, Rockefeller Inst, 40-41; house officer med, Peter Bent Brigham Hosp, Mass, 43-44, asst resident, 44-46, res assoc, 47-49; instr, Harvard Med Sch, 49-51; assoc prof, 51-57, PROF MED & PEDIAT, SCH MED, CHIEF ENDOCRINOL, DEPT MED & DIR METAB RES

INST & GEN CLIN RES CTR, MED CTR, UNIV CALIF, SAN FRANCISCO, 57- *Concurrent Pos:* Consult, NIH, chmn metab study sect, 58-60; consult, US Navy, 57-65, Oak Knoll Hosp, Oakland. *Mem:* AAAS; Am Soc Clin Invest; Asn Am Physicians; Soc Exp Biol & Med; Endocrine Soc. *Res:* Metabolic diseases; pathophysiology of the adrenal cortex, mostly in man; relation of pituitary tropic hormones to activity of target glands; metabolic studies of diabetes mellitus; anti-inflammatory agents used in collagen diseases; antiovulatory hormones from the pineal. *Mailing Add:* Univ Calif Med Ctr San Francisco CA 94143

FORSHEY, CHESTER GENE, b Salem, Ohio, Mar 21, 25; m 56; c 4. POMOLOGY. *Educ:* Ohio State Univ, BS, 50, PhD(hort), 54. *Prof Exp:* Asst, Ohio Agr Exp Sta, 52-54; from asst prof to prof, Hudson Valley Lab, NY State Agr Exp Sta, 54-90, supt, 69-90, emer prof, 90-; RETIRED. *Concurrent Pos:* Mem staff, Rockefeller Found Chilean Agr Prog, 63-64; hon mem fac, Cath Univ Chile & Univ Chile, spec consult Fundacion Chile, 80-83. *Mem:* Am Soc Hort Sci; Am Chem Soc. *Res:* Growth regulator effects and nitrogen and mineral nutrition of fruit plants; apple crop production. *Mailing Add:* Hudson Valley Lab PO Box 727 Highland NY 12528

FORSLUND, DAVID WALLACE, b Ukiah, Calif, Feb 18, 44; m 68; c 2. PLASMA PHYSICS. *Educ:* Univ Santa Clara, BS, 64; Princeton Univ, MA, 67, PhD(astrophys), 69. *Prof Exp:* Fel, Los Alamos Nat Lab, 69-71, mem staff plasma physics, 71-78, dep group leader inertial confinement, 78-81, LAB FEL, LOS ALAMOS NAT LAB, 81-, DEP DIR, ADVAN COMPUT LAB, 88- *Mem:* Fel Am Phys Soc; Am Geophys Union; Am Astron Soc. *Res:* Studying methods of parallel scientific computing using object-oriented techniques in distributed and massively parallel systems as applied to large scale plasma processes. *Mailing Add:* MSB287 Los Alamos Nat Lab PO Box 1663 Los Alamos NM 87545

FORSMAN, EARL N, b Keuterville, Idaho, Oct 27, 36; m 62; c 4. ATOMIC PHYSICS. *Educ:* Gonzaga Univ, BS, 63; Univ Wash, MS, 65, PhD(physics), 70. *Prof Exp:* From asst prof to assoc prof, 70-84, PROF PHYSICS, EASTERN WASH UNIV, 84- *Mem:* Am Asn Physics Teachers. *Res:* Atomic and molecular physics; gas-phase reactions and spectroscopy; neural models. *Mailing Add:* W 7413 Deno Spokane WA 99204

FORSMAN, M(ARION) E(DWIN), b Raymondville, Tex, July 19, 12; m 38; c 4. ELECTRICAL ENGINEERING. *Educ:* Univ Tex, BSEE & MSEE, 40; Iowa State Univ, PhD(elec eng), 54. *Prof Exp:* Supt construct, Am Smelting & Refrigeration Co, Tex & Mex, 40-42; field engr, Fischback & Moore of Tex, Inc, 42-43; res dir, Electronic Chem Eng Co, Calif, 46; res lab analyst, Northrop Aircraft, 46-47; assoc prof, Tulane Univ, 47-51; instr, Iowa State Univ, 51-52; elec engr, Gen Elec Co, Wash, 52-55; asst dir, Eng & Indust Exp Sta, 55-64, asst dir eng admin, 64-65, dir, 65-68, prof elec eng & dir eng external progs, 68-74, emer prof elec eng, Univ Fla, 74-; RETIRED. *Mem:* Am Soc Eng Educ; Inst Elec & Electronics Engrs; Nat Soc Prof Engrs. *Res:* Electrical power transmission; instrumentation and control; nuclear power. *Mailing Add:* 6335 Little Lake Geneva Rd Keystone Heights FL 32656

FORSMAN, WILLIAM C(OMSTOCK), b Grand Rapids, Minn, July 24, 29; m 56; c 4. PHYSICAL CHEMISTRY, CHEMICAL ENGINEERING. *Educ:* Univ Minn, BChE, 52; Univ Pa, PhD(phys chem), 61. *Prof Exp:* Chem engr, Hercules Powder Co, 52-54, 56, res chemist, 61-63; from asst prof to assoc prof, 64-77, PROF CHEM ENG, UNIV PA, 77- *Mem:* Am Chem Soc; Am Phys Soc; Am Inst Chem Engrs. *Res:* Physics and physical chemistry of high-polymer systems; statistical mechanics. *Mailing Add:* 102 S Swarthmore Ave Swarthmore PA 19081

FORSSEN, ERIC ANTON, b Los Angeles, Calif, Apr 6, 51. DRUG DEVELOPMENT, DESIGN RESEARCH. *Educ:* Univ Calif, Los Angeles, BS, 73; Univ Southern Calif, Dr Pharm, 77, PhD(pharmaceut sci), 81. *Prof Exp:* Pharmacist, Hosp Good Samaritan, 77-79; RES SCIENTIST, CANCER CHEMOTHER, CANCER CTR, UNIV SOUTHERN CALIF, 79- *Honors & Awards:* Young Investr Award, Am Heart Asn, 81; Life Sci Res Award, IntraSci Res Found, 80. *Mem:* Am Chem Soc; Am Pharmaceut Asn; AAAS. *Res:* Design of new drug molecules and formulations which have reduced toxicity. *Mailing Add:* 1599 Blvd New Haven CT 06511

FORST, WENDELL, b Trinec, Czech, Sept 28, 26; Can citizen; m 53. PHYSICAL CHEMISTRY, THEORETICAL CHEMISTRY. *Educ:* Prague Tech Univ, BSc, 48; McGill Univ, PhD(phys chem), 55. *Prof Exp:* Res engr, Northeastern Paper Prod, Ltd, Can, 55-56; from asst prof to prof chem, Laval Univ, 56-86; DIR RES, CNRS, FRANCE, 86- *Concurrent Pos:* Res assoc, Univ NC, 58-61; vis, Univ Calif, Berkeley & Free Univ Brussels, 69-70; Commonwealth vis prof, Univ Essex, 75-76; int fel, Stanford Res Inst, Menlo Park, Calif, 83-84; vis prof, Univ Catholique Louvain, Belgium, 88-89. *Mem:* Am Phys Soc. *Res:* Chemical dynamics; theory of unimolecular reactions. *Mailing Add:* Chimie Physique A Univ Bordeaux I Talence Cedex 33405 France

FORSTALL, WALTON, JR, mechanical engineering; deceased, see previous edition for last biography

FORSTER, D LYNN, b Columbus, Ind, Aug 29, 46; m 65; c 2. AGRICULTURAL ECONOMICS. *Educ:* Purdue Univ, BS, 68, MS, 72; Mich State Univ, PhD(agr econ), 74. *Prof Exp:* From asst prof to assoc prof, 74-, PROF AGR ECON, OHIO STATE UNIV. *Mem:* Am Agr Econ Asn. *Res:* Economic analysis of the control of water pollution in agricultural production; development and application of computer decision making models for agriculture; agricultural productivity. *Mailing Add:* Dept Agr Econ Ohio State Univ 2120 Fyffe Rd Columbus OH 43210

FORSTER, DENIS, b Newcastle-on-Tyne, Eng, Feb 28, 41; m 64; c 2. INORGANIC CHEMISTRY, CATALYSIS. *Educ:* Univ London, BSc, 62, PhD(inorg chem), 65. *Prof Exp:* Fel, Princeton Univ, 65-66; res chemist, Monsanto Cent Res Dept, 66-70, group leader homogeneous catalysis, 70-74, sr res group leader, 74-75, fel, 75-80, sr fel, 80-84, DISTINGUISHED FEL, MONSANTO CO, 84-, DIR, CHEM SCI, 85- *Honors & Awards:* Ipatieff Prize, Am Chem Soc, 80; Chem Pioneer Award, Am Inst Chemists, 80; Thomas & Hochwalt Award, 84. *Mem:* Am Chem Soc; Royal Soc Chem. *Res:* Transition metal chemistry; catalysis. *Mailing Add:* 32 Woodcrest Dr St Louis MO 63124

FORSTER, ERIC OTTO, b Lemberg, Oct 24, 18; nat US; m; c 6. PHYSICAL ORGANIC CHEMISTRY. *Educ:* Columbia Univ, BS, 49, MA, 50, PhD(phys org chem), 51. *Prof Exp:* Res chemist, Standard Oil Develop Co, 51-57, from res assoc to sr res assoc, Esso Res & Eng Co, Standard Oil Co of NJ, 57-75, sci adv, Exxon Res & Eng Co, 75-85; RETIRED. *Concurrent Pos:* Lectr, Columbia Univ, 53-67; adj prof, Rutgers Univ, 67-90; bd mem, Nat Acad Sci-Nat Res Coun Conf Elec Insulation & Dielec Phenomena, 72-80; res prof, State Univ NY, Buffalo, 85-86, vis prof, Xi'an Ziatong Univ, 87, Univ Rome, 88, Fed Univ Rio de Janeiro, 89 & 90. *Honors & Awards:* Centennial Medal, Inst Elec & Electronic Engrs, 84, Distinguished Serv Award, 83. *Mem:* AAAS; Am Chem Soc; fel Inst Elec & Electronics Engrs; Am Inst Physics. *Res:* Electrochemistry applied to hydrocarbon systems; electric properties of materials. *Mailing Add:* 1997 Duncan Dr Scotch Plains NJ 07076-2639

FORSTER, FRANCIS MICHAEL, b Cincinnati, Ohio, Feb 14, 12; m 37; c 5. CLINICAL NEUROLOGY. *Educ:* Univ Cincinnati, BS, 35, BM, 36, MD, 37. *Hon Degrees:* LLD, Xavier Univ, 55. *Prof Exp:* Intern, Good Samaritan Hosp, Mass, 37-38; fel psychiat, Pa Hosp, Philadelphia, 38-39; asst neurol, Harvard Med Sch, 39-40; Rockefeller Found fel & res assoc physiol, Sch Med, Yale Univ, 40-41; instr neurol, Sch Med, Boston Univ, 41-43; from asst prof to assoc prof, Jefferson Med Col, 43-50; prof, Med Ctr, Georgetown Univ, 50-58, dean sch med, 53-58; prof nerol, Sch Med, Univ Wis-Madson, 78-83; dir, Epilepsy Ctr, Madison Vet Admin Hosp, 78-83; RETIRED. *Concurrent Pos:* Resident physician, Boston City Hosp, 39-40; specialist, US Vet Admin, 46-; consult, Hosps, Surg Gen, US Air Force, 56- & Surg Gen, US Navy, 58-; vpres, Am Bd Psychiat & Neurol, 59, pres, 59-60. *Mem:* Am Physiol Soc; Am Neurol Asn; Am Asn Neuropath; fel AMA; Am Acad Neurol (vpres, 53-55, pres, 56-58); hon mem Philippine Neurol Soc; Sigma Xi. *Res:* Epilepsy; vascular accidents; multiple sclerosis; muscular dystrophy. *Mailing Add:* 21 Fallen Branch Lane Blue Ash Cincinnati OH 45241

FORSTER, FRED KURT, b Chicago, Ill, Feb 23, 44; m. ULTRASONICS, CARDIOVASCULAR DYNAMICS. *Educ:* Univ Ill, BS, 66; Stanford Univ, MS, 68, PhD(aeronaut & astronaut), 72. *Prof Exp:* Res asst bioeng, Inst Biomed Tech Univ & Swiss Fed Inst Technol, Zürich, 71-72; fel bioeng, Ctr Bioeng, 74-77, res engr, 77-79, res asst prof, 79-84, ASSOC PROF, DEPT MECH ENG, UNIV WASH, 84- *Concurrent Pos:* sr fel, NIH Cardiovascular Technol Traineeship, 74-77; co-investr, NIH prog projs & res grants, 77-79, prin investr, NIH res grants, 80- *Mem:* Inst Elec & Electronics Engrs; Acoustical Soc Am; Am Heart Asn; Cardiovasc Dynamics Soc. *Res:* Medical applications of ultrasound including Doppler and tissue characterization techniques; cardiovascular dynamics including turbulence phenomena in blood; medical instrumentation; acoustics; fluid mechanics. *Mailing Add:* Dept Mech Eng MS FU-10 Univ Wash Seattle WA 98195

FORSTER, HARRIET HERTA, b Vienna, Austria; nat US; m 42, 77. PHYSICS, NUCLEAR STRUCTURE. *Educ:* Univ Calif, MA, 47, PhD(physics), 48. *Prof Exp:* Instr, 48-51, from asst prof to assoc prof, 51-64, prof, 64-88, EMER PROF PHYSICS, UNIV SOUTHERN CALIF, 88- *Concurrent Pos:* Chair, Dept of Physics, VSC, 62-64; asst chair of undergrad instr 75-87. *Mem:* Fel Am Phys Soc. *Res:* Nuclear physics; cosmic rays. *Mailing Add:* 267 S Beloit Ave Los Angeles CA 90049

FORSTER, HUBERT VINCENT, CONTROL OF BREATHING, EXERCISE PHYSIOLOGY. *Educ:* Univ Wis, BS, 62, MS, 65, PhD(biodynamics), 69. *Prof Exp:* PROF PHYSIOL, MED COL WIS, 72- *Mailing Add:* Dept Physiol Med Col Wis PO Box 26509 Milwaukee WI 53226

FORSTER, JULIAN, b New York, NY, Aug 31, 18; m 41; c 2. ENGINEERING. *Educ:* City Univ New York, BS, 40. *Prof Exp:* Sr scientist, US Navy, 41-56; engr & tech mgr, Control & Instrumentation & Safety, Nuclear Power Generating Sta, Gen Elec Corp, 56-80, US rep, 88; CONSULT, QUADREX CORP, 80- *Concurrent Pos:* Mem, Inst Elec & Electronics Engrs, 69-, chmn, Standards Bd, 72, emer mem, 88, life fel, 91. *Honors & Awards:* Centennial Medal, Inst Elec & Electronics Engrs, 84; R F Shea Serv Award, 90. *Mem:* Fel Inst Elec & Electronics Engrs; Power Comput Nuclear Soc. *Res:* Application of computers to electric power generation; in-core neutron monitoring for light water reactors; safety system preparation of standards for emerging technologies in computer hardware and software. *Mailing Add:* 6962 Castle Rock Dr San Jose CA 95120

FORSTER, KURT, b Vienna, Austria, Mar 17, 15; nat US; c 2. PHYSICS. *Educ:* Univ Vienna, PhD(physics), 38. *Prof Exp:* Sr res engr, Jet Propulsion Lab, Calif Inst Technol, 46-47, res fel physics, 47-48; from asst prof to assoc prof, 48-56; PROF ENG, UNIV CALIF, LOS ANGELES, 56- *Concurrent Pos:* Guggenheim fel, 58-59. *Mem:* AAAS; Am Inst Aeronaut & Astronaut; Am Phys Soc. *Res:* Aerodynamics; heat transfer; space dynamics. *Mailing Add:* 3261 Veteran Ave Los Angeles CA 90034

FORSTER, LESLIE STEWART, b Chicago, Ill, May 10, 24; m 46; c 2. PHYSICAL CHEMISTRY, BIOPHYSICAL CHEMISTRY. *Educ:* Univ Calif, BS, 47; Univ Minn, PhD(chem), 51. *Prof Exp:* Fel, Univ Rochester, 51-52; instr chem, Bates Col, 52-54; from asst prof to assoc prof, 55-64, PROF CHEM, UNIV ARIZ, 64- *Concurrent Pos:* NSF sci fac fel, Univ Copenhagen, 61-62; consult, NIH, 63-64; USPHS spec fel, Weizmann Inst, Israel, 68-69. *Mem:* Am Chem Soc; Am Phys Soc; Sigma Xi. *Res:* Luminescence; photophysics of metal complexes; inorganic photochemistry; fluorescence of biomolecules. *Mailing Add:* Dept Chem Univ Ariz Tucson AZ 85721

FORSTER, MICHAEL JAY, b Buffalo, NY, Feb 8, 23; m 48; c 7. POLYMER PHYSICS. *Educ:* Canisius Col, BS, 45; Univ Notre Dame, MS, 50, PhD(physics), 51. *Prof Exp:* Physicist, E I du Pont de Nemours & Co, Inc, 45-46; asst, Univ Notre Dame, 46-51; res physicist, 51-66, group leader, 66-67, mgr textile res, 67-74, RES ASSOC, FIRESTONE TIRE & RUBBER CO, 74- *Mem:* Am Phys Soc; Am Chem Soc; Fiber Soc. *Res:* Fiber structure and properties; fiber morphology; crystalline structure of high polymers. *Mailing Add:* 356 Kimberly Rd Akron OH 44313-4252

FORSTER, ROBERT E, II, b St Davids, Pa, Dec 23, 19; m 47; c 4. RESPIRATORY PHYSIOLOGY. *Educ:* Yale Univ, BS, 41; Univ Pa, MD, 43. *Prof Exp:* Life Ins fel physiol, Harvard Med Sch, 49-50; from asst prof to prof physiol, Dept Physiol & Pharmacol, Grad Sch Med, Univ Pa, 50-59, assoc prof physiol in surg, Sch Med, 57-61, Isaac Ott prof physiol & chmn dept, Grad Sch Med, 59-70, Sch Med, 70-90, PROF PHYSIOL IN SURG, UNIV PA, 61-, PROF PHYSIOL IN PEDIAT, 77-, EMER ISAAC OTT PROF PHYSIOL, 90- *Concurrent Pos:* Lowell M Palmer Sr fel, Grad Sch Med, Univ Pa, 54-56; mem, Cardiovasc Study Sect, NIH, 60-64 & gen clin res ctr comt, 64-71; mem, Nat Adv Heart Coun, 67-71. *Mem:* Nat Acad Sci; Am Physiol Soc (pres, 66-67); Am Soc Clin Invest; Biophys Soc; Soc Gen Physiol. *Res:* Respiratory physiology; gas exchange, rapid exchanges of red cells, tissue oxygenation and temperature regulation. *Mailing Add:* Dept Physiol A-201 Richards Bldg Univ Pa 37th & Hamilton Walk Philadelphia PA 19104-6085

FORSTER, ROY PHILIP, b Milwaukee, Wis, Sept 28, 11; m 35; c 1. PHYSIOLOGY. *Educ:* Marquette Univ, BS, 32; Univ Wis, PhB, 36, PhD(zool), 38. *Hon Degrees:* MA, Dartmouth Col, 48. *Prof Exp:* Asst zool, Marquette Univ, 32-34; asst, Univ Wis, 35-38; from instr to prof biol sci, 38-48, lectr physiol, Med Sch, 64-74, Ira Allen Eastman prof, 64-76, res prof biol sci, 76-82, EMER IRA ALLEN EASTMAN PROF BIOL SCI, DARTMOUTH COL, 76- *Concurrent Pos:* Trustee, Mt Desert Island Biol Lab, 40-79, dir, 40-47, vpres, 61-63, pres, 64-70; sect ed, Biol Abstr, 47-76; Guggenheim fel, Cambridge Univ, 49; mem, Macy Conf Renal Function, 49-53; Rockefeller Found grant, 50-57; Guggenheim fel, Zool Sta, Univ Naples, 56; grant, NIH, 56-82, mem sci rev comt, Health Res Facil, 64-68, mem med biol rev comt & mem ad hoc comt comp pharmacol, 66-67, mem pharmacol-toxicol rev comt, 69-70; vis lectr, Sch Med, George Washington Univ, 59-60 & dir regulatory biol prog, NSF, 59-60; mem Nat Acad Sci-Nat Res Coun comt eval NSF grad fels, 64-65; mem Vet Admin comt res lab sci & med, 64-66; mem coun on circulation, Am Heart Asn; vis prof physiol, Sch Med, Univ Hawaii, 73; adj prof physiol, Dartmouth Med Sch, 74- *Mem:* Fel AAAS; Am Soc Zoologists; Am Physiol Soc; Soc Gen Physiol; Am Soc Nephrology. *Res:* Cellular and comparative physiology of the kidney; renal hemodynamics; membrane transport processes; nitrogen metabolism and excretion. *Mailing Add:* Prof Emer Biol Sci Dartmouth Col Hanover NH 03755

FORSTER, WARREN SCHUMANN, b Denver, Colo, Aug 16, 14; m 41; c 2. ORGANIC CHEMISTRY. *Educ:* Univ Denver, BS, 36; Pa State Univ, MS, 38, PhD(org chem), 40. *Prof Exp:* Asst chem, Pa State Univ, 36-40; res chemist, Bound Brook Lab, Am Cyanamid Co, 40-58, coordr res serv dept, Bus Off, 58-63, tech coordr res & develop dept, 63-81; RETIRED. *Mem:* Am Chem Soc. *Res:* Calorimetry of dimethylamine; aliphatic organic chemistry; fluorescent vat, soluble vat, synthetic fibre dyestuffs and intermediates; optical bleaches; ultraviolet light absorbers; stabilization of plastics. *Mailing Add:* 62 S Alward Ave Basking Ridge NJ 07920-1847

FORSTER, WILLIAM H(ALL), communications, electronics, for more information see previous edition

FORSTER, WILLIAM OWEN, b Dearborn, Mich, July 2, 27; m 48; c 3. ANALYTICAL CHEMISTRY, OCEANOGRAPHY. *Educ:* Mich State Univ, BS, 51, MA, 52; Univ Hawaii, PhD(chem), 66. *Prof Exp:* Anal chemist, Buick Motor Car Co, 50-52; teacher, Mich High Sch, 52-56; instr chem, Henry Ford Community Col, 56-61; instr gen sci, Univ Hawaii, 61-66; asst prof chem occeanog, Ore State Univ, 66-72; MEM STAFF, DIV BIOMED & ENVIRON SCI, ENERGY RES & DEVELOP ADMIN, 72- *Concurrent Pos:* Lectr, Kilolani Planetarium, Bishop Mus, 64-66; Oak Ridge Nat Lab fel, PR Nuclear Ctr, Mayaguez, 69-72, head marine biol prog, 70-72; leader marine prog, Div Biomed & Environ Res, Energy Res & Develop Agency, 72-81; sr officer, Div Nuclear Safety & Environ Protection, Int Atomic Energy Agency, 77-81. *Mem:* AAAS; Am Chem Soc; Am Geophys Union; Am Soc Limnol & Oceanog; Soc Appl Spectros. *Res:* Trace element analysis in sea water, biota and sediments; biogeochemistry of marine environments. *Mailing Add:* General Delivery Waikoloa Village Kamuela HI 96738-9999

FORSTHOEFEL, PAULINUS FREDERICK, b St Sebastian, Ohio, Apr 5, 15. GENETICS. *Educ:* Loyola Univ, Ill, AB, 39; Ohio State Univ, MSc, 51, PhD(zool), 53. *Prof Exp:* Instr math, Loyola Acad, 41-43; instr, Ohio Pvt Sch, 43-44; from instr to prof, 53-81, EMER PROF BIOL, UNIV DETROIT, 82- *Concurrent Pos:* NSF grant, 54-56; USPHS grants, 57-80; judge, Nat Sci Fair, 58 & 68. *Mem:* Fel AAAS; Genetics Soc Am. *Res:* Developmental genetics of mice. *Mailing Add:* 4001 W McNichols Rd Detroit MI 48221

FORSTNER, JANET FERGUSON, b Toronto, Ont, Nov 14, 37; m 60; c 2. MEDICAL RESEARCH. *Educ:* Univ Toronto, BA, 58, PhD(biochem), 71; Univ BC, MD, 62. *Prof Exp:* Intern med, Toronto Gen Hosp, 62-63; resident pediat, Univ Ill Res & Educ Hosp, 63-64; res fel biochem, Boston Univ, 64-66; res assoc med, 71-72, from asst prof to assoc prof biochem, 74-86, ASST PROF MED, TORONTO WESTERN HOSP, UNIV TORONTO, 73-, PROF BIOCHEM, 86- *Concurrent Pos:* Fel, Can Cystic Fibrosis Found, 71-72, chmn med sci adv comt, 87-; Med Res Coun Can scholar, 72-77; investr, Hosp Sick Children, Toronto, 74-; ed, J Biomed Chromatography, 86-; assoc prof biochem, Res Inst, Hosp Sick Children, 74-86, prof, 86- *Mem:* Can Biochem Soc; Asn Women Sci. *Res:* Investigation of the chemical and physical properties and secretion of epithelial mucin glycoproteins from intestine; relevance to cystic fibrosis. *Mailing Add:* Dept Biochem Univ Toronto Fac Med One Kings College Circle Toronto ON M5S 1A8 Can

FORSYTH, BEN RALPH, b New York, NY, Mar 8, 34; m 62; c 3. INTERNAL MEDICINE, INFECTIOUS DISEASES. *Educ:* NY Univ, MD, 57. *Prof Exp:* Intern & asst res, Yale New Haven Med Ctr, 57-60; res fel bact & immunol, Harvard Med Sch & Boston City Hosp, 60-61; sr investr respiratory virus unit, Lab Infectious Dis, Nat Inst Allergy & Infectious Dis, 63-66; assoc prof, 66-71, actg chmn dept med microbiol, 67, dir infectious dis unit, 66-74, PROF MED, COL MED, UNIV VT, 71-, ASSOC PROF MED MICROBIOL, 67-, ASSOC DEAN LONG RANGE PLANNING, DIV HEALTH SCI, 70- *Concurrent Pos:* Sinsheimer Fund fac fel, 66-71; mem res reagents comt, Nat Inst Allergy & Infectious Dis, 66-68, chmn, 68. *Mem:* Am Fedn Clin Res; Am Soc Microbiol; Infectious Dis Soc Am; Soc Exp Biol & Med; Sigma Xi. *Res:* Virology; mycoplasma; epidemiology. *Mailing Add:* Univ Vt 349 Waterman Bldg Burlington VT 05405

FORSYTH, DALE MARVIN, b Charles City, Iowa, Feb 4, 45; m 67; c 2. ANIMAL NUTRITION, ANIMAL SCIENCE. *Educ:* Iowa State Univ, BS, 67; Cornell Univ, PhD(nutrit), 71. *Prof Exp:* Asst prof, 72-78, ASSOC PROF ANIMAL NUTRIT, PURDUE UNIV, 78- *Mem:* Am Soc Animal Sci; AAAS. *Res:* Nutrition of the young pig; utilization of feedstuffs, including high moisture, genetically improved and molded corn; mineral metabolism, especially iron, calcium and fluoride. *Mailing Add:* Dept Animal Sci Purdue Univ Lilly Hall West Lafayette IN 47907

FORSYTH, DOUGLAS JOHN, b Kingston, Ont, Apr 25, 46; m 85; c 2. ECOTOXICOLOGY, BIOACCUMULATION OF POLLUTANTS. *Educ:* Carleton Univ, BSc, 69; Ohio State Univ, MS, 72, PhD(ecol), 80. *Prof Exp:* Pesticides evaluator, 75-81, WILDLIFE TOXICOLOGIST, CAN WILDLIFE SERV, 82- *Mem:* Soc Environ Toxicol & Chem. *Res:* Impact and bioaccumulation of environmental contaminants in ecosystems; effects of pesticides on wildlife; population dynamics and behavior of mammals; environmental physiology. *Mailing Add:* Can Wildlife Serv 115 Perimeter Rd Saskatoon SK S7N 0X4 Can

FORSYTH, ERIC BOYLAND, b Bolton, Eng, Apr 16, 32; m 58; c 2. SUPERCONDUCTIVITY. *Educ:* Manchester Univ, BSc, 53; Toronto Univ, MASc, 61. *Prof Exp:* Elec engr particle accelerator, 60-65, chief elec engr, Alternating Gradient Synchrotron Div, 65-67, dep div head, 67-68, proj mgr advan accelerator div, 71-77, div head advan technol appl, 77-86, CHMN, ACCELERATOR DEVELOP DEPT, BROOKHAVEN NAT LAB, 86- *Mem:* Inst Elec Engrs, UK; fel Inst Elec & Electronics Engrs. *Res:* Development of high power underground transmission systems using superconducting cables; development of particles accelerators. *Mailing Add:* Accelerator Develop Dept Bldg 1005 Brookhaven Nat Lab Upton NY 11973

FORSYTH, FRANK RUSSELL, b Russell, Ont, Mar 15, 22; m 45; c 5. PLANT PHYSIOLOGY. *Educ:* Queen's Univ, Ont, BA, 49; Univ Toronto, PhD(plant physiol), 52. *Prof Exp:* Res officer agr & plant physiologist, Plant Path Sect, Univ Man, 52-59, res officer, 59-62, res officer agr, Pesticide Res Inst, Kentville Res Sta, Can Dept Agr, 62-78; RETIRED. *Mem:* Can Soc Plant Physiol; Agr Inst Can; Am Soc Hort Sci; Can Soc Hort Sci. *Res:* Biochemical aspects of physiological disorders of fruit in storage and controlled atmospheric storage of apples. *Mailing Add:* Box 454 Metcalfe ON K0A 2P0 Can

FORSYTH, J(AMES) S(NEDDON), chemical engineering; deceased, see previous edition for last biography

FORSYTH, JAMES M, b Niagara Falls, NY, Apr 21, 42; m 68. OPTICS. *Educ:* Univ Rochester, BS, 64, PhD(optics), 69. *Prof Exp:* From asst prof to assoc prof optics, Univ Rochester, 68-84; VPRES TECHNOL, HAMPSHIRE INSTRUMENTS INC, 84- *Honors & Awards:* Adolf Lamb Medal, Optical Soc Am, 74. *Mem:* Fel Optical Soc Am. *Res:* X-ray optics; laser plasma generation; x-ray lithography. *Mailing Add:* Hampshire Instruments Inc Ten Carlson Rd Rochester NY 14610

FORSYTH, JANE LOUISE, b Hanover, NH, Nov 9, 21. GLACIAL GEOLOGY, ENVIRONMENTAL GEOLOGY. *Educ:* Smith Col, AB, 43; Univ Cincinnati, MA, 46; Ohio State Univ, PhD(geol), 56. *Prof Exp:* Asst geol, Univ Cincinnati, 43-46; instr geol, Miami Univ, 46-47; asst geol, Univ Calif, Berkeley, 47-48; asst instr, Ohio State Univ, 48-49 & 51-55; Pleistocene geologist, Ohio Geol Surv, 55-65; from asst prof to assoc prof, 65-74, PROF GEOL, BOWLING GREEN STATE UNIV, 74- *Concurrent Pos:* Ed, Ohio J Sci, 64-74; mem, Ohio Natural Area Coun & Nature Conservancy Bd. *Mem:* Geol Soc Am; Nat Asn Geol Teachers; Am Quarternary Asn. *Res:* Pleistocene geology; Wisconsin chronology in Ohio; soils, particularly as related to glacial geology studies in Ohio; geomorphology; human environmental geology; ecology, especially the relationship between plant distribution and geology in the midwest. *Mailing Add:* Dept Geol Bowling Green State Univ Bowling Green OH 43403

FORSYTH, JOHN WILEY, b McKinney, Tex, Mar 18, 13; m 43; c 3. EXPERIMENTAL MORPHOLOGY. *Educ:* Tex Christian Univ, BS, 35, MS, 37; Princeton Univ, PhD(exp morphol), 41. *Prof Exp:* Instr, Tex Christian Univ, 36-38; prof, Presby Col, SC, 41-43; adj prof, Univ SC, 46; from asst prof to assoc prof biol, Tex Christian Univ, 46-50, prof, 50-79; RETIRED. *Mem:* AAAS; Am Soc Ichthyol & Herpet; Soc Syst Zool; Am Soc Zool; Sigma Xi. *Res:* Amphibian limb regeneration; herpetological studies; food habits and distribution in Ft Worth region. *Mailing Add:* 2622 Waits Ave Ft Worth TX 76109

FORSYTH, PAUL FRANCIS, b Ogdensburg, NY, Apr 21, 28; m 51; c 6. INDUSTRIAL CHEMISTRY. *Educ:* Canisius Col, BS, 51, MS, 53. *Prof Exp:* AEC asst, Canisius Col, 51-53; res chemist, Union Carbide Metals Co, 53-61; mat engr, Bell Aerosyst, 61-62; supvry engr measurement sect, 62-69, supvry engr, Prod Develop, 69-74, prin engr, tech br, 74-78, MGR DEVELOP, COATED ABRASIVES DIV, CARBORUNDUM CO, 78- *Mem:* Am Chem Soc. *Res:* Inorganic synthesis; vacuum techniques; analytical instrumentation; rheology; particle size analysis; sandpaper product development and processing; electron beam radiation polymerization. *Mailing Add:* 1937 Saunders Settlement Rd Niagara Falls NY 14304-1063

FORSYTH, PETER ALLAN, b Prince Albert, Sask, Mar 20, 22; m 44; c 2. PHYSICS. *Educ:* Univ Sask, BA, 42-46, MA, 47; McGill Univ, PhD(physics), 51. *Prof Exp:* Sci off radio physics lab, Defense Res Telecommun Estab, Can, 51-53, sect leader, Upper Atmospheric Physics Sect, 53-57, supt, Radio Physics Lab, 57-58; prof physics, Univ Sask, 58-61; head dept, Univ Western Ont, 61-67, prof physics, 61-87, dir, Ctr Radio Sci, 67-87, EMER PROF PHYSICS, UNIV WESTERN ONT, 87- *Mem:* Can Asn Physicists (pres, 79-80); fel Royal Soc Can; fel Can Aeronaut & Space Inst. *Res:* Physics of the upper atmosphere; propagation of radio waves in ionized media; scattering of radio waves by inhomogenous media. *Mailing Add:* One Grosvenor Apt 602 London ON N6A 1Y2 Can

FORSYTH, THOMAS HENRY, b Pikeville, Ky, Nov 8, 42; m 64; c 2. CHEMICAL ENGINEERING, POLYMER SCIENCE. *Educ:* Univ Ky, BS, 64; Va Polytech Inst, MS, 66, PhD(chem eng), 68. *Prof Exp:* Chem engr, Dow Chem Co, Mich, 67-70; from asst prof to assoc prof chem eng, Univ Akron, 70-80, res assoc, Inst Polymer Sci, 76-80; RES ASSOC, B F GOODRICH, 80- *Concurrent Pos:* Lectr, Midland, Mich, 69-70. *Mem:* Am Chem Soc; Am Inst Chem Engrs; Soc Plastic Engrs. *Res:* New plastics and processing of plastics. *Mailing Add:* Box 122 ALTC Avon Lake OH 44012-1195

FORSYTHE, ALAN BARRY, b Brooklyn, NY, Nov 3, 40; m 62; c 2. BIOSTATISTICS. *Educ:* Brooklyn Col, BS, 62; Columbia Univ, MS, 64; Yale Univ, PhD(biomet), 67. *Prof Exp:* Res asst statist anal, Dept Pediat & Cardiac Res, Yale Univ, 64, statistician, Dept Obstet & Gynec, 66; asst res statistician, 67-68, fels, 70-73, SUPVRY STATISTICIAN, DEPT BIOMATH & SCH MED, UNIV CALIF, LOS ANGELES, 68- *Concurrent Pos:* Consult, Div Perinatol, Dept Obstet & Gynec, Med Sch, Univ Southern Calif, 68-75, Southern Calif Permanente Med Group & Kaiser Found Hosps, 75-; assoc ed, Theory & Methods, J Am Statist Asn; mem, President's Task Force Meetings, Am Statist Asn; assoc ed, J Statist Comput & Simulation, 76-, J Biometrics, J Technometrics, 78- *Mem:* AAAS; Am Statist Asn; Biomet Soc. *Res:* Robust estimation and hypothesis testing, especially regression; use of computers for statistical analysis and teaching. *Mailing Add:* Dept Biomath Univ Calif Sch Med Los Angeles CA 90024

FORSYTHE, DENNIS MARTIN, b Toledo, Ohio, June 20, 42; m; c 2. ORNITHOLOGY. *Educ:* Ohio Univ, BS, 64; Utah State Univ, MS, 67; Clemson Univ, PhD(zool), 74. *Prof Exp:* Res supvr upland game, Minn Div Game & Fish, 68-69; from asst prof to assoc prof, 69-80, dir, Vector Biol Res Prog, 80-84, PROF BIOL, THE CITADEL, 80- *Concurrent Pos:* Vis prof biol, Charleston Southern Univ, Charleston, 72-79, prof Activ Continuing Educ Prog, Univ SC, 74-78; res assoc ornith, Charleston Mus, 72-77; res assoc zool, Univ Aberdeen, Scotland, 79; fac grad, Charleston Higher Educ Consortium Marine Sci, 80-; vis prof, Dept Environ Sci, Univ SC, 80-88. *Honors & Awards:* Fel Explorer's Club. *Mem:* Animal Behav Soc; Am Ornithologists Union; Cooper Ornith Soc; Ecol Soc Am. *Res:* Behavior and bioacoustics of birds especially shorebirds; solid waste disposal and bird species associated with the bird-aircraft collison hazard; bird populations and community structure in selected plant communities; Pelagic Bird communities. *Mailing Add:* Dept Biol Citadel Charleston SC 29409

FORSYTHE, HOWARD YOST, JR, b Aiken, SC, Oct 27, 31; m 59; c 4. ENTOMOLOGY. *Educ:* Univ Maine, BS, 58; Cornell Univ, MS, 60, PhD(entom), 62. *Prof Exp:* Asst prof entom, Ohio Agr Exp Sta, 62-66, assoc prof, Ohio Agr Res & Develop Ctr, 66-69; assoc prof, 69-79, PROF ENTOM, UNIV MAINE, ORONO, 79-, CHMN DEPT, 77- *Mem:* Entom Soc Am; Entom Soc Can. *Res:* Control and biology of deciduous fruit tree insects and mites; control and biology of insects on blueberries. *Mailing Add:* Dept Entom Univ Maine Orono ME 04469

FORSYTHE, RANDALL D, b Durand, Wis, Apr 16, 52. PALEOMAGNETICS, GEOLOGY. *Educ:* Lawrence Univ, BA, 74; Columbia Univ, MS, 76, MPhil, 78, PhD(geol), 81. *Prof Exp:* Asst prof geol, Dept Geol Sci, Rutgers Univ, 81-; ASST PROF GEOL, DEPT GEOG & EARTH SCI, UNIV NC, CHARLOTTE. *Concurrent Pos:* Vis res scientist geol, Lamont Doherty Geol Observ, Columbia Univ, 80-; vis prof, Dept Environ Sci, Univ SC, 80-88...

FORSYTHE, RICHARD HAMILTON, BIOCHEMISTRY, NUTRITION. *Educ:* Iowa Univ, BS, 43, PhD(chem), 49. *Prof Exp:* RESEARCHER, CAMPBELL SOUP CO, 73- *Mailing Add:* Dept Animal & Poultry Sci Univ Ark Fayetteville AR 72071

FORSYTHE, WARREN M, soil physics, for more information see previous edition

FORT, RAYMOND CORNELIUS, JR, b Upper Darby, Pa, Mar 28, 38; div; c 2. PHYSICAL ORGANIC CHEMISTRY. *Educ:* Drexel Inst, BS, 61; Princeton Univ, PhD(org chem), 65. *Prof Exp:* Fel chem, Princeton Univ, 64-65; from asst prof to prof chem, Kent State Univ, 65-85; PROF & CHMN DEPT CHEM, UNIV MAINE, 85- *Concurrent Pos:* Consult, Kemin Industs, Eaton, Peabody, Bradford & Veague. *Mem:* Am Chem Soc; Royal Soc Chem. *Res:* Heteroadamantanes; quantum chemistry; computational chemistry. *Mailing Add:* Dept Chem Univ Maine Orono ME 04469

FORT, TOMLINSON, JR, b Sumter, SC, Apr 16, 32; m 56; c 2. SURFACE CHEMISTRY. *Educ:* Univ Ga, BS, 52; Univ Tenn, MS & PhD(surface chem), 57. *Prof Exp:* Stephens res fel, Univ Sydney, 57-58; from res chemist to sr res chemist, E I du Pont de Nemours & Co, Inc, 58-65; from asst prof to prof chem eng sci, Case Western Reserve Univ, 65-73; prof chem eng & chem & head, Dept Chem Eng, Carnegie-Mellon Univ, 73-80; prof chem & chem eng & provost, Univ Mo-Rolla, 80-82; prof chem & mat sci & vpres acad affairs, Calif Polytech State Univ, 82-83, provost, 83-86, prof chem & mat sci, 86-89; CENTENNIAL PROF & CHAIR, DEPT CHEM ENG & PROF MAT SCI, VANDERBILT UNIV, 89- *Concurrent Pos:* Vis prof, Univ Copenhagen, 78 & 80, Nat Univ Mex, 73. *Mem:* Am Chem Soc; Adhesion Soc; Am Inst Chem Engrs; Am Soc Eng Educ; Sigma Xi; Int Asn Colloid & Interface Scientist. *Res:* Adsorption; monolayers and thin films; contact angles and wetting; adhesion; interface thermodynamics in composite materials; composite materials. *Mailing Add:* 845 Otter Creek Rd Nashville TN 37220

FORTE, JOHN GAETANO, b Philadelphia, Pa, Dec 23, 34; m 61; c 3. PHYSIOLOGY. *Educ:* Johns Hopkins Univ, AB, 56; Univ Pa, PhD(physiol), 61. *Prof Exp:* Lab instr biol, Univ Pa, 59 & mammalian physiol, 59-61, instr physiol, 61-62, assoc, 62-64; res assoc biochem, Univ Southern Calif, 64-65; from asst prof to assoc prof, 65-74, chmn, Dept Physiol-Anat, 72-79, Miller res professorship, 81-82, PROF PHYSIOL, UNIV CALIF, BERKELEY, 74- *Concurrent Pos:* Mem physiol study sect, NIH, 74-78, chmn, 76-78, clin sci study sect, 85-; sect ed, Ann Rev Physiol, 81- *Honors & Awards:* Hoffmann LaRoche Prize, Am Physiol Soc, 84; William Beaumont Prize, Am Gastroenterol Asn, 85; Sir Arthur Hurst Lectr, Brit Soc Gastroenterol, 85. *Mem:* Am Physiol Soc; Biophys Soc; Parietal Cell Club; NY Acad Sci. *Res:* Secretory mechanisms in a variety of tissues, particularly glandular systems of gastrointestinal tract. *Mailing Add:* Dept Cell Physiol Univ Calif Life Sci Annex Berkeley CA 94720

FORTE, LEONARD RALPH, b Nashville, Tenn, June 10, 41; m 62; c 3. PHARMACOLOGY. *Educ:* Austin Peay State Col, BS, 63; Vanderbilt Univ, PhD(pharmacol), 69. *Prof Exp:* Res asst pharmacol, Vanderbilt Univ, 63-64, res assoc, 69; from asst prof to assoc prof, 69-81, chmn dept, 81-83, PROF PHARMACOL, UNIV MO-COLUMBIA, 81-, RES CAREER SCIENTIST, VET ADMIN, 88- *Concurrent Pos:* NIH fel, Vanderbilt Univ, 69; NIH career res develop award, 74-79; merit rev bd basic sci res serv, Vet Admin, 78-81; assoc chief staff res, Harry S Truman Vet Admin Hosp, Columbia, 79-81, res pharmacologist, 74-87; vis scientist, Dept Med, Melbourne Univ, Australia, 89-90. *Mem:* AAAS; Am Soc Pharmacol & Exp Therapeut; Endocrine Soc; Am Soc Bone & Mineral Res. *Res:* Renal pharmacology; calcium homeostasis; characterization of kidney plasma membranes in regard to function and molecular action of parathyroid hormone; physiology of vitamin D; regulation of chloride channels by cyclic amp. *Mailing Add:* Dept Pharmacol Univ Mo Columbia MO 65212

FORTES, GEORGE (PETER ALEXANDER), b Omaha, Nebr, Aug 2, 44; Mex citizen; m 75; c 1. ACTIVE & PASSIVE ION TRANSPORT. *Educ:* Univ Nat Autonomous Mex, MD, 68; Univ Pa, PhD(molecular biol), 72. *Prof Exp:* Fel, Johnson Found, Univ Pa, 68-69; res assoc, dept physiol, Sch Med, Yale Univ, 69-72; vis scientist, Inst Animal Physiol, Cambridge, Eng, 72-73; asst prof, 72-79, ASSOC PROF, DEPT BIOL, UNIV CALIF, SAN DIEGO, 79- *Concurrent Pos:* Prin investr, Am Heart Asn, 74-76 & NIH, 76- *Mem:* Biophys Soc; Soc Gen Physiologists. *Res:* Membranes; structure and function of adenosine triphosphate-dependent ion pumps; bioenergetics; fluorescence spectroscopy of membrane proteins; red blood cell membrane properties. *Mailing Add:* Dept Biol Univ Calif San Diego La Jolla CA 92093-0322

FORTIN, ANDRE FRANCIS, b Shawinigan, Que, Dec 29, 51; m 79; c 1. MEATS, BODY COMPOSITION. *Educ:* Macdonald Col, McGill Univ, BSc, 73; Cornell Univ, PhD(nutrit), 77. *Prof Exp:* Teaching asst, Animal Sci Dept, Cornell Univ, 73-76, res asst, 73-77; RES SCIENTIST, ANIMAL RES CTR, AGR CAN, 77- *Concurrent Pos:* Mem, Meat Sci Prog Comt, Am Soc Animal Sci, 82- *Mem:* Am Soc Animal Sci; Am Meat Sci Asn; Can Soc Animal Sci; Agr Inst Can. *Res:* Growth and development of body tissues in beef cattle, swine and sheep; live evaluation of body and carcass composition; hog carcass grading. *Mailing Add:* Cent Exp Farm Bldg 59 Agr Can Ottawa ON K1A 0C6 Can

FORTMAN, JOHN JOSEPH, b Dayton, Ohio, Oct 26, 39; m 68; c 2. INORGANIC CHEMISTRY, CHEMICAL EDUCATION. *Educ:* Univ Dayton, BS, 61; Univ Notre Dame, PhD(phys Inorg chem), 66. *Prof Exp:* Asst prof, 65-69, ASSOC PROF CHEM, WRIGHT STATE UNIV, 69- *Concurrent Pos:* Res analyst, Aerospace Res Labs, Wright-Patterson Air Force Base, 66-70; vis assoc prof chem, Purdue Univ, 73-74. *Mem:* Nat Sci Teachers Asn; Am Chem Soc. *Res:* Electron paramagnetic resonance; nuclear magnetic resonance of inorganic complexes; coordination chemistry; intramolecular rearrangements of chelate complexes; chemical demonstrations; analogies in teaching; course content and organization. *Mailing Add:* Dept Chem Wright State Univ Dayton OH 45435

FORTNER, BRAND I, b Paso Robles, Calif, Nov 22, 55; m; c 3. COMPUTATIONAL SCIENCE, SCIENTIFIC VISUALIZATION. *Educ:* Univ Ill, BA, 77, MAS, 81, PhD, 91. *Prof Exp:* Mgr appln software, Univ Ill, 85-90; CO-FOUNDER & RES & DEVELOP DIR, SPYGLASS, INC, 90- *Mem:* AAAS; Am Phys Soc; Am Astron Soc. *Res:* Scientific software development on the Mac 11. *Mailing Add:* Spyglass Inc 701 Devonshire Dr No C17 Champaign IL 61820

FORTNER, GEORGE WILLIAM, b Middlesboro, Ky, Mar 30, 38; m 77. BIOLOGY, IMMUNOLOGY. *Educ:* Wayne State Univ, BS, 61; Univ Tenn, PhD(microbiol), 73. *Prof Exp:* Res technician microbiol, Oak Ridge Nat Lab, 64-68; investr immunol, Sch Med, Stanford Univ, 73-75; scientist II, Frederick Cancer Res Ctr, 75-77; ASST PROF IMMUNOL, DIV BIOL, KANS STATE UNIV, 77- *Concurrent Pos:* Prin investr grants, Kans State Univ, 77-78, Mid-Am Cancer Ctr, 77-78 & 79-80 & Nat Cancer Inst, 79-83; assoc scientist, Mid-Am Cancer Ctr, 78- *Mem:* Am Soc Microbiol; AAAS. *Res:* Cellular immunology; immunobiology; tumor immunology; host-parasite relationships. *Mailing Add:* Dept Biol Kans State Univ Div Biol Ackert Hall Manhattan KS 66506

FORTNER, JOSEPH GERALD, b Bedford, Ind, May 30, 21; m 47; c 2. SURGERY, ONCOLOGY. *Educ:* Univ Ill, BS, 44, MD, 45; Univ Birmingham, MSc, 65; Am Bd Surg, dipl, FRCS. *Prof Exp:* Intern, St Luke's Hosp, Ill, 45-46; resident path, Charity Hosp, La, 48-49; asst resident surg, Bellevue Hosp, NY, 49-51; asst resident surg, Mem Hosp, 51-52, resident, 52-54; from instr surg to assoc prof clin surg, 54-70, assoc prof surg, 70-72, PROF SURG, MED COL, CORNELL UNIV, 72-; DIR, GEN MOTORS SURG RES LAB, 77-; MEM, SLOAN-KETTERING INST, 85- *Concurrent Pos:* Nat Cancer Inst trainee, 53-54; attend surgeon, Gastric & Mixed Tumor Serv, Mem Hosp, 69-, Chief, Surg Res Serv, 78-; chief, Div Surg Res, Sloan-Kettering Inst, 68-77, chief, Gastric & Mixed Tumor Serv, 70-78; pres, Gen Motors Cancer Res Found. *Honors & Awards:* Alfred P Sloan Award, 63. *Mem:* Am Surg Asn; Soc Univ Surg; Am Asn Cancer Res; fel Am Col Surg; Soc Surg Oncol; fel Hellenic Surg Soc. *Res:* Transplantation; experimental surgery; cancer therapy. *Mailing Add:* Mem Hosp Sloan-Kettering Inst 1275 York Ave New York NY 10021

FORTNER, ROSANNE WHITE, b Logan, WVa, Nov 13, 45; m 66; c 2. MARINE & AQUATIC EDUCATION. *Educ:* WVa Univ, BA, 67; Ore State Univ, MA, 73; Va Polytech Inst & State Univ, EdD, 78. *Prof Exp:* Teacher earth sci, Roanoke County Schs, Va, 67-69 & life sci, 72-76; res asst radiation biol, Ore State Univ, 69-71; teaching assoc sci educ, Va Polytech Inst & State Univ, 76-78; vis asst prof sci educ, 78-79, asst prof, 79-85, ASSOC PROF NATURAL RESOURCES & SCI EDUC, OHIO STATE UNIV, 85- *Concurrent Pos:* Pres, Nat Marine Educ Asn, 88-89; consult ed, J Environ Educ, 82-89; prin investr educ prog, Ohio Sea Grant, 82-; consult, Nat Park Serv, 86-89; bd dir, NAm Asn Environ Educ; ed, The Biosphere; consult, Nat Oceanog & Atmospheric Asn Educ Afairs, 91. *Mem:* Nat Asn Res Sci Teaching; NAm Asn Environ Educ; Nat Marine Educ Asn; Nat Sci Teachers Asn; Am Asn Advan Sci. *Res:* Comparative effectiveness of communications media for presentation of environmental and natural resources messages, including evaluation of television documentaries, use of educational technologies and inservice teacher training techniques; curriculum development in global change education. *Mailing Add:* Ohio State Univ 2021 Coffey Rd 210 Kottman Hall Columbus OH 43210-1085

FORTNER, WENDELL LEE, b Conway, Ark, Nov 17, 35; m 62; c 2. STRATIGRAPHY, HYDROLOGY. *Educ:* Univ Cent Ark, BS, 61. *Prof Exp:* Grad asst, Grad Inst Technol, Univ Ark, 61-62; tech inspector, Martin-Marietta Co, 62-63; chemist, 63-79, CHEM ENGR, PINE BLUFF ARSENAL, US ARMY, 79- *Mem:* Sigma Xi; Am Chem Soc; fel Am Inst Chem Engrs. *Res:* Environmental treatment systems for industrial effluents with continuous chemical and biological monitoring techniques as early warning devices; incineration processes for both bulk chemical and chemical contaminated hardware (fluid-bed, rotary and chain grate); resource recovery operations and techniques. *Mailing Add:* 2113 Mt Vernon Ct Pine Bluff AR 71603

FORTNEY, JUDITH A, b Cheshire, Eng, Jan 28, 38; m 61; c 2. RESEARCH ADMINISTRATION. *Educ:* London Sch Econ BSc, 59; Univ Wis, MS, 63; Duke Univ, PhD(demog), 71. *Prof Exp:* Instr sociol, NC Cent Univ, 64-68; dir res, Low Income Housing Develop Corp, 72-74; DIR, DIV REPRODUCTIVE EPIDEMIOL, FAMILY HEALTH INT, 74- *Concurrent Pos:* Consult, NIH, 85 & WHO, 86-; adj assoc prof, dept epidemiol, Sch Pub Health Univ NC, Chapel Hill, 87- *Mem:* Am Col Epidemiol; Am Publ Health Asn; Soc Epidemiol Res. *Mailing Add:* Family Health Int PO Box 13950 Research Triangle Park NC 27709

FORTNEY, LLOYD RAY, b Enid, Okla, June 22, 36; m 62, 88; c 2. EXPERIMENTAL PHYSICS. *Educ:* NMex State Univ, BS, 58; Univ Wis, PhD(physics), 62. *Prof Exp:* Res assoc high energy physics, Univ Wis, 62-63; res assoc, 63-64, from asst prof to assoc prof, 64-86, dir undergrad studies, 80-83, PROF PHYSICS, DUKE UNIV, 86 - *Mem:* Am Phys Soc. *Res:* Experimental high energy physics. *Mailing Add:* Dept Physics Duke Univ Durham NC 27706

FORTNUM, DONALD HOLLY, b Berlin, Wis, Apr 2, 32; m 58; c 4. PHYSICAL CHEMISTRY. *Educ:* Carroll Col, Wis, 54; Brown Univ, PhD(chem), 58. *Prof Exp:* Instr chem, Providence Col, 56-57; asst prof, Ursinus Col, 58-65; assoc prof, 65-73, PROF CHEM, GETTYSBURG COL, 73- *Mem:* Am Chem Soc. *Res:* Molecular structure; kinetics of inorganic reactions; computer applications in chemistry. *Mailing Add:* Dept Chem Gettysburg Col Gettysburg PA 17325

FORTSON, EDWARD NORVAL, b Atlanta, Ga, June 16, 36; m 60; c 3. LASERS, SPACE-TIME SYMMETRIES. *Educ:* Duke Univ, BS, 57; Harvard Univ, PhD(physics), 64. *Prof Exp:* Res asst prof, 63-65, from asst prof to assoc prof, 66-73, PROF PHYSICS, UNIV WASH, 73- *Concurrent Pos:* Fulbright grant, Fulbright Comn, 65-66; vis prof, Univ Bonn, 65-66; res fel, Univ Oxford, 77, vis fel, 81; Guggenheim fel, Guggenheim Found, 80-81. *Mem:* Am Phys Soc. *Res:* Experimental atomic physics using laser-optical techniques; the study of elementary particle interactions and tests of parity and time-reversal symmetry within atoms. *Mailing Add:* Dept Physics FM-15 Univ Wash Seattle WA 98195

FORTUNA, EDWARD MICHAEL, JR, b Grand Rapid, Mich, Dec 21, 51; m 78; c 2. QUALITY CONTROL, ADHESIVES. *Educ:* Northern Mich Univ, BS, 74. *Prof Exp:* Lab chemist, 77-79, lab supvr, 79-81, LAB MGR, WOLVERINE WORLD WIDE, 81- *Concurrent Pos:* Chmn, Am Standard Testing Methods, 83-, Footwear Indust Am, 87- *Mem:* Am Soc Testing & Mats; Am Soc Qual Control. *Mailing Add:* Qual Control Div Wolverine World Wide Inc 9341 Courtland Dr Rockford MI 49351

FORTUNE, H TERRY, b Ramer, Tenn, Feb 16, 41; m 61; c 2. NUCLEAR PHYSICS. *Educ:* Memphis State Univ, BS, 63; Fla State Univ, PhD(physics), 67. *Prof Exp:* Res assoc nuclear physics, Argonne Nat Lab, 67-69; from asst prof to assoc prof, 69-76, PROF PHYSICS, UNIV PA, 76- *Concurrent Pos:* Vis scientist, Argonne Nat Lab, 74; vis fel, Oxford Univ, 77. *Mem:* Am Phys Soc; Am Asn Physics Teachers. *Res:* Nuclear structure physics; reaction mechanisms; reactions induced by heavy ions. *Mailing Add:* Dept Physics 2E5 DRL E1 Univ Pa 209 S 33rd St Philadelphia PA 19104

FORTUNE, JOANNE ELIZABETH, b Watertown, NY, Jan 24, 45; m 74. REPRODUCTIVE ENDOCRINOLOGY. *Educ:* Col New Rochelle, BA, 67; Cornell Univ, MS, 71, PhD(embryol), 75. *Prof Exp:* Fel reproductive physiol, Dept Obstet/Gynec, Univ Western Ont, 75-76; fel reproduction physiol, Dept Animal Sci, 76-77, from res assoc to sr res assoc, Dept Phys Biol, 77-80, asst prof, 80-86, ASSOC PROF PHYSIOL, SECT PHYSIOL, CORNELL UNIV, 86- *Concurrent Pos:* Co-prin investr, NIH res grant, 78-80, prin investr, 80-; review panel mem, Regulatory Biol Panel, NSF, 81-83; mem bd dirs, Soc Study Reproduction, 83-86. *Mem:* Soc Study Reproduction; Endocrine Soc; Am Soc Zoologists; Soc Develop Biol; Soc Study Fertil; Sigma Xi. *Res:* Ovarian development and function; hormonal control of steroid production by mammalian and amphibian ovarian cells in vitro, using radioimmunoassay, tissue culture, and microsurgical techniques. *Mailing Add:* 823 Vet Res Tower Cornell Univ Ithaca NY 14853

FORWARD, RICHARD BLAIR, JR, b Washington, DC, Jan 13, 43; m 69; c 2. COMPARATIVE PHYSIOLOGY. *Educ:* Stanford Univ, BA, 65; Univ Calif, Santa Barbara, PhD(biol), 69. *Prof Exp:* From res asst biol to teaching asst, Univ Calif, Santa Barbara, 66-69; res assoc biol, Yale Univ, 69-71; asst prof, 71-77, assoc prof, 78-88, PROF ZOOL, DUKE UNIV, 88- *Mem:* Fel AAAS; Am Soc Photobiol; Am Soc Zoologists; Sigma Xi; Am Soc Limnol & Oceanog. *Res:* Photobehavior and photophysiology of zooplankton; biological rhythms. *Mailing Add:* Duke Univ Marine Lab Beaufort NC 28516

FORWARD, ROBERT L, b Geneva, NY, Aug 15, 32; m 54; c 4. EXPERIMENTAL GRAVITATION. *Educ:* Univ Md, BS, 54; Univ Calif, Los Angeles, MS, 58; Univ Md, PhD(physics), 65. *Prof Exp:* Mem tech physics, Hughes Res Labs, 56-57, assoc dept mgr, Theoret Studies Dept, 67-68, mgr, Explor Studies Dept, 68-74, sr scientist, 75-87; OWNER & CHIEF SCIENTIST, FORWARD UNLIMITED, 87- *Concurrent Pos:* Gravity Res Found awards, 62, 63, 64, 65; ed, Interstellar Studies issues, J Brit Interplanetary Soc, 88- *Mem:* Am Phys Soc; sr mem Am Astronaut Soc; assoc fel Inst Aeronaut & Astronaut; fel Brit Interplanetary Soc; Sci Fiction Writers Am. *Res:* Experimental investigations of dynamic gravitational fields, gravitational gradient sensors, tests of general theory of relativity and gravitational radiation; low noise electronics; advanced propulsion. *Mailing Add:* PO Box 2783 Malibu CA 90265-7783

FORYS, LEONARD J, b Buffalo, NY, July 22, 41; m 66, 82; c 2. PROCESSOR PERFORMANCE. *Educ:* Univ Notre Dame, BS, 63; Mass Inst Technol, MS & EE, 65; Univ Calif, Berkeley, PhD(elec eng), 68. *Prof Exp:* Actg asst prof elec eng & comput sci, Univ Calif, Berkeley, 67-68; mem tech staff, Bell Labs, 68-83; DIST MGR, BELLCORE, 84- *Res:* Communication theory; teletraffic theory; queueing theory; computer real time performance. *Mailing Add:* 823 Holmdel Rd Holmdel NJ 07733

FORZIATI, ALPHONSE FRANK, b Boston, Mass, Feb 27, 11; m 45. PHYSICAL CHEMISTRY. *Educ:* Harvard Univ, BA, 32, MA, 34, PhD(phys chem), 39. *Prof Exp:* Lab asst chem, Harvard Univ, 32-33 & Radcliffe Col, 33-38; asst phys chem, Harvard Univ, 39-41; Am Petrol Inst res assoc, Nat Bur Stand, 41-50 & Am Dent Asn res assoc, 50-62; electrochemist, Harry Diamond Lab, 62-63; tech prog mgr, Adv Res Proj Agency, US Dept Defense, 63-66; asst chief res div, Fed Water Pollution Control Admin, US Environ Protection Agency, 66-71, chief measurements & instrumentation br, Div Processes & Effects, Off Res & Monitoring, 71-74, staff scientist sci adv bd, 74-77, dir stratospheric modification res staff, Off Res & Develop, 77-80; RETIRED. *Mem:* AAAS; Am Chem Soc; Electrochem Soc; Soc Appl Spectros; Am Dent Asn. *Res:* Electrochemistry; electrode potentials; fuel cells; high vacuum technique; physical properties of hydrocarbons; precise determination of and correlation with molecular structure; fluorescence and phosphorescence; technical program management; instrumentation; monitoring; photobiology; atmospheric chemistry and physics. *Mailing Add:* 15525 Prince Federick Way Silver Spring MD 20906

FOSBERG, MAYNARD AXEL, b Turlock, Calif, July 7, 19; m 47; c 2. SOIL GENESIS & CLASSIFICATION. *Educ:* Univ Wis, BS, 48, MS, 49, PhD(soils), 63. *Prof Exp:* From asst prof to prof, 58-90, EMER PROF SOIL SCI & SOIL SCIENTIST, UNIV IDAHO, 90- *Mem:* Soil Sci Soc Am; Am Soc Range Mgt; Sigma Xi. *Res:* Soil genesis and classification; soil-plant relationships in range and forest habitats; soil relationship in geology and archeology. *Mailing Add:* Dept Plant Soil & Entom Sci Univ Idaho Moscow ID 83843

FOSBERG, THEODORE M(ICHAEL), b Seattle, Wash, Dec 26, 34; m 58; c 5. CHEMICAL ENGINEERING. *Educ:* Univ Wash, BS, 59, MS, 61, PhD(chem eng), 64. *Prof Exp:* Bioscientist, Boeing Co, 64-71; PROCESS ENGR, RESOURCES CONSERV CO, 71- *Mem:* Am Chem Soc; Am Inst Chem Engrs. *Res:* Interphase mass transfer; life support requirements and systems; human factors; process technology for desalination and brine concentrator systems; evaporator technology. *Mailing Add:* 1913 SW 167th Seattle WA 98166-2755

FOSCANTE, RAYMOND EUGENE, b New York, NY, Mar 24, 42; m 64; c 2. POLYMER CHEMISTRY, TECHNICAL MANAGEMENT. *Educ:* Manhattan Col, BS, 62; Seton Hall Univ, PhD(chem), 66; Univ Mo, JD, 75. *Prof Exp:* From assoc chemist to sr chemist, 69-71, head org & polymer chem sect, Midwest Res Inst, 71-75; dir corp res & develop, 75-81, VPRES, TECH OPER, AMERON PROTECTIVE COATINGS GROUP, AMERON, INC, 81- *Mem:* Am Chem Soc; The Chem Soc. *Res:* Propellant chemistry; structure-property relationships; wastewater treatment; corrosion control; polymer cure chemistry; organic and inorganic coatings; polymer synthesis; carcinogen synthesis; research management; failure analysis. *Mailing Add:* 20547 Manzanita Yorba Linda CA 92686

FOSCHINI, GERARD JOSEPH, b Jersey City, NJ, Feb 28, 40; m 62; c 2. APPLIED MATHEMATICS. *Educ:* Newark Col Eng, BSEE, 61; NY Univ, MEE, 63; Stevens Inst Technol, PhD(math), 67. *Prof Exp:* MEM TECH STAFF, BELL TEL LABS, INC, 61- *Mem:* Math Asn Am. *Res:* Computer communications; communication theory; function theory; stochastic processes. *Mailing Add:* 79 Orchard St South Amboy NJ 08879

FOSCOLOS, ANTHONY E, b Cairo, Egypt, May 15, 30; Can citizen; m 58; c 2. SOIL MINERALOGY, SEDIMENTOLOGY. *Educ:* Univ Thessaloniki, BAgE, 53; Univ Calif, Berkeley, MS, 64, PhD(clay physics, chem), 66. *Prof Exp:* Tech officer, Kopaes Orgn, Greece, 53-64; res asst clay mineral, Univ Calif, Berkeley, 63-66; RES SCIENTIST, III, INST SEDIMENTARY & PETROL GEOL, CAN DEPT ENERGY, MINES & RESOURCES, 66- *Concurrent Pos:* Greek Found Scholars grant, 60; NSF grant, 64; lectr & instr, Dept Archaeol, Univ Calgary, 75- *Mem:* Am Soc Agron; NY Acad Sci; Int Soil Sci Soc; Clay Minerals Soc; Sigma Xi. *Res:* Clay physical chemistry, chemistry and mineralogy. *Mailing Add:* 3407 Varal Rd NW Calgary AB T3A 0A3 Can

FOSDICK, LLOYD D(UDLEY), b New York, NY, Jan 18, 28; m 58. COMPUTER SCIENCE. *Educ:* Univ Chicago, PhB, 46, BS, 48; Purdue Univ, MS, 50, PhD(physics), 53. *Prof Exp:* Systs evaluation, Control Systs Lab, 53-54; assoc head, Comput Div, Midwestern Univs Res Asn, 56-57; res asst prof digital comput lab, Univ Ill, 57-61, res assoc prof physics, 61-64, res prof, 64-70; PROF COMPUT SCI, UNIV COLO, BOULDER, 70- *Concurrent Pos:* Guggenheim fel, 64. *Mem:* Asn Comput Mach; Soc Indust & Appl Math; Sigma Xi. *Res:* Digital computers; statistical mechanics; mathematics of computation. *Mailing Add:* 276 Acorn Lane Boulder CO 80302

FOSDICK, ROGER L(EE), b Pontiac, Ill, Nov 18, 36; m 56; c 2. APPLIED MATHEMATICS, CONTINUUM MECHANICS. *Educ:* Ill Inst Technol, BS, 59; Brown Univ, PhD(appl math), 63. *Prof Exp:* Asst prof mech, Ill Inst Technol, 62-65, assoc prof, 65-69; PROF MECH, UNIV MINN, MINNEAPOLIS, 69- *Concurrent Pos:* Vis prof, Fed Univ, Rio de Janeiro, Brazil, 77; Erskine fel, Univ Canterbury, Christchurch, NZ, 80. *Mem:* Soc Nat Philos (secy, 72-74, chmn, 79-81); Soc for Rheology; Soc Interaction Mech & Math; Sigma Xi. *Res:* Continuum mechanics, especially theory of finite elasticity, theory of finite viscoelasticity, and theory of non-linear fluids; continuum thermomechanics. *Mailing Add:* 122 Aerospace Engr Univ Minn Minneapolis MN 55455

FOSKET, DONALD ELSTON, b Klamath Falls, Ore, July 20, 36; m 87; c 2. PLANT MOLECULAR BIOLOGY, CELL BIOLOGY. *Educ:* Univ Idaho, BA, 58, PhD(plant biol), 64. *Prof Exp:* Res assoc developmental bot, Brookhaven Nat Lab, 64-65; asst prof biol, Mt Holyoke Col, 65-67; NSF fel, Biol Lab, Harvard Univ, 67-69, Bullard fel, 69-70; from asst to assoc prof develop & cell biol, 70-80, PROF BIOL SCI, UNIV CALIF, IRVINE, 80- *Concurrent Pos:* Dir develop biol prog, NSF, 84-85; vis prof, Kyoto Univ, Japan, 87. *Mem:* Am Soc Cell Biol; Soc Develop Biol; fel AAAS; Bot Soc Am; Soc Exp Biol & Med. *Res:* Regulation of cell division and cell differentiation; microtubules. *Mailing Add:* Dept Develop & Cell Biol Univ Calif Irvine CA 92717

FOSNACHT, DONALD RALPH, b Chicago, Ill, Dec 10, 50; m 72; c 2. METALLURGICAL & MINERAL ENGINEERING. *Educ:* MacMurray Col, BS, 73; Columbia Univ, BS, 74, MS, 75; Univ Mo-Rolla, PhD(metall eng), 78. *Prof Exp:* Res engr metal eng, Inland Steel Co, 78-81, sr res eng, 81-85, sect mgr steel & refractures res, 85-87, mgr raw mat & primary processing res, 87-88, mgr steel prods res, 88-90, MGR OPERATING TECHNOL, 80, HOT STRIP MILL, INLAND STEEL CO, 90- *Mem:* Am Inst Mining, Metall & Petrol Engrs; Sigma Xi. *Res:* Raw materials with ironmaking steel making and casting research; ladle and jundish metallurgy; hot metal treatments; ceramic filtering of molten metal, and development of improved refractury practices; steel product development; rolling technology; reheat furnace operations. *Mailing Add:* Inland Steel Co 3210 Watling St East Chicago IN 46312

FOSS, ALAN STUART, b Stamford, Conn, Sept 9, 29. CHEMICAL ENGINEERING. *Educ:* Worcester Polytech Inst, BS, 52; Univ Del, MChE, 54, PhD(chem eng), 57. *Prof Exp:* Sr res engr, Eng Res Lab, E I du Pont de Nemours & Co, 56-61; from asst prof to prof chem eng, 61-73, vchmn dept, 67-69, PROF CHEM ENG, UNIV CALIF, BERKELEY, 73- *Concurrent Pos:* Fel, Univ Trondheim, 69-70. *Mem:* Am Inst Chem Engrs; Sigma Xi. *Res:* Chemical process dynamics and control; computer control; educational software. *Mailing Add:* Dept Chem Eng Univ Calif Berkeley CA 94720

FOSS, DONALD C, b Providence, RI, May 3, 38; m 60; c 4. NUTRITION, PHYSIOLOGY. *Educ:* Univ NH, BS, 60; Univ Wis, MS, 61; Univ Mass, PhD(physiol & nutrit), 66. *Prof Exp:* Fel poultry sci, 66-68, from asst prof to assoc prof, 68-80, PROF ANIMAL SCI, UNIV VT, 80- *Concurrent Pos:* Assoc ed, Poultry Sci; consult poultry mgt, Ecuador & Honduras; chair, Col Agr & Life Scis Recruitment, 80-, Biol Sci Prog, 86- *Mem:* Poultry Sci Asn; Am Soc Photobiol; World's Poultry Sci Asn; Sigma Xi; Soc Exp Biol & Med. *Res:* Effects of light quality on growth reproduction and endocrine balance; role of the pineal gland in mediating light effects and circadian rhythms; nutritional and physiological factors influencing poultry growth and egg production. *Mailing Add:* Animal Sci Biol Lab Univ Vt 655 Spear St Burlington VT 05401

FOSS, FREDERICK WILLIAM, JR, b Lincoln, Mich, June 3, 33; m 57; c 3. INORGANIC CHEMISTRY. *Educ:* Univ Mich, BS, 55; Univ Minn, Minneapolis, MS, 57; Univ of the Pac, PhD(inorg chem), 64. *Prof Exp:* Instr phys sci, 57-63, from asst prof to assoc prof chem, 63-67, head dept, 66-89, PROF CHEM, WINONA STATE UNIV, 67- *Mem:* Am Chem Soc. *Res:* Inorganic polarography in non-aqueous solvents; general chemistry, laboratory approaches; chemical lecture demonstrations. *Mailing Add:* Dept Chem Winona State Univ Winona MN 55987

FOSS, JOHN E, b Whitehall, Wis, May 30, 32; m 53; c 2. SOIL GENESIS, SOIL CLASSIFICATION. *Educ:* Wis State Univ, River Falls, BS, 57; Univ Minn, MS, 59, PhD(soil sci, geol), 65. *Prof Exp:* Asst prof soils, Wis State Univ, River Falls, 60-66; assoc prof soils, Univ Md, 66-73, prof soils, 73-81; PROF & CHMN SOILS, NDAK STATE UNIV, 81-; PLANT & SOIL SCI DEPT, UNIV TENN. *Concurrent Pos:* Consult, Libby, McNeil & Libby Corp, France, 62, Allied Chem, 73-80, Columbia Gas Corp & A D Little, 75 & Dept Interior, 80-81. *Mem:* AAAS; fel Am Soc Agron; Sigma Xi; fel Soil Sci Soc Am. *Res:* Soil development; interpretation and uses of soil surveys; soil-archaeology studies. *Mailing Add:* Plant & Soil Sci Dept Univ Tenn Box 1071 Knoxville TN 37901-1071

FOSS, JOHN F, b Washington, Pa, Mar 24, 38; m 60; c 2. FLUID MECHANICS. *Educ:* Purdue Univ, BS, 61, MS, 62, PhD(mech eng), 65. *Prof Exp:* From asst prof to assoc prof, 65-75, PROF MECH ENG, MICH STATE UNIV, 75- *Concurrent Pos:* Fel, Johns Hopkins Univ, 70-71; Humboldt res fel, Univ Karlsruhe, WGermany, 78-79 & Univ Erlangen-Nurnberg, , 85-86. *Mem:* AAAS; fel Am Soc Mech Engrs; Am Soc Eng Educ; Am Inst Aeronaut & Astronaut; Am Phys Soc. *Res:* Turbulent shear flow and experimental fluid mechanics. *Mailing Add:* Dept Mech Eng Mich State Univ East Lansing MI 48824

FOSS, JOHN G(ERALD), b Brooklyn, NY, Feb 27, 30; m 57; c 3. BIOPHYSICS. *Educ:* Brooklyn Polytech, BS, 51; Univ Conn, MS, 53; Univ Utah, PhD(phys chem), 56. *Prof Exp:* Assoc phys chem, Univ Minn, 56-57 & Univ Ore, 57-61; PROF BIOPHYS, IOWA STATE UNIV, 61- *Concurrent Pos:* Vis prof, Univ Calif, Berkeley, 63-64; liaison scientist, Off Naval Res, London, 70-72. *Res:* Biophysics; spectroscopic methods in biochemistry. *Mailing Add:* RR 1 No 88 Callender IA 50523

FOSS, MARTYN (HENRY), b Salt Lake City, Utah, May 28, 18; m 42; c 6. PHYSICS. *Educ:* Univ Chicago, BS, 38; Carnegie Inst Technol, PhD(physics), 48. *Prof Exp:* Shift supvr lab, Ind Ord Works, 40-41, chem engr, 41-43; mem staff, Manhattan Proj, Del, Ill & Los Alamos, 43-46 & 82-86; res physicist, Carnegie Inst Technol, 48-54; sr physicist, Argonne Nat Lab, 55-59, assoc dir, Particle Accelerator Div, 59-62, sr scientist, 74-82; assoc dir, Nuclear Res Ctr, Carnegie Inst Technol, 62-70, sr physicist, Carnegie-Mellon Univ, 70-74; RETIRED. *Honors & Awards:* Sigma Xi. *Mem:* Fel Am Phys Soc. *Res:* Particle accelerator design: magnets, orbits, and radio frequency. *Mailing Add:* 13734 Aleppo Ave Sun City West AZ 85375

FOSS, ROBERT PAUL, b Chicago, Ill, Aug 22, 37; m 62; c 2. PHYSICAL ORGANIC CHEMISTRY, POLYMER CHEMISTRY. *Educ:* Northwestern Univ, BA, 58; Calif Inst Technol, PhD(chem), 63. *Prof Exp:* NIH fel chem, Lab Chem Biodynamics, Univ Calif, Berkeley, 63-65; res chemist, cent res dept, 65-84, res chemist, Biomedical Prod Dept, 84-86, RES CHEMIST, CENTRAL RES DEPT, E I DU PONT DE NEMOURS & CO, INC, 88- *Mem:* Am Chem Soc; Sigma Xi. *Res:* Photochemistry and energy transfer involving organic and inorganic compounds; polymer chemistry involving general and photopolymerization studies; high energy radiation chemistry of organic and polymer systems; biomaterials; materials for non-linear optics. *Mailing Add:* Cent Res & Develop Dept E328 E I du Pont de Nemours & Co Inc Wilmington DE 19898

FOSSAN, DAVID B, b Faribault, Minn, Aug 23, 34; m 64. PHYSICS. *Educ:* St Olaf Col, BA, 56; Univ Wis, MS, 57, PhD(physics), 60. *Prof Exp:* Asst physics, Univ Wis, 60-61; Ford Found grant, Niels Bohr Inst, Copenhagen, 61-62; mem res staff nuclear physics, Van de Graaff Lab, Lockheed Palo Alto Research Lab, 63-65; from asst prof to assoc prof, 65-72, PROF PHYSICS, STATE UNIV NY STONY BROOK, 72- *Concurrent Pos:* Vis physicist, Lawrence Berkeley Lab, Univ Calif, 78-79; mem prof adv comt, Super Heavy Ion Linear Accelerator, Lawrence Berkeley Lab, Univ Calif, 81- *Mem:* Am Phys Soc. *Res:* Nuclear structure via heavy-ion induced reactions and gamma-ray measurements; excited state lifetimes and static moments of nuclei. *Mailing Add:* Dept Physics State Univ NY Stony Brook NY 11794

FOSSEL, ERIC THOR, b Minneapolis, Minn, Dec 11, 41; m 64, 82; c 2. BIOPHYSICAL CHEMISTRY, CARDIAC METABOLISM. *Educ:* Yale Univ, BA, 64, MS, 66, MPhil, 67; Harvard Univ, PhD(chem), 70. *Prof Exp:* NIH res fel chem, 70-71, NIH res fel biochem, 71-74, from lectr to asst prof biophys, 74-83, asst prof radiol, 83-85, ASSOC PROF RADIOL, SCH MED, HARVARD UNIV, 85- *Concurrent Pos:* Fel, Muscular Dystrophy Asn Am, 72-75; fel, Arthritis Found, 75-78; estab investr, Am Heart Asn, 78-82; res assoc, 83-, dir radiol res, Beth Israel Hosp, 87- *Mem:* Biophys Soc; Am Heart Asn; Asn Univ Radiologists; North Am Soc Cardiac Radiologists; Am Fedn Clin Res; Soc Magnetic Resonance in Med; Soc Magnetic Resonance Imaging. *Res:* Thermal rearrangements; enzyme mechanisms; muscle action mechanism; nuclear magnetic resonance in biological systems; cardiac metabolism; ion transport in cardiac tissue; tumor biophysics; cancer detection. *Mailing Add:* Beth Israel Hosp Dept Radiol 330 Brookline Ave Boston MA 02215

FOSSLAND, ROBERT GERARD, b Winthrop Harbor, Ill, Oct 18, 18; m 60; c 6. REPRODUCTIVE BIOLOGY, EVOLUTIONARY BIOLOGY. *Educ:* Univ Ill, BS, 40; Univ Nebr, PhD(zool), 56. *Prof Exp:* Asst, Univ Nebr, 40-42, from instr to assoc prof dairy sci, Univ Ariz, 45-61; assoc prof zool, Milwaukee-Downer Col, 62; assoc prof biol, 62-65, PROF BIOL, UNIV WIS-EAU CLAIRE, 65- *Res:* Evolutionary biology. *Mailing Add:* 2330 Jordan Ct Eau Claire WI 54701

FOSSLIEN, EGIL, b Norway, Dec 14, 35; US citizen; m 66; c 1. MOLECULAR BIOLOGY, PATHOLOGY. *Educ:* Univ Heidelberg Med Sch, MD, 65; (dissertation in Pediat), 66. *Prof Exp:* Intern, WGer, 66-68; rotating intern, Gunderson Clin & Lacroix Lutheran Hosp, Wis, 68-69; residency path, Univ Chicago, 69-71, instr & trainee, 71-72; asst prof path, Univ Mo, 72-76; assoc prof path, Univ Southern Fla, Tampa, 76-78; assoc prof clin path, Univ Ala, 78-80; assoc prof, 80-84, PROF PATH, UNIV ILL, 84- *Concurrent Pos:* Dir, Hosp Labs, Univ Ill Hosp, 80-86; researcher, Gladstone Found Lab for Cardiovascular Dis, Univ Calif, San Francisco, 87-88. *Mem:* Am Med Asn; Am Asn Path; Am Clin Soc (vpres, 83-84, pres, 84-85); Am Soc Clin Path; Can Asn Path. *Res:* Cardiovascular disease; molecular pathogenesis of atherosclerosis. *Mailing Add:* Dept Path M/C 847 Univ Ill Col Med 1853 W Polk St Chicago IL 60612

FOSSUM, GUILFORD O, b Loma, NDak, Dec 17, 18; m 45; c 4. CIVIL & SANITARY ENGINEERING. *Educ:* Univ NDak, BS, 42; Iowa State Univ, MS, 51. *Prof Exp:* Inspector, US Eng Dept, CZ, 42-43; engr, Grand Forks City Eng Dept, 46; from asst prof to prof civil eng, Univ NDak, 46-; RETIRED. *Honors & Awards:* Arthur Sidney Bedell Award, Water Pollution Control Fedn, 73. *Mem:* Water Pollution Control Fedn; Am Soc Civil Engrs; Am Soc Eng Educ; Nat Soc Prof Engrs. *Res:* Industrial and municipal wastes and water supplies, including treatment methods. *Mailing Add:* 1828 Cottonwood Grand Forks ND 58201

FOSSUM, JERRY G, b Phoenix, Ariz, July 18, 43; m 86; c 2. SEMICONDUCTOR DEVICE PHYSICS, COMPUTER SIMULATION. *Educ:* Univ Ariz, BS, 66, MS, 69, PhD(elec eng), 71. *Prof Exp:* Appl res, Sandia Labs, 71-78; assoc prof, 78-80, PROF, RES/TEACHING, UNIV FLA, 80- *Concurrent Pos:* Consult, Burr-Brown, 70-71, Harris Semiconductor, 84, Tex Instruments, 88- *Mem:* Fel Inst Elec & Electronics Engrs. *Res:* Semiconductor device modeling-simulation; VLSI technology computer-aided design; design for manufacturability. *Mailing Add:* Dept Elect Eng Univ Fla Gainesville FL 32611

FOSSUM, ROBERT MERLE, b Northfield, Minn, May 1, 38; m 60, 79; c 4. MATHEMATICS. *Educ:* St Olaf Col, BA, 59; Univ Mich, AM, 61, PhD(math), 65. *Prof Exp:* From asst prof to assoc prof, 64-72, PROF MATH, UNIV ILL, URBANA, 72- *Concurrent Pos:* Fulbright res grant, Inst Math, Univ Oslo, 67-68, vis prof, 68-69; vis prof, Univ Aarhus, 71-73; ed, Proc Amer Math Soc, 74-78; vis assoc prof, Univ Copenhagen, 76-77; prof, Univ Paris VI, 78-79; gen secy, Inst Algebraic Meditation, 76; assoc secy, Am Math Soc, 84-87. *Mem:* Danish Math Soc; Am Math Soc (secy, 89-); Sigma Xi; Swed Math Soc; Inst Algebraic Meditation. *Res:* Commutative noetherian rings; algebras over commutative noetherian integrally closed integral domains and homological algebra. *Mailing Add:* 1409 W Green Urbana IL 61801

FOSSUM, STEVE P, b Northfield, Minn, Dec 26, 41; div; c 2. SURFACE PHYSICS, VACUUM TECHNOLOGY. *Educ:* St Olaf Col, BA, 63; Univ Wis, Madison, MA, 65, PhD(physics), 70. *Prof Exp:* Assoc prof, 66-75, PROF PHYSICS & CHMN DEPT, UNIV WIS-STOUT, 75- *Concurrent Pos:* Exchange lectr, Newcastle Polytech Inst, Eng, 71-72. *Mem:* Am Phys Soc; Am Asn Physics Teachers. *Res:* Atomic hydrogen reactions on metal surfaces in ultra-high vacuums. *Mailing Add:* Dept Physics Univ Wis-Stout Menomonie WI 54751

FOSSUM, TIMOTHY V, b Northfield, Minn, Oct 29, 42; m 65; c 2. ALGEBRA, COMPUTER SCIENCE. *Educ:* St Olaf Col, BA, 64; Univ Ore, PhD(math), 68. *Prof Exp:* Asst prof math, Univ Utah, 68-74; PROF APPL COMPUT SCI, UNIV WIS-PARKSIDE, 74- *Concurrent Pos:* Vis lectr, Univ Ill, 71-72. *Mem:* Asn Comput Mach. *Res:* Operating systems and computer networks. *Mailing Add:* Dept Appl Comput Sci Univ Wis-Parkside 900 Wood Rd Box 2000 Kenosha WI 53141-2000

FOST, DENNIS L, b Los Angeles, Calif, May 9, 44; m 66; c 3. SURFACTANT CHEMISTRY. *Educ:* Calif State Univ, 67; Purdue Univ, PhD(org chem), 71. *Prof Exp:* Instr org chem, Purdue Univ, 71-77; res chemist, E I du Pont de Nemours & Co, 73-77; res mgr, New Eng Nuclear Corp, 77-81; tech dir, 81-83, VPRES RES & DEVELOP, MONA INDUST INC, 83- *Mem:* Am Chem Soc; Soc Cosmetic Chem; Am Oil Soc; Soc Tribologidt & Lubrication Engrs; Cosmetics, Toileteries Mfrs Asn; Chem Specialities Mfrs Asn. *Res:* Product development and maintenance; technical service for customers and production department; analytical support. *Mailing Add:* Mona Inds Inc 76 E 24th St Paterson NJ 07514-2022

FOSTER, ALBERT EARL, b Madison, SDak, Apr 4, 31; m 52; c 2. BARLEY BREEDING & GENETICS. *Educ:* SDak State Univ, BS, 54, MS, 56, PhD(agron), 58. *Prof Exp:* From asst prof to assoc prof, 58-70, PROF AGRON, NDAK STATE UNIV, 70-, CHMN, DEPT CROP & WEED SCI, 87- *Concurrent Pos:* Consult, San Miguel Corp, 76-, Norteñ, 89- *Mem:* Crop Sci Soc Am; Sigma Xi. *Res:* Developing of malting barley cultivars for North Dakota and adjacent areas; studies of barley genetics. *Mailing Add:* Dept Crop & Weed Sci NDak State Univ Fargo ND 58105-5051

FOSTER, ALFRED FIELD, b Warren, Ohio, Dec 4, 15; m 40; c 2. PHYSICAL CHEMISTRY. *Educ:* Col of Wooster, BA, 38; Ohio State Univ, MA, 40, PhD, 50. *Prof Exp:* Process engr, US Rubber Co, NC, 42-43; asst prof physics, Davidson Col, 43-44; from asst prof to prof chem, Univ Toledo, 46-85, assoc dean col arts & sci, 62-68, assoc dean grad sch, 69-74; RETIRED. *Concurrent Pos:* Res assoc, Ohio State Res Found, 48-50; partic, Argonne Nat Lab Conf, 68, 70. *Mem:* AAAS; fel Am Inst Chemists; Am Chem Soc; Am Crystallog Asn; NY Acad Sci. *Res:* X-ray crystal structure; dipole moments of molecules in solutions; radiochemistry. *Mailing Add:* 2025 Bretton Pl Toledo OH 43606-3318

FOSTER, ARTHUR R(OWE), b Peabody, Mass, Apr 22, 24; m 47; c 2. MECHANICAL ENGINEERING, NUCLEAR ENGINEERING. *Educ:* Tufts Univ, BS, 45; Yale Univ, MEng, 49. *Prof Exp:* Engr, Mat Develop Lab, Pratt & Whitney Div, United Aircraft Corp, 47-48; chmn dept, 61-75, prof mech eng, 49-89, EMER PROF MECH ENG, NORTHEASTERN UNIV, 89- *Concurrent Pos:* Mem, Gen Elec Profs Conf, 55; Latin Am teaching fel, Escuela Politecnica Nacional, Quito, Ecuador, 75-76, vis prof, 84. *Honors & Awards:* Fulbright lectr solar energy, Colombia, 79; Centennial Medal, Am Soc Mech Engrs, 81. *Mem:* Am Soc Mech Engrs; Am Soc Eng Educ. *Res:* Applied thermodynamics; nuclear engineering; solar engineering. *Mailing Add:* Dept Mech Eng Northeastern Univ Boston MA 02115

FOSTER, BILLY GLEN, b Canton, Tex, Mar 6, 32; m 54; c 4. MICROBIOLOGY. *Educ:* NTex State Univ, BS, 55, MS, 62; Univ Iowa, PhD(prev med, environ health), 65. *Prof Exp:* Instr biol, NTex State Univ, 60-62; res assoc microbiol, Univ Iowa, 62-65; from asst prof to assoc prof, 65-85, PROF MICROBIOL, TEX A&M UNIV, 85- *Mem:* AAAS; Am Soc Microbiol; Am Acad Microbiol. *Res:* medical and diagnostic microbiology; epidemiology and serology of infectious disease; humoral and cell medicated immunity; regulation of humoral and cell medicated response; macrophage activation. *Mailing Add:* Dept Biol Tex A&M Univ College Station TX 77843

FOSTER, BRUCE PARKS, b Mussoorie, India, June 27, 25; US citizen; m 47; c 3. EXPERIMENTAL NUCLEAR PHYSICS. *Educ:* Baldwin-Wallace Col, BS, 49; Yale Univ, MS, 51, PhD(physics), 54. *Prof Exp:* Asst prof, 53-56, ASSOC PROF PHYSICS, NTEX STATE UNIV, 56-, 56- *Concurrent Pos:* Fulbright lectr, Peshawar, 60-61; lectr, Forman Christian Col, WPakistan, 61-63; mem vchancellors comt strengthening physics dept, WPakistan Eng Univ, 62-63. *Mem:* Am Phys Soc. *Res:* Investigation and analysis of gamma ray decay schemes in nuclear reactions including the development of computer programs for this purpose; gamma-ray spectroscopy; astronomy. *Mailing Add:* PO Box 2182 Denton TX 76202-2182

FOSTER, C(HARLES) VERNON, heat transfer, for more information see previous edition

FOSTER, CAXTON CROXFORD, b Ft Bragg, NC, Jan 21, 29; m 59; c 5. COMPUTER ARCHITECTURE. *Educ:* Mass Inst Technol, BS, 50; Univ Mich, MSE, 57, PhD(elec eng), 65. *Prof Exp:* Physicist, Tracer Lab, Inc, Boston, Mass, 51-52; engr, Technol Instrument Corp, Acton, Mich, 53-55; res, Mental Health Res Inst, Univ Mich, 55-64; engr, Goodyear Aerospace Corp, 64-65; dir comput ctr, Univ Mass, 65-67, prof comput sci, 67-84; RETIRED. *Concurrent Pos:* Vis lectr, Univ Edinburgh, Scotland, 67-68. *Res:* Design of computer hardware and software. *Mailing Add:* PO Box 488 East Orleans MA 02643-0488

FOSTER, CHARLES DAVID OWEN, b London, Ont, Oct 26, 38; m 60; c 2. ANIMAL ENERGETICS, THERMOREGULATION. *Educ:* Western Ont Univ, BSc, 60, MSc, 62; Univ Wis, PhD(biochem), 67. *Prof Exp:* NIH fel biochem, Princeton Univ, 66-68; asst prof, Univ Ill, Urbana-Champaign, 68-72; from asst res officer to assoc res officer, 72-82, SR RES OFFICER, NAT RES COUN CAN, 83- *Concurrent Pos:* Assoc ed, Can J Physiol Pharmacol, 84- *Mem:* Sigma Xi; Can Physiol Soc; Am Physiol Soc. *Res:* Mechanisms of mammalian energy balance; physiology and biochemistry of brown adipose tissue and nonshivering thermogenesis. *Mailing Add:* Biol Sci Div M54 Nat Res Coun Can Ottawa ON K1C 2M7 Can

FOSTER, CHARLES HENRY WHEELWRIGHT, b Boston, Mass, Mar 18, 27; m 53; c 3. NATURAL RESOURCES POLICY, FORESTRY. *Educ:* Harvard Col, BA, 51; Univ Mich, BS, 53, MS, 56; Johns Hopkins Univ, PhD(geog & environ eng), 69. *Hon Degrees:* DPA, Suffolk Univ, 71; MA, Yale Univ, 76. *Prof Exp:* Comnr, Mass Dept Natural Resources, 59-66; pres, Nature Conservancy, 66-67; chmn, New Eng Natural Resources Ctr, 69-71; secy, Mass Exec Off Environ Affairs, 71-75; prof environ policy, Univ Mass, 75-76; dean, Sch Forestry & Environ Studies, Yale Univ, 76-81. *Concurrent Pos:* Mem adv coun, Pub Land Law Rev Comn, 67-69; Bullard fel, Harvard Univ, 69-70; mem, Environ Adv Bd, Corps Engrs, 70-72, Task Force Land Use Planning, Coun State Govt, 72-73 & Adv Bd Natural Hazards, NSF, 73-74; US comnr, Int Comn NW Atlantic Fisheries, 73-75; sr staff mem natural resources, Arthur D Little, Inc, 75-76; mem & chmn, N Atlantic Region Adv Comn, Nat Park Serv, 75-77; chmn, Appalachian Nat Scenic Trail Adv Coun, Dept Interior, 75-77; mem, Marine Fisheries Adv Comn, Dept Com, 77-79; consult prof human biol, Stanford Univ, 81-83; pres, W Alton Jones Found, 83-84; vis fac mem, Tufts Univ, Clark Univ & Brown Univ, 84-86; Herbert E Kahler res fel, Eastern Nat Park & Monument Asn, 85; res fel & lectr, Harvard Univ, 86. *Mem:* Fel AAAS; Soc Am Foresters. *Res:* Public administration; environmental impact analysis; land use planning; environmental mediation; water resources. *Mailing Add:* 484 Charles River St Needham MA 02192

FOSTER, CHARLES HOWARD, b Greenville, Ky, Sept 8, 47; m 69; c 2. ORGANIC CHEMISTRY. *Educ:* Vanderbilt Univ, BA, 69; Mass Inst Technol, PhD(org chem), 73. *Prof Exp:* Res chemist, 73-75, sr res chemist, 75-84, RES ASSOC, TENN EASTMAN CO, 84- *Mem:* Sigma Xi; Am Chem Soc. *Res:* New synthetic methods; synthesis of natural products; steroid chemistry. *Mailing Add:* 1413 Dobyns Dr Kingsport TN 37664

FOSTER, CHARLES STEPHEN, b Charleston, WVa, May 19, 42; c 1. GENERAL & OCULAR IMMUNOLOGY. *Educ:* Duke Univ, BS, 65 & 68, MD, 69; Am Bd Ophthal, dipl, 76. *Prof Exp:* From instr to asst prof, 69-82, ASSOC PROF OPHTHAL, HARVARD MED SCH, 82-; DIR, HILLES IMMUNOL LAB, MASS EYE & EAR INFIRMARY, 84- *Concurrent Pos:* Instr med, George Washington Univ Med Ctr, 70-72; dir residency training, Mass Eye & Ear Infirmary, 76-81, asst ophthal, 77-82, asst surgeon, 82-85, assoc surgeon, 85-87, surgeon ophthal, 87-, dir, immunol serv, 81; clin asst scientist, dept cornea res, Eye Res Inst Retina Found, 77-81, dir, Immunopath Lab, 77-81, clin assoc scientist, 81-; vis prof, Wills Eye Hosp & Univ Paris, 79, Johns Hopkins Univ, NY Univ, dir rheumatology & Manhattan Eye, Ear & Throat Hosp, 87; mem, Ocular Microbiol & Immunol Group. *Honors & Awards:* Sigma Xi. *Mem:* Am Med Asn; Royal Soc Med; Am Acad Ophthal; Am Asn Ophthal; Contact Lens Soc Am; Corneal Dystrophy Soc; Am Asn Immunol; Am Soc Ophthal & Otolaryngol Allergy; Am Uveitis Soc; Am Fedn Clin Res; Fedn Am Soc Exp Biol; Int Soc Immunopharmacol; Asn Res Vision & Ophthal; Int Soc Eye Res; Asn Eye Res Amsterdam. *Res:* General and ocular immunology; autoimmune ocular disease; corneal transplantation immunology; immune response to ocular herpes simplex infection. *Mailing Add:* Mass Eye & Ear Infirmary 243 Charles St Boston MA 02114

FOSTER, CHARLES THOMAS, JR, b Fremont, Ohio, Aug 30, 49; m 90; c 1. GEOLOGY, PETROLOGY. *Educ:* Univ Calif, Santa Barbara, BA, 71; Johns Hopkins Univ, MA, 74, PhD(geol), 75. *Prof Exp:* Actg asst prof geol, Univ Calif, Los Angeles, 75-77, res fel, 77; asst prof, 78-82, ASSOC PROF GEOL, UNIV IOWA, 82- *Mem:* Mineral Soc Am; Geol Soc Am; Geol Asn Can; Am Geophys Union; Mineral Asn Can; Geochem Soc. *Res:* Igneous and metamorphic petrology; aqueous geochemistry; irreversible thermodynamic processes in geology, geologic mapping, structural geology; computer applications to geology; shape analysis of fossils. *Mailing Add:* Dept Geol Univ Iowa Iowa City IA 52242

FOSTER, DANIEL W, b Marlin, Tex, Mar 4, 30; m 55; c 3. INTERNAL MEDICINE, METABOLISM. *Educ:* Tex Western Col, BA, 51; Univ Tex Southwest Med Sch Dallas, MD, 55. *Prof Exp:* Intern internal med, Parkland Mem Hosp, 55-56, asst resident, 56-58, chief resident, 58-59; fel biochem, Univ Tex Southwestern Med Sch, 59-60; investr, Nat Inst Arthritis & Metab Dis, 60-62; from asst prof to assoc prof, Univ Tex Southwestern Med Sch, 62-69, prof, 69-86, Jan & Henri Bromberg prof, 86-89, CHMN, DEPT INTERNAL MED, UNIV TEX SOUTHWESTERN MED SCH, DALLAS, TEX, 88-, DONALD W SELDIN DISTINGUISHED CHAIR, 89- *Concurrent Pos:* Mem metab study sect, NIH, 68-70, chmn sect, 70-72; assoc ed, J Clin Invest, 72-77 & J Metab, Clin & Exp, 69-87; sr attend physician, Parkland Mem Hosp & Univ Med Ctr, Tex; ed, Diabetes, 78-83; mem, Nat Diabetes Adv Bd, 81-84; chair, sci adv bd, Hartford Found; mem Nat Diabetes & Digestive & Kidney Dis Adv Coun, NIH, 87-90; consult, Vet Admin Hosp, Dallas, Tex, Presby Hosp, Baylor Univ Med Ctr. *Honors & Awards:* Banting Medal, 84; Joslin Medal, 84; Upjohn Award, Am Diabetes Asn, 88. *Mem:* Inst Med-Nat Acad Sci; Am Soc Clin Invest; Am Diabetes Asn; Asn Am Physicians; Am Fedn Clin Res; Am Soc Biol Chemists; AAAS; Am Col Physicians. *Res:* Endocrinology and metabolism; regulation of fatty acid synthesis and ketone body formation; author or co-author of over 100 publications. *Mailing Add:* Dept Internal Med Univ Tex Health Sci Ctr Dallas TX 75235-9030

FOSTER, DAVID BERNARD, b Marine City, Mich, May 28, 14; m 38; c 3. NEUROLOGY. *Educ:* Wayne State Univ, AB, 35; Univ Mich, MD, 38, MS, 41. *Prof Exp:* Instr neurol, Med Sch, Univ Mich, 41-44; dir div neurol & neurosurg & instr neurol, Sch Psychiat, Menninger Found, 47-67, sr physician, Div Neurol & Psychiat, Menninger Clin, 67-72; staff neurologist, Vet Admin Hosp, Topeka, 72-80, actg chief neurol, 81-86; RETIRED. *Concurrent Pos:* Consult, Topeka Vet Admin Hosp, 47-72, actg chief neurol serv, 67; consult, Topeka State Hosp, 47-; lectr, Med Sch, Univ Kans, 49- *Mem:* Am Neurol Asn; AMA; Am Acad Neurologists. *Res:* Clinical neurology and neuropathology. *Mailing Add:* 900 SW 31st No 316 Topeka KS 66611-2196

FOSTER, DONALD MYERS, b Mayersville, Miss, Mar 21, 38; m 64; c 2. CLINICAL CHEMISTRY, BIOCHEMICAL PHARMACOLOGY. *Educ:* Delta State Univ, BSE, 61; Univ Houston, MS, 67, PhD(biol sci), 69. *Prof Exp:* Fel biomath, Univ Mich, Ann Arbor, 69-72; res assoc cation transport ATP-phosphatases, Vanderbilt Univ, Nashville, 72-73; res chemist, Vet Admin Hosp, Minneapolis, 73-76; chemist, OE Teague Vet Ctr, 76-87; assoc prof path, Med Sch, Tex A&M Univ, 81-87; RES CHEMIST, NASA, JOHNSON SPACE CTR, HOUSTON, 87- *Concurrent Pos:* Asst prof lab med & path, Med Sch, Univ Minn, Minneapolis, 75-76; chmn, Tex Sect, Am Chem Soc, 79. *Mem:* Am Chem Soc; Sigma Xi; Am Asn Clin Chem. *Res:* Membrane structure and function as related to actions of drugs and toxic agents; phosphoproteins and mechanisms of active cation transport; clinical chemistry. *Mailing Add:* Dept Chem Univ Mary Hardin UMH-B Sta Box 433 Belton TX 76513-0433

FOSTER, DOUGLAS LAYNE, b Brookings, SDak, Apr 15, 44. REPRODUCTIVE ENDOCRINOLOGY, PHYSIOLOGY. *Educ:* Univ Nebr, BS, 66; Univ Ill, PhD(animal sci), 70. *Prof Exp:* NIH fel child health & human develop, 71-72, from asst prof to assoc prof, 80-87, PROF ANIMAL PHYSIOL, DEPT OBSTET & GYNEC, DIV BIOL SCI, UNIV MICH, ANN ARBOR, 87- *Concurrent Pos:* Nat Inst Child Health & Human Develop grants, 74-94, US Dept Agr grants, 87-92; Tech Adv Comt Sci & Educ, USDA, 85. *Mem:* Soc Study Reproduction; Endocrine Soc; Soc Study Fertility; Int Soc Neuroendocrinol; Soc Gynec Invest. *Res:* Physiology of reproduction; neuroendocrinology; developmental endocrinology; investigation of reproductive neuroendocrine maturation with emphasis on internal and external signals which control onset of puberty in the female; neuroendocrine control of adult ovulatory cycle and seasonality of reproduction. *Mailing Add:* Reproductive Sci Prog Rm 1101 Univ Mich 300 N Ingells Bldg Ann Arbor MI 48109-0404

FOSTER, DUNCAN GRAHAM, JR, b Philadelphia, Pa, Sept 20, 29; m 53; c 4. NUCLEAR PHYSICS, NUCLEAR DATA EVALUATION. *Educ:* Swarthmore Col, 51; Cornell Univ, PhD(exp physics), 56. *Prof Exp:* Res assoc, Pac Northwest Lab, Battelle Mem Inst, 56-69; STAFF MEM, LOS ALAMOS NAT LAB, 69- *Concurrent Pos:* Instr, Ctr Grad Study, Univ Wash, 57-66. *Mem:* Am Phys Soc; Am Nuclear Soc. *Res:* Monoenergetic measurement of neutron age; evaluation of neutron cross sections; fast-neutron total cross sections; medium-energy spallation calculations; post-fission photon and neutron spectra; photon and neutron spectra from actinide decay; neutronic properties of prompt fission fragments. *Mailing Add:* 3153 Villa Los Alamos NM 87544

FOSTER, E(LTON) GORDON, b Milwaukee, Wis, Feb 4, 19; m 41; c 1. CHEMICAL ENGINEERING. *Educ:* Univ Wis, BS, 41, MS, 42, PhD(phys chem & chem eng minor), 44. *Prof Exp:* Asst phys chem, Univ Wis, 41-43; process engr, E I du Pont de Nemours & Co, 44-46, group leader process engr, 46-51; asst prof chem eng, Univ Louisville, 51-52; engr, Shell Oil Co, 52-56, supvr, 56-66, mgr process develop dept, Indust Chem Div, Shell Chem Co, NY, 66-68, head, Process Design-Licensing Dept, Shell Develop Co, Calif, 68-70, mgr process eng-licensing dept, 70-74, sr staff res engr, 74-80, res assoc, 80-81, sr res assoc, Shell Develop Co, 81-87; RETIRED. *Concurrent Pos:* Consult, Shell Oil Co, 88- *Mem:* Am Chem Soc; fel Am Inst Chem Engrs. *Res:* Process engineering; biotechnical processes. *Mailing Add:* 2022 Bruceala Ct Cardiff-By-The-Sea CA 92007-1201

FOSTER, EDWARD STANIFORD, JR, b Port Deposit, Md, June 9, 13; m 36; c 4. APPLIED PHYSICS. *Educ:* Col of Wooster, BA, 35; Wash Univ (St Louis), MS, 37. *Prof Exp:* Field engr, Subterrex-Explor, Geophysicist, 37-39 & Halliburton Oil Well Cementing Co, 39-42; from asst prof to assoc prof, 46-58, PROF ENG PHYSICS UNIV TOLEDO, 58-, PROF PHYSICS, 74- *Concurrent Pos:* Consult, Brush Beryllium Co, 48-58 & Argonne Labs, 64- *Mem:* Am Phys Soc; Am Soc Eng Educ; Am Asn Physics Teachers; Sigma Xi. *Res:* Applied spectroscopy; peaceful uses of nuclear energy; applied meteorology; atmospheric stability; friction. *Mailing Add:* Dept Eng Physics Univ Toledo Toledo OH 43606

FOSTER, EDWIN MICHAEL, b Alba, Tex, Jan 1, 17; m 41; c 1. BACTERIOLOGY. *Educ:* NTex State Teachers Col, BA, 36, MA, 37; Univ Wis, PhD(bact), 40. *Prof Exp:* Instr bact, Univ Wis, 40-41 & Univ Tex, 41-42; from asst prof to assoc prof, 45-52, dir, Food Res Inst, 66-86, PROF BACT, UNIV WIS-MADISON, 52- *Concurrent Pos:* Mem expert comt food hyg, WHO, 67-; mem, Nat Adv Food & Drug Comn, Food & Drug Admin, 72-76; mem, Food & Nutrit Bd, Nat Acad Sci-Nat Res Coun, 73-76. *Honors & Awards:* Nicholas Appert Award Medal, Inst Food Tech, 69. *Mem:* Am Soc Microbiol (secy, 57-61, pres, 69-70); Am Acad Microbiol (secy-treas, 62-65, pres, 66-67); Inst Food Tech. *Res:* Food borne disease; safety of food additives. *Mailing Add:* Food Res Inst Univ Wis 1925 Willow Dr Madison WI 53706

FOSTER, EDWIN POWELL, b Louisville, Ky, May 7, 42; m 66; c 3. STRUCTURAL ENGINEERING, HIGHWAY ACCIDENT RECONSTRUCTION. *Educ:* Vanderbilt Univ, BE, 64, MS, 66, PhD(struct eng), 74. *Prof Exp:* Field engr, Am Bridge Div, US Steel, 64; engr, Brown Eng Co, 65-66; res asst struct eng, Univ Ill, 67-68; teaching asst eng, Vanderbilt Univ, 67-68; assoc prof eng & coordr, Univ Tenn, Nashville, 68-79; PROF ENG & DIR, UNIV TENN, CHATTANOOGA, 79- *Concurrent Pos:* Adj assoc prof, Vanderbilt Univ, 75; res assoc, NASA Langley Res Ctr, 77 & 78; consult, Nashville Bridge Co, John Carpenter & Assoc & Taylor & Cabtree, Architects. *Mem:* Am Soc Civil Engrs; Am Soc Eng Educ; Nat Soc Prof Engrs. *Res:* Structural analysis by the finite element method; analysis and design of geometrically nonlinear cable structures; structural model testing; bridge restoration and structural testing. *Mailing Add:* 507 Highbury Lane Chattanooga TN 37343

FOSTER, ELLIS L(OUIS), b Louisville, Ky, Oct 2, 23; m 50; c 2. METALLURGICAL ENGINEERING. *Educ:* Univ Ky, BS, 50. *Prof Exp:* Prin metallurgist, Battelle Mem Inst, 50-56, asst div chief, 56-66, dept tech adv, 66-71, sr adv, 71-73, prog mgr, Thermal Protection System, 73-75; res staff, Inst for Defense Anal, 75-77, mgr, Defense/Space Syst Mat Requirements Prog Off, 77-87, SR RES ENG, DEFENSE & AEROSPACE MAT GROUP, BATTELLE COLUMBUS LABS, 87- *Mem:* Am Soc Metals; Am Vacuum Soc; Am Inst Aeronaut & Astronaut; Am Inst Mining & Metall Engrs. *Res:* Materials and processing research as it applies to reactive and refractory composite materials for use in nuclear, defense, aerospace, and industrial applications. *Mailing Add:* Battelle Columbus Div 505 King Ave Columbus OH 43201-2693

FOSTER, EUGENE A, b New York, NY, Apr 26, 27; m 52; c 3. PATHOLOGY. *Educ:* Washington Univ, AB, 47, MD, 51. *Prof Exp:* Intern, Salt Lake County Gen Hosp, 51-52; resident path, Peter Bent Brigham Hosp, Boston, 55-58; fel surg path, Barnes Hosp, St Louis, 58-59; from asst prof to prof path, Sch Med, Univ Va, 59-76; PROF PATH, SCH MED, TUFTS UNIV, 76- *Concurrent Pos:* Teaching fel, Harvard Med Sch, 57-58; mem exec comt end results group, Nat Cancer Inst, 62-67, chmn, 65-67; Nat Inst Allergy & Infectious Dis res career develop award, 67-72; res worker, Sir William Dunn Sch Path, Univ Oxford, 67-68. *Mem:* Am Asn Pathologists. *Res:* Natural history and epidemiology of cancer; experimental urinary tract infection; pathogenesis of staphylococcal infections. *Mailing Add:* Dept Path Tufts Univ Sch Med New England Med Ctr 171 Harrison Ave Boston MA 02111

FOSTER, EUGENE L(EWIS), b Clinton, Mass, Oct 9, 22; m 44; c 4. MECHANICAL ENGINEERING. *Educ:* Univ NH, BS, 44, MS, 51; Mass Inst Technol, MechE, 53, ScD, 54. *Prof Exp:* Proj res engr, Procter & Gamble Co, 46; instr mech eng, Univ NH, 47-49; asst prof, Mass Inst Technol, 50-56; pres, Foster-Miller Assocs, Inc, 56-73; PRES, UTD CORP, 76-, CHIEF EXEC OFFICER & CORP RES OFFICER, 80- *Concurrent Pos:* Instr mech, US Mil Acad, 52-56; chmn, Foster-Miller Assocs, Inc, 56- *Mem:* AAAS; Am Soc Eng Educ; Am Soc Mech Engrs; Am Soc Testing & Mat; NY Acad Sci; Sigma Xi. *Res:* Heat transfer; thermodynamics. *Mailing Add:* 3316 Wessynton Way Alexandria VA 22309

FOSTER, GEORGE A, JR, b Stockton, Kans, Aug 31, 38; m 58; c 3. PATENT LIAISON, SCIENCE INFORMATION. *Educ:* Univ Colo, BA, 60; Univ Wis, MS, 62, PhD(biochem), 66. *Prof Exp:* Sr res biochemist, Gen Mills, Inc, 66-69; info scientist, Miles Labs, Inc, 69-76, patent liaison specialist, 76-78, supr stability & automation, 78-81, supvr sci & prod info libr, 81-89; SECT HEAD PATENT INFORMATION, ABBOTT LABS, 89- *Concurrent Pos:* Patent agent, 77- *Mem:* Am Chem Soc. *Res:* Computerized data bases. *Mailing Add:* D441 AP6D Abbott Labs One Abbott Park Rd Abbott Park IL 60064-3500

FOSTER, GEORGE RAINEY, b Fayetteville, Tenn, July 4, 43; m 69; c 2. SOIL EROSION, ENVIRONMENTAL QUALITY. *Educ:* Univ Tenn, Knoxville, BS, 66; Purdue Univ, WLafayette, Ind, MS, 68, PhD(agr eng), 75. *Prof Exp:* Res hydraul engr, USDA-Agr Res Serv, 67-87; PROF & HEAD AGR ENGR, DEPT AGR ENG, UNIV MINN, 87- *Concurrent Pos:* Vis scientist environ sci, Los Alamos Nat Lab, 82-83. *Honors & Awards:* Super Serv Team Award, US Dept Agr, 81, 90. *Mem:* Am Soc Agr Engrs; Soil & Water Conserv Soc; Am Geophys Union; Sigma Xi. *Res:* Processes of soil erosion by water; develop mathematical model of soil erosion that are used by research scientist and action agencies; control research on erosion control methods. *Mailing Add:* Dept Agr Eng Univ Minn 1390 Eckles Ave St Paul MN 55108

FOSTER, GIRAUD VERNAM, b New York, NY, Jan 13, 28; m 52; c 3. ENDOCRINOLOGY. *Educ:* Trinity Col, BS, 52; Univ Md, MD, 56; Univ London, PhD(biochem), 66. *Prof Exp:* Intern, Univ Md Hosp, Baltimore, 56-57, from jr asst resident to sr asst resident med, 57-60; fel physiol chem, Sch Med, Johns Hopkins Univ, 60-62; personal physician to His Majesty Imam Ahmed of Yemen, 62; post med adv to Dept State, Tiaz, Yemen, 62-63; res asst, Dept Med, Royal Postgrad Med Sch, London, 63-64, hon asst lectr, Dept Chem Path, 64-66; hon clin asst med, Hammersmith & St Mark's Hosp, London, 65-68; vis prof med, Univ Ala, Birmingham, 68-72; ASSOC PROF

GYNEC, OBSTET, PHYSIOL & MED, SCH MED, JOHNS HOPKINS UNIV, 72- *Concurrent Pos:* Am Heart Asn grants-in-aid, 66-73 & 74-76; Med Res Coun grants, 68-71; NIH grants-in-aid, 72-76; sr lectr, Dept Chem Path, Wellcome Unit Endocrinol & hon sr lectr chem path, Royal Postgrad Med Sch, London, 70- *Mem:* Soc Endocrinol; Biochem Soc; Endocrine Soc; Am Physiol Soc; Am Soc Biol Chemists. *Res:* Hormonal regulation of calcium metabolism; prolactin secretion. *Mailing Add:* Dept Obstet & Gynec Johns Hopkins Hosp Baltimore MD 21205

FOSTER, HAROLD DOUGLAS, b Tunstall, Eng, Jan 9, 43; m 64, 90; c 3. DISASTER PLANNING, MEDICAL GEOGRAPHY. *Educ:* Univ London, BSc, 64, PhD(geog), 67. *Prof Exp:* From instr to assoc prof, 67-81, PROF GEOG, UNIV VICTORIA, BC, 81- *Concurrent Pos:* Consult, Ottawa Dept Energy, Mines & Resources, UN, NATO & Dept Fisheries & Environ, Prov Govt Ont & BC, 68-69; ed, Western Geog Series, 68- *Mem:* Brit Geog Asn; NY Acad Sci; Can Asn Geog; Am Geog Soc; Am Geophys Union. *Res:* Medical geography; natural hazards; water resources; social significance of geomorphology. *Mailing Add:* Dept Geog Univ Victoria Victoria BC V8W 2Y2 Can

FOSTER, HAROLD MARVIN, b Passaic, NJ, Aug 2, 28; m 54; c 4. INDUSTRIAL CHEMISTRY. *Educ:* Lehigh Univ, BS, 50; Univ Ill, PhD(chem), 53. *Prof Exp:* Res chemist, Am Cyanamid Co, 53-61; sr res chemist & proj leader, Mobil Chem Co, 61-66; res group leader, Colors & Chem Dept, Sherwin-Williams Co, Chicago, 66-70, dir pigments lab, Sherwin-Williams Chem Div, 70-72, sr res assoc, 72-73, com develop specialist, Sherwin-Williams Co, Toledo, 73-78; ASST TECH DIR, PIGMENTS DIV, BORDEN CHEM INC, CINCINNATI, 78- *Mem:* Am Chem Soc. *Res:* Organic synthesis; process research; agricultural chemicals. *Mailing Add:* 909 Valley Rd Carbondale IL 62901-2420

FOSTER, HELEN LAURA, b Adrian, Mich, Dec 15, 19. GEOLOGY, RECONNAISSANCE GEOLOGIC MAPPING ALASKA. *Educ:* Univ Mich, BS, 41, MS, 43, PhD(geol), 46. *Prof Exp:* Asst geol, Univ Mich, 42-44, instr, Wellesley Col, 46-48; geologist, Mil Geol Br, Tokyo, 48-55; chief, Ishigaki Field Party, 55-57 & DC, 57-65; GEOLOGIST, ALASKAN BR, US GEOL SURV, CALIF, 65- *Concurrent Pos:* Geologist, US Geol Surv, Mich, 43-45; instr, Rocky Mt Field Sta, Univ Mich, 47. *Mem:* AAAS; Fel Geol Soc Am; Am Geophys Union; Arctic Inst; Am Asn Petrol Geologists. *Res:* Volcanoes of Japan; geology of Ryukyu Islands, Yukon-Tanana Upland and East Central Alaska. *Mailing Add:* Alaskan Geol Br Geol Surv 345 Middlefield Rd MS 904 Menlo Park CA 94025

FOSTER, HENRY D(ORROH), b Timmonsville, SC, Aug 14, 12; m 42; c 6. CHEMICAL ENGINEERING. *Educ:* Univ SC, BS, 33; Univ Ill, MS, 34, PhD(chem eng), 37. *Prof Exp:* Res chem engr, E I du Pont de Nemours & Co, 37-45, develop chem engr, 45-46, res supvr, 46-47, plant process supvr, Ohio & NJ, 47-48, asst tech supt, NJ & Tex, 48, tech supt, 48-51, asst process mgr, Grasselli Chem Dept, 51-55, process mgr, 55-61, asst mgr plants tech sect, Indust & Biochems Dept, 61-72; RETIRED. *Mem:* Am Chem Soc; Am Inst Chem Engrs. *Res:* Development of chemical manufacturing processes for plastics; insecticides, herbicides, pharmaceuticals, and heavy chemicals; catalytic oxidation in the vapor phase. *Mailing Add:* 1000 Wynnewood Ave Wilmington DE 19803

FOSTER, HENRY LOUIS, b Boston, Mass, Apr 6, 25; m 48; c 3. VETERINARY MEDICINE. *Educ:* Middlesex Col, DVM, 46; dipl, Am Col Lab Animal Med, 61. *Prof Exp:* Consult vet, UNRRA, 46-47; pres dir gen, Soc Des Elevages Charles River France, SA, 64-76, PRES, CHARLES RIVER BREEDING LABS, INC, 47-, CANADA, 70-, ITALY, 71-, CHMN, UK LTD, 70- *Concurrent Pos:* Chmn husb transportation comt, Inst Lab Animal Resources Div, Nat Acad Sci-Nat Res Coun, 58-59, mem prod & stand comts, 58-59, gnotobiotics comt, 59, transportation comt, 60, mem adv coun, 59-64, mem exec comt, 63-64, vpres, Lab Animal Breeders Asn, 58-60, pres, 61-62, secy-treas, 64; mem comt nomenclature randombred animals subcomt gnotobiotes rabbit & rodent procurement stand & stand for qual lab animal feed, Inst Lab Animal Resources, 67; vpres & chmn exec comt, Century Bank & Trust Co, Somerville, Mass, 69-; trustee, Beaver Country Day Sch, 73-75; trustee & fel, Brandeis Univ & mem bd overseers, Rosenstiel Basic Med Sci Res Ctr, 73- *Mem:* Am Vet Med Asn; Am Inst Biol Sci; NY Acad Sci; Am Asn Lab Animal Sci; Am Col Lab Animal Med. *Res:* Commercial production of caesarean-originated, barrier-sustained and gnotobiotic rats, mice, rabbits, hamsters, guinea pigs and non-human primates; production methods; elimination of specific diseases; environmental control. *Mailing Add:* Charles River Labs 251 Ballardvale St North Wilmington MA 01887

FOSTER, HENRY WENDELL, b Pine Bluff, Ark, Sept 8, 33; m 60; c 2. OBSTETRICS & GYNECOLOGY. *Educ:* Morehouse Col, BS, 54; Univ Ark, MD, 58; Am Bd Obstet & Gynec, dipl, 67, 79 & 89. *Prof Exp:* Intern, Receiving Hosp, Detroit, Mich, 58-59; resident, Malden Hosp, Mass, 61-62 & Hubbard Hosp, Nashville, Tenn, 62-65; chief obstet & gynec, John A Andrew Mem Col, 65-73; asst prof, 67-73, prof obstet & gynec & chmn dept, 73-90, DEAN, SCH MED, MEHARRY MED COL, 90-, VPRES HEALTH SERV, 90-; CLIN PROF, DEPT OBSTET & GYNEC, VANDERBILT UNIV, NASHVILLE, TENN, 75- *Concurrent Pos:* Attend, Macon County Hosp & consult staff, Vet Admin Hosp, Tuskegee, Ala; dir, Maternity & Infant Care Proj 556, John A Andrew Mem Hosp, 70-80; mem bd dirs, Planned Parenthood Fedn Am, 75-; exmnr, Am Bd Obstet & Gynec, 76-; mem, Ethics Adv Bd, 77-; mem, Obstet & Gynec Sect, Nat Med Asn, Comt Maternal & Child Health Coun; mem, Nat Adv Child Health & Human Develop Coun, Dept Health & Human Serv, Off Secy, 91. *Mem:* Inst Med-Nat Acad Sci; fel Am Col Obstet & Gynec; AMA; Nat Med Asn; Am Asn Univ Professors; Am Fertil Soc. *Res:* Electron microscopic, findings in placentas of sickle cell anemia patients; nutrition; relationships among intake and toxemia of pregnancy; study of lipids in human amnion and chorion; gestational variation of amniotic fluid phospholipids. *Mailing Add:* Off Dean Sch Med Meharry Med Col 1005 Todd Blvd Nashville TN 37208

FOSTER, IRVING GORDON, b Lynn, Mass, July 15, 12; m 37; c 3. MECHANICS. *Educ:* Va Mil Inst, BS, 35; Univ Wis, PhM, 37; Univ Va, PhD(physics), 48. *Prof Exp:* Res assoc, Weshburn Observ, Univ Wis, 36-37; from instr to prof physics, Va Mil Inst, 37-59; chmn div math & nat sci, Fla Presby Col, 60-73, prof, 60-78, PROF EMER PHYSICS, ECKERD COL, 78- *Concurrent Pos:* Fulbright lectr, Vidyodaya Univ Ceylon; consult, Honeywell, Inc. *Mem:* Am Phys Soc; Am Asn Physics Teachers; Sigma Xi. *Res:* Theory and design of mechanical devices; coherent radiation. *Mailing Add:* Suncoast Manor Apt BS9 6909 Ninth St S St Petersburg FL 33705-6240

FOSTER, J EARL, b New Albany, Ind, Feb 21, 29; m 55; c 4. MECHANICAL ENGINEERING. *Educ:* US Merchant Marine Acad, BS, 50; Univ Iowa, MS, 55, PhD(mech eng), 58. *Prof Exp:* Engr, Collins Radio Co, Iowa, 57-61; prof mech eng, Univ Wyo, 61-68; prof mech, Univ Mo-Rolla, 68-77; LAB MGR, MOTOROLA COMMUN & ELECTRONICS, 77- *Mem:* Am Soc Mech Engrs; Am Soc Eng Educ. *Res:* Vibrations; dynamics. *Mailing Add:* 6451 Turtle Rock Terr Hialeah FL 33014

FOSTER, JAMES HENDERSON, b Chicago, Ill, Feb 21, 46. MATHEMATICS. *Educ:* Northwestern Univ, BA, 68; Univ Southern Calif, PhD(math), 76. *Prof Exp:* Mem tech staff eng, Autonetics Div, Rockwell Int, Inc, 68-70; asst prof math, Univ Santa Clara, 76-80; RES SCIENTIST, LOCKHEED MISSILES & SPACE CO, 80- *Concurrent Pos:* Instr, Univ Calif, Santa Barbara, 76. *Mem:* Am Math Soc; Soc Indust & Appl Math. *Res:* Optimal control theory and differential equations. *Mailing Add:* 4211 Liberty Ogden UT 84403

FOSTER, JAMES HENRY, b New Haven, Conn, June 24, 30; m 55; c 3. SURGERY. *Educ:* Haverford Col, AB, 50; Columbia Univ, MD, 54. *Prof Exp:* Intern surg, Barnes Hosp, St Louis, Mo, 54-55; asst surg resident, Portland Vet Hosp, Ore, 57-59, chief resident combined serv, 60-61, asst chief surg serv, 61-66; asst, Sch Med, Univ Ore, 59-60, from instr to asst prof surg, 61-66; chief surg, Hartford Hosp, 66-78; chmn dept, 78-84, dir surg, univ hosp, 78-86, PROF SURG, SCH MED, UNIV CONN, 72- *Concurrent Pos:* Consult, Rocky Hill Vet Hosp & Newington Childrens Hosp, 66-, Manchester Mem Hosp, 67- & Norwalk Hosp, 74-; assoc prof surg, Sch Med, Univ Conn, 68-73; examr, Am Bd Surg, 74- & dir, 77-83. *Mem:* Soc Surg Alimentary Tract; Am Bd Surg; Am Col Surg; Am Surg Asn; Soc Surg Oncol; Int Soc Surg. *Res:* Benign and malignant liver tumors; portal hypertension. *Mailing Add:* Dept Surg Univ Conn Health Ctr Farmington CT 06032

FOSTER, JOHN EDWARD, b Lesterville, Mo, May 22, 40; c 1. ENTOMOLOGY, PLANT BREEDING. *Educ:* Cent Methodist Col, BA, 64; Univ Mo, MS, 66; Purdue Univ, PhD, 70. *Prof Exp:* Res entom, Agr Res Serv, USDA, 70-76; from asst prof to prof, Agr Exp Sta, Purdue Univ, 70-89; PROF & HEAD, DEPT ENTOMOL, UNIV NEBR, LINCOLN, 90- *Mem:* Entom Soc Am; Sigma Xi; Am Soc Agron. *Res:* Development of insect resistant plants; interrelationship between host plant and insect biotypes; genetics of plant resistance and insect virulence; population management by genetic manipulation; insect genetics. *Mailing Add:* 202 PI Bldg Univ Nebr Lincoln NE 68583-0816

FOSTER, JOHN HOSKINS, surgery; deceased, see previous edition for last biography

FOSTER, JOHN MCGAW, b Philadelphia, Pa, Apr 22, 28; m 51; c 2. BIOCHEMISTRY. *Educ:* Swarthmore Col, BA, 50; Harvard Univ, MA, 53, PhD(biochem), 54. *Prof Exp:* Res assoc biol, Mass Inst Technol, 54-56; asst prof biochem, Sch Med, Boston Univ, 58-67; asst prog dir, Sci Curric Improv Prog, NSF, 67-68, assoc prog dir, 68-69; assoc prof, 69-72, actg dean, Sch Nat Sci, 72-73, PROF BIOL, HAMPSHIRE COL, 72- *Concurrent Pos:* Chief clin chem sect, 406 Med Gen Lab, Japan, 56-57; asst chief, Chem Dept Res & Develop Unit, Fitzsimons Army Hosp, 58; adj prof, Univ Mass, 74- *Mem:* AAAS; Am Inst Biol Sci. *Res:* Biochemical control mechanisms; biosynthesis, structure and control of bacterial photosynthetic membranes. *Mailing Add:* Dept Natural Sci Hampshire Col Amherst MA 01002

FOSTER, JOHN ROBERT, b Washington, DC, Apr 6, 47; m 69; c 2. MARINE BIOLOGY, FISHERIES BIOLOGY. *Educ:* Univ Md, BSc, 69; Univ Toronto, MSc, 71, PhD(behav ecol), 80. *Prof Exp:* Lab technician behav ecol, Dept Zool, Univ Md, 66-69, Univ Toronto, 69-75; environ consult fisheries, Biol Dept, James F MacLaren Ltd, 74-80; environ consult environ protection, Ontario Hydro Ltd, 78-80, fisheries biologist, Res Div, 80-81; RES ED & ED RES COORDR MARINE BIOL, HUNTSMAN MARINE LAB, 81- *Mem:* Animal Behav Soc; Ecol Soc Am; Nat Geog Soc. *Res:* Studies of predator-prey relations of moon snails and grass pickerel; reproductive behavior in sticklebacks; factors influencing fish entrapment; impringement and entrainment at power plants; stream surveys and the life history of hag fish. *Mailing Add:* Dept Mktg Univ Tex El Paso TX 79968

FOSTER, JOHN STUART, JR, b New Haven, Conn, Sept 18, 22; c 4. ENGINEERING MECHANICS. *Educ:* McGill Univ, Can, BS, 48; Univ Calif, Berkeley, PhD(physics), 52. *Hon Degrees:* DSc, Univ Mo, 79. *Prof Exp:* Div leader exp physics, Lawrence Livermore Nat Lab, 52-58, assoc dir, 58-61, dir, 61-65; dir defense res & eng, US Dept Defense, 65-73; vpres energy res & develop, Energy Systs Group, 73-79, vpres sci & technol, 79-88, CONSULT, TRW, INC, 88- *Concurrent Pos:* Chmn, Defense Sci Bd; mem, Air Force Sci Adv Bd, 56, Army Sci Adv Panel, 58, Ballistic Missile Defense Adv Comt, Advan Res Projs Agency, 65 & President's Foreign Intel Adv Bd, 73-90; panel consult, President's Sci Adv Comt, 65. *Honors & Awards:* Ernest Orlando Lawrence Mem Award, Atomic Energy Comn, 60; James Forrestal Mem Award, 69; H H Arnold Trophy, 71; Crowell Medal, 72; Knight Commander's Cross of the Order of Merit, Fed Repub Ger, 74. *Mem:* Nat Acad Eng; Am Defense Preparedness Asn; Nat Security Indust Asn; Am Inst Aeronaut & Astronaut. *Mailing Add:* One Space Park Redondo Beach CA 90278

FOSTER, JOHN WEBSTER, b Concord, NH, Oct 21, 23; m 48, 69; c 4. GEOHYDROLOGY, GROUNDWATER EXPLORATION. *Educ:* Dartmouth Col, AB, 47; Ohio State Univ, MS, 50. *Prof Exp:* Investr, William F Guyton & Assocs, Austin, Tex, 55-57; consult & dir, Asn Stratig Serv, Los Angeles, 57-61; dir, Water Resources, Litton Industs, Beirut, Lebanon, 61-64; proj mgr, Wilson-Murrow Ltd, Jeddah, Saudi Arabia, 64-89; consult geohydrol, Foster & Assocs, Beirut, Lebanon, 70-79; sr vpres H D R Geothermal, H D R Energy Develop Corp, Augusta, Ga, 79-82; ASSOC PROF GEOHYDROL, ILL STATE UNIV, 82- *Concurrent Pos:* Assoc geologist, Ill State Geol Survey, 49-55; actg dept chairperson, Dept Geog-Geol, Ill State Univ, 87-89; geol grad prog coordr, MS in Geohydrol, Ill State Univ, 91. *Honors & Awards:* Exemplary Serv Award, Govt Kuwait, 60. *Mem:* Asn Ground Water Scientists & Engrs; Nat Asn Geol Teachers. *Res:* Author of three United States patents, one on heat extraction from hot dry-rock geothermal source, two on advanced technology downhole instruments; innovative analytical techniques for aquifer evaluations or contamination sources. *Mailing Add:* 1207 Stephens Dr Normal IL 61761

FOSTER, JOYCE GERALDINE, b Farmville, Va, Oct 10, 51. FORAGE LEGUMES, TOXICOLOGY. *Educ:* Longwood Col, BS, 74; Va Tech Inst, MS, 76, PhD(biochem) & PhD(nutrit), 79. *Prof Exp:* Res assoc, dept biochem & nutrit, Va Tech Inst, 79; proj assoc, dept hort, Univ Wis, Madison, 80; res assoc, dept bot, Wash State Univ, 81; RES BIOCHEMIST, LEAD SCIENTIST PLANT RES, USDA-ARS, APPALACHIAN SOIL & WATER CONSERV RES LAB, 82-, ACTG LAB DIR, 86-87. *Concurrent Pos:* Grad res asst, dept biochem & nutrit, Va Tech, 74-77, grad teaching asst, 78; adj prof, dept plant path, physiol & weed sci, Va Polytech Inst & State Univ, 83; vchmn, Am Soc Plant Physiologists, 90-91, chmn, 91-92. *Honors & Awards:* J Shelton Horsley Res Award, Va Acad Sci, 82. *Mem:* Am Chem Soc; Am Soc Plant Physiologists (secy-treas, 89-90); Japanese Soc Plant Physiologists; Am Soc Biochem & Molecular Biol; Am Soc Agron; Scandinavian Soc Plant Physiol. *Res:* Regulatory mechanisms for biochemical constituents affecting legume performance and utilization; the ecological biochemistry of the soil-root interface; the biochemistry of plant adaptation for problem environments; ruminant toxicology; rumen metabolism and ruminant toxicology. *Mailing Add:* Appalachial Soil & Water Conserv Res Lab USDA-ARS Airport Rd PO Box 867 Beckley WV 25802-0867

FOSTER, KEN WOOD, b Trenton, NJ, Apr 12, 50; m 72; c 1. CROP GENETICS, BIOMETRY. *Educ:* Colo State Univ, BS, 72; Univ Calif, Davis, MS, 74, PhD(genetics), 76. *Prof Exp:* Asst geneticist crops, Univ Calif, Riverside, 76-80, ASST GENETICIST CROPS, UNIV CALIF, DAVIS, 80- *Mem:* Crop Sci Soc Am; Am Soc Agron; Am Soc Plant Physiologists; Genetics Soc Am; Sigma Xi. *Res:* Improvement of selection methods for quantitative characters. *Mailing Add:* 5364 Orrville Ave Woodland Hills CA 91364

FOSTER, KENNETH WILLIAM, b Victoria, BC, Jan 8, 44; m 69, 83; c 3. SENSORY PHYSIOLOGY. *Educ:* Univ Victoria, BSc Hons, 65; Calif Inst Technol, PhD(biophysics), 72. *Prof Exp:* Res fel, Biol Div, Calif Inst Technol, 72; res assoc, Dept Molecular, Cellular & Develop Biol, Univ Colo, 72-79; asst prof, Mt Sinai Sch Med, 80-84; ASSOC PROF, SYRACUSE UNIV, 84- *Concurrent Pos:* Lectr biophysics, Dept Molecular, Cellular & Develop Biol, Univ Colo, 79. *Mem:* Biophys Soc; Am Soc Photobiologists; NY Acad Sci; AAAS. *Res:* Spectroscopic, behavioral, electrophysiological, biochemical and genetic analysis of primitive photoreceptors and their associated signaling pathways, primarily using the rhodopsin-based visual system of the alga chlamydomonas. *Mailing Add:* Dept Physics Syracuse Univ Syracuse NY 13244-1130

FOSTER, KENNITH EARL, b Lamesa, Tex, Jan 20, 45; m 67; c 2. ARID LANDS ECOLOGY, AGRICULTURAL ENGINEERING. *Educ:* Tex Tech Univ, BS, 67; Univ Ariz, MS, 69, PhD(watershed mgt), 72. *Prof Exp:* Res asst agr eng, 67-69, res assoc watershed mgt, 70-71, from asst dir to assoc dir, 71-82, ASST PROF & DIR, OFF ARID LANDS STUDIES, UNIV ARIZ, 82- *Mem:* AAAS. *Res:* Possible uses of arid land adapted vegetation; natural resource management using remote sensing; urban water conservation. *Mailing Add:* Off Arid Lands Studies Univ Ariz Tucson AZ 85721

FOSTER, KENT ELLSWORTH, b Evergreen Park, Ill, Nov 30, 43; m 71; c 2. MATHEMATICAL ANALYSIS. *Educ:* St Lawrence Univ, NY, BS, 66; Southern Ill Univ, Carbondale, MS, 71, PhD(math), 75. *Prof Exp:* Asst prof math, Mt Allison Univ, 75-78; asst prof math, Emory Univ, 78-79; ASST PROF COMPUT SCI, WINTHROP COL, 79- *Mem:* Am Math Soc; Sigma Xi. *Res:* Oscillation properties of ordinary differential equations; statistical sampling techniques and simulation. *Mailing Add:* Dept Bus Admin Winthrop Col Rock Hill SC 29733

FOSTER, LEIGH CURTIS, spectroscopy, for more information see previous edition

FOSTER, LORRAINE L, b Los Angeles, Calif, Dec 25, 38; m 59; c 3. MATHEMATICS. *Educ:* Occidental Col, BA, 60; Calif Inst Technol, PhD(math), 64. *Prof Exp:* From asst prof to assoc prof, 64-72, PROF MATH, CALIF STATE UNIV, NORTHRIDGE, 72- *Mem:* Am Math Soc; Math Asn Am; Sigma Xi; Asn Women Math. *Res:* Elementary and algebraic number theory; matrix theories; geometry. *Mailing Add:* 8735 White Oak Ave Northridge CA 91325-3126

FOSTER, M GLENYS, b Cwm Penmachno, Wales; Can citizen. MASS SPECTROMETRY, CHROMATOGRAPHY. *Educ:* Univ London, Eng, BSc, 68; Univ Wales, MSc, 73, PhD, 84. *Prof Exp:* Biochemist, Wellesley Hosp, 74-78; chief, Mass Spectrometry Serv, Lab Serv Br, Ministry Environ, 81-85; LAB MGR, ZENON ENVIRON LABS INC, 85- *Mem:* Am Soc Mass Spectrometry. *Res:* Analytical methodology for the characterization of environmental samples; techniques used include gas chromatography-mass spectrometry with electron impact and chemical ionisation modes detecting both positive and negative ions; comparative studies with liquid chromatography-mass spectrometry. *Mailing Add:* Zenon Environ Labs Inc 5555 N Service Rd Burlington ON L7L 5H7 Can

FOSTER, MARGARET C, b Mar 24, 35. BIOPHYSICS. *Educ:* Univ Richmond, BS, 57; Univ Wis, MS, 60, PhD, 65. *Prof Exp:* Researcher, Europ Orgn Nuclear Res, Geneva, 65-67 & Nuclear Physics Res Lab, Univ Liverpool, Eng, 67-68; asst prof physics, State Univ NY Stony Brook, 68-72, lectr physiol & biophys, 72-73; NIH spec res fel, Univ Calif, San Diego, 73-75, asst res biologist, 75-82; intergovt personnel exchange fel, 80-82, expert & res physicist, Lab Chem Physics, Nat Inst Arthritis, Diabetes, Digestive & Kidney Dis, NIH 82-87; PHYSICIST, BIOMED ENG & INSTRUMENTATION BR, DIV RES SERV, NIH, 87- *Concurrent Pos:* Spec res fel, NIH, 72; acad career develop award, Nat Eye Inst, 75. *Mem:* Am Phys Soc; Biophys Soc; AAAS; Microbeam Anal Soc; Electron Micros Soc Am. *Res:* Biophysical studies of membranes from vertebrate and invertebrate photoreceptors; x-ray microanalysis to determine elemental composition of individual cells and cell organelles. *Mailing Add:* NIH Nat Inst Diabetes & Digestive & Kidney Dis Lab Chem Physics Bldg 2 Rm B2-26 Bethesda MD 20892

FOSTER, MERRILL W, b South Gate, Calif, Mar 18, 39; m 89; c 3. INVERTEBRATE ZOOLOGY, MARINE BIOLOGY. *Educ:* Univ Calif-Berkeley, MA, 64; Harvard Univ, PhD(geol), 70. *Prof Exp:* Teaching asst paleont, Univ Calif-Berkeley, 62-64; teaching fel geol, Harvard Univ, 64-66; from instr to assoc prof, PROF, DEPT GEOL SCI, BRADLEY UNIV, 69- *Concurrent Pos:* Asst investr, sponsored study Antarctic Recent Brachiopods, NSF, 66-69; instr geol & natural hist, Sun Found, 75-90; lead teacher, Funded Field Sci Proj Elem Sch Teacher, NSF, 79-81; pres, Peoria Area Sigma Xi, 83-84; ed, Peoria Acad Sci Proceedings, 84-85; instr, State Funded Sci Concepts, Inst Elem Sch Teachers, 89-90; instr, Sci Lit Coop-Bureau Co, Ill, 90-91; sci-in-residence, East Peoria Sch Dist, 90. *Mem:* Int Paleont Asn; Paleont Soc; Sigma Xi; Paleont Res Inst; Soc Econ Paleontologists & Mineralogists; AAAS. *Res:* The biology of recent brachiopods-particularly those in Antarctica, Subantarctica, Northeast Pacific, Azores, South Atlantic and Extreme South Pacific; the paleobiology of Early Paleozoic (particularly Ordovician) and Pennsylvanian Marine Invertebrates-particularly receptaculitids, soft-bodied Cnidarians, Tullimonstrum, brachiopods, trilobites and echinoderms. *Mailing Add:* Dept Geol Sci Bradley Univ Peoria IL 61625

FOSTER, MICHAEL RALPH, b Indianapolis, Ind, Jan 12, 43; m 66; c 2. FLUID DYNAMICS, AERODYNAMICS. *Educ:* Mass Inst Technol, BS, 65, MS, 66; Calif Inst Technol, PhD(aeronaut eng), 69. *Prof Exp:* Instr math, Mass Inst Technol, 69-70; asst prof, 70-80, assoc prof, 80-86, PROF AERONAUT & ASTRONAUT ENG, OHIO STATE UNIV, 86- *Mem:* Am Phys Soc. *Res:* Theoretical fluid dynamics, especially aspects of geophysical fluid dynamics, rotating and/or stratified fluids, singular perturbation theory; stability theory. *Mailing Add:* Dept Aeronaut & Astronaut Eng 2036 Neil Ave Mall Columbus OH 43210-1276

FOSTER, MICHAEL SIMMLER, b Los Angeles, Calif, Sept 4, 42; c 1. MARINE BIOLOGY, PHYCOLOGY. *Educ:* Stanford Univ, BS, 64, MA, 65; Univ Calif, Santa Barbara, PhD(biol), 72. *Prof Exp:* Res assoc marine bot, Marine Sci Inst, Univ Calif, Santa Barbara, 71-72; asst prof biol sci, Calif State Univ, Hayward, 72-76; assoc prof, 76-81, PROF MARINE SCI, MOSS LANDING MARINE LABS, 81- *Concurrent Pos:* Consult oil pollution effects on marine pop, Off Atty Gen, State of Calif, 74, Environ Protection Agency, 90. *Mem:* AAAS; Phycol Soc Am; Ecol Soc Am; Int Phycol Soc; Brit Phycol Soc. *Res:* Population and community ecology of marine macroalgae and the effects of oil pollution on marine organisms. *Mailing Add:* Moss Landing Marine Labs PO Box 450 Moss Landing CA 95039

FOSTER, NEAL ROBERT, b Nyack, NY, Aug 13, 37; m 80. EMBRYOLOGY, ENDOCRINOLOGY. *Educ:* Cornell Univ, BA, 59, MS, 61, PhD, 67. *Prof Exp:* Teaching asst zool, Cornell Univ, 58-59, curatorial asst fish collection, 59-60, asst vertebrate zool, 60-65; asst cur, Dept Limnol, Acad Natural Sci, Philadelphia, 65-76; proj leader, Physiol & Behav, 77-84, FISH BIOL RES, NAT FISHERIES RES CTR, US FISH & WILDLIFE SERV, GREAT LAKES, 84-; ADJ ASST PROF, SCH NAT RESOURCES, UNIV MICH, ANN ARBOR, 88- *Concurrent Pos:* Pres, Mich Chap, Am Fisheries Soc, 87-88, treas, Early Life Hist Sect, 88- *Honors & Awards:* James W Moffett Award, 86. *Mem:* Am Fisheries Soc; Int Asn Great Lakes Res; Am Soc Ichthyol & Herpetol; AAAS; fel Am Inst Fish Res Biologists. *Res:* Behavioral ecology, morphology and early life history of fishes; aquaculture; plasma hormones and reproductive biology and behavior of lake trout; fish pheromones and chemoreception; Great Lakes fish ecology. *Mailing Add:* US Fish & Wildlife Serv 1451 Green Rd Ann Arbor MI 48105-2899

FOSTER, NEIL WILLIAM, b Guelph, Ont, Nov 7, 45; m 69; c 3. SOIL SCIENCE, FORESTRY. *Educ:* Univ Toronto, BScF, 68; Univ Wash, MSc, 71; Univ Guelph, PhD, 79. *Prof Exp:* Forestry officer res, 68-79, RES SCIENTIST, FORESTRY CAN ONT REGION, 80- *Concurrent Pos:* Assoc ed, Can J Forest Res, 85-90. *Mem:* Soil Sci Soc Am; Can Soc Soil Sci; Can Inst Forestry. *Res:* Soil-plant relationships; biogeochemical cycling of nutrients; nitrogen in forest ecosystems; decomposition of plant materials in forest soils in relation to nutrient availability and biological activity; acid rain effects on forest soils and nutrient cycling in forest ecosystems. *Mailing Add:* Forestry Can Ont Region PO Box 490 Sault Ste Marie ON P6A 5M7 Can

FOSTER, NORMAN CHARLES, b Portland, Ore, June 18, 47; m 85; c 2. SULFONATION. *Educ:* Ore State Univ, BS, 69; Univ Wash, PhD(chem eng), 79. *Prof Exp:* Res engr, Atlantic Richfield Hanford Co, 69-73; res asst, Univ Washington, 73-78; MGR RES & DEVELOP, CHEMITHON CORP, 78- *Concurrent Pos:* Lectr sulfonation, Ctr Prof Advan, 81- *Mem:* Am Oil Chemists Soc. *Res:* Development of equipment and technology for sulfonation of organic raw materials with gaseous sulfur trioxide. *Mailing Add:* 4818 SW Genesee Seattle WA 98116-0572

FOSTER, NORMAN FRANCIS, semiconductor technology, for more information see previous edition

FOSTER, NORMAN GEORGE, b Chicago, Ill, Dec 23, 19; m 44, 85; c 5. PHYSICAL CHEMISTRY, MASS SPECTROMETRY. *Educ:* Univ Chicago, BS, 42; Univ Ark, MS, 54, PhD(phys chem), 55. *Prof Exp:* Control chemist, Sinclair Ref Co, Ind, 42; anal chemist, Dearborn Chem Co, Ill, 43; res chemist, Whiting Res Labs, Standard Oil Co, Ind, 46-51; asst, Univ Ark, 51-55; res chemist, Cities Serv Res & Develop Co, Okla, 55-57; phys chemist, Bartlesville Petrol Res Ctr, US Bur Mines, 57-65, proj leader, 63; assoc prof chem, Tex Woman's Univ, 65-85; RETIRED. *Mem:* Fel AAAS; Am Chem Soc; Am Soc Mass Spectrometry; NY Acad Sci; Fel Am Inst Chem; Coblentz Soc. *Res:* Mass spectrometric fragmentations of organo-sulfur compounds, present in petroleum; fragmentation mechanism studies by utilization of isotopically labeled molecules; characterization of complex mixtures by means of mass spectrometry; structure elucidation utilizing mass spectrometry, nuclear magnetic resonance and infrared. *Mailing Add:* 2719 Robinwood Lane Denton TX 76201

FOSTER, NORMAN HOLLAND, b Iowa City, Iowa, Oct 2, 34; m 56; c 2. GEOLOGY. *Educ:* Univ Iowa, BA, 57, MS, 60; Univ Kans, PhD(geol), 63. *Prof Exp:* From intermediate geologist to geol specialist & explor team supvr oil & gas, Sinclair Oil Corp & Atlantic Richfield Co, 62-69; from dist geologist to vpres geol explor, Trend Explor Ltd, 69-74; vpres geol explor, Filon Explor Corp, 74-79, mem bd dirs, 77-79; INDEPENDENT GEOLOGIST, 79- *Concurrent Pos:* Distinguished lectr, Am Asn Petrol Geologists, 76-77; deleg, US 28th Int Geol Cong, Washington, DC, Nat Acad Sci & US Geol Survey, 89; mem, Explor Comt, Am Petrol Inst, 88-89, Adv Comt, US Geol Survey, Nat Acad Sci & Nat Res Coun, 89-90, US Comt Geol, Nat Acad Sci & Nat Res Coun, 90-; chmn, Finance Comt of the 28th Int Geol Cong, Washington, DC, 89; pres, Am Asn Petrol Geologists, 77, trustee assoc of the Found, 79-, treas & chmn, Ins Comt, 82-84, chmn, Astrogeology Comt, 84-88; hon mem, Rocky Mountain Asn Geol, 83. *Honors & Awards:* A I Levorsen Award, Am Asn Petrol Geologists, 81, Distinguished Serv Award, 85, Spec Award, Co-editorship Treatise Petrol Geol, 91. *Mem:* Am Asn Petrol Geologists (pres, 88-89); fel Geol Soc Am; Soc Econ Paleontologists & Mineralogists; Soc Independent Prof Earth Scientists. *Res:* Oil and gas exploration in Indonesia and in the Rocky Mountain region and other areas of the United States and the world; use of remote sensing and photogeomorphology in oil, gas and mineral exploration; co-ed, treatise of petroleum geology. *Mailing Add:* 1625 Broadway Suite 530 Denver CO 80202

FOSTER, PERRY ALANSON, JR, b Manchester, NH, May 18, 25; m 47; c 5. PHYSICAL CHEMISTRY. *Educ:* St Anselm's Col, BA, 50; Ga Inst Technol, MS, 55. *Prof Exp:* Asst chem, Ga Inst Technol, 50-52, eng exp sta, 52-55; res chemist, Alcoa Labs, Aluminum Co Am, 55-67, sr scientist, 67-76, staff scientist, 76-88; RETIRED. *Concurrent Pos:* Adj prof, Va Polytech Inst & State Univ, 75-78. *Mem:* Res Soc Am. *Res:* High temperature chemistry; molten salts; phase equilibria in non-metallic systems; electrochemical and process development. *Mailing Add:* 295 Ferledge Dr New Kensington PA 15068

FOSTER, RAYMOND ORRVILLE, b Gary, Ind, Sept 25, 21; m 43; c 6. CLINICAL BIOCHEMISTRY, REPRODUCTIVE BIOLOGY. *Educ:* Univ Chicago, BS, 50. *Prof Exp:* Res biochemist, Armour Pharmaceut Co, 50-52; asst tech dir, Biol Testing, Rosner-Hixson Labs, 52-60; dir res clin biochem, Weston Labs, Gen Tire & Rubber Co, 60-73; DIR DIAG, FOSTER RES, 74- *Mem:* Am Chem Soc. *Res:* Basic and clinical research in the development of diagnostic tests. *Mailing Add:* 1202 Clarage St Joliet IL 60436

FOSTER, RICHARD B(ERGERON), b Springdale, Wash, Nov 21, 16; m 55; c 2. ENGINEERING. *Educ:* Univ Calif, BA, 38. *Prof Exp:* Asst supvr, Douglas Aircraft Co, 40-42; dir planning div, Am Aviation, 42-43; chief indust engr, Globe Aircraft Co, 43-44; estimator, Consol Vultee Aircraft Corp, 44; mem staff, Kaiser Shipyards, 44; dist invest chief, US Govt Agencies, 44-47; partner, Foster & Derian, Consults, 47-51, 53-54; exec asst & gen mgr, Marquardt Aircraft, 51-53; DIR STRATEGIC STUDIES CTR, STANFORD RES INST INT, 54- *Mem:* Fel AAAS; Opers Res Soc Am; Sigma Xi; Inst Strategic Studies. *Res:* Philosophy and civil engineering; cost-effectiveness analysis for strategic decisions; values, power and strategy; methods of interdisciplinary research in strategic analysis; comparative strategy. *Mailing Add:* 1101 Arlington Ridge Rd Apt 1204 Arlington VA 22202

FOSTER, RICHARD N(ORMAN), b Cleveland, Ohio, June 10, 41; m 65. CHEMICAL ENGINEERING. *Educ:* Yale Univ, BE, 63, MS, 65, PhD, 66. *Prof Exp:* Tech rep int mkt develop, Union Carbide Corp, 66, int prod-mkt mgr chem mkt develop, 67-69; mgr tech & financial anal, 69-71, DIR TECHNOL MGT GROUP, TECHNOL TRANSFER, FIRE SAFETY, RES PLANNING & EVAL, ABT ASSOCS, 71- *Mem:* AAAS; Am Inst Chemists; Am Chem Soc. *Res:* Diffusion limited nonisothermal chemical reactions; surface diffusion; semiconductor catalysis; plasma physics; technological transfer; research management. *Mailing Add:* 1010 Fifth Ave New York NY 10028

FOSTER, ROBERT EDWARD, II, plant breeding; deceased, see previous edition for last biography

FOSTER, ROBERT H, b Monroe, Ga, Nov 4, 20; m 42; c 4. BIOGEOGRAPHY. *Educ:* Univ Ga, BS, 63; Brigham Young Univ, MS, 66, PhD(bot), 68. *Prof Exp:* assoc prof, 68-80, EMER PROF GEOG, WESTERN KY UNIV, 80-; TEACHER AM HIST & BOT, UTAH TECH COL, PROVO, 83- *Concurrent Pos:* Researcher, threatened & endangered plants, Brigham Young Univ, 80-83. *Mem:* Asn Am Geog; Sigma Xi. *Res:* Plant distribution in the United States intermountain region. *Mailing Add:* 769 West 200 N Provo UT 84601

FOSTER, ROBERT JOE, b Glendale, Calif, June 6, 24; m 51; c 2. BIOLOGICAL CHEMISTRY. *Educ:* Calif Inst Technol, BS, 48, PhD(chem), 52. *Prof Exp:* Res fel biol, Calif Inst Technol, 53-55; asst chemist 55-61, from asst to 55-88, EMER PROF BIOCHEM, WASH STATE UNIV, 88- *Mem:* Am Soc Biol Chem; AAAS; Sigma Xi. *Res:* Mechanisms of enzyme action and specificity; comparative enzymology. *Mailing Add:* Biochem Wash State Univ Pullman WA 99164-4660

FOSTER, ROBERT JOHN, b Cambridge, Mass, Apr 19, 29; m 51; c 2. GEOLOGY. *Educ:* Mass Inst Technol, SB, 51; Univ Wash, MS, 55, PhD(geol), 57. *Prof Exp:* Seismic computer, Geophys Serv, Inc, 51-52; instrumentation engr, Boeing Airplane Co, 53-54; asst, Univ Wash, 55-57; asst prof geol, Mont State Univ, 58-61; from assoc prof to prof, geol dept, 61-88, EMER PROF SCI, SAN JOSE STATE UNIV, 88- *Mem:* AAAS; Geol Soc Am; Am Asn Petrol Geol; Nat Asn Geol Teachers. *Res:* Petrology; structural geology. *Mailing Add:* 3605 Hilltop Rd Soquel CA 95073

FOSTER, ROBIN BRADFORD, b Los Angeles, Calif, Feb 18, 45. TROPICAL PLANT ECOLOGY, FLORISTICS. *Educ:* Dartmouth Col, AB, 66; Duke Univ, PhD(bot), 74. *Prof Exp:* Asst prof biol, Latin Am studies & evolutionary biol, Univ Chicago, 72-80; RES ASSOC, DEPT BOT, FIELD MUS OF NATURAL HIST, 77-; RES ASSOC, SMITHSONIAN TROPICAL RES INST, 78- *Concurrent Pos:* Mem fac, Orgn Trop Studies, 73, 74, 79 & 80; trop forest consult, Panama, Peru & Ecuador, 74-; lectr, Bot Inst, Univ Aarhus, Denmark, 84; res assoc, Mo Bot Garden, 83-; ecol adv, Conserv Int, 87- *Mem:* Ecol Soc Am; British Ecol Soc; Asn Trop Biol; Int Soc Trop Ecol; AAAS. *Res:* Community ecology and floristics; population biology of semelparous plants; systematics and biogeography of neotropical vascular plants; community phenology and reproductive biology in neotropical forests; heterogeneity in tropical vegetation and animal communities; forest regeneration. *Mailing Add:* Dept Bot Field Mus Natural Hist Chicago IL 60605

FOSTER, ROGER SHERMAN, JR, b Washington, DC, Jan 8, 36; m 60; c 4. SURGERY. *Educ:* Haverford Col, AB, 57; Western Reserve Univ, MD, 61; Am Bd Surg, dipl, 69. *Prof Exp:* Res fel, Roswell Park Mem Inst, 66-68; dir, Surg Res Lab, 70-81, dir, Clin Transplant Unit, 72-81, assoc prof, 73-81, PROF SURG, COL MED, UNIV VT, 81-; DIR, VT REGIONAL CANCER CTR, 84- *Concurrent Pos:* Attend surg, Med Ctr Hosp, Vt, 70-; mem, Bd Gov, Am Col Surgeons, 82- *Mem:* Soc Univ Surgeons; fel Am Col Surgeons; Transplantation Soc; Am Soc Transplant Surgeons; Soc Surg Oncol; Am Soc Clin Oncol; Am Surg Asn. *Res:* Clinical trials to compare treatment programs for breast cancer; laboratory and clinical studies of effects of chemotherapy, immunotherapy and surgery on hematopoiesis. *Mailing Add:* Univ Vt Med Ctr Hosp Vt Vt Regional Cancer Ctr Burlington VT 05401

FOSTER, SUSAN J, COMPUTER SCIENCE. *Educ:* Temple Univ, BA, 76, MEd, 79. *Prof Exp:* Admin asst & appln programmer, Computer Concepts Inc, 63-65; appln systs analyst, Computation Ctr, Univ Chicago, 65-68; sr systs analyst, Computer Activ, Temple Univ, 69-77; mgr, Computer & Info Systs, Temple-Woodhaven Ctr, 77-80; systs specialist, Off Comput & Info Serv, Temple Univ, 80-83, dir, 83-89; ASSOC VPRES COMPUT & NETWORK SERV, UNIV DEL, 89- *Concurrent Pos:* Teaching, Computers in Soc, Temple Univ; consult, Peoples Repub China, 87; expert witness, Off Systs Technol, 86; mem, Spec Interest Group Graphics, Asn Comput Mach, Spec Interest Group Univ & Col Computer Serv, Spec Interest Group Computer Uses in Educ. *Mem:* Asn Computer Mach; Inst Elec & Electronics Engrs Comput Soc. *Mailing Add:* Comput & Network Serv Univ Del 192 S Chapel St Newark DE 19716

FOSTER, TERRY LYNN, b Mt Pleasant, Tex, Apr 4, 43; m 64; c 2. MICROBIOLOGY. *Educ:* NTex State Univ, BA, 65, MA, 67; Tex A&M Univ, PhD(microbiol), 73. *Prof Exp:* Assoc prof biol, 67-80, RES PROF, HARDIN-SIMMONS UNIV, 80-, ASSOC DIR, DICKINSON RES CTR, 84-; MGR, NEW REAGENTS, CLIN CHEM DIV, ABBOTT LABS, 90- *Concurrent Pos:* Prin investr, Hardin-Simmons Univ Sci Res Ctr, 72-80, sr res scientist, 80-84; dir, Biomed Div, Fairleigh Dickinson Labs, Inc, 83-87, vpres, 87- *Mem:* Am Soc Microbiol; AAAS. *Res:* Isolation of selected microorganisms from interplanetary spacecraft environments and determination of their response to experimental planetary environments; development and evaluation of new anaerobic procedures and of a new vaccine for bovine brucellosis; evaluation of new clinical diagnostic procedures; evaluation of microbial enhanced oil recovery; development of potential therapeutic for Herpes II; manufacture and marketing of rapid immunodiagnostic kits for medical care. *Mailing Add:* PO Box 167573 Irving TX 75016

FOSTER, THEODORE DEAN, b Plainfield, NJ, July 25, 29. GEOPHYSICAL FLUID DYNAMICS, PHYSICAL OCEANOGRAPHY. *Educ:* Brown Univ, ScB, 52; Univ Colo, MS, 58, MA, 60; Univ Calif, San Diego, PhD(physics), 65. *Prof Exp:* Physicist, Reaction Motors Inc, 52-55; instr phys sci, Univ Colo, 57-60; res physicist, Scripps Inst, Univ Calif, 62-65; asst prof geophys & appl sci, Yale Univ, 65-69; asst res physicist, Scripps Inst Oceanog, 69-71, assoc res oceanogr, 71-77; dean natural sci, 79-80, PROF MARINE SCI, UNIV CALIF, SANTA CRUZ, 77- *Concurrent Pos:* Mem comt polar res, Panel Oceanog, Nat Acad Sci, 71-73; mem working group 38, Sci Comt Oceanic Res, 74-77, convenor, Comt Antarctic Oceanog, 77-82; vis prof, Univ New South Wales, 87. *Mem:* Am Physical Soc; Int Glaciol Soc; Am Meteorol Soc; Philos Sci Asn; Am Geophys Union; Arctic Inst NAm. *Res:* Geophysical fluid dynamics; physical oceanography, especially of polar regions; philosophy of science. *Mailing Add:* Marine Sci Univ Calif Santa Cruz CA 95064

FOSTER, THOMAS SCOTT, b Gloversville, NY, May 25, 47; m 69; c 2. BIOPHARMACEUTICS, CLINICAL PHARMACOLOGY. *Educ:* State Univ NY, Buffalo, BS, 70; Univ Ky, PharmD(pharm), 73. *Prof Exp:* Assoc prof pharm, Col Pharm, Univ Ky, 73; dir biopharmaceut & pharmacokinetics, Drug Prod Eval Unit, 73-80, ASSOC PROF ANESTHESIOL & PHARM, ALBERT B CHANDLER MED CTR, UNIV KY, 80- *Mem:* Am Pharmaceut Asn; Am Soc Hosp Pharmacists; Acad Pharm Pract; Acad Pharmaceut Sci. *Res:* Application of biopharmaceutic and pharmacokinetic principles for adjustment of drug dosage in patients and the critical evaluation of investigational new drugs through design and execution of clinical pharmacology studies. *Mailing Add:* 206 Laburnam Crescent Rochester NY 14620-1838

FOSTER, VIRGINIA, b Joseph, Ore, Feb 4, 14. PHYTOPATHOLOGY. *Educ:* Univ Wash, BS, 49, MS, 50; Ohio State Univ, PhD(plant path, bot), 54. *Prof Exp:* Assoc prof biol, Judson Col, 56-58 & Miss State Col 58-59; assoc prof life sci, La Verne Col, 59-60; asst prof biol sci, Calif Western Univ, 60-61; from teacher to asst prof, Pensacola Jr Col, 61-70, assoc prof biol sci, 70-84; RETIRED. *Mem:* Am Phytopath Soc; Soc Indust Microbiol. *Res:* Physiological mycology; physiological relations between parasitic fungi and plant hosts, especially the production and action of wilting toxins. *Mailing Add:* 9270 Scenic Hwy Pensacola FL 32514-8054

FOSTER, WALTER EDWARD, b Cincinnati, Ohio, Oct 6, 24; m 44; c 2. INDUSTRIAL CHEMISTRY, RESEARCH MANAGEMENT. *Educ:* Univ Cincinnati, ChE, 49, MS, 51, PhD(chem), 53. *Prof Exp:* Res chemist, 53-57, res supvr, 57-61, asst dir, 61-63, dir chem res & develop, 63-65 & pioneer res, 65-68, tech dir res & develop, 68-78, GEN MGR RES & DEVELOP, ETHYL CORP, 78- *Mem:* Am Chem Soc. *Res:* Industrial organic, inorganic and organometallic research. *Mailing Add:* 9116 San Lo Dr Baton Rouge LA 70815-1067

FOSTER, WALTER H, JR, physical chemistry, analytical chemistry, for more information see previous edition

FOSTER, WILFRED JOHN DANIEL, b King William Co, Va, May 20, 37; m 69; c 1. PHYSICAL CHEMISTRY. *Educ:* Va State Col, BS, 60; George Washington Univ, MPhil, 72, PhD(chem), 74. *Prof Exp:* Lab technician, 63-67, from instr to assoc prof, 67-80, PROF CHEM, VA STATE COL, 80- *Mem:* Am Chem Soc. *Res:* Computer assisted investigations of energy transfers that occur when molecules combine, using the Huckle Molecular Orbital Theory. *Mailing Add:* 7807 Millcreek Dr Richmond VA 23235-6741

FOSTER, WILLIAM BURNHAM, b Greenfield, Mass, July 30, 30; m 54; c 1. PARASITOLOGY, INVERTEBRATE ZOOLOGY. *Educ:* Univ Mass, BS, 52, MA, 54; Rice Inst, PhD(parasite physiol), 57. *Prof Exp:* Instr gen biol, invert zool, helminthol & host-parasite relationships, 57-60, from asst prof to assoc prof zool, 64-70, PROF ZOOL, RUTGERS UNIV, 70- *Mem:* Am Soc Trop Med & Hyg; Am Soc Parasitol. *Res:* Physiology and immunology of parasites; ecology; life histories. *Mailing Add:* Dept Biol Rutgers Univ New Brunswick NJ 08903

FOSTER, WILLIAM J, b Seattle, Wash, Mar 09, 59. HORTICULTURE. *Educ:* Va Polytech Inst, BS, 82; Univ Fla, MS, 86. *Prof Exp:* Supt hort, Ornamental Hort Substa, Auburn Univ, 86-89. *Mem:* Int Plant Propagation Soc; Am Soc Hort Sci. *Mailing Add:* Grace-Sierra Hort Prod Co 1001 Yosimete Milpitas CA 95035

FOSTER, WILLIAM RODERICK, physical chemistry, for more information see previous edition

FOSTER, WILLIS ROY, b New Orleans, La, Dec 8, 28; m 57; c 3. MEDICAL SCIENCE, INFORMATION SCIENCE. *Educ:* La State Univ, BA, 50, MS & MD, 57. *Prof Exp:* Res assoc pharmacol, George Washington Univ, 57-58; prof assoc, Smithsonian Sci Info Exchange, Washington, DC, 59-63, assoc dir life sci, 64-71, vpres, 72-76; tech dir, Sci Info Serv, Kappa Systs, Inc, Arlington, Va, 76-78; pres, Advan Concepts for Develop, Inc, Bethesda, 78-83; expert consult, NIH, 83-85, SR STAFF PHYSICIAN, NAT INST DIABETES & DIGESTIVE & KIDNEY DIS, NIH, 85- *Concurrent Pos:* Fel, Johns Hopkins Univ, 58-59. *Res:* Biochemistry; pharmacology; analysis and reporting of research advances and programs in basic and clinical medicine, from the literature, from other experts and from data bases. *Mailing Add:* 6117 Greentree Rd Bethesda MD 20817

FOTH, HENRY DONALD, b Norwalk, Wis, Feb 9, 23; m 48. AGRONOMY, SOIL MORPHOLOGY. *Educ:* Univ Wis, BS, 47, MS, 48; Iowa State Col, PhD, 52. *Prof Exp:* Instr soils, Iowa State Col, 48-52; assoc prof, Agr & Mech Col Tex, 52-55; assoc prof, 55-60, PROF SOILS, MICH STATE UNIV, 60- *Concurrent Pos:* Mem, Int Soil Sci Cong; consult, Encyclopedia Britannica Educ Corp, 74- & USAID, Malawi, 85. *Honors & Awards:* Educ Award, Am Soc Agron, 59; Gustav Oahus Award, Nat Sci Teachers Asn, 74; Ensminger-Interstate Distinguished Teacher Award & Teacher-Fel Award, Nat Asn Cols & Teachers of Agr, 75. *Mem:* Fel AAAS; fel Soil Sci Soc Am; fel Am Soc Agron; Sigma Xi. *Res:* Morphology and genesis of Mollisols; soil fertility and root distribution of field crops; soil geography and land use; teaching programs for soil science. *Mailing Add:* Dept Crop-Soil Sci Mich State Univ East Lansing MI 48823

FOTI, MARGARET A, SCIENTIFIC PUBLICATIONS, MANAGEMENT. *Educ:* Temple Univ, BA, 75, MA, 85. *Prof Exp:* MANAGING ED, AM ASN CANCER RES, PHILADELPHIA, 69-, EXEC DIR, 82-, DIR PUBL, 90- *Concurrent Pos:* Mem, Asn Joint Meeting & Conf, Cambridge, Eng, 81-84, Int Fedn Sci Ed Asn, 81-83, deleg, Peoples Repub China, Coun Biol Ed, 84-88, ideas comt, Women in Cancer Res, 89-, sci publ comt, Am Heart Asn, 90-92; bd dirs & adv group, Int Fedn Sci Ed, 90- *Mem:* Coun Biol Ed; Soc Scholarly Publ; Europ Asn Sci Ed; Am Soc Asn Execs; Int Fedn Sci Ed Asns; Am Asn Cancer Res; AAAS; Coun Eng & Sci Soc Exec. *Res:* Cancer research; timely communication of technical information. *Mailing Add:* 220 Locust St Apt 24A Philadelphia PA 19106

FOTINO, MARILENA, HISTOCOMPATIBILITY. *Educ:* Inst Med & Pharm, MD, 51; Inst Microbiol-Serol, PhD(hemat), 65. *Prof Exp:* DIR IMMUNOGENETICS CTR, ROGOSIN INST, 85- *Mailing Add:* Rogosin Inst 430 E 71st St New York NY 10021

FOTINO, MIRCEA, b Bucharest, Romania, June 6, 27; US citizen; m 69; c 2. BIOPHYSICS, CYTOLOGY. *Educ:* Univ Paris, Lic-es-sci, 51; Univ Calif, Berkeley, PhD(high-energy physics), 58. *Prof Exp:* Res and teaching asst, Univ Calif, Berkeley, 52-58; charge de recherches, Nat Ctr Sci Res, Polytech Sch, Paris, 58-60; res physicist & lectr physics, Univ Calif, Berkeley, 61; res fel physics, Cambridge Electron Accelerator, Harvard Univ, 61-68, res fel biol, 69-70; dir lab high-volt electron micros, 71-85, assoc prof, 71-79, PROF, DEPT MOLECULAR, CELLULAR & DEVELOP BIOL, UNIV COLO, BOULDER, 79- *Concurrent Pos:* Guest, Dept Physics, Mass Inst Technol, 61-68. *Mem:* Am Phys Soc; NY Acad Sci; Electron Micros Soc Am; Biophys Soc. *Res:* Cosmic-ray pi mesons; beam handling and instrumentation for synchrotrons and particle physics; inverse Compton effect and polarized photons; ultrastructure of biological materials by high-voltage electron microscopy; quantitative cryomicroscopy; high-resolution and analytical electron microscopy; physical methods in biology. *Mailing Add:* Dept Molecular Cellular & Develop Biol Univ Colo Boulder CO 80309

FOU, CHENG-MING, b Shanghai, China, May 27, 36; US citizen; m 66; c 2. NUCLEAR PHYSICS, ION-ATOM COLLISION. *Educ:* Nat Taiwan Univ, BS, 56; Univ Munich, dipl physics, 61; Univ Pa, PhD(nuclear exp), 65. *Prof Exp:* Res assoc nuclear struct, Tandem Accelerator Lab, Univ Pa, 65-68; from asst prof to assoc prof, 68-87, PROF PHYSICS, UNIV DEL, 87- *Concurrent Pos:* Vis res prof, Nat Tsing Hua Univ, 74-75; vis scientist, Academia Sinica, 85-86, 89-90. *Mem:* Am Phys Soc. *Res:* Nuclear structure experimental studies using direct reactions and multiparticle final state reactions; particle angular correlation studies for nuclear spectroscopy; inner-shell ionization in ion-atom collision. *Mailing Add:* Dept Physics Univ Del Newark DE 19716

FOUAD, ABDEL AZIZ, b May 10, 28; m; c 2. ENGINEERING. *Educ:* Cairo Univ, BS, 50; State Univ Iowa, MS, 53; Iowa State Univ, PhD(elec eng), 56. *Prof Exp:* Instr, elec eng dept, Iowa State Univ, 54-56; lectr, Ain-Shams Univ, Cairo, Egypt, 56-60; from asst prof to prof, 60-90, DISTINGUISHED PROF, ELEC ENG DEPT, IOWA STATE UNIV, 90- *Concurrent Pos:* Rio Light Co, Rio de Janeiro, Brazil, 60; spec asst, Detroit Edison Co, 61; res engr, Atomics Int, Canoga Park, Ill, 63-64; vis prof, Univ Philippines, 69-71 & Univ Rio, Rio de Janeiro, Brazil, 72; proj mgr, Elec Power Res Inst, Palo Alto, 80-81; consult, Jersey Prop Res Co, 62, Off Educ, 74-75, Brazilian Ministry Planning, 75, Control Data Corp, Minneapolis, 75-79 & Elec Power Res Inst, 84; mem, Nat Res Coun, 75-78. *Mem:* Am Soc Eng Educ; fel Inst Elec & Electronics Engrs. *Res:* Author of 2 books and several journal articles. *Mailing Add:* Elect Eng Dept Iowa State Univ Ames IA 50011

FOUCHER, W(ALTER) D(AVID), b Bennington, Vt, Aug 4, 36; div; c 2. RESEARCH ADMINISTRATION. *Educ:* St Michael's Col, Vt, AB, 58; Univ Vt, MS, 59; Univ Fla, PhD(phys inorg chem), 62. *Prof Exp:* Chemist, Texaco Res Ctr, Texaco Inc, 62-63, sr chemist, 63-67, res chemist, 67-69, group leader, Lubricants Res, 69-75, group leader, Fuels Res, 75-79, technologist, 79-81, technologist sci planning, 81-83, coordr, planning & coordr, employee & pub rels, 85-87, MGR, HUMAN RESOURCES, TEXACO RES & DEVELOP, TEXACO INC, 87- *Concurrent Pos:* Ed, Mid-Hudson Chemist, Am Chem Soc, 63-69; vis instr, State Univ NY Col New Paltz, 64-68; vis lectr, Marist Col, 67-75; ed, Test Tube Texaco, Inc, 80-85. *Mem:* Am Chem Soc; Sigma Xi. *Res:* Preparation catalysts and catalyst supports for hydrocarbon conversion; additives and lubricating oils for internal combustion engines, gasoline and diesel; gasoline additive synthesis and development. *Mailing Add:* Texaco Res & Develop Box 509 Beacon NY 12508

FOUDIN, ARNOLD S, PLANT PATHOGENIC MICROORGANISMS. *Educ:* Univ Ga, PhD(plant path), 75. *Prof Exp:* Postdoctoral, Animal & Plant Health Inspection, Univ Ga, 75-77; plant pathologist, State Mo, USDA, 77, lab chief survey methods develop lab, Univ Mo, Columbia, 77-85, Admin Off, Hyattesville, Md, 85, Plant Protection & Quarantive, 86, DEP DIR BIOTECHNOL PERMITS, BIOLOGICS & ENVIRON PROTECTION DIV, ANIMAL & PLANT HEALTH INSPECTION SERV, USDA, 86- *Res:* Mycotoxin biosynthesis in food stuffs; development of monocloanal antibody serological test kits for the field identification of certain important plant pathogenic microorganisms. *Mailing Add:* Animal & Plant Health Inspection Serv USDA Fed Ctr Bldg Hyattsville MD 20782

FOUGERE, PAUL FRANCIS, b Cambridge, Mass, Feb 29, 32; m 52; c 6. IONOSPHERIC PHYSICS, SIGNAL PROCESSING. *Educ:* Boston Col, BS, 52, MS, 53; Boston Univ, PhD(physics), 65. *Prof Exp:* Physicist, Naval Res Lab, 53-54; tech aircraft engr, Gen Elec Co, Mass, 54-55; PHYSICIST IONOSPHERIC PHYSICS, AIR FORCE GEOPHYS LAB, 55- *Concurrent Pos:* Chmn & ed, Maximum Entropy & Bayesian Methods, Dartmouth, 89. *Mem:* Sr mem Inst Elec & Electronics Engrs; Am Geophys Union; Sigma Xi; NY Acad Sci; Int Sci Radio Union. *Res:* Maximum entropy power spectral analysis: development of method and application to geophysical and ionospheric data sets; space and ionospheric physics; development of methods for computer graphics; maximum entropy power spectrum analysis; theory and analysis of ionospheric scintillation; application of fractal geometry to ionospheric problems; theory and application of ionospheric tomography. *Mailing Add:* Air Force Geophys Lab/LIS Hanscom AFB Bedford MA 01731-5000

FOUGERON, MYRON GEORGE, b Morris, Minn, July 20, 32; m 59; c 2. PHYSIOLOGY, ENDOCRINOLOGY. *Educ:* Sam Houston State Col, BS, 59, MA, 65; Tex A&M Univ, PhD(zool), 67. *Prof Exp:* Teacher jr high sch, Tex, 59-61; partic acad year inst biophys, Col Med, Baylor Univ, 66-67; from asst prof to assoc prof, 67-76, PROF BIOL, KEARNEY STATE COL, 76- *Mem:* Am Inst Biol Sci; Asn Midwestern Col Biol Teachers; Sigma Xi. *Res:* Effect of radiation on placental transport; effect of hypophysectomy on the endocrine glands of the American chameleon, Anolis Carolinensis. *Mailing Add:* Dept Biol Kearney State Col Kearney NE 68849

FOULDS, JOHN DOUGLAS, b Pittsfield, Mass; m 82; c 1. BIOCHEMISTRY, MOLECULAR BIOLOGY. *Educ:* Columbia Col, AB, 60, Columbia Univ, PhD(biochem), 66. *Prof Exp:* Asst prof microbiol, Univ Conn Health Ctr, 68-74; RES BIOCHEMIST, NAT INST ARTHRITIS, METAB & DIGESTIVE DIS, 74- *Concurrent Pos:* USPHS fel, Stanford Univ, 66-68. *Mem:* Am Soc Biol Chem; Am Soc Microbiol. *Res:* Bacterial cell surface components; role of bacterial outer membrane components (proteins, lipids and lipoproteins) in pathogenesis and cellular physiology. *Mailing Add:* Lab Struct Biol Sec Membrane Biol NIH Bldg 8 Rm 106 Bethesda MD 20892

FOULIS, DAVID JAMES, b Hinsdale, Ill, July 26, 30; m 56, 62, 77; c 3. MATHEMATICS, QUANTUM PHYSICS. *Educ:* Univ Miami, BA, 52, MS, 53; Tulane Univ, PhD, 58. *Prof Exp:* Asst prof, Lehigh Univ, 58-59; asst prof, Wayne State Univ, 59-63; prof, Univ Fla, 63-65; PROF MATH, GRAD SCH, UNIV MASS, AMHERST, 65- *Mem:* Am Math Soc; Math Asn Am; Sigma Xi. *Res:* Foundations of statistics; operator algebras; quantum theory; author of 4 publications. *Mailing Add:* Dept Math Univ Mass Amherst MA 01003

FOULK, CLINTON ROSS, b Wichita, Kans, Jan 24, 30; m 58; c 1. COMPUTER SCIENCE, MATHEMATICS. *Educ:* Univ Kans, BA, 51; Univ Ill, Urbana, MA, 58, PhD(math), 63. *Prof Exp:* Res asst comput prog, Digital Comput Lab, Univ Ill, Urbana, 58-62; asst prof math, 63-66, asst prof comput & info sci, 66-68, ASSOC PROF COMPUT & INFO SCI, OHIO STATE UNIV, 68- *Mem:* Asn Comput Mach; Sigma Xi. *Res:* Computer systems programming. *Mailing Add:* 5101 Delancy St Ohio St Univ Columbus OH 43220

FOULKES, ERNEST CHARLES, b Ger, Aug 20, 24; US citizen; m 46; c 4. TOXICOLOGY. *Educ:* Univ Sydney, BSc, 46, MSc, 47; Oxford Univ, DPhil(biochem), 52. *Prof Exp:* Investr, Nat Health Med Res Coun, Australia, 46-49; assoc, May Inst, 52-65; assoc prof, 65-70, PROF ENVIRON HEALTH & PHYSIOL, UNIV CINCINNATI, 70-, ASSOC DIR DEPT ENVIRON HEALTH, 71- *Concurrent Pos:* Estab investr, Am Heart Asn, 56-61; pres adv, Univ Cincinnati, 72-74. *Mem:* Biochem Soc; Biophys Soc; Am Soc Biol Chem; Am Physiol Soc; Soc Exp Biol & Med; Soc Toxicol. *Res:* Action of heme enzymes; active transport; cell permeability, renal physiology and toxicology, heavy metal toxicology. *Mailing Add:* Dept Environ Health Univ Cincinnati Col Med Cincinnati OH 45267-0056

FOULKES, ROBERT HUGH, b Wis, Nov 14, 18; m 44; c 1. EXPERIMENTAL ZOOLOGY. *Educ:* Coe Col, AB, 40; Univ Iowa, MS, 47, PhD(zool), 51. *Prof Exp:* Instr zool, Univ Iowa, 47-51; res fel, Edsel Ford Inst Med Res, 51-55; asst prof biol, Northern Ill Univ, 55-56 & St Louis Univ, 56-59; assoc scientist res div, Dr Salsbury's Labs, 59-61; assoc prof biol, 62-69, prof zool, 69-80, PROF BIOL, UNIV WIS, PLATTEVILLE, 80- *Concurrent Pos:* Independent consult, 61- *Mem:* AAAS; Am Soc Zool; Wildlife Dis Asn; Am Inst Biol Sci. *Res:* Regeneration; neoplasms; histochemistry; toxicology; aerobiology. *Mailing Add:* 826 30th St Rock Island IL 61201

FOULKS, JAMES GRIGSBY, b Bay City, Tex, Sept 18, 16; m 47; c 2. PHARMACOLOGY. *Educ:* Rice Inst, BA, 39; Johns Hopkins Univ, PhD(biol sci), 43; Columbia Univ, MD, 50. *Prof Exp:* Lab asst, Rice Inst, 37-39; lab asst, Univ Rochester, 39-40; lab asst, Johns Hopkins Univ, 40-42; instr anat, Ohio State Univ, 42-43; lab asst pharmacol, Columbia Univ, 50-51; head dept, 51-71, PROF PHARMACOL, UNIV BC, 51- *Concurrent Pos:* Nat Heart Inst fel, Columbia Univ, 50-51. *Mem:* Am Physiol Soc; Soc Exp Biol & Med; Am Soc Pharmacol & Exp Therapeut; Pharmacol Soc Can; Can Physiol Soc. *Res:* Water and electrolyte metabolism; renal and cardiovascular physiology and pharmacology. *Mailing Add:* Dept Pharmacol Fac Med Univ BC Vancouver BC V6T 1W5 Can

FOULSER, DAVID A, b Columbus, Ohio, Apr 10, 33; m 56; c 3. MATHEMATICS. *Educ:* Ohio State Univ, BA, 55; Univ Mich, PhD(math), 63. *Prof Exp:* Instr math, Univ Chicago, 63-65; asst prof, 65-68, assoc prof, 68-74, PROF MATH, UNIV ILL, CHICAGO CIRCLE, 74- *Mem:* Am Math Soc. *Res:* Projective planes; group theory. *Mailing Add:* Dept Math Univ Ill Chicago IL 60680

FOUNDS, HENRY WILLIAM, b Drexel Hill, Pa, Mar 6, 42; m 71; c 4. DIAGNOSTIC SCIENCE, CELL CULTURE TECHNOLOGIES. *Educ:* Villanova Univ, BS, 64; Univ Notre Dame, MS, 68; Rutgers Univ, PhD(microbiol), 81. *Prof Exp:* Instr biol & chem, Bayley-Ellard High Sch, 64-68; assoc prof microbiol, Morris Col, 68-83; mgr, cell cult sect, 83-85, dir res & develop, 85-86, VPRES RES & DEVELOP, VENTREX LABS, 86- *Concurrent Pos:* Consult res & develop, Personal Diagnostics, 81. *Mem:* AAAS; Am Soc Microbiol; Tissue Cult Asn. *Res:* Hybridoma and mammalian cell culture; immunodiagnostic assay development; allergy diagnostics. *Mailing Add:* 13 Hemlock Circle Scarborough ME 04074

FOUNTAIN, JOHN CROTHERS, b Berkeley, Calif, May 8, 47. GEOCHEMISTRY. *Educ:* Calif Polytech State Univ, San Luis Obispo, BS, 70; Univ Calif, Santa Barbara, MA, 73, PhD(geol), 75. *Prof Exp:* Asst prof, 75-80, ASSOC PROF GEOL, STATE UNIV NY, BUFFALO, 80- *Mem:* Geol Soc Am; Geochem Soc. *Res:* Trace element and isotope geochemistry; igneous petrology. *Mailing Add:* Dept Geol SUNY Buffalo Amherst 4240 Ridge Lea Amherst NY 14226

FOUNTAIN, JOHN WILLIAM, b Quincy, Ill, Mar 12, 44. SPACE SCIENCE. *Educ:* Univ Ariz, BS, 66. *Prof Exp:* Res asst, Lunar & Planetary Lab, 65-73, res asst, Lunar & Planetary Lab & Optical Sci Ctr, 73-81, RES ASST, LUNAR & PLANETARY LAB, UNIV ARIZ, 81- *Concurrent Pos:* Co-discoverer, Saturn Satellites Epimetheus & Telesto. *Honors & Awards:* Pub Serv Achievement Award, NASA, 74. *Mem:* Am Astron Soc; AAAS. *Res:* Structure and dynamics of planetary atmospheres; Saturn ring; photometry of planetary atmospheres; astronomical instrumentation; celestial mechanics. *Mailing Add:* Lunar & Planetary Lab Space Sci Bldg Univ Ariz Tucson AZ 85721-5610

FOUNTAIN, LEONARD DU BOIS, b Missouri Valley, Iowa, Jan 25, 29; m 59; c 1. MATHEMATICAL ANALYSIS, APPLIED MATH. *Educ:* Univ Chicago, AB, 50, MS, 53; Univ Nebr, PhD(math), 60. *Prof Exp:* Instr math, Univ Nebr, 58-60; from asst prof to assoc prof, 60-70, PROF MATH, SAN DIEGO STATE UNIV, 70- *Mem:* Math Asn Am. *Res:* Ordinary differential equations; convex functions and their applications to existence theorems for differential equations. *Mailing Add:* Dept Math San Diego State Univ San Diego CA 92182-0314

FOUNTAIN, LEWIS SPENCER, b McCrory, Ark, Oct 18, 17; m 46; c 4. GEOSCIENCE, ACOUSTICS. *Educ:* Trinity Univ, BS, 58, MS, 62. *Prof Exp:* Res engr nondestructive testing, Southwest Res Inst, 58-64, sr res engr, 64-71, actg mgr & sr res engr, 71-75, sr res scientist geosci, 75-85, staff scientist, 85-87; independent consult, 87-89; STAFF SCIENTIST, SOUTHWEST RES INST, 89- *Mem:* Acoust Soc Am; Inst Elec & Electronics Engrs; Am Soc Nondestructive Testing; Soc Explor Geophysicists; Sigma Xi; NY Acad Sci. *Res:* Subsurface cavity detection studies; detection and disposal activities related to clearing hazardous buried ordnance objects; land mine detector evaluations; global soils studies; chemical, physical, electromagnetic and magnetic analysis of soils; magnetic signature measurement and analysis. *Mailing Add:* 127 Postwood San Antonio TX 78228

FOUQUET, JULIE, b Mar 23, 58; m 87. SEMICONDUCTOR LASERS, TIME-RESOLVED PHOTOLUMINESCENCE. *Educ:* Harvard Univ, MS, 80; Stanford Univ, MS, 82, PhD(appl physics), 86. *Prof Exp:* Mem tech staff, 85-88, PROJ LEADER, HEWLETT-PACKARD LABS, 88- *Concurrent Pos:* Assoc ed, Inst Elec & Electronics Engrs, Circuits & Devices Mag, 88-91, ed, 91-; mem, Prog Comt, Lasers & Electro-Optics Soc, 90 & 91, bd gov, 91-93. *Mem:* Am Phys Soc; Inst Elec & Electronics Engrs. *Res:* Optoelectronic devices, including semiconductor lasers; advanced optical measurements on compound semiconductor structures. *Mailing Add:* Hewlett Packard Labs 26M 3500 Deer Creek Rd Palo Alto CA 94304

FOUQUETTE, MARTIN JOHN, JR, b Philadelphia, Pa, June 14, 30; div; c 2. HERPETOLOGY. *Educ:* Univ Tex, BA, 51, MA, 53, PhD(zool), 59. *Prof Exp:* interim asst prof zool, Univ Fla, 59-61; asst prof, Univ Southwestern La, 61-65; ASSOC PROF ZOOL, ARIZ STATE UNIV, 65- *Concurrent Pos:* Prin investr, Nat Sci Found, 63-65, 69-70, assoc dir, Int Biol Prog, 76-78. *Mem:* Am Soc Ichthyologists & Herpetologists; Am Soc Naturalists; Am Soc Zoologists; Sigma Xi; Soc Syst Zool; Soc Study Amphibians & Reptiles. *Res:* Systematics and ecology of anuran amphibians and squamate reptiles; mechanics of speciation; amphibian bioacoustics; anuran sperm morphology. *Mailing Add:* Dept Zool Ariz State Univ Tempe AZ 85287-1501

FOURER, ROBERT, b Philadelphia, Pa, 1950. LINEAR PROGRAMMING. *Educ:* Mass Inst Technol, BS, 72; Stanford Univ, MS(opers res) & MS(statist), 79, PhD(opers res),80. *Prof Exp:* Res analyst, Nat Bur Econ Res, 74-76; asst prof, 79-85, ASSOC PROF NORTHWESTERN UNIV, 85- *Concurrent Pos:* Mem tech staff, AT&T Bell Labs, 85-86. *Mem:* Oper Res Soc Am; Inst Mgt Sci; Math Prog Soc; Asn Comput Mach; Soc Indust & Appl Math. *Res:* Large-scale optimization. *Mailing Add:* Dept Indust Eng Northwestern Univ Evanston IL 60208-3119

FOURNELLE, RAYMOND ALBERT, b St Louis, Mo, Dec 9, 41. PHASE TRANSFORMATIONS, DIFFUSION IN SOLIDS. *Educ:* Univ Mo-Rolla, BS, 64, MS, 68, PhD(metall eng), 71. *Prof Exp:* Res engr, Wood River Res Lab, Shell Oil Co, Ill, 64-66; res assoc, Dept Mat Sci, Northwestern Univ, 71-72; from asst prof to assoc prof, Dept Mech Eng, 72-86, PROF METALL & MAT SCI, DEPT MECH & INDUST ENG, MARQUETTE UNIV, 86- *Concurrent Pos:* Fac res partic, Mat Sci Div, Argonne Nat Lab, 78, Mat Res Dept, Globe Union, Inc, NSF, 79; mem bd rev, Metall Trans, 81-; Fulbright sr res fel, Inst Metall, Univ Stuttgart, Repub of Ger, 83-84, Max Planck Inst Metall, 90-91; Alexander von Humboldt fel, Inst Metall, Univ Stuttgart, Repub of Ger, 85, 86 & 88. *Mem:* Am Soc Metals Int; Metall Soc; Am Ceramics Soc; Electron Microscopy Soc Am; Am Soc Eng Educ; Am Asn Univ Prof. *Res:* Phase transformations in metals and ceramics; defect structure of metals and ceramics; mechanical behavior of materials; electron microscopy; author of 38 publications. *Mailing Add:* Dept Mech & Indust Eng Marquette Univ Milwaukee WI 53233

FOURNELLE, THOMAS ALBERT, b St Paul, Minn, Apr 24, 46; m 78. INFINITE GROUP THEORY, GEOMETRY. *Educ:* St John's Univ, Minn, BS, 69; St Louis Univ, MS, 72; Univ Ill, Urbana, PhD(math), 78. *Prof Exp:* Instr math, Mich State Univ, 78-80; asst prof, Univ Ala, 80-83; asst prof, 83-86, ASSOC PROF MATH, UNIV WIS-PARKSIDE, 86- *Mem:* Sigma Xi; Am Math Soc. *Res:* Infinite group theory; relation of the automorphism group of a group to the structure of a group itself; construction of new groups using generalized wreath product-like constructions, in particular, the construction of locally finite infinite groups and of periodic non-locally finite groups; the structure theory for groups in which rings have been verbally embedded and computer techniques for studying abstract algebraic structures. *Mailing Add:* Dept Math Univ Wis-Parkside Kenosha WI 53141

FOURNEY, M(ICHAEL) E(UGENE), b Blue Jay, WVa, Jan 30, 36. AERONAUTICS, MECHANICAL ENGINEERING. *Educ:* Univ WVa, BS, 58; Calif Inst Technol, MS, 59, PhD(aeronaut), 63. *Prof Exp:* Aerospace engr, Bolkow Entwicklungen KG, Ger, 63-64; from res asst prof to res assoc prof aeronaut & astronaut, Univ Wash, 64-72; assoc prof, 72-75, chmn, 79-83, PROF, DEPT MECH & STRUCTURES, UNIV CALIF, LOS ANGELES, 75- *Concurrent Pos:* Consult, Math Sci Northwest Inc, 60-73; pvt consult, 70-; vis prof, US Mil Acad, West Point, 89-90. *Honors & Awards:* Murray Medal, Soc Exp Mech, 90. *Mem:* Soc Exp Stress Anal (pres, 80-81); Optical Soc Am; fel Soc Exp Mech; fel Am Soc Mech Engrs; fel Am Acad Mech. *Res:* Solid and experimental mechanics. *Mailing Add:* Sch Eng & Appl Sci Univ Calif Los Angeles CA 90024

FOURNIE, JOHN WILLIAM, b St Louis, Mo, March 24, 52; m 77; c 2. FISH & TOXICOLOGIC PATHOLOGY. *Educ:* St Louis Univ, BA, 74, MS, 79; Univ Miss, PhD(biol), 85. *Prof Exp:* Fel, Gulf Coast Res Lab, 85-86; RES AQUATIC BIOLOGIST, US ENVIRON PROTECTION AGENCY, 86- *Mem:* Am Fisheries Soc; Am Soc Parasitologists; Soc Protozoologists; Sigma Xi. *Res:* Fish pathology with special interests in the life history, and infectivity; pathology of parasitic protozoans; experimental carcinogenesis in fishes and invertebrates; the use of small fishes in the development of animal models for carcinogenesis. *Mailing Add:* Path Br USEPA Environ Res Lab Gulf Breeze FL 32561

FOURNIER, JOHN J F, b Can, May 3, 42. MATHEMATICS, ANALYSIS & FUNCTIONAL ANALYSIS. *Educ:* Toledo Univ, BA, 63; Wis Univ, PhD(math), 67. *Prof Exp:* PROF MATH, UNIV BC, 67- *Mem:* Can Math Asn; Am Math Asn. *Mailing Add:* Math Dept Univ BC Vancouver BC V6T 1Y4 Can

FOURNIER, MAURILLE JOSEPH, JR, b Montpelier, Vt, Jan 13, 40; m 64; c 2. MOLECULAR BIOLOGY, BIOCHEMISTRY. *Educ:* Univ Vt, BA, 62; Dartmouth Col, PhD(molecular biol), 68. *Prof Exp:* Res biochemist, Walter Reed Army Inst Res, 68-70; Am Cancer Soc fel, Nat Insts Health, 70-72; from asst prof to assoc prof, 78-84, PROF BIOCHEM, UNIV MASS, AMHERST, 84- *Concurrent Pos:* Mem, Biochem Study Sect, NIH, 83-87. *Mem:* Am Soc Biol Chemists; AAAS; Am Soc Microbiol. *Res:* Genetics and biochemistry of gene expression; biosynthesis, structure and function of novel small RNAs of Escherichia coli and the yeast Saccharomyces cerevisiae, esp the role of yeast small nucleolar RNAs (snoRNA) in ribosome biogenesis; flavivirus molecular biology, creation of recombinant vaccines for dengue and Japanese encephalitis viruses; genetic synthesis of advanced materials; engineering of novel protein-based polymers. *Mailing Add:* Dept Biochem Lederle Grad Res Ctr Univ Mass Amherst MA 01003

FOURNIER, MICHEL, b Montreal, Que, July 27, 54; m 81. IMMUNOTOXICOLOGY. *Educ:* Univ Que, Montreal. BSc, 76; Univ Montreal, MSc, 78; McGill Univ, PhD(exp med), 81. *Prof Exp:* PROF IMMUNOL, UNIV QUE, MONTREAL, 80-, INVESTR, LAB ENVIRON TOXICOL, 85- *Mem:* Can Soc Immunol; Fr Can Asn Advan Sci; NY Acad Sci; Can Health Res. *Res:* Effects of environmental xenobiotics on immune response and consequences on resistance to infections. *Mailing Add:* Dept Biol Sci Univ Que PO Box 8888, Sta A Montreal PQ H3C 3P8 Can

FOURNIER, PIERRE WILLIAM, pathology, for more information see previous edition

FOURNIER, R E KEITH, b Attleboro, Mass, July 26, 49. CELL GENETICS, MOLECULAR BIOLOGY. *Educ:* Providence Col, BS, 71; Princeton Univ, PhD(biochem), 74. *Prof Exp:* Fel, Dept Biol & Human Genetics, Yale Univ, 75-78; asst prof, 78-84, assoc prof, 84-86, prof, Dept Microbiol & Comprehensive Cancer Ctr, Univ Southern Calif, 86-87; FULL MEM, DIV BASIC SCI, HUTCHINSON CANCER RES CTR, 87- *Concurrent Pos:* Adj prof, W Alton Jones Cell Sci Ctr, Lake Placid, NY, 78-82; Jr Fac Res Award, Am Cancer Soc, 80-83, Fac Res Award, 83- *Mem:* Sigma Xi; AAAS. *Res:* Control of eukaryotic gene expression, particularly the genetic and molecular mechanisms which regulate gene activity in higher differentiated mammalian cells. *Mailing Add:* Dept Molecular Med M462 Fred Hutchinson Center 1124 Columbia St Seattle WA 98104

FOURNIER, ROBERT ORVILLE, b San Diego, Calif, Jan 14, 32; m 60; c 2. GEOCHEMISTRY. *Educ:* Harvard Univ, AB, 54; Univ Calif, PhD(geol), 58. *Prof Exp:* Assoc prof lectr geol, George Washington Univ, 58-59; GEOLOGIST, US GEOL SURV, 58- *Concurrent Pos:* Chmn, US orgn comt 2nd UN symp develop & use of geothermal energy, 73-75. *Mem:* Geol Soc Am; Mineral Soc Am; Geochem Soc Japan; Int Asn Geochem Cosmochem; Int Asn Gen Ore Deposits. *Res:* Geologic and geochemical aspects of geothermal energy; geochemistry of hydrothermal solutions and hydrothermal alteration; experimental studies of solution mineral reactions at high temperatures and high pressures. *Mailing Add:* 108 Paloma Rd Portola Valley CA 94025

FOURTNER, CHARLES RUSSELL, b Hollywood, Calif, Oct 3, 44; m 68; c 3. INVERTEBRATE PHYSIOLOGY. *Educ:* Carroll Col, AB, 66; Mich State Univ, MS, 69, PhD(zool), 71. *Prof Exp:* Fel biol, Univ Miami, 71-72; fel physiol, Univ Alta, 73; from asst prof to assoc prof, 74-90, PROF BIOL, STATE UNIV NY, BUFFALO, 90- *Concurrent Pos:* Vis prof, UTMB, Galveston, 85. *Mem:* Soc Neurosci; Am Soc Zoologists; Soc Exp Biol; AAAS. *Res:* Control of ion channel activity in fish intestine and pituitary cells; central nervous control of rhythmic behavior patterns such as locomotion and respiration in annelids. *Mailing Add:* Dept Biol Sci State Univ NY Buffalo NY 14260

FOUSS, JAMES L(AWRENCE), b Warsaw, Ohio, Feb 22, 36; m 57; c 2. AGRICULTURAL ENGINEERING. *Educ:* Ohio State Univ, BAE, 59, MSc, 62, PhD, 71. *Prof Exp:* Res agr engr, USDA, 60-72, lab dir, Coastal Plains Soil & Water Conserv Res Ctr, Agr Res Serv, 72-76; vpres res & sr scientist, res & new prod develop, Hancor, Inc, 76-82; AGR ENG, SOIL & WATER MGT RES, USDA-AGR RES SERVICE, 82-, TEAM LEADER 89- *Concurrent Pos:* Eng consult, USA & Can, 68-75, 82- *Honors & Awards:* Spec Award, USDA, 62, Performance Awards, 68, 72, 89 , Superior Serv Award, 72; Young Designer Award, Am Soc Agr Eng, 72. *Mem:* Am Soc Agr Engrs; Soil Conserv Soc Am; Am Soc Testing & Mat; Am Soc Civil Engrs; Land Improv Contractors Am. *Res:* Development of new equipment and materials for the high-speed and low-cost installation of subsurface drains in agricultural cropland; plastic septic tank for home sewage disposal systems; development of simulation models for automatically controlled soil-water table management systems in humid climate regions, permitting input of daily weather forecast records; development of integrated technology and drainage soil- water, fertilizers and pesticide to reduce and control ground water pollution. *Mailing Add:* Soil & Water Mgt Res USDA-ARS PO Box 25071 LSU Baton Rouge LA 70894-5071

FOUTCH, GARY LYNN, b Poplar Bluff, Mo, Aug 26, 54; m 77; c 3. BIOTECHNOLOGY, ION EXCHANGE. *Educ:* Univ Mo, Rolla, BS, 75, MS, 77, PhD(chem eng), 80. *Prof Exp:* From asst prof to assoc prof, 80-89, PROF CHEM ENG, OKLA STATE UNIV, 89- *Concurrent Pos:* Fel, Jet Propulsion Lab, Calif Inst Technol, 81 & 82, Dow Chem Co, 83, Smith, Kline & French, 85, Phillips Petrol, 87; Fulbright Scholar, Loughborough, Univ UK, 90-91. *Mem:* Am Inst Chem Engrs; Am Chem Soc; Sigma Xi; Nat Soc Prof Engrs. *Res:* Production of specialty bio-chemicals by both biological and thermal processes; reaction kinetics and reactor design; mixed bed ion exchange. *Mailing Add:* Okla State Univ 423 Eng Stillwater OK 74078

FOUTCH, HARLEY WAYNE, b Woodlawn, Ill, Sept 29, 44; m 70. AGRONOMY, PLANT PHYSIOLOGY. *Educ:* Southern Ill Univ, BS, 66, MS, 68; Auburn Univ, PhD(agron), 71. *Prof Exp:* PROF AGR, MIDDLE TENN STATE UNIV, 70-, DEPT CHMN, 80- *Mem:* Am Soc Agron; Weed Sci Soc; Nat Asn Col Teachers Agr. *Res:* Morphological and physiological response of cool season perennial grasses to temperature; forage crop physiology as affected by management practices; incorporation of herbicides by tillage methods. *Mailing Add:* Chmn Agri Dept Box 5 Middle Tenn State Univ Murfreesboro TN 37132

FOUTS, JAMES RALPH, b Macomb, Ill, Aug 8, 29; m 64; c 3. PHARMACOLOGY, TOXICOLOGY. *Educ:* Northwestern Univ, BS, 51, PhD(biochem), 54; Duke Univ, MDiv, 86. *Prof Exp:* Instr & asst chem, Northwestern Univ, 51-54; asst scientist, Nat Heart Inst, 54-56, sr asst scientist, 56; sr res biochemist, Wellcome Res Labs, Burroughs Wellcome & Co, 56-57; from asst prof to prof pharmacol, Col Med, Univ Iowa, 57-70, dir, Oakdale Toxicol Ctr, 68-70; chief pharmacol br, Nat Inst Environ Health Sci, 70-76, sci dir, 76-78, chief lab pharmacol, 78-81, SR EXEC SERV, NAT INST ENVIRON HEALTH SCI, 81-, SR SCI ADV TO DIR, 86- *Concurrent Pos:* Mem, Pharmacol-Toxicol Rev Comt, Nat Inst Gen Med Sci, 64-65 & 67-68; Nat Adv Coun, Nat Inst Environ Health Sci, 66-70, Nat Task Force Res Priorities, 68 & 76; mem comt on anticonvulsant drugs, Nat Inst Neurol Dis & Stroke, 69-75 & 80-83, mem epilepsy adv comt, 72-75 & 80-83; mem sci group on prin of pre-clin testing for drug safety, WHO, Geneva, Switzerland, 66; consult, Dr Salsbury Labs, 60-66, Smith Kline & French Labs, Philadelphia, 66-70, Hoffman La Roche, Nutley, 68-70; Claude Bernard prof, Inst Med & Surg, Univ Montreal, 70; mem comt environ pharmacol, Am Soc Pharmacol & Exp Therapeut, 70-74, chmn, drug metab div, 75-78, mem coun, 81-84; adj prof pharmacol, Sch Med, Univ NC, 70-, adj prof entom & toxicol, Sch Agr & Life Sci, NC State Univ, 71-; mem educ comt, Soc Toxicol, 70-72; mem Sci Adv Bd, Nat Ctr Toxicol Res, Food & Drug Admin/Environ Protection Agency, 72-74; mem drug interactions contract rev, Nat Inst Drug Abuse, 74-77; mem basic pharmacol adv comt, Pharmaceut Mfrs Asn Found, Inc, 73-78; mem & US rep, Sci Comt Prob Environ, Int Coun Sci Unions, 76-80; chmn, Gordon Res Conf Drug Metab, Plymouth, NH, 77 & 78; assoc ed, Pharmacol Rev, 77-88, ed, Chem-Biol Interactions, 73-76, Annual Report Carcinogens, 86-; vis prof, Univ Zurich, 78-79; vis scientist, Swiss Fed Inst Technol, 78-79. *Honors & Awards:* Abel Award, Am Soc Pharmacol & Exp Therapeut, 64. *Mem:* Am Soc Pharmacol & Exp Therapeut; Sigma Xi; Am Asn Cancer Res. *Res:* Drug metabolizing systems and factors affecting them; correlation of cell structure and enzyme activity; developmental pharmacology and toxicology; comparative and marine pharmacology and toxicology; drug interactions; pulmonary toxicology; carcinogenesis. *Mailing Add:* Nat Inst Environ Health Sci PO Box 12233 MDA2 03 Research Triangle Park NC 27709

FOWELL, ANDREW JOHN, b Liverpool, Eng, Sept 27, 36; m 59; c 2. FLUID MECHANICS, ACOUSTICS. *Educ:* Univ Nottingham, BSc, 57, PhD(mech eng), 61. *Prof Exp:* Grad apprentice, English Elec Co, 60-62, develop engr, 62; res scientist, Am Standard Inc, 62-68, supvr acoust, 68-69, mgr adv prod develop, 69-72, mgr fixtures & seats eng, 72-76; chief, Prod Performance Eng Div, Nat Bur Standards, 76-81, chief Fire Safety Eng Div, 81-86, dep dir, Ctr Fire Res, 86-90, CHIEF, FIRE SCI & ENG DIV, NAT INST STANDARDS & TECHNOL, 90- *Mem:* Am Soc Testing & Mat; Sigma Xi; Soc Fire Protection Engrs. *Res:* Building technology; pulsating flow; turbulent diffusion; flow through air moving devices; fluid couplings; air conditioner and plumbing noise; consumer product performance; energy conservation; test methods; product safety; fire. *Mailing Add:* 10128 Gravier Ct Gaithersburg MD 20879

FOWKE, LAWRENCE CARROLL, b Toronto, Ont, June 6, 41; m 62; c 3. PLANT CYTOLOGY. *Educ:* Univ Sask, BA, 63; Carleton Univ, PhD(cell biol), 68. *Prof Exp:* Res fel cell biol, Australian Nat Univ, 68-70; from asst prof to assoc prof, 70-79, PROF BIOL, UNIV SASK, 79- *Concurrent Pos:* Vis scientist, Australian Nat Union, 76-77, Friedrich Miescher Inst, Ciba Geigy, Bd & Inst Physiol Botony, Univ Uppsala, 83-84; affil scientist, Plant Biotechnol Inst, NRC Saskatoon, 88- *Mem:* Can Soc Cell Biol; Am Soc Cell Biol; Int Asn Plant Tissue Cult; Can Soc Plant Physiologists; Can Bot Asn. *Res:* Light and electron microscope studies of plant protoplasts, especially cell wall regeneration, protoplast fusion and the structure and function of coated vesicles. *Mailing Add:* Dept Biol Sci Univ Sask Saskatoon SK S7N 0W0 Can

FOWKES, FREDERICK MAYHEW, b Chicago, Ill, Jan 29, 15; m 37; c 4. CHEMISTRY. *Educ:* Univ Chicago, BS, 36, PhD(chem), 38. *Prof Exp:* Chemist, Nat Aluminate Corp, 37; res chemist, Continental Can Co, 38-42 & Shell Develop Co, 46-52, res supvr, 52-59, spec res chemist, Shell Oil Co, 59-61, res supvr, Shell Develop Co, 61-62; dir res, Sprague Elec Co, Mass, 62-68; chmn dept, 68-81, PROF CHEM, LEHIGH UNIV, 68- *Concurrent Pos:* Exchange chemist, Koninklijke Shell Lab, Amsterdam, 55-56. *Mem:* Am Chem Soc; sr mem Inst Elec & Electronics Eng; Mat Res Soc. *Res:* Charge transfer mechanisms at solid-liquid and solid-solid interfaces; electrokinetic phenomena at surfaces and interfaces; acid-base interactions of polymers, solvents and fillers; dispersion force interactions in solutions and at interfaces. *Mailing Add:* Seeley G Mudd Chem Lehigh Univ Bldg 6 Bethlehem PA 18015-3173

FOWLER, ALAN B, b Denver, Colo, Oct 15, 28; m 50; c 4. SOLID STATE PHYSICS, SURFACE PHYSICS. *Educ:* Rensselaer Polytech Inst, BS, 51, MS, 52; Harvard Univ, PhD(appl physics), 58. *Prof Exp:* IBM FEL, IBM CORP, 58- *Concurrent Pos:* Raytheon Mfg Co, 53-56. *Honors & Awards:* Wetherill Medal, Franklin Inst, 81; Sarnoff Prize, Inst Elec & Electronics Engrs, 87; Buckley Prize, Am Phys Soc, 88; Alexander von Humboldt Prize. *Mem:* Nat Acad Sci; Nat Acad Engrs; fel Am Phys Soc; Inst Elec & Electronics Engrs; AAAS. *Res:* Semiconductor research in surface studies; optical properties of heavily doped crystals and photoconductors; injection lasers and thin film devices; primary contributions in studies of 2-D electrons. *Mailing Add:* 3511 Kamhi Dr Yorktown Heights NY 10598

FOWLER, ARNOLD K, b Exeter, NH, Aug 11, 36; m 59; c 2. ANIMAL PHYSIOLOGY, GENETICS. *Educ:* Univ NH, BS, 58; Univ Conn, MS, 60; Ohio State Univ, PhD(animal physiol), 63. *Prof Exp:* Res scientist virol, US Air Force Sch Aerospace Med, 63-64, res scientist cellular biol, 64-69; asst prof animal sci, Univ NH, 69-70; MEM STAFF, NAT CANCER INST, 70- *Mem:* AAAS; Sigma Xi. *Res:* Mammalian reproductive physiology and genetics; cellular physiology. *Mailing Add:* 7386 Old Lime Ct Middletown MD 21769

FOWLER, AUDREE VERNEE, b Los Angeles, Calif, Oct 7, 33. PROTEIN CHEMISTRY. *Educ:* Univ Calif, Los Angeles, BS, 56, PhD(biochem), 63. *Prof Exp:* Fel, Dept Molecular Biol, Albert Einstein Col Med, 63-65; fel, 65-66, asst, 66-74, assoc res biol chemist, 74-80, RES BIOL CHEMIST, UNIV CALIF, LOS ANGELES, 80- *Mem:* Am Soc Biol Chemists; Sigma Xi. *Res:* Structural studies of protein; sequence, conformation, immunological properties, particularly proteins of the Lac operon. *Mailing Add:* Dept Biol Chem Univ Calif Los Angeles Ctr Health Sci Los Angeles CA 90024

FOWLER, BRUCE ANDREW, b Seattle, Wash, Dec 28, 45; m 68; c 2. TOXICOLOGY, ENVIRONMENTAL BIOLOGY. *Educ:* Univ Wash, BS, 68; Univ Ore, PhD(exp path), 72. *Prof Exp:* Staff fel, 72-74, sr staff fel, 74-77, res biologist environ toxicol, 77-80, sr scientist, Nat Inst Environ Health Sci, 80-87; DIR & PROF PATH, UNIV MD MED SCH, 87-, DIR OFF COLLAB STUDIES ADAPTIVE RESPONSES ESTUARINE SPECIES, 90- *Concurrent Pos:* Chmn, Steering Comt, Res Triangle Environ Metals Group, 73-75; adj asst prof path, Univ NC, Chapel Hill, 74-80, adj assoc prof path, 80-87, adj assoc prof toxicol curric, 83-87; temp adv, WHO, 78-79; mem, working group, Int Agency for Res Against Cancer, 79; chmn, Dahlem Konferenzen Workshop Mechanisms Cell Inquiry, Berlin, WGermany, 85; co-chmn, NY Acad Sci Conf Chem Induced Porphyrinopathies, 86-; res fel, Japanese Soc Prom Sci, 90; mem Md Gov Coun Toxic Substances, 88, chmn, 89-, Nat Acad Sci/Nat Res Coun Toxicol Info Prog Comt, 89-, Comt Toxicol, 89-, Comt Women in Sci & Engr, 91-; chmn Nat Acad Sci/Nat Res Coun Comt Measuring Lead in Critical Pop, 89- *Mem:* NY Acad Sci; AAAS; Am Asn Pathologists; Soc Toxicol; Am Soc Pharmacol Exp Therapeut. *Res:* The ultrastructural/biochemical characterization of mechanisms of cell injury from exposure to trace metals in mammals and marine organisms in relation to intracellular binding of both toxic and essential metals; biochemical mechanisms of metal-induced alterations of cellular heme metabolism; molecular mechanisms of metal-induced alterations in gene expression. *Mailing Add:* Toxicol Prog 660 W Redwood St Baltimore MD 21201

FOWLER, BRUCE WAYNE, b Gadsden, Ala, Dec 10, 48; m; c 1. VECTOR POTENTIAL THEORY. *Educ:* Univ Ala, BS, 70; Univ Ill, Urbana, MS, 72; Univ Ala, Huntsville, PhD(physics), 78; US Army War Col, MEL-1, 91. *Prof Exp:* Syst analyst, Teledyne Brown Eng Co, 72-74; actg tech dir, 85-86, TECH DIR, ADV SYST CONCEPTS OFF, US ARMY MISSILE COMMAND, ARMY DEPT, 86-, PHYSICIST, 74-; ASST PROF PHYSICS, UNIV ALA, HUNTSVILLE, 84- *Concurrent Pos:* instr & mem grad fac, Univ Ala, Huntsville, 79-84; guest prof, Ga Inst Tech, 84- *Mem:* Am Chem Soc; Am Phys Soc; Sigma Xi; Opers Res Soc Am; Asn US Army; Air Defense Artil Asn; Mil Opers Res Soc. *Res:* Development of an empirically consistent methodology of technology forecasting; system concept development and analysis including technical, economic and political extrapolation; mathematical conjugate theories of tactics and strategy. *Mailing Add:* PO Box 279 Lacey's Spring AL 35754-0279

FOWLER, CHARLES A(LBERT), b Centralia, Ill, Dec 17, 20; m 43; c 2. ELECTRONICS. *Educ:* Univ Ill, BS, 42. *Prof Exp:* Mem staff, Radiation Lab, Mass Inst Technol, 42-45; dept head, radar systs, Airborne Instruments Lab, Inc, 45-66; dep dir tactical warfare progs, Off of Dir Defense Res & Eng, Dept Defense, 66-70; vpres & mgr equip develop labs, Equip Div, Raytheon Co, 70-76; vpres, Bedford Opers, Mitre Corp, Mass, 76-80, gen mgr, 80-84, sr vpres, 84-85; PRES, C A FOWLER ASSOCS INC, SUDBURY, MASS, 86- *Concurrent Pos:* Mem, sci adv comt, Defense Intel Agency, 72-, chmn, 76-82; mem, sci adv bd, US Air Force, 73-77; mem, Defense Sci Bd, 71-89, chmn, 84-88, vchmn, 88-89, sr fel, 89- *Mem:* Nat Acad Eng; fel Inst Elec & Electronics Engrs; fel Am Inst Aeronaut & Astronaut; fel AAAS; Asn Old Crows. *Res:* Radar; electronics. *Mailing Add:* 15 Woodberry Rd Sudbury MA 01776

FOWLER, CHARLES ARMAN, JR, b Salt Lake City, Utah, Apr 23, 12; m 34; c 2. MAGNETISM, THIN MAGNETIC FILMS. *Educ:* Univ Utah, AB, 33, MS, 34; Univ Calif, PhD(physics), 40. *Prof Exp:* from instr to asst prof, Univ Calif, Berkeley, 40-46; assoc prof, 46-50, chmn dept, 47-72, prof, 50-77, EMER PROF PHYSICS, POMONA COL, 77- *Concurrent Pos:* NSF sr fel, Univ Grenoble, France, 60-61; mem comt physics fac in cols, Am Inst Physics, 62-65; NSF fac fel, 67-68. *Mem:* Am Phys Soc; Am Asn Physics Teachers; Sigma Xi. *Res:* Magnetism; ferromagnetic domains; magneto-optics; molecular spectroscopy; optics; x-ray diffraction; originated the longitudinal Kerr magneto-optic technique for visualizing ferromagnetic domains (1952); discovered twelve new band systems in the absorption spectra of diatomic alkaline earth fluorides (1941). *Mailing Add:* Millikan Lab Pomona Col Claremont CA 91711

FOWLER, CLARENCE MAXWELL, b Centralia, Ill, Nov 26, 18; m 42; c 1. PHYSICS. *Educ:* Univ Ill, BS, 40; Univ Mich, MS & PhD(physics), 49. *Prof Exp:* Asst wire res, Am Steel & Wire Co, 40-43; prof physics, Kans State Col, 49-57; MEM STAFF, LOS ALAMOS NAT LAB, 57- *Mem:* Fel Am Phys Soc. *Res:* Shock waves; high magnetic fields; explosive energy conversion. *Mailing Add:* 3220 Arizona Los Alamos NM 87544

FOWLER, DONA JANE, b Muncie, Ind, May 8, 28; div; c 2. NEUROENDOCRINOLOGY. *Educ:* Purdue Univ, BS, 55, MS, 62, PhD, 65. *Prof Exp:* Res asst plant physiol, Purdue Univ, 54-55, cardiac res, 56-57; assoc res anal chemist, Eli Lilly Co, 57-60; asst zool & biol, Purdue Univ, 60-62 & physiol & ecol, 62-65; from instr to prof, 65-88, EMER PROF, WESTERN MICH UNIV, 88- *Concurrent Pos:* Guest Scientist, Labs, Genetics, Evolution & Biomet, Nat Ctr Sci Res, Gif-sur-Yvette, France; vis scholar, Biol Dept, Univ Ariz, 80-81 & 87; vis scientist, Argonne Nat labs, 83-; prog dir & reviewer bull, Am Meteorol Soc. *Mem:* AAAS; Am Inst Biol Sci; Am Soc Zool; Int Soc Chronobiol; Am Meteorol Soc; Am Arachnologists; Int Soc Chronobiol; Am Soc Chronobiol; Am Soc Photobiol; Sigma Xi. *Res:* Cellular regulation; environmental factors that influence the regulatory functions of invertebrates, chiefly arachnids, experimental parameters involved, including the analysis of neurosecretions and locomotion as cyclic phenomena; monochromatic light receptors in tissue culture. *Mailing Add:* 8692 N Little Oak Lane Tucson AZ 85704

FOWLER, DONALD PAIGE, b Waterbury, Conn, Nov 26, 32; Can citizen; m 83; c 3. FOREST GENETICS, FORESTRY. *Educ:* Univ NB, BSc, 55; Yale Univ, MS, 56, PhD(forestry), 64. *Prof Exp:* Res scientist forest genetics, Ont Dept Lands & Forests, 56-66; RES SCIENTIST & PROJ LEADER FOREST GENETICS, CAN FORESTRY SERV, 66- *Concurrent Pos:* Sect leader, Can Forestry Serv, 66-; res assoc, Univ NB, 70-; assoc ed, Can J Forest Res, 76-; ed, Silvae Genetica, 80. *Honors & Awards:* Sci Achievement Award, Can Forestry Serv, 81. *Mem:* Can Tree Improv Asn; Sigma Xi; Can Inst Forestry. *Res:* Forest genetics and applied tree improvement with special interest in picea and larix; species hybridization; phylogenetic relationships; population structure and development of breeding strategies. *Mailing Add:* Can Forestry Serv PO Box 4000 Fredericton NB E3B 5P7 Can

FOWLER, EARLE CABELL, b Bowling Green, Ky, June 10, 21; m 50; c 3. ELEMENTARY PARTICLE PHYSICS. *Educ:* Univ Ky, BS, 42; Harvard Univ, AM, 47, PhD(physics), 49. *Prof Exp:* Assoc physicist, Brookhaven Nat Lab, 49-52; from asst prof to assoc prof, Yale Univ, 52-62; prof, Duke Univ, 62-70; prof physics, Purdue Univ, 71-82, head dept, 71-77; CHIEF, FACIL OPERS BR, DIV HIGH ENERGY PHYSICS, DEPT ENERGY, 82- *Concurrent Pos:* Consult, Brookhaven Nat Lab, 52-76; Fulbright lectr, Univ Birmingham, Eng, 58-59; Fulbright scholar, Univ Rome, 67-68; mem bd dirs, Triangle Univ Comput Ctr, 69-71; panel mem, NSF Comput Facil Div, 69-71; chmn exec comt, Fermi Nat Accelerator Lab User's Orgn, 72-73; sr physicist, High Energy Physics Div, US Dept Energy, 80-82. *Mem:* Am Phys Soc. *Res:* Cosmic Rays; high energy particle physics; micrometeorology. *Mailing Add:* ER-223 GTN High Energy Physics US Dept Energy Washington DC 20545

FOWLER, EDWARD HERBERT, b Stoneham, Mass, Oct 25, 36; m 56; c 2. PATHOLOGY. *Educ:* Univ NH, BS, 58; Mich State Univ, MS & DVM, 62; Ohio State Univ, PhD(vet path), 65; Am Col Vet Path, dipl, 67. *Prof Exp:* Asst prof vet path, Ohio State Univ, 66-70, actg asst dean acad affairs, Vet Col, 70; assoc prof oncol, Lab Animal Med & assoc prof path, Sch Med, Univ Rochester, 70-78; chem hyg fel & mgr path & animal care, Carnegie-Mellon Inst Res, 78-79; mgr path & animal care, 80-82, ASSOC DIR, BUSHY RUN RES CTR, UNION CARBIDE CORP, 83- *Concurrent Pos:* Consult, Bur Drugs, Food & Drug Admin, 71-72, Breast Cancer Task Force, NCI, 78-80, NTP-PWG, 84-86. *Mem:* Am Vet Med Asn; Am Soc Vet Clin Path; Soc Toxicol Path; Am Asn Pathol. *Res:* Pathogenesis of animal neoplasms; etiology and pathogenesis of steroid hormone dependent animal and human neoplasms; toxicologic pathology. *Mailing Add:* 6702 Mellon Rd Export PA 15632-8902

FOWLER, ELIZABETH, b Schenectady, NY, Apr 18, 43. PROTEIN CHEMISTRY, CELL BIOLOGY. *Educ:* Cornell Univ, AB, 65; Harvard Univ, PhD(biol chem), 72. *Prof Exp:* Fel genetics, Univ Wis-Madison, 72-76, asst scientist, 76-77; asst prof Bacteria & Immunol, Univ NC, Chapel Hill, 77-83; assoc prof, Microbiol & Immunol, Univ Ala, Mobile, 83-85; STAFF SCIENTIST, CIBA-GEIGY, RESEARCH TRIANGLE PARK, NC, 85- *Mem:* Am Asn Immunologists; Am Soc Cell Biol; AAAS; Entomol Soc Am. *Res:* Genetic engineering of crop plants for insect resistance; insect-selective protein toxins; biologicals for insect control; protein structure and function. *Mailing Add:* Ciba Geigy Biotec Res Unit 111 Briarcliff Durham NC 27707

FOWLER, EMIL EUGENE, b Morgantown, WVa, Sept 15, 23; m 49; c 3. CHEMISTRY. *Educ:* WVa Univ, AB, 46, MS, 47. *Prof Exp:* Chief, Radioisotopes Br, Tenn, 47-56, deputy dir, Isotopes Div, 56, dep asst dir, Div Civilian Appln & Off Indust Develop, 56-59, DIR ISOTOPES DEVELOP, US ATOMIC ENERGY COMN, WASHINGTON, DC, 65- *Concurrent Pos:* Dir chem & indust applns, Int Atomic Energy Agency, Vienna, Austria, 75-77; chief tech adv, UN Develop Prog Indust Proj, Vienna, Austria, 80- *Mem:* AAAS; Am Nuclear Soc; Soc Nuclear Med; Health Phys Soc; Am Inst Chem. *Res:* Accelerating development of widespread applications of radioisotopes and high-intensity radiation; radioisotopes production, process development, pricing and marketing; production and distribution of radioisotopes and isotopes technology training. *Mailing Add:* 5124 Westpath Way Bethesda MD 20816

FOWLER, ERIC BEAUMONT, b Milbank, SDak, May 4, 14; m 42; c 3. SOIL CHEMISTRY. *Educ:* Kans State Univ, BS, 42, MS, 44; Iowa State Col, PhD(chem & physiol bact), 50. *Prof Exp:* Student asst & lab asst bact, Kans State Univ, 39-42; asst soil conserv, Kans State Col, 42-44, instr bact, 44-50, from asst prof to assoc prof, 50- 56; mem staff & alt group leader, Los Alamos Sci Lab, 56-75, mem staff & asst group leader, 75-78, mem staff & prin investr, Soil-Waste Interactions, 78-81; RETIRED. *Concurrent Pos:* Soils element mgr, Nev Appl Ecol Group Energy Res & Develop Admin, 70-78. *Mem:* AAAS; Am Nuclear Soc; Sigma Xi. *Res:* Chemical and physiological bacteriology; disposal of biological and industrial wastes; plant uptake and control of radio isotopes in the environment; radioactive-soil interactions. *Mailing Add:* 2488 45th St Los Alamos NM 87544

FOWLER, FRANK CAVAN, b Kansas City, Mo, June 15, 18; m 43; c 3. CHEMICAL ENGINEERING. *Educ:* Univ Ill, BS, 39; Univ Mich, MS, 40, PhD(chem eng), 43. *Prof Exp:* Chem engr, Phillips Petrol Co, Okla, 43-46; instr univ exten, Univ Okla, 45, from assoc prof to prof chem eng, 46-51; consult chem engr, 51-58; PRES, RES ENGRS, INC, 58- *Concurrent Pos:* Vis prof, Univ Kans, 52-53 & 56-58. *Mem:* Am Chem Soc; Sigma Xi; Am Inst Chem Engrs. *Res:* Fluid flow; heat transfer; distillation; mixing of fluids by successive flow through pipes; process design; thermodynamics. *Mailing Add:* 10000 E 137th St Kansas City MO 64149

FOWLER, FRANK WILSON, b Portland, Maine, May 16, 41; m 63. ORGANIC CHEMISTRY. *Educ:* Univ Colo, PhD(chem), 67. *Prof Exp:* Leverhulme vis fel, Univ EAnglia, Eng, 67-68; asst prof, 68-73, assoc prof, 73-80, PROF CHEM, STATE UNIV NY STONY BROOK, 80- *Res:* Synthesis of interesting and unusual heterocyclic molecules; aromaticity; valence tantomerism; natural products. *Mailing Add:* 12 Point Rd Bellport NY 11713-2611

FOWLER, GERALD ALLAN, b Tacoma, Wash, June 1, 34; m 62; c 2. MICROPALEONTOLOGY, PALEOECOLOGY. *Educ:* Univ Puget Sound, BS,57; Univ Southern Calif, PhD(geol), 65. *Prof Exp:* Asst prof marine geol, Dept Oceanog, Ore State Univ, 64-72; assoc prof earth sci, div sci, 72-, AT DEPT GEOL, UNIV WIS, PARKSIDE. *Mem:* Geol Soc Am; Soc Econ Paleontologists & Mineralogists. *Res:* Interpretation of paleoenvironments and stratigraphy of marine neogene sediments of continental margin of the Northwest United States using benthic foraminifera. *Mailing Add:* Dept Geol Univ Wis Parkside Kenosha WI 53141

FOWLER, GREGORY L, b Wichita, Kans, Aug 19, 34; m 67; c 2. GENETICS. *Educ:* Wichita State Univ, BA, 56, MS, 60; Brown Univ, PhD(genetics), 68. *Prof Exp:* Lectr biol, Wichita State Univ, 59-60; instr, Bethany Col, 60-62; asst prof, George Washington Univ, 67-69; assoc & vis asst prof, Univ Ore, 69-71; asst prof, Univ Dseldorf, 71-74 & Univ Ore, 74-76; ASSOC PROF BIOL, SOUTHERN ORE STATE COL, 76- *Res:* In vitro cytological and biochemical characterization of lampbrush chromosomes in Drosophila hydei testis; light and electron microscope studies of Drosophila spermatogenesis in vitro. *Mailing Add:* Dept Biol Southern Ore State Col Ashland OR 97520r

FOWLER, H(ORATIO) SEYMOUR, b Highland Park, Mich, Mar 1, 19; wid; c 1. BIOLOGY, SCIENCE EDUCATION. *Educ:* Cornell Univ, BS, 41, MS, 46, PhD(sci educ), 51. *Prof Exp:* Teacher pub schs, NY, 46 & 47-49; asst sci educ, Cornell Univ, 49-51; asst prof, Southern Ore Col, 51-52 & biol, Iowa State Teachers Col, 52-57; prof sci educ, 57-83, chmn, sci educ fac, 69-83, coordr, div acad curric & instr, 74-76, EMER PROF NATURE & SCI EDUC, PA STATE UNIV, 83-, DIR, PA CONSERV LAB, 60-; SCI DIR, NAT JR SCI & HUMANITIES SYMP, 84-; SCI DIR, NAT JR SCI & HUMANITIES SYMP, 84- *Concurrent Pos:* Dir, Iowa teachers conserv camp, 52-57; Fulbright lectr, Korea, 68-69; Paul Harris Fel, Rotary Int, 83; sci dir, Nat Jr Sci & Humanities Symp, 84- *Honors & Awards:* Fulbright Lectr, Korea, 68-69. *Mem:* Fel AAAS; Am Nature Study Soc (vpres, 64, pres elect, 66, pres, 67); Nat Asn Res Sci Teaching; hon mem Nat Asn Biol Teachers (vpres, 59); Sigma Xi. *Res:* Science education; conservation education. *Mailing Add:* 1342 Park Hills Ave W State College PA 16803

FOWLER, HOWLAND AUCHINCLOSS, b New York, NY, Jan 25, 30; m 62; c 2. PHYSICS, RESEARCH ADMINISTRATION. *Educ:* Princeton Univ, AB, 52; Brown Univ, MSc, 55, PhD(physics), 57. *Prof Exp:* Nat Res Coun assoc, 57-58, physicist, 58-71, PHYS SCI ADMINR, NAT BUR STANDARDS, 71- *Mem:* Am Phys Soc; Soc Indust & Appl Math; Inst Elec & Electronics Engrs. *Res:* Electron physics; electron scattering; electron-optical technique; far-UV optical constants; application of Josephson effect to voltage measurement; applications of computer graphics to physical problems. *Mailing Add:* 5413 Albemarle St Bethesda MD 20816

FOWLER, IRA, experimental embryology, for more information see previous edition

FOWLER, JAMES A, b New York, NY, Jan 30, 23; m 55; c 2. VERTEBRATE EMBRYOLOGY, COMPUTER MODELING. *Educ:* Princeton Univ, BSE, 44; Columbia Univ, MA, 57, PhD(zool), 61. *Prof Exp:* Power engr, Western Union Tel Co, 47-49; elec engr, St Anthony Mining & Develop Co, 49; lectr embryol, Columbia Univ, 59 & zool, Barnard Col, 59-60; asst dean col arts & sci, 64-69, assoc dean health prof adv, 69-74, asst prof, 61-85, EMER PROF BIOL SCI, STATE UNIV NY STONY BROOK, 85- *Mem:* AAAS; Am Soc Zool; Am Soc Naturalists. *Res:* Interaction between evolution and embryology; population genetics of development and its evolutionary history; theoretical analysis of development. *Mailing Add:* Box L Coraway Rd Setauket NY 11733

FOWLER, JAMES LOWELL, b Stephenville, Tex, Oct 20, 35; m 61; c 2. CROP PHYSIOLOGY. *Educ:* Tex Tech Univ, BS, 62, MS, 66; Tex A&M Univ, PhD(plant physiol), 71. *Prof Exp:* Asst prof, 71-79, ASSOC PROF AGRON, NMEX STATE UNIV, 79- *Mem:* Am Soc Agron; Crop Sci Soc Am. *Res:* Cotton production; plant growth and development; environmental physiology; plant water relations; new crops. *Mailing Add:* NMex State Univ Box 30003 Las Cruces NM 88003-0003

FOWLER, JOANNA S, b Aug 9, 42. ORGANIC CHEMISTRY. *Educ:* Univ SFla, BA, 64; Univ Colo, PhD(chem), 68. *Prof Exp:* Res assoc organothallium chem, Univ East Anglia, Eng, 68-69; res assoc org chem, 69-71, CHEMIST, BROOKHAVEN NAT LAB, 71- *Mem:* Soc Nuclear Med; Am Chem Soc. *Res:* Organic synthesis; reactions of molecular fluorine; design and synthesis of radiopharmaceuticals labeled with short-lived nuclides; mechanisms of drug localization. *Mailing Add:* Dept Chem Brookhaven Nat Lab Upton NY 11973

FOWLER, JOHN ALVIS, b High Point, NC, Oct 17, 21; m; c 2. PSYCHOANALYSIS, PSYCHIATRY. *Educ:* Wake Forest Univ, BS, 43; Bowman Gray Sch Med, MD, 46. *Prof Exp:* Intern, US Naval Hosp, Corpus Christi, Tex, 46-47; resident psychiat, Univ Colo Med Ctr, 49-52; from asst prof to prof,53-88, HEAD DIV, MED CTR, DUKE UNIV, 63-, EMER PROF CHILD PSYCHIAT, 88- *Concurrent Pos:* Resident child psychiat, Univ Colo Med Ctr & Conn Bur Ment Hyg, 51-53; dir, Durham Child Guid Clin, 53-63; training & supv psychoanalyst, Univ NC-Duke Univ Psychoanal Training Prog, 72-; lectr educ, Duke Univ, 73-; Supv child psychoanal, Wash Psychoanal Inst, 75- *Mem:* Am Psychoanal Asn; Asn Child Psychoanal; Am Psychiat Asn; Am Acad Child Psychiat. *Mailing Add:* 2721 Spencer St Durham NC 27705

FOWLER, JOHN RAYFORD, b Winnfield, La, July 11, 43; m 64; c 2. INORGANIC CHEMISTRY, TEXTILE TECHNOLOGY. *Educ:* McMurry Col, BA, 65; Univ Kans, PhD(inorg chem), 69. *Prof Exp:* Teaching asst chem, Univ Kans, 65-68; res chemist, Textile Res Lab, 69-77, res chemist, 77-79, STAFF CHEMIST, SAVANNAH RIVER LAB, E I DU PONT DE NEMOURS & CO, 80- *Concurrent Pos:* Instr, Aiken TEC, SC, 81- *Mem:* Am Chem Soc. *Res:* Transition metal complexes; nonaqueous solvents; cyano complexes; IR spectroscopy; synthetic textile fibers; textile technology; polymer science; nuclear chemistry; nuclear waste management. *Mailing Add:* 849 Magnolia Aiken SC 29801-4907

FOWLER, MALCOLM MCFARLAND, b Houston, Tex, Dec 13, 43; m 73; c 2. NUCLEAR & ANALYTICAL CHEMISTRY. *Educ:* Univ NMex, BS, 66; Wash Univ, St Louis, MA, 67, PhD(chem), 72. *Prof Exp:* Fel nuclear chem, McGill Univ, 72-73; fel heavy ion reactions, Lawrence Berkeley Lab, Berkeley, 73-75; STAFF MEM NUCLEAR CHEM, LOS ALAMOS SCI LAB, 75- *Mem:* Am Chem Soc; Am Phys Soc; Sigma Xi. *Res:* Heavy-ion reactions; instrument development; environmental and atmospheric research. *Mailing Add:* 134 Aztec Los Alamos NM 87544

FOWLER, MICHAEL, b Doncaster, Eng, Apr 30, 38; m 65. THEORETICAL PHYSICS. *Educ:* Cambridge Univ, BA, 59, PhD(field theory), 62. *Prof Exp:* Instr physics, Princeton Univ, 62-63; asst prof, Univ Md, 63-65; asst prof, Univ Toronto, 65-68; assoc prof, 68-73, PROF PHYSICS, UNIV VA, 73- *Mem:* Am Phys Soc. *Res:* Analytic methods in potential theory and perturbation theory; electrons in high magnetic fields in metals. *Mailing Add:* United Technols Res Ctr MS-70 Silver Lane East Hartford CT 06108

FOWLER, MURRAY ELWOOD, b Glendale, Wash, July 17, 28; m 50; c 5. VETERINARY MEDICINE. *Educ:* Utah State Univ, BS, 52; Iowa State Univ, DVM, 55; Am Bd Vet Toxicol, dipl; Am Col Vet Internists, dipl; Am Col Zool Med, dipl, 84. *Prof Exp:* Pvt pract vet med, 55-58; chmn dept med, 73-85, PROF VET MED, UNIV CALIF, DAVIS, 58- *Mem:* Am Vet Med Asn; Am Asn Zoo Vets; Wildlife Dis Asn. *Res:* Clinical toxicology; teaching and research in problems of zoo animal medicine and wildlife diseases. *Mailing Add:* Dept Med Sch Vet Med Univ Calif Davis CA 95616

FOWLER, NOBLE OWEN, b Vicksburg, Miss, July 14, 19; m 42; c 3. INTERNAL MEDICINE, CARDIOVASCULAR DISEASE. *Educ:* Univ Tenn, MD, 41. *Prof Exp:* USPHS fel cardiol, Univ Cincinnati, 48-49, trainee, 49-50, Am Heart Asn res fel & asst prof med, 51-52; asst prof med, State Univ NY, 52-54; assoc prof med & chair cardiovasc res, Emory Univ, 54-57; assoc prof clin med, 57-59, assoc prof med, 59-64, Prof Internal Med, Univ Cincinnati, 64-85, dir div cardiol, 70-86; dir cardiac res lab, Cincinnati Gen Hosp, 64-70; EMER PROF MED & PHARMACOL, UNIV CINCINNATI, 85- *Concurrent Pos:* Consult, Dayton Vet Hosp, 49-52 & Brooklyn Vet Hosp, NY, 54. *Mem:* Fel Am Col Physicians; Am Physiol Soc; Am Clin & Climat Soc; Sigma Xi; fel Am Col Cardiol. *Res:* Physiology and pharmacology of the pulmonary circulation; physiology of regulation of cardiac output; plasma substitutes; pericardial function. *Mailing Add:* 3533 Deep Woods Lane Cincinnati OH 45208

FOWLER, NORMA LEE, b St Louis, Mo, May 19, 52. POPULATION ECOLOGY, POPULATION GENETICS. *Educ:* Univ Chicago, BA, 73; Duke Univ, PhD(bot), 78. *Prof Exp:* Fel, Univ Col NWales, Bangor, UK, 78-79; asst prof, 79-85, ASSOC PROF BOT, UNIV TEX, AUSTIN, 85- *Mem:* Ecol Soc Am; Soc Study Evolution; British Ecol Soc; fel AAAS. *Res:* Herbaceous perennials; plant population dynamics and population regulation, competition, community structure, life histories, and quantitative genetic variation in ecologically relevant characters. *Mailing Add:* Dept Bot Univ Tex Austin TX 78713

FOWLER, RICHARD GILDART, b Albion, Mich, June 13, 16; m 39; c 4. RADIATIVE LIFETIMES, LIGHTNING. *Educ:* Albion Col, AB, 36; Univ Mich, MS, 39, PhD(physics), 41. *Prof Exp:* Asst, Dow Chem Co, Mich, 36-38; asst, Univ Mich, 38-40, res physicist, 41 & 42-46; instr physics, NC State Col, 41-42; from asst prof to prof, 46-61, chmn dept, 55-59 & 66-68, res prof, 61-80, EMER PROF PHYSICS, UNIV OKLA, 80- *Concurrent Pos:* Guggenheim fel, 52; Fulbright lectr, Australia, 63; NATO fel, 70. *Mem:* Fel Inst Physics; AAAS; fel Am Phys Soc. *Res:* Ultraviolet spectrochemical analysis; organic structure determination by infrared spectra; plasma physics and electrically generated shock waves; plasma driven shock tubes; purification of graphite; mechanisms involved in the production of radiation; atomic and molecular lifetimes; electron fluids and lightning. *Mailing Add:* Dept Physics Univ Okla 440 W Brooks Norman OK 73069

FOWLER, ROBERT MCSWAIN, b Georgetown, Tex, Mar 6, 06; m 31; c 3. METALLURGICAL RESEARCH. *Educ:* Southwestern Univ, BA, 27, George Washington Univ, MA, 31. *Prof Exp:* Jr chemist anal, Nat Bur Standards, 27-36; res chemist, Union Carbide Corp, 36-63, US Dept Defense, 63-69 & Cybertech, Inc, 69-71; RETIRED. *Concurrent Pos:* Serv Corps Retired Execs consult, US Small Bus Admin, 71. *Mem:* Am Chem Soc; fel AAAS. *Res:* Analytical and metallurgical research on various industrial problems; exotic metals production; purification and application. *Mailing Add:* 4102 Balcones Dr Austin TX 78731

FOWLER, SCOTT WELLINGTON, b Berkeley, Calif, May 31, 41; m 65; c 2. BIOLOGICAL OCEANOGRAPHY. *Educ:* Univ Calif, Riverside, BA, 64; Ore State Univ, MS, 66, PhD(biol oceanog), 69. *Prof Exp:* Scientist, Battelle-Northwest Labs, 67; Fulbright lectr, Monterrey Inst Technol, Mex, 69-70; SCIENTIST & HEAD RADIOECOL LAB, INT ATOMIC ENERGY AGENCY, INT LAB MARINE RADIOACTIVITY, MONACO, 70- *Concurrent Pos:* UN Environ Prog Consult, Multidisciplinary Mission, Persian Gulf States, 80 & mission leader, Survey Tar & Oil Pollution, Sultanate Oman, 80; pres marine radioactivity comt, Int Comn Sci, Explor Mediterranean, 78-84, 90-; prin investr, subcontract vertical flux of transuranics in the ocean, NSF, 80-89, contract particle flux EROS 2000, EEC, 90-93; consult, UN Environ Prog, Libya, 82, Algeria, 83; UN Environ

Prog consult, multidisciplinary mission, Somalia, 86; mem, Int Union Conserv Nature Comn Ecol, 82- & UN Group Experts Sci Aspects Marine Pollution, 80-84, Int Mussel Watch Comt, 88-, Sci Comt French Ocean Flux Prog, 89- *Mem:* NY Acad Sci; Am Soc Limnol & Oceanog; Sigma Xi; Int Union Radioecologists. *Res:* Transfer of radionuclides through marine food chains, biokinetics of heavy metals and organic pollutants in marine organisms, zooplankton physiology, marine invertebrate physiology and ecology; vertical flux of marine biogenic particulates. *Mailing Add:* Int Atomic Energy Agency Int Lab Marine Radioactivity 19 Ave Des Castellans MC 98000 Monaco

FOWLER, STANLEY D, b Pagosa Springs, Colo, Apr 11, 42; m 68; c 2. EXPERIMENTAL PATHOLOGY, CELL BIOLOGY. *Educ:* Pomona Col, BA, 64; Rockefeller Univ, PhD(cell biol), 69. *Prof Exp:* Teaching fel biochem, Univ Louvain, Belg, 69-71; teaching fel path, Univ Wash Sch Med, 71-73; from asst prof to assoc prof biochem cytol, Rockefeller Univ, 73-82; PROF PATH, SCH MED, UNIV SC, 82-, ASSOC DEAN RES & GRAD STUDIES, 86- *Concurrent Pos:* Standing mem, Path A Study Sect, NIH, 82-86. *Honors & Awards:* Estab Investr Award, Am Heart Asn, 78. *Mem:* Am Soc Biol Chemists; Am Soc Cell Biol; Am Asn Pathologists; Int Acad Path-US Div; Am Heart Asn. *Res:* Pathogenesis of atherosclerosis; identification of cell types in atherosclerosis lesions; mechanisms of intracellular lipid deposition. *Mailing Add:* Dept Path Univ SC Sch Med Columbia SC 29208

FOWLER, STEPHEN C, PSYCHOLOGY. *Prof Exp:* PROF PSYCHOL & PHARMACOL, UNIV MISS, 73- *Mailing Add:* Dept Psychol & Pharmacol Univ Miss University MS 38677

FOWLER, THOMAS KENNETH, b Thomaston, Ga, Mar 27, 31; m 56; c 3. THEORETICAL PHYSICS. *Educ:* Vanderbilt Univ, BE, 53, MS, 55; Univ Wis, PhD(theoret physics), 57. *Prof Exp:* Physicist, Oak Ridge Nat Lab, 57-65 & gen atomic div, Gen Dynamic Corp, 65-67; head plasma physics div, 67, group leader plasma theory, Lawrence Livermore Lab, 67-69, ASSOC DIR CONTROLLED THERMONUCLEAR RES, LAWRENCE LIVERMORE LAB, 70-, DIV LEADER, 69-; CHMN, DEPT NUCLEAR ENG, UNIV CALIF, BERKELEY, 88- *Mem:* Nat Acad Sci; fel Am Phys Soc; Sigma Xi. *Res:* Controlled fusion; plasma and nuclear physics, especially scattering theory. *Mailing Add:* Dept Nuclear Eng 4153 Etcheverry Hall Univ Calif Berkeley CA 94720

FOWLER, TIMOTHY JOHN, b Birmingham, Eng, Jan 11, 38; m 65; c 3. STRUCTURAL ENGINEERING, NONDESTRUCTIVE TESTING. *Educ:* Univ Birmingham, Eng, BSc, 59; Univ London, DIC, 60; Univ Tex, PhD(struct eng), 66. *Prof Exp:* Grad engr civil eng, Binnie, Deacon & Gourley, London, 60-61; design engr civil eng, Binnie & Partners, Kuala Lumpur, Malaya, 62-63; sr engr, eng specialist & prin eng specialist civil, struct & mech eng, Monsanto Co, 65-71, Monsanto fel, 71-75, sr Monsanto fel, 76-83, DISTINGUISHED MONSANTO FEL, MONSANTO CO, ST LOUIS, 83- *Mem:* Am Soc Civil Engrs; Am Soc Mech Engrs; Am Concrete Inst; Inst Civil Engrs. *Res:* Acoustic emission; structural mechanics; structural plastics; refractory concrete. *Mailing Add:* 802 Millfield Court St Louis MO 63017

FOWLER, WALLACE T(HOMAS), b Greenville, Tex, Aug 27, 38; m 68; c 2. AEROSPACE ENGINEERING. *Educ:* Univ Tex, Austin, BA, 60, MS, 61, PhD(eng mech), 65. *Prof Exp:* Asst prof eng mech, 65-67, from asst prof to assoc prof aerospace eng, 67-77, PROF AEROSPACE ENG & ENG MECH, UNIV TEX, AUSTIN, 77-, DIR, BUR ENG TEACHING, COL ENG, OFF DEAN, 80- *Concurrent Pos:* Consult, Manned Spacecraft Ctr, 66- & Gen Dynamics/Ft Worth, 67-69. *Mem:* Am Inst Aeronaut & Astronaut. *Res:* Flight mechanics; numerical optimization; guidance and control. *Mailing Add:* Dept Aerospace Studies Univ Tex Austin Austin TX 78712

FOWLER, WILLIAM ALFRED, b Pittsburgh, Pa, Aug 9, 11; m 40, 89; c 2. NUCLEAR ASTROPHYSICS. *Educ:* Ohio State Univ, BEngPhys, 33; Calif Inst Technol, PhD(physics), 36. *Prof Exp:* Res fel, 36-39, from asst prof to prof, 39-82, EMER PROF PHYSICS, CALIF INST TECHNOL, 82- *Concurrent Pos:* Asst dir res, Nat Defense Res Comt, 41-45; tech observ, Off Field Serv & New Develop Div, Dept War, 44; sci dir proj Vista, Dept Defense, 51-52; Fulbright lectr & Guggenheim fel, Pembroke Col & Cavendish Lab, Cambridge Univ, 54-55, 61-62 & St Johns Col, 61-62; vis prof, Mass Inst Technol, 66; mem, Nat Sci Bd, NSF, 68-74; mem bd dirs, Am Friends Cambridge Univ, 70-78; mem space sci bd, Nat Acad Sci, 70-73 & 77-80, mem coun, 74-77, chmn, Off Phys Sci, 81-84 & mem comt phys sci, math & resources, 81-87; mem gov bd, Am Inst Physics, 74-80; mem, Soc Am Baseball Res, 80- *Honors & Awards:* Nobel Prize for Physics, 83; Lamme Medal, 52; Liege Medal, 5; Barnard Medal, 65; Apollo Achievement Award, NASA, 69; Tom W Bonner Prize, Am Phys Soc, 70; G Unger Vetlesen Prize, 73; Nat Medal of Sci, 74; Eddington Medal, Royal Astron Soc, 78; Bruce Gold Medal, Astron Soc Pac, 79; Sullivant Medal, Ohio State Univ, 85; E A Milne lectr, Milne Soc, 86; William A Fowler Award, Ohio Sect, Am Phys Soc, 86. *Mem:* Nat Acad Sci; AAAS; fel Am Phys Soc (pres, 76); Am Astron Soc; Am Philos Soc; Am Acad Arts & Sci; Royal Soc Arts; Int Astron Union; Asn Royal Astron Soc. *Res:* Studies of nuclear forces and reaction rates; nuclear spectroscopy; structure of light nuclei; thermonuclear sources of stellar energy and element synthesis in stars, supernovae and the early universe; study of general relativistic effects in quasar and pulsar models; nuclear cosmochronology. *Mailing Add:* Kellogg Radiation Lab 106-38 Calif Inst Technol Pasadena CA 91125

FOWLER, WILLIAM MAYO, JR, b Brooklyn, NY, June 16, 26; m 50; c 1. PHYSICAL MEDICINE & REHABILITATION. *Educ:* Springfield Col, BS, 48, MEd, 49; Univ Southern Calif, MD, 57. *Prof Exp:* Instr, dept phys educ, Univ Calif, Los Angeles, 49-52 & United Cerebral Palsy Found grant pediat, 59-60; training grant neurol, NIMH, 60-61; lectr phys med & rehab, Univ Calif, Los Angeles, 63-64, asst prof, 64-68, actg chmn dept, 67, chief rehab med, 67-68, assoc prof phys med & rehab & pediat, 68; assoc prof, 68-72, chmn dept, 68-81, PROF PHYS MED & REHAB, UNIV CALIF, DAVIS, 72- *Concurrent Pos:* Dir phys med & rehab, Harbor Gen Hosp, Torrance, Calif, 65; attend physician phys med, Wadsworth Vet Admin Hosp, Los Angeles, 65-68 & Long Beach Vet Admin Hosp, 65-68; consult phys med & rehab, San Fernando Vet Admin Hosp, 67-68; dir, Off Allied Health Sci, Univ Calif, Davis, 69-71. *Mem:* Acad Phys Med & Rehab (pres, 80-81). *Res:* Muscle biology and muscle disease; work physiology; pediatrics. *Mailing Add:* 44346 N El Macero Dr El Macero CA 95618

FOWLER, WYMAN BEALL, JR, b Scranton, Pa, June 18, 37; m 61; c 4. SOLID STATE PHYSICS. *Educ:* Lehigh Univ, BS, 59; Univ Rochester, PhD(physics), 63. *Prof Exp:* Res assoc physics, Univ Rochester, 63 & Univ Ill, 63-66; assoc prof, 66-69, chmn dept 78-84, 88-91, PROF PHYSICS, LEHIGH UNIV, 69- *Concurrent Pos:* Consult, Argonne Nat Lab, 63-66, Naval Res Lab, 66-81, US Army, ET&D Lab, Ft Monmouth, NJ, 84-87; vis res fel, Oxford Univ, 85; vis prof, Univ Parma, 86. *Honors & Awards:* Eastman Kodak Sci Award, 63. *Mem:* AAAS; Fel Am Phys Soc. *Res:* Solid state theory; electronic properties of insulators and semiconductors; color centers; band structures. *Mailing Add:* Dept Physics No 16 Lehigh Univ Bethlehem PA 18015

FOWLES, GEORGE RICHARD, b Glenwood Springs, Colo, Apr 2, 28; m 54, 88; c 4. HIGH PRESSURE PHYSICS. *Educ:* Stanford Univ, BS, 52, MS, 54, PhD(geophysics), 62. *Prof Exp:* Geophysicist, Phelps Dodge Corp, 54-55; physicist, Poulter Labs, Stanford Res Inst, 55-62, group head shock wave physics, 62-63, div dir, 63-66; assoc prof, 66-73, PROF PHYSICS, WASH STATE UNIV, 73-, CHMN PHYSICS, 84- *Concurrent Pos:* Sr staff scientist, Physics Int Co, 69-70; consult, Nat Mat Adv Bd, Nat Res Coun, 70 & 77-78, Inst Cerac, SA, Switzerland, 72, Lawrence Livermore Lab, Stanford Res Inst & Gen Motors Corp, 71-; Fulbright fel, US Educ Found, NZ, 75; vis prof, Australia Nat Univ, 83. *Mem:* Am Geophys Union; Am Phys Soc. *Res:* Numerical simulation of reactive wave propagation in phase- transfering substances; including such phenomena as explosive boiling in fluids; propagation of martensitic transformations in metals and condensation discontinuities. *Mailing Add:* Dept Physics 2814 Wash State Univ Pullman WA 99164

FOWLES, GRANT ROBERT, b Fairview, Utah, Sept 19, 19; m 42; c 4. QUANTUM OPTICS. *Educ:* Univ Utah, BS, 41; Univ Calif, PhD(physics), 50. *Prof Exp:* From asst prof to prof physics, Univ Utah, 50-90; RETIRED. *Mem:* Am Phys Soc; Optical Soc Am. *Res:* Spectroscopy; metal vapor lasers. *Mailing Add:* 1864 Princeton Ave Salt Lake City UT 84108

FOWLES, PATRICK ERNEST, b Harrow, Eng, Nov 7, 38; m 67; c 2. CHEMICAL ENGINEERING, TRIBOLOGY. *Educ:* Univ London, BSc, 60; Mass Inst Technol, ScD(chem eng, fluid mech), 66. *Prof Exp:* Res chem engr, Mobil Res & Develop Corp, 66-69, sr res engr, Cent Res Div Lab, 69-74, res assoc, 74-77, mgr lubrication & resources sci group, 77-78, asst mgr, Prod Res & Tech Serv Sect, 78-79, mgr, 79-83, sr planning assoc, 83-85, mgr, Lubricants & Additives, 85-88, MGR, PROD RES & TECH SERV DIV, 88- *Mem:* Soc Tribologists & Lubrication Engrs; Am Inst Chem Engrs; Soc Automotive Engrs; Am Soc Testing & Mat. *Res:* Lubricants and fuels product development; tribology; hydrodynamic and elastohydrodynamic lubrication; rheology of lubricants and related fluids; oil shale. *Mailing Add:* Mobil Res & Develop Corp Billingsport Rd Paulsboro NJ 08066

FOWLIS, WILLIAM WEBSTER, b Lanarkshire, Scotland, Jan 21, 37; m 70; c 1. FLUID DYNAMICS CRYSTAL GROWTH, SURFACE TENSION INTERFACE SHAPES. *Educ:* Glasgow Univ, BSc, 59; Univ Durham, PhD(physics), 64. *Prof Exp:* Res staff, Dept Geol & Geophys, Mass Inst Technol, 62-64; res assoc & lectr meteorol, Fla State Univ, 65-68, asst prof & assoc prof meteorol, Geophys Fluid Dynamics Inst, 68-74; assoc prog dir meteorol, Atmospheric Res Sect, NSF, 74-76; res scientist, atmospheric sci div, 76-83, RES SCIENTIST, LOW GRAVITY SCI DIV, MARSHALL SPACE FLIGHT CTR, NASA, 83- *Concurrent Pos:* Assoc prof meteorol, Fla State Univ, 74-76. *Honors & Awards:* Medal Except Sci Achievement, NASA, 82, Cert Recognition, 83. *Mem:* AAAS; Am Geophys Union; Am Meteorol Soc; Am Phys Soc. *Res:* Theoretical fluid dynamical studies of materials processing systems in the low gravity environment of space and in ground-based laboratories; theoretical and experimental studies of laboratory models of planetary atmospheric circulations. *Mailing Add:* 9411 Valley Lane Huntsville AL 35803

FOX, A(RTHUR) GARDNER, b Syracuse, NY, Nov 22, 12; m 38; c 4. ENGINEERING. *Educ:* Mass Inst Technol, BS, MS & EE, 35. *Prof Exp:* Head, Dept Coherent Wave Physics, Bell Tel Labs, 36-78; RETIRED. *Concurrent Pos:* Ed, J Quantum Electronics. *Honors & Awards:* Quantum Electronics Award & Microwave Career Award, Inst Elec & Electronics Engrs, 78, David Sarnoff Award, 79 & Centennial Medal, 84. *Mem:* Fel Inst Elec & Electronics Engrs; fel Optical Soc Am. *Res:* Coherent optical wave techniques; lasers, resonators, optical beams; microwave techniques in centimeter and millimeter wave length ranges; microwave magnetics. *Mailing Add:* 10 Conover Ln Rumson NJ 07760

FOX, ADRIAN SAMUEL, b Chicago, Ill, Apr 3, 36; div; c 2. ORGANIC CHEMISTRY, BIOMEDICAL MATERIALS. *Educ:* Univ Ill, BS, 57; Univ Wash, PhD(org chem), 62. *Prof Exp:* Res fel, Ohio State Univ, 62-63; res chemist, Plastics Div, Union Carbide Corp, NJ, 63-66; staff chemist, Res Dept, Raychem Corp, Calif, 66-68; res chemist, Polymer Res Dept, Pennwalt Chem Corp, 68-70, proj leader, Polymer Res Dept, Pennwalt Corp, 70-76; mgr dent mat res, Myerson Tooth Corp, Cambridge, Mass, 76-77; mgr dent mat res, Howmedica Dent Res, Howmedica Div, Pfizer, Inc, 77-79, mgr Polymer Res, Corp Res & Develop Dept, Howmedica Inc, Groton, Conn, 79-84; mgr tech develop, Advan Polymer Tech Div, Valley Lab Inc, 84-89; vpres, Res & Develop, Bio Polymers Inc, Farmington, Conn, 89-90; MGR, HYDROGEL RES & DEVELOP, NEPERA, INC, HARRIMAN, NY, 90- *Mem:* Am Chem Soc; Soc Biomat; Am Mgt Asn. *Res:* Polymer synthesis and post-polymerization chemistry; medical adhesives; transdermal drug delivery; medical electrodes, wound dressing. *Mailing Add:* Nepera Inc Rte 17 Harriman NY 10926

FOX, ARTHUR CHARLES, b Newark, NJ, Sept 16, 26. MEDICINE, CARDIOLOGY MEDICAL EDUCATION. *Educ:* NY Univ, MD, 48. *Prof Exp:* Intern med, Bellevue Hosp, 48-49, from asst resident to resident, 49-52; prof asst, Div Med Sci, Nat Res Coun, 53-54; asst, 54-56, from instr to assoc prof, 56-68, PROF MED, SCH MED, NY UNIV, 68- *Concurrent Pos:* Vis physician, Bellevue Hosp & Univ Hosp; Nat Heart Inst res fel, 54-56; consult, Vet Admin Hosp, Manhattan; sect chief cardiol, Med Ctr, NY Univ, 69-; gov, Am Col Physicians, NY, 81-85; presidency New York affil, Am Heart Asn, 87-89. *Mem:* Fel Am Col Physicians; fel Am Col Cardiol; AAAS; Sigma Xi; Am Fed Clin Res; Am Heart Asn. *Res:* Cardiology; myocardial metabolism and coronary blood flow; myocardial infarction. *Mailing Add:* 550 First Ave New York NY 10016

FOX, BARBARA SAXTON, b Troy, NY, Oct 22, 57; m 82; c 2. CELLULAR IMMUNOLOGY, T LYMPHOCYTES. *Educ:* Bryn Mawr Col, AB, 78; Mass Inst Technol, PhD(biochem), 83. *Prof Exp:* Postdoctoral fel immunol, NIH, 83-87; ASST PROF, DEPT MED, MICROBIOL & IMMUNOL, BIOL CHEM, SCH MED, UNIV MD, 87- *Mem:* Am Asn Immunol; Am Soc Biochem & Molecular Biol; AAAS. *Res:* Mechanisms of activation and regulation of helper T lymphocytes and their role in directing the outcome of the immune response. *Mailing Add:* Univ Md Med Sch MSTF 8-34 Ten S Pine St Baltimore MD 21201

FOX, BENNETT L, b Chicago, Ill, Aug 13, 38. MATHEMATICS. *Educ:* Univ Mich, BA, 60; Univ Chicago, MSc, 62; Univ Calif, Berkeley, PhD(opers res), 65. *Prof Exp:* Mathematician, Rand Corp, 65-71; mem staff, Dept Info Sci, Univ Montreal, 71-88; PROF MATH, UNIV COLO, DENVER, 88- *Concurrent Pos:* Ford vis prof, Univ Chicago, 71-72. *Mem:* Opers Res Soc Am. *Res:* Reliability theory; dynamic programming; simulation. *Mailing Add:* Dept Math Campus Box 170 Univ Colo Denver PO Box 173364 Denver CO 80217-3364

FOX, BERNARD LAWRENCE, b Canton, Ohio, Aug 18, 40; m 63. ORGANIC CHEMISTRY. *Educ:* John Carroll Univ, BS, 62; Ohio State Univ, PhD(org chem), 66. *Prof Exp:* From asst prof to assoc prof, 66-80, PROF CHEM & DEPT CHMN, UNIV DAYTON, 80- *Concurrent Pos:* Consult, Mat Lab, Wright-Patterson Air Force Base, 67-71; vis scientist, Univ Ariz, 79. *Mem:* Am Chem Soc. *Res:* Synthesis and reactions of azabicycloalkanes; mechanism of hydrogenolysis of benzylamines; mechanism of lithium aluminum hydride reduction of oxazolidines; drug analysis. *Mailing Add:* Dept Chem Univ Dayton Dayton OH 45469-0001

FOX, BRADLEY ALAN, b Circleville, Ohio, Jan 21, 63; m 90. MISFIT DISLOCATION IN SEMICONDUCTORS, POLYSILICON DEPOSITION. *Educ:* Va Polytechnic Inst & State Univ, BS, 85; Univ Va, PhD(mat sci), 90. *Prof Exp:* STAFF ENGR, IBM, 90- *Mem:* Am Soc Metals. *Res:* Epitaxial deposition of both III-V compounds and silicon as well as polysilicon deposition; lattice-mismatch and the dislocations and stresses which result. *Mailing Add:* RD 1 Box 461-7 Richmond VT 05477

FOX, C FRED, b Springfield, Ohio, 37; c 3. GROWTH FACTORS, RECEPTORS. *Educ:* Univ Chicago, PhD(biol chem), 64. *Hon Degrees:* DSc, Wittenberg Univ, 75. *Prof Exp:* PROF MOLECULAR BIOL, UNIV CALIF, LOS ANGELES, 70- *Concurrent Pos:* Pres, Int Genetic Eng Inc, 81-82; dir, Keystone Symp Molecular & Cellular Biol, 72- *Honors & Awards:* Eli Lilly Award, Am Chem Soc, 73. *Mem:* Am Soc Biochem & Molecular Biol; Am Chem Soc; Am Soc Cell Biol; Endocrine Soc; Am Soc Microbiol. *Res:* Hormone receptors; membrane bio-chemistry. *Mailing Add:* Molecular Biol Inst Univ Calif Los Angeles CA 90024

FOX, CARL ALAN, b Waukesha, Wis, Nov 24, 50; m 77; c 3. ARID ECOSYSTEMS ECOLOGY, DENDRO ECOLOGY. *Educ:* Univ Wis-River Falls, BS, 73; Univ Minn, MS, 75; Ariz State Univ, PhD(bot), 80. *Prof Exp:* Res scientist, Southern Calif Edison Co, 79-84, sr res scientist, 84-87; EXEC DIR BIOL SCI, DESERT RES INST, 87- *Concurrent Pos:* Lectr, Univ Calif, Los Angeles, 81-82; tech adv, Univ Calif, Riverside, 82-87, res assoc, 86-87; res adv, Elec Power Res Inst, 83-87; consult, Utility Air Regulatory Group, 87-; bd dirs, World Rainforest Found, 89- *Mem:* AAAS; Air Pollution Control Asn; Am Soc Agron; Ecol Soc Am. *Res:* Develop, conduct and direct ecological research focused on the assessment, interaction and management of biological resources from a local, regional and global perspective. *Mailing Add:* Desert Res Inst Biol Sic Ctr PO Box 60220 Reno NV 89506

FOX, CHARLES JUNIUS, organic chemistry, for more information see previous edition

FOX, CHESTER DAVID, b Albany, NY, Apr 8, 31; m 61; c 1. PHARMACEUTICAL CHEMISTRY. *Educ:* Union Col NY, BS, 58; Univ Md, MS, 63, PhD(indust pharm), 65. *Prof Exp:* Tech adv prod, 64-78, DIR TECH SERV, WYETH-AYERST LABS, INC, DIV AM HOME PROD CORP, 78- *Concurrent Pos:* Chmn mfg pract comt, Pharmaceut Mfrs Asn, 73-76. *Mem:* AAAS; Pharmaceut Mfrs Asn; Am Asn Pharmaceut Scientists; fel Am Inst Chem; Am Inst Chem Engrs. *Res:* Industrial pharmacy; development of automated industrial pharmaceutical production processes; development and evaluation of new pharmaceutical production equipment and methods; solvent recovery systems. *Mailing Add:* Tech Serv Wyeth-Ayerst Labs Inc 64 Maple St Rouses Point NY 12979-1497

FOX, DALE BENNETT, b Sioux Falls, SDak, May 25, 39; m 64; c 3. HETEROGENEOUS CATALYSIS. *Educ:* Hamline Univ, BS, 61; Univ Iowa, MS, 68, PhD(org chem), 69. *Prof Exp:* Assoc chemist, Res Ctr, Marathon Oil Co, Colo, 62-65; res chemist, Mo, 69-71, sr res chemist, 71-75, process specialist, Tex, 75-79, mgr, univ rels & prof employ, 79-82, TECH EMPLOY MGR, MONSANTO C0, 83- *Mem:* Am Chem Soc. *Res:* Catalysis; catalysts and catalytic conversions, especially with hydrocarbons; monomer synthesis; general organic and organometallic synthesis. *Mailing Add:* 1407 Timberwood Lane St Louis MO 63146-5445

FOX, DANIEL WAYNE, plastics chemistry, polymer chemistry; deceased, see previous edition for last biography

FOX, DANNY GENE, b West Unity, Ohio, Mar 18, 40; m 60; c 2. RUMINANT NUTRITION. *Educ:* Ohio State Univ, BS, 62, MS, 68, PhD(ruminant nutrit), 70. *Prof Exp:* Asst prof exten beef nutrit, SDak State Univ, 70-72; asst prof beef nutrit, Cornell Univ, 72-74; prof beef cattle res & exten, Mich State Univ, 74-78; MEM FAC DEPT ANIMAL SCI, CORNELL UNIV, 78- *Concurrent Pos:* Proj leader, USDA Exten Serv Proj Mgt Syst Small Beef Herds, 73- *Mem:* Am Soc Animal Sci. *Res:* Protein and energy requirements of varied cattle types; producing beef on high roughage diets; feeding systems for various cattle types. *Mailing Add:* Morrison Hall Cornell Univ Ithaca NY 14853

FOX, DAVID, b Brooklyn, NY, Sept 8, 20; m 38; c 3. THEORETICAL PHYSICS. *Educ:* Univ Calif, Berkeley, BA, 42, MA, 50, PhD(physics), 52. *Prof Exp:* Lectr physics, Israel Inst Technol, 52-55, sr lectr, 55-56; lectr, Hebrew Univ, 55; res assoc, Inst Optics, Univ Rochester, 56-57, asst prof, 57-59; dean grad sch, 63-66, PROF PHYSICS, STATE UNIV NY STONY BROOK, 59- *Res:* Theoretical solid state physics; statistical theory of eigenvalues; stochastic problems in energy storage. *Mailing Add:* Dept Physics State Univ NY Stony Brook NY 11794-3800

FOX, DAVID WILLIAM, b Dubuque, Iowa, Nov 21, 28; m 50; c 2. MATHEMATICS. *Educ:* Univ Mich, AB, 51, MSE, 52; Univ Md, PhD(math), 58. *Prof Exp:* Sr engr, 53-56, consult, 56-60, mathematician, prin staff, 60-66, SUPVR, APPL MATH RES GROUP, APPL PHYSICS LAB, JOHNS HOPKINS UNIV, 66- *Concurrent Pos:* Res asst prof, Inst Fluid Dynamics & Appl Math, Univ Md, 59-60; consult, Int Div, Battelle Mem Inst, Switz, 60- *Mem:* Am Math Soc. *Res:* Spectral theory of operators and theory of partial differential equations. *Mailing Add:* Comput Sci Univ Minn Minneapolis 136 Lind Hall 207 Church St SE Minneapolis MN 55455-0191

FOX, DONALD A, b Cleveland, Ohio, July 1, 48; m 81; c 2. NEUROBEHAVIORAL TOXICOLOGY & PHARMACOLOGY. *Educ:* Miami Univ, Oxford, Ohio, BS, 70; Univ Cincinnati Med Ctr, PhD(toxicol), 77. *Prof Exp:* Res pharmacologist, US Environ Protection Agency, 76-77; asst prof toxicol & pharmacol, Med Sch, Health Sci Ctr, Houston, 79-82; asst prof, 82-85, ASSOC PROF TOXICOL & PHARMACOL, UNIV HOUSTON, 85- *Concurrent Pos:* Consult, US Environ Protection Agency, 79, Nat Comt Air Quality Control, 80; prin investr grant, Lead Neurotoxicity Visual Syst, Nat Inst Environ Health Sci, 81-84; mem, Comn Vision, Nat Acad Sci, 82-85. *Mem:* Soc Neurosci; Soc Toxicol; Behav Teratology Soc; Asn Res Vision & Ophthalmol; AAAS; Sigma Xi. *Res:* Effects of lead exposure on the developing visual system utilizing electrophysiological and cell biological techniques; visual capacities and the functional bioelectric properties of the scotopic and photopic visual system during development and adulthood. *Mailing Add:* Col Optometry Univ Houston Houston TX 77204-6052

FOX, DONALD LEE, b Wichita, Kans, Sept 13, 43. AIR POLLUTION, ATMOSPHERIC CHEMISTRY. *Educ:* Wichita State Univ, BS, 65; Univ Ariz, PhD(phys chem), 71. *Prof Exp:* Environ Protection Agency fel atmospheric chem, 71-73, lectr air hyg, 73, from asst prof to assoc prof, 74-85, PROF AIR HYG, UNIV NC, CHAPEL HILL, 85-. ASSOC DEAN, 88- *Concurrent Pos:* Consult, indust, govt & Res Triangle Inst, 74- *Mem:* Air Pollution Control Asn; Am Chem Soc. *Res:* Air pollution chemistry; aerosol formation; kinetics modeling; air monitoring and instrumentation; photochemistry applied to air pollution control devices. *Mailing Add:* Environ Sci & Eng Sch Pub Health Univ NC Chapel Hill NC 27514

FOX, DOUGLAS GARY, b New York, NY, July 29, 41; m 64; c 3. AIR POLLUTION, ECOLOGY. *Educ:* Cooper Union, BCE, 63; Princeton Univ, MA & MSE, 65, PhD(fluid mech), 68. *Prof Exp:* Sr staff scientist meteorol, Nat Ctr Atmospheric Res, 68-72; res meteorologist & br chief, Nat Res Ctr, US Environ Protection Agency, 72-74; RES METEOROLOGIST & PROJ LEADER, ROCKY MOUNTAIN FOREST & RANGE EXP STA, US FOREST SERV, 74- *Concurrent Pos:* Affil prof, Princeton Univ, 69-72; mem panel air pollution modeling, NATO, 73-74; sr mem working group 1 air pollution measurement, modeling & methodology, US-USSR Environ Agreement, 73-; fac affil, Colo State Univ, 74-, Mowash Univ, Australia, Camm; ed, Atmospheric Environ, J Air & Waste Mgt Asn; vis fel, Chisholm Inst Technol, Melbourne, Australia; bd dirs, Air & Waste Mgt Asn. *Honors & Awards:* Chief's Stewardship Award, USDA-Forest Serv, 90. *Mem:* AAAS; Am Meteorol Soc; Air & Waste Mgt Asn. *Res:* Interactions between the atmosphere and wildland mangement and resource protection, especially associated with air pollution; mountain meteorology and forest meteorology. *Mailing Add:* Rocky Mountain Forest & Range Sta 240 W Prospect St Ft Collins CO 80526

FOX, EDWARD A(LEXANDER), b New York, NY, Aug 7, 20; m 49; c 3. MECHANICS. *Educ:* Harvard Univ, BS, 41; Columbia Univ, BS, 47, PhD(eng mech), 58. *Prof Exp:* From asst prof to prof, 54-83, EMER PROF MECH, RENSSELAER POLYTECH INST, 83- *Mem:* Am Soc Mech Engrs. *Res:* Mathematical theory of elasticity; stress wave propagation; classical mechanics. *Mailing Add:* 191 Southerland Rd Stephentown NY 12168

FOX, EDWARD A, b Far Rockaway, NY, May 14, 50. COMPUTER SCIENCE. *Educ:* Mass Inst Technol, BS, 71; Cornell Univ, MS, 81, PhD(computer sci), 83. *Prof Exp:* Data process instr, Florence Darlington Tech Col, Florence, SC, 71-72; data processing mgr, Vulcraft Div Nucor Corp, Florence, SC, 72-78; instr, res asst & teaching asst, Dept Computer Sci, Cornell Univ, 78-82; mgr info systs, Int Inst Trop Agr, Nigeria, 82-83; asst prof, 83-88, ASSOC PROF, DEPT COMPUTER SCI, VA POLYTECHNIC INST & STATE UNIV, 88-, ASSOC DIR, RES COMPUT CTR, 90- *Concurrent Pos:* Mem adv bd, Lexical Res Prog, Appl Info Tech Res Ctr, 88-90, tech prog comt, Fifth Annual AI Systs Gov conf, 88-90, Int Conf Multimedia Info, Singapore, 89-91, res comt, Am Soc Info Sci, 89-91, Chem Abstracts Serv Res Coun, 90-91, Press CD-ROM adv panel, Asn Comput Mach, 90-; vchmn, Spec Interest Group Info Retrieval, Asn Comput Mach, 87-91, chmn, 91- *Mailing Add:* Dept Comp Sci Va Polytechnic Inst 562 McBryde Hall Blacksburg VA 24061-0106

FOX, EUGENE N, b Chicago, Ill, Dec 9, 27; m 64; c 2. MEDICAL MICROBIOLOGY. *Educ:* Univ Ill, BS, 49, MS, 50; Western Reserve Univ, PhD(microbiol), 55. *Prof Exp:* Instr microbiol, Western Reserve Univ, 58-60; from asst prof prof microbiol, La Rabida Inst, Univ Chicago, 74-77; dir microbiol res & develop, Cutter Labs Inc, 77-80; PRIN, KENSINGTON CONSULT, 81- *Concurrent Pos:* Sr investr, Arthritis & Rheumatism Found, 59-64; consult, Dept Defense, US Army Med Res & Develop Cmnd, 73-77. *Mem:* Am Soc Microbiol; Am Asn Immunol. *Res:* Immunology and infectious diseases. *Mailing Add:* 235 Willamette Ave Kensington CA 94708

FOX, FRANCIS HENRY, b Clifton Springs, NY, Mar 11, 23; m 46; c 4. VETERINARY MEDICINE. *Educ:* State Univ NY, DVM, 45. *Prof Exp:* Asst mastitis & vet med, Cornell Univ, 45-46; instr vet med & surg, Ohio State Univ, 46-47; from asst prof to assoc prof, 47-53, chmn dept large animal med obstet & surg & dir ambulatory clin, 72-77, PROF VET MED & OBSTET, NY STATE VET COL, CORNELL UNIV, 53- *Mem:* Am Vet Med Asn; Sigma Xi. *Res:* Diseases of large domestic animals, especially the bovine. *Mailing Add:* Vet Col Cornell Univ Ithaca NY 14853-6401

FOX, FREDERICK GLENN, geology; deceased, see previous edition for last biography

FOX, G(EORGE) SIDNEY, b Philadelphia, Pa, Feb 21, 28; m 56, 78; c 2. GEOLOGICAL ENGINEERING. *Educ:* Princeton Univ, BSE, 50. *Prof Exp:* Hydraul engr, US Geol Surv, 51-55; groundwater geologist, 55-67, partner, 67-76, vpres, 76-84, EXEC VPRES, LEGGETTE, BRASHEARS & GRAHAM, INC, 84- *Mem:* Geol Soc Am; Am Water Resources Asn; Am Inst Mining, Metall & Petrol Engrs; Am Geophys Union; Asn Prof Geol Scientists. *Mailing Add:* Leggette Brashears & Graham Inc 72 Danbury Rd Wilton CT 06897

FOX, GEOFFREY CHARLES, b Dunfermline, Scotland, June 7, 44; m 76; c 3. PHYSICS, COMPLEX SYSTEMS. *Educ:* Cambridge Univ, Eng, BA, 64, PhD(physics), 67, MA, 68. *Prof Exp:* Mem physics, Inst Advan Study, 67-68; res scientist physics, Lawrence Berkeley Lab, 68-69; res fel, Peterhouse Col, Cambridge, Eng, 69-70; res fel, 70-71, from asst prof to assoc prof, 71-79, exec officer, 81-83, dean educ comput, 83-86, PROF PHYSICS, CALIF INST TECHNOL, 79-, ASSOC PROVOST COMPUT, 86- *Mem:* Am Phys Soc. *Res:* Computational elementary particle physics; modelling; computer architectures and algorithms; massively parallel computers. *Mailing Add:* Concurrent Computation Prog Calif Tech Mail Code 158-79 Pasadena CA 91125

FOX, GEORGE A, c 4. CIVIL ENGINEERING. *Educ:* Cooper Union Col, BCE, CE, 40; Brooklyn Polytech, MCE, 42. *Prof Exp:* Consult engr in construct, 46-76; chmn Contract Admin Comt, Am Soc Civil Engr, Construct Div, 76-79; pres, NY Bldg Congr, 82-86; PRES & CHIEF ENGR, GROW TUNNELING CORP, 86- *Concurrent Pos:* Chmn, Coun of Pres, NY Bldg affil, 82-; chmn bd trustees, Cooper Union Col, 86-87. *Honors & Awards:* Moles Award, Am Construction Indust, 79; Robling Award, Am Soc Civil Engrs, 88. *Mem:* Nat Acad Eng; Am Soc Civil Engrs. *Res:* Specializing in heavy construction, shafts and tunnels, notably those done using compressed air; design of tunnel linings. *Mailing Add:* Grow Tunneling Corp 71 W 23rd St 3rd Floor New York NY 10010

FOX, GEORGE EDWARD, b Syracuse, NY, Dec 17, 45; m 73; c 2. MICROBIOLOGY. *Educ:* Syracuse Univ, BSChE, 67, PhD(chem eng), 74. *Prof Exp:* Res assoc genetics & develop, Univ Ill, Urbana, 73-77; from asst prof to assoc prof, 77-87, PROF BIOCHEM SCI, UNIV HOUSTON, 87-, DIR INST MOLECULAR BIOL, 88- *Concurrent Pos:* Prin investr, NASA & NSF grants & US Army contract; grants, NIH. *Mem:* Am Soc Microbiol; Am Chem Soc; AAAS. *Res:* RNA structure and function; theoretical biology; microbial phylogeny; molecular evolution. *Mailing Add:* Dept Biochem & Biophys Sci Univ Houston Houston TX 77204-5500

FOX, GERALD, b New York, NY, Apr 23, 23; m 55; c 1. INDUSTRIAL CHEMISTRY. *Educ:* Long Island Univ, BSc, 47. *Prof Exp:* Chemist, Fleischmann Labs, NY, 46-49; chief chemist, Consol Laundries Corp, 49-57; secy & dir res, Gold Par Prod Co, Inc, 57-70 & Gold Par Chem, Inc, 70-74; dir chem res, Chemair Corp Am, 75-80; dir chem res, Am Machinery Corp, 80-82; DIR CHEM RES, HUNTER CHEM & FORMULATING CO, INC, 82- *Concurrent Pos:* Consult electroplaters & mgrs cosmetic mfrs. *Mem:* Am Chem Soc; Soc Cosmetic Chem; fel Am Inst Chem. *Res:* Sanitary chemicals; chemical specialty items; soaps and detergents for laundry, dry cleaning, textile processing, food and beverage, dairy, dishwashing and metal industries; research and development of new cosmetic products; research and development of new post-harvest processing chemical for the fruit and vegetable industry. *Mailing Add:* 714 Village Ln Winter Park FL 32792-3411

FOX, GERARD F, ENGINEERING ADMINISTRATION. *Prof Exp:* Partner, Howard Needles Tammen & Bergendoff; RETIRED. *Mem:* Nat Acad Eng. *Mailing Add:* Three Whitehall Blvd Garden City NY 11530

FOX, HAZEL METZ, nutrition; deceased, see previous edition for last biography

FOX, HERBERT LEON, b Boston, Mass, Jan 6, 30; m 52; c 4. PHYSICS, ECONOMICS. *Educ:* Boston Univ, BA, 58, MA, 62. *Prof Exp:* Proj engr, Farnsworth-Tamari, 57; chief design engr, Tech Prod Co, 55-57; SR PHYSICIST & ECONOMIST, BOLT BERANEK & NEWMAN, 57- *Concurrent Pos:* Res asst, Boston Univ, 63-64; consult comt, Nat Acad Remote Atmospheric Probing; lectr, Dept Econ, Northeastern Univ; consult econ & physics, Idmon, Inc, 75- *Mem:* AAAS; Am Asn Physics Teachers. *Res:* Atmospheric sound propagation; quantum optics; kinetic theory; nonlinear mechanics; development economics; occupational health and safety; environmental impact. *Mailing Add:* 35 Trenor Dr New Rochelle NY 10804

FOX, IRVING, b Hartford, Conn, Dec 19, 12; m 41; c 1. MEDICAL ENTOMOLOGY. *Educ:* George Washington Univ, AB, 37, MA, 38; Iowa State Col, PhD, 40. *Prof Exp:* Asst entom, Iowa State Col, 37-40; instr biol, Univ PR, San Juan, 41-42, from asst prof to prof med entom, 46-83, hon adj prof med entom, 83-90, EMER PROF, SCH MED, UNIV PR, SAN JUAN, 90- *Concurrent Pos:* Collabr, Bur Entom, USDA, 35-42; malaria control officer, US Army, PR, 42-46; prog dir, USPHS grad res training grant, 57-70. *Mem:* Am Soc Parasitol; Entom Soc Am; Am Soc Trop Med & Hyg; Am Mosquito Control Asn; Am Arachnological Soc; fel AAAS. *Res:* Taxonomy, population studies; toxicology, immunology and control of insects and Arachnida such as fleas, mites, ticks, biting flies, mosquitoes and spiders; molluscicides; Tardigrada. *Mailing Add:* Dept Microbiol & Med Zool Univ PR Sch Med GPO Box 5067 San Juan PR 00936

FOX, IRVING HARVEY, b Montreal, Que, Dec 7, 43; m 66; c 3. RHEUMATOLOGY, PURINE METABOLISM. *Educ:* McGill Univ, BSc, 65, MDCM, 67. *Prof Exp:* House officer med, Royal Victoria Hosp, 67-69; fel rheumatology, Duke Univ Med Ctr, 69-72; asst prof med, Univ Toronto, 72-76; assoc prof med, Univ Mich, 76-78, from asst prof to prof biol chem, 76-90, dir, clin res ctr, 77-90, prof med, 78-90, VPRES MED AFFAIRS, UNIV MICH, 90- *Concurrent Pos:* Prof Biol Chem, 84- *Mem:* Am Fedn Clin Res; Am Rheumatism Asn; Am Soc Clin Invest; Am Soc Biol Chemists; Cent Soc Clin Res. *Res:* Regulation of human purine nucleotide degradation in normal and disease states; biotechnology. *Mailing Add:* Biogen 14 Cambridge Ctr Cambridge MA 02142

FOX, J EUGENE, b Anderson, Ind, Aug 7, 34; m 60; c 4. PLANT PHYSIOLOGY, PLANT BIOCHEMISTRY. *Educ:* Ind Univ, BS, 56, PhD(bot), 60. *Prof Exp:* Nat Cancer Inst fel bot, Univ Wis, 60-61; from asst prof to assoc prof, Univ Kans, Lawrence, 61-70, prof biochem & bot, 70-87, chmn dept bot, 74-87; dir, Arco Plant Cell Res Inst; VPRES RES & DEVELOP, BIOTECH GROUP, MILES LABS, 88- *Concurrent Pos:* Asst dean, Col Arts & Sci, Univ Kans, 67-70, dean, 70-72. *Mem:* AAAS; Bot Soc Am; Am Soc Plant Physiol; Scand Soc Plant Physiol; Japanese Soc Plant Physiol. *Res:* Growth and development of plants; biochemistry of plant growth regulators; plant tissue culture. *Mailing Add:* Mallinckrodt Spec Chem Co 675 McDonnell Blvd PO Box 5840 St Louis MO 63134

FOX, JACK JAY, b New York, NY, Dec 21, 16; m 39; c 2. BIOCHEMISTRY. *Educ:* Univ Colo, BA, 39, PhD(biochem), 50. *Prof Exp:* Nat Res Coun fel, Free Univ Brussels, 50-52; Damon Runyon Mem Fund fel, 52-54, asst chem, 54-58, head medicinal chem, 58-60, assoc mem, 60-64, MEM, SLOAN-KETTERING INST CANCER RES, 64-, CHIEF ORG CHEM, 71- *Concurrent Pos:* From asst prof to assoc prof, Sloan-Kettering Div, Cornell Univ, 56-65, prof, 65- *Honors & Awards:* Alfred P Sloan Award Cancer Res, 56; C S Hudson Award Carbohydrate Chem, 77; Papanicolaou Award for Sci Contributions in Cancer, 82; Norlin Award for Sci Achievement, Univ Colo, 84. *Mem:* Am Chem Soc; Am Soc Biol Chem; Am Asn Cancer Res; Int Soc Heterocyclic Chem; Sigma Xi. *Res:* Design and syntheses of potential anti-tumor and anti-viral agents and other medicinals; chemistry of carbohydrates, heterocycles, nucleosides and antibiotics. *Mailing Add:* 424 S Lexington Ave White Plains NY 10606

FOX, JACK LAWRENCE, b Ann Arbor, Mich, Oct 10, 41; m 61; c 2. BIOCHEMISTRY, BIOPHYSICS. *Educ:* Fla State Univ, BA, 62; Univ Ariz, PhD(chem), 66. *Prof Exp:* Guest investr biochem, Rockefeller Univ, 66-68; asst prof, Univ Tex, Austin, 68-74, assoc prof zool, 74-; AT DEPT MOLECULAR BIOL, ABBOTT LABS. *Concurrent Pos:* USPHS fel, 66-68; biol ed, Marcel Dekker, Inc, 73-; Alexander von Humboldt Found fel, 77-78. *Mem:* AAAS; Am Chem Soc; Am Soc Biol Chem; Biophys Soc; Int Union Biochem. *Res:* Enzyme structure and molecular evolution, especially flavoenzymes and phycobiliproteins, chemical and theoretical studies, recombinant DNA. *Mailing Add:* Tex Res Technol Found 14785 Omicron Dr San Antonio TX 78245

FOX, JAMES DAVID, b Gatesville, Tex, May 11, 43; m 66; c 2. ANIMAL GENETICS, STATISTICS. *Educ:* Tex A&M Univ, BS, 65, MS, 67; Ohio State Univ, PhD(animal breeding), 70. *Prof Exp:* Geneticist, DeKalb Agresearch, Inc, 71-72, DIR SWINE GENETIC RES, DEKALB SWINE BREEDERS INC, 72- *Mem:* Am Soc Animal Sci; Biomet Soc. *Res:* Relationship of mature body weight and efficiency of production in livestock; developing swine breeding stock. *Mailing Add:* DeKalb Swine Breeders Inc 3100 Sycamore Rd DeKalb IL 60115

FOX, JAMES GAHAN, b Reno, Nev, Mar 8, 43; m 85; c 2. VETERINARY & COMPARATIVE MEDICINE. *Educ:* Colo State Univ, DVM, 68; Stanford Univ MS, 72; Am Col Lab Animal Med, dipl, 74. *Prof Exp:* Resident vet, Biol Lab Animal Div, US Army Vet Corp, Ft Detrick, 68-70; asst prof & staff vet, Med Ctr, Univ Colo, 73-74; inst vet & dir, Animal Care Facil, 74-75, assoc prof & dir, Div Lab Animal Med, 75-83, PROF & DIR, DIV COMP MED, MASS INST TECHNOL, 83- *Concurrent Pos:* NIH fel lab animal med & med microbiol, Stanford Univ & co-investr, Standard Oil Co, Calif, 70-72; fac affil, Dept Clins & Surg, Colo State Univ, 73-74; dir, Animal Care Facil, Forsyth Dent Ctr, Boston, 74-; prin investr, Animal Res Diag Lab grant, NIH, 75-, Nitrosamine Carcinogenesis grant, 80-, Helicobacter induced gastric dis, 85-, Lab Animal Med postdoctoral training grant, 88- & Animal Res Ctr grant, Nat Cancer Inst, 77-80; spec consult, Brandeis Univ, Forsyth Dent Ctr & Mass Gen Hosp, 76, West Roxbury Vet Admin Hosp, NIH & Nat Cancer Inst, 77, US Environ Protection Agency, 78- & US Food & Drug Admin, 78-81; staff affil, Angell Mem Hosp, Boston, 76-; adj assoc prof comp med, Tufts Univ Sch Vet Med, 81-82 & adj prof, 83-; mem bd dir & exec comt, Mass Soc Med Res, 83-; adj prof, Univ Pa Vet Sch, 90- *Honors & Awards:* Charles River Prize in Lab Animal Med, Am Vet Med Asn, 90. *Mem:* Am Asn Accreditation Lab Animal Care; Am Vet Med Asn; Am Asn Lab Animal Sci; Am Acad Clin Toxicol; Am Col Toxicol; AAAS; Asn Biomed Res; NY Acad Sci; Campylobacter Soc; Mass Soc Med Res. *Res:* Study of infectious, metabolic or inherited diseases in research animals which adversely affect experimental results; development of animal models as analogs of human disease; animal environment and its role in interpretation of research results; pathogenesis of campylobacter infections; study of Helicobacter pylori and gastric disease; development of the ferret as an animal model. *Mailing Add:* Mass Inst Technol Div Comp Med Cambridge MA 02139

FOX, JAY B, JR, b Lincoln, Nebr, July 30, 27; m 52; c 4. BIOCHEMISTRY. *Educ:* Col Puget Sound, BS, 51; Univ Wash, PhD(biochem), 55. *Prof Exp:* Assoc biochemist, Am Meat Inst Found, 55-61, biochemist, 61-64; RES CHEMIST, EASTERN REGION RES CTR, AGR RES SERV, USDA, 64- *Mem:* Am Chem Soc; Inst Food Technologists; Am Inst Chem; Am Meat Sci Asn; Asn Off Anal Chemists. *Res:* Biochemistry of heme pigments; nitrite chemistry. *Mailing Add:* Eastern Regional Res Ctr 600 E Mermaid Lane Philadelphia PA 19118

FOX, JOAN ELIZABETH BOTHWELL, b Newcastle-upon-Tyne, England, Sept 5, 47; m 72; c 2. PLATELETS, CYTOSKELETONS. *Educ:* Southampton Univ, England, BSC, 69; McMaster Univ, Can, PhD(biochem), 76. *Prof Exp:* Teaching fel biochem, McMaster Univ, 76-79 & St Jude Children's Res Hosp, 80-81; staff res investr, 81-84, SCIENTIST & ASST PROF PATH, GLADSTONE FOUND CARDIOVASC DIS, UNIV CALIF, SAN FRANCISCO, 84- *Mem:* Am Soc Cell Biol; Am Heart Asn; Int Soc Thrombosis & Hemostasis; AAAS. *Res:* Biochemistry of platelet with emphasis on the structure and function of the cytoskeleton; investigation of the nature of the interaction between actin filaments and plasma membrane glycoproteins. *Mailing Add:* Gladstone Found Cardiovasc Dis Univ Calif PO Box 40608 San Francisco CA 94140

FOX, JOEL S, b New York, NY, Feb 7, 39; m 63. MECHANICAL ENGINEERING. *Educ:* Polytech Inst Brooklyn, BME, 59, MME, 61, PhD(mech eng), 66. *Prof Exp:* Mem tech staff heat transfer, Hughes Aircraft Co, 61-62; instr mech eng, Polytech Inst Brooklyn, 63-66; asst prof, 66-69, ASSOC PROF MECH ENG, UNIV HAWAII, 69- *Mem:* Am Soc Mech Engrs; Am Soc Eng Educ. *Res:* Electric fields and combustion process interaction; thermal conductivity of polymers; freeze-dry preservation of whole blood. *Mailing Add:* Dept Mech Eng Univ Hawaii Manoa Honolulu HI 96822

FOX, JOHN ARTHUR, b Toronto, Ont, Feb 8, 24; US citizen; m 46; c 4. MECHANICAL & AEROSPACE ENGINEERING. *Educ:* Univ Mich, BS, 48 & 49; Pa State Univ, MS, 50, PhD(aeronaut eng), 60. *Prof Exp:* Asst, Univ Mich, 48-49; instr aeronaut eng, Pa State Univ, 49-60, assoc prof & actg head dept, 60-61, consult, Ord Res Lab, 57-61; assoc prof mech & aerospace sci, Univ Rochester, 61-67; PROF MECH ENG & CHMN DEPT, UNIV MISS, 67- *Concurrent Pos:* Consult, Piper Aircraft Corp & HRB-Singer, Inc, Pa, 61- & Rochester Appl Sci Assocs, NY, 64- *Mem:* AAAS; Am Soc Eng Educ; Am Inst Aeronaut & Astronaut; Sigma Xi. *Res:* Mechanical and aerospace science; dynamics and celestial mechanics; aerodynamics; aerospace structures; magnetohydrodynamics. *Mailing Add:* Dept Mech Eng Univ Miss University MS 38677

FOX, JOHN DAVID, b Huntington, WVa, Dec 8, 29; m 78. NUCLEAR PHYSICS. *Educ:* Mass Inst Technol, SB, 51; Univ Ill, MS, 54, PhD, 60. *Prof Exp:* Asst, Univ Ill, 52-55; asst physicist, Brookhaven Nat Lab, 56-59; from asst prof to assoc prof, 59-65, PROF PHYSICS, FLA STATE UNIV, 65- *Concurrent Pos:* NSF sr fel & guest scientist, Max-Planck Inst Nuclear Physics, Heidelberg, 68-69; sr scientist, Alexander von Humboldt Found, 75; guest prof, Univ Köln; consult, Argonne Nat Labs, 80-87; prog dir, nuclear physics, NSF, 90-91. *Mem:* Fel Am Phys Soc; Sigma Xi; AAAS. *Res:* Experimental nuclear physics; low energy nuclear physics; neutron physics in resonance region; particle detectors and instrumentation; nuclear isomerism; photonuclear phenomena; isobaric analogue resonances; charged particle induced nuclear reactions; superconducting linear accelerator development. *Mailing Add:* Dept Physics Fla State Univ Tallahassee FL 32306

FOX, JOHN FREDERICK, b Lakewood, NJ, Aug 22, 45. ECOLOGY, POPULATION BIOLOGY. *Educ:* Johns Hopkins Univ, AB, 67; Univ Chicago, MS, 69, PhD(biol), 74. *Prof Exp:* Programmer, Syst Eng Lab, 69; res assoc, Univ Wyo, 75-77; ASST PROF, UNIV ALASKA, 77- *Mem:* Ecol Soc Am; Soc Study Evolution; Brit Ecol Soc. *Res:* Population biology and community ecology of terrestrial plants; plant animal interactions; ecological modeling. *Mailing Add:* Inst Arctic Biol 311 Irving Bldg Univ Alaska Fairbanks AK 99775-0180

FOX, JOSEPH M(ICKLE), III, b Philadelphia, Pa, Nov 20, 22; m 48; c 6. REFINERY PROCESS DESIGN, PILOT PLANTS. *Educ:* Princeton Univ, BS, 43, MS, 47. *Prof Exp:* Engr tech serv, Pan Am Refining Corp, 43-45; instr chem eng, Princeton Univ, 46-47; res engr, Res & Develop Dept, M W Kellogg Co Div, Pullman, Inc, 47-52, supvr, 52-58, res assoc, 58-61, pilot plant sect head, 61-66; process develop mgr, 66-76, asst chief process eng, 76-86, CHIEF PROCESS ENG & MGR PROCESS TECHNOL, BECHTEL GROUP RES & DEVELOP DIV, 86- *Concurrent Pos:* Mem, tech data comt, Am Petrol Inst; mem, Panel on Methanol, Calif Energy Comt; consult, Catalytica Assocs Inc; dir, Am Inst Chem Engrs, 76-79, chmn, leg comt, 86-89. *Mem:* Fel Am Inst Chem Engrs; Am Chem Soc. *Res:* Process scale-up; petroleum and petrochemical processing; catalysis; heat and mass transfer; unusual separation techniques; environmental control; synthetic fuels; liquified natural gas; concerned with design, evaluation and scale up of catalytic reactions; methane valorization, particularly as it relates to the development of remote natural gas as a source of liquid fuels. *Mailing Add:* Bechtel Group Res & Develop Div PO Box 3965 San Francisco CA 94119-3965

FOX, JOSEPH S, b Montreal, Que, May 12, 47; m 70; c 3. EXPLORATION GEOLOGY. *Educ:* McGill Univ, BSc, 68, MSc, 70; Cambridge Univ, Eng, PhD(geol), 74. *Prof Exp:* Res scientist, Sask Res Coun, 74-78; proj geologist, Sask Mining Develop Corp, 78-80; res geologist, Teck Corp, 80-85; DIR, MINERAL EXPLOR RES INST, 85- *Concurrent Pos:* Adj prof, McGill Univ, 85- *Mem:* Can Inst Mining Metall. *Res:* Metallogeny of basalt-hosted massive sulphide deposits; geochemistry of gold in the lacuskine environment. *Mailing Add:* Nat Res Coun Can Montreal Rd Bldg M-58 Rm 5107 Ottawa ON K1A 0R6 Can

FOX, KARL RICHARD, b Berkeley, Calif, July 8, 42; c 1. PATHOLOGY, PHYSIOLOGY. *Educ:* Iowa State Univ, BS, 62; Univ Iowa, MD, 67. *Prof Exp:* Fel physiol, Univ Iowa, 67-69; intern path, State Univ NY Upstate Med Ctr, 69-70; NIH fel lab med, Univ Minn, Minneapolis, 70-73; NIH fel & vis asst prof human physiol, Univ Pavia, 73-74; asst prof path, Med Sch, Univ Tex, 74-77; dir hematol, Dept Path, Baylor Univ Med Ctr & asst prof gen path, Baylor Col Dent, 77-81; dir hematol & cytol, Gary Methodist Hosps, 81-; CHIEF, DEPT PATH, ST THERESE MED CTR, WAUKEGAN, ILL. *Concurrent Pos:* USPHS spec fel, 73. *Mem:* AAAS; Biophys Soc; Am Soc Hemat; Am Physiol Soc; Am Soc Clin Pathologists. *Res:* Biophysics of membrane transport, pathology and physiology of hemostasis. *Mailing Add:* Lab Med Allegheny Gen Hosp 320 E North Ave Pittsburgh PA 15212

FOX, KAYE EDWARD, pharmacology; deceased, see previous edition for last biography

FOX, KENNETH, b Highland Park, Mich, Aug 16, 35; m 82; c 3. ATOMIC & MOLECULAR PHYSICS, ASTROPHYSICS. *Educ:* Wayne State Univ, BS, 57; Univ Mich, MS, 58, PhD(physics), 62. *Hon Degrees:* JD, Univ Tenn, 82. *Prof Exp:* Instr physics, Univ Mich, 61-62; vis res physicist, Inst Nuclear Physics Res, Neth, 62-63; vis asst prof physics, Vanderbilt Univ, 63-64; asst prof, Univ Tenn, Knoxville, 64-67; Nat Acad Sci-Nat Res Coun sr res assoc, Jet Propulsion Lab, Calif Inst Technol, 67-69; assoc prof, 69-75, PROF PHYSICS, UNIV TENN, KNOXVILLE, 75- *Concurrent Pos:* Consult, Oak Ridge Nat Lab, 64-70 & Jet Propulsion Lab, Calif Inst Technol, 66-67 & 69-71; mem vis staff, Los Alamos Sci Lab, Univ Calif, 74-77; Fulbright sr lectureship, Univ Dijon & Univ Paris, France, 75; Nat Acad Sci-Nat Res Coun sr res assoc, Goddard Space Flight Ctr, NASA, 77-78, discipline scientist, 85-88. *Mem:* Fel Am Phys Soc; Sigma Xi; NY Acad Sci; Am Astron Soc; Int Astron Union; Am Geophys Union. *Res:* Theoretical and experimental spectroscopy; planetary atmospheres and interstellar gases; observational astronomy; mathematical physics; laser isotope separation; ultra-high resolution spectroscopy. *Mailing Add:* Dept Physics & Astron Univ Tenn Knoxville TN 37996-1200

FOX, KENNETH IAN, b Chicago, Ill, Mar 27, 43; m 75. FOOD MICROBIOLOGY, FOOD SCIENCE. *Educ:* Univ Ill, Urbana, BS, 65; Mich State Univ, MS, 67, PhD(food sci), 71. *Prof Exp:* Res chemist, J R Short Milling Co, 71-78; RES MGR CITRUS MACH DIV, FMC CORP, 78- *Mem:* Am Soc Microbiol; Inst Food Technol. *Res:* Botulism food poisoning; germination of bacterial spores; dry heat resistance of bacterial spores; citrus juice and by-products. *Mailing Add:* FMC PO Box 1708 Lakeland FL 33802

FOX, KEVIN A, b Nashua, NH, Feb 1, 39; m 66; c 3. REPRODUCTIVE BIOLOGY. *Educ:* Univ NH, BA, 60, MS, 62; Univ Vt, PhD(zool), 67. *Prof Exp:* Instr zool, Univ NH, 60-61 & Hebron Acad, 61-64; res technician, Reproductive Physiol, Univ Vt, 65-66; USPHS fel comp path, Penrose Res Lab & Univ Pa, 67-69; res assoc path, Penrose Res Lab, 69-70; from asst prof to assoc prof, State Univ NY Col, Fredonia, 70-77, chmn dept biol, 74-77, assoc acad vpres & dean spec studies, 77-80, prof, grad fac, 77-89, DISTINGUISHED TEACHING PROF, STATE UNIV NY COL, FREDONIA, 89- *Concurrent Pos:* NIMH grants, 70-74. *Mem:* AAAS; Animal Behav Soc; Am Soc Mammalogists; Sigma Xi. *Res:* Reproductive and behavioral physiology; animal behavior. *Mailing Add:* Dept Biol State Univ NY Col Fredonia NY 14063

FOX, LAUREL R, b New York, NY, Jan 5, 46; m 70; c 2. EVOLUTIONARY BIOLOGY. *Educ:* Cornell Univ, BS, 67; Univ Calif, Santa Barbara, MA, 70, PhD(biol), 73. *Prof Exp:* Res fel, Res Sch Biol Sci, Australian Nat Univ, 73-78; asst prof, 78-81, ASSOC PROF, UNIV CALIF, SANTA CRUZ, 81- *Concurrent Pos:* Fel, Am Asn Univ Women, 85-86. *Mem:* Entom Soc Am; Am Soc Naturalist; Int Soc Chem Ecol. *Res:* Plant-animal interactions; three-trophic level interactions. *Mailing Add:* Biol Dept Univ Calif Santa Cruz CA 95064

FOX, MARTIN, b New York, NY, July 25, 29; m 51; c 3. MATHEMATICS, GENERAL. *Educ:* Univ Calif, Berkeley, AB, 51, PhD(statist), 59. *Prof Exp:* From asst prof to assoc prof statist, 59-71, actg chmn dept statist & probability, 66-67, PROF STATIST & PROBABILITY, MICH STATE UNIV, 71- *Concurrent Pos:* Fulbright lectr, Tel-Aviv Univ, 62-63; vis res mem, US Army Math Res Ctr, Univ Wis, 67-68; vis prof, Dept Math, Bowling Green State Univ, 75-76, Tel-Aviv Univ, 83-84, Sichuan Univ, 84. *Mem:* Fel Inst Math Statist (exec secy, 78-81); Am Math Soc; Am Statist Asn; Math Asn Am; Am Assoc Univ Profs. *Res:* Mathematical statistics; probability; game theory. *Mailing Add:* Dept Statist & Probability Mich State Univ East Lansing MI 48824

FOX, MARTIN DALE, b Hackensack, NJ, Oct 18, 46; m 72; c 3. ELECTRICAL ENGINEERING, RADIOLOGY. *Educ:* Cornell Univ, BEE, 69; Duke Univ, PhD(biomed eng), 72; Univ Miami, MD, 83. *Prof Exp:* From asst prof to assoc prof, 72-90, PROF, ELEC & SYST ENG, UNIV CONN, 90- *Concurrent Pos:* Intern in Med, Univ Conn Health Ctr, 83-84, fel in Radiol, 84-85, joint appointment, Dept Radiol, 85- *Mem:* Inst Elec & Electronics Engrs; AMA; Am Inst Ultrasound Med; Radiol Soc NAm; Asn Univ Radiols. *Res:* Diagnostic ultrasound; ultrasound doppler; medical imaging; digital radiography. *Mailing Add:* Dept Elec & Systs Eng Box U-157 Storrs CT 06268

FOX, MARY ELEANOR, b Bellevue, Ky, Aug 14, 19. SOLID STATE PHYSICS. *Educ:* Villa Madonna Col, AB, 42; Cath Univ Am, MS, 44; Univ Cincinnati, PhD(physics), 62. *Prof Exp:* From assoc prof to prof physics, Thomas More Col, 44-86; PHYSICS TEACHER, BISHOP BROFSART HIGH SCH, 86- *Mem:* Am Phys Soc; Am Asn Physics Teachers; Sigma Xi. *Res:* Mathematical and classical physics; nuclear magnetic resonance broad line studies of crystalline materials in relation to motional narrowing; magnetic anisotropy studies in relation to molecular structure. *Mailing Add:* Bishop Brossart HS Grove & Jefferson Alexandria KY 41001

FOX, MARY FRANCES, b Cincinnati, Ohio, June 25, 40; div; c 3. PHYSICAL CHEMISTRY. *Educ:* Mt St Joseph Col, RI, BA, 61; Univ RI, PhD(chem), 76. *Prof Exp:* Res chemist, Div Air Pollution, Robert A Taft Sanit Eng Ctr, 61-62; res assoc, 76-82, RES SCIENTIST OCEANOG, UNIV RI, 82- *Mem:* Am Chem Soc; AAAS; Am Geophys Union. *Res:* Spectroscopy of ionic equilibrium in aqueous solutions; iron species in seawater; fate of chemicals discharged at deepwater dumpsites; chemical distribution in seawater; chemical distributions in Gulf Stream rings. *Mailing Add:* Grad Sch Oceanog Univ RI Kingston RI 02881

FOX, MARYE ANNE, b Canton, Ohio, Dec 9, 47; m 69, 90; c 5. ORGANIC CHEMISTRY, PHYSICAL CHEMISTRY. *Educ:* Notre Dame Col, BS, 69; Cleveland State Univ, MS, 70; Dartmouth Col, PhD(org chem), 74. *Prof Exp:* Instr phys sci, Cuyahoga Community Col, 70-71; fel & res assoc chem, Univ Md, College Park, 74-76; from asst prof to assoc prof, 76-85, prof chem, 85-87, PROF ROWLAND PETTIT CENTENNIAL, UNIV TEX, 87- *Concurrent Pos:* Alfred P Sloan res fel, 78; mem adv panel, Army Basic Res Comt, 85-87; assoc ed, J Am Chem Soc, 86-; dir, Ctr for Fast Kinetics Res; Arthur C Cope Scholar Award, Am Chem Soc, 89; mem, Nat Sci Bd, 91-96. *Honors & Awards:* Garvan Medal, Am Chem Soc, 88. *Mem:* Am Chem Soc; Interam Photochem Soc; Europ Photochem Soc; Electrochem Soc; Am Soc Photobiol. *Res:* Organic photochemistry; electrochemistry; physical organic mechanisms. *Mailing Add:* Dept Chem Univ Tex Austin TX 78712

FOX, MATTIE RAE SPIVEY, nutritional biochemistry; deceased, see previous edition for last biography

FOX, MAURICE SANFORD, b New York, NY, Oct 11, 24; m 55; c 3. GENETICS. *Educ:* Univ Chicago, BS, 44, MS, 51, PhD(chem), 51. *Prof Exp:* Res assoc, Univ Chicago, 51-53; asst, Rockefeller Inst, 53-56, from asst prof to assoc prof, 56-62; from assoc prof to prof genetics, 62-79, LESTER WOLFE PROF MOLECULAR BIOL, MASS INST TECHNOL, 79-, HEAD, DEPT BIOL, 85- *Concurrent Pos:* Nuffield res scholar, 57; Lalor fel award, 59. *Mem:* Inst Med-Nat Acad Sci; fel AAAS. *Res:* Chemical events following high energy nuclear recoil; mutation; continuous culture of microorganisms; biological properties of deoxyribonucleates; microbial genetics; molecular mechanism of genetic recombination; public health and epidemiology. *Mailing Add:* Dept Biol Mass Inst Technol Cambridge MA 02139

FOX, MICHAEL HENRY, b Great Bend, Kans, Mar 19, 46; m 75; c 2. FLOW CYTOMETRY, RADIATION BIOLOGY. *Educ:* McPherson Col, Kans, BS, 68; Kans State Univ, Manhattan, MS 72, PhD(physics), 77. *Prof Exp:* Asst prof physics, McPherson Col, 72-73; fel, 77-79, asst prof, 79-88, ASSOC PROF RADIATION BIOL, COLO STATE UNIV, 88-, CHMN CELL & MOLECULAR BIOL GRAD PROG, 90- *Concurrent Pos:* Prin investr, Nat Cancer Inst & Dept Health & Human Serv, 79-; mem Grad Fac Cell & Molecular Biol, Colo State Univ, 81-; vis scientist, Max Planck Inst Biophys Chem, Göttingen, FRG, 89-90. *Mem:* Soc Anal Cytol; Radiation Res Soc; NAm Hyperthermia Group; Cell Kinetics Soc. *Res:* Cellular mechanisms of hyperthermia and thermotolerance; use of flow cytometry to analyze cell cycle, membrane fluidity, intracellular pH and ion changes induced by hyperthermia; tumor cell kinetics during therapy. *Mailing Add:* Dept Radiol & Health Sci Colo State Univ Ft Collins CO 80523

FOX, MICHAEL JEAN, b Chicago, Ill, Apr 19, 41; div; c 2. COLLOID CHEMISTRY, CORROSION. *Educ:* Southern Ill Univ, BS, 65; Univ Hawaii, PhD(phys chem), 71. *Prof Exp:* Staff mem, Res & Develop, US Gypsum Co, 65-66; researcher biophys, Univ Hawaii, 71-72; sales engr, Brewer Chem Corp, 72-73; staff mem corrosion, Res & Develop, Gen Elec Co, 73-78; res & develop mgr, Elec Power Res Inst, 78-83; MGR, CHEM ENG SERV, APTECH CORP, 83- *Mem:* Fel NDEA; fel NSF. *Res:* Interdisciplined research in surface and colloid chemistry, corrosion, metallurgy and hazardous chemicals. *Mailing Add:* 7490 Stanford Pl Cupertino CA 95014

FOX, MICHAEL WILSON, b Bolton, Eng, Aug 13, 37; m 63; c 3. ETHOLOGY. *Educ:* Royal Vet Col, Univ London, BVetMed & MRCVS, 62; PhD(med), 67, DSc(behav), 76. *Prof Exp:* Fel, Jackson Lab, Bar Harbor, Maine, 62-64; med res assoc brain & behav develop, State Res Hosp, Galesburg, Ill, 64-67; asst prof biol & psychol, Wash Univ, 67-68, assoc prof psychol, 68-76. *Concurrent Pos:* Dir, Inst Study Animal Problems, Humane Soc US. *Honors & Awards:* Felix Wankle Int Prize in Animal Welfare Res, 81. *Mem:* Animal Behav Soc; Am Vet Med Asn; Am Asn Animal Sci; Am Psychol Asn. *Res:* Animal behavior, development of canines, including wolf, coyote, fox and dog; animal rights, philosophy & welfare science; comparative psychopathology. *Mailing Add:* 4912 Sherrier Pl Washington DC 20016

FOX, NEIL STEWART, b Detroit, Mich, July 21, 45. INDUSTRIAL ORGANIC CHEMISTRY. *Educ:* Wayne State Univ, BS, 67; Iowa State Univ, MS, 70, PhD(chem), 74. *Prof Exp:* Sr chemist, 3M Co, 74-77; res scientist, Thiokol-Dynachem Co, 77-78, res supvr, 78-81, group leader & tech res mgr, 81-85; vpres opers, Chemline Indust, 85-87; pres, Armecyn, 88-89; bus dir, Imaging Prod, Inland Specialty Chem, 89-90; PRES, TAIYO AM INC, 90- *Mem:* Am Chem Soc; Sigma Xi; Soc Mfg Eng. *Res:* Process and product development of fine chemicals; UV photopolymerization technology; printed circuit boards; graphic arts clear coatings; dry film photoresists; screen inks, solder and etch resists. *Mailing Add:* 1321 Enterprize Way Carson City NV 89703-3621

FOX, OWEN FORREST, b Madison, Wis, Jan 11, 44; m 87; c 4. BIOCHEMISTRY, MOLECULAR BIOLOGY. *Educ:* Ind Inst Technol, BSci, 67; Univ Mo, PhD(biochem), 71. *Prof Exp:* Res asst protein chem, Max Planck Inst Biochem, 72-75; staff scientist biochem, Okla Med Res Found, 76-80, sr res scientist, 80-87; RES SCIENTIST, PITMAN-MOORE, INC, 87- *Concurrent Pos:* Lectr carbohydrate chem & metab, Univ Okla. *Mem:* Am Chem Soc; NY Acad Sci; AAAS; Sigma Xi. *Res:* Biochemistry, structure and function of glycoproteins in mucous secretions; metabolism and synthesis of glycoproteins in chemical carcinogenesis; immunochemical interactions; autoimmune diseases; bioanalytical chemistry; immunology. *Mailing Add:* 6551 Carlisle Rd Terre Haute IN 47802

FOX, PAUL JEFFREY, b New York, NY, Dec 6, 41. MARINE GEOLOGY, TECTONOPHYSICS. *Educ:* Ohio Wesleyan Univ, BA, 63; Columbia Univ, PhD(geol), 72. *Prof Exp:* Res asst marine geol, Lamont Doherty Geol Observ, 64-72; from asst prof to assoc prof geol, State Univ NY, Albany, 72-81; assoc prof, 81-86, PROF GEOL, GRAD SCH OCEANOG, UNIV RI, 86- *Concurrent Pos:* Vis res assoc, Lamont Doherty Geol Observ, 72-82; mem, Ocean Crust Panel, Deep Earth Sampling Prog, Joint Oceanog Inst, 76-78, chmn, 78-82; panel mem, adv comt ocean sci, NSF, 78-81; steering comt, Conf on Sci Ocean Drilling, 86-87 & Ridge Interdisciplinary Global Exp, 86-; mem, US sci adv comt, Joint Ocean Inst, 87- *Mem:* Fel Geol Soc Am; Am Geophys Union; AAAS; Sigma Xi. *Res:* Structure, composition and evolution of ocean crust; the tectonics of the Mid-Oceanic Ridge System; the geologic history of the Caribbean. *Mailing Add:* Grad Sch Oceanog Univ RI Kingston RI 02881

FOX, PHYLLIS, b Denver, Colo, Mar 13, 23; m 58; c 2. MATHEMATICS, COMPUTER SCIENCE. *Educ:* Wellesley Col, AB, 44; Univ Colo, BS, 48; Mass Inst Technol, MS, 49, ScD(math), 54. *Prof Exp:* Scientist AEC, Courant Inst, NY Univ, 54-58; res assoc, Mass Inst Technol, 58-62; assoc prof math, NJ Inst Technol, 63-67, prof comput sci, 67-73; mem tech staff, 73-84, CONSULT, BELL LABS, 84- *Mem:* Soc Indust & Appl Math; Asn Comput Mach. *Res:* Applications of computers; numerical analysis; applied mathematics; development of mathematical subroutine libraries for numerical computation. *Mailing Add:* 66 Old Short Hills Rd Short Hills NJ 07078

FOX, RICHARD CARR, b Lowell, Mass, Oct 3, 33. VERTEBRATE PALEONTOLOGY. *Educ:* Hamilton Col, AB, 55; State Univ NY, MSc, 57; Univ Kans, PhD(zool), 65. *Prof Exp:* Asst prof, 65-69, PROF GEOL & ZOOL, UNIV ALTA, 69- *Mem:* AAAS; Am Soc Zoologists; Am Soc Ichthyologists & Herpetologists; Am Soc Mammalogists; NY Acad Sci. *Res:* Paleozoic reptiles and the origin of reptiles; late Cretaceous and early Tertiary mammals, lizards, salamanders; community evolution of late Cretaceous vertebrates. *Mailing Add:* Dept Geol Univ Alta Edmonton AB T6G 2M7 Can

FOX, RICHARD HENRY, b Reno, Nev, Nov 1, 38; m 61; c 2. SOIL FERTILITY. *Educ:* Carleton Col, BA, 61; Univ Ariz, MS, 64, PhD(agr chem & soils), 66. *Prof Exp:* Soils adv to Peru, NC State Univ/USAID Agr Mission Peru, 66-69; asst prof soil fertil & dir opers in PR, Trop Soils Proj, Cornell Univ/USAID/Univ PR, 69-74; from asst prof to assoc prof, 75-85, PROF SOIL SCI, PA STATE UNIV, 85- *Mem:* Soil Sci Soc Am; Am Soc Agron; Int Soil Sci Soc; Sigma Xi. *Res:* Developing means to estimate soil nitrogen supply capability; measuring NH3 volatilization losses from surface applied urea-containing fertilizers and developing methods to minimize losses; estimating residual nitrogen from legumes; optimizing nitrogen fertilizer management for corn. *Mailing Add:* Dept Agron Pa State Univ University Park PA 16802

FOX, RICHARD ROMAINE, b New Haven, Conn, Nov 12, 34; m 55, 81; c 6. LABORATORY ANIMAL SCIENCE, GENETICS. *Educ:* Univ Conn, BS, 56; Univ Minn, MS, 58, PhD(animal breeding, genetics), 59. *Prof Exp:* Actg swine herdsman, Univ Conn, 56; asst animal breeding, Univ Minn, 56-59; fel, Jackson Lab, 59-60, from assoc staff scientist to staff scientist, 60-80, sr staff scientist, 80-90, EMER SR STAFF SCIENTIST, JACKSON LAB, 90- *Honors & Awards:* Res Award, Am Asn Lab Animal Sci, 81. *Mem:* AAAS; Am Dairy Sci Asn; Am Soc Animal Sci; Genetics Soc Am; Am Genetic Asn; NY Acad Sci; Sigma Xi. *Res:* Genetics of the rabbit; quantitative genetics; reproductive physiology; germ plasm preservation; genetic monitoring. *Mailing Add:* Jackson Lab Bar Harbor ME 04609

FOX, RICHARD SHIRLEY, b Lexington, KY, Feb 20, 43. INVERTEBRATE ZOOLOGY, MARINE ECOLOGY. *Educ:* Univ Fla, BS, 67, MS, 69; Univ NC, Chapel Hill, PhD(zool),80. *Prof Exp:* Aquatic biologist, Fla Dept Air & Water Pollution Control, 69-71; PROF BIOL, LANDER COL, 77- *Concurrent Pos:* Vis prof, NJ Marine Sci Consortium, 76; res assoc, Baruch Inst Marine Biol, Univ SC, 78-; vis assoc prof zool, Marine Lab, Duke Univ, 88-90; vis prof zool, Harbor Branch Oceanog Inst. *Honors & Awards:* Distinguished Prof Award, Lander Found, 85. *Mem:* Sigma Xi; Am Soc Zoologists; Southeastern Estuarine Res Soc. *Res:* Systematics and ecology of amphipod crustaceans; marine community ecology; shallow water marine benthic invertebrates of the Southeastern United States; invertebrate morphology. *Mailing Add:* Dept Biol Lander Col Greenwood SC 29649

FOX, ROBERT BERNARD, b Boise, Idaho, May 24, 22; m 54. TECHNOLOGY ASSESSMENT. *Educ:* Univ Minn, BChem, 43; Univ Md, PhD(org chem), 59. *Prof Exp:* Chemist, Sharples Chem Inc, 43-44; chemist, 46-54, SECT HEAD, NAVAL RES LAB, 54- *Concurrent Pos:* Secy & mem, Comn Macromolecular Nomenclature, Int Union Pure & Appl Chem, 68-79, secy, Interdiv Comn Nomencalture & Symbols, 79-87; dir, Am Chem Soc, 81-86. *Honors & Awards:* Pure Sci Award, Sigma Xi, 74. *Mem:* Am Chem Soc; Sigma Xi. *Res:* Polymer science, elastomers; interpenetrating networks; polymer photochemistry. *Mailing Add:* 6115 Wiscasset Rd Bethesda MD 20816

FOX, ROBERT DEAN, b Cass City, Mich, Oct 26, 36; m 66; c 6. AGRICULTURAL ENGINEERING, ATMOSPHERIC PHYSICS. *Educ:* Mich State Univ, BS, 57, MS, 58, PhD(agr eng), 68. *Prof Exp:* Instr agr eng & rural electrification, Mich State Univ, 63-64; AGR ENGR, AGR RES SERV, USDA, 68- *Mem:* Am Soc Agr Engrs; Am Meteorol Soc. *Res:* Diffusion of fine particles in air, specifically the effects of turbulent transport of particles in the region of a plant canopy; drift from orchard sprayers. *Mailing Add:* Agr Eng Dept Ohio Agr Res & Develop Ctr Wooster OH 44691

FOX, ROBERT KRIEGBAUM, b Covington, Ohio, Apr 1, 07; m 34; c 3. PHYSICAL INORGANIC CHEMISTRY. *Educ:* Ohio State Univ, BA, 29, MA, 30, PhD(inorg & phys chem), 32. *Prof Exp:* Asst chem, Ohio State Univ, 29-32; from instr to asst prof, Bethany Col, WVA, 32-36; asst prof, Hiram Col, 36-41; partner, Fox Chem Co, Ohio, 41-45; pres, Ind Glass Co, 56-74; pres, 45-76, vpres & treas, Lancaster Colony Corp, 62-83, CHMN, LANCASTER

GLASS CORP, 76- *Mem:* Am Chem Soc; Sigma Xi. *Res:* Catalytic oxidation of acetylene black; solubility of manganese hydroxide; fabric treating compounds; crystal and colored glasses. *Mailing Add:* 1445 Cincinnati-Zanesville Rd SW Lancaster OH 43130

FOX, ROBERT LEE, b Moberly, Mo, May 26, 23; m 48; c 5. AGRONOMY. *Educ:* Univ Mo, BS, 47, MA, 50, PhD(soils), 55. *Prof Exp:* Asst prof agron & asst agronomist, Univ Nebr, 50-56; assoc prof, Univ Ankara, 56-58 & Univ Nebr, 58-61; PROF SOIL SCI & SOIL SCIENTIST, UNIV HAWAII, 61- *Concurrent Pos:* Vis scientist, NC State Univ, 68, Int Trop Agr, 75 & Rurakura Res Ctr, NZ, 82; sr sci fel, NZ; key consult, Foresters Am, Punjab Agr Univ, Ludhiana, India. *Mem:* Fel Am Soc Agron; fel Soil Sci Soc Am. *Res:* Soil fertility and chemistry. *Mailing Add:* 1542 Hauhaku Honolulu HI 96822

FOX, ROBERT WILLIAM, b Montreal, Que, July 1, 34, Can citizen; m 62; c 2. MECHANICAL ENGINEERING, FLUID MECHANICS. *Educ:* Rensselaer Polytech Inst, BS, 55; Univ Colo, MS, 57; Stanford Univ, PhD(mech eng), 61. *Prof Exp:* Instr mech eng, Univ Colo, 55-57; asst, Stanford Univ, 57-60; from asst prof to assoc prof, 60-66, chmn grad prog, 69-71, asst head, Sch Mech Eng, 71-72, asst dean, Eng for Instr, 72-76, actg head, Sch Mech Eng, 75-76, PROF MECH ENG, PURDUE UNIV, 66-, ASSOC HEAD DEPT, 76- *Concurrent Pos:* Vis prof, Bradley Univ, 64-65; consult, Nat Comt Fluid Mech Films, 63- & Owens Corning Fiberglass Corp, Ohio, 63. *Honors & Awards:* M B Scott Award, 83 & 87; Ruth & Joel Spira Award, 87; Solberg Award, 78 & 83. *Mem:* Am Soc Eng Educ; Am Soc Mech Engrs; Sigma Xi. *Res:* Turbulent boundary layers in adverse pressure gradients; diffuser flows. *Mailing Add:* Sch Mech Eng Purdue Univ Lafayette IN 47907

FOX, RONALD FORREST, b Berkeley, Calif, Oct 1, 43; m 69; c 2. THEORETICAL PHYSICS. *Educ:* Reed Col, BA, 64; Rockefeller Univ, PhD(theoret physics), 69. *Prof Exp:* Miller fel theoret physics, Miller Inst Basic Res Sci, Univ Calif, Berkeley, 69-71; from asst prof to assoc prof, Ga Inst Technol, 71-79, asst dir physics, 82-84, assoc dir physics, 86-89, PROF PHYSICS, GA TECH, 79- *Concurrent Pos:* Sloan fel, 74-78; vis prof chem, Univ Calif, Davis, 81; Guggenheim fel, 85; fac assoc, Argonne Nat Lab, Argonne, Ill, 89-91. *Mem:* Am Phys Soc; Sigma Xi; NY Acad Sci. *Res:* Biodynamics; non-equilibrium statistical mechanics; energy conversion; stochastic processes; chaos. *Mailing Add:* Sch Physics Ga Inst Technol Atlanta GA 30332-0430

FOX, RUSSELL ELWELL, b Richmond, Va, Dec 28, 16; m 42; c 3. ATOMIC PHYSICS. *Educ:* Hampden-Sydney Col, BS, 38; Univ Va, MS, 39, PhD(physics), 42. *Prof Exp:* Fel, Westinghouse Elec Corp, 42, res physicist, 42-57, mgr, Physics Dept, 57-64, dir atomic & molecular sci res & develop, 64-69, res dir consumer prods, 69-74, res & develop dir indust prods, Res Lab, 74-82; RETIRED. *Mem:* Fel Am Phys Soc; Sigma Xi. *Res:* Mass spectrometry; research planning and administration. *Mailing Add:* 14 Kenilworth Dr Hampton VA 23666

FOX, SALLY INGERSOLL, b Philadelphia, Pa, Oct 19, 25; m 49; c 3. MICROBIOLOGY. *Educ:* Vassar Col, BA, 46; Columbia Univ, MA, 52; Rensselaer Polytech Inst, PhD(biol), 67. *Prof Exp:* Lectr microbiol, Russell Sage Col, 61; asst microbiol, Rensselaer Polytech Inst, 63-66; from asst prof to prof biol, 67-, EMER PROF BIOL, COL ST ROSE. *Mem:* Fedn Am Scientists; Am Soc Microbiol. *Res:* Microflora of acid mine waters, especially leaching areas of copper mines; toxic effects of Bacillus thesingiensis d-endotoxin on cultured cells and Trichuplusiani larvae; interactions between indigenous yeast and copper oxidizing thiobacilli. *Mailing Add:* 191 Southerland Rd Stephentown NY 12168

FOX, SAMUEL MICKLE, III, b Andalusia, Pa, Feb 13, 23; m 49; c 4. MEDICINE. *Educ:* Haverford Col, BA, 44; Univ Pa, MD, 47; Am Bd Internal Med, dipl, 55. *Prof Exp:* Intern, resident med, Univ Pa Hosp, 47-50; asst instr med, Univ Pa, 48-50; actg chief gastroenterol, Nat Naval Med Ctr, US Navy, Md, 50-51, from mem staff to chief cardiol serv, 53-54, mem staff, Off Naval Res & Med Off, Eastern Atlantic & Mediterranean, Eng, 51-53; head dept clin invest, Naval Med Res Unit, Cairo, Egypt, 54-56; chief cardiol & officer in chg cardiopulmonary res lab, US Naval Hosp, Portsmouth, Va, 56-57; responsible investr sect cardiodyn, Nat Heart Inst, 57-59, co-chief sect, 59-61, asst dir, 61-62; dep chief heart dis & stroke control prog, USPHS, 63-64, chief, 65-70; prof med, Div Cardiol, Sch Med, George Washington Univ, 70-75; PROF MED, DIV CARDIOL, SCH MED, GEORGETOWN UNIV, 75- *Concurrent Pos:* Res fel med & gastroenterol, Hosp Univ Pa, 48-49; res assoc, Res Inst & Abbassia & Embaba Fever Hosps, Cairo, Egypt, 54-56; vis prof, Ein Shams Univ, 55-56; res assoc, Kasr-el-aini, 55-56; consult, WHO, Inter Am Soc Cardiol & Asian-Pac Soc Cardiol; mem NASA res adv comt biotech & human res, 64-68; fel coun clin cardiol, Am Heart Asn; mem, President's Coun Phys Fitness & Sports, 70- *Mem:* Fel AMA; Am Heart Asn; Am Col Physicians; Am Col Cardiol (pres, 72-73); fel Am Col Sports Med (vpres, 76-79). *Res:* Clinical problems of disturbed cardiopulmonary physiology, prevention, management, and rehabilitation. *Mailing Add:* Georgetown Univ 3800 Reservoir Rd NW Washington DC 20007

FOX, SIDNEY WALTER, b Los Angeles, Calif, Mar 24, 12; m 37; c 3. BIOCHEMISTRY. *Educ:* Univ Calif, Los Angeles, BA, 33; Calif Inst Technol, PhD(biochem), 40. *Prof Exp:* Asst, Rockefeller Inst, 34-35 & Calif Inst Technol, 35-37; fel, Univ Mich, 41-42; from asst prof to prof chem, Iowa State Col, 43-55, head chem sect, Exp Sta, 49-55; prof chem, Fla State Univ, 55-64, dir oceanog inst, 55-61 & inst space biosci, 61-64; dir regional lab, Nat Found Cancer Res, 76-84; PROF RES & DIR INST MOLECULAR & CELLULAR EVOLUTION, UNIV MIAMI, 64- *Concurrent Pos:* Consult, AEC, 47-55 & A E Staley Co, 54-60; mem panel selection NSF fels, Nat Acad Sci-Nat Res Coun, 53-55, 60, chmn subcomt nomenclature biochem, Nat Res Coun, 56-57; secy, Nat Comt Biochem, 56-59; subchmn panel selection NSF fac fels, Asn Am Cols, 57-58; Nat Acad Sci deleg, Fourth Int Cong Biochem, Vienna, 58; mem adv panel syst biol, NSF, 58-60 & comn undergrad educ biol sci, 69-71; mem biosci subcomt, NASA, 60-66; USA-USSR Interacad lectr, 69; vpres, Int Soc Study Origin Life, 70-74, dir, Liberty Fund Conf, 81-86; lectr, Chinese Acad Sci, Beijing, 79 & vis investr, Inst Biophysics, 86; AAAS lectr, 79-82; vis prof, Tex A&M Univ, 80; mem, evolutionary biol comt, Nat Asn Biol Teachers, 80-83; Watkins vis prof, Wichita State Univ, 81; vis investr, Max Planck Inst Biochem, Munich, 88. *Honors & Awards:* US Dept Com Medal, 62; Priestman lectr, Univ NB, 64; Daisy lectr, Univ Ill, 70; Iddles Award, 73; Fla Award, Am Chem Soc, 74; Camille & Henry Dreyfus lectr, Pomona Col, 81. *Mem:* AAAS; Geochem Soc; Am Chem Soc; Am Soc Biol Chem; Am Soc Cell Biol; Neurosci Soc. *Res:* Thermal proteins; molecular and cellular evolution; self-sequencing of amino acids and relationship of this process to thermal proteins and their biofunctions; significance to evolutionary biology. *Mailing Add:* Southern Ill Univ Life Sci Bldg II Rm 429 Carbondale IL 62901-8816

FOX, STANLEY FORREST, b Rockford, Ill, Mar 21, 46; m; c 3. EVOLUTIONARY ECOLOGY, HERPETOLOGY. *Educ:* Univ Ill, BS, 68; Yale Univ, MPhil, 69, PhD(ecol & evolution), 73. *Prof Exp:* Res asst forest ecol, Brookhaven Lab, 68; biol teacher, Watertown High Sch, 69-70; fel pop biol, Rockefeller Univ, 73-75; asst prof evolution, Boston Univ, 75-77; asst prof, 77-82, ASSOC PROF EVOLUTION, OKLA STATE UNIV, 82- *Concurrent Pos:* Prin investr, NSF res grants, 67, 71-73, 78-80, 81-83, & 85-86, Am Philos Soc grant, 75, Sigma Xi grants, 71 & 75 & Conacyt res grant, 90; asst cur herpetol, Peabody Mus, Yale Univ, 70, actg cur, 71-73; teaching asst, Yale Univ, 70-71; res assoc, NIH res grant, 75; co-investr, Am Mus Natural Hist & Am Philos res grant, 75; curator herpetology, Okla State Univ, 77-; Fulbright Scholar, Chile, 85. *Mem:* Herpetologists' League; Ecol Soc Am; Soc Study Ecolution; Am Soc Ichthyologists & Herpetologists; Int Soc Behav Ecol. *Res:* Evolutionary ecology of behavior, especially social behavior; ecology of territoriality; population ecology and demography of lizards; natural selection. *Mailing Add:* Dept Zool Okla State Univ Stillwater OK 74078

FOX, STEPHEN SORIN, b New Haven, Conn, July 15, 33; m 59; c 3. PHYSIOLOGICAL PSYCHOLOGY. *Educ:* Univ Pa, BA, 55; Univ Mich, MA, 57, PhD(psychol), 59. *Prof Exp:* Asst psychophysiol, Vision Res Labs, Univ Mich, 55-56; res anatomist, Med Sch, Univ Calif, Los Angeles, 57-59; head animal neurophysiol & behav sect, Schizophrenia & Psychopharmacol Joint Res Proj, Univ Mich, 59-63, assoc res psychobiologist, Ment Health Res Inst, 59-65, asst prof psychol, Univ Mich, 62-65; assoc prof, 65-70, PROF PSYCHOL, UNIV IOWA, 70- *Concurrent Pos:* Res assoc, Ypsilanti State Hosp, Mich, 59-63; investr, Woods Hole Marine Biol Labs, 61- *Mem:* Am Psychol Asn; Biophys Soc; NY Acad Sci; Biofeedback Res Soc; Am Physiol Soc. *Res:* Behavioral neurophysiology; psychobiology; neurophysiological correlates of behavior; electrophysiology; identification of the structural and functional substrate of behavior; neural coding. *Mailing Add:* Dept Psychol Univ Iowa Iowa City IA 52242

FOX, THOMAS ALLEN, b Dover, Ohio, Aug 23, 26; m 53; c 2. REACTOR PHYSICS. *Educ:* Univ Ohio, BS, 49, MS, 51. *Prof Exp:* Instr physics, Ohio Univ, 51-52; from res scientist to head, Exp Reactor Physics Sect, 52-68, Lewis Res Ctr, NASA, 52-68, head, Crit Facil Unit, 68-73, res scientist, Power Amplifier Sect, 73-79, Microwave Amplifier Sect, 80-82; RETIRED. *Concurrent Pos:* Instr, Cayahoga Community Col, 84- *Mem:* Am Nuclear Soc. *Res:* Experimental reactor physics; reactor operations; first analytical studies; reactor critical facilities. *Mailing Add:* 27522 Dunford Rd Westlake OH 44145-5364

FOX, THOMAS DAVID, b Staten Island, NY, Mar 5, 50. MOLECULAR GENETICS, EXTRACHROMOSOMAL GENETICS. *Educ:* Cornell Univ, BS, 71; Harvard Univ, PhD(biochem & molecular biol), 76. *Prof Exp:* Fel biochem, Biocenter, Univ Basel, 76-78, independent res assoc, 79, asst prof, 79-81; asst prof, 81-86, ASSOC PROF GENETICS, CORNELL UNIV, 86- *Concurrent Pos:* Instr, course on molecular genetics of yeast & molecular biol of membranes, Europ Molecular Biol Orgn, Basel, 80; res career develop award, NIH, 83-88, Biochem Study Sect, 87-91. *Mem:* Genetic Soc Am; Am Soc Biochem & Molecular Biol; Am Soc Microbiol. *Res:* Study of gene structure and regulation, especially in the mitochondrial genetic system of yeast. *Mailing Add:* Sect Genetics & Develop Biotechnol Bldg Cornell Univ Ithaca NY 14853-2703

FOX, THOMAS OREN, b Ames, Iowa, Nov 24, 45; m 75. BIOCHEMICAL GENETICS, DEVELOPMENTAL NEUROSCIENCE. *Educ:* Pomona Col, BA, 67; Princeton Univ, PhD(biochem sci), 71. *Prof Exp:* Fel develop biol, 71-73, instr biochem genetics, 73-74, from asst prof to assoc prof neuropath, 74-84, ASSOC PROF NEUROSCI, HARVARD MED SCH, 84-, ASSOC DIR NEUROSCI PROG, HARVARD UNIV, 84- *Concurrent Pos:* Sr res assoc neurosci, Childrens Hosp, Med Ctr, Boston, 79-84; biochemist in neurol, Mass Gen Hosp, 84-, sr biochemist, E K Shriver Ctr, Waltham, Mass, 84- *Mem:* Soc Neurosci; Endocrine Soc; Am Chem Soc; Tissue Culture Asn; Sigma Xi. *Res:* Biochemical genetics and development of sex steroid receptors; biochemical and behavioral investigations of mammalian mutants with syndromes of androgen-resistance; control mechanisms involved in the sexual differentiation of the brain. *Mailing Add:* One Louders Lane Boston MA 02130-3422

FOX, THOMAS WALTON, b Pawtucket, RI, Mar 21, 23; m 48; c 2. ANIMAL GENETICS, ANIMAL BREEDING. *Educ:* Univ Mass, BS, 49, MS, 50; Purdue Univ, PhD(genetics, physiol), 52. *Prof Exp:* Asst poultry, Purdue Univ, 50-52; from instr to asst prof, 52-54, prof poultry & head dept, 54-64, head dept vet & animal sci, 64-79, PROF ANIMAL SCI, UNIV MASS, AMHERST, 79- *Mem:* AAAS; Am Genetics Asn; Poultry Sci Asn; Am Inst Biol Sci; Genetics Soc Am; Sigma Xi. *Res:* Avian genetics and applied breeding; avian physiology. *Mailing Add:* 676 E Pleasant St Amherst MA 01002

FOX, WILLIAM, b New Haven, Conn, Mar 15, 14; m 49; c 2. PHYSICAL & COLLOID CHEMISTRY. *Educ:* City Col, BS, 35; Columbia Univ, MA, 42, PhD(chem), 44. *Prof Exp:* Patrolman, New York Police Dept, 40-45, detective police lab & field command, 45-52, sgt supvr, 52-57, lt, Police Acad & Fac Police Sci, City Col New York, 57-62, lt supvr field command, 62-66, capt, 66-77; ASSOC, OAKLAND RES ASSOCS, 56- *Concurrent Pos:* Phys chemist, Petrol Exp Sta, Bur Mines, Okla, 49; mem, Task Force Forensic Chem, Am Chem Soc, 78-80; invited lectr, 4th Int Conf Colloid Chem, Hungary, 83; nat counr, Am Chem Soc, 77-79, 89-91. *Mem:* Am Chem Soc; AAAS; Sigma Xi; Int Asn Colloid & Interface Scientists. *Res:* Physics and chemistry of fluid interfaces; hydrostatics in various gravitational fields; line of contact of fluid phases; line and edge forces of fluid phases; spontaneous mass movements of fluids; contact angles and force constants of fluid interfaces; surface and interfacial tension; thin film stability; emulsions; colloids; secondary oil recovery; forensic science. *Mailing Add:* Oakland Res Assocs 657 Oakland Ave Staten Island NY 10310

FOX, WILLIAM B, b Clifton, NJ, Aug 12, 28; m 54; c 1. PHYSICAL INORGANIC CHEMISTRY. *Educ:* Univ Ill, BS, 55; Pa State Univ, PhD(inorg chem), 60. *Prof Exp:* Res chemist, Gen Chem Div, Allied Chem Corp, Morristown, 60-61, sr res chemist, 61-62, tech supvr, 62-66, mgr inorg mat res, Indust Chem Res Lab, 66-68, dir inorg res & develop, 68-71; head, Inorg Chem Br, 71-79, asst dir res, Off Undersecy Defense, Res & Eng, 79-81, HEAD, POLYMERIC MATS BR, NAVAL RES LAB, 81- *Concurrent Pos:* Chmn, Fluorine Div, Am Chem Soc, 80. *Mem:* Am Chem Soc; Royal Soc Chem; Sigma Xi. *Res:* Synthesis of fluorocarbon derivatives of oxygen, nitrogen, sulfur, halogens, and metalloids; isolation and spectroscopy of free radicals; chemical laser systems; high temperature fluids, organic and organometallic polymers, coatings and composites. *Mailing Add:* 1813 Edgehill Dr Alexandria VA 22307-1122

FOX, WILLIAM CASSIDY, b Homestead, Fla, Oct 9, 26; m 55; c 1. MATHEMATICS. *Educ:* Grinnell Col, BA, 49; Univ Mich, MA, 50, PhD(math), 55. *Prof Exp:* CLE Moore res instr, Mass Inst Technol, 54-56; asst prof math, Northwestern Univ, 56-60; res assoc, Tulane Univ, 60-61; chair math, 66-67, dir grad prog, 75-78, ASSOC PROF MATH, STATE UNIV NY, STONY BROOK, 61- *Res:* Topology; analytic and harmonic functions; Riemann surfaces. *Mailing Add:* Dept Math State Univ NY Stony Brook NY 11794

FOX, WILLIAM R(OBERT), b Sevierville, Tenn, June 15, 36; m 59; c 3. BIOENGINEERING & BIOMEDICAL ENGINEERING. *Educ:* Univ Tenn, BS, 58, MS, 59; Iowa State Univ, PhD(agr eng, eng mech), 62. *Prof Exp:* Instr agr eng, Iowa State Univ, 60-62; from asst prof to assoc prof, 62-67, PROF AGR & BIOL ENG & HEAD DEPT, MISS STATE UNIV, 67- *Concurrent Pos:* Spec asst, Ala Comn on Higher Educ, 85-86, USDA & SEA Southern Energy Ctr. *Mem:* AAAS; fel Am Soc Agr Engrs; Am Soc Eng Educ; Nat Soc Prof Engrs; Sigma Xi. *Res:* Tillage mechanics and machinery; fundamentals of tillage energy applications; bio-mechanics. *Mailing Add:* Dept Agr & Biol Eng Miss State Univ Box 5465 Mississippi State MS 39762

FOX, WILLIAM TEMPLETON, b Chicago, Ill, Nov 15, 32; m 59; c 1. GEOLOGY. *Educ:* Williams Col, BA, 54; Northwestern Univ, MS, 60, PhD(geol), 61. *Prof Exp:* From instr to assoc prof, 61-72, PROF GEOL, WILLIAMS COL, 72- *Concurrent Pos:* NSF res Grant, 62-64, fel, Stanford Univ, 66-67; Off Naval Res grant, 69-71; geologist, Shell Oil Co, 61; mem coastal res group, Off Naval Res Contract, 72-78; vis prof, Ore State Univ, 73-74. *Mem:* AAAS; Am Asn Petrol Geol; Geol Soc Am; Soc Econ Paleontologists & Mineralogists; Sigma Xi. *Res:* Stratigraphy; sedimentation; paleoecology; computer applications to geology. *Mailing Add:* Dept Geol Williams Col Williamstown MA 01267

FOX, WILLIAM WALTER, JR, b San Diego, Calif, July 18, 45; m 67; c 2. FISHERIES. *Educ:* Univ Miami, BS, 67, MS, 70; Univ Wash, PhD(fisheries), 72. *Prof Exp:* Fisheries biologist pop dynamics, Bur Com Fisheries, Dept Interior, 67-69; fisheries biologist, 72-73, prog leader, 73-75, div chief, Oceanic Fisheries, 75-78, DIR, SOUTHEAST FISHERIES CTR, NAT MARINE FISHERIES SERV, DEPT COM, 78- *Concurrent Pos:* Mem Sci Comt, Int Whaling Comn, 72; chmn, 27th Ann Tuna Conf, 75-76; mem Sci Comt, Int Comn Conserv Atlantic Tunas, 72-; mem Tuna Stock Assessment Panel, Food & Agr Orgn UN, 74-; contrib ed, Fishery Bulletin, Nat Marine Fisheries Serv, 75-78. *Honors & Awards:* W F Thompson Award, Am Inst Fisheries Res Biologists, 71. *Mem:* Am Fisheries Soc; Am Inst Fisheries Res Biologists; Sigma Xi. *Res:* Theory and application of quantitative methods to determine the response of natural populations to controlled and uncontrolled exploitation. *Mailing Add:* 21120 Stonecorp Pl Ashburn VA 22011

FOXHALL, GEORGE FREDERIC, b Worcester, Mass, Feb 20, 39; m 61; c 2. SEMICONDUCTOR PHYSICS. *Educ:* Worcester Polytech Inst, BS, 61; Univ Ill, MS, 62. *Prof Exp:* MEM TECH STAFF, BELL TEL LABS, 61- *Mem:* AAAS; sr mem Inst Elec & Electronics Engrs. *Res:* Semiconductor device physics. *Mailing Add:* Five Plymouth Circle Reading PA 19602

FOXMAN, BRUCE MAYER, b Youngstown, Ohio, Mar 12, 42; m 68; c 2. SOLID STATE CHEMISTRY. *Educ:* Iowa State Univ, BS, 64; Mass Inst Technol, PhD(inorg chem), 68. *Prof Exp:* Res fel chem, Australian Nat Univ, 68-72; from asst prof to assoc prof, 72-85, PROF CHEM, BRANDEIS UNIV, 85- *Concurrent Pos:* Vis prof, T J Watson Res Ctr, IBM CORP, 75; consult, Gen Telephone & Electronics Corp Lab, 81-87. *Mem:* Am Chem Soc; Am Crystallog Asn; Royal Soc Chem. *Res:* Mechanisms of solid-state reactions; engineering and synthesis of reactive single crystals; x-ray diffraction studies. *Mailing Add:* Dept Chem Brandeis Univ PO Box 9110 Waltham MA 02254-9110

FOXWORTHY, BRUCE L, b Spokane, Wash, Dec 30, 25; m 56; c 4. HYDROLOGY. *Educ:* Univ Wash, BS, 50. *Prof Exp:* Geologist, Water Resources Div, US Geol Surv, Wash, 51-59, dist geologist, Ore, 59-64, hydrologist in chg, Long Island, NY, 64-68, sub-dist chief, 68-70, asst dist chief, Wash, 70-73, proj chief, Puget Sound Urban Area Studies, 73-82 & nat water summaries, 83-88; RETIRED. *Honors & Awards:* Common Award, US Dept Interior, 78, Meritorious Serv Award, 82. *Mem:* Am Inst Hydrol. *Res:* Water resources inventories; artificial recharge; hydrology of volcanic terranes; urban hydrology. *Mailing Add:* 601 N Baker Ave 102 East Wenatchee WA 98802

FOXWORTHY, JAMES E(RNEST), b Los Angeles, Calif, Feb 23, 30; m 50; c 6. ENVIRONMENTAL ENGINEERING, OCEANOGRAPHY. *Educ:* Univ Southern Calif, BS, 55, MS, 58, PhD, 65. *Prof Exp:* Instr civil eng, Univ Wis, 55-56; lectr gen eng, Univ Southern Calif, 56-59; from asst prof to assoc prof civil eng & chmn dept, 58-69, dean col eng, 69-73, dean sci & eng, 73-81, PROF CIVIL ENG, LOYOLA MARYMOUNT UNIV, 69-, EXEC VPRES SCI & ENG, 81- *Concurrent Pos:* Consult, WHO, 74- *Mem:* Am Soc Civil Engrs; Am Soc Eng Educ; Am Asn Prof Environ Engrs; AAAS. *Res:* Waste disposal in the marine environment. *Mailing Add:* 27953 Alaflora Dr San Pedro CA 90732

FOXX, RICHARD MICHAEL, b Denver, Colo, Oct 28, 44; m 88; c 1. BEHAVIOR THERAPY, BEHAVIOR ANALYSIS. *Educ:* Univ Calif Riverside, BA, 67; Calif State Univ, MA, 70; Southern Ill Univ, PhD(educ psychol), 71. *Prof Exp:* Res scientist, Anna State Hosp, 70-74; from asst prof to assoc prof psychol, Univ Md, Baltimore, 74-80; RES DIR, ANNA MENTAL HEALTH CTR, 80-; CLIN PROF PSYCHIAT, SOUTHERN ILL UNIV SCH MED, 87- *Concurrent Pos:* Asst prof pediat, Univ Md Med Sch, 74-80; adj prof rehab, Southern Ill Univ, 81-; Erskine fel, Univ Canterbury, 87; mem task force Self Injurious Behav, Asn Advan Behav Ther, 80-81; subcomt mem, Nat Inst Child Health & Human Develop Behav & Social Sci Planning Comt, 84. *Mem:* Fel Am Psychol Asn; fel Am Asn Mental Retardation; Asn Advan Behav Ther; Asn Behav Anal; Behav Ther & Res Soc; fel Am Psychol Soc. *Res:* Developed internationally recognized treatment and training programs for normal and disordered populations; cigarette smoking; language development and social skills; problem solving and aggression. *Mailing Add:* Anna Mental Health Ctr 1000 N Main St Anna IL 62906

FOY, C ALLAN, b Philadelphia, Pa, Jan 27, 44; m 65, 85; c 5. RADAR SYSTEMS. *Educ:* US Naval Acad, Annapolis, Md, BS, 65; Univ Mich, MS, 68, PhD(math), 71. *Prof Exp:* Instr math, US Naval Acad, 77-79; mem res staff, 79-80, asst dir, 80-81, sr prog mgr, 81-82, VPRES, SYST PLANNING CORP, 82- *Mem:* Am Math Soc; Inst Math Statist. *Res:* Radar systems; collection and analysis of radar data; systems level analyses of radar system performance. *Mailing Add:* 2161 Kings Mill Ct Falls Church VA 22043

FOY, CHARLES DALEY, b Buena Vista, Ky, Aug 19, 23; m 50; c 1. SOIL FERTILITY, PLANT NUTRITION. *Educ:* Univ Tenn, BS, 48; Purdue Univ, MS, 53, PhD(soil fertility), 55. *Prof Exp:* Instr vets inst on farm training, Ind 49-51; asst prof agron, Purdue Univ, 55-57; co-op agent agron, Univ Ark, 57-61, soil scientist, Ark, 57-61, soil scientist, Md, 61-72, RES SOIL SCIENTIST, PLANT STRESS LAB & CLIMATE STRESS LAB, AGR RES SERV, USDA, 72- *Honors & Awards:* Environ Qual Award, Am Soc Hort Sci, 74. *Mem:* Fel Am Soc Agron; Soil Sci Soc Am; Crops Soc Am. *Res:* Causes of acid soil infertility; physiological mechanisms for differential tolerances of plant genotypes to high aluminum and manganese and low phosphorus levels in acid soils and mine spoils and low iron availability in calcareous soils; collaboration with plant breeders in tailoring plants for adaptation to soil mineral stresses. *Mailing Add:* Climate Stress Lab Natural Resources Inst Beltsville MD 20705

FOY, CHESTER LARRIMORE, b Dukedom, Tenn, July 8, 28; m 53; c 2. PLANT PHYSIOLOGY. *Educ:* Univ Tenn, BS, 52; Univ Mo, MS, 53; Univ Calif, PhD(plant physiol), 58. *Prof Exp:* Asst instr field crops, Univ Mo, 52-53; asst specialist chem weed control, Dept Bot, Univ Calif, Davis, 53-56, asst plant physiol, 57-58, asst botanist, 58-64, assoc prof bot & assoc botanist, 64-66; assoc prof, 66-68, head dept, 74-80, PROF PLANT PHYSIOL, VA POLYTECH INST & STATE UNIV, 68- *Concurrent Pos:* Nat Acad Sci-Nat Res Coun resident res assoc, Ft Detrick, Md, 64-65; res adv comt, Col Agr & Life Sci, 71. *Mem:* Am Soc Agron; Am Soc Plant Physiol; Weed Sci Soc Am; Plant Growth Regulator Soc; Sigma Xi. *Res:* Agricultural chemicals; chemical weed control; pathways and mechanisms of foliar and root absorption, translocation, accumulation; metabolism, mode of action, selectivity and fate of herbicides and surfactants; physiological, biochemical and morphological changes induced by chemicals; detoxification and other plant protective mechanisms; growth regulators and phytotoxicants; pesticides in the environment. *Mailing Add:* Dept Plant Path Physiol & Weed Sci Va Polytech Inst & State Univ Blacksburg VA 24060-0331

FOY, HJORDIS M, b Stockholm, Sweden, June 28, 26; US citizen; m 56; c 3. INFECTIOUS DISEASES, PREVENTIVE MEDICINE. *Educ:* Karolinska Inst, Sweden, MD, 53; Univ Wash, MS, 67, PhD, 68. *Prof Exp:* Mem staff, Hosp Infectious Dis, Stockholm, 53-55; mem staff, Johns Hopkins Hosp, 56-57 & 58-59; sr fel, Univ Wash, 63-67, from instr to asst prof prev med, 66-70, from asst prof to assoc prof, 70-76, dir prev med training, 83-89, PROF EPIDEMIOL, SCH PUB HEALTH, UNIV WASH, 76- *Concurrent Pos:* Consult epidemiol & biometry training comt, Nat Inst Gen Med, 71-73; consult bact vaccines & toxoids, Bur Biologics, Food & Drug Admin, 72-; consult allergy, immunol res comt, Nat Inst Allergy & Infectious Dis, 75-77; mem, microbiol & infectious dis adv comt, Nat Inst Allergy & Infectious Dis, 78-81; Ministry Pub Health, Thailand, 87-88. *Mem:* Fel Am Col Prev Med; Am Epidemiol Soc; Int Epidemiol Soc; Soc Epidemiol Res; Infectious Dis Soc Am. *Res:* Epidemiology of infectious diseases, in particular Mycoplasma, chlamydia and influenza infections; tropical diseases, dengue, rabies. *Mailing Add:* Dept Epidemiol Univ Wash Sch Pub Health Seattle WA 98195

FOY, ROBERT BASTIAN, b Goodland, Ind, June 14, 28; m 50; c 9. BIOCHEMISTRY, CLINICAL CHEMISTRY. *Educ:* Cent Mich Univ, BS, 50; Mich State Univ, MS, 55, PhD(chem), 60. *Prof Exp:* Technologist, Clin Chem, Hurley Hosp, Flint, Mich, 50-52; technologist, Edward W Sparrow

Hosp, 52-55, tech dir labs, Lab Clin Med, 55-, partner, 75-86; RETIRED. *Concurrent Pos:* Lectr, Mich State Univ, 63-73. *Mem:* AAAS; Am Chem Soc; Am Asn Clin Chemists; Am Soc Clin Pathologists. *Res:* Proteins and enzymes, especially blood coagulation; clinical chemical methodology. *Mailing Add:* 14619 E Horsehead Lake Dr Mecosta MI 49332

FOY, WADE H(AMPTON), b Richmond, Va, Jan 26, 25; m 52; c 2. ELECTRONICS ENGINEERING. *Educ:* US Naval Acad, BS, 46; NC State Col, BEE, 51; Mass Inst Technol, MSEE, 55; Johns Hopkins Univ, DEng(elec eng), 62. *Prof Exp:* Res engr, Martin Co, Md, 56-62; sr res engr, Stanford Res Inst, 62-67, prog mgr, 68-76, asst lab dir, 76-80, STAFF SCIENTIST, SRI INT, 80- *Concurrent Pos:* Lectr, Martin Exten Prog, Drexel Inst Technol, 57-61, adj prof, 61-62; vis prof, Univ Santa Clara, 63-65 & 66-; NSF Fel, 60-61. *Mem:* Sr mem Inst Elec & Electronics Engrs. *Res:* Information theory and statistical communication theory; electronic and control system design; applied mathematics. *Mailing Add:* PO Box 828 Los Altos CA 94022

FOY, WALTER LAWRENCE, b West Springfield, Mass, Aug 6, 21; m 63; c 6. PHYSICAL CHEMISTRY. *Educ:* Am Int Col, BS, 41; Clark Univ, MA, 43, PhD(phys chem), 47. *Prof Exp:* Teacher pub sch, Conn, 41-42; res chemist, E I du Pont de Nemours & Co, 47-52, supvr, 52-55, chief supvr, 55-57, tech supt plant, 57-65, mgr prod res, 65-70, mgr graphic arts & printing res, 70-74, lab dir graphic arts & printing prod res, 74-77 & photopolymer res, 77-82; RETIRED. *Mem:* AAAS; Am Chem Soc; Soc Photog Scientists & Engrs. *Res:* Electrical conductance of solutions; photographic chemistry; electrical conductance of substituted ammonium salts in ethylidene dichloride. *Mailing Add:* 84 Boxwood Terr Red Bank NJ 07701-1188

FOYE, LAURANCE V, JR, b Seattle, Wash, Nov 26, 25; m 51; c 2. MEDICINE. *Educ:* Univ Calif, AB, 49, MD, 52. *Prof Exp:* Asst chief med serv, Vet Admin Hosp, San Francisco, 58-66, chief cancer chemother sect, 60-66; chief cancer ther eval br, Nat Cancer Inst, Md, 66-68, exec secy, Cancer Clin Invest Rev Comt, 66-70, chief clin invests br, 68-70, mem grants assocs bd, NIH, 69-70; dir educ serv & dep asst chief med dir res & educ, 70-73, Vet Admin Cent Off, asst chief med dir acad affairs, 73-74, dep chief med dir, 74-78, dir, Vet Admin Ctr, San Francisco, 78-88; RETIRED. *Concurrent Pos:* Clin instr, Sch Med, Univ Calif, San Francisco, 57-62, asst clin prof, 62-66, res assoc, Cancer Res Inst, 62-66, assoc clin prof, 79-; prin investr, Vet Admin Cancer Chemother Study Group, 61-66; mem adv coun, Nat Heart & Lung Inst, 70-73; mem bd regents, Nat Libr Med, 73-74. *Mem:* Fel Am Col Physicians; Asn Am Med Cols. *Res:* Clinical chemotherapy of cancer; evaluation of drugs and disease response. *Mailing Add:* Univ Calif 125 Cambon Dr No 10-H San Francisco CA 94132

FOYE, WILLIAM OWEN, b Athol, Mass, June 26, 23;; m 76; c 1. MEDICINAL CHEMISTRY. *Educ:* Dartmouth Col, AB, 43; Ind Univ, MA, 44, PhD(org chem), 48. *Prof Exp:* Res chemist, E I du Pont de Nemours & Co, 48-49; asst prof pharmaceut chem, Univ Wis, 49-55; assoc prof chem, 55-63, prof, 63-74, chmn dept, 66-71, dean fac, 70-73, dean grad studies, 73-74, dir res, 64-84, SAWYER PROF PHARMACEUT SCI, MASS COL PHARM, 74- *Concurrent Pos:* vis prof, Univ Cairo, 81; mem, Task Force on Drug Develop, Muscular Dystrophy Asn, 84-89 & US Pharmacopeial Rev Comt, 75-90; chmn, Sect Chem Teachers, Am Asn Col Pharm, 81-82; counr, sect pharmaceut sci, AAAS, 86-90, chair, 91. *Honors & Awards:* Res Achievement Award Pharmaceut & Med Chem, Am Pharmaceut Asn Found, 70; Gold Medal, Kitasato Univ, Tokyo, 85. *Mem:* Fel AAAS; Am Chem Soc; Am Pharmaceut Asn; fel NY Acad Sci; Am Asn Col Pharm; fel Am Asn Pharmaceut Scientists. *Res:* Cancer chemotherapeutic, antibacterial, antihypertensive, radioprotective agents; biological aspects of metal binding; sulfur-containing molecules of medicinal interest. *Mailing Add:* Mass Col Pharm & Allied Health Sci 179 Longwood Ave Boston MA 02115

FOYT, ARTHUR GEORGE, JR, b Austin, Tex, June 17, 37; m 60; c 2. ELECTRON DEVICES. *Educ:* Mass Inst Technol, BSEE & MSEE, 60, ScD(Gunn effect), 65. *Prof Exp:* Mem tech staff, Transistor Develop, Bell Tel Labs, NJ, 60-62; mem tech staff, Gunn Effect & Ion Implatation Technol, Mass Inst Technol, 62-72, group leader, 72-78, mem tech staff, Optical Signal Processing, Lincoln Lab, 78-82; MGR ELECTRONICS RES, UNITED TECHNOL RES CTR, 82- *Mem:* Fel Inst Elec & Electronics Engrs. *Res:* Ion implantation in compound semiconductors; Gunn effect in GaAs and CdTe. *Mailing Add:* United Technol Res Lab Silverlane MS 32 East Hartford CT 06108

FOZARD, JAMES LEONARD, b Detroit, Mich, Aug 23, 30; m 63, 85; c 3. GERONTOLOGY, PSYCHOLOGY. *Educ:* Univ Calif, Santa Barbara, AB, 54; San Diego State Col, MA, 58; Lehigh Univ, PhD(psychol), 61. *Prof Exp:* Asst prof psychol, Colby Col, Maine, 60-63; lectr psychol, Europ Div, Univ Md, 63-64; res fel psychol, Mass Inst Technol, 64-66; asst prof psychol, Southern Methodist Univ, 66-67; res psychologist, 67-76; dir, Geriatric Res Educ Clin Ctr, Vet Admin Outpatient Clin, Boston, 76-78; dir, Patient Treatment Serv, Off Geriat & Extended Care, Vet Admin, Washington, DC, 78-85; ASSOC SCI DIR, BALTIMORE LONGITUDINAL STUDY AGING, GERONT RES CTR, NAT INST AGING, 85- *Concurrent Pos:* Asst prof psychol, dept psychiat, Sch Med, Harvard Univ, 68-78; consult, aging res, Nat Inst Aging, NSF, Admin Aging & NIMH, 70-85 & AMA, 85; ed, psychol sci, J Geront, 72-76; mem, adv coun, Int Asn Perception & Attention, 77-86; chmn, Task Force Aging, Am Psychol Asn for White House Conf Aging, 79-82; co-chmn, Fed Interagency Comt Res Aging, 83-85; Vet Admin rep, Health and Human Serv Task Force Alzheimer's Dis, 84-85; adj prof, Sch Hygiene & Pub Health, Johns Hopkins Univ, 90- *Honors & Awards:* Distinguished Contrib Award, Am Psychol Asn, 84; Commendation Career Contrib Cancer Res, Vet Admin, 85. *Mem:* Fel Am Psychol Asn; fel Geront Soc Am; Human Factors Soc; Int Asn Perception & Attention. *Res:* Memory, vision, interests and abilities in relation to age; Vet Admin long term patient care programs. *Mailing Add:* 16 S Gilmor St Baltimore MD 21223-2455

FOZDAR, BIRENDRA SINGH, biology, horticulture, for more information see previous edition

FOZZARD, GEORGE BROWARD, b Jacksonville, Fla, Apr 13, 39; m 69; c 1. ORGANIC CHEMISTRY. *Educ:* Washington & Lee Univ, BS, 61; Univ NC, PhD(org chem), 67. *Prof Exp:* Res asst org chem, Univ NC, 63-66; sr res chemist, Phillips Petrol Co, 66-; AT CSA INC. *Mem:* Am Chem Soc. *Res:* Conformational analysis; hydrogen bonding; fluorochemicals; laboratory automation; catalytic oxidation. *Mailing Add:* CSA Inc 3043 Faye Rd Jacksonville FL 32226-2336

FOZZARD, HARRY A, b Jacksonville, Fla, Apr 22, 31; m 54; c 2. PHYSIOLOGY, INTERNAL MEDICINE. *Educ:* Wash Univ, MD, 56. *Hon Degrees:* Dr, Univ Uruquay. *Prof Exp:* Intern med, Yale Univ, 56-57; asst, Wash Univ, 59-61, from instr to asst prof, 62-66; prof med & physiol, 66-77, OTHO S A SPRAGUE PROF MED SCI, SCH MED, UNIV CHICAGO, 77- *Concurrent Pos:* Nat Heart Inst fels cardiol, Wash Univ, 61-63 & physiol, Univ Berne, 63-64. *Mem:* Am Fedn Clin Res; Biophys Soc; Am Soc Clin Invest; Am Col Cardiol; Am Physiol Soc. *Res:* Intracellular cardiac electrophysiology, especially excitation-contraction coupling; cardiovascular hemodynamics. *Mailing Add:* Dept Med Box 440 Univ Chicago Chicago IL 60637

FRAAD, LEWIS M, pediatrics; deceased, see previous edition for last biography

FRAAS, ARTHUR P(AUL), b Lakewood, Ohio, Aug 20, 15; m 40; c 2. MECHANICAL ENGINEERING. *Educ:* Case Inst Technol, BS, 38; NY Univ, MS(aeronaut eng), 43. *Prof Exp:* Exp test engr, Wright Aeronaut Corp, 38-40; instr, Aircraft Power Plants, NY Univ, 41-43; exp proj engr, Aircraft Eng Div, Packard Motor Car Co, 43-45; asst prof combustion eng, Case Inst Technol, 45-46; assoc prof, Inst Technol Aeronaut, Brazil, 47-49; mem gen design group, Aircraft Nuclear Propulsion Proj, 50-57, assoc dir, Reactor Div, 57-74, mgr high temperature systs, 74-76, CONSULT, OAK RIDGE NAT LAB, 76- *Concurrent Pos:* Union Carbide Eng fel. *Mem:* Sigma Xi; fel Am Soc Mech Engrs; fel Am Nuclear Soc; assoc fel Am Inst Aeronaut & Astronaut. *Res:* Heat exchangers; combustion engines; fluidized bed combustion, fossil and nuclear power plants. *Mailing Add:* 1040 Scenic Dr Knoxville TN 37919

FRADE, PETER DANIEL, b Highland Park, Mich, Sept 3, 46. ANALYTICAL TOXICOLOGY. *Educ:* Wayne State Univ, BS, 68, MS, 71, PhD(chem), 78. *Prof Exp:* Anal chemist & toxicologist, Dept Path, Div Pharmacol & Toxicol, 68-87, SR CLIN SCIENTIST, DIV CLIN CHEM & PHARMACOL, HENRY FORD HOSP, 87- *Concurrent Pos:* Consult, Inter-Dept, Henry Ford Hosp, 78-; vis scholar, Univ Mich, 80-91; vis lectr, Wayne State Univ, 85; vis scientist, Hypertension Res, Henry Ford Hosp, 86-88; peer reviewer, prof sci journals, 88- *Mem:* Sigma Xi; Am Chem Soc; fel Nat Acad Clin Biochem; AAAS; NY Acad Sci. *Res:* High resolution chromatography: method development for the determination of drugs and their metabolites in biological fluids; fatty acid epoxidation chemistry: synthesis, purification, structure elucidation and biological activity. *Mailing Add:* 20200 Orleans Detroit MI 48203

FRADIN, FRANK YALE, b Chicago, Ill, May 14, 41; m 63; c 3. PHYSICAL METALLURGY, SOLID STATE PHYSICS. *Educ:* Mass Inst Technol, SB, 63; Univ Ill, MS, 64, PhD(metall, physics), 67. *Prof Exp:* SR SCIENTIST & ASSOC LAB DIR PHYS RES, ARGONNE NAT LAB, 67- *Concurrent Pos:* Vis prof, Northern Ill Univ, 69-70, 77-78 & 81-88 & Northwestern Univ, 70. *Mem:* Fel Am Phys Soc; Mat Res Soc. *Res:* Diffusion in solids; nuclear magnetic resonance in alloys and intermetallic compounds; magnetic and transport properties of alloys and intermetallic compounds; superconductivity. *Mailing Add:* Argonne Nat Lab 9700 S Cass Ave Bldg 212 Argonne IL 60439

FRADKIN, CHENG-MEI WANG, genetics; deceased, see previous edition for last biography

FRADKIN, DAVID MILTON, b Los Angeles, Calif, Apr 20, 31; m 59; c 3. THEORETICAL PHYSICS. *Educ:* Univ Calif, Berkeley, BS, 54; Iowa State Univ, PhD(physics), 63. *Prof Exp:* Exploitation engr, Shell Oil Co, 54-56; res assoc theoret physics, Ames Lab, AEC & Iowa State Univ, 63-64; NATO fel, Marconi Inst Physics, Univ Rome, 64-65; from asst prof to assoc prof, 65-75, PROF PHYSICS, WAYNE STATE UNIV, 75-, CHMN DEPT, 81- *Concurrent Pos:* Sr postdoctoral fel, Univ Edinburgh, 77-78. *Mem:* Am Phys Soc. *Res:* Electron polarization operators; coulomb scattering of relativistic electrons; conservation laws and relativistic wave equations; dynamical groups and symmetries; coherent processes in particle beams; particle behavior in intense laser fields. *Mailing Add:* Dept Physics Wayne State Univ Detroit MI 48202

FRADKIN, JUDITH ELAINE, b Oceanside, Calif, Oct 29, 49. ENDOCRINE & METABOLIC RESEARCH. *Educ:* Harvard Col, BA, 71; Univ Calif, San Francisco, MD, 75, Am Bd Internal Med, cert, 78, cert endocrinol & metab, 82. *Prof Exp:* Clin fel med, Harvard Univ, 75-78, fel endocrinol, 78-79; clin assoc, Clin Endocrinol Br, Nat Inst Arthritis, Diabetes & Digestive & Kidney Dis, NIH, 79-82; staff endocrinologist, Naval Hosp, Bethesda, lt comdr, USN & asst prof, Uniformed Serv Sch Health Sci, 82-84; CHIEF, ENDOCRINOL & METAB DIS PROG BR, NAT INST DIABETES & DIGESTIVE & KIDNEY DIS, NIH, 84- *Concurrent Pos:* Med intern, Beth Israel Hosp, 75-76, med jr asst, 76-77, med sr asst, 77-78; consult, Bethesda Naval Hosp; co-chmn, Trans NIH Cystic Fibrosis Coordinating Comt; mem, Interagency Coordinating Comt Human Growth Hormones & Creutzfeldt-Jakob Dis. *Mem:* Endocrine Soc; Am Fedn Clin Res. *Res:* Endocrinology. *Mailing Add:* NIH Westwood Bldg Rm 603 Bethesda MD 20892

FRAENKEL, DAN GABRIEL, b London, Eng, May 6, 37; US citizen; m; c 2. MICROBIOLOGY. *Educ:* Univ Ill, BS, 57; Harvard Univ, PhD(bact), 62. *Prof Exp:* Nat Found fel microbiol, Sch Med, NY Univ, 62-63; res assoc molecular biol, Albert Einstein Col Med, 63-65; from asst prof to assoc prof bact & immunol, 65-73, PROF MICROBIOL & MOLECULAR GENETICS, HARVARD MED SCH, 73- *Res:* Physiology and genetics of carbohydrate metabolism in bacteria and in yeast. *Mailing Add:* Microbiol & Molecular Genetics Harvard Med Sch Boston MA 02115

FRAENKEL, GEORGE KESSLER, b Deal, NJ, July 27, 21; m 51, 67. CHEMICAL PHYSICS. *Educ:* Harvard Univ, BA, 42; Cornell Univ, PhD(chem), 49. *Prof Exp:* Res group leader, Nat Defense Res Comt, Oceanog Inst, Woods Hole, 43-46; from instr to assoc prof chem, Columbia Univ, 49-61, chmn dept, 66-68, dean, Grad Sch Arts & Sci, 68-83, vpres spec proj, 83-86, PROF CHEM, COLUMBIA UNIV, 61-, HIGGINS PROF FAC, GRAD SCH ARTS & SCI, 86- *Concurrent Pos:* Assoc ed, J Chem Physics, 62-64; mem post-doctoral fel comt, Nat Acad Sci-NSF, 64-65; chmn, Gordon Res Conf Magnetic Resonance, 67; trustee, The Walden Sch, 64-66 & Columbia Univ Press, 68-71; mem, Arts Col Adv Coun, Cornell, 64-74, Doctoral Coun, NY State Dept Educ, 68-83; chmn, Comt Policies Grad Educ, Asn Grad Schs, 69-71, mem exec comt, 76-80, pres, 78-79; dir, mem bd & treas, Atran Found, 68-; mem adv bd, Chem Phys Letters, 66-71. *Honors & Awards:* Army-Navy Cert of Appreciation, 48; Off dans l'Ordre des Palmes Academique. *Mem:* Fel AAAS; Am Chem Soc; fel Am Phys Soc; Sigma Xi. *Res:* Electron spin resonance; free radicals. *Mailing Add:* 520 W 14th St Apt 82 Columbia Univ Box 307 Havemeyer New York NY 10025-7811

FRAENKEL, GIDEON, b Frankfurt, Ger, Feb 21, 32; m 61; c 2. CHEMISTRY. *Educ:* Univ Ill, BS, 52; Harvard Univ, MA, 53, PhD(org chem), 57. *Prof Exp:* Res fel chem, Calif Inst Technol, 57-60; from asst prof to assoc prof, 60-67, PROF CHEM, OHIO STATE UNIV, 67- *Concurrent Pos:* Guest prof, Univ Lund, 71; vis prof, Mass Inst Technol, 80-81; Noyes fel, Calif Inst Technol, 59; mem exec comt, Nat Coalition Sci & Technol. *Mem:* Am Chem Soc; Royal Soc Chem; AAAS. *Res:* Reaction mechanisms; nuclear magnetic resonance; organometallic chemistry; kinetics of exchange at equilibrium, reactive intermediates; structure and behavior of carbanions; nuclear magnetic resonance line-shape analysis of dynamic phenomena; mechanism of heavy metal toxicity; generation of carbon atoms. *Mailing Add:* Dept Chem Ohio State Univ 120 W 18th Ave Columbus OH 43210

FRAENKEL, STEPHEN JOSEPH, b Berlin, Ger, Nov 28, 17; US citizen; m 41; c 3. STRUCTURAL MECHANICS, MECHANICS OF MATERIALS. *Educ:* Univ Nebr, BS, 40, MS, 41; Ill Inst Technol, PhD(eng mech), 51. *Prof Exp:* Mgr propulsion & struct res, Ill Inst Technol, 46-55; dir res & develop, Stanray Corp, 55-62; dir eng res, Continental Can Co, 62-64; gen mgr, 64-75, dir res & develop, Container Corp Am, 75-82; pres, Technol Serv Inc, 82-; dir, Technol Commercialization Ctr, Ill Inst Technol, 86-89; RETIRED. *Concurrent Pos:* Consult, Nat Inst Standards & Technol; arbitrator, Am Arbit Asn, 85- *Mem:* Soc Exp Stress Anal; Tech Asn Pulp & Paper Indust; Sigma Xi; Indust Res Inst. *Res:* Physical and mechanical properties and materials; stress analysis; behavior of structural and mechanical systems under dynamic loads; instrumented test procedures; statistical design of experiments. *Mailing Add:* 1252 Spruce St Winnetka IL 60093

FRAENKEL-CONRAT, HEINZ LUDWIG, b Breslau, Ger, July 29, 10; nat US; m 39, 64; c 2. MOLECULAR BIOLOGY. *Educ:* Breslau Univ, MD, 33; Univ Edinburgh, PhD(biochem), 36. *Prof Exp:* Asst, Rockefeller Inst, 36-37; res assoc, Butantan Inst, Sao Paulo, Brazil, 37-38; res assoc inst exp biol, Univ Calif, 38-42; from assoc chemist to chemist, Western Regional Res Lab, Bur Agr & Indust Chem, USDA, 42-50; Rockefeller fel, Eng & Denmark, 51; res biochemist, Virus Lab, 52-58, prof virol, 58-63, prof biochem, 63-81, EMER PROF MOLECULAR CELL BIOL, UNIV CALIF, BERKELEY, 81- *Concurrent Pos:* Guggenheim fel, 63, 67; fac res lectr, Univ Calif, Berkeley, 68; vis prof, Postgrad Med Col, Univ London, 86- *Honors & Awards:* Lasker Award, Am Pub Health Asn; Humbolt Sr US Scientist Award. *Mem:* Nat Acad Sci; Int Soc Toxinology; Am Acad Arts & Sci; Am Chem Soc; Am Soc Biol Chem. *Res:* Chemistry and structure of proteins and nucleic acids, especially viruses, enzymes, hormones, toxins; structural requirements for activity. *Mailing Add:* Stanley Hall Univ Calif Berkeley CA 94720

FRAENKEL-CONRAT, JANE E, b Baltimore, Md, Dec 9, 15. BIOCHEMISTRY. *Educ:* Goucher Col, BA, 36; Univ Calif, PhD(biochem), 42. *Prof Exp:* Asst, Sch Med, Johns Hopkins Univ, 36-37; asst, Inst Exp Biol, Univ Calif, 38-42, res assoc, Radiation Lab, 43, res assoc biochem, Univ, 45-47, res chemist, Dept Home Econ, 48-51; res biochemist, Children's Hosp East Bay, Oakland, Calif, 53-63; clin chemist, Mem Hosp, San Leandro, 63-64; sr scientist, Lockheed Missiles & Space Co, 64; dir lab serv & assoc res scientist, Cancer Res Ctr, Ellis Fischel Hosp, Mo, 64-66, dir res labs, 66-67; lectr biol, Seattle Univ, Pine Lake Campus, 67-69; assoc prof chem, St Mary's Sem Col, 69-73; assoc prof,Univ Maine, 73-77, prof chem, 77-82; RETIRED. *Mem:* AAAS; Am Chem Soc; Am Asn Clin Chemists; Am Asn Allied Health; NY Acad Sci; Sigma Xi. *Res:* Enzyme activity of normal and pathological leukocytes; intermediary protein genetic diseases; metabolism; protein chemistry; development of clinical methods; interpretation of clinical chemistry; effect of irradiation on enzyme activity. *Mailing Add:* Wesley View 64 915 S 216th Des Moines WA 98198

FRAGA, SERAFIN, b Madrid, Spain, Feb 11, 31; m 56; c 3. THEORETICAL CHEMISTRY. *Educ:* Univ Madrid, Lic Sc, 54, DSc(chem), 57. *Hon Degrees:* Dr, Univ Autonoma de Barcelona, 84. *Prof Exp:* Res assoc physics, Univ Chicago, 58-61 & chem, Univ Alta, 61-62; asst prof physics, Royal Mil Col, Que, 62-63; from asst prof chem to assoc prof, 63-69, PROF THEORET CHEM, UNIV ALTA, 69- *Concurrent Pos:* Juan March Found fel, 58-59; fel, Univ Alta, 61-62. *Mem:* Royal Span Soc Physics & Chem. *Res:* Quantum chemistry; theoretical research on atomic and molecular structure. *Mailing Add:* Dept Chem Univ Alta Edmonton AB T6G 2G2 Can

FRAGASZY, RICHARD J, b New York, NY, June 11, 50. GEOTECHNICAL ENGINEERING. *Educ:* Duke Univ, BSE, 72, MS, 74; Univ Calif, Davis, PhD(civil eng), 79. *Prof Exp:* Asst prof, Bucknell Univ, 78-79; ASST PROF CIVIL ENG, SAN DIEGO STATE UNIV, 79- *Mem:* Am Soc Civil Engrs; Int Soc Soil Mech & Found Eng; Sigma Xi; Am Soc Eng Educ. *Res:* Blast-induced liquefaction and use of blasting to compact soils, centrifuge modeling, slope stability, bearing capacity of reinforced earth and dynamic earth pressure. *Mailing Add:* 590 Caldwell Circle Athens GA 30605-3706

FRAHM, CHARLES PETER, b Mason City, Iowa, July 7, 38; m 58; c 2. THEORETICAL PHYSICS. *Educ:* Ga Inst Technol, BS, 61, PhD(high energy physics), 67. *Prof Exp:* Instr physics, Ga Inst Technol, 65-66; res asst, Univ Tel-Aviv, 66-67; asst prof, Ga Inst Technol, 67-68; from asst prof to assoc prof, Ill State Univ, 68-75, prof physics, 75-88; MEM TECH STAFF, ROCKWELL INT, 88- *Concurrent Pos:* Res asst, Advan Res Corp, 64-68; res physicist, Naval Coastal Systs Lab, 72- *Mem:* Am Asn Physics Teachers; Am Phys Soc. *Res:* Theoretical acoustics; magnetics; general relativity. *Mailing Add:* 6537 E Calle Del Norte Anaheim CA 92807

FRAHM, RICHARD R, b Scottsbluff, Nebr, Nov 17, 39; m 61; c 2. POPULATION GENETICS, ANIMAL BREEDING. *Educ:* Univ Nebr, BS, 61; NC State Univ, MS, 63, PhD(genetics), 65. *Prof Exp:* From asst prof to assoc prof, 67-76, prof animal breeding, Okla State Univ, 76-87; PROF & HEAD, ANIMAL SCI DEPT, VA POLYTECH INST & STATE UNIV, 87- *Concurrent Pos:* Chmn, OSU Fac Coun, 81-82; coord breeding & genetics, Regional Res Comt CSRS, USDA. *Mem:* Am Soc Animal Sci; Am Genetics Asn; Am Registry Prof Animal Scientists (secy, 85-); Coun Agr Sci & Technol; Nat Asn Col & Teachers Agr. *Res:* Population genetics and animal breeding research with beef cattle; research designed to elucidate basic quantitative genetic principles and develop procedures to most effectively incorporate them into breeding programs to improve production efficiency. *Mailing Add:* Animal Sci Dept Va Polytech Inst & State Univ Blacksburg VA 24061-0306

FRAIKOR, FREDERICK JOHN, b Duquesne, Pa, Apr 22, 37; m 62; c 3. ELECTRON MICROSCOPY, METALLOGRAPHIC RESEARCH. *Educ:* Carnegie-Mellon Univ, BS, 59; Ohio State Univ, PhD(metall eng), 65. *Prof Exp:* Sr res mgr, Dow Chem, 65-72; dean, Eng Technol, Metrop State Col, 72-74; mgr, Phys Metall Res & Develop, Rockwell Int, 77-80, mgr, Res & Develop Systs & Plans, 80-86, mgr, Future Systs, 86-88, mgr plant training, 88-90; DEP DIR TRAINING, EG&G, INC, 90- *Concurrent Pos:* Contrib, Handbk Comt, Am Soc Metals, 71; vis scientist, Smithsonian Inst, 77; actg exec dir, Colo Advan Technol Inst, 83-86; mem, adv comt, Ctr Indust Training, 90- *Honors & Awards:* Outstanding Young Metallurgist, Am Soc Metals, 72. *Mem:* Am Soc Metals; Sigma Xi; Nat Mgt Asn. *Res:* Electron microscopy and metallographic research on nuclear materials such as beryllium, plutonium, and uranium; metallographic interpretation of archaeological metal artifacts; physical metallurgy of beryllium. *Mailing Add:* EG&G Inc Bldg G14 McIntyre PO Box 464 Golden CO 80402-0464

FRAILEY, DENNIS J(OHN), b Tulsa, Okla, Mar 5, 44; m 69. SOFTWARE SYSTEMS. *Educ:* Univ Notre Dame, BS, 66; Purdue Univ, MS, 68, PhD(comput sci), 71. *Prof Exp:* Res scientist comput sci, Ford Motor Co, 66-71; from asst prof to assoc prof eng & appl sci, Southern Methodist Univ, 71-77; mem tech staff, 77-79, SR MEM TECH STAFF & MGR, TEX INSTRUMENTS, INC, 79- *Concurrent Pos:* Consult, Tex Instruments, Inc, 74-77; adj assoc prof, Univ Tex, Austin, 81-85, adj prof, Southern Methodist Univ, 86- *Mem:* Asn Comput Mach; Inst Elec & Electronics Engrs. *Res:* Compilers; operating systems; data structures; microprogramming; scheduling algorithms; code optimization; real-time computing and associated problems. *Mailing Add:* 3504 Louis Dr PO Box 869305 Ms 8461 Plano TX 75023-1116

FRAIR, KAREN LEE, b Miami, Okla, Jan 26, 45; m 66. EDUCATION, ADMINISTRATION. *Educ:* Univ Tulsa, BS, 67; Univ Okla, MS, 71, PhD(mech eng), 74. *Prof Exp:* Engr & mathematician, Desert Test Ctr, US Army, 67-70; asst prof eng mech, Va Polytech Inst & State Univ, 74-84; prof mech eng, Calif State Univ, Fresno, 84-86, assoc dean eng, 86-89; Am Coun Educ fel, 89-90; PROG DIR, NSF, 90- *Concurrent Pos:* Aero engr, Tinker AFB, 67; fac fel, NASA Langley Res Ctr, 75-77, Eglin AFB, 80-81 & Jet Propulsion Lab, 84. *Mem:* Am Soc Eng Educ; Am Soc Mech Engrs. *Res:* Fluid mechanics, with emphasis on computational methods; technological literacy. *Mailing Add:* NSF Rm 639 1800 G St NW Washington DC 20550

FRAIR, WAYNE, b Pittsburgh, Pa, May 23, 26; m 87. SYSTEMATIC ZOOLOGY, BIOCHEMISTRY. *Educ:* Houghton Col, BA, 50; Wheaton Col, Ill, BS, 51; Univ Mass, MA, 55; Rutgers Univ, PhD(serol), 62. *Prof Exp:* Teacher, Ben Lippen Sch, 51-52; asst, Univ Mass, 52-53 & Brown Univ, 54-55; from instr to assoc prof, 55-67, PROF BIOL, KING'S COL, NY, 67- *Concurrent Pos:* Asst, Rutgers Univ, 59-60. *Mem:* Fel AAAS; Sigma Xi; fel Creation Res Soc (secy, 74-84, vpres, 85-86, pres, 87-); Am Inst Biol Sci; Am Soc Zool; Am Physiol Soc. *Res:* Animal relationships; molecular taxonomy; turtles, with emphasis on biochemical taxonomy; usage of turtle serum proteins for clarifying possible relations among turtles, involving electrophoresis and immunology. *Mailing Add:* Dept Biol King's Col Briarcliff Manor NY 10510

FRAJOLA, WALTER JOSEPH, b Chicago, Ill, Nov 2, 16; m 41; c 2. FORENSIC TOXICOLOGY. *Educ:* Hamline Univ, BS, 38; Univ Ill, MS, 47, PhD(chem), 50. *Prof Exp:* Instr pub sch, SDak, 38-41; instr pub sch, Minn, 41-42; asst chem, Univ Ill, 46-50; from asst prof to assoc prof physiol chem, Col Med, Ohio State Univ, 50-59, asst prof med res, 53-60, assoc prof path, 59-60, dir, H A Hoster Res Lab, 52-60, chief div clin biochem, 59-62, prof path & physiol chem, 60-68, EMER PROF PHYSIOL CHEM, OHIO STATE UNIV, 77-; PRES, WALTER J FRAJOLA PHD CONSULTING, INC, 80- *Concurrent Pos:* Chief scientist, Columbus Div, NAm Rockwell Corp, 62-66; pres, Lab Anal Blood Studies, Inc, 66-71; pres & lab dir, Community Labs of Ohio, 71-79. *Honors & Awards:* B Katchman Award, Ohio Valley Section AACC, 81. *Mem:* Am Chem Soc; Am Asn Clin Chem; Electron Micros Soc Am; NY Acad Sci; Sigma Xi. *Res:* Enzymes; cancer; clinical chemistry. *Mailing Add:* 2558 Onandaga Dr Columbus OH 43221

FRAKER, ANNA CLYDE, b Chuckey, Tenn, June 25, 35. METALLURGICAL & CERAMICS ENGINEERING. *Educ:* Furman Univ, BS, 57; NC State Univ, MS, 61, PhD(ceramic eng), 67. *Prof Exp:* Res asst metall, NC State Univ, 57-62, res assoc, 63-67; METALLURGIST, NAT BUR STANDARDS, 67- *Mem:* Electron Micros Soc Am; Am Soc Metals; Sigma Xi; Soc Biomat. *Res:* Alloy phase studies; interstitial alloys; recrystallization of aluminum; corrosion of titanium and aluminum; thin foil transmission electron microscopy; corrosion; fatigue. *Mailing Add:* 850 Diamond Dr Gaithersburg MD 20878

FRAKER, JOHN RICHARD, b Knoxville, Tenn, Dec 20, 34; m 54; c 3. ENGINEERING MANAGEMENT, INDUSTRIAL ENGINEERING. *Educ:* Univ Tenn, BS, 56, MS, 65; Clemson Univ, PhD(eng mgt), 71. *Prof Exp:* Mfg engr, Westinghouse Elec Corp, 56-57; instr indust eng & opers res, Clemson Univ, 65-72; from asst prof to assoc prof mgt, Western Carolina Univ, 73-75; PROF & CHMN ENG MGT, UNIV DAYTON, 75- *Mem:* Am Inst Indust Engrs; Am Soc Eng Educ; Sigma Xi; Inst Elec & Electronics Engrs. *Res:* Military system simulation; energy and power systems; queueing networks; departure processes. *Mailing Add:* 6150 Ironside Dr Dayton OH 45459-2225

FRAKER, PAMELA JEAN, b Williamsport, Ind, Aug 4, 44. IMMUNOBIOLOGY, TRACE ELEMENT NUTRITION. *Educ:* Purdue Univ, BS, 66; Univ Ill, MS, 68, PhD(microbiol), 71. *Prof Exp:* From asst prof to assoc prof biochem, 73-85, PROF BIOCHEM & MICROBIOL, MICH STATE UNIV, 85- *Honors & Awards:* BioServ Award, Am Inst Nutrit. *Mem:* Am Inst Nutrit; Am Soc Biol Chem; Am Asn Immunol. *Res:* Study of the effects of nutritional deficiency on the immune capacity; modulation of immune function by the opiate-like neuropeptides; role of carnitine and short and long chain fatty acids in the metabolism of lymphocytes and macrophages. *Mailing Add:* Dept Biochem Mich State Univ East Lansing MI 48824

FRAKES, ELIZABETH (MCCUNE), b Huron, Kans, Jan 4, 22; m 65. NUTRITION. *Educ:* Univ Kans, BA, 47; Univ Iowa, MS, 48. *Prof Exp:* Dietician, Watkins Mem Hosp & Student Union, Univ Kans, 48-50; from instr to prof nutrit, Univ Kans Med Ctr, Kansas City, 50-, chmn & dir dept dietetics & nutrit, 72-; RETIRED. *Honors & Awards:* Dietary Prod Found Award, Am Dietetic Asn, 66. *Mem:* Am Dietetic Asn. *Res:* Management phase of dietetics and nutrition. *Mailing Add:* 5214 Yecker Kansas City KS 66104

FRAKES, LAWRENCE AUSTIN, b Pasadena, Calif, Apr 28, 33; m 54; c 1. SEDIMENTOLOGY, MARINE GEOLOGY. *Educ:* Univ Calif, Los Angeles, BA, 57, MA, 59, PhD(geol), 64. *Prof Exp:* Instr geol, Villanova Univ, 59-60; geologist, Pa Geol Surv, 60-62; asst res geologist, Univ Calif, Los Angeles, 64-67, assoc res geologist, 67-69; assoc prof geol, Univ NMex, 69-70, Fla State Univ, 70-73; sr lectr, Monash Univ, 74-75, reader, 76-85; DOUGLAS MAWSON PROF GEOL & GEOPHYS, UNIV ADELAIDE, 85- *Mem:* Fel AAAS; fel Geol Soc Am; Geol Soc Australia. *Res:* Paleoclimatology; Cretaceous climates; atmospheric and oceanic circulation patterns during the late Paleozoic, Mesozoic and Cenozoic; petroleum geology. *Mailing Add:* Dept Geol & Geophys Univ Adelaide Adelaide 5001 Australia

FRAKES, RODNEY VANCE, b Ontario, Ore, July 20, 30; m 52; c 2. AGRONOMY. *Educ:* Ore State Univ, BS, 56, MS, 57; Purdue Univ, PhD(plant breeding), 60. *Prof Exp:* Instr forage crop res, Purdue Univ, 57-60; from asst prof to prof, 60-80, ELIZABETH P RITCHIE DISTINGUISHED PROF PLANT BREEDING, DEPT CROP SCI, 80-, ASSOC DEAN RES, ORE STATE UNIV, 81- *Mem:* Fel Am Soc Agron; fel Crop Sci Soc Am; Soc Res Adminrs. *Res:* Breeding of legumes, and forage and turf grasses. *Mailing Add:* 2625 NW Linnan Circle Corvallis OR 97330

FRAKNOI, ANDREW GABRIEL, b Budapest, Hungary, Aug 24, 48; div. ASTRONOMY, SCIENCE EDUCATION. *Educ:* Harvard Univ, BA, 70; Univ Calif, Berkeley, MA, 72. *Prof Exp:* Instr astron & physics, Canada Col, Calif, 72-78; ED, MERCURY & EXEC OFFICER, ASTRON SOC PAC, 78- *Concurrent Pos:* Instr astron, City Col San Francisco, 74-75; instr astron & physics, San Francisco State Univ, 80-; mem bd dirs, Search for Extraterrestrial Intel Inst; host weekly sci talk show, KGO-FM radio, 82-84; mem, Educ Adv Bd, Am Astron Soc, ed Universe in the Classroom, 84-, assoc ed, The Planetarian Mag, 87-, elected mem-at-large, Astron Sect Comt, AAAS, 89-91. *Mem:* Am Astron Soc; Astron Soc Pac; Am Asn Physics Teachers; Nat Asn Sci Writers. *Res:* Radio astronomy; astronomy education and communicating science to the public; interdisciplinary approaches to science education. *Mailing Add:* Astron Soc Pac 390 Ashton Ave San Francisco CA 94112

FRALEIGH, PETER CHARLES, b New York, NY, Nov 16, 42; m 71; c 2. AQUATIC ECOLOGY. *Educ:* Cornell Univ, BA, 65; Univ Ga, PhD(zool), 71. *Prof Exp:* Res assoc syst sci, Mich State Univ, 71-72; asst prof, 72-80, ASSOC PROF BIOL, UNIV TOLEDO, 80- *Mem:* AAAS; Ecol Soc Am; Am Soc Limnol & Oceanog; Sigma Xi. *Res:* Studies on the environmental regulation of energy flow and community structure in field and laboratory aquatic ecosystems. *Mailing Add:* Dept Biol Univ Toledo Toledo OH 43606

FRALEY, ELWIN E, b Sayre, Pa, May 3, 34; c 6. UROLOGY, PATHOLOGY. *Educ:* Princeton Univ, BA, 57; Harvard Med Sch, MD, 61. *Prof Exp:* Instr surg, Mass Gen Hosp, 66-67; investr urol, NIH, 67-69; PROF UROL, UNIV MINN, MINNEAPOLIS, 69-, CHMN DEPT UROL SURGERY, 80- *Honors & Awards:* Soma Weiss Award, Harvard Med Sch, 63; Clin Res Award, Am Urol Asn, 64, Res Essay Awards, 67. *Mem:* AAAS; Am Soc Exp Path; Am Asn Cancer Res; Am Urol Asn; Soc Univ Urol. *Res:* Nucleic acid and protein metabolism in kidney; relation of viruses to development of human genitourinary neoplasms; human tumor immunology; viral oncology; macromolecular pathology. *Mailing Add:* Univ Minn Hosp 420 Delaware SE Minneapolis MN 55455

FRALEY, ROBERT THOMAS, b Danville, Ill, Jan 25, 53; m 78; c 2. PLANT MOLECULAR BIOLOGY. *Educ:* Univ Ill, BS, 74, MS, 76, PhD(microbiol/biochem), 78. *Prof Exp:* Res assoc biochem, Univ Calif, San Francisco, 79-81; mgr plant molecular biol, 81-85, DIR PLANT SCI, MONSANTO CO, 85- *Concurrent Pos:* Fel, Jane Coffin Childs Mem Fund, 80-81; ed, Plant Molecular Biol J, 81-, Somatic Cell & Molecular Genetics; adj prof, Wash Univ, 81-; mem, Molecular Cytol Study Sect, NIH. *Mem:* Am Chem Soc; fel AAAS. *Res:* Genetic engineering of plants; development of efficient systems for introducing and monitoring the expression of foreign genes in plant cells. *Mailing Add:* Monsanto Co 700 Chesterfield Village Pkwy St Louis MO 63198

FRALICK, RICHARD ALLSTON, b Boston, Mass, July 27, 37; m 65; c 2. MARINE PHYCOLOGY. *Educ:* Salem State Col, BA, 67; Univ NH, MS, 70, PhD(bot), 73. *Prof Exp:* Res asst cryptogamic bot, Farlow Herbarium, Harvard Univ, 61-67; instr biol, Austin Prep Sch, 67-69; teaching asst bot & phycol, Univ NH, 69-73; from asst prof to assoc prof, 73-80, PROF NATURAL SCI, PLYMOUTH STATE COL, 80- *Concurrent Pos:* Field asst marine bot, Harvard Univ & Univ Cent Venezuela Exchange Prog, 63, field coordr, Harvard Univ NSF Exped W Antarctica, 64-65; res asst phytoplankton ecol, Water Resources Res Ctr, Univ NH, 69-70, scientist-aquanaut algal ecol & saturation diving, Tektite Exped US VI, 70 & Flare Exped Fla Keys, 72; guest investr, Environ Syst Lab, Woods Hole Oceanog Inst, 73-76; dir, Plymouth State Col Bermuda Prog; consult, US Dept State USAID Prog-Portugal, 77-80, educ consult Sci, 80-, Azores herbivore proj, 86-; Citation Marine Res, Dept Interior, 70; marine algae consult, UN Develop Prog, Persian Gulf. *Honors & Awards:* Cong Bronze Medal, 65. *Mem:* Sigma Xi; Phycolog Soc Am; Am Polar Soc; AAAS. *Res:* Cultivation of commercially important seaweeds; algal physiological ecology; thermal stress and distribution of marine algae; ecology of nuisance and fouling algae; ecology of Pterocladia and other economically important seaweeds; effects of road-salt on the distribution of marine plants; marine resource management; Persian Gulf seaweeds. *Mailing Add:* Dept Bot Plymouth State Col Plymouth NH 03264

FRAM, HARVEY, b Worcester, Mass, Nov 23, 18; m 44; c 2. FOOD SCIENCE. *Educ:* Univ Mass, BS, 40, MS, 42. *Prof Exp:* Asst, Mass Inst Technol, 42-48; group leader, Nat Dairy Res Labs, 48-60; chief food technologist, B Manischewitz Co, Jersey City, 60-70, tech dir, 70-75; tech dir, Joyce Food Prod, 75-76; CHIEF MICROBIOLOGIST, FITELSON LABS, NEW YORK, 76- *Mem:* Inst Food Technol; Am Asn Cereal Chem. *Res:* Radiation sterilization of foods; cultured dairy products; food preservation, canned, baked and frozen. *Mailing Add:* 100 East Side Pkwy Newton MA 02158

FRAME, HARLAN D, b Muscatine, Iowa, Jan 1, 33; m 60, 76; c 3. INORGANIC CHEMISTRY, PHYSICAL CHEMISTRY. *Educ:* Univ Wichita, BA, 55; Univ Ill, MS, 58, PhD(inorg chem), 59. *Prof Exp:* Assoc chemist, Argonne Nat Lab, 59-69; asst prof, 69-72, ASSOC PROF CHEM, SOUTHWESTERN STATE COL, OKLA, 72- *Mem:* AAAS; Am Chem Soc. *Res:* Synthetic inorganic chemistry employing nonaqueous solvents; inorganic fluorine chemistry; nonaqueous solvents; high temperature reactions; chromatography; microcomputer software. *Mailing Add:* Dept Chem Southwest Okla State Univ Weatherford OK 73096

FRAME, JAMES SUTHERLAND, b New York, NY, Dec 24, 07; m 38; c 4. MATHEMATICS. *Educ:* Harvard Univ, AB, 29, AM, 30, PhD(math), 33. *Prof Exp:* Instr math, Harvard Univ, 30-33; Rogers traveling fel, Univs Harvard, Gottingen & Zurich, 33-34; from instr to asst prof math, Brown Univ, 34-42; assoc prof & chmn dept, Allegheny Col, 42-43; prof math, 43-63, chmn dept, 43-60, proj dir conf bd math sci, 61-62, prof math & eng res, 63-77, EMER PROF MATH & ENG RES, MICH STATE UNIV, 77- *Concurrent Pos:* Mem nat coun, Am Asn Univ Prof, 48-51; mem, Inst Advan Study, 50-51; mem bd gov, Math Asn Am, 50-52 & 58-60; consult, Ford Found, Bangkok, Thailand, 70; consult, Inst fur Quantenchemie, Free Univ Berlin, 72 & 78; vis prof, R W Technische Hochschule, Aachen, WGer, 81; fel, Acad Sr Profs, Eckerd Col, St Petersburg, Fla, 82; mem, People to People Math Deleg to China, 83. *Honors & Awards:* Sr Res Award, Sigma Xi, 52. *Mem:* AAAS; Am Math Soc; Math Asn Am. *Res:* Theory of representations of finite groups; approximations; matrix theory; continued fractions; publisher of approximately 100 papers. *Mailing Add:* 136 Oakland Dr East Lansing MI 48823

FRAME, JOHN W, b Rolla, Mo, Sept 28, 16. METALLURGICAL ENGINEERING. *Educ:* Mo Sch Mines, BS, 37; Lehigh Univ, MS, 38, PhD(metall eng), 61. *Prof Exp:* Mem staff, Metall Dept, Bethlehem Steel Corp, Lackawanna, 38-46, engr res div, 46-56, from supvr to assoc dir res, Phys Metall, 56-64, mgr forming & finishing res, 64-69, mgr prod res, Homer Res Labs, 69-78; RETIRED. *Mem:* Fel Am Soc Metals; Am Inst Mining, Metall & Petrol Engrs; Soc Automotive Engrs. *Res:* Aging, formability and manufacture of sheet steels; corrosion, welding, forming, machining, alloy development, metal physics, fatigue and fracture of steel; organic coating; tin mill products. *Mailing Add:* 831 W Paseo Potrerro Green Valley AZ 85614-1738

FRAME, PAUL S, MEDICINE. *Prof Exp:* MEM MED STAFF, TRI-COUNTY FAMILY MED PROG, DANSVILLE, NY, 76- *Concurrent Pos:* Clin asst prof, Dept Family Med, Sch Med, Univ Rochester, 85- *Mem:* Inst Med-Nat Acad Sci. *Mailing Add:* Tri-County Family Med Prog Box 601 Dansville NY 14437

FRAME, ROBERT ROY, b Oak Park, Ill, Aug 24, 39. INDUSTRIAL ORGANIC CHEMISTRY. *Educ:* Univ Ill, BS, 61, Northwestern Univ, PhD, 65. *Prof Exp:* Fel org chem, Univ Ark, 67-68; vis asst prof, Univ Okla, 68-72; RES CHEMIST, UNIVERSAL OIL PROD CO, 72- *Mem:* Am Chem Soc. *Res:* Catalysis. *Mailing Add:* 725 Juniper Rd Glenview IL 60025-3411

FRAMPTON, ELON WILSON, b New York, NY, July 27, 24; m 58; c 2. MICROBIOLOGY. *Educ:* Syracuse Univ, BA, 49; Northwestern Univ, MS, 51; Univ Ill, PhD(dairy sci, bact), 59. *Prof Exp:* Lab technician, Hosp, Rockefeller Inst, 42-43; microbiologist, Wilson Labs, Ill, 50-54; asst dairy bact, Univ Ill, 54-58; res fel, M D Anderson Hosp & Tumor Inst, Univ Tex, 59-60, res assoc, 60-62, asst radiation biologist, 62-66, assoc biologist, 66-69; ASSOC PROF BIOL SCI, NORTHERN ILL UNIV, 69- *Mem:* AAAS; Am Soc Microbiol; Radiation Res Soc. *Res:* Molecular biology; bacterial physiology. *Mailing Add:* 119 Stoney Creed Rd DeKalb IL 60115

FRAMPTON, PAUL HOWARD, b Kidderminster, Eng, Oct 31, 43. PHYSICS. *Educ:* Univ Oxford, BA, 65, MA, 68, PhD(physics), 68. *Hon Degrees:* DSc, Univ Oxford, 84. *Prof Exp:* Res assoc, Univ Chicago, 68-70; res fel physics, Europ Orgn Nuclear Res, 70-72; vis assoc prof physics, Syracuse Univ, 72-75 & Univ Calif, Los Angeles, 75-77; adj assoc prof, Ohio State Univ, 77-78; vis scholar, Harvard Univ, 78-80; from asst prof to assoc prof, 81-85, PROF PHYSICS, UNIV NC, CHAPEL HILL, 85- *Concurrent Pos:* Assoc, physics dept, Harvard Univ, 81-87; vis prof, Univ Tex, 83; vis prof, Boston Univ, 86-87. *Mem:* Fel Am Phys Soc; fel Inst of Phys (UK); fel AAAS. *Res:* Use of gauge field theories and string theories in unification of fundamental forces. *Mailing Add:* Dept Physics & Astron Univ NC Chapel Hill NC 27599-3255

FRANCAVILLA, THOMAS L(EE), b Buffalo, NY, Dec 3, 39; m 71; c 3. SOLID STATE PHYSICS. *Educ:* Canisius Col, BS, 61; John Carroll Univ, MS, 63; Cath Univ Am, PhD(physics), 73. *Prof Exp:* PHYSICIST, US NAVAL RES LAB, 64- *Mem:* Am Phys Soc; Am Asn Physics Teachers. *Res:* Basic and applied properties of superconducting materials and the phenomenon of superconductivity. *Mailing Add:* US Naval Res Lab Code 6344 4555 Overlook Dr Washington DC 20375

FRANCE, PETER WILLIAM, b Chicago, Ill, Sept 4, 38; m 65; c 2. SOLID STATE PHYSICS, NUCLEAR MAGNETIC RESONANCE. *Educ:* Wayne State Univ, BS, 62, MS, 64, PhD(physics), 67. *Prof Exp:* Sr res physicist, Bendix Res Labs, 68; asst prof, 68-72, ASSOC PROF PHYSICS, UNIV LOUISVILLE, 72- *Mem:* Am Phys Soc; Am Asn Physics Teachers. *Res:* Application of wide line nuclear magnetic resonance of solids, particularly glass; measurement of physical properties of semiconducting and dielectric glasses, such as resistivities, mobilities and Hall constants. *Mailing Add:* Dept Physics Univ Louisville Louisville KY 40292

FRANCE, W(ALTER) DEWAYNE, JR, b New Haven, Conn, Nov 9, 40; m 63; c 3. MATERIALS SCIENCE, FUEL TECHNOLOGY. *Educ:* Yale Univ, BE, 62; Rensselaer Polytech Inst, PhD(mat eng), 66. *Prof Exp:* Asst corrosion res, Rensselaer Polytech Inst, 63-66; assoc sr res chemist, Gen Motors Corp, 66-69, sr res chemist, 69-70, supv specialized activities, 70-73, asst dept head, Anal Chem Dept, 73-85, asst dept head, Fuels & Lubricants Dept, 85-89, DEPT HEAD, ENVIRON SCI DEPT, RES LABS, GEN MOTORS CORP, 89- *Concurrent Pos:* Chmn bd dirs, Am Soc Testing & Mat, 88. *Honors & Awards:* A B Campbell Award, Nat Asn Corrosion Engrs, 70. *Mem:* Nat Asn Corrosion Engrs; fel Am Soc Testing & Mat; fel ASM Int; Am Chem Soc; Soc Automotive Eng; Am Soc Testing & Mat; fel Am Inst Chemists. *Res:* Management of research in such areas as vehicle emissions and pollutant impacts, manufacturing emissions and process waste control and groundwater decontamination. *Mailing Add:* 6455 Bloomfield Glens West Bloomfield MI 48322-2515

FRANCES, SAUL, b Brooklyn, NY, Dec 13, 10; m 41; c 2. BACTERIOLOGY. *Educ:* City Col New York, BS, 35; Univ Minn, MS, 37; Columbia Univ, PhD(bact), 46. *Prof Exp:* Bacteriologist, Skin Res Labs, Minn, 37-39; res bacteriologist & asst to dir, Pease Labs, Inc, NY, 40-43; dir, Coconut Processing Labs, 43-44; bacteriologist, Off Sci Res & Develop Contract, Columbia Univ, 44-46, asst bact, Col Physicians & Surgeons, 46-50; dir, 48-72, EXEC DIR, WELLS LABS, INC, 72- *Concurrent Pos:* Fel, Dazian Found Med Res, 44-46; consult, advert agencies & labs, NY, 44- *Mem:* AAAS; Am Chem Soc; Am Soc Microbiol; Sigma Xi; Soc Indust Microbiol. *Res:* Bacteriology of the mouth; physiology of bacteria; anaerobes; antihistaminics; antibiotics; vaccines; chronic infections; cosmetics; cancer research; toxicology; provides court testimony in cases concerning food products, insecticides, water, and other matters. *Mailing Add:* Claridge House 2 Apt PHAE Verona NJ 07044

FRANCESCHETTI, DONALD RALPH, b Oceanside, NY, Nov 21, 47; m 82. CHEMICAL PHYSICS, SOLID STATE PHYSICS. *Educ:* Brooklyn Col, BS, 69; Princeton Univ, MA, 71, PhD(chem), 74. *Prof Exp:* Res assoc physics, Univ Ill, Urbana-Champaign, 73-75; res assoc, Univ NC, Chapel Hill, 75-77, res asst prof physics, 77-79; from asst prof to assoc prof, 79-86, PROF PHYSICS & CHMN, MEMPHIS STATE UNIV, TENN, 86- *Concurrent Pos:* NSF energy related fel, 75-76; vis lectr, Dept Chem, State Univ, Utrecht, Neth, 82. *Mem:* Am Phys Soc; Am Chem Soc; Electrochem Soc; NY Acad Sci; Hist Sci Soc; Int Soc Solid State Ionics. *Res:* Solid state electrochemistry; impedance spectroscopy. *Mailing Add:* Dept Physics Memphis State Univ Memphis TN 38152

FRANCESCHINI, GUY ARTHUR, b North Adams, Mass, Apr 2, 18; m 47; c 1. METEOROLOGY, OCEANOGRAPHY. *Educ:* Univ Mass, BS, 50; Univ Chicago, SM, 52; Tex A&M Univ, PhD(meteorol), 61. *Prof Exp:* Meteorologist, Nat Adv Comt Aeronaut, 45-47; instr, Univ Chicago, 52; asst prof physics, 52-53, res scientist, Res Found, 52-86, asst prof meteorol, 54-60, from assoc prof to prof, 61-86, EMER PROF OCEANOG & METEOROL, TEX A&M UNIV, 86- *Mem:* Am Meteorol Soc; Am Geophys Union; Sigma Xi. *Res:* Air-sea interactions; solar radiation. *Mailing Add:* Dept Meteorol Tex A&M Univ 5-7671 Rm 1204 O&M Bldg College Station TX 77843

FRANCESCHINI, REMO, b New York, NY, Apr 14, 28; m 53; c 3. FOOD SCIENCE, FOOD TECHNOLOGY. *Educ:* Fordham Univ, BS, 49; Univ Mass, MS, 55, PhD(food sci), 59. *Prof Exp:* Mgr new prod, Libby, McNeill & Libby, 58-61; group leader, 61-63, coordr, 63-64, asst dir res, 64-66, dir prod develop, 66-78, DIR RES & DEVELOP, THOMAS J LIPTON, INC, 78- *Mem:* Inst Food Tech; Am Asn Cereal Chem; Sigma Xi. *Res:* Irradiation of food products; thermal processing; dehydration; utilization of vegetable proteins; new product development; emulsions. *Mailing Add:* Thomas J Lipton Inc 800 Sylvan Ave Englewood Cliffs NJ 07632

FRANCESCONI, RALPH P, b Milford, Mass, Jan 28, 39; m 62; c 3. BIOLOGICAL CHEMISTRY. *Educ:* Amherst Col, AB, 61; Boston Col, MS, 63, PhD(cell biol), 66. *Prof Exp:* USPHS fel biol chem, Harvard Sch Med, 66-68; investr environ biochem, 68-75, res chemist, Heat Res Div, 75-90, CHIEF, COMP PHYSIOL DIV, RES INST ENVIRON MED, US ARMY, NATICK, MA, 90- *Mem:* AAAS; Sigma Xi; Am Physiol Soc; NY Acad Sci. *Res:* Enzyme development and enzyme regulation in mammals; effects of heat stress upon human performance; effects of heat stress on human metabolism and fluid and electrolyte balance. *Mailing Add:* Comp Physiol Div Res Inst Environ Med US Army Natick MA 01760-5007

FRANCH, ROBERT H, b Bessemer, Mich, June 15, 27; m 58; c 7. CARDIOLOGY, CARDIOVASCULAR PHYSIOLOGY. *Educ:* Univ Colo, AB, 48, MD, 52. *Prof Exp:* Nat Heart Inst trainee cardiovasc dis, 55-56; from instr to assoc prof, 56-70, PROF MED, MED SCH, EMORY UNIV, 70-, DIR, CARDIOVASC LAB, UNIV HOSP, 57- *Concurrent Pos:* Fel cardiol, Med Sch, Emory Univ, 56-57; dir cardiac clin, Grady Hosp, 56-57; examr cardiovasc subspecialty bds, Am Bd Internal Med, 61 & 62; mem panel cardiol specialists, State Div Voc Rehab; cardiac consult heart dis control prog, State Dept Pub Health, Ga, 61-; chmn diag treatment facil sect, Div Community Serv, Nat Conf Cardiovasc Dis, 64; fel coun clin cardiol, Am Heart Asn, 64. *Mem:* Fel Am Col Physicians; fel Am Col Cardiol. *Res:* Physiology of the pulmonary vascular bed; anatomy of pulmonary vascular disease; thermodilution technics for measuring flow and ventricular volumes; natural history of small ventricular septal defect; echocardiography in congenital heart disease; flow dynamics of prosthetic heart valves. *Mailing Add:* Emory Univ Clin Emory Univ Sch Med Atlanta GA 30322

FRANCIOSA, JOSEPH ANTHONY, b Easton, Pa, Apr 24, 36; m 73; c 1. CARDIOLOGY. *Educ:* Univ Pa, BA, 58; Univ Rome, Italy, MD, 63; Am Bd Internal Med, cert, 72. *Prof Exp:* From asst prof to assoc prof med, Sch Med, Univ Minn, 74-79; assoc prof, Sch Med, Univ Pa, 79-82; prof med, Univ Ark Med Sci, Little Rock, 82-86; dir cardiovasc renal drugs, ICI Americas Inc, Wilmington, DE, 86-91; EXEC DIR MED AFFAIRS, CIBA GEIGY, 91- *Concurrent Pos:* Adj prof med, Univ Pa, 87- *Mem:* Fel Am Col Physicians; Am Heart Asn; fel Am Col Cardiol; Am Fedn Clin Res; fel Am Col Chest Physicians; Am Soc Clin Pharmacol & Therapeut. *Res:* Cardiovascular clinical pharmacology. *Mailing Add:* Ciba Geigy Corp 556 Morris Ave Summit NJ 07901

FRANCIS, ANTHONY HUSTON, b Philadelphia, Pa, Mar 21, 42; m 65; c 2. CHEMICAL PHYSICS. *Educ:* Yale Univ, BS, 64; Univ Mich, PhD(chem), 69. *Prof Exp:* Asst prof chem, Univ Ill, Chicago Circle, 70-75; from asst prof to assoc prof, 76-85, PROF CHEM, UNIV MICH, 85- *Mem:* Am Chem Soc. *Res:* Optical and magnetic resonance spectroscopy of impurity centers in crystals; vibrational and electronic spectroscopy of surfaces and surface adsorbates; solid state chemistry. *Mailing Add:* Univ Mich Main Campus 1543 Chem Bldg Ann Arbor MI 48109-1055

FRANCIS, AROKIASAMY JOSEPH, b Tiruchirapalli, India, Oct 5, 41; c 3. MICROBIOLOGY, SOIL SCIENCE. *Educ:* Annamalai Univ, BSc, 63, MSc, 65; Cornell Univ, PhD(microbiol), 71. *Prof Exp:* Res assoc microbiol, Cornell Univ, 70-73; microbiologist, Stanford Res Inst, 73-75; assoc scientist, 75-80, MICROBIOLOGIST/SCIENTIST, DEPT APPL SCI, BROOKHAVEN NAT LAB, 80-, GROUP LEADER MICROBIOL, 88- *Concurrent Pos:* Mem tech adv group, Dept Health Serv Nitrogen Pollution, Suffolk County, 75-77; consult on agr & environ microbiol, UN Food & Agr Orgn, 82; consult, Directorate of plant protection, Ministry of Agr, India, 86, Public Serv Elec & Gas Co, NJ, 86-88, Sandia Nat Labs, WIPP prog, 88-; invited partic, Waste Mgt & Environ Restoration Res Plan, Dept Energy, 88, 89 & 90. *Mem:* Am Soc Microbiol; Am Nuclear Soc. *Res:* Anaerobic microbiol transformations of toxic metals, radionuclides and organic compounds in sub-surface environments; microbiology of radioactive wastes, microbial production of sequestering agents of radionuclides and heavy metals, biodegradation of radionuclides and metal organic complexes in the environment, bidegradation of pesticides and organic pollutants; nitrogen transformation and effect of acid rain on soil microbiol process; organic chemistry; geomicrobiology; biogeochemistry; holder of one US patent. *Mailing Add:* Dept Appl Sci Bldg 318 Brookhaven Nat Lab Upton NY 11973

FRANCIS, BETTINA MAGNUS, b Frankfurt, Ger, Sept 3, 43; US citizen; m 67; c 2. DEVELOPMENTAL TOXICOLOGY, NEUROTOXICOLOGY. *Educ:* NY Univ, BA, 65; Univ Mich, MS, 67, PhD(genetics), 71. *Prof Exp:* res assoc environ toxicol, Ill Natural Hist Surv, 75-76; res assoc, 77-80, asst res biologist, 80-81, ASST PROF, INST ENVIRON STUDIES, UNIV ILL, 81- *Mem:* Teratol Soc; AAAS; Sigma Xi; Soc Toxicol; Soc Environ Toxicol & Chem; Int Neurotoxicology Asn. *Res:* Irreversible health effects of xenobiotics, with particular emphasis on developmental and neurological consequences. *Mailing Add:* 801 S Anderson Urbana IL 61801

FRANCIS, CHARLES K, b Newark, NJ, May 24, 39. MEDICINE. *Educ:* Dartmouth Col, BA, 61; Jefferson Med Col, MD, 65. *Prof Exp:* Med intern, Philadelphia Gen Hosp, Pa, 65-66; med resident, Boston City Hosp, Tufts Univ, 69-70, clin fel cardiol, Tufts Circulation Lab, 70-71; clin & res fel cardiol, Mass Gen Hosp, 71-72, sr med resident, 72-73; chief, Cardiac Catheterization Lab, Div Cardiol, Martin Luther King Jr Gen Hosp, Los Angeles, Calif, 73-74, chief, Cardiol Div, 74-77; dir, Cardiol Div, Mt Sinai Hosp, Hartford, Conn, 77-80; co-dir, Cardiac Catheterization, Yale-New Haven Hosp, 80-83, assoc dir, Hypertension Serv, 80-87, dir, Cardiac Catheterization Lab, 83-87; DIR, DEPT MED, HARLEM HOSP CTR, 87-; PROF CLIN MED, COLUMBIA UNIV COL PHYSICIANS & SURGEONS, 87- *Concurrent Pos:* Clin instr med, Sch Med, Tufts Univ,

70-71; teaching fel, Harvard Med Sch, 71-72, clin fel, 72-73; asst prof med, Charles R Drew Postgrad Med Sch & Sch Med, Univ Southern Calif, 73-77; asst prof med & dir, Burgdorf Hypertension Clin, Med Sch, Univ Conn, 77-80; mem cardiac adv comt, Nat Heart, Lung & Blood Inst, NIH, 77-79, Minority Hypertension Proj, 78-80, Nat High Blood Pressure Educ Prog Coord Comt, 89-, study sect A, Prog Proj Peer Rev, 90-; asst prof med, Sch Med, Yale Univ, 80-81, assoc prof, 81-87; mem, Technol Assessment Study Sect, Nat Ctr Health Serv Res, 84-88. *Mem:* Inst Med-Nat Acad Sci; fel Am Col Cardiol; Am Col Physicians; Am Fedn Clin Res; Am Heart Asn. *Res:* Cardiology; hypertension. *Mailing Add:* Dept Med Harlem Hosp Ctr 506 Lenox Ave Rm 14101 New York NY 10037

FRANCIS, CHESTER WAYNE, b Creston, Iowa, Feb 14, 36; m 63; c 1. SOIL SCIENCE, ECOLOGY. *Educ:* Iowa State Col, BS, 58; Univ Wis, MS, 64, PhD(soils), 67. *Prof Exp:* Soil specialist, IRI Res Inst NY, Brazil, 66-68; SOIL CHEMIST, OAK RIDGE NAT LAB, 69- *Mem:* Fel AAAS; Soil Sci Soc Am; Am Soc Agron. *Res:* Soil chemistry, mineralogy and fertility. *Mailing Add:* Oak Ridge Nat Lab PO Box 2008 Oak Ridge TN 37831-6036

FRANCIS, DAVID W, b New York, NY, Oct 31, 36; m 64; c 2. GENETICS, DEVELOPMENTAL BIOLOGY. *Educ:* Harvard Univ, BA, 57; Univ Wis, MS, 59, PhD(bot), 62. *Prof Exp:* NIH fel biol, Princeton Univ, 62-63, res assoc, 63-65; asst prof, Univ BC, 65-69; assoc prof, 69-77, PROF BIOL SCI, UNIV DEL, 77- *Mem:* Soc Develop Biologists; Soc Microbiologists. *Res:* Cell differentiation in protozoa, especially in Acrasiales. *Mailing Add:* Sch Life & Health Sci Univ Del Wolf Hall Newark DE 19716

FRANCIS, DAVID WESSON, b New York, NY, Aug 17, 18; m 50, 77; c 5. POULTRY HUSBANDRY. *Educ:* Rutgers Univ, BS, 41; Univ Del, MS, 52; Univ Md, PhD(poultry husb), 55. *Prof Exp:* Technician, Small Animals, Lederle Labs, 41-42; farmer, Soldier Hill Farm, NJ, 42-50; asst poultry sci, Univ Del, 50-51, res instr, 51-53; teaching asst, poultry dept, Univ MD, 53-55; assoc prof, dept poultry husb, Col A&M Arts, 55-58, prof & head poultry dept, 58-77, prof poultry sci, dept animal & range sci, 77-81, poultry consult, NMex Agric Exten Serv, 81-82, EMER PROF POULTRY SCI, NMEX STATE UNIV, 81- *Concurrent Pos:* Consult poultry mgt & dis, NMex State Dept Agric, 58-72 & NMex State Univ-USAID, Paraguay, 70, 73 & 76; prof poultry & poultry specialist, dept poultry sci, Ore State Univ, 82-84. *Mem:* Fel AAAS; emer mem Poultry Sci Asn; Sigma Xi; World Poultry Sci Asn; Nutrit Today Soc; Am Asn Lab Animal Sci; Am Asn Avian Pathologists. *Res:* Control and treatment of poultry diseases; fertility studies with chickens and turkeys; identification of avian diseases in New Mexico; atmospheric stresses on poultry; salmonella contamination of feeds and feed ingredients; comparison of large commercial egg production units with small research type units. *Mailing Add:* Dept Animal & Range Sci Box 3I NMex State Univ Las Cruces NM 88003

FRANCIS, DONALD PINKSTON, b Los Angeles, Calif, Oct 24, 42; m 76; c 2. MEDICAL EPIDEMIOLOGY, VIROLOGY. *Educ:* Northwestern Univ, MD, 68. *Hon Degrees:* DSc, Harvard, 79. *Prof Exp:* Dept HEW int fel, Childrens Bur, Punjab, India, 68; intern-resident pediat, Los Angeles County, Univ Southern Calif Med Ctr, 69-71; state epidemiologist infectious dis, Rivers State, Nigeria, USAID, 71; epidemic intell serv officer, Ctr Dis Control, 71-73; med officer smallpox, WHO, Sudan, 73 & India, 73-75; infectious dis fel, Channing Lab, Harvard Med Sch, 75-77; CHIEF EPIDEMIOL BR, HEPATITIS LABS DIV, CTR DIS CONTROL, 78- *Concurrent Pos:* Microbiol fel, Harvard Sch Pub Health, 76. *Res:* Hepatitis. *Mailing Add:* Div Oral Dis Ctr Dis Control Atlanta GA 30333

FRANCIS, EUGENE A, b Christiansted, VI, Oct 27, 27; m 55. MATHEMATICS. *Educ:* Inter-Am Univ PR, BA, 49; Columbia Univ, MA, 51. *Prof Exp:* From instr to assoc prof math, 51-67, assoc dir, NSF Math Inst, 59-67, chmn dept math, 67-72, assoc dean studies, 72-74, PROF MATH, UNIV PR, MAYAGUEZ, 67- *Concurrent Pos:* Spec lectr, Inter-Am Univ PR, 51, 56, 58. *Mem:* NY Acad Sci; Nat Coun Teachers Math; AAAS; Math Asn Am; Am Math Soc. *Mailing Add:* Univ PR Mayaguez PR 00708

FRANCIS, FAITH ELLEN, b Batavia, NY, Dec 28, 29. BIOCHEMISTRY, ENDOCRINOLOGY. *Educ:* D'Youville Col, BA, 52; St Louis Univ, PhD(biochem), 57. *Prof Exp:* Res assoc internal med, 57-59, from instr to asst prof, 59-68, asst prof obstet & gynec, 68-72, ASSOC PROF OBSTET & GYNEC, SCH MED, ST LOUIS UNIV, 72-, ASST PROF BIOCHEM, 64- *Mem:* Fel AAAS; Soc Exp Biol & Med; Am Chem Soc; Endocrine Soc; NY Acad Sci; Sigma Xi. *Res:* Metabolism of steroid hormones. *Mailing Add:* Dept Ob/Gyn St Louis Univ Sch Med 3635 Vista Ave PO Box 15250 St Louis MO 63110-0250

FRANCIS, FREDERICK JOHN, b Ottawa, Ont, Oct 9, 21; m 52; c 3. FOOD SCIENCE. *Educ:* Univ Toronto, BA, 46, MA, 48; Univ Mass, PhD, 54. *Prof Exp:* Instr food chem, Univ Toronto, 46-50; lectr hort, Ont Agr Col, 50-54; from asst prof to prof food sci & nutrit, 54-64, Nicolas Appert prof, 64-71, head dept, 71-77, PROF FOOD SCI & NUTRIT, UNIV MASS, AMHERST, 71- *Honors & Awards:* Nicolas Appert Medal, Inst Food Technologists, 79 & Int Award, 82. *Mem:* Fel AAAS; fel Inst Food Technologists (pres, 80-81); Am Soc Hort Sci; Am Chem Soc; NY Acad Sci; Am Inst Nutrit. *Res:* Plant biochemistry, color and pigments; food preservation; food safety. *Mailing Add:* Dept Food Sci & Nutrit Univ Mass Amherst MA 01003

FRANCIS, GARY STUART, b St Paul, Minn, June 5, 43; m 66; c 3. CARDIOLOGY. *Educ:* Creighton Univ, MD, 69. *Prof Exp:* Instr med, Univ Calif San Diego, 76-77; from asst prof to assoc prof, 77-88, PROF MED, UNIV MINN, 88- *Concurrent Pos:* Dir cardiovasc res, Vet Admin Med Ctr, Minneapolis, 77-89, echocardiography lab, 77-88, George E Fahr Cardiac Unit, Univ Minn, 77-79, acute cardiac care, Univ Hosp, Minneapolis, 90-; asst chair, Vet Admin Coop Study on vasodilators in AMI, 79-85; vis prof, Mayo Clin, 89, Univ Calif San Francisco, 91; mem, Coun Circulatory Coun Clin Cardiol. *Mem:* Fel Am Col Physicians; fel Am Heart Asn; Am Fedn Clin Res; Am Phys Soc; Central Soc Clin Res. *Res:* Patients with congestive heart failure; neuroendocrine abnormalities in CHF; molecular biology of myocardial hypertrophy. *Mailing Add:* Box 508 UMHC 420 Delaware St Minneapolis MN 55455

FRANCIS, GEORGE KONRAD, b Warnsdorf, Ger, Mar 7, 39; US citizen; m 67; c 1. MATHEMATICS. *Educ:* Univ Notre Dame, BS, 58, Harvard Univ, MA, 60; Univ Mich, PhD(math), 67. *Prof Exp:* Lectr math, Regis Col, 62-63, Boston Col, 63-64 & Newton Col, 64-65; A H Lloyd fel Univ Mich, 67-68; asst prof, 68-73, ASSOC PROF MATH, UNIV ILL, URBANA, 73- *Mem:* Am Math Soc. *Res:* Differential and combinatorial topology of curves and surfaces; control theory and dynamical systems; confidence bounds in multivariate statistics; biomathematics of morphogenesis and evolution; singularities of maps on manifolds; catastrophe theory; pow dimensional manifolds and cell complexes. *Mailing Add:* Univ Ill 1409 W Green St Urbana IL 61801

FRANCIS, GERALD PETER, b Seattle, Wash, Feb 15, 36; m 64; c 3. MECHANICAL ENGINEERING. *Educ:* Univ Dayton, BME, 58; Cornell Univ, MME, 60, PhD(eng), 65. *Prof Exp:* Instr mech eng, Cornell Univ, 61-64; asst prof, Ga Inst Technol, 64-66; from asst prof to assoc prof, State Univ NY, Buffalo, 66-70, chmn dept, 70-77; prof eng & head dept, US Merchant Marine Acad, 77-80; dean, Col Eng & Math, Univ Vt, 80-85, dean, Div Eng Math & Bus Admin, 81-87, int vpres acad affairs, 85-87, vprovost, 87-90, PROF MECH ENG, UNIV VT, 80-, INT PROVOST, 90- *Concurrent Pos:* NSF grants, 67-70; Am Heart Asn grants, 69-74. *Honors & Awards:* Silver Medal, US Dept Com, 79. *Mem:* Am Soc Mech Engrs; Am Soc Eng Educ; fel Inst Marine Engrs; Nat Soc Prof Engrs; AAAS. *Res:* Turbulent boundary layers; pulsatile blood flow; instrumentation for bio-fluid measurements. *Mailing Add:* 30 Oak Hill Rd Shelburne VT 05482

FRANCIS, HOWARD THOMAS, b St Louis, Mo, Oct 22, 17; m 41; c 2. BIOMEDICAL ENGINEERING. *Educ:* Mt Union Col, BS, 38; Pa State Univ, MS, 40, PhD(phys chem), 42. *Prof Exp:* Mgr electrochem, IIT Res Inst, 42-71; dir biomed eng, Cook County Hosp, 71-85; RETIRED. *Mem:* Am Chem Soc; Electrochem Soc; Nat Asn Corrosion Eng; Asn Advan Med Instrumentation; Am Soc Hosp Engrs. *Res:* Electrical safety; medical equipment testing and maintenance; clinical engineering; automobile battery explosions. *Mailing Add:* 107 Walnut St Park Forest IL 60466-1351

FRANCIS, IVOR STUART, b Napier, NZ, Feb 6, 38. STATISTICS. *Educ:* Univ NZ, BSc, 59, MSc, 60; Harvard Univ, PhD(statist), 66. *Prof Exp:* Asst prof statist, Grad Sch Bus Admin, NY Univ, 65-68; asst prof, 68-74, assoc prof, 74-81, PROF STATIST, CORNELL UNIV, 81- *Concurrent Pos:* Vis prof math, Univ Otago, NZ, 72. *Mem:* Am Statist Asn; Inst Math Statist; Biomet Soc; Royal Statist Soc; Asn Comput Mach. *Res:* Multivariate statistical analysis and statistical computing; evaluation of statistical program packages. *Mailing Add:* PA Consult Group Locked Mail Bag 951 North Sydney NSW 2060 Australia

FRANCIS, JOHN ELBERT, b Kingfisher, Okla, Mar 14, 37; m 62; c 2. HEAT TRANSFER. *Educ:* Univ Okla, BS, 60, MS, 63, PhD(eng sci), 65. *Prof Exp:* Engr, Allis Chalmers Mfg Co, 60; asst prof mech eng, Univ Mo, Rolla, 64-66; from asst prof to prof aerospace, mech & nuclear eng, Univ Okla, 66-90, asst dean grad col, 68-71, actg dean eng, 85, assoc dean eng, 81-88; DEAN ENG & TECHNOL, BRADLEY UNIV, 90- *Honors & Awards:* Ralph Teetor Award, Soc Automotive Engrs, 69. *Mem:* Am Soc Mech Engrs; assoc fel Am Inst Aeronaut & Astronaut; Sigma Xi; NY Acad Sci. *Res:* The use of the scanning infrared camera thermography, in bioengineering and in fatigue studies in metal and composite materials; experimental and analytical solar energy; combined radiative and conductive in insulating materials. *Mailing Add:* 202 E Hanover Pl Peoria IL 61614

FRANCIS, JOHN ELSWORTH, b Toronto, Ont, Jan 25, 32; m 59; c 3. PHARMACEUTICAL CHEMISTRY. *Educ:* Queen's Univ, Ont, BA, 53, MA, 56; Univ NB, PhD(org chem), 58. *Prof Exp:* Jr res chemist div pure chem, Nat Res Coun Can, 54-55; vis scientist, Nat Inst Arthritis & Metab Dis, 58-60; res chemist, J R Geigy AG, Switz, 60-63 & Geigy Chem Corp, 63-67, res assoc, 67-70, sr staff scientist, 70-71, SR RES FEL, CIBA-GEIGY CORP, 72- *Mem:* Am Chem Soc; Int Soc Heterocyclic Chem. *Res:* Aromatic chemistry; organic deuterium compounds; structure elucidation of lycopodium alkaloids; synthesis and selective cleavage of peptides; thiophene derivatives; sulphonamides; nitrofurans; nitrogen heterocyclic compounds; phenethylamines; nucleoside synthesis. *Mailing Add:* 15 Wexford Way Basking Ridge NJ 07920

FRANCIS, LYMAN L(ESLIE), b Alma, Mo, May 17, 20; m 42; c 5. MECHANICAL ENGINEERING. *Educ:* Univ Mo, MS, 50. *Prof Exp:* Design engr, Owen-Ill Glass Co, Ohio, 46-47; jr engr, Toledo Edison Co, 47-48; instr mech eng, Univ Mo, 48-52; assoc prof, Wash State Univ, 52-63; from assoc prof to prof eng, Univ Mo-Rolla, 63-85; RETIRED. *Concurrent Pos:* Chmn coun engrs, Coun for Prof Develop, 77-79; bd dir, Soc Mfr Engrs, 78- *Mem:* Am Soc Eng Educ; Soc Mfg Engrs; Am Soc Mech Engrs. *Res:* Manufacturing engineering, materials, and design. *Mailing Add:* Rt 6 Box 465 Rolla MO 65401

FRANCIS, MARION DAVID, b Campbell River, BC, May 9, 23; m 49; c 2. BONE PHYSIOLOGY, CALCIUM PHOSPHATES. *Educ:* Univ BC, BA, 46, MA, 49; Univ Iowa, PhD(biochem), 53. *Prof Exp:* Chemist, Procter & Gamble Co, 52-76, sr scientist, 76-85; SR SCIENTIST & GROUP LEADER, NORWICH EATON PHARMACEUTICALS INC, 85- *Concurrent Pos:* Consult, Am Dent Asn; guest lectr at numerous Am & Europ Univs; fel USHPS, 51-52; founding mem, Victor Mills Soc, 90. *Mem:* Fel AAAS; fel Am Inst Chem; Soc Nuclear Med; NY Acad Sci; Am Chem Soc; Am Pharmaceut Soc; Am Inst Chemists; Am Asn Pharmaceut Scientists; Am Soc Bone & Mineral Res; Am Col Rheumatology. *Res:* Diagnostic and therapeutic nuclear medicine; calcium and phosphorus chemistry and metabolism of bones and teeth; biochemistry and physiology of the di- and polyphosphonates; arthritis and inflammatory reactions. *Mailing Add:* Norwich Eaton Pharmaceuts Inc Miami Valley Labs Cincinnati OH 45239-8707

FRANCIS, MIKE MCD, b Barbados, WI; Can citizen. BIOENGINEERING, BIOREMEDIATION. *Educ:* Univ King's Col, BSc, 69; Dalhousie Univ, MSc, 72, PhD(microbiol), 76. *Prof Exp:* Res fel & lab instr, Univ Alta, 76-78; res assoc, Univ Calgary, 78-83; RES SCIENTIST, NOVA-HUSKY RES CORP, LTD, 83- *Concurrent Pos:* Lectr, Univ Calgary, 83- *Mem:* Am Soc Microbiologists; Soc Indust Microbiologists; Can Soc Microbiologists; NY Acad Sci. *Res:* Genetics of Streptomyces; regulation and enzymology of secondary metabolism; molecular basis of microbial transport phenomena; physiology of sulfate-reducing bacteria; physiology and metabolism of methylotrophic microorganisms; biodegradation of materials; oil degradation; degradation of toxics. *Mailing Add:* Nova Husky Res Corp Ltd 2928-16 St NE Calgary AB T2E 7K7 Can

FRANCIS, NORMAN, b Rochester, NY, Nov 27, 22; m 47; c 3. THEORETICAL PHYSICS. *Educ:* Univ Rochester, BA, 48, PhD(physics), 52. *Prof Exp:* Res assoc, Ind Univ, 52-55; theoret physicist, 55-64, mgr advan reactor theory, 64-78, CONSULT PHYSICIST, KNOLLS ATOMIC POWER LAB, GEN ELEC CO, 78- *Mem:* Am Phys Soc; fel Am Nuclear Soc. *Res:* Nuclear and reactor theory. *Mailing Add:* 2311 Plum St Schenectady NY 12309

FRANCIS, PETER SCHUYLER, b Youngstown, Ohio, Nov 22, 27; m 51; c 3. PHYSICAL ORGANIC CHEMISTRY. *Educ:* Univ Pa, BA, 51; Univ Del, MS, 53, PhD(org chem), 55. *Prof Exp:* Lab technician chem, Smith Kline & French Labs, 49-51; asst, Univ Del, 52-54; res chemist, Hercules Powder Co, 54-63; sr polymer chemist, North Star Res & Develop Inst, 63-65, dir res polymer chem, 65-67; tech dir chem dept, Franklin Inst Res Labs, 67-69, dir chem dept, 70-71, dir materials & phys sci dept, 71-74; mem bd dir, Germantown Labs, 69-74; dir res, Quaker Chem Corp, 74-76; SR RES CHEMIST, ARCO CHEM CO, 77- *Mem:* Am Chem Soc; Sigma Xi; Am Inst Physics; Soc Plastics Engrs; Soc Automotive Engrs. *Res:* Electrokinetic phenomena; rheology of polymers in bulk and solution; metallizing plastics; polymeric membranes for desalting water; reinforced polymers; flame-retarded polymers; thermoplastics; textile and paper processing. *Mailing Add:* 19 Briarcrest Dr Wallingford PA 19086

FRANCIS, PHILIP HAMILTON, b San Diego, Calif, Apr 13, 38; c 4. ENGINEERING MECHANICS. *Educ:* Calif Polytech State Univ, BS, 59; Univ Iowa, MS, 60, PhD(appl mech), 65; St Mary's Univ, MBA, 72. *Prof Exp:* Stress analyst, Douglas Aircraft Co, Inc, 60-62; sr res engr, Dept Mech Sci, 65-73, mgr, Solid Mech, 73-78, staff engr, Southwest Res Inst, 78-79; dir Indust technol Inst, 84-86; DIR ADV MFG TECH, MOTOROLA INC, 86- *Honors & Awards:* Gustas L Larson Award, Am Soc Mech Engrs, 78. *Mem:* Am Soc Mech Engrs; Am Acad Mech; Sigma Xi; Am Sci Bd. *Res:* Manufacturing technology, applied mechanic, research and development. *Mailing Add:* Square D Co Executive Plaza Palatine TX 60067

FRANCIS, RAY LLEWELLYN, b Detroit, Mich, Feb 24, 21; m 46; c 3. ORGANIC POLYMER CHEMISTRY, ENGINEERING PSYCHOLOGY. *Educ:* Iowa State Univ, BS, 47; Wayne State Univ, MS, 52, MA, 90; Univ Mich, PhD, 75, PD(psychol), 87; Int Inst Advan Studies, PhD(chem eng), 81. *Prof Exp:* Develop engr, Uniroyal, 47-52; process engr, Ternstedt, Gen Motors Corp, 52-53; chem engr, Burroughs Corp, 53-57, sr chem engr, 57-63, sr proj chem engr, 63-69, mgr mat eng lab, 69-87; PSYCHOLOGIST, FAMILY & NEIGHBORHOOD SERV, 90- *Mem:* Tech Asn Pulp & Paper Indust; Soc Mfg Engrs Asn Finishing Processes. *Res:* Materials and processes for electronics and computers. *Mailing Add:* 34051 N Hampshire Livonia MI 48154-2723

FRANCIS, RICHARD L, b Poplar Bluff, Mo, Feb 5, 33; m 65; c 3. MATHEMATICS. *Educ:* Southeast Mo State Col, BS, 55; Univ Mo, MA, 62, EdD(math educ), 65. *Prof Exp:* Instr math, Kemper Sch, 55-61; PROF MATH, SOUTHEAST MO STATE UNIV, 65- *Honors & Awards:* George Polya Award, Math Asn Am. *Mem:* Math Asn Am. *Res:* Mathematics education; number theory. *Mailing Add:* Dept Math Southwest Mo State Univ Cape Girardeau MO 63701

FRANCIS, RICHARD LANE, b Bluefield, WVa, June 9, 38. INDUSTRIAL ENGINEERING, OPERATIONS RESEARCH. *Educ:* Va Polytech Inst, BS, 60; Ga Inst Technol, MS, 62; Northwestern Univ, PhD(indust eng & mgt sci), 67. *Prof Exp:* Systs engr, Eng Res Ctr, Western Elec Co, 62-63; from asst prof to assoc prof indust eng, Ohio State Univ, 66-71; PROF INDUST & SYSTS ENG, UNIV FLA, 71- *Concurrent Pos:* Vis math scientist, Opers Res Div, Nat Bur Standards, 77-78. *Honors & Awards:* Distinguished Serv Award, Am Inst Indust Engrs, 77. *Mem:* Am Inst Indust Engrs; Opers Res Soc Am; Inst Mgt Sci; Sigma Xi. *Res:* Location theory; facility layout; applications of optimization theory. *Mailing Add:* Dept Indust & Syst Eng Univ Fla Gainesville FL 32611

FRANCIS, ROBERT COLGATE, b Pittsfield, Mass, Nov 10, 42; m 61; c 2. BIOMATHEMATICS. *Educ:* Univ Calif, Santa Barbara, BA, 64; Univ Wash, MS, 66, PhD(biomath), 70. *Prof Exp:* Asst prof statist & fishery & wildlife biol, Colo State Univ, 69-71; assoc scientist, 71-74, SR SCIENTIST, INTER-AM TROP TUNA COMN, SCRIPPS INST OCEANOG, 74- *Concurrent Pos:* Math modeler & biometrician, Colo State Univ, 69-71; lectr, Scripps Inst Oceanog, 73- *Mem:* Biomet Soc. *Res:* Population dynamics; estimation; experimental design; modeling of biological systems; computer programming and simulation techniques; application of biomathematics in fisheries and wildlife management. *Mailing Add:* 1790 La Jolla Rancho Rd La Jolla CA 92037

FRANCIS, ROBERT DORL, b West Liberty, Ohio, Sept 28, 20; m 43. MEDICAL MICROBIOLOGY, VIROLOGY. *Educ:* Franklin Col, AB, 42; Univ Chicago, MS, 45; Univ Mich, PhD(microbiol), 55. *Prof Exp:* Anal chemist, Swift & Co, 42-44; prin bacteriologist, Emulsol Corp, 47-49; instr bact, Univ Mich, 50-52; sr asst sanitarian, USPHS, 53-55, sr asst scientist, 55-56, scientist, 56; assoc prof dermat, Sch Med & Dent, 71-73, ASSOC PROF MICROBIOL, MED CTR, UNIV ALA, BIRMINGHAM, 56- *Concurrent Pos:* Consult, Emulsol Corp, 49-52 & Bur Labs, Ala State Health Dept, 66-76. *Mem:* AAAS; Am Soc Microbiol. *Res:* Epidemiology and immunology of psittacosis lymphogranuloma venereum, poliomyelitis and enteroviruses; replication and antigens of herpes viruses; molluscum contagiosum; diagnostic virology. *Mailing Add:* 2508 Belle Terre Dr Birmingham AL 35226

FRANCIS, SAMUEL HOPKINS, b Lancaster, Pa, Jan 27, 43; m 65; c 2. UNDERWATER ACOUSTICS. *Educ:* Yale Univ, BA, 64; Harvard Univ, AM, 68, PhD(physics), 70. *Prof Exp:* mem tech staff, 70-78, tech group supvr, 78-84, tech dept head, 84-87, TECH DIR, AT&T BELL LABS, 87- *Mem:* Am Phys Soc. *Res:* Atmospheric gravity waves and their ionospheric effects; plasma clouds in the ionosphere; flow noise in sonar systems; use of magnetics for locating and tracking undersea cables; turbulent flow noise in moving undersea acoustic sensors. *Mailing Add:* 68 Dale Dr Chatham NJ 07928

FRANCIS, SHARRON H, b Aug 1, 44; m; c 2. PHOSPHO-DIESTERASES, MUSCLE RELAXATION. *Educ:* Vanderbilt Univ, PhD(physiol), 70. *Prof Exp:* Physiologist, Howard Hughes Med Inst, 75-84; ASSOC PROF MOLECULAR PHYSIOL & BIOPHYSICS, VANDERBILT UNIV, 84- *Mem:* Am Soc Biochem & Molecular Biol; AAAS. *Res:* cGMP Kinase. *Mailing Add:* Dept Molecular Physiol & Biophys Vanderbilt Univ Light Hall Rm 767 Nashville TN 37232-0615

FRANCIS, STANLEY ARTHUR, b Moundsville, WVa, Oct 25, 19; m 44; c 3. MOLECULAR SPECTROSCOPY. *Educ:* Ohio Univ, BS, 40; Ohio State Univ, PhD(phys chem), 47. *Prof Exp:* Chemist, Monsanto Chem Co, 44-45; res chemist, Texaco, Inc, 47-63, supvr fundamental res, 63-68, dir, 68-71, asst mgr fundamental res, 71-82; RETIRED. *Mem:* Am Chem Soc. *Res:* Molecular spectroscopy; administration. *Mailing Add:* 335 West Forest Trail Vero Beach FL 32962-4662

FRANCIS, WARREN C(HARLES), chemical engineering, for more information see previous edition

FRANCIS, WILLIAM CONNETT, b Denver, Colo, Dec 23, 22; m 54; c 2. ORGANIC CHEMISTRY. *Educ:* Univ Kans, BS, 47; Ohio State Univ, PhD(org chem), 52; Pepperdine Univ, MBA, 80. *Hon Degrees:* DSc, Cent Methodist Col, 81. *Prof Exp:* Res chemist, Plaskon Div, Libbey-Owens-Ford Glass Co, 47-48; Fulbright grant, Cambridge Univ, 52-54; sr staff mem res dept, Spencer Chem Co, 54-57, group leader org res, 57-60, sect leader, Spencer Chem Div, Gulf Oil Corp, Kans, 60-66, dir res coord chem dept, Tex, 66-69, mgr develop div, 69-71, mgr, Gulf Adhesives Div, 71-73, vpres, Indust & Specialty Chem Div, 73-79, vpres tech & develop specialty chem, Gulf Oil Chem Co, 79-82, gen mgr, Houston Res Ctr, 82-83; RETIRED. *Concurrent Pos:* Chmn, H-F Info Resources, Inc, 84- *Mem:* Am Chem Soc. *Res:* Oxidation of organic compounds; synthesis of fluorinated organic compounds and organic compounds of nitrogen; aminoplast resins; high temperature polymers; organic pesticides. *Mailing Add:* 12126 Broken Bough Houston TX 77024

FRANCIS, WILLIAM PORTER, b St Louis, Mo, Mar 15, 40; m 62; c 3. THEORETICAL PHYSICS, ELEMENTARY PARTICLES. *Educ:* Rensselaer Polytech Inst, BS, 61; Cornell Univ, PhD(physics), 69. *Prof Exp:* Mem tech staff solid state physics, Bell Tel Labs, NJ, 67-68; Nat Res Coun Can fel physics, Univ Windsor, 68-70; asst prof math, 70-77, ASSOC PROF MATH, MICH TECHNOL UNIV, 77- *Mem:* Am Phys Soc; Sigma Xi; Math Assoc Am. *Res:* Gauge theories and the theories of fundamental interactions. *Mailing Add:* 1215 E Fifth Ave Houghton MI 49931

FRANCISCO, JERRY THOMAS, b Huntingdon, Tenn, Dec 18, 32; m 54; c 3. PATHOLOGY. *Educ:* Univ Tenn, MD, 55. *Prof Exp:* Asst instr path, Univ Tenn, 56-57; dir labs, US Naval Hosp, Memphis, 57-59; from instr to assoc prof, 59-67, PROF PATH, INST PATH, UNIV TENN, MEMPHIS, 67- *Concurrent Pos:* Mem staff, John Gaston Hosp, 60-; med examr, Memphis & Shelby County, Tenn, 60-; chief med examr, Tenn, 63-66, forensic path consult for chief med examr, 66-70, chief med examr, 70- *Mem:* Col Am Path. *Res:* Forensic pathology; effect of sickle cell erythrocytes on the in vitro growth characteristics of certain bacteria. *Mailing Add:* Three N Dunlap Univ Tenn Memphis TN 38163

FRANCISCO, JOSEPH SALVADORE, JR, b New Orleans, La, Mar 26, 55. MOLECULAR SPECTROSCOPY. *Educ:* Univ Tex, Austin, BS, 77; Mass Inst Technol, PhD(computer physics), 83. *Prof Exp:* Res fel, Cambridge Univ, Eng, 83-85; provost fel, Mass Inst Technol, 85-86; asst prof, 86-90, ASSOC PROF, WAYNE STATE UNIV, 90- *Concurrent Pos:* Mem, vis comt, Mass Inst Technol Corp, 87-92 & adv comt, Nat Adv Coun Res, Howard Univ, 90-; NSF presidential young investr award, 88; consult, Inst Defense Anal, 88-; Camille & Henry Dreyfus teacher-scholar award, Dreyfus Found, 90; res fel, Alfred P Sloan Found, 90-92; vis prof, Northern Mich Univ, 91; vis assoc, Calif Inst Technol, 91; fac fel, Jet Propulsion Lab, 91. *Mem:* AAAS; Am Phys Soc; Sigma Xi. *Res:* Molecular spectroscopy; photodissociation dynamics; chemical kinetics; atmospheric chemistry; quantum chemistry. *Mailing Add:* Dept Chem 33 Wayne State Univ Detroit MI 48202

FRANCK, RICHARD W, b Germany, May 15, 36; US citizen; m 58; c 2. ORGANIC CHEMISTRY. *Educ:* Amherst Col, AB, 58; Univ Wis, MA, 60; Stanford Univ, PhD(chem), 63. *Prof Exp:* NIH fel chem, Mass Inst Technol, 62-63; from asst prof to prof chem, Fordham Univ, 63-83; PROF CHEM, HUNTER COL, CITY UNIV NEW YORK, 83- *Concurrent Pos:* Vis prof, Univ Calabria, 84. *Mem:* NY Acad Sci; Chem Soc; Am Chem Soc; AAAS; Sigma Xi. *Res:* Modern mechanistic and synthetic organic chemistry; bio-organic chemistry. *Mailing Add:* Five Mary Lane Riverside CT 06878

FRANCK, WALLACE EDMUNDT, b Alexandria, La, Feb 1, 33; m 57; c 2. MATHEMATICAL STATISTICS. *Educ:* La State Univ, BS, 55; Univ NMex, MS, 62, PhD(math), 64. *Prof Exp:* Weapons proj officer, Kirtland AFB, US Air Force, 56-58, programmer math, 58-60; mathematician, US Naval Weapons Eval Facil, 60-62; asst prof, 64-69, ASSOC PROF STATIST, UNIV MO-COLUMBIA, 69- *Mem:* Inst Math Statist. *Res:* Probability; statistics. *Mailing Add:* Dept Statist Univ Mo Columbia MO 65201

FRANCKE, OSCAR F, b Mexico City, Mex, Mar 1, 50. SYSTEMATICS, BIOGEOGRAPHY. *Educ:* Ariz State Univ, BS, 71, PhD(zool), 76. *Prof Exp:* ASST PROF BIOL & ENTOM, TEX TECH UNIV, 76-; VPRES, CROWN CORK DE MEXICO. *Concurrent Pos:* Ed, J Arachnol, 76-; prin investr faunistic surv, Tex Dept Agr, 77-79. *Mem:* Am Arachnol Soc; AAAS; Am Soc Zoologists; Entom Soc Am; Soc Syst Zool; Sigma Xi. *Res:* Phylogeny of Arthropoda, particularly chelicerata; biology of Arachnida; systematics, phylogeny, biology and zoogeography of scorpiones; faunistics of formicidae, ants. *Mailing Add:* Canada 34 Club de Golf Hascenda Tizatan DEZ Aragoza Estabo Mexico

FRANCKE, UTA, b Wiesbaden, Ger, Sept 9, 42; m 67, 86. HUMAN GENETICS, CYTOGENETICS. *Educ:* Univ Munich Med Sch, MD, 67. *Prof Exp:* Asst prof pediat, Univ Calif, San Diego, 73-78; assoc prof, 78-85, PROF HUMAN GENETICS, MED SCH, YALE UNIV, 85- *Concurrent Pos:* Dir med genetics, San Diego Childrens Hosp, 75-78. *Mem:* Am Soc Human Genetics; Soc Pediat Res; Teratology Soc. *Res:* Human cytogenetics including development of a map for chromosome bands; identification and characterization of new chromosomal syndromes; human gene mapping by somatic cell hybridization and chromosomal hybridization; human/mouse comparative chromosome mapping; molecular genetics of muscular dystrophy. *Mailing Add:* 13500 S Fork Lane Los Altos CA 94022

FRANCKO, DAVID ALEX, b Cleveland, Ohio, Aug 15, 52; m 75; c 2. LIMNOLOGY, BOTANY. *Educ:* Kent State Univ, BS, 74, MS, 77; Mich State Univ, PhD(bot), 80. *Prof Exp:* Fel, Kellogg Biol Sta, Mich State Univ, 80-81, instr limnol, 80; from asst prof to prof bot, Okla State Univ, 81-89, dir res, Col Arts & Sci, 87-90; PROF BOT & CHAIR DEPT, MIAMI UNIV, 90- *Concurrent Pos:* Consult, plant biotechnol firms. *Mem:* AAAS; Am Soc Limnol & Oceanog; Int Soc Theoret Appl Limnol; Sigma Xi; Phycological Soc Am. *Res:* Physiological limnology, especially the molecular ecology of aquatic photoautotrophs; cyclic nucleotide research and nutrient dynamics; aquatic toxicology. *Mailing Add:* Dept Bot Miami Univ Oxford OK 45056

FRANCO, NICHOLAS BENJAMIN, b Providence, RI, Apr 26, 38; m 64; c 2. PHYSICAL INORGANIC CHEMISTRY, POLLUTION CHEMISTRY. *Educ:* Providence Col, BS, 59, MS, 62; Univ Miami, PhD(phys chem), 68. *Prof Exp:* Sr res chemist, Olin Corp, 67-69; vpres, treas-mgr water treatment systs, Resource Control, Inc, 69-74; pollution control engr, 74-79, SUPVR RES & DEVELOP, RAW MATS & CHEM PROCESSES, BETHLEHEM STEEL CORP, 79- *Mem:* Water Pollution Control Fedn; Am Chem Soc; Air Pollution Control Asn; Electroplaters Soc. *Res:* Chemistry of fused salts, propellants, coordination compounds and organometallics; homogeneous catalysis by inorganic salts and organometallics; air and water pollution control; minerals processing. *Mailing Add:* 115 W Langhorne Ave Bethlehem PA 18017

FRANCO, VICTOR, b New York, NY, Dec 15, 37; m 83; c 5. THEORETICAL NUCLEAR, PARTICLE & ATOMIC PHYSICS. *Educ:* NY Univ, BS, 58; Harvard Univ, MA, 59, PhD(physics), 64. *Prof Exp:* Instr physics, Mass Inst Technol, 63-64, res assoc, 64-65; fel, Lawrence Radiation Lab, Univ Calif, Berkeley, 65-67 & Los Alamos Sci Lab, 67-69; assoc prof, 69-72, PROF PHYSICS, BROOKLYN COL, 73- *Concurrent Pos:* Consult, Los Alamos Nat Lab; vis scientist, Centre d'Etudes Nucleaives de Saclay, France, 75-76 & 86, Harvard Univ, Cambridge, 83-84, Univ Kurlsruhe, 85; exchange scientist, Inst High Energy Physics, Beijing, 84; prin investr, NSF & NASA res grants; mem fac, New Sch Social Res, 88, 89. *Mem:* Fel Am Phys Soc; Sigma Xi. *Res:* Theoretical medium and high energy nuclear and atomic scattering theory; Glauber theory; diffraction scattering; heavy-ion collisions. *Mailing Add:* Dept Physics Brooklyn Col Brooklyn NY 11210

FRANCOEUR, ROBERT THOMAS, b Detroit, Mich, Oct 18, 31. EMBRYOLOGY, EVOLUTION. *Educ:* Sacred Heart Col, BA, 53; St Vincent Col, MA, 57; Univ Detroit, MS, 62; Univ Del, PhD(biol), 67. *Prof Exp:* Instr biol & theol, Mt St Agnes Col, 61-62; from asst prof to prof embryol, 65-80, PROF BIOL SCI, FAIRLEIGH DICKINSON UNIV, 80- *Honors & Awards:* Buhl Planetarium Award, 60; Educ Found Human Sexuality Award, 78; Fel, Soc for Sci Study Sex. *Mem:* Am Asn Sex Educr & Counr; Groves Conf Marriage & Family; World Future Soc; Soc Sci Study Sex (pres, Eastern region). *Res:* Interdisciplinary research focusing on the future of human sexuality, the impact of non-reproductive sex on social mores, legal codes, religious values, and politico-economic systems; ethical considerations of reproductive technologies; innovative life styles in practice. *Mailing Add:* Dept Biol & Allied Health Sci Fairleigh Dickinson Univ Madison NJ 07940

FRANCO-SAENZ, ROBERTO, b Bogota, Colombia, July 13, 37; div; c 2. ENDOCRINOLOGY, METABOLISM. *Educ:* Nicolas Esguerra Nat Inst, Bogota, BSc, 55; Nat Univ Colombia, MD, 62. *Prof Exp:* Fel metab & genetics, Hosp Univ Pa, 67-69; asst prof med, Nat Univ Colombia, 69-71; asst prof med, endocrinol & metab, 71-75, actg chief, dept endocrinol, 73-75, assoc prof, 75-79, PROF MED, MED COL OHIO, TOLEDO, 79-, CHIEF, DEPT ENDOCRINOL & METAB, 81- *Mem:* Fel Am Col Physicians; Endocrine Soc; Am Fedn Clin Res; Sigma Xi; Cent Soc Clin Res. *Res:* Control mechanisms of Renin-Angiotensin-Aldosterone Axis; atrial natriuretic factor; prostaglandins and hypertension. *Mailing Add:* Dept Med Med Col Ohio Toledo OH 43699

FRANCQ, EDWARD NATHANIEL LLOYD, b Greencastle, Ind, July 5, 34; div; c 2. VERTEBRATE ECOLOGY, MAMMALIAN BEHAVIOR. *Educ:* Univ Md, College Park, BS, 56; Univ Idaho, MS, 62; Pa State Univ, PhD(zool), 67. *Prof Exp:* Instr, 65-67, asst prof, 67-81, ASSOC PROF ZOOL, UNIV NH, 81- *Mem:* AAAS; Am Soc Mammal; Animal Behav Soc. *Res:* Behavior and population ecology of vertebrates, especially small mammals; behavioral responses to environment and stressful situations. *Mailing Add:* Dept Zool Univ NH Durham NH 03824

FRANCY, DAVID BRUCE, b Burbank, Calif, June 26, 32; m 56; c 3. ECOLOGY & CONTROL OF ARBOVIRUSES. *Educ:* Calif State Univ, BA, 59; Univ Utah, MS, 61; Univ Calif, Berkeley, MPH, 63, PhD(epidemiol), 72. *Prof Exp:* Chief, Biol & Lab Units, Arbovirus Ecol Sect, Bur Labs, Ctr Dis Control, 69-74, chief, Arbovirus Ecol Br, Div Vector-borne Infectious Dis, 74-90; EPIDEMIOLOGIST, ANIMAL & PLANT HEALTH INSPECTION SERV-V5, USDA, 90- *Concurrent Pos:* Consult, county, state, nat & int health agencies, 61-90; fac affil, Dept Microbiol, Colo State Univ, Ft Collins, 69-; vis scientist, Dept Virol, State Bact Lab, Stockholm, Sweden, 85. *Mem:* Am Mosquito Control Asn (pres, 76-77); Am Soc Trop Med & Hyg; Entom Soc Am. *Res:* Bovine spongiform encephalopathy; ecology and control of mosquito and tick-borne viruses. *Mailing Add:* 2700 Ringneck Dr Ft Collins CO 80526-2087

FRANDSEN, HENRY, b Chicago, Ill, May 21, 33; m 53; c 3. MATHEMATICS. *Educ:* Univ Ill, BS, 57, MS, 59, PhD(math), 61. *Prof Exp:* Asst math, Univ Ill, 57-61; asst prof, Clark Univ, 61-66; assoc prof, Wheaton Col, 66-67; assoc prof, 67-71, PROF MATH & MATH EDUC, UNIV TENN, KNOXVILLE, 71- *Concurrent Pos:* Mathematician, Mitre Corp, 63-67. *Mem:* Sigma Xi; Am Math Soc; Math Asn Am. *Res:* Algebra; geometry; mathematics education; secondary school teacher training. *Mailing Add:* Dept Math Univ Tenn Knoxville TN 37996-1300

FRANDSEN, JOHN CHRISTIAN, b Salt Lake City, Utah, Aug 25, 33; m 65, 75; c 4. PARASITOLOGY. *Educ:* Univ Utah, BS, 55, MS, 56, PhD, 60. *Prof Exp:* Asst zool, Univ Utah, 55-58, asst NSF & univ res fel, 58-60; res parasitologist, Regional Animal Dis Lab, Agr Res Serv, USDA, 61-66, sr res parasitologist, 66-72, dir, Regional Parasite Res Lab, 72-80, sr res parasitologist, 80-90; ADJ ASSOC PROF ZOOL & ENTOM, SCH VET MED, AUBURN UNIV, 71- *Concurrent Pos:* Vis prof, Sch Vet Med, Tuskegee Inst, 73-; adj instr, Soviet Union State Jr Col, 90- *Mem:* Am Soc Parasitol; Am Asn Vet Parasitologists; Am Soc Trop Med & Hyg; Soc Protozool; Soc Nematologists. *Res:* Applications of histochemistry in parasitology; host-parasite relationship; ecology of endoparasites; biochemistry of parasites; interrelationships of parasitism and nutrition. *Mailing Add:* 637 Cary Dr Auburn AL 36830

FRANDSON, ROWEN DALE, b Fremont, Nebr, July 22, 20; m 50; c 3. ANATOMY. *Educ:* Colo State Univ, BS, 42, DVM, 44, MS, 55. *Prof Exp:* Practicing vet, 47-48; from asst prof to assoc prof, 48-62, PROF ANAT, COLO STATE UNIV, 62- *Mem:* AAAS; World Asn Vet Anat; Am Asn Vet Anat; Am Vet Med Asn; Am Asn Anat. *Res:* Microanatomy of hypothalamic region; application of audio-visual techniques to medical and biological education. *Mailing Add:* Vet Anat Colo State Univ Ft Collins CO 80523

FRANGIONE, BLAS, AMYLOIDOSIS, ALZHEIMERS DISEASE. *Educ:* Cambridge Univ, PhD(molecular biol), 68. *Prof Exp:* PROF PATH, SCH MED, NEW YORK UNIV, 74- *Mailing Add:* Dept Path NY Univ Med Ctr 560 First Ave No TH427 New York NY 10016

FRANGOPOL, DAN MIRCEA, b Bucharest, Romania, July 28, 46; US citizen; m 69; c 2. STRUCTURAL RELIABILITY & OPTIMIZATION, BRIDGE DESIGN & EVALUATION. *Educ:* Inst Civil Eng, Bucharest, Romania, BS, 69; Univ Liege, Belg, PhD(appl sci & struct reliability), 76. *Prof Exp:* Asst prof struct eng, Inst Civil Eng, Bucharest, Romania, 69-74, lectr, 76-79; res struct engr, Univ Liege, Belg, 74-76; proj engr underground struct, A Lipski Consult Engrs, Brussels, Belg, 79-83; assoc prof, 83-88, PROF STRUCT RELIABILITY, UNIV COLO, BOULDER, 88- *Concurrent Pos:* Prin investr, NSF, 86-; consult, Transp Res Bd, 89-91; vis prof, Swiss Fed Inst Technol, Lausanne, Switz & Nat Defense Acad, Yokosuka, Japan, 89; chmn, Subcomt Existing Struct & Subcomt Res Needs, Am Concrete Inst & Transp Res Bd, 89-, Comt Safety of Bldg, Am Soc Civil Engrs, 89-; vis res prof, Univ Waterloo, Ont, Can, 91; mem, numerous tech comts, Am Soc Civil Engrs, Am Concrete Inst & Transp Res Bd; vis sr scientist, Royal Norweg Coun Sci & Indust Res, Univ Trondheim, Norway, 91. *Mem:* Am Soc Civil Engrs; Am Concrete Inst; Earthquake Eng Res Inst; Transp Res Bd; Int Asn Bridge & Struct Eng. *Res:* Academic-oriented research and development in the field of structural reliability, safety evaluation of existing structures and structural optimization; experience in professional structural engineering practice; co-author of a book on reliability of steel structures. *Mailing Add:* Dept Civil Eng Rm OT5-26 Univ Colo CB428 Boulder CO 80309-0428

FRANK, ALAN M, b New York, NY, Dec 22, 44; m 68; c 1. OPTICAL PHYSICS, OPTICAL ENGINEERING. *Educ:* NY Univ, BA, 67; Univ Denver, MS, 69. *Prof Exp:* Staff scientist, optics & detectors, Ball Bros Res Corp, 69-73; tech dir atmospheric physics, Colspan Environ Systs, Inc, 73-74; PHYSICIST & OPTICAL SYST ENGR, LAWRENCE LIVERMORE LAB, UNIV CALIF, 74- *Concurrent Pos:* Consult, Skin & Cancer Groups, NY Univ Hosp & Belvue Med Ctr, 63-67; comptroller, Optical Soc Am & Inst Elec & Electronics Engrs Conf on Lasers & Electrooptics. *Mem:* Optical Soc Am; Am Phys Soc; Inst Elec & Electronics Engrs Quantum Electronics & Appln Soc. *Res:* Development of optical instrumentation and techniques for plasma physics, remote environmental sensing, astronomy and medicine. *Mailing Add:* Lawrence Livermore Nat Lab Box 808 L-281 Livermore CA 94550

FRANK, ALBERT BERNARD, b Fresno, Ohio, July 7, 39; m 62; c 4. PLANT PHYSIOLOGY. *Educ:* Ohio State Univ, BS, 65, MS, 66; NDak State Univ, PhD(agron), 69. *Prof Exp:* Plant Physiologist, 69-82, SUPVRY PLANT PHYSIOLOGIST & RES LEADER, NORTHERN GREAT PLAINS RES CTR, AGR RES SERV, USDA, 82- *Mem:* Am Soc Plant Physiol; Am Soc Agron; Crop Sci Soc Am; Soil Conserv Soc Am; Sigma Xi. *Res:* Forage grass physiology as it relates to management practices and development of selection criteria for varietal improvement of grasses adapted to the northern Great Plains; effect of environmental and genetic factors on plant water relations as related to the water regime of growing plants in the northern Great Plains; mechanisms of drought tolerance in wheat and forages. *Mailing Add:* USDA-ARS Agr Res Box 459 Northern Great Plains Res Ctr Mandan ND 58554

FRANK, ANDREW JULIAN, b Chicago, Ill, Sept 17, 25; m 48; c 2. COMMUNITY HEALTH. *Educ:* Univ Ill, BS, 48, MS, 49, PhD(anal & inorg chem), 51. *Prof Exp:* Asst prof chem, Tulane Univ, 51-52; res chemist raw mat develop lab, AEC, Mass, 52-54; chief anal res sect, Watertown Arsenal, Mass, 54-56; chief metals chem br, Metall Div, Denver Res Inst, 56-60; res chemist, Allis-Chalmers Mfg Co, Wis, 60-62; from assoc prof to prof chem, Western Wash State Col, 62-70, chmn dept, 62-70; exec dir, Comprehensive Health Planning Coun, Whatcom, Skagit, Island & San Juan Counties, State of Wash, 71-76; asst dir admin, Puget Sound Health Systs Agency, 76-79; adminr, Jones, Grey & Bayley, 80-83; proj adminr, Betts Patterson & Mines, Puget Sound, 84-88; adminr, Trial Support Serv, Inc, 88-89; CONSULT, 89- *Concurrent Pos:* Instr, Northeastern Univ, 54-56, Denver Res Inst, 57 & Seattle Community Col, 80. *Mem:* Fedn Am Scientists; Am Chem Soc; Asn Legal Adminrs. *Mailing Add:* 2714 Ninth Ave W Seattle WA 98119-2220

FRANK, ARLEN W(ALKER), b Lima, Peru, Nov 22, 28; US citizen; m 58. ORGANOPHOSPHORUS CHEMISTRY, TEXTILE CHEMISTRY. *Educ:* Acadia Univ, BSc, 50, cert appl sci, 50; McGill Univ, Can, PhD(chem), 54. *Prof Exp:* Fel, Div Pure Chem, Nat Res Coun, Ottawa, 54-55; res chemist, Exp Sta, E I du Pont de Nemours & Co, 56-69 & Res Ctr, Hooker Chem Corp, 59-66; teaching intern org chem, Chem Dept, Ohio Wesleyan Univ, 66-67; RES CHEMIST, USDA, SOUTHERN REGIONAL RES CTR, NEW ORLEANS, 67- *Concurrent Pos:* Actg res leader, 82-85. *Mem:* Am Chem Soc; Am Inst Chem; AAAS; Sigma Xi; Inst Food Technologists. *Res:* Organophosphorus chemistry; flame retardants for cotton textile fabrics; food uses of cotton; soy protein modification. *Mailing Add:* 3812 Croydon St Slidell LA 70458

FRANK, ARNE, b Drammen, Norway, Feb 1, 25; US citizen; m 51, 75; c 2. MECHANICAL ENGINEERING. *Educ:* Purdue Univ, BSME, 50, MSME, 51. *Prof Exp:* Res asst, Purdue Univ, 51-52; sr engr, Carrier Corp, 52-57; engr, Trane Co, 57-65, mgr eng unitary prod, 65-75, vpres compressor develop, 75-82, vpres int eng, 79-88; RETIRED. *Mem:* Am Soc Heating, Refrig & Air-Conditioning Engrs. *Res:* Efficiency optimization in compressors. *Mailing Add:* N1948 Hickory Lane La Crosse WI 54601

FRANK, ARTHUR JESSE, b Cincinnati, Ohio, Jan 17, 45; m 68; c 2. PHOTOELECTROCHEMISTRY, PHOTOCHEMISTRY. *Educ:* Univ Colo, BA, 68; Univ Fla, PhD(chem), 75. *Prof Exp:* Res asst chem, Univ Fla, 70-75; vis scientist, Hahn-Meitner Inst, Berlin, 75-76; res assoc, Chem Biodyn Div, Lawrence Berkeley Lab, 76-78; staff scientist, 78-82, SR SCIENTIST, PHOTOCONVERSION RES BR, SOLAR ENERGY RES INST, 82- *Concurrent Pos:* NRC fel, Proctor & Gamble fel, 74; vis prof, Fed Inst Technol, Lausanne, Switz, 87. *Mem:* Am Chem Soc; Electrochem Soc; Inter-Am Photochem Soc. *Res:* Photoelectrochemistry, photochemistry, solid state semiconductor physics, laser flash photolysis, kinetics, microheterogenous and heterogeneous catalysis, conducting polymers, synthetic membranes. *Mailing Add:* Solar Energy Res Inst 1617 Cole Blvd Golden CO 80401

FRANK, BARRY, b Montreal, Que, Mar 26, 41; m 60; c 2. THEORETICAL SOLID STATE PHYSICS. *Educ:* McGill Univ, BS, 61, MS, 62; Univ BC, PhD(theoret physics), 65. *Prof Exp:* From asst prof to assoc prof, 65-86, PROF PHYSICS, CONCORDIA UNIV, 86- *Concurrent Pos:* Vis prof, Hebrew Univ, 71-72, King's Col, London, 78-79, Weizmann Inst, 86-87 & DTH, Denmark, 87. *Res:* Solid state and statistical physics. *Mailing Add:* Dept Physics Concordia Univ Montreal PQ H3G 1M8 Can

FRANK, BRUCE HILL, b Hartford, Conn, Oct 12, 38; m 61; c 2. BIOCHEMISTRY, BIOPHYSICS. *Educ:* Trinity Col, BS, 60; Northwestern Univ, PhD(phys biochem), 64. *Prof Exp:* RES ADV PHYS BIOCHEM, ELI LILLY & CO, 66-; ASST PROF BIOCHEM, MED SCH, IND UNIV, INDIANAPOLIS, 69- *Concurrent Pos:* NIH fel, Univ Calif, Berkeley, 64-66. *Mem:* AAAS; Am Chem Soc; Am Soc Biol Chem; Am Diabetes Asn; Sigma Xi; Europ Asn Study Diabetes. *Res:* Physical biochemistry; drug design; diabetes; protein chemistry; ultracentrifugation; self-associating systems; products derived using recombinant DNA technology. *Mailing Add:* Lilly Res Labs Lilly Corp Ctr 307 E McCarty St Indianapolis IN 46285

FRANK, CAROLYN, b Princeton, NJ, July 22, 52. TRANSPORTATION RESEARCH, COMPUTER SYSTEMS DEVELOPMENT. *Educ:* Univ Pa, BA, 74, MA, 74, PhD(reg sci), 78. *Prof Exp:* Res fel transp, Univ Pa, 75-76, lectr, econ geog, 76-77; economist consult, Syst Design Concepts, Inc, 77-79; sr economist consult, Evans Econ Inc, 79-81; proj mgr consult, Automated Sci Group, Inc, 81-90; PROG DIR, INFO SYSTS & SERV INC, 90- *Concurrent Pos:* Mem, Transp Res bd & Regional Sci Asn. *Mem:* Am Econ Asn. *Res:* The development of management information database system for research studies of motor vehicle and aircraft accidents; design and develop statistical and computer simulation models of highway and air traffic patterns, accidents and associated human behaviors. *Mailing Add:* 8403 Colesville Rd Suite 750 Silver Springs MD 20910

FRANK, CHARLES EDWARD, b Philipsburg, Pa, May 1, 14; m 39; c 3. ORGANIC CHEMISTRY. *Educ:* Pa State Col, BS, 35; Ohio State Univ, MS, 36, PhD(org chem), 38. *Prof Exp:* Grad asst, Ohio State Univ, 35-37; res chemist exp sta, E I du Pont de Nemours & Co, 38-48; assoc prof grad dept appl sci, Univ Cincinnati, 48-53; head appl res, Res Div, US Indust Chem Co Div, Nat Distillers & Chem Corp, 53-58, asst dir res, 58-75; chem consult, 75-85; RETIRED. *Concurrent Pos:* Sr tech adv, US Environ Protection Agency, 78-83. *Mem:* Am Chem Soc. *Res:* Hydrocarbon oxidation; catalysis; carbon monoxide; polymerization; formaldehyde; sodium. *Mailing Add:* 6887 Kenwood Rd Cincinnati OH 45243

FRANK, DAVID LEWIS, b Kearny, NJ, Apr 4, 43; m 64; c 2. TOPOLOGY. *Educ:* Columbia Univ, BA, 64; Univ Calif, Berkeley, PhD(math), 67. *Prof Exp:* NSF fel, 67-68; Moore instr math, Mass Inst Technol, 68-70; from asst prof to assoc prof, State Univ NY Stony Brook, 70-75; ASSOC PROF MATH, UNIV MINN, MINNEAPOLIS, 75- *Res:* Vector fields on manifolds; differential structures on manifolds. *Mailing Add:* Math 127 Vincent Hall Univ Minn Minneapolis 206 Chruch St SE Minneapolis MN 55455

FRANK, DAVID STANLEY, b Brooklyn, NY, Dec 3, 44; m 67; c 2. BIOCHEMISTRY. *Educ:* Univ Pa, BS, 66; Cornell Univ, MS, 68, PhD(org chem), 71. *Prof Exp:* Instr med, State Univ NY Downstate Med Ctr, 71-72; sr res chemist, Eastman Kodak Co, 72-80, res assoc, 80-81, mkt dir clin chem, 81-84, dir int support serv, clin prod, 85-86, dir worldwide mktg planning, clin prod, 86-88, mgr, Asian, African & Australasian region, clin prod, 88-90, MGR, DISPERSED TESTING DIAGNOSTICS, EASTMAN KODAK CO, 90- *Mem:* AAAS; Am Chem Soc; Am Asn Clin Chem. *Res:* Mechanistic enzymology; bioorganic chemistry; immunochemistry. *Mailing Add:* 127 Warrington Dr Rochester NY 14618-1122

FRANK, DONALD JOSEPH, b Cincinnati, Ohio, Nov 9, 26; m 49; c 4. PEDIATRICS. *Educ:* Univ Cincinnati, MD, 51. *Prof Exp:* From instr to asst prof, 54-67, assoc dir, Div Community Pediat, 67-80, ASSOC PROF PEDIAT, COL MED, UNIV CINCINNATI, 67- *Concurrent Pos:* Pvt pract pediat, 54-64; consult, Margaret Mary Hosp, Batesville, Ind, 54-, Bethesda Hosp, Cincinnati, Ohio, 80- *Mem:* Am Acad Pediat. *Res:* Clinical problems of the newborn; delivery of health care to children; training of pediatric nurse associates. *Mailing Add:* Munson Med Ctr Traverse City MI 49684-2386

FRANK, FLOYD WILLIAM, b Fortuna, Calif, Feb 12, 22; m 48; c 2. VETERINARY MEDICINE, VETERINARY MICROBIOLOGY. *Educ:* Wash State Univ, BS & DVM, 51, PhD(vet sci), 63. *Prof Exp:* Vet, Animal Dis Eradication Br, Agr Res Serv, USDA, Ore, 51-53; vet bact, Wyo State Vet Lab, 53-55; assoc vet, Vet Res Lab, Univ Idaho, 55-63, res vet, 63-67, actg assoc dir agr res, 71, prof vet sci & head dept, 67-84, dean, 74-84, EMER PROF VET MED, UNIV IDAHO, 84- *Concurrent Pos:* NIH trainee, 61-62; mem, Tech Comt Vibriosis Sheep & Tech Comt Urolithiasis Cattle & Sheep; dir, Idaho Vet Med Asn, 64-84; dir, Intermountain Vet Med Asn, 69-73, pres, 73-74; mem, Nat Adv Bd Wild Horses & Burros, 72-76, chmn, 75. *Mem:* Am Vet Med Asn; Am Soc Microbiol; Am Asn Vet Bact; Am Soc Animal Sci. *Res:* Etiology, transmission, treatment and control of vibriosis of sheep; nutritional factors influencing urolithiasis; etiology, transmission and resistance to enteritis in calves and lambs; prophylaxis of enzootic abortion of ewes and epizootic observations regarding tularemia in sheep. *Mailing Add:* Vet Med Univ Idaho Moscow ID 83843

FRANK, FORREST JAY, b Chicago, Ill, Sept 2, 37; m 59; c 1. ORGANIC CHEMISTRY, SCIENCE EDUCATION. *Educ:* Grinnell Col, BA, 59; Purdue Univ, PhD(org chem), 64. *Prof Exp:* Res chemist, Rayonier, Inc, 64-65; chairperson dept, 74-78, ASSOC PROF CHEM, ILL WESLEYAN UNIV, 65- *Mem:* Am Chem Soc; Sigma Xi. *Res:* Fingerprint detection chemical synthesis. *Mailing Add:* Dept Chem Ill Wesleyan Univ Bloomington IL 61702-2900

FRANK, GEORGE BARRY, b Brooklyn, NY, Feb 1, 29; Can citizen; m 51; c 3. PHYSIOLOGY, PHARMACOLOGY. *Educ:* City Col New York, BS, 50; Ohio State Univ, MSc, 52; McGill Univ, PhD(physiol), 56. *Prof Exp:* Lectr physiol, McGill Univ, 56-57; from asst prof to prof pharmacol, Univ Man, 57-65; PROF PHARMACOL, UNIV ALTA, 65- *Concurrent Pos:* USPHS fel, 56-57; USPHS spec fel physiol, Univ Col, Univ London & USPHS spec fel pharmacol, Univ Lund, 63-64; vis prof pharmacol, Univ Geneva, 70-71; vis prof physiol, Sapporo Med Col, 88. *Honors & Awards:* Samuel Racz Medallion, Hungariah Physiol Soc. *Mem:* Am Physiol Soc; Am Soc Pharmacol & Exp Therapeut; Can Col Neuropsychopharmacol; Can Physiol Soc; Pharmacol Soc Can (treas, 78-81); hon mem Hungarian Physiol Soc. *Res:* Electrical and mechanical activities of skeletal muscle; physiology and pharmacology of the central nervous system; pharmacology of general central nervous system depressants and narcotic analgesic drugs. *Mailing Add:* Dept Pharmacol Univ Alta Edmonton AB T6G 2H7 Can

FRANK, GLENN WILLIAM, b Mayfield Heights, Ohio, Jan 13, 28; m 49; c 3. GEOLOGY. *Educ:* Kent State Univ, BS, 51; Univ Maine, MS, 53. *Prof Exp:* From asst prof to assoc prof geol, 53-69, mem fac senate, 69-74, secy, 69-70, chmn, 71-72, asst chmn dept, 69-76, prof, 69-85, DANFORTH ASSOC, KENT STATE UNIV, 71-, EMER PROF GEOL, 85- *Concurrent Pos:* Asst geologist, Maine Geol Surv, 52; consult, Ohio Dept Transp, 73-74; prin investr, NSF grant, 80-82. *Honors & Awards:* Congressional Medal Antarctic Res, 82. *Mem:* Sigma Xi; Am Inst Prof Geol; Geol Soc Am; Nat Asn Geol Teachers; Int Platform Asn. *Mailing Add:* 7446 W Lake Blvd Kent OH 44242

FRANK, H LEE, b Philadelphia, Pa, Sept 9, 41; m 90; c 1. PULMONARY RESEARCH, NEONATOLOGY. *Educ:* Univ Pa, BA, 63; Univ Chicago, MD, 72; Univ Iowa, PhD(pharmacol), 78. *Prof Exp:* Resident, Univ Iowa Hosps & Clins, 72-75; pulmonary res fel, 78-79, from asst prof to assoc prof med, 80-87, PROF MED, UNIV MIAMI SCH MED, 88-, PROF PEDIAT, 90- *Concurrent Pos:* Prin investr, NIH, Nat Heart Lung & Blood Inst, 80-90 & Am Heart Asn, 91-92; Southern Soc Pediat Res young investr award, 84; site visitor, NIH, Nat Heart Lung & Blood Inst, 85, 87 & 89. *Honors & Awards:* Founder's Award Meritorious Res, Southern Soc Pediat Res, 87. *Mem:* Am Thoracic Soc; Am Lung Asn; Am Physiol Soc; Endotoxin Soc; Soc Pediat Res; Undersea & Hyperbaric Med Soc. *Res:* Mechanisms of protection from lung damage due to oxidants, especially hyperoxia; fetal, newborn and adult animal antioxidant defense systems; lung growth and development under normal and hyperoxic conditions. *Mailing Add:* Dept Med & Pediat Pulmonary Div R-120 Univ Miami Sch Med PO Box 016960 Miami FL 33101

FRANK, HENRY SORG, physical chemistry, thermodynamics; deceased, see previous edition for last biography

FRANK, HILMER AARON, b St Paul, Minn, Oct 26, 23; m 53; c 1. FOOD MICROBIOLOGY. *Educ:* Univ Minn, BA, 49; Wash State Univ, MS, 52, PhD(food technol), 54. *Prof Exp:* Asst bact, Wash State Univ, 50-51 & hort, 51-55, Nat Cancer Inst res fel, 55-57; bacteriologist indust test lab, US Naval Shipyard, Pa, 57 & eastern regional res lab, USDA, 57-60; from asst prof to assoc prof, 60-68, PROF FOOD SCI, UNIV HAWAII, 68- *Mem:* Am Soc

Microbiol; Soc Indust Microbiol; Brit Soc Appl Bact; Brit Soc Gen Microbiol; Sigma Xi. *Res:* Food spoilage; spores of anaerobic bacteria; food safety; histamine forming bacteria in fish. *Mailing Add:* Dept Food Sci Univ Hawaii Honolulu HI 96822

FRANK, HOWARD, b New York, NY, June 4, 41; m 65; c 3. ELECTRICAL ENGINEERING. *Educ:* Univ Miami, BS, 62; Northwestern Univ, MS, 64, PhD(elec eng), 65. *Prof Exp:* Asst prof elec eng & comput sci, Univ Calif, Berkeley, 65-69, assoc prof, 67-70; exec vpres, 69-70, PRES COMPUT COMMUN, NETWORK ANALYSIS CORP, 70- *Concurrent Pos:* Vis consult, Exec Off President of US, 68-69. *Honors & Awards:* Leonard G Abraham Award, Inst Elec & Electronics Engrs, 69. *Mem:* Fel Inst Elec & Electronics Engrs; AAAS; Opers Res Soc Am. *Res:* Man-machine communications; computer communications; network analysis, system reliability. *Mailing Add:* 1212 35th St NW Washington DC 20007-3207

FRANK, JAMES RICHARD, b Pittsburgh, Pa, Feb 17, 51; m 80; c 2. ENVIRONMENTAL BIOTECHNOLOGY, MULTIDISCIPLINARY TEAM MANAGEMENT. *Educ:* Cornell Univ, BS, 74; Univ Miami, MS, 78. *Prof Exp:* Scientist, oceanogr, Lamont-Doherty Geol Observ, Columbia Univ, 71-72; from proj mgr to prog mgr, 79-86, MGR, BIOL PROGS, GAS RES INST, 86- *Concurrent Pos:* Planning comt, Tenn Valley Authority, 84-; adj res scientist, Univ Fla, 85-; mem Industr Assocs Prog, Mont State Univ, 87- *Res:* Development, funding, and technical management of the biology-related programs of the US natural gas industry including programs in the production of methane wastewater treatment and chemical production and environmental remediation of industry related wastes. *Mailing Add:* Gas Res Inst Physical Sci Dept 8600 W Bryn Mawr Chicago IL 60631

FRANK, JEAN ANN, b New York, NY, June 17, 29. ORGANIC CHEMISTRY. *Educ:* Hunter Col, BA, 51; Colo State Univ, MA, 58; Utah State Univ, PhD(chem), 64. *Prof Exp:* Geologist, AEC, 51-57; instr chem, Wis State Col, 58-60; res asst, Brandeis Univ, 63-65; from asst prof to prof chem & chmn dept, Am Inst Col, 65-82. *Mem:* Am Chem Soc. *Res:* Chemical kinetics of oxidation-reduction type reactions. *Mailing Add:* 26 Yenom Rd South Dennis MA 02660-3617

FRANK, JOAN PATRICIA, b New York, NY, May 7, 25; wid. RADIOCHEMISTRY. *Educ:* Univ Mich, BS, 45, MS, 46; Univ Southern Calif, PhD(phys chem), 50. *Prof Exp:* Chemist nuclear chem & radiochem, Brookhaven Nat Lab, 51-56; control chemist anal develop, Parke-Davis & Co, 68-72; staff res assoc radiochem, Univ Calif, Irvine, 73-80; CONSULT, 80- *Mem:* Am Phys Soc; Am Inst Physics. *Res:* Experimental and theoretical aspects of radioactive thermal fluorine atom addition to protonated and deuterated ethylenes with particular emphasis on the isotope effects involved. *Mailing Add:* 2514 Vista Del Oro Newport Beach CA 92660

FRANK, JOHN HOWARD, b Stockton-on-Tees, Eng, Apr 13, 42; m 68; c 3. INSECT ECOLOGY. *Educ:* Durham Univ, BSc, 63; Oxford Univ, DPhil(entom), 67. *Prof Exp:* Fel entom, Univ Alta, 66-68; entomologist, Sugar Mfr Asn Jamaica, Ltd, 68-72; entomologist, Fla Med Entom Lab, Fla Div Health, 72-79; assoc prof, 79-87, PROF ENTOM, UNIV FLA, 87- *Mem:* Royal Entom Soc London; Entom Soc Am; Brit Ecol Soc; Coleopterists Soc; Entom Soc Can; Int Org Biol Control. *Res:* Population ecology and biological control of insect pests; ecology of Staphylinidae, particularly of the Caribbean region. *Mailing Add:* Dept Entom Univ Fla Gainesville FL 32611-0740

FRANK, JOHN L, b Lander, Wyo, Aug 12, 35. VECHICLE ENGINEERING. *Educ:* Univ Wyo, BS, 56. *Prof Exp:* Res eng, E I du Pont de Nemours & Co, 56-61; proj eng, Barnes Eng Co, 61-65; head electronics eng, Deere Co, 67-80; dir New Prod Eng, Int Harvester, 80-86; VPRES ENG & RES, CLARK EQUIPMENT CO, 86- *Mem:* Soc Automotive Engrs; Inst Elec & Electronics Engrs. *Res:* Design and development; systems technology; new concepts in material handling. *Mailing Add:* Clark Equip Co 333 W Vine St Suite 500 Lexington KY 40507

FRANK, JOY SOPIS, m 70; c 2. PHYSIOLOGY. *Educ:* Univ Pa, PhD(physiol), 68. *Prof Exp:* PROF MED & PHYSIOL, SCH MED CTR, UNIV CALIF, LOS ANGELES, 88- *Concurrent Pos:* Mem, Cardiol Adv Comt, NIH, 84-87. *Mem:* Am Soc Cell Biol; Biophys Soc; Electron Microscopy Soc Am. *Res:* Ultrastructure of the myocardium; freeze-fracture morphology; excitation-contraction coupling. *Mailing Add:* Dept Med & Physiol Univ Calif Sch Med Los Angeles CA 90024

FRANK, KOZI, applications of probabilistic methods; deceased, see previous edition for last biography

FRANK, LAWRENCE, b Brooklyn, NY, Sept 12, 15; m 40; c 2. MEDICINE. *Educ:* Univ NC, AB, 36; Long Island Col Med, MD, 40; Am Bd Dermat, dipl, 46. *Prof Exp:* Assoc prof, 53-55, PROF DERMAT & DIR DEPT, STATE UNIV NY DOWNSTATE MED CTR, 55-, DIR & HEAD, UNIV HOSP, 68- *Concurrent Pos:* Attend physician, Long Island Col Hosp, 55-, dir dept dermat; attend physician, Kings County Hosp, 57-, dir & head div dermat, 59-; consult, Vet Admin Hosp, Brooklyn, NY, 59-; consult coun drugs, AMA, 59-; consult, Brooklyn Eye & Ear Hosp & Caledonian Hosp. *Mem:* Fel Soc Invest Dermat; fel AMA; fel Am Col Physicians; fel Am Acad Dermat; fel NY Acad Med. *Res:* Dermatology. *Mailing Add:* 125 Argyle Rd Brooklyn NY 11218

FRANK, LEONARD HAROLD, b New York, NY, Jan 24, 30; m 52; c 3. BIOCHEMISTRY. *Educ:* Univ Okla, AB, 50; Johns Hopkins Univ, PhD(biochem), 57. *Prof Exp:* Sr instr biochem, Sch Med, Western Reserve Univ, 57-61; from asst prof to assoc prof, Sch Hyg, Johns Hopkins Univ, 61-69, PROF BIOCHEM, SCH MED, UNIV MD, BALTIMORE, 70- *Mem:* Am Soc Biol Chem. *Res:* Amino acid metabolism and metabolic control; microbial physiology. *Mailing Add:* Dept Biol Chem Univ Md Sch Med 660 W Redwood St Baltimore MD 21201

FRANK, LOUIS ALBERT, b Chicago, Ill, Aug 30, 38; m 60; c 2. PHYSICS, ASTRONOMY. *Educ:* Univ Iowa, BA, 60, MS, 61, PhD(space sci), 64. *Prof Exp:* From asst prof to assoc prof, 64-71, PROF PHYSICS & ASTRON, UNIV IOWA, 71- *Mem:* Fel Am Geophysics Union. *Res:* Measurements of planetary magnetic fields, particles and associated atmospheric phenomena; interplanetary medium; solar phenomena. *Mailing Add:* Dept Physics & Astron Univ Iowa Iowa City IA 52242

FRANK, MARTIN, b Chicago, Ill, Oct 22, 47; m 70. CELL PHYSIOLOGY, CARDIOVASCULAR PHYSIOLOGY. *Educ:* Univ Ill, Urbana, AB, 69, MS, 71, PhD(physiol), 73. *Prof Exp:* Res assoc cell physiol, Mich Cancer Found, 73-74; res assoc pharmacol, Mich State Univ, 74-75; asst prof, George Washington Univ, 75-78, asst prof lectr physiol, 78-; health scientist adminr, physiol study sect, NIH, 78-; EXEC SECY-TREAS, AM PHYSIOL SOC. *Concurrent Pos:* Mem res comt, Am Heart Asn, Nation's Capital Affil, 78- *Mem:* Biophys Soc; Am Heart Asn; Am Physiol Soc; AAAS; Sigma Xi. *Res:* Membrane physiology, particularly with respect to small solute and ion transport mechanisms across cell membranes; ion movements associated with myocardial excitation-contraction coupling; nucleocytoplasmic solute exchange and interactions. *Mailing Add:* Am Physiol Soc 9650 Rockville Pike Bethesda MD 20814

FRANK, MARTIN J, b Detroit, Mich, June 4, 28; m 50; c 4. INTERNAL MEDICINE, CARDIOLOGY. *Educ:* Univ Mich, BS & MD, 53; Am Bd Internal Med, dipl, 62. *Prof Exp:* Intern, Wayne County Gen Hosp, Mich, 53-54, from resident to chief resident med, 56-59; NIH fel cardiol, Wayne State Univ, 59-60; NIH fel, NJ Col Med & Dent, 60-61, from instr to asst prof med, 61-67; assoc prof, 67-69, dir hemodynamic labs, 67-78, PROF MED, MED COL GA, 69-, CHIEF SECT CARDIOL, 78- *Concurrent Pos:* Staff mem, Eugene Talmadge Mem Hosp, Ga, 67; consult, Vet Admin Hosp, 67, Cent State Hosp, 68 & Univ Hosp, 69. *Mem:* Am Heart Asn; Am Fedn Clin Res; fel Am Col Cardiol; Am Col Physicians; Fedn Am Socs Exp Biol. *Res:* Coronary blood flow and myocardial metabolism; left ventricular function in congenital and acquired valvular heart disease; methods for study of circulation by indicator dilution. *Mailing Add:* Dept Med Med Col Ga Hosp & Clins 1120 15th St Augusta GA 30912-3105

FRANK, MAURICE JEROME, b Omaha, Nebr, Dec 20, 42; m 70; c 2. FUNCTIONAL EQUATIONS, PROBABILISTIC GEOMETRY. *Educ:* Univ Chicago, SB, 65; Ill Inst Technol, MS, 69, PhD(math), 72. *Prof Exp:* Asst prof math, Univ Mass, 72-73; lectr math, Univ Wis-Milwaukee, 73-75; asst prof, 76-83, ASSOC PROF MATH, ILL INST TECHNOL, 83-, CHMN, 86- *Mem:* Am Math Soc; Math Asn Am. *Res:* Functional equations, especially the associactivity and related equations, iterative equations and systems; probability, distribution theory, probabilistic geometry. *Mailing Add:* Dept Math Ill Inst Technol Chicago IL 60616

FRANK, MAX, b Detroit, Mich, Feb 25, 27; m 56; c 4. SOLID STATE ELECTRONICS, NUCLEAR RADIATION EFFECTS. *Educ:* Wayne State Univ, BS, 49, MS, 61. *Prof Exp:* Instr elec eng, Wayne State Univ, 49-50; sr staff engr res solid state electronics, Wayne Eng Res Inst, 51-57; electronics staff engr, Defense Eng Div, Chrysler Corp, 57-59; prin engr solid state radiation effects, Res Labs Div, Bendix Corp, 59-79, mgr, Technol Forecasting Exec Off, 79-85; SR ENG SPECIALIST, BUS & TECH PLANNING, LAND SYSTS DIV, GEN DYNAMICS, 85- *Mem:* Sr mem Inst Elec & Electronics Engrs; Sigma Xi. *Res:* Prediction of radiation damage in transistors, diodes and microelectronics; radiation effects on magnetic and insulating materials and electronic components; radiation hardening of electronic systems for nuclear and space radiation environments. *Mailing Add:* 32445 Olde Franklin Dr Farmington Hills MI 48334

FRANK, MICHAEL M, b Brooklyn, NY, Feb 28, 37; m 61; c 3. IMMUNOLOGY. *Educ:* Univ Wis, AB, 56; Harvard Univ, MD, 60; Am Bd Clin Immunol & Allergy, dipl. *Prof Exp:* CHIEF, LAB CLIN INVEST & CLIN DIR, NAT INST ALLERGY & INFECTIOUS DIS, NIH, 76- *Concurrent Pos:* Mem med bd, NIH, 75-, chmn, 78-79. *Mem:* Am Soc Clin Invest (secy-treas, 81-83); Asn Am Physicians; Soc Pediat Res; Asn Am Immunologists; Am Fedn Clin Res. *Res:* Mechanisms of immune damage. *Mailing Add:* NIAID NIH Bldg 10 Rm 11N228 9000 Rockville Pike Bethesda MD 20892

FRANK, MORTON HOWARD, animal physiology, for more information see previous edition

FRANK, NEIL LAVERNE, b Sept 11, 31; US citizen; m 52; c 3. METEOROLOGY. *Educ:* Southwestern Col, BA, 53; Fla State Univ, MS, 59, PhD(meteorol), 70. *Prof Exp:* Forecaster, US Air Force, 53-57; forecaster, Nat Hurricane Ctr, 61-67, hurricane specialist, 68-72, dep dir, 72-74, dir, 74-87; FORECASTER, KHOU-TV CHANNEL 11, 87- *Mem:* Am Meteorol Soc; Sigma Xi; Am Sci Affil. *Res:* Tropical meteorology and hurricanes. *Mailing Add:* KHOU-TV 1945 Allen Pkwy Houston TX 77019

FRANK, OSCAR, b Trieste, Italy, Mar 6, 32; US citizen; m 59; c 2. BIOCHEMISTRY, NUTRITION. *Educ:* Brooklyn Col, BS, 55, MA, 57; NY Univ, PhD(biochem), 61. *Prof Exp:* Assoc chem, Mt Sinai Hosp, New York, 57-60; instr med, Seton Hall Col Med & Dent, 60-64; asst dir vitamin metab, Dept Med Res, Roosevelt Hosp, New York, 64-66; assoc prof, 66-79, PROF MED, UNIV MED & DENT NJ, NEWARK, 79- *Concurrent Pos:* Assoc, Haskins Labs, New York, 56-69. *Mem:* AAAS; Soc Protozool; Am Inst Biol Sci; Am Soc Clin Nutrit. *Res:* Development and application to clinical medicine of techniques; demonstrating metabolic disorders, drug effects, nutrition, medicine, dietetics and vitamin metabolism. *Mailing Add:* 77 Sussex Rd Tenafly NJ 07670

FRANK, PETER WOLFGANG, b Mainz, Germany, Sept 24, 23; nat US; m 46; c 3. ECOLOGY. *Educ:* Earlham Col, AB, 44; Univ Chicago, PhD, 51. *Prof Exp:* Seessel fel, Yale Univ, 51-52; asst prof zool, Univ Mo, 52-57; assoc prof, 57-64, PROF BIOL, UNIV ORE, 64- *Concurrent Pos:* Zool ed, Ecology, 64-70; prog dir gen community ecology, NSF, 76-77; assoc ed, Ann Ref Ecol Syst, 70- *Mem:* Ecol Soc Am; Sigma Xi. *Res:* Population, marine and experimental ecology. *Mailing Add:* Dept Biol Univ Ore Eugene OR 97403

FRANK, RICHARD STEPHEN, b Teaneck, NJ, Sept 7, 40; m 64; c 2. FORENSIC SCIENCE, SCIENCE ADMINISTRATION. *Educ:* Washington Col, BS, 62. *Prof Exp:* High sch teacher chem, physics & gen sci, Chestertown, Md, 62-63; chemist, US Food & Drug Admin, 63-68; forensic chemist, Bur Narcotics & Dangerous Drugs Admin, 68-70, sect chief forensic drug chem, 70-74; actg chief, 74-77, chief, Forensic Sci Sect, 77-91, ASSOC DEP ASST ADMINR, DRUG ENFORCEMENT ADMIN, 91- *Concurrent Pos:* Mem comt revise const, Asn Official Anal Chemists, 74-84; mem, Crime Lab Info Syst Oper Comt & Criminalistics Cert Study Group, Nat Adv Task Force to Chapter on Forensic Sci for Local Govt Police Mgt, 88- *Mem:* Fel Am Acad Forensic Sci (vpres, 85-86, secy, 86-87, pres- elect, 87-88, pres, 88-89); Am Soc Crime Lab Dirs (secy, 80-81); Asn Official Anal Chemists. *Res:* Planning laboratories; standardization of forensic drug analyses; proficiency testing; representative sampling of evidence. *Mailing Add:* Drug Enforcement Admin Forensic Sci Sect Washington DC 20537

FRANK, ROBERT CARL, b Adams, Wis, Aug 12, 27; m 51; c 3. SOLID STATE SCIENCE. *Educ:* St Olaf Col, BA, 50; Wayne State Univ, MA, 52, PhD(physics), 59. *Prof Exp:* Sr res physicist, Res Labs, Gen Motors Corp, 54-64; dir res, Augustana Res Found, 73-85, PROF PHYSICS, AUGUSTANA COL, ILL, 64- *Concurrent Pos:* Vis scientist, Argonne Nat Lab, 82-83. *Mem:* Am Asn Physics Teachers; Am Phys Soc; Minerals Metals & Mat Soc; Sigma Xi. *Res:* Diffusion and trapping of small atom impurities in solids; secondary ion mass spectroscopy experiments and computer simulation of diffusion and trapping. *Mailing Add:* Dept Physics Augustana Col Rock Island IL 61201

FRANK, ROBERT LOEFFLER, b Milwaukee, Wis, Mar 15, 14; m 43; c 2. NUTRITION. *Educ:* Dartmouth Col, AB, 36; Univ Wis, MA, 38, PhD(org chem), 40. *Prof Exp:* Asst chem, Univ Wis, 39-40; du Pont asst, Univ Ill, 40-41, instr org chem, 41-42, assoc, 43-45, asst prof, 45-50; dir res, Edwal Labs, Inc, 50-53; dir res, Ringwood Chem Corp, 53-54, vpres, 54-57; vpres, Morton Int, Inc, 55-71 & Morton-Norwich Prod, Inc, 71-76. *Concurrent Pos:* Consult, 76- *Mem:* Am Chem Soc. *Res:* Synthetic organic chemistry; polymer and synthetic rubber chemistry; salt technology; sodium and potassium metabolism. *Mailing Add:* 68 Stony Ridge Asheville NC 28804-1854

FRANK, ROBERT MCKINLEY, JR, b Newark, NJ, July 29, 32; m 64; c 2. SILVICULTURE, FOREST MANAGEMENT. *Educ:* Pa State Univ, BS, 54, MF, 56. *Prof Exp:* Res forester surv, 57-61, res forester mine spoil reclamation, 61-63, RES FORESTER SPRUCE-FIR SILVICULT, FOREST SERV, USDA, 63- *Concurrent Pos:* Fac assoc, Univ Maine, 72- *Mem:* Soc Am Foresters. *Res:* Silviculture and forest management research in northeastern spruce-balsam fir forest types. *Mailing Add:* 40 Marion Dr Hampden ME 04444

FRANK, ROBERT MORRIS, b New York, NY, Feb 2, 20; c 2. THEORETICAL PHYSICS, DATA PROCESSING. *Educ:* Cornell Univ, PhD(theoret physics), 51. *Prof Exp:* Asst prof theoret physics, Fla State Univ, 50-53; mem staff, Los Alamos Sci Lab, Univ Calif, 53-64; assoc, E H Plesset Assocs, Inc, Calif, 62-68; MEM STAFF, LOS ALAMOS SCI LAB, 68- *Mem:* Am Phys Soc; Asn Comput Mach. *Res:* Logical design of computing machines. *Mailing Add:* 155 Monte Rey Dr S White Rock NM 87544

FRANK, ROBERT NEIL, b Pittsburgh, Pa, May 14, 39; c 4. OPHTHALMOLOGY, VISION RESEARCH. *Educ:* Harvard Col, AB, 61; Yale Univ Sch Med, MD, 66. *Prof Exp:* Intern, Grady Mem Hosp, 66-67 & Emory Univ Sch Med, 66-67; res fel, 67-69, resident asst ophthal, Johns Hopkins Hosp 69-72; sr investr, Nat Eye Inst, Nat Inst Health, 72-76; assoc prof, 76-80, PROF OPHTHAL, WAYNE STATE SCH MED, 80- *Concurrent Pos:* Mem, Visual Scis A-2 Study Sect, US NIH, 79-84 & Policy Adv Comt, Diabetes Control & Complications Trial, 82-, chmn, ed comt, Early Treat Diabetic Retinopathy Study, 81-; mem, Med Sci Rev Comt, Int Juv Diabetes Found, 80-83 & Med Sci Adv Comt, 83-86; chmn, Res Rev Comt, Am Diabetes Asn, 83-86. *Honors & Awards:* Fight for Sight Award, Asn Res Vision & Ophthal, 77. *Mem:* Asn Res Vision & Ophthal; Am Acad Ophthal; AMA; Am Diabetes Asn; AAAS; Int Soc Eye Res. *Res:* Clinical interest in diseases of the retina, in particular diabetic retinopathy; the physiology and pathology of the retinal blood vessels and the retinal pigment epithelium. *Mailing Add:* Kresge Eye Inst Wayne State Sch Med 3994 John St Detroit MI 48201

FRANK, SAMUEL B, dermatology; deceased, see previous edition for last biography

FRANK, SIDNEY RAYMOND, b Minneapolis, Minn, Mar 16, 19; m 50; c 1. METEOROLOGY. *Educ:* Univ Minn, BA, 40; Univ Calif, Los Angeles, MA, 41. *Prof Exp:* Lab asst, Univ Calif, Los Angeles, 41; forecaster, Trans World Airline, Calif, 41-45; res & instr meteorol, Trans World Airline, Kansas City, Mo, 45-52; proj dir, Aerophys Res Found, Calif, 52-56; vpres, Aerometric Res, Inc, 56-68, pres, Aerometric Res Found, 60-68, PRES, SIDNEY R FRANK GROUP & SRF RES INST, 68- *Concurrent Pos:* Ed, J Aeronaut Meteorol, 45-47; lectr, Univ Kansas City, 49; vpres, NAm Weather Consult, 52-68; mem President's Adv Comt on Weather Control, 56-57; pres & dir, Aerometric Res Found, 56-68; assoc ed, J Appl Meteorol, Am Meteorol Soc, 62-67; mem, tech rev comt, USPHS, 63-65; lectr & res assoc, Univ Calif, Riverside, 60-65 & Univ Calif, Santa Barbara, 65-75; trustee, Santa Barbara Community Col Dist, 65-, pres bd, 71-73, 85-87; meteorol coordr Air Pollution Control Inst, Univ Southern Calif, 67-70. *Honors & Awards:* Air Transp Asn Awards, 43-49. *Mem:* AAAS; Air Pollution Control Asn; Am Meteorol Soc; Solar Energy Soc; Am Geophys Union. *Res:* Meso-scale, transport and diffusion; turbulence, synoptic meteorology and air pollution studies; meteorological aspects of air quality in environmental studies; forensic meteorology. *Mailing Add:* 229 El Monte Dr Santa Barbara CA 93109

FRANK, SIMON, b Orange, NJ, 1921; m 45; c 2. INDUSTRIAL CHEMISTRY. *Educ:* Drew Univ, AB, 43; Univ Mich, PhD(chem), 50. *Prof Exp:* From chemist to sr res chemist, 50-60, group leader, 60-67, mgr tech info serv, 67-72, proj leader oil field chem, Stamford Res Lab, 72-76, proj coordr enhanced oil recovery, 76-80, MGR, ENHANCED OIL RECOVERY DEPT & TECH SERV, INDUST PROD DIV, AM CYANAMID CO, 81- *Mem:* Am Chem Soc; Soc Petrol Engrs. *Mailing Add:* 34 Hazelwood Lane Stamford CT 06905-2725

FRANK, STANLEY, b New York, NY, Apr 21, 26; m 60. MATHEMATICS EDUCATION. *Educ:* City Col New York, BS, 50, MA, 53; Univ Fla, PhD(math), 60. *Prof Exp:* Teacher pub sch, NY, 50-53; teacher math & philos & logic, St John's Col, Fla, 60-61; asst prof math & dir math testing, Univ S Fla, 61-62; systs analyst, Mitre Corp, Mass, 62-64, sr statistician, Dynamics Res, 64-65; systs analyst, Gen Elec Co, 65-67; res dir, Booz-Allen Appl Res, 67-68; proj mgr, TRW Systs, 68-70; pres & res dir, Tanglewylde Res Inst, 70; dir, Math & Med Learning Lab, Santa Fe Jr Col, Gainesville, 71-85, at dept tech educ; RETIRED. *Concurrent Pos:* Dir, Div Human Resources, North Fla Conserv Coalition, 81- *Mem:* Am Statist Asn; Math Asn Am; Opers Res Soc Am; Int Blind Writers Educ Asn. *Res:* Medical science; navigation; information storage and retrieval; reliability; logic; autoregressive analysis; management information systems and design of military weapon systems; training of medical personnel; conservation; learning theory; disabilities in reading and quantitative skills. *Mailing Add:* 1240 River St Palatka FL 32177

FRANK, STEVEN NEIL, b Red Oak, Iowa, Feb 15, 47; m 75. ELECTROCHEMISTRY. *Educ:* Colo State Univ, BS, 69; Calif Inst Technol, PhD(electrochem), 74. *Prof Exp:* Assoc electrochem, Calif Inst Technol, 73-74; fel electrochem, Univ Tex, Austin, 74-77; mem tech staff electrochem, Tex Instruments, Inc, 77-80, sr mem tech staff, 80-83, mgr fuel cell develop, 80-83, mgr, Charge Coupled Device Process Technol, 83-86, MGR FOCAL PLANE ARRAY FABRICATION, TEX INSTRUMENTS, INC, 86- *Concurrent Pos:* Robert A Welch fel, Univ Tex, Austin, 74-77. *Mem:* Am Chem Soc; Electrochem Soc; Am Inst Chemists. *Res:* Batteries; fuel cells; electrode kinetics; adsorption at electrode surfaces; interfacial phenomena; silicon device fabrication; infrared focal plane fabrication. *Mailing Add:* Tex Instruments Inc PO Box 655012 MS 492 Dallas TX 75265

FRANK, SYLVAN GERALD, b San Antonio, Tex, Aug 30, 39; m 64; c 3. SURFACE CHEMISTRY, PHARMACEUTICS. *Educ:* Columbia Univ, BS, 62; Univ Mich, MS, 66, PhD(pharmaceut chem), 68. *Prof Exp:* Asst prof pharm, Duquesne Univ, 68-70; from asst prof to assoc prof, 70-83, PROF PHARM, OHIO STATE UNIV, 83- *Concurrent Pos:* Vis prof, Upjohn Co, 72, Swedish Inst Surface Chem, Stockholm, 78-79. *Honors & Awards:* Mead-Johnson Labs, Award 69. *Mem:* Am Pharmaceut Asn; Am Asn Pharmaceut Scientists; Int Asn Colloid & Surface Scientists; Sigma Xi; Royal Soc Chem; Soc Rheology; Am Chem Soc; Controlled Release Soc; Fine Particle Soc. *Res:* Micellar solubilization; emulsions; liquid crystals; application of surface chemistry to biological and pharmaceutical systems; inclusion compounds; colloidal particles. *Mailing Add:* Ohio State Univ Col Pharm 500 W 12th Ave Columbus OH 43210-1291

FRANK, THOMAS PAUL, b Appleton, Wis, July 8, 56; m 78; c 2. SILICON SENSORS, PRESSURE MONITORING. *Educ:* Fox Col, AAS, 77; Milwaukee Sch Eng, BSBE, 80; Ohio State Univ, MBA, 90. *Prof Exp:* Mgr, Ametek Inc, 82-83, dir, 83-85; dir, 85-86, VPRES, MEDEX, INC, 86- *Mem:* Inst Elec & Electronics Engrs; Asn Advan Med Instrumentation. *Res:* Development of silicon based sensors and transducers for biomedical monitoring and industrial process control; support technology in electrostatic bonding; thick film hybrid microelectronic circuit design. *Mailing Add:* 6232 Arapahoe Pl Dublin OH 43017

FRANK, THOMAS STOLLEY, b Milwaukee, Wis, June 26, 31; m 52; c 2. COMPUTER SCIENCE. *Educ:* Lawrence Col, BA, 55; Syracuse Univ, MA, 57, PhD(math), 62. *Prof Exp:* Instr math, Syracuse Univ, 60-62; from asst prof to prof math, 62-77, PROF COMPUT SCI, LE MOYNE COL, NY, 77- *Mem:* Math Asn Am; Sigma Xi; Asn Comput Mach; Inst Elec & Electronic Engrs Comput Soc. *Res:* Hardware systems. *Mailing Add:* Dept Comput Sci Le Moyne Col Syracuse NY 13214

FRANK, VICTOR SAMUEL, b Hartford, Conn, June 18, 19; m 44; c 3. INDUSTRIAL ORGANIC CHEMISTRY, RESEARCH ADMINISTRATION. *Educ:* Mass Inst Technol, BS, 42, PhD(org chem), 48. *Prof Exp:* Chemist, Dewey & Almy Chem Co, 42-46; asst, Mass Inst Technol, 47-48; sr chemist, Merck & Co, Inc, 49-51; group leader org & polymer res, Dewey & Almy Chem Div, 51-54, res dir org chem, 54-62, dir polymer res, Res Div, 62-67, org & polymer res, 67-68, vpes res, res div, W R Grace & Co, 68-83; CONSULT. *Concurrent Pos:* Instr, Northeastern Univ, 52-62. *Mem:* Am Chem Soc; fel Am Inst Chemists. *Mailing Add:* 1104 Lagrande Rd Silver Spring MD 20903-1324

FRANK, WILLIAM BENSON, b Youngstown, Ohio, July 12, 28. INDUSTRIAL CHEMISTRY. *Educ:* Thiel Col, BS, 50. *Prof Exp:* Res engr, Phys Chem Div, 53-67, mgr tech info dept, 67-77, sci assoc, Alcoa Labs, Aluminum Co Am, 77-86; SR ED TECH COMMUN, 86- *Mem:* Am Chem Soc; Electrochem Soc; Am Inst Mining, Metall & Petrol Eng. *Res:* Extractive metallurgy of aluminum. *Mailing Add:* Alcoa Labs Alcoa Center PA 15069

FRANK, WILLIAM CHARLES, b Grand Rapids, Mich, Oct 10, 40; m 64; c 2. ORGANIC CHEMISTRY, PHOTOGRAPHIC CHEMISTRY. *Educ:* Valparaiso Univ, BS, 62; Univ Colo, Boulder, PhD(org chem), 65. *Prof Exp:* Sr chemist, Cent Res Lab, 3M Co, 65-66, sr emulsion chemist, Minn 3M Res Ltd, Eng, 66-67, 3M Italia-Ferrania SpA, Italy, 67-68, res specialist photog chem, Photo Prod Div Lab, 68-71, res & develop supvr, 71-73, prod develop mgr, 73-86, SR RES SPECIALIST, MED IMAGING SYST LAB, 3M CO, 86- *Mem:* Am Chem Soc; Soc Photog Scientists & Engrs. *Res:* Photographic emulsion chemistry; photographic product development; radiographic imaging; phosphors. *Mailing Add:* 3M Co Med Imaging Syst Lab 3M Ctr 235-2B-23 St Paul MN 55144

FRANK, WILSON JAMES, b Kansas City, Mo, June 4, 23. PHYSICS. *Educ:* Univ Ill, BS, 48; Univ Calif, PhD(physics), 53. *Prof Exp:* PHYSICIST, LAWRENCE LIVERMORE LAB, UNIV CALIF, 51- *Mem:* Am Phys Soc; Sigma Xi. *Res:* Nuclear physics. *Mailing Add:* Box 625 Livermore CA 94550

FRANKART, WILLIAM A, b Kansas City, Kans, Feb 25, 43. IMMUNOLOGY. *Educ:* Univ Calif, Davis, PhD(biochem), 73. *Prof Exp:* Fel, Fundacion Bariloche/Argentina, 74-75; SR RES BIOCHEMIST, HOLLISTER-STIER LABS, DIV MILES LABS, 77- *Res:* Allergy products research; parental, topical and diagnostic new products; product development; clinical studies. *Mailing Add:* Hollister-Stier Labs PO Box 3145 Ter Annex Spokane WA 99220

FRANKE, CHARLES H, b Jersey City, NJ, Dec 28, 33; m 59; c 3. MATHEMATICS. *Educ:* Rutgers Univ, AB, 55, PhD(math), 61, MS, 84; Yale Univ, MA, 56. *Prof Exp:* Instr math, Rutgers Univ, 58-62; mem tech staff, Bell Tel Labs, 62-66; prof math, Seton Hall Univ, 66-85, chmn dept, 71-83, coordr prog comput sci, 83-85; PROF & CHAIR DEPT MATH, STATIST & COMPUT SCI, EASTERN KY UNIV, 85- *Mem:* Math Asn Am; Am Math Soc; Sigma Xi; Asn Comput Mach. *Res:* Galois theory of difference fields obtained by adjoining a fundamental system for a linear homogeneous difference equation to a ground field. *Mailing Add:* Math Statist & Comput Sci Eastern Ky Univ Richmond KY 40475

FRANKE, DONALD EDWARD, b Center, Tex, May 6, 37; m 63; c 3. ANIMAL BREEDING, ANIMAL GENETICS. *Educ:* Stephen F Austin State Univ, BS, 61; La State Univ, MS, 65; Tex A&M Univ, PhD(animal breeding), 69. *Prof Exp:* Teacher voc agr, Little Cypress Sch Dist, Ind, 61-63; asst animal breeding, La State Univ & Tex A&M Univ, 63-69; from asst prof to assoc prof animal sci, Univ Fla, 68-76; PROF ANIMAL SCI, LA STATE UNIV, 76- *Mem:* Am Soc Animal Sci; Am Genetic Asn; Biomet Soc; Sigma Xi. *Res:* Breeding systems for beef production; estimation of genetic parameters associated with growth, reproduction, lactation and efficiencies in beef production; systems analyses for estimating efficiency of red meat production. *Mailing Add:* Animal Sci Dept La State Univ Baton Rouge LA 70803

FRANKE, ERNEST A, b Uvalde, Tex, Oct 22, 39; m 64; c 3. ROBOTICS, MACHINE VISION. *Educ:* Tex A&I Univ, BS, 61; MS, 63; Case Western Reserve Univ, PhD(eng), 67. *Prof Exp:* From instr to assoc prof, Tex A&I Univ, 62-70, chmn dept, 69-71, prof eng, 70-79, dean, sch eng, 71-79; vpres res & develop, Alpha Electronics Co, 79-83; sr res engr, 83-85, staff engr, 85-88, MGR MACH PERCEPTION, SOUTHWEST RES INST, 88- *Concurrent Pos:* Consult. *Mem:* Inst Elec & Electronics Engrs; Soc Mfg Engrs; Soc Photo Optic Instrumentation Engrs. *Res:* Development of sensors for machine perception and computer methods for adaptive robotic operations based on sensory input. *Mailing Add:* Southwest Res Inst 6220 Culebra Rd PO Drawer 28510 San Antonio TX 78284

FRANKE, ERNST KARL, biophysics; deceased, see previous edition for last biography

FRANKE, FREDERICK RAHDE, b Pittsburgh, Pa, Oct 14, 18; m 43; c 5. MEDICINE. *Educ:* Univ Pittsburgh, BS, 41, MD, 43; Univ Pa, MSc, 50, DSc(med), 52; Am Bd Internal Med, dipl, 50; Am Bd Cardiovasc Dis, dipl, 62. *Prof Exp:* Instr physiol & pharmacol, 48-49, res assoc appl physiol, 49-53, physician in chg therapeut sect, 53-55, ASST PROF MED, MED SCH, UNIV PITTSBURGH, 55-, CLIN PROF PHARMACOL, SCH PHARM, 67- *Concurrent Pos:* Spec fel cardiol & genetics, Johns Hopkins Hosp, 59-60; chief div med, Western Pa Hosp, Pittsburgh, 63-67, chmn dept cardiovasc dis, 66-, med dir, 67-73. *Mem:* Fel Am Col Physicians; Soc Exp Biol & Med; Am Therapeut Soc; Am Heart Asn. *Res:* Cardiovascular diseases; physiologic and pharmacologic research in clinical diseases. *Mailing Add:* Western Pa Hosp 4800 Friendship Ave Pittsburgh PA 15224

FRANKE, GENE LOUIS, b Baltimore, MD, Feb 13, 51; m 79; c 2. WELDING ENGINEERING, WELD METALLURGY. *Educ:* Worcester Polytech Inst, BS, 73; Univ Ill, MS, 75. *Prof Exp:* METALLURGIST, DAVID TAYLOR RES CTR, USN, 75- *Mem:* Am Welding Soc; Am Soc Metals Int. *Res:* Development of welding consumables for fabrication of the United States Navy's low alloy steels. *Mailing Add:* David Taylor Res Ctr Code 2815 Annapolis MD 21402

FRANKE, JOHN ERWIN, b Belgrade, Minn, Mar 1, 46; m 68. GLOBAL ANALYSIS. *Educ:* Luther Col, Iowa, BA, 68; Northwestern Univ, MS, 71, PhD(math), 73. *Prof Exp:* PROF MATH, NC STATE UNIV, 73- *Concurrent Pos:* Consult, Res Triangle Inst. *Mem:* Am Math Soc; Math Asn Am. *Res:* Global analysis; stability and singularity theory; biomathematics. *Mailing Add:* Dept Math NC State Univ Raleigh NC 27607

FRANKE, MILTON EUGENE, b Springfield, Ill, Apr 7, 31; m 55; c 2. FLUID MECHANICS, PROPULSION. *Educ:* Univ Fla, BME, 52; Univ Minn, MSME, 54; Ohio State Univ, PhD(mech eng), 67. *Prof Exp:* Engr, Westinghouse Elec Corp, 52; proj engr, US Air Force Wright Air Develop Ctr, 54-56, sr proj engr, 56-57; res engr, E I du Pont de Nemours & Co, 57-59; from asst prof to assoc prof, 59-70, PROF, US AIR FORCE INST TECHNOL, 70- *Concurrent Pos:* Chmn, Dayton Sect, Am Soc Mech Engrs, 68-69, mem publ comt, 84-89, mem-at-large bd commun, 86-90, chmn, Dynamic Systs & Control Div, 87-88, chmn, publ comt, 89-90, vpres commun, 90-; vis prof, Technische Hochschule Aachen, Ger, 71; fac coun pres, Air Force Inst Technol, 74-75; assoc ed, J Dynamic Systs, Measurement & Control, Am Soc Mech Engrs, 79-84; mem Liquid Propulsion Tech Comt, Am Inst Aeronaut & Astronaut, 82-85, Air Breathing Propulsion Tech Comt, 86-89. *Honors & Awards:* Charles A Stone Award, Air Force Inst Technol, 86. *Mem:* Am Soc Mech Engrs; Am Soc Eng Educ; assoc fel Am Inst Aeronaut & Astronaut; Instrument Soc Am; Electrostatics Soc Am. *Res:* Fluid transmission lines; heat transfer; fluid mechanics; fluidics; gas dynamics; propulsion; electrostatic effects on heat transfer; fuel systems; aerodynamics; dynamic systems; fluid control; hydraulics; three United States patents. *Mailing Add:* Dept Aeronaut & Astronaut Air Force Inst Technol Wright-Patterson AFB OH 45433-6583

FRANKE, ROBERT G, b Muskegon, Mich, June 27, 33; m 57; c 3. BOTANY. *Educ:* Northern Ill Univ, BS, 56; Northwestern Univ, MS, 61; Univ Tex, PhD(bot), 65. *Prof Exp:* From instr to asst prof natural sci, Mich State Univ, 64-68; assoc prof, 68-73, prof bot & plant path, Iowa State Univ, 73-77; PROF BIOL & HEAD DEPT, UNIV TENN, CHATTANOOGA, 77- *Mem:* Am Inst Biol Sci; Mycol Soc Am; Brit Mycol Soc. *Res:* Taxonomy and natural relationship of Myxomycetes, as disclosed with serological, electrophoretic, and related techniques; ecology of fungi. *Mailing Add:* Col Scis Univ Ark Little Rock AR 72204

FRANKEL, ARTHUR IRVING, b Baltimore, Md, Oct 25, 18; m 42; c 2. TESTICULAR REGULATION, MALE SEX BEHAVIOR. *Educ:* Oberlin Col, Ohio, BA, 40; Univ Mo, MA, 61; Univ Ill, PhD(zool), 66. *Prof Exp:* NIH fel endocrinol, Univ Ill, 65-66; NIH fel biochem, Sch Med, Univ Utah, 66-68; from asst prof to assoc prof, 68-81, PROF ENDOCRINOL, STATE UNIV NY, BINGHAMTON, 81- *Concurrent Pos:* Consult, Biochem Endocrinol Study Sect, NIH, 77-81; NIH prin investr, Nat Inst Aging & Nat Inst Child Health & Human Develop, 69- *Mem:* Endocrine Soc; Soc Study Reproduction; Soc Study Fertil. *Res:* Study of the male animal, specializing in regulation of the testis and prostate; androgen receptors; male sex behavior. *Mailing Add:* Dept Biol State Univ NY Binghamton NY 13901

FRANKEL, EDWIN N, b Alexandria, Egypt, July 3, 28; US citizen; m 50; c 5. BIOCHEMISTRY, ORGANIC CHEMISTRY. *Educ:* Mich State Univ, BS, 50; Univ Calif, Davis, MS, 52, PhD(agr chem), 56. *Prof Exp:* Jr specialist dairy chem, Univ Calif, Davis, 53-56; chemist, Northern Regional Res Lab, USDA, Ill, 56-61; group leader food div, Procter & Gamble Co, Ohio, 61-62; RES LEADER, NORTHERN REGIONAL RES CTR, USDA, 62- *Concurrent Pos:* Res fel, Israel Inst Technol, 66-67; assoc ed, Lipids, Am Oil Chemists' Soc, 66-80; sr vis res fel, Queen Mary Col, Univ London, Eng, 75-76; vis chemist, Univ Calif, Davis, 88- *Honors & Awards:* Superior Serv Award, USDA, 78; A E Bailey Award, N Cent Sect, Am Oil Chemists Soc, 85. *Mem:* Am Oil Chemists' Soc; Am Chem Soc. *Res:* Lipid chemistry; toxidation and of lipids; vegetable oil processing technology. *Mailing Add:* 1669 Colusa Ave Davis CA 95616-3140

FRANKEL, FRED HAROLD, b Benoni, SAfrica, Mar 23, 24; US citizen; m 47; c 3. PSYCHIATRY. *Educ:* Univ Witwatersrand, MB, ChB, 48, dipl psychol med, 52; Am Bd Psychiat & Neurol, dipl, 66. *Prof Exp:* Intern surg & med, Johannesburg Gen Hosp, SAfrica, 48-49; resident psychiat, Tara Hosp Nerv Dis, Johannesburg, 50-52; resident psychiat, Mass Gen Hosp, Boston, 52-53; assoc psychiatrist, 68-69, PSYCHIATRIST, BETH ISRAEL HOSP, BOSTON, 69-, PSYCHIATRIST-IN-CHIEF, 86-; asst psychiat, Mass Gen Hosp, 78-87. *Concurrent Pos:* Consult, Social Serv Dept, Transvaal & Orange Free State Chamber of Mines, 54-60; from second asst neuropsychiatrist to asst, Med Sch, Univ Witwatersrand & Johannesburg Gen Hosp, 54-62; hon psychiatrist, Witwatersrand Jewish Aged Home, 59-62; consult, Mass Gen Hosp, 78-; asst to comnr, Dept Ment Health, Mass, 65-68; assoc prof, Harvard Med Sch, 76-81, prof, 81; med ed, Int J Clin & Exp Hypn, 78- *Mem:* Fel Am Psychiat Asn; fel Royal Soc Med; fel Soc Clin & Exp Hypnosis; fel Am Col Psychiat; Int Soc Hypn (pres, 80-82). *Res:* Psychosomatic medicine; hypnosis. *Mailing Add:* Beth Israel Hosp 330 Brookline Ave Boston MA 02215

FRANKEL, FRED ROBERT, b Baltimore, Md, July 6, 34; m 67; c 2. MOLECULAR BIOLOGY. *Educ:* Pa State Univ, BS, 55, MS, 57; Univ Fla, PhD, 60. *Prof Exp:* Res assoc, Genetics Res Unit, Carnegie Inst, 60-63; from asst prof to assoc prof, 63-76, PROF MICROBIOL, SCH MED, UNIV PA, 76- *Mem:* Am Soc Biol Chem; Am Soc Microbiol. *Res:* Action of steroid hormones in target cell nuclei; role of cytoskeleton in cell structure and movement; mechanism of phagocytosis. *Mailing Add:* Dept Microbiol Sch Med Univ Pa 249 Johnson Pavilion Philadelphia PA 19104

FRANKEL, HARRY MEYER, b Baltimore, Md, Oct 11, 27; m 51; c 2. PHYSIOLOGY. *Educ:* Univ Md, BS, 49; Univ Iowa, PhD(physiol), 58. *Prof Exp:* Physiologist, Med Labs, Chem Warfare Lab, Md, 51-55; asst radiation res lab, Univ Iowa, 55-56; physiologist, Chem Warfare Lab, Army Chem Ctr, 58-60; from asst prof to assoc prof, 60-70, chmn dept, 73-79, PROF PHYSIOL, RUTGERS UNIV, NEW BRUNSWICK, 70- *Mem:* AAAS; Am Physiol Soc; Harvey Soc; Sigma Xi. *Res:* Environmental and respiratory physiology; comparative pharmacology. *Mailing Add:* Dept Physiol Rutgers Univ New Brunswick NJ 08903

FRANKEL, HERBERT, b Harrison, NJ, Feb 12, 14; m 41; c 2. ELECTRICAL ENGINEERING. *Educ:* State Univ NY, BA, 40; Washington Univ, St Louis, BS, 47; Rensselaer Polytech Inst, MEE, 53. *Prof Exp:* Engr, DeLaval Separator Co, NY, 47-48; assoc prof elec eng, Rensselaer Polytech Inst, 48-80; SR ENGR, BENDIX FIELD ENG CORP, 80- *Concurrent Pos:* Consult, Watervliet Arsenal, NY, 56- *Mem:* Inst Elec & Electronics Engrs. *Res:* Circuit theory; nondestructive testing of metals. *Mailing Add:* Bendix Field Eng Corp 9250 Rte 108 Space Flight Ctr Code 543 Greenbelt MD 20902

FRANKEL, JACK WILLIAM, b New York, NY, Feb 15, 25; m 48; c 3. VIROLOGY. *Educ:* Brown Univ, BA, 48; Rutgers Univ, PhD, 51; Am Bd Med Microbiol, dipl. *Prof Exp:* Res fel virol, Pub Health Res Inst, Inc, NY, 51-53; res assoc, Sharp & Dohme Div, Merck & Co, 53-58, res assoc dept microbiol, Sch Med, Temple Univ, 58-59; chief virus diag lab, Dept Res Therapeut, Norristown State Hosp, Pa, 59-61; head virus res, Ciba-Geigy Pharmaceut Co, 61-68; dep dir & dir virus res lab, Life Sci Res Labs, 68-76; prof med microbiol, Col Med, Univ SFla, 76-82, Sci Res Labs, 68-72; PROF MED MICROBIOL, CREIGHTON UNIV SCH MED & UNIV SFLA, 76-, FLA STATE PUB HEALTH LABS, 82- *Concurrent Pos:* Prof, Hunter Col, 61-68; prof, Drew Univ, 63-68; clin prof med microbiol, Col Med, Univ SFla, 72-76; prof, St Petersburg Col, 74-75; assoc ed, Cancer Invest; ed-in-chief, Int J Pub Health. *Mem:* Am Soc Microbiol; Teratology Soc; AAAS; Am Asn Cancer Res; Soc Exp Biol Med. *Res:* Viral oncology. *Mailing Add:* Tampa Lab DHRS Box 2380 Tampa FL 33601

FRANKEL, JOSEPH, b Vienna, Austria, July 30, 35; US citizen; m 61; c 2. DEVELOPMENTAL GENETICS. *Educ:* Cornell Univ, BA, 56; Yale Univ, PhD(zool), 60. *Prof Exp:* NIH fel cell biol, Biol Inst Carlsberg Found, Denmark, 60-62; asst prof, 62-65, assoc prof, 65-71, PROF BIOL, UNIV IOWA, 71- *Concurrent Pos:* NIH res grants, 63-69, 74-80 & 81-94; NSF res grants, 72-74 & 80-81. *Mem:* Soc Protozool; Genetics Soc Am; Soc Develop Biol. *Res:* Development in ciliated protozoa; pattern formation. *Mailing Add:* Dept Biol Univ Iowa Iowa City IA 52242

FRANKEL, JULIUS, b Cernauti, Romania, Dec 10, 35; m 70; c 1. SOLID STATE PHYSICS, ULTRASONICS. *Educ:* Brooklyn Col, BS, 58; Rensselaer Polytech Inst, MS, 65. *Prof Exp:* Res asst physics, NY Univ, 58; grad teaching asst physics, Rensselaer Polytech Inst, 58-60; PHYSICIST, PHYSICS & MATH, WATERVLIET ARSENAL, 60- *Concurrent Pos:* Guest scientist, Max Planck Inst-Festkoerper Forschung, Stuttgart, 82. *Honors & Awards:* Siple Award. *Mem:* Am Phys Soc. *Res:* High pressure; equation of state; ultrasonics. *Mailing Add:* Bldg 15 Benet Lab SMCAR-CCB-RA Watervliet Arsenal Watervliet NY 12189-4050

FRANKEL, LARRY, b New York, NY, July 8, 28; m 50; c 2. INVERTEBRATE PALEONTOLOGY. *Educ:* Brooklyn Col, BSc, 50; Columbia Univ, AM, 52; Univ Nebr, PhD(geol), 56. *Prof Exp:* Asst, Univ Nebr, 52-54; geologist, US AEC, 54-55; from instr to prof geol, 59-77, PROF GEOL & GEOPHYS, UNIV CONN, 77- *Mem:* Geol Soc Am; Am Asn Petrol Geologists. *Res:* Pleistocene Mollusca and geology; foraminifera, estuarine studies. *Mailing Add:* Dept Geol Univ Conn Main Campus U-45 354 Mansfield Rd Storrs CT 06268

FRANKEL, LAWRENCE (STEPHEN), b New York, NY, July 26, 41; m 64; c 1. PHYSICAL CHEMISTRY, ANALYTICAL CHEMISTRY. *Educ:* Hofstra Univ, BA, 63; Univ Mass, PhD(phys chem), 67. *Prof Exp:* RES SECT MGR, ROHM & HAAS CO, 69- *Mem:* Am Chem Soc. *Res:* Transition metal chemistry; kinetics of fast reactions; thermal and spectroscopic characterization of polymers; ion exchange resins and adsorbents; analytical aspects of pollution problems; elastomers. *Mailing Add:* 1110 Delene Rd Jenkintown PA 19046-3019

FRANKEL, MARTIN RICHARD, b Wash, DC, June 16, 43; m 70; c 2. SURVEY SAMPLING DESIGN & THEORY. *Educ:* Univ NC, AB, 65; Univ Mich, MA, 67, PhD(math & sociol), 71. *Prof Exp:* From asst prof to assoc prof, 71-79, PROF STATIST, BERNARD BARUCH COL, CITY UNIV NY, 79- *Concurrent Pos:* Tech dir, Nat Opinion Res Ctr, 74-88, sr statist scientist, 89-91; mem panel & chmn, adv comt US census, Am Statist Asn, 75-81, chmn, Survey Res Sect, 76; mem, Panel Occup Safety & Health Statist, Nat Acad Sci, 85-89; chmn, Res Qual Coun, Advert Res Found, 88- & Standards Comt, Am Asn Pub Opinion Res, 89-90. *Mem:* Fel Am Statist Asn; Royal Statist Soc; Am Asn Pub Opinion Res; Asn Comput Mach. *Res:* Theoretical and practical aspects of the application of statistical sampling techniques to the study of populations. *Mailing Add:* 14 Patricia Lane Cos Cob CT 06807

FRANKEL, MICHAEL JAY, b New York, NY, Feb 20, 47; m 69; c 4. THEORETICAL SOLID STATE PHYSICS, SHOCK WAVE PHYSICS. *Educ:* Yeshiva Univ, BA, 68, MA, 70; NY Univ, PhD(physics), 75. *Prof Exp:* Assoc res scientist theoretical condensed matter physics, City Col, City Univ NY, 75; analyst satellite telemetry, Syst Sci Div, Comput Sci Corp, 75-77; res physicist explosive physics, Energetic Mat Div, Naval Surface Weapons Ctr, US Dept Navy, 77-81; physicist, shock physics div, US Defense Nuclear Agency, 81-85. *Concurrent Pos:* NSF int travel grant & NATO fel, 73, NSF traineeship, NY Univ, 71-74; assoc res scientist & fel, City Univ New York, 75; Legis Cong fel, 85. *Mem:* Am Phys Soc. *Res:* Condensed matter theory; shock wave physics; nuclear weapons effects. *Mailing Add:* 11006 Wheeler Dr Silver Spring MD 20901

FRANKEL, RICHARD BARRY, b St Paul, Minn, June 24, 39; m 60; c 2. PHYSICS, CHEMISTRY. *Educ:* Univ Mo, BS, 61; Univ Calif, Berkeley, PhD(chem), 65. *Prof Exp:* Res asst, Lawrence Radiation Lab, Univ Calif, 62-65; res staff mem, 65-79, SR RES SCIENTIST, NAT MAGNET LAB, MASS INST TECHNOL, 79- *Concurrent Pos:* NATO fel, Munich Tech, 67-68. *Mem:* Fel Am Phys Soc; AAAS; Bioelectromagnetic Soc; Explorers Club. *Res:* Mossbauer spectroscopy; magnetic properties of solids; electronic structure of metalloproteins; magnetism in organisms. *Mailing Add:* 167 Tremont St Cambridge MA 02139

FRANKEL, SHERMAN, b Nov 15, 22; div; c 1. HIGH ENERGY PHYSICS. *Educ:* Brooklyn Col, BA, 43; Univ Ill, MS, 47, PhD(physics), 49. *Prof Exp:* Mem staff, Radiation Lab, Mass Inst Technol, 43-46; from asst prof to assoc prof physics, 52-60, PROF PHYSICS, UNIV PA, 60- *Concurrent Pos:* Assoc ed, Review Sci Instruments, 52-53; Guggenheim fel, 56-57 & 78-79; vis scientist, Niels Bohr Inst, Denmark, 68, CERN, Geneva, Switzerland, 75, CEN de Saclay, France, 79; guest fel, Int Security Arms Control, Stanford Univ Ctr, 87; guest scholar, Brookings Inst, 87. *Mem:* Fel Am Phys Soc. *Res:* Elementary particle and nuclear physics. *Mailing Add:* Dept Physics Univ Pa Philadelphia PA 19104

FRANKEL, SIDNEY, b New York, NY, Oct 6, 10; m 38; c 1. ELECTRONICS. *Educ:* Rensselaer Polytech Inst, EE, 31, MS, 34, PhD(math), 36. *Prof Exp:* Instr math, Rensselaer Polytech Inst, 31-33; jr engr, Eclipse Aviation Corp, 37-38; engr radio transmitter & head dept, Fed Telecommunications Labs, 38-50; assoc head, Microwave Lab, Hughes Aircraft Co, 50-54; head radar & countermeasures dept, Litton Industs, Inc, 54-58; dir eng, Sierra Div, Philco Corp, 58-60; tech consul & mgr, Sidney Frankel & Assocs, 60-78; ASST PROF ELEC ENG, SAN JOSE STATE UNIV, 79- *Mem:* Fel Inst Elec & Electronics Engrs; Sigma Xi. *Res:* Theory of multiconductor transmission lines and cables. *Mailing Add:* 2140 Santa Cruz Ave Apt E203 Menlo Park CA 94025

FRANKEL, THEODORE THOMAS, b Philadelphia, Pa, June 17, 29; m 55; c 3. MATHEMATICS. *Educ:* Univ Calif, AB, 50, PhD, 55. *Prof Exp:* From instr to asst prof math, Stanford Univ, 55-62; assoc prof, Brown Univ, 62-65; PROF MATH, UNIV CALIF, SAN DIEGO, 65- *Concurrent Pos:* Mem, Inst Advan Study, 57-59. *Mem:* Am Math Soc. *Res:* Differential geometry; Morse theory; general relativity. *Mailing Add:* Dept Math Univ Calif at San Diego La Jolla CA 92093

FRANKEL, VICTOR H, b Wilmington, Del, May 14, 25; m 58; c 5. ORTHOPEDIC SURGERY, BIOENGINEERING. *Educ:* Swarthmore Col, BA, 46; Univ Pa, MD, 51; Univ Upsala, MD(orthop surg), 60. *Prof Exp:* Attend orthop surgeon, Hosp Joint Dis, New York, 60-66; assoc prof orthop surg, Case Western Reserve Univ, 66-71, prof orthop surg & bioeng, 69-76; prof & chmn dept orthop surg, Univ Wash, Seattle, 76-81; CHMN, DEPT ORTHOP, HOSP JOINT DIS, NY, 81-, PRES & CHIEF EXEC OFFICER, 87-; PROF, ORTHOP SURG, NEW YORK UNIV, SCH MED, NY, 85- *Concurrent Pos:* Fel, Nelson Nat Found, 58-60; mem comt prosthetics res & develop, Nat Acad Sci, 70- *Mem:* Am Orthop Asn; Asn Bone & Joint Surg; Am Col Surg; AMA; Am Asn Orthop Surgeons. *Res:* Application of engineering techniques to orthopedic surgery; development of biomechanics teaching. *Mailing Add:* Dept Orthop Hosp Joint Dis 301 E 17th St New York NY 10003

FRANKEN, PETER ALDEN, b New York, NY, Nov 10, 28; m 83; c 3. OPTICAL PHYSICS, QUANTUM OPTICS. *Educ:* Columbia Univ, BA, 48, MA, 50, PhD(physics), 52. *Prof Exp:* Lectr math, Columbia Univ, 48-49, asst physics, 50-52; instr, Stanford Univ, 52-56; from asst prof to prof, Univ Mich, Ann Arbor, 56-73; dir, Optical Sci Ctr, 73-83, PROF OPTICAL SCI & PHYSICS, UNIV ARIZ, 73- *Concurrent Pos:* Dep dir, Advan Res Projs Agency, US Dept Defense, Washington, DC, 67-68. *Honors & Awards:* Prize, Am Phys Soc, 67; Wood Prize, Optical Soc Am, 79; Richtmyer Mem Lectr Award, Am Phys Soc-Am Asn Physics Teachers, 88. *Mem:* Fel Am Phys Soc; fel Optical Soc Am (pres, 77); fel AAAS; fel Int Soc Optical Eng; Radiol Soc NAm. *Res:* Experimental atomic and electron physics and imaging technology. *Mailing Add:* Optical Sci Ctr Univ Ariz Tucson AZ 85721

FRANKENBERG, DIRK, b Woodsville, NH, Nov 25, 37; m 60. BIOLOGICAL OCEANOGRAPHY. *Educ:* Dartmouth Col, AB, 59; Emory Univ, MS, 60, PhD(biol), 62. *Prof Exp:* Asst prof zool, Marine Inst, Univ Ga, 62-66 & Univ Del, 66-67; from assoc prof to prof, Univ Ga, 67-74; PROF MARINE SCI & DIR MARINE SCI PROG, UNIV NC, CHAPEL HILL, 74-, CHMN DEPT, 74- *Concurrent Pos:* Dir, Biol Oceanog Prog, NSF, 70-71, dir, Div Ocean Sci, 78-80; mem ocean sci comt, Nat Acad Sci, 73. *Mem:* Ecol Soc Am; Am Inst Biol Sci; Am Soc Limnol & Oceanog. *Res:* Ecology of macro-benthos; oxygen phenomena in estuaries. *Mailing Add:* Dept Marine Sci Univ NC CB 3300 12-5 Venable Hall Chapel Hill NC 27599

FRANKENBERG, JULIAN MYRON, b Chicago, Ill, Jan 18, 38; m 62; c 1. PALEOBOTANY, PLANT MORPHOLOGY. *Educ:* Univ Ill, BS, 61, PhD(paleobot), 68; Univ Minn, MS, 63. *Prof Exp:* Lab asst, Paleobot Lab, Univ Ill, Urbana, 57-61; teaching asst, Univ Minn, 61-63; teaching asst bot & biol, 63-68, actg asst dean, Col Lib Arts & Sci, 68, asst dean, 68-70, DIR & ASST DEAN HEALTH PROF INFO OFF, UNIV ILL, URBANA-CHAMPAIGN, 70- *Concurrent Pos:* Exec dir, Nat Asn Adv for Health Professions (NAAHP, Inc), 79- *Mem:* AAAS; Bot Soc Am; Am Inst Biol Sci; Asn Am Med Cols; Nat Asn Health Professions (pres, 80-82). *Res:* Study of petrified Stigmaria from North America; coal-ball flora of the Pennsylvania period. *Mailing Add:* Health Prof Info Off Univ Ill 901 W Illinois Urbana IL 61801-3028

FRANKENBURG, PETER EDGAR, b Ludwigshafen on Rhine, Ger, Nov 10, 26; nat US; m 54; c 4. ORGANIC CHEMISTRY. *Educ:* Princeton Univ, AB, 49; Univ Rochester, PhD(org chem), 53. *Prof Exp:* SR RES ASSOC, E I DU PONT DE NEMOURS & CO, INC, 53- *Mem:* Sigma Xi. *Res:* Polymers for textile use; hot gas filtration. *Mailing Add:* 2405 Shellpot Dr Wilmington DE 19803

FRANKENFELD, JOHN WILLIAM, b Mesa, Ariz, Oct 29, 32; m 59; c 2. ORGANIC CHEMISTRY. *Educ:* Univ Chicago, BA, 52, SM, 57; Mass Inst Technol, PhD(chem), 62. *Prof Exp:* From res chemist to sr staff chemist, 61-83, RES ASSOC, EXXON RES & ENG CO, 83- *Mem:* Sigma Xi; Am Chem Soc. *Res:* Biological degradation of petroleum hydrocarbons; synthesis of peptides and peptide intermediates; synthetic foods and food additives; synthetic fuels; biological effects of oil pollution. *Mailing Add:* Exxon Res & Eng Co PO Box 51 Linden NJ 07036-0051

FRANKENTHAL, ROBERT PETER, b Berlin, Germany, Sept 11, 30; nat US; m 58; c 2. ELECTROCHEMISTRY, CORROSION. *Educ:* Univ Rochester, BS, 52; Univ Wis, PhD(chem), 56. *Prof Exp:* Res chemist appl res lab, US Steel Corp, 56-60, scientist, Edgar C Bain Lab Fundamental Res, 60-68, sr scientist, 68-72; mem tech staff, 72-83, DISTINGUISHED MEM TECH STAFF, AT&T BELL LABS, 83- *Honors & Awards:* H H Uhlig Award, Corrosion Div, Electrochem Soc, 89. *Mem:* Am Chem Soc; Electrochem Soc (treas, 86-90, vpres, 90-93); Nat Asn Corrosion Eng; Am Vacuum Soc. *Res:* Oxidation and tarnishing of metals and alloys; corrosion of electronic materials and devices; corrosion surface properties of metals and alloys; mechanism of electrode reactions. *Mailing Add:* AT&T Bell Labs Rm 1D-352 Murray Hill NJ 07974-2070

FRANKFATER, ALLEN, b New York, NY, June 23, 41; m 67; c 2. BIOCHEMISTRY. *Educ:* Brooklyn Col, BS, 63; Duke Univ, PhD(biochem), 68. *Prof Exp:* Res assoc biochem, Univ Chicago, 68-70; asst prof, 70-77, ASSOC PROF BIOCHEM, STRITCH SCH MED, LOYOLA UNIV CHICAGO, 77- *Concurrent Pos:* Prin investr, Arthritis Found, 76-77 & Nat Inst Child Health & Human Develop, 77-80. *Mem:* Am Chem Soc; AAAS; Sigma Xi; Am Soc Biol Chemists; Am Soc Microbiol; NY Acad Sci. *Res:* Enzymology; mechanisms of proteolytic enzymes; naturally occurring inhibitors of proteolytic enzymes; intracellular protein degradation and microinjection of cultured cells. *Mailing Add:* Dept Biochem Loyola Med Ctr Maywood IL 60153

FRANKFORTER, WELDON D, b Tobias, Nebr, May 1, 20. GEOLOGY, PALEONTOLOGY. *Educ:* Univ Nebr, BSc, 44, MSc, 49. *Prof Exp:* Dir, Sanford Mus & Planetarium, Cherokee, Iowa, 51-62; asst dir, Grand Rapids Pub Mus, 62-64, dir, 65-88; RETIRED. *Mem:* Fel Geol Soc Am; AAAS; Sigma Xi. *Res:* Author or co-author of more than 30 scientific articles in geology, archeology & paleontology & articles in museum journals. *Mailing Add:* 4856 Fuller Ave SE Kentwood MI 49508

FRANKIE, GORDON WILLIAM, b Albany, Calif, Mar 29, 40; m 69. INSECT ECOLOGY. *Educ:* Univ Calif, Berkeley, BS, 63, PhD(entom), 68. *Prof Exp:* NSF grant & res specialist, Orgn Trop Studies, Inc, Costa Rica, 68-70; from asst prof to assoc prof, Tex A&M Univ, 70-76; assoc prof, 76-81, PROF ENTOM, UNIV CALIF, BERKELEY, 81- *Mem:* AAAS; Soc Study Evolution; Ecol Soc Am; Entom Soc Am. *Res:* Insect-plant relations of insects on ornamental plants; pollination biology in the tropics; biological organization of tropical communities with emphasis on plant reproductive biology; community studies of insects inhabiting insect-induced galls on oak. *Mailing Add:* Dept Entom Univ Calif Berkeley CA 94720

FRANKL, DANIEL RICHARD, b New York, NY, Sept 6, 22; m 51; c 2. SURFACE PHYSICS. *Educ:* Cooper Union, BChE, 43; Columbia Univ, PhD(physics), 53. *Prof Exp:* Process develop engr, US Rubber Co, 43-50; res assoc, Columbia Univ, 51-53; eng specialist, Luminescence, Sylvania Elec Prod, Inc, 53-58, semiconductors, Gen Tel & Electronics Labs, Inc, 58-62; vis prof phys metall, Univ Ill, 62-63; prof physics, Pa State Univ, Univ Park, 63-88; RETIRED. *Concurrent Pos:* Vis sr res assoc, Univ Sussex, Eng, 69-70; vis res chemist, Univ Calif, San Diego, 78-79; vis prof physics, Univ Cambridge, Eng, 86. *Mem:* Fel Am Phys Soc; AAAS; Am Vacuum Soc. *Res:* Atomic beam surface scattering. *Mailing Add:* 236 Davey Lab University Park PA 16802

FRANKL, WILLIAM S, b Philadelphia, Pa, July 15, 28; m 51; c 2. CARDIOVASCULAR DISEASES. *Educ:* Temple Univ, BA, 51, MD, 55, MS, 61; Am Bd Internal Med, dipl, 62 & 74; Cardovasc Dis Bd, dipl, 75. *Prof Exp:* Intern med, Buffalo Gen Hosp, 55-56; resident, Temple Univ, 56-57 & 59-61; instr med, Univ Pa, 61-62; instr, Temple Univ, 62-64, assoc, 64-65, from asst prof to assoc prof, 65-68; physician-in-chief dept med, Springfield Hosp Med Ctr, 68-70; prof & dir, Cardiol Div, Med Col Pa, 70-79; prof med & assoc chief, Cardiol Div, Thomas Jefferson Univ, 79-84; prof med & co-dir, 84-86, DIR, LIKOFF CARDIOVASCULAR INST, HAHNEMANN UNIV, 86-, THOMAS J VISCHER PROF MED & CHMN DEPT MED, 87- *Concurrent Pos:* Res fel cardiol, Univ Pa, 61-62; dir EKG sect, Cardiol Div, Temple Univ, 66-68 & Cardiac Care Unit, 67-68; consult cardiol, Philadelphia Vet Admin Hosp, 70-79; Fogarty Int fel, Univ London & Cardiothoracic Inst, 78-79; fel coun clin cardiol & coun atherosclerosis, Am Heart Asn, pres, Pa Affil, 85-86; gov, Eastern Pa, Am Col Cardiol, 86- *Mem:* Fel Am Col Physicians; fel Am Col Cardiol; Am Soc Clin Pharmacol & Therapeut; NY Acad Sci; Am Fedn Clin Res. *Res:* Cardiac pharmacology, especially beta adrenergic blocking agents; cardiovascular hemodynamics; electrophysiology. *Mailing Add:* 536 Moreno Rd Wynnewood PA 19096

FRANKLE, REVA TREELISKY, b Pittsburgh, Pa; c 1. NUTRITION, PUBLIC HEALTH. *Educ:* Carnegie Mellon Inst, BS, 43; Columbia Univ, MS, 63, EdD(nutrit educ), 75. *Prof Exp:* Metab res dietitian, Columbia Univ Col Physicians & Surgeons, 58-63; pub health nutritionist, New York Health Dept, 63-66, Off Econ Opportunity consult, Proj Headstart, 67; nutrit instr, Teachers Col, Columbia, 69-75; dir nutrit, Weight Watchers Int, 76-88; CONSULT NUTRIT, 88- *Concurrent Pos:* Nutrit coordr, Dept Community Med, Mt Sinai Sch of Med, 67-75; consult pub health & nutrit related fields, var orgn, 58- *Mem:* Am Inst Nutrit; Am Dietetic Asn; Soc Nutrit Educators; Am Pub Health Asn; Am Teachers Prev Med. *Res:* Nutrition in medical education; weight loss and maintenance in ambulatory care settings. *Mailing Add:* 425-D Heritage Hills Somers NY 10589

FRANKLE, WILLIAM ERNEST, b Baldwin, NY, Mar 10, 44; m 68; c 3. ENVIRONMENTAL SCIENCES. *Educ:* Adelphi Univ, BS, 66; Princeton Univ, MS, 69, PhD(chem), 70. *Prof Exp:* sr res assoc paper chem, 70-77, mgr environ protection sci, 77-81, PROJ MGR ENVIRON COMPLIANCE, CORP RES CTR, INT PAPER CO, 81- *Concurrent Pos:* Chem instr, Harriman Col, 75-76. *Mem:* Am Chem Soc; Am Inst Chemists; NY Acad Sci; Am Forest Asn; Tech Asn Pulp & Paper Indust. *Mailing Add:* 14 Wisner Terr Goshen NY 10924

FRANKLIN, ALAN DOUGLAS, b Glenside, Pa, Dec 10, 22; m 43, 60; c 3. SOLID STATE PHYSICS, ARCHAEOMETRY. *Educ:* Princeton Univ, AB, 46, PhD(chem), 49. *Prof Exp:* Sect chief ferromagnetism, Franklin Inst, 49-55; group leader ferroelec, NAT Bur Standards, 55-59, chief mineral prod div, 59-63, theoret physicist & asst to dir inst mat res, 63-67, res chemist, 67-81; lectr, George Wash Univ, 81-82; VIS PROF, UNIV MD, 83- *Honors & Awards:* Gold Medal, Dept Com, 70. *Mem:* Am Chem Soc; fel Am Phys Soc; Sigma Xi; fel Am Ceramic Soc; Soc Archeol Sci. *Res:* Theoretical, experimental studies point defects, ion mobility in crystals; solid electrolyte-electrode interfaces; fuel cell materials; thermoluminescence dating. *Mailing Add:* PO Box 39 Shepherdstown WV 25443

FRANKLIN, ALLAN DAVID, b Brooklyn, NY, Aug 1, 38; div. PHYSICS. *Educ:* Columbia Univ, AB, 59; Cornell Univ, PhD(physics), 65. *Prof Exp:* Res assoc physics, Princeton Univ, 65-66, instr, 66-67; from asst prof to assoc prof, 67-82, PROF PHYSICS, UNIV COLO, BOULDER, 82- *Honors & Awards:* Fel, Am Phys Soc. *Mem:* Am Phys Soc; Hist Sci Soc; Am Asn Physics Teachers; Brit Soc Philos Sci; Philos Sci Asn. *Res:* History and philosophy of science. *Mailing Add:* Univ Colo Campus box 390 Boulder CO 80309

FRANKLIN, BERYL CLETIS, b Colly, Ky, Oct 29, 23; m 47. REPRODUCTIVE PHYSIOLOGY. *Educ:* Ky Wesleyan Col, AB, 48; Univ Ky, MS, 50; Ohio State Univ, PhD, 57. *Prof Exp:* Field rep, Ky Wesleyan Col, 48, instr zool & chem, 49-50; asst zool, Ohio State Univ, 50-52, asst instr, 52-55, supvr grad assts, 54-57, instr, 55-57; instr biol, Del Mar Col, 57-59; asst prof zool, La State Univ, 59-60; from asst prof to assoc prof, 60-67, PROF BIOL, NORTHEAST LA UNIV, 67- *Mem:* AAAS; Am Soc Zool; Sigma Xi. *Res:* Endocrinology; physiological zoology. *Mailing Add:* Dept Biol Northeast La Univ Monroe LA 71209

FRANKLIN, FRED ALDRICH, b Worcester, Mass, July 24, 32. ASTRONOMY. *Educ:* Harvard Univ, AB, 54, MA, 56, PhD, 62. *Prof Exp:* PHYSICIST, SMITHSONIAN ASTROPHYS OBSERV, HARVARD UNIV, 57- *Mem:* Am Astron Soc. *Res:* Photometry and dynamics of planets and satellites. *Mailing Add:* 41 Linnaean St Cambridge MA 02138

FRANKLIN, GENE F(ARTHING), b Banner Elk, NC, July 25, 27; m 52; c 2. ELECTRICAL ENGINEERING. *Educ:* Ga Inst Technol, BEE, 50; Mass Inst Technol, MSEE, 52; Columbia Univ, EngScD, 55. *Prof Exp:* Instr elec eng, Columbia Univ, 52-55; from asst prof to assoc prof, 55-61, PROF ELEC ENG, STANFORD UNIV, 61- *Concurrent Pos:* Consult, industs, 58- *Honors & Awards:* Educ Award, Am Automatic Control Coun, 85; Fel, Inst Elec & Electronics Engrs. *Mem:* Inst Elec & Electronics Engrs; Soc Indust & Appl Math; Sigma Xi. *Res:* Automatic control with emphasis on the digital control of multivariable nonlinear dynamical systems. *Mailing Add:* Dept Elec Eng Stanford Univ Stanford CA 94305

FRANKLIN, GEORGE JOSEPH, b Osmond, Nebr, May 21, 20; m; c 4. GEOLOGY. *Educ:* Ohio State Univ, PhD(geol), 61. *Prof Exp:* Geologist, US Geol Surv, 62-82; GEOLOGIST, US MINERALS MGT SERV, 82- *Mailing Add:* US Mineral Mgt Serv MS 5111 1420 S Clearview Pkwy New Orleans LA 70123-2394

FRANKLIN, JAMES CURRY, b St Charles, Ky, Nov 11, 33; m 55; c 1. ANALYTICAL CHEMISTRY. *Educ:* Western Ky State Univ, BS, 55; Univ Ala, MS, 62; Univ Tenn, PhD(anal chem), 70. *Prof Exp:* Teacher chem, Union County Bd Educ, 57-58 & Lamar County Bd Educ, 58-59; instr, Univ Ala, 59-61; develop chemist, Y-12 Plant, Oak Ridge, Tenn, 61-73; res staff chem, Oak Ridge Nat Lab, 73-78; DEPT HEAD Y-12 LAB, NUCLEAR DIV, UNION CARBIDE, 78- *Mem:* Am Chem Soc; Am Soc Mass Spectrometry; Sigma Xi. *Res:* Spark source mass spectrometry and its application to the trace analysis of energy and environmental related materials. *Mailing Add:* 141 Cumberland View Dr Oak Ridge TN 37830

FRANKLIN, JAMES MCWILLIE, b North Bay, Ont, Nov 9, 42; m 68. ECONOMIC GEOLOGY, STRATIGRAPHY. *Educ:* Carleton Univ, BS, 64, MS, 67; Univ Western Ont, PhD(geol), 70. *Prof Exp:* From asst prof to assoc prof geol, Lakehead Univ, 69-75; RES SCIENTIST, DEPT ENERGY, MINES & RESOURCES, GEOL SURV CAN, 75- *Concurrent Pos:* Geologist, Mattagami Lake Mines Ltd, 71; trip coordr, Int Geol Cong, Montreal, 72. *Mem:* Fel Geol Asn Can; Can Inst Mineral & Metall; Mineral Asn Can. *Res:* Precambrian stratigraphy; chemistry of formation of stratabound copper-zinc massive sulphide deposits in Archean volcanic terrains, effects of metamorphism on these deposits; metallogeny of veins associated with Proterozoic sedimentary sequences; lead isotope studies of Archean and Proterozoic mineral deposits; metallogeny of gold deposits in Archean terrains. *Mailing Add:* Dept Energy Mines & Resources 601 Booth St Ottawa ON K1A 0E8 Can

FRANKLIN, JERROLD, b New York, NY, June 19, 30; m 55; c 2. THEORETICAL PHYSICS. *Educ:* Cooper Union, BEE, 52; Univ Ill, MS, 53, PhD(physics), 56. *Prof Exp:* Asst physics, Univ Ill, 52-56; instr, Columbia Univ, 56-59; asst prof, Brown Univ, 59-64, asst to dean, 62-64; physicist, Lawrence Radiation Lab, Univ Calif, 64-67; assoc prof, 67-70, PROF PHYSICS, TEMPLE UNIV, 70- *Concurrent Pos:* Assoc ed, Math Rev, 62-63; vis prof, Israel Inst Technol, 70-71, 76-77 & Lady Davis fel, 83-84. *Mem:* Am Phys Soc. *Res:* Elementary particle physics; quark models of elementary particles. *Mailing Add:* Dept Physics Temple Univ Philadelphia PA 19122

FRANKLIN, JERRY FOREST, b Waldport, Ore, Oct 27, 36; m 58, 88; c 4. ECOLOGY, FORESTRY. *Educ:* Ore State Univ, BS, 59, MS, 61; Wash State Univ, PhD(bot), 66. *Prof Exp:* Res forester, US Forest Serv, 59-65, plant ecologist, 65-68, prin plant ecologist, 68-73, dept dir coniferous forest biome, 70-73; dir ecosyst anal prog, NSF, 73-75; prof forest sci & bot, Ore State Univ, 76-90; CHIEF PLANT ECOLOGIST, PAC NORTHWEST FOREST & RANGE EXP STA, US FOREST SERV, 75-; BLOEDELL PROF ECOSYSTS, UNIV WASH, 86- *Concurrent Pos:* Bd gov, Nature Conservancy; nat lectr, Sigma Xi; Bullard fel, Harvard Univ, 86. *Honors & Awards:* Arthur S Fleming Award, 72; Barrington Moore Award, Soc Am Foresters, 87. *Mem:* AAAS; Soc Am Foresters; Ecol Soc Am; Brit Ecol Soc; Sigma Xi. *Res:* Forest community ecology and succession; alpine communities; vegetation-soil relationships, especially Abies-Tsuga forest types. *Mailing Add:* Col Forest Resources AR-10 Univ Wash Seattle WA 98195

FRANKLIN, JOEL NICHOLAS, b Chicago, Ill, Apr 4, 30; m 49, 71; c 1. APPLIED MATHEMATICS. *Educ:* Stanford Univ, BS, 50, PhD(math), 53. *Prof Exp:* Instr math, Inst Math Sci, NY Univ, 53-55; asst prof, Univ Wash, 55-56; sr mathematician, Burroughs Corp, 56-57; assoc prof appl mech, 57-65, prof appl sci, 65-70, PROF APPL MATH, CALIF INST TECHNOL, 70- *Concurrent Pos:* Consult, AEC Radiation Lab, Univ Calif, 55, 57 & Jet Propulsion Lab, NASA, 62- *Mem:* Am Math Soc; Math Asn Am; Soc Indust & Appl Math. *Mailing Add:* Appl Math Dept Cal Tech Pasadena CA 91125

FRANKLIN, KENNETH LINN, b Alameda, Calif, Mar 25, 23; m 58; c 3. ASTRONOMY. *Educ:* Univ Calif, AB, 48, PhD(astron), 53. *Prof Exp:* Lab technician, Lick Observ, Calif, 49-50; asst, Leuschner Observ, 53-54; res fel radio astron, Carnegie Inst, 54-56; asst astronr, 56-57, assoc astronr, 58-63, from asst chmn to chmn, 68-74, ASTRONR, AM MUS-HAYDEN PLANETARIUM, 63- *Concurrent Pos:* Adj prof, Cooper Union, 68; vis prof, Rutgers Univ, 68-72; astron ed, World Almanac, 70-; mem comn five, Int Sci Radio Union; consult, Grumman Aircraft Co, Kearfott Div, Gen Precision,

Inc, Eclipse-Pioneer Div, Bendix Corp, AIL Div, Cutler-Hammer Corp, Razdow Labs, Int Tel & Tel, Martin Co, McGraw-Hill Bk Co, Sci Am, NY Times, Time-Life Inc, CBS-TV & NBC-TV. *Mem:* AAAS; Am Astron Soc; Inst Elec & Electronics Engrs; fel Royal Astron Soc; NY Acad Sci. *Res:* Binary stars; galactic structure; radio astronomy; astronomy education. *Mailing Add:* 148070 SW 144th Terr Miami FL 33186

FRANKLIN, LUTHER EDWARD, b Birmingham, Ala, Jan 21, 29; m 48; c 2. DEVELOPMENTAL BIOLOGY. *Educ:* Univ Ala, BS, 53; Fla State Univ, PhD(exp biol), 63. *Prof Exp:* Res assoc, Inst Space Biosci, Fla State Univ, 64 & Inst Molecular Evolution, Univ Miami, 64-65; res assoc & adj asst prof, Delta Regional Primate Res Ctr, Tulane Univ, 65-72, assoc prof, 72; ASSOC PROF BIOL, UNIV HOUSTON, 72- *Mem:* Am Soc Cell Biol; Int Soc Develop Biologists; Soc Study Reprod. *Res:* Morphology and physiology of fertilization; aging of the male reproductive system. *Mailing Add:* Dept Biol Univ Houston University Park 4800 Calhoun Houston TX 77204

FRANKLIN, MARK A, b New York, NY, Apr 25, 40; div; c 2. COMPUTER ARCHITECTURE, COMPUTER PERFORMANCE EVALUATION. *Educ:* Columbia Univ, AB, 61, BSEE, 62, MSEE, 64; Carnegie-Mellon Univ, PhD(elec engr), 70. *Prof Exp:* PROF ELEC ENG & COMPUT SCI, SCH ENG & APPL SCI, WASH UNIV, 70-, DIR, CTR COMPUT SYSTS DESIGN, 81- *Concurrent Pos:* Vchmn, Asn Comput Mach, spec interest group & Comput Archit, 85- *Mem:* Sr mem Inst Elec & Electronics Engrs; Asn Comput Mach. *Res:* Computer architecture, performance evaluation and very large scale integration; parallel processing; computer design automation. *Mailing Add:* Ctr Comput Systs Design Box 1115 Wash Univ St Louis MO 63130

FRANKLIN, MERVYN, b Minehead, Eng, Jan 13, 32; Can citizen; m 63; c 3. MICROBIOLOGY. *Educ:* Univ Reading, BSc, 55 & 56; McGill Univ, PhD(biochem), 59. *Prof Exp:* From demonstr to sr demonstr bact, McGill Univ, 57-59; J C Childs fel microbiol, Western Reserve Univ, 59-60; lectr bact, McGill Univ, 60-62, asst prof, 62-65; asst prof microbiol, New York Med Col, 65-66, assoc prof, 67-69; prof biol & dean fac sci, Univ NB, Fredericton, 69-75, actg vpres, 75-76, acad vpres, 76-78; pres, 78-84, PROF BIOL SCI, UNIV WINDSOR, ONT, 85- *Concurrent Pos:* Assoc ed, Can J Microbiol, 70-73; mem, Sci Coun Can, 70-76; chmn environ coun, Prov of NB, 72-75; mem, Environ Adv Coun Can, 78-81. *Mem:* Can Asn Deans Arts & Sci (chmn 73-74); Can Soc Microbiologists (second vpres, 75-76); Am Soc Microbiol. *Res:* Ecology and physiology of aquatic bacteria. *Mailing Add:* Dept Biol Sci Univ Windsor Windsor ON N9B 3P4 Can

FRANKLIN, MICHAEL R(OGER), b 1944; m; c 2. TOXICOLOGY, BIOCHEMICAL PHARMACOLOGY. *Educ:* Univ Birmingham, Eng, BSc, 66, Univ London, Eng, PhD(biochem), 69. *Prof Exp:* PROF PHARMACOL & TOXICOL, UNIV UTAH, 72- *Mem:* Am Soc Pharmacol & Exp Therapeut; Soc Toxicol. *Res:* Drug metabolising enzymes. *Mailing Add:* Dept Pharmacol-Toxicol Univ Utah 105 Skaggs Hall Salt Lake City UT 84112

FRANKLIN, NAOMI C, b New York, NY, Mar 25, 29. GENETICS, VIROLOGY. *Educ:* Cornell Univ, BS, 50; Yale Univ, PhD(genetics & biochem), 54. *Prof Exp:* Fel genetics & virol, Calif Inst Technol, 54-56; res assoc, Univ Geneva, Switzerland, 57-59, Mass Inst Technol, 59-63; res assoc, Stanford Univ, 63-74, adj prof, 74-79; RES PROF GENETICS & VIROL, UNIV UTAH, 79- *Concurrent Pos:* Prin investr res grants, NSF, 65-; vis res assoc, Univ Paris, 72-73. *Mem:* Genetics Soc Am; Am Soc Microbiol; Sigma Xi; Fedn Am Scientists; AAAS. *Res:* Post-initiation regulation of transcription by termination/antitermination as in bacteriophage lambda; structure and function of the N anti-termination protein and its genetic variants. *Mailing Add:* Dept Biol Univ Utah Salt Lake City UT 84112

FRANKLIN, RALPH E, b Chicago, Ill, Sept 14, 34; m 57, 85; c 3. OIL SHALE. *Educ:* Univ Ark, BS, 55, MS, 56; Ohio State Univ, PhD(agron), 61. *Prof Exp:* From asst prof to prof agron, Ohio State Univ & Ohio Agr Res & Develop Ctr, 61-74; soil scientist, AEC, 72-75 & US Energy Res & Develop Admin, 76-77, prog mgr, US Dept Energy, 78-84; HEAD DEPT AGRON & SOILS, CLEMSON UNIV, 84- *Concurrent Pos:* Adv to Peru, Int Atomic Energy Agency, 66-67; mem, Sci Adv Comt Land & Water Qual Studies, Elec Power Res Inst. *Honors & Awards:* Spec Achievement Awards, US Energy Res & Develop Admin, 75, 76 & 77 & US Dept Energy, 81. *Mem:* AAAS; Soil Sci Soc Am; Am Soc Agron; fel Am Inst Chemists. *Res:* Planning and management of research needed for energy development, expecially oil shale and other sythetic fuel technologies; physical chemistry of soils and plant nutrition; solid waste disposal and mangement; agranomic issues; interactions between plant roots and soil colloids; general plant nutrition; fate of radioactive fallout; environmental consequences of energy development; science and public policy. *Mailing Add:* RD 2 Box 291 Salem SC 29676

FRANKLIN, RENTY BENJAMIN, b Birmingham, Ala, Sept 2, 45; div; c 2. ENDOCRINOLOGY, INTERMEDIARY METABOLISM. *Educ:* Morehouse Col, BS, 66; Atlanta Univ, MS, 67; Howard Univ, PhD(physiol), 72. *Prof Exp:* Instr biol, St Angustine's Col, 67-69; fel physiol, Col Med, Howard Univ, 69-72, from asst prof to assoc prof, 72-80; assoc prof, 80-86, PROF PHYSIOL, SCH DENT, UNIV MD, 86- *Concurrent Pos:* Porter Found res fel pharmacol, Harvard Univ, 74-75; grant reviewer, NSF, 75-77 & Nat Osteop Found, 82-83; prin investr, Nat Inst Arthritis, Diabetes, Digestive & Kidney Dis, 76-79 & NIH, 77-; consult, MBS Res Prog, Howard Univ, 85-86. *Mem:* Am Physiol Soc; AAAS; Sigma Xi; Endocrine Soc. *Res:* Hormonal control of intermediary metabolism; androgenic regulation of enzymatic reactions in prostate epithelial cells; elucidation for unique biochemical markers for neoplastic transformation and for characterization of in vitro cultured epithelial cells. *Mailing Add:* Dept Physiol Sch Dent Univ Md 666 W Baltimore St Baltimore MD 21201

FRANKLIN, RICHARD CRAWFORD, b Spokane, Wash, Nov 11, 15; m 42; c 2. ORGANIC CHEMISTRY. *Educ:* Univ Ill, BA, 37; Univ Wis, PhD(chem), 40. *Prof Exp:* Res chemist, E I du Pont de Nemours & Co, Inc, 40-41 & 46-64, res supvr, 64-78; RETIRED. *Mem:* Am Chem Soc. *Res:* Dyes; anthraquinone colors and intermediates; phthalocyanines; rubber chemicals; hydrogenation and organic chemicals; Azo and basic dyes. *Mailing Add:* 1203 Brook Dr Wilmington DE 19803

FRANKLIN, RICHARD MORRIS, b Medford, Mass, Oct 16, 30; m 58; c 2. MOLECULAR PARASITOLOGY. *Educ:* Tufts Col, BS, 51; Yale Univ, PhD, 54. *Prof Exp:* Am Cancer Soc res fel, Calif Inst Technol, 54-56; res assoc, Max Planck Inst, Tuebingen, Ger, 56-59; asst prof virol, Rockefeller Inst, 59-63; from assoc prof to prof path, Sch Med, Univ Colo, Denver, 63-67; mem, Pub Health Res Inst, City of New York, 67-71; PROF VIROL, UNIV BASEL, 71- *Concurrent Pos:* Fulbright fel, 56; adj prof, Sch Med, NY Univ, 67-71. *Mem:* Swiss Soc Trop Med & Parasitol; Royal Soc Trop Med & Hyg; Brit Soc Immunol; Brit Soc Parasitol. *Res:* Molecular mechanisms of drug resistance in malaria and other parasitic protozoans. *Mailing Add:* Biozentrum Univ Basel Klingelbergstrasse 70 CH-4056 Basel Switzerland

FRANKLIN, ROBERT LOUIS, b Salina, Kans, Feb 21, 35; m 61; c 2. BIOCHEMISTRY, ORGANIC CHEMISTRY. *Educ:* Westmar Col, BA, 63; Mich State Univ, PhD(biochem), 67. *Prof Exp:* From asst prof to assoc prof, 67-78, PROF CHEM, WESTMAR COL, 78- *Mem:* Am Chem Soc; Sigma Xi. *Res:* Synthesis and physiological activity of indolic plant growth regulators; identification and isolation of insect pheromones; solar and wind energy. *Mailing Add:* Rte 5 Box 210-B Le Mars IA 51031

FRANKLIN, ROBERT RAY, b Wetumka, Okla, Jan 20, 28; m 51; c 4. OBSTETRICS & GYNECOLOGY. *Educ:* Univ Tex, BA, 49, MD, 53. *Prof Exp:* Assoc prof, 59-70, PROF OBSTET & GYNEC, COL MED, BAYLOR UNIV, 70- *Concurrent Pos:* Mem attend staff, Woman's Hosp, St Luke's Hosp, Children's Med Ctr Hosp, Texas. *Mem:* AMA; Am Col Obstetricians & Gynecologists; Am Fertility Soc; Am Cancer Soc; Int Soc Adv Humanistic Studies Gynec. *Res:* Endocrinology; bioassay for luteinizing hormone for clinical use; newborn physiology; transitional distress project. *Mailing Add:* Womens Hosp Tex 7550 Fannin Houston TX 77054-1989

FRANKLIN, RONALD, DRUG METABOLISM, MECHANISTIC TOXICOLOGY. *Educ:* St Mary's Med Sch, London, Eng, PhD(biochem), 74. *Prof Exp:* RES SCIENTIST, ELI LILLY & CO, 81- *Res:* Biochemical pharmacology. *Mailing Add:* Dept Drug Deposition Eli Lilly & Co Lilly Corp Ctr Indianapolis IN 46285

FRANKLIN, RUDOLPH MICHAEL, MEDICAL RETINA, UVEITIS. *Educ:* Johns Hopkins Univ, MD, 69. *Prof Exp:* PROF OPHTHAL, LA STATE UNIV EYE CTR, 78- *Mailing Add:* Dept Ophthal LSU Med Ctr Sch Med LSU Eye Ctr 2020 Gravier St New Orleans LA 70112

FRANKLIN, RUDOLPH THOMAS, b Morristown, NJ, Sept 29, 28; m 52; c 5. ENTOMOLOGY. *Educ:* Emory Univ, BA, 50; Univ Ga, MS, 55; Univ Minn, PhD(entom), 64. *Prof Exp:* Forest entomologist, Minn Dept Agr, 58-62; entomologist, Regional Off, Southern Region, US Forest Serv, Ga, 62-63, zone leader entom, NC, 63-65; asst prof entom, 65-70, ASSOC PROF ENTOM & FOREST RESOURCES, UNIV GA, 70- *Mem:* Soc Am Foresters; Entom Soc Am; Entom Soc Can. *Res:* Ecology of forest insects; biology of bark beetles and weevils and hymenopterous parasites of bark beetles. *Mailing Add:* Dept Entom Univ Ga Athens GA 30602

FRANKLIN, SAMUEL GREGG, b Camden, NJ, May 8, 46; div. BIOCHEMISTRY. *Educ:* Rutgers Univ, AB, 68; Thomas Jefferson Univ, PhD(biochem), 73. *Prof Exp:* Res asst biochem, Med Col Ga, 68-69; teaching asst, Thomas Jefferson Univ, 69-73; fel, Inst Cancer Res, Philadelphia, Pa, 72-80, sr res assoc, 80-83; sr res biochemist, Revlon Health Care Group, Tuckahoe, NY, 83-85; sr investr, 85-88, ASST DIR PROTEIN BIOCHEM, SMITH KLINE BEECHAM PHARMACEUTICALS, 88- *Mem:* AAAS; Am Chem Soc. *Res:* The primary structure of histone variants; screening and primary structure of human serum albumin variants; biochemistry of hepatitis B virus binding substance; purification and characterization of tissue plasminogen activator; novel thrombolytics; human immunodeficiency virus related proteins. *Mailing Add:* Smith Kline Beecham Pharmaceuticals 709 Swedleland Rd PO Box 1539 King of Prussia PA 19406-0939

FRANKLIN, STANLEY PHILLIP, b Memphis, Tenn, Aug 14, 31; m 79; c 8. NEUROL NETWORKS, CATAGORICAL TOPOLOGY. *Educ:* Memphis State Univ, BS, 59; Univ Calif, Los Angeles, MA, 62, PhD(math), 63. *Prof Exp:* NSF fel, Univ Wash, 63-64; asst prof math, Univ Fla, 64-65; from asst prof to prof, Carnegie-Mellon Univ, 65-72; chmn dept, 72-84, PROF MATH SCI, MEMPHIS STATE UNIV, 72. *Concurrent Pos:* Indust Trainer & Consult, Artificial Intel. *Mem:* Asn Comput Mach; Am Asn Artificial Intel; Cognitive Sci Soc; Int Soc Neutral Networks. *Res:* Neural networks; mechanisms of mind; formal mathematical theory of neurol networks as a form of massivity parallel computing; categorical topology. *Mailing Add:* Dept Math Sci Memphis State Univ Memphis TN 38152

FRANKLIN, THOMAS CHESTER, b Birmingham, Ala, Feb 5, 23; m 46; c 4. ELECTROCHEM & ANALYTICAL CHEMISTRY. *Educ:* Howard Col, BS, 44; Ohio State Univ, PhD(chem), 51. *Prof Exp:* Instr chem, Howard Col, 46-48; asst phys chem, Ohio State Univ, 48-49, asst instr, 49-50; asst prof chem, Richmond Univ, 51-54; assoc prof, 54-62, PROF CHEM, BAYLOR UNIV, 62- *Concurrent Pos:* Chem res worker, Va Inst Sci Res, 52-53. *Mem:* Electrochem Soc; Am Chem Soc; Am Electroplaters Soc; Int Soc Electrochem. *Res:* Electrochemistry and catalysis. *Mailing Add:* Dept Chem Baylor Univ PO Box 97348 Waco TX 76798-7348

FRANKLIN, THOMAS DOYAL, JR, b Morganton, NC, sept 25, 41; m 64; c 2. PHYSIOLOGY, ORTHOPEDIC RESEARCH. *Educ:* Wake Forest Univ, BS, 63, MS, 67; Univ Ill, Urbana-Champaign, PhD(physiol), 72. *Prof Exp:* Assoc physiologist aerospace med, McDonnell-Douglas Corp, 67-69; res scientist ultrasound, Intersci Res Inst, 69-72; from asst prof to assoc prof radiol, Sch Med, Ind Univ, Indianapolis, 72-90, asst dir, Fortune-Fry Ultrasound Res Labs, 73-80, assoc dir, 80-81, dir, Life Sci Res Div, Indianapolis Ctr Advan Res, Ind Univ, 81-82, pres & exec dir, 82-90; PRES & CHIEF EXEC OFFICER, TEX BACK INST RES FOUND, 90- *Mem:* AAAS; Am Heart Asn; Am Soc Echocardiography; Sigma Xi; Am Inst Ultrasound in Med; Radiol Soc NAm. *Res:* Application of ultrasound to soft tissue visualization for diagnosis; potential therapeutic applications of low intensity ultrasound; non-invasive evaluation of the cardiovascular system; orthopedic research in disc disease; biomechanics of spine; rehabilitation outcome studies. *Mailing Add:* Tex Back Inst Res Found 3801 W 15th St Plano TX 75075-7788

FRANKLIN, URSULA MARTIUS, b Munich, Ger, Sept 16, 21; Can Citizen; m 52; c 2. PHYSICAL METALLURGY. *Educ:* Tech Univ, Berlin, PhD(appl physics), 49. *Prof Exp:* Fel appl physics, Univ Toronto, 50-52; sr res officer mat eng, Ont Res Found, Toronto, 52-67; PROF MAT SCI, UNIV TORONTO, 67-, AFFIL, INST HIST SCI & TECHNOL, 69- *Concurrent Pos:* Res assoc hist technol, Royal Ont Mus, 69-; mem & comt chmn, Sci Coun Can, 74-77; mem coun & exec, Nat Sci & Eng Res Coun, 77- *Mem:* Am Soc Metal; Can Metall Conf; Soc Archaeol Sci. *Res:* Structure of alloys; archeo-metallurgy; study of ancient metals and alloys; origin and development of bronze technology; metals in early China. *Mailing Add:* Dept Metall & Mat Sci Univ Toronto Toronto ON M5S 1A1 Can

FRANKLIN, WILBUR A, b Detroit, Mich, Mar 8, 42; m; c 2. PATHOLOGY. *Educ:* Univ Colo, BA, 64; Northwestern Univ, MD, 68. *Prof Exp:* Intern straight path, Chicago Wesley Mem Hosp, 68-69, resident path, 69-72; asst head, Blood Bank, Nat Naval Med Ctr, Bethesda, Md, 72-73, anat pathologist, 73-74; asst prof path, Med Sch, Georgetown Univ, Washington, DC, 73-74; postdoctoral fel immunol, La Rabida-Univ Chicago Inst, 74-76, from asst prof to assoc prof surg path, 76-89; PROF, DEPT PATH, VCHMN ANAT PATH & DIR, LAB SURG PATH, HEALTH SCI CTR, UNIV COLO, 89- *Concurrent Pos:* Jr fac clin fel, Am Cancer Soc, 79-82; vis res assoc, John Radcliffe Hosp, Oxford, Eng, 83-84. *Mailing Add:* Dept Path Univ Colo Box B216 4200 E Ninth Ave Denver CO 80262

FRANKLIN, WILLIAM ELWOOD, b Washington, DC, Nov 22, 31; m 60. TEXTILE CHEMISTRY, POLYMER CHEMISTRY. *Educ:* Univ Nebr, BSc, 53, MSc, 55; Univ Iowa, PhD(org chem), 59. *Prof Exp:* Asst, Univ Iowa, 55-58; from asst prof to assoc prof chem, Loyola Univ, La, 58-65; RES CHEMIST, SOUTHERN REGIONAL RES CTR, AGR RES SERV, USDA, 65- *Mem:* Am Chem Soc; Am Asn Textile Chemists & Colorists; Sigma Xi. *Res:* Cellulose crosslinking agents; bacterial endotoxins; textile flamability; mass spectrometry byssinosis. *Mailing Add:* 3317 Jefferson Ave New Orleans LA 70125

FRANKLIN, WILLIAM LLOYD, b Santa Monica, Calif, July 31, 41; m 62; c 4. ANIMAL ECOLOGY, MAMMALOGY. *Educ:* Univ Calif, Davis, BS, 64; Humboldt State Univ, MS, 68; Utah State Univ, PhD(wildlife sci), 78. *Prof Exp:* ASST PROF MAMMALIAN WILDLIFE ECOL, DEPT ANIMAL ECOL, IOWA STATE UNIV, 75- *Concurrent Pos:* Prin investr multiple grants, 68-81. *Mem:* Wildlife Soc; Animal Behav Soc; Am Soc Mammalogists; Fauna Preserv Soc. *Res:* Sociobiology and behavioral ecology of mammals; South American wild camelids and ungulate social systems; fish and wildlife sciences. *Mailing Add:* Dept Animal Ecol Iowa State Univ 124 Sci Li Ames IA 50011

FRANKO, BERNARD VINCENT, b West Brownsville, Pa, June 9, 22; m 46; c 9. CARDIOVASCULAR PHARMACOLOGY, AUTONOMIC PHARMACOLOGY. *Educ:* WVa Univ, BS, 54, MS, 55; Med Col Va, PhD(pharmacol), 58. *Prof Exp:* From instr to asst prof, 58-76, ADJ ASST PROF, DEPT PHARMACOL,MED COL, VA COMMONWEALTH UNIV, 76-; pharmacologist, 58-68, assoc dir pharmacol res, 68-71 & 73-76, dir pharmacol res, 71-73, sr monitor qual assurance, 76-78, dir qual assurance, 78-81, MGR, RES COORD & TRAINING SECT, A H ROBINS CO, INC, 81- *Mem:* Fel AAAS; Am Soc Pharmacol & Exp Therapeut; Soc Exp Biol & Med; Int Soc Biochem Pharmacol. *Res:* Autonomic, cardiovascular and diuretic drugs; toxicology. *Mailing Add:* 4012 Patterson Ave Richmond VA 23221

FRANKO-FILIPASIC, BORIVOJ RICHARD SIMON, b Zagreb, Yugoslavia, Jan 5, 22; nat US; m 48; c 3. HIGH PRESSURE EQUIPMENT, CATALYSIS. *Educ:* Northwestern Univ, BS, 43, MS, 51, PhD(chem), 52. *Prof Exp:* Res chemist, Pittsburgh Plate Glass Co, 52-53 & Mathieson Chem Corp, 53-56; supvr org res & develop, FMC Corp, 56-66, mgr process res & develop, 66-84; PRES, B&R FRANKO ASSOCS, INC, 84- *Mem:* Am Chem Soc; Royal Soc Chem; fel Am Inst Chem Eng; Am Soc Test Mat. *Res:* Fats and oils; boron fuels; high pressure reactions; glycerol; flame retardents. *Mailing Add:* 2 Oak Ave Morrisville PA 19067

FRANKOSKI, STANLEY P, b Suffern, NY, May 24, 44; m 68; c 2. PRODUCT RESEARCH & DEVELOPMENT. *Educ:* Villanova Univ, BS, 66; Purdue Univ, MS, 68; Univ Mass, PhD(chem), 72. *Prof Exp:* Res chemist, Gaf Corp, 72-74, sr res chemist, 74-76, mgr anal res & develop, 76-81, mgr quality asst, 81-83, dir Res & Develop, 84. *Mem:* Am Chem Soc; Tech Asn Pulp & Paper Inst. *Res:* Organic and inorganic chemistry; material sciences; identifications; product development; administration. *Mailing Add:* Seven Will Lane West Milford NJ 07480

FRANKS, ALLEN P, b Cleveland, Ohio, Nov 12, 36; m 63; c 2. RUBBER CHEMISTRY. *Educ:* Western Reserve Univ, BS, 59, JD, 63. *Prof Exp:* Phys scientist, Crile Vet Admin Hosp, 59-60; chemist, Western Reserve Univ, 61-63; patent attorney, B F Goodrich Co, 63-65; res chemist, PPG Industs, Inc, 65-66; rubber lab mgr & tech dir, Reichhold Chem, Cuyahoga Falls, 66-76; CONSULT & PRES, IAS INC, 77-; PLANT CHEMIST, CANTON GLOVE, 90- *Concurrent Pos:* Chmn bd dirs, Persephone Found, 76-78; mgr tech sales, Sovereign Chem Co, 81-88; tech serv chemist, Killian Latex, 88-89; chemist, Pioneer Balloon Co, 89-90; secy-treas, Res Div, IAA, 88- *Mem:* Am Chem Soc; NY Acad Sci; fel Am Inst Chemists; AAAS. *Res:* Organic synthesis of antioxidants, antiozonants and nucleotides; bioassay of enzymes and hormones; protective coatings; dispersions and lattices; compounding of rubber latex. *Mailing Add:* 5610 S Main St Akron OH 44319-5158

FRANKS, DAVID A, b Washington, DC, June 24, 29; m 53; c 2. MATHEMATICS. *Educ:* Howard Univ, BS, 51, MS, 52. *Prof Exp:* First lieutenent, US Ordnance Corps, 55-57; MGR, WESTINGHOUSE ELEC CORP, 57- *Mem:* Asn Comput Mach; Am Math Soc; Data Processing Mgt Asn; Inst Elec & Electronics Engrs. *Mailing Add:* 8505 Moon Glass Ct Columbia MD 21045

FRANKS, EDWIN CLARK, b Chagrin Falls, Ohio, Jan 13, 37; m 58. ZOOLOGY. *Educ:* Ohio State Univ, BSc, 58, MSc, 60, PhD(zool, ornith), 65. *Prof Exp:* Lab technician, Battelle Mem Inst, 59-61, biologist, 61-62; USPHS trainee zool, Pa State Univ, 65-66; asst prof, 66-70, assoc prof, 70-77, PROF BIOL, WESTERN ILL UNIV, 77- *Concurrent Pos:* NSF award, Marine Biol Lab, Woods Hole, 65, NSF grant, Bermuda Biol Sta, 72. *Mem:* Inland Bird Banding Asn; Am Ornith Union; Wilson Ornith Soc; Nat Audubon Soc. *Res:* Bird ecology and behavior. *Mailing Add:* Dept Biol Sci Western Ill Univ Macomb IL 61455

FRANKS, JOHN ANTHONY, JR, b Cleveland, Ohio, Nov 20, 35; m 59; c 5. TIN CHEMICALS. *Educ:* Univ Notre Dame, BS, 58; Western Reserve Univ, MS, 61, PhD(org chem), 64. *Prof Exp:* Asst chem, Western Reserve Univ, 59-63, fel org chem, Sch Med, 63-64; sr org chemist, Eli Lilly & Co, 64-66; group leader appln res, color & chem res dept, Sherwin-Williams Co, 66-71; factor rep, Parco Sci Co, 71-72; mgr prod develop, Hull-Smith Chem, Inc, 72-74; sr res chemist, Vulcan Mat Co, 75-80, sr res assoc, 80-87; CHEMIST & BACTERIOLOGIST, DEPT WATER PURIFICATION, TOLEDO, 88- *Mem:* Am Chem Soc; Sigma Xi. *Res:* Investigations on the synthesis and structural elucidations on benzoxazine systems and their derivatives, cyclopentane tetrols plus derivatives and reactions of thebaine type alkaloids; plastisol formulation, extrusion processing, pigments, one package epoxy systems; fungicides and antioxidants; tin chemicals, their chemical and electrochemical processing and applications. *Mailing Add:* 1394 Sanford St Vermillion OH 44089-1521

FRANKS, JOHN JULIAN, b Pueblo, Colo, Apr 9, 29; m 51; c 5. INTERNAL MEDICINE, ANESTHESIOLOGY. *Educ:* Univ Colo, BA, 51, MD, 54. *Prof Exp:* Intern, Cornell Med Div, Bellevue Hosp, 54-55; resident internal med, Med Ctr, Univ Colo, 55-58; dep chief dept pharmacol-biochem, Sch Aviation Med, US Air Force, 58-62; staff hematologist, Lackland Air Force Hosp, 62-63; asst prof, Univ Colo, Denver, 63-68, assoc prof med, Med Ctr, 68-77, assoc dir, Clin Res Ctr, 70-77, dir, 77-80; assoc chief staff, Denver Vet Admin Med Ctr, 68-83; from resident to asst prof, 84-87, PROF & DIR RES, VANDERBILT UNIV MED CTR, 87- *Concurrent Pos:* Res fel med, Sch Med, Harvard Univ & Mass Gen Hosp, 63-64. *Honors & Awards:* Res Career Develop Award, US Pub Health Serv, 63-69. *Mem:* AAAS; Am Physiol Soc; Am Gastroenterol Asn; Am Soc Anesthesiol; Cent Soc Clin Res. *Res:* Plasma protein metabolism; biomathematics; fibrinogen degradation products; action of anesthetic agents. *Mailing Add:* 216 Vaughns Gap Rd Nashville TN 37205-3532

FRANKS, LARRY ALLEN, b Chesterland, Ohio, July 22, 34; m 60; c 2. GENERAL PHYSICS. *Educ:* Hiram Col, AB, 58; Vanderbilt Univ, MS, 60, PhD(physics, math), 68. *Prof Exp:* From scientist to sr scientist, EG&G Inc, 60-68, sci specialist, 68-72; staff scientist, Off Radiation Progs, US Environ Protection Agency, 72-73; sr sci specialist, 73-74, group leader, 74-82, dept mgr, 82-84, ASST PROG MGR, EG&G INC, 84- *Concurrent Pos:* Staff mem, Fed Power Comn-Spec Task Force Natural Gas Technol, 72-73. *Mem:* Am Phys Soc. *Res:* Molecular structure and spectroscopy; development of radiation detection systems; applied optics. *Mailing Add:* 311 Piedont Rd Santa Barbara CA 93105

FRANKS, LEWIS EMBREE, b San Mateo, Calif, Nov 8, 31; m 54; c 3. SIGNAL PROCESSING, COMMUNICATIONS ENGINEERING. *Educ:* Ore State Univ, BS, 52; Stanford Univ, MS, 53, PhD(elec eng), 57. *Prof Exp:* Instr elec eng, Stanford Univ, 57-58; mem tech staff, Bell Labs, Inc, 58-62, supvr data syst, 62-69; assoc prof, 69-71, PROF ELEC ENG, UNIV MASS, AMHERST, 71- *Concurrent Pos:* Adj assoc prof, Columbia Univ, 65; adj prof, Northeast Univ, 67-68; dir, networking & commun res, NSF, 88-90. *Mem:* Fel Inst Elec & Electronics Engrs. *Res:* Theoretical and design aspects of communication systems, signals, and signal processing equipment; circuit theory and design. *Mailing Add:* Juggler Meadow Rd Amherst MA 01002

FRANKS, NEAL EDWARD, b Canton, Ohio, July 24, 36; m 58; c 2. ORGANIC CHEMISTRY, POLYMER CHEMISTRY. *Educ:* Manchester Col, BA, 58; Ohio State Univ, PhD(org chem), 63. *Prof Exp:* Res assoc biochem, Univ Iowa, 65-66; res biochemist enzyme technol, Procter & Gamble Co, 67-72; res scientist & proj leader synthetic fibers, Am Enka Co, 72-77; group leader pulping technol, St Regis Tech Ctr, 77-85; mgr mech pulping, Champion Int Tech Ctr, 85-88; MGR TECH SERV, CHAMPION INT, NEWSPRINT OPER, SHELDON, TX, 88- *Mem:* Can Pulp & Paper Asn; Tech Asn Pulp & Paper Indust. *Res:* Organic and physical chemistry of natural polymers, particularly fibrous materials, with emphasis on their manufacture and modification. *Mailing Add:* Champion Int PO Box 23011 Houston TX 77228

FRANKS, RICHARD LEE, b Portland, Ore, Mar 29, 41; m 66; c 2. COMPUTER PERFORMANCE ANALYSIS. *Educ:* Univ Wash, BSEE, 63; Univ Calif, Berkeley, MS, 69, PhD(elec eng), 70. *Prof Exp:* Mem tech staff, Systs Anal, 70-73, supvr, Systs Anal Group, 73-77, head, Network Mgt Dept,

77-79, HEAD, PERFORMANCE ANALYSIS DEPT, BELL TEL LABS, 79- *Mem:* Inst Elec & Electronics Engrs; Sigma Xi. *Res:* System theory. *Mailing Add:* Decision Support Systems Dept AT&T Bell Labs Rm Ho 1d432 Holmdel NJ 07733

FRANKS, STEPHEN GUEST, b Frankstown, Miss, Apr 15, 46; m 77; c 2. GEOLOGY, PETROLOGY. *Educ:* Millsaps Col, BS, 68; Univ Miss, MS, 70; Case Western Reserve Univ, PhD(geol), 76. *Prof Exp:* Res geologist, Atlantic Richfield Co, 74-77, sr res geologist, Geosci Group, 77-80, res dir rock/fluid systs, Arco Oil & Gas, 80-83, dir, geol serv, Arco Alaska, 83-85, prin geologist, 85-90, EXPLOR ADVISOR, ARCO OIL & GAS, 90- *Honors & Awards:* George C Matson Award, Am Asn Petrol Geol, 78. *Mem:* Am Asn Petrol Geol; Soc Econ Paleontol & Mineral; Clay Min Soc. *Res:* Diagenesis and low-grade metamorphism of clastic sediments, particularly the relationships of tectonism, sedimentation and diagenesis and how these affect the hydrocarbon potential of sedimentary basins. *Mailing Add:* ARCO PO Box 51408 Lafayette IN 70507

FRANS, ROBERT EARL, b Louisville, Nebr, Apr 19, 27; m 49; c 3. AGRONOMY. *Educ:* Univ Nebr, BS, 50; Rutgers Univ, MS, 53; Iowa State Univ, PhD(plant physiol), 55. *Prof Exp:* Asst farm crops, Rutgers Univ, 50-53; asst plant physiol, Iowa State Univ, 53-55; from asst prof to prof, 55-75, DISTINGUISHED PROF AGRON, UNIV ARK, FAYETTEVILLE, 75-, ELMS FARMING-RICHARD S BARNETT JR CHMN, WEED SCI, 80- *Mem:* Weed Sci Soc Am. *Res:* Applied physiological usages of growth-active compounds for herbicidal purposes; mechanisms responsible for plant growth inhibition from biologically active compounds. *Mailing Add:* Dept Agron Univ Ark Main Campus Fayetteville AR 72701

FRANSE, RENARD, b New York, NY, Jan, 17, 48. REAGENT METHODOLOGY DEVELOPMENT, UV-VIS SPECTROPHOTOMETRY SYSTEMS DEVELOPMENT. *Educ:* Acad Aeronaut, AAS, 68; NY Univ, BS, 71; Rutgers Univ, MBA, 84. *Prof Exp:* Sr methodology res engr, Union Carbide Med Prod, 71-76; sr support engr, Honeywell Med Systs, 76-79; staff engr, Burroughs Corp, 79-82; prod planner, Pitney Bowes, 82-84; PROJ MGR, PANASONIC CO, 84- *Mem:* Am Soc Qual Control; Soc Automotive Engrs; Soc Mfg Engrs. *Res:* Clinical methodologies for automated chemistry analyzers; several enzymatic and rate reaction-linear methodologies awarded patents. *Mailing Add:* Four Butterfly Lane Putnam Valley NY 10579

FRANSON, RAYMOND LEE, b Springfield, Mo, Aug 19, 59. INTERACTIONS BETWEEN PLANTS, EFFECT OF MYCORRHIZAE ON PLANTS. *Educ:* Univ Mo, AB, 81; Univ Chicago, MS, 83, PhD(plant & microbial ecol), 89. *Prof Exp:* Biologist, 87-89, PLANT PHYSIOLOGIST, USDA-AGR RES SERV, 89- *Mem:* Ecol Soc Am; Soc Ecol Restoration; AAAS; Scand Soc Plant Physiol. *Res:* Mechanisms by which plants interact; belowground interactions, including the effect of vesicular- arbuscular mycorrhizal fungi on interactions between plants, and the occurence of commensal nutrient interactions between plants. *Mailing Add:* 1931 Glencrest Springfield MO 65804

FRANSON, RICHARD CARL, b Woburn, Mass, Dec 14, 43; m 67; c 3. BIOCHEMISTRY. *Educ:* Univ Mass, BS, 65; Bowman Gray Sch Med, MS, 70, PhD(biochem), 72. *Prof Exp:* Instr exp med, Sch Med, NY Univ, 73-75; asst prof biophys, 75-77, asst prof, 77-79, ASSOC PROF BIOPHYS & BIOCHEM, DEPT BIOPHYS, MED COL VA, 79- *Concurrent Pos:* USPHS fel, Sch Med, NY Univ, 72-75; vis scientist, Univ Utrecht, Neth, 79-80; res career develop award, 80-85. *Mem:* AAAS; Harvey Soc; NY Acad Sci; Sigma Xi; Biophys Soc. *Res:* Cell physiology; leukocyte metabolism and function; metabolism of glycerides and phospholipids; role of lysosomes and lysosomal enzymes in cell physiology; lung phospholipid and surfactant metabolism. *Mailing Add:* 11812 Britain Way Richmond VA 23233-3960

FRANT, MARTIN S, b NY, July 15, 26; m 48; c 2. ANALYTICAL CHEMISTRY. *Educ:* Brooklyn Col, BS, 49; Western Reserve Univ, MS, 51, PhD(org chem), 53. *Prof Exp:* Asst, Western Reserve Univ, 51-52; sr res chemist, Gallowhur Chem Corp, 52-56; sr res assoc, AMP, Inc, 56-62; asst tech dir, Prototech, Inc, 63-64; from asst dir res to dir res, Orion Res Inc, 64-73, vpres res, 73-78; mgr appln serv, anal div, Foxboro Co, 78-84; TECH DIR, ORION RES INC, 84- *Concurrent Pos:* Co-chmn, Gordon Res Conf Electrodeposition, 61. *Honors & Awards:* Speaker Award, Am Chem Soc, 69. *Mem:* Fel AAAS; Am Electroplaters Soc; Am Chem Soc; Electrochem Soc; Sigma Xi. *Res:* Aromatic mercury and quaternary ammonium compounds; corrosion of electroplated surfaces; analytical instrumentation for process control; electroanalytical chemistry, ion-sensing electrodes. *Mailing Add:* 131 Westchester Rd Newton MA 02158

FRANTA, WILLIAM ALFRED, b Ligderwood, NDak, Mar 26, 13; m 40; c 2. TECHNICAL MANAGEMENT, PLASTICS ENGINEERING. *Educ:* Univ NDak, BS, 33, MS, 34; Oxford Univ, AB, 37. *Hon Degrees:* DEng, Univ NDak, 71. *Prof Exp:* Sales engr, Herbert L Brown, Ohio, 37-38; res chemist, Columbia Chem Div, Pittsburgh Plate Glass Co, 38-42; from res chemist to gen lab dir, Polychem Dept, E I du Pont de Nemours & Co, Inc, 46-53, mgr develop & serv, 53-55, mkt mgr, 55-57, prod mgr plastics, 57-60, dir supporting res, res & develop div, 60-64, asst dir, 64, dir res & develop div, dept plastics, 64-78; RETIRED. *Honors & Awards:* Charles D Hurd lectr, Northwestern Univ, 73. *Mem:* Soc Plastics Eng; Am Chem Soc. *Res:* Organic polymers; polymer science; plastics processing and application. *Mailing Add:* Stonegates Box 88 4031 Kennett Pike Greenville DE 19807

FRANTA, WILLIAM ROY, b St Paul, Minn, May 21, 42; m 66. MODELING, LOCAL COMPUTER NETWORKS. *Educ:* Inst Technol, Univ Minn, BMath, 64, MS, 66, PhD(comput sci), 70. *Prof Exp:* From asst prof to assoc prof, 70-81, from assoc dir to co-dir, 72-81, PROF COMPUT SCI & ASSOC DIR, INFO SCI CTR, INST TECHNOL, UNIV MINN, 81- *Concurrent Pos:* Mem affil fac, Tech Univ Berlin, 79-; vis prof tech, Israel Inst Technol, 82; consult, Honeywell, Network Systs Corp, Comput Sci Corp & others; pres, First Midwest Capital Corp, 82-85; treas, ADC Telecommunications, 85-87; vpres, Network Systs Corp, 87- *Mem:* Asn Comput Mach; Inst Elec & Electronics Engrs; Inst Mgt Sci. *Res:* Modeling; simulation; local computer networks; statistical analysis. *Mailing Add:* VPres Engr Network Systems Corp 7000 Boone Ave N Minneapolis MN 55428

FRANTI, CHARLES ELMER, b Ewen, Mich, Apr 29, 33; m 56; c 5. APPLIED STATISTICS. *Educ:* Univ Mich, BS, 55, AM, 60; Mich State Univ, MS, 60; Univ Calif, Berkeley, PhD(biostatist), 67. *Prof Exp:* Instr math, biol & phys educ, Suomi Col, 55-59; instr math, Mich Technol Univ, 57-59 & 60-62; from asst prof to assoc prof, 66-76, from assoc prof to prof community health, 76-85, PROF BIOSTATIST, UNIV CALIF, DAVIS, 76- *Concurrent Pos:* Consult statist, USDA, 72-76 & Univ Alta, Can, 81. *Mem:* Sigma Xi; Am Statist Asn; Nat Coun Teachers Math; Biomet Soc. *Res:* Applied epidemiology; application of statistical methods in biomedical research; statistical methods in medical surveys; zoonotic diseases. *Mailing Add:* EPM Vet Med Univ Calif Davis Davis CA 95616

FRANTSI, CHRISTOPHER, b London, Eng, Oct 13, 42; Can citizen; m 66; c 2. AQUACULTURE, FISH DISEASES. *Educ:* Acadia Univ, BSc, 66; Univ Guelph, MSc, 68, PhD(fisheries), 72. *Prof Exp:* Lectr food microbiol, Univ Guelph, 68; biologist, Gov Can Fisheries, 72-75, adv aquacult, 77-78; biologist, Huntsman Marine Lab, 75-77, teaching supvr aquacult, 78-84; pres, Fundy Isles Marine Ltd, 77-84; MGR, AQUACULT DIV, CONNERS BROS LTD, 84- *Mem:* Am Fisheries Soc; Am Inst Fisheries Res Biologists. *Res:* Fish diseases and virology; aquaculture, management and disease; shellfish depuration. *Mailing Add:* Box 381 St Andrews NB E0G 2X0 Can

FRANTTI, GORDON EARL, b Palmer, Mich, July 28, 28; m 52; c 7. SEISMOLOGY. *Educ:* Mich Technol Univ, BS, 53, MS, 54. *Prof Exp:* Res asst physics, Mich Technol Univ, 53-54; mining engr & geologist, Copper Range Co, 54-55; geologist & geophysicist, Cleveland Cliffs Iron Co, 55-56; geophysicist, US Bur Mines, 56-59; asst prof physics, Mich Technol Univ, 59-60; res assoc, Inst Sci & Technol, Univ Mich, 60-64, lectr geol & mineral, 63-64, assoc res geophysicist, 64-65; assoc prof, 65-82, PROF GEOPHYS & HEAD DEPT GEOL & GEOL ENG, MICH TECH UNIV, 81- *Concurrent Pos:* Consult, US Bur Mines, 59-60, Gen Elec Co, 62-63, var consult activities, 65- *Mem:* Soc Explor Geophys; Am Geophys Union; Seismol Soc Am. *Res:* Mining and rock mechanics; theoretical and applied seismology; blasting and vibration studies; seismic earth noise; ground subsidence. *Mailing Add:* Dept Geol Mich Technol Univ Houghton MI 49931

FRANTZ, ANDREW GIBSON, b New York, NY, May 22, 30. ENDOCRINOLOGY, INTERNAL MED. *Educ:* Harvard Univ, AB, 51; Col Physicians & Surgeons, Columbia Univ, MD, 55. *Prof Exp:* Intern resident med, Presbyterian Hosp, NY, 55-58; fel endocrinol, Columbia Univ, 58-60; from instr to assoc med, Harvard Med Sch, 62-66; from asst to assoc prof, 66-71, PROF MED, COL PHYSICIANS & SURGEONS & CHIEF, DIV ENDOCRINOL, DEPT MED, COLUMBIA UNIV, 71- *Concurrent Pos:* Asst med, Mass Gen Hosp, 62-66; attend physician, Presbyterian Hosp, 66-; assoc ed, Metabolism, 69-; mem med adv bd, Nat Pituitary Agency, 70-73. *Mem:* Endocrine Soc; Am Soc Clin Invest; Int Soc Neuroendocrinol; Am Fedn Clin Res; Asn Am Phys. *Res:* Endocrinology and neuroendocrinology; characterization, assay, regulation of secretion and mode of action of pituitary hormones, especially prolactin; opioid peptides of hypothalamus and pituitary; diagnosis and treatment of pituitary diseases. *Mailing Add:* Dept Med Columbia Univ Col Physicians & Surgeons New York NY 10032

FRANTZ, BERYL MAY, b Cape Town, SAfrica, June 26, 43; US citizen; m 71; c 1. CLINICAL INFORMATION, CLINICAL PHARMACOLOGY. *Educ:* Univ Cape Town, BSc, 63, Hons, 64; Univ London, PhD(chem), 68. *Prof Exp:* Fel biochem, Med Sch, George Washington Univ, 69; res assoc cardiol, Med Col, Cornell Univ, 69-70; fel, Univ Miami, 70-71; proj assoc biochem, Univ Wis-Madison, 71-73; res investr anal chem, 73-78, sr clin info scientist, 81-86, CLIN INFO MGR, SQUIBB INST MED RES, 86- *Mem:* Royal Soc Chem; Am Pharmaceut Asn. *Res:* Data generated from clinical pharmacology studies of new drugs. *Mailing Add:* 135 Hale Dr Princeton NJ 08540

FRANTZ, DANIEL RAYMOND, b Detroit, Mich, Apr 20, 43; m 69; c 3. USER INTERFACE MANAGEMENT SYSTEMS, PRESENTATION SERVICES. *Educ:* Univ Mich, Ann Arbor, BS, 65, PhD(comput sci), 72. *Prof Exp:* Asst prof comput sci, Wayne State Univ, 72-79; CONSULT SOFTWEARE ENGR, DIGITAL EQUIP CORP, 79- *Concurrent Pos:* Mem, Screen Mgt Task Group, Conf Data Systs Lang, 80-84, chair, Form Interface Mgt Syst Comt, 86-; adj prof, Elec Eng Dept, Univ Lowell, 85-86. *Mem:* Asn Comput Mach. *Res:* User interface management systems design and standardization; forms; language design; internationalization; dialogue models; asynchrony; device independence; multiple device types; distributability; mixing media and discrete data in interfaces. *Mailing Add:* Digital Equip Corp ZK02-2/023 110 Spit Brook Rd Nashua NH 03062

FRANTZ, GREGORY CLAYTON, b Palmerton, Pa, Sept 14, 47; m 73; c 1. STRUCTURAL ENGINEERING, REINFORCED CONCRETE. *Educ:* Lehigh Univ, BS, 69; Univ Tex, Austin, MS, 70, PhD(civil eng), 78. *Prof Exp:* AT US Navy Seabees, 71-74; res asst, Univ Tex, Austin, 75-78; asst prof, 78-84, ASSOC PROF CIVIL ENG, UNIV CONN, 84- *Concurrent Pos:* Vis prof, Imp Col Sci & Technol, London, 84. *Mem:* Am Concrete Inst. *Res:* Structural engineering; material properties of concrete; control of cracking in reinforced concrete beams; strength of reinforced concrete structural members. *Mailing Add:* 206 Cedar Swamp Rd Storrs CT 06268

FRANTZ, IVAN DERAY, JR, b Smithville, WVa, Jan 16, 16; m 42; c 5. BIOCHEMISTRY. *Educ:* Duke Univ, AB, 37; Harvard Univ, MD, 41. *Prof Exp:* Intern, Springfield Hosp, Mass, 41-42; asst to assoc med, Harvard Univ, 48-54, tutor biochem, 47-54, instr med, 50-51; CLARK RES PROF, DEPTS MED & BIOCHEM, MED SCH UNIV MINN, MINNEAPOLIS, 54-

Concurrent Pos: Res fel med, Harvard Univ, 46-48; Am Cancer Soc fel, Huntington Mem Labs, Mass Gen Hosp, Boston, 46-47, clin fel med, 46-50; asst physician, Mass Gen Hosp, Boston, 50-54; consult, Mass Inst Technol, 47-54; mem coun arteriosclerosis, Am Heart Asn, chmn, 67. *Mem:* Am Soc Biol Chem; Am Soc Clin Invest; Am Asn Cancer Res; Sigma Xi. *Res:* Bacterial nutrition; enzyme kinetics; protein and lipid metabolism with use of radioactive tracers. *Mailing Add:* Lupid Res Clin Dept Med Univ Minn Hosp Harvard St at E River Rd Box 192 Minneapolis MN 55455

FRANTZ, JOSEPH FOSTER, b McComb, Miss, Feb 21, 33; m 60; c 3. CHEMICAL ENGINEERING. *Educ:* La State Univ, BS, 55, MS, 56, PhD(chem eng), 58. *Prof Exp:* Res chem engr, Monsanto Co, Ark, 58-60, sr chem engr, Mo, 61-62, eng dept, Hydrocarbons Div, 62-65; mgr develop & mkt, South Hampton Co, Houston, 65-68; PRES, THE FRANTZ CO, 68- *Concurrent Pos:* Am Inst Chem Engrs Award, 63-64. *Mem:* Am Inst Chem Engrs; Am Chem Soc. *Res:* Fluidization; thermal cracking of hydrocarbons; marketing. *Mailing Add:* 6410 Tam O'Shanter Houston TX 77036

FRANTZ, WENDELIN R, b Cleveland, Ohio, Apr 28, 29; m 52; c 3. PETROLEUM GEOLOGY, STRATIGRAPHY. *Educ:* Col Wooster, BA, 52; Univ Pittsburgh, MS, 56, PhD(geol), 63. *Prof Exp:* Explor geologist, Sohio Petrol Co, 56-60; instr geol, Capital Univ, 61-64, asst prof, 64-68; PROF GEOG & EARTH SCI & CHMN DEPT, BLOOMSBURG STATE COL, 68- *Mem:* Am Asn Petrol Geol; Nat Asn Geol Teachers; Geol Soc Am. *Res:* Subsurface stratigraphy and sedimentation, particularly the application to exploration for stratigraphic-type hydrocarbon accumulations. *Mailing Add:* Dept Geog & Earth Sci Bloomsburg State Col Bloomsburg PA 17815

FRANTZ, WILLIAM LAWRENCE, b Canton, Ohio, Nov 3, 27; m 50; c 3. ZOOLOGY. *Educ:* Kent State Univ, BSc, 51; Ohio State Univ, MSc, 53, PhD, 57. *Prof Exp:* Asst prof biol, Drake Univ, 57-60; asst prof physiol & pharmacol, 60-74, PROF PHYSIOL, MICH STATE UNIV, 74- *Mem:* Am Soc Zool; Am Physiol Soc. *Res:* Cellular physiology; intermediary metabolism; membrane transport. *Mailing Add:* Dept Physiol Mich State Univ 206 Giltner Hall East Lansing MI 48824

FRANZ, ANSELM, b Schladming, Austria, Jan 21, 00; nat US; m 29; c 2. MECHANICAL & AERONAUTICAL ENGINEERING. *Educ:* Tech Univ Graz, Austria, MS, 24; Tech Univ Berlin, PhD(eng), 39. *Hon Degrees:* Dr, Tech Univ Graz, Austria, 69. *Prof Exp:* Asst prof turbomachinery, Tech Univ Graz, 24-28; res engr, Schwarzkopff Werke, Berlin, 29-35; vpres eng, Junkers Flugzeug & Motorenwerke, Dessau, Germany, 36-45; consult jet propulsion, Wright-Patterson Air Force Base, US Air Force, Dayton, Ohio, 46-50; vpres & asst gen mgr, Avco Lycoming Div, Stratford, Conn, 51-68; CONSULT, AVCO CORP, TEXTRON & LYCOMING CO, 69- *Honors & Awards:* Freudenberg Prize, Lilienthal Soc Aero Res, Germany, 42; Klemin Award, Am Helicopter Soc, 66; Austrian Distinguished Serv Cross for Sci & Art, 1st Class, President of Austria, 78, Austrian Medal of Honor Sci & Art, 84; R Tom Sawyer Award, Am Soc Mech Engrs, 85. *Mem:* Fel Am Soc Mech Engrs; Soc Automotive Engrs; Am Helicopter Soc; assoc fel Am Inst Aeronaut & Astronaut; Ger Acad Aeronaut Sci; NY Acad Sci. *Res:* Aerodynamics; thermodynamics; combustion; jet-propulsion; gas turbine and jet engine development. *Mailing Add:* 488A Commanche Ln Stratford CT 06497

FRANZ, CRAIG JOSEPH, b Baltimore, Md, Apr 12, 53. ECOLOGY, SYSTEMATICS. *Educ:* Bucknell Univ, BA, 75; Drexel Univ, MSc, 77; Univ RI, PhD(biol), 88. *Prof Exp:* Vis res zool, Estacíon de Investigaciones, Fundacfon La Salle de Ciencias Naturales, Venezuela, 86-87; univ fel zool, Univ RI, 87-88; ASST PROF BIOL, LA SALLE UNIV, 88- *Concurrent Pos:* Trustee, Calvert Hall Col, 82-84; mem, Inst Rev Bd, Albert Einstein Med Ctr, 89-; trustee, Villages, Inc, 91-; teaching asst zool, Univ RI. *Mem:* Am Malacological Union; AAAS; Am Soc Zoologists; Ecol Soc Am; Crustacean Soc Am. *Res:* Ecological and systematic work on marine (archaed) gastropods and polyplacophoran molluscs of the Southern Carribean and Western Atlantic. *Mailing Add:* Dept Biol LaSalle Univ 20th St & Olney Ave Philadelphia PA 19141-1199

FRANZ, CURTIS ALLEN, b Gary, Ind, Jan 6, 40; m 69; c 2. ORGANIC CHEMISTRY, PHARMACEUTICAL CHEMISTRY. *Educ:* Purdue Univ, BS, 62; Ind Univ, MS, 65; Univ Iowa, PhD(org chem), 72. *Prof Exp:* Sr chemist, Ott Chem Co, 70-72; process chemist, 73-75, sr process chemist, 75-77, chief chemist, Org Chem Div, 77-82, PROJ LEADER, LEDERLE LABS, AM CYANAMID CO, 82- *Mem:* Am Chem Soc. *Res:* Process development in chemistry, process maintenance and profit improvement. *Mailing Add:* Lederle Labs Div Am Cyanamid Co Bound Brook NJ 08805

FRANZ, DAVID ALAN, b Philadelphia, Pa, Nov 5, 42; m 64; c 2. ANALYTICAL CHEMISTRY. *Educ:* Princeton Univ, AB, 64; Johns Hopkins Univ, MAT, 65; Univ Va, PhD(chem), 70. *Prof Exp:* High sch teacher chem, Baltimore City Pub Schs, 65-66; from asst prof to assoc prof, 70-90, PROF CHEM, LYCOMING COL, 90- *Concurrent Pos:* Vis scholar, dept chem, Univ NMex, 80-81. *Mem:* Am Chem Soc. *Res:* Teaching methods, particularly tested lecture demonstrations; environmental chemical analysis. *Mailing Add:* Dept Chem Lycoming Col Williamsport PA 17701

FRANZ, DONALD NORBERT, b Indianapolis, Ind, Sept 23, 32; div; c 2. NEUROPHARMACOLOGY. *Educ:* Butler Univ, BS, 54, MS, 62; Univ Utah, PhD(pharmacol), 66. *Prof Exp:* Fel vet physiol, Univ Edinburgh, 66-68; from asst prof to assoc prof, 68-85, PROF PHARMACOL, COL MED, UNIV UTAH, 85- *Concurrent Pos:* Prin investr, Nat Heart, Lung & Blood Inst, 79-86; medicolegal consult, 82- *Mem:* AAAS; Am Soc Pharmacol & Exp Therapeut; Soc Neurosci; Am Heart Asn. *Res:* Neurobiology of monoamines in central autonomic transmission; pharmacology of central monoaminergic pathways; mechanisms of action of antihypertensives and opioids. *Mailing Add:* Dept Pharmacol Univ Utah Sch Med Salt Lake City UT 84132

FRANZ, EDGAR ARTHUR, b Staunton, Ill, Dec 9, 19; m 46; c 4. MATHEMATICS. *Educ:* Univ Iowa, BA, 48, MS, 49. *Prof Exp:* From instr to prof math, Culver-Stockton Col, 49-65; PROF MATH, ILL COL, 65-, HITCHCOCK CHAIR MATH, 80- *Mem:* Math Asn Am. *Res:* General mathematics. *Mailing Add:* 348 Sandusky Jacksonville IL 62650

FRANZ, EDMUND C(LARENCE), b Pittsburgh, Pa, July 8, 20; m 49; c 3. PHYSICAL METALLURGY. *Educ:* Carnegie-Mellon Univ, BS, 48; Case Western Reserve Univ, MS, 53. *Prof Exp:* Res metallurgist, Cleveland Res Div, Aluminum Co Am, 48-53, Alcoa Labs, 53-82; RETIRED. *Concurrent Pos:* Consult, E F Metall Consults, 82- *Mem:* Am Soc Metals; Sigma Xi. *Res:* Alloy technology and fabrication of aluminum alloys. *Mailing Add:* 25 Nancy Dr Pittsburgh PA 15235

FRANZ, FRANK ANDREW, b Philadelphia, Pa, Sept 16, 37; m 59; c 1. PHYSICS. *Educ:* Lafayette Col, BS, 59; Univ Ill, MS, 61, PhD(physics), 64. *Prof Exp:* Res assoc physics, Coord Sci Lab, Univ Ill, 64-65; res fel, Swiss Fed Inst Technol, 65-67; from assst prof to assoc prof, 67-74, assoc dean, Col Arts & Sci, 74-76, dean fac, 77-82, PROF PHYSICS, IND UNIV, 74-; AT DEPT PHYSICS, WVA UNIV. *Concurrent Pos:* NSF fel, 65-67; Alfred P Sloan Found fel, 69-71. *Mem:* AAAS; fel Am Phys Soc; Am Asn Phys Teachers. *Res:* Relaxation phenomena in atomic and solid state physics; optical pumping; collisional interactions between atoms in ground and excited states; magnetic resonance in atomic vapors. *Mailing Add:* Two Ridge Place Morgantown WV 26505-3633

FRANZ, GUNTER NORBERT, b Backa Palanka, Yugoslavia, Mar 13, 35; m 65; c 3. PHYSIOLOGY, BIOPHYSICS. *Educ:* Karlsruhe Tech Univ, dipl elec eng, 59; Univ Wash, PhD(physiol, biophys), 68. *Prof Exp:* Asst prof, 68-72, ASSOC PROF PHYSIOL & BIOPHYS, MED CTR, WVA UNIV, 72- *Mem:* AAAS; Inst Elec & Electronics Engrs; Biophys Soc; Am Physiol Soc. *Res:* Theory of voltage and current clamping; lung mechanics. *Mailing Add:* Dept Physiol WVa Univ Health Sci Ctr Morgantown WV 26506

FRANZ, JAMES ALAN, b Seattle, Wash, June 20, 48; m 74; c 2. PHYSICAL ORGANIC CHEMISTRY. *Educ:* Univ Wash, BS, 70; Univ Ill, Champaign-Urbana, PhD(chem), 74. *Prof Exp:* RES SCIENTIST CHEM, PAC NORTHWEST LABS, BATTELLE MEM INST, 74-, GROUP LEADER, CHEM SCI GROUP. *Concurrent Pos:* Lectr chem, Wash State Univ, 75-; mem, Coal Res Adv Comt, Gas Res Inst, 81 & rev comt, Univ Chicago, Argonne Nat Lab, 85-86. *Mem:* Am Chem Soc. *Res:* Applications of C-13 nuclear magnetic resonance to determination of coal structure; kinetics and mechanisms of free radical reactions; laser chemistry; formation, structure and reactivity of coal. *Mailing Add:* 5325 W 25th Ave Kennewick WA 99337

FRANZ, JOHN EDWARD, b Springfield, Ill, Dec 21, 29; m 51; c 4. BIO-ORGANIC CHEMISTRY. *Educ:* Univ Ill, BS, 51; Univ Minn, PhD(chem), 55. *Prof Exp:* Res chemist, Monsanto Agr Co, 55-59, group leader, 59-62, sci fel, 62-75, sr sci fel, 75-80, distinguished sci fel, 80-91; RETIRED. *Honors & Awards:* IR-100 Award, 77; J F Queeny Award, Monsanto, 81; Nat Medal Technol, Washington, DC, 87; Perkin Medal, 90. *Mem:* Am Chem Soc. *Res:* Fundamental organic research; reaction mechanisms; coenzyme A antimetabolites; antiauxin chemistry (isothiazoles, isoxasoles, pyrazoles); plant chemistry; cell membrane chemistry (glyceride and phospholipid syntheses, liposomes); phosphonomethylglycine (glyphosate) chemistry and processes; plant hormone chemistry (abscissic acid analogs, ethylene generators); biorational design of herbicides; nitrile sulfide chemistry; periselective addition reactions of one- and three-dipoles. *Mailing Add:* 9831 Meadowfern Dr St Louis MO 63126

FRANZ, JOHN MATTHIAS, b Oak Park, Ill, May 23, 27; m 51; c 4. CARCINOGENESIS, EXPERT SYSTEMS. *Educ:* Univ Ill, BS, 50; Univ Iowa, MS, 52, PhD(biochem), 55. *Prof Exp:* From instr to asst prof, 55-61, asst chmn dept, 75-76, assoc chmn dept, 76-79, dir undergrad educ, 79-85, ASSOC PROF BIOCHEM, UNIV MO-COLUMBIA, 61- *Concurrent Pos:* Res assoc, Harvard Univ, 65-66. *Mem:* AAAS; Am Chem Soc; Asn Comput Mach; Am Asn Artificial Intel. *Res:* Developmental and comparative aspects of metabolic control; tryptophan metabolism; carcinogenesis; use of computers in biochemistry; expert systems in biochemistry, biology and medicine. *Mailing Add:* Dept Biochem Univ Mo 322 Chemistry Bldg Columbia MO 65211

FRANZ, JUDY R, b Chicago, Ill, May 3, 38; m 59; c 1. ELECTRONIC PROPERTIES DISORDERED MATERIALS. *Educ:* Cornell Univ, BA, 59; Univ Ill, MS, 61, PhD(physics), 65. *Prof Exp:* Res scientist, IBM Res Labs, Switz, 65-67; asst prof physics, Ind Univ, Bloomington, 68-74, assoc prof physics, 74-79, prof physics, 79-87; PROF PHYSICS, WVA UNIV, MORGANTOWN, 87- *Concurrent Pos:* Vis prof, Tech Univ Munich, Germany, 78-79, Cornell Univ, 85-86, 88 & 90; assoc dean, Col Arts & Sci, Ind Univ, 80-82; counr-at-large, Asn Women in Sci, 81-83 & Am Phys Soc, 84-88; chmn, Gordon Res Conf, 85; mem, Div Mat Res Adv Comt, Nat Sci Found, 86-89; mem, gov bd, Am Inst Physics, 87- & exec bd, Coun Sci Soc Presidents, 90. *Mem:* Fel Am Phy Soc fel; Am Asn Adv Sci fel; Am Asn Physics Teachers (vpres, 88, pres-elect, 89, pres, 90); Am Inst Physics; Asn Women in Sci; Sigma Xi. *Res:* Theoretical investigation of the electronic properties of disordered materials, such as liquid metal alloys, doped semiconductors and metals near their critical points; metal-insulator transitions. *Mailing Add:* Dept Phys WVa Univ Morgantown WV 26506

FRANZ, NORMAN CHARLES, b Newark, NJ, June 12, 25; m 49, 81; c 1. WOOD SCIENCE, WOOD TECHNOLOGY. *Educ:* State Univ NY, BS, 48; Univ Mich, MS, 50, PhD(wood sci), 56. *Prof Exp:* Res engr, Eng Res Inst, 50-54; from instr to assoc prof wood tech, Univ Mich, 54-68; prof forestry, Univ BC, 68-; PROF EMER, 86- *Concurrent Pos:* Indust res & develop, expert witness & consult; vis scientist, Commonwealth Sci & Ind Res Orgn, Australia. *Mem:* Soc Wood Sci & Technol; Forest Prod Res Soc; Am Soc Mech Eng; Int Union Forest Res Orgn. *Res:* Forest products engineering; machining of wood and other materials; ultra-high pressure physics and processing; fluid jet cutting; wood machining. *Mailing Add:* Dept Forestry Univ Brit Col 2075 Wesbrook Mall Vancouver BC V6T 1W5 Can

FRANZ, OTTO GUSTAV, b Eggenburg, Austria, Feb 14, 31; m 62; c 3. OBSERVATIONAL ASTRONOMY. *Educ:* Univ Vienna, DPhil, 55. *Prof Exp:* Res asst astron, Vienna Univ Observ, 53-55; res assoc, Dearborn Observ, Northwestern Univ, 55-58; astronr, US Naval Observ, 58-65; ASTRONR, LOWELL OBSERV, 65- *Concurrent Pos:* Adj prof, Ohio State Univ, 68- *Mem:* Am Astron Soc; AAAS; Sigma Xi; Int Astron Union. *Res:* Astrometry and photometry of visual binary stars; photometric investigation of planets and planetary satellites; occultation studies of solar system objects; space telescope astrometry. *Mailing Add:* Lowell Observ Mars Hill Rd 1400 W Flagstaff AZ 86001

FRANZAK, EDMUND GEORGE, b Chicago, Ill, June 27, 30; m 61; c 3. ENVIRONMENT, SAFETY & HEALTH. *Educ:* Fournier Inst Technol, BS, 52; Univ NMex, MS, 55; Northwestern Univ, MS, 57, PhD(physics), 60. *Prof Exp:* Mem staff, Sandia Corp, 52-55; asst prof physics, Univ Mo, 60-61; head neutron tube develop div, Sandia Nat Labs, 62-66, mgr, Electronic Components Dept, 66-74, mgr, Integrated Circuit Design Dept, 74-77, mgr, Measurement Standards Dept, 77-84, mem, mgt staff (cong liaison), 84-89, COORDR, ENVIRON SAFETY & HEALTH, 90- *Mem:* Sr mem Inst Elec & Electronics Engrs. *Res:* Neutron and theoretical solid state physics; electron transport in metals. *Mailing Add:* 7000 Vista Del Arroyo NE Albuquerque NM 87109

FRANZBLAU, CARL, b New York, NY, Sept 26, 34; m 58; c 2. BIOCHEMISTRY. *Educ:* Univ Mich, BS, 56; Albert Einstein Col Med, PhD(biochem), 62. *Hon Degrees:* DSc, Roger Williams Col, 88. *Prof Exp:* From asst prof to assoc prof, 62-71, PROF BIOCHEM, SCH MED, BOSTON UNIV, 71-, CHMN DEPT, 77-, ASSOC DEAN, BIOMED SCI, 89- *Concurrent Pos:* Res collabr, Brookhaven Nat Lab, 62-72; estab investr, Am Heart Asn, 66-71; mem, Nat Heart Lung Blood Adv Comt, NIH, 84-87. *Mem:* Am Chem Soc; Am Soc Biol Chem; NY Acad Sci; Am Heart Asn. *Res:* Chemistry of connective tissue proteins; enzymology; lung diseases; atherosclerosis research. *Mailing Add:* Dept Biochem Boston Univ Sch Med Boston MA 02118

FRANZEN, DOROTHEA SUSANNA, b Emporia, Kans, Mar 17, 12. ZOOLOGY. *Educ:* Bethel Col, AB, 37; Univ Kans, MA, 43, PhD(zool), 46. *Prof Exp:* Instr biol, Cedar Crest Col, 46-47; from asst prof to assoc prof, Washburn Univ, 47-52; from assoc prof to prof, 52-71, George C & Ella Beach Lewis prof, 71-77, EMER PROF BIOL, ILL WESLEYAN UNIV, 77- *Concurrent Pos:* Researcher, Univ Reading, Eng, 59-60. *Mem:* Fel AAAS; Am Soc Zool; Soc Syst Zool; Am Malacol Union. *Res:* Distribution and ecology of Cenozoic terrestrial and freshwater Mollusca; fossil and living Pupillidae in Kansas; Succineidae in North America. *Mailing Add:* Dept Biol Ill Wesleyan Univ PO Box 281 North Newton KS 67117

FRANZEN, HUGO FRIEDRICH, b New York, NY, Aug 27, 34; m 56; c 3. PHYSICAL CHEMISTRY. *Educ:* Univ Calif, Berkeley, BS, 57; Univ Kans, PhD(chem), 62. *Prof Exp:* Fel, Inst Inorg Chem, Univ Stockholm, 63-64; asst chemist, AEC, 64-65, assoc chemist, 65-69, chemist, 69-74, SR CHEMIST, AMES LAB, DEPT ENERGY, 74-; PROF CHEM, IOWA STATE UNIV, 74- *Concurrent Pos:* From instr to assoc prof, Iowa State Univ, 64-74; am ed, J Alloys & Compounds. *Honors & Awards:* Outstanding Res Mat Chem Award, Dept Energy, 86. *Mem:* AAAS; Am Chem Soc. *Res:* High temperature chemistry; thermodynamics; crystallography; heterogeneous equilibrium; vaporization chemistry. *Mailing Add:* Dept Chem Iowa State Univ Ames IA 50011

FRANZEN, JAMES, b Chicago, Ill, Jan 13, 34; m 61; c 5. BIOCHEMISTRY. *Educ:* Wheaton Col, Ill, BS, 55; Univ Ill, MS, 58, PhD(biochem), 60. *Prof Exp:* USPHS fel, Northwestern Univ, 60-61; res assoc biophys chem, Grad Sch Pub Health, 61-62, asst res prof, 62-65, assoc prof biochem, 65-81, PROF BIOCHEM, UNIV PITTSBURGH, 81- *Mem:* Am Chem Soc; Am Soc Biochem & Molecular Biol. *Res:* RNA splicing. *Mailing Add:* Dept Biol Sci Fac Arts & Sci Univ Pittsburgh Pittsburgh PA 15260

FRANZEN, KAY LOUISE, b Minneapolis, Minn, Apr 15, 50. FOOD SCIENCE. *Educ:* Univ Minn, BS, 70, BChem, 71; Cornell Univ, MS, 73, PhD(food chem), 76. *Prof Exp:* Scientist, 75-77, GROUP LEADER RES & DEVELOP, PILLSBURY CO, 77- *Mem:* Inst Food Technologists; AAAS. *Res:* Protein functionality; protein modification; ingredient functionality; cereals; baked goods. *Mailing Add:* 16385 15 Ave N Plymouth MN 55447

FRANZEN, WOLFGANG, b Duesseldorf, Ger, Apr 6, 22; nat US; m 43. OPTICS. *Educ:* Haverford Col, BS, 42; Columbia Univ, MA, 44; Univ Pa, PhD, 49. *Prof Exp:* Instr physics, Princeton Univ, 49-53; asst prof, Univ Rochester, 53-56; sr physicist, Arthur D Little, Inc, 56-61; assoc prof, 61-65, PROF PHYSICS, BOSTON UNIV, 65- *Concurrent Pos:* Exchange fel, Univ Basel, 54-55; NATO sr fel, Toulouse Univ, 69-70; Fulbright lectr, Nat Univ Colombia, 77; vis prof, Fed Inst Technol, Zurich, 83. *Mem:* Fel Am Phys Soc. *Res:* Nuclear reactions and physics; instrumentation; optical pumping; magnetic resonance; electron resonance scattering and polarization; polarized metastable atomic beams. *Mailing Add:* Dept Physics Boston Univ Boston MA 02215

FRANZIN, WILLIAM GILBERT, b Powell River, BC, July 11, 46; m 69; c 1. FISH BIOLOGY. *Educ:* Univ BC, BSc Hons, 67; Univ Man, MSc, 70, PhD(zool), 74. *Prof Exp:* Biologist, Environ Can, Dept Fisheries & Oceans, Can, 73-75, res scientist toxicol, 75-82, res mgt, 82-83, res scientist, Resource Develop Res, 83-89, RES SCIENTIST, FISH HABITAT RES, DEPT FISHERIES & OCEANS, CAN, 89- *Honors & Awards:* Sr Res Award, Sigma Xi, 80. *Mem:* Am Fisheries Soc; Sigma Xi. *Res:* Fish community dynamics in experimental lakes; fish population genetics; zoogeography; invading/colonizing fish species. *Mailing Add:* Dept Fisheries & Oceans 501 University Crescent Winnipeg MB R3T 2N6 Can

FRANZINI, JOSEPH B(ERNARD), civil engineering, for more information see previous edition

FRANZL, ROBERT E, b New York, NY, May 27, 21; m 52; c 2. IMMUNOBIOLOGY, MITOCHONDRIAL RESPIRATION. *Educ:* City Col New York, BS, 41; Columbia Univ, MA, 48, PhD(biochem), 52. *Prof Exp:* Asst org chem & biochem, Columbia Univ, 46-52, res assoc biochem, 53-54; res assoc immunochem, Sloan-Kettering Inst, 54-56; res assoc, Rockefeller Univ, 56-59, asst prof immunol, 59-72; chief immunoneurol res, Vet Admin Hosp, Coatesville, 72-76; RES SCIENTIST, UNIV NEV, RENO, 76- *Concurrent Pos:* Vis res assoc, Brookhaven Nat Labs, 52-53. *Mem:* AAAS; Am Soc Microbiol; Am Chem Soc; Brit Biochem Soc; Am Asn Immunologist. *Res:* Bacterial enzymes; cyclitols; phospholipids; immunochemistry; antibody synthesis; drug addiction; cancer. *Mailing Add:* 17 Shady Rock Ct Round Rock TX 78664-7598

FRANZMEIER, DONALD PAUL, b Greenwood, Wis, May 13, 35; m 60; c 3. SOIL GENESIS, SOIL CLASSIFICATION. *Educ:* Univ Minn, BS, 57, MS, 58; Mich State Univ, PhD(soil sci), 62. *Prof Exp:* Soil scientist, Soil Conserv Serv, USDA, 62-67; assoc prof, 67-76, PROF AGRON, PURDUE UNIV, WEST LAFAYETTE, 76-, SOIL SURV LEADER, PURDUE AGR EXP STA, 70- *Concurrent Pos:* Res consult, Brazil, Portugal & Hungary. *Honors & Awards:* Commendation Award, Soil & Water Conserv Soc. *Mem:* Am Soc Agron; fel Soil Sci Soc Am; Am Asn Quaternary Res; Soil & Water Conserv Soc; Int Soc Soil Sci. *Res:* Soil-geomorphology relations; soil mineralogy, chemistry and physics in relation to the genesis and classification of soils. *Mailing Add:* Dept Agron Purdue Univ West Lafayette IN 47907

FRANZON, PAUL DAMIAN, b Adelaide, Australia, June 4, 61. PACKAGING, VLSI & MICROELECTRONIC SYSTEMS. *Educ:* Univ Adelaide, BS, 83, BE, 84, PhD(elec eng), 89. *Prof Exp:* Engr, Dept Defense, Australia, 84-85; consult, AT&T Bell Labs, 86-87; dir, Communica Pty Ltd, 87-90; ASST PROF COMPUTER ENG, NC STATE UNIV, 89- *Concurrent Pos:* Consult, MCNC, 89-90, Techsearch Int, 90-91; chmn, Inst Elec & Electronic Engrs CHMT Educ Comt, 91- *Mem:* Inst Elec & Electronics Engrs; Asn Comput Mach; Int Soc Hybrid Microelectronics. *Res:* Design techniques and CAD tools for high speed digital systems, particularly package design with multichip modules; design of low power VLSI-based systems. *Mailing Add:* NC State Univ Box 7911 Hillsborough St Raleigh NC 27695-7911

FRANZOSA, EDWARD SYKES, b Boston, Mass, Oct 28, 45. FORENSIC SCIENCE, PHYSICAL CHEMISTRY. *Educ:* Rensselaer Polytech Inst, BS, 67; State Univ NY Binghamton, PhD(phys chem), 75. *Prof Exp:* FORENSIC CHEMIST, DRUG ENFORCEMENT ADMIN, US DEPT JUSTICE, 75- *Concurrent Pos:* Officer & evidence analyst, Whitney Pt Police Dept, NY, 70-75. *Mem:* Am Chem Soc; Am Acad Forensic Sci. *Res:* Pre-resonance raman intensity - quantitative test of theories using vibronically active pyrazine; computerized band contour analysis of laser excited fluorescence of s-tetrazine to determine excited state rotational constants; microscopic and microchemical analysis of illicit and legitimate drug preparations; gas chromatography and mass spectroscopy analysis of synthesis precursors and by-products in clandestinely manufactured drugs. *Mailing Add:* Spec Testing & Res Lab 7704 Old Springhouse Rd McLean VA 22102

FRAPPIER, ARMAND, b Valleyfield, Que, Nov 26, 04; m 29; c 4. BACTERIOLOGY, HYGIENE. *Educ:* Univ Montreal, BA, 24, MD, 30, Lic-es-Sci, 31; Trudeau Sch, dipl, 32. *Hon Degrees:* Dr, Univ Paris, 64, Lavel Univ, 71, Montreal Univ, 76, Nicolas Copernic Acad Med, Cracow, Poland, 78 & Univ Que, 78. *Prof Exp:* Prof, Fac Med, 33-71, emer founder & dean sch hyg, 45-65, dir & founder, Inst Microbiol & Hyg, 38-74, ADV TO DIR, INST ARMAND-FRAPPIER, UNIV QUEBEC, 74- *Concurrent Pos:* Dir clin labs, St Luke's Hosp, 27-43; mem adv comt med res, Nat Res Coun Can, 52-55; mem panel infection & adv comt biol warfare res, Defence Res Bd, Can; mem adv comt, Pub Health Res, Can, 54-60; mem expert comt tuberc, WHO, 53-74. *Honors & Awards:* Companion, Order of Can, 69; Jean Toy Prize, Acad Sci, Inst of France, 71; Marie-Victorin Prize, Que, 79; J L Ldvesque Prize, 83. *Mem:* Fel Am Pub Health Asn; Am Asn Path & Bact; fel Royal Soc Can; Can Physiol Soc; Can Soc Microbiol (pres, 54). *Res:* Allergy and immunization in tuberculosis using Bacillus Calmette-Guerin vaccine; development of scarification test; experimental tuberculosis using radio-isotopes; gas gangrene mechanism; role of bacterial surface washings in production of immunity against whooping cough and other agents; infection promoting factors; poliomyelitis; prevention of leukemia with Bacillus Calmette-Guerin. *Mailing Add:* Inst Armand-Frappier CP 100 Laval-des-Rapides PQ H7V 1B7 Can

FRAREY, MURRAY JAMES, b Midland, Ont, Jan 31, 17; m 54; c 2. GEOLOGY. *Educ:* Univ Western Ont, BA, 40; Univ Mich, MSc, 51, PhD(geol), 54. *Prof Exp:* Geologist, Kerr-Addison Gold Mines, Ltd, 41-43; geologist, Geol Surv Can, 47-85; RETIRED. *Mem:* Geol Asn Can. *Res:* Precambrian geology. *Mailing Add:* 5 Gervin St Nepean ON K2G 0J6 Can

FRASCELLA, DANIEL W, b New Brunswick, NJ, July 6, 34; m 56; c 3. PHYSIOLOGY, BIOCHEMISTRY. *Educ:* Rutgers Univ, BS, 60, MS, 62, PhD(physiol, biochem), 68. *Prof Exp:* Pharmacologist, Wallace Labs, 62-63; asst prof physiol, Rutgers Univ, 66-69; sr res fel, Merck Inst Therapeut Res, 69-70; asst prof physiol, St John's Univ, 70-74; asst dir, 74-75, ASSOC DIR SCI MKT, HOECHST-ROUSSEL PHARMACEUT, INC, AM HOECHST CORP, 74- *Concurrent Pos:* Vis lectr, Dept Natural Sci, Pace Col, 66-67; vis assoc prof biol, City Col New York, 72-74. *Honors & Awards:* Dunning Award, Am Pharmaceut Asn. *Mem:* Am Soc Zoologists; AAAS; NY Acad Sci; Sigma Xi; Am Diabetes Asn; Am Col Clin Pharmacol; Am Soc Hosp Pharmacists. *Res:* Tissue biochemical changes induced by stress and cold exposure; whole body and tissue changes induced by stress in diabetics; dietary effect on tissue enzymes; medical - continuing education program development; new drug development; pharmacology. *Mailing Add:* RD 1 Box 205 Stanton-Lebanon Rd Lebanon NJ 08833

FRASCH, CARL EDWARD, b Albuquerque, NMex, Sept 7, 43; m 63; c 2. MENINGOCOCCAL VIRULENCE. *Educ:* Univ Wash, BS, 67; Univ Minn, PhD(microbiol), 72. *Prof Exp:* post doctoral fel, Rockefeller Univ, New York, NY, 72-75; RES SCIENTIST, OFF BIOLOGICS, FOOD & DRUGS ADMIN, 75- *Concurrent Pos:* Assoc ed, J Immunol, 78-80. *Mem:* Am Soc Microbiol; Harvey Soc. *Res:* Development, evaluation and quality control of vaccines for the prevention of bacterial meningitis; studies on the human immune response to meningococcal antigens and on the mechanisms of meningococcal virulence. *Mailing Add:* Div Bact Prods Off Biologics 8800 Rockville Pike Bethesda MD 20892

FRASCHE, DEAN F, b Council Bluffs, Iowa, Aug 26, 06. GEOLOGY & MINE-GEOLOGY. *Educ:* Univ Wis, PhD(geol), 34. *Prof Exp:* Chmn bd & managing dir, Thailand Explor & Mining Co, Bangkok, 60-71; RETIRED. *Mem:* Am Asn Econ Geol; Geol Soc Am. *Mailing Add:* 36 Park Ave Greenwich CT 06830

FRASCO, DAVID LEE, b Brush, Colo, Apr 8, 31; m 54; c 2. PHYSICAL CHEMISTRY, SPECTROSCOPY. *Educ:* Colo State Col, AB, 53; Wash State Univ, MS, 55, PhD(infrared spectra of solids), 58. *Prof Exp:* PROF CHEM, WHITMAN COL, 58- *Concurrent Pos:* Consult, Pac Northwest Labs, Battelle Mem Inst, 68-78. *Res:* Physical, chemical and electrical properties of thin lipid membranes, especially artificial cell membranes; infrared spectra of molecular solids. *Mailing Add:* 716 Estralla Walla Walla WA 99362

FRASER, ALAN RICHARD, b Arvida, Que, May 22, 44. INORGANIC CHEMISTRY, BIOCHEMISTRY. *Educ:* McGill Univ, BSc, 66; Sir George Williams Univ, PhD(chem), 74. *Prof Exp:* Fel biochem, Rockefeller Univ, 74-76; res assoc, Univ Toronto, 76-78; from asst prof to assoc prof, 78-85, PROF BIOCHEM, UNIV MONCTON, 85-, CHAIRPERSON, 89- *Concurrent Pos:* Med Res Coun Can fel, 74-76; Atkinson Charitable Found fel, 77-78. *Res:* Mitochondrial DNA molluscs and bivalues; biochemistry of marine organisms. *Mailing Add:* Dept Chem & Biochem Univ Moncton Moncton NB E1A 3E9 Can

FRASER, ALBERT DONALD, b Fredericton, NB, Can, Jan 6, 49; m 79; c 1. CLINICAL TOXICOLOGY, FORENSIC TOXICOLOGY. *Educ:* Houghton Col, BA, 71; Boston Univ, PhD(biochem), 76; Univ Toronto, DCC, 78; Am Bd Clin Chem, dipl, 85, Am Bd Forensic Toxicol, 86. *Prof Exp:* Biochemist, Moncton Hosp, 79-81; HEAD, TOXICOL LAB, DIV CLIN CHEM, VICTORIA GEN HOSP, 82- *Concurrent Pos:* Asst prof, Dept Path, Dalhousie Univ, 82-86, assoc prof, 86-; mem, coun, Can Soc Clin Chemists, 85-87. *Honors & Awards:* Med-Chem Award, Can Soc Clin Chemists, 84. *Mem:* Can Soc Clin Chemists; Can Soc Forensic Sci; Am Asn Clin Chem; Am Acad Forensic Sci; Nat Acad Clin Biochem; Int Asn Forensic Toxicologists. *Res:* Pharmacokinetics and new analytical methods for drug analysis; chromatographic studies, especially high performance liquid chromatography; forensic toxicology; drug-drug interactions. *Mailing Add:* Toxicol Lab Victoria Gen Hosp 1278 Tower Rd Halifax NS B3H 2X9 Can

FRASER, ALEX STEWART, b London, UK, Dec 24, 23; m 50; c 6. VISUAL PERCEPTION, GENETICS. *Educ:* Univ Edinburgh, BSc, MSc & PhD, 51. *Prof Exp:* Sr prin res off genetics, Commonwealth Sci & Indust Res Conf, 51-62; prof, Univ Calif, Davis, 62-67; prof genetics, Univ Cincinnati, 67-89; RETIRED. *Res:* Variation of visual perception in humans and the degree to which such variation is genetically determined. *Mailing Add:* 505 Ludlow Apt 13 Cincinnati OH 45220

FRASER, ALISTAIR BISSON, b Rossland, BC, May 31, 39; m 62; c 2. ATMOSPHERIC PHYSICS, OPTICS. *Educ:* Univ BC, BSc, 62; Univ London, DIC, 68, PhD(meteorol), 68. *Prof Exp:* Meteorol officer, Dept Transp, Govt Can, 62-64; from res assoc to asst prof atmospheric sci, Univ Wash, 68-72; assoc prof, 72-79, PROF METEOROL, PA STATE UNIV, 79- *Mem:* Am Meteorol Soc; Optical Soc Am; Royal Meteorol Soc; Can Meteorol Soc. *Res:* The optical properties of the atmosphere with emphasis on image formation and remote sensing. *Mailing Add:* Dept Meteorol Pa State Univ University Park PA 16802

FRASER, ANN DAVINA ELIZABETH, b Wimbledon, Eng, Mar 1, 54; Can citizen. BIOTECHNOLOGY, ANIMAL DISEASES. *Educ:* Queen's Univ, Ont, HBSc, 76; Carleton Univ, Ont, MSc, 78, PhD(microbiol), 82. *Prof Exp:* Res biotechnologist, Borden Co, Ltd, 82-83; RES SCIENTIST, ANIMAL DIS RES INST AGR, CAN, 84. *Concurrent Pos:* Sessional lectr, Carleton Univ, Ottawa, 85- *Mem:* Am Soc Microbiol; Sigma Xi; NY Acad Sci; Am Asn Anal Chemists; Soc Indust Microbiol. *Res:* Studies on the physiology, molecular biology, genetics and biochemistry of food borne pathogenic microorganisms. *Mailing Add:* Animal Dis Res Inst 3851 Fallowfield Rd PO Box 11300 Sta H Nepean ON K2H 8P9 Can

FRASER, BLAIR ALLEN, b Rockford, Ill, July 27, 48; m 74; c 2. BIOCHEMISTRY, IMMUNOLOGY. *Educ:* Hope Col, AB, 70; Pa State Univ, MS, 72, PhD(biochem), 74. *Prof Exp:* Res assoc immunochem, Rockefeller Univ, 74-77; staff fel, Nat Inst Allergy & Infectious Dis, 77-78; sr staff fel, Biochem Br, Div Bact Prods, 78-79 & res chemist, 79-80, RES CHEMIST, DIV BIOCHEM & BIOPHYS, CTR FOR BIOLOGICS EVAL & RES & DIR, CHEM BIOL LAB, DIV BIOCHEM & BIOPHYS, FOOD & DRUG ADMIN, 81- *Concurrent Pos:* Nat res serv award, 75-77; assoc-ed, J Immunol, 78-81; adj prof, Catholic Univ, 80- *Mem:* Sigma Xi; Am Chem Soc; Harvey Soc; Am Soc Biol Chem & Molecular Biol; Am Asn Immunologists; Am Soc Mass Spectrometry. *Res:* Structural and immunochemical studies on complex glycoconjugates of microbial cell surface origin; mass spectrometric methods of structural analysis of biomolecules; peptide structural analysis; neuro-immune functions. *Mailing Add:* Ctr Biol Eval & Res 8800 Rockville Pike Bethesda MD 20892

FRASER, CLARENCE MALCOLM, b Hamiota, Man, June 21, 26; m 55; c 5. VETERINARY MEDICINE. *Educ:* Univ Man, BSA, 49; Univ Toronto, DVM, 54, MVSc, 63. *Prof Exp:* Self employed veterinarian, 54-58; ambulatory clinician vet med, Ont Vet Col, 58-63, prof vet med, 63-65; prof vet med & head dept vet clin studies, Univ Sask, 65-70; assoc ed vet manual, Merck & Co, Inc, 70-75; mgr clin res, Pitman Moore, Inc, 75-76; assoc ed, 76-81, ED, VET MANUAL, MERCK & CO, INC, 82- *Mem:* Am Vet Med Asn; Can Vet Med Asn. *Res:* Veterinary parasitology; infectious and metabolic diseases. *Mailing Add:* Four Old Forge Rd Flemington NJ 08822-9417

FRASER, DAVID ALLISON, b Philadelphia, Pa, Aug 29, 22; m 47; c 2. OCCUPATIONAL HEALTH, ELECTRON MICROSCOPY. *Educ:* Univ Pa, BA, 47; Xavier Univ, Ohio, MS, 57; Univ Cincinnati, ScD(indust health), 61. *Hon Degrees:* DSc, Univ Cincinnati, 61. *Prof Exp:* Res chemist, Chem Mfg & Distrib Co, 47-49; chemist, Div Indust Hyg, USPHS, Washington, DC, 49-52, chief aerosol unit, Div Occup Health Prog, Ohio, 52-61; assoc prof, 61-68, PROF INDUST HEALTH, UNIV NC, CHAPEL HILL, 68- *Concurrent Pos:* Lectr, Univ Cincinnati, 58-61; consult, Div Radiol Health, USPHS, 64-; dir, Occup Safety & Health Educ Resource Ctr, Univ NC, 77-86; mem, Indust Hyg Round-Table, 84. *Honors & Awards:* Cummings Award, Am Indust Hyg Asn, 84. *Mem:* Fel AAAS; Am Chem Soc; Electron Micros Soc Am; Am Indust Hyg Asn; fel Royal Micros Soc. *Res:* Industrial health and toxicology; air sampling techniques; physics and sampling of airborne particulates; optical microscopy. *Mailing Add:* 408 Longleaf Dr Chapel Hill NC 27514-2607

FRASER, DAVID WILLIAM, b Abington, Pa, May 10, 44; m 66; c 2. PUBLIC HEALTH. *Educ:* Haverford Col, BA, 65; Harvard Med Sch, MD, 69. *Hon Degrees:* DSc, Moravian Col, 87. *Prof Exp:* Chief, Spec Pathogens Br, Bacterial Dis Div, Bur Epidemiol, Ctr Dis Control, 75-80; med epidemiol consult, Health Br, Health & Income Maintenance Div, Off Mgt & Budget, 80-81; asst dir med sci, Bacterial Dis Div, Ctr Infectious Dis, Ctrs Dis Control, 81-82; PRES, SWARTHMORE COL, 82- *Concurrent Pos:* Adj prof med, Univ Penn, 83-; mem, Comt Nat Strategy for AIDS, Inst Med, Nat Acad Sci, Immunization Practices Adv Comt Ctr Dis Control. *Honors & Awards:* Richard & Hinda Rosenthal Found Award, Am Col Physicians, 79; John Scott Award, 86. *Mem:* Am Col Physicians; Am Epidemiol Soc; Sigma Xi; Infectious Dis Soc of Am; Am Col Eng. *Res:* Epidemiology of bacterial diseases, with emphasis on Legionnaires diseases and related diseases; bacterial meningitis and other infections of the respiratory tract. *Mailing Add:* 324 Cedar Lane Swarthmore PA 19081

FRASER, (WILLIAM) DEAN, molecular biology; deceased, see previous edition for last biography

FRASER, DONALD, b Toronto, Ont, Feb 14, 21; m 54; c 1. PHYSIOLOGY. *Educ:* Univ Toronto, MD, 44, MA, 46, PhD(physiol), 50. *Prof Exp:* Res assoc, Hosp for Sick Children, 53-58; assoc pediat, 55-58, asst prof pediat & physiol, 58-77, mem fac physiol, 77-80, RES MEM, RES INST, HOSP FOR SICK CHILDREN, 58- *Honors & Awards:* Medal Med, Royal Col Physicians & Surgeons, Can, 56. *Mem:* AAAS; Soc Pediat Res; fel Am Acad Pediat; Can Pediat Soc; Sigma Xi. *Res:* Physiology of bone metabolism as applied to children. *Mailing Add:* 10915 W Santa Fe Dr Sun City AZ 85351

FRASER, DONALD ALEXANDER STUART, b Toronto, Ont, Apr 29, 25. STATISTICS. *Educ:* Univ Toronto, BA, 46, MA, 47; Princeton Univ, MA, 48, PhD(math), 49. *Prof Exp:* Instr math, Princeton Univ, 47-49; from asst prof to prof, 49-77, MEM FAC MATH, UNIV TORONTO, 77- *Mem:* Fel AAAS; fel Am Statist Asn; fel Inst Math Statist; fel Royal Statist Soc; fel Royal Soc Can. *Res:* Mathematical statistics. *Mailing Add:* 4 Old George Pl Toronto ON M4W 1X9 Can

FRASER, DONALD BOYD, b Teaneck, NJ, Nov 15, 30; m 55; c 2. PHYSICAL CHEMISTRY. *Educ:* St Peter's Col, BS, 54; Rutgers Univ, PhD(phys chem), 60. *Prof Exp:* Chemist, Enjay Labs, Esso Res & Eng Co, 60-62; sr develop engr, Celanese Plastics Co, NJ, 62-63, group leader, 63-68; assoc prof, 68-71, PROF CHEM, ESSEX COUNTY COL, 71-, CHMN DEPT, 70- *Mem:* Am Chem Soc. *Mailing Add:* Dept Chem 303 University Ave Newark NJ 07102

FRASER, DONALD C, b New York, NY, Apr 20,41; div; c 2. SCIENCE ADMINISTRATION. *Educ:* Mass Inst Technol, BS, 62, MS, 63, ScD, 67. *Prof Exp:* Tech staff, Mass Inst Technol Instrumentation Lab, 67-69; div leader, 69-81, vpres tech opers, 81-87, EXEC VPRES, C S DRAPER LAB INC, 87- *Concurrent Pos:* Vis prof, Stanford Univ, 70-71; assoc ed, Am Inst Aeronaut & Astronaut, J Spacecraft & Rockets, 70-72, ed-in-chief, 74-78; chmn, Air Force Studies Bd Comt Fault Isolation, 84-; lectr, MIT Aeronaut & Astronaut Dept, 72-; mem, Air Force Studies Bd Comt Advan Avionics; ed-in-chief, J Guidance, Control & Dynamics, Am Inst Aeronaut & Astronaut, 78- *Mem:* Nat Acad Eng; fel Am Inst Aeronaut & Astronaut; fel Am Astronaut Soc. *Res:* Design and prototyping of advanced guidance, control and navigation systems. *Mailing Add:* C S Draper Lab Inc 555 Technol Sq Cambridge MA 02139

FRASER, DOUGLAS FYFE, b Hartford, Conn, June 17, 41. ECOLOGY. *Educ:* Univ Mich, AB, 63; Univ Md, MS, 70, PhD(zool), 74. *Prof Exp:* From asst prof to assoc prof, 74-85, PROF BIOL, SIENA COL, 85- *Concurrent Pos:* Hon vis prof, Univ W I, Trinidad, 89-90. *Mem:* Am Soc Ichthyologists & Herpetologists; AAAS; Am Fisheries Soc; Ecol Soc Am; Am Soc Naturalists. *Res:* Population and community ecology of stream dwelling fish and woodland salamanders; rattlesnake population ecology. *Mailing Add:* Dept Biol Siena Col Loudonville NY 12211

FRASER, FRANK CLARKE, b Norwich, Conn, Mar 29, 20; nat Can; m 48; c 4. MEDICAL GENETICS, TERATOLOGY. *Educ:* Acadia Univ, BSc, 40; McGill Univ, MSc, 41, PhD(genetics), 45, MD, CM, 50. *Hon Degrees:* DSc, Acadia Univ, 67. *Prof Exp:* Demonstr genetics, McGill Univ, 46-50, from lectr to asst prof genetics, 50-55, from asst prof to assoc prof genetics, 55-60, from assoc prof to prof pediat, 60-82, prof med genetics, 60-82, prof clin genetics, Mem Univ Ffld, 82-85, EMER PROF BIOL & PEDIAT, MCGILL UNIV, 85- *Concurrent Pos:* Molson-McConnel res fel, 50-51; dir, dept med genetics, Montreal Children's Hosp, 50-82; clin fel, Royal Victoria Hosp, 50-82; teaching fel, McGill Univ, 52-54; consult, Shriners Hosp Crippled Children, 54-82. *Honors & Awards:* Allen Award, Am Soc Human Genetics, 77; Blacleader Award, Can Med Asn, 68; March of Dimes Award for Contrib Field of Birth Defects; Honors of the Am Cleft Palate Asn, 74; Medal of Honor, Am Asn Plastic Surgeons, 75. *Mem:* Am Soc Human Genetics (vpres, 59, pres, 62); Teratol Soc (pres, 62); Soc Pediat Res; Genetics Soc Can; Royal Soc Can; Can Col Med Geneticists (pres, 80-83). *Res:* Experimental production of congenital defects; inheritance of human diseases; genetic counseling. *Mailing Add:* Dept Biol Stewart Bldg McGill Univ Montreal PQ H3A 1B1 Can

FRASER, GERALD TIMOTHY, b Detroit, Mich, Mar 1, 57; m 89. MOLECULAR SPECTROSCOPY. *Educ:* Loyola Univ, Chicago, BA, 79; Harvard Univ, PhD(phys chem), 85. *Prof Exp:* NAT RES COUN FEL PHYS CHEM, NAT BUR STANDARDS, GAITHERSBURG, MD, 85- *Mem:* Am Phys Soc. *Res:* Molecular spectroscopy; van der Waals complexes; hydrogen bonding. *Mailing Add:* Molecular Physics Div Nat Inst Standards & Technol Gaithersburg MD 20899

FRASER, GORDON SIMON, b Chicago, Ill, Oct 18, 46. SEDIMENTOLOGY. *Educ:* Univ Ill, BS, 68, MS, 70, PhD(geol), 74. *Prof Exp:* Res asst geol, Ill State Geol Surv, 68-74; Sedimentologist, BP-Alaska, Inc, 74-78; SR RES SCIENTIST, IND GEOL SURV, 78-; ASSOC PROF GEOL, IND UNIV, 79- *Mem:* Soc Econ Paleontologists & Mineralogists; Int Asn Sedimentologists; Geol Soc Am; Am Quaternary Asn. *Res:* coastal sedimentation; sedimentology of clastic and carbonate rocks; proglacial sedimentation. *Mailing Add:* 908 Rainier Ct Bloomington IN 47401

FRASER, GRANT ADAM, b New York, NY, Dec 14, 43; m 77; c 2. LATTICE THEORY, UNIVERSAL ALGEBRA. *Educ:* Univ Calif, Los Angeles, AB, 64, MA, 69, PhD(math), 70. *Prof Exp:* Asst prof math, Univ San Francisco, 71-72; from asst prof to assoc prof math, Univ Santa Clara, 72-82; assoc prof, 82-84, PROF MATH & COMPUTER SCI, CALIF STATE UNIV, LOS ANGELES, 84- *Mem:* Am Math Soc; Math Asn Am. *Res:* Investigation of structure of tensor products of lattices, distribute lattices and semilattices; analysis of the word problem in the tensor product of these lattices. *Mailing Add:* Dept Math & Computer Sci Calif State Univ Los Angeles Los Angeles CA 90032

FRASER, HARVEY R(EED), b Elizabeth, Ill, Aug 11, 16; m 40; c 3. MECHANICS. *Educ:* US Mil Acad, BS, 39; Calif Inst Technol, MS, 48; Univ Ill, PhD(fluid mech), 56; Von Karman Inst Fluid Dynamics, Belg, dipl, 61. *Hon Degrees:* DSc, Yankton Col. *Prof Exp:* Commanding officer, Eng Co, US Army, Hawaii, 40-43 & engr combat battalion, Europe, 44-45, opers officer, Oak Ridge, Tenn, 45-46, instr mech, US Mil Acad, 48-50, assoc prof, 50-52, prof, 53-65; dean eng, SDak Sch Mines & Technol, 65-66, pres, 66-75; dean acad affairs, Ore Inst Technol, 75-79; dean acad, Calif Maritime Acad, 79-84; dir, MDU Resources Group Inc, 71-87; RETIRED. *Mem:* Soc Am Mil Engrs; Am Soc Eng Educ; Sigma Xi. *Res:* Fluid mechanics; diffuser flow. *Mailing Add:* 1229 Leisure World Mesa AZ 85206

FRASER, IAN MCLENNAN, b Victoria, Australia, June 21, 27; m 49; c 3. PHARMACOLOGY. *Educ:* Univ Sydney, BSc, 49; Cambridge Univ, PhD(biol), 52. *Prof Exp:* Lectr physiol, New South Wales Univ Technol, 52-53; from instr to assoc prof, 53-67, chmn dept, 67-70, PROF PHARMACOL, SCH MED, LOMA LINDA UNIV, 67-, CHMN DEPT PHYSIOL & PHARMACOL, 70- *Concurrent Pos:* NIH spec fel, 66-67. *Mem:* Am Soc Pharmacol & Exp Therapeut; Mycol Soc Am; Bot Soc Am. *Res:* Chemotherapy; drug metabolism; pharmacogenetics. *Mailing Add:* Dept Physiol & Pharmacol Loma Linda Univ Sch Med Loma Linda CA 92354

FRASER, J(ULIUS) T(HOMAS), b Budapest, Hungary, May 7, 23; US citizen; m 73; c 3. STUDY OF TIME, SCIENCE WRITING. *Educ:* Cooper Union, BEE, 50; Univ Hannover, PhD(philos), 70. *Prof Exp:* Engr, Mackay Radio & Tel Co, NY, 50-53; res engr, Westinghouse Elec Corp, 53-55; staff mem, Res Physics Sect, Kearfott Div, Singer Co, 55-58, sr staff mem, 58-62, sr scientist, 62-71; FOUNDER & ACTG EXEC SECY, INT SOC STUDY TIME, 66- *Concurrent Pos:* Res assoc physics & astron, Mich State Univ, 62-65; vis lectr, Mt Holyoke Col, 67-69; vis prof, Dept Hist, Univ Md, 69-70; assoc prof, Fordham Univ, 71-85. *Mem:* Int Soc Study Time. *Res:* Study of time; philosophy and history of science. *Mailing Add:* Int Soc for the Study of Time PO Box 815 Westport CT 06881-0815

FRASER, JAMES MATTISON, b Bozeman, Mont, July 31, 25; m 51; c 3. PHYSICAL CHEMISTRY, ANALYTICAL CHEMISTRY. *Educ:* Univ Wis, BS, 53, PhD(phys chem), 57. *Prof Exp:* From res chemist to sr res chemist, Res Ctr, Pure Oil Co, 56-60, group supvr phys chem, 60-61, sect supvr instrumental anal, 61-62, from asst dir to dir anal res & serv div, 62-65, supvr spectral anal, 65-69, MGR ANALYTICAL RES & SERV, RES DEPT, UNION OIL CO, CALIF, 69- *Mem:* Am Chem Soc; Soc Appl Spectros. *Res:* Instrumental and chemical analysis and analytical methods development of petroleum and its products. *Mailing Add:* UNOCAL Sci Tech Div 376 S Valencia Ave PO Box 76 Brea CA 92621-6345

FRASER, JAMES MILLAN, b July 5, 25; Can citizen; m 51; c 8. FISH & WILDLIFE SCIENCES. *Educ:* St Francis Xavier Univ, BSc, 50; Univ Toronto, MA, 53. *Prof Exp:* District biologist, Ont Ministry Natural Resources, 53-57, fish & wildlife supvr, 57-60, dir, Harkness Lab Fisheries Res, 64-69, scientist-in-charge brook trout res, 61-89; RETIRED. *Mem:* Am Fisheries Soc; fel Am Inst Fishery Res Biologists. *Res:* Brook trout research program of Ontario. *Mailing Add:* 336 Crosby Ave Richmond Hill ON L4C 2R7 Can

FRASER, JOHN STILES, b Wonsan, Korea, June 23, 21; Can citizen; m 44; c 2. ACCELERATOR PHYSICS. *Educ:* Dalhousie Univ, BSc, 42; McGill Univ, PhD(physics), 49. *Prof Exp:* Asst cyclotron, McGill Univ, 47-49; res physicist nuclear physics, Atomic Energy Can, Ltd, 49-70, sr res officer accelerator physics, 70-81, br head accelerator physics, 78-81; mem staff, Accelerator Technol Div, Los Alamos Nat Lab, 81-87; RETIRED. *Concurrent Pos:* Consult, Triumf Lab, UBC, Vancouver, BC, 87- *Honors & Awards:* IR-100 Award, 87. *Mem:* Fel Am Phys Soc; Can Asn Physicists. *Res:* Accelerator design; data processing; nuclear instrumentation; nuclear fission; nuclear reactions. *Mailing Add:* PO Box 1341 Ganges BC V0S 1E0 Can

FRASER, LEMUEL ANDERSON, b Donora, Pa, June 18, 18; m 42; c 1. ZOOLOGY. *Educ:* Am Univ, BA, 39; Univ Wis, MA, 40, PhD(zool), 44. *Prof Exp:* Asst zool, Univ Wis, 41-44; instr, Univ Tex, 46-48, asst prof, 48-49; from asst prof to assoc prof, 49-59, PROF ZOOL, UNIV WIS-MADISON, 59- *Concurrent Pos:* Chmn dept zool, Univ Wis, 57-62 & 64-67. *Mem:* Am Soc Zoologists; Am Micros Soc. *Res:* Freshwater invertebrates; invertebrate embryology. *Mailing Add:* Prof Zool Birge Hall Rm 258 Madison WI 53706

FRASER, MARGARET SHIRLEY, b Biggar, Sask; US citizen. INORGANIC CHEMISTRY. *Educ:* Univ Alta, BS, 48, MS, 50; La State Univ, PhD(inorg chem), 71. *Prof Exp:* Res assoc biochem, Univ Alta, 50-51; proj asst oncol, McArdle Inst, Univ Wis-Madison, 51-52; instr chem, Univ Wis-Racine Ctr, 52-71; from asst prof to assoc prof chem, Univ Wis-Parkside, 78-83; RETIRED. *Mem:* Am Chem Soc; AAAS. *Res:* Olefin and acetylene reactions with hydridoiridum complexes. *Mailing Add:* 415 Cherry Hill Dr Racine WI 53405

FRASER, NIGEL WILLIAM, b May 14, 47; British citizen. TRANSCRIPTION, VIRAL LATENCY. *Educ:* Univ Aberdeen, Scotland, MSc, 72; Univ Glasgow, PhD(biochem), 75. *Prof Exp:* Sr Fulbright-Hays fel, Rockefeller Univ, 75-78; from asst to assoc prof, Wistar Inst Anat & Biol, 78-88; asst prof, 80, assoc prof, 84-89, PROF, DEPT MICROBIOL, SCH MED, UNIV PA, 89-; PROF, WISTAR INST ANAT & BIOL, 88- *Mem:* Am Soc Microbiol; Am Soc Virol. *Res:* Latency of herpes simplex virus type I in a mouse model system; molecular biology of DNA viruses; transcriptional termination of RNA polymerase II in adenovirus type II. *Mailing Add:* Wistar Inst Anat & Biol 36th St & Spruce Philadelphia PA 19104

FRASER, PETER ARTHUR, b Ancon, Ecuador, Aug 26, 28; nat Can; m 52; c 4. PHYSICS, NUMERICAL ANALYSIS. *Educ:* Univ Western Ont, BSc, 50, PhD(physics), 54; Univ Wis, MS, 52. *Prof Exp:* Asst physics, Univ Wis, 50-52; res assoc, Univ Western Ont, 52-54; Nat Res Coun Can fel, Univ Col, Univ London, 54-56; lectr physics, 56-57, from asst prof to prof physics, 57-70, PROF APPL MATH, UNIV WESTERN ONT, 70- *Mem:* Am Asn Physics Teachers; Brit Inst Physics. *Res:* Low energy atomic collisions. *Mailing Add:* Dept Appl Math Univ Western Ont London ON N6A 5B9 Can

FRASER, ROBERT GORDON, b Winnipeg, Man, June 30, 21; m 45; c 4. MEDICINE. *Educ:* Univ Man, LMCC, 45; FRCP(C), 56. *Prof Exp:* Resident radiol, Royal Victoria Hosp, 48-50; demonstr, 51-54, from lectr to assoc prof, McGill Univ, 54-68, prof radiol, 68-76, chmn dept, 71-76; head, Div Diag Radiol, 76-80, prof diag radiol, 76-88, PROF EMER DIAG RADIOL, MED CTR, UNIV ALA, BIRMINGHAM, 88- *Concurrent Pos:* Fel radiol, Royal Victoria Hosp, 51, clin asst, 52, assoc radiologist, 54-56, radiologist, 57-64, diag radiologist-in-chief, 64-76; consult radiologist, Royal Can Air Force, 54-59, Montreal Children's Hosp, 54-76 & Montreal Neurol Inst, 55-76; adv, Dept Vet Affairs, Can, 70-76; consult, Can Forces Med Coun. *Honors & Awards:* Medal, Am Col Chest Physicians, 72; Gold Medal, Am Roentgen Ray Soc, 89; Gold Medal, Rad Soc NAm, 90. *Mem:* Can Asn Radiol (pres, 70); Fleischner Soc (pres, 70-72); Am Roentgen Ray Soc (1st vpres, 72); Radiol Soc NAm; fel Royal Col Radiol. *Res:* Diagnostic radiology. *Mailing Add:* Dept Radiol Sch Med Univ Ala Univ Sta Birmingham AL 35294

FRASER, ROBERT ROWNTREE, b Ottawa, Ont, Oct 25, 31; m 64; c 4. ORGANIC CHEMISTRY. *Educ:* Univ Western Ont, BSc, 53, MSc, 54; Univ Ill, PhD(org chem), 58. *Prof Exp:* Asst prof org chem, Univ Ottawa, 58-62; sr res chemist, Bristol Labs Div, Bristol-Myers Co, 62-64; assoc prof org chem, 64-70, PROF ORG CHEM, UNIV OTTAWA, 70-, CHMN DEPT, 85- *Mem:* Am Chem Soc; Chem Inst Can. *Res:* Stereochemistry and mechanisms of organic reactions; nuclear magnetic resonance spectroscopy. *Mailing Add:* Dept Chem Univ Ottawa Ottawa ON K1N 9B4 Can

FRASER, ROBERT STEWART, b Nelson, BC, Feb 14, 22; m 49; c 4. INTERNAL MEDICINE, CARDIOLOGY. *Educ:* Univ Alta, BSc, 44, MD, 46, MSc, 50; FRCP(C), 54. *Prof Exp:* Markle scholar, 53-56; Muttart assoc prof med, 55-64, chmn dept, 69-74, prof clin med, 64-78, ASSOC DEAN MED, UNIV ALTA, 78- *Concurrent Pos:* Consult, Charles Camsell Indian Hosp, Edmonton. *Mem:* Can Soc Clin Invest; Can Cardiovasc Soc (pres, 64). *Res:* Clinical cardiology; clinical and experimental hemodynamic studies. *Mailing Add:* 7426 118 A St Univ Alta Edmonton AB T5G 1V5 Can

FRASER, ROBERT STUART, b Hiawatha, Utah, May 4, 23; m 50; c 2. METEOROLOGY. *Educ:* Univ Calif, Los Angeles, BA, 49, MA, 52, PhD(meteorol), 60. *Prof Exp:* Res asst atmospheric radiation, Univ Calif, Los Angeles, 52-58; mem tech staff atmospheric physics, TRW Syst, 58-71; SPACE SCIENTIST, NASA, 71- *Mem:* Am Meteorol Soc; Am Optical Soc. *Res:* Measurement of atmospheric characteristics from satellite; effect of Earth's atmosphere on satellite observations; effect of aerosols on Earth's radiation budget. *Mailing Add:* DFACS US Air Force Acad Colorado Springs CO 80840

FRASER, RONALD CHESTER, b Portage la Prairie, Man, Nov 6, 19; US citizen; m 44; c 4. DEVELOMENTAL BIOLOGY. *Educ:* Univ Minn, BS, 50, MS, 52, PhD(zool), 53. *Prof Exp:* Asst zool, Univ Minn, 50-53; instr biol, Reed Col, 53-54; from asst prof to prof zool, Univ Tenn, Knoxville, 61-82; RETIRED. *Concurrent Pos:* Asst, Univ Minn, 53-; consult, AEC, 54- *Mem:* Am Soc Zool. *Res:* Morphogenesis; mouse tumor growth; hemoglobins and serum proteins of the chick embryo. *Mailing Add:* Rte 13 Box 600 Ft Myers FL 33908

FRASER, THOMAS HUNTER, b Dansville, NY, Mar 19, 48; m 72; c 2. BIOCHEMISTRY, MOLECULAR BIOLOGY. *Educ:* Univ Rochester, BA, 70; Mass Inst Technol, PhD(biochem), 75. *Prof Exp:* Instr microbiol, Mass Inst Technol, 75-76, res assoc biochem, 76; res assoc biochem, Univ Colo, 76-77; res scientist molecular biol, UpJohn Co, 77-81; vpres res & develop, Repligen Corp, 81-83, exec vpres & chief tech officer, 83-89; DIACRIN, INC, 89- *Concurrent Pos:* Damon Runyon-Walter Winchell Cancer Fund fel, 76-77. *Mem:* AAAS. *Res:* Application of molecular biology and recombinant DNA technology to the design, discovery and production of commercially important peptides and proteins. *Mailing Add:* Diacrin Inc 840 Memorial Dr Cambridge MA 02139

FRASER-REID, BERTRAM OLIVER, b Coleyville, Jamaica, Feb 23, 34; Jamaican & Can citizen; m 63; c 2. SYNTHETIC ORGANIC, NATURAL PRODUCTS & CARBOHYDRATE CHEMISTRY. *Educ:* Queen's Univ, BSc, 59, MSc, 61; Univ Alberta, PhD(chem), 64. *Prof Exp:* Asst chem, Imp Col, Univ London, 64-66; prof, Univ Waterloo, Can, 66-80 & Univ Md, 80-82; PROF CHEM, DUKE UNIV, 82-, JAMES B DUKE PROF CHEM, 85- *Concurrent Pos:* Mem med chem study sect A, NIH, 79-83, pharmacol sci review panel, 84-88; consult, Burroughs Wellcome Co, SCM Org Chem, Glaxo. *Honors & Awards:* Merck, Sharp & Dohme Award, Chem Inst Can, 77. *Mem:* Can Inst Chem; Am Chem Soc; Brit Chem Soc. *Res:* Synthesis of organic compounds of pharmacological, biological and theoretical importance from readily available carbohydrate derivatives; influence of electronic effects on reactivities of carbohydrates and carbocycles. *Mailing Add:* Dept Chem Duke Univ Durham NC 27706

FRASER-SMITH, ANTONY CHARLES, b Auckland, NZ, July 7, 38; m 68; c 2. SPACE PHYSICS, GEOPHYSICS. *Educ:* Univ NZ, BS, 59, MS, 61; Univ Auckland, PhD(physics), 66. *Prof Exp:* Lectr physics, Univ Auckland, 61-65; assoc res scientist, Lockheed Missiles & Space Co, Calif, 66-68; res assoc, 68-77, SR RES ASSOC, SPACE, TELECOMMUN & RADIOSCI LAB, STANFORD UNIV, 77- *Concurrent Pos:* vis res prof & chair appl physics, Naval Postgrad Sch, Monterey, Calif, 80. *Mem:* Fel Brit Inst Physics; Am Geophys Union; Am Inst Physics; Int Union Radio Sci; Sr mem Inst Elec Elec Engrs. *Res:* Origin and properties of low-frequency electromagnetic phenomena; use of low frequency electromagnetic waves for communication and studies of geophysical phenomena. *Mailing Add:* Space, Telecommun & Radiosci Lab Stanford Univ Stanford CA 94305

FRASHER, WALLACE G, JR, b Los Angeles, Calif, Dec 2, 20; m 59; c 2. CARDIOVASCULAR PHYSIOLOGY. *Educ:* Univ Southern Calif, AB, 41, MD, 51. *Prof Exp:* Head physician, Los Angeles County Hosp, 55-57; estab investr, Los Angeles County Heart Asn, 58-59; from asst res prof to assoc res prof med, Sch Med Loma Linda Univ, 61-66; from assoc prof to prof physiol, Sch Med Univ Southern Calif, 66-77; DEP EXEC VPRES RES PROG, AM HEART ASN, 77- *Concurrent Pos:* Res fel, Los Angeles County Heart Asn, 57-58; Nat Heart Inst res fel, 60-61, res career develop award, 61-; res fel eng, Calif Inst Technol, 61-63, sr res fel, 63- *Mem:* Am Physiol Soc; Microcirculatory Soc; Soc Rheol; Fedn Am Soc Exp Biol; Sigma Xi. *Res:* Major artery distensibility; flow properties of blood and its constituents in small tubes; microvascular casting. *Mailing Add:* 7320 Greenville Ave Dallas TX 75231

FRASHIER, LOYD DOLA, b Pampa, Tex, Oct 29, 16; m 53; c 3. PHYSICAL CHEMISTRY. *Educ:* Harding Col, BS, 40; Univ Calif, Calif, PhD(phys chem), 49. *Prof Exp:* Chemist, E I du Pont de Nemours & Co, Inc, 41-45; assoc prof chem, Ga Inst Technol, 49-58; chmn dept, 58-78, prof, 58-90, EMER PROF CHEM, PEPPERDINE UNIV, 90- *Mem:* Am Chem Soc. *Res:* Chemical kinetics. *Mailing Add:* 24344 Baxter Ave Malibu CA 90265-4742

FRASIER, GARY WAYNE, b Imperial, Nebr, July 27, 37. RANGELAND HYDROLOGY, WATER HARVESTING. *Educ:* Colo State Univ, BS, 59; Ariz State Univ, MS, 66. *Prof Exp:* Agr engr, USDA, 59-67, res hydraulic engr, Water Conserv Lab, Phoenix, 67-78, res hydraulic eng & res leader, Aridland Watershed Res Ctr, Agr Res Serv, Tucson, Ariz, 78-90, HYDRAULIC ENGR, AGR RES SERV, USDA, FT COLLINS, CO, 90- *Concurrent Pos:* Assoc ed, J Range Mgt, 82-85; ed, Rangelands, Soc Range Mgt, 84- *Mem:* Am Soc Agr Engrs; fel Soc Range Mgt; Soil Conserv Soc Am. *Res:* Water harvesting as a means of water supply for livestock, wildlife and households; growing of arid and semiarid plants by water harvesting and runoff farming techniques. *Mailing Add:* USDA Agr Res Serv Crops Res Lab Colo State Univ 1701 Center Ave Ft Collins CO 80526

FRASIER, S DOUGLAS, b Los Angeles, Calif, Nov 29, 32; m 56; c 3. PEDIATRIC ENDOCRINOLOGY. *Educ:* Univ Calif, Los Angeles, BA, 54, MD, 58. *Prof Exp:* Asst prof pediat, Sch Med, Univ Calif, Los Angeles, 65-67; from asst prof to assoc prof, 67-74, PROF PEDIAT PHYSIOL & BIOPHYSICS, SCH MED, UNIV SOUTHERN CALIF, 74- *Concurrent Pos:* Fel pediat, Sch Med, Univ Calif, Los Angeles, 63-65. *Mem:* Soc Pediat Res; Endocrine Soc; Am Pediat Soc; Am Acad Pediat. *Res:* Growth hormone physiology in children; growth hormone deficiency and therapeutic effects of growth hormone; antigenicity of human growth hormone; endocrine correlates of puberty. *Mailing Add:* USC Sch Med Los Angeles Co USC Med Ctr Pd Pavilion 1129 N State St Los Angeles CA 90033

FRATANTONI, JOSEPH CHARLES, b Brooklyn, NY, May 14, 38; m 65; c 3. HEMATOLOGY. *Educ:* Fordham Univ, BS, 59; Harvard Univ, AM, 61; Cornell Univ, MD, 65. *Prof Exp:* From intern to asst resident med, New York Hosp, Cornell Univ, 65-67; staff assoc, Nat Inst Arthritis & Metab Dis, NIH, 67-69; assoc resident med, New York Hosp, Cornell Univ, 69-71; asst prof med & dir coagulation lab, Sch Med, Georgetown Univ, 71-72; staff physician, Hematol Serv, Clin Ctr, Nat Inst Arthritis & Metab Dis, NIH, 72-74; dir, Thrombosis Prog, Nat Heart Lung & Blood Inst, 74-75, chief, blood dis br, div blood dis & resources, 75-77, chief, Blood Resources Br, 77-78; dir, blood bank prod bur biol, 78-84, chief, Lab Cellular Components, 84-90, CHIEF, LAB CELLULAR HEMATOL, FOOD & DRUG ADMIN, 90-; CLIN PROF, UNIFORMED SERVS UNIV HEALTH SCI, 88- *Concurrent Pos:* Mem exec comt, Coun Thrombosis, Am Heart Asn, 75-77; consult & lectr, Nat Naval Med Ctr, 75-; clin asst prof med & pharmacol, Georgetown Univ, 72-76, clin assoc prof, 76-; clin assoc prof med, Uniformed Servs Univ Health Sci, 77-88. *Mem:* AAAS; Am Fedn Clin Res; Am Soc Hematol; fel Am Col Physicians; Am Asn Blood Banks. *Res:* Hemostasis; platelets; blood substitutes. *Mailing Add:* Ctr Biol Eval & Res 8800 Rockville Pike Bethesda MD 20814

FRATER, ROBERT WILLIAM MAYO, b Cape Town, SAfrica, Nov 12, 28; c 3. THORACIC SURGERY, CARDIOVASCULAR SURGERY. *Educ:* Univ Cape Town, MB, ChB, 52; Univ Minn, MS, 61; FRCS(E), 61. *Prof Exp:* Intern surg, Groote Schuur Hosp, Cape Town, SAfrica, 53-54; Coun Sci & Indust Res fel, Univ Cape Town, 54-55 & Mayo Found, 55-61; Noble Found award, 61; sr lectr thoracic surg, Univ Cape Town, 62-64; from asst prof to assoc prof thoracic surg, 64-71, actg chmn dept surg, 71-76, PROF THORACIC SURG, ALBERT EINSTEIN COL MED, 71-, CHIEF CARDIO-THORACIC SURG, BRONX MUNIC HOSP CTR, 68-, CHIEF CARDIO-THORACIC SURG, MONTEFIORE HOSP & MED CTR, 76- *Concurrent Pos:* Min Heart Asn fel, Univ Minn, 59-60; sr surgeon, Groote Schuur Hosp & Red Cross War Mem Children's Hosp, Cape Town, 62-64; NIH fels, 65-70; Am Heart Asn grant in aid, 67. *Mem:* Am Asn Thoracic Surg. *Res:* Artificial heart valves; membrane oxygenators; right ventricular growth; cardiovascular toxic factor; ventricular compliance; function natural heart valves; criteria for valve testing; intraortic balloon pumping; myocardial protection; endocrine response to cardiac surgery; repair of heart valves. *Mailing Add:* Albert Einstein Col Med Morris Pkwy Eastchester Bronx NY 10461

FRATI, WILLIAM, b New York, NY, Sept 14, 31; m 63; c 3. HIGH ENERGY PHYSICS. *Educ:* Polytech Inst Brooklyn, BS, 52; Columbia Univ, MA, 55, PhD(physics), 60. *Prof Exp:* Res assoc physics, Columbia Univ, 60-61; res assoc, 61-65, res specialist, 66-80, ASSOC PROF PHYSICS, RITTENHOUSE LAB, UNIV PA, 81- *Mem:* Am Phys Soc. *Res:* High energy particle physics. *Mailing Add:* Dept Physics D Rittenhouse Lab Univ Pa Philadelphia PA 19174

FRATIELLO, ANTHONY, b Providence, RI, Mar 16, 36; m 63; c 3. PHYSICAL CHEMISTRY. *Educ:* Providence Col, BSc, 57; Brown Univ, PhD(chem), 62. *Prof Exp:* Fel & mem tech staff, Bell Tel Labs, 62-63; from asst prof to assoc prof chem, 63-69, PROF CHEM, CALIF STATE UNIV, LOS ANGELES, 69- *Concurrent Pos:* Vis fel, Bell Tel Labs, 69-70; NIH res career develop award, 69-73. *Mem:* Am Chem Soc. *Res:* Nuclear magnetic resonance; solution complexes; ion hydration. *Mailing Add:* Dept Chem Los Angeles State Col Los Angeles CA 90032

FRATTALI, VICTOR PAUL, b Scranton, Pa, Mar 23, 38; m 62; c 3. NUTRITIONAL BIOCHEMISTRY, CHEMISTRY. *Educ:* Univ Scranton, BS, 59; Georgetown Univ, PhD(biochem), 65. *Prof Exp:* Chem officer & army res & develop coordr, US Army Missile Command, Ala, 64-66; res chemist, Naval Med Res Inst, Nat Naval Med Ctr, Md, 66-76; nutritionist, 76-80, SUPVRY NUTRITIONIST, CTR FOOD SAFETY & APPL NUTRIT, FOOD & DRUG ADMIN, 80- *Concurrent Pos:* USPHS fel, Naval Med Res Inst, 66, Nat Acad Sci fel, 66-68; res consult, George Washington Univ, 70-75. *Mem:* Am Chem Soc; Am Soc Clin Nutrit; Am Inst Nutrit; Am Diabetes Asn; Soc Exp Biol & Med; Inst Food Technol. *Res:* Impact of federal regulation on nutritional quality of food supply; effect of deep undersea habitation on human nutrient requirements; protein structure and function; protein-protein associating systems; proteinases and proteinase inhibitors. *Mailing Add:* 9404 Ewing Dr Bethesda MD 20817-2436

FRATTINI, PAUL L, b St Louis, Mo, Oct 27, 58; m 82. RHEOLOGY, FLUID MECHANICS. *Educ:* Rensselaer Polytech Inst, BS, 81; Stanford Univ, MS, 82, PhD(chem eng), 85. *Prof Exp:* Process engr agr prod, Monsanto Agr Prod, 81; res assoc chem eng, Stanford Univ, 82-84, teaching asst, 85; postdoctoral fel physiol, Rhein Westfaelia Tech Univ, 85-86; ASST PROF CHEM ENG, CARNEGIE MELLON UNIV, 86- *Concurrent Pos:* Hertz fel, Fannie & John Hertz Found, 81-85; NATO postdoctoral fel, NSF, 85-86; Dreyfus Found teacher-scholar award, 89; counr, NAm Soc Biorheology, 91-93. *Honors & Awards:* Presidential Young Investr Award, NSF, 86. *Mem:* AAAS; Am Chem Soc; Am Inst Chem Engrs; Biomed Eng Soc; Soc Rheology. *Res:* Rheology and microstructural studies of colloidal suspensions and polymeric liquids; instrument development for optical studies in fluid mechanics; biorheology and biophysics of cell deformation; author of various publications. *Mailing Add:* PO Box 44223 Pittsburgh PA 15205

FRAUENFELDER, HANS (EMIL), b Neuhausen, Switz, July 28, 22; nat US; m 50; c 3. PARTICLE PHYSICS, BIOLOGICAL PHYSICS. *Educ:* Swiss Fed Inst Technol, Dipl, 47, Dr sc nat, 50. *Hon Degrees:* DSc, Univ Penn, Tech Univ Munich. *Prof Exp:* Asst physics, Swiss Fed Inst Technol, 46-52; res assoc, 52, from asst prof to assoc prof, 52-58, PROF PHYSICS, UNIV ILL, URBANA, 58- *Concurrent Pos:* Vis scientist, European Orgn Nuclear Res, 58, 59, 63 & 73; consult, Los Alamos Sci Lab. *Mem:* Nat Acad Sci; fel Am Phys Soc; Swiss Phys Soc; fel AAAS; fel NY Acad Sci; fel Am Acad Arts Sci; Acad Leopoldina. *Res:* Nuclear physics; reactions and dynamics of biomolecules. *Mailing Add:* Dept Physics Univ Ill 1110 W Green St Urbana IL 61801

FRAUENGLASS, ELLIOTT, b Hartford, Conn, Aug 7, 34; m 55; c 3. ORGANIC CHEMISTRY. *Educ:* Univ Conn, BS, 56; Cornell Univ, PhD(org chem), 60. *Prof Exp:* Res chemist, Eastman Kodak Co, 60-65; POLYMER CHEMIST, LOCTITE CORP, 65- *Mem:* Am Chem Soc. *Res:* Synthetic and polymer organic chemistry. *Mailing Add:* Loctite Corp 705 N Mountain Rd Newington CT 06111-1411

FRAUENTHAL, JAMES CLAY, b New York, NY, Nov 25, 44; m 69. APPLIED MATHEMATICS, ENGINEERING MECHANICS. *Educ:* Tufts Univ, BS, 66; Harvard Univ, MS, 67, PhD(appl math), 71. *Prof Exp:* Asst prof mech eng, Tufts Univ, 71-73; res assoc pop sci, Harvard Sch Pub Health, 73-75; assoc prof appl math, State Univ NY Stony Brook, 75-82; MEM TECH STAFF, AT&T BELL LABS, 82- *Concurrent Pos:* Prin investr, Alfred P Sloan Found, 75-79; ed, Soc Indust & Appl Math News, 75-84; trustee, Hosp for Joint Dis-Orthop Inst, 76-, Helene Fuld Sch Nursing, 85-; vis assoc prof math, Harvey Mudd Col, 81-82. *Mem:* Resource Modeling Asn (secy, 86-90). *Res:* Population dynamics; structural optimization; mathematical demography; biomathematics; numerical analysis; queueing systems. *Mailing Add:* AT&T Bell Labs Rm 3K-521 Holmdel NJ 07733-1988

FRAUMENI, JOSEPH F, JR, b Boston, Mass, Apr 1, 33. INTERNAL MEDICINE, EPIDEMIOLOGY. *Educ:* Harvard Univ, AB, 54, MScHyg, 65; Duke Univ, MD, 58; Am Bd Internal Med, dipl, 65. *Prof Exp:* Intern med, Johns Hopkins Hosp, 58-59, asst resident, 59-60; sr asst resident, Cornell Second Div, Bellevue Hosp & Mem Ctr, 60-61; chief resident, Mem Sloan-Kettering Cancer Ctr, 61-62; med officer epidemiol, 62-66, head ecol sect, 66-75, from assoc chief to chief environ epidemiol br, 72-82, DIR EPIDEMIOL & BIOSTATIST PROG, NAT CANCER INST, 81- *Concurrent Pos:* Instr, Med Col, Cornell Univ, 61-62; asst med, Peter Bent Brigham Hosp, 64-65; attend physician, Clin Ctr, NIH, 66-; adj prof, Uniformed Serv Univ Health Sci, 80- *Honors & Awards:* Distinguished Serv Medal, Pub Health Serv, 83; Gorgas Medal, Asn Military Surgeons of US, 89. *Mem:* Am Pub Health Asn; Am Col Physicians; Am Epidemiol Soc; Am Asn Cancer Res; Am Col Epidemiol; Am Col Preventive Med. *Res:* Cancer research, especially epidemiological studies; medical oncology. *Mailing Add:* NIH Executive Plaza N Rm 543 Bethesda MD 20892

FRAUNFELDER, FREDERICK THEODOR, b Pasadena, Calif, Aug 16, 34; m 59; c 4. OPHTHALMOLOGY. *Educ:* Univ Ore, BS, 56, MD, 60. *Prof Exp:* Intern, Univ Chicago, 61; resident ophthal, Med Sch, Univ Ore, 64-66; from assoc prof to prof ophthal, Med Sch, Univ Ark, Little Rock, 68-78, chmn dept, 68-78; PROF OPHTHAL & CHMN DEPT, MED SCH, UNIV ORE, 78- *Concurrent Pos:* NIH fel ophthal, Univ Ore, 61-62; NIH spec fel ophthal, Wilmer Inst, Johns Hopkins Univ, 67-68; sabbatical, Eye Inst, Univ London, 74-75; consult, Vet Admin Hosp; trustee, Int Serv for Blind, 70-76. *Mem:* Am Acad Ophthal & Otolaryngol; AMA; Asn Univ Prof Ophthal (secy-treas, 73-74); Asn Res Vision & Ophthal; Am Col Surg. *Res:* External eye disease; ocular cancer. *Mailing Add:* Ore Health Sci Univ Hosp 3181 SW Sam Jackson Portland OR 97201-3098

FRAUNFELTER, GEORGE H, b Hamburg, Pa, June 2, 27; div. GEOLOGY. *Educ:* Lehigh Univ, BA, 48; Univ Mo, AM, 51, PhD(geol), 64. *Prof Exp:* Cur geol, Univ Mo, 51-55; paleontologist, Creole Petrol Corp, 55-58; cur & asst prof geol, 65-70, assoc prof, 70-73, PROF GEOL, SOUTHERN ILL UNIV, CARBONDALE, 73- *Mem:* AAAS; Am Asn Petrol Geologists; Paleont Soc; Paleont Res Inst; Am Inst Prof Geologists; Sigma Xi. *Res:* Ichnofossils (Upper Paleozoic); paleontology and stratigraphy of Middle Devonian of Midwestern United States; lower Pennsylvanian megafaunas (invertebrate). *Mailing Add:* Dept Geol Southern Ill Univ Carbondale Carbondale IL 62901

FRAUTSCHI, STEVEN CLARK, b Madison, Wis, Dec 6, 33; m 67; c 2. ELEMENTARY PARTICLE PHYSICS. *Educ:* Harvard Univ, BA, 54; Stanford Univ, PhD(physics), 58. *Prof Exp:* NSF fel, Yukawa Hall, Kyoto, 58-59; asst physics, Univ Calif, Berkeley, 59-61; asst prof, Cornell Univ, 61-62; from asst prof to assoc prof, 62-66, PROF PHYSICS, CALIF INST TECHNOL, 66- *Concurrent Pos:* Guggenheim fel, 71-72; vis prof, Univ Paris, Orsay, 77-78; Japan Soc Prom Sci fel, 77 & 87. *Mem:* Am Phys Soc. *Res:* Theory of elementary particles. *Mailing Add:* Dept Physics Calif Inst Technol Pasadena CA 91125

FRAUTSCHY, JEFFERY DEAN, b Monroe, Wis, June 22, 19; m 48; c 3. GEOLOGY. *Educ:* Univ Minn, BA, 42. *Prof Exp:* Assoc physicist, Div War Res, Univ Calif, 42-46; assoc, Scripps Inst Oceanog, Univ Calif, San Diego, 46-47; geophysicist, US Geol Surv, 47-49; marine geologist, 49-58, ASST DIR, SCRIPPS INST OCEANOG, UNIV CALIF, SAN DIEGO, 58- *Mem:* Geol Soc Am; Am Geophys Union; Am Asn Petrol Geologists; Soc Econ Paleont & Mineral; Sigma Xi. *Res:* Marine geology; oceanographic field instruments, equipment and methods; oceanographic vessel outfitting and operation; inshore oceanography and shore processes coastline planning; water quality control. *Mailing Add:* 2625 Ellentown Rd La Jolla CA 92037

FRAWLEY, ANTHONY DENIS, b Richmond, Australia, June 10, 49; m 85. HEAVY ION REACTIONS. *Educ:* Melbourne Univ, BSc, 72; Australian Nat Univ, PhD(physics), 77. *Prof Exp:* Res assoc physics, 77-79, res asst prof, 79-81, asst res scientist, 81-86, ASSOC RES SCIENTIST, PHYSICS DEPT, FLA STATE UNIV, TALLAHASSEE, 86- *Mem:* Am Phys Soc. *Res:* Nuclear physics; heavy ion nuclear reaction mechanisms. *Mailing Add:* Dept Physics Fla State Univ Tallahassee FL 32306

FRAWLEY, JOHN PAUL, b Washington, DC, Dec 17, 27; m 53; c 2. TOXICOLOGY, ENVIRONMENTAL SAFTEY. *Educ:* Georgetown Univ, BS & MS, 48, PhD(biochem), 50; Am Bd Indust Hyg, dipl; Am Bd Toxicol, cert. *Prof Exp:* Instr chem, Georgetown Univ, 47-48; res pharmacologist, US Food & Drug Admin, 48-52 & 54-56; lab dir, Surg Res Team in Korea, US Army, 52-54; dir toxicol, Hercules Inc, 56-82, gen mgr, health, environ & safety, 82-89; PRES, HEALTH ENVIRON INT, LTD, 89- *Concurrent Pos:* Lectr, Georgetown Univ, 49-52; mem agr bd, Nat Res Coun, 69-72; mem & chmn bd, Adria Labs Inc, 74-83. *Honors & Awards:* Lea Hitchner Award, 68; Arnold J Lehman Award, Soc Toxicol, 87. *Mem:* Am Chem Soc; Soc Toxicol; Am Soc Pharmacol & Exp Therapeut; Am Indust Hyg Asn; Int Soc Regulatory Toxicol (pres, 88-90); Acad Toxicol Sci (pres, 86-89). *Res:* Toxicity and drugs, cosmetics and food packaging. *Mailing Add:* 111 Chestnut Ave Edgewood Hills Wilmington DE 19809

FRAWLEY, NILE NELSON, b Madison, Wis, Jan 26, 50; m 72; c 3. INDUSTRIAL CHEMISTRY. *Educ:* Hamline Univ, BA, 72; Univ Minn, PhD(anal chem), 77. *Prof Exp:* Teaching assoc, Univ Minn, 72-77; sr res chemist, 77-80, proj leader, 80-84, RESEARCHER LEADER, MICH DIV ANALYTICAL LABS, DOW CHEM CO, 84- *Honors & Awards:* IR 100 Award, 84; Vaaler Award, Polymer Div, Am Chem Soc, 84. *Mem:* Am Chem Soc; Sigma Xi. *Res:* Analytical chemistry of potassium cyanate and its determination in biological materials; analytical methodologies for the analysis of agricultural chemicals and the integration of membrane technology into the practice of analytical chemistry. *Mailing Add:* 4613 Lund Dr Midland MI 48640

FRAWLEY, THOMAS FRANCIS, b Rochester, NY, June 27, 19; m 47; c 3. MEDICINE. *Educ:* Univ Rochester, AB, 41; Univ Buffalo, MD, 44; Am Bd Internal Med, dipl; Am Bd Endocrinol & Metab, dipl; FRCP(I), 82. *Prof Exp:* Res fel med, Sch Med, Univ Buffalo, 47-49; res fel, Harvard Med Sch, 49-51; head sub-dept endocrinol & metab, Albany Med Col, 51-63, prof med, 59-63; chmn dept & prof, 63-74, EMER PROF & CHMN INTERNAL MED, SCH MED, ST LOUIS UNIV, 75-; CHMN GRAD MED EDUC, ST JOHN'S MERCY MED CTR, ST LOUIS, 81- *Concurrent Pos:* Chief endocrine-metab clin, Albany Hosp, 51-63, attend physician, 52-63; consult, Albany Vet Admin Hosp, 51-63; clin investr, Nat Inst Arthritis & Metab Dis, 55-57; physician-in-chief, St Louis Univ Hosps, 63-74; mem sci rev comt, NIH, 70-74. *Honors & Awards:* Laureate Award, Am Col Physicians, 87. *Mem:* Endocrine Soc; Am Diabetes Asn; Asn Am Physicians; fel Am Col Physicians (pres, 81); Am Thyroid Asn; Am Clin & Climatol Asn; Am Fed Clin Res. *Res:* Clinical and basic investigation of endocrine metabolic disorders, particularly adrenal cortical disorders and carbohydrate metabolism. *Mailing Add:* Dept Med St John's Mercy Med Ctr Off Grad Med Educ 615 S New Ballas Rd St Louis MO 63141

FRAWLEY, WILLIAM JAMES, b Cleveland, Ohio, Sept 14, 37; c 3. ARTIFICIAL INTELLIGENCE. *Educ:* John Carroll Univ, BS, 58, MS, 60; Univ Okla, PhD(math), 69. *Prof Exp:* Res engr, Lewis Res Ctr, NASA, Ohio, 58-60; instr math, John Carroll Univ, 60-62; sr res mathematician, 62-65, sr res proj mathematician & sect head, 68-76, PROG LEADER COMPUT INTEL, SCHLUMBERGER-DOLL RES CTR, SCHLUMBERGER LTD, 77- *Concurrent Pos:* Lectr, John Carroll Univ, 58-60; comput use consult, Cleveland Elec Illum Co, 60-62. *Mem:* Am Math Soc; Soc Indust & Appl Math. *Res:* Numerical analysis; differential equations; real time analysis and control; design and implementation of knowledge based systems. *Mailing Add:* 2500 Mystic Valley Pkwy 501 Medford MA 02155

FRAY, ROBERT DUTTON, b Shepherdstown, WVa, Feb 16, 39; m 62. MATHEMATICS. *Educ:* Roanoke Col, BS, 61; Duke Univ, PhD(number theory), 65. *Prof Exp:* Asst prof math, Fla State Univ, 65-71; ASSOC PROF MATH, FURMAN UNIV, 71- *Mem:* Am Math Soc; Math Asn Am. *Res:* Number theory and combinatorial analysis. *Mailing Add:* Dept Math Furman Univ Greenville SC 29613

FRAYER, WARREN EDWARD, b Manchester, Conn, Sept 22, 39; m 60; c 4. FORESTRY. *Educ:* Pa State Univ, BS, 61; Yale Univ, MF, 62, DF, 65. *Prof Exp:* Res forester, US Forest Serv, Upper Darby, Pa, 63-67; assoc prof forest biomet, Dept Forest & Wood Sci, Colo State Univ, 67-75; supvry res forester, US Forest Serv, Washington, DC, 75-76; head, dept forest & wood sci, Colo State Univ, 76-84; DEAN, SCH FORESTRY & WOOD PROD, MICH TECHNOL UNIV, 84- *Concurrent Pos:* Consult fed agencies & companies, 67-; prin invester, various res projs, 68-; vis lectr, Yale Univ, 74. *Mem:* Soc Am Foresters; Forest Prod Res Soc. *Res:* Natural resource sampling methods. *Mailing Add:* Dept Forestry Mich Technol Univ Houghton MI 49931

FRAYSER, KATHERINE REGINA, b Lynchburg, Va, Feb 18, 26. PHYSIOLOGY. *Educ:* Randolph-Macon Woman's Col, AB, 46; Duke Univ, AM, 54, PhD(anat), 60. *Prof Exp:* Instr med, Sch Med, Duke Univ, 60-62; asst prof, Sch Med, Ind Univ, 62-63, asst prof med & physiol, 63-65, from assoc prof to prof, 65-73; prof med res & ophthal, 73-78, PROF MED & PHYSIOL, MED UNIV SC, 78- *Concurrent Pos:* Mem, NIH Physiol Study Sect, 72-76, Clin Trails Rev Panel & Sickle Cell Ctr Rev Panel, 75 & 81 & Pulmonary Acad Awards Rev Panel, 75-76. *Mem:* Am Physiol Soc; Am Asn Anat; Microcirculatory Soc. *Res:* Human and experimental pulmonary physiology and pathophysiology, altitude physiology; microcirculation. *Mailing Add:* Dept med Med Univ SC Col Med 171 Ashley Ave Charleston SC 29425

FRAZEE, JERRY D, b Wichita Falls, Tex, May 26, 29; div; c 4. PHYSICAL CHEMISTRY, BIOCHEMISTRY. *Educ:* Univ Tex, BS, 51, MA, 57, PhD(chem), 59. *Prof Exp:* Vis scientist phys chem, Rohm & Haas Co, Marshall Space Flight Ctr, 59; asst prof chem, La Polytech Inst, 59-60; from chemist to sr chemist, Rocketdyne Div, NAm Aviation, Inc, 60-67; proj leader & supvr instrumental anal sect, Chem Dept, Gulf Oil Corp, 67-68; sr paint chemist, Tex Hwy Dept, 68-86; INSTRUMENTAL ANALYSIS CONSULT, SCI-TECH, AUSTIN, 80-; ASST PROF SCI, AUSTIN COMMUN COL, 87- *Concurrent Pos:* Wine Anal, Cypress Valley Winery, 87; asst prof sci, Concordia Lutheran Col, 90- *Mem:* AAAS; Am Chem Soc; Brit Oil & Colour Chem Asn; Coblentz Soc; Sigma Xi. *Res:* Organic coatings; pyrolysis of hydrocarbons; biochemistry and bacteriology; polymers and resins; spectroscopy; color. *Mailing Add:* 3456 N Hills Dr Austin TX 78731

FRAZER, ALAN, b Philadelphia, Pa, Mar 2, 43; m 68; c 2. PHARMACOLOGY. *Educ:* Philadelphia Col Pharm & Sci, BSc, 64; Univ Pa, PhD(pharmacol), 69. *Prof Exp:* From instr to asst prof, 69-75, ASSOC PROF PHARMACOL, UNIV PA, 75-; RES PHARMACOLOGIST, VET ADMIN HOSP, PHILADELPHIA, 69-, CHIEF, NEUROPSYCHOPHARMACOL UNIT, 81- *Mem:* Am Soc Pharmacol & Exp Therapeut; AAAS. *Res:* Defining the biological basis of affective disorders. *Mailing Add:* Dept Pharm & Psychiat Univ Pa Sch Med Vet Admin Hosp (151 E) University & Woodlands Aves Philadelphia PA 19104

FRAZER, BENJAMIN CHALMERS, b Birmingham, Ala, July 19, 22; m 51; c 3. SOLID STATE PHYSICS. *Educ:* Ala Polytech Inst, BS, 47, MS, 48; Pa State Col, PhD(physics), 52. *Prof Exp:* Res assoc, Brookhaven Nat Lab, 52-55; physicist, Westinghouse Res labs, 55-58; physicist, 58-67, dep chmn dept physics, 67-74, assoc chmn solid state physics, 74-75, sr physicist, Brookhaven Nat Lab, 67-83; phys scientist, 84-85, BR CHIEF, SOLID STATE PHYSICS & MAT CHEM, DIV MAT SCI, US DEPT ENERGY, 86- *Concurrent Pos:* Consult, Westinghouse Res Labs, 55-62; guest scientist, PR Nuclear Ctr, 62-63; assoc ed, J Phys Chem Solids, 64-74; managing ed, The Phys Rev, 74-81. *Mem:* Fel Am Phys Soc; AAAS. *Res:* Structure and dynamics of solids, phase transitions, ferroelectricity, x-ray and neutron scattering, synchrotron radiation. *Mailing Add:* 14 Cherbourg Court Potomac MD 20854

FRAZER, BRYAN DOUGLAS, b Penticton, BC, Feb 17, 42; m 63; c 2. INSECT ECOLOGY. *Educ:* Univ BC, BSc, 65; Univ Calif, Berkeley, PhD(entom), 71. *Prof Exp:* Res officer entom, 65-67, RES SCIENTIST ENTOM, RES BR, AGR CAN, 70- *Concurrent Pos:* Adj prof plant sci, Univ BC, 72- *Mem:* Can Entom Soc; Brit Ecol Soc. *Res:* Population dynamics and biological control of the aphid vectors of plant virus diseases as an approach to understanding and affecting plant virus epidemiology. *Mailing Add:* Agr Can Res Sta 6660 NW Marine Dr Vancouver BC V6T 1X2 Can

FRAZER, J(OHN) RONALD, b Ottawa, Ont, July 17, 23; US citizen; m 48; c 5. INDUSTRIAL ENGINEERING & MANAGEMENT. *Educ:* Clarkson Col Technol, BME, 45; Iowa State Univ, MS, 50, PhD(eng, econ), 54. *Prof Exp:* Instr mech eng, Clarkson Col Technol, 45-46; from instr to asst prof indust eng, Iowa State Univ, 46-53; from asst prof to prof indust eng & chmn dept, 53-67, dean sch bus admin, 67-70, PROF INDUST MGT, CLARKSON COL TECHNOL, 70- *Mem:* Am Inst Indust Engrs; Acad Mgt; Asn Bus Simulation Experiential Learning; Am Inst Decision Sci. *Res:* Cost studies; linear programming; operations research; management simulation. *Mailing Add:* Sch Mgt Clarkson Col 301 Snell Hall Potsdam NY 13699

FRAZER, JACK WINFIELD, b Forest Grove, Ore, Sept 9, 24; m 64; c 3. CHEMISTRY. *Educ:* Hardin-Simmons Univ, BS, 48. *Prof Exp:* Anal chemist, Los Alamos Sci Lab, 48-51, chemist, 51-53; group leader vacuum anal chem, Univ Calif, 53-66, from assoc leader to leader, gen chem div, 66-74, dept head chem & nat sci, 74-78, sr scientist, Lawrence Livermore Lab, 78-83. *Concurrent Pos:* Consult, 57-; consult, Universal Oil Prod, 71-, Merck Inc, 78-87, Oak Ridge Nat Lab, 83-87 & Eastman Kodak, 87-; indust prof chem, Ind Univ, 73-77; vis prof elec eng, Colo State Univ, 75-82; adj prof chem, Univ Ga, 84- *Honors & Awards:* Instrumentation Award, Am Chem Soc, 73. *Mem:* Am Chem Soc; fel Am Soc Testing & Mat; Am Inst Chem Engrs. *Res:* Development of new analytical methods and computer automated instrumentation together with their use in the development of fully computer controlled manufacturing processes, pilot plants and large scale experimental apparatus. *Mailing Add:* PO Box 1420 Tuolumne CA 95379-1417

FRAZER, JOHN P, b Rochester, NY, Sept 14, 14; m 50; c 2. OTOLARYNGOLOGY. *Educ:* Univ Rochester, MD, 39; Am Bd Otolaryngol, dipl, 44. *Prof Exp:* Instr path, Med Col, Cornell Univ, 39-40; instr surg, Long Island Col Med, 40-41; instr otolaryngol, Sch Med, Yale Univ, 41-46 & 47; consult, Tripler Army Hosp, Hawaii Leprosarium & Leahi Tuberc Hosp, Honolulu, 50-63; PROF OTOLARYNGOL, MED CTR, UNIV ROCHESTER, 63- *Mem:* Am Laryngol, Rhinol & Otol Soc; Am Col Surg; Am Broncho-Esophagol Asn; Am Laryngol Asn. *Mailing Add:* 329 Orchard Park Blvd Rochester NY 14609

FRAZER, L NEIL, b Courtenay, Can, Apr 7, 48; m 72; c 3. REFLECTION & REFRACTION SEISMOLOGY, UNDERWATER SOUND. *Educ:* Univ BC, BASc, 70; Princeton Univ, PhD(geophysics), 78. *Prof Exp:* Geophysicist, Gulf Oil Can Ltd, 71-72; res assoc, Dept Geol, Princeton Univ, 77-79; asst seismologist, Hawaii Inst Geophysics, 79-81; from asst prof to assoc prof, 82-89, PROF, DEPT GEOL & GEOPHYS, UNIV HAWAII, 89- *Concurrent Pos:* Consult, Schlumberger-Doll Res, 84-89, Arco Petrol, 88, Sohio Petrol, 85-88, Geo-Pacific Corp, 83- *Mem:* Seismol Soc Am; Soc Exploration Geophysicists; Am Geophys Union; Acoust Soc Am; Royal Astron Soc. *Res:* Theoretical body-wave seismology; inversion of reflection and refraction data for earth structure; petroleum exploration; underwater sound; geophysics. *Mailing Add:* Dept Geol Univ Hawaii Honolulu HI 96822

FRAZER, MARSHALL EVERETT, b Alva, Okla, Jan 19, 44; m 64; c 1. APPLIED PHYSICS. *Educ:* Northwestern State Col, BS, 64; Univ Tex, Austin, PhD(physics), 74. *Prof Exp:* Instr physics, Northwestern State Col, 64; ASST DIR, APPL RES LABS, UNIV TEX, AUSTIN, 65- *Mem:* Inst Elec & Electronics Engrs; Acoust Soc Am; Sigma Xi. *Res:* Underwater acoustics, acoustic scattering, acoustic propagation; communication theory, decision theory, signal processing, artificial intelligence; multidimensional stochastic processes, spatial and temporal properties of stochastic fields. *Mailing Add:* 3010 Honeytree Lane Austin TX 78746

FRAZER, W(ILLIAM) DONALD, b Tampa, Fla, Jan 9, 37; m 61; c 3. COMPUTER SCIENCE. *Educ:* Princeton Univ, BSE, 59; Univ Ill, MS, 61, PhD(elec eng), 63. *Prof Exp:* Res asst, Digital Comput Lab, Univ Ill, 59-61; res staff mem, Thomas J Watson Res Ctr, 63-78, DIR, SYSTS ARCHITECTURE, INFO SYSTS & TECHNOL GROUP, IBM CORP, 79- *Concurrent Pos:* Vis lectr, Princeton Univ, 67-68; adj assoc prof, Courant Inst Math Sci, NY Univ, 70-71. *Mem:* Sigma Xi; Asn Comput Mach. *Res:* Operating systems architecture and design; design and analysis of algorithms for sort, search and optimization; data structures; information theory and coding. *Mailing Add:* Dept 39/180 IBM Corp 1000 Westchester Ave White Plains NY 10604

FRAZER, WILLIAM ROBERT, b Indianapolis, Ind, Aug 6, 33; m 54; c 2. THEORETICAL PHYSICS. *Educ:* Carleton Col, AB, 54; Univ Calif, PhD(physics), 59. *Prof Exp:* Physicist, Lawrence Radiation Lab, Univ Calif, 59; mem physics dept, Inst Advan Study, 59-60; from asst prof to assoc prof physics, 60-67, actg provost, Third Col, 69-70, chmn dept, 75-78, PROF PHYSICS, UNIV CALIF, SAN DIEGO 67-, VPRES, 81- *Concurrent Pos:* Sloan found fel; Guggenheim fel. *Mem:* Am Phys Soc; Fel Am Phys Soc. *Res:* Theoretical physics of the elementary particles. *Mailing Add:* Univ Calif 300 Lakeside Dr Suite 2261 Oakland CA 94612

FRAZIER, A JOEL, b Texarkana, Ark, Dec 16, 38; m 61; c 2. CERAMIC RAW MATERIALS. *Educ:* Univ Ark, BS, 60. *Prof Exp:* Field geologist, WS Dickey Mfg Co, 61-75, chief geologist, 75-85, mgr ceramic technol, 85-87; MGR CERAMIC TECHNOL & CHIEF GEOLOGIST, MISSION CLAY PROD, 87- *Mem:* Am Ceramic Soc. *Res:* Test ceramic raw materials for the structural clay products (brick, tile and sewer pipe); formulate ceramic mixes for clay products; determine C,S & Quartz content of ceramic materials and thermal characteristics. *Mailing Add:* Mission Clay Prod 826 E Fourth St Pittsburg KS 66762

FRAZIER, CLAUDE CLINTON, III, b Hattiesburg, Miss, Aug 27, 48; m 77; c 1. NONLINEAR OPTICAL ORGANIC & ORGANOMETALLIC MATERIALS. *Educ:* Memphis State Univ, BS, 70; Ore Grad Ctr, MS, 72; Calif Inst Technol, PhD(chem), 76. *Prof Exp:* asst prof chem, Univ Minn, Duluth, 78-82; sr scientist, Martin Marietta Labs, Baltimore, Md, 82-88; SCI CONSULT, 88- *Honors & Awards:* Robert Glen Lye Mem Award. *Mem:* Am Chem Soc; Sigma Xi. *Res:* Nonlinear optical organic and organometallic materials; inorganic and organometallic photochemistry. *Mailing Add:* 4718 Hale Haven Dr Ellicott City MD 21043

FRAZIER, DONALD THA, b Martin, Ky, Sept 26, 35; m 56, 84; c 4. PHYSIOLOGY, NEUROPHYSIOLOGY. *Educ:* Univ Ky, BS, 58, MS, 60, PhD, 64. *Prof Exp:* From instr to assoc prof neurophysiol, Sch Med, Univ NMex, 64-69; assoc prof, 69-74, PROF PHYSIOL, MED CTR, UNIV KY, 74-, CHMN DEPT, 80- *Concurrent Pos:* NIH res grants, 66-69, 73-79, 86-89, PI prog proj, 89-94; dir, Grass Found Fel Prog, 69-79; pres, assoc chmn, Dept Physiol; mem study sect, NIH, 84-89. *Mem:* Neurosci Soc; AAAS; Am Physiol Soc; Sigma Xi; Thoracic Soc; Am Heart Asn. *Res:* Neurosciences; nature of respiratory neurons; biophysics of nerve membranes. *Mailing Add:* Dept Physiol Univ Ky Med Ctr Lexington KY 40536

FRAZIER, EDWARD NELSON, b Los Angeles, Calif, Feb 16, 39; m 68; c 2. SPACE PHYSICS, OPTICAL SENSORS. *Educ:* Univ Calif, Los Angeles, BA, 60; Univ Calif, Berkeley, PhD(astron), 66. *Prof Exp:* Analyst, Aeronutronic Corp, 60-61; res asst astrophys, Kitt Peak Nat Observ, 63; teaching asst astron, Univ Calif, Berkeley, 65-66; scientist astrophys, Univ Heidelberg, 66-67; res astronr, Univ Calif, Berkeley, 67-69; staff scientist solar physics, Aerospace Corp, 69-83; SR STAFF ENGR, TRW, 83- *Concurrent Pos:* Adj prof, Calif State Univ, Northridge, 76-82. *Mem:* AAAS; Am Phys Soc; Am Astron Soc; Int Astron Union. *Res:* Mechanical energy transport in solar atmosphere; solar magnetohydrodynamics; properties of interplanetary material; infrared signatures; infrared sensors; sensor and spacecraft technology. *Mailing Add:* 933 Eighth St Manhattan Beach CA 90266

FRAZIER, GEORGE CLARK, JR, b Cumberland, Va, Apr 14, 30; m 53. CHEMICAL ENGINEERING. *Educ:* Va Polytech Inst, BS, 52; Ohio State Univ, MSc, 56; Johns Hopkins Univ, DEng(chem eng), 62. *Prof Exp:* Engr nuclear reactor design, Atomic Power Div, Westinghouse Elec Corp, 56-59; NATO fel, Cambridge Univ, 62-63; asst prof chem eng, Johns Hopkins Univ, 63-68; assoc prof chem & metall eng, 68-77, PROF CHEM ENG, UNIV TENN, KNOXVILLE, 77- *Concurrent Pos:* Consult, Am Potash & Chem Corp, Calif, 64-70, Direct Reduction Corp, 79- & Union Carbide Corp, 77- *Mem:* Am Inst Chem Engrs; Am Chem Soc; Combustion Inst; Sigma Xi. *Res:* Combustion; coal processing; heat and mass transfer; bioprocess engineering. *Mailing Add:* 441 E Hillvale TRN SW Knoxville TN 37919

FRAZIER, HOWARD STANLEY, b Oak Park, Ill, Jan 16, 26; m 50; c 4. MEDICINE. *Educ:* Univ Chicago, PhB, 47; Harvard Univ, MD, 53. *Prof Exp:* Res fel, Harvard Med Sch, 55-56; res fel, Physiol Lab, Cambridge Univ, 56-57; fel cardiol, Western Reserve Univ, 57-58; from asst prof to assoc prof med, Beth Israel Hosp, 65-68; PROF MED, HARVARD UNIV MED SCH, 78- *Mem:* Inst Med-Nat Acad Sci; Am Physiol Soc; Am Soc Clin Invest; Am Soc Nephrol. *Res:* Health services research. *Mailing Add:* Dept Health Policy & Mgt Sch Pub Health 677 Huntington Ave Boston MA 02115

FRAZIER, JAMES LEWIS, b Wadsworth, Ohio, Apr 15, 43; c 1. NEUROBIOLOGY, BEHAVIORAL PHYSIOLOGY. *Educ:* Ohio State Univ, BS, 66, PhD(entom), 70. *Prof Exp:* From asst prof to prof, Miss Agr & Forestry Exp Sta, Miss State Univ, 70-81; sr res biologist, Dupont Exp Sta, E I du Pont de Nemours & Co, Inc, 81-89; PROF & HEAD DEPT ENTOMOL, PA STATE UNIV, 89- *Concurrent Pos:* Vis res scientist, Western Regional Lab, USDA, Albany, Calif, 76 & dept biol sci, Univ Md, Baltimore, 80-81. *Mem:* Sigma Xi; Entom Soc Am; fel Nat Ctr Food & Agr; Asn Chemoreception Sci; Int Soc Chem Ecol. *Res:* Insect behavior, chemoreception, electrophysiology of chemoreceptors. *Mailing Add:* Dept Entom Pa State Univ 501 Agr Sci & Indust Bldg University Park PA 16802

FRAZIER, JOHN MELVIN, b Greensboro, NC, Apr 24, 44; m 65; c 2. IN VITRO TOXICOLOGY, METAL TOXICOLOGY. *Educ:* Johns Hopkins Univ, BS, 66, PhD(physics), 71. *Prof Exp:* Res assoc, 70-73, asst prof environ health sci, 73-82, ASSOC PROF ENVIRON HEALTH SCI, SCH HYG & PUB HEALTH, JOHNS HOPKINS UNIV, 82-; ASSOC DIR, ALTERNATIVES TO ANIMAL TESTING, JOHNS HOPKINS CTR, 85- *Mem:* Sigma Xi; Soc Toxicol. *Res:* Physiological and biochemical control of transition metal metabolism in biological systems with particular interests in the toxicokinetics and biliary excretion of cadmium. *Mailing Add:* 209 Eastspring Rd Lutherville-Timonium MD 21093

FRAZIER, LOY WILLIAM, JR, b Ft Smith, Ark, Aug 14, 38; m 61; c 2. PHYSIOLOGY. *Educ:* Univ Tex, Arlington, BS, 68; Univ Tex Southwestern Med Sch, Dallas, PhD(physiol), 72. *Prof Exp:* From asst prof to assoc prof, 72-78, PROF & CHMN DEPT PHYSIOL, BAYLOR COL DENT, 78-;

PROF, GRAD DIV, BAYLOR UNIV, 73- *Concurrent Pos:* Tex Med Found fel, Baylor Col Dent, 73-74; adj assoc prof, Sch Nursing, Baylor Univ, 74-80; NIH grant, 75-81. *Mem:* Soc Exp Biol & Med; Am Physiol Soc; Int Asn Dent Res. *Res:* Electrolyte transport; acid-base balance. *Mailing Add:* cept Physioliol Baylor Col Dent Dallas TX 75246

FRAZIER, RICHARD H, b Bellevue, Pa, May 29, 1900. CIRCUITRY, ELECTRO-MECHANICAL DEVICES. *Educ:* Mass Inst Technol, SB, 23, SM, 32. *Prof Exp:* Elec engr, Railway & Eng Co, Greensburg, 23-25; prof elec eng, Mass Inst Technol, 25-65; independent consult, 25-82; RETIRED. *Mem:* Fel Inst Elec & Electronic Engrs; Am Soc Eng Educ; Am Asn Univ Prof. *Mailing Add:* Seven Summit Ave Winchester MA 01890

FRAZIER, ROBERT CARL, b Guilford Co, NC, Feb 14, 32; m 53; c 3. MATHEMATICS. *Educ:* Atlantic Christian Col, AB, 53; ECarolina Univ, MA, 59; Univ Ill, MS, 65; Fla State Univ, EdD(math educ), 69. *Prof Exp:* Teacher high sch, NC, 53-54 & 56-59; assoc prof, Atlantic Christian Col, 59-69; PROF MATH, BARTON COL, 69-, CHMN DEPT, 77- *Mem:* Math Asn Am. *Res:* Comparison of methods of teaching mathematical proof; mathematics education. *Mailing Add:* Dept Math Barton Col Wilson NC 27893

FRAZIER, STEPHEN EARL, b Spencer, WVa, Oct 21, 39; m 70; c 2. INORGANIC CHEMISTRY, INDOOR AIR POLLUTION ABATEMENT. *Educ:* Fla Southern Col, BS, 61; Univ Fla, MS, 63, PhD(chem), 65. *Prof Exp:* Res assoc, Case Western Res Univ, 66-67; res chemist res div, W R Grace & Co, 67-69; prof chem, Polk Jr Col, 70-71; indust consult inorg chem, 71; dir res, Rush-Hampton Industs, 71-85; DIR, FRAGRANT AIR PROD, LTD, 87- *Concurrent Pos:* Instr, Baltimore Community Col, 68-69; res assoc, Univ Fla, 70-71; adj instr, Valencia Community Col, 88, Seminole Community Col, 89-, Daytona Beach Community Col, 90- *Mem:* Sigma Xi; NY Acad Sci; Am Chem Soc; Am Soc Heating, Refrig & Air Conditioning Engrs. *Res:* Inorganic phosphorus chemistry; synthesis of organo-silicon chemistry; inorganic polymers; chemistry of chloramine and ammonia; heterogeneous catalysis; inorganic materials science; indoor air pollution abatement; aseptic synthesis. *Mailing Add:* 940-204 Framingham Ct Lake Mary FL 32746

FRAZIER, THOMAS VERNON, b Tonopah, Nev, Feb 25, 21; m 64; c 2. PHYSICS. *Educ:* Univ Calif, Los Angeles, AB, 43, MA, 49, PhD(physics), 52. *Prof Exp:* Asst physics, Univ Calif, Los Angeles, 43-44, 45-49; physicist, Naval Ord Lab, 44-45; instr, 50-52, from asst prof to assoc prof, 52-67, PROF PHYSICS, UNIV NEV, RENO, 67- *Concurrent Pos:* Vis Fulbright lectr, Nat Tsing Hua Univ & Taiwan Prov Norm Univ, 59-61. *Mem:* AAAS; Am Asn Physics Teachers; Am Phys Soc; Acoust Soc Am; Sigma Xi. *Res:* Hearing by bone connection. *Mailing Add:* Dept Physics Univ Nev Reno NV 89557

FRAZIER, TODD MEARL, b Lima, Ohio, Nov 9, 25; m; c 4. PUBLIC HEALTH. *Educ:* Kenyon Col, AB, 49; Johns Hopkins Univ, ScM, 57. *Prof Exp:* Statistician, Army Chem Ctr, Md, 51-54; dir biostatist, Baltimore City Health Dept, 54-63; chief div planning, res & statist, DC Dept Pub Health, 63-68; assoc prof biostatist, Sch Pub Health, Harvard Ctr Community Health & Med Care, 68-78; CHIEF SURVEILLANCE BR, DIV SURVEILLANCE, HAZARD EVAL & FIELD STUDIES, NAT INST OCCUP SAFETY & HEALTH, 78- *Res:* Public health administration, planning, research and statistics; occupational health; health program evaluation. *Mailing Add:* 2164 Cable Car Ct Cincinnati OH 45244

FRAZIER, WILLIAM ALLEN, b Carrizo Springs, Tex, Apr 26, 08; m 35; c 3. OLERICULTURE. *Educ:* Agr & Mech Col Tex, BS, 30; Univ Md, MS, 31, PhD, 33. *Prof Exp:* From instr to assoc prof olericult, Univ Md, 33-37; assoc horticulturist, Exp Sta, Univ Ariz, 37-39; olericulturist & head dept veg crops, Univ Hawaii, 39-49; prof & horticulturist, 49-72, EMER PROF HORT, ORE STATE UNIV, 72- *Mem:* AAAS; Am Soc Hort Sci. *Res:* Breeding vegetable crop plants for disease resistance and improved horticultural characters. *Mailing Add:* 3225 Northwest Crisp Wash Univ Med 660 S Euclid Ave Corvalis OR 97330

FRAZZA, EVERETT JOSEPH, b NJ, Nov 2, 24; m 46; c 4. ORGANIC CHEMISTRY. *Educ:* Univ Md, BS, 50, PhD(org chem), 54. *Prof Exp:* Chemist, Appl Physics Lab, Johns Hopkins Univ, 51-52; res chemist, 54-58, group leader, Indust Chem Dept, 58-60, group leader, Cent Res Div, 60-69, head biotherapeut dept, Lederle Labs, 69-70, dir res & develop, Davis & Geck Dept, 70-75, assoc dir new prod acquisitions, Med Res Div, 76-80, DIR NEW PROD LICENSING, MED GROUP, AM CYANAMID CO, 80- *Concurrent Pos:* Chmn, Gordon Res Conf Biomat, 76. *Mem:* Am Chem Soc (pres, 50); Soc Biomat; AAAS (pres, 65). *Res:* Aliphatic nitriles; petrochemicals; paper chemicals; monomer and polymer synthesis; medical and biological applications of polymers; surgical sutures; hospital specialties. *Mailing Add:* 2135 Van Cortlandt Circle Yorktown Heights NY 10598

FREA, JAMES IRVING, b Sturgeon Bay, Wis, Mar 1, 37; m 59; c 4. BACTERIOLOGY, BIOCHEMISTRY. *Educ:* Univ Wis, BS, 59, MS, 61, PhD(bact, biochem), 63. *Prof Exp:* From asst prof to assoc prof, 65-73, PROF MICROBIOL, OHIO STATE UNIV, 73- *Mem:* AAAS; Sigma Xi; Am Soc Microbiol; Am Chem Soc. *Res:* Interaction of microbes and metals; microbial ecology; biochemical ecology of microorganisms; microbial methane production and oxidation. *Mailing Add:* Dept Microbiol Ohio State Univ Columbus OH 43210

FREADMAN, MARVIN ALAN, b Pittsfield, Mass, Aug 18, 49. BIOLOGY, COMPARATIVE PHYSIOLOGY. *Educ:* Univ Colo, BA, 71; Col William & Mary, MA, 74; Univ Mass, PhD(zool), 78. *Prof Exp:* NIH fel physiol, John B Pierce Found, Yale Univ, 78-80; res assoc zool, Univ Mass, Amherst, 81-; AT DEPT BIOL, CLARK LAB WOODS HOLE OCEANOG INST; INDEPENDENT INVESTR, MARINE BIOL LAB, 88- *Concurrent Pos:* Vis asst fel, John B Pierce Found, Yale Univ, 81- *Mem:* Am Soc Zoologists; Sigma Xi; AAAS. *Res:* Comparative physiology; function of invertebrate respiratory pigments; gill ventilation and swimming in fishes; muscle physiology. *Mailing Add:* 45 Earle St Wellesley MA 02181

FREAR, DONALD STUART, b Wakefield, RI, Sept 5, 29; m 56; c 4. AGRICULTURAL BIOCHEMISTRY. *Educ:* Pa State Univ, BS, 51; Ohio State Univ, MSc, 53, PhD(agr biochem), 55. *Prof Exp:* Res biochemist, Charles F Kettering Found, 55-57; assoc prof to prof biochem, NDak State Univ, 57-76; RES LEADER, AGR RES SERV, USDA, 68- *Concurrent Pos:* Res chemist, Sci & Educ Admin-Agr Res, USDA, 64-68. *Mem:* Am Chem Soc. *Res:* Pesticide metabolism in plants. *Mailing Add:* Met/Rad Res Lab State Univ Sta Fargo ND 58102

FREAS, ALAN D('YARMETT), b Newark, Ohio, Nov 4, 10; m 34; c 3. ENGINEERING. *Educ:* Univ Wis, BS, 33, MS, 47, CE, 52. *Prof Exp:* Instr civil eng, Univ Wis, 36-39; engr, Forest Prods Lab, US Forest Serv, 33-36, 39-42 & 45-56, asst dir forest prods res, US Forest Serv, Washington, DC, 56-58, asst to dir, Forest Prods Lab, 58-67, chief, Div Solid Wood Prod Res, 67-71, asst dir, Timber Utilization Res, 72-74, asst dir, Wood Engr Res, 74-75; ed, Forest Prods Res Soc, 76-77; CONSULT WOOD, 77- *Concurrent Pos:* Asst chief, Wood & Glue Br, Mat Lab, Air Mat Command, Wright Field, Ohio, 42-45. *Mem:* Am Soc Civil Engrs; Forest Prod Res Soc; Am Inst Timber Construct. *Res:* Strength of wood, modified wood and plywood; strength of glued, laminated wood structures; processing and protection of forest products; research administration. *Mailing Add:* 2618 Park Pl Madison WI 53705

FREAS, WILLIAM, b Oakland, Calif, May 14, 46; m 78; c 2. CATECHOLAMINE METABOLISM, VASCULAR CONTRACTILITY. *Educ:* Univ Md, PhD(physiol), 78. *Prof Exp:* RES ASSOC PROF PHYSIOL, UNIFORMED SERV UNIV, 84- *Mem:* Am Physiol Soc. *Res:* Research investigating the effect of oxygen radicals on neurovascular function. *Mailing Add:* Dept Anesthesiol Uniformed Serv Univ Hlth Sci 4301 Jones Bridge Rd Bethesda MD 20814

FREBERG, C(ARL) ROGER, b Hector, Minn, Mar 17, 16; m 41; c 2. MECHANICAL ENGINEERING. *Educ:* Univ Minn, BME, 38, MS, 40; Purdue Univ, PhD(mech eng), 43. *Prof Exp:* Instr mech eng, Univ Minn, 39-40; Purdue Univ, 40-43, asst prof mech eng & aeronaut eng, 43-45; res engr, Carrier Corp, 45-46; dir eng div, Southern Res Inst, 46-49; head equip res, US Naval Civil Eng Lab, 49-52; assoc dir, Borg Warner Res Ctr, 52-57; PROF MECH ENG, UNIV SOUTHERN CALIF, 57- *Concurrent Pos:* Consult, Studebaker Corp, 44, Carrier Corp, 45 & Hughes Aircraft Corp, 59; lectr, NATO, Paris, France, 58. *Mem:* Am Soc Mech Engrs; Am Soc Eng Educ. *Res:* Vibrations; noise; mechanical design; research management; applied mechanics. *Mailing Add:* 846 S Hudson Ave Los Angeles CA 90005-3819

FRECH, ROGER, b Gary, Ind, Mar 26, 41; m; c 2. PHYSICAL CHEMISTRY. *Educ:* Mass Inst Technol, BS, 63; Univ Minn, PhD(phys chem), 68. *Prof Exp:* Res assoc phys chem, Ore State Univ, 68-70, instr, 69-70; from asst prof to assoc prof, 70-79, PROF CHEM, UNIV OKLA, 79- *Concurrent Pos:* Alexander von Humboldt fel, 80-81, 88; guest scientist, Max-Planck Inst, Stuttgart, Ger & Mainz, Ger, 88-89. *Mem:* Am Chem Soc; Am Inst Physics. *Res:* Vibrational spectroscopy of crystals and condensed phases; disordered systems; theory of optical and dielectric constants; fast ion conductivity. *Mailing Add:* Rm 208 620 Parrington Oval Norman OK 73019

FRECHE, JOHN C(HARLES), b Minneapolis, Minn, Apr 29, 23; m 43. METALLURGY, MECHANICAL ENGINEERING. *Educ:* Univ Pittsburgh, BS, 43. *Prof Exp:* Jr engr, Wright Aeronaut Corp, NJ, 43-44; aeronaut res engr, Nat Adv Comt Aeronaut, NASA, 44-48, asst head turbine cooling sect, 49-57, head alloys sect, Lewis Res Ctr, 58-63, chief, Fatigue & Alloys Res Br, 63-71, asst chief mat & struct div, 71-74, assoc chief mat & struct div, Lewis Res Ctr, 74-79, actg chief mats div & struct div, 79-80; RETIRED. *Honors & Awards:* IR-100 Award, Indust Res Mag, 68, 74 & 78; Except Sci Achievement Medal, NASA, 71; Cert Appreciation Tech Contribs Develop & Understanding Alloys, NASA, 80. *Mem:* Fel Am Inst Chem; assoc fel Am Inst Aeronaut & Astronaut; Am Inst Mining, Metall & Petrol Engrs. *Res:* Materials for air breathing and space propulsion systems; nickel and cobalt base alloys; metal fatigue; author of over 100 technical papers on high temp ni-base & co-base, superalloys, air cooled & liquid turbine blades, metal fatigue & ferromagnetic co-base alloys for space power. *Mailing Add:* 20123 Lorain Rd Apt 306 Fairview Park OH 44126

FRECHET, JEAN M J, b Chalon, France, Aug 18, 44; Can citizen; m 67; c 2. ORGANIC POLYMER CHEMISTRY, POLYMER SYNTHESIS. *Educ:* Inst Indust Chem & Physics, France, BS, 67; State Univ NY Col Environ Sci & Forestry, MSc, 69, PhD(chem), 71; Syracuse Univ, PhD(chem), 71. *Prof Exp:* Prof chem, Univ Ottawa, 73-87, vdean grad studies & res, 83-87; PROF POLYMER CHEM, DEPT CHEM, CORNELL UNIV, 87- *Concurrent Pos:* Vis scientist, IBM Res Lab, San Jose, Calif, 79 & 83; consult, Xerox Corp, 80-89, Allied Signal Corp, 84- , Exxon Chem Corp, 88- & E I Du Pont De Nemours & Co, 91-; bd dirs, Ont Ctr Mats Res, 87-; adj prof, Univ Ottawa, 87-; mem, Sci Adv Bd, Bausch & Lomb Corp, 88- *Honors & Awards:* Int Union Pure & Appl Chem Award, 83; A K Doolittle Award, Am Chem Soc, 86. *Mem:* Am Chem Soc; Soc Polymer Sci Japan. *Res:* Polymer chemistry synthesis and characterization; polymers in high technologies; medicine and separation science; blends and composites; photochemistry; polymers for micro- and opto-electronics. *Mailing Add:* 382 Baker Lab Dept Chem Cornell Univ Ithaca NY 14853-1301

FRECHETTE, ALFRED LEO, b Groveton, NH, Aug 22, 09; m 30; c 1. PUBLIC HEALTH. *Educ:* Univ Vt, MD, 34; Harvard Univ, MPH, 39. *Prof Exp:* Instr, Col Med, Univ Vt, 35-36; pvt pract, 36-38; dir venereal dis control, NH Dept Pub Health, 39-42, comnr, 42-45; dir pub health, Brookline, Mass, 45-50; dir health div, United Community Servs Metrop Boston, 50-59; comnr, Mass Dept Pub Health, 59-72; mem staff, Sidney Farber Cancer Ctr, 72-78; COMNR, MASS DEPT PUB HEALTH, 79- *Mem:* Am Pub Health Asn. *Res:* Public health administration. *Mailing Add:* 44 Birney St Boston MA 02115

FRECK, PETER G, b Hinsdale, Ill, Aug 14, 34; m 58; c 3. OPERATIONS RESEARCH, AERONAUTICAL ENGINEERING. *Educ:* Princeton Univ, BSE, 56, MSE, 59; Harvard Univ, MBA, 62. *Prof Exp:* Aerodynamicist, Douglas Aircraft Corp, Calif, 58-60; aeronaut engr, Anal Serv, Inc, Va, 62-64; asst dir, Systs Eval Div, Inst Defense Anal, 64-78; ASSOC TECH DIR, BATTLEFIELD SYSTS, MITRE CORP, 78- *Mem:* Am Inst Aeronaut & Astronaut; Opers Res Soc Am. *Res:* Aircraft performance, stability and control; systems analysis; operations research. *Mailing Add:* MITRE Corp 7525 Colshire Dr McLean VA 22102

FRECKMANN, ROBERT W, b Milwaukee, Wis, Dec 5, 39; m 71. PLANT TAXONOMY. *Educ:* Univ Wis-Milwaukee, BS, 62; Iowa State Univ, PhD(plant taxon), 67. *Prof Exp:* Asst cur bot, Milwaukee Pub Mus, 67-68; from asst prof to assoc prof, 68-79, PROF BIOL, UNIV WIS-STEVENS POINT, 79-, CUR HERBARIUM, MUS NATURAL HIST, 69- *Concurrent Pos:* Wis State Univs res fund grant, 69-71. *Mem:* Am Soc Plant Taxon; Bot Soc Am; Int Asn Plant Taxon. *Res:* Taxonomy and biosystematics of Dichanthelium and Panicum; grasses of Wisconsin; flora of central Wisconsin. *Mailing Add:* Dept Biol Univ Wis Stevens Point WI 54481

FREDEEN, HOWARD T, b MacRorie, Sask, Dec 10, 21; m 53; c 5. ANIMAL GENETICS, BIOMETRY. *Educ:* Univ Sask, BSc, 43; Univ Alta, MSc, 47; Iowa State Col, PhD(genetics), 52. *Prof Exp:* Lectr animal husb, Univ Sask, 43-45; animal husbandman, 47-55, head livestock res, 55-80, head livestock res, res sci animal breeding, meats, Can Dept Agr, 80-84; CONSULT, 84- *Concurrent Pos:* Mem, Quinquennial Rev Conf, Commonwealth Agr Bur, Eng, 55; co-developer Lacombe breed of pigs, 59; mem, Nat Animal Breeding Adv Comt, 59-64; sci ed, Can J Genetics & Cytol, 60-64; sci ed, Can J Animal Sci, 63-67, ed, 67-71; chmn, Can Livestock Geneticists Roundtable, 63-64; chmn, Gen Adv Comn Can ROP for Swine, 70-79; tech adv, Can Pork Coun, Can Charolais Asn & Can Lacombe Breeders Asn; adj prof, Univ Alta, 80- *Honors & Awards:* Pub Serv Merit Award, 69; Cert of Merit, Can Soc Animal Sci, 76; Award of Excellence, Genetics Soc Can, 78. *Mem:* AAAS; NY Acad Sci; Am Soc Animal Sci; Poultry Sci Asn; fel Agr Inst Can; Sigma Xi. *Res:* Genetics of pigs, cattle and poultry, especially factors influencing the effectiveness of selective breeding; meats and carcass evalutation beef and pork. *Mailing Add:* Box 1810 LaCombe AB T0C 1S0 Can

FREDEN, STANLEY CHARLES, b Far Rockaway, NY, Dec 5, 27; div; c 3. SCIENCE ADMINISTRATION, REMOTE SENSING. *Educ:* Univ Calif, Los Angeles, AB, 50, MA, 52, PhD, 56. *Prof Exp:* Asst physics, Univ Calif, Los Angeles, 50-54, assoc, 54-56, instr, 56-57; sr staff physicist, Lawrence Radiation Lab, Livermore, 57-61; sect head, Aerospace Corp, Calif, 61-66, sr staff scientist, 66-68; chief space physics div, Manned Spacecraft Ctr, Tex, 68-70, chief scientist, Lab Meteorol & Earth Sci, 70-74, Lansat Proj scientist & chief Missions Utilization Off, 74-84, asst dir space sta, 84-90, prog scientist, Space Sta Attached Payloads, 88-89, ASST DIR SPACE PROBES, GODDARD SPACE FLIGHT CTR, NASA, 90- *Concurrent Pos:* Guest scientist, Max Planck Inst Extraterrestrial Physics, 64-65. *Honors & Awards:* Except Performance Award, Goddard Space Flight Ctr, 73. *Mem:* fel Am Phys Soc; Am Geophys Union. *Res:* Particles trapped in the earth's magnetic field; experiments performed with space probes and satellites; acquisition of remotely sensed data and their applications to earth sciences and earth resource survey disciplines, including ecology and environmental quality. *Mailing Add:* Code 900 Goddard Space Flight Ctr NASA Greenbelt MD 20771

FREDENBURG, ROBERT LOVE, b Antwerp, NY, Nov 28, 21; m 42; c 5. SCIENCE EDUCATION. *Educ:* Syracuse Univ, AB, 54, MS, 59, PhD(sci educ), 65. *Prof Exp:* Form assoc prof to prof, 59-69, EMER PROF GEOL & PHYS SCI, CALIF STATE UNIV, CHICO. *Res:* General education physical science courses at the college level and student interest in these courses. *Mailing Add:* 2898 Marigold Ave Chico CA 95926

FREDERICK, DANIEL, b Elkhorn, WVa, Apr 29, 25; m 52; c 4. ENGINEERING MECHANICS. *Educ:* Va Polytech Inst, BS, 44, MS, 48; Univ Mich, PhD(eng mech), 55. *Prof Exp:* From instr to prof, 48-74, ALUMNI DISTINGUISHED PROF ENG MECH, VA POLYTECH INST & STATE UNIV, 74-, HEAD DEPT, 70- *Concurrent Pos:* Lectr, Nat Sci Found Adv Mech Inst, Univ Colo, 62-63; consult, Structures Res Div, NASA, Va, Pac Missile Range, Calif, Martin-Marietta Corp, Lord Mfg Co & David Taylor Model Basin; dir conf continuum mech, NSF, 65-70. *Honors & Awards:* Award, Am Soc Civil Engrs, 60; Award, Am Soc Engr Ed. *Mem:* Am Soc Civil Engrs; Am Inst Aeronaut & Astronaut; Am Soc Eng Educ; Am Soc Mech Engrs; Soc Eng Sci. *Res:* Aerodynamic stability of bridges; continuum mechanics; elasticity; plasticity; plates and shells; mechanics of composite materials. *Mailing Add:* Eng Sci/Mech Norris Hall Rm 218 Va Polytech State Univ Blacksburg VA 24061-0219

FREDERICK, DAVID EUGENE, b Chicago, Ill, June 10, 31; div; c 5. REAL-TIME COMPUTER-BASED SYSTEMS, PSYCHOTHERAPY & MEDIATION. *Educ:* Yale Univ, BS, 52; Univ Ill, PhD(physics), 62; Santa Clara Univ, MA, 85. *Prof Exp:* Group leader power plant simulators, Singer Co, 69-71, Harshman Assocs, Inc, 71-72 & Electronic Assocs, Inc, 72-73; consult data acquisition process control systs, 73-77; owner, 77-78, PRES, ARGUS TECHNOL SYSTS, 79- *Mem:* Am Phys Soc; Am Nuclear Soc; Sigma Xi; Inst Elec & Electronics Engrs. *Res:* Medium-energy photonuclear physics research. *Mailing Add:* 506 Kinross Ct Sunnyvale CA 94087

FREDERICK, DEAN KIMBALL, b Providence, RI, Nov 11, 34; m 57; c 4. ELECTRICAL ENGINEERING. *Educ:* Yale Univ, BE, 55; Brown Univ, ScM, 61; Stanford Univ, PhD(elec eng), 64. *Prof Exp:* Engr, New Departure Div, Gen Motors Corp, 57-58; asst prof elec eng, Clarkson Col Technol, 64; ASSOC PROF ELEC ENG, RENSSELAER POLYTECH INST, 64- *Mem:* Inst Elec & Electronics Engrs. *Res:* Automatic control theory; computer-aided design of control systems; dynamic simulation; computer-aided instruction; expert-systems applications. *Mailing Add:* Elec Comput & Systs Eng Dept Rensselaer Polytech Inst Troy NY 12181

FREDERICK, EDWARD C, b Mankato, Minn, Nov 17, 30; m 51; c 5. ANIMAL SCIENCE, PHYSIOLOGY. *Educ:* Univ Minn, BS, 54, MS, 55, PhD(reprod), 58. *Prof Exp:* Assoc prof & dairy specialist, Northwest Sch & Exp Sta, 58-64, assoc prof, Southern Sch & Exp Sta, 64-66, prof animal sci, 66-69, supt, Sch & Sta, 64-69, dir, Col, 69-70, provost, 70-85, PROF ANIMAL SCI, UNIV MINN TECH COL-WASECA, 69-, CHANCELLOR, 85- *Mem:* AAAS; Am Dairy Sci Asn; Am Soc Animal Sci. *Res:* Frozen semen; frequency of semen collection; effects of nutrition on semen production; mastitis; effect of tranquilizers on behavior; fly control studies; dairy-beef feeding trials; high moisture barley studies. *Mailing Add:* Pres/Chancellor Univ Minn Tech Col Waseca Waseca MN 56093

FREDERICK, GEORGE LEONARD, b Kitchener, Ont, July 15, 30; m 50; c 4. PHARMACOLOGY, TOXICOLOGY. *Educ:* Ont Vet Col, DVM, 52; Univ Western Ont, MSc, 53. *Prof Exp:* Res officer, Animal Res Inst, Can Dept Agr, 54-67; sci adv, Bur Sci Adv Serv, 67-73, sci adv, Br Drugs, 73-80, sci adv & sr toxicologist, Bur Human Prescription Drugs, Dept Nat Health & Welfare, Can, 80-87; PRIVATE CONSULT, 87- *Mem:* Can Biochem Soc; Nutrit Soc Can; Brit Soc Study Fertil; Can Vet Med Asn. *Res:* Noninfectious causes and prevention of reproductive failures and neonatal mortality in farm animals; role of vitamin B-12 and iron metabolism; evaluation of drug safety and efficacy. *Mailing Add:* 1808 Juno Ave Ottawa ON K1H 6S7 Can

FREDERICK, JAMES R, b May 29, 46; m; c 1. BIOCHEMISTRY. *Educ:* Wesleyan Univ, Nebr, BS & BA, 68; Univ Pa, MS, 70, PhD(biochem), 75. *Prof Exp:* Vpres, res & develop, Poly-Tek Mfg Corp, 77-80; engr & scientist, Tech Ctr, Aluminium Co Am, 80-82; dir, Spec Res & Serv, Emtech Res Corp, 84-89; VPRES, OMEGA THERMAL TECHNOLOGIES, INC, 89- *Concurrent Pos:* loss control consult, Fidelity Environ Ins Co. *Mem:* Am Mgt Asn; Am Chem Soc; Toastmasters Int; Am Inst Chem Engrs. *Res:* Development of biotechnology for consumer or industrial use; marketing support; management of waste products; toxic waste analysis and management; development of thermal processes which take waste in - no waste out. *Mailing Add:* 97 Streek Rd Medford NJ 08055

FREDERICK, JEANNE M, NEUROLOGICAL TRANSMITTER, DEVELOPMENT. *Educ:* Univ Wis-Madison, PhD(anat), 79. *Prof Exp:* RES ASST PROF, DEPT OPHTHAL, CULLEN EYE INST, BAYLOR COL MED, 83- *Mailing Add:* Dept Ophthal Cullen Eye Inst Baylor Col Med One Baylor Plaza Houston TX 77030

FREDERICK, JOHN EDGAR, b Thursday, WVa, Aug 10, 40; m 70; c 1. PHYSICAL CHEMISTRY. *Educ:* Glenville State Col, BS, 60; Univ Wis, PhD(phys chem), 64. *Prof Exp:* Chemist, Union Carbide Chem Co, 60; fel phys chem, Stanford Res Inst, 64-66; asst prof, 66-72, ASSOC PROF CHEM & POLYMER SCI, UNIV AKRON, 72- *Mem:* Am Chem Soc; Soc Rheol; Am Phys Soc. *Res:* High polymer physics; quasielastic light scattering from polymeric systems; viscoelastic properties of polymeric systems. *Mailing Add:* Dept Polymer Sci Univ of Akron Akron OH 44325-3901

FREDERICK, KENNETH JACOB, b Schenectady, NY, Dec 5, 13; m 36, 78; c 3. PHYSICAL CHEMISTRY. *Educ:* Union Col, BS, 36; Univ Calif, PhD(phys chem), 39. *Prof Exp:* Sr res chemist, Solvay Process Div, Allied Chem & Dye Corp, NY, 39-47; group leader, Paraffine Co, Inc, Calif, 47-49; dir res data opers, Abbott Labs, 49-76; RETIRED. *Mem:* AAAS; Am Inst Chemists; Am Chem Soc. *Res:* Physical chemistry of drugs and drug action; scientific instrumentation and mathematics applied to pharmaceutical research; scientific personnel development and facility planning. *Mailing Add:* 1540 N Greenleaf Lake Forest IL 60045

FREDERICK, LAFAYETTE, b Friarspoint, Miss, Mar, 19, 23; m 50; c 3. MYCOLOGY, PLANT PATHOLOGY. *Educ:* Tuskegee Inst, BS, 43; RI State Col, MS, 50; State Col Wash, PhD(plant path), 52. *Prof Exp:* From assoc prof to prof biol, Southern Univ, 52-62; prof biol, Atlanta Univ, 62-76, chmn dept, 63-76; actg dean, Col Lib Arts, 86-87, PROF BOT & CHMN DEPT, HOWARD UNIV, 76- *Concurrent Pos:* Carnegie res grant, 53; NSF fac fel, Dept Bot, Univ Ill, 60-61; mem, Comn Undergrad Educ Biol, 70-71; mem, Gen Res Support Adv Comt, NIH & Biol Achievement Test Comt, Educ Testing Serv; mem-at-large, Am Inst Biol Sci; mem adv comt, Am Type Culture Mycol. *Mem:* AAAS; Am Phytopathological Soc; Mycol Soc Am; Bot Soc Am; Am Inst Biol Sci; Nat Inst Sci. *Res:* Vascular wilt diseases of plants; developmental studies on Conidia and ascospores; systematics and ecology of Myxomycetes; systematics of imperfect fungi. *Mailing Add:* Dept Bot Howard Univ Washington DC 20059

FREDERICK, LLOYD RANDALL, b Ill, Aug 5, 21; m 43; c 3. SOIL MICROBIOLOGY, SOIL FERTILITY. *Educ:* Univ Nebr, BSc, 43; Rutgers Univ, MSc, 47, PhD(microbiol), 50. *Prof Exp:* Asst microbiologist, Purdue Univ, 49-55; from assoc prof to prof, 55-78, EMER PROF SOILS, IOWA STATE UNIV, 78-; AGR SCIENTIST CONSULT, 90- *Concurrent Pos:* Fulbright res scholar, Ger, 62; consult, W R Grace Inoculant Lab, 67-70, Res Seeds, Inc, 70-74, USDA, 90 & Nat Res Coun-Nat Acad Sci, 91; microbiol specialist, AID, 75-77 & 78-89. *Mem:* AAAS; fel Am Soc Agron; fel Soil Sci Soc Am. *Res:* Transformations of sulfur compounds by soil microorganisms; decomposition of plant residues and chemicals in soil; factors affecting ammonia retention and oxidation; denitrification; effect of temperature on biological soil changes; biological nitrogen fixation; rhizobia in soils and nodules. *Mailing Add:* 119 Eighth St Box 205B Cambridge IA 50046

FREDERICK, MARVIN RAY, organic chemistry, for more information see previous edition

FREDERICK, RAYMOND H, b Johnstown, NY, Oct 14, 23. PHOSPHOR POWDERS, ALUMINUM STRONTIUM CHROMATE PIGMENT. *Educ:* Pratt Inst, NY, BChE, 44. *Prof Exp:* Res engr, Res Div, Reynolds Metals Co, Glen Cove, Long Island, 46-48; qual control engr, Westinghouse Elec, 48-49; res engr, Aluminium Powders & Paste Div, Reynolds Metals Co, Ky, 49-50; lieutenant comdr, Atomic Bomb Test, US Navy, Bikini Islands,

50-52; asst prod supvr, Gen Sales Off, Louisville, Ky, 52-57, REGIONAL SALES ENGR, ALUMINUM POWDERS & PASTE PROD, REYNOLDS METALS CO, KANSAS CITY, MO, 57-, REGIONAL SALES MGR, OVERLAND PARK, KANS, 70- *Concurrent Pos:* Consult aluminum powders and pastes. *Mem:* Am Inst Chemists (treas, 70-76); Nat Paint & Coatings Asn; Fedn Socs Coating Technol. *Res:* Aluminum pigments and powders; maintenance, roof and bridge coatings; general purpose aluminum coatings. *Mailing Add:* 9045 Holly St Kansas City MO 64114

FREDERICK, SUE ELLEN, b Harrisville, WVa, Feb 25, 46; m 71; c 2. ELECTRON MICROSCOPY, MEMBRANES. *Educ:* Glenville State Col, BS, 66; Univ Wis, Madison, PhD(bot), 70. *Prof Exp:* Fel molecular & cell biol, Univ Colo, Boulder, 70-71; instr biochem, Mich State Univ, 71-72; from asst prof to assoc prof biol sci, 73-85, PROF BIOL SCI, MT HOLYOKE COL, 85- *Mem:* Am Soc Cell Biol; Sigma Xi. *Res:* Ultrastructure and cytochemistry of plant organelles; nucleus of plant cells. *Mailing Add:* Dept Biol Sci Mt Holyoke Col South Hadley MA 01075

FREDERICK, WILLIAM GEORGE DEMOTT, b Toledo, Ohio, June 23, 36; m 81; c 3. PHYSICS, MATERIALS SCIENCE. *Educ:* Univ Toledo, BS, 58; Univ Dayton, MS, 68; Univ Cincinnati, PhD(mat sci), 73; Mass Inst Technol, MS, 80. *Prof Exp:* Prof engr magnetic mat, Air Force Mat Lab, 65-68, tech area mgr electromagnetic mat, 68-73, br chief optical mat, 73-79, chief, mat lab plans off, 80-83; ASST DIR SENSOR TECH, OFF SECY DEFENSE/SDIO, 83- *Concurrent Pos:* Sloan fel, Mass Inst Technol, 79-80. *Honors & Awards:* Arthur Flemming Award, 77; Levinstein Award, IRIS Detector Specialty, 89. *Mem:* Am Phys Soc; AAAS. *Res:* Infrared transmitting materials; infrared detector materials; magnetic properties of rare earth-cobalt alloys; electromagnetic properties of solids. *Mailing Add:* 11511 Stonewood Lane Rockville MD 20852

FREDERICKS, CHRISTOPHER M, b Detroit, Mich, May 13, 44; m 71; c 2. PHYSIOLOGY. *Educ:* Oberlin Col, BS, 66; Wayne State Univ, PhD(physiol, pharmacol), 71. *Prof Exp:* Fel pharmacol, Sch Pharm, Ore State Univ, 71-72; asst prof physiol, Sch Med, Wayne State Univ, 72-75, assoc urol, 73-75; from asst prof to assoc prof, 75-84, PROF PHYSIOL, MED UNIV SC, 84- *Mem:* Am Physiol Soc; Soc Study Reprod; Am Fertil Soc. *Res:* Reproductive tract physiology. *Mailing Add:* Dept Physiol Med Univ SC 171 Ashley Ave Charleston SC 29425

FREDERICKS, ROBERT JOSEPH, b New York, NY, Dec 26, 34; m 84. X-RAY CRYSTALLOGRAPHY. *Educ:* Villanova Univ, BS, 57; St Joseph's Col, Pa, MS, 59; Lehigh Univ, PhD(chem), 65. *Prof Exp:* Res chemist, Cent Res Lab, Gen Aniline & Film Corp, 60-65, res specialist, 65-67; res supvr, Cent Res Lab, Allied Chem Corp, 67-72; mgr anal chem, Ethicon Inc, 72-74, dir res serv, 74-76, assoc dir res, 76-78; vpres res & develop, Surgikos, 78-79; vpres res & develop, Johnson & Johnson Dent Prod Co, 79-82; sr vpres & gen mgr, Biosearch Med Prod, 82-85; PRES, ALLEN TRANSL SERV, 85- *Concurrent Pos:* Lectr, Am Chem Soc, 71-73. *Mem:* AAAS; Am Chem Soc; Sigma Xi. *Res:* Research administration; polymer morphology; polymer structure. *Mailing Add:* 16 Butterworth Dr Morristown NJ 07960

FREDERICKS, WILLIAM JOHN, b San Diego, Calif, Sept 18, 24; m 42. PHYSICAL CHEMISTRY. *Educ:* San Diego State Col, BS, 51; Ore State Col, PhD(phys chem), 55. *Prof Exp:* Fulbright fel, Kamerlingh Onnes Lab, Leiden, Holland, 55-56; chemist, Stanford Res Inst, 56-59, chmn, solid state dept, 59-62; from assoc prof to prof, 62-87, EMER PROF CHEM, ORE STATE UNIV, 87-; RES PROF CHEM & MAT SCI, UNIV ALA, HUNTSVILLE, 88- *Concurrent Pos:* Vis acad, Atomic Energy Res Estab, Harwell, Berkshire, UK, 69-70; sr fel, Ctr Chem Physics, Univ Western Ont, 81. *Mem:* AAAS; Am Phys Soc; Am Chem Soc; Am Asn Crystal Growth. *Res:* Solid state chemistry; luminescence; ionic solids; impurity reactions in solids; impurity diffusion; crystal growth and purification; optical properties of solids, light scattering from solids and liquids. *Mailing Add:* Microgravity & Mat Res Ctr Univ Alabama Huntsville AL 35899

FREDERICKSEN, JAMES MONROE, b Blackstone, Va, June 25, 19; m 48; c 3. ORGANIC CHEMISTRY. *Educ:* Univ Richmond, BS, 40; Univ Va, PhD(org chem), 47. *Prof Exp:* Chemist, E R Squibb & Co, NJ, 44; sr chemist, Merck & Co, 47-53; prof chem & chmn dept, Davis & Elkins Col, 53-54 & Hampden-Sydney Col, 54-57; PROF CHEM, DAVIDSON COL, 57- *Mem:* Am Chem Soc. *Res:* Quinolines, thiazoles, isoalloxazines and vitamins; synthetic medicinals for tuberculosis and malaria; local anesthetics; synthetic medicinals in the quinoline, isoalloxazines and thiazole series. *Mailing Add:* Box 111 Davidson NC 28036-0111

FREDERICKSEN, TOMMY L, b Sheboygan, WI, Nov 12, 46. ZOOLOGY. *Educ:* Univ Alta, PhD(zool), 76. *Prof Exp:* Res asst prof, dept microbiol, Sch Med, NY Univ, 80-83; res assoc, dept biochem & orthop surg, Sch Med, Rutgers Univ, 83-84; immunologist, Poultry Res Lab, Agr Res Serv, USDA, 84-85; group leader, Discovery Res, Embrex Inc, 85-90; DIR RES, SPAFAS INC, 90- *Mem:* Am Asn Immunologists; Poultry Sci Asn; World Poultry Sci Asn; NY Acad Sci. *Mailing Add:* SPAFAS Inc 67 Baxter Rd Storrs CT 06268

FREDERICKSON, ARMAN FREDERICK, b Winnipeg, Man, May 5, 18; nat US; m 43; c 5. GEOLOGY, EXPLORATION FOR METALS & PETROLEUM. *Educ:* Univ Wash, BSc, 40; Mont Sch Mines, MSc, 42; Mass Inst Technol, ScD, 47. *Prof Exp:* Miner western US, 36-40; chief geologist, Cornucopia Gold Mines, Ore, 40-41; instr & asst, Mont Sch Mines, 41-42; engr, Alameda Dry Rock Co, Calif, 42; asst, Mass Inst Technol, 42-43 & 46-47; prof geol, Wash Univ, 47-55; supvr geol res, Pan-Am Petrol Corp, Okla, 55-60; prof earth & planetary sci & chmn dept & dir oceanog prog, Univ Pittsburgh, 60-65; vpres & dir res, King Resources Co, 65-69, sr vpres & dir tech progs, 69-71; pres, Sorbotec, Inc, 71-74; PRES, GLOBAL SURV, INC, 74- *Concurrent Pos:* Consult geologist, geochemist & mining engr, Fulbright res prof, Oslo, 51-52; chmn clay mineral comt, Nat Res Coun; contract negotiator petrol & mining progs, Africa, Mid & Far East, Latin & SAm, US, 77-; specialty work in gem mining & explor metals & petrol, Thailand. *Mem:* Fel Geol Soc Am; Soc Econ Geologists; Am Mineral Soc; Am Inst Mining, Metall & Petrol Eng; fel Geol Soc Finland. *Res:* Production of micas from clays; spectrographic, x-ray, diff-thermal and optical research; structure and genesis of ore deposits; isotope and trace geochemistry of sediments; formation fluids; regional stratigraphy and map analysis by computer methods; organization and management of industrial research; mining and petroleum exploration; coal engineering and evaluation; Middle East and North African petroleum exploration; marine geology and geophysics; oceanography. *Mailing Add:* 25221 Perch Dr Dana Point CA 92629

FREDERICKSON, EVAN LLOYD, b Spring Green, Wis, Mar 1, 22; m 46; c 3. ANESTHESIOLOGY. *Educ:* Univ Wis, BS, 47, MD, 50; Univ Iowa, MS, 53. *Prof Exp:* From instr to asst prof anesthesiol, Med Ctr, Univ Kans, 53-56; from asst prof to assoc prof, Univ Wash, 56-59; prof, Med Ctr, Univ Kans, 59-65; PROF ANESTHESIOL, EMORY UNIV, 65- *Honors & Awards:* Frederickson Lectr, Emory Univ. *Mem:* AAAS; Am Soc Anesthesiol; AMA. *Res:* Effect of drugs on cells; cell membranes; mechanisms of anesthesia. *Mailing Add:* Anesthesia Res Emory Univ Atlanta GA 30322

FREDERICKSON, GREG N, b Baltimore, Md, May 20, 47. ANALYSIS OF ALGORITHMS, DATA STRUCTURES. *Educ:* Harvard Univ, AB, 69; Univ Md, MS, 76, PhD(computer sci), 77. *Prof Exp:* Asst prof computer sci, Pa State Univ, 77-82; assoc prof, 82-86, PROF COMPUTER SCI, PURDUE UNIV, 86- *Mem:* Asn Comput Mach; Soc Indust & Appl Math. *Mailing Add:* Dept Computer Sci Purdue Univ West Lafayette IN 47907

FREDERICKSON, RICHARD GORDON, b Port Gamble, Wash, June 19, 44; m 70; c 2. ANATOMY, CELL BIOLOGY. *Educ:* Pac Lutheran Univ, BS, 66; Univ NDak, MS, 68, PhD(anat), 70. *Prof Exp:* From instr to asst prof anat, Univ Mich, 70-73; from asst prof to assoc prof anat, WVa Univ, 73-84; AFFIL ASSOC PROF BIOL STRUCT, UNIV WASH, 85-; res health sci specialist, Vet Admin Med Ctr, Seattle, 85-90; ASSOC PROF BASIC SCI & CHMN DEPT, JOHN BASTYR COL, 91- *Concurrent Pos:* Part-time instr, John Bastys Col Naturopathic Med, Seattle, 86-; part-time instr, Tacoma Community Col, Wash, 86-; part-time instr, Seattle Cent Community Col, 86- *Mem:* Am Soc Cell Biologists; Am Asn Anatomists; Soc Neurosci; Sigma Xi. *Res:* Embryonic growth of connective tissue; fine structure of meninges; analytical electron microscopy of heavy metals in cells and tissues; mesenchymal-epittelial interactions in embryonic and adult tissues; repair myocardium following injury. *Mailing Add:* John Bastyr Col 144 NE 54th St Seattle WA 98105

FREDERIKSE, HANS PIETER ROETERT, b Hague, Neth, July 13, 20; nat US; m 52; c 3. CERAMICS. *Educ:* Univ Leiden, BS, 41, MS, 45, PhD(physics), 50. *Prof Exp:* Vis lectr physics, Purdue Univ, 50-53; physicist, 53-55, chief solid state physics sect, 55-78, chief, Ceramics, Glass & Solid State Sci Div, 78-81, SR MAT SCIENTIST, NAT BUR STANDARDS, 81- *Concurrent Pos:* Fulbright travel grant, 50-52; Solid State Sci Panel, NAS/Nat Res Coun, 56-; Guggenheim fel, 61-62. *Honors & Awards:* Except Serv Award, US Dept Com, 62. *Mem:* Fel Am Phys Soc; Am Ceramic Soc; Fedn Am Scientists; Neth Phys Soc; corresp mem Neth Acad Sci. *Res:* Electrical, optical, and thermal properties of semiconductors, oxides, and nitrides; low temperature physics; properties of thin films. *Mailing Add:* Nat Inst Stand & Tech Mat Bldg Rm A329 Gaithersburg MD 20899

FREDERIKSEN, DIXIE WARD, b San Angelo, Tex, Sept 13, 42; m 63; c 2. PHYSICAL CHEMISTRY, BIOPHYSICS. *Educ:* Tex Technol Col, BA, 63; Wash Univ, AM, 65, PhD(chem), 67. *Prof Exp:* Sr polymer chemist, Eli Lilly & Co, 67; res assoc med, Southwestern Med Sch, Univ Tex, 68-70; res assoc biophys & NIH fel, Kings Col, Univ London, 73-74; asst prof, 74-80, ASSOC PROF BIOCHEM, VANDERBILT UNIV, 80- *Concurrent Pos:* Mellon teacher-scientist fel, Vanderbilt Univ, 74-75; adj assoc prof biol, Rensselear Polytech Inst, 80- *Mem:* Am Chem Soc; Sigma Xi; Biophys Soc; Am Heart Asn; AAAS. *Res:* Physical chemistry of biological macromolecules, muscle proteins, contraction and cell motility. *Mailing Add:* 4500 Price Circle Nashville TN 37205-1320

FREDERIKSEN, RICHARD ALLAN, b Renville, Minn, Aug 9, 33; m 58; c 2. HOST PLANT RESISTANSE, DISEASE MANAGEMENT. *Educ:* Univ Minn, BSc, 55, MS, 57, PhD(plant path), 61. *Prof Exp:* Res plant pathologist, USDA, 56-63; from asst prof to assoc prof, 63-73, PROF PLANT PATH, TEX A&M UNIV, 73- *Concurrent Pos:* Vis prof, Cornell Univ, 85-86. *Mem:* Fel, Am Phytopath Soc, 87; Am Agron Soc; Int Soc Plant Path. *Res:* Devise tactics for disease control in greater cereals; utilizing host resistance as a key strategy and complementing host resistance with management techniques. *Mailing Add:* Dept Plant Path & Microbiol Tex A&M Univ College Station TX 77843

FREDIANI, HAROLD ARTHUR, b New York, NY, Dec 23, 11. ANALYTICAL CHEMISTRY, MICROBIOLOGY. *Educ:* State Univ Iowa, AB, 34, MS, 35; La State Univ, PhD(phys chem), 37. *Prof Exp:* Asst prof anal chem, La State Univ, 37-40; chief chemist reagents, Fisher Sci Co, 40-46; asst dir qual control drugs, Merck & Co, 46-55; asst vpres pharm mfg, Bristol Labs Div, Bristol Myers Co, 55-74; DIR PHARMACEUT COUNR, CAZENOVIA, 74- *Concurrent Pos:* Fulbright prof, USDA, Italy, 51. *Mem:* Am Chem Soc; fel Am Inst Chemists; AAAS; Soc Indust Microbiologists; Am Soc Microbiol. *Res:* Laboratory instrumentation and automation; quality assurance. *Mailing Add:* 1630 Costa St Sun City Ctr FL 33573

FREDIN, LEIF G R, b Malmo, Sweden, Oct 23, 45; US citizen; m 81; c 2. LASER & MATERIALS INTERACTION, LASER BASED ANALYTICAL TECHNOLOGY. *Educ:* Univ Lund, Sweden, BS, 68, PhD(chem), 74. *Prof Exp:* Res assoc thermo chem, Univ Lund, Sweden, 68-80; postdoctoral fel phys chem, Rice Univ, Houston, Tex, 80-82, res assoc, 82-87; SR RES ASSOC LASERS/MAT, HOUSTON ADVAN RES CTR, THE WOODLANDS, TEX, 87- *Concurrent Pos:* Postdoctoral fel phys chem, Univ Calif, Berkeley, 76-79; pres, Spectroscopic Assocs Inc, 85-; sr res assoc, Antropix Corp, 88-; sr scientist, Schmidt Instruments Inc, 90- *Honors*

& Awards: IR-100 Award, Indust Res & Develop Mag, 86. *Mem:* Am Chem Soc; AAAS; Am Ceramic Soc; Int Soc Optical Eng; Soc Appl Spectros. *Res:* High temperature materials studies; properties of ceramics, metals and alloys at high temperature laser/materials interaction studies; laser based analytical techniques development; spectroscopic studies of high temperature development; biomedical applications of lasers. *Mailing Add:* Houston Advan Res Ctr 4802 Res Forest Dr The Woodlands TX 77381

FREDIN, REYNOLD A, b Greenville, Iowa, Feb 13, 23; m 44; c 2. FISHERIES. *Educ:* Iowa State Univ, BS, 48, MS, 49. *Prof Exp:* Assoc fisheries biologist, NC Wildlife Resources Comn, 49-50; fishery res biologist, Bur Com Fisheries, US Fish & Wildlife Serv, 50-70; dir fisheries data & mgt systs div, Northwest Fisheries Ctr, Nat Marine Fisheries Serv, Nat Oceanic & Atmospheric Admin, 70-78; CONSULT FISHERIES, 78- *Concurrent Pos:* Tech adv, US Sect, Int NPac Fisheries Comn, 56-78 & US Sect, Int Pac Halibut Comn, 66-75; lectr, Univ Wash, 69-72. *Honors & Awards:* Silver Medal Award, US Dept Com, 71. *Mem:* Fel Am Inst Fishery Res Biol. *Res:* Population dynamics of North American salmon stocks; biometrics; fisheries and biological statistics of United States salmon stocks; assessment of condition of North Pacific Ocean groundfish stocks; maximum net productivity levels in marine and terrestrial populations of large mammals; management and utilization bottomfish stocks in Alaskan waters. *Mailing Add:* 2328 NE 104th St Seattle WA 98125

FREDKIN, DONALD ROY, b New York, NY, Sept 28, 35. PHYSICS. *Educ:* NY Univ, AB, 56; Princeton Univ, PhD(math, physics), 61. *Prof Exp:* Instr physics, Princeton Univ, 59-61; assoc, 61-63, asst prof, 63-69, ASSOC PROF PHYSICS, UNIV CALIF, SAN DIEGO, 69- *Concurrent Pos:* Consult, Bell Tel Labs, Inc, 60-64 & Aerospace Corp, 62-63; Nat Acad Sci-Nat Res Coun fel, Saclay Nuclear Res Ctr, France, 64-65; Sloan res fel, 64-67. *Mem:* Am Phys Soc; Am Math Soc. *Res:* Solid state and low temperature theoretical physics. *Mailing Add:* Dept Physics Univ Calif La Jolla CA 92093

FREDMAN, MICHAEL LAWRENCE, b Milwaukee, Wis, July 12, 47; m 87. DISCRETE MATHEMATICS, CONCRETE COMPLEXITY THEORY. *Educ:* Calif Inst Technol, BS, 69; Stanford Univ, PhD(comput sci), 72. *Prof Exp:* From instr to asst prof appl math, Mass Inst Technol, 72-75; mem tech staff, Bell Commun Res, 87-89; asst prof, 75-80, PROF COMPUT SCI, UNIV CALIF, SAN DIEGO, 80- *Concurrent Pos:* Prin investr, Nat Sci Found, 76-; consult, Bell Commun Res; ed, J Comput, Soc Indust & Appl Math. *Mem:* Asn Comput Mach; Math Asn Am; Soc Indust & Appl Math. *Res:* Design and analysis of computer algorithms and data structures; emphasis on the design of efficient algorithms and mathematical investigation of the inherent limitations on the possible efficiency of algorithms for solving computational problems. *Mailing Add:* 3436 Syracuse Ave San Diego CA 92122

FREDRICH, AUGUSTINE JOSEPH, b Little Rock, Ark, Sept 12, 39; m 62; c 3. HYDROLOGIC ANALYSIS, WATER RESOURCES POLICY. *Educ:* Univ Ark, BS, 62; Calif State Univ, Sacramento, MS, 72. *Prof Exp:* Hydraulic engr, US Army Eng Dist, Little Rock, 62-66 & Hydrol Eng Ctr, US Army Corps Engrs, 66-68, chief, Res Br, Hydrol Eng Ctr, US Army Corps Engrs, 68-72, sr policy analyst, Off Chief Engrs, 73-75, tech dir, US Army Eng Inst for Water Resources, 75-78, dir, 78-79; fel, Cong of US, 72-73; prof civil eng, Ind State Univ, Evansville, 79-81; CHMN, ENG TECHNOL DIV, UNIV SOUTHERN IND, 81- *Concurrent Pos:* Consult, UNESCO, Porto Alegre, Brazil, 70 & 71. *Mem:* Am Soc Civil Engrs; Am Water Resources Asn; Int Water Resources Asn; Am Soc Eng Educ; Nat Sci Prof Engrs. *Res:* Development of digital simulation models for hydrologic analysis required in planning, design and operation of water resources projects. *Mailing Add:* Dept Eng Technol Univ Southern Ind Evansville IN 47712

FREDRICK, JEROME FREDERICK, b New York, NY, Feb 23, 26; m 46; c 3. BIOCHEMISTRY. *Educ:* City Col New York, BSc, 49; NY Univ, MSc, 51, PhD(biol), 55. *Prof Exp:* Instr biol, City Col New York, 48-49; biochemist, Vet Admin Hosp, New York, 49-51; chemist, US Customs Labs, 51-53; asst dir res, 53-60, DIR BIOCHEM RES, DODGE CHEM CO, 60-, ASSOC PROF BIOCHEM, DODGE INST ADVAN STUDIES, 65- *Concurrent Pos:* Consult, Handbook Biol Data Comt, Nat Acad Sci-Nat Res Coun, 57-60. *Mem:* AAAS; Am Chem Soc; fel NY Acad Sci; fel Am Inst Chemists; Am Inst Biol Sci; Am Inst of Sci & Technol (pres, 81-85); Int Soc Endocytobiology; fel Explorers Club. *Res:* Chelation chemistry; enzymology; chemotherapy of cancer; evolution of isozymes, especially structure and molecular evolution. *Mailing Add:* Dodge Res Labs 3425 Boston Post Rd Bronx NY 10469

FREDRICK, LAURENCE WILLIAM, b Stroudsburg, Pa, Aug 27, 27; m 49; c 3. ASTRONOMY. *Educ:* Swarthmore Col, BA, 52, MA, 54; Univ Pa, PhD(astron), 59. *Prof Exp:* Asst astron, Sproul Observ, 52-59; astronr, Lowell Observ, Ariz, 59-63; prof astron, 63-72, dir, Leander McCormick Observ, 63-79, J D HAMILTON PROF ASTRON, UNIV VA, 72- *Concurrent Pos:* Asst, Flower & Cook Observ, 55-59; mem, Nat Res Coun, 65-68; mem, US Nat Comt, Int Astron Union & Spitzer Comt, Nat Space Bd, 65-71; assoc astronr, Europ Southern Observ, 82-83 & Australia Nat Univ, 91-92; gen secy, Soc Sci Explor, 83- *Honors & Awards:* Fulbright-Hays lectr, Univ Vienna, 72-73. *Mem:* Int Astron Union; Am Astron Soc (secy); Am Meteor Soc (vpres); Royal Astron Soc; Astron Soc Pac; Sigma Xi. *Res:* Binary stars; infrared spectroscopy; image tubes; instrumentation; space astronomy; astrometry. *Mailing Add:* Box 3818 Univ Va Charlottesville VA 22903

FREDRICKS, WALTER WILLIAM, b Philadelphia, Pa, July 19, 35; m 57, 75; c 3. BIOCHEMISTRY, IMMUNOCHEMISTRY. *Educ:* LaSalle Col, BA, 57; Johns Hopkins Univ, PhD(biol), 62. *Prof Exp:* USPHS fel, Nat Heart Inst, 62-64; asst prof physiol, Univ Md, 64-65; asst prof biochem, 65-66, from asst prof to assoc prof, 66- 83, PROF BIOL, MARQUETTE UNIV, 83- *Mem:* Am Soc Biochem & Molecular Biol; Am Asn Immunologists. *Res:* Evolution of the immunoglobulins. *Mailing Add:* Dept Biol Marquette Univ Milwaukee WI 53233

FREDRICKSON, ARNOLD G(ERHARD), b Fairbault, Minn, Apr 11, 32. CHEMICAL ENGINEERING. *Educ:* Univ Minn, BS, 54, MS, 56; Univ Wis, PhD(chem eng, 59. *Prof Exp:* From asst prof to assoc prof, 58-66, PROF CHEM ENG, UNIV MINN, MINNEAPOLIS, 66- *Concurrent Pos:* US ed, Chem Eng Sci, 75-89. *Honors & Awards:* Ford Pharmaceut & Bioeng Div Award, Am Inst Chem Engrs, 88. *Mem:* Am Chem Soc; Am Inst Chem Engrs; Sigma Xi; AAAS. *Res:* Bioengineering. *Mailing Add:* Dept Chem Eng & Mat Sci Univ Minn Minneapolis MN 55455

FREDRICKSON, DONALD SHARP, b Canon City, Colo, Aug 8, 24; m 50; c 2. CARDIOLOGY, MEDICINE. *Educ:* Univ Mich, BS, 46, MD, 49. *Prof Exp:* Intern, Peter Bent Brigham Hosp, Boston, 49-50, asst med, 50-52; res fel med, Mass Gen Hosp, Boston, 52-53; clin assoc, Nat Heart & Lung Inst, 53-55, investr, Lab Cellular Physiol & Metab, 55-61, clin dir inst, 61-66, dir, 66-68, head sect molecular dis, Lab Metab, 62-66, chief molecular dis br, Div Intramural Res, 66, dir, Div Intramural Res, 68-74; pres, Inst Med, Nat Acad Sci, 74-75; dir, NIH, 75-81; scholar in residence, Nat Acad Sci, 81-83; vpres, 83-84, pres & chief exec officer, Howard Hughes Med Inst, 86-87; RESEARCHER, NIH, 87-; PRES, DSF ASSOCS, INC, 88- *Concurrent Pos:* Clin instr med, Sch Med, George Washington Univ, 56-59, spec lectr internal med, 59-; mem cardiovasc study sect, NIH, 59-62; chmn, Coun Arteriosclerosis, Am Heart Asn, 74; chmn, Fed Interagency Comn Recombining DNA Res, 76-81; chmn, Coun Med, Fed Coord Coun Sci Eng Technol, 78-81; chmn, Fed Comn Res Biol Effects Ion Radiation, 78-81; scholar, Nat Libr Med; mem, Europ Community Med Res Prog Eval panel, 90. *Honors & Awards:* Gold Medal Award, Am Col Cardiol, 67; Int Award Heart & Vascular Res, James F Mitchell Found Med Educ & Res, 68; McCollum Award, Am Soc Clin Nutrit & Am Inst Nutrit, 71; Gairdner Award, 78; Lorenzini Medal, 80. *Mem:* Nat Acad Sci; Inst of Med of Nat Acad Sci; Royal Col Physicians (London); fel Am Col Cardiol; fel Am Col Physicians; AAAS; Am Col Arts & Sci; Am Philos Soc. *Mailing Add:* 6615 Bradley Blvd Bethesda MD 20817

FREDRICKSON, JOHN E, b Chicago, Ill, Sept 12, 19; m 46; c 3. PHYSICS. *Educ:* Univ Calif, Berkeley, BS, 43; Univ Southern Calif, MS, 52, PhD(physics), 56. *Prof Exp:* PROF PHYSICS, CALIF STATE UNIV, LONG BEACH, 62- *Mem:* Am Phys Soc; Am Asn Physics Teachers. *Res:* Solid state physics; infrared spectroscopy. *Mailing Add:* 27526 Eastvale Rd Rolling Hills CA 90274

FREDRICKSON, JOHN MURRAY, b Winnipeg, Man, Mar 24, 31; m 56; c 3. OTOLARYNGOLOGY, NEUROPHYSIOLOGY. *Educ:* Univ BC, BA, 53, MD, 57; Am Bd Otolaryngol, dipl, 66; FRCS(C). *Hon Degrees:* MD, Univ Linkoping, Sweden, 75. *Prof Exp:* Instr surg, Univ Chicago, 63-65; asst prof otolaryngol, Med Ctr, Stanford Univ, 65-68; assoc prof otolaryngol, 68-76, ASST PROF PHYSIOL, UNIV TORONTO, 68-, PROF OTOLARYNGOL, 76- *Concurrent Pos:* Vis investr, Univ Freiburg, 64-65; consult, Toronto Gen Hosp, Princess Margaret Cancer Hosp & Hosp for Children. *Honors & Awards:* Res Award, Am Acad Ophthal & Otolaryngol, 64; Hodge Mem Award, Can Otolaryngol Soc, 65. *Mem:* Fel Am Col Surg. *Res:* Vestibular neurophysiology; tissue and organ transplantation. *Mailing Add:* Dept Otolaryngol Wash Univ Med Sch Campus Box 8115 517 Euclid St Louis MO 63110

FREDRICKSON, LEIGH H, b Sioux City, Iowa, Mar 13, 39; m 65; c 2. WILDLIFE ECOLOGY. *Educ:* Iowa State Univ, BS, 61, MS, 63, PhD(zool), 67. *Prof Exp:* Instr zool, Iowa State Univ, 64-65; asst prof biol sci, 67-73, assoc prof wildlife, 73-79, PROF WILDLIFE, SCH FORESTRY, FISHERIES & WILDLIFE, UNIV MO-COLUMBIA, 79-, RUCKER PROF, 87- *Concurrent Pos:* Dir, Gaylord Lab, Univ Mo, Columbia, 67- *Honors & Awards:* E Sydney Stephens Prof Wildlife Award. *Mem:* Wildlife Soc; AAAS; Am Ornith Union; Sigma Xi; Am Inst Biol Sci. *Res:* Behavior and ecology of waterfowl, especially Cairinini and Mergini; marsh and swamp ecology, especially lowland hardwoods; ecology of wintering waterfowl. *Mailing Add:* Gaylord Mem Lab Univ Mo Rte 1 Box 185 Puxico MO 63960

FREDRICKSON, RICHARD WILLIAM, b Blakesburg, Iowa, Apr 28, 19; m 47; c 2. ACAROLOGY, ECOLOGY. *Educ:* Univ Kans, AB, 51, MA, 54, PhD(entom), 61. *Prof Exp:* Lectr zool, Southern Ill Univ, 56-58; lectr biol, Queens Col, NY, 61-62, instr, 62-63; from lectr to asst prof, City Col New York, 63-67; from assoc prof to prof, 67-89, EMER PROF BIOL, ST JOSEPH'S UNIV, PHILADELPHIA, PA, 89. *Mem:* Acarol Soc Am; Am Entomol Soc. *Res:* Zoogeography and systematics of Acarina; ecology and ethology of Arthropoda, particularly Acarina; symbiotic associations of Acarina. *Mailing Add:* Dept Biol St Joseph's Univ 5600 City Ave Philadelphia PA 19131

FREDRIKSSON, KURT A, b Haparanda, Sweden, Apr 9, 26; m 64. GEOCHEMISTRY. *Educ:* Stockholm Inst Technol, Sweden, ChemE, 46; Univ Stockholm, PhD(mineral & geol), 55. *Prof Exp:* Consult engr, Hagconsult Inc, Sweden, 46-52; res asst sediment petrol, Oceanog Inst, Univ Gothenburg, 53-55; res asst meteorite mineral, Univ Stockholm, 55-57; geochemist, Geol Surv Sweden, 58-60; res assoc meteoritics, Univ Calif, San Diego, 60-64; cur-in-chg, 64-67, GEOCHEMIST, DEPT MINERAL SCI, SMITHSONIAN INST, 67- *Concurrent Pos:* Grants, Australian Nat Univ, 61 & NASA, 64; consult econ geol & inorg & org chem, Sweden, 52-60; prin investr lunar samples, Apollo XI-XVII; Fulbright scholar, Max Planck Inst, Ger, 74-75. *Mem:* Geochem Soc; Meteoritical Soc. *Res:* Meteoritics; phase composition in meteorites, particularly chondrites; origin of meteorites and planets; ore mineralogy; electron and ion microprobe analysis. *Mailing Add:* Div Meteorites Smithsonian Inst Washington DC 20560

FREE, ALFRED HENRY, b Bainbridge, Ohio, Apr 11, 13; m; c 9. BIOCHEMISTRY. *Educ:* Miami Univ, AB, 34; Western Reserve Univ, MS, 36, PhD(biochem), 39. *Prof Exp:* Asst biochem, Cleveland Clin, 34-35; instr, Western Reserve Univ, 39-43, asst prof, 43-46; head biochem sect, Ames Res Lab, 46-59, dir, 59-64, dir, Ames Tech Serv, 64-70, vpres tech serv & sci rels, 70-78, SR SCI CONTRACTOR, MILES INC, 78- *Concurrent Pos:* Consult,

Ben Venue Labs, Ohio, 43-46. *Honors & Awards:* Outstanding Serv to Prof, Asn Clin Chemists, 82; Mosher Award, Am Chem Soc, 84. *Mem:* AAAS; Am Soc Biol Chem; Soc Exp Biol & Med; Am Chem Soc; Am Inst Chem; Am Asn Clin Chem; Am Diabetic Asn; Sigma Xi. *Res:* Vitamin and protein nutrition; enzymes and gastrointestinal tract; intestinal absorption; antibiotics; clinical biochemistry and laboratory methodology; radioisotope methodology; metabolism of drugs; urinalysis tests. *Mailing Add:* Miles Inc Diag Div PO Box 70 Elkhart IN 46514

FREE, CHARLES ALFRED, b Cleveland, Ohio, Apr 19, 36; m 61; c 2. BIOCHEMICAL PHARMACOLOGY, ENZYMOLOGY. *Educ:* Purdue Univ, BS, 57; Univ Calif, Los Angeles, PhD(physiol chem), 62. *Prof Exp:* Res fel, Sloan-Kettering Inst Cancer Res, 62-65; sr res scientist, Squibb Inst Med Res, 65-69, sr res investr, 69-82, res fel, 82-90, RES LEADER, SQUIBB INST MED RES, 90- *Concurrent Pos:* USPHS fel, 63-65. *Mem:* AAAS; Am Chem Soc; Sigma Xi; Am Soc Pharmacol Exp Ther. *Res:* Purification and properties of anterior pituitary hormones; cyclic nucleotide-mediated biological systems; immediate hypersensitivity; enzymes and enzyme inhibitors; steroid hormone receptors. *Mailing Add:* Dept Cardiovasc Biochem Bristol-Myer Squibb Pharmaceut Res Inst PO Box 4000 Princeton NJ 08543-4000

FREE, HELEN M, b Feb 20, 23; m 47; c 6. CLINICAL CHEMISTRY. *Educ:* Col Wooster, AB, 44; Cent Mich Univ, MA, 78. *Prof Exp:* Control chemist, Miles-Ames, 44-46, res chemist, Biochem Sect, 46-59, assoc res biochemist & group leader, Ames Res Lab, 59-64, Ames Prod Develop Lab, 64-66, Ames Tech Serv, 66-69, new prod mgr clin test systs, Ames Growth & Develop, 69-74, sr new prod mgr microbiol test systs, 74-76, dir spec test systs, Ames Div, 76-78, dir clin lab, reagents, Res Div, 78-82, PROF RELS, DIAG DIV, MILES INC, 82- *Concurrent Pos:* Adj fac, Ind Univ, South Bend, 76-; mem, bd dirs, Am Chem Soc. *Honors & Awards:* Garvan Medal, Am Chem Soc, 80, Mosher Award, Santa Clara Valley, 82. *Mem:* Am Chem Soc; Am Asn Clin Chem (pres, 90); Am Soc Med Technol; fel Am Inst Chemists; fel AAAS. *Res:* Patents in diagnostics field; author and co-author of over 200 papers and presentations. *Mailing Add:* 3752 E Jackson Blvd Elkhart IN 46516

FREE, JOHN ULRIC, b Homestead, Fla, Sept 4, 41; c 1. SOLID STATE PHYSICS, ACOUSTICS. *Educ:* Eastern Nazarene Col, BS, 64; Mass Inst Technol, PhD(physics), 74. *Prof Exp:* ASSOC PROF PHYSICS, EASTERN NAZARENE COL, 70- *Concurrent Pos:* Educ consult, 74-75; mem fac fel prog, NSF, 75. *Mem:* Am Inst Physics; Am Asn Physics Teachers; Acoust Soc Am; Sigma Xi. *Res:* Investigation of fundamental materials using ultrasonic and hypersonic waves; ultrasonic attenuation in metals at low temperatures used to investigate the electron-photon interaction and the interface between materials. *Mailing Add:* Dept Physics Eng Eastern Nazareth Col 23 E Elm Ave Quincy MA 02170

FREE, JOSEPH CARL, b Cedar City, Utah, May 14, 35; m 55; c 6. MECHANICAL ENGINEERING. *Educ:* Brigham Young Univ, BES, 58; Calif Inst Technol, MS, 61; Mass Inst Technol, PhD(nonlinear models), 68. *Prof Exp:* Res engr, Autonetics Div, NAm Aviation, Inc, 58-61; from instr to asst prof mech eng, Brigham Young Univ, 61-64; teaching asst, Mass Inst Technol, 64-65; assoc prof, 67-77, PROF MECH ENG, BRIGHAM YOUNG UNIV, 77- *Concurrent Pos:* Consult, Rich Bumper Co, 68 & Lawrence Radiation Lab, 68- *Mem:* Am Soc Mech Engrs. *Res:* Nonlinear system modeling; optimum design of engineering systems; efficient computer simulation; automatic controls. *Mailing Add:* 195 W 400 S Orem UT 84057

FREE, MICHAEL JOHN, b Newton Abbot, Eng, Nov 22, 37; m 64; c 2. PHYSIOLOGY. *Educ:* Univ Nottingham, BSc, 64; Ohio State Univ, MSc, 65, PhD(physiol of reprod), 67. *Prof Exp:* Res assoc testis physiol, Animal Reprod Teaching & Res Ctr, Ohio State Univ, 67-69; from asst prof to assoc prof mammalian physiol, Calif State Univ, Hayward, 69-72; SR RES SCIENTIST, BATTELLE, PAC NORTHWEST LAB, 72- *Mem:* Brit Soc Study Fertil; Soc Study Reprod. *Res:* Physiology and biochemistry of the testis and epididymis, particularly effect of temperature and hormones; circulation in vivo metabolism and hormone production and fertility control methods in male and female. *Mailing Add:* 6333 14th Ave NE Seattle WA 98115

FREE, STEPHEN J, b Salt Lake City, Utah, Sept 4, 48; m; c 4. GENETICS, MOLECULAR BIOLOGY. *Educ:* Purdue Univ, BS, 72; Stanford Univ, PhD(genetics), 77. *Prof Exp:* Res assoc, Dept Physiol Chem, Univ Wis, 77-79; ASSOC PROF GENETICS, DEPT BIOL SCI, STATE UNIV NY AT BUFFALO, 79- *Res:* Mechanisms by which fungi adapt to changes in the environment. *Mailing Add:* Cooke Hall Rm 370 Dept Biol Sci State Univ NY Buffalo NY 14260

FREEBERG, FRED E, b Windber, Pa, Nov 5, 37; m 63; c 2. ANALYTICAL CHEMISTRY. *Educ:* Univ Pittsburgh, BS, 59; Pa State Univ, PhD(chem), 65. *Prof Exp:* Asst chemist, Koppers Co, Inc, 59-60; teaching asst, Pa State Univ, 60-65; staff res chemist, 65-80, SECT HEAD, PROCTER & GAMBLE CO, 80- *Mem:* Am Chem Soc; Soc Qual Assurance. *Res:* General use of thermal methods of analysis to obtain thermodynamic constants as well as application to analysis. *Mailing Add:* Ivorydale Tech Ctr Procter & Gamble Co 5299 Spring Grave Ave Cincinnati OH 45217

FREEBERG, JOHN ARTHUR, b Kenosha, Wis, July 2, 32; m 55; c 1. PHYSIOLOGY, PLANT. *Educ:* Harvard Univ, AB, 54, PhD(biol), 57. *Prof Exp:* Asst prof biol, Lehigh Univ, 57-62 & Boston Univ, 62-66; ASSOC PROF BIOL, UNIV MASS, BOSTON, 66- *Mem:* Bot Soc Am. *Res:* Morphogenesis of vascular plants. *Mailing Add:* Dept Biol Univ Mass Boston MA 02125

FREED, ARTHUR NELSON, b Philadelphia, Pa, Mar 14, 52; m 90. RESPIRATORY PHYSIOLOGY, ASTHMA. *Educ:* Ind Univ Pa, BS, 73; Temple Univ, MEd, 75; Univ Fla, PhD(zool), 82. *Prof Exp:* Fel physiol, Univ Fla Col Vet Med, 82-83; fel physiol, Dept Hyg, 83-86, from instr to asst prof physiol, 86-91, ASST PROF MED, DEPT MED, JOHNS HOPKINS UNIV, 90-, ASSOC PROF PHYSIOL, DEPT HYG, 91- *Concurrent Pos:* Prin investr, Nat Heart, Lung & Blood Inst, NIRA, NIH, 86-89, First Award, 87-92. *Mem:* AAAS; Am Physiol Soc; Am Soc Zoologists; Am Thoracic Soc. *Res:* Potential mechanisms underlying airways disease with emphasis on asthma and non-specific airway reactivity. *Mailing Add:* Johns Hopkins Univ 7006 Hygiene 615 N Wolfe St Baltimore MD 21205

FREED, CHARLES, b Budapest, Hungary, Mar 21, 26; US citizen; m 56; c 2. ELECTRICAL ENGINEERING. *Educ:* NY Univ, BEE, 52; Mass Inst Technol, SM, 54, EE, 58. *Prof Exp:* Asst, Res Lab Electronics, Mass Inst Technol, 52-54, Div Indust Co-op, 54-55, Res Lab Electronics, 55-58, consult, Phys Sci Study Comt, 58; sr engr, Res Div, Raytheon Mfg Co, 58-59, Spencer Lab, 59, spec microwave devices oper, 59-62; mem staff, 62-74, SR STAFF, LINCOLN LAB, MASS INST TECHNOL, 78-, DEPT LECTR ELEC ENG & COMPUTER SCI, 69- *Mem:* Fel Inst Elec & Electronics Engrs; Sigma Xi. *Res:* Microwave electronics; electron devices; low noise amplification; parametric devices; vacuum theory and techniques; electron beams; laser design and applications, intensity fluctuations and frequency stability; optical detection; optical radar. *Mailing Add:* 16 Browning Lane RFD-1 Lincoln MA 01773

FREED, JACK H, b Brooklyn, NY, Apr 19, 38; m 61; c 2. CHEMICAL PHYSICS. *Educ:* Yale Univ, BE, 58; Columbia Univ, MA, 59, PhD(chem), 62. *Prof Exp:* NSF & Hon US Ramsay Mem fels, 62-63; from asst prof to assoc prof chem, 63-73, PROF CHEM, CORNELL UNIV, 73- *Concurrent Pos:* Alfred P Sloan Found fel, 66-68; vis scientist, US-Japan Coop Sci Prog, Tokyo Univ, 69-70; sr Weizmann fel, Weizmann Inst Sci, 70-; guest prof, Aarhus Univ, Denmark, 74; assoc ed, J Chem Physics, 76-78; vis prof, Geneva Univ, 77 & Delft Tech Univ, 78; assoc ed, J Phys Chem, 79-83, Chem Physics Lett, 88-; Guggenheim fel, 84-85; guest prof, Ecole Normale Superieure, Paris, 84-85. *Honors & Awards:* Buck-Whitney Award, Am Chem Soc, 81. *Mem:* Fel Am Phys Soc. *Res:* Applications of magnetic resonance to problems of theoretical chemical interest. *Mailing Add:* Baker Lab Chem Cornell Univ Ithaca NY 14853-0001

FREED, JAMES MELVIN, b Enid, Okla, Apr 6, 39; m 69; c 2. CELL PHYSIOLOGY. *Educ:* McPherson Col, BS, 61; Univ Ill, Urbana-Champaign, MS, 63, PhD(physiol), 69. *Prof Exp:* Instr biol, Manchester Col, 63-65; from asst prof to assoc prof zool, 69-87, PROF ZOOL, OHIO WESLEYAN UNIV, 87- *Mem:* AAAS; Am Soc Zool; Sigma Xi; Am Soc Human Genetics. *Res:* Biochemical adaptations in poikilotherms to environmental temperature changes. *Mailing Add:* Dept Zool Ohio Wesleyan Univ Delaware OH 43015-2378

FREED, JEROME JAMES, b New York, NY, May 4, 28; m 51; c 4. CELL BIOLOGY, CELL CULTURE. *Educ:* Yale Univ, BS, 49; Columbia Univ, MA, 51, PhD(zool), 54. *Prof Exp:* Res assoc, 53-62, from asst to assoc mem, 66-78, MEM,, INST CANCER RES, 78- *Mem:* AAAS; Tissue Culture Asn; Am Soc Cell Biol; Am Asn Cancer Res. *Res:* Heredity in cultured cells; genetic events in oncogenic transformation; cell culture methodology. *Mailing Add:* Inst Cancer Res Fox Chase Cancer Ctr Philadelphia PA 19111

FREED, JOHN HOWARD, b New Brighton, Pa, Mar 10, 43; m 71; c 2. TRANSPLANTATION BIOCHEMISTRY, IMMUNOGENETICS. *Educ:* Mass Inst Technol, SB, 65; Stanford Univ, PhD(org chem), 71. *Prof Exp:* Fel immunol, Sch Med, Stanford Univ, 71-73, Albert Einstein Col Med, 73-76; asst prof immunol, Sch Med, Johns Hopkins Univ, 76-; ASSOC PROF, DIV BASIC IMMUNOL, DEPT MED, NAT JEWISH HOSP & RES CTR. *Mem:* Am Asn Immunologists; AAAS. *Res:* Biochemistry of the membrane antigens encoded by the H-2 complex, the major histocompatibility complex of mice: transplant rejection and regulation of the immune response. *Mailing Add:* Dept Med Div Basic Immunol Nat Jewish Ctr Immunol & Resp Med 1400 Jackson St Denver CO 80206

FREED, KARL F, b Brooklyn, NY, Sept 25, 42; m 64; c 2. THEORETICAL CHEMISTRY, PHYSICAL CHEMISTRY. *Educ:* Harvard Univ, AM, 65, PhD(chem physics), 67. *Prof Exp:* NATO fel theoret physics, Univ Manchester, 67-68; from asst prof to assoc prof, 68-76, dir, James Franck Inst, 83-86, PROF CHEM, UNIV CHICAGO, 76- *Concurrent Pos:* Sloan Found res fel, 69-71; Guggenheim Found res fel, 72-73; Dreyfus Found teacher-scholar fel, 72-; sr vis fel, Cavendish Lab, Cambridge, Eng, 72-73; Denkewalter lectr, Loyola Univ Chicago, 76; Phillips lectr, Haverford Col, 76; vis scientist, Centre Nucleaires, Strasbourg, France, 77; adv ed, Chem Physics, 79-, Chemical Rev, 81-83, adv, Chem Physics, 84-; assoc ed, J Chem Physics, 82-84; Hill vis prof, Univ Minn, 84. *Honors & Awards:* Marlow Medal, Faraday Div, The Chem Soc, 73; Am Chem Soc Award Pure Chem, 76. *Mem:* Fel Am Phys Soc; Royal Soc Chem. *Res:* Radiationless processes and photochemistry in polyatomic molecules; electronic structure in disordered systems; statistical mechanics of polymer systems; many-body theory and the electronic structure of molecules; quantum chemistry; polymer physics. *Mailing Add:* James Franck Inst Univ Chicago Chicago IL 60637

FREED, LEONARD ALAN, b Cleveland Ohio, July 17, 47; m 89; c 1. EVOLUTIONARY & BEHAVIORAL ECOLOGY. *Educ:* Northwestern Univ, BA, 69; Univ Iowa, MS, 81, PhD(zool), 81. *Prof Exp:* Res fel, Smithsonian Trop Res Inst, 82-83; ASSOC PROF AVIAN BIOL & VERT ZOOL, UNIV HAWAII, 83- *Honors & Awards:* W C Allee Award, Animal Behav Soc, 81. *Mem:* Ecol Soc Am; Animal Behav Soc; Am Ornithologists' Union; Cooper Orinth Soc; Sigma Xi; AAAS; Soc Conserv Biol; soc Study Evolution. *Res:* Avian life history and mating systems theory; evolution of permanent monogamy and reproductive rate in tropical house wrens; evolution of sexual dichromatism and cavity-nesting in the Hawaii Akepa; population structure of native Hawaiian birds. *Mailing Add:* Dept Zool Univ Hawaii at Manoa 2538 The Mall Honolulu HI 96822

FREED, MEIER EZRA, b Philadelphia, Pa, Oct 20, 25; m 70. ORGANIC CHEMISTRY. *Educ:* Pa State Univ, BSc, 48, MSc, 49; Univ Pa, PhD(chem), 60. *Prof Exp:* res chemist, Wyeth Labs, Inc, 51-87; RETIRED. *Mem:* Am Chem Soc. *Res:* Pharmaceutical chemistry; heterocyclic and alicyclic derivatives; local anesthetics (oxethazine), CNS (iprindole), cardiovascular agents, analgesics (dezocine); 95 patents, 16 publications. *Mailing Add:* 804 Colony Rd Bryn Mawr PA 19010

FREED, MICHAEL ABRAHAM, b Cheltenham, Pa, Dec 26, 53; m 88. NEUROBIOLOGY, VISION. *Educ:* Yale Univ, BA, 76; Univ Pa, PhD(neurobiol), 84. *Prof Exp:* Guest researcher, 85-87, POSTDOCTORAL FEL, NIH, 88- *Mem:* Soc Neurosci; Asn Res Vision & Ophthal. *Res:* Physiological basis of vision; identification in cat retina of specific cell types, their electrical responses and their synaptic connections. *Mailing Add:* NIH Nat Inst Neurol Dis & Stroke Div Intramural Res Lab Neurophysiol Bldg 10 Rm 4D10 Bethesda MD 20892

FREED, MURRAY MONROE, b Paterson, NJ, Oct 9, 24; m 48; c 4. REHABILITATION MEDICINE. *Educ:* Harvard Univ, AB, 48; Boston Univ, MD, 52. *Prof Exp:* Nat Found Infantile Paralysis fel phys med, NY Univ-Bellevue Med Ctr, 53-56; instr phys med & rehab, 56-59, assoc, 59-62, from asst prof to assoc prof, 62-67, PROF REHAB MED & CHMN DEPT, SCH MED, BOSTON UNIV, 67- *Concurrent Pos:* Asst chief rehab med, Univ Hosp, Boston, 56-59, chief, 59-; consult, Vet Admin, 62-74; mem med adv bd, Mass Sect, Nat Multiple Sclerosis Found, 63-80; physician-in-chief phys med & rehab, Boston City Hosp, 64-75, sr vis physician, 75-80; trustee & mem med sci comt, Mass Sect, Arthritis Found, 64-84; deleg, Int Med Soc Paraplegia, 67, 68, 75 & 79; mem med adv bd, Nat Paraplegia Found, 70-81; chmn, City Boston Mayor's Comn Physically Handicapped, 72-82; mem, Mass Gov Comn Employment Handicapped; bd dirs, Am Spinal Injury Asn, 82-88; dir, Am Bd Phys Med & Rehab, 74-87, vchmn, 84-87. *Honors & Awards:* Herbert Talbot Award, Nat Spinal Cord Injury Found, 78; N Neil Pike Prize, 81; Bernard Kutner distinguished lectr, 81; Distinguished Clinician Award, Am Acad Phys Med & Rehab, 88; Award for Distinguished Serv, Am Spinal Injury Asn, 88. *Mem:* Am Acad Phys Med & Rehab (pres, 82-83); Am Cong Rehab Med; Am Col Physicians; Am Spinal Injury Asn; Asn Acad Physiatrists (pres, 71). *Res:* Long term effects of spinal cord trauma; rehabilitation potential of patient with spinal cord trauma; rehabilitation potential of individual with amputation as a result of malignancy. *Mailing Add:* 88 E Newton Boston MA 02118

FREED, NORMAN, b Philadelphia, Pa, June 11, 36; m 60; c 2. THEORETICAL NUCLEAR PHYSICS. *Educ:* Antioch Col, BS, 58; Western Reserve Univ, MS, 61, PhD(theoret physics), 64. *Prof Exp:* Res fel physics, Univ Nebr, 63-64; Nordic Inst Theoret Atomic Physics res fel, Inst Theoret Physics, Univ Lund, Sweden, 64-65; Ford Found fel, Niels Bohr Inst, Univ Copenhagen, 65; from asst prof to assoc prof, 65-82, asst dean, Col Sci, 78-79, PROF PHYSICS, PA STATE UNIV, UNIVERSITY PARK, 82-, ASSOC DEAN, COL SCI, 79- *Concurrent Pos:* Woodrow Wilson fel & Atomic Energy Comn fel, 58; NSF fel, 62-63; Ford Found fel, Niels Bohr Inst, Copenhagen, 64-65; Res Inst Theoret Phys fel, Univ Helsinki, 71; guest scientist & fel, Ctr Nuclear Studies, French Atomic Energy Comn Saclay, Paris, 71-72. *Mem:* AAAS; Am Phys Soc; Sigma Xi; Am Asn Univ Profs. *Res:* Intermediate energy nuclear theory; pion photoproduction; nuclear structure theory. *Mailing Add:* 211 Whitmore Lab Pa State Univ University Park PA 16802

FREED, ROBERT LLOYD, b Allentown, Pa, Feb 13, 50; m. TECHNICAL MANAGEMENT. *Educ:* Drexel Univ, BS, 73; Mass Inst Technol, PhD(mat eng), 78. *Prof Exp:* Res supvr, 78-90, RES MGR, ENG TECHNOL LAB, E I DU PONT DE NEMOURS & CO, INC, 90- *Mem:* Am Soc Metals; Mat Res Soc; Tech Asn Pulp & Paper Indust. *Res:* Materials characterization; mechanics of materials; corrosion, wear and the combined corrosion-wear of materials; thin film technology; ceramics; particle technology. *Mailing Add:* Edge Moore Plant DuPont Chemicals Edge Moor DE 19809

FREED, SIMON, b Lodz, Poland, Nov 11, 99; nat US; m 43. NEUROPHYSIOLOGICAL CHEMISTRY. *Educ:* Mass Inst Technol, SB, 20; Univ Calif, PhD(chem), 27. *Prof Exp:* Instr chem, Univ Calif, 27-28, res assoc, 28-30; Guggenheim fel, Kammerlingh-Onnes Lab, Leiden, 30-31; instr, Univ Chicago, 31-37, asst prof, 37-43 & 45-46; group leader, SAM Lab, Columbia Univ, 43-45; chief chemist, Oak Ridge Nat Lab, 46-49; sr scientist, Brookhaven Nat Lab, 49-65; res prof biochem & neurol, NY Med Col, 67-75; RES COLLABR, BROOKHAVEN NAT LAB, 68-; RES SCIENTIST, HEART DIS RES FOUND, 83- *Concurrent Pos:* Inter-acad exchange, USSR, 66, Czech, 67. *Mem:* AAAS. *Res:* Optical and magnetic properties; symmetry of electric fields about ions in solutions and crystals; chemistry and biochemistry at low temperatures; resolution of states and reactivities in chemistry and biochemistry; endogenous chemistry of animals during learning; physiological chemistry of the nervous system; cryobiochemical analysis of tissue; comparative mnemonic processes. *Mailing Add:* 89 Bedford St New York NY 10014

FREED, VIRGIL HAVEN, b Mendota, Ill, Nov 18, 19; m 44; c 4. BIOCHEMISTRY. *Educ:* Ore State Univ, BS, 43, MS, 48, PhD(chem), 59. *Prof Exp:* Asst prof chem, 44-48, assoc prof chem & farm crops, 48-54, assoc prof chem, 54-59, dir, Environ Health Sci Ctr, 67-81, PROF CHEM, ORE STATE UNIV, 60-, HEAD DEPT AGR CHEM, 61- *Concurrent Pos:* Chmn res comt, USPHS, 47; mem, Western Weed Control Conf, 52; mem, Gordon Res Conf, 58, 65; lectr, Tour Europ Res Ctr, 62; mem, Gov Adv Comt Synthetic Chem in Environ, Ore, 69-; mem toxic substances panel, Nat Acad Sci, 72-73. *Mem:* Fel AAAS; Am Chem Soc; Weed Sci Soc Am. *Res:* Mechanism of action of plant growth regulators and herbicides; physical chemistry of compounds in relation to their biological action. *Mailing Add:* Dept Agr Chem Ore State Univ Corvallis OR 97331

FREEDBERG, ABRAHAM STONE, b Salem, Mass, May 30, 08; m 35; c 2. MEDICINE. *Educ:* Harvard Univ, AB, 29; Univ Chicago, MD, 35; Am Bd Internal Med, dipl, cert Cardiol. *Prof Exp:* House officer med, Mt Sinai Hosp, Ill, 34-35; resident, Cook County Hosp, Ill, 35-36; house officer path, RI Hosp, 36-37; asst med, Beth Israel Hosp, 38-40; res fel, 41-42, asst, 42-46, instr, 46-47, assoc, 47-50, from asst prof to prof, 50-74, EMER PROF MED, HARVARD MED SCH, 74- *Concurrent Pos:* From jr vis physician to physician, Beth Israel Hosp, Boston, 40-69, assoc med res, 40-50, from assoc dir to dir, Cardiol Unit, 50-69; consult & mem thyroid uptake calibration comt, Med Div, Oak Ridge Inst Nuclear Studies, 55-56; Ziskind sr fel, Beth Israel Hosp, Boston, 56; consult metab study sect, USPHS, 56-60; Guggenheim fel, 67-68. *Honors & Awards:* Pioneer Award, Am Nuclear Soc. *Mem:* Am Physiol Soc; Am Soc Clin Invest; Am Heart Asn; Asn Am Physicians; Am Thyroid Asn (first vpres, 65-66). *Res:* Cardiovascular diseases; thyroid in health and disease. *Mailing Add:* 111 Perkins St Jamaica Plain MA 02130

FREEDBERG, IRWIN MARK, b Boston, Mass, July 4, 31; m 54; c 3. DERMATOLOGY. *Educ:* Harvard Univ, MD, 56. *Prof Exp:* Intern, asst resident & res fel med, Beth Israel Hosp, 56-59; asst resident & res fel dermat, Mass Gen Hosp, 59-61; fel biochem, Brandeis Univ, 61-62; from instr to prof, dept dermat, Sch Med, Harvard Univ, 62-77; prof & chmn dept dermat, Johns Hopkins Med Insts, 77-81; PROF & CHMN DEPT DERMAT, SCH MED, NEW YORK UNIV, 81-; DIR DERMAT SERV, BELLEVUE HOSP, NY, 81- *Concurrent Pos:* Fel, Med Found Boston, Inc, 62-65; fac res assoc, Am Cancer Soc, 65-70; John Simon Guggenheim Mem Found fel, dept biophys, Weizmann Inst Sci, Israel, 69-70; ed-in-chief, J Investigative Dermat, 72-77; dir, Am Bd Dermat, 84-; mem, Adv Coun, Nat Inst Arthritis, Diabetes & Digestive & Kidney Dis, 84-; chmn Coun Govt Liaison, Am Acad Dermat, 86- *Mem:* Asn Am Physicians; Am Soc Cell Biol; Soc Investigative Dermat (pres, 81-82); Asn Profs Dermat (pres, 86-); Am Soc Clin Invest; Am Soc Biochemists. *Res:* The role of keratin and keratinization in health and disease. *Mailing Add:* Dept Dermat NY Univ Sch Med 550 First Ave New York NY 10016

FREEDLAND, RICHARD A, b Pittsburgh, Pa, May 9, 31; m 58; c 3. PHYSIOLOGICAL CHEMISTRY. *Educ:* Univ Pittsburgh, BS, 53; Univ Ill, MS, 55; Univ Wis, PhD(biochem), 58. *Prof Exp:* Res assoc pediat, Univ Wis, 58-60; lectr biochem, 60-61, from asst prof to assoc prof physiol chem, 61-69, PROF PHYSIOL CHEM, UNIV CALIF, DAVIS, 69-, CHMN DEPT PHYSIOL SCI, 74- *Concurrent Pos:* Assoc ed, J Biol Chem, 84-88; Fulbright scholar, Australia, 87-88; Wellcome vis prof, Univ Ga, 91. *Mem:* Fel AAAS; Am Soc Biol Chem; Am Inst Nutrit; Soc Exp Biol & Med; Am Physiol Soc. *Res:* Correlation of enzyme activity and productive capacity in isolated cells after nutritional or hormonal treatment of animals, particularly the mechanisms and physiological interpretations; author of one book. *Mailing Add:* Dept Physiol Sci Univ Calif Davis CA 95616

FREEDMAN, AARON DAVID, b Albany, NY, Jan 4, 22; m 48; c 3. INTERNAL MEDICINE, BIOCHEMISTRY. *Educ:* Cornell Univ, AB, 42; Albany Med Col, MD, 45; Columbia Univ, PhD(biochem), 58; Am Bd Internal Med, dipl. *Hon Degrees:* MA, Univ Pa, 73. *Prof Exp:* USPHS fel, Columbia Univ, 54-57, from instr biochem to asst prof med, 58-65; dir dept med, Menorah Med Ctr & dir exp med, Danciger Res Found, 65-69; prof med & assoc dean, Sch Med, Univ Pa, 69-75; dir, Herman Goldman Inst, 75-79, PROF MED, CITY COL NEW YORK, 75- *Concurrent Pos:* Career investr, Health Res Coun, New York, 64-65; clin prof med, Sch Med, Univ Kans, 66-69; exec dir, Grad Hosp, Univ Pa, 72-75; actg vpres health affairs & actg dean, Sophie Davis Sch Biomed Educ, 78-79, dep dean acad affairs, City Univ NY Med Sch, 90- *Mem:* Am Soc Biol Chem; Am Soc Cell Biol; Harvey Soc. *Res:* Intermediary metabolism; medical education; health sciences administration. *Mailing Add:* City Col New York 138th St & Convent Ave New York NY 10031

FREEDMAN, ALFRED MORDECAI, b Albany, NY, Jan 7, 17; m 43; c 2. MEDICAL SCIENCES. *Educ:* Cornell Univ, AB, 37; Univ Minn, MB, 41. *Prof Exp:* Intern, Harlem Hosp, NY, 41-42; asst path, Mt Sinai Hosp, 46; physiologist, Med Div, Army Chem Ctr, Md, 46-48; resident, Psychiat Div, NY Univ-Bellevue Med Ctr, 48-49, asst alienist, 49-50, jr psychiatrist, Children's Serv, 50-51, res fel, Univ-Ctr, 51-52, sr psychiatrist, Children's Serv, 53-54, clin instr psychiat, Col Med, 53-55; from asst prof to assoc prof psychiat, State Univ NY Downstate Med Ctr, 55-60; prof & chmn dept, 60-89, EMER PROF PSYCHIAT & CHMN DEPT, NY MED COL, 89- *Concurrent Pos:* Child psychiatrist, Inst Phys Med & Rehab, Bellevue Hosp, 52-53, clin asst neuropsychiatrist, 54-55; asst, NY Univ Hosp, 53-55; asst pediatrician, Babies Hosp, 53-60; dir pediat psychiat, Kings County Hosp, 55-60; dir psychiat, Flower & Fifth Ave Hosp, Metrop Hosp & Bird S Coler Hosp, New York; pres & chmn, Nat Comm Confidentiality Health Records, Ctr Urban Educ; mem bd dir, Am Bd Psychiat & Neurol; consult, Europ Ment Health Sect WHO; chmn gen comt, World Psychiat Asn; vis prof, Dept Social Med, Harvard Med Sch, 88- *Honors & Awards:* Henry Wisner Award, 64; Samuel Hamitton Award, 72; Cardinal Cooke Medal, 84; Jeanne Knutsen Award, Int Soc Polit Psychol, 89; Wyeth-Ayerst Award, World Psychiat Asn, 89. *Mem:* Fel Am Psychiat Asn (pres, 73-74); fel Am Orthopsychiat Asn; fel NY Acad Sci; Am Psychopath Asn (pres, 71-72); Am Col Neuropsychopharmacol (pres, 72-73); hon fel Royal Col Psychiat UK; hon fel Royal Australia & NZ Col Psychiat. *Res:* Effect of anticholinesterases on central nervous system; clinical and biological aspects of child and adult psychiatry; psychiatric education; narcotic addiction; psychopharmacology and social psychiatry. *Mailing Add:* 1148 Fifth Ave New York NY 10128

FREEDMAN, ALLEN ROY, b Chicago, Ill, Aug 18, 40; m 62; c 2. MATHEMATICS. *Educ:* Univ Calif, Berkeley, AB, 62; Ore State Univ, PhD(math), 65. *Prof Exp:* Asst prof, 65-70, ASSOC PROF MATH, SIMON FRASER UNIV, 70- *Mem:* Am Math Soc. *Res:* Number theory; density theory in additive number theory. *Mailing Add:* Math & Stat Dept Simon Fraser Univ Burnaby BC V5A 1S6 Can

FREEDMAN, ANDREW, b Brooklyn, NY, Apr 14, 51. MATERIALS SCIENCE. *Educ:* Yale Univ, BS, 71; Univ Calif, Berkeley, PhD(chem), 77. *Prof Exp:* Fel, Columbia Univ, 76-79; PRIN RES SCIENTIST, AERODYNE RES, 79- *Mem:* Am Chem Soc; Am Phys Soc; Mat Res Soc. *Res:* Gas phase kinetics and surface studies of electronic material deposition and processing. *Mailing Add:* Aerodyne Res 45 Manning Rd Billerica MA 01821

FREEDMAN, ARTHUR JACOB, b Brooklyn, NY, Dec 10, 24; m 64; c 6. WATER TREATMENT, CORROSION CONTROL. *Educ:* NY Univ, BA, 45, MS, 46, PhD(chem), 48. *Prof Exp:* Res assoc, Mass Inst Technol, 48-49; mem staff, Los Alamos Sci Lab, 49-51; res assoc, Univ NMex, 51-54; from asst proj engr to sr proj supvr, Stand Oil Co, Ind, 54-59; sect head, Nalco Chem Co, 59-66, res coordr, 66-67, tech mgr res, 67-69, tech dir, 69-72, mkt mgr, 72-78, tech dir, 78-81; pres, Arthur Freedman Assocs, Inc, 81-88; EXEC VPRES, THOMAS M LARONGE, INC, 88- *Mem:* AAAS; Am Chem Soc; Nat Asn Corrosion Engrs. *Res:* Phase rule; radiochemistry; kinetics; corrosion; water treatment; petroleum; water and air pollution control. *Mailing Add:* PO Box 338 Califon NJ 07830

FREEDMAN, DANIEL X, b Lafayette, Ind, Aug 17, 21; m 45. PSYCHIATRY. *Educ:* Harvard Univ, BA, 47; Yale Univ, MD, 51. *Hon Degrees:* DSc, Wabash Col, 74, Ind Univ, 82. *Prof Exp:* Resident psychiat, Sch Med, Yale Univ, 52-55, chief, biol sci sect, dept psychiat, dir grad res training prog psychiat & neurobehav sci & attending psychiatrist, Yale-New Haven Community Hosp, 58-66, from asst prof to prof psychiat, 58-66; prof & chmn, Dept Psychiat, 66-83, Louis Block prof biol sci, 69-83, EMER LOUIS BLOCK PROF BIOL SCI, UNIV CHICAGO, 83-; JUDSON BRAUN PROF PSYCHIAT & PHARMACOL, EXEC VCHMN DEPT PSYCHIAT & BIOBEHAV SCI & DIR, DIV ADULT PSYCHIAT, UNIV CALIF, LOS ANGELES, 83- *Concurrent Pos:* Attend psychiatrist, Vet Admin Hosp, West Haven, Conn, 55-66; assoc psychiatrist, Grace-New Haven Community Hosp, 55-66; consult juv courts, 55-57, Fairfield State Hosp, 58-66 & US Dept Army, 65-67; career investr, USPHS, 57-66; consult, NIMH, 60-; chmn panel psychiat drugs efficacy study, Nat Acad Sci-Nat Res Coun, 66-68; mem adv comt, Food & Drug Admin, 67-; dir, Social Sci Res Coun, 67-73; dir, Founds Fund Res in Psychiat, 69-72; chief ed, Arch Gen Psychiat, 70-; mem comt brain sci, Nat Res Coun, 71-73, mem comt probs drug dependence, 71-, Am Psychiat Asn rep to div med sci, 71-73; dir, Drug Abuse Coun, 73; chmn, Pharmacol, Substance Abuse & Environ Toxicol Interdisciplinary Cluster, President's Biomed Res Panel, 75-76; mem selection comt, President's Comn Mental Health, 77 & coordr res task panel, 77-78; mem, Joint Comn Prescription Drug Use, Inc, 77-; mem, expert adv panel mental health, WHO, 80-84; chmn, Bd Sci Counselors, NIMH, 84-86 & Addiction Res Ctr, Nat Inst Drug Abuse, 89-91. *Honors & Awards:* William C Menniger Award, Am Col Physicians, 75; McAlpin Medal for Res Achievement, Nat Ment Health Asn, 79; Paul Hoch Award for Distinguished Serv, Am Col Neuropsychopharmacol, 82; E B Bowis Award, Am Col Psychiatrists, 83; Schilder Lectr, NY Univ, 83. *Mem:* Nat Acad Sci; sr mem Inst Med-Nat Acad Sci; Asn Res Nerv & Ment Dis (pres, 74); fel Am Psychiat Asn (vpres, 75-77, pres, 81-82); fel Am Col Neuropsychopharmacol (pres, 70); fel Am Col Psychiatrists; fel AAAS; AMA; Soc Biol Psychiat (vpres, 83-84); Sigma Xi; Am Soc Pharmacol & Exp Therapeut; Am Psychosomatic Soc; Int Brain Res Orgn; Psychiat Res Soc; Soc Neurosci; Am Med Soc Alcoholism; fel Am Acad Arts & Sci; hon fel Am Asn Psychoanal Physicians; distinguished fel Egyptian Psychiat Asn. *Res:* Psychopharmacology; psychoanalytic, neurophysiologic and social investigation in schizophrenia; central nervous system determinants of allergy; drugs, brain function and behavior; methodology of drug studies; author of numerous publications. *Mailing Add:* Dept Psychiat Neuropsychiat Inst Univ Calif 760 Westwood Plaza Los Angeles CA 90024

FREEDMAN, DANIEL Z, b Hartford, Conn, May 3, 39. THEORETICAL PHYSICS. *Educ:* Wesleyan Univ, BA, 60; Univ Wis, MS, 62, PhD(physics), 64. *Prof Exp:* Res assoc physics, Univ Wis, 64; NSF fel, Imp Col, Univ London, 64-66; res fel, Univ Calif, Berkeley, 65-66, instr, 66-67; mem, Inst Advan Study, Princeton, NJ, 67-68; from asst prof to assoc prof, State Univ NY Stony Brook, 68-74, prof physics, 74-80; PROF APPL MATH, MASS INST TECHNOL. 80- *Concurrent Pos:* Sloan fel, 69-71; vis scientist, Dept Physics, Mass Inst Technol, 70-71; Guggenheim fel, 73-74; vis assoc, Dept Physics, Calif Inst Technol, 77-78. *Mem:* Am Phys Soc. *Res:* Theory of elementary particles and their interactions, particularly as related to the theory of gravitation. *Mailing Add:* Dept Math Rm 2-381 Mass Inst Technol 77 Massachusetts Ave Cambridge MA 02139

FREEDMAN, DAVID A, b Montreal, Que, Mar 5, 38; div; c 2. STATISTICS. *Educ:* McGill Univ, BSc, 58; Princeton Univ, MA, 59, PhD(probability), 60. *Prof Exp:* Can Coun fel, Imp Col, Univ London, 61-62; PROF STATIST, UNIV CALIF, BERKELEY, 62- *Concurrent Pos:* Sloan fel, Univ Calif, Berkeley, 64-66; vis prof, Hebrew Univ Jerusalem, 68-69; consult various industs & govt, 70-73; vis prof, Nat Free Univ Mexico, 73, Venezuelan Inst Sci Invest, 70-71 & 76 & Univ Kuwait, 81. *Mem:* Inst Math Statist; Sigma Xi. *Res:* Ergodic theory; Bayesian inference; Martingales; Markov chains; random distribution functions; Brownian motion and diffusion; representation theorems; empirical histograms. *Mailing Add:* Dept Stat Univ Calif Berkeley Berkeley CA 94720

FREEDMAN, DAVID ASA, b Boston, Mass, May 27, 18; m 47; c 4. NEUROLOGY, PSYCHIATRY. *Educ:* Harvard Univ, AB, 39; Tufts Univ, MD, 43. *Prof Exp:* Rockefeller fel neurol, Columbia Univ, 47-48; fel, Tulane Univ, 49, assoc clin prof neurol, 50-65; assoc prof psychiat & neurol, 65-71, PROF PSYCHIAT, BAYLOR COL MED, 71- *Concurrent Pos:* From instr to training & supv analyst, New Orleans Psychoanal Inst, 54-; clin prof psychiat, Med Sch, La State Univ, 69-73; pres & chmn, Educ Comt, Houston-Galveston Psychoanal Inst, Inc, 78- *Mem:* Asn Res Nerv & Ment Dis; Am Acad Neurol; Am Psychiat Asn; Am Psychoanal Asn; Am Col Psychiat. *Res:* Role of congenital and perinatal sensory deprivations in evolution of psychic structure; an approach to the study of ego development. *Mailing Add:* Meth Hosp Baylor Col Med 6560 Fannin No 950 Houston TX 77030

FREEDMAN, ELI (HANSELL), b Pittsburgh, Pa, Dec 12, 27; m 51; c 2. HIGH TEMPERATURE CHEMISTRY. *Educ:* Carnegie Inst Technol, BS, 47; Cornell Univ, PhD(chem), 52. *Prof Exp:* Res assoc physics, Brown Univ, 52-53, instr chem, 53-56; asst prof, Univ Buffalo, 56-60; res chemist, Chem Br, Ballistic Res Labs, 60-62, chief, Phys Chem Sect, 62-65, chemist, Phys Br, 65-69 & Thermokinetics Group, 69-73, res chemist, Interior Ballistics Div, Ballistic Res Lab, Aberdeen Proving Ground, 73-87; CONSULT, 88- *Concurrent Pos:* Mem, Eve Col Fac, Johns Hopkins Univ, 65-75; vis prof physics, Univ Mo, Kansas City, 85. *Mem:* AAAS; Am Chem Soc; Am Phys Soc. *Res:* Chemical reaction in shock tubes; high temperature chemical kinetics; estimation of thermodynamic properties; computing high-temperature equilibria of real gases. *Mailing Add:* 2411 Diana Rd Baltimore MD 21209-1525

FREEDMAN, GEORGE, b Boston, Mass, Dec 11, 21; m 43; c 2. METALLURGY, PHYSICS. *Educ:* Mass Inst Technol, SB, 43; Boston Univ, MA, 52. *Prof Exp:* Asst metallurgist, Raytheon Co, 43, plant metallurgist, 43-51, head semiconductor eng & develop sect, 51-55, chief engr adv develop, 55-59, chief process engr, 59-60, mgr mat & tech lab, 62-65, group mgr mat & tech eng, 65-69, mgr, New Prod Ctr, Microwave & Power Tube Div, 69-80, dir, New Prod Ctr, 80-87, adv ed solid state technol, 61-72, ed in chief, J Microwave Oven, 87-90, CONSULT, RAYTHEON CO, 87- *Concurrent Pos:* Pres, Tyco Semiconductor Corp, 60-62; lectr dent mat, Harvard Sch Dent Med, 65-80; fel, Int Microwave Inst, 87. *Mem:* Int Microwave Power Inst (chmn, Bd Govs, 76-80). *Res:* Development of new business concepts and organization of new product development and innovation programs. *Mailing Add:* Five Brook Trail Rd Wayland MA 01778

FREEDMAN, HAROLD HERSH, b Malden, Mass, Mar 5, 24; m 51; c 3. ORGANIC CHEMISTRY. *Educ:* Tufts Univ, BS, 49, MS, 50; Boston Univ, PhD(org chem), 56. *Prof Exp:* Chemist, Ionics, Inc, 51-52; res chemist, Dow Chem Co, 56-63, assoc scientist, New England Lab, 63-88. *Concurrent Pos:* Vis scientist, Mass Inst Technol, 80-81; adj prof, Tufts Univ, 85-86. *Mem:* Am Chem Soc; The Chem Soc. *Res:* Mechanism and structure in organic chemistry; organic chemistry of anions in non polar media; enzymes for organic synthesis. *Mailing Add:* 141 Jackson St Newton Ctr Boston MA 02159-1880

FREEDMAN, HENRY HILLEL, b New York, NY, Dec 21, 19; m 48; c 1. IMMUNOLOGY. *Educ:* NY Univ, AB, 48, MS, 50, PhD(physiol), 53. *Prof Exp:* Teaching fel biol, Washington Sq Col, NY Univ, 49-52; res assoc, Princeton Labs, Inc, 52-64; res immunologist, Inst Microbiol, Rutgers Univ, 65; sr res assoc, 65-67, mgr dept microbiol, Warner-Lambert Res Inst, 67-68, dir dept, 68-69, dir dept physiol & microbiol, 69-72, dir div biol res, 72-74; dir biomed res dept, ICI United States Inc, 74-76, gen mgr, Pharmaceut Res & Develop, 76-79, vpres, Pharmaceut Res & Develop, 79-81, vpres, ICI Americas, Inc, 81-85; RETIRED. *Concurrent Pos:* Mem res coun, Univ Del Res Found, 76-79; trustee, Wilmington Med Ctr, 82-85. *Mem:* fel NY Acad Sci; Am Soc Microbiol; Reticuloendothelial Soc (pres, 76); Am Chem Soc; AAAS; Am Physiol Soc. *Res:* Immune mechanisms; non-specific host resistance; bacterial lipopolysacchardies; reticuloendothelial system; delayed hypersensitivity; transplantation immunity; immunosuppression. *Mailing Add:* 138 Valley Rd Princeton NJ 08540

FREEDMAN, HERBERT IRVING, b Winnipeg, Man, Nov 16, 40; m 63; c 4. MATHEMATICAL ANALYSIS, BIOMATHEMATICS. *Educ:* Univ Man, BS, 62; Univ Minn, MA, 64, PhD(math), 67. *Prof Exp:* From asst prof to assoc prof, 67-78, PROF MATH, UNIV ALTA, 78- *Concurrent Pos:* Vis assoc prof math, Univ Minn, 73-74. *Mem:* Am Math Soc; Math Asn Am; Soc Indust & Appl Math; Can Math Soc; Can Appl Math Ser; NY Acad Sci. *Res:* The existence and stability of periodic oscillations and equilibria of ordinary differential equations; applications to mathematical models of species interactions. *Mailing Add:* Dept Math Univ Alta Edmonton AB T6G 2E2 Can

FREEDMAN, JEFFREY CARL, b Brooklyn, NY, Sept 24, 45; m 67; c 2. MEMBRANE TRANSPORT, CELL PHYSIOLOGY. *Educ:* Swarthmore Col, BA, 67; Univ Pa, PhD(biol), 73. *Prof Exp:* Asst prof biol, Reed Col, 73-75; fel, physiol & membrane pathol, Sch Med, Yale Univ, 75-79; ASST PROF PHYSIOL, STATE UNIV NY UPSTATE MED CTR, 79- *Concurrent Pos:* Prin investr, res grant, Nat Inst Gen Med Sci, 80- *Mem:* Biophys Soc; Am Physiol Soc; Soc Gen Physiologists; Red Cell Club; NY Acad Sci. *Res:* Red blood cells; ionic and osmotic equilibria; measurement of membrane potentials with fluorescent dyes; effects of calcium on potassium permeability; reconstitution of membrane transport functions into planar lipid bilayers. *Mailing Add:* Dept Physiol SUNY Health Sci Ctr 750 E Adams St Syracuse NY 13210

FREEDMAN, JEROME, b New York, NY, Aug 16, 16; m 46; c 7. ELECTRICAL ENGINEERING, ELECTRONICS. *Educ:* City Col New York, BS, 38; Polytech Inst Brooklyn, MS, 51. *Prof Exp:* Electronics scientist, Watson Lab, NJ, 46-51; dep chief, Rome Air Develop Ctr, Griffiss Air Force Base, 51-52; staff mem, Mass Inst Technol, 52-53, group leader, 53-55, div head, 55-68, asst dir, Lincoln Lab, 68-88; RETIRED. *Concurrent Pos:* Mem adv comt ballistic missile defense, Adv Res Projs Agency, US Dept Defense, 63-67, chmn, 67-68; mem re-entry progs rev group, US Dept Defense Res & Eng, 63-70; consult, US Arms Control & Disarmament Agency, 65-67 & Inst Defense Anal, 65-; mem avionics panel, Adv Group Aerospace Res & Develop, NATO, 68-79; consult, Joint Chiefs of Staff, 68-72; mem, Army Sci Bd, 78-; consult, Defense Sci Bd, 68-; mem steering comt, Navy Sea LITE Prog, 79-83; assoc ed, J Defense Res. *Honors & Awards:* Editor's Award, Inst Radio Engrs, Inst Elec & Electronics Engrs, 52. *Mem:* AAAS; fel Inst Elec & Electronics Engrs; assoc fel Am Inst Aeronaut & Astronaut. *Res:* Radar system design; signal design theory; microwave technology; systems design and analysis in ballistic missile defense and offense systems; radar system design. *Mailing Add:* Lincoln Lab 244 Wood St PO Box 73 Lexington MA 02173-0073

FREEDMAN, JULES, b Brooklyn, NY, Feb 3, 33; m 65; c 3. MEDICINAL CHEMISTRY. *Educ:* Brooklyn Col Pharm, BS, 57; Univ Mich, PhD(med chem), 62. *Prof Exp:* Sr res chemist, Lakeside Lab, 62-72, group leader cent nervous syst drugs, 72-75; RES CHEMIST, MERRELL-NAT, INC, 75- *Mem:* Am Chem Soc; NY Acad Sci. *Res:* Development of new synthetic substances useful in treatment of disorders of mental function. *Mailing Add:* Merrell Res Ctr 2110 E Galbraith Rd Cincinnati OH 45215

FREEDMAN, LAWRENCE ZELIC, b Gardner, Mass, Sept 4, 19; m 55; c 5. PSYCHIATRY. *Educ:* Tufts Univ, BS, 41, MD, 44; NY Psychoanal Inst, dipl, 50-55. *Prof Exp:* Resident psychiat, New Haven Hosp & Sch Med, Yale Univ, 46-49, assoc clin prof, 49-61; FOUND FUND RES PROF PSYCHIAT, SCH MED, UNIV CHICAGO, 61-, CHMN, INST SOCIAL & BEHAV PATH, 71- *Concurrent Pos:* Vis scholar & sr res assoc, Cambridge Univ, 58-59; fel, Ctr Advan Study Behav Sci, 59-60; fel, Adlai Stevenson Inst Int Affairs, Chicago, 72; vis lectr & mem coop fac, Sch Law, Yale Univ, 49-57, chmn study unit psychiat & law, 53-58; assoc psychiatrist, Grace-New Haven Community Hosp, 49-61; permanent deleg, UN Econ & Social Coun, 49-; psychiat consult, Am Law Inst, 54- & Am Bar Asn, 59-; physician, Billings Hosp, Pritzker Sch Med, Chicago, 61-; med adv bd, Ctr Study Criminal Justice, Sch Law, Univ Chicago, 66-; permanent consult, comt environ design, Nat Housing Ctr, Washington, DC, 66-; psychiatrist, Nat Comn Causes & Prev Violence, 68-70; vis lectr, Child Study Ctr, Yale Univ, 72; vis prof, Univ Tel Aviv, 73. *Mem:* AAAS; Am Psychiat Asn; fel Am Orthopsychiat Asn; NY Acad Sci; Group Advan Psychiat; Int Soc Criminol; Sigma Xi; fel Royal Col Anthrop Great Brit & Ireland. *Res:* Psychosomatic medicine; psychosomatic aspects of obstetrics; psychoanalysis; social psychiatry; neurosis and economic factors; non-conformist behavior and social response; delinquent behavior; sex, aggressive and acquisitive deviant behavior. *Mailing Add:* Univ Chicago Four Winds 5020 S Lake Shore Dr No 3501N Chicago IL 60615

FREEDMAN, LEON DAVID, b Baltimore, Md, July 19, 21; m 45; c 2. ORGANOMETALLIC CHEMISTRY. *Educ:* Johns Hopkins Univ, AB, 41, MA, 47, PhD(org chem), 49. *Prof Exp:* Anal chemist, Johns Hopkins Hosp, USPHS, 41-44, org chemist, Sch Hyg, 46-48; org chemist, Nat Cancer Inst, 48-49 & Univ NC, 49-61; assoc prof, 61-65, PROF CHEM, NC STATE UNIV, 65- *Concurrent Pos:* Dir, Org Electronic Spectral Data, Inc, 62- *Mem:* AAAS; Am Chem Soc. *Res:* Organophosphorus compounds; oxidation-reduction; relationship between chemical structure and biological activity. *Mailing Add:* 2006 Myron Dr Raleigh NC 27607

FREEDMAN, LEWIS SIMON, b Boston, Mass, Mar 21, 36; m 71; c 1. NEUROCHEMISTRY, NEUROPHARMACOLOGY. *Educ:* Harvard Col, AB, 58; Boston Univ, MA, 60; Cornell Univ, PhD(nutrit), 70. *Prof Exp:* NIMH fel neurochem, dept psychiat, NY Univ Med Ctr, 69-72, asst prof neurol & pharmacol, 72-78, res assoc prof neurol & pharmacol, dept neurol & pharmacol, 78-87; RES PROGS DIR, DEPT SURG, SLOAN-KETTERING CANCER CTR, 88- *Concurrent Pos:* NIMH spec fel neurochem, Dept Psychiat, NY Univ Med Ctr, 72-73. *Mem:* Am Soc Pharmacol & Exp Therapeut; Am Asn Cancer Res; Fedn Am Soc Exp Biol. *Res:* Neurotransmitter mechanisms in learning and memory, neonatal undernutrition, neuroblastoma, neurological and psychiatric disorders; oncology; clinical trials. *Mailing Add:* One Fifth Ave New York NY 10003

FREEDMAN, M(ORRIS) DAVID, b Toronto, Ont, May 23, 38; US citizen; m 62; c 2. SOFTWARE ENGINEERING. *Educ:* Univ Toronto, BASc, 60; Univ Ill, Urbana, MS, 62, PhD(elec eng), 65. *Prof Exp:* Res assoc elec eng, Biol Comput Lab, Univ Ill, Urbana, 65-66; sr eng, Bendix Res Labs, 66-69, staff engr, 70-72, sr staff engr, 72-74, prin engr, 74-75, sr prin engr, 75-81; owner & pres, M David Freedman & Assoc Inc, 81-90; SR ENGR, COMPUTER METHODS CORP, 90- *Concurrent Pos:* from adj asst prof to adj assoc prof, Univ Mich, Dearborn, 72-78. *Mem:* Inst Elec & Electronic Engrs. *Res:* Analysis and synthesis of musical tones; digital computer systems-design and applications; pattern recognition-optical character recognition; computer-aided design systems; multiprocessor computer architecture; software for engineering applications and product design; software productivity and productivity tools; human factors; software quality appearance. *Mailing Add:* 6709 Brookeshire West Bloomfield MI 48322

FREEDMAN, MARVIN I, b Boston, Mass, Oct 4, 39; m 66. MATHEMATICS. *Educ:* Mass Inst Technol, BS, 60; Brandeis Univ, MA, 62, PhD(math), 64. *Prof Exp:* Instr math, Univ Calif, Berkeley, 64-66; scientist, Electronic Res Ctr, NASA, 67-70; assoc prof, 70-77, PROF MATH, BOSTON UNIV, 78- *Concurrent Pos:* Vis asst prof, Brown Univ, 68-69. *Honors & Awards:* Co-winner Best Paper Award, Joint Automatic Control Conf, 68. *Mem:* Math Asn Am; Am Math Soc; Soc Indust & Appl Math. *Res:* Stability theory; partial differential equations; harmonic and functional analysis; numerical analysis and system theory; perturbation methods in control; aerodynamics. *Mailing Add:* 22 Woodfield Rd Wellesley MA 02181

FREEDMAN, MELVIN HARRIS, b St John, New Brunswick, Can, July 4, 39; m 63; c 2. PEDIATRIC HEMATOLOGY-ONCOLOGY. *Educ:* Dalhousie Univ, Can, MD, 66; Royal Col Physicians & Surgeons, FRCP, 71; Am Acad Pediat, FAAP, 86. *Prof Exp:* Instr pediat, Univ Southern Calif Sch Med, 69-70; clin teacher, 71-72, from asst prof to assoc prof, 72-86, PROF PEDIAT, UNIV TORONTO, 86- *Concurrent Pos:* Staff hematologist-oncologist, Hosp for Sick Children, 71-79, sr staff hematologist-oncologist, 79-, res proj dir, Res Inst, 72-, dir, Hematopoiesis Res Lab, 72- & Transfusion-Chelation Prog, 81-88; mem, Bd Examrs, Royal Col Physicians & Surgeons, Can, 75-89 & Comt Hemat for Accredition of Training Progs, 86-; mem, Med Adv Comt, Can Red Cross, Toronto Transfusion Serv, 86- *Mem:* Soc Pediat Res; Am Acad Pediat; Am Fedn Clin Res; Am Soc Hemat; Int Soc Exp Hemat; Int Soc Hemat; Can Soc Clin Invest. *Res:* Hematopoiesis and leukemopoiesis in vivo and in vitro; clinical use of biological response modifiers; recombinant growth factor clinically and in vitro on bone marrow. *Mailing Add:* Hosp Sick Children 555 Univ Ave Toronto ON M5G 1X8 Can

FREEDMAN, MICHAEL HARTLEY, b Los Angeles, Calif, Apr 21, 51; m 83; c 3. GEOMETRIC TOPOLOGY. *Educ:* Princeton Univ, PhD(math), 73. *Prof Exp:* Lectr math, Univ Calif, Berkeley, 73-75; PROF MATH, UNIV CALIF, SAN DIEGO, 76-, PROF C L POWELL ENDOWED CHAIR, 85- *Concurrent Pos:* Mem math inst, Inst Advan Study, Princeton, NJ, 75-76; fel, Alfred P Sloan Found, 80-83 & MacArthur Found, 84-89; Humboldt fel, Ger, 88. *Honors & Awards:* Veblen Prize, Am Math Soc, 86; Fields Medal, Int Cong Math, 86; Nat Medal of Sci, 87. *Mem:* Nat Acad Sci; Am Acad Arts & Sci; Am Math Soc; NY Acad Sci. *Res:* Geometric topology. *Mailing Add:* Dept Math 0112 Univ Calif San Diego La Jolla CA 92093-0112

FREEDMAN, MICHAEL LEWIS, b Hartford, Conn, Dec 30, 42. ORAL & PHYSIOLOGICAL MICROBIOLOGY. *Educ:* Brown Univ, AB, 64; State Univ NY Buffalo, PhD(biol), 69. *Prof Exp:* Res assoc, Argonne Nat Lab, 69-71; assoc prof oral diagnosis, Sch Dent Med, Univ Conn, Farmington, 71-; PROF MED, DEPT MED, NY UNIV SCH MED. *Mem:* Am Soc Microbiol; Radiol Res Soc; Sigma Xi. *Res:* Genetics and physiology of cariogenic microorganisms. *Mailing Add:* Oral Radiol Dept Univ Conn Health Ctr Farmington CT 06032

FREEDMAN, PHILIP, b London, Eng, June 25, 26; US citizen; m 54; c 5. MEDICINE. *Educ:* Univ London, MB, BS, 48, MD, 51, FRCP. *Prof Exp:* Bilton Pollard fel, Univ Col Hosp Med Sch, Univ London & Dept Med, Univ Ill, 57-59; first asst physician, St George's Hosp, London, 59-60, clin asst physician, Renal Study Clin, 60-63; assoc prof med, Chicago Med Sch, 63, actg chmn dept, 66-67, chmn, 67-74; chmn dept med, Med Ctr, Mt Sinai Hosp, 66-, prof med, Rush Med Col, 75-; AT DEPT PSYCHOL, UNIV ILL, CHICAGO. *Concurrent Pos:* Consult physician, Woolwich Hosp Group & Redhill Hosp Group, London, 60-63; chief, Chicago Med Sch Serv, Div Med & dir renal unit, Cook County Hosp, 63-66. *Mem:* Am Fedn Clin Res; Am Soc Nephrology; Soc Exp Biol & Med; fel Am Col Physicians; Med Res Soc London; Sigma Xi. *Res:* Renal disease, particularly the pathogenesis of glomerulonephritis and the autoimmune diseases; immunologic aspects of renal and systemic disease. *Mailing Add:* 2808 Knollwood Lane Glenview IL 60025

FREEDMAN, ROBERT RUSSELL, b Philadelphia, Pa, Apr 30, 47; m 81; c 1. BEHAVIORAL MEDICINE, PSYCHOPHYSIOLOGY. *Educ:* Univ Chicago, BA, 69; Univ Mich, MA, 73, PhD(clin psychol), 75. *Prof Exp:* Asst prof, 76-83, assoc prof psychol & psychiat, 83-89, PROF PSYCHOL & PSYCHIAT, WAYNE STATE UNIV, 89-; DIR BEHAV MED, LAFAYETTE CLIN, 75- *Concurrent Pos:* Consult, Nat Inst Drug Abuse, 76-78, Vet Admin, 85-, Nat Inst Mental Health, 86-; mem MHSB Study Sect, Nat Inst Mental Health, 86-; prin investr, Nat Heart, Blood & Lung Inst, 79-, Nat Inst Aging, 87-; chmn publ comt, Biofeedback Soc Am, 84-90; ed, Biofeedback & Self-Regulation, 91- *Honors & Awards:* Res Achievement Award, Biofeedback Soc Am, 88. *Mem:* AAAS; Asn Psychophysiol Study of Sleep; Biofeedback Soc Am; Soc Behav Med; Soc Psychophysiol Res; Acad Behav Med Res. *Res:* Physiological mechanisms of behavioral treatments for cardiovascular diseases; pathophysiology of Raynaud's disease and menopausal hot flashes; regulation of adrenergic receptors; sex differences in adrenergic receptors and in pathophysiological conditions. *Mailing Add:* Behav Med Lab Lafayette Clin 951 E Lafayette Detroit MI 48207

FREEDMAN, ROBERT WAGNER, b Newark, NJ, Feb 15, 15; m 46; c 2. ORGANIC CHEMISTRY. *Educ:* Mass Inst Technol, SB, 38; Polytech Inst Brooklyn, PhD(chem), 51. *Prof Exp:* Res chemist, Battelle Mem Inst, 51-52, Balco Res Corp, 52-54 & Colgate Palmolive Co, 54-57; asst plant mgr, Transition Metals, Inc, 57; sect head anal res, Consol Coal Co, 57-65; PROJ COORDR ANALYTIC RES METHODS, US BUR MINES, 65-, SUPVRY RES CHEMIST, 72- *Mem:* Am Chem Soc; Am Inst Chemists; Royal Soc Chem. *Res:* Gas chromatography. *Mailing Add:* Sherwood Oaks 100 Norman Dr Box 262 Mars PA 16046-8203

FREEDMAN, SAMUEL ORKIN, b Montreal, Que, May 8, 28; m 55; c 4. IMMUNOLOGY. *Educ:* McGill Univ, BSc, 49, MD, CM, 53; FRCP(C), 58. *Prof Exp:* From asst prof to assoc prof, 59-65, dean, Fac Med, 77-81, PROF MED, MCGILL UNIV, 65-, VPRIN, 81- *Concurrent Pos:* Dir div clin immunol & allergy, Montreal Gen Hosp, 65-77; chmn grants panel immunol & transplantation, Med Res Coun Can, 68-73; vis prof, London Univ & Chester Beatty Res Inst, 73-74; chmn grants panel immunol, Nat Cancer Inst Can, 75- *Honors & Awards:* Gairdner Int Award for Med Res, 78. *Mem:* Fel Am Col Physicians; Am Soc Clin Invest; Can Acad Allergy (pres, 68-69); fel Royal Soc Can; Can Med Asn. *Res:* Isolation and identification of human tumor antigens by immunological methods; co-discoverer of the CEA test for cancer; tuberculin hypersensitivity and respiratory allergy. *Mailing Add:* Fac Med McGill Univ 845 Sherbrooke St W Montreal PQ H3A 2T5 Can

FREEDMAN, STEVEN I(RWIN), b New York, NY, June 5, 35; m 58; c 3. MECHANICAL ENGINEERING, ENERGY UTILIZATION. *Educ:* Mass Inst Technol, SB, 56, SM, 57, MechE, 60, PhD(mech eng), 61. *Prof Exp:* From instr to asst prof mech eng, Mass Inst Technol, 59-63; consult engr nuclear space power, Gen Elec Co, 63-68; asst chief engr, Prod Develop Div, Budd Co, 68-70; lectr, Univ Pa, 70-71, chief engr, Nat Ctr Energy Mgt & Power, 71-73; vpres res, IU Energy Systs, 73-74; asst dir combustion & power, Energy Res & Develop Admin, 74-77; asst dir combustion & power, 77-79, chief engr Coal Utilization, Dept Energy, 79-83, EXEC SCIENTIST, GAS RES INST, 83- *Mem:* Am Soc Mech Engrs. *Res:* Energy conversion, heat transfer, power generation, thermodynamics; magnetohydrodynamics; thermoelectricity; thermionics; fluid mechanics. *Mailing Add:* Gas Res Inst 8600 W Bryn Mawr Chicago IL 60631

FREEDMAN, STEVEN LESLIE, b Boston, Mass, Mar 3, 35; m 62. GENERAL MEDICAL SCIENCES, ANIMAL SCIENCE & NUTRITION. *Educ:* Univ NH, BS, 57; Rutgers Univ, PhD(avian physiol), 62. *Prof Exp:* Res assoc, 64-65, asst prof, 65-71, ASSOC PROF ANAT & NEUROBIOL, COL MED, UNIV VT, 71- *Concurrent Pos:* NIH fel anat, Brain Res Inst, Univ Calif, Los Angeles, 62-63 & Col Med, Univ Vt, 63-64. *Mem:* AAAS; Soc Neurosci. *Res:* Neuroanatomical studies of the avian nervous system. *Mailing Add:* Dept Anat & Neurobiol Univ Vt Burlington VT 05405

FREEDMAN, STUART JAY, b Los Angeles, Calif, Jan 13, 44; m 68; c 1. PHYSICS. *Educ:* Univ Calif, Berkeley, BS, 65, PhD(physics), 71. *Prof Exp:* Instr, Princeton Univ, 72-75, lectr, 75-76; ASST PROF PHYSICS, STANFORD UNIV, 76- *Concurrent Pos:* Sloan fel, 78-82. *Mem:* Am Phys Soc. *Res:* Elementary particle and nuclear physics; fundamental interactions and symmetries. *Mailing Add:* Rte 1 Box 90A Cedar Grove NC 27231

FREEDMAN, STUART JAY, b Los Angeles, Calif, Jan 13, 44; m 68; c 1. NUCLEAR STRUCTURE, ASTROPHYSICS. *Educ:* Univ Calif, Berkeley, BS, 65, MA, 67, PhD(physics), 72. *Prof Exp:* Instr physics, Princeton Univ, 72-75, lectr, 75-76; asst prof physics, Stanford Univ, 76-82; physicist, Argonne Nat Lab, 82-87, sr physicist, 87-91; prof physics, Univ Chicago, 87-91; PROF PHYSICS, UNIV CALIF, BERKELEY, 91-, FAC SR PHYSICIST, LAWRENCE BERKELEY LAB, 91- *Concurrent Pos:* Res asst, Univ Calif, Berkeley, 70-72, res physicist, 72; Sloan fel, 78-82; div assoc ed, Phys Rev Lett, 87-90; mem adv comt, Los Alamos Meson Physics Fac Prog, 88- *Mem:* Fel Am Phys Soc. *Res:* Experimental nuclear and particle physics; emphasis on the weak interaction and the other fundamental forces. *Mailing Add:* 5400 S Hyde Park Blvd Chicago IL 60615

FREEDY, AMOS, b Petach-Tikva, Israel, Oct 11, 38; US citizen; m 65; c 2. ARTIFICIAL INTELLIGENCE, ENGINEERING. *Educ:* Univ Calif, Los Angeles, BS, 65, MS, 67, PhD(eng), 69. *Prof Exp:* PRES, PERCEPTRONICS, INC, 70- *Concurrent Pos:* Assoc res engr, Univ Calif, Los Angeles, 70- *Honors & Awards:* Humanitarian Scientist Award, Orgn Rehab Training, 72. *Mem:* Inst Elec & Electronics Engrs; Sigma Xi. *Res:* Artificial limbs control; application of artificial intelligence to man-machine system; computer decision aids; man-computer control systems and problem solving. *Mailing Add:* Perceptronics Inc 21135 Erwin St Woodland Hills CA 91364

FREELAND, FORREST DEAN, JR, hydrology, forestry, for more information see previous edition

FREELAND, GEORGE LOCKWOOD, b New York, NY, Dec 26, 31; m 54; c 2. PETROLEUM GEOLOGY, OCEAN WASTE DISPOSAL. *Educ:* Colo Sch Mines, BS, 54; Lehigh Univ, MS, 54; Rice Univ, PhD(geol), 71. *Prof Exp:* Geologist, Shell Oil Co, 56-66; oceanographer, US Naval Oceanog Office, 66; marine geol, Nat Oceanog & Atmospheric Admin, 70-82; PETROL GEOLOGIST, BISCAYNE EXP INC, 82- *Concurrent Pos:* Bd mem, Fla Bd Prof Geologists, Fla Dept Prof Reg, 88- *Mem:* Fel Geol Soc Am; Am Assoc Petrol Geologists; Am Inst Prof Geologists; Soc Ind Prof Geol; Soc Econ & Paleontol Mineralogists. *Res:* Marine geology, shelf sediment studies, ocean waste disposal, shelf geophysical studies, sediment transport; petroleum geology, sandstone and carbonate studies, reef sediments; plate tectonics, plate motion studies. *Mailing Add:* 710 S Mashta Dr Key Biscayne FL 33149-1737

FREELAND, MAX, b Browning, Mo, Oct 20, 20; div; c 6. ANALYTICAL CHEMISTRY. *Educ:* Northeast Mo State Teachers Col, BS, 41; Iowa State Col, MS, 52, PhD(anal chem), 55. *Prof Exp:* Teacher pub sch, Mo, 46-48; chemist anal method develop, Columbia-Southern Chem Corp, 55-57; from assoc prof to prof chem, Northeast Mo State Univ, 57-84; RETIRED. *Concurrent Pos:* Fel, Tex A&M Univ, 64-65. *Mem:* Am Chem Soc. *Res:* Spectrophotometric analysis; high precision spectrophotometry; water analysis. *Mailing Add:* 810 W LaHarpe St Kirksville MO 63501

FREELAND-GRAVES, JEANNE H, b New Brunswick, NJ, Mar 21, 48; m; c 3. TRACE MINERALS. *Educ:* Rutgers Univ, PhD(nutrit), 75. *Prof Exp:* PROF GRAD NUTRIT, UNIV TEX, 88- *Concurrent Pos:* Nutrit ed specialist, Wheat Indust Coun; Kathyrn Ross Richards Centennial Teaching Fel. *Mem:* Am Inst Nutrit; Soc Nutrit Educ; Am Dietetic Asn. *Mailing Add:* Dept Human Ecol Univ Tex GEA 115 Austin TX 78712

FREELING, MICHAEL, b Ft Wayne, Ind, Jan 14, 45; m 67; c 2. GENETICS. *Educ:* Univ Ore, BA, 68; Ind Univ, PhD(genetics), 73. *Prof Exp:* ASSOC PROF GENETICS, UNIV CALIF, BERKELEY, 73- *Concurrent Pos:* Guggenheim fel, 80-81. *Res:* Gene regulation and development using plant systems. *Mailing Add:* Dept Bot Univ Calif Berkley Berkeley CA 94720

FREEMAN, ALAN R, b Atlantic City, NJ, Jan 16, 37; m 58; c 2. PHYSIOLOGY, PHARMACOLOGY. *Educ:* Philadelphia Col Pharm, BSc, 58; Hahnemann Med Col, PhD(pharmacol), 62. *Prof Exp:* Instr pharmacol, Hahnemann Med Col, 62-63; Univ fel neurophysiol, Med Ctr, Columbia Univ, 63-66; from asst prof to assoc prof physiol, Med Sch, Rutgers Univ, 66-71; assoc prof psychiat & physiol, Med Ctr & dir neurophysiol sect, Inst Psychiat Res, Ind Univ, Indianapolis, 71-74; PROF PHYSIOL & CHMN DEPT, SCH MED, TEMPLE UNIV, 74- *Mem:* AAAS; Biophys Soc; Am Physiol Soc; Brit Soc Gen Physiol; Soc Neurosci; Sigma Xi. *Res:* Application of biophysical techniques to electrophysiological, osmotic and permeability studies as an approach to understanding the physiology and pharmacology of the nervous system at the level of the cellular membrane. *Mailing Add:* Chmn Physiol Dept Temple Univ Med 3420 N Broad St Philadelphia PA 19140

FREEMAN, ALBERT EUGENE, b Lewisburgh, WVa, Mar 16, 31; m 50; c 3. ANIMAL BREEDING. *Educ:* Univ WVa, BS, 52, MS, 54; Cornell Univ PhD(animal breeding), 57. *Prof Exp:* Asst dairy husb, Univ WVa, 52-54; asst animal sci, Cornell Univ, 54-56; from asst prof to assoc prof, 64-65, PROF ANIMAL SCI, IOWA STATE UNIV, 65-, CHARLES F CURTISS DISTINGUISHED PROF AGR, 78- *Concurrent Pos:* Fulbright Hays fel, Fulbright Hays Comn, 75. *Honors & Awards:* Nat Asn Animal Breeders Award, Am Dairy Sci Asn, 75, Borden Award, 82 & J L Lush Award Animal Breeding, 84; Sr Fulbright-Hays Award, 75; Rockefeller Prentice Mem Animal Breeding & Genetics Award, Am Soc Animal Sci, 79. *Mem:* Fel Am Soc Animal Sci; Am Dairy Sci Asn; Am Statist Asn. *Res:* Genetic improvement of domestic animals, dairy cattle, beef cattle, swine, poultry and sheep; genetics of populations; statistical genetics; molecular genetics. *Mailing Add:* Dept Animal Sci Iowa State Univ Ames IA 50011

FREEMAN, ARNOLD I, b Toronto, Ont, Sept 26, 37; m; c 3. PEDIATRICS. *Educ:* Univ Toronto, MD, 62; Am Bd Pediat, dipl, 68. *Prof Exp:* Intern, New Mt Sinai Hosp, Toronto, 62-63; resident pediat, Hosp Sick Children, Toronto, 63-65, resident path, 65-66; resident internal med, Sunnybrook Hosp, 66; instr pediat, Univ Tenn, Memphis, 66-68; res asst prof, State Univ NY Buffalo, 68-73; clinician II, 68-71, assoc chief pediat, 71-76, res assoc prof pediat, 73-80, CHIEF PEDIAT, ROSWELL PARK MEM INST, 76-, PROF PEDIAT, 81- *Concurrent Pos:* Res trainee fel, Lab Virol & Oncol, St Jude Res Hosp, Memphis, Tenn, 66-68; chmn electron micros area, Cancer Core Grant, 73; chmn, Acute Leukemia Group B, Nat Study Chemother Brain Tumors. *Mem:* Am Asn Cancer Res; Am Soc Hemat; Am Soc Clin Oncol; NY Acad Sci. *Res:* Treatment of acute lymphocytic leukemia; toxicity to the central nervous system from treatment for acute lymphocytic leukemia; use of interferon and interferon inducers in cancer and in viral or virally related disease. *Mailing Add:* Childrens Mercy Hosp 2401 Gillham Rd Kansas City MO 64108

FREEMAN, ARTHUR, b Youngstown, Ohio, Jan 12, 25; m 55; c 2. VETERINARY MEDICINE. *Educ:* Ohio State Univ, DVM, 55. *Prof Exp:* Vet pvt pract, Pac Vet Hosp, Wash, 55-56; dir prof rels, Jensen-Salsbery Labs, Inc, Mo, 56-59; ed, J Am Vet Med Asn, 59-72, ed-in-chief, 72-89, asst exec vpres, 77-89; RETIRED. *Mem:* Am Vet Med Asn; Am Asn Equine Practr; Coun Biol Eds (pres, 85-86). *Mailing Add:* 930 N Meacham Rd Schaumburg IL 60196

FREEMAN, ARTHUR JAY, b Lublin, Poland, Feb 6, 30; nat US; m 52; c 4. SOLID STATE PHYSICS. *Educ:* Mass Inst Technol, BS, 52, PhD(physics), 56. *Prof Exp:* Asst physics, Mass Inst Technol, 52-56; instr, Brandeis Univ, 55-56; physicist, Ord Mat Res Off, 56-62; assoc dir & head theoret physics group, Nat Magnet Lab, Mass Inst Technol, 62-67; chmn dept, 67-72, PROF, 67-83, MORRISON PROF PHYSICS, NORTHWESTERN UNIV, EVANSTON, 83- *Concurrent Pos:* Lectr, Northeastern Univ, 59-61; dir, NATO Advan Study Insts, France, 66, Can, 67; Guggenheim fel & Fulbright-Hays fel, Hebrew Univ, Jerusalem, 70-71; ed, Int J Magnetism, 70-74; ed, J Magnet & Magnetic Mat, 75-; Alexander Von Humboldt fel, Tech Univ Munich, 77-78; consult, Argonne Nat Lab, IBM Corp, Nat Magnet Lab, Mass Inst Technol. *Mem:* Fel Am Phys Soc. *Res:* Quantum theory of atoms, molecules and solids; neutron and x-ray scattering; crystalline field theory; electronic band structure of solids; theory of magnetism; hyperfine interactions; theory of electronically driven phase transitions; phonon anomalies in high temperature superconductors; electronic structure of surfaces interfaces, adsorbates on surfaces, and coherent modulated structures; electronic structure of high temperature superconductors. *Mailing Add:* Dept Physics Northwestern Univ Evanston IL 60208

FREEMAN, ARTHUR SCOTT, b Summit, NJ, Apr 11, 55. NEUROPHYSIOLOGY. *Educ:* Univ Colo, BA, 77; Med Col Va, PhD(pharmacol), 82. *Prof Exp:* Postdoctoral fel psychiat, Yale Univ Sch Med, 82-85; head, Extracellular Electrophysiol Sect, Sinai Res Inst, 85-91; ASSOC PROF PSYCHIAT, WAYNE STATE UNIV SCH MED, 91. *Concurrent Pos:* Asst prof, Dept Psychiat, Wayne State Univ Sch Med, 86-91. *Mem:* Soc Neurosci; NY Acad Sci; Am Soc Pharmacol & Exp Therapeut; Int Brain Res Orgn; AAAS. *Res:* Physiological and pharmacological regulation of the electrophysiology of dopamine-containing neurons of the rat midbrain. *Mailing Add:* Hamburger-Jospey Res Bldg 6767 W Outer Dr Detroit MI 48235-2899

FREEMAN, BOB A, b Eastland, Tex, May 7, 26; m 60; c 4. HEALTH EDUCATION ADMINISTRATION, MICROBIOLOGY. *Educ:* Univ Tex, BA, 49, MA, 50, PhD(bact), 54. *Prof Exp:* Instr biol, Agr & Mech Col Tex, 50-51; instr bact, Univ Ark, 54; from instr to asst prof microbiol, Univ Chicago, 54-64; assoc prof, Univ Tenn Ctr Health Sci, 64-66, chmn dept, 70-83, vchancellor acad affairs, 82-88, PROF MICROBIOL, UNIV TENN CTR HEALTH SCI, MEMPHIS, 66-, ALUMNI DISTINGUISHED SERV PROF, 88- *Concurrent Pos:* USPHS grants, Univ Chicago, 56-64; US Army Res grants, Med Univ, Univ Tenn, 65-71, USPHS grant, 71-74, USDA res grant, 79-81; consult, WHO, 68. *Mem:* Am Soc Microbiol; Soc Exp Biol Med. *Res:* Effective immunity and mechanisms of pathogenesis in bacterial diseases; microbial toxins. *Mailing Add:* Alumni Distinguished Serv Prof Memphis TN 38163

FREEMAN, BRUCE ALAN, b Apr 5, 52; m. OXYGEN FREE RADICALS, OXYGEN TOXICITY. *Educ:* Univ Calif, Riverside, PhD(biochem), 78. *Prof Exp:* assoc prof biochem, Univ Ala, Birmingham, 85-, PROF ANESTHESIOL, BIOCHEM & PEDIAT. *Mailing Add:* Dept Anesthesiol Univ Ala, 941 Tinsley Harrison Tower Birmingham AL 35294

FREEMAN, BRUCE L, JR, b Wharton, Tex, Apr 2, 49. EXPLOSIVE PULSED POWER, THERMONUCLEAR PHYSICS. *Educ:* Tex A&M Univ, BS, 70; Univ Calif, Davis, MS, 71, PhD(appl sci), 74. *Prof Exp:* STAFF PHYSICIST, LOS ALAMOS NAT LAB, 74- *Mem:* NY Acad Sci; Inst Elec & Electronics Engrs. *Res:* Application of pulsed power technologies to the achievement of significant fusion energy release from inertial systems; development of increasingly intense pulsed power sources from explosive flux compression generators for fusion and other applications. *Mailing Add:* 1431 11th St Los Alamos NM 87544

FREEMAN, CAROLYN RUTH, b Kettering, Eng, Jan 2, 50; Can & Brit citizen; m 81; c 1. RADIATION ONCOLOGY. *Educ:* London Univ, MB, ChB, 72; FRCP(C). *Prof Exp:* Asst prof, 78-79, ASSOC PROF & CHMN, DEPT RADIATION ONCOL, MCGILL UNIV, 79- *Concurrent Pos:* Asst radiol oncologist, Mont Gen, Jewish Gen, Royal Victoria & Mont Children's Hosp, 78-79, radiation oncologist-in-chief, 79; consult specialist, Queen Elizabeth, Reddy Mem, St Mary's & Mont Chest Hosps, 78- *Mem:* Can Asn Radiologists; Can Oncol Soc. *Res:* Clinical research in treatment of cancer. *Mailing Add:* Dept Therapeau Radiol Montreal Gen Hosp 1650 Cedar Ave Montreal PQ H3G 1A4 Can

FREEMAN, CHARLES EDWARD, JR, b Ironton, Mo, Aug 19, 41; m 63; c 4. PLANT ECOLOGY. *Educ:* Abilene Christian Col, BS, 63; NMex State Univ, MS, 66, PhD(plant ecol), 68. *Prof Exp:* Univ Res Inst grants, 68-70, asst prof bot, 68-79, ASSOC PROF BIOL SCI, UNIV TEX, EL PASO, 74- *Res:* Plant autecology; pollination biology. *Mailing Add:* Dept Biol Sci Univ Tex El Paso El Paso TX 79968

FREEMAN, COLETTE, b Clarence, Iowa, Feb 22, 37. TUMOR BIOLOGY, SCIENCE ADMINISTRATION. *Educ:* Coe Col, BA, 56; Temple Univ, MS, 69, PhD(biochem), 72. *Prof Exp:* Staff fel, Nat Inst Diabetes & Digestive & Kidney Dis, 73-77, sr staff fel, 73-79; prog dir, 79-86, CHIEF CANCER BIOL BR, DIV CANCER BIOL & DIAG, NAT CANCER CTR, NIH, 86- *Mem:* AAAS; Int Soc Differentiation; Women Cancer Res. *Res:* The hormonal regulation of mammary epithelial growth and differentiation in vitro and in vivo; mechanism of insulin action; characterization of insulin-responsive and insulin-refractory tissue; health science administration. *Mailing Add:* Nat Cancer Inst NIH Executive Plaza S Rm 630 Rockville MD 20892

FREEMAN, DAVID HAINES, b Rochester, NY, June 24, 31; m 56; c 4. PHYSICAL CHEMISTRY, ANALYTICAL CHEMISTRY & GEOCHEMISTRY. *Educ:* Univ Rochester, BS, 52; Carnegie Inst Technol, MS, 54; Mass Inst Technol, PhD(chem), 57. *Prof Exp:* Res assoc phys chem, Mass Inst Technol, 57-60; asst prof chem, Wash State Univ, 60-65; res chemist, Anal Chem Div, Nat Bur Standards, Washington, DC, 65, chief, Separation & Purification Sect, 65-74; PROF CHEM, UNIV MD, COLLEGE PARK, 74- *Concurrent Pos:* Chmn, Gordon Res Conf Ion Exchange, 73; lectr, Am Chem Soc, 78-83; vis researcher, Geophys Lab, Carnegie Inst, Wash, 81. *Honors & Awards:* Silver Medal, Dept Com, 69. *Mem:* Am Chem Soc. *Res:* Organic geochemistry; biogeochemical indicators; science of analytical method development; adsorptive materials for chemical separations; marine chemistry; water-sediment partitioning; high performance liquid chromatography; gel permeation chromatography; ion exchange; chromatographic mechanisms; small particle-classification, metrology and microscopy; ultra-pure chemical reagents; issued 1 patent. *Mailing Add:* Dept Chem Univ Md College Park MD 20742

FREEMAN, DWIGHT CARL, b Salt Lake City, Utah, Mar 7, 51; m 76; c 2. PLANT REPRODUCTION, ECOLOGICAL GENETICS. *Educ:* Univ Utah, BS, 73; Brigham Young Univ, MS, 76, PhD(bot), 77. *Prof Exp:* Res geneticist, US Forest Serv, 77; ASST PROF ECOL, WAYNE STATE UNIV, 77- *Res:* The ecology and evolution of dioecious plant species; the evolution and adaptive significance of sexual lability in dioecious plants; the physiological and genetics of sex determination and sexual lability in dioecious plant species. *Mailing Add:* Biology 210 Sci Hall Wayne State Univ Detroit MI 48202

FREEMAN, ERNEST ROBERT, b Brooklyn, NY, Oct 3, 33; m 54; c 4. ELECTROMAGNETICS, COMMUNICATION SYSTEMS. *Educ:* Univ Miami, Coral Gables, Fla, BSEE, 55; George Washington Univ, MEA, 66; Lasalle, LLB, 71. *Hon Degrees:* ScD, London Inst Appl Res, 77. *Prof Exp:* Mem tech staff, Bell Telephone Labs, 59-61; mgr, eng dept, IIT Res Inst, 61-68; PRES, SACHS-FREEMAN ASSOCS, INC, 69- *Concurrent Pos:* Lectr, Ctr Technol, Am Univ, 66-68. *Honors & Awards:* Best Session Award, IEEE, 76. *Mem:* Fel Inst Elec & Electronics Engrs; Asn Fed Commun Consult Engrs. *Res:* Electromagnetic compatibility analysis; communications systems design; computer modeling. *Mailing Add:* 5357 Strathmore Ave Kensington MD 20895

FREEMAN, EVA CAROL, thin film devices, for more information see previous edition

FREEMAN, FILLMORE, b Lexington, Miss, Apr 10, 36; div; c 4. ORGANIC CHEMISTRY. *Educ:* Cent State Univ, BSc, 57; Mich State Univ, PhD(phys org chem), 62. *Prof Exp:* Res chemist, Calif Res Corp, 62-64; NIH fel, Yale Univ, 64-65; from asst prof to assoc prof chem, Calif State Univ, Long Beach, 65-73; vis prof, 73, assoc prof, 73-75, PROF CHEM, UNIV CALIF, IRVINE, 75- *Concurrent Pos:* Vis prof, Univ Paris, 71-72; adj prof chem, Univ Ill, Chicago Circle, 76; vis prof chem, Max Planck Inst Biophys Chem, Gottingen, WGer, 77-78; Fulbright-Hays sr res scholar, 77-78; Alexander von Humboldt Found fel, 77-78 & 79. *Mem:* Am Chem Soc; Royal Soc Chem; Sigma Xi. *Res:* Mechanisms and kinetics of transition metals oxidations; heterocyclic and bio-organic chemistry; organic sulfur compounds. *Mailing Add:* 1396 Morningside Dr Laguna Beach CA 92651

FREEMAN, FRED W, b Logan, Ohio, Aug 27, 24; m 46; c 3. FORESTRY, HORTICULTURE. *Educ:* Mich State Univ, BS, 49. MS, 51, PhD(forestry), 63. *Prof Exp:* Soil scientist, Fed Bur Reclamation, 50-51; ranger, Ohio Div Forestry, 51-53, forester, 53-55; horticulturist, Hidden Lake Gardens, 55-61, asst prof hort, Univ, 61-68, assoc prof hort, Mich State Univ, 68-, cur, Hidden Lake Gardens, 61-86; RETIRED. *Concurrent Pos:* Interchange fel hort, Brit Isles, 63. *Mem:* Am Asn Bot Gardens & Arboretums; Am Soc Hort Sci; Royal Hort Soc. *Res:* Arboretum development and management; establishment and maintenance of both woody and herbaceous plant collections in arboretum and conservatory; educational programs in the natural sciences for the visiting public. *Mailing Add:* 278 N Ridge Rd Brooklyn MI 49230

FREEMAN, GEORGE R(OLAND), b Chattahoochee, Fla, Oct 30, 18; m 43; c 2. AGRICULTURAL ENGINEERING. *Educ:* Univ Fla, BSA, 48, MSA, 57. *Prof Exp:* Asst regional engr, Int Harvester Co, Fla, 48-50; supt field opers, 50-66, asst dir, 66-80, ASST PROF & DIR, AGR EXP STA, UNIV FLA, 80- *Res:* Development of research equipment, tools and buildings for all departments, especially agronomy, animal husbandry, plant pathology and entomology. *Mailing Add:* 2408 NW 13th Pl Gainesville FL 32605

FREEMAN, GORDON RUSSEL, b Hoffer, Sask, Aug 27, 30; m 51; c 2. PHYSICAL CHEMISTRY, RADIATION CHEMISTRY. *Educ:* Univ Sask, BA, 52, MA, 53; McGill Univ, PhD(chem), 55; Oxford Univ, DPhil, 57. *Prof Exp:* Fel, Saclay Ctr Nuclear Studies, France, 57-58; from asst prof to assoc prof, 58-65, chmn div phys chem, 65-75, PROF CHEM, UNIV ALTA, 65-, DIR RADIATION RES CTR, 68- *Mem:* Radiation Res Soc; fel Chem Inst Can; Can Asn Physicists; Am Phys Soc; Epigraphic Soc. *Res:* Reaction kinetics; properties and behavior of electrons in fluids; kinetics of nonhomogeneous processes. *Mailing Add:* Dept Chem Univ Alta Edmonton AB T6G 2G2 Can

FREEMAN, GUSTAVE, b New York, NY, July 2, 09; m 38; c 3. MEDICINE. *Educ:* Brown Univ, PhB, 29; Duke Univ, MD, 34. *Prof Exp:* Instr path, Yale Univ, 34-36; from instr to asst prof med, Univ Chicago, 36-47; mem staff, Peter Bent Brigham Hosp, 49-51; chief clin invest, Med Labs, Army Chem Ctr, 51-53; head clin pharmacol & exp therapeut sect, Nat Serv Ctr, Nat Cancer Inst, 54-56; res assoc, Calif Inst Technol, 56-58; dir, dept med sci, SRI Int, 58-85; RETIRED. *Concurrent Pos:* Res assoc, Childrens Hosp, Boston, 49-51, consult, 51-60; asst prof, Johns Hopkins Univ, 52-56; consult, NIH, 69- & State of Calif, 85; adv, WHO, 76. *Mem:* Am Soc Trop Med & Hyg; Am Soc Hemat; Am Fedn Clin Res; Int Soc Hemat; Am Asn Cancer Res. *Res:* Thrombocytopenic bleeding; clinical investigation; infectious disease; pathology; cholinesterase biology; cancer chemotherapy; viral tumorigenesis in tissue culture; viral hepatitis; environmental medicine experimental emphysema. *Mailing Add:* 1470 Sand Hill Rd Palo Alto CA 94304

FREEMAN, HAROLD ADOLPH, b Wilkes-Barre, Pa, July 12, 09; m 35; c 2. STATISTICS. *Educ:* Mass Inst Technol, BS, 31. *Prof Exp:* Asst statist, 33-35, from instr to assoc prof, 35-50, PROF STATIST, MASS INST TECHNOL, 50- *Concurrent Pos:* Guggenheim Mem fel, Princeton Univ, 52-53; NSF fel, Univ NC, 60-61; consult, Urban Inst, 72- *Mem:* Fel Am Statist Asn (vpres, 48-50); fel Am Soc Qual Control; Inst Math Statist; fel Am Acad Arts & Sci. *Res:* Statistical methods; statistical theory; design of experiments. *Mailing Add:* 185 Hamilton St Cambridge MA 02139

FREEMAN, HAROLD STANLEY, b Raleigh, NC, Apr 7, 51; m 73; c 2. DYESTUFF CHEMISTRY, TEXTILE CHEMISTRY. *Educ:* NC A&T State Univ, Greensboro, BS, 72; NC State Univ, Raleigh, MS, 78, PhD(chem), 81. *Prof Exp:* Res scientist org chem, Burroughs-Wellcome Co, 73-82; from assoc prof to prof textile chem, 82-90, CIBA-GEIGY PROF DYESTUFF CHEM, TEXTILE CHEM DEPT, NC STATE UNIV, 90- *Mem:* Am Asn Textile Chemists & Colorists; Am Chem Soc; Sigma Xi; Nat Tech Asn; Environ Mutagenicity Soc. *Res:* Design, synthesis and analysis of organic dyes and their intermediates; characterization of the photodegradation of dyes in a polymeric medium and the use of the resulting information to enhance the ultra violent stability of synthetic colorants; relationships between chemical structure and the genotoxicity of azo dyes. *Mailing Add:* NC State Univ Box 8301 Raleigh NC 27695-8301

FREEMAN, HARRY CLEVELAND, b Greenfield, NS, Can, Mar 9, 25; m 46; c 5. POLLUTANTS & FISH PHYSIOLOGY, POLLUTANTS STEROID HORMONES. *Educ:* Acadia Univ, BSc, 50, MSc, 52. *Prof Exp:* Asst scientist chem, Fisheries Res Bd Can, 52-57, assoc scientist, 57-62, res scientist 2, 62-78; RES SCIENTIST 3 CHEM, DEPT FISHERIES CAN, 78- *Mem:* Can Inst Chem. *Res:* Devised methods for determining sublethal effects of contaminants in fish. *Mailing Add:* PO Box 100 Queens County Greenfield NS B0T 1E0 Can

FREEMAN, HERBERT, b Frankfurt-am-Main, Ger, Dec 13, 25; US citizen; m 55; c 3. COMPUTER ENGINEERING. *Educ:* Union Univ, NY, BSEE, 46; Columbia Univ, MS, 48, DSc(elec eng), 56. *Prof Exp:* Asst elec eng, Columbia Univ, 46-48; head dept advan studies eng, Sperry Gyroscope Co, 48-58, head dept data processing, 59-60; vis prof elec eng, Mass Inst Technol, 58-59; prof elec eng, NY Univ, 60-75, chmn dept, 68-73; prof comput eng, Rensselaer Polytech Inst, 75-85; STATE NJ PROF COMPUT ENG, RUTGERS UNIV, 85- *Concurrent Pos:* Res grant, Air Force Off Sci Res, 61-78; mem cong comt, Int Fedn Info Processing, 64-65, chmn US comt, cong, 71; NASA res grant, 65-; Nat Sci Found fel, 66-67; vis prof, Swiss Fed Inst Echnol, 66-67; dir Cybex Assoc, Inc; vis prof, Comput Sci Inst, Univ Pisa, Italy, 73 & Stanford Univ, 82. *Mem:* Asn Comput Mach; fel Inst Elec & Electronics Engrs; NY Acad Sci; Pattern Recognition Soc; Soc Info Display. *Res:* Digital computers and control systems; logical design, graphical data processing and pattern recognition; computer image processing. *Mailing Add:* Comput Eng Prog Rutgers Univ PO Box 1390 Piscataway NJ 08855-1390

FREEMAN, HORATIO PUTNAM, b Emmitsburg, Md, Mar 17, 24; m 51; c 3. INORGANIC CHEMISTRY. *Educ:* Dickinson Col, BSc, 47. *Prof Exp:* Res assoc, Bone Char Res Proj, Nat Bur Stand, 48-56; res chemist, Fertilizer Lab, 56-65, RES CHEMIST, SOILS LAB, USDA, 65- *Mem:* Am Chem Soc. *Res:* Cane sugar refining processes; phosphatic fertilizer materials; pesticide chemistry. *Mailing Add:* 10113 Tenbrook Dr Silver Spring MD 20901

FREEMAN, HUGH J, b Edmonton, Alta; c 2. GASTROENTEROLOGY, NUTRITION. *Educ:* Univ Montreal, BSc, 68; Mc Gill Univ, MD, 72. *Prof Exp:* Res fel, Med Res Coun Can, 74-75 & 76-79; from asst prof to assoc prof, 79-86, PROF MED, UNIV BC, 86-, HEAD GASTROENTEROL, 81-; HEAD MED, UNIV HOSP, 88- *Concurrent Pos:* Res scholar, BC Health Care Res Found, 79-83; actg staff, Vancouver Gen Hosp, 79-83, Univ Hosp, 81-; consult staff, Cancer Control Agency BC, 79-, BC Children's Hosp, 86-; chmn, res comt, Can Asn Gastroenterol, 84-87, grants review panel, BC Health Care Res Found & Can Found Ileitis & Colitis, 87-90. *Mem:* Can Found Ileitis & Colitis; Can Asn Gastroenterol; Am Asn Gastroenterol; Can Soc Clin Invest. *Res:* Oncodevelopmental changes in the gastrointestinal tract; nutrient transport; intestinal absorption and malabsorption; inflammatory bowel disease; colonic neoplasia; small intestinal adaptation; celiac sprue and intestinal lymphoma; gastrointestinal endoscopy. *Mailing Add:* Univ Hosp Acute Care Unit F-137 2211 Wesbrook Mall Vancouver BC V6T 1W5 Can

FREEMAN, JAMES HARRISON, b Braddock, Pa, Feb 11, 22; m 44; c 3. ORGANIC CHEMISTRY, POLYMER CHEMISTRY. *Educ:* Juniata Col, BS, 44; Univ Pa, MS, 49, PhD(org chem), 50. *Prof Exp:* Asst instr org chem, Univ Pa, 46-50; res chemist, Westinghouse Res Labs, 50-56, adv chemist, 56-58, mgr org polymer chem sect, Insulation Dept, 58-69, mgr polymers & plastics dept, 69-83; RETIRED. *Mem:* AAAS; Am Chem Soc; Sigma Xi. *Res:* Heterocyclic organic compounds; mechanism of phenolformaldehyde resin formation; paper chromatographic analysis; high temperature application of polymers; thermally stable polymers; flat wiring systems; reinforced composites; adhesives; research managment. *Mailing Add:* 3436 Woodland Dr Murrysville PA 15668

FREEMAN, JAMES J, b Erie, Pa, Mar 11, 40; m 62; c 3. ELECTRICAL ENGINEERING, BIOENGINEERING. *Educ:* Gannon Col, BS, 62; Univ Detroit, MS, 64, MEng, 66, DEng, 68. *Prof Exp:* Instr eng, Univ Detroit, 64-68; NIH Med fel, Carnegie-Mellon Univ, 68-69; from asst prof to assoc prof, Univ Detroit, 69-80, prof & chmn elec eng, 80-82; chmn elec eng, 82-89, ASSOC DEAN, SAN JOSE STATE UNIV 89- *Concurrent Pos:* Systs consult, Pan Aura Div, Parke Davis Corp, 66-69; investr, NSF grant, 70-71; res assoc, Providence Hosp, Detroit 71-; NASA grants, 75-80, 84-; Air Force Off Sci Res grant, 76-78. *Mem:* Am Soc Eng Educ; Inst Elec & Electronics Engrs. *Res:* Electrical and medical engineering with special emphasis in cardiac ultrasonic diagnosis; microprocessors; blood pressure and flow instrumentation. *Mailing Add:* San Jose State San Jose CA 95192

FREEMAN, JEFFREY VANDUYNE, b Orange, NJ, Sept 13, 34; m 56; c 2. ENTOMOLOGY, FORESTRY. *Educ:* State Univ NY Col Environ Sci & Forestry, BS, 57; Rutgers Univ, MS, 60, PhD(entom), 62. *Prof Exp:* Teacher biol, Windham Col, 62-64; from asst prof to assoc prof sci, 64-72, PROF BIOL, CASTLETON STATE COL, 72- *Mem:* Entom Soc Am; Soc Am Foresters; Ecol Soc Am. *Res:* Biology and ecology of horse and deer flies; limnobiological studies, Lake Bomoseen, Vermont and tributaries; big tree survey of Vermont. *Mailing Add:* Box 125 Castleton VT 05735

FREEMAN, JERE EVANS, b Martin, Tenn, Oct 13, 36; m 62; c 4. STRATEGIC PLANNING. *Educ:* Univ Tenn, BS, 58; Univ Ill, MS, 61, PhD(agron), 62; Univ Chicago, MBA, 74. *Prof Exp:* Res asst plant genetics, Univ Ill, 58-62; plant physiologist, CPC Int Inc, 62-67, sect leader res, 67-75, technol forecaster, Off of Pres, 75-76, dir bus environ res, 76-77, cor dir indust res & develop, 77-78, mgr tech planning, CPC NAm, 78-79; vpres corp develop, Gold Kist Inc, 79-85; VPRES, PHILIPPS J HOOK & ASSOCS, INC, 85- *Concurrent Pos:* Mem, adv bd, Sci Res Inst, Atlanta Univ Ctr; bd dir, Ga Freight Bur, Inc; steering comt, Univ Ga Indust Interface Prog; adv bd, Sci Res Inst, Atlanta Univ Ctr; mem, Gov High Technol Adv Comt. *Mem:* Asn Corp Growth; Sigma Xi. *Res:* Genetics, physiology and chemistry of cereal grains. *Mailing Add:* Philipps J Hook & Assocs Inc Suite 300 N 5600 Roswell Rd Atlanta GA 30342

FREEMAN, JEREMIAH PATRICK, b Detroit, Mich, Aug 3, 29; m 53; c 6. ORGANIC CHEMISTRY. *Educ:* Univ Notre Dame, BS, 50; Univ Ill, MS, 51, PhD(chem), 53. *Prof Exp:* Anal chemist, Monsanto Chem Co, 50; asst, Univ Ill, 50-51; sr res chemist, Rohm & Haas Co, 53-57, head org chem group, 57-64; assoc prof, 64-68, chmn dept, 70-79, PROF CHEM, UNIV NOTRE DAME, IND, 68-, ASSOC DEAN, COL SCI, 88- *Concurrent Pos:* Asst head dept, Univ Notre Dame, 65-70; Sloan fel, 66-68; secy, Organic Syntheses, Inc, 79- *Mem:* Am Chem Soc; The Chem Soc. *Res:* Chemistry of oxidized nitrogen compounds; heterocycles as synthetic intermediates. *Mailing Add:* Dept Chem Univ Notre Dame Notre Dame IN 46556

FREEMAN, JOHN A, b Berkeley, Calif, May 7, 38; m 57; c 4. NEUROPHYSIOLOGY, BIOPHYSICS. *Educ:* Trinity Col, BSc, 58; McGill Univ, MD, CM, 62; Mass Inst Technol, PhD(biophys), 67. *Prof Exp:* Lectr physiol, Fac Med, McGill Univ, 61-62; med intern, Wilford Hall US Air Force Hosp, San Antonio, Tex, 62-63; chief in-flight res unit & flight surgeon, US Air Force Sch Aerospace Med, 63-64; fel, Mass Inst Technol, 65-66; fel, Inst Biomed Res, AMA, 67-69; vis scientist, Marine Biol Lab, Woods Hole, 68; res scientist, Aerospace Med Res Lab, Wright-Patterson AFB, 69-71; from asst prof to assoc prof, 71-80, PROF ANAT & OPHTHAL, VANDERBILT UNIV, 80- *Concurrent Pos:* Consult biomed electronics, Technol Inc, Ohio. *Mem:* Inst Elec & Electronics Eng; Soc Neurosci; Am Physiol Soc; NY Acad Sci; Asn Res Vision & Ophthal. *Res:* Neurophysiological mechanisms underlying information processing in the vertebrate visual system; molecular mechanisms associated with nurite growth synapse formation; related problems in biomedical engineering involving electronic design, computer applications and mathematical analysis; design of biomedical instrumentation. *Mailing Add:* Dept Cell Biol & Ophth Vanderbilt Univ Sch Med Nashville TN 37232

FREEMAN, JOHN A, b Daytona Beach, Fla, Aug 13, 48; m 73; c 2. DEVELOPMENTAL BIOLOGY. *Educ:* Fla State Univ, BS, 70; Univ New Orleans, MS, 72; Duke Univ, PhD(zool), 78. *Prof Exp:* Res assoc, Duke Univ Marine Lab, 79-80; from asst prof to assoc prof, 80-90, PROF BIOL, UNIV S ALA, 90- *Mem:* Am Soc Zool; Am Soc Cell Biol; Am Soc Develop Biol; Crustacean Soc. *Mailing Add:* Dept Biol Sci Univ S Ala Mobile AL 36688

FREEMAN, JOHN ALDERMAN, b Raleigh, NC, Aug 27, 17; m 41; c 4. ZOOLOGY. *Educ:* Wake Forest Col, BA, 38, MA, 40; Duke Univ, PhD(zool), 49. *Prof Exp:* Prof natural sci, Louisburg Col, 42; asst prof chem, Wake Forest Col, 42-46; instr zool, Duke Univ, 47-48; asst prof biol, Tulane Univ, 48-52; chmn dept, 62-76, prof biol, Winthrop Col, 52-; RETIRED. *Mem:* Am Inst Biol Sci. *Res:* Human blood phospholipids; tissue enzymes; temperature acclimatization in fish; metabolism of the fish brain; mollusk shell growth; differential longevity of sperm. *Mailing Add:* Island Ford Rd No 4 Brevard NC 28712

FREEMAN, JOHN CLINTON, JR, b Houston, Tex, Aug 7, 20; m 47; c 6. METEOROLOGY, MATHEMATICS. *Educ:* Rice Univ, BA, 41; Calif Inst Technol, MS, 42; Univ Chicago, PhD(meteorol), 52. *Prof Exp:* Asst gas & fluid dynamics, Brown Univ, 46-48; meteorologist, US Weather Bur, 48-49; mem res, Inst Adv Study, 49-50; res assoc meteorol, Univ Chicago, 50-52; sr engr, Cook Res Lab, 51-53; assoc prof meteorol & oceanog, Tex A&M Univ, 52-55; owner, Gulf Consult, 55-62; prof math, Univ Houston, 58; prof, 58-66, PROF METEOROL & DIR INST STORM RES, UNIV ST THOMAS, TEX, 66-; DIR & PRES, WEATHER RES CTR, HOUSTON, TX, 88- *Honors & Awards:* Meisinger Award, Am Meteorol Soc, 51, Citation for Work In Estab of the Tornado Warning Radar Network, 61. *Mem:* AAAS; NY Acad Sci; Meteorol Soc Japan; Am Geophys Union; fel Am Meteorol Soc. *Res:* Fluid dynamics; applied mathematics; meteorology; oceanography. *Mailing Add:* 4404 Mt Vernon St Houston TX 77006

FREEMAN, JOHN DANIEL, b Clarksville, Tenn, Apr 16, 41; m 65; c 2. TAXONOMIC BOTANY. *Educ:* Austin Peay State Univ, BA, 63; Vanderbilt Univ, PhD(biol), 69. *Prof Exp:* Asst prof bot, 68-74, ASSOC PROF BOT, AUBURN UNIV, 74-, CUR HERBARIUM, 68- *Mem:* Am Soc Plant Taxon; Int Asn Plant Taxon. *Res:* Taxonomy and cytogenetics of Trillium; Liliaceae of Alabama; flora of east-central and southeastern Alabama; poisonous plants of Alabama; eastern Asian-eastern North American floristic relationships. *Mailing Add:* Dept Bot & Micro Auburn Univ Auburn AL 36849-5407

FREEMAN, JOHN J, b Washington, DC, Sept 19, 41; m 68; c 2. PHARMACOLOGY. *Educ:* George Washington Univ, BS, 63; Vanderbilt Univ, PhD(pharmacol), 72. *Prof Exp:* Pharmacist, Maxwell & Tennyson, 63-64; fel pharmacol, Univ Calif, Los Angeles, 72-74, res assoc, 74-75; ASST PROF PHARMACOL, UNIV SC, 75- *Res:* Neurotransmitters in both the adrenergic and cholinergic nervous system. *Mailing Add:* Div Pharm Univ SC Col Pharm Columbia SC 29208

FREEMAN, JOHN JEROME, b Arlington Heights, Ill, Sept 28, 33; m 58. PHYSICAL CHEMISTRY. *Educ:* Univ NMex, BSc, 58, PhD(phys chem), 64. *Prof Exp:* Res assoc chem, Brookhaven Nat Lab, 63-66 & Chemstrand Res Ctr, Inc, 66-68; res chemist, 68-80, SCI FEL, MONSANTO CO, 80- *Mem:* Am Chem Soc; Soc Appl Specros. *Res:* Visible absorption and emission spectroscopy of molecules and complexed ions; infrared and Raman spectroscopy; catalyst characterization; polymer characterization; protein characterization. *Mailing Add:* Monsanto Co 700 Chesterfield Village Pkwy St Louis MO 63198

FREEMAN, JOHN MARK, b Brooklyn, NY, Jan 11, 33; m 56; c 3. PEDIATRIC NEUROLOGY. *Educ:* Amherst Col, AB, 54; Johns Hopkins Univ, MD, 58; Am Bd Pediat, dipl, 63; Am Bd Psychiat & Neurol, dipl, 69. *Prof Exp:* From intern to sr resident pediat, Harriet Lane Home, Johns Hopkins Hosp, 61-64; fel pediat neurol, Columbia Presby Med Ctr, New York, 64-66; asst prof pediat & neurol, Sch Med, Stanford Univ, 66-69; PROF PEDIAT & NEUROL, SCH MED, JOHNS HOPKINS UNIV, 69-, DIR PEDIAT-NEUROL SERV & BIRTH DEFECTS TREATMENT CTR, JOHNS HOPKINS HOSP, 69-, DIR PEDIAT EPILEPSY CTR, 74- *Concurrent Pos:* consult, Perinatal Res Br, Nat Inst Neurol Dis & Stroke & Walter Reed Gen Hosp; sci prog comt, Am Acad Ped, 69-75, Common Drugs, 73-79, Fetus & Newborn, 85-90; bd, Epilepsy Asn Maryland, 74-, pres, 77-82; exec coun, Child Neurol Soc, 79-81; vpres, Epilepsy Found Am, 81-; mem sci prog adv comt, NINCDS, 82-88; chair, Pre & Peri Natural Factors Brain Dysfunction, NINCDS-NICHD, 83-84. *Honors & Awards:* Lucy Moses Prize, Columbia Presby Med Ctr, 66. *Mem:* Fel Am Acad Pediat; Am Acad Neurol; Am Pediat Soc; Child Neurol Soc; Asn Prof Child Neurol (pres, 80-82). *Res:* Epilepsy; birth defects; clinical neurology. *Mailing Add:* Meyer 5-109 Johns Hopkins Hosp Baltimore MD 21205

FREEMAN, JOHN PAUL, b Washington, DC, Aug 30, 37. PHOTOGRAPHIC CHEMISTRY. *Educ:* Washington & Lee Univ, BSChem, 59; Univ Wash, MS, 64; Ohio State Univ, PhD(org chem), 70. *Prof Exp:* Index ed, Chem Abstr Serv, Am Chem Soc, Columbus, Ohio, 62-64; sr res chemist, res labs, 71-84, RES ASSOC FOR DEVELOP, EASTMAN GELATINE DIV, EASTMAN KODAK CO, PEABODY, MASS, 85- *Mem:* Am Chem Soc; Soc Photog Scientists & Engrs. *Res:* Sensitizing dye synthesis and behavior in photographic emulsions; mechanism of interaction of organic addenda with photographic systems; photographic gelatin - sensitometric properties and analytical techniques. *Mailing Add:* 27 Corning St Beverly MA 01915-3711

FREEMAN, JOHN RICHARDSON, b Murfreesboro, Tenn, Aug 24, 27; m 64; c 3. HERPETOLOGY. *Educ:* Univ NC, AB, 50; George Peabody Col, MA, 53; Univ Fla, PhD(zool), 63. *Prof Exp:* Instr biol, Jacksonville Jr Col, 53-55; from asst prof to assoc prof & head dept, 59-70, PROF BIOL, UNIV TENN, CHATTANOOGA, 70- *Mem:* Am Soc Ichthyol & Herpet. *Res:* Physiological ecology; respiratory behavior in the salamander Pseudobranchus striatus. *Mailing Add:* Dept Biol Univ Tenn Chattanooga 615 McCallie Ave Chattanooga TN 37403

FREEMAN, JOHN WRIGHT, JR, b Chicago, Ill, July 12, 35; m 57; c 2. SPACE PHYSICS. *Educ:* Beloit Col, BS, 57; Univ Iowa, MS, 61, PhD(physics), 63. *Prof Exp:* Staff scientist, Off Space Sci & Appln, NASA, Washington, DC, 63-64; res assoc space sci, 64-65, from asst prof to assoc prof, 65-72, PROF SPACE PHYSICS & ASTRON, RICE UNIV, 72- *Concurrent Pos:* Consult, NASA, Washington, DC; vis scientist, Royal Inst Technol, Stockholm and Univ Bern, 71-72; prin investr, Apollo Sci Exp; dir, Space Solar Power Res Prog, 77- *Honors & Awards:* Exceptional Sci Achievement Medal, NASA, 73. *Mem:* Am Geophys Union; Int Union Geod & Geophys; Int Asn Geomagnetism & Aeronomy. *Res:* Space plasma physics, particularly magnetospheric modelling; low energy particle measurements and magnetospheric dynamics; lunar exosphere and solar wind interaction with the moon; spacecraft charging; space weather. *Mailing Add:* Dept Space Physics & Astron Rice Univ Houston TX 77251

FREEMAN, JOSEPH THEODORE, b McKeesport, Pa, May 25, 08; m 43; c 1. HISTORY OF AGING, GENERAL MEDICAL HISTORY. *Educ:* Harvard Univ, AB, 30; Thomas Jefferson Univ, MD, 34. *Prof Exp:* Head dept geriat, Doctors' Hosp Philadelphia, 42-48; clin asst & prof med geriat, Med Col Pa, 44-64; SR ATTEND PHYSICIAN INTERNAL MED, DIV INTERNAL MED, LANKENAU HOSP, PHILADELPHIA, 48- *Concurrent Pos:* Lectr geriat, Philadelphia Sch Social Work, 54-59 & Grad Sch Med, Univ Pa, 56-63; vis prof med geront, Med Ctr, Univ Nebr, Omaha, 71; clin lectr geriat, Geront Soc Am, 84. *Honors & Awards:* Strouse Award, Moss Rehab Hosp, 51; Freeman Lectr, Geront Soc, 84. *Mem:* Am Geriat Soc (vpres, 48-52); Geront Soc Am (pres, 61); Int Asn Med Hist; Am Asn Med Hist; Am Col Physicians. *Res:* Geriatrics and gerontology on hip fractures; ascorbic acid metabolism; therapeutics; history of aging; drug principles in aging; gerontism; archives and bibliography of old age. *Mailing Add:* 6013 Walnut St Philadelphia PA 19139

FREEMAN, JULIA BERG, b Moline, Ill, June 7, 40. ARTHRITIS, RHEUMATOLOGY. *Educ:* Radcliffe Col, BA, 62; Ind Univ, PhD(biochem), 66. *Prof Exp:* From asst prof to prof, Dept Pharmacol, Ind Univ Sch Med, 71-85, dir Animal Core, Diabetes Res & Training Ctr & dir grad studies pharmacol, 77-85; prog dir Diabetes Res, Nat Inst Diabetes & Digestive & Kidney Dis, 85-87, PROG DIR, CENTERS PROG, NAT INST ARTHRITIS & MUSCULOSKELETAL & SKIN DIS, NIH, 87- *Concurrent Pos:* Res assoc, Vet Admin Hosp, Indianapolis, 75-85. *Mem:* AAAS; Am Med Writers Asn; Am Soc Pharmacol & Exp Therapeut; Sigma Xi; Soc Exp Biol & Med. *Res:* Biochemistry; pharmacology; over 40 technical publications. *Mailing Add:* Nat Inst Arthritis Musculoskeletal & Skin Dis Prog NIH Westwood Bldg Rm 403 Bethesda MD 20892

FREEMAN, KARL BORUCH, b Toronto, Ont, Jan 21, 34; m 55; c 2. BIOCHEMISTRY. *Educ:* Univ Toronto, BA, 56, PhD(biochem), 59. *Prof Exp:* Res assoc, Univ Toronto, 60-61, Nat Inst Med Res, Eng, 61-63 & Ont Cancer Inst, 64-65; from asst prof to assoc prof, 65-72, chmn res unit, 68-72, chmn dept, 73-79, 82, PROF BIOCHEM, MCMASTER UNIV, 72- *Concurrent Pos:* Vis scientist, London & Edinburgh, 72-73, San Francisco, 80. *Mem:* Am Soc Biol Chem; Can Biochem Soc. *Res:* Biogenesis of mitochondria; import of proteins into mitochondria; molecular biology of brown adipose tissue; uncoupling protein. *Mailing Add:* Dept Biochem McMaster Univ 1200 Main St W Hamilton ON L8N 3Z5 Can

FREEMAN, KENNETH ALFREY, b McAllaster, Kans, June 29, 12; m 38; c 2. ORGANIC CHEMISTRY. *Educ:* Stetson Univ, BS, 34, MS, 35; Georgetown Univ, PhD(biochem), 48. *Prof Exp:* Naval stores asst, Univ Fla, 36-38; high sch teacher, Fla, 38-39; chemist, Food & Drug Admin, USDA, La, 39-40; chemist, Fed Security Agency, Washington, DC, 40-42, chief color certification br, 46-53, chief color certification br, Dept Health, Educ & Welfare, 53-57, dep dir div cosmetics, 57-64, dir div color certification & eval, 64-68, dir div colors & cosmetics, 68-70; independent consult, 70-87; RETIRED. *Concurrent Pos:* Vis lectr, Georgetown Univ, 48-56. *Mem:* Am Chem Soc. *Res:* Terpene chemistry; naval stores; synthetic organic dyes; cosmetics. *Mailing Add:* 6577 Sandspur Lane Ft Myers FL 33919

FREEMAN, LEON DAVID, b Minneapolis, Minn, Oct 21, 20; m 43; c 2. THERAPEUTICS, CLINICAL PHARMACOLOGY. *Educ:* Univ Calif, Los Angeles, BA, 43; Univ Southern Calif, PhD(biochem), 62. *Prof Exp:* Lab supvr chem, US Rubber Co, 43-47; chief chemist, Southern Calif Gland Co, 49-60; res dir, Darwin Labs, 55-60; dir biochem sect, Riker Labs, 60-63; assoc dir clin invest, 63-65; vpres & dir biol res, Calbiochem, 66-71; CHMN, PHARMAQUEST CORP, 75- *Concurrent Pos:* Tech dir Harvard Labs, 55-60; consult, 71- *Mem:* AAAS; Am Chem Soc; Am Soc Clin Pharmacol & Therapeut; Am Acad Dermat; Drug Inf Assoc. *Res:* Chemistry and biology of heparin and other mucopolysaccharides; mechanism of action and metabolism; lipid metabolism and coronary heart disease; pharmaceutical research and clinical investigation; new drug design and human trials; new methods for isolation of heparin. *Mailing Add:* 48 Alta Way Corte Madera CA 94925

FREEMAN, LOUIS BARTON, b New York, NY, May 12, 35; m 62; c 2. REACTOR PHYSICS. *Educ:* Colgate Univ, AB, 55; Harvard Univ, AM, 57; Univ Pittsburgh, PhD(appl math), 65. *Prof Exp:* Assoc scientist, 58-62, scientist, 62-65, sr scientist, 65-71, prin scientist, 71-72, mgr breeder nuclear design, 72-76, ADV SCIENTIST NUCLEAR ENG, BETTIS ATOMIC POWER LAB, WESTINGHOUSE ELEC CORP, 76- *Mem:* Am Nuclear Soc. *Res:* Nuclear reactor calculational methods; analysis of nuclear reactor performance data; nuclear reactor safety calculations. *Mailing Add:* Westinghouse Bettis Atomic Power Lab PO Box 79 West Mifflin PA 15122-0079

FREEMAN, MARC EDWARD, b Philadelphia, Pa, Feb 18, 44; m 70. REPRODUCTIVE ENDOCRINOLOGY. *Educ:* Moravian Col, BS, 65; WVa Univ, MS, 67, PhD(reprod physiol), 70. *Prof Exp:* Res assoc physiol, 70-71, fel, Sch Med, Emory Univ, 71-72; from asst prof to assoc prof, 72-82, assoc chmn, 83-86, PROF BIOL, FLA STATE UNIV, 82- *Concurrent Pos:* Ed, Endocrinol, 88-; mem, Biochem Endo SS, NIH, 88-92. *Mem:* Endocrine Soc; Am Physiol Soc; Soc Study Reprod; Soc Study Fertility; AAAS. *Res:* The regulation of secretion of luteinizing hormone, follicle stimulating hormone and prolactin. *Mailing Add:* Dept Biol Sci Fla State Univ Tallahassee FL 32306

FREEMAN, MAX JAMES, b Columbus, Ohio, Aug 28, 34; m 59; c 3. IMMUNOBIOLOGY, IMMUNOPATHOLOGY. *Educ:* Auburn Univ, DVM, 58; Univ Wis, MS, 60, PhD(vet sci path), 61. *Prof Exp:* Asst prof, Ohio Agr Exp Sta, 61-62; fel microbiol, Western Reserve Univ, 62-63, USPHS fel immunol, 63-64; asst prof microbiol, Univ Kans, 64-67; assoc prof, 67-70, PROF IMMUNOL, PURDUE UNIV, 70- *Mem:* Am Vet Med Asn; Am Soc Microbiol; Am Asn Immunologists; Am Soc Exp Path. *Res:* Antibody biosynthesis; function and heterogeneity; immunologic mechanisms in disease. *Mailing Add:* Dept Vet Microbiol & Path Purdue Univ West Lafayette IN 47907

FREEMAN, MAYNARD LLOYD, b Chicago, Ill, July 19, 50; m 76; c 1. NUCLEAR CARDIOLOGY, DIABETES. *Educ:* Univ Ill, BS, 72; Rush Med Col, MD, 75. *Prof Exp:* Intern med, Wayne State Univ, 75-76; resident med, 78-80, fel, 80-82, ATTEND NUCLEAR MED, USVA HOSP-HINES, 82-, ASST SERV CHIEF, 83- *Concurrent Pos:* Ed, Am J Physiol Imaging, 85- *Mem:* Soc Nuclear Med; Asn Res Vision & Opthal; Asn Engrs Med & Biol. *Res:* Use of computer generated images to study physiology; early diagnosis and therapy of diabetic retinopathy. *Mailing Add:* VA Hosp Loyola Univ Med Sch 5469 Ranier Lisle IL 60532

FREEMAN, MILTON MALCOLM ROLAND, b London, Eng, Apr 23, 34; Can citizen; m; c 3. RESOURCE MANAGEMENT, CULTURAL ANTHROPOLOGY. *Educ:* Univ Reading, BSc, 58; McGill Univ, PhD(physiol ecol), 65. *Prof Exp:* Res asst econ & ecol studies, Arctic Biol Sta, Fisheries Res Bd Can, 61; res demonstr zool, Queen Elizabeth Col, Univ London, 62-64; Dept Northern Affairs & Nat Resources res grant, Northwest Territories & Ottawa, 65-67; asst prof biol, Mem Univ Nfld, 67-68, from asst prof to assoc prof anthrop, 67-72; from assoc prof to prof anthrop, McMaster Univ, 72-80; PROF ANTHROP & HENRY MARSHALL TORY PROF, UNIV ALTA, 82- *Concurrent Pos:* Can Coun Killam res grant, Mem Univ Nfld & Northwest Territories, 69-72; dir, Inuit Land Use & Occupancy Res Proj, 73-75; adj prof environ studies, Univ Waterloo, 77-81; bd dirs, NWT Sci Inst, 85-87; sr res scholar, Boreal Inst Northern Studies, 84-90 & Can Circumpolar Inst; sr sci adv, Dept Northern & Indian Affairs, Can, 79-81. *Mem:* Soc Appl Anthrop Can (pres, 84-85); fel Am Anthrop Asn; fel Arctic Inst NAm; Can Anthrop Soc; fel Soc Appl Anthrop; fel Soc Adv Socio-Econ; Int Asn Study Common Property. *Res:* Human ecological research and public policy issues relating to the Canadian arctic; regulation of whaling (circumpolar, Japan); common property resources management. *Mailing Add:* Dept Anthrop Univ Alberta Edmonton AB T6G 2H4 Can

FREEMAN, MYRON L, b Fergus Falls, Minn, Nov 2, 30; m 57; c 3. PLANT TAXONOMY. *Educ:* Moorhead State Col, BS, 57; Univ NDak, MS, 62; Univ Northern Colo, DAgr(bot), 70. *Prof Exp:* Instr biol, Univ NDak, 60-62; PROF BIOL, DICKINSON STATE COL, 62-, CHMN DEPT SCI & MATH, 77- *Res:* Identification of western North Dakota plants, especially grasses. *Mailing Add:* Dept Math & Sci Dickinson State Univ Dickinson ND 58601

FREEMAN, NEIL JULIAN, b Jersey City, NJ, Nov 25, 39; m; c 2. STRUCTURAL MECHANICS, MOTOR VEHICLE ACCIDENT RECONSTRUCTION. *Educ:* Univ Miami, BS, 62, MS, 65; Northwestern Univ, PhD(theoret & appl mech), 66. *Prof Exp:* From asst prof to assoc prof civil eng, 66-82, CONSULT FORENSIC ENG, UNIV MIAMI, 82- *Concurrent Pos:* Prin investr, NSF res initiation grant, 68-69 & NSF res grant, 70-72; Nat Comt Vehicle Crashworthiness, Dept Transp, 71-72; co dir, Law Ctr Eng Inst, Univ Miami, 85- *Mem:* Am Soc Civil Engrs; Am Asn Automotive Med; Int Asn Accident & Traffic Med; Sigma Xi; Soc Automotive Engrs. *Res:* Analytical studies of load transfer, cracks and stress concentrations in composite materials, within the scope of classical elasticity; mixed boundary value problems in elasticity, thermal conductance; multidisciplinary vehicle accident investigation and reconstruction. *Mailing Add:* 3671 Matheson Ave Miami FL 33133

FREEMAN, PETER KENT, b Modesto, Calif, Nov 25, 31; m 55; c 4. ORGANIC CHEMISTRY. *Educ:* Univ Calif, BS, 53; Univ Colo, PhD(chem), 58. *Prof Exp:* Res technologist, Shell Oil Co, 58; res assoc org chem, Pa State Univ, 58-59; from asst prof to prof chem, Univ Idaho, 59-68; PROF CHEM, ORE STATE UNIV, 68- *Mem:* Am Chem Soc; Sigma Xi. *Res:* Rearrangements; reaction mechanisms; environmental chemistry; photochemistry. *Mailing Add:* Dept Chem Ore State Univ Corvallis OR 97331

FREEMAN, RALPH DAVID, b Cleveland, Ohio, Mar 11, 39. PHYSIOLOGICAL OPTICS, NEUROPHYSIOLOGY. *Educ:* Ohio State Univ, BS, 62, OD, 63; Univ Calif, MS, 66, PhD(biophys), 69. *Prof Exp:* Asst prof, 69-74, assoc prof, 74-80, PROF PHYSIOL OPTICS & OPTOM, UNIV CALIF, BERKELEY, 80- *Concurrent Pos:* Nat Eye Inst grant, Univ Calif, Berkeley, 72-; lectr & consult var univs, 72- *Mem:* AAAS; Am Acad Optom. *Res:* Psychophysics of vision; plasticity of visual system; neurophysiology of visual cortex. *Mailing Add:* Sch Optom Univ Calif Berkeley CA 94720

FREEMAN, RAYMOND, b Long Eaton, Eng, Jan 6, 32; m 58; c 5. PHYSICAL CHEMISTRY. *Educ:* Oxford Univ, BA, 55, MA & DPhil(nuclear magnetic resonance), 57, DSc(chem), 75. *Prof Exp:* Engr, French AEC, 57-59; sr sci officer, Nat Phys Lab, 59-61; res fel nuclear magnetic resonance, Instrument Div, Varian Assocs, 61-62; sr sci officer, Nat Phys Lab, Eng, 62-63; mgr nuclear magnetic resonance basic res, Instrument Div, Varian Assocs, 64-73; univ lectr & fel, Magdalen Col, Oxford Univ, 73-87, Aldrichian praelector chem, 82-87; JOHN HUMPHREY PLUMMER PROF MAGNETIC RESONANCE, CAMBRIDGE, 87- *Concurrent Pos:* Consult, Varian Assocs, Calif, 74-; vis assoc, Calif Inst Technol, 85; Jesus Col fel, 87- *Honors & Awards:* Chem Soc Medal, Theoret Chemistry & Spectros, 78. *Mem:* Fel Royal Soc London. *Res:* Nuclear magnetic resonance spectroscopy; double irradiation techniques and relaxation time studies; two-dimensional Fourier transform spectroscopy; nuclear spin echoes. *Mailing Add:* Dept Chem Jesus Col Lensfield Rd Cambridge CB2 1EW England

FREEMAN, REINO SAMUEL, helminthology, parasitology; deceased, see previous edition for last biography

FREEMAN, RICHARD B, b Allentown, Pa, July 24, 31; m 54; c 3. INTERNAL MEDICINE, NEPHROLOGY. *Educ:* Franklin & Marshall Col, BS, 53; Jefferson Med Col, MD, 57. *Prof Exp:* Intern, Pa Hosp, Philadelphia, 57-58, resident med, 59-61; instr, Med Ctr, Georgetown Univ, 64-67; asst prof, 67-69, ASSOC PROF NEPHROLOGY, MED CTR, UNIV ROCHESTER, 69-, HEAD NEPHROLOGY UNIT, 74- *Concurrent Pos:* Fel hypertension, Pa Hosp, Philadelphia, 58-59; fels nephrology, Med Ctr,

Georgetown Univ, 61, 63-64; consult, Hosp & Med Facil Div, USPHS, 67-74; mem, Nat Adv Comt Artificial Kidney Chronic Uremia Prog, Nat Insts Arthritis & Metab Dis, 69-74; pres, Nat Kidney Found, 70- *Mem:* AAAS; Am Fedn Clin Res; Am Heart Asn; Am Soc Artificial Internal Organs; Am Soc Nephrology. *Res:* Pathogenesis and course of chronic renal disease; performance characteristics of artificial kidneys; cation metabolism. *Mailing Add:* Strong Mem Hosp Sch Med Univ Rochester Med Ctr 601 Elmwood Ave Rochester NY 14642

FREEMAN, RICHARD CARL, b NS, Can, May 8, 29; m 53; c 3. TEXTILE CHEMISTRY. *Educ:* Lowell Technol Inst, BS, 59. *Prof Exp:* From chemist to asst dir res & develop, Joseph Bancroft & Sons Co, 59-70; group mgr wool prod develop, Wool Bur, Inc, Woodbury, 70-73, vpres tech serv & develop, 73-91. *Mem:* Am Asn Textile Chemists & Colorists; Am Textile Mfrs Inst; NY Acad Sci; Knitted Textile Asn. *Mailing Add:* 40 Park Edge Dr Buffalo NY 14225

FREEMAN, RICHARD REILING, b Corpus Christi, Tex, Nov 13, 44; m 67; c 2. X-RAY LITHOGRAPHY, X-RAY SCATTERING. *Educ:* Univ Wash, BS, 67; Harvard Univ, AM, 68, PhD(physics), 73. *Prof Exp:* Res engr, Boeing Co, 68-69 & Dynetics Co, 69-70; consult physicist, Calspan Corp, 72-73; instr, Dept Physics, Mass Inst Technol, 73-76; mem tech staff, 76-81, head electromagnetic phenomena res dept, 81-89, HEAD LITHOGRAPHY RES DEPT, BELL LABS, 89-; AT AT&T BELL LABS, MURRAY HILL, NJ. *Mem:* Sigma Xi; Am Phys Soc; fel Optical Soc Am. *Res:* Atomic physics; nonlinear optics; x-ray physics. *Mailing Add:* One Cherry Rd Holnmdel NJ 07733

FREEMAN, ROBERT, MATHEMATICS. *Educ:* NY Univ, BAE, 47; Univ Calif, Berkeley, PhD(math), 59. *Prof Exp:* Mathematician, Lawrence Radiation Lab, 58-62; from asst prof to assoc prof math, Univ Md, 62-67; ASSOC PROF MATH, UNIV ORE & INST THEORET SCI, 67- *Mem:* NSF grant, 63-65. Mem: Am Math Soc; Math Asn Am. *Res:* Spectral and operation theoretic structure of elliptic boundary value problems. *Mailing Add:* Dept Math Univ Ore Eugene OR 97403

FREEMAN, ROBERT CLARENCE, b Harrellsville, NC, Oct 14, 27; m 54. ORGANIC CHEMISTRY. *Educ:* NC Col, Durham, BS, 50, MS, 52; Wayne State Univ, PhD(chem), 56. *Prof Exp:* Res chemist, Monsanto Chem Co, 56-62; mem fac, Agr & Tech Col NC, 62-64; res chemist, 64-75, SR RES SPECIALIST, MONSANTO CO, 75- *Mem:* Am Chem Soc; AAAS. *Res:* Acyloin condensation; ketenimine chemistry; epoxyether chemistry; reactions of phosphorous compounds with alpha halocarbonyl compounds; reactions of enamines; synthesis of radioactive compounds; synthesis of carbon-13 compounds. *Mailing Add:* 4647 Penrose St Louis MO 63115-2435

FREEMAN, ROBERT DAVID, b Nicholson, Ga, Feb 7, 30; m 57; c 2. CHEMICAL THERMODYNAMICS, HIGH TEMPERATURE CHEMISTRY. *Educ:* N Ga Col, BS, 48; Purdue Univ, MS, 52, PhD(chem), 54. *Prof Exp:* Phys chemist, Goodyear Atomic Corp, 54-55; from asst prof to prof, 55-88, EMER PROF CHEM, OKLA STATE UNIV, 88-; PRES & OWNER, ENTROPOLOGY UNLTD, 90- *Concurrent Pos:* Ed, Bull Chem Thermodynamics, 76-90. *Mem:* AAAS; Am Chem Soc; Int Union Pure & Appl Chem; Am Asn Physics Teachers. *Res:* Vaporization phenomena and chemistry at high temperatures; molecular flow of rarefied gases; thermochemistry; calorimetry; information theory and thermodynamics. *Mailing Add:* Entropology Unltd 116 S Kings St Stillwater OK 74074-2513

FREEMAN, ROBERT GLEN, b Kerrville, Tex, Feb 3, 27; m 50; c 4. PATHOLOGY. *Educ:* Baylor Univ, MD, 49. *Prof Exp:* Instr anat, Col Med, Baylor Univ, 50-52; instr, Med Field Serv Sch, Ft Sam Houston, 52-54; asst prof, Med Br, Univ Tenn, 54-55; resident & asst surg, Vet Admin Hosp, 55-56; resident & instr path, Affil Hosps, Baylor Univ, 57-59, from asst prof to prof, 59-70; prof path & dermat, 70-74, clin prof path & dermat, Univ Tex Health Sci Ctr, Dallas, 74-77, PROF PATH & DERMAT, CHIEF, DIV OF DERMAT PATHOL, UNIV TEX, SOUTHWESTERN MED CTR, DALLAS, 77- *Concurrent Pos:* Consult dermat, Brooke Army Hosp, Ft Sam Houston, Tex. *Mem:* Col Am Path; Am Acad Dermat; Am Soc Dermatopath; Am Dermat Asn; Soc Invest Dermat. *Res:* Dermatopathology; effects of ultraviolet on the skin. *Mailing Add:* Dept Path & Derm Univ Tex Southwestern Med Sch 5323 Harry Hines Blvd Dallas TX 75235

FREEMAN, ROGER DANTE, b New York, NY, Aug 3, 33; m 59; c 3. CHILD PSYCHIATRY. *Educ:* Swarthmore Col, BA, 54; Johns Hopkins Univ, MD, 58; McGill Univ, dipl psychiat, 63. *Prof Exp:* Rotating intern, Montreal Gen Hosp, 58-59, resident gen psychiat, 59-61; fel child psychiat, Child Study Ctr Philadelphia, 61-63; instr psychiat, Jefferson Med Col, 63-64; from instr to asst prof, Sch Med, Temple Univ, 64-70; from assoc prof to prof, 70-83, CLIN PROF PSYCHIAT, UNIV BC, 83- *Concurrent Pos:* Dir psychiat serv, Handicapped Children's Unit, St Christopher's Hosp for Children, Philadelphia, 64-69, dir child psychiat clin, 69-70; Children's Hosp, Vancouver, 70-; consult, US Food & Drug Admin, 72-73; Dept Nat Health & Welfare grant, Univ BC, 72-74; consult, Western Inst Deaf, 74- & Sunny Hill Hosp Children, 75- *Mem:* Am Acad Cerebral Palsy; Can Med Asn; Am Epilepsy Soc. *Res:* Psychiatry of handicapped children and their families; child psychopharmacology. *Mailing Add:* 3507 W 47th Ave Vancouver BC V6N 3N9 Can

FREEMAN, THOMAS EDWARD, b Laurel, Miss, Jan 29, 30; m 53; c 2. PLANT PATHOLOGY. *Educ:* Millsaps Col, BS, 52; La State Univ, MS, 54, PhD(plant path), 56. *Prof Exp:* Asst, La State Univ, 52-56; from asst plant pathologist to assoc plant pathologist, 56-68, PLANT PATHOLOGIST & PROF PLANT PATH, UNIV FLA, 68- *Mem:* Am Phytopath Soc. *Res:* Grass diseases; biological control of weed with plant pathogens. *Mailing Add:* Dept Plant Path Univ Fla Gainesville FL 32611-0513

FREEMAN, THOMAS J, b Miami, Fla, Sept 30, 32; m 55; c 2. GEOLOGY. *Educ:* Univ Ark, BS, 56, MS, 57; Univ Tex, PhD(geol), 62. *Prof Exp:* Jr geologist, Stand Oil Co, Ohio, 57-58; asst prof geol, Univ Mo, 62-63; field geologist, Ark Geol Comn, 63-64; asst prof, 64-67, assoc prof, 67-80, PROF GEOL, UNIV MO-COLUMBIA, 80- *Concurrent Pos:* Vis prof, Univ Madrid, 69-70; leader, Int Field Inst, Spain, 71. *Mem:* Am Asn Petrol Geologists; Geol Soc Am; Soc Econ Paleont & Mineral. *Res:* Economic aspects of the genesis and diagenesis of carbonate rocks. *Mailing Add:* Geol 101 Geol Bldg Univ Mo-Columbia Columbia MO 65211

FREEMAN, THOMAS PATRICK, b Denver, Colo, Aug 14, 38; m 63; c 4. PLANT ANATOMY, CELL ULTRASTRUCTURE. *Educ:* Colo State Univ, BA, 62; Univ Northern Colo, MA, 63; Ariz State Univ, PhD(plant anat), 68. *Prof Exp:* Instr bot, Foothill Col, 63-65; teaching asst, Ariz State Univ, 65-68; from asst prof to assoc prof, 68-78, PROF BOT, NDAK STATE UNIV, 78-, DIR ELECTRON MICROSCOPE LAB, 77- *Mem:* Am Inst Biol Sci; Bot Soc Am; Sigma Xi. *Res:* Developmental anatomy; chloroplast ultrastructure; electron microscopy. *Mailing Add:* Electron Micros Northern Crop Sci Lab NDak State Univ Fargo ND 58102

FREEMAN, VERNE CRAWFORD, b Bentonville, Ind, Dec 25, 00; m 26; c 1. AGRICULTURE. *Educ:* Purdue Univ, BSA, 23, MSA, 26. *Prof Exp:* Teacher pub sch, Ind, 23-25; instr animal husb & asst to dean, 26-35, from asst dean to assoc dean, 35-69, from asst prof to prof animal husb, 38-69, dir resident instr, 58-69, EMER PROF ANIMAL SCI, SCH AGR, PURDUE UNIV, WEST LAFAYETTE, 80-, EMER DIR RESIDENT INSTR, 69- *Concurrent Pos:* Consult, Bd Fundamental Educ. *Mem:* Am Coun Educ; Farm Econ Asn; fel Am Soc Animal Sci. *Res:* Agricultural personnel and guidance programs. *Mailing Add:* 518 Hillcrest Rd West Lafayette IN 47906

FREEMAN, WADE AUSTIN, b Evanston, Ill, Nov 20, 40; m 75; c 2. INORGANIC CHEMISTRY. *Educ:* Univ Ill, BS, 62; Univ Mich, MS, 64, PhD(chem), 67. *Prof Exp:* From instr to asst prof chem, 67-84, asst dean, Col Lib Arts, 69-79, ASSOC PROF CHEM, UNIV ILL, CHICAGO CIRCLE, 84-, DIR FRESHMAN CHEM, 79-, ASST HEAD, 89- *Mem:* Am Chem Soc. *Res:* Determination of molecular structure in relation to circular dichroism of inorganic complex compounds. *Mailing Add:* Dept Chem Box 4348 Univ Ill Chicago Circle Chicago IL 60680

FREEMAN, WALTER JACKSON, III, b Washington, DC, Jan 30, 27; m 52; c 7. NEUROPHYSIOLOGY. *Educ:* Yale Univ, MD, 54. *Prof Exp:* Intern path, Yale Univ, 54-55; intern med, Hosp, Johns Hopkins Univ, 55-56; asst res physiologist, Univ Calif, Los Angeles, 56-59; from asst prof to assoc prof, 59-67, chmn dept, 67-72, PROF PHYSIOL, CALIF, BERKELEY, 67- *Concurrent Pos:* USPHS fel, Univ Calif, Los Angeles, 56-58; Found Fund Res Psychiat & Guggenheim fels, 65-66; titulane de la chaire solvay, Univ Libre, Bruxelles, 74. *Res:* Neurophysiology. *Mailing Add:* Dept Physiol-Anat Univ Calif Berkeley CA 94720

FREEMON, FRANK REED, b Bloomington, Ill, July 18, 38; m 66. NEUROLOGY. *Educ:* Univ Fla, MD, 65; Vanderbilt Univ, MA, 85; Univ Ill, PhD, 92. *Prof Exp:* Clin fel neurol, Univ Fla, 66-69; res fel, Wash Univ, 69-70; asst prof neurol, Med Col Wis, 70-72; assoc prof, 72-78, PROF NEUROL, VANDERBILT UNIV, 78- *Mem:* AAAS; Am Acad Neurol; Am Asn Hist Med. *Res:* Clinical neurology; brain function. *Mailing Add:* 2422 Valleybrook Rd Nashville TN 37215

FREER, RICHARD JOHN, b Poughkeepsie, NY, June 2, 42. PHARMACOLOGY. *Educ:* Marist Col, BA, 64; Columbia Univ, PhD(pharmacol), 69. *Prof Exp:* Instr biochem, Sch Med, Univ Colo, Denver, 69-72; asst prof pharmacol, Univ Conn Health Ctr, Farmington, 72-75; assoc prof, 75-81, PROF PHARMACOL, MED COL VA, 81- *Concurrent Pos:* Estab investr, Am Heart Asn. *Mem:* Am Soc Exp Pharmacol & Ther; Am Heart Asn. *Res:* Synthesis of analogs of vasoactive peptides to study peptide-receptor interactions; study and synthesis of inhibitors of peptide hormones. *Mailing Add:* Dept Pharm & Toxicol Med Col Va Box 524 Richmond VA 23298

FREER, STEPHAN T, b Akron, Ohio, May 14, 33; m 59; c 2. BIOCHEMISTRY, PHYSICAL CHEMISTRY. *Educ:* Univ Calif, Berkeley, AB, 59; Univ Wash, PhD(biochem), 64. *Prof Exp:* Asst specialist chem, Univ Calif, San Diego, 64-67, asst res chemist, 67-72, assoc res chemist, 72-82; scientist, Ma-Com Linkabit, 82-88; SR SCIENTIST, AGOURON PHARMACEUT, 88- *Mem:* AAAS; Am Crystallog Asn. *Res:* Software systems; structure and function of biologically significant macromolecules. *Mailing Add:* 11025 N Torrey Pines Rd La Jolla CA 92037

FREERKS, MARSHALL CORNELIUS, b Wahpeton, NDak, Sept 2, 12; m 41; c 4. ATOMIC PHYSICS. *Educ:* Carleton Col, BA, 39; Univ Minn, PhD(org & phys chem), 49. *Prof Exp:* Org chemist & res group leader, 49-61, scientist, 61-72, SCI FEL, MONSANTO CO, 72- *Mem:* Am Chem Soc; NY Acad Sci. *Res:* Fundamental nature of catalysis as a function of atomic structure; physical nature of atomic structure; nature of the chemical bond. *Mailing Add:* 339 Greenleaf Dr Kirkwood MO 63122

FREERKSEN, DEBORAH LYNNE (CHALMERS), b Calgary, Alta, Can, May 16, 56; US citizen; m 79; c 2. INTERMEDIARY METABOLISM, CELL DIFFERENTIATION. *Educ:* Univ Colo, Boulder, BA, 78; Univ Del, Newark, PhD(biochem), 84. *Prof Exp:* Res Technician, Tufts New Eng Med Ctr, 78-79; res technician, 79-80, res asst, 80-84, res fel, 84-86, BIOCHEMIST, E I DU PONT CO, 87- *Concurrent Pos:* Post doctoral fel, Am Heart Asn, 84-86. *Mem:* Am Chem Soc; AAAS. *Res:* T-cell activation by lymphokines for adoptive immunotherapy; regulation of intermediary metabolism in developing cardiac and skeletal muscle by thyroid hormones and corticosteroids; glycerol phosphate dehydrogenase isoenzymes. *Mailing Add:* 130 DaVinci Ct Hockessin DE 19707

FREERKSEN, ROBERT WAYNE, b Owatonna, Minn, May 19, 52; m 79; c 2. ORGANIC CHEMISTRY. *Educ:* Univ Minn, Minneapolis, BS, 74; Univ Colo, Boulder, PhD(org chem), 78. *Prof Exp:* Res fel chem, Harvard Univ, 78-79; RES CHEMIST, BIOCHEM DEPT, E I DU PONT DE NEMOURS & CO, 79- *Concurrent Pos:* NIH fel. *Mem:* Am Chem Soc. *Res:* Synthesis of biologically active compounds including naturally occuring polyether ionophores and novel steroids; development of synthetic methods. *Mailing Add:* Walkers Mill Bldg-Barley Mill DuPont Co 03470 Wilmington DE 19880

FREESE, ERNST, molecular biology; deceased, see previous edition for last biography

FREESE, KATHERINE, b Freiburg, Ger, Feb 8, 57; US citizen; m 87; c 1. COSMOLOGY. *Educ:* Princeton Univ, BA, 77; Columbia Univ, MA, 81; Univ Chicago, PhD (physics), 84. *Prof Exp:* Postdoctoral physics, Harvard Ctr Astrophys, 84-85, Inst Theoret Physics, Santa Barbara, 85-87 & Univ Calif, Berkeley, 87-88; ASST PROF PHYSICS, MASS INST TECHNOL, 88- *Concurrent Pos:* Sloan Found Award, 89; NSF presidential young investr award, 90. *Mem:* Am Phys Soc. *Res:* Theoretical cosmology; study of the early universe; specialize in research at the interface between astrophysics and particle physics; inflation; dark matter; magnetic monopoles. *Mailing Add:* Dept Physics Rm 6-201 Mass Inst Technol Cambridge MA 02139

FREESE, KENNETH BROOKS, b Leavenworth, Kans, Feb 10, 49; c 1. EXPERIMENTAL PLASMA PHYSICS, TECHNOLOGY TRANSFER. *Educ:* Univ Mich, AB, 70; Dartmouth Col, PhD(physics), 74. *Prof Exp:* Asst physics, Dartmouth Col, 70-74; sr prog officer, Econ Develop & Tourism Dept, State NMex, 83-85; staff mem, Controlled Thermonuclear Res Div, Los Alamos Nat Lab, 74-81, asst admin, 81-83, proj mgr tech develop, Indust & Int Initiatives Off, 83-87, com applications officer, Indust Applications Off, 87-90, actg dep dir, 90; CONSULT, 90- *Concurrent Pos:* Owner, Double Seal Insulated Glass Co, 76-80; loaned exec, State Govt NMex, 83-85. *Mem:* Am Phys Soc; Sigma Xi; Licensing Execs Soc; Optical Soc Am. *Res:* Controlled fusion energy research, plasma physics and plasma diagnostic development; computer data acquisition and handling; solar energy research and development; technology transfer; licensing and contract negotiation; conference organization and management. *Mailing Add:* Indust Appl Off MS-M899 Los Alamos Sci Lab Los Alamos NM 87545

FREESE, RALPH, b Berkeley, Calif, Apr 4, 46. MATHEMATICS. *Educ:* Univ Calif, BA, 68; Calif Tech Inst, PhD(math), 72. *Prof Exp:* PROF MATH, UNIV HAWAII, 72- *Mem:* Am Math Asn. *Mailing Add:* Dept Math Univ Hawaii Honolulu HI 96822

FREESE, RAYMOND WILLIAM, b Foristell, Mo, Dec 17, 34; m 57; c 3. GEOMETRY. *Educ:* Univ Mo, BS, 56 & 58, MA, 58, PhD(math), 61. *Prof Exp:* From asst prof to assoc prof, 61-67, chmn dept, 71-86, PROF MATH, ST LOUIS UNIV, 67- *Mem:* Am Math Soc; Math Asn Am. *Res:* Distance geometry. *Mailing Add:* Dept Math St Louis Univ St Louis MO 63103

FREESE, UWE ERNST, b Bordesholm, WGer, May 11, 25; US citizen; m 61; c 2. OBSTETRICS & GYNECOLOGY. *Educ:* Univ Kiel, MD, 51. *Prof Exp:* From instr to prof, Chicago Lying-in Hosp, Univ Chicago, 71-75, PROF OBSTET & GYNEC & CHMN DEPT, CHICAGO MED SCH-UNIV HEALTH SCI, 75-; DIR OBSTET & GYNEC, COOK COUNTY HOSP, 76- *Mem:* Soc Gynec Invest; Perinatal Res Soc; NY Acad Sci; Am Col Obstetricians & Gynecologists; Sigma Xi; Am Professors Gynec & Obstet; Coun Resident Educ Obstet & Gynec. *Res:* Maternal-fetal medicine; morphophysiology of utero-placental unit. *Mailing Add:* UHS/The Chicago Med Sch 3333 Green Bay Rd North Chicago IL 60064

FREESE, WILLIAM P, II, b Catskill, NY, Aug 20, 34; m 57; c 3. RESEARCH ADMINISTRATION. *Educ:* Lafayette Col, BS, 56; Union Col, MS, 74. *Prof Exp:* Proj chemist, Am Viscose Corp, 56; develop chemist, Hercules Powder Co, 58-62; sr chemist, Huyck Res Ctr, Huyck Corp, 62-68, Bendix Corp, 68-75; asst res scientist, NY State Dept Environ Conserv, 75-76; plant chemist, Albany Plastics Co, 76-77; CHEM SCIENTIST & GROUP LEADER INDUST RESIN, GA-PAC CORP, 77- *Mem:* Am Chem Soc. *Res:* Evaluation of methods to improve the heat resistance, toughness and flexibility of phenolic resins. *Mailing Add:* 3283 Salem Lane Conyers GA 30208-2355

FREESTON, W DENNEY, JR, b Orange, NJ, May 8, 36; m 57; c 3. MECHANICAL ENGINEERING, TEXTILES. *Educ:* Princeton Univ, BS, 57, MS, 58, MA, 59, PhD(mech eng), 61. *Prof Exp:* Asst, Princeton Univ, 57-58; sr res assoc, FRL Res Labs, Inc, Dedham, Mass, 60-65, asst dir appl mech, 65-69, assoc dir, 69-71; prof, French Textile Sch, Ga Inst Technol, 71-80, dir, 76-80, prof & assoc dean eng, 80-90, PROF & DIR CONTINUING EDUC, GA INST TECHNOL, 90- *Concurrent Pos:* Harvard Inst Educ Mgt, 81, chmn, Am Soc Testing Mat, 74-76, mem eng res coun, Am Soc Eng Educ, chmn Nat Acad Eng Comt on technol & int econ & trade issues, 80-83. *Honors & Awards:* Distinguished Achievement Award, Fiber Soc, 69. *Mem:* Fiber Soc (pres, 77); Sigma Xi; Am Soc Eng Educ; Nat Soc Prof Engrs. *Res:* Mechanics of flexible fibrous strucures; high performance materials. *Mailing Add:* 2765 Joel Place NE Atlanta GA 30360

FREETHEY, GEOFFREY WARD, b Denver, Colo, Nov 22, 45; m 70. GROUND-WATER FLOW MODELING, GROUND-WATER MOVEMENT IN CONSOLIDATED-ROCK AQUIFERS. *Educ:* Univ Alaska, BS, 67; Colo State Univ, MS, 69. *Prof Exp:* Hydrogeologist, Conserv & Surv Div, US Geol Surv, Lincoln, Nebr, 72-73, hydrologist, Water Resources Div, Anchorage, Alaska, 73-79, Tucson, Ariz, 79-83, Salt Lake City, Utah, 83-89, SUPVRY HYDROLOGIST, WATER RESOURCES DIV, US GEOL SURV, SALT LAKE CITY, UTAH, 89- *Mem:* Am Inst Prof Geologists; Am Water Resources Asn; Nat Water Well Asn; Asn Arid Lands Studies. *Res:* Regional aquifer-systems analysis; ground-water movement in fractured porous media; ground-water flow modeling; solute-transport modeling. *Mailing Add:* 1539 S 1000 E Salt Lake City UT 84105

FREEZE, ROY ALLAN, b Edmonton, Alta, May 23, 39; m 61; c 4. HYDROLOGY. *Educ:* Queen's Univ, BSc, 61; Univ Calif, Berkeley, MSc, 64, PhD(hydrol), 66. *Prof Exp:* Res scientist, Inland Waters Br, Can Dept Energy, Mines & Resources, 61-69; res staff mem hydrol, Thomas J Watson Res Ctr, IBM Corp, 70-73; from assoc prof to prof geol sci, Univ BC, 73-91; AT R ALLAN FREEZE ENG, INC. *Honors & Awards:* Horton Award, Am Geophys Union, 70 & 72, Macelwane Award, 73; Meinzer Award, Geol Soc Am, 74. *Mem:* Am Geophys Union; Geol Soc Am. *Res:* Computer simulation of regional groundwater flow systems and hydrologic response models. *Mailing Add:* R Allan Freeze Eng Inc 3755 Nico Wynd Dr White Rock BC V4A 5Z4 Can

FREGLY, MELVIN JAMES, b Patton, Pa, May 26, 25; m 56. PHYSIOLOGY. *Educ:* Bucknell Univ, BS & MS, 49; Univ Rochester, PhD(physiol), 52. *Prof Exp:* Instr physiol, Harvard Med Sch, 52-56; from asst prof to assoc prof, 56-65, asst dean grad studies, 67-72, prof physiol, 65-79, GRAD RES PROF, COL MED, UNIV FLA, 79- *Concurrent Pos:* Travel fels, Int Physiol Cong, 56, 59 & 62; mem coun high blood pressure, Am Heart Asn; consult, Strasenburgh Pharmaceut Co, NY, 65-67 & Environ Protection Agency, 76- *Mem:* Soc Exp Biol & Med; Am Physiol Soc; Am Soc Zool; Endocrine Soc; Am Thyroid Asn; Aerospace Med Asn; Can Physiol Soc. *Res:* Cardiovascular hypertension; temperature regulation; behavioral physiology. *Mailing Add:* Dept Physiol Box J-274 JHMHC Univ Fla Col Med Gainesville FL 32610

FREHN, JOHN, b Shippensburg, Pa, Mar 17, 36; m 60; c 2. PHYSIOLOGY. *Educ:* Dickinson Col, BS, 58; Pa State Univ, MS, 60, PhD(zool), 62. *Prof Exp:* From asst prof to assoc prof, 62-70, PROF PHYSIOL, ILL STATE UNIV, 70- *Mem:* AAAS; Am Soc Zoologists; Sigma Xi. *Res:* Cellular metabolism in cold acclimation and hibernation in various species of mammals. *Mailing Add:* Dept Biol Sci Ill State Univ Normal IL 61761

FREI, EMIL, III, b St Louis, Mo, Feb 21, 24; m 48, 87; c 5. CANCER MEDICINE. *Educ:* Yale Univ, MD, 48. *Prof Exp:* Intern, Univ Hosp, St Louis, 48-49, resident path, 52-53, resident internal med, 53-55; head, Chemother Serv, chief, Med Br & assoc sci dir, Nat Cancer Inst, 55-65; assoc dir, Univ Tex M D Anderson Hosp & Tumor Inst, 65-73; DIR & PHYSICIAN-IN-CHIEF, DANA-FARBER CANCER INST, 73-; prof, 80-83, RICHARD & SUSAN SMITH PROF MED, MED SCH, HARVARD UNIV, BOSTON, 83- *Honors & Awards:* Lasker Award, 72; Kettering Award, Gen Motors, 83; Hammer Prize, Pioneering Work Cancer Res, 90. *Mem:* Inst Med-Nat Acad Sci; Am Soc Clin Invest; Am Asn Cancer Res (pres, 71-72); Asn Am Physicians. *Res:* Cancer medicine and chemotherapy; pharmacology. *Mailing Add:* Dana-Farber Cancer Inst 44 Binney St Boston MA 02115

FREI, JAROSLAV VACLAV, b Prague, Czech, Mar 7, 29; Can citizen; m 55; c 5. PATHOLOGY, GASTROENTEROLOGICAL PATHOLOGY. *Educ:* Charles Univ, Prague, MUC, 49; Queen's Univ, Ont, MD, CM, 56; McGill Univ, MSc, 60, PhD(path), 62; FRCP(C), FRCPath. *Prof Exp:* Life Ins Med Res Fund fel, McGill Univ, 59-61, Nat Cancer Inst Can fel, 61-62, asst prof path, 63-66; asst prof, Cancer Res Lab, 66-70, assoc prof, Univ, 70-75, PROF PATH, UNIV WESTERN ONT, 75- *Concurrent Pos:* Nat Cancer Inst Can res assoc, 63-66; Eleanor Roosevelt fel, Med Res Coun Toxicol Unit, UK, 66; pathologist, Univ Hosp, London, 72-; vis scientist, Chester Beatty Res Inst, London, 73-74, 80-81. *Mem:* Am Asn Cancer Res; Am Soc Exp Path; Can Soc Cell Biol; Can Asn Path. *Res:* Experimental carcinogenesis in mice; nitroso compound carcinogenesis; gastroenterological pathology. *Mailing Add:* Dept Path Univ Western Ont London ON N6A 5B7 Can

FREI, SISTER JOHN KAREN, b Jersey City, NJ. EPIPHYTIC ECOLOGY. *Educ:* Douglass Col, BS, 59; Rutgers Univ, MS, 61; Univ Miami, PhD(biol), 72; Barry Univ, MBA, 82. *Prof Exp:* From asst prof to assoc prof biol, 70-75, actg chmn, 73-74, PROF BIOL, BARRY UNIV, 74-, CHMN DEPT, 81-, ASSOC VPRES, ACAD HEALTH & SCI, 85- *Concurrent Pos:* Dir, Biol Dept, Barry Univ, 83-, dean, 84- *Mem:* AAAS; Torrey Bot Soc; Am Asn Univ Professors; Bot Soc Am; Am Orchid Soc. *Res:* Epiphytic ecology; development of models to successfully educate minority students for science. *Mailing Add:* Dept Health Sci Barry Univ 11300 NE Second Ave Miami FL 33161

FREIBERG, SAMUEL ROBERT, b Staten Island, NY, Apr 14, 24; m 49; c 2. PLANT PHYSIOLOGY, BIOLOGY. *Educ:* Rutgers Univ, BSc, 48, PhD(plant physiol), 51. *Prof Exp:* Res assoc hort, Rutgers Univ, 51-52; plant physiologist div trop res, United Fruit Co, Honduras, 52-57, head plant physiol cent res labs, Mass, 58-60, asst dir res, 60-62, dir labs, 62-65; consult, IRI Res Inst, Inc, 65, dir res, 65-79, vpres, 66-79; sr agriculturist, World Bank, Washington, DC, 79-86; AGR CONSULT, 86- *Concurrent Pos:* Vis fel plant physiol, Cornell Univ, 57-58; consult, World Bank, Spain, 71, Indonesia, 72 & 87-88, Malaysia, 74, Mexico, 75, Turkey, 86-91, Pakistan, 88-89, Tanzania, 87 & India, 86; mem bd trustees, IRI Res Inst, Inc, 74-; consult, Int Found Agr Develop, 90-91. *Mem:* AAAS; Am Soc Plant Physiol; Am Soc Hort Sci; Am Soc Agron; Asn Trop Biol. *Res:* Plant nutrition and growth; post harvest fruit physiology; plant, animal and fruit productivity. *Mailing Add:* 8733 Susanna Lane Chevy Chase MD 20815

FREIBERGER, WALTER FREDERICK, b Vienna, Austria, Feb 20, 24; nat US; m 56; c 3. APPLIED MATHEMATICS. *Educ:* Univ Melbourne, BA, 47, MA, 49; Cambridge Univ, PhD(math), 53. *Prof Exp:* Sci res officer, Aeronaut Res Lab, Australian Dept Supply, 47-49, sr sci res officer, 53-54; res assoc, 55-56, from asst prof to assoc prof, 56-64, dir ctr comput & info sci, 68-76, chmn div, 76-82, PROF APPL MATH, BROWN UNIV, 64- *Concurrent Pos:* Tutor, Dept Math, Univ Melbourne, 47-49; Fulbright fel, 55; Guggenheim fel, Inst Math Statist, Univ Stockholm, 62-63; dir comput lab, Brown Univ, 63-68; managing ed, Quart Appl Math, 66-; ed-in-chief, Int Dictionary Appl Math; vis lectr & consult, appl actuarial sci, Bryant Col, 85-; chmn grad prog comt, Brown Univ, 85-88, assoc chmn of the div, 88- *Mem:* Am Math Soc; Asn Comput Mach; Inst Math Statist; Soc Indust & Appl Math. *Res:* Probability and statistics; biostatistics; actuarial mathematics. *Mailing Add:* Div Appl Math Brown Univ 182 George St Providence RI 02906

FREIBURG, RICHARD EIGHME, b Milwaukee, Wis, Apr 2, 23; m 50; c 4. ZOOLOGY, ECOLOGY. *Educ:* Univ Kans, BA, 49, MA, 51; Ore State Col, PhD(zool), 54. *Prof Exp:* Asst instr zool, Univ Kans, 49-51 & Ore State Col, 51-54; asst prof biol, Washburn Univ, Topeka, 54-57; from asst prof to assoc prof, 57-67, PROF BIOL, MACMURRAY COL, 67-, HEAD DEPT, 70- *Mem:* Am Soc Mammal; Am Soc Ichthyologists & Herpetologists; Ecol Soc Am. *Res:* Herpetology; ecological studies on reptiles and amphibians. *Mailing Add:* 1909 Southview Ct Jacksonville IL 62650

FREIDINGER, ROGER MERLIN, b Pekin, Ill, July 26, 47; m 69; c 2. MEDICINAL CHEMISTRY, PEPTIDE CHEMISTRY. *Educ:* Univ Ill, Urbana, BS, 69; Mass Inst Technol, PhD(org chem), 75. *Prof Exp:* Sr res chemist, Merck & Co Inc, 75-80, res fel, 80-82, sr res fel, 82-85, asst dir, 85-87, assoc dir, 87-89, SR SCIENTIST MED CHEM, MERCK & CO INC, 89- *Concurrent Pos:* Consult, NICHD Contraceptive Develop Br, 82, 86, 88 & 89; ad hoc reviewer, 84, mem, NIH Bio-org & Natural Prod Study Sect, 90-, ad hoc, NIH Endocrinol Study Sect, 86; ed adv bd, J Org Chem, 88-; prog comt, Tenth Ann Peptide Symp, 87. *Honors & Awards:* Vincent du Vigneaud Award, Young Investrs Peptide Chem, 86. *Mem:* Am Chem Soc; AAAS. *Res:* Design and synthesis of biologically active peptides; conformationally constrained peptide analogs; nonpeptide ligands for peptide receptors; peptide mimetics; peptide antagonists; CCK/gastrin antagonists; oxytocin antagonists. *Mailing Add:* 744 Newport Ln Lansdale PA 19446

FREIDLINE, CHARLES EUGENE, b San Francisco, Calif, Oct 5, 37; m 60; c 1. INORGANIC CHEMISTRY, ANALYTICAL CHEMISTRY. *Educ:* Westmont Col, BA, 60; Univ Minn, MS, 63, PhD(inorg chem), 66. *Prof Exp:* Assoc prof chem, Cent Methodist Col, 65-82; PROF CHEM, UNION COL, 83- *Concurrent Pos:* Nat Acad Sci grant; fel biochem, Univ Mo, 74-75, Environ Protection Agency, IPA, 80-81. *Mem:* Am Chem Soc. *Res:* Determination of stability constants of complex ions in solution; stabilized complexes of lead IV; analytical chemistry of biogenic amines and amine oxidation products; matrix isolation fourier transform I R. *Mailing Add:* Union Col Chem Sci & Math 5236 Prescott Ave Lincoln NE 68506

FREIER, ESTHER FAY, b Hibbing, Minn, Mar 3, 25. PHYSIOLOGICAL CHEMISTRY. *Educ:* Univ Minn, BS, 46, MS, 56. *Prof Exp:* From instr to assoc prof, 51-68. chemist, univ hosp, 57-68, PROF MED TECHNOL, UNIV MINN, MINNEAPOLIS, 68-, CO-DIR CLIN CHEM, 76- *Concurrent Pos:* Ed staff, Am J Med Technol, 52-72. *Honors & Awards:* G T Evans Award, Acad Clin Lab Physicians & Scientists, 80. *Mem:* AAAS; Am Soc Med Technol; Am Asn Clin Chemists; Acad Clin Lab Physicians & Scientists (secy treas, 76-80). *Res:* Clinical chemistry methodology and quality control; proteins and enzymes. *Mailing Add:* C-292 Mayo Univ Minn Minneapolis MN 55455

FREIER, GEORGE DAVID, b Ellsworth, Wis, Jan 22, 15; m 43; c 2. ATMOSPHERIC PHYSICS. *Educ:* Wis State Col, River Falls, BS, 38; Univ Minn, MA, 44, PhD(physics), 49. *Prof Exp:* Res physicist, Naval Ord Lab, 44-46; from asst prof to assoc prof, 50-67, PROF PHYSICS, UNIV MINN, MINNEAPOLIS, 67- *Mem:* AAAS; Am Phys Soc; Am Asn Physics Teachers; Am Geophys Union; Am Meteorol Soc. *Res:* Atmospheric electricity. *Mailing Add:* 2327 Gordon Ave St Paul MN 55108

FREIER, HERBERT EDWARD, b Delmont, SDak, Mar 19, 21; m 55; c 2. ANALYTICAL CHEMISTRY. *Educ:* Yankton Col, BA, 43; Univ Ill, PhD(org chem), 46. *Prof Exp:* Asst chem, Univ Ill, 43-44, spec asst, 44-46; from asst prof to assoc prof, Univ NDak, 46-50; mem staff, Cent Res, Minn Mining & Mfg Co, 50-54, supvr anal sect, 54-55 & org sect, 55-57, sect leader tech info serv, 57-60, mgr res serv, 60-74, dir anal & properties res lab, 74-85; RETIRED. *Mem:* Am Chem Soc. *Res:* Organic synthesis and analysis. *Mailing Add:* 1661 E Conway St St Paul MN 55106

FREIER, JEROME BERNARD, b New York, NY, May 6, 16. MATHEMATICS. *Educ:* City Col NY, BS, 39; NY Univ, PhD(math), 58. *Prof Exp:* Asst physics, NY Univ, 47-49; instr math, St Peter's Col, 50-53; asst prof, Rensselaer Polytech Inst, 53-65; ASSOC PROF MATH, SOUTHEASTERN MASS UNIV, 65- *Mem:* AAAS; Am Phys Soc; Am Math Soc. *Res:* Analysis; applied mathematics. *Mailing Add:* Dept Math Southeastern Mass Univ North Dartmouth MA 02747

FREIER, PHYLLIS S, b Minneapolis, Minn, Jan 19, 21; m 43; c 2. PHYSICS. *Educ:* Univ Minn, BS, 42, MA, 44, PhD(physics), 50. *Prof Exp:* Physicist, Naval Ord Lab, 44-45; res assoc, 50-70, assoc prof, 70-75, PROF PHYSICS, UNIV MINN, MINNEAPOLIS, 75- *Mem:* AAAS; Am Astron Soc; Am Geophys Union; fel Am Phys Soc. *Res:* Cosmic rays. *Mailing Add:* Sch Phys & Astron Univ of Minn 116 Church St SE Minneapolis MN 55455

FREILICH, GERALD, b Brooklyn, NY, Dec 29, 26; m 53; c 2. MATHEMATICS. *Educ:* City Col New York, BS, 46; Brown Univ, MS, 47, PhD(math), 49. *Prof Exp:* Asst math, Brown Univ, 46-48, instr, 49-50; from instr to prof, City Col New York, 50-71, chmn dept, 66-70; PROF MATH, QUEENS COL, NY, 71- *Mem:* Am Math Soc; Math Asn Am. *Res:* Measure theory; theory of convex sets; operations research. *Mailing Add:* 1619 E 21st St Brooklyn NY 11210

FREILING, EDWARD CLAWSON, b San Francisco, Calif, Aug 11, 22; m 48; c 7. ENVIRONMENTAL SCIENCE. *Educ:* Univ San Francisco, BS, 43; Univ Calif, MS, 47; Stanford Univ, PhD(chem), 51. *Prof Exp:* Assoc prof chem, St Mary's Col, Calif, 49-51; sr investr, US Naval Radiol Defense Lab, 51-68, head, Anal Br, US Naval Surface Weapons Ctr, 69-74, res scientist & chemist, 74-83; software engr, E-OIR Inc, 84-85; TECH DIR, SCI SOFTWARE SERV, 85- *Honors & Awards:* Super Civilian Serv Award, Chief Naval Mat, 68. *Res:* Ion exchange; radiochemical analysis; nuclear detonation phenomena; fused salt chemistry; kinetics; modeling chemical perturbation of the ionosphere; environmental science, baseline studies, impact assessment; software engineering. *Mailing Add:* 7775 Wooden Ct Dublin CA 94568

FREILING, MICHAEL JOSEPH, b San Francisco, Calif, Mar 19, 50; m 81; c 4. ARTIFICIAL INTELLIGENCE, DATABASE SYSTEMS. *Educ:* Univ San Francisco, BS, 72; Mass Inst Technol, PhD(appl math), 77. *Prof Exp:* Luce scholar comput sci, Kyoto Univ, 77-78; asst prof comput sci, Ore State Univ, 78-83; prin scientist, Tektronix, Inc, 84-89; MEM TECH STAFF, WYATT SOFTWARE, 89- *Honors & Awards:* Henry Luce Scholar, 77. *Mem:* Sigma Xi; Am Asn Artificial Intel; Am Finance Asn. *Res:* Tools and methodologies for acquiring and processing knowledge; application of knowledge to solve engineering and business decision-making problems; knowledge engineering; software applications for pension and investment management. *Mailing Add:* 16690 S Glenwood Ct Lake Oswego OR 97034-5034

FREIMAN, CHARLES, b New York, NY, June 17, 32; m 55; c 4. COMPUTER SCIENCE, ELECTRICAL ENGINEERING. *Educ:* Columbia Univ, AB, 54, BS, 55, MS, 56, ScD(eng), 61. *Prof Exp:* Instr elec eng, Columbia Univ, 56-60; mem res staff, Int Bus Mach Thomas J Watson Res Ctr, 60-65, develop engr, Data Systs Div, 65-68, sr mgr, Comput Sci Dept, 68- 73, mgr tech planning staff, Adv Systs Develop Div, 73-75, mgr RAS planning, Data Systs Div, 75-77, sr mgr, Comput Sci Dept, Int Bus Mach Thomas J Watson Res Ctr, 77-84, asst res lab dir & mgr, Comput Sci Inst, Tokyo Res Lab, 85-87, exec asst group dir, Advan Systs, 77-78, mgr knowledge base systs, Enterprise Systs, 88-89, secy, Acad Technol, 89-90; DIR, ENG FOUND, 90- *Concurrent Pos:* Ed, Info Processing, Int Fedn Info Processing, 71. *Mem:* AAAS; Asn Comput Mach; Inst Elec & Electronic Engrs; Int Fedn Info Processing; Sigma Xi; NY Acad Sci; Am Soc Eng Educ. *Res:* Discrete information theory; computer arithmetic; computer system design. *Mailing Add:* Palmer Lane West Pleasantville NY 10570

FREIMAN, DAVID GALLAND, b New York, NY, July 1, 11; m 49; c 2. PATHOLOGY. *Educ:* City Col New York, AB, 30; Long Island Col Med, MD, 35. *Hon Degrees:* AM, Harvard Univ, 62. *Prof Exp:* Intern & resident path, Montefiore Hosp, 38-43; asst pathologist, Mass Gen Hosp, 44-50; from asst prof to assoc prof path, Col Med, Univ Cincinnati, 50-56; pathologist in chief, Beth Israel Hosp, 56-79, from clin prof to Mallinckrodt prof, 56-84, EMER MALLINCKRODT PROF PATH, HARVARD MED SCH, 84-; EMER PATHOLOGIST IN CHIEF, BETH ISRAEL HOSP, 79- *Concurrent Pos:* Instr, Med Sch, Tufts Univ, 47 & 48 & Harvard Med Sch, 49 & 50; attend pathologist, Cincinnati Gen & Drake Mem Hosps, 52-56 & Vet Admin Hosps, Ohio & Ky, 54-56; consult, Vet Admin Hosp, Boston, 62-85, Cambridge Hosp, 68-85 & Children's Hosp Med Ctr, 77-90; lectr path, Simmons Col, 62-78; Kirstein fel med educ, Harvard Univ, 71-72; mem joint fac, Prog Health Sci & Technol, Harvard Univ-Mass Inst Technol, 75-79; prof anat & interim chmn dept, Univ Mass Med Sch, 85-87; spec asst to pres, Beth Israel Hosp, 79- *Mem:* Am Asn Pathologists; Am Soc Clin Path; Histochem Soc; Int Acad Path; Int Soc Thrombosis & Hemostasis; Sigma Xi. *Res:* Histochemistry; pulmonary disease; sarcoidosis; cardiovascular and thromboembolic disease; medical education. *Mailing Add:* Dept Path Beth Israel Hosp 330 Brookline Ave Boston MA 02215

FREIMAN, STEPHEN WEIL, b Alexandria, La, Jan 21, 42; m 69. CERAMICS. *Educ:* Ga Inst Technol, BChE, 63, MS, 66; Univ Fla, PhD(metall, mat eng), 68. *Prof Exp:* Assoc res scientist, IIT Res Inst, 68-71; mem staff, Ocean Technol Div, Naval Res Lab, 71-78; RES SCIENTIST, NAT BUR STANDARDS, 78- *Mem:* AAAS; Am Ceramic Soc; Sigma Xi; Am Soc Testing & Mat. *Res:* Structure and properties of glass ceramics; structure of glass; mechanical properties of glasses and ceramics; fracture mechanics of glass and ceramics; stress corrosion. *Mailing Add:* 7949 Inverness Ridge Rd Rockville MD 20854-4010

FREIMANIS, ATIS K, b Riga, Latvia, Mar 28, 25; US citizen; m 51; c 3. RADIOLOGY. *Educ:* Univ Hamburg, Dr Med, 51; Am Bd Radiol, cert, 58. *Prof Exp:* From instr to prof, Col Med, Ohio State Univ, 58-70; prof radiol & chmn dept, Med Col Ohio, 70-76; prof radiol & chmn dept, Col Med, Ohio State Univ, 76-82; DEPT RADIOL, MICH STATE UNIV, 83- *Concurrent Pos:* Consult, Juv Diag Ctr, 51-70, Brown Vet Admin Hosp, Dayton, 63-70 & 76- & Chillicothe Vet Admin Hosp, 64-70; radiation study sect, NIH, 76-80. *Mem:* Am Col Radiol; Radiol Soc NAm; Asn Univ Radiol; Am Roentgen Ray Soc; Am Inst Ultrasonics in Med; Soc Radiologists Ultrasound. *Res:* Ultrasonic diagnosis of abdominal diseases; teaching programs in radiology. *Mailing Add:* Radiol B220 Clin Ctr Mich State Univ E Lansing MI 48824

FREIMER, EARL HOWARD, b New York, NY, Nov 15, 26; m 48; c 4. MICROBIOLOGY, INFECTIOUS DISEASES. *Educ:* Univ Mich, AB, 48; State Univ NY, MD, 55; Univ Cambridge, MA, 78. *Prof Exp:* Intern, Columbia Med Serv, Bellevue Hosp, 55-56, asst resident, 56-57; res assoc, Rockefeller Inst & asst physician, 57-61; asst prof, Rockefeller Univ & assoc physician, 61-68; PROF INTERNAL MED, DIR DIV INFECTIOUS DIS & PROF MICROBIOL & CHMN DEPT, MED COL OHIO, 68- *Concurrent Pos:* Vis prof, State Univ NY Downstate Med Ctr & vis physician, Kings County Hosp, 67-68; vis prof, Col Med, Pa State Univ, 68-69, Dept Path, Univ Cambridge, Eng, 77-78, Dept Med, John Hopkins Univ, 89; consult, Toledo Hosp, Ohio, 68-, Mercy Hosp, 70- & St Luke's Hosp, 78-; mem, Adv Panel Infectious Dis, US Pharmacopial Conv, 80-; grants, NIH, Schering-Plough Corp, Lilly Res Labs, Merck & Co, 79-85, Squibb, Lederle, Smith Kline & French. *Mem:* AAAS; Am Soc Microbiol; Am Asn Immunol; fel Infectious Dis Soc Am; fel Am Col Cardiol; fel Royal Soc Med. *Res:* Biology of group A streptococcus; pathogenesis of rheumatic fever; bacterial L-forms and protoplasts; bacterial cell walls and membranes; immunochemistry of bacterial antigens; mechanisms of action of antimicrobial agents; immunocytochemistry of cardiac muscle; immunology of connective tissue diseases; biology of the pneumococcus. *Mailing Add:* Dept Microbiol Med Col Ohio C S 10008 Toledo OH 43699

FREIMER, MARSHALL LEONARD, b New York, NY, May 6, 32; m 61; c 2. MATHEMATICS. *Educ:* Harvard Univ, AB, 53, AM, 54, PhD(math), 60. *Prof Exp:* Mem staff, Lincoln Lab, Mass Inst Technol, 57-61; mem tech staff, Inst Naval Studies, 61-63; ASSOC PROF BUS ADMIN, GRAD SCH MGT, UNIV ROCHESTER, 63- *Concurrent Pos:* Ford Found fac fel, 65-66. *Mem:* Inst Math Statist; Math Asn Am. *Res:* Operations research, especially applied probability, and statistics. *Mailing Add:* William E Simon Grad Sch Bus Admin Univ Rochester Rochester NY 14627

FREIMUTH, HENRY CHARLES, b New York, NY, June 24, 12; wid; c 5. CHEMISTRY, FORENSIC SCIENCE. *Educ:* City Col New York, BS, 32; NY Univ, MS, 33, PhD(chem), 38. *Prof Exp:* Asst instr chem, Wash Sq Col, NY Univ, 37-38, asst therapeut, Sch Med, 38-39; from jr chemist to prin chemist & spec agent, Fed Bur Invest, US Dept Justice, Washington, DC, 39-44; toxicologist, Md State Post-Mortem Examr, 44-72; prof, 72-78, chmn, Dept Chem, 77-78, EMER PROF CHEM, LOYOLA COL, 78- *Concurrent Pos:* Instr, Loyola Col, Md, 46-56, prof lectr, 56-68, adj prof, 68-72; from asst prof to assoc prof legal med, Sch Med, Univ Md, 53-72, adj assoc prof pharmacol & toxicol, Sch Pharm, 70-; consult, Baltimore Poison Control Ctr, 56-; assoc div forensic path, Sch Hyg & Pub Health, Johns Hopkins Univ, 65-72. *Honors & Awards:* Med Chem Award, Med Sect, Am Chem Soc, 77; Andrew White Medal, Loyola Col, 78; Alexander Gettler Award, Toxicol Sect Am Acad Forensic Sci, 84. *Mem:* AAAS; Am Chem Soc; Am Acad Forensic Sci; Int Asn Forensic Toxicologists; Sigma Xi. *Res:* Carbon monoxide poisoning; alcoholic intoxication; detection of organic poisons in tissues; drowning tests; asphyxia; boric acid poisoning. *Mailing Add:* 30 D4 North Ridge Rd Cottage No 601 Ellicott City MD 21043

FREINKEL, NORBERT, internal medicine, medical services; deceased, see previous edition for last biography

FREINKEL, RUTH KIMMELSTIEL, b Hamburg, Ger, Dec 26, 26; US citizen; m 55; c 3. MEDICINE. *Educ:* Randolph Macon Col, AB, 48; Duke Univ, MD, 52; Am Bd Dermat, dipl, 61. *Prof Exp:* Res fel biochem, Harvard Med Sch, 54-55; res fel, Cambridge Univ, 55-56; res fel dermat, Harvard Med Sch, 58-60, instr, 60-61, assoc, 61-64, asst prof, 64-66; assoc prof, 66-72, PROF DERMAT, MED SCH, NORTHWESTERN UNIV, CHICAGO, 72- *Mem:* Am Soc Clin Invest; Soc Invest Dermat; Am Dermat Asn; Am Acad Dermat. *Res:* Intermediate and lipid metabolism of skin; pathogenesis of acne. *Mailing Add:* Dept of Dermat Northwestern Univ Med Sch Chicago IL 60611

FREIRE, ERNESTO I, b Lima, Peru, July 28, 49; m 75; c 1. BIOPHYSICS, BIOCHEMISTRY. *Educ:* Univ Cayetano Heredia, Peru, BS, 72, MS, 73; Univ Va, PhD(biophys), 77. *Prof Exp:* Res assoc biophys, Univ Va, 77-78, vis asst prof, 78-81; ASST PROF BIOCHEM, UNIV TENN, KNOXVILLE, 82- *Mem:* Biophys Soc; Am Chem Soc; AAAS. *Res:* Thermodynamics of conformational transitions in proteins, nucleic acids and model membrane systems; organization and function of biological membranes; mechanisms and energetics of protein insertion into membranes; computer modeling of biological structures; scanning calorimetry; isothermal calorimetry. *Mailing Add:* Dept Biochem Johns Hopkins Univ Baltimore MD 21218-2680

FREIREICH, EMIL J, b Chicago, Ill, Mar 16, 27; m 53; c 4. HEMATOLOGY, INTERNAL MEDICINE. *Educ:* Univ Ill, BS, 47, MD, 49; Am Bd Internal Med, dipl, 57. *Hon Degrees:* DSc, Univ Ill, 82. *Prof Exp:* Intern, Cook County Hosp, 49-51; resident internal med, Presby Hosp, Chicago, 51-53; asst med, Boston Univ, 53-55; sr investr, Nat Cancer Inst, 55-65, head leukemia serv, 64-65; asst head dept develop therapeut, 65-72, prof med & chief sect res hemat, Univ Tex M D Anderson Hosp & Tumor Inst, 65-85, head dept develop therapeut, Syst Cancer Ctr, 72-83, prof med, 75-80, chmn dept hemat, 83-85, RUTH HARRIET AINSWORTH PROF, UNIV TEX MED SCH, 80- DIR, ADULT LEUKEMIA RES PROG, 85- *Concurrent Pos:* Asst, Univ Ill, 51-53; res assoc, Evans Mem Hosp, 53-55. *Honors & Awards:* deVillier Award, Leukemia Soc Am, 79; Kettering Prize, General Motors, 83. *Mem:* Fel Am Col Physicians; Am Soc Hemat; Am Asn Cancer Res; Am Soc Clin Invest; Am Soc Clin Oncol; Asn Am Physicians. *Res:* Chemotherapy and natural history of human leukemia. *Mailing Add:* Dept Hemat Univ Tex M D Anderson Cancer Ctr 1515 Holcombe Blvd Houston TX 77030

FREIS, EDWARD DAVID, b Chicago, Ill, May 13, 12; m 34; c 3. MEDICINE. *Educ:* Univ Ariz, BS, 36; Columbia Univ, MD, 40. *Hon Degrees:* DrSc, Georgetown Univ, 85. *Prof Exp:* Intern med, Mass Mem Hosp, Boston, 40-41; sr intern & house physician, Boston City Hosp, 41-42; asst resident, Evans Mem Hosp, 46-47, res fel cardiovasc dis, 47-49; asst chief med, Vet Admin Hosp, 49-54, chief, 54-59; adj clin prof, 49-57, assoc prof, 57-63, PROF MED, SCH MED, GEORGETOWN UNIV, 63-, CHIEF CARDIOVASC RES LAB, UNIV HOSP, 49-; SR MED INVESTR, VET ADMIN HOSP, 59- *Concurrent Pos:* Instr, Sch Med, Boston Univ, 47-49. *Honors & Awards:* Lasker Award, 71; Ciba Award, 81. *Mem:* Am Soc Clin Invest. *Res:* Clinical evaluation and hemodynamic analysis of hypertensive drugs; blood and fluid volume changes in disease; cardiovascular physiology in man. *Mailing Add:* Vet Admin Med Ctr 50 Irving St NW Washington DC 20422

FREIS, ROBERT P, b San Francisco, Calif, Oct 6, 31; m 55; c 2. COMPUTATIONAL ATMOSPHERIC PHYSICS. *Educ:* Univ Calif, Los Angeles, BS, 59; Univ Calif, Berkeley, MS, 64. *Prof Exp:* PHYSICIST, LAWRENCE LIVERMORE NAT LAB, 59- *Res:* Computational plasma physics problems related to the magnetic fusion energy program with emphasis on the development and application of numerical simulation techniques; atmospheric physics and computational problems and atmospheric science relating to the accidental release of toxic and nuclear material. *Mailing Add:* Lawrence Livermore Nat Lab L-262 PO Box 808 Livermore CA 94550

FREISE, EARL J, b Chicago, Ill, Dec 30, 35; m 58; c 4. MATERIALS SCIENCE, METALLURGY. *Educ:* Ill Inst Technol, BS, 58; Northwestern Univ, MS, 59; Cambridge Univ, PhD(metall), 62. *Prof Exp:* From asst prof to assoc prof mat sci, Northwestern Univ, 62-77; adj prof mech eng & dir, Off Res & Prog Develop, Univ NDak, 77-81; asst vchancellor res & prof mech eng, Univ Nebr, Lincoln, 82-87; DIR, OFF SPONSORED RES & INSTR, MAT SCI, CALIF INST TECHNOL, 88- *Mem:* Am Soc Metals; Am Inst Mining, Metall & Petrol Engrs; Am Soc Eng Educ; Asn Univ Technol Managers (secy-treas, 77-79); Nat Coun Univ Res Adminr, (pres 84-85). *Res:* Phase transformations in materials; physical and mechanical properties of high temperature metallic and nonmetallic materials; quantitative x-ray diffraction analysis in crystalline solids. *Mailing Add:* Caltech MC 213-6 1201 E California Blvd Pasadena CA 91125

FREISER, HENRY, b New York, NY, Aug 27, 20; m 42; c 3. ANALYTICAL CHEMISTRY. *Educ:* City Col New York, BS, 41; Duke Univ, MA, 42, PhD(phys chem), 44. *Prof Exp:* Prof anal & phys chem & chmn dept, NDak State Col, 44-45; res fel, Mellon Inst, 45-46; instr anal chem, Univ Pittsburgh, 46-50, from asst prof to assoc prof, 50-58; head dept, 58-67, PROF CHEM, UNIV ARIZ, 58- *Concurrent Pos:* Mem, Comn Equilibrium Data, Anal Div, Int Union Pure & Appl Chem; O M Smith Lectr, 68. *Mem:* Am Chem Soc; Am Soc Testing & Mat; fel The Chem Soc. *Res:* Analytical separations processes and solvent extraction; trace analysis; metal chelates. *Mailing Add:* Dept Chem Univ Ariz Tucson AZ 85721

FREISER, MARVIN JOSEPH, b Brooklyn, NY, Feb 9, 26; m 49; c 1. THEORETICAL PHYSICS. *Educ:* Brooklyn Col, BS, 48; Purdue Univ, MS, 51, PhD(physics), 55. *Prof Exp:* Asst prof physics, Worcester Polytech Inst, 55-56; physicist, Midwestern Univs Res Asn, 56-57; assoc physicist, Res Lab, 57-64, RES PHYSICIST, WATSON RES CTR, IBM CORP, 64- *Mem:* Am Phys Soc. *Res:* Solid state physics; theory of magnetism; statistical mechanics; liquid crystals. *Mailing Add:* 1032 Houston Mill Rd Altanta GA 30329

FREISHEIM, JAMES HAROLD, b Tacoma, Wash, July 19, 37; m 58; c 2. BIOCHEMISTRY. *Educ:* Pac Lutheran Univ, BA, 60; Univ Wash, PhD(biochem), 66. *Prof Exp:* Res assoc biochem, Scripps Clin & Res Found, 66-69; from asst prof to prof biochem, Col Med, Univ Cincinnati, 69-85; PROF BIOCHEM & CHMN DEPT, MED COL OHIO, TOLEDO, 85- *Concurrent Pos:* NIH trainee, 66-67; Am Cancer Soc fel, 67-68. *Mem:* AAAS; Am Soc Biol Chemists; NY Acad Sci; Am Chem Soc; Am Asn Cancer Res. *Res:* Relationship of protein structure to function; molecular properties and mechanism of action of folate-dependent enzymes; mechanism of drug resistance to folate antagonists. *Mailing Add:* Dept Biochem Med Col Ohio C S 10008 Toledo OH 43699

FREITAG, DEAN R(ICHARD), b Ft Dodge, Iowa, Oct 1, 26; m 52; c 4. CIVIL ENGINEERING, SOIL MECHANICS. *Educ:* Iowa State Univ, BS, 49; Harvard Univ, MS, 51; Auburn Univ, PhD,65. *Prof Exp:* Civil engr, Engr Waterways Exp Sta, US Army, 51-55, supv res civil engr, 55-70, asst tech dir, 70-72, tech dir, Cold Regions Res & Eng Lab, 72-81; prof civil eng, Tenn Technol Univ, 81-88; CONSULT, 88- *Concurrent Pos:* Engr, Mat & Res Lab, Calif Div Hwys, 53-54. *Honors & Awards:* US Army Meritorious Civilian Serv Award 70 & 81. *Mem:* Int Soc Terrain-Vehicle Systs; Am Soc Civil Engrs; Am Soc Agr Engrs. *Res:* Mechanics of soil-vehicle systems; soil cutting and excavation; roads and airfields; cold regions engineering. *Mailing Add:* 312 Parragon Rd Cookeville TN 38501

FREITAG, HARLOW, b New York, NY, Apr 17, 36; m; c 1. COMPUTER SCIENCE. *Educ:* NY Univ, AB, 55; Yale Univ, MS, 57, PhD(chem), 59. *Prof Exp:* Assoc, Data Processing Div, Int Bus Mach Corp, 58-59, mgr chem, Advan Systs Develop Div, 59-61, mem staff, Res Div, 61-63, mgr design automation, Comput Systs Dept, Thomas J Watson Res Ctr, 63-67, mgr subsysts & integration, 67-77, sr mgr tech planning, Res Div, 77-80, mem info systs & technol group staff, 80-83; DEP DIR, SUPERCOMPUT RES CTR, 86- *Honors & Awards:* Centennial Medal, Inst Elec & Electronic Engrs. *Mem:* fel Inst Elec & Electronics Engrs. *Res:* Digital computers. *Mailing Add:* Supercomput Res Ctr 17100 Science Dr Bowie MD 20715-4300

FREITAG, JULIA LOUISE, b Allentown, Pa, Nov 29, 27. PREVENTIVE MEDICINE, EPIDEMIOLOGY. *Educ:* Cornell Univ, AB, 49, MD, 53; Harvard Univ, MPH, 57. *Prof Exp:* Epidemiologist, 57-58, asst dir, Off Epidemiol, 58-66, dir, 66-70, dir, Off Med Manpower, 70-75, asst comnr, Health Manpower Group, NY State Dept Health, 75-87, CONSULT, 87- *Concurrent Pos:* Lectr, Rensselaer Polytech Inst, 62-70 & Albany Med Col, Uniion Univ, NY, 70-; mem, NY State Adv Coun Voc Educ, 71; mem, Epidemiol & Biomet Adv Comt, Nat Heart & Lung Inst, 72-74. *Res:* Communicable diseases; population studies and outbreak investigation; genetics; inheritance of blood groups. *Mailing Add:* Unionville Rd Feura Bush Tevra Bush NY 12067

FREITAG, ROBERT FREDERICK, b Jackson, Mich, Jan 20, 20; m 41; c 4. AERONAUTICAL & ASTRONAUTICAL ENGINEERING. *Educ:* Univ Mich, BSE, 41. *Prof Exp:* Aerodyn officer, bur aeronaut, Navy Dept, DC, 42-48; intel officer develop tech intel plans, Off Naval Attache, Eng, 48-49; chief tech systs labs, US Air Force Missile Test Ctr, Cape Canaveral, 49-51; prog plans officer, guided missiles div, bur aeronaut, Navy Dept, 51-53, dir surface launched missiles br, 53-55, dir ballistic missile br, 55, prog officer, off chief naval opers, 55-57, dir plans & requirements, US Navy Pac Missile Range, Calif, 57-59; astronaut officer, bur naval weapons, 59-63; dir launch vehicles & propulsion, NASA, 63, manned space flight ctr develop, 63-70, spec asst to assoc adminr manned space flight, 70-72, dep dir advan progs, Off Manned Space Flight, 73-82, dep dir Space Sta Taskforce, 82-84, dir, Policy & Plans Div, Space Sta Prog Off, Hq, 84-86; AEROSPACE CONSULT, 86- *Concurrent Pos:* Mem subcomt propellers, Nat Adv Comt Aeronaut, 44-46 & spec comt space technol, 58-59; mem, Joint Army-Navy Ballistic Missile Comt, 55-57; mem spec comt adequacy range facilities, Secy Defense, 56-58; mem res adv comt missile & spacecraft aerodyn, NASA, 60-63; mem launch vehicles panel, aeroanut & astronaut coord bd, Joint Defense Dept-NASA,

60-64. *Honors & Awards:* Bronze Medal, Brit Interplanetary Soc, 79; Allan D Emil Mem Medal, Int Acad Astronaut, 85; White House Presidential Meritorious Rank, 87; Int Coop Medal, Am Inst Aeronaut & Astronaut, 90. *Mem:* Fel Am Inst Aeronaut & Astronaut; fel Royal Aeronaut Soc; fel Am Astronaut Soc; mem Int Acad Astronaut; hon mem German Aeronaut & Space Soc; fel Brit Interplanetary Soc. *Res:* Aeronautics and astronautics, especially in aerodynamics, guided missile guidance, rocket propulsion, range testing and instrumentation; spacecraft and space vehicle design. *Mailing Add:* 4110 Mason Ridge Dr Annandale VA 22003-2034

FREIWALD, RONALD CHARLES, b Pittsburgh, Pa, July 21, 43. TOPOLOGY. *Educ:* Washington & Jefferson Col, AB, 65; Univ Rochester, PhD(math), 70. *Prof Exp:* ASSOC PROF MATH, WASHINGTON UNIV, 70- *Mem:* Am Math Soc; Math Asn Am. *Res:* Mapping properties of absolute Borel sets. *Mailing Add:* Dept Math Washington Univ St Louis MO 63130

FRELINGER, ANDREW L, b San Diego, Calif, Jan 15, 53. IMMUNOLOGY. *Educ:* San Diego State Univ, BS, 75; Case Western Reserve Univ, PhD(biol), 84. *Prof Exp:* Postdoctoral fel, Dept Develop Genetics & Anat, Case Western Reserve Univ, 84-86; postdoctoral fel, Dept Immunol, Scripps Clin, 86-88, sr res assoc, Comt Vascular Biol, 88-90; RES STAFF SCIENTIST, DEPT PROTEIN CHEM, BIOGEN INC, 90- *Concurrent Pos:* Res fel award, Am Heart Asn, 87-89; first award, NIH, 90. *Mem:* Am Soc Cell Biol; Am Heart Asn. *Mailing Add:* Biogen Inc 14 Cambridge Ctr Cambridge MA 02142

FRELINGER, JEFFREY, b Brooklyn, NY, July 16, 48; m 70; c 2. IMMUNOGENETICS, MOLECULAR IMMUNOLOGY. *Educ:* Univ Calif, San Diego, BA, 69; Calif Inst Technol, PhD(immunol), 73. *Prof Exp:* Jane Coffin Childs fel immunogenetics, dept human genetics, Univ Mich, 73-75; from asst prof to prof immunogenetics, dept microbiol, Sch Med, Univ Southern Calif, 75-83; PROF, DEPT MICROBIOL & IMMUNOL & LEADER IMMUNOL PROG, LINEBERGER CANCER CTR, UNIV NC, CHAPEL HILL, 83- *Concurrent Pos:* Fac res award, Am Cancer Soc, 78-; mem, Mammalian Genetics Study Sect, NIH, 79-83 & Genetic Basis Dis Comt, 87-91; vis worker, Radiobiol Unit, Med Res Coun Howell, 80; vis prof, Tokyo Univ, 86; vis fel, Trinity Col, Oxford Univ, 89-90; Foggerty sr fel, NIH, 89-90. *Mem:* Sigma Xi; AAAS; Am Asn Immunologists. *Res:* Molecular genetics and function of mouse and human major histocompatibility gene products in the development and regulation of immune responses. *Mailing Add:* CB No 7290 FLOB Univ NC Chapel Hill Chapel Hill NC 27599-7290

FREMLING, CALVIN R, b Brainerd, Minn, Nov 13, 29; m 54; c 1. LIMNOLOGY, ENTOMOLOGY. *Educ:* St Cloud State Col, BS, 51, MS, 55; Iowa State Univ, PhD(zool), 59. *Prof Exp:* Sci teacher high sch, Minn, 51-52; ecologist, Univ Utah, Dugway Proving Ground, 52-54; instr zool & bot, Eveleth Jr Col, 55-56; PROF BIOL, WINONA STATE COL, 59- *Concurrent Pos:* Minn Acad Sci vis scientist, High Schs, 50-; NSF res grant, 61-64 & 69-70; consult, USPHS, 62-, Int Joint Comn, 62, Metrop Structures Can, Montreal, 66-69 & Nasco Inc, Wis, 67-; Am Inst Biol Sci vis scientist, Cols, 63-71; Fed Water Pollution Control Admin grant, 66-70. *Mem:* AAAS; Entom Soc Am; Am Soc Limnol & Oceanog; Nat Asn Biol Teachers; Wildlife Soc; Sigma Xi. *Res:* Ecology of the Mississippi River; biology of Hexagenia mayflies and hydropsychid caddisflies; water pollution and floods. *Mailing Add:* Pasteur Hall Winona State Univ Winona MN 55987

FREMMING, BENJAMIN DEWITT, b Minneapolis, Minn, Oct 27, 24; m 55; c 2. VETERINARY PHYSIOLOGY. *Educ:* Colo State Univ, DVM, 46; Univ Calif, Berkeley, MPH, 52; Am Bd Vet Pub Health, dipl, 54; Am Col Lab Animal Med, dipl, 56. *Prof Exp:* Scientist virol-serol, Walter Reed Res & Grad Ctr, 52-53; chief radiobiol vet res group, Radiobiol Lab, Univ Tex, 53-56; proj mgr physiol life sci sect, Westinghouse Res Develop Ctr, 61-64; scientist adminr, Nat Heart Inst, 65-67; prof path & dir animal care ctr, Univ Tex Med Sch, 67-71, chmn dept lab animal med 68-74, prof lab animal med & anesthesiol, 71-77; prof pharmacol, Schs Med & Pharm & dir, Lab Animal Ctr, 77-86, DIR, CLOUDCROSS RES LAB, 86-, EMER PROF, UNIV MO-KANSAS CITY, 86-; CLIN PROF, OBSTET & GYNEC, UNIV TEX HEALTH SCI CTR, SAN ANTONIO, 87- *Concurrent Pos:* Mem, Inst Lab Animal Resources, Nat Res Coun, 53-58, chmn coun nonhuman primates, 53-58; pres, Am Col Lab Animal Med, 59-62; consult, Off Inspector Gen, US Dept State, 62-64. *Mem:* fel Am Col Pharm & Therapeut; Am Soc Pharm & Exp Therapeut; Sigma Xi; Am Vet Med Asn; Am Asn Lab Animal Sci. *Res:* Radiobiology; epidemiology; experimental surgery; laboratory animal medicine; reproductive physiology research centered around migration of ova through the fallopian tube, capacation and ova transplantation. *Mailing Add:* Cloudcross Res Lab Box 429 Kendela TX 78027-0429

FREMONT, HERBERT IRWIN, mathematics, for more information see previous edition

FREMOUNT, HENRY NEIL, b Easton, Pa, Sept 29, 33; m 57; c 2. BIOLOGY, MEDICAL PARASITOLOGY. *Educ:* East Stroudsburg State Col, BS, 56, MEd, 64; Columbia Univ, MS, 66, DrPH(parasitic dis), 70. *Prof Exp:* Teacher & co-chmn dept sci, Delaware Valley Joint Sch Syst, 56-58; teacher high sch, Belvidere, NJ, 58-65; assoc prof, 66-71, chmn dept, 74-80, PROF BIOL, EAST STROUDSBURG STATE COL, 71- *Concurrent Pos:* Instr, Sch Pub Health, Columbia Univ, 70-75; fel trop med, Sch Med, La State Univ, 70; vis scientist, Gorgas Mem Lab, Panama, 70; Nat Inst Allergy & Infectious Dis res grant, 71-74; adj assoc prof, Inst Pathobiol & Inst Health Sci, Lehigh Univ, 72-76; consult parasitic dis, Sacred Heart Hosp, Allentown, Pa, 74- *Mem:* AAAS; Royal Soc Trop Med & Hyg; Am Soc Trop Med & Hyg; Am Soc Parasitol; Am Soc Clin Path; NY Acad Sci. *Res:* Biology and pathophysiology of malaria; ultrastructural changes in the host red cell as induced by the parasite; ultrastructure. *Mailing Add:* Dept Biol East Stroudsburg Univ East Stroudsburg PA 18301

FREMOUW, EDWARD JOSEPH, b Northfield, Minn, Feb 23, 34; m 60; c 2. AERONOMY, RADIOPHYSICS. *Educ:* Stanford Univ, BS, 57; Univ Alaska, MS, 63, PhD(geophys), 66. *Prof Exp:* Engr, Boeing Co, 57-58; auroral physicist & auroral discipline chief for US Antarctic Res Prog in Antarctica, Arctic Inst NAm, 58-59; engr, Boeing Co, 60; asst elec eng, Univ Alaska, 60-61, asst aeronomy, Geophys Inst, 61-63, res assoc, 63-66, asst prof geophys, Univ, 66-67; physicist, Radio Physics Lab, SRI, Int, 67-70, sr physicist, 70-76, prog mgr, 76-77; staff scientist & vpres, Phys Dynamics Inc, 77-86; SR RES SCIENTIST & PRES, NORTHWEST RES ASSOC INC, 86- *Concurrent Pos:* Consult, Geophys Inst, Univ Alaska, 67-69. *Honors & Awards:* Excellence in Refereeing Award, 84 & 89; Fremouw Peak in Antartica named in honor. *Mem:* Am Geophys Union; Inst Elec & Electronic Engrs; Int Radio Sci Union. *Res:* Ionospheric and auroral physics; satellite communication, surveillance radars, and navigation; radiowave scattering; ionospheric tomography. *Mailing Add:* Northwest Res Assoc Inc PO Box 3027 Bellevue WA 98009

FRENCH, ALAN RAYMOND, b Los Angeles, Calif, Dec 17, 46; m 68; c 2. PHYSIOLOGICAL ECOLOGY, CHRONOBIOLOGY. *Educ:* Univ Calif, Berkeley, BA, 68, Los Angeles, PhD(biol), 75. *Prof Exp:* Actg asst prof biol, Univ Calif, Los Angeles, 75-76; res scientist, Dept Biol Sci, Stanford Univ, 76-77; asst prof biol, Univ Calif, Riverside, 77-84; ASST PROF BIOL, STATE UNIV NY, BINGHAMTON, 84- *Mem:* Ecol Soc Am; Am Soc Mammalogists; Int Hibernation Soc. *Res:* Vertebrate physiological ecology; timing and energetics of mammalian hibernation; environmental control of reproduction and dormancy; physiology of circadian and circannual rhythms. *Mailing Add:* Dept Biol State Univ NY Binghamton Binghamton NY 13901

FRENCH, ALEXANDER MURDOCH, plant pathology; deceased, see previous edition for last biography

FRENCH, ALFRED DEXTER, b Boston, Mass, June 27, 43; div. PHYSICAL CHEMISTRY. *Educ:* Iowa State Univ, BS, 65; Ariz State Univ, PhD(phys chem), 71. *Prof Exp:* Res chemist, Northern Lab, USDA, 65-66, Nat Res Coun fel, Southern Lab, 71-73, RES CHEMIST, SOUTHERN LAB, USDA, 73- *Honors & Awards:* ARS Fel. *Mem:* Am Chem Soc; Am Crystallog Asn. *Res:* Structure of polymers, especially polysaccharides. *Mailing Add:* 201 Central Ave Jefferson LA 70121

FRENCH, ALLEN LEE, b East Grand Rapids, Mich, Jan 19, 39; m 64; c 8. ENTOMOLOGY, ANIMAL PARASITOLOGY. *Educ:* Mich State Univ, BS, 62, MS, 64, PhD(entom), 68. *Prof Exp:* Asst entom, Mich State Univ, 62-68; head entom & parasitol res, 68-75, COORDR BIO-HEALTH RES & REGULATORY COMPLIANCE, MOORMAN MFG CO, 75- *Mem:* Entom Soc Am; Sigma Xi. *Res:* Evaluation of control procedures for pests and parasites of livestock. *Mailing Add:* Res Dept Moorman Mfg Co Quincy IL 62301

FRENCH, BEVAN MEREDITH, b East Orange, NJ, Mar 8, 37; m 67; c 3. GEOCHEMISTRY, ASTROGEOLOGY. *Educ:* Dartmouth Col, AB, 58; Calif Inst Technol, MS, 60; Johns Hopkins Univ, PhD(geol), 64. *Prof Exp:* Nat Acad Sci-Nat Res Coun resident res assoc geochem, Goddard Space Flight Ctr, NASA, 64-65, aerospace technologist, 65-72; prog dir geochem, NSF, 72-75; discipline scientist planetary mat, 75-84, advan progs scientist, 84-89, DISCIPLINE SCIENTIST, SPEC PROJS, NASA, 89- *Concurrent Pos:* Vis prof, Dartmouth Col, 68; co-investr, Apollo XI, XII & XIV Lunar Samples; vis res scientist, Univ Pretoria, SAfrica, 81-82. *Mem:* Meteoritical Soc. *Res:* Shock metamorphism of natural materials; geology of terrestrial meteorite impact craters; mineralogy and shock metamorphism of lunar samples; equilibrium relations in natural and artificial solid-gas systems; experimental synthesis and stability studies of carbonate minerals; sedimentary and metamorphosed iron formations. *Mailing Add:* 7408 Wyndale Lane Chevy Chase MD 20815

FRENCH, CHARLES STACY, b Lowell, Mass, Dec 13, 07; m 38; c 2. PLANT PHYSIOLOGY. *Educ:* Harvard Univ, SB, 30, MA, 32, PhD(biol), 34. *Hon Degrees:* PhD, Univ Goteborg, Sweden, 74. *Prof Exp:* Asst gen physiol, Radcliffe Col, 30-31 & Harvard Univ, 31-33; res fel biol, Calif Inst Technol, 34-35; guest worker, Kaiser Wilhelm Inst, 35-36; Austin teaching fel biochem, Harvard Med Sch, 36-38; instr chem, Univ Chicago, 38-41; from asst prof to assoc prof bot, Univ Minn, 41-47; dir dept plant biol, 47-73, DIR EMER, CARNEGIE INST, 73-, PROF BIOL BY COURTESY, 64- *Honors & Awards:* Award of Merit, Bot Soc Am, 73. *Mem:* Nat Acad Sci; AAAS; Bot Soc Am; Am Soc Biol Chem; Am Acad Arts & Sci; Am Soc Plant Physiologists. *Res:* Cellular respiration; photosynthesis of purple bacteria, leaves and algae; characteristics, spectroscopy and functions of plant pigments; construction of spectroscopic equipment. *Mailing Add:* Dept Plant Biol Carnegie Inst Stanford CA 94305

FRENCH, DAVID MILTON, b Alexandria, Va, July 11, 14; c 3. POLYMER CHEMISTRY. *Educ:* Univ Va, BS, 36, PhD(chem), 40. *Prof Exp:* Chemist, US Rubber Co, 40-46; res engr, Eng Exp Sta, Univ Fla, 46-48, Nat Bur Stand, 49-50, Acme Backing Corp, 53-56 & Wyandotte Chem Corp, 56-59; br head, US Naval Ord Sta, 62-74, group leader, Naval Surface Weapons Ctr, 74-78; CONSULT, 78- *Mem:* Am Chem Soc; AAAS. *Res:* Characterization of prepolymers; physical chemistry of polymers; degradation of crosslinked polymers; polymer network topology; emulsion polymerization; methods and materials of the coatings, adhesives, and solid propellant industries. *Mailing Add:* 703 S Fairfax St Alexandria VA 22314

FRENCH, DAVID N(ICHOLS), b Newton, Mass, Jan 24, 36; m 60; c 4. PHYSICAL METALLURGY, MATERIALS SCIENCE. *Educ:* Mass Inst Technol, BS, 58, MS, 59, ScD(metall), 62. *Prof Exp:* Res metallurgist, Linde Div, Union Carbide Corp, 62-63; mem tech staff, Ingersoll-Rand Res Ctr, 63-68; phys metallurgist, Abex Corp Res Ctr, Mahwah, 68-73; dir corp qual assurance, Riley Stoker Corp, 73-82; vpres qual assurance, Leighton Indust, 82-83; staff metallurgist, D G Peterson & Assocs, 83-84; PRES, D N FRENCH, INC, METALLURGISTS, 84- *Mem:* Am Soc Metals; Am Inst

Mining, Metall & Petrol Engrs; Nat Asn Corrosion Engrs. *Res:* Mechanical properties and solidification of sea ice; heat flow and transfer; mechanical behavior, x-ray stress analysis and manufacture of composites; ash corrosion; materials for high temperature environments. *Mailing Add:* One Lancaster Rd Northborough MA 01532

FRENCH, DAVID W, b Mason City, Iowa, Nov 10, 21; m 44; c 3. FOREST PATHOLOGY. *Educ:* Univ Minn, BS, 43, MS, 49, PhD, 52. *Prof Exp:* From instr to assoc prof, 50-63, head dept, 79-85, PROF PLANT PATH, UNIV MINN, ST PAUL, 63- *Mem:* Am Phytopath Soc; Forest Prod Res Soc; Int Soc Arboriculture; Mycol Soc Am; Am Forestry Asn. *Res:* Products pathology; mycology. *Mailing Add:* 3569 Baltic Ave St Paul MN 55122

FRENCH, EDWARD P(ERRY), b Boise, Idaho, Aug 9, 24; m 49; c 2. SOLAR ENGINEERING, SOLAR POWER. *Educ:* Stanford Univ, BS, 48, MS, 50, PhD(metall), 53. *Prof Exp:* Group supvr ramjet engines, Marquardt Aircraft Co, 52-58; mem tech staff, Rockwell Int, Seal Beach, Calif, 58-85; CONSULT, 85- *Concurrent Pos:* Instr, eng exten, Univ Calif, Los Angeles, 57 & 62, lectr, col eng, 65-66. *Mem:* Am Inst Aeronaut & Astronaut; Sigma Xi; Int Solar Energy Soc. *Res:* Aircraft and space propulsion; aerothermochemistry; thermodynamics; solar engineering. *Mailing Add:* 208 Northrop Place Santa Cruz CA 95060

FRENCH, ERNEST W(EBSTER), b Osnabrock, NDak, July 19, 29; m 54; c 2. AGRICULTURAL ENGINEERING. *Educ:* NDak State Univ, BS, 51, MS, 56. *Prof Exp:* Salesman, Standard Oil Co, 54-55; asst agr engr, 57-59, asst prof agr eng, 59-60, SUPT, WILLISTON BR EXP STA, N DAK STATE UNIV, 60- *Mem:* Am Soc Agr Engrs; Soil & Water Conserv Soc. *Res:* Soil and water conservation research and application. *Mailing Add:* Williston Res Ctr NDak State Univ Box 1445 Williston ND 58802-1445

FRENCH, FRANCIS WILLIAM, b Brooklyn, NY, Dec 28, 27; m 67; c 5. AEROSPACE ENGINEERING, APPLIED MECHANICS. *Educ:* Polytech Inst Brooklyn, BS, 51, MS, 56, PhD(aeronaut eng), 59. *Prof Exp:* Struct draftsman, Grumman Aircraft, 51-52; stress analyst, Aerophys Lab, NAm Aviation, Inc, 52-53; res asst aerospace eng & appl mech, Polytech Inst Brooklyn, 53-55, res assoc, 55-59; sr engr, Technik, Inc, 59-60; sr staff mem, Allied Res Assocs, 60-61, chief appl mech group, 61-62; mem tech staff, Space Systs Dept, Mitre Corp, Mass, 62-65; sr consult engr, Sci Satellite Proj Off, Avco Corp, 65-67, prin res engr, Avco Everett Res Lab, 67-70, sr consult scientist, Avco Systs Div, 70-71, prin res engr, Avco Everett Res Lab, 71-73; PRIN ENGR, W J SCHAFER ASSOCS, 73- *Concurrent Pos:* Assoc ed, J Spacecraft & Rockets. *Mem:* Sr mem Am Astronaut Soc; assoc fel Am Inst Aeronauts & Astronauts. *Res:* Laser systems; manned and unmanned spacecraft; structural mechanics. *Mailing Add:* VP Govt & Cust Rel Space Transp Syst United Technologies 1825 L St NW Suite 700 Washington DC 20006

FRENCH, FRANK ELWOOD, JR, b Lubbock, Tex, Feb 20, 35; m 58; c 2. ENTOMOLOGY, PARASITOLOGY. *Educ:* Tex Tech Col, BS, 57; Iowa State Univ, MS, 58, PhD(entom), 62. *Prof Exp:* Res assoc med entom, Queen's Univ, Ont, 66-68; from asst prof to assoc prof, 68-88, PROF BIOL, GA SOUTHERN UNIV, 88- *Mem:* Entom Soc Am. *Res:* Insecta and Acarina of medical and veterinary importance, ectoparasites of mammals; horse flies (Tabanidae) distribution and associated spiroplasms, population dynamics; Demodex hair follicle mites; life history, host aquisition, host's pathologic response. *Mailing Add:* Dept Biol Ga Southern Univ Statesboro GA 30460-8042

FRENCH, GORDON NICHOLS, b Washington, DC, Mar 14, 19; m 42; c 4. INTERNAL MEDICINE. *Educ:* Yale Univ, AB, 41; Tufts Univ, MD, 44. *Prof Exp:* Intern path, Boston City Hosp, 44-45; resident med, Vet Admin Hosp, Boston, 47-48; Bellevue Hosp, New York, 48-49; physician, Green Mountain Clin, Vt, 49-53; instr med, Sch Med, Tufts Univ, 54-59; asst physician, Harvard Univ, 55-59, res assoc, Sch Pub Health, 57-59; dir med, Misericordia Hosp, 59-67; asst prof clin med & assoc dean, Sch Med, Univ Pa, 67-72; DIR MED, NEW BRUNSWICK AFFIL HOSPS, 72- *Concurrent Pos:* Fel pharmacol, Harvard Univ, 46-47; consult, Lemmuel Shattuck Hosp, Boston, 55-59. *Mem:* Fel Am Col Physicians. *Res:* Cardiovascular physiology; ventricular pressure; volume relationships and oxygen consumption in fibrillation and arrest. *Mailing Add:* Med Col Penn PO Box 150 Unionville PA 19375

FRENCH, J(OHN) BARRY, b Toronto, Ont, Aug 22, 31; m 51; c 4. AEROSPACE ENGINEERING. *Educ:* Univ Toronto, BSc, 55, PhD(low density plasmas), 61; Univ Birmingham, MSc, 57. *Prof Exp:* Lectr aerospace eng sci, Univ Toronto, 61-62, from asst prof to assoc prof, 62-68, prof aerospace eng sci, 68-76, asst dean grad studies, 83-86; CORP SCI CONSULT, SCIEX INC, 76- *Concurrent Pos:* Mem bd, Univ Toronto Innovations Found, 82-87, Sci Coun Can, 87- & Can Genetic Dis Network, 90- *Honors & Awards:* Barringer Award, Spectros Soc Can; Bell Can-Forum Award, 90. *Mem:* Fel Can Aeronaut & Space Inst; fel Royal Soc Arts; fel Royal Soc Can. *Res:* Development of instrumentation for trace gas analysis; application studies in environmental, medical, military and other areas; quadrupole mass spectroscopy; atmospheric pressure chemical ionization; ion mobility applications; molecular beams; space simulation. *Mailing Add:* Inst Aerospace Studies 4925 Dufferin St Downsview ON M3H 5T6 Can

FRENCH, JAMES C, b Detroit, Mich, Apr 25, 30; m 56; c 1. ORGANIC CHEMISTRY. *Educ:* Wayne State Univ, BS, 51, PhD(chem), 54. *Prof Exp:* USPHS fel, Harvard Univ, 54-56; RES CHEMIST, PARKE, DAVIS & CO, 56- *Mem:* Am Chem Soc; AAAS; Am Soc Pharmacog; Japan Antibiotics Res Asn. *Res:* Isolation and characterization of antibiotics; peptide synthesis. *Mailing Add:* 3150 Rumsey Dr Ann Arbor MI 48105-3291

FRENCH, JAMES EDWIN, b Chicago, Ill, Jan 11, 42. CERAMIC COMPOSITES, MAGNETIC MATERIALS. *Educ:* Knox Col, BA, 64; Stanford Univ, PhD(chem), 68. *Prof Exp:* NIH fel inorg chem, Mass Inst Technol, 68-69; res chemist, sr res chemist & res scientist, Hercules, Inc, 69-89; PRIN TECH SPECIALIST, MCDONNELL DOUGLAS, 89- *Mem:* Am Chem Soc; Am Ceramic Soc. *Res:* Transition metal research, electrochemistry; high temperature polymers; magnetic materials; ceramic composites; current research centers on ceramic composites, coatings, magnetic polymeric and tailored inorganic materials for advanced aircraft applications. *Mailing Add:* 1719 Big Horn Basin Dr Ballwin MO 63011-4821

FRENCH, JEAN GILVEY, infectious disease, population genetics, for more information see previous edition

FRENCH, JEPTHA VICTOR, b Cripple Creek, Colo, Dec 16, 36; m 62; c 2. ENTOMOLOGY, VIROLOGY. *Educ:* Colo State Univ, BSc, 60, MSc, 62; Mich State Univ, PhD(entom), 73. *Prof Exp:* Res assoc, Univ Calif, Riverside, 62-69; fel entom, Mich State Univ, East Lansing, 69-73; asst prof, 73-77, assoc prof, 77, PROF ENTOM, CITRUS CTR, TEX A&I UNIV, 77- *Concurrent Pos:* Fel, Mich State Univ, 69-73; consult, S Tex Citrus Growers, 73- *Mem:* Entom Soc Am; Am Registry Prof Entomologists; Int Org Citrus Virologists; Int Soc Citricult. *Res:* Insect transmission of plant pathogenic viruses; biological and chemical control of arthropods attacking citrus. *Mailing Add:* Dept Entom Tex A&I Univ Citrus Ctr Weslaco TX 78501

FRENCH, JOHN DONALD, b New Orleans, La, Feb 19, 23; m 48; c 4. PHYSICS. *Educ:* La State Univ, BS, 48, MS, 52, PhD(physics), 58. *Prof Exp:* Physicist, US Geol Surv, 53-55; asst prof, 58-63, ASSOC PROF PHYSICS, AUBURN UNIV, 63- *Mem:* Am Phys Soc; Sigma Xi. *Res:* Nuclear physics. *Mailing Add:* 389 Clubhouse Dr Pl Lakewood Villas Gulf Shores AL 36542

FRENCH, JOHN DOUGLAS, neurosurgery; deceased, see previous edition for last biography

FRENCH, JOSEPH H, b Toledo, Ohio, July 4, 28; c 4. PEDIATRICS, NEUROLOGY. *Educ:* Ohio State Univ, BA, 50, MD, 54; Am Bd Pediat, dipl, 60; Am Bd Psychiat & Neurol, dipl neurol, 65. *Prof Exp:* Asst prof pediat & neurol, Sch Med, Univ Colo, 61-64; from asst prof to assoc prof, 64-75, PROF PEDIAT & NEUROL, ALBERT EINSTEIN COL MED, 75-; ASST DEAN EDUC, MONTEFIORE HOSP & MED CTR, 70- *Concurrent Pos:* John Hay Whitney Found fel, 54-55; Nat Inst Neurol Dis & Blindness sr clin trainee, 57-61; assoc attend, Montefiore Hosp & Med Ctr, 64-; assoc vis pediatrician, Morrisania City Hosp, 64-; asst vis neurologist, Bronx Munic Hosp, 64-; consult, Jewish Mem Hosp, 66-; Commonwealth Fund fel, 70. *Mem:* Fel Am Acad Pediat; fel Am Acad Neurol; Soc Pediat Res. *Res:* Neurochemistry and clinical pediatric neurology. *Mailing Add:* NYS Inst Basic Res Dev Disabil SUNY Hlth Sci Ctr 1050 Forest Hill Rd Staten Island NY 10314

FRENCH, JUDSON CULL, b Washington, DC, Sept 30, 22; m 51; c 1. ELECTRONICS & ELECTROMAGNETICS. *Educ:* Am Univ, BS, 43; Harvard Univ, MS, 49. *Prof Exp:* Instr physics, Johns Hopkins Univ, 43-44; instr, George Washington Univ, 44-47; proj leader gaseous electronics, Nat Bur Standards, 48-50, group leader, 51-56, group leader solid state devices, 54-64, from asst chief to chief, Electron Devices Sect, 64-73, chief, Electronic Technol Div, 73-78,; DIR, CTR ELECTRONICS & ELEC ENG, NAT INST STANDARDS & TECHNOL, 78- *Honors & Awards:* Silver Medal, Dept Com, 64 & Gold Medal, 78; Edward Bennett Rosa Award, Nat Bur Standards, 71. *Mem:* Nat Acad Eng; fel Inst Elec & Electronic Engrs; Am Soc Testing & Mat; Am Phys Soc. *Res:* Gaseous electronics and semiconductors; microwave gas switching tubes; measurements and research on semiconductor devices and materials. *Mailing Add:* Ctr for Electronics & Elec Eng Nat Inst Standards & Technol Gaithersburg MD 20899

FRENCH, KENNETH EDWARD, b Elyria, Ohio, Apr 16, 29; m; c 3. AEROSPACE ENGINEERING, MECHANICAL ENGINEERING. *Educ:* Calif Inst Technol, BS, 53; Mass Inst Technol, SM, 57; Univ Santa Clara, Engr, 68. *Prof Exp:* Chief engr, Irving Air Chute Co, Inc, 57-58; staff engr, Lockheed Missiles & Space Co, 58-90; RETIRED. *Concurrent Pos:* Expert examr mech eng, Calif State Bd Regist Prof Engrs, 67-73; lectr, San Jose State Univ, 74-80. *Mem:* Am Soc Mech Engrs; Am Inst Aeronaut & Astronaut. *Res:* Parachute system research, development and design; design of Discoverer parachute system. *Mailing Add:* PO Box 5988 Huntsville AL 35814

FRENCH, LARRY J, b Freedom, NY, Feb 19, 40; m 64; c 2. INTEGRATED CIRCUITS. *Educ:* Pratt Inst, BSEE, 62; Polytech Inst Brooklyn, MSEE, 64; Columbia Univ, PhD(elec eng), 71. *Prof Exp:* Mem tech staff, RCA Labs, 62-71, mgr design auto, Solid State Tech Ctr, 71-76, dir, 76-80, vpres, 80-84; corp officer & vpres tech, 84-91, CORP OFFICER & SR TECH OFFICER, NA PHILIPS, 91- *Concurrent Pos:* Chief exec officer & pres, Magnascreen. *Res:* High definition television; interactive graphics, data bases, memories, MOS devices and processes; computer aided design; computer aided manufacturing; solid state ballasts. *Mailing Add:* 631-B Coppermine Rd Princeton NJ 08540

FRENCH, LYLE ALBERT, b Worthing, SDak, Mar 26, 15; m 41; c 3. NEUROSURGERY. *Educ:* Univ Minn, BS, 36, MB, 39, MD, 40, MS, 46, PhD(neurosurg), 47; Am Bd Neurol Surg, dipl, 48. *Prof Exp:* From instr to prof, Univ Minn, 47-85, dir dept, 60-72, vpres health sci, 70-83, EMER PROF NEUROSURG, MED SCH, UNIV MINN, MINNEAPOLIS, 85- *Mem:* AMA; Am Asn Neurol Surgeons; Neurosurg Soc Am; Am Surg Soc. *Res:* Peripheral nerve injuries; cerebral edema; brain tumors in children. *Mailing Add:* Univ Minn Hosps Minneapolis MN 55455

FRENCH, RICHARD COLLINS, b Camden, NJ, Dec 11, 22. PLANT BIOCHEMISTRY. *Educ:* Rutgers Univ, BS, 47, MS, 48; Purdue Univ, PhD(plant physiol), 53. *Prof Exp:* Plant physiologist biol warfare labs, Ft Detrick, 53-57; plant physiologist biol sci br agr mkt serv, USDA, 57-62; plant physiologist biol labs crops div, US Army, 62-68, plant physiologist, Plant Sci Lab, 68-71; PLANT PHYSIOLOGIST, PLANT DIS RES LAB, USDA, 71- *Mem:* Fel Am Inst Chemists; Am Soc Plant Physiol; Bot Soc Am; Mycol Soc Am; Am Soc Hort Sci; Sigma Xi. *Res:* Seed and spore physiology. *Mailing Add:* Bldg 1301 Ft Detrick Plant Dis Res Lab USDA Frederick MD 21701

FRENCH, ROBERT DEXTER, b Springfield, Mass, Oct 26, 39; m 65; c 2. PHYSICAL METALLURGY, SURFACE PHYSICS. *Educ:* Northeastern Univ, BSc, 62, MSc, 64; Brown Univ, PhD(eng), 67. *Prof Exp:* Res asst crystall growth, Northeastern Univ, 62-64; asst phys metall, Brown Univ, 64-67, instr, 67, res assoc, 67-68; metallurgist, 68-77, group leader, 73-78, br chief, 78-79, div chief, 79-81, chief, metals & ceramics lab, 81-84, GEN MGR, ARMY MAT & MECH RES CTR, 84- *Mem:* Am Inst Mining & Metall Engrs; Am Soc Metals; Am Crystallog Soc; Metall Soc; Sigma Xi. *Res:* Study of metal structure and metal surfaces through electron and field ion microscopy; diffusion of interstitials; creep and stress-rupture in single and two-phase alloys. *Mailing Add:* 40 Reed St Lexington MA 02173

FRENCH, ROBERT LEONARD, b Camarillo, Calif, June 13, 33; m 55; c 3. PROJECT MANAGEMENT. *Educ:* Univ Calif, Los Angeles, BS, 56; Univ Southern Calif, MS, 66. *Prof Exp:* Supvr, Rocketdyne Div, NAm Rockwell, 56-69; mem tech staff, 69-79, tech mgr, Salton Sea solar pond exp, 79-81, DEP PROJ MGR SOLAR PONDS, JET PROPULSION LAB, CALIF INST TECHNOL, 81- *Mem:* Int Solar Energy Soc. *Res:* System design and analysis of salt gradient solar pond power plants; thermodynamic and hydrodynamic behavior, water clarity and measurement and zone boundry control. *Mailing Add:* R L French & Assoc 3815 Lisbon St Suite 201 Ft Worth TX 76107

FRENCH, WALTER RUSSELL, JR, b Inman, Nebr, Sept 29, 23; m 45; c 3. PHYSICS. *Educ:* Nebr Wesleyan Univ, AB, 48; Univ Iowa, MA, 50; Univ Nebr, PhD(physics), 57. *Prof Exp:* Instr physics, Nebr State Teachers Col, Peru, 50-51; from asst prof to prof, 51-62, head dept, 52-85, E C AMES DISTINGUISHED PROF PHYSICS, NEBR WESLEYAN UNIV, 62- *Concurrent Pos:* NSF fac fel, Univ Wis, 65-66; dir & lectr, NSF-in-Serv Insts, 60-63, dir, 64-65; co-dir, Nebr Acad Sci Vis Scientist Prog, 56-64; consult, Oak Ridge Inst Nuclear Studies, AEC, 62-; participant, Oak Ridge Nat Labs, 63; mem, Cottrell Grants Adv Comt, Res Corp, 71-76; engr instrument design group, Lawrence Berkeley Lab, Univ Calif, 76-77 & 78; vis prof, Dept Radiation Biol & Biophys, Univ Rochester, 79-85. *Mem:* Am Phys Soc; Am Asn Physics Teachers. *Res:* X-ray fluorescence applications; experimental nuclear physics. *Mailing Add:* Dept Physics Nebr Wesleyan Univ 5000 St Paul Ave Lincoln NE 68504

FRENCH, WILBUR LILE, b Hammond, Ind, Mar 16, 29; m 54; c 3. GENETICS. *Educ:* Univ Ill, BS, 56, MS, 57, PhD(cytol, genetics), 62. *Prof Exp:* Instr biol, Eastern Ill Univ, 58-59; res assoc genetics, Univ Ill, 62-63; USPHS fel, Univ Calif, Riverside, 63-64 & Univ Mainz, 64-65; asst prof, 65-70, assoc prof, 70-80, EMER PROF GENETICS, LA STATE UNIV, BATON ROUGE, 80- *Mem:* AAAS; Genetic Soc Am; Entomol Soc Am; Sigma Xi. *Res:* Basic molecular mechanisms of life and aging. *Mailing Add:* 1943 Nicholson Baton Rouge LA 70803

FRENCH, WILLIAM EDWIN, b Jackson, Mich, Nov 17, 36; m 58; c 2. ENVIRONMENTAL MANAGEMENT, MARINE GEOLOGY. *Educ:* Univ Mich, BS, 58, MS, 60, PhD(oceanog), 65. *Prof Exp:* Asst res geologist, Great Lakes Res Div, Univ Mich, 59-66; oceanogr, Marine Geophys Surv, US Navy Oceanog Off, DC, 66-68; asst prof geol, Hope Col, 68-72; sr staff consult & vpres logistics, Environ Consult, Inc, 72-80; sr staff consult, WAPORA, Inc, 80-81; SR STAFF CONSULT, COASTAL ECOSYSTEM MGT, INC, 81-; SR GEOLOGIST, ENVIRON ASSOC, INC, 89- *Mem:* Am Soc Limnol & Oceanog; Soc Econ Paleont & Mineral; Geol Soc Am. *Res:* Measurement of sedimentary activity and near-bottom current regime; lake and estuary circulation; shoreline stabilization; groundwater hydrology of large surface mines. *Mailing Add:* 914 Wayside Way Richardson TX 75080-4017

FRENCH, WILLIAM GEORGE, b Seattle, Wash, Jan 23, 43; m 65; c 2. PHYSICAL CHEMISTRY, GLASS TECHNOLOGY. *Educ:* Univ Calif, Riverside, BA, 65; Univ Wis, PhD(phys chem), 69. *Prof Exp:* Mem tech staff chem, Bell Labs, 69-79; sr res specialist, Minn, 79-; AT OPTICAL TECHNOL CTR, 3M CO, CALIF. *Mem:* Am Chem Soc; Am Ceramic Soc; Optical Soc Am. *Res:* Physics and chemistry of glass; optical waveguide fibers; high purity materials. *Mailing Add:* Bldg 260-5A-11 3M Co 3M Ctr St Paul MN 55144-1000

FRENCH, WILLIAM STANLEY, b Colfax, Iowa, Sept 17, 40; m 68; c 6. EXPLORATION GEOPHYSICS. *Educ:* Iowa State Univ, BS, 62; Ore State Univ, PhD(geophys), 70. *Prof Exp:* Asst phys, Ore State Univ, 62-64, asst geophysics, 64-68; from geophysicist to sr res geophysicist, Gulf Res & Develop Co, Gulf Oil Corp, 68-74; assoc prof geophys, Sch of Oceanog, Ore State Univ, 74-75; res supvr, Tulsa Res Ctr, Amoco Prod Co, Standard Oil Co, Ind, 75-77, head Dept Geophys, New Orleans Region, 77-81; PRES, TENSOR GEOPHYS SERV CORP, 81-; CONSULT PROF PHYSICS, UNIV NEW ORLEANS, 83- *Concurrent Pos:* Mem, res comt, Soc Explor Geophysicists, 73-77, distinguished lectr, 89, pres elect to pres, 90-92; ed, Geophys Soc Tulsa, 77; data processing ed, Geophysics, 82-83. *Mem:* Soc Explor Geophys (vpres, 87-88); Am Geophys Union; Seismol Soc Am; Europ Asn Explor Geophysicists. *Res:* Application of mathematical theory of elastic wave propagation to problems in earthquake seismology and petroleum exploration; oceanographic seismology; gravity and magnetic exploration; ultrasonic modeling; three-dimensional seismic exploration methods. *Mailing Add:* Tensor Geophys Serv Corp 3510 N Causeway Blvd 501 Metairie LA 70002

FRENGER, PAUL F, b Houston, Tex, May 9, 46; m 79; c 1. NEURAL NETWORKS-COMPUTERS, MEDICAL DATABASES. *Educ:* Rice Univ, Houston, BA, 68; Univ Tex, San Antonio, MD, 74. *Prof Exp:* Course supvr, USAF Physician Asst Sch, 76-78; spec consult, Med Networks, 79-82; dir, Med Prod, Microprocessor Labs, Inc, 83; chief med officer, Mediclinic, Inc, 84-86; computer consult, Working Hypothesis, Inc, 86-87; programmer, Telescan, Inc, 87-89; clin physician, McCarty Indust Clin, 89-90; CLIN PHYSICIAN, FT BEND FAMILY HEALTH CTR, 90-; CHIEF MED OFFICER, HEALTH TESTING INC, 91- *Mem:* Asn Comput Mach; Inst Elec & Electronic Engrs; Am Med Informatics Asn; Int Neural Network Soc. *Res:* Bioengineering; computing; medical databases; neural networks; Forth programming language; microcomputer applications. *Mailing Add:* PO Box 820506 Houston TX 77282

FRENKEL, ALBERT W, b Berlin, Ger, Jan 1, 19; nat US; m 48; c 3. PLANT PHYSIOLOGY. *Educ:* Univ Calif, BA, 39, PhD(plant physiol), 42. *Prof Exp:* Asst, Univ Calif, 39-42 & Calif Inst Technol, 42-44; assoc radiol, Univ Rochester, US Army, Manhattan Dist & AEC Contract, 45-47; from asst prof to assoc prof bot, 47-89, head dept, 71-75, EMER PROF BOT, UNIV MINN, ST PAUL, 89- *Concurrent Pos:* Res fel, Mass Gen Hosp, 54; vis scientist, Hopkins Marine Sta, Stanford Univ, 67-68. *Mem:* Fel AAAS; Am Soc Plant Physiol; Am Chem Soc; Am Soc Microbiol; Am Soc Biol Chem; Sigma Xi; NY Acad Sci. *Res:* Photobiology; photochemistry; photosynthesis; regulation of microbial metabolism; microbiol ecology. *Mailing Add:* Dept Plant Biol Col Biol Sci 220 Biol Sci Ctr Univ of Minn St Paul MN 55108

FRENKEL, EUGENE PHILLIP, b Detroit, Mich, Aug 27, 29; m 58; c 2. INTERNAL MEDICINE, HEMATOLOGY & MEDICAL ONCOLOGY. *Educ:* Wayne State Univ, BS, 49; Univ Mich, MD, 53; Am Bd Internal Med, cert, 62, cert hematol, 72 & cert med oncol, 73. *Prof Exp:* Intern, Wayne County Gen Hosp, Mich, 53-54; resident, Boston City Hosp, 54-55; resident, Med Ctr, Univ Mich, 57-59, res assoc hemat, 59-60, instr, 60-62; from asst prof to assoc prof, 62-69, Am Cancer Soc prof clin oncol, 73-81, PROF INTERNAL MED & RADIOL, UNIV TEX, SOUTHWESTERN MED CTR, DALLAS, 69-, CHIEF, DIV HEMAT-ONCOL, 62- *Concurrent Pos:* Consult & assoc chief nuclear med, Dallas Vet Admin Hosp, 62-; consult, St Paul Hosp, 64-, Baylor Univ Med Ctr, Presby Hosp & Brooke Army Hosp, 65-; Bd Gov Am Joint Comn Cancer, Am Bd Internal Med, 80-87; Emma Freeman prof res radiation ther, Am Cancer Soc, 78-90; Raymond D & Patsy R Nasher Distinguished Prof, Cancer Res, Univ Tex, Southwestern Med Ctr, Dallas, 90- *Mem:* Fel Am Col Physicians; Am Soc Hemat; Am Asn Cancer Res; Am Soc Clin Oncol; Asn Am Physicians; Am Soc Clin Invest; Am Soc Biol Chemists; Am Asn Cancer Educ. *Res:* Cancer chemotherapy; vitamin B-12 and folic acid metabolism; clinical and theraputic aspects of malignant brain tumors and lymphoma. *Mailing Add:* Dept Internal Med Univ Tex Southwestern Med Ctr Dallas TX 75235-8852

FRENKEL, GERALD DANIEL, b New York, NY, Dec 29, 44; m 70; c 3. NUCLEIC ACID ENZYMOLOGY, SELENIUM & MERCURY BIOCHEMISTRY. *Educ:* Columbia Univ, BA, 66; Harvard Univ, PhD(biol chem), 71. *Prof Exp:* Fel virol, Weizmann Inst, 71-73; res assoc, Princeton Univ, 73-75; asst prof microbiol, Albany Med Col, 75-79; res scientist biochem, NY State Dept Health, 79-83; assoc prof, 83-90, PROF BIOCHEM, RUTGERS UNIV, 90- *Concurrent Pos:* Prin investr, NIH & Am Cancer Soc grants, 76-; adj assoc prof, State Univ NY, Albany, 79-83. *Mem:* Am Soc Biochem & Molecular Biol. *Res:* Cytotoxic effects of mercury and selenium compounds are studied in cells in culture and in various in vitro systems. *Mailing Add:* Dept Biol Sci Rutgers Univ 101 Warren St Newark NJ 07102

FRENKEL, JACOB KARL, b Darmstadt, Ger, Feb 16, 21; nat US; m 54; c 3. PATHOLOGY. *Educ:* Univ Calif, AB, 42, MD, 46, PhD(comp path), 48. *Prof Exp:* Asst zool, Univ Calif, 40-42 & 43, asst bact, 42-43, asst anat, 43-44, intern path, Univ Hosp, 46-47, clin asst, Med Sch, 47-48; pathologist, Rocky Mountain Lab, USPHS, 48-50; instr, Med Sch, Univ Tenn, 51-52; from asst prof to assoc prof, 52-60, PROF PATH, SCH MED, UNIV KANS, 60- *Concurrent Pos:* Researcher, Path Lab, NIH, 50-51; Fulbright fel & vis prof, Nat Univ Mex, 63-64; mem, Subcomt Comp Path, Div Med Sci, Nat Res Coun, 65-71; vis prof, Medellin, Colombia, 67 & San Jose, Costa Rica, 71; sci panel mem, Lunar Receiving Lab, 69-72. *Honors & Awards:* Humboldt Sr Scientist Award, Univ Bonn, Fed Repub Ger, 77; Tinker Award for Res in Cent Am, 83-84. *Mem:* Am Soc Parasitol; Soc Exp Biol & Med; Am Asn Pathologists; Infectious Dis Soc Am; Am Soc Trop Med & Hyg; Sigma Xi. *Res:* Pathogenesis of infection with obligate intracellular organisms; adrenal infection and necrosis; effects of corticoids on immunity and hypersensitivity; cellular immunity; toxoplasmosis. *Mailing Add:* Dept Path & Oncol Univ Kans Med Ctr Kansas City KS 66103

FRENKEL, KRYSTYNA, b USSR, Mar 2, 41; US citizen; m 77. CARCINOGENESIS, OXYGEN RADICALS. *Educ:* Warsaw Univ, Poland, MS, 64; NY Univ, PhD(biochem), 74. *Prof Exp:* Asst res scientist org chem, Warsaw Univ, 64-66, sr asst res scientist, 66-68; asst res scientist chem carcinogenesis, Dept Environ Med, Med Ctr, NY Univ, 69-74; fel, Inst Cancer Res Columbia Univ Physicians & Surgeons, 74-76, staff assoc, 76-77; assoc res scientist, 77-81, res asst prof, 81-82, asst prof, Dept Path, Med Ctr, 82-88, asst prof, Dept Environ Med, 84-88, ASSOC PROF ENVIRON MED & PATHOL, NY UNIV, 88- *Concurrent Pos:* Instr, Dept Org Chem, Warsaw Univ, 64-68. *Mem:* Am Chem Soc; Am Asn Cancer Res; NY Acad Sci; AAAS; Soc Free Radical Res. *Res:* Modification of nucleic acids by chemical and physical carcinogens such as active oxygen species generated by ionizing radiation or by tumor promoter-activated neutrophils; structural and conformation characterization of the products; excision repair mechanisms in normal human cells and repair deficient cells. *Mailing Add:* Dept Environ Med 550 First Ave New York NY 10016

FRENKEL, NIZA B, b Tel Aviv, Israel, June 3, 47; m 68; c 2. VIROLOGY. *Educ:* Univ Chicago, MSc, 70, PhD(virol), 72. *Prof Exp:* Fel genetics, Weizmann Inst Sci, Israel, 72-73; instr virol, 73-74, asst prof biol, 74-80, ASSOC PROF BIOL, DEVELOP BIOL, VIROL & GENETICS, UNIV

CHICAGO, 81- *Concurrent Pos:* Prin investr, grants, NIH & NSF, 78- *Mem:* Am Soc Microbiol; Am Soc Virol. *Res:* Herpes simplex virus DNA replication and expression; viral mediated cell transformation; eukaryotic gene amplification; viral vectors. *Mailing Add:* 1363 E Park Pl Chicago IL 60637

FRENKEL, RENE A, b Santiago, Chile, Sept 1, 32; m 59; c 4. BIOCHEMISTRY. *Educ:* Univ Chile, BS & MS, 56; Cornell Univ, PhD(biochem), 64. *Prof Exp:* First asst biochem, Univ Chile, 5657; res asst, Sloan-Kettering Inst Cancer Res, 57-60; res assoc, Univ Pa, 67-68, asst prof, 68; asst prof, 68-73, ASSOC PROF BIOCHEM, UNIV TEX SOUTHWESTERN MED CTR, DALLAS, 73- *Concurrent Pos:* Johnson Res Found fel biophys, Univ Pa, 64-67. *Mem:* AAAS; Am Soc Biol Chemists. *Res:* Metabolic control; enzyme kinetics and interactions; integrated metabolic sequences. *Mailing Add:* Dept Biochem Univ Tex Health Sci Ctr Dallas TX 75235

FRENKIEL, RICHARD H, SYSTEMS ENGINEERING, CELLULAR RADIO PRODUCT DEVELOPMENT. *Educ:* Tufts Univ, BS, 63; Rutgers Univ, MS, 65. *Prof Exp:* Engr, supvr, & head, Mobile Radio Syst Eng Dept, 65-77, head, Mobile Syst Eng Dept, 77-83, HEAD, CORDLESS TEL DEVELOP DEPT, AT&T BELL LABS, 83- *Honors & Awards:* Alexander Graham Bell Medal, Inst Elec & Electronics Engrs, 87. *Mem:* Fel Inst Elec & Electronics Engrs. *Mailing Add:* Dept Head Product Develop AT&T Bell Labs Crawford Corner Rd Holmdel NJ 07733

FRENKLACH, MICHAEL Y, b Moscow, USSR, Oct 31, 47; US citizen; m 81; c 2. COMBUSTION CHEMISTRY, POWDER NUCLEATION. *Educ:* Mendeleyev Inst Chem Technol, MSc, 69; Hebrew Univ, PhD(phys chem), 76. *Prof Exp:* Res assoc combustion, McGill Univ, Montreal, Can, 76-78; postdoctoral assoc chem, Mass Inst Technol, 78-79; from asst prof to assoc prof chem eng, La State Univ, 79-85; PROF FUEL SCI, PA STATE UNIV, 85- *Concurrent Pos:* Outstanding jr fac award, Arco Oil & Gas Co, 81; fac res award, La State Univ, 83; vis prof, Heidelberg Univ, Fed Repub Ger, 85-86. *Mem:* AAAS; Am Chem Soc; Combustion Inst; Mat Res Soc. *Res:* Experimental and modeling studies of reactive systems at high temperatures including: combustion chemistry, powder nucleation and deposition of diamond thin films. *Mailing Add:* 202 Acad Projs Pa State Univ University Park PA 16802

FRENSDORFF, H KARL, b Hannover, Ger, Apr 7, 22; nat US. PHYSICAL CHEMISTRY, POLYMER CHEMISTRY. *Educ:* Rensselaer Polytech Inst, BS, 49; Princeton Univ, AM, 51, PhD(phys chem), 52. *Prof Exp:* Res staff, E I du Pont de Nemours & Co, Inc, 52-84; ed, J Macromolecular Sci & Chem, 84-89; RETIRED. *Concurrent Pos:* Assoc ed, Rubber Chem & Technol, 72-74, 77-84, ed, 75-77. *Mem:* Am Chem Soc. *Res:* Structure and properties of high polymers; polymerization statistics; sorption and diffusion; macrocyclic polyethers. *Mailing Add:* Apt 403 2304 Riddle Ave Wilmington DE 19806

FRENSLEY, WILLIAM ROBERT, b Wichita, Kans. SOLID STATE PHYSICS, SEMICONDUCTORS. *Educ:* Calif Inst Technol, BS, 73; Univ Colo, PhD(physics), 76. *Prof Exp:* Researcher elec eng, Univ Calif, Santa Barbara, 76-77; mem tech staff, Cent Res Labs, Tex Instruments, Inc, 77-90; PROF ELEC ENG, UNIV TEX DALLAS, 90- *Mem:* Am Phys Soc; Inst Elec & Electronics Engrs. *Res:* Semiconductor device physics, particularly gallium arsenide field-effect transistors, heterojunctions and quantum devices. *Mailing Add:* Univ Texas Dallas PO Box 830688 BE28 Richardson TX 75083-0688

FRENSTER, JOHN H, b Chicago, Ill, Oct 14, 28; m 58; c 3. CELL BIOLOGY, ONCOLOGY. *Educ:* Univ Ill, BS, 50, MD, 54. *Prof Exp:* Intern, Cook County Hosp, Chicago, 55; resident med, Res & Educ Hosps, Univ Ill, 55-58, res fel hemat, 56-57; Am Cancer Soc fel & guest investr cell biol, Rockefeller Inst, 58-60; cell radiobiologist, Walter Reed Army Inst Res, Washington, DC, 60-62; asst prof, Rockefeller Inst, 62-66; chief oncol, Santa Clara Valley Med Ctr, San Jose, 72-89; CLIN ASSOC PROF MED, SCH MED, STANFORD UNIV, 66- *Concurrent Pos:* Asst hemat, Johns Hopkins Hosp, Md, 60-62; USPHS res career develop award, 62-67. *Mem:* Fel Am Col Physicians; Biophys Soc; Am Soc Hemat; Am Asn Cancer Res; Am Soc Cell Biol. *Res:* Structure and function of the cell nucleus; clinical and biological aspects of leukemia and lymphomas; control of RNA synthesis; human tumor immunotherapy; quantitation of health or disease within individual human systems. *Mailing Add:* Stanford Univ 247 Stockbridge Ave Atherton CA 94017-5446

FRENZEL, HUGH N, petroleum geology; deceased, see previous edition for last biography

FRENZEL, LOUIS DANIEL, JR, b San Antonio, Tex, Aug 22, 20; m 45; c 2. BIOLOGY, ECOLOGY. *Educ:* NTex State Col, BS, 47, MS, 48; Univ Minn, PhD(wildlife mgt), 57. *Prof Exp:* Instr biol, Ely Jr Col, Minn, 48-54; asst, Univ Minn, 54-56; prof, Macalester Col, 57-69; PROF ENTOM, FISHERIES & WILDLIFE, UNIV MINN, ST PAUL, 69- *Mem:* Wildlife Soc; Am Soc Mammal; Ecol Soc Am. *Res:* Ecology of terrestrial vertebrates, especially of avian and mammalian populations; field and general biology; wildlife ecology; ecological studies and management of American Bald Eagles. *Mailing Add:* 1506 Crawford Ave St Paul MN 55113

FRENZEL, LYDIA ANN MELCHER, b Victoria, Tex, July 26, 44; m 73. RISK MANAGEMENT & ASSESSMENT, CHEMICAL MARKETING. *Educ:* Univ Tex Austin, BS, 66, DPhil, 71. *Prof Exp:* Asst prof chem, La State Univ, New Orleans, 72-77; dir res, Baker Sand Control, Subsid Baker Int, 77-78; dir res & develop & secy, Coastal Sci Assocs, Inc, 78-87; DIR RES & DEVELOP, AM LIGNITE PROD CO, SUBSID ABB, 87- *Concurrent Pos:* Consult, Southern Imp Corp, Shell Oil Co & Richmond Tank Co, 74-83; mem, La State Task Force Small Bus Innovation, 80; tech adv, CCI Serv Inc, Woodlands, Tex, 87- *Mem:* Am Chem Soc; Fedn Soc Coatings & Technol; Steel Structures Painting Coun; Soc Petrol Engrs; Sigma Xi. *Res:* Chemical product development; marketing; regulatory compliance; public relations; coatings systems; corrosion; water jetting; surface contamination; author of various publications. *Mailing Add:* Am Lignite Prod Co 4655 Coal Mine Rd Ione CA 95640

FRENZEN, CHRISTOPHER LEE, b Chicago, Ill, July 30, 54. APPLIED MATHEMATICS, FLUIDS. *Educ:* Univ Chicago, AB, 76; Calif Inst Technol, MS, 78; Univ Wash, PhD(appl math), 82. *Prof Exp:* Teaching fel math, Univ BC, 82-84; ASST PROF MATH, SOUTHERN METHODIST UNIV, 84- *Mem:* Soc Indust & Appl Math. *Res:* Nonlinear waves; asymptotic and perturbation methods; asymptotic expansion of integrals; mathematical biology. *Mailing Add:* Dept Math Southern Methodist Univ Dallas TX 75275

FRENZEN, PAUL, b Oak Park, Ill, Sept 4, 24; m 51; c 4. METEOROLOGY, FLUID DYNAMICS. *Educ:* Univ Chicago, BS, 49, MS, 51, PhD(meteorol), 64. *Prof Exp:* Weather forecaster, US Army Air Force, 44-46; res assoc fluid dynamics, Hydro Lab, Univ Chicago, 51-56; res assoc meteorol, Argonne Nat Lab, 56-59, assoc meteorologist, 59-78, assoc div dir, 76-83, sr meteorologist, Environ Res Div, 78-88; RES PROF GEOPHYS SCI, UNIV CHICAGO, 85- *Concurrent Pos:* Vis scientist atmospheric physics, Commmonwealth Sci & Indust Res Orgn, Melbourne, 62-63 & 68-69, environ mech, Canberra, 85; mem, Comt Turbulence & Boundary Layers, Am Meteorol Soc, 84- *Mem:* Am Meteorol Soc; Royal Meteorol Soc; Sigma Xi. *Res:* Atmospheric turbulence; diffusion in air and water; turbulence instrumentation. *Mailing Add:* 1034 N Damen Ave Chicago IL 60622

FRERE, MAURICE HERBERT, b Sheridan, Wyo, Sept 8, 32; m 57; c 3. FORAGE & LIVESTOCK MODELS. *Educ:* Univ Wyo, BS, 54, MS, 58; Univ Md, PhD(soils), 62. *Prof Exp:* Res soil scientist, Soil & Water Conserv Div, 58-75, dir, Southern Great Plains Res Watershed, Sci & Educ Admin-Agr Res, 75-79, asst regional adminr, 79-85, DIR, S PIEDMONT CONSERV RES CTR, AGR RES SERV, USDA, 85- *Mem:* AAAS; Am Soc Agron; Soil Sci Soc Am; Am Chem Soc; Soil & Water Conserv Soc. *Res:* Systems analysis of soil-water-plant-animal relations. *Mailing Add:* S Piedmont Ctr Box 555 Watkinsville GA 30677

FRERICHS, WILLIAM EDWARD, b Des Moines, Iowa, Mar 30, 39; m 63; c 2. MICROPALEONTOLOGY. *Educ:* Iowa State Univ, BS & MS, 63; Univ Southern Calif, PhD(geol), 67. *Prof Exp:* Res geologist, Esso Prod Res Co, 66-68; from asst prof to assoc prof, 68-75, PROF GEOL, UNIV WYO, 75- *Concurrent Pos:* NSF grant, 70-72. *Mem:* AAAS. *Res:* Foraminiferal paleoecology; geopolarity history of the Cretaceous; Cretaceous biostratigraphy; vertical movements related to continental drift. *Mailing Add:* Dept Geol Univ Wyo Box 3006 Laramie WY 82071

FRERKING, MARGARET ANN, b Jamaica, NY, July 15, 50. ASTROPHYSICS, MILLIMETER AND SUBMILLIMETER INSTRUMENTATION. *Educ:* Mass Inst Technol, BSc, 72, PhD(physics), 77. *Prof Exp:* Consult infrared receivers, Laser Analystics Inc, 76-77; fel radio astron, Bell Labs, 77-80; MEM TECH STAFF MILLIMETER & SUBMILLIMETER WAVE INSTRUMENTATION, JET PROPULSION LABS, 80- *Honors & Awards:* Sigma Xi. *Mem:* Am Astron Soc. *Res:* Millimeter wave and submillimeter wave instrumentation for atmospheric and astrophysical research. *Mailing Add:* Tech Staff Jet Propulsion Lab 4800 Oak Grove Dr Ms 168 318 Pasadena CA 91109

FRERMAN, FRANK EDWARD, b Louisville, Ky, Feb 7, 42; m 66; c 3. BIOCHEMISTRY, MICROBIOLOGY. *Educ:* Bellarmine Col, BA, 64; Univ Ky, PhD(biochem), 68. *Prof Exp:* Fel biochem, Johns Hopkins Univ, 68-70; from asst prof to assoc prof microbiol, Med Col Wis, 75-87; B F STALINSKI LAB, COLO UNIV HEALTH SCI CTR. *Res:* Enzymology; lipid metabolism. *Mailing Add:* B F Stalinski Lab Colo Univ Health Sci Ctr Eighth & Colorado Denver CO 80262

FRESCO, JACQUES ROBERT, b New York, NY, May 30, 28; m 57; c 3. BIOCHEMISTRY, MOLECULAR BIOLOGY. *Educ:* NY Univ, BA, 47, MS, 49, PhD(biochem), 53. *Hon Degrees:* MD, Univ Göteborg, Sweden, 79. *Prof Exp:* Asst, Lebanon Hosp, NY, 47-48; instr chem, Col Med, NY Univ, 53-54, instr pharmacol, 54-56; res fel chem, Harvard Univ, 56-60; from asst prof to assoc prof, Princeton Univ, 60-65, prof chem, 65-70, chmn dept biochem sci, 74-80, prof biochem sci, 70-90, PROF MOLECULAR BIOL, PRINCETON UNIV, 90-, PFEIFFER PROF LIFE SCI, 77- *Concurrent Pos:* Fel, Sloan-Kettering Inst Cancer Res & USPHS, 52-54; tutor biochem sci, Harvard Univ, 57-60; Lalor Found fel, Cavendish Lab, Cambridge, Eng & Inst Biol Physico-Chim, Paris, 57; estab investr, Am Heart Asn, 58-63; Guggenheim Found fel, Med Res Coun Lab Molecular Biol, Cambridge, Eng, 69-70; vis prof, Hebrew Univ, Jerusalem, 73; chmn, Sci Adv Comt, BioTech Int, 81-86. *Honors & Awards:* Am Scientist Award, AAAS, 62. *Mem:* Am Chem Soc; Am Soc Biol Chem. *Res:* Biochemistry of nucleic acids; molecular genetics; conformation of biomacromolecules; mechanisms of mutation; base pairing; protein-nucleic acid recognition; biophysics. *Mailing Add:* Dept Molecular Biol Princeton Univ Lewis Thomas Princeton NJ 08544

FRESCO, JAMES MARTIN, b Yonkers, NY, Sept 29, 26. ANALYTICAL CHEMISTRY. *Educ:* NY Univ, AB, 49; Polytech Inst Brooklyn, MS, 56; Univ Ariz, PhD(chem), 61. *Prof Exp:* Chemist, Health & Safety Div, NY Opers Off, US AEC, 49-56; res assoc chem, Univ Ariz, 61-62; asst prof, Univ Nev, 62-64; chemist, US Naval Radiol Defense Lab, 64; asst prof, Tex Technol Col, 64-65; asst prof, 65-69, ASSOC PROF CHEM, MCGILL UNIV, 69- *Mem:* Am Chem Soc; The Chem Soc; Chem Inst Can. *Res:* Radiochemistry; infrared spectroscopy; chemistry of coordination compounds; chemical separation processes. *Mailing Add:* Dept Chem McGill Univ 801 Sherbrooke St W Montreal PQ H3A 2K6 Can

FRESCURA, BERT LOUIS, b Glendale, Calif, Aug 13, 36; m; c 3. MICROELECTRONICS, SOLID STATE PHYSICS. *Educ:* Univ Calif, Los Angeles, BS, 59; Ore State Univ, MS, 63; Stanford Univ, PhD(elec eng), 75. *Prof Exp:* Commun officer, US Air Force, 59-62; engr integrated circuits, Fairchild Semiconductor Res & Develop Labs, 63-67; mem tech staff integrated circuits, Leds & Lasers, Palo Alto, 67-81, res & develop sect mgr, Microwave Semiconductor Div, San Jose, 81-83, DESIGN ENGR, CMOS ASICS, CTR, HEWLETT-PACKARD, SANTA CLARA, CALIF, 83- *Mem:* Inst Elec & Electronics Engrs; Soc Info Display. *Res:* Bipolar and metal-oxide-silicon integrated circuits; large area led display; double-heterojunction lasers; dynamic model operations section power transistors. *Mailing Add:* 10100 Phar Lap Dr Cupertino CA 95014

FRESH, JAMES W, b Toccoa, Ga, Jan 9, 26; m 57; c 2. PATHOLOGY. *Educ:* Lenoir-Rhyne Col, BA, 49; Univ NC, MPH, 50, MD, 57. *Prof Exp:* Instr physiol, Univ NC, 52-54, instr path, 57-58; pathologist, Marine Corps Air Facil, US Navy, New River, NC, 58-59 & Naval Hosp, St Albans, NY, 59-62, head path, Naval Med Res Unit 2, Taipei, Taiwan, 62-69, cmndg officer, Naval Med Res Unit 1 & Head path sect, Naval Biomed Res Lab, 69-71, exec officer, Naval Med Res Unit 3, 71-75, chief path, Naval Northwest Regional Med Ctr, Jacksonville, Fla, 75-76; RETIRED. *Concurrent Pos:* Vis prof, Col Med, Taiwan Nat Univ, 62-69, Nat Defense Med Col, Taiwan, 65-68 & Taipei Med Col, 65-69; consult, Chinese Navy. *Mem:* Am Soc Clin Path; Col Am Path; Asn Mil Surgeons US; AMA; Am Pub Health Asn; Sigma Xi; Int Acad Path. *Res:* Clinical research on malignancies, infectious diseases, nutritional deficiencies and coronary artery disease. *Mailing Add:* 7409 Fleming Island Dr Green Cove Springs FL 32043

FRESIA, ELMO JAMES, b Pittsfield, Mass, Sept 10, 31; m 58; c 3. ELECTROCHEMISTRY, OXIDE FILMS. *Educ:* Univ Mass, BS, 53; Univ Conn, MS, 59. *Prof Exp:* Proj leader high reliability develop, 59-62, sect head, Solid State Capacitors Res & Develop, 62-69, proj mgr, Electrochem Prod Develop Lab, 69-78, sect head, Device Develop, & Res, Develop & Eng, 78-85, sr res eng, 85-87, CONSULT, SPRAGUE ELEC CO, 87- *Concurrent Pos:* Chmn Components Comt, Electronic Components Conf. *Mem:* Sigma Xi. *Res:* Preparation and structure determination of ternary oxides; development of solid state capacitors, electrolytic type; dielectric oxide films; corrosion of metals. *Mailing Add:* Longview Terr Williamstown PA 01267

FRESQUEZ, CATALINA LOURDES, b Socorro, Tex, Feb 11, 37. GENETICS. *Educ:* Incarnate Word Col, BA, 63; Univ Tex, Austin, MA, 69, PhD(zool, genetics), 76. *Prof Exp:* Teacher math, Archbishop Chapelle High Sch, 63-66, chmn dept, 65-66; from instr to assoc prof, 69-85, chmn div natural sci, 78-79 & 85-87, PROF BIOL, INCARNATE WORD COL, 85- *Concurrent Pos:* Prin investr, Minority Biomed Support Prog, 76-, choice reviewer, 80-, Nat Sci Found, 88; fel, Nat Sci Found, 64-65, Nat Inst Health, 71-75 & Ford Found, 75-76. *Mem:* Am Soc Cell Biol; Genetics Soc Am; AAAS. *Res:* Regulatory mechanisms; developmental biology; RNA puff clones; hormone activated sequences. *Mailing Add:* Incarnate Word Col 4301 Broadway San Antonio TX 78209

FRESTON, JAMES W, b Mt Pleasant, Utah, July 20, 36; m 56; c 4. INTERNAL MEDICINE, GASTROENTEROLOGY. *Educ:* Univ Utah, MD, 61, PhD(med), 67. *Prof Exp:* From asst prof to prof med & pharmacol, Col Med, Univ Utah, 67-80, chief, Div Clin Pharmacol, 69-80, chmn, Div Gastroenterol, 70-80; PROF & CHMN, DEPT MED, UNIV CONN, 80- *Concurrent Pos:* Nat Inst Arthritis & Metab Dis fel, Dept Med, Royal Free Hosp, London, Eng, 65-67; staff physician, Med Ctr, Univ Utah, 67-; clin investr, Vet Admin Hosp, Salt Lake City, Utah, 67-69; Burroughs-Wellcome scholar, 69; vis prof, Univ Man, 70. *Mem:* AAAS; AMA; Am Asn Study Liver Dis; Am Col Physicians; Am Soc Internal Med. *Res:* Gastrointestinal pharmacology and drug toxicity; peptic ulcer therapy. *Mailing Add:* Chmn Med Univ Conn Health Ctr Farmington CT 06032

FRETER, KURT RUDOLF, b Hamburg, Ger, Jan 26, 29; Can citizen; m 59; c 4. ORGANIC CHEMISTRY. *Educ:* Univ Frankfurt, Hauptdiplom, 53, Dr res nat, 55. *Prof Exp:* Vis scientist, NIH, 56-57; group leader med chem, C H Boehringer, Ger, 57-64; head Dept Chem, Pharm Res Cen Ltd, 64-77; head dept chem, Boehringer Ingelheim Ltd, 77-88; RETIRED. *Concurrent Pos:* Adj prof, Wesleyan Univ, Middletown, Conn, 77-89. *Mem:* Am Chem Soc; Chem Inst Can; Soc Ger Chem. *Res:* Peptides; Heterocyclic Chemistry. *Mailing Add:* 906 Lucas Ave Victoria BC V8X 4M7 Can

FRETER, ROLF GUSTAV, b Hamburg, Ger, Jan 5, 26; US citizen; m 55; c 4. MEDICAL BACTERIOLOGY, IMMUNOLOGY. *Educ:* Univ Frankfurt, PhD(bact), 51. *Prof Exp:* Res assoc bact, Biol Budesanstalt, Braunschweig, Ger, 51; instr, Loyola Univ, Ill, 54-57; assoc prof microbiol, Jefferson Med Col, 57-65; PROF MICROBIOL, UNIV MICH, ANN ARBOR, 65- *Concurrent Pos:* Logan fel & res assoc, Univ Chicago, 52-54. *Mem:* AAAS; Am Soc Microbiol; fel Am Acad Microbiol; Am Asn Immunologists. *Res:* Oral enteric vaccines; ecology of normal enteric flora of man; bacterial toxins and enzymes; experimental enteric animal infections; anaerobic bacteria; experimental and mathematical models of intestinal microecology. *Mailing Add:* Dept Microbiol 6734 Med Sci Bldg II Univ Mich Ann Arbor MI 48109-0620

FRETTER, WILLIAM BACHE, b Pasadena, Calif, Sept 28, 16; m 39; c 3. PHYSICS. *Educ:* Univ Calif, AB, 37, PhD(physics), 46. *Prof Exp:* Asst physics, Univ Calif, 37-41; res assoc, Mass Inst Technol, 41; res engr, Westinghouse Elec Co, Pa, 41-44; res engr, Manhattan Dist Proj, Radiation Lab, 45-46, from instr to assoc prof physics, 46-55, dean col lett & sci, 62-67, PROF PHYSICS, UNIV CALIF, BERKELEY, 55- *Concurrent Pos:* Fulbright res scholar, France, 52-53 & 60-61; Guggenheim fel, 60-61. *Honors & Awards:* Chevalier, Legion of Honor, France. *Mem:* Fel Am Phys Soc. *Res:* High power microwaves; isotope separation; cosmic rays; mass of cosmic ray mesotrons; penetrating showers, heavy mesons and hyperons; elementary particle physics. *Mailing Add:* 130 Condor Ct Bodega CA 94922

FRETWELL, LYMAN JEFFERSON, JR, b Rockford, Ill, Oct 8, 34; m 72; c 5. SIGNAL PROCESSING, SYSTEM PERFORMANCE ANALYSIS. *Educ:* Calif Inst Technol, BS, 56, PhD(physics), 67. *Prof Exp:* Mem tech staff, 66-68, supvr, 68-88, DISTINGUISHED MEM TECH STAFF, AT&T BELL LABS, 88- *Mem:* AAAS; Am Phys Soc. *Res:* Algorithm and system design for ocean acoustic signal processing systems; system performance evaluation; adaptive signal processing; statistical modeling techniques; physics and system implications of ocean acoustics. *Mailing Add:* Bell Labs Whippany NJ 07981

FRETZ, THOMAS ALVIN, b Buffalo, NY, Oct 9, 42; m 66; c 2. ORNAMENTAL HORTICULTURE. *Educ:* Univ Md, BS, 64; Univ Del, MS, 66, PhD(plant sci), 70. *Prof Exp:* Asst prof hort, Ga Sta, Univ Ga, 69-72; from asst prof to assoc prof hort, Ohio State Univ, 72-81; PROF HORT & HEAD DEPT, VA POLYTECH INST & STATE UNIV, 81- *Mem:* Am Soc Hort Sci; Int Plant Propagators Soc; Am Hort Soc; Agron Soc Am; Crop Sci Soc; Sigma Xi. *Res:* Chemotaxonomy of ornamental plants; propagation of ornamental plants and utilization of hardwood bark as a growth media for plants. *Mailing Add:* Dept Hort Va Polytech Inst Blacksburg VA 24061-0327

FREUD, GEZA, mathematical analysis, for more information see previous edition

FREUD, PAUL J, b Mineola, NY, Sept 22, 38; m 61; c 3. SOLID STATE PHYSICS. *Educ:* Dartmouth Col, BA, 60; Rutgers Univ, PhD(physics), 64. *Prof Exp:* Sr scientist, Columbus Lab, Battelle Mem Inst, 64-73; prin scientist, 73-80, CORP SCIENTIST, LEEDS & NORTHRUP CORP, 80- *Mem:* Am Phys Soc; Int Soc Hybrid Microelectronics; Am Vacuum Soc. *Res:* Solid state and thin film device research and development; microelectronic sensor development; optical, fiber optical and integrated optical device and sensor development; transport proporties of thin film metals; semiconductors and alloys. *Mailing Add:* 3113 Cloverly Dr Furlong PA 18925

FREUDENSTEIN, FERDINAND, b Ger, May 12, 26; nat US; m 59. MECHANICAL ENGINEERING. *Educ:* Harvard Univ, MS, 48; Columbia Univ, PhD(mech eng), 54. *Prof Exp:* Develop engr, Am Optical Co, NY, 48-50; mem tech staff, Bell Tel Labs, Inc, 54; from asst prof to assoc prof, 54-70, chmn dept, 58-64, PROF MECH ENG, COLUMBIA UNIV, 70- *Concurrent Pos:* Guggenheim fels, 61-62 & 67-68; indust consult. *Honors & Awards:* Machine Design Award, Am Soc Mech Engrs, 72. *Mem:* Nat Acad Eng; fel Am Soc Mech Engrs. *Res:* Engineering design; mechanisms; kinematic analysis and synthesis. *Mailing Add:* Dept Mech Eng 220 Mudd Bldg Columbia Univ 500 W 120th St New York NY 10027

FREUDENTHAL, HUGO DAVID, b Brooklyn, NY, Mar 29, 30; m 55; c 2. MARINE SCIENCES, HYDROLOGY & WATER RESOURCES. *Educ:* Columbia Univ, BS, 53, MS, 55; NY Univ, PhD(protozool), 59. *Prof Exp:* Vpres, H2M Corp, Environ Engrs & Scientists, 73-85; assoc prof biol, 59-65, chmn dept, 65-73, PROF GRAD MARINE & ENVIRON SCI, POST COL, LONG ISLAND UNIV, 65-; PRES, FREUDENTHAL & ELKOWITZ CONSULT GROUP, 88- *Concurrent Pos:* Assoc dir, Living Foraminifera Lab, Am Mus Natural Hist, 61-69; mgr life sci, Fairfield Repub Div, Fairchild Industs, Inc, 66-73; res assoc, Nat Bur Standards, 70-72. *Mem:* AAAS; Soc Protozoologists; Am Soc Limnol & Oceanog; Am Micros Soc; Am Soc Microbiol; Sigma Xi. *Res:* Physiology and ecology of plantonic Foraminifera; calcification and pressure studies on marine microorganisms; ecology and taxonomy of zooxanthellae; photomicrography of planktonic organisms; microbiology of polluted waters; aerospace life sciences; environmental engineering and impact studies. *Mailing Add:* 13 Iroquios Pl Massapequa NY 11758

FREUDENTHAL, PETER, b New York, NY, Aug 12, 34; m 56; c 2. ENVIRONMENTAL HEALTH, METEOROLOGY. *Educ:* NY Univ, BA, 54, MS, 63, PhD(environ health sci), 70; Columbia Univ, BS, 60. *Prof Exp:* Meteorologist, Air Weather Serv, 55-57; asst res scientist, Inst Environ Med, NY Univ Med Ctr, 63-66; phys scientist, Environ Studies Div, Health & Safety Lab, AEC, 66-70; chief, Air Qual Control Eng, 70-75, dir air & noise progs, 75-83, DIR AIR & LAND USE, CONSOL EDISON CO, NY, 83- *Concurrent Pos:* Asst res scientist, Inst Environ Med, Med Ctr, NY Univ, 66-70; adj assoc prof, Long Island Univ, 68-75; vchmn, Environ Sci Comn, NY Acad Sci, 80-81, chmn, 81-82; chmn, Consol Edison Radiation Safety Subcomt, 85- *Mem:* AAAS; Am Meteorol Soc; Air Pollution Control Asn; NY Acad Sci. *Res:* Air pollution; epidemiology; control techniques; cooling tower environmental effects; radioactive fallout transport and deposition; aerosol sizing and deposition; dispersion modeling; risk analysis. *Mailing Add:* 1226 York Ave New York NY 10021

FREUDENTHAL, RALPH IRA, b New York, NY, Aug 27, 40. PESTICIDE & REGULATORY TOXICOLOGY. *Educ:* NY Univ, BS, 63; State Univ NY Buffalo, PhD(biochem, pharmacol), 69. *Prof Exp:* Biochem pharmacologist, Res Triangle Inst, 69-73; assoc mgr path, pharmacol/toxicol & animal resources sect, Columbus Labs, Battelle Mem Inst, 73-77; dir, Dept Toxicol & Environ Health Ctr, 77-84, dir prod safety, occup med, toxicol & regulatory affairs, 84-88, SR CONSULT TOXICOL, PROD SAFETY & REGULATORY AFFAIRS, STAUFFER CHEM CO, 88-; PROF, SCH PHARM, UNIV CONN, 79- *Concurrent Pos:* Assoc ed, J Toxicol & Environ Health. *Mem:* AAAS; Int Soc Biochem Pharmacol; NY Acad Sci; Am Soc Pharmacol & Exp Therapeut; Soc Toxicol. *Res:* Drug-enzyme interactions; drug metabolism with emphasis on steroidal agents; whole body autoradiography; mechanisms in carcinogenesis; molecular toxicology; pesticide toxicity; product safety. *Mailing Add:* 20 Linda Lane Westport CT 06880

FREUND, GERHARD, b Frankfurt, Ger, Apr 21, 26; US citizen; m 55; c 2. INTERNAL MEDICINE, ENDOCRINOLOGY. *Educ:* Univ Frankfurt, MD, 51; McGill Univ, MS, 57. *Prof Exp:* Resident internal med, Augustana Hosp, Chicago, 52-53; resident, Res & Educ Hosp, Univ Ill, 53-55; clin instr internal med, Col Med, Univ Ill, 57-60, asst clin prof, 60-63; res assoc endocrinol, 63-64, asst prof, 64-70, assoc prof internal med, 70-75, PROF INTERNAL MED, COL MED, UNIV FLA, 75-; CHIEF ENDOCRINOL, VET ADMIN HSOP, GAINESVILLE, 67- *Concurrent Pos:* Res fel endocrinol, McGill Univ, 55-57. *Mem:* AAAS; fel Am Col Physicians; Soc Neurosci; Sigma Xi; Soc Biol Psychiat. *Res:* Effects of aging, chronic ethanol ingestion and acetaldehyde metabolism on brain structure and synaptic function in man and experimental animals. *Mailing Add:* Dept Med Box J-277 Univ Fla Col Med Gainesville FL 32610

FREUND, HARRY, b Tulsa, Okla, Nov 21, 17; m 45; c 3. ANALYTICAL CHEMISTRY. *Educ:* City Col New York, BS, 40; Univ Mich, MS, 41, PhD(chem), 45. *Prof Exp:* Asst, Anal Lab, Dept Eng Res, Univ Mich, 41-42, res assoc, 42-45, res chemist, 45-47; instr, 47-49, from asst prof to assoc prof, 47-60, prof, 60-80, EMER PROF CHEM, ORE STATE UNIV, 80- *Concurrent Pos:* Chemist, 48- *Honors & Awards:* NY Cocoa Exchange Award, 40. *Mem:* Am Chem Soc. *Res:* Inorganic chemistry; instrumental analysis; chemical instrumentation. *Mailing Add:* 1320 NW Forest Dr Corvallis OR 97701

FREUND, HOWARD JOHN, b Philadelphia, Pa, Dec 27, 46; m 71; c 2. CHEMICAL KINETICS. *Educ:* Pa State Univ, BS, 68; Cornell Univ, PhD(chem), 75. *Prof Exp:* Fel, Fuel Sci Dept, Pa State Univ, 75-76; SR STAFF CHEMIST, EXXON RES & ENG CO, 76- *Mem:* Am Chem Soc; Combustion Inst. *Res:* Reaction kinetics and modeling of hydrocarbons and carbonaceous materials. *Mailing Add:* Exxon Prod Res Co PO Box 2189 Rm N-225 Houston TX 77252-2189

FREUND, JACK, b New York, NY, Nov 19, 17; m 46; c 3. CLINICAL PHARMACOLOGY. *Educ:* NY Univ, BS, 37; Univ Va, 41-42; Med Col Va, MD, 46. *Prof Exp:* Intern, Hosps, Med Col Va, 46-47; resident path, Beth Israel Hosp, New York, 47-48; asst therapeut, Dept Med, NY Univ-Bellevue Med Ctr, 49; resident med, Vet Admin Hosp, Bronx, 49-50, sr resident chest & cardiol, 50-51; lectr pharmacol & assoc med, 51-58, clin assoc med, 58-61, from asst clin prof to assoc clin prof med, 61-70, ASST PROF PHARMACOL, MED COL VA, 58-, CLIN PROF MED, 70- *Concurrent Pos:* Asst dir clin res, A H Robins Co, Inc, 55-58, dir, 58-60, med dir, 60-62, vpres & med dir, 62-64, vpres res, 64-74, sr vpres res & develop, 74-78; consult pharmaceut indust, 78- *Mem:* Fel Am Col Physicians; fel Am Soc Clin Pharmacol & Therapeut; fel Am Col Cardiol; Soc Toxicol. *Res:* Clinical evaluation of drugs; peripheral vascular disease. *Mailing Add:* 310 Old Bridge Lane Richmond VA 23229

FREUND, JOHN ERNST, b Berlin, Ger, Aug 6, 21; nat US; m 49; c 2. MATHEMATICAL STATISTICS. *Educ:* Univ Calif, Los Angeles, BA, 43, MA, 44; Univ Pittsburgh, PhD(math), 52. *Prof Exp:* Prof math, Alfred Univ, 46-54; prof statist, Va Polytech Inst, 54-57; prof math, 57-71, EMER PROF MATH, ARIZ STATE UNIV, 71- *Mem:* Am Math Soc; Am Statist Asn; Math Asn Am; Inst Math Statist. *Mailing Add:* 7035 N 69th Place Scottsdale AZ 85253

FREUND, LAMBERT BEN, b McHenry, Ill, Nov 23, 42; m 65; c 3. APPLIED MECHANICS, MATERIALS SCIENCE. *Educ:* Univ Ill, Urbana-Champaign, BS, 64, MS, 65; Northwestern Univ, PhD(appl mech), 67. *Prof Exp:* Fel mat, Brown Univ, 67-69, from asst prof to assoc prof, 69-75, chmn, Div Eng, 79-83, PROF ENG, 75-, HENRY LEDYARD GODDARD UNIV PROF, BROWN UNIV, 88- *Concurrent Pos:* Consult, Am Iron & Steel Inst, 72-78; vis prof, Stanford Univ, 74-75; vis scholar, Harvard Univ, 83-84; ed, J Appl Mech, Am Soc Mech Engrs, 83-88; mem, US Nat Comt Theoret & Appl Mech, 85-; mem, Gen Assembly, Int Union Theoret & Appl Mech, 86-; gen ed, Cambridge Monogr Mech & Appl Math, 89-; mem ed adv comt, Aeta Mechanica Sinica, 90- *Honors & Awards:* Henry Hess Award, Am Soc Mech Engrs, 74; George R Irwin Medal, Am Soc Testing & Mat, 87. *Mem:* Fel Am Soc Mech Engrs; Am Geophys Univ; fel Am Acad Mech. *Res:* Mechanics of solids; fracture mechanics; stress waves in solids; theoretical seismology; mechanics of thin film structures and microelectronic devices. *Mailing Add:* Div Eng Brown Univ Providence RI 02912

FREUND, MATTHEW J, b New York, NY, Aug 3, 28; m 52; c 2. PHYSIOLOGY, PHARMACOLOGY. *Educ:* NY Univ, BA, 48; Univ Nebr, MSc, 50; Rutgers Univ, PhD(physiol), 57. *Prof Exp:* Asst dairy husb, Rutgers Univ, 55-56; asst anat, Med Col, Cornell Univ, 56-58; instr physiol & pharmacol, 58-60, from asst prof to assoc prof, 60-67, prof pharmacol, New York Med Col, 68-78, assoc prof obstet & gynec, 64-78, res prof urol, 73-78; prof physiol, Southern Ill Univ, 78-, assoc prof obstet & gynec, 64-, res prof urol, 73-, chmn dept physiol, 80-; PROF OBSTET & GYNEC, UNIV MED & DENT NJ-SCH MED. *Concurrent Pos:* Health Res Coun City New York res grant & career scientist, 62- *Mem:* AAAS; Am Physiol Soc; Am Soc Pharmacol & Exp Therapeut; Am Fertil Soc; Am Soc Cell Biol. *Res:* Physiology and pharmacology of reproduction fertility and infertility; cryobiology. *Mailing Add:* 316 Hialeah Dr Cherry Hill NJ 08002

FREUND, PETER GEORGE OLIVER, b Temesvar, Rumania, Sept 7, 36; m 63. THEORETICAL PHYSICS. *Educ:* Polytech Inst Temesvar, dipl eng, 58; Univ Vienna, PhD(physics), 60. *Prof Exp:* Res assoc physics, Inst Theoret Physics, Univ Vienna, 61; res assoc, Inst Theoret Physics, Univ Geneva, 61-62; res assoc, Enrico Fermi Inst Nuclear Studies, 62-64, from asst prof to assoc prof, Inst & Univ, 64-74, PROF PHYSICS, UNIV CHICAGO & ENRICO FERMI INST NUCLEAR STUDIES, 74- *Concurrent Pos:* Mem, Inst Advan Study, Princeton, NJ, 64-65. *Mem:* Fel Am Phys Soc; Austrian Phys Soc; Ital Phys Soc. *Res:* Quantum field theory; dispersion theory; symmetries of elementary particle interactions. *Mailing Add:* Enrico Fermi Inst 5630 S Ellis Chicago IL 60637

FREUND, PETER RICHARD, b Baltimore, Md, Oct 27, 35; m 59; c 3. FOOD SCIENCE, MICROBIOLOGY. *Educ:* Univ Wis, BS, 57, MS, 58, PhD(food sci), 69. *Prof Exp:* Sr scientist nutrit prod develop, Mead Johnson & Co, 61-65; prod develop mgr candy res, M&M/Mars, 65-67; dir qual control bakery prod mfg, Chapman & Smith Div, DCA Food Industs, 67-68; corp dir qual control & res multi-food mfg, Jewett & Sherman Co, 69-73; tech mgr prod develop, Universal Foods Corp, 73-75, dir tech develop multi-food res, 75-79, mgr res & develop, 79-81; asst to the pres, Multi-Natural Food & Agr Ingred Mfg, 81-84, MGR PROD APPLN, CHR HANSEN'S LAB, INC, 84- *Mem:* Inst Food Technologists. *Res:* Research and development of microbiol cultures and fermentation products, naturally derived enzymes and natural colors and flavors. *Mailing Add:* 10522 Culpepper Ct Seattle WA 98177

FREUND, RICHARD A, engineering, statistics; deceased, see previous edition for last biography

FREUND, ROBERT M, b New York, NY, Nov 3, 53. MATHEMATICAL PROGRAMMING, LINEAR SYSTEM. *Educ:* Princeton Univ, BA, 75; Stanford Univ, MS, 79, PhD(opers res), 80. *Prof Exp:* ASSOC PROF, MASS INST TECHNOL, 83- *Concurrent Pos:* Consult assoc, ICF, Inc, 80-83. *Mem:* Oper Res Soc Am; Inst Management Sci; Math Asn Am. *Res:* Optimization Theory and algorithms; linear and nonlinear programming. *Mailing Add:* Mass Inst Technol 50 Memorial Dr Cambridge MA 02139

FREUND, ROBERT STANLEY, b Newark, NJ, Jan 26, 39; m 62; c 2. SEMICONDUCTOR PROCESSING. *Educ:* Wesleyan Univ, BA, 60; Harvard Univ, MA, 62, PhD(chem physics), 65. *Prof Exp:* Res fel chem, Harvard Univ, 65-66; mem tech staff, AT&T Bell Labs, 66-76, head dept environ chem, 76-79, head dept chem Kinetics Res, 79-87, head dept plasma processing res, 87-90, HEAD, POWER SYSTS RES DEPT, AT&T BELL LABS, 90- *Concurrent Pos:* Mem comt, Atomic & Molecular Scis, Nat Res Coun, 76-82; mem, Sci Adv Comt, Gov NJ, 79-83; exec comt, Gaseous Elect Conference, 83-85; alt coun, div physical chem, Am Chem Soc, 85-87; exec comt, Atomic Molecular & Optical Physics, Am Physical Soc, 87-90; local co-chmn, 16th Int Conf Physics Electronic & Atomic Collisions, 89; chmn, Int Soc Optical Eng conf Multichamer & In-situ Processing Electronic Mat, 89. *Mem:* Int Elec & Electronic Engrs; Am Chem Soc; Am Vacuum Soc. *Res:* Molecular spectroscopy; molecular dissociation; high rydberg states; electron collisions with molecules; double resonance and anticrossings; plasma processing of semi conductors. *Mailing Add:* AT&T Bell Labs Rm 1D-256 Murray Hill NJ 07974

FREUND, ROLAND WILHELM, b Schweinfurt, Ger, Aug 1, 55; m 79. NUMERICAL ANALYSIS, SCIENTIFIC COMPUTING. *Educ:* Univ Wuerzburg, Ger, dipl, 82, PhD(math), 83. *Prof Exp:* Res asst math, Inst Appl Math, Univ Wuerzburg, Ger, 82-84, asst prof, 85-90; RES SCIENTIST, RES INST ADVAN COMPUTER SCI, NASA AMES RES CTR, 88- *Concurrent Pos:* Vis res assoc, Computer Sci Dept, Stanford Univ, 85-86; ed, J Numerical Linear Algebra with Applications, 90-; ed, J Numerical Anal, Soc Indust & Appl Math, 91- *Mem:* Soc Indust & Appl Math. *Res:* Numerical linear algebra; sparse matrix computation; iterative solution of large linear systems; structured matrices; algorithms for massively parallel machines; constructive approximation theory and its application in iterative matrix computations. *Mailing Add:* Res Inst Advan Computer Sci Mail Stop Ellis St NASA Ames Res Ctr Moffett Field CA 94035

FREUND, RUDOLF JAKOB, b Kiel, Ger, Mar 3, 27; nat US; m 48; c 3. STATISTICS. *Educ:* Univ Chicago, MA, 51; NC State Univ, PhD(exp statist), 55. *Prof Exp:* From asst prof to assoc prof statist, Va Polytech Inst, 55-62; assoc prof, 62-74, assoc dir inst statist, 62-77, PROF STATIST, TEX A&M UNIV, 74- *Concurrent Pos:* Vis res scholar, Univ Okla, 60. *Mem:* Fel Am Statist Asn. *Res:* Use of computers in statistics; use of statistics and mathematics in economics; use of linear models. *Mailing Add:* Dept Stat Tex A&M Univ College Station TX 77843-3142

FREUND, THOMAS STEVEN, b New York, NY, Jan 11, 44; m 66; c 2. BIOCHEMISTRY. *Educ:* Lehigh Univ, BS, 65, PhD(biochem), 69. *Prof Exp:* Teaching asst chem, Lehigh Univ, 65-66, res asst marine sci & biochem, 66-69; NIH trainee ophthal, Col Physicians & Surgeons, Columbia Univ, 69-70; res assoc oral biol, Sch Dent Med, Univ Conn, 70-75; vis asst prof biol, Trinity Col, Conn, 74-75; from asst prof to assoc prof, 75-84, PROF BIOCHEM, SCH DENT, FAIRLEIGH DICKINSON UNIV, 84-, CHMN, DIV HUMAN BIOCHEM, 85- *Concurrent Pos:* Vis scientist, Nat Inst Med Health Res, Paris, France, 75 & 82; biohazard safety coordinator, Fairleigh Dickinson Univ; pres, Ringwood Bd Health, NJ. *Mem:* AAAS; NY Acad Sci; Am Chem Soc; Int Asn Den Res; Am Asn Den Res. *Res:* Proteins, their structure and function; hormonal regulation; cellular metal ion requirements and transport; microbial biochemistry, mineral nutrition and development, calcification, and biohazard safety. *Mailing Add:* Dept Biochem Fairleigh Dickinson Univ Sch Dent Hackensack NJ 07601

FREUNDLICH, MARTIN, b New York, NY, Dec 15, 30; m 52; c 4. BIOCHEMISTRY, MOLECULAR BIOLOGY. *Educ:* Brooklyn Col, BA, 55; Long Island Univ, MS, 57; Univ Minn, PhD(microbiol), 61. *Prof Exp:* USPHS fel, Cold Spring Harbor Lab Quant Biol, 61-64; asst prof microbiol, Dartmouth Med Sch, 64-66; from asst prof to assoc prof, 66-84, actg chmn dept, 74-75, 86-89, PROF BIOCHEM, STATE UNIV NY, STONYBROOK, 85- *Concurrent Pos:* USPHS res grant, 64-67, 69-72 & 73-94 & career develop award, 70-75; Lederle med fel,65-66; ed, J Bact, 66-70; NY State Res Found grant, 68-69. *Mem:* Am Chem Soc; Am Soc Microbiol; Am Soc Biol Chem. *Res:* Mechanism and interaction of genetic regulatory elements in bacteria. *Mailing Add:* Dept Biochem State Univ NY Stony Brook NY 11794-5215

FREUNDLICH, MARTIN M, b Goerlitz, Ger, Nov 23, 05; nat US; m 47; c 1. ELECTRONICS. *Educ:* Tech Univ, Berlin, ME, 29, PhD(electronics), 33. *Prof Exp:* Asst, Tech Univ, Berlin, 33-34 & Royal Tech Col, Glasgow, Scotland, 35; mem staff, Pye Radio Ltd, Eng, 35-36, Columbia Broadcasting Syst, 36-44 & 45-49 & NAm Phillips, 44-45; asst supv eng consult, AIL Div, Eaton Corp, 49-70; consult, 70-88; RETIRED. *Concurrent Pos:* Adj asst prof, Queensborough Community Col, City Univ New York, 72-78. *Mem:* AAAS; Am Astron Soc; Inst Elec & Electronics Engrs. *Res:* Tube development, oscillographs for lightning investigations, electron microscopes; television tubes, properties of phosphors; storage tubes; color television systems; materials in space; space lubrication; vacuum technology; exobiology. *Mailing Add:* 16 Suydam Dr Melville NY 11747

FREVEL, LUDO KARL, b Frankfurt am Main, Ger, May 31, 10; nat US; m 37; c 3. CHEMISTRY. *Educ:* Johns Hopkins Univ, PhD(chem), 34. *Prof Exp:* Nat res fel chem, Calif Inst Technol, 34-36; res chemist, Dow Chem Co, 36-74; FEL CHEM, JOHNS HOPKINS UNIV, 74-; CHMN, INT CTR DIFFRACTION DATA, 90- *Concurrent Pos:* Lab dir chem physics, Dow Chem Co, 69-72; vis res prof mat sci, Pa State Univ, 75. *Honors & Awards:* J D Hanawalt Award, 83. *Mem:* AAAS; Am Chem Soc; Crystallog Asn. *Res:* X-ray studies of substances under pressure; crystal structure; identification of compounds by x-ray and electron diffraction methods; catalysis; iterated functions. *Mailing Add:* 1205 W Park Dr Midland MI 48640

FREVERT, RICHARD KELLER, agricultural engineering; deceased, see previous edition for last biography

FREY, BRUCE EDWARD, b New York, NY, Jan 29, 45; m 79; c 2. BIOLOGICAL OCEANOGRAPHY. *Educ:* Cornell Univ, BS, 67; Ore State Univ, MS, 74, PhD(oceanog), 77, Ore Health Sci Univ, MD, 88. *Prof Exp:* res assoc, Ore State Univ, 77-79, asst prof oceanog, 79-84; resident, Good Samaritan Hosp, Portland, Ore, 88-89; RADIATION ONCOLOGIST, ORE HEALTH SCI UNIV, 89- *Concurrent Pos:* Prin investr, Environ Protection Agency grant, 77-79 & Columbia River data develop grant, 79-84. *Mem:* Am Soc Therapeut Radiol & Oncol; AAAS. *Res:* Plankton ecology and physiology, phytoplankton-nutrient interactions; radiation oncology. *Mailing Add:* 6047 SW Brugger St Portland OR 97219

FREY, CARL, b New York, NY; m 55; c 3. ENGINEERING. *Educ:* NY Univ, BME, 49; Columbia Univ, MA, 51. *Prof Exp:* Asst exec secy, Eng Manpower Comn, 60, exec secy, 61-64; secy, Engrs Joint Coun, 65-67, exec dir, 67-79; exec dir, Am Asn Eng Socs, 79-83; CONSULT, ENG UTILIZATION, 83- *Concurrent Pos:* Adv, US Dept Labor, 62; mem comt specialized personnel, Off Emergency Planning, Off of the President, 62-64, mem exec res, adv panel select comt govt res, House of Rep, 64-; treas, US Comt Large Dams & US Nat Comt, World Power Conf, 65-; Founder, US/Egyptian Alliance Eng Corp, 79-82, dir, eng utilization study, 83- *Mem:* AAAS; Am Soc Eng Educ; Am Inst Aeronaut & Astronaut. *Res:* Policies in field of engineering and scientific manpower; professional society; engineering utilization. *Mailing Add:* 19 Harding Lane Westport CT 06880

FREY, CHARLES FREDERICK, b New York, NY, Nov 15, 29; m 57; c 5. SURGERY. *Educ:* Amherst Col, BA, 51; Cornell Univ, MD, 55. *Prof Exp:* Instr, New York Hosp, Med Sch, Cornell Univ, 63; from instr to prof surg, Med Ctr, Univ Mich, Ann Arbor, 64-76; PROF & EXEC VCHMN, DEPT SURG, UNIV CALIF, DAVIS, 76- *Concurrent Pos:* Fel, Dept Surg, Univ Mich, 64-65; attend physician surg serv, Vet Admin Hosp, Ann Arbor & Wayne County Gen Hosp, Eloise, 65-71; surg consult, Student Health Serv, 67-70; pres, Univ Asn Emergency Med Serv, 70-71; chmn, Mich Emergency Serv Health Coun, consult to ad hoc adv group on emergency med serv, Vet Admin, 71-72; consult ed, J Am Col Emergency Physicians, 71-74; surg consult, Vet Admin Hosp, 71-76; dir surg, Wayne County Gen Hosp, Eloise, 71-76; mem, Ad Hoc Comt Emergency Med Commun, Nat Acad Sci, 71-72; consult, Ann Arbor Vet Hosp, 71-76; consult & mem, Bd Dirs, Govr's Hwy Safety Res Inst, 73-76; mem, Comt Regional Emergency Med Commun Systs, Nat Acad Sci, Nat Res Coun, 73-82, chmn, Mich Comt Trauma, 70-74, regional dir, Area V Comt Trauma, 74-76; chief surg, Vet Admin Hosp, 76-80, consult to dir surg, Cent Off, 77-80. *Mem:* Fel Am Col Surgeons; Soc Surg Alimentary Tract; Am Surg Asn; fel Am Asn Surgeons Trauma; Soc Univ Surgeons; Nat Coun Trauma. *Res:* Pancreatic and biliary pathophysiology; trauma and emergency health service systems. *Mailing Add:* Dept Surg Sch Med Univ Calif 4301 X St Sacramento CA 95817

FREY, CHRIS(TIAN) M(ILLER), b Cumberland, Md, Feb 26, 23; m 43; c 2. MECHANICAL ENGINEERING. *Educ:* Univ Md, BS, 51. *Prof Exp:* Supvr gen design group, Allegany Ballistics Lab, 51-53, design res group, 53-59; chief engr, 59-67, MGR RES & ADVAN TECHNOL DEPT, UNITED TECHNOL CTR, 67- *Concurrent Pos:* Chmn, Missile Booster Mat Comt, 56-59; consult mat adv bd, Nat Acad Sci, 58-60; consult, Nat Aeronaut & Space Admin, 59-60; lectr, Stanford Univ. *Mem:* Am Soc Mech Engrs; Am Soc Testing & Mat; Am Ord Asn. *Res:* Rocket propulsion; high temperature materials; high strength materials, including fiberglass and boron structures; unique fabrication techniques and rocket design. *Mailing Add:* Res & Advan Technol Dept 1890 Newcastle Dr Los Altos CA 94024-6937

FREY, DAVID ALLEN, b Mendota, Ill, Nov 26, 35; m 57; c 2. ORGANIC CHEMISTRY, POLYMER CHEMISTRY. *Educ:* Monmouth Col, Ill, BA, 57; Univ Iowa, MS, 59, PhD(org chem), 61. *Prof Exp:* Fel high temperature polymers under C S Marvel, Univ Ariz, 61-62; res chemist, Phillips Petrol Co, 62-64; res chemist, Morton Thiokol Inc, 64-69, res supvr, 69-75, mgr polymer res, 75-78, qual assurance mgr, Morton Chem Co Div, qual assurance mgr, spec chem group, 88-90, QUAL ASSURANCE MGR, MORTON INT, INC, 91- *Mem:* Am Chem Soc; Am Soc Qual Control. *Res:* Inter-intra polymerizations, high temperature polymers and emulsion and solution polymerizations using free radical and ionic catalysts; quality assurance; precautionary labeling. *Mailing Add:* Morton Int Inc 1275 Lake Ave Woodstock IL 60098-7499

FREY, DAVID GROVER, b Hartford, Wis, Oct 10, 15; m 48; c 3. LIMNOLOGY, CHYDORID CLADOCERA. *Educ:* Univ Wis, BA, 36, MA, 38, PhD(zool), 40. *Prof Exp:* Asst, Conserv Dept, Wis, 35-40; jr aquatic biologist, US Fish & Wildlife Serv, Wash, 40-42, asst aquatic biologist, Md, 42-43, assoc aquatic biologist, 43-45; assoc prof zool, Univ NC, 46-50; assoc prof, Ind Univ, Bloomington, 50-55, prof zool, 55-86; RETIRED. *Concurrent Pos:* Fulbright fel, Austria, 53-54, Ireland, 85 & Guggenheim fel, Austria, 53-54; ed J, Am Soc Limnol & Oceanog & aquatic ed J, Ecol Soc Am; Nat Acad Sci-Nat Res Coun Exchange to Soviet Union, 62; Ford Found consult, Mindanao State Univ, 67-68. *Honors & Awards:* Einar Naumann-August Thienemann Medal, Int Asn Limnol, 80; Distinguished Serv Award, Ecol Soc Am, 83. *Mem:* Am Soc Limnol & Oceanog (pres, 55); Am Micros Soc (vpres, 70); Ecol Soc Am; Am Quaternary Asn (pres, 73-74); Int Asn Limnol (vpres, 80-87, pres 87-90); foreign mem Royal Danish Acad Sci & Lett. *Res:* Developmental history of lakes and micropaleontology of freshwater deposits; systematics, ecology and evolution of chydorid cladocera. *Mailing Add:* 2625 S Smith Rd Bloomington IN 47401

FREY, DENNIS FREDERICK, b Chickasha, Okla, Apr 1, 41; m 62; c 2. ANIMAL BEHAVIOR. *Educ:* Okla State Univ, BS, 63, PhD(zool), 70; Va State Col, MS, 67. *Prof Exp:* PROF BIOL SCI, CALIF POLYTECH STATE UNIV, 70- *Concurrent Pos:* Res grant, NIMH, 71-72 & Calif Polytech State Univ, 72-73; researcher, Animal Behav Res Group, Dept Zool, Oxford Univ, 78-79. *Mem:* Animal Behav Soc; Am Soc Zoologists. *Res:* Social Behavior in fishes with particular interest in determinants and consequences of dominance phenomena; time sharing in fish behavior. *Mailing Add:* Dept Biol Sci Calif Polytech State Univ San Luis Obispo CA 93407

FREY, DONALD N(ELSON), b St Louis, Mo, Mar 13, 23; m 42; c 6. INFORMATION SYSTEMS, INNOVATION. *Educ:* Univ Mich, BS, 47, PhD(metall), 50. *Prof Exp:* Asst eng, Univ Mich, 47-48, res assoc, 48-49, instr metall eng, Col Eng, 49-50, asst prof chem & metall eng, 50-51; assoc dir, Sci Lab, Ford Motor Co, 51-57; asst chief engr, Ford Div, 57-61, prod planning mgr, 61, asst gen mgr, 62-65, vpres & gen mgr, 65-67, vpres prod develop, 67-78; pres, Gen Cable Corp, 68-71; chmn & chief exec officer, Bell & Howell Co, 71-88; PROF INDUST ENG & MGT SCI, NORTHWESTERN UNIV, 88- *Honors & Awards:* Nat Medal Technol, 90. *Mem:* Nat Acad Eng; fel Am Soc Metals; fel AAAS. *Res:* Metallurgy of high temperature alloys; general metallurgy including iron and steel making; general automotive engineering; information systems; innovation; technical management. *Mailing Add:* Technol Inst Northwestern Univ Evanston IL 60208

FREY, DOUGLAS D, b Lincoln, Nebr, Sept 15, 55. CHEMICAL ENGINEERING. *Educ:* Willamette Univ, BA, 78; Stanford Univ, BS, 78; Univ Calif, Berkeley, MS, 80, PhD(chem eng), 84. *Prof Exp:* Asst prof, 84-90, ASSOC PROF CHEM ENG, YALE UNIV, 90- *Concurrent Pos:* Consult, Teltech Resource Network Inc, 84- *Honors & Awards:* Henry Ford II Award, Stanford Univ, 78. *Mem:* Am Inst Chem Engrs; Am Chem Soc; Adsorption Soc. *Res:* Fundamental phenomena underlying separation and purification processes and related systems. *Mailing Add:* 570 Prospect St No 1 New Haven CT 06511

FREY, ELMER JACOB, b Buffalo, NY, Jan, 3, 18; m 45, 63; c 2. SYSTEMS ENGINEERING, INERTIAL INSTRUMENTATION. *Educ:* City Col New York, BS, 37; NY Univ, MS, 40; Mass Inst Technol, PhD(math), 49. *Prof Exp:* Teacher, Pub Sch, NY, 40-41; instr math, Mass Inst Technol, 46-49, mathematician, Instrumentation Lab, 49-53, group leader, 53-55, asst dir, 55-57, dep assoc dir, 57-69, lectr aeronaut & astronaut, 62-72, assoc dir, Measurement Systs Lab, 65-70, mem staff, Ctr Space Res, 70-72; eng mgr satellite systs, Fairchild Space & Electronics Co, 72-75; consult systs eng, Frey Assocs, Inc, 75-78, EMRAY INT, INC, 79- *Concurrent Pos:* Mem, US Defense Dept Ad Hoc Comt Inertial Guid, 63; vis prof, Spec Sch Aeronaut, Paris, 63-64; consult engr, France, 63-64; consult, US Air Force Sci Adv Bd, 63-65, chief scientist, 64-65; NATO Adv Group Aerospace Res & Develop, 64 & 69; US Dept Transp lectr, Univ Naples, Cath Univ Louvain & Nat Sch Advan Aeronaut Studies, Paris, 69-70. *Mem:* Am Geophys Union; assoc fel Am Inst Aeronaut & Astronaut; Soc Indust & Appl Math. *Res:* Computation systems and borehole instrumentation; railroad automation; transportation safety and cost benefit analysis. *Mailing Add:* Emray Int Inc 270 Chestnut Hill Rd RD 4 Amherst NH 03031

FREY, FREDERICK AUGUST, b Milwaukee, Wis, Apr 1, 38; m 72; c 1. GEOCHEMISTRY. *Educ:* Univ Wis, BS, 60, PhD(phys chem), 67. *Prof Exp:* Res chemist, Hercules Chem Co, 60-61; asst prof, 66-73, assoc prof, 73-80, PROF GEOCHEM, MASS INST TECHNOL, 80- *Concurrent Pos:* Ed, Geochimica et Comochimica Acta. *Honors & Awards:* VGP Bowen Award, Am Geophys Union, 87. *Mem:* Geol Soc Am; Am Geophys Union; Geochem Soc. *Res:* Elemental distribution in geologic systems; origin and evolution of volcanic rocks. *Mailing Add:* Dept Earth Atmospheric & Planetary Sci Mass Inst Technol Rm 54-1220 Cambridge MA 02139

FREY, FREDERICK WOLFF, JR, b New Orleans, La, Sept 30, 30; m 52; c 5. INORGANIC CHEMISTRY. *Educ:* Loyola Univ, La, BS, 50; Tulane Univ, MS, 52, PhD(chem), 54. *Prof Exp:* Chemist, Mallinckrodt Chem Works, 54-55; supvr chem res, 55-73, asst dir indust chem res, 73-85, assoc dir math appl, 85-87, ASSOC DIR, ADMIN, ETHYL CORP, 87- *Mem:* Am Chem Soc. *Res:* Metal hydride chemistry; organometallics. *Mailing Add:* 1469 Crescent Dr Baton Rouge LA 70806

FREY, HENRY RICHARD, b New York, NY, July 16, 32; m 60; c 3. PHYSICAL OCEANOGRAPHY, OCEAN ENGINEERING. *Educ:* Queen's Col, BS, 60; NY Univ, MS, 66, PhD(oceanog), 71. *Prof Exp:* Group leader underwater technol, Uniroyal Res Ctr, Uniroyal Inc, 60-66, prog mgr ocean sci & eng, Technol Transfer Div, 66-67; sr res scientist oceanog, Sch Eng & Sci, NY Univ, 67-73 & NY Inst Ocean Resources, 73-74; res assoc prof, Polytech Inst NY, 73-76; tech adv, 77-80, CHIEF MARINE ENVIRON SERVS DIV, NAT OCEANIC & ATMOSPHERIC ADMIN, NAT OCEAN SURV, 80- *Concurrent Pos:* Chmn, Uniroyal Ocean Eng Group, Uniroyal Inc, 60-67; dir ocean eng prog, NY Univ, 71-73 & Polytech Inst NY, 73-76; consult, Ministry of Mining & Nat Resources, Jamaica, Chesapeake Res Consortium & Alpine Geophys Assocs, 76-77. *Mem:* Am Geophys Union; Marine Technol Soc; Sigma Xi. *Res:* Continental shelf and estuarine physical oceanography, especially circulation, mixing, wave climate and transports. *Mailing Add:* 9914 Thornwood Rd Kensington MD 20895-4230

FREY, JAMES R, b De Young, Pa, Feb 27, 32; m 53; c 2. BACTERIOLOGY, IMMUNOLOGY. *Educ:* Defiance Col, BS, 52; Miami Univ, Ohio, MA, 57; Mich State Univ, PhD(virol), 61. *Prof Exp:* Asst bact, Tulane Univ, 55-58; asst bact, Mich State Univ, 58-59, virol, 59-61; from asst prof to assoc prof, 61-70, head dept, 68-78, div chmn Natural Systs Studies, 69-78, PROF BIOL, DEFIANCE COL, 70- *Concurrent Pos:* Vis scientist, NSF, 62-64. *Mem:* AAAS; Am Pub Health Asn; Am Soc Microbiol; Am Inst Biol Sci. *Res:* Quantitative study of the enteric viruses in sewage; immuno-genetics approach to study of gene differences. *Mailing Add:* Dept Math & Sci Defiance Col Defiance OH 43512

FREY, JEFFREY, b New York, NY, Aug 27, 39. ELECTRICAL ENGINEERING, MATERIALS SCIENCE. *Educ:* Cornell Univ, BEE, 60; Univ Calif, Berkeley, MS, 63, PhD(elec eng), 65. *Prof Exp:* Mem tech staff, microwave semiconductor devices, Watkins-Johnson Co, Calif, 65-66; NATO fel, Rutherford High Energy Lab, Eng, 66-67; res assoc ion implantation, UK Atomic Energy Res Estab, Harwell, 67-69; from asst prof to assoc prof, 70-79, prof elec eng, Cornell Univ, 79-87; PROF, UNIV MD, 87- *Concurrent Pos:* Japan Soc Promotion Sci fel, Univ Tokyo, 76-77; consult, US Air Force, Carborundum Co & Westinghouse Elec Corp, 77-; mgr device physics & actg mgr advan lithography, Advan Tech Ctr, Signetics Corp, 80-81; vis prof, Univ Tokyo, 84-85. *Honors & Awards:* Paul Rappaport Award, IEEE, 83. *Mem:* Inst Elec & Electronics Engrs. *Res:* Microwave semiconductor devices; microwave integrated circuits; semiconductor materials and devices. *Mailing Add:* Dept Elec Eng Univ Md Col Park MD 20742

FREY, JOHN ERHART, b Chicago, Ill, May 6, 30; m 57; c 4. INORGANIC CHEMISTRY. *Educ:* Northwestern Univ, BS, 52; Univ Ill, MS, 55; Univ Chicago, PhD(inorg chem), 56. *Prof Exp:* Res chemist, Univ Chicago, 56-57; from instr to asst prof chem, Bowdoin Col, 57-60; asst prof, Ill Inst Technol, 60-63 & Western Mich Univ, 63-66; assoc prof, 66-71, asst dean common learning, 69-73, PROF CHEM, NORTHERN MICH UNIV, 71- *Mem:* Am Chem Soc; Sigma Xi. *Res:* Boron hydrides and halides; charge-transfer complexes. *Mailing Add:* Dept Chem Sch Arts & Sci Northern Mich Univ Marquette MI 49855

FREY, KENNETH JOHN, b Charlotte, Mich, Mar 23, 23; m 45; c 3. PLANT BREEDING, AGRONOMY. *Educ:* Mich State Univ, BS, 44, MS, 45; Iowa State Univ, PhD, 48. *Prof Exp:* Asst, Mich State Univ, 44-45, asst prof farm crops, 48-53; from assoc prof to prof agron, 53-70, asst dean grad col, 67-70, actg vpres res & dean grad col, 70-71, C F CURTIS DISTINGUISHED PROF AGR, IOWA STATE UNIV, 70- *Concurrent Pos:* Fulbright scholar, Australia, 68, Yugoslavia, 77. *Mem:* Fel AAAS; fel Am Soc Agron; Genetics Soc Am; Am Soc Agron (pres); Crop Sci Soc Am (pres). *Res:* Plant breeding methodology for self-pollinated crops; biochemistry of cereal grains. *Mailing Add:* Dept Agron Iowa State Col Ames IA 50012

FREY, MARY ANNE BASSETT, b Washington, DC, Dec 15, 34; c 3. MEDICAL PHYSIOLOGY. *Educ:* George Washington Univ, BA, 70, PhD(physiol), 75. *Prof Exp:* Lectr physiol, Montgomery Col, 72-75; Nat Heart & Lung Inst fel, Sch Med, George Washington Univ, 75-76; asst prof physiol, Wright State Univ, 76-80, assoc prof, 80-82; TECH MGR, KENNEDY SPACE CTR, BIONETICS CORP, FLA, 82- *Concurrent Pos:* Consult, Presidents Adv Coun Mgt Improv, 72-73 & US Nat Olympic Men's Volleyball Team; mem, NIH Gen Med Sci Review Comt Access to Res Careers Prog. *Mem:* Asn Women Sci; AAAS; Am Heart Asn; Sigma Xi; Am Physiol Soc. *Res:* Cardiovascular physiology, neural control of the circulation, exercise and stress physiology. *Mailing Add:* Lockheed Eng Serv No 600 600 Maryland Ave SW Washington DC 20024

FREY, MAURICE G, b Cincinnati, Ohio, July 26, 13; m 45; c 2. STRUCTURAL GEOLOGY, PETROLEUM GEOLOGY. *Educ:* Univ Cincinnati, BS, 36; Univ Minn, MS, 37, PhD(geol), 39. *Prof Exp:* Party chief gravity meter crew, Chevron Oil Co Div, Stand Oil Co Calif, 39-41, seismologist, 42-45, supvr geophys, 45-46, dist geologist, 46-47, explor res supvr, 47-48; assoc prof geol, Univ Cincinnati, 48-52; mem staff geol & geophys explor, Chevron Oil Co Div, Standard Oil Co Calif, 52-54, chief geologist, 55-67, asst to vpres, 67-71, consult geologist environ affairs, 71-78; CONSULT, GEOL & GEOPHYS, 78-; PROF, DEPT GEO & GEOPHYS, UNIV NEW ORLEANS, 78- *Concurrent Pos:* Adj prof earth sci & geophys, Univ New Orleans, 78- *Mem:* Am Asn Petrol Geol; Soc Explor Geophys; fel Geol Soc Am. *Res:* Exploration for oil and gas in relation to origin and development of salt domes; oil in marine environment. *Mailing Add:* Dept Earth Sci Univ New Orleans New Orleans LA 70148

FREY, MERWIN LESTER, b Manhattan, Kans, Apr 20, 32; m 61; c 2. VETERINARY MICROBIOLOGY. *Educ:* Kans State Univ, BS & DVM, 56; Univ Wis, MS, 61, PhD(vet sci), 66. *Prof Exp:* Res asst vet sci, Univ Wis, 59-61, proj assoc, 61-64; NIH trainee, 64-66; from asst prof to prof vet microbiol, Iowa State Univ, 66-73; prof & head dept vet parasitol, microbiol & pub health, Col Vet Med, Okla State Univ, 73-77; prof, Dept Vet Sci, Univ Nebr-Lincoln, 77-87, emer prof, 87-; USDA NAT VET SERV LAB, AMES. *Mem:* Am Vet Comput Soc; Am Vet Med Asn; Am Soc Microbiol; Am Asn Vet Lab Diagnosticians. *Res:* Infectious diseases of domestic animals, especially mycoplasmas and viruses as causes of pneumonia in food animals. *Mailing Add:* USDA Nat Vet Serv Lab 13th St & Dayton Rd Ames IA 50010

FREY, NICHOLAS MARTIN, b Earlham, Iowa, Apr 14, 48; m 71. CROP PHYSIOLOGY. *Educ:* Iowa State Univ, BS, 70; Univ Minn, PhD(plant physiol), 74. *Prof Exp:* Physiologist, Pioneer Hi-Bred Int, Inc, 74-80; coordr, Johnston Res, 80-82, dir, biotechnol res, 82-87, coordr, 83-88, group dir, Plant Breeding Div, 84-88, dir, technol acquisition & develop, 88-89, PROD DEVELOP MGR, SPECIALTY PLANT PROD DIV, PIONEER HI-BRED INT INC, 89- *Mem:* Crop Sci Soc Am; Am Soc Agron; Am Soc Plant Physiologists; fel AAAS. *Res:* Research management in biotechnology and other departments supporting plant breeding. *Mailing Add:* Pioneer Hi-Bred Int Inc 317 Sixth Ave Suite 740 Des Moines IA 50309

FREY, PERRY ALLEN, b Plain City, Ohio, Nov 14, 35; m 61; c 2. BIOCHEMISTRY. *Educ:* Ohio State Univ, BS, 59; Brandeis Univ, PhD(biochem), 68. *Prof Exp:* Chemist, USPHS, 60-64; NIH res fel chem, Harvard Univ, 67-68; from asst prof to prof chem, Ohio State Univ, 69-81; PROF BIOCHEM, UNIV WIS-MADISON, 81- *Mem:* Am Chem Soc; Am Soc Biol Chemists; Protein Soc; AAAS. *Res:* Chemical mechanism of action of enzymes and coenzymes. *Mailing Add:* Inst Enzyme Res Univ of Wis Madison WI 53706

FREY, SHELDON ELLSWORTH, b Wheelerville, Pa, Apr 29, 21; m 47; c 2. ORGANIC CHEMISTRY. *Educ:* Pa State Col, BS, 43; Univ Tenn, MS, 50; Univ Ill, PhD(chem), 53. *Prof Exp:* Jr engr mass spectrom, Kellex Corp, NY, 43-44, jr engr vacuum test, Tenn, 44; from tech supvr to assoc chemist process control, Carbide & Carbon Chem Div, Union Carbide & Carbon Corp, 44-50; res chemist, E I du Pont de Nemours & Co, 53-62; coordr opers improv, Allied Chem Corp, 62-72; RES CHEMIST & ASST MGR, MONROE CHEM CO, 73- *Mem:* Am Chem Soc. *Res:* Fluorocarbon polymers; organic chemistry of plastics; elastomers; polymers; hindered ketones; products of formaldehyde and hydroxybenzoic acid; fluorinated hydrocarbons; tetraethyl lead; dispersed dyes; cumene hydroperoxide; phenol; phthalic anhydride; benzaldehyde. *Mailing Add:* 404 Kings Hwy Moorestown NJ 08057-2771

FREY, TERRENCE G, b Feb 6, 48; wid; c 3. STRUCTURAL BIOLOGY. *Educ:* Ohio State Univ, BS, 70; Univ Calif, Los Angeles, PhD(biochem), 75. *Prof Exp:* from res asst prof to res assoc prof biochem & biophys, Univ Pa, 78-86; assoc prof, 86-88, PROF BIOL, SAN DIEGO STATE UNIV, 88- *Concurrent Pos:* res asst, Dept Anat, Duke Univ Med Ctr, 77-78 & Microbiol, Biocenter, Univ Basel, Switz, 75-77. *Mem:* Am Soc Biochem & Molecular Biol; Electron Micros Soc; Sigma Xi; Am Biophys Soc. *Res:* Protein structure and function; electron microscopy membrane structure. *Mailing Add:* Dept Biol Col Sci San Diego State Univ San Diego CA 92182-0057

FREY, THOMAS G, b Eugene, Ore, Sept 24, 43; m 75; c 2. ORGANIC CHEMISTRY. *Educ:* Univ Ore, BA, 65; Univ Idaho, PhD(org chem), 71. *Prof Exp:* From asst prof to assoc prof, 70-80, PROF CHEM, CALIF STATE POLYTECH UNIV, SAN LUIS OBISPO, 80- *Concurrent Pos:* Conserv & restaration studies, Hearst San Simeon State Hist Monument; proj dir grant, State of Calif, 87; instr chem ceramic glazes & chem glass for studio glass makers, Calif Poly Art Dept; vis prof, Univ Ill, Urbana, 82-83, Cornell Univ, Ithaca, 90-91. *Mem:* Am Chem Soc. *Res:* Study of base catalyzed nucleophilic additions to activated acetylenes. *Mailing Add:* Dept Chem Calif State Polytech Univ San Luis Obispo CA 93407

FREY, WILLIAM ADRIAN, b York, Pa, Jan 1, 51; m 72; c 1. DIAGNOSTICS, INFECTIOUS DISEASES. *Educ:* King's Col, BS, 72; Pa State Univ, PhD(org chem), 76. *Prof Exp:* Res fel biol chem, Harvard Med Sch, 77-80; res biochemist, E I du Pont de Nemours & Co Inc, 80-84, res supvr, 84-86, mkt mgr, 87-89, planning mgr, 90, OPERS MGR EUROPE, E I DU PONT DE NEMOURS & CO, INC, 91- *Concurrent Pos:* Assoc staff med, Peter Bent Brigham Hosp, Boston, 77-80; fel, Am Cancer Soc, 77-78; traineeship, NIH, 78-80. *Mem:* Am Chem Soc; AAAS; NY Acad Sci; Am Asn Clin Chemists. *Res:* Diagnostics; infectious diseases; immunochemistry. *Mailing Add:* 34 Raphael Rd Hockessin DE 19707

FREY, WILLIAM CARL, b Newark, NJ, Aug 21, 23; m 47; c 2. STATISTICS, CHEMISTRY. *Educ:* Upsala Col, BA, 43; Rutgers Univ, MS, 54. *Prof Exp:* Control chemist, Bristol-Myers Co, 43-45, asst to purchasing agent, 45-47, supvr anal unit, Control Dept, 47-53, supvr statist unit, 53-57, head statist serv dept, 57-66, dir tech serv, 66-68, dir qual assurance & regulatory conformance, 68-70, tech dir, 70-81, vpres tech servs, Bristol- Myers Co, 81-; RETIRED. *Concurrent Pos:* Adj instr grad sch, Rutgers Univ, 58-; chmn, Ellis R Ott Found. *Honors & Awards:* Ellis R Ott Award, 76. *Mem:* Am Statist Asn; fel Am Soc Qual Control. *Res:* Applied statistics; application of statistical methods in the treatment of experimental data, control of quality, production improvement and in assorted management functions. *Mailing Add:* Five Cedar Ave Linden NJ 07207

FREY, WILLIAM FRANCIS, b Bristol, VA, Nov 23, 33; m 55; c 3. EXPERIMENTAL PHYSICS. *Educ:* King Col, AB, 55; Vanderbilt Univ, MS, 57, PhD(physics), 60. *Prof Exp:* From asst prof to assoc prof, 60-83, PROF PHYSICS, DAVIDSON COL, 83- *Concurrent Pos:* Consult, Health Physics Div, Oak Ridge Nat Lab, 73-83; alt comnr, SE Compact for Low-Level Radioactive Waste, 85- *Res:* Beta-ray spectroscopy; internal conversion coefficients; photoelectric angular distributions; fluorescent yields; multi-photon absorption processes in atoms using tunable dye laser. *Mailing Add:* Dept Physics Davidson Col Davidson NC 28036

FREY, WILLIAM HOWARD, II, b Atlanta, Ga, Nov 19, 47; m 77; c 3. NEUROCHEMISTRY, BEHAVIORAL PSYCHOLOGY. *Educ:* Wash Univ, St Louis, BA, 69; Case Western Reserve Univ, PhD(biochem), 74. *Prof Exp:* ASST PROF PSYCHIAT, ST PAUL-RAMSEY MED CTR, UNIV MINN, 77- , DIR, PSYCHIAT RES LABS, 80- *Concurrent Pos:* Adv, Geriat Res, Educ & Clin Ctr, Vet Admin Ctr, Minneapolis, 80-; res dir,Alzheimers Treatment & Res Ctr & Dry Eye &Tear Res Ctr,80- *Mem:* Am Soc Biol Chemists. *Res:* Human brain biochemistry and neurochemistry of Alzheimer's disease; human emotional crying and tears. *Mailing Add:* Psychiat Res Labs St Paul-Ramsey Med Ctr 640 Jackson St St Paul MN 55101

FREYBERGER, WILFRED L(AWSON), b Newark, NJ, Feb 28, 28; m 51; c 3. MINERAL ENGINEERING, SURFACE CHEMISTRY. *Educ:* Mass Inst Technol, SB, 47, ScD(metall), 55. *Prof Exp:* Res engr, NJ Zinc Co, Pa, 55-60; res metallurgist, Am Cyanamid Co, Conn, 60-64; assoc prof metall & res engr, 64-68, dir, Inst Mineral Res, 70-80, PROF METALL ENG, MICH TECHNOL UNIV, 68- *Mem:* AAAS; Am Inst Mining, Metall & Petrol Engrs; Sigma Xi. *Res:* Flotation collector chemistry; flotation kinetics; developmental studies in the practice of mineral engineering. *Mailing Add:* 718 Cedar Bluff Dr Houghton MI 49931

FREYBURGER, WALTER ALFRED, b Philadelphia, Pa, May 14, 20; m 47; c 5. PHARMACOLOGY. *Educ:* Bucknell Univ, BS, 42; Univ Mich, PhD(pharmacol), 51. *Prof Exp:* Asst pharm, Univ Mich, 46-50; pharmacologist, 50-57, sect head cardiovasc-renal pharmacol, 57-68, MGR CARDIOVASC DIS RES, UPJOHN CO, 68- *Concurrent Pos:* Mem, Coun High Blood Pressure Res & Circulation, Am Heart Asn. *Mem:* Am Soc Pharmacol & Exp Therapeut. *Res:* Cardiovascular and autonomic nervous system physiology and pharmacology. *Mailing Add:* 200 S Tara Taveres FL 32778

FREYD, PETER JOHN, b Evanston, Ill, Feb 5, 36; m 57; c 2. MATHEMATICS. *Educ:* Brown Univ, AB, 58; Princeton Univ, PhD(math), 60; Univ Pa, MA, 71. *Prof Exp:* Consult, Batton, Barton, Durstine & Osborn, 58-60; Ritt instr math, Columbia Univ, 60-62; from asst prof to assoc prof, 62-68, chmn grad group math, 84-87, PROF MATH, UNIV PA, 68-, PROF COMPUT SCI, 87- *Concurrent Pos:* Assoc chmn dept math, Univ Pa, 64-73; vis prof, Pahlavi Univ, Iran, 68-69, Univ Mex, 74, Univ Chicago, 80, Univ Louvain, France, 81, Univ Sydney, Australia, 85 & Univ Milan, Italy, 86; managing ed, J Pure & Appl Sci, 70-; Fulbright sr fel, Univ New S Wales, Australia, 71; prin lectr, Nat Sci Bd Res Seminar, Univ Montreal, 73; fel, St John's Col, Cambridge, Eng, 80-81. *Mem:* Am Math Soc; Math Asn Am. *Res:*

Categorical algebra & topos theory; theory of functors; Abelian categories and their embeddings; representation in Abelian categories; existence of adjoints; applications to relative homological algebra, stable homotopy, proof theory, model theory and foundations; theoretical computer science. *Mailing Add:* 2020 1/2 Addison St Philadelphia PA 19146

FREYER, GUSTAV JOHN, b Ancon, CZ, Nov 11, 31; US citizen; m 57; c 2. SPACE PHYSICS. *Educ:* US Mil Acad, West Point, BS, 54; Air Force Inst Technol, MS, 60; Pa State Univ, PhD(physics), 69. *Prof Exp:* Assoc prof physics, Air Force Inst Technol, 67-71; chief scientist, Air Force Weapons Lab, 73-74, comdr, 75-77; SYST SURVIVABILITY MGR, DENVER DIV, MARTIN MARIETTA AEROSPACE, 77- *Mem:* Am Geophys Union; Sigma Xi. *Res:* Aurora; magnetospheric physics; nuclear weapon effects. *Mailing Add:* 4537 Mockingbird St Las Cruces NM 88001

FREYERMUTH, HARLAN BENJAMIN, b Muscatine, Iowa, Sept 15, 17; m 46; c 1. ORGANIC CHEMISTRY. *Educ:* Univ Iowa, BA, 38, MS, 40, PhD(org chem), 42. *Prof Exp:* Asst, Univ Iowa, 40-42; tech assoc, GAF Corp, 42-69, mgr col rels & tech employment, 69-76; chemist & consult, J T Baker Chem Co, 70-88; RETIRED. *Mem:* Am Chem Soc. *Res:* Organic research in dyes and dye intermediates; formation and properties of uretidinedione; textile finishes; catalytic hydrogenation; diisocyanates; optical bleaching agents; ultra violet absorber; research and process development in inorganic chemicals. *Mailing Add:* 3385 Allen St Easton PA 18042

FREYGANG, WALTER HENRY, JR, b Jersey City, NJ, Dec 27, 24; m; c 3. PHYSIOLOGY. *Educ:* Stevens Inst Technol, ME, 45; Univ Pa, MD, 49. *Prof Exp:* Intern & asst med resident, Bellevue Hosp, NY, 49-51; asst resident neurol, Columbia-Presby Med Ctr, 51-52, asst neurologist, 52; neurophysiologist, NIH, 52-59; vis scientist, Cambridge Univ, 59; guest p of, Univ Heidelberg, 67-69; chief, Sect Membrane Physiol, NIMH, 60-67, mem staff, Div Biol & Biochem Res, 69-72; RETIRED. *Concurrent Pos:* Vis fel neurol & Nat Found Infantile Paralysis fel, Columbia Univ, 52; clin prof neurol, Georgetown Univ; mem, Marine Biol Lab, Woods Hole, Mass. *Honors & Awards:* Commendation Medal, USPHS. *Mem:* AAAS; Soc Gen Physiol; Biophys Soc; Am Physiol Soc; Asn Res Nerv & Ment Dis; Sigma Xi. *Res:* Muscle; central nervous system; peripheral nerve. *Mailing Add:* 6247 29th St NW Washington DC 20015

FREYMANN, JOHN GORDON, b Omaha, Nebr, Apr 9, 22; m 50; c 4. MEDICINE. *Educ:* Yale Univ, BS, 44; Harvard Univ, MD, 46; Am Bd Internal Med, dipl, 55, dipl oncol, 75. *Hon Degrees:* DSc, Univ Nebr, 84. *Prof Exp:* Intern med, Mass Gen Hosp, 46-47, asst resident med, 50-51; asst, Harvard Med Sch, 54-56, instr, 56-60; asst prof med, Sch Med, Tufts Univ & dir med educ, Mem Hosp, Worcester, Mass, 59-65; gen dir, Boston Hosp Women & lectr prev med, Harvard Med Sch, 65-69; dir educ, Hartford Hosp, 69-75; pres, Nat Fund Med Educ, 75-87, PROF FAMILY PRACT, SCH MED & DENT, UNIV CONN, 69- *Concurrent Pos:* Fel internal med, Mayo Found, Minn, 49-50; res fel, Huntington Lab, 51-53; Damon Runyon fel, Harvard Med Sch, 53-54; asst, Mass Gen Hosp, 54-60, clin assoc, 61-69; pres, Educ Comn Foreign Med Grads, 70-76. *Mem:* Fel Am Col Physicians; Asn Hosp Med Educ (vpres, 70-72); Asn Am Med Cols; Soc Med Adminrs (pres, 79-81). *Res:* Chemotherapy of cancer. *Mailing Add:* Dept Family Med Univ Conn Health Ctr Farmington CT 06032

FREYMANN, MOYE WICKS, b Omaha, Nebr, Sept 2, 25; m 56; c 3. PUBLIC HEALTH. *Educ:* Yale Univ, BS, 45; Johns Hopkins Univ, MD, 48; Harvard Univ, MPH, 56, DrPH, 60. *Prof Exp:* Intern internal med, Univ Hosp, Yale Univ, 49-51; health officer, US Tech Coop Mission, Iran, 52-55; chief consult health & family planning, India Off, Ford Found, 57-66; dir, Carolina Pop Ctr, 66-74, PROF HEALTH ADMIN, SCH PUB HEALTH, UNIV NC, CHAPEL HILL, 66-; SPEC ASST TO ASST SECY HEALTH & SCI AFFAIRS, US HEALTH & SERV, 71- *Concurrent Pos:* Fel virol, Med Sch, Yale Univ, 48-49; USPHS trainee, 51; mem adv comt pop dynamics, Pan-Am Health Orgn; consult, WHO; spec asst to asst secy health & sci affairs, Dept Health, Educ & Welfare, 71-; consult, UN Develop; consult, US Govt Agency Int Develop, NIH, govts of Egypt, Mexico & various UN orgn; med dir, US Public Health Serv Reserve; vis sr scientist, Battela Mem Inst, 81-; bd dir, US Planned Parenthoood Fedn. *Mem:* AAAS; fel Am Pub Health Asn; fel Am Sociol Asn; Soc Int Develop; Int Union for Sci Study of Pop. *Res:* International development theory and practice; population dynamics; public health policy and administration. *Mailing Add:* Univ North Carolina Pub HHL Univ North Carolina at Chapel Hill Rosenau Hall Chapel Hill NC 27514

FREYMUTH, PETER, b Warmbrunn, Ger, Dec 4, 36; m 65; c 2. AEROSPACE ENGINEERING. *Educ:* Berlin Tech Univ, MPhysics, 62, DrEng, 65. *Prof Exp:* Res assoc aerodynamics, inst turbulence res, Ger Exp Estab Air & Space Res, Berlin, 62-65; res assoc aerospace eng, 66-67, from asst prof to assoc prof, 67-81, PROF AEROSPACE ENG SCI, UNIV COLO, BOULDER, 81- *Honors & Awards:* Erich Trefftz Award, Ger Soc Flight Sci, 66. *Mem:* Am Phys Soc; Am Inst Aeronaut & Astronaut; Sigma Xi. *Res:* Aerospace engineering sciences, especially hydrodynamic stability, theory and design of hot-wire anemometers; experimental investigation of turbulent flows; flow visualization; unsteady flow. *Mailing Add:* Dept Aerospace Eng Sci Univ Colo Campus Box 429 Boulder CO 80309-0429

FREYRE, RAOUL MANUEL, b Gibara, Cuba, Jan 18, 31. PHYSICS, MATHEMATICS. *Educ:* Inst Holquin, Cuba, BS, 49; Univ Havana, PhD, 55. *Prof Exp:* Asst physics, Inst Holquin, Cuba, 53-54, asst prof math, 54-55, prof & head dept, 55-57; from instr to asst prof physics, NC State Col, 57-63; res physicist, Nat Co, Mass, 63; head dept physics, Col Adv Sci, Canaan, NH, 63-64, dean, 64; asst prof math & physics, Lowell Technol Inst, 64-66; assoc prof, 66, PROF MATH, BOSTON STATE COL, 67- *Mem:* Soc Indust & Appl Math; Am Math Soc; Math Asn Am; Sigma Xi. *Res:* Lasers. *Mailing Add:* 48 Ships Way Bourne MA 02532

FREYTAG, PAUL HAROLD, b Laramie, Wyo, Dec 3, 34; m 67; c 2. SYSTEMATIC ENTOMOLOGY. *Educ:* Univ Wyo, BS, 56; Ohio State Univ, MSc, 60, PhD(entom), 63. *Prof Exp:* Res assoc entom, Ohio State Univ, 63-64, asst prof, 64-66; asst prof, Ark State Col, 66-67; assoc prof, 67-82, PROF, UNIV KY, 82- *Concurrent Pos:* NSF grant, 64-72. *Mem:* AAAS; Entom Soc Am. *Res:* Systematics of leafhoppers (Cicadellidae) and their parasites. *Mailing Add:* Dept Entom Univ Ky Lexington KY 40546-0091

FREYTAG, SVEND O, MOLECULAR GENETICS, GENE REGULATION. *Educ:* Case Western Univ, PhD(biochem), 81- *Prof Exp:* ASST PROF BIOCHEM & MOLECULAR BIOL, SCH MED, UNIV MICH, 84- *Res:* Cellular differentiation. *Mailing Add:* Dept Biol & Chem Univ Mich Med Sch Med Sci Bldg 1 Rm 4311 Ann Arbor MI 48109

FREZON, SHERWOOD EARL, b Highland Park, Mich, Nov 28, 21. ECONOMIC GEOLOGY. *Educ:* Univ Mich, BS, 50, MS, 63. *Prof Exp:* GEOLOGIST, US GEOL SURV, 51- *Mem:* Am Asn Petrol Geol; Geol Soc Am; Paleont Soc; Soc Econ Paleont & Mineral. *Res:* Paleozoic stratigraphic studies in Oklahoma and Arkansas; domestic oil and gas resource appraisal. *Mailing Add:* 410 Queen St Denver CO 80226

FRIAR, BILLY W(ADE), b Rose Hill, Va, July 18, 31. MECHANICAL ENGINEERING. *Educ:* Berea Col, AB, 53; Va Polytech Inst, BS, 58; Ohio State Univ, MSc, 59, PhD(mech eng), 70. *Prof Exp:* Asst prof mech eng, Va Polytech Inst, 60-62; ASST PROF ENG, WRIGHT STATE UNIV, 70- *Concurrent Pos:* Mech engr, Babcock & Wilcox Co, 60; facilities eng sect, US Army Chem Ctr, 61. *Mem:* AAAS; Sigma Xi. *Res:* Thermodynamics; measurement of gas density; ionization of gases by beta-particle radiation. *Mailing Add:* Mech Syst Eng Dept Wright State Univ Dayton OH 45435

FRIAR, JAMES LEWIS, b Mansfield, Ohio, June 26, 40; m 62; c 2. NUCLEAR PHYSICS, THEORETICAL PHYSICS. *Educ:* Case Inst Technol, BS, 62; Stanford Univ, PhD(physics), 67. *Prof Exp:* NATO fel nuclear physics, Europ Orgn Nuclear Res, 67-68; fel, Univ Wash, 68-70; fel, Mass Inst Technol, 70-72; asst prof, Brown Univ, 72-76; MEM STAFF INTERMEDIATE ENERGY NUCLEAR PHYSICS, LOS ALAMOS NAT LAB, 76- *Mem:* Am Phys Soc. *Res:* Electromagnetic nuclear physics; few body problem. *Mailing Add:* 460 Cheryl Los Alamos NM 87544

FRIAR, ROBERT EDSEL, b Warren, Ind, Dec 30, 33; m 61; c 3. HUMAN SEXUALITY, BIRTH CONTROL. *Educ:* Purdue Univ, BS, 56, MS, 59, PhD(physiol), 68. *Prof Exp:* Instr high sch, Ind, 56-58, 59-64; from asst prof to assoc prof physiol, 67-74, head dept biol sci, 71-80, PROF PHYSIOL, FERRIS STATE COL, 74- *Concurrent Pos:* Fulbright scholar, Univ Khartoum, Sudan, 85-87; mem bd dir, Soc Sci Study Sex, 90-93. *Mem:* Am Physiol Soc; Soc Study Reprod; Soc Sci Study Sex; AAAS; Human Anat & Physiol Soc. *Res:* Affinity of various estrogenic compounds for the estrogen receptor; human physiology; effectiveness of various methods of birth control; improved methods of teaching. *Mailing Add:* Dept Biol Ferris State Univ Big Rapids MI 49307

FRIARS, GERALD W, b Sussex, NB, Apr 26, 29; m 54; c 3. POULTRY BREEDING. *Educ:* McGill Univ, BSc, 51; Purdue Univ, MSc, 55, PhD(genetics), 61. *Prof Exp:* Asst genetics, Ont Agr Col, 51-55, lectr, 55-61, res scientist, 61-63; from asst prof to prof genetics, Univ Guelph, 63-85; CHIEF SCIENTIST, ATLANTIC SALMON RES INST, 85-; RES ASSOC, UNIV NB, 85- *Concurrent Pos:* Nat Res Coun Can grants quantum genetics, 64-; res grants, Can Dept Agr, Ont Turkey Bd; consult geneticist, Peels Poultry Farm Ltd, Shaver Poultry Breeding Farms & King Cole Ducks; mem, NB Inst Agrologists. *Mem:* Biomet Soc; Poultry Sci Asn; Genetics Soc Can; Sigma Xi; Aquacult Asn Can. *Res:* Quantitative genetics of economic and physiological traits of chickens and turkeys; quantitative and population genetics of Tribolium castaneum used as a pilot organism; quantitative genetics of fish. *Mailing Add:* Atlantic Salmon Res Inst PO Box 429 St Andrews NB E0G 2X0 Can

FRIAUF, ROBERT J, b Pittsburgh, Pa, Mar 31, 26; m 49; c 3. SOLID STATE PHYSICS. *Educ:* Duke Univ, BS, 47; Univ Chicago, SM, 51, PhD(physics), 53. *Prof Exp:* From asst prof to assoc prof, 53-64, PROF PHYSICS, UNIV KANS, 64- *Concurrent Pos:* Fulbright fel, Univ Stuttgart, 65-66; Argonne Univs Asn distinguished award, Argonne Nat Lab, 78-79. *Mem:* Am Chem Soc; fel Am Phys Soc; Am Asn Physics Teachers; Sigma Xi; Soc Imaging Sci & Technol; Am Asn Univ Professors. *Res:* Ionic conductivity, diffusion and color centers in ionic crystals; theory of correlation effects for diffusion and of electronic structure of point defects in insulators; molecular dynamics modeling of insulator surface-plasma interactions. *Mailing Add:* Dept Physics Univ Kans Lawrence KS 66045

FRIBERG, EMIL EDWARDS, b Wichita Falls, Tex, Apr 11, 35; m 57; c 3. ENERGY USE ANALYSIS, ENERGY CONSERVATION. *Educ:* Univ Tex, Austin, BSME, 58. *Prof Exp:* Engr, Tex Elec Serv Co, 58-64, eng consult, 64-69; prin, Cowan, Love & Jackson, Inc, 69-71; secy-treas, Love, Jackson & Friberg, Inc, 71-73; PRES, FRIBERG ASSOCS INC, 73- *Honors & Awards:* Distinguished Serv Award, Am Soc Heating, Refrig & Air Conditioning Engrs. *Mem:* Fel Am Soc Heating, Refrig & Air Conditioning Engrs (dir & regional chmn VIII, 75-78, dir-at-large, 79-82); Am Soc Mech Engrs; Soc Petrol Engrs, Am Inst Mining, Metall & Petrol Engrs; Nat Soc Prof Engrs; Am Consult Engrs Coun; Consult Engrs Coun Tex (pres, 79-80). *Res:* Energy analysis for buildings and energy conservation; computer programs for calculating energy use; building energy management systems. *Mailing Add:* 3406 Woodford Dr Arlington TX 76013

FRIBERG, JAMES FREDERICK, b Florence, Wis, Aug 27, 43; m 68; c 2. GEOLOGY, SEDIMENTOLOGY. *Educ:* Univ Wis-Milwaukee, BS, 65; Ind Univ, MA, 67, PhD(geol), 70. *Prof Exp:* Res geologist, 70-75, sr res geologist, 75-78, res assoc petrol explor, 78-79, SUPVR, SEDIMENTARY GEOL RES, UNION OIL CO CALIF RES CTR, BREA, 80- *Mem:* Am Asn Petrol Geologists; Soc Econ Paleontologists & Mineralogists. *Res:* Sandstone depositional models; diagenesis of sandstones; exploration methodology for stratigraphic traps; tectonic and sedimentological evolution of fore-arc basins. *Mailing Add:* 20061 Canyon Dr Yorba Linda CA 92686

FRIBERG, LAVERNE MARVIN, b Siren, Wis, Mar 11, 49; m 71; c 2. IGNEOUS & METAMORPHIC PETROLOGY, MICROCOMPUTER APPLICATIONS. *Educ:* Univ Wis, River Falls, BS, 71; Ind Univ, MS, 74, PhD(geol), 76. *Prof Exp:* Asst prof, 76-81, ASSOC PROF PETROL, UNIV AKRON, 81- *Mem:* Mineral Soc Am; Geochem Soc; Mineral Asn Can; Geol Soc Am; Sigma Xi. *Res:* Igneous and metamorphic petrology; utilizing petrography, chemical and cathodoluminescence data in research activities. *Mailing Add:* Dept Geol Univ Akron Akron OH 44325-4101

FRIBERG, STIG E, b Tjärstao, Sweden, Feb 8, 30; US citizen; m 67; c 2. EMULSIONS, FOAMS. *Educ:* Univ Stockholm, Sweden, BS, 58, DSci(chem), 66, PhD(chem), 67. *Prof Exp:* Teaching asst chem, Univ Stockholm, Sweden, 57-62; res scientist, Res Inst Nat Defense, Stockholm, 66-69; head, Lab Surface Chem, Stockholm, 68-69; dir, Swedish Inst Surface Chem, Stockholm, 69-76; prof & chair, Univ MBS, Rolla, 76-79, Curtors prof, 79-87; PROF & CHAIR, CLARKSON UNIV, 87- *Concurrent Pos:* Adj prof chem, Chem Ctr, Univ Lund, Swe, 73-76; ed, Jour Dispersion Sci & Technol, 85- *Honors & Awards:* Kendall Award, Am Chem Soc, 85. *Mem:* Am Inst Chemists fel; Royal Swedish Acad Eng Sci; Am Chem Soc; Am Assoc Adv Sci. *Res:* application of amphiphilic association structures to the properties of emulsions, foam and microemulsions; reactions in microemulsions. *Mailing Add:* Dept Chem Clarkson Col Potsdam NY 13676

FRIBOURG, HENRY AUGUST, b Paris, France, Mar 10, 29; nat US; m 56; c 2. CROP ECOLOGY. *Educ:* Univ Wis, BS, 49; Cornell Univ, MS, 51; Iowa State Univ, PhD(agron), 54. *Prof Exp:* From asst agronomist to assoc agronomist forage crop prod res, 56-70, PROF PLANT & SOIL SCI, UNIV TENN, KNOXVILLE, 70- *Concurrent Pos:* Interpreter, Int Grassland Cong, Pa State Univ, 52; chief interpreter, Inter-Am Meeting Livestock Prod, Brazil, 52 & food & agr meeting on exten methods, Caribbean Area, Jamaica, 54; pres, Southern Pasture Forage Crops Improv Conf & Southern Appalachian Sci Fair; vis scientist, Tenn Acad of Sci, 65-; sr lectr & Fulbright-Hays lectr, Ataturk Univ, 73-74; assoc ed, Crop Sci, 85-89 & J Prod Agr, 85-91. *Mem:* Fel Am Soc Agron; Am Meteorol Soc. *Res:* Digital computer use in agronomic research; forage crop ecology and management; crop climatology. *Mailing Add:* Dept Plant & Soil Sci Univ Tenn Knoxville TN 37901-1031

FRIBOURGH, JAMES H, b Sioux City, Iowa, June 10, 26; m 55; c 3. ZOOLOGY, SCIENCE EDUCATION. *Educ:* Univ Iowa, BA & MS, 49, PhD(zool, sci educ), 57. *Hon Degrees:* LHD, Morningside Col, 89. *Prof Exp:* Instr biol, 49-56, assoc prof, 57-59, prof & chmn dept, 59-69, vpres acad affairs, 69-70, vchancellor acad affairs, 70-72, interim chancellor, 72-82, exec vchancellor, 73-82, provost, 82-84, DISTINGUISHED PROF, UNIV ARK, LITTLE ROCK, 84- *Concurrent Pos:* Res biologist, Fish Farming Exp Sta, US Dept Interior, 60-; vis scientist, NSF Vis Scientist Prog, Ark Acad Sci; consult, Radioisotope Serv, Vet Admin Hosp, Little Rock, 60- *Mem:* Fel AAAS; fel Am Inst Fisheries Res Biologists; Am Fisheries Soc; Sigma Xi; Electron Micros Soc Am; Am Soc Swed Engrs; fel Col Preceptors, London, Eng. *Res:* Fisheries biology and culture; experimental embryology. *Mailing Add:* Univ Ark 33rd & University Ave Little Rock AR 72204

FRICK, JOHN P, b Kansas City, Mo, Jan 20, 44; m 67; c 2. CORROSION, QUALITY ASSURANCE. *Educ:* Univ Kans, BS, 66; Pa State Univ, MS, 69, PhD(metall), 72. *Prof Exp:* Metallurgist, Wean United, 72-74; res metallurgist, Youngstown Sheet & Tube Co, 74-77; res scientist, Dresser Indust, 78-79; SR ENGR ADV, MOBIL OIL CORP, 79- *Concurrent Pos:* Ed, Woldman's Eng Alloys. *Mem:* Am Soc Metals; Am Inst Mining, Metall & Petrol Engrs; Nat Asn Corrosion Engrs. *Res:* Corrosion phenomena with particular emphasis on environmental cracking; tribological and metallurgical phenomena. *Mailing Add:* 6924 Middle Cove Dr Dallas TX 75248

FRICK, NEIL HUNTINGTON, b Rockville Centre, NY, July 14, 33; m 60; c 4. PHYSICAL CHEMISTRY, POLYMER CHEMISTRY. *Educ:* Col Wooster, BA, 60; Princeton Univ, MA, 62, PhD(phys chem), 64. *Prof Exp:* Sr res chemist, 64-67, from res assoc to sr res assoc, 67-72, asst dir indust coatings develop, 72-73, mgr coil, appliance & container coatings develop, 73-79, mgr indust coating res, 79-84, dir, 84-86, VPRES RES & DEVELOP, PPG INDUSTS, INC, 87- *Mem:* Am Chem Soc; Fedn Socs Coatings Technol; Sigma Xi. *Res:* Polymer characterization; electrophoretic deposition of organic coatings; development of new coatings for the coil appliance and container segments of the marketplace; involves the polymer properties, application properties, final film properties and colloidal behavior of systems; research administration, technical management. *Mailing Add:* 4028 Ewalt Rd Gibsonia PA 15044

FRICK, OSCAR L, b New York, NY, Mar 12, 23; m 54. IMMUNOLOGY, PEDIATRICS. *Educ:* Cornell Univ, AB, 44, MD, 46; Univ Pa, MMedSci, 60; Stanford Univ, PhD(microbiol), 64. *Prof Exp:* Assoc prof, 64-73, PROF PEDIAT, SCH MED, UNIV CALIF, SAN FRANCISCO, 73- *Concurrent Pos:* Mem, Sub-bd allergy, Am Bd Pediat, 67-73; secy-gen, Int Asn Allergol & Clin Immunol, 85-94. *Honors & Awards:* Bret Ratner Award, Am Acad Pediat, 82. *Mem:* Am Asn Immunologists; Am Acad Allergy & Immunol (pres, 77-78); Am Acad Pediat; Int Asn Allergol; Am Thoracic Soc. *Res:* Hypersensitivity, especially immediate type related to immunoglobulins and allergy; adaptation of clinical situations to laboratory evaluation. *Mailing Add:* Dept Pediat Sch Med Univ Calif Box 0546 San Francisco CA 94143

FRICK, PIETER A, b Heidelberg, Transvaal, SAfrica, Feb 11, 42; US citizen; m 67; c 1. COMPUTER ARCHITECTURE. *Educ:* Univ Stellenbosch, SAfrica, BSc, 64, MEng, 66; Imperial Col Sci & Technol, UK, DIC, 71; Univ London, PhD(control systs), 72. *Prof Exp:* Tech officer elec eng, SAfrican Air Force, 64-67; res engr, SAfrican Iron & Steel Corp, 67-68; lectr & scholar syst sci, syst sci dept, Univ Calif, Los Angeles, 71-72; from asst prof to assoc prof elec eng, Ore State Univ, 73-80; prof & dept head elec eng, Portland State Univ, 80-87; DEAN, COL ENGR, 88- *Concurrent Pos:* Vis prof, Inst di Automatica, Univ Rome, 76 & Ore Grad Ctr, 79-80; lectr, Univ Pretoria, SAfrica, 78; head, comput sci dept, Portland State Univ, 84-87. *Mem:* Soc Indust & Appl Math; sr mem, Inst Elec & Electronic Engrs; Nat Security Indust Asn; Asn Comput Mach. *Res:* Development of new computer architectures that are efficient for on-line real time control systems and their implementation in silicon; industrial robots, vision and computer systems; computer control systems. *Mailing Add:* Col Eng & Appl Sci Colo Univ PO Box 7150 Colo Springs CO 80933-7150

FRICKE, ARTHUR LEE, b Huntington, WVa, Mar 6, 34; m 54; c 3. CHEMICAL ENGINEERING. *Educ:* Univ Cincinnati, ChE, 57; Univ Wis, MS, 59, PhD(chem eng), 61. *Prof Exp:* Instr chem eng, Univ Wis, Madison, 59-60; res engr, Shell Develop Co, Calif, 61-63, group leader mech eng res, 63-65, asst dept mgr mkt develop prod, 65-66, sr technologist, process develop, NY, 66-67; from asst prof to assoc prof chem eng, Va Polytech Inst & State Univ, 67-76; PROF CHEM ENG & CHMN DEPT, UNIV MAINE, 76- *Concurrent Pos:* Vpres develop, Polytron Corp, Va, 68-69. *Mem:* Am Chem Soc; Am Inst Chem Engrs; Soc Plastics Engrs; Tech Asn Pulp & Paper Indust. *Res:* Polymer foaming; mass and heat transfer in polymers; phase change in polymers; composities; economic analysis. *Mailing Add:* 1502 NW 110th Terr Gainesville FL 32606-5466

FRICKE, EDWIN FRANCIS, b Mackay, Idaho, July 25, 10; m 42; c 5. NUCLEAR PHYSICS, CHEMICAL PHYSICS. *Educ:* Univ Idaho, BS, 35; Univ Calif, Los Angeles, MA, 37, PhD(physics), 40. *Prof Exp:* Design engr, Stone & Webster Engr Corp, Mass, 40-43; engr, Manhattan Proj, Kellex Corp, NY, 43-44; sr engr, Sylvania Elec Prod, Inc, 44-45; phys chemist, Gen Chem Co, 45-46; physicist, Repub Aviation Corp, 46-49; nuclear physicist & staff engr, Argonne Nat Lab, 50-56; sr nuclear physicist, Nuclear Prod Div, ACF Industs, 56-59; sr develop engr & nuclear physicist, Repub Aviation Corp, 59-65; res scientist, Bell Aerosysts Co, 65-66; sr staff scientist, Sanders Assocs, Nashua, 66-68; prin engr, Jackson-Moreland Engrs, 68-70; nuclear eng consult, 70-80; RETIRED. *Concurrent Pos:* Instr, Fournier Inst Technol, 50-55. *Mem:* AAAS; Am Chem Soc; Am Nuclear Soc; Sigma Xi; NY Acad Sci. *Res:* Adsorption of sound in five triatomic gases; statistical thermodynamics applied to chemical kinetics. *Mailing Add:* County Rd Merrimack NH 03054

FRICKE, GERD, b Magdeburg Ger, Mar 28, 46; m; c 2. MATHEMATICS ANALYSIS & FUNCTIONAL ANALYSIS. *Educ:* Kansas Univ, MA, 69; Kent State Univ, PhD(math), 71. *Prof Exp:* PROF MATH, WRIGHT STATE UNIV, 73- *Concurrent Pos:* Mem, IBM. *Mem:* Math Asn Am; Sigma Xi. *Res:* Combinatorics and finite mathematics. *Mailing Add:* Dept Math & Statist Wright State Uni Dayton OH 45435

FRICKE, GORDON HUGH, b Buffalo, NY, Apr 18, 37; m 60; c 2. ANALYTICAL CHEMISTRY, CHEMOMETRICS. *Educ:* Goshen Col, BA, 64; State Univ NY Binghamton, MA, 66; Clarkson Col Technol, PhD(chem), 71. *Prof Exp:* Res assoc, State Univ NY Buffalo, 70-71; res fel, Wright State Univ, 71-72; asst prof, 72-75, ASSOC PROF CHEM, IND UNIV-PURDUE UNIV, INDIANAPOLIS, 75- *Concurrent Pos:* Indust fel, Dow Chem Co, 80-81, consult, 81- *Mem:* Sigma Xi. *Res:* experimental optimization techniques; expert systems and chemometrics. *Mailing Add:* Dept Chem Ind Univ-Purdue Univ 1125 E 38th St Indianapolis IN 46205-2810

FRICKE, MARTIN PAUL, b Franklin, Pa, May 18, 37; m 59. NUCLEAR PHYSICS. *Educ:* Drexel Inst, BS, 61; Univ Minn, MS, 64, PhD(physics & math), 67. *Prof Exp:* Res asst physics, Univ Minn, 61-64; fel, Univ Mich, 67-68; staff physicist, Defense Sci Dept, Gulf Gen Atomic Inc, 68-70, staff physicist & cross sect group leader, Gulf Radiation Technol Div, Gulf Energy & Environ Systs Co, San Diego, 70-74; div mgr, Sci Appl, Inc, 74-76, asst vpres, 76-77, vpres, 77- 79, corp vpres, 79-; sr vpres, Titan Systs Inc, 84-87, exec vpres, Titan Technol, 87-89, SR VPRES, THE TITAN CORP, SAN DIEGO, 89- *Concurrent Pos:* Assoc res scientist, Res Lab, Honeywell Corp, Minn, 62-64. *Mem:* Fel Am Phys Soc. *Res:* Experimental and theoretical research in nuclear reactions. *Mailing Add:* Titan Corp PO Box 12139 La Jolla CA 92037

FRICKE, ROBERT F, biochemistry, toxicology, for more information see previous edition

FRICKE, WILLIAM G(EORGE), JR, b Pittsburgh, Pa, May 10, 26; m 48; c 4. METALLURGY. *Educ:* Pa State Univ, BS, 50, MS, 51; Univ Pittsburgh, PhD(metall eng), 61. *Prof Exp:* Res metallurgist, Alcoa Labs, Aluminum Co Am, 52-67, sr res engr, 67-73, group leader, 73-85, sr sci assoc, 85-90, FEL, ALCOA LABS, ALUMINUM CO AM, 90- *Honors & Awards:* Templin Award, Am Soc Testing & Mat, 55. *Mem:* Am Soc Metals Int; Am Inst Mining, Metall & Petrol Engrs; Int Soc Stereology; Am Soc Testing & Mat; Microbeam Anal Soc; Sigma Xi. *Res:* Physical metallurgy and metallography; mechanisms of metal fatigue; deformation and plastic flow; electron microprobe analysis; electron microscopy; auger analysis; crystallographic texture and diffusion. *Mailing Add:* Alloy Technol Div Alcoa Labs Aluminum Co Am Alcoa Center PA 15069

FRICKEN, RAYMOND LEE, b New Orleans, La, June 25, 37. HIGH ENERGY PHYSICS, RESEARCH ADMINISTRATION. *Educ:* Loyola Univ, La, BS, 59; La State Univ, PhD(physics), 63. *Prof Exp:* Physicist, Div High Energy Physics, US AEC, 63-74, Energy Res & Develop Admin, 74-77, US Dept Energy, 77-81, chief prog opers br, 81-87, actg dir, 87-88, EXEC OFFICER, SSC DIV, US DEPT ENERGY, 88- *Mem:* Am Phys Soc. *Res:* Interactions of primary cosmic rays; high energy particle accelerators. *Mailing Add:* Off SSC US Dept Energy ER-90 Washington DC 20545

FRICKEY, PAUL HENRY, b Syracuse, NY, Nov 14, 31; m 56; c 2. MICROBIOLOGY, VIROLOGY. *Educ:* Syracuse Univ, BS, 54, MS, 57; Univ Rochester, PhD(microbiol), 63. *Prof Exp:* Res virologist, Biol Process & Prod Improv Dept, Lederle Labs Div, Am Cyanamid Co, 62-64, res virologist, Virus & Rickettsial Res Sect, 64-70; dir prod & develop, Flow Labs, Inc, 70-73; RES ASSOC, EASTMAN KODAK CO, 73- *Mem:* AAAS; Am Soc Microbiol; Sigma Xi. *Res:* Factors affecting multiplication of various human respiratory viruses in primate and avian tissues and the development of vaccines from such viruses; identification and anti-viral activity of antibodies in nasal secretions using immunological techniques; development of viral diagnostics. *Mailing Add:* 49 Country Club Dr Rochester NY 14618

FRIDAY, ELBERT W, JR, b DeQueen, Ark, July 13, 39; m 59; c 2. GENERAL ATMOSPHERIC SCIENCES. *Educ:* Univ Okla, BS, 61, MS, 67, PhD(meteorol), 69. *Prof Exp:* Weather Officer, US Airforce, 61-81; dir, Environ & Life Sci, Dept Defense, 78-81; dep dir, 81-87, DIR, NAT WEATHER SERV, 87- *Mem:* Fel Am Meteorol Soc; Sigma Xi; Nat Weather Asn. *Mailing Add:* 1325 East-West Hwy Rm 18130 Silver Spring MD 20910

FRIDINGER, TOMAS LEE, b Washington, DC, Dec 21, 40; m 63; c 2. PESTICIDE CHEMISTRY, ORGANIC CHEMISTRY. *Educ:* Col William & Mary, BS, 62; Univ Md, PhD(org chem), 67. *Prof Exp:* Sr res chemist, Minn Mining & Mfg Co, 67-70, res specialist, 70-72, supvr synthesis & process chem, 72-74, mgr chem & regulatory affairs, 74-75, tech mgr, 75-79, LAB MGR, 3M CO, 79- *Mem:* Am Chem Soc; Weed Sci Soc Am. *Res:* Synthetic approaches to azirinones; synthesis, formulation and screening of potential agrichemicals, herbicides, plant growth regulators, insecticides, plant disease control agents; agrichemical process, metabolism and environmental chemistry; Environ Protection Agency registration of agrichemicals. *Mailing Add:* 270-2N-03 3M Co St Paul MN 55144

FRIDLAND, ARNOLD, b Antwerp, Belg; Can citizen; m 63; c 2. BIOCHEMICAL PHARMACOLOGY. *Educ:* Univ Montreal, BS, 63, MS, 64; McGill Univ, PhD(biochem), 68. *Prof Exp:* Fel oncol, McArdle Lab Cancer Res, 68-71; asst mem, 71-76, ASSOC MEM PHARMACOL, ST JUDE CHILDREN'S RES HOSP, 76- *Mem:* Am Chem Soc; Am Asn Cancer Res; Am Soc Biol Chemists. *Res:* Molecular mechanism of action of antimetabolites; DNA replication in eukaryotic cells. *Mailing Add:* Dept Pharm St Jude Children's Res Hosp 332 N Lauderdale PO Box 318 Memphis TN 38101

FRIDLEY, ROBERT B, b Burns, Ore, June 6, 34; m 55; c 3. RESEARCH & TEACHING ADMINISTRATION. *Educ:* Univ Calif, Berkeley, BS, 56; Univ Calif, Davis, MS, 60; Mich State Univ, PhD(agr eng), 73. *Hon Degrees:* Dr, Universidad Politécnica de Madrid, 88. *Prof Exp:* Prof agr eng, Univ Calif, Davis, 69-78, actg assoc dean, Col Eng, 72, chmn dept, 74-76; mgr dept res & develop, Weyerhaeuser Co, 77-85; dir aquacult & fisheries prog, 85-90, EXEC ASSOC DEAN, COL AGR & ENVIRON SCI, UNIV CALIF, DAVIS, 89- *Concurrent Pos:* Asst specialist, Univ Calif, Davis, 56-60, from asst prof to assoc prof, 61-69. *Honors & Awards:* Charles G Woodbury Award, Am Soc Hort Sci, 66; Young Researcher Award for Eng Achievement, Am Soc Agr Engrs, 72-, Presidential Distinguished Serv Award, 88. *Mem:* Nat Acad Eng; Am Fisheries Soc; Am Soc Eng Educ; fel Am Soc Agr Eng; World Aquacult Soc. *Res:* Engineering research, design and development of production systems in aquaculture, horticulture and silviculture, particularly as related to ocean ranching, fruit harvest mechanization and reforestation. *Mailing Add:* Univ Calif 228 Mrak Hall Davis CA 95616

FRIDLUND, PAUL RUSSELL, b Minneapolis, Minn, Jan 3, 20; m 49; c 4. PLANT PATHOLOGY. *Educ:* Augsburg Col, BA, 42; Univ Minn, MS, 52, PhD(plant path), 54. *Prof Exp:* Asst, Univ Minn, 46-49; supvr sect plant path, State Dept Agr, Minn, 49-55; assoc plant pathologist, Irrigation Exp Sta, 55-66, PLANT PATHOLOGIST, IRRIGATED AGR RES & EXTEN CTR, WASH STATE UNIV, 66- *Concurrent Pos:* Nat Acad Sci exchange scientist, Romania, 68, 77, 84 & 86; consult, de la Vince de Bergerac, 85. *Mem:* Am Pomol Soc; Am Phytopath Soc. *Res:* Virus diseases of deciduous fruit trees; thermotherapy of virus diseased plants. *Mailing Add:* Irrigated Agr Res & Exten Ctr Wash State Univ Prosser WA 99350

FRIDOVICH, IRWIN, b New York, NY, Aug 2, 29; m 52; c 2. BIOCHEMISTRY, ENZYMOLOGY. *Educ:* City Col New York, BS, 51; Duke Univ, PhD(biochem), 55. *Hon Degrees:* Dr, l'Université Rene Descartes, Paris, France, 80. *Prof Exp:* Instr, 56-58, from asst prof to assoc prof, 61-71, dir grad stud, 65-67, prof biochem, 71-76, JAMES B DUKE PROF BIOCHEM, DUKE UNIV, 76- *Concurrent Pos:* NIH res fel, 55-56; NIH res career develop award, 59-69; mem, Biochem Study Sect, NIH, 68-71; mem, Nat Bd Med Examrs, 73-76; mem, adv comt biochem & chem carcinogenesis & study sect, Am Cancer Soc, 80-84; mem, Sci Adv Comm, Sandoz Found, 87- *Honors & Awards:* Cressy A Morrison Award in Sci, NY Acad Sci, 84; Herty Medal, Am Chem Soc, 80; Townsend Harris Medal, 90. *Mem:* Nat Acad Sci; Am Chem Soc; Am Acad Arts & Sci; Am Soc Biol Chemists (pres-elect 81-82, pres, 82-83). *Res:* Generation and scavenging of free radicals in biological systems. *Mailing Add:* Dept Biochem Duke Univ Med Ctr Durham NC 27710

FRIDOVICH-KEIL, JUDITH LISA, b Durham, NC, Feb 8, 61; m 84. CELL CYCLE REGULATION, REGULATION OF GENE EXPRESSION. *Educ:* Princeton Univ, AB, 83; Mass Inst Technol, PhD(biol), 88. *Prof Exp:* TEACHING ASST & INSTR GENETICS, HARVARD UNIV EXTEN SCH, 88- *Concurrent Pos:* Postdoctoral res fel cell growth & regulation, Dana-Farber Cancer Inst & Harvard Med Sch, 88-90, training fel clin molecular genetics, 88-91. *Mem:* Am Soc Cell Biol; Am Soc Human Genetics; AAAS. *Res:* Molecular aspects of cell growth and regulation; molecular aspects of human genetic disease. *Mailing Add:* Dana-Farber Cancer Inst Rm D810A 44 Binney St Boston MA 02115

FRIDY, JOHN ALBERT, b Lancaster, Pa, Sept 30, 37; m 71; c 4. MATHEMATICS. *Educ:* Pa State Univ, BS, 59, MA, 61; Univ NC, PhD(math), 64. *Prof Exp:* Asst prof math, Rutgers Univ, 64-66; assoc prof, 67-77, PROF MATH SCI, KENT STATE UNIV, 77- *Concurrent Pos:* NSF res grant, 65-67. *Mem:* Am Math Soc; Math Asn Am. *Res:* Summability theory; number theory. *Mailing Add:* Dept Math Kent State Univ Kent OH 44242

FRIEBELE, EDWARD JOSEPH, b New York, NY, Oct 15, 46; m 70; c 2. SOLID STATE PHYSICS, ELECTRICAL ENGINEERING. *Educ:* Davidson Col, BS, 68; Vanderbilt Univ, MS, 70, PhD(solid state physics, elec eng), 73. *Prof Exp:* Nat Res Coun/Naval Res Lab fel, 73-75, prog mgr, Defense Advan Res Projs Agency, 80-85, RES PHYSICIST SOLID STATE PHYSICS, NAVAL RES LAB, 75- *Honors & Awards:* George Morey Award for Glass Res, Am Ceramic Soc. *Mem:* Am Phys Soc; Inst Elec & Electronics Engrs; fel Am Ceramic Soc; Sigma Xi; Soc Photog & Instrumentation Engrs. *Res:* Radiation effects in optical fiber waveguides; radiation-induced defect centers in glasses; fiber optic sensors; preparation of radiation resistant and polarization holding optical fibers; guided wave nonlinear optical effects. *Mailing Add:* Code 6505 Naval Res Lab Washington DC 20375-5000

FRIED, BERNARD, b New York, NY, Aug 17, 33; m 69; c 1. PARASITOLOGY. *Educ:* NY Univ, AB, 54; Univ NH, MS, 56; Univ Conn, PhD(zool), 61. *Prof Exp:* Asst zool, Univ NH, 54-56; res technician, Archbold Biol Sta, Fla, 57; instr parasitol, Sch Med, Yale Univ, 59; asst zool, Univ Conn, 59-61; NIH res fel parasitol, Emory Univ, 61-63; from asst prof to assoc prof, 63-70, actg head dept, 70-71 & 81-82, KREIDER PROF BIOL, LAFAYETTE COL, 75- *Concurrent Pos:* Consult thin layer chromatography, Kontes Glass Co, Vineland, NJ, 76 & Ctr Prof Advan, East Brunswick, NJ, 77; assoc ed, Am Soc Parisitol; co-prin investr, NIH grant. *Mem:* Am Soc Parasitol; Am Micros Soc. *Res:* Biology and physiology of endoparasitic trematodes; applications of thin-layer chromatography to biological research. *Mailing Add:* Dept Biol Lafayette Col Easton PA 18042

FRIED, BURTON DAVID, b Chicago, Ill, Dec 14, 25; m 47; c 2. PLASMA PHYSICS. *Educ:* Ill Inst Technol, BS, 47; Univ Chicago, MS, 50, PhD(physics), 52. *Prof Exp:* Instr physics, Ill Inst Technol, 48-52; res physicist, Lawrence Berkeley Lab, Univ Calif, 52-54; mem sr staff, TRW Space Tech Labs, Inc, 54-60, dir res Thompson-Ramo-Wooldridge Comput Div, 60-62; sr staff physicist, TRW Systs Group, 62-86; assoc prof, 63, PROF PHYSICS, UNIV CALIF, LOS ANGELES, 64- *Concurrent Pos:* Consult, Lawrence Livermore Nat Lab & Sandia Nat Lab. *Mem:* Fel Am Phys Soc. *Res:* Physics of elementary particles; quantum field theory; theory of ballistic missile trajectories; on-line computing; magnetohydrodynamics; theoretical plasma physics; controlled fusion. *Mailing Add:* Dept Physics Univ Calif Los Angeles CA 90024

FRIED, DAVID L, b Brooklyn, NY, Apr 13, 33; div; c 3. OPTICAL PHYSICS. *Educ:* Rutgers Univ, BA, 57, MS, 59, PhD(physics), 62. *Prof Exp:* Engr, Astro-Electronic Div, Radio Corp Am, 57-59; sr tech specialist, Space & Info Systs Div, NAm Rockwell Corp, 61-66, sr tech specialist, Autonetics Div, 66-67, chief, Sensor Technol Sect, Electrooptical Lab, 67-68, mgr, Avionic Systs Sect, 69-70; owner, 70-77, PRES, OPTICAL SCI CO, 77- *Concurrent Pos:* Consult mem, Army Sci Bd, 68-86; assoc ed, J Optical Soc Am, 68-82 & Optics Letters 77-79. *Mem:* Am Phys Soc; fel Optical Soc Am; Inst Elec & Electronics Engr. *Res:* Quantum field theory; optical propagation in stochastic media; electrooptic and mechanooptic interactions; signal processing and noise theory; infrared systems; adaptive optics. *Mailing Add:* PO Box 1329 Placentia CA 92670

FRIED, ERWIN, b Vienna, Austria, Aug 24, 22; US citizen; m 47; c 2. HEAT TRANSFER, FLUID MECHANICS. *Educ:* Columbia Univ, BSME, 52; Union Col, MS, 58. *Prof Exp:* Develop engr, Gen Eng Lab, 52-61, consult engr, Space Div, 61-71, CONSULT HEAT TRANSFER & FLUID MECH, MACH APPARATUR OPERS, GEN ELEC CO, 71- *Concurrent Pos:* Thermophys comt, Am Inst Aeronaut & Astronaut, 69-71; chmn, Heat Transfer Div, Am Soc Med Engrs, 77-78; conf chmn, 21st Nat Heat Transfer Conf, Seattle, Wash, 83- *Honors & Awards:* Centennial Medal, Am Soc Mech Engrs, 82. *Mem:* Fel Am Soc Mech Engrs; assoc fel Am Inst Aeronaut & Astronaut. *Res:* Contact heat transfer, performed experimental and analytical work; thermal hydraulics as applied to steam power plants and fluid systems. *Mailing Add:* 12 Oakleaf Dr Clifton Park NY 12065

FRIED, GEORGE H, b New York, NY, Apr 16, 26; m 54; c 3. PHYSIOLOGY. *Educ:* Brooklyn Col, BA, 47; Univ Tenn, MS, 49, PhD(zool), 52. *Prof Exp:* Fel physiol, NY Univ, 53-54; lectr, 57-64, from asst prof to assoc prof, 64-71, chmn, dept, 83-87, PROF BIOL, BROOKLYN COL, 71- *Concurrent Pos:* Res assoc, Levy Found, Beth Israel Hosp, 54-68; lectr, Dent Col, NY Univ, 58-61. *Mem:* Fel AAAS; Am Soc Exp Path; Am Physiol Soc. *Res:* Metabolic studies in obesity; cellular physiology; tissue enzyme levels as a reflection of difference in metabolic intensity in animals of different sizes; enzymatic basis for psychopharmacology; histochemical studies with tetrazolium salts. *Mailing Add:* Dept Biol Brooklyn Col Brooklyn NY 11210

FRIED, HERBERT MARTIN, b New York, NY, Sept 22, 29; m 52; c 3. THEORETICAL PHYSICS. *Educ:* Brooklyn Col, BS, 50; Univ Conn, MS, 52; Stanford Univ, PhD(physics), 57. *Prof Exp:* NSF fel physics, Univ Paris, 57-58; res lectr, Univ Calif, Los Angeles, 58-63; vis mem, Courant Inst Math Sci, NY Univ, 63-64; from asst prof to assoc prof, 64-69, PROF PHYSICS, BROWN UNIV, 69- *Concurrent Pos:* Consult, Rand Corp, 58- *Mem:* Am Phys Soc. *Res:* Quantum field theory; strong-coupling approximation. *Mailing Add:* Dept Physics Brown Univ Providence RI 02912

FRIED, HOWARD MARK, YEAST GENETICS. *Educ:* Cornell Univ, PhD(biochem), 78. *Prof Exp:* ASST PROF BIOCHEM, UNIV NC, 82- *Mailing Add:* Dept Biochem Univ North Carolina CB 7260 335 FLOB Chapel Hill NC 27599

FRIED, JERROLD, b New York, NY, Mar 3, 37; m 65; c 2. BIOPHYSICS. *Educ:* Calif Inst Technol, BS, 58; Stanford Univ, MS, 60, PhD(biophys), 64. *Prof Exp:* Res assoc biol, Hunter Col, 64-65; res assoc, 65-67, assoc, 68-76, assoc mem, Dept Hematopoietic Cell Kinetics, Sloan-Kettering Inst Cancer Res, 76-87; HEALTH SCIENTIST ADMIN, DIV RES GRANTS, NIH, 87- *Concurrent Pos:* Muscular Dystrophy Asn Am, Inc fel, 65; asst prof, Grad Sch Med Sci, Cornell Univ, 68-76, assoc prof, 76-87; Leukemia Soc Am spec fel,

71-73. *Mem:* Cell Kinetics Soc; Soc Anal Cytol; Biophys Soc; Am Asn Cancer Res; Am Soc Hemat. *Res:* Cell proliferation kinetics; mathematical and computer models of cell populations; chemotherapy of acute leukemia; flow cytometry; antigenic modulation; bone marrow purging of tumor cells. *Mailing Add:* 7301 Durbin Terr Bethesda MD 20817-6127

FRIED, JOEL ROBERT, b Memphis, Tenn, Dec 9, 46; m 69; c 2. POLYMER CHEMISTRY. *Educ:* Rensselaer Polytech Inst, BS, 68 & 71, ME, 72; Univ Mass, MS, 75, PhD(polymer sci), 76. *Prof Exp:* Assoc staff mem, Gen Elec, 72-73; sr res engr, Monsanto Co, 76-78; from asst prof to assoc prof chem eng, 78-90, dir, chem eng grad studies, 86-90, DIR, POLYMER RES CTR, UNIV CINCINNATI, 89-, PROF CHEM ENG, 90- *Concurrent Pos:* Consult, Monsanto Co, 79-81, Res Dynamics, 83-84 & Borden Chem, 85; dir chem eng labs, Univ Cincinnati, 82-84; pres, Polymer Res Assocs, Inc, 83-; vis fel, Sci & Eng Res Coun, UK, 85. *Mem:* Am Chem Soc; Am Inst Chem Engrs; Am Phys Soc; Soc Plastics Engrs; Soc Rheology. *Res:* Properties of polymer blends; computer modelling of polymer properties; permeation in polymer systems; polymer composites. *Mailing Add:* Dept Chem Eng Univ Cincinnati Cincinnati OH 45221-0171

FRIED, JOHN, b Leipzig, Ger, Oct 7, 29; US citizen; m 55; c 3. ORGANIC CHEMISTRY, MEDICINAL CHEMISTRY. *Educ:* Cornell Univ, AB, 51, PhD(org chem), 55. *Prof Exp:* Fel, Columbia Univ, 55-56; sect head steroid chem, Merck & Co, NJ, 56-64; dept head, Syntex Inst Steroid Chem, 64-65, assoc dir, 65-67, vpres & dir, Inst Org Chem, 67-75, exec vpres, Syntex Res, 75-76, PRES, SYNTEX RES, 76- *Concurrent Pos:* Dir, Syntex Corp, 82-, sr vpres, 81-85, vchmn bd, 85- *Mem:* Am Chem Soc. *Res:* Organic synthesis; medicinal chemistry; chemistry of natural products. *Mailing Add:* Syntex Res 1238 Martin Ave Palo Alto CA 94301

FRIED, JOHN H, b Linz, Austria, Oct 9, 24; m 51; c 1. MICROBIOLOGY, CHEMISTRY. *Educ:* Univ Conn, BS, 49; Syracuse Univ, MS, 51. *Prof Exp:* Microbiologist, Stauffer Chem Co, 51-57; MICROBIOLOGIST, PFIZER INC, 71- *Mem:* Am Soc Microbiol; Am Chem Soc; Soc Indust Microbiol. *Res:* Microbiologically derived products; development of commercial fermentation processes. *Mailing Add:* Pfizer Cent Res Eastern Point Rd Groton CT 06340

FRIED, JOSEF, b Przemysl, Poland, July 21, 14; nat US; m 39; c 1. ORGANIC CHEMISTRY. *Educ:* Columbia Univ, PhD(org chem), 41. *Prof Exp:* Eli Lilly fel, Columbia Univ, 40-42; res chemist, Givaudan Res Inst, NY, 43; head dept, Squibb Inst Med Res, NJ, 44-59, dir, Div Org Chem, 59-63; chmn, Dept Chem, 77-79, LOUIS BLOCK PROF, DEPTS CHEM & BIOCHEM, & PROF, BEN MAY INST, UNIV CHICAGO, 63- *Concurrent Pos:* Chmn, Gordon Res Conf Steroids & Related Natural Prod, 55-57, Conf Med Chem, 81; Knapp Mem lectr, Univ Wis, 58; mem med chem study sect, NIH, 63-67 & 68-72 & comt arrangements, Laurentian Hormone Conf, 64-71; bd ed, J Biol Chem, 78-88. *Honors & Awards:* Med Chem Award, Am Chem Soc, 74. *Mem:* Nat Acad Sci; fel AAAS; Am Chem Soc; Am Soc Biol Chem; fel NY Acad Sci; Am Acad Arts & Sci. *Res:* Chemistry of steroids; prostaglandins. *Mailing Add:* Dept Chem Univ Chicago Chicago IL 60637

FRIED, MAURICE, b New York, NY, Nov 6, 20; m 43; c 5. PLANT NUTRITION, AGRICULTURE. *Educ:* Cornell Univ, BS, 41, MS, 45; Purdue Univ, PhD(soil chem), 48. *Prof Exp:* Asst soil chem, Cornell Univ, 44-45; instr soils, Purdue Univ, 47-48; prin res scientist, Bur Plant Indust, Soils & Agr Eng, USDA, 48-53 & Agr Res Serv, 53-61; head agr sect, Int Atomic Energy Agency, 61-64; dir, Joint Food & Agr Orgn-Int Atomic Energy Agency Div, Atomic Energy in Agr, 64-83; CONSULT, NAT ACAD SCI, 84- *Concurrent Pos:* Vis res soil scientist & lectr soil sci, Univ Calif, 74-75, Storer lectr, 84. *Honors & Awards:* Outstanding Performance, USDA, 59. *Mem:* fel Am Soc Agron; Am Soc Plant Physiol; Am Chem Soc; NY Acad Sci; Soil Sci Soc Am. *Res:* Poor crop growth in acid soils; sufficiency of sulfur for plant growth; radioactive tracers in soil and mineral nutrition investigations; nitrogen cycle; denitrification; isotopes in soil; mineral nutrition and fertilizer investigations. *Mailing Add:* 130 Aspen Lane Scientists Cliffs Port Republic MD 20676

FRIED, MELVIN, b Brooklyn, NY, May 28, 24; m 47; c 3. BIOCHEMISTRY. *Educ:* Univ Fla, BS, 48, MS, 49; Yale Univ, PhD(biochem), 52. *Prof Exp:* Instr biochem, Washington Univ, 53-56; from asst prof to prof, 56-85, asst dean grad educ, 72-81, EMER PROF BIOCHEM, COL MED, UNIV FLA, 85- *Concurrent Pos:* Childs Fund fel med res, Cambridge Univ, 52-53; NIH sr res fel, 57-62; vis res prof, Inst Biol Chem, Univ Aix-Marseille, 68-69; vis prof, Univ WI, Kingston, Jamaica, 85-89. *Mem:* Am Soc Biol Chemists; Am Chem Soc; Brit Biochem Soc; Soc Exp Biol & Med; Sigma Xi. *Res:* Proteolytic enzymes; protein biosynthesis; nucleic acid chemistry; metal-peptide complexes; chemistry and metabolism of serum lipoproteins; marine biochemistry. *Mailing Add:* Dept Biochem & Molecular Biol Box J-245 Univ Fla Col Med Gainesville FL 32610

FRIED, VICTOR A, PROTEOLYTIC PROCESSING, UBIQUITINATION. *Educ:* Univ Ore, PhD(biochem), 70. *Prof Exp:* ASSOC DIR BIOCHEM & DIR PROTEIN STRUCT FACIL, ST JUDE CHILDREN'S RES HOSP, 84- *Mailing Add:* 1727 Bryn Mawr Circle Memphis TN 38138

FRIED, VOJTECH, b Lozin, Czech, Aug 27, 21; m 51; c 1. PHYSICAL CHEMISTRY. *Educ:* Univ Chem Technol, Prague, 51, DrTSc, 53, DrChSc, 57, DrPhysCh, 63. *Prof Exp:* From instr to assoc prof phys chem, Univ Chem Technol, Prague, 50-64; PROF CHEM, BROOKLYN COL, 65- *Concurrent Pos:* Vis prof, Arya Mehr Univ Technol, Tehran, Iran, 74-75; Masahiro Yorizane chair professorship, Hiroshima Univ, Japan, 75; vis prof, Jiao Tong Univ, China, 81. *Honors & Awards:* Czech State Prize in Sci, 63. *Mem:* Am Chem Soc; Czech Chem Soc; fel Japanese Soc Prom Sci. *Res:* Thermodynamic and statistical theories of solutions of nonelectrolytes; equations of state. *Mailing Add:* The Meadows 3014 Rosemead Sarasota FL 34235

FRIED, WALTER, b Frauenkirchen, Austria, Mar 21, 35; US citizen; m 65; c 2. INTERNAL MEDICINE, HEMATOLOGY. *Educ:* Univ Chicago, BA, 54, MD, 58. *Prof Exp:* Clin investr, West Side Vet Admin Hosp, Chicago, 65-68; dir hemat labs, Presby St Luke's Hosp, 68-71; dir hemat unit, Univ Ill Hosps, 71-76; dir Div Hemat & Oncol, Michael Reese Hosp, 76-81, actg chmn, dept med, 80-82; ASSOC DEAN, MED SCI & SERV, RUSH PRESBY ST LUKE'S HOSP, 82- *Concurrent Pos:* From assoc prof to prof med, Univ Ill, 67-76; consult staff, Presby St Luke's Hosp, Chicago, 71-82; mem hemat study sect, NIH, 76-80 & 83-87; prof med, Univ Chicago, 76-82 & Rush Med Col, 82- *Mem:* AAAS; Am Soc Clin Invest; Int Soc Exp Hemat (treas, 70-78, pres, 83); Am Soc Hemat; Am Fedn Clin Res; Sigma Xi. *Res:* Studies on the regulation of erythropoiesis; studies on the regulation of hematopoietic; stem cell kinetics. *Mailing Add:* 2714 Tennyson Highland Park IL 60035

FRIED, WALTER RUDOLF, b Vienna, Austria, April 10, 23; US citizen; m 50; c 3. NAVIGATION SYSTEMS, COMMUNICATION SYSTEMS. *Educ:* Univ Cincinnati, BS, 48; Ohio State Univ, MS, 53. *Prof Exp:* Sr engr electronic navig, Wright Air Develop Ctr, 48-58; sect mgr doppler navig, Gen Precision Inc, 58-64; chief scientist radar & navig, Rockwell Int, 64-75; SR SCIENTIST COMMUN, HUGHES AIRCRAFT CO, 75- *Concurrent Pos:* Lectr, Univ Calif, Irvine, 68, 79-81 & 88; deleg, Int Civil Aviation Orgn, United Nations, 57-58. *Mem:* Fel Inst Elec & Electronic Engrs; Inst Navig. *Res:* Early development of doppler velocity and altitude measuring radars and systems for aircraft and helicopters; development of relative navigation systems; research and development of integrated communication and navigation systems. *Mailing Add:* Communications Systems Div Hughes Aircraft Co Bldg 675 C212 PO Box 3310 Fullerton CA 92634

FRIEDBERG, CARL E, PHYSICS. *Educ:* Harvard Univ, BA, 64; Princeton Univ, MA & PhD(physics), 69. *Prof Exp:* Res assoc fel physics, Lawrence Berkeley Lab, Univ Calif, 71-76; asst prof radiation ther & physicist, Mass Gen Hosp, Harvard Med Sch, 76-79 & Mem Sloan Kettering Cancer Ctr, Cornell Med Sch, 79-81; pres, In House Systs, 80-83; pres, Rocket Sci Inc, 87-91; PRES, SEAPORT SYSTS INC, 83- & COMET & CO, 91- *Mem:* Am Physics Soc; NY Acad Sci; Asn Comput Mach; Digital Equip Computer Users Soc. *Res:* Real time and commercial computing; high energy experimental physics; medical physics. *Mailing Add:* Comet & Co 165 William St New York NY 10038

FRIEDBERG, ERROL CLIVE, b Johannesburg, SAfrica, Oct 2, 37; m 61; c 2. BIOCHEMISTRY, PATHOLOGY. *Educ:* Univ Witwatersrand, BSc, 58, MB & BCh, 61; Educ Coun Foreign Med Grad, cert, 69. *Prof Exp:* Intern med & surg, King Edward VIII Hosp, Univ Natal, SAfrica, 62-63; registr path, Univ Witwatersrand, 63-65; resident path, Cleveland Metrop Gen Hosp & Sch Med, Case Western Reserve Univ, 65-66; res investr, Walter Reed Army Inst Res, 69-70; ASSOC PROF PATH, SCH MED, STANFORD UNIV, 70- *Concurrent Pos:* Fel path, Cleveland Metrop Gen Hosp & Sch Med, Case Western Reserve Univ, 65-66; fel biochem, Sch Med, Case Western Reserve Univ, 66-68; AEC, NIH & Am Cancer Soc grants; USPHS res develop career award, 74-79; Josiah Macy Fac Scholar, 78-79; mem, Path B Study Sect, NIH, 78-82. *Mem:* AMA; Biophys Soc; NY Acad Sci; Radiation Res Soc; Am Chem Soc. *Res:* Enzymes in DNA metabolism; repair of radiation damage to DNA; role of DNA repair in carcinogenesis. *Mailing Add:* Dept Path Stanford Univ Sch Med Stanford CA 94305

FRIEDBERG, FELIX, b Copenhagen, Denmark, Apr 3, 21; nat US; m 71. BIOCHEMISTRY. *Educ:* Univ Denver, BS, 44; Univ Calif, PhD(biochem), 47. *Prof Exp:* Asst biochem, Univ Calif, 44-46; from instr to assoc prof, 48-61, PROF BIOCHEM, COL MED, HOWARD UNIV, 61- *Concurrent Pos:* USPHS sr fel, 47-48; vis lectr, Cath Univ Am, 50-52; Lederle med fac award, 57; Commonwealth Fund fel, Howard Univ, 62-63. *Mem:* Biophys Soc; Am Soc Biol Chemists. *Res:* Protein metabolism and structure; total amino acid sequence of bacterial alpha and beta amylases; human calmodulin gene sequence. *Mailing Add:* Dept Biochem Howard Univ Col Med Washington DC 20059

FRIEDBERG, MICHAEL, b New York, NY, Aug 7, 39. TOPOLOGY. *Educ:* Univ Miami, BS, 61; La State Univ, PhD(math), 65; Univ Houston, MBA, 84. *Prof Exp:* Asst prof math, Emory Univ, 65-66; asst prof math, Univ Tex, 66-67; from asst to assoc prof, 67-76, PROF MATH, UNIV HOUSTON, TEX, 76- *Mem:* Am Math Soc; Math Asn Am. *Mailing Add:* Univ Houston University Park Cullen Blvd Houston TX 77004

FRIEDBERG, RICHARD MICHAEL, b New York, NY, Oct 8, 35; m 63. THEORETICAL PHYSICS. *Educ:* Harvard Univ, BA, 56; Columbia Univ, MA, 61, PhD(physics), 62. *Prof Exp:* Mem, Inst Adv Study, 62-64; from asst prof to assoc prof, 64-77, PROF PHYSICS, BARNARD COL, COLUMBIA UNIV, 77- *Concurrent Pos:* Sloan res fel, 61-65; visitor, Europ Orgn Nuclear Res, Geneva, Switz, 64-65. *Res:* Elementary particles; mathematical logic; artificial intelligence; quantum optics; statistical mechanics. *Mailing Add:* Dept Physics Barnard Col Columbia Univ New York NY 10027-6598

FRIEDBERG, SIMEON ADLOW, b Pittsburgh, Pa, July 7, 25; m 50; c 3. LOW TEMPERATURE PHYSICS, MAGNETISM. *Educ:* Harvard Univ, AB, 47; Carnegie Inst Technol, MS, 48, DSc(physics), 51. *Prof Exp:* Fulbright grant, State Univ Leiden, 51-52; res physicist, 52-53, from asst prof to assoc prof, 53-62, chmn dept, 73-80, PROF PHYSICS, CARNEGIE-MELLON UNIV, 62- *Concurrent Pos:* Alfred P Sloan res fel, 57-61; Guggenheim fel, 65-66. *Mem:* Fel Am Phys Soc; fel AAAS; Sigma Xi. *Res:* Solid state and low temperature physics; thermal, magnetic and transport properties; cooperative behavior in magnetic crystals. *Mailing Add:* Dept Physics Carnegie-Mellon Univ Pittsburgh PA 15213

FRIEDBERG, STEPHEN HOWARD, b Malden, Mass, July 29, 40; m 66; c 3. MATHEMATICAL ANALYSIS. *Educ:* Boston Univ, BA, 62; Northwestern Univ, MS, 64, PhD(math), 67. *Prof Exp:* Instr math, Mass Inst Technol, 67-69; staff mathematician, Comt Undergrad Prog Math, NSF,

69-70; assoc prof, 70-80, PROF MATH, ILL STATE UNIV, NORMAL, 80- *Concurrent Pos:* Lectr math, Royal Holloway Col, Univ London, 74-75. *Mem:* Math Asn Am; Am Math Soc; Am Statist Asn. *Res:* Harmonic analysis of locally compact abelian groups, particularly closed ideals in group algebras. *Mailing Add:* Dept Math Ill State Univ Normal IL 61761

FRIEDBERG, WALLACE, b New York, NY, Apr 12, 27; m 57; c 3. RADIOBIOLOGY. *Educ:* Hope Col, AB, 49; Mich State Univ, MS, 51, PhD(physiol), 53. *Prof Exp:* Asst, Mich State Univ, 49-53; res assoc, Children's Hosp Philadelphia, Univ Pa, 54-55; res assoc, Biol Div, Oak Ridge Nat Lab, 55-56, from assoc biologist to biologist, 58-60; from asst prof to assoc prof, 61-69, prof parasitol & lab pract, 69-74, PROF RES BIOCHEM & MOLECULAR BIOL, HEALTH SCI CTR, UNIV OKLA, 69-; SUPVR, RADIOBIOL RES, AVIATION TOXICOL LAB, CIVIL AEROMED INST, FED AVIATION ADMIN, 60- *Concurrent Pos:* NIH res fels, Ind Univ, 53-54 & Oak Ridge Nat Lab, 57-58; assoc biol, Oklahoma City Univ, 68-70 & 74-75; vis investr, Oak Ridge Nat Lab, 69-79; adj prof zool, Univ Okla, 71- *Mem:* Am Chem Soc; Soc Exp Biol & Med; Am Physiol Soc; Radiation Res Soc; Bioelectromagnetics Soc. *Res:* Biological effects of radiation. *Mailing Add:* 7805 NW 26th Bethany OK 73008

FRIEDE, REINHARD L, b Jaegerndorf, Czech, May 12, 26; US citizen; m 53; c 2. NEUROPATHOLOGY. *Educ:* Univ Vienna, MD, 51; Am Bd Path, dipl, 63. *Prof Exp:* Resident, City Hosp, St Poelten, Austria, 51-52; intern, Neurol Inst & Clin, Univ Vienna, 53; mem staff, Clin Neurosurg, Univ Freiburg, 53-57; with civil serv, Wright Air Develop Ctr, Ohio, 57-59; instr psychiat, Univ Mich, 59-60, asst prof histochem, 60-62, assoc prof path, 62-65; prof neuropath, Inst Path, Case Western Reserve Univ, 65-75 & Inst Path, Univ Zurich, Switz, 75-81; PROF NEUROPATH, UNIV GOTTINGEN, GERMANY, 81- *Concurrent Pos:* Adv bds, Acta Neuropath, J Neuropath Exp Neurol, Clin Neuropath, Neuropath & Appl Neurobiol. *Honors & Awards:* Civil Service Award, 61. *Mem:* Am Asn Neuropath; German Asn Neuropath; Swiss Asn Neuropath. *Res:* Histochemistry of nervous system; experimental pathology; developmental neuropathology; experimental neuropathology; chemical cytology of nervous system; myelin developmental neuropathology. *Mailing Add:* Univ Gottingen Robert-Koch-Strasse 40 Gottingen D 3400 Germany

FRIEDEL, ARTHUR W, b Pittsburgh, Pa, Nov 14, 37. INORGANIC CHEMISTRY. *Educ:* Univ Pittsburgh, BS, 59, MEd, 63; Ohio State Univ, PhD(sci educ), 68. *Prof Exp:* Teacher pub schs, Pa, 59-63; asst sci educ, Ohio State Univ, 64-65 & chem, 65-67; from instr to asst prof, 67-72, ASSOC PROF CHEM, IND UNIV-PURDUE UNIV, FT WAYNE, 72- *Concurrent Pos:* Dir, Northeastern Ind Regional Sci Fair, 67-91; exchange teacher, Fulbright-Hays Act, Trent Polytech, Nottingham, Eng, 75-76. *Mem:* Am Chem Soc; Nat Sci Teachers Asn; Nat Asn Res Sci Teaching. *Res:* Chemical education; problem solving. *Mailing Add:* Dept Chem Ind Univ-Purdue Univ Ft Wayne IN 46805

FRIEDELL, GILBERT H, b Minneapolis, Minn, Feb 28, 27; m 50; c 5. MEDICINE, PATHOLOGY. *Educ:* Univ Minn, BS, 47, MB, 49, MD, 50; Am Bd Path, dipl, 55. *Prof Exp:* Intern, Minneapolis Gen Hosp, 49-50; asst resident, Mallory Inst Path, 50-52; resident, Salem Hosp, 53-54 & Pondville Hosp, 54-55; from asst to assoc, Mass Mem Hosps, Boston, 57-62; pathologist, New Eng Deaconess Hosp & res assoc, Cancer Res Inst, Boston, 62-67; assoc pathologist, Mallory Inst Path, Boston City Hosp, 67-69; chief dept path, St Vincent Hosp, 69-78, med dir, 78-82; prof path, Univ Mass, 70-82; dir, Markey Cancer Ctr, 83-90, PROF PATH, UNIV KY, 83-, AIR, CANCER CONTROL, MARKEY CANCER CTR, 90- *Concurrent Pos:* Am Cancer Soc fel path, Free Hosp Women, 52-53; teaching fel, Harvard Med Sch, 52-53, asst, 57-61, instr, 62-67, lectr, 67-69; USPHS spec fel, Strangeways Res Lab, Eng, 61-62; instr, Boston Univ, 58-61, assoc prof, Sch Med, 67-69; prof path, Sch Med, Univ Mass, Worcester, 70-73; dir, NCI Nat Bladder Cancer Proj, 71-84. *Mem:* Am Asn Cancer Res; Am Soc Clin Oncol. *Res:* Clinical and experimental studies of tumor-host relationships with special reference to breast, cervix and bladder cancer. *Mailing Add:* Markey Cancer Ctr 800 Rose St Rm 170 Lexington KY 40536-0093

FRIEDELL, HYMER LOUIS, b St Petersburg, Russia, Feb 6, 11; nat US; m 35; c 3. RADIOLOGY. *Educ:* Univ Minn, BS, 31, MB, 35, MD, 36, PhD(radiol), 40. *Prof Exp:* Instr radiol, Univ Calif, 41-42; chmn dept, 46-78, prof, 78-80, EMER PROF RADIOL, SCH MED, CASE WESTERN RESERVE UNIV, 80- *Concurrent Pos:* Nat Cancer Inst fel, Chicago Tumor Inst, Mem Hosp, New York, 39-40 & Univ Calif Hosp, 40-41; consult, Nat Adv Comt Aeronaut; chmn, Radiation Study Sect, NIH; chmn, Comt Allocation of Isotopes for Human Use & mem, Reactor Safeguard Comt, AEC; chmn, Subcomt Radiobiol & mem, Comt Radiol, Nat Res Coun; mem, Subcomt Permissable External Dose, Nat Bur Stand; mem, Coun Exec Bd, Argonne Nat Lab, Cent Adv Comt, Radioisotope Sect, Res & Educ Serv, US Vet Admin, Res & Develop Bd, Joint Panel Med Aspects Atomic Warfare & Vis Comt for Med Dept, Brookhaven Nat Lab; partic, Int Cong Radiol; vpres, Nat Coun Radiation Protection & Measurements, 77-83. *Honors & Awards:* Hon Mem, Nat Coun Radiation Protection & Measurement. *Mem:* AAAS; Am Radium Soc; fel Am Col Radiol; Radiation Res Soc (past pres); Radiation Soc NAm; Sigma Xi. *Res:* Biological effects of radioisotopes; use of radiation and radioisotopes for therapy and diagnosis; radiation protection and radiation hazards. *Mailing Add:* Case Western Reserve Univ 2119 Abington Rd Cleveland OH 44106

FRIEDELL, JOHN C, b Dubuque, Iowa, Nov 2, 29. PURE MATHEMATICS. *Educ:* Loras Col, BA, 51; Gregorian Univ, STL, 55; Univ Iowa, MS, 58; Catholic Univ, PhD(math), 62. *Prof Exp:* From instr to assoc prof, 57-75, PROF MATH, LORAS COL, 75- *Mem:* Math Asn Am. *Res:* Linear and topological algebra; Banach algebras. *Mailing Add:* Dept Math Loras Col 1450 Alta Vista Dubuque IA 52001

FRIEDEN, BERNARD ROY, b Brooklyn, NY, Sept 10, 36; m 62; c 3. OPTICAL PHYSICS. *Educ:* Brooklyn Col, BS, 57; Univ Pa, MS, 59; Univ Rochester, PhD(optics), 66. *Prof Exp:* Res asst optics, Univ Rochester, 62-63; from asst prof to assoc prof, 66-74, PROF OPTICAL SCI, UNIV ARIZ, 74- *Concurrent Pos:* Consult, US Navy, Mass, 73, E I du Pont de Nemours & Co, Inc, 74, 78, 87, Can Ctr Remote Sensing, 76, Kitt Peak Nat Observ, 76-80, Aerospace Corp, 83-, Aberdeen Proving Ground, 83-; assoc ed, Optical Soc Am, 78-82; vis prof, Univ Groningen, Netherlands, 83. *Mem:* Fel AAAS; fel Optical Soc Am; fel Soc Photo-optical Instrumentation Engrs. *Res:* Image restoration and enhancement; probability and statistics; Fischer information and the foundation of physics; synthetic aperture radar; apodizing, super-resolving pupils; Shannon information theory; numerical, statistical analysis. *Mailing Add:* Optical Sci Ctr Univ Ariz Tucson AZ 85721

FRIEDEN, CARL, b New Rochelle, NY, Dec 31, 28; m 53; c 3. BIOCHEMISTRY. *Educ:* Carleton Col, BA, 51; Univ Wis, PhD(chem), 55. *Prof Exp:* Fel, 55-57, from instr to assoc prof, 57-67, PROF BIOCHEM, WASH UNIV, 67- *Honors & Awards:* Am Chem Soc Award, 76. *Mem:* Nat Acad Sci; AAAS; Am Chem Soc; Am Soc Biochem & Molecular Biol; Am Soc Cell Biol. *Res:* Enzymes and enzyme kinetics; physical chemistry of proteins; actin polymerization; protein folding. *Mailing Add:* Dept Biochem & Molecular Biophysics Box 8231 Wash Univ Sch Med St Louis MO 63110

FRIEDEN, EARL, b Norfolk, VA, Dec 31, 21; m 42; c 2. BIOCHEMISTRY. *Educ:* Univ Calif, Los Angeles, BA, 43; Univ Southern Calif, MS, 47, PhD(biochem), 49. *Prof Exp:* Lab supvr anal chem, Rubber Reserve Co, US Rubber Co, 43-45; res biochemist, Bedwell Labs, 45-47, instr, Univ Southern Calif, 48; from asst prof to assoc prof, 49-57, chmn dept, 62-68, PROF CHEM, FLA STATE UNIV, 57-, R O LAWTON DISTINGUISHED PROF, 69- *Concurrent Pos:* USPHS spec fel, Carlsberg Labs, Denmark, 57-58 & USPHS fel, Univ Calif, La Jolla, 71; vis prof, Univ Southern Calif, 63; ed, Biochem of Elements, 71- & Biol Trace Element Res, 78. *Honors & Awards:* Lalor Award, Inst Enzyme Res, Univ Wis, 55; Fla Award, Am Chem Soc, 68. *Mem:* Am Soc Biol Chem; Am Chem Soc. *Res:* Chemistry and mechanism of enzymes; copper and iron metalloproteins and metalloenzymes; biochemistry of amphibian metamorphosis; role of thyroid hormones in vitro and in vivo systems; author of numerous publications. *Mailing Add:* Dept Chem Fla State Univ Tallahassee FL 32306

FRIEDENBERG, RICHARD M, b New York, NY, 26; m; c 3. RADIOLOGY. *Educ:* Columbia Univ, AB, 46; State Univ NY, MD, 49. *Prof Exp:* Asst prof radiol, Albert Einstein Col Med, 55-66, assoc clin prof, 66-68; prof radiol & chmn dept, New York Med Col, 68-80; chmn, Dept Radiol, Westchester Med Ctr, 74-80; PROF RADIOL & CHMN, DEPT RADIOL SCI, UNIV CALIF, IRVINE, 80- *Concurrent Pos:* Dir & chmn dept radiol, Bronx-Lebanon Hosp Ctr, 57-68. *Mem:* Fel Am Col Radiol; Radiol Soc NAm; Am Roentgen Ray Soc; fel NY Acad Sci. *Mailing Add:* Irving Med Ctr Univ Calif 101 City Dr S Orange CA 92668

FRIEDENSTEIN, HANNA, b Vienna, Austria; nat US; m 89. CHEMISTRY, INFORMATION SCIENCE. *Educ:* Univ London, BSc, 41; Simmons Col, MS, 50. *Prof Exp:* Res chemist, Phillips Elec Ltd, Eng, 42-46; asst intel officer, Brit Oxygen Co, 46-47; res librn, Cabot Corp, 47-57, head tech info serv, 57-68, mgr, Info Ctr, 68-81; RETIRED. *Mem:* Am Chem Soc; Spec Libr Asn. *Res:* Chemical documentation; communication and retrieval of information; library science. *Mailing Add:* 1010 Waltham St Apt B341 Lexington MA 02173-8044

FRIEDER, GIDEON, b Zvolen, Czech, Sept 30, 37; US citizen; m 60; c 3. COMPUTER SCIENCES. *Educ:* Israel Inst Technol, BS, 59, MS, 61, DSc(quantum physics), 67. *Prof Exp:* Res assoc numerical anal, Sci Dept, Israeli Ministry Defense & Prog, 59-63, res group mgr, 64-67, dep mgr comput sci, 68-69, mgr, 69-70; staff mem, IBM Sci Ctr, 73-75; from asst prof to assoc prof, State Univ NY Buffalo, 70-73, assoc prof comput sci, 75-80, prof, 80-81; prof & chmn, Dept Comput & Commun Sci, Univ Mich, Ann Arbor, 81-83, Div Comput Sci & Eng, 84-87; DEAN, SCH COMPUT & INFO SCI, SYRACUSE, 87. *Concurrent Pos:* Consult microprog, comp archit, syst design & intel property litigation, 70-; nat lectr, Asn Comput Mach, 73 & 77-79. *Mem:* Asn Comput Mach; Inst Elec & Electronic Engrs Comput Soc. *Res:* Distributed architectures of computing system, and their operating systems; nonbinary computers, especially the ternary computer; tomography and computer structures for supporting it; intermediate machines for high-level languages; computer vision via rapid surface recognition. *Mailing Add:* Sch Comput & Info Sci Ctr Sci & Technol Syracuse Univ Syracuse NY 13244-4100

FRIEDERICI, HARTMANN H R, b Asuncion, Paraguay, Jan 25, 27; US citizen; m 58; c 3. MEDICINE, PATHOLOGY. *Educ:* Goethe Col, Paraguay, BS, 46; Univ La Plata, MD, 53. *Prof Exp:* Asst path, Univ Bonn, 55-56; resident, Univ Ill Col Med, 57-59; from instr to assoc prof, Univ Ill Col Med, 60-69; prof biol sci, Col Arts & Sci, 70-80, PROF PATH, SCH MED, NORTHWESTERN UNIV, 69-; HEAD DEPT PATH, EVANSTON HOSP, 71- *Concurrent Pos:* Assoc pathologist, Res & Educ Hosp, 61-69. *Mem:* AAAS; Int Acad Path; Am Asn Pathologists; Am Soc Cell Biol; Sigma Xi. *Res:* Study of ultrastructure of blood and lymphatic capillaries under physiologic and pathologic circumstances in man and experimental animals; cell surfaces. *Mailing Add:* Dept Path Evanston Hosp 2650 Ridge Ave Evanston IL 60201

FRIEDEWALD, WILLIAM THOMAS, b New York, NY, Mar 7, 39; m 67; c 3. MEDICINE. *Educ:* Univ Notre Dame, BS, 60; Yale Univ, MD, 63. *Prof Exp:* Intern med, Grace-New Haven Hosp, 63-63, Jr resident, 63-64; res assoc virol, Nat Inst Allergy & Infectious Dis, NIH, 65-67; sr resident med, Yale-New Haven Hosp, 67-68; fel statist, Stanford Univ, 68-69; med officer Epidemiol, Med, Statist, Nat Heart, Lung & Blood Inst, NIH, 69-73, br chief, 73-79, assoc dir, 79-84, dir Div Epidemiol & Clin Appln, Med, Statist, 84-86, assoc dir Off Dis Prev, 86-89; CHIEF MED DIR, METROP LIFE INS CO, 89- *Concurrent Pos:* Adj assoc prof, Uniformed Serv, Univ Health Sci, 82-

Honors & Awards: Pub Health Serv Commendation Medal, Nat Heart, Lung & Blood Inst, NIH, 76. *Mem:* Am Fedn Clin Res; Am Heart Asn; Am Epidemiol Soc; Am Asn Physicians; Soc Clin Trials. *Res:* Cardiovascular disease, prevention and treatments; major clinical trials and epidemiology. *Mailing Add:* Metrop Life Ins Co One Madison Ave Area 16X New York NY 10010

FRIEDHOFF, ARNOLD JEROME, b Johnstown, Pa, Dec 26, 23; m 46; c 3. PSYCHIATRY. *Educ:* Univ Pa, MD, 47. *Prof Exp:* Intern psychiat, WPa Hosp, 47-48; jr staff psychiatrist, Mayview State Hosp, Pa, 48-51; res psychiat div, Bellevue Hosp & Clin, Med Ctr, 53-54, clin asst psychiat, Col Med, 55-56, clin instr, 56-57, from instr to asst prof, 57-66, PROF PSYCHIAT, COL MED, NY UNIV, 66-, DIR, CTR STUDY PSYCHOTIC DIS, 64- *Concurrent Pos:* Jr psychiatrist, NY Univ Hosp & Clin, 54-57, asst vis neuropsychiatrist, 57-, head psychopharmacol res unit, 56-; dir, Millhauser Labs; mem, Clin Proj Rev Comt, NIMH, 70-74; pres, Soc Biol Psychiat, 80-81; mem, Nat Adv Ment Health Coun, 87- *Mem:* Harvey Soc; Asn Res Nerv & Ment Dis; Am Psychiat Asn; Am Psychopath Asn (pres); Am Col Neuropsychopharmacol (vpres, 77-78, pres, 79-80); Royal Col Psychiatrists (Gt Brit); Nat Alliance Res Schizophrenia & Depression. *Res:* Neurochemistry; neuropharmacology. *Mailing Add:* 1382 Lexington Ave New York NY 10028

FRIEDKIN, MORRIS ENTON, b Kansas City, Mo, Dec 30, 18; m 43; c 3. BIOCHEMISTRY. *Educ:* Iowa State Col, MS, 41; Univ Chicago, PhD(biochem), 48. *Prof Exp:* Chemist penicillin proj, North Regional Res Lab, USDA, 41-45; USPHS fel, Copenhagen Univ, 48-49; from instr to assoc prof pharmacol, Sch Med, Wash Univ, St Louis, 49-58; prof & chmn dept, Sch Med, Tufts Univ, 58-69; provost, Revelle Col, 74-78, PROF BIOL & MEM FAC, SCH MED, UNIV CALIF, SAN DIEGO, 69- *Mem:* Nat Acad Sci; Am Soc Biol Chem; Am Acad Arts & Sci. *Res:* Enzymology; nucleic acid metabolism; biochemical pharmacology. *Mailing Add:* Dept Biol MC 0601 Univ Calif San Diego La Jolla CA 92093

FRIEDL, FRANK EDWARD, b Stewart, Minn, May 29, 31; m 56; c 4. ANIMAL PHYSIOLOGY, PARASITOLOGY. *Educ:* Univ Minn, BA, 52, PhD(zool), 58. *Prof Exp:* Asst zool, Univ Minn, 52-58; res fel, USPHS, Rockefeller Inst, 58-60; from asst prof to assoc prof prof, 60-72, PROF, DEPT BIOL, UNIV SFLA, 72- *Mem:* AAAS; Am Soc Parasitol; Am Soc Zool. *Res:* Comparative physiology and biochemistry; oxygen uptake and metabolism; nitrogen metabolism in molluscs. *Mailing Add:* Dept Biol Univ SFla Tampa FL 33620

FRIEDLAENDER, FRITZ J(OSEF), b Freiburg, Ger, May 7, 25; nat US; m 69; c 2. MAGNETICS. *Educ:* Carnegie Inst Technol, BS, 51, MS, 52, PhD(elec eng), 55. *Prof Exp:* Asst elec eng, Carnegie Inst Technol, 51-54; asst prof, Columbia Univ, 54-55; from asst prof to assoc prof, 55-62, PROF ELEC ENG, PURDUE UNIV, 62- *Concurrent Pos:* Consult, Gen Elec Co, Ind, 56-58, Components Corp, Ill, 59-61, Lawrence Radiation Lab, Univ Calif, 67-69 & P R Mallory & Co, 74-78, Oak Ridge Nat Lab, 79-82; guest prof, Max-Planck Inst Metal Res, 64-65, Inst fur Werkstoffe der Elektrotechnik, Ruhr Univ, WGer, 72-73, Nagoya Univ, Japan, 80 & Univ Regensburg, WGer, 81-82; rev ed, IEEE Trans on Magnetics, 65-67; co-ed, Magnetic Separation News, 83-; Meyerhoff vis prof, Weizmann Inst Sci, Israel, 90. *Honors & Awards:* Humboldt Award, 72; Centennial Medal, Inst Elec & Electronic Engrs, 84, Magnetics Soc Achievement Award, 86. *Mem:* Am Phys Soc; Am Soc Eng Educ; fel Inst Elec & Electronic Engrs Magnetics Soc (vpres, 75-76, pres, 77-78); Swiss Elec Eng Soc. *Res:* Magnetics, magnetic devices and memories; high gradient magnetic separation; applications of magnetic separation to medicine, biotechnology, and mineral beneficiation; magnetic wall dynamics; vertical bloch lines; dynamics; microwave ferrites. *Mailing Add:* Sch Elec Eng Purdue Univ West Lafayette IN 47907

FRIEDLAENDER, JONATHAN SCOTT, b New Orleans, La, Aug 24, 40; m 71; c 2. PHYSICAL ANTHROPOLOGY. *Educ:* Harvard Univ, AB, 62, PhD(anthrop), 69. *Prof Exp:* Pop Coun del demog & pop genetics, Univ Wis-Madison, 68-69, asst prof anthrop, 69-70; from asst prof to assoc prof anthrop, Harvard Univ, 70-76; assoc prof , 76-83, PROF ANTHROPOL, TEMPLE UNIV, 83- *Mem:* AAAS; Am Asn Phys Anthrop; Human Biol Coun; Brit Soc Study Human Biol; Pop Coun; Sigma Xi. *Res:* Studies of human physical variation over population boundaries; populations on Bougainville Island, Territory of New Guinea; epidemiology of hypertension; hypercholesterolemia; diabetes melitus in the Solomons. *Mailing Add:* Dept Anthropol Temple Univ Philadelphia PA 19122

FRIEDLAND, BEATRICE L, b New York, NY, Feb 12, 14; m 39; c 2. BIOLOGY. *Educ:* NY Univ, BA, 35, MS, 38, PhD(biol), 43. *Prof Exp:* Instr biol, NY Univ, 43-46 & Hunter Col, 47-51; from instr to prof, 51-79, chmn dept, 55-71, EMER PROF BIOL, STERN COL, YESHIVA UNIV, 79- *Res:* Embryological genetics; action of genes at specific times in development. *Mailing Add:* One Fifth Ave New York NY 10003

FRIEDLAND, BERNARD, b Brooklyn, NY, May 25, 30; m 59; c 3. ELECTRICAL ENGINEERING, APPLIED MATHEMATICS. *Educ:* Columbia Univ, AB, 52, BS, 53, MS, 54, PhD(elec eng), 57. *Prof Exp:* From instr to asst prof elec eng, Columbia Univ, 54-61; head, Control Systs Lab, Melpar, Inc, 61-62; mgr systs res, Kearfott Guidance & Navigation Corp, 62-90; DISTINGUISHED PROF, ELEC ENG, NJ INST TECHNOL, 90- *Concurrent Pos:* Adj prof, Columbia Univ, 67-72, NY Univ, 70-73 & Polytech Inst NY, 74-90. *Honors & Awards:* Oldenburger Medal, Am Soc Mech Engrs, 80. *Mem:* Fel Am Soc Mech Engrs; fel Inst Elec & Electronic Engrs; assoc fel Am Inst Aeronaut & Astronaut. *Res:* Modern control theory and application. *Mailing Add:* 36 Dartmouth Rd West Orange NJ 07052

FRIEDLAND, DANIEL, b Columbus, Ohio, Apr 5, 16; m 40; c 2. CHEMICAL ENGINEERING. *Educ:* City Col New York, BChE, 37; Polytech Inst Brooklyn, MChE, 45, DChE, 48. *Prof Exp:* Asst, M W Kellogg Co, 37; chem engr, Cities Serv Oil Co, New York, 38-46; vpres, Truland Chem Co, 46-59; vpres, Trubek Labs, East Rutherford, 59-64, exec vpres, Trubek Chem Co, 64-65; exec vpres, UOP Chem Co, 65-66, vpres mkt, UOP Chem Div, 66-77, pres, Org Chem Div, Crompton & Knowles, 77-78; CONSULT, 78- *Mem:* Am Chem Soc; Am Inst Chem Engrs; Am Inst Chem; Commercial Chem Develop Asn. *Res:* Distillation; extraction; unit operations of chemical engineering; petroleum refining. *Mailing Add:* 1082 Lakeside Blvd Boca Raton FL 33434

FRIEDLAND, FRITZ, b Berlin, Ger, Jan 2, 10; nat US; m 38. MEDICINE. *Educ:* Univ Berlin, MD, 34. *Prof Exp:* Chief phys med, Warren City Hosp, Ohio, 43-44; chief phys med & rehab serv, Vet Admin Hosp, Framingham, 46-53 & Boston, 53-73; asst clin prof, 54-61, assoc prof, 61-73, clin prof phys & rehab med, Sch Med, Tufts Univ, 73-85; RETIRED. *Concurrent Pos:* Consult, Lemuel Shattuck Hosp; mem assoc staff, New Eng Med Ctr Hosps; consult physiatrist, Spaulding Rehab Hosp, Boston; chief phys med serv, Leonard Morse Hosp, Natick, Mass, 73-81. *Mem:* Fel Am Col Physicians; Am Cong Rehab Med. *Res:* Ultrasound in medicine; medical rehabilitation of paraplegics, amputees, arthritic and poliomyelitic patients, cerebrovascular accidents. *Mailing Add:* 707 S Gulfstream Ave Sarasota FL 34236-7701

FRIEDLAND, JOAN MARTHA, b Binghamton, NY, Feb 6, 36. BIOCHEMISTRY, OCCUPATIONAL HEALTH. *Educ:* Cornell Univ, BS, 59; Univ Ill, MS, 64, PhD(biochem), 68 Univ Calif, Los Angeles, MPH, 87- *Hon Degrees:* Environ Health, MS, Hunter Col, 85, Univ Col Los Angeles, MPH, 87. *Prof Exp:* Res asst, Johns Hopkins Univ, 59-62; res asst, Biol Labs, Harvard Univ, 66-67; res assoc enzym & neurochem, Kingsbrook Jewish Med Ctr, Brooklyn, 68-73; asst physiol & biophys, Mt Sinai Sch Med, 73-74; res assoc, 74-76, asst prof, Dept Med, State Univ NY Downstate Med Ctr, 76-85, fel Nat Inst Environ Health Sci, UCLA, 85-87; EPIDEMIOLOGIST, NAT INST OCCUP SAFETY & HEALTH, CINCINNATI, OHIO, 88- *Concurrent Pos:* Res fel, Harvard Univ, 68. *Mem:* Am Chem Soc; Am Public Health Asn. *Res:* Bacterial bioluminescent enzymes; subunit structure; biochemistry of lipid storage diseases, hexosaminidase; isozyme studies on Tay Sachs disease and enzymes; studies in mucopolysaccharidosis; prenatal diagnosis of Tay Sachs disease; Angiotensin converting enzyme in Sarcoidosis and induction of Angiotensin converting enzyme in rabbit and human macrophages; access to prenatal diagnostic procedures for poor women; AIDS education; biological markers for susceptibility to disease; screening and notification activity. *Mailing Add:* 5911 E Woodmount Ave Cincinnati OH 45213-2009

FRIEDLAND, MELVYN, b Aug 24, 32; m 61; c 2. BIOCHEMISTRY. *Educ:* Univ Calif, Los Angeles, BS, 54; Univ Southern Calif, PhD(biochem), 63. *Prof Exp:* Clin chemist, Bio-Sci Labs, Los Angeles, 63-67; CLIN CHEMIST, UPJOHN CO, 67- *Mem:* AAAS; Am Asn Clin Chem; Am Chem Soc. *Res:* Nucleic acid antimetabolites; development of new tests for clinical chemistry; management of laboratory operations. *Mailing Add:* Upjohn Co 7000 Portage Rd Kalamazoo MI 49001

FRIEDLAND, STEPHEN SHOLOM, medical physics; deceased, see previous edition for last biography

FRIEDLAND, WALDO CHARLES, b Menasha, Wis, Dec 18, 23; m 46; c 6. CHEMISTRY. *Educ:* Iowa State Col, BS, 48, PhD(chem), 51. *Prof Exp:* Res microbiologist, 51-59, sect mgr, 59-63, asst to dir develop, 63-67, dir develop, 67-81, dir licensing, Abbott Labs, 81-82; RETIRED. *Mem:* AAAS; Am Chem Soc. *Res:* Antibiotics and microbial products. *Mailing Add:* 1020 Gracewood Dr Libertyville IL 60048

FRIEDLANDER, ALAN L, b Chicago, Ill, Aug 31, 36; m 58; c 3. ELECTRICAL ENGINEERING. *Educ:* Ill Inst Technol, BS, 58; Case Inst Technol, MS, 63. *Prof Exp:* Aeronaut res engr, Lewis Res Ctr, NASA, 58-63; sr engr, IIT Res Inst, 63-72; SR RES ENGR, SCI APPLN, INC, 72- *Mem:* Am Inst Aeronaut & Astronaut; Sigma Xi; Am Astronaut Soc. *Res:* Theoretical analysis and design of space vehicle guidance and control systems; astronautical engineering; trajectory and space mission analysis. *Mailing Add:* 5041 Wright Terr Skokie IL 60077

FRIEDLANDER, CARL B, b Chicago, Ill, May 7, 49; m 70; c 2. OPERATING SYSTEMS, PARALLEL PROCESSING. *Educ:* NC State Univ, BS, 74; Mich State Univ, MS, 79; Wayne State Univ, PhD(comput sci), 82. *Prof Exp:* Lectr comput sci, Wayne State Univ, 80-81, from instr to asst prof, 81-85; sr staff mem, BDM Corp, 85-89; VPRES, ISX CORP, 89- *Concurrent Pos:* Nat lectr, Asn Comput Mach, 84-86. *Mem:* Sigma Xi; Asn Comput Mach; Inst Elec & Electronics Engrs; Spec Interest Group Comput & Handicapped (secy-treas, 85-87). *Res:* Distribution of operating systems for parallel processing; computer languages for specifying parallel programs in an efficient and correct notation. *Mailing Add:* 3271 Rosehill Circle Thousand Oaks CA 91360

FRIEDLANDER, ERIC MARK, b Santurce, PR, Jan 7, 44; m 68. ALGEBRAIC TOPOLOGY, ALGEBRAIC GEOMETRY. *Educ:* Swarthmore Col, BA, 65; Mass Inst Technol, PhD(math), 70. *Prof Exp:* instr, Princeton Univ, 70-71, lectr, 71-72, asst prof, 72-75; assoc prof, 75-80, PROF MATH, 80-, CHMN, NORTHWESTERN UNIV, 87- *Concurrent Pos:* Res fel, US-France Exchange Scientists, 73-74; vis sr lectr, Sci Res Coun, UK, 77-78; vis mem, Inst Advan Study, Princeton, 81, 85-86; ed, J Pure & Appl Algebra, K-Theory, Oxford Math Monogr. *Mem:* Sigma Xi; Am Math Soc. *Res:* Techniques from algebraic geometry and topology to exploit newly discovered relationships to obtain results in algebraic geometry, algebraic topology, algebraic K-theory and group theory. *Mailing Add:* Dept Math Northwestern Univ Evanston IL 60208

FRIEDLANDER, GERHART, b Munich, Ger, July 28, 16; nat US; m 41, 83; c 2. NUCLEAR CHEMISTRY. *Educ:* Univ Calif, BS, 39, PhD(radiochem), 42. *Prof Exp:* Asst chem, Univ Calif, 42; instr, Univ Idaho, 42-43; chemist & group leader, Los Alamos Sci Lab, 43-46; res assoc, Res Lab, Gen Elec Co, NY, 46-48; vis lectr chem & physics, Wash Univ, St Louis, 48; chemist,

Brookhaven Nat Lab, 48-52, sr chemist, 52-81, chmn dept chem, 68-77, consult, 81-89, SR CHEMIST, BROOKHAVEN NAT LAB, 89- *Concurrent Pos:* Chmn panel future nuclear sci, Nat Acad Sci, 75-77; Humboldt award, 78-79 & 87; comn phys sci, math & resources, Nat Res Coun, 82-90. *Honors & Awards:* Nuclear Applns in Chem Award, Am Chem Soc, 67. *Mem:* Nat Acad Sci; AAAS; fel Am Phys Soc; Am Chem Soc; Am Acad Arts & Sci. *Res:* Nuclear reactions; solar neutrinos; properties of radioactive nuclides; cluster impact fusion. *Mailing Add:* Chem Dept Brookhaven Nat Lab Upton NY 11973

FRIEDLANDER, HENRY Z, b New York, NY; c 2. ORGANIC CHEMISTRY. *Educ:* Oberlin Col, AB, 48; Univ Ill, MS, 49, PhD(chem), 52; Fordham Law Sch, JD, 74. *Prof Exp:* Res assoc, Case Inst Technol, 48; asst, Univ Ill, 49-52; chemist, Stamford Labs, Am Cyanamid Co, 52-56; sr chemist, Johnson & Johnson, 57-58; sr scientist, Springdale Labs, Am Mach & Found Co, 59-64; res mgr, Dorr-Oliver, Inc, 64-66; mem staff, Union Carbide Corp Res, 66-78 & Stauffer Chem Co, 78-83; CONSULT, LACKENBACH, SIEGEL ET AL, SCARSDALE, NY, 85- *Concurrent Pos:* Nat Counr, Am Chem Soc. *Mem:* Am Chem Soc. *Res:* Polymerization; synthetic membranes; law. *Mailing Add:* 85 Riverside Ave Stamford CT 06905

FRIEDLANDER, HERBERT NORMAN, b Chicago Heights, Ill, Mar 12, 22; m 43; c 3. PHYSICAL ORGANIC CHEMISTRY, RESEARCH ADMINISTRATION. *Educ:* Univ Chicago, SB, 42, PhD(phys org chem), 47. *Prof Exp:* Lab asst org chem, Univ Chicago, 42-44, jr chemist, Metall Labs, 44, res corp fel & asst, 47-48; jr scientist, Los Alamos Labs, NMex, 44-46; proj chemist, Res Dept, Standard Oil Co, Ind, 48-56, group leader, Res Dept, 57-58, sect head, 59-60; res assoc, Res Dept, Amoco Chem Corp, 61-62; mgr polymer sci, Basic Res Dept, Chemstrand Res Ctr, Inc, 62-65; dir res, Monsanto Co, 65-66, dir new prod res & develop, 66-67, assoc dir cent res dept, 67-68, vpres & dir tech opers, Chemstrand Res Ctr, Inc, 68-74 & Monsanto Triangle Park Develop Ctr, Inc, 74-82, mgr technol assessment, Monsanto Res Corp/Mound, 82-84; PRES, FRIEDLANDER CONSULTS, INC, 85- *Concurrent Pos:* Consult, Nat Coun Res & Develop, State of Israel, 62 & 64; adj assoc prof, Dept Textile Chem, Sch Textiles, NC State Univ, 72-78, adj prof, 81-82; consult dir indust rels, dept chem, Univ Calif, San Diego, 85- *Mem:* AAAS; Am Chem Soc. *Res:* Reaction of organic free radicals; slow neutron cross sections; atoms and free radicals in solution; chemicals from petroleum; catalysis and polymerization; stereo-regulated polymerization with solid catalysts; structure and properties of fiber-forming polymers; high performance fibers and fabrics for industrial applications; research management and technology transfer. *Mailing Add:* 1620 B Royal Palm Dr S Gulfport FL 33707-3830

FRIEDLANDER, IRA RAY, b Harvey, Ill, Apr 10, 53; m 84; c 2. CELLULAR CARDIAC ELECTROPHYSIOLOGY. *Educ:* Univ Chicago, BA, 75; Univ NC, Chapel Hill, MD, 79; Am Bd Internal Med, diplomate, 84. *Prof Exp:* Resident med, 79-82, RES FEL CARDIOL, UNIV CHICAGO HOSP & CLIN, 82- *Concurrent Pos:* Mem, Coun Sci Affairs, AMA, 82-, Diag & Therapeut Technol Assesment subcomt, 83-, Nat Heart, Lung, & Blood Inst comt cholestoral educ, 85- *Mem:* Sigma Xi; AAAS; AMA; Am Soc Internal Med; Am Col Cardiol. *Res:* Cellular cardiac electrophysiology; ion selective micro electrodes. *Mailing Add:* 4923 S Kimbark Ave Chicago IL 60615-2954

FRIEDLANDER, JOHN BENJAMIN, b Toronto, Ont, Oct 4, 41; m 74. NUMBER THEORY. *Educ:* Univ Toronto, BSc, 65; Univ Waterloo, MA, 66; Pa State Univ, PhD(number theory), 72. *Prof Exp:* Vis mem math, Inst Advan Study, 72-74; lectr, Mass Inst Technol, 74-76; vis prof, Scuola Normale Superiore, Pisa, 76-77; from asst prof to assoc prof, 77-82, chmn dept, 87-91, PROF MATH, UNIV TORONTO, 82- *Concurrent Pos:* Vis prof, Univ Ill, Urbana, 79-80; vis mem math, Inst Advan Study, 83-84; res prof, MSRI, Berkeley, 91-92. *Honors & Awards:* FRSCan, 88. *Mem:* Am Math Soc; fel Royal Soc Can. *Res:* Sieve methods; quadratic residues; class numbers; distribution of primes; exponential sums; modular forms. *Mailing Add:* Phys Sci Group Univ Toronto Scarborough ON M1C 1A4 Can

FRIEDLANDER, MARTIN, VISUAL SCIENCE, PROTEIN BIOCHEMISTRY. *Educ:* Univ Chicago, PhD(develop biol anat), 76. *Prof Exp:* Asst prof cell biol, Rockefeller Univ, 83-86; CONSULT, 86- *Res:* Topogenesis. *Mailing Add:* Dept Ophth Juels Stein Eye Inst UCLA Sch Med 800 Westwood Plaza Los Angeles CA 90024

FRIEDLANDER, MICHAEL J, b Miami, Fla, Jan 30, 50; c 3. NEUROSCIENCES, ANIMAL PHYSIOLOGY. *Educ:* Fla State Univ, BS, 72; Univ Ill, MS, 74, PhD(physiol), 77. *Prof Exp:* NIH fel physiol, Univ Ill, 74-77 & Univ Va, 77-79; res asst prof anat & asst prof neurobiol, State Univ NY Stony Brook, 79-80; from asst prof to assoc prof, 80-87, PROF PHYSIOL & BIOPHYSICS, UNIV ALA, BIRMINGHAM, 87-, DIR NEUROBIOL RES CTR, 87- *Concurrent Pos:* Co-investr res proj, Nat Eye Inst, 79-80, prin investr, Develop Structure & Function Vis Syst, 81-84; prin investr, NATO Collaborative Res, 83-87, Nat Sci Found, Develop Vis Syst, 85-, Effects of Vis Deprivation on Genicolocortical Pathway, 84-, co-investr, Sloan Found Comput Modelling Award, 82, Fogerty Sr Int Res Fel for Australia, 88. *Honors & Awards:* Sloan Young Neuroscientist Award. *Mem:* Soc Neurosci; Sigma Xi; Asn Res Vis & Ophthal; AAAS. *Res:* Structural basis of function of individual mammalian brain cells involved in processing visual information in the normal adult brain and during postnatal development chemical communication. *Mailing Add:* Physiol-Biophys Dept Univ Ala 401 Volker Hall Birmingham AL 35294

FRIEDLANDER, MICHAEL WULF, b Cape Town, SAfrica, 28. PHYSICS, ASTROPHYSICS. *Educ:* Univ Cape Town, SAfrica, BSc, 48, MSc, 50; Univ Bristol, Eng, PhD(physics), 55. *Prof Exp:* Jr lectr physics, Univ Cape Town, SAfrica, 51-52; res assoc, Univ Bristol, Eng, 54-56; from asst prof to assoc prof, 56-67, PROF PHYSICS, WASHINGTON UNIV, ST LOUIS, 67- *Concurrent Pos:* Guggenheim fel, 62-63. *Mem:* Am Astron Soc; AAAS; Am Physics Soc; Int Astron Union; Am Asn Univ Professors (second vpres, 78-80). *Res:* Cosmic rays; infrared astronomy; archaeoastronomy. *Mailing Add:* Dept Physics Washington Univ St Louis MO 63130

FRIEDLANDER, SHELDON K, b New York, NY, Nov 17, 27; m 58; c 4. CHEMICAL & ENVIRONMENTAL ENGINEERING. *Educ:* Columbia Univ, BS, 49; Mass Inst Technol, MS, 51; Univ Ill, PhD(chem eng), 54. *Prof Exp:* Asst prof chem eng, Columbia Univ, 54-57; from asst prof to prof, Johns Hopkins Univ, 57-64; prof chem & environ eng, Calif Inst Technol, 64-78; prof eng & vchmn chem eng, Univ Calif, Los Angeles, 78-84, Parsons prof chem eng & chmn dept, 84-88; CONSULT, 88- *Concurrent Pos:* Fulbright scholar, France, 60-61; indust & govt consult; mem environ sci & eng study sect, US Pub Health Serv, 65-68; Guggenheim fel, 69-70; chmn panel on abatement particulate emissions, Nat Res Coun, 70-72; mem technol assessment & pollution control adv comt, Environ Protection Agency, 76-79, chmn clean air sci adv comt; mem, Environ Studies Bd, Nat Resource Coun, 77-80. *Honors & Awards:* Colburn Award, Am Inst Chem Engrs, 59; Walker Award, Am Inst Chem Engrs, 79; Sr US Scientist Award-Humboldt Award, 84; Fuchs Mem Award, 90. *Mem:* Nat Acad Eng; Am Inst Chem Engrs; Am Chem Soc; Am Asn Aerosol Res. *Res:* Aerosol dynamics; diffusion and interfacial transfer; air pollution; particle technology. *Mailing Add:* Dept Chem Eng Univ Calif Los Angeles 5531 Boelter Hall Los Angeles CA 90024

FRIEDLANDER, SUSAN JEAN, b London, Eng, Jan 26, 46; m 68. APPLIED MATHEMATICS, GEOPHYSICAL FLUID DYNAMICS. *Educ:* Univ London, BS, 67; Mass Inst Technol, MS, 70; Princeton Univ, PhD(fluid dynamics), 72. *Prof Exp:* Vis mem math, Courant Inst, NY Univ, 72-74; instr, Princeton Univ, 74-75; from asst prof to assoc prof, 75-89, PROF MATH, UNIV ILL, CHICAGO, 89- *Concurrent Pos:* Kennedy Mem Fel, 67-69; consult, Goddard Inst Space Studies, 74-75 & Math Res Ctr, Univ Wis, 80; vis lectr, Math Inst, Oxford Univ, 77-78; res assoc, Univ Calif, Berkeley, 81; assoc prof, Univ Paris, 82-83; elect mem-at-large, Coun Am Math Soc, 83-85; mem, ETH, Zurich, 86. *Mem:* Am Math Soc; Soc Indust Appl Math; Am Geophys Soc. *Res:* The application of mathematical techniques to problems in geophysical fluid dynamics. *Mailing Add:* Dept Math Univ Ill Chicago Circle Chicago IL 60680

FRIEDLANDER, WALTER JAY, b Los Angeles, Calif, June 6, 19; m 76; c 3. NEUROLOGY. *Educ:* Univ Calif, MD, 45; Am Bd Psychiat & Neurol, dipl. *Prof Exp:* Chief EEG lab, NIH, 52-53; clin instr med, Sch Med, Stanford Univ, 53-55, asst clin prof neurol, 55-56; asst prof, Sch Med, Boston Univ, 57-61; from asst prof to prof, Albany Med Col, 61-66; prof neurol, Creighton Univ Col Med, 72-, dir, Clin Neurol Info Ctr, 72-80; prof neurol, Col Med, Univ Nebr, 66-81, dir, Ctr Humanities & Med, 75-80, prof med jurisp & humanities, 80-81, EMER PROF NEUROL, COL MED, UNIV NEBR, 81- *Mem:* sr mem Am Acad Neurol; fel Am Col Physicians; Am Assoc Hist Med. *Res:* History of medicine. *Mailing Add:* Dept Prev & Societal Med Univ Nebr Col Med Omaha NE 68105

FRIEDLANDER, WILLIAM SHEFFIELD, b Evanston, Ill, Feb 17, 30; m 51; c 5. ORGANIC CHEMISTRY. *Educ:* Dartmouth Col, AB, 51; Univ Ill, MS, 52, PhD, 54. *Prof Exp:* Sr chemist & group supvr, 54-61, asst mgr, Adv Res Proj Agency Proj, 61-62, head, Polymer Sect, 62-65, dir, Contract Res Lab, 65-67, tech dir, New Bus Ventures Div, 67-73, tech dir, Duplicating Prods Div, 73-75, tech planning mgr, Graphic Systs Group, 3M Co, 75-76, managing dir, Minn 3M Res Ltd, Harlow, Eng, 76-81, DIR TECHNOL EVAL, INDUST & CONSUMER SECTOR, 3M CTR, 81- *Mem:* Am Chem Soc; AAAS; Soc Adv Educ; Royal Soc Chem. *Res:* Synthetic organic chemistry; free radical reactions; chemistry of high energy propellants and rocket fuels; polymer science; fluorochemicals; graphic sciences; photographic science; new product development. *Mailing Add:* 3M Co 3M Ctr 220-9E-09 St Paul MN 55144-0001

FRIEDLER, ROBERT M, b Budapest, Hungary, Jan 7, 36; US citizen; m; c 4. NEPHROLOGY. *Educ:* Univ Chile, Baccalaureat, 53, MD, 60; Am Bd Internal Med, dipl, 73. *Prof Exp:* Asst anat, Sch Dent, Univ Chile, Santiago, 54-55, asst physiol, Inst Physiol, Med Sch, 55-56, instr med, 62-63, assoc med, 66-69; rotating intern, Salvador Hosp, Santiago, Chile, 59-60; fel, CIBA, 61 & Med Soc Santiago, 61-62; res fel kidney & electrolytes, Thorndike Mem Lab, Boston City Hosp, Harvard Med Sch, 63-66, clin fel med & nephrology, II-IV Harvard Med Serv, 65-66; med intern, St Elizabeth's Hosp, Boston, Mass, 69-70, resident med, 70-71; asst chief, Nephrology Sect, Wadsworth Vet Admin Hosp, Los Angeles, Calif, 72-74, dir, Renal Res & Electrolyte Lab, 73-74; asst prof med, Sch Med, Univ Calif, Los Angeles, 72-74, assoc prof, Santa Clara, 74-81; res physiologist, Cedars-Sinai Med Ctr, Los Angeles, Calif, 73-81; assoc dir, 81-89, PROF MED, DIV NEPHROLOGY, BONE & MINERAL METAB, UNIV KY, LEXINGTON, 81-; CHIEF NEPHROLOGY, VET ADMIN MED CTR, LEXINGTON, KY, 89- *Concurrent Pos:* Chmn, Res Comt, Kidney Found Southern Calif, 75; asst ed, J Mineral & Electrolyte Metab, 75-81; sabbatical, Div Nephrology, Univ Calif, Los Angeles, 88. *Mem:* Am Fedn Clin Res; Am Soc Nephrology; Int Soc Nephrology; Am Physiol Soc; Am Soc Bone & Mineral Res; NY Acad Sci. *Res:* Sodium handling by the kidney; renal osteodystrophy; experimental aluminum intoxication in uremia. *Mailing Add:* Div Nephrology Bone & Mineral Metab Univ Ky Col Med 800 Rose St Lexington KY 40536-0084

FRIEDLY, JOHN C, b Glen Dale, WVa, Feb 28, 38; m 62; c 4. CHEMICAL ENGINEERING. *Educ:* Carnegie Inst Technol, BS, 60; Univ Calif, Berkeley, PhD(chem eng), 64. *Prof Exp:* Chem engr, Gen Elec Res Lab, 64-67; asst prof chem eng, Johns Hopkins Univ, 67-68; from asst prof to assoc prof, Univ Rochester, 68-81, assoc dean, 79-81, chmn chem eng, 81-90, PROF CHEM ENG, UNIV ROCHESTER, 81- *Concurrent Pos:* NATO fel, Univ Oxford, 75-76; vis prof, US Geol Surv, Menlo Park, Calif, 90-91. *Mem:* Am Inst Chem Engrs; Am Chem Soc; Am Soc Eng Educ. *Res:* Dynamics of chemical processes; automatic control as applied to the process industries; heat transfer; groundwater transport. *Mailing Add:* Dept Chem Eng Univ Rochester Rochester NY 14627

FRIEDMAN, ABRAHAM S(OLOMON), b Brooklyn, NY, Oct 25, 21; m 52; c 4. SCIENCE POLICY, PHYSICAL CHEMISTRY. *Educ:* Brooklyn Col, BA, 43; Ohio State Univ, PhD(phys chem), 50. *Prof Exp:* Chem engr, Manhattan Proj, Decatur, Ill, 44-45; res chemist, Manhattan Proj, Metall Lab,

Univ Chicago, 45-46; res assoc, Ohio State Univ, 50-51; Fulbright res fel, Van der Waals Lab, Amsterdam Univ, 51-52; phys chemist, Nat Bur Standards, 52-56; chemist, US AEC, 56-61, sr chemist, 61-62, sci rep, Paris, 62-65, dep asst dir res chem, Washington, DC, 65-66, dep dir int affairs, 66-71, dir int affairs, 72-75; actg asst adminr int affairs, US Energy Res & Develop Admin, 75; sci counr, Dept State, Mex, 75-77, Bonn, 77-78, Paris, 78-82, Washington, DC, 83; res prof, dept physics, Cath Univ, 84-85; CONSULT, INT SCI & TECHNOL, 84- *Concurrent Pos:* Alt US rep, Int Atomic Energy Agency Gen Conf, 74; OAS Interamer Nuclear Energy Comn, 70-74, US-Iran Joint Nuclear Energy Comt, 70-74; mem, US-USSR Joint Comt Sci & Tech Coop, 72-74; vis prof, Ctr Inst Sci & Tech Policy, George Wash Univ, 90. *Mem:* Am Nuclear Soc. *Res:* Nuclear policy, arms control, technology transfer, chemical physics, energy technology. *Mailing Add:* 6305 Phyllis Lane Bethesda MD 20817

FRIEDMAN, ALAN E, b New Castle, Pa, Apr 18, 45; div; c 2. TOXICOLOGY, INFORMATION PROCESSING. *Educ:* Pa State Univ, BS, 66; State Univ NY, Buffalo, PhD(med chem), 73. *Prof Exp:* Asst chem, Res Triangle Inst, 72-74; assoc chem, Sinai Hosp, Baltimore, 74-76; asst state toxicol, Off Chief Med Examr, State Md, 76-77; proj mgr, Tracor Jitco Inc, 77-79; clin proj mgr, Alpha Therapeut Corp, 79-86; PROJ MGR, AMGEN, 86- *Concurrent Pos:* Instr, Essex Community Col, 77-79. *Mem:* Am Chem Soc. *Res:* Study the safety and effectiveness of new therapeutic agents in animal and human clinical trials. *Mailing Add:* c/o AmGen 1900 Oak Terr Lane Thousand Oaks CA 91320

FRIEDMAN, ALAN HERBERT, b New York, NY, Sept 24, 37; m 60; c 4. OPHTHALMOLOGY. *Educ:* Cornell Univ, BA, 59; NY Univ, MD, 63. *Prof Exp:* Intern med, Bellevue Hosp, New York, NY, 63-64; resident ophthal, Med Ctr, NY Univ, 66-70; from asst prof to assoc prof ophthal, Albert Einstein Col Med, 70-77, asst prof path, 73-77; CLIN PROF OPHTHAL & ATTEND OPHTHAL PATH, MT SINAI SCH MED, 77- *Concurrent Pos:* Res fel, Royal Postgrad Med Sch London, 72. *Mem:* Am Ophthal Soc; Asn Res Vision & Ophthal; Ophthal Soc UK; French Ophthal Soc; Royal Soc Med. *Res:* Ocular pathology and histochemistry. *Mailing Add:* Mt Sinai Hosp 888 Park Ave New York NY 10029

FRIEDMAN, ALEXANDER HERBERT, b Yonkers, NY, July 26, 25; m 61; c 2. PHARMACOLOGY. *Educ:* NY Univ, BA, 48; Univ Ill, MS, 56, PhD(pharmacol), 59. *Prof Exp:* Asst pharmacol, Yale Univ, 53; asst, Col Med, Univ Ill, 55-59; from instr to asst prof, Col Med, Univ Wis, 59-65, actg chmn dept, 63-64; assoc prof, 68-78, PROF PHARMACOL, STRITCH SCH MED, LOYOLA UNIV CHICAGO, 80- *Concurrent Pos:* Lederle med fac award, 66-69; adj prof ophalmol, Stritch Sch Med, Loyola Univ. *Honors & Awards:* Sigma Xi Res Prize, 58. *Mem:* AAAS; Int Soc Chronobiol; Am Soc Photobiol; hon mem NY Acad Sci; Am Soc Pharmacol & Exp Therapeut; Asn Res Vision & Ophthalmol. *Res:* Neuropharmacology; tremor and rigidity-Parkinsonism; autonomic nervous system pharmacology; muscle spindles; modulator effects of drugs and circadian rhythms of biogenic amines as a basis for drug action; role of neurotransmitters in retinal coding; physiochemical factors affecting membrane and signal transduction; non-invasive screening for open angle glaucoma. *Mailing Add:* Dept Pharmacol & Therapeut Loyola Univ Stritch Sch of Med Maywood IL 60153

FRIEDMAN, ARNOLD CARL, b Bronx, NY, Nov 17, 51; m 75; c 2. RADIOLOGY, CROSS-SECTION IMAGING. *Educ:* Cornell Univ, BA, 72; Albert Einstein Col Med, MD, 75. *Prof Exp:* Chief gastrointestinal radiol, Walter Reed Army Med Ctr, 80-83; dir genitourinary-computed tomography & gastrointestinal-genitourinary, George Washington Univ Hosp, 83-84; dir computed tomography-ultrasound-magnetic resonance imaging, 84-85, DIR ABDOMINAL IMAGING, TEMPLE UNIV HOSP, 85- *Concurrent Pos:* From instr to asst prof, Uniformed Serv Univ Health Sci, 79-83; asst prof lectr, George Washington Univ, 79-83, assoc prof radiol, 83-84; assoc prof, Temple Univ Hosp, 84-87, prof, 88-; ed, Williams & Wilkins, 87, Mosby, 89; lectr, Radiol Soc NAm, 88. *Mem:* Am Col Radiol; Asn Univ Radiologists; Radiol Soc NAm; Soc Gastrointest Radiologists; Am Inst Ultrasound Med; Am Roentgen Ray Soc. *Res:* Evaluation of new imaging modalities in the abdomen, chest and pelvis. *Mailing Add:* 524 Hoffman Dr Bryn Mawr PA 19010

FRIEDMAN, ARTHUR DANIEL, b New York, NY, Apr 4, 40; m 68; c 2. FAULT-TOLERANT COMPUTER DESIGN & SWITCHING THEORY. *Educ:* Columbia Univ, BA, 61, BS, 62, MS, 63, PhD(elec eng), 65. *Prof Exp:* Mem tech staff, Bell Telephone Labs, 65-72; assoc prof elec eng, Univ Southern Calif, 72-77; PROF ELEC ENG & COMPUT SCI, GEORGE WASHINGTON UNIV, 77- *Mem:* Fel Inst Elec & Electronics Engrs; Asn Comput Mach. *Res:* Writer and researcher in areas of fault-tolerant computer design, switching theory and automatic theory. *Mailing Add:* Dept Elec Eng & Comput Sci George Washington Univ Washington DC 20052

FRIEDMAN, AVNER, b Israel, Nov 19, 32; US citizen; c 4. DIFFERENTIAL EQUATIONS. *Educ:* Hebrew Univ, MSc, 54, PhD, 56. *Prof Exp:* Res assoc, Univ Kans, 56-57; lectr, Ind Univ, 57-58; asst prof, Univ Calif, Berkeley, 58-59; vis assoc prof, Stanford Univ, 61-62; prof, Northwestern Univ, 62-82, Noyes Prof Math, 84-86; Duncan distinguished prof math, Purdue Univ, 85-87; assoc prof, 59-61, DIR INST MATH & ITS APPL & PROF, UNIV MINN, 87- *Concurrent Pos:* Ed bd, Procedures AMS, 62-65, J Differential Equations, 69-, SIAM J Control, 70-86, Math Oper Res, 76-82, Commun in Partial Differential Equations, 76-, J Nonlinear Anal, Theoret Math, Appl, 76-83, Stochastic Anal & Appl, 83-, J Math Anal & its Appl, 86-; Sloan fel, 62-65, Guggenheim fel, 66-67, vis fel, Oxford Univ, 82-87. *Honors & Awards:* Stampacchia Prize, 82. *Mem:* AAAS. *Res:* Stochastic Control; free boundary problems; differential games; nonlinear partial differential equations. *Mailing Add:* Inst Math & Its Application Univ Minn 514 Vincent Hall 206 Church St SE Minneapolis MN 55455

FRIEDMAN, BEN I, b Cincinnati, Ohio, Oct 18, 26; m 54; c 1. NUCLEAR MEDICINE, INTERNAL MEDICINE. *Educ:* Univ Cincinnati, MD, 48. *Prof Exp:* Instr med, Col Med, Univ Cincinnati, 55-59, asst clin prof, 59-62, from asst prof to assoc prof, 62-68, asst prof radiol, 64-68, assoc dir nuclear med, 66-68; prof radiol, Col Med, Univ Tenn, Memphis, 68-73, prof med, 68-73, prof nuclear med & chmn dept & dir, Div Radiation Sci, 73-77; MEM STAFF, DEPT NUCLEAR MED, MORTON PLANT HOSP, 77- *Concurrent Pos:* Fel hemat & nutrit, Cincinnati Gen Hosp, 53-55; clinician, Outpatient Dept, Cincinnati Gen Hosp, 55-68, from asst attend physician to attend physician, 56-68, asst chief clinician, Hemat Clin, 56-68; attend physician, City Memphis Hosps, 68-77, Doctors Hosp & LeBonheur Childrens Hosp, Memphis. *Mem:* Am Col Nuclear Physicians; AMA; Soc Nuclear Med. *Res:* Hematology; radioisotopes; radiobiology. *Mailing Add:* Four Belleview Blvd No 601 Belleair FL 34616

FRIEDMAN, BERNARD SAMUEL, b Chicago, Ill, Jan 4, 07; m 38; c 2. PETROLEUM CHEMISTRY. *Educ:* Univ Ill, AB, 30, PhD(org chem), 36. *Prof Exp:* Teacher pub sch, Ill, 30-33; asst instr chem, Univ Ill, 33-36; res chemist, Universal Oil Prods Co, Ill, 36-45; tech dir, Reyam Plastic Prods Co, 45-47; dir chem lab, QM Res & Develop Labs, 47-48; assoc dir org res div, Sinclair Res Labs, Inc, 48-59, res assoc, 59-69; CONSULT, 69- *Concurrent Pos:* Mem, Chicago Bd Educ, 62-77; prof lectr, Univ Chicago, 69-73. *Honors & Awards:* Honor Scroll, Am Inst Chem, 59. *Mem:* Am Chem Soc (pres, 74); Am Inst Chem Engrs; fel Am Inst Chem. *Res:* Petroleum and petrochemical catalytic chemistry. *Mailing Add:* 4800 S Chicago Beach Dr Apt 1616 N Chicago IL 60615

FRIEDMAN, CHARLES NATHANIEL, b New York, NY, June 10, 46; div; c 2. MATHEMATICS. *Educ:* Cornell Univ, AB, 68; Princeton Univ, PhD(math), 71. *Prof Exp:* C L E Moore instr math, Mass Inst Technol, 71-73; asst prof, 73-78, ASSOC PROF MATH, UNIV TEX, AUSTIN, 78- *Mem:* Am Math Soc. *Res:* Mathematical physics; measurement theory in quantum mechanics; asymptotics of partial differential equations; product integrals. *Mailing Add:* Dept Math Univ Tex Austin TX 78712

FRIEDMAN, CONSTANCE LIVINGSTONE, b Montreal, Que, July 30, 20; m 40. ANATOMY. *Educ:* McGill Univ, BSc, 41, MSc, 42, PhD(anat), 48. *Prof Exp:* Demonstr histol & biochem, McGill Univ, 43-45, res asst & assoc, 45-50; res assoc prof anat, 50-85, EMER PROF ANAT, UNIV BC, 86- *Mem:* Can Asn Anat. *Res:* Hypertension; endocrinology; aging. *Mailing Add:* Dept Anat Univ BC Vancouver BC V6T 1W5 Can

FRIEDMAN, DANIEL LESTER, b Cleveland, Ohio, Sept 25, 36; m 81; c 4. CANCER, AFFECTIVE DISORDERS. *Educ:* Western Reserve Univ, BA, 58, MD & PhD(pharmacol), 65. *Prof Exp:* Fel oncol, McArdel Lab, Univ Wis, 65-67; asst prof, 67-72, ASSOC PROF, VANDERBILT UNIV, 72-; RESIDENT PSYCHIAT, 88- *Concurrent Pos:* Mem sci adv panel, NSF, 78-80. *Mem:* Am Soc Cell Biol; Am Soc Biol Chemists; Sigma Xi. *Res:* Biochemistry of growth and cell division; biochemistry and genetics of psychiatric disorders. *Mailing Add:* Box 1694 Sta B Vanderbilt Univ Nashville TN 37203

FRIEDMAN, DON GENE, b Long Beach, Calif, May 26, 25; m 49; c 6. APPLIED METEOROLOGY, APPLIED SEISMOLOGY. *Educ:* Univ Calif, Los Angeles, BA, 50; Mass Inst Technol, MS, 51, ScD(meteorol), 54. *Prof Exp:* Asst meteorol, Mass Inst Technol, 50-54; fel math statist, Univ Chicago, 54-55; res assoc appl meteorol & opers res, 55-60, asst dir res, 60-66, assoc dir res, 66-76, DIR, CORP RES DIV, TRAVELERS INS CO, 76- *Concurrent Pos:* Consult, Dept Housing & Urban Develop, 66-67, NSF, Assesment Res Natural Hazards Proj, Univ Colo, 73-75, NSF, 75-78, UNESCO, UN Disaster Relief Prog, UN Develop Prog & UN Environ Prog, 78-85; mem, Nat Res Coun, 81-82, mem, Indust Res Adv Coun, 83- *Honors & Awards:* Nat Award for Outstanding Contrib to the Advance of Appl Meteorol, Am Meteorol Soc, 76. *Mem:* Am Meteorol Soc; Am Seismol Soc; AAAS; Am Geophys Union; Earthquake Eng Res. *Res:* Computer simulation used to generate, superimpose and interact geographical severity patterns of natural hazards (storms, floods, earthquakes) on computerized mapping of population; property to better understand mechanism that produces disasters and how to mitigate their detrimental effects. *Mailing Add:* 99 Knollwood Rd No NT Newington CT 06111

FRIEDMAN, EDWARD ALAN, b Bayonne, NJ, Sept 29, 35; m 63; c 2. TECHNICAL & ACADEMIC ADMINISTRATION. *Educ:* Mass Inst Technol, BS, 57; Columbia Univ, PhD(physics), 63. *Hon Degrees:* MEng, Stevens Inst Tech, 83. *Prof Exp:* From asst prof to assoc prof physics, Stevens Inst Technol, 63-80, dean col, 73-83, vpres acad affairs, 83-85, PROF MGT, STEVENS INST TECHNOL, 80-, DIR, CTR IMPROVED ENG & SCI EDUC, 88- *Concurrent Pos:* US Agency Int Develop contract loan, Kabul, Afghanistan, 65-67; chief of party, US Eng Team, Fac Eng, Kabul Univ, 70-73; chmn, Coun for Understanding Technol in Human Affairs, 79-82; mem bd dir, Asn Independent Col & Univ, NJ, 78-83; sr vpres, Afghanistan Relief Comt, 80- *Honors & Awards:* Res Award, Stevens Inst Technol, 70. *Mem:* AAAS; Am Phys Soc; NY Acad Sci; Am Soc Eng Educ. *Res:* Mossbauer measurements of hyperfine fields; inelastic neutron scattering from magnetic materials; philosophy of technology; studies on role of computers in education and in business; full-text information retrieval. *Mailing Add:* Stevens Inst Technol Hoboken NJ 07030

FRIEDMAN, EILEEN ANNE, US citizen. CELL BIOLOGY. *Educ:* NY Univ, AB, 67; Johns Hopkins Univ, PhD(molecular biol), 72. *Prof Exp:* Fel cell biol, Albert Einstein Col Med, 74-78; ASSOC GASTROENTEROL, MEM SLOAN-KETTERING CANCER CTR, 78- *Concurrent Pos:* Nat Cancer Inst fel, Albert Einstein Col, 75-77. *Mem:* Am Soc Cell Biol; Am Asn Cancer Researchers. *Res:* Cell biology; differentiates; DNA transfection; molecular biology. *Mailing Add:* Mem Sloan-Kettering Cancer Ctr 1275 York Ave Box 244 New York NY 10021

FRIEDMAN, EITAN, PSYCHOLOGY. *Prof Exp:* PROF PSYCHOL & PHARMACOL & DIR, DIV NEUROCHEM, MED COL, PA, 86- *Mailing Add:* Dept Psychol & Pharmacol Med Col Pa EPPI 3200 Henry Ave Rm 737 Philadelphia PA 19129

FRIEDMAN, ELI A, b New York, NY, Apr 9, 33; m 57; c 3. INTERNAL MEDICINE, IMMUNOLOGY. *Educ:* DS, Maduraikamaraj Univ, India, Long Island Univ; DSc, Univ Maduri, India. *Hon Degrees:* DS, Maduraikamaraj Univ, India. *Prof Exp:* Intern med, Peter Bent Brigham Hosp, Boston, 57-58, sr resident, 60-61; instr, Emory Univ, 61-63; from asst prof to assoc prof, 63-71, PROF MED, STATE UNIV NY DOWNSTATE MED CTR, 71- *Concurrent Pos:* Res fel nephrol, Peter Bent Brigham Hosp, Boston, 58-60; consult, Vet Admin & USPHS, 64; coordr regional med prog, USPHS, 76- *Honors & Awards:* Hoenig Award, New York Kidney Found. *Mem:* Fel Am Col Physicians; Transplantation Soc; Am Soc Nephrology; Am Fedn Clin Res; Am Soc Artificial Internal Organs; Asn Am Physicians; Int Soc Artificial Organs. *Res:* Development and application of artificial kidneys; effect and synergistic relationship of immunosuppressive drugs. *Mailing Add:* Dept Med State Univ NY Downstate Med Ctr Brooklyn NY 11203

FRIEDMAN, EMANUEL A, b New York, NY, June 9, 26; m 48; c 3. GYNECOLOGY. *Educ:* Brooklyn Col, AB, 47; Columbia Univ, MD, 51, MedScD(physiol), 59; Am Bd Obstet & Gynec, dipl. *Hon Degrees:* AM, Harvard Univ, 69. *Prof Exp:* Intern path, Cornell Med Div, Bellevue Hosp, New York, 51-52; resident obstet & gynec, Columbia-Presby Med Ctr, 52-57; instr, Fac Med, Columbia Univ, 57-59, assoc, 59-60, from asst prof to assoc prof, 60-63; prof & chmn dept, Chicago Med Sch, 63-69; chmn, Div Obstet & Gynec, Michael Reese Hosp & Med Ctr, 63-69; PROF OBSTET & GYNEC, HARVARD MED SCH, 69-; CHMN DEPT OBSTET & GYNEC, BETH ISRAEL HOSP, BOSTON, 69- *Concurrent Pos:* Asst, Columbia-Presby Med Ctr, 57-59, from asst attend to assoc attend, 59-63. *Honors & Awards:* Joseph Mather Smith Award, Columbia Univ, 58, Commemorative Silver Award, 67. *Mem:* AAAS; fel Am Col Obstet & Gynec; fel Am Col Surg; Soc Exp Biol & Med; fel Int Col Surg. *Res:* Physiology and pathophysiology of human labor phenomena; lactation and milk ejection; placental physiology. *Mailing Add:* Harvard Med Sch Beth Israel Hosp 330 Brookline Ave Boston MA 02215

FRIEDMAN, EMIL MARTIN, b Brooklyn, NY, Feb 15, 48; m 74; c 3. RUBBER CHEMISTRY, POLYMER PHYSICS. *Educ:* Mass Inst Technol, SB, 68; Princeton Univ, MA, 70, PhD(phys chem), 73; Case Western Reserve Univ, MS, 87. *Prof Exp:* Vis scientist, Polymer Sci & Eng, Univ Mass, 73-74; sr chemist, 74-83, ASSOC SCIENTIST, GOODYEAR TIRE & RUBBER CO, 83- *Mem:* Am Chem Soc; Am Statist Asn; Am Soc Qual Control. *Res:* Physical properties of uncured rubber; structure-properties relationships; mathematical modeling of chemical processes; experimental design and statistics. *Mailing Add:* Goodyear Tech Ctr 410D PO Box 3531 Akron OH 44309-3531

FRIEDMAN, EMILY PERLINSKI, computer science, for more information see previous edition

FRIEDMAN, EPHRAIM, b Belvedere, Calif, Jan 1, 30; m 54; c 4. OPHTHALMOLOGY. *Educ:* Univ Calif, Los Angeles, BA, 50; Univ Calif, San Francisco, MD, 54. *Prof Exp:* Intern, San Francisco City & County Hosps, 54-55; resident ophthal, Mass Eye & Ear Infirmary, 59-61; instr, Howe Lab Ophthal, 61-64; prof ophthal, Sch Med, Boston Univ, 65-74, dean, 71-74; PROF OPHTHAL & DEAN, ALBERT EINSTEIN COL MED, 74- *Mem:* Asn Res Vision & Ophthal. *Res:* Physiology of the circulation of blood in the eye and related clinical problems. *Mailing Add:* Harvard Med Sch Mass Eye & Ear Infirmary 243 Charles St Boston MA 02114

FRIEDMAN, FRANK L, b Baltimore, Md, Dec 9, 42. SOFTWARE SYSTEMS. *Educ:* Antioch Col, BA, 65; Johns Hopkins Univ, MS, 69; Purdue Univ, MA, 72, PhD(computer sci), 74. *Prof Exp:* From asst prof to assoc prof, 74-91, PROF COMPUTER & INFO SCI, TEMPLE UNIV, 91- *Concurrent Pos:* Vis scientist, Computer Sci, 84-85 & Software Eng Inst, 89-90; chmn conf bd, Asn Comput Mach, 86- *Mem:* Asn Comput Mach. *Mailing Add:* Dept Computer & Info Sci Rm 303 Temple Univ Computer Ctr Bldg Philadelphia PA 19122

FRIEDMAN, FRED JAY, b New York, NY, June 28, 51; div; c 1. RELATIONAL DATABASES, ADA. *Educ:* Mich State Univ, BS, 71; Yale Univ, MS, 73, MPh, 74. *Prof Exp:* Consult, Naval Underwater Sound Lab & Xerox, 73-74, Advan Comput Techniques, 74, Gen Elec Co, 74-75; mem staff, Inco, Inc, 75-80, dir, Adv Syst Res, 80-81, corp sr scientist, 81-83; OWNER, RACOM COMPUTER PROFESSIONALS, 83- *Res:* Relational data bases, specifically user-friendly interfaces to back-end data base machines; compilers; computer networks; text editors; software productivity techniques and metrics; advanced Ada technology. *Mailing Add:* 7321 Franklin Rd Annandale VA 22003-1620

FRIEDMAN, FRED K, b Montreal, Can, Oct 23, 52. CHEMISTRY. *Educ:* City Col NY, BS, 73; Columbia Univ, MA, 75, PhD(chem), 79. *Prof Exp:* SUPVR RES, CHEM LAB MOLECULAR CARCINOGENESIS, NAT CANCER INST, NIH, 82- *Mem:* Am Soc Biochem & Molecular Biol; Biophys Soc; AAAS; Am Chem Soc; Sigma Xi. *Mailing Add:* Lab Molecular Carcinogen Nat Cancer Inst NIH Bldg 37 Rm 3E24 Bethesda MD 20892

FRIEDMAN, GARY DAVID, b Cleveland, Ohio, Mar 8, 34; m 58; c 3. CHRONIC DISEASE EPIDEMIOLOGY. *Educ:* Univ Chicago, BS, 56, MD, 59; Harvard Univ, SM, 65; Am Bd Internal Med, cert, 66. *Prof Exp:* From intern to asst resident, Harvard Med Serv, Boston City Hosp, 59-61; asst resident, Univ Hosp, Cleveland, 61-62; med officer, Heart Dis Epidemiol Study, US Pub Health Serv, Mass, 62-66, chief, Epidemiol Unit, Epidemiol Field & Training Sta, Heart Dis Control Prog, 66-68; sr epidemiologist, 68-76, ASST DIR EPIDEMIOL & BIOSTATIST, DIV RES, KAISER-PERMANENTE MED CARE PROG, 76- *Concurrent Pos:* From res fel to res assoc, Dept Prev Med, Med Sch, Harvard Univ, 62-66; asst, Dept Ambulatory & Community Med, Univ Calif, San Francisco, 67-80, assoc clin prof, 80-; lectr, Sch Pub Health, Univ Calif, Berkeley, 68-; prin investr res grants & contracts, 71-; mem, US-USSR Working Group Sudden Cardia Death, Nat Heart, Lung and Blood Inst, NIH, 75-82, mem epidemiol & dis control study sect, 82-86, prev serv task force, 84-; mem sci rev panel, Calif Air Resources Bd; outstanding investr grant, Nat Cancer Inst, 89-96. *Mem:* Fel Am Col Physicians; Am Epidemiol Soc; Soc Epidemiol Res; Int Epidemiol Asn; fel Am Heart Asn; Int Soc Twin Studies; Am Pub Health Asn; Am Soc Prev Oncol. *Res:* Epidemiology of chronic diseases, focusing especially on cardiovascular diseases, cancer, and the effects of smoking, alcohol and medicinal drugs; author of over 190 publications. *Mailing Add:* 3451 Piedmont Ave Oakland CA 94611

FRIEDMAN, GEORGE J(ERRY), b New York, NY, Mar 22, 28; m 53; c 3. SYSTEMS ANALYSIS, APPLIED MATHEMATICS. *Educ:* Univ Calif, Berkeley, BS, 49; Univ Calif, Los Angeles, MS, 56, PhD(eng & appl sci), 67. *Prof Exp:* Mech eng assoc, pub utility systs anal, Los Angeles Dept Water & Power, 49-56; res & develop engr, Servomechanisms, Inc, Calif, 56-60; res scientist & mgr adv systs, 60-80, Vpres Tech & Electron Systs Group, 80-84, PRES ENG, NORTHROP, CENTURY CITY, CALIF, 84- *Concurrent Pos:* Lectr, Univ Calif, Los Angeles, 57, Calif State Univ, Northridge, 83- *Honors & Awards:* W R G Baker Award, Inst Elec & Electronics Engrs, 70. *Mem:* Soc Gen Systs Res; fel Inst Elec & Electronics Engrs; Am Inst Aeronaut & Astronaut. *Res:* Operations analysis; constraint theory applied to the analysis of complex, multidimensional math models; computer simulation of evolutionary processes; automata theory; systems analysis methodology; artificial intelligence and multidimensional kinematics. *Mailing Add:* VPres Northrop Corp 1840 Century Park E Los Angeles CA 90067

FRIEDMAN, GERALD MANFRED, b Berlin, Ger, July 23, 21; nat US; m 48; c 5. SEDIMENTOLOGY, SEDIMENTARY PETROLOGY. *Educ:* Univ London, BSc, 45; Columbia Univ, MA, 50, PhD(geol), 52; DrSci, Univ London, 77. *Hon Degrees:* Dr, Univ Heidelberg, 86. *Prof Exp:* Sr chemist, J Lyons & Co, 45-46; anal chemist, E R Squibb & Sons, 47-49; asst geol, Columbia Univ, 50; from instr to asst prof, Univ Cincinnati, 50-54; consult geologist, Sault Ste Marie, Can, 54-56; sr res geologist, Amoco Prod Co, 56-60, res assoc & supvr sedimentary geology res, 60-64; prof, 64-84, EMER PROF GEOL, RENSSELAER POLYTECH INST, 84-; prof, 85-88, DISTINGUISHED PROF GEOL, BROOKLYN COL & GRAD SCH & UNIV CTR, CITY UNIV NEW YORK, 88- *Concurrent Pos:* Sect ed, Chem Abstr, 62-69; Fulbright sr vis lectr, Hebrew Univ, 64; ed, J Sedimentary Petrol, 64-70, Northeastern Geol, 79- & Earth Sci Hist, 82-, Carbonates & Evaporites, 86-; lecturer, Oil & Gas Consult Int, 68-; pres, Northeastern Sci Found, Inc, 79-; vis scientist, Geol Surv, Israel, 67-73 & 78. *Honors & Awards:* Distinguished lectr. Am Asn Petrol Geol, 72-73, Distinguished Serv Award, 88. *Mem:* Asn Earth Sci Ed (vpres, 70-71, pres, 71-72); hon mem Soc Econ Paleontologists & Mineralogists (vpres, 70-71, pres, 74-75); hon mem Int Asn Sedimentologists (vpres, 71-75, pres, 75-78); fel AAAS (counr, 79-80); Nat Asn Geol Teachers (treas, 51-55); hon mem Am Asn Petrol Geol (vpres, 84-85); Earth Sci Soc (pres, 82-86); fel Mineralogical Soc Am. *Res:* Carbonate sedimentology; chemistry of sedimentation; petrology and sedimentology of clastic and carbonate sediments. *Mailing Add:* Dept Geol Brooklyn Col Brooklyn NY 11210

FRIEDMAN, HAROLD BERTRAND, b Montgomery, Ala, Oct 13, 04; m 53; c 3. PHYSICAL CHEMISTRY, INDUSTRIAL CHEMISTRY. *Educ:* Univ Ala, AB, 23; Univ Va, PhD(chem), 27. *Prof Exp:* Teacher pub sch, Ala, 23-24; instr phys chem, Univ Maine, 27-28; asst chem, Columbia Univ, 28; from asst prof to assoc prof, Ga Tech, 29-42; chem dir, 45-63, VPRES, ZEP MFG CO, 63- *Concurrent Pos:* Consult chem, 74- *Mem:* Emer Am Chem Soc; fel Am Inst Chem; Sigma Xi. *Res:* Solid-gas catalysis; neutral salt action; reaction velocity and kinetics; activity coefficients of electrolytes; history of chemistry; soaps, waxes and detergents. *Mailing Add:* Zep Mfg Co 1310 Seaboard Industrial Blvd NW Atlanta GA 30301

FRIEDMAN, HAROLD LEO, b New York, NY, Mar 24, 23; m 45; c 2. PHYSICAL CHEMISTRY, CHEMICAL PHYSICS. *Educ:* Univ Chicago, BS, 47, PhD(chem), 49. *Prof Exp:* From instr to assoc prof chem, Univ Southern Calif, 49-59; adv chemist, Res Ctr, Int Bus Mach Corp, 59-65; PROF CHEM, STATE UNIV NY, STONY BROOK, 65- *Concurrent Pos:* Guggenheim fel, 57-58; Alfred P Sloan res fel, 59-61; adj prof, Polytech Inst Brooklyn, 64-65. *Honors & Awards:* Robinson Medal, Faraday Soc. *Mem:* Am Chem Soc; Am Inst Physics; Royal Soc Chem; AAAS; Sigma Xi; fel Am Phys Soc. *Res:* Equilibrium and dynamic properties of liquid solutions. *Mailing Add:* Dept Chem State Univ NY Stony Brook NY 11794-3400

FRIEDMAN, HARRIS LEONARD, organic chemistry, for more information see previous edition

FRIEDMAN, HARVEY MARTIN, b Chicago, Ill, Sept 23, 48; m 73. MATHEMATICAL LOGIC. *Educ:* Mass Inst Technol, PhD(math), 67. *Prof Exp:* From asst prof to assoc prof logic, Stanford Univ, 67-73; prof math, State Univ NY Buffalo, 73-77; PROF MATH, OHIO STATE UNIV, 77-, PROF PHILOS & COMPUT SCI, 87- & DIR PROG FOUND STUDIES, 87- *Concurrent Pos:* Vis assoc prof math, Univ Wis-Madison, 70-71, vis prof, State Univ NY Buffalo, 72-73; prin investr, NSF Grants, 67-; vis scientist, IBM, 83 & 84; consult, Bell Labs, 84; Guggenheim fel, Ohio State Univ, 86 & 87. *Honors & Awards:* Alan T Waterman Award, NSF, 84. *Mem:* Am Math Soc; Asn Symbolic Logic. *Res:* Mathematical logic and the foundation of mathematics; theoret computer science; foundational studies. *Mailing Add:* Dept Math Ohio State Univ 231 W 18th Ave Columbus OH 43210

FRIEDMAN, HARVEY PAUL, b New York, NY, Dec 31, 35; m 70. EMBRYOLOGY, IMMUNOLOGY. *Educ:* City Col New York, BS, 57; Univ Kans, PhD(anat), 63. *Prof Exp:* Bacteriologist, Sloan-Kettering Inst, 57-58; asst prof microbiol, Meharry Med Col, 65-66; res assoc immunol, Univ Kans, 66-68; asst prof, 68-71, ASSOC PROF, DEPT BIOL, UNIV MO-ST LOUIS, 71- *Concurrent Pos:* NIH fel, Case Western Reserve Univ, 63-65; Nat Inst Neurol Dis & Blindness grant; consult, Vet Admin Hosp, Topeka, Kans. *Mem:* AAAS; Am Soc Neurochem; Soc Study Develop. *Res:* Immunological investigation of embryonic development; in vitro synthesis of antibody. *Mailing Add:* Dept Biol Univ Mo 8001 Natural Bridge Rd St Louis MO 63121

FRIEDMAN, HELEN LOWENTHAL, b New York, NY, Jan 25, 24; m 46; c 2. PHYSICAL OPTICS, MEDICAL PHYSICS. *Educ:* Hunter Col, BA, 44; Purdue Univ, MS, 46; Columbia Univ, PhD(physics), 51. *Prof Exp:* Asst physics, Purdue Univ, 44-46; asst, Columbia Univ, 49-51; lectr, Hunter Col, 52-53; instr, Queen's Col, NY, 53-55, lectr, 58-61; asst prof, 61-68, ASSOC PROF PHYSICS, HOFSTRA UNIV, 68- *Mem:* AAAS; Am Phys Soc; Optical Soc Am; Am Asn Physics Teachers; Am Asn Physicists Med. *Res:* Physical optics. *Mailing Add:* 242-23 54th Ave Douglaston NY 11362

FRIEDMAN, HENRY DAVID, mathematics, for more information see previous edition

FRIEDMAN, HERBERT, b New York, NY, June 21, 16; m 40; c 2. PHYSICS, ASTRONOMY. *Educ:* Brooklyn Col, BS, 36; Johns Hopkins Univ, PhD(physics), 40. *Hon Degrees:* DSc, Univ Tubingen, 77, Univ Mich, 79. *Prof Exp:* Physicist, Metall Div, US Naval Res Lab, 40-41, head, Electron Optics Br, 41-58, supt, Astron & Astrophys Div, 58-63, Space Sci Div, 63-80, EMER CHIEF SCIENTIST, E O HULBERT CTR SPACE RES, US NAVAL RES LAB, 80- *Concurrent Pos:* Adj prof, Univ Md, 61-83; mem Space Sci Bd, Nat Acad Sci, 62-75, adv coun, Dept Geol Sci, Princeton Univ, 67-76, vis comt, Div Phys Sci, Chicago Univ, 74-80, gov bd, Nat Res Coun, 79-86, bd trustees, Univ Res Assoc, 85; chmn, Comt Space Res Working Group II for Int Year Quiet Sun, 62-65; vchmn, Inter-Union Comt on Ionosphere, 63-66; mem exec comt, Int Asn Geomagnetism & Aeronomy, 63-; pres, Inter-Union Comn on Solar-Terrestrial Physics, 63-66 & spec comt, Int Coun Sci Unions, 66-74; mem bd trustees, Assocs Univs, Inc, 66-70; vis prof, Yale Univ, 66-68; mem bd trustees, Assoc Univs, Inc, 66-70; mem exec comt, Div Phys Sci, Nat Res Coun-NAS, 69-, adv panel atmospheric sci & chmn comt solar-terrestrial res, Geophys Res Bd; mem gen adv comt to US Atomic Energy Comn, 69-74; pres comn 48 on high energy astrophys, Int Astron Union, 70-; mem, President's Sci Adv Comt, 71-73; mem, Space Prog Adv Coun, Nat Aeronaut & Space Admin, 74-77; adj prof, Univ Pa, 74-; mem ad hoc adv group on sci progs, Presidents's Sci Adv, NSF, 75-76; adv bd, Off Phys Sci, Nat Res Coun, 75-, chmn Geophys Res Bd, 76-; chmn, X-Ray Astron Comt, Univs Res Asn, Inc, 76-81, vis comt, Dept Astron, Harvard Univ, 81-86; mem coun Nat Acad Sci, 80-83, chmn, Assembly Math & Phys Sci Nat Res Coun, 80-83, chmn, Comn Phys Sci, Math & Resources, 83-86; hon mem, Spec Comt Solar Terrestrial Physics, 82, Comt Space Res, Inter-Coun Sci Unions, 71-75 & 82- *Honors & Awards:* Annual Award, Soc Appl Spectros, 57; Janssen Medal, Fr Photog Soc, 62; Space Sci Award, Am Inst Aeronaut & Astronaut, 63, Dryden Res Award, 73; Capt Robert Dexter Conrad Award, Off Naval Res, 64; Eddington Medal, Royal Astron Soc, 64; President's Award Distinguished Fed Serv, 67; Nat Medal Sci, 68 & NASA Medal Excep Sci Achievement, 70 & 78; Albert A Michelson Medal, Franklin Inst, 72; Lovelace Award, Am Astron Soc & Dryden Res Award, Am Inst Aeronauts & Astronauts, 73; William Bowie Medal, Am Geophys Union, 81; Henry Norris Russell Award, Am Astron Soc, 83; Navy Award Distinquished Sci Achievement, 62. *Mem:* Nat Acad Sci; fel AAAS (vpres elect, 71, vpres, 72); fel Am Acad Arts & Sci; fel Am Phys Soc; fel Optical Soc Am; fel Am Inst Aeronaut & Astronaut; fel Int Acad Astron; Am Philos Soc; Int Acad Astronaut; Am Philos Soc. *Res:* X-ray spectroscopy and diffraction; electron diffraction and microscopy; nucleonics; upper atmosphere research; electron optics; high energy astronomy; radio astronomy. *Mailing Add:* Code 4190 Space Sci Div Naval Res Lab Washington DC 20375

FRIEDMAN, HERBERT ALTER, b Brooklyn, NY, May 9, 37. PHARMACEUTICAL CHEMISTRY. *Educ:* Yeshiva Univ, BA, 58; Polytech Inst Brooklyn, PhD(chem), 64. *Prof Exp:* From instr pharmacol to res specialist, Univ Pa Sch Med, 67-76; asst prof pharmacol, Jefferson Med Col, Thomas Jefferson Univ, 76-81; chem fac, West Chester Univ, 81-85; sci fac, Tower Hill Sch, 85-86; CRIMINALIST, PA STATE POLICE, 88- *Concurrent Pos:* Fel, Sloan-Kettering Inst Cancer Res, 64-67. *Mem:* Am Chem Soc; Am Soc Pharmacol & Exp Therapeut; Sigma Xi. *Res:* Structure-activity relationships; medicinal chemistry; molecular aspects of interactions with muscle; mass spectrometry in pharmacology. *Mailing Add:* Seven Orchard Rd Merion Station PA 19066-1816

FRIEDMAN, HERMAN, b Philadelphia, Pa, Sept 22, 31; m 58. MICROBIOLOGY, IMMUNOLOGY. *Educ:* Temple Univ, AB, 53, AM, 55; Hahnemann Med Col, PhD(microbiol), 57; Am Bd Microbiol, dipl. *Prof Exp:* Instr microbiol, Hahnemann Med Col, 57-58; res biochemist & chief allergy dept, Vet Admin Hosp, Pittsburgh, 58-59; from asst prof to prof microbiol, Sch Med, Temple Univ, 70-78; head dept, Albert Einstein Med Ctr, 59-78; PROF & CHMN DEPT MICROBIOL, COL MED, UNIV SOUTH FLA, 78- *Concurrent Pos:* Am Heart Found fel, Childrens Hosp, Philadelphia & Hahnemann Med Col, 57-58; NIH & NSF grants, 60-; Am Cancer Soc grants, 65-68 & 72-73; chmn, Comt Immunol, Am Bd Med Microbiol, 67-79; deleg & session chmn, Int Cong Microbiol, Moscow, 66 & Mexico City, 70, Cong Bact, Jerusalem, 73, Immunol, Brighton, Eng, 74, Sydney, Australia, 78 & Transplant Cong, Jerusalem, 74; deleg & session chmn, Virol Cong, Madrd, 75, New York City, 76 & Reticuloendothelial Socs Cong, Panplona, Spain, 75 & Jerusalem, 78; chmn, Int Conf, NY Acad Sci, 71, 73, 74, 75 & 79; consult, Miles Labs, Damon Labs, Smith Kline & French Labs & Wyeth Labs. *Mem:* AAAS; Am Soc Microbiol; fel NY Acad Sci; Brit Soc Gen Microbiol; Am Asn Immunologists; Sigma Xi. *Res:* Bacterial physiology; tumor virus immunology and oncology; nucleic acid synthesis; immunology and immunochemistry-antibody formation; hypersensitivity and allergy. *Mailing Add:* Col Med Univ SFla 12901 N Bruce B Downs Blvd MDC Box 10 Tampa FL 33612

FRIEDMAN, HOWARD STEPHEN, b Elizabeth, NJ, Mar 31, 48; m 72. ORGANIC CHEMISTRY. *Educ:* NY Univ, AB, 70, MS, 72, PhD(org chem), 76. *Prof Exp:* Scholar & lectr chem, Univ Mich, 75-77; sr res chemist, Goodyear Tire & Rubber Co, 77-79; sr res chemist, Ferro Corp, Bedford, Ohio, 80-82; sr chemist, Velsicol Chem Corp, 82-85, corp scientist 86-87; proj leader, 88-90, REGULATORY SCIENTIST, UNIROYAL CHEM CORP, MIDDLEBURY, CT, 90- *Mem:* Am Chem Soc; Royal Soc Chem; Sigma Xi. *Res:* Flame retardant, specialty, agrichemical pharmaceutical chemicals; plastic polymerization inhibitors. *Mailing Add:* Uniroyal Chem Co World Hq Middlebury CT 06749

FRIEDMAN, IRVING, b New York, NY, Jan 12, 20; m 46. GEOCHEMISTRY. *Educ:* Mont State Col, BS, 42; State Col, Wash, MS, 44; Univ Chicago, PhD(geochem), 50. *Prof Exp:* Chemist, US Naval Res Lab, 46-48; asst geol, Univ Chicago, 48-50, res assoc, Enrico Fermi Inst Nuclear Studies, 50-52; GEOCHEMIST, US GEOL SURV, 52- *Concurrent Pos:* Adv, Int Atomic Energy Agency, Vienna, 62-; adj prof, Univ Pa, Dept Geol, 69-; distinguished vis prof, Quartenary Res Inst, Univ Wash, 77. *Mem:* Fel Geol Soc Am; fel Am Geophys Union; Am Chem Soc; AAAS. *Res:* Phase equilibria in hydrous silicate systems at high temperature and pressure; abundance of stable isotopes applied to earth sciences; volcanology. *Mailing Add:* Five Hedge Lane Merrick NY 11566

FRIEDMAN, IRWIN, b New York, NY, Dec 15, 29; m 53; c 3. MEDICINE. *Educ:* Union Col, BS, 51; NY Univ, MD, 55. *Prof Exp:* Dir Pulmonary Function Lab, 63, RESPIRATORY SERV, BUFFALO GEN HOSP, 68-; ASSOC PROF INTERNAL MED & PHARMACOL, STATE UNIV NY, BUFFALO, 72- *Concurrent Pos:* Nat Heart Inst fel, 60-61; asst prof internal med, State Univ NY Buffalo, 66-72; pvt practr, 61. *Mem:* Am Thoracic Soc; Am Heart Asn; AMA; Am Col Physicians; Am Col Chest Physicians. *Res:* Clinical pulmonary physiology. *Mailing Add:* Buffalo Gen Hosp 85 High St Buffalo NY 14203

FRIEDMAN, JACK P, b New York, NY, Sept 4, 39; wid; c 2. ENGINEERING PHYSICS, APPLIED MATHEMATICS. *Educ:* Rensselaer Polytech Inst, BS, 60; Univ Chicago, MS, 61, PhD(physics), 65. *Prof Exp:* ENGR COMPUT SCI, APPL MATH & ENG PHYSICS, GEN ELEC KNOLLS ATOMIC POWERLAB, 65- *Mem:* Am Phys Soc. *Res:* Heat transfer; applied mathematics; computer science; fluid mechanics. *Mailing Add:* 2251 Berkely Ave Schenectady NY 12309

FRIEDMAN, JEROME ISAAC, b Chicago, Ill, Mar 28, 30; m 56; c 4. PHYSICS. *Educ:* Univ Chicago, AB, 50, MS, 53, PhD(physics), 56. *Prof Exp:* Res assoc physics, Univ Chicago, 56-57 & Stanford Univ, 57-60; from asst prof to prof physics, Mass Inst Technol, 60-83, dir, Lab Nuclear Sci, 80-83, head physics dept, 83-88; CONSULT, 88- *Concurrent Pos:* Bd mem & vpres, Univ Res Asn, 77-83; mem, High Energy Physics Adv Panel, Dept Energy, 88-; chmn, Sci Policy Comt, Superconducting Super Collider, 89- *Honors & Awards:* Co-recipient Nobel Prize Physics, 90; Co-recipient Panofsky Prize, Am Phys Soc, 88. *Mem:* Fel Am Phys Soc; Am Acad Arts & Sci. *Res:* High energy physics; elementary particles. *Mailing Add:* Dept Physics 24-512 Mass Inst Technol Cambridge MA 02139

FRIEDMAN, JOEL MITCHEL, b Brooklyn, NY, Aug 13, 47; m 71; c 2. CHEMICAL PHYSICS, BIOPHYSICS. *Educ:* Brooklyn Col, BS, 69; Univ Pa, MD, 76, PhD(chem), 75. *Prof Exp:* Fel, 75-77, MEM TECH STAFF, BELL LABS, 77- *Concurrent Pos:* Adj prof chem, Univ NMex, 84- *Mem:* Biophys Soc; fel Am Phys Soc; Am Chem Soc. *Res:* Experimental and theoretical study of resonant light scattering from molecular systems; application of resonant light scattering to reveal the molecular mechanisms of biological activity. *Mailing Add:* Chem Dept NY Univ New York NY 10003

FRIEDMAN, JOYCE BARBARA, b Washington, DC, Jan 5, 28. COMPUTER SCIENCES. *Educ:* Wellesley Col, BA, 49; Radcliffe Col, AM, 52; Harvard Univ, PhD(appl math), 65. *Prof Exp:* Res assoc, Logistics Res Proj, George Washington Univ, 50-51; mathematician, Weapons Syst Eval Group, Off Secy Defense, 52-54 & ACF Indust Inc, 54-56; sr mathematician, Tech Oper Inc, 56-60; mem tech staff, MITRE Corp, 60-65; asst prof comput sci, Stanford Univ, 65-68; assoc prof, 68-71, PROF COMPUT & COMMUN SCI, UNIV MICH, 71- *Concurrent Pos:* Vis scientist, IBM Watson Res Ctr, 70-71; consult, Stanford Res Inst, 74-76. *Mem:* Asn Comput Linguistics (vpres, 70, pres, 71); Asn Comput Mach; Asn Symbolic Logic; Am Math Soc. *Res:* Computational linguistics; computer models of linguistic theories; computer aids to linguistic research; speech understanding systems. *Mailing Add:* 221 Mt Auburn St Cambridge MA 02138

FRIEDMAN, JULES DANIEL, b Poughkeepsie, NY, Oct 24, 28; m; c 3. GEOLOGY. *Educ:* Cornell Univ, AB, 50; Yale Univ, MS, 52, PhD(geol), 58. *Prof Exp:* Asst dept geol, Yale Univ, 50-53; chief, Remote Sensing Sect, 83-85; geologist, DC, 53-69, geologist, Regional Geophys, DC, 69-74, geologist, Theoret & Appl Geophys, Colo, 74-81, GEOLOGIST, GEOPHYS & REMOTE SENSING, US GEOL SURV, COLO, 81- *Concurrent Pos:* Mem panel geodesy & cartography, Comt Polar Res, Nat Acad Sci-Nat Res Coun; secy, exec comt, Front Range Br, Am Geophys Union; ed trans of USSR Acad Sci, Earth Sci Sect (John Wiley). *Mem:* Fel Geol Soc Am; Am Geophys Union. *Res:* Geophysics; remote sensing; volcanology; structural geology; geomorphology; application of aerial and satellite infrared thermography to volcanic geology, heat flow, geothermal sources and geomorphology, especially neovolcanic zone of Iceland; Mono and Inyo Basins, California; Cascade Range, especially Mt St Helens and Lassen volcanic region; lineament analyses; the development of practical infrared-image analysis method for rapid assessment and surveillance of volcanic heat flow, energy partition, and total energy yield for inactive and eruptive periods; geophysics and geochemistry of Shawangunk, New York ore deposits. *Mailing Add:* US Geol Surv Dept Interior Denver CO 80225

FRIEDMAN, JULIUS JAY, b Brooklyn, NY, Mar 6, 26; c 2. MICROCIRCULATION. *Educ:* Tulane Univ, BS, 49, MS, 51, PhD(physiol), 53. *Prof Exp:* Res physiologist, Biophys Lab, Tulane Univ, 53-58; from asst prof to assoc prof, 58-71, PROF PHYSIOL & BIOPHYSICS, SCH MED, IND UNIV, INDIANAPOLIS, 71- *Concurrent Pos:* Lederle Med Fac fel, 60-63; Alexander von Humboldt Sr Scientist Award, 82-83. *Mem:* Microcirculatory Soc; Am Asn Univ Profs; Am Physiol Soc; Am Inst Biol Sci. *Res:* Tissue blood volumes and hematocrits; hemorrhagic and traumatic shock; peripheral vascular dynamics; microvascular function; trans-capillary exchange; capillary permeability; capillary regulation; interstitial transport. *Mailing Add:* Dept Physiol Ind Univ Med Ctr Indianapolis IN 46223

FRIEDMAN, KENNETH JOSEPH, b Brooklyn, NY, July 28, 43; m 69. PHYSIOLOGY, BIOPHYSICS. *Educ:* Lawrence Col, BA, 64; State Univ NY Stony Brook, PhD(biol sci), 69. *Prof Exp:* Lectr, State Univ NY Stony Brook, 68-69; NIH trainee, Dept Zool, Univ Calif, Los Angeles, 69-70, NIH fel, 70-72, staff fel, 72-75; asst prof, 75-82, ASSOC PROF PHYSIOL, COL MED & DENT, NJ MED SCH, 82- *Mem:* AAAS; Am Physiol Soc; Biophys Soc. *Res:* Role of lipids in determining electrical and transport properties of the cell membrane; neurophysiology and electrophysiology. *Mailing Add:* Dept Physiol NJ Med Sch 100 Bergen St Newark NJ 07103

FRIEDMAN, LAWRENCE, MEDICAL RESEARCH. *Prof Exp:* ASSOC DIR, CLIN APPLN & PREV PROG, NAT HEART, LUNG & BLOOD INST, NIH, 86- *Mailing Add:* NIH Nat Heart Lung Blood Inst Clin Appln Prog Fed Bldg Rm 5C01 7550 Wisconsin Ave Bethesda MD 20892

FRIEDMAN, LAWRENCE BOYD, b Minneapolis, Minn, May 9, 39; m 61; c 2. INORGANIC CHEMISTRY. *Educ:* Univ Minn, Duluth, BA, 61; Harvard Univ, MA, 63, PhD(chem), 66. *Prof Exp:* Asst prof chem, Oakland Univ, 66-68; lab mgr, Polaroid Corp, 74-80, sr scientist, 80-83, qual mgr, 84-87; asst chmn, Dept Chem, Univ Pittsburgh, 87-89; ASSOC DIR, BECKMAN CTR HIST CHEM, UNIV PA, 89- *Mem:* Am Chem Soc; Am Crystallog Asn. *Res:* Boron hydride chemistry; x-ray crystallography; analytical techniques; photographic science. *Mailing Add:* 520 Spring Mill Rd Villanova PA 19085

FRIEDMAN, LAWRENCE DAVID, b Newark, NJ, Aug 25, 32; m 60; c 3. GENETICS. *Educ:* Rutgers Univ, BA, 54; Northwestern Univ, MS, 55; Univ Wis, PhD(zool, genetics), 60. *Prof Exp:* Asst genetics, Northwestern Univ, 54-55; asst zool, Univ Wis, 55-57, asst genetics, 57-60, proj assoc, 60-61; asst prof biol, Hiram Col, 61-66; ASSOC PROF BIOL, UNIV MO-ST LOUIS, 66- *Concurrent Pos:* Vis prof, Univ Newcastle, England, 72-73. *Mem:* AAAS; Genetics Soc Am; Am Soc Zoologists; Am Inst Biol Sci. *Res:* Genetics of Drosophila. *Mailing Add:* Dept Biol Univ Mo St Louis MO 63121

FRIEDMAN, LEONARD, b Brooklyn, NY, Jan 27, 29; m 64. BIOCHEMICAL TOXICOLOGY. *Educ:* NY Univ, BA, 51; Rutgers Univ, MS, 53, PhD(biochem), 59. *Prof Exp:* Fel, Iowa State Univ, 59-61; res chemist biochem, NIH, 61-62; res chemist, 62-80, SUPV RES CHEMIST, DIV TOXICOL, FOOD & DRUG ADMIN, 80- *Mem:* AAAS; Soc Toxicol; Am Chem Soc; Sigma Xi; Asn Off Anal Chemists; Am Soc Pharmacol & Exp Therapeut. *Res:* Chemical, physical and enzymic nature of proteolytic enzymes, especially rennin and proteinase; metabolism of malaria; effect of nutrition on enzymes and metabolism of tissues; biological and biochemical effects of natural and environmental toxicants on in vitro and in vivo mammalian systems; cell proliferation and carcinogens. *Mailing Add:* Div Toxicol Food & Drug Admin 8501 Muirkirk Rd Laurel MD 20708

FRIEDMAN, LESTER, organic chemistry, for more information see previous edition

FRIEDMAN, LEWIS, b Spring Lake, NJ, Aug 8, 22; m 48; c 3. PHYSICAL CHEMISTRY. *Educ:* Lehigh Univ, AB, 43, AM, 45; Princeton Univ, PhD(chem), 47. *Prof Exp:* Inst Nuclear Studies fel, Univ Chicago, 47-48; from assoc chemist to chemist, 48-64, SR CHEMIST, BROOKHAVEN NAT LAB, 64- *Concurrent Pos:* Guest scientist, Found Fundamental Res Matter Lab Mass Separation, Amsterdam, Neth, 60-61. *Res:* Electron and ion impact phenomena; high sensitivity and high molecular weight mass spectrometry; chemical studies with isotopes. *Mailing Add:* Brookhaven Nat Lab Upton NY 11973

FRIEDMAN, LIONEL ROBERT, b Philadelphia, Pa, May 8, 33; m 56; c 3. SOLID STATE PHYSICS, OPTICS. *Educ:* Swarthmore Col, BS, 55; Univ Pittsburgh, PhD(physics), 61. *Prof Exp:* Scientist, Westinghouse Atomic Power Div, 55-61; res assoc solid state physics, Univ Pittsburgh, 61-62; mem tech staff, RCA Labs, 62-72; sr res assoc, Sch Math & Physics, Univ East Anglia, Norwich, UK, 72-73; lectr, Univ St Andrews, 73-74, reader, dept theoret physics, 74-80; mem tech staff, GTE Labs, 80-85; PROF, WORCESTER POLYTECH INST, 85- *Concurrent Pos:* Instr, Temple Univ, 64-66; Sci Res Coun sr vis fel, Cavendish Lab, Cambridge Univ, 71-72; contrib lectr physics & technol amorphous mat course, sponsored by Univ Edinburgh, Glasgow, Dundee, Heriot-Watt & St Andrews; mem sci & organizing comt, 7th Int Conf Amorphous & Liquid Semiconductors, Edinburgh, 77; co-dir, Scottish Univs Summer Sch Physics, 78. *Mem:* Am Phys Soc; Sigma Xi; Int Soc Optical Eng; Inst Elec & Electronic Engrs. *Res:* Transport phenomena in solids; small polaron theory; organic semiconduction-Hall effect; amorphous semiconductors; metal-insulator transition; quantum wells and superlattices; guided-wave optics. *Mailing Add:* 49 Winfield Rd Holden MA 01520-2442

FRIEDMAN, M H, ENTERO-ENTERIC REFLEXES, CARDIAC REFLEXES IN EXERCISE. *Educ:* McGill Univ, Montreal, PhD(physiol), 37. *Prof Exp:* Prof & chmn dept, Thomas Jefferson Univ, 57-74, emer prof physiol, 74-86; CONSULT, 86- *Concurrent Pos:* Vis prof physiol, Philadelphia Col Osteop Med, 74-86. *Res:* Gastrointestinal physiology; cardiac dynamics of exercise. *Mailing Add:* 2420 Greenhill Rd Lansdowne PA 19050

FRIEDMAN, MARC MITCHELL, NEUTROPHIL ACTIVATION, MAST CELL SECRETION. *Educ:* Johns Hopkins Univ, PhD(cell biol), 77. *Prof Exp:* RES ASST CELL BIOL, MED CTR, GEORGETOWN UNIV, 80- *Res:* Cell biology. *Mailing Add:* 17455 MacDuff Ave Olney MD 20832

FRIEDMAN, MARVIN HAROLD, b New York, NY, July 20, 23; m; c 1. PHYSICS. *Educ:* City Col, BS, 43; Univ Ill, MS, 49, PhD(physics), 52. *Prof Exp:* NSF fel, Cornell Univ, 52-53; res assoc physics, Columbia Univ, 53-55; asst prof, Mass Inst Technol, 55-61; assoc prof, 61-66, PROF PHYSICS, NORTHEASTERN UNIV, 66- *Mem:* Fel Am Phys Soc. *Res:* Elementary particle theory; condensed matter theory. *Mailing Add:* Dept Physics Northeastern Univ Boston MA 02115

FRIEDMAN, MATTHEW JOEL, b Newark, NJ, Mar 10, 40; c 4. PSYCHIATRY, PHARMACOLOGY. *Educ:* Dartmouth Col, AB, 61; Albert Einstein Col Med, PhD(pharmacol), 67; Univ Ky, MD, 69. *Prof Exp:* Intern, Univ Ky Hosp, 69-70; resident psychiat, Mass Gen Hosp, Boston, 70-72; resident, Dartmouth-Hitchcock Ment Health Ctr, 72-73; PROF PSYCHIAT & PHARMACOL, DARTMOUTH MED SCH, 73- *Concurrent Pos:* Chief psychiat, Vet Admin Hosp, White River Junction, Vt, 73- *Mem:* Am Psychiat Asn; Soc Biol Psychiat; Physicians Soc Responsibility. *Res:* Biological basis of affective disorders; hypertension and depression; physical dependence and tolerance; clinical psychopharmacology; blood levels of psychotropic drugs and clinical response; tryptophan metabolism during alcoholism and the abstinent state; post-traumatic stress disorder; factors affecting utilization of psychiatric outpatient treatment. *Mailing Add:* Dartmouth Med Sch Vet Admin Hosp White River Junction VT 05001

FRIEDMAN, MELVIN, b West Orange, NJ, Nov 14, 30; m 54; c 2. ROCK MECHANICS, STRUCTURAL GEOLOGY. *Educ:* Rutgers Univ, BS, 52, MS, 54; Rice Univ, PhD, 61. *Prof Exp:* Res geologist & sect leader, Shell Develop Co, 54-67; assoc prof, 67-69, dir, Ctr Tectonophysics, 79-82, PROF GEOL, TEX A&M UNIV, 69-, DEAN, COL GEOSCI, 83- *Concurrent Pos:* Field geologist, Newfoundland Geol Surv, 52, Bear Creek Mining Co, 53; consult, numerous companies; co-ed-in-chief, Technophysics, 80- *Honors & Awards:* Res Award, Intersoc Comt Rock Mech, 69. *Mem:* Fel AAAS; fel Geol Soc Am; Am Geophys Union. *Res:* Dynamic analysis of tectonic structures through a knowledge of the physical and mechanical properties of minerals and rocks. *Mailing Add:* Col Geosci Tex A&M Univ College Station TX 77843

FRIEDMAN, MENDEL, b Pultusk, Poland, Feb 13, 33; US citizen; m 57; c 3. FOOD CHEMISTRY, FOOD SAFETY. *Educ:* Univ Ill, BS(pharm), 54; Univ Chicago, MS, 58, PhD(chem), 62. *Prof Exp:* Res assoc, Univ Wis, 61-62; res chemist, Northern Regional Res Lab, 62-69, ACTG RES LEADER, FOOD SAFETY RES UNIT, WESTERN REGIONAL RES CTR, AGR RES SERV, USDA, ALBANY, CALIF, 69- *Concurrent Pos:* Instr, Bradley Univ, 66-67 & Eureka Col, 67-69; fel, Intra-Sci Found Protein Chem, 71-75; lectr, Dept Nutrit Sci, Univ Calif, Berkeley, 82-87. *Mem:* Am Chem Soc; fel AAAS; Am Soc Biol Chemists; Am Inst Nutrit; Fel Agr & Food Chem Div, Am Chem Soc. *Res:* Protein chemistry and biochemistry; food chemistry; food safety and nutrition. *Mailing Add:* 3896 Paseo Grande Moraga CA 94556

FRIEDMAN, MICHAEL A, b Port Arthur, Tex, Aug 10, 43; m; c 3. CANCER RESEARCH. *Educ:* Tulane Univ, BA, 65; Univ Tex, MD, 69; Am Bd Internal Med, dipl, 74, depl med oncol, 75. *Prof Exp:* Clin assoc, Cancer Ther Eval Prog, Nat Cancer Inst, 70-72; sr asst resident med, Stanford Univ Hosp, 72-73, postdoctoral fel med oncol, 73-75; from asst prof to assoc prof med, Univ Calif, San Francisco, 75-83, interim dir, Cancer Res Inst, 81-83, assoc dir, 81-82; chief, Clin Invest Br, 83-88, ASSOC DIR CANCER THER EVAL PROG & ACTG ASSOC DIR, RADIATION RES PROG, DIV CANCER TREAT, NAT CANCER INST, 88- *Concurrent Pos:* Mem, Study Sect Regional Coop Cancer Groups, NIH, 81-83, Oncol Comt Med Knowledge, Self Assessment Prog, Am Col Physicians, 86-, Comt Comprehensive Ctr, Nat Cancer Adv Bd, 87-89, Ad Hoc Comt Clin Trials, Am Soc Clin Oncologists, 89-, Comt Patient Care & Res, Am Col Surgeons, Comn Cancer, 89- *Mem:* Am Asn Cancer Res; Am Soc Clin Oncol; Am Fedn Clin Res; Am Asn Cancer Educ. *Res:* Cancer therapy; oncology. *Mailing Add:* Cancer Ther Eval Prog Div Camcer Treat Nat Cancer Inst Health Executive Plaza N Rm 742 Bethesda MD 20892

FRIEDMAN, MICHAEL E, b Bronx, NY, Aug 17, 37; m 64; c 1. BIOPHYSICAL CHEMISTRY, ANALYTICAL CHEMISTRY. *Prof Exp:* Univ Pa, AB, 59; Polytech Inst Brooklyn, MS, 63; Cornell Univ, PhD(biophys chem), 66. Prof Exp: Technician chem, Pack Med Group, 58-60; staff fel, NIH, 66-68; from asst prof to assoc prof, 68-84, PROF CHEM, AUBURN UNIV, 84- *Mem:* Am Chem Soc. *Res:* Structural studies of biological polymers and their relationship to the body. *Mailing Add:* Dept Chem Auburn Univ Auburn AL 36849

FRIEDMAN, MISCHA ELLIOT, b Worcester, Mass, Nov 7, 22; m 56; c 2. BACTERIOLOGY. *Educ:* Univ Mass, BS, 48; Univ Ill, MS, 49, PhD(dairy sci), 53. *Prof Exp:* Res assoc dairy sci, Univ Ill, 53; microbiologist, Med Bact Div, US Army Chem Corps, Biol Labs, Ft Detrick, 53-70; exec secy, Allergy & Immunol Study Sect, NIH, 70-76, chief clin sci rev sect, Sci Rev Br, 76-84, chief referral & rev br, Div Res Grants, 84-88, assoc dir referral & rev, 87-88; CONSULT, 88- *Concurrent Pos:* Secy Army res & study fel & vis lectr, Hadassah Med Sch, Hebrew Univ, Israel, 64-65; vis lectr, Hadassah Med Sch, Hebrew Univ, Israel, 64-65. *Honors & Awards:* Dirs Award, NIH, 79 & 87. *Res:* Bacterial nutrition and metabolism; nitrogen metabolism; nutritional inhibitors; bacteriophage; protein synthesis; immunology; science administration; research grants; science writing; training. *Mailing Add:* 314 W College Terr Frederick MD 21701

FRIEDMAN, MITCHELL, b 1944; m; c 2. PULMONARY DISEASE, CRITICAL CARE. *Educ:* Univ Miami, MD, 69. *Prof Exp:* ASSOC PROF MED & DIR CRITICAL CARE, DEPT MED, UNIV NC, 76- *Honors & Awards:* Pulmunary Young Investr Award, NIH. *Mem:* Fel Am Col Phys; fel Am Col Chest Phys; Am Osteop Asn. *Res:* Pulmonary endothelial cell biology. *Mailing Add:* Dept Med Univ NC CH 724 Clin Sci Bldg 229-H Chapel Hill NC 27514

FRIEDMAN, MORTON (BENJAMIN), b Bayonne, NJ, Mar 14, 28; m 56; c 2. ENGINEERING. *Educ:* NY Univ, BAero Eng, 48, MAero Eng, 50, DEng Sc, 53. *Prof Exp:* Engr aerodyn, Cornell Aeronaut Lab, Inc, 48; res assoc, NY Univ, 53-55; appl mech, Inst Flight Structures, 55-56, asst prof, 56-61, assoc prof, 61-76, PROF AERODYN & APPL MATH, COLUMBIA UNIV, 76- *Concurrent Pos:* Field scholar, Found Instrumentation Educ & Res, 60; Fulbright lectr, Netherlands, 62-63. *Mem:* AAAS. *Res:* Gas dynamics; shock diffraction; acoustics; viscous fluids; applied mathematics and mechanics. *Mailing Add:* MVDD Bldg Rm 630 Columbia Univ New York NY 10027

FRIEDMAN, MORTON HAROLD, b New York, NY, June 18, 35; m 61; c 3. HEMODYNAMICS, BIOLOGICAL TRANSPORT. *Educ:* Cornell Univ, BChE, 57; Univ Mich, MS, 58, PhD(chem eng), 61. *Prof Exp:* Sr chem engr, Cent Res Labs, Minn Mining & Mfg Co, 60-65; sr staff engr, Appl Physics Lab, 65-68, prog coordr, Biomed Progs Off, 74-77, supvr theoret probs res, 75-77, dep dir biomed progs, 77-81, mem prin prof staff, Appl Physics Lab, Johns Hopkins Univ, 68-88, assoc prof ophthal, Sch Med, 71-88, assoc prof biomed eng, Sch Med, 77-88, chief scientist biomed progs, Applied Physics Lab, 82-88, PROF BIOMED & CHEM ENG, COL ENG, 88-; PROF PATH, COL MED, OHIO STATE UNIV, 88- *Concurrent Pos:* Vis scholar, Stanford Univ Med Ctr, 76; ed, J Biomech Eng, 84-90, Ann Biomed Eng, 85-; chmn, Bioeng Div, Am Soc Mech Engrs, 89-90. *Honors & Awards:* Nat Capital Award, DC Coun Eng & Archit Soc, 70. *Mem:* AAAS; Biomed Eng Soc (dir, 85-88, pres, 88-89); Am Inst Chem Engrs; Am Soc Mech Engrs; NAm Soc Biorheology; Am Heart Asn. *Res:* Arterial fluid mechanics; angiographic image processing; physiological transport. *Mailing Add:* Biomed Engr Ctr Ohio State Univ Columbus OH 43210-1002

FRIEDMAN, MORTON HENRY, b Uniontown, Pa, Apr 16, 38; m; c 2. HUMAN ANATOMY, MICROSCOPIC ANATOMY. *Educ:* Washington & Jefferson Col, AB, 60; Hofstra Univ, MA, 64; Univ Tenn, PhD(anat), 69. *Prof Exp:* From instr to asst prof, 69-73, ASSOC PROF ANAT, SCH MED, W VA UNIV, 73- *Mem:* NY Acad Sci; AAAS; Am Asn Anatomists; Electron Micros Soc Am; Am Soc Cell Biol. *Res:* Electron microscopy; microanatomy. *Mailing Add:* Dept Anat WVa Univ Sch Med Health Sci Ctr N Morgantown WV 26506

FRIEDMAN, NATHAN BARUCH, b New York, NY, Jan 30, 11; m 42, 60; c 6. PATHOLOGY. *Educ:* Harvard Univ, BS, 30; Cornell Univ, MD, 34. *Prof Exp:* Resident pneumonia serv, Harlem Hosp, NY, 34-35; intern, Montefiore Hosp, New York, 35-37; resident path, Univ Chicago Clins, 38-39; instr, Sch Med, Stanford Univ, 41-42; sr pathologist, Army Inst Path, 46-47; dir labs, Cedars of Lebanon Hosp, 48-71; SR CONSULT, DIV LABS, CEDARS-SINAI MED CTR, 72- *Concurrent Pos:* Littauer fel, Harvard Med Sch, 39-40; prof lectr, George Washington Univ, 47; from assoc clin prof to clin prof, Sch Med, Univ Southern Calif, 52- *Mem:* Endocrine Soc; AMA; Am Asn Pathologists; Am Asn Cancer Res; Am Soc Clin Path. *Res:* Pathology of endocrine glands, genitourinary organs and tumors; radiation reactions. *Mailing Add:* Dept Pathol Cedars-Sinai Med Ctr 8700 Beverly Blvd Los Angeles CA 90048

FRIEDMAN, ORRIE MAX, b Grenfell, Sask, June 6, 15; nat US; m 59; c 4. RESEARCH ADMINISTRATION, BIO-ORGANIC CHEMISTRY. *Educ:* Univ Man, BSc, 35; McGill Univ, BSc, 41, PhD(chem), 44. *Prof Exp:* Jr res chemist, Nat Res Coun Can, 44-46; res fel chem, Harvard Univ, 46-49; res assoc, Harvard Med Sch, 49-52, asst prof chem, 52-53; from assoc prof to prof, Brandeis Univ, 53-62, adj res prof, 62-66; dir, Technol Int Ltd, 75-78; pres & sci dir, 62-82, CHMN, COLLAB RES, INC, 82-, CEO, 86- *Concurrent Pos:* Assoc, Beth Israel Hosp, 49-54, trustee, Worcester Found Exp Biol; consult, Harvard Med Sch, 53-54 & 56-57; spec consult, Nat Cancer Inst, NIH, 65-70; dir, United Chem Co, Ltd, 60-84; bd gov, Technion, Israel Inst Technol; mem corp, Dana Farber Cancer Ctr, Boston Mus Sci; trustee, Beth Israel Hosp, Boston, Worcester Fedn Exp Biol, Barnett Inst, Northeastern Univ. *Mem:* Fel AAAS; Am Chem Soc; Am Asn Cancer Res; fel Chem Soc London; NY Acad Sci. *Res:* Chemistry of high explosives; cancer chemotherapy, psychopharmacology and the chemistry of nucelic acids. *Mailing Add:* 49 Warren St Brookline MA 02146

FRIEDMAN, PAUL, b Brooklyn, NY, Oct 12, 31; m 54; c 2. ORGANIC CHEMISTRY. *Educ:* City Col NY, BS, 53; Brooklyn Col, MA, 57; Stevens Inst Technol, PhD(chem), 63. *Prof Exp:* Sr res chemist, Evans Res & Develop Corp, 55-60; instr chem, Newark Col Eng, 60-61; res assoc, Stevens Inst Technol, 61-63; NSF res assoc, Univ Southern Calif, 63-64; from asst prof to assoc prof, 64-70, PROF CHEM, PRATT INST, 70-, CHMN DEPT MATH & SCI, 75- *Concurrent Pos:* Vis sr fel, Princeton Univ, 83 & 86. *Mem:* Fel Royal Soc Chem; Am Chem Soc; Sigma Xi; NY Acad Sci; AAAS. *Res:* Heterocyclic chemistry; non benzenoid aromatics; photochemistry; quantum organic and physical organic chemistry; medicinal chemistry. *Mailing Add:* Dept Chem Pratt Inst Brooklyn NY 11205

FRIEDMAN, PAUL J, b New York, NY, Jan 20, 37; m 60; c 4. RADIOLOGY, EDUCATIONAL ADMINISTRATION. *Educ:* Univ Wis, BS, 55; Yale Univ, MD, 60. *Prof Exp:* Chief radiol, US Naval Submarine Med Ctr, Conn, 64-66; from asst prof to assoc prof radiol, Univ Calif, San Diego, 68-75, chmn fac, Sch Med, 79-80, assoc dean acad affairs, 82-88, PROF RADIOL, UNIV CALIF, SAN DIEGO, 75-, DEAN ACAD AFFAIRS, SCH MED, 89- *Concurrent Pos:* James Picker Found advan fel acad radiol, 66-68 & scholar radiol res, 68-69; Markle scholar acad med, 69-74; vchmn, Dept Radiol, Univ Calif, San Diego, 78-82; subcomt chest x-ray epidemiol studies, Am Thoracic Soc, 75-78, comt lung cancer, 79-82; B Reader, Nat Inst Occup Safety & Health, 82-86, 88- *Mem:* Fleischner Soc; Asn Univ Radiologists; fel Am Col Radiol; Am Thoracic Soc; fel Am Col Chest Physicians. *Res:* Pulmonary diseases; lung cancer staging; radiologic-pathologic correlation; pulmonary radiology; chest computed tomography; research ethics; faculty demographics. *Mailing Add:* Sch Med M-002 Univ Calif San Diego La Jolla CA 92093

FRIEDMAN, RAYMOND, b Portsmouth, Va, Feb 9, 22. FIRE RESEARCH, COMBUSTION SCIENCE. *Educ:* Va Polytech Inst & State Univ, BS, 42; Univ Wis, PhD(chem eng), 48. *Prof Exp:* Res scientist, Westinghouse Elec Corp, 43-46, sr res scientist, 48-55; res mgr, Atlantic Res Corp, 55-62, vpres, 62-69; vpres, Factory Mutual Res Corp, 69-87; CONSULT, 87- *Concurrent Pos:* Vchmn, Int Asn Fire Safety Sci, 87. *Mem:* Combustion Inst (pres, 78-82); Am Chem Soc; Soc Fire Protection Engrs; Int Asn Fire Safety Sci; Nat Fire Protection Asn. *Res:* Combustion phemonena; fire detection, spread and suppression; solid-propellant rocket combustion; gaseous flames; chemical themodynamics and kinetics. *Mailing Add:* 77 Florence St Newton MA 02167

FRIEDMAN, RICHARD M, b Cleveland, Ohio, Aug 25, 30; m 52; c 4. SPACE PHYSICS. *Educ:* Case Inst Technol, BS, 52; Stanford Univ, MS, 54, PhD(physics), 56. *Prof Exp:* Res scientist, Res Lab, Lockheed Missiles & Space Co, 56-61; mgr nuclear instrumentation, Vela Prog, Aerospace Corp, 61-62, proj leader, infrared measurements, Space Physics Lab, 62-63, staff scientist, 63-66, head space tech support dept, 66-67; proj mgr, TRW Defense & Space Systs Group, 67-69, proj mgr, Viking Lander Biol Instrument, 69-73, advan studies mgr, 73-77, dir, Systs Eng Defense Systs, 77-79, asst proj mgr, Defense Support Prog, 79-87, DIV TECH STAFF MGR, APPL TECHNOL DIV TRW SPACE & TECHNOL GROUP, 87- *Mem:* Am Phys Soc; Am Geophys Union; Am Inst Aeronaut & Astronaut. *Res:* Development of sensor systems for space applications; measurement of solar x-rays; earth's background radiance and nuclear phenomena. *Mailing Add:* Appl Technol Div TRW One Space Park Redondo Beach CA 90278

FRIEDMAN, ROBERT BERNARD, b Chicago, Ill, June 9, 38; m 69; c 5. BIOCHEMISTRY, CARBOHYDRATE CHEMISTRY. *Educ:* Northwestern Univ, PhB, 62; Univ Ill, PhD(biochem), 69. *Prof Exp:* Res assoc, Sch Med, Tufts Univ, 68-70; asst prof chem, Boston Univ, 70-71; res fel biochem, Eunice Kennedy Shriver Ctr Ment Retardation, Mass, 71-76; asst prof biochem, Med Sch, Northwestern Univ, 76-79; res biochemist, Vet Admin Lakeside Hosp, 76-79; sect leader, 79-80, ASSOC DIR RES, RES AM MAIZE PROD, 88- *Concurrent Pos:* Asst prof chem, Salem State Col, 73-74. *Mem:* AAAS; Sigma Xi; Am Chem Soc; Soc Complex Carbohydrates; Controlled Release Soc. *Res:* Study of the structure of glycoproteins and starch polysaccharides; enzymology; synthetic and analytic carbohydrate chemistry. *Mailing Add:* 6654 N Mozart Chicago IL 60645-4308

FRIEDMAN, ROBERT DAVID, b New York, NY, Mar 24, 35; c 2. MEDICAL GENETICS, DENTISTRY. *Educ:* Brooklyn Col, BA, 56; NY Univ, DDS, 60; Brandeis Univ, MA, 68; Ind Univ, PhD(med genetics), 72. *Prof Exp:* Asst prof biochem, Sch Dent, Temple Univ, 72-77; ASST PROF ORAL MED, SCH DENT, UNIV PA, 78- *Concurrent Pos:* NIH spec fel med genetics, Ind Univ, 71-72. *Mem:* AAAS; Am Soc Human Genetics; Am Genetics Soc; Sigma Xi. *Res:* Genetic and biochemical study of the polymorphic proteins in human saliva. *Mailing Add:* 2132 E Cumberland St Philadelphia PA 19125

FRIEDMAN, ROBERT HAROLD, b Sioux City, Iowa, Jan 11, 24; m 45, 66; c 4. PHYSICAL CHEMISTRY. *Educ:* Univ Chicago, PhB, 47, BS, 49; Univ Tex, PhD, 57. *Prof Exp:* Sr res engr, Humble Oil & Ref Co, 56-64; res scientist, Tidewater Oil Co, 64-69; res scientist, 69-76, SR RES SCIENTIST, GETTY OIL CO, 72- *Mem:* Am Phys Soc; Am Chem Soc; Am Inst Mining, Metall & Petrol Eng. *Res:* Theoretical chemistry; operations research; oil recovery from subsurface formations. *Mailing Add:* 6015 Volkeith Houston TX 77096-3832

FRIEDMAN, ROBERT MORRIS, b New York, NY, Nov 21, 32; m 57; c 2. PATHOLOGY, VIROLOGY. *Educ:* Cornell Univ, BA, 54; NY Univ, MD, 58. *Prof Exp:* Intern Med, Mt Sinai Hosp, New York, 58-59; investr virol, Div Biol Stand, NIH, 59-61, pathologist, Clin Ctr, 61-63; vis scientist virol, Nat Inst Med Res, London, Eng, 63-64; investr, Nat Cancer Inst, 64-70, pathologist, 65-70, chief molecular path sect, Lab Molecular Biol, 70-74; chief, Lab Exp Path, Nat Inst Arthritis, Metab & Digestive Dis, 74-81; PROF & CHMN, DEPT PATH, UNIFORMED SERV UNIV HEALTH SCI, 81- *Concurrent Pos:* Consult, Antiviral Chemother Prog, Nat Inst Allergy & Infectious Dis; McLaughlin lectr, Univ Tex Med Br, Galveston, 71; vis scientist, Biochem Dept, Nat Inst Med Res, London, Eng, 71-73; med dir, USPHS. *Mem:* Am Soc Microbiol; Am Soc Immunologists; Am Soc Exp Path. *Res:* Experimental pathology; virology; interferon studies; immunology. *Mailing Add:* Uniformed Serv Univ Health Sci 4301 Jones Bridge Rd Bethesda MD 20814

FRIEDMAN, RONALD MARVIN, b Brooklyn, NY, Apr 26, 30; c 2. CELL BIOLOGY. *Educ:* Columbia Univ, BS, 60; NY Univ, MS, 67, PhD(cell biol), 76. *Prof Exp:* Fel biochem, Columbia Univ, 75-76 & Yale Univ, 77-78; vis fel, Princeton Univ, 78-79; vis scientist enzym, NY State Inst Basic Res, 79-81; fel genetics, Albert Einstein Col Med, 81-82; res assoc, dept endocrinol, Sloan Kettering Mem Hosp, NY, 82-83; sr res scientist hematol, City Univ NY, 84-86; sr res assoc biochem, Roswell Park Mem Inst, Buffalo, NY, 86-87; res assoc, infectious diseases, Harvard Med Sch, Boston Mass, 87-89; VIS FEL, NUCLEIC ACID RES, LEWIS THOMAS LABS, PRINCETON UNIV, NJ, 89- *Concurrent Pos:* Scientific adv, Knights Template Res Found, 82-84. *Mem:* Sigma Xi; Am Soc Cell Biol; Harvey Soc; NY Acad Sci; Fedn Am Socs Exp Biol. *Res:* Cell culture in serum free medium; transfer of genes from one cell type to another (transfection); cell transformation, its transition point and specific proteins involved in this transition; biochemistry of platelets and mechanisms with respect to their role in metastasis, pancreatic tumor antigens there biochemistry; research nucleic acid synthetic polymers; relationship of ph and conc of synthetic acid polymer to thulium transitions. *Mailing Add:* Lewis Thomas Labs Princeton Univ Princeton NJ 08544

FRIEDMAN, RUTH T, b Sisseton, SDak, Nov 23, 36. NUTRITION PHYSIOLOGY. *Educ:* Med Col Va, Va Commonwealth Univ, PhD(physiol), 68. *Prof Exp:* prof biol sci, Mary Washington Col, 75-84; RETIRED. *Mem:* Am Physiol Soc. *Mailing Add:* 204 Kent Ave Fredericksburg VA 22405

FRIEDMAN, SAMUEL ARTHUR, b Brooklyn, NY, Jan 21, 27; m 88; c 2. COAL GEOLOGY. *Educ:* Brooklyn Col, BS, 50; Ohio State Univ, MS, 52. *Prof Exp:* Geologist, Ind Geol Surv, 52-67; geologist, US Bur Mines, 67-71; SR COAL GEOLOGIST, OKLA GEOL SURV, 71- *Concurrent Pos:* Adj prof geol, Grad Fac, Univ Okla, 74-; chmn, coal geol div, Geol Soc Am, 77-78; vpres, Energy Minerals Div, Am Asn Petrol Geologists, 80-81, 89-90, pres, 90-91. *Mem:* Fel Geol Soc Am; Soc Sedimentary Geologists; Am Asn Petrol Geologists; Am Inst Prof Geologists. *Res:* Coal geology, coal resources and net recoverable reserves; middle Pennsylvanian lithostratigraphy of Indiana and Oklahoma; low-sulfur coal deposits and coalbed methane resources; depositional environments of coal; distribution of trace elements in coal. *Mailing Add:* Okla Geol Surv 100 E Boyd St Norman OK 73019-0628

FRIEDMAN, SELWYN MARVIN, b New York, NY, May 17, 29; m 72. BIOCHEMICAL GENETICS. *Educ:* Univ Mich, BS, 51; Purdue Univ, MS, 53, PhD(bact), 61. *Prof Exp:* Fel biochem, Western Reserve Univ, 60-61; fel cell biol, Albert Einstein Col Med, 62-63; res fel, Med Col Physicians & Surgeons, Columbia Univ, 63-66; asst prof, 66-69, ASSOC PROF BIOL, HUNTER COL, 69- *Concurrent Pos:* Ed, Biochem Thermophily, 78; external evaluator, grad prog biol, William Paterson Col, 87; vice-chair, microbiol sect, NY Acad Sci, 88. *Mem:* AAAS; Am Soc Microbiol; NY Acad Sci. *Res:* Protein synthesis and the genetic code; physiology of thermophilic bacteria. *Mailing Add:* Biol Sci NY Acad Sci Two E 63rd St New York NY 10021

FRIEDMAN, SEYMOUR K, b New York, NY, July 1, 28; m 56; c 4. ORGANIC CHEMISTRY, COLLOID CHEMISTRY. *Educ:* City Col NY, BS, 48. *Prof Exp:* Group leader chem, Cent Res, Stauffer Chem Co, 53-56; asst mgr develop, Emulsol Div, 57-60, mgr prod appln & planning, Detergent Div, 61-63, CORP DIR COM DEVELOP, 64-76, PROD MGR INDUST SURFACTANTS, WITCO CHEM CO, 76- *Mem:* Com Chem Develop Asn. *Res:* Information techniques; research planning. *Mailing Add:* 60 E Eighth St Apt 14J New York NY 10003-6514

FRIEDMAN, SIDNEY, b Union City, NJ, Jan 24, 26; m 57; c 3. ORGANIC CHEMISTRY. *Educ:* Purdue Univ, BS, 49; Harvard Univ, AM & PhD(org chem), 53. *Prof Exp:* Fel petrol, Mellon Inst, 53-54; res chemist, US Bur Mines, 55-75; res chemist, US Energy Res & Develop Admin, 75-77, br chief, 77-87, SR TECH ADV, US DEPT ENERGY, 87- *Concurrent Pos:* Lectr, Duquesne Univ, 56-57. *Honors & Awards:* Bituminous Coal Res Award, 69, 77. *Mem:* Am Chem Soc. *Res:* Organometallic catalysis; organic reaction mechanisms; organic spectroscopy; gas chromatography; origin, structure and reactivity of coal; coal sulfur chemistry and desulfurization; microbiology of coal. *Mailing Add:* Pittsburgh Energy Technol Ctr PO Box 10940 Pittsburgh PA 15236-0940

FRIEDMAN, STANLEY, b New York, NY, Dec 11, 25; m; c 4. INSECT PHYSIOLOGY. *Educ:* Univ Ill, BA, 48; Johns Hopkins Univ, PhD(biol), 52. *Prof Exp:* Res assoc entom, Univ Ill, 52-56; biochemist, NIH, 56-58; from asst prof to assoc prof entom, Purdue Univ, 58-64; assoc prof, Univ Ill, Urbana, 64-67, PROF ENTOM, 67-, HEAD DEPT, 75-, ASSOC DIR, SCH LIFE SCI, UNIV ILL, URBANA, 89- *Mem:* Am Soc Zool; Am Soc Biol Chem; Entom Soc Am; fel AAAS. *Res:* Biochemistry and physiology of insects. *Mailing Add:* Dept Entom Univ Ill 505 S Goodwin Ave Urbana IL 61801

FRIEDMAN, STEPHEN BURT, b Amsterdam, NY, Mar 23, 31; m 64; c 2. MICROBIOLOGY. *Educ:* Univ Rochester, BA, 53; Syracuse Univ, MS, 55; Univ Ill, PhD(microbiol), 62. *Prof Exp:* NIH fel microbiol, State Univ, Belgium, 62-64; res assoc molecular biol, Cold Spring Harbor Lab Quant Biol, 64-66; ASSOC PROF BIOL, WESTERN MICH UNIV, 66- *Concurrent Pos:* Europ Molecular Biol Orgn fel, Belgium, 72. *Mem:* AAAS; Am Soc Microbiol; Am Inst Biol Sci; NY Acad Sci; Sigma Xi. *Res:* Bacterial and phage regulation; active transport; membrane structure and function; bacterial isozyme variation and population genetics. *Mailing Add:* Dept Biomed Sci Western Mich Univ Kalamazoo MI 49008

FRIEDMAN, SYDNEY MURRAY, b Montreal, Que, Feb 17, 16; m 40. MEDICINE. *Educ:* McGill Univ, BA, 38, MD & CM, 40, MSc, 41, PhD(renal physiol), 46. *Prof Exp:* Demonstr histol, McGill Univ, 40-41, from asst prof to assoc prof anat, 44-50; prof anat & head dept, 50-81, prof, 81-85, EMER PROF ANAT, UNIV BC, 86- *Concurrent Pos:* Asst physician, Royal Victoria Hosp, 48-50; Pfizer traveling fel, Clin Res Inst, Montreal, 71; vis lectr, Can Cardiovasc Soc, 71; fel, Coun High Blood Pressure Res. *Honors & Awards:* Premier Award Res in Aging, Ciba Found, 55; J C B Grant Award, Can Asn Anatomists; Distinguished Achievement Award, Can Hypertension Soc, 87. *Mem:* Am Physiol Soc; fel Royal Soc Can; Can Asn Anatomists (vpres, 62-64, pres, 65-66); Am Asn Anatomists; fel Coun High Blood Pressure Res. *Res:* Cardiovascular-renal physiology; endocrinology; hormonal hypertension; aging. *Mailing Add:* Dept Anat Univ BC Vancouver BC V6T 1W5 Can

FRIEDMAN, THEA MARLA, MOLECULAR BIOLOGY. *Educ:* Univ Pa, PhD(anat), 85. *Prof Exp:* ASST MOLECULAR BIOL, FOX-CHASE CANCER CTR, 86- *Mailing Add:* 20 Tunbridge Rd Cherry Hill NJ 08003

FRIEDMAN, THOMAS BAER, b Detroit, Mich, Dec 9, 44; m 70; c 2. HUMAN GENETICS, DEVELOPMENTAL GENETICS. *Educ:* Univ Mich, BS, 66, PhD(biol), 71. *Prof Exp:* Staff fel biochem, NIH, 71-74; asst prof develop biol & genetics, Oakland Univ, 74-77; ASSOC PROF GENETICS, MICH STATE UNIV, 78-, AT DEPT ZOOL. *Mem:* Genetics Soc Am; AAAS; Soc Develop Biol. *Res:* Galactose metabolism in cultured human cells; regulation of urate oxidase and adenide phosphoribosyl transferase during development of Drosophila Melanogaster. *Mailing Add:* Dept Zool 203 Natural Sci Bldg Mich State Univ East Lansing MI 48824

FRIEDMAN, WILLIAM ALBERT, b Chicago, Ill, May 29, 38; m 61. PHYSICS. *Educ:* Cornell Univ, BEP, 61; Mass Inst Technol, PhD(physics), 66. *Prof Exp:* NSF fel, Niels Bohr Inst, Copenhagen Univ, Denmark, 66-67; instr, Princeton Univ, 67-70; from asst prof to assoc prof, 70-80, PROF PHYSICS, UNIV WIS-MADISON, 80- *Concurrent Pos:* Vis prof physics, Mich State Univ, 81-82, 84-85. *Mem:* Am Phys Soc. *Res:* Theoretical nuclear physics, including nuclear reactions and nuclear structure. *Mailing Add:* Dept Physics Univ Wis Madison WI 53706

FRIEDMAN, WILLIAM FOSTER, b New York, NY, July 24, 36; c 2. PEDIATRIC CARDIOLOGY. *Educ:* Columbia Univ, AB, 57; State Univ NY, MD, 61. *Prof Exp:* Intern pediat, Harriet Lane Home, Johns Hopkins Hosp, 61-62, asst & sr resident, 62-64; clin assoc, Cardiol Br & pediat consult, Clin Surg, Nat Heart Inst, 64-66, sr investr & pediat cardiologist, Cardiol Br, 66-68; from asst prof to assoc prof pediat, Sch Med, Univ Calif, San Diego, 68-73, prof, 73-80, chief pediat cardiol, 68-80; PROF PEDIAT CARDIOL & EXEC CHMN DEPT PEDIAT, UNIV CALIF, LOS ANGELES, SCH MED, 80- *Concurrent Pos:* Nat Heart Inst Res Career Develop Award; chmn, Sub-board Pediat Cardiol, Am Bd Pediat, 80-86; chmn, Pediat Cardiol Comt, Am Col Cardiol, 86-; cardiol adv comt, Nat Heart Lung & Blood Inst, 87-; gov, Am Col Cardiol, 88-; J H Nicholson endowed chair, Pediat Cardiol, UCLA, 88- *Mem:* Am Soc Clin Invest; Am Pediat Soc; Am Heart Asn; Soc Nuclear Med; Soc Pediat Res. *Res:* Physiology, pharmacology and biochemistry of the developing heart. *Mailing Add:* Dept Pediat MDCC 22-412 Univ Calif Los Angeles Sch Med Los Angeles CA 90024

FRIEDMAN, YOCHANAN, b Chicago, Ill, Feb 25, 45; m 76; c 2. ENDOCRINOLOGY, THYROIDOLOGY. *Educ:* Roosevelt Univ, BS, 67; Univ Ill Med Ctr, PhD(biochem), 73. *Prof Exp:* Instr biochem & immunol, Univ Ill Med Ctr, 72-73; sci investr endocrinol, 73-80, DIR, RADIOIMMUNOASSAY LAB, COOK COUNTY HOSP, CHICAGO, 80- *Mem:* Endocrine Soc; Am Thyroid Asn; Am Fed Clin Res. *Res:* Regulation at the cellular level of thyroid physiology; cyclic adenosine monophosphate adenylate cyclase system. *Mailing Add:* 6742 N Whipple Chicago IL 60612

FRIEDMANN, E(MERICH) IMRE, b Budapest, Hungary, Dec 20, 21. PHYCOLOGY, MICROBIAL ECOLOGY. *Educ:* Sch Agr, Kolozsvar, Hungary, BSc, 43; Sch Agr, Magyarovar, MSc, 49; Univ Vienna, DrPhil(bot), 51. *Prof Exp:* Instr bot, Hebrew Univ, Israel, 52-56, lectr, 57-61, sr lectr, 61-66, assoc prof, 66-67; assoc prof biol, Queen's Univ, Ont, 67-68; assoc prof, 68-76, PROF BIOL SCI, FLA STATE UNIV, 76-; DIR, POLAR DESERT RES CTR, 85- *Concurrent Pos:* Res fel, Univ Manchester, 56-58; Dept Sci & Indust Res sr vis res fel, Univ Leeds, 59; res assoc, Queens Col, NY, 65-66; vis assoc prof, Fla State Univ, 66-67; vis prof, Univ Vienna, 75, prof, Nanjing Univ, Peoples Repub China, 87- *Mem:* Am Soc Microbiol; Linn Soc London; Phycol Soc Am; Soc Phycol France; Phycol Soc India; Royal Micros Soc; Int Phycol Soc. *Res:* Microbial ecology of the Antarctic desert; desert microorganisms and cave microorganisms; life history, sexuality, cytology and ecology of marine algae; fine structure of gamete fusion in algae; Phycomycetes parasitic on algae; experimental taxonomy of cyanobacteria. *Mailing Add:* Dept Biol Sci Fla State Univ Tallahassee FL 32306

FRIEDMANN, GERHART B, b Mannheim, Ger, Jan 10, 29; m 59. MEDICAL PHYSICS, OPTICS. *Educ:* Univ Madras, BSc, 49, MA, 51; Univ BC, PhD(physics), 58. *Prof Exp:* Res asst cosmic ray physics, Tata Inst Fundamental Res, 51-54; spec lectr physics, Univ Victoria, BC, 58-62, from asst prof to assoc prof, 62-90, emer assoc prof, 90-; RETIRED. *Concurrent Pos:* Grant radiation physics, BC Cancer Clin, 58-59; physicist, Victoria Cancer Clin, 58-76; radiol health officer, Royal Jubilee Hosp, 63-85. *Mem:* Can Asn Physicists. *Res:* Biological and medical physics. *Mailing Add:* 2957 Sea Point Dr Victoria BC V8N 1T1 Can

FRIEDMANN, HERBERT CLAUS, b Mannheim, Ger, June 19, 27; nat US; m 61; c 2. BIOCHEMISTRY, MICROBIOLOGY. *Educ:* Univ Madras, India, BSc, 47, MSc, 51; Univ Chicago, PhD(biochem), 58. *Prof Exp:* Chemist, Allergic Asthma Enquiry, Govt of Madras, India, 51-54; asst, Univ Chicago, 55-58, res assoc, 58-59; fel, Damon Runyon, McCollum-Pratt Inst, Johns Hopkins Univ, 59-60; res assoc, Dept Physiol, 60-64, asst prof biochem, 64-69, ASSOC PROF BIOCHEM, UNIV CHICAGO, 69- *Concurrent Pos:* Fulbright fel, Philipps Univ, Marburg, Ger, 85-86. *Mem:* Am Soc Biol Chem; Am Soc Microbiol. *Res:* Enzymes of intermediary metabolism; flavoproteins; vitamin B12; amino acids; porphyrins; delta-aminolevulinic acid. *Mailing Add:* Dept Biochem & Molecular Biol Univ Chicago 920 E 58th St Chicago IL 60637

FRIEDMANN, PAUL, b Vienna, Austria, Dec 2, 33; US citizen; m 62; c 2. SURGERY. *Educ:* Univ Pa, AB, 55; Harvard Univ, MD, 59. *Prof Exp:* Asst chief surg, 68-71, CHMN DIV SURG, BAY STATE MED CTR, SPRINGFIELD, 71- *Concurrent Pos:* From asst clin prof to clin prof, Sch Med, Tufts Univ, 72-85, prof, 85- *Mem:* Am Col Surgeons. *Res:* Clinical research in vascular surgery, alternative techniques in bypass grafting; surgical oncology. *Mailing Add:* 449 Waukena Ave Oceanside NY 11572

FRIEDMANN, PERETZ PETER, b Timisoara, Rumania, Nov 18, 38; US citizen; m 64. AEROELASTICITY, STRUCTURAL DYNAMICS. *Educ:* Israel Inst Technol, BS, 61, MS, 68; Mass Inst Technol, DSc, 72. *Prof Exp:* Eng off, Air Force, Israel Defence Forces, 61-65; sr engr structures, Israel Aircraft Indust, 65-69; res asst aeroelasticity, Mass Inst Technol, 69-72; asst prof, 72-77, assoc prof, 77-80, PROF MECH, AEROSPACE & NUCLEAR ENG DEPT, UNIV CALIF, LOS ANGELES, 80-, CHMN MECH, AEROSPACE & NUCLEAR ENG DEPT, 88- *Honors & Awards:* Recipient, Struct & Mat Award, Am Soc Mech Engrs, 84. *Mem:* Fel Am Inst Aeronaut & Astronaut; Am Soc Mech Engrs; Am Helicopter Soc; Am Acad Mech; Soc Eng Sci. *Res:* Rotary-wing aeroelasticity; jet engine blade aeroelastic problems; fluid-structure interaction problems; structural dynamics; unsteady aerodynamics; numerical methods; active control of aeroelastic problems. *Mailing Add:* 46-147N Eng IV Univ Calif Los Angeles CA 90024

FRIEDMANN, THEODORE, b Vienna, Austria, June 16, 35; m 65; c 2. MEDICINE, GENETICS. *Educ:* Univ Pa, BA, 56, MD, 60; Am Bd Pediat, cert, 68; Am Bd Med Genetics, cert, 87. *Prof Exp:* From asst prof pediat to assoc prof pediat, 70-82, PROF PEDIAT, UNIV CALIF, SAN DIEGO, 82-; DIR, AUGORON INST, LA JOLA, 82- *Concurrent Pos:* mem Congressional Biomed Ethic Adv comt, 82-, Exp Virol Study Sect, NIH, 85-; res fac lectr, Univ Clif, San Diego, 88. *Mem:* Soc Pediat Res. *Res:* The development of techniques for gene therapy for human disease. *Mailing Add:* Dept Pediat Univ Calif Ctr Molecular Genetics La Jolla CA 92093

FRIEDRICH, BENJAMIN C, b Fond du Lac, Wis, Feb 2, 29; div; c 2. SCIENCE EDUCATION. *Educ:* St Cloud State Col, BS, 54; Univ Ind, MS, 57; Pa State Univ, DEd, 61. *Prof Exp:* Instr chem, Luther Jr Col, 54-56; instr sci educ, Ind Univ, 57-59; asst prof, Northeastern Univ, 61-66; from asst prof to assoc prof, Dept Geosci, 66-81, PROF ASTRON,JERSEY STATE COL, 81- *Concurrent Pos:* Chmn dept geosci, Jersey City State Col, 69-73. *Mem:* AAAS. *Mailing Add:* Dept Geosci Jersey City State Col Jersey City NJ 07305

FRIEDRICH, BRUCE H, b Clinton, Okla, Oct 20, 36; m 63; c 1. PHYSICAL CHEMISTRY. *Educ:* Univ Iowa, MS, 61, PhD(chem), 63. *Prof Exp:* Fel phys chem, Univ Calif, Berkeley, 62-63; asst prof chem, Gustavus Adolphus Col, 63-66; from asst prof to assoc prof, 66-76, PROF CHEM, UNIV IOWA, 76- *Mem:* AAAS; Am Chem Soc; Am Phys Soc. *Res:* Infrared spectra of molecular crystals; spectra of donor-acceptor complexes. *Mailing Add:* Dept Chem Univ Iowa Iowa City IA 52242-1000

FRIEDRICH, DONALD MARTIN, b Mich, Jan 3, 44. MOLECULAR SPECTROSCOPY. *Educ:* Univ Mich, BSci, 66; Cornell Univ, PhD(chem), 73. *Prof Exp:* Res assoc, Wayne State Univ, 73-75; from asst prof to assoc prof chem, Hope Col, 75-85; sr res scientist, 85-89, TECH GROUP LEADER, PAC NORTHWEST LAB, 90- *Concurrent Pos:* Vis Prof, Univ Calif, Berkeley, 82-83. *Mem:* Am Chem Soc; Am Phys Soc; AAAS. *Res:* Molecular science by optical spectroscopics; Raman, non-linear optical techniques; ultraviolet vibronic excitation & emission spectroscopics; structure and dynamics in molecular clusters. *Mailing Add:* K2-18 Pac Northwest Lab Battelle Blvd Richland WA 99352

FRIEDRICH, EDWIN CARL, b Woodbury, NJ, Jan 15, 36; m 67; c 2. ORGANIC CHEMISTRY. *Educ:* Univ Ill, Urbana, BS, 57; Univ Calif, Los Angeles, PhD(chem), 61. *Prof Exp:* Res assoc chem, Univ Calif, Los Angeles, 61-62; res chemist, Calif Res Corp, 62-64; fel chem, Mass Inst Technol, 64-65; from asst prof to assoc prof, 65-76, PROF CHEM, UNIV CALIF, DAVIS, 76- *Mem:* Am Chem Soc. *Res:* Kinetics, product studies, salt and solvent effects in carbonium ion reactions of cyclopropylcarbinyl and bridged bicyclic systems; free-radical brominations; cyclopropylcarbinyl-allylcarbinyl radical rearrangements; organotin and organozinc reaction mechanisms and synthesis. *Mailing Add:* Dept Chem Univ Calif Davis CA 95616

FRIEDRICH, JAMES WAYNE, b Boonville, Mo, Dec 16, 52. PLANT BREEDING & GENETICS, AGRONOMY. *Educ:* Univ Mo, BS, 74; Univ Wis, MS, 75, PhD(agron), 78. *Prof Exp:* Plant physiologist, Univ Calif, Davis, 78-80; tech supvr agr res, Allied Chem Corp, 80-84; PRES, MAIZE GENETIC RESOURCES, INC. *Mem:* Am Soc Plant Physiologists; Am Soc Agron; Crop Sci Soc Am. *Res:* Nitrogen and sulfur metabolism in plants; mineral nutrition. *Mailing Add:* 2630 Davis St Raleigh NC 27608

FRIEDRICH, LOUIS ELBERT, b Wilmington, Del, Sept 7, 41; m 65. ORGANIC CHEMISTRY. *Educ:* Mass Inst Technol, BS, 63; Univ Calif, Berkeley, PhD(chem), 66. *Prof Exp:* NSF fel chem, Yale Univ, 66-67; from asst prof to assoc prof chem, Univ Rochester, 67-80; sr res scientist, 80-85, RES ASSOC, EASTMAN KODAK, 85- *Res:* Determine mechanisms of organic reactions, using both statistically designed experiments and theoretical techniques. *Mailing Add:* Res Lab Bldg 82-D 7th Floor Eastman Kodak Rochester NY 14650-2102

FRIEDRICH, OTTO MARTIN, JR, b Austin, Tex, Jan 29, 39. ELECTRICAL ENGINEERING, ELECTROOPTICS. *Educ:* Univ Tex, Austin, BS, 61, MS, 62, PhD(elec eng), 65. *Prof Exp:* Asst dir electronics res, 71-75, RES ENGR, UNIV TEX, AUSTIN, 65- *Concurrent Pos:* Consult eng, Indust & Govt, 65- *Honors & Awards:* Dr D S Draiper Award, Instrument Soc Am, 72; Mills Dean III Award, Int Instrumentation Symposium, 90. *Mem:* AAAS; Inst Elec & Electronic Engrs; Instrument Soc Am; Am Phys Soc; Optical Soc Am. *Res:* Electronics; lasers; holography; signal and data processing. *Mailing Add:* 1125 Shady Lane Austin TX 78721

FRIEL, DANIEL DENWOOD, b Queenstown, Md, Aug 11, 20; m 43; c 4. CHEMICAL ENGINEERING, PHYSICS. *Educ:* Johns Hopkins Univ, BsChE, 42. *Prof Exp:* Chemist, E I du Pont de Nemours & Co, Inc, 42-43; optics supvr, Manhattan Proj, E I du Pont de Nemours & Co, Inc, Univ Chicago, 43-44 & Manhattan Proj-Hanford Ord Works, 44-45; supvr, Appl Physics Lab, E I du Pont de Nemours & Co, Inc, 45-54, res mgr lab, 54-57, asst lab dir, 57-59, mgr, eng prod, 59-60, mgr, invest develop dept, 60-69; pres, Holotron Corp, 69-71; mgr, Riston Div, E I Du Pont De Nemours & Co, Inc, 71-73, dir, Electronic Prods, 73-75, dir, instrument prods, 75-83; PRES, EDGECRAFT CORP, 84- *Mem:* Optical Soc Am; Am Phys Soc; Instrument Soc Am. *Res:* Instrumentation. *Mailing Add:* PO Box 4319 Wilmington DE 19807

FRIEL, LEROY LAWRENCE, b Marlinton, WVa, Feb 3, 38; m 64; c 1. STRUCTURAL DESIGN, FOUNDATION DESIGN. *Educ:* WVa Univ, BS, 62, MS, 63; Ga Inst Technol, PhD(civil eng), 73. *Prof Exp:* Bridge engr, WVa Highway Comn, 63-65; asst prof struct & found, WVa Inst Technol, 65-67; found engr, A C Ackenheil, 67; asst prof struct & found, Univ Maine, 70-76; assoc prof, 76-79, prof & head eng sci, 79-90, PROF, MONT TECH, BUTTE, 90- *Concurrent Pos:* Prof struct, Metrop Univ, Mex City, 83. *Mem:* Am Soc Civil Engrs; Am Soc Eng Educrs. *Res:* The strength of hydraulic fill; optimum design of structures; instructional improvement. *Mailing Add:* Dept Eng Sci Mont Tech Butte MT 59701

FRIEMAN, EDWARD ALLAN, b New York, NY, Jan 19, 26; m 49, 67; c 5. PLASMA PHYSICS. *Educ:* Columbia Univ, BS, 45; Polytech Inst Brooklyn, MS, 48, PhD(physics), 51. *Prof Exp:* Instr physics, Polytech Inst Brooklyn, 45-52, res assoc, 47-49; res assoc, Proj Matterhorn, Princeton Univ, 52-53, head theoret div, 53-64, prof astron sci, 61-79, assoc dir, plasma physics lab, 64-79; dir, Energy Res, Dept Energy, 79-81; exec vpres, Sci Appln Inc, 81-86; DIR, SCRIPPS INST OCEANOG, 86-; VCHANC MARINES SCI, UNIV CALIF, SAN DIEGO, 86- *Concurrent Pos:* Consult, Lawrence Radiation Lab, Univ Calif, 53-57 & Los Alamos Sci Lab, 53-64; mem res adv comt nuclear energy processes, NASA, 59-61; John Simon Guggenheim Mem Found fel, 70; vchmn, White House Sci Coun, 81-89; mem, Defense Sci Bd, 84-90; sr consult, 87-; chmn, supercollider site eval comt, NRC, 87-88; mem & secy Energy Adv Bd. *Honors & Awards:* Richtmyer Award, Am Soc Physics Teachers-Am Phys Soc, 83. *Mem:* Nat Acad Sci; AAAS; Am Philos Soc; fel Am Phys Soc. *Res:* Theoretical plasma physics; hydromagnetics; hydrodynamics stability; astrophysics. *Mailing Add:* Scripps Inst Oceanog Dir Off 0210 Univ Calif-San Diego La Jolla CA 92093-0210

FRIEND, CYNTHIA MARIE, b Hastings, Nebr, Mar 16, 55; m 80; c 2. PHYSICAL CHEMISTRY. *Educ:* Univ Calif, Davis, BS, 77; Univ Calif, Berkeley, PhD(chem), 81. *Prof Exp:* fel, Stanford Univ, 81-82; from asst prof to assoc prof, 82-88, MORRIS KAHN ASSOC PROF, HARVARD UNIV, 88- *Concurrent Pos:* Vis researcher, Gen Motors Res, 81; res collabr, Nat Synchrotron, Brookhaven Nat Labs, 83-; fac develop award, IBM, 83-85; Presidential Young Investr Award, NSF, 85- *Mem:* Am Chem Soc; Am Phys Soc; Am Vacuum Soc. *Res:* Surface chemistry relating to catalysis and metal deposition on semiconductors; relating surface electronic structure to selectivity and reactivity; using surface chemical and spectroscopic methods to characterize the effect of nonmetallic, electronegative adlayers on the chemistry of organic molecules absorbed on molybdenum and tungsten surfaces; reactions of organic bases on silicon. *Mailing Add:* Dept Chem Harvard Univ 12 Oxford St Cambridge MA 02138-2902

FRIEND, DANIEL S, b Passaic, NJ, Nov 20, 33; c 3. EXPERIMENTAL PATHOLOGY, CELL BIOLOGY. *Educ:* NY Univ, BA, 57; State Univ NY Downstate Med Ctr, MD, 61. *Prof Exp:* Intern, Boston City Hosp, 61-62, resident path, 62-63; res fel anat, Harvard Med Sch, 63-65; res fel, 65-66, lectr, 66-67, from asst prof to assoc prof, 67-78, PROF PATH & VCHMN DEPT, UNIV CALIF, SAN FRANCISCO, 78- *Concurrent Pos:* USPHS career develop award, 67-88 & res grant, 68-88. *Mem:* AAAS; Am Soc Cell Biol(sect 88-91); Soc Study Reproduction; Am Asn Pathologists; Am Asn Anatomists. *Res:* Electron microscopy; structure and function of cell organelles; sperm and male reproductive tract; cell junctions; cytochemistry. *Mailing Add:* Dept Path 501 HSW Univ Calif 501 HSW Box 0506 San Francisco CA 94143

FRIEND, JAMES PHILIP, b Hartford, Conn, Nov 30, 29; m 55; c 2. ATMOSPHERIC CHEMISTRY. *Educ:* Mass Inst Technol, SB, 51; Columbia Univ, MA, 53, PhD(chem), 56. *Prof Exp:* Asst chem, Columbia Univ, 53, 55-56; proj engr, Perkin-Elmer Corp, 56-57; sr res scientist, Isotopes, Inc, 57-64, sr scientific adv, 64-67; assoc prof atmospheric chem, NY Univ, 67-72, prof, 72-73; R S HANSON PROF ATMOSPHERIC CHEM, DREXEL UNIV, 73- *Concurrent Pos:* Independent consult; mem, Climatic Impact Comt, Nat Acad Sci, 72-75 & Panel Atmospheric Chem Climatic Impact Comt, 75-; mem, biol effects of increased solar ultraviolet radiation comt, Nat Acad Sci, 81-82, chem & physics of ozone depletion comt, 81-82 & atmospheric effects nuclear explosions comt, 83-85. *Mem:* AAAS; Am Geophys Union; Am Meteorol Soc; Am Chem Soc; Sigma Xi. *Res:* Atmospheric chemistry and radioactivity; atmospheric diffusion; air pollution chemistry; global cycles and geochemistry of trace materials in the atmosphere; aerosol formation from gas phase reactions; atmospheric chemistry of volcanic plumes; nuclear war effects on the atmosphere. *Mailing Add:* 108 Righters Ferry Rd Bala-Cynwyd PA 19004

FRIEND, JONATHON D, US citizen. VETERINARY MEDICINE. *Educ:* Kans State Univ, DVM, 45, MS, 59; Okla State Univ, BS, 49. *Prof Exp:* From asst prof to assoc prof vet anat, 48-58, PROF PHYSIOL SCI, OKLA STATE UNIV, 58- *Mem:* Am Asn Vet Anat; Am Vet Med Asn. *Res:* Regional innervation in the bovine. *Mailing Add:* 4824 Country Club Ct Stillwater OK 74074

FRIEND, JUDITH H, b Attleboro, Mass, May 14, 41. CORNEAL BIOLOGY, OPHTHALMOLOGY. *Educ:* Harvard Univ, MA, 66. *Prof Exp:* RES COORDR & CLIN ASST PROF OPHTHAL, EYE & EAR INST, PITTSBURGH, 85- *Mem:* Asn Res Vision & Ophthal; Am Soc Cell Biol; Soc Res Adminr. *Mailing Add:* Eye & Ear Hosp Pittsburgh 230 Lathrop St Pittsburgh PA 15213

FRIEND, MILTON, b Malden, Mass, Dec 11, 35; m 64; c 2. WILDLIFE DISEASE INVESTIGATIONS, EPIZOOTIOLOGY. *Educ:* Univ Maine, Orono, BS, 58; Univ Mass, MS, 65; Univ Wis-Madison, PhD(vet sci & wildlife ecol), 71. *Prof Exp:* Wildlife biologist upland game birds, NY State Conserv Dept, 61 & wildlife dis, 61-66; wildlife biologist pesticide res, Denver Wildlife Res Ctr, 71-72, chief, Sect Pesticide Res, 73-75, DIR WILDLIFE DIS, NAT WILDLIFE HEALTH RES CTR, US FISH & WILDLIFE SERV, 75-; ADJ PROF, UNIV WIS-MADISON, 87- *Concurrent Pos:* Adj asst prof, Univ Wis-Madison, 75-87; Robert & Virginia Rausch Fund vis scientist, Univ Sask, 86; mem US deleg, Environ & Ecol Subgroup, Ninth Indo-NS Sci & Technol Subcomn Meeting, New Delhi, India, 87; mem Wis Dept Nat Res Wildlife Dis Rev Team, 87-88. *Mem:* Wildlife Dis Asn (secy, 71-73, pres, 74-75); Wildlife Soc; Am Asn Vet Lab Diagnosticians; US Animal Health Asn; AAAS; Sigma Xi. *Res:* Eye lens weights for wildlife age determination; environmental containment-infectious disease interactions; lead poisoning; diseases of wildlife; disease control; epidemiology. *Mailing Add:* Nat Wildlife Health Res Ctr 6006 Schroeder Rd Madison WI 53711

FRIEND, PATRIC LEE, b Iron River, Mich, Sept 4, 38; m 62; c 5. APPLIED & ENVIRONMENTAL MICROBIOLOGY. *Educ:* Northern Mich Col, BS, 61; Northwestern Univ, PhD(microbiol), 65. *Prof Exp:* Res asst microbiol, Med Sch, Northwestern Univ, 65-66; asst prof, Col Med, Univ Cincinnati, 66-73; RES DEPT, BETZ LABS INC, TREVOSE, PA, 73- *Mem:* NY Acad Sci; Am Soc Microbiol; Soc Indust Microbiol. *Res:* Aquatic and environmental microbiology, biodegradation and ecological relationships. *Mailing Add:* 9669 Grogans Mill Rd The Woodlands TX 77387

FRIEND, WILLIAM GEORGE, b Toronto, Ont, July 25, 28; m 53. INSECT PHYSIOLOGY. *Educ:* McGill Univ, BSc, 50; Cornell Univ, PhD(insect physiol), 54. *Prof Exp:* Entomologist, Can Dept Agr, 50-59; from asst prof to assoc prof, 59-66, PROF ZOOL, UNIV TORONTO, 66- *Mem:* Entom Soc Am; Entom Soc Can; Can Biochem Soc; Nutrit Soc Can. *Res:* Insect nutrition and biochemistry. *Mailing Add:* Univ Toronto St George Campus Toronto ON M5S 1A1 Can

FRIERSON, WILLIAM JOE, b Batesville, Ark, July 8, 07; m 30; c 2. ANALYTICAL CHEMISTRY. *Educ:* Ark Col, AB, 27; Emory Univ, MS, 28; Cornell Univ, PhD(inorg chem), 36. *Prof Exp:* Asst chem, Ark Col, 26-27; from asst prof to assoc prof, Hampden-Sydney Col, 28-44; prof, Birmingham-Southern Col, 44-46; prof, 46-69, William Rand Kenan, Jr prof, 69-75, EMER PROF CHEM AGNES SCOTT COL, 75- *Concurrent Pos:* Fel, Cornell Univ, 35-36; Prof, Univ Calif, Berkeley, 68- *Mem:* Am Chem Soc. *Res:* Chemistry of inorganic nitrogen compounds; organic reagents in analytical chemistry; boiling points of pure compounds under varying pressures; paper chromatography of inorganic ions. *Mailing Add:* 689 Kenilworth Circle Stone Mountain GA 30083

FRIES, CARA ROSENDALE, b Toledo, Ohio, Feb 11, 42. COMPARATIVE PATHOBIOLOGY. *Educ:* Univ Del, BA, 66, MS, 69, PhD(comp immunobiol), 77. *Prof Exp:* Instr microbiol & physiol, 68-71, res assoc, 77-79, scientist, 79-85, ASST PROF, UNIV DEL, 85- *Mem:* AAAS; Am Soc Zoologists; Sigma Xi; Am Soc Microbiol. *Res:* Immunology; microbiology; ultrastructure of marine animals; paleoparasitology. *Mailing Add:* Sch Life & Health Sci Univ Del Newark DE 19711

FRIES, DAVID SAMUEL, b Manassas, Va, June 22, 45; m 65; c 3. MEDICINAL CHEMISTRY. *Educ:* Bridgewater Col, BA, 68; Va Commonwealth Univ, PhD(med chem), 71. *Prof Exp:* Fel med chem, Univ Minn, 71-72; asst prof chem, Winona State Col, 73; PROF MED CHEM, UNIV OF THE PAC, 73- *Concurrent Pos:* Vis res prof med chem, Univ Groningen, Netherlands, 84-85, Ger Cancer Res Ctr, Heidelberg, Ger, 89-90. *Mem:* Am Chem Soc; Am Asn Col Pharm; Fedn Int Pharmaceuts; Sigma Xi. *Res:* The design and synthesis of narcotic antagonist; elucidation of narcotic-receptor topography; enzyme inhibitors of polyamine metabolism; neurotoxicology. *Mailing Add:* Sch Pharm Univ the Pac Stockton CA 95211

FRIES, JAMES ANDREW, b St Louis, Mo, June 25, 43; m 61; c 2. PHYSICAL CHEMISTRY, ENVIRONMENTAL CHEMISTRY. *Educ:* Univ SDak, BSEd, 65; Univ Iowa, MS, 68, PhD(phys chem), 69. *Prof Exp:* Asst phys chem, Univ Iowa, 65-69; from asst prof to assoc prof chem, 69-75, head dept, 69-78, prof chem, 75-78, ASST to PRES & DIR DEVELOP, NORTHERN STATE COL, 78- *Mem:* Am Chem Soc. *Res:* High temperature mass spectrometric studies of the vaporization and thermodynamics of the lanthanide metal sulfides; water quality studies. *Mailing Add:* Pres Chancellor Col St Michaels Dr Santa Fe NM 87501

FRIES, RALPH JAY, b Lancaster, Pa, Oct 22, 30; m 50; c 3. PHYSICAL CHEMISTRY. *Educ:* Pa State Univ, BSc, 52; Univ Pittsburgh, MLitt, 58, PhD(chem), 59. *Prof Exp:* Res assoc phys chem, Mellon Inst, 55-58; mem staff, Los Alamos Nat Lab, 58-73, assoc group leader, Group L-4, 74-77, group leader, Group L-7, 77-81, PROG MGR, LASER-FUSION TARGET FABRICATION PROG, GROUP L-4, LASER DIV, LOS ALAMOS NAT LAB, 73-, GROUP LEADER, GROUP CMB-10, 81-, PROG MGR, LASER ISOTOPE SEPARATION, 83- *Concurrent Pos:* Lectr, Univ Pittsburgh, 56-57; vis scientist, Ispra Lab, Europ Atomic Energy Community, Italy, 67-68; vis fel, Inst Laser Eng, Osaka Univ, Japan, 81; proj leader, Cognetive Systs Engr Group A-6, 88- *Mem:* Sigma Xi; AAAS; Am Phys Soc; Am Vacuum Soc. *Res:* Application of computers to technical education and training; surface and high temperature chemistry; physical measurements; diffusion permeation; coatings; microfabrication; laser-fusion target fabrication. *Mailing Add:* MS-M997 PO Box 1663 Los Alamos Nat Lab Los Alamos NM 87545

FRIESEL, DENNIS LANE, b Chicago, Ill, July 24, 42; m 64; c 2. PARTICLE ACCELERATOR DESIGN PHYSICS. *Educ:* St Procopius Col, BS, 64; Univ Notre Dame, PhD(physics), 70. *Prof Exp:* Res assoc, dept physics, 70-72, staff physicist, 72-87, HEAD, ACCELERATOR DIV, CYCLOTRON FACIL, IND UNIV, 88- *Mem:* Am Phys Soc; Sigma Xi. *Res:* Beam development and improved performance of Indiana University Cyclotron Facility 200 mega electron volts separated sector cyclotron; design and development of electron cooled storage ring using internal targets. *Mailing Add:* Ind Univ Cyclotron Facil Milo B Sampson Lane Bloomington IN 47401

FRIESEM, ALBERT ASHER, b Haifa, Israel, Jan 18, 36; US citizen; m 56; c 3. APPLIED PHYSICS, ELECTROOPTICS. *Educ:* Univ Mich, BS, 58, PhD(elec optics), 68; Wayne State Univ, MS, 61. *Prof Exp:* Engr, Bell Aircraft Co, NY, 58-59; engr, Res Labs, Bendix Corp, Mich, 59-63; res assoc, Radar & Optics Lab, Inst Sci & Technol, Univ, Mich, 63-66, assoc res engr, 66-68, res engr, 68-69; prin res engr, Electro Optics Ctr, Radiation Inc, 69-73; assoc prof, 73-77, PROF, WEIZMANN INST SCI, 77- *Concurrent Pos:* Vis scholar, Univ Mich, 69-71; sr vis scientist, Weizmann Inst Sci, 72-73, chmn sci coun, 82-84, chmn prof coun, 88-90, dept head, 87-90; sr res engr, Environ Res Inst Mich, 81-82; vis prof, Univ Mich, 81-82 & Univ Neuchatel, 87. *Mem:* Sr mem Inst Elec & Electronic Engrs; fel Optical Soc Am; Sigma Xi. *Res:* Optical Computing; optical image processing; wavefront reconstruction; holographic applications; electro-optic devices; optical memories; optical displays; optical fibers. *Mailing Add:* Dept Electronics Weizmann Inst Sci Rehovot 76100 Israel

FRIESEN, BENJAMIN S, b Garden City, Kans, Mar 24, 28; m 53; c 4. RADIATION BIOPHYSICS. *Educ:* Univ Kans, BA, 52, MA, 54; Iowa State Univ, PhD(biophys), 59. *Prof Exp:* USPHS fel, 59-60; from asst prof to prof radiation biophys, 60-85, PROF BIOCHEM, UNIV KANS, 85- *Concurrent Pos:* Dir, Radiation Safety Serv. *Mem:* Radiation Res Soc; Health Phys Soc. *Res:* Radiation safety. *Mailing Add:* Res Health & Safety 217 Burt Hall Univ Kans Lawrence KS 66045

FRIESEN, DONALD KENT, b Morrison, Ill, Mar 31, 41. ALGEBRA. *Educ:* Knox Col, BA, 63; Dartmouth Col, MA, 65, PhD(math), 66. *Prof Exp:* Res instr math, Dartmouth Col, 66-67; from asst prof to assoc prof math, Univ Ill, Urbana, 74-88; ASSOC PROF COMPUT SCI, TEX A&M, 88- *Mem:* Math Asn Am. *Res:* Finite subgroups of orthogonal groups and related problems in theory of finite groups. *Mailing Add:* Dept Comp Sci Tex A&M Univ College Station TX 77843

FRIESEN, EARL WAYNE, particle physics; deceased, see previous edition for last biography

FRIESEN, HENRY GEORGE, b Morden, Man, July 31, 34; m 67; c 2. ENDOCRINOLOGY. *Educ:* Univ Man, MSc & MD, 68. *Prof Exp:* Intern, Winnipeg Gen Hosp, 58-59, asst resident, 59-60; res fel endocrinol, New Eng Ctr Hosp, Boston, 60-61 & 62-63; asst res, Royal Victoria Hosp, Montreal, 61-62 & 63-65; from asst prof med to assoc prof, McGill Univ, 65-71, prof exp med, 72-73; PROF MED & PHYSIOL & HEAD DEPT PHYSIOL, UNIV MAN, 73- *Concurrent Pos:* Asst prof med, Tufts Univ Sch Med, 65-66; assoc, Med Res Coun Can, 68-73; chmn & mem, Med Res Coun Can comts, 70-; assoc ed, Can J Physiol & Pharmacol, 74-78; mem organizing comt, VI Int Endocrine Cong, Hamburg, 75; mem NIH task forces, 75-77 & 78; Sandoz lectr, Can Soc Endocrinol & Metab, 78. *Honors & Awards:* Eli Lilly Award, Endocrine Soc, 74; Gairdner Found Award, Toronto, 77. *Mem:* Can Soc Endocrinol & Metab (pres, 74); Int Soc Neuroendocrinol; fel Can Soc Clin Invest (pres, 78); Royal Soc Can; fel Royal Col Physicians & Surgeons Can. *Res:* Endocrinology, prolactin, placental lactogen and their receptors; growth factors. *Mailing Add:* Dept Physiol 770 Bannatyne Ave Winnipeg MB R3T 2N2 Can

FRIESEN, JAMES DONALD, b Rosthern, Sask, Nov 4, 35; m 58; c 3. MOLECULAR BIOLOGY, MICROBIOLOGY. *Educ:* Univ Sask, BA, 56, MA, 58; Univ Toronto, PhD(med biophysics), 62. *Prof Exp:* Nat Cancer Inst Can fel, Inst Microbiol, Copenhagen Univ, 62-64; vis asst prof physics, Kans State Univ, 64-65, asst prof, 65-67, assoc prof biol, 67-68; assoc prof biol, York Univ, 69-74, prof, 74-81; CHMN, DEPT MED GENETICS, UNIV TORONTO, 81- *Mem:* Am Soc Microbiol. *Res:* Genetics and regulation of transposition and translation in microorganisms. *Mailing Add:* Hosp for Sick Children 555 University Ave Toronto ON M5G 1X8 Can

FRIESEN, RHINEHART F, b Gretna, Man, Jan 6, 14; m 44; c 4. OBSTETRICS & GYNECOLOGY. *Educ:* Univ Man, MD, 44; FRCPS(C), 57. *Prof Exp:* Demonstr obstet & gynec, 58-65, asst pediat, 59-65, lectr & sr res asst, 65-67, asst prof, 67-78, ASSOC PROF OBSTET & GYNEC, UNIV MAN, 78- *Concurrent Pos:* Asst obstetrician & gynecologist, Health Sci Ctr, 59; consult, Man Rehab Hosp. *Mem:* Soc Obstetricians & Gynecologists Can; Can Med Asn; fel Am Col Obstet & Gynec. *Res:* Obstetrics and gynecology; perinatal mortality; fetal transfusions in Rh-sensitized mothers. *Mailing Add:* 45 Wilton St Winnipeg MB R3M 3B3 Can

FRIESEN, STANLEY RICHARD, b Roshern, Sask, Sept 8, 18; US citizen; m 42; c 4. SURGERY. *Educ:* Univ Kans, AB, 40, MD, 43; Univ Minn, PhD(surg), 49; Am Bd Surg, dipl, 50. *Prof Exp:* From asst prof to assoc prof surg & oncol, 49-59, PROF SURG, MED SCH, UNIV KANS, 59- *Concurrent Pos:* Consult, Vet Admin Hosp, Kansas City. *Mem:* AAAS; Soc Exp Biol & Med; Soc Univ Surgeons; Am Surg Asn; fel Am Col Surg. *Res:* Surgical endocrinology and gastroeneterology, specifically acid-peptic ulceration and multiple endocrine adenomatosis. *Mailing Add:* Dept Surg A 407 Col Health Univ Kans 39th St & Rainbow Blvd Kansas City KS 66103

FRIESEN, WOLFGANG OTTO, b Ger, 42; US citizen. NEUROBIOLOGY, ELECTROPHYSIOLOGY. *Educ:* Bethel Col, Kans, AB, 64; Univ Calif, Berkeley, MA, 66; Univ Calif, San Diego, PhD(neurosci), 74. *Prof Exp:* Res physicist pulmonary physiol, Cardiovasc Res Inst, Univ Calif, San Francisco, 69-70; res biologist neurophysiol, Univ Calif, Berkeley, 74-77; asst prof biol & neurobiol, 77-82, assoc prof, 82-88, PROF BIOL, UNIV VA, 88- *Concurrent Pos:* NIH fel molecular biol, Univ Calif, Berkeley, 75-77; NIH grant, 78-81, prin investr, 84-, NSF grant, 81- *Mem:* AAAS; Soc Neurosci. *Res:* Physiological studies and modeling analysis of the neural basis of animal movements; neuronal cell culture. *Mailing Add:* Dept Biol Gilmer Hall Univ Va Charlottesville VA 22901

FRIESER, RUDOLF GRUENSPAN, b Vienna, Austria, Apr 20, 20; US citizen; m 55; c 3. INORGANIC CHEMISTRY, SURFACE CHEMISTRY. *Educ:* Columbia Univ, BS, 50; Brooklyn Polytech Inst, MS, 58. *Prof Exp:* Chem supvr, Path Lab Dr Block, NY, 47-51; anal chemist, Fisher Sci Corp, 51-52; sr chemist, Res Lab, Interchem Corp, 52-58; engr semiconductor chem, RCA Corp, NJ, 58-60; mem tech staff, Bell Tel Lab, 60-65; sr res scientist, Res & Develop Ctr, Sprague Elec Co, 65-68; adv chemist, IBM Corp, Fishkill, 68-85; ADJ PROF, CHEM DEPT, UNIV NC, CHARLOTTE, 85- *Concurrent Pos:* Instr gen chem, N Adams State Col, 66-67; mem bd dir, Electrochem Soc, 78-80; div ed, J Electrochem Soc, 81-88. *Mem:* Am Chem Soc; Electrochem Soc; Sigma Xi. *Res:* Characterization and deposition of insulator thin films by chemical vapor deposition, plasma deposition; chemical and plasma etching of semiconductors and insulators; surface chemistry of semiconductors; metals and insulators. *Mailing Add:* 5211 Basswood Dr Lakeshore Estates Concord NC 28025

FRIESINGER, GOTTLIEB CHRISTIAN, b Zanesville, Ohio, July 4, 29; m 52; c 4. MEDICINE, PHYSIOLOGY. *Educ:* Muskingum Col, BS, 51; Johns Hopkins Univ, MD, 55; Am Bd Internal Med, dipl, 65. *Prof Exp:* Intern, Osler Med Serv, Johns Hopkins Univ Hosp, 55-56, asst resident med, 56-57, 59-60, chief resident, 62, from instr to assoc prof med, Sch Med, John Hopkins Univ, 63-71; PROF MED & DIR DIV CARDIOL, VANDERBILT UNIV, 71- *Concurrent Pos:* Fel med, Cardiovasc Div, Johns Hopkins Univ, 60-62, Clayton scholar, 63-71; dir myocardial infarction unit, Johns Hopkins Univ, 68-71; mem, Coun Circulation, Am Heart Asn, 66. *Mem:* Am Soc Clin Invest; Am Fedn Clin Res; Am Physiol Soc; Am Col Phys; Am Col Cardiol. *Res:* Applied cardiovascular physiology, especially ischemic heart disease including acute myocardial infarction. *Mailing Add:* Dept Med Rm CC-2218 Vanderbilt Med Ctr N Vanderbilt Univ Nashville TN 37232

FRIESS, SEYMOUR LOUIS, b Detroit, Mich, July 1, 22; m 53; c 2. ORGANIC CHEMISTRY. *Educ:* Univ Calif, Los Angeles, AB, 43, MA, 44, PhD(chem), 47. *Prof Exp:* Res chemist, Manhattan Eng Dist, Oak Ridge, 44-45; instr chem, Univ Calif, Los Angeles, 47-48; instr, Univ Rochester, 48-51; res chemist, US Naval Med Res Inst, 51-59, head phys biochem div, 59-68, actg dir, Physiol Sci Dept, 67-68, dir, 68-70, dir, Environ Biosci Dept, 70-80; CONSULT TOXICOL, ARLINGTON, VA, 80- *Concurrent Pos:*

Mem comt on toxicol, Nat Res Coun, 67-76 & toxicol info prog comt, 72-75; adj prof pharmacol, Med Sch, Uniformed Serv Univ, 77-; managing dir, Drill, Friess, Hays, Loomis & Shaffer, Inc. *Mem:* Am Chem Soc; Undersea Med Soc; Soc Toxicol. *Res:* Enzymatic topography; kinetics and catalysis; cholinesterase and conduction in nerve; mechanisms of toxic interactions in tissues; hyperbaric pharmacology; marine toxins; chemical hazard assessment and chemical safety evaluation; toxicological mechanisms; pharmacology and toxicology. *Mailing Add:* 6522 Lone Oak Ct Bethesda MD 20817-1644

FRIGYESI, TAMAS L, b Budapest, Hungary, June 7, 27; US citizen. NEUROBIOLOGY. *Educ:* Univ Budapest, MD, 51. *Prof Exp:* Asst neurol, Columbia Univ, 65-67; asst prof anat, Albert Einstein Col Med, 67-69; assoc prof physiol, Col Med & Dent NJ, 69-70; assoc prof neurol, Columbia Univ, 70-72; mem staff, Neurosurg Clin, Kanton Hosp, Univ Zurich, 72-76; prof physiol & neurol, Sch Med, Tex Tech Univ, Lubbock, 76-79; DIR, EPID RES INST, 79- *Concurrent Pos:* Nat Inst Neurol Dis & Blindness spec fel, 62-65; vis prof, Col Med & Dent NJ, Newark, 70-72; res prof physiol, New York Med Col, 72- *Honors & Awards:* Semmelweis Award, Am-Hungarian Med Asn, 70. *Mem:* Am Physiol Soc; Am Asn Anat; Am Electroencephalog Soc; Soc Neurosci. *Res:* Neurobiology of central integration of sensorimotor activities; synaptic organizations in basal ganglia-diencephalon functional linkages; epilepsy. *Mailing Add:* 2322 17th St Lubbock TX 79401

FRIIHAUF, EDWARD JOE, b Cleveland, Ohio, Apr 29, 36; m 58; c 3. TRIBOLOGY. *Educ:* Kent State Univ, BS, 58; Univ Ill, Urbana, MS, 60, PhD(chem), 61. *Prof Exp:* Chemist, J T Baker Chem Co, 61-62; chemist, Lubrizol Corp, 62-81. *Mem:* Am Chem Soc; Soc Lubrication Engrs. *Res:* Lubricants and related additives. *Mailing Add:* 12 Sabin St Montpelier VT 05602

FRIMPTER, GEORGE W, b Haverstraw, NY, Mar 17, 28; m 51; c 6. MEDICINE. *Educ:* Williams Col, BA, 48; Cornell Univ, MD, 52. *Prof Exp:* Estab investr, Am Heart Asn, 64-69; from asst prof to assoc prof med, Med Col, Cornell Univ, 61-69; prof med, Univ Tex Med Sch, San Antonio, 69-77; PROF MED & ASSOC CHIEF OF STAFF, AMB CARE, AUDIE MURPHY VA HOSP, 77- *Concurrent Pos:* Res fel med, USPHS, 58-59; sr res fel, NY Heart Asn, 59-64. *Mem:* Am Fedn Clin Res; Am Soc Clin Invest; fel Am Col Physicians; Aerospace Med Asn. *Res:* Internal medicine; metabolic aspects of kidney disease, especially errors of amino acid metabolism in various inherited conditions. *Mailing Add:* Univ Tex Health Sci Ctr 7400 Merton Minter Blvd San Antonio TX 78284

FRIMPTER, MICHAEL HOWARD, b New York, NY, Dec 10, 34; m 62; c 3. GEOLOGY, HYDROLOGY. *Educ:* Williams Col, BA, 57; Boston Univ, MA, 61, PhD(geol), 67. *Prof Exp:* Geologist water resources div, US Geol Surv, 63-68, hydrologist, 68-69; asst prof geol, Wis State Univ, 69-71; hydrologist, 71-79, MASS SUBDIST CHIEF WATER RESOURCES DIV, US GEOL SURV, 79- *Mem:* Geol Soc Am; Nat Waterwell Asn; New Eng Waterworks Asn. *Res:* Geology of the Hudson Highlands; hydrogeology and water chemistry; geochemistry of surface and ground water; glacial geology and ground water resources, New York, New England and Massachusetts. *Mailing Add:* Water Resources Div 28 Lord Rd Suite 280 Marlboro MA 01752

FRINDT, ROBERT FREDERICK, b Edmonton, Alta, March 8, 39; m 63; c 2. ENGINEERING PHYSICS. *Educ:* Univ Alta, BSc, 60; Univ Cambridge, PhD(physics), 63. *Prof Exp:* Sr student 1851, Cavendish Lab, Univ Cambridge, 63-64; asst res officer, Physics Div, Nat Res Coun Can, 64-65; asst prof, Simon Fraser Univ, 65-67, assoc prof, 67-78, actg dir, Energy Res Inst, 80-81, PROF PHYSICS, SIMON FRASER UNIV, 78- *Concurrent Pos:* Mem, Energy Res Eval Comt, BC Sci Coun, 80-89 & Electronics & Commun Res Eval Comt, 89-91; Nat Sci & Eng Res Coun Can grant, Selection Comt, 90-91. *Mem:* Can Asn Physicists. *Res:* Physical properties of layered materials and composite systems. *Mailing Add:* Physics Dept Simon Fraser Univ Burnaby BC V5A 1S6 Can

FRINGS, CHRISTOPHER STANTON, b Birmingham, Ala, Aug 10, 40; m 65; c 2. CLINICAL CHEMISTRY, TOXICOLOGY. *Educ:* Univ Ala, BS, 61; Purdue Univ, PhD(chem), 66. *Prof Exp:* Res assoc biochem, Mayo Clin & Mayo Grad Sch Med, 66-67; SR VPRES TECH SERV & DIR CLIN CHEM & TOXICOL, MED LAB ASSOCS & CUNNINGHAM PATH ASSOCS, 67- *Mem:* Am Asn Clin Chem; Am Chem Soc; Am Acad Clin Toxicol. *Res:* Clinical laboratory toxicology and effective therapeutic drug monitoring. *Mailing Add:* 633 Winwood Dr Birmingham AL 35226-2837

FRINK, CHARLES RICHARD, b Keene, NH, Sept 26, 31; m 53; c 3. SOIL CHEMISTRY. *Educ:* Cornell Univ, BS, 53, PhD(soil chem), 60; Univ Calif, Berkeley, MS, 57. *Prof Exp:* From asst soil chemist to soil chemist, 60-70, CHIEF, DEPT SOIL & WATER, CONN AGR EXP STA, 70-, VDIR, 72- *Mem:* Fel AAAS; Am Chem Soc; Soil Sci Soc Am; fel Am Inst Chemists; NY Acad Sci. *Res:* Aluminum chemistry and clay mineralogy in acid soils; nutrient cycles in soil, water and lake sediments; toxic organic wastes; agricultural production. *Mailing Add:* Conn Agr Exp Sta PO Box 1106 New Haven CT 06504

FRINK, DONALD W, b Madison, Ohio, Apr 25, 33; m 55; c 3. MECHANICAL ENGINEERING. *Educ:* Ohio State Univ, BME, 56, MME, 57. *Prof Exp:* Res engr, 56-60, proj leader, 60-62, proj dir, 62-63, GROUP DIR, BATTELLE MEM INST, 63- *Mem:* Nat Soc Prof Engrs; Am Soc Mech Engrs. *Res:* Kinematics; mechanism; dynamics and statistics; mechanism synthesis and analysis. *Mailing Add:* 2309 Johnston Rd Columbus OH 43220

FRIOU, GEORGE JACOB, b Brooklyn, NY, Oct 5, 19; div; c 4. IMMUNOLOGY, RHEUMATOLOGY. *Educ:* Cornell Univ, BS, 40, MD, 44; Am Bd Internal Med, dipl, 51. *Prof Exp:* Intern internal med, New Haven Hosp, Yale Univ, 44-45, asst resident, 45-46, res fel, 48-49, chief resident, 49-50, instr, Sch Med, 49-50, from clin instr to clin asst prof, 50-58; asst prof med, 58-60, assoc prof med & microbiol, Sch Med, Univ Okla, 60-64; assoc prof med, Sch Med, Univ Southern Calif, 64-68, prof, 68-78, chief immunol & rheumatic dis sect, 64-78; PROF MED, SCH MED, UNIV CALIF, IRVINE, 78-, DIR DIV IMMUNOL & RHEUMATIC DIS, 78- *Mem:* Am Soc Clin Invest; Am Asn Immunologists; Am Rheumatism Asn; fel Am Col Physicians; Am Fedn Clin Res. *Res:* Rheumatic diseases; immunology. *Mailing Add:* Dept Med Med Sci One Univ Calif Sch Med Irvine CA 92717

FRIPP, ARCHIBALD LINLEY, b Columbia, SC, Jan 15, 39; m 63; c 3. MATERIALS SCIENCE, ELECTRICAL ENGINEERING. *Educ:* Univ SC, BS, 66; Univ Va, MS, 69, PhD(mat sci), 74. *Prof Exp:* STAFF MEM PHYSICS, NASA-LANGLEY RES CTR, HAMPTON, 66- *Mem:* Am Asn Crystal Growth; Electrochem Soc; Inst Elec & Electronics Engrs. *Res:* Crystal growth of compound semiconductor materials, which are used to build infrared detector arrays. *Mailing Add:* Langley Res Ctr Hampton VA 23665

FRISANCHO, A ROBERTO, b Peru, Feb 4, 39. BIOLOGY. *Educ:* Nat Univ Cuzco, Peru, PH, 62; Pa State Univ, MA, 66, PhD(bioanthrop), 69. *Prof Exp:* Asst prof anthrop, Univ Mich, 68-73, assoc prof anthrop & res sci, Ctr Human Growth, 73; prin investr, Nat Heart, Lung & Blood Inst, NIH, 71-73; PROF ANTHROP, CTR HUMAN GROWTH & DEVELOP, UNIV MICH, 76- *Concurrent Pos:* Prin investr adolescent pregnancy & birthweight, NSF, 80-84; mem, Amt Stand Growth & Nutrit Status, Maternal Child & Health Bur, 89-91. *Mem:* Am Asn Phys Anthrop; Am Soc Clin Nutrit; Am Human Biol Coun; Am Inst Nutrit; Soc Int Nutrit. *Mailing Add:* Ctr Human Growth & Develop Univ Mich 300 N Ingalls Ann Arbor MI 48109

FRISBIE, RAYMOND EDWARD, b Barstow, Calif, Apr 14, 45; m 72. ENTOMOLOGY. *Educ:* Univ Calif, Riverside, BS, 67, MS, 69, PhD(entom), 72. *Prof Exp:* Res asst entom, Univ Calif, Riverside, 67-72; exten entomologist & pest mgt, 72-75, exten entomologist & pest mgt leader, Tex Agr Exten Serv, 75-77, MEM FAC ENTOM, TEX A&M UNIV, 77-, INTEGRATED PEST MGT COORDR, 79- *Mem:* Entom Soc Am. *Res:* Pest management program designed to reduce production costs to cotton and grain sorghum farmers by using integrated technology to reduce pest populations. *Mailing Add:* Systs Build Rm 312 Tex A&M Univ College Station TX 77840

FRISBY, JAMES CURTIS, b Bethany, Mo, Oct 22, 30; m 69. AGRICULTURAL ENGINEERING. *Educ:* Univ Mo, BS, 52 & 56; Iowa State Univ, MS, 63, PhD(agr eng), 65. *Prof Exp:* Classroom instr math sci, ed & training dept, Caterpillar Tractor Co, Ill, 56-57, tech writer, serv dept, 57-58, mkt analyst, engine div, 58-60; asst mgr, farm serv dept, Iowa State Univ, 61-63, instr agr eng, 63-66; from asst prof to assoc prof, 66-74, PROF AGR ENG, UNIV MO-COLUMBIA, 74-, DEPT CHAIR, 89- *Mem:* Am Soc Agr Engrs; Am Soc Eng Educ; Nat Asn Col & Teachers Agr; Sigma Xi; Nat Soc Prof Engrs. *Res:* Application of operations research techniques to machine systems used for agricultural enterprises. *Mailing Add:* Dept Agr Eng 214 Agr Eng Bldg Univ Mo Columbia MO 65211

FRISCH, ALFRED SHELBY, b San Diego, Calif, July 1, 35; m 62; c 2. OCEANOGRAPHY, ATMOSPHERIC SCIENCES. *Educ:* San Diego State Univ, BA, 57; Univ Wash, MS, 64, PhD(geophys), 70. *Prof Exp:* Thermodyn engr mass spectros, Convair Astronaut, 57-59; physicist nuclear physics, Edgerton, Germeshausenire & Griar, Inc, 61-63; oceanogr phys oceanog, Appl Physics Lab, Univ Wash, 65-67; PHYSICIST ATMOSPHERIC SCI & OCEANOG, WAVE PROPAGATION LAB, NAT OCEANIC & ATMOSPHERIC ADMIN, 70- *Honors & Awards:* Group Award, Nat Oceanic & Atmospheric Admin, 75. *Mem:* Am Geophys Union. *Res:* Atmospheric turbulence and air motion using acoustic echo sounding and doppler radars; physical oceanography using high frequency doppler radar for ocean current measurement. *Mailing Add:* 690 Tenth St Boulder CO 80302

FRISCH, HARRY LLOYD, b Vienna, Austria, Nov 13, 28; nat US; div; c 2. THEORETICAL CHEMISTRY. *Educ:* Williams Col, Mass, AB, 47; Polytech Inst Brooklyn, PhD(phys chem), 52. *Prof Exp:* Res assoc phys chem, Polytech Inst Brooklyn, 51-52; res assoc physics, Syracuse Univ, 52-54; from instr to asst prof chem, Univ Southern Calif, 54-56; 56; mem tech staff, Bell Labs, 56-67; assoc dean, Col Arts & prof chem, 67-78, DISTINGUISHED PROF CHEM, STATE UNIV NY ALBANY, 78- *Concurrent Pos:* Vis assoc prof, Yeshiva Univ, 62-; vis mem, Courant Inst Math Sci, NY Univ, 64-65 & 85; assoc ed, J Chem Physics, 64-66; mem adv bd, J Adhesion, 70-80, J Phys Chem, 76-78, J Polymer Sci, 78-, J Colloid & Interface Sci, 80-83; Jolliot Curie Prof, Paris, 79; sr Humboldt fel, 87. *Honors & Awards:* Boris Pregel Medal, NY Acad Sci, 73. *Mem:* Am Chem Soc; fel Am Phys Soc. *Res:* Statistical mechanics and kinetic theory; colloid and high polymer chemistry; solid state chemistry and physics. *Mailing Add:* Dept Chem State Univ NY Albany NY 12222

FRISCH, HENRY JONATHAN, b Los Alamos, NMex, Aug 21, 44; m 70; c 2. EXPERIMENTAL HIGH ENERGY PHYSICS. *Educ:* Harvard Univ, BA, 66; Univ Calif, Berkeley, PhD(physics), 71. *Prof Exp:* From instr to assoc prof, 71-83, PROF PHYSICS, UNIV CHICAGO, 84- *Honors & Awards:* Fel, Am Phys Soc, 87. *Mem:* Am Phys Soc. *Res:* Direct lepton and dilepton production and high transverse momentum particle production; very high energy particle collisions; monopole searches. *Mailing Add:* EFI HEP 320 5640 Ellis Ave Chicago IL 60637

FRISCH, I(VAN) T, b Budapest, Hungary, Sept 21, 37; US citizen; m 62. ELECTRICAL ENGINEERING. *Educ:* Queens Col, NY, BS, 58; Columbia Univ, BS, 58, MS, 59, PhD(elec eng), 62. *Prof Exp:* Asst prof elec eng, Univ Calif, Berkeley, 62-65 & 66-68, assoc prof, 68-69; Ford Found resident eng pract, Bell Tel Labs, 65-66; Guggenheim fel, 69, sr vpres & gen mgr, Network Anal Corp, 71-; AT CONTEL INF SYSTS. *Concurrent Pos:* Consult, Collins Radio Co, Inc, 62-; off emergency planning, Exec Off President, 68-; ed-in-chief, Networks; vis prof, State Univ NY, 73-74; adj prof, Columbia Univ, 74-75. *Mem:* Fel Inst Elec & Electronics Engrs. *Mailing Add:* Polytech Univ 333 Jay St Rm 321 Brooklyn NY 11201

FRISCH, JOSEPH, b Vienna, Austria, Apr 21, 21; nat US; m 57; c 3. MECHANICAL ENGINEERING. *Educ:* Duke Univ, BSME, 46; Univ Calif, MS, 50. *Prof Exp:* Sr engr, dept pub works, Baltimore, 47; from asst prof eng design to prof mech eng, 47-63, asst dir, inst eng res, 63-66, chmn dept mech design, 66-70, assoc dean, 73-76, PROF MECH ENG, UNIV CALIF, BERKELEY, 63- *Concurrent Pos:* A E Taylor distinguished prof, Univ Birmingham, Eng, 70. *Mem:* Fel Am Soc Mech Engrs. *Res:* Mechanical engineering design; materials behavior and processing; computer control. *Mailing Add:* Dept Mech Eng Col Eng Univ Calif Berkeley CA 94720

FRISCH, KURT CHARLES, b Vienna, Austria, Jan 15, 18; nat US; m 46; c 3. CHEMISTRY. *Educ:* Realgymnasium, Austria, BS, 35; Univ Vienna, MA, 38; Columbia Univ, MA, 41, PhD(org chem), 44. *Prof Exp:* Anal & res chemist, Am Dietaids Co, NY, 41; res chemist, Gen Elec Co, 44-52; asst mgr res, E F Houghton & Co, 52-56; dir polymer res & develop, Wyandotte Chem Corp, 56-68; PROF POLYMER ENG & CHEM & DIR POLYMER INST, UNIV DETROIT, 68- *Mem:* Am Chem Soc; NY Acad Sci; fel Am Inst Chem. *Res:* Polymer research on polyurethanes; silicones; phenolics; vinyls; organic synthetic and application research on textile and paper chemicals; synthetic lubricants; antimalarials. *Mailing Add:* 17986 Parke Lane Grosse Ile MI 48138

FRISCH, P DOUGLAS, b Tiffin, Ohio, Aug 24, 45; m 67; c 2. INORGANIC CHEMISTRY. *Educ:* Ohio Univ, BS, 67; Univ Wis-Madison, PhD(inorg chem), 72. *Prof Exp:* Teaching asst, Univ Wis, 67-70, res assoc, 70-72; Sci Res Coun fel & staff tutor chem, Univ Sheffield, Eng, 72-74; asst prof chem, Univ Maine, Orono, 74-79; res eng specialist, Monsanto Co, St Louis, 79-81; STAFF CHEMIST, EXXON CHEM CO, LINDEN, NJ, 81- *Mem:* Am Chem Soc; Sigma Xi. *Res:* Synthetic inorganic and organometallic chemistry; solid state and solution structural determination; homogeneous and heterogeneous catalysis; new product development. *Mailing Add:* 4006 Pecan Park Lane Kingwood TX 77345

FRISCH, ROSE EPSTEIN, b New York, NY, July 7, 18; m 40; c 2. REPRODUCTIVE BIOLOGY, POPULATION SCIENCES. *Educ:* Smith Col, BA, 39; Columbia Univ, MA, 40; Univ Wis, PhD(physiol genetics), 43. *Prof Exp:* Res fel biol, 53-54, res assoc, Ctr Pop Studies, 70-75, lectr, Sch Pub Health, 75-84, MEM, CTR POP STUDIES, HARVARD UNIV, 75-, ASSOC PROF POP SCI, SCH PUB HEALTH, 84- *Concurrent Pos:* Guggenheim Mem Found fel, 75-76; NIH res grant award, 88-91; Nat Lectr, Sigma Xi 89-90. *Mem:* Soc Study Reproduction; Soc Study Fertil; Pop Asn Am; Soc Study Social Biol; Endocrine Soc; AAAS. *Res:* Determinants of puberty and reproductive ability; effects of environmental factors such as nutrition and physicial exercise on female fertility, adeposity and fertility; exercise and cancer risk; population dynamics. *Mailing Add:* Harvard Ctr Pop Studies Nine Bow St Cambridge MA 02138

FRISCHER, HENRI, b Brussels, Belg, Jan 15, 34; US citizen; m 61; c 3. INTERNAL MEDICINE, GENETICS. *Educ:* Santo Domingo Univ, MD, 58; Univ Chicago, PhD(genetics), 65. *Prof Exp:* Intern, Michael Reese Hosp, Chicago, 58-59, from resident to chief resident internal med, 59-62; from instr to asst prof med, Univ Chicago, 64-71; assoc prof hemat, 71-77, DIR, SECT BLOOD GENETICS & PHARMACOGENETICS, DEPT PHARMACOL, RUSH UNIV, 71-, PROF MED & PHARMACOL, RUSH MED COL, 77- *Concurrent Pos:* USPHS spec fel, 65-66; sr attend physician, Rush-Presby St Luke's Med Ctr, 77- *Mem:* AAAS; Am Soc Human Genetics. *Res:* Biochemical genetics; hereditary hemolytic anemias; glucose-6-phosphate dehydroglucose deficiency; disorders and regulation of pentose phosphate shunt; carbohydrate metabolism. *Mailing Add:* Rush Med Col Rush Univ 1753 W Congress Pkwy Chicago IL 60612

FRISCHMUTH, ROBERT WELLINGTON, b Cleveland, Ohio, Jan 28, 40. CHEMICAL ENGINEERING. *Educ:* Case Inst Technol, BS, 62; Northwestern Univ, MS, 63. *Prof Exp:* Engr, NASA Lewis Res Ctr, 63-68 & Shell Develop Co, 68-74; group leader, 74-77, mgr res, Occidental Res Corp, 77-83; ELECTRIC POWER RES INST, 85- *Mem:* Am Inst Chem Engrs; Soc Mining Engrs; Am Chem Soc; Am Soc Mech Eng. *Res:* Coal research; chemicals from coal; management of research; power generation. *Mailing Add:* 661 Wingate Dr Sunnyvale CA 94087-2456

FRISCO, L(OUIS) J(OSEPH), b Patchogue, NY, Aug 21, 23; m 50; c 2. ELECTRICAL ENGINEERING. *Educ:* Johns Hopkins Univ, BE, 49, MScE, 53. *Prof Exp:* Asst res contract dir, Dielec Lab, Johns Hopkins Univ, 50-58, res contract dir, 58-64; prog mgr dielec mat, Adv Tech Labs, Gen Elec Co, NY, 64-65; mgr, 65-66, dir, Tech Serv Labs, 66-69, dir mfg, 70-77, tech dir, Wire & Cable Div, 77-79, GEN MGR, WIRE & CABLE DIV & CORP DIR PROD REV, RAYCHEM CORP, MENLO PARK, 79- *Concurrent Pos:* Mem conf elec insulation, Nat Acad Sci-Nat Res Coun; US deleg, Int Electrotechnol Comn. *Mem:* Inst Elec & Electronics Engrs; Am Soc Testing & Mat. *Res:* Electrical insulation and dielectric phenomena. *Mailing Add:* Tech Dir Electronics Sector 1018 Earlington Ct Sunnyvale CA 94087

FRISELL, WILHELM RICHARD, b Two Harbors, Minn, Apr 27, 20; m 48; c 2. ENZYMOLOGY, METABOLISM. *Educ:* St Olaf Col, BA, 42; Johns Hopkins Univ, MA, 43, PhD(org chem), 46. *Prof Exp:* Jr instr chem, Johns Hopkins Univ, 42-46, res assoc, 46-47, instr, 47-51; from asst prof to prof biochem, Sch Med, Univ Colo, 51-69, assoc dean, Grad Sch, 59-69; prof biochem & chmn dept, NJ Med Sch, Univ Med & Dent, NJ, 69-76; prof biochem, chmn dept & assoc dean basic sci & grad affairs, 76-90, EMER PROF BIOCHEM, EAST CAROLINA UNIV SCH MED, NC, 90- *Concurrent Pos:* Am Scand Found fel, Univ Uppsala, 49-50; chmn, Int Res Fel Comt, Fogarty Int Ctr, NIH, 62-71 & 81, mem, 62-71 & 81- *Mem:* Fel AAAS; Am Chem Soc; Am Soc Microbiol; Harvey Soc; Am Soc Biol Chemists; Soc Exp Biol Med; Sigma Xi. *Res:* One-carbon metabolism; flavins and flavoenzymes; mitochondrial structure and function; microbiol metabolism; muscle metabolism. *Mailing Add:* Dept Biochem East Carolina Univ Med Sch Greenville NC 27858-4354

FRISHBERG, CAROL, b Brooklyn, NY, Sept 11, 47; m 73; c 3. QUANTUM CHEMISTRY. *Educ:* Brooklyn Col, BS, 68; City Univ NY, PhD(phys chem), 75. *Prof Exp:* Res assoc chem, Univ NC, Chapel Hill, 75-76; from asst prof to assoc prof, 77-86, PROF CHEM, RAMAPO STATE COL, NJ, 86- *Concurrent Pos:* Vis prof, Ctr Advan Studies, Univ NMex, 85- *Mem:* Am Chem Soc; Am Phys Soc; Sigma Xi; AAAS. *Res:* Electronic structure of atoms and molecules. *Mailing Add:* Sch Theoret & Appl Sci Ramapo State Col Mahwah NJ 07430

FRISHKOPF, LAWRENCE SAMUEL, b Philadelphia, Pa, June 26, 30; m 60; c 3. BIOPHYSICS. *Educ:* Univ Pa, AB, 51; Mass Inst Technol, PhD(physics), 56. *Prof Exp:* Mem res staff commun biophys, Mass Inst Technol, 55-57; NIH fel biophys, Rockefeller Inst, 57-58; mem tech res staff, Bell Labs, 59-68; PROF ELEC ENG, MASS INST TECHNOL, 68- *Mem:* AAAS; fel Acoust Soc Am; Soc Neurosci. *Res:* Physiology and anatomy of acousticolateralis system; relation to perception and behavior; physiology and anatomy of sensory systems; animal communication. *Mailing Add:* Dept Elec Eng 36-824 & Comp Sci Mass Inst Technol Cambridge MA 02139

FRISHMAN, AUSTIN MICHAEL, b Brooklyn, NY, May 28, 40; m 62; c 2. ENTOMOLOGY. *Educ:* Cornell Univ, BS, 62, MS, 64; Purdue Univ, PhD(entom), 68. *Prof Exp:* Res asst livestock entom, Cornell Univ, 62-64; exten asst entom, Purdue Univ, 64-67, instr pest control & entom, 67-68; from asst prof to assoc prof, 68-75, PROF ENTOM, BOT & BIOL, STATE UNIV NY AGR & TECH COL FARMINGDALE, 75- *Concurrent Pos:* Pvt consult, 68-; NSF res grant & consult, Huntington Comput Proj, Polytech Inst Brooklyn, 68-69; training dir, Copesan Serv, Inc, 69-74 & Southern Mill Creek Prod, Inc, 74-; NSF res grant comput, 70-73. *Mem:* Entom Soc Am. *Res:* Structural and industrial control of pests; extermination; pest control; livestock and medical entomology; extension entomology. *Mailing Add:* 20 Miller Rd 30 Miller Rd Farmingdale NY 11735

FRISHMAN, DANIEL, b Brooklyn, NY, Oct 19, 19; m 42; c 4. DEVELOPMENT & PROCESSING OF POLYMERIC MATERIALS. *Educ:* Brooklyn Col, BA, 40; Cath Univ, MS, 47; Georgetown Univ, PhD, 50. *Prof Exp:* Anal chemist, Navy Yard, Washington, DC, 40-42; asst, Cath Univ, 42-43; res assoc, Textile Found, Nat Bur Standards, 43-44; tech adv, Fercleve Corp, Tenn, 44-45; res assoc, Harris Res Labs, 45-49, proj leader, 51-55; dir res, A Hollander & Son, NJ, 49-51; dir res & develop, Malden Mills, Mass, 55-62; pres & dir res, Fibresearch Corp, 62-65; tech adv, Reid-Meredith Inc, Lawrence, 65-69; pres, 74-89, CHMN BD & CONSULT, AKKO INC, 89- *Concurrent Pos:* Consult, 69- *Mem:* Am Soc Plastics Engrs. *Res:* Lead-acid storage batteries; thermal diffusion of uranium isotopes; chemistry and physics of textile fibers, animal fibers, fur and leather; coated and pile fabrics; plastics; synthetic furs; blood chemistry. *Mailing Add:* 105 Paradise Harbour, Apt 204 North Palm Beach FL 33408-5038

FRISHMAN, FRED, b New York, NY, Aug 4, 23; m 48; c 4. ENGINEERING & BUSINESS STATISTICS. *Educ:* City Col New York, BBA, 47; George Washington Univ, AB, 56, MA, 57, PhD, 71. *Prof Exp:* Enumerator, US Bur Census, NY, 48; statistician, NY Bd Educ, 48-49; eng statistician, Naval Inspector Ord, 49-51, Bur Ord, Dept Navy, 51-54, head appl math & statist group, US Naval Propellant Plant, 54-60; chief math br, Off Chief Res & Develop, Hq, Dept Army, 60-73; dir, Math Div, Army Res Off, Durham, 73-74; chief, Math Statist Br, US Internal Revenue Serv, 74-79; exec secy comt appl & theoret statist, Nat Res Coun-Nat Acad Sci, 79-83; assoc prof math, Weber Col, Babson Park, Fla, 83-84; ASSOC STATIST, UNIV CENT FLA, ORLANDO, 84-; CONSULT, LONGWOOD, FLA, 83- *Concurrent Pos:* Assoc statist, Col Gen Studies, George Washington Univ, 55-59, lectr, 60-64, asst prof lectr, 66-73, assoc prof lectr, 74-77, prof lectr, 77-80, adj prof, 80-83; chief, Math Br, Army Res Off, London, 71-72. *Mem:* Math Asn Am; Inst Math Statist; fel Am Statist Asn; Int Asn Statist Phys Sci; Int Statist Inst; Sigma Xi. *Res:* Statistical design of experiments; development of sampling techniques; applications of statistics in the physical sciences. *Mailing Add:* 446 Stanton Pl Longwood FL 32779

FRISHMAN, LAURA J(EAN), b Washington, DC, Dec 21, 47; m 72. RETINAL NEUROBIOLOGY & PROCESSING, ELECTRORETINOGRAM. *Educ:* Vassar Col, BA, 69; Univ Pittsburgh, MS, 75, PhD(psychobiol), 79. *Prof Exp:* Fel, biomed eng, Northwestern Univ, 79-83; asst res physiol, Univ Calif, San Francisco, 83-90, adj asst prof ophthal, 87-90; ASSOC PROF OPTOM & PHYSIOL, COL OPTOM, UNIV HOUSTON, 90- *Concurrent Pos:* Prin investr, 86- *Honors & Awards:* Basic Res First Prize, Italian Opthamol Soc, 90. *Mem:* Asn Res Vision & Opthal; Neurosci Soc; Int Soc Clin Electrophysiol; Sigma Xi; AAAS. *Res:* Neurophysiology and nueropharmacology of retinal processing; retinal neurocircuitry; intraretinal origins of the clinically relevant electroretinogram. *Mailing Add:* Col Optom Univ Houston 4901 Calhoun Rd Houston TX 77204-6052

FRISHMAN, WILLIAM HOWARD, b Bronx, NY, Nov 9, 46; m 71; c 3. CARDIOLOGY & CRITICAL CARE MEDICINE, PHARMACOLOGY. *Educ:* Boston Univ, BA, 69, MD, 69. *Prof Exp:* Intern med, Montefiore Hosp, 69-70, resident, 70-71; resident, Bronx Munic Hosp, 71-72; fel cardiol, Cornell-NY Hosp, 72-74; instr med, Cornell Med Sch, 74-76; from asst prof to assoc prof, 76-85, PROF MED, ALBERT EINSTEINOL COL MED, 85-, PROF EPIDEMIOL & SOCIAL MED, 88- *Concurrent Pos:* Chief cardiol, US Walson Army Hosp, 74-76; prin investr, Bronx Aging Study, NIH & Bronx Longitudinal Aging Study, 80; dir med, Hosp Albert Einstein Col Med, 82-; consult, Nat Conf Standards & Guidelines Cardiopulminary Resuscitation & Emer Cardiac Care, 85-; Dir med, Hosp Albert Einstein Col Med, 82- *Mem:* Am Col Physicians; Am Col Cardiol; Am Heart Asn; Am Col Chest Physicians; Am Col Clin Pharmacol; Am Fedn Clin Res. *Res:* Investigation of new cardiovascular drugs; cardiovascular epidemiology; medical education research; author and editor of books on cardiovascular pharmacology. *Mailing Add:* Albert Einstein Med Col Albert Einstein Med Col Hosp 1825 Eastchester Rd Bronx NY 10461

FRISILLO, ALBERT LAWRENCE, b Rome, NY, July 24, 43; m 68; c 3. GEOPHYSICS. *Educ:* Utica Col, BS, 65; Univ Dayton, MS, 67; Pa State Univ, PhD(geophys), 72. *Prof Exp:* Nat Res Coun res assoc, Johnson Spacecraft Ctr, Houston, 72-74; RES SCIENTIST, AMOCO PROD CO RES CTR, 74- *Mem:* Am Geophys Union; Sigma Xi. *Res:* Physical properties of geologic materials as related to the development or improvement of geophysical exploration techniques. *Mailing Add:* 7822 Park Ave Broken Arrow OK 74012

FRISINGER, H HOWARD, II, b Ann Arbor, Mich, Feb 28, 33; m 54; c 4. MATHEMATICS, METEOROLOGY. *Educ:* Univ Mich, Ann Arbor, BS & BBA, 56, MS, 61, DEduc(math), 64. *Prof Exp:* assoc prof, 64-76, PROF MATH, COLO STATE UNIV, 76- *Concurrent Pos:* Fac improv comt grant, 64-65. *Mem:* Hist Sci Soc; Am Meteorol Soc; Math Asn Am. *Res:* History of science, especially contributions of mathematics and mathematicians to the development of the science of meteorology. *Mailing Add:* Dept Math Colo State Univ Ft Collins CO 80523

FRISK, GEORGE VLADIMIR, b Schenectady, NY, Apr 3, 46; m 74; c 2. OCEAN ACOUSTICS. *Educ:* Univ Rochester, BA, 67; Brown Univ, ScM, 69; Cath Univ Am, PhD(physics), 75. *Prof Exp:* Physicist acoust, Naval Res Lab, 68-77; asst scientist, 77-81, ASSOC SCIENTIST OCEAN ENG, WOODS HOLE OCEANOG INST, 81- *Mem:* Acoust Soc Am; Sigma Xi. *Res:* Acoustic propagation, reflection, and scattering in the ocean and sea floor; acoustic surface waves; scattering theory of waves; computational physics. *Mailing Add:* Bigelow 208 Woods Hole Oceanog Inst Woods Hole MA 02543

FRISKEN, WILLIAM ROSS, b Hamilton, Ont, May 29, 33; m 56; c 3. PARTICLE PHYSICS. *Educ:* Queen's Univ, Ont, BSc, 56, MSc, 57; Univ Birmingham, PhD(physics), 60. *Prof Exp:* Teaching fel physics, McGill Univ, 60-62, asst prof, 62-64; assoc physicist, Exp Planning Div, Accelerator Dept, Brookhaven Nat Lab, NY, 64-66; assoc prof physics, Case Western Reserve Univ, 66-71; assoc prof, 71-74, PROF PHYSICS, YORK UNIV, 74- *Mem:* Am Phys Soc; Inst Particle Physics; Can Asn Physicists. *Res:* High energy particle physics experiments, particularly with electron-positron, electron-proton, and proton-proton colliders; design of experimental facilities for new accelerators, storage rings. *Mailing Add:* Dept Physics York Univ 4700 Keele St Toronto ON M3J 1P3 Can

FRISSEL, HARRY FREDERICK, b Alphen aan de Rhine, Netherlands, Mar 29, 20; US citizen; m 43; c 3. NUCLEAR PHYSICS. *Educ:* Hope Col BA, 42; Iowa State Univ, MS, 43, PhD(physics), 54. *Prof Exp:* Res physicist, Cornell Aeronaut Lab, 43-48; chmn dept, 63-75, PROF PHYSICS, HOPE COL, 48- *Concurrent Pos:* Mem, Denver Conf Undergrad Physics, 61- *Mem:* Am Asn Physics Teachers; Sigma Xi; Am Sci Affiliation. *Res:* Area of Walsh-Hadamard transforms and sinc and cosinc transforms. *Mailing Add:* Dept Physics Hope Col Holland MI 49423

FRISTEDT, BERT, b Minnesota, April 8,37. MATHEMATICS. *Educ:* Univ Minn, BA, 59; Mass Inst Technol, PhD(math), 63. *Prof Exp:* PROF MATH, UNIV MINN, 63- *Mailing Add:* Math Dept Univ Minn Sch Math Minneapolis MN 55455

FRISTROM, DIANNE, b Sydney, Australia, Dec 18, 40; m 66; c 2. ELECTRON MICROSCOPY, IMMUNOCYTOCHEMISTRY. *Educ:* Univ New S Wales, BSc, 62, MSc, 65; Univ Sydney, dipl educ, 63; Univ Calif, Berkeley, PhD(genetics), 69. *Prof Exp:* Asst res geneticist, 76-83, ASSOC RES GENETICIST, DEPT GENETICS, UNIV CALIF, BERKELEY, 83- *Concurrent Pos:* Fulbright fel, 65; NIH fel, 69. *Mem:* Genetics Soc Am; Soc Cell Biol. *Res:* Cellular and molecular basis of morphogenesis during insect metamorphosis; processes of cell rearrangement and cell shape changes; synthesis and extracellular assembly of cuticle by insect cells. *Mailing Add:* Dept Genetics Univ Calif 345 Mulford Hall Berkeley CA 94720

FRISTROM, JAMES W, b Chicago, Ill, July 7, 36; m 66. GENETICS. *Educ:* Reed Col, BA, 59; Rockefeller Univ, PhD(genetics), 64. *Prof Exp:* NSF fel, Biol Div, Calif Inst Technol, 64-65; from asst prof to assoc prof, 65-75, PROF GENETICS, UNIV CALIF, BERKELEY, 75- *Concurrent Pos:* Vis lectr, Univ Sydney, 69; UNESCO expert insect genetics, Biol Res Ctr, Hungarian Acad Sci, 74; Guggenheim fel, 76-77; vis researcher, Genetics Res Labs, CSIRO, Sydney, 76-77. *Res:* Developmental genetics; chemical analysis of morphological mutants and biochemistry of imaginal discs in Drosophila Melanogaster. *Mailing Add:* Dept Genetics Univ Calif Berkeley CA 94720

FRISTROM, ROBERT MAURICE, b Portland, Ore, May 26, 22; m 57; c 1. CHEMISTRY, PHYSICS. *Educ:* Reed Col, AB, 43; Univ Ore, AM, 45; Stanford Univ, PhD(chem), 48. *Prof Exp:* Res fel chem, Harvard Univ, 48-51; trustee prod res comt, 74-79, PRIN MEM STAFF CHEM & PHYSICS, APPL PHYSICS LAB, JOHNS HOPKINS UNIV, 51- *Concurrent Pos:* Parson's fel chem eng, Johns Hopkins Univ, 59-60, lectr, 60-61; ed, Fire Res Abstr & Rev, 65 & 78; mem comt fire res, Nat Acad Sci, 70; prin investr fire probs, NSF grant, 70; vis prof, Inst Phys Chem, Univ Gottingen, 73-74, vis prof mech eng, Stanford Univ, 77, Springer vis prof mech eng, Univ Calif, Berkeley, 80 & vis prof chem eng, Univ Calif, Los Angeles, 85. *Honors & Awards:* Humboldt Prize, Humboldt Found WGer, 73; Silver Combustion Medal, Combustion Inst, 66. *Mem:* Am Chem Soc; Am Phys Soc; Combustion Inst. *Res:* Microwave spectroscopy; combustion; molecular beams; chemical kinetics; fire research. *Mailing Add:* Appl Physics Lab Johns Hopkins Univ Laurel MD 20723-6099

FRITCHLE, FRANK PAUL, b Chicago, Ill, Aug 27, 22; m 49; c 6. PHYSICS. *Educ:* De Paul Univ, BS, 49, MS, 50; Univ Santa Clara, MBA, 65. *Prof Exp:* Physicist instrument design, Cent Sci Co, 50-53; asst to dir res, 54; group leader prod & equip design, Helipot Div, 55-58, chief equip design engr, 58-59, mgr prod develop, Spinco Div, 59-60, MGR ENG, SPINCO DIV, BECKMAN INSTRUMENTS, INC, 60- *Mem:* AAAS; Am Phys Soc; Inst Elec & Electronics Engrs. *Res:* Design and development of scientific instruments; electronics; precision mechanics; optics; servo systems; precision component design including potentiometers. *Mailing Add:* 1163 Pomegranate Ct Sunnyvale CA 94087

FRITCHMAN, HARRY KIER, II, b Portland, Ore, Sept 11, 23; m 55; c 3. INVERTEBRATE ZOOLOGY. *Educ:* Univ Calif, BA, 48, MA, 51, PhD, 53. *Prof Exp:* Actg instr biol, Univ Calif, 53; actg instr zool, Univ Wash, 54; instr, 54-67, chmn dept biol, 68-72, PROF ZOOL, BOISE STATE UNIV, 67- *Concurrent Pos:* Consult, Am Inst Biol Sci Film Series, 60; lectr, NSF Marine Inst, Ore, 62-64. *Res:* Natural history of marine and fresh water invertebrates; aquaculture of Gammarus. *Mailing Add:* 1131 Michigan Ave Boise ID 83706

FRITSCH, ARNOLD RUDOLPH, b Passaic, NJ, Mar 28, 32; m 53; c 4. NUCLEAR CHEMISTRY. *Educ:* Univ Rochester, BS, 53; Univ Calif, PhD(chem), 57. *Prof Exp:* Sr engr, Westinghouse Elec Corp, 56-59; br chief, Div Int Affairs, US Atomic Energy Comn, 59-61, spec asst to chmn, 61-68; mgr tech eval dept, Gulf Gen Atomic Inc, Calif, 68-71, pres, Gulf United Nuclear Fuels, Inc, 71-74, dir nuclear coord, Gulf Oil Corp, 74-84; CONSULT, 84- *Honors & Awards:* Arthur S Flemming Award, 67. *Mem:* Am Phys Soc; Am Chem Soc; Nuclear Soc; Am Inst Chemists. *Res:* Nuclear fuel; reactor technology; nuclear decontamination and decommissioning. *Mailing Add:* 707 Amberson Ave Pittsburgh PA 15232

FRITSCH, CARL WALTER, b New York, NY, July 21, 28; m 54; c 2. LIPIDS. *Educ:* Blackburn Col, AB, 52; Ohio State Univ, PhD(agr biochem), 55. *Prof Exp:* Sect leader, Dept Food Processing, 55-67, res assoc, Cereal Develop Dept, 67-76, RES ASSOC CORP RES, JFB TECH CTR, GEN MILLS, INC, 76- *Mem:* Am Oil Chemists' Soc. *Res:* Fats and oils, antioxidants; food deterioration; nutrition. *Mailing Add:* Gen Mills Inc JFB Tech Ctr 9000 Plymouth Ave N Minneapolis MN 55427

FRITSCH, CHARLES A(NTHONY), b Maysville, Ky, Mar 9, 36; m 63; c 5. MECHANICAL ENGINEERING. *Educ:* Univ Dayton, BME, 58; Purdue Univ, MS, 60, PhD(mech eng), 62. *Prof Exp:* Mem tech staff, 61-65, SUPVR, BELL LABS, INC, 65- *Mem:* Am Soc Mech Engrs. *Res:* Thermodynamics and heat transfer; computer aided design; computer integrated manufactoring. *Mailing Add:* 6249 Sleepy Hollow Dr New Albany OH 43054

FRITSCH, EDWARD FRANCIS, b Pittsburgh, Pa, June 1, 50; m 71; c 2. MOLECULAR CLONING, GENE EXPRESSION. *Educ:* Mass Inst Technol, BS, 72; Univ Wis, PhD(molecular biol), 77. *Prof Exp:* Fel, Univ Southern Calif, 77-78, Calif Inst Technol, 78-80; asst prof biochem, Mich State Univ, 80-82; SR SCIENTIST, GENETICS INST, 82- *Concurrent Pos:* Instr, Cold Spring Harbor Lab, 80-83; consult, Genetics Inst, 81-82. *Res:* Molecular biology of gene expression; isolation and characterization of eukaryotic genes. *Mailing Add:* Genetics Inst 87 Cambridge Pk Dr Cambridge MA 02140

FRITSCH, KLAUS, b Mannheim, WGer, Sept 26, 41; m 65; c 2. ELECTRONICS, ACOUSTICS. *Educ:* Georgetown Univ, BS, 61; Mass Inst Technol, MS, 65; Cath Univ Am, PhD(physics), 68. *Prof Exp:* From asst prof to assoc prof, 67-72, PROF PHYSICS, JOHN CARROLL UNIV, 77-, CHMN PHYSICS, 89- *Concurrent Pos:* Guest prof, Univ Saarland, WGer, 79-80. *Mem:* Inst Elec & Electronic Engrs; Am Phys Soc. *Res:* Fiber optic sensors; applications of linear and digital integrated circuits. *Mailing Add:* John Carroll Univ Cleveland OH 44118

FRITSCHE, HERBERT AHART, JR, b Houston, Tex, Nov 30, 41; m 69; c 2. CLINICAL CHEMISTRY, ANALYTICAL CHEMISTRY. *Educ:* Univ Houston, BS, 63; Tex A&M Univ, MS, 65, PhD(chem), 68. *Prof Exp:* Asst chief chem div, First US Army Med Lab, Ft Meade, Md, 67-69; CHIEF CLIN CHEM, M D ANDERSON HOSP & TUMOR INST, 69- *Concurrent Pos:* Assoc biochemist & assoc prof biochem, Univ Tex Syst Cancer Ctr, M D Anderson Hosp, 69-; fac mem, Grad Sch, Univ Tex, Houston, 73- *Mem:* Am Asn Clin Chemists. *Res:* Electroanalytical techniques; polarography; immunochemistry; biochemical markers of cancer; automation; computers in the clinical laboratory; alkaline and acid phosphatases; serum isoenzymes; immunoglobulins; carcinoembryonic proteins. *Mailing Add:* Lab Med Rm C3013 M D Anderson Hosp Houston TX 77030

FRITSCHE, RICHARD T, b Dallas, Tex, Jan 29, 36; m 59; c 2. MATHEMATICS. *Educ:* St Louis Univ, BS, 57, MS, 59; Univ Ariz, PhD(math), 67. *Prof Exp:* Instr math, Univ Ariz, 60-61; from instr to asst prof, Univ Dallas, 61-68; ASSOC PROF MATH, NORTHEAST LA UNIV, 68- *Mem:* Am Math Soc; Math Asn Am. *Res:* Topological structures for probabilistic metric spaces. *Mailing Add:* Dept Math Northeast La Univ Monroe LA 71209

FRITSCHEN, LEO J, b Salina, Kans, Sept 14, 30; m 53; c 7. METEOROLOGY. *Educ:* Kans State Col, BS, 52; Kans State Univ, MS, 58; Iowa State Univ, PhD(agr climat), 60. *Prof Exp:* Res meteorologist, US Water Conserv Lab, USDA, 60-66; assoc prof, 66-73, chmn div biol sci, 73-79, PROF FOREST METEOROL, COL FORESTRY, UNIV WASH, 73-, ADJ PROF ATMOSPHERIC SCI, 73- *Mem:* Am Meteorol Soc; Am Soc Agron; Am Soil Sci Soc. *Res:* Agricultural climatology; micrometeorological research and instrumentation. *Mailing Add:* Dept Forest Univ Wash Seattle WA 98195

FRITSCHEN, ROBERT DAVID, b Mitchell, SDak, Nov 27, 35; m 59; c 2. ANIMAL SCIENCE, BIOLOGY. *Educ:* SDak State Univ, BS, 61, MS, 63. *Prof Exp:* County exten agent agr, 63-65, from asst prof to assoc prof, 65-76, prof animal sci, 76-84, DIR, PANHANDLE RES & EXTENSION CTR, UNIV NEBR, 84- *Honors & Awards:* Sci Extension Award, Am Soc Animals. *Mem:* Am Regist Cert Animal Scientists; Am Soc Animal Sci; Am Mgt Asn. *Res:* Design and management of swine facilities to spare energy; study of factors influencing swine behavior; effect of floor materials on foot and leg lesions. *Mailing Add:* Box 466 Aurora NE 68818

FRITSCHY, J(OHN) MELVIN, b Tucson, Ariz, Aug 23, 21; m 50; c 2. METALLURGICAL ENGINEERING. *Educ:* Univ Ariz, BS, 44. *Prof Exp:* Asst test engr, Phelps Dodge Corp, 44-45, test engr, 45-46; jr chem engr flotation res, Potash Co Am, 46-51, chem engr, 51-56, plant res & develop engr, 56-69, process engr, 69-82. *Concurrent Pos:* Self employed, 85- *Mem:* Am Inst Mining, Metall & Petrol Engrs. *Res:* Mineral beneficiation, especially of potash ores by flotation process. *Mailing Add:* 1118 Thomas Carlsbad NM 88220

FRITTON, DANIEL DALE, b Cheyenne Wells, Colo, Oct 1, 42; m 65; c 7. SOIL PHYSICS. *Educ:* Colo State Univ, BS, 64; Iowa State Univ, MS, 66, PhD(soil physics, agr climat), 68. *Prof Exp:* Res assoc soil physics, Iowa State Univ, 66-68; asst prof, Cornell Univ, 68-71; from asst prof to assoc prof, 71-81, PROF SOIL PHYSICS, PA STATE UNIV, 81- *Concurrent Pos:* Interim head, Dept Agron, Pa State Univ, 85. *Mem:* Soil Sci Soc Am; Am Soc Agron. *Res:* Soil compaction; physical aspects of waste disposal in soil. *Mailing Add:* Dept Agron Pa State Univ University Park PA 16802

FRITTS, HAROLD CLARK, b Rochester, NY, Dec 17, 28; m 82; c 2. PLANT ECOLOGY, DENDROCHRONOLOGY. *Educ:* Oberlin Col, AB, 51; Ohio State Univ, MSc, 53, PhD(bot), 56. *Prof Exp:* Asst prof bot, Eastern Ill Univ, 56-60; from asst prof to assoc prof, 60-69, PROF DENDROCHRONOL, LAB TREE RING RES, UNIV ARIZ, 69- *Concurrent Pos:* Guggenheim fel, 68-69; mem panel climatic variation, US Comt Global Atmospheric Res Prog, Nat Acad Sci, 72-74; organizer & dir, Int Tree-Ring Data Bank, 75-, quaternary res, ed adv bd, 77-82; mem staff, Nat Defense Univ, 78; mem adv comt on paleoclimat, Climate Dynamics Prog, NSF, 78-81; mem organizing group, Int Conf Dendroclimat, Eng, 78-80; fac, North Atlantic Treaty Orgn Advan Study Inst, Italy, 80; coun rep & chmn paleoecol sect, Ecol Soc Am, 84 & Dendrochronological Modeling, 90. *Honors & Awards:* Outstanding Bioclimat Achievement Award, Am Meterol Soc, 82; Dendrochronological Award of Appreciation, 90. *Mem:* AAAS; Ecol Soc Am; Tree Ring Soc; Am Meterol Soc; Am Quaternary Asn; Am Inst Biol Sci; Int Asn Ecol; Int Soc Ecol Modeling. *Res:* Tree growth and forest tree physiology; climatology; microenvironment; dendrochronology and ecology. *Mailing Add:* Lab Tree Ring Res Univ Ariz Tucson AZ 85721

FRITTS, HARRY WASHINGTON, JR, b Rockwood, Tenn, Oct 4, 21; m 49; c 3. MEDICINE. *Educ:* Mass Inst Technol, BS, 43; Boston Univ, MD, 51; Am Bd Internal Med, dipl. *Prof Exp:* Mem res staff, Mass Inst Technol, 46-47; instr elec eng, Northeastern Univ, 47-51; assoc, Col Physicians & Surgeons, Columbia Univ, 56-57, from asst prof to Dickinson W Richards prof med, 57-73; prof med & chmn dept, sch med, 73-87, EDMUND D PELLEGRINO PROF MED, STATE UNIV NY STONY BROOK, 87-88. *Concurrent Pos:* Vis physician & consult, Manhattan Vet Admin Hosp, 57-68; vis physician, Bellevue Hosp, New York, 57-68 & Presby Hosp, 61-73; Guggenheim fel, 59; mem, Bd Dirs & Adv Coun Res, NY Heart Asn; mem, Physiol Study Sect & Cardiovasc Training Comt, USPHS; assoc ed, J Clin Invest; Am Bd of Internal Med, 76-79; coun, Nat Heart, Lung & Blood Inst, 76-80; vis prof, Univ London, 82; William Harris vis prof, Nat Med Sch, Taiwan, 87-88. *Mem:* Fel Am Col Physicians; Am Physiol Soc; Am Soc Clin Invest; Asn Am Physicians; Am Clin & Climat Soc. *Res:* Diseases of the heart and lungs. *Mailing Add:* Dept Med Sch Med State Univ NY Stony Brook Stony Brook NY 11794

FRITTS, ROBERT WASHBURN, b Rochester, NY, Oct 26, 24; m 47; c 3. PHYSICS. *Educ:* Oberlin Col, AB, 47; Northwestern Univ, MS, 48, PhD(physics), 50. *Prof Exp:* Asst dir res, Milwaukee Gas Specialty Co, Wis, 50-57; supvr thermoelectricity mat group, Minn Mining & Mfg Co, 57-59, mgr thermoelectricity proj, 59-66, sr res specialist energy conversion, 66-68; tech develop mgr, Nat Advert Co, 68-75; mgr, archit murals & com display proj, Graphic Technol Sector Div, 3M Co, 75-83; OWNER, ST CROIX GRAPHIC ASSOCS, STILLWATER, MN, PORTABLE DISPLAY & GRAPHICS PROD & SALES, CONSULT, LARGE FORMAT GRAPHICS, 83- *Mem:* Am Phys Soc. *Res:* Semiconductor research on intermetallic compounds; thermoelectric effects and transport phenomena; thermoelectric generator devices and thermoelectric power supply systems; traffic control devices and systems; optical-electronic color imaging systems; uses for large format graphics; technical management. *Mailing Add:* 1575 N 2 St Stillwater MN 55082

FRITTS, STEVEN HUGH, b Fayetteville, Ark, July 14, 48; m 69; c 3. ECOLOGY OF WOLVES. *Educ:* Univ Ark, Ba, 70, MS, 72; Univ Minn, PhD(ecol), 79. *Prof Exp:* Wildlife res biologist, 79-84, LEADER ECOL SECT, ENDANGERED SPECIES RES BR, PATUXENT WILDLIFE RES CTR, US FISH & WILDLIFE SERV, 84- *Mem:* Sigma Xi; Am Soc Mammalogists; Wildlife Soc. *Res:* Distribution and numbers of wolves (Canis lupus) in north-central and northwestern Minnesota; nature and extent of wolf depredations on domestic animals in Minnesota. *Mailing Add:* Fish & Wildlife Enhancement Fed Bldg US Courthouse 301 S Park Box 10023 Helena MT 59626

FRITTS-WILLIAMS, MARY LOUISE MONICA, b Detroit, Mich, Apr 16, 40; m 68; c 1. ANATOMY, EMBRYOLOGY. *Educ:* Univ Detroit, BS, 62, MS, 64; Wayne State Univ, PhD(embryol, anat), 71. *Prof Exp:* Instr microbiol & embryol, Mercy Col, Mich, 64-67, asst prof, 67-68; asst prof microbiol, Univ Detroit, 68-69; ASST PROF ANAT, MED SCH, WAYNE STATE UNIV, 71-, ASST PROF MORTUARY SCI, 76-, CHMN MORTUARY SCI. *Concurrent Pos:* Dir, Pathologists' Assts Prog. *Mem:* AAAS; Genetics Soc Am; Am Soc Microbiol; Soc Develop Biol. *Res:* Primordial germ cells; gonadal development; tissue culture. *Mailing Add:* Dept Anat Wayne State Univ Detroit MI 48202

FRITZ, GEORGE JOHN, b New York, NY, Feb 19, 27; m 57. PLANT PHYSIOLOGY. *Educ:* Mont State Univ, BS, 49; Univ Calif, MS, 51; Purdue Univ, PhD(bot), 54. *Prof Exp:* Fel plant physiol, Duke Univ, 54-55; asst prof bot, Pa State Univ, 55-60; ASSOC PROF BOT, UNIV FLA, 60- *Concurrent Pos:* AEC res grant, 57, 61-68; Sigma Xi grant, 58; NSF grant, 62; ed-in-chief, What's New in Plant Physiol, 69- *Mem:* AAAS; Am Soc Plant Physiol; Scand Soc Plant Physiol. *Res:* Oxygen fixation by plants. *Mailing Add:* 202 NW 20th Terr Gainesville FL 32603

FRITZ, GEORGE RICHARD, JR, b Detroit, Mich, Mar 28, 32; m 60; c 3. PHYSIOLOGY, ENDOCRINOLOGY. *Educ:* Mich State Univ, BS, 54, MS, 58; Univ Ill, PhD(dairy sci), 63. *Prof Exp:* Instr, Sch Med, Univ Pittsburgh, 65-66, res assoc, 66-67, res asst prof physiol, 67-83, from asst dir to assoc dir, Primate Res Lab, 67-83; gen mgr, Federated Med Resources, 87-88; CONSULT, 88- *Concurrent Pos:* NIH fel physiol, Sch Med, Univ Pittsburgh, 62-65. *Honors & Awards:* Pac Coast Fertility Soc Award, 71; Donna Paich Award, Three Rivers Br Am Asn Lab Animal Sci, 86. *Mem:* Am Soc Animal Sci; Brit Soc Study Fertil; Am Asn Lab Animal Sci; Soc Study Reprod. *Res:* Physiology of reproduction, especially regulation of gonadal function and control of parturition. *Mailing Add:* 4294 Skyvue Dr Murrysville PA 15668

FRITZ, GEORGIA T(HOMAS), b Allentown, Pa, Mar 7, 33; m 57; c 3. WASTE MANAGEMENT, LIQUID CHROMATOGRAPHY. *Educ:* Ursinus Col, BS, 55; Univ NMex, MS, 77. *Prof Exp:* Asst chem, Cornell Univ, 55-61; res asst anal chem, Molecular Biol Group, Los Alamos Nat Lab, 62-69, staff mem, 69-70, 72-75, staff mem anal chem, Explosives Technol Group, 75-87, staff mem hazardous waste mgt, 88-90, STAFF MEM, ANALYTICAL CHEM GROUP, LOS ALAMOS NAT LAB, 91- *Concurrent Pos:* Founder & bd mem, Los Alamos Women Sci, 79-88; bd mem, NMex Network Women Sci & Eng, 79-81, 86-; chair, task group, Am Soc Testing & Mat, 79-; treas, Cen NMex sect Am Chem Soc, 90- *Honors & Awards:* Award of Merit, Am Soc Testing & Mat, 91. *Mem:* Am Chem Soc; Am Soc Testing & Mat; AAAS. *Res:* General analytical chemistry; chromatographic and thermal analysis of organic high explosives; thermal analysis of oil shales; hazardous waste management; writer and editor on standard operating procedures. *Mailing Add:* Los Alamos Nat Lab PO Box 1663 MD G740 Los Alamos NM 87545

FRITZ, GILBERT GEIGER, b Washington, DC, Apr 16, 42; m 71. X-RAY ASTRONOMY, X-RAY INSTRUMENTATION. *Educ:* Johns Hopkins Univ, BA, 64, MA, 65. *Prof Exp:* Physicist, 65-72, SUPVRY ASTROPHYSICIST, US NAVAL RES LAB, 72- *Mem:* AAAS; Sigma Xi. *Res:* X-ray pulsars; soft x-ray background; supernova remnants; x-ray binaries; spectroscopy of x-ray sources; proportional detectors; scintillators; x-ray collimators. *Mailing Add:* Naval Res Lab Code 4120 Washington DC 20375

FRITZ, HERBERT IRA, b New York, NY, May 31, 35; m 62; c 2. NUTRITION, BIOCHEMISTRY. *Educ:* Univ Calif, Davis, AB, 58, PhD(nutrit), 64. *Prof Exp:* Res asst nutrit, Univ Calif, Davis, 58-64; NIH fel, Univ Pa, 64-66; from asst prof to assoc prof biol sci, 66-74, ASSOC PROF BIOL CHEM, SCH MED, WRIGHT STATE UNIV, 74- *Mem:* AAAS; Teratol Soc; Am Soc Cell Biol; Soc Study Reprod; Sigma Xi. *Res:* Embryo nutrition; effect of vitamins on development; reproduction of American marsupials; comparative and experimental teratology; fetal alcohol syndrome and hyperactivity. *Mailing Add:* Biol Chem Dept Wright State Univ Dayton OH 45431

FRITZ, IRVING BAMDAS, b Rocky Mount, NC, Feb 11, 27; m 50, 72; c 5. REPRODUCTIVE BIOLOGY, BIOCHEMISTRY. *Educ:* Med Col Va, DDS, 48; Univ Chicago, PhD(physiol), 51. *Prof Exp:* Asst dir dept metab & endocrine res, Med Res Inst, Michael Reese Hosp, 55-56; from asst prof to prof physiol, Univ Mich, Ann Arbor, 56-68; chmn dept, 68-78, PROF MED RES, BANTING & BEST DEPT MED RES, BEST INST, UNIV TORONTO, 68-, DISTINGUISHED PROF, 84- *Concurrent Pos:* USPHS fel, Inst Med Physiol, Copenhagen, 53-55; Guggenheim fel, 78-79; vis res scientist, Int Animal Physiol & Genetics Res, Cambridge Res Stat, Cambridge, UK, 91- *Honors & Awards:* Gairdner Award, 80. *Mem:* Am Physiol Soc; Am Biochem Soc; Am Soc Cell Biologists; Endocrine Soc; Soc Study Reprod. *Res:* Spermatogenesis; developmental biology; carnitine; cell-cell interactions. *Mailing Add:* Banting & Best Dept Med Univ Toronto Best Inst 112 Col St Toronto ON M5G 1L6 Can

FRITZ, JAMES JOHN, b Sunbury, Pa, Sept 21, 20; m 43; c 4. PHYSICAL CHEMISTRY. *Educ:* Pa State Univ, BS, 39; Univ Chicago, MS, 40; Univ Calif, PhD(chem), 48. *Prof Exp:* Asst chem, Univ Calif, 43-48; res assoc phys chem, Ohio State Univ, 48-49; from asst prof to prof, 49-85, EMER PROF CHEM, PA STATE UNIV, 85- *Concurrent Pos:* Guggenheim fel, Oxford Univ, 61-62. *Mem:* Am Chem Soc; Sigma Xi. *Res:* Low temperature phenomena; solutions of electrolytes. *Mailing Add:* 624 S Ditmar St Oceanside CA 92054

FRITZ, JAMES SHERWOOD, b Decatur, Ill, July 20, 24; m 49, 89; c 4. ANALYTICAL CHEMISTRY. *Educ:* James Millikin Univ, BS, 46; Univ Ill, MS, 46, PhD(chem), 48. *Prof Exp:* Asst prof chem, Wayne State Univ, 48-51; from asst prof to assoc prof, 51-60, PROF CHEM, IOWA STATE UNIV, 60- *Honors & Awards:* Chromatography Award, Am Chem Soc, 76, Anal Chem Award, 85; First Int Chromotography Forum Award, 88; Dal Nogare Award, 91. *Res:* Gas, liquid and ion-exchange chromatography; solid-phase extraction, chelating and ion-exchange resins. *Mailing Add:* Dept Chem Iowa State Univ Ames IA 50011

FRITZ, JOSEPH N, b Klein, Mont, Dec 27, 31; m 57; c 3. SOLID STATE PHYSICS. *Educ:* Mont State Col, BS, 53; Cornell Univ, PhD(physics), 61. *Prof Exp:* STAFF MEM, LOS ALAMOS SCI LAB, 61- *Mem:* Am Phys Soc. *Res:* Theoretical solid state physics; solid state physics at high and very high pressures. *Mailing Add:* Los Alamos Nat Lab M-6 J970 Los Alamos NM 87545

FRITZ, K(ENNETH) E(ARL), b Monroeville, Ohio, Oct 17, 18; m 40; c 2. MATERIALS ENGINEERING. *Educ:* Ohio State Univ, BS, 42; Univ Wis, MS, 51. *Prof Exp:* Metall engr, Naval Res Lab, 42-45 & Bucyrus-Erie Co, 45-55; metall engr locomotive & car equip dept, 55-56, mat & process lab, 56-59 & gas turbine dept, 69-73, MGR STRUCT MAT APPLICATIONS, GAS TURBINE DIV, GEN ELEC CO, 73- *Mem:* Am Soc Metals. *Res:* Ultrahigh strength materials; statistical evaluation of data; precipitation hardening process; high temperature properties of materials. *Mailing Add:* RR 1 No 414A Pattersonville NY 12137

FRITZ, KATHERINE ELIZABETH, b Omaha, Nebr, June 24, 18; m 40; c 2. PATHOLOGY, IMMUNOLOGY. *Educ:* Univ Omaha, BA, 39; Albany Med Col, MS, 61, PhD(path), 66. *Prof Exp:* Res assoc, 61-63, from instr to asst prof, 66-74, assoc prof path, 74-81, RES PROF PATH, ALBANY MED COL, 81- *Concurrent Pos:* Res biologist, Vet Admin Hosp, Albany, NY, 66-; fel, Coun Arteriosclerosis, Am Heart Asn, 71. *Mem:* Am Soc Cell Biol; Electron Micros Soc Am; Am Soc Exp Path. *Res:* Mechanisms influencing the regression of atherosclerotic lesions with emphasis on changes in hydrolytic enzymatic activities and/or phagocytic activities of lesion cells as studied biochemically and histochemically. *Mailing Add:* Vet Admin Hosp Albany NY 12208

FRITZ, LAWRENCE WILLIAM, b Oceanside, NY, Sept 1, 37; m 59; c 3. PHOTOGRAMMETRY, GEODESY. *Educ:* Lafayette Col, AB, 59; Ohio State Univ, MS, 67. *Prof Exp:* Cartogr, US Coast & Geod Surv, 61-67; res phys scientist photogram, Nat Oceanic & Atmospheric Admin, 67-80; SR POLICY ANALYSIS SPACE POLICY, OFF SCI & TECHNOL POLICY, EXEC OFF PRES, 80- *Concurrent Pos:* Eng reports ed, Photogram Eng & Remote Sensing, 69-73; vis lectr, Va Polytech Inst & State Univ, 81- *Mem:* Am Soc Photogram; AAAS; Int Soc Photogram & Remote Sensing. *Res:* Photogrammetric system for the densification of geodetic networks; precision calibration techniques for mapping cameras and photogrammetric comparators used in the sciences of geodesy and cartography; author or coauthor of more than 30 technical publications. *Mailing Add:* 1104 Barlett Rd Wayne PA 19087

FRITZ, MADELEINE ALBERTA, paleontology; deceased, see previous edition for last biography

FRITZ, MARC ANTHONY, b Detroit, Mich, Aug 8, 51; m 73. OBSTETRICS & GYNECOLOGY. *Educ:* US Air Force Acad, BS, 73; Tulane Univ Sch, MD, 77. *Prof Exp:* Residency obstet & gynec, Sch Med, Wright State Univ, 77-81; INSTR OBSTET & GYNEC, ORE HEALTH SCI UNIV & FEL, ORE REGIONAL PRIMATE RES CTR, 81- *Mem:* Jr fel Am Col Obstet & Gynec; Am Fertil Soc. *Res:* Investigation of the endocrine vascular relationships of the fetoplacental unit. *Mailing Add:* SGHO-USAF Med Ctr Wright Patterson Wright Patterson AFB OH 45433

FRITZ, MICHAEL E, b Boston, Mass, Feb 26, 38; m 60; c 3. PERIODONTOLOGY, CELL PHYSIOLOGY. *Educ:* Univ Pa, DDS, 63, cert periodont & MS, 65, PhD(physiol), 67. *Prof Exp:* Instr periodont, Sch Dent, Univ Pa, 63-67; from assoc prof to prof, 67-80, chmn dept, 67-82, dean, 82-85, CHARLES HOWARD CANDLER PROF PERIODONT, SCH DENT, EMORY UNIV, 80. *Concurrent Pos:* Fel physiol, Sch Med, Univ Pa, 63-67; Nat Inst Dent Res grant, 68-; consult, US Vet Admin Hosp, Atlanta, Ga, 67-78 & Ft Benning, Ga, 70-78; Fogarty Int fel, 73-74; assoc ed, J Periodont, 89. *Mem:* AAAS; Int Asn Dent Res; Am Dent Asn; Am Acad Periodont; Fedn Am Socs Exp Biol; fel Am Col Dent; Am Acad Implant Dent; Am Col Dentists. *Res:* Physiology of salivary glands; bone pathology; virus infection of cells; clinical dentistry and dental implants. *Mailing Add:* Dept Periodont Emory Univ Sch Dent Atlanta GA 30322

FRITZ, PAUL JOHN, b Belleville, Ill, May 17, 29; m 56; c 4. BIOCHEMISTRY. *Educ:* Washington Univ, AB, 51; Auburn Univ, PhD(biochem), 62. *Prof Exp:* Instr chem, Auburn Univ, 60-62, from asst prof to assoc prof pharmacol, 64-66; asst prof biochem & sr investr, Molecular Biol Lab, Med Ctr, Univ Ala, Birmingham, 6669; ASSOC PROF PHARMACOL, MILTON S HERSHEY MED CTR, PA STATE UNIV, 69- *Concurrent Pos:* USPHS fel biochem, Biol Div, Oak Ridge Nat Lab, 62-64; vis prof pharmacol, Samford Univ, 68-69. *Mem:* AAAS; Am Chem Soc; Am Soc Pharmacol & Exp Therapeut; Am Soc Biol Chemists; Sigma Xi. *Res:* Enzymatic control of metabolism; mechanisms of enzyme action; physical and chemical properties of enzymes; factors controlling synthesis and degradation of enzymes. *Mailing Add:* Food Sci Dept 111 Borland Lab Pa State Univ University Park PA 16802

FRITZ, PETER, b Stuttgart, Ger, Mar 18, 37; m 66; c 2. GEOLOGY, GEOCHEMISTRY. *Educ:* Univ Stuttgart, dipl, 62, Dr rer nat, 65. *Prof Exp:* NATO res fel, Univ Paris, 65-66; fel, Univ Alta, 66-68, res assoc isotope geol, 68-71; prof geochem, Univ Waterloo, 71-85, chmn dept earth sci, 80-85; DIR, INST HYDROL, GSF, WGER, 87- *Mem:* Geochem Soc; Geol Asn Can; Int Asn Geochem & Cosmochem. *Res:* Geochemistry of stable isotopes in groundwater, tritium and carbon-14 dating; quarternary paleoenvironments; geochemistry in nuclear waste disposal studies; sulfur cycle in soils and ground water. *Mailing Add:* Inst Hydrol GSF Engolstaedterland St No 1 Neuerberg D8040 Germany

FRITZ, RICHARD BLAIR, b Washington, DC, Jan 23, 36; m 88; c 2. ATMOSPHERIC PHYSICS, IONOSPHERIC PHYSICS. *Educ:* Knox Col, AB, 58; Cornell Univ, MS, 61. *Prof Exp:* Physicist, Gen Elec Res Labs, NY, 58; res asst physics & astron, Cornell Univ, 58-61; PHYSICIST, BOULDER LABS, US DEPT COM, 64- *Concurrent Pos:* US Army, 62-64. *Mem:* Am Geophys Union; Air & Waste Mgt Asn; AAAS; Sigma Xi. *Res:* air pollution meteorology; application of optical propagation techniques for remote sensing of atmospheric conditions such as wind and rain; ionospheric monitoring by satellite radio beacon techniques. *Mailing Add:* NOAA-ERL-WPL 325 Broadway Boulder CO 80303-3328

FRITZ, ROBERT B, b Milwaukee, Wis, Nov 16, 37. MICROBIOLOGY. *Educ:* Bowdoin Col, AB, 59; Univ Maine, MS, 64; Duke Univ, PhD(immunol), 67. *Prof Exp:* Postdoctoral, Duke Univ, 67-69; from asst prof to assoc prof microbiol, Emory Univ, 69-89; PROF MICROBIOL, DEPT MICROBIOL, MED COL WIS, 89- *Mem:* AAAS; Am Asn Immunologists. *Mailing Add:* Dept Microbiol Med Col Wis 8701 Watertown Plank Rd Milwaukee WI 53226

FRITZ, ROBERT J(ACOB), b New Orleans, La, June 5, 23; m 47; c 4. MECHANICAL & NUCLEAR ENGINEERING. *Educ:* Tulane Univ, La, BEE, 44; Union Univ, NY, MS, 50, MME, 52; Rensselaer Polytech Inst, DEng(mech eng), 70. *Prof Exp:* With test prog, Knolls Atomic Power Lab, Gen Elec Co, 46-47, heat transfer engr, 47-52, mech analyst, 52-53, supvr mech anal, 53-55, sr specialist, 55, supvr eng eval, 55-57, consult eng heat transfer & mech, 57-61, supvr & mgr, Reactor Core Heat Transfer Sect, 61-63, consult engr, rotating equip & mech, 63-66, struct mech & dynamics, 66-87; RETIRED. *Mem:* Fel Am Soc Mech Engrs; Sigma Xi. *Res:* Flow vibrations and mechanics of nuclear reactors and power systems. *Mailing Add:* 2511 Whamer Lane Schenectady NY 12309

FRITZ, ROY FREDOLIN, b Oakland, Calif, Nov 15, 15; m 39; c 2. ENTOMOLOGY, EPIDEMIOLOGY. *Educ:* Kans State Col, BS, 37, MS, 39; Univ Calif, MPH, 48. *Prof Exp:* Asst entomologist, Agr Exp Sta, Kans State Col, 39-42; asst entomologist, Malaria Control in War Areas, USPHS, 42-43, dist entomologist, 43-44, asst chief entom div, 44-46, typhus control entomologist, 46-47, resident, Univ Calif, 47-49, resident rep commun dis ctr, Region X, 50, chief mosquito control & invest, Eng Br, 51-52, chief vector-borne dis unit, Epidemiol Br, 52-55, actg chief, Surveillance Sect, Ga, 54-55; malariologist adv, Int Coop Admin, Mex Govt, 55-57; malariologist & chief, Malaria Eradication Br, US AID, 58-63; secy, Report Expert Comt Malaria, 62, scientist, Prog & Planning, Malaria Eradication Div, 63-66, scientist, Anopheline Res, Vector Biol & Control, 66-72, MEM EXPERT PANEL MALARIA, WHO, 60- *Concurrent Pos:* Consult, US AID, 74-75. *Mem:* Sigma Xi; Am Soc Prof Biologists (pres, 48); Entom Soc Am; Am Mosquito Control Asn; fel Am Pub Health Asn. *Res:* Medical entomology; epidemiology of vector-borne diseases; ecology; biological control; environmental entomology. *Mailing Add:* 11112 Nocturne Ct Sun City AZ 85351

FRITZ, SIGMUND, b Brooklyn, NY, June 9, 14; m 46; c 2. METEOROLOGY. *Educ:* Brooklyn Col, BS, 34; Mass Inst Technol, MS, 41, ScD, 53. *Prof Exp:* Observer, US Weather Bur, 37-42, meteorologist, Washington, DC, 46-67; meteorologist, Nat Environ Satellite Serv, Nat Oceanic & Atmospheric Admin, 67-75; prof meteorol, Univ Md, 75-84, sr res assoc, 84-89; CHIEF SPACE SCIENTIST, MENTOR TECHNOLOGIES INC, 90- *Concurrent Pos:* Pres, Int Asn Meteorol & Atmospheric Physics, 71-75. *Honors & Awards:* Gold Medal, Dept Com, 65. *Mem:* Fel Am Meteorol Soc; fel Am Geophys Union. *Res:* Solar radiation; albedo of ground; albedo and absorption; atmospheric ozone measurements; meteorological satellites; climate. *Mailing Add:* Mentor Technologies Inc 12750 Twin Brook Pkwy Rockville MD 20852

FRITZ, THOMAS EDWARD, b Detroit, Mich, May 24, 33; m 66; c 3. PATHOLOGY, RADIOBIOLOGY. *Educ:* Mich State Univ, BS, 55, DVM, 57; Univ Ill, MS, 60. *Prof Exp:* Instr path, Col Vet Med, Univ Ill, 57-62; supvr, Diag Res Lab, Ill State Dept Agr & Univ Ill, 62-63; assoc pathologist, 63-72, asst div dir animal facil, 76-78, PATHOLOGIST, DIV BIOL & MED RES, ARGONNE NAT LAB, 72-, GROUP LEADER RADIATION TOXICITY, 77-, ASSOC DIV DIR, DIV BIOL & MED RES, 79- *Concurrent Pos:* Clin assoc prof path, Loyola Univ, 75-; assoc prof biol, Northern Ill Univ, 78- *Mem:* AAAS; Am Vet Med Asn; Int Soc Exp Hemat; Am Asn Lab Animal Sci; Radiation Res Soc. *Res:* Comparative pathology of diseases of domestic and laboratory animals; pathology of radiation and isotope toxicity; experimental hematology and leukemogenesis; biological effects of ionizing radiation. *Mailing Add:* Div Biol Environ & Med Res Bldg D-202 Argonne Nat Lab Argonne IL 60439-4833

FRITZ, WILLIAM HAROLD, b Cathlamet, Wash, Aug 24, 28; m 58; c 1. PALEONTOLOGY. *Educ:* Univ Wash, BS, 52, MS, 58, PhD(geol), 60. *Prof Exp:* NSF fel, 60-61; explor geologist, Shell Oil Co, 61-64; CAMBRIAN PALEONTOLOGIST, GEOL SURV CAN, 64- *Concurrent Pos:* Res scientist, NSF, 61- *Mem:* Am Paleont Soc; Brit Paleont Asn. *Res:* Structure and stratigraphy; Cambrian paleontology. *Mailing Add:* Geol Surv Can 601 Booth St Ottawa ON K1A 0E8 Can

FRITZ, WILLIAM J, b Vancouver, Wash, Mar 31, 53; m 86; c 1. VOLCANICLASTIC SEDIMENTATION, PALEOECOLOGY. *Educ:* Walla Walla Col, BS, 75, MS, 77; Univ Mont, PhD(geol), 80. *Prof Exp:* Grad teaching asst paleont & biol labs, Walla Walla Col, 75-77; head grad teaching asst geol labs, Univ Mont, 77-80; petrol geologist oil & gas explor, Amoco Prod Co-Standard Oil, 80-81; asst prof, 81-87, ASSOC PROF GEOL, GA STATE UNIV, 87- *Concurrent Pos:* Petrol consult, Davis Oil Co, 81-82. *Mem:* Geol Soc Am; Am Asn Petrol Geologists; Soc Econ Paleontologists & Mineralogists; Sigma Xi; Int Asn Sedimentologists. *Res:* Influence of explosive volcanism on terrestial sedimentation; ordovician volcaniclastic rocks of North Wales and Ireland; volcanic controls on sedimentation; tertiary sedimentology, paleoecology, and recent depositional environments including the Eocene Yellowstone fossil forest covering both the paleobotany and sedimentology. *Mailing Add:* Dept Geol Ga State Univ Atlanta GA 30303

FRITZE, CURTIS W(ILLIAM), b LeSueur, Minn, June 2, 22; m 48; c 3. ELECTRONICS & ELECTRICAL ENGINEERING. *Educ:* Univ Minn, BSEE, 47. *Prof Exp:* Res & design engr, Minn Electronics Corp, 47-50; sr engr, Eng Res Assoc, Inc, 50-54; eng dept mgr, Remington Rand Univac, 54-57, dir spec prod eng, 57-59; vpres & gen mgr, Monarch Electronics Co, 59-60; systs specialist, corp staff mkt, Control Data Corp, 60-62, asst to vpres indust group, 62-65, dir planning, indust & govt group, 65-66, dir corp planning, 66-68, vpres, 68-74, vpres space comput systs, 74-76; vpres opers & admin, Control Data Worldtech, 76-85; PRES, CURTIS, INC, 85- *Concurrent Pos:* Consult, Lab Psychol Hyg, 47-57; mem bd dir, Am Nat Standards Inst, 69-73; mem bd trustees, Sci Mus Minn, 68-80. *Res:* Technological and corporate planning systems; personal computer accessory boards; technical management. *Mailing Add:* Ten Anemone Circle St Paul MN 55127

FRITZELL, ERIK KENNETH, b Grand Forks, NDak, Oct 30, 46; m 73; c 2. WILDLIFE BIOLOGY, ECOLOGY. *Educ:* Univ NDak, BS, 68; Southern Ill Univ, MS, 72; Univ Minn, PhD(wildlife), 76. *Prof Exp:* Res assoc ecol, Delta Waterfowl Res Sta, 68-72 & Northern Prairie Wildlife Res Ctr, US Fish & Wildlife, 72-76; lectr wildlife resource, McGill Univ, 76-77; PROF, FISHERIES & WILDLIFE, UNIV MO, COLUMBIA, 78-, PROG LEADER, 84- *Mem:* Wildlife Soc; Am Soc Mammalogists. *Res:* Mammalian and avian ecology; sociobiology of carnivores; wetland ecology. *Mailing Add:* Dept Fisheries & Wildlife Univ Mo Columbia MO 65211

FRITZLEN, GLENN A, b Indianapolis, Ind, Apr 25, 19. MATERIALS SCIENCE ENGINEERING, METALLURGY & PHYSICAL METALLURGICAL ENGINEERING. *Educ:* Purdue Univ, BS, 41. *Prof Exp:* Engr & tech mgr, Haynes Int, 80-83; PROF & COORDR MECH ENG, COMPUTER INTEGRATED MFG & INDUST TECHNOL, PURDUE UNIV, 85- *Mem:* Fel Am Soc Metals; sr mem Metals Soc. *Res:* Developed nickel and cobalt alloys for heat corrosion and wear resistance mostly for rocketry materials. *Mailing Add:* Purdue Univ 300 S Washington St Kokomo IN 46902

FRITZSCHE, ALFRED KEITH, b Hamilton, Ohio, Jan 3, 43; m; c 2. MEMBRANE TRANSPORT. *Educ:* Miami Univ, Ohio, BS, 64; Case Inst Tech, MS, 67; Univ Mass, PhD(polymer sci), 72. *Prof Exp:* Latin Am teaching fel, Univ Simon Bolivar, Venezuela, 72-73; sr res chemist, Monsanto Co, 74-79, res specialist, 79-89; proj mgr, Romicon, Inc, 89-90; MGR, MEMBRANE RES & DEVELOP, BLOKEN SEPARATIONS, 90- *Concurrent Pos:* At Gen Elec Res & Develop Ctr, NY, 67-69. *Mem:* Am Chem Soc; AAAS; Am Inst Chem Engrs; Nat Am Membrane Soc. *Res:* Polymer crystallization kinetics; membrane transport of gases; epoxy sealant technology; co-inventer of prism alpria oxygen/nitrogen separation membrane for controlled atmospheres; ultrafilterations membrane technology. *Mailing Add:* 33 Douglas Rd Dracut MA 01826-4214

FRITZSCHE, HELLMUT, b Berlin, Ger, Feb 20, 27; US citizen; m 52; c 4. SOLID STATE PHYSICS. *Educ:* Univ Gottingen, dipl, 52; Purdue Univ, PhD(physics), 54. *Hon Degrees:* DSc, Purdue Univ, 88. *Prof Exp:* Asst prof physics, Purdue Univ, 55-57; from asst prof to assoc prof, 57-63, dir, Mat Res Lab, 73-77, chmn dept, 77-86, PROF PHYSICS, UNIV CHICAGO, 63-; VPRES & DIR, ENERGY CONVERSION DEVICES INC, 67- *Concurrent Pos:* Mem, adv bd, Encycl Britannica, 67-; mem solid state sci panel, Nat Res Coun-Nat Acad Sci, 74-; hon prof, Shanghai Inst Ceramics, 85, Nanjing Univ, 86, Beijing Univ Aeronaut & Astronaut, 88. *Honors & Awards:* Alexander von Humboldt Sr Scientist Award, 85; Oliver E Buckley Condensed Matter Physics Prize, 89. *Mem:* Fel NY Acad Sci; Fel Am Phys Soc; Am Arbitration Asn; hon mem Chinese Acad Sci. *Res:* Electronic and optical properties of amorphous and crystalline semiconductors and metals; low temperature physics. *Mailing Add:* James Franck Inst Univ Chicago 5640 Ellis Ave Chicago IL 60637

FRIZ, CARL T, b Elmhurst, Ill, July 9, 27; m 63. ANATOMY, BIOCHEMISTRY. *Educ:* Univ Ill, BS, 51, MS, 52; Univ Minn, PhD(anat), 59. *Prof Exp:* USPHS fel biol, Carlsberg Lab, Copenhagen, 59-62; instr anat, Univ Minn, 62-64; asst prof, 64-71, assoc prof, 71-76, PROF ANAT, UNIV BC, 76- *Mem:* Soc Protozool; Am Asn Anat. *Res:* Phylogenetic analysis using molecular characters of the free-living amoeba. *Mailing Add:* Dept Anat Univ BC Med Sch Vancouver BC V6T 1W5 Can

FRIZZELL, LEON ALBERT, b West Stewartstown, NH, Sept 12, 47; m 68; c 1. ELECTRICAL ENGINEERING, BIOENGINEERING. *Educ:* Univ NH, BS, 69; Univ Rochester, MS, 71, PhD(elec eng), 76. *Prof Exp:* Vis asst prof, 75-78, asst prof elec eng & bioeng, 78-83, ASSOC PROF ELEC & COMPUT ENG & BIOENG, UNIV ILL, URBANA, 83- *Mem:* AAAS; Inst Elec & Electronics Engrs; fel Acoust Soc Am; fel Am Inst Ultrasound Med. *Res:* Ultrasonic biophysics, bioengineering and dosimetry. *Mailing Add:* Bioacoust Res Lab Univ Ill 1406 W Green St Urbana IL 61801

FRIZZERA, GLAUCO, b Gondar, Ethiopia, Oct 26, 39; nat US; div; c 2. EXPERIMENTAL BIOLOGY. *Educ:* Univ Bologna, Italy, MD, 64; Am Bd Path, dipl, 88. *Prof Exp:* Resident path, Inst Path Anat, Univ Bologna, Italy, 65-69, asst prof path, 70-77; from asst prof to prof lab med & path, Univ Minn, Minneapolis, 77-89; CHMN, DEPT HEMATOPATH, ARMED FORCES INST PATH, WASHINGTON, DC, 89- *Concurrent Pos:* Brit Coun scholar, Dept Path, Charing Cross Med Sch, Univ London, Gt Brit, 69; USPHS trainee hematopath, Univ Chicago, Ill, 72-74; hon instr path, Univ Chicago, Ill, 73-74; staff pathologist surg path, Univ Minn, 77-89, hematopathologist, 77-89, sr staff pathologist autopsy path, 77-89; mem, Dept Lab Med & Path, Grand Rounds Comt, Univ Minn, 78-79, chmn, 79-81, mem, Residency Comt, 79-81, co-dir, Hematopath Sect, Path Course, 82-86; consult, Hennepin County Med Ctr, Minneapolis, Minn, 80-89, Vet Admin Hosp, Minneapolis, Minn, 81-89; mem, Nat Path Panel Lymphoma Clin Studies, 80-87; head, Second Regional Ctr Lymphoma Clin Studies, Cancer & Leukemia, Group B, 80- & vchmn, Hemat Subcomt, Path Comt, 80- *Mem:* Int Acad Path; Soc Hematopath; Am Asn Pathologists. *Mailing Add:* Dept Hematopath Armed Forces Inst Path 6825 16th St NW Washington DC 20306-6000

FROBISH, LOWELL T, b Flanagan, Ill, Aug 3, 40; m 60; c 2. ANIMAL NUTRITION, BIOCHEMISTRY. *Educ:* Univ Ill, Urbana, BS, 62; Iowa State Univ, MS, 64, PhD(animal nutrit), 67. *Prof Exp:* Assoc, Iowa State Univ, 66-67; res animal husbandman, Swine Res Br, animal Sci Res Div, Agr Res Serv, USDA, 67-72, chief, Non-Ruminant Animal Nutrit Lab, Animal Sci Inst, 72-81; HEAD, DEPT ANIMAL SCI, CLEMSON UNIV, 81- *Mem:* Am Soc Animal Sci. *Res:* Utilization of energy by gravid animals and the neonatal pig. *Mailing Add:* 1153 Owens Rd Auburn AL 36830-2517

FRODEY, RAY CHARLES, b Pittsburgh, Pa, Sept 6, 23; m 44; c 3. FOOD SCIENCE & NUTRITION, SCIENCE ADMINISTRATION. *Educ:* Mass Inst Technol, BS & MS, 49. *Prof Exp:* Chemist, Gerber Prod Co, 49-51, sr researcher, 51-56, dir new prod, 56-64, vpres res & qual control, 64-85; RETIRED. *Mem:* Inst Food Technologists; Sigma Xi; AAAS. *Res:* All aspects of food research and engineering including food biochemistry and nutrition, microbiology, packaging, product development and process engineering. *Mailing Add:* 4345 Chippewa Tr Fremont MI 49412

FRODYMA, MICHAEL MITCHELL, b Holyoke, Mass, Mar 3, 20; m 51; c 1. ANALYTICAL CHEMISTRY. *Educ:* Univ Mass, BS, 42; Columbia Univ, AM, 47; Univ Hawaii, MS, 49; George Washington Univ, PhD(chem), 52. *Prof Exp:* Asst chem, Univ Hawaii, 47-49, from asst prof to prof, 52-67; prog dir, 65-66, PROG DIR, POSTDOCTORAL PROGS, DIV RES CAREER DEVELOP, SCI FOUND, 67- *Concurrent Pos:* Exchange asst prof, Vassar Col, 57-58, lectr, 58; fac fel, NSF, 58-61. *Mem:* AAAS; Am Inst Chemists; Am Chem Soc; Royal Soc Chem. *Res:* Microchemistry; oceanographic chemistry; organic polarography; analytical aspects of spectral reflectance. *Mailing Add:* Div Res Career Develop Nat Sci Found Washington DC 20550

FROEDE, HARRY CURT, b Cortez, Colo, Sept 9, 34; m 64; c 3. PHARMACOLOGY, BIOCHEMISTRY. *Educ:* Univ Colo, BS, 58, MS, 61; Washington Univ, PhD(pharmacol), 65. *Prof Exp:* Asst prof pharmacol & toxicol, Sch Pharm, Univ Colo, Boulder, 70-77, res assoc, dept chem, 77-85; RETIRED. *Concurrent Pos:* Fel pharmacol, Washington Univ, 65-66 & Sch Med, Stanford Univ, 66-69. *Res:* Enzyme mechanisms and control of enzyme activity. *Mailing Add:* 7831 Raven Ct Boulder CO 80303

FROEHLICH, FRITZ EDGAR, b Ger, Nov 12, 25; nat US; m 49; c 3. PHYSICS. *Educ:* Syracuse Univ, BS, 50, MS, 52, PhD(physics), 55. *Prof Exp:* Asst physics, Syracuse Univ, 50-53, asst instr, 53-54; mem tech staff, Bell Tel Labs, 54-56, supvr commun tech, 56-63, head, Wideband Data & Spec Systs Dept, 63-66, Data Systs Dept, 66-67 & Tel Technol Dept, 67-75, head, New Sta Serv Dept, 75-78, Bus Terminal Dept, 78-81 & Adv Terminal Systs Dept, 81-83; head univ rels, 83-87, PROF TELECOMMUNICATIONS, UNIV PITTSBURGH, 87- *Concurrent Pos:* Instr, Utica Col, 52-54. *Mem:* AAAS; fel Inst Elec & Electronic Eng; NY Acad Sci; Sigma Xi; Inst Elec & Electronic Engrs. *Res:* Techniques of digital and voice communication. *Mailing Add:* SLIS Bldg Rm 743 Univ Pittsburgh Pittsburgh PA 15260

FROEHLICH, JEFFREY PAUL, b Milwaukee, Wis, Mar 31, 43; m 65; c 1. BIOPHYSICS. *Educ:* Univ Wis-Madison, BS, 65; Univ Chicago, MD, 69. *Prof Exp:* Fel cancer res, Ben May Lab Cancer Res, Univ Chicago, 65-68, fel biophys, 70-72; USPHS officer, 72-85, CHIEF, SECT MEMBRANE BIOL, NAT INST AGING, NIH, 85- *Concurrent Pos:* Adj prof, dept biol chem, Univ Md. *Mem:* Am Soc Biol Chemists; Biophys Soc. *Res:* Kinetic properties of cellular cation pumps and channels. *Mailing Add:* Nat Inst Aging Francis Scott Key Med Ctr 4940 Eastern Ave Baltimore MD 21224

FROEHLICH, LUZ, b Manila, Philippines, Feb 19, 28; US citizen; m 56; c 2. SCIENCE ADMINISTRATION, SCIENCE POLICY. *Educ:* Univ Philippines, MD, 53; Am Bd Path, cert, 61. *Prof Exp:* From resident to chief resident path, Buffalo Gen Hosp, 56-60; asst attend pathologist & actg head, Dept Path, Buffalo Children's Hosp, 61-63; med officer, Sect Path, Perinatal Res Br, Nat Inst Neurol Dis & Stroke, NIH, 63-71, from asst dir to assoc dir clin progs, Nat Inst Allergy & Infectious Dis, 71-76, chief, Prog & Proj Rev Br, 76-90, DEP DIR, DIV EXTRAMURAL ACTIV, NAT INST ALLERGY & INFECTIOUS DIS, NIH, 83- *Concurrent Pos:* Sr cancer res scientist, Roswell Park Mem Inst, Buffalo, 60-62; exec secy, Allergy & Immunol Res Comt, 71-76; actg dir, Extramural Activ Prog, Nat Inst Allergy & Infectious Dis, 81-82, Drug Enforcement Admin, 90-91. *Mem:* Am Soc Microbiol; Teratology Soc; Soc Pediat Path; Am Med Women's Asn. *Res:* Teratology; perinatal infections; perinatal and pediatric pathology; placental pathology; immunologic and infectious diseases; AIDS in women and minorities. *Mailing Add:* NIH Nat Inst Allergy & Infectious Dis Prog & Proj Rev Br Westwood Bldg Rm 703 5333 Westbard Ave Bethesda MD 20892

FROELICH, ALBERT JOSEPH, b Cleveland, Ohio, Aug 2, 29; m 58; c 2. GEOLOGY. *Educ:* Ohio State Univ, BS, 52, MS, 53. *Prof Exp:* Geologist, US Geol Surv, 53-54; geologist-party chief, San Jose Oil Co, Inc, 56-60; sr geologist, United Canso Oil & Gas, Inc, 60-65; chief geologist, Magellan Petrol Corp, 65-67, geol consult, 67-68; geologist, Ky, 68-71, geologist, Baltimore-Washington Urban Area Proj, Washington, DC, 71-74, chief, Fairfax County Environ Geol Proj, 74-79, chief, Culpeper Basin Environ Geol & Hydrol Proj, 79-85, CHIEF, EARLY MESOZOIC EVOL EASTERN US, US GEOL SURV, 85- *Mem:* Am Asn Petrol Geol; Geol Soc Am; Am Geophys Union. *Res:* Geology, particularly international petroleum and mineral exploration in Philippines, western and northern Canada, North Africa and Australia; environmental, geophysical and geohydrological studies; geochemistry of mafic igneous rocks, especially PGE's. *Mailing Add:* STOP 926 Nat Ctr US Geol Surv Reston VA 22092

FROELICH, ERNEST, b Vienna, Austria, Dec 7, 12; nat US; m 47; c 1. MICROBIOLOGY, CHEMOTHERAPY. *Educ:* Univ Zagreb, DVM, 37. *Prof Exp:* Vet asst, Vet Inst, Belgrade, 38-39; head vet, Vet Serum Labs, Zemun, 39-40; head vet, Biol Control Labs, 40-41; UNRRA fel, 46-47; res assoc microbiol & chemother, Sterling-Winthrop Res Inst 48-50, head, Chemother Lab, 50-62, dir res & tech serv, Vet Dept, 62-78, consult, 78-79; RETIRED. *Mem:* Am Soc Microbiologists; Am Vet Med Asn; NY Acad Sci. *Res:* Chemotherapy of bacterial, viral and rickettsial diseases; immunological studies of diseases of domestic animals. *Mailing Add:* 174 Tampa Ave Albany NY 12208

FROELICH, PHILIP NISSEN, b Winston-Salem, NC, Apr 12, 46; m 71, 85; c 1. MARINE GEOCHEMISTRY, CHEMICAL OCEANOGRAPHY. *Educ:* Duke Univ, BS, 70; Univ PR, MS, 73; Univ RI, PhD(oceanog), 79. *Prof Exp:* Assoc prof oceanog, Fla State Univ, 79-85; SR RES SCIENTIST, LAMONT-DOHERTY GEOL OBSERV & ADJ PROF, DEPT GEOL SCI, COLUMBIA UNIV, 85- *Mem:* AAAS; Am Geophys Union; Am Soc Limnol & Oceanog; Geol Soc Am; Geochem Soc. *Res:* Marine nutrient geochemistry (phosphorus, nitrogen, carbon, silicon); estuarine biogeochemistry; marine geochemistry of intersitial fluids (nutrients, gases and trace metals); germanium geochemistry and paleoceanography; phosphorites. *Mailing Add:* Lamont-Doherty Geol Observ Columbia Univ Palisades NY 10964

FROELICH, ROBERT EARL, b St Louis, Mo, July 24, 29; m 54; c 2. PSYCHIATRY. *Educ:* Wash Univ, AB, 51, MD, 55. *Prof Exp:* Resident, Med Col Ga, 58-61; asst prof psychiat, Univ Mo, 61-67; asst prof, Univ Southern Calif, 67-68; assoc prof, Univ Mo, 68-69; from assoc prof to prof psychiat & behav sci, Univ Okla, 69-74; asst dean, 74-76, PROF PSYCHIAT & CHMN DEPT, SCH PRIMARY MED CARE, UNIV ALA, HUNTSVILLE, 74- *Mem:* Health Sci Commun Asn; AMA; Am Psychiat Asn; Asn Educ Commun & Technol. *Res:* Education methods and self instructional packages. *Mailing Add:* Sch Primary Med Care Univ Ala 201 Governors Dr Huntsville AL 35801

FROEMKE, JON, b Sioux Falls, SDak, June 23, 41; m 63; c 1. MATHEMATICS. *Educ:* Univ Nebr, Lincoln, BA, 62, MA, 63; Univ Calif, Berkeley, PhD(math), 67. *Prof Exp:* Asst prof, 67-71, ASSOC PROF MATH, OAKLAND UNIV, 71- *Mem:* AAAS; Am Math Soc; Math Asn Am. *Res:* General algebraic systems. *Mailing Add:* Dept Math Sci Oakland Univ Rochester MI 48309

FROEMSDORF, DONALD HOPE, b Cape Girardeau, Mo, Mar 4, 34; m 54. ORGANIC CHEMISTRY. *Educ:* Southeast Mo State Univ, BS, 55; Iowa State Univ, PhD(org chem), 59. *Prof Exp:* Asst proj chemist, Res & Develop Dept, Standard Oil Co of Ind, 59-60; assoc prof to prof chem, 66-70, chmn div sci & math, 70-75, DEAN, COL SCI, SOUTHEAST MO STATE UNIV, 75- *Mem:* AAAS; Am Chem Soc; The Chem Soc; Sigma Xi. *Res:* Physical organic chemistry; organic reaction mechanisms, especially alpha and beta elimination reactions; structure and properties of high polymers. *Mailing Add:* Col Sci & Tech Southeast Mo State Univ Cape Girardeau MO 63701

FROES, FRANCIS HERBERT (SAM), b Liverpool, Eng, Apr 21, 40; US citizen; m 65; c 2. PROCESSING MICROSTRUCTURE MECHANICAL PROPERTY RELATIONSHIPS, APPLICATION OF METALS IN AEROSPACE SYSTEMS. *Educ:* Univ Liverpool, BS, 62; Univ Sheffield, MS, 63, PhD(phys metall), 67. *Prof Exp:* Engr, Crucible Steel Co, 67-75, mgr, 75-78; sr scientist, USAF Mat Lab, 78-81, tech area mgr, 81-87, br chief, 87-90; AT INST MAT & ADVAN PROCESSES, UNIV IDAHO, 90- *Concurrent Pos:* Adj prof, Univ Dayton, 83-88, Wright State Univ, 84-88; adv bd, Int J Powder Metals, 86-88, J Metals, 85-88. *Mem:* Am Soc Metals Int; Metal Powder Indust Fed; Am Inst Metall Engrs Metal Soc; Soc Am Metall & Petrol Engrs. *Res:* Area of the light metals aluminum, titanium, magnesium and beryllium; monolitnic and composite materials and ranges from basic research to applications in US Air Force systems. *Mailing Add:* Inst Mat & Advan Processes Univ Idaho Mines Bldg Moscow ID 83843

FROESE, ALISON BARBARA, b Dauphin, Man, Can, Aug 27, 45; m 69; c 2. RESPIRATORY PHYSIOLOGY, NEONATAL LUNG DISEASE. *Educ:* Univ Man, BSc & MD, 68, Royal Col Physicians & Surgeons, Can, FRCP(C), 73. *Prof Exp:* Investr, Res Inst Hosp Sick Children, Toronto, 74-81; assoc prof anesthesia, Univ Toronto, 80-81; asst prof physiol, 81-86, assoc prof anesthesia, 81-90, ASSOC PROF PEDIAT, QUEEN'S UNIV, 84-, ASSOC PROF PHYSIOL, 86-, PROF ANESTHESIA, 90- *Concurrent Pos:* Vis prof, Univ Man, 78 & 82, Univ Wash, 83, Mayo Clin, Univ Rochester, 83, Univ Ottawa, 84, Dartmouth Univ, 84; consult, NIH steering Comt, HFO Trial for Infants with Respiratory Distress Syndrome, 84-85; vis prof, Univ Ottawa, 87, Columbia Univ, 87, McGill Univ, 87, Univ Sask, 88, Yale Univ, Univ Conn & Johns Hopkins Univ, 89; ed, J Appl Physiol, 84-87; Queen's Nat scholar, 88-91. *Mem:* Am Soc Anesthesiol; Am Physiol Soc; Int Anesthesia Res Soc; Can Anaesthetists Soc; Can Thoracic Soc; Christian Med Dent Soc. *Res:* Introduced high frequency oscillatory ventilation as an experimental form of artificial ventilation; human and animal-based investigations into its therapeutic role, focusing on mechanisms of ventilation-related lung injury; influence of anesthesia on chest wall mechanics. *Mailing Add:* Dept Anesthesia Kingston Gen Hosp 76 Stuart St Kingston ON K7L 2V7 Can

FROESE, ARNOLD, b Halbstadt, Ukraine, Mar 18, 34; Can citizen; m 69; c 3. IMMUNOCHEMISTRY. *Educ:* Univ Western Ont, BS, 57; McGill Univ, PhD(chem), 63. *Prof Exp:* Lectr, McGill Univ, 63-64; res assoc immunochem, Max Planck Inst Phys Chem, 64-67 & McGill Univ, 67-69; assoc prof, 69-79, PROF IMMUNOL, UNIV MAN, 79- *Concurrent Pos:* Med Res Coun Can scholar & res grant, 68-73. *Mem:* Chem Inst Can; Can Biochem Soc; Can Soc Immunol (secy-treas, 77-83, vpres, 87-89, pres, 89-91); Am Asn Immunologists. *Res:* Kinetics of antibody-hapten reactions; structure and function of immunoglobulins; receptors on mast cells; mast cell differentiation. *Mailing Add:* Dept Immunol Univ Man Winnipeg MB R3E 0W3 Can

FROESE, GERD, b Saporoshje, USSR, May 18, 26; Can citizen; m 53; c 3. BIOPHYSICS. *Educ:* Univ Sask, 54; Univ Western Ont, MSc, 56, PhD(biophys), 58. *Prof Exp:* Physicist, Manitoba Cancer Treatment & Res Found, 60-; AT DEPT ANAT & RADIOL, UNIV MAN. *Concurrent Pos:* Donner res fel, Max Planck Inst Biophys, Ger, 58-59 & res unit radiobiol, Brit Empire Cancer Campaign, Middlesex, 59-60. *Mem:* Radiation Res Soc; Can Soc Immunol; Brit Asn Radiation Res. *Res:* Mammalian temperature regulation; radiation physics; radiobiology and cellular biology; experimental tumor immunology; hyperthermia. *Mailing Add:* Dept Med Physics Man Cancer Found Winnipeg MB R3E 0V9 Can

FROGEL, JAY ALBERT, b New York, NY, Apr 28, 44. INFRARED PHOTOMETRY & SPECTROSCOPY. *Educ:* Harvard Univ, BA, 66; Calif Inst Technol, PhD(astron), 71. *Prof Exp:* Res fel & lectr astron, Harvard Univ, 71-75; staff mem astron, Cerro Tololo Inter-Am Observ, 75-; DEPT ASTRON, OHIO STATE UNIV. *Mem:* Am Astron Soc; Astron Soc Pac; Int Astron Union. *Res:* The stellar content of clusters and galaxies; infrared emission from H II regions and galactic nuclei. *Mailing Add:* Dept Astron Ohio State Univ 174 W 18th Ave Columbus OH 43210

FROHLICH, EDWARD DAVID, b New York, NY, Sept 10, 31; m 59; c 3. INTERNAL MEDICINE. *Educ:* Washington & Jefferson Col, AB, 52; Univ Md, MD, 56; Northwestern Univ, MS, 63. *Prof Exp:* Intern med, DC Gen Hosp, 56, resident internal med, 57-58; resident, Georgetown Univ Hosp, 59-60; clin investr internal med & cardiovasc res, Vet Admin Res Hosp, Northwestern Univ, 62-64; mem staff, Res Div, Cleveland Clin, Ohio, 64-69; assoc prof, Med Ctr, Univ Okla, 69-71, assoc prof pharmacol & prof med & physiol & biophys, 71-76, George Lynn Cross res prof, 75-76; VPRES EDUC & RES, 76-, ALTON OCHSNER DISTINGUISHED SCIENTIST & VPRES ACAD AFFAIRS, ALTON OCHSNER MED FOUND & MEM, SECT HYPERTENSIVE DIS, OCHSNER CLIN, 85- *Concurrent Pos:* Fel cardiovasc res, Georgetown Univ, 58-59; mem, Med Adv Bds, Coun on Circulation, Am Heart Asn, 64-, Coun High Blood Pressue Res, 68, Coun Basic Sci, 69; ed-in-chief, J Lab & Clin Med, 74-76; chmn, Coun High Blood Pressure Res, 88- *Mem:* Fel AAAS; Am Soc Clin Pharmacol & Therapeut (pres, 73-74); fel Am Col Cardiol; fel Am Col Physicians; Am Soc Pharmacol & Exp Therapeut; Am Physiol Soc; Am Soc Clin Invest; Am Heart Asn; Am Asn Physicians. *Res:* Cardiovascular rescarch; hypertension; clinical pharmacology; pathophysiological mechanisms underlying hypertensive diseases and elucidation of antihypertensive drug actions. *Mailing Add:* Vpres Acad Affairs Alton Ochsner Med Found1516 Jefferson Hury New Orleans LA 70121

FROHLICH, GERHARD J, b Simmern, Ger, Feb 20, 29; US citizen; m 58; c 2. CHEMICAL ENGINEERING, PHYSICAL CHEMISTRY. *Educ:* Darmstadt Tech Univ, dipl chem, 53; Polytech Inst Brooklyn, MChE, 54, DChE, 57. *Prof Exp:* Chem engr, Vulcan-Cincinnati, Inc, Ohio, 55-59; sr chem engr, St Paul Ammonia Prod, Inc, Minn, 59-61, chief process engr, 61-62, mgr res & develop, 62-64; sr chem engr, 64-70, mgr chem eng develop, 70-77, GEN MGR & CORP ENGR, HOFFMANN-LA ROCHE, INC, NUTLEY, 77- *Mem:* Am Inst Chem Engrs; Am Chem Soc; Electrochem Soc; Faraday Soc; Nat Soc Prof Engrs; Sigma Xi. *Res:* Chemical process design and development; electrochemical engineering; heterogeneous catalysis; liquid-liquid extraction. *Mailing Add:* 669A Mountain Rd Kinnelon NJ 07405

FROHLICH, JIRI J, b Nespeky, Czech, Aug 12, 42; Can citizen; m 70; c 2. CHOLESTEROL. *Educ:* Charles Univ, Prague, MD, 65; FRCP(C), 74. *Prof Exp:* Teaching fel, Charles Univ, Prague, 66-68; res fel, McGill Univ, Montreal, 68-70; res fel, Univ BC, Vancouver, 73-74, clin asst prof path, 76, from asst prof to assoc prof, 77-86, PROF PATH, UNIV BC, VANCOUVER, 87-; ASSOC DIR LAB MED, BIOCHEM, UNIV HOSP-SHAUGHNESSY SITE, VANCOUVER, 90- *Concurrent Pos:* MRC res fel, Can, 69-70 & 73-74; prin investr grants, Vancouver Found, 74, BC MRF, 76, Can Heart Fund, 77-78, MRC, 78-79, Health & Welfare Can, 78-82, BC Heart Foundn, 79-82, BCHCRF, 79-81; corresp mem, Comt Med Biochem, Royal Col Physicians Surgeons, 75-77, core mem, 77-82; invited lectr, Congress Lab Med, Winnipeg, 78; mem, Res Comt, VGH, 79-80, Med Adv Comt, BC Heart Fund, 80-84, Comt Med Biochem, Royal Col Physicians & Surgeons, 89; consult, Cardiovasc Res Inst, Univ Calif, San Francisco, 81-82; lectr, FEBS workshop on LCAT deficiency, Edinburgh, 81, Cardiovasc Res Inst, Univ Calif, San Francisco, BMC Urinalysis Course & BC-CSLT meeting, 82, CSCC-Alta Div & NIH, 83 Hammersmith Hosp, London, St Mary's Hosp Med Sch, London, Inst Med Sci, Univ Toronto, BMC seminar, Toronto & numerous on lipidemia, BC, 84; vchmn, Med Adv Comt, BC Heart Found, 82-83, chmn bd dir, 84; co-chmn, Int Conf Lipoprotein Deficiency, Vancouver & Sem Hyperlipidemia, Inter-Am Cong Cardiol, Vancouver, 85; chmn, Med Biochem Sect, RCPS Meeting, 85; BSCC award, Can Soc Clin Chemists, 85-86; annual lectureship, Man Soc Clin Chemists, 88, Ont Soc Clin Chemists, 89; chmn bd, Can Cholesterol Res Found; exec bd mem, Can Lipoprotein Conf; coun mem, Can Asn Med Biochem & Can Atherosclerosis Soc. *Mem:* Royal Col Physicians & Surgeons; Can Soc Clin Invest (past pres); Can Cholesterol Res Found; Can Asn Med Biochem; Can Atherosclerosis Soc. *Res:* Cholesterol transport and the clinical, biochemical and genetic factors of inborn deficiencies of high density lipoprotein metabolism focusing on tissue removal of cholesterol in these patients. *Mailing Add:* Univ Hosp-Shaughnessy Site 4500 Oak St Vancouver BC V6H 3N1 Can

FROHLIGER, JOHN OWEN, b Indianapolis, Ind, Dec 21, 30; m 52; c 3. ANALYTICAL CHEMISTRY. *Educ:* Purdue Univ, BS, 57; Univ Iowa, MS, 59, PhD(anal chem), 61. *Prof Exp:* Asst prof chem, Duquesne Univ, 61-67; from asst prof to assoc prof indust health & air chem, Grad Sch Pub Health, Univ Pittsburgh, 67-80, actg chmn, Dept Occup Health, 73-75; PRES, INDUST HYG ASSOCS, 80- *Mem:* AAAS; Am Chem Soc; Air Pollution Control Asn; Am Pub Health Asn. *Res:* Analysis of chemical compounds in the environment; liquid chromatography; acid nature of precipitation. *Mailing Add:* 4527 Clairton Blvd Pittsburgh PA 15236-2111

FROHMAN, CHARLES EDWARD, b Columbus, Ind, Oct 25, 21; m 46; c 4. BIOCHEMISTRY, PHYSIOLOGY. *Educ:* Ind Univ, Bloomington, BS, 47, AM, 48; Wayne State Univ, PhD(physiol chem), 51. *Prof Exp:* Res assoc effect radiation, Kresge Eye Inst, Detroit, 51-52; res assoc, Dept Physiol Chem, Wayne State Univ, 52-55, instr, 55-57, dir dept biochem, Lafayette Clin, 55-86, from asst prof to assoc prof physiol chem, 67-72, PROF BIOCHEM, DEPT PSYCHIAT, WAYNE STATE UNIV, 72- *Mem:* Soc Biol Psychiat; Asn Anal Chemists; Am Asn Biol Chemists. *Res:* Mental illness of the brain. *Mailing Add:* 1380 Brys Dr Grosse Pointe Woods MI 48236

FROHMAN, LAWRENCE ASHER, b Detroit, Mich, Jan 26, 35. ENDOCRINOLOGY. *Educ:* Univ Mich, Ann Arbor, MD, 58. *Prof Exp:* From asst prof to assoc prof med, State Univ NY Buffalo, 65-73; prof med, sch med, Univ Chicago, 73-81; dir div endocrinol & metab, Michael Reese Med Ctr, 73-81; dir, clin res ctr, 86-90, PROF MED & DIR, DIV ENDOCRINOL METAB, SCH MED, UNIV CINCINNATI, 81- *Concurrent Pos:* Mem endocrinol study sect, NIH, 72-76; mem adv comt to assoc dir diabetes, endocrinol & metab, NIH, 78-82; mem, Vet Admin Merit Review Bd Endocrinol, 79-82; mem, endocrinol & metab adv comt, Food & Drug Admin, 82-85. *Mem:* Endocrine Soc; Am Diabetes Asn; Am Soc Clin

Invest; Int Soc Neuroendocrinol; Asn Am Physicians. *Res:* Neuroendocrine regulation of anterior pituitary function as related to growth hormone and prolactin secretion; neuropharmacology of hypothalamic releasing and inhibiting hormones; molecular heterogeneity of peptide hormones; neural control of carbohydrate metabolism. *Mailing Add:* Div Endocrinol Univ Cincinnati Med Ctr 231 Bethesda Ave Cincinnati OH 45267

FROHMBERG, RICHARD P, b Newburgh Heights, Ohio, Mar 30, 20; m 45; c 4. ENGINEERING, MATERIALS SCIENCE. *Educ:* Case Inst Technol, BS, 42, MS, 48, PhD(phys metall), 54. *Prof Exp:* Instr phys metall, Case Inst Technol, 48-51, res assoc metals, 51-55; chief prod develop lab, Rocketdyne Div, N Am Aviation, Inc, 55-60, chief mat res, 60-70; mgr eng, Bendix Corp, 70-87; RETIRED. *Honors & Awards:* Henry Marion Howe Award, 55. *Mem:* AAAS; fel Am Inst Aeronaut & Astronaut; fel Am Soc Metals. *Res:* High temperature materials sciences, including physical metallurgy, fabrication, ceramics and polymer science. *Mailing Add:* 11019 W 99th Pl Overland Park KS 66214

FROHNSDORFF, GEOFFREY JAMES CARL, b London, Eng, Feb 4, 28; US citizen; div; c 3. PHYSICAL CHEMISTRY, MATERIALS SCIENCE. *Educ:* Univ St Andrews, BSc, 53; Lehigh Univ, MS, 56; Univ London, PhD(phys chem) & DIC, 59. *Prof Exp:* Res fel chem, Royal Mil Col, Ont, 59-60; mgr res & develop, Am Cement Corp, 60-70; group leader, Gillette Res Inst, Inc, 70-73; chief, Mat & Composites Sect, 73-80, sr scientist, Structures & Mat Div, 80-81, CHIEF BUILDING MAT DIV, BLDG & FIRE RES LAB, NAT INST STANDARDS & TECHNOL, 81- *Concurrent Pos:* Chmn, comt res cement & concrete, Transp Res Bd, Nat Acad Sci-Nat Res Coun, 83-86. *Honors & Awards:* P H Bates Award, Am Soc Testing Mat, 78; Delmar Bloem Award, Am Concrete Inst, 88; Silver Medal, Dept Com, 82. *Mem:* Fel Am Ceramic Soc (vpres, 81-82); Int Union Mat & Structures Res & Testing Labs; Am Concrete Inst; Am Soc Testing & Mat; fel Am Concrete Inst. *Res:* Performance and durability of building materials; relationships between composition, structure and engineering performance of materials; surface chemistry; cement chemistry; surface coatings; adhesion; corrosion; preservation technology; resource recovery; materials conservation. *Mailing Add:* Bldg & Fire Res Lab Nat Inst Standards & Technol Gaithersburg MD 20899

FROHRIB, DARRELL A, b Oshkosh, Wis, June 25, 30; m 55; c 3. MECHANICAL & BIOMEDICAL ENGINEERING. *Educ:* Mass Inst Technol, SB, 52, SM, 53; Univ Minn, Minneapolis, PhD(mech), 66. *Prof Exp:* Engr, Sperry Gyroscope Co, NY, 53-59; lectr mech eng, Univ Minn-Minneapolis, 59-66, asst prof, 66-68, dir design ctr, 71-89, dir grad educ in Biomed Eng, 78-81, ASSOC PROF MECH ENG, UNIV MINN-MINNEAPOLIS, 68-, DIR GRAD EDUC MECH ENG, 88- *Concurrent Pos:* Am Petrol Soc fel, 63-65; Fulbright-Hays sr fel, India, 71; consult, Am Med Systs, 76-79, Theradyne Corp, 79-80 & Daromed Corp, 86-89. *Mem:* Am Soc Mech Engrs; Am Soc Eng Educ. *Res:* Shock and vibration; engineering design; dynamics; mathematical modeling of regional development systems; automation; urological system implant design; orthopaedic long bone healing prostheses; wheelchair optimization; flywheel energy conservation and retrieval. *Mailing Add:* 325 B Mech Engr Bldg Univ Minn Minneapolis MN 55455

FROILAND, SVEN GORDON, b Astoria, SDak, May 4, 22; m 43; c 6. BIOLOGY, ECOLOGY. *Educ:* SDak State Univ, BS, 43; Univ Colo, MA, 51, PhD, 57. *Hon Degrees:* DH, Luther Col, 78. *Prof Exp:* Instr biol, Augustana Col, 46-49 & Univ Colo, 49-53; assoc prof, 53-57, chmn, Dept Biol, 53-70 & Div Natural Sci, 59-76, PROF BIOL, AUGUSTANA COL, 57-, FELLOWS CHAIR NATURAL SCI, 80- *Concurrent Pos:* Vis scholar, Univ Ariz, 70-71. *Mem:* AAAS; Ecol Soc Am; Am Inst Biol Sci; Nat Asn Biol Teachers; Soc Study Evolution; Sigma Xi. *Res:* Morphology of genus betula; taxonomy and ecology of genus salix in Black Hills of South Dakota; natural history of Black Hills. *Mailing Add:* 4808 Arden Sioux Falls SD 57103

FROILAND, THOMAS GORDON, b Henderson, Ky, Dec 9, 43; m 66; c 2. DEVELOPMENTAL GENETICS. *Educ:* Augustana Col, BA, 65; Univ Nebr, MS, 67, PhD(zool), 70. *Prof Exp:* Instr biol, Kans State Univ, 70-71; asst prof, Wayne State Col, 71-72; from asst prof to assoc prof, 72-81, actg dept head, 78-79, PROF BIOL, NORTHERN MICH UNIV, 81-, DEPT HEAD, 79- *Concurrent Pos:* Consult, Upper Peninsula Med Educ Prog, Mich State Univ, 75 & 80. *Mem:* Int Pigment Cell Soc; Sigma Xi. *Res:* Control of pigmentation in genetic systems; ultrastructural relationships of microtubules to pigment; sex reversal, growth and development in lake trout; copper metabolism in mice. *Mailing Add:* Dept Biol Northern Mich Univ Marquette MI 49855-5341

FROJMOVIC, MAURICE MONY, b Brussels, Belg, Feb 2, 43; Can citizen; m 66; c 3. CELL ADHESION, BLOOD CELL AGGREGATION. *Educ:* McGill Univ, PhD(org chem), 67. *Prof Exp:* Nat Res Coun-NATO fel, Chem Biodynamics Lab, 66-68, asst prof, 68-76, ASSOC PROF PHYSIOL, McGILL UNIV, 76- *Concurrent Pos:* vis scintist, Can Heart Found; polymer dept, Weizman Res Inst, Israel, 77-78. *Mem:* Int Soc Hemostasis & Thrombosis; Am Heart Asn; Can Physiol Soc. *Res:* Model cell and particle studies of cell adhesions; platelet structure and function in health and disease; regulation of cell aggregation in variable flow regimes; blood cell adhesion molecules (integrins); flow cytometric dynamic studies of adhesion receptors. *Mailing Add:* Dept Physiol McIntyre Med Bldg 3655 Drummond Montreal PQ H3G 1Y6 Can

FROLANDER, HERBERT FARLEY, b Providence, RI, Sept 29, 22; m 50; c 4. BIOLOGICAL OCEANOGRAPHY. *Educ:* RI Col Educ, EdB, 46; Brown Univ, ScM, 50, PhD(biol), 55. *Prof Exp:* Instr oceanog, Univ Wash, 52-56, asst prof, 57-59; assoc prof, 59-65, asst dean grad sch, 66-67, actg chmn dept oceanog, 67-68, coordr marine sci & technol prog, 68-73, PROF OCEANOG, ORE STATE UNIV, 65- *Concurrent Pos:* Pres-elect, Asn Sea Grant Prog Insts, 70-71, pres, 71- *Mem:* Am Soc Limnol & Oceanog. *Res:* Marine zooplankton; population dynamics. *Mailing Add:* 1210 NW Alder Creek Dr Corvallis OR 97330

FROLIK, CHARLES ALAN, b New London, Wis, Mar 11, 45; m 69; c 2. BONE BIOLOGY, OSTEOPOROSIS. *Educ:* Mich State Univ, BS, 67; Univ Wis, PhD(biochem), 72. *Prof Exp:* Postdoctoral, Univ Wis, 72-75; sr staff scientist, NIH, 75-84; RES SCIENTIST, LILLY RES LABS, ELI LILLY & CO, 84- *Mem:* Am Soc Biochem & Molecular Biol; AAAS; NY Acad Sci. *Res:* Bone biology and osteoporosis; factors and mechanisms involved in bone formation and resorption, especially growth factors and steroid hormones. *Mailing Add:* Lilly Corp Ctr Indianapolis IN 46285

FROLIK, ELVIN FRANK, b DeWitt, Nebr, June 9, 09; m 38; c 3. AGRONOMY. *Educ:* Univ Nebr, Lincoln, BS, 30, MS, 32; Univ Minn, PhD, 48. *Prof Exp:* Exten agronomist, 36-45, from assoc prof to prof agron 46-75 chmn dept agron, 52-55, dir agr exp sta, 55-60, dean col agr, 60-73, EMER PROF AGRON, UNIV NEBR, LINCOLN, 75- *Concurrent Pos:* Consult, Develop & Resources Corp, Iran, 73-75, USAID, 76- & USDA, 78- *Mem:* AAAS; Am Soc Agron; Soil Conserv Soc Am; Int Crop Improv Asn (pres, 45-47). *Res:* Maize genetics; radiation genetics of crop plants. *Mailing Add:* 2317 The Knolls Lincoln NE 68512

FROM, ARTHUR HARVEY LEIGH, b South Bend, Ind, Oct 1, 36; m 62. CARDIOVASCULAR PHARMACOLOGY, CLINICAL CARDIOLOGY. *Educ:* Ind Univ, AB, 58, MA, 60, MD(physiol), 61. *Prof Exp:* from asst prof to assoc prof, 69-88, PROF MED, MED SCH, UNIV MINN, 88-; STAFF CARDIOLOGIST & DIR ECG LAB, MINN VET ADMIN MED CTR, 76- *Mem:* AAAS; Am Fedn Clin Res; Am Heart Asn; Int Soc Heart Res; Soc Gen Physiologists; Am Physiol Soc. *Res:* Molecular pharmacology of disitalis; NMR studies of myocardial bioenergetics. *Mailing Add:* Cardiol Sect Vet Admin Med Ctr 54th St & 48th Ave S Minneapolis MN 55417

FROM, CHARLES A(UGUSTUS), JR, b Chicago, Ill, Apr 12, 15; m 40; c 4. CHEMICAL ENGINEERING. *Educ:* Univ Mich, BS, 36. *Prof Exp:* Anal chemist, E I du Pont de Nemours & Co, 36-37; develop engr raw mat, AT&T Technol, 37-42; plastics develop, Powder Finishing, Nucleonics & Plastics Develop & Eng Labs, 46-55, dept chief, 55-59, non-metallic raw mat develop, 59-61, org finishing, insulating & encapsulation develop, 61-75, chief, Dept Mat, 75-80; RETIRED. *Concurrent Pos:* Chmn, Environ Comn, Elmhurst, Ill, 73- *Mem:* Am Chem Soc; Am Inst Chem Engrs; Soc Plastics Engrs; Am Watchmakers Inst; Nat Asn Watch & Clock Collectors; Tel Pioneers Am. *Res:* Plastics; non-metallic raw materials; organic finishing; mechanical finishing; adhesives. *Mailing Add:* 241 Kenmore Ave Elmhurst IL 60126

FROMAGEOT, HENRI PIERRE-MARCEL, b Paris, France, Jan 25, 37; US citizen; m 63; c 3. BIOCHEMISTRY. *Educ:* Nat Sch Advan Chem, Paris, dipl eng, 59; Sorbonne, BS, 60; Cambridge Univ, PhD(chem), 66. *Prof Exp:* Res assoc biochem, Rockefeller Univ, 66-69; staff biochem, Gen Elec Co Res & Develop Ctr, 69-78; sr res assoc, 78-81, SR RES & DEVELOP SCIENTIST, INT PAPER CO, CORP RES CTR, 81- *Mem:* Ger Chem Soc; AAAS. *Res:* Materials for wound healing. *Mailing Add:* Int Paper Corp Res Ctr Long Meadow Rd Tuxedo NY 10987

FROMAN, DAROL KENNETH, b Harrington, Wash, Oct 23, 06; m 31, 79; c 2. NUCLEAR PHYSICS. *Educ:* Univ Alta, BSc, 26, MSc, 27; Univ Chicago, PhD(physics), 30. *Hon Degrees:* LLD, Univ Alta, 64. *Prof Exp:* Asst physics, Univ Chicago, 29-30; lectr, Univ Alta, 30-31; lectr, Macdonald Col, McGill Univ, 31-36, asst prof, 36-39 & McGill Univ, 39-41; prof, Univ Denver, 41-42; group leader, Radio & Sound Lab, US Navy, San Diego, 42; res assoc, Manhattan Dist Proj, Metall Lab, Univ Chicago, 42-43; group leader, Los Alamos Sci Labs, 43-45, div leader, 45-48, asst dir weapons develop, 49-51, tech assoc dir, 51-62; RETIRED. *Concurrent Pos:* Dir, Mt Evans High Altitude Lab, 41; sci dir, Atomic Weapons Tests, Eniwetok, 48; consult prof, Univ NMex, 54-61; mem, Sci Adv Comt Ballistic Missiles, Secy Defense, 55-60, Sci Directorate, Douglas Aircraft Co, 63-69 & Gen Adv Comt, AEC, 64- 66; dir develop, Espanola Hosp, 67-70; chmn bd, First Nat Bank Rio Arriba, 71-78; consult, 62-75. *Mem:* Fel Am Phys Soc; fel Am Nuclear Soc; NY Acad Sci. *Res:* Supersonics; x-rays; cosmic rays; electronics applied to nuclear physics. *Mailing Add:* 250 E Alameda No 438 Santa Fe NM 87501

FROMAN, SEYMOUR, b New York, NY, May 19, 20; m 46; c 3. MICROBIOLOGY. *Educ:* NY Univ, BA, 46; Univ Calif, Los Angeles, MA, 49, PhD, 52. *Prof Exp:* Asst clin prof, 64-65, ASSOC PROF-IN-RESIDENCE MICROBIOL & IMMUNOL, SCH MED, UNIV CALIF, LOS ANGELES, 65-; DIR, MICROBIOL RES LAB, OLIVE VIEW HOSP, 64- *Concurrent Pos:* Consult, Tuberc Unit, WHO & Clin Lab, Univ Calif, Los Angeles. *Mem:* Am Soc Microbiologists; Am Thoracic Soc; fel Am Pub Health Asn; NY Acad Sci; Int Union Against Tuberc; Sigma Xi. *Res:* Microbiology of tuberculosis. *Mailing Add:* Olive View Med Ctr 14445 Olive View Dr Sylmar CA 91342

FROMHOLD, ALBERT THOMAS, JR, b Birmingham, Ala, Nov 25, 35; m 60; c 2. SOLID STATE PHYSICS. *Educ:* Auburn Univ, BEngPhys, 57, MS, 58; Cornell Univ, PhD(eng physics), 61. *Prof Exp:* Mem res staff, Sandia Corp, 61-65; alumni assoc res prof, 65-69, PROF PHYSICS, AUBURN UNIV, 70- *Concurrent Pos:* Vis scientist, Nat Bur Standards, 69-70, Hokkaido Univ, Sapporo, Japan, 79 & Oak Ridge Nat Lab, 80; res fel, Japan Soc Prom Sci, 79; NSF Prof Develop Award, 80. *Honors & Awards:* Jessie Beams Research Award, Am Phys Soc, 81. *Mem:* Am Phys Soc. *Res:* Theory of oxide film formation; optical properties of metals; solid state transport. *Mailing Add:* Dept Physics Auburn Univ Auburn AL 36849

FROMKIN, VICTORIA A, b Passaic, NJ, May 16, 23; m 48; c 1. LINGUISTICS, NEUROLINGUISTICS. *Educ:* Univ Calif, Berkeley, BA, 44; Univ Calif, Los Angeles, MA, 63, PhD(ling), 65. *Prof Exp:* Res linguist, 65-66, from asst prof to assoc prof, 66-72, chmn dept, 72-76, dean, grad div, 79-89 & vchancellor, grad progs, 80-89, PROF LING, UNIV CALIF, LOS ANGELES, 72- *Concurrent Pos:* Vis prof ling, Univ Stockholm, 77 & Cambridge Univ, 78; NIH Commun Sci Study Sect, 80-82, NIH Sensory Disorders & Lang Study Sect, 82-84; vis fel phonetics & ling, Wolfson Col,

Oxford Univ, 83, 87; chmn, Coun Grad Deans, Univ Calif, 85-86; Ida Beam Prof, Univ Iowa, 85; Cecil & Ida Green prof, Univ BC, 86; NRC Comt Basic Res Behav & Social Scis, 82-88; pres, Asn Grad Schs AAU, 88; US deleg & exec bd, Int Permanent Comt Linguists; NSF Rev Panel Ling, 73-75, NSF Adv Panel Fac Awards Women Sci & Eng, 90-91. *Honors & Awards:* Harvey L Eby Award for Distinguished Grad Teaching, 74. *Mem:* Ling Soc Am, (secy-treas, 79-84, pres, 85); fel Acoust Soc Am; fel NY Acad Sci; Int Neuropsychol Soc; Am Asn Phonetic Sci; Acad Aphasia; fel Am Psych Soc. *Res:* Brain mechanisms underlying language and cognition, using data from normal spontaneous speech errors, aphasia patients and normal experimental speech production and perception studies. *Mailing Add:* Dept Ling Univ Calif Los Angeles CA 90024-1543

FROMM, DAVID, b New York, NY, Jan 21, 39; m 61; c 3. ION TRANSPORT, INTRACELLULAR PH REGULATION. *Educ:* Univ Calif, BS, 60, MD, 64. *Prof Exp:* Intern surg, Univ Calif, 64-65, resident, 65-70, chief resident, 70-71; res fel surg, Harvard Med Sch, 67-69, from asst prof to assoc prof, 73-88; surgeon, Walter Reed Army Inst Res, 71-73; prof & chmn surg, State Univ NY, 78-88; PENBERTLY PROF & CHMN SURG, WAYNE STATE UNIV, 88- *Concurrent Pos:* Res career development award, NIH, 75-78, mem study sect, 91-; gov, Am Col Surgeons, 77-83; bd ed, Proc of Soc for Exp Biol & Med, 85-91; mem, Coop Studies Eval Comt, Vet Admin, 86-90; chief surg, Harper Hosp, 88-; surgeon-in-chief, Detroit Med Ctr, 88- *Honors & Awards:* Meyer O Cantor Award, Int Col Surgeons, 89. *Mem:* Am Physiol Soc; Am Gastroenterol Asn; Am Surg Asn; Soc Univ Surgeons; Asn Acad Surg; Am Fedn Clin Res. *Res:* Characterization of intestinal ion transport; ion transport by healthy and damaged gastric mucosa; regulation of intracellular PH by gastricmucosal surface cells. *Mailing Add:* Dept Surg 6-C Univ Health Ctr 4201 St Antoine Detroit MI 48201

FROMM, ELI, b Niedaltdorf, Ger, May 7, 39; US citizen; m 62; c 3. ELECTRICAL ENGINEERING, PHYSIOLOGY. *Educ:* Drexel Univ, BSEE, 62, MS, 64; Jefferson Med Col, PhD(physiol, biomed eng), 67. *Prof Exp:* Engr, Missile & Space Div, Gen Elec Co, 62-63; engr, Appl Physics Labs, E I du Pont de Nemours & Co, 63; teaching asst physics, Drexel Univ, 63-64, from asst prof to assoc prof biol sci, Biomed Eng Inst, 70-80, assoc dean, 87-89, interim dean engr, 89-90, PROF ELEC & COMPUTER ENGR, DREXEL UNIV, 80-, VPROVOST RES & GRAD STUDIES, 90- *Concurrent Pos:* Lalor Found grant, 68-70; NSF grant, 69-71, 83-84 & 88-93; NIH grant, 69-76; consult, var indust co, 68-; lectr, Dept Obstet & Gynec, Sch Med, Univ Pa, 73-; assoc dir, Biomed & Eng Sci Prog, Drexel Univ, 73-74; cong fel & staff mem, Comt Sci & Technol, US House Rep, Wash, DC, 80-81; dir, Bioeng Prog, NSF, 83-84; mem, bd dirs & tech act bd, Inst Elec & Electronic Engrs, ed, act bd. *Honors & Awards:* Centennial Medal, Inst Elec & Electronic Engrs. *Mem:* Fel Inst Elec & Electronic Engrs; Eng Med & Biol Soc (secy-treas, 73-75, vpres, 76-77, pres, 78-81). *Res:* Physiologic function by telemetry; engineering education; rhythmic responses of physiologic and biophysical parameters; development of transduction and transmission techniques for biologic variables. *Mailing Add:* Vprovost Res & Grad Studies Drexel Univ Philadelphia PA 19104

FROMM, GERHARD HERMANN, b Königsberg, Ger, Sept 7, 31; US citizen; m 73; c 2. NEUROLOGY, NEUROPHARMACOLOGY. *Educ:* Univ PR, BS, 49; Jefferson Med Col, MD, 53. *Prof Exp:* Fel neurophysiol, Univ Freiburg, Ger, 59-60 & Kungl, Veterinärhögskolan, Stockholm, Sweden, 60-61; from instr to assoc prof neurol, Sch Med, Tulane Univ, 61-68; assoc prof, 68-81, PROF NEUROL, SCH MED, UNIV PITTSBURGH, 81- *Concurrent Pos:* Psychiat resident, Tulane Univ & Vet Admin Hosp, New Orleans, 54-55; neurol resident, Tulane Univ & Charity Hosp, New Orleans, 55-56, chief resident, 58-59; staff neurologist, US Naval Hosp, St Albans, NY, 56-68; attend neurologist, Charity Hosp, New Orleans, La, 61-68, Presby-Univ Hosp & Children's Hosp, Pittsburgh, Pa, 68. *Mem:* Am Neurol Asn; Soc Neurosci; Int Asn Study Pain; Am Epilepsy Soc; Am Acad Neurol. *Res:* Development of new laboratory model for testing antiepileptic and antineuralgic drugs which more accurately predict their clinical spectrum of activity; discovery of new drugs for treatment of petit mal epilepsy and trigeminal neuralgia. *Mailing Add:* Dept Neurol Univ Pittsburgh Scaife Hall Rm 325 Pittsburgh PA 15261

FROMM, HANS, b Hagenow, Germany, Aug 1, 39; m 68; c 2. GASTROENTEROLOGY. *Educ:* Univ Freiburg, DrMed, 65. *Prof Exp:* Intern, Mem Hosp, Worcester, Mass, 66-67; resident, Lemuel Shattuck Hosp, Boston, Mass, 67-68; resident, Albany Med Col, 68-69, res fel gastroenterol, 69-70; res fel, Gastroenterol Res Unit, Mayo Clin, 70-71; res asst, Div Gastroenterol, Med Hochsch Hannover, 71-74, pvt lectr, 74-75; asst head gastroenterol & nutrit unit, Montefiore Hosp, 75-78; from asst prof to assoc prof, 75-84, PROF MED, UNIV PITTSBURGH, 84-; DIV DIR & CHIEF, GASTROENTEROL, GEORGE WASHINGTON UNIV, 84- *Concurrent Pos:* Assoc head gastroenterol & nutrit unit, Montefiore Hosp, Pittsburgh, 78-81; head, gastroenterol Unit, Montefiore Hosp, 81-84. *Honors & Awards:* Res Career Develop Award, NIH, 77. *Mem:* Am Asn Study Liver Dis; Am Gastroenterol Asn; European Soc Clin Invest; Am Fedn Clin Res; AAAS; Am Soc Clin Investigators. *Res:* Bile acid metabolism in cholelithiasis and in inflammatory bowel disease in man; development of new methods of treatment for cholelithiasis and inflammatory bowel disease. *Mailing Add:* Div Gastroenterol HB Burns Mem Bldg Suite 5405N George Washington Univ Med Ctr 2150 Pennsylvania Ave NW Washington DC 20037

FROMM, HERBERT JEROME, b New York, NY, Apr 5, 29; m 64; c 4. BIOCHEMISTRY. *Educ:* Mich State Univ, BS, 50; Loyola Univ, Ill, PhD(biochem), 54. *Prof Exp:* From asst prof to prof biochem, Sch Med, Univ NDak, 54-66; PROF BIOCHEM, IOWA STATE UNIV, 66- *Mem:* Am Chem Soc; Am Soc Biol Chemists. *Res:* Enzyme chemistry. *Mailing Add:* Dept Biochem & Biophys Iowa State Univ Ames IA 50011-0061

FROMM, PAUL OLIVER, b Ramsey, Ill, Dec 2, 23; m 47; c 2. COMPARATIVE PHYSIOLOGY. *Educ:* Univ Ill, Urbana, BS, 49, MS, 51, PhD(physiol), 54. *Prof Exp:* From instr to prof, 54-86, EMER PROF PHYSIOL, MICH STATE UNIV, 86- *Concurrent Pos:* Fulbright res grant, Mus Oceanog, Monaco, 63-64; vis prof, Ariz State Univ, 71, adj prof physiol, 78 & 79. *Mem:* Am Physiol Soc; N Am Benthological Soc; Am Soc Zoologists; Soc Exp Biol & Med. *Res:* Toxic effect of water soluble pollutants on freshwater fish; transfer of materials across isolated-perfused gills of fishes; comparative physiology of vertebrate eyes. *Mailing Add:* 6741 S Lake St Rte 1 Pentwater MI 49449

FROMMER, GABRIEL PAUL, b Budapest, Hungary, Apr 27, 36; US citizen; m 58; c 2. NEUROPSYCHOLOGY, NEUROSCIENCES. *Educ:* Oberlin Col, BA, 57; Brown Univ, ScM, 59, PhD(psychol), 61. *Prof Exp:* Sr asst scientist, USPHS, 60-62; USPHS fel, Yale Univ, 62-64; from asst prof to assoc prof, 64-79, PROF PSYCHOL, IND UNIV, BLOOMINGTON, 79- *Concurrent Pos:* Vis assoc res scientist physiol, Univ Mich, 73-74 & 75-76, USPHS spec fel, 75-76. *Mem:* Am Psychol Asn; Soc Neurosci; Psychonomic Soc; Sigma Xi; AAAS; Am Physiol Soc; Am Psychol Soc. *Res:* Neural mechanisms of somesthesis; neuropsychology. *Mailing Add:* Dept Psychol Ind Univ Bloomington IN 47405-1301

FROMMER, JACK, b New Haven, Conn, Jan 21, 18; m 44; c 4. ANATOMY. *Educ:* Univ Conn, BS, 40, MS, 46; Brown Univ, PhD(biol), 54. *Prof Exp:* Instr zool, Univ Conn, 47-48; asst res physiol, Yale Univ, 48-49; instr zool, Univ Conn, 49-50; from instr to assoc prof, 53-75, PROF ANAT, MED SCH, TUFTS UNIV, 75- *Mem:* AAAS; Am Asn Anatomists. *Res:* Temporomandibular joint; mandible development; cleft palate; osteoclasts; teratology; radioautography. *Mailing Add:* 39 Whittemore Rd Framingham MA 01701

FROMMER, PETER LESLIE, b Budapest, Hungary, Feb 13, 32; US citizen; m 53; c 4. CARDIOLOGY, BIOMEDICAL ENGINEERING. *Educ:* Univ Cincinnati, EE, 54; Harvard Med Sch, MD, 58. *Prof Exp:* Intern, Med Ctr, Univ Cincinnati, 58-59, asst resident med, 61-63; jr investr, Lab Tech Develop, 59-61, sr investr & attend physician, Cardiol Br, 63-66, asst chief, Myocardial Infarction Br, 66-67, chief, Myocardial Infraction Br, 67-72, assoc dir cardiol, Div Heart & Vascular Dir, 72-78, DEP DIR, NAT HEART, LUNG & BLOOD INST, NIH, 78- *Concurrent Pos:* Mem, Joint US Comt Eng Med & Biol, Nat Res Coun, 60-65, chmn comt, 64, mem, US Nat Comt Eng Med & Biol, 66-68; mem, Admin Coun, Int Fedn Med & Biol Eng, 65, vpres, 69. *Mem:* Fel Am Col Cardiol; fel Am Col Physicians; Am Fedn Clin Res; Am Heart Asn; Am Physiol Soc; Am Soc Artificial Internal Organs. *Res:* Cardiology and cardiovascular physiology; instrumentation; research planning and adminstration. *Mailing Add:* Nat Heart Lung & Blood Inst NIH Bethesda MD 20892

FROMMHOLD, LOTHAR WERNER, b Wurzburg, Ger, Apr 20, 30; m 58; c 2. ATOMIC & MOLECULAR PHYSICS. *Educ:* Univ Hamburg, Dipl, 56, PhD(gas discharges), 61, Dr habil, 64. *Prof Exp:* Res assoc physics, Univ Hamburg, 56-63, instr, 61-63, asst prof, 64; res assoc & instr, Univ Pittsburgh, 64-66, res prof, 66; assoc prof, 66-69, PROF PHYSICS, UNIV TEX, AUSTIN, 69- *Concurrent Pos:* Fulbright travel grant, Univ Pittsburgh, 64, res fel, 64-66. *Mem:* Fel Am Phys Soc. *Res:* Gaseous electronics; atomic and molecular physics; plasma diagnostics; physical chemistry; laser Raman spectroscopy. *Mailing Add:* Dept Physics Univ Tex Austin TX 78712

FROMOVITZ, STAN, b Lodz, Poland, May 31, 36; Can citizen. OPERATIONS RESEARCH, APPLIED STATISTICS. *Educ:* Univ Toronto, BASc, 60, MA, 61; Stanford Univ, PhD(statist), 65. *Prof Exp:* Mathematician, Shell Develop Co, 64-67; asst prof statist, Grad Sch Bus, Univ Santa Clara, 67-71; ASSOC PROF MGT SCI, UNIV MD, 71- *Mem:* Opers Res Soc Am; Inst Mgt Sci; Sigma Xi. *Res:* Mathematical programming; inventory theory; optimization algorithms; applied operations research. *Mailing Add:* Col Bus & Mgt Univ Md College Park MD 20742

FROMOWITZ, FRANK B, SURGICAL PATHOLOGY. *Educ:* Jefferson Med Col, MD, 76. *Prof Exp:* ASSOC PROF PATH, STATE UNIV NY, STONY BROOK, 82- *Res:* Hematopathology. *Mailing Add:* Dept Path State Univ NY Sch Med Univ Hosp Z-756 Stony Brook NY 11794

FRONCZEK, FRANK ROLF, b Lake Charles, La, Dec 17, 48; m 78; c 1. DEPARTMENTAL CRYSTALLOGRAPHY. *Educ:* La State Univ, BS, 70; Calif Inst Tech, PhD, 75. *Prof Exp:* Postdoctoral, dept chem, Univ Calif, Berkeley, 75-76; vis asst prof, 76-80, RES ASSOC CHEM, LA STATE UNIV, 80- *Concurrent Pos:* Adj prof, Univ S Fla, Tampa, 87- *Mem:* Am Crystallographic Asn; Am Chem Soc; Sigma Xi. *Res:* X-ray crystallography; structure of macrocycles and metal ion complexes of macrocyclic ligands; structure of plant natural products. *Mailing Add:* Dept Chem La State Univ Baton Rouge LA 70803-1804

FRONDEL, CLIFFORD, b New York, NY, Jan 8, 07; m 41; c 2. MINERALOGY. *Educ:* Colo Sch Mines, Geol Eng, 29; Columbia Univ, AM, 36; Mass Inst Technol, PhD(crystallog), 39. *Prof Exp:* Res assoc mineral, Harvard Univ, 39-46, assoc prof, 46-54, prof, 54-77, chmn dept geol sci, 65-69; RETIRED. *Concurrent Pos:* Sr physicist, US War Dept, 42-43; res dir, Reeves Sound Labs & Reeves-Ely Labs, 43-45; corr mem, Am Mus Nat Hist & Hist Mus Vienna; pres, Crystal Res, Inc, 49-72- *Honors & Awards:* Becke Medal, Austrian Mineral Asn, 58; Roebling Medal, Am Mineral Soc, 64; Boricky Medal, 68. *Mem:* Fel Am Mineral Soc (pres, 56); fel Geol Soc Am; Am Acad Arts & Sci; Geochem Soc; Austrian Acad Sci; Sigma Xi. *Res:* Descriptive mineralogy; crystal synthesis; lunar geochemistry. *Mailing Add:* 20 Beatrice Circle Belmont MA 02178

FRONEK, ARNOST, b Topolcany, Czech, Aug 8, 23; m 57; c 2. PHYSIOLOGY. *Educ:* Charles Univ, Prague, MD, 49; Czech Acad Sci, Prague, cert, 55. *Prof Exp:* Intern med, County Hosp, Most, Czech, 49-51; intern, Inst Exp Physiol, Charles Univ, Prague, 51-52; vis asst prof, Bockus

Res Inst, Univ Pa, 62-63, asst prof, 65-68; from assoc prof to prof aerospace & mech eng sci, 68-73, PROF BIOENG, SCH MED, UNIV CALIF, SAN DIEGO, 73-, PROF SURG, 76- *Concurrent Pos:* Res fel cardiovasc physiol, Inst Cardiovasc Res, Prague, 52-64. *Mem:* Czech Soc Cardiol (hon secy, 61-64); Czech Soc Physiol; Czech Soc Med Electronics; Sigma Xi. *Res:* Cardiovascular physiology; local thermodilution; cardiac energetics. *Mailing Add:* Dept Surg & Bioeng M-043 Univ Calif La Jolla CA 92093

FRONEK, DONALD KAREL, b Sunnyside, Wash, July 11, 37; m 59; c 2. DIGITAL COMPUTERS, ELECTRICAL ENGINEERING. *Educ:* Wash State Univ, BA, 60, BS, 64; Univ Idaho, MS, 69, PhD(elec eng), 73. *Prof Exp:* Proj officer elec, US Army, 60-62; elec engr, Gen Elec Co, 64-66, Hughes Aircraft Co, Calif, 66-67; instr, Univ Idaho, 68-73; asst prof, Univ Ala, Huntsville, 73-76; prof elec, Univ Nev, Reno, 76-84; PROF & DEPT HEAD, ELEC ENG DEPT, LA TECH UNIV, RUTAN, LA, 85- *Concurrent Pos:* Prin investr, NSF, 74-75, Army Res Orgn, 75-76 & 78-79, NASA, 75-76; consult, Battelle Durham, SC, 74-76 & 78-88. *Mem:* Sigma Xi; Inst Elec & Electronics Engrs. *Res:* Microprocessors; digital switching theory; digital electronics; thick film electronics; image processing articfical intellingence; optical processing; pattern recognition. *Mailing Add:* 15 Camino del Senador Tijeras NM 87059

FRONEK, KITTY, b Grenoble, France, Mar 23, 25; m 50; c 2. HUMAN PHYSIOLOGY. *Educ:* Charles Univ Prague, MD, 51; Czech Acad Sci, CSc(physiol), 60. *Prof Exp:* Instr pharmacol, Med Sch, Charles Univ, Prague, 50-51; sr investr physiol & clin res, Inst Cardiovasc Res, 51-60; head physiol div, Nuclear Lab, Inst Clin & Exp Surg, 60-64; res assoc cardiovasc res, Med Sch, Temple Univ, 65-68; RES PHYSIOL, SCH MED, UNIV CALIF, SAN DIEGO, 68- *Concurrent Pos:* Head physiol lab, Episcopal Hosp, Philadelphia, Pa, 67-68; consult, HRA Health Serv Res, Rockville, Md,73-74; mem, NIH, cardiovasc-renal study sect, 75-79. *Mem:* Am Soc Physiol; Am Heart Asn. *Res:* Control and regulation of the cardiovascular system under physiological and pathophysiological conditions; relationship between cardiac output, blood pressure and blood flow in several peripheral regions; pathophysiology of atheriosclerosis. *Mailing Add:* Res Physiol Univ Calif San Diego CA 92093

FRONING, GLENN WESLEY, b Gray Summit, Mo, Sept 8, 30; m 62; c 2. BIOCHEMISTRY. *Educ:* Univ Mo, BS, 53, MS, 57; Univ Minn, PhD(food tech), 61. *Prof Exp:* Asst prof food sci, Rutgers Univ, 61-63; asst prof poultry sci, Univ Conn, 63-66; prof poultry & wildlife sci, 66-77, chmn dept, 72-77, prof animal sci, 77-84, PROF FOOD SCI & TECHNOL UNIV NEBR, LINCOLN, 84- *Honors & Awards:* Res Award, Nat Turkey Fedn, 72 & Am Egg Bd, 88. *Mem:* Poultry Sci Asn (secy-treas, 80-83, 2nd vpres, 89-90, 1st vpres, 90-91); Inst Food Technol; AAAS; Am Chem Soc. *Res:* Egg white proteins; color and textural properties of poultry meat and poultry meat myoglobin; functional properties of eggs; supercritical fluid extraction of lipids and cholesterol from animal products. *Mailing Add:* Dept Food Sci & Technol Univ Nebr Lincoln NE 68583-0919

FROSCH, ROBERT ALAN, b New York, NY, May 22, 28; m 57; c 2. AUTOMOBILE MANUFACTURE. *Educ:* Columbia Univ, AB, 47, MA, 49, PhD(theoret physics), 52. *Prof Exp:* Res scientist theoret physics, Hudson Labs, Columbia Univ, 51-53, from asst dir to dir, Theoret Div, 53-63; dir nuclear test detection, Adv Res Proj Agency, Off Secy Defense, 63-65, dep dir, 65-66; asst secy Navy for res & develop, Dept Navy, 66-73; asst exec dir, UN Environ Prog, 73-75; assoc dir, Woods Hole Oceanog Inst, 75-77; adminr, NASA, 77-81; pres, Am Asn Eng Soc, 81-82; VPRES, RES LABS, GEN MOTORS CORP, 82- *Concurrent Pos:* Chmn, Interagency Comt Oceanog, 66-67 & US del, Intergovt Oceanog Comt, UNESCO, Paris, 67 & 70; mem, Interagency Comt Marine Res, Educ & Facil, 67-69 & Vis Comt Earth Sci, Mass Inst Technol; Trustee Woods Hole Oceanog Inst, 82-88. *Honors & Awards:* Arthur S Flemming Award, 66. *Mem:* Nat Acad Engrs; fel AAAS; fel Acoust Soc Am; Am Phys Soc; fel Inst Elec & Electronic Engrs; Am Acad Arts & Sci. *Res:* Theoretical physics; acoustical oceanography; seismology; system analysis and design; marine physics. *Mailing Add:* Gen Motors Res Labs 30500 Mound Rd Warren MI 48090-9055

FROSCH, WILLIAM ARTHUR, b New York, NY, June 24, 32; m 53; c 2. PSYCHIATRY, PSYCHOANALYSIS. *Educ:* Columbia Col, AB, 53; NY Univ, MD, 57; NY Psychoanal Inst, cert, 65. *Prof Exp:* From intern to chief resident psychiat, Bellevue Hosp, New York, 57-61; from teaching asst to prof psychiat, Sch Med, NY Univ, 59-74, asst dean, 61-63, dir grad educ, 72-74; PROF PSYCHIAT & VCHMN, NY HOSP-CORNELL UNIV MED CTR, 75-, MED DIR, PAYNE WHITNEY CLINIC, 78- *Concurrent Pos:* USPHS fel psychiat, Bellevue Hosp, New York, 58-60; from asst dir to assoc dir, Psychiat Div, Bellevue Hosp, 67-72. *Mem:* AAAS; fel Am Psychiat Asn; Asn Res Nerv & Ment Dis. *Res:* Drug abuse; psychopharmacology. *Mailing Add:* Cornell Univ Med Col NY Hosp 525 E 68th St New York NY 10021

FROSETH, JOHN ALLEN, b Eau Claire, Wis, July 18, 42; m 61; c 2. INTERNATIONAL ANIMAL AGRICULTURE. *Educ:* Wis State Col, BS, 64; Purdue Univ, MS, 66, PhD(animal nutrit), 70. *Prof Exp:* Instr animal sci, Purdue Univ & Nat Univ South, Arg, 67-69; asst prof & asst animal scientist, 69-74, assoc prof & assoc animal scientist, 74-82, PROF ANIMAL SCI & ANIMAL SCIENTIST, WASH STATE UNIV, 82- *Concurrent Pos:* Vis prof, Mich State Univ, 81-82, Eastern Island Univ, Indonesia, 84 & 86, Sichuan Agr Univ, China, 86. *Mem:* Am Soc Animal Sci; Am Inst Nutrit. *Res:* Barley and other Northwest grains and by-product feeds for swine; fatty acid metabolism in neonatal pigs. *Mailing Add:* Dept Animal Sci Wash State Univ Pullman WA 99164-6320

FROST, ALBERT D(ENVER), b Mass, Feb 12, 22; m 56; c 4. PHYSICAL INSTRUMENTATION. *Educ:* Tufts Univ, BS, 44; Harvard Univ, AM, 47; Mass Inst Technol, ScD(appl physics), 52. *Prof Exp:* From instr to asst prof physics, Tufts Univ, 47-57; assoc prof elec eng, 57-61, PROF ELEC ENG, UNIV NH, 61-, DIR ANTENNA SYSTS LAB, 58- *Concurrent Pos:* Res assoc acoust lab, Mass Inst Technol, 49-53; engr, Bolt Beranek & Newman, 51-53; consult, Arthur D Little, Inc, 56-57 & Stanford Res Labs, 68-71; asst dir res, Transistor Appln Inc, 59-61; staff scientist, Inst Defense Anal, 63-64; vis prof, Jodrell Bank Radio Astron Lab, Univ Manchester, 64-65, physics dept, Univ Sheffield, 71-72 & Oxford Univ, Res Lab Archaeometry, 78-79; univ fel, US Dept Transp, 81; consult scientist, radio navigation systs, Transp Systs Ctr, Dept Transp, 81-; vis prof, Kings Col, London, 89-90. *Mem:* Acoust Soc Am; Soc Archaeol Sci. *Res:* Physical acoustics; photoacoustic examination of archaeological materials; radio navigation systems; very low frequency radio propagation; precision time transfer; GPS applications. *Mailing Add:* Dept Elec Eng Antenna Systs Lab Univ NH Durham NH 03824

FROST, ARTHUR ATWATER, b Onarga, Ill, Aug 3, 09; m 34; c 3. PHYSICAL CHEMISTRY. *Educ:* Univ Calif, BS, 31; Princeton Univ, PhD(phys chem), 34. *Prof Exp:* Res fel, Harvard Univ, 34-36; instr, 36-42, from asst prof to assoc prof, 42-54, chmn dept, 57-62, prof, 54-76, EMER PROF CHEM, NORTHWESTERN UNIV, 76- *Concurrent Pos:* Adj prof, Northern Ariz Univ, 76- *Mem:* AAAS; Am Chem Soc; Am Phys Soc. *Res:* Molecular quantum mechanics. *Mailing Add:* 195 Eagle Lane Sedona AZ 86336

FROST, BARRIE JAMES, b Nelson, NZ, June 3, 39; Can/NZ citizen; m 65; c 3. VISION, SENSORY PROCESSES. *Educ:* Univ NZ, BA, 61; Univ Canterbury, MA, 64; Dalhousie Univ, PhD(psychol), 67. *Prof Exp:* Asst res physiologist, Univ Calif, Berkeley, 68-69; asst prof, 68-72, assoc prof, 72-78, PROF PSYCHOL, QUEEN'S UNIV, KINGSTON, 78- *Concurrent Pos:* Vis prof physiol, John Curtin Sch Med Res, Australian Nat Univ, 75-76; mem, Grant Selection Comt, Nat Sci & Eng Res Coun, 79-82; prin investr grants, Med Res Coun, Nat Sci & Eng Res Coun & Nat Health & Welfare. *Mem:* fel Can Psychol Asn; AAAS; Asn Res Vision & Ophthal. *Res:* Single cell recording studies of visual systems; autoradiography development of tactile vocoders for deaf. *Mailing Add:* Dept Biol Sci Queens Univ Kingston ON K7L 3N6 Can

FROST, BRIAN R T, b London, Eng, Sept 6, 26; US citizen; m 54; c 2. METALLURGY. *Educ:* Univ Birmingham, BSc, 47, PhD(metall), 49; Univ Chicago, MBA, 74. *Prof Exp:* Scientist metall, UK Atomic Energy Authority, 49-69; assoc dir metall, Argonne Nat Lab, 69-73, dir Mat Sci Div, 73-84, dir tech transfer, 85-90, SR TECH ADV, ARGONNE NAT LAB, 90- *Concurrent Pos:* Vis lectr, Queen Mary Col, Univ London, 59-63; vis prof, Northwestern Univ, 76-77; mem, Nat Mat Adv Bd, 82-88. *Mem:* Fel Am Nuclear Soc; fel Am Soc Metals; Am Inst Metall Engrs; fel Brit Inst Metals; Am Ceramics Soc; Mats Res Soc. *Res:* Radiation effects; corrosion and mechanical properties of materials for nuclear fission and fusion reactors; materials for non-nuclear energy systems; technology transfer. *Mailing Add:* Argonne Nat Lab 9700 S Cass Ave Argonne IL 60439

FROST, BRUCE WESLEY, b Brunswick, Maine, Mar 8, 41; m 63; c 2. BIOLOGICAL OCEANOGRAPHY. *Educ:* Bowdoin Col, BA, 63; Univ Calif, San Diego, PhD(oceanog), 69. *Prof Exp:* Asst prof, 69-75, assoc prof, 75-80, PROF OCEANOG, UNIV WASH, 80- *Concurrent Pos:* NSF res grant, 70- *Mem:* AAAS; Am Soc Limnol & Oceanog; Soc Syst Zool; Ecol Soc Am; Sigma Xi. *Res:* Ecology and systematics of marine zooplankton. *Mailing Add:* Sch Oceanog WB10 Univ Wash Seattle WA 98195

FROST, DAVID, b Brooklyn, NY, Dec 19, 25; m 46; c 2. SCIENCE EDITOR, SCIENCE WRITER. *Educ:* City Col New York, BS, 46, MS, 49; NY Univ, MS, 52, PhD, 60. *Prof Exp:* Tutor biol, City Col New York, 46-49; instr, pvt sch, 49-52; from instr to asst prof biol sci, Rutgers Univ, 52-59; sci ed, Squibb Inst Med Res, 59-75; CONSULT ED, 75- *Concurrent Pos:* Adj prof, Rutgers Univ, 53-78. *Mem:* Coun Biol Ed. *Res:* Genetic control of axolotl pigmentation; inhibition of lens regeneration in the newt; thin-layer chromatography of antibiotics; writings cover new drugs, surgery, medicine, chemistry, forestry, biology, optometry and environmental sciences. *Mailing Add:* 1229 E Seventh St Plainfield NJ 07062

FROST, DAVID CREGREEN, b Ormskirk, Eng, Mar 27, 29; m 60. PHYSICAL CHEMISTRY. *Educ:* Univ Liverpool, Eng, BSc, 53, PhD(chem), 58, DSc, 79. *Prof Exp:* PROF CHEM, UNIV BC, 59- *Concurrent Pos:* Guggenheim fel, 80. *Mem:* Fel Am Chem Soc; fel Can Inst Chem; fel Royal Soc Chem, UK. *Res:* U-V and X-ray photoelectron spectroscopy; ionization and dissociation of molecules by electron and photon impact in the gaseous and solid phases. *Mailing Add:* Dept Chem Univ BC 2075 Westbrook Pl Vancouver BC V6T 1W5 Can

FROST, E A M, b Glasgow, Scotland, Oct 29, 38; US citizen; wid; c 4. ANESTHESIOLOGY. *Educ:* Glasgow Univ, MBChB, 61; Am Bd Anesthesiol, dipl, 67. *Prof Exp:* PROF ANESTHESIOL, ALBERT EINSTEIN COL MED, 80- *Mailing Add:* Dept Anesthesiology Albert Einstein Col Medicine Yeshiva Univ 1300 Morris Park Ave Bronx NY 10461

FROST, H(AROLD) BONNELL, b Ft Washakie, Wyo, Dec 3, 23; m 46; c 4. COMPUTER SCIENCE, ELECTRICAL ENGINEERING. *Educ:* Univ Nebr, BS, 48; Mass Inst Technol, SM, 50, ScD(elec eng), 54. *Prof Exp:* Asst, Lincoln Lab, Mass Inst Technol, 48-54; mem tech staff, Bell Tel Labs, NJ, 54-63, Pa, 63-80, Colo, 80-84; RETIRED. *Mem:* AAAS; sr mem Inst Elec & Electronics Engrs; Sigma Xi. *Res:* Engineering applications of digital computers. *Mailing Add:* 5559 Cott Dr Longmont CO 80503-8604

FROST, HAROLD MAURICE, III, b New Haven, Conn, June 4, 42; m 64; c 1. DIELECTRICS, EFFECTS OF NUCLEAR RADIATION ON MATERIALS. *Educ:* Univ Vt, BA, 64, MS, 69, PhD(physics), 74. *Prof Exp:* Scientist biophys, Microbiol Lab, White Sands Missile Range, NMex, 67; Nat Res Coun res assoc ultrasonics, Microwave Physics Lab, Air Force Cambridge Res Labs, 74-75; res physicist ultrasonics, Bur Radiol Health, US Dept HEW, 75-77; proj scientist, Nondestructive Testing & Mat Sci Res, Pa State Univ, 77-81; physicist, E H V Weidmann Indust, Inc, St Johnsbury, Vt, 81-84; STAFF MEM PHYSICIST, MAT SCI & TECHNOL DIV, LOS

ALAMOS NAT LAB, LOS ALAMOS, NM, 84- *Concurrent Pos:* Consult, 77-; Int ed, Ultrasonics, 77-81. *Mem:* Inst Elec & Electronics Engrs; Sigma Xi; Am Ceramic Soc; Am Phys Soc; Mat Res Soc; AAAS. *Res:* Measurement of dielectric and ultrasonic properties of metals, polymers, ceramics and composites, from audio to millimeter-wave frequencies; microstructural characterization of materials; mathematical analysis; ultrasonic transducers, nondestructive evaluation; effects of high power ultrasonics and microwaves on materials; coordinating compliance with environment, safety and health federal regulations in research and development; project research and development management. *Mailing Add:* 98 Royal Crest Los Alamos NM 87544-3207

FROST, HERBERT HAMILTON, b New York, NY, Jan 22, 17; m 48; c 5. VERTEBRATE ZOOLOGY. *Educ:* Brigham Young Univ, BA, 41, MS, 47; Cornell Univ, PhD, 55. *Prof Exp:* Teacher & head dept biol sci, Ricks Col, 47-60, chmn div math & natural sci, 55-60; assoc prof zool, Brigham Young Univ, 60-67, prof zool, 67-; RETIRED. *Honors & Awards:* Comstock Award, Cornell Univ. *Mem:* AAAS; Sigma Xi; Audubon Soc. *Res:* Ornithology; natural history; conservation. *Mailing Add:* 461 East 350 South Orem UT 84058

FROST, JACKIE GENE, b Feb 18, 37; US citizen; m 63; c 2. CHEMISTRY, POLAROGRAPHY. *Educ:* Eastern Ill Univ, BS, 58; Duke Univ, PhD(polarog), 62. *Prof Exp:* Res assoc, Univ NC, 62-63; develop chemist anal instrumentation, 63-66, res chemist, 66-69, sr res chemist, 69-77, res supvr, 77-83, RES ASSOC, HALLIBURTON CO, 83- *Mem:* Am Chem Soc; Nat Asn Corrosion Engrs. *Res:* Complex ion chemistry; potentiometry; steel corrosion in alkaline chelating media; steel passivation; stability of metal complex ions; corrosion specialist. *Mailing Add:* Halliburton Serv Drawer 1431 Duncan OK 73536-0439

FROST, JOHN KINGSBURY, image analysis, environmental & industrial epidemiology; deceased, see previous edition for last biography

FROST, KENNETH ALMERON, JR, b San Francisco, Calif, Dec 12, 44; m 69; c 2. ORGANIC CHEMISTRY, FUEL SCIENCE. *Educ:* Ind Univ, AB, 66; Univ Calif, Davis, PhD(chem), 71. *Prof Exp:* Res chemist, 71-75, SR RES CHEMIST, CHEVRON RES CO, STANDARD OIL CALIF, 75-, SR RES ASSOC, 80- *Concurrent Pos:* Res unit mgr, Chevron Res & Technol Co. *Mem:* Am Chem Soc; Soc Automotive Engrs. *Res:* Fuel and lubricant additives; lubrication. *Mailing Add:* Chevron Res & Technol Co 100 Chevron Way Richmond CA 94802

FROST, L(ESLIE) S(WIFT), b Pittsburgh, Pa, May 7, 22; m 47; c 2. ELECTRICAL ENGINEERING. *Educ:* Carnegie Inst Technol, BS, 46, MS, 48, DSc(elec eng), 49. *Prof Exp:* Res engr physics, 49-73, ADV ENGR, WESTINGHOUSE RES LABS, 73- *Mem:* Am Phys Soc; Am Inst Elec & Electronics Engrs; Fedn Am Soc Exp Biol. *Res:* Studies of low pressure helium positive column; interpretation of electron transport data; metal vapor arc lamps; experimental and theoretical studies of gas-blast circuit breaker arcs; arc and plasma physics studies. *Mailing Add:* 423 Beulah Rd Pittsburgh PA 15235

FROST, LAWRENCE WILLIAM, b Fredonia, NY, Dec 2, 20; m 42; c 2. POLYMER CHEMISTRY. *Educ:* Allegheny Col, AB, 41; Purdue Univ, PhD(org chem), 44. *Prof Exp:* Res Found fel, Purdue Univ, 44-46; res chemist, Westinghouse Elec Corp, 46-59, fel chemist, 59-73, adv chemist, 73-83; RETIRED. *Mem:* AAAS; Am Chem Soc. *Res:* Organic fluorine and organo silicon compounds; aromatic polyimides; high temperature electrical varnishes and wire enamels; laminates. *Mailing Add:* 3265 Chalmers Ave Murrysville PA 15668

FROST, PAUL D(AVIS), physical metallurgy, process metallurgy, for more information see previous edition

FROST, ROBERT EDWIN, b Gowanda, NY, Feb 1, 32; m 58; c 3. INORGANIC CHEMISTRY. *Educ:* Allegheny Col, BS, 53; Harvard Univ, AM, 55, PhD(chem), 57. *Prof Exp:* Res chemist, Res Ctr, B F Goodrich Co, 57-61; assoc prof, 61-64, PROF CHEM, STATE UNIV NY ALBANY, 64- *Concurrent Pos:* Vis lectr, Univ Ill, 65-66. *Mem:* Am Chem Soc. *Res:* Transition metal complexes; organometallic chemistry. *Mailing Add:* Dept Chem State Univ NY Albany NY 12222

FROST, ROBERT HARTWIG, b Riverside, Calif, July 2, 17; m 44; c 2. EXPERIMENTAL PHYSICS. *Educ:* Univ Calif, AB, 39, MA, 45, PhD(physics), 47. *Prof Exp:* Asst physics, Univ Calif, 41-45; asst prof, Univ Mo, 47-53; from asst prof to prof, 53-83, head dept, 71-83, EMER PROF PHYSICS CALIF POLYTECH STATE UNIV, SAN LUIS OBISPO, 83- *Mem:* AAAS; Am Asn Physics Teachers; Am Phys Soc; Sigma Xi. *Res:* Specific ionization of electrons. *Mailing Add:* 215 Longview Lane San Luis Obispo CA 93405

FROST, ROBERT T, b Towson, Md, Sept, 14, 24; m 48; c 5. PHYSICS. *Educ:* Johns Hopkins Univ, AB, 49, PhD, 53. *Prof Exp:* Jr instr physics, Johns Hopkins Univ, 49-51; res assoc, Knolls Atomic Power Lab, 53-55, mgr critical assemblies groups, 55-57, adv reactor physics, 57-59, adv concepts, 59-61, consult, 61-62, mgr space physics, 62-70, mgr earth orbital opers, Space Sci Lab, 70-80, CONSULT ENG, GEN ELEC CO, 80- *Concurrent Pos:* Mem space & atmos physics comt, Am Inst Aeronaut & Astronaut, chmn space processing tech comt. *Mem:* Am Phys Soc; Am Geophys Union. *Res:* Reactions of light nuclei; nuclear reactors; space physics; development of airborne and satellite experiments for remote sensing of earth and atmosphere, electromagnetic levitation and zero gravity metals solidification studies. *Mailing Add:* PO Box 269 Accomac VA 23301

FROST, STANLEY H, geology, paleontology, for more information see previous edition

FROST, TERRENCE PARKER, b Concord, NH, Apr 29, 21; m 57; c 2. WATER POLLUTION. *Educ:* Univ NH, BS, 42, MS, 55. *Prof Exp:* Water pollution biologist, 47-67, chief aquatic biologist & dir permits & enforcement planning, 67-81, chief aquatic biologist, 67-82, dir biosyst & basin planning, 73-77, lab dir, 81-82, CONSULT WATER POLLUTION BIOLOGIST, 82- *Concurrent Pos:* Mem, NH Pesticides Control Bd, 65-79, Gov Adv Comt Oceanog, 66-70, NH dredge bd, 67-69, NH Gov Citizens Task Force Subcomt Indust & Domestic Wastes, 69-70, NH Water Resource Res Ctr Adv Bd, 69-82, NH Bulk Power Site Eval Comt, 71-82, NH Wetlands Bd, 79- & Vt Yankee Nuclear Power Co Tech Adv Bd, 73-; ed, J NH Water Works Asn, 71-; lectr water resource mgt, New Eng Col, 71-78. *Mem:* AAAS; Water Pollution Control Fedn; Am Soc Limnol & Oceanog; Am Fisheries Soc; Nat Environ Health Asn. *Res:* Travel of underground pollution plumes; deep-lake homogenization for algae control; nutrient removal by duckweed harvest; pollutant removal efficiency of spray irrigation systems on forestlands; acid rainfall; phosphorus rainfall. *Mailing Add:* 37 Clinton St Concord NH 03301

FROST, WALTER, b Edmonton, Alta, Apr 6, 35; m 62; c 2. MECHANICAL ENGINEERING, HEAT TRANSFER. *Educ:* Univ Wash, Seattle, BS, 61, MS, 63, PhD(mech eng), 65. *Prof Exp:* Assoc prof mech eng, 65-76, PROF AVIATION SYSTS, SPACE INST, UNIV TENN, 76- *Mem:* Am Soc Mech Engrs; Am Inst Aeronaut & Astronaut; Sigma Xi. *Res:* Boiling heat transfer and two phase flow; radiation and convection heat transfer from finned surfaces; natural convection from finite horizontal surfaces; heat transfer in porous media. *Mailing Add:* 217 Lakewood Dr Tullahoma TN 37388

FROSTICK, FREDERICK CHARLES, JR, organic chemistry, for more information see previous edition

FROST-MASON, SALLY KAY, b New York, NY, May 29, 50; m 90. DEVELOPMENTAL BIOLOGY, PIGMENT CELL BIOLOGY. *Educ:* Univ Ky, BA, 72; Purdue Univ, MS, 74; Univ Ariz, PhD(cellular, molecular & develop biol), 78. *Prof Exp:* Postdoctoral res assoc develop biol, Ind Univ, Bloomington, 78-80; from asst prof to assoc prof, 80-91, PROF BIOL, DEPT PHYSIOL & CELL BIOL, UNIV KANS, 91-, ASSOC DEAN, DIV DEAN NATURAL SCI & MATH, COL LIB ARTS & SCI, 90- *Concurrent Pos:* Prin investr, NSF, NIH & Wesley grants, 81-; actg chairperson, Dept Physiol & Cell Biol, Univ Kans, 86-89. *Mem:* AAAS; Soc Develop Biol; Am Soc Zool; Am Soc Cell Biol; Int Pigment Cell Soc. *Res:* Developmental biology and genetics of pigment cell differentiation from the vertebrate embryonic neural crest; mechanisms underlying embryonic cell differentiation and pattern formation. *Mailing Add:* Dept Physiol & Cell Biol Univ Kans Lawrence KS 66045

FROTHINGHAM, THOMAS ELIOT, b Boston, Mass, June 21, 26; c 4. PEDIATRICS, PUBLIC HEALTH. *Educ:* Harvard Med Sch, MD, 51. *Prof Exp:* Asst prof epidemiol, Tulane Univ, 59-60; assoc mem, Pub Health Res Inst, City New York, 60-61; from asst prof to assoc prof trop pub health, Harvard Univ, 61-69; mem staff, Corvallis Clin, 69-73; dir area health educ prog, 73-80, dir, Div Gen Pediat, 73-88, PROF PEDIAT & COMMUNITY & FAMILY MED, MED CTR, DUKE UNIV, 73- *Concurrent Pos:* Teaching fel pediat, Harvard Univ, 56-57, res fel, 57-58. *Mem:* Am Soc Trop Med & Hyg; Am Acad Pediat. *Res:* Infectious diseases; clinical, laboratory and public health aspects; child health; clinical and public health aspects. *Mailing Add:* Box 3937 Duke Univ Med Ctr Durham NC 27710

FROUNFELKER, ROBERT E, b West Allis, Wis, Apr 20, 25; m 45; c 4. ENGINEERING SCIENCE. *Educ:* Marquette Univ, BME, 45, MS, 48 & 63. *Prof Exp:* Asst physics, Marquette Univ, 46-48, from instr to asst prof math, 48-63; assoc prof, 63-66, PROF ENG SCI, TENN TECHNOL UNIV, 66- *Concurrent Pos:* Consult engr, Bayley Blower Co, 58-63, Sheldons Eng Ltd, 59-63 & Clarage Fan Co, 59-63; comput consult, Johnson Serv Co, 60-63. *Res:* Materials and computer science; computer techniques for rating capacities of air moving equipment; determination of the cohesive and surface energies of ionic crystals. *Mailing Add:* Dept Mech Eng Tenn Technol Univ Cookeville TN 38505

FROYD, JAMES DONALD, b Brooklyn, NY, May 25, 39; m 63; c 3. PLANT PATHOLOGY. *Educ:* Denison Univ, BS, 61; Univ Minn, MS, 64, PhD(plant path), 67. *Prof Exp:* Res asst plant dis, Univ Minn, St Paul, 61-65, exten plant pathologist, 65-67; sr plant pathologist, Eli Lilly & Co, 67-85; RES FEL, AM CYANAMID, 85- *Mem:* Am Phytopath Soc; Int Soc Plant Pathologists; Sigma Xi; Soc Nematologists. *Res:* Ecology of the dissemination of forest pathogens; diseases of ornamentals; chemical control of plant diseases; systemic fungicides; discovery of new fungicides. *Mailing Add:* Am Cyanamid Res & Develop PO Box 400 Princeton NJ 08543

FROYEN, SVERRE, b Oslo, Norway, June 25, 51; m 84; c 2. PHYSICS. *Educ:* NTH, Trondheim, Norway, Siv Ing, 76; Stanford Univ, PhD(appl physics), 81. *Prof Exp:* Fel physics, Univ Calif, Berkeley, 80-83; fel physics, Nordita, Copenhagen, 83-85; staff scientist, 85-90, SR SCIENTIST, SOLAR ENERGY RES INST, 90- *Mem:* Am Phys Soc. *Res:* Theoretical solid state physics; electronic structure and structural properties of semiconductors and metals. *Mailing Add:* Solar Energy Res Inst 1617 Cole Blvd Golden CO 80401

FRUCHTER, JONATHAN S, b San Antonio, Tex, June 5, 45; m 73; c 2. ENVIRONMENTAL CHEMISTRY, GEOCHEMISTRY. *Educ:* Univ Tex, Austin, BS, 66; Univ Calif, San Diego, PhD(chem), 71. *Prof Exp:* Res asst chem, Univ Calif, San Diego, 66-71; res assoc geochem, Univ Ore, 71-74; sr res scientist environ chem & biochem, 74-79, mgr, Geochem Sect, 80-87, STAFF SCIENTIST, PAC NORTHWEST DIV, BATTELLE MEM INST, 88- *Mem:* Am Chem Soc; Geol Soc Am; AAAS. *Res:* Environmental and analytical chemistry of products and effluents from synthetic fossil fuel plants, especially oil shale; trace elements in igneous rocks and meteorites; interactions of cosmic rays with matter, nuclear waste migration, toxic waste migration; chemistry of coal combustion wastes. *Mailing Add:* Pac Northwest Div PO Box 999 Richland WA 99352

FRUEH, ALFRED JOSEPH, JR, b Passaic, NJ, Sept 2, 19; m 43; c 3. CRYSTALLOGRAPHY, MINERALOGY. *Educ:* Mass Inst Technol, BS, 42, MS, 47, PhD(mineral), 49. *Prof Exp:* Res assoc, Univ Chicago, 49-52, asst prof mineral & crystallog, 52-58; vis prof, Univ Oslo, 58-59; from asst prof to prof crystallog, McGill Univ, 59-69; chmn dept, 69-78, prof, 69-87, EMER PROF GEOL, UNIV CONN, 87- *Concurrent Pos:* Chmn, Teaching Comn, Int Union Crystallog, 60-66, Int Union Crystallog rep, Int Coun Sci Unions, 63-66; co-ed, Structures Reports, 60-63 & 69-75; chmn & organizer, Int Mineral Asn, 62 Symp Sulfide Mineralogy; assoc ed, Am Mineralogist, 62-65; mem adv comt crystallog, Nat Res Coun Can, 66-69. *Mem:* Mineral Soc Am; Geol Soc Am; Am Crystallog Asn; Fr Soc Mineral & Crystallog; Brit Mineral Soc. *Res:* X-ray crystallography; mineral structures; solid state physics and chemistry applied to minerals. *Mailing Add:* 23 Bundy Lane Storrs CT 06268

FRUEHAN, RICHARD J, b Scranton, Pa, Feb 22, 42; m 61; c 3. PHYSICAL CHEMISTRY. *Educ:* Univ Pa, BS, 63, PhD(metall eng). *Prof Exp:* Res consult, US Steel Res, 66-80; PROF METALL ENG, CARNEGIE MELLON UNIV, 80- *Concurrent Pos:* Assoc ed, Metall Trans, 80-; dir, Ctr Iron & Steelmaking Res, 85- *Honors & Awards:* Hunt Medal, Am Inst Metall Engrs, 70 & 81, Chipman Medal, 81; Mathewson Gold Medal, Metall Soc, 89. *Mem:* Am Inst Metall Engrs (pres, 90-91). *Res:* Physical chemistry of metallurgical reactions primarily in the area of steelmaking and other metals production. *Mailing Add:* MEMS Dept Carnegie Mellon Univ Pittsburgh PA 15213

FRUHMAN, GEORGE JOSHUA, b Boston, Mass, Oct 17, 24. ANATOMY. *Educ:* NY Univ, AB, 48, MS, 52, PhD, 54. *Prof Exp:* Asst prof biol, Stern Col Women, Yeshiva Univ, 55-58; asst prof, 56-70, ASSOC PROF ANAT, ALBERT EINSTEIN COL MED, 70- *Concurrent Pos:* Fel, Damon Runyon Found, NY Univ, 54-56; lectr, NY Univ. *Mem:* Fel NY Acad Sci; Am Asn Anatomists; Am Soc Zoologists; Reticuloendothelial Soc; Soc Exp Biol & Med. *Res:* Experimental hematology; histology. *Mailing Add:* Dept Anat Albert Einstein Col Med 1300 Morris Park Ave New York NY 10461

FRUMERMAN, ROBERT, b Rochester, Pa, Aug 14, 24; m 55; c 2. COAL CONVERSION TECHNOLOGY, CHEMICAL PROCESS DEVELOPMENT. *Educ:* Univ Pittsburgh, BS, 47; Carnegie-Mellon Univ, MS, 55. *Prof Exp:* PRES & CHIEF CONSULT ENGR, FRUMERMAN ASSOCS INC, 62- *Mem:* Fel Am Inst Chem Engrs; Air Pollution Control Asn; Am Asn Cost Engrs. *Res:* Coal gasification systems; cleanup of gas from coal, cold and hot; research and development management methodology; energy & environmental systems. *Mailing Add:* Frumerman Assocs Inc 218 S Trenton Ave Pittsburgh PA 15221

FRUMKES, THOMAS EUGENE, b Rochester, NY, July 25, 41. NEUROSCIENCES, PHYSIOLOGICAL OPTICS. *Educ:* Cornell Univ, AB, 63; Syracuse Univ, PhD(physiol psychol), 67. *Prof Exp:* Asst, Syracuse Univ, 63-64; NIH fel, Queens Col, NY, 67-69, lectr psychol, 68-69, from asst prof to assoc prof, 69-78, PROF PSYCHOL, QUEENS COL, NY, 78-, PROF BIOL & NEUROBIOL, 87- *Concurrent Pos:* Res assoc, Dept Neurosurg, Mt Sinai Hosp, City Univ NY, 69-70, Dept Ophthal, Col Physicians & Surgeons, Columbia Univ, 70-71, Dept Physiol, State Univ NY, Buffalo, 76-77, Max-Plank Inst, 83-84, Moorefields Eye Hosp, 87, NIH, 90-91. *Mem:* AAAS; Optical Soc Am; Asn Res Vision & Ophthal; NY Acad Sci; Soc Neurosci; Sigma Xi. *Res:* Vision; neurophysiology; psychophysics; sensory processes; central nervous system. *Mailing Add:* Dept Psychol Queens Col Flushing NY 11367

FRUMP, JOHN ADAMS, b Butterfield, Ark, Oct 1, 21; m 46; c 3. ORGANIC CHEMISTRY. *Educ:* Ind Univ, BS, 50. *Prof Exp:* Chemist, Chrysler Corp, 50-51; plant chemist, Autolite Corp, 51-52; res chemist, Int Minerals & Chem Corp, 52-85; RETIRED. *Mem:* Am Chem Soc. *Res:* Oxazoline and oxazolidine chemistry; emulsion chemistry; nitroparaffin chemistry; process development for recovery of organic compounds. *Mailing Add:* RR 21 Box 701 Terre Haute IN 47802

FRUSHOUR, BRUCE GEORGE, b Frederick, Md, Sept 27, 47; m 71; c 2. MACROMOLECULAR SCIENCE, PHYSICAL CHEMISTRY. *Educ:* Juniata Col, BS, 69; Case Western Reserve Univ, MS, 72, PhD(macromolecular sci), 74. *Prof Exp:* Sr res chemist, Monsanto Polymer Prod Co, 74-77, res specialist, Textiles Div, 77-80, sr res specialist, 80-84, SCI FEL, MONSANTO POLYMER PROD CO, 84- *Concurrent Pos:* Adj assoc prof, Sch Textiles, NC State Univ, 79- *Mem:* Am Chem Soc; Am Phys Soc. *Res:* Structure of acrylic and nylon fibers; mechanical properties of plastics; thermal analysis; high-performance thermoplastics. *Mailing Add:* Monsanto Co 730 Worchester St Indian Orchard MA 01151

FRUTHALER, GEORGE JAMES, JR, b New Orleans, La, Aug 8, 25; m 48; c 3. MEDICINE. *Educ:* Tulane Univ, BS, 46, MD, 48; Am Bd Pediat, dipl, 54, 80, 86. *Prof Exp:* Intern, Charity Hosp, La, 48-49 & pediat, Vanderbilt Univ Hosp, 49-50; from instr to assoc prof, 54-68, PROF CLIN PEDIAT, TULANE UNIV, 68-; MEM PEDIAT STAFF, OCHSNER CLIN, 53- *Concurrent Pos:* Ochsner Med Found fel, 50-51 & 53; asst vis physician, Charity Hosp, 50-68 & vis physician, 68-; mem, Am Bd Pediat, 66-72. *Mem:* AMA; Am Acad Pediat. *Res:* Pediatrics; allergy. *Mailing Add:* Ochsner Clin 1514 Jefferson Hwy New Orleans LA 70121

FRUTIGER, ROBERT LESTER, b Harrisburg, Pa, Sept 22, 46; m 72; c 2. MECHANICAL ENGINEERING, SOLID MECHANICS. *Educ:* Univ Pittsburgh, BS, 69, MS, 72, PhD(mech eng), 75. *Prof Exp:* Asst instr, Univ Pittsburgh, 70-75; mem res staff plastics processing, Eng Res Ctr, Western Elec Co, 75-79; SR RES ENGR, ENG MECH, GEN MOTORS RES LABS, 79- *Mem:* Am Soc Mech Engrs; Soc Rheology; Sigma Xi. *Res:* Viscoelasticity; finite elasticity; continuum mechanics; polymer processing. *Mailing Add:* Eng Mech Dept Gen Motors Res Labs Warren MI 48090

FRUTON, JOSEPH STEWART, b Czestochowa, Poland, May 14, 12; nat US; m 36. BIOCHEMISTRY, HISTORY OF SCIENCE. *Educ:* Columbia Univ, BA, 31, PhD(biochem), 34. *Hon Degrees:* MA, Yale Univ, 50; DSc, Rockefeller Univ, 76. *Prof Exp:* Asst biochem, Col Physicians & Surgeons, Columbia Univ, 33-34; fel chem, Rockefeller Inst, 34-35, asst, 35-38, assoc, 38-45; assoc prof physiol chem, 45-50, prof biochem, 50-57, chmn dept, 51-67, Eugene Higgins prof biochem, 57-82, dir div sci, 59-62, prof hist med, 80-82, EMER PROF BIOCHEM & HIST MED, YALE UNIV, 82- *Concurrent Pos:* Chmn panel enzymes, Comt Growth, Nat Res Coun, 46-49, mem biochem, Adv Comt, Chem Biol Coord Ctr, 46-51 & Div Chem & Chem Technol, 50-52, fel bd, 53-60, chmn, 58-60, mem, Exec Comt, Div Med Sci, 60-63; spec fel, Rockefeller Found, 48; consult, Anna Fuller Fund, 51-72; Harvey lectr, 55; Dakin lectr, 62; Commonwealth Fund fel, 62-63; mem, Int Union Pure & Appl Chem-Int Union Biochem, 64-69; vis prof, Rockefeller Univ, 68-69; Sarton Lectr, 76 & Xerox, 77; Guggenheim Found fel, 83-84; Benjamin Franklin fel, Royal Soc Arts; exec secy, Corp Presidential Search Comt, Yale Univ, 85-86. *Honors & Awards:* Lilly Award, Am Chem Soc, 44; Pfizer Award, Hist Sci Soc, 73; Lewis Award, Am Philos Soc, 90. *Mem:* Nat Acad Sci; AAAS; Am Chem Soc; NY Acad Sci (vpres, 45-46); Am Soc Biol Chem; Am Philos Soc; Hist Sci Soc; Am Acad Arts Sci. *Res:* Chemistry and metabolism of amino acids, peptides and proteins; enzymes; history of science. *Mailing Add:* 123 York St New Haven CT 06511

FRY, ALBERT JOSEPH, b Philadelphia, Pa, May 12, 37; m 66; c 3. ORGANIC ELECTROCHEMISTRY. *Educ:* Univ Mich, BS, 58; Univ Wis, PhD(org chem), 63. *Prof Exp:* Res fel chem, Calif Inst Technol, 63-64; res fel, 64-65, from asst prof to assoc prof, 65-77, PROF CHEM, WESLEYAN UNIV, 77- *Mem:* Am Chem Soc; Royal Soc Chem; Int Electrochem Soc. *Res:* Nonbenzenoid aromatic species; synthetic and mechanistic organic electrochemistry. *Mailing Add:* Hall-Atwater Lab Chem Wesleyan Univ Middletown CT 06457

FRY, ANNE EVANS, b Philadelphia, Pa, Sept 11, 39. ZOOLOGY, DEVELOPMENTAL BIOLOGY. *Educ:* Mt Holyoke Col, AB, 61; Univ Iowa, MS, 63; Univ Mass, PhD(zool), 69. *Prof Exp:* Instr biol, Carleton Col, 63-65; from asst prof to assoc prof, 69-80, PROF BIOL SCI, OHIO WESLEYAN UNIV, 80- *Mem:* AAAS; Am Inst Biol Sci; Am Soc Zoologists; Soc Develop Biol; Sigma Xi. *Res:* Amphibian metamorphosis. *Mailing Add:* Dept Zool Ohio Wesleyan Univ 70 S Henry St Delaware OH 43015

FRY, ARTHUR JAMES, b Dodson, Mont, Mar 10, 21; m 47; c 3. ORGANIC REACTION MECHANISMS. *Educ:* Mont State Col, BS, 43; Univ Calif, Berkeley, PhD(chem), 51. *Prof Exp:* Assoc chemist, Oak Ridge Nat Lab, 46-48; from asst prof to prof, 51-84, chmn dept, 56-57 & 64-67, UNIV PROF CHEM, UNIV ARK, FAYETTEVILLE, 84- *Concurrent Pos:* Vis prof, Univ Auckland, 69-70 & 78, Univ Adelaide, 70, Monash Univ, 77, Univ Sheffield, 88, NC State Univ, 88. *Honors & Awards:* Southwest Region Chemist Award, Am Chem Soc, 85. *Mem:* AAAS; Am Chem Soc; The Chem Soc. *Res:* Organic reaction mechanisms; heavy atom isotope effects on rates and mechanisms of chemical reactions; tracer studies of mechanisms of acid catalyzed ketone rearrangements; tracer and isotope effect of organic reaction mechanisms. *Mailing Add:* Dept Chem Chem Bldg Univ Ark Fayetteville AR 72701

FRY, CLEOTA GAGE, b Shoshone, Idaho, Dec 30, 10. MATHEMATICS. *Educ:* Reed Col, BA, 33; Purdue Univ, MS, 36, PhD(math), 39. *Prof Exp:* Asst instr math, 39-40, instr physics, 41-45, from instr to asst prof math, 45-55, asst to dean sch sci, 52-61, assoc prof, 55-77, EMER ASSOC PROF MATH, PURDUE UNIV, 77- *Mem:* Am Phys Soc; Am Math Soc; Sigma Xi. *Res:* Mathematical analysis in field of complex variables; acoustics; theoretical physics; calculation of atomic form factors; applied mathematics. *Mailing Add:* Dept Math Purdue Univ W Lafayette IN 47907

FRY, DAVID LLOYD GEORGE, b Detroit, Mich, Sept 22, 18; m 42; c 3. PHYSICS. *Educ:* Kalamazoo Col, BA, 40; Ohio State Univ, MS, 42. *Prof Exp:* Asst, Kalamazoo Col, 39-40; res physicist, Gen Motors Corp, 42-48, group leader spectros, 48-52, supvr spectros, 52-56, supvr chem physics, 56-65, supvr chem physics & magnetics, 65-76, dept res scientist, Res Labs, 76-80; RETIRED. *Mem:* Fel AAAS; Am Phys Soc; fel Optical Soc Am. *Res:* Emission and absorption spectroscopy in ultraviolet, visible and infrared; mass spectroscopy; diffusion of gases in solids; nuclear magnetic resonance; internal friction; electro-optics; magnetics; atmopsheric physics; combustion physics. *Mailing Add:* 685 Princeton Rd Berkley MI 48072-3055

FRY, DONALD LEWIS, b Des Moines, Iowa, Dec 29, 24; m 57; c 4. PHYSIOLOGY. *Educ:* Harvard Univ, MD, 49. *Prof Exp:* Intern med, Univ Minn Hosps, 49-50, res med, 50-52, res fel, Heart Hosp, 52-53; independent investr, Nat Heart Inst, NIH, 53-61, head, Sect Clin Biophysics, Nat Heart & Lung Inst, 61-71, head, Sect Exp Atherosclerosis, 71-75, chief, Lab Exp Atherosclerosis, Nat Heart & Lung Inst, 75-80; PROF MED PATH & DIR, LAB EXP ATHEROSCLEROSIS, COL MED, OHIO STATE UNIV, COLUMBUS, 80- *Concurrent Pos:* Fel, Var Club Heart Hosp, 52-53. *Mem:* AAAS; Am Physiol Soc; Biophys Soc; Am Soc Clin Invest; NY Acad Sci. *Res:* Physiology and biophysics of the blood-vascular interface, with particular emphasis on mechanochemical events associated with endothelial degeneration, transvascular transport and the genesis of atherosclerosis. *Mailing Add:* Dept Med & Pathol Lab Exp Atherosclerosis 2025 Wiseman Hall Ohio State Univ Col Med 400 W 12th Ave Columbus OH 43210

FRY, EDWARD IRAD, b US, Jan 7, 24; m 50. BIOLOGICAL ANTHROPOLOGY. *Educ:* Univ Tex, BA, 49, MA, 50; Harvard Univ, PhD(anthrop), 58. *Prof Exp:* Off coordr, Values Study, Harvard Univ, 50-51; asst prof anthrop & tech dir, Anthropomet Proj, Antioch Col, 55-56; from instr to assoc prof anthrop, Univ Nebr, Lincoln, 56-66; prof anthrop, Southern Methodist Univ, 66-87; CONSULT, 87- *Concurrent Pos:* Fulbright fel, Univ NZ, Cook Islands, 53-54; field consult, Carnegie Inst, Mex, 55; Fulbright res fel, Univ Hong Kong, 63-64; chmn sect H, AAAS, 80-81. *Mem:* AAAS; Am Anthrop Asn; Am Asn Phys Anthropologists (pres, 73-75); Soc Study Human Biol. *Res:* Human biology, especially growth, skeletal age and bone; Polynesia; United States; Southeast Asia. *Mailing Add:* Dept Anthrop Southern Methodist Univ Dallas TX 75275

FRY, EDWARD STRAUSS, b Meadville, Pa, July 27, 40; m 64; c 3. PHYSICS, ATOMIC & MOLECULAR PHYSICS. *Educ:* Univ Mich, BS, 62, MS, 63, PhD(physics), 69. *Prof Exp:* From asst prof to assoc prof physics, Tex A&M Univ, 69-77; vis assoc prof, Univ Mich, 77-79; assoc prof, 79-86, PROF PHYSICS, TEX A&M UNIV, 86- *Mem:* Optical Soc Am; Am Phys Soc. *Res:* Experimental atomic physics; foundations of quantum mechanics; tests for parity violation; polarized light scattering studies. *Mailing Add:* Dept Physics Tex A&M Univ College Station TX 77843

FRY, FRANCIS J(OHN), b Johnstown, Pa, Apr 2, 20; m 46; c 9. ELECTRICAL ENGINEERING. *Educ:* Pa State Univ, BS, 40; Univ Pittsburgh, MS, 46. *Prof Exp:* Design engr, Westinghouse Elec Corp, 40-46; res assoc elec eng, Univ Ill, Urbana, 46-50, res asst prof, 50-57, res assoc prof, 57-68, assoc prof, 68-72; assoc dir, 72-78, DIR, ULTRASOUND RES LABS, INDIANAPOLIS CTR ADVAN RES, 78-; ASSOC PROF SURG, IND SCH MED, 72- *Concurrent Pos:* Vpres, Intersci Res Inst, 57-68, pres, 68-70, secy & chmn bd dirs, 70-72; vis assoc prof elec eng, Univ Ill, Urbana, 72- *Honors & Awards:* Pioneer Award, Am Inst Ultrasound Med, 81. *Mem:* Am Inst Ultrasound Med; Am Soc Artificial Internal Organs; Neurosci Soc; Biomed Eng Soc; Acoust Soc Am; Sigma Xi. *Res:* Application of ultrasound to medical and biological problems; artificial hearts, blood flow; effects of ultrasound on biological systems; ultrasonic soft tissue visualization and tissue modifying systems; neuroscience, quantitative aspects of structural organization in brain; relations between structure and function. *Mailing Add:* 414 Spring Lake Blvd NW Port Charlotte FL 33952-6429

FRY, FREDERICK ERNEST JOSEPH, ecology; deceased, see previous edition for last biography

FRY, HAROLD, mechanical engineering, for more information see previous edition

FRY, JACK L, b Thomas, Okla, Dec 24, 30; m 52; c 3. ACADEMIC ADMINISTRATION, POULTRY SCIENCE. *Educ:* Okla State Univ, BS, 52, MS, 56; Purdue Univ, PhD(food technol), 59. *Prof Exp:* Asst prof, Kans State Univ, 59-64; assoc prof poultry sci, 64-70, fac develop grant, 70, PROF POULTRY SCI, UNIV FLA, 70-, ASST DEAN COL AGR, 75- *Honors & Awards:* Inst Am Poultry Industs Award, 69. *Mem:* Poultry Sci Asn; World Poultry Sci Asn; Inst Food Technologists. *Mailing Add:* Col Agr Univ Fla 2001 McCarty Hall Gainesville FL 32611

FRY, JAMES LESLIE, b Fostoria, Ohio, May 24, 41; m 64; c 1. ORGANIC & STRUCTURAL CHEMISTRY, COMPUTATIONAL CHEMISTRY. *Educ:* Bowling Green State Univ, BS, 63; Mich State Univ, PhD(org chem), 67. *Prof Exp:* NIH fel org chem, Princeton Univ, 67-69; from asst prof to assoc prof, 69-78, PROF CHEM, UNIV TOLEDO, 78-, CHMN, 87- *Concurrent Pos:* Vis scholar, Univ Sci & Technol, Languedos, France, 82-83. *Honors & Awards:* Outstanding Res Award, Sigma Xi, 88. *Mem:* Am Chem Soc; The Chem Soc; Sigma Xi; Int Union Pure & Appl Chem; Am Asn Univ Prof. *Res:* Kinetic isotope effects; asymmetric syntheses; reaction mechanisms; synthetic methodology; organosilicon chemistry; computational chemistry; surface chemistry. *Mailing Add:* Dept Chem Univ Toledo Toledo OH 43606

FRY, JAMES N, b Philadelphia, Pa, Aug 6, 52. THEORETICAL COSMOLOGY. *Educ:* Cornell Univ, BS, 74; Princeton Univ, MA, 76, PhD(physics), 79. *Prof Exp:* McCormick fel, Univ Chicago, 79-81, res assoc theoret astrophys, 81-82, sr res assoc, 82-83; vis asst res scientist, Inst Theoret Physics, 84; asst prof, 83-89, ASSOC PROF PHYSICS, UNIV FLA, 89- *Concurrent Pos:* Vis asst res scientist, Inst Theoret Physics, Santa Barbara, Calif, 84; vis scholar, Inst Astron, Cambridge, UK, 82. *Mem:* Am Astron Soc; Am Phys Soc; Astron Soc Pac. *Res:* Nature of the large scale structure of the universe; implications of grand unified theories in the early universe; statistical treatment of the clustering of galaxies. *Mailing Add:* Dept Physics 215 Williamson Hall Univ Fla Gainesville FL 32611

FRY, JAMES PALMER, b Detroit, Mich, May 2, 39; m 61; c 6. INTELLIGENT SYSTEMS. *Educ:* Univ Mich, BSE, 62, MSE, 63 & 74. *Prof Exp:* Engr, E I du Pont de Nemours, 62-63; res engr, Boeing Co, 64; mem tech staff, Mitre Corp, 65-69, head tech transfer, 69-70; sr res assoc, dept indust opers eng, Univ Mich, 70-75; asst res scientist, 75-77, dir, info syst res group, 77-84 & comput serv, 79-82, assoc res scientist, 82-85, ASST PROF COMPUT INFO SYST, GRAD SCH BUS, UNIV MICH, 85- *Concurrent Pos:* Nat lectr, Asn Comput Mach, 69-70; consult, ADP Corp, Dana Corp & Standard Oil. *Mem:* Asn Comput Mach. *Res:* Effective design of decision support, expert and information systems, both centralized and distributed; development of model-based information systems; strategic models for information systems. *Mailing Add:* Dept Comput Info Syst Univ Mich Main Campus Ann Arbor MI 48109-1234

FRY, JOHN CRAIG, b Salem, Ore, Dec 11, 26; m 49; c 3. MARINE SCIENCE, SCIENCE POLICY. *Educ:* US Naval Acad, BS, 47; Univ Calif, MS, 52. *Prof Exp:* Prog officer acoust, Naval Underwater Sound Lab, US Navy, 56-59, proj officer, Cruiser Destroyer Flotilla Three, 60-62, head, Requirements Br, Oceanog Div, Off Chief Naval Opers, 62-65; tech asst oceanog, Off Sci & Technol, Exec Off President, 65-66, sr staff mem, Nat Coun Marine Resources & Eng Develop, 67-69; vpres, Ocean Data Systs, Inc, 69-71; dep dir, Off Sci & Technol, AID, 71-76; dir, Off Bilateral & Multilateral Sci & Technol Affairs, Dept State, 76-80; personal rank minister-counr, US Mission Europ Communities, Brussels, 81-84, Am Embassy, Stockholm, 82, counr, Sci Technol Affairs, 84-86; SR RES FEL, NAT DEFENSE UNIV, WASHINGTON, DC, 87- *Mem:* Sigma Xi. *Res:* Science technology for international development. *Mailing Add:* 173 Prince George St Annapolis MD 21401

FRY, JOHN SEDGWICK, b Philadelphia, Pa, Oct 22, 29; m 51; c 2. ORGANIC CHEMISTRY. *Educ:* Duke Univ, BS, 51, MA, 55. *Prof Exp:* Proj scientist, Plastics Div, 55-68, res scientist, res & develop dept, 68-83, SR DEVELOP SCIENTIST, UNION CARBIDE CORP, 83- *Res:* Polymers and resins and derived products. *Mailing Add:* 14 Westbrook Ave Somerville NJ 08876-4815

FRY, LOUIS RUMMEL, b New York, NY, Aug 8, 28; m 65. ORTHOPEDIC SURGERY. *Educ:* Denison Univ, BA, 51; Temple Univ, MD, 55; Am Bd Orthop Surg, dipl. *Prof Exp:* Intern, Sch Med, Temple Univ, 56; instr anat, Med Sch, Univ Mich, 57-58, resident, orthop, 61, instr orthop surg, 61-62; from instr to assoc prof orthop surg, Sch Med, Univ Wash, CLIN CONSULT, SCHS MED, UNIV ORE, PORTLAND & UNIV WASH, 75- *Concurrent Pos:* NIH training grant, 64-65; adj prof, Pac Univ, Forest Grove, Ore, 77-; mem staff, Tuality Community Hosp, chief staff, 79-80. *Mem:* AAAS; fel Am Col Surg; fel Am Acad Orthop Surg; Orthop Res Soc; AMA. *Res:* Cell morphology; cartilage; ultrastructure; bone ultrastructure. *Mailing Add:* Univ Ore Tuolity Comm Hosp 349 SE Seventh St Hillsboro OR 97123

FRY, PEGGY CROOKE, b Conroe, Tex, Oct 24, 28; m 50. NUTRITION. *Educ:* Univ Tex, BS, 49; Harvard Univ, MPH, 55; Univ Nebr, Lincoln, PhD(nutrit), 67. *Prof Exp:* Am Dietetic Asn admin dietetic intern, Aetna Life Ins Co, Hartford, Conn, 49-50; asst dietitian, Rice Univ, 50; asst admin dietitian, Mt Auburn Hosp, Cambridge, Mass, 51; nutritionist, Forsyth Dent Infirmary for Children, Boston, 51-53; org chemist, Charles F Kettering Found, 55-56; from instr to asst prof food & nutrit, Univ Nebr, Lincoln, 56-63; ASST PROF PEDIAT & CHIEF NUTRIT SERV, CHILDREN & YOUTH PROJ, UNIV TEX SOUTHWESTERN MED SCH DALLAS, 68- *Concurrent Pos:* Fulbright Fel to NZ & Cook Islands, 53-54; Nutrit consult, Yellow Springs Med Clin, Ohio, 55-56; consult, Interdept Comt Nutrit Nat Defense, NIH, 62-63, Inst Cross-Cult Res, Washington, DC, 68-, Ctr Res & Educ, Denver, Colo, 72- & Diet, Nutrit & Cancer Prog, Nat Cancer Inst, NIH, 75-77; res fel, Anat Dept, Univ Hong Kong Med Sch, Hong Kong, 63-64; adj prof, Anthrop Dept, Southern Methodist Univ, 78- *Honors & Awards:* Borden Award, Univ Tex, 49; Effie I Raitt Award, Am Home Econ Asn, 53. *Mem:* Am Inst Nutrit; Am Dietetic Asn. *Res:* Childhood obesity; utilization of amino acids from plant proteins; methodology for determining nutritional status of children; effectiveness of various methods for improving the iron, protein and vitamin A nutriture of preschoolers. *Mailing Add:* 3004 Fondren Dr Dallas TX 75205-1916

FRY, RICHARD JEREMY MICHAEL, b Dublin, Ireland, July 8, 25; m 56; c 3. PHYSIOLOGY, RADIOBIOLOGY. *Educ:* Univ Dublin, BA, 46, MB, BCh & BAO, 49, MD, 62. *Prof Exp:* Lectr physiol, Univ Dublin, 52-59; resident res assoc radiobiol, Argonne Nat Lab, 59-61; lectr physiol, Univ Dublin, 61-63; assoc scientist, Argonne Nat Lab, 63-70, sr scientist radiobiol, 70-77; SECT HEAD, CANCER & TOXICOL SECT, BIOL DIV, OAK RIDGE NAT LAB, 77- *Concurrent Pos:* Prof radiol, Univ Chicago, 70-77. *Mem:* Radiation Res Soc; Am Asn Cancer Res; Am Soc Photobiol; Cell Kinetics Soc; Sigma Xi. *Res:* Cell proliferation and radiation effects; ionizing and ultraviolet radiation carcinogenesis. *Mailing Add:* Biol Div PO Box Y 9211 Oak Ridge TN 37830

FRY, THOMAS R, b Franklin, Pa, Dec 10, 23; m 44; c 3. ELECTRICAL UTILITY ENGINEERING. *Educ:* Univ Pittsburgh, BS, 49. *Prof Exp:* Test engr, Porcelain Prods, Inc, WVa, 49-51, chief engr, Parkersburg & Carey plants, 51-59; chief engr, A B Chance Co, 59-67, mgr eng, Utility Systs Div, 67-71; consult, Black & Veatch, Consult Engrs, 71-74; mgr prod develop, Fargo Mfg Co, 74-77; mgr eng, South Miss Elec Power Asn, 77-80, dir fuels, 80-; RETIRED. *Concurrent Pos:* Consult, Monongahela Power Co, WVa, 53, Am Casualty Co, 53-58 & Puritan Pottery Co, Pa, 57. *Mem:* Nat Soc Prof Engrs; sr mem Inst Elec & Electronics Engrs. *Res:* Generation and transmission of electrical energy. *Mailing Add:* 113 Belmont Ave Hattiesburg MS 39402

FRY, WAYNE LYLE, b Inwood, Iowa, Oct 6, 22. PALEOBOTANY. *Educ:* Univ Minn, BSc, 47; Cornell Univ, PhD(bot geol), 53. *Prof Exp:* Asst bot, Univ Minn, 47-49; from asst to instr, Cornell Univ, 49-53; geologist, Geol Surv Can, 53-57; asst prof paleont, Univ, 57-61, fac fel, 59-62, dir mus & chmn dept, 69-70, ASSOC PROF PALEONT, UNIV CALIF, BERKELEY, 61-, CUR PALEOBOT, MUS PALEONT, 57- *Concurrent Pos:* Dir, Summer Inst Earth Sci, 65-67 & 69-; assoc prog dir & foreign affairs coordr, Pre-Col Div, NSF, 67-69; ed in chief, J Paleont, 71-77. *Mem:* Fel Geol Soc Am; fel Bot Soc Am; fel Paleont Soc Am. *Res:* Carboniferous and Devonian petrifications; Mesozoic and Tertiary floras of British Columbia; Jurassic floras of western North America. *Mailing Add:* Mus Paleont Univ Calif Three Earth Sci Berkeley CA 94720

FRY, WILLIAM EARL, b Lincoln, Nebr, July 6, 44; m 70; c 2. PLANT PATHOLOGY, EPIDEMIOLOGY. *Educ:* Nebr Wesleyan Univ, BA, 66; Cornell Univ, PhD(plant path), 70. *Prof Exp:* Asst prof biol, Cent Conn State Col, 70-71; from asst prof to assoc prof, 71-84, PROF PLANT PATH, CORNELL UNIV, 84-, CHMN, 81- *Mem:* AAAS; Am Phytopath Soc; Int Soc Plant Path; Potato Asn Am. *Res:* Biological and physical factors affecting plant disease epidemiology and control; genetics. *Mailing Add:* 821 Bostwick Rd Ithaca NY 14850

FRY, WILLIAM FREDERICK, b Carlisle, Iowa, Dec 16, 21; m 43; c 2. EXPERIMENTAL PHYSICS. *Educ:* Iowa State Col, BS, 43, PhD(physics), 51. *Prof Exp:* Physicist, Naval Res Lab, 43-47; with AEC, Chicago, 51-52; asst prof, 52-58, prof, 58-88, HILLDALE PROF PHYSICS, UNIV WIS-MADISON, 88- *Concurrent Pos:* Fulbright lectr, Italy, 56-57; Guggenheim fel, 56-57; Hilldale fel, Univ Wis, 87; lectr, hist phys sci, Florence, Italy, 88. *Mem:* Fel Am Phys Soc. *Res:* Elementary particle physics using bubble chambers; acoustics of violin family; astrophysics; ultra high energy gamma ray astrophysics. *Mailing Add:* Dept Physics Univ Wis Madison WI 53706

FRY, WILLIAM JAMES, b Ann Arbor, Mich, Mar 21, 28; m 49; c 2. SURGERY. *Educ:* Univ Mich, MD, 52; Am Bd Surg, dipl, 60. *Prof Exp:* From instr to prof, Med Sch, Univ Mich, Ann Arbor, 59-74, asst surg, Dept Postgrad Med, 60-66, chief res surg serv, Med Ctr, 64-67, head sect gen surg, 67-76, Frederick A Coller prof surg, 74-76; LEE HUDSON-ROBERT PENN PROF SURG & CHMN DEPT, UNIV TEX SOUTHWESTERN MED SCH, 76- *Concurrent Pos:* Attend physician, Ann Arbor Vet Admin Hosp, 60-61, chief surg, 61-64, consult, 64-76; consult, Wayne Co Gen Hosp,

67-76, Vet Admin Hosp, Dallas, 76- & Med Ctr, Baylor Univ, 76-; mem sr med staff, Parkland Mem Hosp, 76-; USPHS grant. *Mem:* Am Surg Asn; Soc Univ Surg; fel Am Col Surg; Soc Surg Alimentary Tract; Soc Vascular Surg. *Res:* Vascular physiology. *Mailing Add:* Univ Tex Health Sci Ctr Parkland Mem Hosp 5301 E Huron River Rd Ann Arbor MI 48106

FRYBERGER, DAVID, b Duluth, Minn, Feb 22, 31; m 57; c 1. HIGH ENERGY PHYSICS. *Educ:* Yale Univ, BE, 53; Ill Inst Technol, MS, 62; Univ Chicago, PhD(physics), 67. *Prof Exp:* Res engr elec eng, Ill Inst Technol Res Inst, 59-67; ENG PHYSICIST PARTICLE PHYSICS, STANFORD LINEAR ACCELERATOR CTR, 67- *Concurrent Pos:* Hertz fel, Fannie & John Hertz Found, 66-67. *Mem:* Am Phys Soc. *Res:* Elementary particle physics. *Mailing Add:* 2361 Middlefield Rd Palo Alto CA 94301

FRYDENDALL, MERRILL J, b Portis, Kans, Mar 28, 34; m 59; c 2. VERTEBRATE ECOLOGY, ANIMAL BEHAVIOR. *Educ:* Ft Hays Kans State Univ, BS, 56, MS, 60; Utah State Univ, PhD(zool), 67. *Prof Exp:* High sch instr, Kans, 56-57 & 60-62; instr ornith, Utah State Univ, 65; PROF BIOL, MANKATO STATE UNIV, 66- *Mem:* Am Ornithologists Union; Animal Behav Soc; Cooper Ornith Soc; Sigma Xi; Audubon Soc. *Res:* Vertebrate ecology and behavior, especially activity patterns and utilization by mammals and aves; auditory communication especially in aves. *Mailing Add:* Dept Biol Mankato State Univ Mankato MN 56001

FRYE, ALVA L(EONARD), b Gray, Okla, July 19, 22; m 43; c 2. CHEMICAL ENGINEERING. *Educ:* Iowa State Col, BS, 43. *Prof Exp:* Control chemist, Shell Chem Co, Tex, 43-44; process engr, Nat Synthetic Rubber Corp, Ky, 44-47; sect leader, Minn Mining & Mfg Co, 47-57, mgr, cent res pilot plant, 57-62, tech dir paper prod div, 62-68; vpres res & commercial develop, Inmont Corp, Aladdin Indust, Inc, 69-70, dir res & develop, 70-71, vpres res & develop, 71-87; PRES, TECH TRAN CONF, INC, 87- *Concurrent Pos:* Dir, ALH, Inc. *Mem:* Am Chem Soc; Am Inst Chem Engrs; Tech Asn Pulp & Paper Indust; Int Platform Asn; Indust Res Inst. *Res:* Synthetic polymers; process development; fluoro-chemical and pilot plant; specialty papers and new products. *Mailing Add:* Tech Tran Conf Inc 325 Plus Park Blvd Suite 108 Nashville TN 37217

FRYE, C(LIFTON) G(EORGE), b Port Clinton, Ohio, Dec 13, 18; m 43; c 2. CHEMICAL ENGINEERING. *Educ:* Univ Cincinnati, ChE, 41; Mass Inst Technol, MS, 47. *Prof Exp:* Sr chem engr petrochem res, Pan Am Petrol Co, 47-58; sr res eng petrol ref res, Standard Oil Co, Ind, 58-61; res assoc, Amoco Oil Co, 61-74, dir, 74-79, mgr res & develop, 79-83; RETIRED. *Mem:* Am Meteorol Soc; Am Inst Chem Engrs; Am Chem Soc. *Res:* Application of thermodynamics and kinetics to the development of new catalytic processes. *Mailing Add:* 2334 N Carriage Lane Port Clinton OH 43452-2915

FRYE, CECIL LEONARD, b Dearborn, Mich, Apr 3, 28; m 52; c 4. ORGANIC CHEMISTRY. *Educ:* Univ Mich, BS, 50, MS, 51; Pa State Univ, PhD(org chem), 60. *Prof Exp:* Res chemist, 51-56, proj leader, Res Dept, 60-63, lab supvr, 63-70, mgr, Sealants Res Unit, 70-72, mgr bio-sci chem res, 73-74, scientist, corp res, 75-80, SCIENTIST, HEALTH & ENVIRON SCI DEPT, DOW CORNING CORP, 80- *Mem:* Am Chem Soc; Sigma Xi; AAAS; Soc Environ Toxicol & Chem. *Res:* Silicon stereochemistry; extra-coordinate silicon compounds; polysiloxanes; environmental chemistry of silicones. *Mailing Add:* Dow Corning Corp Midland MI 48686-0994

FRYE, CHARLES ISAAC, b Peterboro, NH, Dec 15, 35; m 65; c 3. COAL GEOLOGY, STRATIGRAPHY. *Educ:* Univ NH, BA, 58; Univ Mass, Amherst, MS, 60; Univ NDak, PhD(geol), 67. *Prof Exp:* Asst prof, Muskingum Col, 65-70, chmn, Dept Geol-Geog, 69-74, assoc prof geol, 70-79; geologic consult, 79-81, ASSOC PROF GEOL, NW MO STATE UNIV, 81-, CHMN, 90- *Concurrent Pos:* Spec consult, Ohio Environ Protection Agency, 75-78; consult, Div Reclamation, Ohio Dept Natural Resources, 80-81. *Mem:* Geol Soc Am; Am Asn Petrol Geol; Am Inst Prof Geologists; Nat Asn Geol Teachers. *Res:* Stratigraphy and sedimentation of Hell Creek formation and related beds, Montana, North Dakota, South Dakota and of the Conemaugh formation in southeastern Ohio; coal geology, southeast Ohio; reclamation of abandoned coal mine lands in Ohio; rare plant (orchid) ecology; environmental geology; groundwater quality Northwest Missouri. *Mailing Add:* Dept Geol-Geog NW Mo State Univ Maryville MO 64468

FRYE, GERALD DALTON, b Winchester, Va, Aug 27, 50; m 71; c 1. NEUROPHARMACOLOGY, PHARMACOLOGY. *Educ:* Va Polytech Inst & State Univ, BS, 72; Univ NC, PhD(pharmacol), 77. *Prof Exp:* Res scientist, neuropharmacol, Biol Sci Res Ctr, Univ NC, Chapel Hill, 77-79, pharmacologist, Ctr For Alcohol Studies & res asst prof dept psychiat, 79-83; asst prof, 83-87, ASSOC PROF, DEPT MED PHARMACOL & TOXICOL, TEX A&M UNIV, 87- *Concurrent Pos:* Fel, Nat Inst Alcoholism & Alcohol Abuse, 77-79, 87-92, Res Scientist Awardee, 87-92. *Mem:* Soc Neurosci; Res Soc Alcoholism; fel Nat Inst Alcoholism & Alcohol Abuse; Am Soc Pharmacol & Exp Therapeut. *Res:* Neuropsychopharmacology of psychotropic drugs; neuropharmacology of ethanol tolerance and physical dependence; central nervous system mechanisms of GABAergic adaptation. *Mailing Add:* Dept Med Pharmacol & Toxicol Tex A&M Univ College Sta TX 77843-1114

FRYE, GLENN MCKINLEY, JR, b Ithaca, Mich, Apr 20, 26; m 48; c 3. PHYSICS. *Educ:* Univ Mich, BSE, 46, MS, 47, PhD(physics), 50. *Prof Exp:* Mem staff, Physics Div, Los Alamos Sci Lab, 51-58; physicist, Physics Br, Res Div, AEC, 58-60; assoc prof, 60-66, PROF PHYSICS, CASE WESTERN RESERVE UNIV, 66- *Concurrent Pos:* Guest lectr, Univ Minn, 57; prof, Univ NMex, 58. *Mem:* Am Phys Soc; AAAS; Am Astron Soc; Int Astron Union. *Res:* Gamma-ray astronomy; high energy astrophysics; neutral cosmic ray secondaries in the atmosphere; cosmic rays. *Mailing Add:* Physics Dept ROC Bldg Case Western Reserve Univ University Circle Cleveland OH 44106

FRYE, HERSCHEL GORDON, b Long Beach, Calif, Apr 6, 20; m 40; c 2. ANALYTICAL CHEMISTRY. *Educ:* Col of the Pac, AB, 47, MA, 49; Univ Ore, PhD(chem), 57. *Prof Exp:* Chemist, Permanente Metals Corp, 42-44; chief chemist, Kaiser Magnesium Co, 51-52; head sci dept, High Sch, 52-54; from asst prof to assoc prof, 56-62, PROF CHEM, UNIV OF THE PAC, 62- *Concurrent Pos:* Res Corp grants, 58 & 60; Am Philos Soc grants, 59, 63 & 64; gen partner, Anal Consults, 77- *Mem:* AAAS; Am Chem Soc; NY Acad Sci; Consult Chemists Asn; Am Acad Forensic Sci. *Res:* Absorption spectrophotometry; forensic analytical chemistry; synthesis of coordination compounds. *Mailing Add:* Dept Chem Univ of the Pac Stockton CA 95211

FRYE, JAMES SAYLER, b Washington, DC, Dec 4, 45; m 67; c 3. NUCLEAR MAGNETIC RESONANCE SPECTROSCOPY. *Educ:* Reed Col, BA, 67; Wash State Univ, PhD(chem), 75. *Prof Exp:* Res chemist, Univ Calif, Davis, 74-76; asst prof chem, Whitman Col, 76-77 & Reed Col, 77-78; res assoc chem, Colo State Univ, 78-90; APPLNS SCIENTIST, CHEMAGNETICS INC, 90- *Mem:* Am Chem Soc. *Res:* Nuclear magnetic resonance of organic and inorganic compounds in solution and solid phases. *Mailing Add:* Chemagnetics Inc Ft Collins CO 80525

FRYE, JENNINGS BRYAN, JR, b Springhill, La, Sept 3, 18; m 40; c 4. DAIRY SCIENCE. *Educ:* La State Univ, BS, 35; Iowa State Univ, MS, 40, PhD(dairy cattle nutrit), 45. *Prof Exp:* Assoc prof dairy sci, Univ Ga, 45-48; prof & head dept dairy sci, La State Univ, Baton Rouge, 48-; RETIRED. *Concurrent Pos:* Spec dairy indust consult, Venezuela, 53, spec study, Dairy Indust, 57; Breeding Res, Cuba, 58. *Mem:* Am Forage & Grassland Coun; Am Dairy Sci Asn. *Res:* Dairy cattle management, especially in tropical and subtropical environments; influence of soybeans on the flavor of milk, cream and butter and on the body and texture of butter; bloat in dairy cattle; silage, including use of various preservatives and methods of self feeding; feeding green forage to cattle. *Mailing Add:* 551 Leeward Dr Baton Rouge LA 70808

FRYE, JOHN H, JR, b Birmingham, Ala, Oct 1, 08; m 35; c 3. PHYSICAL METALLURGY. *Educ:* Howard Col, AB, 30; Lehigh Univ, MS, 34; Oxford Univ, DPhil(phys sci), 42. *Prof Exp:* Engr, Am Cast Iron Pipe Co, Ala, 34-35; instr metall, Lehigh Univ, 35-37, from asst prof to assoc prof, 40-44; res engr, Bethlehem Steel Co, 44-48; dir metall div, Oak Ridge Nat Lab, 48-62, dir, Metals & Ceramics Div, 62-73; PROF METALL ENG, UNIV ALA, 73- *Concurrent Pos:* Lectr, Grad Sch, Univ Tenn, 50; adj prof, Col Eng, Univ Ala, 64-66. *Honors & Awards:* Lincoln Gold Medal, Am Welding Soc, 43. *Mem:* Fel AAAS; fel Am Soc Metals; fel Metall Soc. *Res:* Rates of reaction solids; phase equilibria; nuclear materials. *Mailing Add:* Dept Metall & Mat Eng PO Box 870202 University AL 35487

FRYE, KEITH, b Columbus, Ohio, July 17, 35; wid. GEOCHEMISTRY, MINERALOGY. *Educ:* Oberlin Col, AB, 57; Univ Minn, MS, 59; Pa State Univ, PhD(geochem), 65. *Prof Exp:* Chemist, Hyman Labs, Inc, 61-62; asst prof geol, Univ Ga, 65-67; assoc prof geosci, Old Dominion Univ, 67-87; SELF EMPLOYED, 87- *Mem:* AAAS; Am Geophys Union; fel Geol Soc Am; Mineral Soc Am; Mineral Asn Can; Min Soc Gt Brit. *Res:* Geology and petrology of the central Virginia Blue Ridge; Appalachian geology; distribution of trace elements and of minor minerals in granites; interaction between man and his geological environment; writer and editor in geology and mineralogy. *Mailing Add:* Tyro VA 22976

FRYE, ROBERT BRUCE, b Wilmington, Del, Apr 10, 49; m 74; c 3. ORGANIC & SILICONE CHEMISTRY. *Educ:* Univ Del, BS, 71; Mass Inst Technol, PhD(org chem), 76. *Prof Exp:* Sr res chemist dyestuffs, GAF Corp, 76-77; prod develop chemist silicones, Gen Elec Co, 77-80, res specialist silicones, 81-85, mgr prod res, 85-90; MGR, NORYL PROD TECHNOL, 90- *Mem:* Am Chem Soc; Sigma Xi. *Res:* Silicone copolymers; flame retardants. *Mailing Add:* Gen Elec Plastics One Noryl Ave Selkirk NY 12158

FRYE, WILBUR WAYNE, b Finger, Tenn, Aug 6, 33; m 57; c 2. SOIL FERTILITY, SOIL MANAGEMENT & CONSERVATION. *Educ:* Univ Tenn, Knoxville, BS, 61, MS, 64; Va Polytech Inst & State Univ, PhD, 69. *Prof Exp:* Instr agron, Tenn Technol Univ, 63-64, asst prof, 64-67, prof & chmn dept, 70-74; assoc prof, 74-83, PROF AGRON, UNIV KY, 83- *Concurrent Pos:* Danforth assoc, Danforth Found, 81-86. *Mem:* Am Soc Agron; Soil Sci Soc Am; Int Soc Soil Sci; fel Soil & Water Conserv Soc; Coun Agr Sci & Technol; World Asn Soil & Water Conserv. *Res:* Use of biologically fixed nitrogen for crop production, improving nitrogen fertilizer efficiency and effects of soil erosion on soil properties; soil productivity and surface water quality. *Mailing Add:* Dept Agron Univ Ky Lexington KY 40546-0091

FRYE, WILLIAM EMERSON, b Detroit, Mich, June 20, 17; m 42; c 2. SPACE PHYSICS, OPTICS. *Educ:* Univ Ill, AB, 37, MS, 38; Univ Chicago, PhD(physics), 41. *Prof Exp:* Group leader, Airborne Radio Div, Naval Res Lab, Wash, 42- 45, asst sect head, 45-46; group leader, Aerophysics Lab, NAm Aviation, Inc, Calif, 46-48; physicist, Rand Corp, 48-56; mgr dept, 56-59, consult scientist, 59-62, 64-68, sr mem res lab, 68-81, STAFF SCIENTIST, LOCKHEED MISSILE & SPACE CO, 81- *Concurrent Pos:* Mem adv comt control, guid & navig, NASA, 60-64. *Mem:* Am Phys Soc; Am Inst Aeronaut & Astronaut; Am Astronaut Soc. *Res:* Space systems and scientific instrumentation. *Mailing Add:* 536 Lincoln Ave Palo Alto CA 94301

FRYER, CHARLES W, b Springfield, Mo, May 4, 28; m 53, 76; c 4. GEMOLOGY. *Educ:* Gemol Inst Am, grad gemol, 63. *Prof Exp:* Plant mgr, C Holle Glass Co, 62-64; mgr, R&B Artcraft Co, 64-66; instr, 66-67, lab supvr gem identification, 67-76, DIR GIA-GEM TRADE LAB, WESTERN DIV, GEMOL INST AM, 76- *Concurrent Pos:* Ed, Gem Trade Lab, Lab Notes, Gem & Gemology; cert gemologist, Am Gem Soc. *Mem:* Fel Gemol Asn Gt Brit. *Res:* X-ray powder diffraction identification of gem materials and their inclusions; optical and physical identification of gem materials and their substitutes; x-ray analysis of pearls to determine natural or cultured origin. *Mailing Add:* 1153 Stanford Santa Monica CA 90403

FRYER, ELSIE BETH, b Kopperl, Tex, June 25, 25; m 66. NUTRITION. *Educ:* Univ NMex, BS, 45; Ohio State Univ, MS, 49; Mich State Univ, PhD(foods & nutrit), 59. *Prof Exp:* High sch teacher, NMex, 45-48; asst prof foods & nutrit, SDak State Col, 49-56; assoc prof, 59-75, PROF NUTRIT, KANS STATE UNIV, 75- *Mem:* Am Dietetic Asn; Am Asn Cereal Chemists; Am Home Econ Asn; Am Inst Nutrit; Inst Food Technologists. *Res:* Food habit and nutritional status surveys of women and children; quality of cereal proteins; human metabolic studies concerned with weight reduction, lipid metabolism and nitrogen balance. *Mailing Add:* Dept Foods & Nutrit Justin Hall Kans State Univ Manhattan KS 66506

FRYER, HOLLY CLAIRE, b Carlton, Ore, Dec 6, 08; m 34, 66; c 2. BIOLOGICAL STATISTICS. *Educ:* Univ Ore, BS, 31; Ore State Col, MS, 33; Iowa State Col, PhD(statist), 40. *Prof Exp:* Instr math, Iowa State Univ, 37-40; asst prof, Kans State Univ, 40-42, assoc prof & statistician exp sta, 42-44; assoc res mathematician, Columbia Univ, 44-45; prof math, 45-59, head statist dept, 59-75, dir statist lab, 46-75, statistician, exp sta, 40-79, prof statist, 45-79, EMER PROF STATIST, KANS STATE UNIV, 79- *Mem:* Biomet Soc; fel Am Statist Asn; Inst Math Statist; fel Royal Statist Soc; Sigma Xi. *Res:* Designing biometric experiments; mathematical statistics. *Mailing Add:* Dept Statist Kans State Univ Manhattan KS 66506

FRYER, JOHN LOUIS, b Ft Worth, Tex, July 4, 29; m 52; c 4. MICROBIOLOGY, FISHERIES. *Educ:* Ore State Univ, BS, 56, MS, 57, PhD(microbiol), 64. *Prof Exp:* Instr fisheries, Ore State Univ, 57-58; fish pathologist, Res Div, Ore Fish Comn, 58-60, state fisheries pathologist, 60-64; from asst prof to assoc prof, 64-73, PROF MICROBIOL & FISHERIES, ORE STATE UNIV, 73-, CHMN MICROBIOL, 77- *Mem:* Am Soc Microbiol; Am Inst Fishery Res Biol; Am Fisheries Soc; Wildlife Dis Asn; Sigma Xi; Am Acad Microbiol. *Res:* Pathogenic microbiology, virology and immunology in relation to infectious diseases of cold blooded animals; methods for prevention, detection and control of diseases in populations of fishes; fish pathology. *Mailing Add:* Dept Microbiol Ore State Univ Corvallis OR 97331-3804

FRYER, MINOT PACKER, b Conn, Mar 16, 25; m 42; c 2. SURGERY. *Educ:* Brown Univ, AB, 37; Johns Hopkins Univ, MD, 40. *Hon Degrees:* DSc, Brown Univ, 71. *Prof Exp:* assoc prof, 57-67, prof, 67-, EMER PROF CLIN SURG, MED SCH & MAXILLOFACIAL SURG, DENT SCH, WASH UNIV. *Concurrent Pos:* Asst surgeon, Barnes & St Louis Children's Hosps, 48-; mem staff, De Paul Hosp, 48-; consult, Vet Admin Hosp, 49-; vchmn, Am Bd Plastic Surg, 67-68. *Mem:* Am Soc Plastic & Reconstruct Surg; Am Asn Plastic Surg (vpres, 66-67, pres, 67-68); fel Am Asn Surg of Trauma; fel Am Col Surg; Soc Head & Neck Surg. *Res:* Plastic and reconstructive surgery. *Mailing Add:* PO Box 3907 Evansville IN 47737

FRYER, RODNEY IAN, b London, Eng, Mar 13, 30; US citizen; m 54; c 4. SYNTHETIC ORGANIC CHEMISTRY, MEDICINAL CHEMISTRY. *Educ:* Univ Calif, Los Angeles, BS, 57; Univ Manchester, PhD(chem), 60. *Prof Exp:* Sr chemist pharmaceut res, Hoffmann-La Roche, Inc, 60-63, asst group chief, 63-65, group chief pharmaceut res, 65-69, sect chief, Res Dept, 70-78, dir medicinal chem, 79-85; distinguished vis prof, 85-86, PROF (II) CHEM, RUTGERS, STATE UNIV NJ, NEWARK, 86- *Concurrent Pos:* Consult, 85- *Mem:* AAAS; Am Chem Soc; Royal Soc Chem; Int Soc Heterocyclic Chem; NY Acad Sci. *Res:* Synthesis and transformations of heterocyclic compounds of potential medicinal interest. *Mailing Add:* Five Eton Dr N Caldwell NJ 07006

FRYKLUND, VERNE CHARLES, JR, b Greeley, Colo, July 2, 20; m 43; c 4. GEOLOGY. *Educ:* Univ Minn, BS, 44, MS, 47, PhD(geol), 49. *Prof Exp:* Field asst, US Geol Surv, 43, jr geologist, 43-45; from instr to asst prof geol, Univ Idaho, 48-51; geologist, US Geol Surv, 51-62; chief lunar & planetary sci br, Manned Space Sci Div, NASA, 62-64; chief mil geol br, US Geol Surv, 64-66; prog mgr, Off Secy Defense, 66-68, dep dir, Nuclear Monitoring Res Off, 68-72, dir technol, Assessments Off, Defense Advan Res Projs Agency, 72-74; CONSULT GEOLOGIST, 74- *Mem:* Fel Geol Soc Am; Soc Econ Geol; Am Inst Mining, Metall & Petrol Eng. *Res:* Management; application of geology to national defense matters; arms control problems; economic geology. *Mailing Add:* 6805 Broyhill St McLean VA 22101

FRYLING, ROBERT HOWARD, b Danville, Pa, Nov 17, 21; m 84; c 2. MATHEMATICS. *Educ:* Gettysburg Col, BA, 43; Univ Pittsburgh, MS, 51, PhD(math), 72. *Prof Exp:* Instr math, 47-50, dean of men, 52-56, asst prof math, 58-69, assoc prof, 69-77, chmn dept, 74-79, PROF MATH, GETTYSBURG COL, 77- *Mem:* Math Asn Am. *Res:* Theory of random variable centers. *Mailing Add:* 473 Custer Dr Lake Heritage Gettysburg PA 17325

FRYREAR, DONALD W, b Haxtum, Colo, Dec 8, 36; m 56; c 2. AGRICULTURAL ENGINEERING. *Educ:* Colo State Univ, BS, 59; Kans State Univ, MS, 62. *Prof Exp:* Agr engr, Cent Great Plains Field Sta, USDA, 59-60, agr engr wind erosion lab, 60-62, agr engr, Blackland Conserv Exp Sta, 62-63, res agr engr, 63-65, SUPT & RES AGR ENGR, BIG SPRING FIELD STA, AGR RES SERV, USDA, 65- *Concurrent Pos:* Group mem, World Meteorol Orgn, 76-; consult, UNESCO, 81 & 83; US Dryland Cropping Rep to US-Indo Wkshp, Jodhpur, 84. *Mem:* Am Soc Agr Engrs; Nat Soc Prof Engrs; Sigma Xi; fel Soil Conserv Soc Am. *Res:* Dryland water conservation in Colorado; application of annual barriers for wind erosion control; land modifications for controlling erosion; development of equipment for measuring erosion rates; wind erosion and moisture conservation on sandy soils in a semiarid environment; design and construction of six wind tunnels for erosion research. *Mailing Add:* Agr Res Serv USDA Big Spring Field Sta Big Spring TX 79720

FRYSTAK, RONALD WAYNE, b Chicago, Ill, Sept 25, 41; m 65; c 1. PHYSICAL CHEMISTRY, ENVIRONMENTAL CHEMISTRY. *Educ:* Millikin Univ, BS, 63; Univ Hawaii, PhD(phys chem), 68. *Prof Exp:* Asst prof, 68-77, ASSOC PROF CHEM, HAWAII LOA COL, 77- *Mem:* Am Chem Soc. *Res:* Physical interpretation of the He-He interaction through the use of the associated density matrices; saving of Hawaiian beaches in case an accidental oil spillage occurs. *Mailing Add:* Dept Math & Sci Hawaii Loa Col 45-045 Kamehameha Hwy Kaneohe HI 96744

FRYXELL, FRITIOF MELVIN, b Moline, Ill, Apr 27, 00; m 28; c 1. GEOLOGY. *Educ:* Augustana Col, Ill, AB, 22; Univ Ill, MA, 23; Univ Chicago, PhD(geol), 29. *Hon Degrees:* ScD, Wittenberg Univ, 60 & Upsala Col, 60; LLD, Univ Wyo, 79. *Prof Exp:* From asst prof to prof geol, 23-73, cur mus, 29-73, chmn sci div, 46-51, DIR RES FOUND, AUGUSTANA COL, ILL, 47- *Concurrent Pos:* Asst, Univ Chicago, 27-28; naturalist, Grand Teton Nat Park, Wyo, 29-34, geologist, Mus Planning Staff, Nat Park Serv, 35-37; sr geologist, Philippine Govt, 39-40; trustee, Putnam Mus, 41-; geologist, Mil Geol Unit, US Geol Surv, 42-46; Am-Scand traveling fel, 48; lit exec, Francois E Matthes, 49-65; dir, Am Geol Inst, 50-51; Guggenheim Found fel, 54; NSF res grant, Augustana Col, 59-61 & 64. *Honors & Awards:* Miner Award, 53. *Mem:* Fel AAAS; fel Geol Soc Am; fel Am Geog Soc; hon mem Nat Asn Geol Teachers (pres, 38); hon mem Am Alpine Club. *Res:* Geomorphology; glacial and military geology; history of geology. *Mailing Add:* 1331 42nd Ave Rock Island IL 61201

FRYXELL, GRETA ALBRECHT, b Princeton, Ill, Nov 21, 26; m 47; c 3. MARINE PHYTOPLANKTON. *Educ:* Augustana Col, BA, 48; Tex A&M Univ, MEd, 69, PhD(oceanog), 75. *Prof Exp:* Teacher math, Davenport Sch Syst, Iowa, 48-49; teacher sci, Ames Independent Sch Dist, Iowa, 49-52; res assoc, 71-77, from asst res scientist to assoc res scientist, 77-80, from asst prof to assoc prof, 80-86, PROF OCEANOG, TEX A&M UNIV, 86- *Concurrent Pos:* Mem, Int Comt Standardize Diatom Terminology, 74-76, Int Diatom Comt, 74-76 & mem arrangements comt, Fifth Symp Living & Fossil Diatoms, Antwerp, 78; convenor, Polar Diatom Colloquium, 86. *Mem:* Bot Soc Am; Brit Phycol Soc; Int Phycol Soc; Phycol Soc Am; Am Soc Limnol & Oceanog; Am Soc Plant Taxonomists; fel Am Asn Univ Women. *Res:* Comparative morphology; systematics; distribution and place in marine ecosystem of phytoplankton, principally diatoms and coccolithophorids; light and electron microscopy on field samples, living cultures and marine sediments. *Mailing Add:* Dept Oceanog Tex A&M Univ College Station TX 77843-3146

FRYXELL, PAUL ARNOLD, b Moline, Ill, Feb 2, 27; m 47; c 3. SYSTEMATIC BOTANY. *Educ:* Augustana Col, AB, 49; Iowa State Univ, MS, 51, PhD(genetics), 55. *Prof Exp:* Asst agron, Agr Exp Sta, NMex, 52-55; asst prof bot, Univ Wichita, 55-57; from geneticist to prin res geneticist, 57-75, CHIEF RES BOTANIST, AGR RES SERV, USDA, 75- *Concurrent Pos:* Mem grad fac, Tex A&M Univ, Univ Tex; ed, Brittonia, 71-74; dep adminr rep, Plant Germplasm Coord Comt, Sci Educ Admin, 73-82; mem adv panel, NSF, 80-83; vis scientist, Univ NAC AUT Mex, 81. *Honors & Awards:* Cotton Genetics Res Award, 67; H A Gleason Award, 89. *Mem:* Fel AAAS; Am Soc Plant Taxonomists (pres, 83-84); Bot Soc Am; Soc Bot Mexico; Int Asn Plant Taxon; Soc Econ Bot (pres, 88-89). *Res:* Systematics of Malvaceae, especially of the Neotropics, involving revisionary studies of selected genera; floristic treatments of significant regions; experimental and evolutionary studies where relevant. *Mailing Add:* Dept Soil & Crop Sci Tex A&M Univ College Station TX 77843-2474

FRYXELL, ROBERT EDWARD, b Moline, Ill, Mar 24, 23; m 61; c 3. PHYSICAL INORGANIC CHEMISTRY. *Educ:* Augustana Col, BA, 44; Univ Chicago, MS, 49, PhD(chem), 50. *Prof Exp:* Chemist, Manhattan Proj, Metall Lab, Univ Chicago, 43-44 & Los Alamos Sci Lab, Univ Calif, 44-46; anal chemist, Inst Study Metals, Univ Chicago, 46-50; chemist, M&P Lab, Transformer Div, 50-59, prin engr, Aircraft Nuclear Propulsion Dept, 59-61 & Nuclear Mat & Propulsion Oper, 61-64, mgr high temperature fuels res, 64-69, SR PHYS CHEMIST, AIRCRAFT ENGINE GROUP, GEN ELEC CO, 69- *Mem:* AAAS; Am Chem Soc. *Res:* Separation procedures in analytical chemistry; electrochemistry; reactions of non metallic impurities in metals; high temperature reactions; corrosion. *Mailing Add:* 8430 Old Hickory Dr Cincinnati OH 45243

FRYXELL, RONALD C, b Moline, Ill, Aug 31, 38; m 59; c 2. MATHEMATICS. *Educ:* Augustana Col, Ill, AB, 60; Wash State Univ, MA, 62, PhD(math), 64. *Prof Exp:* From asst prof to assoc prof, 64-77, PROF MATH, ALBION COL, 77-, CHMN, 85- *Mem:* Asn Comput Mach; Math Asn Am; Inst Elec Electronic Engr Comput Soc. *Res:* Finite geometry, particularly finite planes of square order; autonomous robots-navigation. *Mailing Add:* Dept Math Albion Col 701 E Porter Albion MI 49224

FTACLAS, CHRIST, b New York, NY. COSMOLOGY, GALAXY STRUCTURE. *Educ:* City Col NY, BS, 65, MA, 71; City Univ NY, PhD(physics), 78. *Prof Exp:* ASST PROF ASTRON, UNIV PA, 78- *Concurrent Pos:* Proj engr, Leesona Mooos Lab, Great Neck Long Island. *Mem:* Am Phys Soc; Am Astron Soc; Int Soc Gen Relativity & Gravitation; NY Acad Sci. *Res:* Relativistic cosmology galaxies; clusters of galaxies; physical cosmology. *Mailing Add:* Space Sci Progs MS813 Perkin-Elmer 100 Wooster Heights Rd Danbury CT 06810

FTHENAKIS, EMANUEL JOHN, b Salonica, Greece, Jan 30, 28; US citizen; c 2. SYSTEMS DESIGN & SYSTEMS SCIENCE. *Educ:* Tech Univ Athens, Dipl, 51; Columbia Univ, MS, 54. *Prof Exp:* Mem tech staff, Bell Telephone Lab, 53-57; dir eng, Gen Elec Co, 57-62; vpres & gen mgr, Space & Re-entry Div, Philco-Ford Co, 62-69; pres aerospace & optic div, ITT, 69-70; dir, info system, space & electronics div, 71, corp vpres, 71-79, sr vpres, 79-85, pres & chief oper officer, 85, CHMN & CHIEF EXEC OFFICER, FAIRCHILD INDUST INC, 86- *Concurrent Pos:* Adj prof, Dept Elect Eng, Univ Md, 81-84. *Mem:* Fel Inst Elec & Electronic Engrs; assoc fel Am Inst Aeronaut & Astronaut. *Res:* Digital system for space and avionics; advanced digital communications systems include the development of advance system hardware and software elements. *Mailing Add:* Fairchild Indust Inc 300 W Service Rd PO Box 10803 Chantilly VA 22021

FU, CHENG-TZE, b China, Oct 12, 49; Can citizen; m 75; c 3. HEAVY OILS & OIL SAND BITUMEN. *Educ:* Chun-Yuan Univ, BE, 72; Univ Ottawa, MS, 78, PhD(chem eng), 83. *Prof Exp:* RES OFFICER THERMODYN, ALTA RES COUN, 82- *Mem:* Am Inst Chem Engrs; Soc Petrol Engrs; Can Soc Chem Eng; Can Inst Chem. *Res:* Vapour-liquid equilibrium properties for natural gases, heavy oils and bitumens systems at field conditions; calculations by an equation of state related with these thermodynamic properties. *Mailing Add:* Alta Res Coun (OSRD) 1021 Hayter Rd Edmonton AB T6H 5X2 Can

FU, HUI-HSING, b Hou-Pei, China, Dec 5, 47; m 70; c 1. NUCLEAR ENGINEERING, SOLID STATE PHYSICS. *Educ:* Nat Tsing Hua Univ, Taiwan, BS, 69; Ohio State Univ, MS, 72, PhD(physics), 74. *Prof Exp:* Res assoc physics, Univ Ill, Urbana-Champaign, 74-76; asst prof physics, Ill State Univ, 76-79; engr, 80-81, SR ENGR, GEN PUB UTILITIES, 81- *Res:* Nuclear fuel management; low temperature physics; many body theory. *Mailing Add:* Gen Pub Utilities 110 Interpace Pkwy Parsippany NJ 07054

FU, JERRY HUI MING, b Hupeh, China, July 15, 32; US citizen; m 64; c 2. PLASMA PHYSICS, SPACE PHYSICS. *Educ:* Univ Taiwan, BS, 56; Northwestern Univ, MS, 61; Univ Mich, PhD(aerospace sci), 67. *Prof Exp:* Fel inst fluid dynamics & appl math, Univ Md, 67-68, res assoc, 67-70; SR SCIENTIST, EG&G, INC, 70- *Mem:* Am Phys Soc; Am Geophys Union. *Res:* Shock wave in collisionless plasma and microstructure of the earth's bow shock; surface potential distribution near a satellite in interplanetary space. *Mailing Add:* 387 Manhattan Loop Los Alamos NM 87544

FU, KUAN-CHEN, b Nanking, China, Feb 2, 33; m 63; c 2. CIVIL & STRUCTURAL ENGINEERING. *Educ:* Taiwan Prov Col Eng, BS, 55; Univ Notre Dame, MS, 59, PhD(civil eng), 67. *Prof Exp:* Struct engr, Chas Cole & Sons, 59-63; from asst prof to assoc prof civil eng, 67-78, PROF CIVIL ENG, UNIV TOLEDO, 78- *Concurrent Pos:* Prin investr, NSF res grant, 69-71, NSF US-China Int res grant, 74-75 & NASA res grant, 83-85; chair prof civil eng, Nat Cheng-Kung Univ, Taiwan, 74-75. *Mem:* Am Soc Civil Engrs. *Res:* Optimization of structural configurations under deterministic and probabilistic loading conditions; discrete structural optimization using complex-simplex method; finite element analysis of a large wind turbine blade under thermal load; integral equation method and boundary element method on the analysis of plates and shells. *Mailing Add:* Dept Civil Eng Univ Toledo 2801 W Bancroft Toledo OH 43606

FU, LEE LUENG, b Taipei, Taiwan, Oct 10, 50; m 77; c 1. PHYSICAL OCEANOGRAPHY, REMOTE SENSING. *Educ:* Nat Taiwan Univ, BS, 72; Mass Inst Technol, PhD(oceanog), 80. *Prof Exp:* Res assoc, Mass Inst Technol, 80; RES SCIENTIST, JET PROPULSION LAB, CALIF INST TECHNOL, 80-, TECH GROUP SUPVR, 86-, TOPEX/POSEIDON PROJ SCIENTIST, 88- *Mem:* Am Geophys Union; Am Meteorol Soc; Sigma Xi. *Res:* Ocean circulation; geophysical fluid dynamics; geophysical data analysis; oceanic remote sensing with microwave sensors. *Mailing Add:* 300-323 Jet Propulsion Lab 4800 Oak Grove Dr Pasadena CA 91109

FU, LI-SHENG WILLIAM, b China; US citizen. ENGINEERING MECHANICS. *Educ:* Nat Taiwan Univ, BSc, 62; Northwestern Univ, MSc, 65, PhD(theoret & appl mech), 67. *Prof Exp:* Asst prof, 67-71, ASSOC PROF ENG MECH, OHIO STATE UNIV, 71- *Concurrent Pos:* Prin investr, NSF, 69-72 & NASA, 79-82. *Mem:* Am Acad Mech; Am Soc Mech Engrs; Am Soc Testing & Mat; Am Soc Eng Educ; Sigma Xi. *Res:* Fracture and fatigue analysis; crack initiation and growth at elevated temperature; elastic and plastic analysis; nondestructive testing of fracture toughness. *Mailing Add:* Dept Eng Mech 155 W Woodruff Ave Columbus OH 43210

FU, LORRAINE SHAO-YEN, b China, Jan 7, 39; US citizen; m 67; c 1. MATHEMATICS. *Educ:* Hunter Col, BA, 60; NY Univ, MS, 63; Polytech Inst Brooklyn, PhD(math), 72. *Prof Exp:* Lectr, Hunter Col, 64-67 & 70-71, asst prof math, 71-76; asst prof, 77-80, ASSOC PROF MATH, PACE UNIV, 80- *Concurrent Pos:* Mem, New York City Community Sch Bd, 75-77. *Mem:* Sigma Xi; Am Math Soc; Math Asn Am; Soc Indust & Appl Math. *Res:* Inverse problems. *Mailing Add:* 2277 85th St Apt 3 Brooklyn NY 11214-3305

FU, PETER PI-CHENG, b Canton, China, June 12, 41. TOXICOLOGY RESEARCH. *Educ:* Nat Taiwan Normal Univ, BS, 65; Ill Inst Technol, MS, 69, PhD(chem), 73. *Prof Exp:* Assoc prof, Univ Ark, Little Rock, 79-82; RES CHEMIST, NAT CTR TOXICOL RES, US FOOD & DRUG ADMIN, 79-, DIR TOXICOL, ASIAN TOXICOL RES PROG, 86- *Concurrent Pos:* Adj prof chem, Univ Ark, Little Rock, 82- adj prof toxicol, Univ Ark Med Sci, 88-; adj prof chem, Providence Col, Taiwan, 84-; consult, UN Develop Prog, Chinese Acad Prev Med, & IEHE, China, 87. *Mem:* Am Chem Soc; Am Assoc Cancer Res; Int Soc Study Xenobiotics; Soc Chinese Bioscientist Am. *Res:* Chemical carniogenesis and chemistry of polycrylic aromatic hydrocarbons, nitroarenes, aromatic amines and nitrofurans. *Mailing Add:* 26 Kingspark Rd Little Rock AR 72207-4814

FU, SHOU-CHENG JOSEPH, b Beijing, China, Mar 19, 24; nat US; m 51; c 3. BIOCHEMISTRY. *Educ:* Cath Univ Beijing, China, BS, 41, MS, 44; Johns Hopkins Univ, PhD(chem), 49. *Prof Exp:* Asst chem, Cath Univ Beijing, 42-44; jr instr, Johns Hopkins Univ, 47-49; vis scientist, Nat Cancer Inst, 51-54; chief enzyme & bio-org chem lab, Children's Cancer Res Found, Boston, 56-66; prof chem & chmn chem bd, Chinese Univ Hong Kong, 66-70, vis prof ophthal, Col Physicians & Surgeons, Columbia Univ, 70-71; from asst dean to dean, Grad Sch Biomed Sci, 75-79, PROF BIOCHEM, UNIV MED & DENT NJ, 71- *Concurrent Pos:* Res fel, Nat Cancer Inst, 49-51; Bissing fel from Johns Hopkins Univ to Univ Col, Univ London, 55; res assoc, Children's Hosp Med Ctr & Harvard Med Sch, 56-66; univ dean sci fac, Chinese Univ Hong Kong, 67-69; hon prof & acad consult, Inner-Mongolia (AR) Med Col, Huthot, Peoples Repub China, 88- *Mem:* Fel AAAS; Am Asn Cancer Res; Am Soc Biol Chem; NY Acad Sci; fel Royal Soc Chem; Sigma Xi. *Res:* Organic chemistry; proteins and enzymes; chemical kinetics and reaction mechanism; cancer chemotherapy and prevention; lens development, aging and cataractogenesis. *Mailing Add:* Dept Biochem & Molecular Biol NJ Med Sch & Grad Sch Biomed Sci Univ Med & Dent NJ Newark NJ 07103

FU, SHU MAN, b Canton, China, July 29, 42; US citizen; m; c 2. MICROBIOLOGY & IMMUNOLOGY. *Educ:* Dickinson Col, BS, 65; Stanford Univ, MD, 70; Rockefeller Univ, PhD, 75; Am Bd Internal Med, dipl, 76 & 86. *Prof Exp:* Intern & resident, Med Ctr, Stanford Univ, 70-71; asst resident physician, Rockefeller Univ, 73-75, from asst prof to assoc prof, 75-82; Okla Med Res Found prof pediat, Dept Med & Pediat, Sch Med, Univ Okla, 82-85, Okla Med Res Found prof med, Dept Med, Microbiol & Immunol, 85-88; PROF MED & MICROBIOL, SCH MED, UNIV VA, 88- *Concurrent Pos:* Asst resident physician, NY Hosp, 75-76; assoc physician, Hosp Rockefeller Univ, 75-78, physician, 78-82; asst ed, J Exp Med, 77-82, adv ed, 82-; assoc ed, J Immunol Methods, 81- & J Immunol, 84-, sect ed, 88-; mem, Cancer Res Prog, Okla Med Res Found, 82-85, mem & head, Immunol Prog, 85-88; attend physician, Okla Mem Hosp, Okla Children's Mem Hosp & Presby Hosp, 82-88 & Univ Va Hosps, 88-; mem, Immunol Sci Study Sect, NIH, 83-87, Adv Comt Res Etiology, Diag, Natural Hist, Prev & Ther Mult Sclerosis, 85-89, Fel Comt, Arthritis Found, 86-89 & Immunol & Microbiol Res Study Comt, Am Heart Asn, 90-; adj prof microbiol & immunol, Sch Med, Univ Okla, 85-88. *Mem:* Am Soc Clin Invest; Am Asn Immunologists; Am Col Rheumatology; Am Soc Microbiol; Clin Immunol Soc; AAAS; Am Fedn Clin Res; Am Rheumatology Soc. *Res:* Author of more than 150 technical publications. *Mailing Add:* Dept Med & Microbiol Sch Med Univ Va Box 412 Charlottesville VA 22908

FU, WALLACE YAMTAK, b Macao, China, June 10, 43; US citizen; m 70. ORGANIC CHEMISTRY. *Educ:* St Johns Univ, BS, 67; Marquette Univ, PhD(org chem), 73. *Prof Exp:* Res assoc chem, Cornell Univ, 73-75; chemist, 75-80, PROJ SCIENTIST, UNION CARBIDE CORP, 80- *Mem:* Am Chem Soc; Sigma Xi; Weed Sci Soc Am. *Res:* Organic photochemistry, organic synthesis and pesticide process development. *Mailing Add:* 910 Grove St Chapel Hill NC 27514

FU, WEI-NING, b Huang Hsien, China, Feb 24, 25; m 55; c 3. CYTOGENETICS. *Educ:* Nat Honan Univ, China, BS, 49; Okla State Univ, MS, 60, PhD(plant breeding, genetics), 63. *Prof Exp:* Teacher high sch, Taiwan, 49-51; assoc agronomist, Taiwan Tobacco Res Inst, 52-57; from asst prof to assoc prof, 63-75, PROF GENETICS & CYTOL, CENT CONN STATE UNIV, 75- *Concurrent Pos:* fel med, Sch Med, Johns Hopkins Univ, 75 & Weslyan Univ, 84. *Mem:* AAAS; Am Genetic Asn; Am Soc Human Genetics; Am Hort Soc. *Res:* Human cytogenetics; molecular genetics; oriental vegatables; work on satellite DNA of Drosophila melanogaster; study the function of the junction between satellite DNA and non-satellite DNA in heterochromatin. *Mailing Add:* Dept Biol Sci Cent Conn State Univ New Britain CT 06050

FU, YAOTIAN, theoretical solid state physics, for more information see previous edition

FU, YUAN C(HIN), b Formosa, Feb 16, 30; US citizen; m 60; c 2. FUELS ENGINEERING, CHEMICAL ENGINEERING. *Educ:* Nat Taiwan Univ, BS, 53; Univ Utah, PhD(fuels eng), 61. *Prof Exp:* Res engr, Union Indust Res Inst, Formosa, 53-56; res engr, Phillips Petrol Co, 61-64; res assoc, Univ Southern Calif, 64-65; res chemist, US Bur Mines, 65-75; res engr, ERDA, 75-77 & Pittsburgh Energy Technol Ctr, US Dept Energy, 77-89; PROF, MURORAN INST TECHNOL, JAPAN, 89- *Concurrent Pos:* Instr, Univ Pittsburgh, 67-69. *Honors & Awards:* Bituminous Coal Res Award, Am Chem Soc, 68. *Mem:* Am Chem Soc; Chem Soc Japan; Catalysis Soc. *Res:* Coal liquefaction; coal and coal-water mixture combustion technology; synfuel technology; flue gas cleanup. *Mailing Add:* Dept Chem Eng Muroran Inst Technol Muroran 050 Japan

FU, YUN-LUNG, b China, June 12, 42; US citizen; m 70; c 2. MATERIAL SCIENCE, ORGANIC & POLYMER CHEMISTRY. *Educ:* Nat Taiwan Univ, BSc, 65; Syracuse Univ, PhD(chem), 72. *Prof Exp:* Res assoc, Wood Chem Lab, Univ Mont, 72-73; cancer res scientist, Roswell Park Mem Inst, 73-77; res assoc, Int Paper Co, 77-78; res chemist, Am Cyanamid Co, 78-80, sr res chemist, 80-87; pres & consult, YLF Int Corp, 87-89; SR SCIENTIST, KOCH INDUSTS, INC, 89- *Mem:* Am Chem Soc; AAAS; Nat Asn Corrosion Engrs. *Res:* Organic synthesis; polymer synthesis; wood, carbohydrate, cellulose, starch and polysaccharide chemistry; pyrolysis; paper chemicals; mining chemicals; asphalt; coatings; materials science. *Mailing Add:* 8831 Shadowridge Wichita KS 67226

FUBINI, EUGENE G(HIRON), b Turin, Italy, Apr 19, 13; nat US; m 45; c 6. ENGINEERING. *Educ:* Univ Rome, Dr Physics, 33. *Hon Degrees:* LLD, Polytech Inst Brooklyn, 68; DEng, Pratt Inst, 67; ScD, Rensselaer Polytech Inst. *Prof Exp:* Res assoc, Nat Inst Electrotechnics, Rome, 35-38; engr in chg microwave & int broadcasting, CBS, Inc, 38-42; res assoc develop electronic countermeasures, Radio Res Lab, Harvard Univ, 42-44; mem staff, Airborne Instruments Lab, 45-61, from div head to vpres, 60-61; asst secy defense & dept dir, Defense Res & Eng, 61-65; vpres & group exec, IBM Corp, 65-69; PRES, E G FUBINI CONSULTS LTD, 69- *Concurrent Pos:* Spec lectr, Harvard Univ, 56; mem, President's Sci Adv Comt, 57-61, 69-71, Adv Coun Advan Sci Res & Develop, NY State Univ, 58-, Panel Sci Adv Bd, Nat Security Agency, 58-61; chmn, Electromagnetic Warfare Adv Group, Air Res & Develop Command, 58-61; mem, Adv Group Spec Projs, Dept Defense, 58-61, President's Comn Law Enforcement, 65-67, Defense Intel Agency, 65-, chmn, 65-70; trustee, Urban Inst, 65 & 82; mem, Defense Sci Bd, 66-, Am Newspaper Publn Asn, 69-78 & Defense Spec Projs Group, 70-; dem bd dirs, Volunteers Int Tech Assistance, Inc, 69-; dir, Tex Instruments Inc, 69-83; consult, IBM Corp, 69-; vis comt, Comput Ctr, Harvard Univ, Sch Eng, Stanford Univ, & George Washington Univ. *Mem:* Nat Acad Eng; fel Inst Elec & Electronics Engrs; NY Acad Sci. *Res:* Advanced development of special electronic devices; nonconventional antennas and microwave devices; radar; radar countermeasures; computers. *Mailing Add:* 2706 N Quincy Arlington VA 22207

FUCCI, DONALD JAMES, b New Kensington, Pa, July 17, 41; m 66; c 2. SPEECH PATHOLOGY. *Educ:* Univ Pittsburgh, BA, 62, MS, 64; Purdue Univ, PhD(speech path), 68. *Prof Exp:* PROF SPEECH PATH, OHIO UNIV, 68- *Mem:* Fel Am Speech & Hearing Asn; Acoust Soc Am. *Res:* Oral vibrotactile sensitivity and perception as related to area of speech and hearing. *Mailing Add:* Sch Hearing & Speech Sci Lindley Hall 219 Ohio Univ Athens OH 45701

FUCHIGAMI, LESLIE H, b Lanai City, Hawaii, June 11, 42; m 63; c 5. HORTICULTURE, PLANT PHYSIOLOGY. *Educ:* Univ Hawaii, BS, 64; Univ Minn, St Paul, MS, 66, PhD(plant physiol), 70. *Prof Exp:* Assoc prof hort, Ore State Univ, 70-81; prof hort, Univ Hawaii, 81-; AT DEPT HORT, ORE STATE UNIV. *Concurrent Pos:* Chmn Dept Hort, Univ Hawaii, Honolulu; consult, Nursery Crops; actg asst dir, Col Trop Agr & Human Resources, Univ Hawaii, Honolulu. *Mem:* Fel Am Soc Hort Sci. *Res:* Environmental, physiological and chemical control of cold acclimation in Cornus stolonifera; Meloidogyne hapla versus Treflan; rooting and establishment of Douglas fir; storage of ornamentals; forcing of colored lilies; root regeneration of nursery crops and maturity, dormancy, hardiness, storage and defoliation of deciduous plants; physiology of tropical fruit crops. *Mailing Add:* Dept Hort Ore State Univ Corvallis OR 97331

FUCHS, ALBERT FREDERICK, b Philadelphia, Pa, Feb 15, 38; m 63; c 2. BIOMEDICAL ENGINEERING. *Educ:* Drexel Inst, BSEE, 60, MS, 61; Johns Hopkins Univ, PhD(biomed eng), 66. *Prof Exp:* PROF PHYSIOL & BIOPHYS, DEPT PHYSIOL & BIOPHYS & MEM CORE STAFF, REGIONAL PRIMATE RES CTR, UNIV WASH, 69- *Concurrent Pos:* Nat Inst Combat Blindness fel, Johns Hopkins Univ, 66-67; NSF fel, Univ Freiburg, 67-68. *Honors & Awards:* Samuel A Talbot Award, Biomed Eng Group, Inst Elec & Electronics Engrs, 69. *Mem:* Soc Neurosci; assoc Soc Res Vision & Ophthal. *Res:* Oculomotor system of primates. *Mailing Add:* Dept Physiol & Biophys Univ Wash Sch Med Seattle WA 98195

FUCHS, ANNA-RIITTA, b Helsinki, Finland; m 48; c 4. REPRODUCTIVE PHYSIOLOGY. *Educ:* Univ Helsinki, Phil Cand, 48, Phil Mag(chem), 50; Univ Copenhagen, DSc, 78. *Prof Exp:* Fel reproduction, Dept Embryol, Carnegie Inst, Wash, DC, 50- 51; res assoc, Dept Physiol, Univ Copenhagen, 52-54, fel bact, Dept Pub Health, 54-56, res assoc reproduction, Dept Physiol, 57-62, univ adj, 62-65; res assoc reproduction, Biomed Div, Population Coun, NY, 65-70, staff scientist, 70-77; assoc prof, 77-86, PROF REPRODUCTIVE BIOL IN OBSTET & GYNEC & IN DEPT PHYSIOL & BIOPHYSICS, CORNELL UNIV MED COL, 86- *Concurrent Pos:* Adj assoc prof reproductive biol in obstet & gynec, Med Col, Cornell Univ, 71-77, fac mem, Grad Sch Med Sci, 73-; vis prof reproduction, Fac Med, Fed Univ Bahia, Brazil, 67- 68 & Med Sch, Chulalongkorn Univ, Thailand, 72-73. *Mem:* Endocrine Soc; Soc Gynec Invest; Soc Study Reproduction; Soc Study Fertil; fel NY Acad Sci; Asn Women Sci; hon mem, Finnish Gynec Asn; AAAS. *Res:* Endocrinology of gestation and the onset of parturition; uterine physiology; prostaglandins uterine and gonadal function; oxytocin and vasopressin in uterine and gonadal function; hormonal and molecular mechanisms regulating length of gestation, onset of labor, and the prevention of preterm birth. *Mailing Add:* Dept Obstet & Gynec S-412 1300 York Ave New York NY 10021

FUCHS, ELAINE V, b Hinsdale, Ill, May 5, 50; m. GENE EXPRESSION, DIFFERENTIATION. *Educ:* Univ Ill, Champaign-Urbana, BS, 72; Princeton Univ, PhD(biochem), 77. *Prof Exp:* Res fel biochem, Mass Inst Technol, 77-80; asst prof biochem, 80-85, assoc prof molecular & cell biol & biochem, 85-88, PROF MOLECULAR GENETICS, MOLECULAR & CELL BIOL & BIOCHEM, HOWARD HUGHES MED INST, UNIV CHICAGO, 88-, INVESTR, 88- *Concurrent Pos:* Andrew Mellon fel, Univ Chicago, 80; Searle Scholar Award, Searle Corp, 81; career develop award, NIH, 82; Pres Young Investr Award, Nat Sci Found, 85-90. *Honors & Awards:* R R Bensely Award, Am Asn Anatomists, 88. *Mem:* Am Soc Cell Biol; Am Soc Biol Chem; Soc Investigative Dermat. *Res:* Molecular biology of gene expression in differentiating human epidermal cells and in transgenic mouse epidermis; biochemical changes in the cytoskeletal architecture during differentiation; effect of vitamin A and growth factors on gene expression in normal and cancerous epithelial cells. *Mailing Add:* Dept Molecular Genetics Howard Hughes Med Inst Univ Chicago 5841 S Maryland Rm N314 Chicago IL 60637

FUCHS, EWALD FRANZ, b Munderkingen, Ger, Dec 5, 39; m 77. ELECTRICAL ENGINEERING, MATHEMATICS. *Educ:* Univ Stuttgart, BS, 63, dipl eng, 67; Univ Colo, Boulder, PhD(elec eng), 70. *Prof Exp:* Res engr, Siemens AG, Erlangen, Ger, 67; teaching asst, Univ Colo, Boulder, 67-69, res asst, 69-70, res assoc, 70-71, asst prof elec eng, 71; sr engr power indust, Kraftwerk Union AG, Muelheim-Ruhr, Ger, 71-77; ASSOC PROF ELEC ENG, UNIV COLO, BOULDER, 77- *Mem:* Inst Elec & Electronics Engrs; Verein Deutscher Elektrotechniker; Sigma Xi. *Res:* Investigation of harmonic effects in power systems; rotating machine optimization and design from a field modeling standpoint; finite difference and finite element formulations. *Mailing Add:* 1030 Rosehill Dr Boulder CO 80302-7148

FUCHS, FRANKLIN, b New York, NY, July 1, 31; m 57; c 2. PHYSIOLOGY. *Educ:* Univ Mich, AB, 53; Tufts Univ, PhD(physiol), 62. *Prof Exp:* From instr to asst prof, 64-71, assoc prof physiol, 71-80, PROF SCH MED, UNIV PITTSBURGH, 80- *Concurrent Pos:* Res fel physiol, Sch Med, Tufts Univ, 58-62; USPHS fel, Inst Neurophysiol, Copenhagen, 62-64; Lederle med fac award, 67-70; mem, Physiol Study Sect, NIH, 79-83; vis investr, Boston Biomed Res Inst, 88. *Mem:* AAAS; Soc Gen Physiol; Biophys Soc; Am Physiol Soc. *Res:* Physiology and biochemistry of muscle. *Mailing Add:* Dept Physiol Univ Pittsburgh Sch Med Pittsburgh PA 15261

FUCHS, FRITZ FRIEDRICH, b Frederiksberg, Denmark, Nov 27, 18; m 48; c 4. OBSTETRICS & GYNECOLOGY. *Educ:* Copenhagen Univ, MD, 45, DrMedSci, 57. *Prof Exp:* Second asst surgeon, Kommunehosp, Copenhagen, 53-55; second asst obstetrician & gynecologist, Rigshosp, 55-56, first asst obstetrician & gynecologist, 56-58; gynecologist-in-chief, Kommunehosp, 58-64; Given Found prof, Cornell Univ, 65-77, dept chmn, 65-78, Harold & Percy Uris prof reproductive biol, 77-80, prof obstet & gynec, 65-89, EMER PROF OBSTET & GYNEC, CORNELL UNIV, 89- *Concurrent Pos:* Mem ed staff, J Danish Med Assoc, 53-61; assoc prof, Copenhagen Univ, 56-64; assoc ed & founder, Danish Med Bull, 62-64; consult, Rockerfeller Univ, 68-78; consult, WHO, Thailand, 72-73; obstetrician & gynecologist-in-chief, NY Hosp, 65-78, attend obstet & gynecologist, 79; Am ed, J Obstet Gynecol, 87- *Mem:* Fel Am Gynec Soc; fel Am Col Obstet & Gynec; Am Fertil Soc; Endocrine Soc; AAAS. *Res:* Permeability of placenta; endocrinology of pregnancy; biology of reproduction; to prevent birth defects by antenatal diagnosis of genetic malformations and by prevention of preterm birth; preterm labor, dysmenorrhea, amniocentesis for antenatal diagnosis. *Mailing Add:* Dept Obstet & Gynec Cornell Univ Med Col 1300 York Ave New York NY 10021

FUCHS, H(ENRY) O(TTEN), mechanical engineering; deceased, see previous edition for last biography

FUCHS, HENRY, b Tokaj, Hungary, Jan 20, 48; US citizen. COMPUTER SCIENCE. *Educ:* Univ Calif, Santa Cruz, BA, 70; Univ Utah, PhD(comput sci), 75. *Prof Exp:* Res asst & teaching fel, dept comput sci, Univ Utah, 70-74; asst prof math sci, Univ Tex, Dallas, 75-78, comput sci coordr, 77-78; assoc prof comput sci, 78-83, PROF COMPUT SCI, UNIV NC, CHAPEL HILL, 83- *Mem:* Asn Comput Mach; Sigma Xi; Inst Elect & Electronics Engrs. *Res:* Computer graphics; microelectronic design; man-machine systems; medical image processing. *Mailing Add:* Dept Comput Sci Sitterson Hall Box 3175 Univ NC Chapel Hill NC 27599-3175

FUCHS, JACOB, b New York, NY, May 7, 23; m 46; c 2. ANALYTICAL CHEMISTRY, SPECTROCHEMISTRY. *Educ:* NY Univ, AB, 44; Univ Ill, MS, 47, PhD(chem), 50. *Prof Exp:* Asst chem, Univ Ill, 48-52; from asst prof to assoc prof, 52-59, exec officer dept, 61-75, PROF CHEM, ARIZ STATE UNIV, 59-, DIR, MOD INDUST SPECTROS, 56- *Concurrent Pos:* vis prof chem, Univ Colo, Boulder, 65. *Mem:* Am Chem Soc; Soc Appl Spectros (pres elect, 84, pres, 85); fel Am Inst Chemists. *Res:* X-ray diffraction and spectroscopy; ultraviolet and infrared spectroscopy. *Mailing Add:* Dept Chem Ariz State Univ Tempe AZ 85287-1604

FUCHS, JAMES ALLEN, b Ballinger, Tex, July 5, 43; m 66; c 2. BIOCHEMISTRY. *Educ:* Tex A&M Univ, BS, 65, MS, 67, PhD(genetics), 70. *Prof Exp:* NSF fel biochem, Univ Enzyme Inst, Copenhagen, 70-71; from asst prof to assoc prof, 72-87, PROF BIOCHEM, UNIV MINN, ST PAUL, 87- *Concurrent Pos:* NIH fel, Nobel Inst Biochem, Sweden. *Mem:* Am Soc Microbiol; AAAS; Am Soc Biochem & Molecular Biol. *Res:* Function and regulation of deoxynucleotide metabolism in Escherichia coli; protein engineering. *Mailing Add:* 140 Gortner Lab Univ Minn St Paul MN 55108

FUCHS, JAMES CLAIBORNE ALLRED, b Coronado, Calif, June 13, 38; m 65; c 1. VASCULAR SURGERY. *Educ:* Princeton Univ, AB, 60; Johns Hopkins Univ, MD, 64. *Prof Exp:* Resident, Duke Univ Med Ctr, 64-66; surgeon, US Pub Health Serv, 66-68; resident, 68-72, asst prof, 74-77, assoc prof, 78-81, PROF SURG, DUKE UNIV MED CTR, 81- *Concurrent Pos:* Chief, Surg Serv, Vet Admin Hosp, 80- *Mem:* Southern Surg Asn; Soc Univ Surgeons; Soc Vascular Surg; Int Cardiovascular Soc; Am Col Surgeons. *Res:* Vascular surgery; the evolution and treatment of atherosclerotic vascular disease; evaluation of postoperative vacular morphology. *Mailing Add:* Duke Univ Med Ctr PO Box 3351 Durham NC 27710

FUCHS, JULIUS JAKOB, b Monzingen, Ger, Feb 12, 27; US citizen; m 66; c 2. ORGANIC CHEMISTRY. *Educ:* Univ Mainz, PhD, 52. *Prof Exp:* From chemist to res chemist, 53-68, res assoc, 68-79, RES FEL ORG CHEM, E I DU PONT DE NEMOURS & CO, 79- *Concurrent Pos:* Fel, Swiss Fed Inst Technol, 52-53. *Res:* Economical synthesis of biologically active organic compounds. *Mailing Add:* 1104 Greenway Rd Wilmington DE 19803-3518

FUCHS, LASZLO, b Budapest, Hungary, June 24, 24; m 74; c 2. MATHEMATICS. *Educ:* Univ Budapest, MS & PhD(math), 47. *Prof Exp:* Asst math, Eotvos Lorand Univ, Budapest, 49-52, docent, 52-54, prof, 54-68; PROF MATH, TULANE UNIV, 68- *Concurrent Pos:* Vis prof, Tulane Univ, 61-62, Univ New South Wales, 65, Univ Miami, 66-68, Univ de Montpellier, 68, Univ Ariz, 72, Univ di Padova, 77 & 81, Univ West Australia, 80, Univ Essen, 82, Univ Orange Free State, 83 & Bar Ilan Univ, 84; W R Irby prof, Tulane Univ, 79. *Mem:* Am Math Soc; Math Asn Am; Hungarian Math Soc(treas, 51-63, secy gen, 63-66). *Res:* Abstract algebra, particularly commutative groups, module theory and partially ordered algebraic structures. *Mailing Add:* Dept Math Tulane Univ New Orleans LA 70118

FUCHS, MORTON S, b New York, NY, Nov 16, 32; m 60; c 2. INSECT ENDOCRINOLOGY. *Educ:* Mich State Univ, BS, 58, MS, 60, PhD(biochem), 67. *Prof Exp:* From asst prof to assoc prof biol, 66-69, chmn dept microbiol, 81-84, PROF BIOL, UNIV NOTRE DAME, 74-, CHMN DEPT BIOL SCI, 85- *Concurrent Pos:* Nat Commun Dis Control Ctr grant, 68-71 & NIH grants, 71-; mem, NIH study sect, TMP, 84-87. *Mem:* AAAS; Entom Soc Am; Am Soc Zool. *Res:* Biochemical and genetic control of differentiation; reproductive physiology of insects. *Mailing Add:* Dept Biol Sci Univ Notre Dame Notre Dame IN 46556

FUCHS, NORMAN H, b Newark, NJ, Aug 2, 38; m 59; c 2. THEORETICAL PHYSICS. *Educ:* Carnegie Inst Technol, BS, 59; Mass Inst Technol, PhD(physics), 64. *Prof Exp:* Res assoc physics, Mass Inst Technol, 64; res investr, Univ Pa, 64-66; from asst prof to assoc prof, 66-80, PROF PHYSICS, PURDUE UNIV, 80- *Mem:* Am Phys Soc; Am Asn Physics Teachers. *Res:* Symmetries and their breakdown; elementary particle physics. *Mailing Add:* Dept Physics Purdue Univ West Lafayette IN 47907

FUCHS, RICHARD, b Baltimore, Md, Dec 29, 26. ORGANIC CHEMISTRY. *Educ:* Cornell Univ, AB, 49; Univ Kans, PhD(chem), 53. *Prof Exp:* Asst instr chem, Univ Kans, 49-53; instr math, Calif Inst Technol, from asst prof to assoc prof, 55-66, prof, 66-; assoc prof, 63-72, PROF CHEM, UNIV HOUSTON, 72-; CHMN, DEPT AVIATION, CENT CAMPUS, OHIO UNIV. *Concurrent Pos:* From instr to asst prof, Univ Tex, 55-63. *Mem:* Am Chem Soc. *Res:* Solvation effects on rates and equilibria; thermochemistry of strained molecules. *Mailing Add:* 10119 Holly Springs Dr Houston TX 77042-1525

FUCHS, RICHARD E(ARL), b Milan, Tenn, July 26, 36; m 61; c 2. CHEMICAL ENGINEERING. *Educ:* Univ Tenn, BS, 58; La State Univ, MS, 62, PhD(chem eng), 64. *Prof Exp:* Chem engr, Esso Res Labs, 58-61; asst prof chem eng, Univ Miss, 64; sr res engr, comput processes, Olinkraft, Inc, 64-72; mgr pulp mill systs, Indust Nucleonics Corp, 72-75; develop mgr, Paper Div, Gulf States Paper Corp, 75-77; sr tech serv engr, Mill Div, 77-82, MGR PROCESS CONTROL, MANVILLE FOREST PROD CORP, INC, 82- *Mem:* Tech Asn Pulp & Paper Indust. *Res:* Decolorization of pulp mill bleaching effluents; forest products; process control. *Mailing Add:* Manville Forest Prod Corp PO Box 488 West Monroe LA 71291

FUCHS, ROBERT L, b Bay Shore, NY, Dec 7, 29; m 86; c 2. GEOLOGY. *Educ:* Cornell Univ, BA, 51; Univ Ill, MS, 52. *Prof Exp:* Area explorer, Mobil Oil, NY, staff geologist, jr geologist, 52-65; dir prog support, Flow Labs Inc, 65-66; pres, founder, dir, Con-Serv Corp & Automation Inst, 66-69; sr vpres, Intercontinental Energy Corp, 69-71; dir, Canyon Resources Corp, 79-90; dir, Sundance Oil Co, 82-84; pres, founder, dir, Geosystems Corp, 71-86; dir, Sheffield Explor Co Inc, 81-90; MANAGING DIR, FIRST FAIRFIELD INVESTMENT CO, 87- *Concurrent Pos:* Pres, Geol Soc Am Found, 87- *Honors & Awards:* Distinguished Serv Award, Am Asn Petrol Geologists. *Mem:* Am Asn Petrol Geologist (pres, 81-82); fel Geol Soc Am. *Mailing Add:* First Fairfield Investment Co 1675 Larimer St Suite 310 Denver CO 80202

FUCHS, ROLAND JOHN, b Yonkers, NY, Jan 15, 33; m 57; c 3. EDUCATION ADMINISTRATION. *Educ:* Columbia Univ, BA, 54; Clark Univ, MA, 57, PhD(geog), 59. *Prof Exp:* Vis assoc prof, Clark Univ, 63-64; asst prof, Univ Hawaii, chmn & assoc prof, 64-68, asst & assoc dean, Col Arts & Sci, 65-67, asst to pres, Int Prog, 86, CHMN & PROF GEOG, UNIV HAWAII, 68-; VRECTOR, UN UNIV, 87- *Concurrent Pos:* Res fel, Fac Geog, Moscow State Univ, 60-61; res assoc prof geog, Clark Univ, 63-64; Fulbright res scholar, Tribhuvan Univ, Nepal, 65; chmn, US Nat Comt-Int Geog Union, Nat Acad Sci-Nat Res Coun, 72-80, mem bd sci & technol in develop & chmn adv comt, Int Coun Sci Unions, 80-85; mem bd dirs, Am Asn Advan Slavic Studies, 76-81; mem, Coun Pac Sci Asn, 80- & US Nat Comt for Pac Basin Econ Coop, 85-; adj res assoc, E W Ctr, 80- *Honors & Awards:* Honors Award, Asn Am Geogr, 87. *Mem:* Asn Am Geogr; AAAS; Int Geog Union (vpres, 88, pres, 88-); Inst of British Geogr; Asn of Japanese Geogr. *Res:* Urbanization and population distribution policies in socialist and developing countries; land use and environmental issues in development. *Mailing Add:* UN Univ 15-1 Shibuya 2-chome Shibuya-ku Japan Tokyo 150 Japan

FUCHS, RONALD, b Los Angeles, Calif, Jan 27, 32; m 63; c 2. SOLID STATE PHYSICS. *Educ:* Calif Inst Technol, BS, 54; Univ Ill, PhD(physics), 57. *Prof Exp:* Fulbright fel, Stuttgart Tech Univ, 57-59; physicist, Div Sponsored Res, Lab Insulation Res, Mass Inst Technol, 59-61; from asst prof to assoc prof, 61-74, PROF PHYSICS, IOWA STATE UNIV, 74- *Mem:* Am Phys Soc. *Res:* Theoretical solid state physics; optical properties. *Mailing Add:* Dept Physics Iowa State Univ Ames IA 50011

FUCHS, VICTOR ROBERT, b New York City, NY, Jan 31, 24; m 48; c 4. MEDICAL SCIENCES. *Educ:* NY Univ, BS, 47; Columbia Univ, MA, 51, PhD(econ), 55. *Prof Exp:* Assoc prof econ, NY Univ, 59-60; prog assoc econ, Ford Found, 60-62; prof community med, Mt Sinai Sch Med & prof econ, City Univ New York Grad Ctr, 68-74; vpres, 68-78, RES ASSOC, NAT BUR ECON RES, 62-; HENRY J KAISER JR PROF, STANFORD UNIV, 88-, PROF ECON, DEPT ECON & DEPT HEALTH RES & POLICY, 74- *Concurrent Pos:* Mem, Pres Comt Mental Retardation, 68-71; mem, bd dirs, Prin Financial Group, 81-90; distinguished investr award, Asn Health Serv, 88; res prize, Baxter Found Health Serv, 91. *Mem:* Inst Med-Nat Acad Sci; fel Am Acad Arts & Sci; Nat Acad Social Ins; Am Philos Soc; fel Am Econ Asn. *Res:* Economic aspects of health, medical care, family, work, and dimensions of human behavior and social institutions in post-industrial society; women's quest for economic equality; well-being of children. *Mailing Add:* 204 Junipero Serra Blvd Stanford CA 94305-8715

FUCHS, VLADIMIR, b Czech, Oct 14, 35; Can citizen; m 67; c 2. THEORY OF MODE COUPLING, PLASMA HEATING & CURRENT GENERATION. *Educ:* Charles Univ, Prague, Czech, MSc, 61, RNDr, 66, PhD(physics), 68. *Prof Exp:* Asst prof physics, Charles Univ, Prague, 64-68; mem res staff, Inst Res Hydro, Que, 68-89; MEM RES STAFF, CAN CTR MAGNETIC FUSION, 89- *Concurrent Pos:* Vis scientist, Plasma Fusion Ctr, Mass Inst Technol, 79-80, Culham Lab, UK Atomic Energy Authority, 81 & 90-91. *Honors & Awards:* Ayrton Premium, Inst Elec Engrs, UK, 74. *Mem:* Fel Am Phys Soc; Can Asn Physicists. *Res:* Interaction of waves with plasmas; wave coupling and mode conversions; plasma heating and current generation with waves. *Mailing Add:* Can Ctr Magnetic Fusion 1804 Montee Ste-Julie Varennes PQ J3X 1S1 Can

FUCHS, W KENT, b Elk City, Okla, Nov 3, 54; m 81; c 3. FAULT-TOLERANT COMPUTING. *Educ:* Duke Univ, BSE, 77; Univ Ill, MS, 82, PhD(elec eng), 85; Trinity Evangelical Divinity Sch, MDiv, 84. *Prof Exp:* asst prof comput eng, 85-89, ASSOC PROF, ELEC & COMPUTER ENG, UNIV ILL, URBANA, 89- *Mem:* Asn Comput Mach; Inst Elec & Electronics Engrs. *Res:* Highly reliable parallel computer architectures and VLSI systems; computer aided design. *Mailing Add:* Coord Sci Lab Univ Ill 1101 W Springfield Ave Urbana IL 61801

FUCHS, WALTER, b Munich, Germany, Dec 29, 32; US citizen. EXPERT SYSTEMS, MATHEMATICAL MODELLING. *Educ:* Munich Tech Univ, dipl, 61; Carnegie Mellon Univ, MS, 64, PhD(physics), 68. *Prof Exp:* Asst prof physics, Univ Toledo, 67-68, res assoc, Cornell Univ, 68-70; res assoc, Cornell Univ, 68-70; phys scientist, Physics & Comput Sci, US Dept Energy, 70-89; RETIRED. *Concurrent Pos:* Tech consult, US Dept of Defense, Picatinny Arsenal, 70-72. *Mem:* AAAS; Am Phys Soc. *Res:* Coal conversion, coal-to-oil, coal-to-gas; instrumentation and control of process facilities; automatic data processing; mathematical modeling of conversion processes; solid state physics; computer science; artificial intelligence. *Mailing Add:* US Dept Energy PO Box 10940 Pittsburgh PA 15236

FUCHS, WOLFGANG HEINRICH, b Munich, Ger, May 19, 15; nat US; m 43; c 3. PURE MATHEMATICS. *Educ:* Cambridge Univ, BA, 36, PhD(math), 41. *Prof Exp:* Teacher, Brit Univs, 40-49; from assoc prof to prof, 50-85, EMER PROF MATH, CORNELL UNIV, 85- *Concurrent Pos:* Vis assoc prof math, Cornell Univ, 48-49. *Res:* Theory of functions. *Mailing Add:* Dept Math White Hall Cornell Univ Ithaca NY 14853-7901

FUCHSMAN, CHARLES H(ERMAN), b New York, NY, June 12, 17; m 37; c 3. CHEMISTRY. *Educ:* City Col New York, BS, 36; Western Reserve Univ, PhD(org chem), 65. *Prof Exp:* Technologist, US Bur Mines, 42-45; res chemist potash refining, Int Minerals & Chem Corp, 45-51; res group leader, Micro-Pilot Plant Develop, Columbia Southern Chem Corp, 51-56; dir res & develop, Ferro Chem Corp, 56-66, dir res & develop, 66-69, assoc dir res org chem, Ferro Chem Div, 69-72; from assoc prof to prof & dir, Ctr Environ Studies, Bemidji State Univ, 72-82; CONSULT, 72- *Concurrent Pos:* Consult, 72- *Mem:* Phytochem Soc NAm; AAAS; Am Chem Soc; NY Acad Sci; Int Peat Soc. *Res:* Organic chemical synthesis; polymer chemistry; history of organic chemistry; chemistry of peat; environmental chemistry; natural products chemistry. *Mailing Add:* 3006 Cedar Lane NW Bemidji MN 56601-4202

FUCHSMAN, WILLIAM HARVEY, b New York, NY, June 2, 41; m 64; c 3. BIOINORGANIC CHEMISTRY. *Educ:* Harvard Univ, BA, 63; Johns Hopkins Univ, PhD(biochem), 67. *Prof Exp:* Asst prof chem, Univ SFla, 67-68; fel, E I du Pont de Nemours & Co, 68-70; from asst prof to assoc prof, 70-83, PROF CHEM, OBERLIN COL, 83- *Mem:* AAAS; Am Chem Soc; Crop Sci Soc Am. *Res:* Chemistry of porphyrins, metalloporphyrins and hemeproteins; nitrogen fixation. *Mailing Add:* Dept Chem Oberlin Col Oberlin OH 44074-1083

FUCIK, EDWARD MONTFORD, b Chicago, Ill, Jan 25, 14; m 43; c 3. HYDROELECTRIC ENGINEERING. *Educ:* Princeton Univ, BSCE, 35; Harvard Univ, MS, 37. *Prof Exp:* Mem staff, Harza, Consult Engrs, Chicago, 38-40; found engr, Panama Canal, 40-42; mem staff, Harza Eng Co, 45-69, pres, 69-74, chmn bd, 69-77, emer chmn, 77-85; RETIRED. *Concurrent Pos:* Mem consult bd, Kalabagh Dam, Pakistan; mem bd mgrs, Highland Park Hosp, Ill, 66-69; mem adv coun, Ill Inst Technol, 71-75. *Honors & Awards:* Thomas Fitch Rowland Prize, Am Soc Civil Engrs, 53. *Mem:* Nat Acad Eng; Am Soc Civil Engrs; Soc Am Mil Engrs; Nat Soc Prof Engrs. *Mailing Add:* 57 S Deer Park Dr Highland Park IL 60035

FUCIK, JOHN EDWARD, b Waukegan, Ill, May 30, 28; m 56; c 3. HORTICULTURE, PLANT PHYSIOLOGY. *Educ:* Univ Ill, BS, 49, MS, 57, PhD(hort), 63. *Prof Exp:* Asst prod mgr, Pfister Assoc Growers, Inc, 49-51; supvr & teacher hort, Univ Ill, 57-65; PROF AGR CITRUS CTR, TEX A&I UNIV, 65- *Concurrent Pos:* Citrus consult, AID, Jordan. *Mem:* Am Soc Hort Sci; Int Soc Citricult; Plant Growth Regulation Soc Am; Int Soc Hort Sci. *Res:* Physiology of horticultural tree crops; tree-soil-environment relationships; native plants. *Mailing Add:* Texas A&I Citrus Ctr PO Box 1150 Weslaco TX 78596

FUDA, MICHAEL GEORGE, b Albany, NY, Oct 31, 38; m 62; c 3. THEORETICAL NUCLEAR PHYSICS, RELATIVISTIC QUANTUM MECHANICS. *Educ:* Rensselaer Polytech Inst, BS, 60, PhD(physics), 67. *Prof Exp:* Res physicist, Knolls Atomic Power Lab, 62-64; from asst prof, to assoc prof, 67-78, PROF PHYSICS, STATE UNIV NY BUFFALO, 78- *Mem:* Am Phys Soc. *Res:* Scattering theory with applications to nuclear physics; theory of three-particle systems; relativistic quantum mechanics. *Mailing Add:* Dept Physics State Univ NY Buffalo NY 14260

FUDALI, ROBERT F, b Minneapolis, Minn, July 7, 33; m 56; c 3. GEOCHEMISTRY, PETROLOGY. *Educ:* Univ Minn, Minneapolis, BA, 56; Pa State Univ, PhD(geochem), 60. *Prof Exp:* Res fel, Pa State Univ, 60-62; staff scientist, Bellcomm Inc, 62-66; RES SCIENTIST, DEPT MINERAL SCI, SMITHSONIAN INST, 66- *Mem:* Fel Meteoritical Soc; Geochem Soc; fel Mineral Soc Am; fel AAAS. *Res:* Genesis of igneous rocks and meteorites; genesis of lunar rocks and landforms; role of large cometary and meteorite impacts in the development of planetary surfaces. *Mailing Add:* Dept Mineral Sci Stop 119 Smithsonian Inst Washington DC 20560

FUDENBERG, H HUGH, b New York, NY, Oct 24, 28; m 55; c 4. HEMATOLOGY, IMMUNOLOGY. *Educ:* Univ Calif, Los Angeles, AB, 49; Univ Chicago, MD, 53; Boston Univ, MA, 57. *Prof Exp:* Intern med, Univ Utah, 53-54; asst resident med, Mt Sinai Hosp, New York, 56-57 & Peter Bent Brigham Hosp, Boston, 57-58; res assoc immunol, Rockefeller Inst, 58-60; from asst prof to prof med, Sch Med, Univ Calif, San Francisco, 66-75, chief hemat unit, 62-75; chmn dept basic & clin immunol, Med Univ SC, 75-89; RETIRED. *Concurrent Pos:* Fel hemat, Sch Med, Tufts Univ, 54-56; mem, Expert Comt Immunol, WHO, 62-82; prof bact & immunol, Univ Calif, Berkeley, 66-75; chief ed, J Chem Immunol, 74-86. *Honors & Awards:* Pasteur Medal, Inst Pasteur, Paris, 62; Robert A Cooke Mem Medal, Am Acad Allergy, 67; Petrov Medal, Cancer, USSR, 79. *Mem:* Fel AAAS; Am Asn Immunologists; Am Soc Human Genetics; Am Soc Clin Invest; Asn Am Physicians. *Res:* Properties of related proteins; genetic control of normal antibody synthesis and genetically determined immunologic aberrations predisposing to disease; neuro immunology. *Mailing Add:* Dept Microbiol Med Univ SC Charleston SC 29425

FUELBERG, HENRY ERNEST, b Brenham, Tex, May 13, 48. METEOROLOGY. *Educ:* Tex A&M Univ, BS, 70, MS, 71, PhD(meteorol), 76. *Prof Exp:* From res asst to res assoc meteorol, Tex A&M Univ, 74-77; from asst prof to assoc prof meteorol, St Louis Univ, 77-85; ASSOC PROF METEOROL, FLA STATE UNIV, 85- *Concurrent Pos:* Prin investr, NASA grants, 78- *Mem:* Am Meteorol Soc. *Res:* Synoptic meteorology; atmospheric energetics; satellite meteorology. *Mailing Add:* Dept Meteorol Fla State Univ Tallahassee FL 32306-3034

FUELLING, CLINTON PAUL, b Decatur, Ind, Dec 8, 37; m 58; c 2. NUMERICAL APPROXIMATION, PERFORMANCE MEASUREMENT. *Educ:* Ball State Univ, BS, 60; Univ Chicago, MS, 62; Am Univ, PhD(math), 72. *Prof Exp:* Mathematician, Defense Intel Agency, 65-68; sr analyst opers res, Gen Res Agency, 68-73; mem fac, 73-83, PROF COMPUT SCI, BALL STATE UNIV, 83- *Mem:* Asn Comput Machinery; Math Asn Am; Inst Elec & Electronics Engrs; Soc Indus & Appl Math. *Res:* Numerical methods of approximation and performance of computer systems. *Mailing Add:* 2905 W Twickingham Dr Muncie IN 47304

FUENNING, SAMUEL ISAIAH, b Ft Morgan, Colo, Sept 20, 16; m 44; c 5. PREVENTIVE MEDICINE, ATHLETIC MEDICINE. *Educ:* Univ Nebr, BSc, 40, MS, 41, MD, 45. *Prof Exp:* Asst bot, 40-41, assoc prof pub health & assoc cellular res, 50-62, prev med, 62-74, dir, Nebr Ctr Health Educ, 74-77, coordr res & develop, 77-79, PROF PREV MED, UNIV NEBR-LINCOLN, 74-, FAMILY WELLNESS SPECIALIST, 78- *Concurrent Pos:* Chmn, Nebr Voluntary Adv Comn Selective Serv System; med adv, Nebr Govr's Coun Fitness; chmn, Nat Conf Health Col Community, Boston, 70 & Nebr Conf Educ Health & Fitness, 64, 68, 70 & 72; mem, Vis Comt Overseers, Harvard Univ, 70-76; med dir, Univ Nebr Health Serv & Health Ctr, 47-74; med dir res, Athletic Med, 77-89, Sports Med, 89; pres, Nebr Interagency Health Coun, 80-; co-chmn, Gov Coun Physical Fitness & Sports, 80- *Honors & Awards:* Hitchcock Award, Am Col Health Asn, 64, Ruth Boynton Award, 69. *Mem:* Am Col Preventive Med; AMA; Am Col Health Asn (pres, 60-61); Am Public Health Asn; Soc Prospective Med; Sigma Xi. *Res:* Clinical medicine in the college age group; stress; fitness and mental health; sports medicine. *Mailing Add:* 5100 Grandview Lane Lincoln NE 68521

FUENTES, RICARDO, JR, b McAllen, Tex, July 28, 48; m 70; c 1. ZIEGLER-NATTA CATALYSIS, ANIONIC POLYMERIZATION. *Educ:* Pan Am Univ, BS, 70; Univ Tex, Austin, PhD(chem), 75. *Prof Exp:* Res assoc biochem, Mich State Univ, 75-77; RES MGR, DOW CHEM CO, 77- *Concurrent Pos:* Technol mgr, Pexco Polymers. *Mem:* Am Chem Soc. *Res:* Organometallic chemistry of Ziegler-Natta catalysts; anionic polymerization; styrene block copolymers product and process research. *Mailing Add:* Dow Chem Res & Develop PO Box 400 Plaquemine LA 70765-0400

FUERNISS, STEPHEN JOSEPH, b York, Nebr, July 17, 45. ORGANIC CHEMISTRY. *Educ:* Col St Thomas, BA, 67; Univ Nebr, PhD(org chem), 75. *Prof Exp:* Fel, Brandeis Univ, 75-76, Ohio State Univ, 76-77; res chemist, E I du Pont de Nemours & Co, Inc, 77-85; ADV CHEMIST, IBM CORP, 85- *Mem:* Am Chem Soc; Sigma Xi; Soc Appl Spectros. *Res:* Photochemistry; polymer chemistry. *Mailing Add:* Boswell Hill Rd Box 415 A Endicott NY 13760

FUERST, ADOLPH, b Linz, Austria, Sept 11, 25; US citizen; m 51; c 2. PAPER & INK RELATIONSHIPS, RHEOLOGICAL TESTING. *Educ:* City Univ New York, BS, 49, MA, 57. *Prof Exp:* Chem lab technician, Mt Sinai Hosp, 49-50; asst prod chemist, US Vitamin Corp, 50-52; pigment chemist, Ansbacher-Siegle Corp, 52-55; pigment & dyestuff chemist, Gen Aniline & Film Corp, 55-58; group leader, Sinclair & Valentine Co, 58-69, sr group leader, 69-72, res & develop mgr, 72-74; mgr spec serv, GRAPHIC ARTS LABS, SUN CHEM CORP, 74-86; CONSULT, 86- *Mem:* Am Chem Soc; fel Am Inst Chemists. *Res:* Development of paper and ink tests for predicting on-press performance; rheological testing of ink; organic pigment synthesis; pigment dispersion studies. *Mailing Add:* 266 Marlborough Rd Brooklyn NY 11226-4512

FUERST, CARLTON DWIGHT, b Nagoya, Japan, Apr 19, 55; US citizen; m 85. MAGNETIC MATERIALS. *Educ:* Col William & Mary, BS, 77; Univ Pa, PhD(physics), 83. *Prof Exp:* STAFF RES SCIENTIST, GEN MOTORS RES LABS, 83- *Mem:* Am Phys Soc. *Res:* Rare earth-transition metal materials, especially permanent magnets produced by rapid solidification and hot forming of same. *Mailing Add:* Physics Dept Gen Motors Res Labs Mound Rd Box 9055 Warren MI 48090-9055

FUERST, PAUL ANTHONY, b Rockville Centre, NY, June 15, 48; m 70. POPULATION GENETICS, MOLECULAR GENETICS. *Educ:* Manhattan Col, BA, 70; Brown Univ, ScM, 72, PhD(biol), 75. *Prof Exp:* Res assoc human genetics, Univ Tex Health Sci Ctr Houston, 75-77; asst prof biol, Univ Houston, 76-77; sr res assoc human genetics, Univ Tex Health Sci Ctr, Houston, 77-80; asst prof, 80-86, ASSOC PROF MOLECULAR GENETICS, OHIO STATE UNIV, 86- *Concurrent Pos:* Vis scientist, Nat Inst Genetics, Mishima, Japan, 83, 85; dir, Prog Molecular Cell Develop Biol, Ohio State Univ, 88- *Mem:* Am Genetics Asn; Genetics Soc Am; Am Soc Human Genetics; Sigma Xi; Am Soc Naturalists. *Res:* Population genetics especially theoretical genetics including statistical aspects of protein and nucleic acid polymorphism in natural populations; evolutionary genetics of intracellular bacteria in genus Rickettsia and subcellular organelles. *Mailing Add:* Dept Molecular Genetics Ohio State Univ 484 W 12th Ave Columbus OH 43210

FUERST, ROBERT, b Vienna, Austria, Jan 12, 21; nat US; m 46; c 2. GENETICS, MICROBIOLOGY. *Educ:* Univ Houston, BS, 44; Univ Tex, MA, 48, PhD(zool), 55. *Prof Exp:* Res scientist biochem, M D Anderson Hosp & Tumor Inst, 49-52, res scientist, Genetics Found, 52-55, asst biologist chg microbiol sect, 55-57; assoc prof, 57-64, head microbiol res, 57-87, PROF BIOL, TEX WOMAN'S UNIV, 64- *Concurrent Pos:* Asst prof genetics, Univ Tex, 55-57. *Mem:* AAAS; Am Soc Human Genetics; Soc Indust Microbiol; Genetics Soc Am; Sigma Xi. *Res:* Microbial genetics; radiation; biology; author of over one hundred publications and several textbooks of microbiology in several languages, also lab manuals in microbiology. *Mailing Add:* TWU Sta Tex Woman's Univ Biol Dept Denton TX 76204

FUERSTENAU, D(OUGLAS) W(INSTON), b Hazel, SDak, Dec 6, 28; m 53; c 3. MINERAL PROCESSING. *Educ:* SDak Sch Mines & Technol, BS, 49; Mont Sch Mines, MS, 50; Mass Inst Technol, ScD(metall), 53. *Hon Degrees:* Dr, Univ Liege, Belg, 89. *Prof Exp:* Asst prof mineral eng, Mass Inst Technol, 53-56; res engr, Union Carbide Metals Co, 56-58; mgr mineral eng, Kaiser Aluminum & Chem Corp, 58-59; from assoc prof to prof metall, 59-86, chmn dept mat sci & mineral eng, 70-78, P MALOZEMOFF PROF MINERAL ENG, UNIV CALIF, BERKELEY, 87- *Concurrent Pos:* Mem, Bd Mineral Resources, Nat Resource Coun, 75-77; Humboldt Sr Scientist award, Fed Repub Germany, 84. *Honors & Awards:* Hardy Gold Medal, Am Inst Mining, Metall & Petrol Engrs, 57, Raymond Award, 61, Richards Award, 76 & Gaudin Award, 78; Aplan Award, Eng Found, 90. *Mem:* Nat Acad Eng; Am Chem Soc; hon mem Am Inst Mining, Metall & Petrol Engrs; Am Inst Chem Engrs. *Res:* Mineral processing; applied colloid and surface chemistry; properties of particulate materials. *Mailing Add:* Dept Mat Sci & Mineral Eng Univ Calif Berkeley CA 94720

FUERSTENAU, M(AURICE) C(LARK), b Watertown, SDak, June 6, 33; m 53; c 4. GEOLOGICAL ENGINEERING, METALLURGY. *Educ:* SDak Sch Mines & Technol, BS, 55; Mass Inst Technol, MS, 57, ScD(metall), 61. *Prof Exp:* Res assoc metall, Mass Inst Technol, 60-61; res engr, NMex Bur Mines, 61-63; from asst prof to assoc prof, Colo Sch Mines, 63-68; from assoc prof to prof, Univ Utah, 68-70; prof metall eng & head dept, SDak Sch Mines & Technol, 70-87, interim vpres, 87-88; ECHO BAY MINES DISTINGUISHED PROF, UNIV NEV, RENO, 88- *Honors & Awards:* Arthur F Taggart Award, Soc Mining Engrs, 78; Antoine M Gaudin Award, Soc Mining Engrs, 79; Robert H Richards Award, Am Inst Mining, Metall & Petrol Engrs, 82. *Mem:* Nat Acad Eng; Am Inst Mining, Metall & Petrol Engrs (vpres, 83); Soc Mining Engrs (pres, 82). *Res:* Surface chemistry and adsorption phenomena; kinetics and reaction mechanisms involved in hydrometallurgy; mechanisms involved in froth flotation; fundamentals of sedimentation. *Mailing Add:* Dept Chem & Metall Eng Univ Nev Reno NV 89557

FUESS, FREDERICK WILLIAM, III, b Syracuse, NY, Nov 5, 27; m 52; c 2. AGRONOMY. *Educ:* Cornell Univ, BS, 52, MEd, 55; Mich State Univ, PhD, 63. *Prof Exp:* Asst prof agron, State Univ NY Agr & Tech Inst, Morrisville, 55-60; asst, Mich State Univ, 60-63; from asst prof to prof plant & soil sci, Ill State Univ, 63-90; RETIRED. *Concurrent Pos:* Book reviewer, AAAS, 70; Nat Asn Col Teachers Agr publ comt. *Mem:* Am Soc Agron; Crop Sci Soc Am; AAAS; Sigma Xi; Nat Asn Col Teachers Agr. *Res:* Crop physiology and management; soybean production; hydroponic crop production. *Mailing Add:* 1609 N Linden St Normal IL 61761

FUGATE, JOSEPH B, b Kansas, Sept 28, 33. MATHEMATICS. *Educ:* Univ Kans, BA, 56; Univ Iowa, PhD(math), 64. *Prof Exp:* PROF MATH, UNIV KY, 64- *Mem:* Am Math Asn; Math Asn Am. *Mailing Add:* Math Dept Univ Ky Lexington KY 40506

FUGATE, KEARBY JOE, b Dallas, Tex, Aug 6, 34; m 62; c 2. MICROBIOLOGY, BIOCHEMISTRY. *Educ:* Baylor Univ, BA, 56, MS, 63, PhD(biochem), 67. *Prof Exp:* Res asst microbiol, J K & Susie Wadley Res Inst, 60-63; asst prof, Dent Sch, Baylor Univ, 64-67; BIOCHEMIST, USPHS, 67- *Concurrent Pos:* Br chief hematol & path, Div Clin Lab Devices, Bur Med Devices. *Mem:* AAAS; Am Soc Microbiol; Am Pub Health Asn; fel Am Inst Chemists; Sigma Xi; Am Soc Clin Pathologists. *Res:* Immunochemistry of bacterial cell wall constituents; immunochemistry of antigen-antibody interactions; immunology of murine leukemia-virus transmission; immunochemistry of tumor-associated markers; software systems. *Mailing Add:* 12860 Whitefur Lane Herndon VA 22070

FUGELSO, LEIF ERIK, b Minot, NDak, Oct 29, 35; m; c 4. GEOPHYSICS, APPLIED MATHEMATICS. *Educ:* Univ Chicago, SM, 63, PhD(geophys), 73. *Prof Exp:* Res engr, Mech Res Div, Am Mach & Foundry Co, 59-63 & Gen Am Transp Corp, 63-75; seismologist, Portland Cement Asn, 75; STAFF MEM, LOS ALAMOS SCI LAB, 75- *Res:* Seismology; structural dynamics; terminal ballistics; shock waves; weapons and blast effects; explosive safety. *Mailing Add:* Los Alamos Nat Lab PO Box 1663 MS J576 Los Alamos NM 87544

FUGET, CHARLES ROBERT, b Rochester, Pa, Dec 15, 29; m 56; c 1. PHYSICAL CHEMISTRY, EDUCATIONAL ADMINISTRATION. *Educ:* Geneva Col, BS, 51; Pa State Univ, MS, 53, PhD(phys chem), 56. *Hon Degrees:* LHD, Hahnemann Univ; DSc, Geneva Col. *Prof Exp:* Res chemist, Esso Res & Eng Co, NJ, 55-56; from asst prof to assoc prof chem, Geneva Col, 56-63; res chemist, Callery Chem Co, 57-59; prof sci, State Univ NY Col Buffalo, 63-64; prof physics & chmn dept, Geneva Col, 64-71; dir, Div Natural Sci & Math, Indiana Univ Pa, 71-73, prof physics, 71-77, assoc dean, Col Natural Sci & Math, 73-76, dean, 77-88; DEP SECY & COMNR HIGHER EDUC, DEPT EDUC, COMMONWEALTH, PA, 88- *Mem:* Am Chem Soc; Am Phys Soc. *Res:* Thermodynamic properties; reaction calorimetry; electrochemistry. *Mailing Add:* 4414 D Ontario Dr Harrisburg PA 17111-8065

FUGGER, JOSEPH, b Jesenice, Yugoslavia, Feb 11, 21; nat US; m 53; c 7. INDUSTRIAL CHEMISTRY. *Educ:* Univ Pittsburgh, BS, 46, MS, 48, PhD(chem), 52. *Prof Exp:* Res intern, Northern Regional Lab, USDA, 49-50; asst prof res, Antioch Col, 52-53; res chemist, Buckeye Cellulose Corp, 54-55; fel chem, Harvard Univ, 55-56; res group leader, Thiokol Chem Corp, 57-58; res specialist gas dynamics, Boeing Airplane Co, 58-59; staff scientist res & opers, Aeronutronic Div, Ford Motor Co, 59-60; res scientist, Missiles & Space Div, Lockheed Aircraft Corp, 61-62; res group leader, Westreco, Inc, Nestle Int, 63-64, head basic res sect, 64-67; consult chemist, Hops Extract

Corp Am, Wash, 69-70; consult chemist, 70-75; PRES, FUGGER-TECH CONSULTS, 75- *Concurrent Pos:* Consult, appln of foreign chem tech info US indust. *Res:* Organic synthesis; macromolecular systems; beverage and food; solid propellants; materials science. *Mailing Add:* 739 Greenleaf Dr Monroeville PA 15146

FUGO, NICHOLAS WILLIAM, b Syracuse, NY, Sept 15, 13; m 40; c 1. OBSTETRICS & GYNECOLOGY. *Educ:* Syracuse Univ, AB, 35; Univ Iowa, MS, 37, PhD(endocrinol), 40; Univ Chicago, MD, 50; Am Bd Obstet & Gynec, dipl. *Prof Exp:* Res assoc zool, Univ Iowa, 39-40, instr pharmacol, Med Labs, 40-43; res assoc obstet & gynec, Univ Chicago, 46-52, from instr to assoc prof, 52-60; from prof to emer prof obstet & gynec, Sch Med, WVa Univ, 60-85, Margaret Higgins Sanger chair family planning & reproductive physiol, 67-80; RETIRED. *Mem:* Soc Exp Biol & Med; Am Soc Pharmacol & Exp Therapeut; Am Fertil Soc; Am Fedn Clin Res; fel Am Col Obstet & Gynec. *Res:* Endocrinology of sex; human reproduction. *Mailing Add:* Dept Obstet & Gynec WVa Med Ctr PO Box 4127 Morgantown WV 26505

FUHLHAGE, DONALD WAYNE, b Virgil, Kans, Sept 2, 31; m 61; c 2. ORGANIC CHEMISTRY. *Educ:* Univ Kans, BS, 53, PhD(org chem), 58. *Prof Exp:* Sr res chemist, Tidewater Oil Co, Calif, 58-62, res assoc, Pa, 62-63; supvr process develop, 63-70, sect leader process develop & residue anal, 70-75, SUPVR SYNTHESIS & PILOT PLANT, THOMPSON-HAYWARD CHEM CO, HARRISON & CROSSFIELDS GROUP, 75- *Mem:* Am Chem Soc. *Res:* Pyrrole chemistry; synthesis of pesticides; liquid chromatography; surfactant synthesis; environmental analysis. *Mailing Add:* PO Box 2383 Kansas City KS 66110

FUHR, IRVIN, b Sharon, Pa, Jan 16, 13; m 40; c 2. BIOPHYSICAL CHEMISTRY. *Educ:* Univ Wis, BS, 37, MS, 39, PhD(biochem), 42. *Prof Exp:* Toxicologist, Army Chem Ctr, US War Dept, Md, 43-48; exec secy, Biochem Study Sect, Div Res Grants, NIH, 48-55, exec secy, Biophysics & Biophys Chem Study Sect, 55-78. *Mem:* Biophys Soc. *Res:* Toxicology of drugs and chemical warfare agents; vitamin and mineral metabolism; anemia and rickets; development of experimental diets and chemical methods; development of rodenticide cartridge. *Mailing Add:* 12925 Crisfield Rd Silver Spring MD 20906

FUHR, JEFFREY ROBERT, b Baltimore, Md, Oct 27, 46; m 72; c 2. ATOMIC SPECTRAL LINE SHAPES, ATOMIC TRANSITION PROBABILITIES. *Educ:* Earlham Col, AB, 68; Purdue Univ, MS, 70. *Prof Exp:* PHYSICIST, NAT INST STANDARDS & TECHNOL, 70- *Mem:* Am Phys Soc. *Res:* Evaluation and compilation of data on atomic transition probabilities for publication in journals and handbooks; annotated bibliographies; availability and reliability of atomic data. *Mailing Add:* Nat Inst Standards & Technol Physics Bldg Rm A267 Gaithersburg MD 20899

FUHR, JOSEPH ERNEST, b New York, NY, June 30, 36; m 61; c 3. CELL PHYSIOLOGY, HEMATOLOGY. *Educ:* Le Moyne Col, AB, 60; LI Univ, MS, 62; St John's Univ, PhD(biol), 68. *Prof Exp:* Res assoc hemat, Albert Einstein Col Med, 66-68; hemat trainee, Col Physicians & Surgeons, Columbia Univ, 68-70, instr human genetics & develop, 70-71; asst res prof, 71-75, assoc prof med biol, 78-81, ASSOC RES PROF CELL PHYSIOL & HEMAT, MEM RES CTR, UNIV TENN, KNOXVILLE, 75-, PROF MED BIOL & DIR MEM RES CTR, 81- *Concurrent Pos:* Adj assoc prof biol, St John's Univ, NY, 70-71; Leukemia Soc Am spec fel, 70-72; mem ad hoc study sect, Nat Heart, Lung & Blood Inst, 75-76; res career develop award, Nat Inst Arthritis, Metab & Digestive Dis, 76-81; consult sickle cell, Nat Heart, Lung & Blood Int, 81-82. *Mem:* AAAS; Am Soc Hemat; Soc Exp Biol & Med; Int Soc Exp Hemat. *Res:* Erythropoiesis and hemoglobin synthesis; control of protein synthesis. *Mailing Add:* 623 Broome Rd Knoxville TN 37909

FUHRIMAN, D(EAN) K(ENNETH), b Ridgedale, Idaho, June 6, 18; m 41; c 5. CIVIL ENGINEERING. *Educ:* Utah State Col, BS, 41, MS, 50; Univ Wis, PhD(civil eng), 52. *Prof Exp:* Dept asst, exp sta, Utah State Col, 37-41; jr engr, US Corps Engrs, 41; instr, civil eng, Utah State Col, 41-43; asst irrig engr, soil conserv serv, USDA, 46-48, agr engr, PR, 48-50; assoc prof civil eng, Colo Agr & Mech Col, 51-52; assoc prof irrig & drainage engr, Utah State Agr Col, 52-54; assoc prof civil eng, 54-56, chmn dept, 58-61, prof, 56-83, EMER PROF CIVIL ENG, BRIGHAM YOUNG UNIV, 83- *Concurrent Pos:* Pres & prin engr, Fuhriman, Rollins & Co, 56-66; pres, Fuhriman, Barton & Assoc, Consult Engrs, 70-81; tech adv, US Water Pollution Control Admin, Washington, DC, 67-68 & Indust Pollution Control Res Div, Environ Protection Admin, 74-75; mem, US comt Int Comn Irrig, Drainage & Flood Control; pres, Dean K Fuhriman & Assocs, Eng Consults. *Mem:* Am Soc Civil Engrs; Am Soc Eng Educ. *Res:* Water resources engineering; irrigation; drainage; agricultural engineering; environmental engineering. *Mailing Add:* 1457 Cherry Lane Provo UT 84604

FUHRKEN, GEBHARD, b Frankfurt, Ger, Feb 10, 30. MATHEMATICS. *Educ:* Univ Calif, Berkeley, PhD(math), 62. *Prof Exp:* Asst prof, 62-71, ASSOC PROF MATH, UNIV MINN, MINNEAPOLIS, 71- *Mem:* Am Math Soc; Asn Symbolic Logic; Math Asn Am. *Res:* Mathematical logic; theory of models. *Mailing Add:* Sch Math Univ Minn Minneapolis MN 55455

FUHRMAN, ALBERT WILLIAM, b Brooklyn, NY, Jan 13, 21; m 43; c 2. ORGANIC CHEMISTRY, PROCESS DEVELOPMENT. *Educ:* City Col New York, BS, 42; Purdue Univ, MS, 44, PhD(org chem), 48. *Prof Exp:* Res chemist chem div, Glenn L Martin Co, 48-50; res & develop chemist, Naugatuck Chem Div, US Rubber Co, 50-55; plant mgr, Plastics & Chem Div, Great Am Plastics Co, 55-59, vpres, Great Am Chem Corp, 59-81; RETIRED. *Concurrent Pos:* Consult, 81-82. *Mem:* Am Chem Soc. *Res:* Ketene reactions; vinyl resins; vinyl polymerization; reaction of ketene with certain carbonyl compounds; vinyl compounding. *Mailing Add:* 6847-12 Caminito Mundo San Diego CA 92119

FUHRMAN, FREDERICK ALEXANDER, b Coquille, Ore, Aug 13, 15; m 42. MEDICAL PHYSIOLOGY, TOXINOLOGY. *Educ:* Ore State Univ, BS, 37, MS, 39; Stanford Univ, PhD(physiol), 43. *Prof Exp:* Res assoc physiol, 41-44, from instr to prof, 44-59, dir, Fleischmann Labs Med Sci, 59-70, prof exp med, 59-72, prof, 72-81, EMER PROF PHYSIOL, STANFORD UNIV, 81- *Concurrent Pos:* Guggenheim fel, Copenhagen Univ, 51-52; NSF fel, Donner Lab, Univ Calif, 58-59; Commonwealth Fund fel, 65-66; vis investr, Zoophys Lab, Copenhagen Univ, 49, 51-52 & 58-59; consult, Tech Info Div, Libr Cong, 56; vis investr, Hopkins Marine Sta, 71-79, affil prof, 73-79. *Mem:* Fel AAAS; Am Physiol Soc; Am Soc Pharmacol & Exp Therapeut; Soc Gen Physiol; Soc Toxinol. *Res:* Pharmacology of marine and amphibian toxins; temperature and drug action; experimental frostbite; tissue changes following ischemia; ion transport; drugs and tissue respiration. *Mailing Add:* 2445 Sharon Oaks Dr Menlo Park CA 94025

FUHRMAN, JED ALAN, b New York, NY, Oct 25, 56. MICROBIAL ECOLOGY, BIOLOGICAL OCEANOGRAPHY. *Educ:* Mass Inst Technol, SB, 77; Scripps Inst Oceanog, Univ Calif, San Diego, PhD(oceanog), 81. *Prof Exp:* ASST PROF MARINE SCI, MARINE SCI RES CTR, STATE UNIV NY, STONY BROOK, 81-; ASSOC PROF, BIOL SCI, UNIV SOUTHERN CALIF, 88- *Concurrent Pos:* Res collabr, chem dept, Brookhaven Nat Lab, 85-; assoc prof, State Univ NY, Stony Brook, 86-88; rep, AAAS & Am Soc Limnol & Oceanog, Sect G, 84- *Honors & Awards:* Antarctic Serv Medal, NSF, 79. *Mem:* Am Soc Limnol & Oceanog; AAAS; Am Soc Microbiol; Sigma Xi; Oceanog Soc. *Res:* Microbial ecology of aquatic and marine habitats; roles of bacteria and protozoa in pelagic ecosystems; influence of chemical and physical factors; microbially-mediated biodegradation; marine viruses; molecular biological approaches to these problems. *Mailing Add:* Dept Biol Sci Univ Southern Calif Los Angeles CA 90089-0371

FUHRMAN, ROBERT ALEXANDER, b Detroit, Mich, Feb 23, 25; m 49; c 3. AEROSPACE ENGINEERING & TECHNOLOGY. *Educ:* Univ Mich, BS, 45; Univ Md, MS, 52. *Prof Exp:* Proj engr, Naval Air Test Ctr, Patuxent River, Md, 46-53; chief tech eng, Ryan Aeronaut Co, San Diego, 53-58; vpres & asst gen mgr, Missile Systs Div, Lockheed Missiles & Space Co, 58-69, vpres & gen mgr, 69, exec vpres, 73-76, pres, 76-84; vpres, Lockheed Corp, Burbank, Calif, 69-76 & sr vpres, 76-83, pres, Lockheed Ga Co, Marietta, 70-71, Lockheed Calif Co, Burbank, 71-73 & Space & Electronics Systs Group, Lockheed Missiles, 83-85; chmn, 79-90, vchmn chief oper officer, 88-91, SR ADV, LOCKHEED MISSILES & SPACE CO, 91-; PRES, LOCKHEED CORP, 86- *Concurrent Pos:* Chmn bd, Ventura Mfg Co, 70-71; dir, Charles Stark Draper Lab, Cambridge, Mass, 87- *Honors & Awards:* John J Montgomery, Nat Mgt Asn, 64; Award, Soc Mfg Engrs, 73. *Mem:* Nat Acad Eng; fel Am Inst Aeronaut & Astronaut (pres-elect, 91-); Nat Aeronaut Asn; Inst Elec & Electronics Engrs; Am Astron Soc; Am Defense Preparedness Asn. *Mailing Add:* PO Box 9 Pebble Beach CA 93953

FUHS, ALLEN E(UGENE), b Laramie, Wyo, Aug 11, 27; m 51; c 1. MECHANICAL ENGINEERING. *Educ:* Univ NMex, BS, 51; Calif Inst Technol, MS, 55, PhD(mech eng), 58. *Prof Exp:* Asst, Calif Inst Technol, 54-55, lectr jet propulsion, 57-58; asst prof mech eng, Northwestern Univ, 58-59; mem tech staff, Phys Res Lab, Space Technol Labs, Thompson Ramo Wooldridge, Inc, 59-60; staff scientist, Plasma Res Lab, Aerospace Corp, 60-66; prof aeronaut, Naval Postgrad Sch, 66-68, chmn dept, 67-68; chief scientist, Air Force Aero Propulsion Lab, Wright-Patterson AFB, Ohio, 68-70; prof,70-74, chmn, dept mech eng, 75-78, chmn space systs, 82-87, DISTINGUIISHED PROF AERONAUT, NAVAL POSTGRAD SCH, 74-; CHIEF CONSULT, MONTEREY CONSULT SERUS, 88- *Concurrent Pos:* Private consult, 56-; Joint Inst Lab Astrophys vis fel & assoc prof, Univ Colo, 64-65. *Honors & Awards:* Ralph R Teetor Award, Soc Automotive Engrs, 81. *Mem:* Fel Am Inst Aeronaut & Astronaut; Am Phys Soc; fel Am Soc Mech Engrs; Optical Soc Am; Soc Automotive Engrs; Am Inst Astronaut & Aeronaut (pres-elect 85, pres, 86). *Res:* Magnetohydrodynamics; spectroscopy; combustion; jet propulsion; re-entry physics; instrumentation for high speed plasma flow; aircraft gas turbines; combustion; high energy lasers. *Mailing Add:* 25932 Carmel Knoll Carmel CA 93923

FUJI, HIROSHI, b Sept 26, 30; Japanese citizen; m; c 2. CANCER. *Educ:* Osaka Med Col, MD, 55; Kyoto Univ, DMSc, 60. *Prof Exp:* Res mem, Med Gen Lab, Kanagawaken, Japan, 60-62; USPHS fel, Wistar Inst, Univ Pa, 62-63; res assoc microbiol, Sch Med, Univ Pittsburgh, 63-66; mem guest fac, Sch Med, Univ Frankfurt & Paul Ehrlich Inst, Ger, 66-67; res asst prof & Henry C & Bertha H Buswell fel, Dept Microbiol, Sch Med, State Univ NY, Buffalo, 68-72; SR CANCER RES SCIENTIST, ROSWELL PARK MEM INST, 72- *Mem:* Am Asn Cancer Res; Am Asn Immunol. *Res:* Tumor immunology; immunopathology; formation of allo antibodies in mouse; cellular immunology; plaque-forming cells; methodology and theory. *Mailing Add:* 72 Farber Lane Williamsville NY 14221

FUJII, JACK K, b Phoenix, Ariz, June 9, 1940; m 67; c 4. APICULTURE. *Educ:* Univ Calif, Berkeley, BS, 63; Univ Hawaii, Manoa, MS, 68, PhD(entom), 75. *Prof Exp:* State forest entomologist, Hawaii State Div Forestry, 74-76; ASSOC PROF ENTOM & DEAN COL AGR, COL AGR, UNIV HAWAII, HILO, 76- *Mem:* Entom Soc Am; Soc Invertebrate Path; Am Inst Biol Sci. *Res:* Biology, ecology and control of termites; insect pathology. *Mailing Add:* Col Agr Univ Hawaii Hilo HI 96720

FUJIKAWA, NORMA SUTTON, b Albany, NY, Oct 19, 28; m 68. ANALYTICAL CHEMISTRY. *Educ:* Rutgers Univ, AB, 52; Univ Southern Calif, MS, 68. *Prof Exp:* Chemist, Mallinckrodt Chem Works, NJ, 52-56; assoc scientist, Walter Kidde Nuclear Labs, NY, 56-58; res engr, Rocketdyne Div, NAm Aviation, Inc, 58-62, sr res engr, 62-64, prin scientist, 64-70, mem tech staff, NAm Rockwell Corp, 70-72, supvr, 72-76, MGR, ROCKETDYNE DIV, ROCKWELL INT, CANOGA PARK, 76- *Mem:* Am Chem Soc; MENSA Soc. *Res:* Analytical chemistry of explosives and rocket propellants; propellant and pollution technology; aerospace operations safety investigations; environmental chemistry. *Mailing Add:* 21900 Marylee St Unit 285 Woodland Hills CA 91367

FUJIMOTO, ATSUKO ONO, b Japan, 35; US citizen; m 59; c 1. PEDIATRICS, BIOCHEMISTRY. *Educ:* Int Christian Univ, Tokyo, BA, 58; Univ Calif, Los Angeles, PhD(chem), 63, MD, 69. *Prof Exp:* Asst biochem, Univ Calif, Los Angeles, 58-62, res biochemist VI, 62-65; from intern to resident, 69-72, ASST PROF PEDIAT, SCH MED, UNIV SOUTHERN CALIF, 74- *Concurrent Pos:* Fulbright scholar. *Mem:* AAAS; Am Soc Human Genetics; Soc Pediat Res; Sigma Xi; Am Pediat Soc. *Res:* Diagnosis, genetic counseling and prenatal diagnosis of genetic disorders; biochemical disorders. *Mailing Add:* 1070 Vista Grande Dr Pacific Palisades CA 90272

FUJIMOTO, GEORGE IWAO, b Seattle, Wash, July 1, 20; m 49; c 3. BIOCHEMISTRY. *Educ:* Harvard Univ, BA, 42; Univ Mich, MS, 45, PhD(chem), 47. *Prof Exp:* Fel chem, Calif Inst Technol, 47-49; asst res prof biochem, Sch Med, Univ Utah, 49-55; assoc prof biochem, Albert Einstein Col Med, 55-85; vis scientist, Univ Calif, San Diego, 85-87; RETIRED. *Mem:* AAAS; Soc Study Reprod; Am Soc Biol Chem; Endocrine Soc. *Res:* Reproduction biochemistry and endocrinology. *Mailing Add:* 8720 Glenwick Lane La Jolla CA 92037

FUJIMOTO, GORDON TAKEO, b Chicago, Ill, Jun 17, 52. MATERIALS SCIENCE. *Educ:* Univ Ill, Urbana, BS(chem eng) & BS(chem), 74, Northwestern Univ, Evanston, 79. *Prof Exp:* Nat Res Cour res assoc, Naval Res Lab, 79-81; res assoc, Nat Bur Standards, Gaithersburg, Md, 81-84; ASSOC MAT SCIENTIST, MIDWEST RES INST, 85- *Mem:* Am Chem Soc; Am Inst Chem Engrs. *Res:* Chemical reation dynamics; gas phase kinetics; photochemistry; vibrational energy transfer; gas surface reactions. *Mailing Add:* Surface Res & Appln 8330 Melrose Dr Lenexa KS 66214-1630

FUJIMOTO, JAMES G, b Chicago, Ill, Sept 28, 57. LASERS & QUANTUM ELECTRONICS, ULTRAFAST PHENOMENA. *Educ:* Mass Inst Technol, SB, 79, SMEE, 81, PhD(elec eng), 85. *Prof Exp:* Asst prof, 85-88, ASSOC PROF ELEC ENG, MASS INST TECHNOL, 88- *Concurrent Pos:* NSF presidential young investr, 86; vis lectr ophthal, Harvard Med Sch, 87-; travelling lectr, Inst Elec & Electronics Engrs Lasers & Electro-Optics Soc, 90. *Honors & Awards:* William Baker Award, Nat Acad Sci, 90. *Mem:* AAAS; Optical Soc Am; Inst Elec & Electronics Engrs; Am Phys Soc. *Res:* Lasers and quantum electronics; studies of ultrafast phenomena; picosecond and femtosecond optics; laser medicine and laser surgery. *Mailing Add:* Dept Elec Eng & Computer Sci Mass Inst Technol Cambridge MA 02139

FUJIMOTO, JAMES MASAO, b Vacaville, Calif, May 10, 28; c 2. PHARMACOLOGY. *Educ:* Univ Calif, AB, 51, MS, 53, PhD, 56. *Prof Exp:* Asst pharmacol, Univ Calif, 51-55; from instr to prof, Sch Med, Tulane Univ, 56-68, actg chmn dept, 64-65 & 67-68; PROF PHARMACOL, MED COL WIS, 68- *Concurrent Pos:* Res career scientist, Wood Vet Admin Ctr, 78- *Mem:* Soc Exp Biol & Med; Am Soc Pharmacol & Exp Therapeut; Soc Toxicol; Sigma Xi. *Res:* Drug metabolism; toxicology; opioids. *Mailing Add:* Pharmacol Toxicol Dept Med Col Wis PO Box 26509 Milwaukee WI 53226

FUJIMOTO, MINORU, b Takasago, Japan, Feb 11, 26; m 55; c 4. PHASE TRANSITIONS. *Educ:* Osaka Univ, BSc, 48; Univ Southampton, PhD(physics), 59. *Prof Exp:* Res assoc physics, Univ Md, 59-61; asst prof, Clark Univ, 61-63 & Univ Man, 63-67; assoc prof, 67-78, PROF PHYSICS, UNIV GUELPH, 78- *Mem:* Am Phys Soc; Can Assoc Physics. *Res:* Structural phase transitions. *Mailing Add:* Dept Physics Univ Guelph Guelph ON N1G 2W1 Can

FUJIMOTO, SHIGEYOSHI, b Toyky, Japan, Sept 5, 36; m 72; c 3. TUMOR IMMUNOLOGY, IMMUNOTHERAPY. *Educ:* Chiba Univ, MD, 63, PhD(med), 68. *Prof Exp:* Lectr path & immunol, Dept Path, Sch Med, Chiba Univ, 68-70, from asst prof to assoc prof immunol, Div Immunol, 75-80; postdoctoral fel immunol, Dept Immunol, Fac Med, Univ Man, 70-72, asst prof immunol, 72-75; PROF & HEAD IMMUNOL, DEPT IMMUNOL, KOCHI MED SCH, 80- *Mem:* Can Soc Immunol; Am Asn Immunologists; NY Acad Sci. *Res:* Molecular and cellular mechanisms of immune response to tumor; development of specific immunotherapy against cancer. *Mailing Add:* Kohasu Okoh-cho Nankoku-shi Kochi 783 Japan

FUJIMOTO, WILFRED Y, b Hilo, Hawaii, July 15, 40; m; c 2. MEDICINE, ENDOCRINOLOGY. *Educ:* Johns Hopkins Univ, BA, 62, MD, 65; Am Bd Internal Med, dipl internal med, endocrinol & metab, 75. *Prof Exp:* Intern & asst resident med, Presby Hosp, Columbia-Presby Med Ctr, 65-67; clin assoc, Sect Human Biochem Genetics, Nat Inst Arthritis & Metab Dis, NIH, 67-69; asst resident, 69-70, from instr to assoc prof, 70-83, PROF, DEPT MED, SCH MED, UNIV WASH, SEATTLE, 83- *Concurrent Pos:* Sr fel, Dept Med, Sch Med, Univ Wash, Seattle, 70-71, dir, Core Tissue Cult Facil, Diabetes-Endocrinol Res Ctr, 77-; attend physician, Med Serv, Harborview Med Ctr, Seattle, 71-76, Univ Hosp, 71- & USPHS Hosp, 74-76; Nat Inst Arthritis, Metab & Digestive Dis res career develop award, 71-76; lectr, numerous US & foreign univs, 72-91; consult, numerous adv comts, NIH, 75-91; assoc med staff, Harborview Med Ctr, 76-; assoc prog dir, Clin Res Ctr, Univ Hosp, Seattle, 76-; assoc ed, Diabetes, 86; mem, Steering Comt, Joint Japan-US Training Prog Diabetes Epidemiol, 89- *Mem:* AAAS; fel Am Col Physicians; Am Diabetes Asn; Am Fedn Clin Res; Am Inst Nutrit; Am Soc Clin Nutrit; Endocrine Soc; Soc Exp Biol & Med; Tissue Cult Asn. *Res:* Author of 236 technical publications. *Mailing Add:* Dept Med Div Metab Endocrinol & Nutrit Univ Wash RG-26 Seattle WA 98195

FUJIMURA, OSAMU, b Tokyo, Japan, Aug 29, 27; m; c 5. EXPERIMENTAL PHYSICS, LINGUISTICS. *Educ:* Univ Tokyo, BS, 52, DSc(physics), 62. *Prof Exp:* Speech res mem, Mass Inst Technol, 58-61, Royal Inst Technol, Sweden, 63-65; prof speech sci & ling, Univ Tokyo, 65-73; head dept ling & artificial intelligence res, AT&T Bell Labs, 73-88; res prof, Hahnemann Med Univ, 76-81; CONSULT, 81- *Mem:* Acoust Soc Am; Inst Elec & Electronic Engrs; Ling Soc Am; Acoust Soc Japan; NY Acad Sci; Asn Comput Ling; AAAS. *Res:* Studies in physical, physiological, psychological and linguistic aspects of speech phenomena. *Mailing Add:* 1070 Carmack Rd Pressey Hall Rm 110 Columbus OH 43210-1002

FUJIMURA, ROBERT, b Seattle, Wash, July 28, 33; m 62; c 3. SCIENCE POLICY. *Educ:* Univ Wash, BS, 56; Univ Wis, MS, 59, PhD(biochem), 61. *Prof Exp:* NIH fel biochem, Inst Protein Res, Osaka, Japan, 61-62; fel biophys, Univ Wis, 63; MEM STAFF BIOCHEM, BIOL DIV, OAK RIDGE NAT LAB, 63- *Concurrent Pos:* Adj prof, Univ Tenn; res fel, Japan Soc Promotion Sci, 81; foreign com serv officer biotechnol, Am Embassy, Tokyo, 85-86; genetic sci task force, Social & Ethical Impact Gene Manipulation, United Methodist Church, 89-91. *Mem:* Sigma Xi; Am Soc Biolchem & Molecular Biol; AAAS. *Res:* Interaction of bacteriophages with their hosts; biosynthesis of small bacteriophages and their nucleic acids; bacteriophage T5 DNA repair and replication; mechanisms of phage T5 DNA polymerase and chemical mutagenesis in mouse embryos; DNA-protein and protein-protein interactions. *Mailing Add:* Biol Div PO Box 2009 Oak Ridge Nat Lab Oak Ridge TN 37831-8077

FUJINAMI, ROBERT S, b Salt Lake City, Utah, Dec 8, 49. VIRAL PATHOGENESIS, AUTOIMMUNITY. *Educ:* Univ Utah, BA, 72; Northwestern Univ, PhD(immunol), 77. *Prof Exp:* Assoc prof path, Univ Calif, San Diego, 85-90; res fel, 77-80, res assoc, 80-81, asst mem, 81-85, VIS INVESTR, SCRIPPS CLIN & RES FOUND, 85-; PROF, UNIV UTAH, SALT LAKE CITY, 90- *Concurrent Pos:* Instr med microbiol, Northwestern Univ, 73-76; res immunopathologist, Univ Calif, San Diego, 80-82; mem, NIH site visit team, Neurol Inst, 87, exp immunol study sect, 88-92, spec rev study sect, 88; mem, Nat Multiple Sclerosis Study Sect, 90-92, sci adv, SLC, Nat Multiple Sclerosis Soc chap; fel, NIH, 77-79, new investr award, 81-84, res grant, 86-89; mem, Nat Multiple Sclerosis Soc, 85-88, Harry Weaver Neurosci Award, 82-86; Javits Neurosci Award, NIH, 89- *Mem:* Am Asn Immunologists; Am Asn Pathologists; Am Soc Microbiol; Am Soc Virol. *Res:* Understanding how viruses initiate and cause disease in man, particularly of the central nervous system; virus induced autoimmunity including immune responses to self components; viral pathogenesis. *Mailing Add:* Dept Neurol 3R210 Sch Med Univ Utah Salt Lake City UT 84132

FUJIOKA, ROGER SADAO, b May 11, 38; US citizen; m 66. VIROLOGY, WATER POLLUTION. *Educ:* Univ Hawaii, BS, 60, MS, 66; Univ Mich, PhD(virol), 70. *Prof Exp:* Clin lab officer med technol, US Army, 60-63; res asst microbiol, Univ Hawaii, Honolulu, 63-66; fel virol, Baylor Col Med, 70-71; from asst researcher to assoc researcher, 72-85, RESEARCHER, UNIV HAWAII, HONOLULU, 85-, MEM GRAD FAC, 76-, PROF, SCH PUB HEALTH. *Mem:* AAAS; Am Soc Microbiol; Water Pollution Control Fed; Am Water Works Asn. *Res:* Public health consequences of human viruses in the water environment; technology to improve the disinfection of viruses in water and the mechanism by which viruses are inactivated by disinfectants; use of bacteria to assess hygienic quality of water. *Mailing Add:* Dept Microbiol Univ Hawaii Honolulu HI 96822

FUJITA, DONALD J, b Hunt, Idaho, July 2, 43. VIROLOGY, MICROBIOLOGY. *Educ:* Reed Col, BA, 65; Univ Chicago, PhD(microbiol), 71. *Prof Exp:* Fel virol, Sch Med, Stanford Univ, 71-73; vis scientist virol, Sch Med, Univ Southern Calif, 73-74; fel molecular virol, Sch Med, Univ Calif, San Francisco, 74-77; from asst prof to assoc prof, Cancer Res Lab & Dept Biochem, Univ Western Ont, 85-88; DEPT MED BIOCHEM, UNIV CALGARY HEALTH SCI CTR, 88- *Concurrent Pos:* Damon Runyon Found Cancer Res fel, 71-73; Leukemia Soc Am fel, 73-75; trainee, NIH, 75-77; Nat Cancer Inst Can res grants, 78-81 & 81-; Med Res Coun Can res grants, 79-81 & 81-; Leukemia Res Fund grant, 79-86; Crusade Against Leukemia res grant, 79-80; sci officer, Res Grant Rev Panel A, Nat Cancer Inst Can, 82. *Mem:* Am Soc Microbiol; AAAS. *Res:* Cellular and viral oncogenes; biochemical and genetic studies on avian RNA tumor viruses; viral gene expression, structure and evolution; cell transformation. *Mailing Add:* Dept Med Biochem Univ Calgary Health Sci Ctr 300 Hospital Dr NW Calgary AB T2N 4N1 Can

FUJITA, SHIGEJI, b Oita City, Japan, May 15, 29; m 58; c 4. THEORETICAL PHYSICS. *Educ:* Kyushu Univ, BS, 53. Univ Md, PhD(physics), 60. *Prof Exp:* Res asst physics, Kyushu Univ, 53-56 & Univ Md, 56-58; res assoc, Northwestern Univ, 58-60; sr res assoc phys chem, Univ Brussels, 60-65; vis assoc prof physics, Univ Ore, 65-66; assoc prof, 66-68, PROF PHYSICS, STATE UNIV NY BUFFALO, 68- *Concurrent Pos:* Vis assoc prof, Pa State, 64-65; vis prof, Ctr Math Res, Univ Montreal, 72-73, Nat Univ Mexico, 79 & Univ Zurich, 81. *Mem:* Am Phys Soc; Phys Soc Japan. *Res:* Statistical mechanical theory of reduced density mechanics and distribution functions; applied to transport phenomena; in gases, liquids and solids. *Mailing Add:* Physics Dept State Univ NY Buffalo Amherst NY 14260

FUJITA, TETSUYA T, b Japan, Oct 23, 20; m 48; c 1. METEOROLOGY. *Educ:* Meiji Inst Technol, Japan, BS, 43; Univ Tokyo, DSc(meteorol), 53. *Prof Exp:* Asst Meiji Inst Technol, Japan, 43, asst prof, 43-50; asst prof, Kyushu Inst Technol, 50-53; vis prof, 53-56, dir mesometeorol res, 56-61, assoc prof, 61-69, PROF METEOROL, UNIV CHICAGO, 69- *Concurrent Pos:* Consult, Ill State Water Surv, 54-58. *Honors & Awards:* Okada Award, 59; Kamura Award, 65. *Mem:* Am Meteorol Soc; Am Geophys Union; Optical Soc Am; Am Soc Photogram; Meteorol Soc Japan; Sigma Xi. *Res:* Mesometeorology; severe local storms; aerial photogrammetry; satellite meteorology. *Mailing Add:* 5727 S Maryland Ave Chicago IL 60637

FUJIWARA, HIDEO, b Nara, Japan, Feb 9, 46; m 69; c 2. DESIGN & TEST OF COMPUTERS, FAULT TOLERANT COMPUTING. *Educ:* Osaka Univ, BE, 69, ME, 71, PhD(electronic eng), 74. *Prof Exp:* Asst prof electronic eng, Osaka Univ, 74-85; assoc prof electronic eng, 85-90, PROF COMPUTER SCI, MEIJI UNIV, 90- *Concurrent Pos:* Vis asst prof, Univ Waterloo, Can, 81; vis assoc prof, McGill Univ, Can, 84; ed, Inst Elec & Electronic Engrs Design & Test Computers, 86-91, Inst Electronics Info & Commun Engrs Trans Info & Systs & J Electronic Testing, 90- *Mem:* Fel Inst Elec & Electronic Engrs; Inst Electronics Info & Commun Engrs; Info Processing Soc Japan. *Res:* Computer design and test; test generation; fault simulation; design for testability; built-in-self-test; parallel processing; neural networks; computational complexity; fault-tolerant computing. *Mailing Add:* Dept Computer Sci Meiji Univ/Fac Eng 1-1-1 Higashi-Mita Tama-Ku Kawasaki 214 Japan

FUJIWARA, KEIGI, b Nita-cho, Shimane, Japan, May 16, 44; m 69; c 1. CELL BIOLOGY. *Educ:* Int Christian Univ, Tokyo, BA, 68; Univ Pa, PhD(biol), 74. *Prof Exp:* Asst biol, Int Christian Univ, 68-69; res fel cell biol, 74-77, asst prof anat, 77-83, assoc prof anat & cellular biol, Harvard Med Sch, 83-86; AT DEPT STRUCTURAL ANAL, NAT CARDIOVASC CTR, OSAKA, JAPAN. *Mem:* Am Soc Cell Biol; AAAS; NY Acad Sci; Am Asn Anatomists. *Res:* Morphological and biochemical analyses of the cytoskeleton of eukaryotic non-muscle cells. *Mailing Add:* Dept Structural Anal Nat Cardiovasc Ctr Res Inst 5-125 Fujishirodai Suita Osaka 565 Japan

FUKA, LOUIS RICHARD, b New York City, NY, Dec 19, 37; m 59; c 4. VIBRATION, WAVE PROPAGATION. *Educ:* St Louis Univ, BS, 59; Univ Mo-Rolla, MS, 63; Univ Tex, Austin, PhD(eng), 71. *Prof Exp:* Engr, US Army Corps Engrs, 59-61; engr & scientist specialist, McDonnell Douglas Missiles & Space Systs, 63-67; teaching asst eng mech, Univ Tex, Austin, 69; res engr & scientist, Eng Mech Res Lab, 69-71; staff mem, Los Alamos Nat Lab, 71-80; PRES, LOMAR RES & DEVELOP, 80- *Concurrent Pos:* Consult, Douglas Aircraft Co, 66, Atomic Energy Comn, 72-74; prin investr, Nuclear Regulatory Comn, 72-74; lectr, Am Mgr Asn, 73-74; vis assoc prof, Univ Hawaii, 78-79; Univ Tex Eng Found fel; prog dir mech technol, Univ NMex-Los Alamos, 86-87; Middle East adv, 90-91. *Mem:* Nat Soc Prof Engrs. *Res:* Analog digital minicomputer; vibration, shock, wave propagaton, dynamic structural analysis and nuclear weapons effects; United States and foreign patents. *Mailing Add:* 2073 North Rd Los Alamos NM 87544

FUKAI, JUNICHIRO, b Noda, Japan, Mar 6, 38; m 64; c 2. PLASMA PHYSICS. *Educ:* Waseda Univ, Japan, BEngr, 61; Univ Denver, MS, 66; Univ Tenn, Knoxville, PhD(physics), 70. *Prof Exp:* Staff engr design, Tokyo Shibaura Elec Co, 61-64; res assoc physics, Univ Tenn, Knoxville, 70-72; res assoc appl physics, Yale Univ, 72-74; asst prof, 74-78, ASSOC PROF PHYSICS, AUBURN UNIV, 78- *Mem:* Am Phys Soc. *Res:* Theoretical investigations on plasma instabilities and vacuum spark discharges-nonlinear instabilities in a beam plasma and dynamics of plasma discharges; fundamental physics. *Mailing Add:* Dept Physics Auburn Univ Auburn AL 36830

FUKUDA, MINORU, b Taipei, Taiwan, July 6, 45; m 70; c 2. CANCER RESEARCH, MEMBRANE BIOCHEMISTRY. *Educ:* Univ Tokyo, Japan, BS, 68, MS, 70, PhD(biochem), 73. *Prof Exp:* Res assoc biochem, Dept Pharmaceut Sci, Univ Tokyo, 73-75; assoc biochem, Sch Med, Yale Univ, 75-77; assoc cancer res, Fred Hutchinson Cancer Res Ctr, 77-81; asst prof pathobiochem, Sch Pub Health, Univ Wash, 80-81; STAFF SCIENTIST CANCER BIOL, LA JOLLA CANCER RES FOUND, 82-, DIR, LAB CARBOHYDRATE CHEM, 84-, PROG DIR, GLYCOBIOL & CHEM, 88- *Mem:* Am Soc Cell Biol; Am Soc Biol Chemists; NY Acad Sci; AAAS; Am Asn Cancer Res. *Res:* Biochemistry of cell surface glycoproteins in normal and cancer cell differentiation; hematopoietic cells (leukemic cells); epithelial cells (gastrointestinal carcinoma); lysosomal membrane glycoproteins; carbohydrate-protein interaction. *Mailing Add:* La Jolla Cancer Res Found Cancer Res Ctr 10901 N Torrey Pines Rd La Jolla CA 92037

FUKUHARA, FRANCIS M, b Seattle Wash, Jan 30, 25; m 53; c 4. FISHERIES. *Educ:* Univ Wash, BS, 55, PhD(fisheries), 71. *Prof Exp:* Fishery res biologist, Bur Com Fisheries, Nat Marine Fisheries Serv, 52-66, asst dir, Biol Lab, 66-71, dir, Div Resource Ecol & Fish Mgt, 71-87. *Concurrent Pos:* Scientist mem comt biol & res, Int N Pac Fisheries Comn, 69-; affil prof, Univ Wash, 73-80; nat resources consult, 80-87. *Mem:* Am Inst Fishery Res Biologists. *Res:* Ecosystem and population dynamics of demersal fish and shellfish communities of the east Bering Sea and east North Pacific Ocean; development of biological bases for management of single species and multiple species complexes. *Mailing Add:* 320 36th Ave Seattle WA 98112

FUKUI, H(ATSUAKI), b Yokohama, Japan, Dec 14, 27; m 54; c 2. ELECTRICAL ENGINEERING. *Educ:* Miyakojima Tech Col, Grad, 49; Osaka Univ, DEng, 61. *Prof Exp:* Res assoc, Osaka City Univ, 49-54; engr, Shimada Phys & Chem Indust Co, 54-55; sr engr, semiconductor div, Sony Corp, 55-59, supvr, 59-61, mgr eng div, 61-62; mem tech staff, Bell Tel Labs, 62-69, supvr, 69-73, mem tech staff, 73-81, supvr, 81-83, SUPVR, AT&T BELL LABS, MURRAY HILL, NJ, 84- *Concurrent Pos:* Lectr, Tokyo Metrop Univ, 62; asst to chmn, Sony Corp, Tokyo, 73; vpres, Sony Corp Am, New York, 73. *Honors & Awards:* Inada Prize, Inst Elec Eng, Japan, 59; Microwave Prize, Inst Elec & Electronics Engrs, 80. *Mem:* Fel Inst Elec & Electronics Engrs. *Res:* Microwave semiconductor devices: GaAs MESFETs, bipolar transistors, tunnel diodes, avalanche diodes and bulk-effect devices; microwave electron tubes; microwave circuits; image pickup and display devices; videophone subsystems; solid-state consumer electronics; semiconductor lasers; lightwave technology. *Mailing Add:* 53 Drum Hill Dr Summit NJ 07901

FUKUNAGA, KEINOSUKE, b Japan, July 23, 30; m 60; c 2. ELECTRICAL ENGINEERING. *Educ:* Kyoto Univ, BS, 53, PhD(elec eng), 62; Univ Pa, MS, 59. *Prof Exp:* Res engr, res lab, Mitsubishi Elec Co, Amagasaki-shi, Japan, 53-57, sect head res comput control, 59-65, develop of comput, Kamakura-shi, 65-66; asst instr elec eng, Univ Pa, 57-59; assoc prof, 66-73, PROF ELEC ENG, PURDUE UNIV, 73- *Concurrent Pos:* Consult, Corp & Govt Agencies. *Mem:* Fel Inst Elec & Electronics Engrs. *Res:* Pattern recognition; decision making processes; statistical data analysis. *Mailing Add:* Dept Elec Eng Purdue Univ Lafayette IN 47907

FUKUNAGA, TADAMICHI, b Japan, Mar 21, 31; m 57; c 2. ORGANIC CHEMISTRY, RESEARCH ADMINISTRATION. *Educ:* Osaka Univ, BS, 53, MS, 55; Ohio State Univ, PhD(org chem), 59. *Prof Exp:* Fel, Harvard Univ, 59-62; res staff, E I du Pont de Nemours Co, Inc, 62-65; assoc prof, Tokyo Univ, 65-66; RES STAFF, E I DU PONT DE NEMOURS & CO, INC, 66- *Concurrent Pos:* Vis prof, Univ Ill, Urbana-Champaign, 79; mem, adv bd, J Org Chem, 80-84. *Mem:* Am Chem Soc; Chem Soc Japan. *Res:* Synthetic, physical and theoretical organic chemistry. *Mailing Add:* Du Pont Exp Sta PO Box 80328 Wilmington DE 19880-0328

FUKUSHIMA, DAVID KENZO, b Fresno, Calif, Aug 24, 17; m 42; c 2. BIOCHEMISTRY. *Educ:* Whittier Col, BA, 39; Univ Calif, Los Angeles, MA, 43; Univ Rochester, PhD(org chem), 46. *Prof Exp:* Asst, Sloan-Kettering Inst Cancer Res, 47-51, assoc, 51-60, mem, 60-63; sr investr, Montefiore Hosp & Med Ctr, New York, 63-77, dir, Inst Steroid Res, 77-71; RETIRED. *Concurrent Pos:* Fel, Sloan-Kettering Inst, 46-47; prof biochem, Albert Einstein Col Med, 66-81. *Mem:* Am Chem Soc; Endocrine Soc; Am Soc Biol Chemists; Brit Soc Endocrinol. *Res:* Biochemistry of steroid hormones in man; steroid chemistry. *Mailing Add:* 1430 N Sharpless St LaHabra CA 90631

FUKUSHIMA, EIICHI, b Tokyo, Japan, June 3, 36; US citizen; m 63; c 3. NUCLEAR MAGNETIC RESONANCE, MULTIPHASE FLOW. *Educ:* Univ Chicago, AB & SB, 57; Dartmouth Col, MA, 59; Univ Wash, PhD(physics), 67. *Prof Exp:* Staff mem, Los Alamos Sci Lab, 67-85; SCIENTIST, LOVELACE MED FOUND, 85- *Concurrent Pos:* Vis prof, Univ Fla, 75-76. *Mem:* Am Phys Soc; Am Asn Physics Teachers. *Res:* Nuclear magnetic resonance studies of fluid flow with emphasis on biological flow, multiphase flow, and instrumentation. *Mailing Add:* 2425 Ridgecrest Dr SE Albuquerque NM 87108

FUKUSHIMA, TOSHIYUKI, b Tacoma, Wash, Sept 26, 21; m 51; c 6. MECHANICAL ENGINEERING. *Educ:* Swarthmore Col, BS, 51; Univ Pa, MS, 59. *Prof Exp:* Engr, Ultra-Mechanisms, Inc, 53, Prewitt Aircraft Corp, 53-54 & Thermal Res & Eng Corp, 54-56; instr mech eng, Swarthmore Col, 56-61; asst prof, Drexel Inst Technol, 61-65; sr engr theoret aerodyn, Dynasci Corp, 65-67; sr engr aerodynamics res, Vertol Div, Boeing Co, Pa, 67-71; TEACHER MATH, DELAWARE COUNTY AREA VOC-TECH SCHS, 71- *Concurrent Pos:* Lectr, Widener Univ, 65- *Res:* Fluid mechanics and heat transfer; thermodynamics and magnetohydrodynamics as applied to energy conversion; helicopter rotor aerodynamics. *Mailing Add:* 218 Lafayette Ave Swarthmore PA 19081

FUKUTA, NORIHIKO, b Tokoname, Japan, May 11, 31; m 66. CLOUD PHYSICS, ATMOSPHERIC SCIENCES. *Educ:* Nagoya Univ, BSc, 54, MSc, 56, PhD(phys chem), 59. *Prof Exp:* Res asst chemist, Nagoya Univ, 59-61; vis res fel, Commonwealth Sci & Indust Res Orgn, Sydney, Australia, 62-64; head microphys lab, Meteorol Res Inc, Calif, 66-67, dir microphys lab, 67-68; head, Cloud Physics Lab, Denver Res Inst, 68-77; prof environ eng, Col Eng, Univ Denver, 68-75; PROF METEOROL, UNIV UTAH, 77- *Concurrent Pos:* Vis scientist, Imp Col, London, 61-62; assoc ed, J Appl Meteorol, 72-79 & J Atmospheric Sci, 76-; adj prof physics, Univ Denver, 75-77; sr academian, Environ Res Lab, Nat Oceanic & Atmospheric Admin, 83-84; consult, Govt Agencies & Pvt Sectors. *Mem:* Am Meteorol Soc; Meteorol Soc Japan; Am Geophys Union; Sigma Xi; Am Phys Soc; Weather Modification Asn. *Res:* Ice nucleation; organic ice nuclei; development of non-AgI cloud seeding generators; convective cloud seeding; ice crystal growth; hydrometeor growth kinetics; supercooled fog clearing; development of cloud condensation and ice nucleus spectrometers. *Mailing Add:* Dept Meteorol Univ Utah Salt Lake City UT 84112

FUKUTO, TETSUO ROY, b Los Angeles, Calif, Dec 15, 23; m 53; c 4. ORGANIC CHEMISTRY. *Educ:* Univ Minn, BS, 46; Univ Calif, Los Angeles, PhD, 50. *Prof Exp:* Res fel, Univ Ill, 50-51; develop chemist, Aerojet Eng Corp, 51-52; from asst insect toxicologist to assoc insect toxicologist, 52-63, PROF ENTOM & CHEM & INSECT TOXICOLOGIST, CITRUS RES CTR & AGR EXP STA, UNIV CALIF, RIVERSIDE, 63-, HEAD DIV TOXICOL & PHYSIOL, 74- *Concurrent Pos:* Consult toxicol study sect, USPHS, 62-66. *Mem:* Am Chem Soc; Entom Soc Am; Sigma Xi. *Res:* Chemistry and mode of action of insecticides. *Mailing Add:* Div Toxicol & Physiol Univ Calif PO Box 112 Riverside CA 92521

FUKUYAMA, KIMIE, b Tokyo, Japan, Dec 11, 27. DERMATOLOGY. *Educ:* Tokyo Women's Med Col, MD, 49, PhD, 64; Univ Mich, MS, 58. *Prof Exp:* Intern, Tokyo Med Sch, 49-50, resident & asst dermat, 50-56; res assoc, Univ Mich, 58-61; asst, Tokyo Women's Med Col, 61-63, lectr, 64-65; lectr, 65-67, from resident asst prof to resident assoc prof, 67-78, RESIDENT PROF, SCH MED, UNIV CALIF, SAN FRANCISCO, 78- *Mem:* Soc Invest Dermat; AMA. *Mailing Add:* Dept Dermat Univ Calif San Francisco CA 94143

FUKUYAMA, THOMAS T, b Juneau, Alaska, Dec 30, 27; m 60; c 3. MICROBIOLOGY, BIOCHEMISTRY. *Educ:* Univ Wash, BS, 50, MS, 53; Univ Pa, PhD(microbiol), 61. *Prof Exp:* Res assoc, Sch Med, 63-64, from instr to asst prof, 64-69, asst prof, Sch Pharm, 69-71, chmn, Grad Educ & Res Coun, 73-76, ASSOC PROF MICROBIOL, SCH PHARM, UNIV SOUTHERN CALIF, 71-, BIOPHARMACY AREA COORDR, 77-, LEVEL II COORDR, 77- *Concurrent Pos:* Res fel microbiol, Harvard Med Sch, 60-63; USPHS fel, 61-63 & res grant, 66; Univ Southern Calif gen res support grant, 66; coordr grad progs comt, Univ Southern Calif, 75- *Mem:* NY Acad Sci; Sigma Xi; AAAS; Am Soc Microbiol. *Res:* Microbial physiology; enzyme action and regulation; mechanisms of action purine nucleoside analogs. *Mailing Add:* 3985 Prospect Ave Los Angeles CA 90027

FUKUYAMA, TOHRU, b Anjo, Japan, Aug 9, 48; m 77; c 3. ORGANIC CHEMISTRY. *Educ:* Nagoya Univ, BA, 71, MA, 73; Harvard Univ, PhD(chem), 77. *Prof Exp:* Fel org chem, Harvard Univ, 77-78; from asst prof to assoc prof, 78-88, PROF ORG CHEM, RICE UNIV, 88- *Mem:* Am Chem Soc. *Res:* Total synthesis of complex natural products of biological importance. *Mailing Add:* PO Box 1892 Houston TX 77251-1892

FULBRIGHT, DENNIS WAYNE, b Lynwood, Calif, Aug 20, 52; c 2. BIOLOGICAL CONTROL OF PLANT PATHOGENS, WHEAT PATHOLOGY. *Educ:* Whittier Col, AB, 74; Univ Calif, Riverside, PhD(plant path), 79. *Prof Exp:* From asst prof to assoc prof, 79-90, PROF PLANT PATH, MICH STATE UNIV, 90- *Concurrent Pos:* Vis scientist, Roche Inst Molecular Biol, 89. *Mem:* Am Phytopath Soc; Am Soc Microbiol; Sigma Xi. *Res:* Attenuated forms of the chestnut blight pathogen and the survival of the American chestnut tree; molecular genetics of double-stranded RNA viruses of fungi. *Mailing Add:* Dept Bot & Plant Path Mich State Univ East Lansing MI 48824-1312

FULBRIGHT, HARRY WILKS, b Springfield, Mo, Sept 19, 18; m 44. NUCLEAR PHYSICS. *Educ:* Wash Univ, AB, 40, MS, 42, PhD(physics), 44. *Prof Exp:* Physicist in charge cyclotron, Wash Univ, 42-44; physicist & group leader Manhattan dist proj, Calif, Los Alamos, 44-46; asst prof physics, Princeton Univ, 46-50; from asst prof to assoc prof, 50-56, PROF PHYSICS, UNIV ROCHESTER, 56- *Concurrent Pos:* Consult med sch, Univ Rochester, 54-57; Fulbright & Guggenheim fel, Inst Theoret Physics, Copenhagen Univ, 56-57; consult, Gen Atomic Co, 58-60 & Tropel, Inc, 61-64; vis prof, Univ Louis Pasteur, Strasbourg, France, 74-75; vis prof, Punjab Univ, Chandigarh, India, 75. *Mem:* Fel Am Phys Soc; Sigma Xi. *Res:* Nuclear reactions and structure; nuclear instrumentation. *Mailing Add:* 245 Castle Rd Rochester NY 14623

FULBRIGHT, TIMOTHY EDWARD, b Abilene, Tex, Sept 24, 53; m 78; c 1. PLANT ECOPHYSIOLOGY, DEER HABITAT MANAGEMENT. *Educ:* Abilene Christian Univ, BS, 76, MS, 78; Colo State Univ, PhD(range sci), 81. *Prof Exp:* Asst prof, 81-85, ASSOC PROF RANGE MGT, TEX A&I UNIV, 85- *Concurrent Pos:* Assoc ed, J Range Mgt, 89-; consult, Hwy Div, Off Atty Gen Tex & Frost Bank, 90. *Mem:* Soc Range Mgt; Wildlife Soc. *Res:* Shrub ecophysiology; mechanisms of woody plant succession; wildlife habitat improvement; improving white-tailed deer habitat. *Mailing Add:* 1002 S 23rd Kingsville TX 78363

FULCHER, WILLIAM ERNEST, b Eden, NC, Aug 21, 31; m 56; c 3. BOTANY, PLANT MORPHOLOGY. *Educ:* NC State Univ, BS, 53; Appalachian State Univ, MA, 60; Univ NC, PhD(bot), 71. *Prof Exp:* Teacher biol, Morehead High Sch, Eden, NC, 53-54, 56-62; instr mil subj, Signal Sch, Ft Monmouth, NJ, 54-56; from instr to assoc prof, 62-81, PROF BIOL, GUILFORD COL, 81-, CHMN DEPT, 76- *Mem:* Am Fern Soc; Royal Hort Soc. *Res:* Plant anatomy; economic botany. *Mailing Add:* Dept Biol Guilford Col Greensboro NC 27410

FULCO, ARMAND J, b Los Angeles, Calif, Apr 3, 32; m 55; c 4. CYTOCHROME ENZYMES, BACTERIAL LIPIDS. *Educ:* Univ Calif, Los Angeles, BS, 57, PhD(physiol chem & lipid biosynthesis), 60, Harvard Univ, Cambridge, MA. *Prof Exp:* Postdoctoral fel, Univ Calif Los Angeles, 60-61, Harvard Univ, 61-63; asst res biochemist, dept biophys & nuclear med, 63-65, asst prof, dept biol chem, 65-70, assoc prof, 70-76, PROF BIOL CHEM & PRIN INVESTR, LAB BIOMED & ENVIRON SCI, UNIV CALIF, LOS ANGELES, 76- *Concurrent Pos:* USPHS res fel, Lipid Labs, Univ Calif, Los Angeles, 60-61, consult biochemist, Vet Admin, Los Angles, 68-; NIH res fel chem, Harvard Univ, 61-63; NIH res grants, 70-73, 73-78, 77-82, 78-81 & 82-86, 86-91. *Mem:* Sigma Xi; AAAS; Am Chem Soc; Am Oil Chemists Soc; Am Soc Biochem & Molecular Biol; Am Soc Microbiol. *Res:* Pathways, mechanisms and enzymology of membrane lipid biosynthesis; comparative biochemistry; cytochrome P-450 enzymes in bacteria; biosynthetic control mechanisms. *Mailing Add:* Dept BIOL CHEM Sch Med Univ Calif Los Angeles CA 90024-1737

FULCO, JOSE ROQUE, b Buenos Aires, Arg, Dec 5, 27; m 49; c 5. PHYSICS. *Educ:* Argentine Army Tech Sch, CE, 55; Univ Buenos Aires, PhD(physics), 62. *Prof Exp:* Argentine Army vis fel physics, Lawrence Radiation Lab, Calif, 57-59; prof, Univ Buenos Aires, 59-62; asst res physicist, Univ Calif, San Diego, 62-64; assoc prof, 64-69, assoc dir educ abroad prog, 70-76, PROF PHYSICS, UNIV CALIF, SANTA BARBARA, 69-, CHMN DEPT, 78- *Concurrent Pos:* Prof, La Plata, 62. *Mem:* Fel Am Phys Soc. *Res:* Theoretical high energy physics. *Mailing Add:* Dept Physics Univ Calif Santa Barbara CA 93106

FULDA, MYRON OSCAR, b New York, NY, Mar 18, 30; m 61; c 3. ANALYTICAL CHEMISTRY. *Educ:* Iowa State Col, BS, 53, MS, 55. *Prof Exp:* Chemist, 55-64, res chemist, 64-70, sr chemist, 70-78, SR RES CHEMIST, E I DU PONT DE NEMOURS & CO, INC, 78- *Mem:* Am Chem Soc; Am Assoc Textile Colorists & Chem. *Res:* Non-aqueous, complexiometric and electrometric titrimetry; radiochemistry; analytical chemistry of textile fibers; dyeability of acrylic fibers; analysis of air and water pollutants, personnel exposure. *Mailing Add:* 2106 Burns Lane Camden SC 29020

FULDE, ROLAND CHARLES, b Chicago, Ill, Nov 23, 26; m 52; c 2. FOOD SCIENCE. *Educ:* Mich State Univ, BS, 48, MS, 49, PhD, 53. *Prof Exp:* Res bacteriologist, Swift & Co, 53-62; div head, 62-80, MGR, TECHNOL DEVELOP, CAMPBELL SOUP CO, 80- *Mem:* Inst Food Technol; Am Chem Soc. *Res:* Industrial biochemical processes. *Mailing Add:* 78 Knollwood Dr Cherry Hill NJ 08002-1400

FULEKI, TIBOR, b Budapest, Hungary, June 28, 31; Can citizen; c 3. FOOD SCIENCE, BIOCHEMISTRY. *Educ:* Col Hort & Viticult, Budapest, BSc, 53; McGill Univ, MSc, 61; Univ Mass, Amherst, PhD(food chem), 67. *Prof Exp:* Res scientist, Res Sta, Agr Can, NS, 61-67 & Res Labs, Health & Welfare Can, Ont, 67-68; RES SCIENTIST, HORT PROD LAB, HORT RES INST ONT, 68- *Mem:* Can Soc Oenol; Can Soc Color; Groupe Polyphenols; Phytochem Soc N Am. *Res:* Chemical composition of fruits, vegetables and their products; tristimulus colorimetry of horticultural crops; anthocyanin composition of economic plants; chemistry and technology of juice making; food analysis. *Mailing Add:* Hort Prod Lab Hort Res Inst Ont Vineland Station ON L0R 2E0 Can

FULFORD, MARGARET HANNAH, b Cincinnati, Ohio, June 14, 04. BOTANY. *Educ:* Univ Cincinnati, BA, 26, BE, 27, MA, 28; Yale Univ, PhD(bot), 35. *Prof Exp:* Instr bot & cur, 27-40, from asst prof to prof, 40-76, EMER PROF BOT, UNIV CINCINNATI, 76- *Concurrent Pos:* Guggenheim fel, Harvard Univ, Yale Univ & NY Bot Garden, 41; mem staff biol sta, Univ Mich, 47-53; fel grad sch, Univ Cincinnati, 58. *Honors & Awards:* George Rieveschl Jr Award, 82. *Mem:* AAAS; Bot Soc Am; Am Soc Plant Taxon; Am Bryol & Lichenological Soc; hon mem Latin Am Bryol Soc; Brit Bryol Soc. *Res:* Hepatics; leafy hepatics of Latin America; sporeling development and regeneration of hepatics; nutrient studies in a leafy hepatic; leafy hepaticae of Mexico and Jamaica. *Mailing Add:* Edgecliff Apts 1008 2200 Victory Pkwy Cincinnati OH 45206

FULGER, CHARLES V, b Budapest, Hungary, Sept 27, 32; US citizen; m 71. FOOD CHEMISTRY, FOOD TECHNOLOGY. *Educ:* Univ Budapest, BSc, 56; Univ Melbourne, MSc, 60; Univ Mass, PhD(food sci), 67. *Prof Exp:* Anal chemist, Univ Melbourne, 57-60; asst biochem, Univ Calif, 60-63; res fel food res, Univ Mass, 63-67; from proj leader to dir res labs, Kellogg Co, 67-77; corp res mgr, Gen Foods, 77-; MGR FOOD SCI & ENG, MCCORMICK & CO. *Mem:* Am Cereal Chem Soc; Am Chem Soc; Inst Food Technologists. *Res:* Stereochemistry of lipid reactions; experimental design of food systems; cereal chemistry and technology; modification of seed proteins. *Mailing Add:* McCormick & Co 202 Wight Ave Hunt Valley MD 21031-1051

FULGHUM, ROBERT SCHMIDT, b Washington, DC, Mar 3, 29; m 53; c 3. BACTERIOLOGY. *Educ:* Roanoke Col, BS, 54; Va Polytech Inst, MS, 59, PhD(bact), 65. *Prof Exp:* Res asst biol, Va Agr Exp Sta, 56-60; from instr to asst prof biol, Susquehanna Univ, 60-64; from asst prof to assoc prof bact, NDak State Univ, 64-68; asst prof oral biol, Col Dent, Univ Ky, 68-71; dir anaerobic prod, Robbin Labs Div, Scott Labs, 71-72; actg chmn, 74-76, assoc prof, 72-90, PROF & ASST CHMN, DEPT MICROBIOL & IMMUNOL, SCH MED, EAST CAROLINA UNIV, 90- *Concurrent Pos:* Nat Inst Dent Res grant, 71-72; Deafness Res Found grant, 79-81, 82-85, 85-87; mem, organizing comt, Educ Strategies Workshop, 89-90. *Mem:* Am Soc Microbiol; Brit Soc Gen Microbiol; Sigma Xi. *Res:* Anaerobic bacteriology; anaerobic bacteria of medical importance; rumen bacteriology; cecal bacteriology; oral bacteriology; Methodology in anaerobic bacteriology; bacterial nutrition; bacterial flora-host animal ecology; animal models for polymicrobic infection; aerobic and anaerobic bacteriology of otitis media. *Mailing Add:* Dept Microbiol & Immunol Sch Med E Carolina Univ Greenville NC 27858-4354

FULGINITI, VINCENT ANTHONY, b Philadelphia, Pa, Aug 8, 31; m; c 3. PEDIATRICS, IMMUNOLOGY. *Educ:* Temple Univ, AB, 53, MD, 57, MS, 61. *Prof Exp:* Asst pediat & chief resident, St Christopher's Hosp, Philadelphia, 60-61; from instr to assoc prof, Univ Colo Med Ctr, Denver, 62-68; prof pediat & head dept, Col Med Univ Ariz, 69-85, vdean, 85-87, actg dean, 87-88; DEAN, SCH MED, TULANE UNIV, 89- *Concurrent Pos:* Fulbright scholar; fel pediat infectious dis, Univ Colo Med Ctr, Denver; Waldo E Nelson lectr, 59; consult, Tucson Med Ctr, Kino Community Hosp & Davis-Monthan Hosp; Markle scholar, 64-69; ed, Am J Dis Child, 83- *Honors & Awards:* Ross Award, Western Soc Pediat Res, 65. *Mem:* AAAS; Infectious Dis Soc Am; Am Soc Microbiol; Am Acad Pediat; AMA; Am Pediat Soc; Am Pub Health Asn; Pan Am Health Orgn. *Res:* Pediatric virology; measles virus, immunization and reaction; smallpox vaccination; varicella vaccination; immunodeficienoic. *Mailing Add:* Deans Off Tulane Univ Sch Med New Orleans LA 70112

FULKERSON, JOHN FREDERICK, b Los Angeles, Calif, Feb 20, 22; m 53; c 2. MICROBIOLOGY, PLANT PATHOLOGY. *Educ:* Western Reserve Univ, BS, 49; NC State Col, MS, 51, PhD(plant path), 57. *Prof Exp:* Asst biol, Western Reserve Univ, 41-42, 46-49 & NC State Col, 49-51; assoc entomologist & plant pathologist, WVa Univ, 51-53; asst biol, NC State Col, 53-56; res pathologist, 56-60, plant pathologist & prin scientist, USDA, 60-88; RETIRED. *Concurrent Pos:* Nat Inst Pub Affairs-Ford Found fel, Ind Univ, 65-66. *Mem:* Fel AAAS; fel Am Phytopath Soc. *Res:* Phytobacteriology; microbial physiology; disease physiology. *Mailing Add:* 207 Lexington Dr Silver Spring MD 20901

FULKERSON, WILLIAM, b Baltimore, Md, May 26, 35; m 57. RESEARCH MANAGEMENT. *Educ:* Rice Univ, BA, 57, PhD(chem eng), 62. *Prof Exp:* Group leader, Metals & Ceramics Div, 69-70, prog mgr, 72-75, sect head, Energy Div, 74-75, DIR ENERGY DIV, OAK RIDGE NAT LAB, 75- *Mem:* AAAS; Sigma Xi. *Res:* Management, research and development of energy conservation; electric power systems; environmental effects of energy facilities, programs and data systems. *Mailing Add:* Rte 2 Box 393 Wheat Rd Lenoir City TN 37771

FULKS, WATSON, b Ark, Jan 24, 19; m 43; c 3. MATHEMATICS. *Educ:* Ark State Teachers Col, BS, 40; Univ Ark, MS, 41; Univ Minn, PhD(math), 49. *Prof Exp:* Instr math, Univ Ark, 41-44; from instr to assoc prof, Univ Minn, 45-60; prof, Ore State Col, 60-63; PROF MATH, UNIV COLO, BOULDER, 63- *Concurrent Pos:* Asst, Calif Inst Technol, 49-50. *Mem:* AAAS; Am Math Soc; Math Asn Am. *Res:* Partial differential equations; asymptotics. *Mailing Add:* Dept Math Univ Colo Campus Box 426 Boulder CO 80309

FULLAGAR, PAUL DAVID, b Ft Edward, NY, Dec 19, 38; m 59; c 2. GEOCHRONOLOGY. *Educ:* Columbia Univ, AB, 60; Univ Ill, PhD(geol), 63. *Prof Exp:* Asst prof geol, Old Dominion Col, 63-67; from asst prof to assoc prof, 67-74, PROF GEOL, UNIV NC, CHAPEL HILL, 74-, CHMN DEPT, 79- *Concurrent Pos:* Vis fel, Wolfson Col, Cambridge Univ, Eng, 84-85. *Mem:* Geochem Soc; fel Geol Soc Am; Am Geophys Union; AAAS; Nat Asn Geol Teachers; Am Asn Univ Prof. *Res:* Rubidium-strontium geochronology; evolution orogenic belts; lead isotopic studies. *Mailing Add:* Dept Geol CB 3315 Univ NC Chapel Hill NC 27599-3315

FULLAM, HAROLD THOMAS, b Tacoma, Wash, June 4, 27. CHEMICAL ENGINEERING. *Educ:* Univ Wash, Seattle, BS, 51, MS, 52, PhD(chem eng), 56. *Prof Exp:* Design engr, Standard Oil Co Calif, 52; res engr, Martinez Refinery, Shell Oil Co Calif, 56-57; group leader, Richmond Res Lab, Stauffer Chem Co, 57-63; sr nuclear engr, Hanford Atomic Prod Opers, Gen Elec Co, 63-65; SR NUCLEAR ENGR, PAC NORTHWEST LAB, BATTELLE MEM INST, 65- *Mem:* Am Chem Soc; Electrochem Soc; Am Ord Asn. *Res:* Fused salt electrochemistry; inorganic process development; high temperature thermodynamics, actinide chemistry and processing. *Mailing Add:* 1500 Torthay Pl Richland WA 99352

FULLER, BARBARA FINK BROCKWAY, b Chicago, Ill, Feb 29, 36; m 59, 80; c 2. VERTEBRATE ZOOLOGY, PHYSIOLOGY. *Educ:* Univ Calif, Santa Barbara, BS, 56; Cornell Univ, MS, 58, PhD, 62, RN, 81. *Prof Exp:* Instr biol, Western Reserve Univ, 61-63, NSF instnl grant, 62-63; asst prof zool,

Ohio State Univ, 64-68; vis lectr biol, 68-69, lectr, 69-70, assoc prof, 70-73, PROF PHYSIOL, SCH NURSING, UNIV COLO, DENVER, 73- *Concurrent Pos:* NSF res grant, 65-78; isntnl grants, Sch Nursing, Univ Colo, 78-88; family nurse practr, 81-; NIH res grants, 86-90, 90- 95. *Mem:* NY Acad Sci; AAAS; Am Nursing Asn; Soc Psychophysiol Res; Int Soc Human Ethology. *Res:* Social Behavior and reproductive physiology; nursing intervention and patient stress; vocal indices of stress; infant pain: behavioral characteristics and nursing cues used in assessment. *Mailing Add:* Sch Nursing Univ Colo 4200 E Ninth Ave Denver CO 80262

FULLER, BENJAMIN FRANKLIN, JR, b St Paul, Minn, Aug 7, 22; m 45; c 3. MEDICINE. *Educ:* Univ Minn, BA, 43, MD, 45, MS, 50. *Prof Exp:* Clin asst prof internal med, 53-66, from asst prof to assoc prof, 66-70, dir div family med & community health, 66-68, head dept family pract & community health, 68-70, prof med, Med Ctr, 70-76, head sect primary care, Dept Internal Med, 72-76, CLIN PROF MED, UNIV MINN, MINNEAPOLIS, 76- *Mem:* Fel Am Col Physicians; Sigma Xi. *Res:* Peripheral vascular disease. *Mailing Add:* 2555 East County Rd E White Bear Lake MN 55110

FULLER, CHRISTOPHER ROBERT, b Crystal Brook, S Australia, Feb 17, 53. ACTIVE CONTROLS, INTELLIGENT SYSTEMS. *Educ:* Univ Adelaide, Australia, BEng, 73, PhD(acoust), 79. *Prof Exp:* Engr, Australia Dept Defence, 74-76; res fel, Inst Sound & Vibration Res, Southampton Univ, UK, 78-81; NRC scientist, NASA Langley Res Ctr, 81-83; PROF MECH ENG, VA POLYTECH INST & STATE UNIV, 83- *Concurrent Pos:* Consult, 76-78 & 83-; mem, Tech Comt Exp Acoustics & Instrumentation, Am Soc Mech Engrs, 87-; mem, AHS Tech Comt on Acoust, 89-, Acoust Soc Am Tech Comt on Structural Acoust & Vibration, 89-; chmn, Inst Noise Control Eng Tech Comt, Active Control, 91-; assoc tech ed, Am Soc Mech Engr J Vibration of Acoust, 91- *Mem:* Acoust Soc Am; fel Am Inst Aeronaut & Astronaut; Sigma Xi. *Res:* Active control of structurally radiated sound by adaptive structures; developing and applying smart systems for identification of acoustic signatures. *Mailing Add:* Dept Mech Eng Va Polytech Inst & State Univ Blacksburg VA 24061

FULLER, DEREK JOSEPH HAGGARD, b London, Eng, June 17, 17. MATHEMATICAL ANALYSIS. *Educ:* Univ SAfrica, BSc, 48, BA, 53, BSc, 56, MSc, 60; Univ Witwatersrand, BSc, 50; Univ Calif, Los Angeles, MA & PhD(math), 63. *Prof Exp:* Lectr math, Pius XII Col, Roma, Basutoland, 50-59 & 63; from lectr to sr lectr, Univ Basutoland, Bechuanaland Protectorate & Swaziland, 64-65; from asst prof to prof, 65-85, actg chmn dept, 66-67, chmn dept, 67-76, EMER PROF MATH, CREIGHTON UNIV, 85- *Res:* Analytic functions on Riemann surfaces. *Mailing Add:* 5050 Grover St No 10 Omaha NE 68106-3846

FULLER, DUDLEY D(EAN), b Woodhaven, NY, Feb 8, 13; m 45; c 1. MECHANICAL ENGINEERING. *Educ:* City Col New York, BME, 41; Columbia Univ, MS, 45. *Prof Exp:* Instr mech eng, City Col New York, 41-44; from instr to assoc prof, 43-54, chmn dept, 64-70, prof, 54-81, STEVENS EMER PROF MECH ENG, COLUMBIA UNIV, 81- *Concurrent Pos:* Prin engr, Franklin Inst, 54-69; mem bearing panel, high-temperature bearings, Nat Acad Sci, 59, Inst Defense Anal, 62 & proj forecast, US Army-Air Force, 63; chmn, Int Gas Bearing Symposium, Off Naval Res, 59. *Honors & Awards:* Award, Am Soc Lubrication Engrs, 57; Richards Mem Award, 62 & Mayo D Hersey Tribology Award, 74, Am Soc Mech Engrs; Tribology Gold Medal, Inst Mech Engrs, London, 78. *Mem:* Hon mem Am Soc Lubrication Engrs; fel Am Soc Mech Engrs; Soc Automotive Engrs; inst fel Franklin Inst; Brit Inst Mech Engrs. *Res:* Lubrication and bearing design; gas bearing research. *Mailing Add:* Box 219 Lake Hill NY 12448

FULLER, ELLEN ONEIL, b Providence, RI; m 46; c 4. CARDIOVASCULAR PHYSIOLOGY. *Educ:* Med Col Ga, BS, 60; Emory Univ, MN, 60, MSc, 64, PhD(physiol), 68. *Prof Exp:* Assoc prof health sci, Ga State Univ, 68-70; asst prof, Emory Univ, 70-80, assoc prof physiol, 80-; AT DEPT PHYSIOL, SCH MED, UNIV PA. *Mem:* Sigma Xi. *Res:* Hemodynamics of the uterine circulation; effect of exercise-training on heart. *Mailing Add:* RD 2 Box 352 Phoenixville PA 19460

FULLER, EUGENE GEORGE, b Reno, Nev, May 12, 38; m 66; c 2. EMBRYOLOGY, HISTOLOGY. *Educ:* Univ Nevada, BS, 60, MS, 63; Ore State Univ, PhD(zool), 66. *Prof Exp:* Instr zool, Ore State Univ, 66-67; from asst prof to assoc prof, 67-77, PROF ZOOL, BOISE STATE UNIV, 77- *Mem:* AAAS; Am Soc Zoologists; Sigma Xi. *Res:* Histochemistry of the crustacean antennal gland in relation to its function; histochemistry, histology and morphogenesis of the rodent placenta. *Mailing Add:* Dept Biol Sci Boise State Univ Boise ID 83725

FULLER, EVERETT G(LADDING), b Springfield, Mass, Dec 6, 20; m 47; c 3. NUCLEAR PHYSICS. *Educ:* Amherst Col, BS, 42; Univ Ill, MA, 47, PhD(physics), 50. *Prof Exp:* Asst physics, Univ Ill, 49-50; physicist, 50-61, section chief, 61-77, dir, Photonuclear Data Ctr, 77-81, PHYSICIST, NAT BUR STANDARDS, 81- *Concurrent Pos:* Guest, Inst Theoret Physics, Copenhagen, 64-65; vis guest prof, Inst Theoret Physics, Frankfurt, 67; com sci fel, US Dept Com, 70-71; res assoc prof, Dept Radiol, Sch Med, Univ Md, 71-74; mem, Int Adv Comt to the Pres, Max Planck-Gesellschaft, Ger, 75-80. *Mem:* Am Phys Soc. *Res:* Photonuclear interaction; interaction of electromagnetic radiation with nuclei. *Mailing Add:* 9016 Honeybee Lane Bethesda MD 20817

FULLER, EVERETT J, b Twin Falls, Idaho, July 3, 29; m 52; c 4. PHYSICAL CHEMISTRY. *Educ:* Idaho State Col, BS, 51; Univ Utah, PhD(chem), 60. *Prof Exp:* Chemist, Atomic Energy Div, Am Cyanamid Co, 51-53 & Phillips Petrol Co, 53; from chemist to sr chemist, 60-67, res assoc, Exxon Res & Eng Co, Exxon Corp, NJ, 60-86; CONSULT, 86- *Mem:* Am Chem Soc; Am Water Works Asn; AAAS; Soc Mining Engrs. *Res:* Enzyme model systems; theoretical statistical mechanics; radiochemistry; separations. *Mailing Add:* 666 Long Hill Rd Gillette NJ 07933

FULLER, FORST DONALD, b Stroh, Ind, Apr 25, 16; m 37; c 2. PHYSIOLOGY. *Educ:* DePauw Univ, BA, 38; Purdue Univ, MS, 41, PhD(zool), 52. *Prof Exp:* Physiologist, Armour Labs, Ill, 42-43; instr zool, Purdue Univ, 46-47; from instr to assoc prof zool, DePauw Univ, 47-58, prof, 58-81, head dept, 73-81, EMER PROF ZOOL, DEPAUW UNIV, 81- *Concurrent Pos:* Adv premed prog, DePauw Univ, 73- *Res:* Vertebrate and invertebrate physiology. *Mailing Add:* 1005 Albin Pond Rd Greencastle IN 46135-9220

FULLER, FRANCIS BROCK, b Eugene, Ore, July 8, 27; m 57; c 1. MATHEMATICS. *Educ:* Princeton Univ, AB, 49, PhD, 52. *Prof Exp:* Instr math, Princeton Univ, 51-52; instr, 52-55, from asst prof to assoc prof, 55-66, PROF MATH, CALIF INST TECHNOL, 66- *Concurrent Pos:* Fulbright res scholar, Univ Strasbourg, 67-68. *Mem:* Am Math Soc. *Res:* Applications of algebraic topology, especially fixed point theory; global theory of ordinary differential equations; mechanics of DNA molecules. *Mailing Add:* 527 Pepeekeo Pl Honolulu HI 96825

FULLER, GERALD ARTHUR, b Sask, Sept 29, 39; m 68; c 3. ENVIRONMENTAL & ECONOMIC CONSEQUENCES OF AGRICULTURAL DRAINAGE. *Educ:* Univ Sask, BEng, 62, MSc, 66; Univ Waterloo, PhD(civil eng), 74; Univ Regina, cert admin, 88. *Prof Exp:* Lab instr eng, Univ Sask, 65-66; water supply engr, Can Dept Agr, 62-65 & 70-73; civil engr, Sask Dept Environ, 65-68; assoc prof eng, Univ Regina, 73-79, dept head, 76-80, asst dean, 78-81, PROF ENG, UNIV REGINA, 79- *Concurrent Pos:* Mem, Sask Water Appeal Bd, 82-88, Sask Forest Mgt Task Force, 84-85 & Sask Water Studies Inst, 84-87. *Mem:* Can Water Resources Asn; Int Asn Irrig & Drainage. *Res:* Environmental and economic consequences of agricultural drainage; estimation of streamflow data for ungauged streams; use of wind power in remote areas; use of computers for teaching purposes. *Mailing Add:* Eng Dept Univ Regina Regina SK S4S 0A2 Can

FULLER, GERALD M, b Beaumont, Tex, Sept 13, 35. CELL BIOLOGY. *Educ:* Univ Austin, BA, 58; Sam Houston State Univ, MA, 60; Univ Calif, San Diego, PhD, 68. *Prof Exp:* Instr, Biol Dept, Sam Houston State Univ, Huntsville, Tex, 60-61; asst prof, Univ San Diego, La Jolla, 63-65; from asst prof to prof, Dept Human Biol Chem & Genetics & from adj asst prof to adj prof, Marine Biomed Inst, Univ Tex Med Br, Galveston, Tex, 70-85, dir, Grad Prog Human Genetics & Cell Biol, 81-85; PROF DEPT CELL BIOL & ANAT, UNIV ALA, BIRMINGHAM, 86- *Concurrent Pos:* Numerous grants, NIH, Cystic Fibrosis Res, 77-95; mem ad hoc comt, Heart Lung & Blood Res Comt, NIH, Coun Thrombosis, Am Heart Asn, 79-82. *Mem:* Am Soc Biochem & Molecular Biol; Am Soc Cell Biol (treas, 83-); Am Chem Soc; Am Heart Asn; Sigma Xi; Am Soc Hemat; AAAS. *Res:* Cystic fibrosis; biology; plasma protein biocynthesis; hepatocyte-monocyte interation in acute inflammation. *Mailing Add:* Dept Cell Biol & Anat Univ Ala Birmingham BHS 670 UAB Sta Birmingham AL 35294

FULLER, GLENN, b Lancaster, Calif, May 18, 29; m 54; c 2. ORGANIC CHEMISTRY, BIOCHEMISTRY. *Educ:* Stanford Univ, BS, 50; Univ Ill, MS, 51, PhD(chem), 53. *Prof Exp:* Chemist, Shell Develop Co, 53-64; asst prof chem, Mills Col, 64-65; res chemist, 65-71, chief, Fruit Lab, 71-72, RES LEADER PLANT DEVELOP QUAL, WESTERN REGIONAL RES CTR, USDA, 72- *Mem:* AAAS; Am Chem Soc; Inst Food Technol; Am Oil Chemists' Soc; Am Soc Plant Physiologists; Scand Soc Plant Physiol. *Res:* Chemistry of fats and oils; chemistry of free radicals; antioxidants; lubricants; preservation of fruit and fruit products; plant hormones; aflatoxins. *Mailing Add:* 1060 Cragmont Ave Berkeley CA 94708-1434

FULLER, HAROLD Q, b Waynetown, Ind, Apr 21, 07; m 32; c 3. PHYSICS. *Educ:* Wabash Col, AB, 28; Univ Ill, AM, 30, PhD(physics), 32. *Hon Degrees:* DSc, Wabash Col. *Prof Exp:* Res physics, Univ Ill, 32-33; asst prof math & physics, Ill Col, 33-37; instr physics, Univ Ill, 37-38; from asst prof to prof, Albion Col, 38-47; chmn dept physics, 48-70, actg dean sch sci, 66-67, dean col arts & sci, 70-72, prof 47-77, EMER PROF PHYSICS, UNIV MO, ROLLA, 77-, EMER DEAN, COL ARTS & SCI, 72- *Concurrent Pos:* Sr physicist, Process Improv Div, Tenn Eastman Corp, 44-45. *Mem:* AAAS; Am Phys Soc; Am Soc Eng Educ; Am Asn Physics Teachers. *Res:* Spectroscopy; photochemistry; soft x-rays. *Mailing Add:* Univ Mo Rolla Univ Mo Rolla MO 65401

FULLER, HAROLD WAYNE, b Chicago, Ill, Sept 4, 25; m 52; c 5. SPECTROSCOPY, ISOTOPES. *Educ:* Kalamazoo Col, Mich, BA, 50. *Prof Exp:* Lab technician II biochem, Univ Fla, Hague, 63-64; teacher, Gilcrest County Bd Pl, Trenton, Fla, 63-64 & Putnam County Bd Pl, Palatka, 65-66; chemist I biochem, Univ Fla, Gainesville, 66-79; CHEMIST III ANALYTICAL CHEM, FLA DEPT TRANSP, GAINESVILLE, 79- *Mem:* Fel Am Inst Chem; Royal Soc Chem; Soc Appl Spectros. *Res:* Cement and paint analysis, using atomic absorption; environmental waters trace mineral analysis for pesticides; enzyme activity. *Mailing Add:* 901 NW 37th Terr Gainesville FL 32607

FULLER, HENRY LESTER, b Andreasky, Alaska, June 25, 16; m 46; c 2. POULTRY NUTRITION. *Educ:* State Col Wash, BS, 40; Iowa State Col, MS, 41; Purdue Univ, PhD(poultry nutrit), 51. *Prof Exp:* Salesman livestock feeds, Hales & Hunter Co, 46-48; dir educ sales, Feed Mill Div, Glidden Co, 48-49; assoc prof, 51-57, prof, 57-86, EMER PROF POULTRY NUTRIT, UNIV GA, 86- *Concurrent Pos:* Indust consult. *Mem:* Poultry Sci Asn; Am Inst Nutrit; World Poultry Sci Asn; Fedn Am Soc Exp Biol; Sigma Xi. *Res:* Effects of restricted feeding of young chickens on subsequent egg production characterisitics and longevity; nutrition requirements of chickens; effect of diet on heat stress in chickens. *Mailing Add:* Dept Poultry Sci Univ Ga Athens GA 30602

FULLER, JACKSON FRANKLIN, b Salt Lake City, Utah, Oct 9, 20; m 42; c 4. ELECTRICAL ENGINEERING, ENERGY CONVERSION. *Educ:* Univ Colo, BSEE, 44. *Prof Exp:* Systs eng elec, Gen Elec Co, 45-69; PROF ELEC, UNIV COLO, 69- *Concurrent Pos:* Consult, US Govt, Pvt Indust, State & Local Govt, 53- *Honors & Awards:* Chas A Coffin Award, Gen Elec Co, 53; Power Educator Award, Edison Elec Inst, 77. *Mem:* Inst Elec & Electronics Engrs; Am Soc Eng Educ. *Res:* Generation; transmission and distribution of electrical energy; energy conversion from nuclear and fossil sources; air quality control from power plants. *Mailing Add:* Dept Elec Eng Campus Code 425 Colo Univ Boulder CO 80309-0425

FULLER, JAMES OSBORN, b Chaumont, NY, Aug 14, 12; m 39, 74, 82; c 4. GEOLOGY. *Educ:* Lehigh Univ, AB, 34; Columbia Univ, PhD(geol), 41. *Hon Degrees:* LLD, Lehigh Univ, 72. *Prof Exp:* Asst petrol, Columbia Univ, 36-37, asst econ geol, 37-39; asst prof geol, Mt Union Col, 39-41; instr, Ohio State Univ, 41-43; asst prof, WVa Univ, 43-44; asst geologist, US Geol Surv, Va, 44-45; asst prof geol, WVa Univ, 45-46; assoc prof, Ohio State Univ, 46-48, prof, 48-67, acting dean col arts & sci, 51-52, assoc dean, 52-57, secy adv res coun, 55-57, dean col arts & sci, 57-67; pres, Fairleigh Dickinson Univ, 67-74, distinguished prof marine geol, 74; dir part-time & evening prog & spec asst to pres, 74-75, dean arts, 75- 76, prof, 74-83, EMER PROF GEOL, OHIO STATE UNIV, 83- *Concurrent Pos:* Consult, Nat Sea Grant Adv Panel, 71-76 & Nat Sea Grant Recert Panel, 88; trustee, secy & treas, Int Ctr Preserv Wild Animals, 84- *Mem:* AAAS; Am Asn Petrol Geol; fel Geol Soc Am; Sigma Xi. *Res:* Economic and petroleum geology; field mapping of Sharon conglomerate and Lee County, Virginia, oil field; ore petrography and polishing; water resources of Ohio. *Mailing Add:* 3928 Fairlington Dr Upper Arlington OH 43220

FULLER, KENT RALPH, b Northfield, Minn, Feb 15, 38; m 58; c 1. MATHEMATICS. *Educ:* Mankato State Col, BS, 60, MS, 62; Univ Ore, MA, 65, PhD(math), 67. *Prof Exp:* From asst prof to assoc prof, 67-75, PROF MATH, UNIV IOWA, 75- *Concurrent Pos:* Ed, J Algebra, 83-; vis prof, Univ Hawaii, 79-80; vis scholar, Aarhus Univ, 77, Fern Universitat, 82, Univ Leicester, 86, Univ Calgary, 87, Univ Marcia, 88. *Mem:* Am Math Soc. *Res:* Algebra, especially ring theory. *Mailing Add:* Dept Math Univ Iowa Iowa City IA 52242

FULLER, LEONARD EUGENE, b Casper, Wyo, July 25, 19; m 43. MATHEMATICS. *Educ:* Univ Wyo, BA, 41; Univ Wis, MS, 47, PhD(math), 50. *Prof Exp:* Asst, Univ Wis, 46-50, instr, 50-51; mathematician, Goodyear Aircraft Corp, 51-52; from asst prof to prof, 52-84, EMER PROF MATH, KANS STATE UNIV, 84- *Mem:* Am Math Soc; Math Asn Am; Sigma Xi. *Res:* Applications of matrix theory. *Mailing Add:* 404 Ehlers Rd Manhattan KS 66502-4533

FULLER, LYNN FENTON, b Schenectady, NY, Dec 8, 47; m 70; c 2. MICROELECTRONICS MANUFACTURING, VERY LARGE-SCALE INTEGRATION DESIGN. *Educ:* Rochester Inst Technol, BS, 70, MS, 73; State Univ NY, Buffalo, PhD(elec eng), 79. *Prof Exp:* PROF ENG & HEAD MICROELEC ENG, ROCHESTER INST TECHNOL, 70- *Concurrent Pos:* Vis prof eng, Univ Hawaii, 79-80. *Mem:* Sr mem Inst Elec & Electronics Engrs; Inst Elec & Electronics Engrs Electron Devices Soc; Int Soc Hybrid Microelectronics. *Res:* Microelectronics; solid state devices. *Mailing Add:* Ctr Microelec & Computer Eng Rochester Inst Technol One Lomb Memorial Dr Rochester NY 14623

FULLER, MARIAN JANE, b Scottsbluff, Nebr, Jan 9, 40. PLANT TAXONOMY, GENETICS. *Educ:* Univ Nebr, Lincoln, BS, 62, PhD(plant taxon), 67. *Prof Exp:* From asst prof to assoc prof, 67-78, PROF BIOL, MURRAY STATE UNIV, 78- *Mem:* Am Soc Plant Taxon; Int Asn Plant Taxon; Bot Soc Am. *Res:* Relationships among members of the genus Carduus in Nebras; relationships among members of the genus Goodyera; Orchidaceae of Kentucky; flora of West Kentucky. *Mailing Add:* Dept Biol Murray State Univ Murry KY 42071

FULLER, MARK ROY, b Minneapolis, Minn, Sept 8, 46; m 70; c 2. RAPTOR ECOLOGY, PREDATOR-PREY RELATIONSHIPS. *Educ:* Colo State Univ, BS, 68; Cent Wash State Univ MS, 71; Univ Minn, PhD(ecol & behav biol), 79. *Prof Exp:* Res assoc, Bell Mus Nat Hist & Dept Vet Biol, Univ Minn, 76-78; RES BIOLOGIST, US FISH & WILDLIFE SERV, PATUXENT WILDLIFE RES CTR, 78- *Mem:* Am Ornithologists Union; Ecol Soc Am; Sigma Xi; Int Soc Chronobiol; Animal Behav Soc. *Res:* Habitat use by birds of prey including minimum area requirements and influence of land-use on behavior, density, productivity; raptor population biology, assessment of population status; eco-physiology and predator behavior. *Mailing Add:* Patuxent Wildlife Res Ctr Laurel MD 20708

FULLER, MARTIN EMIL, b Inglewood, Calif, Sept 15, 30; m 61; c 2. ENVIRONMENTAL CHEMISTRY. *Educ:* Univ Calif, Los Angeles, BS, 52; Mass Inst Technol, PhD(chem), 56. *Prof Exp:* Asst phys chem, Mass Inst Technol, 52-53; asst prof chem, Pomona Col, 56-61; res asst, Univ Fla, 61-63; assoc prof, Colo Sch Mines, 63-68 & Prescott Col, 68-70; teacher, Colo Rocky Mountain Sch, 70-71; assoc prof chem, Ft Lewis, Colo, 71-73; dir phys chem, Lee Pharmaceut, 73-80, assoc dir res & develop, 80-82; AT DEPT CHEM, AUSTIN COL. *Mem:* Am Chem Soc; Am Phys Soc; AAAS; Sigma Xi. *Res:* Composite polymer systems; dielectric properties of polypolar polymers and solutions; instrumental methods of analysis; science education; environmental chemistry. *Mailing Add:* Dept Chem Austin Col Sherman TX 75091-1177

FULLER, MELVIN STUART, b Livermore Falls, Maine, May 5, 31; m 55; c 3. BOTANY, MYCOLOGY. *Educ:* Univ Maine, BS, 53; Univ Nebr, MS, 55; Univ Calif, PhD, 59; Brown Univ, MS(ad eundem), 63. *Prof Exp:* Instr bot, Brown Univ, 59-60, from asst prof to assoc prof, Brown Univ, 60-64; asst prof, Univ Calif, Berkeley, 64-66, assoc prof & dir electron microscope lab, 66-68; head dept, 68-73, 86-89, PROF BOT, UNIV GA, 73- *Concurrent Pos:* Consult ed, McGraw-Hill. *Mem:* Mycol Soc Am (vpres, 73-74, pres elect, 74-75, pres, 75-76); Phytopath Soc Am. *Res:* Biology of the fungi; aquatic Phycomycetes; mitosis in fungi; fungal motile cells; mechanism of action of fungicides. *Mailing Add:* Dept Bot Univ Ga Athens GA 30602

FULLER, MICHAEL D, b London, Eng, June 10, 34. GEOPHYSICS. *Educ:* Cambridge Univ, BA, 58, PhD(geophys), 61. *Prof Exp:* Fel geophys, Scripps Inst, Calif, 61-62; res geophysicist, Gulf Res & Develop Co, Pa, 62-64; assoc prof geophys, Univ Pittsburgh, 65-69, prof, 69-74; PROF GEOPHYS, UNIV CALIF, SANTA BARBARA, 74- *Concurrent Pos:* Mem, NASA Lunar Sci Comts, 69-72 & NSF Earth Sci Panel, 77-79. *Mem:* Fel Am Geophys Union. *Res:* Physical process of magnetization of rocks in nature; paleomagnetic records of secular variation and field reversals; lunar magnetism; seismomagnetism; interpretation of geological structure from aeromagnetic surveys; tectonics of SE Asia. *Mailing Add:* Dept Geol Univ Calif Santa Barbara CA 93106

FULLER, MILTON E, physical chemistry; deceased, see previous edition for last biography

FULLER, PETER MCAFEE, b Grand Rapids, Mich, Jan 24, 43; m 70; c 1. NEUROANATOMY. *Educ:* Olivet Col, BA, 66; Univ Va, PhD(anat), 74. *Prof Exp:* Res assoc, Highway Safety Res Inst, 67-70; asst prof, 74-80, ASSOC PROF ANAT, SCH MED, UNIV LOUISVILLE, 80- *Concurrent Pos:* Consult, Div Prod Safety, Nat Bur Stand, 72-74. *Mem:* Am Asn Anatomists; Am Asn Automotive Med; Aerospace Med Asn; Soc Neurosci. *Res:* Neuroanatomical studies of the vestibular and trigeminal sensory systems in various vertebrate animals in an attempt to understand more fully the sensory input to the central nervous system; accident reconstruction and head injury studies. *Mailing Add:* Dept Anat Health Sci Ctr Univ Louisville Louisville KY 40292

FULLER, RAY W, b Dongola, Ill, Dec 16, 35; m 56; c 2. BIOCHEMISTRY. *Educ:* Southern Ill Univ, BA, 57, MA, 58; Purdue Univ, PhD(biochem), 61. *Hon Degrees:* ScD, Purdue Univ, 90. *Prof Exp:* Dir biochem res lab, Ft Wayne State Sch, Ind, 61-63; sr pharmacologist, 63-67, res scientist, 67-68, head, Dept Metab Res, 68-71, res assoc, 71-75, res adv, 76-89, LILLY RES FEL, ELI LILLY & CO, 89- *Concurrent Pos:* Adj assoc prof biochem, Sch Med, Ind Univ, 74-84, adj prof neurobiol, 85-; vis lectr, Mass Inst Technol, 76-; mem, Basic Psychopharmacol Adv Comt, Pharmaceut Mfrs Asn Found, 78-82; pres, Catecholamine Club, 79-80; mem, Basic Psychopharmacol & Neuropsychol Res Rev Comt, Nat Inst Mental Health, 80-84; mem, NASA Space Adaptation Syndrome Steering Comt, 82-84. *Mem:* AAAS; Am Chem Soc; Am Soc Biol Chemists; Am Soc Pharmacol & Exp Therapeut; Am Soc Neurochem; Soc Neurosci; Endocrine Soc; NY Acad Sci. *Res:* Biochemical mechanism of drug action; neuropsychopharmacology; brain biochemistry; neuroendocrinology; biogenic amine and amino acid metabolism. *Mailing Add:* Res Labs Eli Lilly & Co Indianapolis IN 46285

FULLER, RICHARD H, engineering education; deceased, see previous edition for last biography

FULLER, RICHARD KENNETH, b Chicago, Ill, Apr 2, 35; m 60; c 2. GASTROENTEROLOGY. *Educ:* Monmouth Col, BA, 57; Case Western Univ, MD, 62, MS, 71. *Prof Exp:* Asst prof biomet, Case Western Univ, 72-90, from instr to assoc prof med, 82-90; DIR, DIV CLIN & PREV RES, NAT INST ALCOHOL ABUSE & ALCOHOLISM, 90- *Mem:* Am Gastroentorol Asn; Am Col Physicians; AAAS. *Res:* Evaluation of treatments for pancreatitis, ascites, hepatic cirrhosis and alcoholism. *Mailing Add:* 5600 Fishers Lane Rm 14C10 Rockville MD 20817

FULLER, RICHARD M, b Crawfordsville, Ind, July 23, 33; m 55; c 3. SOLID STATE PHYSICS. *Educ:* DePauw Univ, BA, 55; Univ Minn, MA, 60; Mich State Univ, PhD(physics), 65. *Prof Exp:* Instr physics, Alma Col, Mich, 59-68, from asst prof to assoc prof, 61-68; assoc prof, 68-70, dean, 90-91, PROF PHYSICS, GUSTAVUS ADOLPHUS COL, 68- *Concurrent Pos:* Lectr, Mich State Univ, 65 & 68, Univ Mo-Rolla, 66-67 & People's Repub China, 88-89 & 91; co-auth, Physics, Including Human Applns, 78. *Mem:* AAAS; Am Asn Physics Teachers; Am Phys Soc; Sigma Xi. *Res:* Infrared physics; optical and electrical properties of solids particularly ionic crystals and bone; history and philosophy of physics. *Mailing Add:* 723 Upper Johnson Circle St Peter MN 56082

FULLER, ROBERT GOHL, b Crawfordsville, Ind, June 7, 35; m 61; c 3. SOLID STATE PHYSICS. *Educ:* Univ Mo, Rolla, BS, 57; Univ Ill, Urbana, MS, 58, PhD(physics), 65. *Prof Exp:* Teacher high sch, Burma, 59-61; res asst physics, Univ Ill, Urbana, 61-65; Nat Res Coun-Nat Acad Sci res fel, Naval Res Lab, 65-67, physicist, 67-69; assoc prof, 69-76, PROF PHYSICS, UNIV NEBR, LINCOLN, 76- *Concurrent Pos:* Vis physicist orthop res, Vet Admin Hosp, Syracuse, NY, 75; vis prof, Univ Calif, Berkeley & Lawrence Hall Sci, 76-77. *Mem:* Am Asn Physics Teachers (vpres, 78, pres-elect, 79, pres, 80-81); Am Phys Soc; Nat Sci Teachers Asn. *Res:* Diffusion, pulse radiolysis, and electrical conductivity of ionic solids; photoconductivity in bone and tendon; physics teaching and Piagetian psychology; faculty development; interactive videodiscs in physics teaching. *Mailing Add:* Dept Physics Univ Nebr Lincoln NE 68588

FULLER, ROY JOSEPH, b Little Rock, Ark, June 19, 39. MATHEMATICS. *Educ:* Univ Ark, BSEE, 61, MS, 63; Princeton Univ, PhD(math), 67. *Prof Exp:* asst prof, 67-80, ASSOC PROF MATH, UNIV ARK, FAYETTEVILLE, 80- *Mem:* Am Math Soc. *Res:* Non-Abelian Gaussian sums and their connection with Hecke operators; estimation of damage to aircraft due to fragmenting-warhead projectiles. *Mailing Add:* Dept Comput Sci Univ Ark Main Campus Fayetteville AR 72701

FULLER, RUFUS CLINTON, b Providence, RI, Mar 5, 25; m 46; c 4. MICROBIOLOGY, BIOCHEMISTRY. *Educ:* Brown Univ, AB, 45; Amherst Col, AM, 48; Stanford Univ, PhD(biol), 52. *Hon Degrees:* MA, Dartmouth Col, 61. *Prof Exp:* Teaching asst biol, Brown Univ, 45-46 & Amherst Col, 46-48; teaching asst, Stanford Univ, 48-50, asst microbiol, 51-52; res microbiologist, Lawrence Radiation Lab, Univ Calif, 52-55; assoc plant physiologist, Brookhaven Nat Lab, 55-58, plant biochemist, 58-59; prof microbiol & chmn dept, Dartmouth Med Sch, 60-65; prof biomed sci & dir grad sch biomed sci, Univ Tenn, Oak Ridge Nat Lab, 66-71; chmn dept, 71-77, PROF BIOCHEM, UNIV MASS, AMHERST, 71- *Concurrent Pos:* Vis prof life sci, Univ Calif, Riverside, 66; consult biol div, Oak Ridge Nat Lab, 66-72; Sigma Xi sr fel, NSF, Oxford Univ, Eng, 59-60; mem med scientists training comt, Nat Inst Gen Med Sci, 63-65; mem cell biol study sect, NIH, 65-69; mem microbiol training comt, Nat Inst Gen Med Sci, 69-74, consult minority access to res careers training comt, 74-79; vis prof, Swiss Tech Univ, Zurich, 84-85. *Honors & Awards:* US Sr Scientist Award, Alexander von Humboldt Found, Freiburg Univ, Ger, 77-78; Chancelors Medal, Univ MA. *Mem:* AAAS; Am Soc Plant Physiol; Am Soc Microbiol; Am Soc Biol Chem; Am Soc Cell Biol. *Res:* Comparative biochemistry of photosynthesis; cellular biochemistry as related to cellular structures; developmental aspects of microbial ultrastructure; biosynthesis and biodegredation of microbial polymers; membrane structure and function. *Mailing Add:* Dept Biochem Univ Mass Amherst MA 01003

FULLER, SAMUEL HENRY, b Detroit, Mich, June 1, 46; m; c 3. COMPUTER SCIENCE, ELECTRICAL ENGINEERING. *Educ:* Univ Mich, BS, 68; Stanford Univ, MS, 69, PhD(elec eng & comput sci), 72. *Prof Exp:* Asst prof comput sci, Carnegie Mellon Univ, 72-74, assoc prof, 74-78; VPRES RES & ARCHIT, DIGITAL EQUIP CORP, 78- *Concurrent Pos:* Consult, Naval Res Lab, 73-77, US Army, Ft Monmouth, 75-77. *Mem:* Nat Acad Eng; Asn Comput Mach; Inst Elec & Electronics Engrs. *Res:* Computer architecture; multiple processor structures and performance evaluation; design and evaluation of two experimental multiprocessors; developed techniques for evaluating alternative computer architectures. *Mailing Add:* Res & Archit Digital Equip Corp 146 Main St M1012-1-T7 Maynard MA 01754

FULLER, STEPHEN WILLIAM, b Portchester, NY, Mar 14, 45; m 74; c 2. BOTANY, PHYCOLOGY. *Educ:* Cornell Univ, BS, 67; Univ NH, PhD(bot), 71. *Prof Exp:* Res fel phycol, NY Ocean Sci Lab, 71-72; from asst prof to assoc prof, 72-84, chmn, 76-88 PROF BIOL MARY WASH COL, 84- *Concurrent Pos:* Consult, floral invest, wetlands delineation. *Mem:* Atlantic Estuarine Res Soc; Soc Wetlands Scientists. *Res:* Marine phycology. *Mailing Add:* Dept Biol Mary Washington Col Fredericksburg VA 22401-5358

FULLER, THOMAS CHARLES, b Evanston, Ill, Aug 2, 18; m 45; c 2. WEED SCIENCE. *Educ:* Northwestern Univ, BS, 40; Univ NMex, MS, 42; Univ Chicago, PhD(cytol), 47. *Prof Exp:* Asst bot & zool, Univ NMex, 40-41; instr bot, RI State Col, 46-48; prof, Hanover Col, 48-49; asst prof, Univ Southern Calif, 49-53; from jr plant pathologist to asst plant pathologist, Calif Dept Agr, 53-57, botanist, 57-81; RETIRED. *Mem:* Asian-Pac Weed Sci Soc. *Res:* Cytology; mitosis; effects of several sulfa compounds on nuclear and cell division; diseases and culture of ornamentals; distribution of weeds in California, poisonous plants of California. *Mailing Add:* 171 Wescott Way Sacramento CA 95864

FULLER, WALLACE HAMILTON, b Old Hamilton, Alaska, Apr 15, 15; m 39; c 1. BIOCHEMISTRY, BACTERIOLOGY. *Educ:* State Col Wash, BS, 38, MS, 39; Iowa State Col, PhD(soils chem), 42. *Prof Exp:* Res assoc, State Col Wash, 37-39; soil scientist, Soil Conserv Serv, USDA, 39-40; res assoc, Iowa State Col, 40-45; bacteriologist bur plant indust & agr eng, USDA, 45-47, soil scientist, 47-48; biochemist & assoc prof, 48-56, head dept agr chem & soils, 56-72, PROF CHEM & BIOCHEMIST, AGR RES STA, SOILS, WATER & ENG DEPT, UNIV ARIZ, 56- *Mem:* AAAS; Am Soc Agron. *Res:* Soil microbiology; biodegradation of wastes; chemistry of soil organic matter; soil fertilizers; plant nutrition; plant decomposition; soil phosphorus; plant decomposition and nutrient release; waste disposal; metal migration through soils; water quality; author various books and articles. *Mailing Add:* 5674 W Flying Circle Tucson AZ 85713

FULLER, WAYNE ARTHUR, b Brooks, Iowa, June 15, 31; m 56; c 2. STATISTICS. *Educ:* Iowa State Univ, BS, 55, MS, 57, PhD(agr econ), 59. *Prof Exp:* asst agr econ, 55-56, res assoc, 56-59, from asst prof to prof, 59-83, DISTINGUISHED PROF STATIST, IOWA STATE UNIV, 83- *Concurrent Pos:* NSF fel, 64-65. *Mem:* Int Statist Inst; Am Statist Asn; Inst Math Statist; Economet Soc; Biometric Soc. *Res:* Survey sampling, time series, measurement error, estimation and econometrics. *Mailing Add:* Dept Statist Iowa State Univ Ames IA 50011

FULLER, WENDY WEBB, b Washington, DC, Feb 3, 52. EXPERIMENTAL SOLID STATE PHYSICS, EXPERIMENTAL LOW TEMPERATURE PHYSICS. *Educ:* Univ Md, BS, 75; Univ Calif, Los Angeles, MS, 76, PhD(physics), 80. *Prof Exp:* Fel, 80, RES PHYSICIST, NAVAL RES LAB, 80- *Mem:* Am Phys Soc; AAAS. *Res:* Electronic transport properties in low dimensional materials and superconductivity in bulk materials and thin films. *Mailing Add:* Code 6344 Naval Res Lab Washington DC 20375-5000

FULLER, WILLIAM ALBERT, b Moosomin, Sask, May 10, 24; m 47; c 4. VERTEBRATE ECOLOGY. *Educ:* Univ Sask, BA, 46, MA, 47; Univ Wis, PhD(ecol), 57. *Prof Exp:* Mammalogist, Can Wildlife Serv, 47-59; from asst prof to prof, Univ Alta, 59-84, chmn dept, 69-74, emer prof zool, 84-; RETIRED. *Concurrent Pos:* Mem exec bd, Int Union Conserv Nature & Natural Resources, 64-70. *Mem:* Am Soc Mammal; Arctic Inst NAm. *Res:* Ecology of mammals, especially in the taiga in winter. *Mailing Add:* Box 672 Athabasca AB T0B 0B0 Can

FULLER, WILLIAM RICHARD, b Indianapolis, Ind, Oct 27, 20; m 43; c 3. MATHEMATICS. *Educ:* Butler Univ, BS, 48; Purdue Univ, MS, 51, PhD(math), 57. *Prof Exp:* Instr math, Butler Univ, 48-49; asst, Purdue Univ, 49-51; mathematician, US Naval Ord Plant, Ind, 51-54; instr math, Purdue Univ, 54-57, asst prof, 57-59, from assoc prof & actg head, Dept Math & Statist to prof & actg head, Div Math Sci, 59-61 & 63-64, assoc dean, Sch Sci, 65-76, prof math, 60-91, EMER PROF MATH, PURDUE UNIV, 91- *Concurrent Pos:* Consult, US Naval Avionics Facil, 58-61, Radio Corp Am Serv Co, Patrick AFB, Fla, 59 & Gen Elec Co, 60-61; chancellor, Purdue Univ, N Cent Campus, 79-82. *Mem:* Am Math Soc; Math Asn Am; Am Soc Eng Educ; AAAS. *Res:* Nonlinear differential equations; computers in undergraduate education; Fortran for use in calculus. *Mailing Add:* Dept Math Math Sci Bldg Purdue Univ West Lafayette IN 47907-1395

FULLERTON, CHARLES MICHAEL, b Oklahoma City, Okla, Mar 10, 32; m 54; c 3. ATMOSPHERIC PHYSICS. *Educ:* Univ Okla, BS, 54; NMex Inst Mining & Technol, MSc, 64, PhD(geophys), 66. *Prof Exp:* Instr physics & math, Col St Joseph, NMex, 57-61; from asst prof physics & asst geophysicist to assoc prof physics & assoc geophysicist, 66-74, actg dir, Observ, 69-71, actg provost, Hilo Col, 74, dir observ, 71-83, PROF PHYSICS & METEOROLOGIST, CLOUD PHYSICS OBSERV, UNIV HAWAII, HILO, 74-, DEAN COL ARTS & SCI, 84- *Concurrent Pos:* Prin investr, US Dept Interior Off Water Resources res grant, 71-76. *Mem:* Am Geophys Union; Am Meteorol Soc. *Res:* Precipitation physics; solar coronal emission; electrical effects and coalescence processes in warm rain; ice nuclei concentrations in relation to volcanic activity; microbarometric oscillations; distribution of high intensity rainfall. *Mailing Add:* Col Arts & Sci Univ Hawaii at Hilo Hilo HI 96720-4091

FULLERTON, DAVID STANLEY, b Norwalk, Ohio, Mar 30, 41. QUATERNARY GEOLOGY. *Educ:* Waynesburg Col, BS, 63; Yale Univ, MS, 65; Princeton Univ, AM, 68, PhD(geol & geophys sci), 71. *Prof Exp:* Instr geol, NY Univ, 69-71, asst prof, 71-72; geologist, Ohio State Geol Surv, 72-74; GEOLOGIST, US GEOL SURV, 74- *Res:* Quaternary glaciation and environments in the Montana Plains; holocene environmental changes in the Great Plains; quaternary stratigraphy and chronology in northeast United States. *Mailing Add:* 12022 W Louisiana Ave Denver CO 80228

FULLERTON, DWIGHT STORY, b Oakland, Calif, June 9, 43; m 64; c 2. MEDICINAL CHEMISTRY. *Educ:* Ore State Univ, BS(chem) & BS(pharm), 67; Univ Calif, Berkeley, PhD(org chem), 71. *Prof Exp:* NIH fel, Univ Va, 71-72; asst prof med chem, Col Pharm, Univ Minn, Minneapolis, 72-76; ASSOC PROF MED CHEM, SCH PHARM, ORE STATE UNIV, 76- *Mem:* Acad Pharmaceut Sci; Am Chem Soc; Sigma Xi; Am Pharmaceut Asn. *Res:* Structure-activity studies of digitalis analogs and of natural product tumor inhibitors; novel synthetic methods. *Mailing Add:* Dean Col Pharm Univ Utah Salt Lake City UT 84112

FULLERTON, H(ERBERT) P(ALMER), b Philadelphia, Pa, Aug 14, 12; m 34; c 2. MECHANICAL ENGINEERING. *Educ:* Univ Pa, BS, 33, ME, 45; Univ Buffalo, MA, 44. *Prof Exp:* With Gen Elec Co, 33-36 & Am Eng Co, 36-41; assoc prof eng, Univ Buffalo, 41-45; assoc prof mach design, Univ Va, 45-48; develop & design engr, Gen Elec Co, 48-62, proj engr, 62-68, consult engr, 68-75; RETIRED. *Concurrent Pos:* Adj prof, Drexel Inst, 48-64. *Mem:* Am Soc Mech Engrs; Am Soc Eng Educ; Sigma Xi. *Res:* Mathematical analysis of engineering and business systems. *Mailing Add:* Eight E Franklin St Media PA 19063

FULLERTON, LARRY WAYNE, b Fayetteville, Ark, Dec 11, 50; m 72; c 2. RADAR, COMMUNICATIONS. *Educ:* Univ Ark, Fayetteville, BSEE, 75. *Prof Exp:* Chief engr, Laser Commun, Inc, 75-77; sr engr, Telecommun Div, ITT, 77-79; sr systs engr, Computer Sci Corp, 79-80; systs engr, Tex Instruments, 80-81; chief engr, Mid-South Technol, 81-87; VPRES ENG, TIME DOMAIN SYSTS, INC, 87- *Res:* New technology of ultra wide band electromagnetics; radar; communications; security; awarded four patents. *Mailing Add:* 4825 University Sq Suite 3 Huntsville AL 35816

FULLERTON, RALPH O, b Benwood, WVa, May 22, 31; m 52; c 1. REMOTE SENSING, LATIN AMERICAN GEOGRAPHY. *Educ:* Univ Ky, BS, 60; Ind Univ, MS, 62, PhD(geog & admin), 71. *Prof Exp:* Teacher geog, Valley High Sch, Louisville, Ky, 60-63; from asst prof to assoc prof, 63-71, PROF GEOG & CHAIR DEPT, MID TENN STATE UNIV, 71- *Concurrent Pos:* Vis prof, US Army Command & Gen Staff Col, 70-75; mem, Nat Place Names Comn, 76-80; consult, Rutherford County & Murfreesboro Planning Comns, Tenn, 80-; mem, geog info systs comt, State Tenn, 83- *Mem:* Am Soc Photogram & Remote Sensing; Asn Am Geogrs. *Res:* Political geography of Latin America, especially Mexico; application of remote sensing to resource management problems of Tennessee; examination of the Usumacinta River of Mexico by satellite imagery. *Mailing Add:* 803 N Rutherford Blvd Murfreesboro TN 37130

FULLHART, LAWRENCE, JR, b Mystic, Iowa, Sept 14, 20; m 42; c 2. CHEMISTRY. *Educ:* William Jewell Col, AB, 42; Iowa State Col, PhD(org chem), 46. *Prof Exp:* Asst, William Jewell Col, 41-42; chemist, 46-57, tech rep, 57-62, develop specialist, 62-70, mkt res rep, 70-72, mkt res specialist, 72-78, mgr mkt res prog, E I Du Pont De Nemour & Co, Inc, 78-82; RETIRED. *Concurrent Pos:* Chem consult, 82- *Mem:* AAAS; Am Chem Soc; NY Acad Sci; Chem Mkt Res Asn. *Res:* Organic sulfur compounds; synthetic antimalarial and antitubercular compounds; synthetic vitamin D; nutritional biochemistry; amino acids and proteins; polymer dispersions; new product opportunities, marketing concepts and strategies. *Mailing Add:* 112 Neptune Dr Newark DE 19711-3011

FULLING, STEPHEN ALBERT, b Evansville, Ind, Apr 29, 45. THEORETICAL PHYSICS, MATHEMATICAL ANALYSIS. *Educ:* Harvard Col, AB, 67; Princeton Univ, MA, 69, PhD(physics), 72. *Prof Exp:* Fel & lectr physics, Univ Wis-Milwaukee, 72-74; res asst appl math, King's Col, Univ London, 74-76; from asst prof to assoc prof, 76-84, PROF MATH, TEX A&M UNIV, 84- *Concurrent Pos:* Prin investr, NSF res grant, 77-86. *Mem:* AAAS; Am Asn Physics Teachers; Am Phys Soc; Math Asn Am. *Res:* Quantum field theory; general relativity; spectral and asymptotic analysis of differential operators. *Mailing Add:* Dept Math Tex A&M Univ Col Sta TX 77843

FULLMAN, R(OBERT) L(OUIS), b Sewickley, Pa, Sept 13, 22; m 44; c 2. METALLURGY, MATERIALS SCIENCE. *Educ:* Yale Univ, BEng, 43, DEng(metall), 50. *Prof Exp:* Instr metall, New Haven YMCA Jr Col, 47-48; res assoc, 48-55, mgr mat & processes studies, 55-59, mgr metal studies, 60-63, mgr fuel cell studies, 64-65, mgr properties br, 65-68, mgr planning & resources, mat sci & eng, 69-72, metallurgist, Res & Develop Ctr, Gen Elec Co, 72-83; CONSULT, 83- *Concurrent Pos:* Vis lectr, Rensselaer Polytech Inst, 51-56, adj prof, 56-65; secy treas bd gov, Acta Metallurgica, 65- *Honors & Awards:* Geisler Mem Award, Am Soc Metals, 55. *Mem:* Fel Am Soc Metals; Am Inst Mining, Metall & Petrol Engrs. *Res:* Deformation of metals; interfacial energies in solids; crystal growth; origin of microstructures; recrystallization and grain growth; relationships between microstructure and properties of metals. *Mailing Add:* 206 Jamaica Way Punta Gorda FL 33950

FULLMER, GEORGE CLINTON, b Tucson, Ariz, Feb 17, 22; m 77; c 2. NUCLEAR ENGINEERING, PHYSICS. *Educ:* Univ Wash, BS, 47. *Prof Exp:* Physicist & supvr pile physics, Hanford Atomic Prod Oper, Gen Elec Co, 47-56, mgr oper physics, 56-65; mgr instrumentation, Pac Northwest Labs, Battelle Mem Inst, 65-66; mgr reactor physics, Hanford Atomic Prod Oper, Gen Elec Co & Douglas United Nuclear, 66-71; mgr core mgt eng, Nuclear Energy Eng, Gen Elec Co, 71-76, prog mgt core mgt serv, 76-82; RETIRED. *Mem:* Am Phys Soc; Am Nuclear Soc. *Res:* Pioneered development and application of techniques for dynamic flux distribution control; operator physics training; international monitoring of shutdown production reactors, and multi-BWR power reactor core management. *Mailing Add:* 1940 Noel Dr Los Altos CA 94024

FULLMER, HAROLD MILTON, b Gary, Ind, July 9, 18; m 42, 87; c 2. EXPERIMENTAL PATHOLOGY. *Educ:* Ind Univ, BS, 42, DDS, 44; Am Bd Oral Path, dipl. *Hon Degrees:* Dr, Univ Athens, Greece, 81. *Prof Exp:* Intern & resident, Charity Hosp, New Orleans, 46-48; assoc prof gen & oral path & head dept path, Sch Dent, Loyola Univ, La, 48-53; prin investr lab histol & path, Nat Inst Dent Res, 53-64, chief sect histochem, 65-70, chief exp path br, 68-70; prof path & dent, assoc dean Sch Dent & dir Inst Dent Res, Med Ctr, 70-87, EMER PROF, UNIV ALA, BIRMINGHAM, 87- *Concurrent Pos:* Mem dent study sect, NIH, 66-70 & dent caries adv comt, 75-79; assoc ed, Oral Surg, Oral Med & Oral Path, 68-76; ed-in-chief, J Oral Path, 72-89; assoc ed, J Cutaneous Path, 73-86; ed bd, Tissue Reactions, 76-88; trustee, Biol Stain Comn, 77-85; mem Grants & Allocations Comt, Am Fund Dent Health, 77-83; Vet Admin Med Res Career Develop Comt, 77-81; mem bd dirs, Nat Soc Med Res, 77-; consult ed, Gerodont, 81- *Honors & Awards:* Isaac Schour Award, Int Asn Dent Res, 73. *Mem:* Fel AAAS (secy, Sect R, 79-86); Histochem Soc; Am Asn Dent Res (vpres, 74-75, pres-elect, 75-76, pres, 76-77); Int Asn Dent Res (vpres, 74-75, pres-elect, 75-76, pres, 76-77); Int Asn Oral Pathologists (pres, 78-81); Am Acad Oral Path (vpres, 84-85, pres elect, 85-86 & pres, 86-87); hon mem Brit Soc Oral Pathologists. *Res:* Histochemistry and microchemistry of connective tissues, bones and teeth; mechanisms of staining reactions; cytological changes with development and age. *Mailing Add:* 3514 Bethune Dr Birmingham AL 35223

FULLMER, JUNE ZIMMERMAN, b Peoria, Ill, Dec 16, 20; m 53. HISTORY OF SCIENCE. *Educ:* Ill Inst Technol, BS, 43, MS, 45; Bryn Mawr Col, PhD, 48. *Prof Exp:* Instr chem, Hood Col, 45-46; asst prof, Pa Col Women, 49-52; res chemist metall eng, Carnegie Inst Technol, 53-54; vis asst prof chem, Newcomb Col, Tulane, 56-57, asst prof, 57-59, assoc prof & head dept, 59-64; vis assoc prof, Ohio Wesleyan Univ, 64-65; from adj assoc prof to prof hist, 66-84, EMER PROF HIST, OHIO STATE UNIV, 84- *Concurrent Pos:* Berliner fel, Oxford Univ, 48-49; Am Coun Learned Socs fel, 60-61; Guggenheim fel, 63-64; hon fel, Univ Wis, 63-64; consult, Battelle Mem Inst, 72-73; chmn, Hist Chem Div, Am Chem Soc, 70-71; mem comt hist geol, Nat Res Coun, 76-79 & 80-83. *Mem:* Fel AAAS; Hist Sci Soc. *Res:* Nineteenth century physical sciences; Humphry Davy; scientific biographies; social relations of science. *Mailing Add:* 781 Latham Ct Columbus OH 43214

FULLWOOD, RALPH ROY, b Hereford, Tex, Sept 16, 28; m 74; c 3. NUCLEAR ENGINEERING, PHYSICS. *Educ:* Tex Technol Col, BS, 52; Harvard Univ, AM, 55; Rensselaer Polytech Inst, PhD(nuclear eng), 65. *Prof Exp:* Physicist, Knolls Atomic Power Lab, 56-57; res assoc physics, Univ Pa, 57-60; staff mem nuclear eng, Rensselaer Polytech Inst, 60-65, mem fac, 65-66; mem staff physics, Los Alamos Sci Lab, 66-72; mgr, 72-79, ASST VPRES, SCI APPLN INC, 80- *Mem:* Inst Elec & Electronics Engrs; Am Nuclear Soc; Am Phys Soc; Inst Nuclear Mat Mgrs; Int Solar Energy Soc. *Res:* Probabilistic safety analysis applied to nuclear reactor and nuclear fuel cycle; cross section measurements; nuclear spectrum measurements; statistical modeling of failure. *Mailing Add:* Tech Adv Brookhaven Nat Labs Bldg 130 Upton NY 11973

FULMER, CHARLES V, b Council Bluffs, Iowa, Nov 15, 20; m 47; c 4. ENVIRONMENTAL GEOLOGY, GEOLOGICAL ENGINEERING. *Educ:* Univ Wash, BS & MS, 47; Univ Calif, Berkeley, PhD(geol paleont), 56. *Prof Exp:* Geologist, Standard Oil Co, Calif, 51-53, field geologist geol mapping, 53-56, stratigrapher, Wash & Ore, 56-59 & Alaska, 59-60, div stratigrapher, Wash, Ore & Alaska, 60-62; res engr appl res, Boeing Co, 62-64, space explor, 64-72; EARTH SCIENTIST, DEPT LAND USE MGT, KING COUNTY, WASH, 72- *Concurrent Pos:* Consult geologist, 85-91. *Mem:* Geol Soc Am; Am Asn Petrol Geol; Asn Eng Geol. *Res:* Tertiary stratigraphy of Washington and Oregon; space exploration; environmental analysis for land use management. *Mailing Add:* 6174 NE 187th Pl Seattle WA 98155

FULMER, GLENN ELTON, b Istanbul, Turkey, Aug 3, 28; US citizen; m 52; c 4. POLYMER PHYSICS, POLYMER CHEMISTRY. *Educ:* Oberlin Col, BA, 49; Johns Hopkins Univ, MA, 56. *Prof Exp:* Chemist, Nalco Chem Co, 49-50; sr engr, Martin Marietta Corp, 56-58; RES ASSOC, RES DIV, W R GRACE & CO, 58- *Concurrent Pos:* Adj prof, Antioch Col, 73-79. *Mem:* Soc Rheol; Am Soc Testing & Mat; Am Chem Soc. *Res:* Rheological research; failure of materials and development of superior materials; testing of nuclear fuel elements and nuclear fallout; evaluation of ion exchange resings. *Mailing Add:* 11505 Crows Nest Clarksville MD 21029

FULMER, HUGH SCOTT, b Syracuse, NY, June 18, 28; m 52; c 3. INTERNAL MEDICINE, PREVENTIVE MEDICINE. *Educ:* Syracuse Univ, AB, 48; State Univ NY, MD, 51; Harvard Univ, MPH, 61. *Prof Exp:* Instr internal med, Sch Med, State Univ NY Syracuse, 57; res assoc pub health & prev med, Cornell Univ, 58-60; from asst prof to prof community med, Col Med, Univ KY, 60-68; tech rep health progs, Peace Corps, Malaysia, 68-69; head, Dept Community & Family Med, Col Med, Univ Mass, 69-77, prof, Community & Family Med, 69-83, assoc dean, Clin Educ & Primary Care, 75-79, prof med, Dept Med & dir, sect Gen Med, 78-83; dir, Ambulatory & Community Servs, 83-88, DIR, COMMUNITY ORIENTED PRIMARY CARE, CARNEY HOSP, BOSTON, MASS, 88-; PROF CLIN MED, SOCIOL MED SCI, COMMUNITY MED & PUB HEALTH, SCH MED & PUB HEALTH, BOSTON UNIV, 83- *Concurrent Pos:* Bd Regents, Am Col Prev Med, 88- *Mem:* AMA; Am Pub Health Asn; AAAS; Asn Teachers Prev Med (pres, 75-76); Am Inst Community Health (vpres, 78-80); Am Col Prev Med. *Res:* Epidemiology of non-infectious disease; medical care. *Mailing Add:* 2100 Dorchester Ave Boston MA 02124

FULMER, RICHARD W, b Istanbul, Turkey, Jan 23, 30; US citizen; m 47; c 4. AGRICULTURAL & FOOD CHEMISTRY, BIOCHEMISTRY. *Educ:* De Pauw Univ, BA, 52; Univ Ill, PhD(chem), 55. *Prof Exp:* Res scientist, Gen Mills Chem, Inc, 55-64; res mgr chem, 64-67, asst vpres res, 67-85, VPRES RES, CARGILL INC, 85- *Mem:* Am Chem Soc; Am Oil Chemists Soc; Agr Res Inst. *Res:* Polymers; nitrogen compounds; biological research management in polymers, biochemicals, foods, feeds and seeds. *Mailing Add:* 36045 Fannon SE Delano MN 55328

FULMOR, WILLIAM, organic chemistry, spectroscopy; deceased, see previous edition for last biography

FULOP, MILFORD, b New York, NY, Nov 7, 27; m 57; c 2. INTERNAL MEDICINE. *Educ:* Columbia Univ, AB, 46, MD, 49. *Prof Exp:* From asst prof to assoc prof, 56-68, actg chmn dept, 75-80, PROF MED, ALBERT EINSTEIN COL MED, 68-; DIR MED SERV, BRONX MUNIC HOSP CTR, 70- *Mem:* Asn Am Physicians; Am Fedn Clin Res; Am Col Physicians. *Res:* Renal diseases; metabolic disorders. *Mailing Add:* Dept Med Albert Einstein Col Med Bronx NY 10461

FULP, RONALD OWEN, b Trinity, NC, June 29, 36; m 59; c 3. ALGEBRA. *Educ:* Wake Forest Col, BS, 58; Univ NC, Chapel Hill, MA, 61; Auburn Univ, PhD(math), 65. *Prof Exp:* Instr math, Univ NC, 58-61; instr, Auburn Univ, 61-62; asst prof, Ga State Col, 63-65; from asst prof to assoc prof, Univ Houston, 65-69; assoc prof, 69-76, PROF MATH, NC STATE UNIV, 76- *Res:* Geometry of fiber bundles and applications to gauge theory, gauge theory as used in both particle physics and in gravitational theories; Hamiltonian mechanics and symplectic geometry. *Mailing Add:* Dept Math NC State Univ Raleigh Main Campus Box 8205 Raleigh NC 27695-8205

FULTON, CHANDLER MONTGOMERY, b Cleveland, Ohio, Apr 17, 34; m 55, 81; c 3. DEVELOPMENTAL BIOLOGY, CELL BIOLOGY. *Educ:* Brown Univ, BA, 56; Rockefeller Univ, PhD(biol), 60. *Prof Exp:* From instr to assoc prof, 60-76, chmn dept, 81-84, PROF BIOL, BRANDEIS UNIV, 76-,. *Concurrent Pos:* NSF grants, 60-; mem, NSF Adv Panel Develop Biol, 74-77; NIH grant, 78- *Mem:* Soc Develop Biol; Am Soc Cell Biol; Soc Protozool. *Res:* Analysis of molecular and cellular events of cell differentiation and cell organelle morphogenesis using the amebo-flagellate Naegleria gruberi; gene expression. *Mailing Add:* Dept Biol Brandeis Univ Waltham MA 02254

FULTON, GEORGE P(EARMAN), b Milton, Mass, June 3, 14; m 42, 70; c 4. PHYSIOLOGY. *Educ:* Boston Univ, BS, 36, MA, 38, PhD(physiol), 41. *Prof Exp:* Instr biol, Boston Univ, 41-42; res physiologist, Arthur D Little, Inc, Mass, 46-47; from asst prof to prof, Boston Univ, 47-60, head dept, 56-74, Shields Warren prof biol, 60-76; asst dir Health Affairs, SC Comn Higher Educ, 77-84; EMER PROF BIOL, BOSTON UNIV, 76- *Concurrent Pos:* Vis prof, Sch Med, Stanford Univ, 58-59; mem staff, Childrens Cancer Res Found, Boston, 64-72; consult biophys lab, Aeromed Lab, Wright-Patterson AFB; consult div nursing, USPHS; mem adv bd, Sea Farms Found; mem sci adv bd, New Eng Aquarium; founding ed, Microvascular Res; mem, Marine Adv Bd, SC Wildlife & Marine Resources Comn & SC Nutrit Coun; founding dir, SC Hall Sci & Technol, 78-87. *Mem:* Fel Am Acad Arts & Sci; Am Physiol Soc; Soc Exp Biol & Med; Am Asn Anat; Gerontol Soc; Am Microcirculation Soc. *Res:* Blood capillary circulation; innervation of blood vessels; skin temperature and heat tolerance work; topical effects of irritant chemicals on skin; thromboembolism; vascular effects of irradiation; petechial formation; self-help among senior citizens. *Mailing Add:* 4740 Cedar Springs Rd Columbia SC 29206

FULTON, J(AMES) C(ALVIN), metallurgy; deceased, see previous edition for last biography

FULTON, JAMES W(ILLIAM), b Oklahoma City, Okla, Dec 2, 28; m 62; c 2. CHEMICAL ENGINEERING. *Educ:* Harvard Univ, BA, 51; Univ Okla, MS, 61, PhD(chem eng), 64. *Prof Exp:* Serv engr oil well treating, Dowell, Inc, 55-56; process engr, Monsanto Chem Co, 57; instr chem eng, Univ Okla, 58-59; asst prof, Okla State Univ, 63-66; group supvr, 66-77, DESIGN ENGR, MONSANTO CO, 77- *Mem:* Am Inst Chem Engrs. *Res:* Reaction kinetics and catalysis; heterogeneous catalysts; chemical reactor design; biochemical process design; heat and mass transfer in chemical reactors; distillation; applied statistics; economic evaluations. *Mailing Add:* Dist Mgr 2574 Woodberry Dr Winston-Salem NC 27106

FULTON, JOHN DAVID, b Norton, Va, Dec 4, 37; m 59; c 1. MATHEMATICS. *Educ:* NC State Univ, BS, 60, MS, 63, PhD(math), 65. *Prof Exp:* Instr math, NC State Univ, 65-66; res assoc, Math Div, Oak Ridge Nat Lab, 66-67; from asst prof to assoc prof, 67-76, PROF MATH, CLEMSON UNIV, 76-, HEAD DEPT MATH SCI, 78- *Mem:* Am Math Soc; Math Asn Am; Soc Indust & Appl Math. *Res:* Combinatorial mathematics. *Mailing Add:* Col Arts & Sci Univ WFla Pensacola FL 32514

FULTON, JOSEPH PATTON, b Princeton, Ind, July 1, 17; m 45; c 2. PLANT PATHOLOGY. *Educ:* Wabash Col, AB, 39; Univ Ill, MA, 41, PhD(plant path), 47. *Prof Exp:* Asst bot, Univ Ill, 39-42 & 46-47; from asst prof to assoc prof, 47-54, PROF PLANT PATH, UNIV ARK, FAYETTEVILLE, 54- *Concurrent Pos:* Head dept plant path, Univ Ark, Fayetteville, 59-64. *Mem:* AAAS; Am Phytopath Soc (pres, 71). *Res:* Diseases of small fruit; virus diseases; beetle vector of virus diseases. *Mailing Add:* Box 645 Crystal City TX 78839

FULTON, PAUL F(RANKLIN), b Irwin, Pa, Mar 28, 16; m 48. PETROLEUM ENGINEERING. *Educ:* Univ Pittsburgh, BS, 38, MS, 51; Pa State Univ, PhD, 64. *Prof Exp:* From instr to assoc prof, 47-56, assoc chmn dept, 65-80, PROF PETROL ENG, UNIV PITTSBURGH, 56-, UNDERGRAD PROG COORDR, DEPT CHEM & PETROL ENG & DIR PETROL ENG CURRIC, 77- *Concurrent Pos:* Res engr, Gulf Res & Develop Co, 49-59; NSF sci fac fel, 60-61. *Mem:* Am Inst Mining, Metall & Petrol Engrs; Sigma Xi. *Res:* Petroleum reservoir engineering; core analysis research; thermal methods of oil recovery; role of wettability in oil recovery. *Mailing Add:* 145 Woodshire Rd Pittsburgh PA 15215

FULTON, ROBERT E(ARLE), b Dothan, Ala, Jan 23, 31; m 53; c 3. STRUCTURAL ENGINEERING, APPLIED MECHANICS. *Educ:* Auburn Univ, BSc, 53; Univ Ill, Urbana, MS, 58, PhD(civil eng), 60. *Prof Exp:* Struct designer, Chicago Bridge & Iron Co, Ala, 53-54; asst prof civil eng, Univ Ill, Urbana, 60-62; aerospace engr, 62-65, head, Automated Methods Sect, 65-76, MGR, INTEGRATED PROG AEROSPACE VEHICLE DESIGN PROJ, LANGLEY RES CTR, NASA, 76- *Concurrent Pos:* Instr, Va Polytech Inst, 62-65 & Univ Va, 62-68; adj prof lectr, George Washington Univ, 68-; adj prof, Old Dominion Univ, 70 & NC State Univ, 71. *Mem:* Am Soc Mech Engrs; Am Inst Aeronaut & Astronaut; Int Asn Shell Struct; Am Soc Civil Engrs; Nat Comput Graphics Asn. *Res:* Structural mechanics including shell structures, stability theory, vibrations and numerical methods; use of computers for analysis and design of complex structures occurring in aerospace and civil engineering; author of over one hundred technical publications and presentations. *Mailing Add:* Geo W Woodruff Sch Mech Eng Ga Inst Technol Atlanta GA 30332-0405

FULTON, ROBERT JOHN, b Birtle, Man, May 27, 37; m 63; c 3. ENGINEERING. *Educ:* Univ Man, BSc, 59; Northwestern Univ, PhD(geol), 63. *Prof Exp:* Geologist, 63-67, res scientist quaternary geol, 67-73, head, Regional Surficial Geol Mapping, 73-87, HEAD, WESTERN REGION SECT, QUATERNARY GEOL SUBDIV, 87- *Mem:* Geol Asn Can; Geol Soc Am; Can Asn Quaternary Res; Am Asn Quaternary Res. *Res:* Mapping of Quaternary deposits; Quaternary history; delta and lake sedimentation. *Mailing Add:* Geol Surv Can 601 Booth St Ottawa ON K1A 0E8 Can

FULTON, ROBERT LESTER, b Weymouth, Mass, May 13, 35; m 65. CHEMICAL PHYSICS. *Educ:* Brown Univ, ScB, 57; Harvard Univ, AM, 60, PhD(chem physics), 64. *Prof Exp:* Res fel chem, Harvard Univ, 64-65; from asst prof to assoc prof, 65-75, PROF CHEM, FLA STATE UNIV, 75- *Mem:* Am Phys Soc. *Res:* Relaxation phenomenon; light scattering by macromolecules; macroscopic quantum electrodynamics; crystal optics; theory of dielectrics. *Mailing Add:* Dept Chem Fla State Univ Tallahassee FL 32306

FULTON, ROBERT WATT, b Sistersville, WVa, Jan 29, 14; m 74. PLANT PATHOLOGY. *Educ:* Wabash Col, AB, 35; Univ Wis, PhD, 40. *Prof Exp:* Instr bot, Wabash Col, 35-37; from asst to asst prof hort, Univ Wis-Madison, 37-55, from asst prof to prof plant path, 55-84; RETIRED. *Mem:* AAAS; fel Am Phytopath Soc. *Res:* Viruses and virus diseases of plants. *Mailing Add:* 5060 Sunrise Trail Middleton WI 53562

FULTON, ROBERT WESLEY, b Blackwell, Okla, Oct 19, 42; m 75; c 2. VETERINARY VIROLOGY, VETERINARY MICROBIOLOGY. *Educ:* Okla State Univ, BS, 64, DVM, 66; Wash State Univ, MS, 72; Univ Mo-Columbia, PhD(microbiol), 75; Am Col Vet Microbiologists, dipl, 76. *Prof Exp:* Base vet, USAF Vet Serv, 66-70; resident vet clins, Wash State Univ, 70-72; res assoc virol, Univ Mo-Columbia, 72-75; prof vet virol, Sch Vet Med, La State Univ, Baton Rouge, 75-82; PROF & HEAD, DEPT PARASITOL & PUB HEALTH, OKLA STATE UNIV, STILLWATER, 82- *Mem:* Sigma Xi; Am Vet Med Asn; Am Soc Microbiol. *Res:* Interferon; host defense mechanisms; bovine infectious diseases. *Mailing Add:* Dept Vet Parasitol Microbiol & Pub Health Col Vet Med Okla State Univ Stillwater OK 74078

FULTON, THOMAS, b Budapest, Hungary, Nov 19, 27; nat US; m 52; c 2. THEORETICAL PHYSICS. *Educ:* Harvard Univ, BA, 50, MA, 51, PhD, 54. *Prof Exp:* Mem & Jewett fel, Inst Adv Study, 54-55, NSF fel, 55-56; from asst prof physics to assoc prof, 56-64, PROF PHYSICS, JOHNS HOPKINS UNIV, 64- *Concurrent Pos:* Consult, Brookhaven Lab, 54, Res Inst Advan Study, Inc, 54-55; vis scientist, Univ Calif, Berkeley, 59, Stanford Univ, 67, Argonne Lab, 68, Brandeis Inst, 62, Fulbright sr res scholar & Guggenheim fel, Inst Theoret Physics, Vienna, 64-65; Aspen Inst, 66-67, Aspen Ctr Phys, 68, 71, 85; assoc ed, J Math Physics, 68-71; Europ Orgn Nuclear Res, 69-70, 76-77, 83-84; Int Ctr Theoret Physics, Trieste, 86, ITP, Santa Barbara, 88. *Mem:* Am Phys Soc; Fedn Am Scientists. *Res:* Quantum theory of fields; elementary particle physics; scattering theory; QED applied to atomic theory. *Mailing Add:* Dept Physics Johns Hopkins Univ Baltimore MD 21218

FULTON, WINSTON CORDELL, b Minto, NB, Nov 19, 43. SYSTEMS ANALYSIS. *Educ:* Univ NB, BSc, 72; Mich State Univ, MS, 75, PhD(entom), 78. *Prof Exp:* Asst prof plant path, Mich State Univ, 78-80; ASST PROF ENTOM, PURDUE UNIV, 80- *Mem:* Entom Soc Am; Am Phytopath Soc; AAAS. *Res:* Modelling the effects of Diabrotica damage to roots on grain yield in maize with emphasis on using plant water relations parameters as measures of damage. *Mailing Add:* 1512 Summit Dr West Lafayette IN 47907

FULTYN, ROBERT VICTOR, b Chicago, Ill, Nov 8, 33; m 55; c 1. COMPUTER SCIENCES. *Educ:* Northwestern Univ, BS, 54, MS, 55; Harvard Univ, MS, 58, ScD(indust hyg), 61. *Prof Exp:* STAFF MEM, LOS ALAMOS SCI LAB, UNIV CALIF, 63- *Mem:* Sigma Xi. *Res:* Industrial research toward improving manufacturing processes. *Mailing Add:* 38 Northwood Dr Nashua NH 03063

FULTZ, BRENT T, b Troy, NY, Feb 24, 55; m 84; c 2. STATISTICAL KINETIC THEORY, MOSSBAUER & ELECTRON SPECTROSCOPIES. *Educ:* Mass Inst Technol, BS, 75; Univ Calif, Berkeley, MS, 78, PhD(eng sci), 82. *Prof Exp:* Staff scientist, Lawrence Berkeley Lab, 82-85; asst prof, 85-90, ASSOC PROF MAT SCI, CALIF INST TECHNOL, 91- *Concurrent Pos:* Consult, Contact Prod Pomona, 86-; prin investr, Dept Energy, 86-; IBM fac develop award, 86, 87; NSF presidential young investr, 88. *Mem:* Metall Soc; Electron Micros Soc Am; Mat Res Soc; Am Soc Metals. *Res:* Materials science, physical science: local arrangements of atoms and how they move in nonequilibrium materials. *Mailing Add:* Eng & Appl Sci Calif Inst Technol Pasadena CA 91125

FULTZ, DAVE, b Chicago, Ill, Aug 12, 21; m 46; c 3. DYNAMIC METEOROLOGY, FLUID MECHANICS. *Educ:* Univ Chicago, SB, 41, PhD(meteorol), 47. *Prof Exp:* Asst, US Weather Bur, Chicago Sta, 42; res assoc, Univ Chicago & PR, 42-44; opers analyst, US Air Force, 45; res assoc, 46-47, instr meteorol, 47-48, from asst prof to assoc prof, 48-60, PROF METEOROL, UNIV CHICAGO, 60-, HEAD HYDRODYN LABS, 46- *Concurrent Pos:* Guggenheim fel, 50-51; sr fel, NSF, 57-58; mem sci adv bd, USAF, 59-63; Nat Comt Fluid Mech Films, 62-71; res grants adv comt, Air Pollution Control Off, 70-71. *Honors & Awards:* Meisinger Award, Am Meteorol Soc, 51, Rossby Res Medal, 67. *Mem:* Nat Acad Sci; fel Am Meteorol Soc; fel Am Geophys Union; fel AAAS. *Res:* Geophysical experimental fluid mechanics; convectional flows in stationary and rotating systems; synoptic study of the upper air, cloud forms; upper air trajectories in weather forecasting. *Mailing Add:* Dept Geophys Sci Univ Chicago Chicago IL 60637

FULTZ, R PAUL, b Staunton, Va, Oct 5, 23; m 46; c 4. PUBLIC HEALTH, PEDODONTICS. *Educ:* Med Col Va, DDS, 51; Univ Tex, MPH, 70. *Prof Exp:* Pvt pract pedodont, 53-69; assoc prof prev med, Univ Tex Dent Br Houston, 70-71; prof community health & prev dent & chmn dept, Baylor Col Dent, 71-; RETIRED. *Concurrent Pos:* Consult, Region VI Head Start Prog, 72-, Region VI Nat Health Serv Corps, 72- & Tex State Dept Health, 75-; pres, Morcheron Remounts Inc, 77- *Mem:* Am Asn Dent Schs; Am Dent Asn; Am Pub Health Asn; Am Asn Pub Health Dentists. *Res:* Preventive dentistry; dental services delivery. *Mailing Add:* Rte 1 Box 1 Brownsboro TX 75756

FUNCK, DENNIS LIGHT, b Palmyra, Pa, Nov 23, 26; m 49; c 2. POLYMER CHEMISTRY. *Educ:* Lebanon Valley Col, BS, 49; Univ Del, MS, 50, PhD(org chem), 52. *Prof Exp:* Res Chemist process develop, E I du Pont de Nemours & Co, 52-62, sr res chemist prod develop, 62-73, res assoc, 73-83, sr res assoc, 84-86; RETIRED. *Mem:* Am Chem Soc; Sigma Xi; Soc Plastics Eng. *Res:* Chromic acid oxidation of secondary and tertiary alcohols; free radical chemistry; polymer process and product development, synthesis and oxidation; acoustics; radiation; applications scouting. *Mailing Add:* 104 Monticello Rd Fairfax Wilmington DE 19803

FUNCK, LARRY LEHMAN, b Hershey, Pa, Dec 25, 42; m 64; c 2. INORGANIC CHEMISTRY. *Educ:* Lebanon Valley Col, BS, 64; Lehigh Univ, PhD(inorg chem), 69. *Prof Exp:* Asst prof, 69-77, assoc prof, 77-80, PROF CHEM & SCI COORDR, WHEATON COL, ILL, 80- *Mem:* Am Chem Soc. *Res:* Visible and ultraviolet spectra of transition metal complexes; solution studies of complex equilibria. *Mailing Add:* Dept Chem Wheaton Col Wheaton IL 60187-4295

FUNDERBURGH, JAMES LOUIS, b Manhattan, Kans, June 30, 45; m 65; c 3. CORNEAL BIOCHEMISTRY, CARBOHYDRATE BIOCHEMISTRY. *Educ:* Univ Tex, Austin, BA, 65; Univ Minn, MS, 69, PhD(biochem), 73. *Prof Exp:* Assoc, Dept Path, Univ Geneva, Switz, 73-75; sr fel, Ophthal Dept, Univ Wash, 73-78; res assoc, Swed Hosp Med Ctr, Seattle, 79-81; ASST SCIENTIST, DIV BIOL, KANS STATE UNIV, MANHATTAN, 81- *Mem:* Asn Res Vision & Ophthal. *Res:* Cellular-extracellular matrix interaction; biochemistry of proteoglycans; corneal biochemistry; early events in herpes virus infection; biochemistry of translation; arterial biochemistry. *Mailing Add:* Dept Biol Kans State Univ Manhattan KS 66506

FUNG, ADRIAN K, b Liuchow, Kwangsi, Chin, Dec 25, 36; m 66; c 3. ELECTROMAGNETIC WAVES. *Educ:* Nat Cheng Kung Univ, BSc, 58; Brown Univ, MSc, 61; Univ Kans, PhD(elec eng), 65. *Prof Exp:* Prof elec eng, Univ Kans, 72-; PROF ELEC ENG, UNIV TEX, ARLINGTON. *Concurrent Pos:* NSF grants, 75-77 & 80-82; US Army Res Off grants, 77-79 & 80-; mem, US Comn F, Int Sci Radio Union. *Honors & Awards:* Distinguished Achievement Award, Inst Elec & Electronics Engrs, 89. *Mem:* Sigma Xi; fel Inst Elec & Electronics Engrs. *Res:* Remote sensing of sea, land and vegetations; wave scattering emission and propagation from inhomogeneous media. *Mailing Add:* Dept Elec Eng Univ Tex Arlington TX 76019

FUNG, DANIEL YEE CHAK, b Hong Kong, May 15, 42; m 68; c 1. MICROBIOLOGY, FOOD TECHNOLOGY. *Educ:* Int Christian Univ, Tokyo, BA, 65; Univ NC, Chapel Hill, MSPH, 67; Iowa State Univ, PhD(food technol), 69. *Prof Exp:* Asst prof & dir microbiol, Pa State Univ, 69-78; from asst prof to assoc prof, 78-85, PROF, DEPT ANIMAL SCI INDUST, KANS STATE UNIV, 85-, CHMN, FOOD SCI GRAD PROG, 79- *Concurrent Pos:* NSF instnl grant & Agr Exp Sta res grant, Pa State Univ, 70-78, Kans State Univ, 78-; Am Pub Health Asn res grants, 74-76; mem Local Course Improv Proj, NSF,81-83; pres, Kans City Sect, Inst Food Technologists, 83; USDA Fel grant, 84-88. *Mem:* Am Soc Microbiol; Inst Food Technologists; Int Asn Milk Food & Environ Sanitarians; Nat Environ Health Asn; Sigma Xi; Am

Meat Sci Asn; fel Am Acad Microbiol; fel Am Acad Microbiol Res. *Res:* Development of miniaturized microbiological methods for diagnostic microbiology; enterotoxigenesis of Staphylococcus aureus; detection of Salmonella; effects of modern food processing on bacteria survival in foods; silage microbiology; effects of food additives on microbial survival in foods. *Mailing Add:* Kans State Univ Call Hall Manhattan KS 66506

FUNG, HENRY C, JR, b San Francisco, Calif, Feb 5, 39; m 61; c 3. IMMUNOLOGY. *Educ:* Univ Calif, BA, 59; San Francisco State Col, MA, 62; Wash State Univ, PhD(bact), 66. *Prof Exp:* From asst prof to assoc prof, 66-75, PROF MICROBIOL, CALIF STATE UNIV, LONG BEACH, 75- *Concurrent Pos:* Collab scientist, City of Hope Nat Med Ctr; consult, Concept Media, 75; chmn, Am Soc Microbiol-BET comt on Undergrad & Grad Educ, 82-90; Am Pub Health Asn rep to Am Bd Med Microbiol, 87-92. *Mem:* AAAS; Am Pub Health Asn; Am Soc Microbiol; Tissue Cult Asn; NY Acad Sci; Sigma Xi. *Res:* Cancer immunology; cellular immunity; antigenic mosaic of microorganisms influenced by their environment; immunobiology. *Mailing Add:* Dept Microbiol Calif State Univ Long Beach CA 90840

FUNG, HO-LEUNG, b Hong Kong, Nov 17, 43; m 70; c 2. PHARMACOLOGY. *Educ:* Victorian Col Pharm, Australia, Cert, 66; Univ Kans, PhD(pharmaceut), 70. *Prof Exp:* From asst prof to assoc prof, 70-80, CHMN DEPT, STATE UNIV NY, BUFFALO, 78-, PROF PHARMACEUT, 80- *Honors & Awards:* Res Achievement Award, Am Asn Pharmaceut Scientists, 88; Merit Award, Nat Heart, Lung & Blood Inst, NIH, 88. *Mem:* Fel, AAAS; fel Acad Pharmaceut Sci; fel Am Asn Pharmaceut Scientists. *Res:* Drug delivery in the human body; pharmacokinetics and pharmacology of organic nitrates; cardiovascular pharmacology. *Mailing Add:* Dept Pharmaceut Sch Pharm State Univ NY Cooke Hall Buffalo NY 14260

FUNG, HONPONG, b Hong Kong, July 29, 20; nat US; m 47; c 2. CIVIL ENGINEERING. *Educ:* Lingnan Univ, BSc, 42; Iowa State Univ, MSc, 48, PhD, 56. *Prof Exp:* Asst mechs surv, Nat Chaio-tung Univ, China, 42-46; res assoc, Eng Exp Sta, Iowa State Univ, 51-55, asst prof res, 55-63, asst prof hwy transp, 55-76, ASSOC PROF RES, BITUMINOUS RES LAB, IOWA STATE UNIV, 63-, ASSOC PROF CIVIL ENG, 76- *Concurrent Pos:* Mem, Hwy Res Bd, 46- *Mem:* Asn Asphalt Paving Technol; Am Road Builders Asn; Sigma Xi. *Res:* Planning of highway transportation; bituminous paving materials; technology of asphalt pavement construction and quality control of the bituminous paving materials. *Mailing Add:* Box 1231 ISU Sta Iowa State Univ Ames IA 50011

FUNG, K Y, b Dec 20, 48; m; c 2. AERODYNAMICS, COMPUTATIONAL FLUID MECHANICS. *Educ:* Nat Taiwan Univ, BS, 72; Cornell Univ, PhD(aerospace eng), 76. *Prof Exp:* Res assoc, 76-79, asst prof, 79-85, ASSOC PROF AEROSPACE & MECH ENG, UNIV ARIZ, TUCSON, 85- *Concurrent Pos:* Vis scientist aerospace & mech eng, Ger Res Inst Space & Aerodyn, Gottingen, WGer, 81; Alexander von Humboldt Found fel, W Ger, 81; consult, Univ Ariz & Zonatech, 83- *Mem:* Am Inst Aeronaut & Astronaut. *Res:* Steady and unsteady aerodynamics; efficient computational methods for the design and analysis of flow over aircraft. *Mailing Add:* Dept Aerospace & Mech Eng Col Eng & Mines Univ Ariz Tucson AZ 85721

FUNG, LESLIE WO-MEI, b Nanking, China, Sept 23, 46; US citizen. MOLECULAR BIOPHYSICS. *Educ:* Univ Calif, Berkeley, BSc, 68; Mass Inst Technol, PhD(phys chem), 71. *Prof Exp:* Res fel phys chem, Rice Univ, 71-72; lectr chem, Chinese Univ Hong Kong, 72-73; res assoc biophys, Univ Pittsburgh, 73-77; assoc prof chem, Wayne State Univ, 77-82; AT DEPT CHEM, LOYOLA UNIV, CHICAGO. *Honors & Awards:* NIH Nat Res Serv Award, 75-77. *Mem:* AAAS; Am Biophys Soc; Am Chem Soc; Asn Women Sci; Sigma Xi. *Res:* Studies of structural-functional relationship of biological systems. *Mailing Add:* Dept Chem Flanner Hall Loyola Univ Chicago 6525 N Sheridan Rd Chicago IL 60626

FUNG, SHUN CHONG, b Kwongtung, China, Jan 28, 43; US citizen; m 73; c 2. CATALYSIS, REACTION KINETICS. *Educ:* Univ Calif, Berkeley, BS, 65; Univ Ill, Urbana, MS, 67, PhD(chem eng), 69. *Prof Exp:* ENG ASSOC, EXXON RES & ENG CO, 69- *Mem:* Am Chem Soc; NAm Catalysis Soc. *Res:* Designing and characterizing new catalysts for petroleum and chemical processes; the understanding of the chemistry of the catalytic reaction via kinetics and mechanism studies of the reaction of interest. *Mailing Add:* Clinton Twp Rte 22E Annandale NJ 08801

FUNG, STEVEN, b New Westminster, BC, Oct 31, 51; Can citizen. SYNTHETIC ORGANIC CHEMISTRY, MEDICAL CHEMISTRY. *Educ:* Univ BC, BSc, 73; Univ Alta, PhD(org chem), 78. *Prof Exp:* Fel, Zoecon Corp, 78-80; sr sci, Ayerst Labs, 80-; AT APPL BIOSYSTS. *Mem:* Chem Inst Can; Am Chem Soc. *Res:* Synthetic organic and medicinal chemical research. *Mailing Add:* Appl Biosysts 850 Lincoln Ctr Dr Foster City CA 94404-1155

FUNG, SUI-AN, b Chekiang, China, Dec 18, 22; c 5. MECHANICAL ENGINEERING. *Educ:* Nat Cent Univ NanKing, BS, 48; Univ Rochester, MS, 56; Cornell Univ, PhD, 65. *Prof Exp:* Mech engr, supvr & mgr, Signal Equip Co, 48-52; asst prof mech eng, Univ Evansville, 56-59; assoc prof, Tex Tech Univ, 59-62; mem staff, Space Div, Rockwell Int, 65-; prof, South Bay Univ, 79-; RETIRED. *Concurrent Pos:* Consult engr, RCA Whirlpool Refrig, Inc, 59; NSF scholar, 66. *Mem:* AAAS; Am Soc Mech Engrs; Am Soc Eng Educ; Am Inst Aeronaut & Astronaut. *Res:* Thermodynamic properties; heat transfer; fluid mechanics; applied mathematics and low-g propellant behavior. *Mailing Add:* 2822 Castle Heights Los Angeles CA 90034

FUNG, YUAN-CHENG B(ERTAAM), b China, Sept 15, 19; m 49; c 2. BIOENGINEERING, STRUCTURAL DYNAMICS. *Educ:* Nat Cent Univ, China, BS, 41, MS, 43; Calif Inst Technol, PhD(aeronaut), 48. *Prof Exp:* Instr mech, Nat Cent Univ, China, 41-43, Bur Aeronaut Res, 43-45; from asst prof to prof aeronaut, Calif Inst Technol, 50-66; PROF BIOENG & APPL MECH, UNIV CALIF, SAN DIEGO, 66- *Concurrent Pos:* Ed, J Biomech Eng, 79-83 & J Biorheology, 78-88. *Honors & Awards:* Landis Award for Res, Microcirculatory Soc, 75; Von Karman Medal, Am Soc Civil Engrs, 76; Lissner Award, Am Soc Mech Engrs, 78, W R Warner Reed Medal, 84; Poiseville Medal, Int Soc Biorheology, 86. *Mem:* Nat Acad Eng; Inst Med-Nat Acad Sci; Am Soc Mech Engrs; Am Physiol Soc; Int Soc Biorheology (vpres, 74-); Microcirculatory Soc; Am Inst Aeronaut & Astronaut. *Res:* Physiology; circulation; respiration; elasticity; aeroelasticity; dynamics; author or coauthor of over 250 scientific papers and five books. *Mailing Add:* Dept Ames/Bioeng Univ Calif San Diego R-012 La Jolla CA 92093-0412

FUNK, CYRIL REED, JR, b Richmond, Utah, Sept 20, 28; m 51; c 3. PLANT BREEDING. *Educ:* Utah State Univ, BS, 52, MS, 55; Rutgers Univ, PhD(plant breeding), 62. *Prof Exp:* Instr farm crops, 56-61, from asst res prof to res prof, 61-81, PROF II TURFGRASS BREEDING, RUTGERS UNIV, 81- *Mem:* Am Soc Agron; Am Genetic Asn; AAAS; Sigma Xi; fel Crop Sci Soc Am. *Res:* Field corn breeding, production and management; ecology of crop mixtures; seed quality; turfgrass breeding and genetics; apomixis; author or co author of many scientific publications. *Mailing Add:* Crop Sci Dept Cook Col Rutgers Univ New Brunswick NJ 08903

FUNK, DAVID CROZIER, b Wilmington, Del, Sept 24, 22; m 47; c 4. CARDIOLOGY. *Educ:* Emory Univ, BS, 43, MD, 46. *Prof Exp:* Intern, Hosps, 46-47, resident, 49-52, from clin instr to clin assoc prof, Col Med, 52-70, ASSOC PROF INTERNAL MED, COL MED, UNIV IOWA, 70- *Concurrent Pos:* Clin investr, Vet Admin, 57-60; trainee, Cardiovascular Res Inst, Univ Calif, San Francisco, 59-60; chief cardiovasc lab, Vet Admin Hosp, Iowa City, 57-; fel coun cardiol, Am Heart Asn. *Mem:* AAAS; fel Am Col Physicians; fel Am Col Cardiol. *Res:* Cardiovascular and pulmonary physiology. *Mailing Add:* Vet Admin Hosp Iowa City IA 52240

FUNK, DAVID TRUMAN, b Greensburg, Ind, Oct 17, 29; m 56; c 1. FOREST GENETICS. *Educ:* Purdue Univ, BS, 51, MS, 56; Mich State Univ, PhD, 71. *Prof Exp:* Res forester, Cent States Forest Exp Sta, 56-65, plant geneticist, 65-70, prin plant geneticist, NCent Forest Exp Sta, 70-79, ASST DIR, NORTHEAST FOREST EXP STA, US FOREST SERV, 79- *Concurrent Pos:* Chmn, Cent States Forest Tree Improv Conf, 68-71; co-chmn working party breeding nut species, Int Union Forestry Res Orgns, 73-81. *Mem:* AAAS; Soc Am Foresters; Fedn Am Scientists. *Res:* Strip-mine reclamation and reforestation research; population genetics; hardwood tree breeding. *Mailing Add:* 123 Mill Rd Durham NH 03824

FUNK, EMERSON GORNFLOW, JR, b Highland Park, Mich, Jan 27, 31; m 53; c 3. NUCLEAR PHYSICS, ENVIRONMENTAL RADIOACTIVITY. *Educ:* Wayne State Univ, AB, 53; Univ Mich, MA, 54, PhD(physics), 58. *Prof Exp:* Instr physics, Univ Mich, 58; from asst prof to assoc prof, 58-71, PROF PHYSICS, UNIV NOTRE DAME, 71- *Mem:* Am Phys Soc; Am Asn Physics Teachers; Health Physics Soc; AAAS. *Res:* environmental radioactivity. *Mailing Add:* Dept Physics Univ Notre Dame Notre Dame IN 46556

FUNK, GLENN ALBERT, b Highland Park, Mich, Feb 28, 42; m 66; c 2. MEDICAL VIROLOGY, MEDICAL MICROBIOLOGY. *Educ:* Univ Calif, Berkeley, BA, 63; Univ Calif, Los Angeles, MS, 65; Stanford Univ, PhD(med microbiol), 75. *Prof Exp:* Ment health trainee med microbiol, NIH, 63-65; dir, Clin Virol Lab, Naval Med Res Unit 2, Taipei, Taiwan, 66-69, mem staff res microbiol, Naval Biosci Lab, Oakland, Calif, 70; res assoc med virol, Baylor Col Med, 74-76; asst prof, 76-81, ASSOC PROF MICROBIOL & MED VIROL, SAN JOSE STATE UNIV, 81-; MGR RES, INT DIAGNOSTIC TECHNOL, SANTA CLARA, CALIF, 81- *Concurrent Pos:* Asst, Sch Med, Stanford Univ, 72-74, USPHS fel med microbiol, 69-74; fel, Baylor Col Med, 75-76; consult, Microbiol Assoc, Rockville, Md. *Mem:* Am Soc Microbiol; Soc Wine Educrs. *Res:* Biochemical and biophysical characterization of animal viruses; virology and immunology of hepatitis; biomedical instrumentation and computer interfacing; viral immunodiagnosis. *Mailing Add:* GE Govt Serv PO Box 138 Moffett Field CA 94035

FUNK, JAMES ELLIS, b Cincinnati, Ohio, Nov 8, 32; m 55; c 5. MECHANICAL & CHEMICAL ENGINEERING. *Educ:* Univ Cincinnati, ChemE, 55; Univ Pittsburgh, MS, 58, PhD(chem eng), 60. *Prof Exp:* Sr engr, Bettis Atomic Power Lab, Westinghouse Elec Corp, 55-63; group leader, Allison Div, Gen Motors Corp, 63-64; from assoc prof to prof mech eng & assoc dean grad progs, Univ Ky, 64-70; dir tech activities, indust group, Combustion Eng, Inc, 70-71; dean, 71-79, assoc vpres, Acad Affairs & coordr, Energy Res Educ, 79-81, PROF MECH ENG, COL ENG, UNIV KY, 81-; DIR, ADV TECH CTR, ALLIS-CHALMERS CORP, 81- *Concurrent Pos:* Consult, NSF & Nat Bur Standards. *Mem:* AAAS; Am Inst Chem Engrs. *Res:* Chromatography; systems analysis; thermochemical hydrogen production; thermodynamics. *Mailing Add:* Dept Mech Eng Univ Ky Lexington KY 40506

FUNK, JOHN LEON, b Coschocton, Ohio, Nov 15, 09; m 33; c 1. ECOLOGY, FISH BIOLOGY. *Educ:* Kent State Univ, BS, 32; Univ Mich, MA, 39. *Prof Exp:* Aquatic biologist, Inst Fisheries Res, Mich Dept Conserv, 40-44 & Wash State Pollution Control Comn, 44-45; fishery biologist, Mo Dept Conserv, 45-52, supvr steam res, 52-59, supt fisheries res & training, 59-72, staff specialist, 72-74; ecol consult, 74-89; RETIRED. *Mem:* Am Fisheries Soc; emer fel Am Inst Fishery Res Biol. *Res:* Warm water stream fisheries; ecology of fishes of warm water streams; migration of fishes; populations and ecology of streams and associated fauna. *Mailing Add:* 2917 W Rollins Rd D-13 Columbia MO 65203

FUNK, RICHARD CULLEN, b Omaha, Nebr, Sept 22, 34; m 57; c 4. ACAROLOGY, INVERTEBRATE ZOOLOGY. *Educ:* Colo State Univ, BS, 57, MS, 60; Univ Kans, PhD(entom), 68. *Prof Exp:* From asst prof to assoc prof, 65-78, PROF ZOOL, 78-, CHMN, DEPT PROF STUDIES, 85-, CHMN, DEPT ZOOL, EASTERN ILL UNIV, 88- *Concurrent Pos:* Chmn, Health Prof Studies, 85- *Mem:* Acarological Soc Am; Am Arachnological Soc; Entom Soc Am. *Res:* Systematics, zoogeography and biology of celaenopsoid, fedrizzioid and other taxa of phoretic mites. *Mailing Add:* Dept Zool Eastern Ill Univ Charleston IL 61920

FUNK, WILLIAM HENRY, b Ephraim, Utah, June 10, 33; wid; c 1. LIMNOLOGY, SANITARY BIOLOGY. *Educ:* Univ Utah, BS, 55, MS, 63, PhD(zool), 66. *Prof Exp:* Instr high schs, Utah, 57-63; assoc prof sanit sci, 66-75, chair, environ sci prog, 77-80, PROF CIVIL & ENVIRON ENG, WASH STATE UNIV, 75- *Concurrent Pos:* Consult lake asns & indust on water qual; consult, US Civil Serv Comn, 73-75; nat dir, Pac Northwest Pollution Control Asn, 78-81, pres, 83-84; pres, NAm Lake Mgt Soc, 84-85; chair, Nat Asn Water Inst Dirs, 85-87. *Honors & Awards:* Arthur Sidney Bedell Award , Water Pollution Control. *Mem:* Water Pollution Control Fedn; Am Soc Limnol & Oceanog; Am Micros Soc; Sigma Xi; NAm Lake Mgt Soc; Nat Asn Water Inst Dirs. *Res:* Nature and causes of eutrophication of lakes and reservoirs; development of multidisciplinary research and inter-university cooperation on water research. *Mailing Add:* State of Wash Water Res Ctr Wash State Univ Pullman WA 99164-3002

FUNKE, PHILLIP T, b Bend, Ore, Nov 1, 32; m 70. MASS SPECTROMETRY, ORGANIC CHEMISTRY. *Educ:* Univ Puget Sound, BS, 54; Stevens Inst Technol, MS, 56, PhD(phys chem), 62. *Prof Exp:* Instr chem, Newark Col, Eng, 60-63; res scientist mass spectrometry, Stevens Inst Technol, 63-70; res investr, Squibb Inst Med Res, E R Squibb & Sons, 70-72, res fel mass spectrometry, 72-82, sect head, molecular spectros & microanal, 82-90; ASSOC DIR ANALYTICAL RES, BRISTOLOMYERS SQUIBB, 91- *Concurrent Pos:* Vis lectr, Stevens Inst Technol, 63-70; consult, Sandoz Pharmaceut, Inc, 66-69. *Mem:* Am Chem Soc; Am Soc Mass Spectrometry; AAAS. *Res:* Organic structure determination by mass spectrometry; ion-molecule reactions; analysis of thermal degradation products of polymers; gas chromatography; quantitative analysis of drugs in body fluids; positive and negative chemical ionization. *Mailing Add:* Bristol-Myers Squibb PO Box 4000 Rte 206 & Provinceline Rd Princeton NJ 08543-4000

FUNKENBUSCH, WALTER W, b Canton, Mo, Feb 13, 18; m 45; c 4. MATHEMATICS. *Educ:* Culver-Stockton Col, BA, 40; Ore State Col, MS, 50. *Prof Exp:* Asst math, Ore State Col, 40-41, inst, 42-44; seismograph comput, W Geophys Co, 44; instr math, Mich Col Mining & Technol, 44-45 & 45-49; from asst prof to assoc prof, 49-67, PROF MATH, MICH TECHNOL UNIV, 67- *Concurrent Pos:* Instr, Mich State Col, 45; Proj dir undergrad partic, NSF, 59-60. *Mem:* Math Asn Am; Sigma Xi. *Res:* Probability and game theory. *Mailing Add:* 510 W Jackie Ave Houghton MI 49931

FUNKHOUSER, EDWARD ALLEN, b Trenton, NJ, Sept 30, 45; m 68; c 2. PLANT PHYSIOLOGY, PLANT BIOCHEMISTRY. *Educ:* Del Valley Col, BS, 67; Rutgers Univ, MS, 69, PhD(plant physiol), 72. *Prof Exp:* Instr biol, Del Valley Col Sci & Agr, 69-72; res assoc plant biochem, State Univ NY, Buffalo, 72-74; sci asst plant physiol, Vennesland res dept, Max Planck Soc Advan Sci, 74-76; from asst prof to assoc prof plant physiol, Dept Plant Sci, 76-83, ASSOC PROF BIOCHEM & PLANT PHYSIOL, DEPT BIOCHEM & BIOPHYS, TEX A&M UNIV, 83-, ASSOC HEAD UNDERGRAD EDUC, DEPT BIOCHEM & BIOPHYS, 90- *Mem:* AAAS; Am Soc Plant Physiologists; Japanese Soc Plant Physiologists; Sigma Xi; Am Soc Biochem & Molecular Biol; Scand Soc Plant Physiol. *Res:* Plant biochemistry and molecular biology; control of nitrate utilization in plants and factors which control nitrate reductase activity; effects of drought-stress on enzyme induction and activity. *Mailing Add:* Dept Biochem & Biophys Tex A&M Univ College Station TX 77843-2128

FUNKHOUSER, JANE D, b Ark, July 10, 39. BIOCHEMISTRY. *Educ:* Univ Ark, BS, 70, PhD(biochem), 75. *Prof Exp:* Instr & res assoc, Dept Pediat, Univ SAla Col Med, 75-77, asst prof pediat, 77-82, assoc prof biochem, 82-89, PROF BIOCHEM, UNIV S ALA COL MED, 89- *Concurrent Pos:* Prin investr, NIH, 86-92 & 89-94, co-investr, 88-91; spec study sect, Oncogenes & Growth in Fetal Lung, NIH, 89, study sect, Respiratory & Physiol, 91- *Mem:* Am Soc Biochem & Molecular Biol. *Res:* Biochemistry of developmental processes; lung development; pulmonary surfactant system; phospholipid biochemistry. *Mailing Add:* Dept Biochem Univ Southern Ala Col Med MSB 2152 307 University Ave Mobile AL 36688

FUNKHOUSER, JOHN TOWER, b Paterson, NJ, Dec 19, 28; m 55; c 3. ENVIRONMENTAL SCIENCES. *Educ:* Princeton Univ, AB, 50; Mass Inst Technol, PhD(chem), 54. *Prof Exp:* Chemist, E I du Pont de Nemours & Co, 54-62; chemist, Arthur D Little Inc, 62-63, group leader anal res, 63-69, dir, Environ Sci Res & Develop Div, Cambridge, 71-73, mgr consult, 73-89; dir, Ctr Environ Assurance, 78-84; RETIRED. *Mem:* AAAS; Am Chem Soc. *Res:* Analytical chemistry. *Mailing Add:* 505 Via Media Palos Verdes Estates CA 90274

FUNSTEN, HERBERT OLIVER, III, b Princeton, NJ, May 29, 62; m 89; c 1. LOW ENERGY ATOMIC INTERACTIONS. *Educ:* Wash & Lee Univ, BS, 84; Univ Va, MEP, 86, PhD(eng physics), 90. *Prof Exp:* POSTDOCTORAL FEL, LOS ALAMOS NAT LAB, 90- *Mem:* Am Phys Soc. *Res:* Ion beam-induced deposition of thin films; ion beam interactions with solids; secondary ion emission from insulators. *Mailing Add:* SST-8 Los Alamos Nat Lab MS D-438 Los Alamos NM 87545

FUNT, B LIONEL, b Jan 20, 24; Can citizen; m 47; c 3. POLYMER CHEMISTRY. *Educ:* Dalhousie Univ, BSc, 44, MSc, 46; McGill Univ, PhD(phys chem), 49. *Prof Exp:* Lectr chem, Dalhousie Univ, 46-47; from asst prof to prof, Univ Man, 50-68, dean grad studies, 64-68; dean sci, 68-71, prof, 68-89, EMER PROF CHEM, SIMON FRASER UNIV, 89-, DIR SCI PROGS, HARBOR CTR, 89- *Concurrent Pos:* Res consult, Nuclear Enterprises, 54-76, dir, UK, 56-76, Can, 60-, US, 68-76; mem bd dirs, Asn Univs & Cols Can, 66-69 & Univs Grants Comn Man, 67-69; mem bd gov, Simon Fraser Univ, 75-78. *Honors & Awards:* Macromolecular Sci & Eng Award, Chem Inst Can, 91. *Mem:* Fel Chem Inst Can. *Res:* Kinetics of polymerization; electrically initiated polymerizations; electrically conducting polymers. *Mailing Add:* Dept Chem Simon Fraser Univ Burnaby BC V5A 1S6 Can

FUOSS, RAYMOND MATTHEW, physical chemistry; deceased, see previous edition for last biography

FUQUA, MARY ELIZABETH, b Dresden, Tenn, Sept 12, 22. EDUCATIONAL ADMINISTRATION, NUTRITION. *Educ:* Univ Tenn, BS, 44; Ohio State Univ, MS, 48, PhD(food & nutrit), 52. *Prof Exp:* Therapeut dietician, Med Ctr, Univ Ind, 45-57; instr foods & nutrit, Ohio State Univ, 48-49; asst prof, Univ Ill, 51-53; assoc prof, Pa State Univ, 53-63; head dept, Purdue Univ, 63-66, assoc dean exten, 73-78, prof foods & nutrit, 63-87, assoc dean, 78-87; RETIRED. *Mem:* Am Dietetic Asn; Sigma Xi; Am Inst Nutrit. *Res:* Calcium metabolism. *Mailing Add:* Dean-Dir Continuing Educ North Adams State Col Church St North Adams MA 01247

FUQUAY, JAMES JENKINS, b Montesano, Wash, May 18, 24; m 46; c 4. METEOROLOGY, CLIMATOLOGY. *Educ:* Univ Wash, BS, 50, MS, 52. *Prof Exp:* Jr res meteorologist atmospheric diffusion, Univ Wash, 50-51, res meteorologist, 51-52; res meteorologist, Atmospheric Physics Oper, Hanford Labs, Gen Elec Co, 52-60, mgr, 60-65; Mgr, Atmospheric Physics Sect, Physics & Instruments Dept, Pac Northwest Lab, Battelle Mem Inst, 65-66 & Environ 7 Radiol Sci Dept, 66-68, assoc mgr, Environ & Life Sci Div, 68-74, asst dir prog support, 75-80, asst dir admin, 85-86; RETIRED. *Mem:* Am Meteorol Soc; Am Geophys Union; Air Pollution Control Asn; Sigma Xi. *Res:* Micrometeorology; microclimatology; application of meteorology in industrial plant operation and dispersing of stack effluents. *Mailing Add:* 314 Fuller St Richland WA 99352

FUQUAY, JOHN WADE, b Burlington, NC, July 31, 33; m 70; c 2. REPRODUCTIVE PHYSIOLOGY, ENVIRONMENTAL PHYSIOLOGY. *Educ:* NC State Univ, BS, 55, MS, 66; Pa State Univ, PhD(dairy sci), 69. *Prof Exp:* Herd mgr dairy cattle, Fuquay's Jersey Farm, 59-64; asst dairy cattle physiol, NC State Univ, 64-66; asst dairy cattle nutrit, Pa State Univ, 66-69; from asst prof to assoc prof, 69-79, PROF DAIRY CATTLE PHYSIOL, MISS STATE UNIV, 79-, COORDR, GRAD PROG ANIMAL PHYSIOL, 87- *Concurrent Pos:* Vis prof, Univ Calif, Davis, 79, 85-86; chmn, Am Dairy Sci Asn Educ Comn, 81-82; CSRS rev panel, Clemson Univ, 84, Univ Arkansas, 84; consult, USAID, Pakistan, 90; assoc ed, J Dairy Sci, 91- *Mem:* Am Dairy Sci Asn; AAAS; Am Soc Animal Sci; Soc Study Reproduction. *Res:* Dairy cattle, especially endocrine regulation of estrous cycle and fertility; environmental stress or effects on physiology and performance. *Mailing Add:* Dept Dairy Sci Drawer Miss State Univ Mississippi State MS 39762

FURBISH, FRANCIS SCOTT, b Portland, Maine, Mar 25, 40; m 69; c 1. BIOCHEMISTRY. *Educ:* Univ Maine, BS, 63; Iowa State Univ, PhD(biochem), 69; Northeastern Univ, MBA, 90. *Prof Exp:* Chief chem sect, US Army, Valley Forge Gen Hosp, 69-72; guest researcher biochem, NIH, 75-76, sr staff fel biochem, 76-84; res scientist, Genzyme Corp, 84-85, sr scientist biochem, 85-86, mfg mgr, 86-88, govt contracts mgr, 89, EDUC PROGS MGR, GENZYME CORP, 90- *Mem:* AAAS; Am Chem Soc; NY Acad Sci; Sigma Xi; Am Soc Biochem & Molecular Biol; Am Asn Clin Chem. *Res:* Enzyme replacement therapy; protein purification and modification; glycoprotein receptors in mammalian systems; analytical biochemistry; coenzymes, enzymology. *Mailing Add:* Genzyme Corp One Kendall Sq Cambridge MA 02139

FURBY, NEAL WASHBURN, b Hanford, Calif, Apr 20, 12; m 37; c 3. CHEMISTRY. *Educ:* Univ Calif, BS, 39. *Prof Exp:* From res chemist to sr res chemist, Chevron Res Co, 39-58, supvry res chemist, 58-64, sr res assoc, 64-77; VPRES, INT LUBRICATION CONSULTS, 77- *Mem:* Am Chem Soc; Am Soc Test & Mat; Am Soc Lubrication Eng. *Res:* Petroleum chemistry; lubricating oils; lubricants; hydraulic fluids; engine oils; synthetic fluids. *Mailing Add:* 1985 Paquita Dr Carpinteria CA 93013-3026

FURCHGOTT, ROBERT FRANCIS, b Charleston, SC, June 4, 16; m 41; c 3. MEMBRANE RECEPTORS, VASCULAR PHARMACOLOGY. *Educ:* Univ NC, BS, 37; Northwestern Univ, PhD(biochem), 40. *Hon Degrees:* DM, Autonomous Univ, Madrid, 84, Univ Lund, 84; DSc, Univ NC, Chapel Hill, 89. *Prof Exp:* Res fel med, Med Col, Cornell Univ, 40-43, res assoc, 43-47, instr physiol, 43-48, asst prof med biochem, 47-49; from asst prof to assoc prof pharmacol, Med Sch, Washington Univ, 49-56; chmn dept, 56-83, prof, 56-88, UNIV DISTINGUISHED PROF, HEALTH SCI CTR, BROOKLYN, STATE UNIV NY, 88-, EMER PROF PHARMACOL, 90- *Concurrent Pos:* Mem pharmacol training comt, USPHS, 61-64, mem pharmacol-toxicol rev comt, 65-68; Commonwealth fel, 62-63; vis prof, Univ Geneva, 62-63; mem bd sci coun, Nat Heart Inst, 64-68 York City Health Res Coun, 65-70; USPHS spec fel, 71-72; vis prof, Univ Calif, San Diego, 71-72; vis prof, Med Univ SC, 80 & Univ Calif, 80; adj prof pharmacol, Sch Med Univ Miami, 89- *Honors & Awards:* Goodman & Gilman Award, Am Soc Pharmacol & Exp Therapeut; Res Achievement Award, Am Heart Asn, 90; Bristol-Myers Squibb Award for Achievement in Cardiovasc Res, 91. *Mem:* Nat Acad Sci; AAAS; Am Chem Soc; Am Soc Biochem; Am Soc Pharmacol & Exp Therapeut (pres, 71-72); Harvey Soc; hon mem Polish Physiol Soc; Soc Gen Physiologists; Sigma Xi. *Res:* Red cell structure; circulatory shock and hypertension; pharmacology and biochemistry of cardiac and smooth muscle; adrenergic mechanisms; theory of drug-receptor interactions; mechanisms of vasodilitation; endothelium-dependent relaxation of blood vessels. *Mailing Add:* Health Sci Ctr Brooklyn State Univ NY Brooklyn NY 11203

FURCINITTI, PAUL STEPHEN, b Milford, Mass, Mar 8, 49; m 78. IMAGE PROCESSING, STRUCTURAL BIOLOGY. *Educ:* Worcester Polytech Inst, BS, 71; Univ NH, MS, 74, PhD(physics), 75. *Prof Exp:* Res assoc laser physics, Worcester Polytech Inst, 76; res assoc radiation biol, Pa State Univ, 76-78; res assoc, Carrcinogenesis Risk Assessment, Div Health & Safety Res, Oak Ridge Nat Lab, 78-79; res assoc, radiation biophys, Columbia Univ, 79-82; res collabr, Brookhaven Nat Lab, 79-82, sr res assoc, 82-85, asst biophysicist, 85-87, assoc biophysicist, 87-88; asst res biophysicist, Univ Mich, 88-90; SR RES ASSOC, UNIV COLO, 90- *Mem:* AAAS; Sigma Xi; Optical Soc Am; Electron Micros Soc Am. *Res:* Cellular radiation biology; analytical cell electrophoresis; laser light scattering; laser theory; chemical carcinogenesis; image processing of electron micrographs; structural biology. *Mailing Add:* 6988 Hunter Pl Boulder CO 80301

FURDYNA, JACEK K, b Kamionka, Poland, Sept 6, 33; US citizen; m 60; c 4. SOLID STATE PHYSICS. *Educ:* Loyola Univ, Chicago, BS, 55; Northwestern Univ, PhD(physics), 60. *Prof Exp:* Res assoc, Northwestern Univ, 60-62; res physicist, Francis Bitter Nat Magnet Lab, Mass Inst Technol, 62-66; assoc prof, 66-72, PROF PHYSICS, PURDUE UNIV, 72- *Mem:* Am Phys Soc. *Res:* Physics of semiconductors and metals; galvano-magnetic phenomena; magneto-optical phenomena at microwave frequencies; plasma effect in solids. *Mailing Add:* Dept Physics Univ Notre Dame Notre Dame IN 46556

FURESZ, JOHN, b Miskolc, Hungary, Oct 13, 27; Can citizen; m 77; c 3. VIROLOGY. *Educ:* Univ Budapest, MD, 51 & McGill Univ, 61. *Prof Exp:* Med officer virol, State Inst Hyg, Budapest, 51-56; biologist, Lab of Hyg, Ottawa, 56-64; med officer, Can Commun Dis Ctr, 64-73; dir, Bur Viral Dis, 73-74, DIR BUR BIOL, DRUGS DIRECTORATE, 74- *Honors & Awards:* Merit Award Outstanding Sci Achievement, Health & Welfare, Can, 91. *Mem:* Fel Am Acad Microbiol; Can Soc Microbiol; Can Asn Med Microbiol; Am Soc Microbiol; Int Asn Microbiol Socs. *Res:* Antigenic structure of influenza viruses; genetic markers of polioviruses; measles, German measles and mumps vaccines in the laboratory and field trials; slow virus infections. *Mailing Add:* Bur Biol Drugs Directorate Tunney's Pasture Ottawa ON K1A 0L2 Can

FUREY, ROBERT LAWRENCE, b Canton, Ohio, June 25, 41; m 64; c 2. FUEL CHEMISTRY, VEHICLE EMISSIONS. *Educ:* Kent State Univ, BS, 63, PhD(org chem), 67. *Prof Exp:* Res chemist, Res Labs, Edgewood Arsenal, Md, 67-69; assoc sr res chemist, 69-73, sr res chemist fuels, 73-80, staff res scientist, 80-85, SR STAFF RES SCIENTIST, GEN MOTORS RES LABS, 85- *Mem:* Am Chem Soc; Soc Automotive Eng; Am Soc Testing Mat. *Res:* Effects of conventional and alternative automotive fuels and fuel additives on vehicle emissions and durability. *Mailing Add:* Fuels & Lubricants Dept Gen Motors Res Labs Warren MI 48090-9055

FURFARI, FRANK A, b Morgantown, WVa, Feb 14, 15; m 42; c 3. INDUSTRIAL APPLICATIONS OF ELECTRICAL ENGINEERING. *Educ:* WVa Univ, BSSE, 38. *Prof Exp:* Serv mgr, Indust Equip & Serv Group, Westinghouse Corp, 40-80; CONSULT, 80- *Concurrent Pos:* Mem bd dir, Inst Elec & Electronics Engrs, pres, Indust Appl Soc, 76. *Honors & Awards:* Outstanding Achievement Award, Inst Elec & Electronics Engrs Indust Appl Soc; Centennial Medal, Inst Elec & Electronics Engrs. *Mem:* fel Inst Elec & Electronics Engrs. *Mailing Add:* 117 Washington Rd Pittsburgh PA 15221

FURFINE, CHARLES STUART, b St Louis, Mo, Apr 26, 36; m 58; c 2. BIOLOGICAL CHEMISTRY. *Educ:* Wash Univ, St Louis, BA, 57, PhD(biochem), 62. *Prof Exp:* Res fel biochem, Albert Einstein Col Med, 62-63, from instr to asst prof, 63-68; asst prof biochem, Georgetown Univ, 68-75; CHEMIST, FOOD & DRUG ADMIN, 75- *Concurrent Pos:* NIH grant, 65-72. *Mem:* Am Chem Soc. *Res:* Effect of substrates on the physical chemical parameters of proteins; nature of allosteric effects and its relation to phenomena observed in vivo. *Mailing Add:* 10416 Crossing Creek Rd Potomac MD 20854-4201

FURGASON, ROBERT ROY, b Spokane, Wash, Aug 2, 35; m 64. CHEMICAL ENGINEERING. *Educ:* Univ Idaho, BS, 56, MS, 58; Northwestern Univ, PhD(chem eng), 61. *Prof Exp:* From instr to assoc prof chem eng, 57-67, chmn dept, 63-74, dean, Col Eng, 74-78, PROF CHEM ENG, UNIV IDAHO, 67-, VPRES ACAD AFFAIRS & RES, 78- *Concurrent Pos:* Consult, Minute Maid Corp, 58-59, J R Simplot Co, 63-; NSF grant, 64-; develop consult, B F Goodrich Chem Co, 69-70, continuing consult, 70- *Mem:* Am Inst Chem Engrs; Am Chem Soc; Am Soc Eng Educ; Sigma Xi. *Res:* Heat transfer in chemically reacting systems, especially nitrogen tetroxide-nitrogen dioxide and ozone-oxygen decomposing systems; physical properties of reacting systems. *Mailing Add:* Corpus Christi State Univ 6300 Ocean Dr Corpus Christi TX 78412

FURGIUELE, ANGELO RALPH, b Washington, Pa, Apr 14, 29; m 52; c 2. PHARMACOLOGY. *Educ:* Duquesne Univ, BS, 51; Univ Pittsburgh, MS, 59, PhD(pharmacol), 62. *Prof Exp:* Sr res scientist, Squibb Inst Med Res, 62-64, sci coordr biol, Squibb-Int Div, Olin Mathieson Chem Corp, 64-67, dir biol res, E R Squibb & Sons Inc, 67-72; DIR, INT NEW PROD DEVELOP, MCNEIL LABS INC, 72- *Mem:* AAAS; NY Acad Sci; Am Pharmaceut Asn; Am Soc Pharmacol & Exp Therapeut; Am Soc Toxicol. *Res:* Central nervous system pharmacology; toxicology; new product planning, development, marketing; product acquisition/licensing; research management. *Mailing Add:* 16 High Gate Lane Blue Bell PA 19422

FURIE, BRUCE, b Coral Gables, Fla, Apr 17, 44; m 66; c 2. PROTEIN BIOCHEMISTRY, HEMATOLOGY. *Educ:* Princeton Univ, AB, 66; Univ Pa, MD(med), 70. *Prof Exp:* Res, Hosp Univ Pa, 70-72; res assoc protein chem, NIH, 72-74; from asst prof to assoc prof, 75-85, PROF MED & BIOCHEM, TUFTS UNIV SCH MED, 85-; DIR, CTR RES & CHIEF, DIV HEMAT & ONCOL, NEW ENG MED CTR, 86- *Concurrent Pos:* Chief, Coagulation Unit, New Eng Med Ctr, 75-; consult, NIH, 88-; mem, Am Bd Internal Med. *Honors & Awards:* Dameshek, Am Soc Hemat, 84. *Mem:* Asn Am Physicians; Am Soc Clin Invest; Am Soc Hemat; Am Soc Biochem & Molecular Biol; Am Col Physicians. *Res:* Molecular basis of blood coagulation; role of vitamin K in synthesis of blood clotting proteins; cell adhesion molecules. *Mailing Add:* Dept Med Div Hemat Oncol New Eng Med Ctr Hemostasis & Thrombosis Res Boston MA 02111

FURLAN, VALENTIN, b Toulouse, France, Apr 24, 42; Can citizen. ENDOMYCORRHIZAE, PLANT SYMBIOSIS. *Educ:* Univ Montreal, BSc, 71; Univ Laval, Que, DSc, 76. *Prof Exp:* Res assoc, fac forestry, Univ Laval, Que, 77-82; RES SCIENTIST, AGR CAN, SAINTE-FOY, QUE, 82- *Concurrent Pos:* Consult, Premier Peat Moss, Ltd, 83-86. *Res:* Ecology, physiology, biochemistry, biological interactions and applied research on endomycorrhizae on agricultural, horticultural and tree crops. *Mailing Add:* Res Sta Agr Can 2560 Hochelaga Blvd Sainte-Foy PQ G1V 2J3 Can

FURLANETTO, RICHARD W, b Chicago, Ill, Oct 23, 45; m; c 2. PEDIATRICS. *Educ:* Northwestern Univ, BA, 67; Univ Chicago, PhD(biochem), 72, MD, 73; Am Bd Pediat, cert, 83, pediat endocrin, 83. *Prof Exp:* Resident pediat, Wyler Children's Hosp, Univ Chicago, 73-75; fel, pediat endocrin, Univ NC, Chapel Hill, 75-77; asst prof, Dept Pediat & Human Genetics, Univ Tex Med Br, Galveston, 77-80, med staff, 77-82; assoc prof, Dept Pediat, Children's Hosp Philadelphia, Univ Pa, 82-88, med staff, 82-88; PROF PEDIAT ENDOCRINOL & MEM MED STAFF, UNIV ROCHESTER, NY, 89- *Concurrent Pos:* Prin investr, numerous grants, 77-93. *Mem:* AAAS; Am Soc Cell Biol; Endocrine Soc; Soc Pediat Res; Sigma Xi. *Res:* Pediatrics; endocrinology; numerous technical publications. *Mailing Add:* Sch Med & Dent Univ Rochester 601 Elmwood Ave Box 777 Rochester NY 14642

FURLONG, IRA E, b June 2, 31; US citizen. GEOLOGY. *Educ:* Boston Univ, AB, 53, MS, 54, PhD(geol), 60. *Prof Exp:* Instr geol, Marshall Univ, 58-59; lectr, Boston Univ, 59-60; from instr to assoc prof, 60-69, PROF GEOL, BRIDGEWATER STATE COL, 69- *Mem:* AAAS; Asn Am Geographers; Geol Soc Am; Nat Asn Geol Teachers (vpres, 65, pres, 66). *Res:* Genesis of granite plutons and their relationship on resulting geomorphology. *Mailing Add:* Dept Earth Sci Bridgewater State Col Bridgewater MA 02324

FURLONG, NORMAN BURR, JR, b Norwalk, Ohio, Jan 6, 31; m 56; c 3. BIOCHEMISTRY. *Educ:* Southern Methodist Univ, AB & BS, 52; Stanford Univ, MS, 54; Univ Tex, PhD(chem), 60. *Prof Exp:* USPHS fels, Biol Div, Oak Ridge Nat Lab, 60-61 & Univ Tex M D Anderson Hosp & Tumor Inst, 61-62; asst biochemist, 62-66, asst prof, 64-66, assoc dean curric develop, 68-73, assoc prof biochem, Grad Sch Biomed Sci & Assoc Biochemist, Univ Tex M D Anderson Hosp & Tumor Inst, 66-85; SCI TEACHER, KINKAID SCH, 85- *Mem:* AAAS; Am Asn Cancer Res; Sigma Xi. *Res:* Biosynthesis of DNA, enzymology and mechanism of DNA replication; tumor-host relationships; oligonucleotide biochemistry; nonaqueous subcellular separation methods; mechanisms of carcinogenesis; mechanisms of anti-tumor agents; biomolecular information theory; interdisciplinary curricula in the sciences. *Mailing Add:* Kinkaid Sch Kinkaid Dr Houston TX 77024

FURLONG, RICHARD W, b Norwalk, Ohio, Mar 30, 29; m 51; c 2. STRUCTURAL ENGINEERING. *Educ:* Southern Methodist Univ, BS, 52; Wash Univ, St Louis, MS, 57; Univ Tex, PhD(struct), 63. *Prof Exp:* Engr, McDonnell Aircraft Corp, Mo, 52-53, Petrol Equip Co, 53-55 & F Ray Martin, Inc, 55-58; from asst prof civil eng & struct to assoc prof civil eng, 58-71, PROF CIVIL ENG, UNIV TEX, AUSTIN, 71- *Concurrent Pos:* Lectr, Washington Univ, 54-56; Erskine fel, Univ Canterbury, NZ, 73; mem, Struct Stability Res Coun; vis prof, Univ Toronto, 78. *Mem:* Am Soc Civil Engrs; Nat Soc Prof Engrs; fel Am Concrete Inst; Sigma Xi; Can Soc Civil Engrs. *Res:* Analysis and design of structures, design aids for reinforced concrete and limit analysis of continuous frames; composite steel-concrete construction; design of metal and concrete structures. *Mailing Add:* 4103 Greystone Dr Austin TX 78731-1301

FURLONG, ROBERT B, b Malone, NY, Jan 19, 34; m 60; c 2. MINERALOGY, GEOCHEMISTRY. *Educ:* Harpur Col, BA, 62; Univ Ill, Urbana, MS, 65, PhD(clay mineral), 67. *Prof Exp:* From asst prof to assoc prof, 66-81, PROF GEOL, WAYNE STATE UNIV, 81-, CHMN DEPT, 74- *Mem:* Clay Minerals Soc; Mineral Soc Am; Geochem Soc. *Res:* Use of electron microscopy and electron diffraction to study high temperature changes in the clay minerals; clay mineralogy of deep sea sediment; heavy metal contamination in lake waters; erosion of airless planetary bodies and asteroids by meteorite impact (with Luciano Ronca); computer sciences; geochemical exploration for petroleum; location of natural resources by remote sensing techniques. *Mailing Add:* Dept Geol Wayne State Univ Detroit MI 48202

FURMAN, DEANE PHILIP, b Richardton, NDak, June 4, 15; m 38; c 3. ENTOMOLOGY, PARASITOLOGY. *Educ:* Univ Calif, BS, 37, PhD(med entom), 42. *Prof Exp:* Entomologist, USPHS, 46; asst prof parasitol & asst entomologist, 46-52, assoc prof & assoc entomologist, 52-58, chmn, Div Parasitol, 63-72, chmn, Div Entom & Parasitol, 73-75, prof & entomologist, Exp Sta, 58-82, assoc dean, Col Nat Resources, 84-87, EMER PROF ENTOM, UNIV CALIF, BERKELEY, 82- *Concurrent Pos:* NIH spec fel, 64-65; guest investr, US Naval Med Res Unit-3, Egypt, 64-65; chmn, Interdisciplinary Grad Group Parasitol. *Mem:* AAAS; Am Soc Parasitol; Entom Soc Am; Am Soc Trop Med & Hyg; Wildlife Dis Asn. *Res:* Control of parasitic arthropods; systematics of parasitic mites and ticks; biology of helminths; arthropod vectors of diseases and parasites of man and animals. *Mailing Add:* Div Entom & Parasitol Univ Calif Berkeley CA 94720

FURMAN, ROBERT HOWARD, b Schenectady, NY, Oct 23, 18; m 45; c 4. RESEARCH ADMINISTRATION. *Educ:* Union Col, NY, AB, 40; Yale Univ, MD, 43. *Prof Exp:* Asst med, Sch Med, Yale Univ, 44-45, from intern to asst resident physician, 44-45; instr physiol & asst med, Sch Med, Vanderbilt Univ, 46-48, from asst resident physician to resident physician, 48-50, from instr to asst prof med, 49-52; from assoc prof to prof res med, Sch Med, Univ Okla, 52-70, head cardiovasc sect, Okla Med Res Found, 52-70, assoc dir res, 58-70; exec dir clin res, Eli Lilly & Co, 70-73, vpres, Lilly Res Labs, 73-76, VPRES, CORP MED AFFAIRS, ELI LILLY & CO, 76-; PROF MED, SCH MED, IND UNIV, INDIANAPOLIS, 70- *Concurrent Pos:* Lilly fel, Sch Med, Vanderbilt Univ, 46-47, Nat Res Coun fel med sci, 47-48; mem cardiovasc study sect, NIH, 60-63; mem heart spec proj comt, Adv Heart Coun, Coun Arteriosclerosis, Am Heart Asn. *Mem:* Endocrine Soc; Soc Exp Biol & Med; Am Physiol Soc; Cent Soc Clin Res; Am Heart Asn; Sigma Xi. *Res:* Nutritional-endocrinologic-lipid metabolic interrelationships in atherogenesis; clinical pharmacology research. *Mailing Add:* PO Box 4878 Scottsdale AZ 85261

FURMAN, SEYMOUR, b New York, NY, July 12, 31; m 57; c 3. SURGERY. *Educ:* State Univ NY, MD, 55. *Prof Exp:* Intern med, Montefiore Hosp & Med Ctr, NY, 55-56, resident surg, 56-60; resident thoracic surg, Baylor Univ, 62-63; adj attend surgeon, Montefiore Hosp & Med Ctr, 63; from asst prof to assoc prof, 68-77, PROF SURG, ALBERT EINSTEIN COL MED, 77- *Concurrent Pos:* Instr, Dept Surg, Baylor Univ, 63; dir, Biomed Div, F Dickinson Labs, Inc, 83-; assoc dir, Hardin-Simmons Univ, Dickinson Res Ctr, 84- *Mem:* AMA; Am Soc Artificial Internal Organs; fel Am Col Surg; Asn Advan Med Instrumentation; Am Heart Asn. *Res:* Cardiothoracic surgery; cardiac pacemaker; cardiac support systems; manufacture and marketing of rapid immunodiagnostic kits for medical care. *Mailing Add:* Albert Einstein Col Med Montefiore Hosp Med Ctr 111 E 210th St Bronx NY 10467

FURMANSKI, PHILIP, b July 26, 46. EXPERIMENTAL BIOLOGY. *Educ:* Temple Univ, BA, 66, PhD(microbiol), 69. *Prof Exp:* Res assoc & instr, Dept Microbiol, Dartmouth Med Sch, 70-72; res scientist, Dept Biol, Mich Cancer Found, 72-73, chief, Lab Cell Biol, 74-81, chmn dept biol, 78-81, assoc mem, 80-81; sci dir, AMC Cancer Res Ctr, 82-90; PROF & CHMN, DEPT BIOL, NY UNIV, 90- *Concurrent Pos:* Adj assoc prof, Dept Immunol & Microbiol, Wayne State Univ Sch Med, 79-81; full mem, Univ Colo Cancer Ctr, 87-90; adj prof, Dept Med, Univ Colo Sch Med, 88-90; mem, Path B Study Sect, Div Res Grants, NIH, 88-, Resources & Repositories Tech Rev Group, Div Extramural Affairs, Nat Cancer Inst, 82-85, adv bd, Human Tumor Cell Bank, 85-, RNA Tumor Virus Working Group, WHO/FAO Int Prog Comp Virol; chmn, Cancer Biol & Immunol Contracts Rev Comt, Nat Cancer Inst, 85-87. *Res:* Biology; leukemia virus; immunology; microbiology. *Mailing Add:* Dept Biol NY Univ 100 Main Bldg Washington Sq New York NY 10003

FURNAS, DAVID WILLIAM, b Caldwell, Idaho, Apr 1, 31; m 56; c 3. PLASTIC SURGERY, RECONSTRUCTIVE SURGERY. *Educ:* Univ Calif, AB, 52, MD, 55, MS, 57; Am Bd Surg, dipl, 66; Am Bd Plastic Surg, dipl, 65. *Prof Exp:* Asst pharmacol, Univ Calif, 55, from intern to asst resident surg, Univ Hosp, 55-57; asst resident psychiat, Langley Porter Neuropsychiat Inst, 57-60; resident surg, Gorgas Hosp, CZ, 60-61; resident plastic surg, NY Hosp-Cornell Univ, 61-63; registr, Glasgow Royal Infirmary, 63-64; assoc prof surg, Univ Iowa, 64-69; from assoc prof to prof, 69-80, CLIN PROF, DIV PLASTIC SURG, UNIV CALIF, IRVINE, 80-, CHIEF, DIV PLASTIC SURG, 69- *Concurrent Pos:* NIMH fel psychiat, Langley Porter Neuropsychiat Inst, 57-60; surgeon, African Med & Res Found, Kenya, 72-73; bd dir, Am Bd Plastic Surg, 79; with Balakbayan Med Mission, Mindanao, Philippines, 80, 81. *Honors & Awards:* Res Award, Am Soc Surg, Hand, 70 & Am Soc Plastic & Reconstructive Surg, 87. *Mem:* Soc Maxillofacial Surg; Asn Surg EAfrica; Am Soc Surg of the Hand; Am Soc Plastic & Reconstruct Surg; Am Asn Plastic Surg; fel Am Col Surgeons; fel Royal Col Surgeons; fel Royal Geog Soc Can; Am Bd Plastic Surg; AMA. *Res:* Surgery and growth of the facial skeleton; microsurgery; nasoorbital surgery; hand surgery; surgery in the tropics; surgical anatomy; experimental transplantation surgery; technology and psychology of esthetic surgery of the face; traditional and tribal surgery. *Mailing Add:* Univ Calif Div Plastic Surg Irvine Med Ctr 101 City Dr S Orange CA 92668

FURNER, RAYMOND LYNN, b Parkersburg, WVa, May 19, 43. PHARMACOLOGY, CHEMOTHERAPY. *Educ:* WVa Univ, BA, 65, MS, 67, PhD(pharmacol), 68. *Prof Exp:* Alexander von Humboldt Stiftung res assoc pharmacol, Univ Tubingen, 69; Nat Acad Sci res assoc, Ames Res Ctr, NASA, Moffett Field, Calif, 70-71; sr pharmacologist, Southern Res Inst, 71-79; DIR GAS CHROMATOGRAPHY-MASS SPECTROMETRY LAB, NEUROPSYCHIAT RES PROG, UNIV ALA, BIRMINGHAM, 81- *Concurrent Pos:* Assoc prof pharmacol, Univ Ala, Birmingham, 73- *Mem:* Am Soc Pharmacol & Exp Therapeut; Sigma Xi; AAAS; Am Asn Cancer Res; Am Soc Mass Spectrometry. *Res:* Absorption, distribution, metabolism, excretion and pharmacokinetics of chemotherapeutics used in the treatment of cancer; analytical biochemistry; diagnosis of depression and schizophrenia. *Mailing Add:* Neuropsychiat Res Prog Univ Ala Birmingham AL 35294

FURNIVAL, GEORGE MASON, b Johnson City, Tenn, May 1, 25; m 46; c 2. FORESTRY. *Educ:* Univ Ga, BSF, 48; Duke Univ, MF, 52, DF, 57. *Prof Exp:* Res forester, Miss State Col, 48-50 & US Forest Serv, 52-55; instr forest mensuration, Sch Forestry, Yale Univ, 55-57, from asst prof to assoc prof, 57-64; dir biomet studies, US Forest Serv, 64-65; PROF FOREST BIOMET, YALE UNIV, 65-, WEYERHAUSER PROF FOREST MGT, 71-, DIR GRAD STUDIES, 77- *Res:* Application of statistical methods in forestry. *Mailing Add:* Dept Forest Mgt Yale Univ 23 Marsh New Haven CT 06520

FURR, AARON KEITH, b Salisbury, NC, Mar 5, 32; m 58; c 4. LABORATORY SAFETY, HEALTH PHYSICS. *Educ:* Catawba Col, BA, 54; Emory Univ, MS, 55; Duke Univ, PhD(nuclear physics), 62. *Prof Exp:* From asst prof to assoc prof physics, Va Polytech Inst & State Univ, 60-70, prof physics & nuclear eng, 70-71, dir, Nuclear Lab, 72-75, PROF NUCLEAR SCI & ENG, VA POLYTECH INST & STATE UNIV, 71-, DIR ENVIRON HEALTH SERV, 75- *Mem:* Nat Fire Protection Asn; Health Phys Soc; Sigma Xi. *Res:* Nuclear level measurements; neutron spectroscopy; activation analysis; indoor air quality. *Mailing Add:* Environ Health & Safety Va Polytech Inst & State Univ Blacksburg VA 24061

FURRER, JOHN D, b Walton, Nebr, Jan 23, 20; m 51; c 4. AGRONOMY. *Educ:* Univ Nebr, BS, 47, MS, 52. *Prof Exp:* Exten agronomist, 47-64, exten specialist & assoc prof, 65-70, PESTICIDE SPECIALIST & PROF AGRON, UNIV NEBR, LINCOLN, 70- *Mem:* Weed Sci Soc Am. *Res:* Herbicides for perennial weed control; pre-emergence herbicides for annual weed control in corn, sorghum and soybeans and for lawn weed control. *Mailing Add:* Dept Agron E Campus Univ Nebr Lincoln NE 68583

FURROW, STANLEY DONALD, b Bangor, Maine, Mar 6, 34; m 61; c 2. PHYSICAL CHEMISTRY. *Educ:* Univ Maine, BS, 56, MS, 62, PhD(chem), 65. *Prof Exp:* Instr chem, Univ Maine, 63-64; res asst, Exeter Univ, 65-66; asst prof chem, Univ Maine, 66-69; asst prof, 69-82, ASSOC PROF CHEM, PA STATE UNIV, 82- *Mem:* Am Chem Soc. *Res:* Non-electrolyte solution thermodynamics; calorimetry; oscillating chemical reactions. *Mailing Add:* Dept Chem Berks Campus Pa State Univ PO Box 7009 Reading PA 19610-6009

FURRY, BENJAMIN K, b Wadsworth, Ohio, Nov 21, 23; m 45; c 3. RUBBER CHEMISTRY. *Educ:* Muskingum Col, BS, 44. *Prof Exp:* Lab technician adhesives, Firestone Tire & Rubber Co, 44-46; asst chemist, Seiberling Latex Prod, 46-51, chief chemist latex polymers, 51-67,vpres res & develop, 68-73; develop scientist, B F Goodrich Chem Co, 67-68; vpres res, MCM Hosp Supplies Inc, El Reno, 73-79; tech dir, Latex Industs, Chippewa Lake, Ohio, 79-82; CONSULT, 82- *Mem:* Am Chem Soc. *Res:* Dipping compounds in natural and synthetic lactices; foam polymers; dipping plastisols; catheters and hospital devices. *Mailing Add:* 751 W St Wadsworth OH 44281

FURRY, RONALD B(AY), b Niagara Falls, NY, Oct 22, 31; m 53; c 3. AGRICULTURAL ENGINEERING. *Educ:* Cornell Univ, BS, 53, MS, 55; Iowa State Univ, PhD, 65. *Prof Exp:* Engr drawing & descriptive geometry, 53-56, asst prof agr exten eng, 56-60; from asst prof to assoc prof, 60-72, teaching & dir NY State Planning Serv, 60-63, grad field rep & coordr grad instr, 69-73, PROF AGR ENG, CORNELL UNIV, 72- *Concurrent Pos:* NSF sci fac fel, 63-65. *Mem:* Am Soc Agr Engrs; Am Soc Eng Educ. *Res:* Plant structures and environments; controlled atmosphere storage of fruits and vegetables; similitude methodology. *Mailing Add:* Riley-Robb Hall Cornell Univ Ithaca NY 14853

FURSE, CLARE TAYLOR, b Salt Lake City, Utah, May 18, 31; m 55; c 5. ANALYTICAL CHEMISTRY. *Educ:* Univ Utah, BS, 57, PhD(anal chem), 61. *Prof Exp:* Res chemist, Esso Res & Eng Co, 61-64; assoc prof, 64-74, PROF CHEM, MERCER UNIV, 74- *Mem:* Am Chem Soc. *Res:* Square-wave polarography; electrode kinetics; ultraviolet and fluorescence spectroscopy; separation and identification of polynuclear aromatics. *Mailing Add:* Dept Chem Mercer Univ Macon GA 31207-0003

FURSHPAN, EDWIN JEAN, b Hartford, Conn, Apr 18, 28; m 57, 85; c 3. NEUROBIOLOGY. *Educ:* Univ Conn, BS, 50; Calif Inst Technol, PhD(animal physiol), 55. *Hon Degrees:* MS, Harvard Med Sch, 67. *Prof Exp:* Fel & hon asst, Univ Col, Univ London, 55-58; instr neurophysiol & neuropharmacol, Harvard Med Sch, 58-59, assoc, 60-62, asst prof, 62-66, from asst prof to prof neurobiol, 66-88, ROBERT HENRY PFEIFFER PROF NEUROBIOL, HARVARD MED SCH, 88- *Concurrent Pos:* Sister Kenny Found scholar, 58-62; NIH career develop award, 63-73. *Honors & Awards:* Thomas Hunt Morgan Award, 55. *Mem:* Nat Acad Sci; Soc Neurosci; Am Acad Arts & Sci; hon mem Harvey Soc; Am Physiol Soc; Tissue Cult Soc; assoc Neurosci Res Prog. *Res:* Electrophysiology and chemistry of excitable cells; cell interaction; development of neuronal properties; control of transmitter choice in neurons; central nervous system neurons in cell culture. *Mailing Add:* Dept Neurobiol Harvard Med Sch Boston MA 02115

FURST, ARTHUR, b Minneapolis, Minn, Dec 25, 14; m 40; c 4. CANCER, TOXICOLOGY. *Educ:* Univ Calif, Los Angeles, AB, 37, MA, 40; Stanford Univ, PhD(chem), 48. *Hon Degrees:* ScD, Univ San Francisco, 83. *Prof Exp:* Asst, Univ Calif, Los Angeles, 38-39; teacher, Pac Mil Acad, 39-40; teacher petrol inspection & org chem, Univ Calif, 41-45; teacher chem, City Col of San Francisco, 40-47; from asst prof to assoc prof, Univ San Francisco, 47-52; from assoc prof to prof med chem, Sch Med, Stanford Univ, 52-61; prof chem & dir, Inst Chem Biol, Univ San Francisco, 61-80, dean, Grad Div, 76-80; CONSULT TOXICOL, 81- *Concurrent Pos:* Res assoc, Mt Zion Hosp, 50-82; clin prof, Col Physicians & Surgeons, Columbia Univ, 69-70; vis fel, Battelle Res Ctr, 74; grants, Res Corp, USPHS & Am Cancer Soc. *Honors & Awards:* Klaus Schwarz Medal, 86. *Mem:* AAAS; Am Chem Soc; Am Asn Cancer Res; Am Soc Pharmacol & Exp Therapeut; Soc Toxicol; Am Col Toxicol (pres, 85). *Res:* Synthesis of possible growth inhibitors; chemotherapy of cancer and virus; carcinogenesis; metals and hydrocarbons. *Mailing Add:* 3736 La Calle Ct Palo Alto CA 94306

FURST, MERRICK LEE, b New York, NY, Jan 18, 55; m; c 2. COMPUTER THEORY & SOFTWARE. *Educ:* Bucknell Univ, BS & MS, 76; Cornell Univ, MS & PhD(comput sci), 80. *Prof Exp:* ASSOC PROF COMPUT SCI, CARNEGIE-MELLON UNIV, 80-, ASSOC DEAN, SCH COMPUTER SCI, 88- *Mem:* Am Math Soc; Asn Comput Mach. *Res:* Algorithm design and computational complexity. *Mailing Add:* Dept Comput Sci Carnegie-Mellon Univ Pittsburgh PA 15213

FURST, MILTON, b New York, NY, Sept 10, 21; m 45; c 2. PHYSICS. *Educ:* City Col New York, BS, 42; NY Univ, BS, 48, PhD, 52. *Prof Exp:* Physicist, NY Naval Shipyard, 47-50; asst physics, NY Univ, 50-52; from instr to assoc prof, 55-67, PROF PHYSICS, HUNTER COL, 67- *Concurrent Pos:* Res assoc, NY Univ, 52-63, res scientist, 63-67; consult & researcher, Saclay Nuclear Res Ctr, France, 61-62; vis res prof, Oakland Univ, 74-75. *Mem:* AAAS; Am Phys Soc. *Res:* Fluorescence and energy transfer in liquid organic systems under high energy and light excitations. *Mailing Add:* 145 W 67th St New York NY 10023

FURST, ULRICH RICHARD, b Vienna, Austria, Jan 18, 13; nat US; m 44; c 2. PHYSICS. *Educ:* Inst Tech, Vienna, EE, 35, PhD(physics), 38. *Prof Exp:* Engr, Keystone Mfg Co, Mass, 38-40 & Rehtron Corp, Ill, 40-41; electronic engr, Offner Electronics, 41-42; chief electronic engr, Russell Elec Co, 42-46; pres, Furst Electronics, Inc, 46-56; chief electronics engr, Elec Eye Equip Co, 56-58; chief missile opers unit, Bomarc Proj, Aero-Space Div, Boeing Co, 58-62; sect mgr, Syst Eng Dept, Missile Systs Div, Hughes Aircraft Co, 62-73, sr staff engr, Electro-Optical & Data Systs Group, 73-83; RETIRED. *Concurrent Pos:* Lectr, Northwestern Univ, 46-49. *Mem:* Sr mem Inst Elec & Electronics Engrs; Sigma Xi. *Res:* Military electronics systems. *Mailing Add:* 3620 Weslin Ave Sherman Oaks CA 91423

FURSTENBERG, HILLEL, MATHEMATICS. *Prof Exp:* PROF, INST MATH, HEBREW UNIV, 75- *Mem:* Nat Acad Sci. *Mailing Add:* Inst Math Hebrew Univ Girat Ram Jerusalem 91904 Israel

FURTADO, DOLORES, b West Warwick, RI, July 4, 38. MEDICAL MICROBIOLOGY. *Educ:* Cornell Univ, BS, 60; Univ Mich, MS, 63, PhD(bact), 66. *Prof Exp:* From asst prof to assoc prof, 70-82, PROF MICROBIOL, UNIV KANS MED SCH, 82- *Concurrent Pos:* NIH fel, Guy's Hosp Med Sch, London, 66-67 & Yale Univ Sch Med, 67-70; fel, Nat Kidney Found, 69; coun, Am Soc Microbiol, 87-91; vchmn microbiol, Univ Kans Sch Med, 85-88. *Mem:* Am Soc Microbiol; Am Soc Nephrol; Int Soc Nephrol; Sigma Xi; Am Asn Univ Professors; Infectious Dis Soc Am; NY Acad Sci. *Res:* Pathogenesis of bacterial infections; experimental urinary tract infections; host-parasite interactions during bactermia; asymptomatic bacterial infections. *Mailing Add:* Dept Microbiol Univ Kans Med Ctr 39th & Rainbow Blvd Kansas City KS 66103

FURTADO, VICTOR CUNHA, b Elizabeth, NJ, Mar 21, 37; wid; c 4. INDUSTRIAL HYGIENE, HEALTH PHYSICS. *Educ:* Newark Col Eng, BSCE, 58; Univ Calif, Berkeley, MBiorad, 63; NY Univ, PhD(civil eng, environ health sci), 71. *Prof Exp:* Sanit & indust hyg engr, 810th Med Group, Fairchild AFB, Wash, USAF, 58-61, health physicist, 392nd Med Group, Vandenberg AFB, Calif, 63-64 & 6595th Aerospace Test Wing, 64-65, bioenviron engr, Aerospace Med Div, Brooks AFB, Tex, 65-66, Health Physicist, Radiol Health Lab, Wright-Patterson AFB, 69-72, bioenviron engr, 1st Med Serv Wing, Clark AB, Philippines, 72-75, Sch Aerospace Med, 75-78, sr bioenviron eng, Off Air Force Surg Gen, Brooks AFB, Tex, 78-83, assoc chief, Biomed Sci Corps, 79-83; DIR ENVIRON QUALITY, PAC GAS & ELEC CO, SAN FRANCISCO, CALIF, 83- *Mem:* Health Physics Soc; Am Conf Govt Indust Hygienists. *Res:* Evaluation of low levels of iodine-131 in nuclear power reactor environments. *Mailing Add:* 677 Tampico Dr Walnut Creek CA 94598

FURTAK, THOMAS ELTON, b Ord, Nebr, May 23, 49; m 71; c 3. SURFACE PHYSICS. *Educ:* Univ Nebr, Lincoln, BS, 71; Iowa State Univ, PhD(solid state physics), 75. *Prof Exp:* Res fel photoelectrochem physics, Ames Lab, ERDA, 75-77; leader electrochem group, Solid State Physics Div, Ames Lab, Dept Energy, 77-80; assoc prof physics, Rensselaer Polytech Inst, 80-86; PROF PHYSICS, COLO SCH MINES, 86- *Mem:* Am Phys Soc; Am Chem Soc; Electrochem Soc; Sigma Xi; Optical Soc Am; Mat Res Soc. *Res:* Physics of interfaces, particularly the solid and aqueous-solution interface; monochromatic light as a quantum probe in the surface; second harmonic generation; electrochemical modulation spectroscopy; Raman spectroscopy; photoelectrocatalysis; ellipsometry; x-ray absorption at interfaces. *Mailing Add:* Colo Sch Mines Golden CO 80401

FURTER, W(ILLIAM) F(REDERICK), b North Bay, Ont, Apr 5, 31; m 66; c 3. CHEMICAL ENGINEERING. *Educ:* Royal Mil Col Can, RMC, 53; Univ Toronto, BASc, 54; Mass Inst Technol, SM, 55; Univ Toronto, PhD(chem eng), 58; Nat Defence Col Can, NDC, 70. *Prof Exp:* Asst instr, Mass Inst Technol, 54-55 & Univ Toronto, 55-58; res engr & sr tech investr, Res & Develop Dept, Du Pont of Can Ltd, 58-60; from asst prof to assoc prof, 60-66, prof-in-charge chem eng div, Dept Chem & Chem Eng, 60-80, secy, Grad Sch, 67-80, actg dean grad studies & res, 78-79, dean, Can Forces Mil Col & chmn, Exten Div, 80-84, PROF CHEM ENG, ROYAL MIL COL CAN, 66-, DEAN GRAD STUDIES & RES, 84- *Concurrent Pos:* Spec lectr, Royal Mil Col Can, 58-60; consult engr, 60-; res grant, Defence Res Bd Can, 63-, NSF & Petrol Res Fund; Dominion scholar; consult engr, Hexcel Corp, Calif, Air Liquide Ltd, Can & Union Carbide Corp, WVa. *Honors & Awards:* Eng Inst Can Prize; Bronze Medal, Gov Gen Can, Silver Medal; Silver Medal, Lt Gov, Ontario. *Mem:* Fel Chem Inst Can; Can Soc Chem Engrs; Can Nuclear Asn; Am Nuclear Soc. *Res:* Vapor-liquid phase equilibria; solution thermodynamics; extractive distillation; gas-liquid tray design; packed tower design; nuclear chemical engineering; engineering economics and administration. *Mailing Add:* Dept Chem Eng Royal Mil Col Can Kingston ON K7K 5L0 Can

FURTH, DAVID GEORGE, b Cleveland, Ohio, May 10, 45. BIO-SYSTEMATICS. *Educ:* Miami Univ, BA, 67; Ohio State Univ, MS, 69; Cornell Univ, PhD(entom), 76. *Prof Exp:* Teaching asst zool, Ohio State Univ, 67-69; nursery inspector plant pests, Ohio Dept Agr, 70; teaching asst entom, Cornell Univ, 70-72; cur insects entom, Tel Aviv Univ, 72-74; lectr biol, Yale Univ, 76-77; fel entom, Hebrew Univ Jerusalem, 77-79; consult, aquatic entom, Nature Reserve Authority Israel, 80-81; collection mgr curentom, Peabody Mus Natural Hist, Yale Univ, 81-89; CURATORIAL ASSOC ENTOM, MUS COMP ZOOL, HARVARD UNIV, 89- *Concurrent Pos:* Fulbright-Hays fel, 72-74. *Mem:* Sigma Xi; Entom Soc Am; Coleopterist's Soc; Xerces Soc (treas); Entom Soc Israel. *Res:* Systematics; food plant ecology; biogeography; evolution of Leaf Beetles (Chrysomelidae); general biogeography of the Middle East; faunistics and ecology of aquatic insects, especially Hemiptera; insect co-evolution with plants, including ferns; invertebrate conservation. *Mailing Add:* Mus Comp Zool Harvard Univ Cambridge MA 02138

FURTH, EUGENE DAVID, b Philadelphia, Pa, Jan 25, 29; m 52; c 2. MEDICINE, ENDOCRINOLOGY. *Educ:* Wesleyan Univ, AB, 50; Cornell Univ, MD, 54; Am Bd Internal Med, dipl. *Prof Exp:* Asst prof med & radiobiol, Col Med, Cornell Univ, 63-67, asst dir radioisotope lab, NY Hosp, 62-67; from assoc prof to prof med, Albany Med Col, 67-76; PROF & CHMN, DEPT MED, SCH MED, ECAROLINA UNIV, 76- *Concurrent Pos:* Hartford Geriat Fel, Johns Hopkins Med Sch, 86-87. *Mem:* Am Fedn Clin Res; Am Thyroid Asn; Endocrine Soc; AAAS; Am Geriat Soc; Am Diabetes Asn. *Res:* Thyroid gland physiology and pathophysiology; glucose metabolism. *Mailing Add:* Dept Med ECarolina Univ Greenville NC 27858-4354

FURTH, HAROLD PAUL, b Vienna, Austria, Jan 13, 30; nat US; div; c 1. PHYSICS. *Educ:* Harvard Univ, AB, 51, PhD, 60. *Prof Exp:* Physicist, Lawrence Radiation Lab, Univ Calif, Berkeley, 56-67; co-head, Exp Div, 67-78, assoc dir & head res dept, 78-80, prog dir, 80-81, dir, plasma physics lab, 81-90, PROF ASTROPHYS SCI, PRINCETON UNIV, 67- *Concurrent Pos:* Assoc ed, Rev of Modern Physics, 75-80. *Honors & Awards:* E O Lawrence Mem Award, AEC, 74; James Clerk Maxwell Prize in Plasma Physics, Am Phys Soc, 83; Joseph Priestley Award, Dickinson Col, 85. *Mem:* Am Acad Arts & Sci; fel Am Phys Soc. *Res:* Plasma physics; controlled thermonuclear research, particularly toroidal confinement experiments, theory of nonideal magneto-hydrodynamic stability, optimization of magnetic confinement configurations, design of toroidal magnetic fusion reactors. *Mailing Add:* Plasma Physics Lab Princeton Univ Princeton NJ 08543

FURTH, JOHN J, b Philadelphia, Pa, Jan 25, 29; m 59; c 3. BIOCHEMISTRY, PATHOLOGY. *Educ:* Cornell Univ, BA, 50. *Prof Exp:* Intern med path, Cornell Univ, 58-59; resident path, NY Univ, 59-60; fel biochem, Sch Med, NY Univ, 60-62; assoc, 62-65, from asst prof to assoc prof, 65-79, PROF PATH, SCH MED, UNIV PA, 79- *Concurrent Pos:* USPHS fel, 60-62, res career develop award, 62-71; mem path B study sect, NIH, 73-77; Eleanor Roosevelt fel, Am Cancer Soc-Int Union Against Cancer, 77-78; vis prof biochem, Aberdeen Univ, 77-78; assoc ed, Cancer Res, 77-84. *Honors & Awards:* Roche Award, 56. *Mem:* Am Soc Biochem & Molecular Biol; Am Asn Cancer Res. *Res:* Enzymic synthesis of RNA; structure of the 5s ribosomal RNA gene(s) and regulation of its transcription; regulation of collage symthesis in normal, senescent and transformed fibroblasts. *Mailing Add:* Dept Path Sch Med Univ Pa Philadelphia PA 19104-6082

FURTHMAYR, HEINZ, b Linz, Austria, May 5, 41. PATHOLOGY. *Educ:* Univ Vienna, MD, 65. *Prof Exp:* Instr anat, Univ Vienna, 62-65, res asst, Dept Immunol, 67-69; intern clin path, Hanuschkrankenhaus, Vienna, 65-66; intern med, LandeskranKenhaus Mistelbach, Austria, 66-67; res assoc biochem, Max Planck Inst Biochem, Munich, Ger, 69-72; res assoc path, Sch Med, Yale Univ, 72-76, asst resident, 75-76, from asst prof to prof path, 76-89, dir grad studies, 82-88; PROF, DEPT PATH, STANFORD UNIV, 89- *Concurrent Pos:* Vis fel, CSSR Acad Sci, Prague & Hosp St Louis, Paris, 68; Am Cancer Soc res award, 78; Mem, Molecular Biol Study Sect, 79, Path A Study Sect, 80-84 & 88, Prog Comt, Am Asn Pathologists, 80-82 & 84-86, Meritorious Awards Comt, 82-84 & Spec Study Sect, Nat Inst Diabetes & Digestive Kidney Dis, 90. *Mem:* Am Asn Pathologists; Am Soc Cell Biol; Asn Univ Pathologists. *Res:* Structure and function of basement membranes; cell-matrix interaction, gene regulation, differentiation; role of extracellular matrix in development and organogenesis; genetic defects involving extracellular matrix components; molecular biology of filopodia; author of more than 150 technical publications. *Mailing Add:* Dept Path R204 Med Ctr Stanford Univ 300 Pasteur Dr Stanford CA 94305

FURUKAWA, DAVID HIROSHI, b San Pedro, Calif, Mar 26, 38; m 64; c 5. CHEMICAL ENGINEERING. *Educ:* Univ Colo, BSChE, 60. *Prof Exp:* Chem engr saline water demineralization sect, Bur Reclamation, 60-64, supvr sect, 64-66, head sect, 66-69; mgr res & develop, Havens Int, 69-70, mgr com develop dept, Calgon-Havens Systs, 70-73; mgr com develop, UOP, Inc, Fluid Systs Div, 73-76; consult chem engr, Furukawa & Assocs, 76-78; prin chem engr, Boyle Eng Corp, 78-81; DIR MKT, RESOURCES CONSERV CO, 81- *Mem:* Am Inst Chem Engrs; Am Chem Soc; Am Water Works Asn; Water Pollution Control Fedn; Int Desalination & Environ Asn. *Res:* Desalting, particularly reverse osmosis but also electrodialysis and transport depletion; characterization of permselective and semipermeable membranes. *Mailing Add:* 13511 Willow Run Road Poway CA 92064

FURUKAWA, GEORGE TADAHARU, b Calif, May 25, 21; m 51; c 2. THERMODYNAMICS. *Educ:* Cent Col, Mo, AB, 43; Univ Wis, PhD(chem), 48. *Prof Exp:* Instr chem & physics, Cent Col, Mo, 43-45; physicist, Bur Standards, 48-85; METROLOGIST CONSULT, 85- *Concurrent Pos:* Prof lectr, George Washington Univ, 64-68. *Mem:* Am Chem Soc; Am Phys Soc; Calorimetry Conf (secy-treas, 60-63); Sigma Xi. *Res:* Low temperature heat capacity; vapor pressure; latent heats; surface tension; gas heat capacity; heterogeneous phase equilibria; temperature scale and thermometry; investigation of platinum resistance thermometry and thermometric fixed points. *Mailing Add:* 1712 Evelyn Dr Rockville MD 20852

FURUKAWA, TOSHIHARU, b Japan, Mar 22, 48; m 77. MATERIALS SCIENCE, MATERIALS ENGINEERING. *Educ:* Osaka Univ, Japan, BS, 70; Pa State Univ, MS, 73, PhD(solid state sci), 77. *Prof Exp:* Proj assoc mat sci, Pa State Univ, 77-80; MEM STAFF, IBM CORP, BURLINGTON, VT, 80- *Mem:* Am Ceramic Soc; Electrochem Soc; Inst Elec & Electronics Engrs. *Res:* Silicon semiconductor physics, semiconductor process engineering; characterizations and engineering applications of refractory, luminescence and semiconducting materials; surface and interfacial science; interactions of organic and inorganic materials; structure and crystallization of amorphous materials. *Mailing Add:* IBM Essex Junction VT 05452

FURUMOTO, AUGUSTINE S, b Honolulu, Hawaii, Aug 12, 27; m 65; c 2. SEISMOLOGY, GEOLOGY. *Educ:* Univ Dayton, BS, 49; Univ Tokyo, MSc, 55; St Louis Univ, PhD(geophys), 61. *Prof Exp:* Asst prof geophys & seismol, 61-67, assoc prof geophys, 67-71, PROF GEOPHYS & SEISMOLOGIST, INST GEOPHYS, UNIV HAWAII, HONOLULU, 71- *Concurrent Pos:* Long term vis Japan, NSF, 89. *Mem:* AAAS; Am Geophys Union; Seismol Soc Am; Nat Sci Teachers Asn; Sigma Xi. *Res:* Seismicity of Hawaii; earthquake hazards mitigation; volcanology. *Mailing Add:* Univ Hawaii Inst Geophys 2525 Correa Rd Honolulu HI 96822

FURUMOTO, HORACE WATARU, b Honolulu, Hawaii, Dec 13, 31; m 59; c 2. PHYSICS, ATOMIC & MOLECULAR PHYSICS. *Educ:* Calif Inst Technol, BSc, 55; Ohio State Univ, PhD(physics), 63. *Prof Exp:* Mem staff physics, Avco Systs Div, Avco Corp, 63-66 & NASA, 66-70; mem staff elec eng, US Dept Transp, 70-72; mem staff physics, Avco Everett Res Lab, Inc,

72-77; PRES, CANDELA LASER CORP, WAYLAND, MASS, 77- *Concurrent Pos:* Vis lectr, Harvard Med Sch. *Mem:* AAAS; Am Phys Soc; Optical Soc Am; Inst Elec & Electronics Engrs; Am Soc Laser Med & Surg; Laser Inst Am. *Res:* Laser research; laser applications; tunable dye lasers; medical lasers. *Mailing Add:* 14 Woodridge Rd Wellesley MA 02181

FURUMOTO, WARREN AKIRA, b Honolulu, Hawaii, Dec 17, 34; m 62. PLANT SCIENCE. *Educ:* Calif Inst Technol, BS, 57; Univ Calif, Los Angeles, PhD(bot sci), 61. *Prof Exp:* Instr bot, Univ Chicago, 60-62; from asst prof to assoc prof, 62-70, PROF BIOL, CALIF STATE UNIV, NORTHRIDGE, 70- *Concurrent Pos:* NSF res grants, 60-65. *Mem:* AAAS. *Res:* Studies on the infective process of tobacco mosaic virus. *Mailing Add:* Dept Biol Calif State Univ Northridge CA 91330

FURUSAWA, EIICHI, b Japan, Jan 25, 28; m; c 1. CHEMOTHERAPY, VIROLOGY. *Educ:* Osaka Univ, MD, 54, PhD, 59. *Prof Exp:* Res assoc, Res Inst Microbial Dis, Osaka Univ, 55-59; fel microbiol, Stanford Univ, 59-61, res assoc, 61-64; assoc pharmacologist, Pac Biomed Ctr, 64-65, assoc prof, 65-69, PROF PHARMACOL, SCH MED, UNIV HAWAII, MANOA, 69- *Concurrent Pos:* Fel microbiol, Columbia Univ, 60; NIH res grant, 67-81 & 84-87; Leukemia Soc res grant, 70-71; AMA res grant, 71-75; Am Cancer Soc res grant, 71-78; Univ Hawaii Found res grant, 81-88. *Mem:* Soc Exp Biol & Med; Am Asn Cancer Res; Am Soc Microbiol. *Res:* Virus and cancer chemotherapy and search for antiviral and anticancer agents from natural products. *Mailing Add:* Sch Med Univ Hawaii Honolulu HI 96822

FURUTA, TOKUJI, environmental horticulture; deceased, see previous edition for last biography

FURUTO, DONALD K, b Los Angeles, Calif, Mar 23, 48; m 83; c 1. CONNECTIVE TISSUE, PROTEINS. *Educ:* Univ Calif, Los Angeles, BA(hist) & BA(zool), 70; Univ Southern Calif, PhD(biochem), 77; Birmingham Theol Sem, MDiv, 85. *Prof Exp:* Res asst dent biochem, Sch Dent, Univ Southern Calif, 72-76; teaching fel, 77-80, res fel, 81-83, lectr, Sch Dent, 84 & 86, RES ASSOC BIOCHEM, INST DENT RES, UNIV ALA, BIRMINGHAM, 83- *Concurrent Pos:* Ed secy, Gustav Fischer, Stuttgart-NY, 81-; Nat Res Sci award, NIH, 77-80. *Res:* Connective tissue protein biochemistry and molecular biology. *Mailing Add:* 404 Oak Glen Lane Birmingham AL 35244

FUSARO, BERNARD A, b Charleston, WVa, Aug 9, 24; m 66; c 1. MATHEMATICS. *Educ:* Swarthmore Col, BA, 50; Columbia Univ, MA, 54; Univ Md, PhD(math), 65. *Prof Exp:* Instr physics & math, Ripon Col, 51-52; asst math, Columbia Univ, 52-54; instr, Middlebury Col, 54-57 & Univ Md, 57-61; vis asst prof, Univ Okla, 61-62; asst prof to assoc prof, Univ SFla, 62-67; prof & chmn dept, Queens Col, NC, 67-69, Dana prof, 69-74; prof math & chmn dept, Salisbury State Col, Md, 74-82; PROF MATH, SALISBURY STATE UNIV, 82- *Concurrent Pos:* Lectr, NSF Inst, Univ Okla, 61-62; Fulbright prof, Nat Taiwan Univ, 70-71; vis adj prof, Dept Environ Eng Sci, Univ Fla, 80-81; vis prof, US Mil Acad, 87-88. *Mem:* Am Math Soc; Math Asn Am; Soc Indust & Appl Math. *Res:* Linear and hyperbolic second order differential equations, particularly the generalized Euler-Poisson-Darboux equation; harmonic Riemannian spaces. *Mailing Add:* Dept Math Sci Salisbury State Univ Salisbury MD 21801

FUSARO, CRAIG ALLEN, b San Jose, Calif, July 19, 48. POPULATION BIOLOGY. *Educ:* San Jose State Univ, BA, 70; Univ Calif, Santa Barbara, MA, 73, PhD(biol), 77. *Prof Exp:* Assoc scientist, Ecomar, Inc, 78-83; INSTR BIOL, SANTA BARBARA CITY COL, 81-; DIR JOINT OIL/FISHERIES LIAISON OFF, SANTA BARBARA, 83- *Res:* Population ecology of marine intertidal crustaceans; community ecology of marine subtidal kelp beds and reefs; fate-and-effects studies of exploratory drilling fluids discharge; marine fisheries operations. *Mailing Add:* Dept Biol Sci Santa Barbara City Col 721 Cliff Dr Santa Barbara CA 93109

FUSARO, RAMON MICHAEL, b Brooklyn, NY, Mar 6, 27; m 51; c 2. DERMATOLOGY. *Educ:* Univ Minn, BA, 49, BS, 51, MD, 53, MS, 58, PhD, 65. *Prof Exp:* Intern, Minneapolis Gen Hosp, Minn, 53-54; resident, Minneapolis Gen Hosp-Univ Minn Hosp, 54-57; from instr to assoc prof dermat, Med Sch, Univ Minn, 57-70; prof & chmn dept, Med Ctr, Univ Nebr-Omaha, 70-82; prof dermat & chmn dept, Creighton Univ & dir dermat prog, Creighton-Nebr Univs Health Found, 75-82; RETIRED. *Mem:* Am Acad Dermat; Soc Invest Dermat; Sigma Xi. *Res:* Clinical dermatology; carbohydrate metabolism; photosensitivity, immunology; hereditary malignant melanoma. *Mailing Add:* 600 S 42nd St Univ Nebr Med Omaha NE 68198-4360

FUSCALDO, ANTHONY ALFRED, b New York, NY, Nov 11, 39; m 63; c 2. VIROLOGY, GENETICS. *Educ:* St John's Univ, NY, BS, 61, MS, 63; Ind Univ, Bloomington, PhD(microbiol), 67. *Prof Exp:* Prin investr viral genetics, US Army Biol Labs, 67-71; res assoc, Merrell Nat Labs Div, Richardson-Merrell, Inc, 71-74; asst prof & head res labs, Dept Med Hemat & Oncol & asst prof, Dept Biol Chem, Hahnemann Med Col & Hosp, 74-80, prof & adminr, Dept Hemat & Oncol, 75-80; regional safety dir, Ecol & Environ, 80-81; health & safety dir, Roy F Weston, 81-82; PRES, PHOENIX SAFETY ASSOCS, 82- *Mem:* AAAS; Tissue Cult Asn; Soc Occup & Environ Health; Am Soc Microbiol. *Res:* Biochemical and biophysical analysis of viruses; viral replication, morphogenesis and genetics; virus purification; isolation and characterization of oncornaviruses from human tissue specimens; tissue culture growth of human leukemic cells. *Mailing Add:* 1220 Valley Forge Rd Suite 33 & 34 Valley Forge PA 19481

FUSCALDO, KATHRYN ELIZABETH, b New York, NY, Jan 4, 31. GENETICS, MEDICAL CYTOGENETICS. *Educ:* Queens Col, NY, BS, 52; Hofstra Col, MA, 55; Mich State Univ, PhD, 60. *Prof Exp:* Jr bacteriologist immunol, NY State Dept Health, 52-53; asst, Carnegie Inst, 55-56; asst genetics, Mich State Univ, 56-60; asst prof, St John's Univ, NY, 60-63; asst prof genetics, 63-65, assoc prof med, 74-76, assoc prof anat, 65-77, res assoc prof microbiol, 66-75, PROF MED ONCOL, HEMAT & PATH, HAHNEMANN MED COL, 78-, ASSOC DIR & ADMINR CANCER INST, 73- & DIR MED GENETICS, 73- *Concurrent Pos:* Instr, Hofstra Col, 55-56; guest investr, Carnegie Inst, 61-63; vis lectr, Med Col Pa, 71-74. *Mem:* Am Asn Cancer Inst; Am Asn Cancer Res; AAAS; Genetics Soc Am; Am Soc Microbiol; Am Soc Cell Biol; Am Soc Human Genetics; Am Asn Univ Women. *Res:* Cytogenetics of myeloproliferative diseases; biochemical genetics; oncogenetics; immunogenetics. *Mailing Add:* 954 Conestoga Rd Rosemont PA 19010

FUSCO, CARMEN LOUISE, b Syracuse, NY; m 65; c 1. NUTRITIONAL ASPECTS OF AGING. *Educ:* Syracuse Univ, BS, 55; NY Med Col, MS, 86. *Prof Exp:* Instr pharmacol fundamentals metrics, NY Hosp Sch Nursing, Cornell Univ, 57-63; instr commun fundamentals, Sch Nursing, Syracuse Univ, 63-64; instr nutrit, 86-89, ASSOC PROF NUTRIT, NY MED COL, 90- *Concurrent Pos:* Res scientist, Grad Sch Health Sci, NY Med Col, 84-86; lectr, Continuing Educ Physicians, St John's Riverside Hosp, 90 & Med Students, NY Med, 90. *Mem:* NY Acad Sci; AAAS; Am Age Asn. *Res:* Nutritional aspects of aging; metabolism of riboflavin in hypothyroidism. *Mailing Add:* Seven Brook Rd Bronxville NY 10708

FUSCO, GABRIEL CARMINE, b Pittsburgh, Pa, Nov 11, 36; m 60; c 3. ORGANIC CHEMISTRY. *Educ:* Duquesne Univ, BS, 58, MS, 60; Univ Colo, PhD(org chem), 65. *Prof Exp:* Res chemist, Jackson Lab, E I du Pont de Nemours & Co, NJ, 65-67; asst prof, 67-68, assoc prof, 68-71, PROF CHEM, CALIF STATE COL, 71- *Mem:* Am Chem Soc; The Chem Soc. *Res:* Carbonium ion rearrangements; addition and elimination reactions; substitution reactions of saturated carbons. *Mailing Add:* Dept Chem Calif State Col California PA 15419

FUSCO, MADELINE M, b Waterbury, Conn, Nov 7, 24. PHYSIOLOGY. *Educ:* Ohio State Univ, BS, 48, MS, 49; Univ Pa, PhD(physiol), 59. *Prof Exp:* Vis lectr physiol, Goucher Col, 49-50, instr, 50-52; asst instr, Univ Pa, 52-55; instr, Vassar Col, 54-55; from instr to assoc prof, Univ Mich, 59-67; assoc prof, Med Col Pa, 67-71; assoc dean, 77-79, dean, 79-81, PROF ANAT SCI, SCH BASIC SCI STATE UNIV NY, STONY BROOK, 71- *Mem:* AAAS; Am Physiol Soc; Sigma Xi. *Res:* Animal calorimetry; temperature regulation, especially in hypothalamus; temperature regulation; neural control of energy exchange; behavior; neurophysiology. *Mailing Add:* Dept Anat Health Sci State Univ NY Stony Brook NY 11794

FUSCO, ROBERT ANGELO, b Middletown, NY, Aug 22, 41; m 64; c 2. ENTOMOLOGY. *Educ:* Univ Ky, BS, 64, MS, 67; Pa State Univ, PhD(entom), 71. *Prof Exp:* Entomologist, Md State Bd Agr, 71-72; ENTOMOLOGIST, PA DEPT ENVIRON RESOURCES, 72- *Mem:* Sigma Xi; Entom Soc Am; Soc Am Foresters. *Res:* Parasitoid-host interactions; parasitoid behavior; parasitoid biology; laboratory rearing and colonization techniques. *Mailing Add:* Pa Bur Forestry 34 Airport Dr Middletown PA 17057

FUSELER, JOHN WILLIAM, b Columbia, SC, May 3, 43; m 73. CELL BIOLOGY. *Educ:* Ga Inst Technol, BS, 67; Univ Pa, PhD(biol), 73. *Prof Exp:* Res assoc cell biol, Univ Pa, 73-75; SR RES INVESTR CELL BIOL, UNIV TEX MED BR, 75-; MEM FAC CELL BIOL, UNIV TEX HEALTH SCI CTR, DALLAS, 78- *Mem:* Am Soc Cell Biol; Am Inst Biol Sci; Bot Soc Am. *Res:* Mechanism of Mitosis in mammalian tissue culture cells; effect of temperature and various drugs on the microtubule assembly-disassembly process involved in chromosome movement. *Mailing Add:* 5660 S Lakeshore Dr Apt 215 Shreveport LA 71119

FUSHTEY, STEPHEN GEORGE, b Wasel, Alta, Sept 17, 24; m 51; c 4. PLANT PATHOLOGY. *Educ:* Univ Alta, BSc, 47, MSc, 50; Univ London, PhD(plant path), 53. *Prof Exp:* Asst plant pathologist, Plant Path Lab, Can Dept Agr, 47-51; lectr bot & plant path, Univ Guelph, 53-57, from asst prof to assoc prof, 64-80; RES SCIENTIEST, AGR CAN, 80- *Mem:* Am Phytopath Soc; Soc Nematol; Can Phytopath Soc; Agr Inst Can. *Res:* Diseases of cereal crops; nematology; turf grass diseases. *Mailing Add:* Agr Can Res Sta PO Box 1000 Agassiz BC V0M 1A0 Can

FUSON, ERNEST WAYNE, b Cawood, Ky, Oct 13, 47; m 69; c 2. IMMUNOLOGY. *Educ:* Lee Col, Tenn, BS, 69; Univ Tenn, Knoxville, MS, 72, PhD(zool), 75. *Prof Exp:* From asst prof to assoc prof med biol, Mem Res Ctr, Univ Tenn, Knoxville, 78-85; DIR IMMUNODIAGNOSTICS SECT, BLOUNT MEM HOSP, MARYVILLE, TENN, 85-; DIR LABS, THOMPSON CANCER SURVIVAL CTR, KNOXVILLE, TENN, 85- *Concurrent Pos:* Fel tumor immunol, Dept Surg, Univ Ala, Birmingham, 75-78; Nat Res Serv Award, Nat Cancer Inst-NIH, 77-78. *Mem:* Am Asn Immunologists; Int Soc Exp Hemat; Soc Exp Biol & Med; Am Soc Microbiol; Am Soc Histocompatabiltiy & Immunogenetics. *Res:* Immune response, cellular and humoral to tumor induction and development; mechanisms of tissue damage in autoimmune diseases; mechanisms of leukocyte-mediated cytotoxicity, including target recognition; effector cell activation and suppression; mechanisms of cytolysis. *Mailing Add:* Blount Mem Hosp 907 Lamar Alexander Pkwy Maryville TN 37801

FUSON, NELSON, b Canton, China, Sept 4, 13; US citizen; m 45; c 2. MOLECULAR SPECTROSCOPY. *Educ:* Col Emporia, AB, 34; Univ Kans, MA, 35; Univ Mich, PhD(physics), 39. *Prof Exp:* Lab instr physics, Univ Mich, 35-37; instr, Rutgers Univ, 38-41; res physicist, Off Sci Res & Develop, Univ Mich, 45; res assoc, Rockefeller Found Proj, Johns Hopkins Univ, 46-48; asst prof physics, Howard Univ, 48-49; from assoc prof to prof physics, Fisk Univ, 49-83, dir, Fisk Infrared Spectros Inst, 50-83, Latin Am, San Paulo, Brazil, 65, co-dir, Fisk Infrared Spectros Res Lab, 52-69; RETIRED. *Concurrent Pos:* Consult, AMP, Inc, Pa, 55-57; res assoc, Univ Bordeaux, 56-57, vis prof, Fac of Sci, 57-59; vis prof, Vanderbilt Univ, 60; pres, Bd Mgt, Coblentz Soc, 66-68; dir coop study, Nashville Univ Ctr Coun, 69-70, exec secy, 70-72; co-prin investr, Fisk Univ/Vanderbilt Univ physics consortium, NSF res prog, 80-83. *Honors & Awards:* Pegram Medal, Am Phys Soc, 75.

Mem: Fel AAAS; Am Soc Physics Teachers; Am Phys Soc; Coblentz Soc. *Res:* Applications of infrared spectroscopy to problems in chemical physics such as inter-molecular interactions, carcinogenic properties, inorganic ions in alkali halide matrices; biophysics; teaching of infrared spectroscopy short courses. *Mailing Add:* 1803 Morena St Nashville TN 37208

FUSON, ROBERT L, b Indianapolis, Ind, Mar 12, 32; m 59; c 2. THORACIC SURGERY, CARDIOVASCULAR SURGERY. *Educ:* DePauw Univ, MA, 56, MA, 57; Ind Univ, MD, 61; Am Bd Surg, dipl, 70. *Prof Exp:* Vpres res, Hemathermatrol Corp, 58-62; resident, Med Ctr, Duke Univ, 62-69; mgr surg prod res & develop dept, Ethicon, Inc, NJ, 69-73, dir prod develop, 73-75; SR VPRES, ZIMMER INC, 75- *Concurrent Pos:* Intern, Med Ctr, Ind Univ, 61-62; asst prof surg, Rutgers Univ, 71-75. *Mem:* Am Soc Artificial Internal Organs; Sigma Xi. *Res:* Development of biomedical devices; clinical and experimental surgical research; medical product regulatory and quality affairs. *Mailing Add:* PO Box 708 Warsaw IN 46581-0708

FUSON, ROGER BAKER, b Hazard, Ky, Mar 7, 16; m 48. IMMUNOLOGY, ANATOMY. *Educ:* Univ Ky, BS, 39; Univ Utah, BS, 51, MS, 52, PhD(anat, microbiol), 58. *Prof Exp:* Res bacteriologist, Vet Admin Hosp, Salt Lake City, 52-58; assoc dir exp med lab, Deaconess Hosp, Great Falls, Mont, 58-61; scientist adminr res grants br, Nat Inst Gen Med Sci, 61-63, head predoctoral fels sect, 63-74, fels officer, 74-82; RETIRED. *Mem:* AAAS; Am Soc Cell Biol; Am Soc Microbiol; NY Acad Sci; Am Asn Anat. *Res:* Inhibition of hyaluronidase activity by blood sera as a function of neoplasia and cytotoxins elicited by transplanted tumors and normal tissues. *Mailing Add:* Two Azalea St Scientists Cliffs Port Republic MD 20676

FUSSELL, CATHARINE PUGH, b Philadelphia, Pa, July 13, 19. BOTANY, MEIOSIS. *Educ:* Colby Col, AB, 41; Cornell Univ, MS, 58; Columbia Univ, PhD(cell biol), 66. *Prof Exp:* Admin asst, Shipping & Purchasing Dept, Am Friends Serv Comt, Philadelphia, 47-55; res asst biol, Brookhaven Nat Lab, 57-60; res assoc, Inst Cancer Res, Philadelphia, 66-67; NIH res fel, Fels Res Inst, Sch Med, Temple Univ, 67-68; from asst prof to assoc prof, McKeesport Campus, Pa State Univ, 68-77, assoc prof biol, Ogontz Campus, 77-88, res scientist, 88-89; VIS SCHOLAR, SCH MED, UNIV PA, 90- *Concurrent Pos:* NSF instnl res grant, Pa State Univ, 69-70 & small cols prog, 81; vis scientist, Med Res Coun, Clin & Pop Cytogenetics Unit, Western Gen Hosp, Edinburgh, Scotland, UK, 75-76; assoc investr, Exp Marine Bot Prog, Marine Biol Lab, Woods Hole, Mass, 78, prin investr, 81, libr reader. *Mem:* AAAS; Am Soc Cell Biol; Bot Soc Am; Genetics Soc Am. *Res:* Chromosome position in interphase nuclei in relation to differentiation and function; chromosome arrangements leading into meiosis; mechanics of meiotic synapsis; meiosis and the cytoskelton. *Mailing Add:* 7807 Spring Ave Elkins Park PA 19117

FUSTER, JOAQUIN MARIA, b Barcelona, Spain, Aug 17, 30; m 57; c 3. NEUROPHYSIOLOGY. *Educ:* Univ Barcelona, MD, 53; Univ Granada, PhD, 67. *Prof Exp:* Intern psychiat, Sch Med, Univ Barcelona, 52-53; asst resident, Neuropsychiat Clin, Innsbruck Univ, 54; asst resident, Inst Prev Psychiat, Univ Barcelona, 55-56; from assoc res psychiatrist to res psychiatrist, 57-67, PROF PSYCHIAT, MED CTR, UNIV CALIF, LOS ANGELES, 67- *Concurrent Pos:* Balmes fel, Neuropsychiat Clin, Innsbruck Univ, 54; Del Amo fel, 56; NIMH career develop award, 60-70, career scientist award, 70- *Mem:* AAAS; NY Acad Sci; Soc Neurosci; Am Psychiat Asn. *Res:* Neurophysiological basis of behavior; biological psychiatry. *Mailing Add:* Dept Psychiat Univ Calif Los Angeles Med Ctr 405 Hilgard Ave Los Angeles CA 90024

FUTCH, ARCHER HAMNER, b Monroe NC, Mar 21, 25; m 53; c 3. PHYSICS. *Educ:* Univ NC, BS, 49, MS, 51; Univ Md, PhD(physics), 56. *Prof Exp:* Physicist, E I du Pont de Nemours & Co, 55-58; PHYSICIST, LAWRENCE RADIATION LAB, 59- *Mem:* Am Phys Soc; Am Nuclear Soc; Sigma Xi. *Res:* Atomic and nuclear physics; controlled fusion research including plasma diagnostics and numerical computations. *Mailing Add:* 1252 Westbrook Pl Livermore CA 94550

FUTCH, DAVID GARDNER, b Schofield Barracks, Hawaii, Aug 31, 32; m 67; c 1. GENETICS, EVOLUTION. *Educ:* Univ NC, BA, 55; Univ Tex, MA, 61, PhD(zool), 64. *Prof Exp:* Res fel genetics, Calif Inst Technol, 64-65, Inst Animal Genetics, Edinburgh, 65-67 & City of Hope Med Ctr, 67; asst prof, 67-72, ASSOC PROF BIOL, SAN DIEGO STATE UNIV, 72- *Mem:* Genetics Soc Am; Soc Study Evolution; Am Genetic Asn. *Res:* Population genetics and evolution; genetic and cytological studies of evolution in Drosophila; genetics of parthenogenesis and sexual isolation in Drosophila. *Mailing Add:* Dept Biol San Diego State Univ San Diego CA 92182

FUTCHER, ANTHONY GRAHAM, b Rugby, Eng, Jan 18, 41; m 64; c 2. BIOSYSTEMATICS, NATURAL HISTORY. *Educ:* Columbia Union Col, BA, 62; Loma Linda Univ, PhD(biol), 74. *Prof Exp:* High sch teacher, Md, 62-63; from instr to assoc prof, 66-80, PROF BIOL, COLUMBIA UNION COL, 80- *Mem:* Asn Field Ornithologists; Am Ornithologists Union; Am Soc Mammalogists; Eastern Bird-Banding Asn. *Res:* Mammalian and avian biosystematics and karyology; avian populations and distribution utilizing bird-banding data; longevity of banded birds. *Mailing Add:* Dept Biol Columbia Union Col Takoma Park MD 20912-7796

FUTCHER, PALMER HOWARD, b Baltimore, Md, Sept 13, 10; wid; c 2. MEDICINE. *Educ:* Harvard Univ, AB, 32; Johns Hopkins Univ, MD, 36. *Prof Exp:* Intern, Johns Hopkins Hosp, 36-39; fel & asst res physician, Rockefeller Inst Hosp, 39-41; resident, Johns Hopkins Hosp, 41; asst prof med, Sch Med, Washington Univ, 46-48; assoc prof, Johns Hopkins Univ, 48-66, dir personnel health clin, Hopkins Med Insts, 62-66; clin prof med, Sch Med, Univ Pa, 89-; RETIRED. *Concurrent Pos:* Exec dir, Am Bd Internal Med, 67-75. *Mem:* Am Soc Clin Invest; Endocrine Soc; Am Diabetes Asn; AMA; Am Col Physicians; Am Clin Climat Asn. *Res:* Evaluation of clinical competence of physicians, clinical medicine. *Mailing Add:* 273 S Third St Philadelphia PA 19106

FUTRELL, J WILLIAM, b Shreveport, La, Jan 13, 41; m 65; c 2. PLASTIC SURGERY. *Educ:* Duke Univ, AB, 63, MD, 67. *Prof Exp:* From asst prof to assoc prof, 75-79, PROF PLASTIC SURG, UNIV PITTSBURGH, 79- *Res:* Tissue expansion; lymphatic system. *Mailing Add:* Presby-Univ Hosp 1117 Scaife Hall Univ Pittsburgh Med Ctr Pittsburgh PA 15261

FUTRELL, JEAN H, b Dry Prong, La, Oct 20, 33; m; c 3. ANALYTICAL CHEMISTRY. *Educ:* La Polytech Inst, BS, 55; Univ Calif, PhD(chem), 58. *Prof Exp:* Res chemist, Humble Oil & Refining Co, 58-59; sr res chemist & group leader, Aero-Space Res Labs, Ohio, 61-68; prof chem, Univ Utah, 68-86; WILLIS F HARRINGTON, PROF & CHAIR, CHEM & BIOCHEM, UNIV DEL, 86- *Concurrent Pos:* Sloan fel, 68-72; NIH Career Development Award, 69-74; Fulbright fel, Austria, 80-81. *Mem:* Fel AAAS; Am Chem Soc; Am Phys Soc; Am Soc Mass Spectros (pres, 76-78). *Res:* Chemical kinetics; mass spectrometry; ion-molecule reactions; tandem mass spectrometry. *Mailing Add:* Dept Chem & Biochem Univ Del Newark DE 19716

FUTRELL, MARY FELTNER, b Cadiz, Ky, Jan 5, 24; m 47; c 2. BIOCHEMISTRY, HUMAN NUTRITION. *Educ:* Austin Peay State Univ, BS, 44; Univ Wis, MS, 49, PhD(human nutrit), 52. *Prof Exp:* Grade sch teacher, Tenn, 42-43; high sch teacher, Tenn, 44-46 & Ky, 47-49; res asst nutrit, Univ Wis, 48-52; res assoc biochem, Tex A&M Univ, 52-54, asst prof home econ, 54-56; lectr, Ahmadu Bello Univ, Nigeria, 64-66; prof nutrit, 67-80, PROF HOME ECON, MISS STATE UNIV, 80- *Mem:* Fel Am Inst Chemists; Am Home Econ Asn; Am Dietetic Asn. *Res:* Amino acids in self-selected diets; ascorbic acid requirements; nutritional status of preschool children. *Mailing Add:* Dept Home Econ Miss State Univ Mississippi State MS 39762

FUTRELLE, ROBERT PEEL, b Washington, DC, Apr 23, 37; m 62; c 3. ARTIFICIAL INTELLIGENCE, THEORY OF SCIENTIFIC RESEARCH. *Educ:* Mass Inst Technol, SB, 59, PhD(physics), 66. *Prof Exp:* Mem tech staff theoret physics, Sci Ctr, NAm Rockwell Corp, 65-69, staff engr optical physics, Electrooptical Lab, Autonetics Div, 69-71; vis fel, Joint Inst Lab Astrophys, Univ Colo-Nat Bur Standards, 71-72; res assoc biophysics & theoret biol, Univ Chicago, 72-75; asst prof genetics & develop, Univ Ill, 75-85; ASSOC PROF COMPUT SCI, NORTHEASTERN UNIV, 86-, HEAD, BIOL KNOWLDEGE LAB, 89- *Concurrent Pos:* Chmn, Gordon Res Conf Atomic & Molecular Interactions, 70-71; NSF Biol Instrumentation Panel, 83-88; biomed comput coun, Asn Comput Mach, 85-87; NIH spec fel, 72-73. *Mem:* AAAS; Asn Comput Mach; Inst Elec & Electronics Eng; Am Asn Artificial Intel; NY Acad Sci; Sigma Xi. *Res:* Artificial intelligence to understand the research process, especially in biology; machine understanding of technical writing and graphics; computer vision; expert knowledge assistant for scientists; models of language acquisition; biophysics of chemosensing and chemotaxis. *Mailing Add:* Col Comput Sci Northeastern Univ 360 Huntington Ave Boston MA 02115

FUTUYMA, DOUGLAS JOEL, b New York, NY, Apr 24, 42. EVOLUTIONARY BIOLOGY, POPULATION BIOLOGY. *Educ:* Cornell Univ, BS, 63; Univ Mich, MS, 66, PhD(zool), 69. *Prof Exp:* from asst prof to assoc prof, 69-83, PROF ECOL & EVOLUTION, STATE UNIV NY, STONY BROOK, 83- *Concurrent Pos:* Mem fac, Orgn Trop Studies, 66, 69, 71, 74, 81 & 84; Ed, Evolution, 81-83. *Mem:* Soc Study Evolution (pres, 87); Genetic Soc Am; Ecol Soc Am; Am Soc Naturalists; Soc Syst Zool. *Res:* Genetic, phylogenetic, and ecological aspects of the evolution of associations of insects and plants; evolution of insect diet. *Mailing Add:* Dept Ecol & Evolution State Univ NY Stony Brook NY 11794-5245

FUXA, JAMES RODERICK, b Lincoln, Nebr, Jan 26, 49; m 72; c 2. ENTOMOLOGY, INSECT PATHOLOGY. *Educ:* Univ Nebr, BS, 71; Ore State Univ, MS, 75; NC State Univ, PhD(entom), 78. *Prof Exp:* From asst prof to assoc prof entom, 78-86, interim dept head, 90-91, PROF ENTOM, LA STATE UNIV, BATON ROUGE, 86- *Concurrent Pos:* AAAS/EPA Sci Eng fel, 87. *Mem:* Soc Invert Path; Entom Soc Am; Sigma Xi; AAAS; Int Orgn Biol Control. *Res:* Insect pathology, primarily epizootiology and microbial control of lepidopterous pests of field crops. *Mailing Add:* Dept Entom La State Univ Baton Rouge LA 70803-1710

FUZEK, JOHN FRANK, b Knoxville, Tenn, Dec 21, 21; m 43; c 3. FIBER SCIENCE. *Educ:* Univ Tenn, BS, 43, MS, 45, PhD(phys chem), 47. *Prof Exp:* Chem eng aide, Tenn Valley Authority, 40-42, chemist, Hercules Powder Co, Wilmington, 43-44; Off Naval Res fel, Tenn, 47-48; res chemist, Beaunit Fibers Div, Beaunit Corp, 48-55, head res physics, 55-66; sr res chemist, chem div, Eastman Kodak Co, 66-70, res assoc, 70-86; CONSULT, 87- *Honors & Awards:* Oak Ridge Inst Res Award, 49; Fiber Soc Lectr, 80-81; Cert of Appreciation, Am Soc Testing & Mat, 83, Merit Award, 90. *Mem:* Fel Am Soc Testing & Mat; Am Chem Soc; Fiber Soc; fel Am Inst Chem; Am Asn Textile Chem & Colorists; Sigma Xi. *Res:* Catalysis; kinetics; adsorption; cellulose chemistry; physical chemistry of polymers; x-ray diffraction of polymers; analytical instrumentation; fiber science; textile comfort; moisture in textiles. *Mailing Add:* 4603 Mitchell Rd Kingsport TN 37664

FYE, ROBERT EATON, b Cresco, Iowa, Jan 19, 24. BIOMATHEMATICS, AGRONOMY. *Educ:* Iowa State Col, BS, 49; Wash State Univ, MS, 51; Univ Wis, PhD, 54. *Prof Exp:* Asst entomologist, NMex State Univ, 54-55; entomologist, Entom Res Div, Agr Res Serv, USDA, 55-59; res officer, Can Dept Forestry, 60-65; entomologist, Cotton Insect Biol Control Lab, Ariz-NMex Area, USDA, 65-77, res entomologist, Yakima Agr Res Lab, Wash-Ore Area, Western Region, 77-84, at Crop Sci Res Lab, Mid S Area, Agr Res Serv, 84-86; RETIRED. *Concurrent Pos:* Assoc prof, Univ Ariz, 67-77; vis scientist, Cotton Res Unit, Commonwealth Sci & Indust Res Orgn, Australia, 74-75. *Mem:* Entom Soc Am; Ecol Soc Am. *Res:* Biology and ecology of cotton insects, pear psylla, lygus bugs, pollinators, native bees and wasps; predation; insect population dynamics; bioclimatology. *Mailing Add:* 104 Edgewood Dr Starkville MS 39759

FYFE, I(AN) MILLAR, b Glen Ridge, NJ, Nov 13, 25; m 51; c 1. AERONAUTICAL ENGINEERING, APPLIED MECHANICS. *Educ:* Royal Col Sci & Technol, ARTC, 51; Univ Del, MME, 54; Stanford Univ, PhD(mech), 58. *Prof Exp:* Res engr dynamics, Boeing Co, 57-60; from asst prof to assoc prof aeronaut, 60-69, PROF AERONAUT & ASTRONAUT, UNIV WASH, 69- *Concurrent Pos:* NATO sr fel, 68. *Mem:* Am Soc Mech Engrs; Am Inst Aeronaut & Astronaut; Soc Exp Stress Anal. *Res:* Wave propagation in solids and fluids; impact dynamics. *Mailing Add:* Dept Aeronaut Univ Wash Seattle WA 98195

FYFE, JAMES ARTHUR, b Woodstock, Ill, Mar 16, 41; m 64; c 3. MECHANISMS OF ANTI-HERPES COMPOUNDS. *Educ:* Carleton Col, BA, 63; Univ Chicago, PhD(biochem), 68. *Prof Exp:* Asst prof biochem & chem, Trinity Univ, Tex, 68-70; RES BIOCHEMIST, WELLCOME RES LABS, BURROUGHS WELLCOME CO, 70- *Mem:* Am Chem Soc. *Res:* Mechanisms of action and substrate-inhibitor specificities of purine and pyrimidine metabolizing enzymes; enzymology of antiviral chemotherapy. *Mailing Add:* RFD 6 Box 195/A1 Durham NC 27703-9802

FYFE, RICHARD ROSS, b Binghamton, NY, Nov 19, 41; m 66; c 2. CHEMICAL ENGINEERING, ELECTROCHEMISTRY. *Educ:* Lafayette Col, BS & BA, 64; Columbia Univ, MS, 66, DEngSc(chem eng), 69. *Prof Exp:* Chem engr, Alcorn Combustion Co, 68-69; res chem engr, Picatinny Arsenal, US Army, 69-72; staff chemist, Lawrence Livermore Lab, Univ Calif, 72-74; res scientist, Union Camp Corp, NJ, 74-79; sr res engr, Koppers Co, Inc, 79-89; SR DEVELOP ENGR, HOECHST CELANESE CORP, 89- *Honors & Awards:* Sigma Xi. *Mem:* Am Inst Chem Engrs; Am Chem Soc. *Res:* Colloid chemistry; kinetics; electrostatics; shock waves; metastable and explosive materials characterization; surface chemistry; tall oil and fatty acid processes; terpene processes; coal tar chemical processes; copper and arsenic compounds; economics and process design. *Mailing Add:* 4245 Snowbird St Corpus Christi TX 78413-4402

FYFE, WILLIAM SEFTON, b Ashburton, NZ, June 4, 27; Can citizen; m 52; c 3. GEOLOGY. *Educ:* Univ Otago, BSc, 48, MSc, 49, PhD(chem), 52. *Hon Degrees:* DSc, mem, Lisbon, Lakehead. *Prof Exp:* Prof geol, Univ Calif, Berkeley, 58-66 & Manchester, UK, 66-72; prof & chmn geol, 72-86, dean sci, 86-90, PROF GEOL, UNIV WESTERN ONT, CAN, 91- *Honors & Awards:* Logan Medal, Geol Asn Can, 81; Arthur Holmes Medal, EUGS; Day Medal, Geol Soc Am, 90. *Mem:* Geol Soc Am; Minerals Soc Am; Royal Soc London; Geol Asn Can; Chem Soc UK; fel Am Acad Arts & Sci; Royal Soc NZ; hon mem Acad Sci Brazil; fel Explorer's Club; hon fel Geol Soc UK. *Res:* Geology; geochemistry; environmental science. *Mailing Add:* Dept Geol Univ Western Ont London ON N6A 5B7 Can

FYFFE, DAVID EUGENE, b Washington, Ind, June 29, 25; m 65; c 3. INDUSTRIAL ENGINEERING, OPERATIONS RESEARCH. *Educ:* Purdue Univ, BSME, 50, MSIE, 55; Northwestern Univ, PhD, 64. *Prof Exp:* Mgr qual control, Appliance Motor Dept, Gen Elec Co, 57-61; from assoc prof to prof indust eng, Ga Inst Technol, 64-86; PSI INC, 86- *Concurrent Pos:* Consult indust eng & mgt. *Mem:* Am Inst Indust Engrs; Am Soc Eng Educ; Am Soc Qual Control; Inst Mgt Sci. *Res:* Manufacturing systems; project feasibility analysis; quality control; inventory control. *Mailing Add:* 21 N Main St Suite 207 Alpharetta GA 30201

FYHRIE, DAVID PAUL, b Spokane, Wash, May 31, 55; m 87; c 2. BIOMECHANICS. *Educ:* Gonzaga Univ, BS, 77; Stanford Univ, Spokane, MS, 78, PhD(mech eng), 86. *Prof Exp:* Design engr, Westinghouse Marine Div, Sunnyvale, Calif, 78; biomed engr, Rehab Res & Develop Ctr, Palo Alto, Va, 79-80, 81-87; student res asst, Stanford Univ, Calif, 80-81, teaching asst, 82-83; res assoc, Mech Eng Dept, Wayne State Univ, Detroit, Mich, 87-88; assoc staff investr, 88-89, HEAD, SECT BIOMECH, BONE & JOINT CTR, HENRY FORD HOSP, DETROIT, MICH, 90-, ASST DIR RES, 90- *Mem:* Orthop Res Soc; Am Soc Biomech. *Res:* Investigate Wolff's Law of trabecular bone by calculating 3D stress state of bone and relating to biological activity of tissues; determine loading effect on bone pattern in growth. *Mailing Add:* Henry Ford Hosp 2799 W Grand Blvd E&R-2015 Detroit MI 48202

FYLES, JOHN GLADSTONE, b Vancouver, BC, Feb 27, 23; m 50; c 3. ENVIRONMENTAL GEOLOGY, ENVIRONMENTAL MANAGEMENT. *Educ:* Univ BC, BASc, 46, MASc, 50; Ohio State Univ, PhD(geol), 56. *Prof Exp:* Tech officer, Dept Energy, Mines & Resources, Geol Surv Can, 50-56, geologist, 56-68, chief terrain sci div, 68-73, environ eng adv, 73-77; dir gen, Northern Pipelines Br, Dept Indian Affairs & Northern Develop, 77-79; chief geologist & dep dir gen, 79-89, EMER SCIENTIST, GEOL SURV CAN, 89- *Concurrent Pos:* coord for Dept Energy, Mines & Resources, Environ-Social Prog, Northern Pipelines of Task Force on Northern Oil Develop, 71-74, head pipeline appln assessment group, 74; head inquiry appraisal team, Mackenzie Valley Pipeline Inquiry, 75-77. *Mem:* Geol Soc Am. *Res:* Pleistocene geology; geomorphology; engineering geology. *Mailing Add:* Geol Surv Can 601 Booth St Ottawa ON K2A 3K3 Can

FYMAT, ALAIN L, b Casablanca, Morocco, Dec 7, 38; m 60; c 2. METEOROLOGY, PLANETARY ATMOSPHERES. *Educ:* Nat Super Sch Meteorol, Paris, BA, 59; Sorbonne, MA, 60; Univ Bordeaux, MS, 63; Univ Calif, Los Angeles, PhD(meteorol), 67. *Prof Exp:* Staff engr, Weather Bur Morocco, 60-63; res meteorologist, Univ Calif, Los Angeles, 64-67; sr res scientist, Calif Inst Technol, 67-70, mem tech staff, Space Sci Div, Jet Propulsion Lab, 67-82; SCIENTIST, AEROJET ELECTROSYSTS, CALIF, 82- *Concurrent Pos:* Asst prof & lectr meteorol, Univ Calif, Los Angeles, 67-70; prof geol, Univ Southern Calif, 70; vis prof physics, Univ Lille, 70-71; adv, Int Radiation Comn, 73-; assoc ed, Int J Appl Math & Comput, 74-88; vis prof physics, Univ Leningrad, 78; vis prof atmospheric sci, Univ Calif, Los Angeles, 79. *Honors & Awards:* NASA Awards for Sci Inventions, 71 & 75. *Mem:* Am Inst Physics; Am Astron Soc; Optical Soc Am; Am Geophys Union; Am Meteorol Soc. *Res:* Hydrogen lyman alpha geocorona; radiative transfer in planetary atmospheres; interferometric polarimetry; spectral line formation in scattering atmospheres; integral equations theory; differential equations theory; system identification; optimization; environmental pollution; societal aspects and implications, space defense systems; electro optical sensors. *Mailing Add:* PO Box 90891 Pasadena CA 91109-0891

FYSTROM, DELL O, b Minneapolis, Minn, Aug 29, 37; div; c 3. ATOMIC PHYSICS. *Educ:* St Olaf Col, BA, 59; Univ Colo, Boulder, PhD(physics), 69. *Prof Exp:* Chmn dept, 70-73 & 82-91, ASSOC PROF PHYSICS, UNIV WIS-LA CROSSE, 69- *Concurrent Pos:* Chmn dept physics, Haile Sellassie Univ, 74-75. *Mem:* Am Asn Physics Teachers; Acoust Soc Am; Soc Exp Mech. *Res:* Magnetic moment; precision measurements; proton moment; nuclear magnetons. *Mailing Add:* Dept Physics Univ Wis La Crosse WI 54601

FYTELSON, MILTON, b Bridgeport, Conn, Nov 15, 17; m 46; c 2. CHEMISTRY. *Educ:* Yale Univ, BS, 37, PhD(org chem), 41. *Prof Exp:* Pfizer fel, Columbia Univ, 41-42; res chemist, Yale Univ, 42-43; chemist, Am Quinine Co, Bogota, Colombia, SAm, 43; fel, Mellon Inst, 44-45; res chemist, E I du Pont de Nemours & Co, 46-58; group leader, Toms River Chem Corp, 58-67; mgr process develop, Otto B May, Inc, 67-68; pres, Fytelson & Assocs, Chem & Eng Consult, 68-74; plant mgr, Sandoz Colors & Chem, 74-79, asst to vpres mfg, 79-85; PRES, FYTELSON & ASSOCS, 85- *Concurrent Pos:* Adj instr, Orgn Chem Environ Sci, Belmont Abbey Col, 86-88; treas, c/p sect, Am Chem Soc, 89-91, counr, 91- *Mem:* Am Chem Soc. *Res:* Synthesis of vitamin A; synthetic lubricants; synthesis and applications of organic pigments; metallo-organic complexes; process development and engineering; sulfur and chromium chemistry; manufacture dyes, chemicals and intermediates. *Mailing Add:* 3026 Willowbrae Rd Charlotte NC 28226-3044